Jane's

Space Directory 2005-2006

Edited by David Baker

Twenty-first Edition

Total number of entries 2,902 New and updated entries 1,698

Visit jsd.janes.com and view the list of latest updates that have been added to the online version of *Jane's Space Directory* subsequent to this print edition.

ISBN 0 7106 2706 8

In the US and its dependencies
Jane's Information Group Inc, 110 N Royal Street, Suite 200, Alexandria, Virginia 22314, US

Printed in the United Kingdom

EDITORIAL AND ADMINISTRATION

Director: Ian Kay, e-mail: Ian.Kay@janes.com

New Media Publishing Director: Sean Howe, e-mail: Sean.Howe@janes.com

Publisher: Sara Morgan, e-mail: Sara.Morgan@janes.com

Content Services Director: Anita Slade, e-mail: Anita.Slade@janes.com

Content Systems Manager: Jo Agius, e-mail: Jo.Agius@janes.com

Pre-Press Manager: Christopher Morris, e-mail: Christopher.Morris@janes.com

Team Leader: Melanie Rovery, e-mail: Melanie.Rovery@janes.com

Content Editor: Rebecca Davies, e-mail: Rebecca.Davies@janes.com

Production Controller: Laura-Jane Walker, e-mail: Laura-JaneWalker@janes.com

Content Update: Jacqui Beard, Information Collection Team Leader
Tel: +44 (0) 20 8700 3808 Fax: +44 (0) 20 8700 3959
e-mail: yearbook@janes.com

Jane's Information Group Limited, Sentinel House, 163 Brighton Road, Coulsdon, Surrey CR5 2YH, UK
Tel: +44 (0) 20 8700 3700 Fax: +44 (0) 20 8700 3900

SALES OFFICES

Send Europe and Africa enquiries to: *Mike Gwynn – Head of Information Sales*
Jane's Information Group Limited, Sentinel House, 163 Brighton Road, Coulsdon, Surrey CR5 2YH, UK
Tel: +44 (0) 20 8700 3750 Fax: +44 (0) 20 8700 3751
e-mail: customer.servicesuk@janes.com

Send US/Latin America/Canada enquiries to: *Greg Wallis – VP Information Sales*
Jane's Information Group Inc, 110 N Royal Street, Suite 200, Alexandria, Virginia 22314, USA
Tel: (+1 703) 683 37 00 Fax: (+1 703) 836 02 97 Telex: 6819193
Tel: (+1 800) 824 07 68 Fax: (+1 800) 836 02 97
e-mail: customer.servicesus@janes.com

Send Asia enquiries to: *David Fisher – Group Business Manager*
Jane's Information Group Asia, 78 Shenton Way, #10-02, Singapore 079120, Singapore
Tel: (+65) 63 25 08 66 Fax: (+65) 62 26 11 85
e-mail: asiapacific@janes.com

Send Australia/New Zealand enquiries to: *Russel Smith – Business Manager*
Jane's Information Group, PO Box 3502, Rozelle Delivery Centre, New South Wales 2039, Australia
Tel: (+61 2) 85 87 79 00 Fax: (+61 2) 85 87 79 01
e-mail: oceania@janes.com

Send Middle East enquiries to: *Ali Abdellatif Siali – Regional Sales Manager*
Jane's Information Group, PO Box 502138, Dubai, United Arab Emirates
Tel: (+971 4) 390 23 35 Tel: (+971 4) 390 23 36 Fax: (+971 4) 390 88 48
e-mail: customer.servicesmideast@janes.com

Send Japan enquiries to: *Norihisa Fukuyama – Information Consultant*
Jane's Information Group, Palaceside Building, 5F, 1-1-1, Hitotsubashi, Chiyoda-ku, Tokyo 100-0003, Japan
Tel: (+81 3) 52 18 76 82 Fax: (+81 3) 52 22 12 80
e-mail: japan@janes.com

Send India enquiries to: *T C Martin – Information Consultant*
Jane's Information Group, PO Box 3806, New Delhi 110049, India
Tel/Fax: (+91 11) 26 51 61 05
e-mail: india@janes.com

Send China enquiries to: *Jia Mao – Business Consultant for People's Republic of China*
Jane's Information Group, 78 Shenton Way, #10-02, Singapore 079120, Singapore
Tel/Fax: (+86 0) 10 84 47 27 56
e-mail: china@janes.com

ADVERTISEMENT SALES OFFICES

(Head Office)
Jane's Information Group
Sentinel House, 163 Brighton Road,
Coulsdon, Surrey CR5 2YH, UK
Tel: (+44 20) 87 00 37 00
Fax: (+44 20) 87 00 38 59/37 44
e-mail: defadsales@janes.com

Richard West, Senior Key Accounts Manager
Tel: (+44 1892) 72 55 80 Fax: (+44 1892) 72 55 81
e-mail: richard.west@janes.com

Nicky Eakins, Advertising Sales Executive
Tel: (+44 20) 87 00 38 53 Fax: (+44 20) 87 00 38 59/37 44
e-mail: nicky.eakins@janes.com

(US/Canada office)
Jane's Information Group
110 N Royal Street, Suite 200,
Alexandria, Virginia 22314, USA
Tel: (+1 703) 683 37 00
Fax: (+1 703) 836 55 37
e-mail: defadsales@janes.com

USA and Canada
Katie Taplett, US Advertising Sales Director
Tel: (+1 703) 683 37 00 Fax: (+1 703) 836 55 37
e-mail: katie.taplett@janes.com

Northern US and Eastern Canada
Linda Hewish, Northeast Region Advertising Sales Manager
Tel: (+1 703) 683 37 00 Fax: (+1 703) 836 55 37
e-mail: linda.hewish@janes.com

Southeastern US
Kristin D Schulze, Advertising Sales Manager
PO Box 270190, Tampa, Florida 33688-0190, USA
Tel: (+1 813) 961 81 32 Fax: (+1 813) 961 96 42
e-mail: Kristin.Schulze@janes.com

Western US and Western Canada
Richard L Ayer
127 Avenida del Mar, Suite 2A, San Clemente, California 92672, USA
Tel: (+1 949) 366 84 55 Fax: (+1 949) 366 92 89
e-mail: ayercomm@earthlink.com

Australia: *Richard West* (see UK Head Office)

Benelux: *Nicky Eakins* (see UK Head Office)

Brazil: *Katie Taplett* (see US address)

Eastern Europe (excl. Poland): MCW Media & Consulting Wehrstedt
Dr Uwe H Wehrstedt
Hagenbreite 9, D-06463 Ermsleben, Germany
Tel: (+49) 07 00/WEHRSTEDT / (+49 03) 47 43/620 90
Fax: (+49 03) 47 43/620 91
e-mail: info@Wehrstedt.org

France: Patrice Février
BP 418, 35 avenue MacMahon,
F-75824 Paris Cedex 17, France
Tel: (+33 1) 45 72 33 11 Fax: (+33 1) 45 72 17 95
e-mail: patrice.fevrier@wanadoo.fr

Germany and Austria: *MCW Media & Consulting Wehrstedt* (see Eastern Europe)

Greece: *Nicky Eakins* (see UK Head Office)

Hong Kong: *Nicky Eakins* (see UK Head Office)

India: *Nicky Eakins* (see UK Head Office)

Iran: Eideh Info Company
Ali Jahangard
19 4th Street, Ghaem Magham Avenue, Tehran, Iran
Tel: (+98 21) 873 59 23
e-mail: eideh@mavara.com

Israel: Oreet – International Media
15 Kinneret Street, IL-51201 Bene Berak, Israel
Tel: (+972 3) 570 65 27 Fax: (+972 3) 570 65 27
e-mail: admin@oreet-marcom.com
Defence: Liat Shaham
e-mail: liat_s@oreet-marcom.com

Italy and Switzerland: Ediconsult Internazionale Srl
Piazza Fontane Marose 3, I-16123 Genoa, Italy
Tel: (+39 010) 58 36 84 Fax: (+39 010) 56 65 78
e-mail: genova@ediconsult.com

Middle East: *Nicky Eakins* (see UK Head Office)

Pakistan: *Nicky Eakins* (see UK Head Office)

Poland: *Nicky Eakins* (see UK Head Office)

Russia: Vladimir N Usov, PO Box 98, 622018 Nizhniy Tagil, Sverdlovsk Region, Russian Federation
Tel/Fax: (+7 3435) 23 02 68
e-mail: uvn125@uraltelecom.ru

Scandinavia: The Falsten Partnership
PO Box 27, Portslade, East Sussex BN41 2XA, UK
Tel: (+44 1273) 77 10 20 Fax: (+ 44 1273) 77 00 70
e-mail: sales@falsten.com

Singapore: *Richard West/Nicky Eakins* (see UK Head Office)

South Africa: *Richard West* (see UK Head Office)

South Korea: Infonet Group Inc
Sanbu Rennaissance Tower 902, 456 Gongdukdong, Mapogu, Seoul, South Korea
Contact: Mr Jongseog Lee
Tel: (+82 2) 716 99 22
Fax: (+82 2) 716 95 31
e-mail: jslee@infonetgroup.co.kr

Spain: Via Exclusivas SL
e-mail: viaexclusivas@viaexclusivas.com

Turkey: *Richard West* (see UK Head Office)

ADVERTISING COPY
Linda Letori (Jane's UK Head Office)
Tel: (+44 20) 87 00 37 42 Fax: (+44 20) 87 00 38 59/37 44
e-mail: linda.letori@janes.com

For North America, South America and Caribbean only:
Lia Johns (Jane's US/Canada Office)
Tel: (+1 703) 683 37 00 Fax: (+1 703) 836 55 37
e-mail: lia.johns@janes.com

Contents

Jane's Space Directory website: jsd.janes.com

Front cover caption: ESA's ATV is designed to supplement the lifting capacity of reusable systems for servicing and logistics supply to the International Space Station.

NEW/1121434

How to Use *Jane's Space Directory*

As with last year's edition, this year the content is structured around the categorisation of functions, beginning with a description of government and non-government space programmes and organisations providing detailed information on relevant activities. This is followed by a directory of the global space industry which serves the aforementioned programmes. These two sections provide the core from which all the technology, research, development and operational activity flows and includes information about trends and objectives, budgets, programme and project prioritisation and about centres of application. These two sections form the institutional backbone to government and non-government space activity.

The second main section of the book covers civilian operations involving launch sites, launch vehicles, launch vehicle propulsion and in-orbit satellite activity, planetary and scientific operations and human space flight including space stations and resupply vehicles. The opening part catalogues all known world space centres including launch sites, training and tracking facilities, research and development facilities and space monitoring sites. It continues with a survey and technical description of the world's orbital and suborbital launch vehicles followed by a section on propulsion. Civilian satellites follow with information on design types, pedigree, operations history and specifications. Planetary and space science spacecraft follow in a similar format. Human space flight serves to connect major goals outlined in earlier sections with realisation of national or international objectives.

The third main section groups all military operations by programme category and then by country but cross-reference with the launch vehicles section is necessary for determining the characteristics of launchers and propulsion systems where they apply. Categories of military satellites are those exclusively dedicated to military support functions but do not include those commercial or civilian satellites from which some defence application is obtained. Nor do they include the military flights, or defence related experiments, conducted by manned vehicles, details of which can be found in the appropriate section.

The fourth main section is that devoted to chronicling the contractors, to providing tables of data and lists of information for reference purposes in a series of space logs enabling cross-reference through the index.

One major addition introduced in 2000 is the comprehensive list of every satellite, spacecraft and payload launched since the first artificial satellite orbited the earth on 4 October 1957. This is intended to stand as a benchmark for reference purposes and as a precedent for adding additional information about the status of historic payloads still in orbit in subsequent editions. This year, due to the ever growing size of the publication, the world launch log from 1957-1994 can be found online only, at jsd.janes.com.

Any update to the content of this product will appear online as it occurs (see jsd.janes.com for the additional benefits of an online subscription to *Jane's Space Directory* and details of our free online trial) and will be incorporated annually into future print editions.

To help users evaluate the data of this edition, the following identifiers have been used:

- ***VERIFIED*** The editor has made a detailed examination of the entry's content and checked it's relevancy and accuracy for publication in the new edition to the best of his ability.
- ***UPDATED*** During the verification process, significant changes to content have been made to reflect the latest position known to Jane's at the time of publication.
- ***NEW ENTRY*** Information on new equipment and/or appearing for the first time in the title.
- ***NEW*** New images are identified as ***NEW*** and some are followed by a seven-digit number for ease of identification by our image library.

A full list of all entries indicating their current status is provided in the index.

Total number of entries **2,902** New and updated entries **1,698**

Visit jsd.janes.com and view the list of latest updates that have been added to the online version of *Jane's Space Directory* subsequent to this print edition.

British Library Cataloguing-in-Publication Data.
A catalogue record for this book is available from the British Library.

Printed and bound in Great Britain by Biddles Ltd, Guildford and Kings Lynn

Jane's Libraries

To assist your information gathering and to save you money, Jane's has grouped some related subject matter together to form 'ready-made' libraries, which you can access in whichever way suits you best – online, on CD-ROM, via Jane's EIS or through Jane's Data Service.

The entire contents of each library can be cross-searched, to ensure you find every reference to the subjects you are looking for. All Jane's libraries are updated according to the delivery service you choose and can stand alone or be networked throughout your organisation.

Jane's Defence Equipment Library

Aero-Engines
Air-Launched Weapons
Aircraft Upgrades
All the World's Aircraft
Ammunition Handbook
Armour and Artillery
Armour and Artillery Upgrades
Avionics
C4I Systems
Electro-Optic Systems
Explosive Ordnance Disposal
Fighting Ships
Infantry Weapons
Land-Based Air Defence
Military Communications
Military Vehicles and Logistics
Mines and Mine Clearance
Naval Weapon Systems
Nuclear, Biological and Chemical Defence
Radar and Electronic Warfare Systems
Strategic Weapon Systems
Underwater Warfare Systems
Unmanned Aerial Vehicles and Targets

Jane's Defence Magazines Library

Defence Industry
Defence Weekly
Foreign Report
Intelligence Digest
Intelligence Review
International Defence Review
Islamic Affairs Analyst
Missiles and Rockets
Navy International
Terrorism and Security Monitor

Jane's Market Intelligence Library

Aircraft Component Manufacturers
All the World's Aircraft
Defence Industry
Defence Weekly
Electronic Mission Aircraft
Fighting Ships
Helicopter Markets and Systems
International ABC Aerospace Directory
International Defence Directory
Marine Propulsion
Naval Construction and Retrofit Markets
Police and Security Equipment
Simulation and Training Systems
Space Directory
Underwater Technology
World Armies
World Defence Industry

Jane's Security Library

Amphibious and Special Forces
Chemical-Biological Defense Guidebook
Facility Security
Fighting Ships
Intelligence Digest
Intelligence Review
Intelligence Watch Report
Islamic Affairs Analyst
Police and Security Equipment
Police Review
Terrorism & Security Monitor
Terrorism Watch Report
World Air Forces
World Armies
World Insurgency and Terrorism

Jane's Sentinel Library

Central Africa
Central America and the Caribbean
Central Europe and the Baltic States
China and Northeast Asia
Eastern Mediterranean
North Africa
North America
Oceania
Russia and the CIS
South America
South Asia
Southeast Asia
Southern Africa
The Balkans
The Gulf States
West Africa
Western Europe

Jane's Transport Library

Aero-Engines
Air Traffic Control
Aircraft Component Manufacturers
Aircraft Upgrades
Airport Review
Airports and Handling Agents –
  Central and Latin America (inc. the Caribbean)
  Europe
  Far East, Asia and Australasia
  Middle East and Africa
  United States and Canada
Airports, Equipment and Services
All the World's Aircraft
Avionics
High-Speed Marine Transportation
Marine Propulsion
Merchant Ships
Naval Construction and Retrofit Markets
Simulation and Training Systems
Transport Finance
Urban Transport Systems
World Airlines
World Railways

Alphabetical list of advertisers

Jane's Users' Charter

This publication is brought to you by Jane's Information Group, a global company with more than 100 years of innovation and an unrivalled reputation for impartiality, accuracy and authority.

Our collection and output of information and images is not dictated by any political or commercial affiliation. Our reportage is undertaken without fear of, or favour from, any government, alliance, state or corporation.

We publish information that is collected overtly from unclassified sources, although much could be regarded as extremely sensitive or not publicly accessible.

Our validation and analysis aims to eradicate misinformation or disinformation as well as factual errors; our objective is always to produce the most accurate and authoritative data.

In the event of any significant inaccuracies, we undertake to draw these to the readers' attention to preserve the highly valued relationship of trust and credibility with our customers worldwide.

If you believe that these policies have been breached by this title, you are invited to contact the editor.

A copy of Jane's Information Group's Code of Conduct for its editorial teams is available from the publisher.

www.janes.com

Jane's

Intelligence and Insight You Can Trust

Executive Overview

The events of the last year have been one of the most significant for the health and prosperity of the global space industry and portend great things for the prime manufacturing companies, for its contractor teams and for the ancillary industries. Good news too for investors as the privatisation of the industry continues apace. Privatisation too is working for the major national and international service and carrier industries with a veritable plethora of denationalised organisations. Concerns such as Intelsat and Inmarsat, each now fully privatised, are showing their mettle against open competition. Freed from the institutionalised constraints that bound them to their former mandates, these giant conveyors of international traffic have shown mature and competent ability to sustain their original traffic custom and to extend into new and innovative technologies and applications for better market orientation. They also have a more flexible and timely response to customer needs. We can expect to see rapid response reaction to market needs and better proactive display of technology applications.

Control of government programmes and projects in the major space-fairing countries has undergone a major transformation. In China, continued and sustained resolve to challenge the big space powers in human space flight and rocket technology underpins a paced initiative at further human flights toward permanent orbital operations and, eventually, a lunar base. China continues to develop propulsion technologies for advanced launch vehicle systems and for new and more powerful vehicles. It is also investing in the development of hybrid rocket systems and new forms of air-breathing propulsion. Japan has transformed its research and development divisions with the first full year of activity with JAXA, the replacement government body for the former National Space Development Agency and associated science and aerospace engineering research facilities. This should cut costs and enhance efficiency as well as consolidating future project development and programme applications. Japan has yet to see the launch of its experiment module to the International Space Station but is already looking to join the US in its new Space Exploration Initiative.

European Space activities reached a new level of maturity with the success of its first major planetary mission - Mars Express. It continues to provide valuable data and to complement the existing fleet of NASA planetary missions still in operation (Mars Global Surveyor and Mars Odyssey in orbit and Spirit and Opportunity landers on the surface). Vital supplementary detail provided by the unique suite of scientific equipment on board Mars Express enhances the opportunity for further planetary missions and for the future development of planetary spacecraft. In launcher technology Europe is maintaining its hold on commercial satellite orders, with Ariane 5 now secure as a dominant player in global launcher markets. Matched in capability by only the US Delta 4 and Atlas 5 variants, Ariane 5 is proving a perfect match for required payload lifting requirements. Despite early pessimism about the payload lifting capability (said by some to have been compromised by its original design mandate to lift the now cancelled Hermes spaceplane) the programme continues to impress. Satellite development continues to address the changing demands of an applications market. Supplementary to the US Global Positioning System, and its designers hope it will eventually be the successor, the Galileo navigation satellite programme anticipates launches in 2005 and 2006 for a test period prior to operational capability with four satellites in 2008 and full deployment of 30 satellites thereafter.

Europe's Galileo programme is in the vanguard of a new generation of satellite applications shaped to government requirements for supporting new national requirements. Galileo will address not only the traditional market currently filled by the US GPS system, but it will also allow Europe-wide management of traffic systems, road tolling, security operations and the monitoring of individual vehicles on roads and highways. Satellites of this type are filling a unique niche, independent of the traditional sectors. They are now playing a part in the political and administrative management of land, sea and air based traffic operations and will form an integral component in Europe's next political experiment - the Europe-wide control of transportation, traffic management and security measures across national boundaries in a seamless monitoring operation. The European Union has expressed an interest in folding the European Space Agency (ESA) into its own administrative envelope and bringing Europe's space related community under EU management for the first time. ESA, traditionally and historically outside the political and economic union of member EU states, is likely to have its future role transformed by expediency and rationalisation. It is also likely to change ultimately because of the events of 11 September 2001 when the hijacking of four airliners resulted in the death of 3,000 people and changed the way countries view security and border protection. Satellites can greatly assist with security and anti-terrorist operations throughout Europe and the need for a military support role in Europe has never been greater with, at present, only France and Germany, showing strong inclination to expand ESA's mandate into these areas.

Russia continues to apply almost 50 years of space engineering and technology with little real prospect of new development projects. This is due largely to financial constraints and the lack of inertia at a political level. There is greater interest in spending national resources on defence and security related priorities and seemingly little enthusiasm for government

NASA's new Aqua hydrology survey satellite successfully launched by Delta 2 on 4 May 2002 0576699

Parts of the Shuttle Columbia recovered and examined for any clues to why the orbiter broke up during re-entry on 1 February 2003 0576704

A candidate design concept for NASA's proposed Orbital Space Plane currently receiving scrutiny as Congressional interrogators debate the future of human space flight
0576702

money going to new launch vehicle or scientific research programmes. Russia has several new concepts and designs for improved launch vehicles. There are some projects in private development that promise great improvements over the current generation of Russian launchers but there is little prospect of these reaching the customer. The Russian government is, however, consolidating its inventory of military satellites and preserving the existing infrastructure with improvements and upgrades. Russia has a family of launchers capable of supporting requirements for the next several decades. Only if a major breakthrough occurs in cost/kg to orbit by new launcher designs outside Russia will the existing inventory become uneconomic. Russia too has been the mainstay of the human presence aboard the International Space Station since the grounding of NASA's three remaining Shuttle orbiters. Following the tragic accident to Columbia in February 2003 the Space Station has been maintained by a succession of two-man crews ferried back and forth by Russia's able Soyuz spacecraft and kept replenished by its unmanned derivative, the Progress supply tanker. This has placed pressure on the partnership due to Russia's claim for financial compensation, which has not been possible due to a veto by the US State Department over human rights issues and the transfer of funds to the Russian government.

US commitment to space activities and funding of both military and civil programmes continues to show optimistic progress. Spending on military space activities again exceeds that of NASA and the prolific array of new programmes responds to the new expansion of space-related defence applications. Operations by air and ground forces in Afghanistan built confidence in a new generation of weapon and support systems operated

Looking to match the interplanetary aspirations of the historic space-faring superpowers, the European Space Agency Venus Express 0576701

For the time being the International Space Station is managed by a two-person contingency crew with only Russia's Soyuz able to launch crewmembers to and from the pressure modules until the Shuttle is returned to flight 0576700

Folded for launch, NASA's 2003 Mars lander and roving vehicle folded for encapsulation prior to launch 0576706

from or through satellites in space. These developments have resulted in a new requirement for sustained space-based assets in time of conflict. The Pentagon is keen not to pursue a "blinkers-on" approach, believing that while terrorism and conventional warfare dominates the attention of the military at present, it may not always be that way. The need to maintain a vital nuclear deterrent force is crucial and the support infrastructure for that relies heavily on space assets. Accordingly, and in pursuit of long held belief in independent launch insurance, the Pentagon is funding development of a new launch system capable of doubling as a weapon. The Affordable Responsive Spacelift system would have a turnaround time of 24 to 48 hours and lower launch costs to about one-third of their present value for loads up to 4.5 tonnes. The Pentagon wants the system in place by 2018, presently conceived as a two-stage launcher, the first stage reusable and the second stage expendable. At present the Defence Department is developing a series of small satellite launchers capable of lifting 1,000 kg to orbit and these may form the second stage of the partially reusable spacelift vehicle.

NASA is transforming itself from primarily a space station maintenance and cargo truck delivery organisation back into a research and development agency, pushing forward the exploration and utilisation of space. Approximately 25 per cent of NASA's budget will be redirected from the Space Station and the Shuttle into the development of a successor to the Shuttle, the Crew Exploration Vehicle (CEV). The CEV will be funded as Shuttle operations are shut down when full assembly of the initial Space Station configuration is complete and when all the requirements under the partnership agreement have been fulfilled. US participation in the Station will be limited to experiments and activities directly in support of future human space flight operations, where the ambition is to return to the moon. The aim is to support a permanent human presence on the moon and to develop hardware and mission capabilities commensurate with an early establishment of Mars colonies for scientists and explorers. NASA cannot continue to fly the Shuttle beyond about 2010 for the same reason that it cannot be upgraded or modified: the system is simply too vulnerable and the risk of another orbiter loss is too high. NASA has not sought, nor will it support, the acquisition of funds to carry out improvements beyond the essential steps necessary to maintain as safe a vehicle fleet as possible until its remaining obligations are discharged.

NASA's MER Mars spacecraft compared to the first rover, Mars Pathfinder, which landed on the surface in July 1997 0576705

Under the Space Exploration Initiative, NASA opens a path to exploration in a programme remarkably similar to that originally envisaged for Apollo in 1959-60 when a successor to the tiny one-man Mercury spacecraft was conceived at NASA's Langley Research Centre by Max Faget and his team. Named Apollo, the project was hijacked by the White House when President Kennedy gave it a new and accelerated role: the landing of two astronauts on the moon by the end of the 1960s. Apollo had been intended as an earth-orbit operations support vehicle, one that could also be used for circumlunar trips and, with appropriate additional stages and landing gear, put down to the surface of the moon. The specification and mission requirement for NASA's new CEV concept is very close to that original requirement for Apollo 45 years ago. Everything else, between Apollo and the present, has been a digression from the original concept of NASA's exploration mission, framed by political expediency and Cold War challenges. In returning to old visionary concepts NASA has engaged the interest of the European Space Agency, which sees its aspirations too as following the NASA programme. Funding for the new NASA role will be masterminded by the incoming administrator, Dr Michael Griffin. In the first funded year of the new space initiative, Congress voted the fully requested amount, but this should not be taken as a green light for full steam ahead. Congress is going to put up some tough fights for NASA in the years to come. If NASA loses another Shuttle orbiter, the entire blueprint may change.

In the commercial sector the shift from big, government funded, satellites to smaller privately owned systems is pointing a way to the future few would have anticipated a decade ago. A large proportion of defence imagery is now purchased on the open market from suppliers handling data from non-government satellites. This trend is shifting in to defence communications. In turn, defence satellites are being "privatised" to save money and to expand the user base, thereby lowering costs.

Events of the last year have consolidated a robust space industry with solid growth in launch vehicle markets and in satellite manufacturing. A price war on launch prices has lowered profits and the US manufacturers are only surviving due to assured procurement from government agencies. This will temper expansion of fleets but will ultimately benefit the industry by pruning out weak elements. Elsewhere, the return from commercial sales of imagery has not produced the levels forecast a decade ago, but here too, thinning in the market has helped focus commercial success on this with refined product and add-on services tailored to the market. In all sectors, commercialisation is happening not in new but existing space market sectors. This bodes well for the future and a market that shows all indications of a healthy future.

Acknowledgements

As the editor I would like to thank all those people in the space industry worldwide who have helped make this edition of *Jane's Space Directory* a reality. I would also like to thank Simon Michell, my Managing Editor until December 2004, and New Media Publishing Director Sean Howe. Appreciation too for Rebecca Davies and all the production team at *Jane's*, together with essential IT support and back-up from many who work to make this product a success. So many individuals have helped continuously throughout the preparation and production phase. To you all, my heartfelt thank's.

David Baker

Schooled and educated in the UK and the US, Dr David Baker has had a lifelong career in space programmes involving government and industry activities around the world. Active from the early days of the Apollo Programme in which he was involved with future mission planning and in the formative studies for reusable Shuttle and space station concepts, he founded a London based consulting company in the early 1980s supporting the global commercialisation of space. During that time Dr Baker discussed launch vehicle commercialisation in Moscow and Beijing assisting US and European efforts to link Russia and China in an expanding effort to broaden the competition for placing satellites in orbit. He has also worked extensively in India and countries throughout the Far East helping develop space applications in earth-based activities such as agriculture, weather forecasting and education. He has also been involved in planning the military application of space based capabilities and in analysis of existing or potential threats to in-orbit infrastructure and national security. David maintains an academic interest in space education and in management practices applicable to non-space based industries. He has written more than 60 books and has made numerous appearances on TV.

Units, Abbreviations and Acronyms

The metric system of measurement is employed throughout the Directory. Times and dates are quoted in Greenwich Mean Time (GMT) unless otherwise noted. Shuttle missions, in addition to their official STS designations, are also allocated 'SM' numbers to indicate the actual flight sequence. Abbreviations and acronyms are usually explained in the text; the most common are listed below, along with some unit conversions not given elsewhere.

μg:	microgravity
μm:	10^{-6} m (micron)
Å:	Angstrom, 10^{-10} m
ACS:	Attitude Control System
A/D:	Analogue/Digital
AF:	Air Force
AFB:	Air Force Base
AIT:	Assembly, Integration and Testing
AKM:	Apogee Kick Motor
AO:	Announcement of Opportunity
AOCS:	Attitude and Orbit Control System
AS:	Air Station
ASEAN:	Association of South East Asia Nations
ASIC:	Application Specific Integrated Circuit
atm:	Earth atmospheric pressure (1.0336 kg/cm^2; 14.7 lb/in^2)
AU:	Astronomical Unit (149.598 million km) or ESA Accounting Unit (equivalent to the ECU) billion: 1,000 million
BOL/EOL:	Beginning Of Life/End Of Life
C^3:	Command, Control and Communications
C^4:	Command, Control, Communications and Computers
CAD:	Computer-Aided Design
CCT:	Computer-Compatible Tape
CDR:	Critical Design Review
cleanliness:	the Directory adopts the US system specified by the numerical limit of 0.5 particles per ft^3 (0.028 m^3). Thus a class 10,000 environment has fewer than 10,000 particles of 0.5 in 1 ft^3 (0.028 m^3). Most satellite and launcher integration and testing facilities are class 100,000. Class 10 and class 100 facilities are used for assembly of sensitive items such as gyros
cm:	centimetre (1 inch = 2.54 cm)
CNC:	Computer Numerical Control
Conus:	Continental United States
COTS:	Commercial-Off-The-Shelf
CPV:	Common Pressure Vessel
DAMA:	Demand Assigned Multiple Access
DBS:	Direct Broadcast Satellite
DGPS:	Differential GPS
DoF:	Degrees of Freedom
DRAM:	Dynamic Random Access Memory
DTH:	Direct To Home
ECLSS:	Environment Control and Life Support System
ECU:	European Currency Unit
EGSE:	Electrical Ground Support Equipment
EIRP:	Equivalent Isotropically Radiated Power
elint:	electronic intelligence
EM:	Engineering Model
EPC:	Electronic Power Conditioner
ESA:	European Space Agency
eV:	electron Volt
fl:	focal length
FM:	Flight Model
FoV:	Field of View
FSD:	Full Scale Development
FWHM:	Full Width Half Maximum
FY:	Fiscal Year
g:	gravitational acceleration (9.806 m/s)
GDP:	Gross Domestic Product
GEO:	Geostationary Orbit; altitude 35,786 km, 0° inclination, period 23.934 h
GH_2:	Gaseous Hydrogen
GIS:	Geographic Information System
GMDSS:	Global Maritime Distress and Safety System
GMT:	Greenwich Mean Time (=UT, Universal Time)
GN&C:	Guidance, Navigation and Control
GNP:	Gross National Product
GOX:	Gaseous Oxygen
GPS:	Global Positioning System
GSE:	Ground Support Equipment
GSM:	Groupe Spéciale Mobile
GSO:	Geosynchronous Orbit
GTO:	Geostationary Transfer Orbit
HES:	Health, Environment and Safety
HGA:	High Gain Antenna
HOOD:	Hierarchical Object Oriented Design
HPA:	High-Power Amplifier
HRVIR:	High-Resolution Visible Infra-Red
IC:	Integrated Circuit
ICBM:	InterContinental Ballistic Missile (range >5,500 km)
IFOG:	Interferometric Fibre Optic Gyro
IMU:	Inertial Measurement Unit
INS:	Inertial Navigation System
INU:	Inertial Navigation Unit
IOC:	Initial Operational Capability
IRBM:	Intermediate Range Ballistic Missile (range 500-5,500 km)
IRU:	Inertial Reference Unit
ISDN:	Integrated Services Digital Network
km:	kilometre (0.6214 mile)
L:	Launched
LEO:	Low Earth Orbit, typically up to 1,500 km
LEOP:	Launch and Early Orbit Phase
LH_2:	Liquid Hydrogen
LHCP:	Left-Hand Circular Polarisation
LiOH:	Lithium Hydroxide
LITVC:	Liquid Injection Thrust Vector Control
LNA:	Low-Noise Amplifier
LOX:	Liquid Oxygen
m:	metre (39.37 inch)
MAU:	Member Accounting Unit
MEOP:	Maximum Expected Operating Pressure
MGSE:	Mechanical Ground Support Equipment
micron:	10^{-6} m
MIR:	Mid Infra-Red
MMH:	Monomethyl Hydrazine
MMIC:	Monolithic Microwave Integrated Circuit
MON:	Mixed Oxides of Nitrogen
MoU:	Memorandum of Understanding
MPD:	Magnetoplasmadynamic
MW:	Momentum Wheel
N:	Newton (0.225 lbf)
NASA:	National Aeronautics and Space Administration (US)
Ni/Cd:	Nickel Cadmium
Ni/H_2:	Nickel Hydrogen
NiMH:	Nickel Metal Hydride
NIR:	Near Infra-Red
NSSK:	North-South Station-Keeping
NTO, N_2O_4:	Nitrogen Tetroxide
OBDH:	Onboard Data Handling
OEM:	Original Equipment Manufacturer
PAN:	Panchromatic
PDR:	Preliminary Design Review
PI:	Principal Investigator
PKM:	Perigee Kick Motor
PMD:	Propellant Management Device
ppm:	parts per million
PROM:	Programmable Read-Only Memory
QM:	Qualification Model
RAM:	Random Access Memory
RCS:	Reaction Control System
RDT&E:	Research, Development, Test and Evaluation
RF:	Radio Frequency
RHCP:	Right Hand Circular Polarisation
RLG:	Ring Laser Gyro
RLV:	Reusable Launch Vehicle
RW:	Reaction Wheel
SAR:	Synthetic Aperture Radar
SCADA:	Supervisory Control and Data Acquisition
SCOE:	Special Check-Out Equipment
SEP:	Spherical Error Probability
SEU:	Single Event Upset
SGLS:	US DoD's Space Ground Link Subsystem
sigint:	signals intelligence
SL:	Sea Level
SLBM:	Submarine-Launched Ballistic Missile
SNG:	Satellite News Gathering
SOS:	Sapphire On Silicon
SSPA:	Solid-State Power Amplifier
SSR:	Solid-State Recorder
STDN:	NASA's Spacecraft Tracking and Data Network
SW:	Space Wing
SWIR:	Short Wave Infra-Red
t:	metric tonne (1,000 kg; 0.984 ton; 1.102 US ton)
TC:	Telecommand
TCR:	Telemetry, Command and Ranging
TT&C:	Tracking, Telemetry and Command
TVC:	Thrust Vector Control
TVRO:	TV Receive Only
TWTA:	Travelling Wave Tube Amplifier
UTM:	Universal Transverse Mercator grid projection
VHSIC:	Very High-Speed Integrated Circuit
VLBI:	Very Long Baseline Interferometry
VNIR:	Very Near Infra-Red
VTIR:	Visible/Thermal Infra-Red
W:	Watt
WWW:	World Wide Web (Internet)
XS:	multispectral

GOVERNMENT AND NON-GOVERNMENT SPACE PROGRAMMES

GOVERNMENT AND NON-GOVERNMENT SPACE PROGRAMMES

Argentina

Current Status
Despite stringent financial controls the Argentine government continues to maintain interest in space project applications. Nahuel 1 was launched by Ariane 44L on 30 January 1997. Negotiations continue for Nahuel 2, a geostationary communications satellite scheduled for launch after 2005. The SAC-C remote sensing satellite was launched by Delta 7320 from Vandenburg Air Force Base on 21 November 2000. Development of the SAOCOM-A/B-1 vehicle is in development while MiniSat 1.1 is under way as a co-operative project with Spain. SAC-D is delayed while SABIA, a co-operative venture with Brazil, is expected to reach orbit in 2004. SAC-D scientists held their first co-ordination meeting in March 2003.

Background
Argentina's modest space activities emphasise applications and space science, although now the country has broadened the charter to consider developing an indigenous orbital launcher. Until May of 1991 the air force executed overall authority for all space research projects. Then the Argentine government transferred the CNIE National Commission for Space Research into the civil CONAE National Commission for Space Activities, under direct supervision of the country's President. CONAE's strategic programme covering 1995–2006 was approved by President Menem 28 November 1994, projecting expenditure of US$701 million 1995–2006. This figure still reflects significant contributions from unspecified third parties. It is reviewed and extended every two years so that the horizon is always a decade away.

The current plan states these objectives:

Establish a ground station for access to data from international and domestic remote sensing satellites.

Continue the SAC satellite series for remote sensing, carrying national instruments and science, of which SAC-C was launched in November 2000.

Develop the SAOCOM Satellites for Observation and Communications.

Reactivate the old Condor 2 missile facility at Falda del Carmen as the Teófilo Tabanera Space Center, adding laboratories for integrating/testing satellites and payloads. The Mario Gulich Institute for Advanced Space Studies will be located at the centre providing courses tailored to satisfying the needs of CONAE and the space community.

CONAE and its predecessor have worked on the **SAC** Satelite de Aplicaciones Cientificas (Scientific Applications Satellite) project since 1981. Two further SAC missions are planned with NASA, the SAC-C Earth observation (multispectral medium resolution) satellite having been launched in November 2000.

An agreement was signed with Italy in October 1992 to begin discussions on space co-operation, emphasising communications and Earth observation.

Intelsat and Inmarsat are accessed primarily through the Balcarce station near Buenos Aires. CNT signed a 24-year concession in May 1993 with a consortium headed by DASA for the Nahuel S American satellite communications system.

Impsat was formed in 1988 to develop and operate a regional satellite-based network and is now the leading provider of satellite communications in Latin America, operating 2,500 VSATs by 1995. A US$30 million contract was signed February 1995 with Hughes Network Systems for a further 2,000 VSATs and five shared hubs. The Lusat packet radio satellite was provided for Amsat-Argentina and launched by Ariane in January 1990 as one of four Microsats.

Argentina is negotiating on the Nahuelsat SA commercial geostationary communications satellite for launch on a date yet to be determined.

Reports emerged during 1989 suggesting that Argentina was co-operating with Iraq and Egypt (Egypt actually withdrew in 1988) on the development of a 900 km range Condor 2 missile that could also form the basis of a space launcher. The single stage Condor 1, capable of lofting a 400 kg payload to 70 km, began development 1977–78 as a sounding rocket with assistance from Germany's MBB. A parallel design entered service 1990 as the Alacran missile, capable of delivering 500 kg over 300 km. The military government approved Condor 2 in late 1982 after the Falklands defeat as a two-stage solid-propellant missile capable of hitting the islands with a 500 kg payload.

Eventually, the air force built a US$400 million installation under a mountain at Falda del Carmen in Cordoba province in accordance with plans prepared by MBB. President Menem announced the project's deactivation in April 1990 as part of defence budget cuts, but apparently all material was handed over to CONAE and the technology was then transferred to Spain's Capricornio launcher programme.

In July 1995, the Asociacion Argentina de Tecnologia Espacial, AATE, obtained a reservation to fly a Getaway Special payload aboard a United States shuttle flight. The formal agreement between NASA and AATE was signed 6 August 1996. The payload called Proyecto PADE (Paquete Argentino de Experimentos) comprises seven educational experiments, designed and manufactured by Argentine universities and scientific institutions.

VERIFIED

Australia

Current Status
The Optus C1 (Optus and Defence C1) was launched by Ariane 5 on 11 June 2003. It was designed to operate in Ku-band for commercial services for Singtel Optus and in UHF, X- and Ka-bands for the Australian Department of Defence. Singtel Optus has overall prime contractor role to the Australian Department of Defence for the military communications payload and the satellite was built by Mitsubishi Electric. The spacecraft bus is a Space Systems/Loral FS1300 while Raytheon was appointed military UHF payload subsystem contractor. In 2004 the Aries (Australian Resource Information and Environment Satellite) 1 remote sensing satellite is expected to be launched with data from 7-day revisit cycles provided by CSIRO.

Background
Australia originally ran its space activities through the NSP, which was established in 1985 to provide a base for the country's participation in the world space industry. Shortly thereafter, the government formed the Australian Space Office in 1987 as the focus for space-related activities.

The Federal government established the Australian Space Council in 1993 as a result of the 1992 review of the National Space Programme by the Curtis committee. The Curtis report recommended an integrated NSP with designated programme funding and the new council. A major task was the formulation of a five-year rolling plan to set new directions for the NSP and determine priorities. Bureaucratically the ASC reported to the Federal Minister for Science & Technology within the Industry, Science & Tourism portfolio. Then, as part of the 1996/1997 Science and Technology budget announced by the Minister for Science, Peter McGauran MP, on 20 August 1996 the Commonwealth Scientific and Industrial Research Organisation (CSIRO), Australia's largest R&D body, got new authority to undertake space projects on behalf of the Australian government. The former Australian Space Office and Australian Space Council were terminated.

The new space plan, the first controlled by the new government, elected 2 March 1996, made a commitment to a viable space research programme and fostering applications. To carry this out ASC recommended the government place priority on the development and demonstration of light launch capabilities, funding the launch of two rockets from Australia within development agreements with private interests with confirmed capital. The launches would create the centrepiece for building other Plan elements. They could be the launch vehicles for carrying Australian-built demonstration satellites, leading to the development of Australian Earth observation and communications payloads. Australia seeks to secure a significant share of Asian space business in launches, small satellites and space-based services, beginning with launches first from Woomera then Darwin and Cape York.

Studies have been undertaken under the Earth observation programme on the need for Australia to develop an Earth observation data network, the development of value-added services and the development of training programmes. Other studies may be undertaken on the development of sensors such as SAR and an imaging spectrometer.

ASO operated with a budget of A$3.2 million 1987–88, A$5.4 million 1988–89, A$4.4 million 1989–90, A$6.299 million 1990–91, A$6.179 million 1991–92, A$6.989 million 1992–93, A$7.653 million 1993–94, A$11.3 million 1994–95 and A$9.0 million 1995–96 (fiscal year begins 1 July).

Foremost among the new space projects the CSIRO plans to introduce are programmes to place one or more micro-satellites into orbit in late 2001. Known as Aries, the satellites will return to image the same ground track every seven days. CSIRO is heading this project with universities, industry and other parties. A proposal to establish a Co-operative Research Centre to conduct this programme is being considered by the Australian government.

A range of other organisations undertakes space-related science and applications programmes, including the Australian Centre for Remote Sensing (ACRES), Commonwealth Industrial & Scientific Organisation (CSIRO), and the Bureau of Meteorology; commercial satellite communications are handled by Telstra and Optus.

Although space activities are necessarily modest, a population of 18 million dispersed over a 7.68 million km^2 land mass, coupled with a dependence on natural resources and agriculture, provides the impetus for development of space applications. Remote sensing plays a significant role in national space activities. Landsat, Spot, NOAA, ERS and JERS data are received by ACRES and the Bureau of Meteorology operates meteorological satellite reception sites at Melbourne, Perth, Darwin and Casey (Antarctica). Some 70 per cent of Australia's overseas telecommunications traffic is carried by Telstra through Intelsat satellites, and Optus's own spacecraft carry a major part of domestic services, including routes to Papua New Guinea, New Zealand and deeper into the southwest Pacific.

Two satellites were launched by Shuttle in 1985, followed by a third in 1987, to improve domestic communications and supply radio/TV to 16 million people over sparsely populated areas. The largest single application is the Homestead and Community Broadcasting Satellite Service (HACBSS), provided by the Australian Broadcasting Corp. By January 1986, 650,000 Australians in remote areas were able to receive radio/TV for the first time. The second of two new-generation satellites was launched by China on a CZ-2E from Xichang in August 1994, capable of carrying mobile communications services. L-band satellite phone services were inaugurated 31 August 1994. Before then, only 5 per cent of the country had phone coverage (80 per cent of the population), the remainder relied on UHF radio. A third-generation series, Optus C, began with the launch of C1 in 2003 and will continue with the C2 at a later date. The Optus C series satellite bus is being assembled by Space Systems Loral with the communications payload provided by Mitsubishi Electric Corporation.

The Australian Fiducial Network of GPS Reference Stations was established in 1994. There are 14 sites, including a site at Cocos Island (South Pacific), Macquarie Island, Wellington (New Zealand) (South Pacific) and three in Antarctica plus the mainland sites, for satellite integrity monitoring, scientific studies and to provide the precise framework for surveying and mapping. A Norfolk Island (South Pacific) site is imminent. The network also broadcasts error corrections to provide high-accuracy navigation for mobile users.

Space tracking activities are based at the Canberra Deep Space Communication Complex at Tidbinbilla. Tidbinbilla was commissioned in 1965 and employs 150 staff. CDSCC routinely supports about 25 missions, including Voyager and Galileo. The 64 m antenna of the main Deep Space Station, DSS 43, was enlarged to 70 m in 1989 (and is active 80 per cent of the time); the new 34 m DSS 34 became operational in 1996. The two other major stations installed in the Australian Capital Territory during the 1960s, Honeysuckle Creek and Orroral Valley, were closed and their residual functions transferred to CDSCC. Orroral's antenna was donated to the University of Tasmania for radio astronomy, and Honeysuckle Creek's DSS 45 was re-erected at CDSCC to track Earth-orbiting satellites. Stations are also operated for the US agency at Yathagarra in western Australia (Mobile Laser Ranging Facility) and for ESA at Gnangara near Perth. Radio astronomy facilities include the 64 m Parkes telescope and the Australia Telescope at Culgoora, completed in 1988. Parkes was upgraded for linking directly with CDSCC to support Galileo's Jupiter operations in 1996.

A meteorological satellite command/data acquisition station is operated at Perth. Earth station sites are also provided for US military satellites, including the DSP early warning facility at Nurrangar, the signals intelligence satellite station at Pine Gap (near Alice Springs), and the Harold E Holt Naval Communications Station at North West Cape, western Australia, working through DSCS satellites. Harold E Holt remained a Joint Facility until it transferred to Australian control in May 1999, although the US Navy continues to have access. DSCS is used by other military installations, mainly the Australian Defence Signals Directorate in Melbourne. One Ku-band and one L-band transponder are also leased aboard Optus for military communications. DSP left Nurrangar in 2000 as the US consolidates all data reception/processing in a single domestic site. An Australian imaging surveillance system is projected for 2004, possibly using commercial imagery or a purchased dedicated satellite, to cover Australia and the surrounding seas.

UPDATED

Austria

Current Status
Austria elected to participate in the European programme for Life and Physical Sciences and Applications utilising the International Space Station (ELIPS) and Aurora (Space Exploration Programme). Austrian Space Agency is also a member of the Galileo satellite navigation programme. Austria has sought a stronger participation in this endeavour but has little industrial infrastructure to substantiate a beneficial return and so has implemented a research orientated participation through its universities and academic institutions. These research results have provided an important overview of national capabilities and in addition to assisting, define the ways Austria can play a more active role in Galileo, this is also helping restructure the manner in which the country's national capabilities can best be applied to other ESA activities.

Background
Austria began participating in the European Space Agency's activities in 1975 and on 1 January 1987 became a full member. ESA accounts for almost all of the national space expenditure. The 2000 contribution is €31.2 million, 1.4 per cent of total member contributions compared with €30.3 million (1.3 per cent) for 1999. The 1999 contribution is €30.3 million, 1.3 per cent of members' contributions. Former years' contributions were: 1996 €31.4 million; 1995 30.6; 1994 26.7; 1993 25.9; 1992 25.4; 1991 18.8; 1990 16.2; 1989 12.5. It had contributed a total of Sch3.340 million to the agency by September 1996, receiving Sch2.562 million in contracts in return. Its contribution to ESA's optional activities is frozen 1994-97 at Sch250 million annually. Austria's principal interests within ESA are Artemis, DRS (1.5 per cent), ERS 1/2 (0.73/1 per cent), Ariane 5 (0.4 per cent), Envisat (1.0 per cent), Meteosat Second Generation (2.3 per cent), EOPP (2-17 per cent) and FESTIP (2.31 per cent). The Space Research Institute contributed two experiments to the four-satellite Cluster project.

Austria solidified its growing programme by the signing in November 1988 of the protocol for the 'AustroMir' mission. The neo-commercial agreement with Glavkosmos called for a payment of Sch85 million (about US$6 million) plus research equipment provided for the mission. Total cost to Austria was Sch200 million. Joanneum Research in Graz provided management, after the selection of 14 experiments (10 biomedical, three technological and one remote sensing using Mir's MKF-6MA multispectral camera). The science research required 42 h of in-orbit time.

Austrian Franz Viehböck flew the successful Soyuz-TM 13 mission 2–10 October 1991.

Previous Soviet collaboration includes magnetometers carried by the Venera 13/14 and Vega 1/2 Venus/Halley's Comet probes, and similarly modest contributions to the international Phobos and Interball projects. Two ion emitter modules from Joanneum Research and the Research Centre Siebersdorf are flying on Japan's Geotail to cancel electrostatic charging; the design was tested during AustroMir. Interball, ESA's Cluster and Germany's Equator-S carry further modules. Manufacturing and testing support have been provided to Interball's two Magion Czech sub-satellites since 1993. France's Mars balloon includes an Austrian magnetometer, should Russian funding permit the flight.

Austria participates in Intelsat (holding a 0.41 per cent investment share in 1996) and Eutelsat (0.61 per cent 1996). In January 1994 it became EUMETSAT's 17th member through its National Meteorological Service (Zentralanstalt für Meteorologie und Geodynamik).

UPDATED

Belgium

Current Status
Organised under the auspices of the Federal Office of Scientific, Technical and Cultural Affairs, the OSTC controls space activities in several discrete disciplines including space aeronomy. An effort to expand aeronomy activity and integrate with other national science objectives is underway.

Background
Belgium is a federal state comprising the Flanders, Walloon and Brussels economic regions and the Flemish, French and German cultural communities. Space activities at the national level are funded and managed by the federal Ministry of Science Policy through the SSTC Services Fédéraux des Affaires Scientifiques, Techniques et Culturelles (Office for Science, Technology and Cultural Affairs). The three major regions participate separately. For example, the Walloon region contributed to France's Ramses electrophoresis experiment flown on Shuttle in 1994.

National space activities are almost all centred on the European Space Agency, to which it contributes more than 90 per cent of its space budget. The 2000 contributions is €110.5 million, or 5.1 per cent of total member contributions, compared with a 1999 contribution of €122.1 million. Space expenditure totals about 0.08 per cent of GNP. Belgium is the fifth largest contributor, after France, Germany, Italy and UK, to ESA overall (5.3 per cent of Member State income for 1999) and about the fourth largest contributor to major programmes: 2.9 per cent science 1996-2000, 6 per cent Ariane 5 (0.2 per cent Ariane 5E upgrade), 5 per cent Columbus, 4.0 per cent DRS, 4.5 per cent Envisat.

One of ESA's major ground stations is located at Redu in Belgium, providing TT&C facilities for Eutelsat's satellites. Redu has been operated and maintained by CISET since 1982 and its tasks include launch and early orbit phase, maintenance and control of satellites in orbit and during the pre-operational test phase, and command and control functions during operations. Belgium is leading the proposal for the IRIS (Intercontinental Retrieval of Information via Satellite) satellite system for commercial messaging and localisation. The LLMS Little LEO Messaging System development project is funded within ESA's communications programme. SAIT Systems is system designer and plans to establish a subsidiary for operating IRIS; Alcatel Bell Telephone is building the payload. The first launch was of an integral package on Russia's Resurs-O1 in 1997.

There is increasing emphasis on remote sensing observations and applications. Belgium is a 4 per cent shareholder in France's Spot remote sensing satellite system and will contribute 4 per cent of the cost for Spot 5 and another 10 per cent to the development of the 'Vegetation' instruments for Spot 4/5. Belgium is a candidate for Vegetation's data processing centre. Mir flies the major MIRAS spectrometer.

Belgium also provides 4 per cent to ESA's Metop costs and 2.7 per cent to the EUMETSATs, 3 per cent to ESA's Meteosat Second Generation costs and 2.7 per cent to the Meteosat Transition Programme.

The first Belgian in space was Dr Dirk Frimout, flying as a Payload Specialist aboard Shuttle STS-45 in March 1992. Physician Marianne Merchez was selected in May 1992 by ESA as one of six new astronauts. Belgium is participating in the International Space Station through its ESA membership.

Belgacom manages national participation in Intelsat, Inmarsat and Eutelsat. Belgacom operates the V-STAR 1.2 to 2.4 m VSAT service, operating from hubs in Liedekerke via Eutelsat for Europe and in Lessive via Intelsat for Africa, the Middle East and North and South America.

Belgian Space Expenditure (BFr million)

Year	ESA	National	Total
1989	3,453.0	368.0	3,821.0
1990	3,928.0	347.6	4,275.6
1991	4,111.7	380.0	4,491.7
1992	4,959.0	408.6	5,367.6
1993	5,750.4	430.3	6,180.7
1994	5,906.2	456.8	6,363.0
1995	6,107.5	396.9	6,783.5
1996	6,180.8	383.0	5,587.7
1997	5,319.6	388.0	5,707.6
1998	4,991.4	1,160.0	6,151.4
1999	5,942.4	1,170.0	6,212.3
2000	5,647.6	1,210.2	5,647.6

UPDATED

Brazil

Current Status
Brazil has continued to expand its communications coverage with the launch of Brasilsat B4 in August 2000 and followed this with the launch of Loral Skynet Telstar 14/Estrela do Sul 1 by Delta 4 on 11 January 2004. This satellite is manufactured by Space Systems Loral and, until it was launched, services to Brazil and surrounding areas were handled by Brazil 1 (T) launched in 1985 and moved from 106.8° W to 63.0° W in August 2000. Loral Skynet do Sul 1 introduced Ku-band to Brazil and carries 51 transponders. Brazil has concluded several co-operative and joint venture initiatives with financially strapped Argentina and has sought partners for several projects subsequently put on hold. On 22 August 2003 the VLS-1 launch vehicle exploded on its launch pad killing 21 people on the ground. This is the third failure in three attempts at development of the VLS launch vehicle. The ignition system is suspected as the cause of the accident.

Background
As a developing country with a land area of 8.512 million km^2, Brazil's space activities emphasise the development of remote sensing and communications programmes in conjunction with a move towards an independent launch capability. Brazil desires an indigenous domestic and commercial satellite infrastructure after 2002. The federal government passed the law February 1994 creating the Brazilian Space Agency (AEB) as the successor to the inter-ministerial COBAE Commission for Space Activities. The civil AEB, reporting directly to the President, is now in charge of co-ordinating and planning space programmes. The existing governmental space organisations remain linked to their ministries, but they now function as the operational arms of AEB.

On 8 December 1994, the government approved the revision of the PNDAE National Policy for the Development of Space Activities proposed by the AEB's General Council, setting the main objectives for Brazil's space activities. It promotes the development of technologies and capabilities relevant to Brazil's problems, and the preparation of industry to become competitive in the space goods and services markets. The Brazilian government is particularly interested in space surveillance of the Amazon, patrolling the land borders and the Atlantic coastal area, cataloguing and monitoring natural resources, planning and inspecting land use, forecasting agricultural harvests, weather and climate.

The ambitious MECB (Missão Espacial Completa Brasileira) 'All-Brazilian Space Programme' began with government approval in 1978. It began in earnest in 1980 to design, build and launch four satellites, two 115 kg SCD (Satelites de Coleta de Dados) satellites to relay environmental information from ground data collection platforms, and two 170 kg SSR (Satelites de Sensoriamento Remoto) to provide 200 m resolution Earth imagery from CCD cameras. A third SCD was added in February 1994 and a fourth in 1995.

MECB's charter required the development of the VLS orbital launcher, similar in performance to India's ASLV (first launched 1987), and associated launch sites and tracking/data network. Commercial US Pegasus launched the first data relay satellite February 1993. The first imaging craft was planned for 2001 on VLS, followed by a second satellite in 2002, but problems with the launcher delayed these flights. Total MECB cost was estimated at more than US$1 billion at 1988 rates, divided between US$280 million for the satellites and more than US$700 million for the launcher which is now under test.

In February 1988, the National Institute for Space Research, the national agency responsible for applications satellites, committed US$45 million to the joint development of the China/Brazil Earth Resources Satellite (CBERS), carrying sensors to return 20 to 260 m resolution imagery. Brazil will pay US$15 million for two CZ-4 launches, according to the contract signed November 1993, in return for which China guarantees a 100 per cent return on the commercialisation of Brazilian high-tech products in China, such as aerospace items, through a trade agreement. CBERS-1 was launched on 14 October 1999 and Brazil will build, integrate and test CBERS 2 for launch in 2001.

INPE's 1996 CBERS budget is US$17 million; 1995 24; 1994 10; 1993 21. Brazil will provide the HSB humidity sensor for NASA's EOS PM 1 satellite. CNES and AEB agreed in May 1996 to launch a joint 80 kg science satellite in 1998.

Technology transfer constraints on European and US companies helped to slow launcher development. Arianespace's successful bid for Brasilsat second generation launch services during 1989 was reported to include proposals for transferring satellite thruster, gyros, other satellite technology and Ariane Viking rocket motor technology. However, the Missile Technology Control Regime signed by the G7 group of the world's seven most industrialised nations prevented technology transfer. As a result, VLS underwent significant redesign. Brazil became part of the MTCR in October 1995, agreeing not to develop missiles capable of delivering 500 kg warheads over 300 km, or exporting technology to others. The first orbital launch attempt of VLS-1 failed on 2 November 1997 when the SCD2A satellite was destroyed.

INPE is responsible for the satellites and their ground system, projecting a cost of US$295 million at 1988 rates. US$190 million had been expended to end-1995 (US$150 million by December 1991 when SCD 1 was completed; US$3 million in 1994 plus US$5 million in 1995). The Satellite Control Centre was inaugurated September 1989 at INPE's São José dos Campos site. Cuiaba is the primary reception station for the data collection and remote sensing satellites, while INPE's Cachoeira Paulista site 80 km northeast of São Paulo handles the environmental data processing and dissemination.

UPDATED

Bulgaria

Current Status

A government policy statement issued in early 2000 indicated an intention to expand space sciences activity and to increase co-operation with ESA and the Russian Space Agency. Problems with financial restrictions have prevented further involvement sought by Bulgaria. Nevertheless, further co-operation has been established with Romania, Russia and the National Aerospace Agency of the Republic of Kazakhstan. Bulgaria has also offered to act as broker for non-European countries seeking active participation in the space science programmes of ESA and Russia. Several potentially lucrative projects have been developed and Russia is known to be interested in obtaining services and joint projects with Bulgarian space assets and science capabilities.

Background

The Bulgarian Aerospace Agency (BASA) was formed in December 1993 as the national body for the development and application of aerospace research and technology. Bulgaria was one of the most active Interkosmos members, and the only one to have undertaken two manned visiting missions to USSR space stations. The Space Research Institute of the Academy of Sciences is a BASA member and acts as the national centre for space science activities. Twelve ionospheric, geodetic and upper atmosphere experiments were contributed to Interkosmos 22 (Bulgaria 1300), the Fregat TV/spectrometer was provided for Phobos, and scientists are participating in the Aktivny, Apex, Interball, Spektrum-X/γ and Mars-96 projects.

The first space station mission was aborted in April 1979 when Soyuz 33, carrying researcher Georgi Ivanov (real surname was Kakalov, a name which was deemed inappropriate for a cosmonaut), suffered a main engine failure before it could dock with Salyut 7. Ivanov's back-up, Alexander Alexandrov, flew to Mir aboard Soyuz-TM 5 in June 1988 for the Shipka 47-experiment research programme. The CCD-based Rozhen telescope and camera system was demonstrated successfully and left in position for future crews; a version of it was to fly as Spektrum-X's star tracker (subsequently cancelled) and it is offered for other collaborative projects. SRI also provided the Spectrum-256 imaging spectrometer, the Zora equipment for cosmonaut psycho/physiological monitoring and the Svet greenhouse, still operational aboard Mir. The new-generation Svet 2 is being developed, and Zora's Neurolab-B medical system successor will be launched to Mir in 1996. SRI signed a distribution agreement with Spot Image in April 1993.

Bulgaria is a member of the Intersputnik, Inmarsat and Eutelsat (joining 1995) satellite communications organisations and joined INTELSAT on 15 May 1996.

UPDATED

Canada

Current Status

Developed by the Canadian Space Agency, Dynacon Enterprises Ltd, the University of Toronto and the University of British Columbia, Canada's first microsatellite and its first space telescope was launched in 2003. Called MOST (Microvariability and Oscillations of STars) it was assigned launch by Rockot. Also launched in 2003 is the Atmospheric Chemistry Experiment abroad SCISAT-1 launched by NASA. It is Canada's first science satellite since 1971.

Following the loss of the Shuttle Columbia on 1 February 2003, NASA's STS-115 mission carrying Canadian astronaut Steve MacLean has been postponed from its planned launch date of 23 May 2003.

Background

Canada's space programme is governed by the Long Term Space Plan. A new LTSP was released 3 June 1994 covering 1995–2004, adding (after a 12 per cent cut in 1995) C$670 million to the C$1.7 billion remaining from the previous plan and focusing on Earth observations, Space Station, science and communications. Many feared that election of a new government in October 1993 and a large national deficit would lead to a reduction in Canada's Space Station commitment and loss of manned flights. But the Mobile Servicing System will be completed and there are sufficient funds for one astronaut a year on Shuttle.

Manned funding is C$498 million 1995 to 2004. The C$637 million for Earth observation is the single largest element, including C$240 million for a second Radarsat scheduled for launch in 2005. The C$271 million for space science includes new 250 kg SciSat satellites, a co-operative venture with the US. The C$334 million for communications includes C$160 million for an advanced satellite to demonstrate Ka-band high-rate links for fixed and mobile terminals. The Communications Research Centre is looking for a private partner to at least equal that investment. Adding in C$271 million for technology, C$236 million for infrastructure and a C$124 million reserve brings the total for 1995–2004 to C$2,371 million.

The Canadian Space Agency oversees the Space Station programme, space science, space technology development, the astronaut programme and Radarsat. It is also responsible for overseeing national space policy.

Industry sales, 70 per cent of them export, topped C$353 million in 1990, about the same in 1991. Sales of goods and services in 1995 reached C$1 billion, while employing more than 5,000. The Space Station represents Canada's largest single expenditure, with the Mobile Servicing System accounting for most. The two primary satellite programs initiated under the previous LTSP are Radarsat 1 (remote sensing) and MSAT (mobile communications). Canada signed a 10-year renewal of an ESA agreement as an associate member in May 1989, contributing about 4 per cent (C$6 million) of the general ESA budget and an additional C$7.4 million for specific European programmes such as ERS. Since 1991, it has contributed a half share on the GNP scale employed to calculate the contributions of full member states. The latest accord, signed 21 March 1991, called for C$130 million over the following five years, with specific interest in ERS 2, Envisat, the ASTP Advanced Systems & Technology Programme and the Artemis Data Relay & Technology Mission. 1996's contribution was 0.7 per cent of the total from member states, or C$32.8 million (1995 29.4; 1994 25.2; 1993 30.6). In 2000, Canada contributed €14.8 million, or 0.7 per cent of total ESA member contributions.

SovCan Star Satellite Communications Inc was formed in 1990, 50 per cent owned by Canada's COM DEV, Spar Aerospace, Canadian Satellite Communications and General Discovery, and 50 per cent by Russia's NPO PM and Intersputnik. The goal is to build five GEO telecommunications satellites, with NPO PM providing the launcher and bus and Canada the payload.

Canada was one of the four founding members of the Cospas-Sarsat satellite-based search and rescue system in 1979. Teleglobe Canada represents it in Intelsat and Inmarsat.

Telesat Canada, jointly owned by the government and telephone/telegraph companies (the government's 53.7 per cent interest was sold to the Alouette Telecommunications Inc consortium for C$154.8 million in March 1992), became the world's first operators of domestic satellites with the launch of Anik A1 in 1972. With Anik C2 launched by Shuttle STS-7 in 1983, Telesat Canada pioneered five-channel TV direct to US subscribers' homes in November 1983. TMI Communications began operating the MSAT system following April 1996's launch. Canada is a prime mover for development of land mobile systems because of its geography.

Years of close collaboration with NASA, and the development at a cost of C$100 million of the Remote Manipulator System for the Space Shuttle, made Canada the largest national contributor to the Space Transportation System outside the USA. In return, in April 1980 NASA ordered three more RMS sets for the next three Orbiters at a cost of US$63 million.

Canada was an observer member of ESA from 1975, and in 1978 signed a co-operation agreement under which it would pay 1 per cent of the net fixed common costs of ESA's general budget. This agreement was renewed in 1988 for a further 10 years, with Canada's contribution to ESA's general budget growing to about 4 per cent in 1990. During 1981–85 Canada spent a total of

Headquarters building of the Canadian Space Agency 0572494

Canada's M-Sat telecommunications satellite in assembly and checkout stages 0572495

C$476 million on space activities, including five Anik launches. The hybrid Anik E appeared on Ariane in April and September 1991 and will serve well into the next century. The next generation Anik began with the launch of F1 in November 2000, an HS-702 bus carrying 36 C-band and 48 Ku-band transponders, followed by F2 in mid-2002.

In December 1987, Canada became the first foreign partner to reach agreement with NASA and State Dept officials on the text of agreements covering participation in Space Station. Canada is providing the dextrous manipulators, collectively known as the Mobile Servicing System (MSS), to play a critical role in Space Station assembly, maintenance and servicing. Spar was awarded the prime C$195 million second phase contract in March 1991 (the first phase was worth C$50 million). Canada will have access to 3 per cent of station facilities, which includes instrument space, power, data handling and astronaut servicing time. The MSS includes the Space Station Remote Manipulator System (SSRMS), or Canadarm 2, and its Mobile Remote Servicer Base System (MBS), a mobile platform to support the arm. Canada will also provide the Special Purpose Dextrous Manipulator (SPDM), a dual armed robot that works in conjunction with Canadarm 2. The SSRMS was delivered to the Space Station in April 2001 on the STS-100/6-A mission. The Space Vision System is being developed in co-operation with NASA for aiding astronaut RMS control by providing synthetic visual clues. Canada has interests in materials and life sciences projects within the station, in addition to attached payloads for materials, Earth observations, astronomy and space physics.

Canada was among a number of countries to sign collaborative space agreements with the Soviet Union in the 1987/88 period. After a four-day meeting in Moscow between representatives of the National Research Council of Canada and the Soviet Academy of Sciences, the parties signed a joint protocol. Canada agreed to participate in the international Interball magnetospheric research project, and to undertake joint research in solar/terrestrial physics, high-energy astrophysics, sub-mm/mm astronomy, radio astronomy, materials processing, space biology/medicine. A University of Calgary ultra-violet auroral imager is aboard Interball as part of this agreement, following up versions on Sweden's Viking/Freja satellites. Two experiments, including a radiation detector, flew in September 1989 aboard the Cosmos 2044 biosatellite.

The MIM Microgravity Vibration Isolation Mount 6-DOF magnetic levitator for isolating experiments from onboard disturbances is flying on Mir's Spektr module, joined in 1996 by the NFE Nuclear Fragmentation Experiment for measuring neutron and charged particle-induced events inside Mir. A CSA agreement with the Russian Space Agency on space co-operation was signed 21 May 1993. A joint Space Panel with Japan was established in 1989 to co-ordinate co-operative activities, and established working groups in remote sensing, materials science and space science. A Thermal Plasma Analyser, similar to a Freja instrument, is carried by Japan's 1998 Mars orbiter Planet-B. COM DEV is prime under CSA's C$3.7 million contract.

NASA recognised Canada's contributions to the Shuttle programme by providing three flights for Payload Specialists. The first, Marc Garneau, a naval communications and electronic warfare expert, flew on STS-41G in October 1984. The other two flights, which aimed to test the SVS Space Vision System to increase the usefulness of the RMS, and study the disorientation experienced when waking from sleep in zero-*g*, were scheduled for 1986, but postponed indefinitely after the Challenger accident. The National Research Council had reduced 4,300 applicants to a team of six. In January 1989, NASA announced the selection of Drs Roberta Bondar and Kenneth Money as candidates from Canada's cadre of six astronauts for the single life sciences PS post aboard the first Shuttle International Microgravity Laboratory mission in January 1992. Bondar was selected in January 1990 to fly the mission; Bondar and Money left the corps in 1992. Steve MacLean tested SVS aboard STS-52 in October 1992. Bob Thirsk flew as a PS aboard STS-78's Life & Microgravity Spacelab in June 1996. The sixth member of the original group, Bjarni Tryggvason, flew as PS aboard STS-85's earth observation mission in August 1997. A further four astronauts were selected mid-1992 from 5,330 applicants to cope with Shuttle and Space Station demands: Major Chris Hadfield (AF pilot), AF Captain Michael McKay (electrical/computer engineer), Julie Payette (computer engineer) and Dafydd Rhys Williams (physician). Garneau and Hadfield joined NASA's Shuttle Mission Specialist training group in August 1992; Williams joined 1995's MS group. They are paid by CSA but are part of NASA's corps. Hadfield flew as an MS on the second Mir docking mission, STS-74, in November 1995; space vision experiments were again conducted. Garneau flew on STS-77 in May 1996 and Williams flew aboard STS-90 on a 15-day 21-hour mission in April 1998. Payette was launched aboard STS-96 in May 1999. Garneau flew again on STS-97 in November-December 2000 followed by Chris Hadfield aboard STS-100 in April-May 2001. Supporting the installation of the Canadarm 2 remote robotic arm, Hadfield became the first Canadian to perform an EVA.

Fiscal Year	**1995–96**	**1996–97**	**1997–98**
Earth Observation	58.857	86.117	80.017
Manned Flight	139.494	85.900	40.485
Space Science	32.803	36.330	36.845
Communications	8.778	17.495	33.495
Space Technology	50.688	47.576	42.654
Space Awareness	1.432	1.417	1.014
David Florida	7.996	8.505	10.138
Central Services	22.910	14.896	15.003
Contingency	0	13.000	12.800
Total	**322.958**	**311.236**	**272.451**

All figures are C$ million.
CSA accounts for about 70 per cent of all federal government space expenditure.

Canadian Space Expenditure (C$ millions)

Year	**ESA**	**National**	**Total**
1993	34.0	400.0	434.0
1994	23.3	385.0	408.3
1995	30.6	350.3	381.0
1996	33.8	356.1	390.0
1997	24.1	375.9	400.0
1998	22.6	377.4	400.0
1999	22.0	378.0	400.0
2000	22.0	378.0	400.0

UPDATED

Chile

Current Status

Chilean sources say the government is interested in revitalising former plans for participation in manned space flight, through the NASA-led International Space Station. During 2000, these asperations were tempered by recommendations for emphasis on scientific research with unmanned satellites and remote sensing. In 2001 and 2002 continued dialogue sought ways to expand Chile's involvement with space science and engineering and plans were formulated around telecommunications satellite systems serving the South American continent. In early 2003, the Chilean government again made robust approaches through its relevant science and technology directorates that it should be allowed stronger participation in several international space related endeavours, although these have been tempered by an increasing desire to align space-related science research with earth and atmospheric studies. Participation in the ISS has been less enthusiastically promoted by Chile since the limitations on ISS science activity predicated by the grounding of the Shuttle fleet following the loss of Columbia on 1 February 2003. Chile has focused its limited efforts on space science and space applications and there are signs that the government is prepared to re-examine the resources it is prepared to commit.

Background

Space affairs are conducted under the auspices of the CAE Committee for Space Matters, created in 1980, as part of the Ministry of Defence. The Committee proposes to the President specific space budgets and projects. The Space Division of the Fuerza Aérea de Chile (FACH, Chilean Air Force) launched the 50 kg FASat-Alfa, as a passenger payload in 1995, into polar orbit aboard a Ukrainian/Russian Tsyklon.

A £3 million contract was signed with the UK's Surrey Satellite Technology Ltd in May 1994 for the satellite, ground station and on-the-job training for eight FACH engineers (launch cost was additional). The satellite was built at Surrey by a Chilean team working alongside SSTL staff. The programme includes sending students to gain Master's Degrees on Surrey University's satellite engineering course. Unfortunately, a pyrotechnic problem meant that FASat-Alfa could not be separated from Sich. FACH thus signed a £1.3 million contract with SSTL for the identical FASat-Bravo launched in July 1998. Educational activities using the satellite will help to create space awareness in the country. FACH plans the similar FASat-Beta (built in Chile). Two others will follow to form a constellation providing LEO communications services. Work is also under way on the FASat-Gamma mini-satellite for remote sensing from 2003. The expertise derived from these projects is expected to lead to a GEO satellite around 2010.

The CEE Centro de Estudios Espaciales was established in 1959 to provide TT&C services to NASA from the Peldehue station some 40 km north of Santiago. The station still operates. CEE also designs/builds GOES WEFAX/VISSR stations for the military, but they might also become commercially available. A GPS station is operated as part of the Topex-Poseidon project. It offers a certified Cospas/Sarsat terminal. Chile was the first South American country to establish a Cospas/Sarsat ground station. The country does not possess Landsat or Spot ground stations but Landsat coverage is available via the Brazil, Argentina and Ecuador facilities.

In conjunction with CEE, Chile's Amsat-CE amateur satellite organisation, formed in April 1992, is working on the Cesar 1 satellite at the University of La Frontera in Temuco for amateur digital voice links. The 10 kg Microsat design is being adapted under agreement with Amsat North America and could include a GPS receiver. Uplink will be in L-band and downlink in S-band. FACH supplied SSTL software for Cesar 1.

Chile's signatory agency to Intelsat and Inmarsat is Entel (Empressa Nacional de Telecomunicaciones), with 1.51 per cent and 0.065 per cent interests respectively. The private CTC Compañia de Telefonos contracted with Scientific-Atlanta of the US in 1989 to build a US$29 million domestic telephone and data network, creating the world's largest digital satellite communications network. Scientific-Atlanta installed two 6 m Earth stations, five 7 m regional stations, an 11 m hub at Santiago, 35 4.5 m remote stations and more than 100 1.8 m VSAT stations for interactive data transmission. Chilesat and Satel Comunicaciones SA (a partnership of Entel and a Comsat subsidiary) provide domestic and international Intelsat services from their respective Intelsat A stations at Melipilla and Santiago.

Following a March 1985 agreement between Chile's Director General of Civil Aviation and NASA, a Shuttle contingency landing site was built on Easter Island, situated 3,700 km due west of Chile's Pacific coast and suitable for an aborted Shuttle launch from Vandenberg AFB in California. As a result of this agreement, there was a possibility of a Chilean national flying on the US Shuttle. FACH, in 1993, expressed interest in a US or Russian manned flight as part of an increased involvement in space development; there is no plan at the moment for such a flight.

UPDATED

China

Current Status

China has taken a major step toward a manned space flight capability with the flight test of the Project 921 Shenzhou spacecraft in November 1999. A second Shenzhou spacecraft was launched on 9 January 2001 carrying a monkey, dog and rabbit in a test of life support systems on a seven day flight. Launched on 25 March 2002, Shenzhou 3 was a further qualification test of the spacecraft and launcher followed by Shenzhou 4 on 29 December 2003. The first Chinese manned space flight took place in mid-October 2003 and is part of a long-term programme to expand China's space technology programme. China has already publicised ambitious plans for long duration manned space flight and bases on the moon. In October 2003 the Chinese government announced that it had authorised the formal start to its unmanned lunar satellite programme and that a lunar reconnaissance mission would be launched around 2005. China has stated its interest in developing an earth orbiting laboratory and a manned lunar landing programme. The Chinese government has benefited from the incorporation of Hong Kong and its commercial satellite assets.

Background

China's growing space programme aims in the near-term to operate polar and geostationary meteorological satellites, polar Landsat-type Earth resources satellites, geostationary communications satellites and recoverable Earth resources/microgravity capsules, in addition to marketing its range of CZ/Long March launchers. The most powerful of its launchers provides a capability equivalent to Europe's Ariane 4 and Russia's Proton. Longer term ambitions have been acknowledged to include manned flights and space stations by the middle of next century. Future CZ versions are being studied to provide a LEO capability of at least 20 tonnes within the next decade.

The space agency proposed a manned project within the 1996–2000 five-year plan and this resulted in an unmanned test launch in 1999. The two-man capsule was launched unmanned on a stretched CZ-2E from Jiuquan. Russian Soyuz technology in life support and cosmonaut training was required for this project. The programme now has high level backing. A 10-year shuttle programme was claimed in late 1992 to be under way, with work beginning on a new launch site near Jiuquan.

Studies have been conducted of a Global Mobile Satellite Information System (GMSIS) which would provide personal hand-held communications via 18 to 24 satellites in medium orbits. Three or four of the 500 kg satellites could be delivered by a single CZ-3A. Also under study is the SSRSS Small Satellite Remote Sensing System to use frequent revisits for monitoring environmental changes, particularly natural disasters. Seven satellites in 773 km near-polar orbits would observe a particular site twice daily. Several of the 250 kg satellites could be launched by one CZ-4. China is offering this as a collaborative project for the Asia-Pacific region. The Twin Star proposal would use two satellites some 40° apart in the GEO arc to provide positioning, messaging and timing services, after the fashion of OmniTracs. A lunar orbiter was mentioned in 1995 as possible for about 2002 but this did not materialise and projected dates for launch are deferred.

Estimating China's space expenditure is difficult and figures have only recently begun to appear. An official report in September 1994 claimed that almost ¥15 billion of space goods were produced in 1993, a 35 per cent increase over 1992. Sales increased 32 per cent to ¥15.3 billion (about US$1.7 billion), for a profit of ¥540 million. A civil annual budget of ¥1.4 to 1.5 billion (US$168-180 million) has been claimed for R&D and rocket and satellite production, but excluding launch operations and tracking, which are performed by a separate organisation (see below). By contrast, one US analysis in 1993 estimated US$1.35 billion annual expenditure on all programmes. With the return of Hong Kong to Chinese rule in July 1997, the communications satellite companies in the former-British colony automatically became Chinese.

The Commission of Science, Technology & Industry ultimately directs most civil and military space programmes for national defence. Civil space programmes fall under the aegis of the China Aerospace Corp and the China National Space Administration, the latter being the facet for foreign contact. Both were formed in June 1993 from their dissolved predecessor Ministry of Aerospace Industry, created in 1988 from the merging of the Ministry of Astronautics Industry and the Ministry of Aeronautics Industry. Commercial activities are directed through CASC's China Great Wall Industry Corporation. The Ministry of Posts & Telecommunications is the national signatory to Intelsat, and the Beijing Marine Communications & Navigation Co to Inmarsat. MPT's 'Chinasat' China Telecommunications Broadcast Satellite Co is the state operator of telecommunications satellites.

The principal space and military launcher manufacturing facility is the Beijing Wan Yuan Industry Corporation, founded in 1957 and now called the China Academy of Launch Vehicle Technology (CALT) near the town of Nan Yuan some 15 km south of the capital. Here, the smaller CZ-2 family and cryogenic upper stages are manufactured in a complex of 13 research institutes and seven factories. The 27,000 employees include 9,000 engineers/technicians. Its Shanxi Liquid Rocket Engine Co designs/builds the hypergolic engines. Production of the CZ-3 launchers' lower stages and of the entire CZ-4/2D is undertaken at the Shanghai Bureau of Astronautics (SBA), founded in 1961 and now also known as the Shanghai Academy of Space Technology (SAST), in 17 institutes and 11 factories. It employs 30,000, including 8,000 technical staff. Spacecraft production is the responsibility of the Chinese Academy of Space Technology. Founded in February 1968, its 14 research institutes and factories employ 11,000, including 6,000 technicians. SBA, CAST and CALT all report to CASC. The Shanghai Institute of Satellite Engineering (part of CAST until 1993, when it came under SBA/SAST) builds the FY meteorological satellites with funds and specifications from the State Meteorology Administration and the Shanghai Institute of Technical Physics provides the imaging system. The Xian Institute of Space Radio Technology is similarly providing the 20 m resolution CCD imaging sensors for the CBERS remote sensing satellite. The Institute of Space Medico-Engineering (ISME), founded in April 1968, studies principally the medical, physiological and psychological aspects of manned space flight, including the development of life support systems.

The China Satellite Launch and TT&C General organisation, with more than 20,000 personnel, operates the launch sites and is responsible for all TT&C services through to the conclusion of a mission. It operates under the Commission of Science, Technology & Industry for National Defence. China maintains three orbital lauhch sites, the most recent identified in 1988. Jiuquan in the Gobi desert was the starting point for China's first satellite and has two pads for CZ-1/2 launches into 57 to 70° orbits. Xichang became operational in 1984 for GEO missions with the CZ-3 and a third site at Taiyuan south of Beijing is utilised for Sun-synchronous meteorological and Earth resources launches with the CZ-4.

Launch and satellite TT&C is provided by six fixed, three mobile and three ship ground stations, headquartered at Xian City in Shaanxi Province. Ground stations at Beijing, Guangzhou and Urumqi feed NOAA and FY meteorological satellite data to the National Satellite Meteorological Centre in the capital. A Landsat Earth resources station has been operating since December 1986, provided by Hughes STX of the US, with the dish at Miyun 100 km northeast of Beijing and the processing facilities in northwest Beijing itself. Facilities were upgraded in 1993 to add Landsat 6 ETM, ERS and JERS 1.

China adopted a 'Twelve year Development Plan of Science and Technology' incorporating rocket propulsion in 1958. In 1965, the Academy of Sciences formulated a satellite programme that resulted in the first successful launch in 1970. China was the fifth nation after the USSR, USA, France and Japan to achieve national launcher capability. The ability to recover payloads from orbit was added in November 1975 (FSW capsules are now offered commercially) and the then only cryogenic upper stage outside of the US and European Space Agency carried the first Chinese GEO communications satellite in 1984.

Development of China's long-range military rockets and space launchers is largely credited to Dr Chien Hsue-shen, expelled from the US in August 1955 as a result of Senator Joseph McCarthy's anti-Communist campaigns. Chien had taken a PhD at Toronto University and then worked as a research engineer at the California Institute of Technology until ordered to return to mainland China. He visited the US in November 1972 heading a team of Chinese scientists. Ren Xinmin, chairman of the Ministry of Astronautics' scientific committee in the 1980s, also returned to China after completing his studies abroad and participated in developing plans for the Earth resources, meteorological and communications programmes now under way.

There were rumours of a Chinese launch attempt in November 1969, but these appear to have related to posters appearing anticipating China's first satellite launch rather than actually celebrating a launch attempt. The first Chinese satellite was Dong Fang Hong, launched on 24 April 1970. While the satellite became infamous for its playing of the Chinese anthem 'The East is Red' (after which the satellite was named), examination of the payload suggests that it carried communications transponders similar to the United States Telstar 1 and 2. This raises the possibility that it might have had a limited communications repeater role.

Shi Jian 1 was a science payload (carrying a magnetometer, cosmic ray and solar X-ray detectors) launched a year later using a modified version of the DFH-1 satellite: this time solar cells were carried and thermal control was achieved by using louvres on the satellite. The triple payload which comprised Shi Jian 2, 2A and 2B was a science mission, the latter satellite being a balloon attached to a heavy mass which quickly decayed from orbit. Shi Jian 3 is a designator used for an abandoned early-1980s meteorological satellite programme and was never used for an orbital mission. Shi Jian 4 – along with KF 1 – was a science payload for the maiden flight of the CZ-3A launch vehicle: KF 1 (the acronym has never been explained by the Chinese) appears to have been an active satellite, but its purpose has never been explained. It is believed that a Shi Jian 5 satellite might be being developed, but few details are available: the satellite is said to be intended to prove a design of a planned satellite but which could have other applications.

The series of Ji Shu Shiyan Weixing satellites has never been discussed in Chinese literature, but it is thought that they comprise a series of Elint satellites, which were never followed up as an operational programme. Since Mao Zedong primarily backed the Shanghai space activities, which were responsible for this programme, it is possible that when he died so did the political support for this programme.

The most numerous satellites in the Chinese programme have been the Fanhui Shi Weixing recoverable satellites, with three main series having flown. The FSW-0 had 10 launches, one launch failure and nine recoveries (lifetimes three days initially, then five days). FSW-1 had five launches, no launch failures, four recoveries (lifetimes typically eight days) and FSW-2 had three launches, no launch failures, three recoveries (lifetimes 15–16 days). The series of successful consecutive recoveries starting in 1975 and ending in 1992 is remarkable, with China being the third nation to demonstrate this ability.

The FSW satellites were originally developed for a photoreconnaissance role, but after a break in launches from 1978–1982 the satellites were modified for a dual-role photoreconnaissance and remote sensing role. They have also been used to carry both domestic and foreign commercial microgravity and biomedical experiments. The final FSW-0 satellite, launched in 1987, carried experiments for MATRA Espace, the second FSW-1 (1988) carried COSIMA 1 for INTOSPACE in the (then) Federal Republic of Germany. COSIMA 2 had been planned for an FSW-1 in August 1989, but the flight was cancelled - in part for political reasons as a reaction against the events in Tiananmen Square in June that year: the experiment actually flew on a Russian Resurs-F1 satellite, launched in September 1989. No details appear to be available, but the fourth FSW-1 is said to have carried a small Japanese 710°C semiconductor furnace.

The next group of applications satellites to begin flying was for communications. The first launch in January 1984 failed to place its DFH-2 satellite into geosynchronous orbit due to a malfunction of the new liquid oxygen/liquid hydrogen third stage of the launch vehicle. But only three months later a test satellite was placed into geosynchronous orbit and it continued to operate over 125° E until June 1988 when it was boosted into a new orbit slightly higher than the geosynchronous orbit band. A third satellite was launched in February 1986 and this entered geosynchronous orbit over 103° E where it operated until around September 1990: at this time the satellite might have simply run out of station-keeping propellant and 'died' on station.

In 1988, an improved series of communications satellite was introduced, DFH-2A. The launches in March and December 1988 and February 1990 placed satellites in orbit over 87.5° E, 110.5° E and 98° E respectively: all three satellites were still operating as of April 1997. A launch in December 1991 left the final DFH-2A stranded in a low orbit after the launch vehicle's third stage failed once more.

The Chinese have developed, in collaboration with Germany, the new DFH-3 series of communications satellites. The first launch (Zhongxing 6) in November 1994 was initially successful, but problems with the satellite's attitude control system meant that by the time a geosynchronous drift orbit was reached there was not sufficient propellant left to reach a geosynchronous orbit and perform the necessary station-keeping manoeuvres. As a result the satellite was abandoned.

As a stop-gap measure the Chinese purchased a Hughes HS-276 satellite and launched it in August 1996 as Zhongxing 7: however a third re-ignition failure of the CZ-3 launch vehicle's third stage left the satellite stranded in a low orbit.

The third series of Chinese applications satellites has been used for meteorology and remote sensing. Two Feng Yun-1 satellites have been launched to Sun-synchronous orbit, but each suffered in-orbit problems. The first, launched in 1988, only operated for 39 days. The second, launched in September 1990, suffered a loss of attitude control in February 1991, but this was solved after the Chinese spent 50 days trouble-shooting the problem. Later, radiation problems again meant that control of the satellite was lost.

The FY-2 series of meteorological satellites is intended for launch to geosynchronous orbit. The first was lost in April 1994 when there was an explosion during its propellant loading before being mated with the launch vehicle: a replacement flew in June 1997.

China began offering commercial launches following the successful positioning of its first GEO satellite in 1984. The issue of a *Long March 3 User's Manual* in 1985 began the appearance of detailed information on most of the launcher family. GWIC indicated in 1986 that up to a dozen departures a year could be handled from the two operational sites, but that at least half would be required for domestic purposes. Although none of the early agreements was converted to a firm contract, they did demonstrate China's willingness to provide undertakings, in order to satisfy western concerns over the potential for technology transfer, that customers' spacecraft would be admitted and processed without inspection. Upgraded handling and pad facilities were introduced to meet international standards, in addition to construction of a second Xichang pad for the new CZ-2E. The first firm launch contract was signed in November 1988 with Hughes for Australia's Optus B 1/2 satellites. AsiaSat became the first to depart, aboard a CZ-3 in April 1990 at a cost of US$30 million. These successes and international concern over CZ pricing policies produced an agreement in December 1988 that China would carry no more than nine foreign satellites by end-1994 and at costs compatible with the international launcher market. It has been reported that China has consistently underbid US counterparts by at least 30 per cent, although comparisons are difficult. A new agreement was reached in January 1995, allowing 11 further GEO satellites by 31 December 2001 (in addition to those signed up under the previous accord). The total may be increased if the annual global requirement is at least 20 satellites, or if western vehicles cannot accommodate the market. Prices must be comparable with the international market and certainly within 15 per cent. LEO launches must also be on a cost par, although no percentage limit is given. There has been notable success in signing up commercial payloads, including Intelsat, Globalstar, Mabuhay, APStar and EchoStar, although mission failures have cast doubts on reliability.

China's first 24-transponder DFH-3 three-axis satellite at last appeared in 1994 to help satisfy the country's burgeoning telecommunications demands, although it had to be abandoned in orbit after its leaking attitude control system exhausted the propellant. Its replacement was launched in May 1997. The first successful demonstration communications satellite was placed in GEO during 1984, a second in 1986, followed by three DFH-2A versions in 1988/90. DASA and CASC established the EurasSpace GmbH joint venture in 1994 to help produce China's domestic satellites and to bid for international contracts. A Local User Terminal and Mission Control Centre, supplied by France's CEIS TM under a US$1.5 million contract, became operational in 1995 in Beijing for the Cospas-Sarsat international search/rescue system.

Collaboration with Brazil has developed over recent years: joint development of the US$150 million CBERS (China/Brazil Earth Resources Satellite) was finalised in 1988. The potential exists for construction of a Brazilian CZ launch site near the equator for GEO missions. A similar Indonesian facility, possibly on Biak Island, was considered in 1987.

Discussions began in 1994 with Canada on the Radarsat programme as microwave imaging would be useful over China's persistently cloudy south. Meteorological satellites have proved difficult to establish: the first two FY-1 polar craft had only short lives and the first FY-2 GEO imager was lost in 1994's ground accident. The replacement was launched in June 1997. As a result, China has continued to rely on NOAA and Japanese GMS satellites and its own ground radar. The country's vast expanse and its susceptibility to natural disasters provides an impetus to spaceborne monitoring. China is estimated to lose ¥50 billion annually from natural disasters, a cost expected to increase.

The first commercial flight was AsiaSat 1 in April 1990, the satellite being successfully launched. The major Chinese success was in winning the launch contract for the two AUSSAT-B satellites to be flown for AUSSAT in Australia: these satellites were the latest-generation Hughes HS-601 satellites. By the time that the launches took place the AUSSAT company had become Optus. The first Optus-B launch in August 1992 was successful, but the malfunctions which surrounded the second Optus-B launch in December that year have never been explained although the Chinese state that their launch vehicle performed flawlessly. However, the Australians stuck with the Chinese and the replacement Optus-B 3 satellite was successfully launched in August 1994. The previous month APStar 1 had been successfully launched.

In 1995, the Chinese suffered a major failure when the CZ-2E launch vehicle with APStar 2 (a further HS-601 satellite) exploded only seconds after launch: as with Optus-B 2 the western media blamed the Chinese launch vehicle for the failure. At the end of the year the first two flights of the CZ-2E with the Chinese new third stage EPKM successfully placed commercial payloads into orbit: AsiaSat 2 in November and EchoStar 1 in December.

Just as the Chinese commercial launch programme was recovering, the maiden flight of the CZ-3 in January 1984 failed when the third-stage engine malfunctioned: it was supposed to re-ignite to take the satellite out of a low Earth orbit and into a geosynchronous transfer orbit. The chamber pressure rose to 90 per cent of the normal value, maintained this for 3 seconds and then declined to zero with a resulting loss of thrust. The problem had its roots in the weightless behaviour of the propellants (it was the first time that a Chinese rocket engine was planned to shut down and then re-ignite in orbit) and modifications were made to the next flight vehicle, successfully launched the following April.

Following a string of successful launches the eighth CZ-3 launch vehicle resulted in a failure in December 1991. As with the maiden flight, the problem appeared when the time came to re-ignite the third stage. 58 seconds after re-ignition the chamber pressure began to drop gradually and failed completely 77 seconds later. Post- failure analysis showed that the problem had been due to leakage of the helium pressurant supply.

The eleventh CZ-3 was the next one to suffer a third-stage failure. 261 seconds after the re-start of the engine what was described as an 'abnormal phenomena' appeared in the engine control gas line system and this caused the engine's thrust to drop, resulting in shut-down some 48 seconds early. After a failure of the first orbited launch attempt with the CZ-3B in February 1996 the second succeeded in August 1997.

Turning to the maiden flight of the CZ-2E vehicle, this carried two payloads. The basic launch vehicle worked as planned and the Pakistani Badr 1 was deployed after parking orbit injection. However, problems arose with the new PKM third-stage which was intended to place a mockup of the AUSSAT-B (later renamed Optus-B) satellite into a 24°, 340 to 25,740 km orbit. Although the PKM/AUSSAT-B model assembly reached parking orbit it was never tracked in orbit by US sensors and it seems possible that when the time came for the PKM burn the assembly was wrongly-oriented and the rocket stage brought itself and the payload out of orbit. Chinese literature has stated that the failure was due to an attitude control and orientation problem with the second stage of the CZ-2E, which would suggest that the scenario just presented is a reasonable one.

The next two failures of the CZ-2E have never been explained in public to everyone's satisfaction, with the Chinese and Hughes (who built the satellites) initially blaming each other. These two failures involved Optus-B 2 and APStar 2, both Hughes HS-601 satellites launched on a standard two-stage CZ-2E but with a United States STAR-63F third stage added as the perigee kick motor.

Video coverage shows that about 70 seconds after the launch of Optus-B 2 a cloud of gas could be seen emerging from the payload shroud, although the launch vehicle continued in flight. The CZ-2E successfully launched the STAR-63F/Optus-B assembly into LEO – except that after orbital injection no contact would be made with the satellite. After the launch, when recovering the payload shroud, downrange debris from the outer parts of the satellite were also found and, it was therefore obvious that the satellite had exploded within the payload shroud at the time that the gas cloud was seen shortly after launch. The Chinese themselves have said in their official User's Manual for the CZ-3A/3B/3C launch vehicles that, in the case of the Optus-B 2 launch 'the launch vehicle performed well in all aspects, although the spacecraft mission was not completed due to an explosion in the spacecraft'. Failure analysis indicated that the explosion had been caused by a failure in the payload shroud itself.

Since the mid-1980s it has been known that the Chinese were working on a launch vehicle which would be in a similar class to the United States Saturn-1/1B, although no design details have been officially published. In 1992, a paper entitled 'A Modular Space Transportation System' was presented at the International Astronautical

Federation. It described a new two-stage launch vehicle using liquid oxygen and kerosene in the first stage and liquid oxygen/liquid hydrogen in the second stage. This basic vehicle would be capable of placing 11 tonnes into a 60°, 300 to 500 km orbit.

The two-stage launch vehicle was described as the basis for a new family of launch vehicles, based upon the clustering concept. The baseline vehicle supplemented by two strap-on boosters only slightly modified from the first stage would be capable of putting 22 tonnes into LEO, while a third modification would be capable of orbiting 38 tonnes (see above illustration). The largest variant would have six strap-on boosters clustered and would be capable of placing 70 tonnes into LEO.

To date the Chinese have not used LOX/kerosene on any of their launch vehicles or missiles and it is therefore interesting that they have shown an interest in purchasing Russian engines using these propellants. Originally the Chinese wanted to buy an RD-171 engine (used on the Zenit first stage) for examination but the Russians refused the sale. The Chinese were able to purchase some (the number is not known – it might only be one) RD-120 class engines which are used on the second stage of Zenit. It is known that a modification of the RD-120, the RD-120K will be used on either the strap-on boosters or the core of the new Russian Soyuz-2 vehicle and therefore the engine has been modified for first stage applications. If the 1992 studies have any official basis, then one might speculate that four uprated RD-120K class engines – designed and manufactured in China – could be clustered for the baseline launch vehicle's first stage.

In 1992, the Chinese stated that a new launch complex was being built as part of the Jiuquan launch site and images from photoreconnaissance satellites show that its size indicates that it is to support far larger launch vehicles than those currently operating.

The only reason for developing launch vehicles, which are destined to place payloads of 20 tonnes or more into LEO, is in support of a manned space programme. Such a programme has been in an 'on-off-on' situation since the mid-1970s as the Chinese government placed a higher financial priority in the development of 'practical' satellites which could be used for reconnaissance, remote sensing and communications. At the end of the 1970s photographs of apparent Chinese astronaut trainees appeared in the press and similarly different photographs from later experienced were published in 1988.

The Chinese have a programme designated 'Project 921' for the launch of a manned spacecraft using a Chinese launch vehicle. In March 1995, ITAR-TASS stated that the Chinese were interested in purchasing hardware which could be used in a manned programme. The areas of interest were said to be "systems of emergency rescue and thermal control, as well as other 'joints and units' including a docking system". A complete Soyuz life-support system was said to have been purchased along with a spacesuit. It was stated that the Chinese were planning a manned mission around 2000 to 2002 with an orbital station planned for around 2015.

Visitors to the Russian cosmonaut training centre at Zvezdny Gorodok in July 1996 were told that a group of Chinese astronauts would be arriving in October to undergo basic training for space missions. The number was initially hinted to be up to 10, but in December 1996 only two Chinese trainees were named by the Russians: Wu Tse and Li Tsinlung, said to have been selected for the Chinese national astronaut team in March 1995. There were other Chinese at Zvezdny Gorodok, but it was not clear whether these were also astronaut candidates or simply trainers. While it was thought that a Russian-Chinese manned flight to Mir might be in the planning stage, it would appear that the Chinese simply want their astronauts to go through basic training and then return to China for final training using Chinese hardware for a two-manned flight in 2000/2001. Of course, simply the experiences of using Russian equipment and going through the training process will be of major benefit to the Chinese when they start to train their own astronauts from the initial selection all the way through to orbital flights.

On 19 November 1999, 'Project 921' emerged to public view when a CZ-2F launched the unmanned Shenzhou spacecraft from the Jiuquan launch site. Shenzhou was the name chosen by president Jiang Zemin for China's first manned spacecraft design which closely resembles the Russian Soyuz spacecraft. Shenzhou comprised a forward orbital module, a re-entry capsule and an aft service module. Three more spacecraft of this type were launched by 2003 with the first manned flight having taken place in mid-October 2003.

UPDATED

Czech Republic

Current Status

Prior to its split to become the Czech and Slovak Republics on 1 January 1993, Czechoslovakia had a proud and distinguished record as a major participant in Soviet-led Eastern Bloc space activities. However, after the collapse of communism as the ruling authority, funding difficulties, uncertainty over Interkosmos projects and preliminary discussions with the European Space Agency, all contributed to the lack of focus of future plans. However, the Czech Republic signed a draft co-operative agreement with ESA in November 1995 and is an active participant in ESA's Cluster programme. The Czech Republic maintains a Council for Co-operation with ESA. The Czech Republic is developing a space science research programme compatible with its new funding profiles and with regard to opportunities arising from a broader marketing of skills and capabilities to an international market.

In May 1992, The Ministry of Economy, the national signatory to Inmarsat and Sprava radiokomunikaci Praha became the Intelsat signatory. The Czech Republic are now debating Eutelsat membership and the country is also an Intersputnik member (operating two Earth stations). Czech scientists continue to make significant contributions to international space science missions. Recent involvement includes instrumentation for Phobos and Magion sub-satellites for the Aktivny, APEX and Interball projects. Under the agreement providing for Interkosmos member countries to participate in manned space flights, military pilot Vladimir Remek became the first non-US/Soviet citizen to enter space. As a researcher on the Soyuz 28 mission in 1978 he spent a week aboard Salyut 6.

In addition to the Magion satellites, Czech instruments for Interkosmos have included a laser reflector for geodetic ranging measurements and a micrometeoroid detector. Prognoz was provided with fields and particles instrumentation and Prognoz 10 in 1985 carried a six-channel Czech radiometer. Czechoslovakia also contributed the steerable camera platforms to the Vega probes that intercepted Halley's Comet in 1986. A derived version, the ASP-GM, was provided for mounting instruments on Mir's Kvant 2 module, docked to the space station from December 1989. The development of a micro-accelerometer, possibly for an aeronomy Magion, continues. It was tested successfully aboard Mir in 1992 and returned to Earth. Czech government has discussed the possibility of a natural astronaut flying to the ISS.

Acadamies and observatories in the Czech Republic are active in tracking near-to-Earth asteroids. Supporting grants from US and western European universities have helped intergrate Czech research into a global effort. Czech Republic has organised international seminars for astronomy specialists seeking involvement with NEO research. The Czech government has also expanded its financial support for materials processing experiments in space and is actively exploring opportunities for this research aboard the International Space Station through the ESA partnership.

UPDATED

Denmark

Current Status

In 1996, the Danish Research Advisory Board for Space was established following new Danish legislation on research advisory services. The main task of the Danish Advisory Board for Space is to provide the Minister for Research and Information Technology with independent and qualified advice within the space research area, while improving the possibilities of creating cohesion in Danish space research. As part of its task, the Board follows all developments within the field of space research. Denmark has extended these analyses to include major ESA programmes and special projects which it defines as pertinent to its indigenous capabilities and seeks to incorporate new ways in which it can participate more fully in international ventures out of area.

Denmark has been a member of ESA since the Agency's creation. The country's contribution to ESA's budget for 2000 is about 75 to 80 per cent of the domestic annual space budget and accounts for 1.1 per cent of the total ESA budget, its latest annual contribution being €24.8 million. The national space budget for 1997 was DKr241 million, 1996 – DKr246.8 million, 1995 – DKr231 million, 1994 – DKr253.3 million, 1993 – DKr201.2 million and 1992 DKr230 million. Around DKr20 million is spent on non-ESA activities. The Danish Space Research Institute is under the Ministry of Research and Information Technology. It handles basic space research tasks and experimentation and assists companies and research institutes with new and innovative development projects based on national or international interests.

Denmark participates through the Space Research Institute in Russia's Granat and Spektrum-X astronomical missions. Ørsted is the first national satellite. The country also participates in Intelsat, Inmarsat and Eutelsat through its Post & Telegraph Administration. Has developed a telecommunications experimental technology programme improving pocket switching opportunities.

As part of the Danish Small Satellite Programme, the country finished the Ørsted in 1995 and stored it as the US schedule slipped. It was successfully launched on 23 February 1999. The Romer satellite, another DSSP project, carries x-ray and gamma-ray instruments for astronomical observations in these portions of the spectrum. Detailed definitions began in May 2000. A third DSSP project was submitted by the Danish Meteorological Institute with instruments designed to measure the emission of visual light and x-rays in association with large thunderstorms. A second DSSP project from the Danish Meteorological Institute known as FACE-IT (Faraday Current Meter) is designed to measure the structural components of the magnetsphere.

Denmark Space Expenditure (DKr millions)

Year	ESA	National	Total
1993	190.5	28.0	218.5
1994	173.4	30.0	203.4
1995	187.4	30.0	217.5
1996	173.8	30.0	203.8
1997	170.3	30.0	200.3
1998	176.9	30.0	206.9
1999	186.0	30.0	216.0
2000	186.0	30.0	216.0

UPDATED

Finland

Current Status

The Finnish government has carried out a study to explore the potential advantages of a modest expansion in space funding, following several years of static growth. Some conclusions have been reached that would bring commercial interests to government programmes. Some approaches have been made to private companies to engage interest in co-operative ventures. Significant progress has been made with developing nationwide utilisations of remote servicing data through US and European distribution channels including government and commercial outlets. Increased participation in microgravity research. Expressed interest in astrobiology. The Finnish government has received representation from several ESA countries for co-operative ventures, joint programmes and development projects utilising skills from Finnish companies and government departments. Finland is conducting discussions on the possible application of unique test requirements for spacecraft components applied to ESA projects and is developing joint agreements with non-ESA countries about the possibility of conducting test programmes for commercial clients.

Background

Space activities began in the early 1980s, when Finland put instruments aboard the Soviet Phobos Mars probes, handled by the Geophysics Dept of the Finnish Meteorological Institute. Space activities now come under the principal authority of the Ministry of Trade & Industry (KTM). TEKES acts as the co-ordinator and main funding agency.

The Finnish Space Committee was established by the government in 1985 as a policy advisory body and transferred from the Ministry of Communications to KTM in 1989. The Ministry of Education supports space science through the Academy of Finland, and CIS collaboration takes place through a special commission supported by the Academy of Finland. The Finnish Space Committee presented a National Strategy for Space Research and Development in Finland in 1994 and this was updated in 1996 to cover the years 1996–1998. It made several recommendations. First, Finnish space activities should concentrate on ESA programmes, with special emphasis on the sectors, which would best impact Finnish industries. Second, Finland should undertake bilateral programmes as well as promoting national activities. Third, in the areas of remote sensing and satellite geodesy, Finland should place emphasis on the development of operational applications. In space science, Finland should focus on the scientific utilisation of instruments, with Finland continuing to participate in the development of instruments for ESA scientific satellites. Finally, in the fields of satellite communications and navigation, the emphasis should be on industry and the development of key technologies, primarily within ESA programmes to ensure and promote profitability in the field, to expand the markets and to enhance the effectiveness of all Finnish industries.

Two technology programmes began in 1996, jointly funded at FMk8 million per annum: Space 2000 and the RSTP Remote Sensing Technology Programme. Space 2000 aims at creating a network of companies and institutes capable of handling large space projects. Software, structures, electronics and reliability engineering are being emphasised. RSTP will last for five years, covering: remote sensing technology and research projects of international significance; innovative projects demonstrating applications to end-users; implementation of operational remote sensing methods; commercial applications.

Finland contributed the ERNE particle analyser to ESA's Soho and the University of Oulo is co-investigator in the EFW Electric Fields and Waves instrument on the failed Cluster (also on NASA's Polar). Sodankylä Geophysics Observatory has provided several magnetometers for NASA sounding rockets. It has industrial involvement in the ESA's Huygens portion of Cassini; the VTT Technical Research Centre, the University of Oulu and FMI are involved in both elements. Finnish scientists are collaborating on the Russian-led Spektrum-X (X-ray detectors, developed from versions provided for Mars Observer), the failed Mars-96 (FMI's Metegg meteorology sensors), Radioastron (22 GHz receivers) and Interball missions, and contributed the Aspera plasma spectrometer to Phobos. There is involvement with Sweden's Freja and Odin.

FMI is the national signatory to EUMETSAT, with Finland's contribution paid via FMI's budget. Data from polar meteorological satellites have been used since 1969 by FMI: a Primary Data User Station for Meteosat was installed in 1986 and a receiving station for full resolution digital NOAA data was installed during 1987. The National Land Survey (Maanmittauslaitos) is the National Point of Contact for ESA/ESRIN. Finland is involved in the GOMOS ozone monitor for ESA's Envisat 1 polar platform (1.20 per cent interest); Metop 1 platform involvement is 0.70 per cent. It is contributing 1.35 per cent to Meteosat Second Generation and 1.82 per cent to ERS 2 phase E. ERS provides ice data for the Baltic Sea. The Finnish Geodetic Institute participates in the CIGNET Co-operative International GPS Network and the DORIS/Spot programmes.

Outside of ESA, Finland's major space involvement has been in the Scandinavian Tele-X direct broadcast satellite, for which the Nokia group provided the ground control centre. There is a 0.40 per cent commitment to ESA's Data Relay Satellite programme, plus participation in ARTES 1 (1.71 per cent + 1.60 per cent extension), ARTES 5 (4.05 per cent), ARTES 7 (1.50 per cent) and ASTP 4 (1.90 per cent). Telecom Finland is the national signatory to Intelsat, Eutelsat and Inmarsat. Sweden's Telia AB and Telecom Finland International jointly own/operate the Tanum Teleport, on Sweden's west coast, with five Intelsat antennas. Inmarsat communications are routed through Norway's Eik station.

The Space Committee comprises a chairman, vice chairman and 10 members, with the executive working group as the sole subordinated body. The chairman is the national representative on ESA's Council. A new Committee was appointed on 1 April 1995 to serve into 1998.

1996 funding for space R&D was FMk158 million, including contributions to ESA. It remained flat through 1999. The country became an Associate Member of ESA for a five-year period from 1 January 1987 (extended in 1991 to end-1994), contributing 0.81 per cent of the General Budget and expressing the intention of eventually becoming a full member. The agreement for full membership from 1 January 1995 was signed at the ESA Council Meeting on 22 March 1994 and ratified by Finland's parliament in December 1994. It began participating in the agency's science programme on a full GNP level in 1990 and in 1996 contributed AU17.5 million (FMk10.0 million; 0.68 per cent) to ESA's total budget. In 1999, it contributed €14 million, or 0.6 per cent of ESA's budget and, in 2000, contributes €14.3 million or 0.7 per cent of total member contributions. 1988 FMk2.2 million (0.16 per cent); 1989 2.9 (0.2 per cent), 1990 5.9 (0.28 per cent), 1991 7.0 (0.28 per cent), 1992 8.2 (0.30 per cent), 1993 9.5 (0.30 per cent), 1994 11 (0.38 per cent), 1995 14.5 (0.54 per cent).

Finnish Space Expenditure (FMk millions)

Year	ESA	National	Total
1993	57.3	78.7	136.0
1994	71.5	67.0	138.5
1995	94.2	66.4	160.6
1996	102.1	66.4	168.5
1997	79.2	40.3	119.5
1998	83.2	40.9	124.2
1999	86.2	41.6	127.8
2000	85.0	41.6	126.6

UPDATED

France

Current Status

France is Western Europe's major space power and, with Germany and Italy, provides the principal impetus behind ESA. Its expansion into all areas of space activity now includes strong military surveillance interests, exemplified by the launch of Europe's first independent reconnaissance satellite, Helios 1A, in 1995.

France was the third nation after the USSR and the US to achieve national launcher capability and now possesses a significant telecommunications and Earth observation satellite manufacturing capacity, principally through Aerospatiale and Matra Marconi Space. As elsewhere, the French space industry is undergoing consolidation to meet the new financial climate and manufacturing overcapacity. MMS in 1994 acquired the space interests of British Aerospace. CNES (Centre National d'Etudes Spatiales) is the agency responsible for prosecuting French space activities and has been instrumental in establishing a number of companies and GIEs (Groupements d'Intérêt Economique) to commercialise and promote these activities, including Arianespace (Ariane launcher family) and Spot Image (Earth observation).

December 1995's Franco-German summit in Baden-Baden between Chancellor Helmut Kohl and President Jacques Chirac yielded an accord on military satellite programmes previously propounded by France. It opened the door for the long-planned merger of the Aerospatiale and DASA satellite divisions. Aerospatiale Chairman Louis Gallois and DASA President Manfred Bischoff signed an MoU on 18 December 1995 that produced the European Space Industries in 1996. Headquartered in Munich, ESI is the largest European space company (with subsidiaries ESI-France and ESI-Deutschland) and the world's fifth largest, with 2,600 personnel and more than US$1 billion annual sales. Originally, it was planned that France and Germany would contribute equally overall to the FFr11.5 billion Helios 2 optical/IR and FFr13 billion Horus radar satellite projects, while expressing the hope that they will become pan-European.

In April 1998, Germany formally ended its participation in the Helios programme, prompting French reluctance to continue in the Horus programme. In the wake of this blow, France remains committed to the Helios.

On 24 October 2000, CNES and NASA signed a statement of intent to co-operate in the exploration of Mars. The statement envisages an initial validation mission in 2007 followed by sample return flights in 2009, 2011 or 2014. Scientists from the USA and France will also work together in 2005 on the NASA Mars orbiter. CNES plans to launch a Mars orbiter in 2007.

France is responsible for the successful series of Ariane commercial launchers, providing 46 per cent of the funding for the €6.5 billion Ariane 5/5E development programmes, in return for prime contractorship. France leads European initiatives for new launch vehicle initiatives and in conducting studies including reusable, semi-reusable and expendable systems. The timeframe envisages formulation of concepts and opportunities over the period 2002–2006 followed by formal moves to

France has consistently underpinned the development of Europe's highly successful Ariane satellite launcher, paying more than 60 per cent of total development cost since the programme began 30 years ago. (ESA)
0572487

have ESA proceed with the next generation launch vehicle for operations beginning 2012. Until Russia became part of the international Space Station project and NASA began flying joint Shuttle-Mir missions, France was the leading collaborator with Russia. The country provided an extensive range of instrumentation to Earth-orbiting and deep space vehicles, including instruments for Mars-96 and the primary Sigma X/gamma-ray telescope of the Granat observatory. Annual agency-level meetings are held alternately in France/Russia to review current co-operation and consider proposals for future missions.

French AF pilot Jean-Loup Chrétien flew twice to Soviet space stations, in December 1988 performing the first non-US/USSR EVA. As part of the accord on long-term co-operative activities signed with Glavcosmos in Moscow December 1989, CNES/Licensintorg formalised an agreement for a French cosmonaut to spend 12 days aboard Mir in July-August 1992. Lt Col Michel Tognini's 'Antares' visit cost France FFr72.3 million, less than the quoted rate because of the close collaborative ties. CNES/NPO Energia signed a memorandum on 29 July 1992 for four further missions, in 1993, 1996, 1998 and 2000 to exploit Mir's facilities. Lt Col Jean-Pierre Haignere's 'Altair' visit used Soyuz-TM 17 in July 1993. Altair back-up Dr Claudie André-Deshays flew August 1996's 'Cassiopeia' mission. FFr100 million was invested in the experiments programme. FFr126 million was paid jointly for the 1993/96 missions, plus equipment which remained aboard Mir for Leopold Eyharts' planned visit in August 1997 on Soyuz-TM 27 (Pegasus mission). That mission was delayed until January 1998 by problems aboard Mir. CNES astronaut Jean-Francois Clervoy was selected in May 1992 by ESA as one of six new astronauts; he flew as a NASA Mission Specialist on STS-66 in November 1994. Dr Jean-Jacques Favier (materials science) and Michel Viso (veterinary science) were selected as CNES astronauts in 1985, followed by test pilot Philippe Perrin in 1990. Favier flew as a payload specialist aboard STS-78's Life and Microgravity Spacelab in June 1996 and Viso was due on 1998's Neurolab Shuttle. Chrétien and Tognini qualified as NASA Mission Specialists in 1995; Chrétien flew aboard the STS-86 mission in September 1997 and Tognini was scheduled to fly on the STS-93 mission in July 1999. France also pioneered the flight of μg payloads aboard China's recoverable spacecraft. On smaller projects, in November 1999 India and France signed an agreement to build and launch a scientific satellite named Megha Tropiques as continuation of a co-operative endeavour that began in 1972.

The space industry accounted for FFr18.3 billion of all military and civil aerospace sales of FFr100.6 billion in 1995 (17.5/105.5 1994; 16.7/110 1993). Comparison with 1989's 11.4/112.3 highlights the increasing importance of space. 12,000 are employed full time on space activities, including 8,700 in industry. 1996's military space budget was proposed at FFr4.6 billion; 1995's was FFr4.1 billion, 1994 4.1, 1993 3.9. While 1992's overall military budget remained static at about FFr103 billion, its space element rose from 1991's FFr3 billion (itself a 50 per cent increase over 1989) to FFr3.5 billion for 1992; 1995's was then projected at FFr5.5 billion and 2000's at FFr8 billion. It was announced 26 February 1993 that CNES had responsibility for design and development of military space projects, under the Ministry of Defence. France's fiscal year 2000 contribution to the ESA budget amounted to €634.0 million, 29.4 per cent of total member contributions.

CNES Budget Authority (FFr millions)

	ESA	Total
1984	1,793.0	4,610.1
1985	1,740.2	4,858.3
1986	2,209.8	5,686.9
1987	2,436.2	6,076.1
1988	2,657.9	6,660.5
1989	3,243.0	8,323.7
1990	3,614.1	9,528.1
1991	4,151.2	10,054.3
1992	4,725.0	10,623.1
1993	4,947.0	11,016.0
1994	4,860.0	11,774.7
1995	4,776.3	11,782.5
1996*	5,190.0	12,182.0
1997	4,166.4	12,536.0
1998	3,892.7	12,799.0
1999	3,935.7	12,960.6
2000	4,329.3	11,635.0

*provisional. Note: the contribution to ESA grew by 55 per cent in real terms 1987–92. It is capped at the current real level until 2000 in order to maintain national activities.

VERIFIED

Gabon

Current Status

Gabon has assumed a central role in co-ordinating satellite remote sensing of Africa through the Food and Agriculture Organisation of the United Nations. It has co-ordinated data from synthetic aperture radar and optical measurements and images through US and European government and commercial projects. Progress made with integrating data from several sources and different programmes providing optical and radar images. Gabon faces financial limits imposed by aggressive government funding constraints and from foreign investors anxious to see stability without excessive government financing of non-essential services and government run programmes.

The Republic of Gabon is served by the 'Equasat' C-band domestic satellite communications network provided on a turnkey basis by Scientific-Atlanta of the US. A 72 MHz hemispheric transponder was purchased aboard Intelsat's 1°W satellite in October 1986. Elements of this initial network include an 18 m International IDR Intermediate Data Rate gateway station in Franceville, a 15 m hub station in Libreville, eight 11 m stations, two 7 m stations and a single 7 m transportable. These stations provide >400 telephony channels and two TV channels. The IDR link from Franceville to Paris provides additional international connectivity. Completed in 1995, the system covers 36 stations: nine for telephony and TV distribution, two for TV receive and telephony, and 25 TVRO. These support two 17.5 MHz TV, two FM radio and about 370 telephone channels. The country is a signatory to Intelsat and Inmarsat through its TIG Télécommunications Internationales Gabonaises.

VERIFIED

Germany

Current Status

Pressure on government to reduce expenditure has raised questions about Germany's degree of participation in ESA programmes. Opposition from certain political elements to the extant level of financial involvement has reduced support for expanded ESA involvement. The government is keen to extend co-operation with foreign countries. Increased opposition to expanded development of microgravity experiments for the International Space Station has resulted from increased costs for research and development and a lack of full support from the US Bush administration. Germany has effectively mobilised an all-European effort to maximise microgravity research on the ISS and to carry out similar work on independent Shuttle missions. Concerns were expressed prior to the Columbia disaster regarding the paucity of scheduled science activities with the down-sized station complement of three astronauts. These concerns and financial pressures have eliminated possibilities put forward by proponents of extended human space flight initiatives for an increased commitment to ISS activities. The German government has shown strong support for ESA activities but is increasingly concerned about the apparent waste of resources in the ISS endeavour. Toward this end Germany had shown strong intellectual support for ESA's future programmes including the Aurora and Spaceplane propositions but financial and economic problems for the country have prevented full participation.

Background

The federal government has been supporting space research and technology since 1962. In February 1976, the government decided that there would be no further national or even bilateral research satellite projects such as Symphonie. Consequently, two-thirds of the 1976–79 space budget of DM2.2 billion would be allocated to ESA. However, activities such as the Franco-German TDF/TV-Sat direct broadcast satellite programme were subsequently undertaken. While France focused on the Ariane launcher, West Germany became the largest contributor (53.3 per cent) to Spacelab, developed by prime contractor ERNO and companies from 10 European countries for the US Space Shuttle. In addition, DM40 million was expended over the eight years to 1984 on contributions by Dornier and MBB to Ariane. 1983's budget totalled DM770 million.

The DARA German Space Agency, created in 1989, is in full charge of implementing the space programme, except for the hypersonic technology programme, which remains embedded within the BMFT Ministry for Research & Technology's aeronautics activities. Federal expenditure over 1985-96 totalled DM18.9 billion. 1999 federal funding for ESA space activities amounts to €570.1 million and ESA funding for 2000 is €555.3 or 25.7 per cent of total member contributions. The federal states, universities, industry and the German Aerospace Establishment (DLR) made minor additional contributions. Germany is the largest contributor to ESA's mandatory programme (25 per cent in 1999) and second only to France in the total budget, including the optional activities. National budgetary restrictions arising from the worldwide recession and unification mean that Germany has reduced its ESA payments and cut back on national activities.

At the end of the Cold War, West Germany had to fold a modest but vigorous East German space programme into its own. The space involvement of the former East Germany was restricted to co-operation with the USSR and other eastern European countries through its membership of Interkosmos, established in 1967. Beginning with a UV photometer aboard Interkosmos 1 in 1969, about 175 items of scientific equipment from institutes and industry flew on more than 40 Soviet spacecraft. The MKF-6 multispectral film camera, first flown aboard Soyuz 22 in 1976, was a permanent fixture on the Salyut 6/7 space stations and was attached to Mir in December 1989 as part of Kvant 2. The camera was operated aboard Salyut 6 in August 1978 by East German guest cosmonaut Sigmund Jähn, launched in Soyuz 31. His flight covered more than 20 experiments in remote sensing, materials science and biomedical research. IR Fourier spectrometers performed thermal mapping of Venus from Venera 15/16 and similar instruments flew on three Meteor Earth weather satellites. Image processing and magnetometers were contributed to Vega and Phobos and VNIR spectrometers to two Interkosmos satellites plus Salyut 7 and Mir for atmospheric remote sensing. Plasma probes were developed for several Interkosmos missions in addition to geophysical and meteorological rockets.

East German participation was principally through the Institut für Kosmosforschung (IKF, Space Research Institute), which specialised in spectrometric remote sensing, opto-electronic systems, extraterrestrial physics and the development and testing of space research equipment. It was absorbed by DLR in 1991, which continues the international projects. At that time, a programme of planned IKF international co-operation existed. It included Mir Priroda module (MOS Multispectral Opto-electronic Scanner), Mars-96 (WAOSS Wide-Angle Opto-electronic Stereo Scanner, Orbiter Magnetometer, IR spectrometer, image processing; balloon magnetometer), Regatta (magnetospheric plasma diagnostics), Sodart (Spektrum-X/γ high resolution X-ray spectroscopy), APEX Active Plasma Experiment mission (launch preparation, mission assistance, telemetry acquisition), Cassini (cosmic dust analyser; data compression), Okean (data acquisition from Okean satellites), Coronas solar observations (main station for telemetry data acquisition) and Beacon (satellite beacon network).

Germany, in 1994, made its first strong declaration of commitment to military space activities. December 1995's Franco-German summit in Baden-Baden between Chancellor Helmut Kohl and President Jacques Chirac yielded an accord on military satellite programmes previously propounded by France. It opened the door for the long-planned merger of the Aerospatiale and DASA satellite divisions. Aerospatiale Chairman Louis Gallois and DASA President Manfred Bischoff signed an MoU 18 December 1995 that formed the European Space Industries in 1996.

Germany shocked its French partners in April 1998 when it withdrew from the Helios spy satellite programme. France had the role of prime contractor for the Helios and Germany the Horus as part of their joint military space programme. France expressed reservations about continuing in the Horus programme, but Germany has reasserted its commitment to the project.

German involvement in the Helios 2 programme appeared to be in doubt at the beginning of December 1996. Since the satellite was not capable of seeing through cloud cover, Germany saw it as being of limited use and with German budget limitations, funding for the programme appeared to be uncertain.

Two major ESA centres are located in Germany. The ESOC European Space Operations Centre for TT&C of satellites during launch and orbital operations is housed at Darmstadt, heading a network of seven stations worldwide. The European Astronauts Centre was established in 1990 in Cologne as the home base of ESA's astronauts. EAC hosts the nearby CTC Crew Training Complex, where the bulk of Columbus training will be done. Germany's commitment to manned flight is reflected by the GSOC German Space Operations Centre satellite and manned mission control facility at Oberpfaffenhofen, providing the first manned capability outside of the US/Russia and already experienced with the Spacelab D1/D2 and EuroMir 95 missions. It was recently expanded at a cost of DM533 million (national/ESA funding) to cope with European space station operations. EUMETSAT headquarters are sited in Darmstadt and, until December 1995, Meteosat activities were controlled through ESOC's Odenwald station; EUMETSAT then assumed control via its new Fucino station in Italy.

Manned and microgravity programmes have been emphasised since the early 1970s. It has extensive manufacturing and manned flight experience with the Spacelab vehicle and the Spacelab D1/D2 missions, devoted primarily to life sciences and other µg activities. COF's ECU658.5 million fixed price construction contract was signed by ESA and DASA 28 March 1996. Test pilot Maj Klaus-Dietrich Flade flew March 1992's eight-day Soyuz-TM 14 mission to Mir, with physicist Reinhold Ewald as back-up, conducting 14 experiments emphasising biomedical research. DARA provided programme management, while DLR was responsible for the astronauts. A DARA/RKA agreement was signed 9 December 1995 and Ewald flew to Mir aboard TM25 in February 1997. National payload specialists flew the Spacelab D1 (Reinhard Furrer and Ernst Messerschmid. Furrer died on 9 September 1995 in a plane crash) and D2 (Ulrich Walter and Hans William Schlegel) missions. AF test pilot Thomas Reiter was selected in May 1992 by ESA as one of six new astronauts, flying ESA's EuroMir-95 Soyuz-TM 22 mission September 1995 – February 1996. Dr Ulf Merbold has been an ESA astronaut since 1977, flying Spacelab missions 1983/92 and Soyuz-TM 20 to Mir in October 1994 for EuroMir-94.

The former DASA provided ESA's Eureca free-flyer and operates the Texus, Mini-Texus and Maxus sounding rocket µg programmes. The recoverable SPAS platform, flown five times on Shuttle by end-1995, is available for µg payloads but it is now principally the Astro-SPAS astronomical observatory. The Express project in January 1995 launched Germany's reusable capsule on Japan's M-3SII but it was delivered into an untenable orbit; the cost renders future missions unlikely. Instead, Germany continues to lead Europe's utilisation of Russia's Photon µg craft. DASA Jena-Optronik and Kayser-Threde have developed the piggyback MIRCA Micro Re-entry Capsule which was due to debut on Photon in late 1996 for re-entry investigations.

In satellite construction, MBB was one of five Eurosatellite partners that produced the TV-Sat, TDF and Tele-X direct broadcast satellites and developed the national DFS-Kopernikus craft with Dornier. Dornier was also prime for ESA's Cluster satellites (Germany provided two instruments), following its lead role for the ERS 1/2 microwave remote sensing satellites, the Ulysses solar probe and the international Rosat X-ray observatory.

After termination of the former East Germany's membership of Interkosmos, Germany included those projects in its co-operative programme with Russia. Including Mars-96 (involvement in 14 experiments), the programme covers about 40 space science and 10 Earth observation projects. In 1996, DASA split into two new organisations: Dornier Satellitensysteme GmbH and Daimler-Benz Aerospace-Space Infrastructure Division.

To meet space requirements, the federal cabinet decided in April 1989 to create a new structure, broken into three organisations. These are:

A Cabinet Committee on Space Activities, chaired by the Federal Chancellor and with the Federal Minister for Research & Technology taking the lead rôle; A State Secretaries' Committee on Space, chaired by the Secretary of State in the Federal Ministry for Research & Technology; DARA GmbH, the German Space Agency.

The Cabinet Committee is responsible for advising the Federal government on space policy decisions, in addition to fundamental strategic and planning aspects and on the allocation of the space budget. Funds are channelled to DARA through the budgets of the responsible ministries. The committee comprises: Federal Chancellor (chairman), head of the Federal Chancellory, Foreign Minister, Minister of Finance, Minister of Economics, Defence Minister, Minister of Transport, Minister for Research & Technology and Minister of Posts & Telecommunications.

The State Secretaries' Committee on Space prepares the decisions to be taken by the Cabinet Committee, co-ordinates the relevant Federal ministries and defines goals and commissions for DARA. Its members are the Secretaries of State of the ministries represented on the Cabinet Committee. DARA is the central management organisation for German space activities, headed by a Director General and two Managing Directors. Since June 1990 it has been legally responsible for drafting space planning and issuing the programmes, managing them and awarding contracts to science and industry. DARA also represents German interests at international level, most notably ESA.

The increasingly ambitious nature of future projects, with a return to national satellites and the selection in December 1982 of two national astronauts, reflected a growing commitment to space. The Spacelab D1 mission, costing DM565 million, was devoted to materials/life sciences; Spacelab D2 in 1993 cost DM890 million. Microgravity research in the Texus sounding rocket programme began in 1978. DARA, in 1994, withdrew from funding Maxus, preferring the cheaper Texus instead. Launched science missions include ESA's Exosat (MBB was prime), NASA's Galileo mission to Jupiter (MBB provided the propulsion module), AMPTE, Hipparcos, Hubble, ERS and Rosat (Dornier prime). Earth observation instruments include the SCIAMACHY spectrometer for ESA's Envisat, the Millimetre Atmospheric Sounder flying on Atlas Shuttle missions and the major XSAR X-band synthetic aperture radar as part of Shuttle's Space Radar Lab.

Federal Ministries' expenditure (DM millions)

	ESA	Nat'l	BMV	BMPT	Total
1985	414.1	416.4	36.5	545	1,412.0
1986	559.2	358.3	43.9	480	1,441.4
1987	639.6	418.8	35.0	385	1,478.4
1988	644.1	462.4	31.5	323	1,461.0
1989	712.9	506.5	17.4	570	1,806.8
1990	839.0	576.5	25.0	404	1,844.5
1991	964.0	580.0	30.0	–	1,574.0
1992	1,133.0	604.0	30.0	–	1,767.0
1993	1,202.0	580.0	30.0	–	1,812.0
1994	1,092.0	478.0	30.0	–	1,600.0
1995	1,072.0	300.0	30.0	–	1,402.0
1996	1,003.0	300.0	–	–	1,303.0
1997	977.8	272.0	–	–	1,250.0
1998	1,095.0	300.0	–	–	1,385.0
1999	1,095.0	270.0	–	–	1,365.0

BMFT = Bundesministerium für Forschung und Technologie (Federal Ministry for Research & Technology); its figures exclude the German National Hypersonic Technology Programme. BMV = Bundesministerium für Verkehr (Federal Ministry for Transportation); mainly contributions to EUMETSAT. BMPT = Bundesministerium für Post und Telekommunikation (Federal Ministry for Postal Services & Telecommunications); expenditures for satellite broadcasting.

UPDATED

Greece

Current Status

The first national involvement in space activities was initiated on 27 September 1989 by the signing of an MoU in Athens between Moscow's Institute of Space Research (IKI) and the Crete Research Centre covering participation in the Mars-96 project. The government and ESA signed a co-operative agreement in July 1994. Greece is a member of Intelsat, Inmarsat and Eutelsat through the signatory Hellenic Telecommunications Organisation (OTE). An Inmarsat A/C/B/M station covering the Indian and East Atlantic Oceans is operated at Thermopylae. ESA is providing consultancy to the Ministry of Transport and Communications on the proposed Hellas-Sat national telecom satellite system.

Greece has increased its utilisation of remote sensing and has begun studies of active participation in ESA earth observation programmes. The Institute of Ionospheric and Space Research (IISR), founded in 1955 and given its present name in 1990 is now located to a university campus north of Athens. Funded by several branches of the Greek government, the IISR conducts a wide range of atmospheric and near-Earth space observations. Modest expansion of facilities, personnel and equipment resulting in enhanced interpretation skills used by a wider range of government agencies. Remote sensing has become a major government priority within its science support programmes for earth observation and analysis with several different government departments committed to using this technology.

VERIFIED

Hungary

Current Status

The Hungarian Space Office has managed all national space activities under the Hungarian Space Board since January 1992. As the umbrella authority, the Hungarian Space Organisation was founded in January 1992, supervised since 1994 by the minister of Transport, Communication and Water Management. Membership of ESA is a possibility for the future; formal meetings were held with the European Space Agency in July 1990. Hungary is a member of Intersputnik (single Earth station operating via the 14° W Gorizont) and joined Eutelsat in 1993 and Intelsat in January 1994 (two 18 m Intelsat A stations). Israel Aircraft Industries plans a second Amos satellite by 1999, owned by Magyarsat Ltd in Hungary, a 50/50 joint venture by IAI and Antenna Hungaria, the country's public broadcasting monopoly and the second largest telecommunications service provider. For this CERES Central European Communications Satellite to proceed, it required a commitment from the Hungarian government to guarantee two transponders, which was agreed 30 April 1995. Lack of financial support caused a collapse in this project and a reduction in opportunities for transponder utilisation.

As part of the Interkosmos organisation, Hungary has contributed to several Russian-led science missions. Hungary provided elements of Vega's TV system and the controlling electronics for the Phobos long-lived lander, in addition to involvement in Aktivny, Apex and Interball. Lt Col Bertalan Farkas of the Hungarian Air Force flew to the Salyut 6 space station aboard Soyuz 39 in May 1980 under Interkosmos; radiation dosimeters carried on the mission were subsequently adopted for Salyut 7 and Mir space station monitoring. A space co-operation agreement was signed with India on 27 October 1995, emphasising Earth observation and space physics.

Hungary is seeking co-operation with ESA and in 2000 makes a contribution of €0.3 million in areas of earth science, physical science and general industrial co-operation projects. Studies have now begun into co-operating with ESA on human space flight activities through the International Space Station programme. Several unspecified research laboratories have participated in tests to evaluate different ways in which they can make a meaningful scientific contribution to the ISS but these efforts have been suspended pending a decision on the science activity aboard the ISS when Shuttle flights resume.

VERIFIED

India

Current Status

Resolved to extend its involvement with space science the Indian cabinet has approved a plan to send an unmanned spacecraft to lunar orbit by 2008. The government has approved a fund of US$83 million and plans to place a 400 kg spacecraft in moon orbit following launch by a PSLV in order to extend India's involvement in planetary research, creating say Indian scientists, a new cadre of specialists in planetary research. Noting the strident progress being made by leading space countries such as China, the Indian government is determined to maintain a positive growth in space research, science, exploration and engineering development for extended participation among the space faring countries of the world. The Indian government has sustained a major growth in national space activities and has also increased its commitment to remote sensing (see relevant section) and expansion of telecommunications services. India is also addressing the needs of the military with concerted efforts to seek a means of providing secure communications links through a dedicated military satellite system. The first launch of the GSLV took place on 18 April 2001 with the launch of GSAT-1. The second development flight was successfully conducted on 8 May 2003 when GSLV-D2 launched the GSAT-2 satellite.

Background

India, with more than three-quarters of its 850 million population dependent on agriculture, concentrates its space development activities on applications satellites and autonomy in launch systems. The two prime aims involve operational space-based remote sensing and communications systems: meteorological and Earth resources images are routinely returned from space and 701 TV and 182 radio stations are linked into the space network. The US-built Insat 1 geostationary communications and meteorological satellite family was superseded by indigenous Insat 2s in 1992 and the new IRS 1C, launched in December 1995, is the world's most advanced civil remote sensing satellite.

India was the seventh nation to achieve orbit capability, in July 1980. By late 1985, Insat 1B, which was launched in August 1983, had extended TV coverage from 20 to 70 per cent of the population, with the goal of 90 per cent now satisfied. Following the successful experiments with educational TV from NASA's ATS 6 in 1975, more than 35,000 of India's 520,000 villages have been equipped to receive Insat services. Professor Rao, who became chairman of ISRO in October 1984, commented at the time that space technology had given India the opportunity to convert backwardness into an asset: developing countries could bypass the intermediate technology stage and leapfrog into a high technology era.

Like France, India has benefited from simultaneous co-operation with the CIS/USSR, ESA and US. After the loss of Insat 1A, initial problems with Insat 1B were overcome, and this Ford-built series should have been completed with the Shuttle launch of Insat 1C, accompanied by an ISRO scientist or engineer, in mid-1986. Thus, India would have joined France as the only country to have had specialists aboard both USSR and US spacecraft, but 1986's Shuttle accident required a switch to Ariane. Maj Rakesh Sharma of the Indian Air Force, after two years training, took part in the successful seven-day Soyuz-T11/Salyut 7 mission in April 1984. ISRO and the Russian Space Agency signed a general agreement continuing space co-operation on 23 December 1994.

Two commercial agreements reached in 1995 demonstrate that India has advanced to the front rank of the world's space nations. While development of the US Landsat remote sensing series has stumbled along, EOSAT plans to market imagery and data from the 14 satellites that India expects to fly by 2005. Over 1991–1997 there were four satellites launched towards polar orbit. They are not only placing India in a strong position to satisfy its own applications needs, but also to dominate a significant portion of the global commercial market. The IRS series is now also moving from land applications to environmental monitoring, with an emphasis on oceanography. Additional remote sensing satellites planned by India include CartoSat (IRS-P5), ResourceSat 1 (IRS-P6) launched in 2003, ResourceSat 2 (IRS-P7) and CartoSat 2.

The second notable development, India's agreement with Intelsat meant the European consortium will lease 11 transponders on Insat 2E for 10 years beginning in early 1998 at a cost of around US$100 million. It was noted in February 1996 that the system already generates revenues of Rs4 billion annually.

Launcher and propulsion programmes represent half of space expenditure, which runs at some US$300 million. The PSLV Polar Satellite Launch Vehicle, fully successful on its second attempt in October 1994, provides a capability of placing one tonne-class IRS satellites into Sun-synchronous orbits and is now offered commercially through Antrix Corp. PSLV offers only 450 kg GTO capability but an upgrading into the Geostationary SLV is under way to satisfy 2.5 tonne-class GTO requirements by the end of 2000. Notably, GSLV incorporates a cryogenic KVD-1 engine acquired from Russia. The same engine will fly on Proton. India's first attempt at firing a 10 kN LOX/LH_2 engine ended with an explosion at ignition in July 1993. LH_2 became available in the country only after a US company built a plant at end-1992. The first seven GSLVs will use the Russian engine to allow time for India's own to be developed by 2003-2005. GSLV will also create a planetary capability: Mars, Mercury and Venus missions are under consideration. An advanced GSLV will be powered by a solid propellant first stage and a cryogenic upper stage and will be capable of placing payloads weighing up to 3 tonnes in geostationary transfer orbit.

1988 saw the successful orbiting of IRS 1A under a semi-commercial agreement with Glavcosmos – the first of its kind – and an agreement was reached in November 1988 for IRS 1B to be carried by another Vostok in August 1991 at a cost of about Rs220 million. The agreement of January 1991 added IRS 1C in December 1995 for Rs500 million. Insat 1D departed 12 June 1990 aboard a Delta 2.

HECL Hughes Escorts Communications Ltd, a subsidiary of Hughes Network Systems of the US, began providing India's first private satellite shared hub service in January 1995 using Insat 2 leased capacity. VSAT-based wide area networks in India are expected to grow rapidly. Afro-Asian Satellite Communications Ltd (ASC), part of India's Essel Group, signed a contract with Hughes Space & Communications International in January 1995 for two satellites to provide GSM-compatible mobile telephone links from 1998, initially covering an area bounded by India, Middle East, Singapore and Moscow. The value was not disclosed but system cost has been estimated at US$700 million. First launch was planned for 2001, but the programme was subsequently cancelled.

ISRO and Malaysia's Maxstar consortium signed an agreement in January 1995 for development of two 50 kg satellites for launches in 1997 and 2000 as part of Malaysia's space capability development programme.

Growth in India's space budget has been unprecedented, and unparalleled by any other country, rising by 91 per cent in the last four years alone. Major areas of growth are in satellite operations and space applications, each of which greatly benefit local and regional development. The fiscal 2000–2001 budget for the year ending March 31 2001, of 20.192 billion rupees, is equivalent to approximately US$459 million. The 17 per cent increase in the '00-01' space budget is set against a general government spending increase of 7 per cent for this year.

UPDATED

Indonesia

Current Status

Indonesia is undertaking a modest programme of remote sensing and atmospheric probing, but its emphasis is on satellite-based communications linking the country's 6,064 inhabited islands, with their 182 million population, spanning some 5,000 km along the equator. The first launch in July 1976 resulted in the system, operated by Perumtel (the private PT Satelit Palapa Indonesia replaced what had become known as PT Telkom), being declared operational the following month. The second generation was introduced in June 1983 by Palapa B1 and the Palapa C third generation appeared in 1996 to expand services to China, India, Japan and Pakistan, for example. CI launched 1 February on a US Atlas-2AS and C2 16 May on an Ariane-44L. The private PT Satelindo was set up in 1993 to finance and operate the Palapa C system; it also took over the Palapa Bs, leaving PTTelkom with the terrestrial telecommunications network. Palapa C1 was launched on 31 January 1996, followed by Palapa C2 on 15 May 1996. The private Indostar was launched in November 1997 to provide up to 40TV channels; CD-quality will be added to Indostar 2, a launch data for which has yet to be specified. Indostar 1 is known as Cakrawartha-1, the only S-band satellite operating over Asia. Three more Indostar satellites are planned. National TV cannot currently be received by 32 per cent (57 million) and radio by 20 per cent (36 million). Indonesia is leading the US$650 million Garuda/ACeS Asia Cellular Satellite System project to provide personal mobile services throughout Asia. Palapa CI and C2 were launched in January and May 1996, respectively.

The LAPAN space agency maintains ground stations for reception of Earth resources and meteorological satellite imagery. Indonesia was the 11th country to build a Landsat station, in 1981. A Cospas-Sarsat satellite-based search/rescue system Local User Terminal (from Canada's CAL Corp) was installed in Ambon and one with an Operations Control Console in Jakarta in 1991. In conjunction with the Netherlands' NIVR, LAPAN studied the TERS Tropical Earth Resources Satellite to provide equatorial multispectral coverage from a 1,680 km, 10° inclination orbit, employing a cloud sensor to direct the CCD-based imager towards clear regions. An equatorial launch site for handling China's CZ-3 was considered in 1987, possibly located on Biak Island but plans for this were subsequently abandoned. China subsequently reopened approaches to Indonesia and coupled offers for trade deals with further consideration of a near equatorial launch site facilitating a commensurate increase in lifting performance.

International

European Space Agency

Current Status

ESA has consolidated several separate programmes and is preparing to move into the operational phase of human space flight activity with the Columbus module for the International Space Station. These plans have been placed temporarily on hold due to the loss of the NASA Shuttle Columbia and the grounding of remaining orbiters. Pressures from member countries to reduce costs continue to apply restrictions to the development of new systems. Some initiatives are developing liaison with private companies and organisations. ESA is planning a strategic blueprint for the exploration of the solar system and is presently evolving a series of missions which would support a general objective of putting humans on Mars by the middle of the century.

Background

Eleven states created ESA in 1975 to replace the European Launcher Development Organisation (ELDO) and European Space Research Organisation (ESRO). The states were Belgium, Denmark, France, West Germany, Ireland, Italy, Netherlands, Spain, Sweden, Switzerland and the UK. Austria and Norway, with Finland as an associate member (full member from January 1995) and Canada as a co-operating state subsequently joined the charter members. In 2000, Portugal became the 15th

India's budget summary (Rs million)

	89–90	90–91	91–92	92–93	93–94	*94–95	*95–96	97–98	98–99	99–00	00–01
Rocket development	1,474.1	1,812.9	2,057.9	2,680.4	2,967.2	3,678.2	3,868.2	5,005	5,670	5,548	7,744
Satellite development	1,389.2	1,180.0	1,432.4	645.2	326.8	358.5	327.5	2,226	6,473	6,898	7,244
Insat/IRS ops**	481.7	216.0	360.8	819.1	2,481.9	2,701.7	3,925.7	1,503	2,008	1,873	2,658
Space applications	450.7	444.5	493.2	508.8	658.6	649.3	610.6	947	1,055	1,404	1,453
Space sciences	84.8	94.7	140.4	132.7	136.2	146.2	141.9	296	297	398	391
Administration	105.1	114.2	116.2	122.9	382.7	216.1	293.7	566	521	1,138	702
Totals	**3,985.6**	**3,862.2**	**4,601.0**	**4,909.2**	**6,953.4**	**7,750.0**	**9,167.6**	**10,543**	**16,024**	**17,259**	**20,192**

Includes National Remote Sensing Agency and Physical Research Lab expenditure.
*1995-96 figures are estimates; 1994–95 revised.
** IRS included from 1992–93.

member of ESA. Member states contribute to the mandatory general and science budgets (€2,553.2 million approved for 1996–2000 (1995 rates)) on a gross national product basis, and for the remainder receive contracts in proportion to their funding.

ESA's future was mapped out in November 1987 when the ministers responsible for space matters in its (then) 13 member states met in The Hague, Netherlands and approved programmes into the next century. Total budget for this 1988–2000 long-term plan was estimated in 1987 at AU32 million, but the member states suggested a pruning of 15 to 20 per cent at Germany's instigation. Essentially all programmes were approved, with the Columbus space station element and the Hermes manned spaceplane being phased so that after a Phase 1 of three years the member states could review their status.

At the ministerial meeting in The Hague in November 1987 the Hermes spaceplane was approved along with Columbus and Ariane-5. Hermes was the most technically-demanding of the three and a two-phase approach was adopted: Phase 1 began in January 1988 and it was to have ended in December 1989 but was delayed until the end of 1992. At the end of Phase 1 ESA's planning was revised and called for Hermes' maiden flight in 2002, followed by its manned debut in 2003 and its first operational servicing mission in 2004.

Programme cost was estimated at AU7,320 million (1990 rates; AU6,222 million 1986 rates), a 40.5 per cent increase over the AU4,429 million (1986 rates) envelope established at The Hague ministerial meeting in 1987. 17.5 per cent of the increase was credited to technical changes and 23 per cent to the four-year programme stretch-out. AU1,167 million had been expended to end-1992 (1991 rate). However, ESA's need to reduce funding commitments produced November 1992's Granada decision to cut Hermes back to a technology programme, principally on joint ESA/Russian spaceplane studies. The funded 1993–95 reorientation period began studying three strategic options: co-operation with Russia, co-operation with the US and an autonomous European scenario. These covered system studies primarily directed towards definition of an ESA/Russian vehicle and development of critical technologies based on the Hermes definition. A detailed definition study of an Assured Crew Return Vehicle for Space Station, detailed definition studies and pre-development of an Automated Transfer Vehicle (ATV) also entered the mix of options.

These were reworked as MSTP (Manned Space Transportation Programme) and approved by ESA's Council in February 1994. Development of a joint spaceplane with Russia was no longer an option. Full development required approval by the ministerial meeting in October 1995 when ATV was fully approved and CTV began a three-year study. ATV could perform logistics and propellant resupply to ISS beginning February 2003. It would also have performed the re-boost function, to raise the orbit of ISS, which is continuously decreased by the residual atmospheric drag. The October 1995 ministerial meeting approved €667 million (1995 rate) for the ATV for 1996–2004. The November 1991 Munich ministerial meeting reconsidered the plan, accepting a revised 1992–2005 plan as a strategic framework for ESA's activities. Cost of the 14-year plan was projected at AU39.1 billion (US$45.0 billion) at 1990 rates. The expected Phase 2 approval for Hermes (launch delayed to 2002), Columbus (its free-flyer element's launch delayed to 2003) and DRS was deferred a year, to a third ministerial meeting in Spain. The Munich meeting did approve POEM 1 start (to be reviewed in Spain) and a 5 per cent reduction in 1992's budget. 1992's Granada meeting approved the 1993–2000 long-term programme, costing AU22.7 billion at 1991 rates (reduced from Munich's planned AU25.8 billion for the same period).

The Columbus Orbital Facility (COF) module was approved, but with a 5 per cent cost reduction and 5 per cent financial under-subscription. The Columbus Free-Flyer module was cancelled. However, continuing financial pressures and the major reorientation of the international space station again required significant changes in February 1994 of ESA's long-term plan.

These programmes were funded only for 1994–95 (ERA was fully approved September 1994); the objective being to make a single programme out of the two proposals and then submit the single programme in 1995 at the ministerial meeting in October 1995. The resulting merged programme was called the European Participation in the International Space Station, wherein COF and ATV were fully approved and CTV began a three-year study phase. COF has been significantly reduced until it is about the size of Spacelab. The Toulouse meeting approved €2,651.2 million (1995 rate) for the whole programme in 1996–2004, with Germany contributing 41.0 per cent, France 27.6 per cent, Italy 18.9 per cent, Belgium 3.0 per cent, Switzerland 2.5 per cent, Spain 2.0 per cent, Denmark 1.17 per cent, Netherlands 0.94 per cent and Norway 0.45 per cent. ESA will have access to 5.3 per cent of ISS resources in exchange for ATV services including re-boost.

ATV is now scheduled for launch sometime in 2004 with Columbus in May 2005. ESA is participating in development of the US-led X-38 International SpaceStations Crew Return Vehicle (CRV) and is responsible for 15 X-38 subsystems or major elements. It is expected to fly for the first time in February 2002 when it will be carried into orbit aboard the NASA Shuttle for an unmanned return to earth demonstration. The ESA CRV is expected to be operational in 2005.

ESA permanent staff totals 1,820 (as of 1 January 1997), with 349 based at the Paris headquarters (including Washington, Brussels, Moscow, Kourou and Toulouse). Three centres provide support. The ESTEC European Space Research & Technology Centre in Netherlands designs, develops and tests spacecraft and payloads (1,078 personnel). The ESOC European Space Operations Centre, staff 269, in Darmstadt, Germany is responsible for spacecraft operations. And the ESRIN European Space Research Institute at Frascati, Italy with a staff of 147 houses ESA's Information Retrieval Service and the system for collecting, pre-processing and distributing data from remote sensing satellites. The European Astronauts Centre, with 15 personnel, in Cologne is a centre of expertise for ESA's manned space activities and acts as the home base for all ESA astronauts.

The ESA programme plan provides the following status:

Science

Annual funding will remain at a constant €347 million from 1996 to 2000, yielding a reduction in real terms of 15 per cent. This requires some of the later missions to be delayed and some to be placed on indefinite hold. The Science Programme Committee in February 1996 ordered a 10 per cent reduction in the budgets of new missions.

Horizon 2000 comprises four major Cornerstone missions interspersed with smaller projects into the next century.

The loss of the Cluster payloads on the Ariane 501 mission meant ESA needed supplemental funding for a replacement mission launched on two Russian launchers in 2000. The X-ray astronomy cornerstone is the X-ray Multi-Mirror (XMM) mission launched in December 1999 by Ariane 5. The primordial body's cornerstone (Rosetta) was selected in 1993 to fly in January 2003, leaving the sub-millimetre astronomy cornerstone (FIRST Far Infra-Red/Sub-mm Space Telescope) as the fourth, in 2007. Horizon 2000 also includes medium and small space science projects selected as funding allows. The Giotto Comet Halley fly-by probe of 1986 intercepted a second comet in July 1992 and is now dormant.

ESA shut down the International Ultra-violet Explorer (IUE) in 1996 to save money. The Hipparcos astrometry satellite was orbited in 1989 (mission completed 1993), followed in 1990 by the Hubble SpaceTelescope (15 per cent contribution to NASA project, covering solar array and Faint Object Camera) and the Ulysses solar polar probe. The Infra-red Space Observatory departed in November 1995, the Cassini Saturn orbiter collaboration with NASA for which ESA is providing the HuygensTitan atmospheric entry probe, the Integral (International Gamma-Ray Laboratory) in 2001 and the M3/M4 medium missions for 2004/2007. Budget cuts mean that M3 is deferred a year while its cost is studied and M4 is shelved for the moment.

Preparatory work began in 1993 on the follow-up Horizon 2000 Plus programme to cover new missions 2007–2016. This new science long-term plan was endorsed by October 1995's ministerial meeting. The call for mission concepts was made 29 June 1993 and 110 were received by the 15 October 1993 deadline. the special Survey Committee presented its report October 1994. With funding remaining at 1995's level, two Cornerstone missions (maximum ECU625 million; 1993 rate) will be complemented by four medium (ECU345 million) plus small missions and Space Station (possibly a high-energy astrophysics instrument). One Cornerstone will establish a Mercury orbiter for an intensive geophysical and magnetospheric survey. The second will be an interferometric observatory, either astrometric or IR imaging. The astrometric would improve on Hipparcos by a factor of 100, to 10 μarcsec, to ascertain the distances, motions and luminosity of tens of millions of Milky Way stars, and the mass distribution of nearby galaxies. The IR would search for Jupiter-like planets and brown dwarf companions of other stars.

ESA participation in Hubble should continue, currently planned through 2002. Annual funding increases of 5 per cent for 2001–2005 are required (but look unlikely in the new financial climate) for a third Cornerstone and augmented technology development. The gravitational waves Cornerstone (the LISA Large Interferometer Space Antenna was studied for M3 but proved far too expensive) in 2017 would explore the very early phases of the Universe and observe massive black holes and their coalescence. Technology studies could lead to X-ray and IR observatories beyond 2017.

Earth observation

Meteosat operations continue with development/procurement of the MSG second generation under way. ESA has budget authority for ERS 1/2 microwave all-weather remote sensing satellites, the first Polar Platform, Envisat 1, launched in 2001 and the second PPF, Metop 1, was approved in 1996 in reduced form with launch scheduled for 2005.

Microgravity

ESA facilities aboard Columbus, Eureca (stored), Spacelab, Mir, Photon, Bion and sounding rockets such as Texus and Maxus.

ESA's microgravity programme had performed 343 experiments through August 1995, including 183 on manned missions. The first phase of the materials science, fluid physics and life sciences μg research programme ran from January 1982 to 1985 at a cost of AU44 million (1993 rates, AU28 million 1983 rate) and comprised two principal elements: Spacelab experimentation and a sounding rocket programme. Three multi-user facilities were provided for Spacelab: the Biorack unit for investigating bacteria, cells, insects and plants (flown Spacelab D1), the Fluid Physics Module to study hydrodynamic phenomena of floating liquid zones (flown Spacelab 1/D1) and the Sled facility to investigate the human's vestibular system (flown D1). Eureca was added to ESA's overall programme in 1983 as a self-contained programme element and contributed significantly to μg research (see the separate entry above).

The current Phase 2 (known as EMIR 1 since 1993) began in 1985 to accomplish several goals. First, it was meant to maximise use of Spacelab and its multi-user facilities. Secondly, it needed to develop and fly the new Eureca payload facilities and exploit facilities developed nationally for the Spacelab 1/D1 and for the sounding rocket programme. Finally, it had to design, develop and launch new multi-user facilities for all μg disciplines; and provide flight opportunities for existing or new individual experiments. It therefore re-flew Biorack (Spacelab IML-1/2) and developed three fluid physics facilities (Advanced Fluid Physics Module – flown on Spacelab D2, The Bubble, Drop and Particle Unit – flown IML-2/LMS and the Critical Point Facility – flown IML-1/2). The Anthrorack human physiology module (flown Spacelab D2), the Advanced Protein Crystallisation Facility (flown Spacehab 1/IML-2/LMS, STS-72) and the materials science Advanced Gradient Heating Facility (flown LMS Spacelab); further flight was expected on 1998's Neurolab Spacelab. Total Phase 2 cost envelope is AU450.6 million at 1983 rates.

Apart from the short duration flight opportunities on sounding rockets, all other elements required flights of either Spacelab or Eureca. However, Challenger's loss in January 1986 halted all Shuttle flights for more than two years. As a result, Phase 2, conceived as a four-year programme, was stretched into 1997 to cover the delayed Shuttle activities. This extension was performed in two steps: the first was approved in 1988 and the second in 1991. Besides increasing the short duration opportunities, in order to maintain programme momentum, using drop towers, aircraft flights, Get Away Specials and sounding rockets, it emphasised two further broad objectives. The first further utilises Spacelab. Secondly, it sought to develop other research outlets independent of Shuttle/Spacelab: small payloads are flying on other retrievable carriers (such as Biobox and Biopan on Russian capsules) and the EuroMir missions. For example, the Fluid Physics Facility was planned for Photon in 1998 and Biorack was due to fly three times on Shuttle/Mir (STS-76, March 1996; STS-81 December 1996; STS-84, May 1997). In May 1993, the Microgravity Programme Board approved re-orientation of the extended Phase 2 activities towards basic research under the EMIR 1 European Microgravity Research Programme. EMIR 2 is essentially a utilisation period covering 1996-2001, costing AU146 million (1995 rates) and approved July 1996.

Phase A studies for Columbus payloads began 1988/89: Biolab, Fluid Science Laboratory, High Temperature Materials Processing Laboratory and an Anthrolab. Protein and vapour crystal growth units, a Combustion Facility and a Biotechnology Facility were also earmarked for further study. It is now planned to develop/build the multi-user facilities for Space Station under a programme separate from EMIR starting 1997: MFC Microgravity Facilities for Columbus. €206.7 million (1995 rate) was approved October 1995 for 1997–2003. Planned are Biolab, Fluid Science Laboratory, Physiology Modules, New Physiology Facility, Low Gradient Furnace, and Solidification and Quench Furnace. Low Gradient Furnace will fly aboard ISS's US Laboratory module; the others on ESA's COF.

Communications

ESA plans to maintain one Marecs as Inmarsat back-up plus one for experiments, keep three ECS satellites in service with Eutelsat and operate Artemis to demonstrate future data relay, land mobile and other innovative communications technologies. Launched in July 2001, Artemis suffered from a second-stage Ariane 5 failure which resulted in it being left in the wrong orbit. Onboard propulsion was required to lift it to a geostationary position.

Navigation

The GNSS Geostationary Navigation Satellite Service will be provided through Inmarsat 3's navigation transponders. Preparatory studies for the GNSS2 second-generation system (2005) are under way. ESA/EU/Eurocontrol are co-operating on the €150 million 1995–2000 programme. On 7 December 1999, ESA signed a contract for the GalileoSat study, the Agency's contribution to the Galileo European global navigation satellite programme, which is expected to comprise at least 21 satellites in medium orbit at 24,000 km altitude possibly complemented by geostationary orbit satellites at 36,000 km altitude. Galileo is expected to start operations in late 2005 or 2006 and become fully operational in 2008.

Space transportation

ESA and CNES, with Aerospatiale as industrial architect, were responsible for developing the Ariane 1–5 range of orbital launchers. The vehicles are now procured, marketed and launched by Arianespace. The six Ariane 4 versions will be phased out from 2002 as the Ariane 5 takes up more of the launch manifest.

Ariane 4 became Europe's primary launcher in 1989. ESA is developing Ariane 5 under a €6,572 million (1995 rate, total including preparatory programme) programme to expand capacity. A further €1,713.4 million was approved in October 1995, principally to develop the more capable 5E version. The Kourou base in French Guiana will continue as the launch site. ESA began FESTIP studies in 1994 of future reusable systems, and they have focused on reusable systems for post-Ariane 5 generation launchers.

Manned space transportation

ESA's Manned Space Transportation Programme is developing support systems for Space Station, including the ATV Automated Transfer Vehicle and ERA External Robotic Arm.

The CTV is intended for ISS support and possibly crew rescue. Delivered by an Ariane-5/ATV, the 10 tonne ballistic vehicle could carry a four person crew plus 400 kg in the 51.6°, 400 km orbit. The first manned launch is due around 2002 and the cost is more than €2.5 billion. The October 1995 ministerial meeting approved €52.3 million (1995 rate) for the 1996–1998 definition study, with full approval to be discussed at a further ministerial meeting in late 1997. The 2,665 kg, 2.8 m diameter, 2.04 m high Atmospheric Re-entry Demonstrator (ARD) flew on Ariane 503, splashing down close to 5° N, 148° W in the Pacific after one circuit and an apogee of 875 km. It is ESA's first re-entry craft, a 0.7-scale Apollo capsule (L/D = 0.3) with 200 sensors to monitor aerodynamic, aerothermodyanamic, thermal protection and blackout conditions. A Sextant Avionique NSS100-P GPS receiver is included to improve landing accuracy and study signal penetration during blackout (as will a TDRS S-band link). Protected by Aerospatiale's Aleastrasyl heatshield and Norcoat 622-50FI on the aft body. The electrical and control systems (including 7 × 400 N hydrazine thrusters) were derived from Ariane 5. Poser comes from two 28 V Spot 4 Ni/Cd batteries plus one pyrotechnics battery. Two 1 Gbit SSRs. The €30 million contract was signed with Aerospatiale on 30 September 1994; some 25 per cent of the contract value went to Belgium (SABCA supplied the aluminium structure).

As part of the 'early deliveries', ESA will also develop a robotic arm, ERA (European Robotic Arm). ERA is a derivative of the HERA (Hermes Robotic Arm). It will be used on the Russian segment of ISS to support assembly and servicing tasks such as the installation and exchange of external equipment and also external inspection. The 10 m long symmetrical arm will have two 'wrists' and be attached at either end, thus being able to relocate itself, operating with either wrist acting as the manipulator's base. The €145 million project was approved in 1994 and has Fokker space of the Netherlands (which is the major contributor to the project) leading an international consortium from Belgium, Denmark, Germany, Italy, Netherlands, Sweden and Switzerland. Launch delivery is anticipated for 2004.

Columbus

Orbital Facility to be attached to Space Station at a date to be determined when Shuttle flights resume. €1,234.1 million allocated 1996–2004, plus €206.7 million 1996–2003 for µg equipment.

On 11 and 12 May 1999, the Council of Ministers met in Brussels to confirm the main objectives of the ESA Long Term Plan for the period 1999–2006, incorporating new programmes spanning the fields of space science, access to space (launchers), space applications (Earth observation, telecommunications, navigation and multimedia), exploitation of the International Space Station and advanced technology. See the table for optional programmes.

On top of these optional programmes, the Ministers approved the Level of Resources for the period 1999–2003 amounting to €2.17 billion of which €1.85 billion is for Scientific Programme and the remaining amount for the General Budget. The Level of Resources sees a constant level for the Mandatory Activities (General Budget and Science) at €531.2 million per annum based on the 1998 budget.

Science Programme

Based on the long-term plans Horizon 2000 and Horizon 2000 Plus. The programme was drafted in a consensual manner with the involvement of the whole European community (in the following, brackets indicate actual or projected launch dates as of July 2001).

Four ESA space science missions are in orbit in 1999:

- Ulysses, launched October 1990, exploring the heliosphere;
- the Hubble Space Telescope, launched in April 1990, a joint ESA-NASA mission;
- SOHO (part of Cornerstone), the Solar & Heliospheric Observatory launched December 1995;
- Huygens/Cassini mission to Saturn and Pluto launched October 1997 in collaboration with NASA.

Five spacecraft are under development:

- XMM, X-ray Multi-mirror Mission, the second cornerstone of Horizons 2000 launched in December 1999;
- Cluster II, part of first Cornerstone, four-satellite magnetospheric mission for launch in July and August 2000;
- INTEGRAL, International Gamma Ray Laboratory, to be launched April 2002;
- Rosetta, third Cornerstone, for analysis of in-situ comet, to be launched January 2003;
- FIRST/Planck, combination fourth Cornerstone and smaller mission, submillimetre launch Febuary 2007.
- SMART-1, advanced technology Moon mission for launch 2002;
- MiniSTEP, mission to test Equivalence Principle, launched 2003 in collaboration with NASA;
- four candidates for launch of next Cornerstone 10 years from now: Mercury mission, advanced interferometric astrometry mission (GAIA), interferometric mission to find terrestrial extra-solar planets (IRSI), mission for detection of gravitational waves (LISA).

At its November 1998 meeting the Science Planning Committee worked on two priorities defined by the scientific community:

- Mars Express (to be launched in a launch window beginning May 2003) unanimously approved at a cost of €150 million under the condition that adequate resources are available and that there is no impact on already approved missions. This mission subsequently included the UK Open University Beagle 2 lander, first as a payload and then as a separate spacecraft.
- the decision was taken to ensure the presence of the European astronomical community on board the HST mission, the plan being to link the HST and Next Generation Space Telescope (NGST) in negotiations with NASA. Collaboration in these two missions should not cost more than €175 million.

The overall science budget requirements for 1993–2003 amount to €1.85 billion, contributions shared on basis of GNP of each member state.

Optional programmes for the ESA

	million Euro (€)
1. Navigation/GalileoSat (€40 million until end-2000 + €460 million until end-2005)	500
2. Telecommunications/Preliminary studies – ARTES 1 – (2000–2005)	50
3. Telecommunications/multimedia – ARTES 3 – (1999–2002)	309
4. Earth Observation/Living Planet programme (1999–2002)	759
5. International Space Station/Exploitation (2000–2001)	344
6. Microgravity follow-on Programme (2000–2003)	98.4
7. Ariane 5 Plus (step 2 until end-2001)	462
8. Vega (step 2 until end-2002)	317
9. Ariane 5 ARTA (extension of two years – 2001 & 2002)	161
10. Ariane 5 Infrastructure (extension of one year – 2001)	26
11. Future Launchers Technology (until end-2001)	70
12. CSG Korou (extension of one year – 2001)	87

ESA spacecraft Mars Express in final assembly and checkout (ESA) 0572491

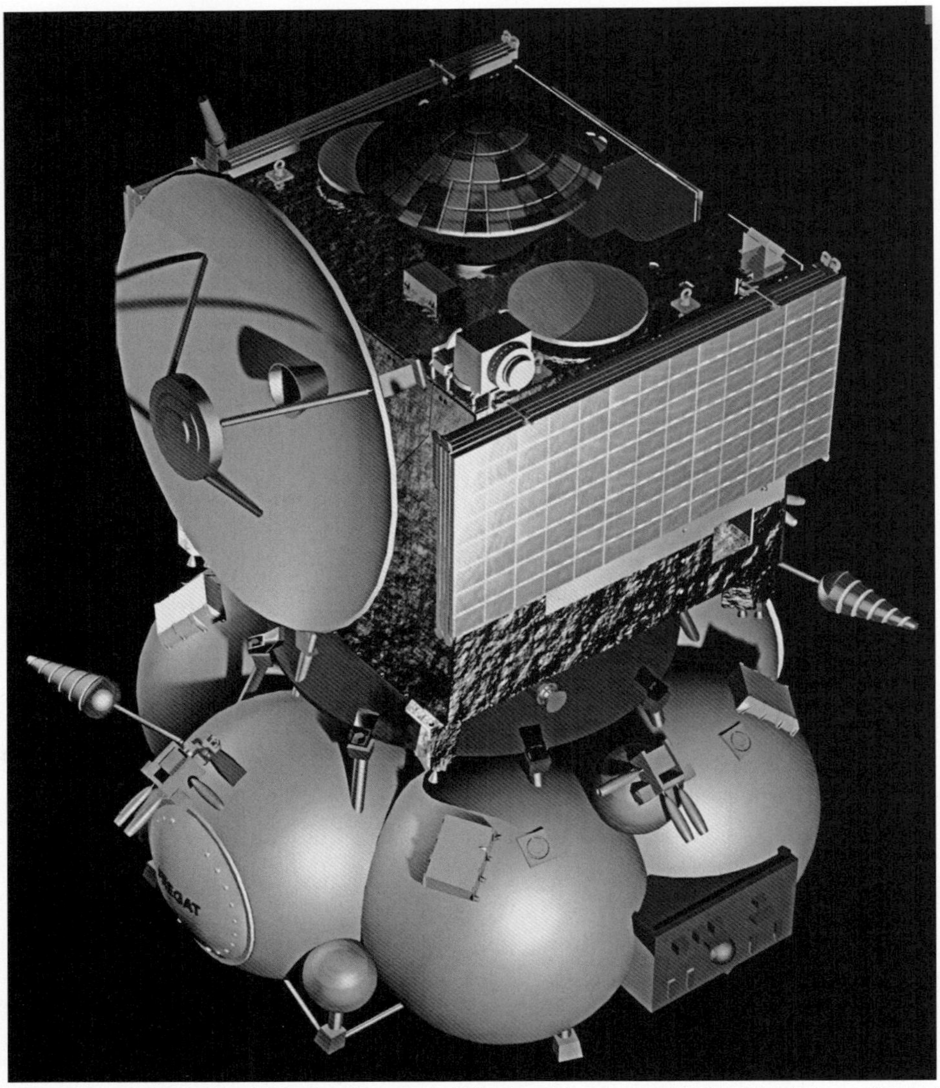

Mars spacecraft attached to the Russian Fregat stage indicative of the co-operation that exists between ESA and the Russian Space Agency (ESA) 0572490

Earth Observation programme
Better known now as the Living Planet Programme, earth observation programmes comprise two main components:

- Earth Explorer Component, including the definition, development, launch and operations of Earth Explorer (Core and Opportunity) missions covering Earth's interior, oceans, atmosphere, cryosphere and land surface;
- Development and Exploitation Component including instrument pre-development for Earth Explorer and Earth Watch; definition of Earth Watch type missions and the preparation of detailed programme proposals for optional Earth Watch type programmes; mission exploitation/market development.

The Living Planet Programme is an optional programme implemented in five-year successive periods, its activities starting under the Earth Observation Preparatory Programme (EOPP) funded in June 1998 as a transitional measure with the following commitment:

- continuation of ERS operations beyond January for three years;
- one Earth Explorer Core Mission;
- two Earth Explorer Opportunity Missions.

The start of the Earth Watch missions includes instrument development, technology development and market development. A new decision will be made in 2001 to include the development of a second Earth Explorer Core Mission in order to accommodate the objectives of the first five years of the Earth Observation Envelope Programme.

Telecommunication Programmes
The new ESA Telecommunications Programme is based on a portfolio of activities that reflect the need to adapt swiftly to market conditions, the main instrument being the ARTES (Advanced Research in Telecommunications Systems) programme consisting of several elements covering different activities and recently extended to include the period 2001–2005. ARTES 1 is intended to carry out preparatory activities in order to identify new mission opportunities and to identify and study major opportunities for partnership programmes. ARTES element 3 is directly aimed at supporting privately led European Multimedia Satellite Systems to offer industry support in the start-up phase of new projects with which there is difficulty in finding finance from the market.

Satellite Navigation Programme
The Global Navigation Satellite System (GNSS), called Galileo, will become a central feature of all European transport systems, and the European Union, in close co-operation with ESA, has defined a strategy which includes:

- the implementation of a navigation satellite constellation and associated ground infrastructure providing worldwide coverage compatible with GPS;
- international co-operation with the US to ensure system compatibility and with Russia to benefit from GLONASS know-how;
- an approach in phases including a definition phase until 2000, a development/validation

ESA's budget by %

By country	1991	1992	1993	1994	1995	1996	1997	1999	2000
Austria	0.76	0.93	0.92	1.05	1.14	1.04	1.10	1.30	1.4
Belgium	4.04	4.25	4.49	4.90	5.90	5.51	5.70	5.30	5.1
Denmark	0.80	0.80	0.85	0.85	0.94	0.92	1.00	1.10	1.1
Finland	0.28	0.30	0.30	0.42	0.54	0.60	0.50	0.60	0.7
France	24.58	25.66	27.48	26.07	23.92	26.59	26.80	29.00	29.4
Germany	20.53	20.54	21.09	19.10	19.01	20.11	20.10	25.00	25.7
Ireland	0.19	0.21	0.18	0.20	0.19	0.20	0.20	0.30	0.3
Italy	15.07	14.52	15.96	14.61	14.37	14.23	11.90	13.70	14.0
Netherlands	2.19	2.15	2.54	2.58	2.97	3.56	3.50	3.60	3.1
Norway	0.65	0.70	0.70	0.77	0.80	0.83	0.80	0.90	1.0
Spain	3.95	3.99	3.96	4.04	3.93	4.10	3.70	4.50	4.0
Sweden	1.87	1.89	2.08	2.29	2.29	2.39	2.20	2.60	2.6
Switzerland	1.96	2.02	2.13	2.38	2.62	2.61	2.70	3.30	3.3
United Kingdom	5.12	5.04	5.84	6.50	6.47	6.35	6.30	8.00	7.4
Canada	0.48	0.59	0.70	0.66	0.67	0.66	0.50	0.60	0.7
Other income	17.51	16.42	10.78	13.57	14.36	10.29	13.00	0.20	0.2
By activity	**1991**	**1992**	**1993**	**1994**	**1995**	**1996**	**1997**	**1999**	**2000**
General Budget	5.94	5.86	5.96	6.72	6.35	6.22	7.10	6.30	6.3
Associated to GB	3.19	3.41	3.34	4.03	4.03	3.96	4.40	4.10	4.3
Science	9.62	9.74	10.44	13.08	12.80	12.85	14.50	13.40	13.2
Earth Observation	7.43	6.81	9.65	15.96	18.63	18.59	18.80	24.30	18.1
Microgravity	2.26	2.49	3.12	3.56	3.47	3.30	4.10	3.70	3.8
Telecommunications	9.25	8.64	9.48	10.63	10.86	10.81	7.70	7.80	6.5
Manned[1]	12.16	12.16	14.06	5.14	10.04	9.74	13.30	16.40	18.0
Launchers[2]	47.23	46.94	40.54	37.73	29.90	32.04	25.70	18.40	19.6
TDP 1/2 & Prodex[3]	0.73	0.73	0.50	0.83	1.16	1.19	1.50	1.50	1.8
Third Party Programmes	2.18	3.22	2.91	2.37	2.76	1.22	2.70	3.90	2.8
Transformation Programme	–	–	–	–	–	–	0.20	0.10	–

1997 total AU2,417.3 million from Members AU2,104.1 million. 1996's budget totalled AU2,699.5 million, including AU2,294.7 million from Member States. ESA's financial year runs 1 January to 31 December. For accounting purposes, the agency employs the Accounting Unit, which is taken as the average value of each national currency, based on the ECU (€), in June of the previous year. 1 AU = US$1.33 for 1996 figures (1.18 1995; 1.19 1994; 1.30 1993; 1.15 1992; 1.22 1991).
The columns may not total 100 per cent due to rounding.

Notes
[1] Space Stations & Platforms' until 1995.
[2] Space Transportation Systems' until 1995.
[3] this element includes Hipparcos operations in 1991-92, and GSTP from 1994. Previous budget totals were: 1983 977, 1984 1,124,1985 1,139, 1986 1,457, 1987 1,605, 1988 1,882, 1989 2,030.5, 1990 2,070.4, 1991 2,552, 1992 2,731.6, 1993 2,817.3, 1994 2,531.3, 1995 2,679.4 MAU.

phase financed mainly through public funding and completed by 2006 and an operational phase involving private funding leading to full operational capability of the Galileo system by early 2008.

ESA's GalileoSat Programme involves the definition, development and validation phases and if private funding is forthcoming this will be added to the €500 million equally divided between the EU and ESA.

Manned Space Flight & Microgravity Programmes
Exploitation of the International Space Station is aimed at developing European operational capabilities for long-term manned space exploration, to build up the necessary know-how to master operations with complex manned outposts in space and to encourage support of the ISS by the European user community. Technical content includes operations and maintenance of the ESA/ISS elements and ground segments (Columbus pressurised module, Automated Transfer Vehicle, data management system. European robotic arm and microgravity facilities for Columbus as well as respective ATV and Columbus control centres. Other commitments include the fulfilment of ISS common-systems operation, including Ariane 5/ATV transportation, refuel and reboost services to the ISS and support and co-ordination of ISS utilisation by European users (payload integration, logistics, data transmission and astronauts).

Exploitation of the ISS is divided into steps: Step 1, early activities in 2000–2001 for €344 million including ATV preparation and development, Ariane 5 in support of a first operational ATV launch in 2004, NASA reimbursable services in the form of payments for ISS utilisation and bartering these services in exchange for development of the Crew Return Vehicle (CRV); Step 2, (2002–2004 will launch in 2007) €336 million completing the Toulouse commitment. Subsequently the CRV was cancelled and the Bush administration proposed through the FY2004 budget an Orbital Space Plane to sustain that role and add functions.

The microgravity research programme aims to continue with the technological development of microgravity research, implement microgravity missions involving sounding rockets, parabolic flights and Spacehab missions, support early utilisation of the ISS and continue to upgrade existing facilities and to develop new ones. The future microgravity programme is structured in two slices; slice 1 (2000–2003) with €98 million covering precursor missions such as those with sounding rockets and Spacehab flights, early ISS utilisation and new microgravity facilities; slice 2 (2002–2004) with €145 million for decision in 2001. Further details can be found in the section on human space flight.

Launcher Programmes
The launcher development programme aims to acquire new types of launch vehicle to orbit larger telecommunications satellites, satellite constellations and small payloads. Strategy is based around improvements to Ariane 5, to develop complementary launch capabilities and to prepare the technology for future reusable launchers to maintain Europe's strong lead in launch services for global customers.

The Ariane 5 Plus programme aims to increase the performance of the Ariane 5 in a stepped sequence. This includes the Ariane 5V version in 2002, the Ariane 5C-A programme providing a 9-tonne GTO capability in 2002 followed by Ariane 5C-B with an 11-tonne GTO capability in 2006, thus doubling the present Ariane 5 lift capability. The programme has a total budget of €1.064 billion broken down into three steps. The first step for €108 million was decided in June 1998 and is ongoing, the second step is funded at €462 million to cover development until the end of 2001 and the third step involving €484 million covers the period 2002–2006.

In the Vega small launcher development programme with a total multiyear funding of €370 million agreed in June 1998, ESA planned to develop a low-cost launcher for payloads around 1-tonne into LEO. Uncertainties about a solid market for Vega have led to ESA withdrawing from the programme. In the Future Launchers Technology Programme ESA is analysing a variety of optional launch vehicle concepts for cost-effective and reusable space transportation with a decision required around 2007 for implementation of a development programme. Additional activity is proceeding on maintaining and improving the level of reliability for Ariane 5 with a view to developing increased efficiency, recovery and reuse of solid-propellant boosters and enhanced mathematical models.

UPDATED

Inter-Agency Solar Terrestrial Physics Programme

Current Status
The Inter-Agency Solar Terrestrial Physics Programme is a 12-satellite project involving NASA, ESA, ISAS and Russia's Institute of Space Research. IASTP is undertaking detailed investigation of the Sun, Earth's space environment and Sun-Earth interaction. The Inter-Agency Consultative Group (IACG), selected Solar Terrestrial Science as the next discipline area for multilateral co-operation following the highly successful Comet Halley collaboration. IASTP, started in 1977 with planned completion in 2001 but sustained on a year by year basis, involves more than 100 universities, research laboratories and major contractors in 16 countries. Informal approaches have been made by several countries in the Middle East for participation but nothing has been concluded to date.

ESA's Soho and Cluster contributions, pursued in collaboration with NASA as the STSP Solar-Terrestrial Science Programme, comprise five which were to have been satellites launched in 1995/6. NASA includes its involvement in Soho, Cluster and Geotail under the ISTP International Solar Terrestrial Physics programme. IKI planned to fly two Regatta satellites 1993-96, the second closely complementing Cluster, but severe financial constraints forced indefinite postponement. Launched in 1995 and 1996, the four-satellite Interball is contributing, however. The US Global Geospace Science (GGS) element is designed to measure the terrestrial portion of the solar-terrestrial connection. Its Wind (launched November 1994) and Polar (February 1996) are complemented by ISAS' Geotail (July 1992). Wind is using lunar swing-bys to hold its apogee on Earth's dayside to survey the upstream region out to 250 Earth radii. After two years, it will enter a small halo orbit around the Sun-Earth L1 Lagrangian point to act as the solar wind monitor for the other craft.

From these measurements, models will trace the energetic plasma particles and fields in the solar wind through acceleration and storage in the Earth's magnetosphere, to their ultimate deposition in the atmosphere. In the polar regions, this energy deposition equals the solar radiant energy input, and therefore has a significant influence on the atmospheric energy balance. The results should make it possible to predict disruptive effects on the ground and in space.

NASA budgeted US$18.6 million in FY88 to start development of Wind/Polar (rising to 64.4 FY89, 57.6 FY90, 96.6 FY91, 75.3 FY92, 72.6 FY93, 21.6 FY94, 40.0 FY95, 0 FY96) and provide instruments for Geotail, Soho and Cluster. Russia's KONUS v-burst detector was delivered 1992 to fly aboard Wind, the first CIS instrument on a US satellite. GE Astro Space was awarded the prime Polar/Wind contracts in January 1989, ending at US$175 million. Total NASA development cost for GGS, Geotail, Soho and Cluster is estimated at US$732 million. Wind/Polar design/construction difficulties resulted in launch delays, leaving Geotail as the first in orbit. Wind was finally launched by Delta 7925 from Cape Canaveral on 1 November 1994. Followed by Polar launched in February 1996, Soho was launched to solar orbit in November 1995 followed by Cluster in July 2000.

VERIFIED

Interboll

Current Status
Originally part of the Interkosmos programme with flights planned for the 1980s. A major part of the Inter-Agency Solar Terrestrial Physics programme, Interboll is a 20-nation collaborative project to investigate the interaction of solar wind with Earth's magnetosphere and its acceleration down into the auroral oval. Experiments developed as part of the Interboll programme were flown on other Interkosmos missions. Satellites based on the new Prognoz-M2 design are operating in two distinct orbits, one monitoring the magnetotail processes in conjunction with its S2-T sub-satellite, and the second at a higher inclination for auroral observations with its S2-A companion. The Czech Institute of Atmospheric Physics' Upper Atmosphere Dept provides the Magion sub-satellites. The lower craft includes a Canadian UV imager first flown aboard Sweden's Viking auroral research satellite in 1986. Relikt 2, should it appear, would provide additional fields/particles data deep in the magnetotail from its halo orbit around the L2 libration point 1.5 million km on the anti-Sun side from Earth. Interball is part of the Inter-Agency Consultative Group plan. The magnetospheric tail pair were launched on 2 August 1995 followed by the auroral pair on 29 August 1996.

Interkosmos

Current Status
Interkosmos, formally the Council for International Co-operation in the Studies and Uses of Outer Space, was responsible for collaborative science projects among its 10 member states and with other nations. Based at the Academy of Sciences, it was closed in 1994 with the launch of Interkosmos 26 and the projects continued under IKI, IZMIRAN and other institutes. Current activities include the Aktivny, APEX, Interball and Relikt science satellite

RFAS/USSR-led International Earth-orbit Science Missions

Project	Launch Date	Spacecraft	Orbit	Objective
Aktivny-IK	September 1989	AUOS parent + Magion sub-satellite	Low Earth Orbit	Study the effects of ULF waves on the magnetosphere; Interkosmos project; parent still operational
Granat	December 1989	Astron-type	Prognoz-type	Study of gamma/X-ray radiation with France/Denmark; still operational
Gamma	July 1990	Progress-type	Low Earth Orbit	Study of gamma/X-ray radiation in co-operation with France
APEX	December 1991	AUOS parent + Magion sub-satellite	Low Earth orbit	Electron and plasma probing of magnetosphere; Interkosmos project; parent still operational
Coronas-I	March 1994	AUOS-SM	Low Earth Orbit	Helioseismology and solar activity; Interkosmos project, still operational
Interball	August 1995	Prognoz-M2	Prognoz + intermediate	Study of magnetosphere and plasmasphere in 14-nation
	August 1996	+ Magion sub-satellites	Molniya-type	co-operation; one Prognoz/Magion pair in each orbit type
Priroda	April 1996	Mir module	Low Earth Orbit	Mir remote sensing module, including France's Alissa radar
Bion 11	December 1996	Vostok-type	Low Earth Orbit	Life sciences research on living organisms, including 2 monkeys; NASA involvement
Relikt 2	?	Prognoz-M2	L2 halo orbit	Study of Universe's microwave background
Coronas-F	1997?	AUOS-SM	Prognoz-type	Helioseismology and solar activity; Interkosmos project
Bion 12	1998	Vostok-type	Low Earth Orbit	Life sciences research on living organisms, including 2 monkeys; NASA involvement
Spektrum-X/v	1998	Spektrum-type	Prognoz	High-energy astronomical observations
Radioastron (Spektrum-R)	2000?	Spektrum-type	Elliptical 24h	VLBI study of cm-radio emissions with 10 m diameter radio telescope in conjunction with 70 m ground-based telescopes
Spektrum-UV	2002?	Spektrum-type	Elliptical synchronous?	Study of ultraviolet sources with 170 cm diameter telescope

projects. There was no joint Interkosmos fund: Russia offered its technology and launchers free of charge and each country bore its own costs. No new missions have recently been announced and existing projects continue to suffer delays. Several efforts have been made to revitalise the Interkosmos programme and member countries have generated several initiatives which could form the basis for integrated space science research programmes between Interkosmos countries and ESA member states.

Interkosmos was created in May 1966 by the Soviet-affiliated countries of CMEA (Committee for Mutual Economic Assistance): Bulgaria, Cuba, Czechoslovakia, GDR, Hungary, Mongolia, Poland, Romania and USSR; Vietnam was added in 1979 and the UK participated with the AUOS satellite launched on 2 March 1994. Areas of interest covered space physics, meteorology, communications, biology and medicine, with Earth resources added in 1975. Specific emphasis has been made of fields/particles studies of solar radiation and its interaction with Earth's magnetosphere. The first launch, Cosmos 261 in 1968 from Plesetsk, studied air density and polar auroras. Each member financed its own contributions, with the CIS/USSR supplying the basic satellite and the launch; scientific results were the common property of all Interkosmos members.

The first programme ran 1967–80, with Czechoslovakia a noted contributor to every unmanned mission. This culminated in its provision on Interkosmos 18 of the Magion sub-satellite, the first non-Soviet satellite in the series, Magion 2 in 1989–90 and #3 in 1991 (#4/5 are flying within Interball). The second programme, for 1980-85, saw a move towards Earth surface studies, with the ocean dynamics investigations of #20/21. By October 1984, when the 17th Interkosmos conference was held in Berlin, 11 high-altitude Vertikal research rockets and 22 Interkosmos satellites had been launched, the last in August 1981.

The outline of the programme issued for 1986–90 noted there would be economic use of the results already achieved, but made no mention of any new launches. However, as a member of the Inter-Agency Consultative Group, Interkosmos was closely involved in the Vega missions and the international collaboration on the 1985/86 studies of Halley's Comet, and a major contributor to 1988/89's Phobos mission. In April 1985, Prognoz 10 (Intershock), one of the solar radiation series, was co-designated Interkosmos 23. Aktivny, APEX and Interball subsequently appeared as major projects and in October 1987 Interkosmos announced proposals for two Coronas observatories that would monitor the Sun during its period of maximum activity in the early 1990s. The SAS aeronomy satellites were described in 1989 (but later cancelled) and agreement was near on the Regatta contribution (now postponed indefinitely) to the Inter-Agency Solar-Terrestrial Physics programme. During 1988, a second Bulgarian cosmonaut, Alexander Alexandrov, made a Soyuz flight, apparently in compensation for the failed Soyuz 33 docking with Salyut 6 in 1979.

With the exception of Interkosmos 6, a Vostok-type, and #15, an AUOS Automatic Unified Orbital Station prototype, #1-16 were light research satellites. Weighing 320 to 375 kg, they were based around a standard ellipsoidal pressurised shell. Czech scientists made contributions to all these satellites and provided the orientation system for the more advanced 422 kg AUOS, introduced in operational form with Interkosmos 17. Features included onboard data processing and provision for a sub-satellite. APEX is flying the last AUOS-Z model (Z=Zemlia for Earth, denoting three-axis pointing about z-axis as vertical). Coronas-I introduced the AUOS-SM in 1994; the Prognoz-M2 debuted on Interball in 1995. The last Interkosmos mission, Interkosmos 26 was launched byTsyklon 3 from Plesetsk on 2 March 1994.

Orbiting Satellites Carrying Amateur Radio (OSCAR)

Current Status

The most recent satellite in this series was Oscar 50, also known as SaudiSat-1C launched by a Dnepr from Baikonur cosmodrome to a low earth orbit on 20 December 2002.

Background

The Orbiting Satellites Carrying Amateur Radio (Oscar) series of small satellites was initiated for radio amateurs to experience satellite tracking and participate in radio propagation experiments. The World Administrative Radio Conference allocated frequencies for the Amateur Satellite Service, including 29 MHz (10 m), 145 MHz (2 m), 435 MHz (70 cm), 1,270 MHz (24 cm) and 2,400 MHz (13 cm). Transmitting low-powered signals, initially battery-operated and offering short lives, the satellites have become increasingly sophisticated. They have served school science groups, provided emergency communications for disaster relief, acted as technology demonstrators and transmitted Earth imagery. The UK's UoSAT series alone has involved hundreds of schools and thousands of groups worldwide in activities using simple antennas, receivers and personal computers. The work ranges from telemetry transmission only, through amateur radio communications, to advanced experiments such as testing indium phosphide solar cells and transputers, and monitoring radiation effects on electronics – satellite engineering for a fraction of the time and cost of the more-advanced conventional satellites. Somewhat confusingly, US military Transit satellites also carry Oscar designations.

The first amateur satellite was launched on 12 December 1961 and designated OSCAR 1. More than 570 amateurs in 28 countries forwarded observations to the Project OSCAR data reduction centre before the satellite burned up in the atmosphere 22 days later. OSCAR 1 led to the creation of the Amateur Satellite Corporation (AMSAT) in 1969.

OSCAR satellites are separated into three specific categories: low earth orbit voice, high-elliptical orbit and digital satellites.

VERIFIED

Iran

Current Status

Iran is a signatory to the Intelsat and Inmarsat organisations, and has hopes for its own domestic satellite communications system. In 2003 Iran made several approaches through unidentified third countries to major satellite manufacturers including the national laboratories of India and China to seek possible co-operation with other countries in the development of an Iranian communication satellite system and has expressed interest in a hybrid design similar to the first-generation Insat satellites procured from (the then) Ford Aerospace. An Iran Space Administration was believed to be in the process of formation, but little has been heard recently. Iranian P&T has contracted with Alcatel for development of Zohreh communications satellite. During early 2001 unconfirmed reports indicated that this programme may be considerably delayed due to financial restrictions. Iran has retained a strong interest in procuring a domestic satellite but is also interested in developing capabilities for military surveillance systems procured perhaps with a foreign partner. In the aftermath of the regime change in Iraq, Iran is believed to have procured the services of specialist satellite consultants for the conceptualisation and design of an independent communications system satisfying military requirements. Iran is also known to have studied the possibility of procuring a multipurpose satellite capable of providing telecommunications services for the army and medium-resolution images for intelligence gathering among border states.

UPDATED

Iraq

Current Status

Following the removal of Saddam Hussein from power all space based telecommunications projects involving deals with outside countries have been stopped and a new strategy for restoring development in these areas is presently being defined between the United States and the United Nations. With the transfer of power to the Iraqi interim government, all future decisions regarding telecommunications and space projects will depend upon new strategies for economic re-development and direction. Prior to the attack on Iraq it was believed the regime was planning to use satellites from other countries to combine telecommunications and intelligence gathering. Prior to the 1990 invasion of Kuwait it was reported that Iraq was developing a satellite launcher. The 24 m 48 tonne Tamouz 1 vehicle launched from the Al Anbar facility 80 km west of Baghdad on 5 December 1989 was claimed by the Iraqis as a satellite launcher test. Only the first of the three stages was live, reportedly attaining 12 km altitude. Stage 1 appears to comprise five clustered Al Abbas versions of the SS-1 Scud missile, topped by a single Al Abbas stage 2 and a short stage 3. A range of 2,000 km was claimed for the Al Aabed (The Worshipper) missile version. A 1989 report suggested that Brazil was to provide an SCD environmental platform data collection satellite. Technical assistance from China and France was also mentioned. Close ties with the Soviet Union suggested involvement and that associations has survived the collapse of communism in Russia. It is possible that plans revealed during 1990 to build a 1 m-bore 'supergun' under Project Babylon by the Brussels-based Space Research Corp included the firing of small payloads into low Earth orbit. No further developments have been confirmed.

Iraq has reportedly opened up discussions with Russia about access to rocket and satellite technology. In 1999, Iraq was reported to have signed a deal with NPO Mashinostroyenia for high- and medium-resolution satellite images of Iraq's immediate neighbours. It was stated that these were to be used for missile targeting and guidance plans. Since then Iraq has acquired images from most outside, commercial, sources via third parties.

Iraq is a signatory to Intelsat and Inmarsat.

UPDATED

Ireland

Current Status

Ireland's national space activities are managed by the Department of Enterprise & Employment, and directed almost exclusively through ESA. Forbairt manages the technological aspects of the Irish programme. Contributions to ESA were €6.7 million in 2000, €5.7 million in 1999 and €5.0 million (US$5.9 million) for each of 1995 and 1996, €5.1 million (US$6.1 million) in 1994, I£5.2 million (€5.0 million) 1993, 4.4 (5.8) 1992, 3.6 (4.8) 1991, 3.3 (4.2) 1990, 3.0 (3.9) 1989, 1.7 (2.4) 1988 and 1987, and 1.6 (2.2) 1986, amounting to 0.2 per cent of ESA's budget, the smallest contribution of any member. 0.2 per cent is contributed to the Ariane 5 programme. Ireland has been a member of ESA since 1975 and in recent years several companies in Ireland have secured prime contracts for major international projects including the Space Station. Expanded payload participation through research activities and universities in Ireland as well as parallel organisations in ESA.

Ireland's first ESA facility, the Microelectronics Technology Support Laboratory, was established in 1988 at the National Microelectronics Research Centre in Cork. MTSL provides support in semiconductor technologies, particularly microwave, opto-electronic, power, digital and analogue functions. Ireland is the sixth country to host an ESA facility. SpaceTechnology (Ireland) Ltd, was established in 1987 following the successful provision of space radiation monitors to the Giotto, Vega and Phobos international missions.

Telecom Eireann is the national signatory to Intelsat and Eutelsat. A commercial satellite communications system was announced in March 1986 by Atlantic Satellites Ltd, 80 per cent owned by Hughes Communications Inc, to use the 31° W slot allocated to Ireland. HCI later withdrew and the project fell into abeyance.

At the end of the 1990s, discussions were re-opened but a decision has been postponed. The Irish government has reopened discussions with unnamed scientific research establishments regarding utilisation of existing earth imaging and remote sensing in efficient management of coastal and inshore areas in estuaries and rivers. Ireland has opened new links to other ESA countries in an effort to establish a new and innovative way of utilising international assets and the remote sensing sector is one which is being used to demonstrate improvements in dialogue and operating mechanisms.

Irish Space Expenditure (I£ million)

Year	ESA	Total
1993	4.0	4.0
1994	3.4	3.4
1995	3.6	3.6
1996	4.8	4.8
1997	10.0	10.0
1998	4.3	4.3
1999	4.3	4.3
2000	5.1	5.0

UPDATED

Israel

Current Status

Israel has concluded an agreement with the government of Taiwan for the use of its EROS-1 imaging satellite by Taiwan and for Taiwan to control the operational use of the satellite when it comes within 1,000 km of a Taiwanese ground station. This agreement allows Taiwan to use the Israeli satellite rather than procure images from the Ikonos satellite system. With an eye to sales as a stimulus for the development of major projects, Israel has secured an agreement with Singapore for the development of a surveillance and intelligence gathering satellite estimated to cost in excess of US$1 billion. Israel continues to receive assistance from the USA and has a programme involving Israeli astronauts. The first, Col Ramon, was killed when the Shuttle Columbia broke up over the USA during re-entry on 1 February 2003. The government has since reaffirmed its intention to fly more of its personnel in space.

Background

Israel became the eighth nation to launch an indigenous satellite when a three-stage Shavit solid-propellant booster placed the Ofeq 1 demonstration satellite into LEO from the Palmachim Air Force Base south of Tel Aviv on 19 September 1988. The launcher is believed to have been developed from the Jericho 2 missile. The satellite was developed by Israel Aircraft Industries under the auspices of the Israeli space agency as a precursor to more sophisticated science and communications carriers.

US President Clinton in January 1994 decided to allow Israeli companies to compete in the US on civil space activities. Reconnaissance platforms are also a possibility. The defence minister hinted in March 1991 at such a project and there have been rumours that at least one has flown. Reconnaissance satellites can be expected to play a major role in the Middle East strategic balance in the next decade. Since the surprise attacks that began the 1973 Yom Kippur War, Israeli intelligence analysts have given high priority to developing an independent source of space-based intelligence. It was charged that immediately before that war the US withheld critical satellite intelligence information on Arab offensive military formations. After the 1990–91 Gulf War, defence minister Moshe Arens explicitly and publicly declared Israel's intention of launching an indigenous reconnaissance satellite.

The Technion Institute of Technology, has proposed three or four small satellites at 7,000 km for continuous early warning coverage, although the Knesset claimed little official interest. Ofeq 3 is primarily an imaging satellite, with UV/visible sensors. If its technical capabilities are similar to those planned for South Africa's Greensat, it is providing 2 m resolution. There has clearly been strong collaboration between Israel and South Africa on launchers and satellites.

The partnership between IAI and Core Software Technology of Pasadena, California, planned a commercial 250 kg remote sensing satellite for launch by Russian Molniya in July 1997 to return 2 m-resolution imagery. CST is responsible for the ground segment. The satellite is based on Ofeq 3. The first improved version was planned for 1998, offering 1 m PAN plus 5 m XS. At least seven regional partners are each required to invest US$60 to 80 million, the cost of flying one satellite. Each would operate a dedicated satellite within its region, with IAI/CST controlling it beyond.

ISA chairman Professor Yuval Ne'eman noted in 1990 that the Ofeq programme began in 1983 and had then cost US$150 million; a principal aim was development of a national space infrastructure. The US$200 million commercial domestic/subregional Amos 1 telecommunications satellite developed by IAI appeared in 1996 on Ariane. Amos 2 was launched in 2003 with up to 16 Ku-band transponders.

Shavit was proposed unsuccessfully in 1990 and 1994 as NASA's Meteor and Ultralight launcher, respectively. IAI is working on the NEXT commercial version of Shavit (Shavit 2), stretching stages 1/2 and adding a bipropellant stage 4. Unconfirmed reports suggest that a 4,800 km range Jericho 3 is under development, based on NEXT. Early in 1999 the Israeli government let it be known that a Defence Ministry Committee has backed development of a heavy lift satellite launcher. Called Star-460, this vehicle would be capable of lifting 28 tonnes to low Earth orbit and 12 tonnes to geosynchronous orbit, more than any other launcher currently in operational use.

MBT Systems & Space Technology, part of IAI's Electronics Division, is Israel's major space centre, responsible for developing Ofeq and Amos. Its Space Technology Center includes 3,400 m^2 of satellite integration/test facilities and a 10 m S/X-band ground station. The station's Image Processing Center handles the received Spot imagery (and presumably Ofeq 3); upgrading to handle other satellites is planned. Raw/processed data are transferred to ISA for marketing by ICTAF.

Elop, Israel's largest electro-optics company, has developed the Earth Resources Monitoring System for small satellites as part of a major company effort to lead large scale high-tech R&D programmes for remote sensing, navigation sensors (TechSat) and astronomy (see TAUVEX). The pushbroom linear CCD camera would provide 16 m resolution from 400 km but increased resolution is possible. A similar camera is aboard Ofeq 3.

UPDATED

Italy

Current Status

Italy continues to consolidate its space industry and has begun negotiations with NASA over the possible development and assembly of a habitation module for the International Space Station, an element originally to have been built by NASA but cancelled by the incoming Bush administration. Sustaining active development and production of unmanned logistical ISS support modules. President Bush announced that NASA will be directed to focus all research aboard the ISS toward space exploration endeavours rather than basic scientific research. Habitation module studies have been switched to support that objective.

Background

The Italian government established a National Space Programme (PSN) in 1979 based on a plan that was to be updated every five years. The Interministerial Committee for Economic Planning (CIPE) initially entrusted the overall management and administration of the programme to the CNR and then to ASI. Early in 1984, Italy joined West Germany in proposing that ESA should authorise development of the Columbus space station facility, operating either independently or with the US, but leading to a manned free-flying European laboratory by 1999. Columbus draws upon the technology developed by Aeritalia in building the Spacelab pressurised modules and by West Germany with Spacelab, Eureca and SPAS. Italy was instrumental in the October 1991 Munich ESA ministerial meeting in calling for the Hermes and Columbus programmes to be stretched out to reduce annual funding requirements.

The creation of the ASI national space agency in 1988 to manage the National Space Plan was accompanied by allocation of an Lit800 billion annual budget for the first three years, more than doubling 1985's Lit362.6 billion space funding. The next five-year plan was submitted in December 1990 and approved in July 1991 by the CIPE Interministerial Committee for Economic Planning, which directs government spending, totalling Lit6,260 billion. More than Lit1,300 billion for each year 1992–94, following Lit1,029 billion 1990 and Lit1,211 billion 1991. In fact, because of government financial difficulties, only Lit800 billion (US$470 million at 1993 rates) was provided for each of 1992–94, with Lit850 billion planned for 1995. Acquiring hard figures is difficult in the current uncertain climate. The previous plan concluded with Lit772 billion in 1988 and Lit742 billion in 1989.

Italy is ESA's third largest contributor, accounting for 14 per cent of member states' contributions in 2000 and 13.7 per cent of member states' 1999 contributions and encompassing a 15.0 per cent interest in Ariane 5 (plus 7 per cent in Ariane 5 future developments), 19 per cent in Columbus, 11 per cent in Envisat and 45 per cent in the Data Relay Satellite. Expenditure was previously divided almost equally between national/bilateral programmes, but the current plan has increased ESA's element to about two-thirds. Italy's financial problems meant that ESA had to take out a Lit189 billion loan on its behalf to help pay its ESA contributions 1994–96. In 1996, the Italian civil space budget was equivalent to US$600 million.

Industry is dominated by the Space Division of Alenia Aerospazio (previously Alenia Spazio SpA which is now incorporated within Finmeccanica). The significant national interest in Columbus arises from Aeritalia's expertise with pressurised modules for manned spacecraft, beginning with Spacelab, while Selenia headed communications programmes such as Italsat. Italy was a major driving force behind Olympus, contributing 31.5 per cent, and is taking the lead role in DRS. National propulsion capabilities are represented by BPD, responsible for the IRIS upper stage motor, the Ariane 4 solid propellant strap-ons and development in co-operation with France's SEP of Ariane 5's large boosters.

The last five-year plan approved development of the San Marco Scout launcher and BPD was awarded ASI's industrialization contract in 1991 aiming for a 1995 maiden flight. However, the University of Rome proposed a separate San Marco Scout programme closely based on the US Scout, prompting legal wrangling. BPD is continuing development of the current Vega design using internal funding plus Ministry of Research support for development of the main Zefiro solid motor. Should government support ever be provided, the qualification flight would be in the next decade. The San Marco site is now regarded as too costly to maintain, and Vega's principal sites would be Kourou and Vandenberg AFB using transportable equipment. Discussions are under way for commonality with Spain's Capricornio and with CNES's ESL. Development cost would be about US$200 million.

The five-year plan also approved a new contribution to Space Station outside of Italy's ESA involvement. NASA/ASI signed an MoU 6 December 1991 for three Alenia Mini Pressurised Logistics Modules to deliver Space Station supplies aboard Shuttle. Cost is about US$300 million. The agreement also covers a Mini-Lab, currently under definition. MPLM will also form the basis for ESA's Columbus module. In another bilateral project with NASA, Italy is providing a significant contribution to the Cassini Saturn orbiter. Alenia Spazio is prime for the high gain antenna, the radar mapper's RF element and parts of the RF subsystem, and Officine Galileo for the VIMS Visual IR Mapping Spectrometer visible channel and stellar reference unit. Italy might also build a Ka-band station on its soil to enhance gravitational wave experiments.

An Italian plasma physics specialist, Dr Franco Malerba (back-up Dr Umberto Guidoni) flew Shuttle's Tethered Satellite mission in August 1992. Guidoni flew the repeat TSS 1R aboard STS-75 in February 1996. AF test pilot Maurizio Cheli was selected in May 1992 by ESA as one of six new astronauts and also debuted on STS-75. A mission to Mir has been discussed with Russia. ASI proposed a corps of 10 to 15 astronauts to satisfy these and future collaborative ventures, working from the National Centre for Space Training & Experimentation, but such a group is now unlikely because of financial constraints.

Telespazio was responsible for developing the Argo system for the Civil Defence Department to provide national emergency communications and datalinks through Eutelsat within 4 h of a major disaster. An inter-service study for a PSM Piano Spaziale Militare was completed in 1990 but funding restrictions render its future uncertain. In addition to participation in France's Helios observation satellite and the proposed Sicral military/civil emergency communications satellite (more likely now a package aboard a civil satellite), it recommended a Helios successor, an electronic intelligence satellite, a navigation system, and communications and data relay satellites, all with international co-operation.

Italian Space Expenditure (Lit million)

Year	ESA	National	Total
1993	693.4	150.0	843.4
1994	660.7	150.0	810.7
1995	747.0	195.0	943.0
1996	686.2	63.74	750.0
1997	686.2	263.7	950.0
1998	572.5	355.0	868.0
1999	522.7	348.5	871.2

UPDATED

Japan

Current Status

From 1 October 2003, the Institute of Space and Astronautical Science (ISAS), the National Aerospace Laboratory of Japan (NAL) and the National Space Development of Japan (NASDA) will merge into a single body known from that date as the Japanese Aerospace Exploration Agency. The president of the new body, bearing the English acronym JAXA, is Shuichiro Yamanouchi of NASDA who was appointed to this position on 20 May 2003. ISAS had been responsible for space science and planetary exploration, NASDA had been responsible for launch vehicles, satellites and human space flight programmes, NAL had been responsbile for general aviation and space research and development. Exactly 45 years after the founding the US aerospace research and development agency NASA, Japan is following suit with a reorganisation designed to merge aviation and space technology into a single funding entity.

Japan's decision to halt operations with the HII launcher and proceed with activities using the HIIA, and a commitment to convert the HOPE spaceplane concept into a fully fledged Shuttle, points Japan firmly toward an expanded space transportation programme. The Japanese Experiment Module was scheduled for launch to the international Space Station in May 2004, but has been indefinately postponed following the grounding of the Shuttle fleet. NASDA conducted the first H2A flight on 29 August 2001. The launch of Kibo, Japanese Experiment Module, for attachment to the ISS is on hold pending return to flight of the NASA Shuttle. NASDA personnel levels continue to move upwards, increasing from 1,038 in 1997 to 1,079 in 1999 and 1,090 in 2001. ISAS maintains a staff of about 300 people.

Background

Japan was the fourth country after the USSR, USA and France to achieve national satellite launcher capability and the third into GEO. It now stands only third to the USA and Russia in space expenditure, at about US$2.5 billion state spending, or 0.04 per cent of GDP. The Space Activities Commission in January 1996 released a new 15-year plan, proposing to more than double spending to ¥7,000 billion (US$68 billion) over 1996–2010, emphasising Earth observation, Space Station, launcher and spaceplane activities, and increased deep space ventures. The influential Federation of Economic Organisations in mid-1993 recommended more than doubling such spending to ¥500 billion annually. In the light of significant financial difficulties these plans were deemed unrealistic and Japan's space budget is unlikely to increase.

The space programme has grown significantly since the early 1970s, aiming for independent launcher and applications satellite capabilities from modest beginnings. The H2, for example, became the first wholly-indigenous NASDA launcher when it debuted successfully in 1994 and an important member of the international launcher scene. Development of the Upgraded H2A began FY96 with a flight debut in August 2001 followed by the second flight in February 2002, the third flight in September 2002 and the fourth in December 2002. The fifth flight in March 2003 carried a Japanese military reconnaissance satellite but the sixth flight launched November 2003 failed when two of the strap-on boosters failed to separate after burnout. The H2A has been developed as a family capable of delivering 4 tonnes GEO and 20 tonnes LEO. Unit cost is about ¥9 billion.

Japan is aiming ultimately for a national manned space presence, gaining experience via Spacelab and JEM. These also support the growing μg interest, which includes the Space Flyer Unit, USERS and Spacelab. A national space station and a lunar base have been identified as specific goals for the next several decades, supported by LEO servicing and space transfer vehicles.

Until 1989's launches of Superbird (Space Communications Corp) and JCSat (Japan Communications Satellite Co, now Japan Satellite Systems), telecommunications satellites had been state-owned and operated through the Telecommunications Advancement Organisation of Japan (TAO). The CS-3/BS-3 series and, now, NTT's NStar provide national telecommunications and direct TV services. US pressure helped to open the follow-on programmes to foreign bidders, resulting in both NStar and the BSAT DTH satellites of Broadcasting Satellite System Corp being built in the US. NASDA continues to develop and operate the ETS series of technology demonstration satellites.

Earth observations are well established in Japan: GMS provides routine global meteorological imaging from GEO, the MOS Marine Observation Satellite system was operational 1987–96 and was joined by the synthetic aperture radar of JERS in 1992. The Science & Technology Agency has undertaken an extensive study of remote sensing requirements through 2010 to improve environmental monitoring and research, while complementing other nations' programmes. The inclusion of military imaging satellites in the new National Defence Outline was under consideration in 1994, with reported support from the Defence Agency. The US has traditionally served Japan's intelligence requirements for satellite imagery, with the drawback of limited access.

Although it operates from a modest budget, ISAS maintains an impressive science and launcher programme. The two Comet Halley probes of 1985–6 demonstrated a deep space capability and in March 1990 ISAS established the first lunar orbiter since 1976. The agency was granted permission during 1989 to develop a larger M-5 all-solid launcher which made its first flight in February 1997 carrying the Haruka radio telescope that will test techniques for deploying large space structures before performing a radio astronomy mission. The second M-5 carried the Mars orbiter Nozomi launched in July 1998 and future vehicles are assigned to carry Lunar-A, a combined moon orbiter/penetrometer mission. NASDA has also developed the all-solid vehicle, the J1, to carry LEO payloads too small for the expensive HIIA. ISAS is also researching reusable launch systems and has completed several flights with its experimental Reusable Test Vehicle.

Independent of the national programme, journalist Toyohiro Akiyama visited the Soviet Mir space station during December 1990 under a commercial agreement between Glavcosmos/NPO Energia and Japan's TBS television network.

Space budget summary (¥ billion)

JFY	NASDA	ISAS	Total
1974	47.56	5.08	57.47
1975	61.42	6.54	77.04
1976	72.65	7.14	88.01
1977	79.17	8.42	94.91
1978	79.41	9.23	95.96
1979	81.84	10.28	99.46
1980	83.73	10.37	102.01
1981	85.36	11.24	105.00
1982	86.38	12.93	108.47
1983	86.07	15.18	113.39
1984	84.36	15.83	112.95
1985	88.86	10.96	112.55
1986	90.66	12.38	117.38
1987	92.648	11.825	121.924
1988	96.534	19.790	141.781
1989	106.76	20.79	154.99
1990	116.293	18.022	162.160
1991	128.313	20.841	177.658
1992	140.789	20.87	189.66
1993	152.184	20.646	201.700
1994	164.348	21.414	217.502
1995	172.398	21.414	228.812
1996	182.2	22.072	231.245
1997	180.3	21.970	202.27
1998	185.1	22.440	207.54
1999	191.9	–	–
2000	200.2	–	–
2001	200.8	–	–

April 1996 exchange rate was ¥110 per US$. Figures are for Japan's Fiscal Year of 1 April – 31 March. They exclude private sector spending on production items such as telecommunications satellites and industry-funded R&D. The total does not include non-government contributions, such as from NTT, to NASDA's overall budget since 1979. Some three-quarters of government funding is routed through the Science & Technology Agency, with the remainder through the ministries of Education, International Trade & Industry, Transport, and Post and Telecommunications. ISAS budget for 1997 was not available as this went to press.

UPDATED

Korea, South

Current Status

South Korea seeks to build its second-generation satellite system with a large proportion of the technology from domestic sources. Financial problems have delayed these plans, however. The country's programme to develop space technology began with the launches of two small satellites in 1992/3. An agreement worth £2.7 million (US$4.6 million; actual cost was less than £2.4 million) was signed in 1990 by KAIST with the UK's University of Surrey and associated Surrey Satellite Technology Ltd covering five years training, one Kitsat micro-satellite, parts for a second and a ground station. Students took the university's MSc in Satellite Communications & Spacecraft Engineering and the UoSAT 5 project was used for training. The second satellite was built in Korea at a total project cost of less than US$5 million. The indigenous Kitsat 3 was launched on 26 May 1999 by India's PSLV 2.

China and South Korea expected to reach an agreement in 1994 on a US$24 million project to launch the CKETS Chinese-Korean Engineering Test Satellite medium-class LEO satellite in 1998 to provide communications and data collection from environmental platforms, but the project was cancelled. Instead, KARI selected TRW Space & Electronics Group in 1994 in competition with Lockheed and Matra Marconi to jointly develop the Korea Multipurpose Satellite for launch in 1999 carrying science and communications payloads.

KARI launched the first indigenous sounding rocket (KSR 1, the Korean Sounding Rocket 1) on 4 June 1993 to 39 km altitude to study the ozone layer. The one-stage 6.7 m long 41.3 cm diameter 1.1 tonne rocket was unguided. KSR 2, with the same payload, attained 55 km on 1 September 1993. Development of a two-stage vehicle is under way. This has been delayed due to financial restrictions on government spending. South Korea contracted TRW to build an ocean colour sensor for its Kompsat multipurpose satellite assembled at the Korean Aerospace Research Institute. Kompsat was launched by Taurus from Vandenberg AFB on 21 December 1999. Based on research carried out in Korea by the KARI further studies of multipurpose platforms have been conducted and several new designs have been proposed for evaluation. The development of multispectral imaging and remote sensing satellites are considered an important part of Korea's future space project planning and several government institutions have been asked to recommend technology support programmes leading to an indigenous effort. South Korea has successfully integrated civil interests with those related to national defence and security and is planning further use of existing satellites in space by downlinking or leasing services.

UPDATED

Luxembourg

Current Status

Luxembourg's space activities are focused on the Astra TV distribution satellite system that began operational broadcasting to Western Europe in February 1989 using a bought-in satellite. Twelve Astra satellites have been launched to date (seven at 19.2° E and five at 28.2° E) and Astra has filed for a further eight slots to accommodate expanding market needs. In 2000, a third orbital position was opened at 23.5° E. The government is also a signatory to Intelsat and Eutelsat, but not Inmarsat. Although there is no national space programme and limited industrial capacity, Switzerland is a member of the European Space Agency and, in 2000, contributed the sum of €71.9 million, or 3.3 per cent of all members' contributions. Luxembourg has sought participation in ESA telecommunications satellite technology research. The possibility has been mooted of Luxembourg hosting a research and development organisation. This would link pan-European countries, irrespective of other trade or economic alliances, in a grouping to support scientific research linking commercial interests with space projects. The intention would be to support companies with commercial science, technology or engineering experiments likely to benefit from international finance or multi-national support from private organisations and companies outside the region.

VERIFIED

Malaysia

Current Status

The government, in 1994, identified three principal strategies for developing national space technology capabilities; involvement of the private sector in foreign space programmes, establishing formal links between Malaysian institutes with foreign space agencies and fostering closer regional governmental relationships. The country is in the process of formulating a more comprehensive national space policy and programmes. National telecommunications policy is viewed as a catalyst to Malaysia achieving the status of an industrialised nation by 2020. Also, Malaysia is interested in taking a lead role in development of South East Asia communications systems development.

The telecommunications company Binariang launched the MEASAT 1 (Malaysian-East Asian Satellite) telecommunications satellite to provide domestic and regional services. MEASAT 1 was launched on 12 January 1996 by Ariane-44L, MEASAT 2 followed on 13 November 1996. Binariang signed a MoU with Arianespace in January 1995 for launching a 50 kg Malaysian educational and science micro-satellite. This is being developed with India's assistance, under an agreement signed on 3 August 1995 between the Maxstar (Malaysian Space and Telecommunications Research) consortium and ISRO. The payload includes amateur communications (voice/data), a CCD Earth observation camera and a GPS receiver. The cost for two or three ground stations will be M$45 million to M$50 million, funded by the government and private sector. The consortium includes universities, government agencies and private companies such as Binariang and Telekom Malaysia. MEASAT 3 and 4 are destined for location at 91.5° E and 148° E respectively. Measat 3 is scheduled for launch by International launch Services on a Proton vehicle in 2005.

The MACRES Malaysian Centre for Remote Sensing is being upgraded and operates a major ground receiving station. Meteorological satellite data are already received. Limited expansion of remote sensing data. Malaysia is expanding its contacts with the Indian space industry and with government agencies in India and in other countries.

UPDATED

Mexico

Current Status

Eighteen million out of a population of 75 million had no access to telephones or TV in 1982 when Mexico ordered two Hughes HS-376 satellites to provide national telephone, TV, data and facsimile services. To launch the satellites, Mexico agreed with NASA to use the Shuttle if Mexican payload specialist, Rodolfo Neri Vela, could fly aboard the Shuttle for the second launch in November 1985. Morelos 1 was launched on June 18 1985, and retired in March 1995 and Morelos 2, launched in November 1985 aboard STS-61B, had to be placed in a drift orbit moving over two years to its final position because financing problems and the September 1985 earthquake delayed completion of the ground segment. A request for proposals for a replacement was released in 1996. Hughes also won the contract in 1991 for two second generation Solidaridad satellites, based on its HS-601 platform and adding L-band transponders for mobile communications. First launch was made in November 1993, with Solidaridad 2 launched in October 1994, followed by Morelos 3 in December 1999, all three by Ariane 4. Morelos 4 (SatMex 6) is planned for launch in 2005. Mexico has suffered from failures with Solidaridad and from lack of investment in future replacement systems.

Telecomunicaciones de Mexico (Telecomm Mexico) was created in January 1990 to assume responsibility for the country's telecommunications systems; SCT now functions as a regulating body. Telecomm was privatised in 1996. Mexico is also a member of Intelsat and joined Inmarsat in January 1994, with Telecomm as signatory.

As part of its winning Solidaridad bid, Arianespace included the future launch of a Mexican micro-satellite and training for technicians with the French/German space agencies. The separate 12 kg Unamsat 1 was built by the UNAM Autonomous University of Mexico using Amsat-NA's Microsat design, but was lost in the 28 March 1995 launch failure of Russia's Start from Plesetsk. The main mission was meteor sounding by detecting echoes from 60 W 1 to 10 ms pulses transmitted every 1 to 10 seconds on 40.097 MHz. Amateur store/forward links were to be provided on 2 m/70 cm up/down. A replacement is expected.

UPDATED

Netherlands

Current Status

The Netherlands' space activities were undertaken principally within Europe's ESRO and ELDO organisations until 1970. In 1969, however, the government decided to fund the Astronomical Netherlands Satellite (ANS) at a cost of NLG76 million, with NASA help. Instrumentation was also provided to OGO 5, Solar Max, ESRO 2/4, TD 1, COS B, ISEE 2, Exosat and Ulysses. In 1976, the government approved the pioneering Dutch/US/UK IRAS infra-red radiation observatory satellite (launched 1983), but then withdrew from large national space activities in favour of increased ESA participation. In 1999, the Netherlands' participation in ESA was 3.6 per cent, accounting for 69 per cent of its national space budget. As a full ESA member, the Netherlands contributes to the Mandatory Programme on a GNP basis (about 4.5 per cent; for 1972–91 it was 2.74 per cent). In 2000 the Netherlands contributes €67.5 million to ESA, 3.1 per cent of total members' contributions.

Some 80 per cent of the Netherlands' 2001 space budget was devoted to ESA, largely via the ministries of Economic Affairs (60 per cent) and Education and Science (20 per cent). The contribution is relatively modest, but the country hosts the agency's primary ESTEC technical R&D centre at Noordwijk. The Interdepartmental Committee on Space Research and Technology is responsible for formulating space policy. Space activity management is exercised by the NIVR, Netherlands Agency for Aerospace Programmes, a semi-governmental institute. Space science is controlled through SRON Space Research Organisation Netherlands. Industrial co-operation is formalised through the NISO Netherlands Industrial Space Organisation. Added emphasis on collection and archiving of space science and astrophysical data.

The first Dutch satellite was the ANS astronomical platform launched in 1974, followed by significant participation in IRAS and a 10 per cent interest in the SAX satellite. The SAX bilateral MoU, worth NLG65 million to Dutch companies, was signed on 29 June 1990 and accounts for the attitude control system and X-ray telescope; Fokker Space also provided the solar array. SAX was launched on 30 April 1996 and is operating successfully. The government has declared Earth observation to be a top priority; the SCIAMACHY instrument is being developed with Germany for Envisat under a MoU signed in 1994.

ESA, in 1994, approved the €180 million ERA European Robotic Arm for installation on Russia's segment of Space Station. The Netherlands has a 59 per cent share, making it the largest Dutch space project during 1995–2000. Fokker Space leads the industrial consortium.

Fokker Space represents the Netherlands principal industrial space capability. It is 50 per cent owned by Fokker NV and 50 per cent by Ultra Centrifuge Nederland NV. Its solar arrays are included on Inmarsat 2, Telecom 2, SAX, Soho, Hispasat, Artemis, Hot Bird 2-4, DRS and Envisat and its RARA retractable/retrievable array flew on Eureca. It provided ISO's attitude/orbit control subsystem and was responsible for the integration and testing of the ERS 1/2 payloads.

Through the Dutch Telecommunications Administration and the Dutch Meteorological Institute, the Netherlands is a member of Intelsat (1.1 per cent), Eutelsat (7.4 per cent), Inmarsat (3.8 per cent) and EUMETSAT (4.2 per cent).

Netherlands Space Expenditure (DFl millions)

Year	ESA	National	Total
1993	167.6	48.0	215.6
1994	146.6	50.0	196.6
1995	176.4	43.2	217.6
1996	189.3	45.0	234.3
1997	186.6	56.4	243.1
1998	170.1	90.9	261.1
1999	174.1	101.3	275.4
2000	148.7	88.1	236.9

VERIFIED

Netherlands space budget (million guilders)

	1990	1991	1992	1993	1994	1995	1996	1997
ESA contribution	145	146	153	160	150	175	200	190
NIVR technology programme	13	13	13	14	14	14	14	15
SRON space research	19	19	20	20	20	20	20	20
SAX & SCIAMACHY*	10	12	14	28	34	34	21	15

ESA Participation	
General & Science	4.5%
Telecommunications	2%
Micro-*g* & Earth Observation	4%
Ariane 5	2.1%
Manned Systems	7%
ERA	59%

* SCIAMACHY funding began 1992; SAX funding until 1996.

New Zealand

Current Status

There is no single space agency or co-ordinating group in New Zealand, but the New Zealand Spaceflight Association was formed in May 1977 to support and encourage public interest in space activities. The Spaceflight Association publishes a bi-monthly magazine. It had as its patron the late Dr William Pickering, former Director of the NASA Jet Propulsion Laboratory. Telecom Corp of New Zealand Ltd is the Intelsat and Inmarsat signatory, operating Intelsat stations at Warkworth, Wellington, Rangiora, Chatham Islands and Scott Base (Antarctica). It also leases capacity on Australia's Optus. MetService in Wellington receives and distributes NOAA and GMS satellite imagery and products. Landcare and its Dept of Scientific & Industrial Research predecessor has undertaken Earth resources satellite data processing since 1973. It maintains a large archive of satellite data of New Zealand and holds licences to distribute Spot and Landsat data. A Cospas/Sarsat ground station (CAL Corp's ATLUT-320) was installed in Wellington in late 1991 for the Ministry of Transport. New Zealand receives transmissions from the Australian Optus B series telecommunications satellite. New Zealand has participated in discussions with Micronesian island groups about the supply of weather and remote-sensing images. This has stimulated interest in New Zealand serving as the focus for processing activities in the region and for operating as a terminal for disseminating regional geographical, atmospheric and hydrological information.

UPDATED

Nigeria

Current Status

The Space Research Centre of the University of Nigeria was founded in 1972 but funds have since ceased and the facility in now entirely dependent on charitable donations. Nigeria has considered the procurement of a national communications satellite system, with three spacecraft located at 20/16/14° E employing C-band frequencies. Preliminary discussions have been held with several satellite manufacturers but no further news has been forthcoming. Meanwhile, Nigeria leases two transponders from Intelsat, with 26 ground stations, for telecommunications and TV services. Studies were also made for the establishment of a Landsat and Spot data receiving station. In February 1988, Nigeria became the 54th member of Inmarsat; Nitel is also the national Intelsat signatory. UN supported aid programmes include information provided to Nigeria about the value and effectiveness of satellite imagery and specialist advice has educated users on analysis and interpretation. Nigeria continues to anticipate a future need for a dedicated remote sensing satellite and has received technical representatives from industrial countries to explain the advantages of a multipurpose satellite integrating remote sensing and atmospheric observation combining surface and atmosphere measurements for improved weather and land-use applications.

VERIFIED

Norway

Current Status

Through the Norwegian Space Centre at Norsk Romsenter, a national earth observation resource, data from SAR and optical imaging satellites are fed to Norwegian Navy and Coastguard users as well as Norwegian Pollution Control Authority. On 16th June 1999 the Svalbard Satellite Station was opened allowing continuous reception of polar orbiting satellites on respective orbits. Norway has plans to launch a Thor 4 telecommunications satellite but excellent performance of Thor 1, 2 and 3 launched 1990, 1997 and 1998 has obviated the need for this in the near future. Thor 1, Telenor's first TV and communications satellite, was de-orbited on 7 January 2003, having been purchased from BskyB in July 1992. Norway has increased resources for utilisation of polar orbiting satellite data covering northern latitudes. Norway is pioneering several innovative applications including the first use of satellites to track porpoises in harbour areas. Set up in 1999 the activity is operated by researchers from the Institute of Marine Research and is designed to locate porpoises caught in salmon nets so that they can freed before they die. Scientists also use this method to determine where porpoises go when they leave the area during the Arctic winter.

Background

Norway became an associate member of ESA from 1981, in 1985 sharing a 1.8 per cent contribution to the general budget with Austria, Canada, Finland, Portugal and Turkey. 1985 government direct space expenditure totalled NKr39.4 million, almost equally divided between ESA and national activities. It became a full member from January 1987, contributing 0.7 per cent of the general budget and established NSC as the national space agency five months later. Direct space expenditure more than doubled from its 1986 level of NKr50.6 million (27 to ESA).

The public sector funding of space activities in 1996 amounted to NKr461 million (1995 – NKr430 million). Of this, ESA programmes accounted for 43 per cent, the balance being divided between universities/colleges, Telenor Research, Norwegian Space Centre national programmes, administration, Norwegian Defence Research Establishment and the Research Council of Norway. Space-related sales totalled NKr3.2 billion in 1996, NKr2.6 billion in 1995, and NKr2.7 billion in 1994. More than a third of government space expenditure is devoted to ESA programmes, for which Norway's €22.2 million represents some 0.9 per cent of members 1996 contributions. The Norwegian Space Centre's 1997 budget is NKr304 million (1996 – NKr310 million) which includes the ESA funding. Contracts in return from ESA totalled NKr1,320 million by the end of 1996.

The Andøya Rocket Range is a major European facility, launching primarily sounding rockets undertaking scientific investigations at high latitudes. The commercial Polar Satellite Launch Service is planned for delivering 250 kg satellites into polar orbit, but a suitable launch vehicle remains to be found. The Tromsø Satellite Station reflects strong Norwegian interest in Earth observation and was upgraded in 1991 as a near-real-time (30 minutes) processing and distribution centre for ESA's ERS radar satellite; Radarsat and JERS 1 were added later. The Swedish Space Corp acquired a half interest in 1995.

The country's principal space emphasis is on communications: its status as Inmarsat's fourth largest member, with 6.9 per cent in 1996, complements the size of its maritime fleet and the importance of the major Earth station at Eik. Norwegian firms are among international leaders in supplying coastal Earth stations and ship terminals. Telenor Satellite Services is the national signatory to Intelsat and Eutelsat and is among the largest VSAT and gateway providers in Europe. In addition, Telenor (Norwegian Telecom until January 1995) 'through its subsidiary Telenor Satellite Services' is a member of the Skyphone consortium providing aeronautical communications services through the Inmarsat system. NT bought the UK Marcopolo 2 satellite in 1992 and moved it to 1° W as Thor 1, to provide direct-to-home services in conjunction with an Intelsat at 1° W and Tele-X at 5° E; Germany's TV-Sat 2 has been leased at 1° W since March 1995. 35 channels were planned by 1998 by investing NKr1.5 billion in two new satellites. Hughes built Thor 2 for launch on 21 May 1997 and the larger Thor 2B contract went to Hughes, being launched as Thor 3 on 9 June 1998.

The navigation and positioning sector was identified in 1992 as a new focus. The Norwegian Mapping Authority in 1995 completed a network of 11 GPS reference stations providing 5 m navigation accuracy to mobile users. Comparing the known locations with GPS-derived positions allows GPS error corrections to be broadcast on AM/FM radio to users. Norway participates in the European tri-partite project EC-NOS.

The Norwegian Meteorological Institute is the EUMETSAT signatory.

The NDRE Norwegian Defence Research Establishment is leading work on the national NISSE Norwegian Ionospheric Small Satellite Experiment satellite on behalf of the space physics community. Funds come from the Research Council of Norway. Further development is pending due to lack of national funding. Cost would be NKr83 million, including launch.

The national plan approved by Parliament as a step towards ESA membership proposed that by 2000 Norway should equal an average small ESA country in space activity. This required funding of NKr220 million annually (1987 rates); 70 per cent would be channelled through ESA, and 30 per cent through national programmes. Special projects such as Tele-X, the direct broadcast satellite in which Norway originally shared a 15 per cent interest with Sweden (82 per cent) and Finland (3 per cent), would be treated on a case-by-case basis. In fiscal year 2000 Norway has made a €2.1 million contribution to ESA, 1.0 per cent of all members' contributions. At the November 1987 ESA ministers' meeting Norway indicated that it would make a 0.4 per cent contribution to each of the Ariane 5 (increased in 1991 to 0.6 per cent) and Columbus projects. It subsequently committed 0.2 per cent to Hermes but withdrew from the re-oriented programme starting 1993. The first ESA scientific project in which Norway could participate fully was Cluster/Soho, followed by Cassini/Huygens. Norway is fully participating in the replacement Cluster-2 programme following the loss of the original satellites aboard Ariane 501. Before that, its ESA activities had concentrated on remote sensing, contributing to Operational Meteosat, ERS and EOPP. ERS and Polar Platform are seen as important steps towards future operational remote sensing systems and it is intended to participate directly in both their utilisation and their ground segments.

Norwegian Space Expenditure (NKr million)

Year	ESA	National	Total
1993	164.3	60.0	224.3
1994	173.4	60.0	233.4
1995	182.5	60.0	242.5
1996	186.0	60.0	246.0
1997	176.0	60.0	236.0
1998	177.7	60.0	237.7
1999	177.0	60.0	237.0
2000	170.1	60.0	230.1

UPDATED

Summary of Norway's ESA Contributions 1991–1997 (NKr million)

Year	1991	1992	1993	1994	1995	1996	1997
Science	29.4	33.4	36.9	39.3	41.4	41.5	45.0
Telecommunications	23.8	23.5	27.4	23.0	22.1	21.2	14.4
Earth observation	19.4	24.8	30.1	38.2	43.8	48.3	49.7
Int'l Space Station	9.1	9.6	9.9	4.5	9.2	6.7	8.3
Space transportation	36.9	37.0	38.4	46.0	34.7	40.0	27.5
Microgravity	1.5	1.8	2.4	2.5	2.6	2.5	0.9
General budget	26.4	32.9	39.3	20.9	25.2	23.9	30.4
Other	–	–	–	10.5	12.1	13.9	14.5

Pakistan

Current Status

The development of a space infrastructure started with the space policy of the Space Research Council and continued with the Space Upper Atmosphere Research Commission (SUPARCO) formed in 1964. Promoting the development of sounding rockets and satallite applications, it is a low key organisation only loosely associated with space research and the Clinton administrator's sanctions imposed in 1998 restricted activities clearly linked to the development of missiles. Peaceful applications of space activities has focused on the design and development of low-cost Badr ('Full Moon') LEO satellites for communications, remote sensing and science applications. Badr 2 is part of a programme to develop a remote sensing satellite system, although the cost could require an international partner. Pakistan had planned to have the Badr C system launched in 2000 but the project has been postponed. Remote sensing applications occupy a central position in Suparco's activities. The agency's Islamabad station began reception of Landsat, Spot and NOAA imagery in 1988. This coverage has been expanded to include imported satellite data from commercial high-resolution earth imaging data.

Suparco has also studied a 'Pakstar' Radio Determination Satellite Service based on Geostar's principles. TV and telecommunications capacity is leased on AsiaSat 1. A Local User Terminal & Mission Control Centre for the Cospas/Sarsat distress system was commissioned during 1990 at Lahore. The country is a member of Inmarsat, Intelsat, ICAO, Cospar/ICSU, Cospas/Sarsat, the UN Committee on the Peaceful Uses of Outer Space (Copuos), the ITU, IAF and WMO.

Another long range goal is to develop a small satellite launcher, but western technology transfer restrictions are slowing progress.

The government sold one of its GEO slots to Alcatel Espace for a commercial telecommunications satellite. Pakistan has withdrawn from a domestic system and even the Nahuel approach of licensing the location while acting as an anchor for national use. There are currently no communications services for more than 70 per cent of the rural and remote population. Out of a country of more than 100 million people, three quarters of the 2 million working telephones serve 15 major cities. Pakistan has been approached by several consortia with proposals to begin discussions about a national satellite system. This would be funded by outside agencies and leased to the government for a sum less than the savings, which would be made through greater efficiency and the economic use of digital technology.

VERIFIED

Poland

Current Status

The country is a member of Inmarsat and Intersputnik, and became Eutelsat's 27th member in 1990 and Intelsat's 130th in January 1994. Polish Telecom is the national signatory. Inmarsat Atlantic and Indian ocean Earth stations operate at Psary, 200 km south of Warsaw, and a 13 m station become operational in 1993 for digital telephony as the Eutelsat system's 19th such station.

The Polspace consortium was formed in 1993 by research centres and businesses to promote and participate in space projects. Poland was a leading member of the Russian-led Interkosmos space science organisation, participating in the Phobos, Vega, Aktivny, Apex, Gamma, Interball, Biokosmos, Mars-96 and Interkosmos missions. Polish researcher Miroslav Hermaszewski visited the Salyut 6 space station in June-July 1978 under the Interkosmos program as part of the Soyuz 30 crew. Discussions began with ESA in 1990 on closer co-operation, with a view to possible membership. A general co-operative agreement was signed 28 January 1994. The Academy of Science's Committee for Space Research co-ordinates space science activities for execution by the Space Research Centre.

Poland has already selected several ESA programmes in which it has plans to propose participation and is a strong participant in ESA's Integral satellite mission designed to detect and measure intergalactic gamma radiation. Astrophysics study groups have assembled a set of experimental research programmes for follow-on ESA participation and some government support has been pledged for formal proposals. It is reported that several links have been established with ESA research institutes with a view to expanding the national commitment to astrophysics and space science research projects at an international level. A policy paper from the Polish government supports the view that co-operation with ESA and other international bodies is to be preferred over co-operative ventures with individual countries.

Poland has increased its participation in remote sensing, weather and earth resource analysis programmes and has focused support among non-ESA European countries for expanded use of existing resources and space-based assets.

UPDATED

Portugal

Current Status

Portugal hosted the June 2002 European conference on Jupiter exploration after Galileo and Cassini. This important event builds on previous international conferences discussing options for the exploration of Jupiter in the next decade. Portugal has maintained close links to NASA's Jupiter Icy Moons Orbiter (JIMO) proposal and has begun discussions within Europe regarding further co-operation on the exploration of the Jovian system.

Background

In the first half of 2000 Portugal became the 15th member of the European Space Agency with a contribution of €1 million. The first Portuguese satellite, PoSAT 1, was launched in 1993. A Portuguese team, within a technology transfer programme between the UK and Portugal, built it. The primary objective is to stimulate a national space industry and generate a nucleus of engineers with first-hand expertise for possible future satellite projects. The PoSAT consortium is led by INETI National Institute of Engineering & Industrial Technology and includes EFACEC, Alcatel Portugal, OGMA Oficinas Gerais de Material Aeronautico, CPRM Companhia Portuguesa de Rádio Marconi, IST Instituto Superior Técnico, UBI Universidade de Beira Interior and CEDINTEC. Overall cost was about £5 million, funded by the Ministry of Industry through the PEDIP programme. Each consortium member supported its own manpower. The programme included on the job training for 11 engineers at Surrey Satellite Technology Ltd in the UK, who also provided the ground station in Portugal. The Portuguese Planetary and Space Science community, or CEPt, has been formed to formalise contact with ESA and NASA. On 1 January 2001, Portugal became a member of the European Southern Observatory. Portugal has expressed interest in organising a European technologies and capabilities conference uniting all ESA countries. This has been supported by several ESA countries although a formal position has yet to be composed which would allow preparations to begin.

UPDATED

Romania

Current Status

Romania has recently increased its space activities although overall funding has remained static. Expectations of greater participation with former Soviet-controlled East European states have not been fulfilled.

Background

Formed in 1991, the Romanian Space Agency has extended studies in space physics and has begun studies into the utilisation of microsatellites. Romanian government has expressed interest in remote-sensing analysis and interpretation and is active in providing facilities for East European countries to join proposed organisations for disseminating optical and radar images of the earth. This has developed to a level of interim planning and will possibly result in a greater participation in international telecommunications and remote sensing projects similar to its involvement in telemedicine tests with Italy. A member of Interkosmos and Intersputnik since their inceptions, Romania, in 1990, joined Eutelsat (28th member state), Intelsat (119th) and Inmarsat (60th) through its Ministry of Communications signatory. It was the first East European country to be a signatory to Intelsat. Its initial use of the Eutelsat system covers international telephony links with six other member countries; an Earth station built by Satellite Transmission Systems of Hauppauge, New York at a cost of about US$3 million became operational in early 1991 near Bucharest. The country has no Intersputnik Earth stations and it declared in 1990 that it would not take part in the organisation's future programmes (although it remains a member). Membership of the European Space Agency has also been mentioned as a possibility; a basic co-operative agreement was signed on 11 December 1992.

Space contributions had been modest and all have been made through Interkosmos but that has changed in recent years. Participation in the Biokosmos, Apex, Interball and Interkosmos 6/17/18/20/21 science missions has been acknowledged but was at a low level. Romania was the ninth/last country to fly a cosmonaut to a Soviet space station under the 1976 Interkosmos agreement: Dumitru Prunariu spent a week aboard Salyut 6 in May 1981 and performed a brief programme of experiments, primarily biomedical studies.

The Romanian Space Industry has been formed to strengthen links between scientific research and industry and to develop a microsatellite for remote sensing and educational purposes. Since 1998, Romania has participated in telemedicine transmissions in co-operation with Italy. These have been extended recently to include areas of the Balkans.

UPDATED

Russian Federation

Current Status

Because the Shuttle programme is recovering from the loss of Columbia on 1 February 2003, Russia has since provided the only means of access to the International Space Station. While NASA expects to resume ISS support flights by the end of 2005, the decision to terminate Shuttle operations around 2010, the gap before the next US human space flight vehicle is ready (the Crew Exploration Vehicle) allocates to Russia a continuing role in support of the ISS unimagined at the outset. This elevates Russian importance in providing logistical and supply flights. Continued fiscal restraint prevent full utilisation of indigenous Russian expertise and capabilities in both industry and research institutes. Total number of payloads lofted has fallen dramatically since the early 1990s with more than half comprising orders from foreign clients for launching non-Russian satellites. There is now virtually no money for government planetary and space science research although considerable exchange takes place at an international level in data interpretation and analysis. Russian launch vehicles continue to account for a credible proportion of launch traffic in an over-subscribed launcher market.

Russia's historic emphasis on human space flight has left a legacy of technical capability and operational competence which now underpins the International Space Station. Since the loss of NASA's Shuttle Columbia and the grounding of the remaining three orbiters, Russia has been the only country capable of maintaining the ISS at a caretaker level. Progress cargo-tankers will continue to provide consumables and supplies while Soyuz TM will deliver and return to earth two-man housekeeping crews until the Shuttle can be returned to flight. The funds for this additional activity has been made available by the European Space Agency who have relinquished seats reserved aboard Soyuz ferry vehicles for European astronauts until the Shuttle orbiters resume their role. The Russian Space Agency has taken over some of the funding for continued support of the ISS which, since the Columbia loss has been maintained wholly by Russian launch vehicles and spacecraft. They will be required to continue this burden until at least late 2005 when the Shuttle should be restored to support for the ISS following two qualification flights earlier that year.

A major success for Russia's propulsion industry, the NPO-Energomash RD-180 now powers Atlas 3 and 5 series launchers providing efficiency and power levels unavailable with indigenous US motors. First stage sea-level thrust is raised from 2,120 kN of the MA-5A package for Atlas 2 to 3,827 kN with the RD-180.

Background

Although launches have fallen to about a third of their Soviet levels, the up to 180 satellites operating at any one time is unchanged and reflects improving longevity. Much of the apparent robustness can be attributed to the momentum remaining from the Soviet era, but that is now clearly dissipating. Civil and military agencies chronically complain of inadequate budget allocations, being paid late and only partially. Western currency from commercial contracts now plays a crucial role. Visiting missions to the Mir space station in 1995 brought in some Rb350 billion, highlighting why Russia has decided to try and keep Mir operational through to at least 2000. The income is essential for paying for launch vehicles, which are now hand built to specific order. Momentum was being maintained partly by drawing launchers from the strategic stockpile, but without replacing them. Likewise, much of the rocket engine production capacity has been shut down or, in some cases, dismantled.

At least the high-profile manned programme has the power and the currency to buy vehicles on a reasonably regular basis, but less glamorous projects such as the Photon microgravity and Resurs-F observation missions struggle for funding. February 1995's Photon, for example, was postponed from September 1994 because of problems in paying the rocket supplier and the VKS Military Space Forces for launch services. Photon and its Bion life sciences equivalent would probably not survive without European and US involvement. There was only one Resurs-F mission in 1994–95 because of the high price of launchers and none in 1996. The improved F1M has been ready since 1994 but there is no flight funding. Science missions are particularly vulnerable: no new programmes have started in recent years, or are likely to, and it has been a struggle to maintain operations of some already flying. The loss of Mars 96 has only made matters worse. French payments were required to keep the Granat observatory in service and Franco-German contributions were crucial for Mars 96. Future planetary involvement may depend on NASA missions.

A 12-point space pact was signed in Minsk, Belarus 30 December 1991 by heads of all members of the new Commonwealth of Independent States (CIS), except for Ukraine and Moldova. It proposed the creation of an ESA-like structure, financed proportionately depending on level of participation, while allowing independent space programmes and agencies. An Interstate Space Council would control CIS activities, while military space operations continue under central control of the joint strategic forces. The Prime Ministers voted on 13 November 1992 to form the council, although its value is not easy to determine. Russia's determination to go it alone does sideline it, though. Most importantly, the pact agreed that Baikonur/Tyuratam would provide free access for other states. Although Russia holds up to 90 per cent of space industry capability, the loss of the Baikonur launch site in Kazakhstan would have been a major blow.

Russia is determined to eliminate its dependence on former USSR countries for any space technologies, materials and capabilities. Major areas include Ukrainian electronics for launchers and spacecraft and the Yevpatoria ground station. Proposals for a new cosmodrome at the decommissioned missile site at Svobodny-18 near the Chinese border, were widely brandished, probably mainly as a strong negotiating ploy against Kazakhstan, although President Yeltsin signed a decree in March 1996 approving its creation. The cost would be high but Kazakhstan could not then dictate terms. First orbital launch from Svobodny came in March 1997. For example, helicopter crews for recovering manned Soyuz and unmanned capsules cannot leave their base on the border until the Kazakh government has approved their flight plans. They also have to carry cash to pay for refuelling.

A positive note among all the negative developments is the completion of the Glonass satellite navigation system. Under deployment since 1982, the launch of triple satellites in December 1995 brought the system up to its full strength of 24 operating satellites for the first time. It provides better accuracy for civil users than GPS (50 m *vs.* 100 m), and combined receivers are now beginning to appear. Russia hoped to create a unified navigation network of differential stations providing at least 10 m accuracy nationwide by 2000 but prospects for keeping Glonass at this level looked bleak in early 1999.

The federal civil budget for 1996 was Rb2,219.5 billion, including Rb719.9 billion for Baikonur, although the RKA Russian Space Agency did not receive it all. RKA head Yuri Koptev claimed in March 1995 that RKA received only 51 per cent of 1994's allocated budget. 1995's budget was Rb1,895 billion after RKA and VKS had each requested about Rb2,900 billion. VKS claimed in January 1996 that it received only half of that request. RKK Energia, Russia's largest space company and responsible for the manned programme, complained in February 1995 that it had been allocated only 2 per cent of the required resources but the following month was guaranteed the funding by Parliament. It is doubtful whether manned activities would survive without integration into international missions. RKA took the lead role in developing a seven-year civil space plan adopted by Parliament in December 1993. This emphasises expansion of telecommunications, such as a 20-fold increase in telephone lines, Earth observation, upgrading the Proton and Soyuz launch vehicles and, above all, continuing the manned programme with Mir and involvement in the international Space Station. Koptev has even floated the idea of Russia becoming an associate member of ESA. Nonetheless, it is still an impressive achievement when the civil space agency's budget is less than that of India's.

Russia's fight to enter the commercial launch market lasted 14 years before Proton successfully dispatched Luxembourg's Astra 1F in April 1996. Inmarsat was the first to sign up to Proton, in 1992 and Proton launched Inmarsat-3 in September 1996 and once that credibility barrier had been breached and Lockheed Martin gave its stamp of approval by jointly marketing the vehicle, there has been marked success. The first accord with the US stipulated that only eight GEO/GTO commercial launches could be conducted through 2000 (limited to two in any 12 months). Although not specifically a Proton agreement, in practice it covered principally that vehicle's GEO version. Part of the deal required prices within 7.5 per cent of the western equivalents. The agreement excluded Proton's earlier Inmarsat contract; LEO and other non-GEO/GTO launches would be dealt with on a case-by-case basis. A more relaxed agreement was signed on 30 January 1996, allowing 16 launches (plus another four should market conditions allow) within 15 per cent of western prices.

Khrunichev is commercially the most successful Russian space company, its employees earning the highest rates in the industry (Rb800,000 monthly, as opposed to Rb230,000 in the TsUP manned control centre). Its resources are allowing it to develop the Breeze-M storable stage and a cryogenic stage for Proton to handle the next generation of large satellites. In fact, Khrunichev plans to begin flight-testing the new Angara vehicle in about 2002 to replace Proton. Not only would it improve Proton's cost and performance, it would free Russia of Proton's dependence on Kazakhstan. Khrunichev's smaller Rockot is being marketed jointly with DASA and can be expected to be successful with smaller payloads. Payloads are also appearing on AKO Polyot's Cosmos vehicle and a sea-launched version of Zenit is being developed with Boeing.

While there were calls during 1990 for a strong central space agency, the power of Glavcosmos was declining. Organisations such as RKK Energia were unleashed to offer their particular expertise directly; Energia first took over responsibility from Glavcosmos for commercial manned missions and in late 1991 assumed management of Mir, reporting to the Russian Ministry of Industry. Meanwhile, the Council of Ministers' main space planner had thrown doubt on a Mir successor, condemning national space stations as too expensive. The Russian Federation established its own Ministry of Communications, Information and Space (under Minister Vladimir Bulgak) in 1990 and paid the Ministry of Defence for launching the first of three dedicated Gorizont communications satellites. The Interkosmos international science collaboration programme, based at the Academy, ended in 1994 and the projects were distributed among institutes such as IKI.

RUSSIA/USSR PAYLOADS 1963–2002

	65	66	67	68	69	70	71	72	73	74	75	76	77	78	79	80	81	82	83	84	85	86	87	88	89	90	91	92	93	94	95	96	97	98	99	00	01	02	Total
2 Luna (Lunik)	4	5	-	1	1	2	2	1	1	2	-	1	-	-	-	-	-	-	-	-	-	-	-	-	-	-	-	-	-	-	-	-	-	-	-	-	-	-	20
3 Vostok/khod	1	-	-	-	-	-	-	-	-	-	-	-	-	-	-	-	-	-	-	-	-	-	-	-	-	-	-	-	-	-	-	-	-	-	-	-	-	-	1
4 Cosmos	52	34	61	64	55	72	81	72	85	74	85	101	86	96	79	88	94	97	94	94	99	96	97	79	68	66	54	55	38	38	21	10	10	17	3	7	9	5	2,336
5 Venera	3	-	2	-	2	2	-	2	-	-	2	-	-	2	-	-	2	-	2	2	-	-	-	-	-	-	-	-	-	-	-	-	-	-	-	-	-	-	21
6 Mars	-	-	-	-	-	-	2	-	4	-	-	-	-	-	-	-	-	-	-	-	-	-	-	-	-	-	-	-	-	-	-	1	-	-	-	-	-	-	7
7 Polyot	-	-	-	-	-	-	-	-	-	-	-	-	-	-	-	-	-	-	-	-	-	-	-	-	-	-	-	-	-	-	-	-	-	-	-	-	-	-	0
8 Elektron	-	-	-	-	-	-	-	-	-	-	-	-	-	-	-	-	-	-	-	-	-	-	-	-	-	-	-	-	-	-	-	-	-	-	-	-	-	-	0
9 Zond	1	-	-	3	1	1	-	-	-	-	-	-	-	-	-	-	-	-	-	-	-	-	-	-	-	-	-	-	-	-	-	-	-	-	-	-	-	-	6
10 Molniya	2	2	3	3	2	5	3	6	8	7	10	7	6	6	5	4	8	5	7	4	8	7	1	7	4	6	5	4	5	2	1	2	1	2	1	-	2	-	161
11 Proton	2	1	-	1	-	-	-	-	-	-	-	-	-	-	-	-	-	-	-	-	-	-	-	-	-	-	-	-	-	-	-	-	-	-	-	-	-	-	4
12 Soyuz	-	-	1	2	5	1	2	-	2	3	4	3	3	5	4	6	3	3	2	3	2	2	3	3	1	3	2	2	2	3	2	2	1	2	1	2	2	2	89
13 Meteor	-	-	-	-	2	4	4	3	2	5	4	3	4	-	3	2	2	2	1	1	3	1	2	2	2	2	2	-	1	1	-	-	-	-	-	-	1	-	59
14 Intercosmos	-	-	-	-	2	2	1	3	2	2	2	2	1	2	2	-	2	-	-	-	-	-	-	-	2	-	1	-	-	-	2	-	-	-	-	-	-	-	28
15 No desig	-	2	-	-	-	-	-	-	-	-	-	-	-	-	-	-	-	-	-	-	-	-	-	-	-	-	-	-	-	-	-	-	-	-	1	-	-	-	3
16 Salyut	-	-	-	-	-	-	1	-	1	2	-	1	1	-	-	-	-	1	-	-	-	-	-	-	-	-	-	-	-	-	-	-	-	-	-	-	-	-	7
17 Aureole	-	-	-	-	-	-	1	-	1	-	-	-	-	-	-	-	1	-	-	-	-	-	-	-	-	-	-	-	-	-	-	-	-	-	-	-	-	-	3
18 Prognoz	-	-	-	-	-	-	-	2	1	-	1	1	1	1	-	1	-	-	1	-	1	-	-	-	-	-	-	-	-	-	-	-	-	-	-	-	-	-	10
19 For other Countries	-	-	-	-	-	-	-	1	-	-	2	-	1	-	1	-	1	-	-	-	-	-	-	1	-	-	2	-	1	1	8	4	19	15	33	35	8	18	151
20 Raduga	-	-	-	-	-	-	-	-	-	-	1	1	1	1	1	2	3	1	2	2	2	2	2	1	3	3	2	-	2	3	-	1	-	-	1	-	1	-	38
21 Ekran	-	-	-	-	-	-	-	-	-	-	-	1	1	-	2	2	1	2	2	2	1	1	2	2	-	-	-	1	-	-	-	-	-	-	-	-	1	-	21
22 Progress	-	-	-	-	-	-	-	-	-	-	-	-	-	4	3	4	1	4	2	5	1	2	7	6	4	4	5	5	5	5	5	3	4	3	2	4	6	3	97
23 Radio	-	-	-	-	-	-	-	-	-	-	-	-	-	2	-	-	6	-	-	-	-	-	-	-	-	-	-	-	-	1	-	-	-	-	-	-	-	-	9
24 Gorizont	-	-	-	-	-	-	-	-	-	-	-	-	-	1	2	1	-	2	2	2	1	2	1	2	3	3	2	3	2	1	-	2	-	-	-	1	-	-	33
25 Iskra	-	-	-	-	-	-	-	-	-	-	-	-	-	-	-	-	1	2	-	-	-	-	-	-	-	-	-	-	-	-	-	-	-	-	-	-	-	-	3
26 Astron	-	-	-	-	-	-	-	-	-	-	-	-	-	-	-	-	-	-	1	-	-	-	-	-	-	-	-	-	-	-	-	-	-	-	-	-	-	-	1
27 Mir/module	-	-	-	-	-	-	-	-	-	-	-	-	-	-	-	-	-	-	-	-	-	1	1	-	1	1	-	-	-	-	1	1	-	-	-	-	-	-	6
28 Phobos	-	-	-	-	-	-	-	-	-	-	-	-	-	-	-	-	-	-	-	-	-	-	-	2	-	-	-	-	-	-	-	2	-	-	-	-	-	-	4
29 Photon	-	-	-	-	-	-	-	-	-	-	-	-	-	-	-	-	-	-	-	-	-	-	-	1	1	1	1	1	-	1	1	-	1	-	-	-	-	-	8
30 Okean	-	-	-	-	-	-	-	-	-	-	-	-	-	-	-	-	-	-	-	-	-	-	-	1	-	1	1	-	-	1	1	-	-	-	1	-	-	-	6
31 Buran	-	-	-	-	-	-	-	-	-	-	-	-	-	-	-	-	-	-	-	-	-	-	-	1	-	-	-	-	-	-	-	-	-	-	-	-	-	-	1
32 Resurs-F	-	-	-	-	-	-	-	-	-	-	-	-	-	-	-	-	-	-	-	-	-	-	-	-	5	4	4	3	3	-	1	-	1	-	1	-	-	-	22
33 Pion	-	-	-	-	-	-	-	-	-	-	-	-	-	-	-	-	-	-	-	-	-	-	-	-	4	-	-	2	-	-	-	-	-	-	-	-	-	-	6
34 Nadezhda	-	-	-	-	-	-	-	-	-	-	-	-	-	-	-	-	-	-	-	-	-	-	-	-	1	1	1	-	-	1	-	-	-	-	-	1	-	1	6
35 Granat	-	-	-	-	-	-	-	-	-	-	-	-	-	-	-	-	-	-	-	-	-	-	-	-	1	-	-	-	-	-	-	-	-	-	-	-	-	-	1
36 Gamma	-	-	-	-	-	-	-	-	-	-	-	-	-	-	-	-	-	-	-	-	-	-	-	-	-	1	-	-	-	-	-	-	-	-	-	-	-	-	1
37 Informator	-	-	-	-	-	-	-	-	-	-	-	-	-	-	-	-	-	-	-	-	-	-	-	-	-	-	1	-	-	-	-	-	-	-	-	-	-	-	1
38 Almaz	-	-	-	-	-	-	-	-	-	-	-	-	-	-	-	-	-	-	-	-	-	-	-	-	-	-	1	-	-	-	-	-	-	-	-	-	-	-	1
39 Mak	-	-	-	-	-	-	-	-	-	-	-	-	-	-	-	-	-	-	-	-	-	-	-	-	-	-	1	1	-	-	-	-	-	-	-	-	-	-	2
40 Resurs 500	-	-	-	-	-	-	-	-	-	-	-	-	-	-	-	-	-	-	-	-	-	-	-	-	-	-	-	1	-	-	-	-	-	-	-	-	-	-	1
41 Gals	-	-	-	-	-	-	-	-	-	-	-	-	-	-	-	-	-	-	-	-	-	-	-	-	-	-	-	-	-	1	1	-	-	-	-	-	-	-	2
42 GEO-IK	-	-	-	-	-	-	-	-	-	-	-	-	-	-	-	-	-	-	-	-	-	-	-	-	-	-	-	-	-	1	-	-	-	-	-	-	-	-	1
43 Express	-	-	-	-	-	-	-	-	-	-	-	-	-	-	-	-	-	-	-	-	-	-	-	-	-	-	-	-	-	1	-	1	-	-	-	2	-	-	4
44 Coronas	-	-	-	-	-	-	-	-	-	-	-	-	-	-	-	-	-	-	-	-	-	-	-	-	-	-	-	-	-	1	-	-	-	-	-	-	1	-	2
45 Luch	-	-	-	-	-	-	-	-	-	-	-	-	-	-	-	-	-	-	-	-	-	-	-	-	-	-	-	-	-	1	1	-	-	-	-	-	-	-	2
46 Resurs-O	-	-	-	-	-	-	-	-	-	-	-	-	-	-	-	-	-	-	-	-	-	-	-	-	-	-	-	-	-	1	-	-	-	1	-	-	-	-	2
47 Elektro	-	-	-	-	-	-	-	-	-	-	-	-	-	-	-	-	-	-	-	-	-	-	-	-	-	-	-	-	-	1	-	-	-	-	-	-	-	-	1
48 Tsikada	-	-	-	-	-	-	-	-	-	-	-	-	-	-	-	-	-	-	-	-	-	-	-	-	-	-	-	-	-	-	1	-	-	-	-	-	-	-	1
49 EKA	-	-	-	-	-	-	-	-	-	-	-	-	-	-	-	-	-	-	-	-	-	-	-	-	-	-	-	-	-	-	1	-	-	-	-	-	-	-	1
50 Gonets-10	-	-	-	-	-	-	-	-	-	-	-	-	-	-	-	-	-	-	-	-	-	-	-	-	-	-	-	-	-	-	-	3	-	-	-	-	3	-	6
51 Bion	-	-	-	-	-	-	-	-	-	-	-	-	-	-	-	-	-	-	-	-	-	-	-	-	-	-	-	-	-	-	-	1	-	-	-	-	-	-	1
52 Bankir	-	-	-	-	-	-	-	-	-	-	-	-	-	-	-	-	-	-	-	-	-	-	-	-	-	-	-	-	-	-	-	-	1	-	-	-	-	-	1
53 Tubsat	-	-	-	-	-	-	-	-	-	-	-	-	-	-	-	-	-	-	-	-	-	-	-	-	-	-	-	-	-	-	-	-	-	2	-	-	-	-	2
54 ISS module	-	-	-	-	-	-	-	-	-	-	-	-	-	-	-	-	-	-	-	-	-	-	-	-	-	-	-	-	-	-	-	-	-	1	-	1	-	-	2
55 Yamal	-	-	-	-	-	-	-	-	-	-	-	-	-	-	-	-	-	-	-	-	-	-	-	-	-	-	-	-	-	-	-	-	-	-	2	-	-	-	2
56 LMI	-	-	-	-	-	-	-	-	-	-	-	-	-	-	-	-	-	-	-	-	-	-	-	-	-	-	-	-	-	-	-	-	-	-	1	-	-	-	1
57 Globus	-	-	-	-	-	-	-	-	-	-	-	-	-	-	-	-	-	-	-	-	-	-	-	-	-	-	-	-	-	-	-	-	-	-	-	1	-	-	1
58 Strela	-	-	-	-	-	-	-	-	-	-	-	-	-	-	-	-	-	-	-	-	-	-	-	-	-	-	-	-	-	-	-	-	-	-	-	-	-	2	2
59 Mozhayets	-	-	-	-	-	-	-	-	-	-	-	-	-	-	-	-	-	-	-	-	-	-	-	-	-	-	-	-	-	-	-	-	-	-	-	-	-	1	1
60 Glonass	-	-	-	-	-	-	-	-	-	-	-	-	-	-	-	-	-	-	-	-	-	-	-	-	-	-	-	-	-	-	-	-	-	-	-	-	-	3	3
Totals	**65**	**44**	**67**	**74**	**70**	**89**	**97**	**90**	**107**	**95**	**111**	**121**	**105**	**120**	**102**	**110**	**125**	**119**	**116**	**115**	**118**	**114**	**116**	**108**	**100**	**96**	**85**	**78**	**59**	**65**	**47**	**33**	**38**	**43**	**47**	**54**	**34**	**35**	**3,212**

Order was slow to emerge following 1991's collapse of the USSR and the subsequent struggle to establish a formal structure inevitably left the space program in an organisational limbo. The Soviet Ministry of Machine Building (MOM), for more than a quarter of a century the principal governmental body with direct oversight of space and missile activities, was disbanded after August 1991's failed coup. The RKA, created by President Yeltsin's decree in February 1992, rapidly established itself under Koptev, formerly MOM deputy minister, as the principal civil authority. Russia's parliament formalised its existence in 1993.

Even before 1991's events, Western perception of the Soviet space programme had undergone a remarkable transformation. Contrasting with the traditional view of a large and vigorous monolithic program, public criticism of space expenditure appeared and the concerted response revealed that Soviet space planners were having to justify and fight for their funding. As part of the reaction, detailed budget figures were released for the first time. They revealed that 1989's total space budget amounted to Rb6.9 billion, or almost 1.5 per cent of the state total. Koptev in late 1992 noted that 1989 represents the peak of funding. More than half, Rb3.9 billion, was allocated to military activities, Rb1.3 billion to development of reusable space systems (principally Energia/Buran) and Rb1.7 billion to the remaining civil programme of applications satellites and manned missions. 1988's figure for the last category was Rb1.3 billion. 1990's budget reportedly fell to Rb6.3 billion (including Rb220 million for the manned programme), and 1991's was Rb5.8 billion. The Academy of Sciences 1991 allocation was Rb300 million. Officials in April 1991 revealed that Rb31.6 billion had been expended 1986–1990, with Rb34.5-36 billion (including Rb15 billion for civil) requested for 1991–95.

Other quoted figures included Rb1,471 million for the whole Mir space station programme up to April 1989 and Rb14 billion over the 13 years of the Energia/Buran programmes. Meteorological satellites were claimed to return Rb500-700 million on their investment annually, supported in 1988 by Rb500 million from communications satellites. Overall, it was estimated that such projects with economic applications had returned some Rb12 billion in 33 years.

Planned investment for the 1991–95 period was some Rb6 billion, comprising Rb1 billion for development of new commercial satellites, Rb2 billion for science research and Rb3 billion for manned flights and new hardware.

The military has publicly criticised the power RKA has acquired, complaining that civil programmes are diverting funding from military efforts. 1993's civil appropriation in December 1993 was Rb164.8 billion (pegged to July 1993's rate, about US$130 million), only half of the 1980s' level after allowing for inflation. That sum breaks down as Rb126 billion for R&D, Rb7.8 billion for maintaining Baikonur, Rb21 billion for purchasing spacecraft production and Rb10 billion for capital construction. 1994's simultaneous allocation was Rb274.8 billion (July 1993 rate), breaking down as 167, 16, 81.8, 10, respectively.

1995's draft allocation to begin 1 April was reported in March 1995 to be Rb2,970 billion and Rb2,889 billion for military (inflation makes it difficult to compare on a year by year basis). Koptev has said that he wants civil budgets to outstrip the military's. Both have suffered, most dramatically from inflation, since the Rb6.9 billion peak in 1989. Civil space funding fell 10 to 12 per cent over 1990–92, while military development funding dropped 30 per cent and production funding 45 per cent for the same period.

Mir's visiting missions are offered through RKK Energia (through Glavcosmos until 1990) at a cost of US$12 million to US$15 million. Austria in November 1988 became the first to sign an agreement, flying in October 1991. Austria paid Russia US$6 million plus the contribution of research apparatus. That mission was preceded by Japanese and UK visits, and it was followed (July 1992) by France's third trip to a Russian station and Germany's first (March 1992). CNES/Energia signed a memorandum on 29 July 1992 for four further missions, in July 1993, August 1996 (Cassiopeia), 1998 and 2000 to exploit Mir's facilities. FFr126 million was paid jointly for the 1993/96 missions, plus equipment. RKA/DARA signed an agreement in December 1995 for a second German visit, aboard TM25 in February 1997. ESA's two visits were the 30-day EuroMir-94 from October 1994 and the six-month EuroMir-95 from September 1995, paying €45 million for the pair. NASA paid more than US$300 million for its Shuttle visits.

Once it became clear in the 1980s that technology transfer restrictions imposed by the US had effectively strangled Soviet attempts at launcher marketing, the USSR began more vigorously pushing its microgravity, remote sensing, communications and manned space facilities. Commercial accommodation aboard the Photon μg and Resurs-F Earth resources craft are regularly leased. The drive for hard currency is providing an international impetus towards higher resolution Earth imaging, a move the US military has traditionally resisted. Some material has been available since the late 1980s, for example through Soyuzkarta, but its use has been hampered by a rigid and unresponsive system. But in 1993 Sovinformsputnik (whose dominant shareholder is the Central specialised Design Bureau, builders of Resurs-F and military systems) became licensed to sell data from some film cameras of military satellites. Likewise, Priroda in 1993 established the WorldMap consortium with the UK's JEBCO Information Services to sell digitised versions of its Resurs-F imagery. Its archives, holding two million negatives, are being digitised and added to a database to facilitate user searches. Priroda/WorldMap advertised three 1994 Resurs-F flights in advance and invited requests for imaging targets. However, no missions appeared because of a dearth of launch vehicles. Both organisations can provide 2 m resolution (and greater if allowed), exceeding western commercial capability. Priroda plans to distribute current digital 3 m resolution panchromatic imagery from a military satellite, presumably the 5th generation photo-reconnaissance design.

Mars-96's November 1996 departure could be the last launch of a Russian planetary probe for many years. Any talk of manned Mars expeditions is long ended. Funding for science missions has become so tight that external contributions have been required. CNES in early 1992 began contributing FFr1.5 million every six months to keep the Yevpatoria station open to control the joint Granat astronomical satellite.

Participation in the International Space Station should ensure the survival of Russia's manned programme and is providing a major source of hard currency. The international partners on 7 December 1993, formally invited Russia to join the project. The intention is to integrate Russian core elements and fly modified Soyuz-TM craft as attached rescue vehicles and uprated Progress as ferries. Russia will be paid upwards of US$600 million for its contributions, including the US$190 million agreed with Khrunichev in February 1995 for the Salyut. NASA Administrator Dan Goldin and RKA General Director Yuri Koptev signed an additional US$400 million USFY94-97 protocol on 16 December 1993 (with the final agreement signed 21 June 1994). It included US$305 million for up to 10 Shuttle Mir 1 visits encompassing six NASA astronauts spending a total of 24 months aboard. The visits existed primarily for life/μg sciences research, crew exchanges, cargo delivery and experiment return.

UPDATED

South Africa

Current Status

South Africa has enthusiastically embraced the SAFARI 2000 science project involving a global team of scientists, a large number of aircraft and several NASA spacecraft. SAFARI 2000 began in August 1999 as a three year project to study the environment of Southern Africa enlarging understanding of a wide range of phenomena including regional crop productivity and global climate change. An important part of SAFARI 2000 is an operational test of the Terra satellite. This met outstanding success and operators have instigated a successor programme yet to be defined.

Background

The country's overt involvement in space has been modest, principally comprising the Satellite Applications Centre. Recent years saw attempts to find commercial applications of defence technologies, but ultimately without success. Denel (Pty) Ltd, formed in 1992 to take on the mantle of South Africa's defence manufacturing from Armscor and expand into the commercial sector, began promoting industrial space capabilities. Although the company is government-funded, projects must show commercial promise for them to be continued.

The government agency charged with developing plans, CSIR, considered proposals in 1988 for a national space programme emphasising communications and remote sensing, but including the possibility of developing an indigenous orbital launcher. The Minister of Trade and Industries announced the formation of a South African Space Committee in 1990. Its brief included the monitoring of trends in space technology applications and encouraging appropriate national utilisation. Committee representation covered CSIR, Dept of Posts and Telecommunications, the Weather Bureau, Dept of Trade and Industries, Industrial Development Corp, and private and official broadcast agencies and academic interests. However, the Committee was disbanded in 1991. The government in 1993 passed legislation to establish a Council for Space Affairs to manage and control national space activities, reporting to the Dept of Trade and Industries, but it did not appear.

Denel's Houwteq division, specialising in missiles and high tech systems, spearheaded the design of the Greensat Earth observation satellite. However, Denel announced in October 1994 that Greensat was being cancelled because of the lack of interest and that Houwteq's space activities had ended.

Denel had planned a three-stage all-solid launcher for delivering small payloads to LEO from 1995, but announced in June 1993, after expending US$55 million, that it would be uneconomic. Termination removed a major obstacle to South Africa's membership of the MTCR Missile Technology Control Regime, which became effective in 1995. Denel had two motors which would still be static tested. Three successful test flights were made from the Overberg Test Range (part of Denel) near Bredasdorp north of Cape Town: June 1989's flew only stage 1 and stages 1/2 were tested July 1989 and November 1990. 1989's flight was originally associated with Israel's Jericho 2 missile or Shavit space launcher; there seems to be some collaboration. Israel was apparently closely linked with South Africa's space efforts.

Of the Denel group, Somchem in Somerset West near Cape Town specialises in propellants and propulsion systems, Kentron in guided weapons and systems, and Eloptro in electro-optical systems.

Telkom SA is the national signatory to Intelsat and Inmarsat (joined April 1994). Studies have been made of a domestic 'Syncom' system. The Satellite Applications Centre (part of the Council for Scientific & Industrial Research, now known only as CSIR) receives, processes and markets ERS, Landsat, Spot and NOAA data and provides TT&C services to foreign space agencies. For example, it currently operates as part of France's 2 GHz TT&C network.

In the late 1990s, South Africa began to formulate a strategy for expanded participation in remote sensing studies. These have resulted in several proposals for different levels of interpretation with other African countries as South Africa itself seeks a central role in coordinating regional earth resource studies. South Africa has been at the forefront of pan-African discussions about the need for a satellite programme capable of providing continental databases serving the specific needs of individual countries.

VERIFIED

Spain

Current Status

Although the majority of Spain's space activities continue to be routed through ESA, there is a growing national programme. Spain launched a domestic telecommunications satellite service in 1992 with Hispasat 1A followed by Hispasat 1B in 1993 and Hispasat 1C in Feburary 2000. Hispasat 1D comprises a Alcatel Space Industries Spacebus 3000 scheduled for launch by Atles 2AS through the International Launch Services organisation. Ariane launched a micro-satellite in 1995, a small platform flew in 1997 and a micro-launcher is being developed.

Hispasat SA was created in 1989 to operate the domestic satellite system. The satellites also provide secure government military links at X-band, complementing Spain's 7 per cent participation in France's Helios 1 reconnaissance satellite project.

The nine-nation Western European Union inaugurated a satellite data interpretation and training centre in 1993 at Spain's Torrejon Air Base. Minisat has been developed as a national entry into the space market and studies began in 1991 of the three-stage Capricornio, capable of placing about 100 kg in orbit from Spain. Pedro Duque was selected in May 1992 by ESA as Spain's first astronaut; he was back-up for October 1994's Mir visit. Discussions for a national visiting mission to Russia's Mir space station were conducted in 1989, but no accord was reached. Spain has attempted to find a way to send a Spanish national to the ISS and has proposed several science experiment packages in support.

2000 ESA contribution is €87.1 million, or 4.0 per cent of total member contributions, 1999 ESA contribution was €102.1 million or 4.5 per cent of the agency's total budget. 1995 3.9 per cent (AU105.4 million), 1994 4.7 per cent (AU102.2 million), 1993 4.0 per cent (AU97.8 million), 1992 4.0 per cent (AU113.1 million), 1991 4.1 per cent (AU96.3 million); 1990 4.0 per cent (AU71.6 million); 1989 3.8 per cent (AU57.3 million). Space expenditure more than doubled over 1984–87 to about Pta6.5 billion; by 1991 the ESA contribution was Pta12,680 million. Involvement in ESA programmes has been modest, taking small but significant roles in Ariane 5, Envisat, Polar Platform, Data Relay Satellite; it is now contributing 2 per cent to the Columbus Orbital Facility and the Automated Transfer Vehicle.

National space planning and co-ordination is undertaken by the Ministry of Industry & Energy's Centre for the Development of Industrial Technology (CDTI), which in 1986 replaced the Ministry of Defence's Comisión Nacional de Investigación del Espacio (CONIE). INTA (see below) provides R&D and its INSA subsidiary operates ESA and NASA tracking stations at Villafranca (IUE control centre) and Madrid. Madrid's 70 m, two 34 m and 26 m antennas are part of NASA's Deep Space Network for controlling spacecraft such as Voyager and Pioneer. A renewed space co-operation agreement with the US was signed 11 July 1991, including the continued use of emergency Shuttle landing sites (Moron and Zaragoza). The Instituto Nacional de Meteorología is the national signatory to EUMETSAT, and Telefónica to Inmarsat, Eutelsat and Intelsat.

Spanish Space Expenditure (Pta million)

Year	ESA	National	Total
1993	11,913.7	3,870.5	15,784.2
1994	14,843.5	4,510.0	19,353.4
1995	16,776.3	4,770.5	21,546.8
1996	18,163.8	4,857.2	23,021.0
1997	13,711.9	4,809.8	18,521.7
1998	14,886.8	4,906.2	19,793.0
1999	15,141.1	4,991.6	20,132.7
2000	14,492.2	4,991.6	19,483.8

VERIFIED

Sweden

Current Status

A flurry of launches took place in the first few months of 2001, including the Odin space observatory by Start-1 launcher on 20 February and the Maxus 4 sounding rocket on 29 April.

Sweden has been an ESA member since 1975, accounting for 2.6 per cent of members' contributions in 2000 and 1999, 2.7 per cent in 1996, 2.7 per cent 1995, 2.6 per cent 1994, 2.3 per cent 1993, 2.2 per cent 1992, 2.3 per cent 1991, 2.3 per cent 1990, 2.5 per cent 1989, 2.5 per cent 1988 and 2.2 per cent 1987. Sweden's contributions for 2000 amounted to €55.2 million. Bilateral co-operation includes France (Spot) and Germany (Sanger). An MoU was signed with Germany in 1990 agreeing to share hypersonic information; Sweden expended SKr30 million to end-1992 on hypersonic work in Germany's Phase 1 programme. It has also made contributions to CIS/USSR space projects, including Prognoz and Interkosmos. Astrid 1 was the first western satellite launched on a Cosmos-3M. Physicist Dr Christer Fuglesang was selected in May 1992 by ESA as one of six new astronauts: he was back-up for Mir's Soyuz-TM 22 mission in September 1995.

Sweden joined the German Texus µg sounding rocket programme in 1977 and 10 years later began its own Maser, offering one launch annually. The former DASA and the Swedish Space Corp developed the Maxus Castor 4-based vehicle to extend µg time to more than 14 minutes. Sweden's guidance control system ensures the vehicle and payload descend within Esrange. The maiden launch failed in May 1991; number 2 was successful 8 November 1992 and number 3 28 November 1995; number 4 was planned for April 1998.

In April 1979, the government embarked upon a national space programme to enable its industries to be more competitive in the international space communications market. The scientific Viking and the Tele-X direct broadcast satellite were the first results. Telia AB (previously Swedish Telecom), the national signatory to Intelsat, Inmarsat and Eutelsat, operates as a private company but is wholly government-owned, under the Ministry for Transport & Communications.

Sweden's space budget comprises two primary elements: national and ESA/bilateral. The 1997 budget total was SKr548 million. Within its SKr573 million ESA contribution for 18 months in 1995 and 96 (559 1994/95; 556 1993/94; 469 1992/93; 414 1991/92; 418 1990/91; 280 1989/90), it participates in the mandatory basic and scientific program, Ariane 5, telecommunications, remote sensing (including ERS) and µg activities. National interests cover space science, remote sensing, Esrange management and industrial development. Scientific research is traditionally focused on satellite and sounding rocket magnetospheric/ionospheric investigations, including modest but important collaboration on CIS missions. Upgrades to sounding rocket launch facilities and planned improvements to tracking and recovery.

The other national activity aims to enhance industry's capabilities through satellite projects such as Viking, Freja and Odin. Saab Ericsson Space and Volvo represent the primary national industrial space capability. Three DBS channels from Sweden's Tele-X satellite were augmented in 1994 with the purchase of the UK's Marcopolo 1. Tele-X's 1997 Sirius replacement was contracted in 1995.

The Swedish Space Coporation employs about 300 people. Established in 1972, it provides for the primary technical services on behalf of the Swedish Space Board.

Sweden's Space Budget (SKr million)

National Activities	1994/95	1995/96	1997
Management	13	20	14
Basic activities (Esrange)	13	22	15
Space science (including µg)	51	72	46
Remote sensing	18	26	18
Industrial development	47	47	28
National total	142	187	121
International Activities	**1994/95**	**1995/96**	**1997**
Basic activities	64	89	62
Space science	99	129	82
Space transportation	160	235	146
Manned programme	14	40	14
Telecommunications	61	61	22
Remote sensing	238	226	184
Microgravity	38	49	38
International total	674	829	548
Total	**816**	**1,016**	**669**

The annual accounting period was originally 1 July to the following 30 June, but it was decided that the accounting period should be adjusted to begin on 1 January: as a result the 1995/96 figures cover the 18 months period of 1 July 1995 to 31 December 1996. The totals for earlier years are 1993/94 – SKr733 million, 1992/93 – SKr662 million, 1991–92 – SKr589 million, 1990/91 – SKr561 million.

Swedish Space Expenditure (SKr million)

Year	ESA	National	Total
1993	436.1	180.0	616.1
1994	540.5	180.0	720.5
1995	590.2	130.0	770.2
1996	660.0	130.0	790.0
1997	503.8	150.0	653.8
1998	517.5	150.0	667.5
1999	515.0	150.0	665.0
2000	474.1	150.0	624.1

VERIFIED

Switzerland

Current Status

Switzerland's space involvement is almost totally routed through ESA – the 2000 contribution of €71.9 million accounting for 3.3 per cent of the members' contributions. Although modest, funding for ESA more than tripled during the period 1984 to 1994 and has remained almost stable since then: SFr27.5 million in 1984, growing to SFr114 million in 1994, SFr110.8 million in 1995, SFr111.7 million in 1996, SFr113.4 million in 1997 and SFr347 million planned for 1998–2000.

The Federal Council (government) determines the national space policy under the advice of a 20-member Federal Space Affairs Commission (CFAS). Science projects fall under the Federal Office of Education and Science in the Federal Department of the Interior and are carried out by the universities and federal institutes of technology with funding from the Swiss Academy of Sciences' Space Research Commission, the Swiss National Science Foundation and the various Cantons. Switzerland is a signatory to Intelsat, Eutelsat and Inmarsat through the PTT General Directorate of Posts, Telephones and Telecommunications: it is also a member of Eumetsat.

Since 1988, Switzerland has been contributing 4 per cent to all of the new Earth observation and microgravity development programmes. These programmes, together with the mandatory ESA science programme, give Swiss scientists the opportunity when selected to fly experiments on ESA missions which are mainly funded through the ESA Prodex (SFr7 million per year). In the field of launchers Switzerland participates with 2 per cent of the main Ariane-5 development, 2.5 per cent to the Ariane-5 infrastructure programme, 0.65 per cent for Ariane-5 research and technology and 0.38 per cent at the Ariane-5 evolution programme. At the ESA ministerial meeting in October 1995 Switzerland agreed to contribute 2.5 per cent to the European participation in the International Space Station programme, with its main interest being the Automated Transfer Vehicle. Swiss participation in telecommunications development programmes has been limited to a 2 per cent share in the Artemis programme, but following a decision in 1995, future funding of technology and telecommunications programmes has been stabilised at SFr9 million.

No national programme is being pursued. However, the Bern-based International Space Science Institute, co-funded by Switzerland and ESA, was created in 1995 to fully exploit space missions in a multi-disciplinary environment – in particular, the solar-terrestrial physics missions presently co-ordinated by ESA, ISAS, IKI and NASA. The Observatory of Geneva holds the recently opened ESA Integral Science Data, which will serve as an access point for the astronomical community.

More than 30 Swiss companies are directly involved in space; products and on-going developments include satellite structures, Ariane fairings, MGSE, pointing mechanisms, cryogenic valves, advanced motors, sensors, bioreactors, atomic clocks, electronic units, EGSE and dedicated space science instruments. ESA has nominated the Observatory of Neuchâtel as its external laboratory for time and frequency. Swiss scientists are actively involved in Ulysses, HST, ISO, Soho and XMM. There is particular interest in ERS SAR activities, as well as data interpretation from Spot, Landsat and NOAA. EuroMir, Shuttle, Spacelaboratory and sounding rockets are employed for life sciences research. The Space Biology Group was set up in 1977 to conduct basic research using Shuttle and Spacelab missions. This group is now involved in supporting ESA life science activities on the International Space Station and has considerably expanded its involvement in the selection and definition of life science tasking.

Claude Nicollier, from Vevey, flew as the first non-American Mission Specialist aboard 1992's Shuttle mission STS-46. He was responsible for deploying ESA's Eureca platform. He flew again on STS-61 (December 1993 – first Hubble servicing mission) and STS-75 (February 1996 – second Tethered Satellite mission).

Switzerland hosts the International Space Science Institute, an interdisciplinary body where scientists from many countries can come together for workshops and work groups in a variety of fields related to space research.

Swiss Space Expenditure (SFr million)

Year	ESA	National	Total
1993	110.7	3.7	114.4
1994	105.9	3.5	109.4
1995	113.4	3.3	116.7
1996	110.9	3.3	114.2
1997	104.6	3.0	107.7
1998	117.9	3.2	121.2
1999	117.4	3.2	120.6
2000	117.2	3.2	120.5

VERIFIED

Syria

Current Status

The country is a member of Intersputnik, joining in October 1991, and Intelsat but not of Inmarsat or Eutelsat. On 22 July 1987, Lt. Col. Mohammed Faris, 36, a Syrian Air Force pilot, represented his country on the Soviets' 12th international mission to a space station. Launched on Soyuz-TM 3, Faris was aboard Mir for six days. In addition to participating in materials processing experiments, he conducted photographic surveys of Syria to help the search for water and mineral resources.

China, North Korea and Russia have all been associated with the transfer of Gallactic missile technology to Syria, among other countries, on the pretext that they could assist with development of indigenous satellite launches. Syria has frequently sought to acquire missile technology on the pretext of developing a satellite launcher. Syria also obtained information about manufacture of propulsion systems and engines from existing designs. Some sources report Syria's willingness to downgrade its military rocket development in the wake of the coalition attack on Iraq and the collapse of Saddam Hussein's regime in that country believing that a compliance with US requests will allow it access to technology from previously prohibited sources.

VERIFIED

Taiwan

Current Status

Taiwan has been using imagery from the Ikonos satellite for monitoring Chinese missile developments and deployments but under an agreement with the Israeli government now uses that country's EROS-1 satellite with 2 m resolution. Taiwan has control over the satellite when it flies to within 1,000 km of the island's satellite ground station. Acquisition of commercial remote sensing data has accelerated since late 2002 and Taiwan regularly uses satellite imagery for defence purposes.

Taiwan has scheduled the launch of two telecommunications satellites, Destiny 1 and Destiny 2, for launch in 2001 and 2002 respectively. The satellites and their launch dates have been delayed. Taiwan has selected Orbital Sciences to build a fleet of six microsatellites to study Earth's atmosphere, weather and climate in a programme known as the Republic of China Satellite-3/Constellation Observing System for Meteorology, Ionosphere and climate, or ROCSAT-3/Cosmic. Each satellite will weigh 68 kg and are scheduled for launch in 2005.

The US has helped Taiwan build a US$6.5 million ground receiving station for data collected over mainland China. Taiwan has expanded its acquisition of data and earth imaging data as well as SAR and other radar facilities and capabilities.

Background

Taiwan authorised an NT$13.6 billion (about US$540 million) 15-year programme in December 1990 to begin development of space technologies. The premier also appointed a five-member Space Programme Planning Committee, which identified feasible goals and developed a long-range strategic plan by August 1991. The National Space Programme Office, under the auspices of the National Science Council, was established in 1992. The principal goal is to develop an indigenous space technology capability. The programme is creating a cadre of 300 personnel capable of developing future space missions.

The basic plan includes:

the purchase and operation of a remote sensing satellite ground station (operations began 1993);

the establishment of a small satellite development and operations capability, constructing three Rocsat satellites for launch at five-year intervals by existing foreign vehicles.

TRW was awarded the ROCSAT-1 bus development contract. It worked alongside Taiwanese companies to help develop critical technologies and transfer capabilities: Taiwan performed the systems engineering, integration and testing, while TRW designed, manufactured and tested the bus. A team led by AlliedSignal Technical Services Group was awarded a contract in January 1995 worth potentially US$31.6 million for the ground system. France's Intespace was awarded a 1993 contract to assist in building an integration/test centre, including space simulation, vibration, acoustic and metrology facilities. ROCSAT-1 was launched in January 1999.

The Directorate General of Telecommunications and Singapore Telecom in March 1996 jointly ordered the ST-1 satellite from MMS to provide regional broadcasting and telecommunications services from Singapore's 88° E in 1998. The country is not a member of Intelsat or Inmarsat.

UPDATED

Thailand

Current Status

Shinawatra Computer & Communications Group, the largest group of Thai companies operating telecommunications and broadcasting service concessions, was selected in 1991 for a 30-year satellite concession to lease transponders to government and private sector users, with excess capacity offered to neighbouring countries. The Hughes HS-376 Thaicom 1 was launched successfully 18 December 1993; number 2 followed 8 October 1994. The larger Thaicom 3 was launched 16 April 1997 by an Ariane-4, to be followed by the duplicate number 4 scheduled for launch in 2004. The first pair offers up to 25 DTH channels; their successors' high power Ku-band payload will be principally devoted to DTH. The launch of Thaicom 5 has yet to be arranged. Both satellites will carry C-band, Ku-band and Ka-band capability. Thaicom 4 will operate from 120° east.

Thailand joined Inmarsat in January 1995, through its signatory Post & Telegraph Dept. The first Inmarsat land Earth station is expected at Nonthaburi. The country has been a member of Intelsat since 1966. Shinawatra Satellite and the Communications Authority of Thailand are jointly investing 5 per cent (US$65.4 million) in ICO Global Communications. Population is about 55 million.

The Thai MicroSatellite Co signed a £3 million contract 28 November 1995 with the UK's Surrey Satellite Technology Ltd for a 50 kg micro-satellite. The 18-month technology transfer programme involves a team of Thai engineers working alongside SSTL staff as TMSAT is built. It carries Earth imaging cameras and communications payloads, controlled from a station installed by SSTL in Bangkok. Its GPS receiver was developed by SSTL under ESA contract to provide position and attitude data. TMSAT was launched by Zenit in July 1998, the first commercial microsatellite to display 100 m image resolution capability. Thailand acquires commercial and government funded remote sensing data for interpolation with TMSAT.

China's Xinhua news agency claimed November 1993 that Thailand's Ministry of Transport & Communications planned to establish a satellite technology centre with Chinese assistance. It further claimed there was already an agreement for help in building and launching two small weather and science satellites. The Thailand Remote Sensing Centre operates a Landsat/Spot/MOS/ERS/JERS receiving station in Bangkok. In a region plagued by dense cloud cover up to six months annually, there is interest in microwave data from satellites such as ERS.

On 31 March 1997, Radio Thailand carried a report that there were plans to develop military satellites. It was stated that the Military Satellite for Defence Project was important because the cost would not be too great and it was essential for modern warfare. Two systems were noted as forming the project: one would be a satellite to be used for telecommunications and command, while the other would be for military surveillance and intelligence purposes. No schedules, budget levels, and so on, were given in the broadcast, but three steps in the programme were planned: establishment of ground stations, acquisition of high resolution surveillance satellites and finally the acquisition of a telecommunications satellite. It was stated that the first stage would not require a large budget, while the second and third stages would depend upon the country's economic situation. These plans have been temporarily suspended due to economic difficulties.

UPDATED

Turkey

Current Status

Turkey began operating a domestic satellite system in 1994 for TV distribution (including Europe), data and voice services. Its first satellite was lost in Ariane's January 1994 failure but the second was successful in August 1994. Insurance funded 1996's replacement launched on an Ariane-4, 9 July 1996. Turksat IC, is still operating and as of June 2004 was at 42° E.

The country is a member of Intelsat, Eutelsat and Inmarsat, operating two stations at Ata to provide Inmarsat coverage of the Indian Ocean and Atlantic Ocean regions. A ground station for Landsat 6 imagery was planned near Ankara, but none has been built since the launch failure. Recently, Turkey has started using Ikonos imaging through a regional affiliate, IntaSpace Systems Inc. Turkey has also expressed an interest in handling regional distribution of images obtained through commercial organisations and this has been extended to include other aspects of general space applications data, retrieval from space-based assets and distribution to national, regional and international operators embracing remote sensing, earth studies, climatology, oceanography, atmospheric sciences and space sciences.

UPDATED

Ukraine

Current Status

The European Union and Ukraine signed an agreement on co-operation in the creation of an all-European global navigation satellite system. Further co-operation led to integration of national interests into a policy planning document which led to the originating standard for the European Galileo satellite system. Ukraine, the second major space force to emerge from the break-up of the Soviet Union, is pursuing an independent policy but it keenly monitors international projects for opportunities to participate. The future of its major space enterprise, NPO Yuzhnoye, was uncertain. After it appeared the state considered halting production of the Zenit orbital launcher. As commercial markets appeared in the early 1990s, though, the pressure relaxed and now the Zenit's future is assured. The first commercial success came with the May 1995 selection by Space Systems/Loral for three 1998 launches to carry 36 US Globalstar satellites. Boeing, too, issued a vote of confidence in the Zenit when it selected the booster for the Sea Launch venture. This projects calls for three-stage Zenits to be shipped to the US, assembled and then towed into international waters for near-Equator launches. The world's largest commercial telecommunications satellite manufacturer, Hughes Space & Communications Co (now Boeing Satellite Systems), placed a firm order in December 1995 with Sea Launch for 10 launches beginning in 1998. At the end of 1998, however, political forces associated with technology transfer questions prevented Sea Launch from inaugurating its service until October 1999, launching DirecTV-1R seven months after the first test flight. Since then a further six launches have taken place of which the first, in March 2000, was a failure.

The Ukraine and US signed a launch agreement on 14 December 1995 that allowed Ukraine to launch 16 western satellites through to 2000. Five were allowed on purely Ukrainian vehicles (six if the market grows) and 11 on joint US/Ukrainian ventures, effectively Sea Launch.

The future of Ukraine's Tsyklon launcher is less assured – its launch rate is falling under funding limitations and the loss of suppliers for the ageing avionics. Tsyklon production appears to have halted; it has been suggested that some of Russia's Tsyklon-M strategic stockpile may be converted into the three-stage version. 1995's sole Tsyklon delivered the national Sich ocean radar satellite, controlled by Russia's Military Space Forces until Yevpatoria took it over from October 1995. A Ukrainian spaceport is not likely because of the investment required, although polar orbits would be accessible for southern departures from the Black Sea coast. The new generation Sich/Okean is Ukraine's major national satellite programme, expanding to provide extensive remote sensing capabilities. 1998's Sich 2 included a synthetic aperture radar with 10 m resolution. The Ukrainian government has expressed concerns at the reduction in importance for its national launcher programmes and has sought ways to co-operate with foreign operators to develop a new generation of launchers.

Ukraine President Leonid Kravchuk announced on 2 March 1992 the creation of the Ukrainian Space Agency. The Commission for Space Research was also established, under the Academy of Sciences. Presidents Leonid Kuchma and Bill Clinton establishing direct contact between their space agencies for the first time signed a five-year agreement November 1994. Ukrainian cosmonaut Leonid Kadenyuk flew on Shuttle STS-87 in November 1997 as a consequence. Kazakhstan issued a decree the previous September setting up the Kazakhstan Space Research Agency;

cosmonaut Toktar Aubakirov became its General Director in 1993 (when it became the National Aerospace Agency). He stepped down in May 1994 after his election as a parliamentary deputy. Alsultan Kalybayev replaced him. Azerbaijan formed its own ANAKA aerospace agency in February 1992. The Uzbek State Agency for Space Research (Uzbekkosmos) followed in March 1993, headed by Kamol Muminov. Tajikistan approved a plan in January 1994 to set up a space research centre under its Academy of Sciences.

UPDATED

United Kingdom

Current Status

Currently, the UK spends approximately £170 million on civil space, down considerably from the mid-1990s, of which about two-thirds is invested in ESA. BNSC has co-ordination of British interests in the Galileo European Global Positioning System, scheduled to be operational in 2008. The UK government continues to seek ways of reducing space expenditure and is keen to engage industry and non-government organisations in co-operative ventures. The general decline in financial commitments continues and fails to match that of other ESA member states. Seeking ways to reduce the cost of government funded programmes, the UK Ministry of Defence has undertaken to lease defence communications services from the Skynet 5 series satellites due for launch from 2005. Shamed into support for the Beagle 2 Mars lander carried by the ESA Mars Express, the UK government, in 2000, provided the same funds for Mars exploration which encouraged commercial participation. Beagle 2 was lost during the final descent to the surface of Mars in December 2003 and an investigation into the failure was begun in February 2004. The report indicated several key areas where the spacecraft could have failed and was unable to provide a definitive conclusion due to lack of evidence.

Background

The United Kingdom's space programme is brought together by BNSC, the British National Space Centre. Two-thirds of the UK's annual space budget is devoted to ESA programmes, with the balance supporting national R&D activity through the universities, research centres and companies.

Historically, the UK was the sixth nation to launch its own satellite into orbit when Prospero flew aboard the fourth Black Arrow launch vehicle (designated R-3) on 28 October 1971: however, this was the final Black Arrow launch to be funded – planned further launches had been cancelled in July 1971. The first two Black Arrow launches had been sub-orbital missions (R-0 on 28 June 1969 failed, R-1 on 4 March 1970 with the X-1 payload was successful), while the third had been the first orbital attempt (2 September 1970 with the Orba/X-2 payload) failed to reach orbit. A sad comment on the programme was made in the United States Library of Congress report 'Worldwide Space Activities' in the introduction to the Black Arrow discussion: 'The only continuous thread which runs through the history of the British space programme has been its lack of continuity'.

MMS continues to market the successful series of Skylark sounding rockets, first launched in 1957 and the backbone of European Texus microgravity launches. MMS has worked with ESA partners on possible two-stage to orbit systems using the Russian Antonov An-225 and interim Hotol spaceplane, with results showing a medium term feasibility. UK interest in reusable launchers is now led by Reaction Engines, with its Skylon vehicle using a derivative of the Hotol dual-mode engine with a new airframe design: one of the people behind Reaction Engines is Alan Bond, who was responsible for the original Hotol concept.

The first Briton in space, food scientist Helen Sharman, flew to Russia's Mir orbital station aboard Soyuz-TM 12 in May 1991 under the commercial Project Juno. British-born Michael Foale (now joint US/UK nationality), a career astronaut rather than a one-off commercial flier, was selected in 1989 as a NASA mission specialist: his first flight was aboard STS-45 in 1992 and when flying STS-63 in 1995 he became the first British-born person to perform an EVA. He flew to Mir in May 1997 aboard the STS-84 mission and returned to Earth in October 1997 aboard STS-86.

By early 2001, the UK had flown 22 indigenously-built national satellites: one launched by Black Arrow, 13 by the US and seven by Ariane. The total includes the five UoSATs of the University of Surrey. These are in addition to craft produced for international and commercial programmes, such as Giotto and Inmarsat. Prospero remains the only UK satellite launched on a UK vehicle. Like the original US/USSR programmes, the national space effort depended initially on the conversion of a ballistic missile into a launcher. This was the Hawker Siddeley Blue Streak, finally cancelled as a missile in 1960 though continued until 1973 as first stage for ELDO's Europa. A more modest attempt at a national launch vehicle was announced in September 1964, developing a small three-stage vehicle from the successful Westland Black Knight research rocket. Construction of three Black Arrows was ordered in March 1967. Success in October 1971 with only the second complete vehicle was contrasted by the programme's previous cancellation in July 1971. The final Black Arrow now resides in a museum. With Prospero, the UK became the sixth nation to achieve orbital capability – after the USSR, USA, France, Japan and China.

Following up its success with the Ariel science series, Britain provided the ground control for the three-nation Infra-red Astronomical Satellite, launched January 1983. RAL also provided the UKs satellite to 1984's AMPTE investigation of the magnetosphere.

By early 1984, NASA had agreed to carry a UK payload specialist on each of the Skynet 4 Shuttle launches due in 1986. Sq-Ldr Nigel Wood was chosen, from four candidates who were given six months' training at an 'expenses only' cost of about US$150,000 each, as prime candidate for the first mission. *Challenger*'s loss in January 1986 resulted in the Skynets transferring to expendable launchers and the team disbanding. NASA's overtures about its readiness to carry a British PS in 1985 met with little response, and the opportunity to have a British scientist accompany the ambitious solar X-ray telescope aboard the Spacelab 2 mission in July was missed. The agreement for the commercially-sponsored non-government Mir Project Juno visiting mission to the Soviet Mir space station, including 26 UK experiments, was signed with the Soviets 29 June 1989. Sponsorship target was £16 million, about £9 million of which was to be paid to the Soviets. The City of London Moscow Narodny Bank (the only wholly owned Soviet state bank outside of the USSR) managed the project through its Antequera Ltd company. Scientist Helen Sharman and Army Air Corps pilot Major Tim Mace, selected from 13,000 applicants, began training in Star Town in November 1989. Juno was re-oriented during 1990 when it became clear that sponsorship was insufficient; the Soviets agreed to continue the flight at below commercial rates and to provide the experiments.

UK space priorities were re-cast in 1996, following long consultation between the government, industry and academia. In a new 'Forward Plan' for space, announced by the then Space Minister, Ian Taylor MP in July 1996, some new aims were set for BNSC and for the UK in space. These aims built on past UK activities to provide a coherent programme for the future and BNSC is charged with updating them annually. These aims are:

Foster the development of a competitive Earth/observation industry.
Develop supportive technology and R&D programmes driven by future needs.
Maintain continuity of global datasets to meteorology and climate monitoring.
Establish a recognised means for procuring Earth observation services, and microgravity.
To exploit, sustain and develop UK science expertise and the links and mutual support which science, Earth observation and industrial applications share.
Support a strong ESA science programme.
Pursue opportunities for bilateral activity and a national 'small missions' programme; and maintain interest in ESA microgravity options.
Encourage a new European approach to spectrum authorising and regulations.
Promote international standards and industrial access.
Take significant roles in Europa's GNSS (Global Navigation Satellite System) programme.
Aid competitiveness through strong partnerships with industry (including R&D and technology demonstrators).
Urge the commercialising of international satellite organisations; and help the development of multimedia services by seeking government uses for them.
Help Europe optimise use of its own launcher services and those provided elsewhere and seek ways of reducing costs while stimulating availability.
Encourage military/civil space synergy; support UK industrial competitiveness.
Utilise and develop technology and technology transfer.
Encourage synergy between ESA (as a provider) and the European Union (as a user) of space services and facilities.
Encourage more co-funding with industry; and heighten awareness and applications of space in society.

Recent BNSC studies have indicated that the UK has a thriving space industry sector with more than 450 organisations, a turnover of £850 million, a workforce of some 7,000 and markets throughout the world.

BNSC itself was established in 1985 to bring together the space interests of government department and research councils, optimise value for money and provide the UK with a coherent position in European and World space activities. The government through the BNSC partnership, initially focused the UK space effort in satellite communications.

With major participation through BNSC in infrastructure projects, the government enabled and encouraged industrial investment and a sound commercial position for the UK in telecommunications before, in the late 1980s, moving to meet the challenge of Earth observation.

UK contributions to international Earth observation programmes – especially through ESA – are considerable. They include two major instruments on both ESA's ERS (European Remote Sensing) satellites, a 22.5 per cent stake in the next-generation Polar Platform and a 17.5 per cent stake in ENVISAT and its ground segment. ENVISAT is the large Earth observation satellite launched by ESA in 2001. The stake in ENVISAT was worth up to £250 million in the UK over 10 years in the form of support of industrial involvement in the platform's development, and in instrument development. The UK provides two key instruments – ASAR (Advanced Synthetic Aperture Radar) and AATSR (Advanced Along-Track Scanning Radiometer).

The ERS missions (ERS 2 is still active) have justified the UK's funding share of £130 million, returning a steady stream of data about the Earth's oceans, ice, land and atmosphere. Key instruments – AMI (SAR) and ATSR, both forerunners of ENVISAT instruments – were provided from the UK. ERS has enabled scientists to derive a series of new technologies. In 1995, when both ERS 1 and ERS 2 were in operation in tandem, joint interpretation of the data led to interferometric techniques capable of detecting surface movements of around 2 cm. At this level of resolution scientists can identify areas of subsidence and ground heave, thus making it possible to predict and monitor the

An artists impression of the Beagle 2 lander on the surface of Mars. Following separation from the Mars Express spacecraft Beagle failed to communicate with earth after its planned touchdown time on 25 December 2003 (ESA)
0572500

UK SPACE BUDGET (£ million)

	1994–95	1995–96	1996–97	1997–98	1998–99	1999–00	2000–01
DTI	98	100	104	103	88	91	91
PPARC	48	51	51	43	48	44	45
Met Office	22	21	24	15	23	27	25
NERC	12	12	12	13	11	11	8
MoD	7	8	7	7	7	6	6
DoE/DoT/DETR	7	3	3	3	17	1	3
Total	**194**	**195**	**201**	**183**	**195**	**181**	**178**
Earth Obs	104	94	107	96	100	89	78
Science	47	53	53	43	50	47	48
Telecoms	16	17	14	12	8	11	14
Satnav	5	4	2	2	16	5	3
Technology	5	3	3	4	2	9	15
Launchers	4	8	5	5	4	5	5
ESA General	13	17	18	21	14	14	14
Total	**194**	**195**	**201**	**183**	**195**	**181**	**178**
ESA	127	131	137	133	139	114	118
Non-ESA	67	64	64	50	56	67	60
Total	**194**	**195**	**201**	**183**	**195**	**181**	**178**

(All figures rounded to the nearest million.)

effects of eruptions, earthquakes and flooding with unprecedented accuracy.

ERS data are acquired, processed and distributed by the £20 million Earth Observation Data Centre at Farnborough, one of four ESA Processing and Archiving Facilities (PAF). UK PAF operations are managed under ESA contract by the National Remote Sensing Centre Ltd.

The UK is providing instruments for other international Earth observing missions: HIRDLS (High Resolution Dynamic Limb Sounder) on NASA's EOS and AMSU (Advanced Microwave Humidity Sounder) on later NOAAs. BNSC's partner NERC is part-funding a Global Earth Radiation budget instrument for EUMETSAT's second generation METEOSAT.

Communications remain important commercially to the UK, which is committed to four key ESA programmes. The programmes are the DRTM (Data Relay and Technology Mission, 6.7 per cent to Artemis and 0.5 per cent to the Data Relay Satellite), EMS (European Mobile System) on ITALSAT 2, ARTES Element 4 (competitively improvement programme) and RTES Element 9 (GNSS).

Two Hughes-built satellites, Marcopolos 1 and 2, were launched for the commercial British Satellite Broadcasting (BSB) company and became operational in 1990, providing direct broadcast channels. In late 1990, BSB merged with Sky Television (to become BSkyB). As a result in December 1993 Marcopolo 1 was sold to Nordiska Satellit AB in Sweden who operated it under the new name Sirius 1, while in 1992, Norwegian Telecom (now Telenor Satellite Services AS) had purchased Marcopolo 2 and operated it re-named to Thor 1.

Industrial involvement in satellite navigation has been supported with good effect – Racal's Global Skyfix network now has some 60 reference points worldwide, manned around the clock, providing positioning information to a wide spectrum of users. The UK is a strong advocate through ESA of the EGNOS civil overlay satellite communications positioning system and the planned GNSS programme as a whole.

The UK's Skynet satellite communications system remains the leading one outside the USA and CIS committed to military communications. Plans for Skynet 5 have matured toward the assignment of launch dates beginning in 2006. The UK with British Telecom as a signatory participates in Inmarsat (9.36 per cent), Intelsat (9.19 per cent) and Eutelsat (21.67 per cent) and the UK hosts the headquarters for the Maritime Communications Agency.

While UK science ambitions suffered a serious setback, in June 1996, with the uninsured loss of the Cluster solar monitoring missions on Ariane 501, current missions with UK participation such as ESA's Ulysses, ISO and SOHO, the ESA-NASA Hubble Space Telescope and the US/German/UK ROSAT and future missions with a UK contribution such as Russia's Spektrum-RG, ESA's Rosetta cometary probe and the NASA-ESA Cassini-Huygens mission to Saturn proceeded according to plan. BNSC worked to ensure ESA's smoothest adoption of a new space science strategy, Horizon 2000+, which commences in 2006.

The loss of Ariane 501, carrying Cluster, was immediately subjected to an ESA investigation, from which it was clear that no UK components were at fault. Two UK companies contribute to Ariane 5 and the UK has a commitment of £2.8 million over three years to the new launch vehicle's infrastructure programme. The UK continues its parallel commitment to the highly-successful Ariane 4 launch vehicle family, with various companies supplying equipment for the vehicle: MMS supplies the SPELDA payload units and GEC-Marconi Avionics the ring laser guidance units.

The telecommunications expertise developed by British Aerospace and Matra Marconi Space UK has provided the greatest return on space investment. BAe/MMS France in the Satcom International partnership jointly developed the successful Eurostar platform. Marconi Space and Matra Espace merged in 1990 to create Matra Marconi Space to compete in the changing European market against strong aggregations such as Deutsche Aerospace. MMS acquired BAe Space Systems in 1994 for £56 million, at the time, creating Europe's largest space hardware group and the world's third (after Hughes and Lockheed Martin), with annual sales of more than US$1 billion. BAe was once Europe's leader in telecommunications satellite manufacturing and perhaps a factor in its decline was the government's late 1980s switch from satellite communications to remote sensing. BAe acted as prime contractor for the £34 million Giotto probe, which encountered Halley's Comet in March 1986, and ESA's experimental Olympus communications satellite. The company also provided 20 Spacelab pallets in return for Britain's 6 per cent contribution to Spacelab.

In response to a forecast that a world remote sensing market worth £250 million a year would develop by 1990, Britain invested £40 million 1983–86 in a joint science and industry programme covering data acquisition, processing, dissemination, interpretation and forecasting. The RAE (now DRA) National Remote Sensing Centre at Farnborough was established in 1980 and joined a decade later by the Earth Observation Data Centre, initially to process, archive and distribute ERS data. MMS (ex-BAe) now leads ESA's Polar Platform.

UPDATED

United States

Current Status

In January 2004 President Bush announced that NASA was to be directed to retire the Shuttle by 2010, reduce work on the Space Station, develop a new generation of lunar exploration vehicles by 2008 and return humans to the moon soon after 2015. Originally, NASA faced several major changes under the new Bush administration when it came to power in January 2001 with strict budget constraints and a potential shift towards a more research and development role and less of an operations orientated organisation. The Bush administration ordered a cut of US$4 billion in FY02-06 ISS funding, at least half of which has been achieved by cancelling the Propulsion Module, Habitation Module and Crew Return Vehicle. With the successful launch of Mars Odyssey in April 2001, NASA recommenced planetary mission flights following the loss of Mars Climate Orbiter and Mars Polar Lander in 2000. The Mars Odyssey was launched in 2001 and two landers were launched in 2003, successfully reaching the surface of Mars. NASA continues to grow its Earth Science Enterprise series of environmental and earth observation missions. The Bush administration endorsed a FY02 budget request of US$14.5 billion, an increase of 2 per cent over FY01 with a US$15 billion budget for FY03 and a US$15.5 billion budget for FY04. To fund the new Bush initiative, NASA has been given a 5.6 per cent budget increase for FY2004 with an average 5 per cent per annum through 2010 when the NASA budget is scheduled to return to a level paced to inflation.

Commercial activities continue to grow in remote sensing and earth resource data but the surfeit of capacity in the launcher market threatens high returns for operators as prices continue to fall under competition.

Robust development of a National Missile Defense system promises to increase the role for the military in space and funds for Pentagon space activity continues to increase, now about 10 per cent higher than NASA.

Background

NASA hoped that the recommendations of 1990's Advisory Committee on the Future of US Space Policy (generally known as the Augustine Committee after its Chairman) would set the tone for the decade. It concluded that the agency's budget should grow 10 per cent annually in real terms, attaining US$30 billion (FY90 rates) by 2000. This is equivalent to 0.4 per cent GDP, or about half of peak spending during the 1960s. It quickly became clear, however, that civil and military budgets would remain flat at best for the foreseeable future, together accounting for 2.4 per cent of the federal budget. In fact, NASA's FY95 appropriation, at US$14.46 billion, reflected the first reduction in two decades when inflation is taken into account. FY96 fell to US$13.8 billion, but that budget was not settled well into 1996 because of disputes between Congress and the White House. 1995's proposed plan through FY99 would see a continuing decline, providing only US$12.6 billion in FY99 (FY94 rate) and saving US$5 billion over those years. There was even grimmer news for NASA when the FY97 US$13.8 billion budget was released in March 1996. President Clinton plans to take out another US$3.3 billion as part of his drive to balance the federal budget by 2002, leaving NASA with only US$11.6 billion by Fiscal 2000. The conclusion is that NASA must reinvent itself, divesting itself of responsibility for operations and returning to cutting-edge projects.

US federal space expenditure of some US$30 billion annually was, until the early 1990s, challenged only by that of the USSR with three or four times the annual launch rate of the United States. The Department of Defense and the civil National Aeronautics & Space Administration together account for about 99 per cent, with military expenditure standing at around half. Commercial space sales have been increasing at 20 per cent annually during the 1990s, largely because of satellite communications. The US government projects that commercial sales will surpass military and government activities by 2010, partly because of the downturn in those latter areas. Satellite communications continue to represent by far the largest element, accounting for 65 per cent. Space export earnings fell to US$590 million in 1995 from US$750 million in 1994 (US$554 million in 1993). Employment in the missile/space sector fell to 153,000 by 1997, from 157,000 in 1995, 165,000 in 1994 and 176,000 in 1993.

The major manned programme for the next decade will be the low Earth orbit International Space Station, costing more than US$60 billion to establish and with significant international contributions. The arrival of a new administration and concern over costs resulted in President Clinton in February 1993 ordering yet another restructuring to cut cost to completion to US$9 billion (US$19 billion was achieved; US$9 billion has already been expended). Annual spending is capped at US$2.1 billion in 1993 dollars.

Station *Freedom*, as such, ceased to exist in 1993. The programme and its management were overhauled and by the end of 1993, Russia had agreed to become a fully-fledged international partner. NASA Administrator Dan Goldin and Russian Space Agency director general Yuri Koptev signed a US$400 million USFY94-97 final agreement on 21 June 1994. It included US$305 million for up to 10 Shuttle Mir visits encompassing six NASA astronauts spending a total of 24 months aboard the Mir Station, primarily for life and µg sciences research, crew exchanges, cargo delivery and experiment return. Starting with STS-76/SMM-03 a long-term US presence was established on Mir when Shannon Lucid was left on board the station when the shuttle returned to Earth. Her stay on Mir ended when STS-79/SMM-04 docked to leave John Blaha in her place. STS-81/SMM-05 (January 1997) returned Blaha to

US space budget summary (then-year $ million)

FY Authority	NASA	DoD	DoC	DoE	Total
61	926.0	813.9	–	67.7	1,808.2
62	1,796.8	1,298.2	50.7	147.8	3,294.8
63	3,626.0	1,549.9	43.2	213.9	5,434.5
64	5,016.3	1,599.3	2.8	210.0	6,831.4
65	5,137.6	1,573.9	12.2	228.6	6,955.5
66	5,064.5	1,688.8	26.5	186.8	6,969.8
67	4,830.2	1,663.6	29.3	183.6	6,709.5
68	4,430.0	1,921.8	28.1	145.1	6,528.9
69	3,882.0	2,013.0	20.0	118.0	5,975.8
70	3,547.0	1,678.4	8.0	102.8	5,340.5
71	3,101.3	1,512.3	27.4	94.8	4,740.9
72	3,071.0	1,407.0	31.3	55.2	4,574.7
73	3,093.2	1,623.0	39.7	54.2	4,824.9
74	2,758.5	1,766.0	60.2	41.7	4,640.3
75	2,915.3	1,892.4	64.4	29.6	4,914.3
76	3,225.4	1,983.3	71.5	23.3	5,319.9
TQ	849.2	460.4	22.2	4.6	1,340.5
77	3,440.2	2,411.9	90.8	21.7	5,982.8
78	3,662.9	2,738.3	102.8	34.4	6,518.2
79	4,030.4	3,035.6	98.4	58.6	7,243.5
80	4,680.4	3,848.4	92.6	39.6	8,688.8
81	4,992.4	4,827.7	87.0	40.5	9,977.8
82	5,527.6	6,678.7	144.5	60.6	12,440.7
83	6,327.9	9,018.9	177.8	38.9	15,588.5
84	6,648.3	10,194.9	236.0	34.1	17,135.7
85	6,924.9	12,767.9	422.9	34.0	20,166.5
86	7,165.0	14,126.0	308.9	34.6	21,659.4
87	9,809.0	16,286.8	277.9	47.6	26,447.9
88	8,302.4	17,678.7	351.5	240.8	26,606.7
89	10,097.5	17,906.1	301.3	97.2	28,443.3
90	12,141.6	15,616.0	243.0	78.6	28,140.0
91	13,036.0	14,181.0	251.0	251.0	27,779.0
92	13,199.0	15,023.0	327.0	223.0	28,845.0
93	13,064.0	14,106.0	324.0	165.0	27,729.0
94	13,022.0	13,166.0	312.0	74.0	26,649.0
95	12,543.0	10,644.0	352.0	60.0	23,676.0
96	12,569	11,514	472	46	24,601
97	12,457	11,727	448	35	24,667
98	12,321	12,359	456	63	25,199
99	12,459	13,203	575	102	26,599

FY Outlay	NASA	DoD	DoC	DoE	Total
61	694	710	–	64	1,468
62	1,226	1,029	1	130	2,387
63	2,517	1,368	12	181	4,079
64	4,131	1,564	12	220	5,930
65	5,035	1,592	24	232	6,886
66	5,858	1,637	28	188	7,719
67	5,337	1,673	39	184	7,237
68	4,595	1,890	29	147	6,667
69	4,078	2,095	31	118	6,326
70	3,565	1,756	24	103	5,453
71	3,171	1,693	30	97	4,999
72	3,195	1,470	37	60	4,772
73	3,069	1,557	29	51	4,719
74	2,960	1,777	64	39	4,854
75	2,951	1,831	64	34	4,891
76	3,336	1,864	71	26	5,314
TQ	869	458	23	8	1,361
77	3,600	1,833	87	22	5,559
78	3,582	2,457	101	29	6,188
79	3,743.9	2,891.8	97.4	54.7	6,808.3
80	4,340.1	3,162.3	88.7	48.8	7,667.7
81	4,877.1	4,130.5	81.0	46.9	9,165.5
82	5,463.3	4771.5	142.4	59.5	10,466.2
83	6,100.9	6,246.7	178.0	39.6	12,590.4
84	6,461.4	8,000.2	208.7	33.4	14,726.1
85	6,607.4	10,441.3	155.4	34.0	17,254.8
86	6,756.0	11,448.5	316.9	34.7	18,581.0
87	7,254.0	14,264.3	261.9	37.4	21,843.9
88	8,450.5	14,397.4	333.9	199.1	23,414.1
89	10,195	14,504	306	97	25,143
90	12,292	12,962	279	79	25,671
91	13,351	14,432	266	251	28,360
92	12,838	14,437	298	223	27,865
93	13,092	13,779	295	165	27,398
94	12,363	10,973	297	83.0	23,790
95	12,593	11,494.0	330	70.0	25,564.0

These figures are taken from NASA's *Aeronautics & Space Report of the President*. 'Outlay' represents actual spending; Fiscal Year (FY) runs 1 October – 30 September. TQ = transitional quarter. NASA's exclude non-space activities. DoD = Dept of Defense; DoE = Dept of Energy; DoC = Dept of Commerce. The total includes relevant expenditures of the departments of Interior, Agriculture and others.

NASA budget summary US$ millions (FY98 is request)

	FY97	FY98
Human Space Flight	5,674.8	5,326.5
Science, Aeronautics and Technology	5,453.1	5,642.0
Mission Support	2,564.3	2,513.2
Total Budget Authority	13,709.2	13,500.0
Human Space Flight	5,674.8	5,326.5
Space Station	2,148.6	2,121.3
Development	1,766.3	1,386.1
Operations	177.6	490.1
Research Programme	204.7	245.1
US/Russian Co-Operative Programme		100.0
Space Shuttle	3,150.9	2,977.8
Shuttle Operations	2,514.9	2,494.4
Safety/Performance Upgrades	636.0	483.4
Payload and Utilisation Operations	275.3	227.4
Spacelab	50.3	14.2
Payload Processing And Support	41.7	51.6
Advanced Projects	34.7	58.7
Engineering And Technical Base	148.6	102.9
Science, Aeronautics and Technology	5,453.1	5,642.0
Space Science	1,969.3	2,043.8
Advanced X-Ray Astrophysics		
Facility	178.6	92.2
Cassini	89.6	9.0
Relativity Mission (Gp-B)	59.6	45.6
Space Infra-Red Telescope Facility (Sirtf)		81.4
Thermosphere, Ionosphere, Mesosphere Energetics And Dynamics (TIMED)	18.2	48.2
Payloads	16.9	12.3
Explorer Development	125.0	142.7
Mars Surveyor Programme	90.0	139.7
Discovery	76.8	106.5
New Millennium	48.6	75.7
Advanced Space Technology	132.0	151.2
Mission Operations and Data Analysis	583.3	507.4
Supporting Research and Technology	246.0	311.2
Suborbital Programme	64.1	84.4
Life and Microgravity Sciences and Applications	243.7	214.2
Life Sciences	97.4	85.5
Microgravity Science Research	105.3	101.4
Aerospace Medicine	3.8	7.5
Shuttle/Spacelab Payload Mission Management and Integration	24.2	6.9

	FY97	FY98
Space Product Development	13.0	12.9
Mission To Planet Earth	1,361.6	1,417.3
Earth Observing System (Eos)	586.7	679.7
Eos Data Information System	254.6	244.7
Earth Probes Development	57.2	40.7
Science	295.4	271.1
Operations, Data Retrieval And Storage	78.0	54.2
Global Observations To Benefit	5.0	5.0
The Environment (GLOBE)		
Launch Services	84.7	121.9
Aeronautics And Space Transportation Technology	1,339.5	1,469.5
Aeronautical Research And Technology	844.2	920.1
Research and Technology Base	404.2	418.3
Focused Programmes	440.0	501.8
High Performance Computing and Communications	23.3	45.7
High Speed Research	243.1	245.0
Advanced Subsonic Technology	173.6	211.1
Advanced Space Transportation	336.7	396.6
Reusable Launch Vehicle	283.5	353.5
Advanced Space Transportation Technology	53.2	43.1
Commercial Technology Programme	158.6	152.8
Commercial Technology Programmes	25.8	20.0
Technology Transfer Agents	7.8	7.8
Small Business Innovative Research Programmes	125.0	125.0
Mission Communication Services	418.6	400.8
Ground Network	245.6	224.7
Mission Control And Data Systems	147.1	145.0
Space Network Customer Services	25.9	31.1
Academic Programmes	120.4	96.4
Education Programmes	65.6	55.5
Minority University Research And Education	54.8	40.9
Mission Support	2,564.3	2,513.2
Safety, Reliability And Quality Assurance	38.8	37.8
Space Communication Services	277.7	245.7
Space Network	185.1	161.2
Telecommunications	92.6	84.5
Research And Programme Management	2,092.5	2,070.3
Construction Of Facilities	155.3	159.4
Inspector General	17.0	18.3

Earth and Jerry Linenger, who was in turn replaced by Michael Foale in May 1997, took his place. It was courtesy of these residences on Mir that, on 22 March 1997 NASA was able to claim that there had been a continuous American presence in orbit for a full year. By contrast, the Russians far surpassed this mark since there had been a permanent Russian presence in orbit since September 1989. In fact, apart from a short break during April-September 1989, the Russians celebrated 10 years of residencies abroad Mir in February 1997. Additional shuttle flights to Mir included STS-86 in September 1997 leaving astronaut Daniel Wolfe and returning (in early October) with Foale, STS-89 in January 1998 leaving astronaut Andy Thomas and STS-91 in June 1998 returning him to Earth. These flights concluded the Shuttle-Mir programme of visits.

President Bush announced on 20 July 1989, the 20th anniversary of the first manned lunar landing, that the next major US goal after Space Station would be a permanent presence on the Moon before beginning manned Mars exploration in 2019. This US$400 billion 30-year Space Exploration Initiative was supported in spirit by the Augustine review, but it depended on funding availability and was not driven by dates. Congress repeatedly denied funding and NASA closed its Office of Exploration in early 1993. The prospect of manned missions to the Moon and beyond has receded well into the future.

NASA Budget Summary FY99-01 (US$ million)

	1999	2000	2001*
International Space Station	2,304.7	5,467.7	5,499.9
Human Space Flight	–	5,467.7	5,499.9
Launch Vehicles and Payloads Ops	3,175.3	–	–
Science, Aeronautics & Technology	5,653.9	5,580.9	5,929.4
Mission Support	2,511.1	2,532.2	2,584.0
Inspector General	20.0	20.0	22.0
Total Budget Authority	13,665.0	13,600.8	14,035.3

* Budget request for financial year beginning 1 October 2000.

Solar system exploration has arguably suffered most from launcher and budgetary difficulties. The continual drain to fund the space station has turned the tide against the funding of deep space missions. It is now impossible to fund ventures akin to the highly successful Magellan Venus radar mapper and the Galileo Jupiter orbiter. CRAF was sacrificed to save 1997's associated Cassini Saturn orbiter but, at a development cost of US$1.4 billion, that mission survived because it was so far down the road. Termination would have been a major blow to ESA's enthusiasm for joint ventures with NASA. The European agency suffered grievously in the 1980s when NASA unilaterally cancelled its half of the Ulysses mission.

Instead, the emphasis is on frequent and affordable science missions with highly focused objectives: numerous micro-spacecraft carrying

NASA Budget Breakdown

	FY2000	FY2001
Human space flight	***5,467.7***	***5,499.9***
Space stations	**2,323.1**	**2,114.5**
Vehicle	890.1	442.6
Operations capability	763.6	826.5
Research	394.4	455.4
Russian program assurance	200.0	300.0
Crew return vehicle	75.0	90.0
Space Flight Operations (Space Shuttle)	**2,979.5**	**3,165.7**
Safety and Performance Upgrades	488.8	–
(Construction of Facilities included)	(11.0)	–
Shuttle Operations	2,490.7	–
Flight Hardware	–	2,005.9
Ground Operations	–	551.8
Flight Operations	–	273.6
Program Integration	–	334.4
(Construction of Facilities included)	–	(15.5)
Payload and Utilisation Operations	**165.1**	–
Payload Carriers and Support	49.3	–
Expendable Launch Vehicle Mission Support	30.6	–
(Construction of Facilities included)	(2.2)	–
Engineering and Technical Base	85.2	–
Payload and ELV Support	–	**90.2**
Payload Carriers and Support	–	57.0
Expendable Launch Vehicle Mission Support	–	33.2
(Construction of Facilities included)	–	(0.8)
Investment and Support	–	**129.5**
Rocket Propulsion Test Support	–	28.0
Additional Funding for Academic Programs	–	8.0
Technology and Commercialisation	–	20.0
Engineering and Technical Base	–	73.5
Science, Aeronautics and Technology	***5,580.9***	***5,929.4***
Space Science	**2,192.8**	**2,398.8**
Chandra X-Ray Observatory	4.1	–
Relativity Mission (GP-B)	49.9	13.8
HST Development	160.1	168.1
Stratospheric Observatory For Infra-red Astronomy (SOFIA)	39.0	33.9
Space Infra-Red Telescope Facility (SIRTF)	123.4	117.6
Thermosphere, Ionosphere, Mesosphere Energetics and Dynamics (Timed)	27.5	–
Payloads	13.6	7.1
Explorer Development	122.3	138.8
Mars Surveyor Program	248.4	326.7
Discovery	154.8	196.8
Mission Operations	75.4	80.0
Supporting Research and Technology	1,179.3	1,302.8
Additional Funding for Academic Programs	(10.2)	13.2
Undistributed Reduction	–5.0	–
Life and Microgravity Sciences and Applications	**274.7**	**302.4**
Advanced Human Support Technology	30.2	30.9
Biomedical Research and Countermeasures	57.2	76.9
(Construction of Facilities)	(9.0)	(8.5)
Fundamental Biology	38.2	39.2
Microgravity Research	108.8	129.3
Health Research	8.7	11.3
Space Products Development	14.4	13.6
Mission Integration	17.2	0.2
Additional Funding for Academic Programs	(1.0)	1.0
Earth Science	**1,443.4**	**1,405.8**
Earth Observing System (EOS)	575.4	–
Earth Observing System Data Information System	261.9	–
Earth Probes	157.4	–
Applied Research and Data Analysis	436.5	–
Global Observations to Benefit the Environment	5.0	–
Construction of Facilities	1.0	–
Shuttle radar Topography mission	6.2	–
Major Development	–	**819.5**
Earth Observing System	–	447.1

	FY2000	FY2001
Earth Observing System Data Information System	–	252.0
Earth Probes (includes SRTM)	–	120.4
Research and Techology	–	**533.3**
Earth Science Program Science	–	353.2
Applications, Commercialisation and Education	–	69.2
Technology Infusion	–	110.9
Mission Operations	–	**42.7**
Additional Funding for Academic Programs	(7.3)	10.3
Aerospace Technology	**1,124.9**	**1,193.0**
Research and Technology Base	**581.7**	**539.4**
Information Technology	77.9	115.8
Vehicle System Technology	157.5	136.3
Propulsion and Power	76.3	75.5
Flight Research	70.7	81.0
Operations Systems	17.0	17.5
Rotorcraft	26.9	26.7
Space Transfer & Launch Technology	136.5	70.6
Construction of Facilities	11.7	16.0
Additional Funding for Academic Programs	7.2	(11.2)
Focused Programs	**403.2**	**507.4**
High-Performance Computing and Communications	24.2	24.2
Aviation System Capacity	62.9	59.2
Aviation Safety	64.4	70.0
Ultra Efficient Engine Technology	68.3	35.0
Small Air Transport System	–	9.0
Quiet Aircraft Technology Program	–	20.0
2nd Generation RLV Focused	–	235.0
Pathfinder Technology Demonstrations	31.3	37.1
X-33 Advanced Technology Demonstrator	111.6	–
X-34 Technology Demonstration	40.5	17.9
Additional Funding for Academic Programs	**(7.2)**	**11.2**
Commercial Technology Program	**140.0**	**135.0**
Commerical Programs	35.0	29.2
Technology Transfer Agents	7.4	5.8
Small Business Innovative Research Programs	97.6	100.0
Mission Communication Services	**406.3**	*
Ground Networks	162.0	–
Mission Control and Data Systems	233.8	–
Space Network Customer Services	10.5	–
Space Operations	–	**529.4**
Operations	–	329.8
Mission and Data Service Upgrades	–	106.2
TDRSS Replenishment Project	–	55.0
Technology	–	38.4
Academic Programs	**138.8**	**100.0**
Education Programs	85.0	54.1
Minority University Research and Education	53.8	45.9
(Additional funding for Academic Programs not included above)	(28.8)	(43.7)
*Funding has been transferred to Space Operations		
Mission Support	***2,532.2***	***2,584.0***
Safety, Mission Assurance, Engineering and Advanced Concepts	**43.0**	**47.5**
Space Communication services	**89.7**	*
Space Network	36.1	–
Telecommunications	53.6	–
Research and Program Management	**2,217.6**	**2,290.6**
Construction of Facilities	**181.9**	**245.9**
Inspector General	***20.0***	***22.0***

* Funding has been transferred to SAT/Space Operations.

NATIONAL AERONAUTICS AND SPACE ADMINISTRATION CONSTRUCTION OF FACILITIES BY APPROPRIATION FY 2001 BUDGET SUMMARY

Human Space Flight	**13.2**	**16.3**
Space Flight Operations (Space Shuttle)	**11.0**	**15.5**
Repair Payload Changeout Room, Pad B (KSC)	–	2.3
Repair and Upgrade Substations 20A/20B (MAF)	–	1.8
Refurbish Elevator Controls, Vehicle Assembly Building (KSC)	2.3	–
Restore Pad Surfaces and Slopes, Pad B (KSC)	1.8	–
Rehabilitate 480V Electrical Distribution System, ET Manufacturing Bldg (MAF)	1.8	–
Minor Revitalisation of Facilities at Various Locations, Not in excess of US$1,500,000 per project	2.4	8.6
Facility Planning and Design	2.7	2.8
Payload and ELV Support	**2.2**	**0.8**
Minor Revitalisation of Facilities at Various Locations, Not in excess of US$1,500,000 per project	2.2	0.8
Science, Aeronautics and Technology	**24.2**	**37.5**
Space Science	2.5	13.0
Construct Laboratory for In-Situ Microbiology (JPL)	–	13.0
Construct Optical Interferometry Development Laboratory (JPL)	2.5	–
Life and Microgravity Science and Applications	**9.0**	**8.5**
Construct Booster Applications Facility, Brookhaven National Laboratory, Phase 4	9.0	8.5
Earth Science	**1.0**	–
Restore Meteorological Development Laboratory (GSFC)	1.0	–
Aero-space Technology	**11.7**	**16.0**
Replace Fan Blades, National Full-scale Aerodynamic Complex, Phase 3 (ARC)	3.4	6.0
Replace Main Drive for 14 × 22 ft Subsonic Tunnel (LaRC)	7.3	–
Construct Rocket-Based Combined Cycle (RBCC) Facility (Phase 1)	1.0	10.0
NASA Employee Levels by Facility		
Johnson Space Centre	2,926	3,030
Kennedy Space Centre	1,806	1,825
Marshall Space Flight Centre	2,651	2,758
Stennis Space Centre	272	280
Ames Research Centre	1,457	1,486
Dryden Flight Research Centre	634	634
Langley Research Centre	2,382	2,387
Glenn Research Centre	1,983	1,972
Goddard Space Flight Centre	3,282	3,282
Headquarters	1,020	1,087
Total, full-time permanent workyears	**18,413**	**18,741**
Inspector General	**210**	**213**

advanced miniaturised instruments returning a continuous information flow. NASA unveiled the New Millennium programme in its FY96 budget request for the identification, development and flight validation of key technologies so that 21st Century science missions can take advantage of them without the risks inherent in their first use. Spending will be within US$50 million each year. The 486 kg spacecraft Deep Space 1 was launched in October 1998 to investigate an asteroid and comet. There are also Earth observing equivalents, demonstrating new technologies for Mission to Planet Earth.

NASA's Discovery series of rapid turnaround missions, each costing less than US$150 million, began with the NEAR asteroid mission departure in February 1996 and the Mars Pathfinder lander in December 1996. Pathfinder and Surveyor's 1998 and 2001 follow-up landers are all that remains of the MESUR mission that was to build a network of 12 small surface stations on Mars between 1999 and 2003, but NASA was committed to a Mars launch at every planetary opportunity (25 months) although plans have been drastically altered following the loss of the two spacecraft launched to Mars in 1998.

NASA advertised that the Hubble Space Telescope would provide a tenfold increase in observing power for optical astronomers after a four year launch delay, but embarrassing shortcomings lay unrevealed until it attained orbit. However, a series of servicing flights with the shuttle in 1993, 1997, 1999 and 2002 corrected instrument inadequacies, and performance was restored to original expectations and enhanced and upgraded. Additional servicing missions provide the opportunity to replace instruments and reboost the altitude before the final closeout mission in 2010.

The next 'Great Observatory', the Compton Gamma Ray Observatory, followed in April 1991 but the third, the 15 tonne Advanced X-ray Astrophysics Facility re-named the Chandra X-Ray Observatory, was delayed beyond its planned launch date and placed in orbit by Columbia in shuttle mission STS-93 in July 1999.

The military reversed its earlier course and shifted payloads from Shuttle to expendable launchers after the Challenger disaster of January 1986. The military decision provided the stimulus for vigorous commercial marketing of Atlas, Delta and Titan launchers. These vehicles, with their origins in the 1950s, complement the revised Shuttle programme, but will probably be dropped in favour of the EELV early in the next century.

Commercial enterprises have found survival more difficult, but OSC's Pegasus and Lockheed Martin's LMLV have emerged as strong contenders. Lockheed Martin has also been a prime mover in the recent international marketing success of Russia's Proton. Many critics including the US's own Defense Intelligence Agency have stated the US launch capability is out of date, inefficient and expensive. Numerous studies over recent years have recommended radical programmes, including the NLS National Launch System, the air force's Spacelifter and the more exotic single stage to orbit plans of the National AeroSpace Plane and Delta Clipper. DoD, in 1993, opted for gradual updating of existing vehicles, while NASA remained interested in SSTO, but the White House OSTP Office of Science & Technology Policy in September 1993 began the studies anew. The three broad options covered by OSTP were updating existing vehicles, developing a new family of launchers based on current technology, and developing SSTO. NASA envisaged a fully recoverable manned Shuttle 2 by 2010, capable of delivering 11 tonnes to Space Station. The approach is similar to the original Shuttle requirements, providing full re-usability, carrying modest payloads on brief missions and offering rapid turnaround as a Space Station support vehicle.

The Space Transportation Policy, signed by President Clinton 5 August 1994, set the pattern for US launch vehicle development well into the next century. It gave the DoD responsibility for upgrading existing medium and heavy ELVs and NASA the role of developing a reusable next generation after 2000. After several years of deliberation the DoD announced in October 1998 that it had awarded contracts to Boeing Expendable Launch Systems and Lockheed Martin Astronautics for a modern and flexible series of launches in the Evolved Expendable Launch Vehicle (EELV) programme. This will build an existing Delta Atlas and Titan to develop new variants supposedly capable of cutting launch costs by 25–50 per cent over the next 20 years.

NASA, in January 1995, invited industry to submit competing designs for two reusable launch vehicles. The goal is for a payload cost of US$500/kg instead of the current US$900-3,600/kg. Seven proposals were evaluated for the two experimental vehicles: the X-33 single-stage-to-orbit (SSTO) reusable launch vehicle (RLV) leading to development of a definitive vehicle which would eventually replace the Shuttle and on which NASA was committed to spending US$660 million; and the X-34, a much smaller launcher, partially reusable, on which NASA would spend about US$70 million. NASA selected an Orbital Sciences Corporation (OSC)/Rockwell team to develop the two-stage X-34 aiming for a suborbital flight in 1997 and orbital flights beginning a year later. The team formed American Space Lines as an 'Arianespace' look-alike for commercial development of the small launcher market but in early 1996 the team pulled out believing the costs could not produce a viable commercial return. NASA re-contracted with OSC for a single-stage suborbital demonstrator, still designated X-34. Capable of a maximum speed of Mach 8 the X-34 will demonstrate all-weather, pilotless landings as a technology precursor to fully reusable launch systems of the future. X-33 finalists McDonnell Douglas, Rockwell and Lockheed competed for the precursor RLV work, the former submitting its DC-X Delta Clipper vertical-take-off-and-landing (VTOL) vehicle with which it had been conducting flight trials as a private venture. In July 1996, Lockheed Martin announced it had been selected by NASA to build the X-33, a suborbital horizontal-take-off-and-landing precursor to a full-scale Shuttle replacement called Venture Star which was to rely on composites and integrated design for lightness and a linear aerospike engine for propulsion. After several delays the X-33 was expected to begin flight trials no earlier than 2002 and demonstrate RLV technologies to a maximum speed of Mach 12. However, early in 2001 the incoming administration of G W Bush cancelled both X-33 and X-34 and ordered NASA to re-focus on less ambitious technology for launch vehicle developments.

The three principal communications satellite manufacturers, Hughes (the world's largest manufacturer, with about a third of the market), Lockheed Martin and Space Systems/Loral, hold relatively healthy order books. The US accounted for more than 70 per cent (worth more than US$10 billion) of the satellites constructed in the 1990s, with an increasing proportion of non-domestic contracts. Commercial geostationary telecommunications systems have, of course, been long established, but low Earth orbit exploitation is imminent. The 66-satellite Iridium constellation had been expected to go online in late 1998 but a series of financial problems pushed Iridium into chapter 11 in 1999. It had been projected that mobile satellite communications alone would be worth US$20 billion annually worldwide by 2000, including US$11 billion for the satellites, ground segment and launches. A rescue plan was put in place and military contracts helped save the programme for a much reduced service from the optimistic commercial expectations first mooted at inception.

Military space programmes accounted for about US$13 billion of FY99's US$263 billion defence budget, sustaining steady growth from US$10.6 billion in 1995 but still only a fraction of the US$17 billion it had in 1988 (in then-year dollars), today equivalent to US$25 billion. The US military satellite communications network is extensive, employing DSCS, Fleetsatcom and Leasat craft plus the recent additions of UFO and Milstar. Much of the general traffic is being shifted to commercial transponders on long term leases as a means of reducing the estimated US$30 billion cost of replacing the current military systems.

The DoD's Navstar Global Positioning System (GPS) is providing a base for commercial receiver systems. They initially offered 16 m accuracy navigation information to civil users but the signals have been deliberately degraded since mid-1991. Nevertheless, it is a rapidly growing market and the constellation of 24 satellites attained full strength in 1994. The US GPS Industry Council projected a worldwide market for GPS receivers of US$8.47 billion by 2000, from US$867 million in 1994.

The US has long maintained the world's most comprehensive military surveillance and monitoring capability. The LEO and geosynchronous networks of imaging reconnaissance, electronic intelligence and early warning satellites were particularly affected by the launcher problems of 1986–87, but the first US imaging radar satellite was finally launched by the Shuttle in late 1988 and several reconnaissance satellites have departed since. However, it seems that such programmes have suffered deep cuts in recent years, reflected in the decreased requirements for USAF Titan

4 launches through the 1990s. Current expenditure is estimated at around US$5 billion annually, although outgoing CIA Director James Woolsey in January 1995 noted to a Senate committee that a radical restructuring would cut the number of future intelligence satellites by almost half. The four major companies building these systems (Lockheed Martin, Rockwell, TRW and Hughes) are competing for fewer contracts and having to diversify.

The Strategic Defense Initiative in 1993 experienced a fundamental restructuring. The Defense Secretary in May 1993 declared an end to the Star Wars era and renamed SDIO as the Ballistic Missile Defense Organisation. Highest priority is now Theater Missile Defense, in recognition of DoD's focus on regional conflicts and experience in Desert Storm.

The US maintains a double system of civil meteorological satellites providing coverage from both GEO and polar craft. Ideally, two of each type are in service at any one time but the GEO system operated a single vehicle since 1989 until a replacement from the next generation was launched in April 1994. These GOES satellites complement the similar European Meteosat, Japanese GMS and CIS Elektro craft for synoptic coverage. The civil polar satellites are matched by the DMSP for military applications; their merging is planned to orbit the first satellite in about 2006, halving the operational fleet to two and saving at least US$1 billion over 15 years.

Commercialisation of the Landsat remote sensing satellite system has been far from an unqualified success, partly because of uncertain funding for continued operations with the current Landsat 4/5 and for fabrication of successors Landsat 6/7. Landsat 6's loss in 1993 was a serious blow and Landsat 7, which became entangled in protracted wrangling with the military, was not launched until 1999. Instead, image distributors are increasingly looking towards Indian and Russian sources. The first private commercial system licensed by the Dept of Commerce is EarthWatch's system offering 3 m resolution.

The military's exclusive right to high resolution imagery was removed in March 1994 down to at least 1 m by the Clinton administration. The Dept of Commerce, as the licensing agency, retains the power to order shut down in times of crisis. The same decision also places tight restrictions on the export of remote sensing technology. DoC estimated in 1994 that the US$400 million annual market for remote sensing data would grow to US$2 billion by 2000. In reality it has exceeded that, although this market is now shared among several different projects in many countries.

President Clinton signed the executive order in February 1995 for the public to be given access by August 1996 at the National Archives to more than 800,000 images from some reconnaissance satellites flown 1960-72. Impetus has come from US companies looking for new sources of revenue in the post-Cold War world and from Russian sales of digitised film imagery down to 2 m from both civil and military missions.

NASA's Upper Atmosphere Research Satellite began operating in 1991 as the first Mission to Planet Earth satellite and the largest ever flown for atmospheric research. One goal is a better understanding of stratospheric ozone depletion. The US$21 billion EOS Earth Observing System programme running through 2019 is a major inter-agency effort to understand global environmental interactions, humanity's influence and develop techniques for predicting changes.

Congress has granted NASA authorisation for US$14.184 billion for FY01, US$149 million more than the President requested (see chart), with US$14.625 billion authorised for FY02, an increase of US$160 million on the President's request. Congress has agreed that manpower reductions, particularly at the Kennedy Space Center, have been too high in recent years and that a modest increase is essential to prevent increased safety risks.

FY2002 Budget Request

NASA's budget has traditionally been divided into three significant appropriations lines: human space flight; science, aeronautics and technology; and mission support. From FY2002 the budget condenses three separate lines of appropriation into two, merging costs traditionally within mission support into the other two. The following summarises the new structure for FY01-06 (in millions US$):

	01	02	03	04	05	06
Human Space Flight	**7,163.4**	**7,296.0**	**6,881.0**	**6,545.0**	**6,439.0**	**6,494.0**
ISS	2,112.9	2,087.4	1,817.5	1,509.1	1,349.3	1,389.0
Shuttle	3,118.8	3,283.8	3,218.9	3,253.3	3,213.5	3,228.0
Payload and ELV Support	90	91.3	92.5	100.0	104.7	111.6
Investments and Support	1,272.5	1,303.5	1,333.5	1,348.1	1,381.7	1,420.6
Safety and Engineering	47.4	47.8	47.8	48.0	48.0	48.0
Science, Aeronautics and Technology	**7,066.9**	**7,191.7**	**8,079.8**	**8,812.9**	**9,193.9**	**9,552.0**
Space Science	2,624.7	2,786.4	3,144.2	3,560.5	3,897.5	4,008.1
Biological and Physical Research	378.8	360.9	380.7	402.6	405.6	419.4
Earth Science	1,716.2	1,515.0	1,587.4	1,571.0	1,572.9	1,578.7
AeroTechnology	2,214.5	2,375.7	2,823.8	3,135.1	3,174.2	3,402.1
Academic Programmes	132.7	153.7	143.7	143.7	143.7	143.7
Inspector General	**22.9**	**23.7**	**24.6**	**25.5**	**26.5**	**27.4**
Total	**14,253.2**	**14,511.4**	**14,985.4**	**15,383.4**	**15,659.4**	**16,073.4**

UPDATED

GOVERNMENT AND NON-GOVERNMENT INSTITUTIONAL ORGANISATIONS

Military
Civil

MILITARY

Poland

Military Institute of Aviation Medicine

Current Status
The Institute has played a major part in determining physical requirements for space flight has specialised in high altitude/high-*g* research. Polish cosmonaut M. Hermaszewski made the second flight of a non-Russian aboard a Soviet Soyuz spacecraft when he was launched to the Salyut 6 space station in June 1978. The institute is researching into astronaut fitness during long flights, μg effects on muscles and inner organs, and the influence of gravity changes on mammalian reproductive systems. Poland is working with human factor data provided both by NASA and the Russian Space Agency and has carried out similar work for the European Space Agency with a view to participating in the International Space Station as biomedical observers. In recent years the institute has applied theoretical research to general health care and has received international funding for this activity. Increased interest in biomedical experiments in space has created a proposal for stronger representation to ESA for discussions concerning joint ventures.

UPDATED

United States

11th Space Warning Squadron

Current Status
An element of the 21st Space Wing 11th SWS at Falcon AFB, Colorado, operated the ALERT Attack and Launch Early Reporting to Theatre system but has now been disbanded.

VERIFIED

21st Space Wing

Current Status
The 21st Space Wing was activated 15 May 1992, formed from the assets of 1st Space Wing and 3rd Space Support Wing, to operate and maintain a system of space/land-based sensors to detect/track ballistic missile launches, detect space launches, and provide data on foreign ballistic missile nuclear detonations. Space-based early warning is provided by the DSP satellites, the catalyst of the US early warning system as it is the first to detect missile launches. 21st SW DSP squadrons send crucial launch and nuclear detonation reports to the NORAD and USSPACECOM centres at Cheyenne Mountain Air Station.

The 5th Space Warning Squadron at Woomera Air Station in Australia is unique among wing units because it is assigned with the Australian 1st Joint Communications Squadron at the Joint Defense Facility, Nurrangar for DSP. The 11th Space Warning Squadron was an element of the 21st SW but has now been disbanded.

The wing's ground-based sensors until April 1995 comprised BMEWS and SLBM. The two BMEWS radar units are the 12th SWS at Thule AB in Greenland and the 13th SWS at Clear AS in Alaska. SLBM warning system units are the 6th SWS at Cape Cod AS, Massachusetts, 7th SWS at Beale AFB, California. These sensors also provide essential tracking reports on space objects. The 73rd Space Group at Falcon AFB was absorbed 29 April 1995, along with its eight space tracking squadrons.

As host wing for the Peterson Complex, the 21st provides complete base support services for Peterson AFB and Cheyenne Mountain AS, plus some support functions for Falcon AFB. Personnel total 4,127 military, 846 AF civilian and 3,100 contractor employees in nine countries, through five groups in 33 squadrons at 20 locations around the world. Catalogues more than 9,500 manmade objects in space from low to geosynchronous orbit.

30th Space Wing

Current Status
The Western Space and Missile Center in 1991 was re-designated the 30th Space Wing. An element of the 14th Air Force. Elements of AF Space Command, the east and west launch ranges and bases were managed by AF Systems Commands' Space Systems Div until 1 October 1990. 30SW comprises four groups: 30th Operations Group, 30th Logistics Group, 30th Medical Group and 30th Support Group. 30OG operates the Western Range. The 30th Range Squadron operates and maintains the Western Range and the 2nd Space Launch Squadron is responsible for small/medium launch operations (including Atlas II, Delta 2, Pegasus and Taurus 2 Titan 2 and Titan 4). The 576th Flight Test Squadron handles ballistic operations. The 30th Space Wing also handles Titan IV and II, Atlas E, AMROC, and ICBM test programmes. 30th Space Wing also supports test launches for the Ballistic Missile Defense programme. On 6 December 1999 the 30th Space Wing was awarded the USAF Outstanding Unit Award.

45th Space Wing

Current Status
The 45th Space Wing (45 SW) designation in November 1991 replaced the Eastern Space and Missile Center organisation, established 1 October 1979. Its origins date back to 1950 with the Army/Air Force establishment of the Joint Long Range Proving Ground. The Air Force assumed full control in 1951, redesignating it the Air Force Missile Test Center. It was renamed the Air Force Eastern Test Range in 1964. A unit of Air Force Space Command, it is the host organisation for Cape Canaveral AS, Patrick AFB and the two island stations (Ascension and Antigua) of the Eastern Range (ER). The eastern and western launch ranges and bases were managed by AF Systems Commands' Space Systems Division until 1 October 1990. The Wing comprises four major Groups: Operations; Support; Logistics; and Medical.

The Operations Group provides operational control over all aspects of DoD space launches. The 1st Space Launch Squadron was activated 1 October 1990 to assume all responsibility for Delta operations; the 3rd Space Launch Squadron activated 3 August 1992 to take over Atlas work; the 5th SLS activated 14 April 1994 to oversee all Titan operations. Each squadron exercises field operations management/control of all DoD satellites launched from CCAS.

A single story 11,800 m^2 Range Operations Control Center became operational in March 1995 as the ER control facility under a US$134 million Harris Corporation contract. The goal is to allow flights every 12 hours. Operationally, it serves the ER by processing worldwide radar metric data, generating inter-range vectors to aid in acquisition, calculating state vectors for navigation uplink, displaying telemetry and orbital navigation information and functioning as LRCC Lead Range Control Center to co-ordinate radar tracking and retrieval.

Precision tracking is provided by modified AN/FPQ radars and the Multiple Object Tracking Radar. 45th Space Wing adopted the Department of Defense Manned Space Flight Support Office in October 1999.

50th Space Wing

Current Status
The host unit at Schriever AFB, the 50th Space Wing is a component of Air Force Space Command at 18 km-distant Peterson AFB. The wing was originally established 8 July 1985 as the 2nd SW and then redesignated 30 January 1992. The facility at Falcon was built under the air force programme known as the CSOC Consolidated Space Operations Center. Its mission is to command and control operational DoD satellites and to manage the worldwide AFSCN AF Satellite Control Network.

Previously, the Air Force Space and Missile Systems Center's Consolidated Space Test Center (CSTC) in California was the prime control facility. Its satellites include DSP, DMSP, GPS, DSCS, NATO 3, NATO 4/Skynet 4, Fltsatcom, IUS, UHF Follow-On and Milstar.

The wing has 3,300 military and civil personnel assigned to its 45 units at 16 locations worldwide. It comprises four groups: 50th Operations Group, 50th Maintenance Group and 50th Mission Support Group at Falcon AFB. The 50th Operations Group commands/controls the satellites, trains space operations crews and provides operational support and evaluation for managing the satellite operations centres and ground stations. The group's nine Space Operations Squadrons are:

1st SOPS support during launch, early orbit, mission and anomaly phases of GPS and DSP. This includes routinely operating older, more troublesome, satellites and eventually de-orbiting them. Once satellites are checked out, day to day control is turned over to the other SOPS.

2nd SOPS day to day GPS command/control. The Navstar GPS Master Control Station took command of the first GPS operational satellite following launch 14 February 1989. Det 1 at Cape Canaveral AFS operates a ground antenna for prelaunch compatibility testing and on-orbit control.

3rd SOPS day to day command/control of 14 communications satellites, including DSCS 3, Fltsatcom and UFO (Fltsatcom transferred to NAVSOC at Point Mugu on 19 June 1996, to be followed by UFO in 1999). The operators also provide LEOP support, trend analysis and anomaly resolution, which they also do for Milstar (before handing over to 4th SOPS). A total of 20 DSCS, FLTSAT and NATO 3 satellites was transferred 11 July 1991 from CSTC to what was then CSOC.

4th SOPS responsible for day to day operations of Milstar, including satellite command/control, communications management and ground segment maintenance.

5th SOPS at Onizuka AS in California to plan/conduct launch and on-orbit operations for several DoD, allied and commercial missions, including IUS for NASA/DoD, NATO 4/Skynet 4 and DSCS 3. It also supports Shuttle and Titan 4 missions. Responsible for day to day operations of NATO 3 and DSCS 3. Staffed by Lockheed Martin personnel.

6th SOPS located at Offutt AFB in Nebraska for day to day command/control of DMSP (will relocate to Falcon). CSOC assumed DMSP control 1 April 1987. Detachment 1 at Fairchild AFB in Washington is responsible for emergency command/control antenna operations; it will close September 1997 and Vandenberg will provide TT&C station.

21st SOPS operates and maintains operating and logistical support to the USAF Satellite Control Network.

22nd SOPS develops, publishes and executes the network tasking order and co-ordinates launch and on-orbit operations of over 100 DoD satellites.

23rd SOPS provides command and control capability.

The 50th Space Wing has almost 4,000 personnel.

750th Space Group

Current Status
The 750th Space Group at Onizuka AS is responsible for the daily operation of AFSCN, which AFSPC took over in late 1987. The network comprises nine tracking stations: 23rd SOPS, New

Boston AFS (New Hampshire), Det 1 Vandenberg AFB (California), Det 2 Diego Garcia, Chagos Archipelago, Det 3 Thule AFB (Greenland), Det 4 Kaena Point (Hawaii), Det 5 Andersen AFB (Guam), TCS Oakhanger (UK). Two resource control complexes, the 21st SOPS at Onizuka AS and 22nd SOPS at Falcon AFB, schedule satellite contacts for the tracking stations.

Some 400 contacts of 518 minutes each are scheduled daily for about 100 satellites. Camp Parks at Pleasanton in California is responsible for satellite on-orbit testing. The Transportable Vehicle Checkout Facility at Cape Canaveral AS is responsible for pre-launch compatibility testing. AFSCN was recently upgraded, automating many operations. Automation has emphasised the interaction of inter-service data systems. Satellites were controlled from the 1960s under the CDS Current Data System from CSTC, using a 26 m roll of graph paper specifying the contact schedule of each ground station over seven days. The computerised CCS Command and Control System took over at 50th SW and CSTC in spring 1992.

US Army, Space and Missile Defense Command (USASMDC)

Current Status

USASMDC develops space products for army users and supports theatre missile defence projects for BMDO. FY95 budget was US$363.2 million. The US Army Space Command is its operational arm. ARSPACE interfaces with industry and other DoD agencies to identify and demonstrate space technologies useful to Army warfighters through the ASEDP Army Space Exploitation and Demonstration Program. USASDC has been instrumental in developing concepts for the proposed National Missile Defence screen against ballistic missile attack. In November 2002 a laser successfully destroyed an artillery projectile in flight. Part of the Tactical High Energy Laser programme, this followed 25 successful intercepts of Katyusha rockets.

The US Army has been involved with launching satellites since the late 1950s. Here a Juno 2 derivative of the Jupiter missile is prepared for launch (DoD)
0572492

Army Space Command (forward) is the operational arm of the US Army Space and Missile Defense Command. Command of the DSCS System Operations Centers and management of the joint tactical use of DSCS satellites is a primary ARSPACE mission. The Joint Staff assigned ARSPACE the operation of five worldwide DSCSOCs, three worldwide AN/MSQ-114s and three worldwide Regional Space Support Centres.

ARSPACE operation of the JTAGS Joint Tactical Ground Stations in Europe and Korea provides theatre Commanders with a transportable system for processing/disseminating near-real-time warning of theatre ballistic missile launches. Using DSP data, JTAGS can process in stereo, improving the sensor's capability. JTAGS ties directly to theatre communications systems to transmit launch point information, impact area prediction, time of flight and positional information. ARSPACE also provides oversight of space surveillance operations at the US Army Kwajalein Atoll. ARSST Army Space Support Teams have deployed with units such as the 10th Mountain Division worldwide, providing weather, terrain analysis, mapping, imagery, communications and early warning information.

US Strategic Command

Current Status

In October 2002 Secretary of Defense Rumsfeld merged US Space Command with US Strategic Command. A congressionally mandated 15 per cent staff reduction for the FY2000 was reduced on appeal to 7.5 per cent. Reaffirmation of the primary role of Galactic missile naming was reaffirmed through the incoming Bush administration. The Space-Based Infared System (SBIRS) will be crucial to that task. SBIRS is scheduled for launch in 2006.

Background

The unified USSPACECOM, comprising the Army, Naval and AF Space Commands, was authorised by President Reagan in November 1984 and activated in September 1985. Its singular responsibility is to ensure the US has continuous access to and the use of space. Army, Naval and AF space operations had previously functioned separately since the DoD's space programme began in the 1950s. USSPACECOM conducts joint space operations in accordance with Unified Command Plan assigned missions in space forces support, space force enhancement, space force application and space force control. Staff totals about 647 military and 139 civilian.

AFSPC, formed September 1982, is the largest component, with 20,000 military and 4,800 civilian personnel. The FY97 budget was US$1.7 billion; FY96 budget US$1.6 billion; FY95 US$2.84 billion, FY94 US$2.8 billion, FY93 US$1.97 billion, FY92 US$1.5 billion. The Army Space Agency was activated August 1986 and redesignated Army Space Command (ARSPACE) April 1988. It employs 380 military and 89 civilian personnel. FY95 budget was US$47.8 million. Naval Space Command, formed in 1983, is headquartered at Dahlgren, Virginia and is staffed with 408 military and 280 civilian personnel.

AFSPC provides forces for: strategic and tactical ballistic missile warning, space control, satellite operations, DoD satellite launchings and, since 1 July 1993, ICBM missile operations. The 14th AF, activated 1 July 1993, is the day-to-day manager of AFSPC's space forces. The 20th AF has daily operational control over ICBMs. AFSPC's 50th Space Wing (which in 1992 took over the CSOC Consolidated Space Operations Center) at Falcon AFB is the focal point for daily command/control of more than 90 operational satellites.

The 28 ground stations of the AF Satellite Control Network (AFSCN) make an average of 11,485 monthly satellite contacts. Responsibility for the space launch bases at Cape Canaveral (45th SW) and Vandenberg (30th Space Wing) was transferred from Air Force Systems Command to Space Command in October 1990. Its Space Launch Squadrons are responsible for Delta, Titan, Atlas and IUS launch operations. Air Force Space Command assumed AFSATCOM responsibility 1 June 1992 from US Strategic Command at Offutt AFB, Nebraska.

Through its component commands, US Space Command operates a worldwide network of missile warning sensors providing tactical warning and attack assessment of ICBM and SLBM launches posing potential threats against North America, and provides tactical warning of ballistic missile attacks to US Commanders worldwide. These are the responsibility of 21st Space Wing (formed May 1992 from 1st Space Wing and 3rd Space Support Wing), which is also the host wing for the Peterson complex. Data are relayed via satellite almost instantaneously to four primary locations: the command's Missile Warning Center (Cheyenne Mountain AS), Strategic Command HQ (Offutt AFB, Nebraska), National Military Command Center (Washington DC) and Alternate NMCC (Maryland). The data are also employed by the NORAD/USSPACECOM Command Center in Cheyenne Mountain Operations Center. NORAD is a bi-national US-Canadian command charged with safeguarding the sovereign airspace of the two countries. The missile warning sensors are also used for space surveillance. Together with dedicated space surveillance sensors (operated by the 21st Space Wing at Peterson AFB, NAVSPACECOM at Dahlgren and contributing ARSPACE and civil assets), they form the Space Surveillance Network, which provides timely and accurate detection, tracking and identification of space objects and events. The network makes 50,000 to 80,000 daily observations for the command's Space Control Center in Cheyenne Mountain and maintains a catalogue of more than 8,500 orbiting objects.

US Space Command is integral to studies of possible threats from theatre ballistic missiles and intercontinental ballistic missiles. Key to monitoring efforts are the Tactical Information Broadcast Service and the Tactical Related Applications Data Dissemination System.

CIVIL

Argentina

Asociacion Argentina de Tecnologia Espacial (AATE)

Current Status
AATE campaigned to develop a set of experiments that could be lifted into space in NASA's Getaway Special programme and brought together a wide range of potential experiments from several research institutes in Argentina. Eventually, it obtained a reservation for a Getaway Special aboard a US shuttle flight which took place in December 2001 as Getaway Special G-761 aboard STS-108. Known as Paquete Argentino de Experimentos (PADE) experiments are being used to study fluid movement, the reaction of seeds exposed to space and microgravity.

UPDATED

Centro de Investigaciones Aplicades

Current Status
This was originally the Instituto de Investigaciones Tecnológicas de la Fuerza Aérea and was part of the air force. In 1991, it was renamed Centro de Investigaciones Aplicades (Centre for Applied Research) and is part of the Instituto Universitario Aeronautico (University Institute of Aeronautics). The centre designed a range of very small satellites for multipurpose use and application with the capability for near-earth or deep space applications. However, none of these have been produced due to limited financial resources. Its responsibilities are to co-ordinate the research, projects and developments in the fields of space and aeronautical work. It was responsible for Victor 1 (MuSAT before launch) which was launched by the Russians aboard a Molniya-M in August 1996. The Centre is now researching the ways to apply its theoretical and practical work to economic development of technical and engineering designs. Funding has been reduced in fiscal year 2001–2002.

VERIFIED

Comision Nacional de Actividades Espaciales (CONAE)

Current Status
The CNIE Comision Nacional de Investigaciones Espaciales (National Commission for Space Research) was transferred from the air force in May 1991 to become the civil CONAE, under the direct supervision of Argentina's President. CNIE directed Argentina's space science activities from the 1960s. It began with development of a two-stage Castor sounding rocket able to carry small payloads up to 500 km, and in 1981 embarked on the SAC science satellite project. The San Miguel Centre houses the SAC-B Mission Operations Control Centre and TT&C station. Intentionally, SAC-B remained attached to the Pegasus third stage following launch on 4 November 1996. CONAE has conducted expensive studies of promising space technology initiatives with commercial potential. Negotiations have been completed with representatives of the Ukraine where discussions have been undertaken on the use of reciprocal launch sites. Lack of funding and a shift in government priorities have held up negotiations as Argentina wrestles with economic problems.

UPDATED

Comision Nacional de Telecomunicaciones

Current Status
CNT is the national signatory to Intelsat and Inmarsat. Intelsat services are routed through the Balcarce ground stations, also expected to provide Inmarsat services. CNT provides oversight of telecom satellite operations. CNT has recently performed an analysis of multisatellite tracking capability. CNT has also conducted studies on national satellite systems and has performed analysis on regional systems. CNT has provided in-depth analysis on an autonomous satellite for Argentina with broadband capability for regional commercial purposes, but the proposed project has not received government funding. During 2003 CNT conducted inter-regional discussions with interested parties in neighbouring countries with a view to organising a South American satellite system but with little positive response. Some interest in setting up a leased system was expressed by European satellite manufacturers but these discussions failed to produce a workable solution.

UPDATED

Instituto de Astronomia y Fisica del Espacio

Current Status
The IAFE Institute of Astronomy and Space Physics is an institute of Argentina's National Research and Technology Council (CONICET) of SECYT. It is the Principal Investigator institute of SAC-B's HXRS instrument. SAC-B remained attached to the third stage of the Pegasus following launch on 4 November 1996 together with the HETE payload. The IAFE raised plans for a science satellite for aeronomy and space physics studies and has conducted discussions with other countries in South America regarding international participation. This has attracted interest from some groups in the Middle East although no agreements have been reached. Further possibilities have been suggested whereby the National Research and Technology Council would serve as a focus for a new South American effort to focus research on future astronomy research.

UPDATED

Australia

Australian Centre for Remote Sensing

Current Status
Established in 1979, ACRES is a unit within the Australian Surveying and Land Information Group (AUSLIG). It is the data acquisition, archiving, processing and distribution facility for ERS, Landsat, Spot and NOAA satellite imagery together with radar data from Russia and Canada. ERS/JERSS data have also been acquired, archived and processed since 1992. ACRES has facilities at Canberra, the home of the Data Processing Facility, and Alice Springs, where the Data Acquisition Facility is located. ACRES ground station coverage includes all of Australia, Papua New Guinea and parts of Indonesia. ACRES also operates the Tasmanian Earth Resources Satellite (TERSS), which extends coverage across the Southern Ocean to Antarctica (including New Zealand). BAE Systems is contracted to provide operations and maintenance services. Increasingly complex requirements are being met by expanded data-acquisition capabilities and by new software capabilities for Australian and non-Australian data requirements.

Processed data are used to generate remote sensing products on either magnetic or photographic media. These products are used for agriculture monitoring, mapping, land use assessment, environmental impact monitoring, valuable resource exploration, and regional and urban planning. A digital quick-look on-line browse facility was introduced in 1996. Studies have been carried out to investigate the possibility of the ACRES conducting global survey analyses.

UPDATED

Australian GNSS Co-ordination Committee

Current Status
The Federal Government has recognised a need for effective co-ordination of GNSS implementation in Australia. The Hon John Anderson MP, Deputy Prime Minister and Minister for Transport and Regional Services, established the Australian GNSS Co-ordination Committee (AGCC) in May 2000. The AGCC reports to the Minister for Transport and Regional Services, and to the Australian Transport Council (Federal and State Transport Ministers).

The role of the AGCC is to consider and develop mechanisms to co-ordinate all land, sea and air aspects of GNSS; promote the safe and effective utilisation and development of GNSS in Australia; and co-ordinate national security issues, the application of augmentation systems, and the national use of GNSS in other relevant applications.

VERIFIED

Centre for Remote Sensing and GIS

Current Status
Founded in 1981, at the University of New South Wales, the centre is Australia's pre-eminent remote sensing and GIS educational institution. Facilities include a dedicated image analysis laboratory for undergraduate and masters students, and an advanced research laboratory.

The Centre is also engaged in global climate change research. The Centre has conducted studies on the integration between remote sensing and ground truth data for integrated climate change models.

CRS provides regional and offshore environmental data and pollution monitoring analyses and has links to several regional universities and research institutes.

VERIFIED

Commonwealth GPS Group

Current Status
The Commonwealth GPS Group has now been disbanded, its duties lengthened to the Australian GNSS Co-ordinates Committee (see entry).

The GPS Group was established in September 1992 by the Commonwealth Spatial Data Committee as a forum for Commonwealth government civil and defence agencies. Issues cover co-ordination and standardisation of base stations, development of a consistent Commonwealth approach to commercial opportunities and co-ordination of integrity monitoring. The GPS Group has been studying the impact of a separate European GPS system and of problems posed by integration of various different systems operated by a variety of international groups.

VERIFIED

CSIRO Earth Observation Centre

Current Status

The centre, collocated with COSSA, was established by CSIRO in 1995 to centralise its collection of remote sensing data; data product management, including core-level processing. It performs strategic research for advancing the scientific base of remote sensing and is conducting research into global climate change. CSIRO is a major source of theoretical and applications analysis worldwide. Developed interactive optical/non-optical interactive data formats using both imaging and SAR data that are integrated for supporting analysis. The centre has expanded its data handling and archiving capabilities.

VERIFIED

CSIRO Office of Space Science and Applications (OSSA)

Current Status

CSIRO is Australia's primary body for industrial and scientific R&D and is responsible through its OSSA for space science activities. Its annual budget is about A$700 million, of which space-related expenditure is about A$27 million (funding 240 full-time-equivalent staff out of 7,100). The 35 divisions include radio physics, atmosphere, oceans, applied physics and materials science. The collocated Earth Observation Centre was established in 1995 to centralise its remote sensing activities. These have been expanded to handle global data and resgional interpretation for government use. Private companies pay for data analysis and for information sub-sets.

CSIRO hosts the Australia Telescope National Facility, which operates a synthesised array of six 22 m antennas at the Paul Wild Solar Observatory (Narrabri, New South Wales), a 22 m antenna at Coonabarabran (New South Wales) and the Australian National Radio Astronomy Observatory at Parkes, New South Wales. Parkes' 64 m dish is used for supporting deep space missions such as NASA's Voyager. ATNF budget is A$13.4 million; 145 staff. It will play a critical role in southern hemisphere observations for Russia's Radioastron and Japan's VSOP radio astronomy satellites.

CSIRO had its role within Australian space re-defined as a result of the major re-organisation ordered in August 1996. With the demise of the Australian Space Office and Australian Space Council, CSIRO has taken over the responsibility of undertaking space projects on behalf of the Australian government. It is leading the project for the planned FEDSAT programme, intended to launch a micro-satellite into orbit. At the same time it was also announced that the activities of the NASA tracking station at Tidbinbilla would be transferred to CSIRO as a result of CSIRO's experience in radio astronomy and space physics.

OSSA identifies and co-ordinates space-related R&D in more than 20 of CSIRO's divisions. It provides international liaison for CSIRO space interests, manages multidivision projects, provides technical assistance to users of space-related technologies, negotiates commercial agreements and joint ventures, and operates an airborne data collection and processing facility.

CSIRO has also participated in the technology for the planned military utilisation of the Orion series.

VERIFIED

Land Information Centre (LIC)

Current Status

A division of the Department of Conservation and Land Management, LIC's Remote Sensing Unit provides image processing and analysis services for air and space-based systems as well as commercial US and European products, particularly to New South Wales state government agencies. Optical digital images and SAR data sensing Far East land conservation organisations. The centre plays a major role in the integration of data from ground based sensors and airborne platforms as well as untended DCPs. Its activities have been enlarged as a result of expanded national and regional capabilities. A greater focus on international cooperation has widened the scope of the LIC and several countries in Europe have exchanged results in pursuit of global studies.

UPDATED

Optus Networks Pty Ltd

Current Status

Optus was formed in 1992 on changes to the Australian telecommunications industry with the framework for *carriers* and *service providers*. With the privatisation of the national carrier, Aussat was transformed into Optus to become a second carrier to Telstra. Aussat, Australia's national satellite operator, was established by the government in late 1981 to own/operate the Aussat system. The company was legislated under the 1984 Aussat Act and was 75 per cent owned by the Commonwealth of Australia, with the remaining portion allocated to the Australian Telecommunications Commission (Telecom Australia). Aussat A1 satellite was launched in August 1985 followed by A2 in November 1985 and A3 in September 1987. The government announced plans in November 1990 to de-regulate the telecommunications industry, selling Aussat to establish a second carrier to compete with the merged OTC/Telecom Australia. It immediately removed Telecom Australia's 25 per cent and lifted all legislative restrictions imposed upon Aussat's operation. The successful bidder was announced 19 November 1991: Optus Communications Pty Ltd., a 51 per cent Australian-owned company with BellSouth of Atlanta and Cable and Wireless of the UK holding the remainder, acquired Aussat for A$800 million. The government paid out Aussat's debt and terminated the leverage lease financing arrangements for the satellites at a total cost of about A$740 million. The new ownership arrangements came into operation 31 January 1992.

The Australian element, Optus Pty Ltd, has shareholders Mayne Nickless, AMP, National Mutual and other institutional investors whose interests are managed by the AIDC Telecommunications Fund Management Pty Ltd. Optus B1 was launched in August 1992 followed by B2 in December 1992 and B3 in August 1994. B1 and B3 are expected to operate until 2006 and 2009, respectively, after which they will be operated in inclined orbits to conserve propellant. A3 is already in an inclined orbit.

The first Optus C satellite, C 1, was launched in 2003 and is operated by SingTel and the Australian Ministry of defence. Built by Space Systems/Loral it has a design life of 15 years.

UPDATED

Remote Sensing Applications Centre (RSAC)

Current Status

As part of the Department of Land Administration (DOLA), RSAC is the major government distributor of Landsat, Spot and airborne scanner digital and photographic data in Western Australia. Services include full image processing, image writing, geo-rectification to AMG (Australian Map Grid) and bureau facilities from Landsat, Spot and other remote sensing systems including optical, non-optical and radar data. RSAC has coordinated data from several national sources to improve geodetic survey analysis. With the demise of several national remote-sensing satellite projects the RSAC has placed emphasis on gathering data from existing in-orbit assets and integrating data into information obtained from land, sea and air instruments. RSAC has conducted studies of regional changes in geology and geomorphology and has acted as a consultant to major international organisations concerning exploitation of Australia's land resources.

UPDATED

Satellite Communications Research Centre (SCRC)

Current Status

The Australian Space Office established SCRC (as the Australian Space Centre for Signal Processing) to undertake strategic R&D of space-related subsystems, including: mobile terminal modems for the Mobilesat L-band service, supported by Optus. It also performs high-speed satellite communication at up to 155.52 Mbit/s, supported by Intelsat; and remote-control and telemetry systems. The 155.52 Mbit/s encoder/decoder was developed for use via Intelsat to deliver such services as up to 10 HDTV channels (or 50 regular channels) bundled for simultaneous transmission over a single transponder. It will also be used for Asynchronous Transfer Mode transmissions for high-speed datalinks. Satellite demonstrations were performed in 1995, 1996 and 1997. Additional research and technology experiments conducted to extend mobile telephone applications. The Centre has conducted tests with military communication bands for the Optus C system. It is also developing new specifications for future systems with hybrid civil and military functions.

UPDATED

Space Policy Unit (SPU)

Current Status

The Space Policy Unit (SPU) was established as part of the re-organisation of the Australian space effort in 1996 and it operates within the Department of Industry, Science and Tourism to maintain an overview of national and international space activities and to set policy in the national interest. The SPU provides advice on space policy and facilitates commercial space activities and interaction with other governments beyond the role and responsibilities of individual organisations currently involved in space research. Integrated studies on national versus international projects and space programmes involving general Australian and South East Asian interests. The SPU has been heavily involved in determining the advantages in utilising Optus satellites for military communications and has drawn up protocols whereby Optus channels can be leased to other users. It has also been asked by the government to make recommendations on levels of involvement with space science and applications programmes and with international projects and programmes such as the ISS.

UPDATED

Telstra Corp Ltd

Current Status

Telstra is Australia's government-owned domestic and international telecommunications company, formed in 1992 by the merger of OTC and Telecom Australia. It is Australia's signatory to Intelsat (2.82 per cent in 1996) and Inmarsat (1.55 per cent in 1996). Telstra's net profit for 1999 was A$3,486 million versus A$3,004 million for 1998. Total revenue for 1999 was A$18,218 million, an increase of A$916 million over 1998. With 1994–95 revenues of A$14.08 billion (1993–94: A$13.4 billion; 1992–93: A$12.6 billion) on assets of A$24.08 billion, Telstra is ranked in the world's top 20 telecommunications companies. Its Sydney International Telecommunications Centre is a major satellite Earth station, with a number of antennas: SYD1A 18 m limited motion, 183° E, primarily for cable restoration, SYD2A 32 m full motion, 174° E, services include IDR, IBS, SCPC, DAMA and itinerate video, SYD3A 18 m limited motion, 180° E, primarily for full time video leases and itinerate video, SYD4A 18 m limited motion, 177° E, IDR, IBS, itinerate video; PAS 2 9-1 m limited motion, PAS 2, video leases and pay TV video for domestic distribution, AsiaSat 2 9.1 m limited motion for pay TV domestic distribution. Telstra's Perth ITC (Western Australia) is a major satellite communication and tracking centre. It encompasses 32 m Intelsat relocated from Ceduna SA; 7.5 m submarine cable restoration; two 18 m Intelsat and a 27.5 m relocated from Moree NSW (OTC's station opened there in 1969); two 18 m Inmarsat; 20 m TT&C for Intelsat satellites over the Indian and Pacific oceans; four 8 m Intelsat ranging; 7.5 m linking to the Intelsat control centre in Washington DC. A 15 m S/X-band full track antenna, relocated from Carnarvon (Western Australia) in 1987, provides LEOP services for ESA. Telstra is a member of the Satellite Aircom consortium providing aeronautical communications services through the Inmarsat system. Telstra has begun re-engineering studies for broadband services from Intelsat and Inmarsat. Advancing beyond that, Telestra is researching dedicated satellites or constellations that can facilitate the next generation of telecommunications systems.

VERIFIED

Austria

Austrian Space Agency

Current Status

The purpose of ASA is the following:

- Organisation and management of programme contributions on ESA, bilateral and international level
- Promotion and support of Austrian science and industry
- Supporting and advising the Federal Ministry of Transport, Innovation and Technology in carrying out the Austrian space policy
- Co-ordination of national space activities and liaison management with foreign space agencies
- Organisation and management of space-related events and education and training activities
- Public relations.

The Austrian Space Agency, ASA, established by the Federal authorities in Vienna in 1972, serves as a focal point for the co-ordination of space activities in Austria and is the Austrian link to international space activities. ASA assists the Federal Ministry for Transport, Innovation and Technology in co-ordinating space activities and is responsible for establishing the programmatic framework and strategic guidance to promote industry and scientific institutions to play a significant role in international space-related undertakings. It has placed particular emphasis on the established teams in Austria dedicated to remote sensing to integrate international data to the benefit of the country. It has also examined the possibility of expanding broadband services through satellite systems.

To fulfill these objectives ASA is supported by the Government of Austria, represented by the Federal Ministry of Transport, Innovation and Technology and by the following Austrian government bodies, leading space companies and institutes:

- The City of Graz
- The Federal Economic Chamber
- The Austrian Research Centres Seibersdorf
- Joanneum Research
- Plansee AG
- Austrian Aerospace
- Siemens Austria.

The Austrian government has conducted a study of national and pan-European needs and measured the strengths and capabilities of Austria's space related resources with a view to adjusting financial participation depending on the perceived returns to the economy.

VERIFIED

Austrospace

Current Status

Austrospace, the Association of Austrian Space Industries, was created in 1991 to promote the views of its space industry members and it now serves as a strong link between Austrian companies and foreign governments. Austrospace has developed a wide range of proposals for atmospheric physics experiments using sounding rockets and balloon experiments. A revised examination of Austrian space policy has been conducted representing industry within the national interest. Space project direction has been formulated under the national policy of the government.

VERIFIED

Joanneum Research

Current Status

Joanneum is one of the Austrian Space Agency's supporting centres and was responsible for managing 1991's AustroMir mission. Its Institute of Applied Systems Technology undertakes satellite-based communications R&D. The organisation began with a high percentage of space professionals with expertise in telecommunications and broadcast systems from both the terrestrial and space segment envionments. Other specialisations include: remote sensing software and analysis; computer vision systems for autonomous navigation; low temperature materials testing. Joanneum Research is Austria's largest provisionally owned research institution and has more than 300 skilled employees. Research has been conducted on the use of Austria's space resources in encouraging ISS participation at national level.

VERIFIED

Space Research Institute (SRI)

Current Status

SRI is Austria's principal space science research facility, with emphasis on near-Earth auroral, ionospheric and magnetospheric investigations (including rocket, balloon and radar probing) and deep space fields/particles monitoring with magnetometers and particle detectors aboard ESA Space lab and Huygens and Soviet/RFAS Venera, Vega, Phobos, Interball and Mars-96/98 craft. It has strong connections throughout Europe and is a major contributor to international projects both inside and outside Europe. The Institute provided two experiments for ESA's Cluster mission and for the replacement set of Cluster spacecraft. Graz also provides one of the world's reference geodesy stations. Its laser unit provides ±8 mm satellite ranging accuracy. This has been proposed as an adjunct to a rendezvous and docking range rate and docking system. It has been considered for applications with a new generation of autonomous servicing systems.

UPDATED

Belgium

Belgacom

Current Status

Belgacom manages national participation in Intelsat, Inmarsat and Eutelsat. Eutelsat's space segment is controlled from ESA's station at Redu in Belgium. Belgacom operates the V-STAR 1.2 to 2.4 m VSAT service, operating hubs in Liedekerke via Eutelsat for Europe and in Lessive via Intelsat for Africa, Middle East and North/South America. Also operates the Belgacom news and information services via the internet and has developed new techniques for data transfer satellite-to-satellite. Belgacom has also developed more efficient mechanisms for controlling integrated communication relays by converging Intelsat and Inmarsat through Eutelsat networks. Continued studies on advanced telecommunications relay for enhanced broadcasting services. Belgacom is developing new technologies for expanded capabilities with international telecommunications services.

UPDATED

Belgian Institute of Space Aeronomy (BISA)

Current Status

Created in 1964, BISA's space physics and aeronomy research includes solar wind interaction with Earth, terrestrial and planetary magnetospheres, and the physics/chemistry of ionosphere and upper atmospheres. Its infra-red grille spectrometer flew on the Spacelab 1 (1983) and Atlas 1 (1992) missions to observe trace constituents in the high atmosphere. In its improved MIRAS version, it was launched to Mir in 1995 aboard the Spektr module and attached externally by EVA in July 1995. BISA has since worked with Russia on space physics experiments for the Russian ISS elements. Portable experiments have been designed for transfer through Progress cargo tankers. Adaptations of these design concepts have been applied to potential applications in other programmes.

BISA contributed to SPICAM (atmospheric spectrometer) and MAREMF (magnetometer) on Russia's Mars-96. Work is under way on instrumentation for the International Space Station. BISA is part of the Prime Ministers Services through the Federal office for Scientific, Technical and Cultural Affairs.

VERIFIED

Centre Spatial de Liège (CSL)

Current Status

CSL (IAL Space until April 1992) is an autonomous unit within the University of Liège and, along with Intespace (France) and IABG (Germany), is one of the Co-ordinated European Test Facilities group under ESTEC. The test facilities are based around a 480 m^2 class 10,000 cleanroom and include three thermal vacuum optical test facilities. The 150 m^3 5 m diameter horizontal 33 tonne Focal 5 chamber provides a 10^{-9} atmosphere vacuum and 10 temperature settings over 268 to +70°C and optical bench vibration levels of 10^5 *g* on a 350 tonne seismic block. There is also an associated 30 m^2 class 100 laminar flow clean zone. In the same area is the Focal 2 chamber, with identical performance but 21 m^3/2 m diameter. The 1.2 m^3 Focal 1.5 provides a similar environment but with a 1.5 m vertical diameter and 80 m^2 class 10,000 cleanroom and 4 m^2 class 100 laminar flow clean zone. CSL undertakes testing/calibration of camera tubes, image intensifiers, CCDs, photomultipliers and photon counting systems, in addition to the design/manufacture of meteorological instruments. Has developed new systems for multifunctional satellite concepts involving platforms capable of conducting remote sensing surveys and data store and dump. CSI is also involved in graduate and post-graduate studies and supports the development of new instruments for weather satellites. Test facilities leased to other government agencies and to international groups involved in Belgian space research activities. CSL has developed several new ways of converging data sets within a common set of data-gathering instruments.

UPDATED

Office for Science, Technology and Cultural Affairs (OSTC)

Current Status

OSTC (formerly the Science Policy Office), under the authority of the Minister of Science Policy, manages Belgium's space programme, representing the national interest within ESA and non-ESA projects such as Spot and SAR data acquired commercially. OSTC has provided links between Belgian companies and other European space-related organisations as well as with non-European institutions regarding international co-operation in manufacturing and research.

Has provided administration links to non-ESA companies and organisations for joint projects and research activities, connecting technology centres to universities for interdisciplinary programmes. Provides a forum for linking national capabilities with international projects and for connecting research institutes with international organisations.

UPDATED

Royal Meteorological Institute of Belgium (RMIB)

Current Status

RMIB represents Belgium in EUMETSAT and houses the national Space Remote Operations Center, from where scientists can control their experiments aboard Space Shuttle. RMIB's Solcon instrument has flown four times on Shuttle (Spacelab 1, Atlas 1–3) to measure the solar constant, as did its SOVA on ESA's Eureca 1992–93. The Virgo absolute radiometer is now flying on ESA's Soho. Negotiations are under way for Solcon to be included in NASA's Hitchhiker programme, and SOVA (as SOVIM) is under consideration for ESA's Solar payload on Space Station. RMIB is responsible for the optics element of the GERBE Global Earth Radiation Budget Experiment for the Meteosat Second Generation. RMIB also help plan suites of experiments for international weather watch programmes by satellite. RMIB is interested in meteorological observations from Shuttle and other NASA research satellites and has conducted integration tasks.

UPDATED

Brazil

Agencia Espacial Brasileira (AEB)

Current Status

The formation of a co-ordinating body for Brazil's space interests followed several ministries creating their own, duplicating, space departments, each of which lacked singly the resources to focus research or co-ordinate national programmes. AEB became operational in March 1994 to co-ordinate and plan Brazil's space programme, replacing the inter-ministerial COBAE Commission for Space Activities. Brazil's President signed a decree 5 December 1991 nominating a commission representing relevant ministries to discuss the creation of the agency. AEB is now involved in re-structuring Brazil's remote sensing programme around increasingly stringent financial constraints and possible ventures with other countries. AEB has agreed to combine resources with other South American countries in the field of atmospheric, meteorological and climatological sciences. This has been extended to near-earth space science studies and support for international initiatives involving Brazilian university and research institutes.

UPDATED

Andrade Gutierrez Quimica Ltda

Current Status

AGQuimica's plant at Jacarei-São Paulo can manufacture up to 100 tonnes of ammonium perchlorate annually to US DoD MIL-A-192B standards for solid rocket propellants. Applications include Brazil's Sonda sounding rockets and VLS orbital launcher and design studies have been completed on propellant combinations delivering a higher ISP. Technical studies leading to research into new chemical combinations and improved quality control standards. The Ministry of Aeronautics' Centro Tecnico Aerospacial initiated research on AP production technology in 1973, and in late 1984, the Andrade Gutierrez Group was selected for the industrial phase. AGQuimica was founded in 1985. Plant start-up was achieved in late 1986 and during June 1987 the product was certified as compatible with international standards. Production rates have been reduced and other commercial applications have been found in the non-aerospace sector. In the wake of new safety criteria set up as a result of a number of accidents and at least one disaster, protocols emplaced to inhibit such accidents have restrained opportunities that were under way to export propellant combinations.

UPDATED

Centro Tecnico Aeroespacial (CTA)

Current Status

The Institute of Aeronautics and Space (previously the Institute of Space Activities) is one of the five technical divisions of the Ministry of Aeronautics' CTA (Aerospace Technical Centre), which has a permanent staff of about 4,500. IAE is responsible for developing the Sonda sounding rockets and VLS orbital launcher for the national space programme, and for operating the Alcantara equatorial launch site. CTA has taken a leading role in attempting to organise a contingent of developing nations for participation in the International Space Station. Serious financial constraints applied to Brazil, national space endeavours has discouraged the use of resources to accomplish this feat. Nevertheless, further efforts have been made from external interest sources and although there is little enthusiasm internally for this activity, discussions have been held regarding some development in this area. CTA has also promoted the use of sounding rockets for a broad-based programme of technology research.

The Ministry created the GETEPE Executive Group for Space Studies and Projects in June 1964, replaced in 1971 by the Institute of Space Activities. During 2003 the CTA negotiated possible export opportunities with foreign governments, some of which had been holding discussions via third parties for substantial quantities of solid propellants.

UPDATED

Empresa Brasileira de Telecomunicacões

Current Status

Embratel, with a staffing level of just over 9,000, operates the national Brasilsat satellite system and works through the Intelsat system via its Tangua ground station for international links. It is also Brazil's national Inmarsat signatory. Brazil has examined options for future utilisation of mobile and data forward transfer systems. Discussions have been held with other countries in South America about an international venture serving mobile users and paging operators. Continuous modification and refinement of adaptable models for future expansion have stalled due to financial constraints but new telecommunications opportunities have been mooted with the prospect of international co-operation.

Embratel was created in 1965 as the operational arm of the Ministry of Communication's Telecomunicacoes Brasileiras (Telebras) component, and is responsible for the domestic Brazilian communications network and links through the Intelsat and Inmarsat systems. Embratel began developing the ground element for a satellite-based domestic system by employing a single leased Intelsat transponder in 1974, rising to seven by 1982, with 21 Earth stations providing telephony, telex and three TV channels.

In June 1982, Embratel awarded Canada's Spar Aerospace a US$125 million contract for two C-band Hughes HS-376 satellites similar to the Canadian Anik Ds, built under license, to create the national Sistema Brasilero de Telecomunicacoes por Satelite (SBTS, or Brasilsat). Attempts to procure a domestic system were initiated during 1975, but were cancelled after two years because of economic considerations. Ariane successfully launched BRASILSAT A1/A2 in 1985–86, working by end-1987 through about 100 Earth stations and with A2 acting as an in-orbit spare.

The ground segment comprises 53 principal stations (one TV distribution, 35 basic telecommunications, seven TVRO and ten 48/64 kbits/s private data) and more than 11,000 private dishes. 1988 use was 54 per cent of capacity, compared to 1984's projection of 83 per cent. With the system fully operational, Brazil's position as Intelsat's 8th largest user at 3.04 per cent in 1985 fell to 15th with 1.24 per cent in 1986. Transponder leasing to neighbouring countries was considered. There was a possibility of a Brasilsat A3 being commissioned for an early-1990s launch to 61°W to extend system lifetime to century's end. However, in early 1989, RfPs were released for a second-generation system offering C- + X-band links on each of two 28-transponder satellites. Hughes was awarded the contract in April 1990. Brazilsat B1 was launched in August 1994 followed by B2 in March 1995. B3 was ordered December 1995 under a US$70 million agreement and launched in February 1998. Deleting the X-band payload allows it to be optimised for C-band. In 1999 Embratel set up a dedicated unit to operate the Brazilian Satellite Telecommunications System. B4 was launched in August 2000.

The satellites are controlled from the Satellite System Operational Centre (SSOC), located in Guaratiba near Rio de Janeiro, comprising the Spacecraft Control Centre (SCC) and the Communications Operations Control Centre (COCC). SCC provides TT&C commands through 14.2 m and 6.0 m antennas, and the COCC primary traffic routing through a 16.2 m dish.

UPDATED

Instituto de Aeronautica e Espaco (IAE)

Current Status

IAE is responsible for developing the Sonda sounding rockets and VLS orbital launcher for the national space programme. IAE is also responsible for VLS guidance equipment and has conducted tests with advanced navigation and guidance equipment. IAE has conducted an intensive investigation of the VLS assembly and launch preparation safety criteria, examining a wide range of potential fail points from sneak circuits to spontaneous combustion of propellants. This technology has been adapted for other aerospace applications. Guidance equipment is being developed for other sounding rockets.

VERIFIED

Instituto Nacional de Pesquisas Espaciais (INPE)

Current Status

The INPE National Institute for Space Research was created in 1971 under the Ministry of Science and Technology to succeed the National Space Activities Commission (created 1961) and be responsible for the development of the ground and space segments of applications satellite programmes. As such, it is developing the five MECB satellites and the Brazilian contribution to the joint CBERS Sino-Brazil remote sensing satellite. INPE has expanded remote sensing activities and integrates satellite operations with ground truth data. Its 1996 budget is US$100 million (1995 US$80 million; 1994 US$40 million), excluding personnel costs. Personnel total has decreased from just over 1,000 to a little more than 900 in 2001. INPE has its headquarters at São José dos Campos (23° S/46° W, where the CRC Satellite Tracking and Control Centre is sited), with additional facilities in Cachoeira Paulista (23° S/45° W, 80 km north of São Paulo, where the mission centre for SCD satellite data collection is located), Cuiaba (15° S/56° W, where it operates Brazil's Spot, Landsat and ERS station), Natal (5° S/35° W; where the owner Ministry of Aeronautics operates a small sounding rocket launching station) and Fortaleza (3° S/38° W).

INPE's interests include spacecraft integration/testing, propulsion (including bipropellant and monopropellant catalytic thrusters), space materials, solar cells, sensors, computer sciences, space and atmospheric sciences, meteorology and remote sensing. Science scanning technology development.

INPE's 10,000 m^2 LTI Laboratory for Test and Integration opened in 1988 as the largest such facility in the southern hemisphere: 1,600 m^2 work hall, 450 m^2 integration room, control room, 13 and 80 kN shakers, 22 m^3 thermal vacuum chamber and electromagnetic test chambers; Intespace will complete an acoustic chamber in 1997. CBERS 2 will be assembled here, and Brasilsat B final assembly/testing was undertaken beginning March 1994. All system qualification testing of Argentina's SAC-B satellite was undertaken here. Contact with the satellite was maintained through CONAE's ground station at San Miguel.

INPE also operates Tiros/GOES ground stations, in addition to the Local User Terminal in São Paulo for the international Cospas-Sarsat distress system. A 14.2 m VLBI antenna began operating 1993 at Eusebio in northeastern Brazil to link South America geodetically with other continents as part of the International Interferometry Survey, co-ordinated by the International Earth Rotation Service. It also has a GPS station to compare results with VLBI-derived parameters. INPE is in the vanguard of an international move to expand VLBI operations still further.

UPDATED

Bulgaria

Navigation Maritime Bulgare Ltd

Current Status

NMB has participated in studies to determine levels of integration that ESA might provide to non-ESA countries in the Galileo programme, an enquiry stimulated by media announcements of China's participation in the European satellite navigation service. The company has a long history of planning systems for maritime operations and for communications requirements and has formulated strategies for integrating national efforts to support the interlinked infrastructure. The company is Bulgaria's Inmarsat signatory and has also participated in studies relating to the use of competing national systems but has taken steps to participate in regional and global maritime communications at an international level. Government restructuring has given NMB a greater role in planning future communication needs for Bulgaria.

UPDATED

Canada

Alliance for Marine Remote Sensing (AMRS)

Current Status

Founded by a small group of marine remote-sensing enthusiasts in 1990, AMRS is a 'not-for-profit' association designed to develop and promote marine applications of remote-sensing technologies. Significant steps have been taken to integrate optical and radar data. It publishes the *Backscatter* newsletter three times yearly. The association has also supported the use of global data for climate analysis. This has been distributed to science research establishments and used to develop new remote sensing applications. AMRS has also developed international networks for information dissemination and for integrating national programmes into international projects. AMRS maintains a network of 10,000 contacts, focuses on practical applications and develops applications.

UPDATED

Canada Centre For Remote Sensing (CCRS)

Current Status

Central organisation for the development and promotion of remote sensing technology and applications, including reception and processing of Spot, Landsat, NOAA, MOS, ERS, JERS and Radarsat data. Canada's government has imposed financial constraints on the country's space research budget and expanded initiatives sought by institutes and universities have been unable to participate in several international opportunities. CCRS has provided detailed technical guidance from representatives and consultants on the future direction of remote sensing and has advised the government on Canada's role in that process.

The CCRS was established in 1971 with a mandate to improve remote sensing technology, facilitate the acquisition and dissemination of remotely sensed data and to work with Canadian industry involved in remote sensing. It has general responsibility for remote sensing research and development within the government. CCRS employs a total of about 110 government staff, with a further 110 on contract from industry working on various developmental and/or technology transfer projects.

The centre has full-remote sensing satellite reception capabilities, via ground stations at Prince Albert (Saskatchewan) and Gatineau (Quebec), including Landsat, Spot, NOAA, MOS 1, ERS 1/2, JERS and Radarsat, under the auspices of its Data Acquisition Division. The Methods and Systems Division is developing methods to correct and produce satellite and other data, as well as the international standards associated with such products. It also contains a group focusing on the development of image processing algorithms, and methods to incorporate artificial intelligence and expert systems into image interpretation. The Applications Division is active in developing applications of remote sensing in agriculture, forestry, geology, ice, hydrology, cartography and environmental monitoring. The division is also working on GIS and remote sensing for regional resource management. The Center's major current focus is in radar technology and applications, for ERS 1/2, Radarsat and other planned radar satellites.

The Centre has a number of state-of-the-art systems for the collection, production, correction and analysis of data from ground measurement systems, as well as data from a wide variety of airborne and spaceborne remote sensors. Geocoding is available to superimpose satellite data on to map projections of the National Topographic System of Canada. Geocoding also facilitates comparison of images of the same site on different dates, possibly from different satellites or in combination with aircraft information.

CCRS is developing a component of the Canadian Geospatial Data Infrastructure known as the Canadian Earth Observation Network. CCRS also maintains a strong tutorial emphasis and carries out promotional and educational programmes as well as archiving a large database from which students, teachers and scientists can access a wide range of information.

UPDATED

Canadian Space Agency

Current Status

Canada has proposed a wide ranging series of space exploration programmes and is keenly pursuing possible involvement with the new NASA initiative for the colonisation of the moon and the human exploration of Mars. Budget constraints have limited the opportunities for planetary research. New international ventures are sought for active participation in global environmental monitoring. CSA is responsible for the management, planning and policy development of the Canadian space programme, including co-ordination of the space activities of other agencies of the federal government. It was created in March 1989, drawing together the space activities of the Ministry of State for Science, the Dept of Communications, the Dept of Energy, Mines and Resources, and the National Research Council. It accounts for some 70 per cent of federal space expenditure. The President reports to the Minister of State for Science. Parliament, on 14 December 1990, approved legislation creating the agency, allowing it to present its own

Canada has built decades of space engineering expertise on a wide range of space-related products, not least of which is the Shuttle remote manipulator system, seen here in its refined and redesigned form as the Space Station Remote Manipulator System. 0572493

budget and appropriate its own funds. In a recent summary of its priorities and programmes the CSA places completion of Radarsat 2 development as its primary strategic outcome for economic benefit and the launch and operation of SCISAT-1 for understanding the environment.

CSA relocated to the new C$70 million 26,600 m^2 facility at the 0.41 km^2 St Hubert site in 1993 where it maintains a personnel total of 350. The David Florida Laboratory, a portion of the Space Science Programme and a liaison office remain in Ottawa. The CSA also employs some 225 service contractors and 50 students. The Space Operations Support Center has been operational for supporting space station operations for over a year at St Hubert. These have extended through the operational deployment of Canadarm 2, the remote manipulator system permanently attached to the ISS. This is capable of joint activity with the shuttle orbiter-attached Canadarm, when the Shuttle is docked to the station. Canadarm 2 was launched to the ISS on 12 July 2001.

CSA's programmes branch develops new programmes, passing them to the Engineering branch for implementation (for example, Engineering now has Radarsat and Space Station). The Operations branch handles on-orbit operations.

UPDATED

Provincial Remote Sensing Office

Current Status

PRSO, established in 1973, is part of the Information Resources Division of the Ontario Ministry of Natural Resources, a provincial government agency. PRSO works with government, industry and end-user client groups to facilitate the development/application of remote sensing techniques for natural resources management. Data from air/space platforms are used to update natural resource inventories, conduct environmental studies, geological exploration and mapping. PRSO has developed new software interactive programmes for integrating optical and radar data. Canada seeks to expand its environmental monitoring activity through the PRSO and is developing new methods of integrating data from land, sea and air sources. These include integrating buoys, data collection platforms and instrumented balloons for meteorological, atmospheric and climatological data gathering.

The centre operates an aircraft modularly equipped for imaging sensors and electronic navigation. Experience includes analogue/digital analysis for processing imaging sensor technologies; aerial photography, thermal infra-red scanning, SAR, multispectral scanners and imaging spectrographs. The technology transfer programme includes training for domestic and foreign scientists. PRSO also provides information on services available from province's remote sensing companies. A new function is the development of corporate data standards by the Data Standards Secretariat.

VERIFIED

Teleglobe Canada

Current Status

Teleglobe Canada, a member of Teleglobe Inc located in Reston, Virginia., USA, is the national signatory to Intelsat and Inmarsat as the country's international telecom carrier via satellite and undersea cable reaching customers in more than 240 countries and territories. It is the world's fifth largest international communications service provider and is a member of the Satellite Aircom consortium providing aeronautical voice/data services through the Inmarsat system. Teleglobe has conducted telecommunications tests with airline operators and with manufacturers to develop new and efficient ground-to-air communications systems. In 1997, conforming to the deregulation of telecommunications services stipulated by the World Trade Organisation, the Canadian government ended Teleglobe Canada's exclusive mandate to handle international traffic. In June 1997 Teleglobe Inc listed its stock on the New York Stock Exchange as well as the stock exchanges in Montreal and Toronto. Teleglobe expansion continued as it rapidly developed new undersea cable and satellite links. In February 2000 Bell Canada Enterprises (BCE) announced its intention to purchase all remaining outstanding shares from its existing 23 per cent share and the transaction closed on 1 November.

UPDATED

Telesat Canada

Current Status

A private company, incorporated by an act of Parliament in 1969, was originally 53.7 per cent owned by the Canadian government and the remainder by a consortium of common (telephone) carriers and CNCP. In 1992, the government sold its shares to Alouette Telecommunications Inc, a consortium of Canadian telephone companies and Spar Aerospace Ltd. In 1998, BCE Inc, increased its majority stake in Telesat to 100 per cent by acquiring all outstanding common shares in Alouette Telecommunications Inc.

In 2003 revenue was posted of C$344.5 million with net earnings of C$70.2 million compared to C$326.8 million and net earnings of C$54. 1 million in 2002. In 2000, reflected earnings of C$47.4 million, following 1999's C$31.0 million and 1998's C$28.9 million. 2000 operating revenue was C$272.4 million (1999 – C$241.9 million; 1998 – C$218.1 million). Canadian Broadcast Services has consistently contributed about half of total revenues; telecommunications services contribute approximately 25 per cent; carrier services, 15 per cent; and international consulting, 10 per cent.

Telesat is currently operating eight satellites. Aniks E1, E2, F1 and Nimiq are owned and operated by the Company. The Anik FSS satellites are used for broadcast and business telecommunications services. Telesat's Nimiq DBS satellite is used in the delivery of Direct-to-Home services. Telesat operates MSAT-1 (on behalf of Loral Spacecom Corp), and the Rock and Roll satellites (on behalf of XM Satellite Radio Inc).

Telesat has two additional satellites: Anik F2, which included C-, Ku- and Ka-band payloads, manufactured by Boeing Satellite Systems and was launched in 2004. Nimiq 2, the company's second DBS system was launched on 29 December 2002 featuring 32 Ku-band transponders and a small Ka-band payload.

Telesat's headquarters are located in Ottawa, Ontario (Canada), where the majority of its 500 employees are located. The Company has Teleports in major Canadian cities, and offices throughout the Americas.

In 2001, Telesat acquired 100 per cent common shares ownership of Infosat Communications Inc. Infosat is a full-service provider of satellite-based voice, fax, paging and data communications.

Telesat is wholly owned by BCE Inc, a leading communications company.

UPDATED

China

Beijing Marine Communications and Navigation Co

Current Status

In addition to being China's Inmarsat signatory, MCN also leases/sells Inmarsat A/C equipment. Some 70 per cent of sales are to state ministries. Rental customers are mainly tourist groups and foreign oil and mineral extraction companies performing exploratory work. Intelsat/Inmarsat interlinks provided for dual use operators. MCN is expanding its service capabilities and has successfully demonstrated out-of-region telecommunications capabilities in terrestrial and non-terrestrial links and has demonstrated ways in which cross frequency networks are both feasible and practical.

China Aerospace Science and Technology Corporation

Current Status

The China Aerospace Science and Technology Corporation (CASC) specialises in the development and manufacture of spacecraft and launch vehicles as well as a variety of different ballistic missiles. It ranks high as a world-class provider of satellites, launchers and propellant and has strengths in applications, automated control and systems integration. The CATC has been formed out of the China Aerospace Corporation (CAC) which dates back to October 1956 and the establishment of Military Academy 5 which specialised in rockets and ballistic missiles. Renamed Military Academy 7 in 1964 it was subsequently known as the Ministry of Aerospace Industry from which grew a variety of research and development institutes. A series of facilities numbered 061–068 possessed a wide range of research, development and manufacturing capabilities and although set up to comprise third-tier manufacturing centres they quickly became first-line plants. The CAC was set up in 1993 incorporating the No 5 Research Academy of the Ministry of National Defence, the 7th Ministry of Machine Building Industry, the Ministry of Astronautics Industry and the Ministry of Aerospace Industry.

On 1 July 1999, the State Council approved reforms which affected the top 10 of China's Defence and Technology Corporations which brought about the establishment of the CASC from the former CAC in an effort to make it more competitive. With some 270,000 employees the former CAC is divided into the China Aerospace Science and Technology Corporation and the China Aerospace Machinery and Electronics Corporation. More than 130 separate organisations are subordinate to the CASC and these include five research academies known as the Chinese Academy of Space Technology (CAST), the Shanghai Academy of Space Flight Technology (SAST), the Chinese Academy of Space Electronics Technology (CASET) and the Academy of Space Chemical Propulsion Technology. The CASC also presides over two large research and manufacturing facilities: the Sichuan Space Industry Corporation and the Xian Space Science and Technology Industry Corporation. It also controls a number of factories and companies. The CASC has approximately 115,000 employees including more than 43,000 technical staff, 1,300 researchers and 22 academicians from the Chinese Academy of Sciences and the Chinese Academy of Engineering. The CASC has primary responsibility for the development and manufacture of launch vehicles as well as ballistic missiles and the provision of commerical launch services. CASTC has played a major role in the adaptation of CZ launches for human space flight and in the development of the Shenzhou manned spacecraft which carried the first Chinese astronaut into orbit during late 2003.

CASC subsidiaries include:

Beijing Computer Technology and Application Institute (computer systems, software, CAD and CAD-CAM concepts);

Beijing Institute of Aerodynamics (applied technology research institute with wind tunnels and computing centres);

Beijing Institute of Radio Measurement (tracking and instrumentation);

Beijing Institute of Radio Meteorology and Measurement (designs and manufactures equipment and calibrates time and frequency systems);

Beijing Institute of Remote Sensing Equipment (research non optical and electro-optical systems, radar and remote control);

Beijing Institute of Space Automation Control (control systems, launch systems, monitoring and command systems);

Beijing Institute of Telemetry Technology (research into telemetry and tracking but also produces ground stations, GPS receivers and control equipment);

Beijing Simulation Centre (a national laboratory for space and civilian simulation requirements);

China Aerospace Holding Ltd (combines technology with trade, finance and industry for manufacturing high-technology products);

China National Instruments Import and Export Corporation (computers, satellites and communications equipment);

China Precision Machinery Import-Export Corporation (established 1980 to export ballistic missiles, solid propellant rocket motors and tactical missiles);
Chongqing Space Electromechanical Design Academy (research on robotics, precision machines and automation equipment);
Harbin Fenghua Machine Factory (machines products for the space programme);
Nanjing Chenguang Machinery Manufacture (stainless steel products, bellows and expansion joints); Polytechnologies (reports to the PLA General Staff Equipment Department);
Shanghai Institute of Electronic Communication Equipment Engineering (radar and electronic communication equipment);
Shanghai Institute of Space Power Source (development and manufacture of power systems for satellites and launch vehicles);
Shanghai Institute of Spaceflight Automatic Control Equipment (attitude and orbit control systems); Shanghai Institute of TT&C and Telecommunications (research, development and manufacture of electronic equipment for launch vehicles and satellites);
Shaanxi Liquid Rocket Engine Company (produces first and second stage engines for LM-2, LM-3 and LM-4);
Shenyang Xinle Precision Machinery Company (precision mechanical treatment and electronic technology).

UPDATED

China Telecommunications Broadcast Satellite Co

Current Status

ChinaSat has organised and set up China Space Moblie Satellite Telecommunications Co Ltd for operating global satellite mobile telecommunications services and Beijing Spacenet Information Telecommunications Pte Ltd for operating satellite broadcast services. These plans have run into problems likely to delay the initiative, not least because of uncertainty in technical standards.

Background

China Telecommunications Broadcast Satellite Co or 'ChinaSat' is the state satellite operator and a wholly owned subsidiary of the Ministry of Posts and Telecommunications. Capacity is also provided by AsiaSat, which is partly owned by China. ChinaSat was established under the Ministry of Radio, Film and Television in 1983 to provide satellite based TV services to rural China.

Official estimates claimed more than 40,000 TVROs by 1993. The government in October 1993 announced restrictions on the manufacture, sale and installation of terminals, but it remained to be seen how strictly the regulations would be enforced in such a huge market. A 1994 estimate claimed 500,000 home dishes. There were 2,600 VSATs in operation by 1992, rising to more than 8,000 by 2002. Hughes Network Systems in 1993 provided 300 VSATs to the Bank of China, working through AsiaSat 1. MPT lost its monopoly in January 1994.

MPT is China's signatory to Intelsat, and the Beijing Marine Communications and Navigation Co to Inmarsat.

Preparations for the STTW-T1 included a 1983 agreement to employ Italy's Sirio 1 for communications trials, and a US$20 million March 1984 contract to Spar Aerospace for 26 Earth stations. 15 were in operation by October 1986 at Lhasa, Tibet, Guangzhou and Hohhot, linking off-shore oil/gas production facilities by voice/data. Spar received a further US$36 million contract mid-1991 for 13 m Intelsat A stations at 10 new locations and enhancements at seven existing sites. Spar signed a letter of intent 8 November 1994 with CASC, China National Railway Signal and Communications Corp, and Space Wit Communications Co Ltd to develop Earth stations and terminal networks for voice/data. A new Shanghai station opened August 1992 to provide European links via Indian Ocean satellites. A new Intelsat TT&C and Monitoring station was inaugurated at Beijing in October 1992 as part of the organisation's worldwide network. Scientific-Atlanta in September 1995 was awarded a US$3.7 million contract by MPT for an emergency satellite communication network to support relief operations. The initial phase of the Skylinx DDS Digital DAMA telephony network became operational in early 1996 using AsiaSat 2 Ku-band capacity. It employs a 4.5 m master station in Shanghai working with 3.6 m stations in seven provinces and 37 mobile 1.2/1.8 m VSATs. The service will eventually be expanded to include a 3.6 m station in all 22 provinces, with at least five mobile units per province. A US$9.9 million contract for the Ministry of Foreign Trade and Economic Co-operation installed hubs in Beijing and Shenzhen working to almost 60 VSATs in trading locations for financial and customs communications.

Studies are under way of a GMSIS Global Mobile Satellite Information System, which would provide personal hand-held communications via 18 to 24 satellites in medium orbits. Three or four of the 500 kg satellites could be delivered by a single CZ3A. The Twin Star proposal would use two satellites some 40° apart in the GEO arc to provide positioning, messaging and timing services, after the fashion of OmniTracs and EutelTracs.

Chinese Society of Astronautics (CSA)

Current Status

Long term goals and national priorities in space have led the CSA to significant new roles in communicating objectives and research activities to a national and international audience. The affirmed aim from China's government to operate an indigenously developed space station is reshaping its interface with space agencies in Europe. China seeks to participate in several international space ventures, acquiring technology and a commitment to launch satellites on Long March vehicles in return. The CSA assumes in a responsible role in these objectives and the success of China's first human space flight during October 2003 has reinvigorated those plans. CSA was established in October 1979 to promote the development of space science, technology and professional training. It has extended its sphere of interest to include liaison with members of the former Soviet Union and has helped enlarge contact with technical companies in Russia for development of China's human space flight programme and for eventual participation in the International Space Station. There have been limits to the degree of co-operation and to the levels of technical support provided.

Czech Republic

Commission on Space Activity

Current Status

The 25-man commission was created in 1990 within the Academy of Science to make recommendations on space activities. It had a proud heritage in that government space activity supported selection and preparation of the first non-Soviet cosmonaut to fly a Soyuz into space. That event took place in June 1978 when Vladimir Remek was launched to the Salyut 6 space station aboard Soyuz 28. It officially ceased to exist 31 December 1992 but in practice remains the point of contact for international co-operation. Recently it has been the subject of possible revival to bolster Czech space programmes and in 2000, the Czech government consulted its members on fuller participation on ESA activities. The Czech government turned down a request for increased funding essential to support a plan put forward by the commission for greater participation in ESA and non-ESA research projects. In 2003 the Czech government made several approaches to former Russian contacts regarding international programmes in which the country may play a leading role, these being mainly scientific and applications. The core of Czech space activity, however, remains the indigenous and autonomous research made available through the extensive work conducted during the period of communist control.

UPDATED

Denmark

Danish Research Advisory Board for Space

Current Status

The Danish Research Advisory Board for Space provides the Minister for Research and Information Technology with independent and qualified advice in space research and applications and has conducted analysis of ESA programmes to advise on opportunities for Danish participation. It plays an important role attracting external funds for academic research in physics and atmospheric science. Instrumental in framing national policy, integrating national capabilities into international projects within ESA and forming policy for future initiatives.

VERIFIED

Dansk Rumforkningsinstitut

Current Status

The Danish Space Research Institute (Dansk Rumforkningsinstitut) is a government-funded research organisation, founded in 1966, under the direct supervision of the Research Directorate of the Ministry of Education and Research. The early development period was characterised by a high level of interaction with the nascent European wide space organisations, such as ESRO in particular, with a consequent level of knowledge and expertise that placed the Institute to make a major impact on developing national space science and space physics research programmes with considerable progress in foreign agreements for joint research. Considerable effort has been taken to provide educational seminars and to extend post-graduate level courses. Its principal areas of interest are space astronomy, cosmic rays and space plasma physics. Four wide angle x-ray transient and gamma burst detectors which form the Watch all-sky monitor are flying on the CIS Granat satellite and a single Watch unit was provided to ESA for Eureca's 1992 mission. Two large area x-ray telescopes, four imaging detectors and a Bragg crystal spectrometer (Xspect) are being produced as part of the RFAS/Danish Sodart telescope for the Spektrum-X satellite in 1998. Ground support equipment was supplied for the Isophot instrument on ESA's Infra-red Space Observatory in co-operation with Heidelberg's Max Planck Institute of Astronomy, and DSRI is a member of the Hipparcos Northern Data Consortium. The Swedish Viking magnetospheric mission carried a wave experiment built at DSRI; current activities include participation in the Wave and Magnetometer experiments on ESA's Cluster. A series of plasma physics sounding rocket flights in co-operation with Scandinavian countries and NASA is also under way. DSRI is part of Denmark's Ørsted satellite project which was successfully launched on 23 February 1999.

DSRI receives a line item budget voted each year by parliament as part of the national budget. This includes salary, operating and development costs. Additional support is obtained from the Danish Space Board on a project by project basis. DRSI is headed by a Director, who reports to the Research Directorate on financial and personnel matters and to an external Institute Board for science/programme affairs. Otherwise, the Institute is autonomous. 1996's budget is DKr26.1 million; 1995 DKr19.3 million; 1994 DKr19.4 million; 1993 DKr24 million; 1992 DKr22 million; 1991 DKr22.544 million; 1990 DKr21.844 million; 1989 DKr21.335 million. This supports a fixed staff of about 40, half of whom are physicists and engineers. Included in the total is funding from the Danish Space Board to support seven engineers, scientists and technicians, and a development programme for thin foil x-ray optics. This has now been approved. Further work has begun on components for high-resolution spectrophotometric instruments.

UPDATED

Ecuador

Centro de Levantamientos Integrados de Recursos Naturales por Sensores Remotos (CLIRSEN)

Current Status
The CLIRSEN Centro de Levantamientos Integrados de Recursos Naturales por Sensores Remotos (Centre for Integrated Surveys of Natural Resources through Remote Sensors) was created in December 1977 to be responsible for remote sensing and the cataloguing of natural resources. The organisation had a strong base in aerial surveying and in bringing together ground observations and airborne data retrieval, setting it on course for a determined effort to integrate these fields of work with space-based data retrieval, analysis and interpretation. Driven by national priorities for economic development and optimum utilisation of natural resources, government agencies use the CLIRSEN on a demand basis. It operates the Cotopaxi ground station, some 60 km south of Quito, which began taking imagery from Landsat in June 1991 as the system's 18th station; Spot and ERS were added a few years later. The Centre now works with radar and SAR data and integrates air and spaceborne sources. The 2,500 km^2 coverage of 25 countries stretches from Mexico's Yucatan Peninsula to Antofagasta in Chile. Products include 23 × 21 cm film, 72 × 97 cm paper, quick look, microfiche (84 scenes each) and digital tapes (6,250/1,600 bpi). Processing levels: bulk, georeferenced, geocoded and user-specified highlighting. CLIRSEN is currently working with several commercial remote sensing companies in South America. There have been several attempts to integrate natural resource data across Central America but this initiative has met little enthusiasm from neighbouring countries.

UPDATED

Finland

Academy of Finland

Current Status
The Academy is the central government agency for science policy, planning and administration, responsible to the Ministry of Education. It has formed research policy resulting in the development of several small satellite programmes. Formed in 1970, it comprises four research councils (Natural Sciences and Technology, Environment and Natural Resources, Culture and Society and Research Council for Health) and the board. The Academy has responded to a government initiative calling for greater participation in scientific research with a study of ESA projects and how Finnish institutes can contribute. This has been extended to a range of international projects and the integration of these with national objectives. Specialist sub-groups formed to report to various government departments on opportunities and capabilities.

VERIFIED

Finnish Geodetic Institute

Current Status
Space-related activities are carried by two of the Institute's four departments. The Department of Geodesy and Geodynamics operates a satellite laser ranger at the Metsähovi Space Geodetic Station and permanent GPS stations for the International Geodynamics Service. The Dept of Photogrammetry and Remote Sensing studies/processes remote sensing data from aircraft and satellites such as Landsat and Spot but also high resolution data. The department is also developing integrated concepts for converging optical and radar data. The annual budget is €2.9 million, of which €1.0 million is devoted to remote sensing, space geodesy and geodetic astronomy. The Geodetic Institute has also participated in a new standards body set up to help standardise scales and options for geodetic surveys at an internatioal levels.

UPDATED

Finnish Meteorological Institute

Current Status
The Finnish Meteorological Institute (FMI) is a government research and service institute, with a staff of 500. It builds on a long history of meteorological studies and significant achievements in environmental and meteorological studies. The FMI is Finland's signatory to EUMETSAT with the national contribution paid from FMI's budget. FMI's research programme covers a wide range of meteorological and atmospheric topics, as well as geomagnetism, and space physics.

The Geophysical Research department at FMI (FMI/GEO) is a research unit with 56 staff. FMI/GEO has a research programme on the physics of Sun-Earth interaction, middle atmosphere, and planetary physics.

Of the main fields of research, the physics of Sun-Earth interactions has longest traditions. FMI is running a unique ground-based network for research on aurora and other space weather effects in Northern Scandinavia. The results are best used in combination with modeling and simultaneous measurements on magnetospheric satellites. FMI/GEO hosts the world's most advanced computational model for the behavior of the magnetosphere. The long-term goal is to develop methods for predicting space weather events.

In middle atmosphere research, FMI/GEO is specialised in retrieval of ozone and other constituents of the atmosphere, using data from instruments that use the UV to visible wavelength range. Methods developed at FMI/GEO are used in two current satellite programs: Envisat of ESA and the Swedish-led Odin mission.

The third major field of research is planetary exploration. FMI/GEO has been actively involved in developing small meteorological stations for Mars exploration, and is currently working on a next-generation lander as a joint Finnish-Russian project. This work has not been followed by a specific programme plan and there is little future potential for a joint Russo-Finnish venture. Research applications are directed towards comparative planetology, especially mesoscale modeling of meteorological phenomena.

UPDATED

Finnish Space Committee

Current Status
The Space Committee comprises a chairman, vice-chairman and 10 members, with the executive working group as the sole subordinated body. The chairman is the national representative on ESA's Council. A new committee was appointed 1 April 1995 and this produced the 2001–2004 space plan. Space Committee formulated the long-range plan in response to a government initiative and worked with the Academy of Finland to achieve that. Work has included formulating international ventures which focus skills and research experience in Finland's academic and research institutes.

VERIFIED

Helsinki University of Technology, Space Technology Laboratory

Current Status
HUT is Finland's oldest and largest technical university specialising in basic research and development and in technology applications, securing patents which are sold to companies for commercial application. STL main research area is airborne and microwave remote sensing, including instrumentation, measurements and development of geophysical inversion algorithms. It also runs space technology courses and proposed the 50 kg HUTSat designed to provide amateur satellite communications and incorporate a particle detector. It is also involved in planning a new 10-year Finnish space programme. This has strong government support and is focused on possible research activities and on environmental studies from satellites.

VERIFIED

National Land Survey of Finland (MAANMITT AUSLAITOS)

Current Status
SIC imports satellite images and SAR data offers related services: initial processing and distribution of RS data, archives, technical services, user studies and short tailor-made courses. A major task is to support Finnish industry and the research sector in the use of satellite imagery. SIC has a wide base of research and analysis experience with specialists in a wide range of fields related to remote sensing and earth observation. Extended geographic and geophysical surveys conducted by surface exploration on data from satellites. It is the National Point of Contact for ESA/ESRIN. SIC is distributor of satellite data products in Finland under agreement with Sweden's SSC Satellitbild and Eurimage. SIC adds analysis and interpretation capabilities to a prospectus of services tailored to individual requirements. National Land Survey staff 2,120. SIC was established in April 1995 and has a staff of 10.

UPDATED

Technology Development Centre (TEKES)

Current Status
TEKES was established in 1983 by the Ministry of Trade and Industry to co-ordinate and finance applied technical and industrial research and development. Formed from a base of expertise in technology transfer and venture capitalist promotions. It is responsible for implementing, co-ordinating and financing national space projects and Finnish participation in international programmes, including ESA. TEKES is the largest source of funding for Finland's space projects, accounting for approximately two-thirds of Finnish space R + D. Covering the period 2001–2004, the space research programme Antares aims to extend Finland's participation in related programmes through training, international networking and benefit applications. The 2002 budget is FMk112 million, 2001 – FMk113 million, 2000 – FMk114 million: 1999 – FMk110 million, 1998 – FMk108 million, 1997 FMk110 million, 1996 – FMk104 million, 1995 – FMk106 million; 1994 – FMk95 million and 1993 – FMk85 million.

VERIFIED

France

Centre National d'Études Spatiales

Current Status
CNES was established in 1962 with a staff of 27 to execute the national space programme. Organized since April 1993 under the regulating authority of three ministries (Defence; Industry,

Posts, Telecommunications and Foreign Trade; Higher Education and Research), CNES is headed by a President and a Director General. It has an 18-member Conseil d'Administration, with six members elected by agency employees. In 1993 it was also allocated responsibility for design/development of military space projects, under the Ministry of Defense. Personnel total 2,488, with 9 per cent located at headquarters in Paris, 10 per cent in Evry, 68 per cent in Toulouse and 13 per cent at the CSG Ariane launch site. The 2000 CNES budget is FFr12,309 million including a state subsidiary of FFr9,265 million.

CNES was instrumental in creating eight subsidiaries and four GIE Groupements d'Intérêt Economique to exploit and market its space activities. Arianespace produces and markets Ariane launchers and in conjunction with a US subsidiary, Arianespace Inc, S3R offers insurance as an Arianespace subsidiary. Intespace performs space environment tests and maintains facilities for the tests. Spot Image promotes commercialisation of Spot satellite data, in conjunction with US subsidiary Spot Image Corp, Australian Spot Imaging Service, Swedish SSC Satellitbild and Singapore's Spot Asia CLS. For commercialisation of Argos location and data collection services CNES uses US subsidiary Service Argos Inc and North American CLS. Novespace promotes space technology transfer, including microgravity applications. SCOT Conseil is an Earth observation consultancy. SIMKO provides urban development of Kourou town. MEDES-IMPS researches space medicine and physiology. Satel Conseil is a space telecommunications consultancy. (GIE) Prospace promotes French space sector products and services, (GIE) GDTA gives remote sensing training and (GIE) DERSI develops relations between French and Russian space industries.

CNES does not maintain in-house space science laboratories but supports a number of principal French research facilities. For example, a CESR/SAp collaboration provided the Sigma telescope for Russia's Granat observatory through CNES. The laboratories are: Institut d'Astrophysique Spatiale (IAS, Institute of Space Astrophysics at Verrières-le-Buisson), Laboratoire de Météorologie Dynamique (LMD, Dynamic Meteorology Lab, Paris and Palaiseau), Département de Recherche Spatiale de l'Observatoire de Meudon (DESPA, Meudon Observatory Space Research Dept), Service d'Astrophysique du Département de Physique Générale du CEA (astrophysics section of CEA atomic energy authority, at Saclay), Laboratoire d'Astronomie Spatiale (LAS, Space Astronomy Lab, Marseille), Centre de Recherche en Physique de l'Environnement terrestre et planétaire (CRPE, Centre for Research into the Physics of the Terrestrial and Planetary Environment, Issy-les-Moulineaux), Centre d'Etude Spatiale des Rayonnements (CESR, Centre for Space Investigation of Radiation, Toulouse), Groupe de Recherches de Géodésie Spatiale (GRGS, Space Geodesy Research Group, Grasse and Toulouse), Laboratoire de Physique et Chimie de l'Environnement (LPCE, Environmental Physics and Chemistry Lab, Orléans-la-Source), Laboratoire d'Etude de la Solidification et Cristallogénèse (SESC, Solidification and Crystallogenesis Study Lab, Grenoble), Laboratoire de Physiologie Neurosensorielle (Lab of Neurophysiology, Paris), Laboratoire de Biophysique Médicale (Lab of Medical Biophysics, Tours), Service d'Aéronomie CNRS (Verrières-le-Buisson), Service de Physique de l'Etat Condensé (CEA lab, Gif sur Yvette), Institut de Physique du Globe (IPG, Global Physics Institute, Paris), Laboratoire d'Etudes et de Recherches en Télédétection (LERTS, Toulouse), Laboratoire de Biologie Médicale (Toulouse), Laboratoire de Cytologie et Morphogénèse Expérimentale (Paris), Laboratoire de Biophysique Appliquée (Fontenay-aux-Roses) and Centre d'Etudes et de Recherche de Médecine Aérospatiale (CERMA, Brétigny-sur-Orge). CESR, LPCE and CRPE contributed instruments to the Russian-led Interball; and LMD provides the Scarab Earth radiation budget radiometer for various satellites.

Sustained support for continued development of Ariane 5 supporting technologies and for new launcher concepts. Has played a major role in resolving the technical problems with the Ariane 5 ESC-A variant.

VERIFIED

Centre Spatial d'Evry

Current Status

Evry is CNES' launcher development centre, responsible for the Ariane series for the European Space Agency. It has a mandate to continue development of the Ariane 5 and had authority to provide technical direction for the FESTIP and FTLP activities. Responsibility for upper stage development and for propulsion systems analysis. Instrumental in resolving Ariane 5 failure determination. On the continuing search for a reusable launch vehicle Evry has played a very minor role in providing consultative recommendations on the impact a reusable fly-back booster may have on launch complex design and resource requirements. No formal programme has been authorised for such a vehicle, but consistent interest from government agencies across Europe have encouraged a sustained study and analysis programme.

UPDATED

Centre Spatial de Toulouse (CST)

Current Status

CST at Toulouse is the agency's principal engineering and technology facility and exploits the operational systems. It has developed several engineering activities designed to provide research, evaluation and test of small and medium satellites. CST has conducted research activities in the design and development of small to medium-class satellites. It developed nanotechnology concepts for the micro-miniaturisation of satellite and spacecraft systems for application in small or medium size bus designs. CST also manages the balloon launching sites at Aire-sur-l'Adour and Gap-Tallard. The centre comprises 20 laboratories, including the spacecraft test facility managed by CNES subsidiary Intespace. The operation centres located at Toulouse include: two satellite launch/early orbital phase control rooms (second opened 1991), Sarsat-Cospas control/mission centre; the Telecom 1/2 dedicated control centre; the TDF 1/2 dedicated control centre; the Spot 1-3 control centre and the Spot imagery rectification centre; the Helios 1/Spot 4 Fresnel building; the computer centre; and the 2 GHz network control centre, with stations at Issus Aussaguel (15 km southwest of Toulouse), Kourou, Hartebeesthoek (South Africa) and Kerguelen Island (3,000 km southeast of South Africa). CST has conducted integrated mission control studies for multinational projects.

VERIFIED

Etat Major des Armées

Current Status

The Armed Forces Headquarters is responsible for developing France's multiyear military space plan and is also involved in planning military space surveillance systems. It has had a significant role in the Syracuse military communications system. Optional projects have been analysed for their viability, including electronic surveillance satellite systems and advanced imaging reconnaissance systems. It has organised a military space projects development plan under, the aegis of France's government, and formulated plans for technical and operational activities necessary for expanded satellite programmes driven by perceived national security needs. Several studies have been conducted for third parties interested in participating projects and co-operative ventures.

UPDATED

Geosys

Current Status

Geosys was established in 1987. From 1990 to 1996 it was a member of the SGS group (Société Générale de Surveillance) and the Caisse des Dépôts et Consignations (Deposit and Consignment Office) and has three shareholders: ITH (Damien Lepoutre) 60 per cent; IRDI (Investment Company) 35 per cent and Geosys employees 5 per cent. Personnel total more than 40. It compiles regular crop inventories and prepares yield forecasts using remote sensing data, including Spot, NOAA AVHHR and Landsat TM. It also provides geographic information to the agribusiness and environment sector. It develops/sells GIS handling remote sensing images and vector cartographic data. Geosys is developing adaptive software programmes for integrated satellite observation systems. These have application for developing countries. The company has expanded from Geosys (France) to include Geosys Inc in the US for geographic information and Geosys Data for catalogues data available to clients.

UPDATED

Groupement pour le Développement de la Télédetection Aérospatiale (GDTA)

Current Status

The Groupement pour le Développement de la Télédetection Aérospatiale is a GIE (Groupement d'Intérêt Economique) of CNES, the Institut Géographique National (IGN), Bureau de Recherches Géologiques et Minières (BRGM) and Institut Francais de Recherche pour l'Exploitation de la Mer (IFREMER). The principal aim is to promote the development of remote sensing activities and the use of satellite imagery (in particular, Spot products). Personnel total about 30; annual turnover is about FFr30 million. GDTA also has access to the resources and personnel of its four member organisations.

GDTA offers remote sensing training, from introductory to advanced levels. It provides technology transfer services abroad for the French Co-operation and Foreign Affairs Ministries and, in addition to remote sensing training, it provides advisory services for establishing remote sensing centres. GDTA also organises airborne campaigns to test sensors planned for eventual satellite applications. It is currently working on high-resolution European imaging radars. Development of analytical models for synergistic interpolation of imagery from optical and radio sources. It has integrated the skills and capabilities of its four member bodies and provides services to government departments and other companies in France. Conducts analysis and makes recommendations on equipment and future requirements.

VERIFIED

Météo-France

Current Status

Coming under the authority of the Ministry of Equipment, Transportation and Sea and responsible for providing space-based meteorological services, Météo-France acted as the national representative in establishing EUMETSAT and in the implementation of the Meteosat operational satellite system. Specialised expertise in meteorological systems and survey requirements. Météo-France also provides aeronautical weather charts. Météo-France has also participated in the development of instrument specifications for the METOP ESA programme. Provides government advisory bodies with detailed information on national requirements and co-ordinates effort into international ventures for global meteorological support.

Its dedicated Centre de Météorologie Spatiale in Lannion receives, processes and relays meteorological satellite data.

UPDATED

Ministère de la Défense

Current Status

This ministry is the customer through the Délégation Générale pour l'Armement (DGA, military procurement agency) and CNES for the Syracuse military communication element of the Telecom satellite system and the Helios reconnaissance satellite. Has assembled a portfolio of missions capable of giving France's defence interests independent access to surveillance and reconnaissance information. Alcatel Space won the contract to build the new Syracuse 3 military communications system, the contract being signed on 30 November 2000. The Syracuse 3 series comprises three satellites all to be launched by Ariane 5. The ministry has evaluated other defence related satellite projects and is examining optional spy and reconnaissance concepts.

UPDATED

Services de Consultance en Observation de la Terre (SCOT CONSEIL)

Current Status

SCOT (Services de Consultance en Observation de la Terre) was established in 1987 as a CNES affiliate to act as a consultancy and promoter of satellite-based remote sensing data, drawing particularly on Spot-derived imagery. Staff totals about 35, including nine engineers. French banks account for 25 per cent shareholding. The company provides a spectrum of services, from user advice on specifications for Earth stations, processing centres and software to overseeing complete remote sensing programmes. It has conducted analytical surveys of national user requirements, customer needs and the available technology fit with new satellite designs. It heads a team of contractors selected by the Commission of the European Communities to develop an operational system to produce rapid estimates of areas under cultivation and potential yields of major annual crops at European scale. Since the mid-1990s SCOT has integrated surveys in east European countries with those in western Europe. Dedicated satellite data convergence software taking images from optical instruments and data from SAR.

VERIFIED

Société Nationale des Poudres et Explosifs (SNPE)

Current Status

The Société Nationale des Poudres et Explosifs SNPE Group, created as a state-owned company in 1971 and now with more than 5,500 personnel and FFr5.3 billion 2002 turnover (FFr120 million profit), manufactures liquid rocket propellants, with emphasis on UDMH, and develops/produces solid propellants. Has conducted research into exotic and tripropellant combinations as well as hybrid solid-liquid propellant combinations for exotic and specific propulsion requirements. Along with BPD Difesa e Spazio it participates in the Regulus partnership. The company provides the propellant for Ariane 5's P230 solid boosters using facilities at Kourou constructed by subsidiary SNPE Ingénierie. SNPE maintains facilities at Saint-Médard en Jalles near Bordeaux (solid propellant grains), Toulouse (UDMH, AP), Toulon (pyrotechnics) and Le Bouchet near Paris (HQ and R&D). The Toulouse site manages the production of ammonium perchlorate with a capacity of 6,000 tonnes annually to ensure Ariane 5 supplies. SNPE formed a second partnership with BPD, Société Européenne de perchlorate d'ammonium (EUPERA; SNPE 66 per cent, BPD 34 per cent), to handle production. Toulouse also provides the UDMH for Ariane 4 and the MMH for Ariane 5's upper stage. Subsidiary Pyromeca manufactures detonating cords and flexible linear cutting charges and ultra-fast pyrotechnic valves for Ariane 4 release from the launch pad. Propellant management test rig and flow evaluation equipment capable of handling rapid fuelling and dump capability.

In cryogenic propulsion, SNPE provides the main igniter and turbine starter grain for Ariane 4's HM-7B stage 3 engine and the shaped ducts for the Ariane 5 Vulcain igniter and starter.

UPDATED

Télédiffusion de France (TDF)

Current Status

TDF, created in 1975 as a limited company with the French government as majority shareholder (and a France Telecom subsidiary since 1988), operates 10,000 TV transmitters/repeaters, 1,500 radio transmitters and 10,500 km of multi-channel microwave links, handling broadcasting for all public/private French TV channels and the principal public/private radio networks. Personnel total about 3,750, including 2,400 engineers and technical staff.

CNES/TDF undertook a study of a national DBS system, following the signing of the WARC '77 agreements, to provide high quality TV and CD-quality radio. In October 1979, France/Germany agreed to joint development of respective national DBS systems, signing the accord April 1980. The Eurosatellite consortium was awarded the space segment contract in 1984, basing the design on Aerospatiale's Spacebus 300 platform. Aerospatiale acted as prime for TDF 1/2 and MBB for TV-Sat 1/2. TDF 1 was launched in October 1998. The decision to build TDF 2 was taken by the French government in December 1984. There was concern in 1986 that the programme would be cancelled because of developing competition from medium-power systems such as Astra. By then FFr1.5 billion of the FFr3.5 billion allocated to TDF had been expended and the French Prime Minister ruled that it should continue in order to promote D2-MAC as the standard for HDTV in the 1990s. D2-MAC Paquet specifications were adopted by France/Germany in June 1985. FFr550 million was paid to Arianespace for the launches. TDF 2 was launched by Ariane 44L in July 1990. Although designed for a life of 8 years the satellite was still operating in late 2002.

TDF underwent minor modifications after TV-Sat 1 was written off when one solar array failed to deploy in November 1987 and thruster overheating was discovered during orbital manoeuvres. TDF was to provide five TV channels from 230 W TWTAs to 45 cm diameter receivers but at the time of launch none had been allocated to users. The failures of two tubes aboard each satellite reduced the number of available transponders from 12 to eight, prompting the 1990 decision to provide only four TV and three digital radio channels in order to maintain full redundancy. At the same time, it was decided there would be no TDF 3 and that the future system would lease Eutelsat from the mid-1990s. By 1992, it was reported that only some 35,000 locations were equipped with TDF dishes. Annual transponder leasing cost was reduced from FFr80 million to FFr35 million. System cost was not recouped.

VERIFIED

Germany

Aachen Centre for Solidification in Space

Current Status

The Aachen Centre for Solidification in Space was established at the town's Institute of Technology in 1986 to provide a link between μg technology and potential users. Research topics include containerless processing, metal-ceramic composites, high-temperature superconductors, directional solidification, model systems and numerical simulation. Has developed unique furnace design for pressure containment and high-temperature environments. Equipment includes Bridgman-type furnaces for ground-based research accompanying space experiments, containerless processing facilities (EM levitation and a droptube), devices for corrosion investigations and wettability experiments, test setups for determination of mechanical and magnetic properties and microstructure characterisation (SEM, EDX, DTA/DSC, and so on). From the mid-1990s work has focused on microgravity products for the ISS research programmes. From this 1 g experimental analogues for zero-*g* on-orbit experiments has emerged. Development of small carry-on research experiments aimed at commercial users for ISS or Shuttle mid-locker.

VERIFIED

Bundesminister für Forschung

Current Status

The Federal Ministry for Research and Technology is responsible for Germany's space research policy. This included negotiation with foreign governments for international projects and programmes. Also established operating guidelines for international projects and collaborative programmes. Following the restructuring of national space administration, responsibility for planning, programming, management and external representation was transferred to DARA. A considerable portion of its work and responsibility now focuses on development, Germany's long term space business interests and in pointing the government toward major initiatives that benefit the country and the economy. The broad mandate held by the Ministry is said by critics to be too broad to adequately serve the needs of the space community on Germany but the effectiveness with which it links all areas of responsibility is especially evident in space engineering and technology ventures.

UPDATED

Deutsche Agentur für Raumfahrtangele genheiten

Current Status

DARA was founded and began operating in July 1989. Its legal status is that of a private company with limited liability (GmbH) owned and financed by the Federal Government. Personnel total varies between 200 and 250. The Management Board comprises three Managing Directors, each responsible for a specific sector covering individual departments: scientific utilisation, Earth observation, application utilisation, orbital infrastructure, transport systems, ground infrastructure and administration. A Supervisory Board of members from Federal Ministries, science and industry provides guidance on policy. An Advisory Board of high-ranking representatives from science and industry provides scientific and economic input. There are also several Advisory Committees on the different fields of science and technology. DARA organises and co-ordinates inter-disciplinary science tasks and manages policy meetings for integrating various experiments from a wide range of sources. Acts as a representative body for industrial interests at national and international conventions and seminars and formulates national directions in space policy which it recommends to the German government.

Since June 1990, when the 'Law Governing the Transfer of Responsibilities with Regard to Space Activities' was adopted by the German parliament, DARA has had government authority for the space programme, particularly in awarding contracts and representing Germany at the international level. For the first time, domestic space legislation stipulates an obligation for German authorities to transfer responsibilities to a single central body. This applies primarily to BMFT, but also to other governmental authorities with vested interests in space. DARA is in full charge of implementing the German Space Programme, except for the hypersonic technology programme, which remains embedded within BMFT's aeronautics activities. DARA has represented Germany since 1 December 1989 in the various bodies of ESA (Council, Committee and Programme Boards) and other international organisations, in addition to being responsible for multilateral/bilateral agreements.

VERIFIED

Deutsche Forschungsanstalt für Luft und Raumfahrt

Current Status

In close co-operation with DARA, the German Aerospace Research Establishment's responsibility is to execute the research element of the German space programme. It was formed in 1969 and employs a total staff of about 4,500, including 1,500 scientists in five research areas: Telecommunications Technology and Remote Sensing, Materials and Structures, Energetics, Fluid Mechanics, and Flight Mechanics/Guidance and Control. The annual budget is about DM700 million. Research centres are located in Berlin, Braunschweig, Göttingen, Köln, Lampoldshausen, Oberpfaffenhofen and Stuttgart, with local branches in Hamburg, Bonn, Trauen, Neustrelitz and Weilheim.

Test facilities include subsonic-to-hypersonic wind tunnels, rocket propulsion test stands, spacecraft TT&C stations and space simulators. Space mission operations, particularly those for Spacelab, Columbus and Eureca, are centred at the Köln-Porz and Oberpfaffenhofen sites. The Crew Training Complex (for ESA's astronaut cadre) and Microgravity User Support Centre in Köln-Porz, and the German Space Operations Centre (GSOC), Manned Space Laboratories Control Centre, User Data Centre and Automation In Orbit Centre at Oberpfaffenhofen have performed manned/unmanned missions for more than 20 years. Change options for development of on-orbit mission tasks and payload monitoring. This allows crew elements to control general modular housekeeping studies and simultaneously track progress with experiments.

The DFD German Remote Sensing Data Centre is the national PAF Processing and Archiving Facility for ERS and other satellites. Köln is the home base for federal astronauts and seat of the EAC European Astronaut Centre of ESA. The Materials and Structures department includes the Institute of Space Simulation, which operates the Microgravity User Support Centre and was most recently primarily concerned with preparing users, astronaut crew and equipment for Spacelab D2. Under the Department of Energetics, the Space Propulsion Division at Lampoldshausen operates test facilities for rocket motors such as Ariane 5's Vulcain. A new stand was completed in 1995 under a bilateral programme with France.

VERIFIED

Deutsche Telekom AG

Current Status

Deutsche Telekom has separated itself from its DeTeSat satellite communications activities. T-Systems, however, is continuing to operate the Business Line Media Broadcast satellite TV business with the Usingen earth station, along with the Raisiting earth station, which is part of the Service Line Network Services, where the satellite part is handled within International Carrier Services. Other business activities include T-Mobile and T-Online.

UPDATED

DLR Microgravity User Support Centre (MUSC)

Current Status

MUSC, based at the German Aerospace Research Establishment's Köln centre, supported the preparation, operation and utilisation of materials and life science experiments on Spacelab, Texus and Mir and will provide similar experiments on the International Space Station. Has developed a new generation of rapid test and evaluation procedures for multitasked experiment containers. Generates modular and integrated experiment suites according to science and engineering tasks. Accepts both materials science and life science equipment interfaces. It provides user information by seminars, databases and a library, technical support for experiment and experiment facility preparation, operational support for interactive experiment operation (telescience), and scientific support by consultancy and measurement/diagnosis equipment for quick-look analysis. MUSC has also embarked on studies to integrate microgravity research programmes from a variety of international projects.

Facilities: ground versions of Spacelab's Anthrorack, Biolabor, Holop, MEDEA and TEMPUS units, Eureca's core payload and Mir's CSK/TITUS furnace. There is also an experiment control room for remote-control operations.

VERIFIED

Institut für Raumfahrtsysteme

Current Status

The Space Systems Institute (IRS) was founded in 1970 and plasma thruster development has always been a principal interest. Facilities allow MW-class testing at 0.5–2 g/s propellant flow rates (argon) under selected 10^{-6}–10^{-3} atmosphere conditions. 0.5–1 MW MPD stationary thrusters of different geometries have been developed, mainly under AFOSR grants. Based on this experience, IRS has designed and built a range of thermal arcjet thrusters. Current arcjet projects include: Atos 600 to 700 W ammonia, Artus 2 1 to 2 kW NSSK hydrazine (with DASA; DARA contract); 10 kW and 100 kW devices. Atos is a simplified version of Artus 2 for flight on Amsat-Deutschland's amateur satellite aboard Ariane 502. The satellite was launched by 502 in October 1997.

Four plasma wind tunnels are used to investigate re-entry vehicle thermal protection materials and to validate aerothermodynamic CFD codes. Two have MPD plasma generators (0.10 to 1 MW) and are especially suited for high specific enthalpy (up to 150 MJ/kg) and low total pressure conditions. One tunnel has an inductive plasma generator, specially used for catalycity investigations, and the fourth has a thermal plasma generator for higher total pressure and low specific enthalpy areas. Sustained research into plasma generators for research and test activity with the facility available for hire by research institutes and government organisations and agencies.

IRS' Mission and System Analysis division studies development and numerical simulation and design tools for space transportation systems, mission and system optimisation, and performance assessment of air-breathing launchers. The Space Technology and Utilization division encompasses space station design, numerical flow field and simulation methods, and space systems safety. Developed software fail-safe checkout systems technology for mapping fault tree analysis.

VERIFIED

ZARM

Current Status

ZARM (Zentrum für angewandte Raumfahrttechnologie und Mikrogravitation) was established in September 1985 in the Faculty for Production Technology at the University of Bremen. ZARM consists of eight departments covering hydrodynamic stability, space technology, rotating fluids, aerodynamics, interface phenomena, gravitation physics, ferrofluids and combustion. μg research interests include fluid mechanics, combustion and thermodynamics, and technical assistance in its industrial exploitation. ZARM's DM30 million 146 m Drop Tower Bremen was commissioned in 1990, providing 10^{-5} *g* for up to 5 seconds. ZARM-FAB mbH (ZARM Fallturm Betriebsgesellschaft mbH: ZARM-Drop Tower Operation and Service Co) co-ordinates drop tower operations. ZARM-Lab GmbH provides technology/science support for users of Columbus and other flight facilities. ZARM-Förderverein eV (ZARM Promoting Association) is a non-profit organisation to support space and μg activities. Conducts new and off-nominal test evaluations and has created new rationales for generating accurate results. Has designed an automated, multifunctional, tribology research facility for encapsulation in existing test structures on manned habitats such as the ISS and research tests have resulted in a new series of proposed experiments involving suborbital rocket flights.

ZARM has played a substantial role in developing microgravity experiments for the ISS.

A 0.25 m hypersonic wind tunnel (ZARM-HHK) became operational in 1993 within the Hypersonic Technology research division. ZARM was also responsible for the university's Bremsat Shuttle satellite.

The 3.5 m diameter 110 m long drop shaft provides a 4.74 second free-fall through a 10^{-5} atmosphere vacuum (pumping requires 1½ h). The pressurised capsule is decelerated over 6 m by styropore granules. A larger capsule became operational in 1995.

Specifications

Capsule accommodation: 700 mm diameter, 1,200 mm high; 150 kg mass; 28 V 10 A (100 A short term) power; 1.6 Mbit/s telemetry; peak deceleration 350 m/s^2 (design must allow for safety factor of 2).

UPDATED

Hungary

Hungarian Satellite

Current Status

HUNSAT is the signatory to Intersputnik, Intelsat and Eutelsat. Two 18 Intelsat A stations are operated: Balaton-2 (Atlantic services) and Balaton-3 (Indian Ocean). Hunsat is conducting active negotiations with other East European states regarding inter-service exchanges and has developed a telecommunications traffic relay service as a model for further co-operation with former Eastern Bloc states.

VERIFIED

Hungarian Space Office

Current Status

The new space research structure was formally created 1 January 1992 on the advice of the Scientific Council for Space Research. Before then, the Interkosmos Council managed participation in Interkosmos and related space science projects, such as Phobos and Interball.

It incorporates five main research areas: space-earth system; space physics; lifesciences; space technologies; space technology-International co-operation with ESA includes participation in Rosetta and Cluster. Co-operation with NASA includes development of the electronics for Cassini and the Pille dosimeter for the ISS. Has proposed a wide range of new and innovative devices for small satellites and long duration deep-space missions. During 2003 several attempts were made, apparently unsuccessfully, to export research and development technologies and space-based applications developed during the early 1990s as a result of restructuring of national science and technology assets after the collapse of Soviet-led rule.

UPDATED

KFKI Research Institute for Particle and Nuclear Physics (RIPNP)

Current Status

The five institutes previously part of the KFKI (Kozponti Fizikai Kutato Intezete, or Central Research Institute for Physics) became independent legal entities on 1 January 1992 belonging to the Academy of Sciences. RIPNP, who designed the TV camera system and two plasma instruments (Tünde and Plazmag) for Vega, was involved in five plasma detectors on Phobos (in addition to its lander's central control/data acquisition computer), and dosimeters on Salyut 6 (1980) and Challenger (1984). Dosimeters were also provided for Salyat 7 and Mir. Current projects include Spektrum-X/γ's (now Spektrum-RG) data acquisition system/onboard science computer, Kvant 2's Microsvit image processing system and some onboard software for the CAPS particle analyser and MAG magnetometer on Cassini and co-investigator for the Rapid and Lion particle detectors of ESA's Cluster and Soho, respectively. The Institute is participating in ESA's Rosetta mission to Comet Wirtanen: it developed the onboard electronics and ground support equipment of the RPC plasma experiment on the orbiter. The Institute also produced the onboard computer and software for the lander. Rosetta was scheduled for launch in January 2003 with encounter at Wirtanen in November 2011 but a failure in an

Ariane 5 upper stage grounded the launcher. The mission was redefined for a rendezvous with comet 67P/Churyumov-Gerasimenko in February 2004, 10 years after launch.

VERIFIED

Remote Sensing Centre of the Institute for Geodesy, Cartography and Remote Sensing (FÖMI RSC)

Current Status

FÖMI RSC has two principal interests: R&D of remote sensing techniques for environmental protection and agriculture; distribution, processing and utilisation of remote sensing data for users. Services cover soil mapping, monitoring of crop cultivation, updating of topographical maps using Spot and Cosmos imagery, and digitally enhanced space photomaps from Spot data. The centre integrates a diverse range of products from several separate sources and provides recommendations for development on a wide range of technical solutions. New arrangements have been concluded for the dissemination of refined and processed data capabilities with analytical and photo-interpretation capabilities.

The centre is participating in the CORINE Community-wide Co-ordination of Information on the Environment programme. ESA is co-operating in upgrading facilities for ERS and NOAA AVHRR data preprocessing and for co-ordination of Envisat data. FÖMI RSC are developing data integration techniques for optical and radar data. This includes ERS, Envisat and NOAA/VHRR.

UPDATED

India

Indian National Satellite System

Current Status

The Insat system uniquely provides geostationary platforms for simultaneous domestic communications and Earth observation functions. The four first-generation Insat 1 satellites were all US-built, but the advanced Insat 2 and Insat 3 series is indigenously produced. The Insat 2 series were launched between July 1992 and April 1999 while the first Insat 3 was launched in March 2000. The Insat system is a joint venture of India's Department of Space, Dept of Telecommunications, India Meteorological Dept, All India Radio and Doordarshan. Overall co-ordination and management rests with the inter-ministerial Insat Coordination Committee (ICC); the DOS has direct responsibility for establishing and operating the space segment. Insat 3B was launched in March 2000 followed by Insat 3C in January 2002, both launched by Ariane 5. They have design lifetimes of 14 years. These were followed by Insat 3A on 10 April 2003 and Insat 3E on 28 September 2003 both launched by Ariane 5 and will be followed in 2004–2005 by Insat 3D, an advanced meteorological satellite.

The satellites are handled from the Insat Master Control Facility (MCF) at Hassan, Karnataka through two Satellite Control Earth Stations (one with a 14 m fully steerable antenna and the other with a 7.5 m limited steering dish), one additional 14 m fully steerable antenna and an Insat 1 Satellite Control Centre (SCC) with associated TT&C equipment, on-orbit checkout equipment, computer facilities and auxiliary power services. MCF was upgraded with the addition of the Insat 2 SCC and two associated 11 m Satellite Control Earth Stations.

UPDATED

Indian Space Research Organisation (ISRO)

Current Status

ISRO has notched up a creditable series of flights and anticipates a robust period for development of both satellite and launch vehicle technologies. The INSAT 3A multipurpose satellite was launched by Ariane 5 on 10 April 2003 and will augment the present INSAT capacity for telecommunication and broadcasting as well as providing meteorological services along with INSAT 2E and KALPANA 1. The second development test of the GSLV launch vehicle took place on 8 May 2003 from the Satish Dhawan Space Centre, revalidating vehicle systems and carrying the 1.8 tonne GSAT-2 satellite into orbit. Success with the PSLV launcher saw the KALPANA-1 satellite orbited on 12 September 2002 by PSLV-C4 followed by Resourcesat-1 (IRS-P6) by PSLV-C5 on 17 October 2003. ISRO was established in 1969 as India's primary space R&D organisation, responsible for developing launcher and propulsion systems, launch sites, satellites and their tracking networks. This was followed in 1972 by the Space Commission and Department of Space. Personnel totals rose from 13,488 in 1986 to 16,800 in 1996 and stood at 16,400 in 2002, accommodated in eight main centres and units:

Vikram Sarabhai Space Centre (VSSC). ISRO's single largest facility (4,700 personnel), near Trivandrum, providing the technology base for launcher and propulsion development.

Liquid Propulsion Systems Centre (LPSC). Development wings in Bangalore and Trivandrum are supported by major test facilities at Mahendragiri for a wide spectrum of liquid motors, from reaction control system thrusters to the 720 kN Vikas and cryogenic engines. Personnel total 1,600.

ISRO Satellite Centre (ISAC). ISRO's lead centre for the design, fabrication and testing of science, technology and applications satellites. Staff strength is 2,400.

ISRO Inertial Systems Unit (IISU). Provides inertial systems/components for satellites and launchers.

SHAR Centre. ISRO's orbital launch site and largest solid motor production and test facility with 2,300 employees.

ISRO Telemetry, Tracking and Command Network (Istrac). Headquartered in Bangalore, Istrac operates a network of ground stations to provide TTC support for launcher and satellite operations. Personnel total is 460.

Space Applications Centre (SAC). Located at Ahmedabad, SAC is ISRO's applications R&D centre, including communications, remote sensing and geodesy. For example, it developed IRS 1C's cameras. 2,000 personnel.

Development and Educational Communications Unit (DECU) at Ahmedabad.

Insat Master Control Facility (MCF) at Hassan, 180 km from Bangalore. 295 personnel.

National Remote Sensing Agency Ground facilities for the reception, processing and dissemination of IRS, Landsat and ERS data, as well as the development of remote sensing techniques and applications.

Regional Remote Sensing Service Centre located at Bangalore, Naqpur, Kharagpur, Jodhpur and Dehradun.

Physical Research Laboratory covering research into astronomy and astrophysics, planetary atmospheres and aeronomy.

National Mesosphere/Stratosphere Troposphere Radar Facility at Gadankinear Tirupati.

UPDATED

National Remote Sensing Agency (NRSA)

Current Status

The NRSA was created in 1974 as a focus for remote sensing activities and since 1979 has provided India's primary Earth station, at Shadnagar 55 km south of Hyderabad, for reception of NOAA, Landsat, ERS and IRS imagery. It also operates the IIRS training centre. The NRSA Data Centre (NDC) and satellite earth station in Hyderabad disseminates photographic and digital products from IRS, Landsat, ERS and NOAA, in addition to maintaining archives of acquired data. There are 1,500–2,000 data users, concerned with vegetation, waste land, ground water and forest mapping in addition to agricultural applications. Ground truth correlation integrates data from satellites and from sensor platforms at sea, on land and in the air. The NRSA has developed a dual-track capability for analysis and interpretation of space data and for the development of new and improved tools for imaging and SAR data.

The NRSA was set up as an autonomous organisation under the Department of Space and embraces remote-sensing applications in water resources, agriculture, soil and land degradation, mineral exploration, ground water targeting, geomorphological mapping, coastal and ocean resources monitoring, environment, geology, ecology and forest mapping, land use and land cover mapping and urban area studies and large-scale mapping activity. The agency is a contributing element to the National Natural Resources Management System (NNRMS), which includes five Regional Remote Sensing Service Centers (RRSSC) at Bangalore, Dehra Dun, Nagpur, Jodhpur and Kharagpur. The first, at Bangalore, was inaugurated in August 1987. Each RRSSC provides facilities for digital analysis of satellite data, complemented by photo processing and visual interpretation laboratories. The system is configured around a VAX 11/780 with VIPS 32 image analysis software (provided by France's SEP) and three Numelec PC 2001 image displays. The system can handle digital classifications for the equivalent of 1½ LISS 2 (IRS) scenes each day, or three Landsat MSS, or three-quarters of a Landsat TM, or three Spot images. Cost/frame to users of processed IRS data is about Rs 2,300 for LISS 1 on 24 cm false colour composite paper print.

UPDATED

Physical Research Laboratory (PRL)

Current Status

India's principal centre for space science activities, covering ground-based telescopes and satellite/sounding rocket instrumentation as well as physics experiments carried aboard foreign satellites. Latterly an enhanced emphasis on international ventures with Europe, Russia and the USA. Astrophysical experiments and a continuing programme of co-operation with Russia on space science experiments. The PRL has a wide range of test and laboratory facilities and is linked with other national assets to provide a national resource which conducts co-operative programmes with foreign equivalents.

VERIFIED

Space Commission/ Department of Space

Current Status

The Dept of Space (DOS) implements the policies framed by the Space Commission through ISRO, the National Remote Sensing Agency, the Physical Research Laboratory and other agencies. As an enabling agency it co-ordinates activities between various government departments and provides advisory information from which the Space Commission frames its policy. Increasingly, throughout the 1990s, it has played a leading role in determining the balance between scientific and commercial space and the application of space capabilities and assets for defence and national security considerations. Now, in a new century, it has added responsibility for harnessing India's assets in space-related research and development and organising project themes for recommendations to the government regarding capabilities and possible objectives for government agencies and non-government research facilities. The DOS has also conducted studies on the advisability or otherwise of linking national objectives with other space programmes in other countries.

VERIFIED

Videsh Sanchar Nigam Ltd (VSNL)

Current Status
VSNL is India's national signatory to Intelsat and Inmarsat and conducts research into new applications including broadband and USAT services. It has been instrumental in exploring the synergy between global and regional systems. LEO and MEO systems have been examined to detemine optimum roles for mobile telephone systems. Internal research programmes have identified several unique areas where VSNL could benefit from stronger international co-operation to import technology trends and export consulting and engineering fabrication techniques. In 1986 it was incorporated to take control of India's Overseas Communication Services. In February 2002, the Indian government released 26 per cent of VSNL's equity to the control of TATAs which assumed responsibility for administrative control.

UPDATED

VSCC Liquid Propulsion Systems Centre (LPSC)

Current Status
VSCC is responsible for the development of launcher liquid and cryogenic propulsion stages and auxiliary propulsion for launch vehicles and spacecraft. Main research activities cover avionics, aeronautics, materials and mechanical engineering, solid propulsion and composites, propellants, polymers and chemicals. It also conducts research on systems reliability and computer integration. It is growing in importance as main liquid engines are introduced to India's orbital launchers and with the increasing size of indigenous satellites. Current main projects cover the 720 kN Vikas PSLV stage 2 engine, PSLV's stage 4 dual 7.5 kN system, Insat 2's 440 N Liquid Apogee Motor and unified network of 22 N ACS thrusters, and the 680 kN Vikas GSLV strap-on. Seven Russian 76 kN cryogenic engines are being provided for the initial GSLVs the first of which flew on 18 April 2001, but India's own engine will complete development in 2003–5. India signed a deal for the Russian engines in 1990 incorporating technical support on cryogenic development. Facilities development of cryogenic handling structures assisted by Russian aid.

VSSC's test facilities are sited southeast of Trivandrum at Mahendragiri. Vikas' Principal Test Stand (PTS) was commissioned during 1987 and used in January 1988 for the engine's first full-duration 150-second firing. Altitude facilities are also available for PSLV's 7.5 kN motor, Insat 2's LAM and smaller thrusters. Cryogenic engine and stage facilities were commissioned in 1997. The site for liquid propulsion and cryogenic stages is located at Thiravananthapuram while test facilities are located at Mahendragiri in Tamil Nadu. An ammonium perchlorate experimental plant is operated by VSSC at Aluva.

UPDATED

VSSC Solid Propulsion Group

Current Status
India's first 75 mm diameter solid motor was produced in 1967, followed by a 125 to 560 mm range for sounding rocket applications and, in the early 1970s, motors for the SLV-3 satellite launcher. Work on the most powerful so far began in 1984: the 2.8 m diameter 3,500 kN thrust solid powers PSLV's first stage. Although the Solid Propulsion Group is an element of VSSC, the primary Solid Propellant Space Booster Plant (SPROB) is sited at SHAR Centre on Sriharikota Island, along with the Vehicle Assembly Static Test and Evaluation Complex (VAST) for solids. At VSSC the motor cases are produced in the Mechanical Engineering Facility and the propellant binders in the centre's Propellant Fuel Complex. Hybrid propellant combinations studied for possible adoption at primary stage rocket motors for future launches. VSSC has begun development of hybrid systems for advanced high-energy propulsion requirements.

UPDATED

Indonesia

National Institute of Aeronautics and Space (LAPAN)

Current Status
LAPAN (Lembaga Penerbangan dan Antariksa Nasional), founded in 1963, is a non-departmental government agency and is directly responsible to the President. Its chief activities are remote sensing and probing of the upper atmosphere by balloon and rocket sondes. LAPAN pioneered methods of seeding rain clouds and used meteorological satellite data to co-ordinate rain-making activities to water rice fields. It also performs telecommunications research, notably on signal attenuation caused by heavy tropical rainfall. First and second stage solid propellant sounding rockets designed and built by LAPAN are launched up to 100 km from the range at Pameungpeuk some 150 km south of Bandung. The agency maintains Earth stations at Jakarta and Biak Island (Irian Jaya) for reception of GMS, NOAA and Landsat MSS/TM imagery. The Multimission Remote Sensing Satellite Ground Station was commissioned in October 1993 at Pare-Pare on Sulawesi island to receive Spot, ERS and Landsat 5 (Landsat 6 ETM was also to have been included); Japan's JERS was added December 1995 and the facility is being upgraded to take advantage of new commercial remote sensing systems. LAPAN has secured optical and radar imaging to integrate with sounding rocket data for national and regional evaluations.

UPDATED

PT Indosat

Current Status
Indosat announced 1 June 2001, that it had completed the acquisition of PT Birngraha Telekomindo for US$372 million, a holding company whose only asset is a 45 per cent share in Satelindo. Plans to expand international transponder sales and regional telecommunications development. At the end of 2002 the government of Indonesia undertook a 41.94 per cent divestment of its shares in Indosat to Singapore Technologies Telemedia Pte Ltd through the holding company Indonesia Communications Ltd. PT Indosat has extended its activities in mobile and cellular telephone systems and for 2003 declared a 48 per cent divided on net profits 19 per cent of revenue.

Background
PT Indonesian Satellite Corporation Tbk. was established in 1967 as a foreign investment company to provide international telecommunications services in Indonesia commencing its operations in 1969 with the inauguration of the Jatiluhur earth station. In 1980 the government of Indonesia acquired all the shares of Indosat which then became a State Owned Enterprise (SOE). The Indonesian Satellite Corp is the national signatory to Intelsat, which the country joined in 1964 as one of the first members, and to Inmarsat, which it joined in October 1986 as the 48th member, with a 0.385 per cent interest. The government in 1994 sold 35 per cent of its interest. Indosat is responsible for overseas communications links, accessing both the Indian and Pacific satellites through its Jatiluhur gateway. It secured a competitively-bid contract in 1983 to provide TT&C services to Intelsat, previously handled by Australia's OTC from its Carnarvon station. Apart from satellite links, Indosat is responsible for national participation in international submarine cables in the region. Talks to merge PT Indosat with PT Telkommunikasi Indonesia began in late 1999.

UPDATED

International

ArabSat

Current Status
The Arab Satellite Communications Organization was established in 1976 to meet the increasing communications needs of its members. There are 21 members (with percentage shares): Algeria (1.72), Bahrain (2.45), Egypt (1.59), Iraq (1.90), Djibouti (0.12), Jordan (4.05), Kuwait (14.59), Lebanon (3.83), Libya (11.28), Mauritania (0.27), Morocco (0.61), Oman (1.23), Palestine (0.25), Qatar (9.81), Saudi Arabia (36.66), Somalia (0.24), Sudan (0.27), Syria (2.08), Tunisia (0.74), United Arab Emirates (4.66), Yemen (1.65). The first two satellites were launched in 1985 (both removed from service in 1992), and the third/last first generation in 1992, extending operations until the second generation (ArabSat 2A) could appear in July 1996, launched by Ariane 44L followed by ArabSat 2B in November 1996 and ArabSat 3A in February 1999 which was co-located with ArabSat 2A at 26° E. Planning strategic expansion using existing on-orbit infrastructure through optimised transponder usage. Paid capital is worth about US$163 million.

The original operational plan was to carry all traffic on one satellite and to hold the second as an in-orbit spare, but increasing demand required distribution between both. At the beginning of 1992, C-band transponder usage was: nine providing more than 3,500 regional telephone channels, two regional TV, 3¼ Saudi Arabia telecommunications, one Morocco telecommunications, one Oman telecommunications, one Mauritania telecommunications, one ASBU Arab States Broadcasting Union TV, one CNN (US) and 12 India telecommunications. India leased its 12 on 1B for two years beginning 1 October 1989, extending the lease (then on 1C) in 1991 until September 1992. Saudi Arabia leases one S-band.

UPDATED

Argos

Current Status
Argos is a joint programme of CNES, NASA and NOAA, operated under an MoU signed in 1974 and extended in 1986. It provides global data telemetry and geo-positioning services. Presently a candidate for expanded global environmental data monitoring activity with potential for integrating advanced position fixing. It was first developed under a Memorandum of Understanding between Centre National d'Etudes Spatiales (CNES), NASA in the USA and NOAA also in the USA. System operation and promotion is the responsibility of CLS, established in April 1986 as a subsidiary of CNES (55 per cent), IFREMER (French Marine Institute; 15 per cent) and a pool of banks. CNES operates the French Global Processing Centre (FRGPS) in Toulouse and handles all user relations outside North America. US subsidiary Service Argos Inc is responsible for all US/Canadian users; it operates the identical USGPC, which provides redundancy. North American CLS is concerned with developing new services and products. The Australasian regional processing centre began operations in December 1988.

The system consists of independent user Platform Transmitter Terminals (PTTs, down to 25 g and the size of a matchbox), two NOAA polar satellites with Argos Data Collection and Location System packages to receive PTT messages on a random access basis for separation, time coding, formatting and retransmission to ground stations, and the ground stations and two Global Processing Centers in Toulouse and Landover, where data are retrieved, processed and distributed to users. Location of PTTs can be determined to within typically 150 m, if required, by Doppler techniques.

The PTTs transmit data from up to 32 sensors encoded in 8-bit words on 401.650 MHz with typically 200 mW power at intervals of 60 to 120 seconds for location PTTs and >200 s for data collection-only platforms. Message length is 360 to 920 ms and contains the PTT identification number.

The principal Argos system processes and disseminates space and terrestrial environmental data received by dedicated packages aboard US NOAA polar meteorological satellite from fixed/mobile platforms anywhere in the world. More than 5,000 transmitters are operating; the past five years has seen a 15 per cent growth in activity. More than 23,000 transmitters have used the system.

The NOAA Argos packages receive all messages within a 5,000 km diameter visibility circle at any instant; four can be processed simultaneously by NOAA A-J and eight will be handled by the third generation system aboard NOAA K and its successors from 1996. NOAA 11 can handle data at 960 bit/s, NOAA J 1,200 bit/s and NOAA K-N 2,560 bit/s. NOAA K-N increase bandwidth from 24 kHz to 80 kHz. The data are formatted/stored, then dumped each time the satellite moves within reach of one of the three ground stations. VHF/S-band transmitters also perform real-time relay for any user station within the visibility circle. Current onboard packages can process up to 1,400 data-only PTTs, or 415 requiring the location service, but from NOAA K on the capacity will be quadrupled.

Under an agreement signed in April 1990 with EUMETSAT, CLS manages all non-meteorological applications using the Meteosat data collection function. The MAEDS Multisatellite Applications Extended Dissemination Service adds Meteosat and GOES to the existing Argos system to provide high data collection capacity (649 byte messages up to 24 times daily), real-time access to data, and continuous coverage from the Pacific to the Urals. The first MAEDS element became operational in October 1990, followed in 1991 by the GOES satellites covering America and the Pacific. CLS also operates the Doris location system and is involved in the Starsys system.

Global Processing Centers NOAA's tape recorders are read out every 100 minutes over one of the Gilmore Creek (Alaska), Wallops Island (Virginia) and Lannion (France) stations. All of the satellite's data are routed to NOAA's National Environment Satellite, Data and Information Service (NESDIS) in Maryland, where the Argos component is extracted and relayed to CLS in Toulouse and Landover. Here, PTT locations are determined and the sensor data processed. About two-thirds of the processed messages are provided to users within 3 hours of the uplink. Location is derived from measuring the Doppler shift in the carrier frequency of received messages during the 10 minute overflight. An accuracy of 150 m is typical but for PTTs at fixed positions or on slow-moving carriers 100 m is possible. The transmit frequency must remain stable throughout the pass. Each frequency measurement yields a set of possible positions and with a known elevation only four messages are required on each pass to generate a unique solution. The satellites' own positions are known to within 300 m at any instant from a reference system of 11 beacons worldwide at precisely known geodesic positions.

VERIFIED

ASETA

Current Status

ASETA (Asociacion de Empresas Estatales de Telecomunicaciones del Acuerdo Subregional Andino) comprises the five telecom enterprises of the Andean Group: ENTEL-SAM (Bolivia), TELECOM (Colombia), EMETEL SA (Ecuador), Telefonica del Peru SA (Peru) and CANTV (Venezuela). The Group represents a population of 120 million over 4.8 million km^2. ASETA was established in 1974 to develop telecom services facilitating regional integration of the Andean countries. It is now developing the Regional Telecommunications Master Plan in co-operation with the ITU, principally focusing on establishing and exploiting the SAT Andean Telecommunication System. SAT's main aim is building the Corredor Andino Digital, the Andean information and telecom highway comprising terrestrial, satellite and submarine route through the five countries. The fibre optic terrestrial route is under construction by each of the five enterprises in their own territories. The submarine route will be developed through the Panamerican Pacific Fiber Optic System Agreement, in which the five enterprises will invest about 40 per cent of the total cost.

Until then, Intelsat in 1991 offered eight 36 MHz leases until 2010, in addition to the 11 36 MHz units already in use. Each country contracts directly with Intelsat; the initial leases began operating in May 1995 on Intelsat 705 at 50° W.

ASETA earlier planned a dedicated satellite system. Respective ministers signed the agreement creating the OATS Andean Satellite Telecommunications Organization serving the interests of Bolivia, Colombia, Ecuador, Peru and Venezuela in November 1988. Peru, Colombia and Venezuela were each to provide 28 per cent of the US$900 million Condor program (US$220 million space segment), with Ecuador and Bolivia contributing 7.7 per cent. Four GEO slots at 106°, 72°, 77.5° and 89° W were allocated by the ITU. A two-satellite system was projected for 1993–94. All five, with a total population of 120 million covering 4.74 million km^2, operate Intelsat A stations and had been considering a regional satellite system since 1974. Chile withdrew from ASETA in 1977. Three participants had to ratify the OATS agreement for it to be activated, but economic constraints instead dictated service leasing. The launch slots have since been cancelled and ASETA restricted its operations to utilisation of existing capacity. This has been modified to keep open the opportunity for redeveloping plans for launch constellations.

VERIFIED

Committee on Earth Observation Satellites (CEOS)

Current Status

CEOS was created at the suggestion of the G7 Economic Summit in 1984 and now encompasses all of the world's civil space agencies responsible for Earth observation space programmes, along with agencies that receive and process data acquired remotely from space. Integrates international programmes for national agencies and provides new incentives through regional development projects. It provides a forum to exchange information on programmes, plans and requirements. Activities include: completion of an initial assessment of overlaps and gaps in current/planned space missions to meet observing requirements of the major international operational and research programmes in environment, climate and global change; development of remote sensing training materials for use in developing countries; consideration of what role CEOS should play in defining/implementing a long term global observing system for the next decade and beyond. As a service of the CEOS, the International Directory Network is an effort to assist researchers in locating information on available data sets and provides free on-line access to information on earth sciences, space physics, solar physics, planetary science and astronomy and astrophysics. The plenary session meets once a year and brings together members and associates to discuss matters of global interest. CEOS work in developing countries to encourage the maximum use of indigenous capabilities and Working Groups exist to co-ordinate activities with associates and aid development organisations.

UPDATED

COSPAR

Current Status

The Committee on Space Research was established by the International Council of Scientific Unions (ICSU), now known as the International Council for Science, in October 1958 to continue the co-operative programmes of rocket and satellite research undertaken during the International Geophysical Year of 1957–58. Membership comprises National Academies of Science or equivalent, and International Scientific Unions adhering to ICSU. Associate Membership is open to any bonafide scientist conducting space research. Operating under ICSU rules, COSPAR is a non-political organisation for promoting international co-operation through meetings and publication of their proceedings. Publications include: *Advances in Space Research* (the Committee's flagship journal in which the proceedings of the Biennial COSPAR Scientific Assemblies are published – two volumes, 24 issues per year), the trimester *COSPAR Information Bulletin*, the irregular *COSPAR Colloquia Series*, and the *Directory of Organizations and Associates*, every two years. Co-ordinating centrally funded programmes for member states. Provides funding integration for national ventures. COSPAR organises workshops and conferences and serves as a sponsor for congress meetings at various locations around the world.

VERIFIED

Cospas-Sarsat

Current Status

Cospos-Sarsat plans to phase out emergency beacons operating in the 121.5/243 mHz range in favour of the 406 mHz beacons which provide more accurate and reliable alert data. Currently more than 225,000 406 mHz distress beacons are in use. After 2009, users will have only this frequency available. However, expanded applications are promised to ensure technology available for potential interface with Galileo satellite navigation system for accurate position fixing information modulated on to the carrier signal.

Background

As of December 1999 there were eight satellites operational including Cospas 4, 6, 8, 9, Sarsat 3, 4, 6 and 9. Cospas-Sarsat is an international satellite-based system for search/rescue, established in 1979 by Canada, France, US and USSR (membership now assumed by Russia), based on the detection of distress beacons by four polar satellites. It is capable of locating an activated beacon to within 2 km and was credited with helping to save more than 11,500 lives by October 2000. The system was governed by an MoU of October 1984 until the International Cospas-Sarsat Program Agreement was signed by the four states on 1 July 1988 in Paris.

Distress signals are relayed by the satellites to Local User Terminals for processing to determine beacon location. The information is passed to a Mission Control Centre to alert the rescue authorities. Some 600,000 121.5 MHz beacons are operational, but these are restricted to real-time relay by the satellite and therefore require an LUT to be within range (about 2,500 km) for position determination to within 10 to 15 km. The 406.025 MHz units include user identification codes in the message but they still operate on the Doppler shift concept: this provides two locations for each distress signal, an ambiguity resolved by allowing for the Earth's rotation.

If the beacon's frequency stability is sufficient, as with the 406 MHz devices, the true solution is determined on a single satellite pass. The 406 MHz information is not only relayed in real-time but also time-tagged and stored for dumping as each LUT comes into view. This frequency therefore provides a global service with an average waiting time of 44 min, with 94 per cent detected in more than 90 minutes (results from 1990 exercise). All satellite downlinks operate at 1.5445 GHz. By end-1995, more than 119,000 406 MHz beacons had been produced and distributed in more than 130 countries.

The 406 MHz beacons emit a 5 W RF burst of about 500 m/s duration every 50 seconds. The improved frequency stability 10^{-9}/min (a factor of two relaxation was permitted until 1991) assures more precise location accuracy (90 per cent within 5 km), while the high peak power increases the probability of detection on a single pass to more than 98 per cent. Typical delay before detection is 1 to 2 hours near the equator (less at the poles).

Each satellite can handle 90 simultaneous signals. The digital encoded message conveys the country of origin and identification of the vessel or aircraft. Position information from systems such as GPS and Glonass will be encoded in the messages of new 406 MHz beacons from January 1997. This will be particularly useful for the emerging geostationary satellite relays, which at the moment cannot provide the position element. Depending on the beacon type (maritime, airborne or land), it can be activated manually or automatically by shock or immersion. An optional homing device is

Cospas-Sarsat satellite launches

	Launch	Orbit	Payload
C1383[1]	29 June 1982	989 × 1,028 km, 83°	Cospas 1
C1447[2]	24 March 1983	959 × 1,013 km, 83°	Cospas 2
NOAA 8[3]	28 March 1983	803 × 825 km, 98.7°	Sarsat 1
C1574[4]	21 June 1984	965 × 1,005 km, 83°	Cospas 3
NOAA 9[5]	12 December 1984	841 × 862 km, 99°	Sarsat 2
NOAA 10[6]	17 September 1986	808 × 826 km, 98.7°	Sarsat 3
NOAA 11	24 September 1988	849 × 865 km, 98.9°	Sarsat 4
Nadezhda 1[7]	4 July 1989	960 × 1,014 km, 83.0°	Cospas 4
Nadezhda 2[8]	27 February 1990	956 × 1,021 km, 83.0°	Cospas 5
Nadezhda 3[9]	12 March 1991	958 × 1,018 km, 82.9°	Cospas 6
NOAA 13[10]	9 August 1993	850 × 863 km, 98.9°	Sarsat 5
Nadezhda 4[11]	14 July 1994	954 × 1,005 km, 82.9°	Cospas 7
NOAA 14[12]	30 December 1994	848 × 863 km, 98.9°	Sarsat 6
NOAA 15	13 May 1998	808 × 823 km, 98.7°	Sarsat 7
NOAA 16	September 2000		Sarsat 8

GOES 7 (May 1986), 8 (April 1994) and (9 May 1995), Insat 2A (July 1992) and 2B (July 1993), Cosmos 2054 Luch (December 1989), Luch 1 (December 1994) and GMS 5 (March 1995) carried 406 MHz relays into GEO and remain in experimental use.

Notes:
[1] decommissioned March 1988
[2] decommissioned December 1989
[3] decommissioned December 1985
[4] decommissioned June 1990
[5] 406 MHz local/global processor mode operational, global mode reactivated October 1991
[6] 406 MHz global/local modes not operating since September 1988
[7] limited availability in Southern Hemisphere
[8] decommissioned February 1996
[9] started S hemisphere operations December 1991
[10] satellite contact lost 21 August 1993 while SAR testing under way
[11] acting as backup to Nadezhda 3 in same orbit
[12] global mode failed soon after launch.

usually collocated with the 406 MHz beacon for the SAR services but it is not a system specification. Beacons are required to function within a –20/55°C thermal range for a minimum 24 hours and automatically limit any inadvertent continuous transmission to more than 45 seconds.

Mission Control Centres have been established in each country operating at least one LUT to disseminate information to the appropriate Rescue Co-ordination Centres. By end-1995 there were 33 LUTs in 21 countries, with others planned.

The USSR began deploying the space segment with the launch of Cosmos 1383 in 1982. Designated Cospas 1 ('Space System for Search of Vessels in Distress'), the 121.5 MHz band remained operational until Mar 1988, with 406 MHz utilized primarily for interference monitoring. Cosmos 1447 (1983), Cosmos 1574 (1984), Nadezhda 1 (1989), Nadezhda 2 (1990), Nadezhda 3 (1991) and Nadezhda 4 (1994, in the same orbit as Cospas 5 as a spare) adopted the roles of Cospas 2–7, with one more satellite ready for launch as required. The other relays to bring the system up to its full operational complement of four are provided by packages on the US NOAA meteorological satellites. Sarsat 1 (S&R Satellite-aided Tracking) was implemented with NOAA 8 in 1983. NOAA 9-11 brought the system to full strength, with further NOAAs planned for launch as required for meteorological operations. NOAA 14 (Sarsat 6) appeared December 1994 (#12 does not carry the Sarsat package; #13 failed 1993); NOAA 15 launched in 1998.

An operational geostationary system is under consideration following successful tests of a relay aboard NOAA's GOES 7 meteorological satellite, launched May 1986 to 75° W GEO. The NASA/CNES/DOC demonstration, working through the Washington, Lannion and Ottawa stations, showed that most beacons would be detected within 10 minutes. GEO satellites, however, cannot pinpoint a source but must rely on position information encoded in distress signals. GOES 7 detected its first real mayday 15 June 1988, 75 minutes before the LEO satellites responded. 406 MHz GEO relays are included aboard NOAA's three GOES- Next (first, GOES 8, in April 1994; GOES 9 May 1995) and India's Insat 2 satellites (2A 1992; 2B 1993; 2E planned 1997). Russia's Luch 1 (December 1994) carries a 406 MHz repeater (1989's Cosmos 2054 Luch carried an experimental relay). Japan's GMS-5 began providing an experimental relay in 1995. A payload was flown on the Meteosat Second Generation (MSG) geosynchronous-orbit meteorological satellite operated by EUMETSAT from 2003. A 406 MHz GEOSAR demonstration and evaluation is under way. Inmarsat also introduced the Inmarsat E distress service in 1992, using beacons operating at 1.6 GHz L-band through its GEO satellites.

UPDATED

Eucosat

Current Status

Eucosat (European Control by Satellite) is a non-profit organisation promoting joint civil/military use of satellites for treaty compliance, crisis management and environmental monitoring. It is primarily functioning as an intermediary body, linking politicians, industrial leaders, government officials, military decision-makers, scientists and engineers. Created in 1990, its corporate members include DASA, Aerospatiale, CNES, Matra Marconi Space, Alcatel Espace, Alenia Spazio, OHB-System, Fokker Space and Systems, and National Remote Sensing Centre Ltd (UK). It is financed by corporate contributions, member fees and independent research studies. Increasing emphasis on integrating civil and military satellites for arms reduction and UN compliance monitoring. It has conducted analysis of existing commercial and civil government space projects to ascertain their suitability for combined civil-military applications and has made a study of these areas for ESA.

UPDATED

European Centre for Space Law (ECSL)

Current Status

ECSL was established in 1989 at the initiative and under the auspices of ESA. About 450 people are involved. Its main objective is to develop/improve the knowledge in Europe of the law of space activities. It also promotes European activities beyond Europe and building a unique position for Europe in space law practice, teaching and publications. ECSL has set up, with ESA's support, the ESALEX space law database, containing basic texts of space law, ESA basic texts, statutes of other international organisations in full text, as well as bibliographical files. ECLS also initiates research: the first concerned the legal protection of remote sensing data; the second concerns intellectual property rights in outer space. A one-day forum is held annually plus the two week ECSL Summer Course on Space Law and Policy. It also organises, with IISL, the Space Law Moot Court Competition. The *ECSL News* is issued quarterly. Now promotes international law integration with regard to consumer access to privately funded remote sensing satellite projects and has diversified its business attention into new areas of international business and risk assessment through legal constraints. The ECSL is also responsible for reporting potential loopholes in space law to the United Nations and is regularly consulted by operators and satellite owners.

UPDATED

European Commission

Current Status

The Commission is the executive arm of the European Union. It is Europe's largest purchaser of satellite imagery as a result of the Union's agricultural and environmental interests. Its 1995–98 budget for Earth observation-related activities is ECU270 million; it spends some ECU15 million annually on satellite imagery. It is providing half of the development funding for Spot 4's Vegetation instrument. The commission has expressed interest in combining space policy planning at the European Space Agency with its own technology development programme and has requested ESA to study policy changes that would be necessary should the Commission invoke powers over the ESA on decisions regarding finance and programmes selection. Has issued asset of convergence strategies for integrating ESA policy frameworks into the EU planning initiatives. The EU is increasingly interested in space based capabilities for European operating integration in transport, security, tracking, navigation and communications. There are indications that it seeks to embrace the activities of the European Space Agency and utilise its resources to EU requirements involving national government operating plans.

UPDATED

Eurospace

Current Status

Founded in 1961, Eurospace is the Organisation of the European Space Industry with a mandate to perform a forum for professional consultation and to serve as a promotional tool for the industry. Its members comprise the 46 major industrial space companies from 13 countries in Western Europe representing 90 per cent of the total turnover of the European space industry. The association is responsible for the promotion of European Space activity. It represents European interests on international markets and promotes government and non-government projects where they benefit member organisations. It is the spokesman of industry to the European Space Agency (ESA), in particular with regard to future space programmes and industrial policy matters. Its relationship with the European Union is continuously developing, in particular with the General Directorates directly or indirectly involved in space applications and support to industry. Eurospace is taking an increasingly important part in representing the interests of respective companies in member states. Eurospace maintains a global watch on world wide satellite and space-vehicle data providing European industry sources with a compendium of intelligence information about competitors.

VERIFIED

EUMETSAT

Current Status

The first Meteosat Second Generation (MSG) satellite was successfully launched on 28 August 2002, by Ariane 5 from Kourou, French Guiana. MSG satellites, of which there will be four in all, are manufactured by Alcatel Space Industries. The four MSG satellites are expected to remain operational until at least 2015 when the Meteosat Third Generation (MTG) satellites will be introduced. MTG preparation started in late 2000 in co-operation with ESA following the decision of the EUMETSAT Council to proceed with a Post-MSG User Consultation Process aimed at capturing foreseeable needs of the user community in the period 2015–2025. Developed in co-operation

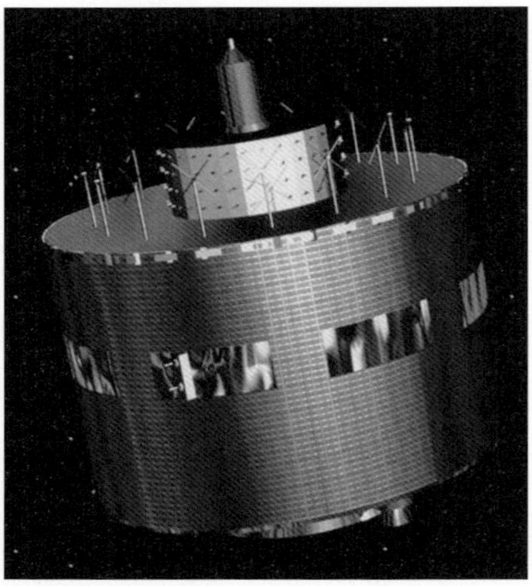

MSG Meteosat satellite of the type used by Eumetsat (Eumetsat) ***NEW***/1047513

with ESA, Metop 1 will be a polar orbiting satellite providing 'morning' coverage and is scheduled for launch in 2005. It is to be followed by Metop-2 in 2006 and Metop-3 in 2007, the three satellites observing meteorological, climatological and environmental features of the world transmitting regional data to user stations throughout the world with stored global data sent down to central facilities in Europe for further processing and distribution.

Background

EUMETSAT is an intergovernmental organisation created through an international convention agreed by 18 member states: Austria, Belgium, Denmark, Finland, France, Germany, Greece, Ireland, Italy, Luxembourg, the Netherlands, Norway, Portugal, Spain, Sweden, Switzerland, Turkey and the UK. It also has eight co-operating states: Croatia, the Czech Republic, Hungary, Poland, Romania, Republic of Serbia and Montenegro, Slovak Republic and Slovenia. The EUMETSAT Convention came into force 19 June 1986, with the organisation assuming overall and financial control of the MOP Meteosat Operational Program on 12 January 1987. Its headquarters were, and remain, at Darmstadt, Germany. A Council providing one seat per Member State, with four subsidiary bodies controls EUMETSAT: Administrative and Finance Group; Scientific and Technical Group; Policy Advisory Committee; Working Group on Charging Policy. The Council represents the National Meteorological Services of 17 European states. Contributions to programmes beginning with MTP are based on GNP (percentage) and for 2002 were the following: Austria 2.25, Belgium 2.79, Denmark 1.87, Finland 1.39, France 15.81, Germany 22.73, Greece 1.38, Ireland 0.90, Italy 12.78, Luxembourg 0.21, Netherlands 4.33, Norway 2.72, Portugal 1.22, Spain 6.47, Sweden 2.63, Switzerland 3.08, Turkey 2.22 and UK 16.23. Cooperating states include Hungary (0.47 per cent), Croatia (0.22 per cent), Poland (1.70 per cent) and the Slovak Republic (0.22 per cent). Total planned expenditure for 2003 is €279.45 million.

The majority of staff nationalities are German (27 per cent), French (20 per cent), British (16 per cent) and Italian (10 per cent). Overall, Germany contributes 25.25 per cent of the total Eumetsat budget followed by France (16.58 per cent), the UK (13.51 per cent), Italy (12.66 per cent), Ireland (6.15 per cent) and the Netherlands (4.34 per cent). Others contribute to the balance.

The Meteosat Operational Program, MOP, began 23 November 1983 under ESA's auspices and ended 30 November 1995. Overall responsibility was transferred to EUMETSAT in January 1987 and in-orbit control delegated to ESA. Total cost for the three-satellite series was estimated in 1982 terms at ECU378 million (revised to ECU721 million, or US$901 million, in 1995 terms), funded by Member States (percentage): Belgium 4.4, Denmark 0.58, Finland 0.35, France 25.60, Germany 26.39, Greece 0.30, Ireland 0.11, Italy 12.00, Netherlands 3.00, Norway 0.50, Portugal 0.30, Spain 5.24, Sweden 0.93, Switzerland 3.03, Turkey 0.50 and UK 16.76 (0.01 per cent not covered).

Aerospatiale was awarded an ECU139.1 million contract in May 1984 for three flight models and one spare. MOP's three primary missions were: Earth imaging, dissemination of image and other meteorological data, and data collection/distribution; with two secondary objectives of meteorological processing and data archiving/retrieval. Real-time image data (1,686.833 MHz) and up to 66 channels of DCP data (1,675.181 to 1,675.381 MHz) are transmitted by Meteosat at 333 kbit/s to EUMETSAT's Primary Ground Station in Fucino for relay to the Mission Control Centre in Darmstadt.

The images are processed into a range of formats (for example, Europe-only or 24 frames covering most of the full disc), with lat/long grids and coastlines added for transmission up to Meteosat (channel 1 2,101.5 MHz; channel 2 2,105.0 MHz) and relay to users on two channels). Channel 1 (1,691.0 MHz) operates on conventional analogue WEFAX, compatible with other GEO meteorological satellites and NOAA's Automatic Picture Transmission service, and available to simpler Secondary Data User Stations. Channel 2 (1,694.5 MHz) provides high resolution digital transmissions to Primary Data User Stations. Some of NOAA's GOES images are also available via Meteosat using the Lannion, northwest France relay station. Three other channels (1,695.725, 1,695.756, 1,695.787 MHz) are primarily concerned with the relaying of digitised facsimile and selected conventional met observations to Africa, as the Meteorological Data Dissemination (MDD) service. In 1996 DCP messages received by Meteosat totalled 3,311,644, of which 1.1 million bulletins were distributed to meteorological services around the world.

The MARF Meteorological Archive and Retrieval Facility is the single repository for all Meteosat image data and derived products acquired since 1978. Live data are continuously archived and historic data are being copied. All data are recorded on 6.6 Gbyte optical disks. Online access the MARF's catalogue is available. During MOP, ESOC produced more than 1.1 million images and gathered some 40,000 tapes' worth of data, some dating back to Meteosat 1. This valuable archive is being transferred to MARF.

A new ground segment was developed by EUMETSAT to take full control of all satellites from 1 December 1995. It includes the MCC Mission Control Centre in Darmstadt and Primary/Secondary Ground Stations, replacing the system previously operated from ESOC. The PGS Primary Ground Station is at Fucino, Italy and includes facilities for operational support of two satellites. A high speed link connecting PGS/MCC transmits data, telemetry and satellite commands. The back-up station is at Weilheim, Germany.

The MCC is located at the new ECU32 million EUMETSAT HQ totalling 5,000 m^2 in Darmstadt, opened in June 1995. The centre comprises four main facilities, each being developed by industry under direct contracts from EUMETSAT. The Core Facility is the MCC's central element and includes mission management, satellite control, ground segment control and pre-processing of satellite data. It also manages the dissemination of data and meteorological products to the user community. The MPEF Meteorological Products Extraction Facility completes full data processing and generates a range of meteorological products for the end-users. The MARF Meteorological Archive and Retrieval Facility allows users to access historical data from the satellites. The USDF User Station Display Facility provides real-time visibility of the end products to operations staff within the MCC.

A technical means of controlling access to Meteosat HRI High Resolution Image data was implemented 4 September 1995. This encrypts the data normally received by PDUSs; analogue (WEFAX) transmissions are not encrypted. Over the past decade there has been a rapid increase in the value-added commercial activity by the private sector, based on meteorological forecasts and data provided by meteorological services, which have also been faced with increasing pressure to recover their costs by charging for services. Additionally, they bear the cost of establishing and maintaining the space and ground observing systems. Exchange mechanisms are being established for meteorological satellite data, taking into account their ownership, value and the benefit generated for the user. As a first step, all users of PDUS were requested to register their use. Regular test transmission of encrypted data began in 1994. Decryption requires an MKU Meteosat Key Unit from EUMETSAT and some adaptation to the PDUS for MKU interfacing. However, there is a range of exceptions. For example, HRI images at 00/06/12/18 GMT remain unencrypted, and the 3-hourly encrypted data are free to all National Meteorological Services. In general, the NMS of countries with a GNP per capita greater than US$2,000 have free access to hourly and half-hourly data for internal use; wealthier countries pay according to their GNPs. Educational and science programs have free access to all data. At the beginning of 1996 there were more than 300 PDUSs and almost 2,000 SDUSs.

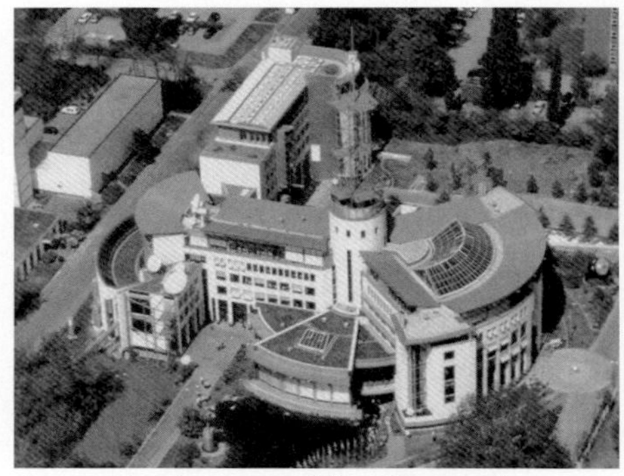

Eumetsat headquarters buildings (Eumetsat) ***NEW***/1047514

EUMETSAT budget (ECU million)	1994	1995	1996	1997
MOP	30.5	27.3	–	–
MTP	59.4	55.4	31.0	34.916
MSG	42.0	57.2	86.6	132.507
EPS polar	13.8	11.3	10.8	63.657
General Budget	15.0	15.6	22.4	12.738
Total	160.7	166.8	150.8	243.818

Previous years: 1993 – 131.916; 1992 – 91.720; 1991 – 64.847; 1990 – 50.841; 1989 – 42.8; 1988 – 59.25; 1987 – 62.763. Staff total 141 by end-1996.

An independent study published in 1993 concluded that five principal European industries (energy, agriculture and fisheries, construction and manufacturing, transport and services) benefited by ECU130 million annually as a result of Meteosat's products.

Meteosat Second Generation's mission is defined as:

1. Basic multispectral imagery. Improved spatial, temporal and spectral resolution, for nowcasting and shortrange forecasting.
2. High resolution imagery. AVHRR-type spatial resolution in visible band, for mesoscale convective monitoring over Europe.
3. Air mass analysis. Water/CO_2 absorption channels for nowcasting.
4. Meteorological product extraction. Improved wind and temperature data.
5. Support of climate and environment monitoring.
6. Continuity of data collection/dissemination.
7. Secondary mission for the relay of SAR distress signals and observation of the global Earth radiation budget.

UPDATED

Eutelsat

Current Status

On 2 July 2001, Eutelsat completed corporate restructuring to become a private company under French law. Eutelsat currently has a fleet of 20 satellites including five Hot Bird, five W series, three Eutelsat II, SESAT, two Eurobird, three Atlantic Bird and the new e-Bird launched 23 September 2003. Eutelsat reports interest in expanding national, regional, inter-regional and international services through a new generation of satellites. The new e-Bird satellite attempts to provide a valuable niche service through broadband services delivered by satellite to European customers. With 22 Ku-band transponders it will deliver two-way broadband internet services from 33° east, sufficiently close to Eurobird 1 to enable bundled digital TV and broadband services that can be received by small domestic antennas. Eutelsat also leases Telstar 12, Express A3 and Telecom 2D and anticipates the launch of W3A to 7° east in 2004, Express AM1 to 40° east in 2004, Hot Bird 7A to 13° east in 2005 and Hot Bird 8° to 13° east in 2006.

Background

The European Telecommunications Satellite Organization, with 245 staff, is an inter-governmental entity established to operate Europe's regional satellite system for 47 member states. 1996 profit ECU80 million; 1995's profit was ECU67 million (1994 51; 1993 33.6; 1992 30; 1991 21.7; 1990 14.'; 1989 13.9; 1988 3.3) on income from satellite operations of 1996 income ECU292 million (1995 275; 1994 260; 1993 244; 1992 231; 1991 183; 1990 122; 1989 100; 1988 70). 70 per cent of revenue is generated by audiovisual services (TV and radio channel delivery to cable headends and domestic satellite reception systems, SNG, outside broadcasts, European Broadcasting Union programme exchanges) and the rest by telephony, business and land mobile services. 1991 saw the commercial introduction of the EutelTracs two-way messaging and position-reporting land mobile service through the Alcatel-Qualcomm joint venture.

Eutelsat tariffs are designed to cover the satellite transponder alone and do not include charges for uplinks and other services on the ground. The ECU2.8 million annual tariff for a pre-emptible narrowband transponder is based on a three-year contract. For a non pre-emptible service over three years, the charge is ECU3.8 million annually, falling to ECU3.4 million for a five year contract for at least five transponders. The five-year tariff is fixed for the first three years and then revised based on the consumer price index for all European Union members (if the result is a tariff increase, the signatory has the right to terminate the agreement). Other options, for example, provide for one year leases and temporary leases.

VERIFIED

Inmarsat

Current Status

Astrium has been awarded a contract to build the Inmarsat 4 generation satellite series which will comprise three satellites. The satellites are primarily aimed at supporting the market for broadband GAN services at speeds up to 432 kbits/sec expanding ten-fold the capacity of the present generation. Inmarsat has decided to retire its highly successful Inmarsat A service on 31 December 2007 as it migrates its customers to the new digital services such as GAN and Fleet. Existing users will have the opportunity to make the transition to Inmarsat Fleet F77 which, in addition to GMDSS facilities provides customers with near global high-speed communications from a compact digital terminal. The decision to shut down Inmarsat A services follows a period of 12 years since the last type approval for a new maritime Inmarsat A terminal was requested (in 1991).

Background

The International Mobile Satellite Organization is a 79 national shareholder, commercial co-operative operating a satellite system to provide telephone, telex, data and fax services to the shipping, aviation, offshore, land mobile and remote industries. It also provides the secretariat for the Cospas/Sarsat search/rescue system. More than 70,000 Inmarsat terminals have been commissioned since global operations began 1 February 1982. In 1999, Inmarsat became the first government-to-government organisation to be privatised and expects to go to public offering in 2001.

Most current users operate Inmarsat A ship Earth stations (SES) or their transportable derivatives. The maritime terminals feature driven 0.85 to 1.2 m parabolic dishes housed in radomes. The Inmarsat B system is the updated digital equivalent of Inmarsat-A and is expected to dominate the market for this class of terminal in the future. The Inmarsat C range of terminals, typically about 30 cm high, 20 cm and, handle data at 600 bit/s. Three types of aircraft terminals are available: Aero-C for low rate store/forward; Aero-L for flight deck and operations datalinks; Aero-H high rate data and one to six voice channels. More than 700 aeronautical terminals had been commissioned by January 1996. Inmarsat 3's advent will allow Aero-I (Intermediate gain) to be introduced in 1997, offering 4.8 kbit/s telephony, 2.4 kbit/s fax/data within the satellites' spot beams (1,200 or 600 bit/s global). Four-channel terminals will weigh only 50 kg, costing US$50,000-100,000. The first voice test, from the ground, was made October 1995 using Inmarsat 2-F4 AOR-W, and prototype flight trials began February 1996. The Inmarsat E service was introduced in February 1992 for maritime distress beacons, although the first was not available until end-1992. Using GPS to generate position data, the beacon transmits a coded 1.6 MHz message for routeing to the appropriate rescue co-ordination centre.

Inmarsat B/M systems became available on a global basis in 1993. B offers services similar to A (which it will eventually replace) but its all-digital nature makes more efficient use of the space segment, offering fax rates up to 9.6 kbit/s and data up to 16 kbit/s. M lightweight (briefcase-sized) mobile terminals offer 4.2 kbit/s telephony and 2.4 kbit/s data/fax services. Mini-M terminals working through Inmarsat 3 satellites were introduced in 1996. A global personal messager, Inmarsat D, became operational in 1997.

Land Earth Stations (LESs; also known as Coast Earth Stations in maritime circles and as Ground Earth Stations for aeronautical applications) link the satellites and international telecommunications networks, and are usually owned/operated by their host Signatories. Stations communicate with the satellites in four ocean regions over C-band while mobile users operate in the L-band. Until 1990's Inmarsat 2 debut, the organisation operated with leased satellites, listed in the separate table. The second generation was followed by the five Inmarsat 3 series launched between 1996 and 1998, providing multiple spot beams. Inmarsat 3 was built by the then Lockheed Martin AstroSpace (now Lockheed Martin Missiles & Space) with the then European Matra Marconi Space (Astrium) developing and building the communication payload.

Inmarsat's constitution also covers the provision of navigation services, and could lead to a global civil system. Inmarsat 3's transponder is the first step, augmenting GPS with ground-derived integrity information, additional ranging signals and wide-area differential corrections. The Inmarsat 4 GEO series, expected to appear in 2005, will support the new Broadband Global Area Network (B-GAN) to deliver internet and inranet content and solutions, video on demand, videoconferencing, fax, e-mail, telephone and LAN access at speeds up to 432 kbps almost anywhere in the world. The Inmarsat 3 integrity monitoring network would allow a reduction in clock stability and data storage. The next step would add these payloads to intermediate orbit satellites (as was planned on the ICO satellites, but Inmarsat's Council in March 1996 rejected the lease proposal). 15 of these teamed with six to eight GEO satellites, although still intended primarily to augment GPS, could provide a 2D service should GPS be removed. A full civil system would require another 15 dedicated satellites in intermediate paths.

Inmarsat has strenuously marketed new business opportunities including leased transponders for military communications, open channels carrying un-encrypted or uncoded signals embracing voice, data and TV.

UPDATED

Intelsat Ltd

Current Status

In March 2004 Intelsat acquired four operational satellites from Loral giving immediate entry to the North American video and corporate data markets. On 15 June 2004 the second Intelsat 10 satellite was successfully launched by Proton from Baikonur. In 2002 Intelsat had total assets of US$3.965 billion and 850 staff members. Intelsat completed its transformation from a treaty-based organisation to a project held company on 18 July 2001. The new company, Intelsat Ltd, was created with over 200 shareholders comprised of companies from more than 145 countries. In 2000, Intelsat had revenues of US$1.1 billion and a net income of US$504 million. In 2001, revenues were again US$1.1 billion with profit of US$499 million. In 2002, Intelsat reported revenues of US$992 million and profit of only US$274 million. In June 2000, the first Intelsat 1X series satellite was launched in a campaign which put a total seven of this type in orbit by 15 February 2003, increasing the fleet to 34 satellites and increasing capacity by more than 40 per cent. Intelsat has yet to schedule its initial public offering of stock.

Background

Established on 20 August 1964, the original treaty-based company known as the International Telecommunications Satellite Organisation (Intelsat) was a not-for-profit commercial co-operative of more than 140 member nations.

Together, they own/operate a global system of communications satellites serving the entire world. Through its network of 20 satellites in the Atlantic, Pacific and Indian oceans and Asian Pacific regions, Intelsat provides international telecom services to more than 180 countries, territories and dependencies, and domestic telecom services to 40 nations. More than 900 Earth station antennas link users via 2,700 pathways. Intelsat carries more than half of all international telephone calls and almost all transoceanic TV. International financial networks, multinational corporations, international news services, and TV/radio broadcasters rely on the system for day-to-day global communications. Full-time Intelsat international traffic exceeds 120,000 derived channels of full-time service. In addition, the system carries more than 60,000 h of occasional use television annually, and 100 full-time leases, three of which are designated to support United Nations peacekeeping operations. Revenue was projected to grow to US$1.28 billion by 2002 but, as reported above, revenue was down by US$288 million on that expectation in response to overall weakness in telecommunications markets. 1997 revenue US$962 million; 1996 revenue US$911 million; 1995 revenue was US$805 million; 1994 US$706 million; 1993 US$658 million; 1992 US$622 million; 1991 US$563.4 million; 1990 US$498.6 million; 1989 US$614 million; 1988 US$614 million; 1987 US$519 million. Full-time channel use has grown from 4,258 in 1970, through 13,368 in 1975, to 133,000 in 1992 and 121,000 in 1993.

Ownership/investment is shared among the members according to their respective use of the system. Investment shares determine each signatory's percentage of the total contribution needed to finance capital expenditures. Revenues derive primarily from utilisation charges and, after deduction of operating costs, are redistributed to the signatories in proportion to their investment share as repayment of capital and compensation for use of capital. Non-member users pay the same rates because non-discriminatory access is a basic tenet of the organisation. All net revenues are reinvested into the organisation.

Intelsat operates under a four-tier structure. The Assembly of Parties comprises representatives of all of the governments who signed the Intelsat Agreement that entered into force 12 February 1973. This group considers resolutions and recommendations (essentially from the Board of Governors) affecting the long-term objectives. The Meeting of Signatories is composed of representatives of Signatories to the Operating Agreement who meet to consider issues related to the financial, technical and operational aspects of the system.

Both the Assembly of Parties and the Meeting of Signatories operate under a one nation/one vote rule, where an affirmative vote by at least two-thirds of the parties present is required on substantive issues and a majority vote is required on procedural matters. Representation on the Board of Governors, which functions as the managing directorate, is determined according to a standard minimum investment share established by the Meeting of Signatories. The Board of Governors, composed of representatives of Signatories whose investment shares either individually or as a group meet or exceed the minimum share requirement (currently 1.90598 per cent), is responsible for all major decisions regarding the establishment and operation of the Intelsat system. 27 members of the Board of Governors represent 107 of the Signatories.

Four countries joined Intelsat in 1996: Tajikistan (22 February), Bosnia and Herzegovina (6 March), Bulgaria (15 May) and Equatorial Guinea (11 December).

Earth station antennas accessing Intelsat's satellites are owned/operated by the host telecom organisations; 14 categories of Earth station standards are used. Co-ordination/control of services are the responsibility of the Satellite Control Center (SCC) and the Intelsat Operations Center (IOC) at the organisation's HQ. The IOC is the focal point for the communications networks and handles more than 100 TV transmission, antenna verification and other tests daily. The SCC provides TT&C services and pointing data for transmissions to Earth stations. Six TT&C stations totalling 54 antennas are operated around the world, collocated with Std A stations at Fucino (Italy), Paumalu (Hawaii), Perth (Australia), Raisting (Germany), Beijing (China) and Clarksburg (Maryland, US). A major upgrading of the system was completed October 1992. To cope with the Intelsat 7 generation and its successors, two antennas were added to Fucino and one each to Raisting and Beijing.

UPDATED

International Space University (ISU)

Current Status

ISU is an international, interdisciplinary educational institution based in France providing courses and qualifications for the space sector, principally through three academic programmes: a two-month Summer Session Programme and two 11-month Masters' Programmes (Master of Space Studies and Master of Space Management). The Summer Session Programme is hosted by an institution in a different country each year. The Master of Space Studies is conducted at the Strasbourg Central Campus and in industry, government, university and research institutions around the world for a 12-week professional placement period. In both, all major space-related disciplines, technical and non-technical, and their interactions are studied. Students from many countries and cultures work together to produce a Team Project. The MSS also includes an individual project. Fees for: the SSP FFr70,000 (all include food, accommodation, transport during the session); MSS: FFr130,000 tuition fees (not inclusive of living expenses). The business development unit is responsible for generating interest among companies in recruiting students. The ISU now works with affiliates in 14 countries. It links its own courses with those of affiliate institutions, universities and colleges.

UPDATED

International Telecommunications Union (ITU)

Current Status

The ITU, comprising 189 member states, is the international organisation responsible for the regulation, development and planning of telecom systems worldwide and for establishing equipment and systems operating standards. It was established as the ITU in 1934 from former bodies dating back to 1865. The Union consists of eight organs, including the Radio Regulations Board (formerly the International Frequency Registration Board until March 1993), which maintains a Master Register of more than 1 million radio frequency assignments. The ITU's main source of funding is from member states' contributions which account for 66.1 per cent of the total budget with a further 17 per cent from sale of publications, project execution and satellite notifications.

Frequency band allocations for satellite communication systems were established initially by 1963's Extraordinary Administrative Radio Conference. The regulations for space radiocommunications were revised in 1971 by the World Administrative Radio Conference for Space Telecommunications, allocating frequency bands for the first time to the broadcasting services. This was followed in 1977 by the World Broadcasting Satellite Administrative Radio Conference, in 1979 by the World Administrative Radio Conference, which completely revised the Radio Regulations, in 1983 by the Regional Administrative Radio Conference for the Planning of the Broadcasting Satellite Service in the Americas, in 1988 by the World Administrative Radio Conference on the Use of the Geostationary Satellite Orbit and the Planning of the Space Services Utilising It, and in 1992 by a WARC for dealing with frequency allocations in certain parts of the spectrum (WARC 92), which partially revised the table of frequency allocations to accommodate growing demands for new space services, satellite HDTV and satellite sound broadcasting. WRC 95 considered MSS frequencies. WRC 97 addressed BSS for Europe, Africa and Asia Pacific, science and Earth observation satellites, inter-satellite links and MSS. Subsequent WARC decisions provide a regulatory frame for expanded telecommunications services. The ITU has developed a new set of strategic plans for integrating the increasing diversity of telecommunications systems and to adapt member countries to the new and more open environment involving formerly politically closed states.

UPDATED

Intersputnik

Current Status

Intersputnik operates 3 Gorizont satellites at 50° E, 130° E and 142.5° E, 2 Express satellites at 14° W and 80° E and LM-1 at 75° E. Express 3A will operate from 11° W. Intersputnik membership has now grown to 24 countries with the latest LMI-1 satellite manufactured under the management of the Lockheed Martin Intersputnik joint venture. In July 2002 Intersputnik and Eutelsat signed a distribution agreement allowing Intersputnik to sell Eutelsat's satellite capacity. LMI-1 commenced service in November 1999 from 75° E over the Indian Ocean providing C-band and Ku-band services to Eastern Europe, Asia, Africa and Australia.

Background

Intersputnik was established in 1971 to offer satellite-based TV, radio, telephony and datalinks. Membership is Afghanistan, Belarus, Bulgaria, Cuba, Czech Republic, Georgia, Germany, Hungary, India, Kazakhstan, Kyrgyzstan, Laos, Mongolia, Poland, Republic of Korea, Republic of Nicaragua, Romania, Russia, Syria, Tajikistan, Turkmenistan, Ukraine, Vietnam and Yemen. A Board composed of one representative from each member governs the organisation; sessions are held at least annually. The Directorate is a permanent executive and administrative body, headed by a Director General. Financial activities are controlled by an Auditing Committee elected for a three-year term. On joining Intersputnik, a member makes a minimum 1 per cent contribution to the Statutory Fund; profits are distributed in proportion to these contributions.

As of 1 January 1996, Intersputnik provided its services through 26 36 MHz C-band transponders on Express 1 at 14° W and eight Gorizont satellites at 40° E, 70° E, 80° E, 85° E, 96.5° E, 103° E, 130° E and 142.5° E, plus two 27 MHz Ku-band transponders on Gals 1 at 70° E. Four transponders (#7/8/10 at 14° W + #8 at 80° E) provide voice/data using digital IDR technology. The others are used for TV, three of which (#7/9 14° W + #10 80° E) are for occasional use. The even-numbered transponders (except for #6) provide minimum 31.0 dBW EIRP hemispheric coverage, the odd numbers are global, at 28.0 dBW minimum. 14° W/80° E are the principal Intersputnik locations.

C-band Earth stations in the system utilise 12 m diameter main dishes of more than 31 dB/K G/T at 3.800 GHz 5° elevation. Transmitter output power is 1 to 3 kW. The stations operate at 3.700 to 4.150 GHz rx, 6.025-6.475 GHz transmit. Ku-band 10 m dishes provide 38.0 dB/K. The network comprises 42 standard Earth stations. TV can be carried in PAL, SECAM and NTSC formats. 6.5 to 7.5 m C-band and Ku-band stations also operate. 2.0 to 3.0 and 3.5 to 4.5 m are used for telephony, fax, telex, data, videoconferencing (3.5–4.5 m only) and TV/radio reception.

Bulgaria, Hungary, German Democratic Republic, Cuba, Mongolia, Poland, Romania, Czechoslovakia and USSR signed the Intersputnik agreement 15 November 1971. The first installations to enter service were those of the USSR and Cuba's Caribe ground station, used for TV transmissions via Molniya 1 of Leonid Brezhnev's 1973 Cuban visit. The system formally became operational in 1974; by 1980 there were eight stations and 11 member countries. By end-1984, the organisation was claiming that 60 per cent of all TV transmissions between member states was being relayed on its channels. Germany's DBP Telkom inherited the GDR membership in German unification. In 2004, Intersputnik has 24 member countries.

Intersputnik has approved the development of a series of small/medium satellites designated Intersputnik-100 m. Development of micro-dish antennas for special user needs and selected frequency allocations. Frames user needs into operational specifications.

UPDATED

RASCOM

Current Status

In March 2001, Cape Town Telkom said it would not be investing in the African satellite initiative, believing it to be too costly and incapable of meeting the needs of the business community. As of that date, although 44 countries had signed up to the RASCOM system only half the required investment for the planned August 2003 launch date. Plans are under review and schedules are being re-structured to reflect a possible shift towards high-power mobile telecommunications systems serviced through MEO. These plans have been shelved due to lack of confidence in business and revenue projections.

Background

RASCOM (Regional African Satellite Communications) held its first Assembly of Parties in September 1993 and the meeting of Board of Directors at its Abidjan HQ. A meeting of all the RASCOM countries in October 1993 was held at Intelsat HQ to discuss the pooling of transponders. The agreed upon pooling implementation began January 1994. It was decided in October 1994 to consolidate all services on Intelsat 804 at 21.5° W in 1996. RASCOM's ultimate objective is a dedicated regional satellite system.

The International Telecommunications Union began a feasibility study in March 1987 of a RASCOM system. The first global study of this type in Africa, it analysed the continent's telecom requirements and the terrestrial and satellite options to meet them in both urban and rural/isolated areas. The final report was delivered in late 1990 and considered by a meeting of ministers in Nigeria in February 1991. It was decided to proceed to a transitional stage during which the ITU would be the executing agency under the supervision of a Committee of experts from 15 African countries. The report on this transitional stage, including all elements for the setting up of the RASCOM operating organisation, was submitted to the second Conference of Telecommunication Ministers (25–27 May 1992, Abidjan. The RASCOM Convention was signed at Abidjan by 41 countries (minimum investment is US$50,000), envisaging the initial pooling of the 14 36 MHz Intelsat domestic transponders as a first step towards the ultimate objective of a dedicated African satellite. 35 African countries, UNDP, ITU, the government of Italy and the UNCTADA II Resource Mobilization Committee contributed/pledged towards financing the project activities.

One option proposed for the dedicated system calls for three satellites, including one spare. Cost would be US$1.3 billion: US$500 million for the satellites and US$800 million for the 50 countries' ground segment.

UPDATED

Iran

Telecommunications Company of Iran

Current Status

The Thurya Satellite Telecommunications Company based in Abu Dhabi has signed a service provider agreement with TCI. Iran has also opened a new government investigation into national satellite and rocket support capabilities with a view to determining future requirements.

Background

The Telecommunications Company of Iran (TCI) is the national signatory to Intelsat and Inmarsat. Plans for a domestic satellite system under the Shah's regime were subsequently revived, with four slots for Zohreh (Venus) satellites at 34° E, 26° E, 41° E and 47° E confirmed at 1988's ITU meeting (Iran now also holds slots at 59/61.5° E).

An RfP was expected in 1987 to be released by end-1988 for two 24-transponder C-band satellites for launch by Ariane. System cost was estimated in 1987 to be US$600 million. The Soviet Union was reported at one stage to be interested in providing both the satellites and associated launch services. The contract was expected to favour European companies as a US satellite would be unlikely to receive an export license from the State Department. Specifications were presented in March 1989 to invited companies in Rome, calling for two satellites (one a ground spare) featuring 20 Ku-band transponders each (six as back-up) with 54 MHz channel bandwidth. Inclusion of up to 1,500 1.8 m ground terminals was expected in the US$700–800 million project. Bids were due in by end-September 1989. Two responses were reported but rejected.

A further request was issued in early 1991. The Alcatel Espace/Aerospatiale team was announced as winner June 1992 of the US$350 million contract, required to deliver two 1,850 kg satellites with 10-year lives to 26/34° E for TCI in mid/end-1995 by Ariane. However, the contract still has not been signed, reportedly because of disagreement within Iran on whether such a system is necessary. 14 Ku-band 14.0 to 14.5/10.95 to 11.7 GHz up/down transponders would provide EIRP >50 dBW, allowing small stations for TVRO and rural telephone and business services. Four would be switchable to a steerable spot antenna. The L-band payload would provide 100 voice channels of Inmarsat-standard for mobile users. An X-band payload might be included for government/military links. Two TT&C stations would each work via two 7.5 m antennas; one station would include the 5.5 m In-Orbit Test antenna for monitoring and calibrating the payload. Reportedly talking to Alcatel about the purchase of an Iranian communications satellite named Zohreh 1 a launch date of 2002 was suggested but that proved both premature and optimistic. Parallel studies and use evaluation have re-structured planning towards a different system.

TCI's Inmarsat A station at Boumehen in southwest Iran, working through Inmarsat's Indian Ocean satellite, added Inmarsat C services in 1994.

UPDATED

Ireland

Team Aer Lingus

Current Status
Aer Lingus, wholly owned by the Irish government, provides propellant filters for Ariane 4's stage 1 tank, telemetry equipment containers for Ariane 4's stage 3 and supports for the Ariane 5 Vulcain main engine. Studied high-grade telemetry structures configuration for post Ariane concepts. Space personnel totals about 10, with an annual space revenue of about I£1.8 million.

UPDATED

Israel

Asher Space Research Institute

Current Status
The institute was established in 1984 to provide expertise in space sciences and engineering, including astrophysics, spacecraft propulsion, control, materials, structural design and remote sensing. It employs 28 full time engineers and 25 academic members drawn from the aerospace engineering and physics faculties. Its TechSat 1 satellite was launched on 28 March 1995 along with two other satellites by Start launcher. The Gurwin-II Techsat satellite was launched from Baikonur by Zenit 2 launcher on 10 July 1998. The Institute continues to carry out research on physics experiments in planning for a new generation of Israeli science satellites. These are envisaged as multipurpose platforms for diverse users and experimenters. Integrated space support systems packages would ensure maximum efficiency for minimum mass/volume allowed. Several potential missions have been defined from new nanotech designs for ultra-low mass properties. Research programmes include broadband laser inter-satellite technology, hyperspectral data acquisition and retrieval for micro-satellite remote sensing, integrated optical inertial space navigation systems, strap-down stellar star-track sensors, magnetic attitude control systems for small satellites and diagnostic techniques for plasma thrusters in flight.

UPDATED

Interdisciplinary Center for Technological Analysis and Forecasting (ICTAF)

Current Status
The Space and Remote Sensing Division at Tel-Aviv University was established in 1985; it has conducted technology surveys in several space-related fields, such as materials, solar cells, bondings and coating, but its principal emphasis is on promoting remote sensing applications within Israel. The Division has conducted pioneering work in multispectral sensor development for hybrid tasks and has applied these throughout the range of civil and military products. The Centre has a strong space technology and applications theme with a diverse range of skill areas in its personnel, facilities and areas of research leading to extensive involvement in pioneering activity related to satellite and space vehicle research and development. ICTAF is the national distributor for Space Imaging, Space Imaging Eurasia, Radarsat International, Eurimage and Russian imagery and products. It was appointed in 1991 as distributor for Israel Space Agency's receiving station. It is also involved in several applications projects: limnology, urban studies, geology and agriculture, soil and forestry mapping. Experience includes thermal mapping for industrial applications. ICTAF has reinvigorated its work on applications resulting from the new generation of high-resolution remote sensing satellites.

This division also carries out literature surveys on a broad range of space related technology subjects.

UPDATED

Israel Military Industries Ltd (IMI)

Current Status
IMI is a government-owned company comprising a HQ group and four operating groups of 10 divisions and 18 plants. There are about 4,250 personnel. The Rocket Systems Division (RSD) of the Systems Group develops/manufactures solid rocket motors, weapon systems and metal/composite material products. For example, it provides the propulsion systems for the Gabriel surface-to-surface and Shafrir air-to-air missiles. Conducts basic R&D in exotic propellant combinations and mixed liquid/solid combinations for hybrid motors. In the space field, it developed/produces the stage 1/2 motors for the Shavit/NEXT launchers. These are HTPB cylindrical grains in filament wound composite cases. Steering options are jet vane, LITVC and flexseal. The motors are marketed as the ASTM family in the US by Atlantic Research Corporation. RSD is deeply involved in both Israel's space programme and the SD10 Arrow. The propulsion systems and the pyrotechnics for these programmes are IMI developed and produced. IMI has researched technology requirements for NEXT launcher derivatives using HTPB grains. No development commitment has yet been made and plans for further development have been shelved.

UPDATED

Israel Space Agency (ISA)

Current Status
ISA was established in 1983 within the Ministry of Science and Technology to direct national space research and R&D activities. The Ofeq programme is conducted under its auspices. The ISA has broad responsibilities for both civil and military technologies but only on the same basis as that of NASA in the USA whereby common technologies and experiments are frequently hosted by the civilian agency. However, unlike US civil and military space organisations, the ISA conducts research and technology studies directly applicable to both interests. This has proved an effective use of limited resources. R and D includes space physics research, space systems engineering and nano-technologies. Liaison with other government agencies for policy planning. ISA actively conducts international projects and provides a point of convergence for military and non-military projects.

VERIFIED

National Committee for Space Research

Current Status
An element of the Israeli Academy of Sciences and Humanities formed in 1960, it advises the government, ISA and universities on space-related research projects and has played an important role in the development of remote serving capabilities. Has no perceived roles and responsibilities in military space objectives. Liaison with the Israel Space Agency converging science objectives with operational opportunities such as integrated mission objectives on multipurpose payload packages. Has taken a leading role in making recommendations on national and international space programmes and has assisted the government to converge separate strands of research into focused projects involving national scientific and academic resources.

VERIFIED

Remote Sensing Laboratory

Current Status
Established in 1988, at Ben-Gurion University of the Negev, this BGU/IDR research facility specialises in the analysis of arid/semi-arid regions using satellite data. Data input from many parts of the world have contributed to a major and globally endorsed inventory of desert information and in studies in the desertification of arid regions. Information from ground measurement is correlated with data (imaging and radar) obtained by satellite. This is analysed by several universities and research institutes. Facilities cover an ERDAS IMAGINE image processor, an ARC/INFO GIS system and a NOAA/AVHRR receiving station. Current activities include: software development for automatic examination of faults, cracks and dykes in satellite images (such features often indicate water). Automatic extraction of surface features such as drainage networks, lineaments and roads from satellite images and DEMs; monitoring environmental trends in the Negev from satellite imagery; studying how dust and aerosols degrade satellite images; looking for water stress in plants before it is apparent on the ground. The facility works with existing medium and low-resolution imagery and helps develop new high-resolution systems. Recent developments include increased emphasis on intergrating optical and radar generated data from satellites. The facilities have several applications for remote sensing correlation studies and for military applications.

UPDATED

Italy

Agenzia Spaziale Italiana (ASI)

Current Status
ASI was formally established 25 August 1988 under MURST to manage the PSN national space plan (Piano Spaziale Nazionale), previously directed from within the National Research Council (CNR, Consiglio Nazionale delle Ricerche). ASI is responsible for Italy's ESA involvement and manages the four

principal national satellite projects: Italsat, Tethered Satellite, SAX and Lageos 2.

The ASI has been instrumental in establishing a strong contribution to the European effort supporting the NASA led International Space Station. Negotiations with NASA over the possible development of a habitation module for the ISS began when the Bush administration cancelled the US element in early 2001. Little progress has been made on this and further discussion will await resolution of the changing nature of the ISS programme and the future of the Shuttle.

The agency's annual budget is described at the beginning of the national entry. The Cabinet elects chairman of the Board, proposed by the MURST Minister. The DG is appointed by MURST after consultations with the Board of Directors. Board Members, elected by the Cabinet, sit for five years.

The CGS at Matera is part of the worldwide integrated geodetic network. Returns are regularly made using the principal laser ranging satellites, Lageos 1/2, Starlette, Ajisai and Stella. AlliedSignal Technical Services Corp of the US installed a Mobile Laser Ranging System in 1994. The station includes a 20 m VLBI dish. ASI's Milo base in Sicily (38°01′ north/12°35′ east) launches stratospheric balloons along the Mediterranean to Spain's west coast.

VERIFIED

Directorate of International Affairs

Current Status

MURST is the parent body of the ASI space agency. Its Office for Space Affairs co-ordinates the government's space interests. It has been given stronger powers to liaise between ASI departments, commercial organisations and foreign governments on major space related ventures. Extended powers provide ample scope for recommending policy changes and for liaising through appropriate embassies for future co-operation. Has taken a stronger role in representing space-related companies and organisations in other countries and in directing foreign interest to appropriate companies and organisations.

UPDATED

MARS Centre

Current Status

The Microgravity Advanced Research and Support Centre was created in 1988 by Alenia and the University of Naples as a user support and operations centre for space experimentation, with specific expertise in fluid physics. Employs about 30 personnel. The Centre hosts a fluid physics lab for experiment preparations, and a telescience room for the remote control and operation of space experiments. For example, the Telescience Control Room controlled an experiment during 1996's Life and Microgravity Sciences Spacelab mission. Can support telepresence requirements on demand. MARS has shifted its effort from Spacelab, now completed, to supporting Columbus, the ESA Space Station module. From this has emerged new initiatives for autonomous Italian initiatives to ESA. Several specialised experiments suites have been proposed and considerable diversity in materials processing tasks have been supported. MARS publishes the *Microgravity Quarterly* and *Space Technology* journals.

VERIFIED

Nuova Telespazio

Current Status

Nuova Telespazio, 50/50 owned by Telecom Italia and STET, is Italy's international satellite communication carrier under the Ministry of Post and Telecommunications. Telecom Italia is the national signatory to Intelsat, Inmarsat and Eutelsat. Staff total averages a little over 800 personnel. Its primary Earth station is the Piero Fanti Space Centre at Fucino, complemented by the Lario facility adjacent to Lake Como and the Scanzano site near Palermo that became operational during 1989. Fucino is a major European TT&C facility: its antennas include 17/27.4/32 m Intelsat Atlantic, 29.5 m Intelsat Indian, 13 m Inmarsat Atlantic, 4/8 m Eutelsat, 11 m Olympus, 10 m Marisat, 7.5 m LASSO, 10 m Landsat and 7.7/9/13.4/16.5/32 m in-orbit test dishes. Lario maintains two 32 m dishes for servicing Intelsat Atlantic/Indian Ocean satellites and a 17 m antenna. Telespazio also receives, processes and distributes Landsat, Spot, ERS and MOS data from Fucino as Italy's National Remote Sensing Centre and the national point of contact for ESA's ESRIN Earth observation data, and operates the Matera laser station for geodetic ranging measurements with the Lageos satellite. The company acquired a 2.3 per cent interest in Spot Image in 1990.

Telespazio was responsible for developing the Argo system for the Civil Defense Department to provide emergency communications and datalinks through Eutelsat within 4 hours of a major disaster. Working through a 9 m Fucino antenna, Rome's Civil Defense HQ control centre has 12 2.2 m Ku-band mobile stations ready on lorries and helicopters to provide telephone, fax and TV links. Three helicopters are equipped for airborne TV. 110 1.8 to 2.5 m fixed terminals provide environmental data (seismometric, volcanic and hydrological, for example) to help disaster forecasting. Nuova Telespazio has conducted studies of integrated data processing networks collectively bringing together current and future technologies including the new European space-based positioning system and datalinks through existing satellite systems, applications extending to a wide range of uses including traffic logging, road toll charging and the monitoring of individuals for national security or policing purposes.

In addition to being the national signatory to Intelsat, Inmarsat and Eutelsat and Italy's exclusive carrier for space communications, Nuova Telespazio also receives, processes and distributes Landsat, Spot, ERS and MOS imagery via its major Fucino facility as Italy's National Remote Sensing Centre and the national point of contact for ESA/ESRIN. NT is setting up a depository of remote sensing data and meteorological information from which future national policy decisions can be made regarding further development of Italy's earth observation programme can be framed. In 1991 Telespazio introduced the Vsat network for the Argo emergency communications unit of the Department of Civil Defence and later extended this to embrace other users, all operated through the Fucino earth station. As part of a 1996 ESA contract, Telespazio led a team of five European companies in a system definition architecture for the European GNSS2. In 2000, Telespazio established E-Geos, a joint venture with the Italian Space Agency for Italy's first national earth observation centre. In 2001 Telespazio managed the early orbit phase of the SICRAL military telecommunications satellite.

UPDATED

Japan

Earth Observation Centre (EOC)

Current Status

NASDA's 115,000 m^2 EOC was established in October 1978 to receive and process Landsat, Spot and MOS data for distribution as CCT and prints through the Remote Sensing Technology Centre (RESTEC). It is part of NASDA's Office of Earth Observation Systems and it performs MOS/JERS 1/ADEOS mission management, archiving and processing, plus Landsat, Spot and ERS archiving/processing. EOC also processes high-resolution imagery and has been instrumental in developing sensors for advanced remote sensing satellites. EOC's two 10 m parabolic dishes channel data to high-density digital tape recorders, with associated quick-look units. It can handle 30 SAR and 30 OPS scenes daily from JERS 1. EOC has provided guidance on Japan's next generation weather and remote serving satellites. This is framing a new policy strategy for research programmes in remote sensing and in co-ordinating greatly different data sets to more efficiently accommodate national needs. The facility will remain as an integrated processing centre after the convergence of NASDA and the ISAS into a new national aerospace research agency.

VERIFIED

Institute for Unmanned Space Experiment Free Flyer (USEF)

Current Status

USEF was established in 1986 to promote development and utilisation of unmanned space systems. Company products were stimulated by US interest in free-flyers for manned space stations assembled through a series of modules, co-orbiting microgravity structures achieving lower levels of vibration and accelerations than possible through manned, pressurised, modules. The SFU Space Flyer Unit carried nine experiments on its first flight before retrieval by NASA's Shuttle in January 1996. The Express and Space Robotics projects began in 1992. The Users project began in 1995 and the technology has been applied to the development of concepts for unmanned platforms. This in turn has provided further research and development opportunities for advanced SFU design concepts.

USEF is a non-profit organisation under the direction of the Ministry of International Trade and Industry and endowed by 13 companies: Fujitsu, Hitachi, IHI, Kawasaki Heavy Industries, MELCO, Mitsubishi Heavy Industries, Mitsubishi Precision, NEC, Nippon Electronics Development, Denso, Nissan Motors, Sharp and Toshiba.

VERIFIED

Institute of Space and Astronautical Science (ISAS)

Current Status

On 1 October 2003 the ISAS ceased to exist. Its activities, along with the National Space Development Agency (NASDA) and the National Aerospace Laboratory (NAL), were absorbed into the new combining organisation known as the Japan Aerospace Exploration Agency (JAX). The ISAS flagship mission MUSES-C to an asteroid was delayed from its planned launch date of December 2002 and the spacecraft was eventually launched on 9 May 2003. Renamed Hayabusa, it will rendezvous with the asteroid Itokawa from where it is expected to retrieve a small sandwich and return to earth.

Background

ISAS was established 14 April 1981 by reorganisation of the Institute of Space and Aeronautical Science, at the University of Tokyo. It was responsible for launching Japan's first satellite in 1970. The university began developing the all-solid M orbital launcher in April 1963 after opening the Kagoshima Space Centre in 1962. The original ISAS was not formed within the University until 1 April 1964. ISAS is one of the National Inter-University Research Institutes answerable to Mombusho (Ministry of Education, Science and Culture) and is responsible for Japan's science satellites, launching every one to two years. It was restricted to all-solid vehicles of 1.4 m diameter but permission for the 2.5 m M-5 was received in 1989 for modest deep space missions from 1996. The Sakigake and Suisei Halley's Comet probes of 1985–86 provided Japan's first experience beyond Earth orbit. ISAS is conducting tests of a vertical liftoff and landing vehicle as part of the Removable Rocket Vehicle Test (RVT) programme.

ISAS is headed by a Director-General advised on policy and administrative matters by a 19-member Board of Councillors and on research planning by a 21-member (11 from within ISAS itself) Advisory Council for Research and Management. Mombusho appoints all members. ISAS' organisational structure incorporates two Technical Divisions (Space Operation and Engineering Support) and nine Research Divisions responsible for 49 Research Sections. The Office of Project Co-ordination oversees all project activities. The Research Divisions provide the core of each Project Team; the Research Sections concentrate on specific fields and are headed by a Professor, an Associate Professor and two Research Associates. Personnel total about 300 with an additional 175 graduate and research students. In 1998 ISAS had a budget of ¥29.7 billion and that level has been maintained.

ISAS operates seven facilities:

1. Kagoshima Space Centre (KSC), initial use 1962, launch centre for sounding rockets and M orbital vehicles; 349 launches to end-1995. There is also a 20 m space tracking dish at Kagoshima.
2. Noshiro Testing Centre (NTC), established 1961 and located on Asanai Beach for static firing of large solids and research/static tests of a LOX/LH_2 engine and a turbo-ramjet.
3. Usuda Deep Space Centre (UDSC), 64 m S/X-band deep space TT&C station. Constructed for Comet Halley probes of 1985–86, began operations October 1984.
4. Space Utilization Research Centre (SURC), established 1988 at Sagamihara campus.
5. Centre for Planning and Information Systems (PLAIN), established 1993 at Sagamihara campus.
6. Centre for Advanced Spacecraft Technology, established 1995 at Sagamihara campus.
7. Sanriku Balloon Centre, launch site for 500 kg payloads on 200,000 m^3 helium balloons to 40 km altitude over the Pacific.

UPDATED

Japan Meteorological Agency (JMA)

Current Status

Renamed the Japan Meteorology Agency when it became an organ of the Ministry of Transport in 1956, the TMA can trace its origins to the National Meteorological Service set up in 1875. The JMA is responsible for providing space-based meteorological services and undertaking sounding rocket investigations. The Meteorological Satellite Centre, established April 1977 near Tokyo and attached to JMA is now situated in the centre of Tokyo and undertakes satellite control, image data acquisition through its Command and Data Acquisition Station (CDAS) at Hatoyama 35 km northwest of Tokyo, and data analysis for the GMS satellites. JMA played a leading role in the conceptualisation, design and development plans for the first Japanese meteorological satellites and has played a vital role in subsequent years developing and refining successors to the first satellites. NASDA procured/launched the satellites under JMA funding and requirements. Raw imagery, data collection platform signals and spacecraft telemetry are received on the two CDAS 18 m Cassegrain dishes; they are then relayed by 2.0 GHz microwave link to the Data Processing Centre (DPC). The DPC maintains four mainframe computers: two M-360R (24 Mbit each) for spacecraft control/operations and two M-380S (32 Mbit each) for image processing and subsequent extraction of parameters such as sea temperature and cloud top height. The smaller machines can assume some of the 380's processing duties in the event of failure. The Stretched-VISSR and WEFAX data are passed to the CDAS for uplinking and broadcast to users: Medium Data Utilization Stations (MDUS) can receive S-VISSR but the Smaller Data Utilization Stations (SDUS) receive the lower-resolution (7 to 8.5 km) WEFAX. Two TARS Turn-Around Ranging Stations are operated on Ishigaki Island (Japan) and Crib Point (Australia) in conjunction with CDAS for satellite orbit determination and prediction. The JMA has organised Japan's participation in the global weather watch involving ESA and NASA satellites. This is revising plans for new generations of global weather satellites and Japanese commitment to the international weather watch. JMA has also developed plans for a new regional weather watch incorporating most of the meteorological research and analysis branches of the Japanese government and remote data platforms throughout the hemisphere.

UPDATED

Japan Space Utilisation Promotion Centre (JSUP)

Current Status

JSUP was founded in February 1986 to promote industrial exploitation of the space environment and has since provided industry and manufacturing companies with introductions for joint ventures. JSUP has provided policy direction for the shifting changes in Japan's government space structures. Major policy recommendations formulated for changes in the balance of government and privately funded ventures. JSUP is supported by the Ministry of Education, Culture, Sports, Science and Technology and by the Ministry of Economy, Trade and Industry. Utilisation promotion includes aircraft parabolic flights and the µg drop test facilities in Hokkaido and Gifu (MGLAB: Microgravity Laboratory of Japan). JSUP has a long standing collaborative agreement with INTOSPACE in Germany. Experience with international programmes has been used as a template for a broader analysis of Japan's technology base in space-related activities and this has been fed into the analysis of non-space related companies for broader participation across a wider base. JSUP revenue for fixed year 2001 was ¥2.3 billion.

UPDATED

Microgravity Laboratory of Japan (MGLAB)

Current Status

MGLAB began operations in 1993, providing a 1.50 m diameter drop tube with 100 m free-drop zone. The 900 mm diameter 2.280 m capsule is decelerated over 50 m and provides two payload racks to users. Test equipment has been utilised for development of microgravity equipment for the JEM. This has expanded to include offers for Japan's metallurgical industries to invest in new research programmes for materials science. Under government reorganisation of nationally funded research projects MGLAB has developed a set of design initiatives which may expand Japan's ability to carry out materials processing tests in space. Stalled plans to get JEM operational due largely to the hiatus in Shuttle missions has encouraged MGLAB to develop new concepts for man-tended and unmanned platform research lifting total reliance from the manned elements formerly believed to be in the vanguard of this research.

UPDATED

Ministry of International Trade and Industry (MITI)

Current Status

MITI is Japan's third largest government funding agency for space activities, after the Science and Technology Agency and the Ministry of Education. It was important in the unification of Japan's aerospace development organisations (primarily NASDA and ISAS), which took place on 1 October 2003 when JAXA (Japan Aerospace Exploration Agency) was formed. MITI also supports foreign participation in aerospace projects. MITI is leading the USERS Unmanned Space Experiment Re-entry System project and provided most of the Space Flyer Unit's funding. It was also instrumental in establishing the USEF Institute for the Unmanned Space Experiment Free Flyer. MITI's Agency of Industrial Science and Technology (AIST) operates several institutes with space interests: ETL Electrotechnical Lab (advanced technologies, such as robotics, space power and remote sensing), MEL Mechanical Engineering Lab (R&D such as attitude control for large space structures) and GSJ Geological Survey of Japan (geological applications for remote sensing). MITI is now studying potential applications of Japan's involvement with the International Space Station and is applying development experience with the Unmanned Space Experiment Free Flyer to the design of unmanned free flying platforms. MITI has taken a leading role in integrating electrical and mechanical disciplines in potential space applications.

VERIFIED

Ministry of Posts and Telecommunications – Communications Policy Bureau

Current Status

The Ministry of Posts and Telecommunications' Communications Policy Bureau is responsible for space-based communications systems, supervising NASDA and the Telecommunications Advancement Organisation of Japan, the latter established in August 1979 to operate national communications satellites. The Ministry also provides analyses of potential telecommunication paths for future development and conducts recommendations made to other government departments on policy and initiatives. The Telecommunications Bureau supervises the Nippon Telegraph and Telephone Corp and the Kokusai Denshin Denwa Co, respectively the provider of domestic satellite communication links and the Japanese signatory to the Intelsat and Inmarsat operating agreements. Recommendations on technology development have provided guidance for industry and manufacturers alike. New guidelines for international telecommunications strategies involving greater participation from private companies and less from government, the latter only reserving exclusivity over regulatory standards. Ministry is developing new criteria for national and international regulatory standards proposed by Japan and forwarded to government for it to be submitted to the ITU.

UPDATED

Ministry of Posts and Telecommunications

Current Status

Until 1989's launches of Superbird (Space Communications Corporation) and JCSat (Japan Communications Satellite Company), Japanese telecom satellites were state-owned and operated through the Telecommunications Advancement Organisation of Japan (TAO). National space agency NASDA was responsible for procurement and launch of the CS and BS series before handing them over to TAO for operational use. The Japan Broadcasting Corp (NHK: Nippon Hoso Kyokai) and industry-owned Japan Satellite Broadcasting Inc (JSB, created 1984 by the Ministry of Posts and Telecommunications) broadcast through BS-3 transponders.

As a result of US pressure, the 'BS-4' generation was opened to foreign bidders. Hughes Space and Communications was selected in December 1993 by the Broadcasting Satellite System Corp of NHK, JSB and others to negotiate for two BSATs to be launched 1997/98. Nippon Telegraph and Telephone provided communications services through CS-3, but NTT's own two NStars were built by SS/L and launched in 1995, the open market competition resulting from US pressure against a government-funded CS-4 closed to foreign bidders. Satellite Japan Corp (Sajac) planned to launch its Sajac Hughes HS-601 satellites in 1994 but financial problems forced it to merge with JCSat in 1993, creating Japan Satellite Systems (JSAT); JCSat 3 began commercial services in November 1995 and JCSat 4 was ordered in February 1996 and launched in February 1997. As part of the BS-4 generations BSAT-1a was launched in April 1997 and BSAT-1b in April 1998 followed by BSAT-2a in March 2001 and BSAT-2b in July 2001.

NASDA continues to develop and operate the ETS series of technology demonstration satellites. ETS 6 appeared in 1994 incorporating inter-satellite data relay and mobile communications packages, but a propulsion problem left it in a low orbit and allowed only a truncated programme. A Data Relay and Test Satellite system will build on ETS 6 and COMETS (1997) experience. The OICETS Optical Inter-orbit Communications Engineering Test Satellite is planned to test laser links with ESA's Artemis. The Ministry of Transport plans two GEO MT-Sats to augment GPS signals and air traffic control links for improving aircraft navigation and control; they will also replace the GMS meteorological satellite. See the Earth Observation section for further information. ETS 8 in 2002 will demonstrate a cellular phone system that can cope with Japan's mountainous geography. MPT reports that cellular telephone subscriptions have increased from 242,000 in 1989 to 41.5 million in 1999 and 52 million by April 2000.

MPT's Communications Policy Bureau is responsible for Japan's R&D and utilisation policy in space communications in CRL is a leading research laboratory in the field of satellite communications and broadcasting technologies. These have been extended to new technologies and to adopting innovative techniques to improve the efficiency of Japan's telecommunication services.

VERIFIED

National Aerospace Laboratory (NAL)

Current Status

The activities of the former National Aerospace Laboratory have now been absorbed into the new Japan Aerospace Exploration Agency (JAXA) formed on 1 October 2003. Under this new government agency, which also replaces NASDA, the activities of the NAL are focused in much the same way the NACA was transformed into NASA. The NAL was Japan's national aerospace technology research centre; it was founded in 1958. It provides the lead on aerospace plane activities. Research staff total approximately 300 people. The NAL is conducting entry dynamics research on possible Japanese spaceplane configurations and has developed new control techniques for unmanned systems. Developing hybrid concepts involving low-lift, semi-ballistic, re-entry vehicles capable of carrying humans or cargo in an automated system. Substantial re-examination of priorities and possible mission requirements has brought a change to the way NAL has conducted its research and this has resulted in less emphasis on a fully aerodynamic, high-lift, configuration.

UPDATED

National Space Development Agency of Japan (NASDA)

Current Status

On 1 October 2003 NASDA, NAL and ISAS merged to form the Japan Aerospace Exploration Agency (JAXA). The following information is retained in this volume only to supply historic material relevant to the former NASDA. The National Space Development Agency of Japan (NASDA) was established in October 1969 to act as a nucleus for implementing space development and promoting space utilisation exclusively for peaceful purposes. It is responsible for the following tasks based upon the Japanese Space Development Programme: development of satellites (including space experiments and space station operations) and launch vehicles, and their launching and tracking; and development of methods, facilities and equipment required for satellite development.

NASDA's first Engineering and Test Satellite was successfully launched in September 1975 using an N1 rocket, based upon the McDonnell Douglas Delta vehicle first stage and built under license. This was followed by the N2 vehicle and later by the H1, the latter being retired in 1992. These three launch vehicles used United States Castor strap-on boosters and Japanese second stages. The first NASDA launch vehicle to be based upon wholly-Japanese stages is the H2, first flown in 1994. The H2 has now been succeeded by the H2A. The all-solid propellant J1 vehicle was first flown in February 1996 and carries small payloads. NASDA successfully launched the H2A No 1 on 29 August 2001 and H2A No 2 on 4 February 2002.

NASDA's personnel totals have increased only slightly to 1,090 in 2002 from the 1982 level of 904 (1969 – 151), and they are accommodated in six principal facilities:

1. Headquarters: accounting for about one-third of the total employees.
2. Tsukuba Space Center: established in 1972, it now employs approximately half of NASDA personnel.
3. Tanegashima Space Center: NASDA's largest facility.
4. Earth Observation Center (EOC), set-up in 1978 as a NASDA field centre to establish and develop satellite remote sensing technology. EOC receives and processes data from Earth-observing satellites for a variety of applications and research.
5. Katsuura, Masuda and Okinawa Tracking and Data Acquisition Stations: used for feeding data to the Tsukuba Tracking and Control Center. Co-operation with other domestic and foreign stations are arranged as necessary for launch and flight support.
6. Kakuda Propulsion Center: Japan's largest propulsion test centre. KPC has two main test stands: High Altitude Test (HAT) facility to simulate upper mainstage engine firing in the space environment; Propellant Feed Systems Test (PFST) facility to evaluate the cryogenic liquid hydrogen turbo pump performance of LE-7A.

NASDA budgets have increased from ¥91 billion in 1980 to ¥136 billion in 1990 and ¥204.03 billion for 2002.

UPDATED

Remote Sensing Technology Centre (RESTEC)

Current Status

The Remote Sensing Technology Centre of Japan disseminates remote sensing data received/processed by NASDA's Earth Observation Centre. Radarsat data were added in May 1996. RESTEC was established in July 1975 to conduct R&D into remote sensing techniques and applications, and operates FACOM M1400/10 K-650, VAX 4000 and Sun 4/630 computer systems. RESTEC also integrates maritime and land use management techniques. Research into new operating capabilities for existing data sets involving high-speed processing for reducing image and SAR data interpretation time. RESTEC has diversified into integrating marine and land use data for national and foreign customers. RESTEC has taken a leading role in recommending strategic changes to the way the remote-sensing assets of Japan are organised and how they can be more effectively locked in to international organisations.

UPDATED

Science and Technology Agency (STA)

Current Status

In addition to being the SAC secretariat, the STA's research and development bureau is responsible for planning basic policies on space development, international co-operation in space utilisation and acting as NASDA's parent body. It is also responsible for the National Aerospace Laboratory (NAL), where research on basic space technology is undertaken. As the central space body, it was responsible for ¥231 billion FY96 funding; it is responsible for ¥244 billion for FY97 funding. The STA has presently presided over a reappraisal of the Hope-X spaceplane, which is planned for launch by H2A in 2004. More recently, plans for human spaceplanes have been tempered by priorities related to commercial activities. Japan is re-examining the role of spaceplanes in future space policy issues and concern over reliance on US heavy lift vehicles has stimulated a re-examination of independent ISS access. The STA has responsibilities for guidance and policy to the newly formed Japan Aerospace Exploration Agency, which, on 1 October 2003, merged the collective activities of the former National Space Development Agency, Institute of Space Astronautical Sciences and the National Aerospace Laboratory.

VERIFIED

Space Activities Commission (SAC)

Current Status

The SAC, composed of four commissioners and chaired by the Minister for Science and Technology, was established in May 1968 to centralise space activities under a systematic programme. The 'Outline of Japan's Space Development Policy was issued in March 1978 as a guideline for space developments over the next 15 years; a review is produced about every five years (the latest in 2002). The STA performs the Commission's secretariat functions. Interim reviews have reduced aspirations in accord with reduced financial projections. Redirection of national direction in space activities to include defence interests and national security issues. However, this cabinet level policy body has approved formal development of an improved H-2A launch vehicle which will double LEO and GTO performance from about 2008. Not yet awarded a formal designation, the improved H-2A will have two LE-7A cryogenic first stage motors instead of one, widening the first stage from 4 to 5 metres, while retaining the same LE-5B upper stage as the standard H-2A. This will prepare Japan's national heavy launcher to conduct unmanned automated resupply missions to the Kibo pressurised laboratory module to be attached to the International Space Station, giving it a lift capacity equal to the Atlas 2AS and the Delta 4 series launchers. Plans for the next generation of Japanese launch vehicles are being formulated at the Space Activities Commission and lessons from the partly successful H2 programme have been fed into a new model for Japanese commercial and economic development, criteria new to the assessment and ones likely to result in a significantly changed policy.

UPDATED

Space Activities Promotion Council (SAPC)

Current Status

The SAPC was formed in 1968 as part of Keidanren (Federation of Economic Organizations) to promote space activities. Membership stands at more than 100 companies and trade associations; most major space-related Japanese companies are represented. The SAPC has become a major marketing support structure for Japanese companies. International representation for Japanese companies at conventions andexhibitions. Representation through embassies. A concerted effort by the Japanese government to commercialise many space projects, hitherto run by government funds alone, has given the SAPC a greater role in shifting activity to long term objectives for services provided by launch-vehicle operators and satellite manufacturers. SAPC has developed a new working dialogue with the Space Activities Commission to integrate a new national strategy for space, following the merging of activities previously organised under NASDA, ISAS and the NRL.

UPDATED

Telecommunications Advancement Organisation of Japan (TAO)

Current Status

The Telecommunications Satellite Corp of Japan (TSCJ) was established in August 1979 to operate the CS and BS satellite series procured and launched by NASDA. The name changed to TAO in October 1992 as it introduced new services such as R&D. Japan's first two operational communications satellites, CS-2a/2b, each with six Ka-band and two C-band transponders, were launched in 1983 principally for maintaining communications with the more scattered islands of the Japanese archipelago. Both were still operational when they were replaced by CS-3a/3b in 1988. NTT's NStar took over in 1995. Now operated by TAO, NStar A was launched in August 1995 followed by NStar B in February 1996, both built by SpaceSystems Loral, and NStar C in July 2002, a satellite built by Lockheed Martin Commercial Space Systems. The BS-2 series introduced DBS services in 1984 using 100+W transponders but two of BS-2a's three amplifiers failed soon after launch on 23 January 1984. BS-2b brought the system to full capacity in 1986 but a BS-2x was purchased from GE Astro Space by NHK for an Ariane launch to ensure system continuity before the first of the BS-3 series could be launched in 1990. However, BS-2x was lost in the Ariane failure of 22 February 1990; the replacement BS-3H was then lost in its April 1991 launch. Both would have been controlled through TSCJ. BS-3a, launched 28 August 1990, replaced BS-2b but it operates on marginal power. 3b completed the system in 1991, but NHK procured BS-3N in late 1992 (launched July 1994) to guarantee services. The 'BS-4' generation is operated by B-SAT through its BSAT-1 series, BSAT-1a launched April 1997 and BSAT-1b launched April 1998. Manufactured by Orbital Sciences, BSAT-2a was launched by Ariane 5 on 8 March 2001 followed by BSAT-2b on 12 July 2001, the latter placed in an incorrect orbit due to a terminal stage failure.

Research and development activity includes the test and evaluation of new fail-safe operating concepts.

VERIFIED

Korea, South

South Korea Telecom (SKT)

Current Status

SKT was 80 per cent government-owned in 1995, reduced to 51 per cent in 1996. It began operating the Kumsan coast Earth station in March 1991 for Pacific Inmarsat A services; Indian Ocean services were added 1993. Inmarsat C was added to both in 1994. Korea is a leading maritime nation, with 960 ocean-going ships and 550 deep sea fishing vessels carrying 130 Inmarsat terminals. SKT is the Inmarsat and Intelsat signatory. By December 1999, SK Telecom broke the 10 million mark for cellular service subscribers and sales for the year totalled 4.28 trillion Nwon, a 21 per cent increase over 1998. SKT is evaluating benefits it might obtain from a national satellite service, possibly involving one or more other nations in the region. Plans to form a regional telecommunications service are dependent upon other countries participating and providing financial contributions. New studies into regional development have explored options for South Korea in opening new agreements for a combined or joint venture absorbing these aspirations.

UPDATED

Libya

Libyan Centre for Remote Sensing and Space Science (LCRSSS)

Current Status

LCRSSS, administered under the Secretariat of Scientific Research, signed a data distribution agreement in 1994 with Spot Image. The two also completed a study using Spot imagery on the impact of urban growth on agricultural land around Tripoli. LCRSSS also handles data from commercial satellite data distributors. LCRSSS supports government efforts to expand Libya's use of satellite data for economic enhancement through improved land resource evaluation and providing inventories. The Libyan initiative to secure agreements with other countries for remote sensing analysis and interpretation has been supported by the LCRSSS through other government departments and agencies but no known agreements have been reached. The impact of recent political changes to the relationship with European countries has stimulated new initiatives between LCRSSS and companies in France, Germany and the UK for enhanced photo-analysis and interpretation.

UPDATED

Mexico

Secrétaria de Comunicaciones y Transportes (SCT)

Current Status

Until 1990, SCT was responsible through the Departamento Espacial of the DGT Dirección General de Telecomunicaciones for Mexico's telecom systems. Telecomm, an independent non-stock company owned by the government, was created in January 1990 to assume control, with SCT operating as a regulating body. It is Mexico's Intelsat (Earth stations at Tulancingo and Cuncun) and Inmarsat signatory. SCT provides feeder services for business services from peripheral countries and receives technical support from the US. Operates as consultant to regional telecommunications entities and sometimes makes recommendations to the Mexican government about future telecommunications policy.

By 2001, SCT was operating 282 Earth stations, including 191 for TV distribution, 11 for transmitting TV, telephony and data, 15 data, 9 telegraphy, 26 rural telephony, 8 mobile and 186 VSATs. SCT has begun an upgrade and improvement programme to increase operability.

UPDATED

Morocco

Centre Royal de Télédétection Spatiale du Maroc (CRTS)

Current Status

CRTS was established in 1989 to promote, develop and co-ordinate satellite-based remote sensing for the country. It has specific responsibility for the archiving and distribution of satellite image data, and has distribution agreements with Spot Image for Spot, and Eurimage for Landsat, NOAA and ERS. CRTS is also acquiring Eurimage data and has begun to work with high-resolution imagery. Morocco's CNTS Comité National de Télédétection Spatiale was set up in 1993, including representatives of ministerial departments involved in remote sensing and GIS. CRTS has recently been expanded by a department dedicated to SAR data. Development of image and SAR data in banks and archives available to government users, and on licence to commercial organisations. Morocco has placed increasing emphasis on the use of satellite information for national and economic development and the CRTS has been tasked with conducting an in-depth analysis of the country's capabilities and in its ability to increase the value of remote sensing for this purpose.

The Technical University of Berlin is providing its Tubsat C micro-satellite platform for a space mission carrying an Earth imaging camera provided commercially by the UK's Rutherford Appleton Lab. The CCD camera, returning 250 m resolution from 800 km, was developed for Pakistan's Badr 2 satellite.

UPDATED

Netherlands

National Aerospace Laboratory (NLR)

Current Status

The **National Aerospace Laboratory (NLR)** is the central institute for aerospace research in the Netherlands. Since 1937 NLR had been an independent non-profit organisation that provides technological support to aerospace industries, operators, authorities and international organisations concerned with aviation and spaceflight, all over the world. NLR participates in programmes of the European Space Agency (ESA) and the European Union. NLR closely co-operates with national industrial partners in space-related projects supported by the Netherlands Agency for Aerospace Programme (NIVR).

NLR employs a staff of about 900 in two main establishments, one in Amsterdam and one in the Noordoostpolder, Flevoland. Over two-thirds of the staff are graduates from universities or technical colleges. NLR holds an ISO 9001; 2000/AQAP-110 quality assurance certificate and several calibration accreditations.

NLR owns several wind tunnels, including a transonic (DNW-HST) and a supersonic (DNW-SST) wind tunnel extensively used in test programmes for the development of the ARIANE launchers, ARD and CTV/CRV capsules and the X-38, as well as heat protection systems. Both of these wind tunnels are capable of being used for the validation of the aerodynamic performance of future space transport systems. NLR's wind tunnels are operated by the German-Dutch Wind Tunnels organisation, DNW, along with similar German facilities. NLR develops wind tunnel models and test equipment.

NLR operates facilities for research in the area of structures and materials used in space projects. Expertise in the area of loads and use monitoring is combined with structural response monitoring and materials characterisation to perform damage tolerance assessment. Advanced dynamic and non-linear analysis and test systems are applied to solving problems associated with design and verification of spacecraft structures, including fatigue and crack growth analysis. Structural optimisation with regard to strength, active vibration control or stability performance is also executed. Materials science and engineering projects include the evaluation of properties of metallic, composite and hybrid materials, also at elevated temperatures, and the testing of full-size structures.

NLR's extensive computer network, featuring a 64 GFlops NEC SX-5/8 supercomputer, is used for theoretical research and simulation, especially in the fields of computational fluid dynamics, computational solid mechanics, robotics, computational electromagnetics, and structural design and analysis, among other things.

NLR's moving-base research flight simulators can be configured not only for transport aircraft, fighter aircraft and helicopters but also for shuttle or spaceplane-type vehicles.

Test and verification equipment is available for use in the development of space avionics systems. Various environmental testing facilities and an antenna test range are available.

Spaceflight specialisations include the simulation and testing of satellite subsystems, the development of two-phase flow heat control systems, and research on liquid dynamics problems in space. In addition, NLR develops electronic control systems for biological experiments in micro-gravity. NLR activities in the area of satellite navigation are focused on validation and validation. To support mission preparation, execution and evaluation, NLR works in the area of space robotics, telescience, and utilisation support.

NLR is developing operational systems for the management of natural resources using remote sensing data. NLR acts as National Point of Contact for the dissemination of remote sensing data, and has co-developed a mobile receiving station. To enable Dutch users to access a wide range of earth observation products and to allow Dutch service providers to make their products and services available, NLR is working on the development of a Netherlands Earth Observation Network (NEONET), in correct with the European CEO initiative. Optical and microwave remote systems are developed, and operated from NLR's Cessna Citation and Fairchild Metro research aircraft. NLR is one of the developers and the ESA focal point of the NASA crack growth program NASGRO.

VERIFIED

Netherlands Agency for Aerospace Programmes

Current Status
The Nederlands Instituut voor Vliegtuigontwikkeling en Ruimtevaart (NIVR) founded in 1946, is a semi-governmental non-profit agency with the statutory objective of promoting industrial aerospace activities, including space since 1969. This includes promoting international ventures through Netherlands industry and serving to link individual companies with similar organisations in other countries. Following on from ANS and IRAS, the current major bilateral programmes are SAX and SCIAMACHY. NIVR is responsible for SAX programme management, co-operating with SRON and Italy's ASI space agency. SCIAMACHY, which measures a large number of trace gases and constituents in the stratosphere and higher troposphere from ESA's Envisat, is a joint project with Germany's DARA, with a contribution from Belgium. Dutch interest totals NLG90 million. NIVR has conducted a semi-official inventory of national resources to gather enhanced opportunities for international projects.

NIVR is conducting a multiyear space technology programme (NRT) to stimulate national technology developments. This includes the solar array development efforts of Fokker Space. Another example is the technological study by Bradford Engineering which led to the development of different types of gloveboxes for use in Spacelab, Mir, space shuttle and International Space Station. These have been grouped in size and volume proportions most sought by potential experimenters following a survey of researchers and scientists.

VERIFIED

Netherlands Industrial Space Organisation (NISO)

Current Status
Initiated in 1986, NISO was officially established in December 1989. It comprises Dutch industries, research organisations and users in the space field with the aims of: defining common interests and strategies; promoting industrial involvement; stimulating the use/application of space facilities and products by industrial enterprises and institutes; broadening the social base of space; and promoting technology spin-offs. The Institute has helped formulate government policy and has been instrumental in aligning industry with international requirements and expectations through ESA and with other international organisations outside Europe. Membership is 33. NISO has latterly provided data and analysis to the government for determination of appropriate levels of space spending. Also for directional policy on the effective use of existing resources, the prospective economic and national value to new projects and international ventures proposed by companies from the Netherlands.

UPDATED

Netherlands Remote Sensing Board

Current Status
Beleidscommissie Remote Sensing (BCRS) is the co-ordinating body for national remote sensing activities which began its programme of research and inventorying before any involvement with satellite data. The BCRS Program Bureau is located at the Survey Department of the Ministry of Transport, Public Works and Water Management at Delft and has been examining the opportunities afforded by ESA's Envisat programme. Through ESA's remote sensing programmes and stimulated by European use of global satellite data BCRS has considerably expanded space derived data analysis and interpretation. Evaluation of technology trends matching impending capabilities to future requirements helps frame government plans for enhanced environmental monitoring. Services available to other EU organisations.

VERIFIED

Observatory for Satellite Geodesy

Current Status
Operated by the Faculty of Geodesy at the Delft University of Technology and currently working on international programmes of geodesic survey including ESA geodesic programmes. Recommendations to ESA on geodesic requirements for implementation in international surveys. The company has conducted a wide range of national and international scientific co-operation and liaison activities involving universities and research institutes in the Netherlands and other European countries.

VERIFIED

Prins Maurits Laboratory (PML) TNO

Current Status
PML is one of the three defence research institutes of the Netherlands Organization for Applied Scientific Research (TNO). PML has about 270 employees in two research and development departments that includes propulsion technology. It developed the gas generator igniter, thrust chamber igniter and turbine pump starter for Ariane 5's Vulcain cryogenic engine, in addition to undertaking research and development of high-performance solid-propellants based on phase-stabilised ammonium nitrate and hydrazinium nitroformate for future spacecraft. Facilities include an indoor test stand for solids, a connected pipe facility for ramjets, mixers for propellant and ramjet fuel manufacturing, and equipment for determining the chemical, mechanical and burning properties of propellants. The laboratory seeks to perform, and conduct, ground breaking research into new and exotic propellants that show results in tests commensurate with substantial improvements in launch vehicle performance. Research into advanced propellant combinations and tri-propellants supporting high performance and high Isp propulsion concepts. Has specialised in matching new and innovative propulsion concepts with projected launcher and in-space applications.

UPDATED

PTT Telecom Netherlands

Current Status
PTT Telecom Netherlands is the national signatory to Intelsat, Inmarsat and Eutelsat. Its principal Earth station is at Burum in the north, with smaller stations at Amsterdam, Rotterdam and Nederhorst den Berg. In addition to the antennas listed below, two Inmarsat A/C land Earth stations were installed at Burum to support mobile communications over the Atlantic/Indian regions from 1991. Extensive evaluation of improved efficiency and continuity of service relying on terrestrial back-up.

The dates given below indicate year of installation. Burum 1 28.5 m, 4/6 GHz Intelsat A, AOR, 1973; Burum 2 32.0 m 4/6 GHz Intelsat A, AOR, 1978; Burum 3/4 32.0 m 4/6 GHz Intelsat A, IOR, 1984; Burum 6/7 9.0 m ECS-SMS/IBS, 1988; Burum 8 9.0 m 11/14 GHz ECS 1988. Amsterdam 3 9.0 m 12/14 GHz ECS-SMS/IBS, 1988; Rotterdam 1 6 m 12/14 GHz SMS, 1985; Nederhorst den Berg 1 11 m 11/14 GHz Eutelsat, 1984; Nederhorst den Berg 2 4.2 m 11/14 GHz Eutelsat, 1989.

VERIFIED

Space Research Organisation Netherlands

Current Status
The Stichting Ruimteonderzoek Nederland (SRON) was founded 10 June 1983 within the Netherlands Organization for Scientific Research (NWO) in The Hague, by which it is financed, to manage/execute space science projects. SRON comprises two laboratories, in Utrecht and Groningen, with specific interest in x/υ-ray and IR/sub-mm astronomy. The Institute acts as the national agency for space research and as the national point of contact for ESA programmes. SRON operates in an advisory role for government and also conducts analysis of capabilities and makes recommendations to the government regarding future programmes. Current and recent projects include detectors for NASA's Compton Observatory and AXAF, two wide-field cameras for SAX, the short wavelength spectrometer for ESA's ISO, the SCIAMACHY instrument on ESA's ENVISAT and a reflection grating spectrometer for ESA's XMM cornerstone mission. SRON is responsible for the development of a Heterodyne Instrument for Far Infrared (HIFI) on the Herschel observatory was due to be launched in 1997. SRON also manages Dutch μg, radiation, exobiology and Earth-oriented research. Some 10 per cent of its G20 million annual budget derives from commissioned research. This is set to increase with an involvement in major international research programmes sought by SRON.

VERIFIED

Technisch Physische Dienst TNO-TU DELFT

Current Status
TPD, the TNO Institute of Applied Physics, is an element of Netherlands Organization for Applied Scientific Research (TNO). Space personnel total 24. TPD began space instrumentation research and development in 1964 with the S59 UV stellar spectrometer for Europe's TD-1A astronomy satellite. It now specialises in attitude sensors, science instrumentation and Earth observation sensors. For example, the department provided Giotto's star mapper and Eureca's Sun acquisition sensors. Science instruments cover the wavelength range from hard X-ray up to the far-IR/mm. Recent projects include Hipparcos' modulation grid and refocusing mechanism, and ISO's cryogenic Short Wavelength Spectrometer. Remote sensing activities are concentrating on instruments for Meteosat Second Generation, Envisat and Metop, including optical systems, detection techniques, focal plane layout and precision mechanisms. Examples are the MIPAS cryogenic focal plane assembly (4 to 15 μm) and SCIAMACHY's optical bench. It was intensively involved in the development of the GOME Global Ozone Monitoring Experiment for ERS 2. TPD supplied the onboard calibration unit and performed the whole instrument's ground calibration under ESA contract. For that, TPD developed a calibration

facility unique in Europe; it will also be used for the calibration of SCIAMACHY and other atmospheric science instruments, such as MERIS. Development of new sensor instruments and research into alternative sensor suite packages for remote sensing, meteorological and earth observation payload requirements. Has taken a lead role in projecting a new generation of requirements for next-generation atmospheric sciences satellites.

UPDATED

Norway

Norsk Romsenter

Current Status

The Norwegian Space Centre is an independent organisation established by the Ministry of Industry 5 June 1987 as the immediate successor to NTNF, Space Activity Division. Including daughter companies there are 75 personnel. Its mandate includes compiling and submitting to the Ministry recommendations for co-ordinated long-term programmes. In particular, it represents Norway in ESA and co-ordinates the space-related activities of Norwegian companies, universities and research institutes. Maintains policy planning and future mission recommendations for government. The Andoya Rocket Range and Tromso Satellite Station are both limited companies, with the Norwegian Space Centre having 100 per cent and 50 per cent interests. The organisation also conducts research into international projects and programmes likely to benefit from Norwegian science. Norsk Romsenter has been integrated with several levels of government in an attempt to maximise the potential economic benefit of space-related activities.

UPDATED

Norwegian Meteorological Institute

Current Status

The Norwegian Meteorological Institute (DNMI) is the national signatory to EUMETSAT. It provides operational meteorological services to national users, including reception, processing and dissemination of NOAA and Meteosat imagery/data products. Co-ordinates land-, air- and space-based data gathering assets. The DNMI has extended its coverage of data sources and expanded its inventory of activities, providing marine and coastal waters survey analysis and marine life monitoring processes integrated with atmospheric studies.

VERIFIED

Telenor Satellite Services AS

Current Status

TNSS, a Telenor subsidiary, is the national signatory to Intelsat, Inmarsat and Eutelsat. The country is Inmarsat's fourth largest member, with 6.8 per cent, complementing the size of its maritime fleet and emphasising the importance of the major Earth station at Eik. TNSS is among the largest VSAT and gateway providers in Europe, with an installed base of 2,700 terminals by 2002. Telenor is a member of the Skyphone consortium providing aeronautical communications services through the Inmarsat system. Has developed innovative concepts for integrating space-based and terrestrial communications systems. Telenor is advancing the concept of satellite linked broadband transmissions to and from major hub sectors allowing high data rates within cable links at terrestrial level.

VERIFIED

Pakistan

Space and Atmospheric Research Centre

Current Status

Through SARC, SUPARCO has embarked on a programme designed to significantly improve the use of satellite data for civilian and military applications, enhancing utilisation and effectiveness of the financial outlay. Expanded funding base for utilising space assets in national resource and security considerations in so far as they pertain to enhanced weather prediction. SUPARCO has expanded its research programmes designed to accommodate expanded government requirements for civil and military space activities.

Background

Located near the University of Karachi on a 0.2 km^2 site, Sparcent is the main establishment of the Space and Upper Atmosphere (SUPARCO) agency and encompasses Satellite Communications, Remote Sensing Applications, Agriculture and Landuse Applications, Remote Sensing Operations, Space Science and Ionospheric Research divisions. The campus houses a tracking and telemetry facility, which served as the prime TT&C station for Badr 1, an online information retrieval facility, Local User Terminal data collection platforms operating via the Argos system, an APT station and a NOAA TOVS acquisition and processing ground station to extract vertical atmospheric profiles for 45 parameters (such as water content and T) up to 35 km. Suparco operates:

The Remote Sensing Applications Centre (Resacent) functions under the Remote Sensing Applications Division as the national centre responsible for remote sensing data acquisition and applications co-ordination.

Suparco Plant, Karachi: manufacture of propellant and sounding rockets. Static test facilities to handle full size motors are maintained.

Flight Test Range, Miani Beach: Suparco's 200 hectare launch site 50 km northwest of Karachi at about 24.5° N/66.5° E overlooking the Arabian Sea equipped with modern facilities. Several Shahpar 7 m long 50 cm diameter sounding rockets are launched annually, capable of delivering 55 kg to 450 km.

Instrumentation Laboratories, Karachi: provision of flight, tracking, telemetry and sensing instrumentation for sounding rockets.

Satellite Ground Station, Islamabad: acquisition of Landsat TM/MSS, Spot HRV (10 m Cassegrain dual feed) and NOAA AVHRR (2.5 m parabolic) imagery; operational since May 1989. Facilities include data reception, acquisition and pre-processing incorporating high density digital tape recorders and two VAX 11/780 computers, 5210 FPS Array Processors and I^{2}S Image Terminals. EOSAT signed an agreement in 1991 for marketing rights to the Landsat data. SGS covers about 20 countries, including most of Central Asia. In addition to standard products, CD-ROMs will soon be available.

The Pakistan Meteorological Dept operates four NOAA APT stations at Karachi, Islamabad, Lahore and Peshawar.

Space Applications and Research Centre (Sparc), Lahore: established near the University of the Punjab, this Centre will develop into one of Suparco's main facilities for satellite development and space applications. It is currently engaged in satellite tracking, telemetry and reception, processing and analysis of science data. A Cospas/Sarsat Local User Terminal and Mission Control Centre were commissioned in April 1990.

Computer Centre, Karachi: computing facilities in support of Suparco's activities.

Aerospace Institute, Islamabad: the institute aims to satisfy Suparco's growing need for qualified manpower in space science and technology. In addition to regular postgraduate courses, it provides short preparatory courses for scientists and engineers travelling abroad for advanced training. Construction completed in 1995.

UPDATED

Space and Upper Atmosphere (SUPARCO)

Current Status

Suparco is the national space agency for execution of space technology, science and applications research programmes. It was established in 1961 under the chairmanship of Professor Abdus Salam and re-organised in 1981 as an autonomous commission under the aegis of the 10-man Space Research Council, which is responsible for formulating national space policy, headed by the Prime Minister. The Executive Committee of the SRC, headed by the Federal Minister for Finance, monitors Suparco on behalf of the SRC. Suparco is headquartered at Karachi. Suparco has been given responsibility for integrating civilian and military space applications and for directing optimum ways in which national sercurity can be enhanced. These have received subsidies to increase their effectiveness and to provide enhanced availability of space assets for earth observation. Pakistan is progressing with a study to determine optimum ways in which space assets and space data, from other national satellite programmes, can be used in serving the economic and security interests of the country and Suparco is central to that effort.

VERIFIED

Poland

Copernicus Astronomical Centre

Current Status

The centre specialises in astrophysics studies. The CAC played a major role in forming ESA astronomical and space astronomy plans. It is currently participating in ESA's Integral mission and is researching new ESA-led initiatives. Integral encountered technical problems in some instrument engineering and the previously anticipated launch date of April 2002 was deferred with launch eventually taking place on 17 October 2002 by Proton from Baikonur.

VERIFIED

Institute of Aviation

Current Status

The institute provides R&D in space science instrumentation, including spectrum analysers, telemetry systems and GPS receivers. Conducted essential research in micro-instrumentation for robotic spacecraft and advanced satellites with high levels of self-modulation and control. It has contributed instruments to Interkosmos 19, Vega and Phobos and most recently spectrum analysers for the ASPI and POLRAD instruments aboard Interball. The Institute has begun development of instruments for remote serving satellites and is marketing some innovative technologies for export to Russian companies. Contributions to Russia's PAKAT remote sensing investigations programmes designed to examine earth surface morphologies. The Institute has brought together several distinct technologies each capable of separate investigations.

UPDATED

Institute of Geodesy and Cartography

Current Status

The only national state-owned geodetic research institute (a principal aim is development/promotion of space technology for geodesy and remote sensing) the Insitute of Geodesy and Cartography has had a long and prestigious background dating back to the International Geophysical Year of 1957–58 and has been the focus for international projects involving Poland's national and government resources. About 40 of the 120 staff are engaged in space research. The geodetic observatory in Borowa Gora is constantly monitored by GPS to detect its geodynamic behaviour within research programs such as the International Geodynamic Service and EUROPROBE. The institute is developing systems to update topographic maps using satellite imagery and to incorporate remote sensing data in GIS. This includes interpretation of satellite images to monitor lake water pollution, forests, crop forecasting, etc. It is a Spot Image distributor. The institute is preparing recommendations on requirements for a new generation of European remote serving satellites and has developed new services for bidding on RFPs. The IGC pursues imaging and radar data to expand environmental monitoring and pollution clean-up activities.

VERIFIED

Institute of Meteorology and Water Management

Current Status

The Institute's satellite centre handles NOAA and Meteosat digital and analogue data as well as archive data from concurrent programmes. Compiled historic weather data from NOAA satellites and provides national research centres with meteorological analysis for forecast and climatic analysis of trends impacting national interests. Processing of NOAA TOVS data for atmosphere temperature, moisture and ozone vertical profiles and of NOAA AVHRR for sea surface temperature, soil temperature and moisture, cloud classification, Baltic Sea ice, and vegetation index changes. Has begun integrating air and satellite data sets for Baltic sea life health studies. Indices thus developed are primarily a base for environmental pollution mapping and migration paths.

VERIFIED

Polspace Ltd

Current Status

Polspace represents almost the entire national potential in space-related technology. It inherits a rich legacy from the former days of the communist rule and has gathered together a credible and highly experienced base of expertise and capabilities. Membership includes SRC, Institute of Aviation, Warsaw University of Technology, Institute of Geodesy and Cartography, Institute of Telecommunications, Telecommunications Research Institute, Institute of Automation and Measurements, Institute of Precise Mechanics, Machine Tools Research and Design Centre, Polish Aviation Factory PZL-Mielec, Diginet Ltd, MAG Ltd, Milemi Ltd and NAVI Ltd. Polspace makes recommendations on national space policy and issues related to international co-operation. Polspace extends representation for Polish companies abroad and assists foreign companies in joint ventures. Polspace has begun discussions with ESA and the RSA over possible co-operations at an industrial level. Polspace has conducted negotiations with industry participants in an effort to increase its access to foreign technology ventures.

UPDATED

Space Research Centre

Current Status

The SRC was established in 1977 to conduct and co-ordinate national space research and to develop international programmes. 1995 budget was Zl 53.1 billion (1994 Zl 41.4 billion) with a staff total of about 145. Staff levels are now believed to have been substantially reduced. Principal interests are solar system and space physics, geodynamics and satellite geodesy and remote sensing. Current projects are Interball (plasma waves), Coronas (solar X-ray spectroscopy), Mars-96 (plasma waves, atmosphere infra-red spectrometry), Relikt 2 (neutral ions, plasma waves), Cassini (Titan's surface temperature), ISTOK 1 (atmosphere infra-red spectrometry from Mir's Priroda module) and VIZIR (advanced Mir TV). Research into ESA driven atmospheric and astrophysical research projects. The organisation has been completely re-evaluated to emphasise the need for research in areas which can benefit the nation and its technology transfer programme.

VERIFIED

Telekomunikacja Polska SA

Current Status

Polish Telecom is the national signatory to Inmarsat, Eutelsat, Intelsat and Intersputnik and has sought integration of national and regional systems. PT has performed infrastructure upgrades to improve operating capabilities. The PT has begun discussions about leasing purchased capacity to businesses and industries.

UPDATED

Portugal

Instituto Nacional de Engenharia e Tecnologia Industrial (INETI)

Current Status

INETI operates within the Ministry of Industry and has a staff of about 1,200 as of July 2001, specialising in biotechnology, materials, energy, opto-electronics and computer sciences. Space activities include remote sensing and cartography (which have also involved the Institute of Information Technology). INETI has also acted as a defence research lab in opto-electronics and space. Off shore marine surveys utilising available real-time and radar data together with sensors on ESA's Envisat. Extended use of this technology for international projects and ESA land resource inventory support using European and US satellites for the database.

UPDATED

Russian Federation

Association for the Advancement of Space Science and Technology (ASCONT)

Current Status

ASCONT was founded in 1992 by IKI, Lavotchkin, NIIKP, Lunokhod and NII Radio to promote Russia's science and technology capabilities but has largely ceased to be influential in national space policy since the collapse of the Soviet Union. Was cited by Russian trade organisations as an instrument for multinational co-operation and joint ventures. Elements of the old organisation have continued to exist under a different organisational structure and it is uncertain which departments have survived and which have been absorbed by other government agencies and affiliates.

UPDATED

AKO Polyot

Current Status

Polyot designs and builds the Cosmos 3M satellite launcher (marketed by the Cosmos USA joint venture) and has been instrumental in the development of several specialised spacecraft types. It has built the Glonass, Tsikada, Cospas, Parus, Strela, Gonets and Informator satellites in conjunction with designer NPO PM. It built the RD-170 engine for NPO Energomash, but production has halted and the plant is now dismantled.

Polyot was created in 1941 to build aircraft. It employed 20,000 as of 1994 and approximately 10,500 by 2003. Production of the An-74 cargo aircraft resulted in the PO designation changing to AKO.

VERIFIED

Astelit

Current Status

NPO Astra and Italy's Nuova Telespazio worked together on the Astelit joint venture, founded in February 1992, and providing Russia with reliable links into the international telephone network using a Raduga satellite. Studies on the utilisation of other satellites have been conducted. Efforts have been made to select alternative satellites for interconnects between national and regional assets. The organisation has failed to attract investment in further expansion of the concept and there has been no additional development. Contact with Telespazio continues to progress and the links on existing programmes are retained.

UPDATED

Central Specialised Design Bureau, Progress Division

Current Status

CSDB (TsSKB) was founded in 1959 as a division of Korolev's OKB-1 Moscow design bureau to provide design support for launcher manufacturing at the 'Progress' factory, where the SL-4 Soyuz and SL-6 Molniya vehicles and satellites are now built. It is the parent of KB Photon. Employees, including KB Photon, totalled less than 2,500 in early 2003. CSDB has reduced manufacturing levels responding to fluctuations in demand. Cost effective measures introduced to eliminate outages of resources and personnel. Its future appears assured as a principal supplier of launchers and satellites to the Russian Ministry of Defence.

VERIFIED

Centre for Space Biotechnology

Current Status
The Institute of Biorganic Chemistry, a part of the Academy of Sciences research microgravity and provides equipment for research. Has conducted experimental tests with a large, medium and small biorack systems for application to the ISS, the Russian ISS pressure modules or the mid-deck lockers of the Shuttle. Partners ESA bio-physical institutes engaged in multinational research with a view to developing new channels of scientific activity. The Centre has produced several experiments carried aboard the Mir space station and has commenced development of new generation experiments for the ISS. These have been directed toward research activities contracted to the Centre for Space Biotechnology by industrial aerospace companies interested in the renewal of an autonomous Russian human space flight programme. Some studies have been conducted on the experiments that might be carried aboard an independent Russian earth orbiting man-tended platform.

UPDATED

ESKOS SA

Current Status
ESKOS was registered in July 1992 in Paris to promote the products and services of Russia's space industry in the post-Soviet era. In this role it had the capacity to represent Russian industrial capabilities but not to conduct or close deals between Russian an foreign clients. Proactive marketing in conferences and seminars linking Russia's space companies with foreign organisations and companies. Major shareholders include the COSMOS concern (leading Russian space enterprises and institutes) of the Russian Space Agency and the COSMOS International Centre for Advanced Studies based at the Moscow Aviation Inst. ESKOS signed a co-operation agreement with the Russian Space Agency 10 May 1993. ESKOS has been largely subsumed into other organisations although some elements remain for representation of space industry interests to foreign governments.

VERIFIED

AO Gazcom

Current Status
RKK Energia and Gazprom, Russia's state natural gas monopoly (94 per cent of Russia's production, 21 per cent of world's), created Gazcom in 1992 to develop the Yamal telecom satellite system for supporting transcontinental transport and communications through remote station relays and terminals. Gazcom is the owner, operator and service provider. Gazprom uses the satellites to communicate with its gas fields, particularly in Siberia. Gazprom were already using 30 stations operating through Gorizont by 1995, with another 75 over 1995–96. By June 2004 there were more than 120 earth stations operating through three teleports. Satellite control is maintained through a mission complex in Moscow. Yamal 1 was launched on 6 September 1999 along with Yamal 2. The system and satellite design are also being offered commercially and talks have been held with Russian and non-Russian companies. Energia, Gazcom and Loral Corp formed a joint venture 9 February 1996. Some of the communications payload is being bought from Space Systems/ Loral, and Gazcom are jointly marketing the design with SS/L. RKK Energia's advanced design includes Xe ion thrusters, CPV Ni/H_2 battery and unpressurised compartments. Low cost nano-satellites conceived and proposed for small budget operators. New specifications for next-generation Gazprom requirements.

UPDATED

Glavcosmos

Current Status
Glavcosmos USSR was established in 1985 under the Ministry of Machine Building (MOM) as the prime authority for implementing co-operative agreements, including commercial utilisation of Soviet systems and flying foreign cosmonauts aboard Soviet spacecraft. Responsibility for manned missions was subsequently transferred to RKK Energia. It acts as the technical and management organisation for commercial space systems. Glavcosmos was instrumental in starting the discussions resulting in the commercial availability of Soviet launch vehicles, first contacts being made with Inmarsat. Intermediaries in helping set up financial arrangements for joint ventures, sale of Russian space services or funding of private participants as cosmonauts on Soyuz flights to the ISS.

Specific deals are negotiated with the Licentsintorg foreign trade agency. Glavcosmos (dropping the USSR in late 1991) now comes under the Department of Machine Building within the Russian Ministry of Industry. Following the creation of RKA, it now appears more concerned with converting military technology to civil applications. It is responsible for the sale of the cryogenic engine to India for its ESLV programme and for promoting the sale of small launcher technology. It has also conducted several negotiations with undisclosed countries for rocket motor technology and launch vehicle lower stage configurations.

VERIFIED

GNPP Kvant

Current Status
Set up in 1987 Kvant builds solar cells and arrays for space applications.The origins of the company go back decades and has been involved in power production systems since the late 1950s.They have been involved in producing solar cell arrays for almost every Russian satellite and spacecraft so equipped. They conducted research into new forms of deployment for array configurations and were instrumental in developing the arrays for the early Soyuz spacecraft concept. Scientific facilities include electron microscopy, micro x-ray analysis, secondary ion spectroscopy and photo cathode electro-fluorimetric analysis. Expanding research in large solar array configurations for high-power requirements. Developed concepts for unfurlable antenna arrays as a spin-off application. Kvant has conducted several engineering studies for companies in India.

UPDATED

Institute of Biomedical Problems (IBMP)

Current Status
IBMP, founded in 1963, is the Russian Academy of Science's premier life sciences research centre. The Institute conducted extensive research into biophysical reactions to weightlessness and conducted several ground-based research programmes based on results fed back from the cosmonaut team. It is responsible for cosmonaut medical care under the Principal Medical Commission, chaired by Grigoriev and the head of the Russian AF's medical section. Poliakov is head of the physician-cosmonauts group, which is based at the institute. He set a space duration record on his return to Earth in March 1995. The Institute has conducted extensive studies into the medical effects of long-duration space flight and has performed several experiments into ways in which the adverse effects can be countered.

IBMP also heads the international Bion biocosmos missions, maintaining a control room on the premises. IBMP has participated in formulating life science experiments aboard the International Space Station and is reportedly at the forefront of developing a new EVA suit for Russian cosmonauts. Spasmodic liaison with NASA Johnson Space Center on proposal to define and design a candidate suite for all ISS EVA operations.

UPDATED

Izmiran

Current Status
The Institute of Terrestrial Magnetism, Ionosphere and Radio Wave Propagation of the Academy of Sciences is Russia's principal investigator of the magnetosphere, ionosphere, Sun and Sun-Earth phenomena. It operated the Coronas-I solar observatory launched in 1994 from Plesetsk, the first combined Russian/Ukraine helioseismology satellite. It was to be followed by Coronas F originally planned for launch in 1997 and then for 2001. Izmiran also provides instruments, particularly fields and particles measurement devices for satellites and is developing devices which it believes it can sell to ESA countries. Proposes major set of small scale upper atmosphere sounding rocket programmes to accommodate incremental payload packages. Has proposed a series of internationally funded near-earth space physics experiments but has not yet received approval.

UPDATED

KB Arsenal

Current Status
Collocated with PO Arsenal, historically a major producer of artillery, KB Arsenal provides the bus for the military Eorsats to TsNPO Kometa. They now offer science and commercial satellite design and are developing a new family of bus designs applicable to civil and military applications. Conceptual work on secondary military piggy-back payloads operating as autonomous vehicles and hybrid satellite and spacecraft bus configurations which, due to a downturn in global market uptake, have not yet succeeded in orders. Much of this work has been abandoned in favour of more traditional engineering contracts.

UPDATED

KB Khimautomatiki

Current Status
The bureau of Semyin A Kosberg (1903–65) was formed in 1941, and began investigating liquid propellant rocket propulsion for aircraft in the early 1950s. It later became responsible for some Soviet space launcher upper stages. In 1974 it was reorganised as KB Khimautomatiki. Developed hybrid propulsion systems using mixed solid-liquid propellant combinations, tri-propellant systems and long duration motors using high temperature burning cycles.

Responsible for Energia's RD-0120 cryogenic core engine, Proton's stage 2/3 engines and Soyuz/Molniya stage 2 engines. Formed from old Kosberg upper stage propulsion bureau. Hybrid propellant development. New chemical research facilities for supporting launch vehicle and rocket motor development.

VERIFIED

KB Khimmach

Current Status
A bureau headed by Alexei M Isayev was established in 1944 for research into storable liquid propellant rocket engines. Whereas Glushko's GDL concentrated on major launcher engines and Kosberg's bureau on upper stage propulsion systems, Isayev's produced smaller spacecraft engines, including manned, Earth orbit and deep space systems. A number of designations have

been adopted but 'KTDU' (Korrektyroushchaya-Tormorznaya Dvigatelnaya Ustanovka, or Corrective-Braking Rocket Motor) is most frequent, with the others variations on this theme.

Information released in early 1991 indicated the bureau is part of NPO Soyuz in Kaliningrad. It is responsible for the 73.58 kN cryogenic engine of the new Proton KM version, originally designed for the L-3M manned lunar mission, and sold to India for its GSLV vehicle. A newly-developed DMT-600 600 N NTO/UDMH thruster was referred to in 1994. Khimmach developed all of the engines for the Heavy Cosmos TKS transport craft. Continued with research into new and improved forms of hybrid propulsion systems for special requirements set out by the Ministry of Defence. Produces the 11D49 engine for the second stage of the Cosmos 3 launch vehicle.

Starting in 1971 and at the instigation of Isayev, the bureau started the development of low-thrust engines, which could be used for attitude control of spacecraft. A series of 11 engines with thrusts up to 2,206 kN (225 kg) and eight with thrusts up to 49 kN (5 kg) were developed. They are capable of burns lasting from hundredths of a second through to hundreds and thousands of seconds. Considerable progress has been made with this technology since the mid-1990s. Aspirations to extend research activities into ion propulsion induced some early efforts in this area but these have been halted.

UPDATED

KB Kolomna

Current Status

Kolomna specialises in surface-air missile warheads and anti-tank guided missiles, but also offers the solid-propellant Sphere geophysical research rocket to deliver 230 kg payloads up to 330 km. Sphere employs guidance systems recovered from SS-23 missiles and adapted for science research purposes. Kolomna has successfully applied SS-23 guidance technology to other, unspecified, research rocket projects. Development of new sounding rocket concepts for marketing to scientific and research organisations, institutes and universities. It has concentrated on developing ideas for a new family of sounding rockets and suborbital high-altitude boosters. There is also some cross development of civil and science rockets with short range military missiles.

UPDATED

KB MAKEYEV

Current Status

V P Makeyev (1925–1985) founded the bureau in 1955 as an offshoot of Korolev's bureau and produced Russia's first ballistic missile to use storable propellants, the R11, a development of the German Wasserfall SAM. It developed the SS-N-6 Zyb, SS-N-8 Vysota, SS-N-18 Volna, SS-N-20 and SS-N-23 Shtil SLBM. KB Makeyev was involved in the development of the Surf/Priboy satellite launcher under an agreement between the US Sea launch organisation and Russia's Ramcon Association. The bureau was responsible for several unsuccessful designs and its research department developed a wide range of propellant combinations, chemical matrices and motor designs that characterised rocket development in Russia during the 1960s.

The Makeyev bureau offers satellite launch services using its converted SLBMs. A Volna was launched 7 June 1995 from the 'Kalmar' Delta 4 boat submerged in the Barents Sea on a 20 minute suborbital µg flight carrying a 120 kg experiment from Germany's ZARM (electrically-driven thermal convection in a fluid shell between two concentric spheres). Recovery was 5,600 km downrange on Kamchatka. There were apparently three SLBM demonstrations 1991–93. A Shtil 2N orbital demonstration launch was reportedly planned carrying a 100 kg Izmiran satellite. Major activity by the end of the 1990s was conversion of SLBMs into potential satellite launches.

Discussions with potential customers for modified Surf/Priboy variants. A wide range of options continues to be developed with few indications of solid customer orders.

UPDATED

KB Photon

Current Status

KB Photon designs and builds a range of spacecraft derived from the original Vostok as a result of taking over Korolev's Zenit photo-reconnaissance satellite programme in 1964. This covers the Resurs-F Earth observation film camera satellite, the Photon µg vehicle, the international Bion biocosmos, and the Zenit Gen 3, Yantar Gen 4 and Kuban Gen 6 photo-reconnaissance. Kuban derivatives will apparently replace Resurs-F, Photon and Bion. It operates under the Central specialised Design Bureau. The satellites are built at the Progress factory in Samara and the technology has been supplied to other countries for substantial retrofit of mission packages launched by Russia. Adapted launcher interface module capable of housing very small secondary payloads for clusters of low-cost operators. The Photon microgravity spacecraft has a mass of 6,200 kg and provides a continuous 400 W (700 W maximum for 90 minutes in any 24 hr period) of electrical power for science experiments. Photon spacecraft are placed in modestly eccentric orbits to minimise perturbation. Photon has carried Zona 1, Zona 4, Zona 4M, Splav 2 and Konstantina 2 electric furnaces and the Kashtan electrophoresis unit. Orbital lifetime is usually 16 days with a total payload capacity of 404 kg. Prototype Photon spacecraft were launched as Cosmos vehicles 1645 in April 1985, Cosmos 1744 in May 1986 and Cosmos 1841 in 1987 but since 1988 the spacecraft has been referred to by the Russians as Photon with Photon 4 being launched on 12 April that year.

UPDATED

KB Salyut

Current Status

KB Salyut's history can be traced back to 1951 as Vladimir Myasischev's (died 14 October 1978) OKB-23 bureau to develop strategic bombers. The first was the M4 Bison, still in service. A modification has transported Buran and Energia's core. OKB-23 created the first spaceplane designed in the Soviet Union, a single-seat vehicle with a faceted, trapesoidal wing, presented in 1957 as an alternative to ballistic spacecraft such as Russia's Vostok design. In 1959, Vladimir Chelomei took over part of the aircraft bureau (including the Khrunichev factory) to develop ICBMs. It produced the SS-11 and its follow-on derivative the SS-19 and Proton. Chelomei's OKB-52 was responsible for designing the Almaz military space station and its three-man ferry craft. It performed the detailed design work on Salyut and Mir under NPO Energia's general direction; it also built the FGB and Service Module for the International Space Station. It provided the re-entry vehicle for the German-Japanese Express µg project. In 1976 the Design Bureau was separated from Chelomei and became a branch of NPO Energia run by Valentin Glushko. In 1985 the Bureau left Energia and became independent. In 1994 Salyut KB merged with Khrunichev Machine Building Plant into the Khrunichev State Space Science and Production Centre (GKNPTs). Dmitri A Polyukhin was KB Salyut's General Designer until his death on 7 September 1993 when he was succeeded by Anatoliy K Nedayuoda. KB Salyat has developed cryogenic kick stages for India's GSLV delivered through Glavkosmos. Further development of this stage as a potential export opportunity has failed to attract interest but the design technology has been separated and applied to other propulsion systems in different designs sold to unnamed third parties.

VERIFIED

Keldysh Institute of Applied Mathematics

Current Status

IPM was founded in 1953 as the Division of Applied Mathematics of the Institute of Mathematics of the USSR to perform mathematical research and modelling for the nuclear and rocket industries. It was later promoted to a separate Institute of Applied Mathematics subordinate to the Soviet Academy of Sciences from which emerged the Institute of Space Research (IKI) in 1965. General directions of research include mathematical modelling of sophisticated phenomena and structures including applications, theory of numerical methods, theoretical and applied celestial mechanics, non-linear analysis, systems software development, applied systems software and pattern recognition. It's Ballistic Centre provides navigational computing expertise for manned and unmanned spacecraft. It also conducts mathematical analysis of planetary body trajectories and designs guidance systems for deep space vehicles. Has conducted tests to determine optional re-entry paths for planetary probes traversing at off-nominal angles and flight paths for optimised science data sensor instrumentation. Has applied designed computer models for ballistic trajectories applicable to fast entering atmospheric probes with military applications as well as planetary.

UPDATED

Keldysh Research Centre

Current Status

The Keldysh Research Centre – previously known as RNII, NII-1 and then NIITP – was established in 1933 as the first institute for research and development of liquid and solid propellant rocket engines. It currently develops, manufactures and tests advanced prototype of rocket engines, space power systems, high-energy beam generators and particle accelerators of various types. Its products have been used in the development of launch vehicles, spacecraft and orbital stations. Before and during the Second World War the centre concentrated on ground-based and air-launched weapons, the major achievement at that time being the development and manufacture of the first batch of Katyusha missiles, for which the centre received the Order of the Red Star. During its history the centre boasted having M V Keldysh, S P Korolyov and V P Glushko working within its organisation.

In rocket motor development, KRC has developed staged combustion cycle motors and pioneered work on highly reliable engines of this type for Proton, Zenit and Energia launchers. Hybrid rocket motors are being developed utilising solid fuel and liquid or gaseous oxidiser intermediate in Isp between solid rocket motors and cryogenic, high-energy, motors. New moveable nozzles are being designed in the Nozzle Differential Facility as are ultralight uncooled composite nozzles reducing engine weight by 12 to 15 per cent.

In general, space activity KRC has researched energy transport in space and from space to earth and into solar dynamics power systems on a Brayton cycle gas turbine with 10 kW output power. Also, research into large solar electric propulsion systems for powering satellites and orbital transfer vehicles. Also support systems include the droplet-coolant radiator for powerplant weight reduction applications.

KRC also conducts research into electron beam generators of 35–1,000 kW, 30–300 KeV, negative ions producing a beam current of 1 mA, 1–20 KeV.

VERIFIED

Khrunichev State Research and Production Space Centre

Current Status

The history of GKNPTs Khrunichev began in 1916 when the factory was established in Fili for building cars by the Russian-Baltic Carriage Factory Joint Stock Co. It began aircraft construction in the 1920s, initially leased to Junkers. In 1927 Plant 22 was established to build metal aircraft for domestic use including types designed by Tupolev, Arkhangelskiy and Petlyakov. In 1941 after the outbreak of war with Germany Plant 22 was moved to Kazan where it remained and is now known as the Kayan Aviation Production Association. In Moscow, on the original site of Plant 22, a new plant, designated No 23, was set up to build bombers, including the turbojet Tu-12 and Tu-14 of 1947–49. In 1951 Plant 23 built Myasishchev bombers from which date it became OKB-23 under the Moscow Council of People's Economy. The plant was renamed after the late Minister of Aviation Industry, Mikhail Vasilievich Khrunichev, on 3 July 1961. On 30 October 1960 OKB-23 had been transferred to OKB-52 headed by V N Chelomei for the production of rockets and missiles and, later in the decade, the development of the Almaz spacecraft. In 1965 the Khrunichev Plant came under the Ministry of General Machine-building of the USSR but after this structure was disbanded in 1991 it became part of the Kompomash association. In 1997 it reported to the Ministry of Economy and, in May 1998, control was transferred to the Russian Space Agency. This move encouraged research and test activity in hybrid propulsion systems. In 1994 Khrunichev merged with KB Salyut to become the M V Khrunichev State Space Science and Production Centre (GKNPTs) after which it received a role in development of spacecraft and launchers. At this point Khrunichev accessed several technologies for exotic propellants and built its own research facilities for high-pressure propellant and combustion systems. These funded new research and development structures for potential competition with parallel organisations. Serial Proton production has been undertaken since the 1960s, followed by the SS-19 missile (now available as the Rockot satellite launcher) and space station modules (Salyut, Mir, Almaz, Kvant, Kristall, Priroda, Spektr, FGB and Service Module), these latter elements for the ISS. Khrunichev is actively developing the Angara series of modular launch vehicles and the Baikal reusable launch vehicle which could fly on Angara stages. The ambitious Baikal incorporates a deployable wing for lifting entry followed by deployable air-breathing engines for controlled flight in the atmosphere preceding touchdown.

UPDATED

Koskon

Current Status

Koskon (Space Conversion) Global Space Communication System is the creation of a 15-member consortium involving the former Soviet PTT, Metallurgy, Electronics and Health Ministries, and the Moscow Aviation Institute; industrial contractor is AKO Polyot. The proposed Koston system would require 32-45 Informator class satellites. The 860 kg satellites built by NPO Polyot would be placed in four 1,100 km 83° orbits, offering 4.8 kbit/s real-time and store/forward voice/data traffic to fixed and mobile terminals. User up: 320.8322 and 1,642.5 to 1,634.4 MHz; down: 290–320 and 1,555 to 1,555.9 MHz. Intersatellite links would be at S-band (2,056 to 2,060 and 2,2402,244 MHz). Feeder links would be at C-band. Polyot's Informator 1 launched on 29 January 1991 was a Koskon demonstrator. None of these plans were realised due to financial difficulties and political difficulties as the USSR collapsed but several initiating bodies assembled for the initial proposal leading to Koskon have continued to campaign for development of the system.

VERIFIED

Lebedev Physics Institute Astro Space Centre

Current Status

The P N Lebedev Physical Institute of the Russian Academy of Sciences was the leading research institute for physics research when a department of IKI separated and became an affiliate with the name Astro Space Centre in 1991. The Astro Space Centre is leading the Spektrum-R/Radioastron VLBI project. It has diversified and conducted several research activities with ESA and with China and has sought agreements with states of the former USSR and with potential markets in the Latin American countries. The Centre continues its research into high-energy physics and astro-related technologies.

UPDATED

Lyulka-Saturn OAO

Current Status

Set up in 1944 as a gas turbine design bureau under Arkhip M Lulka, the bureau specialises in military aircraft engines. It also provides turbomachinery and gas generators for space applications, including Buran's gas generators. The 400 kN LOX/LH_2 D57 engine was developed in the 1960s to provide the lunar N1 launcher's third and fourth stage. Lunar project work was cancelled in 1972 and sometime in that decade the bureau was reorganised into the Saturn Science and Production Association (NPO). After privatisation in the 1990s it became the A Lyulka-Saturn Joint-Stock Company. Lyulka teamed with Aerojet Propulsion in 1993 to improve and market the engine. During this period the company successfully tested combinations of propellant that it had been unable to apply to proven designs and this stimulated dialogue between Russian and American rocket propulsion engineers.

NPO cryogenic engine, which has been described in the 1990s as the D-57, would appear to be the 11D57 gimballing engine, which was developed during the early 1970s for use on the Block S of the modified N-1 launch vehicle, which would have been flown for the dual-launch L-3M manned lunar missions. The bureau was also tasked to develop the 11D54 fixed-chamber cryogenic engine with the same thrust level of 40 tonnes (390 kN). This engine would have been used in a cluster of six to eight on a modified Block V of the N-1 vehicle. The propulsion system was marketed as a potential export for interested clients and some derivatives were tested in support of that. Following the cancellation of Buran and the lack of support for the Energia launcher, Lyulka-Saturn OAO seems to have had no space business but has retained its skill base funded by other activities.

VERIFIED

Ministry for Posts and Telecommunications

Current Status

The Ministry for Posts and Telecommunications is responsible for space-based communications services (joining Eutelsat in July 1994). The Russian Federation established its own Ministry of Communications, Information and Space in 1990, and paid the Ministry of Defence for launching the first of three dedicated Gorizont communications satellites (the third appeared April 1992). The Ministry has developed plans for an integrated civil and military communications satellite network for submission to the Russian government but this is no more than a conceptual study. Research and evaluation for an improved and efficient mobile telephone and paging system. As of mid-2003 none of these aspirations had been possible due to technical and financial limitations but the range of services has grown within the existing capabilities of the unmodified Gorizont system.

VERIFIED

Morsviasputnik

Current Status

Morsviasputnik is a department of Morflot, Russia's merchant marine, and is the national Russian signatory to Inmarsat, accounting for about 4 per cent. The department was responsible for developing communications planning for the Soviet merchant fleet and as such built up the Inmarsat element with headquarters in London, the UK being the introductory signatory. Initially, during the 1980s, it played a modest role in supporting the offer to use Proton launch vehicles for lifting Inmarsat satellites into orbit. A key element in determining the level of support for maritime telecommunications by satellite and a major influence in the origin and evolution of Intelsat. Morsviasputnik has extended its remit to include navigation and interactive communication and telecommunication services seeking to expand the convergence of space-accessed assets.

UPDATED

Moscow Aviation Institute (MAI)

Current Status

MAI is one of the largest engineering and technical universities in Russia, with nine colleges or departments, all with applied technology orientation. MAI was founded in 1930 on the basis of the air mechanics faculty of Moscow's Higher Technical School. It educates specialists for the aerospace industry in 10 faculties: aircraft construction, engines, automatic flight control systems, radio electronics, business/management, aerospace, robotic and intelligent systems for aerospace and military craft, applied mathematics, applied mechanics, humanitarian. Affiliates are located in Arzamas, Zhukovsky, Kaliningrad, Tyuratam and Moscow's radio higher school. Initiated engineering analysis involving exotic materials for high-thermal loads and a broad range of temperatures. There are about 1,500 faculty members, 5,700 other staff and more than 10,000 students. Its extensive cosmonautics section includes flight-standard spacecraft, including the N1 manned lunar lander. Extensive analyses and surveys of future space missions provide innovative project development opportunities but lacking the funding to conduct operating evaluations. Some work has been conducted on behalf of manufacturing industries regarding human factors requirements for air-manoeuvrable vehicles such as those associated with recoverable launch vehicles flown back down to the surface of the earth.

UPDATED

Moscow Institute of Thermal Technology (MITT)

Current Status

MITT was founded in 1948 by Alexander Nadiradze's (1914–87) design bureau and was responsible for the design and development of the RS-12/SS-13 silo-based, RS-14/SS-16, RSD-10/SS-20 and RS-12M/SS-25 road mobile and RS-22/SS-24 rail mobile solid propellant ICBMs. Its Science and Technical Centre Complex was created in 1991 to develop the Start orbital launchers based on the SS-20/SS-25; the Start-1/Start test launches were made 25 March 1993/28 March 1995 from Plesetsk. Responsible for enhanced, boosted, propellants capable of propelling high-acceleration rockets and ballistic missiles. In 1998 the MITT was subordinated to the Russian Space Agency but its expertise remains intact and it has played a leading role in developing the Start launcher. The company has designed a new range of small launchers based on Start technology but lack of financial support from potential investors and a general glut in the launch vehicle market has conspired to prevent further development.

VERIFIED

N D Kuznetsov Company

Current Status

Nikolai Dmitrievich Kuznetsov (24 June 1911–30 July 1995) established an aircraft engine design and development bureau in April 1946 at Kuibyschev (now Samara). Kuznetsov headed the company from 1949 to 1994 where he was succeeded by Yevgeniy Gritsenko. Otherwise known as OKB-276, the bureau became Russia's largest producer of aircraft engines and, in the late 1950s, diversified into liquid rocket motors. Information finally released in 1989 revealed that the bureau had been assigned the responsibility for the design and development for engines for Korolyov's four-stage N-1 launch vehicle. Previously the bureau had developed engines planned for use on Korolyov's R-9 missile, but politics ensured that the missile used other engines. Began research into long life propellants for extended deep storage of rocket motors and boost stages. After 1976 the company reverted to the development of advanced aircraft engines.

In July 1993, Aerojet signed a teaming agreement with N D Kuznetsov SSC to market these engines within the United States. Two NK-33 engines were delivered to Aerojet in 1995. One of them was used by Aerojet for five firings which totalled 410 seconds operating time within a 57 to 114 per cent thrust range which followed the proposed Atlas-2AR launch profile. The engines were under consideration to power the first stage of the Atlas-2AR,but they lost out to the Energomash RD-180 engine. This propulsion system was applied to the Atlas 3A, the first of which was launched successfully on 24 May 2000 and to the Atlas 5 which made its first flight on 21 August 2002.

The United States Kistler Aerospace Corporation planned to use NK-33 engines on its reusable K-1 launch vehicles: three would be on the first stage and one on the second stage. Kuznetsov has begun research into new propulsion technologies for extended reusable rocket motors such as those used on long-life assignments with recoverable and re-usable launch vehicles.

UPDATED

NII Argon

Current Status

Argon designs onboard computers, including those carried by the manned Soyuz and Mir and by modules attached to the Mir common block. Argon has begun development of secondary computer systems for modules that may be attached to the Russian elements of the ISS or flown separately but under the control authority of the ISS. This work has been discontinued and computer design technologies have been redirected to the civil, non-space-related industries. Also supplies computer equipment for the Yamal communications satellite programme. These have been extended to incorporate higher use demand due to the significant increase in earth station links with the national satellite system.

UPDATED

NII Grafit

Current Status

Grafit works on applied research and development of graphite materials including carbon-carbon composites, leading to brakes, nose and leading edges for the now cancelled Buran reusable shuttle. Grafit has applied its research base to other re-entry requirements for Russian spacecraft returning from orbit or deep space. Designed unique carbon-carbon layup techniques for high-velocity entry conditions (high-press/high-temperature environments) required by fast entry probes in planetary atmospheres. The organisation has continued to develop the technology for high-thermal loadings and rocket-boosted entry vehicles for accelerated atmospheric survival.

VERIFIED

NII KHIMMASH

Current Status

NII Khimmash was founded in 1948 as a facility of NII-88 and conducted its first firing in December 1949. Known as Branch 2 it was located 15 km north of Zagorsk and, in 1961, separated from its parent body to become NII-229. From 1965 it was under the Ministry of General Machine Building but when that organisation was disbanded in 1991 it was embraced by the Neptun corporation until absorbed into the Russian Space Agency in April 1992. From meagre beginnings as Russia's oldest motor test facility dating back to the R-7 ballistic missile, it is Russia's largest rocket engine test facilities, covering 25 km^2 and capable of testing complete stages up to 9 m diameter and 12 MN. Unsuccessful efforts to gain political approval for a new generation of single-chamber rocket motors operating on cryogenic propellants. The major test stands are, Stand V1: 100 kN cryogenic; V2: 2 positions for 100 to 2,000 kN cryogenic (such as RD-0120); V3: stages to 490 kN. Stand 102: 11.8 MN LOX/kerosene. Six thermal vacuum chambers of 0.3 to 100 m^3 can simulate missions of up to nine months for 5 N to 6 kN engines. 900 m^3 and 8,300 m^3 chambers test complete satellites. Employees total 3,800. SEP heads a European team that began testing the RD-0120 at the site in 1995. NII Khimmash is developing several initiatives for foreign participation with the SEP tests stimulating new joint deals and international ventures.

VERIFIED

NII Kosmisheskovo Priborostroenye

Current Status

NIIKP was responsible for Glonass, Tsikada, GEO-IK, remote sensing and meteorological satellite payloads, including the MSU imagers aboard Resurs-O. It provided the 60 laser reflectors for Germany's GFZ 1 laser geodetic satellite. NIIKP has developed payload packages of military/civil applications to remote sensing satellites. Hybrid payload packages including optical imaging and radar for manoeuvrable satellites. Has developed new mechanisms for integrating multifunctional experiment and science equipment sensor packages into lightweight satellites and spacecraft.

VERIFIED

NII MACHINOSTROENYE R&DIME (Research and Development Institute of Mechanical Engineering)

Current Status

NII MACHINOSTROENYE R&DIME (Scientific Research Institute of Mechanical Engineering) was established in 1958 as part of NII Thermal Processes, but became independent in the early 1980s. Development and production of 0.8 to 400 N spacecraft thrusters (implying connection with old Isayev bureau), including Soyuz, Progress and Mir. Development and small batch production of engines up to 6 kN can be accommodated. Pioneering work on nano-thrust propulsion systems for long duration firing and low acceleration. Test facilities can handle hydrogen (up to 3 kN), methane and kerosene fuels. The Elektro Earth observation satellite carries 16 DEN-16 electrothermal thrusters. Note: LTRE means Low Thrust Rocket Engine, or RDMT in Russian (Raketnay Dvigatel Maloi Tyagi). Employs fewer than 2,000 people and further work is based more on continued production of extant hardware than development of new technologies. Diversified into applications within defence industries. Has engineered new variants of existing small scale engines and has completed tests with hybrid reaction control systems.

VERIFIED

NII Parachutostroenye

Current Status

The NIIPS Institute of Parachute Construction provides recovery parachutes for Soyuz, Buran, Energia boosters and Progress return capsule. Developed new lock-up control links to satisfy safety evaluation recommendations. It is also supplying the recovery parachutes for ESA's Ariane 5 and has conducted experiments with the recovery of very heavy booster stages. New, phased deceleration deployment for masses in excess of 25 tonnes. Experimental development of very heavy load retardation in support of a new generation of Russian launch vehicles proposed by Energomash but not funded. Work has begun on deceleration systems for recoverable launch vehicle stages and on re-shaped parafoils. Some manufacturers would like to develop land recovery for reusable launch vehicle stages.

UPDATED

NII PM (Prikladnoi Mekhaniki)

Current Status

NII Applied Mechanics is a major designer and developer of launcher and spacecraft navigation instruments, including laser gyros and has produced guidance equipment for planetary spacecraft. Developed research programmes and technology efforts resulting in radical changes to guidance platform design during the 1970s and 1980s. Ring gyro systems for deepspace operations and stop-restart functions following dormancy. NII PM has conducted research into integrated guidance and navigation systems for deep space systems.

VERIFIED

NII Radio

Current Status

Established in the 1930s, NIIR is the main research centre of the State Committee on Communications and Information of the Russian Federation subordinated to the State Committee on Communications and Informatisation of the Russian Federation. Controls standards on frequencies and makes appropriate recommendations on boundaries within which to allocate band spread. NII Radio has carried out tests with unusual frequency distribution networks design but these are related to civilian applications outside the space industry. Research projects paid for by industrial manufacturers have been carried out over the last several years.

UPDATED

NII Radio

Current Status

Payloads for Ekran, Gals, Gelikon and Raduga telecom satellites. Conducted research and development on new and advanced technology research projects including a telecommunications research satellite project which never received approval. Part of the Informcosmos consortium. Participated in development of Orbita, Moskva, and Moskva-Globalnaya networks. Has developed terrestrial and hybrid networks for several former states of the USSR. Continues to upgrade and improve the Ekran and Gals systems.

VERIFIED

NII THERMAL PROCESSES NII TP

Current Status

NII THERMAL PROCESSES NII TP performs research in all forms of launcher and space propulsion, including ramjet, electric and nuclear. BMDO took delivery in October 1993 of a T-100 1.5 kW Hall-type Stationary Plasma Thruster for evaluation. N_2O_4/UDMH thrusters include 10 N and 1,000 N models. NII TP developed the Phobos spacecraft's network of hydrazine thrusters for attitude control and minor orbital modifications. 28 (24 × 50 N, 4 × 10 N) were mounted on four spherical tanks. A set of 0.5 N hydrazine microthrusters of an unusual design was also included to avoid the need for separate N_2 cold gas jets; the propellant was burned internally and cooled inside an expansion tank before passing to the thrusters. Successfully completed extended life cycle tests showing operability after lengthy period of dormancy in cryogenic environment. NIITP has attempted to engage with other European countries to enhance the development of microthruster technology. Responding to a direction of effort in the United States toward nuclear propulsion for deep space transportation, NII TP has attempted to link its own (largely historic) development of nuclear and nuclear-electric in an effort to combine research and provide a base for international co-operation.

High thermal demand propulsion systems are being researched and some limited testing carried out.

UPDATED

NPO AP

Current Status
NPO Automatics and Instrument Engineering originated in 1946 as the Department of Automation in the NII-885 department of the Ministry of Communication Means. In 1963 the Ministry of Industry of Communications Means separated to become an independent entity. NPO AP was run by Nikolai created in 1978 as part of NII AP to develop control, navigation and guidance systems for launchers and spacecraft. NPO AP was run by Nikolai A Pilyugin (1908–82; Lapigin succeeded him in 1962) and provided the flight control systems for the Vostok launcher family, Proton (including Block D/DM), N1, Zenit and Buran. In 1978 it formally received its present title and in 1994 it was absorbed by the Russian Space Agency. Areas of diversification included automated control systems, robotics and 'intelligent' machine technology. Its capabilities remain intact and have been dispersed into other departments. Prior to dispersing the assets of the organisation, NPO AP developed automated flight control systems, that were designed to remotely control spacecraft and space vehicles originally designed for man-tended operations and this technology was adapted to automotive engineering processes in ground facilities unrelated to the space programme.

VERIFIED

NPO Astra

Current Status
Responsible for most of the previous Soviet satellite communication ground segment, including the Moskva and Orbita systems. Built the military links to connect communications elements during former USSR. Astra and Italy's Nuova Telespazio are working together in the Astelit joint venture, providing Russia with reliable links into the international telephone network using a Raduga satellite from early 1992. Fixed, mobile and portable systems providing flexibility in civil and military applications. Several mobile ground stations sold to undisclosed customers. NPO Astra have attempted to expand this into other networks with little result to date. Some efforts have been made to integrate Yamal, Moskva and Orbita requirements and capabilities.

UPDATED

NPO Cryogenmash

Current Status
Design and manufacture of ground-based cryogenic equipment; supply of cryogenic fuel to space programme. Design concepts for upper stage pumping systems on long duration missions configured with small, insulated, cryo tanks. This NPO has provided cryo-storage equipment to several companies outside Russia. New research has concluded that development of storable cryogenic propellants are not only possible but feasible; confirmation of this has been sought by companies outside Russia and several tests have been conducted in which the company has demonstrated this possibility. Cryogenmash has contracts or working relationships with companies in 30 countries around the world. The company estimates annual turnover of about US$200 million.

UPDATED

NPO Elas

Current Status
Elas builds communications and imaging sensor payloads. It is providing the payloads for the Kupon and Courier communications satellites; it is probably responsible for the military Geizer satellite's payload and has developed advanced military communications payloads for future milsats. Elas conducted extensive research into the development of commercial by-products from its work on military communications satellite systems and has derived some business from that transfer, although more revenue accrued from diverse civilian applications than work for the military. Expanded sensor development for multispectral instruments on remote sensing satellites.

VERIFIED

NPO Energomash

Current Status
Set up in 1946 as OKB-456 of the People's Comissariate of Aviation Industry (NKAP), the company was at first responsible for manufacturing rocket motors. In the early 1950s OKB-456 was transferred to the Ministry of Defence Industry. Chief Designer Uasiliy P Mishin was replaced in 1974 by Valentin P Glushko who had begun his career at the Gas Dynamics Laboratory. The Gas Dynamics Laboratory OKB DL was created 1928 in Leningrad (St Petersburg) for research into and production of liquid propellant and electric rocket engines. Valentin P Glushko was director until his death in January 1989. It became responsible for the majority of space launcher lower stage engines, notably the Vostok and Proton families. GDL merged with Sergei Korolev's bureau in May 1974 to form NPO Energia (now RKK Energia). It left Energia after Glushko's death, becoming NPO Energomash under Viktor Radovski. Energomash now reports directly to RKA, employing about 6,800. Two test stands can handle 10 MN engines. An agreement was signed with P&W in October 1992 providing exclusive US marketing rights on Energomash products and technology. To date NPO Energomash has developed almost 60 separate types of rocket engine and has indirectly contributed to the propulsion studies completed by US companies of rocket motors for the NASA Space Launch Initiative. Energomash has offices and facilities in Saint Petersburg, Ekaterinburg and Volgodonsk.

UPDATED

NPO Geofizika

Current Status
One of the older organisations responsible for pioneering work on the development of spacecraft and satellite sensors for attitude stabilisation and orientation. Now subordinated to the Russian Space Agency, Geofizika is Russia's primary designer and producer of spaceborne attitude control sensors, optical instrumentation, spectrophotometers, radiometers, spectrometers, interferometers, star/Sun/Earth simulators, star simulator calibration, astronaut training simulators and high-precision optical angle encoders, plus test and calibration equipment. Geofizika provides opto-mechanical instruments for general use outside the space programme. New forms of non-optical imaging through cadmium light source radiation and reflectance measurement.

VERIFIED

NPO Iskra

Current Status
Iskra ('Spark') was created in 1978 as part of the Makayev Design Bureau (formed in 1955), a leading developer of solid propellant missiles. Produced separation, ullage and retrofire solid propellant rocket motors and developed stage separation sequencers for ballistic missiles. It specialises in solid motors and gas generators; for example, it is responsible for the Soyuz launch escape tower. Throughout the 1990s NPO Iskra diversified into a wide range of non-space related applications. These included solid propellants for gas ignition systems on large boilers, propulsion systems for high-speed subsurface missiles and fast accelerating rocket motors. After the disbandment of MOM in 1991 Iskra was embraced by the Komposmash corporation and then reported to the Ministry of Industry of Russian Federation, then to the Committee on Defence Branches of Industry, Ministry of Defence, and finally back to the re-establishment of MOM.

VERIFIED

NPO Istochnik

Current Status
Istochnik began as a power source company developing long life electrical power storage systems for maritime requirements and diversified into space projects. Supplying batteries for space applications. It has also diversified into other non-space-related industries. Developed high-density accumulators for submarines and other long life requirements. The company has very little participation in space-related industries.

UPDATED

NPO Istok

Current Status
Istok State Scientific Production Corporation is Russia's only producer of TWTA travelling wave tube assemblies for communications satellites. Solid state power amplifiers for high-power (+250 W) transmission on short and long duration satellites. It also makes solar panels for geostationary satellites and for the Mir space station. NPO Istok is part of the Moscow industrial complex. Has developed TWTA for high-power data relay and satellite-to-satellite links to a conceptual project not yet funded. Some of this technology has been applied to advanced projects in former Soviet states.

UPDATED

NPO Kompozit

Current Status
Advanced metallic, non-metallic and composite material research in Russia, including Buran thermal protection, takes place at Kompozit. Research work has been expanded in high thermal stress requirements for multi-use re-entry systems for large structures. Other products include microgravity research equipment, including semiconductor and optical materials units. Spacecraft and launcher instrumentation. Thermal protection research provided base for development of ceramic insulation for boilers and gas turbines. Employment stood at 10,000 in 1992. NPO Kompozit has completed joint ventures with Canada, Germany, Italy and Switzerland.

UPDATED

NPO Kraznaya Zvezda (Red Star)

Current Status
Established in 1972 Red Star specialises in nuclear power systems. It developed the 2 kW Bouk thermionic reactor for the Rorsat programme, and the twice-flown (Kosmos 1818/1867) Topaz 1 5 to 6 kW unit. Redstar has developed plans for the design of high power reactors on long duration planetary space flights and has conceptually speculated on the requirements of a nuclear power source for planetary surface basis. Limited research and development on a small nuclear power source for remote regions of earth. Brayton cycle converters. The company has also developed commercial applications and has specialised in heat-pipe technologies. Developed high-thermal transfer systems for oil driven powerplants. Space-related applications have continued despite little work other than theoretical research in that area.

VERIFIED

NPO LAVOCHKIN

Current Status

NPOL specialises in planetary and deep space craft, science satellites, including Luna, Venera, Zond 1-3, Mars, Phobos, Spektrum, Astron and Granat. The Babakin centre, which emerged from Korolev's OKB-1, is solely responsible for planetary probes and science satellites until its associated NPO Lavochkin was identified openly in 1989. Georgi Babakin (1914–1971) established his own OKB in 1965 and at the same time took over the satellite arm of OKB Lavochkin (founded 1937 as an aircraft manufacturer). Lunar and deep space work was transferred from Korolev's OKB-1 in 1965. Sergei Kryukov was chief designer August 1971 to December 1977. Vyacheslav M Kovtunenko then took over (died 11 July 1995, aged 73), succeeded in turn by Stanislav D Kulikov from 1996. Lavochkin was embraced by the Russian Space Agency from 1994. Specialises in autonomous operation on pre-programmed command and control systems with software uplink changes.

Lavochkin is providing the Kupon telecom satellite for Global Information Systems. It provides the bus for the Oko and Prognoz early warning satellites to TsNPO Kometa, and Fregat as a launcher upper stage. Has made industrial connections with companies in the Ukraine and with several ESA industrial consortia with a view to international cooperation and technical deals.

VERIFIED

NPO Mashinostroyenia

Current Status

Originates from the Design Bureau Factory No.51 set up in October 1944 as Special Design Group (SKG) under the Ministry of Aviation Industry and headed by Vladimir N Chelomei to develop a naval cruise missile. Reorganised into OKB-52 in 1955 and from 1958 expanded to ballistic missile development absorbing several major manufacturing plants. Developed anti-missile system 1963–64 until Krunichev dismissed and developed lunar fly-by mission using LK spacecraft and UR-500 Proton. In 1965 the OKB was placed under the administration of the new Ministry of General Machine-building. Vladimir Chelomei's bureau was divided after his death in December 1984 into KB Salyut and NPO Machinostroenye under Gerbert A Yefremou. Almaz radar satellites, and probably others, are built at the Khrunichev factory. It operates the Almaz foreign trade company to market Almaz remote sensing data. Technology acquired in SAR development has been used in development projects for a proposed new generation of remote sensing satellites. Has made a major contribution to the development of military SAR systems and platforms. In 2000 NPO Mashinostroyenia had sales of US$24.2 million, an increase of 46 per cent over 1999. The company states that where once all its business came from the Russian government, now only 10 per cent originates from that source and 75 per cent of its income is generated from work with the Indian government on missiles and rockets.

UPDATED

NPO Molniya

Current Status

By merging several design bureau the Molniya Scientific Production Association was set up in 1976 specifically to develop the Buran reusable spaceplane. As such it was responsible for its aerodynamic design, flew BOR Cosmos subscale missions, and conducted hypersonic vehicle research. Dr Gleb E Lozino-Lozinsky was Director until February 1993. Employment declined from 5,000 in 1992 to 3,500 in 1995 and to 2,800 in 2000. Responsible for proposed MAKS concept whereby a reusable orbiter is carried aloft on the back of an Antonov An-225 from where it is released for flight to orbit. Design teams from NPO Molniya continued to refine spaceplane concepts but have no realistic funding agent for tests. Molniya are responding to renewed interest from the Russian government in cost-effective launch vehicle programmes through enhanced reusability and recovery techniques.

UPDATED

NPO Orion

Current Status

Established in 1985, Orion is a major designer and developer of detectors, focal plane arrays, cryogenic systems, thermal imagers and processing electronics. Orion has participated in several joint ventures with France. Since early 1990s agreements on co-operation in developing portable stereo receivers permanently tuned to Europe Plus, NPO Orion has diversified and reduced in size. From 2000, large plane detector arrays for solar cell power systems in terrestrial application. New markets are sought from former states of the USSR and some technology development has resulted in a new and diverse range of small-scaled systems and components.

VERIFIED

NPO Planeta

Current Status

Formed in 1989, Planeta was absorbed, in 1991 by Roskomgidromet, the Federal Service for Hydrometeorology and Environmental Monitoring, to manage Earth observations and resources programmes, emphasising meteorological applications following the break-up of the USSR. Specialised not only in space-based systems or data collection but in ground truth and in correlating ground, air and space-based surveys with strengths in water runoff studies and hydrometeorological phenomena. Collected and collated remote sensing information from European and US sources for large scale environmental monitoring. Responsible for Meteor, Resurs 01, Okean 01 and Elektro satellites and for commercial distribution of products. Elements of NPO Planeta were dispensed into Scientific and Research Centre of Space Hydrometeorology and the Scientific and Research Centre for Studies of Natural Resources in the mid-1990s.

VERIFIED

NPO PM (Prikladnoi Mekhaniki)

Current Status

Set up in 1959 as a separate branch of the OKB-1, the M F Reshetnyov Science and Production Association of Applied Mechanics (NPO PM) was dedicated to serial production of ballistic missiles. Production on licence from major missile design bureau.

Since 1961 it switched exclusively to satellites and had been responsible for 900 satellites by the end of 2002 (13 in 1995), including the Express, Gals, Ekran, Gorizont, Raduga and most other major telecom satellites. Designed large military consats offered as a constellation of high-powered geostationary satellites. Its other responsibilities included the Strela 3 military sextets, Glonass and Cosmos navigation and GEO-IK geodetic satellites (the NIIKP Institute of Space Device Engineering provides the payloads). Also responsible for Molniya satellites. It remains a state enterprise, reporting to RKA, and has not yet decided if it will privatise. Employees total about 4,900 in 2000. Its business has been divided equally between civil and military, but the military element is now sharply declining. As a result, it is moving into other areas, such as ground antennas and mobile antenna structures with autonomous power sources. Space vehicle assembly has stabilised and is no longer in decline.

Since January 1994, NPO-PM controls the Gals and Express satellites itself through its Persei subsidiary, instead of relying on the military. TT&C stations in Krasnoyarsk and Gouss Khroustalni allow control over 20° W-154° E GEO. Its Mercury subsidiary offers transponder leasing. From inception NPO PM has been led by Mikhail F Reshetnev (died 26 January 1996) succeeded by his first Deputy Albert Eavrilovich Kozlov. It developed Yangel's R14/SS-5 Skean missile into the Cosmos space launcher. NPO PM made a commercial breakthrough in July 1995 when Eutelsat selected it to provide Sesat (Siberia Europe Satellite). The satellite bus has been adapted for other applications and missions.

UPDATED

NPO Precision Instruments

Current Status

Set up in 1963 as Department 4 of NII-885 the Scientific Research Institute of Precision Instruments NPO TP specialises in manned and unmanned spacecraft payloads and electronics, including control systems, telemetry, reception stations and communications. The organisation served as one of the most important suppliers and design bureau for small instrumentation and for subsystems and components. NPO Precision Instruments provided guidance systems for SS-11 to SS-19 ICBMs, Buran orbiter, Luna, Mars, Venera and Vega spacecraft and SS-25/Start 1/Start launchers. It developed the Salyut/Mir Igla and Kurs rendezvous and docking systems and the Vostok telecommand system. Prior to the collapse of the USSR, developed a set of sensor suites for autonomous planetary explorers. It is part of the Smolsat consortium developing the Gonets satellite system. Prime contractor is Applied Machanics.

VERIFIED

NPO Scientific Centre (NPO-NTs)

Current Status

NTs is an affiliate of the electronics industry ministry. It has a long history of developing special electronic and electrical equipment for the Ministry of defence. Electronic components and μg research equipment, including Mir's Gallar semiconductor furnace. Components also provided for ISS FGB elements. Manufacture of sub-surface marine exploration devices with robotic assembly of unmanned platforms, company says, incorporates technology for planetary exploration. NT has diversified into exotic microelectronic components, development and testing.

UPDATED

NPO Soyuz

Current Status

Established in 1947 as NII-125 the Soyuz NPO operates a major production and R&D facility for double base and composite solid propellants. It is closely connected with STC Complex, indicating that it provides the propellant for the Start launcher. Research into long life monopropellants and hybrid, high-performance, solids for use in terrestrial and space applications. Work has reportedly diversified into the production of solid propellants for rockets and missiles and for foreign companies on demand. The company is interested in developing hybrid propellant combinations involving a solid base with liquid oxidiser intrusion during primary combustion followed by secondary burn when the solid is consumed creating an integrated propellant-staging concept in one continuous burn cycle. Employs approximately 3,600 people.

UPDATED

NPO Start

Current Status

Start is a commercial effort to build rocket motor test equipment and stands, for small and medium size satellite launchers. Based on technologies and engineering development funded prior to the collapse of the USSR. Applied technologies to non-military, non-space requirements. Start has been commissioned to plan ballistic missile launch sites and to administer the development of plans for launch complexes and essential infrastructure. This has been extended to the development of remotely operated launch sites for small and medium launchers.

UPDATED

NPO Stekloplastik

Current Status

Supplier of the composite materials employed on Buran and producer of thermoplastic support structures. Ceramic and thermoplastic products for heat-soak conditions in turbines, boilers and heat pipes. Diversified into thermal protection and lightweight structures for satellites, spacecraft, and robotic machines for inaccessible areas in hazard regions on earth.

UPDATED

NPO Tekhnomash

Current Status

Set up in 1938 Tekhnomash is a commercial concern that provides space engineering research, development and consultancy. It created several technical applications for utilisation in the defence and space industries. Ceramic products for Mir space station, Energia launcher and Buran orbiter. Branches in Omsk, Tomsk, Krasnoyarsk, Kuibyshev (Samara), Dniepropetrovsk, Kharkov, Kiev, St Petersburg, Orenburg and Urga. It has 1,900 employees, about half of whom are graduate engineers. In 1990 it was the founding concern of Kompomash. A broad base of engineering and manufacturing technologies underpins a major capability for converting government work into commercially viable endeavours.

VERIFIED

NPO Tekhnologiya

Current Status

Manufactured high-temperature resistant materials for the gas and power production industry and diversified into the ceramics industry for coatings and thermal protection in industrial applications. A supplier of advanced materials for Buran and other reusable vehicles including hot cast ceramics and composite materials since utilised on space structures. Converted almost completely to non-space materials development and production.

VERIFIED

NPO Trud

Current Status

The former Kuznetsov aircraft engine bureau developed engines for the abandoned N1 lunar launcher. They are now offered commercially. Aerojet markets them in the US. Nikolai D Kuznetsov was General Director until 1994, aged 83. Developed smaller versions of N-1 propulsion units for upper stage applications and some were offered to other countries in the1980s. Now operates test stands for medium to large rocket motors. The marketing responsibility is no longer carried out and Trud has been reapplied to general commercial marketing activities for a variety of aerospace contractors and subcontractors. Some work has been conducted on decommissioning old ICBM silos and missile test stands.

UPDATED

NPO Vega-M

Current Status

Headed by the MNIIP Moscow Science and Research Institute of Instrument Engineering, Vega provides space SAR radars, including those of the Almaz remote sensing satellites. Formerly developed technologies for military reconnaissance and surveillance systems. Pursued development of SAR and other space-based radar systems for the Ministry of Defence. This work led to more advanced SAR designs for Almaz vehicles. Developed advanced radar surveillance equipment for military applications to domestic clients. The company has developed various technology applications which has added revenue value greater than that accruing from space business.

UPDATED

NPP Zvezda

Current Status

Previously known as the Zvezda bureau, NPP Zvezda is a joint-stock company. Since 1952 it has been involved in the development of ejection seats and from the early 1960s has specialised in the life support systems for all manned spacecraft as well as EVA and pressure suits for space missions dating back to Vostok 1 and the SK-1 suit. It was responsible for the Krechet 'Gyrfalcon' suit planned for use on the lunar surface in the late 1960s/early 1970s. From this evolved the Orlan family of EVA suits which have seen service on the Salyut orbital stations and currently Mir. The bureau also produced unflown manned manoeuvring units for use in the mid-1960s and the Ikar 'space chair' tested outside Mir. Among its other activities, it designed Buran ejection seats, the design of which was submitted for possible use aboard Hermes. Dornier/Zvezda led ESA's EVA 2000 studies based upon Mir's Orlan-DMA EVA suit, updated with western technology. Zvezda suits fly on board the International Space Station. Zvezda has extensive test facilities including a Mach 2.2 wind tunnel. Company has developed Mach-speed ejection seats and capsule systems for safe recovery of crew members. Tests demonstrate the viability of Mach 6 ejection, an application proposed for manual rocket launches. Continues to conduct research into hypersonic aerospace vehicles and the problems facing pilots and crewmembers of high-*g* vehicles.

UPDATED

OKB Fakel

Current Status

Fakel is a propulsion laboratory which was established in the Academy of Sciences in 1955 and re-organised in 1972 as OKB Fakel ('Torch'), specialising in spacecraft attitude control thrusters, ion engines and plasma sources. The first EPS flew in 1972 aboard a Meteor-1 satellite and further models was subsequently flown on the later Meteor generation satellites, including the Meteor-Priroda series. Research and development work on Stationary Plasma Thrusters (SPT) started at Fakel in 1964. Starting in 1982, EPS units based upon STP-70 and the K-10 hydrazine thermal-catalytic thrusters (TCT) have flown on communications satellites such as Geizer and Luch (including flights within the Cosmos programme). In 1994, a third generation EPS was introduced, SPT-100 aboard Gals and Ekspress satellites. In 1992, Fakel together with Space Systems/Loral and the Moscow Scientific Research Institute of Applied Mechanics and Electrodynamics (RIAME) established a joint venture - International Space Technology Inc (ISTI) - for promoting, marketing and selling the EPS outside Russia. Subsequently, SEP and ARC joined the company. In 1996 the SPT-100 was certified in accordance with western standards: 7,440 hours of firing tests were successfully conducted at Fakel and 5,000 hours were accomplished at the Jet Propulsion Laboratory in Pasadena. Fakel proposed a joint venture through a commercial company whereby the SPT-100 concept could be applied to US projects for space applications.

Its SPT-70 and SPT-100 (Stationary Plasma Thruster) Hall electric thrusters have flown on more than 50 Meteor polar meteorological satellites since 1972 to provide orbit control. The numerical designators indicate beam diameter in mm. SPTs employ a DC-gas discharge in an annular chamber, in which a radial magnetic field traps the electrons. These Hall currents ionise the Xe propellant, the ions of which are accelerated inside a quasi-neutral plasma, without grids, and by the discharge voltage itself. The advantages are their rugged simplicity (no grid system or high voltage supply), but they provide lower efficiencies, lower exhaust velocities and increased Xe consumption than ion thrusters such as the UK-10. Thrusters utilised on Zord-3, Meteor and Meteor-Priroda satellites, Luch, Gals, Express, Yamal, Arkos and Kupon satellites. In 1994 OKB Fakel became part of the Russian Space Agency. Co-operative exchange system with ESA and has extended several international opportunities to countries in the Far East and South East Asia.

UPDATED

OKB Moskovskaya Energetitsheski Institut

Current Status

This institute builds solid-state recorders for space applications and has diversified into non-space related markets during the last decade. Has developed data storage systems to the rigorous specifications of space projects and dual-tasked for defence applications. Has diversified using space-based technology for commercial, domestic and some foreign business.

VERIFIED

PO Mayak

Current Status

Established in 1948 to produce plutonium for nuclear weapons and located in Chelyabinsk at intersecting water courses where waste was distributed. Employed approximately 12,000 at peak, approximately 3,500 in 2000. Mayak produces radioisotopes for all applications, including space and nuclear weapons. It is a potential supplier of material for future US RTGs. Users included the Lunokhod lunar rovers. PO Mayak has sustained a central role in Russia's nuclear defence industry. Space products are distinctly peripheral. Potential for new generation of space nuclear power systems has resulted in several initiatives made by research scientists employed at the company and these have been evaluated for civil and military applications for which a decision is pending. PO Mayak has specialised in decommissioning activity and is selling its consultative services to other countries.

UPDATED

PO Samara Frunze Engine Building

Current Status

Originated by the French in Moscow in 1912, becoming Aviation Plant 24 named after M V Frunze and from 1932 produced first indigenously manufactured aircraft engines. Redirected for serial production of R-7 ICBM engines in 1957. One of Russia's principal aerospace engine production plants, Samara builds the Soyuz/Molniya RD-107/108 engines for NPO Energomash. After privatisation, became Open Joint-Stock Company named Motorostroitel. Manufacture of propulsion components for the Ministry of Defence and contracts let through the Russian Space Agency. Demonstrated innovative methods of assembling propulsion systems for low cost production methods. This has extended to a new series of small upper-stage propulsion systems for vernier and attitude control applications.

UPDATED

PPO Motorostroitel

Current Status

Originated as PO Samara Frunze Engine Building bureau and now carrying the former mantle as one of Russia's leading aircraft engine and gearbox plants, Perm Motorostroitel builds Proton's RD-253 stage 1 engines for NPO Energomash. Engine production tied to Energomesh contracts. Considerable effort applied through design and testing for propulsion system improvements and enhanced efficiency in manufacture through better production techniques and changes to component breakdown for more efficient assembly. Diversification fails to find non-space applications for the technology.

VERIFIED

Priroda

Current Status

Set up in 1975 State Centre Priroda is the primary civil remote sensing acquisition and processing organisation in Russia, under the administration of Roskartografia, the Federal Service of Geodesy and Cartography. Materials include geographic, geophysical and cartographic information allowing Priroda to conduct geomorphological analyses on demand from Russian agencies. It also carried a large archive of demographic information. Archives almost all the former USSR and supports government as well as commercial customers. A major competitor to Soyuzkarta with products from Rusars remote sensing satellites. It operates a large cartographic division that employs satellite data to create cartographic and GIS products. New imagery is collected during Resurs-F missions. It created the WorldMap consortium in 1993 to market digitised versions of its archive of 2 million film images beginning in the late 1990s. This has expanded to include product from the new generation of high-resolution satellites.

VERIFIED

RKK Energia

Current Status

Set up in 1956 as a long-range missile development department within NII-88 it was tasked with launching an artificial earth satellite under the jurisdiction of Sergei Korolev. It quickly became responsible for all manned spacecraft, planetary flights and orbiting space stations. RKK Energia is Russia's largest space company, responsible for the design, development and manufacture of manned and man-related vehicles: Soyuz, Progress, Mir, Buran, and Energia. Its personnel total 19,500, with 1,000 permanently sited at Baikonur. President Yeltsin signed an order in early 1994 to sell 49 per cent of the company with the government is retaining its 51 per cent interest for at least 3 years. The name became RKK Energia, or Energia Space Rocket Corp.

Energia's predecessor, NPO Energia was formed May 1974 by merger of Sergei Korolev's bureau and the Gas Dynamics Laboratory OKB of Valentin Glushko. Korolev's OKB had been renamed Central Construction Bureau of Experimental Machine Building (TsKBEM) after his death in January 1966. GDL left Energia after Glushko's death in January 1989, becoming NPO Energomash. Mir management and foreign cosmonaut missions are handled by Energia, reporting to the Russian Ministry of Industry.

Soyuz and Progress craft are assembled in the KIS Integration and Test Facility, and shipped to Baikonur in rail containers. KIS is also used for Proton stage 4. As of September 1995, KIS also housed Buran, Mir core, Kvant 1 and Kristall high fidelity mockups, and a Space Station ACRV Soyuz model. Energia has developed plans for an advanced manned Mars mission using elements which it would like to test and evaluate on the International Space Station. These plans envisage some level of international co-operation and limited numbers of hardware elements have been built but only small-scale engineering fixtures for test.

Energia negotiates for non-Russian trade deals and the export of Russian space technology and has a broad base of expertise for dealing with foreign companies.

UPDATED

Russian Academy of Sciences (RAS)

Current Status

The Council on Space of the (then) Soviet Academy of Sciences was set up in 1960 under the directorship of M V Keldysk. After the breakup of the USSR the Russian Academy of Sciences established a Space Council in March 1992 to direct fundamental space research, select missions for proposal to RKA and organise international missions. The council comprises eight sections: plasma physics and Sun/Earth interactions, solar system studies, astrophysics, life sciences, materials, legal and social, Earth sciences and international co-operation (Interkosmos until 1994). Formulates space science requirements for government evaluation. Defines future programme recommendations and connects research institutes to the national space policy. Estimated 1996 personnel strength of 135,000 now approximately 110,000. RAS has directed much of its formerly military orientated research into specialised projects which integrate research programmes capable of attracting foreign interest and through programmes capable of joint ventures.

VERIFIED

Russian Space Agency

Current Status

RSA was created by President Yeltsin's decree in February 1992 to draw up and manage Russia's civil space programme. Its director, Yuri Koptev, was formerly MOM deputy minister and an NPO Lavotchkin engineer. It has 300 HQ employees and a further 600 eslewhere. The nine principal divisions are: state programmes, manned projects and launch facilities, implementation of state programmes, science and commercial satellites, international; ground infrastructure, external relations and legal affairs and resources and business affairs. All rocket engine organisations, such as Energomash and NII TP, now come under RSA. In 1999, the Aviation Industry Ministry became part of the State Committee on the Defence Industry and then under the Defence Industry Ministry before becoming part of the Russian Space Agency.

Controls the Russian side of the International Space Station partnership, negotiates with the other ISS partners and serves as the government agent on the procurement of Soyuz and Progress vehicles. The RSA had developed strategies for maintaining the ISS during the hiatus in Shuttle operations pending a return to flight no earlier than September 2003. NASA has decided to fly two Shuttle missions to demonstrate its compliance with the recommendations of the Accident Investigation Board which would put the earliest resumption of full ISS support at early 2005. As of June 2004 those plans had slipped considerably and the RSS now anticipates maintaining the ISS until late 2005 following the first Shuttle flight earlier that year.

UPDATED

SKAS (Moscow Institute of Radio Links)

Current Status

SKAS builds communications payloads and ground terminals. Has developed military UHF communications transceivers for the Ministry of Defence and produced ground terminals for military communications systems with switchable links to national nets. Only a small percentage of business related to space activities.

VERIFIED

Smolsat

Current Status

The Smolsat consortium includes NPO PM, NPO Precision Instruments, Selkhoz Bank and Moscow's Soyuzmedinform Programme Management (connected with the Ministry of Health). It plans the Gonets satellite communication constellation working through hand-held terminals and has proposed different LEO and broadband systems which fail to attract interest only because of parlous funding. Has evaluated a pan-European plan for involving Western Europe in a mobile telephone and messaging service. Designed internal government department communications systems to operate through information transfer systems.

VERIFIED

Sovinformsputnik Association

Current Status

Companies and organisations responsible for military Earth observation satellites formed Sovinformsputnik in 1991 with authority to distribute high-resolution images of foreign territory. Much of the inventory of images came from early military reconnaissance satellite systems and was a direct marketing attempt to provide high resolution imagery in competition with the European Spot Image and US Landsat systems with much lower resolution. The majority shareholder is the State Research & Production Space Rocket Center "TsSKB-Progress", builders of "Resurs-F", "Kometa" series mapping satellites and other Remote Sensing satellites to acquire 1 and 2 m resolution data, as well as the stereoscopic 10 m resolution images over the different territories. Other shareholders include "Krasnogorski Zavod" joint stock company and "Priroda" Center. Sovinformsputnik has the licensed right to distribute products/services from all these satellites. Multi-year collected archives of space imagery stretch back to 1981. Sovinformsputnik works directly with Russian Industry and Russian Aviation-Space Agency. All three organisations are pooling resources to develop a new and advanced earth resource and remote-sensing package for adaptable interface with several bus designs already in production in Russia.

Sovinformsputnik Profile (products and services): space data acquisition, DEMs and DTMs generation, aero- and space-data orthorectification, GIS layers creation, 3-D modeling, 3-D animation, maps and image scanning and digitalising, software designing for data processing, training, topographic and thematic maps creation, space data/value-added products distribution worldwide.

UPDATED

Sovzond

Current Status

The Sovzond consortium of NPP VNIIEM, NPO Planeta and RNIIKP was created in December 1992 to market remote sensing expertise and to sell high-resolution images and interpretation capabilities. At the time these included Russian companies and government organisations, processing data and images from national satellite assets, but increasingly this has embraced international data acquisition sets and information from ESA satellites in a cooperative exchange which significantly broadened the product base. An agreement with Sweden's SSC Satellitbild is allowing reception, processing and marketing of imagery from November 1994's Resurs-O1 directly at Kiruna. Services are provided to commercial outlets outside the Russian Federation. Sovzond has approached ESA on possible co-operation. Marketing initiatives now embrace countries in South East Asia with offers of brokerage for commercial remote sensing data.

VERIFIED

Soyuzkarta

Current Status

Soyuzkarta originated in 1985 and was established in 1987 as a Soviet foreign trade association affiliate of Priroda to market high-resolution (2 m) aerial/space-based photographic, cartographic and geodetic products. In 1990, it undertook agreements with several western companies for the distribution of high resolution images. In 1992, it was transformed into a joint stock company; Kosmokarta is the space division. 5 m resolution colour film imagery is available from the KFA-1000 camera, 6 to 8 m b/w spectral from the MK-4 and 15 to 30 m colour film from the Kate-200, carried by Mir and Resurs-F. This imagery is now available in digitised form direct from Priroda and its WorldMap consortium. Lower altitude coverage is provided by An-30 and Tu-134 aircraft. Extensive ground-truth and environmental data has built one of the world's largest earth observation archives. Soyuzkarta served as a marketing node for Russian radar imagery and set up franchised subsidiaries in Europe and the United States to handle that data in addition to digital and wet-film products.

VERIFIED

Space Research Institute (IKI)

Current Status

IKI, part of the Russian Academy of Sciences since its creation in 1965, is Russia's premier space science centre and leads missions such as Spektrum and Relikt 2. Bureaux in Frunze and Tarusa support it. Employment declined from around 1,600 at the collapse of the USSR in 1991 to 1,300 in 1994 and about 750 today. Responsible for international co-operation between numerous establishments in Europe and the USA. Conducts high-level academic discussions on international space science policy and research trends. Has continued to provide a focus for Russian research and technology development programmes related to space activities. Emphasis on obtaining joint ventures and co-operative projects with foreign organisations. Since 2001 IKI has developed several long-term strategies for resurring the unmanned exploration of the solar system and is working several concepts for deep space missions to the outer solar system and flight beyond the heliosphere to interstellar space. This is one of several long-term mission goals and objectives being worked as theoretical possibilities for government guidance.

UPDATED

Splav Technical Centre

Current Status

The sole Russian centre dedicated to µg research and payloads, it provides the Zona, Konstanta and Splav furnaces for manned Mir and unmanned Photon missions and is developing research experiments for the Russian elements of the International Space Station. Co-operating with east European countries on microgravity experiments and has developed strong links with several other technical research centres in former states of the USSR. Splav has broadened its activity to include microgravity payload development, linking research initiatives and manufacturing companies with opportunities to fly their experiments in space.

VERIFIED

TsAGI

Current Status

The Central Aerohydrodynamics Institute (TsAEI), was formed in 1918 in Moscow under the leadership of N I Zhukovsky. It maintains aircraft and satellite test facilities employing 6,400 personnel. TsAGI worked with British Aerospace on interim Hotol research. In 1993, it worked with ESA on development of a Buran real-time entry flight simulator. Buran pilots trained on TsAGI's 6-DOF simulator. Has conducted extensive research into hypersonic flight and conducted tests with scramjet engine designs. Some co-operation with the French. Emphasis in space related research has focused on the potential development of trans-atmospheric vehicles, a technology which has historically interested Russian engineers and their former Soviet colleagues and some effort has been made in recent years to interest European companies in an agreement to focus research on hypersonic sramjet technology. The Russians are building a credible research base on these applications and have attracted the interest of Indian companies. Refocusing on exoatmospheric boundary vehicles (capable of flying just inside or outside the earth's atmosphere) TsAGI has designed several hypersonic vehicle configurations for scramjet applications, extending original work conducted during the 1960s and again during the 1980s.

UPDATED

TsNII Kometa

Current Status

Kometa is responsible for several high value tactical and strategic space systems, including the ASAT anti-satellite, Oko/Prognoz early warning, Eorsat and Rorsat. TsNII evolved from the KB-1 in the late 1950s which specialised in anti-satellite and ocean reconnaissance and became an independent bureau under A I Savin. It also played a leading role in monitoring the development of US nuclear weapons and in the development of space-based early warning systems. Charged with projecting an enhanced ABM deployment concept. Has shadowed US ABM technologies and tested warhead entry trajectories to evade ABM missiles using space-based sensors. Has evaluated new methods of in-orbit detection, examination and analysis of in-orbit satellites as a simple inspection capability.

VERIFIED

TsNII Mashinostroenye

Current Status

TsNIIMash central research establishment of Russia's Ministry of General Machine Building (previously the USSR's MOM) is now part of RKA. The TsUP Space Operations Centre is one of its departments. Originally an artillery factory, it became NII-88 (Science Research Institute 88) in May 1946 and, with the help of interned German V2 specialists, under Korolev developed the first Soviet rockets. Korolev's OKB-1 branched off from NII-88 in August 1956. Utkin was General Director of NPO Yuzhnoye until he moved to TsNIIMash in 1990. TsNIIMash-Export was created in 1991 to market TsNIIMash capabilities. It is working with the US BMDO on TAL (thruster with anode layer) D-55 1.5 kW and TAL D-100 5 kW Hall-type Stationary Plasma Thrusters. TsNIIMash employs about 5,750 people. Has provided research development work for the Russian Ministry of Defence on ABM technologies. Has developed new methods of creating high-energy plasma directors which are mobile and could theoretically be used in orbiting platforms. Several direct applications into civilian, earth-based, activity has resulted in some commercial success for projects that had little direct application in space projects.

UPDATED

Ural Electrochemical Integrated Plant (UEIP)

Current Status

UEIP is a major supplier of enriched uranium for nuclear power plants. Although apparently no Russian fuel cell has flown in space, UEIP has developed at least two. The Volna cell, employing circulating KOH as electrolyte, was developed for the manned lunar programme. The Photon fuel cell, immobilising the KOH in an asbestos matrix, was developed and qualified for future Buran missions. UEIP successfully marketed the concept to west European companies and conducted demonstrations to several European government research establishments. ESTEC ran a two-week continuous test campaign on Photon in 1992, attaining 12 kW at 400 A. Photon comprises eight 32-cell stacks connected in parallel, with 176 cm^2 active cell area. UEIP's fuel cell interests now focus on large plants for terrestrial applications. Extensive marketing to European and US companies seeking joint venture for operational deployment of UEIP fuel cell systems. Re-examining the application of fuel cells in space for government research programmes developing technology for export or licence manufacture.

VERIFIED

VNII Elektromekaniki (VNIIEM)

Current Status

Founded in 1941 to build electrical equipment with post-war emphasis on the development of military equipment including missile electronics and guidance systems, VNIIEM is responsible for the Elektro, Resurs-O and Meteor Earth observation satellites. It contributes instruments to other projects, including Almaz and Mir. VNIIEM has diversified into non-space related applications and in 1991 teamed with General Electric to produce x-ray computer tomography scanners for medical purposes. Performing detailed evaluation of large earth environment monitoring platforms similar to those envisaged by NASA's Mission to Planet Earth programmes of the early 1990s. Has worked with several companies in western Europe to determine common specifications for earth observation instruments. VNIIEM employs about 2,100 people.

UPDATED

VNII TransMash

Current Status

Established in 1930 TransMash specialises in the design and fabrication of transport vehicles and tanks. In the space field, it developed the Mars-98 rover (Lavotchkin integrates the payload) and was responsible for the Lunokhods, small Mars 3 rover and the Phobos hopper. It has experience in the design and initial breadboard testing of systems layouts for semi-automated planetary exploration vehicles, capable of significantly closing the gap between the conventional rolling rover, an independently self-controlled robotic systems, with the capability to climb, roll across rough but level surfaces and crossing small ravines. Continues to research the technology for advanced planetary rovers and remotely controlled vehicles. Some work in this direction has been contracted by Energia and Khrunichev. Employees total 1,750.

UPDATED

Singapore

Centre for Remote Imaging, Sensing and Processing (CRISP)

Current Status

Funded by the National Science and Technology Board, and located at the National University of Singapore, CRISP was set up in late 1992 and its ground station became operational in September 1995 to receive Spot and ERS data. Radarsat was added January 1996 and CRISP is now extending its access to the new generation of Ikonis imaging satellites. The 12.8 m antenna (1°17'30" north/103°47'00" east, in campus, 2 km from the processing site) and related ground station facilities were built under a US$7.3 million contract by US companies Datron and International Imaging Systems Inc. Imaging standards convergence programming is opening up options for integrating dissimilar data sets. Analysis and interpretation for earth observation specialists in optical and radar imaging. CRISP has established several international links and provides detailed analysis and interpretation for countries throughout the region.

The data processing facility features Silicon Graphics computers running under Unix. The main computer can process 10 ERS SER and 25 Spot scenes in an 8 hour shift. Ocer 1996–97, CRISP is concentrating its research effort on forestry, oceanography and advanced data processing techniques (stereo, SAR interferometry, radargrammetry and optical/radar fusion).

UPDATED

South Africa

Satellite Applications Centre (SAC)

Current Status

As the leading Earth observation centre in southern Africa, the SAC receives, processes, archives and distributes data and map format products from LANDSAT, SPOT, NOAA and ERS payloads. During 1996, SAC signed an agreement with Core Software Technology in the US, under the terms of which SAC will receive and distribute high-resolution images from future EROS satellites. Starting 1997, SAC also distributes data from Canada's RADARSAT. Further negotiations are under way with several agencies to broaden the range of SAC's products so as to ensure its customers an appropriate choice of spectral and spatial characteristics. Integration of radar and imaging data promises to serve the needs of land resource specialists in fossil fuel and mineral searches. The Satellite Applications Centre (SAC) operates as part of CSIR, South Africa, and has two main areas of business:

Earth observation: the reception, processing, archiving and commercialisation of image data from a range of remote sensing satellites. Converts to specified formats for users including government agencies. Routinely supplies and processes requests to and from the defence ministry and other government agencies responsible for elements of South Africa's national security programme. Also assists with guidance optimum utilisation of global earth resource data.

Tracking, telemetry and command (TT&C) services to international satellites operators and launch agencies. Compiles user requirements documents.

Located at 25.88° S, 27.70° E, SAC's footprint reaches northwards to 5° S, and it is therefore ideally located to serve any clients with an interest in and over the southern African sub-continent.

The infrastructure in place includes extensive buildings and secure power, reception and transmission facilities covering the L-, S- and X- bands and state-of-the-art communications links and powerful hardware and software for data processing. A team of 60 engineers, technicians and support personnel staffs the Centre. Turnover of the SAC is expected to be in excess of US$5 million during the FY97–98.

The SAC archive contains digital data from 1972 and it remains up-to-date with the daily addition of data. Training opportunities are organised twice a year and are typically designed and presented jointly by SAC and selected partners.

The first SAC TT&C services were for NASA during the 1960s through to the mid-1970s. At present SAC's services are available on a contract basis to launch agencies and the owners and operators of satellites. These services are rendered by an experienced team 24 hours a day, throughout the year. In this context SAC is part of the 2 GHz network of the French National Space Agency, CNES. Remote-sensing applications and their potential for improving South Africa's national security considerations are highlighted in recent government reports.

UPDATED

University of Stellenbosch

Current Status

Sunsat was announced June 1991 to promote development of industrial and educational space engineering. It is part of a programme to establish an internationally recognised satellite engineering activity at Stellenbosch, started with the Grinaker company's sponsorship of a Chair of Satellite Systems. The control centre was sited at the University over the five-year life. Stellenbosch already has an Oscar ground station and a 4 m antenna. Students are supported by major electronics companies: Altech and Siemens of Johannesburg, Grinaker Electronics of Pretoria, Plessey SA of Cape Town, Analysis Management and Services of Stellenbosch, and First National Bank of Johannesburg. Cost is estimated at US$800,000 and NASA provides the launch free in exchange for the GPS receiver and laser reflector payloads.

Sunsat was originally planned for Ariane's Helios 1A launch but the US$400,000 cost was prohibitive. Instead, it replaces the balance mass for Denmark's Ørsted on Delta's Argos launch. The satellite was successfully launched on 23 February 1999. Breadboard prototypes were completed January 1992 to March 1993; the engineering was completed April 1994. Development of design concepts for follow-on types stimulated further student participation and a small government research grant.

Successor proposals formulated for stimulating low-cost university participation.

VERIFIED

Spain

Centro para el Desarollo Tecnologico Industrial (CDTI)

Current Status

The Ministry of Industry and Energy created the Centre for the Development of Industrial Technology (Centro para el Desarollo Tecnologico Industrial). To implement April 1986's 'Law of Science' policy, which identified space as one of 14 areas requiring improved official support, the military-based CONIE was dissolved by a law decree and the official Spanish delegation to ESA was entrusted to CDTI. It drew together 45 companies and 11 research institutes to compete for ESA and other space-related business. CDTI has played a major role in organising national space capabilities and in bringing together several separate companies for common marketing initiatives. Joint ventures have been struck with several companies in ESA countries. CDTI routinely garners support from industry and aerospace companies for technical concepts with which to secure ESA contracts within the appropriate allocations. Expanded efforts to link national and international projects and joint endeavours through national and ESA programmes. The CDTI is playing an important role in defining new ways in which Spain can participate in other European technology ventures, many outside the space-related signature industries.

UPDATED

Instituto Nacional de Técnica Aeroespacial (INTA)

Current Status

The defence ministry established the Instituto Nacional de Técnica Aeroespacial (INTA) in 1942. It remains an autonomous R&D body reporting to the Ministry of Defence. Personnel total about 1,050 in four divisions, namely, Materials and Structures, Aerodynamics, Energy and Propulsion, and Avionics. It also operates the Aeronautical/Space Documentation and Information Centre (CIDAE). In the space field, it provides management and technical expertise and is currently managing the Minisat and Capricornio programmes. It was responsible for Spain's first satellite (Intasat), conducts INTA 100/300 sounding rocket firings, provides technical management for Spain's Helios contribution, developed S/Ku-band TT&C antennas used on satellites such as Olympus, Hipparcos, Telecom 2, Eureca and Hispasat, and operates the ESA Spasolab Laboratory for testing and certifying photovoltaic cells. INTA's INSA subsidiary operates and maintains the Villafranca tracking sites for ESA and the Madrid site for NASA. Its own 10 m S-band antenna at Maspalomas receives Landsat and Spot imagery, which it distributes as ESRIN's National Point of Contact. Special applications to government, defence and national security needs. Co-ordinates the use of earth observation images for military applications and has conducted studies of commercial applications for civil satellites.

VERIFIED

Telefónica de España SA

Current Status

The Compania Telefónica Nacional de España is the national signatory to Intelsat, Eutelsat and Inmarsat. It also holds 25 per cent of Hispasat SA, operators of the national satellite system. Telefónica's Eutelsat Earth station at Guadalajara, east of Madrid, operates an 18 m Ku-band antenna for handling TDMA and TV traffic. Telefonica Sistemas de Satelites was created as a subsidiary to provide business communications. Development of SSTDMA and advanced broadband capabilities within the national terrestrial infrastructure. Supporting ground station development for broadband services and developing a set of specifications for the implementation of earth station upgrades.

VERIFIED

Sweden

Swedish Institute of Space Physics

Current Status

The Swedish Institute of Space Physics (Institutet för rymdfysik, IRF) was established in 1957. It has participated in more than 40 rocket and 25 satellite investigations since 1964, primarily targeted at magnetospheric and ionospheric studies. IRF provided Promics particle detectors for Prognoz 7/8, hot plasma and low-frequency wave experiments onboard the Swedish satellites Viking and Freja, the Aspera hot ion composition spectrometres onboard both Phobos spacecraft, the Promics-3 hot plasma detectors for both Interball orbiters, the Electric Fields and Waves instrument for Cluster and Cluster-2, the Aspera-C ion/neutral particle imager for Mars-96, the 0.5 eV-40keV/q ion spectrograph on Japan's Nozomi/Planet-B Mars orbiter launched in July 1998, and in November 2000 launched its own nanosatellite Munin with MEDUSA, a combined electron and ion spectrometer, and DINA, a neutral particle detector. IRTF assembled instruments for two major ESA missions: an energetic atoms analyser Aspera-3 for Mars Express launched in June 2003, and an ion composition analyser RPC-ICA for the spacecraft Rosetta launched to the comet Churyumov-Gerasimenko in February 2004. The institute has carried out several analyses on possible future science packages for comet landers and sample return missions.

The research facility's head office is at Kiruna and is primarily concerned with upper atmosphere phenomena, the ionosphere and the planetary magnetosphere. Since 1996 IRF has also had an atmospheric research programme in Kiruna specialising in the physics of the middle and upper atmosphere. IRF's headquarters in Kiruna and its division in Uppsala focus on studies of the ionosphere and magnetosphere. It also has researchers based in Umeå and Lund specialising respectively in the propagation of infrasonic waves and solar terrestrial physics.

UPDATED

Swedish National Space Board (Rymdstyrelsen)

Current Status

The SNSB was created in 1972 under the Ministry of Industry as the central governmental agency responsible for all publicly-funded national and international space programmes, including co-ordination of remote sensing activities. Three advisory committees cover science, remote sensing and industrial policy. Executive management of space projects, including the Esrange sounding rocket range and satellite ground facilities, is assigned to the SSC state-owned limited corporation. Has limited responsibility for conducting policy shaping and serves as a general advisory body to the Swedish government.

UPDATED

Swedish Space Corporation (SSC)

Current Status
Formed in 1961, the Swedish Space Corporation is a government-owned limited company under the Ministry of Industry and Trade, responsible for the technical implementation of Sweden's space and remote sensing programmes, including feasibility studies, operational applications, systems engineering and research and development. The SSC supports the Swedish Space Board with technical interpretation. The SSC has a 50 per cent share in the Nordic Satellite Company, NSAB, which provides capacity for broadcasting and other television and radio broadcasting, as well as the transmission of data using the SIRIUS satellite system (comprising the Sirius 1 and Tele-X satellites located at 5° E). The SSC is responsible for the operation and procurement of NSAB satellites. It also operates the Esrange sounding rocket range and satellite control and data reception facilities. It acquired a 50 per cent interest in Norway's Tromsø station in 1995. The total number of employees is 300.

SSC offers μg flight services within the Maser (Material Science Experiment Rockets) programme, in addition to design/manufacture of specific experiment modules. By using an improved Black Brant 9, 7 minutes μg can be provided for a 375 kg experiment module. Skylark 7 can provide 6 minutes for 370 kg. Launches are made almost annually. Modules are interchangeable between Maser, Texus and MiniTexus with slight modifications. Maser 1 flew at Esrange 19 March 1987; #2 29 February 1988; #3 10 April 1989; #4 29 March 1990; #5 9 April 1992; #6 4 November 1993 (Skylark 7); #7 (Skylark 7) 3 May 1996, with ESA as the principal customer.

SSC and DASA/ERNO jointly developed the Maxus suborbital vehicle to provide 14 minutes μg. Also special payload support systems for remote payloads and experiment control options during microgravity flight.

VERIFIED

Telia AB

Current Status
Telia AB (previously Swedish Telecom), the national signatory to Intelsat, Inmarsat and Eutelsat, operates as a private company but is wholly government-owned, under the Ministry for Transport and Communications. Eutelsat is accessed through the station at Ågesta 10 km south of Stockholm. Telia AB and Telecom Finland International jointly own and operate the Tanum Teleport, on Sweden's west coast, with five Intelsat antennas. Inmarsat communications are routed through Norway's Eik station. In November 2000 Telia added to its broadband services logging 22 providers of internet-based services. Additional providers were onloaded during 2001 and 2002 with significant increase in the use of multichannel access. Relay co-ordination unit for space/terrestrial feeds in and out of Sweden.

UPDATED

Switzerland

International Space Science Institute

Current Status
ISSI began operating in 1995 with a nucleus of six to eight staff and 10 to 30 research fellows, research associates and visiting scientists. In 2003 it consisted of 12 staff including three directors, one honorary director, three staff scientists and five staff members. Much of the early period was spent developing potentially co-operative ventures and joint endeavours with existing space science teams and in liaising with foreign groups who had already been working with Swiss scientists; Switzerland has had a long history of involvement with space science and is most famous for the Solar Wind Collector experiment carried to the lunar surface by the first manned moon landing, Apollo 11, in July 1969. Integrates universities research programmes. Wide range of specialised activities. Responsive to initiatives from abroad.

UPDATED

Taiwan

Centre for Space and Remote Sensing Research

Current Status
CSRSR, part of the National Central University, signed a contract 27 April 1993 with Spot Image and began receiving Spot imagery directly through the new station, 50 km west of Taipei, on the Chung-Li university campus (first image 25 May; contract came into formal effect July). This image bank and the resulting archive underpins further data acquisitions. CSRSR built an archived data source from which further interpretation and analysis was conducted on a demand basis.

It signed a contract 15 December 1994 with EOSAT to receive and process Landsat 5 data. This added dimensional values to the SPOT data and was used to expand applications resulting in a major national effort to increase environmental issues. It signed an agreement 10 May 1994 with ESA to receive and process ERS SAR data. Annual operating budget is FFr24 million; 48 staff operate it 12 hours daily to generate standard products. The centre's Meteorological Satellite Laboratory receives and processes GMS and NOAA AVHRR images. Other facilities, such as digital image processing, digital photogrammetry and GIS are used to generate value-added products. CSRSR has been a commercial Spot distributor since 1988 and has begun to provide image interpretation for several government departments and agencies. Obtaining Ikonos digital imaging data and providing radar imaging interpolation.

UPDATED

Thailand

Post and Telegraph Department

Current Status
Thailand joined Inmarsat in January 1995, through its signatory Post and Telegraph Dept. The first Inmarsat land Earth station opened in 1996 at Nonthaburi. Considering establishment of separate services for regional telecoms providers. Has proposed a new regional telecommunications hub relieving some traffic from Singapore and has developed efficient route links to demonstrate that the country has the knowledge and claims it has the technology to absorb overload and to serve as a contingency back-up in the event of outage or failure in Singapore. P&TD has been asked to formulate a new telecommunication strategy for Thailand and is liaising with regional partners to integrate objectives.

UPDATED

Thai Remote Sensing Center (TRSC)

Current Status
The TRSC is a division of the National Research Council and has operated the Thailand Landsat Station (TLS) since November 1981. The facility was upgraded to receive Landsat TM and Spot HRV in 1987 with the assistance of the Canadian International Development Agency, added MOS capability in 1988, and ERS/JERS in 1993. Working with regional resource organisations through Swedish development funding. New and expanded capabilities due to software and computer hardware upgrades. Integrated with other regional centres throughout the area to provide a large range of government and commercial remote sensing data products and linking with other countries in the sector to provide local area data and ground truth information. Thailand has developed a wide range of regional and national assessments that it currently uses for analyses and consultative arrangements with friendly states in the area. This work is particularly important for supporting national security and defence requirements.

UPDATED

Turkey

Turkish Telekom

Current Status
Turkey has been a signatory to Intelsat since 1968 and in November 1989 became the 58th member of Inmarsat. The Ata station 37 km south of Ankara became operational in July 1989 to handle Inmarsat A communications; it added Inmarsat C in 1994. COMSAT Corp and IDB both operate Inmarsat A stations in Anatolia for the Indian Ocean region. Turkey has held discussions on a regional relay role for foreign operators. Major upgrade of ground receiving stations and control systems providing opportunities for receiving multichannel satellite relays from several separate operators simultaneously. Also conducting a study of requirements for incorporating military communications satellite nets through commercial operators. Turkey is interested in developing contacts and joint activity with ESA for telecommunications and technology development.

UPDATED

Ukraine

Marlin-Yug Ltd

Current Status
Formed in 1990 as a manufacturer of Scientific equipment, Marlin Yug produces environmental units returning data through the international Argos satellite system. The Berkut balloon unit allows the paths of stratospheric balloons to be tracked. Ten Marlin SVP-B drifters are operating in several international ventures and include operations from the Black Sea. Marlin-Yug has been approved by the Data Buoy Co-operation Panel (DBCP) of the World Meteorological Organisation (WMO) and Intergovernmental Oceanographic Commission (IOC), UNESCO, as a manufacturer of data platforms. Marlin SVPB drifters are certified safe to deploy by aircraft of the US Naval Oceanographic Office. A diving drifter transmits ocean vertical profile parameters and is tracked for surface currents. Ground tests conducted for VEGA and Mars-96 mission. Conducting analysis of engineering requirements for habitation in hostile planetary environments. Minimal effort made to develop plans for proposing airborne balloons as candidates for atmospheric floaters on

distant planetary objectives. Some work has been undertaken to combine atmospheric data collection with ocean data and tracking sensors and data collection platform information relay.

UPDATED

NPO Khartron/ Elektropribor

Current Status
With facilities in Kiev, Moscow and Zhaporje, Khartron supplies onboard spacecraft computers, including Mir and Buran and it provides Zenit SL-16 and Cosmos 3 SL-8 avionics and has developed guidance equipment for autonomous control of launchers. Has conducted software analysis for trajectory changes supporting off-nominal mission requirements. Guidance equipment upgrades and complete replacement of control computer systems. The company has started a new open architecture plug-in module for launch vehicle applications. Intention is to sell the unit to a variety of LV designers, manufacturers and operators.

UPDATED

NPO Musson

Current Status
Worked co-operatively with Russia's telecommunications and navigation agencies to develop maritime telecommunications satellite requirements. Musson handles all maritime and airborne satellite communications terminals, including Inmarsat A maritime in HF and UHF communications. Musson employs about 3,800 people. Also builds GLONASS and Navstar satellite navigation receivers. Has unsuccessfully attempted to export these receivers outside the bloc of former Warsaw Pact/USSR. Very little of the legacy remains in place today.

VERIFIED

NPO Yuzhnoye

Current Status
Based at the Yuzhmash factory, Yuzhnoye is the world's largest integrated missile production facility. NPO Yuzhnoye (formerly OKB-586) is a major launcher, missile and satellite design, development and manufacture bureau. Formed in 1954 under the directorship of M K Yangel until his death in 1971, NPO Yuzhnoye has a 15 per cent interest in Sea launch. Origin of the company goes back to 1951 when what was a large automobile factory was transformed into a missile factory in Dniepropetrovsk and it developed and tested the R-1 (SS-1), R-2 (SS-2) and R-5M (SS-3) missiles and was made responsible for serial production. It is responsible for the SL-11/14 Tsyklon and SL-16 Zenit space launchers, the SS-4 Sandal (basis of SL-7), SS-5 Skean (basis of SL-8 Cosmos), SS-9 Scarp (basis of SL-11/14 Tsyklon), SS-18 Satan (now offered as a satellite launcher) and SS-N-5 Sark missiles, the Okean-O/Sich satellites, Tselina global Elints, and about 400 Cosmos and Interkosmos satellites (including those based on the AUOS bus family). It also probably handles the military Rorsats. Employees totalled about 39,000 (including 14,000 on R&D), with 5,800 involved in space work by 1991. Facilities total about 185,000 m^2, including titanium forging and propellant tank fabrication (also undertaken for other organisations). Test facilities now open for leasing to other Ukrainian companies.

Yuzhmash was producing automobiles and aircraft at the end of the Second World War. In 1951, Stalin allocated it to Sergei Korolev. In about 1954, Korolev's deputy Mikhail K Yangel (25 October 1911 – 25 October 1971) assumed control for production of the SS-6 Sapwood, which formed the basis of the current SL-4 Soyuz and SL-6 Molniya launchers (later switched to the Central specialised Design Bureau). OKB Yangel was also responsible for the manned lunar lander's propulsion system (led by Boris Gubanov, who later became responsible for the Energia launcher), and proposed the R56 manned lunar launcher. NPO Energia allocated it in 1974. These were developed in parallel as the SL-16 Zenit 2. NPO Yuzhnoye was formed out of OKB Yangel in 1974.

UPDATED

PO Monolith

Current Status
PO Monolith provided research programmes and technical direction for manufacturing strapdown guidance platforms and built new integrated trajectory and attitude stabilisation control programmes for computerised and electromechanical operation. Monolith builds control and guidance systems for rockets and ballistic missiles. Exports limited number of strap-down gyro systems and some components. PO Monlith has expanded research and development to auto-adjusted inertial strap-down reference systems which have long life, extended dormancy and high-thermal survival specifications.

UPDATED

Ukrainian Space Agency

Current Status
The agency and commission were established in November 1992 to manage and supervise national space activities while licensing national companies. The Agency has extended negotiations with several foreign governments and an agreement with NASA in 1994 resulted in a Ukrainian cosmonaut flying on a shuttle mission in 1997. USA has been gradually moving toward a policy alignment to ESA and NASA organisations in an attempt to increase participation in space science and engineering projects. The USA has attempted to forge a unique set of skill capabilities for harnessing national assets from government and semi-private sectors. These have increased contacts and projects participation with several institutes, universities and research centres outside the country.

UPDATED

United Kingdom

ASTOS

Current Status
The Association of Specialist Technical Organisations for Space (ASTOS) was formed in 1988 by companies providing specialist manufacturing skills and services to the space industry. It represents the interests of smaller specialist companies competing against larger companies and more well established names. Has secured UK and continental European contracts and joint ventures associated with national and ESA projects. ASTOS has successfully negotiated positions between small companies in the UK and their larger counterparts within ESA. ASTOS has also been consulted by UK government on critical issues related to the support needed for minor participants. Contacts with committees and working groups of BNSC, BT, RAeS and SERC. Membership comprises Aegis Systems, Aetheric Engineering, Anvil Technology, Flow Line Communications, Go-Sat International, Moreton Hall Associates, Space Innovations, Set Resource, Spur Electron, Storm Integration, TRL Technology and Westover Flight Technology.

VERIFIED

British Association of Remote Sensing Companies (BARSC)

Current Status
BARSC was established in 1985 as the trade association of UK remote sensing companies. It aims to ensure that the interests of its members are represented on national, international and government committees and to provide a forum for discussion within UK government departments, such as the British National Space Centre, the Department of the Environment and the Ministry of Defence. For example, BARSC plays a prominent role in the UK's all-party Parliamentary Space Committee. There are 19 member companies. It holds open meetings five or six times annually, in addition to one-day workshops. Prime lobbying body for remote sensing companies with strong representation into government. It also conducts review, analysis and advisory activities for government departments, industry participants and specialist organisations. BARSC provides two industrial members to the all-party Parliamentary Space Committee, set up in 1989.

UPDATED

British National Space Centre (BNSC)

Current Status
BNSC, created in 1985, focuses the civil space interests of the Department of Trade and Industry (its parent body), the Cabinet Office, the Defence Evaluation and Research Agency, the Department of the Environment, Food and Rural Affairs, the Department of Transport, the Foreign and Commonwealth Office, The Meteorological Office, the Ministry of Defence, the National Environment Research Council, the Natural Environment Research Council, the Office of Science and Technology, the Particle Physics and Astronomy Research Council and the Rutherford Appleton Laboratory. A staff of 235 operates from the London HQ and technical centres such as the Central Laboratory of the Research Councils. Released in August 2001, an independent report into space expenditure in the UK endorsed the perception of a gradual decline in government spending with a 20 per cent fall during the tenure of the Blair administration. Secured UK government funding and backing for Open University Beagle 2 Mars lander in conjunction with ESA's Mars Express.

UPDATED

British Telecommunications (BT)

Current Status
BT has significant investment in three satellite consortia: INTELSAT, EUTELSAT and Inmarsat. Each consortium was established through an inter-governmental treaty as a co-operative but with a requirement to function under accepted commercial principles. BT is designated by the UK government to act as the signatory to each consortium and as a consequence, BT represents UK interests at consortia management board meetings.

INTELSAT and EUTELSAT were established respectively with a view to providing public telecommunications globally, and within Europe. Inmarsat was set up to serve the maritime community, a mission which was subsequently expanded to include aeronautical and land mobile services.

The UK's investment in the satellite consortia, including BT's, is as follows: EUTELSAT - 22.6 per cent (largest investor), INTELSAT – 9.2 per cent (second largest investor) and Inmarsat – 9.3 per cent (second largest investor). BT also operates an advisory service for regional and international operators as well as local service providers.

BT provides a wide range of satellite services, accessing satellite consortia capacity and private satellite systems such as PanAmSat, Orion and Astra SES to provide:

- Public Switched Telecommunications Network Services including voice, data and facsimile.
- Multimedia including Turbo-Internet.
- Global mobile satellite services including the Skyphone aeronautical system, maritime communications and land mobile services. BT recently introduced the 'Mobiq' terminal which is smaller than a laptop PC and can provide global direct-dial phone fax and data.
- Entertainment TV and radio. Contribution and distribution TV services as well as CD quality radio.
- Satellite newsgathering and special events coverage.
- In the UK, BT operates earth stations at Goonhilly, Madley, London Teleport and Aberdeen. BT also operates earth station facilities in several European countries providing uplink services for TV broadcasters. BT Broadcast Services operates a fleet of transportable earth stations, which provide uplink services for special event coverage (for example, major sporting events). A BT earth station opened in New Zealand in spring 1997, providing a range of BT Inmarsat services via the Inmarsat satellite system.

BT has researched integrated communications technologies linking several separate satellite constellations simultaneously. Laser communications have been evaluated by BT, along with other national PTT agencies. BT has conducted development programmes to provide multiplexed access routes for broadband and increased data handling capabilities from its development centre at Martlesham Heath in Suffolk. New technology needs for satellite based broadband delivery services are in development, together with wireless transmissions for high-bandwidth services.

UPDATED

Central Laboratory of the Research Councils

Current Status

The Central Laboratory is the focus for the UK's space science projects and directs both ESA and bilateral activities, such as the current Spectrum-X astronomical satellite collaboration with Russia. It was created 1 April 1995 as an independent entity from the merged Daresbury and Rutherford Appleton Laboratories, reporting to the Office of Science and Technology. Its space funding is from PPARC. Technology applications, particulary those from RAL continue. Sustained efforts have continued to propose new and innovative techniques for radical science experiments, representation being made to the UK government and to ESA. Several new initiatives have been expressed through BNSC and to European laboratories via the national organisations responsible for international dialogue.

UPDATED

DERA, Culham

Current Status

Formerly known as the Royal Aircraft Establishment, and DERA since 1995, the Culham facility provides: programme technical lead, thruster design/manufacture, testing, electronics and hollow cathode development. Culham: thruster testing (including diagnostic life tests), plasma physics, ion beam extraction modelling, UK-25 development/testing. Astrium (EADS): development of the UK-10 system for Artemis, including the power conditioning and control equipment and the propellant supply/monitoring equipment and the manufacture and qualification of the flight hardware for Artemis. An element of AEA Technology, Culham houses Europe's largest Xe propellant ion thruster test chamber: 5.8 m long, 1.3 m diameter, equipped with helium cryopumps for handling the inert gas. During 1996, the then MMS commissioned a new test facility, capable of handling two 25 mN thrusters simultaneously.

Ion thruster development in the UK began in the late 1960s at RAE/Culham, concentrating on a 10 cm diameter thruster with a nominal 10 mN thrust and mercury propellant. The T4A engineering model achieved very high efficiency, stable operation and a long operational lifetime. The propellant was changed to Xe in 1985 and new versions of the T5 flight model thruster have been manufactured/tested in collaboration between DRA, Culham and Matra Marconi Space. 0.2-70 mN thrust has been demonstrated; qualification testing at 18 mN is under way at MMS and DRA, specifically for operational NSSK application on ESA's Artemis. This will be extended to 25 mN during 1997.

A major experimental programme was completed in 1995 at the Aerospace Corp in Los Angeles, using USAF funding and a thruster supplied by DRA. This concentrated on a detailed characterisation of the ion beam. A lower level effort continues at Aerospace funded by DRA, EADS and the USAF.

The thruster comprises a cylindrical discharge chamber closed at one end by a soft iron backplate and with a set of closely-spaced grids at the other. A magnetic field is generated by six equispaced peripheral solenoids. The field lines inside the chamber link an inner cylindrical soft iron pole and a larger diameter outer pole. Propellant gas is introduced through the axial hollow cathode and a bypass distributor on the backplate. A DC discharge is set up between the cathode and the cylindrical anode. This ionises the gas, the efficiency being enhanced by the magnetic field and the correct design of the inner pole/baffle disc arrangement. The positive ions are extracted and accelerated by a high electric field between the grids, attaining typically 30 to 60 km/s. A triple grid design is used to minimise damage caused by the impact of charge-exchange ions. Electrons from an external cathode neutralise the ion beam's charge, and an earthed screen to prevent those electrons from reaching other parts of the device surrounds the whole thruster.

MMS UK is responsible for the electronics and propellant feed systems, and commercial exploitation of the fully qualified operational system. Philips Components Ltd previously manufactured the hollow cathode and neutraliser; the work has now been transferred to DRA. DRA/Culham in 1994 produced a low cost version of the complete system for experimental applications. This was accepted for flight on The Johns Hopkins University's NEPSTP Nuclear Electric Propulsion Space Test Program, since cancelled, which was intended to test the Russian Topaz 2 nuclear reactor. Other flight opportunities are being sought.

ESA's Artemis will carry two 18 mN versions. Another application may be ESA's Gravity Explorer Mission; DRA/Dornier studied the possibility in 1995/6 with encouraging results. DRA's STRV-1A launched in June 1994, included an experiment to allow the hollow cathode assembly to demonstrate spacecraft electrostatic discharging. Although the cathode assembly operated correctly, charging was not observed. The nominal operating parameters given below are in 18/25 mN order.

Design of a laboratory model 25 cm thruster began in early 1986 using scaling laws formulated during the UK-10 programme. Culham testing began October 1986; nominal thrust was 200 mN (highest obtained was 316 mN). Following the successful conclusion of the lab model work, an engineering model was manufactured at the beginning of 1989 and completed a comprehensive test/evaluation programme in 1992. Included were studies of life-limiting factors. Investigation of other propellants was completed in 1995 at Southampton University (although an ideal inert propellant, xenon is costly). Development of a power conditioning system was undertaken at Birmingham University and high current hollow cathodes were developed by DRA/Philips Components. Further work on the ESA-XX hybrid design, using features from this thruster and the German RIT-35, is under way with ESA funding. The first prototype of this thruster, intended for interplanetary missions, has been successfully tested at the University of Giessen in Germany.

VERIFIED

DERA – Space Department

Current Status

DERA is a government-owned UK research and development organisation providing independent services to civil and military organisations and industries world-wide. DERA draws upon well-equipped test and production facilities, and state-of-the-art satellite control and ground receiving stations. DERA is involved in most space activities ranging from concept and feasibility studies through to project management, design and hardware construction. Space Department also supports the satellite communications work of the Strategic Communications and Networks Department at DERA Defford.

Activities include analysis of future satellite and ground system requirements, procurement support, post launch mission analysis and resolution of spacecraft operating anomalies. Experience in the UK Skynet military communications programme is used to support overseas customers in the initial studies and procurement of new civil and military satellite systems. DERA provides expertise and operational guidance on control systems and ground station electronics architecture.

In the last two decades DERA has had a responsibility to develop remote sensing applications. Capabilities include SAR and visible/infra-red instrument design, and research into the exploitation of satellite remote sensing data for both civil and military use. Capabilities include digital imaging, radar and wet film photoanalysis and interpretation. Two ground stations for operational reception of meteorological data and high resolution radar data and near real-time delivery of application-specific products support the activities. DERA teams with industry to promote and develop commercially self-sustaining satellite remote sensing markets. Examples of collaboration includes Radarsolutions, a consortium to receive and distribute RADARSAT data across the British Isles and Denmark. DERA works with privately run and publicly owned remote sensing organisations supporting data interpretation and analysis.

DERA's research, development and consultancy is offered on all aspects of space technology, including spacecraft subsystems, attitude and orbit control systems, power systems, onboard data handling, materials and structures, thermal modelling and space debris. In addition, DERA is an acknowledged leader in the field of space environmental effects and ion propulsion. DERA has worked on Ion Thruster technology for more than 20 years, leading the development of the T5 Thruster which is employed in the UK10 Ion Propulsion system. Considerable research work conducted at Culham laboratory on electric propulsion technology and engineering.

DERA also maintains a comprehensive range of test facilities available to industry and academia for qualification testing of space hardware, instruments, components and satellites.

DERA runs a micro-satellite development programme offering a complete service to organisations wishing to take advantage of the comprehensive DERA space capability. This includes complete systems, satellite design and predictive modelling, comprehensive testing, manufacture and the complete provision of ground segment facilities.

Facilities

West Freugh satellite ground station
Acquisition of radar data
ERS-1,2; RADARSAT
Reception downlink at X band
Lasham satellite ground station
Acquisition of meteorological data:
NOAA, Meteosat, GOES
COSPAS/SARSAT data dissemination
L&S band TT&C
Image Data Facility
Image and data transcription facilities
Space Test Facility
Thermal vacuum chambers
Solar simulator
Vibration testing
Mass properties facility
Solar cell measurement and testing.

VERIFIED

Earth Observation Data Centre (EODC)

Current Status

EODC archives, processes and distributes data and products from ERS and other remote sensing satellites. One of four ESA off-line ERS Processing and Archiving Facilities (PAFs) was established at EODC. National Remote Sensing Centre Ltd manages its operations under contract to ESA. Feeds all government departments and national research and academic institutions with remote sensing and earth imaging data. It has been involved in producing policy recommendations about future directions for remote sensing programmes and international projects.

UPDATED

Ministry of Defence, Meteorological Office

Current Status

An Executive Agency within the Ministry of Defence, the department undertakes R&D into the use of satellite instruments for meteorological studies and forecasting of weather systems. The work falls into two principal categories: instrument development and imagery. The department built the SSU Stratospheric Sounding Unit currently flying on NOAA, the Focal Plane Assembly for ERS' Along-Track Scanning Radiometer and the AMSU-B humidity sounders for NOAA K-M.

Meteorologically useful products are derived from AVHRR and Meteosat data for research and operational purposes. TOVS sounding data have been used for many years and R&D work is under way for ATOVS. Images are employed for studies of atmospheric systems and for short-period weather forecasting.

Remote sensing consultancy services are offered through Metstar Consultants, a Met Office business unit. Instrument thermal vacuum calibration is undertaken for industry and space agencies. Testing is planned of MMS, MIMR, MWR; SSU, AMSU-B, MIR, MARSS, AMSU-A have been tested. The MoD Meteorology Office is working when requested, on co-operating with the US DoD on plans to integrate civil and military weather satellites. Input to ESA for Meteosat programmes recommendations on future meteorological aims and objectives. Input parallel to that of other member states of global weather watch. MoD Met Office has developed defence requirement weather forecasts using civilian satellites and this is integrated with US Department of Defence practice to develop a single system serving both user communities.

VERIFIED

Particle Physics and Astronomy Research Centre (PPARC)

Current Status

PPARC is the primary funding body for space science in the UK (mainly conducted in the universities and CCLRC), the second largest partner in BNSC and provider of much of the UK's subscription to ESA's science budget. PPARC is responsible for setting guidelines and research programmes in astronomy and in selecting objectives for the strategic exploration of the solar system. PPARC also contributes money for UK telescopes overseas including the Isaac Newton Group on La Palma, telescopes of the Joint Astronomy Group on Hawaii, the Anglo-Australian telescope in Australia, the Gemini telescopes in Chile and Hawaii and the UK Astronomy Technology Centre at the Royal Observatory in Edinburgh. PPARC also provides the UK's subscription to the European Southern Observatory (ESO) and it's suite of telescopes in Chile. PPARC maintains its strong position as the second largest spender in civil space activities in the UK. It continues to be regarded as one of the central research facilities in Europe's inventory of physics and astronomy research centres. Only the DTI has a larger budget.

UPDATED

Queen Mary and Westfield College

Current Status

QMW's Compact Antenna Test Range was opened in November 1991 to measure the radiation characteristics of spacecraft antennas up to 200 GHz; it is one of several such facilities in Europe. Single offset antenna 5.4 m focal length, 8–20, 26–40, 90, 150 and 180 GHz; test antenna mounted in class 10,000 tent on rail-mounted turntable, capacity 1 tonne, 360° azimuth motion, ±90° elevation. Test and measurement equipment includes HP Series 9000 model 310 desktop computer and model 375 UNIX workstation connected to an Ethernet LAN for fast pattern processing and efficient file transfer; receiver SA 1771 or HP 8510B. CATR was designed specifically for RF testing with instrument contractor Matra Marconi Space of the UK's AMSU-B Advanced Microwave Sounding Unit contribution to NOAA's Tiros-Next series. Testing of an EM and three FMs was successfully completed summer 1993. Microwave experiment antenna designs. Has recently expanded research in SHF frequencies and has pioneered work on shaped antennas.

A new far-field mm-wave anechoic chamber for feed measurements became operational in early 1995. The receiving system is based on a LHe-cooled InSb detector providing 100 to 600 GHz coverage. The CATR and QMW's 6 m far-field Anechoic Chamber are available for commercial use. New enhanced data collection support for external customers subsidised by indigenous requirements.

UPDATED

Satellite Navigation Group

Current Status

Formerly known as the UK Civil Satnav Group the Satellite Navigation Group was established by RIN in 1988, and represents UK interests on the US Civil GPS Interface Committee. Membership is 254 corporate/individual members. Services include the organisation of meetings and conferences and the publication of a newsletter. It also acts as a UK node for dissemination of material from the US Coast Guard Navigation Information Service. Serves as a focus for users, manufacturers and suppliers on representation to government and to the international controlling authority. It is playing a leading role in preparing for the new European Galileo space based navigation system and has made constructive recommendations to the authorising bodies throughout the European Union regarding the several diverse applications for navigation satellites and coupled transponder devices.

UPDATED

UK Astronomy Technology Centre (UKATC)

Current Status

Hosted by the Royal Observatory, Edinburgh, the UK Astronomy and Technology Centre is funded by the Particle Physics and Astronomy Research Council. It was established in 1998 following the restructuring of the two PPARC funded Royal Observatories: the Royal Observatory, Edinburgh, and the Royal Greenwich Observatory, Cambridge. The UKATC is the UK's national centre for the design and production of state-of-the-art astronomical technology and develops instruments in-house and participates as a partner in various national and international consortia. The centre has particular expertise in infra-red and submillimeter instrumentation for both ground and space based astronomy. As a consequence the ATC has considerable experience of cryogenic mechanisms and the application of infra-red array detector technology.

ATC facilities and core competences include CADCAM, optical design, an optical test laboratory, a detector test facility and clean room, software development environments, cryogenic and vacuum facilities and telescope simulators. The ATC was responsible for the Michelle mid-infra-red imager/spectrograph and the Uist near infra-red spectrometer/camera recently delivered to the UK Infra-red Telescope in Hawaii. It also delivered the Submillimeter Common User Bolometer Array (SCUBA) for the James Clerk Maxwell Telescope and is now at work on its successor, SCUBA-2. The ATC is also involved in the NAOMI adaptive optics system for the William Herschel Telescope in La Palma and the Gemini Multi-Object Spectrograph project.

The UKATC is the management organisation for the design, construction and commissioning of the VISTA telescope, a 4 m wide field survey telescope intended to operate in both visible and infra-red. It will be constructed at the European Southern Observatory site near Cerro Paranal in Chile. ATC staff are also working on the proposed very large ground-based telescopes such as the proposed European 100 m OWL telescope.

Space activities carried out on the site began in the 1960s and 1970s, with work on Skylark sounding rockets and early European astronomy satellites such as the S2/68 experiment for TD-1. The staff was involved in the optics for the ISOCAM infra-red camera orbited in 1995 on the ESA Infra-red Space Observatory. Current projects include a contribution to the SPIRE instrument for the ESA cornerstone mission Herschel (formally FIRST) and study work related to instruments for the Next Generation Telescope.

The UKATC collaborates with Sira and University College London's Mullard Space Science Laboratory in the UK Smart Optics Faraday Partnership established in June 2001. The partnership aims to create new and effective links between industry and the academic world in the field of smart optics. In addition to its technology work, a programme of astronomical research is carried out by the ATC project science group which comprises approximately 12 scientists.

VERIFIED

University of Dundee Centre for Coastal Zone Research

Current Status

The centre was established in 1995 jointly by Dundee University's Dept of Applied Physics and Electronic and Mechanical Engineering and Cray's Space and Defense Division. Projected revenue after four years was about £6 million. The main focus is research that directly involves operational end-users of information services, and emphasises the integration of satellite data with information derived from GIS and models of the physical, chemical and biological processes in the coastal regions. Cray will lead development of the operational systems arising from the research. Considerable coastal, environmental and pollution related data for scientific analysis and a data base applicable to local and national utilisation.

The initial programme includes work analysing methods of automatically detecting and classifying algal blooms and monitoring coastal ships. Future activities will cover the potential users of satellite data for flood monitoring and related disaster monitoring applications, and methods of using SAR interferometry for monitoring small scale changes in coastal defence systems.

Significant emphasis has been placed on the dual use of air and satellite systems for accurate monitoring of coastal erosion and unique software programmes for optical and SAR imaging have been developed. The remit has been extended to the study of estuarine levels and ice inundation studies.

UPDATED

University of Dundee, Satellite Station

Current Status

University of Dundee operates the Satellite Station, a NOAA HRPT data acquisition, archiving and distribution service funded by the Natural Environment Research Council. Data set consists of daily recordings from NOAA since October 1978. Data directly applicable to wide user access for academic study and for planning environmental and climate research. More recently, the Station was upgraded to take SeaWiFS data from SeaStar, specialised ground stations are manufactured for use with GMS S-VISSR, NOAA HRPT and Meteosat PDUS/WEFAX/MDD/DCP.

VERIFIED

United States

Center for Commercial Space Communications

Current Status
CCSC was established in 1991 by Virginia Tech, partly under initial funding from the state's Center for Innovative Technology, as an interdisciplinary university centre to develop, manage and undertake research in commercial aspects of space communications. LEO satellites are of particular interest. Also developed hybrid communications systems and researched the potential market value of ultra-high bandwidth capability using satellites for nodal connections. CCSC is associated with Virginia Tech's Satellite Communications Group, which has operated for two decades and received more than US$6 million in federal/corporate funding for space communications research. It has designed/built receivers for NASA's ACTS programme developed by NASA's Lewis Research Centre and launched by Shuttle in September 1993. CCSC has conducted work to support the prolific programme proposals from LEO companies. Diverse interactive telecommunications research linked to results from consultation with users and operators. The CCSC attempted to link government based research projects with several new and innovative commercial activities.

UPDATED

Center for Satellite and Hybrid Communication Networks

Current Status
The Center at the University of Maryland Institute for Systems Research was established in October 1991 as one of NASA's Centers for the Commercial Development of Space. Its primary goal is development and commercialisation of advanced space-based communications, emphasising hybrid satellite/terrestrial networks. Specialist areas cover teleconferencing, DBS, HDTV, database access, supercomputing linkage, satellite digital cellular networks, narrowband mobile networks and personal communications. Extensive analysis of advanced communication services included (SHF and laser technology) in a comprehensive research effort. Work has been conducted on advance integrated telecommunications services involving space- and earth-based systems. Work also includes narrowband and broadband switching elements developed in concept for global satellite capabilities. New technologies sought from both commercial research facilities and government laboratories are being united in a concerted effort to integrate development projects.

UPDATED

Civil GPS Services

Current Status
NIS (until 1995 the GPS Information Center) is operated by the US Coast Guard within the Dept of Transportation as part of the Civil GPS Service to provide basic information and as a point of contact for the civil GPS user. It co-ordinates and manages the Civil GPS Service Interface Committee as part of the Department of Transportation's programme. It has a manned watch 24 hours a day. The Bulletin Board is a constantly updated computer service that can be accessed via a modem. The Internet service has been available since July 1995. The other three CGS elements are: CGS Interface Committee (identifies civil GPS technical needs); Differential GPS (operational in 1996, providing 10 m accuracy to US maritime users); PPS Program Office (under formation, allowing qualified civil users access to the PPS signal). NIS gathers, processes and disseminates timely GPS, DGPS, Omega and Loran-C status and other navigation information to users. NIS has performed integrated navigation system evaluation studies examining the application of several different satellite constellations. Consultation body for integrating civil and military considerations for continued management of the Navstar GPS and for international consultation with the European Galileo fixing system.

UPDATED

Department of Commerce, Office of Space Commercialization

Current Status
The Office of Space Commercialization (OSC) was established 2 December 1988 to co-ordinate space-related issues and programmes within the Department. OSC's Director is responsible for advising the Secretary on all space issues. Input was sought from all areas of the US government and from other agencies around the country in efforts to shape space activity for policy issues. Early efforts were focused on stimulating legislative frameworks for commercial endeavours but these were opposed by NASA. The mission statement defines the goal of OSC as one to foster an economic and policy environment that ensures the growth and international competitiveness of the US commercial space industry. OSC conducts activities in three primary areas: policy development; market analysis; outreach and education. The OSC plays a leading role in space based weather forecasting through the National Oceanic and Atmospheric Administration. New areas of potential commercialisation include government services given or provided as part of a national public service.

UPDATED

Department of Defense, Defense Advanced Research Projects Agency (DARPA)

Current Status
DARPA was established in February 1958 as the Advanced Research Project Agency, partially as a response to Sputnik, and on 23 March 1972 the name was changed to Defense Advanced Research Projects Agency and was established as a separate agency under the DoD. It operates as the Dept of Defense's central R&D organisation for maintaining technical development and proof of concept demonstrations, pursuing them through other DoD and US government agencies. For example, it funded development of the Pegasus and Taurus launch vehicles, several small satellites and numerous advanced technologies for satellite subsystems and payloads. Consultation role in future launch and space transportation systems decision making at NASA and the Department of Defense.

However, DARPA in 1994 officially withdrew from space activities, returning to structure the Discoverer II programme at the end of that decade, a joint initiative with the US Air Force and the National Reconnaissance Office to develop an affordable space-based radar for tactical geolocation. DARPA has conducted a research programme into lightweight, compressible technologies that offer the potential for developing and deploying extremely large antennas in space. The Innovative Space Based Radar Antenna Technology (ISAT) programme assesses feasible and affordable designs for space-based radar systems. The Office of Director is supported by a number of Special Assistants, five management offices and six programme offices: Defense Sciences; Electronic Technology; Information Systems; Information Technology; Sensor Technology; and Tactical Technology. DARPA budgets have increased from US$1.887 billion in FY1999 to US$1.951 in FY 2001.

VERIFIED

Department of Defense, Defense Information Systems Agency (DISA)

Current Status
DISA is a Department of Defense (DoD) combat support agency whose main objective is to anticipate and respond to the needs of its customers, the warfighters, by providing them with seamless, end-to-end, innovative and integrated information services, which provide a fused picture of the battlefield. Core mission areas incorporate the Defense Information System, Defense Message System, Global Combat Support System and the Global Command and Control System. It is responsible for planning, developing and supporting command, control, communications, computers and intelligence (C^4 I) and information systems that serve the needs of the National Command Authorities (NCA) under all conditions of peace and war. It provides guidance and support on technical and operational C^3 and information systems issues and coordinates DOD planning and policy for the integration of C^4I systems and the insertion of C^4 I for the Warrior (C^4 IFTW) leading edge technologies into the Defense Information Infrastructure (DII). DISC has worked with DARPA on geolocation systems. Creative concept evaluation and analysis for potential new systems capabilities in supporting defence requirements in space. Application of new technologies to evolving defence strategies.

Department of Defense, National Geospatial-Intelligence Agency (NIMA)

Current Status
In 2003 the National Imaging and Mapping Agency was renamed the National Geospatial-Intelligence Agency (NGA) in response to changing requirements and a shift in the agency's role and application by defence and relevant civilian authorities. The NIMA National Imaging and Mapping Agency was created 1 October 1996 to co-ordinate imagery collection, processing, exploitation, analysis and distribution, absorbing NPIC, CIO and the Defense Mapping Agency. It is a DoD combat support agency formed by consolidating the Defense Mapping Agency, the Central Imagery Office, the Defense Dissemination Program Office and the National Photographic Interpretation Center. NRO underwent internal reorganisation in 1992, apparently opting for multi-agency staffing of its various branches so that, for example, the CIA and USAF no longer compete on imaging satellite projects. This degree of interaction stems on a long-standing difficulty with polarising the needs of specific services within the Department of Defense. NGA is one of the US Intelligence Community members.

Department of Defense, National Reconnaissance Office (NRO)

Current Status
A member of the US Intelligence Community of 14 agencies and bodies, NRO was formally established 25 August 1960, although its existence was not officially acknowledged by DoD until 18 September 1992, to co-ordinate overhead reconnaissance operations, including those from satellites. The annual budget is estimated to be in excess of US$10 billion, although outgoing CIA Director James Woolsey in January 1995 noted to a Senate committee that a radical restructuring would cut the number of future intelligence satellites by almost half. Recent reorganisation has established functional lines of intelligence gathering involving

signals, imagery and ocean surveillance. About 4,000 staff. The US$100 million HQ was completed in 1996 near Dulles airport; NRO was previously housed in the Pentagon.

It receives its budget through the National Reconnaissance Program portion of the National Foreign Intelligence Program. The office reports to the Secretary of Defense and the Director of Central Intelligence controls allocation of NRO assets. Until 1992, the inter-agency Comirex Committee on Imagery Requirements and Exploitation met daily to review requests and direct targeting. That role was assumed by the DoD Central Imagery Office, created in June 1992. Imagery is relayed to Fort Belvoir in Virginia, from where it is passed to the National Photographic Interpretation Centre in Washington DC (established 1961 and run by the CIA for all the intelligence community) and other users. In February 1995 the NRO declassified the Corona programme and 800,000 images taken between 1960 and 1972 were transferred to the National Archives and Records Administration.

The Assistant Secretary of the Air Force serves as the Director of the NRO and reports to the Secretary of Defence who has ultimate responsibility for the NRO.

With the advent of commercial remote sensing images on a reliable and routine basis the NRO has tasking assignments influenced by the use of some commercial data at high-resolution for general observation.

Department of Defense, Office of Space Acquisition and Technology

Current Status

Military space policy and acquisition responsibilities were consolidated in December 1994 into a single new 'space architect office': Space Acquisition and Technology Programs (SA&TP), headed by a Deputy Defense Undersecretary (DUSD-Space, Robert Davis). Air Force Major Gen Robert Dickman was appointed as the Space Architect 15 September 1995. The new Office has authority over all military space activities and organisations, with a goal of integrating space capabilities into a unified approach. The Joint Space Management Board was activated December 1995 to ensure an integrated approach for the first time between DoD/NRO. In 1996, the JSMB created the National Strategic Space Master Plan to cover US$130 billion of spending over 10 years. While this plan has removed the core architecture, several changes have taken place in response to changing priorities in defense and security strategies under the Bush administration.

Military activities were previously drawn together in 1985 by the activation of a unified US Space Command at Peterson AFB in Colorado. Collocated are the USAF Space Command and US Army Space Command; the Naval Space Command is sited separately in Dahlgren, Virginia. Space Command provides missile and satellite tracking information, operational launch services at Canaveral and Vandenberg and command/control of operational DoD satellites. Management, operation, TT&C and execution of R&D space systems is allocated to the AF Space and Missiles Systems Center in California. Presently handling analysis and evaluation of space transportation and launch vehicle planning policy related to the new generation of Delta and Atlas heavy lift vehicles. Also developing rationales for integrating defence procurement programmes with NASA technology development initiatives.

Department of Energy, Office of Space and Defense Power Systems

Current Status

DoE developed the radioisotope thermoelectric generators (RTGs) for the Cassini mission (US$45 million FY95) launched in 1997. Funding cuts mean that it might be the last unless NASA or other agency provides support; FY96's request excluded all space reactor work. More recent studies have been conducted for the DoD on high output nuclear reactors for classified defence missions. DoE's Office of Energy Research is supporting the Alpha Magnetic Spectrometer for Space Station. The Office of Intelligence and National Security also provides space sensors for treaty verification, nuclear proliferation detection and environmental protection; it funded the Alexis project. See the Los Alamos National Laboratory entry in the World Space Centres section. NASA has revitalised the potential for space nuclear power systems by re-opening a programme for nuclear powered electric propulsion. NASA is also resurrecting plans for space based power systems for a variety of spacecraft and missions.

Department of the Navy, Office of Naval Research

Current Status

NRL was established in July 1923 as the Naval Experimental and Research Laboratory. Now an element of ONR, it is the navy's corporate laboratory. Staff total about 3,500, mostly civilian, in Washington DC, Stennis Space Center (Bay St Louis, Missouri), Monterey (California) and Orlando (Florida). NRL's Naval Center for Space Technology (NCST) maintains space technology expertise through two departments: Space Systems Development and Spacecraft Engineering. The Remote Sensing Division (RSD) and the Space Science Division (SSD) come under NRL's Ocean and Atmospheric Science and Technology Directorate.

NRL directed the Vanguard programme of the mid-1950s, America's first publicly revealed satellite programme, and since 1960 has developed more than 80 satellites, most recently the Clementine technology demonstrator lunar/asteroid probe for BMDO launched in January 1994. It has flown numerous payloads and experiments, such as the Hercules unit carried by STS-53 in December 1992 to allow astronauts to photograph surface features while automatically recording positions within 2 km. It flew again on STS-56 in April 1993 and on STS-70 in July 1995; no other flights are yet manifested but the ONR is actively pursuing several candidate experiments for the International Space Station. The HTSSE High Temperature Superconducting Space Experiment is aboard STP's Argos satellite.

NCST facilities include cleanrooms, anechoic RF chambers, shock/vibration chambers, an acoustic reverberation chamber, thermal/vacuum chambers and long term testing of satellite clock time/frequency standards.

RSD undertakes a broad programme in sensing applications over frequencies from ultra-violet to radio. Sensor systems include RAR Real Aperture Radar, scatterometers, lidars, optical/radio interferometers and passive microwave imagers. Its middle atmosphere sensors include the POAM Polar Ozone and Aerosol Monitor on Spot 3, the MAS Millimetre-wave Atmospheric Sounder on Shuttle Atlas missions (three flown), and a suite of water vapour and ozone monitors as part of the Network for Detection of Stratospheric Change. RSD's facilities include digital image processing, a tactical environmental visualisation centre, an aerosol measurement facility, 25.6/25.9 m radio antennas at Maryland Point Observatory and an optical interferometer at Mount Wilson Observatory for monitoring background environmental emissions at high angular resolution.

SSD maintains facilities for designing, constructing, assembling, calibrating and analyzing space experiments, principally upper atmosphere UV sensing, solar atmosphere spectrometry and celestial radiation over UV to cosmic rays. For example, it provided the SUSIM Solar UV Spectral Irradiance Monitor for UARS and Atlas, the RAIDS Remote Atmospheric and Ionospheric Detection System to measure airglow, the OSSE Oriented Scintillation Spectrometer Experiment gamma instrument on NASA's Compton Observatory, solar x-ray measurements using a Bragg crystal on Japan's Yohkoh, and the LASCO Large Angle Spectrometric Coronograph on Soho. SSD provided three of the main instruments (USA, HIRAAS, GIMI) on Argos.

Definitive trade studies on the role of space technology and systems design for continuing studies of the space environment. Space engineering analysis.

VERIFIED

Federal Aviation Administration, Office of Commercial Space Transportation

Current Status

The Commercial Space Launch Act of 1984 and Executive Order 12465 gave the Department of Transportation the authority to regulate US commercial ELV activities. Orbital/suborbital firings are listed in OCST's quarterly Commercial Launch Manifest. The first licensed orbital launch was of the BSB satellite in August 1989 by Delta. Licenses are granted based on launch providers presenting evidence they are in compliance with all safety regulations and other requirements, and have sufficient insurance or financial resources to cover any probable losses from a launch mishap. The office is also responsible for the regulation of any future commercial launch sites and to oversee commercial re-entry vehicles. It is also the compliance monitor for the Chinese and Russian commercial launch agreements. The Space Systems Development Division studies environmental impacts of new launch sites, develops long range commercial launch forecasts and develops space transportation policies. Evaluated privately owned rocket launcher projects, licensing entrepreneurial ventures, launch forecasts compiled for government agencies analysis of future requirements.

The COMSTAC Commercial Space Transportation's Advisory Committee provides policy guidance. OCST, previously reporting directly to the Office of the Secretary, transferred intact 16 November 1995 to the Federal Aviation Administration. 28 staff.

VERIFIED

Federal Communications Commission (FCC)

Current Status

The FCC is the regulatory body for interstate and international communications by radio, TV, wire, satellite and cable. Satellite proposals are submitted to the commission for licensing consideration. The International Bureau has oversight of all satellite services. Five Commissioners, appointed for five-year terms direct the FCC. There are six operating bureaus: Mass Media, Common Carrier, Wireless Telecommunications, Compliance and Information, International, and Cable Services. FCC frequently advises on the integrated networks operating national and non-US service providers in and out of domestic systems. FCC conducts evaluation of communications satellite payload technologies and provides forecasts of frequency requirement. It also carries out new research into federally supported broadband services.

UPDATED

Florida Space Authority

Current Status

Created in 1989 by Florida's Governor and Legislature, and governed by a board of supervisors, the Spaceport Authority aims to foster development of Florida's space enterprise, including industry, research and education. The Authority is authorised to issue industrial development bonds and its legislation provides statewide tax incentives for space business, and tax exemptions for launch vehicles, payloads and propellants. It has financed the development of the US$30 million Apollo-Saturn 5 Center museum for NASA, and built a US$35 million 6,094 m^2 Titan 4 solid rocket motor warehouse/support complex for the air force in northeast Florida, both completed in 1996.

Canaveral's Launch Complexes 46 and 20 have been modified by the Authority for suborbital rockets and small orbital vehicles beginning 1997, including AF Minuteman and those employing Thiokol's Castor 120 solid motor, such as Taurus and LMLV.

In November 2000 Spaceport Florida Authority selected Lockheed Martine Technical Operations to maintain and support Launch Complexes 46 and 20. Other facilities under the auspices of the SFA include LC 41, LC 37, Hangar AM, Space Station Commerce Park, Florida Space Research Institute and the Space Operations Control Center.

International Space Research Park is being developed by the Authority and NASA's KSC. It frequently hosts visits from foreign students and teachers while playing a major role in disseminating information about Florida's space related activities.

VERIFIED

Geosat Committee

Current Status

The Committee was formed in 1976 as a non-profit educational organisation dedicated to the civil use of remote sensing technology for geological and environmental applications. Geosat four-point programme embraces public education through workshops and lectures, utilisation of remote serving through member activities, dissemination of remote serving products and sponsored research. Integrated inter-agency requirements for remote sensing and geodynamic investigations co-ordinating NASA, NOAA and the Department of the Interior. Geosat has more than 45 corporate members and 18 alumni/consultant chapter members, including Spot Image, MacDonald Dettwiler, EOSAT, Eurimage, ARCO, Earthsat, Amoco and Texaco. Research projects have been conducted with NASA, US Geological Survey, Australia, France, Germany, Japan and UK. Plays key roles in national and international forums and is an important policy making instrument for industrial participants and for international consensus on standards and specifications.

VERIFIED

Lockheed Martin Global Telecommunications

Current Status

Following two years of negotiations Lockheed Martin merged with the Comsat Corporation to form Lockheed Martin Global Telecommunications in 2000 for US$2.7 billion. COMSAT is the US signatory to Intelsat (19.1 per cent contribution) and Inmarsat (23.0 per cent). Units include COMSAT General, operators of Comstar satellites, and COMSAT Labs. COMSAT Corp was created in 1963 following passage of the Communications Satellite Act, signed into law by President Kennedy in late 1962. COMSAT acted as technical manager for Intelsat until 1979, forming COMSAT Laboratories in 1967 for technology R&D. COMSAT General was established in 1973 to branch out into domestic satellite communication.

The shareholder-owned Communications Satellite Corp was established in February 1963 to carry out a US government mandate for a global commercial satellite communication system in co-operation with other countries.

The World Systems division provides satellite communication services to US international carriers and other customers through Intelsat. COMSAT Mobile Communications provides satellite and land Earth station services for shipping, offshore oil platforms and international land mobile users through Inmarsat. As part of the Skyways Alliance, it is providing aeronautical communications through the Inmarsat system. Through COMSAT, flexible route-paths for terrestrial link hook-up with satellite systems or constellations.

The COMSAT General Corporation subsidiary was established in 1973 as a satellite operator: it manages the five satellites of the Marisat, Comstar and SBS 2 networks. COMSAT Video Enterprises owns and operates a satellite-based network distributing entertainment and videoconference services to the US lodging industry. COMSAT Laboratories perform communications R&D for its parent and other organisations, such as advanced technologies for NASA's ACTS, low-cost antennas, and advanced transmission systems. It developed ACTS' NASA ground station and the master control station. The Laboratory employs some 320 of COMSAT's 1,500 personnel. Radiation Systems Inc was acquired in 1994 and merged with COMSAT Technology Services to create COMSAT RSI, able to provide more than 80 per cent of complete satellite communication ground systems.

The Corporation operates Earth stations at Santa Paula, California and Southbury, Connecticut, each with two 12.8 m dishes and a 10.4 m non-tracking TT&C antenna.

National Aeronautics and Space Administration, Ames Research Center

Current Status

NASA ARC continues to operate ER-2 monitoring aircraft for rapid response to natural disasters and in the aftermath of the changes being brought about by the Bush Initiative for a reinvigorated programme of space exploration the centre is active in support for the Mars Exploration Rovers "Spirit" and "Opportunity".

Background

Ames Research Center was established in 1939 as the second facility of the National Advisory Committee on Aeronautics. Ames became part of NASA when the agency was created in 1958. Ames is located in the heart of California's Silicon Valley, about 10 miles north of San Jose in the San Francisco Bay area.

Ames is a NASA leader in information technology and has additional core competencies in biotechnology and nanotechnology. Ames pursues NASA missions in aerospace operations systems and astrobiology. The centre exercises scientific leadership in fundamental space biology and key disciplines of NASA's multifaceted astrobiology thrust. Ames also performs key work in support of NASA missions in computational systems, aviation capacity, air traffic management, thermal protection systems and a host of other areas.

The grand challenge of NASA's mission to explore space and study the origin and role of life in the universe is driving the agency's focus on the technology triad of information technology, biotechnology and nanotechnology. These arenas are widely accepted as the most likely source of breakthrough in the next decade. Within NASA, only Ames has the integrated research environment needed to exploit both the crossover potential and the individual field of breakthroughs of the technology triad to meet the agency's mission needs.

NASA's bold missions in space exploration and aeronautics will require advances in many areas of science and technology, particularly the enabling technologies of computer science. The information sciences team at Ames is a world-class creator of innovative, intelligent, high-performance, and reliable information technologies that enable space and aeronautics missions. Ames specialises in research and development of methods, technologies and processes in autonomous reasoning, human-centred computing and intelligent data understanding for broad application to NASA mission requirements.

Ames is leading NASA's research in astrobiology, the study of the origin, evolution, distribution and destiny of life in the universe. Astrobiology represents a synthesis of disciplines from astronomy to zoology, ecology to molecular biology, and geology to genomics. Ames hosts the NASA Astrobiology Institute, which conducts and integrates groundbreaking research in the astrobiology sciences.

Ames is spearheading agency efforts in fundamental biology research and related technology development. Ames leads research into the effects of space on the evolution, development and function of living organisms, and applies space technology to detect global ecological changes. Ames supports such research both on Earth and in space.

The goal of Ames' life sciences research is to understand the role and influence of gravity on living systems, from cells in culture to physiological studies in animals and humans. Through a better understanding of fundamental biology will come knowledge useful for the development of countermeasures to the deleterious effects of weightlessness and for the maintenance of human health on Earth. The life sciences research team at Ames facilitates the use of the space shuttle and the International Space Station to reach NASA experiment objectives. The research team also conducts extensive ground-based studies utilising a suite of unique gravitational research facilities and advanced 3-D imagery technologies.

Ames' Space Station Biological Research Project supports fundamental biology research goals by enabling long-term space science research in relevant science disciplines. The research program will initially emphasise microbiology and cell-culture research using equipment developed for use on the International Space Station.

Ames is at the forefront in development of automated aerospace operations for applications, from decision tools for air traffic management (used by the Federal Aviation Administration) to autonomous reasoning control of spacecraft in unpredictable environments. Ames works with the FAA and industry to meet the nation's expanding air transportation demands by developing and demonstrating pioneering concepts and technologies to improve safety, increase efficiency and enhance environmental protection. Ames is striving to provide the research the technology products needed to improve aviation safety and expand capacity.

The research centre also plays a major role in efforts to enhance the capacity and productivity of the national airspace system via revolutionary operations systems and vehicle concepts.

Ames lead the Agency in Aviation Systems Technology Advanced Research, which will accelerate the development of selected air traffic management technologies identified by industry and the FAA to improve the capacity and reliability of the current system. Ames will also conduct the fundamental research to support the air transportation system of the future.

VERIFIED

National Aeronautics and Space Administration, Dryden Flight Research Center

Current Status

DFRF merged in 1981 with NASA Ames but again became a fully independent centre 1 October 1994. FY97 budget was US$146 million. Some 500 civil service and 500 contractors are employed. It is located 130 km north of Los Angeles on the edge of the 114 km^2 Rogers Dry Lake at the south end of an 800 km high-speed flight corridor at the USAF Edwards AFB. It has 51,800 km^2 of restricted airspace to conduct aeronautical research with high performance aircraft such as the SR-71, F- 16XL and the X-36 tail-less fighter.

The Apollo Lunar Landing Research Vehicle and Lifting Body family operated out of Dryden in the 1960s-70s. The 8 to 17 km runways on the dry lakebed and Edwards' main 4,600 m concrete runway are now used as Shuttle landing sites when Cape Canaveral conditions are unacceptable, additionally supporting the ferry flights to NASA's Florida launch site.

The OSC/Hercules Pegasus small orbital launch missions were based initially at the site, released over the Pacific from the same USAF/Boeing B-52 employed for X15 hypersonic research flights up to 1968.

Dryden was involved in the development of the X-33 and X-34 reusable launchers until their cancellation by the Bush administration in 2001.

First drop test for the unpowered Boeing X-40A developed for the US Air Force as a quick reaction orbital vehicle, 14 March 2001 (NASA/DFRC)
0569808

NASA DFRC is also supporting development of the X-38 crew return vehicle which completed its third and final free flight on 30 March 2000. Currently working the X-37 approach and landing test vehicle, X-43 hypersonic research aircraft and associated support for the Columbia Accident Investigation Board.

National Aeronautics and Space Administration, Goddard Space Flight Center

Current Status

GSFC was established in May 1959 around a core of 157 transferred from NRL's Vanguard team. It is the only US national facility that can develop, fabricate, test, launch and analyse data from its own space science missions. FY98 budget was US$3,472 million; for FY97 it was US$3,495 million; FY93-96 budgets each about US$2.5 billion; FY92 US$2,459 million; FY91 US$2,379 million; FY90 US$1,849 million. Total workforce is about 12,000, including more than 3,400 civil servants and 8,500 contract personnel. Goddard has managed the development of more than160 satellites for NASA and NOAA, including the Explorer series, COBE, Compton Observatory, Solar Max, ERBS, Spartan and UARS. Hubble Space Telescope science operations are controlled by the ST ScI atThe Johns Hopkins University. GSFC directs NASA's Delta launch activities.

Goddard directs the operation of NASA's Spaceflight Tracking and Data Network. STDN's Space Network (SN) segment can provide 85 per cent coverage for satellites orbiting up to 1,200 km using six TDRS Tracking and Data Relay Satellites. Each TDRS can simultaneously handle multiple users at up to 300 Mbit/s. A network of fully automated computer-controlled equipment and 18.3 m ground antennas controls the TDRS constellation under Goddard's Network Control Center. The White Sands Complex in New Mexico comprises two terminals, each responsible for three TDRS. STDN's Ground Network (GN) comprises three facilities with S-band (9 m) and C- band tracking systems. Located at Ponce de Leon and Merritt Island (Florida) and Bermuda Island (UK), they are used primarily by GSFC to support all Kennedy Space Center launch operations, including Shuttle and expendables. Three further tracking stations in Canberra (Australia), Goldstone (California) and Madrid (Spain) are operated separately by JPL. Although closely aligned with Goddard's STDN for supplementing Shuttle and selected Earth satellites, these Deep Space Network (DSN) stations are principally dedicated to JPL managed projects such as Galileo and Voyager.

The 8,082 m^2 US$16 million Spacecraft Systems Development and Integration Facility opened in June 1990 for handling Shuttle payloads of up to 27 tonnes. The 1,161 m^2 class 10,000 laminar flow cleanroom is one of the world's largest and can accommodate two full size Shuttle payloads (up to 27,216 kg, 4.572 m diameter) simultaneously. Other facilities include: Large Area Pulsed Solar Simulator (test complete panels), Space Simulation Test Facility, Vibration Facility, Battery Test Facility, High Voltage Test Facility, Magnetic Field Component Test Facility (calibrate/align magnetometers), Spacecraft Magnetic Test Facility (entire craft and sounding rockets), High Capacity Centrifuge, Acoustic Facility, Electromagnetic Interference Facility, Static/Dynamic Balance Facility, Optical Thin Film Deposition Facility, Material Properties and Analysis Laboratories.

NASA's Goddard runs the Getaway Special Program (GAS) which accommodates small, autonomous payloads in Shuttle's cargo bay utilising standardised hardware. Hitchhiker offers two carriers providing capabilities beyond GAS. Hitchhiker Junior entered service 1995, using HH avionics and GAS cans but offering satellite ejection (previously from GAS) and limited Orbiter power and services such as pointing.

The canister is an aluminium cylinder provided with a standard experiment mounting plate which, while it may not be altered, does carry provisions for a variety of attachments. GAS can be evacuated and/or pressurised and includes an insulated exterior on the bottom/sides for passive thermal control (an insulated top end cap is available). Operations are independent of the Shuttle Orbiter other than three on/off controls activated by the crew. The experimenter is responsible for providing electrical power, heating/cooling and data acquisition systems. A Motorised Door Assembly is available (satellite ejection no longer is).

A GAS Bridge capable of holding 12 canisters was first flown aboard 61C in January 1986; second flight was STS-40 in June 1991. Payload access is not available following installation in the Orbiter at KSC some two to three months before launch. NASA classifies payloads within three categories: educational (class 1), foreign/commercial (2) and government (3). Within each group, payloads are flown on a first-come-first-served basis. The agency must be provided with detailed information no later than 18 months before proposed launch to ensure adherence to safety regulations. Shipping to KSC is required three to four months before launch. 53 GAS payloads had been flown on the first 24 Shuttle missions before *Challenger's* loss in January 1986, including 13 on the 61C mission alone. The next mission was not until 1991's STS-40, carrying the GAS Bridge with 12 cans. STS-42 added 10 in January 1992; STS-45 flew #78 March 1992, STS-47 #79-87 September 1992, STS-57 #88-97 June 1993, STS-60 #98-101 February 1994, STS-59 #102-104, STS-64 September 1994 (10 GAS on Bridge), STS-68 September 1994 (3 GAS), STS-67 March 1995 (2 GAS), STS-69 September 1995 (4 GAS), STS-72 January 1996 (5 GAS), STS-76 March 1996 (1 GAS), STS-77 May 1996 (12 GAS on Bridge); STS-80 November 1996 (1 GAS); STS-85 August 1997 (2 GAS); STS-87 November 1997 (1 GAS); STS-89 January 1998 (4 GAS); STS-90 April 1998 (3 GAS); STS-91 June 1998 (8 GAS); STS-95 October 1998 (1 GAS); STS-88 December 1998 (1 GAS). Fewer opportunities are now available as the Space Station era begins.

The cost is US$3,000, US$5,000, US$10,000 US Educational; US$8,000, US$14,000, US$27,000 others; for the three size categories listed below; US$500 non-refundable earnest money required.

The Hitchhiker HH-G/HH-M comprise mechanical mounting structures and an avionics unit connecting to the Orbiter's electrical systems, providing up to six payloads with power, real-time telemetry and crew/ground commands. The canisters are mechanically very similar to GAS cans. HH-G is a side-mount version carrying up to three canisters or 63.5 × 99 cm vertical mounting plates holding up to 272 kg in total. It is normally installed in KSC's OPF at the forward starboard side. The HHM bridge is installed either in the OPF or on the pad, and can accommodate 907 kg in up to 11 canisters or 71 × 71 cm plates (six on each side, one reserved for avionics); four 71 × 91.4 cm top plates are also available. Both HH versions offer two 0.142 m^3 50.2 cm diameter 71 cm high canister types: sealed with dry N_2 or air, 90.7 kg capacity; with a Hitchhiker Motorized Door Assembly, 77.1 kg capacity. Customer hardware must be delivered about 6 months before flight; launch occurs typically 8 weeks after Orbiter integration. Cost is $1.108 million at FY90 rates. First HH mission was HH-G1 on STS-61C in January 1986; #2 (STP 1) STS-39 April 1991; #3 STS-52 October 1992; #4 STS-53 December 1992; #5 (DXS) STS- 54 January 1993; #6 STS-60 February 1994; #7 STS-69 September 1995; #8 STS-85 August 1997; #9 STS-95 October 1998. 36 Hitchhiker experiments have flown to date.

Another programme Hitchhiker Junior (HH-J) was introduced in 1995 as an expanded GAS. Up to 100 W can be provided from the Orbiter, totalling up to 4 kWh per mission. The HH avionics report T/P, battery voltage/current, door status and commanded relay status to a laptop computer in the cabin. Real-time ground command or data are not provided. HH-J can eject a satellite up to 68 kg 48 cm diameter at 0.6-1.2 m/s from a 23.8 cm diameter marmon plate interface.

As GSFC enters a new decade focus centres on NASA's Earth science initiative and series of environment observation and monitoring missions including the Earth observing Satellites, the first of which was launched on 21 November 2000. GSFC is currently working up the Thermosphere, Ionosphere, Mesophere, Energetics and Dynamics (TIMED) mission to study a region of earth's atmosphere 60 to180 km above the earth's surface. GSFC developed the GALEX (Galaxy Evolution Explorer) satellite successfully launched in 2003 and the SWIFT explorer launched in 2004.

UPDATED

National Aeronautics and Space Administration, Headquarters

Current Status

NASA has been tasked with carrying out a new Space Exploration Initiative which was announced by President Bush in January 2004 following almost two years of deliberation over future space programme options. NASA was formally established 1 October 1958 to plan and execute the US civil space programme. It comprises seven principal offices and about a dozen major centres and facilities employing 18,741 civil servants, including about 1,100 headquarters staff. FY2000 budget was US$14.035 billion with US$14,253 billion authorised for FY2001, the last budget sent to Congress by the Clinton administration. The first budget request by the Bush administration allows NASA US$14.511 billion for FY2002. The FY2003

NASA has completed its Great Observatories programme with the SIRTF Spitzer telescope (NASA)
NEW/1047540

NASA plans to retire the Shuttle in 2010 and develop a new class of vehicles allowing a return to the moon and the beginning of human Mars visits (NASA) 0572488

Technicians test components on the Mars Exploration Rover launched in 2003 (JPL) ***NEW***/1047539

NASA envisions a mission to Mars for human colonisation some time before the middle of this century (NASA) ***NEW***/1047544

The Spitzer telescope launched in 2003 (NASA) ***NEW***/1047541

budget provided US$15 billion and for FY2004 the proposed budget is US$15,469 billion, an increase of 3.1 per cent over the preceding year. Prior to the Columbia disaster, an event which may result in major budget changes for human space flight, the White House projects NASA budgets increasing by 3.7 to 3.8 per cent over the succeeding three years. The White House in January 1995 directed NASA to cut the budget to US$12.6 billion (FY94 rate) by 2000, saving US$5 billion over those years. President Clinton's March 1996 FY97 request plans were to take out another US$3.3 billion, leaving NASA with only US$11.6 billion by FY2000. Recent budget increases have restored the NASA budget to one which keeps pace with inflation and the new Space Exploration Initiative announced by President Bush in January 2004 is pledged to increase the budget by 5 per cent a year until FY2010 whereafter it will be pegged at the rate of inflation. The FY2005 budget request of US$16.244 billion is 5.6 per cent higher than the US$15.378 billion finally appropriated for FY2004 and is consistent with the projections of the new Initiative. Initial Congressional Hearings in February 2004 indicated support from both Houses and across the political divide. To accomplish the broad objectives of the SEI, NASA budget projections incorporate a growth of 4.7 per cent in FY2006, 4.8 per cent in FY2007, and 1 per cent in FY2008.

Headquarters offices are:

NASA Office of Space Flight which is responsible for development of Space Station and large propulsion systems; operation of Shuttle, Spacelab and Station through its major Kennedy (as the US manned launch site), Marshall, Johnson and Stennis centres.

Office of Space Science, which conducts, unmanned space activities directed at planetary and astronomical investigations and the science elements of manned missions, through the Jet Propulsion Laboratory and Goddard centres.

Office of Aerospace Technology exercising research and development programmes and directs the Ames, Langley, Lewis and Marshall research centres.

Office of Space Communications which is responsible for all launcher, satellite and deep space probe tracking, and NASA's communications/data systems.

Office of Biological and Physical Research, newly formed in 2000 to strengthen this area of research for Space Station and Shuttle.

Office of Earth Science, encompassing missions such as Earth Observing System, Earth Probes and UARS to study changes in the global environment.

Office of Policy and Plans, pursuing defined goals leading to effective utilisation of resources for strategic objectives.

Office of Safety and Mission Assurance to enhance the success of NASA activities, lower risk and improve safety.

UPDATED

National Aeronautics and Space Administration, Jet Propulsion Laboratory

Current Status

JPL is a government-owned facility operated by the California Institute of Technology under contract since 1959 to NASA. The space agency provides about 90 per cent of the budget, with the remainder from the departments of Defense and Energy. Work force totals 4,500, mostly from the reduction in Galileo and Cassini activities. FY95 total operating budget was US$1.25 billion; FY94 US$1.1 billion. JPL is responsible for most of NASA's deep space missions, including Voyager, Galileo, Cassini, Mars Surveyor and Mars Pathfinder. It manages NASA's portion of the joint Ulysses solar probe, the US/French Topex/Poseidon oceanographic satellite project, the SIR Spaceborne Imaging Radar (Shuttle flights 1994) and developed the Wide Field/Planetary Camera for Hubble.

JPL developed and manages NASA's worldwide Deep Space Network for providing links with spacecraft above 10,000 km. DSN comprises three complexes near Canberra (Australia), Goldstone (California, opened 1958) and Madrid (Spain). There is also a launch support facility at NASA Kennedy, and network control and spacecraft compatibility test facilities at JPL. DSN can supplement Goddard's tracking system for Shuttle and selected Earth satellites. The typical DSN Deep Space Communications Complex, operated by the host government under agreement, comprises: 70 m antenna (S/X-band downlink; 20 kW and 400 kW S-band uplink); 34 m standard antenna (full downlink; 20 kW S band uplink); 34 m high efficiency system (full downlink; X-band uplink) and 26 m antenna (S-band up/down). Goldstone also has a prototype 34 m beam waveguide X-band antenna; the other locations will have it from 1996. To boost data reception rates, the Goldstone complex can be arrayed with the 27 25 m Very Large Array dishes of the US National Radio Astronomy Observatory in New Mexico, and Canberra linked with the 64 m Parkes radio telescope.

Spacecraft assembly/test facilities include an 890 m^2 high bay area, class 100,000, and a 16.8 m very high bay class 10,000 area of 390 m^2. A flight system testbed opened in 1993 for development-phase integration/testing of small spacecraft prototypes and subsystems, in a 150 m^2 lab near the high bays. Environmental test facilities provide vibration and spin balancing. The 6.1 m diameter ×7.63 m (working volume) space/solar simulator was built in 1961 and reopened in early 1994 after refurbishment. It provides 5×10^{-6} torr, –196°C to +93°C, solar intensity to 2.7 Suns (5.6 m diameter beam) or 12 Suns (2.2 m beam). A 3 m simulator, added in 1965, and a 93 m^2 class 10,000 assembly room supplement it.

Mission data facilities include the MIPL Multimission Image Processing Laboratory based on the digital image processing technology begun at JPL in the early 1960s. The Planetary Data System archives and distributes (principally via CDROM) digital data from planetary missions, ground observations and lab measurements. here are more than 190 titles since inception in 1991.

JPL has been a key element in re-structuring the Mars exploration programme following the loss of Mars Polar Lander and Mars Climate Orbiter in 1999 successfully conducting the Mass Odyssey mission in 2001, the first in a restructured Mass exploration mission sequence. Successfully reaching Mars orbit, it began operations in March 2002 and continues to operate. Launched on 10 June 2003, Mars Exploration Rover A was followed by MER B on 8 July 2003. Both rovers landed in January 2004.

NASA's first planetary launch, Mariner 1, lifts off from Cape Canaveral on 22 July 1962. Intended for a fly-by of Venus, the spacecraft was destroyed when the launch vehicle failed due to an error in flight guidance equations (NASA/JPL)

NEW/0569815

UPDATED

National Aeronautics and Space Administration, John H Glenn Research Center (formerly Lewis Research Center)

Current Status

Founded in 1941, the John H Glenn Research Center (formerly LeRC) is NASA's lead centre for research, technology and development of aircraft propulsion, space propulsion, space power and satellite communication. As such, it oversees the ACTS Advanced Communications Technology Satellite project by operating the Master Control Station. It is the lead centre for the propulsion technology development for NASA's High Speed Research Program. The Center is also the lead for development of ion propulsion and is participating in the Prometheus programme. The Microgravity Materials Science Laboratory is housed at the centre for qualifying space experiments, with facilities including a 130 m (1 second) µg drop tower, furnaces and a crystal growth system. The center also specialises in turbomachinery and to developing new and innovative technology for computational, analytical and experimental research. The center manages NASA's Atlas and Centaur launch activities. GRC has been lead centre for the design and development of the electrical power system for the International Space Station. Staff total 3,200 people including civil service employees and support services contractors. FY2005 budget proposes expenditure of US$106 million at the facility in G&A.

In 1999 the Lewis Research Center was renamed the John H. Glenn Research Center in honour of the first American to orbit the earth.

National Aeronautics and Space Administration, Johnson Space Center

Current Status

Johnson Space Center (JSC) was opened in 1962 as NASA's centre for design, development and testing of manned spacecraft. Its many capabilities include management of the Space Shuttle Program and the International Space Station Program, studies in advanced human support technology, biomedical research and countermeasures, space medicine, space operations/communications management, Extra Vehicular Activity (EVA) and curatorial care and study of Lunar and Planetary materials. JSC is also home to NASA's astronaut corps, where they train for space flight and support NASA's human space flight programs. The staff is comprised of 2,932 civil servants and approximately 13,782 contractors. The Center is divided into directorates responsible for specific functions: Administration, Flight Crew Operations, Mission Operations, Engineering, Space/Life Sciences and Information Systems. JSC also manages the White Sands Test Facility in New Mexico, a pre-eminent resource for testing and evaluating potentially hazardous materials, space flight components and rocket propulsion.

Most of JSC's 100 buildings on the 6.5 km^2 Houston site provide office space and laboratories. Its unique research facilities support life, planetary and Earth sciences, artificial intelligence and lunar sample analysis. Specialised training facilities include the Shuttle simulators, Shuttle Orbiter trainer in the Space Vehicle Mock-up Facility, Precision Air-Bearing Facility and Space Station mock-ups, Space Environment Simulations and acoustic test facilities.

NASA's Mission Control Center (MCC) has controlled all NASA manned flights since Gemini 4 in 1965. The Flight Control Room (FCR), on the third floor is now an historical landmark. A new wing was added to facilitate new technology for Mission Control – adding two new FCRs, one for the Shuttle and one for the International Space Station (ISS) to facilitate new technology for Mission Control. The new workstations are more efficient requiring less

NASA's Johnson Space Center has directed all US manned space flights from the early Gemini missions in 1965 including historic moon walks in the early 1970s (NASA) ***NEW***/1047482

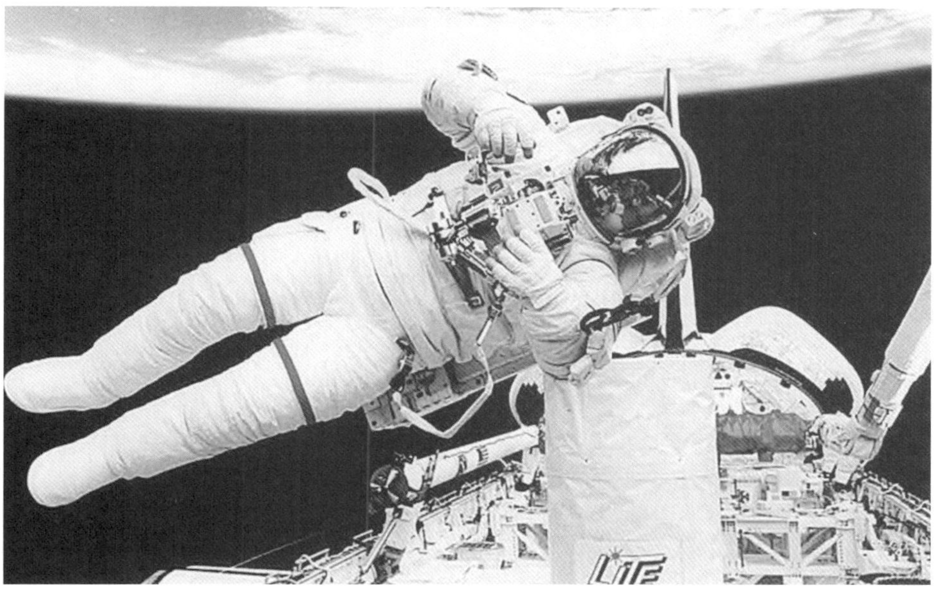

JSC in-house mission support includes crew equipment and aids such as this SAFER system for EVAs (NASA) ***NEW***/1047488

JSC has overall responsibility for the ISS seen here in its May 2004 configuration of partial assembly (NASA) ***NEW***/1047537

manpower to control the combined Shuttle and Station operations.

The Jake Garn Training Facility houses three high fidelity simulators: the Shuttle and Mission Simulator. The motion flight deck simulator uses hydraulic actuators for 6 Degrees Of Freedom (DOF) to simulate launch, landing and other dynamic manoeuvers. A mid-deck might be added for entire crews to train together. Integrated simulations with MCC begin some 10 weeks before a flight, concluding with a fifth ascent/entry session two days before the crew flies to Kennedy Space Center (KSC). The two fixed-base simulators provide on-orbit operations training. The detailed aft flight decks cover activities such as Remote Manipulator System (RMS), EVA, rendezvous and payload deployment. Experiments, meals and housekeeping chores are practised in the mid-deck. There are also high-fidelity ISS simulators.

The Space Vehicle Mock-up Facility houses two high-fidelity Orbiter forward sections (flight deck and mid-deck), which can be rotated into launch attitude. Astronauts use it to learn the cabin's layout and engineers make fit-checks of new equipment. A separate full Fuselage Trainer cargo bay mock-up allows a fit-check of payload mock-ups. The Manipulator Development Facility trains Mission Specialists for RMS operations.

The Neutral Buoyancy Laboratory houses a 23.47 million litre, 31 × 10 × 61.59 × 12.20 m pool capable of accommodating a full-scale Orbiter payload bay and working RMS for crews to simulate microgravity. Space station assembly missions are also simulated in the NBL.

The Lunar Sample Building holds most of Apollo's Lunar samples, as well as meteorite and cosmic dust samples. JSC is the designated curatorial center for all of NASA's extraterrestrial samples.

UPDATED

National Aeronautics and Space Administration, Langley Research Center

Current Status

LaRC continues to focus on four primary areas of research and investigation: aeronautics; earth science; space technology; structures and materials. Aeronautics and space technology merged together in the science aerospace vehicle systems.

Background

LaRC is primarily a research centre for advanced aerospace technology. Major research fields include aerodynamics, materials, structures, flight controls, information systems, acoustics, aeroelasticity, atmospheric sciences and non-destructive evaluation. 40 per cent of the work supports space activities, including technology for advanced space transportation, large space structures and the Earth Observing System as part of NASA's Mission to Planet Earth. Langley covers 3.2 km^2 in its West area, plus 0.08 km^2 on its East side, has more than 220 buildings and a replacement value on today's market of US$2.1 billion. LaRC employs about 3,800 civil service and contractor employees.

The centre was established in 1917 as the first research laboratory of the National Advisory Committee for Aeronautics. It became one of the four original NASA facilities (with Ames, Edwards, and Lewis) when the agency was created in 1958. The US manned space programme began there in 1959 as the Space Task Group, which completed its transfer to new dedicated facilities in Houston in mid-1962. LaRC was responsible for the highly successful Lunar Orbiters and Viking Mars orbiters/landers.

Langley's LITE Lidar In-Space Technology Experiment during STS-64 in September 1994 detected stratospheric and tropospheric aerosols, probed the planetary boundary layer and measured cloud top heights. The 1 m diameter telescope with three-wavelength Nd: YAG laser is a testbed for future operational spaceborne lidars. No further Shuttle flights are planned, but a smaller satellite version is being designed. The LASE Lidar Atmospheric Sensing Experiment, the first autonomous aircraft-based lidar to measure water vapour, completed final validation studies aboard NASA's ER 2 in September 1995.

The centre manages the HALOE Halogen Occultation Experiment aboard UARS to monitor the vertical distributions of ozone and key trace gases. The MAPS Measurement of Air Pollution from Satellites instrument has already flown four times on Shuttle to measure the global distribution of carbon monoxide in the free troposphere; it will operate for one year aboard Mir from June 1996. Langley's Earth Radiation Budget Experiment continues operations on the ERBS satellite (which also carries LaRC's SAGE instrument) and NOAA 9/10. A set of instruments is being developed for EOS missions. CERES, the ERBE follow-on, flew on TRMM in 1997 and SAGE III on Russia's Meteor3M in 1998.

The remote sensing office specialises in developing new technologies for future systems, including solid-state lasers, detectors, lightweight optics, deployable telescopes, miniature sensors and onboard data processing. LaRC leads NASA in developing high energy, high efficiency, long-life solid-state lasers for advanced lidar and other applications requiring stable laser sources. Together with NASA Stennis, it is exploring commercial markets for remote sensing technology. LaRC is an EOS Distributed Data Active Archive Center, which carries data from 18 major NASA atmospheric field missions and satellites.

LaRC provided several experiments during the 1990s for Russia's Mir station. The MEEP Mir Environmental Effects Payload was attached to the docking module by STS-76 in March 1996, where it remained until late 1997 collecting micrometeoroid particles and the effects of strikes on Station Alpha materials. The Enhanced Dynamic Load Sensors set up has carried Mir's Priroda module to measure the effects of crew movement. The MIDAS Materials in Devices As Superconductors experiment remained aboard for several months after delivery by STS-79 in August 1996. Four circuit boards, including one with Russian samples, measured effects on the electrical and magnetic properties of high temperature superconductor materials.

LaRC had the lead in developing the composite primary structures and the advanced composite thermal protection systems for the now cancelled X-33 and X-34. It is helping the industry teams to develop reusable cryogenic tanks and advanced propulsion, in addition to vehicle systems analysis, aerodynamic and aerothermodynamic testing/analysis, and flight control development. It performed wind tunnel testing of an X-34 model mated to a Boeing 737. In 2000 Langley Research Center began its Revolutionary Concepts (REVCON) programme participating with industry in the development of high-risk but radical technologies for atmospheric flight.

LaRC supports the Virginia Air and Space Center, Hampton, Virginia., with IMAX theatre.

Rocket engines developed at the Marshall Space Flight Center and elsewhere, tested in giant cells like these, were the lifeblood of the US civilian launch industry since the early 1960s (MSFC) ***NEW***/1047489

National Aeronautics and Space Administration, Marshall Space Flight Center

Current Status

MSFC is rescoping its primary mandate to develop propulsion systems and launch vehicles with direction from the Bush administration toward a more long-term goal to consolidate a cost effective infrastructure for space transportation. The Center hopes to play a significant role in the new exploration initiative toward which the Bush Administration has directed NASA in a new set of human and robotic goals announced January 2004. Predicated to a large extent by the need to retire the Shuttle, these objectives appeal to the core duties of the MSFC and for which the facility was originally established as a NASA Center.

Background

Marshall is one of the largest of NASA's 10 field centres, accounting for about a quarter of the agency's budget. Employees total approximately 2,760 plus 3,000 contractors. It is the lead NASA centre for research, technology maturation, design, development and integration of space transportation and propulsion systems. This includes both reusable and expendable launch vehicles, as well as vehicles for orbital transfer and deep space missions. The centre has an Agency-assigned Center of Excellence role for space propulsion.

Marshall is also lead centre for NASA's Microgravity Research Program with responsibilities for planning and direction; resource management; assessment; and definition, development, integration and flight of microgravity science and space processing experiments and facility payloads; as well as programme outreach and education activities.

Marshall developed and continues to provide the Space Shuttle's main engines, solid rocket boosters and external tank.

Other key programme assignments include the Space Launch Initiative which focuses on reducing the cost of Earth-to-orbit transportation and the Reusable Launch Vehicle Technology Program which is demonstrating technologies needed for commercial development of a low-cost, next-generation launch system. Both are based on the now defunct Advanced Space Transportation Program which incorporated the X-33 and X-34 programmes cancelled in 2001 by the Bush administration.

Marshall provides support to the International Space Station Program office with engineering personnel and facilities to develop pressurised modules, support equipment and assigned payloads. Marshall has lead responsibility for the Advanced X-ray Astrophysics Facility for development, launch and mission operations phases of the X-Ray Observatory program and for archiving the mission and science data. Marshall manages the Gravity Probe-B and other smaller scientific payloads and instruments and conducts research in high-energy astrophysics, solar magnetic fields, and low- energy space plasma physics. Marshall also supports the Mission to Planet Earth Program through research at the Global Hydrology and Climate Center in Huntsville. Since the Astro-1 mission in 1990, Spacelab science activities have been controlled from Marshall's Spacelab Mission Operations Control Center.

Marshall is supporting the Advanced Space Transportation Programme involving several promising technologies for future space transportation systems. These include the Rocket Based Combined Cycle engine, an air-breathing propulsion system.

Marshall was established in 1960 around a nucleus headed by Dr Wernher von Braun, the team which, as part of the US Army Ballistic Missile Agency, orbited Explorer 1. Marshall became NASA's lead centre for Apollo's Saturn launchers and developed the Skylab space station. It is located on a 7.3 km^2 tract within the US Army's Redstone Arsenal in Huntsville, Alabama. Also operated by Marshall is the Michoud Assembly Facility in Louisiana, occupying a 3.37 km^2 site where Shuttle external tanks are manufactured and shipped to Florida by barge.

UPDATED

National Aeronautics and Space Administration, Stennis Space Center

Current Status

Previously known as the National Space Technology Laboratory, SSC is NASA's primary large propulsion test facility, including development

A technician prepares an SSME for tests (NASA) 0572501

Local crowds gather to experience the exhilaration of a live Shuttle engine test (NASA) 0572502

testing of the Space Shuttle SSME main engines. The NASA-related workforce in FY2002 totalled 288. Stennis is also NASA's lead centre for the commercialisation of remote sensing technology and for earth system science. Thirty other federal and state agencies occupy the centre, including the Naval Meteorology and Oceanographic Command, bringing the workforce to 4,600.

Total land area is 562 km^2, of which 55.8 km^2 constitute the operational base and the remainder is held as an acoustic buffer zone. The site was selected in 1961 to accommodate Apollo's F1 and J2 engines and Saturn 5 stages, testing a total of 27 stages. All main Shuttle engines undergo acceptance testing at Stennis before flight. The first SSME test was made during June 1975. Two stands handle single SSME trials. A1 evaluates sea level performance. A2 simulates firing up to 16.5 km altitude, operating with the engine 18° from the vertical to mimic the installation on the launch pad. A1/A2 were both built in the 1960s for Saturn 5 stage 2 testing and can each handle 4.89 MN, 10.07 m diameter. Stand B1 was modified to test Boeing's RS-68 engine that powers the Delta 4 launch vehicle. The other side of the B complex tested is the Main Propulsion Test Article where three-engine SSME cluster testing was performed; it originally handled Saturn 5's stage 1 and as such can cope with 48.9 MN, 10.07 m diameter.

The three-cell E-1 Test Facility can handle engines using LH_2 or hydrocarbons, solid or hybrid, high flow rates, high pressure cryogenics and ultra-high pressure gases. The state of the art complex is being used for testing hardware for future launch vehicles. Testing is underway supporting NASA's Space Launch Initiative engine concepts

The E-2 Test Facility supports component testing similar to E-1 but at about 1/10 scale. The E-3 facility supports even smaller scale tests utilizing hydrogen peroxide, hydrocarbon or liquid oxygen propellants.

Stennis plays a key role in NASA's Earth science applications program within the Earth Science Enterprise and is home to the Earth Science Applications Directorate. Earth science applications use NASA's unique Earth science research results, data, remote sensing and other technical capabilities to solve practical problems and allow managers to make more informed decisions. The Earth Science Applications Directorate office at Stennis leads the effort on three key issues of national importance: agricultural competitiveness for improved and efficient agricultural; coastal management which examine the effects of natural and man-made changes on coastal ecosystems; and community growth which help communities plan for the future.

National Aeronautics and Space Administration, Wallops Flight Facility

Current Status

Under direction from the NASA GSFC, Wallops has superceded its 1997 Mission 2000 Implementation Plan with Mission 2005 capturing unique partnerships with other NASA centres and to meet national defence goals through the Navy's AEGIS system.

Background

Established in 1945 under NACA auspices, NASA's sounding rocket programme is based at the Wallops Flight Facility in Virginia using vehicles from commercial sources or developed by the agency. Some 25 launches are made each year from a range of sites, including Wallops itself, Poker Flat and White Sands. NASA currently employs 15 sounding rocket types (Aries use ended 1993), including several utilising Bristol Aerospace's Black Brant 5 motors. Extensive use is made of military surplus motors and all are unguided. Saab's S19 Boost Guidance System can be added to the Nike Black Brant and Black Brant 5/9/10: the gyro platform controls aerodynamic canards, decreasing impact dispersion by a factor of 5/10, allowing higher apogees and launches in higher winds.

Now a part of Goddard and on Virginia's east coast, Wallops became the third US orbital site on 16 February 1961 with the launch of the Explorer 9 balloon by the all-solid Scout. 19 vehicles achieved orbit from Wallops by the end of 1995 (plus three failures), the most recent in 1985. Although the facility is still available, Scout, its only orbital vehicle, retired in 1994. Orbital activity resumed October 1995, but the commercial Conestoga failed. Wallops also provided control for the B52/Pegasus launch of Brazil's SCD February 1993 (the B-52 took off from Florida). One air-launched Pegasus carried from Wallops by Lockheed L-1011: SAC-B/HETE (Argentina/NASA; November 1996).

Standard orbital inclination is 37.7°. Wallops' major activity now is as the base for NASA's sounding rocket programme. The first firing made from Wallops was a Tiamat on 4 July 1945 and since then more than 15,000 rockets have been dispatched. Current vehicles include Super Arcas, Black Brant, Taurus-Tomahawk, Taurus-Orion and Terrier-Malemute. Some 30 launches are made annually (32 in 1995, 33 in 1994, 26 in 1993; 16 in 1996), including Wallops-managed firings from the Poker Flat Research Range in Alaska and the White Sands Missile Range in New Mexico. 1995's schedule included six from Woomera. 1994's had more than 30 from Brazil. Wallops' own total was 807 firings by end-1996.

Wallops supported the X-33 programme expecting to provide mobile telemetry, command uplink and flight termination systems, of which since the X-33 has been cancelled there is now no need.

The Wallops Orbital Tracking Station was established in 1986 to provide NASA tracking support for projects such as IUE, IMP 8, Nimbus 7 and COBE. High-speed data transfer to Goddard at Greenbelt is provided by a satellite link. WFF utilises telemetry receiving and command antennas ranging over 2.4 to 18 m for satellite and sounding rocket support. The radars include an 8.8 m parabolic FPS 6 radar that can track objects out to 60,000 km with a precision of ±3 m and 9 cm/s range-rate; peak output 2.5 MW. Wallops also houses NOAA's Command/Data Acquisition Station of nine 7.3 to 26 m antennas to track, command and take data from the NOAA and GOES meteorological satellites.

National Oceanic and Atmospheric Administration

Current Status

NOAA is currently second only to NASA as the agency most closely focused of all US government agencies on environmental monitoring and climate change. Hurricane forecasts have now shifted from three to five days and, with NASA, major new initiatives focus on marine monitoring.

Background

Operators of the NOAA polar and GOES geostationary meteorological satellite systems, and federal overseer until October 1992 of the commercial Landsat 4/5 remote sensing satellite system (a role it will repeat for Landsat 7). As an element of the Department of Commerce, NOAA is the US signatory to the Cospas-Sarsat agreement. Its Space Environment Laboratory in Boulder, Colorado provides real-time monitoring and forecasting services of solar disturbances and their effects on Earth's environment.

Satellite data are used to monitor weather conditions, provide data for issuing warnings of severe weather, prepare charts and coastal maps, and improve assessment and conservation of

Artist's impression of Aqua deploying in orbit (NASA) ***NEW***/1047511

Increasingly, organisations like the NOAA collect and assimilate data from NASA earth science and observation missions such as Aqua, see here (NASA) ***NEW***/1047497

marine life. They are also employed to assess the impact of natural factors and human activities on global food and fuel supplies and on environmental quality. NOAA agencies participating directly in space projects include NESDIS, the Systems Acquisition Office (SAO), the National Weather Service, the National Marine Fisheries Service, the Office of Oceanic and Atmospheric Research, and the National Ocean Service. In co-ordination and co-operation with other appropriate organisations, SAO has 'responsibility for designated major systems acquisition', such as the GOES I-M series. NOAA provides government authority for the Landsat 4/5 Earth resources satellites, now operated commercially by EOSAT. Landsat 7 operations come under NOAA. The NOAA is integral to the convergence of civil and military meteorological services now that the DoD has decided to obtain its weather information from civil satellites.

NOAA has approximately 12,500 personnel at 375 locations and an annual budget of about US$2.5 billion.

UPDATED

NOAA, National Centers for Environmental Prediction, Space Environment Center

Current Status

Changes in the discrimination and budgeting of national meteorological and environmental satellites has unified the separate civilian and defence systems. NESDIS manages the NOAA polar-orbiting and GOES geostationary meteorological satellite systems, gathering, archiving and distributing environmental data. NESDIS' FY92 budget was US$343.9 million; US$329.5 million FY93; US$349.5 million FY94; US$387.4 million FY95; US$472 million FY96 and US$532 million requested FY97. NOAA was established within the Department of Commerce in 1970.

NOAA operates the geostationary GOES satellites, shown here being assembled at Loral Space Systems (NOAA) ***NEW***/1047491

SEC provides real-time space environment monitoring and forecasting services, develops techniques for forecasting solar disturbances and their effects on the Earth's environment, and researches in solar-terrestrial physics. The data and forecast products are of interest to satellite operators, HF radio users, those concerned with radiation damage to passengers in high-flying aircraft or spacecraft, radio navigation users and operators of electric power distribution networks.

SEC receives some 1,500 data streams daily on the state of the space environment from space platforms, such as Tiros and GOES, and ground-based observatories, such as the National Solar Observatory, USGS's Intermagnet Network and USAF's SEON. The more important databases and indices are available for public access by computer networks. Much of this data is fed to the Department of Defense space surveillance networks and to NASA flight operations centres for such diverse applications as orbit decay prediction to a daily atmosphere balance assessment for Shuttle re-entry.

UPDATED

Office of Science and Technology Policy

Current Status

The Office of Science and Technology Policy was formed in 1976 to provide the President with policy advice and to co-ordinate national investment in science and technology. OSTP has crafted policy for the continuation of the Landsat remote serving programme. The Senior Interagency Group (SIG) on Space was replaced in 1989 by the National Space Council to co-ordinate and monitor national space policies and strategies. NSC was chaired by the Vice-President and included 10 cabinet-level officials: NASA Administrator, director of Office of Management and Budget, Secretaries of State, Defense, Commerce, Energy, Transportation and Treasury, CIA director, presidential Chief of Staff, and the President's assistants for National Security Affairs and for Science and Technology. The Executive Secretary led an 11-member policy staff. The new President Bill Clinton replaced NSC in February 1993 with a science and technology council chaired by Vice-President Al Gore. The Bush administraiton has recently reaffirmed its commitment to the OSTP. Bush administration has worked the OSTP in formulating direction and policy for NASA and DoD space planning initiatives.

Ohio State University Center for Mapping

Current Status

The Center, established in 1986, is a NASA commercial space centre, concentrating on real-time satellite mapping, which makes use of data as soon as they are transmitted from satellites. R&D projects concentrate on commercial products for mapping, remote sensing and GIS user communities. Special software systems developed for rapid image interpretation and analysis for specialised user requirements and unique tasking from government agencies and commercial organisations, research institutes and commercial clients. The Center's current research project is the development of a fully-digital, real-time Airborne Integrated Mapping System (AIMS). It combines positioning sensors, GPS and INS, with digital imagery. AIMS has many applications from traditional large-scale topographic mapping to military reconnaissance, determining the extent of an oil spill, corridor surveys such as rail and highway construction and management, drug interdiction efforts and emergency response deployment. Earlier research included the development of the GPSVan, which can map highway and transportation infrastructure at normal traffic speed. The Center is particularly interested in the development of computer-aided "D-to-D" (data to decision) systems which provide the basis for new integrated approaches to mapping and environmental monitoring. The Center has conducted integration functions linking optical and radar derived images from US and non-US, applying them to ground truth solutions. By connecting mapping and imaging systems in an integrated loop with GPS positioning, the OSUCM has developed tri-axis co-ordinate tracking with geographic overlay.

VERIFIED

Pennsylvania State University, Center for Cell Research

Current Status

The Center for Cell Research (CCR) was established in 1987 as one of NASA's Centers for the Commercial Development of Space (CCDS), focusing on commercial product oriented projects: physiological testing, bioseparations and illumination. NASA withdrew funding in December 1993, but the Center continues operations. Space-based physiological testing involves space-flown animals, tissues and/or cells as pharmaceutical test subjects or in discovery research for health problems that occur on Earth. CCR flew the first commercial proprietary physiological test aboard Shuttle in October 1990.

The CCR bioseparation programmes provide continuous flow electrophoresis and aqueous two-phase partitioning. The USCEPS United States Commercial Electrophoresis Program in Space is the first to couple industry access to space-based systems with ground-based testing and product trial separations. This service enables industrial clients to evaluate, before flight, how much improvement going into space will produce. The team began testing an electrophoresis unit in 1995; was certified in 1996 and became available to commercial interests in 1997. The CCR has also conducted preliminary studies of Space Station support packages for electrophoresis experiments in the international laboratory

Developed autonomous cell research packages for modular applications in Shuttle mid-deck area. Experiment rocks can carry several devices from different investigators. The CCR continued to develop experiment racks for man-tended and autonomous space applications with a view to exporting the product to other operators outside the United States.

UPDATED

United States Geological Service, Eros Data Center

Current Status

Opened in the early 1970s, the US Geological Survey's Earth Resources Observation System Data Center (EDC) is responsible for managing a global archive to ensure long-term preservation of satellite remote sensing data. EDC will manage land processes data for the EOS programme; a 5,600 m^2 extension was completed in April 1996 to handle EOS processing/distribution facilities. It is the world's single largest repository of remote sensing data, with more than three million images. EDC maintains the national archive of Landsat imagery; its name derives from the Department of Interior's plan for an EROS Earth Resources Observation Satellite, which evolved into Landsat. A field site for the USGS National Mapping Program, EDC also manages other forms of Earth data, including more than nine million aerial mapping photographs covering the US. The EDC has also integrated products from other imaging systems to serve the needs of its prime customers and these systems include commercial image enhancement companies. The EDC employs about 600 government and contractor personnel. Comprehensive data integration programme for special modelling, making images available for analysis and interpretation configured for the needs of individual researchers or users.

VERIFIED

United States GPS Industry Council

Current Status

USGIC represents the interests of GPS manufacturers and users. It is instrumental in advising on successor systems and in liaising with the government on protecting the business opportunities of US companies in foreign markets. Advises government agencies on technical standards and requirements for licensing receivers and has conducted an analysis of the impact of new competitive systems such as China's aspiring programme and the European Galileo system. This work has followed through on the recommendations of US government agencies that favour some co-operation with competitive systems. Technology research has started on integrating US Navstar and non-US systems.

UPDATED

University of Alabama, Center for Macromolecular Crystallography

Current Status

The CMC was established in 1985 as one of NASA's CSC Commercial Space Centers, focusing on space-based growth of biological crystals and their applications. The centre's Protein Crystal Growth experiments have made numerous Shuttle flights. Government funds have been withdrawn for work that had started on experiments for the International Space Station and funding sought from commercial enterprises and research projects at national and international level. CMC has been developing several crystal growth experiment techniques for application as a template model to both Shuttle and ISS applications.

VERIFIED

University of Alabama, Consortium for Materials Development in Space

Current Status

UA Huntsville's Consortium for Materials Development in Space was established in 1985 as a Center for the Commercial Development of Space (CCDS) through NASA's Office of Commercial Programs to focus on investigations in space to develop new materials and processes. This approach includes commercial materials development that benefits from the space environment, commercial applications of the physical chemistry occurring at the surface of a new material, and how materials are transported to it, and frequent space experiments.

UAH flew six experiments under a US$1 million contract aboard Space Services Inc's first Starfire 1 sub-orbital rocket 29 March 1989 as the Consort 1 mission. Each was developed by one or more principal investigators from the University, industry or both: de-mixing of immiscible polymers, powdered metal sintering, electrodeposition, convective mixing of elastomer modified epoxy resins, foam formation, and operation of the Materials Dispersion Apparatus. Consort 2, funded by a US$1.1 million grant from the Office of Commercial Programs, failed 15 November 1989 because of stage 1 guidance problems. The 455 kg payload was recovered intact and was reflown as Consort 3 on 17 May 1990, with 10 of the 12 experiments successful. Consort 4 flew successfully 16 November 1991 carrying 12 experiments; NASA provided the funding of about US$1.9 million. EER Systems' Space Services Div was awarded a contract worth up to US$9.8 million for up to seven further Consorts through 1996. Consort 5 flew 10 September 1992 but stage 2's burn ended 3 seconds early when the casing burnt through and µg data were not obtained because of the attitude perturbations. Consort 6 flew successfully 19 February 1993 with seven experiments, attaining 301 km. The US$2 million funding was provided from the insurance following C5's partial failure. Conquest 1, the first of a new series, was launched 3 April 1996 carrying eight experiments on Starfire at White Sands Missile Range: providing 7 minutes µg for the 293 kg payload, attaining 320 km. One experiment planned to extrude a 6 m foam beam into space, but it instead formed a large ball. Conquest 2 is not yet planned.

In co-operation with industrial partners the CMDS flew materials and product development experiments on STS-79 with the Extreme Temperature Translating Furnace in Spacehab 5, and on STS-80 with biological materials in the Shuttle mid-deck lockers. Shannon Lucid using the Russian Optizone furnace processed metal sintering samples aboard the Mir complex. Several Russian companies had discussed co-operative ventures for experiments on board the Russian elements of the International Space Station.

The Consortium has also worked with Russian scientists on the development of an international telemedicine capability. New standards were developed and significant progress made toward a fully integrated system capable of global operations across different satellite systems and networks. Interest in this has been expressed by several European countries. Pursuing this work, CMDS had proposed a continuing line of materials processing experiments for Shuttle applications leading to ISS development.

VERIFIED

University of Colorado, Bioserve Space Technologies

Current Status

The University of Colorado, Department of Aerospace Engineering runs BioServe in co-operation with Kansas State University, Manhattan, Kansas, after NASA established it in October 1987 as one of NASA's CCDS (Centers for the Commercial Development of Space). It focuses on space-based bioprocessing and biomedical testing, CELSS technologies and apparatus development. Almost 50 private sector concerns participate in the CCDS. It is responsible for fabricating and qualifying its own apparatus. Development efforts focus on providing a suite of generic life sciences space facilities, including fluids mixing, micro-organisms culture and plant growth: CGBA Commercial Generic Bioprocessing Apparatus (flown STS-50 June 1992, STS-53 January 1993, STS-57 June 1993, STS-60 February 1994, STS-62 March 1994, STS-63 February 1995, STS-69 September 1995, STS-73 October 1995), AGBA Autonomous Generic Bioprocessing Apparatus, A-MASS Animal Module for Autonomous Space Support and P-MASS Plant Module for Autonomous Space Support. BST has developed several research projects for International Space Station application. These were integrated with experimenters from the STS flights in an attempt to ensure continuity for long term ISS research. These plans have been suspended since the loss of Columbia on 1 February 2003. However, ongoing research into space bioprocessing has allowed investigation into the use of automated experiment and instrument equipment on untended platforms. It has also allowed investigation into the ability of astronauts to periodically visit uninhabited modules.

VERIFIED

University of Colorado, National Snow and Ice Data Center

Current Status
The NSIDC archives and distributes analysis of snow and ice data from satellites, including NOAA and DMSP, and maintains an archive of satellite and *in situ* data. It is affiliated with the National Oceanic and Atmospheric Administration's National Geophysical Data Center. Currently receiving non-US data sets to augment domestic products. Incorporated radar data interpolated with imaging information for snow density measurements. Specialist advice provided on new ways of measuring ice melt parameters and water run-off values.

UPDATED

University of Houston, Space Vacuum Epitaxy Center (SVEC)

Current Status
SVEC was established in 1985 as one of NASA's Centers for the Commercial Development of Space, focusing on space-based thin-film growth using Molecular Beam Epitaxy (MBE) and Chemical Beam Epitaxy (CBE). More than 60 personnel are located at SVEC within 1,000 m^2 of laboratory space, including class 10/1,000/10,000 cleanrooms. A spin-off company, Applied Optoelectronics Inc, was set up in Sugarland, Texas, for manufacturing mid-IR optoelectronics. SVEC has planned and advised other companies on international projects involving US, European and Russian consortia. Constructed a consortium for advising agencies on epitaxial experiments and recommended a unified programme. Has diversified work into other ground-based thin-film epitaxial applications.

UPDATED

University of New Mexico, Earth Data Analysis Center

Current Status
EDAC was established in 1964 to transfer aerospace technology. The organisation gathered a wide range of technologies and applied them to other disciplines. It is incorporated industry-wide developments and technology applications from civil and military space projects. Since 1974, it has focused on observations using spatial and spectral analysis, GIS development and GPS applications. EDAC has four service areas: Image Processing; GIS Development; Geographic and Electronic Data Services; Training. These are supported by collocated organisations, including the NM Natural Heritage Program, National Park Service (Southwest GIS Center), US Fish and Wildlife Service, US Geological Survey Geographical Names Information System, and Autometric Inc. Information services include access through Internet and others to provide metadata and digital image transfers. Create educational forums for gathering universities research programmes. Information held by the Center distributed to users. Through strategic partnerships with industry and government, EDAC employs state of the practice hardware and software systems for data processing, modelling, visualisation and environmental simulations. Company activity has diversified greatly since it was formed, responding to changes in the need for specific applications. Much work has been done on integrating COTS components into high-grade systems with demanding specifications, building failure-rate trends and reliability statistics. Product bases now include non-US remote sensing and imaging projects. The centre employs 10 permanent staff.

UPDATED

University of Tennessee Center for Space Transportation

Current Status
The CSTAR Center for Space Transportation and Applied Research at The University of Tennessee Space Institute headed the original three-flight programmes of the Comet retrievable payload sponsored by NASA. Westinghouse Electric Corporation provided the service module and systems engineering, EER Systems the Conestoga launcher (contract worth about US$45 million), Space Industries (contract US$15 million) the recovery capsule, orbital operations and payload integration. The Conestoga concept dates back to the early 1980s when it was proposed by Space Services Inc led by ex-Mercury astronaut Deke Slayton. EER assumed responsibility for the programme in late 1994. NASA's FY94 budget allocated US$14.5 million, leaving a shortfall of US$6 million required for the launch. Commercial customers provided the remainder, so NASA's US$14.5 million was released by the 28 March 1995 contract, which allocated the agency half of the capacity. FY93 US$23 million; FY92 US$18 million; FY91 US$10.5 million. The US$85.1 million contract signed in January 1990 called for three Comet flights: #2 September 1995 and #3 March 1997. NASA modified it to a single mission costing US$65.8 million, after projecting that three missions would cost more than US$158 million. That contract did not cover the companies' investments, but the commercial Westar (Westinghouse Space Transportation and Recovery) service was also offered, charging $35 million for each mission or about US$200,000/kg payload. Westinghouse projected 10 annual missions by 2000. Westinghouse, France's Novespace and Arianespace in May 1993 began marketing the service to European users for launch on Ariane. Cost would be US$68,000/kg payload. A suitable Ariane slot would be available about once annually from 1995. However, Westinghouse withdrew from Comet/Westar in 1994.

VERIFIED

Venezuela

Centre for Digital Image Processing (CPDI)

Current Status
The CPDI section of Venezuela's Fundacion Instituto de Ingenieria was established in August 1983 and was appointed in 1988 as national representative for distribution of EOSAT's Landsat products/services. These products have been supplemented with data from commercial remote sensing images and from processed Landsat data formats. A large database of historic images is available for global users, but emphasis on regional issues related to conservation and government involvement in resource analysis, has stimulated a focus on local needs and customer requirements. More than 50 trained professionals provide advisory and technical services, and conduct studies in remote sensing applications, GIS and software development for special applications. CPDI also works with images from Spot Image and has recently absorbed data from Russian and commercial image processing distribution outlets. Has recently conducted extended negotiations for regional co-operation with new image processing technology pooled by several participants from different countries.

UPDATED

SPACE INDUSTRY

COMMUNICATION SATELLITES, COMPONENTS AND SERVICES

Australia

Satellite Aircom Consortium

Current Status
MITEC emerged from an R&D Centre of the University of Queensland Electrical Engineering Department in 1987 with 15 staff and an A$700,000 turnover; staffing level is 120. It is listed on the Australian Stock Exchange. Major space interests cover satellite and terrestrial communications systems and microwave components. It has designed and manufactured space-qualified transmit and receive equipment for Optus B (three 30 GHz beacons) and Russia's RadioAstron (L-band cryogenic amplifiers and converters). For ground station applications, it produces a range of frequency conversion (up/down) equipment for L, C, X and Ku bands. SSPAs operate up to 70 W Ku band. Low-noise L to Ku amplifiers provide 0.6 to 1.8 dB noise figures. Planar antennas are manufactured for mobile applications and low-noise converters for weather satellite ground stations. MITEC also produced equipment for CSIRO to receive signals from Japan's MOS satellites in Australia. It also developed microwave antennas for multipurpose applications and has developed SSPAs for specific customers with power levels of up to 120 W. It is currently researching new technologies for high power amplifiers for DBS satellite systems.

UPDATED

Belgium

Sirius Communications NV

Current Status
Founded in 1996, Sirius Communications emerged from IMEC, Europe's largest microelectronics research institute as a specialist company looking to develop new electronic systems support capabilities for the commercial satellite industry.

Sirius Communications is producing state-of-the-art digital chips and development systems for CDMA/spread spectrum-based satellite communications. The company supplies its new transceiver chips for a target market in the VSAT and portable voice/data satellite terminals for communications, positioning and messaging communication with its new transceiver chips. About 75 per cent of the company's current revenue comes from satellite communication hardware and services sales and it is involved in W-CDMA (UMTS) developments.

Sirius also provides baseband ICs for broadband CDMA-based satellite links including Internet connections. The chips support both LEO and GEO applications while supplying a worldwide customer base. For specific applications Sirius offers a custom design service for CDMA and spread spectrum users.

One of Sirius' principal products is the ASTRA chip. The Astra is a 3.3V 100-PQFP full-duplex CDMA/spread spectrum transceiver chip. It supports two-parallel QPSK or a four-parallel BPSK channel communication, with spreading code lengths up to 1,023 bits. The spreading codes are fully flexible and all variants of PSK modulation are supported. Data rates from a 50 bit/s up to 150 kbit/s are possible. The Astra has on-chip Direct Digital Synthesis up- and down-conversion. The on-ship receiver has a dual demodulator architecture for operation in S-CDMA (Synchronous CDMA) and QS-CDMA (Quasi-Synchronous CDMA) satellite networks.

The ASTRA Development System including graphical receiver analysis software allows field trials early in the product development cycle. The development boards serve as design examples, which can be customised and miniaturised in the terminal product. FES functions can be executed on plug-in FPGA based 'Gemini' boards available from Sirius.

The DIRAC chip combines a flexible CDMA/spread spectrum receiver chain with an on-chip 32-bit ARM microcontroller core. Sirius Communications works with the customer from the design phases through to the silicon etch for individual or volume manufacture.

In 1999, Sirius produced the CDMAX, the first programmable WCDMAIP core for 3G communications.

In May 2001, Agilent Technologies Inc announced it had signed an agreement to acquire all the issued share capital of Sirius Communications NV. Sirius Communications continues to operate its research and development department under Agilent. In 2003 Agilant extended its global penetration by concluding a deal with Navair to resell electronic test equipment for the wireless, broadcast and avionics industries. This has been extended to acquire new technologies and to extend and expand market potential.

UPDATED

India

Agrani Satellite Services (ASC)

Current Status
ASC and Hughes Space & Communications Company announced a contract 19 January 1995 for two Agrani satellites to provide GSM-compatible mobile telephone links from 1998, covering at least an area bounded by India, Middle East, Singapore and Moscow. ASC was part of India's Essel Group. The turnkey contract with Hughes Space & Communications International called for the first launch December 1997, followed by the second six months later (possibly covering Africa). The value was not disclosed but system cost has been estimated at US$700 million. HSC would provide the satellite, Hughes Network Systems the four gateways and the network control centres (the primary gateways) in Bombay and Gibraltar. Likely slots were 46° E and 53° E, provided by Gibraltar. Each HS-601 HP would have been able to handle up to 16,000 simultaneous calls using about 300 L-band spot beams and 30 kHz channels. 10 to15 W SSPAs would generate approximately 1 kW RF power. Each spot could be adjusted from 30 to 50 dBW. The dual transmit/receive antenna is 12 m in diameter. The Si solar wings would provide 5 kW; design life was 12 years. Unusually, it would be allowed to drift ±6° N/south as mobile users do not require standard NSSK. Several delays postponed the launch of Agrani 1 to 2002, the satellite now being built by Lockheed Martin Telecommunications.

In November 2000 Agrani Satellite Services switched plans and cancelled its plans for a mobile satellite system, opting instead for a conventional DTH broadcast system. Agrani Satellite Services Ltd, an ASC enterprise, signed a deal with Alcatel for a DTH. Agrani is essentially a C-band and Ku-band satellite with C-band for Indian and Pan-Asian coverage and Ku-band for India.

UPDATED

Indonesia

PT MediaCitra (Cakrawartha 1 Satellite)

Current Status
PT. MediaCitra, part of the Indodivision Group of Companies founded in 1993, commissioned the Indostar satellite series of which only one, Indostar 1 renamed Cakrawartha in 1997 has been launched. Indostar aimed to provide the world's first DBS dedicated to radio/TV for a single nation Cakrawartha 1 was originally known as Indostar until renamed in November 1997. The first of four satellites carries five S-band transponders for digital TV, using 8:1 compression for 49 channels. This is the first satellite to use S-band frequencies for commercial communications. The second will add L-band for CD-quality radio. The US$100 million contract for the first satellite and its launch was signed with CTA International 8 December 1993. Funding is provided by: 40 per cent private consortium, 30 per cent PT Amcol Graha Ltd (a Jakarta electronics company) and 30 per cent PT Bimantra Citra. The Cakrawartha 1 DBS system costs US$271 million. PT MediaCitra Indostar is the satellite company of the Indovision group. Indovision is a multichannel pay television company. National TV cannot be received by 32 per cent (57 million) and radio by 20 per cent (36 million). The commercial Cakrawartha system will be received via US$100 hand-held radios, US$100 analogue receivers and US$500 digital satellite decoders with satellite dishes less than 1 m in diameter. It is designed to guarantee quality reception within Indonesia, whatever the weather conditions. With a mass of 1,385 kg at launch and 803 kg in geostationary orbit, Indostar 1 was launched by Ariane 4 on 12 November 1997 to a slot at 96.7° E but has since been moved to 108° E. The satellite is designed for a life of 7 years and carries 12 years propellant.

VERIFIED

PT Pasifik Satelit Nusantara (PSN)

Current Status
PT Pasifik Satelit Nusantara is the first private telecommunications satellite company in Indonesia. Shareholders include PT Telekomunikasi Indonesia, PT Elektrindo Nusantara, Hughes Space and Communications and Telesat Canada. Palapa B1, retired by PT Telkom at 118° E, was sold to PSN in 1991 to provide Pacific Rim services from mid-1992. PSN's teleport at Bataan opened in early 1994. Still operated by PT Telkom, PSN's major shareholder, B1's inclination was allowed to increase to extend its life for three years. Projected revenue for that period was US$3 million. It began drifting in October 1995. B2P was replaced by C1 in February 1996 and transferred to PSN. It is expected the two other Palapa Bs (and then the Palapa Cs) will be sold to PSN as they near the end of their lives. PSN already owns and operates the six extended C-band transponders aboard each Palapa C. Discussions have been held with several regional telecommunications companies in an attempt to link business traffic interest throughout the region but this has had little response.

PSN led the US$650 million ACeS Asean Cellular Satellite System with the PLDT Philippines Long Distance Telephone Co and Jasmine International Public Company Ltd (Thailand). An agreement was signed with Lockheed Martin in

May 1995 for exclusive negotiations; the contract was announced 6 July 1995. The Garuda 1 satellite (named after a mythological Indonesian bird) launched on 12 February 2000 can handle 10,000 simultaneous calls from the 2 million subscribers via the 150-beam phased array antenna. It was launched by Proton K. Subsequent problems with the satellite's L-band system reduced operating capability. PSN also operates Agila 2 and Palapa C1.

PSN's shareholders are PT Telkom (40 per cent), PT Electrindo Nusantara (30 per cent), Hughes Aircraft (10 per cent), Telesat Canada (10 per cent) and private investors.

UPDATED

PT Satelindo

Current Status

PT Satelit Palapa Indonesia was created 29 Jan 1993 to finance and operate the US$240 million new Palapa C1 and C2 satellites. Shareholders were PT Bimagraha Telekomindo (60 per cent), PT Telkom (30 per cent) and PT Indosat (10 per cent) until the DeTeMobil subsidiary of Deutsche Telekom agreed in March 1995 to pay US$566 million for a 25 per cent interest. PT Satelindo will also take over the Palapa Bs from PT Telkom. PT Telkom released the Palapa C RfP 10 August 1992, with bids due 10 October 1992. Hughes' HS-601 was selected and the contract signed 15 April 1993. The winner was required to return at least 20 per cent of the value to Indonesian industry. Satelindo is planning the Palapa D series. Satelindo is now 22.5 per cent owned by Indosat. PT Nusa Era Persada Jaya (NEPJ), a data and voice communication internet based company was founded in January 2000 as a subsidiary company of PT Satelindo with 99.6 per cent shares held by Satelindo and 0.4 per cent by Antariksa.

VERIFIED

PT TELKOM (PT Telekomunikasi Indonesia)

Current Status

Perumtel (Perum Telekomunikasi) became a Limited Company on 24 September 1991, changing its name to PT Telkom and in 1996 undergoing substantial re-structuring. It originally operated the whole satellite network and continues to be responsible for Palapa series satellites. It is instrumental in developing the Palapa D programme and has developed integration concepts for linking with other Asian satellite operators. Has continued to expand on-line services for new transponder operators and has created new interlink connections for regional operators. This has been extended to potential coverage for regional users.

UPDATED

International

NEW ICO

Current Status

ICO Global Communications was constituted in January 1995 to provide global hand-held telephone services using a medium orbit satellite system. Four million subscribers were projected by 2010 and original plans anticipated operations starting in 2000. First launch was scheduled for second half of 1998, with services beginning in late 1999. Cost was projected at some US$3 billion. The 47 investors provided US$1.5 billion total capitalisation, including Inmarsat's own US$150 million. At least 70 per cent was to have been owned by Inmarsat and its signatories. (ICO allows current Inmarsat signatories to invest as desired, instead of in proportion to their Inmarsat interests; maximum is US$103 million.) The rest would be financed by outside strategic investors and bank debt.

ICO targets hand-held phones retailing at US$1,500 for dual-mode models (accessing cellular when in range) and US$1,000 for satellite only. Usage cost was to be average US$2 per minute. It would also provide fax, 4.8 kbit/s data and alerting services. The satellites would relay calls to SAN Satellite Access Nodes within their view. SANS are interconnected terrestrially to form the ICONet, linked through gateways owned and operated by third parties to the public switched networks. Each satellite will provide 4,500 simultaneous calls or equivalent via 163 beams. A 53 kg navigation payload provided by Hughes under a US$105 million contract was planned to augment GPS but Inmarsat in March 1996 voted against its inclusion.

ICO required a constellation of 10 satellites plus two spares in two planes at 10,355 km. An order for 12 Hughes HS601 satellites, worth US$1.4 billion, was announced 20 July 1995; the final contract was signed 5 October 1995. An additional US$925 million launch services contract was signed 7 December. Hughes Electronics Corp also became a strategic investor (US$93.8 million). This was to have been the first HS-601 application in medium orbits; the design will employ GaAs cells, direct injection by the launch vehicle, an enhanced thermal radiator and a modified attitude control system.

Due to financial difficulties and the slow acquisition of business contracts ICO Global Communications entered Chapter 11 bankruptcy protection in August 1999 nine months after the launch of the first of 10 satellites aboard a Sea Launch vehicle. The 2,750 kg HS 601 satellite was to have been placed in a medium orbit of 10,390 km by direct injection with an inclination of 45°. Five satellites were to have been positioned in each of two orthogonal planes providing overlapping coverage of the earth, orbiting every six hours. In October 1999, telecommunications entrepreneur Craig McCaw and his affiliates, Teledesic and Eagle River Investments LLC agreed to lead a group of investors that provided up to US$1.2 billion to ICO to enable it to emerge from bankruptcy thus allowing the launch of the first satellite to proceed. On 28 February 2000 the Sea Launch vehicle was launched but suffered an anomaly several minutes into flight and both launch vehicle and satellites were destroyed.

As a result of the new financial backing the company that emerged from Chapter 11, on 17 May 2000, is known as New ICO and will consist of the same operational suite of 10 satellites in medium orbit with two built as spares. New ICO corporate aspirations are modified and services enhanced to make the system more attractive for a wider range of users. These include: maritime markets in commercial shipping; fishing and recreational vessels; transportation for freight companies that carry goods over distances greater than 480 km; government services such as data, encryption, restricted user groups and global coverage; the oil and gas industry linking professionals in remote locations; individuals and small businesses where there is a need for fixed and mobile networks. The New ICO satellites will use the 2 GHz band for service links and 5/7 GHz for feeder links and the system is expected to begin operations in 2003 with a full constellation of satellites. These will operate in medium earth orbit at an altitude of 10,390 km divided equally between two planes each 45° to the equator. Two of the 12 satellites will operate as spares. On 13 September 2000, Hughes Space and Communications announced that it would build an additional three HS 601 satellites for New ICO, bringing the total to 12, and modify the 11 under construction.

The system will use a bent-pipe architecture carrying signals between user equipment and a ground station and will communicate with terrestrial networks through the ICONET systems which will comprise a high-bandwidth global Internet Protocol network. The New ICO air interface will use a communications protocol similar to that employed by terrestrial cellular systems which allows user equipment to communicate with networks. The air interface is expected to be developed to allow next generation terrestrial wireless applications to operate through New ICO satellites at 144 kbps. Interworking functions will be provided via the GSM global mobile system enabling the company to deliver its services, including cellular links, to all customers whose home networks are based on this system without the need to apply technology upgrades. New ICO expects to support services across packet-mode and ISP operations supporting data packets and other types of internet protocol service.

The first satellite for New ICO was successfully launched on 19 June 2001, passing orbit tests with exceptional success in September 2001. ICO 3–15 satellites have no presently scheduled launch date.

Orion Atlantic LP

Current Status

Orion Atlantic has established a high-power Ku-band trans-Atlantic system optimised for small diameter VSATs and rooftop-to-rooftop communications supporting integrated voice, data and video applications. The structure of what is now known as Orion Atlantic LP in the project was revealed in January 1992. Seven companies joined Orion Network Systems to provide US$90 million equity for shares totalling 66.66 per cent: British Aerospace (UK; 25 per cent), COM DEV (Canada; 4.17 per cent), Martin Marietta (US; 8.33 per cent), Kingston Communications (UK; 4.17 per cent), Matra Hachette (France; 8.33 per cent), Nissho Iwai (Japan; 8.33 per cent), STET (Italy; 8.33 per cent). In addition to the equity investment, an international syndicate of banks, led by Chase Manhattan, provided the partnership a credit facility of US$250 million.

Orion Network Systems (ONS) was formed in 1987 as a telecommunications carrier providing international transmission facilities and services to corporations and US inter-exchange carriers. The OrionNet subsidiary provides digital private line and switched transmission services from its US gateway locations to overseas locations. ONS's subsidiary Orion Satellite Corp is the General Partner and management arm of Orion Atlantic. Its Orion Atlantic Network Services and Orion Atlantic Satellite Services carry out the marketing and sales activities for the partnership.

BAe and Orion Satellite Corporation signed a US$360 million contract on 1 September 1989 for the in-orbit delivery of two satellites. It marked the first US commercial sale by Europe. Construction began 7 January 1992. Under the turnkey contract, Martin Marietta Commercial Launch Services provided launch and insurance services. The FCC granted final authority to proceed 28 June 1991. Orion 1 was launched by Atlas 2A on 29 November 1994. The all-Ku Orion 2 was ordered by Orion Atlantic 24 October 1995 and launched by Ariane 44 LP on 19 October 1999. Total cost, including launch and insurance, is US$265 million, underwritten by bank commitments. The Eurostar 2000+ design is similar to Orion 1 but the higher power transponders allow additional services such as DTH. The coverage will overlap with Orion 1's in eastern US and Europe. Onboard switching will allow beam connectivity reconfiguration according to customer demand.

The company (ONS, its subsidiaries and Orion Atlantic) has returned annual losses through 1999. Launched 5 May 1999, by the second Delta III launcher Orion 3 was a total loss due to second stage failure. Orion 1 and 2 were subsequently sold to the Loral Orion Network Systems company and re-named Telstar 11 and 12 respectively.

SES Global

Current Status

In Novermber 2001 SES Americom and SES Astra combined to form a new global satellite company, SES Global S.A. with 41 satellites covering 95 per cent of the world's population. SES Americom operates 17 satellites while SES Astra has 13 satellites. It also partly owns SIRIUS, AsiaSat, Americom Asia-Pacific, Nahuelsat and Star One. On 13 January 2003 SES Global announced a strategic relationship with US based satellite operator Foundation Telecommunications Inc allowing cable system operators to deliver broadband Internet connections to customers. The system will enable Internet delivery at up to 8 mbps and will be transmitted through combination C-band and Ku-band on SES Americom's AMC-4 satellite.

Israel

Israel Aircraft Industries Ltd

Current Status

IAI began the Amos programme in 1990, with full go-ahead formally in January 1992, to develop, build, operate and market a medium-class GEO satellite, principally for subregional TV broadcasting. IAI retains ownership of Amos 1 and SpaceCom was created in 1993 to sell capacity. SpaceCom's equal partners are IAI, Gilat Communication Engineering, General Satellite Services Co and Mer Services Group Ltd. Amos is built from flight-proven hardware to minimise risk and only the FM was assembled. Cost through launch in May 1996, plus operating costs over 10 years, is projected at US$190 million to US$200 million; adding interest yields a total cost of US$270 million. US$100 million was provided by Israeli banks, guaranteed by the government. The project began when the government guaranteed three transponders over the 12-year life (replacing the three previously leased from Intelsat). A Portugal feed was included early in the programme under a letter of intent, but it is unlikely to be used. Transponder cost is about US$3 million annually non-preemptible. It was hoped that Arab users would be signed up before launch as the majority of the 250 million people within the 700,000 km^2 Middle East beam are Arab-speakers.

IAI plans the launch of the Amos 2/CERES Central European Regional Satellite in 2001–2002. For the project to proceed, it required a commitment from the Hungarian government to guarantee two transponders, agreed 30 April 1995. The US$150 million financing was acquired in 1996. The satellite is owned by Magyarsat in Hungary, a 50/50 joint venture by IAI and Antenna Hungaria, and will be collocated with Amos 1 with eight active 72 MHz transponders (plus two back-ups). The Israel beam will provide Amos 1 back-up (there will be no Portugal beam). IAI concluded other contracts in 1996.

IAI is marketing the approximately 1 tonne design for domestic and subregional communications. The modular approach allows integration of other payloads, and the low mass offers more launch slots. Several different operating concepts have been developed for applications as diverse as mobile telephone relay and data relay services covering national territories.

Italy

Alenia Spazio

Current Status

Alenia Spazio, the major italian space company, is involved in most international projects managed by ASI (Agenzia Spaziale Italiana) and ESA (European Space Agency) and in cooperation with NASA. Italsat, an advanced telecommunications satellite system and Artemis, which will allow in-orbit experimentation and qualification of innovative technologies; it is prime contractor for the coordination and definition of the Galileo project, leading a team of forty European companies and has guided a consortium for the development of Sicral, the first national system for military satellite telecommunications. It has had a key role in commercial telecommunications programmes like Atlantic Bird 1 (for television broadcasting, telephony and fast Internet) and Globalstar, one of the first international mobile communications system, with responsibility for integrating all 72 satellites that make up the constellation. In the multimedia field has developed the SkyPlexNet technology, the satellite platform that allows the various operators to access the satellite autonomously and is working at EuroSkyWay, the first European broadband satellite network for interactive multimedia services. EuroSkyWay is scheduled for introduction in 2005 using 90 cm diameter dish transceivers providing dedicated broadband services.

Japan

Broadcasting Satellite System Corporation

Current Status

B-SAT was incorporated 13 April 1993 as a consortium of NHK, WOWOW, five other broadcasters and eight banks to procure and operate the 'BS-4' generation of broadcasting satellites for launches in 1997–98. These were re-designated in the B-Sat series, B-AT-1A launched in April 1997 and B-AT-1B launched April 1998. The government is expected to allocate four DBS channels, including HDTV. Hughes Space and Communications International Inc was selected in December 1993 (signed June 1994) for on-orbit delivery, in competition with Space Systems/Loral and Martin Marietta. B-SAT 2B was launched by Ariane 5 on 12 July 2001 but an upper stage failure left the satellite stranded in low earth orbit. B-SAT increased operating revenues in 2000, 2001 and 2002.

Japan Satellite Systems, Inc (JSS Inc)

Current Status

In October 2003 JSS Inc had an operating constellation of nine satellites including JCST 1B at 150° east, JCSAT 2A at 154° east, JCSAT3 at 128° east, JCSAT 4A at 124° east, JCSAT 110 at 110° east, N-STARa at 132° east, N-STARb at 136° east and Horizons 1 at 127° east. Horizons 1 was launched on 1 October 2003 by Sea Launch and is based on a Boeing 601HP with a Ku-band payload providing 24 × 36 MHz (108 W) transponders serving North America and Hawaii which, together with JCSAT 2A, will provide a seamless network from the US east coast to Asia and Oceania. Japan Communications Satellite Company Inc (JCSat) was formed in April 1985 to provide the first commercial Japanese satellite communication system. Satellite Japan Corp (Sajac, licensed in May 1991 to launch two satellites in 1994/5) was forced by financing problems to merge with JCSat in 1993, creating JSAT. Hughes Communications sold its 30 per cent equity stake in JCSat and Sajac's Nissho Iwai Corp and Sumitomo Corp major shareholders joined JCSat's Mitsui and ITOCHU. Interests are now: ITOCHU (28.5 per cent), Mitsui (24.5 per cent), Sumitomo (23.5 per cent), Nissho Iwai (23.5 per cent). ¥17 billion revenue was projected for the year ending 31 March 1996 and ¥20 billion for the following year, but has yet to return a profit. JCSAT 1 was launched in March 1989 followed by JCSAT 2 in January 1990, JCSAT 3 in August 1995, JCSAT 4 in February 1997, JCSAT 5 in December 1997 and JCSAT 6 in February 1999. JCSAT 8 was launched in March 2002 and JCSAT 9 will combine commercial and mobile telephone applications.

About 65 per cent of the 64 transponders on the first two satellites are leased; there are 55,000 subscribers to the five TV channels. In addition to network and cable TV distribution, Japan Satellite Communications Network Corp (JSNet) employs JCSat to provide VSAT business data services. Three transponders are used for DTH video services and two carry 12 stereo channels of high-quality Digital Audio Broadcast programming. Other full period lease customers provide time-shared and other value-added services for third parties. Annual transponder charge is ¥300 million to ¥650 million, depending on leasing conditions for a full period, full bandwidth transponder; it also leases partial transponders on a full period, power and bandwidth basis. From autumn 1992, JCSat has provided a full transponder time-share service (15-minute increments).

JSAT started digital multichannel broadcasting named 'PerfecTV!' with more than 70 television channels and 100 radio channels in October 1996 using JCSAT 3. In December 1996, there were about 100,000 subscribers to PerfecTV! and more than 300,000 were acquired by the end of FY96 (March 1997).

Kokusai Denshin Denwa Company Ltd

Current Status

KDD is the Japanese international telecommunications carrier and the signatory to Intelsat and Inmarsat, operating through the Ibaraki and Yamaguchi Earth stations. With COMSAT of the US, it is providing aeronautical communications through the Inmarsat system and has conducted negotiations on integrating other carriers in the region.

KDD has developed special carrier services technology capabilities for multiplexed operations. KDD is also active in plans to develop a network of new earth stations in south-east Asia and has produced network operations segments that integrate national and international carriers.

Nippon Hoso Kyokai (NHK)

Current Status

Major shareholder in Broadcasting Satellite System Corporation (BSSC) but has reduced its holding in recent years. Active in the development of new DTH services. Has been instrumental in developing new policy controls and brought unique analytical capabilities to the BSSC.

Space Communications Corporation of Tokyo (SCC)

Current Status

In November 2002, SCC and SES Americom announced an agreement for multiple-year satellite transponder utilisation of Americom's AMC-13, launched in 2003. SCC owns and operates four Superbird satellites.

Established 22 March 1985, SCC is wholly Japanese-owned by the Tokyo-based Mitsubishi Electric Corp (MELCO, 18.94 per cent), its general trading affiliate Mitsubishi Corp (28.41 per cent), Mitsubishi Heavy Industries (10.0 per cent), The Mitsubishi Bank (5.0 per cent), Mitsubishi Trust and Banking Corp (5.0 per cent), Mitsubishi Estate Co (5.0 per cent), The Tokyo Marine and Fire Insurance Co (5.0 per cent), The Meiji Mutual Insurance Co (5.0 per cent), Kirin Brewery Co (5.0 per cent), and 20 other companies of the Mitsubishi Group (12.65 per cent); share capital totals ¥60 billion. Revenue for the year ended 31 March 1995 was ¥16.4 billion; it was hoped the company would break even in 1995. SCC was established in March 1985 and provides Japan's second commercial satellite communication system (see JSAT); personnel total about 230. Two Superbird satellites were planned, built by Space Systems/Loral and based on the FS-1300 platform. Programme cost was ¥70 billion (US$636 million), covering two satellites, launch services and insurance at premiums of about 20 per cent. Launch of Superbird A 5 June 1989 by Ariane to 158° E was successful but most of its stationkeeping oxidiser was lost in December 1990 and commercial operations were ended; some customers were transferred to the rival JCSat. The insurance claim was reportedly US$170 million. Superbird B, delayed from a December 1989 launch because of transponder problems, was lost in the Ariane 4 accident of 22 February 1990. Insurance cover was US$94.3 million. Launch contract for the replacement Superbird B1 was signed with Arianespace in November 1990; SS/L delivered the satellite within 24 months and the satellite was launched on 27 February 1992. Superbird C was ordered from Hughes in March 1995 to expand coverage to the Asia Pacific region, including HDTV and 150 Mbit/s data and, following launch on 28 July 1997, is located at 144° E with Superbird B1 at 162° E and Superbird A at 158° E. Cost with launch and insurance is about ¥35 billion (US$412 million). Also built by Hughes, Superbird 4 was launched by Ariane 44 LP on 18 February 2000. Built by Aerospatiale, Superbird 5 (known as N-SAT-110) was launched by Ariane 42L on 6 October 2000. The current constellation consists of Superbird A, Superbird B2, Superbird C, Superbird D, Superbird A2 and SN4. The SN4 satellite operates from 172° east and will be used until the launch of AMC-13.

UPDATED

Telecommunications Advancement Organisation of Japan

The Telecommunications Satellite Corporation of Japan (TSCJ) was established in August 1979 to operate the CS and BS satellite series procured and launched by NASDA. The name changed to TAO in October 1992 as it introduced new services such as R&D. Japan's first two operational communications satellites, CS-2a and 2b, each with six Ka-band and two C-band transponders, were launched in 1983 principally for maintaining communications with the more scattered islands of the Japanese archipelago. Both were still operational when they were replaced by CS-3a and 3b in 1988. NTT's NStar took over in 1995. The BS-2 series introduced DBS services in 1984 using 100+W transponders but two of BS-2a's three amplifiers failed soon after launch on 23 January 1984. BS-2b brought the system to full capacity in 1986 but a BS-2x was purchased from GE Astro Space by NHK for an Ariane launch to ensure system continuity before the first of the BS-3 series could be launched in 1990. However, BS-2x was lost in the Ariane failure of 22 February 1990; the replacement BS-3H was then lost in its April 1991 launch. Both would have been controlled through TSCJ. BS-3a, launched 28 August 1990, replaced BS-2b but it operated on marginal power and has been retired. BS- 3b completed the system in 1991 but NHK procured BS-3N in late 1992 (launched July 1994) to guarantee services. BSAT 1a was launched on 16 April 1997 followed by BSAT 1b on 29 April 1998. BSAT-2 series satellites are built by Orbital Sciences Corporation and provide 4-channel digital TV transmission through 180 W ku-band channels. BSAT-2A was launched on 8 March 2001 by Ariane 5 and placed in geosynchronous orbit at 110° E. BSAT-2B was lost in an Ariane 5 launch vehicle failure on 12 July 2001. Replacement satellite BSAT-2C was launched on 11 June 2003.

UPDATED

Luxembourg

SES Astra

Current Status

Astra is the satellite operating company providing SES Global with broadband satellite communication solutions. It consists of 13 satellites providing more than 1, 100 TV and radio channels to 91 million European homes. As of October 2003 the constellation comprised of: Astra 1B, 1C, 1E, 1F, 1G, 1H and 2C at 19.2° east; Astra 3A and 1D at 23.5° east/24.2° east; Astra 2A, 2B and 2D at 28.2° east; and Astra 1A at 5.2° east.

SES was incorporated as a private company in March 1985 to establish a medium-power satellite system for TV distribution from one of the GEO slots assigned to the Grand Duchy of Luxembourg. It operates under a franchise extending to 2010 from the Duchy (which retains a 20 per cent interest). Germany's Deutsche Telekom acquired a 16.67 per cent interest in June 1994. By end-1995 the system was reaching 61.33 million homes (39.60 million cable; 21.73 million SMATV/DTH). 1995 profit was Lfr2,800 million from sales of Lfr10,300 million.

A contract was signed with RCA-Astro Electronics in October 1985 for a single series 4000 satellite operating in the FSS range. Europe's first private satellite began operational programme transmissions 1 February 1989. SES then acquired Satcom K3, previously owned by Crimson Satellite Associates. The 5000-class satellite was in an advanced stage of completion and required only modifications to its transmission characteristics. Transmitting on adjacent frequencies and collocated with Astra 1A from 1991, it doubled the number of channels to 32. Rather than using its more powerful tubes to increase EIRP, the 52 dBW coverage is extended. Hughes was awarded the contract of about US$428 million in December 1990 for two further satellites, based on the HS-601 platform, and their launches on Ariane 4 for collocation at 19.2° E in a 140 km cube with 1A and 1B. 1C operates below 1A's frequencies and acts primarily as its back-up, replacing transponders if they fail. 1D acts as 1B and 1C's back-up but its frequencies at the very bottom of the FSS range require receiver retrofitting. It can also switch to BSS frequencies that can be employed for HDTV. Astra 1E was approved in 1992 to provide digital DTH programming. An ECU220 million loan facility was contracted in February 1993 to finance it, bringing total SES capital investment to ECU1 billion. Astra 1F was approved in 1993, also for 19.2° E and digital DTH; the enhanced 1G was ordered January 1995 and the identical 1H in July 1995 to back up 1E to 1G. Astra 1H was launched in June 1999 introducting Ka-band services followed by Astra 1K in 2001 and Astra 1L in 2002–2003. An ECU175 million loan facility was contracted September 1995; ECU360 million was added July 1996. Total investment through 1996 was ECU1,250 million. The first two satellites of the second-generation, replacing the first and expanding into other business areas from the new 28.2° E slot were ordered in June 1996.

Astra's Betzdort control facility (NASA) 0517408

Astra 2A launched in August 1998 followed by Astra 2B in September 2000 and Astra 2D in December 2000. Astra 2C was launched in June 2001. Astra 3A was launched on 29 March 2002. In September 2003 SES Astra announced the extension of one-way broadband services to Hungary, Poland, Ireland, Czech Republic and Algeria, making it the first operator to carry broadband capability to the African continent.

Luxembourg has filed for eight further 10.70 to 12.75 GHz slots: 24.2, 26.2, 28.2, 31.5, 35.5, 37.5, 41.2, 43.2° E.

SES reported net income margin of 32.9 per cent in six months ended 30 June 2001 with operating profit up 4.9 per cent. SES employs 465 people.

Malaysia

Binariang SDN BHD

Current Status

Binariang signed an MoU in November 1991 with Hughes for two HS-376 satellites to provide domestic and regional services. The first MEASAT (Malaysian-East Asian Satellite) launch was then planned for 1994 aboard Ariane. Before MEASAT appeared, Malaysia spent some US$30 million annually on Intelsat and Palapa leases. A contract was not signed with Hughes Communications International Inc until 17 May 1994 following a four-month international competition. The agreement covered one satellite to be launched by late 1995, an option for a second (authorised January 1995), the ground control station and training for the Malaysian operators. This HS-376 is the first to carry GaAs solar cells (increased output), a lightweight shaped antenna (improves gain and eliminates multiple feedhorns) and a bipropellant stationkeeping propulsion system. Binariang also holds a licence for international gateway services. MEASAT 1 was launched by Ariane 44L in January 1996 to 91.5° E followed by MEASAT 2 to 148° E in November 1996 on another Ariane 44L. Plans for MEASAT 3 envisage C-band services to a wide area from central Africa to Australia and Japan, with Ku-band coverage from Pakistan to Taiwan. Binariang selected Boeing Satellite Systems to build the satellite around a Boeing 601 HP bus and will employ 24 C-band and 24 Ku-band transponders, each providing 36 MHz of bandwidth over a 15-year service life, using a Proton Breeze M launcher from ILS to put MEASAT 3 in orbit during 2005, co-located with MEASAT 1 at 91.5° E. The MEASAT 1 replacement is planned for launch no later than the fourth quarter of 2007. Binariang is also planning a regional expansion which could see a need for additional capacity.

UPDATED

Norway

Telenor Satellite Services AS

Current Status

Telenor became a listed company in December 2000. Financial highlights for 2001 showed a 22.5 per cent increase over 2000. Shares grew by 0.5 per cent against a European index down by 28 per cent.

TNSS, a Telenor subsidiary, is the national signatory to Intelsat, Inmarsat and Eutelsat. The country is Inmarsat's fourth largest member, with 6.8 per cent in 1996, complementing the size of its maritime fleet and emphasising the importance of the major Earth station at Eik. TNSS is among the largest VSAT and gateway providers in Europe, with an installed base of 1,000 terminals by end-1995. Telenor is a member of the Skyphone consortium providing aeronautical communications services through the Inmarsat system. The Mobiq system provides satellite telephone communications to and from receivers at sea, in the air or on land. The NORSAT Sealink is a permanent line for data transmissions incorporating mobile sea sites and one or more land-based site. GAN provides high-speed access via Inmarsat's new M4 satellite service. Telenor now offers its E-Mail Advanced optimised communications service designed to reduce costs and improve transmission efficiency. Wholly owned by Telenor, Satellite Services supplies satellite-based voice and data services for use on land, in shipping and the offshore industry. Also wholly owned by Telenor, Satellite Networks supplies solutions and VSAT technology to the business market and national and international organisations.

Telenor, known as Norwegian Telecom until 1 January 1995, bought the UK's Marcopolo 2 satellite in 1992 for a reported £25 million to provide DTH services from 0.8° W to a potential six million homes. Germany's TV-Sat 2 was added on lease in March 1995 at 0.6° W for the rest of its life. In addition, Telenor leases transponders on an Intelsat at 1° W, employing the Scandinavian spot beam at 50 W per channel. A Nordic channel was previously provided from Tele-X at 5° E. Approximately 20 TV channels (analogue and digital) are currently transmitted from TNSS's 1° W transponders. 35 channels were planned by 1998 by investing NKr1.5 billion in two new satellites. The Thor 2A contract with Hughes was announced in November 1995, with a launch in 1997, with 15 transponders. The larger Thor 3 was launched in June 1998. Equipped with Ku-band transponders, Thor 4 is currently in the planning stage. Telenor has 2,300 employees.

UPDATED

Singapore

Government Sponsored Organisations

Current Status
Nanyan Technological University and the UK's University of Surrey signed an MoU in September 1995 for joint research and training in satellite engineering, communications satellites and, specifically, LEO technology. This will enable NTU to help Singapore develop its own LEO satellite industry. NTU's Merlion initiative is acquiring LEO engineering and applications expertise, at first operating a Surrey-provided ground station and, in March 1996, signing a £1.35 million contract for Surrey's UoSAT 12 to carry NTU's communications payload, funded by the National Science and Technology Board of Singapore. NTU's package comprises: bent pipe L/S-band up/down transponder; L/S-band processing transponder; propagation beacons; processor for various modulation/coding methods and protocols. The ground station was upgraded to operate at these frequencies. Experiments include real-time mobile communications and satellite linking of vehicles in GPS-based fleet tracking/control. UoSAT 12 was launched from Baikonur cosmodrome on 21 April 1999 by SS-18 Dnepr. It was proposed as a follow-up to fly a satellite in equatorial LEO; five would provide continuous coverage for Singapore. Singapore has continued to study multisatellite constellations for improving traffic monitoring and toll logging, the concept being pursued by European manufacturers.

UPDATED

Singapore Telecommunications Ltd

Current Status
Matra Marconi Space was selected by Singapore Telecom and Taiwan's Directorate General of Telecommunications in March 1996 for the ST-1 satellite to be delivered in-orbit within 24 months and ground stations in Singapore and Taipei. Total project cost is about US$240 million. The Eurostar 2000+ satellite bus, approximately 3,000 kg and approximately 12-year life, is positioned at Singapore's 88° E slot to provide regional broadcast and telecom services using 14 C-band and 16 Ku-band (about eight serving India) transponders serving eastern China, Indonesia, Malaysia, the Philippines, Singapore and Taiwan. The Ariane first quarter 1998 launch contract was signed 30 July 1996. Singapore Telecom also has a 14 per cent interest in AMSC with a design life of 12 years. ST-1 was eventually launched on 25 August 1998 joining Apstar 1, launched in 1994, and Apstar 3R, launched in 1998, in providing SingTel services. Plans for a ST-2 and ST-3 have been postponed and there is no possibility that they will be resurrected.

VERIFIED

Sweden

Swedish Space Corporation

Current Status
The Nordic countries of Denmark, Finland, Iceland, Norway and Sweden began studying a joint regional satellite system in the mid-1970s. From the beginning, TV direct broadcasting via satellite was regarded as a major role for such a system.

Tele-X is based on the Spacebus 300 platform (Aerospatiale) 0517013

In 1977, the World Administrative Radio Conference on broadcast satellite services allocated eight channels to the Nordic group as an entity (excluding Iceland), plus another 17 to the individual countries (including five for Iceland). What became the Tele-X project assumed its current form in 1982, when it was envisaged as providing both DBS and business communications services within Scandinavia, and also as a means of stimulating the space engineering capabilities of these countries. The 'X' represented Experimental and it was planned to have only one satellite in orbit. It was to provide two high-power DBS TV channels (plus one back-up,which might be used operationally) and two plus one video/data channels.

By then, Denmark and Iceland had withdrawn from the project, though Denmark had earlier expressed interest in a beam covering southeast Greenland. An agreement between Sweden and Norway was concluded in 1983, giving them 85 per cent and 15 per cent interests, respectively, in NSAB (Nordiska Satellit AB). 3 per cent of Sweden's share of cost/work in the project (but not in NSAB) was later assigned to Finland in a separate bilateral agreement.

Taking advantage of the Franco-German TDF/TV-Sat programme, initiated in 1980, the Swedish Space Corp ordered a similar satellite based on the Spacebus 300 platform. The contract was signed with both the Eurosatellite consortium and one of its members, Aerospatiale, specifying the latter as prime contractor. NSAB was to be proprietor of the system, while SSC was to be the procurement and executive agency. The satellite was named Tele-X and after launch in April 1989, ownership and operational responsibility was to pass to a PTT-owned consortium, Notelsat (50/50 per cent Sweden/Norway). Shortly before launch, Norway withdrew from NSAB, trading its 15 per centshare for a lifetime lease of one channel. NSAB is now a 50/50 partnership between SSC and Teracom, a major distributor of TV/radio programming. SSC became wholly responsible for operating the system. System cost is SKr1.25 billionat fixed 1982 rates, now about SKr1.8 billion, or US$280 million. All three DBS TV channels are occupied: the Norwegian public service transmits in D-MAC, and TV4 (from September 1990) and TV5 Nordic (from November 1991) operate under PAL. Originally named Marco Polo when launched by B Sky B in 1989, the satellite was acquired by NSAB at the end of 1993 and renamed Sirius 1. At the beginning of 2000 Sirius 1 was moved from 5° E to 13° W and renamed Sirius W. It is expected to remain in service until 2006. It was followed by Sirius 2 in November 1997 and Sirius 3 in October 1998. Sirius 2 was sent into orbit as GE-1E. All three Sirius satellites are owned by the Nordic Satellite Company, NSAB. In October 2000 the SSC became 50 per cent owner of the NSAB, the balance owned by SES/ASTRA. Launched in July 2001, the multipurpose Artemis satellite was left stranded in a low orbit following the failure of the Ariane 5 upper stage. On-board propulsion was used to boost the satellite to geostationary altitude in 2002. Germany, Sweden and Norway have signed a contract for fixed cost rocket and balloon activities at the Esrange facility.

UPDATED

Thailand

Asian Broadcasting and Communications Network Company Ltd

Current Status
ABCN was established in 1994 as a subsidiary of Thailand's The M Group Plc to develop joint ventures with regional governments for satellite systems. Concepts included DBS, data relay, telecommunications and television relay. A contract was signed November 1995 with the Lao People's Democratic Republic to jointly manage and operate a system from January 1998 at 116° E; the concession extends 30 years. ABCN holds 80 per cent of the shares in the LaoStar Company set up with Laos. These developments have been stalled by lack of finance and political difficulties in Laos.

Shinawatra Satellite Public Company Ltd

Current Status
The government invited tenders in 1987 for a domestic satellite system; the winner would receive a 30 year concession to lease transponders to government and private sector users, with excess capacity offered to neighbouring countries. Thailand's population is about 55 million. Shinawatra Computer & Communications Group (SC&C; chairman Dr Taksin Shinawatra), the largest group of Thai companies operating telecommunications and broadcasting service concessions, was selected in 1991 and given a 30-year concession. It has an eight-year monopoly to supply satellite capacity to domestic users (and was required to pay existing users of other satellites to transfer to Thaicom). It began formal negotiations June 1991 with Hughes, signing the contract 8 October 1991 at around US$100 million for two satellites, ground equipment and training. Deliveries were due within 24 to 28 months. The first launch contract was signed with Arianespace 6 December 1991. As lighter versions of Hughes' HS-376 bus, Thaicom was the second to come within Ariane's 1 tonne SDS (Spelda Dedicated Satellite) class. Shinawatra signed a letter of intent 1 November 1993 to negotiate with Hughes for a third satellite but Aerospatiale's Spacebus 3000 platform was selected December 1994 for the more powerful Thaicom 3; negotiations for Thaicom 4 continued. Total cost for Thaicom 3 on station is estimated at US$240 million. 1994 revenue was Bt371.60 million, for a Bt156.50 million loss.

Thailand previously leased 2¼ transponders aboard AsiaSat 1 for telecom services. Thaicom 1 was intended to be placed at 101° E and Thaicom 2 at 78.5° E but an agreement was reached with AsiaSat September 1993 to co-locate at the more westerly slot to avoid interference with AsiaSat 2 at

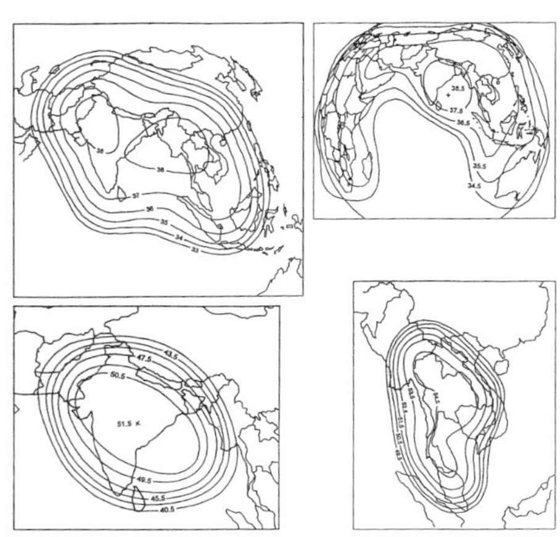

Thaicom 3 footprints (dBW). Top left: C-band India/ Indochina; top right: C-band hemispherical; bottom left: Ku-band steerable (India); bottom right: Ku-band Thailand (Shinawatra Satellite Public Co Ltd) 0517386

100.5° E. Thaicom 1 was launched on 17 December 1993 followed by Thaicom 2 on 2 October 1994 and Thaicom 3 on 16 April 1997. With 25 C-band and 14 Ku-band transponders serving users in Africa, Asia, Australia and Europe, Thaicom 3 is co-located with Thaicom 2 at 78.5° east.

SSPC has ordered a broadband satellite known as iPSTAR-1 from SS/ Loral for launch in 2003 by Arianespace. When launched its 45 Gbps capability will equal that of all existing satellites serving Asia today.

United States

AT&T Alascom

Current Status

AT&T Alascom got out of the satellite business in 1997 but its Aurora satellites continue to operate with other providers.

AT&T Alascom, a subsidiary of AT&T, employed the Aurora satellite to carry domestic telephone, the Alaskan Television Network and emergency messages. Aurora 1 was launched in 1982 as RCA Satcom 5, the first RCA 'Advanced Satcom' C-band spacecraft to employ SSPAs in its 24 active transponders. AT&T Alascom bought the satellite for US$88 million and leased back four transponders to RCA; GE Americom provided TT&C services. GE's Satcom 1R at 139° W and then C-1 at 137° W acted as in-orbit back-ups. Customers had to be moved to C-1 in late 1990 when Aurora suffered service interruptions because of attitude control problems, forcing retirement in 1991 (the satellite is still held inclined 4.0° at 105° W. Aurora 2 (Satcom C5) was launched in 1991 to 139° W as replacement, incorporating 24 active 11W C-band SSPAs. Although it employs basically the same GE 3000 bus, it offers a lifetime of 12 years, an improved redundancy scheme allowing instantaneous payload reconfiguration, Ni/H_2 batteries, larger hydrazine tanks and electro-thermal thrusters. AT&T Alascom signed a US$100 million agreement with GE Americom in spring 1990 to procure two-thirds of the satellite's capacity. Since then, AT&T Alascom has bought two more transponders, leaving GE Americom with a quarter ownership. Telstar 6 replaced it at 129° W in 1999 and is currently at 93° W.

Boeing Satellite Systems (Hughes Space and Communications Company)

Current Status

By the end of 2002, BSS (formerly HSCC) had launched 204 satellites and spacecraft since its first (Syncom 1) in February 1963.

In October 2000 The Boeing Company acquired three units within Hughes Electronics Corporation: Hughes Space and Communications Company, Hughes Electron Dynamics, and Spectrolab Inc.

HSC, created in August 1970 around the Space Systems Division, has been responsible for 198 launched satellites and probes (as of June 2001), including more than 40 per cent of all the world's commercial communications satellites. Its 100th telecommunications satellite was launched in December 1993 and its 100th commercial telecommunications satellite in December 1994. 1991 to 96 sales each totalled about US$2 billion per annum (1989: US$1.7 billion; 1988 US$1.4 billion); it has 7,200 employees. Hughes' first major space projects were the Surveyor lunar soft lander and the Syncom geosynchronous communications demonstrator. The company was responsible for Early Bird (Intelsat 1) and the four Intelsat 2s. Its USAF Tacsat tactical communications craft in 1969 demonstrated technology for future generations, introduced to the international market by Intelsat 4/4A. HS 333 became the first generation of domestic telecommunications satellites, with new technology shaping the beam footprint to the host country's requirements. Anik A1 in 1972 was the first domestic GEO satellite. That HS 333 was joined by Canada's Anik A2/A3, Westar 1–3 and Palapa A1/A2; Westar 3 was the last to leave service, in 1990.

Hughes also provided the three Marisats to COMSAT General for the first maritime service in 1976. 22 years later they were still in service. It then provided the payload to British Aerospace for the second-generation Inmarsat. The Comstar design for COMSAT introduced the use of dual polarisation to double capacity, preceding development of the most successful commercial satellite yet: the HS 376, first launched in November 1980. By February 1997, 50 had been ordered for domestic and regional systems on five continents. The advent of the Shuttle resulted in the wide-body communications satellite: five Leasat HS 381 models were built to provide US Navy links. A blending of the wide-body approach with the deployable functions of the 376 led to the development, in the early 1980s, of the international Intelsat 6 (five ordered) and the third-generation HS 393 domestic platform (three flown). Australia's Optus B1 introduced the modular three-axis HS 601 design in August 1992. The first HS 601 HP (high performance) version was ordered in January 1995 for Luxembourg's Astra 1G, offering increased power, enlarged payload and ion propulsion. The first 601 HP launch was PAS 5. HS 702 was unveiled in October 1995, the mass savings increasing the transponder number by 50 per cent. 1998's Galaxy 1 is the first user.

In addition, Hughes was responsible for five ATS Applications Technology Satellites, numbers 1 and 3 of which remain functional after more than 28 years. It also built 1975's OSO 8 solar observatory, the five Pioneer Venus probes, Magellan's radar payload, Galileo's atmospheric probe, the second-generation GOES and Japan's five HS 378 GMS meteorological satellites. Further information is included in the appropriate sections of *Jane's Space Directory*. HSC's Springback 20.4 kg 4.9 × 6.7 m graphite mesh antennas were introduced on AMSC 1 in 1995. The antennas are rolled into a 5 m-high cone for launch, taking advantage of the normally unused space in the forward fairing. It is assembled on a large graphite mandrel and then baked. Hughes also designs, develops, builds and commissions telemetry, command and satellite control ground terminal facilities on a worldwide basis.

Facilities: HSC has been delivering one satellite per month (on average) for more than three years. As of May 1999 there are 37 satellites ordered to be built and launched before 2001. The smaller spinner requires less than 18 months to build. In 1992, Hughes began streamlining operations so that even an HS 601 or HS 702 from initial contract to launch site delivery requires 24 months or less. Spacecraft are assembled into a full-up configuration and tested for flight readiness in the company's Systems Integration Activity area. The Systems Test and Engineering Laboratory undertakes systems level integration and testing.

Hughes 376

Until HS 601, the HS 376 was numerically the most successful commercial communications satellite, with 55 ordered for 19 domestic and regional communications systems since it was introduced in 1980 to replace the HS-333. As of February 1997, 35 of the 53 orbited since SBS 1 remained operational. The total includes the two Canadian Anik Ds and the Brazilian Brasilsat-A series, built by Spar Aerospace employing licensed HS 376 technology. With a mission life of up to 12 years, it can accommodate up to 24 transponders. Mexico's Morelos was the first to provide for the simultaneous use of C and Ku-band channels, and was the first Hughes model to incorporate a planar array, for receiving the four Ku-band channels. The UK's BSB satellites, numbers 31 and 32 ordered, were the first to use the 376 for DBS applications.

Based on decades of experience in building satellites, the HS376 provided commercial and government operators with a reliable platform for wide ranging telecommunications and TV broadcasting services from the late 1970s (Boeing)
0572486

They carried a larger solar drum for increased power and unusually created the 110 W DBS output by linking paired 55 W transponders. They were later sold in-orbit to companies in Sweden and Norway. Malaysia's MEASATs are the first to employ GaAs cells and a shaped antenna to eliminate multiple feedhorns.

The enhanced 376W wide body version almost doubles solar power generation and expands the propulsion capability by adopting some HS-393 features. Hughes was awarded a US$175 million contract in April 1990 for the first two 376Ws as Brasilsat's second generation: Brasilsat B3 was ordered in 1995. The first 376L lightweight version, with 439 kg dry mass, was ordered by Thailand in 1991; the two-satellite contract was worth about US$100 million.

Total ordered: 57 by October 2001 (SBS (5); Anik C (3); Anik D (2); Telstar 3 (3); Brasilsat A (2); Brasilsat B (4); Ausat A (3); Morelos (2); Westar (3); Palapa-B (4); Galaxy (8); Marcopolo (2); Thaicom (2); APSTAR 1, 1A (2); MEASAT (2); BSAT-1 (2); ChinaSat 7 (1); Thor II (1); Thor III (1); Sirius 3 (1); Bonum-1 (1); Astra 2D (1)); Astra 3A (1); e-BIRD (1)
Total launched: 56 by end of 2002 (BSAT 1A/1B, Brasilsat B3, Thor 2A; Galaxy 1R lost in launch failure)
Operating: 37 by end of 2000
Launch mass: about 1,200 kg (excluding perigee stage); 376W 1,750 kg; 376L 1,080 kg
On-orbit mass: about 700 kg BOL; 376W 1,052 kg BOL; 376L 630 kg BOL
Deployed dimensions: 376/376L 2.16 m diameter, 6.6 m height (2.84 m stowed); 376W 3.65 m diameter, 8.3 m height (3.43 m stowed)
Communications section reliability: more than 0.97 at 10 years
Communications payload mass: about 160 kg
Mission life: up to 12 years, depending on application
Structure: cylindrical conventional design in three sections: bus (including AOCS and so on), solar array and despun payload/antenna platform. The antenna typically incorporates a single despun 1.83 m diameter reflector. The lower telescoping concentric solar array is stowed around the upper body for launch
Power: Si cells mounted on cylindrical surface providing 19.7 mW/cm^2, 1,000 W BOL, 900 W EOL (376W: 1,400 W BOL; 376L: 800/670 W BOL/EOL) at solstice; two Ni/Cd or NiH_2 batteries (376L: 52 Ah) provide 100 per cent eclipse protection. BSB's bottom drum was lengthened 25 cm for 1,100 W BOL. MEASAT is the first to employ GaAs, providing 1.2 kW to the payload
Pointing: antenna accuracy of 0.05° in azimuth and elevation via RF beacon
AOCS: spin-stabilised at 50 rpm by four 22.2 N mono thrusters (two axial plus two radial) supplied with up to 210 kg hydrazine from four peripheral tanks; GEO insertion provided by solid propellant Thiokol Star 30 AKM. MEASAT uses bipropellant HS 601 N_2O_4/MMH attitude thrusters plus Star 30. The HS 376W adopts HS 393 elements: a central Kaiser Marquardt R-4D 490 liquid thruster for GEO insertion plus four 22.2 N ARC attitude thrusters
Thermal systems: quartz mirror radiator band encircling upper body provides primary heat rejection path; area 4.0 m^2

Hughes 601
HS 601 is the company's first three-axis design, accommodating up to three bands simultaneously for high-power services such as land mobile, large area DTH and HDTV. The first commercial contract was for Australia's Optus B; Hughes invested US$100 million in platform development. 10 are for USN under a US$1.8 billion contract, including launches. The Solidaridad contract is worth US$183.47 million. The ¥30 billion JCSat 3 contract was signed 29 October 1993. JCSat ordered numbers 4, 5 and 6 in 1996 and 1997. Modular construction reduces usual production time by about a year. The control system incorporates MIL STD 1750A microprocessor (16k RAM). As part of its on-orbit delivery packages, Hughes paid US$128.8 million in early 1992 to cover eight satellites for approximately US$300 million insurance through brokers Marsh and McLennan Aviation and C T Bowring: two Optus, three UFO, three Galaxy. Luxembourg's SES in January 1995 ordered the first HS 601 HP (high performance) version for its Astra 1G. It subsequently ordered two more in 1996 - Astros 1H and 2A. The twin four-panel GaAs wings provide 8 kW BOL, the enlarged payload can carry 32 active 100 W TWTAs, the XIPS xenon ion propulsion system provides NSSK and lightweight contoured reflectors reduces mass. NASA on 23 February 1995 selected medium power HS 601s for its three TDRS H-J under a US$481.6 million fixed-price contract. First delivery is due 1999. Japan's Space Communications Corp ordered the 42nd HS 601 2 March 1995 for its Superbird C, with Mitsubishi Electric providing the cell panels for the solar wings. PanAmSat ordered an HS 601 HP 10 March 1995 for PAS 5 due for launch in 1997. An order of 12 ICO Global Communications satellites, worth approximately US$1.3 billion, was announced 20 July 1995; the final contract was signed 6 October 1995. This makes the HS 601 the most numerous model with 65 ordered as of February 1997. An additional US$925 million launch services contract for the ICO satellites was signed 8 December. Hughes Electronics Corporation also became a strategic investor, with a US$93.8 million investment. This will be the first HS 601 application in intermediate orbits; the design will employ GaAs cells, a simplified orbit adjust system (the launcher will be responsible for direct placement), an enhanced thermal radiator and a modified attitude control system.

Total ordered: 84 ordered (52 launched) through 2002 (HS601: Optus B (3); UHF (11); MSAT (2); Galaxy (7); DBS (3); Solidarad (2); Astra 1 (4); Palapa C (1); JCSAT (5); Apstar 2 (1); TDRS (3); Superbird (2); GOES (2); DirecTV (3). HS-601HP: Superbird 4 (1); Galaxy (10); Astra 1/2 (4); AsiaSat (3); SATMEX (1); Orion (1); DirecTV (2); HS 601MEO: ICO (15); HS GEM: Thuraya (2))
Total launched: 52 through end of 2002
Number operating: 49 through to end of 2000 (Optus B2, PAS 3, APStar 2 lost during launch; UFO 1 stranded in low orbit by launcher; AsiaSat 3; Galaxy 10; Orion 3)
On-orbit mass: 1,582 kg on-station BOL (Optus B), 1,925 kg ICO
Deployed dimensions: 0.227 m^3 body, solar arrays span 18.3–26 m depending on three- or four-panel wings (satellite envelope 229 × 254 × 254 cm at launch)
Communication section reliability: not available
Communication payload mass: not available
Communication payload power: 3–4.8 kW RF
Design life: 10–15 years, depending on applications
Structure: conventional aluminium structure; payload mounted on north/south panels and Earth-facing floor; east/west faces carry hardpoints for antenna mountings, fixed antenna mountings on Earth-facing wall
Power: twin solar wings of three or four 2.16 × 2.54 m panels carrying K43/4 large area Si cells on Kevlar substrate generate up to 6 kW. HS-601 HP employs twin four-panel GaAs wings for 8 kW BOL. NiH_2 batteries provide 100 per cent eclipse protection
Pointing: 0.025°
AOCS: provided by unified 12 × 22 North Atlantic Research Corporation and single 490 N Kaiser Marquardt R-4D bipropellant thrusters incorporating onboard control processor, Sun and Barnes Earth sensors and two 61N ms two-axis gimballed momentum bias wheels. 1,658 kg NTO/MMH in four spheres. Solar tacking techniques reduce conventional ΔV requirements. Spin-stabilised in transfer orbit. When required, Thiokol's 4.7 tonne Star 63 provides perigee kick. HS-601 HP carries Hughes' XIPS xenon ion propulsion subsystems for NSSK. Comprises four 13 cm 18 mN thrusters (two primary plus two back-up), power supply (439 W input each thruster off 29 to 34V DC bus) and a propellant storage/control unit. Total XIPS package mass 68 kg, for a 400 kg saving. In normal NSSK operation, the thrusters fire for 2 to 3 hours daily.

Hughes 702
HS 702 performance spans and extends HS 601's range. A payload module tailored to customer preference, and able to carry 90 high-power transponders, (72 plus spares) mounts to a common bus module by four bolts and six electrical connectors. Production cycle is about 24 months.
Total ordered: 20 (6 launched) through 2002 (HS702: Galaxy (3); XM-1; XM-2; XM-3; Anik F1; Anik F2 Spaceway (3); Wideband Gapfiller (6); NSS-8 (3))
Total launched: 3 through end of 2000
Number operating: 3 through end of 2000
On-orbit mass: up to 5,200 kg at launch
Deployed dimensions: solar arrays span 40 m
Communication section reliability: not available
Communication payload mass: up to 1,200 kg
Communication payload power: 10 kW RF
Design life: 15 years
Structure: conventional aluminium structure; payload module bolted to common bus. East/west faces carry hardpoints for antennas up to 272 cm, fixed-antenna mountings on Earth-facing wall
Power: twin four-panel GaAs wings for 10 kW BOL, growth to 15 kW planned. NiH_2 batteries provide 100 per cent eclipse protection
Pointing: 0.025° being achieved
AOCS: provided by bipropellant (1,700 kg NTO/MMH) thrusters incorporating onboard control processor, Earth sensors and MWs. Hughes' XIPS xenon ion propulsion subsystem for NSSK: four 25 cm thrusters (two primary plus two back-up) consuming only 5 kg per year. XIPS can also perform final orbit insertion.

VERIFIED

Constellation Communications, Inc

Current Status
CCI formed in 1991. A January 1995 Memorandum of Understanding between CCI (Bell Atlantic is the major shareholder) and Brazil's Telebras could develop the ECCO satellite constellation to provide voice, data, position location and other mobile services for rural and remote areas poorly covered by terrestrial systems. Eleven 425 kg low-cost satellites in 2,000 km, 0° orbits (plus a spare) would yield continuous coverage between 23° N/south. System cost: less than US$500 million. Later, five satellites (plus a spare) in each of seven 62° planes would extend coverage to 71° N/south. Studies have also been conducted on providing similar services riding on other LEO constellations. In 1998 Orbital Sciences Corporation reached an agreement for the manufacture of 12 satellites, the first of 24, then scheduled for launch in 2001 as part of a revised programme of equatorial satellites. At the time OSC was to have been a major investor, providing up to US$150 million in equity capital and vendor financing. The proposal has failed to achieve market viability and the financing has not been forthcoming.

DirecTV, Inc

Current Status
DirecTV was established by Hughes Electronics Corporation in 1992 to operate a domestic US DTH system using Ku-band HS-601s; it became an independent subsidiary in November 1993. DBS 1 was the 100th Hughes-designed telecom satellite to be launched. A 3,720 m^2 programming uplink facility was added to Hughes Communications Inc's Castle Rock (Colorado) station, working via 13 m antennas, capable of delivering 216 simultaneous channels. DirecTV has an FCC licence to co-locate up to four HS-601 satellites around 101° W offering a total of 27 transponders.

The demise of the Sky Cable consortium of Hughes, Cablevision Systems, NBC and The News Corporation was followed in June 1991 by HCI's creation of a Direct Broadcast Services business unit. Hughes unveiled its DirecTV direct-to-home distribution system at the same time and simultaneously sold five of the 16 transponders aboard the first satellite to the USSB United States Satellite Broadcasting Inc subsidiary of Hubbard Broadcasting for US$100 million plus. The channels are BSS 24, 26, 28, 30, 32. DirecTV owns all 16 on DBS 2. The two satellites will deliver 175 channels. Each transponder can provide four video channels using digital compression, or eight films. The venture requires three million subscribers to

DirecTV can provide 175 channels directly to 45 cm home antennas (Hughes) 0517410

recover its US$750 million cost (one million was achieved as at 2 November 1995).

The current inventory includes DBS I (DirecTV-1) launched December 1993 with 3 × 240 W transponders at 110° west, DBS 2 (DirecTV-2) in August 1994 and DBS 3 (DirecTV-3) in June 1995 each with 8 × 240 W transponders at 101° west, DirecTV1R in October 1999 with 16 × 200 W transponders at 101° west and DirecTV4S in November 2001 with 11 × 200 W at 119 deg west.

VERIFIED

EchoStar Communications Corporation

Current Status

EchoStar Communications Corporation is the parent company of DISH network and currently employs approximately 15,000 people. EchoStar Satellite Corporation signed a contract with the then Martin Marietta Astro Space in October 1992 for one 7000-series satellite, plus options on six others, with launch operations and support for 100 US DTH channels via 45 cm dishes. One million receivers were projected after one year. EchoStar is 100 per cent owned by EchoStar Communications Corporation, which had 1995 revenues of US$163.9 million (1994: US$190.6 million), for a US$11 million loss. The company is programming some of its own capacity and leases some channels on a non-common carrier basis. EchoStar's own 5,580 m^2 programme uplink facility is in Cheyenne, Wyoming.

The second satellite could use one transponder to provide 30 radio broadcast channels. Sister company Sky Highway Radio Corporation filed an FCC application 15 December 1992. EchoStar has an 80 per cent interest in the E-Sat Inc joint venture with DBS Industries Inc to launch six LEO data relay satellites serving mainly electric and gas utilities. Cost is estimated at less than US$70 million.

Directsat merged with EchoStar in January 1994 and has FCC approval to merge their DBS licences for 119° W. Directsat had been allocated 10 BSS channels at 119° W by the FCC in December 1993. EchoStar made the US$52.3 million winning bid at the FCC's 26 January 1996 auction for the 24-channel slot at 148° W. EchoStar filed an application before the FCC's 29 September 1995 deadline to build a Ka-band broadband system to deliver high-rate data and video services via small terminals.

EchoStar currently operates six satellites designated EchoStar 1–6 launched between December 1995 and July 2000. In February 2000 Echostar announced plans for three more satellites to serve DISH network subsidiaries. EchoStar 7 and 8 are advanced DBS satellites while EchoStar 9 will be a hybrid Ku/Ka band satellite for business customers. Launched on 21 February 2002, EchoStar 7operates on GEO from 119° W while EchoStar 8, launched on 21 August 2002 by Proton, operates from 110° W. EchoStar 9 was launched by Zenit-3SL on 7 August 2003 and operates from its planned location at 121° west.

Globalstar Limited Partnership

Current Status

Loral Corporation and Qualcomm Inc in 1991 formed LQSS Loral Qualcomm Satellite Services Inc. Loral owns directly and indirectly approximately 34 per cent of the outstanding equity of Globalstar. The company filed an FCC application in June 1991 for the US$2 billion Globalstar mobile communications system. The licence was awarded 31 January 1995. The name was changed to Globalstar LP in March 1994 to reflect expansion of the partnership: Loral, Qualcomm, AirTouch Communications, Alcatel, Finmeccanica, DACOM, DASA, France Telecom, Hyundai, Space Systems/Loral and Vodafone. SS/L's fixed-price contract for 56 satellites is US$1.3 billion; each will cost US$13.1 million plus US$1.6 million in-orbit incentives.

Globalstar is a worldwide, low-earth orbit satellite services system that will provide low-cost, high-quality telephony and other digital telecommunications services, such as data transmission, paging, facsimile and position location. Globalstar will provide access to existing terrestrial cellular telephone services as it extends, enhances, and operates with public land mobile networks, public switched telephone networks and government and private networks. Users of Globalstar will make or receive calls using hand-held or vehicle-mounted terminals similar to today's cellular phones that will be able to switch from conventional cellular telephony to satellite telephony as required. In remote areas with little or no existing wireline telephony, subscribers will communicate through fixed-site telephones, similar to phone booths or ordinary wireline telephones.

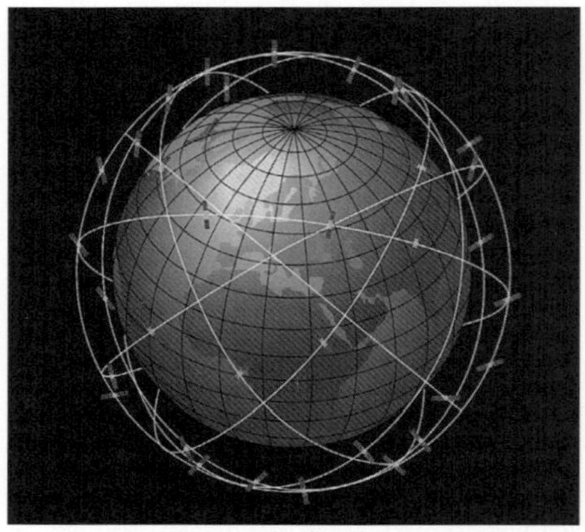

The planned 48-satellite Globalstar constellation, which was to be complete by early 1999 (Globalstar) 0005570

Globalstar's designated service providers, a worldwide network of regional and local telecommunications service providers that includes its strategic partners, will sell access to the service and seek to obtain all necessary local regulatory approvals. Currently, Globalstar has service provider agreements in more than 100 countries, accounting for approximately 88 per cent of the world's population. The service providers - AirTouch Communications, Dacom/Hyundai, France Telecom/Alcatel, Vodafone, Elsag Bailey and Loral - each have the exclusive right to offer Globalstar service in its operating areas. The countries in which Globalstar has most recently established service provider agreements include Brazil, Canada, China, India, Mexico. The first cluster of four Globalstar satellites was launched by Delta 2 in February 1998 followed by a second cluster of four in April 1998. Additional Globalstar clusters were launched by Soyuz-U/Ikar. Twelve satellites were lost in launch vehicle failure.

On 8 February 2000 a Delta 2 launch vehicle placed two Globalstar satellites in orbit, the last of 52 including four spares. Full commercial service was launched in the USA on 29 February 2000. By mid-January 2001 with fewer than 32,000 subscribers Globalstar suspended debt repayment to save money and maintain operations. Delta restructuring was announced in February 2002 as the company sought protection from creditors by filing for Chapter 11 in the US Bankruptcy Court. First Quarter 2002 service revenues were up 75 per cent over the previous quarter with 69,000 subscribers and a reported net loss at US$129 million, down 30 per cent. Globalstar now claims to be the most widely used handheld satellite telephone service with expansion into Europe and the Middle East during 2003.

Iridium Satellite LLC

Current Status

Commercial service commenced in March 2001 following a contract from the US Department of Defence which rescued the system from Motorola. It gives unlimited air time to 20,000 government users for mobile voice, data and paging. The original US$72 million two-year contract has been renewed. The constellation consists of 66 operational satellites and 14 spares operating from six polar planes, each plane hosting 11 mission satellites at an altitude of 780 km.

Motorola unveiled its Iridium mobile communications system proposal in June 1990 after 2½ years of internal work. Iridium planned a system of 66 small LEO satellites by end-1998 providing global pocket mobile digital 4.8 kbit/s telephone services via onboard switching and inter-satellite links for US$3 per minute to the user. Handsets will retail at about US$3,000. Cost was projected at US$3.45 billion for establishing the operational system of 66 satellites and 14 in-orbit spares by 1998, followed by another US$2.8 billion over the next five years for operations and maintenance. 800,000 users are required within five years for viability. Motorola projects 1.824 million by 2001, broken down as 42 per cent pager, 42 per cent portable phone and 7 per cent mobile. After 10 years, in 2006, it projects 3.224 million. Viability is driven by the terminal.

Iridium, Inc was incorporated 14 June 1991 to own and operate the system, with Motorola eventually holding 15 per cent. Owner organisations are: Motorola, Lockheed Martin, Iridium Africa Corporation, Iridium Canada Inc, Iridium India Telecom Private Ltd, Iridium Middle East Corporation, Iridium SudAmerica Corporation, Nippon Iridium Corporation, Korea Mobile Telecommunications Corporation, Raytheon, Sprint Corporation, STET (Italy), Khrunichev (Russia), China Great Wall Industry Corporation, Pacific Electric Wire & Cable Co (Taiwan), Thai Satellite Telecommunications Corporation and Vebacom GmbH (Germany). A total of US$1.888 billion was raised by early 1996; the remainder will come from debt financing.

The proposal was submitted to the FCC 3 December 1990 and an experimental licence was awarded August 1992 for five satellites. The full licence was awarded 31 January 1995. Launches were to have begun by end-1996. Lockheed Space Systems Division (now Lockheed Martin) was awarded the US$700 million contract in August 1993 for 120 spacecraft by 2003. 79 are required for the initial system; the remaining 41 will be launched as required.The bus is built in Nashua, New Hampshire at Lockheed Martin Sanders and integrated with Motorola's communications module in Motorola's final assembly facility in Chandler. The production line system there can handle five satellites simultaneously; the goal is for final assembly to require 21 days. Satellite CDR was held June 1992. The power and control systems and structure are similar to Lockheed's venerable Agena upper stage. Volume production allows use of GaAs solar arrays. In February 1993, Motorola announced that it had signed a contract (subject to US government approval) with Khrunichev Enterprises for three Proton launches of seven satellites each 1996 to 98; Khrunichev invested US$40 million in return. A similar agreement was signed with GWIC April 1993 for 10 launches of two satellites each 1996 to 2002 on CZ-2C. Eight Delta 2s were ordered to handle five satellites each. For replacements, five Delta 7420 launches were ordered December 1995, each handling three satellites. Each satellite will de-orbit at EOL to avoid pollution.

Each satellite operates 48 L-band cells 670 km across using FDMA/TDMA. Calls will be routed to the Earth gateways and among the space segment at Ka-band. The user's 0.7 kg 0.6 W handset will automatically select a terrestrial cellular network if available and cheaper. The network will maintain a log of each handset's last known position: each device will determine its location via GPS or Doppler shift and relay it to a visible satellite for passing on to the nearest gateway. Each gateway will comprise three terminals dispersed sufficiently to avoid weather problems affecting links. Scientific-Atlanta is providing 14 terminals under

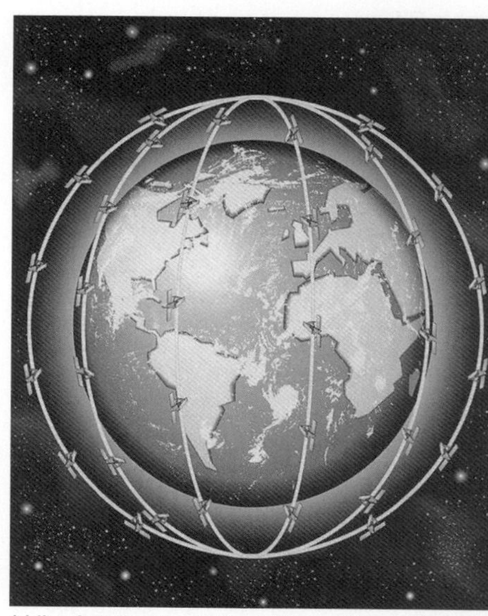

Iridium's network of 66 satellites will provide global pocket mobile telephone services via the L/S-band phased array antennas. Iridium is the 77th element, reflecting the original constellation size 0517224

US$22.6 million contracts for the System Control Segment.

The first cluster of five Iridium satellites was launched by Delta 2 in May 1997 and it took less than two years to orbit 66 operational satellites for the prime constellation. Seven were lost in launch vehicle failures but by February 1999 there were a total of 79 satellites in orbit of which 13 were spares. Predicted subscriber levels were far higher than actually achieved, however, and by 1 April 1999, only 10,294 subscribers had signed up to the Iridium system, far short of the 500,000–600,000 needed to begin paying debt on the US$4.85 billion programme. In the first three months of 1999 the company announced revenue of US$1.4 million and a net loss of US$505 million bringing down stock from US$72 in 1998 to US$15 by late April 1999, casting doubts on the viability of high-end satellite telephone services. Analysts did, however, point to the high pricing of Iridium, which (with 1.5 billion minutes of capacity per year) has to charge US$3–8 a minute for calls. It needs to recoup at the rate of US$1.4 a minute just to break even. Competing systems such as Globalstar, with 7 billion minutes of capacity, can wholesale at 21 cents while ICO, with 5 billion minutes, can do it for 23 cents. Iridium's telephone handset price of US$2,700 was set to plummet during 1999, a year in which the company still expected to sign up a total 500,000 subscribers. On 13 August 1999, Iridum filed for chapter II bankruptcy protection. In the same month the company filed a bankruptcy petition.

In November 2000 the US Bankruptcy Court approved the sale of Iridium assets to Iridium Satellite LLC, which has contracted with The Boeing Company to operate and maintain the satellite constellation. On 6 December 2000 the US Department of Defence announced that it had agreed to contract Iridium Satellite LLC for 24 months of satellite communication services, a contract worth $72 million, with options which, if exercised, would increase the value to US$252 million securing services through 2007. Iridium LLC provides remote communications solutions for heavy construction, defence and government agencies, emergency services, maritime operators, mining operators, forestry agencies, the oil and gas industry, aviation (commercial, corporate and leisure) and leisure and adventure pursuits.

As of December 2000, Iridium Satellite LLC executives believed 30,000 people still owned an Iridium telephone with about 7,000 continuing to use the service.

Loral Space and Communications Ltd

Current Status

Loral Space & Communications, with headquarters in New York City, was spun off from Loral Corporation in April 1996 as part of the sale of its defense electronics and systems integration businesses to Lockheed Martin Corporation.

On a pro forma basis, combining the results of the operations under its management, the company's revenue for what was a nine-month fiscal year, ending 31 December 1996, was US$950 million. Effective 1 January 1997, Loral reported on a 12-month calendar year basis.

In March 1997, Loral completed its acquisition of Skynet, a leading domestic satellite service provider, from AT&T. Loral Skynet advances Loral's strategy of becoming a global provider of satellite-based communications services and will complement its existing satellite manufacturing capabilities. The Telstar fleet lies at the heart of the Loral Skynet system of satellites and comprises Telstar 4, 5, 6, 7, 8, 10, 11, 12 and 13 in addition to Satmex 5, Satmex 6, Solidaridad 2, Europe Star 1, Brasil 1 and Estela do Sul 1.

Loral Skynet owns 60 major US Earth stations and operates a US$100 million satellite monitor and control facility in Hawley, Pennsylvania.

A major subsidiary of Loral Space & Communications, Space Systems/Loral has a strong pedigree of more than 40 years standing beginning in 1957 as the Western Development Laboratories division of Philco. As a division of Ford Aerospace, Philco was acquired by the Loral Corporates in October 1990 and renamed as the satellite portion of SS/L.

Other business units include Loral CyberStar and Satélites Mexicanos.

MCI/News Corporation

Current Status

MCI Communications Corporation acquired the rights to the 110° W DBS slot at the FCC's 25 January 1996 auction with a bid of US$682.5 million. MCI, America's second largest long-distance phone carrier and News Corporation will use the slot in an equal partnership to deliver more than 100 DTH and data channels via 45 cm dishes. The partners awarded Space Systems and Loral a US$400 million contract in March 1996 for the on-orbit delivery of two satellites, each with 32 Ku-band transponders: fourth quarter 1997 on Proton and second quarter 1998 on Atlas 2AS. In November 1998 News Corporation and MCI Worldcomm transferred to EchoStar Communications Corporation a licence for 28 DBS frequency channels at 110° W. Following the merger between MCI and Worldcom plans were shelved but options retained. Restructured development plans has been proposed and market opportunities assessed pending further decisions. Several optional proposals have been put forward and rejected. No definitive agreement has been reached.

Orbcomm LLC

Current Status

Orbcomm now has 30 satellites in LEO providing narrow-band two-way digital messaging, data communications and geo-positioning services. In December 2002 Orbcomm completed a deal with AZERTIA to provide data communication services for monitoring motor vehicles in Europe and the Americas.

OSC's Orbital Communications Corporation subsidiary, created at the beginning of 1990, filed an FCC application 28 February 1990 for a constellation of LEO satellites for two-way communications and geolocation services. Orbcomm will provide low-cost alphanumeric data communications and position determination (375 m) for emergency assistance, data acquisition and messaging services using pocket portable and mobile subscriber terminals. The system is optimised for shorter messages of 6 to 250 characters. OSC and Canada's Teleglobe Inc signed an agreement 26 July 1993 for the joint financing and operation of the system. Teleglobe provided US$10 million of the US$55 million cost of the two first phase satellites. For the second phase of 34 more (including eight ground spare), it has increased its equity share in Orbcomm Global LP to 50 per cent. Teleglobe is contributing US$85 million and OSC US$75 million.

The system comprised 28 43 kg satellites by 1998 at 775 km altitude; 24 at 45° inclination (eight in each of three planes) and two at 70° for high-latitude coverage. Total cost is projected at US$220 million, with four launches aboard OSC's own Pegasus and one on Taurus, plus a set of eight ground spares intended for Pegasus. The first two satellites, in polar orbit, appeared successfully 3 April 1995. Both soon developed receiver problems. Number 2's problems in this respect were solved by a self-activated reset 13 May (subsequent satellites carry a separate receiver to allow reset by ground command). Number 1's subscriber receiver was freed 18 May by sending commands correcting an apparent software blockage; the subscriber transmitter was restored 13 June by commanding battery drain, which triggered a computer reset. Number 2 relayed the system's inaugural message 25 May to the Arcade gateway from a production prototype Panasonic personal communicator at Dulles. Both completed acceptance tests 7 July 1995. A message from an Elisra EL-2000 Data Communicator near the Dead Sea in Israel was successfully routed 25 July 1995 to Dulles via the Georgia gateway. The satellites began providing an initial service 1 February 1996, by which time approximately 300,000 messages had already been sent.

Each satellite carries a GPS receiver for the 375 m position-fixing service using Doppler. OSC's own Magellan Systems Corporation will build hybrid GPS/Orbcomm communicator units from late 1996.

Orbcomm was granted an experimental licence April 1992 for two satellites, plus 1,000 user terminals, to demonstrate the system and initiate revenue services. The full licence was granted October 1994, the first for an LEO mobile constellation. The first two satellites provide the polar coverage for the operational system. The 14.5 kg CDS Capabilities Demonstration Satellite was launched 9 February 1993 into 750 km 25° by Pegasus as a pathfinder to demonstrate services via prototype hand-held communicators built by various electronics manufacturers. Planned for a six-month life, it operated for two years.

Orbcomm seeks initially to serve the US market, where it foresees a total of 10 to 20 million users, beginning with 150,000 the first full year of constellation operations and rising to 5.2 million after seven years. Emergency services are expected to provide 75 to 85 per cent of the market. The company does not seek to offer cross-border services, but it is licensing affiliated companies and PTTs in other countries. Three grades of terminals are envisaged: US$100, 5 W power, signalling and data transmit only; US$150 adding position-fixing; US$400 alphanumeric messaging. The first type requires a US$30 to US$50 annual service charge, with the others supported by US$35 monthly plus usage-linked charges.

Each satellite carries 17 data processors and seven antennas, allowing the relay of 50,000 messages hourly in each direction, or more than five million two-way daily for the 28-satellite system. Four regional gateways will cover the US: Oscilla Georgia, Arcade New York, St Johns Arizona and Washington State, with the Network Control Center in Dulles Virginia backed up in Boulder Colorado. The orbits provide 70° N/south coverage with 5,000 km diameter footprint, allowing users to access the system for more than 95 per cent of the time. Polar region accessibility will be 14 minutes every half hour.

The first launch of the operational Orbcomm satellites took place in December 1997 when a cluster of eight was carried into orbit by a Pegasus XL. By February 1999, 24 satellites had been placed in orbit and the system had three sets of eight satellites in each orbit plane, thus completing the first operational phase.

In January 2000 OSC and Teleglobe announced plans to restructure the partnership and increase equity by $100 million. The full constellation is now expected to comprise 38 satellites. In August 2001, Orbcomm announced that Garmin International Inc will make satellite-based weather available on select Garmin avionics via the Orbcomm system. Volvo Trucks selected Orbcomm to provide two-way communications for a tracking and monitoring solution developed by Volvo Trucks North America for use on heavy vehicles. GE TIP uses Orbcomm for asset tracking its trailers for location of tethered or untethered trailers anywhere in the USA.

PanAmSat Corporation

Current Status

PanAmSat is 81 per cent owned by Hughes Electronics Corporation. The most recent satellites launched were Galaxy 12 on 8 April 2003 and Galaxy 13/Horizons 1 on 1 October 2003. Galaxy 13 is the C-band component of the payload and Horizons 1 is the Ku-band. Satellites are currently operating at 70 per cent of capacity. On 16 May 1997, the merger between PanAmSat Corporation and the Galaxy Satellite Services was completed: the newly formed company would retain the name PanAmSat Corporation. The merger had been announced on 20 September 1996 by Hughes Electronics Corporation and PanAmSat Corporation, with the PanAmSat stockholders receiving an aggregate of US$1.5 billion in cash and 28.5 per cent of the new company.

The original PanAmSat had been established in 1984; 1995's revenue was US$116.2 million (1994: US$63.7 million). The company was granted an FCC licence in 1985 and took over the partially-completed ASC 3 satellite, surplus to Contel ASC requirements, to provide the first private international satellite communication links outside of Intelsat via PanAmSat's Miami International Gateway at Homestead, south of the city. PAS 1 was launched by the maiden Ariane 4 demonstration mission at a reduced fee of US$8 million; insurance coverage was US$40 million at a 25 per cent premium. Prevented by the FCC from connecting with public switched networks, the service is intended primarily for TV programming of US or Mexican origin, and business data, video and voice traffic. PAS 2 appeared in July 1994 to serve the Asia-Pacific region, and by November 1995 had 17 full-time programmers. Hundreds of customers worldwide use the first two satellites; broadcast services represent 67 per cent of revenue. There are also teleports in Napa California, Homestead Florida and Atlanta Georgia.

PanAmSat is the first private sector operator of a global system, with more than 20 satellites over the Pacific, Atlantic and Indian Ocean regions. Hughes was selected in 1991 for a contract worth about US$300 million to provide four of its HS-601 models for PAS 2 to 4; the fourth was to be held as a ground spare. Each carries cross-strapped C and Ku-band 54 MHz transponders optimised for TV programmers and corporate telecommunications. The launch contract went to Ariane in June 1992, in return for which Credit Lyonnais arranged French export loans covering 85 per cent of the US$240 million launch costs. To finance the US$700 million expansion, 50 per cent of PanAmSat's equity was acquired by Mexican media company Televisa SA in December 1992 for US$200 million. US$440 million was raised from a bond sale, completed August 1993. The 2,985 kg PAS 3, intended for 43° W to provide DTH services to Latin America from June 1995, was lost in Ariane's 1 December 1994 failure. Insurance was reportedly US$214 million at 17.5 per cent. The ground spare was thus converted for launch as PAS 3R to begin the Latin America service in first quarter 1996. PAS 3R was launched by Ariane 44L in January 1996. PAS 5 was ordered from Hughes Space & Communications in March 1995 and PAS 6 from Space Systems/Loral in August 1994 to expand the DTH Latin America capacity to approximately 120 digital channels per customer (approximately 500 throughout Latin America). PAS 5 was launched by Proton from Baikonur in August 1997, PAS 6 on an Ariane 44P in the same month followed by Galaxy 8-I in December that year. US$262 million was raised April 1995 in a preferred stock offering to finance PAS 5/6. PAS 7/8 were ordered from SS/L in April 1996. FCC filings were made in September 1995 for the PAS 10/11 Atlantic Ka-band satellites, costing US$409 million through the first year.

In 1998, PanAmSat announced its intention to double its worldwide transmission capabilities by the end of 1999 with the launch of nine additional satellites, launches in 1998 included (in order) Galaxy 10, PAS 7, PAS 8 and PAS 6B.

By September 2003, PanAmSat had 22 satellites in orbit (SBS 6; Galaxy IR, Galaxy 3C; Galaxy 3R; Galaxy 4R, Galaxy 5, Galaxy 8I; Galaxy 9, Galaxy 10R; Galaxy 11; Galaxy 12; Galaxy 13/Horizons 1; PAS 1R; PAS 2; PAS 3; PAS 4; PAS 6; PAS 6B; PAS 7; PAS 8; PAS 9; PAS 10). Of the total, eight were covering the Atlantic Ocean Region, three the Indian Ocean Region, two the Pacific Ocean Region and 11 the USA.

PanAmSat increased its capacity to more than 900 transponders with the launch of Galaxy 3C in June 2002 and PAS 10 in March 2001. PanAmSat has more than 800 employees, assets totalling US$6.1 billion and a 2001 revenue of US$870.1 million. Total revenues for 2002 were US$812.3 million. Revenues for the first three quarters of 2003 are reported as US$613.4 million compared to US$615.5 million for the first nine months of 2002.

UPDATED

SES Americom

Current Status

SES Americom was combined with SES Astra in November 2001 forming a new global satellite company, SES Global S.A. The new company has a fleet of 17 satellites. Formerly known as GE Americom Communications Inc and RCA Americom before that, SES Americom operates AAP-1 (formerly GE-1A), AMC-1 (GE-1), AMC-2 (GE-2), AMC-3 (GE-3), AMC-4 (GE-4), AMC-5 (GE-5), AMC-6 (GE-6), AMC-7 (GE-7), AMC-8 (GE-8), AMC-9 Gstar 4 (formerly GE Gstar 4), Satcom C1 (GE Satcom C1), Satcom C3 (GE Satcom C3), Satcom C4 (GE Satcom C4), Sirius 2 (GE - 1E), Spacenet 4 (GE Spacenet 4), TDRS-5 and TDRS-6. Transoceanic connectivity is enabled by the recently acquired capacity of Columbia Communications. Yet to be launched are AMC-10 to AMC-15 all of which are scheduled for launch in 2004.

RCA Americom was created in 1976 as a wholly owned RCA subsidiary to own and operate the parent company's domestic system. Following GE's 1987 acquisition, GE Americom began operating as part of GE's Communications & Services Group.

Space Systems/Loral

Current Status

Space Systems/Loral is a subsidiary of Loral Space and Communications and its satellites have delivered a total of more than 1,000 years on on-orbit service, the majority of that time from tri-axis stabilised 1300 series satellites. Space Systems/Loral (SS/L) was formed out of the Loral Corporation Space Division when the parent company took over Ford Aerospace Corporation in October 1990 in an acquisition worth US$715 million. Since then, Space Systems/Loral has become the world's fastest growing commercial producer of communications and weather satellites and key subsystems by answering customers' needs for rapid delivery of reliable, state-of-the-art spacecraft. The company is a full service manufacturer that develops, produces and delivers satellites, payloads and complete satellite systems, while offering risk management, insurance, mission control and launch services around the world. Global markets range from broadcast and cable television to cellular mobile communications and direct-to-home TV, on to video conferencing, environmental monitoring, educational and scientific data networks and global positioning.

Employees total 3,500. Loral Space & Communications Ltd, SS/L's parent company, made a strategic decision in 1996 to increase its ownership in SS/L to 100 per cent by acquiring the 18.3 per cent interest held by the Lehman Partnerships in August 1996, and in February 1997, by acquiring the remaining 49 per cent minority interest held by SS/L's four European aerospace partners - Aerospatiale, Alcatel, DASA and Finmeccanica. SS/L maintains a close working relationship in the marketing and manufacture of satellites with these four companies.

SS/L offers satellites based on its standard three-axis spacecraft. By early 1997, the number of satellites in this family of spacecraft that have been built or ordered totalled 59, including 22 satellites in production. The company has supplied this type of spacecraft to many significant customers including Intelsat, Nippon Telegraph & Telephone, NASA, TCI, PanAmSat, the Japanese Ministry of Transport, APT Satellite, Skynet, Mabuhay Philippines Satellite, MCI, Asia Broadcasting and Communications Network, Pasifik Satelit Nusantara and Chinasat. In February 1997, the satellites built by SS/L accumulated more than 600 years of service in space. This landmark of longevity and reliability represents the combined success of 82 communications and weather satellites built during the past three and a half decades.

Loral applied to the FCC in May 1995 for the CyberStar satellite to cover North America from 1998. This broadband data communications satellite would use Ka-band to provide business and consumer users with high-quality video and high-speed data services via small, low-cost antennas. Later in 1995, Loral amended its application, increasing the number of satellites for worldwide coverage. Loral also received orbital assignments from the FCC for this international coverage. Space Systems/Loral CyberStar provides access to US-based internet content to more than 173 ISPs in more than 32 countries.

Space Systems/Loral is the prime contractor for the space segment for the Globalstar 48 low-earth orbit satellite-based system. The digital telecommunications system that SS/L and its partners are constructing is designed to enable low-cost, high-quality wireless voice telephony and data services such as paging, facsimile and position location in virtually every populated area of the world, including areas currently underserved or not served by existing wireline and cellular telecommunications systems.

Space Systems/Loral and Russia's OKB Fakel in 1992 established the International Space Technology, Inc. (ISTI) joint venture to market electric thrusters. Using technology licensed by Trimble, SS/L is developing the GPS *Tensor* TM solid-state, lightweight GPS spacecraft attitude sensor. It is a spin-off of work performed with Stanford University researchers on NASA's Gravity Probe B. SS/L is subcontracted by Rocketdyne Division of Rockwell International Corporation to build Space Station Alpha's Ni/H_2 battery system.

The company features the FS-1300 platform and since 1957 SS/L and its predecessors have built, or been awarded contracts for, more than 200 satellites with 134 launched by September 2002 amassing more than 800 years of cumulative in-orbit time. The FS-1300 is in a continuous state of evolution and can provide satellite power from 5 kW to 12 kW electrical energy or, in expanded variants, from 12 kW to 18 kW. The mass of the largest option is almost 6,700 kg. SS/L is offering the 20.20™, providing greater communications capacity by integrating new technologies into proven elements of the 1300 platform. Total satellite power ranges from 17–30 kW and the platform can accommodate up to 150 transponders with a total launch mass of 8,500 kg.

FS-1300 platform
Total ordered: 50 (including Superbird, GOES, Intelsat VII, N-Star, Tempo, PanAmSat, MTSAT, APStar, TELSTAR, Mabuhay, Sky, L-Star, M²A. Chinasat)
Total launched: 27 (including Superbird, GOES, Intelsat VII/VIIA, N-Star, Tempo)
Total currently operating: 23
Launch mass: up to 6,200 kg, depending on *payload:* EOL mass up to 2,900 kg
Deployed dimensions: 25.7 m solar array span, 7.5 m high
Availability: production cycle 18 months
Platform reliability: better than 0.875 EOL
Communications payload power: up to 19,000 W, depending on payload mass
In-orbit lifetime: designed for more than 15 years
Power: 100% eclipse capability
Pointing: antenna pointing accuracy is less than 0.10°
Attitude control: a microprocessor-based momentum bias system controls the satellite and auxiliary functions. The processor can be pre-programmed with more than a month of manoeuvres; execution is by either automatic or manual control
Propulsion: a single-integrated bipropellant subsystem provides both apogee (400 N) and on-station manoeuvring
Thermal: a passive thermal subsystem includes heat pipes, and heaters for some components. Large N/S surfaces covered with internal heat pipes allow maximum payload mounting flexibility
Telemetry, Tracking and Control: custom capability tailored to each programme in any frequencyband.

VERIFIED

Tempo DBS, Inc

Current Status

Tempo awarded Space Systems/Loral a US$400 million contract July 1994 for two high-power DTH satellites and their launches. Tempo is a subsidiary of Tele-Communications, Inc, the largest cable TV operator in the US. The payload carries 34 Ku-band transponders and is switchable to 200 W transponders, optimised for 120 to 150 digitally compressed channels to home dishes. PrimeStar, owned by six cable TV companies (including 22 per cent by TCI) and GE Americom, lease capacity on the satellites, although operation remains with Tempo. It already provided DTH programming on Americom's FSS Satcom K-1. Tempo filed for launch on an Atlas 2A, which took place on 8 March 1997 but the satellite suffered from power fluctuations and was not accepted by the owners. Insurance paid US$21 million. A replacement Tempo satellite was launched 7 May 2002 on a Proton launch vehicle. The satellite is now operated by DirecTV as DirecTV-5. Tempo has let a request for bids on the development of a successor for the DirecTV-5 satellite.

UPDATED

WorldSpace, Inc

Current Status

WorldSpace was founded in 1990 with audio services to Africa and the Middle East beginning in October 1999 via AfriStar. It's initial function was to use direct audio broadcasting to help stop the spread of AIDS in Africa but its function today is to provide a wider range of services to the people of Africa and Asia.

WorldSpace's subsidiary AfriSpace filed an FCC request in July 1990 to launch a single AfriStar satellite to provide digital audio broadcasting to the African-Arabian region. The licence was awarded 21 June 1991. AfriStar was launched by Ariane 44L on 28 August 1998 and has a lifetime of 15 years. The plan is for three L-band satellites, each beaming up to 288 radio channels to Africa, Middle East, Asia, the Caribbean and South and Central America. The CaribSpace licence was granted by Trinidad & Tobago in December 1992 for AmeriStar, and in September 1995 Australia notified the AsiaStar satellite for AsiaSpace. A contract was signed 21 January 1995 with Alcatel Espace to build, launch and insure the three satellites (plus a ground spare) for US$600 million. Alcatel Espace signed the Ariane launch contract 18 March 1996. More than US$750 million financing was secured, with investment banker Morgan Stanley of New York serving as financial advisor, by a group of international investors.

AsiaStar was launched by Ariane 5 on 20 March 2000. The launch of AmeriStar has been postponed to a future date. Each satellite is capable of transmitting 40 audio channels.

The WorldSpace receiver, cost US$150 to US$200 initially, will have several stationary and mobile models, including solar powered and serial data ports for faxcasting. Sales are projected at 20 million annually within five years of the full service. The radio chipset will also be integrated onto computer interface cards for data downloads and audio programmes via PC.

EARTH OBSERVATION FACILITIES AND EQUIPMENT

Argentina

Aeroterra SA

Current Status
Aeroterra has been involved in remote sensing since 1973 for the evaluation, cataloguing and development of natural resources. Thematics maps drawing on Landsat, Skylab and Spot data cover approximately 7,000,000 km^2 of Argentina, Bolivia, Brazil, Paraguay and Uruguay. Images contributed to a major environmental survey which contributed to the release of UN funds for regional development programmes. It is a representative of EOSAT, ESRI and ERDAS and provides systems and services in GIS, image processing, consulting, training and education. Its Training Centre opened in 1992 for remote sensing and GIS courses. Aeroterra has exclusive agreements to distribute Radarsat data in Argentina and Uruguay. Integrated software packages procured for interleaving with natural resource modelling. Co-operative agreements with other South American companies have been sought in attempts to provide a larger base of data including SAR as well as optical imaging. Several private companies have made use of Aeroterra data for managing regulated revenue analysis surveys. Funding has been significantly reduced for 2002–2005. The research activities have been completely re-scoped in favour of existing deals and no funds have been available for deals which would have significantly increased the capabilities of Aeroterra.

UPDATED

Australia

Earth Resource Mapping Pty Ltd (ERM)

Current Status
ERM specialises in integrated mapping software. Its latest ER Mapper is designed principally to register, process and merge remotely sensed geoscience data along with geographic and other grid overlay information of the user's choice. Development was backed by Sun Microsystems and Hitachi Data Systems as a standard package requiring no customising or programming knowledge. It can handle complete datasets such as full Landsat TM and Spot scenes, and large magnetic datasets. The latest release includes 3-D visualisation and hardcopy production and direct editing of ARC/INFO coverages.

ER Mapper hosts a list server for geophysical and mapping specialists around the world and markets a wide range of products for amateurs and scientists. It also processes radar from satellite and interpolates this with optical imagery from a variety of satellite sources. ERM has integrated pan-Asian environmental studies data into a new and unique signature base for controlling earth resource and environmental systems data used as a control standard for earth resource studies. Mapping software has applications for environmental pollution and aberrant marine and air toxins monitoring through integrated data from a wide variety of sources. This can also be used for extrapolating threat patterns for living organisms.

VERIFIED

Geoimage Pty Ltd

Current Status
An independent, privately owned Australian image-processing and remote sensing consultancy with offices in Brisbane, Darwin and Perth, Australia.

Geoimage specialises in the sales and processing of remote sensing imagery and other raster data for the mineral and petroleum exploration industries.

In addition, Geoimage undertake land use, land cover and vegetation interpretation of remotely sensed data and GIS integration. Geoimage is an authorised reseller for satellite data including LANDSAT, SPOT, IKONOS and RADARSAT data. In addition, Geoimage have a number of regional image mosaics at various resolutions over a number of international regions.

VERIFIED

Moonraker Technology Pty Ltd

Current Status
MRT offers products/services for oceanographic and environmental science researchers, including the MRT-112 drifting buoy for use with the Argos satellite data relay system. Used extensively for tracking whales and large marine mammals. The streamlined cylindrical hull fully encloses the device, apart from the antenna, and provides standard sea surface temperature, battery level and drogue loss indication sensors, with air temperature and atmospheric pressure sensors as options. Activation is effected by removal of a magnet; the MRT-201 hand-held set confirms operation before deployment. Battery packs for 6- and 12-month lives. Other products include marine animal trackers rated to 1,500 m depth. MRT has undertaken design work on integrated data collection and distribution system incorporating platforms on land and at sea. Advanced store and forward dump facility for satellite surveillance and monitoring. MRT has upgraded for adverse weather and sea state conditions.

VERIFIED

Spot Imaging Services Pty Ltd (SIS)

Current Status
SIS was formed in 1988 as a subsidiary of France's Spot Image to market Spot data in Australia and the South Pacific region. Revenues for 1996 were approximately US$900,000. Derived products based on Spot images, such as Digital Terrain Models, can be provided in addition to other services such as image processing for geological, cartographic and civil engineering applications. Regional imagery is received directly at the Australian Centre for Remote Sensing. The company also offers other remote sensing data and serves as a distribution centre for Earth data analysis and interpretation. Spot Imaging Services also market interpretation analyses of Spot data integrated with other data from commercial remote sensing satellites by request from customers.

VERIFIED

Austria

Geospace

Current Status
Geospace is the authorised distributor of Spot Image, Landsat, IRS, ERS, NOAA, Meteosat and products supplied through Eurimage, EOSAT, EUROMAP and Radarsat data for Austria. It also represents the ESA Earthnet National Point of Contact and distributes ERS data. It provides consulting services, evaluation of remote sensing data and satellite-based cartography and is equipped for all image processing and production of colour transparencies from processed data on colour laser plotter. The company has published atlases of Austria and Germany based on Landsat TM and Spot imagery at 1:200,000 and 1:220,000, respectively. It is leading the ISY Global Satellite Image Mapping Project and Geospace has extended its products to include Envisat data from the ESA remote sensing satellite. Geospace has also negotiated archive access for customers integrating radar imagery. Geospace works with the Austrian Remote Sensing Data Centre to advise on data utilisation including products from Russia.

VERIFIED

Belgium

CLEO

Current Status
The Cluster for Earth Observation (CLEO) was created in 1995 to promote the value-added capabilities of Alcatel Bell, SAIT Systems and Trasys. Trasys Space is responsible for management operations at the image-processing and archiving centre for Spot 4's Vegetation instrument. CLEO conducted detailed analysis of land-use mapping and charge state evaluation. This data was delivered to field users for applications in the next generation of Earth resource satellites. Trasys signed a Radarsat data distribution agreement in April 1996 and has since extended that to include data from ESA environmental satellites. Data from US and Russian satellites are integrated into general environmental data on demand or for specific user needs. Selected applications can now be administered direct from user base interfaces and used to promote the multisource applications for independent analysis. There has been a successful effort at integrating data sets and in providing new ways in which users and customers can receive tailored interpretation and analysis reports.

UPDATED

Eurosense

Current Status
Satellite remote sensing activities include digital image treatment and interpretation for Earth sciences, forestry, agriculture and environment, in addition to thematic map production. Eurosense specialise in remote sensing, photogrammetry, GIS, cartography and hydrography. It also provides aerial photography, laser altimetry and multispectral/TIR scanning. Eurosense has integrated air and space derived imagery

for optimised data interpretation. Customer tailored analyses are available utilising these assets. Eurosense was the first company to produce digital colour orthophotomaps through its unique Eudicort® system. Provides detailed image and data retrieval across a wide range of products and can build image convergence from several separate products for specific user needs. Eurosense has several unique asset enhancement programmes and is focused on photomap products with a variety of plug-in applications.

UPDATED

Canada

Array Systems Computing Inc (ASC)

Current Status
ASC specialises in remote sensing, defence and security software development and systems integration. In remote sensing, it focuses on interpreting and processing captured satellite data in real time.

Radarsat Fastscan Processor, FASTSCAN, processes real-time data and produces real-time imagery for viewing on the Moving Window Display. It supports full resolution ScanSAR processing, fast Data Quality Reports and Catalogues, comprehensive data products and a Raw Data Interface. Radarsat Image Data Calibration Workstations provide operational radiometric and geometric calibration and verification methodologies, to assess and maintain end-to-end SAR image quality. Meteorological Applications Processing System (MAPS), ground stations receive, process and distribute HRPT data from polar and GEO satellites. It combines real-time data reception with on-the-fly image production. Automatic or on-demand data distribution can be made to remote sites.

ASC has specialised in SAR interpretation with innovative computerised data handling now involving European, Russian and Japanese satellites. Adapted to future requirements for advanced SAR missions, the SAR interpretation methods developed for these existing missions have eased the process of data extraction and analysis and interpretation for the next generation of radarsats anticipated for the coming decade.

ASC has developed the Hydros SAR processor for the Canadian Space Agency. Hydros is a co-operative venture between the Massachusetts Institute of Technology, NASA's Jet Propulsion Laboratory, the Goddard Space Flight Center and the Canadian Space Agency. The satellite instrument package includes radar and radiometer at work in a low frequency microwave range via a rotating antenna to measure soil moisture content. Hydros will be launched to an altitude of 670 km in December 2009. The Hydros satellite will make the first global measurement of soil moisture and freeze/thaw fields.

UPDATED

Barringer Research Ltd

Current Status
Development of techniques and instrumentation for remote sensing and pollution monitoring. These include GASPEC, a gas filter spectrometer used by two NASA programmes on Space Shuttle, and laser fluorosensors for oceanography and maritime pollution measurements. Barringer also produces field sensors for monitoring pollutants and toxins, equipment which has been applied to the detection of harmful substances used by terrorists. Barringer has provided instrument designs to manufacturers of coastal pollution monitoring equipment receiving signals from remote sensing satellites. Development of multisensor platform receivers has begun with a view to producing monitors capable of receiving data from land, sea and air sensors. Convergent radar and optical imagery through sensor packages. Barringer Research Ltd is a part of the Smiths Detection group that also includes Graseby Dynamics and the Environmental Technologies Group (ETG).

UPDATED

The Bercha Group

Current Status
Set up in 1975, Bercha Group of companies actively competes in engineering, risk analysis and geomatics and specialises in the acquisition, processing and analysis of remote sensing data, including satellite and airborne data. Key companies in the Bercha Group include Bercha Engineering Ltd, Bercha International Inc and Bercha (Malaysia) Sdn. Bhd. Capabilities include digital elevation modelling and specialised SAR processing. Registered Spot distributor. Provides integration of optical and digital imagery and interpretation of geomorphological formations according to indices. Also distributes Radarsat interpretation tools and is integrating Radarsat and Envisat data for correlation with US and Russian images.

The Group also specialises in consulting, engineering, risk analysis and surveillance. Key companies include Bercha Engineering Ltd, Bercha International Inc and Bercha (Malaysia) Sdn. Bhd.

UPDATED

Dendron Resource Surveys Inc

Current Status
Dendron is a forest resources consulting company specialising in the application of remote sensing technologies, including airborne and spaceborne image analysis. Conducted analysis of regional surveys utilising mixed air and space derived data sets integrated with ground truth data for land-based correlation, also with radar imagery input. The company provides GIS services for the input, manipulation and output of large sets of geographically referenced data. Dendron has specialised in cost effective management solutions for forest management and land usage using aerial and satellite data. DRS has extended this work into the integration of optical and radar imaging with management models for cultivation and inventory projects. Specialist areas now include forest inventorying and forest management, mapping, geomatics (for forestry) and environmental management. Dendron has produced forest inventory maps for areas in Alberta, British Colombia, Maine, New Brunswick, New Hampshire, Northwest Territories, Ontario, Prince Edward island (Canada and New York State), Quebec and Yukon.

UPDATED

Eidetic Digital Imaging Ltd

Current Status
The company was established in 1985 to develop modestly priced, stand-alone, PC-based, digital image analysis systems for remotely sensed data obtained from both satellite and airborne sensors. EARTHSCOPE is a complete set of Windows™-based programs for performing image enhancements, classifications, geometric corrections, false- and true-colour 24-bit image display, and utility functions including TIFF and BMP input and output. TEXALL is a program which transcribes image data from Exabyte 8 mm tapes in any format to the PC's hard disk. EarthScenes is a new image processing system which is the software component of Geo/SAT 2, an inexpensive, integrated, remote sensing educational package consisting of image data, teachers' lesson modules and software. It is a new version of the original Geo/SAT package, a co-operative project of Eidetic, the National Council for Geographic Education (NCGE), and EOSAT. It develops unique educational earth science related software programmes and provides packages that tailor specified tutorial requirements and integrates them into both existing and customer-built products.

VERIFIED

Gregory Geoscience Ltd

Current Status
Incorporated in 1973, Gregory Geoscience is also marketed under the name GeoInsight Corporation by which its web site is identified and specialises in remote sensing analysis, conducting geological surveys and geophysical analysis. GG develops municipal databases for GIS to include georeferenced maps, conducts agricultural soil surveys and provides very high resolution scanning of satellite and high altitude imagery. About 20 per cent of income stems from cartographic products, 35 per cent from land and geographic information processing, 35 per cent from remote sensing and 10 per cent from consulting. It has developed the Procom-2 system for optical transfer and integration of diverse data sets from satellite and airborne imagery to a common map scale which is then digitally recorded and entered into a GIS, image analysis or computer mapping system. Company capabilities include very high-resolution scanning and digital analysis of land use change, resource mapping and environmental monitoring. Activities have included the extensive use of satellite and airborne imagery for detecting changes on 1:50,000 maps, mapping peat bogs and wetlands from Landsat, and undertaking a national inventory of mine wastes from satellite coverage. Gregory Geoscience also has expertise in the integration of geographic databases with municipal applications, soil analysis mapping, environmental risk to biodiversity mapping and analysis. Geoscience has already developed analytical tools for coupling imaging data with nadir radar and SAR. GG Ltd has exported products to the application of non-US user packages incorporating linear transformation databases. Has integrated major archival sources for optical and radar data synergistically interleaved with ground truth and remotely derived data base systems.

UPDATED

Horler Information Inc

Current Status
Founded in 1984, HII is a consulting company specialising in the technologies and applications of remote sensing and GIS. In 1990 Horler was licensed to operate the Remote Sensing Online Retrieval System on a commercial basis. Recent projects are the development of an Earth observation strategy for the Canadian Space Agency and an assessment of the geological applications of Radarsat. HII has developed applications packages integrating global data from government and non-government, national and international projects. Specialist areas include GIS ground truth solutions, forest and wetland data interpretation and geomapping analysis. HII has expanded space-based data archive materials to open analytical tools for geomapping and environmental studies. This is sought not only from Canadian sources but the international fraternity involved in all aspects and applications of remote sensing.

UPDATED

Intermap Technologies Corporation

Current Status
Intermap specialises in airborne SAR and other remote sensing acquisition, processing and analysis services, in addition to GIS, mapping

and digital ortho-imaging. Activities now include Radarsat data with processing applied also to a broad base of data from international satellite projects including those from Europe and Russia.

Intermap specialise in photogrammetric mapping and airborne radar providing digital elevation models and digital terrain models for a wide range of GIS, GPS and 3D visualisation applications. Intermap have performed surveys and mapping projects in 85 countries. Intermap is assembling complete national coverage of the USA, Japan and countries in western Europe for Digital Elevation Mapping (DEM), a market estimated at US$500 million for 2005.

UPDATED

Macdonald Dettwiler

Current Status

In 2001 MDA filed total revenues of C$481.275 million compared with C$401.125 million in 2000. The former majority shareholder Orbital Sciences Corporation reduced from 67 per cent in January 2000 to zero in July 2001.

MDA specialises in computer-based systems for aerospace, defence and resource management applications, exporting 75 per cent of its products. In November 1995 it became part of the Orbital Sciences Corporation group. Company employs 1,800 people worldwide.

MDA's product line includes: space-qualified software; MIL-SPEC (including DOD-STD-2167A) systems and software; turnkey remote sensing satellite ground stations; image mapping system for generating maps from digital images; optical/radar image analysis systems for space/airborne sensor data; meteorological data analysis systems. For Canada's Space Station Mobile Servicing System, it is contributing to development of software, data processing subsystems and artificial intelligence applications.

MDA is a leading supplier of ground receiving/processing systems for remote sensing satellites, working as prime contractor on 18 turnkey satellite ground stations worldwide and as major subcontractor on 10. It provides systems for all optical and SAR missions. Its SAR team was the first to digitally process SAR data and to develop a commercial digital processor, and is the world's leading supplier of spaceborne SAR processing systems. MDA is active in the development of analytical software for Radarsat data.

In 2001, company was selected to build the first QuickBird ground station for receiving and processing high resolution imagery from QuickBird satellites. Has contributed information about optimised hybrid satellite technologies and software options for optimised systems. The company is processing high resolution images for use with several companies and organisations related to security concerns and for organisations involved in planning regional ultra-large development schemes.

UPDATED

RADARSAT International, Inc

Current Status

RADARSAT International (RSI, Richmond, BC, Canada), a subsidiary of MacDonald Dettwiler, has been providing Earth-observation data and derived information products, as well as leading-edge services to our global clients since the company was established in 1989.

RSI's broad range of products are derived from the majority of commercially-available Earth-observation satellites operating (or to be operating) in the Visible Infra-Red (VIR) spectrum and Synthetic Apeture Radar (SAR): RADARSAT-1, LANDSAT 5, LANDSAT 7, IKONOS, IRS, ERS-1, ERS-2, QuickBird, ENVISAT and RADARSAT-2. These products include geocorrected imagery, digital elevation models, and application-specific products such as land classification maps and ocean oil seeps vectors. The hallmark of RSI's comprehensive range of services has been our pioneering work in near-realtime delivery to clients with time-critical operations such as disaster management and ship navigation. Other services include training, custom product generation, and GIS project implementation. Also, oil basin screening, offshore and continental oil exploration, geological mapping, monitoring illegal fishing activity, cartographic mapping, DEMs, mosaics, radarmaps, and so forth.

VERIFIED

Satlantic

Current Status

Specialises in real-time monitoring of agricultural products, marine measurements and surveillance systems. Analysis systems for SAR; receiving, processing, analysis stations for NOAA, SeaWiFS, Radarsat, Spot; marine Inmarsat C/M systems. Produces analytical tools for ESA remote sensing programmes and for Radarsat/Envisat correlation. Integrates ground, air and space derived data sets for specific user requirements and for new satellite and remote sensing equipment requirements. The company has integrated optical imaging and radar data sets for national and area mapping and geometics applications.

UPDATED

Seimac Limited

Current Status

Seimacs Satellite Asset Tracker Argos PTT provides intermodal container tracking, automatic utility meter reading and environmental applications such as oceanographic research. Positioning accuracy with GPS is less than 100 m (typically 20 m) and without less than 1 km (typically 350 m). Antenna 3 × 13 × 38 cm, electronics 13 × 18 × 19 cm; rechargeable battery (45 day capacity) 8 × 18 × 19 cm; operating tolerance –30 to +60°C. Seimac's 3.2 × 5 × 13 cm 250 g Smart Cat III Argos PTT is small enough for animal tracking. It is commonly used in compact air-deployable buoys and underwater mooring markers. It includes complete data acquisition and logging capability and an 8-bit analogue-to-digital converter. Seimac equipment has also been adapted for Russian Glonass system and Seimac is developing adaptive electronic packages for new navigation systems. Seimac is developing integrated nets for multiple satellite constellation sourcing to integrate Navstar, Glonass and Galileo systems. Seimac is the prime supplier of weather services to Canada's offshore oil industry using weather maps, radar images and numerical model outputs.

VERIFIED

Czech Republic

GISAT

Current Status

GISAT was founded as a remote sensing and GIS company in 1990 and is the first to carry out this activity in the Czech Republic. Cooperative ventures through Intersputnik countries and with western European distribution organisations to improve access and dissemination inside the Czech Republic. It is a Spot Image, EOSAT and Eurimage distributor within the Czech Republic. Also distributes optical and radar images from Russian satellites. It is involved in the CORINE Land Cover and the agricultural MARS programmes of the European Community and utilises data from ESA programmes. GISAT provides mosaics from satellite images of the Czech and Slovak Republics. GISAT is combining several data sets to optimise real and false-colour images and for data required by radarsat operators and data users.

UPDATED

Denmark

Kampsax Geoplan

Current Status

Founded in 1917, the company specialises in digital and analytical mapping from remote sensing images and aerial photography. It is a distributor of Spot and Landsat data products in Denmark under agreement with SSC Satellitbild and Eurimage, respectively. Kampsax provides interpolated data sets from different satellite programmes and has expanded its archived database for European clients and customers. Also provides Envisat support for ESA and now integrates radar data from air and space subsets. Kampsax provides approximately 25,000 new aerial photographs each year to clients and customers. Kampsax employs about 500 people.

VERIFIED

Ecuador

CLIRSEN

Current Status

The CLIRSEN Centro de Levantamientos Integrados de Recursos Naturales por Sensores Remotos (Centre for Integrated Surveys of Natural Resources through Remote Sensors) was created in December 1977 to be responsible for remote sensing and the cataloguing of natural resources. It operates the Cotopaxi ground station, some 60 km south of Quito, which began taking imagery from Landsat in June 1991 as the system's 18th station; Spot and ERS were added later. The 2,500 km^2 coverage of 25 countries stretches from Mexico's Yucatan Peninsula to Antofagasta in Chile. Products include 23 × 21 cm film, 72 × 97 cm paper, quick look, microfiche (84 scenes each) and digital tapes (6,250/1,600 bpi). Processing levels: bulk, georeferenced, geocoded and user-specified highlighting. CLIRSEN configured the facilities to handle ESA Envisat data and to integrate other remote sensing satellites for environmental monitoring and has absorbed radar data into its product stream. CLIRSEN have specialised in providing imagery for remote sensing analysis and geographic data throughout Central America and Northern South America. It has specialised in regional mapping and resource analysis together with environmental pollution monitoring from global space-based assets including commercial imagery and government programmes. CLIRSEN have also conducted data interpolation analysis for specialised applications and are capable of transforming 3D terrain and topography formulations.

UPDATED

Egypt

Eight M Ltd

Current Status

Eight M Ltd specialises in environmental and remote sensing services, offering state-of-the-art digital image processing and GIS projects, consultancies and workshops. The company has a long record of specialised equipment for the earth observation industry and these have been applied to space-based data sources. It represents EOSAT (Landsat products), Eurimage, Radarsat and Earth Resources Data Analysis Systems (ERDAS; image processing, GIS software packages) and Cirrus laser film writers. Adaptive hardware for converting film imagery to digital format and converting optical imagery to GIS formats. Eight M Ltd has also processed platform data from European and Russian satellites. The company had attempted to provide exclusive regional access and has made considerable progress toward generating a unique set of regional geomorphology maps based on acquired data from in-orbit satellites.

UPDATED

France

EADS Cap Systèmes

Current Status
EADS Cap Systèmes is the European leader in Communication, Control, Command and Intelligence systems and the world leader in Earth observation ground stations for optical and radar imaging including SAR. Matra Cap Systèmes offers multisatellite Earth observation ground stations for SPOT, HELIOS, LANDSAT, ERS, RADARSAT, CBERS, either fixed (in Toulouse, Sweden, US, Pakistan, Brazil) or mobile (US Air Force, IOSAT Canada). The company is also working on several demonstration Pilot projects using Earth observation for EEC (image data servers, agriculture calamities information server). Provides custom built software solutions for multispectral remote sensing data. Also supports military applications. Several R&D projects are run by the company for the European Space Agency in new archiving technologies, information client/server software, SAR interferometry applications, etc. Products are offered by EADS Cap Systèmes for image processing and applications development (Multiscope) and digital cartographic systems (Prodigeo, T10) which are used in several remote sensing and mapping centres in the world.

VERIFIED

Fleximage

Current Status
Now a subsidiary of the European Aeronautics Defence and Space Company (EADS), Fleximage was established in 1989 to specialise in computer-aided photo-interpretation and geographic databases. Fleximage has developed new software for standard databases allowing interpolation from several separate sources (satellites or ground stations) and these have been added to the existing inventory of services. Aerospatiale holds 78 per cent. The extraction of thematic maps from satellite imagery is based on the OCAPI (Operational Computer-Aided Photo-Interpretation) software, initially developed for the military to exploit Spot data. Also, GEOCAPI, a customised software package for information management.

VERIFIED

Geosys

Current Status
Geosys was established in 1987; it is a member of the SGS group (Société Générale de Surveillance) and the Caisse des Dépôts et Consignations (Deposit & Consignment Office). Personnel total on average 42. It compiles regular crop inventories and prepares yield forecasts using remote sensing data, including Spot, NOAA AVHHR and Landsat TM. It also provides geographic information to the agribusiness and environment sector. It develops and sells GIS-handling remote sensing images and vector cartographic data together with ground truth companions integrated with satellite data. Recent additions include agricultural analysis and precision farming support from satellite data. Crop maps, field maps, land parcel characteristics and optional sampling maps are offered to assist with agricultural management. Produces specific analysis of regional areas and integrates land use data for environmental surveys in addition to ground data and atmospheric integration.

VERIFIED

ISTAR

Current Status
Imagerie Stéréo Appliquée au Relief was established in 1988 to market Digital Terrain Models and orthophotos computed by an automatic relief plotting software, for use with Spot stereopairs. DTMs are generated with a 20 m grid cell and an average altimetric accuracy of 10 m. The company offers VUE3D software products for interpreting DTMs and orthophotos; they can create composite images providing 3-D views and calculate successive image sequences for animation purposes. ISTAR is also a consultant in satellite imagery, geometry and relief mapping utilising different sensors. Also sets up orientation packages for multidisciplinary applications and tutorials for estimating new use applicants.

ISTAR's high-resolution mapping products are available for over 180 cities in Europe and North America, and medium resolution geo data covers another 2,000 cities and regions in 100 countries. ISTAR has expanded its historic database and has produced an integrated set of reference matrices for engineering applications using a geodetic overlay.

UPDATED

Spot Image SA

Current Status
To date more than 9 million images have been archived. Spot 1 was launched on 22 February 1986 followed by Spot 2 on 22 January 1990, Spot 3 on 26 September 1993 and Spot 4 on 24 March 1998. Spot 3 ceased functioning in 1997. Spot 5 was launched by Ariane 4 on 3 May 2002 and began operations on 12 July 2002.

Spot Image, with CNES and Matra as the majority shareholders, was established in 1982 to distribute Earth remote sensing data and their products from the Spot satellites, the first of which was launched by Ariane into polar orbit in February 1986, followed by Spot 2 in January 1990, Spot 3 in September 1993 and Spot 4 in 1998. The two main image receiving stations (Toulouse and Kiruna) are designed to receive/store 250,000 images annually.

UPDATED

Germany

Energia Deutschland GmbH

Current Status
The company is a partnership between Germany's Kayser-Threde and Russia's RKK Energia, principally to exploit an exclusive right to distribute Earth observation data from Mir. A 120 × 120 km 7.5 m resolution colour image from Mir's two KFA-1000 cameras (Priroda 5 package) costs DM1,900. Company has directed its activities toward ESA images and Landsat data and has adapted to the absence of new Mir data, since the deorbiting of Mir, by utilising the extensive backlog of images. The company is a subsidiary of Kayses-Threde. Large but selective backlog from the Russia archive has provided product for integrating later images from ESA or Landsat.

VERIFIED

GAF mbH

Current Status
The Company for Applied Remote Sensing (Gesellschaft für Angewandte Fernerkundung mbH) specialises in multidisciplinary applications of remote sensing and image processing techniques. It is a distributor of Radarsat, EOSAT (Landsat, IRS 1C), Spot Image (Spot), Soyuzkarta (KFA-1000) satellite data products. It directly receives, processes and archives IRS 1C data via DLR's Neustrelitz ground station under a contract signed with EOSAT 1 February 1996. Can collect integration signal sources and direct to appropriate tracking and data extraction equipment for parallel analysis. GAF and Spain's Minas de Almadèn y Arrayanes SA created Télédéteccion Aplicada SA (TdA) in 1990 as a joint venture specialising in remote sensing services in Spain and Latin America. Developed optical and radar imaging integration processing which allows users to interpolate data on simultaneous overlay. GAF has expanded its activities to include Envisat data. Now also processes Radarsat and some Japanese data but also preparing to offer consultancy services. GAP has provided new services to the applications and ancillary industries.

UPDATED

VCS Nachrichtentechnik GmbH

Current Status
VCS has specialised in radio production and has applied skills to the space sector. VCS Engineering has provided ESA with new protocols for Internet standard 'differentiated services'. VCS has operated in the space ground system market for 22 years, specialising in satellite ground stations, image workstations, satellite communication, meteorological applications and task force engineering and consulting. ESA and EUMETSAT contracts include development of the whole ground infrastructure for the Meteosat MDD mission, GOES/Meteosat relay and projects on data encryption and compression and significant parts of the core mission control centre for the Meteosat Transition Programme. Has advised on regional meteorological data distribution networks and extended this to international activities linking several separate ground stations and data collections facilities for interactive dialogue and large volume data dissemination.

VERIFIED

India

Speck Systems PVT Ltd

Current Status
Since 1984 Speck Systems has specialised in the design and manufacture of scientific equipment, the testing of precision, electro-optical instruments for remote sensing and software solutions. Manufacturers and suppliers of digital image processing, GIS and photogrammetry systems, digital film recorders and scanners. Production job services in topographic mapping, map digitisation, image analysis and GIS. Also produces analytical tools for Radarsat and for Envisat plus images from Russian and Japanese satellites. In 1995 the company implemented software photogrammetry systems for strategic and tactical applications. In 1999 SSL began operations from Botswana, opening European offices in Germany during 2000 and in the US during 2001. Plans for further expansion have slowed future investment through a policy of consolidation.

UPDATED

Indonesia

PT Indica Dharma Consulting Services

Current Status

Established in 1991, the company distributes data from the Thailand Remote Sensing Centre, LAPAN's Pare-Pare station, Australia's ACRES and FR mapper software distributor. The focus is on the mapping of forestry concession areas, land use, coastal zones, plus training. The company also takes images and data from Landsat and Spot Image satellites and has developed techniques for integrating radar and optical imagery. Several contracts have been acquired throughout the region but general marketing on a commercial basis has not been successful. The company distributes Landsat data products and performs integration with other data sets from other remote sensing programmes. Provides full spectrum imagery in optical and radar bands for data analysis and interpretation or onward provision of raw data. Has created a unique product for regional co-ordination projects integrating local and area requirements from existing satellite imagery or data archive.

UPDATED

International

CLS Argos

Current Status

CLS provides specialised expertise in space based oceanography and operates the Argos system and the Doris system. Doris calculates satellite orbits and is responsible for control of the Topex/Poseidon mission. Until 1998 Doris operated in delayed time only, but with the launch of the DIODE navigator on SPOT-4 it can now provide data in real time to within a few metres. Collecte Localisation Satellites Argos (CLS Argos) is operating/developing several satellite-based location and data collection systems. The principal Argos system processes and disseminates space and terrestrial environmental data received by dedicated packages aboard US NOAA polar meteorological satellites from fixed/mobile platforms anywhere in the world. Approximately 5,000 transmitters are operating. The past five years have seen a 15 per cent growth in activity (approximately 20,000 transmitters have used the system). Argos is a joint programme of CNES, NASA and NOAA, operated under a Memorandum of Understanding signed in 1974 and extended in 1986. System operation and promotion is the responsibility of CLS, established in April 1986 as a subsidiary of CNES (55 per cent), IFREMER (French Marine Institute; 15 per cent) and a pool of banks. CNES operates the French Global Processing Centre (FRGPS) in Toulouse and handles all user relations outside North America. US subsidiary Service Argos Inc is responsible for all US/Canadian users; it operates the identical USGPC, which provides redundancy. North American CLS is concerned with developing new services and products. The Australasian regional processing centre began operations in December 1988.

The system consists of independent user Platform Transmitter Terminals (PTTs, down to 25 g and the size of a matchbox); two NOAA polar satellites with Argos Data Collection & Location System packages to receive PTT messages on a random access basis for separation, time coding, formatting and retransmission to ground stations; and the ground stations and two Global Processing Centres in Toulouse and Landover where data is retrieved, processed and distributed to users. Location of PTTs can be determined to within typically 150 m, if required, by Doppler techniques.

The PTTs transmit data from up to 32 sensors encoded in 8-bit words on 401.650 MHz with typically 200 mW power at intervals of 60 to 120 seconds for location PTTs and approximately 200 seconds for data collection-only platforms. Message length is 360 to 920 ms and contains the PTT identification number.

The NOAA Argos packages receive all messages within a 5,000 km diameter visibility circle at any instant; four can be processed simultaneously by NOAA A-J and eight will be handled by the third-generation system aboard NOAA K and its successors from 1996. NOAA 11 can handle data at 960 bit/s, NOAA J 1,200 bit/s and NOAA K-N 2,560 bit/s. NOAA K-N increases bandwidth from 24 to 80 kHz. The data is formatted/stored, then dumped each time the satellite moves within reach of one of the three ground stations. VHF/S-band transmitters also perform real-time relay for any user station within the visibility circle. Current onboard packages can process up to 1,400 data-only PTTs, or 415 requiring the location service, but from NOAA K on the capacity will be quadrupled. The Argos 3 generation was defined by the end of the 1990s for post-2000 satellites. Japan's 1999 ADEOS 2 will carry an Argos package allowing users to command their PTTs.

Under an agreement signed in April 1990 with EUMETSAT, CLS manages all non-meteorological applications using the Meteosat data collection function. The Multisatellite Applications Extended Dissemination Service (MAEDS) adds Meteosat and GOES to the existing Argos system to provide high data collection capacity (649 byte messages up to 24 times daily), real-time access to data and continuous coverage from the Pacific to the Urals. The first MAEDS element became operational in October 1990, followed in 1991 by the GOES satellites covering America and the Pacific. CLS also operates the Doris location system and is involved in the Starsys system. These systems have been given integration capability through newly developed software provided by a subcontractor.

Global Processing Centres NOAA's tape recorders are read out every 100 minutes over one of the Gilmore Creek (Alaska), Wallops Island (Virginia) and Lannion (France) stations. All of the satellite's data are routed to NOAA's National Environment Satellite, Data and Information Service (NESDIS) in Maryland, where the Argos component is extracted and relayed to CLS in Toulouse and Landover. Here, PTT locations are determined and the sensor data processed. About two-thirds of the processed messages are provided to users within 3 hours of the uplink. Location is derived from measuring the Doppler shift in the carrier frequency of received messages during the 10-minute overflight. An accuracy of 150 m is typical but for PTTs at fixed positions or on slow-moving carriers 100 m is possible. The transmit frequency must remain stable throughout the pass. Each frequency measurement yields a set of possible positions and with a known elevation only four messages are required on each pass to generate a unique solution. The satellites' own positions are known to within 300 m at any instant from a reference system of 11 beacons worldwide at precisely known geodesic positions.

VERIFIED

Eurimage S.c.r.l.

Current Status

Established in 1989 to supply remote sensing products and services, Eurimage's shareholders are: British Aerospace (UK), Dornier (Germany), SSC Satellitbild (Sweden) and Telespazio (Italy). Data is distributed from Landsat, MOS, NOAA/AVHRR, JERS and ERS. Revenue for 1995 was €5.4 million (1994 €5.15 million, 1993 €5 million, 1992 €3.3 million). Products cover digital and

Satellite images of remote regions of the Earth are a vital tool in assessing the impact of climate change on fragile eco-environments such as Svalbard, a part of Spitbergen north of Norway (NASA) ***NEW***/1044057

photographic formats of large areas or small regions: raw data, standard products, geocoded and enhanced products. The rush service offers near-real-time cloud cover assessment (within a few hours of acquisition) and enables the final processed product to be dispatched to the end user within three days. Eurimage now offers European and Mediterranean countries products from Envisat, ERS, Ikonas, IRS, Landsat, Quickbird, and Radarsat as well as NOAA/AVHRR, JERS-1 and Russian camera systems.

UPDATED

RESURS-F WorldMap joint venture

Current Status

The WorldMap consortium was created in 1993 by JEBCO, Priroda and Rosvneshgeo. Priroda is Russia's primary non-military remote sensing acquisition and processing organisation. It also operates a large cartographic division that employs satellite data to create cartographic and GIS products. Priroda is under the administration of Roskartografia, the Federal Service of Geodesy & Cartography. Rosvneshgeo is a private Russian company affiliated with Roskomnedra, the Russian State Committee of Geology in charge of mineral and hydrocarbon exploration. WorldMap also provides SAR data from Russian and Canadian projects and handles the interpolation of data between optical imaging and radar measurements. The activities have extended to added data from daily marine platforms correlated with satellite images and SAR. The consortium has maintained a large archive of historic information and mapping products. It also provides analysis of specific geographic regions and assists with determination of geologic and environmental aspects. WorldMap has contributed to the further development of satellite systems, technical capabilities and improved photo-interpretation.

Priroda collects new imagery each year using the Resurs-F spacecraft. Each F1 satellite carries two KFA-1000 plus three Kate-200 cameras; F1M three KFA-3000 plus one Kate-200; F2 one MK-4 and F3 two KFA-3000s. Priroda's archives contain some two million XS/PAN scenes at 2 to 20 m resolution covering most of the globe. The data collected for the KFA-1000, MK-4 and Kate-200 film camera systems are now available digitally in a bit sequential raster format. Some 500 scenes monthly are being digitised by two Vexcel VX3000 enhanced scanners linked to Sun Microsystems SPARC stations and output on to 8 mm tape. Cost per archive scene to users is: US$3,000 KFA-1000; US$4,000 KFA-3000; US$3,000/US$4,500 1/3-bands MK-4; US$1,600/2,400 1/3-band Kate-200. Prices drop rapidly for large purchases. JEBCO is the sole distributor of digital data derived from the original Resurs-F satellite negatives. The consortium accepts requests for targets to be included in upcoming flights. Priroda plans to distribute digital 3 m resolution panchromatic imagery from a military satellite, presumably the fifth-generation photoreconnaissance model, which had Cosmos 2320 in operation at the beginning of 1996.

Sovinformsputnik (whose dominant shareholder is the Central Specialized Design Bureau, builders of Resurs-F) is licensed to sell data from the KVR-1000, TK-350 and DD-5 camera systems of military satellites. These negatives are also being digitised and Priroda has a sales agreement with Sovinformsputnik. Orders can be directed through the WorldMap consortium.

The KFA-1000 has a focal length of 1,000 mm, 0.57 to 0.68, 0.67 to 0.81 μm, 5 m resolution, 80 km swath width (160 km by two cameras, 240 km by three), frame 300 × 300 mm (optimal enlargement ×10), altitude 220 to 280 km scales 1:220,000 to 1:280,000, stereo overlap 60 per cent, 1,800 frames per camera.

The MK-4 has a focal length of 300 mm, four filters for usually 0.635 to 0.690, 0.810 to 0.900, 0.515 to 0.565, 0.580 to 0.800 μm; 0.460 to 0.505, 0.400 to 0.700 μm also available. From 220 to 270 km altitude, 5 to 8 m resolution, 130 to 160 km frame width, scale 1:730,000 to 1:900,000, stereo overlap 60 per cent, frame size 18 × 18 cm, 2,500 frames per camera. Film improvement planned for F2M in 1997 to increase resolution 1.4 times.

The Kate-200 has a focal length of 200 mm, 0.5 to 0.6, 0.6 to 0.7, 0.7 to 0.85 μm, 15 to 30 m resolution, 180 km swath width, 18 × 18 cm frame size, overlap 20, 60 or 80 per cent, 2,500 frames per camera.

The KFA-3000 is a high-resolution camera flying aboard Resurs-F3/F1M. Focal length is 3000 mm, panchromatic, 2 to 3 m resolution (1.6 m from 180 km), frame 300 × 300 mm, 21 × 21 km, scale 1:70,000.

The KVR-1000 is a panoramic low-angle camera flying aboard Kometa since 1983, usually in conjunction with TK-350. Focal length is 1,000 mm, 0.56 to 0.8 μm, 75 cm resolution (but available commercially only down to 2 m), frame 18 × 72 cm (37 × 165 km; cut into three), scale 1:220,000 but can be enlarged to 1:10,000 without loss of detail. Images nested within TK-350's.

The TK-350 is a topographic camera flying aboard Kometa since 1983. Focal length is 350 mm, 0.50 to 0.68 μm, 10 m resolution, frame 30 × 45 cm, 200 × 300 km, scale 1:660,000 (can be enlarged to 1:50,000 without significant loss of detail), 60 to 80 per cent stereo overlap.

The DD-5 is a high-resolution data degraded to 2 m resolution in the digitising process. DD-5 (DD means digital data) does not refer to a specific camera system but is a catch-all designation for digitised data from 'conversion' military satellites.

VERIFIED

Italy

Alenia Aerospazio

Current Status

As co-contractor with Dornier, it developed and integrated the RF subsystem for the German-Italian X-SAR, flown twice in 1994 as part of the Space Radar Laboratory. It developed the RF project for the 12 m planar antenna and was responsible for its integration and testing. It was also responsible for the design and development of the ground segment Mission Planning and Operation System (MPOS). Alenia Spazio led the design and development of ERS 1 and 2's Radar Altimeter and is studying a second-generation instrument for Envisat. It is developing ESA's Multifrequency Imaging Microwave Radiometer (MIMR), with a 1.6 m antenna, planned for Metop 1. Has produced components for ESA Envisat data processing hardware. Furthermore it is involved in the realisation of second-generation METEOSAT satellite, in the HELIOS national security satellite programme and in the Cosmo Skymed programme, an environmental observation system for the Mediterranean basin. This has stimulated work to put together a radar-based Earth observation system proposal incorporating sensors from optical to non-optical imaging and radar generated data sets.

UPDATED

Japan

NEC

Current Status

NEC's space sensors include the MultiSpectral Electronic Self-Scanning Radiometer (MESSR) flown on MOS, the Optical Sensor (OPS) on JERS 1, the Ocean Colour Temperature Scanner (OCTS) on ADEOS and the Visible and Near-Infra-Red (VNIR) Radiometer of the Advanced Spaceborne Thermal Emission and Reflection (ASTER) Radiometer for NASA's EOS AM1. See the project entries in the space segment section for instrument details. Developed nano-technology for the microsat programme aimed at launching large numbers of surveillance and remote sensing sensors in a national emergency. NEC also supplies the ground systems for receiving/processing Earth observation satellite data. NEC is now developing Earth observation sensors that may be used in carry-on packages to the Japanese Experiment Module, part of the International Space Station. NEC has discussed production of sensors for European Earth observation and remote sensing programmes and has developed a new range of sensor packages for non-space applications. These now include sensors for battlefield application with potential for use with imaging and spectral discrimination.

UPDATED

Nissho Iwai Corporation

Current Status

Nissho Iwai Corporation is a 'Sogo Shosha' trading company with 180 offices worldwide. It is a partner in the JCSat Japan Satellite Systems company owner and operator of telecommunications satellites. The Space Project Office was established in 1985 and now distributes Spot and Landsat data to Japanese users. It also offers Terra-Mar's Image Processing and GIS workstations. The Corporation also handles SPOT and Ikonas data and is integrating radar and SAR images. NIC is developing software options capable of integrating data from large area optical systems and fine scale radar sets.

In 2003 Nissho Iwai Corporation reported total trading transactions of US$43.4 billion with a gross trading profit of US$2.1 billion.

UPDATED

Kenya

Teleos International BV

Current Status

Teleos is a partnership between EOSAT and Nuova Telespazio of Rome, Italy, to provide East and South Africa with remote sensing and GIS applications services in co-operation with Nairobi's Regional Centre for Services in Surveying, Mapping & Remote Sensing (RCSSMRS). Teleos' portable ground station was installed in the grounds of RCSSMRS in September 1994, initially receiving Landsat 5 TM imagery and upgraded in 1995 to add Spot, ERS and IRS. Integrating SAR data and high-resolution images for regional multisourced analysis. Product development in complex GIS and remote sensing interpolations. The partnership has incoporated high-resolution imagery Ikonos class satellites and has adapted applications services to accommodate the growing availability of non-military high-resolution earth observation images. Is converging separate data sets and plans to utilise new generation commercial high-resolution earth observing systems with ground truth and airborne data acquisition. The company has opened up full data archives for both optical imaging and radar satellite data.

UPDATED

Korea, South

Chang Woo, Inc

Current Status

The company is a national provider of remote sensing satellite data products: LANDSAT, ERS and Russian KVR-1000 (under agreement with EOSAT), Spot and ERS (Spot Image), Spot, ERS and JERS (Korea Advanced Institute of Science and Technology, a Korean ground station), RADARSAT (Radarsat International) and Russian Resurs-F (Soyuzkarta). It provides turnkey direct reception ground stations for local markets under agreements with MacDonald Dettwiler and Associates for high-resolution Earth observation satellites like EarthWatch's EarlyBird and QuickBird and OrbImage's OrbView. It has provided a direct reception ground station for Spot, ERS and JERS and has been undertaking a contract for OrbView, RADARSAT and IRS-1C downlink to Korea. It provides ground station facilities for GMS, NOAA and SeaStar and can also provide uplink ground stations or customised facilities for the ground links to Earth observation satellites. Chang Woo is also managing SAR data. Currently working to provide access to long run archive data for studies of spatial and temporal changes. Chang Woo Inc is also available for analysis of integrated air-space data and has performed several specialised studies covering integrated multisensor data sets. Chang Woo has developed several unique products for integrating various multimedia data sets and customer demand has stimulated the production of elevation and 3-D image presentation plus formats for individual customer requirements. Company has developed new processing formats for digital readout and these have become standard for use by national defence and security forces.

UPDATED

New Zealand

Meteorological Service of New Zealand Ltd

Current Status

Evolving in 1991 from a government department into a major state enterprise, MetService receives and processes NOAA HRPT and GMS S-VISSR data streams. Image data are rectified on to standard map projections, Argos transmitters located. Sea surface temperature is retrieved from NOAA AVHRR radiances. Integrates satellite and sea level data and provides data from Japanese environmental satellites. Continues to extend data evaluation and analysis from meteorological sources around the world. The Meteorological Service of New Zealand is an active component of the World Meteorological Organisation and provides daily updates on sea states and weather hazards as well as basic weather forecasts.

VERIFIED

Sirtrack Limited

Current Status

A wholly owned subsidiary of Landcare Research, Sirtrack packages satellite transmitters working through the Argos system for wildlife and marine applications. For example, the Prestel unit is used to locate lost seismic streamers: if the water pressure exceeds a preset limit (indicating sinking) over a set period, the transmitter package is released. Argos can output the location directly to the host vessel via Inmarsat. Upgrades include precise position location equipment. Sirtrack has developed a transponder package for multisatellite operation. Currently working to achieve marine life tracking capabilities through satellite relay. Extended applications for environmental and marine biology preservation projects and programmes. Sirtrack is well placed to take a leading role in the use of satellite information for monitoring marine life and for ocean biota forms.

UPDATED

Norway

NORUT Information Technology Ltd

Current Status

NORUT Informasjonsteknologi is a Norwegian institute specialising in earth observation and information and communication technology. NORUT Information Technology Ltd is one of five research institutes in the NORUT Group Ltd. NORUT's main activities are satellite remote sensing, image processing, GIS, high-speed data communication, electronic data interchange, and object-oriented methods for data modelling and system development. NORUT has developed analytical math models to handle data from different providers. Developed high-speed data base with off-site fast retrieval for customers in country and abroad. 40 staff.

VERIFIED

Spain

AURENSA

Current Status

Auxiliar de Recursos y Energia SA (AURENSA) is an Earth resources company divided into: remote sensing; geological engineering; drilling engineering. It has represented Spot Image interests in Spain since 1988 and is integrating ESA environmental programmes for clients. It has provided coverage for clients using radar as well as optical imagery. AURENSA presently represents SPOT, IRS Ikonas and Radarsat and provide GIS, environmental engineering studies, geological engineering surveys and evaluations of potential drill sites. The company has grown since the early 1990s with integration of a wide range of space-based capabilities and renewed inclusion of ground derived and air-based data capture. Substantial ground-truth data incorporated into data obtained through ground, air and space based imaging and radar data.

VERIFIED

United Kingdom

EADS-Astrium

Current Status

On 24 October 2003 EADS-Astrium announced the award of the Skynet 5 contract by Paradigm Secure Communications for the UK MOD's next generation Skynet 5 satellite system. Under this contract Astrium will build two Skynet 5 satellites and under the Skynet 5 Private Finance Initiative (PFI) the MOD will procure satellite communications services from Paradigm over the next 15 years. In order to provide a seamless transition, Astrium-built Skynet 4 satellites, the last of which was launched in February 2001, will be operated and managed by Paradigm and will include Skynet 5A and 5B scheduled for launch in 2006 and 2007. Astrium was formed in May 2000 by the merger of Matra Marconi Space in France and the UK and the space division of DaimlerChrysler Aerospace in Germany. EADS-Astrium (UK) is providing instruments for several Earth observation satellites. A £9.5 million contract was awarded by the UK's Meteorological Office in March 1989 for development and fabrication of the Advanced Microwave Sounding Unit-B (AMSU-B) for NOAA's Tiros-Next series to return atmospheric humidity information. The company designed and built UARS' Improved Stratosphere and Mesospheric Sounder (ISAMS) for Oxford University/Rutherford Appleton Laboratory to measure gas species and aerosol levels in the upper atmosphere. The company provided the Digital Electronics Unit for the Along-Track Scanning Radiometer (ATSR) for determining sea surface temperatures and atmospheric water vapour by infra-red radiometry on ERS 1. It developed ATSR 2 for ERS 2 and the Advanced ATSR for Envisat 1.

Astrium offers a Scanning Mechanism for high-accuracy platform pointing of remote sensing instruments. It comprises a scan platform mounted on bearings within a structural housing and driven through a gearing system by an electric motor. A closed-loop control system works off an Inductosyn position pick-off attached to the rotating platform, using an electronic processor to drive a pre-determined scan pattern. The mechanism can incorporate momentum compensation and bearing offload during launch if required, and offers both direct drive and gear drive versions. It employs dry lubrication throughout and offers extended operational lifetime (approximately 100 million revolutions). It is used on AMSU-B and derivatives are baselined for Metop 1's MIMR and Envisat 1's GOMOS steering front assembly. MMS also has an active support programme for atmospheric sounding rockets.

Specifications

Size: 300 mm diameter, 180 mm high, 6 kg
Performance (5 kg load mass, 1 A current): 0.01° dynamic accuracy (3σ) Earth viewing, 0.005° positional knowledge, 360° scan range
Power: 20 W
Momentum compensation: better than 1%
Life: 100 million revolutions.

VERIFIED

Earth Observation Sciences Ltd

Current Status

EOS was established in 1985 and now has a staff of more than 60 people providing expertise in science, technology and applications of Earth observation and environmental data management: systems design/development, consultancy, data management and data and information processing. It is involved in ERS, EOSDIS, ADEOS and CEO. As part of the Orbital Sciences Corporation family, EOS has global interests in high-resolution imagery and data interpretation requirements and has specialised in formatting overlays for land use mapping integrating satellite data in optical and non optical portions of the spectrum. EOS has broadened its activities to include environmental consulting services and environmental support software. The company has extensive archives on digitised wet-film imagery and can interpolate digital and radar data sets.

UPDATED

EDS Ltd

Current Status

The company's space business included a £2.5 million contract to supply the SAR processor for the UK Earth Observation Data Centre to process ERS data. In addition, the PulSAR high-performance SAR processing workbench has been developed to provide high-quality SAR imagery with rapid throughput on low-cost hardware. It supplied the SARCALQ image QA and calibration system for ESA's Earthnet and is undertaking projects in the interpretation of Earth observation data using neural nets, the development of integrated GIS, and the development of a system for determining terrain heights from SAR imagery. EDS is active in systems development and integration and in managing earth resource data networks. EDS has integrated SAR data exchange formats for radar data and SAR images. It analyses potential SAR applications and mission requirements. The company produced new and innovative designs for a radical fold-array design concept with a view to space vehicle applications where either limited space or light weight are criteria.

VERIFIED

HTS Development Ltd

Current Status

Formerly known as Hunting Technical Services, HTS Development Ltd offers GIS services, remote sensing analysis, land information systems management, mapping and training courses.

Studies and analyses of airborne, satellite optical and microwave remote sensing data for assessment of all renewable and non-renewable natural resources. Colour Fire 240 precision filmwriter for high-quality photographic hardcopy production. Data analysis using ERDAS image processing and ARC/INFO GIS systems; data acquisition and procurement; consultancy and training. Company offers field verification, systems research, systems definition and training programmes. Has developed interactive computational programmes for integrating optical and radar imaging systems through interpretive analysis set up through a client's software systems.

VERIFIED

Nigel Press Associates Ltd

Current Status
NPA has been operating as a remote sensing consultancy since 1972, concentrating on the digital and photographic enhancement of satellite imagery, particularly for petroleum exploration. It is leading a consortium developing new applications of ERS data for offshore oil exploration. The NPA Data Centre holds one of the largest independent collections of Landsat imagery outside of the USA. NPA, designated an official UK Spot image distributor, offers production and interpretive facilities using in-house equipment for computer and photographic image processing. NPA handles Ikonos, Landsat, SPOT image and other satellite data products for a wide range of applications and requirements. NPA also provides Ikonos images, radar interferometry data and GIS technology. Conducts offshore basin screening supporting oil-seep detection methods, thermal basin screening to map on-shore oil seepage and specialises in civil engineering and environmental applications. NPA also provides surface mapping and structural displacement analysis through three classes of SAR interferometry which it markets under the name InSAR and includes permanent scatter interferometry and corner reflector interferometry.

VERIFIED

Space Technology Systems Ltd

Current Status
STS was established in 1958 and now specialises in systems for receiving and processing data and imagery from Earth observation satellites, and meteorological and oceanographic instruments, data buoys and data collection platforms using satellite telemetry links. STS has made considerable progress for customers in developing data interpretation assets for specific requirements dictated by user needs. Has conducted offshore data surveys involving ships and aircraft for environmental specialists in local authorities. Extensive facilities for integrating meteorological oceanographic and hydrological data and for applying multispectral and radar imaging to specific user requirements.

VERIFIED

Timestep Weather Systems Ltd

Current Status
Timestep's PROsat for Windows PC-based systems receive images from polar and geostationary meteorological satellites. Geostationary images can be received, displayed and animated automatically in full colour with up to 1,000 frames per sequence, as well as 3-D projection, image processing and country outlines on Meteosat images. Polar features include visable and infra-red images viewable simultaneously, multispectral colour, temperature readout with no calibration needed, continuous Autosave with Auto Scheduling to receive all passes without user intervention, integrated Tracking/Prediction functions and many more. The company also offers equipment for NOAA HRPT (High Resolution Picture Transmission) images and CHRPT from Feng Yun 1C and successor satellites. The company supplies equipment throughout the world to government, educational and private users.

VERIFIED

United States

EarthSat (EarthSatellite Corporation)

Current Status
Specialises in image processing, weather, environment and geology. EarthSat has developed the first off-the-shelf medium resolution land cover database in a programme known as GeoCover LC. EarthSat, founded in 1969, utilises remote sensing data from satellites and aircraft and provides value-added services for satellite imagery. It provides image processing and interpretation services and environmental, geologic and GIS consulting. EarthSat is a distributor for all civil satellites, including AVHRR, ERSI, GOES, JERS, Landsat, Radarsat, Russian and Spot. Company also provides higher technology solutions for integrated data returns and analysis on a case-by-case basis. EarthSat provides agricultural and weather information to more than 200,000 antennas each day.

VERIFIED

Intergraph Mapping and Geospatial Solutions

Current Status
Intergraph provides interactive computer graphics systems for GIS, mapping and Earth sciences applications. Produces GeoMedia WebMap, a tool for accessing GIS web data in visualisation form. Assists distribution to customers via Internet. Image Station Imager (ISI) is a workstation software product providing multispectral digital image enhancement and analysis. MGE Base Imager and MGE Advanced Imager are Windows-based software products providing multispectral digital imaging enhancement and analysis. Intergraph has sales and support offices in 63 countries. It also provides military GIS solutions, faster processing and access to imagery platforms for the intelligence analyst. The company has conducted analysis of geospectral radiance mapping for Earth heat sensor plots. This has applications for scientific and military purposes and has found use in hybridised requirements developed with US and non-US companies.

UPDATED

Lockheed Martin IR Imaging Systems

Current Status
Formerly a Loral company, IR Imaging Systems is developing the Atmospheric IR Sounder (AIRS) for JPL under a US$145 million contract for the National Center for Atmospheric Research. IR Imaging Systems also participated in developing the imager and sounder instruments for GOES-Next, for which Space Systems/Loral was prime. IRIS is offering an advanced sounder for proposed successors to the GOES-Next. Developing advanced optical sensors in near IR for post-METOP requirements and has designed integrated sensor packages for future meteorological satellites.

VERIFIED

Morcom International, Inc

Current Status
Morcom acquired Alden Electronics, Inc in 2001 and continues to market Alden products. Alden Electronics specialises in communication systems and weather data display terminals for transmitting, recording and display of meteorological data. The WeatherWorks 2000 provides weather information in the form of NEXRAD, DIFAX, GRIB, satellite and lightning images or text. Operating on a 486 PC with a high-resolution graphics option for pan, zoom, loop, rotation, colour and storage functions. Images can be captured and saved in GIF or TIFF formats. The APTS/WEFAX System 6A is a 486 PC-based system to acquire, process, store and display images from polar and GEO satellites. The 6A, running on UNIX, allows simultaneous ingest and data manipulation, image rotation, time lapse zooming, automatic gridding with latitude and longitude insertion, geopolitical boundary generation and superimposition. Has produced updated image interpolation electronics for 486 minimum operation. Interpolation requirements can be tailored to customer use and needs.

Alden also provides the 9315 CTP Thermal Imaging Printer producing 256 shades of grey per pixel and 2,048 pixels per line for high-resolution photographic quality images.

Alden Weather Systems' products include the AE-900 Nortex Receiver and the survivable Satfind 406 EPIRB.

UPDATED

Northrop Grumman Corporation

Current Status
Aerojet has sold its Electronic and Information Systems business to Northrop Grumman Corporation, effective 22 October 2001. In December 2002 Northrop Grumman acquired TRW Space and Electronics. These acquisitions have boosted the technical base for remote sensing equipment and instrumentation.

The Aerojet plant specialises in electro-optic sensor systems. It was selected for a US$45 million contract by NASA Goddard in September 1993 for the Advanced Microwave Sounding Unit (AMSU) for NASA's EOS-PM1; US$82.4 million was added November 1995 for four further units (one will fly on Europe's Metop 1). For the new USAF DMSP D3 meteorological satellites, it was awarded a US$62.1 million contract in 1989 for three Special Sensor Microwave Imager Sounders (SSMIS). A US$7.9 million option was exercised in January 1990 for a fourth sensor. The 57 kg, 70 W sensor will provide 19 channels over 19 to 183 GHz for observations from 30 km to ground level, including rainfall to an accuracy of 5 mm/h and ocean windspeeds to within 2 m/s. Temperature profiles up to 73 km will also be generated.

The first SSMIS was delivered in October 1997. The first DMSP 5D-3 platform carrying the SSMIS was delivered in 2001.

Aerojet is one of two contractors awarded a US$4 million contract by NASA's Office of Earth Science for formulating an Advanced Technology Microwave Sounder. On 21 December 2000 Aerojet won a US$206.6 million order to build the ATMS. Aerojet has also developed a refined, less costly and lighter version of unspecified applications.

The plant also provides the imaging systems for the DSP early warning satellites. Each of the infra-red array's 2,000 PbS cells views a region on Earth less than 3 km across. The last two spacecraft of the second generation were retrofitted with generation 3 sensors, incorporating 6,000 elements. Aerojet delivered the last DSP sensor in August 1998, ending 28 years of delivering IR products to the USAF.

For TRW space capabilities see separate entry.

VERIFIED

Santa Barbara Remote Sensing Group

Current Status
As one of the companies embraced by Hughes Electro-Optical Systems, SBRS is now part of Boeing Satellite Systems.

Raytheon's SBRS specialises in remote sensing instruments for Earth, meteorological and planetary surface observations. It has specialised in-house team base with connections to a wide range of nationwide facilities and research centres; it also has international links for global consulting and analysis. Its engineering organisation was restructured in 1995 to be an integral part of the Hughes Electro-Optical Systems engineering centre in Los Angeles, reporting to the Space Electro-Optics Business Unit. By January 1997, 69 instruments designed and developed by SBRS since its founding 30 years ago had been launched. There were no failures, on average operating for 2.5 times their design lives, totalling 289 instrument-years. It provided Galileo's Photopolarimeter Radiometer, the Thematic Mapper and Multispectral Scanners for Landsat 4/5 and the Enhanced Thematic Mapper for Landsat 6, in addition to the main instruments for GOES and GMS. It built the Thermal Emission Spectrometer for NASA's Mars Observer. TES 2, completed in 1996, is aboard Mars Global Surveyor. The Visible Infra-Red Scanner (VIRS) delivered to NASA in 1996 for the Tropical Rainfall Measurement System (TRMM) is undergoing its integration and test cycle. The MODIS instrument for the EOS AM platform and the Enhanced Thematic Mapper Plus for LandSat 7 were completed and delivered in 1997. Developed by OSC, the SeaStar spacecraft, launched by an extended Pegasus launcher on 1 August 1997, carried the SeaWifs experiment, its sole payload, built by SBRC.

Developing synergistic technologies for sea-state satellite concept proposed to a NASA/NOAA environmental sciences group.

UPDATED

SeaSpace Corporation

Current Status
Founded on the work begun by scientists at the Scripps Institution of Oceangraphy in the mid-1970s, SeaSpace provides complete reception and processing systems for full resolution data from all sensors on polar and GEO satellites: NOAA HRPT (1 km AVHRR, TOVS, Argos), SeaStar (1 km SeaWiFS), DMSP RTD (0.5 km OLS, SSM/I, SSM/T-1, T-2), GOES GVAR (imager, sounder), GMS SVISSR imager, OceanSat-1, Meteosat PDUS imager. TeraScan software includes 340 UNIX-based functions for multisensor data fusion and analysis. TeraScan systems are operating in over 30 countries and at more than 350 academic, commercial, government and military installations worldwide. SeaSpace is a wholly owned subsidiary of the Allied Defense Group Inc and also produces the TeraScan software solutions programme. Current satellites in the TerraScan system include NOAA, FY-1C, Orbview 2, DMSP, Coriolis, GOES, Meteosat, GMS, FY-2, Terra, Aqua, IRS P4, ADEOS 2, Landsat 7, Rdarsat, ERS 2, Metop, Aura and NPOESS.

VERIFIED

Space Instruments, Inc

Current Status
SI specialises in design and studies of visible, infra-red and microwave sensors, their in-flight calibration, data handling, mission planning, control and detailed algorithms for signal processing. It built the Infra-red Spectral Imaging Radiometer (ISIR) for NASA Goddard, subsequently flown aboard Shuttle Discovery on STS-85 in August 1997 and the Total and UV Irradiance Radiometer (TUVIR) for the Naval Research Laboratory carried by MightySat 2 launched in December 1998. SI also manufactures a line of high-precision cavity radiometers for use as absolute radiometric standards, FireMapper Airborne Imaging System, ForestMapper Airborne Imaging System and a fire detection device with a thermal imaging radiometer. It has conducted miniaturisation designs for nanosat integration with existing operational systems. Space Instruments has designed in-space calibration devices for optimised data collection through sensors likely to degrade or deform with use.

UPDATED

Space Liaison and Imaging Corporation (SLIC)

Current Status
SLIC is a technical and marketing consulting firm, with offices in San Diego and Moscow. It represents Sovinformsputnik in the worldwide distribution of high-resolution satellite imagery from Russian military satellites. Most of these images were originally marketed through separate Russian government agencies and comprise the first generation products of former Soviet spy satellites with resolution down to 5–10 m. Provided archive of 1970s and 1980s Earth imagery for data bank users seeking comparative changes analyses. Marketed temporal data studies analysing selected areas to monitor surface changes over several decades. Attempts to diversify into SAR data dissemination. The corporation has conducted international liaison activities with government agencies in Russia, Western Europe and the United States, bringing together dissemination outlets for integrating data and images from a variety of different sources.

UPDATED

Spot Image Corporation

Current Status
Established in 1982, the company is the exclusive US distributor for Spot products and services. DigitalGlobe in Colorado is the Spot Image Corporation's sole distributor of Spot Image data to the US Department of Defense. Spot's standard product covers 60 × 60 km; scenes are available on CCT, 8 mm tape, QIC, 4 mm and CD-ROM or as photographic products (prints or transparencies) at a variety of scales and processing levels. A computerised catalogue maintains a record of all imagery. The corporation's services include image processing and production, applications and technical support. The image bank now includes more than eight million images.

VERIFIED

Telonics Inc

Current Status
Evolving from a partnership formed in the late 1960's to accomodate the needs of the wildlife research community, Telonics Inc became a privately held corporation in 1978. The company name derives from 'TELemetry electONICS consultants'. Telonics specialises in the design and manufacture of electronic environmental monitoring equipment, including radio telemetry for the wildlife field and related areas of environmental research and remote sensing systems. These include Earth stations with 1.2 to 1.73 m antenna/positioner systems for NOAA AVHRR/HRPT and TIP/TOVS, SeaWiFs, D-band Local User Terminals for Argos downlink processing, certified miniaturised satellite transmitters (PTTs) for oceanographic (low-cost drifters, temperature, depth profile chains and Argos PTTs with GPS positioning capability) and animal (birds; land and marine animals) research. Certified GOES Data Collection Platforms are built for stand-alone operation and for interfacing to dataloggers. Other products include PC-based orbital prediction and graphic display software with multiple propagator models, 1.0/1.2/1.4/1.8/2.4 m tracking antenna receiving systems for D/S/L-band applications, including NOAA HRPT receiving systems and SGLS full duplex command and control systems.

The ST-10 Argos PTT is designed for smaller mammals and birds. One bird version is less than 75 g. The electronics are available in two versions: 4.6 × 4.1 × 1.15 cm flat and 2.4 cm diameter, 4.6 cm high cylindrical. The more powerful ST-9 PTT provides 1 W transmit.

Also provides GPS collars for animal tracking applications and ST-14 units for meterological and oceangraphic applications. Telonics now develops and markets products for four functional operator and user categories: Earth stations, intrusion detection, oceanography and wildlife tracking.

UPDATED

Terra-Mar

Current Status
Specialist in developing software and integration of computer systems for GIS and raster image processing systems. Also for integrating separate software sets for different on-orbit systems converging at the user point. Terra-Mar is co-operating with EOSAT in a study of the economics of GIS for resource management and environmental applications. It is also a Landsat and Spot satellite data product distributor and is working closer with ESA on Envisat in utilising existing data sets and in developing software systems for future satellites. Terra-Mar has diversified into geotechnical engineering, construction materials testing, environmental engineering and drilling. Currently employs about 200 at its seven facilities.

UPDATED

Wirin & Associates

Current Status
The company specialised in the distribution of Russian satellite imagery, specifically 2 m resolution optical (KVR-1000), 10 m resolution stereo pairs (TK-350), KFA-1000 and MK-4 and is working with European consulting companies to integrate digital data and wet-film imagery. The franchise to disseminate Russian wet-film archive products was granted in 1992, one of several distributed to companies in the US and other countries. Sought higher levels of interaction between Russian satellite data and US and European remote sensing satellite information with a view to creating a world standard. Has made available radar and SAR data.

VERIFIED

EARTH OBSERVATION SATELLITES, SYSTEMS AND SERVICES

International

European Association of Remote Sensing Laboratories (EARSeL)

Current Status
EARSeL is a scientific network of European remote sensing institutes belonging both to academia and the commercial and industrial sectors. It is unique in that it represents the interests of these institutes rather than individuals. There are approximately 250 member laboratories in Albania, Algeria, Austria, Belgium, Brazil, Canada, Croatia, Czech Republic, Denmark, Finland, France, Germany, Greece, Hungary, Iceland, Ireland, Italy, Malta, Moldavia, Norway, Poland, Portugal, Russia, Slovenia, Spain, Sweden, Switzerland, Turkey, Ukraine and the UK. EARSeL is chaired by Dr Eberhard Parlow, University of Basel, with vice-chair Prof Rudi Goosens, University of Ghent. The secretary general is Dr Raimer Reuter, university of Oldenburg,

EARSeL was founded in 1977 under the auspices of the European Space Agency, the Council of Europe and the European Commission and is run by a Council of elected national representatives from each country where there are member laboratories and an executive Bureau elected by the Council. The main scientific efforts of EARSeL are concentrated in the Special Interest Groups (SIGs) which form the foundation of the activities of EARSeL. They encourage co-operation and foster innovative applications of remote sensing. The SIGs organise workshops and specialist meetings, the reports and proceedings of which are published, the conclusions and specific recommendations presented to the sponsoring agencies and other relevant institutions including International Society of Phorogrammetry and Remote Sensing, EUROGI, OEEPE, Eurogeographics and EARSEC.

VERIFIED

United States

Orbital Sciences Corporation/Orbimage

Current Status
OrbView-3 was launched by Orbital Sciences Corporation on a Pegasus rocket on 26 June 2003 and offers 1 m resolution panchromatic and 4 m multispectral digital imagery. In September 2004 OSC was awarded a US$500 million contract by the National Imaging and Mapping Agency for the high resolution Orbimage 5 satellite. Known as the NextView/Orbimage 5, the satellite will join Orbview 3 in early 2007 and provide 0.41 m resolution from its orbiting altitude of 660 km.

Background
Orbital Sciences Corporation's Orbimage subsidiary was formed in 1992 to build and operate OSC's commercial Earth observation system, and to market the company's remote sensing products and services.

Orbital Sciences Corporation's Eyeglass International was formed in 1994 to build and operate the Eyeglass satellite system and to market imagery products and services. Orbimage in 1995 decided instead to pursue the smaller, lower cost OrbView satellite derived from its MicroLab design. The cost of deploying one satellite and establishing the ground system is less than US$100 million. OSC's MacDonald Dettwiler Associates is managing the ground segment requirements. Saudi Arabia's EIRAD Co signed in June 1994 to become a major investor (operating a receive station); DoC approval was granted 5 June 1995. EIRAD took a 20 per cent stake in Orbimage; US regulations impose a 25 per cent limit. OrbView 1 was launched in 1995 followed by OrbView 2 in 1997 and a second-generation system. These consist of OrbView 3 launched 26 June 2003 and OrbView 4 launched September 2001 but lost in a launcher failure. OrbView 3 and 4 offer 1 m panchromatic and 4 m multispectral digital imagery. OrbView 4 would have offered the first hyperspectral imagery.

US distributor of Radarsat-2 data planned for launch in 2005. National Space Program Office of Taiwan selected OSC to build the spacecraft for the ROCSAT-3/COSMIC programme which envisages a constellation of six remote sensing microsatellites to collect atmospheric data for weather prediction and for ionosphere, climate and gravity research.

Specifications
Resolution: 1 to 2 m PAN (4 to 8 km swath), 4 m XS (8 km swath)
Launch: launched in the fourth quarter of 1997 by Pegasus XL
Lifetime: 3 years
Mass: 146 kg
Orbit: planned 460 km polar, 10.30 equatorial crossing, revisit less than 3 days
Spacecraft: 96.5 cm diameter; disc about 60 cm high. Although there will be some onboard storage, most imagery will be downlinked in real time
Sensor: bands (μm): PAN 0.50–0.90; XS 0.45–0.52, 0.52–0.60, 0.63–0.69, 0.76–0.90.
Orbimage has filed for financial restructuring and sought protection from bankruptcy under Chapter 11 in September 2001.

UPDATED

Space Imaging, Inc

Current Status
Responding to a new policy directive from the White House issued May 2003, Space Imaging, Inc has teamed with Lockheed Martin Corporation and Raytheon Company as principal subcontractors bidding on NextView, the National Imagery and Mapping Agency (NIMA) competitive contract for sustained availability of high-resolution imagery. NIMA is already a major customer for Space Imaging.

On 19 January 2001, Space Imaging announced that it had been granted a licence by the National Oceanic and Atmospheric Administration to launch and operate a remote sensing satellite with 0.5 m resolution in panchromatic black and white and 2 m multispectral. Launch is scheduled for 2005. In March 2001, Eastman Kodak announced it had struck a deal with SI for it to be the master reseller of Kodak Earth Imaging Products which includes CITIPIX aerial imagery of 95 major US metropolitan areas, 7,000 cities and 600 counties in the US and Canada. In October 2001, SI announced it had signed a deal with Navigation Technologies Corporation making it master reseller of NAVTECH® and NAVSTREETS® data.

Space Imaging's satellite system was initiated by Lockheed Corporation as the Commercial Remote Sensing System in April 1991. In November 1996, Space Imaging acquired Earth Observation Satellite Co (EOSAT), which in 1995 experienced a US$3 million loss on revenues of US$30 million. Founded in 1985 to take over the commercial marketing of Landsat imagery and eventually the satellites themselves. EOSAT had a worldwide distribution network and access to imagery from several leading space-based imaging systems. These included the Indian government's Remote Sensing System (IRS), the Canadian Space Agency's Radarsat, the Landsat series satellites, Japan's JERS satellite system and the European Space Agency's ERS synthetic-aperture satellite. India's IRS-1C has a 5 m resolution capability.

Space Imaging aspires to becoming the world's leading distributor of Earth observation and remote sensing imagery and to revolutionise the way people use imagery to conduct business in the 21st century. It gathers, processes and produces information about the Earth's natural and man-made features with an unequalled combination of resolution, accuracy, information, content, speed of delivery and price. Space Imaging clients are in a wide variety of markets including environmental monitoring, mapping agricultural monitoring, urban planning, oil and gas exploration, disaster assessment, forestry management, telecommunications, real estate, insurance, entertainment and media. The market for high (1m) resolution imagery worldwide has been estimated at US$3 billion to US$5 billion over the next five to six years.

Space Imaging's imagery and information products are stored in a digital archive, which houses space-based and aerial imagery from multiple sources. Users can search the CARTERRA Archive using a standard Web browser; search for Ikonos, Landsat and IRS imagery in a point-and-click environment; and browse the imagery and accompanying geographic information available online. The catalogue is updated with new imagery several times each day.

Lockheed was granted a licence on 22 April 1994 by the Department of Commerce to operate a private, 1 m resolution, Earth imaging system under Presidential Directive 23 March 1994. Lockheed formed Space Imaging in May 1994 to handle the project; it became an incorporated company in December 1994. The cost for two satellites in orbit, two launchers, a primary ground station, two supporting stations, all image processing hard/software and five years of operation was projected at about US$500 million.

The Ikonos satellite routinely collects 1 m PAN and 4 m XS digital imagery, plus 1 m PAN-sharpened or colour enhanced imagery. Mapping North America's largest 100 cities is an early priority. Products include: radiometrically corrected imagery, geometrically corrected imagery, ortho-rectified imagery, digital terrain models and multispectral ratio images. Space Imaging is establishing a network of global regional affiliates that will take customer orders, task the satellite to collect requested imagery and market imagery in their regions. The ground segment covers a primary station in Thornton and two support stations in Alaska and southeast US. The regional affiliates will also operate their own stations. Tasking commands can be modified up to 10 minutes before satellite contact, accommodating last-minute priorities or changing weather conditions. Each station can download images in real time while in contact. Satellite PDR was completed 16 June 1995. A failure involving the Lockheed Martin Astronautics Athena II launcher during the launch attempt on 27 April 1999, destroyed the Ikonos 1 satellite when it failed to reach orbit. Space Imaging launched a replacement on another Athena II in September 1999.

Specifications
Resolution: 1 m Pan, 4 m XS. It will be able to image between ±85° latitude and register any point within the image to less than 12 m horizontally. With inclusion of ground control points, products will have less than 2 m horizontal accuracy
Design life: 7 years
Mass: less than 900 kg at launch, including 50 kg hydrazine
Orbit: 680 km circular in 98.20° Sun-synchronous, descending node between 10.00 and 14.00 local time, providing same-scene access every three days or fewer at better than 1 m ground sample distance

Spacecraft: 1.5 m diameter hexagonal bus, 2 m height. High-accuracy 3-axis pointing control and knowledge by CT-601 star trackers, rate gyros, thermal stability and refined scanning laws. Hydrazine thrusters. Slew rate sufficient for both wide-area monoscopic and same-pass stereo collecting. Two-axis high-gain X-band antenna downlinks wideband imagery at 320 Mbit/s; stored in 64 Gbit SSR. 1,500 W from three deployed GaAs/Ge panels plus bus panel; 50 Ah Ni/H_2 battery
Sensor: totally reflecting optical telescope with 0.7 m aperture primary mirror mated to a digital focal plane array, which can generate 6,500 lines/s PAN. 11 km nadir swath. Spectral ranges (μm): Band 1 0.45–0.52, Band 2 0.52–0.60; Band 3 0.63–0.69; Band 4 0.76–0.90; PAN 0.50–0.90. PAN utilises 12 μm-pixel pushbroom linear CCDs; XS 48 μm-pixel four pushbroom linear CCDs. Low-loss data compression on original 11-bit data.

VERIFIED

GENERAL

Australia

Asia Pacific Aerospace Consultants Pty Ltd (APAC)

Current Status
APAC is a commercial space consultancy specialising in the Asia Pacific region. Its expertise covers satellite-based communications, launch services, remote sensing and related areas. It provides support for commercial space companies interested in growing their business in the Asia Pacific, and to companies located in that region that are interested in growing global space business. APAC generally works at the strategic business planning level. Its services include identifying market trends, undertaking market research, locating and researching prime marketing targets, identifying critical success factors, and advising on objectives and strategies. Assessing the risks of space-related investments, contributing to feasibility studies, advising on acquisitions, and identifying and making contact with potential investors and customers are all services APAC offers. The company primarily serves the region from Japan and Korea to Australia and from Thailand to Indonesia with clients also in Europe and the USA. It has also been called upon to advise companies and government departments on the applicability of space technologies to non-aerospace related industries and to policy-making decisions related to high technology requirements.

UPDATED

Auspace Limited

Current Status
Auspace was established in 1983, and in 1986 it became a subsidiary of Matra MarconiSpace, with capabilities and experience in satellite system engineering and R&D in space systems, ground communications satellites and electro-optics.

Current and recent projects include: the design and building of the Endeavour experimental ultra-violet imaging system to space-qualify a large format photo counting array detector for use in satellite astronomy. The focal plane array for the ETS 2 along-track scanning radiometer; major contractor for Envisat's Advanced Along-Track Scanning Radiometer (infra-red/visible focal plane array, foreoptics, instrument electronics unit and signal pre-amplifier unit); an airborne prototype of a passive spaceborne atmospheric pressure scanner for CSIRO; a high-performance multispectral scanner for airborne applications; conceptual study of the ARIES satellite to have provided hyperspectral imaging of the Earth in the visible and short-wave infra-red regions of the spectrum for applications in the mining, environmental and agricultural areas. Auspace supplied a system to the Defence Science and Technology Organisation to provide rapid access to SAR satellite images and to assist them in an exercise to investigate feature extraction of SAR images over oceans.

In 1994, Auspace sold its GPS business to SIGTEC and evolved a remote sensing programme called Aries (Australian Resource Information and Environment Satellite) with a scheduled launch date of 2001. This programme has now been cancelled.

Auspace is a core partner of the Cooperative Research Centre for Satellite Systems which developed the FedSat microsatellite launched by Japan's H2A on 14 December 2002.

VERIFIED

Australian Space Insurance Group (ASIG)

Current Status
ASIG provides capacity for the worldwide space risk insurance market and represents national interest in brokerage and underwriting with co-brokerage arrangements in London, New York and Tokyo. Launch risk, on-orbit insurance, life risk and revenue loss coverage. It has bundled categories of space related insurance into higher risk packages for marine and aviation sectors.

VERIFIED

British Aerospace Australia (BAeA)

Current Status
British Aerospace Australia is one of Australia's leading defence and aerospace companies – having supported Australia's Defence Forces for nearly 50 years.

With more than 1,600 employees located at the company's 25 sites around Australia and overseas, British Aerospace Australia is the country's second largest defence employer.

The company's strong commitment to meeting Australian Defence Force (ADF) requirements has led to a broad range of capabilities in the support of land, sea and air defence systems including: aircraft systems integration, military vehicles, missiles, decoys and targets, electronic warfare, electro-optics and communication, command and control.

Under contract to the Australian Space Office, British Aerospace Australia manages and operates the Moblas 5 facility located at Yaragadee in Western Australia. Staffed by five technicians, the station provides satellite laser ranging data for investigation of earth geodynamics.

British Aerospace Australia has managed and operated the Canberra Deep Space Communication Complex (CDSCC) on a turnkey basis for the Australian Space Office since 1990. In 1996 the Australian government transferred management of the facility from the Department of Industry, Science and Technology to CSIRO. In 1998 the management contract was transferred to BAeA.

The company employs more than 140 engineers, technicians and administrative personnel who carry out the entire CDSCC operation. Their technical activities provide a link between the ground control and processing environment and various Earth orbital and deep space spacecraft. In addition British Aerospace Australia provides a broad range of support functions.

On behalf of the Australian Surveying and Land Information Group (AUSLIG), British Aerospace Australia provides data acquisition, processing and distribution services for the Australian Centre for Remote Sensing (ACRES). Located at the Data Acquisition Facility in Alice Springs and the Data Processing Facility in Canberra, British Aerospace Australia employs approximately 35 personnel. These personnel are responsible for the operation of a complex manufacturing system which generates highly customised photographic imagery and digital data products by using data received from orbiting remote sensing satellites. These products are sold through a network of Australian distributors and satellite operations outside Australia. Some work has been completed on integrated data packets involving optical imaging and radar.

UPDATED

Electro Optic Systems Pty Ltd (EOS)

Current Status
EOS specialises in sighting and surveillance applications in the aerospace and defence markets including satellite laser ranging and tracking. It has developed high-power, eye-safe lasers for communications links to satellites and for satellite-to-satellite links. Technology for optical transmission links developed. Has provided laser tracking telescopes at six sites for Project Keystone measuring crustal deformation in the Tokyo, Japan region. EOS has applied this design to a new technology incorporating path-adaptive optics for field use. These technologies have been adopted for land-based optical sighting devices.

UPDATED

Hawker de Havilland Ltd

Current Status
HDH's space activities focus on structural, mechanical and thermal systems. HDH has conducted research projects which aim to unite the alloy components of conventional satellite within the same stress envelope as the carbon fibre structures preferred for light weight, strength and long life. It has studied small satellites and worked with BAeA on feasibility studies for the Southern Launch Vehicle. HDH developed a synergistic multi-material based thermal reaction structure designed to conform to temperature profiles. HDH has also developed lightsat technologies applicable to small, low cost, environmental research satellites. HDH provides low-cost composite structures for aerospace applications, has developed carbon fibre elements for ultra lightweight specifications.

UPDATED

KEL Aerospace Pty Ltd

Current Status
KEL specialises in HF radar, advanced HF components, electrical transient monitoring, science instrumentation and commercial information systems. Its ERS Frame Synchroniser is a combined high/low rate frame format synchroniser for data processing ground stations. Produces IPS-71 Advanced Digital Ionosonde, a ground-based radar system that measures the characteristics and movement of the Ionosphere. Specialises in OTH radar for both Ionospheric and surface wave propagation. Has been given government quality award for upgrades to the JINDALEE OTH radar system at Alice Springs in Central Australia.

VERIFIED

Austria

ACT Hochleistungskunst-stofftechnik GmbH

Current Status
ACT provides development, manufacture and assembly of lightweight structures and components, mainly using composites.

The company has specialised in significant applications of new and innovative technology and has pioneered many structures and materials (composite and alloy) which have been used not only in the aerospace industry but in several commercial sectors also. It provided the composite struts for ISO's sunshade support, the first satellite composite hardware designed and manufactured nationally. High-grade hybrid composites for extendible booms, support struts, BAPTA rings and ribs on furlable arrays. It provided the solar array yoke for Orion and Tempo. ACT has developed designs for new lightweight solar array support structures. Proposing compatible frame ultralight support frame for large deployable antennas. Has responded to enquiries for very large deployable array structures where rigidity and low weight are primary requirements packaged in a compact volume with long duration storage capability.

UPDATED

Andritz AG

Current Status
Andritz fabricates the rings for attaching Ariane 5's solid boosters to the cryogenic core and is developing support rings for the advanced Ariane 5. Andritz has been approached regarding lightweight stage separation truss elements for an undisclosed customer. Work has diversified to a limited extent but the company has applied fabrication techniques for the attachment rings to its other business.

UPDATED

Austrian Aerospace GmbH

Current Status
Founded in 1997, Austrian Aerospace GmbH was created by merging Schrack Aerospace and OR5, both companies active since 1983 focusing on electronics, mechanics and thermal hardware. The merger was driven to some extent by the market trend which made commercial sense to integrate and maximise the inherent potential in each separate leg of the combined operation. Austrian Aerospace is a manufacturer of electronics equipment and subsystems for communications, Earth observation and science satellites. It specialises in the design/development of onboard Earth observation instruments, subsystems for optical intersatellite links and satellite control/monitoring electronics. Also developed an interactive and semi-automated satellite control and response logic for ground contingency recovery procedures designed to prevent catastrophic collapse in a sudden and unpredicted satellite failure. In the ground segment, it is developing satellite checkout equipment and has conducted studies in satellite datalink relay units.

UPDATED

Austrian Research Centres Seibersdorf

Current Status
With a staff of 500 across the country, Seibersdorf is Austria's largest non-university research centre. The federal government holds 50.5 per cent of the shares. Activities are focused in five divisions: instrumentation and information technology; process and environmental technologies; engineering; life sciences; and systems research (technology, economy and environment). Seibersdorf holds a long-term contract as an ESTEC test house for space materials, qualifying advanced materials and developing space-specific tests, for example, mass loss, tribology and cold welding in the space environment. Developing remote sensing suite for follow-on Envisat proposal. Seibersdorf further participates with ESA material technology programmes like FESTIP or ALSCAP. ARC provides assistance to ESTEC through the Space Materials Testhouse supporting welding and soldering technology, simulated space environment testing and characterisation of materials in space. Seibersdorf provided support for ESA's Giotto and developed an ion emitter for controlling electrostatic charging of Geotail, Interball and Cluster. Other work includes a micro-analysis station for materials testing in space, investigation of µgravity effects on the human immune system and genetic material. The Centre has conducted studies of integrated biomedical and biophysiological research linking microgravity and 1 *g* research in a unique interpolation of separate disciplines based on space medical research.

VERIFIED

Böhler Edelstahl GmbH

Current Status
Böhler is one of the world's largest producers of special steel, including materials for satellite components since 1982 and Böhler Edelstahl was established in 1991 as a subsidiary of Böhler GmbH. The P802 Invar and P812 Super-Invar steels were employed in the Rosat X-ray astronomy satellite's mirror system to ensure accurate alignment and dimensional stability. Nickel-based alloys are supplied for Ariane and new lightweight alloys are being developed for structural elements of satellites and for lightweight antenna structures. Has been asked to develop new integrated structures combining load bearing, circumferential, support elements for hybrid materials including reinforced carbon and steel alloys. Currently employs about 2,100 people and has an average turnover of €450 million.

UPDATED

Fischer Advanced Composite Components – FACC

Current Status
Fischer Advanced Composite Components (FACC) was established in 1989 to centralise Fischer GmbH's composite component activities. The company made an early name in alloying and composite integration welds and in new fabrication and formulation processes which were applied directly to space engineering requirements and to the commercial non-aerospace market. In 1991 Salinen AG took over 50 per cent of FACC's shares. An annual research and development budget of US$2 million supports production of components for space applications including filament winding for high tensile booms. Specialises in aircraft components and aircraft interiors, basic engineering solutions and CAD/CAM in 3D applications. Composite structures for lightweight planetary spacecraft landing gear in development. Has been asked to bid on large composite matching structures with a metallic core base and on hybrid structures for high load bearing specifications.

UPDATED

Österreichische Raumfahrt- und Systemtechnik GmbH (ORS)

Current Status
ORS is owned by Saab Ericsson Space (74.9 per cent) and Dornier (25.1 per cent). Sales in 1996 were Sch98 million (1995 Sch130 million); 61 staff. Flight and ground support equipment covers ERS, ISO, Ariane 5, Columbus, Soho, Cluster, Huygens and Artemis. It designs, develops and manufactures spacecraft structures, mechanisms and thermal hardware and MGSE. For Italsat, ORS supplied the MGSE and a full-size engineering structure manufactured from aluminium honeycomb. It is developing MGSE and thermal hardware for Envisat 1 instruments. The company provided ISO's aluminium honeycomb sunshield and its support. The company has also developed high-stress, low-mass, cantilevered composite material based space structures for unique applications where high-density packaging and long life are essential.

Other fields concern mainly ECLSS (hygiene, food, water management) and space biology (design of biological and µgravity payloads, utilisation studies). ORS has developed autonomously operated closed-loop ECLSS systems for space station applications and open-loop systems for small resupply vehicles. In addition, ORS is engaged in space robotics (end-effectors, sensors) and the design of advanced composite material and high-accuracy structures. ORS has contributed to the development of environmental control systems in the ESA Columbus laboratory. Developing advanced food/hygiene/contaminant containment facilities for human habitations in space.

VERIFIED

Plansee AG

Current Status
Plansee is a leading producer of refractory metals (tungsten, molybdenum, tantalum, niobium, chromium and their alloys, heavy metals, tungsten-copper alloys, copper-chromium). It is a manufacturer of components for high-temperature applications, and offers a range of products and services for space industries including ceramic thermal protection for re-entry vehicles. Specialises in high-performance materials and innovative tooling systems for cutting and wear protection. Can produce hybrid structures for unique specifications derived for clients as needed. In January 2003, Plansee acquired the California, US-based Polese Co Inc, and now employs 500 people at five locations in the US. Plansee group now employs 6,400 employees in 43 production and sales subsidiaries worldwide.

UPDATED

Siemens AG Österreich

Current Status
Siemens Austria specialises in communications and data technology, power engineering and automation, and programme and system development. Space activities include onboard processing, application of expert systems and object-oriented software for ground systems, telescience software and data processing, EGSE and software development for ESRIN, ESOC and ESTEC. The company has developed satellite imaging technology applications and is designing interactive optical software control systems for adaptive lenses. Providing EGSE and mission control system for ESA's Herschel Plank programme. Siemens AG has developed new methods of telerobotic support software and has been approached by ESA to examine telepresence technology capable of duplicating human actions independent of direct mechanical control. Specialised integration capabilities have now been developed for combining microtechnology and new open architecture designs for complex systems designed and developed for small installation volumes.

UPDATED

Steyr-Daimler-Puch Fahrzeugtechnik AGd Co KG (SDPF)

Current Status
In its former business structures the company that now comprises the base of SDPF has been producing specialised motor vehicles since 1900 and in 1928 merged with Daimler of Austria. SDPF developed and produces mechanical components for Ariane 5,

such as cryogenic fuel and pressurisation lines, and booster attachment elements. Machining of fuel valve casings and pressurisation manifolds. Moulding of hard foam insulation for feed lines. Developed flexible high-temperature resistant bellows and thermal protection sleeves for long-run piping and conduits. Manufacture of XMM mirror shell containers. Participation in ESA's FESTIP study and the Austro/Swiss ALPSAT project. Development of high-pressure gas and liquid vessels for satellites. Concept development for large capacity vessels applicable to propellant systems for satellite and pressurant vessels for launch vehicle upper stages.

UPDATED

Germany

Anite Systems GmbH

Current Status
Previously Cray Systems GmbH, the company is a subsidiary of Anite Systems Ltd of the UK. The company is a specialised spin-off from the computer design and manufacturing company. It specialises in the design and development of software and systems for ground segments including multi-object tracking. Developing scan-track technology for multisatellite data downloading from LEO. This technology adaptation has been added to the product range and has improved multi-object tracking tasks. AS diversified into the production of multisourced tracking for active and repeater transponders.

UPDATED

Autoflug GmbH & Co

Current Status
Autoflug GmbH is one of four member companies in the Autoflug Gruppe GmbH and specialises in rescue and safety technology, fuel measuring and control systems, gyroscope-based sensors and simulators. Development and manufacture of parachute recovery systems and shock-absorbing landing bags, flotation and location systems for launch vehicles and components with specialisation in capsule recovery. Systems applicable to entry descent requirements for soft landing planetary spacecraft. Autoflug have developed designs for the recovery of exceptionally heavy loads, either large capsules recovered from atmospheric descent or oversized booster stages for launch vehicles. The company has designed and promoted descent retardation systems for planetary landers and for heavy weights dropped by parachute in Earth's atmosphere. Some of these designs are incorporating hybrid retro-rockets and inflatable impact absorbing landing bags of the type first used in planetary exploration by Mars Pathfinder.

UPDATED

Bankhaus Reuschel & Co

Current Status
A private bank based in Munich, Bankhaus Reuschel provides space project financing support. The bank has played a leading role in several international ventures throughout Europe. Involvement includes Eutelsat, EUMETSAT and Ariane 5 with secondary financial interests in partnerships. Skills in developing privately funded international ventures in satellite and space vehicle applications to commercial activities. Has provided analytical surveys for ESA on consequences of redirected funding from the EU on space project management. Also specialised capabilities for integrating national and international ventures with cross-currency accounting and project/programme management standards to full accounting levels.

UPDATED

Carl Zeiss

Current Status
Carl Zeiss provides optical equipment with separate divisions providing instruments for astronomy, industrial meterology, microscopy, opthalmology, photogrammetry, semiconductor technology, surgical operations and surveying. Carl Zeiss specialises, in the space field, in high-quality optical systems such as telescopes and microscopes. It will provide Abrixas's X-ray optics system. Hipparcos ROSAT and SELEX and ESTEC's large space simulator employ the company's optics. Research into satellite imaging optics technology. New optical folding mirror mechanisms for remote sensing satellites and planetary spacecraft. CZ has developed folding optic arrays for adaptive lens systems on telescopes for Earth observation and astronomical purposes. Has designed adaptive optical mirrors for large observational devices.

UPDATED

Computer Anwendung Für Management GmbH

Current Status
A subsidiary of France's CISI, CAM's space expertise includes control centre software for controlling satellites such as TV-Sat, DFS Kopernikus and Eutelsat, computer facilities for the Columbus payload operations centre, and development of image processing systems for Earth observation satellites. CAM has provided systems integration software for the Columbus programme and new control logic for multiple payload information processing. It has proposed a multisatellite control system for co-ordinating remote sensing tasks with Earth-based command and control facilities. CAM has developed very advanced command and control systems capable of being semi-autonomously operated by preset instructions.

VERIFIED

Drägerwerk AG

Current Status
Designs and manufactures pressure control systems and environmental support system hardware for a wide range of industries. Has developed high-pressure cleaning systems for high-risk industries. Drägerwerk specialises in gaseous control systems and gas sensors. These include systems for Spacelab and Columbus ECLSS and for onboard scientific experiments. Equipment developed, manufactured and qualified covers vessels, valves, gas and air filters, servicing units, carbon monoxide removal systems, contamination monitoring and control systems, and water servicing units. Has contracts to develop miniaturised systems for personal life-support equipment. Developing enhanced efficiency, low cost, life-support systems for personal or habitat use. Further analyses have applied existing ECLSS technologies to extended duration missions and to demanding new requirements generated by specifications driven by as diverse a range of capabilities as long duration missions and mini-shuttle reusable earth-to-orbit transportation systems.

VERIFIED

EADS (DaimlerChrysler Aerospace AG (DASA))

Current Status
Early in 2003, the UK BAE Systems gave up its 25 per cent share in Astrium, ceding to EADS total ownership of that company. This has had an effect on EADS-Germany by the potential business interests that will accrue.

Background
Cleared by the European Commission in May 2000 for DASA, Aerospatiale Matra and CASA to converge and establish a new company, the European Aeronautic Defence and Space Company (EADS) was formed on 10 July 2000 with shares offered on stock exchanges in Frankfurt, Madrid and Paris.

DaimlerChrysler created DASA (known until 1 January 1995 as Deutsche Aerospace) in 1989 from the country's four major aerospace companies: MBB, Dornier, MTU and TST. Their resources were reorganised in 1991 into four operating groups: space, aircraft, propulsion and defence/civil applications. MBB/TST were absorbed into DASA in 1992; Dornier retains its identity. 1995's DM15 billion revenue produced a DM4.3 billion loss (1994: DM438 million loss on DM17.39 billion; 1993: DM694 million loss on DM18.6 billion). Workforce totals 75,000, from 86,000 in 1994. The Space Systems Group's 1995 revenue totalled DM1.9 billion (1994 DM1.354 billion; 1993 DM1.4 billion; 1992 DM1.91 billion); personnel total 4,200. SSG comprises two divisions: Satellites Systems (Dornier) and Space Infrastructure; they are specified in greater detail below. DASA in 1992 took a 12.25 per cent interest in Space Systems/Loral of the US. Following reorganisation in 1996 the DASA Space Systems Group was split into Dornier Satellitensysteme GmbH and DaimlerChrysler Aerospace – Space Infrastructure Division.

The Space Infrastructure Division is responsible for infrastructure programmes such as Columbus; DASA is leader (the €658.5 million fixed price construction contract was signed with ESA 28 March 1996). The Bremen division was prime contractor for Spacelab, including design, development, integration, qualification and operational support, and for ESA's Eureca retrievable μg carrier. It also conducts the Texus and Express μg research programmes. It is a major contributor to the Ariane programme, including prime for stage 2 integration. The company is prime for the Astro-SPAS reusable platform.

The former Space Transportation Systems and Propulsion division was divided in 1993 into two centres of excellence, one for the Satellite Systems division and the other for the Space Infrastructure division. Strategic responsibility for satellite systems was transferred in 1991 by DASA to Dornier, while production of subsystems remains in Ottobrunn. The former MBB was prime for Exosat and provided Galileo's propulsion module. It undertook 40 per cent of the work on Rosat, provided the Imaging Compton Telescope and part of the Energetic Gamma Ray Experimental Telescope for NASA's Compton Observatory, and developed the Infra-red Space Observatory's payload module. Extensive communications satellite involvement includes DFS (platform, integration/testing), TV-Sat (system management, integration/test, AOCS, propulsion, antenna systems, TT&C), TDF (AOCS, propulsion), and Eutelsat 2 and Turksat (AOCS, TT&C, solar array); the Spacebus platform is offered in association with partner Aerospatiale. Earth stations have been provided to Spain, Belgium, Australia and Italy, in addition to a Landsat terminal in Italy and the ERS antenna in Antarctica. Far field and near field ranges are operated for the testing of antenna systems. The division works with France's SEP on Ariane engine development. The division also develops and manufactures opto-electronic sensors and cameras, such as the MOMS-02 package which flew on Russia's Priroda Mir module.

VERIFIED

Elbe Space & Technology Dresden GmbH & Co KG (ESTD)

Current Status
ESTD was formed in 1992 to market RFAS products and services, finance enhancement of Russian systems for the Western market and act as a contractor in Western projects. Its STIR (Science and Technology International of Russia) subsidiary is headquartered in Moscow. ESTD offers satellite-based navigation and communications systems

such as Elekon, GPS/Glonass receivers, Start launch services, IL-76 parabolic flights and software development. ESTD developed adaptive navigation ground receivers compatible with Navstar GPS and Glonass satellite. Also provides support to GPS marketing companies and operates as distributor for integration software. Elbe is working on a new generation of adaptive global positioning system technologies for the European Galileo system.

VERIFIED

Elektronik- und Luftfahrtgeräte GmbH

Current Status
Elekluft is now absorbed into the EADS organisation and its space activities have been integrated into the space business units of EADS. Elekluft, wholly owned by DaimlerChrysler Aerospace, provides system support for ESOC and EUMETSAT and general support for ESTEC, DLR and the Ariane Kourou base. Advanced system support technologies have been developed for increasingly demanding operating requirements of a new generation of space science and meteorological programmes. Space personnel total 75.

UPDATED

ESG Elektronik System GmbH

Current Status
Originating in 1963 as Flug-Elektronik-Gesellschaft mbH, ESG was founded in 1967 and merged in 1992. Principal space activities cover system design and analysis, system simulation, information management, logistics, operations support, artificial intelligence and expert systems, and product/quality assurance. Contracts have included Anthrorack failure analysis for Spacelab D2 and logistic concepts for Columbus in addition to internal racking design. ESG have conducted analysis of privatised space station module concepts, interaction with government funded stations and logistics/resupply flight requirements. Studies have been carried out linking automation control systems, advanced high-speed computers and information management systems to control and operate new automated space systems involving several separate satellites and controlling ground segments. Recent activity includes telepresence and telerobotic studies with breadboard mockup of conceptual systems involving automated and human-tended operations.

UPDATED

Kayser-Threde GmbH (KT)

Current Status
Kayser-Threde is one of seven companies in the SEKAS group and is active in telematics, process control, data management, Earth observation, scientific payloads and specialised optical systems. KT provides expertise in the design, build and integration of space experiment payloads, covering sounding rockets, Spacelab, SPAS, Photon, Shuttle and re-entry capsules. Sales in 1995 were DM51.5million (DM80million including subsidiaries); staff totals about 240 people. Experience covers: 43 sounding rocket payloads (7 Aries, 33 Texus, 3 Maxus), vestibular experiments on Spacelab 1 and Mir-92, venous pressure measurements on Spacelab 1/D1, high-precision thermostat on Eureca 1, Spacelab D1/D2, Cryostat protein growth facility on IML-1, main contractor for Astro-SPAS' Orfeus telescope and Holop holographic lab for Spacelab D2. It was system integrator for all German experiments on Mir-92. The Energia Deutschland GmbH partnership with NPO Energia exploits an exclusive right to distribute Earth observation data from Mir.

KT built Temisat (plus a spare) under a US$10 million contract to Telespazio for piggyback launch on 31 August 1993 with a Meteor 2 aboard a Tsyklon, demonstrating relay from environmental data collection platforms. Temisat was the first of KT's Blackbird family of micro-satellites, although no others are under development. KT also acted as launch agent for the Technical University of Berlin's Tubsat B in January 1994. It built the GFZ 1 laser geodetic satellite for Germany's National Centre of Earth Sciences for an April 1995 launch on a Russian Progress Mir station ferry.

KT developed a range of DGPS stations for DARA and is involved in the pre-development of an advanced combined GPS/Glonass receiver for ESA. The company has been marketing Glonass receivers since June 1991. It provided the MOMSNAV orbit/attitude determination package for Germany's MOMS-02P imaging system on Mir's Priroda module. This uses two GPS receivers and a high-precision gyro package for 2 m/3 arcsec accuracy. A Precise Range & Range Rate Experiment (PRARE) package was provided under DARA contract for January 1994's Meteor 3 meteorological satellites. KT developed the transputer-based multiprocessor for Soho's SUMER instrument and the camera control processor for Mars-96's High Resolution Stereo Camera.

KT developed the Biopan biological experiment carrier for the exterior of Photon and Resurs descent capsules under ESA contract. The hinged lid on the 37 cm diameter 22 cm high pan-shaped container opens through 180° to expose samples mounted on the lid and fixed bottom plate, totaling 910 cm^2 for experiments. Its own 2 Mbyte microcontroller commands the power supply for the hinge drive and heaters (maintaining 0–37°C) and performs temperature sensor switching. Solar aspect angle and 0.25–4.5 μm ultra-violet radiation are also monitored. A 3.8 cm thick layer of glass fibre and phenol provides re-entry protection. The test flight was made aboard Photon 5 in October 1992. ESA has approved two further flights; first was June 1994 on Photon 6 carrying six experiments, using the refurbished ground qualification unit. The next was in September 1996.

The company signed an agreement with Glavcosmos in December 1987 for options on Photon and Resurs missions, acting as marketing agent. Four missions flew 1989 to 1991, carrying Intospace's Cosima 2–4, Casimir and LZZ experiments. A further contract was signed in autumn 1991 for up to six flights. The next crystal growth experiments flew on Photon 6 in June 1994. Photon 7 in February 1995 carried seven KT materials processing experiments under DARA contract in the Zona 4M and Konstanta 2M furnaces. MIRKA (Micro Re-entry Capsule) was launched aboard the Photon 8 satellite in October 1997. The 150 kg MIRKA was mounted on the outside of the Photon descent capsule where it remained until after the main spacecraft's de-orbit manoeuvre had taken place. It came down 110 km southeast of Orsk on 23 October 1997.

Kayser-Threde was awarded the contract from DARA to be responsible for the logistical and technical preparation for the MIR-97 visit of Reinhold Ewald to Mir – a role which the company had played for the MIR-92 mission. Progress-M33 launched in November 1996 carried 158 kg of German scientific equipment to Mir in advance of the Soyuz-TM 25 launch. Kayser-Threde was responsible for a MIR-97 experiment involving real-time transmission between Mir and the German control centre at Oberpfaffenhofen. The company is now participating in the Columbus experiment racking and experiment schedule. Development work has been carried out on long-term science experiment support aboard the International Space Station with a view to producing cost effective logistics and resupply operations. Kayser-Threde is a partner in the Rapideye commercial satellite-based geo-information service set up in 1998 with products and services for agriculture, cartography, natural disaster assessment and 3D visualisation and so forth. The concept envisages five Earth orbiting imaging satellites with S-band TT&C and X-band data downloads. From an evenly distributed polar orbit the five satellites provide multispectral images with a resolution of about 6.5 metres and an imaging area of 80 × 1,500 km.

UPDATED

MAN Gutehoffnungshutte AG

Current Status
MAN Gutehoffnungshutte AG (MAN GHH) is a subsidiary of the MAN Group and specialises in cages and ore skips as well as hoisting equipment and lifting gear. It also serves space industry in the design and building of ground installations for the aerospace industry, parabolic antennas and mobile bridge systems. The company is also active as a consultancy within these fields. Within the space-related fields, MAN GHH has been involved with: rocket launching facilities, test stands for rocket motors, buildings for satellite preparation, research and communications antennas and ground stations for telecommunications and research.

In 1976, MAN GHH provided and installed the launch table for the first Ariane launch complex, ELA 1 at Kourou. In 1981, the company was commissioned to build the main components for ELA 2 to be used by the Ariane-3 and Ariane-4 launch vehicles. In 1988, MAN GHH was appointed the general contractor to build the third launch pad at Kourou, ELA 3 for Ariane-5. Also at Kourou, MAN GHH was responsible for the satellite preparation buildings S1B and S3B, as well as the satellite encapsulation bay which is part of the final assembly building of Ariane-5.

MAN GHH has also planned and built stands for the testing of both solid and liquid propulsion systems, including the equipment required for the test operations such as holding devices, cranes, propellant supply systems and safety systems. Studies on the plausibility of using these stands and test facilities for large propulsion systems have been conducted with a view of offering their availability for new propulsion systems and development projects which may be required for Ariane 5 replacement generations of satellite launch vehicles.

The Vulcain engines for Ariane-3 and Ariane-4 are tested on the PF 20 test stand, built by MAN GHH and located at the SEP test site in Vernon, France. At Kourou MAN GHH constructed the BEAP test stand for the solid-propellant boosters for Ariane-5.

MAN GHH has supplied fixed, transportable and mobile ground stations for satellite communications, TT&C and geodesy.

Conducted extensive launch support and infrastructure needs for new advanced semi-reusable launch vehicles for ESA. Progressing those studies has generated new ways in which ground handling and servicing costs can be cut through the use of common hardware support equipment.

UPDATED

Munich Reinsurance Company

Current Status
Munich Reinsurance (Münchener Rückversicherungs-Gesellschaft) is one of the largest risk carriers for commercial space projects, providing 10 to15 per cent of the world's coverage. A team of specialists, set up at the end of the 1970s, concentrates exclusively on the insurance and reinsurance of space risks. The company is represented in 32 countries with 48 offices. Munich Reinsurance employs about 5,000 people in total, with a specialised unit staffed with permanent specialists in risk assessment and insurance.

MR works with lead underwriters in London, New York and Tokyo on high-risk satellite and launch/life policies as well as ground support insurance. Specialises in finite risk reinsurance and alternative risk transfer.

UPDATED

Satellite Operational Services GmbH (SOS)

Current Status
SOS, founded in 1983 and now with more than 50 staff, provides expertise in mission operations for manned and unmanned programmes, space flight dynamics and ground station operations. The company participated in Spacelab D1/D2, Helios, TV-Sat, DFS Kopernikus, Eutelsat, Galileo, Rosat, Giotto, IRS and AMPTE. Its current principal customer is the DLR Aerospace Research Establishment at Oberpfaffenhofen. SOS' capabilities include defining and developing satellite operational requirements, training spacecraft operators, control

centre software definition and acceptance testing, satellite in-orbit testing, spacecraft command operations, communications networking, ground data/tracking station operations, orbit and attitude determination, and manoeuvre planning. Human space flight tracking, ground and on-orbit checkout operations, high dynamic flight crisis management and operations scheduling. SOS has performed mission planning analysis and execution for multinational space operations and trajectory co-ordination calling for the use of several separate hardware sets integrated in a single utilisation. SOS can adapt conventional trajectory analysis and spacecraft/satellite tracking sets into a single integrated element specifically tailored to customer requirements and specifications.

VERIFIED

Thysen Krupp VDM GmbH

Current Status

High-performance aerospace materials: nickel-based alloys, cobalt alloys and special stainless steels. It provides the cobalt-based Conicro 5010W-alloy 25 for the nozzle extension of Ariane 5's upper stage engine. Ariane 4 carried VDM steel propellant tanks. Thysen Krupp is developing new alloys for high-temperature, telescopic engine expansion chambers. Development is also underway on high temperature alloys for hypersonic scramjet designs and for long burn rocket motors. Thysen Krupp is also experimenting with heat-protection metals for heat-sink (versus ablative) thermal protection systems. Now developing long-life reusable thermal protection systems with high resilience and stress resistance factors capable of all-weather use for launch and recovery systems, launch vehicles, shuttle-type vehicles and recovery pods or modules.

VERIFIED

W L Gore & Associates GmbH

Current Status

Gore manufactures space-qualified wires, cables, waveguides and data lines. These include wires and cables with/without drainwire qualified to Spacelab SLP 2110 and ESA/SCC detail specification 3901/009, 017, 018, 019, 021. Applications include ERS, SAX, ISO, PPF/Envisat, XMM, GlobalStar and ERA. Also included are high-voltage cables for TWTs and dielectric waveguides 26.5–110 GHz. Customers in Europe and the US and with links to potential clients in Japan. Also developing high-voltage supply cables for high-power demand applications such as radar satellites. Integrating existing wire products with new specifications from advanced Earth observation satellites. Presently has 6,000 associates in 45 locations around the world with revenues of about US$1.4 billion worldwide.

UPDATED

Zeppelin Technologie GmbH

Current Status

Now owned by MAN Technologie AG and a subsidiary of Zeppelin GmbH. Zeppelin manufactures propellant and pressure tanks for all kinds of space applications, as well as individual parts like bulkheads, parts, flanges and pipes. Designing tanks for ISS logistics module also for advanced launch vehicles and for high-pressure gas and liquid containers for very long duration mission requirements. Also examining the feasibility of inflatable, long duration, high-pressure dewars for encapsulation during launch and deployment in orbit or on escape trajectories for application in small satellites and space vehicles occupying small volumes.

VERIFIED

International

ASTRIUM (Matra Marconi Space)

Current Status

Early in 2003 BAE Systems ceded its 25 per cent shareholding in Astrium asserting that space was not part of the company's 'core' business. No capital gain was made. Earlier BAE Systems had agreed to sell its share in Astrium for €165 million and a deterioration in company standing resulted in it abandoning further involvement in space projects. This gives EADS total ownership of Astrium, which employs about 8,000 people.

Background

Astrium is a joint venture owned by EADS (75 per cent) and BAE Systems (25 per cent). Astrium employs about 8,000 people in France, Germany, UK and Spain.

Matra Hachette SA and GEC Marconi plc in 1990 merged the activities of their respective Matra Espace and Marconi Space Systems companies under Matra Marconi Space. Matra Hachette holds 51 per cent and GEC Marconi 49 per cent. Matra brought assets worth £10.7 million to the partnership and GEC £8.7 million; total workforce was 3,200 and 1993 turnover was FFr5,167 million. Workforce had increased to 4,000 by early 2000. MMS acquired British Aerospace Space Systems in 1994 for £56 million, creating Europe's largest space company. BAeSS 1993 sales were £151.6 million, for a £5 million profit. At the time of the purchase, its backlog stood at £512 million (MMS' own was FFr11.5 billion) and the workforce at 1,050. MMS' 1995 turnover was FFr6.7 billion (plus FFr9 billion backlog); 4,400 personnel. MMS is Europe's first fully integrated space company and its largest spacecraft manufacturer. With five major UK/France sites, it can undertake all major applications of space technology: Earth observation, science, civil/military communications, and space transportation and manned space systems.

MMS France is concerned with the design, development and manufacture, as prime contractor, of satellites and spacecraft, platforms, payloads and onboard electronic equipment. The company is prime contractor for Telecom 1/2, Hispasat, Nilesat, Hot Bird 2–5, Spot 1–4, Helios, Soho and Ariane's equipment bay; it is the main subcontractor for Eutelsat 2, Inmarsat 3 and Koreasat. MMS will also play a major role in the Columbus data management and AOCS.

MMS provides attitude control, navigation and guidance equipment for telecom, Earth observation and science satellites, launch vehicles and sounding rockets. Other onboard equipment includes stabilisation systems, Sun sensors, data processors and gyro packages. In the ground-based area, it provides control equipment, Earth stations, automatic test equipment, simulators and image processing systems. It holds 35 per cent interest in Spot Image and 45 per cent in NRSC Ltd.

MMS UK specialises in the design and manufacture of payloads for communications, remote sensing and science applications, together with the associated ground support and control equipment, and user terminals. Expertise also encompasses structures, thermal control, power, propulsion, AOCS, TT&C and mechanisms. Extensive facilities include a fully equipped manufacturing complex, comprehensive AIT resources and high-technology research and development facilities. These include a fully-screened indoor near/far-field antenna test range, one of Europe's largest.

Marconi was prime contractor for Skynet 2 and Ariel 5/6, and provided much of the onboard equipment for the highly successful Marecs, Exosat and IUE satellites. Recent work includes the communications payloads on Skynet 4 and NATO 4, the active microwave system on ERS 1/2 and onboard electronics equipment of the Meteosat series, including the attitude measurement system, AOCS, imaging electronics and communications packages. BAe participated in approximately 100 satellite projects over three decades. It developed Ariane's Spelda carrier, Spacelab's pallets and the Skylark sounding rocket (see the Launchers section). It primed most of Europe's science satellites, including Giotto, provided Hubble's solar arrays and Photon Detector Assembly. Bristol continues as ESA's prime for Polar Platform (first launch as Envisat 1 in 1998). It developed the ISAMS multichannel infra-red instrument for NASA's Upper Atmosphere Research Satellite and the passive microwave AMSU-B sensor for NOAA's Tiros, in addition to providing the Digital Electronics Unit for the UK's Along-Track Scanning Radiometer on ERS.

The company's products span radio, radar and analogue/digital processing applications for civil/military satellite systems, covering space bands less than 0.2–183 GHz, and extending to optical frequencies for intersatellite links. The subsystems encompass: UHF/SHF/EHF transponders, transmitters, frequency converters, receivers, filters and multiplexers, beacons, local oscillators, launcher and satellite command equipment, modulators/demodulators, encoders/decoders, meteorological radiometers, SARs, TT&C systems, onboard processors and memories, optical communications systems.

Earth Station terminals are available in 1 to 19 m diameter antenna sizes. Military stations include: fixed stations with 7 to 13 m antennas, 6.4 m transportable stations, 1.7/4.5 m vehicle-mounted terminals, 1.06/1.2/1.8 m SCOT ship terminals with fully stabilised antennas (more than 70 ordered; the MoD ordered an Enhanced SCOT in March 1990), MASTER aircraft terminals, 12 kg Manpack man-portable SHF terminal and TT&C stations for satellite station keeping. Civil stations include: Inmarsat transportable, Apollo, TVRO stations (fixed and for ships), TV news gathering, meteorological data receiving, satellite measuring stations (antennas up to 19 m), and aircraft data relay. The company provides a turnkey service.

Ground Support Equipment. Up to 45 GHz, covering high power amplifiers, receivers, frequency converters, modems, TDMA systems, filters and multiplexers, and processors.

Satellite Ground Control Facilities. MMS UK offers equipment for the monitoring/control of civil/military systems, including launcher command, TT&C links, security/safety systems and satellite simulators. Most recently, it provided Skynet 4's Satellite Operational Facility at RAF Oakhanger.

Research/Technology. Current programmes include onboard processing, Ni/H_2 batteries, ion propulsion, GaAs solar cells, phased array antennas, laser communications, and space-based surveillance systems.

In Toulouse the company performs commercial satellite integration: 1,332 m^2 10 m high High Bay and 1,190 m^2 3 m high Low Bay, both class 100,000; two 100 m^2 class 100 laminar flow areas. Alignment area on seismic block with computer-driven table (accurate to 1 arcsec). It also conducts Ariane vehicle equipment bay integration: three 400 m^2 areas of 3/6/12 m high; class 100,000 available; two seismic blocks for inertial platform calibration and military satellite integration: three 400 m^2 areas of 3/9/12 m high; class 100,000 specified (10,000 measured); 120 m^2 class 100 10 m high laminar flow area.

The Toulouse antenna range has a 400 m slant range, class 100,000 under 14 m radome, 8 tonne positioner, 70 m tower for transmit/rx measurements.

The company's Portsmouth facilities include two principal class 100,000 cleanroom complexes. Building 43's 41 × 30 m includes a 32 × 16 m satellite assembly/integration area with 10 m hook height/3,000 kg crane capacity. Accessed through a 5 × 6 m high door, the conditions are controlled to within 20±2°C and 35 to 60 per cent relative humidity. A 10 m^2 seismic block provides a position for optical measurements on spacecraft critical reference axes. The building also houses a reception area for loading/unloading, clean store, EGSE room, EMC chamber suite, thermal chamber, mass properties area and antenna test range, all maintained to class 100,000 conditions.

The 11 × 13 m (7.3 m high) EMC facility of five rooms, with 5 × 5 m access door, provides better than 80 dB magnetic screening 14 kHz–10 MHz, and better than 120 dB electric field screening 14 kHz–10 MHz, 115 dB 0.10–1 GHz and 126 dB 1–40 GHz.

The 5 × 5 × 5 m thermal chamber can handle heat dissipation of up to 5 kW from equipment under test and control conditions over –40/+110°C with 2°C accuracy. Data recording is performed on 16 channels.

The mass properties measurement area provides mass determination up to 5,000 kg (or 500 kg in 100 gm increments), moment of inertia measurement up to 3,000 kg using Schenck M1, M4 and M6 machines, centre of gravity location for subjects up to 4,000 kg using Schenck WS12 and WM50/6 machines; and spin balancing using Schenck E5 (capacity 500 kg, up to 300 rpm) and E6 (400 kg, 300 rpm) units.

Building 43 also houses Portsmouth's large Near-Field Far-Field antenna test range. A monolithic foundation supports both the elevation/roll-over-azimuth antenna and test positioner and the source/probe tower, track mounted for variable length. Loads up to 4,000 kg can be accommodated. The 15 × 15 × 31 m chamber offer a 5 m diameter quiet zone, 0.25–60 GHz, better than –40 dB reflectivity at 1 GHz rising to better than –55 dB at 7 GHz, 100 dB magnetic screening at 200 kHz (110 dB at 1 MHz) and 130 dB electrical screening 1 kHz–50 MHz (110 dB over 0.050–1 GHz, 100 dB over 16–18 GHz, 80 dB over 18–60 GHz). Positioner tower capacity 1,600 kg, positional accuracy 1 mm and pointing accuracy 0.03°. The range operates primarily in the spherical near-field mode. The radiated field is sampled in the antenna near-field and the far-field patterns calculated. Determination of gain, radiation patterns, cross polar response and boresight location are undertaken.

Building 48 provides a class 100,000 35 × 12.5 m assembly/integration area with optical block and houses a clean store and EGSE operating room. In addition, the building includes 250 m^2 class 10,000 and 1,300 m^2 class 100,000 manufacturing facilities with class 100 laminar flow workstations. Other Portsmouth facilities cover three thermal-vacuum chambers in building 23A, each offering a 1,250 × 1,250 mm cylindrical internal working volume and less than 10^{-5} torr over –65/+100°C. An LN_2 turbomolecular pump can achieve 10^{-6} torr within 4 hours. The shrouds can also be operated at cryogenic temps with LN_2 (–196°C). 16 auxiliary T-measuring channels are recorded on a data logger for each chamber. A range of climatic chambers is available, providing up to –70/+175°C.

Building 17 houses seven vibrators installed in three rooms and controlled from a common position. Drop, centrifuge and shock test equipment is also housed in the building. The four principal vibrators are: Ling 980 (5–2,000 Hz, 150.52 kN sine peak, ±12 mm max displacement), Ling 964 (5–2,000 Hz, 75.26 kN, ±12 mm), Derritron VP2500 (5–2,100 Hz, 111.2 kN, ±12 mm) and Derritron VP1200 (5–3,000 Hz, 55.6 kN, ±12 mm). The 1.2 m centrifuge achieves 60 *g* on 50 kg test items (up to 0.5 × 0.5 × 0.5 m) and is equipped with 60 sliprings for functional testing. The Portsmouth site does not offer acoustic test facilities but the high intensity acoustic unit in GEC's Mechanical Engineering Lab at Whetstone has been extended for spacecraft testing.

Building 36's Hydrazine Propulsion and Test Facility houses a 3.4 m diameter × 6.7 m long 55 m^3 vacuum chamber with associated pumps and exhaust scrubber, providing an initial chamber pressure of 10^{-2} torr for hot firing tests. T controlled to –15/+60°C. The unit is connected to a hydrazine cart (weighing accuracy 50 gm, pressure up to 35 atmosphere) and an oxidiser cart handling nitrogen tetroxide, nitric oxide or MON. Exhaust is scrubbed by recirculated dilute hydrochloric acid to remove ammonia and unburned hydrazine. Chamber instrumentation includes P/T sensors, flowmeters and thrust transducers associated with a 24-channel data acquisition system. The control system permits thruster pulse duration programming. The remaining 200 m^2 of building 36's class 100,000 cleanroom is used for propulsion subsystem assembly and test, including a class 100 area for detailed assembly work and automatic particle counting. An x-ray room caters for pipework joint inspection by weld radiography, and flow rate/pressure drop tests using GN_2 are conducted in the area. A 1.5 × 1.5 m vacuum chamber for propane tests is also available.

The Stevenage facilities covering 9 hectares, has approximately 6,000 m^2 manufacturing, assembly and test floor area, and a new three-storey office building with 7,000 m^2 for management, admin, design and engineering staff.

The Satellite Assembly Hall measures 30 × 24 m, 11 m high, maintained to Fed-Std-209B class 100,000, air-conditioned to 21±3°C, 35 to 55 per cent relative humidity. Access by 7 × 5 m roller shutter doors. Crane with 10 tonne capacity and 9.7 m hook height, movable over 90 per cent floor area. Two 7.6 × 7.6 m areas can be lowered to provide 16.7 m crane height. Four assembly bays capable of simultaneous accommodation of four satellites and associated checkout equipment. Adjacent 5 × 35 m EGSE control area.

The Microwave/EMC Test Chamber. 12 m high, 10 × 14 m, accessed directly from the Assembly Hall by airlock and protected corridor, and two adjacent 4 m high 6 × 8 m screened rooms. All controlled to class 100,000. Anechoic lining ensures a return loss of greater than 30 dB at 500 MHz and greater than 50 dB at 5 GHz, and can handle an input power of 0.3 W/m^2. A central ceiling access panel allows antennas to be mounted above the spacecraft to excite payload transponders. Chamber screening effectiveness is better than 80 dB in the H field (14 kHz–10 MHz) and better than 115 dB in the E field (14 kHz–40 GHz). T controlled to 5 to 50°C (±3°C); relative humidity 30 to 60 per cent.

Further test facilities include a 7.7 × 18.3 m, 7.3 m high Antenna Anechoic/EMC Chamber, with cylindrical near/far-field range capabilities, and a 2.4 × 2.4 m High Power RF Vacuum Chamber, with up to 3.5 kW at 8 to 18 GHz and 600 W L-band. The Environmental Test Area contains six Climatic Chambers, six Thermal Vacuum Chambers and 40/160 kN shakers. An ultrasound non-destructive test unit allows the bonding of panels and the largest (4 m diameter) honeycomb panels for Ariane 4's Spelda, to be checked. Manufacturing areas include a 150 m^2 class 100 horizontal laminar flow Precision Products Clean Room with associated cleaning/test facilities and panel bonding, antenna lay-up, satellite assembly, electronics assembly and Spelda assembly areas totalling 4,700 m^2. A six-axis CNC carbon fibre filament winding machine capable of 2.4 m diameter × 6 m and its adjacent curing facility (three autoclaves; largest 3.5 m diameter × 5.5 m) and four ovens.

The Thermal Blanket and Secondary Surface Mirror Assembly Facility is a 160 m^2 controlled to Std-209B class 100,000, 21±5°C, less than 50 per cent relative humidity conditions. Complete blanket manufacturing facility for spacecraft applications.

In Bristol the Halley Building (20X) houses the majority of the executive, project management, engineering and drawing office staff on four floors totalling 4,800 m^2 and accommodating 340. Building 20Y houses the site's manufacturing and AIT facilities, including a single story AIT Assembly Hall with pit and lift, capable of accommodating the largest spacecraft currently envisaged in Europe (equivalent to full STS cargo bay), together with an associated two-storey manufacturing, test and office facility.

The Main Assembly Hall, on the ground floor has a working area 589 m^2, access 5 × 6 m high door from haulingway (airlock), head room 12.5 m to crane hook (10 tonne limit and full traverse facilities). 10 × 10 m hydraulic lift platform capable of lowering a six-tonne load 7.93 m below normal floor level. Whole area protects against RF interference up to 60 dB, 0.1–18 GHz. Air conditioning 20±3°C, 50±5 per cent relative humidity, class 100,000 with cross flow ventilation (not laminar) at eight changes per hour. Floor is conductive and walls/ceiling are RF shielded with cleanroom linings.

The Detail Manufacturing/Assembly Area is 562 m^2, air conditioned (warm air only) 20±5°C with a workshop area of 116 m^2 for producing spacecraft detail parts to flight standard. Equipped with small lathe (Harison M250), guillotine, bench drill, bench drill with compound table, bench folder, finger press, 2× finishers, grinder, large folder, fly press with associated equipment and various power tools such as jigsaw. The RCS assembly/test area: 116 m^2, controlled to class 100,000 Fed-Std-209B, air conditioned to 20±3°C, 50 per cent ±5 per cent RH, warning system with interlocks for identification of x-ray in use. The Ultrasonic cleaning facility: includes de-ionised water plant for water rinse and acid holding tank for the passivation and pickling of stainless steels. The paint spray facility with DeVilbis Turbo-Dynaclean Spray Booth (24 m^2) with 2.5 m wide double door access to booth and curing oven.

Spacelink Europe

Current Status

Spacelink was established in 1991 by JRA (UK), Novespace (France), MST Aerospace (Germany) and D'Appolonia (Italy) to co-ordinate ESA's technology spin-off activities. JRA covers UK/Scandinavia and acts as ESA's programme co-ordinator. The network of technology transfer agents spans 17 European countries and Canada. Spacelink produces the annual TEST (Transferable European Space Technologies) catalogue of technologies from space programmes that are available for licence or further development for non-space uses. More than 800 inter-company introductions have been made, the majority of them transnational. Approximately 100 negotiations between space/non-space companies are underway at any one time. Organisation has geared up to onboard government and industry customers for the International Space Station. Has conducted introductory negotiations with the Russians with respect to joint ventures and international co-operation. Provides industry links between European companies and other ISS partners (USA, Russia, Canada and Japan). Spacelink Europe has extended discussions and negotiations with countries around the world and is a primary link in the integration of technology transfer activities between ESA and other non-ESA European countries.

Israel

Amsat-Israel

Current Status

Amsat-IL is collaborating on the TechSat project, which includes an amateur digital packet communications unit. Also researching air-space interactive communications packages. Has conducted amateur satellite relay support links to astronauts aboard the International Space Station. Capable of interlinking several separate packet systems and researching the concept of a new unified telecommunications system netting several separate existing systems.

Elbit Ltd

Current Status

Elbit began its space participation in September 1989 with the award by General Dynamics Space Systems of a contract for avionics boxes for the Atlas launch vehicle. It provided TechSat 1's OBC. It is one of Israel's largest electronics systems houses and is developing control systems for future Israeli satellites including communications and data handling and advanced reconnaissance and surveillance systems. Provides high-resolution panchromatic and multispectral earth observation systems (civil and military), science payloads and a wide range of optical equipment for space and ground-based systems. Elbit Ltd has developed a wide range of satellite and space vehicle programmes for foreign clients and agencies.

ELTA Electronics Industries Ltd

Current Status

Established in 1967, ELTA is part of Israeli Aircraft Industries' Electronics Division and as such participates in the development and fabrication of Israeli satellites including remote sensing and surveillance satellites. Products cover military electronics systems, radars, active/passive early warning, secure communications, computers, signal processing, microwave components, tubes and antennas, high-voltage power supplies and microelectronics. New high-speed processing and control systems for advanced surveillance satellites. Has conducted early design analysis on integrated civil/military earth surveillance and remote sensing technologies and on multifunctional instrument platforms.

Currently employs 2,500 people with annual sales of approximately US$500 million.

Israel Aircraft Industries Ltd (IAI)/MBT Systems & Space Technology (MBT)

Current Status

Launched Eros A1 earth resource observation satellite from Svolodny, Russia on 5 December 2001. An element of IAI's Electronics Division, MBT is Israel's major space centre. It was responsible for developing Israel's first satellite, Ofeq, its all-solid Shavit launcher, and the Amos commercial telecom satellite. MBT may provide some Ellipsat hardware and act as the local service provider. MBT is developing electronic control systems for autonomous housekeeping.

MBT's Space Technology Centre includes the Satellite Integration and Test Centre an S/X-band TT&C station (10 m S/X-band antenna and 2 m S-band back-up), laboratories and fabrication facilities. The ground station's Image Processing Centre handles the received Spot imagery; upgrading to handle other satellites is planned. The satellite centre covers 3,400 m^2, including 2,000 m^2 class 100,000 and a class 10,000 30 m^2 portable laminar flow booth. Class 100 can also be provided.

Part of IAI's Electronics Division, MLM System Engineering and Integration specialises in telemetry systems, communications, C^{3}I and image processing. It is leading the NEXT project, to develop a commercial derivative of Israel's Shavit satellite launcher. MLM provided the solar panels for Technion's TechSat launched in 1997. Also builds the Amos series of lightweight communication satellites. Colocated at 4° W the satellites deliver satellite broadcast solutions for DBS providers, cable TV companies and TV broadcasters.

An element of IAI's Electronics Division, TAMAM specialises in inertial navigation and reference systems, inertial sensors and stabilised platforms.

Soreq Nuclear Research Centre (NRC)

Current Status

The Soreq NRC space evnvironment section focuses on the space environmental aspects of materials, electronic components and sytems, as well as on related subjects (for example, contamination control, safety and reliability of space systems, and so on). It has been participating in all the Israeli space projects, including the Offeq program, AMOS communication satellites, EROS observation satellites, Techsat Gurwin II students satellite, TAUVEX UV telescope, the space shuttle MEIDEX experioment (to be operated by an Israeli astronaut) and other space programs. Nuclear test facilities utilised for non-nuclear activities. The space environment section consists of three groups:

1. electronics in space
2. spacecraft materials and contamination
3. computations and modeling.

Tadiran Spectralink Ltd

Current Status

Tadiran is Israel's largest electronics company. In the space field, it provides end-to-end TT&C systems qualified in the Ofeq programme and is developing communications systems for a new and more advanced platform and satellite support technology. It provided receivers and transmitters for TechSat. Elisra is a subsidiary. It produces ground-to-space automated command systems and satellite integration software. It produces conceptualisation designs for advanced integrated satellite technologies.

VERIFIED

Italy

Alenia Spazio SpA

Current Status

Alenia Spazio SpA, a FINMECCANICA company, is Italy's major space company, born from the merger of two major national space industries. Since the formation of EADS, Alenia has assumed an important role as an independent supplier. Selenia Spazio SpA and Aeritalia – Space Systems group. Alenia Spazio employs 3,000 people at seven plants in Rome, Turin, L'Aquila, Milan and Naples equipped with 16,000 m^2 of clean rooms and research laboratories. It also co-ordinates the operations of subsidiaries Laben, SSI Space Software Italia and Quadrics.

Company expertise covers complete space systems, satellites for telecommunications, remote sensing, meteorology and science applications, manned systems and space infrastructures, transport, launch and re-entry systems, telemetry/control stations and specialist space software. In telecommunications, it is prime for the national Italsat and for ESA's Artemis and Data Relay Satellite. It integrated Globalstar's 56 satellites and contributed to the construction of Argentina's Nahuelsat.

It was prime contractor for the Lageos 2 geodetic satellite launched October 1992 by Shuttle. An Italian industrial team led by the company developed the IRIS (Italian Research Interim Stage), a low-cost spinning solid upper stage for deploying lightweight satellites from Shuttle. First use was with Lageos 2. Other Shuttle launches of ASI programmes with Alenia Spazio as prime are X-SAR (X-band Synthetic Aperture Radar) for Earth observation (two flights in 1994) and the Tethered Satellite System for electrodynamic experiments (responsible for the satellite and the Core Equipment experiment). It was prime for the SAX x-ray satellite. In addition to X-SAR, it is developing other Earth observation instruments.

Following its role as a major industrial contractor for Spacelab (for which it built the pressurised module), Alenia Spazio is involved in the Columbus Orbital Facility, with particular responsibility for development and integration of the thermal-mechanical system. It is also responsible for the structure and thermal control. It similarly built Spacehab's pressurised modules. It will build three Mini Pressurised Logistics Modules to deliver Space Station supplies aboard Shuttle; the PDR was completed in October 1993. Significant roles in ESA programmes include: ERS 1/2, Meteosat (development and manufacture of onboard electronics systems), Hipparcos, Eureca, Envisat and Cassini-Huygens. It is undertaking studies on behalf of ASI for the Carina µg capsule (see the Microgravity section).

Alenia Spazio is currently working on:

EDR: The European Drawer Rack (EDR) is one of the facilities of the Columbus Laboratory and it will be part of the Columbus launch configuration. The EDR is a multi-user facility, capable of simultaneously accommodating up to seven experiment modules for different experimental fields and disciplines. Delivery of the Flight Unit to the customer European Space Agency (ESA) is planned in 2003.

FSL: The Fluid Science Laboratory (FSL) is a flexible, highly modular facility for the Columbus Laboratory. The laboratory will support scientific microgravity research in fluid physics and, in general, provides all typical on-board services and resources. As specific support for fluid science, it is equipped with dedicated visual and interferometric optical instrumentation, supported by very powerful video image management and processing. The FSL project is continuing, having completed the Critical Design Review and the Engineering Model integration, the delivery of the Flight Unit to ESA was made in 2003.

SOLAR/EXPORT Payload (SEP): SOLAR and EXPORT are two ISS external payloads, developed under ESA contract. They will be flown (2005) to the Columbus Exposed Payload Facility (EPF). SOLAR will integrate three solar Sun observation instruments. They will measure the solar radiance with very high accuracy. EXPORT will integrate two experiments: SPOrt and EXPOSE.

COARSE POINTING DEVICE (CPD): The CPD is a two degree-of-freedom pointing system, designed to accommodate payloads on ISS external locations (Columbus EPF/EXPRESS Pallet). This allows, for ISS-based payloads, an observation time of about 20 minutes per observation orbit. The CPD is developed under ESA contract. The first Flight Unit was delivered in 2003. Possible design upgradings are being investigated for future applications.

Sky Polarization Observatory (SPOrt): Designed specifically to respond to a call for proposals from ESA to put scientific experiments on board the International Space Station (ISS) during Early Opportunity Phase. SPOrt is an astrophysical observatory whose main aim is to measure the Sky Polarisation in an unexplored microwave frequency range (20–90 GHz). In this range we expect to detect both polarised radiation coming from our galaxy and some sign of polarisation of the Cosmic Microwave Background (CMB). SPOrt is the most powerful microwave polarimeter for scientific investigations in space.

HEXAPOD (Pointing System): The HEXAPOD Pointing System is designed to accommodate payloads on ISS external locations (EXPRESS Pallet). It will provide a nadir pointing platform to the NASA Stratospheric Aeroso and Gas-Experiment. HEXAPOD is developed under ESA contract. The Flight Unit will be delivered in 2002.

UPDATED

Assicurazioni Generali SpA

Current Status

The company provides satellite, launch and space risk assessment and insurance. It is the world's largest space insurance underwriter, with US$125 million underwriting capacity. Generali expects ROEV (return on embedded value) to be in excess of 10 per cent over the period 2001 to 2005. Financial revenues of €191 billion at the beginning of 2002. It has historically conducted risk analysis for global projects and spreads analysis over a wide acutarial base throughout its own team of specialists in spacecraft, satellite, launch vehicle and operator activities.

Cap Gemini SpA

Current Status

Cap Gemini SpA specialises in onboard software, control centres, data communications, test systems and data processing for European and US satellites together with integration software for adopting different rational standards to a common bus. Has developed software for Leonardo and communications software for logistics modules in support of ISS operations. Can provide software for automated control and housekeeping systems in unmanned or manned space vehicles. Specialised skills in bus software integration. The company has developed a new range of software for the integration of separate control suites for man-tended platforms and unmanned science laboratories.

UPDATED

Carlo Gavazzi Space SpA (CGS)

Current Status

CGS specialises in facilities and payloads for µg applications, onboard instrumentation control electronics, micro-satellites, ground support equipment and ground segment. Principal projects include: EGSE for Cassini, Polar Platform and ERS; Tethered Satellite reflight control units; Spacelab (bubble drop/particle unit); µg instruments for Maxus and Neurolab; feasibility studies for Columbus Anthrolab, Fluid Science and Biology labs; onboard digital image processors for

Columbus and Maxus. High-speed data recorder integration incorporating electronic tracking and monitoring of bubble drop/particle unit. CGS has new development models for integrated drop facilities simulation tasking. It is a member of the Europayload Consortium (see Microgravity section) and has approximately 80 staff.

UPDATED

CISI AID

Current Status
The company is a subsidiary of France's CISI Ingénierie, created in 1987 to focus on the software capabilities of the CISI group. CISI has developed software sets for a wide range of applications in the aero, marine and transport markets, providing solutions to high volume data handling. It specialises in the space field in science payload data processing, mission analysis, orbit determination, software for satellite control centres, telemetry analysis, remote sensing and image processing. Has created software package for integrating optical and non-optical data sets. Creates software solutions to electro-mechanical actuation systems and for processing information feedback from devices into analytical data sets for health and usage monitoring of unattended space systems. Has also extended this work into software suites for control of man-tended platforms and stable, automated, systems.

UPDATED

Dataspazio SpA

Current Status
Dataspazio was established in 1988 by Telespazio (49 per cent) and Datamat (51 per cent) to design/develop systems and software for space applications: satellite Operational Control Centres (OCCs), data processing facilities, satellite simulators and flight dynamics models and software. National programmes include OCCs for Sirio 1/2, Italsat and SAX, and the satellite simulator for SAX. European programmes include the payload data handling subsystem for polar platforms, Olympus data gathering centre, and Columbus user information systems, and crew training centre which is developing a crew interface simulator. New applications for advanced human operations in space including flight simulator and mission operations software. Dataspazio has conducted extensive research on navigation satellite software requirements and has been involved in developing the avionics package for equipment supporting the Galileo navigation satellite programme. Dataspazio continues to research new and innovative ways of developing integrating software and computer simulations for navigation systems interacting with the space-based segment.

VERIFIED

Elettronica SpA

Current Status
Elettronica produces telecommunications equipment for analogue and digital TV broadcasting, terrestrial transmitters and transposers in VHF, UHF and MMDS bands. Space activities cover microwave/mm-wave technologies, analogue/digital electronics, infra-red sensors and optical instrumentation for the space segment, and small/mobile stations, GPS receivers, HPAs and high power/low noise solid-state components for the ground segment. Participating in Columbus programme and developing packages for successor operations requirements compatible with multimodule operations. Development projects indicate potential applications for visual monitoring through video and digital TV systems on remote observing stations interacting with space-based imaging satellites.

VERIFIED

FIAR SpA

Current Status
Fabbrica Italiana Apparecchiature Radioelettriche (FIAR) is primarily concerned with the design, development and production of onboard electronic equipment and small satellite communication terminals. It designs and manufactures radars, transmitters, sonar systems and altimeters for the defence industry. For satellites, FIAR provides full power systems (including batteries), regulators and ancillary equipment, as well as designing and supplying GaAs solar arrays. RF equipment, such as telecom transponders, TWTAs (such as the 135 W 12 GHz version for DBS satellites and 80 W 12 GHz for FSS), SSPAs, new high-efficiency RF equipment for extended life and durability, frequency generators, oscillators and microwave components. In the Earth station field, FIAR manufactures Prodat, a portable L-band terminal for messaging and position reporting. Has carried out tests on intra-satellite links and is researching the possibility of auto-frequency switching for multiple users operating different clusters of satellites in different regional zones (Atlantic, Indian or Pacific ocean regions).

FIAR and COMAU formed Tecnospazio in 1986 for space automation and robotics activities. Electronic systems account for about 84 per cent of annual revenues, space systems 14 per cent and environmental support systems 2 per cent. Efforts have been made to integrate separate field functions for space-based applications, utilising strengths in electronics and environmental systems.

UPDATED

INTECS Sistemi SpA

Current Status
INTECS provides software solutions for the aerospace industry, the transport market, telecommunications specialists and the public sector. Has produced computer programmes capable of interconnecting various control systems for large switching requirements appropriate to managing automated subsystems. INTECS specialises in software technologies, such as artificial intelligence, test/control systems, command/control systems, GIS and real-time embedded systems. Their staff totals about 190.

UPDATED

Laben SpA

Current Status
Laben SpA now includes Proel Technolgie which specialises in electron generation and plasma sources. Laben, a company of the Alenia Finmeccanica Group, specialises in electronic systems for space applications and is a leading member of the Star Consortium for ESA programmes. Personnel total about 400. The company's space activities include the design of onboard computers on chips, electronic data handling, control systems and ground checkout equipment for spacecraft and ground stations. It also provides instrumentation for science, remote sensing and μg satellites. AOCS sensors/actuators and GPS receivers for attitude/orbit determination. The company maintains extensive manufacturing facilities; its thick film hybrid circuit manufacturing procedures are certified by ESA. Products are manufactured for spacecraft and satellite bus configurations, on-board payloads, ground applications, navigation, electric propulsion and electron beam curing. Laben also provides high density processing capability on thin circuit designs and can adapt low volume installations for specilised satellites and spacecraft.

VERIFIED

Microtecnica-Torino SpA (MT)

Current Status
MT was established in 1929 and is now a business unit of the Hamilton Standard Group: 2001 space revenue €5.77 million; 2000 space revenue €5.84 million; staffing level is 800. Its main business areas include environmental and thermal control, drive/actuation, power control and generation, and electronic control equipment. Can provide electromechanical actuators with software controlled management systems. Has developed innovative thermal control systems integrated with environmental control mechanisms. Projects include the spin table for Italy's IRIS upper stage, Ariane 5's Vulcain main engine check and helium control valves boxes, Ariane 5's main stage pressurisation system and helium control valve boxes, Columbus main active thermal control system (ATCS) equipment/GSE, and Mini Pressurised Logistics Module main ATCS and ECS equipment. Contributions since 1974, when the company became active in the space field, include Spacelab, Eureca (ACTS elements) and Lageos 2 (structure).

VERIFIED

Space Software Italia (SSI)

Current Status
SSI was established in 1988 for the design/development of system and equipment software. The company performed critical examinations of existing software sets and played a major role in helping to develop new programmes and logic routines for advanced space missions. Has also developed software under contract for non space-related industries. Alenia Spazio owns 92.4 per cent and Alenia-Finmeccanica 7.6 per cent. 65 employees. Occupies 4,500 sq m office and facilities area. Provides services in consultancy and training, software design and development, COTS, mission support and operational services management. Specialisations cover onboard systems (system and mission management, data processing and communications), mission control and payload operation control systems (mission management and analysis, data processing, mission planning, management and analysis, system simulation, C^3 data processing and test/validation) and services such as logistics support. Space contracts include Columbus advanced flight software prototyping and Central Mission Control Centre detailed definition; XSAR Mission Planning & Operation Software; SAX software ICD for all spacecraft subsystems and payload; Envisat 1 Radar Altimeter EGSE software design/development. Developing software for data distribution system. Operates a software development environment with fast Earthnet technology, advanced WAN for LAN connection.

VERIFIED

Tecnodata Italia Srl

Current Status
A member of the Serco Group since 1998. The main space specialisations are computer services, communications services (VSAT provision and operation; ground station operations, maintenance, scheduling) and software development, including remote sensing and satellite simulation. Principal contracts include software services for ESRIN, ESOC and ESTEC; maintenance and operation of ESOC's Control Centre and Odenwald ground station; software maintenance and systems support for ESA's IRS; ESTEC satellite ground station operations, maintenance and scheduling. Major subcontractor to ESA on the International Space Station. Developing integration software for dedicated

hardware sets to allow several separate satellite or ground systems to operate in synergy through a centralised control and co-ordination centre. It has also conducted demonstration tests with a closed-loop control system for man-tended and automated systems.

UPDATED

Tecnospazio SpA

Current Status
A Galileo Avionica company, Tecnospazio was formed in 1987 by Comau and FIAR to pursue space automation and robotics activities; it also employs the expertise of those companies. Participated in Columbus onboard control logic. Provides the Video Camera Assembly for the inside of the Columbus Orbital Facility and will conduct EUROPA (External Use of Robotics Payloads Automation) aboard the ISS. Developed software controlled automation and robotic systems for advanced telepresence operations for unmanned earth orbit or deep space applications ranging from remotely monitored human space systems to robotic spacecraft in the far reaches of the solar system. Designed and manufactures microgravity payloads and subsystems for application to parobolic flights, sounding rockets and on-orbit facilities. 1998 space revenue Lit8' billion; 1997 space revenue Lit7.9 billion; 1996 space revenue Lit6.2 billion. 1995 space revenue was Lit5.5 billion; 1994 Lit5 billion, 1993 Lit5 billion, 1992 Lit6.4 billion, 1991 Lit5 billion, 1990 Lit3.3 billion, 1989 Lit1.6 billion. In 1995, it participated in the Robotic Joint Control experiment during EuroMir 95.

VERIFIED

VITROCISET SpA

Current Status
Utilising permanent facilities in Belgium, Germany, Holland and French Guyana, VITROCISET specialises in integrated logistics support, launch campaigns/operations, software engineering (space/ground), information systems/databases, and remote sensing and image processing. For example, it provides operation, maintenance, technical assistance and engineering support at ESOC and ESTEC, operation and maintenance services at Kourou, and software engineering support at ESOC and ESRIN. Also contributes to the continuous upgrading and site re-configuration analyses for improved site operability and efficiency. Recent activities include design/development of Helios ground segment elements, integrated logistic support for Italy's Helios facilities, design/development of cartography image handling systems, software design/development of Artemis' Control Centre for LEOP and routine operations, and software design/development of generic spacecraft Integrated Central System (SCOS II). Developing software integration programmes for multisatellite command and control operations to minimise manpower requirements in satellite and spacecraft control facilities. Major national contracts include maintenance/operation of ASI's Trapani-Milo stratospheric balloon launch base, and technical assistance to National EGSE for TSS, IRIS-Lageos, SAX, Italsat F2. Provides SCOS 2000, the ATV simulator airborne support equipment, satellite navigation and a wide range of tools for earth observation.

VERIFIED

Japan

Fuji Heavy Industries Limited (FHI)

Current Status
FHI participated in studies of Japan's HOPE unmanned shuttle and the initial design of a horizontal takeoff and landing spaceplane. It built the two ALFLEX models first flown in 1996. Redirected efforts to changed requirements statement for fly-back recovery vehicle and now developing concepts for ISS rescue vehicles. FHI has researched possible variants of the HOPE concept with conceptual designs similar to the revised NASA re-entry vehicle being considered as an ISS rescue vehicle. These studies have now shifted emphasis as Japan's plans for orbital space planes have changed. Early plans for a reusable Hermes-like shuttlecraft have been moderated in favour of continued research for technology development with the eventual goal of a reusable, affordable, system for spaceplane applications. This has resulted in renewed interest in a reusable delivery system capable of carrying astronauts if necessary, but this is not crucial to the craft's operation. Technology development supported by FHI includes wind tunnel, CFD, composites and structures and materials science and engineering.

UPDATED

Fujitsu Limited

Current Status
The company produces communications satellites equipment, including small Earth stations, spaceborne electronics such as telemetry systems and MOS-1's visible/thermal infra-red radiometer, and data processing systems such as Yohkoh's onboard data processing computer. Also produces components for scientific experiments and modular engineering packages designed to support in-space payload operations. It undertook biological and materials processing experiments aboard Mir in early 1992; its equipment was launched aboard Progress-M 11 on 25 January and returned aboard Soyuz-TM 13 on 25 March. It is a major shareholder of the H2 launcher and H2A Rocket Systems Corporation. Presently also conducting studies on advanced re-usable launch systems and support infrastructure. Conducting system integration study for hybrid launch systems and unconventional launch concepts associated with H2 successor. It has participated in the conceptualisation of large heavy-lift launch vehicles for H2A derived lifters and for new generations of high performance, commercial, launch vehicles.

UPDATED

High-Reliability Components Corporation

Current Status
High-Reliability Components Corporation was established in 1988 to provide testing, inspection, centralised procurement, distribution and storage of space components. Has played a crucial role in increasing safety aspects, successfully weeding out flawed components and increasing subcontractor standards. Has also conducted simulation trials and materials tests. Main activities are QA, vendor quality surveillance, radiation testing, failure analysis, preparation of specifications and selection of space components. HRCC has performed predictive failure analysis models of critical components in sealed elements and demonstrated high reliability monitoring standards using existing software and hardware. The company has been involved with the H-2, H2A, JEMLISS, ADEUS, COMETS, ETS-VII, ADEOS 2, OICETS and DRTS programmes. Facilities include the Tsukuba Test Laboratory and the Minami-suna Storage Facility.

UPDATED

IHI Company Ltd

Current Status
Ishikawajima-Harima Heavy Industries provides the LOX and LH_2 turbopumps for Japan's H2 launcher's cryogenic engines, in addition to stage 2's hydrazine Reaction Control System modules. Other propulsion interests include scramjet and turboramjet systems. IHI also working on high-temperature/high-pressure propulsion system turboprop designs. Working possible solutions to challenging new specifications which may drive next-generation launch vehicle systems incorporating hybrid propulsion units from air-breathers to mixed cycle engines.

In materials processing, IHI developed systems and elements for the First Material Processing Test (FMPT) for the Shuttle Spacelab-J mission (flown September 1992), including an acoustic levitation furnace, continuous heating furnace and a gradient heating furnace. In addition, the company is actively involved in the development of some instruments for the USERS Unmanned Space Experiment Re-entry System. It is responsible for JEM system engineering and integration support as well as software integration. IHI currently owns 102 Japanese companies and 16 shipyards as well as 43 overseas subsidiaries and joint ventures.

VERIFIED

ITOCHU Corporation

Current Status
Formerly C Itoh & Co, ITOCHU is one of Japan's leading sogo shosha (general trading companies), overseeing more than 870 consolidated subsidiaries and associated companies. In the space field, it provides Earth stations, launch services and satellite control systems as well as integration systems and software. Has developed new methods of integrating add-on elements to existing infrastructure through new computer controlled processing and management systems. It holds a 28 per cent interest in Japan Satellite Systems (JSAT), which now operates three telecom satellites; see the Communications section for further information. Currently employs 4,580 people. Trading transactions have been falling over the last five years, being US$85.5 billion for 2002.

VERIFIED

The Japan Amateur Radio League, Inc (JARL)

Current Status
JARL developed three Fuji Oscar amateur radio satellites, launched piggyback by NASDA's H1/H2 and H2A vehicles. JAS-1 (Fuji FO-12) was launched in August 1986 followed by JAS-1G (FO-20) in February 1990 and JAS-2 (FO-29) in August 1996. Further satellites in this series are in development and JARL is seeking international support for a major expansion of the radio satellite programme involving users from around the world.

VERIFIED

Japan Manned Space Systems (JAMSS) Corporation

Current Status
JAMSS was created in 1990 by 14 companies, headed by MHI, IHI, NEC and JGC, to support NASDA for safety and product assurance oversight. It also supports NASDA for JEM utilisation, operation and astronaut training in addition to fail-safe evaluation. Conducts training activities for operational routines, contingency activities to save experiment equipment and time-saving flight rule interpolation tasks. Evaluating JEM follow-on project and possible free-flying laboratory module. Capitalised at ¥445 million with annual revenue (2000) of ¥3.36 billion. Employs 140 people.

VERIFIED

Kawasaki Heavy Industries Ltd (KHI)

Current Status
KHI was involved in the construction of launch facilities for the N, H-1, H2 and H2A at NASDA's Tanegashima Space Centre. It has participated in system development of the berthing mechanism, airlock and environmental control systems for Space Station's Japanese Experiment Module (JEM) since 1985. It has been involved with the HOPE project and spaceplane research since 1986. It is system integrator of the ALFLEX Auto Landing Flight Experiment and rear body integrator for the HYFLEX Hypersonic Flight Experiment. Both are key steps in developing HOPE and possible crew return vehicle for the ISS.

KHI has developed and manufactured the experimental Geodetic Satellite launched by H1 in 1986.

Space robotics and docking mechanisms with advanced proximity sensors are being studied on ground testbeds for future in-orbit servicing. KHI developed the docking systems for 1997's ETS-7. In 1994, it completed the WETS Weightless Environment Training System neutral buoyancy facility at NASDA's Tsukuba Space Center for crew basic training, procedure development for on-orbit rack replacement and JEM system development. It was contracted in 1994 for a crew Isolation Chamber at Tsukuba completed in 1996. In 2000, KHI developed and manufactured the ablative heat shield and shell structure for the re-entry module of the Unmanned Space Experiment Recovery System. Adapted this technology to unmanned recovery systems designs and to potential thermal protection systems for unmanned or manned reusable space vehicles.

UPDATED

Kyokuto Boeki Kaisha, Ltd (KBK)

Current Status
Established in 1947, KBK handles rocket test stands, load cells, flexures, space retroreflectors, lasers and GPS receivers. FY96 annual turnover was ¥182.2 million (FY95 – ¥176.5 million, FY94 – ¥170.6 million), employees 594 (1995 – 621) 2001, 460 employees. Evaluating launch complex expansion at Tanegashima facility and improved access transportation. Designing enhanced pod support facilities for H2A at Tanegashima. KBK has also conducted studies of efficiency improvements to existing test stands at Tanegashima and Kagoshima with a view to reducing costs and improving efficiency, prerequisites, say Japanese space officials, to more comprehensive out-sourcing for maintenance, repair, refurbishment and operational costs.

VERIFIED

Mitsubishi Corporation

Current Status
The Mitsubishi Corporation comprises a number of companies engaged in space activities. The prime manufacturers are Mitsubishi Electric Corporation, Mitsubishi Heavy Industries, Ltd, Mitsubishi Precision Co Ltd, and Mitsubishi Space Software Co Ltd. The first satellite in which the Mitsubishi Electric Corp was involved was the CS domestic communications satellites in the mid 1970s. It was also responsible for the CS-2 and CS-3 series of the 1980s.

Mitsubishi Corporation handles the import and export sales of space hardware, software and services for launch vehicles, satellites, the International Space Station and ground stations. Business creation and capital investment in space-related application such as Space Communications Corporation (satellite communication operator), Japan Space Imaging (provider of commercial high-resolution digital satellite imagery) and Galaxy Express (commercial launch service provider). Mitsubishi is conducting extensive studies on reusable launch systems and on recoverable space vehicles capable of carrying humans but not exclusively for manned flight.

UPDATED

Mitsubishi Electric Corporation (MELCO)

Current Status
Established in 1921, MELCO provides telecom and Earth observation satellites, remote sensing systems and satellite communication systems. It acted as prime for Japan's CS telecom satellites, the ETS-2/4/5 engineering test satellites, the JERS 1 synthetic aperture radar Earth observation satellite and the ISAS/USEF Space Flyer Unit. It is working with Space Systems/Loral and Aerospatiale as contractors for the Intelsat 7 series and with Lockheed Martin Astro Space on Inmarsat 3. Its xenon ion thruster and inter-satellite communications system flew on ETS-6. MELCO owns the Space Communications Corp of Japan, operators of the Superbird communications satellite system. Subsystem expertise covers AOCS (electronics, wheel drive assembly, Earth/Sun sensors, nutation dampers), communications/data handling (transponders, telemetry/command units, central processing unit), electrical power (battery assemblies, converters, shunt, control units), antennas (shaped beam, multibeam, phased array and omni antennas, feeds, multiplexers), solar arrays (GFRP, KFRP, CFRP substrates, semi-rigid/rigid, solar cells modules, GaAs cells), propulsion (mono hydrazine thrusters, ion and MPD arc jet engines), thermal (louvre, heat pipe, shield, multilayer insulation) and structures.

MELCO provided the microwave scanning radiometer for the MOS satellites and developed the SAR for JERS 1. The company also offers turnkey Earth stations, small Earth stations for SNG, VSAT and USAT, and was responsible for ISAS' 64 m Usuda deep space tracking facility. It is a major supplier of mobile Earth terminals for the US MSAT system.

The class IB spacecraft manufacture/test facility at Kamakura can fabricate/test subsystems and components, and integrate/test the entire spacecraft system in one building. The Satellite Integration and Test Facility includes a centre of gravity machine, mass measuring machine, dynamic balancing machine and a data acquisition system. The Subsystem Test Facility houses an anechoic chamber, antenna range test tower, EMC test system, Sun/Earth simulators, three-axis air bearing and gimballed servo tables, infra-red reflectometer and three-axis co-ordinates measuring machine. The environment test facility accommodates a space simulation chamber, spin table, vibration machine, rotary accelerator, shock test machine and outgassing = measuring machine. The manufacturing facility allows fabrication of solar cell modules and heat pipes. MELCO employs 116,000 people in its main organisation and subsidiary companies. Net sales for the first nine months of 2002 were reported as US$27.3 billion with equity of US$4 billion.

VERIFIED

Mitsubishi Precision Co Ltd (MPC)

Current Status
Core business develops and manufactures pilot and driver-driven simulators, also navigation and guidance equipment. Formed in 1962.

MPC specialises in inertial reference systems, RW/MW assemblies, and electronics. IRUs, RWAs and MWAs for ETS-5/6/7, JERS-1, SFU, ADEOS-1/2, COMETS and OICETS. ING and RWA for M-3S and M-5. E-PKG and DIU for H1/2, H2A and JEM. Applications for new control core equipment and simulated contingency operations planning for unmanned and human space flight missions.

Net sales for 2000 to 2001 financial year reported at US$180 million with shareholders equity of US$28 million.

VERIFIED

Mitsubishi Space Software Co Ltd

Current Status
A subsidiary of Mitsubishi Heavy Industries and founded in 1992, the company is responsible for the software of the H2 and H2A launcher. This technology has also been applied to unattended orbital platform control systems. Developing control logic for possible applications to future launch vehicles. Directing some low-level research effort at improvements to self-analysis and health monitoring for risk mitigation in launch vehicle and propulsion operations with a view to improving on-time launch records.

Capital of ¥445 million. Revenue for 2000 reported at ¥3.36 billion. Employs 134 people.

UPDATED

Mitsui & Co Ltd

Current Status
Established in 1947, Mitsui has capital of ¥192 billion and employs more than 10,000 people of which 3,500 are overseas in 43 subsidiaries with facilities in 88 countries. Hardware, software and services for satellites, launchers and Space Station. Has committed some modest resources to integrated launch vehicle automation processes sought by national and company efforts to reduce costs, limit risks, remove excess monitoring and control equipment installation expansion. This work continues with emphasis on the new generation of reusable spacecraft sought by the new Japanses aerospace agency JAXA. The company holds a 24.5 per cent interest in Japan Satellite Systems (covered in the Communications section), which began operating its first satellite in 1988.

UPDATED

NEC Corporation

Current Status
NEC is a major contractor for satellites, rocket borne equipment and ground systems. It has acted as prime for major applications satellites: meteorological (GMS series), Earth observation (MOS series), direct broadcast (BS-3 series) and amateur radio communications (JAS series). Its role as system integrator for Japan's science/test engineering satellites includes: Ohsumi, ETS 1, MS-T1-5, ExosA-D, AstroA-E, Planet-A/B, Muses-A/B, Geotail, Solar-A, Debut, COMETS, OICETS and Lunar-A, totalling more than 50 satellites by 2000. In addition, more than 100 other satellites employ NEC's highly reliable bus equipment, communications and special-purpose payloads and other space-related products.

NEC's Space Systems Development Division makes mission payloads including communications transponders, communications antennas and space sensors; TT&C subsystems, including transponders, omni antennas, databus and data processors; AOCS, including attitude sensors/control electronics; power subsystems, including solar array paddles; structural subsystems; thermal control subsystems. Space Station equipment, μg experimental equipment and Get Away Special. Rocket borne equipment, including telemetry transmitters, radar transponders and command destruct receivers. Ground systems, including TT&C, remote sensing stations and meteorological satellites data receive stations.

The Guidance/Electro-Optical makes guidance computers for H2 and H2A launch vehicle; SAR high-speed image processing system.

The Overseas Microwave/Communications satellites Systems makes Intelsat/Inmarsat/domsat Earth stations and VSAT/MSAT terminals.

Sales for 1998 ¥4.97 billion, for 1999 ¥4.815 billion, sales for 2000 ¥5.21 billion, sales for 2001 ¥5.59 billion, sales for 2002 ¥5.21 billion. Total assets for 2002 ¥5 trillion.

VERIFIED

NIPPI Corporation

Current Status
NIPPI Corporation space-related structures include: M launch vehicle tail fuselage and fins; H2 and H2A booster nosecones and aft skirts; satellite extendible masts (including Akebono and Geotail science booms); extensor for extending motor nozzle; paddles, antennas and satellite structures. Studying extendable nozzle technology and high-temperature liquid cooled skirts. Also integrating steel and composite structures for high-temperature variable conditions in vacuum conditions.

VERIFIED

Nippon Sanso Corporation

Current Status
Space simulation chambers, propellant supply systems for cryogenic engines, hydrogen liquefiers, cryogenic storage tanks, and high-vacuum equipment. It provided Japan's largest solar simulation chamber, the 13 m diameter unit at NASDA's Tsukuba centre. Also Japan's largest is the 1,200 litre/hr liquid hydrogen production facility it built for NASDA. It manufactured the μg drop capsule for the Japan Microgravity Centre and supports microgravity research for JEM. Additional facilities for Japanese microgravity research proposed by Nippon Sanso enabling advanced experiments on autonomous Japanese platform. Nippon Sanso is developing new high performance microgravity simulation facilities and is developing new solar simulation chambers for NASDA. The company is also developing advanced simulation chambers for high-pressure conditions and for simulating several different, varying, space environment conditions for fault prediction and failure propagation prediction.

VERIFIED

OKI Electric Industry Co Ltd

Current Status
Specialises in multimedia networks and services. Developing new concepts for integrating information systems supporting new services options involving multiplexed media inputs for wideband users and high traffic rates on multifrequency gates. Testing new optional frequency diverters for use in advanced communication satellites.

Satellite Earth station and command, control and communications systems for launch vehicles including H2 family. Currently working H2A CCC equipment. Reported 2001 sales increase of 10.5 per cent compared to 2000. Sales in 2001 ¥740 billion (US$5.97 billion). Operating income in 2001 ¥28.3 billion (US$228 million).

UPDATED

Sumitomo Heavy Industries Ltd

Current Status
Specialises in environmental systems, semiconductor and liquid equipment, medical instruments, lasers and materials processing. Extended development testing of a new generation of integrated checkout facilities for spacecraft and launch vehicles.

First Japanese manufacturer of liquid helium dewar. Rocket launch structures and ground base transportation systems; cryogenic technology, including helium liquefaction plants. In July 1994, Sumitomo Corporation acquired a 20 per cent interest in Hamilton Standard Space Systems ISO14000 certification in 1998. Net sales for 2002 reported at ¥517 billion with net income of ¥1.65 billion. Total assets of ¥634 billion. Has conducted extensive study of improved and expanded launch site facilities and support infrastructure for a (potential) heavyweight launch vehicle.

UPDATED

Taisei Corporation

Current Status
Founded in 1873, Tasei Corporation is active in building construction, civil engineering, real estate and engineering. Has designed new floating launch vehicle stabilisation devices.

The Space Technology Section specialises in facilities for space-related ground experiments and the planning and construction of launch bases at Kagoshima and Tanegoshima. Developing earthquake-resistant propellant and explosives facilities. Also leading research into new ways of containing the blast of hazardous facilities in the event of failure or accident.

Taisei employs more than 10,000. Net sales of ¥1.75 billion in 2002 with operating income of ¥66 million.

UPDATED

Toshiba Corporation

Current Status
Spacecraft systems and components (command receivers, star sensors, attitude control electronics, Ni/H_2 batteries), manipulator subsystem for Space Station Japanese Experiment Module, satellite ground systems (Inmarsat ship Earth stations, tracking/control equipment), satellite systems software and high-speed 32-bit Image Processor System. The company was prime contractor for NASDA's ETS 6 and was prime for BS-2. It is prime for NASDA's Precipitation Radar on the US/Japan TRMM mission and is developing advanced follow-on systems for later satellites of this type. Toshiba has conducted preliminary design details for an integrated satellite system using a highly integrated suite of science instruments for earth science research and development.

Corporation sales for financial year 2001 (ended 31 March 2002) reported as ¥394 billion with net income of ¥254 billion. Employs 176,398 people (as of March 2002).

VERIFIED

Netherlands

Brandt Fijnmechanische Industrie BV (BFI)

Current Status
BFI specialises in low-quantity, high-precision machining of components and tools in steels, titanium, aluminium, super-alloys and ceramics. Space involvement includes Exosat, Solar Max, IRAS, Hipparcos and Ariane 4 and 5. Developing new heat resistant materials for high temperature structures and ceramic coatings for heat furnaces used in processing exotic materials and compound molecular structures. BFI has developed a new process for integrating different materials chemistries with hybrid composite materials in complex shapes and structures. The objective is to develop structures for some of the more demanding requirements posed by long-life satellites and spacecraft designed for extreme thermal environments.

UPDATED

CCM

Current Status
Centrum voor Constructie en Mechatronica (CCM) specialises in technical development projects. It designed, developed and built Spacelab biological experiment units that were later modified for sounding rocket application. CCM has designed a protein crystallisation module with Fokker Space and Comprimo to accommodate 400 samples. Parallel development of ground-based applications has resulted in further development of bioracks for future space-based requirements and these have been selected for further analysis as potential commercial products for non-government users and experimenters. CCM has been working on Columbus racks for the ISS and on the design of specialised rack and equipment containers for specific customer adaptation. Diversification has allowed CCM to market space products into earth-based laboratory and test station facilities.

VERIFIED

DEP BV (Delft Electronische Producten)

Current Status
Set up in the 1990s to serve the defence industry, DEP manufactures hybrid photo-diodes for use in high-energy physics research. Products capable of surviving harsh environments with components rated to high-*g* load tolerances, high thermal stress environments and extended radiation saturation.

DEP/Scientific specialises in the design, development and manufacture of customised high-performance low light level imagers and detectors. The range includes image intensifiers, intensified CCDs, (imaging) photon counters, hybrid photodiodes and related electronics. Detection capabilities can reach down to photon-counting level. The company provided VUV 1,200–1,800 Å auroral imagers for the Canadian contribution to Freja and Interball; Solar Blind Imaging Photon Detectors for Israel's TAUVEX UV telescope aboard Russia's Spektrum-X/v also developing the Imaging Photon Detectors for XMM's optical monitor (in co-operation with University College London). Supplies earth-based observation for single pixel detection. Continues to develop optical systems for next-generation telescopes and in developing integration optics for adaptive telescopes.

VERIFIED

DINFA BV

Current Status
Manufacture of high-precision parts and assembly of components, instruments and special tooling. Assembles components for satellite as well as launcher markets. DINFA has developed markets for precision assembly in the nanotechnology sector and is designing nano-tech automated production concepts. Also manufactures shock resistant support structures for small scale engineering requirements in low-volume voids and high-density packaging. Specialising in advanced specification components, structures and polymer-based composites. A member of the Netherlands Aerospace Group.

UPDATED

Dreves Engineering BV

Current Status
Space mechanisms and structures, mechanical devices and electromechanical sampling applications for space- and earth-based applications. Project involvement includes a drill study for ESA's Rosetta comet nucleus sample return mission, design and development of the

gearbox for the Hera robot arm joints, and a feasibility study of Fokker's Star mast deployment system. Remotely controlled autonomous robot activated by external sensors developed for an ESA study on long life planetary exploration vehicles. Continuing studies on deep drilling applications for planetary probes. Developing mass surface penetration devices for ESA sample retrieved missions.

VERIFIED

Drukker International BV

Current Status
Drukker is a diamond company that has specialised since 1967 in science and technology applications. It provided the diamond windows for the infra-red instruments carried by the Pioneer-Venus and Galileo atmospheric probes, the thin diamond bolometer substrate for COBE's FIRAS instrument, and the diamond substrate for a Nimbus 6 long wavelength pass filter. It also supplies diamond anvils for high-pressure research and heat sinks as semiconductor laser substrates or for high power microwave devices. Has developed penetrative diamond aperture products enabling scientists to install micro-engineering detection equipment. Has developed and produced diamond substrates for long wavelength meteorological applications and has researched long exposure opportunities for high-vibration environments. Also provides new multi-frequency optical apertures for high-stress environments and high temperature regimes.

UPDATED

Dutch Space

Current Status
Previously known as Fokker Space BV, Dutch Space is the primary outlet for the Netherlands in European space activity. Fokker Space was created in January 1996 from Fokker Space & Systems, with equal shareholders Fokker NV (part of DASA since 1992) and the state-owned Ultra Centrifuge Nederland NV. Fokker developed Ariane 4's carbon fibre interstage 2/3 and stage 3's engine frame. For DASA/ERNO, it built Ariane 4 booster nosecones. It was contracted for six test and three flight models of Ariane 5's 1.6 tonne aluminium engine thrust frame and subsequent contracts have supported ongoing production of Ariane 5 for commercial and government customers.

Over the last 30 years the company has built up a comprehensive package of expertises, services and products and nowadays is an important supplier of space subsystems. The company has about 300 employees and a turnover of €100 million.

Dutch Space is an independent manufacturer of solar arrays for spacecraft. The solar array program ranges from telecommunication and science to earth observation applications. The solar arrays are based on the Advanced Rigid Array (ARA) concept enabling maximal heritage and flexibility derived from a broad expertise basis.

Dutch Space is also a specialist in advanced robotics technology. The major activity is the current development and production of the ERA (European Robotic Arm), an 11 m long, standalone space robot for the International Space Station.

In the field of earth observation instruments Dutch Space has gained considerable experience. Sciamachy, flying on Envisat, is already known to be a world-class trace-gas monitoring system. A derivative, called OMI (Ozone Monitoring Instrument) will fly on the US platform EOS-CHEM.

The expertise in real-time simulation is demonstrated in the EuroSim program package. This tool is used in-house, but is also licensed to parties that wish to develop their own simulation application. Dutch Space also uses its experience in engineering and simulation for defence applications and has successfully converged several separate technology developments from national research institutions contributing to an integrated research programme with potential for civil and military applications.

VERIFIED

Genius Klinkenberg BV

Current Status
Space-related activities include design/manufacturer of Ariane 5 and advanced Ariane STEP ground equipment and transport systems, and manufacture of aluminium parts for launchers and satellites. Designed multivehicle launch pod adaptations to convert dedicated firing platforms into adaptive, multipurpose ranges. Capable of designing and constructing launch pads and support facilities including full GSE and fluid/electrical interfaces for small or medium launchers. The company has conducted integrated launch structures and firing pad designs for new generations of launcher.

UPDATED

Hymec BV

Current Status
A Neways company, Hymec was founded in 1982 to design/produce thick film hybrid circuits. The 1,000 m^2 class 100,000 + 200 m^2 class 10,000 cleanrooms and class 100 laminar flow bench production facility is certified by ESA for the supply of thick film hybrid microcircuits; contributions have been made to SAX, Soho, Cluster, Ulysses, Sciamachy. Production technology includes plated-through and multilayer screening (up to eight conductor layers), abrasive and laser resistor trimming (passive/active), chip/wire-bonding, reflow soldering, packaging (hermetic, ceramic or conformal coating) and high-reliability screening (MIl-STD-883). The facility is equipped to produce power-hybrids on aluminium nitride and direct bonded copper. Contracts secured for development research on thin-film technologies. Researching hybrid circuit migration and data export potential within chips and circuits. Employs 60 people in the Netherlands and 60 in China.

VERIFIED

ICT Aerospace Group

Current Status
The 12 employees of the Aerospace Group operate within ICT Automatisering Deventer BV, a working company of the ICT Beheer BV, which has more than 4,000 employees. As a systems house, ICT's main space activities cover EGSE, languages and development environment, human computer interfacing, telescience/μg utilisation and simulation, remote sensing, radar system development and transputer technology. Telepresence remotely programmable logic for extended science tasking and zero-*g* simuation exercises. Integration studies on optical and microwave data convergence. ICT has integrated hardened circuit components into a shock-resistant design for long-duration missions.

UPDATED

Mechanical Engineering & Consulting (MECON) BV

Current Status
MECON provides feasibility studies, project management, technical design and detail engineering. Space involvement includes the glovebox camera housing for Bradford Engineering and a science instrument accommodation study for ESA/TNO-TPD. Has conducted fit-check analysis for micro-boxes designed to provide utilities for experiments. These have extended to lockers and experiment pods for interchangeable racks on man-tended or unmanned platforms. Staff totalled approximately 45 people in 2002.

UPDATED

Netherlands Aerospace Group (NAG)

Current Status
The NAG is the national association for the Dutch aerospace sector, representing 50 companies and organisations engaged in R&D, engineering, manufacturing and maintenance. Activities of the NAG include collective participation at airshows, promotion of member capabilities, organisation of trade missions, workshops and seminars, as well as being the national point of contact. Has been an effective lobbying channel for national projects and the stimulation of new small research and development activities and for effectively bringing together the separate elements of the spare-related industries in the Netherlands. Space manufacturing members comprise Fokker Elmo BV, KLM Cargo Aerospace Logistics, Schreiner Components, Ten Cate Advanced Composites BV, TNO Defensieonderzoek FEL and Urenco Nederland BV.

VERIFIED

Origin/Nieuwegein BV

Current Status
Information handling and automation consultancy, project management and software. Specialists in electronic control and software interaction systems with automation interface if necessary. Experience includes spacecraft command/control, simulators, database management systems, science data processing and man/machine interface techniques. Part of the Grace Consortium. Hybrid electronic switching elements developed for high-stress environments. Perfecting software for multiple science experiment tasking where simultaneous data recording is necessary. Conducts analytical evaluations of automated man-machine interface hardware with software interpolation.

UPDATED

Polymarin BV

Current Status
Design, development and fabrication of high-performance components of reinforced plastics, including the nosecones for Ariane 4's liquid strap-on boosters. Studies have also been made of a carbon fibre satellite antenna and for Ariane 5's booster recovery system. Studies continuing in FESTIP follow-on programme. Also developing capabilities to produce composite structures for lightweight planetary landers. High temperature resistant maleable products with embedded instrumentation. Intelligent structures research for sustained shape and form memory re-programmed remotely through thermal modelling at the molecular level. Has advanced this activity into remotely programmable and artificially deformable materials.

UPDATED

Ramaer Connection Technology BV

Current Status
Printed circuit board production for aerospace and high-technology applications in addition to manufacture of flat-board IC boards. Specialised products for hardned radiation proof environments pertaining to high-risk environments for satellite and spacecraft flight trajectories. Development of hardened IC boards and semiconductors. Researched into radiation resistant electronic components and has developed high-temperature resistant control boards.

UPDATED

RDM Technology BV

Current Status
Founded in 1902 as a major shipbuilding company, RDM reorganised in x1995 into three separate companies: RDM Technology BV; RDM Submarines BV and RDM BV. Diversification tripled the size of the company and acquisitions formed RDM Holding.

RDM's space interests include machining of frame body rings and final machining of assembled structures for Ariane 4/5 and FESTIP applications. It has participated in cost reduction exercises for efficiency improvements for machining and so forth. Also develops new and more efficient methods of preforming structures from advanced composite-alloy fusing.

Employs 2,000 people with annual turnover of US$400 million.

UPDATED

Satellite Services BV

Current Status
Established in 1984, Satellite Services BV specialises in satellite data communications, GSE, real-time interface, chip and board level electronics and the design and test of payloads.

Telemetry simulators and data transfer interfaces; telemetry, telecommand and analysis systems in addition to ground antenna applications. Space-based antenna structure possible. Capable of setting up complete end to end systems layout for medium- or high-capacity communication systems through dedicated satellites systems, constellations of common-owned networks and hybrid assemblies.

VERIFIED

Stork NV

Current Status
Stork NV is involved in five market areas: textile printing; poultry and food processing; industrial components; technical services and aerospace.

Services include consultancy, feasibility studies, project management, management consultancy, conceptual and basic design engineering, procurement, construction, HES (health, environment and safety) engineering, quality and product assurance, commissioning and start-up and operator training. The SE&C Group employs more than 17,000 personnel in more than 35 subsidiary offices worldwide. Turnover in 2001 was a reported €2.2 billion with a net loss of €8 million.

Regarding space technology, Stork contributed to the design and construction of ESTEC's Large Space Simulator, acoustic chamber and hydraulic shaker building. R&D is performed on life support systems for short and long duration manned flights and in microgravity research (biological and materials research, space medicine, experiment hardware development, analytical/sensor technology). Current and recent projects include membrane bioreactor systems, biological air filtration units, membrane CO_2 absorption technology, sensor and microsystems technology, a facility for protein crystallisation and a High Performance Capillary Electrophoresis (HPCE) facility for space applications (prototypes have been tested in the Crew Work Station Testbed and the Columbus Automated Testbed at ESTEC); a new HPCE facility based on microsystems technology is under development. SE&C is management consultant to ESA's RADIUS project on zeolite research in space, executed by an international consortium of universities and private industries; responsible for project management, quality assurance, FMECA and safety assessment of biological experiments in the EUROMIR-95 and EUROMIR-E missions; responsible for reliability and safety assessment of ERA's mechanical joints; and participant in the Dutch Utilisation Centre, an infrastructure aimed at microgravity research user support. Stork has developed new biological test and environmental containment chambers for untended operation in orbital facilities.

VERIFIED

Norway

Alcatel Space Norway

Current Status
A subsidiary of Alcatel, ASN (formerly AME Space AS) designs, develops and produces electronics hardware for space applications. Staff total 110. Capabilities include space-qualified thin film hybrid components, space-qualified Surface Acoustic Wave (SAW) components (including bandpass filters, chirp filters and delay lines for applications over 10 MHz to 1 GHz), and onboard signal processing equipment for advanced payloads of communications and remote sensing satellites. Recent contracts include: SAW filter modules for ICO, 900 Globalstar filter modules, IF channel filters for Artemis, SAW bandpass filters for Hot Bird, new generation chirp generator for the Advanced Radar Altimeter on Envisat 1, and development/supply of Ariane 5 electronics (such as solid state power controllers, hybrid circuits for attitude control system, and sensor interface units for cryogenic tank level monitoring). Refined and advanced circuits for complex re-startable stage motors. Development programme for new multi-stage launcher control systems integrated into sensor detector software. New test facilities for an expanded research and development programme.

The company has a class 100 cleanroom for SAW device manufacturing, class 1000–10,000 for thin film hybrids, class 100,000 for PCBs. Total cleanrooms cover 350 m^2.

UPDATED

CAP Gemini Ernst & Young Norway A/S

Current Status
Part of the CAP Gemini Ernst & Young Group, the company specialises in: test/checkout systems such as EGSE, SCOE and simulators; space communication networks and data dissemination systems; telemetry/telecommand systems based on Packet TM/TC and CCSDS standards; Earth observation systems (including GIS); onboard data management systems, in particular onboard software; testing/verification of critical software. Development of responsive electronic systems for multiple satellite constellations. Management and financial programming studies and consultative services and conducts specific telecommunications concepts and data management system layouts. Employs approximately 40 people.

UPDATED

Det Norske Veritas Industry A/S

Current Status
DNV is a specialist risk management company with 300 offices in 100 countries. It serves the maritime industry, oil and gas industries and general industrial sectors. It has focused on several accidents and analysed the reasons and potential preventative directions and as such the company is highly experienced in hardware, software and management weaknesses.

DNV Industry's Space Department specialises in R&D for product assurance, structural analysis, noise/vibration analysis, fracture/fatigue analysis, RAMSES analysis, thermal analysis, weight optimisation, human engineering and expert systems. Space contracts include a safety study of the Andøya Rocket Range and a study of in-orbit non-destructive testing together with simulated mechanical stress. More than 1,800 employees.

VERIFIED

EIDEL (Eidsvoll Electronics A/S)

Current Status
EIDEL was established in 1966 for electronics R&D and production in communications and telemetry for ground based systems. Early development systems for telecommunications support linking ground and orbital systems. Its major activity is based on PCM (Pulse Code Modulation) telemetry encoders, decoders and simulators. EIDEL has developed a HAVE QUICK compatible time distribution and communication control system. Produces EE240 PCM encoder, EE235 solid state receiver and EE315 and EE350 PCM decoders with EE200 software programming tool.

ESTEC contracted an Ion Propulsion Diagnostic Package for Artemis, comprising sensors on the satellite's outer surface to monitor possible interactions from the ion thrusters. Metallic deposition and charged particles are the main concern. Research development programmes for deposition placing. IPDP's general core can be configured for other sensor types and satellites. EIDEL can provide exterior mounted sensors for effects monitoring and contamination assessment.

VERIFIED

Kongsberg Gruppen A/S

Current Status
The Kongsberg Group is Norway's largest defence products manufacturer, established as an independent company in 1987; 51 per cent is owned by the Ministry of Industry. Its space-related activities are managed by the Space and Avionics Dept of Kongsberg Defence and Aerospace and began with development of booster attach and release mechanisms for Ariane 5: structures and mechanisms; SCOE and simulators; electro-optical systems; ground stations for Earth observation satellites (Kongsberg Spacetec). Principal space contracts cover development of Ariane 5's booster attach/separation system; SCOEs and simulators for Soho, Huygens and Ariane 5; MIPAS optical differential system. Set up in 2002, Kongsberg Satellite Services employs 45 people.

Konsberg Informasionskontroll A/S or IK is a subsidiary of Kongsberg Defense. Its 1994 space turnover was NKr15 million out of a total NKr40 million. Principal space capabilities cover system development and integration, software engineering computer systems, signal processing and electronics. Main products include software and computer systems for processing and operational analysis of Earth observation data, and EGSE. Contracts include SAR processing systems for ERS and Radarsat, a system for automatic ship detection in SAR images, VICOS Verification Integration Check-Out Software for Columbus Ground System Software, and EGSE (hard/software) for Soho. Kongsberg has developed new checkout and housekeeping system software package for ISS modules based on a pattern originally designed for scientific satellites.

Kongsberg Spacetec is an information technology company, a subsidiary of Kongsberg Defence and Aerospace, providing system engineering services in electronics, computers and software. 75 staff. Located in the Tromsø Satellite Station, it maintains the station under contract to the Norwegian Space Centre. It offers Earth observation ground stations, meteorological systems for Meteosat PDUS/DCP and NOAA AVHRR systems, and value added products and services based on satellite data, including vegetation maps and pollution monitoring. MEOS (Multimission Earth Observation System) is Spacetec's UNIX-based concept for satellite data acquisition and processing. Increased applications for scientific tasks in addition to support of national Earth observation and remote sensing programmes.

UPDATED

Nordic VLSI

Current Status

A global semiconductor company, Nordic VLSI specialises in microchip design solutions for high-speed data converters and short-range radio communication.

Nordic VLSI designs ASICs and ASIC-based electronic systems. With EADS, it has designed a radiation tolerant signal processing video front end ASIC for space applications. Conducted research on high EMP impact on protected IC boards. Developed high radiation saturation-tolerant boards. New range of radiation hardened IC components for deep space missions to radiation environments. VLSI has conducted preliminary designs of a stress-tolerant range of components with applications in control systems for recoverable spacecraft.

UPDATED

Norsk Elektro Optikk A/S

Current Status

Established in 1985 Norsk Elektro Optikk AS employs 18 people.

The company, formed by engineers from the Norwegian Defense Research Establishment, provides expertise in electro-optics, laser technology, spectroscopy and fibre optic applications. Space experience includes development of the optical path difference measurement sensor for the MIPAS instrument on Envisat 1. It has produced an advanced version of the Envisat model for future generation earth observation and remote sensing satellites including those for meteorological applications. Also provides high-resolution imaging interferometer, high-spectral resolution sounder and airborne spectral imager. The company continues to develop concepts for successor generations of Earth observation and remote sensing satellites.

UPDATED

Norwegian Industrial Forum For Space Activities (NIFRO)

Current Status

Established in 1986, NIFRO is a grouping of 19 Norwegian R&D and manufacturing companies and research institutes promoting members' participation in space activities. The board of directors comprises senior executives from Alcatel Space Norway, Kongsberg Spacetec AS, SINTEF Group, the Andoya Rocket Range and Telenor Broadband Services AS. Areas of interest include: ground segment and support data systems, space electronics, materials technology, manufacturing techniques, space transportation technology, remote sensing, telecommunications and quality assurance/control, consulting. Integrated remote sensing data interpretation from multispectral and non-optical sources. Also, integrated data sets packaged for international projects where customers deal in various technical standards and specifications. Increasingly playing a part in representing Norway's space industry at international and commercial forums.

UPDATED

Prototech A/S

Current Status

Prototech was established as a wholly owned limited company in 1988 by the Christian Michelsen Institute (CMI), restructuring its Mechanical Engineering Section. Specialises in oil and gas, energy, space, medical, industrial and defence engineering, analysis and manufacturing. Space industry applications include rocket payload instruments and deployment systems, satellite components/systems, high-pressure gas supply systems for materials science experiments, and custom-designed scientific instrumentation for research institutes and universities. Prototech and its predecessor have designed/built more than 100 payloads for Norway's own sounding rocket programme. The company was responsible for the design, construction, integration and testing of payloads for Hero, Poleward-Leap, Mapwine, Bugatti, Lomas, Madame, Maimik mother/daughter, Mac-Epsilon, Need and Pulsar II. Payloads have been launched from Andøya in northern Norway, Huelva in Spain and Søndre Strømsfjord in Greenland.

Main space contracts include the gas storage module for Spacelab's Advanced Gradient Heating Facility, design/manufacture of a highly serviceable gas supply system for a Columbus high-temperature materials processing laboratory, engineering/flight models of electronic boxes for an ERS 1/2 altimeter chirp generator, as were similar structures for Soho/Cluster, Radarsat and Artemis, a fluid physics experiment for Texus, linear drive mechanisms for Tiros, and design of sensor/electronic boxes for the IMPS experiments of the Norwegian Defense Research Establishment. Company facilities were inherited from CMI: NC/CNC milling machines and benches, 2D/3D CAD/CAM systems, a measurement laboratory, a rocket payload testing laboratory, vibration and environmental test facilities, and high-precision jig-boring machine/tools; a cleanroom. All equipment/facilities of CMI are available to Prototech. Fluid physics laboratory test facilities for component and experiment systems checkout and test bay integration and multithematic analysis.

Scheduled for delivery in 2003, Prototech is providing advanced lightweight equipment housings for the communications payload on AMC 15 and 16 satellites.

VERIFIED

Raufoss Technology and Industrial Management A/S

Current Status

A subsidiary of Raufoss AS, Raufoss Technology and Industrial Management is a 53 per cent government-owned company mainly involved in pyrotechnics and solid propellants, lightweight metals and composite materials. Space-related contracts include latch valves, antenna pointing and solar array mechanisms for Tele-X, and the separation motors for Ariane 5's P230 boosters. Development work on integration of composite frame technology for release mechanisms controlling separation sequences for large, high-mass, structures.

VERIFIED

SINTEF Group

Current Status

The SINTEF Group (Foundation of Scientific and Industrial Research at the Norwegian Institute of Technology) is a non-profit research organisation. The staff of approximately 1,600 undertakes almost 3,000 projects annually. SINTEF consists of eight research institutes (in applied chemistry, applied mathematics, civil and environmental engineering, electronics and cybernetics, industrial management, materials technology, telecom and informatics and medical sciences) and four research companies. Operating revenues of Nkr1,700 million. Space-related expertise covers human engineering and ergonomics, technical operations and logistics, materials and mechanical engineering, and human safety and reliability. The Neutral Buoyancy Facility (Ocean Basin) at MARINTEK is available for simulation and astronaut training, such as EVA, docking, rendezvous and logistics support as well as simulated EVA repair tasks.

SINTEF DELAB is an R&D institute specialising in information technologies based on computer science and telematics, telecommunications, acoustics and solid-state electronics. Some 20 of the 150 staff work in the space branch, although space-related technical work is also undertaken at other branches. The personnel at DELAB and the Dept of Electrical Engineering & Computer Science, totalling more than 350, make this the largest information technology research centre in Scandinavia. Principal specialisations include: space/ground electronics applying SAW, digital and electrical signal processing; space/ground antennas (particularly reflector antennas and feed systems); distributed databases; knowledge-based systems; Si and GaAs custom-designed ICs; labs for solid-state technology, surface analysis, verification, validation and testing. Specialises also in solid state mapping for interactive chip designs and for multiplexed solutions.

VERIFIED

South Africa

Denel (Pty) Ltd

Current Status

Denel was formed in 1992 to take on the mantle of South Africa's defence manufacturing from Armscor and expand into the commercial sector. Of the group, Somchem in Somerset West near Cape Town specialises in propellants and propulsion systems, Kentron in guided weapons and systems, Eloptro in electro-optical systems, and OTR (Overberg Test Range) flight tests and launch services. Participated in SA launch vehicle programme. Has conducted satellite applications studies for civil and military activities and studies involving South African companies for potential involvement in US and European plans for a new generation of small satellites. Can provide specialist support technologies for small and medium satellite design, design consultancy for basic satellite bus design and systems integration, analysis of launch vehicle design and small satellite test and integration requirements. Denel has conducted several studies related to increased participation in international space activities focusing on scientific and engineering capabilities inherent within the organisation. Denel has proposed a South African satellite network supporting regional communications needs.

UPDATED

Spain

Canava Electronica SA

Current Status

The company specialises in printed circuit boards. Experience includes Hispasat, Shuttle payloads, ISS control boards for Columbus and ground stations. PCB design for advanced experiment control applications. ISS panel boards and filters. Printed boards for new telecommunications satellite concepts requiring open-architecture design concepts. New specification standards applied to all products for ESA and NASA integration through non-US programmes.

UPDATED

Cray Sistemas SA

Current Status

A subsidiary company of Cray Systems Ltd of the UK, specialists in the design and development of software and systems for space applications and ground segment applications. Also for integrating ground and space-based computational requirements. Services available for processing high-value data from different sources in remote sensing and earth management services. Superfast processing and high-information load capacity addressed by coupled systems and integrated hardware sets working through common modules or connecting nodes for applications in very sophisticated multi-experiment integration in space station modules and space based facilities such as ISS elements, advanced observatories and so on. Has conceptualised semi-autonomous command and control systems for unattended satellites and spacecraft.

UPDATED

CRISA

Current Status

Founded in 1985, CRISA specialises in flight electronics, ground support equipment and software engineering, emphasising control and processing units, power units, communications networks, onboard software and video communications. It has produced unique tailored software for mechanical systems and electromechanical control units. It has also pioneered new manufacturing techniques developed to reduce cost and improve reliability. Shareholders are 50 per cent EADS France and 50 per cent Abengoa; turnover Pta2,350 million, staff about 180. Processing units: instrument control units for ASCAT, MIPAS and GOMOS on Envisat and Metop; ASCAT digital processing unit; Columbus vital telecommand unit. Power electronics: Hispasat, Soho and Artemis payload converters; Soho and Ariane 5 electronic unit converters; primary power systems such as battery regulator unit, batteries charge/discharge units, shunt regulators, power distribution units and pyro units. Control electronics: Ariane 5 sequential electronics, sight change mirror mechanism driving electronics, Soho antenna pointing mechanism electronics, deployment control unit, Asar ICUIF 2 board, ISO valve actuation unit. Software engineering: audio/video communications (Central Payload Video System based on 4 Gbit/s data transmission technology); optical communications; control centres; ground stations (baseband equipment, station management, video/audio, monitoring and control functions), HELIOS main spanish control and programming centre. EGSE: power (Helios, Spot 4), simulation subsystem (OCOEs), flight unit test benches (Ariane 5, Artemis) and RF (ERS, Hispasat).

UPDATED

EADS-CASA Espacio

Current Status

EADS-CASA Espacio, Spain's leading space sector company, began operations in 1966 as Construcciones Aeronáuticas SA (CASA). Current activities include: launchers (Ariane 5), science satellites (MINISAT, XMM, Soho, Huygens), Earth observation satellites (Envisat, Helios II, Meteosat Second Generation), telecommunications satellites (Artemis, Hotbird, WorldStar, Sesat, Hispasat), research and technology (dual gridded antenna, CFRP rings, ASYRIO), robotics (ROPODA, SRT) and simulations software (Space Vision Device, VISTA). Specialisations include: launcher structures; satellite service module structural, thermal and harness subsystems and integration; spacecraft platforms; high-stability structures for precision optical systems; multibeam reflector antennas from L- to Ka-bands and feed systems; phased array antennas (L- to C-bands); waveguide antennas, mechanisms and associated electronics; power distribution networks; robotics and 3-D simulation software and virtual reality.

For Ariane 5, CASA developed the inner structures of the upper composite: payload adaptors, upper stage propulsive structure and equipment bay. As an Arianespace shareholder, the Space Division produces elements for each Ariane 4, including the equipment bay structure, stage 1 and intertank skirts, and most of the payload adaptors. CASA provided ERS' wind scatterometer antenna and its radar altimeter reflector.

For national entities such as INTA, CASA has developed the Minisat platform for small missions. For SEPI, CASA has developed the Teleoperated Mobile Robot for use in nuclear power plants. It develops/manufactures elements of the Helios military satellites for the Ministry of Defense. International participation include the European Community's ADONNIS, PROMENVIR, PAGE-IN and MULTICUBE, aimed at developing new technologies.

CASA's Space Division is sited in the Barajas Centre, covering 23,250 m^2 of manufacturing and assembly area. Facilities include 700 m^2 of class 100,000 cleanrooms, including 300 m^2 for satellite integration and 100 m^2 for electronics and harness activities, in conjunction with two class 10,000 laminar flow cabins. Test facilities include type LD2 200 N (modal survey), LDS 18 kN and LE 90 kN shakers, a –60/120°C, 10^{-3} atmospheres Sapratin thermal chamber, a –70/125°C Tenney thermal chamber and optical measurement system chamber.

Collaborates on Atlas 5 launcher and the Soyuz-Starsem launcher. Prime contractor on Minisat 01 for ESA. CASA Espacio has been instrumental in promoting the need for a lightweight space bus for multipurpose satellites. It has also developed a new carbon composite unfurlable antenna structure for application to existing satellite bus designs from other manufacturers. Developing articulated remote manipulator systems for remotely controlled and tele-robotic systems.

UPDATED

GMV, SA

Current Status

GMV is one of three subsidiaries of the holding company Grupo MV, a privately owned company founded in 1984. Set up in 1992 it provides engineering services and turnkey systems development, maintenance and operations. The Technical Department comprises four divisions: Flight Dynamics and Avionics; Operations; Simulation; and Software Engineering. GMV was designated a Centre of Excellence in Orbital Mechanics by ESA in 1989. It is participating in most major ESA and Spanish national programmes, and is providing engineering services and ground control, and data processing systems for other space customers, including Eutelsat, EUMETSAT, CNES and WEU. Provides service on Envisat programme. Capabilities include consultancy and advisory functions in addition to test and checkout through leased facilities or checkout at client venue using GMV staff and equipment. Has performed several consultative analyses for governments and aerospace corporations in Spain and in other countries.

In October 2002, GMV joined the Galileo Sistemas y Servicios along with AENA, CASA, Hispasat, INDRA and Sener.

UPDATED

Iberica del Espacio SA

Current Status

Iberica del Espacio SA was created in September 1989 as a 50/50 subsidiary by France's SEP and Spain's Empresarios Agrupados, a grouping of EPTISA, TRSA and GHESA, specialising in defence and engineering for fossil and nuclear energy production. The company has a total of 55 personnel, with about 50 engineers.

Expertise covers flight components and propulsion, including significant participation in Vulcain engineering; ground segment (design, supply, installation and commissioning); services and engineering analysis (including RAMS Reliability, Availability, Maintainability and Safety studies), and flight data analysis. Specialities include the design and layout of hybrid launch complex engineering requirements where a wide range of different launch vehicle requirements accommodate large and small vehicles. Designs and produces improved value outlet sleeves for high-performance propulsion systems. Also new ground handling interface designs for automated checkout procedures.

UPDATED

Indra Espacio SA

Current Status

Subsidiary of Indra (51 per cent) and Alcatel Space Industries (49 per cent).

Principal space specialisations are: spacecraft control systems (control centres, TT&C stations, OBDH, EGSE, in-orbit test equipment); satellite communications (ground stations, VSATs, network management systems); Earth observation systems (data reception stations, image processing centres, archiving/retrieval systems), GIS applications, Ikonos, Eurimage, SAR processing and risk management. Interactive risk assessment software modes selected for calibrating risk factors in new technology initiatives without actuarial history of failure probability. Products include ground stations for meteorological satellite data reception for the Spanish Meteorological Institute. The complete satellite control system was supplied for the Hispasat system, and TT & C stations were provided for ESA at Maspalomas, Kourou, Villafrance, Malindi, Redu and Michelstadt. INDRA participated in the Helios Image Reception Centre and Image Processing Centre. The company has developed new and innovative tools for integrating space and airborne data retrieval systems.

UPDATED

INSA

Current Status

Ingeniería y Servicios Aeroespaciales SA was set up in July 1992 as a government company under INTA (Instituto Nacional de Técnica Aerospacial) as Spain's major space services company. In the field of aerospace engineering INSA's activity is oriented towards the development of Earth observation services (for example, the FUEGO programme under contract with EC Forestry Observation for ESA). Advanced multispectral image discrimination processing and software patches for different earth resource programming. The company also performs specialised technical and managerial services in the development of aerospace systems (ESA's SMOL, INTA's Capricorno launcher). Business development is oriented to provide commercial services based upon space platforms. Continuous development of space-based, ground-based and interactive discrimination processing for land, air and space-based imaging platforms. New software integrations for existing operator systems and multi-sourced input options.

UPDATED

NTE SA

Current Status

Formed in 1987 as an engineering and software company, NTE's space activities are focused on µgravity, life sciences and life support, with emphasis in thermal design, signal processing electronics, power electronics, image processing, telescience and biology payload hard/software (particularly freezers and coolers). Has conducted research on development of science support hardware/software. ESA has awarded a contract for a feasibility study on pulsed laser evaporation for improved cell biology research. NTE has shown capabilities in integrated systems utilising common control mechanisms and electronic relay systems coupled to software driven control modules. It is working on designs for autonomous life science payload modules adapted to human space flight standards and specifications requirements.

UPDATED

Sener Ingenieria y Sistemas, SA (SIS)

Current Status

SIS specialises in space mechanisms, structures and associated electronics. Activities include the optical path and refocusing mechanism of HST's Faint Object Camera, Ulysses' mechanisms subsystems, mechanisms and structures on Columbus (mainly airlocks and docking equipment), Soho antenna pointing mechanism, Cluster mechanisms and booms, GOMOS instrument electronics, Minisat attitude control system, Helios baffle mechanisms and image processing software, and Meteosat Second Generation's SEVIRI calibration unit, mechanism and pyros subsystem. SIS is also involved in Artemis's Silex payload and Envisat's MERIS and MIPAS instruments. Proposed instrumentation for second generation remote-sensing satellites. Developed remote manipulator systems for nanomodules and small systems packages for interchangeability on unmanned and remotely controlled platforms. It has produced a series of launcher facilities, such as Ariane's umbilical tower and arms, and ground instrumentation, and provided the Spacelab ground integration system and the planar scanner for ERS. Has designed heavy launcher fluids and gases support infrastructure and launch pad umbilical connectors and flexible conduit layouts. Designed flexible composite booms.

UPDATED

Tecsidel SA

Current Status

Formed in 1980 as a specialist company operating in telecommunications, information systems, logistics and mass transport systems. The company has applied mathematical models used in mass transport systems to a wide range of related problem-solving analyses, including the use of software systems for manned habitats in space; this has led to several contracts dealing with man-machine interface in human-tended systems such as orbital facilities. Tecsidel is part of the GSI Group of France and more directly of the Advanced Technology Division, specialising in information processing systems, data communications networks and embedded systems. Graduate personnel total about 45. Space involvement includes studies for ESA on vision processing for in-orbit servicing, Columbus software replaceable units and operating system, and architecture for a distributed real-time operating system. Integrated control systems for telerobotic operations. The company has proposed a family of automated checkout systems involving minimal hardware and software changes.

UPDATED

Universidad Politecnica de Madrid

Current Status

CIDE/UPM is a centre grouping several laboratories of the Faculties of Aeronautical Engineering, Telecommunications Engineering and Naval Engineering. Space activities include small satellite design/construction, spacecraft thermal control and structural analysis, real-time signal processing and communications, µg fluid sciences/materials, and hypersonic aerodynamics. Has researched flight control systems for hypersonic vehicles and manoeuvrable unmanned spacecraft. Designed micro-satellite thermal control systems. Its UPM-Sat 1 micro-satellite was launched in July 1995. Co-ordinated studies from several colleges and universities into telecommunications systems technology. Remote-sensing technology research. Integration studies for new and innovative proposals requesting support for hybrid nano-satellite design concepts and technologies research support.

UPDATED

Sweden

ABB HAFO AB

Current Status

Mechanical and electronic design and construction, spacecraft structures, sensor technology, microwave antennas and semi-rigid antenna structures. Hybrid structural components utilising matrix composite or carbon-fibre structure for aerospace applications including flexible lock struts and folding antenna ribs. Development of thin-mast structures for high tensile specifications. Product research into new hybrid structures for large arrays controlled through semi-autonomous control systems and MEMS technology.

VERIFIED

FFV Aerotech AB

Current Status

FFV Aerotech, a member of the Celsius Group, comprises three units: avionics, aircraft systems and subsidiaries FFV Test Systems and FFV AvioComp. Employees total about 1,800. It concentrates on aviation maintenance and consulting, but in 1989 it was awarded contracts for AIT of Sweden's Freja scientific satellite, which was launched October 1992 by China's CZ-2C. Development and manufacture of Freja's 450 MHz uplink antennas, EGSE, flight harness and thermal blankets were also undertaken by the Avionics Division. Within the Group, composite structure items for space applications are developed and manufactured at FFV Applied Composites AB in Linköping. High-strength composite matrix structures for heat-soaked antennas and locking frames with long life in high-thermal or stress-related loadings. Development has been completed on a thin-spar umbrella concept for large telecommunications satellite antennas. Has developed a new passive thermal control system for small satellites.

UPDATED

Saab Ericsson Space

Current Status

Saab Ericsson (SE) was founded 1 January 1992 by the merger of Saab Space AB and the Space Department of Ericsson Radar Electronics AB. Saab Combitech AB owns 60 per cent and Ericsson Microwave Systems 40 per cent. The company supplies launcher and spacecraft onboard subsystems and equipment, including fault-tolerant computers, data handling systems, antenna systems, microwave electronics, satellite separation systems and guidance systems. Each Ariane 5 carries two new-generation parallel guidance and navigation computers; the equivalent continue to be supplied for Ariane 4. The Columbus Polar Platform employs SE's Cassegrain antenna (including DRS pointing control electronics) and an OBC similar to the Spot 4 fault tolerant concept. For Envisat, SE will deliver control units for two instruments, central RF electronics and a radar signal processing subsystem for the ASAR radar. Meteosat Second Generation will carry a data handling system, UFH receiver/transmitter and TT&C antennas. Thaicom 3, Agila, Nahuel, Sirius 2 and Eutelsat W24 will have SE's data handling systems. Earlier deliveries include computers and microwave antennas for ERS 1/2, computers for Spot 1–3, data handling system, antennas, RF distribution network and central software for Soho, and attitude control computer, antennas and cover release mechanism for ISO. Data handling systems and antennas were supplied for Eutelsat 2, Arabsat 2 andTurksat. Employs 678 people of which 85 per cent are qualified engineers.

Sales in 2001 reported SEK800 million, of which over 90 per cent on export to some 15 countries in Africa, Europe and the USA.

VERIFIED

Technology Trade International (TTI)

Current Status

Initially, TTI's business was to locate and develop possibilities of industrial cooperation between the RFAS space industry and companies/industry with related activities in the rest of the world.

Over the years several important agreements of co-operation and worldwide representation have been reached with major space technology companies of the RFAS. In particular TTI is the representative of the Cosmos Group on the international market. TTI markets the products and services of the Cosos Group and its associated companies including NPO Energomash and Central Specialized Design Bureau (CSDB).

TTI is also the international business representative for STC COMPLEX, the company converting missile technology into the low orbit launch vehicles START and START-1. TTI has linked several separate companies with partners outside Sweden to propose further endeavours based on the experience with START. It has developed market analyses ranging from recruiting investors to funded development of new initiatives outside the areas of interest displayed by venture capitalists.

Recently TTI has taken additional steps to consolidate the international space co-operation through agreements with interested parties all over the world and in particular in Europe, Asia and Australia. This has among other things resulted in development of new software for improvement of existing Russian satellite hardware.

To adequately meet the demands of the challenging period ahead TTI is enhancing its resources in the space technology field to better serve its customers and partners and also provides advisory services in the satellite and launch vehicle fields. Several major studies have been completed on the integration of national technologies underpinning TT international projects. TTI has assessed potential for growth in Sweden's space related industries to project possible areas of co-operation with ESA.

VERIFIED

Switzerland

Aerospace Engineering Office (AEO)

Current Status

AEO provides engineering consulting services to the aerospace industry, including studies of software and light structures and finite element analyses. It has pioneered the integration of predictability studies to integrate random motion factors into new structures with advanced composites bringing long life, greater operating flexibility and tighter operating margins and specifications. Its activities have included static tests of the Ariane fairing, and OTS/ECS solar panel structural analysis. AEO also developed a kinematic model under NASA contract of the Shuttle remote manipulator and ISS telerobotic systems. Have analysed advanced space robotics systems for manned and unmanned vehicles. AEO has conducted tests on new fairing concepts for Soyuz launchers. Finite element analysis on optional fairing configurations. Has also conducted studies into composite beam construction and application with a view to the large diameter antenna market.

UPDATED

APCO Technologies SA

Current Status

Created in 1992, APCO Technologies specialises in the design and manufacture of metallic flight structures, flight mechanisms, MGSE and special electromechanical test equipment. Designs control board systems for integrated space related operations. Conducts feasibility and conceptual studies; mechanical, dynamic and thermal finite element analysis; detailed and manufacturing design; development; manufacture; functional testing and high precision dimensional control. APCO has conducted finite element analysis of long extendible structures manufactured in composite and hybrid material combinations with varying thermal and load requirements. Recent space work includes XMM Mirror Module support structures, ISO payload module outer vessel,

Huygens back cover, Ariane 5 fairing containers, MGSE for Envisat's ASAR, GOMOS and MERIS instruments, and the two main elevator platforms of Kourou's ELA-3 BAF for Ariane 5 final assembly, including the lift system.

UPDATED

Compagnie Industrielle Radioelectrique SA (CIR)

Current Status
CIR belongs to France's Sextant Avionique. Space turnover is about SFr20 million; personnel total about 110. CIR has been active since the 1960s providing TC/TM systems, satellite timing systems and EGSE test/simulation systems. It has developed the image processing system for Envisat's MERIS instrument and for Meteosat Second Generation. Contracts include EGSE equipment for Hipparcos, ECS, Spacelab, ERS and Columbus; time bases for satellite tracking stations; timing equipment for Vulcain engine test facilities; and Anthrorack's conditioning amplifier subsystem. Has designed a biorack control system for ISS applications and is a participant in the ESA Huygens probe programme, providing the payload interface simulator. It has developed enhanced complex interactive simulation and control systems for a suite of separate science experiments multiplexed through a single data processor. These have potential application in deep-space missions with requirements for extensive autonomous capabilities.

UPDATED

Contraves

Current Status
Contraves' Space Division is part of Oerlikon-Bührle Holding Ltd and is responsible for that organisation's space activities. Provides special services in fields of engineering, design, structural analysis, thermal analysis, materials and processes, optics, electronics, computer aided engineering and testing. The company provided the spacecraft structures for ESRO 1, GEOS, Giotto, ISEE 2, Exosat, Giotto and Ulysses, in addition to deployment mechanisms for scientific instruments, antennas and solar arrays. Other space interests include the production of Ariane 4/5 payload fairings, thermal control and EODP. It also provides fairings for Lockheed Martin's commercial Titan 3 and Titan 4. Has available designs for lightweight oversize fairings and payload protection envelopes, also for massive payloads with load support structures as an integral part of the fairing and enclosure. Contraves has developed fairing and payload shroud designs capable of addressing very large volume requirements and specifications for oversize structures.

UPDATED

Centre Suisse d'Electronique et de Microtechnique SA (CSEM)

Current Status
CSEM (Swiss Centre for Electronics and Microtechnology) is a privately held company which provides research and development and small quantity production in systems design and engineering (including wavefront engineering), instrumentation (based on microsystems), data communications, automation and robotics subsystems, smart structures, optical systems and tribological coatings. 1999 space turnover was about US$5.4 million. Work for ESA includes a micro-accelerometer (Olympus, Spacelab D2, Shuttle GAS June 1991), lubrication for Exosat, Meteosat and Hipparcos, hand controllers for Hermes, mechanisms and subsystems for optical calibration of Earth observation instruments, and XMM satellite as well as opto-mechanical systems for laser pointing. Designed switch configurations for manned space vehicles. Has developed a series of switch controls for pre-set and semi-autonomous requirements. CSEM also designs integrated systems connecting microsystems with control nodes in electronic and software links.

CSEM employs 257 people over two thirds of whom hold an academic degree. CSEM has generated 10 independent spin-off and start-up companies which collectively employ a further 350 people. In 2001, CSEM reported consolidated income above €36 million.

UPDATED

Leica AG

Current Status
Optics and sensors, particularly infra-red detectors, for the space and remote sensing industries including photo-analytical devices. Has conducted research into adaptive optics capable of distortion compensation. Corrective optics research for adaptive mirrors utilised in remote sensing and astronomical research programmes for next generation systems. Major research programmes into multi-frequency optics requirements for Earth observation and remote sensing applications. Customers now located in every space-faring country.

UPDATED

Mecanex SA

Current Status
Mecanex is a private company founded in 1959, specialising in the development and manufacture of electromechanical devices, high-performance mechanisms and micro-systems. The company now trades in space, defence and manufacturing industry. The first space products were provided to ESA in 1975, the year it was formed. In the space field, these include slipring brush assemblies for signal/power transmission in solar array drive mechanisms (TV-Sat, Tele-X, TDF, Olympus, Spot, ERS, Eutelsat 2, Turksat, Arabsat, Nahuel), Coarse Pointing Mechanism Assembly for SILEX (Spot 4, Artemis, DRS), Space Bioreactor (Spacelab IMI-2), valves and filters mechanisms (Huygens, ACP-Cassini), and actuator units for ERA's end effector. ISS telerobotic design upgrades proposed and utilised as the basis for new actuator elements. Also specialises in nano-technology for microbot applications in space engineering requirements for small satellites and microcraft designed for solar system exploration. This work has generated a series of nano-tech devices applicable to DCP and sensor-bot requirements supporting new and stringent capabilities.

UPDATED

Oscilloquartz SA

Current Status
A member of the Swatch Group, Oscilloquartz specialises in sync solutions for turnkey synchronisation projects in PDH, SDH and mobile telecom networks. Manufacturer of ultra-stable and very low noise frequency sources, from quartz crystal oscillators to cesium frequency standards. Oscilloquartz also designs/supplies complete integrated systems for the generation and distribution of precise frequency and time references required for satellite communications, time transfer and satellite tracking applications, and equipment for synchronisation of telecommunications networks. Components in ground timing devices for GPS and for integrating different timing systems for commonality of use. Clock and timer applications as well as sequencing and phased cycling requirements. Advanced timing systems for test purposes and special customer requirements.

UPDATED

RST Raumfahrt Systemtechnik AG

Current Status
From its offices in Switzerland and Germany, RST specialises in the design and development of synthetic aperture radar (SAR), altimeter and ground penetration radars. Provides consultancy, engineering services and space products: space systems, space-related ground systems for Earth observation, navigation and communications. Studies completed on integrated earth observation systems. In 1994 it introduced a gyro-stabilised (less than 40 arcmin accuracy elevation/azimuth) 1.2 m antenna for shipboard reception of 1 to 13 GHz satellite TV. In 1996 a new forward-looking CW radar was introduced. RST is the radar system architect for the German SAR-Lupe satellite reconnaissance system and has developed the wideband altimeter for ASIRAS, an ESA project. RST is capable of providing high-level digitised imagery from SAR systems and can integrate altimetry and SAR swaths. Also new SAR applications for small satellites with limited power capabilities.

UPDATED

RUAG Aerospace

Current Status
Initially known as the Swiss Aircraft and Systems Company.

Formed in 1996 from defence companies owned by the Swiss government, it was converted into a public corporation as a subsidiary of the RUAG SUISSE Holding. In 2001, it changed its name to RUAG Aerospace. The company has provided structural elements, instrumentation and integration of Ariane 4 and 5 payload fairings, active cooling systems for Spacelab experiments, structural elements for ESA's Biorack, and satellite containers. Elements for ISS biological experiment racks and micro-containers for small, isolated sample analysis and long-term experiment incubation for ISS and small experiment volumes. Has also developed glove-box type experiment containers with integrated control and power/data transfer capability. Employs 1,700 people.

UPDATED

Vibro-Meter SA

Current Status
A Meggitt company since 1998, Vibro-Meter was forwarded in 1952 to supply instrumentation systems to aerospace, marine and industrial companies. In the 1970s Vibro-Meter diversified into the nuclear industry and in 1987 began designing measuring systems for the space industry.

Pressure and vibration sensors for space applications. Sensors for Ariane 5's Vulcain engine monitor absolute vibration and dynamic pressure of the engine's gas generator and output nozzle; displacement, relative and absolute vibration, rotational speed and dynamic pressure of Vulcain's turbopumps; absolute vibration of the solid boosters. Sensor tolerance is –253 to 780°C. Shock accelerometers are also provided for Ariane 5 pyrotechnics. Applications include satellite and spacecraft boom, landing leg and deploy truss systems for any space vehicle concept. High-resolution standard proximity sensor for ESA is under development. Ariane-4 displacement measuring system for the LOX turbopump which has been applied to marine applications.

UPDATED

Thailand

Thai Micro-Satellite Co Ltd (TMSC)

Current Status
TMSC signed a £3 million contract on 28 November 1995 with the UK's Surrey Satellite Technology Ltd for a 50 kg micro-satellite named TMSat and launched by Zenit in 1998. It was renamed Thai-Puhtt and produced imaging data approximately equal in resolution to the Landsat MSS imager. The 18-month technology transfer programme involves a team of Thai engineers working alongside SSTL staff as TMSAT is built. It carries Earth-imaging cameras and communications payloads, controlled from a station installed by SSTL in Bangkok. Its GPS receiver was developed by SSTL under an ESA contract to provide position and attitude data. Proposals for successor satellites have been completed.

United Kingdom

Advanced Products (Seals & Gaskets) Ltd

Current Status
Advanced Products is a subsidiary of the Advanced Products Company of Connecticut and specialises in the manufacture of seals for extreme environmental conditions, including rocket engines, fuel systems, and cryogenic and vacuum systems. Provides pressure vessel bladders and has developed flexible steel alloy bellows for propulsion systems. Positive expulsion technology for cryogenic applications developed for launch vehicles. AP has developed new flexible long-life fluid and gas conduits capable of resisting extreme temperatures.

AEA Technology

Introduction
AEA Technology Space and ESTL (the European Space Tribology Laboratory) are business units of AEA Technology plc, one of the World's leading innovation businesses, with a turnover in the FY2000–2001 of £379 million. Space has been a significant activity for AEA Technology since the late 1960s, when a process of diversification beyond the traditional field of nuclear technology research and development commenced. In 1972 AEA Technology was awarded a contract by ESRO (now ESA) to construct, equip and operate a specialist tribology laboratory for space applications. Thus the European Space Tribology Laboratory (ESTL) was created. Since these early days of the European space industry, AEA Technology's work has ranged from research and development through to the manufacture of flight hardware and the provision of testing and other services to industry. Approximately 40 personnel work in AEA Technology Space and ESTL.

Industrial Space Activities
Although far from exhaustive, the current range of AEA Technology's space products includes:

- Lithium-ion Batteries – a new type of battery technology used to store surplus energy generated by solar arays for use when spacecraft are in eclipse.
- Radiometric Calibration Systems – precision reference sources flown on spacecraft to calibrate optical images.
- Cell Bypass Protection Devices – devices attached to battery cells that automatically detect open circuit failures and provide a short circuit bypass.
- Helical Antennas – used to provide communications between the spacecraft and the ground segment at UHF and VHF frequencies.
- Advanced Tribology – providing a centre of excellence in tribology with focussed research and development on advanced materials and processes, also training and design tools.
- Space Lubricants and Bearings – specification, procurement and space-compatible lubrication of moving parts such as bearings and gears.
- Space Mechanisms – design, manufacture and qualification of spacecraft mechanisms, independent evaluation and thermal-vacuum testing of mechanisms.

Resources and collaboration
The space activities of AEA Technology are undertaken from two sites in the United Kingdom, Culham Science Centre near Oxford and Birchwood Park near Manchester.

At Culham, space work is undertaken within its 300 m^2 test facility which includes 120 m^2 class 100,000 cleanroom and 20 m^2 class 100 cleanroom. Test equipment includes Ø800 mm class 100 thermal-vacuum chamber, Ø1.5 m thermal-vacuum chamber, two portable thermal chambers, eight environmental chambers (–70 to 180°C) and three low-temperature chambers (–30 to 0°C). Other facilities include six Clive Hurley thermal control plates, 49 channels of battery testers, up to 7.5 kW, 75 V, 256 channels ofcell testers, Lithium-ion cell Lot Acceptance Test (LAT) facility, precisiontemperature calibration bridge and reference standards, ESA qualified soldier station, crimping station, etc, blackbody emissivity calibration facility, residual Gas Analysers and leak testers, particle meter, PFO meter.

At the ESTL facility in Birchwood Park, operations take place within 320 m^2 of class 10,000 cleanroom that includes the following specialist facility: a vibration test facility, eleven thermal-vacuum chambers (diameters 150 mm to 1 m and temperatures –150 to +150°C), three physical vapour deposition coating chambers, three tribometers, a mechanism centrifuge test facility, precision metrology equipment for bearing and mechanism inspection, an XRF film thickness measures facility, optical microscopes and digital imaging equipment, a large component ultrasonic cleaner and a cyrogenic test facility for bearings and linear slides.

AEA Technology has been involved in most European missions over the last 30 years. Thus, AEA Technology has in one form or another, worked with virtually all of the main players in the European space industry over the past three decades, during which time both formal and informal collaborations have been developed. Recently, in the field of space mechanisms, the successful partnership between AEA Technology and Sula Systems balances the design expertise of Sula with the assembly and test experience of AEA Technology. In the field of Lithium-ion batteries, the partnership between AEA Technology and COM DEV is central to the exploitation of the European lead in Lithium-ion technology in the global space market.

Alcoa International Ltd

Current Status
High-technology sheets, plates, extrusions and forgings in aluminium, titanium and other metals for the aerospace industry. Provides numerous specialised products in all forms of fabrication for European launch vehicles and satellites. Develops materials unique to space application and has produced new matrix alloys for specific requirements and high-end specifications. Employs 142,000 in 37 countries.

Analytical Graphics

Current Status
A UK-based subsidiary of the United States Analytical Graphics, established in 1996. Markets AG's software packages for orbital and satellite analysis to UK/European clients, involved with education facilities. Developer of the Satellite Tool Kit (STK) system for analysis software. Interface software available for STK operation in conjunction with real-time data sets. Convergent programming linking database to real-time downloads. Representationin France, Germany, Netherlands and the US.

Anvil Technology Ltd

Current Status
Anvil specialises in software development for spacecraft control, to ESA Software Engineering Standard PSS-05-0. Working relationship with University of Hertfordshire. International availability for multitasked software and software solutions to customer specifications; programmes written to requirements for specified tasks. Also substantial large-scale, real-time and database development. Set standards for environmental control data processing and earth based satellite remote-sensing information.

Asdor Filters

Current Status
Asdor specialises in the design and manufacture of filtration systems for aerospace applications. Fluid and gas filtration for liquid and gas delivery systems. High-velocity flow channels incorporating metering devices and flow control valves, bifurcated conduits and tubing. Miniaturised systems for folded piping designs. Asdor is a member of the Howden Aircontrol group.

BAJ Coatings Limited

Current Status
A Meggitt Aerospace company, BAJ provides high-performance surface engineered coatings, applied by the Tribomet, HVOF and plasma spraying processes, to resist wear, corrosion, erosion and oxidation at elevated temperatures. Provides thermal and electro-deposited coatings. Research into high-temperature gradient coatings and surface coverings. Has conducted developmental experiments into electroplating deposition techniques.

BP Chemicals Ltd

Current Status
BP's Aerospace Composites group was formed from the amalgamation of several existing composite companies. It offers design, development manufacturing and repair capability in composite components and structures for satellites. Also design configurations for high-rigidity composite frame structures and low-tensile antenna structures. BP has focused its business interests in exploration and production, retail, chemicals, gas and power, and solar energy. BP Chemicals has also researched new and innovative products manufactured from synthetic copies of natural compounds. Currently employs more than 100,000 people.

CT Bowring Space Projects Ltd

Current Status
Now totally absorbed into Marsh & McLennan Companies (MMC), (see US Space Industry), the company provides space insurance brokerage services, develops risk management and insurance programmes for risks of pre-launch, launch, in-orbit operations, manufacturers, product exposures and third-party liability. One of the oldest participants in international space insurance. Provides risk analysis and failure assessment.

Commercial Space Technologies Ltd (CST)

Current Status
CST is an independent UK technical consultancy, founded in 1983, comprising individual technical specialists and support staff. Work has been undertaken for clients in launchers, propulsion, spacecraft, communications, remote sensing, μgravity and biotechnology. Current specialisations include small launcher and communication satellite services and insurance and brokerage of 'piggy-back' launch services. Direct contacts exist with Russian manufacturers, research institutes, government organisations and universities. Specialises in Russian space technology and services. Provided consultative services for launch preparation for Chilean FASAT-BETA and Thailand's TMSAT (1998). CST is marketing piggy-back payload opportunities with Russia's Tsyklon, Rockot, Cosmos, Dnepr or Shtil.

Cray UK Ltd

Current Status
Cray UK, a wholly-owned subsidiary of Cray Inc, is a software and systems house with an annual turnover of approximately £92 million generated by more than 700 staff. Of this, space projects employ 135 and contribute revenue of £10.2 million in the following strategic areas: spacecraft control centres; spacecraft simulators; mission planning/evaluation tools; ground station systems; terrestrial communications systems; Earth observation data processing centres; information systems; software products and system studies. Cray Inc was formed in March 2000 with the merger of Cray Research and Tera Computer Company.

As prime contractor, Cray UK has designed/developed software systems for 10 spacecraft control centres over the past decade. Five are operational for Ulysses, ECS, Rosat, DFS and ERS. In collaboration with Germany's DLR, it has developed an Advanced Generic Control Centre. It has been used to date for seven satellites and has been installed for Deutsche Telekom for operation of their telecommunication satellites. Cray Systems designed/developed, on behalf of ESA, the £3 million ERS Mission Management and Control Centre – the largest ESA spacecraft control centre to date. It has recently developed the control centres for ESA's Pastel and Cluster missions, and is now developing the requirements for ESA's Envisat control centre.

The company was awarded the development contracts for eight of the last 11 simulators commissioned by ESA, including Giotto, Meteosat P2, Hipparcos, ERS, ISO and Cluster. It also provides spacecraft simulation consultancy to prime contractors in Italy and Canada.

Software systems and engineering services have been provided for ESA's Estrack packet-switching network since 1976. The company's software systems have been installed at all ESA ground stations. A new generation of station computer, the Monitor and Control Module and Front End Controller, is under development.

Since 1988, Cray UK has provided engineering services for ESANET, covering strategic planning, product assessment and procurement, installation and pre-operational support for new facilities.

A £4.5 million contract for the Western European Union for a turnkey image-processing facility has been completed. The Dundee Centre for Coastal Zone Research has been established with Dundee University.

Software engineering services have been provided to ESRIN's Information Systems Division since 1993, particularly on the CUIS and Cluster projects, as part of a five-year frame contract.

Satcom Earth station products have been developed for Inmarsat, including: Inmarsat M/B Control/Signalling Component, Production/Factory Test Unit, and Earth Station Test Set for Inmarsat type approval. Cray also provides CLEO (for spacecraft monitoring/control), MMOrbit and VISIM (for spacecraft simulation and operations visualisation and analysis).

EASAMS Limited

Current Status
EASAMS specialises in business systems, defence consultancy, software and computer services. It has provided onboard guidance and control software for all Ariane versions to date, including: flight software; studies, mission planning and algorithm development; ground integration support software for the launcher electrical systems; launcher simulation for real-time testing, studies, and post-flight analysis of telemetry; and integrated vehicle checkout software at Kourou. EASAMS is contracted to provide individual Flight Programs for each Ariane 4 and 5 flight. Improved and upgraded software programme following the failure of the first Ariane 5. Designs software-hardware interactive programmes and equipment for complex flight operations logic. Adaptive avionics and flight control programme integration software for advanced guidance equations.

EDS Ltd

Current Status
Founded in 1962 as Electronic Data Systems, EDS employs 138,000 people in 60 countries with more than 35,000 clients.

The EDS Group employs 8,450 people in the UK. Its space business centres on consultancy, systems development, systems integration, systems management and process management. This included a £2.5 million contract to supply the synthetic-aperture radar processor for the UK Earth Observation Data Centre to process ERS data. In addition, the PulSAR high performance SAR processing workbench has been developed to provide high-quality SAR imagery with rapid throughput on low-cost hardware. ESA projects include: supply of the SARCALQ image QA and calibration system for Earthnet; development of expert systems demonstrators and assessment of multimedia systems for the agency's Information Retrieval Service; and leading an international consortium which is investigating the application of formal methods to space software development for ESTEC. Other projects include: interpretation of Earth observation data using neural nets; development of integrated GIS; development of a system for determining terrain heights from SAR imagery; assessment of radiation effects on spaceborne instruments; simulation of space-based instruments and systems; and finite element analysis for space-based structures. Integration software for optical and non-optical earth observing systems and for space-based astronomy observatories.

Electron Tubes Ltd

Current Status
Established in 1953, Electron Tubes is a major manufacturer of very-low-light-level photodetectors, particularly for space activities. Products include light detector assemblies, x-ray detectors, voltage dividers, photomultipliers, housings, power supplies and accessories. The management buy-out from THORN EMI was completed in January 1994. Current and recent space projects include: image intensifiers for the Faint Object Camera in NASA's Hubble Space Telescope; photomultipliers for the NASA Compton Observatory's Compton Imaging Telescope, Burst/Transient Source experiment and the Gamma Ray Telescope; photomultipliers for ESA's Hipparcos astrometry project; photomultipliers for Ulysses (taking a Cerenkov detector into solar polar orbit); and photomultipliers for Italy's SAX astronomical x-ray detector. Space-related facilities include simulation thermal, vacuum and vibration test equipment, and expertise in encapsulation techniques. Qualification programmes and quality control procedures. Customers include NASA, ESA, MMS, SERC, Aeritalia and AEG.

Empcom Systems Ltd (ESL)

Current Status
ESL provides consultancy, contract negotiation and subcontractor monitoring services to the European space, telecommunications and satellite remote sensing industries. Specification writing and review to ESA and NASA standards. Directs small scale multidisciplinary projects for UK space related companies and for European companies in joint endeavours. ESL has built a database of system capabilities and company skill centres with a compendium of staff and professional assets.

Encore Real Time Computing Ltd

Current Status
Part of Encore Real Time Computing, Inc, the company supplies mission-critical computer systems. ERTC Inc was set up in 1961 as Systems Engineering Laboratories, acquired by Gould Inc in 1980 to become Gould Computer Systems Division and established as Encore Computer Corp in 1983. The Infinity R/T real-time and fault-tolerant systems are used for simulation, control and data-acquisition systems. The Infinity SP Enterprise-wide Storage Systems provide high-capacity storage and sharing of data between mainframe, UNIX and LAN systems. Employs 4 people.

ESYS PLC

Current Status
ESYS is an independent public limited company, registered in the UK and based in Guildford. Founded in 1990, ESYS business is divided between four consulting practices serving Space, Telecoms, Defence and Education and Research. Each of these practices offers a portfolio of consulting services, namely strategy and policy, business assessment, project management, technology evaluations and system. ESYS has worked on many of the European space missions of the past ten years, including the Galileo and EGNOS Navigation systems, the SPOT remote sensing satellite, the Envisat environmental monitoring missions and Meteosat weather system. The work has covered both the technology and business aspects of space systems. ESYS has worked for the UK Ministry of Defence on space programmes such as Skynet 4 and Skynet 5 and on wider defence issues associated with telecommunications and battlefield digitisation. ESYS has also worked for the major UK primes including BAE Systems, Astrium, General Dynamics and QinetiQ. ESYS provides market analysis and systems integration consultancy to the satellite communications sector and telecommunications sector. Working with agencies such as the DTI and European Space Agency, ESYS has brought together partners in a number of sectors to create multidisciplinary projects. ESYS has worked with manufacturers such as Astrium, Alcatel and Hughes Network Systems and service providers such as Eutelsat, Astra, Globalstar and New ICO. Technology plays an increasing role in education and research. The perception is that more could be achieved from a better understanding of how technology is applied. ESYS can draw on skills in programme management, evaluation, technology transfer and business analysis to help create this better understanding. ESYS has provided independent evaluation of the effectiveness a number of national research programmes for agencies such as JISC, NERC and BNSC.

GEC – Marconi Materials Technology Limited (MMT)

Current Status
A Technology Centre within the GEC-Marconi Engineering and Technology business group.

Established 1940 to research the applications of new electronic materials. The company traded from its Caswell site which saw significant expansion during 1980s with additions of 5,000 sq m facility for advanced microelectronics and materials research. Other facilities set up at Towcester and Borehamwood.

The company's Caswell facility specialises in the manufacture of GaAs, opto-electronic and pyro-electric sensor products. The GaAs range includes MMIC, WLAN, SART, amplifiers and switches, discrete FETs and a foundry service offering F20 FET and F40 HEMT technologies. The F20 MMIC process is undergoing Capability Approval for ESA. Opto-electronic capacity includes DFB lasers, OEICs, specialist LEDs and GaAs modulators, operating at DC-50 GHz. The company's Towcester facility specialises in radar-absorbent materials and radar-transparent products for endo/exo-atmospheric applications (including anechoic chambers and screened rooms), equipment testing and microwave antenna beam shaping. The transparent high-performance low-loss resin systems (RP13) and lightweight foam materials (P10 range) are designed specifically for radomes. New high-intensity screening materials have been applied to these designs to render them impervious to high heat or vibration stresses.

GEC Plessey Semiconductors (GPS)

Current Status
The company was formed in 1990 by the merger of Marconi Electronic Devices and Plessey Semiconductors. In 1998 the General Electric Company plc sold GEC Plessey Semiconductors to Mitel Corporation of Canada for US$225 million in cash. It is Europe's largest manufacturer of integrated circuits for space and defence applications. It specialises in VLSI CMOS Silicon on Sapphire (SOS) for high performance and radiation hardness. The SOS range covers advanced monolithic MIL-STD-1750A microprocessors and peripherals, static RAMS to 256 kbit, CMOS logic and MIL-STD-1553B databus products. Semi-custom gate arrays and standard cells to 40k gates are available. Self-certification to MIL-STD-883; ESA 9000 Capability Approval achieved 1993. At the same Lincoln facility, GPS manufactures thick/thin-film hybrid circuits to space quality levels.

Graseby Dynamics Ltd

Current Status
Part of Smiths Detection, a group under Smiths Group PLC, Graseby Dynamics is a research and development and production specialist in trace-vapour detection and monitoring, using gas chromatography and ion-mobility spectrometry. Specialists in explosives vapour detectors and highly sensitive particulate detectors for plastic explosives and drugs. Space capabilities include the detection and monitoring of hydrazine, total hydrocarbons and volatile organic compounds for flight and ground operations. The company's hydrazine monitor/data-logger was used on Shuttle STS-37/49 and was planned for reflight in 1997. Graseby developed air-quality monitors for Shuttle (completed 1996) and Space Station (completed 1997) installation.

Hunterskil Howard plc

Current Status
A member of the MPS Group International, Hunterskil Howard plc is a communications and marketing company with about 50 employees.

Space project consultants, including μgravity, component engineering and RF systems. Has conducted studies on μgravity field equipment programmes and has projected software integration regimes for high-density data absorption and applications.

IGG Component Technology Ltd

Current Status
IGG is the oldest independent European company devoted solely to parts engineering and procurement for the space and defence industries. IGG has a subsidiary in Philadelphia, USA. Both serve more than 90 equipment manufacturers in 24 countries. The service includes procurement of individual line items to total project requirements, specification writing, part type reduction, evaluation, constructional analysis, destructive physical analysis and screening. Project involvement includes most ESA craft and satellites such as Insat 2, Telecom 2, Intelsat 7 and Italsat. Facilities include a Class 10,000 cleanroom, automatic thermal shock chamber and full screening facilities. Staff of more than 50.

IGG's Spartan Microwave specialises in the design/manufacture of multichip microwave components using GaAs and/or Si technology for space and defence applications.

Skills include obsolescence management, destructive physical analysis, failure analysis, product evaluation and electronic component kitting. Performs full system integration recommendations and analysis planning for full system structures together with specialised operability analysis and line predictability with potential failure mode evaluation.

IMI Titanium Ltd

Current Status
A wholly owned subsidiary of IMI plc, IMI Titanium is Europe's largest producer of semi-finished titanium mill and stock products in titanium and titanium alloys. Applications development embraces hafnium, niobium, superconductors, zirconium and its alloys. The company holds approvals on testing/analysis from standards authorities such as the UK Ministry of Defence and the Civil Aviation Authority. Production facilities encompass computer monitored vacuum arc melting furnaces capable of producing ingots of up to 17 tonnes, hydraulic forging presses and machining, sawing and cutting equipment. Mechanical, chemical, ultra-sonic, corrosion testing and metallography are conducted throughout the manufacturing operations. Recent acquisitions have diversified the business base to include service valves, fluid power automation, climate control, building products and services and dispensing machines.

International Space Brokers Ltd

Current Status
ISB are insurance brokers specialising in all aspects of prelaunch, launch, liability and in-orbit coverage, in association with International Space Brokers Ltd (UK) and European Space Brokers (France). Has successfully applied global risk management analysis in failure analysis assessment.

Irvin-GQ

Current Status
Previously Irvin Aerospace Ltd, the Irvin-GQ was, until June 2001, a member of Hunting plc and comprised: Irvin Aerospace Inc; Irvin Aerospace Ltd; Irvin Aerospace SpA; and Irvin Aerospace Canada Ltd. Each has a development and manufacturing organisation responsible for a wide range of parachutes and associated equipment. Irvin was acquired by Alchemy Partners in June 2001 to be merged with GQ Parachutes in a deal worth US$17 million. The Irvin Group is a leader in parachute design/manufacture for space projects; for example, recovery systems for heavy vehicles such as boosters and retarders for slow descents and landings on planets and moons. Completed projects include: Ariane 1 Stage 1 recovery parachutes, Pioneer-Venus descent parachute, Atlas MA3 main thrust stage recovery parachutes, and Shuttle landing-brake parachute. IRVIN was involved in the Cassini-Huygens Titan probe deployment parachute and a comprehensive software development tool for ESA to evaluate space-related retarder systems. The Group has one of the world's largest research and development divisions for parachute work, with specialist expertise in materials and the prediction and analysis of system performance using advanced computer modelling techniques. Irvin has conducted analysis of the use of high-temperature resistant materials, steel alloys and carbon-based derivative designs for retardation in extreme temperature and pressure environments for planetary research and for high velocity retardation in low pressure environments. Recent research programmes include the behaviour and performance of materials in the space environment. Also studied extreme environment parachute recovery systems applicable to hot and dense planetary atmospheres.

J S Chinn Engineering Co Ltd

Current Status
A part of the JS Chinn Group of Companies, Chinn Engineering has developed a specialised range of services in systems design, test, modification and repair and environmental engineering. Manufacturer of satellite ground support equipment, transportation packages, and special access and lifting equipment. The company provided the Skynet 4 integration trolley. Metal spinning and flow forming up to a diameter of 2.4 m; maximum mass for single unit: 20 tonnes. Heavy satellite trolley and fine-adjustment handling rigs available. A new range of specialist equipment handling trolleys.

Johnson & Higgins Ltd

Current Status
Space insurance broker based in New York specialising in pre-launch and launch insurance, satellite in-orbit and legal liability insurance. Completed a strategic business combine with Marsh and McLennan Companies, Inc., in March 1997. Marsh and McLennan have more than 80,000 employees worldwide. J&H has extended space coverage to include business risks and life coverage for revenues.

JRA Technology Ltd

Current Status
JRA was established in 1988 to provide marketing and management support within the UK/European aerospace and defence community. As a founder member of the Spacelink Europe consortium, JRA co-ordinates the activities of ESA's technology spin-off programme (while also covering UK/Scandinavia) through a network of 'technology transfer' companies spanning 13 European countries. The company has a 24-hour turnaround interrogation service for all major technical, patent and commercial UK/European databases. Provides technology audit, assessment and marketing services. Capable of integrating multimedia data sets with high-capacity handling options. JRA is conducting a study for ESA examining the degree of competitiveness of the European space industry in the global marketplace. It is also working with ESA's ESTEC to set up spin-off companies.

Kalgard (UK) Ltd

Current Status
The company is a licensee of the Kalgard Company of South Africa. Facilities are available for the application of military standard, resin-based coatings: corrosion-resistant (aluminium-filled resin and aluminium-filled PTFE) low friction coatings; dry film lubricants (MOS_2 loaded); high-temperature-resistant coatings; forming and forging lubricants; industrial non-stick coatings; cadmium coatings; DEF STAN 03–28 PHI. Specialised coatings for the defence and aerospace industry including RFI and EMI shielding. Capable of high heat load structures providing rigid, high-impact, survivability. Has pioneered the development of flexible flat sheet structures for pressure and thermal accommodations with vacuum and high-pressure loading restraints.

KG Coating Limited

Current Status
KG Coating (formerly Scanwel Scientific) is a specialist manufacturer of advanced technology metal deposition equipment. Applications range from large-scale protective coating of structural components to special multilayer conductive and insulating coatings for pressure and temperature sensing devices. High electrostatic resistance coatings for protection from high intensity currents and fields. Ion Vapour Deposition services are available for aluminium and cadmium: For aluminium, services include either barrel coating or jigging of components to MIL-C-83488, DEF STAN 03–28/1; capacity 99 cm (length) × 35 cm (diameter); For cadmium, either barrel coating or the individual jigging of components is available to MIL-C-8837, DTD 940, DEF STAN 03–28/2; capacity 61 cm (length) × 30 cm (diameter). Also provides corrosion inhibitors for the defence and aerospace industries, metal finishing and polishing services. KG has developed special thermal insulation for low temperature resistant coatings.

Kinesix (Europe) Ltd

Current Status
Founded in January 1990, the company manages all aspects of the pan-European markets of 'Sammi', with technical support offices in London, Paris and Rome. Kinesix (UK) began operating in 1994. 'Sammi' is an X-Windows-based tool-set that enables users to build graphical user interfaces for real-time distributed applications. It is designed to run in high-performance mission-critical environments, where the real-time display of high data volumes is essential. 'Sammi' is used at all NASA sites in the US and has been chosen by ESA for Columbus Space Station onboard and ground control systems. It is used extensively by defence contractors, including IBM/Loral, for satellite and telemetry systems. Kinesix also provide Sendera, a graphical user interface tool, and Sendera Lite. The Sammi Application Development Kit is a COTS product. Upgraded versions available and tailored software integrations optional extras.

Ling Dynamic Systems

Current Status
Set up in the early 1980s, LDS helped develop the digital switching amplifier for vibration testing. LDS is a member of the SPX corporation which has an annual turnover of US$5.2 billion. The company specialises in manufacturing, installing and commissioning vibration test systems for analysis and test applications. The force range is 8 N-300 kN. The company is a leading supplier of vibration test systems for satellite integration facilities. Test systems also available for high-mass simulator beds and for high-frequency vibration on small components.

Logica UK Ltd

Current Status
Logica supplies computer-based systems for all aspects of space-related activity and undertakes associated studies and consultancy.

Onboard software: design/development for Huygens; definition of Rosetta autonomy requirements; support for ESA Polar Platform; software environment for ESA 31750 and 32-bit processors. Selected by ESA to develop Phase B of the Envisat ground segment. Also developed Envisat on-board software for Astrium. Is producing interactive software for autonomous rendezvous and docking systems.

Employs more than 6,000 staff with offices in 23 countries.

LRE Relays and Electronics Ltd

Current Status
Part of the LI Group, the company provides relays and contactors for power switching in aerospace, military and industrial applications. Satellite switching capabilities on high-sensor rate systems. Has developed new multifunctional switch relays for selective discrimination designed to satisfy fault tree criteria. New switching control systems for multiple sensor packages.

Marshall Aerospace

Current Status
Marshall is a leading European aeronautical engineering organisation. It has manufactured a number of instrumentation systems for space application, including the Space Sled used aboard Spacelab. In 1993, it completed the conversion, for Orbital Sciences Corporation, of an ex-Air Canada Lockheed L-1011 Tristar to 'belly-carry' the air-launched Pegasus XL orbital vehicle. Continues to maintain a capability to provide space-qualified engineering services. Has been approached to carry out modifications for air-launched rocket systems.

Martin-Baker Aircraft Company Ltd

Current Status
Founded in 1929, the privately owned Martin-Baker specialises in aircraft ejection seats and helicopter crashworthy seats. The company has produced over 67,800 ejection seats that have saved a total of over 6,800 aircrew lives in 90 air forces. Latest developments include ejection seats for EF 2000, Pilatus MKII (JPATS) and the new Joint Strike Fighter programme. Martin-Baker crashworthy seats have been selected for both the military and the civil market that includes programmes such as the Eurocopter Tiger and Sikorsky S-92 Helibus. As well as a wide range of safety equipment that includes automatic lifejacket and liferaft inflation units, the company has designed, developed and produced the Descent Control System for the Huygens space probe that will soft-land on Titan, one of Saturn's moons. Martin-Baker is a partner in the Beagle 2 lander for ESA's Mars Express mission which was launched on 2 June 2003.

Meggitt Aerospace Avica

Current Status
Specialists in high-performance ducting and ducting components for space vehicles, including gimbal joints, sliding joints, flanges/clamps, metallic hoses and thermal insulation. Avica provides the fuel feed lines and exhaust ducts for Ariane 5's Vulcain engines. It provided the titanium flexible hoses for Cassini's hydrazine system. Also controls and bifurcated conduits for gas or liquid flow in propulsion and expulsion systems.

Mitsubishi Electric UK Ltd (MEUK)

Current Status
MEUK commenced business in 1972 and was established in 1978 as a wholly owned subsidiary of Japan's Mitsubishi Electric Corporation (MELCO). The Electronics Division was created in 1981 to offer expertise in spacecraft, satellite communications, sensor and image-processing systems, drawing on the extensive space activities of its parent company. MEUK is now involved with mobile telephone assembly, project engineering, semiconductors and environmental control systems. In 1996 Mitsubishi Electric Europe BV was formed as a wholly owned subsidiary of the Japanese parent.

Mitsubishi parent has a global turnover of US$30 billion employing more than 110,000 people in 30 countries.

Molniya Space Consultancy

Current Status
The consultancy specialises in information on the Chinese and RFAS space programmes. Additional expertise covers worldwide launch vehicles, technology and space history. It compiles the monthly *Worldwide Satellite Launches* newsletter. The MSC was set up in 1991 by its sole proprietor and has expanded its range of services to include analytical studies of launch vehicle, satellite and spacecraft operations.

Moreton Hall Associates (MHA)

Current Status
Moreton Hall Associates (MHA) incorporated Eurobird Ltd with a turnover of about £0.3 million and provides space consultancy services, including system engineering, insurance, tender preparation, launch and satellite statistics, procurement and illustration services. It is a member of the ASTOS Association of Specialist Technical Organisations for Space. Has conducted studies for ESA on integration engineering concepts involving the use of financial and insurance businesses to underpin risk assessment. Company has pioneered the interaction of government and private partner association, or partnership, in development of new high-risk ventures unlikely to be funded by government resources alone.

Normalair-Garrett Ltd

Current Status
Normalair-Garrett Ltd (NGL), part of Westland Technologies Ltd, designs and manufactures precision products, including environmental control and life-support systems, pneumatic and hydraulic solenoid valves, investment castings, actuators, severe environment recorders and electronic controllers. Design of advanced ECS elements for upgrades to ISS, also improved sensor links. Has developed advanced ECS concepts for efficient mechanisms essential to new generations of long duration environmental control system concepts. Continued research into advanced autonomous ECS designs has developed into a new series of technology enhancements which could significantly improve performance for long-duration systems.

Parker Hannifin PLC

Current Status
The Parker Hannifin Seal Group is part of Parker Hannifin Corporation, the world's largest manufacturer of hydraulic and pneumatic equipment,with more than 75 years experience. Worldwide personnel number approximately 32,000, of which around 4,500 are in Europe and 1,000 in the UK. The Seal Group was established in Europe in 1968 and, with the acquisition of the German-based Pradifa Company in 1979, became a major European supplier of high-quality O-rings and packings. Specialist supplier of flight control activators, electrohydraulic servoactuators, hydraulic pumps, pneumatic controls, heat exchangers, fluid metering devices, atomisation devices and fuel systems. For Spacelab, Spacehab and COF, Gask-O-Seals of up to 4 m diameter are installed between the elements of the pressurised laboratory. For Ariane 4, O-rings of almost 4 m inner diameter are applied to seal the 8,200 litre water tanks used for cooling the first stage gas generators. These special-compound seals are manufactured as endless units, with no joints or connection spots, for very narrow tolerances. For Ariane 5, Parker Pradifa worked with MAN Technologie GmbH to research different sealing compounds on the housings for the solid-propellant P230 boosters. Has provided additional seal products for rocket motor assemblies.

Plowshare Technology Ltd

Current Status
Plowshare is the European marketing representative for the Cosmos USA joint venture between AKO Polyot and Assured Space Access, which offers Polyot's Cosmos 3 satellite launcher. It has studied brokerage for Russia/China launch services and retained consultative capacities with organisations linking trading companies around the world.

Quality CNC Engineering Ltd

Current Status
A subcontract engineering company specialising in high-precision components using DNC/CNC. Examples of precision machining from solid raw materials, to flight qualification, include antenna arrays, filter bodies and LNA/SSPA housings. New products include integrally stiffened carbon fibre antenna array systems and spin-linked lightweight unfurlable arrays. Has conducted precision machinery examples for new SSPA designs. Extended product range includes antenna systems.

Queensgate Instruments Ltd

Current Status
Queensgate Instruments Ltd is a member of Sitam Instruments Ltd and provides nano-positioning systems for OEM development and automation applications. Also developing innovative lasers and interferometers.

The company is engaged in the development/manufacture of space-qualified piezoelectric translators and high-precision capacitance sensors for positioning and position measurement to less than 10^{-9} m. It also manufactures high-stability Fabry-Perot etalons (including cryogenic) for remote sensing applications. Space experience includes: high-stiffness piezoelectric translators for several pointing mechanisms in Japan's OICETS; development of the Silex Point Ahead Assembly; a scanning interferometer mirror for the WINDII instrument on UARS; mirror control actuators on Soho's Luminosity Oscillations Imager; and capacitance displacement sensors for Columbus module for the International Space Station. Has extended tests with these systems to allow compatibility with remotely operated pointing systems.

Rose Bearings Limited

Current Status
The company is a manufacturer of precision rod end and spherical bearings for the aerospace industry, supplying spherical bearings and self-lubricating bushes for Ariane 4/5. Rose became a wholly owned subsidiary of the Minebea Company of Japan in 1988. Company successfully bid on new bearings for H2A launcher. For aerospace applications, Rose manufactures self lubricating bearings, metal to metal bearings and self lubricating bushes. The company has developed a new form of sealed load bearing roller mount for heavy loads.

Rosemount Limited

Current Status
Established in the early 1960s, Rosemount is a subsidiary company of Rosemount in the USA, a company owned by Emerson Corporate. Its instrumentation includes pressure sensors and air data probes on the Space Shuttle, and surface and immersion temperature sensors for launchers and satellites. Quantity and pressure sensors for liquid and gaseous dewars. Has produced a diverse range of pressure transmitters, mass flow transmitters, temperature transmitters and associated accessories.

SciSys

Current Status
SciSys has secured a €2 million contract to develop the ground segment for the first satellite in ESA's Galileo global navigation system which, unlike US and Russian equivalent systems, will be under civilian control. Formed in 1980, SciSys offers software development and consultancy services. Now part of the CODA SciSys group. Staffing is 120, with 1996 turnover being £14.5 million (1995: £13.5 million). The company builds the ENVISAT platform and payload simulator in the UK and is engaged in specification and development of the ENVISAT PDS Centre Monitoring and Control and Common Services Facilities, specification and design of the Master Control Centre for the EGNOS satellite navigation system and development of a Small Low-cost Integrated Kernel (SLIK) for small satellite control. SS has also developed control systems for remote sensing and earth observation solutions.

The company offers a set of real-time software packages, including the Kernel control system – an integrated program suite providing satellite monitoring and control. Used on Skynet 4, the Hispasat Spacecraft Control Centre and the Meteosat Transition Programme. Facilities include telemetry reception and monitoring, telecommand generation and orbit control. ISO9001 accredited.

Sedgwick Space Services

Current Status
Sedgwick provides all aspects of space insurance brokering, specialising in technical, legal, insurance and risk management on behalf of manufacturers, owners and operators. It provides the design and implementation of insurance, self-insurance and captive programmes in respect of build, ground property, pre-launch and in-orbit operations. Insurance also available for loss of revenue and failure to provide services to customers. Advises on risk analysis and failure probability assessment connected with launch vehicles and satellites.

Serco Europe Limited

Current Status
Serco Europe Limited are an international management contractor providing comprehensive engineering and support services across a wide range of applications, including scientific, spacecraft engineering, IT and space project management services. Serco have some 34,000 staff in 36 countries, with over 4,800 people in the science and technology sector, 330 of which are currently working on contracts for ESA. They provide support to a number of contracts with major clients throughout Europe, including EUMETSAT, CERN and the European Commission. They have also developed management criteria for large project control systems at a systems level and have applied these rationales to the aerospace sector.

Serco provide task management, operational and logistic support to the three services of the UK Ministry of Defence. They have two specialist operating companies employing some 5,000 people at 120 locations. Defence contracts account for 25 per cent of their business.

Serco Justice offer a range of services to civil, criminal justice and security organisations throughout the UK and Europe. Services include the provision of communication encryption devices together with the design, build, maintenance and operation of demand management centers.

Smith System Engineering Ltd

Current Status
Smith System Engineering is an independent company of consulting system engineers, offering expertise in: remote sensing; environmental issues; onboard data processing; advanced computing and specialist information technology; C^3; radar, sonar and electro-optics. Staff level is about 1,800, employed in Guildford (UK) and in offices in Brussels and Tokyo. Key space projects include: a review of satellite Earth observation data for climate change studies within the UK; a three-year project on the extraction of useful mesoscale ocean features from microwave satellite data; and a two-year project to demonstrate the utility of satellite data in coastal zone management. Has conducted studies on integrated remote sensing data sets for optical and radar sensors. Combining land, sea and air environmental atmosphere and climate data to correlate satellite data for earth science users. Smith System has conducted inventory analysis for land management and land use planning and agrarian development programmes using land, sea and air data platforms and satellites for bulk observation.

Space Innovations Ltd (SIL)

Current Status
Space Innovations Ltd ceased trading through voluntary liquidation on 7 July 2000. A major technology developed by SIL, Spread Spectrum/Digital Communications, was bought by the SEA Group and adapted for ESA's SMART 1 Ka-band transponder experiment.

Background
Space Innovations Ltd (SIL) was established in 1983 and was previously Satellites International Ltd. Capabilities range from feasibility studies and research and development, to build, production and systems integration of space and space/ground equipment. Products include: small satellite busses; onboard data-handling systems; low-cost Sun/Earth sensors; star mappers; magnetorquers; magnetometers and associated attitude-control systems; power conditioning systems and batteries; radiation dose measuring instrumentation (its modular dosimeter operated on Meteosat 3 from 1988); space/ground data-compression systems; EGSE; radiometer instruments; telemetry systems and S-band transmitters/receivers and antennas for space/ground applications; and variable-rate/spread spectrum modems applicable to LEO and deep

space missions. SIL is involved in: Soho; Meteosat; ERS; STRV-1a/b; DRS; Artemis Envisat; Ørsted (S-band transceiver) and Badr-B (S-band transceiver, OBDH, magnetometer, magnetorquers, power conditioning, batteries, attitude sensing system).

SIL now offers two families of semi-standardised platforms: MicroSIL for 40 to 80 kg micro-satellites, and MiniSIL for 100 to 350 kg mini-satellites.

MicroSIL is available in three basic forms: MicroSIL-G gravity-gradient stabilised; MicroSIL-S spin-stabilised; MicroSIL-3 three-axis stabilised. Subsystems are densely packed within 170 × 470 × 470 mm on an equipment shelf.

SIL developed plans for rockets and launch vehicles.

Specifications

MicroSIL

Size: 490 × 490 × 510 mm; 40–80 kg

Payload (kg, for 55 kg total): 13 -G; 18 -S; 14 -3

OBDH: DHS-848B data-handling system, dual T805 or SPARC processors, 1 Mbit program ROM, 48 Mbit SRAM data memory; memory expandable in 32 Mbit modules

Attitude control: two DSS-256 2-axis digital Sun sensors with 0.5° accuracy (options to 0.125°). MFM-3L 3-axis fluxgate magnetometer ±50,000 nT. Control by two MTR-25 magnetorquers. MicroSIL-G nadir pointing ±4° using 6 m boom with 4 kg tip mass; MicroSIL-S 1–10 rpm spin, 0.5° knowledge; MicroSIL-3 pointing within 0.5° using redundant MWs

Communications: two STX-91B 1W S-band transmitters; BPSK at 1 kbit/s to 1 Mbit/s

Power: 20 37-cell strings of 20 × 40 mm GaAs cells + 2 × 4 Ah Ni/Cd batteries. Provides 15–20 W for -G in sunlight (orbit dependent), 20 W for -S/-3

MiniSIL

There are two basic types: MiniSIL-P for narrow fairing Pegasus-class vehicles; MiniSIL-L for larger fairings such as LMLV, Delta and Ariane Each is then available in a double-height version (2P/2L) and in both spin and three-axis designs. They employ a simple primary structure of a thrust tube with 937 mm V-band interfaces fore and aft. The tube is stiffened by longitudinal stringers and shear panels, to which the eight vertical sides are attached, with GaAs cells.

Size (P/2P/L/2L): 1.10 × 0.64 m/1.10 × 1.28 m/1.60 × 0.64 m/1.60 × 1.28 m; 100–150 kg/150–200 kg/ 150–250 kg/200–350 kg

Payload (P/2P/L/2L): 40–55 kg/65–85 kg/65–120 kg/ 90–180 kg

OBDH: DHS-8240B data-handling system, triple T805 or SPARC processors, 3 Mbit program ROM, 240 Mbit SRAM data memory; optional mass memory

Attitude control: six DSS-256 2-axis digital Sun sensors with 0.5° accuracy (options to 0.125°), MFM-3L 3-axis fluxgate magnetometer ±50,000 nT + optional RLG pack, GPS, static Earth horizon sensor or star mapper. Control by six MTR-35 magnetorquers; optional RWs + cold gas thrusters

Communications: two STX-91B 1W S-band transmitters; BPSK at 1 kbit/s to 1 Mbit/s

Power: N × 37-cell strings of 20 × 40 mm GaAs cells + 7 Ah Ni/Cd battery (redundant optional). Provides up to 55W/135W/80W/185 W (P/2P/L/2L).

Spacesense Consultants

Current Status

Spacesense offers worldwide information, marketing data/analysis, and advisory services to organisations active in space technology and its application. The Spacesense Environmental Research Group (SERG) was set up in 1994 by Imperial College, NRSC Ltd, Phoenix Systems and Spacesense for satellite environmental monitoring consultancy, research and development. Has played significant role in developing market survey for Envisat and follow-on projects. Conducted detailed study of national and international environmental surveys integrating earth- and space-based studies.

Spectrogon UK Ltd

Current Status

Formed in 1996, Spectrogon UK Ltd is a wholly owned subsidiary of Spectrogon AB, Taby, Sweden, which also has offices in the USA. Spectrogon UK Ltd has purchased the Diffraction Grating Department of Tayside Optical Technology Ltd, and therefore inherits the equipment and expertise to manufacture high-quality Ion-Etched Diffraction Gratings. Space Optics previously manufactured by this group include the variable line density master grating for ESA's XMM project. Key elements of technical expertise are: holographic optical elements (particularly diffraction gratings); ion etching (or milling) of holographic patterns to give high damage thresholds; and hydrocarbon-free metal coatings suitable for UHV applications. The Spectrogon group also produces Replica Holographic Gratings and Optical Interference Filters for a range of wavelengths from ultra-violet to infra-red. Customers include ESTEC, LLNL and NASA. Optical coatings for UV solar telescope mirrors developed also. Recent investment increase funded a new manufacturing facility in Arninge for gratings, increasing production. Has developed new differential gratings production capability with improved production standards.

Spemco Group

Current Status

Originally known as Argos Electronic Circuits, Spemco Group bought the assets of Argos when it entered receivership and combined it with Southern Circuits to set up the current company in the late 1990s. The company is engaged in printed circuit manufacture, including double sided PTH, multilayer, FR4 and polyamide, to full ESA standards. Have diversified into unique tailored product provision.

Spur Electron Limited

Current Status

Established in 1982, Spur Electron offers a total parts procurement service in addition to specialist engineering support, software development and technical publishing expertise. It also specialises in product assurance activities, and parts, materials and process support for space hardware components. Current activities include Columbus (product assurance database, central parts procurement database); Houwteq Greensat (total parts procurement activity); ESA/ESTEC (radiation effects database); UK Defence Research Agency (data validation unit support; STRV satellite procurement support) and UK MoD (reliability study of generic components). Advises on large scale logistical and maintainability component procurement, management and cargo marketing. Customers include Alcatel. CNES, Daimler Benz, Denel, DERA, ISRO and Matra Marconi.

Surrey Satellite Technology Ltd (SSTL)

Current Status

University of Surrey began its micro-satellite research with the UoSAT programme of 1979. UoSAT-1 was launched by Delta 2910 from Vandenberg AFB in October 1981.

SSTL was formed in 1985 by the University of Surrey to support technology transfer to industry and commercial development of research activities, in addition to undertaking industrial contracts. SSTL is located within the Surrey Space Centre supporting 35 post-graduate students and research fellows. The company's expertise includes designing, fabricating and operating low-cost (around £2 million each) micro-satellites: UoSAT 1–12, Kitsat 1/2, S80/T, PoSAT 1, HealthSat 2, Cerise, FASat satellites, Thai-Phutt, TiungSat-1, SNAP-1, PicoSAT, Clementine and Tsinghua-1 launched 1984–99. SSTL is located within the purpose-built 1,500 m^2 Surrey Space Centre, operating as an autonomous business with management and administration separate from the University. There are three varieties of cleanrooms (which can accommodate AIT of five micro-satellites simultaneously): 25 m^2 class 10,000 instrument assembly, 80 m^2 flight subsystem/ payload assembly and 80 m^2 flight model AIT. The 300 m^2 R&D laboratories cover RF and ground station support, OBDH and digital systems, power and sensors, and an MCAD/ECAD design office. SSTL's Mission Operations & Control Groundstation was improved and relocated to CSER in March 1994. The three independent tracking antennas (VHF, UHF, S-band) frequently track as many as eight satellites, supporting more than 110 daily passes.

Research studies cover: power systems; OBCs and data-handling networks; modulation/ demodulation schemes; RF transmitters/receivers; signal processing; attitude determination; stabilisation and control; imaging; radiation effects on components; failure-resilient architectures; VLSI subsystem design and store/forward techniques from LEO satellites. CSER is developing a 350 N hybrid motor. Using hydrogen peroxide as oxidiser and a solid fuel, it was successfully tested at Royal Ordnance's Westcott facility in February 1995. Under an ESA contract, SSTL is developing a GPS receiver for providing micro-satellite position and attitude; it will be carried by FASat-Bravo and TMSAT.

The modular multipurpose bus forming the core of the UoSat-12 mini-satellite provides S-band communications and enhanced OBDH for 180 kg payloads. The first will carry RWs, a 40 m resolution Earth camera and propulsion system for orbit control.

SSTL contracts have included supply of UoSAT ground stations to South Korea, Pakistan and Portugal; ESTEC spacecraft studies; mission studies for the Swedish Space Corporation; licensing of UoSAT ground station software; manufacture of flight hardware and launch and integration services on UoSAT missions; and supply of platforms for integration of dedicated payloads for commercial customers Matra and Aerospatiale (S80/T and Cerise). SSTL signed an agreement with Arianespace in August 1988 to act as the prime customer for the Ariane Structure for Auxiliary Payloads (ASAP) project. Of the six spacecraft slots on the first ASAP mission in January 1990, two were filled by UoSAT D/E, and four by micro-satellites from Amsat of North America. As a result, SSTL undertook a number of contracts for integration and launch services for experiments from ESTEC, RAE and Volunteers in Technical Assistance (VITA, US). In July 1990, SSTL signed a further Ariane agreement for UoSAT 5 launch aboard ERS 1 s vehicle in 1991. The SSTL-built Kitsat 1 was launched in 1992 (SSTL supplied parts for the South Koreans to build 1993's Kitsat 2 based on SSTL's design), PoSAT 1 in 1993 and Healthsat 2 in 1993. A £3 million contract was signed in May 1994 with the Chilean Air Force for the FASat-Alfa satellite, ground station and training. The £1.3 million contract was signed in January 1996 for the identical FASat-Bravo. A £3 million contract was signed in November 1995 with the Thai MicroSatellite Company for the first national Thai micro-satellite, TMSAT, in 1997. At the beginning of 1996, contracts in progress totalled £11 million. A £1.35 million contract was signed in March 1996 for a communications payload aboard UoSAT 2. Photographs of SSTL-built satellites are included in the South Korea, Portugal, UK, and Oscar entries.

The £6 million demonstration mission of SSTL's multipurpose MiniBus platform will be carried free on the fourth and last test launch of Russia's Rockot. MiniBus can support 50 to 180 kg LEO payloads, including SSTL standard module boxes and a payload bay for heavier instruments or those with nadir-pointing apertures. Payload data will be distributed via the Internet.

During 1996, SSTL became part of a team selected by ESA to conduct a Lunar orbiter Mission study including the Technical University of Munich and the Swedish Institute of Space Physics. Known as the Lunar Academic Research Satellite (LunARSat) the study sought to encourage young pupils in schools and universities in a belief that they could participate in useful science. In October 1998, SSTL announced that it had broken into China's tightly controlled internal satellite industry with formation of a collaborative venture company in Beijing to develop micro-satellites. The first £3 million contract had been signed between SSTL and Tsinghua for a 50 kg micro-satellite to be called Tsinghua-1, the demonstrator for a constellation of seven micro-satellites which will provide daily worldwide high-resolution imagery for disaster monitoring. Launched 29 June 2000 by COSMOS.

First in a constellation of five disaster monitoring satellites, AlSAT-1 was successfully launched by Cosmos 3M from the Plesetsk Cosmodrome on 28 November 2002. This is Algeria's first satellite, joined in orbit by four more disaster monitoring satellites in 2003. SSTL has achieved a placing in the Sunday Times Tech Track 100, companies with the fastest growing technology. It has seen an increase in sales from £5.4 million in 2000 to £12.2 million in 2002.

Specifications

UoSat Bus

Configuration: octagonal prism, 1.1 m (diameter) × 80 cm (height), accommodating standard SSTL module boxes
Mass: 300 kg on-orbit (180 kg payload)
Power supply: 9 GaAs panels provide 150 W orbit average processed power; 21 Ah Ni/Cd battery supports eclipse operations
OBCs: 186 and 386 computer operating under a multitasking system provide data handling. Specialist computing resources provided by TMS320C32 and Transputer-T800 systems. Standard SSTL TM/TC protocol. Inherent in the store/forward architecture is a 72 h autonomy period with automatic fault reporting and data downloading
Attitude control: 3-axis by MWs, magnetorquers and cold gas thrusters
Communications: data rates of 9.6–76.8 kbit/s on 400 MHz using redundant UHF and 1 Mbit/s on redundant S-band. 9.6 kbit/s VHF command uplink
TT&C: by SSTL's Mission Control and Operations Centre.

MiniSat Bus

Total mass: 50 kg, including launcher fittings
Configuration: 600 × 345 × 345 mm box-shaped body with solar panels on four sides, 6 m boom deployed from top face for gravity gradient stabilisation. Each housekeeping system or payload is housed in a standard module box, mass-produced by a computer-controlled process. The boxes are then stacked to form the main spacecraft structure; their number can be varied to meet mission requirements
Power supply: 4 GaAs panels totalling 4,992 cm^2 providing 18 W (orbit average processed); 6 Ah Ni/Cd battery supporting eclipse operations
OBC/Data Handling: the 80C186 payload experiment computer is used for high-level data manipulation. 80C188 version provides secondary OBC. On UoSAT 5 the payload Transputers are employed for high-performance data processing and attitude control
Attitude control: Earth pointing is maintained by combination of passive gravity gradient stabilisation and active magnetorquing; Sun/Earth horizon sensors and 3-axis fluxgate magnetometer with ±2 nT resolution over ±64 μT range provide attitude information. Gravity gradient supplied by 3 kg tip mass on 6 m boom deployed from top face, magnetorquing by six 150-turn aluminium wire coils (one on each spacecraft facet)
Communications: data rate of 9.6 or 38.4 kbit/s, 3-channel receiver (2 communications, 1 command), redundant transmitter switchable to 1/2.5/10 W RF output, 4-element turnstile VHF antenna, one monopole UHF antenna.

Systems Engineering & Assessment Ltd (SEA)

Current Status

Systems Engineering & Assessment Ltd (SEA) is an independent space systems software/hardware development company with professional staff of more than 200 with annual turnover of approximately £13 million. It provides consultancy services on feasibility, systems studies and soft/hardware engineering skills. Remote sensing expertise emphasises active sensors such as SAR for ground-based systems. For satellite communications and satellite navigation, it is active in bespoke ground segment systems and hardware design/development. SEA develops and markets menu-driven software products, including the Space Systems Analysis Toolset, a modular suite of Earth observation mission-analysis software tools, SAR processing solutions and NextPerf, the SAR Performance Analysis and Prediction Tool for the next generation of advanced SARs. Provides SAR interactive software for integration with optical image overlay. Current projects include involvement with Mars Express/Beagle 2, Herschel/Planck, METOP, SMART-1 and Skynet 5.

T S Space Systems

Current Status

Set up in the early 1980s, T S Space Systems provides outgassing monitoring equipment to meet ESA and NASA standards.

This contract research and development company specialises in designing, building and operating: space-simulation equipment for testing satellite components/materials and ion thrusters; solar cell characterisation equipment; spectral response; close-match simulators; proton irradiation chambers. Expertise includes high-vacuum equipment and cryogenics (pumping/shrouds, computer control/data acquisition). Can provide simulation equipment for high-pressure cryogenic vessels, high-temperature flow systems and oscillation and POGO suppression systems. Current projects include Beagle 2 and a large scale ultraviolet soak system. The company has developed higher levels of toxic sniffing and can provide sensory monitors for complete environmental systems.

UK Industrial Space Committee (UKISC)

Current Status

Formed in 1975, UKISC, the national trade association of the UK space industry, is sponsored jointly by the Society of British Aerospace Companies (SBAS) and the Federation of the Electronics Industry (FEI). UKISC represents the collective interests of member companies in increasing space and space-related business and their share of the market. Has developed programme and project co-ordination strategies for specific industrial sectors and for unique member interests. Membership stands at 18 companies, representing over three-quarters of the total turnover (£750 million) and employees of the British space industry (more than 6,000), and includes Matra Marconi Space, BT, Logica and Vega.

VEGA Group Plc

Current Status

Formed in 1978, VEGA is an independent systems engineering and consultancy group with offices in the UK, Germany and the USA and approximately 500 people. 1995/1996 turnover was £16 million (profit £3.5 million) increasing to £40.2 million in 1999/2000. VEGA is organised into several markets including space, government, defence, financial services, transport, utilities and health care.

VEGA specialises in spacecraft operations from the design study phase, through mission planning, to orbital operations. Since forming in 1978, it has been involved in operations/planning for all of ESA's science and applications missions, and has supported Inmarsat 2/3, Eutelsat 2, Skynet 4, NATO 4, Italsat, SAX and Tethered Satellite. It also produces operations-related software such as spacecraft simulators, data-analysis software and training packages. In January 2003, VEGA announced contracts worth €3 million for work on the EUMETSAT programme. Has played an influential role in developing the next generation of meteorological sensor packages for ESA.

Willis Corroon

Current Status

Operating within the Willis Corroon Group's Aerospace division, Willis Corroon Inspace provides full space insurance services, including launch, in-orbit, products liability and third-party legal liability. Customers include AsiaSat, NHK and NASDA. Willis has been involved with space insurance for more than 20 years and secured brokeage on Australia's AUSSAT system in the early 1980s when the space brokerage was under Willis Faber & Dumas before it became Willis Corroon.

United States

Advanced Products Company

Current Status

Seals for extreme environmental conditions, including rocket engines, fuel systems and cryogenic and vacuum systems. Also for high-temperature vacuum environments. Products include gaskets for high-performance cryogenic propulsion systems. The company has tested shock-absorbent seals and connector bellows for high-temperature environments. Long term storage vessels and dewars for small spacecraft and low usage requirements, long duration inert storage and surge-use after extended storage in hostile environments.

UPDATED

AEC-ABLE Engineering Company Inc

Current Status

Founded in 1975, and acquired in 1999 by Pressure Systems Inc, ABLE provides deployable structures and special-purpose space mechanisms. In addition to deployable coilable and articulated lattice masts and tubular columns, ABLE provides: composite structures; honeycomb panels; wire deployers; solar array deployment and tracking/drive mechanisms; payload separation devices; brushless DC motors; linear/rotary drive systems; hinge devices; structural and fluid quick-connect; atomic oxygen protection and thermal control blankets. Composite, high tensile, umbrella antenna structures extending from compact folded configuration to full deployment of up to 100 m. ABLE is providing the bearing assembly and solar array deployer for Space Station. The failed Meteor 1 satellite in 1995 carried ABLE's SCARLET solar wing, with four concentrator panels and two Si planar panels. New panel assemblies from twin Si arrays. Other space projects include Deep Space 1, Mars Surveyor, Cassini, SRTM and Galileo.
ABLE offers a range of solar arrays, including:
The UltraFlex Solar Array which provides 142 W/kg, 10.6 kg at 1,500 W BOL with Si; 166 W/kg, 9.0 kg at 1,500 W BOL with GaAs/Ge cells. 0.0794 m^3 stowed volume for 1,500 W wing.
The PUMA Solar Array which provides 47 W/kg for 1,300 W BOL; self-synchronising deployment, high stiffness. Provided for MSTI 3.
The deployable masts and columns include:
The CoilABLE Boom which is a triangular cross section with three continuous longerons, or rectangular cross section with four, for use with any of three deployment systems: nut; lanyard; carousel. Booms up to 45 m have been deployed in space. A Nut-Deployed CoilABLE boom was first used to deploy the large solar array on Shuttle STS-41D in September 1984. Others include LACE, UARS and Galileo.
The ABLE Articulated Retractable Mast or AARM employing four articulated longerons for greater bending strength and stiffness than above. Deployed by Nut Deployer. The FASTMast is the strongest version, stowing in 2 to 5 per cent of its deployed length. Side length 18 to 274 cm. Italy's TSS Tethered Satellite employs FASTMast, as does ISSA for solar array deployment.
The Tubular Masts which complement the lattice structures consist of a line of tubular masts of 0.6 to 10 cm diameter, including BESTMast and STALK. Flights include six on Transit (SOOS) and a pair for Cornell University's WISP II.

VERIFIED

AeroAstro Corporation

Current Status

AeroAstro was formed in 1988 to design and build small low-cost space systems and components. 1994 revenue was US$3.5 million. There are about 50 personnel. Its first major contract was spacecraft development, test and payload integration and ground station for the Los Alamos National Laboratory Alexis satellite. MIT's 1996 HETE satellite and ground station was developed under a similar contract, awarded in February 1991. The company built Boston University's Terriers satellite to study the upper atmosphere and ionosphere. Terriers was launched in May 1999 and is part of the Student Explorer Demonstration Initiative sponsored by the Universities Space Research Association. AeroAstro is planning a low-cost small satellite launcher. In September 2001, AeroAstro was contracted to develop the DoD Space Test Programme (STP) Satellite Mission 1 (STPSat-1) for the USAF Space Command. A successful Preliminary Design Review was held in July 2002. STPSat-1 will offer more than 200 W of orbit power with pointing accuracy better than 0.1°. Launch is planned for 2005–2006 on Delta 4.

AeroAstro provided the 1 Gbit solid-state digital mass memory system for the NASA/Los Alamos MOXE x-ray detector on Russia's Spektrum-X satellite in 1998. The same mission carries a 0.4 Gbit memory for the Canadian Space Agency's Euvita experiment.

The company also designs and fabricates spacecraft RF equipment, torque coils, Sun sensors (0.1° accuracy over 10°), solar panels, integrated control systems, radiation-hardened flight computers, software, ground stations and other elements for small satellite projects. It provided Ni/Cd batteries for Sweden's 1992 Freja satellite under a US$340,000 contract.

Co-developed with the USAF, AeroAstro's Bitsy micro-satellite provides an autonomous three-axis satellite weighing only 1 kg. Its simple, single circuit board design is easily tailored to specific mission requirements, such as store/forward, asset tracking, and hosting remote sensing instruments. Cost is US$100,000 for the basic satellite, or less than US$1 million for a turnkey system, including launch. Bitsy can be operated by the customer's PC or Macintosh-based ground station, or via AeroAstro on Internet. The debut Bitsy will determine its attitude using a star camera, image the Earth and transmit the compressed image.

Specifications

Availability: 6 months after order
Mass: 1.0 kg
Size: 5 × 15 × 15 cm with attached solar panel
AOCS: 0.5 mrad (2σ) using star camera, 50 mrad (2σ) attitude control by cold gas thrusters, 1 mrad/s stability (target, orbit dependent)
Power: 2 to 10W (increase optional), 8V unregulated; 4 Wh lithium ion battery
Communications: 8 Mbit minimum, up to 1 Gbit, 5 Mbit/s payload raw data max output; UHF or S-band, 2 kbit/s up/down (10 to 100 kbit/s down option).

VERIFIED

The Aerospace Corporation

Current Status

The Aerospace Corporation was founded in 1960 and operates as a private, non-profit federally funded research and development centre dedicated to national security needs, specialising in space systems and related technologies, and providing general system engineering and integration. It provided detailed analytical surveys on defence procurement requirements to meet perceived threats from anti-satellite systems and was instrumental in conducting background research for the Saint programme. Played a major role in the development of the Navstar GPS and was awarded the Robert J. Collier Award for this in 1992. The Aerospace Corporation provides the Space and Missile Systems Center of the Air Force Space Command with risk reduction analysis in launch vehicle operations. Technology development for picosatellites, remote sensing and hyperspectral imaging. Aerospace provides technical oversight for the national security programme, including launchers and satellites such as Milstar, DSCS, GPS, DMSP, DSP and future missile launch detection. For example, it wrote the DSCS II programme plan and provided technical oversight from 1965's study contract through the continuing operations phase. The Aerospace Corporation has full technical support facilities and tracking and control facilities for civilian and military satellite constellations. The organisation is funded from more than 60 programme budgets. More than 3,000 personnel include 66 per cent technical staff. Reported revenue of US$475.5 million for 2000.

UPDATED

Aerospace Industries Association of America Inc (AIA)

Current Status

The Aerospace Industries Association of America,Inc, based in Washington, D.C., is the trade association representing the nation's manufacturers of commercial, military and business aircraft, helicopters, aircraft engines, missiles, spacecraft, materials and related components and equipment.

Founded in 1919 as the Aircraft Industries Association of America and changed to its present title in 1959, AIA speaks aggressively and effectively to convey industry goals and accomplishments and to voice common concerns to Congress relevant federal agencies, international organisations, news media and the American public. The association provides a forum for industry and government to exchange views and resolve issues on non-competitive matters that relate to the industry. It provides a valuable annual analysis and evaluation of the industry, its resources, its assets and its capabilities. It also evaluates the business relationship between different elements of the industry.

UPDATED

Allied Aerospace

Current Status

Formerly known as Dynamic Engineering Inc, Allied Aerospace provides design, analysis and manufacturing support for research and development hardware: mechanical design (100 per cent CAD); electronic design, engineering static, dynamic and thermal analyses, structural dynamics and aeroelasticity; conventional and CNC milling, turning, wire EDM; MIL-Q-9858/MIL-I-45208 Quality Assurance programme, control system hard/software development. Computational CAD/CAM interactions with CFD equations provides three-dimensional stress modelling for hypothetical configurations. Produces CAD/CAM simulation of mechanical design options and presents total system software analysis and requirements evaluation. Has produced interactive CAD/CAM modelling software for unique wire and flow route programming. Staff total 466.

AlliedSignal Aerospace Hardware Product Group

Current Status

AlliedSignal Aerospace Product Group has facilities in the USA and Germany. AlliedSignal has merged with Honeywell, and is now part of that company along with AirResearch, Bendix/King, Garrett, Laseref, Pegasus, Primus/Epic Press. The Group employs 49,000 people.

ASAC's principal interests cover auxiliary power units, environmental control systems, launch vehicle and satellite systems (through Aerospace Equipment Systems, see below), propulsion systems, engine controls, avionics, guidance systems, secure communications and technical services. Combined resources of ASA were deployed to manage project requirements analysis after the terrorist attacks on New York and Washington, DC, on 11 September 2001. Diversification plans have encouraged a higher level of research into new non-space related applications.

AlliedSignal Hardware Product Group/ Environmental Control Systems (ECS)

Current Status

Previously named AiResearch, Aerospace Equipment Systems, the company is involved in environmental control systems. AlliedSignal's heat exchangers, pumps, valves, sensors, fans controllers, blowers, smoke detectors and actuators are being combined into Space Station systems. These include the CO_2 Removal Assembly and the Solar Dynamic Power Array. Developing product range of ECS requirements in large and small ECS systems. Long duration closed-loop ECS integrated with passive regeneration systems for extended life orbital or stored environmental requirements.

AlliedSignal Hardware Product Group/Power Management and Generation Systems (PMGS)

Current Status

PMGS specialises in the design, development and manufacture of components and systems for launch vehicles and satellites. These include: valves; actuators; turbopumps; turboalternators; cold gas and advanced monopropellant propulsion; propellant management systems; flight control actuation; power generation; dynamic power systems and attitude control systems. Current programmes include: Atlas; Centaur RL10A-4/E-1; LLV; Vega; Star 37/48; X33; EELV; Castor 4B/120; Gravity Probe B and military projects such as Theater Missile Defense. The NASA Solar Dynamic Ground Test Demonstrator and the NASA/RKA Flight Test Demonstrator are covered in the Thermal and Power industry section. Power and energy efficient management systems for next-generation launches and reusable transportation systems. Studies on advanced PMG systems for high-power requirements in orbiting systems. Requirements include independence of operation, high thermal loads and low accessibility. The company has developed new methods of power conversion and storage for electric propulsion and for large power sources.

AlliedSignal Hardware Product Group/ Technical Services Corporation

Current Status

The corporation is a technical services unit of AlliedSignal Aerospace, providing systems management for space tracking and data acquisition, systems engineering/design, laser tracking systems, VLBI systems and communication equipment operations. It operates/maintains part or all of three key NASA Goddard networks: the Ground Network, Space Network (including the White Sands TDRSS ground terminal) and the NASCOM NASA Communications Network. In addition, it operates/maintains GSFC's five Payload Operations Control Centers and provides: engineering support to the Data Technology Division; computer operations and maintenance to the Flight Dynamics Facility; and all elements of support to the Information Processing Division in the data reduction of spacecraft telemetry data and delivery to the principal investigators. For NASA's Crustal Dynamics Program, it provides technical support for the Satellite Laser Ranging element, including operation/maintenance of fixed/transportable laser ranging stations, engineering development to improve the accuracy of ranging observations, and precision satellite orbit determination. It is designing, installing and implementing a state-of-the-art satellite laser ranging system for the Italian space agency at the Matera Geophysical Observatory. The VLBI effort includes the design, construction, transportation and operation of hydrogen maser frequency standards, cryogenically cooled low-noise microwave amplifiers and microwave water-vapour radiometers.

At NASA Johnson, Technical Services Corporation is a member of the STSOC Space Transportation Systems Operations Contract team. Functions cover project management, maintenance/operations, flight preparation, production and mission operations involving the Mission Control Center, Shuttle Mission Simulator, Shuttle Avionics Integration Laboratory, Software Production Facility, Central Computing Facility, and the Mockup and Integration Laboratory.

For JPL, the corporation operates/maintains the Deep Space Network elements in Pasadena and Goldstone, providing tracking, data acquisition and network support for manned and unmanned missions. Integrated earth-space systems for information processing through common nodes.

Advanced studies on integrated ground and space-based communications and data management systems, laser communications and new data distribution technologies. Conducts systems integration designs for frequency-shift scan and search systems.

AMP NETCONNECT

Current Status
AMP is a strategic part of Tyco Electronics and specialises in manufacturing electrical and electronic interconnection systems and components for the aerospace industry. Tyco Electronics operates in more than 100 countries and reported 2001 revenues in excess of US$36 billion with more than 250,000 products across the aerospace, automotive and engineering industries. Products include packaging systems, cable and board assemblies, shell circular metal, hermetic, rack and panel, fibre optic, coaxial and printed circuit connectors, in addition to solderless terminals, switches and EMP/EMI filters. Provides protected high energy screened circuits and cabling for unusual environments and extreme conditions of stress, pressure and heat.

Analytical Graphics, Inc

Current Status
Co-developer of the Satellite Tool Kit family, an interactive graphical software package to access, manage, display and manipulate data. Aerospace systems can be analysed graphically and numerically. Data are generated using the provided Two-Body Keplerian, Space Command SGP4/SDP4 (which accepts two-line mean element sets) J2 or J4 propagators. Programmer's Library provides integration of STK functions into an existing environment. Visualisation Option permits analysis in 3-D. Upgrades extended to improve 3-D from graphic software improvements. AGI began operating in 1990 and has 115 employees and has delivered STK products to 28,000 customers in the satellite industry. New product development has provided 3-D projections in software capable of holographic projection.

Applied Technologies Associates

Current Status
Systems and software engineering, command/control engineering, automatic data processing, and continuous support for the USAF Satellite Control Network. The company's Real-Time Orbit Determination system provides real-time satellite orbit information autonomously to the user. Integrated software for object tracking and analysis with discrimination capability for divergent trajectories. Provides assistance in developing secondary, advanced successors. Has developed complex orbit interaction and position determination plots extrapolated from orbital data sets reflecting a historic trend.

ARAC

Current Status
ARAC, a former NASA Industrial Applications Centre, focuses on the utilisation of existing technology for resolving industrial technological needs. Strengths included technology transfer skills in personnel and practices and had good links throughout industry with connections to international research institutes and agencies. It accesses government laboratories, industry, universities and research facilities and hundreds of technical databases. Twenty in-house technical staff analyse and apply information and prepare a proprietary report according to a client-defined technological requirement. Can collate various databases and extract for analysis several separate research sources. Provides technical solutions by combining processes in parallel industries and determining an appropriate application. It is the US 'Technomart' co-ordinator. A 'Technomart' event takes place in Seoul, South Korea, every year. ARAC is also involved in meeting management, marketing and design, and is a consultant in technology transfer. ARAC conducts international as well as national searches and surveys and tailors customer contracts according to individual client requirements.

VERIFIED

Ardak Corporation

Current Status
In addition to serving intelligence and UAV markets, Ardak provides tailored analyses of US government space markets, including launch, ground and customer-specified market niches. Analyses of companies providing space-related products/services are performed, especially competitions and teaming arrangements, merger/acquisition services, and evaluation of independent research and development programmes. Arranges analytical teams for unique client requirements and can set up links with financial consultants or venture capitalists. Active in specialised markets such as remote sensing. Ardak can also provide detailed analyses on national and international ventures, programmes and teaming arrangements where complex legal, financial and contractual arrangements are essential to teaming criteria. Ties defence and non-defence related programmes and technologies into working packages for third party users.

UPDATED

Arrowhead Products

Current Status
In addition to serving the aircraft industry, Arrowhead Products designs, develops and manufactures pneumatic and cryogenic ducting systems, metal bellows and other flexible components, elastomeric and composite structures, and integral thermal insulation for engines and spacecraft. Experience includes: Shuttle Orbiter LOX/LH_2 main propulsion feedlines and manifolds; Shuttle External Tank LOX/LH_2 feedlines, LH_2 recirculation line and GOX/GH_2 pressurisation system ducting; argon-jacketed, vacuum-jacketed and polyurethane foam insulation for cryogenic systems to −251°C, as on Shuttle ET; polyurethane-based high temperature resistant cryogenic sheathing. Next-generation cryogenic and high-pressure launch vehicle and satellite applications. Arrowhead has conducted tests of a special shock-resistant composite structure for cryogenic storage vessels and applied this to insulation problems which have specific application to the construction industry.

VERIFIED

Astro Aerospace Corporation (AAC)

Current Status
Astro began in 1958 as Astro Research Corporation formed by Swiss born scientists Dr Hans Schuerch and developed concepts for inflatable space stations, filament wound structures and heliogyro structures. In 1972 the company was acquired by Canada's Spar Aerospace Ltd and shifted focus from research to development of flight hardware. In 1999 Astro Aerospace became a wholly owned subsidiary of TRW Inc. In 2002 with the merger of TRW and Northrop Grumman, Astro became a subsidiary of Northrop Grumman Space Technology. Astro Aerospace specialises in spacecraft deployable systems, including the three-longeron Astromast folding truss, STEM (Storable Tubular Extendable Member), telescopic deployable booms, specialised solar array developments, large deployable mesh antennas (6 m diameter Astromesh) and deployable support structures for radar systems. Examples are Astromasts for Mars Observer instrument booms, the Milstar solar array deployment booms, Hubble Space Telescope solar array deployment booms, the Astro Edge solar arrays for NASA's Clark satellite, and the support structures for Seasat and Radarsat SARs. Astro is designing and building the Mobile Transporter for Space Station. Personnel total about 100. AAC conducting research on ultra-lightweight composite structures for large diameter antennas also capable of supporting modest *g*-loads when deployed. AAC supplied the Gamma Ray Spectrometer boom for the NASA 2001 Mars Odyssey spacecraft deployed on 5 June 2002. AstroMesh reflectors, capable of apertures up to 20 m and frequencies up to Ka band, are offered. In June 2002 the shuttle Endeavour delivered the Mobile Base System to the ISS. AAC has designed unique folding arrays, long-life deployment systems with in-built memory, extendible structures for large arrays and high-stress arrays for in-orbit manoeuvring.

UPDATED

Avibank Manufacturing, Inc

Current Status
Now a subsidiary of SPS Technologies Inc, set up for business in 1903, Avibank was founded in the 1940s and specialises in fasteners, devices and multicomponent fastening systems. Products range from quick-release pins and accessories to special bolts, latches and keepers, adjustable-diameter fasteners, hold-open rods, electronic hardware, panel fasteners, grounding hardware and threaded inserts. Ball-lok quick-release pins are provided for space missions. The Truss-lok mechanical joint mechanism has been developed for assembling erectable space structures and attaching equipment to the International Space Station's truss to which solar cell arrays and radiators will be attached. Truss-lok variant available for high-*g* loads and for inertia damping on high-mass structures or modules manoeuvred at off-nominal angles with transverse loads and *g* forces. Orders for 2001 reported at US$914 million with earnings of US$51.1 million.

Ball Aerospace & Technologies Corporation (BATC)

Current Status
Ball Aerospace has been awarded a system architecture contract for the next generation GOES-R system planned for launch in 2012. Ball will develop end-to-end configuration for the satellites and associated ground system for the NOAA. A subsidiary of The Ball Corporation, BATC reported 1998 sales of US$363 million. The company was founded in 1956 and comprises the aerospace systems and telecommunications products divisions. It was prime contractor for NASA's ERBS satellite, SDI's Relay Mirror Experiment, and the CRRES science satellite. Ball was awarded the US$46 million contract in 1992 to build the Geosat Follow-On radar altimeter satellite for the US Navy. It provided technical assistance to Spain's small satellite programme under a four-year agreement signed in October 1992.

In 1996, Ball Aerospace Australia was set up with operations in Brisbane, Canberra, Sydney, Melbourne and Adelaide, Australia. BATC built two instruments for the Chandra X-ray observatory. Past programmes include Brilliant Pebbles (1988), Compton GRO (1991), Chandra X-ray observatory (1999), COSTAR for HST (1993), CRRES (1990), Darpasat (1994), Geosat Follow-on (1998), GHRS (1990), HST (1990), IRAS (1983), MTI (2000), NICMOS for HST (1997), OSO (1962), OSSE (1991), QuickSCAT (1999), Radarsat (1995),

Ball built the Combined Radiation Release Effects Satellite as a joint NASA/USAF endeavour (NASA) ***NEW***/1047493

RME (1990),SIR-C (1994), SBUV-2 (1996), SRTM (2000), STIS (1997), and SWAS (1998). Current programmes include CALIPSO, CloudSat, Deep Impact, ICESat, JWST, Kepler, Mars Rovers, NPOESS, QuickBird, SAGE III, SIRTF and THEL.

BATC employs 2,200 people and reported sales of US$363 million for 2000. Ball Corporation reported sales of US$3.9 billion in 2002.

UPDATED

BDM Federal

Current Status

A subsidiary of BDM International, BDM Federal, provides space systems engineering and services in military space and C^4, Ballistic Missile Defence, national security and civil space applications with engineering and analysis on NMD and derivative applications (anti-satellite). It developed missile defence risk probability analysis for system trade-off studies. It supports NASA's RSPAC Remote Sensing Public Access Center. It has diversified into IT and computer systems for civilian agencies and private industry and has developed new analytical tools for statistical evaluation of threat scenarios and probabilities.

VERIFIED

Bergen Cable Technologies Inc

Current Status

A subsidiary of Leggett and Platt Inc, Bergen Cable Technologies was formed in the late 1940s to produce mechanical cables and assemblies for the aerospace industry. It produced articulation systems and mechanical cable extension/retraction devices with high tensile loads. Users include Space Shuttle and International Space Station. Company has developed composite based articulation systems for large structures.

UPDATED

BFGoodrich Aerospace

Current Status

The Military Aircraft section of the Aircraft Wheels/Brakes unit is responsible for the Space Shuttle's wheels and brakes. The new carbon-carbon brakes were introduced on Discovery in April 1990. The Carbon Products unit manufactures the materials. Super-Temp develops high-temperature composites for space, aviation, electronics and nuclear applications. Aircraft Integrated Systems provides measurement, sensing and control systems for manned spacecraft, launchers and strategic missiles. Shuttle carries three systems: point level sensors to monitor pre-launch loading/unloading operations; propellant measurement during orbital manoeuvring and re-entry; cryogenics measurement in orbit. Strategic missile sensors measure the burn rate of solid-motor insulation. On expendable launchers, it provides multi-channel excitation and signal conditioning for instrumentation sensors and destruct control systems. Lighting Systems provided the approach systems for the nine prime and back-up Shuttle landing sites, plus other airborne lighting equipment. Rosemount Aerospace is a division of BFGoodrich Aerospace. The company was a prime partner in the X-33 programme cancelled in 2001. BFGoodrich purchased Raytheon Optical Systems Inc in December 2000. Goodrich, the parent company, reported 2001 sales of US$4.184 billion and net income of US$289.2 million.

Boeing Aerospace Operations, Inc (BAO)

Current Status

BAO's 4,460 m^2 Flight Equipment Processing Facility near NASA, Johnson, houses a laboratory dedicated to processing and providing astronaut equipment for each Shuttle mission. EVA suits, photographic/TV equipment (including 16/35/70 mm film cameras), crew communications items, tools, batteries, food, clothes and personal items are stored, assembled, tested and readied for flight. Equipment is packaged and shipped to the Cape in isopods for installation in the Orbiter; items are checked and refurbished after the flight. Now supporting ISS work. Subcontractor interest in utilising BAO facilities for product evaluation and testing.

Boeing Autometric, Inc

Current Status

In August 2000, The Boeing Company completed acquisition of Autometric Inc. Revenues for 1999 declared at US$80 million with 500 employees. The Omni space system planning software of Autometric's Space Technology and Applications Division provides interactive 4-D whole-Earth modelling and simulation. Provides data production software tools for NASA's SRTM. Has developed new software handling tools for configuring different packets and integrating off-board systems.

Boeing Commercial Space Company (BCSC)

Current Status

Boeing Commercial Space has been disbanded, its activities distributed into other Boeing activities. BCSC was formed in July 1987 to pursue commercial space ventures. Its flagship programme is the Sea Launch venture to launch

Zenit GTO payloads from international waters. BCSC applied for a licence in January 1996 to operate the Resource 21 four-satellite constellation, beginning 1999, for crop monitoring. This project has since been abandoned. It has been resurrected in concept by other companies previously consulted on the idea. Other activities include development of a vapour-transport crystal-growth furnace for Shuttle flights under a Joint Endeavour Agreement with NASA, and participation in NASA Centers for the Commercial Development of Space. The furnace flew first aboard STS-52 Oct 1992.

Boeing Integrated Defence Systems

Current Status
A Boeing Company business element that combines communication networks, air, land and sea-based platforms for global military, government and commercial customers. Embracing the space business acquired from Rockwell International in 1996, then renamed Boeing North American Inc, IDS is under contract to the United Space alliance for Shuttle support operations and is the contractor for Boeing work on the International Space Station. IDS has a turnover of more than US$23 billion and employs 78,000 people. IDS elements specific to space systems are NASA Systems division based at Houston, Texas, Space and Intelligence Systems division at Seal Beach, California and Launch and Satellite Systems division also located at Seal Beach. Products and services include the Delta family of launch vehicles, the Global Positioning System Navstar navigation satellites, the International Space Station, the Shuttle and the Sea Launch Company, LCC.

Background
Boeing signed an agreement on 1 August 1996 to acquire Rockwell's space and defence businesses, which would be renamed Boeing North American Inc, part of Boeing's Defense & Space Group. 1995's combined space revenue was approximately US$3.5 billion. The acquisition would provide Boeing with Rockwell's Shuttle operations and satellite manufacturing expertise (particularly GPS satellite contracts), and Rocketdyne's engine production.

Boeing's Missiles/Space Division, created in May 1991, operates within the Space Systems Group, formally established to centralise all space activities. Boeing Defense & Space Group was named on 17 August 1993 as prime contractor for the revised Space Station. The contract was signed with NASA on 13 January 1995 at US$5.63 billion. Other current projects include the Inertial Upper Stage, covered in detail in the Launchers/Propulsion section.

Booz Allen Hamilton Inc

Current Status
Booz Allen Hamilton has staff of 13,000 in offices throughout the world with sales of US$2.2 billion in fiscal year ending 31 March 2003. Booz Allen has over 1,000 staff specialising in helping clients develop space programmes in public and private sectors using in-house staff comprising space professionals from business and industry, business strategists, design and manufacturing engineers, software engineers, integration specialists and manned and unmanned vehicle operators. Services span concept definition, development, production, operation, support and data exploitation. Major programmes supported include Space Station (as part of teams performing systems engineering and integration, and training systems development), AF Space Command and NASA's Space Network. Supports Shuttle operations analyses.

Brush Wellman Applications Development Center

Current Status
Supports aerospace, automotive, oil, gas, mining, telecommunications and undersea and marine markets. Development Center continues to apply new materials technology to user requirements. Diverse product range and production capabilities for customer needs and to tailored specifications. Design, development and fabrication of spacecraft structures, including solar array booms, antenna booms, central cylinders, optical support structures, thrust cone adaptors, radiators and heat sinks. Antenna struts were provided for Galileo, sunshields and radiators for Voyager and the solar array support structure for Magellan. Produces alloy, ceramic copper beryllium and beryllium products for high-end technology applications with annual sales of US$4,557 million for 1999; an 11 per cent increase over 1998. Produces brush engineered bronze products and Brush Alloy 25, a heat treatable metal.

California Space Institute

Current Status
Research and development for space science programmes. Created by the University of California Regents in 1980. Over 120 affiliated research staff on five campuses in earth science, space science and space-related engineering and technology. Disseminates technical data sets including global climate and radiation measurements from Scripps Institute of Oceanography. CSI has an outreach centre under the University of California and has conducted extensive development of intra-state research and development programmes and has played a major role in integrating California's science and technology applications programmes.

Calspan SRL Corporation

Current Status
SRL was formed in 1982 to design, market and operate commercial space-based facilities and to provide commercial services to government and industrial users. It merged with Arvin Industries' Calspan R&DTE subsidiary in 1992; Arvin took a 70 per cent interest. The creation of Space Business International was announced in June 1993 by SIII, OHB-System GmbH and Novespace to co-operate on space contracts. The activities carried out framed the next several years by integrating all disciplines of science with technology ventures and exotic projects. That work now subsumed into one or other of the six dimensions covering aeronautics, human systems technology, information systems and electronic warfare technology, chemical defence and demilitarisation, automated inspection systems or transportation research. Operate stockpile laboratories and hazardous materials processing facilities. A member of the NBC Industry Group. Conducts tests on protection from radioactive sources carried into space.

Celestis Inc

Current Status
Celestis Inc is a company formed to launch human ashes into orbit. In 2001, the cost for flying each person's phial of ashes (the mass of each set of remains was limited to 7 g) was US$5,300. Other services offered include a lunar orbit or impact option for US$12,500 or a Voyager service to deep space for US$12,500 also. The first launch took place on 21 April 1997 with 24 phials being attached to the third stage of the Pegasus-XL launch vehicle, entering a 151°, 554 to 582 km orbit. Among those carried into orbit were Gene Roddenberry ('Star Trek' TV series creator), Timothy Leary (1960s 'icon'), Gerard K O'Neill (who established the modern concept of orbiting space colonies), Todd Hawley (founder of the International Space University) and Krafft Ehricke (engineer with the German V-2 and American Apollo programmes). Celestis 1 re-entered earth's atmosphere on 20 May 2002 after completing 28,132 orbits of the earth and reached the surface northeast of Papua New Guinea. Further launches aboard Pegasus-XL and Taurus vehicles followed in February 1998 and December 1999. Celestis also provided the capsule carrying a token portion of the cremated remains of Eugene Shoemaker on board the Lunar Prospector spacecraft, which impacted the moon in July 1999, 30 years after the first manned landing on the moon. Dr Shoemaker had played an instrumental role in training Apollo astronauts. Celestis has reached an agreement with SpaceQuest Ltd of Fairfax, Va., to market Celestis services for launch on a series of Dnepr launcher from Baikonur, the first planned for a piggy-back flight with the launch of LatinSat in May 2004. Named Kosmos 1, the remains will be placed in a sun-synchronous orbit at 700 km altitude with an orbital life of approximately 140 years.

Circle Seal Controls, Inc

Current Status
A division of Circor International Inc, Circle Seal Controls was set up in 1974 under the name James-Pond-Clark and is now involved in the development and manufacture of manifold systems, control valves, relief valves, check valves, solenoid valves, regulators, flow limiters, relief valves and shutoff valves for use in rockets, spacecraft and space-related programmes. It has also produced high-temperature gas conduits for cryogenic dewars and special flow control equipment for high-pressure systems in low-temperature environments. Experience covers Shuttle, Titan 2 and Apollo. Aktomatic 6000 series solenoid valves used in the Crawler Transporter used to move the NASA Shuttle to the launch pad. ISO 9001 qualified.

CMC Electronics Cincinnati

Current Status
Formerly known as Cincinnati Electronics Corporation, CMC Electronics Cincinnati (now owned by CMC Electronics Inc) specialises in high-technology electronics and engineering services. The parent company was set up in 1903. The company's main space product lines are the design/manufacture of spacecraft TT&C command systems, high-rate data transmitters, range safety flight termination systems for launchers, and TDRSS telemetry transmitters. The company also provides command encoders, message-format generators, frequency synthesisers and automated test equipment. Spacecraft command systems are offered in SGLS, STDN and TDRSS formats. Range safety flight-termination systems are available in both the standard IRIG 3/4 tone formats as well as the secure, Hi-Alphabet, format. CMC Electronics Cincinnati is the leading producer of command destruct receivers, with equipment qualified for Shuttle, Atlas, Delta, Titan, Centaur and LMLV-1, plus numerous tactical missiles and ICBMs such as Minuteman and Peacekeeper. Extensive supplier for commercial Atlas and Delta derivatives as well as Titan government launches. Has examined the automated checkout and ascent control requirements of autonomous unmanned launch vehicles minimising the number of launch support personnel.

Computer Sciences Corporation (CSC)

Current Status
Formed in 1959, CSC developed the SAP assembly programme and became a public company in 1963. CSC introduced its STEPP Change management programme at the UK based Thales Defence Information Systems.

In December 2000, Computer Sciences Corporation completed acquisition of Mynd Corporation. CSC provides tracking, data acquisition, processing, analysis and software support to US space programmes. The company's Applied Technology Division was selected in May 1994 for a US$1 billion eight-year contract to provide computer support to NASA Marshall. FY93 saw a US$566 million, six-year USAF contract for range technical services and a US$200 million 71/2 year contract to operate/maintain NASA Langley's Central Scientific Computer Complex. Space Software Italia was established by CSC/Aeritalia. US government contacts now account for about 25 per cent of business. Now provides strategic and IT consultancy services. Annual revenues in excess of US$11.4 billion with 64,000 employees worldwide.

Conatec, Incorporated

Current Status
Conatec was established in 1984 to provide space engineering and technical services. It has a broad and diverse skill base and can provide end-to-end technical solutions from an in-house resource base with access to specialised facilities for engineering and breadboard testing in a variety of specialised disciplines. Support has been provided to Texus, TDRSS and BMDO's target programmes. Also provided technical analysis for NMD concepts. Specialising in suborbital launch services as a peripheral to main engineering business. Licensed to use White Sands, New Mexico, range

COSMIC

Current Status
Established as NASA's Software Technology Transfer Center distributing computer programs originally developed by NASA and its contractors for the US space programme. Run by the University of Georgia Research Foundation which terminated its NASA-related activity in 1998, all Open Channel Software now distributed through that organisation with some work carried out for specified NASA field centres.

CTA Incorporated

Current Status
Formed in 1979, Computer Technology Associates Incorporated has a current staff of over 1,150 employees and 1996 revenue of US$`80 million. CTA has completed more than US$1.7 billion of IT projects for over 700 clients. CTA specialises in the engineering, design and development of complex space and communications systems, and information technology applications and services. Their principal efforts are directed toward the innovative applications of new technologies in software, computer and information sciences and the design and development of small satellites, and associated ground-based command, control, communications and intelligence systems.

CTA provides skilled information technology services to commercial clients and state/federal government agencies. CTA's information technology business base includes engineering and acquisition support contracts which address programmes and systems through the entire life cycle of requirements: specifications, design, development, integration, test, training, and maintenance. CTA's government clients include the US Air Force, NASA, NOAA, DOJ, Treasury, the US Navy, Department of Administration, Department of Health and Environment, Department of Ageing and the FAA. Their activities include C^3I, avionics, space systems, air traffic control, and automated enterprise management technology applications. CTA participates in the systems engineering of major large-scale systems such as the Air Force Integrated Tactical Warning/Attack Assessment (ITW/AA) System, NASA's Earth Observing Systems, the FAA Advanced Automation System, and the independent test of such systems as the Air Force Peace Shield Air Defense System, the NASA Hubble Space Telescope Ground System and the FAA Advanced Air Traffic Control System.

CTA performs the operation and maintenance of GSA's Information management systems providing support for a wide variety of governmental users. This work includes state-of-the-art Information Technology services including client-server implementations, network engineering and development of relational base applications. CTA has completed a Year 2000 date conversion for the Bureau of Veteran Affairs, Austin Center and was selected by Nebraska and Kansas to perform Year 2000 conversions. CTA performs avionics integration for the Naval Air Warfare Center (NAWC) at China Lake and provided avionics software support for Operation Desert Storm. CTA develops range instrumentation for NAWC test ranges under the Range Instrumentation Development contract. CTA has a growing commercial IT business. Customers include Reynolds Metals, United Services Automobile Association and Applied Materials.

CTA International (CTAI), Inc markets, engineers and integrates turnkey communications and broadcast satellite systems for International markets. CTAI is under contract to deliver INDOSTAR-1™, an innovative digital direct-to-home television broadcasting satellite and ground system, to PT MediaCitra Indostar (MCI) and to operate and maintain the satellite broadcast system.

CTA Special Payloads Group supports spacecraft system design, hardware, software, test, and flight operations on Goddard Space Flight Center's Small Explorer Program, Space Shuttle Small Payloads Project, and Spartan Shuttle Missions.

Clients include the FBI, DoD, USAF, USN, NASA, the FAA as well as US state governments.

CTA Space Systems (CTA/SS)

Current Status
CTA Space Systems was acquired by Orbital Sciences Corporation in 1997 but this entry is retained to provide a record of projects still continuing. Activities are much the same utilising existing skills and capabilities.

A world leader in manufacturing small satellite systems, CTA Space Systems (CTA/SS) has developed, manufactured, fielded and operated nationally important surveillance and communications systems since 1980. Since 1984, the company has designed and built 27 satellites, and has an additional seven satellites in various stages of development. The company's spacecraft have varied widely in capability and have ranged in weight from 150 to 2,700 lbs. CTA/SS satellites currently under construction include the NASA Small Spacecraft Technology Initiative (SSTI) 'Clark' spacecraft; INDOSTAR, a private commercial geostationary direct broadcast mini-satellite; the EarthWatch (formerly World View) EarlyBird commercial remote sensing spacecraft and several DOD experimental mini-satellites. CTA/SS also builds satellite command and control ground systems, meteorological satellite data ingest systems, airborne and air-dropped remote sensor systems, and communications systems.

DSI and DSI's sister company, International Technologies Inc, were acquired by CTA/SS in mid-1992. 1992 and 1991 revenues were each about US$25 million, with 200 personnel. DSI built 27 satellites, launched between 1990 and 1994; six more were under construction for 1996/1997 launches, including NASA's Clark and EarthWatch's Early Bird. The company's first satellite, Global Low Orbiting Message Relay (GLOMR) was released from a Shuttle GAS canister in 1985. The first Pegasus air-launched booster in April 1990 released a DSI satellite for US Navy communications relay. Three 68 kg Maestro satellites, collectively the USAF P87-2 Stacksat, were launched by Atlas 28E on 11 April 1990. This US$11 million Multiple Autonomous Experimental Spacecraft for Telecommunications, Recording and Observations contract started in August 1987 as an Office of Naval Research and DoD Space Test Program project. The company received a US$7.4 million contract in August 1988 from DARPA for seven mini-satellites. The 21.8 kg micro-satellites were carried by the second Pegasus on 17 July 1991, to demonstrate UHF tactical voice/data links. Two 68 kg MACSAT Multiple Access Communications Satellites were launched on 9 May 1990 aboard a single Scout. 1991's Shuttle STS-39 observed propellant samples released from DSI's three 80 kg CRO subsatellites. DSI built the 90 kg ISES Independent Space Experiment System launched by Scout on 25 June 1991 for communications experiments. The Radcal radar calibration satellite was launched by Scout on 29 June 1993. STP's STEP M1/M3 were lost in the June 1994 and June 1995 launch failures, and GEMstar 1 on 15 August 1995 on LMLV 1. REX 2 was launched on 9 March 1996 for STP. DSI built Meteor's service module under contract to Westinghouse.

Dallas Remote Imaging Group (DRIG)

Current Status
DRIG is interested in satellite tracking, telemetry decoding, observation of new launches and digital processing of satellite imagery. It provides consulting and educational services on weather satellite imagery, satellite communications and tracking, image acquisition and processing for educational purposes. The bulletin board makes available satellite information and images from the Okean and Sich satellites and from other observation satellites. From the outset DRIG had a strong international base of contacts and affiliations. There are approximately 18,500 members worldwide, with more than 110 actively tracking satellites daily.

Decca Valves

Current Status
As the fluid control division of Parker-Hannitin Corporation, Decca provides hydraulic swivels and titanium control valves for space applications including satellites, launch vehicles and ground tent rigs and support systems. Products include motion control systems, high stress relief and turn valves for bifurcated pipes and steel lined propellant conduits.

Draper Laboratory

Current Status
Founded in 1930 by Dr Charles Stark Draper, Draper Laboratory Inc continues to build its reputation as a nationally recognised space avionics design and development organisation for manned and unmanned space programmes. This is based upon years of experience in overall guidance, navigation and control system design and supporting sensor, actuator and data-management system design. Draper designed the navigation and control system for the Apollo spacecraft and provided ground breaking developments for celestial navigation, providing the Apollo programme with solutions to one of its most daunting challenges. Draper continues to explore and develop new designs for advanced inertial systems, attitude control and energy storage, and fibre optic gyro designs in support of advanced space systems. Draper has become the expert in on-orbit flight control and powered flight guidance design. Currently it is involved in support of manned and unmanned systems for both commercial and government programmes. It is under contract to the NASA Johnson Space Center to provide guidance, navigation and control support to the shuttle programme. It redesigned the shuttle on-orbit flight control system for docking with large structures and supported X-33, X-34 and X-38 programmes. Technology applications entered to microelectromechanical systems, modelling and simulation, control systems, information systems and reliability models. Draper Laboratory employs more than 750 engineers, scientists and technicians.

Eagle Aerospace, Inc

Current Status
Aerospace systems engineering and programme management. Acquired by investor group headed by Dr H. Dean Cubley in 1996. Recent activities include Space Station development and telecommunications. Systems studies and telecommunications concept evaluation are recent activities and the organisation has conducted several analytical evaluations for government agencies and for other commercial companies.

Eastman Kodak Company

Current Status
Supplier of image capture, storage, processing and printing components and subsystems, employing photographic, optical, electro-optical and electronic imaging technologies. Develops/manufactures custom imaging products and subsystems, including high-performance electronic cameras, imaging workstations, photographic processing equipment and large ultra-lightweight space-qualified optical components and subsystems. This unit is a major subcontractor to TRW for integration of NASA's AXAF telescope system. It is providing the camera payload for Space Imaging's commercial remote sensing satellite. Other current space projects include NASA's Far-Ultra-violet Spectroscopic Explorer (FUSE), a sub-arcsec star-sensing telescope and a custom digital camera now flying on Space Shuttle. It was also a bidder for the Hubble Space Telescope's mirrors and, in fact, cut/polished the back-up primary mirror. The company is believed to provide optical systems for reconnaissance satellites. It also provided the imaging packages for NASA's five Lunar Orbiters in the 1960s.

The company reported sales in 2001 of US$13.2 billion.

VERIFIED

Eaton Corporation

Current Status
Eaton's Engineered Sensors Division supplies four types of pressure transducers to the Space Shuttle programme: 41SG51 hydrogen/oxygen fuel-cell supply tanks, 27.2 to 82 atmospheres, –87 to +107°C, 48 mV full scale output; 41SG78 Kaiser Marquardt thruster (44 per Orbiter) chamber pressure, 13.6 atmospheres, –1 to +149°C, 30 mV full scale output; 41SG85 LH_2/LOX main engines (19 per engine), 20.4 to 646 atmospheres, –54 to +74°C; 41SG275 main engine hydrogen supply, 340 atmospheres, –107 to +38°C. In the aerospace industry Eaton is a supplier of hydraulic, electro-hydraulic pump and generator products and integrated systems; electric motors; aircraft flap and slat systems; fluid conveyance products and systems; nose wheel steering systems; pressure sensors and fluid monitoring products and systems. Eaton serves commercial and military aviation, aerospace, military weapons, marine and off-road markets worldwide.

The division is fully certified to NHB5300. Eaton has 51,000 employees and sells products in more than 50 countries with 2002 sales of US$7.2 billion.

EG&G Florida, Inc

Current Status
The company was established in 1982 when NASA awarded KSC's base operations a 10-year US$1.6 billion contract. NASA announced on 1 October 1993 that EG&G Florida had won the second 10-year US$1.7 billion contract. It provides operations and maintenance, engineering, computer services, astronaut rescue teams, medical services, security and other support services. EG&G developed a liquid-air pack for the astronaut rescue team and a state-of-the-art meteorological system for the Shuttle Landing Facility. It also prepared design and modification packages for the structures and support systems at KSC. Company has diversified into technology applications in multidisciplinary industries but retains a diversity of specialised skills and levels of expertise relevant to several space related functions and requirements. EG&G has annual sales of more than US$2 billion and employs almost 13,000 people worldwide.

Encore Computer Corporation

Current Status
Originally Systems Engineering Laboratories (SEL) and then Gould/Computer Systems Inc, established in 1983, Encore is a major computer supplier for range/tracking operations and supplies special antenna designs. It has conducted tests of integrated computerised range and antenna farm checkout systems for automated launch and clearance systems. Provides upgrades to existing data acquisition systems including LCRS with Linux code. Contracted to provide analytical tools for new antenna configurations of a non-linear conceptual structure.

Engineered Systems Co (ESCO)

Current Status
ESCO was created by the merger of All American Engineering and the Advanced Development and Engineering Center. AAE was selected in 1987 to provide up to seven Space Shuttle Orbiter Arresting Systems (SOAS) for vehicle braking in the event of a Shuttle landing on an emergency airfield. Three were installed in 1988 in time for flight programme resumption following the Challenger disaster in 1986, a fourth was ordered in October 1988, and a fifth in March 1989 for installation at Hickam AFB, Hawaii. SOAS is a transportable arresting gear designed to capture an Orbiter should a landing on a short runway be necessary. The vehicle would engage a net (height: 13.7 m) across the runway, pulling a nylon tape from each Model 44 Water Twister energy-absorber tape drum and generating 0.75 *g* (maximum) braking. The 36-element net is a scaled-up version of the company's design used with military aircraft and would stop an Orbiter within 250 m without structural damage. The technique was demonstrated by towing Enterprise into a net at Washington DC's Dulles Airport. Technology applications extend to airport installed aircraft restraint systems. ESCO has conducted engineering studies on high velocity retardation systems and braking capture nets for projectiles. Developed versions have been tested in reduced scale for arresting large objects ejected from primary structures, and this has been found to have applications to planetary spacecraft where elements are ejected but contained within arrestor systems.

Entran Devices, Inc

Current Status
Founded in 1970, Entran specialises in miniature and compact static/dynamic measurement devices. In addition to extensive utilisation on missile and aircraft projects, the company's devices have been employed aboard Shuttle, Spacelab, Deep Space Probe and Mars Lander. Also Intelsat 5, Spacelab and Ariane 4 and 5. Entran has offices in England, France, Germany, Italy and Sweden. Entran have developed a fine-scale dynamic measurement device for shock frequency discrimination. The company has expanded its range of measurement devices to include nanotechnology which allows monitoring to take place in very small mechanical systems remotely operated and in alien environments.

Essex Corporation

Current Status
Essex provides patented ImSyn™ processes for the reconstruction of SAR images, onboard spacecraft/other platforms and as workstation accelerators. Essex supplies custom design and development of trainers and mockups, provides quality training to mission specialists, logistics support and technology publications support. Provides integration of SAR and digital imaging data. Has refined interlocked system for connecting digitised images to a common format for optical data comparison. Essex has patented the Virtual OC-768 for long-haul fibre optic networks, a system to increase CDMA availability and a SAR processor operating at 1.6 teraflops per second. Essex reported revenues of US$2.642 million for 2001, an increase of 71 per cent over 2000, revenues of US$4.506 million for 2002 and a revenue of US$11.22 million for the first nine months of 2003.

Fiberite, Inc

Current Status
Fiberite specialises in glass and graphite fibre combined with epoxy and cyanate resins for the aerospace industry, including the Shuttle Orbiter's main doors and OMS pods, the Hubble Space Telescope's main support structure, and rocket-motor exit cones (including IUS, Ariane and Shuttle). Fiberite has designed special large diameter furlable antenna for satellite applications. It also developed unique compacted antenna structures with low weight and very high folded density. It has marketed three products to the airliner industry and is a member of the Brakenet network for automotive applications.

FLIR Systems

Current Status
Founded in 1978, the company was known as Inframetrics before changing its name to FLIR Systems, specialising in infra-red camera technology and measurement systems. The company's radiometer flew aboard Shuttle STS-50 in June 1992 to map oil surface temperature flows as part of a microgravity experiment. IR equipment has also been provided for ground truth data acquisition using technology built upon its space product range.

Furon

Current Status
Advanced composite components for aerospace and defence applications. Furon has provided components for classified satellite programmes and insulation material for the Atlas launch vehicle and its derivatives. Capabilities include twin sheet and pressure forming of thermoplastic materials, compression moulding, oven and autoclave cure of structural composites, and mechanical assembly. Furon has pioneered new methods of fabricating high-strength fibre structures for high-temperature and high-stress environments. Furon has developed special composite material for high-stress structures designed for large antenna requirements. Also new and innovative materials for classified orbit missions of long duration and formed composites in a hybrid matrix for special requirements.

G&H Technology, Inc

Current Status
Non-explosive actuators for satellite and space probe applications, including solar array, antenna, cable tension and connector releases and experiment jettison. Separation devices and pyrotechnic ignites. Developed new latch-actuated position fixing poles and connectors for high-*g* structures. Has provided coupled end-effector connectors for high mass structures and has the capacity to tailor actuator charge shaping for a wide range of customer requirements to individually tailored specifications.

General Research Corporation

Current Status
GRC specialises in decision support and information systems, including space systems and operation support, experiment design, sensor performance modelling/analysis, cost engineering and estimation, systems engineering, advanced database technology, and prototyping and evaluation of distributed real-time simulations. GRC developed the Integrated Technology Assessment System, a regional database of international spacecraft components. Variations are used by the Space Station programme and

the National Facilities Study. Active in heuristic methods of artificial intelligence, information storage and retrieval and software performance analysis. GRC have conducted research analyses on telerobotic and telepresence systems ntegrated with man-tended and human habitation systems.

Gulton Data Systems

Current Status
Acquired by Goodrich in 1997, Gulton provides spacecraft and launch-vehicle command, processing and telemetry systems. Recent products include control systems for Space Station. Telemetry system designed for Delta 2. Comprehensive facilities include design, manufacture and environmental testing for individual components or full-up systems. Revising assembly configuration models for ISS module schedule based on deletion of the habitation module. Performs optional contingency paths relating to mass distribution and alternative assembly and configuration control paths related to independent ISS and with various docked configurations.

Hamilton Sundstrand Space Systems International, Inc (HSSSI)

Current Status
HSSSI is a subsidiary of United Technologies Corporation. HSSSI's space interests focus on environmental control, thermal control and life support systems. Japan's Sumitomo Corporation acquired a 20 per cent interest in July 1994. Hamilton Standard provided Apollo's lunar suit life-support system and the Lunar Module's environmental and thermal control system; it continues to manufacture environmental control systems for the Space Shuttle Orbiter and provide the Shuttle EMU space suit. The current EMU contract calls for strengthening the design to extend its life, reduce on-orbit maintenance and improve its capability for Space Station assembly. SSI and Russian suit maker Zvezda are collaborating on development of common components between the EMU and Russia's Orlan suit for Space Station. HSSSI provides on-orbit replaceable units for EMU including the displays and controls module, the primary life support system, the secondary oxygen pack and enhancements such as the heated gloves, improved hard upper torso, EMU battery, arms and legs.

SSI provides seven major Orbiter systems: Regenerable CO_2 Removal System; Atmosphere Revitalisation System (provides shirtsleeve environment, avionics bays cooling air and a water coolant loop to collect and transport cabin heat to the external Freon Coolant Loop); FCL (collects heat from the ARS and other equipment in the Orbiter mid-body/aft areas and delivers it to one of three heat rejection devices); Flash Evaporator System (rejects all FCL heat during ascent/entry when Orbiter approximately 30 km; supplements radiators on-orbit and rejects excess water collected onboard); Water Spray Boiler (cools the hydraulic fluid and lubricating oil during launch/descent); Ambient Temperature Catalytic Oxidiser (converts CO to CO_2); Waste Collector Subsystem (collects/stores crew solid/liquid body wastes).

For Space Station, SSI is providing: temperature- and humidity-control systems to circulate air between the modules, remove airborne particulates and cool/dehumidify cabin air; similar components also circulate air to cool avionics and other heat loads; waste water processor, storing it for drinking; active thermal control systems, including pumps, heat exchangers and cold plates for the external loop (as part of Rocketdyne's team). Elements are being provided for the modules of Japan, ESA and Italy.

SSI's Hamilton Standard Management Services Inc subsidiary in Houston provides engineering support at NASA Johnson for its EMU and Shuttle hardware, and furnishes EVA tools and training aids. A separate Space and Sea Systems Group focuses on military and commercial applications of proton exchange membrane SPE electrolysers, undersea naval auxiliary equipment, and aircraft cabin oxygen generation and humidification equipment.

HSSSI has created more than 20 different portable life support system designs, 32 EVA space suit designs and 160 space suit assemblies providing more than 34,000 h of manned pressurised service. HSSSI employs approximately 850 people.

Harris Corporation

Current Status
Harris Corporation is an international communications equipment company focused on providing product, system and service solutions for commercial and government customers. The company's five operating divisions serve markets for microwave, broadcast, network support, rf communications and government systems. Harris has sales and service facilities in more than 90 countries.

Harris GCSD conducts advanced research studies, develops prototypes, and produces and supports state-of-the-art, highly reliable communications and information systems that solve the mission-critical communications challenges of its military and government customers, as well as provides the technology base for the company's diverse commercial businesses.

Harris has more than 10,000 employees including 4,000 scientists and engineers in 150 countries. Reported fiscal year 2001 sales of US$1.95 billion.

Hernandez Engineering Inc

Current Status
HEI provides industrial and occupational safety and specialises in manned space engineering and technical services, including mission operations, payload integration, safety, reliability, quality assurance, technical information and training analysis/implementation. HEI has conducted engineering compatibility studies on integrating mission elements manufactured by separate countries using the ISS as a baseline. Extended engineering studies on human habitat systems and long duration planetary habitation modules. Has performed simulated mission management operations for long duration human space flight and has simulated deep space planetary missions. Contracts are held with NASA Goddard, Johnson and Kennedy, DLR, ESTEC and Alenia. Regular contact with ESA centres.

Honeywell International Inc

Current Status
The present company was formed in late 1999 with the merger of AlliedSignal and Honeywell Inc, but it can trace its origins to 1885. Aerospace Solutions and services has sales exceeding US$10.5 billion and employs 49,000 people at 20 sites. Honeywell International Inc serves the space industry through Honeywell Aerospace and Honeywell Specialty Materials.

Under Aerospace Services, the space related activities of the Defense and Space division concentrate on: inertial systems; onboard processors; altitude and control; control-moment gyroscopes; reaction wheel and momentum wheel assemblies; communications/data handling; and isolation/pointing. This includes the development and spacecraft application of fibre optic rotation sensors to supplement the presently manufactured ring laser and dry tuned gyro inertial measurement units. Integrated satellite attitude-control systems and their components are also available, including star trackers, reaction wheels and control-moment gyroscopic effectors, phased array momentum control systems and their operational software.

Data-management systems, in selectable configurations, employing fibre-optic networks, flat-panel displays and signal processors, are ready for spacecraft incorporation and are being provided for Space Station. Integrated components are also available to include optical mass storage, upgraded Shuttle standard multiplexers/demultiplexers, network and bus interface units and onboard data processors.

Honeywell, under contract to the USAF Phillips Laboratory, developed the GVSC Generic VHSIC Spaceborne Computer, which has a high-performance general purpose 1750A CPU, a local-memory controller and 256 + 8 static RAM. A 32-bit processor and single-board computer is in production and has been ordered for various military and commercial space platforms. All elements are radiation hardened. HII produces radiation-hardened computers for meteorological satellite imaging sensor and inertial navigation units for Atlas 5.

Advanced payload pointing systems with magnetic and passive isolation have been developed to attain sub-arcsecond accuracy with low bandwidth isolation. To support a wide range of satellite pointing applications, high-gain modular antenna pointing systems, solar array drives with integrated power transfer, gimbal actuators and spin/de-spin payload control systems are being manufactured. On Shuttle, Honeywell provides the main engine controllers, hand controllers, flight control system, radar altimeter and autoland software. It is subcontracted (US$70 million) by Rockwell for the Crew Multifunction Electronic Display System (MEDS 'glass cockpit') to replace the Orbiter's old cockpit displays; this was completed in 1999 and is installed in all orbiters on a rotation basis.

HII provides range safety tracking systems for the Kodiak launch complex and the company has developed a new range of optical and radar tracking instruments for fast accelerating rockets and missiles.

Hughes STX Corporation

Current Status
Hughes STX Corporation (HSTX) is a wholly owned subsidiary of Hughes Aircraft Company and operates as part of Hughes Technical Services Company. Its primary focus is science, engineering and software support services in: astronomy, Earth and atmospheric sciences; science data management, satellite ground data systems; aerospace engineering and computer networking. For example, HSTX developed the current Landsat ground processing facility for EOSAT. Also, improved multisatellite data compression fixtures, including hardware and software, for remote sensing experimenters. Until 1995 it was co-owner of Starsys Global Positioning Corporation and has now been absorbed by Boeing.

ILC Dover, Inc (ILC)

Current Status
Formed in 1947, ILC specialises in the development of high-technology soft goods materials and the design of inflatable structures for unique applications. It has designed and manufactured space suits for Apollo, Skylab and Shuttle, and has developed new and more flexible suits for International Space Station. ILC also conducts experimental research programmes on integrated materials combinations for advanced suits offering greater dexterity and long life. It has also provided Mars Pathfinder's lander airbag system and a similar system for the two Mars landers launched in the 2002 launch window. ILC also provides inflatable structures for space and lightweight gossamer structures such as solar cell arrays. Also rigidised composite materials and deployment control mechanisms.

Interferometrics Inc

Current Status
The company was founded in 1982 for research, scientific investigation, technical analysis and the design/construction of hardware for NASA, DoD, the radio-astronomy community and commercial space telecommunications companies. Conducted

pioneering development work which was applied to a wide range of products for various US government agencies. More than half of the engineering staff have advanced degrees and 83 per cent of the science staff have PhDs. From June 1988 to August 1989, it provided engineering support to the Radio Amateur Satellite Corp (Amsat) for the development of a micro-satellite in return for the commercial rights to this 'Eyesat' technology. The first four were launched in January 1990 aboard Ariane. The fifth, the 10.5 kg Eyesat 1, was launched on 25 September 1993 with France's Spot 3 into a 791 × 806 km, 98.6° orbit to demonstrate worldwide commercial digital store/forward communications. This was the first commercially licensed low earth orbit satellite in the US. The company is funding the project itself, controlling the satellite from its Chantilly facility.

The Eyesat design is configured as an expandable modular stack and can support a range of LEO missions, from store/forward to specialised science applications, and be launched as a secondary payload on most vehicles. The specially developed general purpose electronic bus interfaces with the command and mission payload equipment, controlled by a 16-bit microprocessor with 18 Mbyte of program/data storage memory. Advanced solar cells and load-side power management optimises power usage. The RF transmission system allows the use of discrete output levels under computer control. Conducted development work on small to medium satellites and has developed some nanosat design concepts.

In 1999, Interferometrics Inc formed Transmitter Location Systems, the only company in the world known to be working exclusively on solving satellite interference problems. TLS has produced the Model 2000 receiving station. Customers include PanAmSat, Intelsat, Eutelsat, Loral Skynet, Telesat, NOAA and SES.

Interferometrics provides information on the spectral irradiance of the Sun as well as imaging the solar disc.

Specifications

Configuration: 23 cm cube carrying no deployables. Each major subsystem is housed in one of five aluminium-frame modules bolted together
Power: 2 x 4 cm Si cells on each face provide 5–7 W orbit average, supported by 6.5 Ah Ni/Cd battery
AOCS: passive magnetic stabilisation with solar spin; magnets maintain alignment with local geomagnetic field
Computer: CMOS V-53 microprocessor, 16 Mbyte RAM data memory, 2 Mbyte RAM EDAC program memory
Communications: primary commercial payload comprises two 300–9,600 baud UHF transmitters, output of each controllable to over 4 W; omni 4-element turnstile antenna. 5-channel VHF receiver, 300–9,600 baud; omni whip antenna. Frequencies not available.
Experimental amateur payload: one 300–9,600 baud UHF transmit (436.7975 MHz), output controllable to 0.5 or 1.5 W; omni whip antenna. 2-channel VHF rx (145.85 MHz), 300–9,600 baud; omni whip antenna. It may be configured for a single analogue channel.

International Space Brokers (ISB)

Current Status

ISB are insurance brokers specialising in all aspects of prelaunch, launch, liability and in-orbit coverage, in association with International Space Brokers Ltd (UK) and European Space Brokers (France). The brokerage house is one of the oldest in the business of space insurance and has played a prominent role in shaping the development of this as a business.

International Technology Underwriters (Intech)

Current Status

Underwrites insurance with manufacturers, launch vehicle providers and operators, satellite owners and operators. Policies indemnify against physical loss or damage to the vehicle and satellite during launch and satellite operations. Coverage can apply to: launch-vehicle performance, satellite initial operations, launch-vehicle and satellite initial operations and satellite on-orbit operations as well as total life cover and insurance against loss of revenue. Intech began space coverage in the 1980s and was one of the leading underwriters in developing space insurance for commercial satellites during the early years of the shuttle and the Ariane launch vehicle programme. Intech is instrumental in advising government, corporate and private sectors on risk assessment and management.

Intraspace Corporation

Current Status

Intraspace, founded in 1986, has developed designs for low cost commercial communication satellites known as the T-SAT (the 'T' borrowed from the analogy with the low cost model T Ford automobile). Intraspace has facilities located at North Salt Lake City, Utah. No orders have yet been received. Intraspace fabricates composite metal parts under the name Dunmon. It has proposed a new manufacturing method for lightweight satellites and spacecraft.

ISSI

Current Status

ISSI (previously International Space Systems, Inc) is a space technology R&D company, providing support in mission planning, mission operations and systems analysis. Project participation covers Spacelab, Shuttle, Galileo and Israel's Ofeq satellite. ISSI has conducted mission operations analysis for planetary spacecraft and Mars lander options and has performed scenario simulations for alternate Mars exploration strategies. Software packages and services include: Mission Planning and Orbit Definition and Design (mission planning; orbital analysis and requirements definition; orbit lifetime forecasting and impact predictions; launch-vehicle performance and dispersion analysis), Mission Control and Operations (mission control centre design and requirements definition; flight/mission operations; expert systems software for orbital analysis/control; expert systems software for timeline generation; simulation/analysis software), Satellite System/Subsystem Design and Analysis (payload systems design and tradeoff analysis; systems engineering; satellite systems design/evaluation; satellite guidance/control design/analysis; electrical power system design/analysis; communications balance analysis; simulation/analysis software), and Integration and Testing Equipment and Support (systems integration support/testing; automated test equipment). Active in NASA's Intelligence Synthesis Environment/Collaborative Engineering Environment (ISE/CEE) through an internal product called RECIPE, advanced launch vehicle analysis and nuclear air-breathing rockets concepts. Has structured analytical software programmes for integrating competing design configuration analysis for advanced propulsion and aerothermal, aerodynamic and space based systems.

ITT Industries Inc

Current Status

Specialised in fluids technology, ITT Industries gets more than half its sales from pumps, mixers, shock-absorbers and friction pads. Also focused on defence electronics. The Aerospace and Communications Division provides voice and data communications, voice recognition, and meteorological and navigation spaceborne payloads. Also provides communications and telemetry sets for hardware and software configurations. It is responsible for the payload of GPS Navstar Block 2R under contract to Lockheed Martin Astro Space; the Imaging Radiometer and Vertical Sounder instruments of GOES 8 to 12 for Space Systems/Loral; AVHRR radiometers and HIRS sounder for NOAA polar as well as DMSP satellite designs for future applications.

Johnson & Higgins

Current Status

Established in 1981 to provide launch and transponder insurance packages and risk management expertise. Co-brokerages in London, Paris and Tokyo. Established an office in Malaysia in 1982 and has since expanded representation in every major developed country. Covers all aspects of space related insurance including, launch, life and revenue streams.

Kaiser Electro-Optics, Inc (KEO)

Current Status

Set up as the optical department for Kaiser Electronics in 1983, KEO specialises in refractive and reflective optical and electro-optical assemblies for space, surface, airborne and underwater military and commercial applications. Expertise covers production, assembly and testing of spherical, flat, cylindrical and aspheric optics and systems for visible to infra-red wavelengths. Electro-optical systems for high revolution imaging cameras and CCD elements in visible and and non-visible portions of the spectrum. The company delivered in 1984 the first of 12 Cassegrain telescopes for DMSP's Operational Linescan System imagers. The 20 cm f/1.0 primary and hyperboloid secondary nickel-coated beryllium mirrors were mounted in a beryllium structure. Eight 15 cm aperture telescopes (f/0.5 parabola primary and hyperbolic secondary) have been delivered for NOAA's High Resolution Infra-Red Sounder, in addition to 20 cm Cassegrain systems for NOAA's Advanced Very High Resolution Radiometer. Has also produced the High Resolution Infrared Radiation Sounder for NOAA, the Block 5 Telescope for the DMSP programme, the NASA Lightning Imaging Sensor, NASA Lightning Mapper Sensor, the NPOESS Visible/Infrared Imager/Radiometer Suit, the JPL MUSES-CN Nanorover Lens and the 92 mm and 200 mm StarTracker lenses. Acquired by Rockwell Collins in November 2000.

Kaman Aerospace

Current Status

Kaman Aerospace is the largest subsidiary of Kaman Corporation. Kaman began as a helicopter manufacturer and its Aerospace division was the historic core of the company which now specialises in aerostructures, aircraft bearings, fuzing systems and electro-optical systems. It developed the US$86 million Wavefront Control Experiment for SDI's Starlab mission, designed to demonstrate image-distortion compensation techniques employing a flexible mirror (the mission was cancelled). It has developed the 'Pamela' segmented-optics mirror system for DARPA, designed/developed a tracking/pointing system for Lawrence Livermore National Laboratory's KESTREL programme, and completed design efforts on NASA's Space Debris Warning Sensor system. EODC is participating in USAF Philips Laboratory's High Altitude Balloon Experiment (HABE), incorporating a precision pointing/tracking system for detecting/tracking ICBMs. EODC has developed several versions of the Magic Lantern lidar for detecting, classifying and locating floating/submerged objects for naval defence.

Kaman reported revenue of US$877 million for 2001 with earnings of US$11.7 million.

Kaman Sciences Corporation

Current Status

A wholly owned subsidiary of Kaman Corporation, Kaman Sciences Corporation was sold to ITT Industries in 1997. It provides services and products in information technology, consulting and technical assistance, military system effects and analysis, and hardware design, test and evaluation. Specific expertise includes software development, operations and maintenance; surveillance and intelligence systems O&M; modelling, simulation and virtual systems development; information network security; information analysis centres; facilities management; astrodynamics and space mission analysis; open source information analysis; EME assistance; advanced weapon system performance assessments and threat modelling; systems testing; photonics and smart sensor systems. Coupled sensor development for automated data transfer.

Kaman Sciences is the prime contractor for Cheyenne Mountain Complex software support, warning systems software support and NORAD/US Space Command/AF Command C^4 support services programmes. It has established the Space Environment and Orbital Debris Resource Center to promote the study of space environmental issues. The Center includes 15 million satellite orbital element data sets dating from the early 1960s. Kaman Sciences also develops specialised software such as the commercial Kaman Satellite Analysis Tool (KSAT), a high-fidelity 3-D animated visual modelling tool for the analysis of orbiting bodies. Interactive software set for three-dimensional modelling of high-revolution images.

Kamatics Corporation

Current Status
The company is a wholly owned subsidiary of Kaman Corporation. Founded in 1996 to apply proprietary technology for solving complex operational problems, Kamatics specialises in high-performance mechanical devices. Also provides rapid response fluid and gas control units systemised into an integrated control network. It develops and manufactures self-lubricated slide bearings used on a range of satellites, spacecraft and launch vehicles, specifically bearing and power take-off assemblies.

KKI

Current Status
Founded in 1984, KKI develops satellite and in orbit operations simulation software: Orion for Windows, Orbit II Plus for Macintosh and PCOrbit for DOS. It supports space-based wide area surveillance, mission planning, visibility studies, communications studies, constellation design, ASAT or threat studies, and production support. Software simulator packages for integrated communications tracking and security services. KKI has evolved an NMD applicable software for adapting ballistic and warhead targeting to anti-satellite requirement and this has resulted in several applications being adapted for operator training activities.

KPMG

Current Status
Established in 1867, KPMG has provided support to the space industry since 1966 and in 1986 formalised a specialised Commercial Space Group. KPMG was one of the founding instruments of finance and insurance for the nascent commercial space industry during the late 1960s. It provides financial and business advisory services to governments and aerospace companies related to telecommunications, space transportation/infrastructure, space-related research and services. Capabilities cover strategic planning, market research, international joint ventures, technology commercialisation, productivity improvements and cost analysis. The company operates 672 offices in 135 countries. Total revenues reported at £1,018 million for 2002 versus £1,023 million for 2001.

L-3 Communications Analysis Corporation

Current Status
A wholly owned subsidiary of L-3 Communications. Formerly known as Space Applications Corporation, L-3 Communications Analytics Corporation specialises in systems engineering, spacecraft command/control, space-qualified hardware and mission planning tools. It developed the station acquisition, orbit/attitude control, command generation and status telemetry monitoring software for GPS and CRRES. It provided experiment control software for the ultra-violet plume instrument flying on SDI's LACE satellite and the Drop Physics Module for Shuttle's US Microgravity Laboratory. The company is also providing systems engineering and technical assistance to the AF Satellite Control Network. SAC has conducted space warfare studies. Also conducting studies on integrated NMD/ASAT applications. Employs more than 400 people. Reported annual revenue of approximately US$70 million.

Launchspace, Inc

Current Status
Launchspace provides courses and seminars for the professional space community and a range of magazine products. During 1999–2001, Launchspace extended electronic information services to web-based subscription services. Launchspace Training, a division of Launchspace Inc, offers public and client-site professional development and continuing education programmes. Courses embrace science, engineering, operations, management and planning. Courses also include conceptual strategies, business planning, financial management of space-related companies and organisations and how to prepare proposals and bids.

L'Garde, Inc

Current Status
Founded in 1971, L'Garde specialises in lightweight and gossamer space structures. It produces a variety of space targets and RV decoys and simulated warheads. It built the inflatable parabolic antenna (diameter: 14 m) flown as a Spartan payload on Shuttle STS-77 in May 1996. It produced the world's first inflatably deployed and supported photovoltaic array, and is developing technology for power densities of 150 W/kg for more than 1 kW. Has manufactured more than 150 structures launched into orbit. Has developed designs for very large ridged deployable solar arrays and micro satellite array systems where weight allocation is at a premium. L'Garde has many specialised design concepts for undisclosed customers in the United States which provide unique support for telecommunications, data retrieval and power conversion through pv cells.

Lockheed Martin

Current Status
The world's two largest space companies, Lockheed and Martin Marietta, formally merged on 15 March 1995. Combined space sales in 1993 were US$6.7 billion, with 33,000 space employees. The companies' overall 1994 sales totalled US$22.9 billion, with 170,000 employees. In 1998, sales were US$26.26 billion compared with US$28.07 billion for 1997. Excluding the sale of divested businesses, 1997 net sales would have been US$26.14 billion. At the end of 1998, sales backlog was US$45.3 billion. Lockheed Martin had 1999 sales exceeding US$25 billion and employs approximately 140,000 people. LMC sales for 2000 exceeded US$25 billion and the company had 130,000 employees. The former Lockheed Missiles and Space Company, Martin Marietta Astronautics now come under the Space Systems Company; Lockheed Engineering and Sciences Company, Martin Marietta Space Systems and Lockheed Space Operations Company under the Information and Technology Services Sector; and Sanders under the Electronics Sector. Two of the four satellite plants will eventually be closed: most of the East Windsor, New Jersey and Valley Forge, Pennsylvania capabilities are transferring to Sunnyvale, California, although existing commercial, Landsat 7 and EOS AM 1 contracts were completed there. Denver, Colorado, will retain its military and interplanetary projects. All space launcher activities are being consolidated in Denver at Lockheed Martin Astronautics. Space Systems Company now incorporates Lockheed Martin International Launch Services Space Imaging Inc, and United Space Alliance.

Lockheed Martin in April 1996 acquired most of Loral Corporation for US$9.1 billion and took a 20 per cent interest in the remaining Loral Space and Communications Ltd. Total annual sales are expected to be US$30 billion. The former Loral companies are initially grouped into a sixth sector, Tactical Systems.

At the time of the Lockheed-Martin merger, Martin Marietta's Space Group comprised Astronautics, Manned Space Systems and Astro Space. The Martin Marietta Corporation was established in 1961 by the merger of the Glenn L Martin Company and the America-Marietta Corporation. General Electric's aerospace businesses were combined in April 1993 with Martin Marietta for US$3.05 billion, establishing the world's largest aerospace electronics company, employing 97,000. GE Astro Space Division became Martin Marietta Astro Space.

Lockheed Martin reported operating revenues of US$23.99 billion for 2001 (down 2 per cent on 2000) and a profit loss of US$1.046 billion (up more than 100 per cent on 2000). Employs 126,000 people.

Lockheed Martin Space Operations

Current Status
LMSO and contractors employ 7,800 people and comprises several divisions handling: the Consolidated Space Operations contract; Science, Engineering, Analysis and Test (SEAT); Enterprise Solutions; Integrated Technology Solutions; and Stennis Programs.

The former Loral Corporation elements acquired in 1996 are initially grouped under Lockheed Martin's sixth business sector of Tactical Systems, comprising 30 operating units and facilities employing 35,000. 1995 sales totalled US$6.2 billion. Loral in 1992 acquired the missile division of LTV Aerospace & Defense, builder of the Scout orbital launcher, in a deal worth US$260 million. IBM Federal Systems was acquired in March 1994 for US$1.5 billion. Loral was responsible for building NASA's new Mission Control Center in Houston for Space Station, in addition to modernising the existing Shuttle facility. On Space Station, it is handling systems engineering and integration for propulsion, communications, tracking and information, plus the integration/validation of the flight software and attached payload integration. Ford (then Loral, now Lockheed Martin) has been prime contractor since 1983 for development of a centralised C^3I capability for NORAD's Space Defense Operations Center. It has also provided approximately 200 large satellite ground terminals around the world. The company operated the USAF Satellite Control Network from the 1950s until September 1995 (when Lockheed Martin took over the new contract). The former Loral AeroSys, now Lockheed Martin Space Mission Systems, provides satellite ground systems and operations, including control centres for Hubble, NOAA and commercial Japanese satellites. Expertise covers information processing and data storage systems, integration and engineering services for space/ground support systems, and space segment engineering and operations, including launch, on-orbit support and payload support.

Lockheed Martin Space Systems Company

Current Status
Lockheed Martin Space Systems Company is responsible for the design, production and integration of launch vehicles and systems, spacecraft for telecommunications, remote sensing and space science and for missile systems for defence and strategic missions. It also comprises three joint ventures: United Space Alliance, Space Imaging and International Launch Services.

Background
Comprises Astronautics Operations, Management and Data Systems, Michoud Operations, and

Missile and Space Operations. Commercial Space Systems designs and manufactures a variety of commercial satellites including the A2100 modular bus design. Joint ventures include United Space Alliance, Space Imaging and International Launch Services.

Lockheed Martin Space Systems Company is responsible for satellite telecommunications and navigation, remote sensing and space science, defence systems and strategic systems. Formerly, Lockheed Missiles & Space Company became the Lockheed Martin Missiles & Space Company in March 1995, part of Lockheed Martin's Space & Strategic Missiles sector. LMSC's Astronautics division was folded into the Space Systems and Missile Systems divisions at the end of 1990. LMSC was systems integrator and main contractor for NASA's Hubble Space Telescope, responsible to NASA for the structure and support; it designed the shroud that encloses the telescope assembly and the scientific instruments. It is building the Iridium satellites for the LEO system. It developed and manufactures the silica fibre insulation tiles for the fleet of Shuttle Orbiters. A range of solar-array designs has been produced for NASA, including an experimental 32 m panel tested aboard Shuttle.

Astronautics Division has designed, built and flown more than 25 space science payloads, including three on Solar Max, particle spectrometers on CRRES, a soft x-ray telescope on Japan's Solar-A and two major instruments on UARS. For IASTP, the Research and Development Division was involved in Soho and Polar satellite instruments. It is the programme integrator for, and is building, the TRACE Small Explorer and is providing the cryogenic cooler for the WIRE Small Explorer.

In military work, Lockheed has been responsible for more satellites than any other company in the West. For example, it is prime contractor for the Milstar satellite to provide highly jam-resistant strategic and tactical communications. The company is a major BMDO/SDI contractor; it was awarded the US$350 million to US$400 million development contract in November 1985 for the Exoatmospheric Re-entry Interceptor System. It is competing for the follow-on early warning system to replace DSP. LMSC was also responsible for the US Air Force's Agena, the 'most-flown' western upper stage. LMMSC comprises Lockheed Martin Astronautics (including International Launch Services, a joint venture with Khrunichev Energia International), Lockheed Martin Michoud Space Systems, Lockheed Martin Missiles and Space (incorporating Communications and Power, Advanced Technology Center and the Commercial Satellite Center), Space Imaging Inc and United Space Alliance (a joint venture with Boeing).

Lockheed Martin Space Systems Company – Astronautics Operations

Current Status

Astronautics Operations is one of the largest operating elements of Lockheed Martin's Space Systems Company and is headquartered at the Deer Creek facility about 20 miles from Denver, Colorado. Originating with the Glenn L. Martin Co. missile and space activities began in the mid-1950s with development of the Titan ICBM and associated launch facilities. It designs, develops, tests and manufactures a variety of advanced technology systems for space and defence among which are planetary spacecraft and other space systems, space launch systems and ground systems. The facility houses some of the most sophisticated production, space simulation and test facilities in the world, including 70 modern laboratories as well as computerised engineering design, manufacturing and administrative buildings. Astronautics also operates the Titan, Atlas and Athena launch vehicle programmes at both Vandenberg Air Force Base, California, and Cape Canaveral, Florida. In January 1999, it employed almost 10,000 people (7,100 at Deer Creek) and had an order backlog of US$8.8 billion. Astronautics is involved in two major core businesses: Space Transportation and Space Systems. Astronautics Operations also has facilities at Harlingen, Texas, Huntsville Alabama, Vardenberg Air Force Base, California, and Cape Canaveral Air Force Station, Florida.

Space Transportation includes Space Launch Systems which is responsible for the Titan II and Titan IV, the Athena I and II, the MultiService Launch System (MSLS) and the Target Delivery System. Advanced Launch Systems is responsible for Atlas and Centaur launchers including development of the Atlas V in co-operation with the USAF Evolved Expendable Launch Vehicle (EELV) programme and Reusable Launch Systems which includes the Crew Return Vehicle (CRV) for use at the International Space Station as an emergency lifeboat.

Space Systems includes Flight Systems, which is responsible for the Mars Global Surveyor spacecraft, the first of a new generation of robot explorers in the Mars Surveyor programme. This also includes Mars Climate Orbiter and Mars Polar Lander which were launched in December 1998 and January 1999 respectively. The company is also building two spacecraft, the Mars Odyssey spacecraft sent to Mars in 2001 and a lander for launch in 2003 respectively, and is studying designs and building one lander for two Surveyor programme missions in 2005. Flight Systems also built the propulsion module for NASA's Cassini spacecraft launched to Saturn in October 1997 for a rendezvous with the planet in June 2004. It is also an industrial partner on Stardust which was launched in February 1999 with the objective of collecting, and returning to Earth in January 2006, sample material from the comet Wild 2. Flight Systems is an industrial partner with JPL in the Genesis programme and after launch in January 2001 gathered substantial quantities of charged particles from the solar wind for return to earth in August 2003 with which scientists will sample primordial material released from the Sun. Flight Systems is also a partner on teams selected by NASA in 1998 to study Aladdin, a spacecraft to sample the Mars moons Phobos and Deimos, and VESPER, which would study weather and climate patterns in the atmosphere on Venus.

Special Programs and Defense Systems includes various classified space and ground support systems.

Loral Space & Communications Ltd

Current Status

On 15 July 2003 Loral Space & Communications announced that it had filed for reorganisation under Chapter 11 of US Bankruptcy law. It also announced that it had reached a definitive agreement to sell its six North American telecommunications satellites to Intelsat Ltd for up to US$1.1 billion in cash. Loral expected to use US$959 million of this to pay off its outstanding secured bank debt. Satellites sold included Telstars 4, 5, 6 and 7 currently in orbit and Telstars 13 and 18 which were scheduled for launch later in 2003 and early in 2004, respectively. The Loral fleet retained included Telsar 10, 11 and 12 in orbit and Telstar 18/Apstar V and Telstar 14/Estrela do Sul to be launched. Loral would continue to own and operate Space Systems/Loral. On 22 September 2003 Loral Skynet, a subsidiary of LS&C declared the Telstar 4 satellite a total loss following a short circuit in the primary power bus three days earlier. Nearly all Telstar 4 customers were shifted to Telsar 5, 6 and 7 satellites. Telstar 4 was insured for US$141 million and under the agreement to sell six satellites to Intelsat Ltd the purchase price would be adjusted less the receipt of the insured sum. At 89° east, Telstar 4 would be replaced by Telstar 8 scheduled for launch in mid-2004 and carrying a Ka-band payload in addition to C- and Ku-band. Telstar 4 was launched in 1995. Registered as a Bermuda Company, Loral Space & Communications is organised in three operating business segments: satellite manufacturing and technology (Space Systems/Loral); fixed satellite services (Loral Skynet, Loral CyberStar, Loral Skynet do Brazil ltda, Satmex, Europe*Star Limited) through the Loral Global Alliance; Data Services providing managed communications networks and Internet and Intranet services through Loral CyberStar. In addition a subsidiary of Loral acts as the general manager of Globalstar of which Loral owns 39 per cent. Loral Space & Communications, with headquarters in New York City, was 'spun off' from Loral Corporation in April 1996 as part of the sale of its defence electronics and systems integration businesses to Lockheed Martin Corporation.

On a pro forma basis, combining the results of the operations under its management, the company's revenue for what was a nine-month fiscal year, ending 31 December 1996, was US$950 million. Effective 1 January 1997 Loral reported on a 12-month calendar year basis. Combined revenues for 2000 were US$1.224 billion compared with US$1.457 billion in 1999 and US$1.301 billion in 1998.

Space Systems/Loral (SS/L), a wholly owned subsidiary of Loral, is one of the world's leading manufacturers of space systems. Globalstar, a system of low-earth-orbiting (LEO) satellites supports a digital telephone service to hand-held and fixed terminals worldwide. Loral and its partners act as Globalstar service providers in Brazil, Canada and Mexico, and with Qualcomm, will hold the exclusive rights to provide an in-flight phone service using Globalstar in the United States.

Loral completed its acquisition of Skynet, a leading domestic satellite services provider, from AT&T in March 1997. Loral Skynet advances Loral's strategy of becoming a global provider of satellite-based communications services, and will complement its existing satellite manufacturing capabilities.

Loral is pursuing additional satellite-based communications service opportunities, including CyberStar, designed to provide interactive, broadband, multimedia data transmission. Loral is also developing a location information service that is designed to improve the accuracy of GPS receivers, and is pursuing opportunities in partnership with others to offer domestic and international direct-to-home services.

In March 1998, Loral completed the acquisition of Orion Network Systems, an Internet access, corporate data networking and satellite services company. Loral and its partner Telefonica Autrey, S.A. de C.V. own a 75 per cent share in Satélites Mexicanos, S.A. de C.V. (SatMex) consisting of three operating satellites: Morelos II, Solidaridad 1 and 2 and SatMex 5 soon to be launched. In March 1999, Loral purchased Satmex.

By 2002, Loral had in place a robust communications network of more than 65 in-orbit geostationary and low-earth orbit satellites providing mobile and fixed telephone services, Internet access, high-speed video and data transmissions, direct-to-home broadcasting and broadband on demand. These plans have been suspended.

Loral and its corporate predecessors have built, or been awarded contracts for, more than 220 satellites, of which 31 have been for Intelsat. Loral reported operating revenues of US$1.07 billion for 2001, down 13 per cent on 2000, and employs 3,000 people.

Lord Mechanical Products

Current Status

Isolators, couplings and elastomeric bearings for mounting vibration, noise and shock-sensitive equipment. Ground-based vibration platforms and rigid grounding test rigs for dynamic and high to low frequency oscillations incorporating resonance sensors with tests programmable through automated or selectable programming. Also supplies solid state electronics equipment for ground, air and space applications.

Malin Space Science Systems (MSSS)

Current Status

Company founder, Dr Malin, was principal investigator for the Mars Observer Camera, lost in August 1993's failure, and is principal investigator for the replacement camera on 1996's Mars Global Surveyor, and on the new cameras for 1998's Mars Surveyor orbiter and lander. MSSS is involved in developing new cameras for future Mars missions, and in developing data processing and archiving technologies. MSSS supported operations of the Mars Global Surveyor camera, the Mars 98 Orbiter colour imager, the Mars 98 landed descent imager and MSSS built the Visible Imager for the THEMIS instrument on the Mars 2001 Odyssey Orbiter. MSSS is a principal investigator on the Mars 2003 lander/rover missions involving spacecraft named Spirit and Opportunity and have conducted extensive processing, analysis and interpretation for MGS, Mars Odyssey and the two rovers. It is also designing the optical equipment for the Mars orbiter due for launch in 2005 and for imaging systems for the landers scheduled for 2007 and 2009.

Marsh & McLennan Space Projects

Current Status
Provides space insurance brokerage services, develops risk management and insurance programmes for risks of pre-launch, launch, in-orbit operations, manufacturers, product exposures and third-party liability.

The parent company Marsh & McLennan reports annual revenues exceeding US$10 billion and employs 59,000 people.

McDonnell Douglas Aerospace

Current Status
McDonnell Douglas Aerospace has been incorporated into the Boeing Company. MDA was Product Group 1 contractor for Space Station, responsible for building the integrated truss, mobile transporter, data-handling systems, crew health-care and monitoring equipment. MDA plays a multiple role with the production of Shuttle hardware, flight training and mission programming and payload integration. This includes building the aft/fore skirts, frustrum and other structures for the solid-fuel rocket boosters, and the aft propulsion system incorporating an orbital manoeuvring system and a reaction control system. MDA also checks and installs all military Shuttle payloads. Spacehab's Shuttle pressurised module was fabricated by the company and it has provided mission support on each flight including the fated STS-107 when orbiter Columbia was destroyed.

McDonnell Douglas built the Delta launch vehicle, with approximately 230 satellites placed in orbit over 35 years. MDA has worked on an SSTO single-stage-to-orbit launcher for BMDO; the experimental DC-X one-third scale version of this Delta Clipper underwent flight testing in 1993.

Meade Instruments Corporation

Current Status
Questar, incorporated in 1951, manufactures space-qualified telescopes, bore-sighted scopes for optical tracking, long-distance microscopes and non-contact measurement systems for laboratory applications, in addition to a range of catadioptric telescopes. All optics are diffraction limited and hand crafted. Now known as the Meade Corporation, the company manufactured the first telescope to fly in space (Gemini, 1965) and is currently involved in several space projects. Founded in 1972, Meade was made public in 1997. Questar has been approached by NASA for light, compact telescopes for hand use aboard the ISS. Meade is active in researching new forms of small collapsible telescopes for space use and has adapted several lightweight elements that would have application for personnel use aboard the ISS.

VERIFIED

Metalmart, International Inc

Current Status
Supplier of speciality metals and alloys to the aerospace industry, including high-temperature materials for thermal insulation of systems aboard the Space Shuttle Orbiter. Metalmart was incorporated in 1974 and now employs more than 60 personnel, with an annual sales volume of approximately US$17 million. Provides aluminium honeycomb, magnesium, non-standard and exotic metals and hybrid products. Specialises in metal products fashioned from billets and prepared to fine tolerances for unique customer requirements. The company is the world's largest stocking distributor of sheet and plate magnesium, currently supplying Boeing, Hughes, Lockheed Martin and other satellite manufacturers.

Microcosm, Inc

Current Status
Established in 1984, Microcosm developed a pressure-fed LOX/kerosene 22.2 kN engine with a claimed manufacturing cost of less than US$5,000 each, excluding the injector. There are only 38 parts, full duration 200 seconds, ablative film-cooled; performance is comparable to solid rockets, although key specifications remain confidential. A 200-second run was successful on 20 November 1995. Assisted by USAF Phillips Laboratory funding (US$3.1 million through September 1996). Microcosm proposed the Scorpius family of vehicles based on the same engine; SR-S sounding rocket (724 kg GLOW, 50 kg to 225 km; US$99,000 recurring cost); SR-1 sounding rocket (400 kg to 190 km; US$295,000); Mini-Lift (100 kg into LEO; US$875,000); Liberty Light Lift (71.7 tonnes GLOW, four stages, seven propulsion pods totalling 49 engines, 1 tonne into LEO; US$1.8 million); Exodus Medium Lift (6,800 kg LEO; US$8.5 million). Non-recurring development cost of the vehicles through Liberty is less than US$30 million (FY96). Development halted due to funding problems and a surfeit of systems in the launch vehicle market. Provides mission operating software, analysis and design together with hybrid engineering solutions. Technology development conducted during the formative evolution of this concept has been fed across into other technology applications and the company is developing new and efficient versions of the propulsion system.

MTS Systems Corporation

Current Status
MTS, founded in 1966, provides aerospace testing needs, including: full scale structural; assembly and substructure; component; materials properties determination; strength/durability of parts; quality control. Analysis of multi-faceted materials test results for composites. Produces long-life motors, customised engineering, software and automation systems and sensors. Supports component and subassembly testing for airframes and propulsion systems and ground vibration testing for damage tolerance design verification, certification tests and air loads calibration testing.

NERAC, Incorporated

Current Status
NERAC is a not-for-profit problem-solving and technology-transfer centre serving US industry since 1966. Has conducted analysis of international ventures and joint ventures involving private companies and foreign governments. NERAC has conducted detailed analyses of international space projects as they pertain to fiscal and programme management, their impact on timescales and the overall scope of the project. Resources include a large staff of technical specialists; technical/scientific resources providing global coverage of historic/current developments; US government technology and procurement opportunities; US and European patents; Reference Plus expert and consultant matching; and document retrieval. Specialisation in analysis of foreign trade stimulated by technical developments. NERAC has also completed detailed analyses of multinational space science projects proposed by small countries unable to achieve appropriate funding allocations and with the desire to spread financial load and risk.

Northrop Grumman Electronic Systems Sector Space Systems

Current Status
Embracesthreetechnologysectors:EO/IR(Electronic Observation/Infra-Red); RF (Radio Frequency); and C[4] ISR Systems.

Westinghouse Electric Corporation's Electronic Systems unit was acquired by Northrop Grumman in 1996 for US$3 billion and renamed the Electronic Sensors and Systems Division. Principal interests include radars and electronics for combat aircraft and battlespace management systems, as well as for military space and underseas systems. Provided electronic equipment for space-based warfighting simulations. A leading supplier of air traffic control radars to the Federal Aviation Administration and to countries in Europe, the Middle East, Africa, Asia and South America. Produces nationwide air defence systems, tactical communication equipment, anti-submarine warfare systems, undersea vehicles, mine countermeasures, marine propulsion, and shipboard instrumentation. Development of systems for space-based radar and TRAM 2, a transmit/receive antenna module for space-based RF applications. Also responsible for development of transceiver systems for high threat yield satellite and spacecraft systems.

Northrop Grumman Space Technology (formerly TRW Space & Electronics)

Current Status
Northrop Grumman SpaceTechnology is the former TRW Space & Electronics before its acquisition by Northrop Grumman in December 2002. Northrop Grumman ST now comprises the space interests of Northrop Grumman and includes electronic systems, information technology, mission systems and integrated systems.

Background
TRW's Space & Electronics Group came into being on 1 January 1993 as a merging of the Space & Technology and Electronic Systems groups; personnel total 7,000. Under the Aerospace and Information Systems group, space business is conducted under the TRW Space Systems division embracing command and control. It has built approximately 185 satellites, built/integrated approximately 130 payloads, developed approximately 200 space instruments and integrated some 550 experiments into spacecraft. Military programmes include DSP, DSCS II, Fltsatcom, Milstar's payload and many covert activities such as KH11. DSCS covers 16 satellites, launched from 1971 and designed to operate into the 1990s. UHF military communications satellites are provided by the US Navy's Fltsatcom; TRW was the lead contractor for NASA's TDRS satellites, the seventh and last of which was launched in July 1995.

TRW provided three High Energy Astronomical Observatories in the 1970s to investigate x-ray sources; these were followed by the Compton Observatory in 1991. NASA selected the company in August 1988 as AXAF prime contractor. Launch took place by Shuttle on 23 July 1999 with the AXAF renamed Chandra. TRW is also responsible for the Aqua and Aura satellites for NASA's Earth Observing System programme. TRW signed a US$5.48 million contract on 17 April 1990 for the US Air Force's STEP Space Test Experiment Platform. STEP is designed to produce satellites of less than 450 kg, with lives of one to three years using small ground stations. Options provide for a further 11 satellites; the first three are to be followed by five more (the fourth was contracted mid-1992 for US$14.5 million). The platform can provide 300 W power and be three-axis or gravity gradient stabilised. Debut was on Taurus in March 1994 with the TAOS Technology for Autonomous Operational Survivability payload but STEP-2 was lost in a launch vehicle failure.

TRW is marketing this Eagle/STEP design: it was selected in April 1994 for Taiwan's Rocsat 1

(US$61 million contract) and in September 1994 for South Korea's Kompsat (US$75 million contract signed in March 1995). TRW was selected in June 1991 for NASA's US$29.3 million 1994 TOMS Earth Probes contract. TRW built the spacecraft and integrated the instrument. NASA awarded the US$398.7 million contract on 15 September 1995 for the EOS Earth Observing System Common Spacecraft, with its AB1200 bus derived from Odyssey's AB940 bus. The initial contract covered PM 1 and Chem 1; these satellites have now been renamed Aqua and Aura respectively. The launch of Aqua took place in 2002 with Aura following in 2003.

In the ground terminal field, the GEODSS Ground-based Electro-Optical Deep Space Surveillance system was developed by TRW Systems Integration Group for USAF. It designs/develops space communications and processing payloads; programmes include Milstar, TDRSS, DSP and SMTS. The division also handles design, development and production of the ground data-handling systems for these payloads.

TRW makes satellite bus systems for the USAF Step programme. Each system comprises three modules: the standardised self-sustaining core, and the payload and propulsion modules that can be tailored to mission requirements. Each module is a 12-sided prism (diameter: 96.5 cm) of monocoque aluminium construction.

The primary or Core Module has a mass of 91.6 kg and a height of 29 cm.

Specifications

Attitude control: 3-axis (gravity gradient or spin available); scan wheels provide horizon scanners and RWs for pitch bias momentum control system. 3-axis magnetometer + magnetorquers unload wheels and control yaw. 0.5° knowledge/control; 1°/s drift rate.

Power: six deployable panels + body-mounted cells provide 110 W orbit average (75 W for payload). Articulated array offers 350 W orbit average. Five Ni/Cd batteries provide 330 Wh storage. 18 VDC bus.

Communications: UHF BPSK or FSK rx, 10 W transmit, 1/125 kbit/s up/down; 5 W S-band transponder, 1/256 kbit/s up/down.

Digital Processing System: 80C86 digital processor/controller, 200 kips; 512 kbyte RAM + 512 kbyte EPROM; 8.2 Mbyte data storage, 256 kbyte command storage, mass memory options.

The Payload Module has a height of 36.8 cm, and a payload volume 0.255 m^3, body solar cells add 15 W orbit average. Modules are stackable to accommodate long items such as telescopes.

The Propulsion Module has a height of 58.4 cm, monopropellant hydrazine (Olin Aerospace 178 N MR-107 + four 5 N MRE-1), body-mounted solar cells.

Northrop Grumman Space Technology and its heritage company TRW has built more than 130 satellites and has reported annual sales of US$2.02 billion for 2001. Employs 8,600 people.

UPDATED

Northstar Aerospace

Current Status

Northstar Aerospace, formerly known as Derlan Inc, provides machined components, fabricated systems and launch vehicle ground-handling systems for solid and liquid propellant launchers. Northstar also provides heavy structures handling fixtures for large stage elements. The company has produced forgings for test stands and grounded thrust structures. The company has annual turnover of over US$180 million.

Oceaneering Space Systems (OSS)

Current Status

OSS is one of the application arms of Oceaneering International, the world's largest publicly owned underwater services company. Founded in 1965, it began serving NASA in 1987 and in 1988 founded its Space Systems division. In May 1993 it acquired ILC Space Systems. Expertise covers thermal and acoustic protection, EVA astronauts' tools and hardware, robotics, life-support systems and neutral-buoyancy training. The range includes thermal insulation blankets, ablative cork insulation, power tools, foot-restraint working platforms, lights, tethers and other tools and containers. OSS is designing the nitrox breathing system for the new buoyancy facility at NASA Johnson.

OSS is contracted for the design, fabrication, test and integration of Space Station's food and science refrigerators and freezers (0.57 m^3 each, −15 to +4°C). Work on other galley components, including oven, drink dispenser, trash compactor and wardroom, is planned. OSS is furnishing Station's EVA support tools and containers. These include a stowage unit mounted on the airlock which will house about 75 subassemblies such as electromechanical tools and crew translation and worksite restraint equipment, much of which is to be robotically compatible. OSS is responsible for the contract from concept design to flight qualification. OSS has also provided breadboard test platforms for space station systems. OSS has designed the Robot-Tank Inspection Fund Effector for the US Department of Energy and has proposed applications as a robotic manipulator.

OSS occupies a 5,900 m^2 building adjacent to NASA Johnson. It houses a 507 m^2 precision machine shop for fabrication of prototype items and space hardware, capable of handling materials ranging from aluminium and exotic alloys to plastics such as Lexan, Teflon and Vespel. Soft goods are produced in two areas: a 176 m^2 area is oriented towards Shuttle items, fabric bags, containers and other items using conventional materials; the 502 m^2 area is isolated to handle more unusual materials such as quartz, ceramic and carbon cloth. OSS maintains a laminar flow tunnel (14.6 m along flow, 7.3 m wide) for work requiring strict cleanliness. There is access to a 4.6 m (depth) × 7.3 m (diameter) neutral-buoyancy facility at Oceaneering's Park Ten Facilities, 68 km NW. OSS also supports EVA and suit testing in Johnson's neutral-buoyancy tank and KC-135. At NASA Marshall, OSS provides neutral-buoyancy simulations, including Shuttle suit-support and general underwater operations.

Oceaneering International currently operates in 54 locations in 23 countries. Reported sales of US$500 million in 2001. Employs 3,200 people. Space Systems reported earnings of US$20 million in 2000. ISO 9001 certified.

Optical Coating Laboratory, Inc (OCLI)

Current Status

Optical Coating Laboratory is a wholly owned subsidiary of JDS Uniphase Corporation which also owns Flex Products Inc which manufactures optically variable pigments used on about 65 currencies worldwide to inhibit counterfeiting. The company has extended this to incorporate more sophisticated forms of counterfeiting.

OCLI, founded in 1949, is the world's largest independent producer of thin-film-coated optical products. Thin-film-coating deposition methods such as plasma plating and MetaMode (Metal Mode Reactive Sputtering) have been developed from OCLI's theoretical research. It has provided anti-glare coatings for the windows of every manned US spacecraft, in addition to coatings for optical instrumentation on Voyager, GOES and UARS. It has coated AXAF's eight x-ray mirrors. OCLI also provides solar-cell covers, thermal-control mirrors and instrumentation filters.

OCLI provides satellite temperature control equipment and satellite and spacecraft instrumentation. Research in laser technology supports satellite laser communication concepts and this has resulted in tests with point-to-point laser communications equipment.

Orbital Sciences Corporation (OSC)

Current Status

OSC has business interests in satellite and space systems, launch vehicles, space technical services and transportation management systems.

Revenue for 2003 was reported at US$581.5 million, an increase of 5 per cent over the US$551.6 million for 2002. At US$35.6 million, 2003 operating income was up 22 per cent on 2002 (US$29 million). The firm order backlog for 2003 reached US$1 billion with a total backlog of US$2.68 billion. OSC reported a revenue of US$459.7 million in 1999 with net loss of US$121.9, a revenue of US$379.5 million in 2000 with net loss of US$278.2 million, and revenue of US$415.2 million in 2001 with net loss of US$18.9 million. It acquired Fairchild Space & Defense Corporation from Matra in August 1994, Magellan Systems Corporation in December 1994, Perkin-Elmer's Applied Science Operation in September 1993 and MacDonald Dettwiler & Associates in November 1995. OSC was established in 1982 to develop/market Space Shuttle upper stages, resulting in the Transfer Orbit Stage. The Pegasus air-launched small orbital delivery system was developed in partnership with Hercules Aerospace, and OSC was contracted in 1989 by ARPA for the Taurus rapid-response orbital launcher based on Pegasus. Space Data Corporation of Tempe, Arizona became a subsidiary in November 1988, extending OSC's capabilities to a wide range of sounding rockets: it has manufactured/launched approximately 600 sub-orbital boosters in 35 different configurations, in addition to producing up to 1,000 meteorological rockets annually. OSC acquired CTA Space Systems in 1997 (see separate entry) adding greatly to its capabilities. OSC employs about 5,000 people worldwide.

OSC's Orbital Communications Corp (Orbcomm) subsidiary plans a constellation of LEO satellites for two-way commercial communications services. First launch was successful in April 1995. By early 1999, 28 Orbcomm satellites were in orbit and during the first quarter total subscriber units in service doubled from 32,000 to 63,000 units. Total of 35 satellites on orbit by July 2000.

On 15 September 2000, Orbcomm filed for bankruptcy protection under Chapter 11. As of December 2001, Orbcomm had strengthened its business around the world, securing Hitachi as a customer in the Asia-Oceania region.

The Orbital Imaging Corporation (Orbimage) is building the OrbView Earth-imaging system. OrbView-1 was launched on 3 April 1995 and was followed by OrbView-2 in 1997. OSC was selected by NASA in 1991 to deliver multispectral ocean colour data to NASA investigators for five years as part of the agency's Earth Probes programme. This is the first time the US government has purchased global environmental data from a privately designed/operated remote sensing satellite. Orbimage's SeaStar satellite, based on its PegaStar bus, carried the SeaWiFS Sea-viewing Wide Field Sensor in late 1996.

In early 1999, OSC formed the Orbital Navigation Company (Orbnav) to develop, operate and market satellite aided automotive guidance and related value-added information services for the rental car, private car, commercial vehicle and emergency vehicles markets. Orbnav plans to leverage Orbcomm's satellite data network and is currently focusing on the rental car market through a joint venture with the Hertz Corporation known as the 'Neverlost®' system.

OSC has developed/integrated several Shuttle payloads, including high-precision two-axis pointing/tracking experiments such as the USAF Cirris 1A. It has also designed, built and installed more than 60 major telemetry/tracking stations in 10 countries.

The OSC Pomona division specialises in imagers and sounders for infra-red to x-ray remote sensing. It is fabricating four Total Ozone Mapping Spectrometers under a US$20 million contract for NASA. To ensure continuity of ozone data, the agency negotiated for the engineering model of Nimbus 7's TOMS to be launched on a CIS/USSR Meteor in 1991 and new models were planned for NASA's own Earth Probe in 1996, Japan's ADEOS in 1996 and Russia's Meteor 3M in 2000.

OSC is studying the Orbital Space Plane under NASA's Space Launch Initiative programme and has chosen the HI-10 Lifting Body configuration as a baseline. OSC is building NASA's Dawn spacecraft designed to orbit asteroids Vesta and Ceres after launch in 2006. On 6 February 2003, OSC launched its first Taurus-Lite, boost vehicle for the Ground-Based Midcourse Defense System.

Picolab

OSC offers PicoLab for ultra-small orbital missions, including instrument development, component proof of concept and space and Earth sciences. Weighing up to 22 kg, PicoLab is carried in the avionics section of OSC's Pegasus or Taurus launchers and can be ejected or remain attached. The firm fixed price cost is quoted at less than US$3 million, which includes satellite, launch

Launch of the first OSC Taurus vehicle (OSC) 0572497

and one year of operations. Cost can be reduced to US$1 million if the customer provides its own control/data collection. Launch would be between 9 and 18 months after contract signing. OSC is targeting industry, universities and NASA field centres.
Design life: 3 years
Size: 20.3 cm (diameter), 15 cm (height), 45.7 cm solar array span; 22 kg, including 2.7 kg payload (4,900 cm^3)
Attitude control options: ±0.5° Sun-pointing or inertial spin; ±5° nadir pointing; ±20° passive magnetic
Power: 10–25 Wh orbit average to spacecraft (5–10 W payload), 3.3–15 V regulated bus; deployable or body solar arrays
Payload data: more than 10 Mbyte per day per station, 64 Mbyte (expandable) storage, 32 bit microprocessor payload interface.

MicroLab
MicroLab employs the MicroStar platform design developed for the Orbcomm satellites. MicroLab 1 was launched successfully in April 1995 carrying a NASA lightning detector. Total contract was less than US$7 million.

Design life: 3 years
Internal size: 96.5 cm (diameter), 30.5 cm (height), 45 kg payload (0.17 m^3)
Attitude control options: nadir pointing ±0.5° or Sun-pointing, spin-stabilised to ±2°
Power: 95 Wh orbit average to payload, 28 V unregulated bus; two deployable articulated solar discs
Payload data: 2.0 Mbit/s max, 512 Mbit (expandable) storage.

PegaStar
The SeaStar and APEX satellites both employ OSC's PegaStar bus. It incorporates launcher/spacecraft avionics functions into a single vehicle, yielding greater power/mass for mission experiments. Cost less than US$20 million; lead time 18 to 24 months. Payload accommodation up to 91 kg/0.454 m^3. The all-aluminum primary structure mates directly to the Pegasus stage 3 motor, and houses launch-vehicle guidance, telemetry/control systems as well as spacecraft housekeeping functions. The power system is shared by the satellite/launcher, incorporating two 6 Ah Ni/H_2 high-energy density batteries (10 Ahr on SeaStar). Four 55.9 × 152.4 cm rigid solar arrays provide 500 W peak BOL (300 W to payload). ACS is provided by a momentum-biased 3-axis system of magnetorquers, magnetometer, Sun sensor and constant-speed momentum wheel to provide Sun pointing to 0.5° in two axes. Control about the Sun line is held to 5.0° with very low angular rates. Four 22.2 N hydrazine thrusters provide initial orbit insertion, drag make-up and precise orbit maintenance; hydrazine capacity 68 kg. A GPS receiver obtains position information to within 100 m independent of ground systems. The multiprocessor satellite computer controls all housekeeping functions; an advanced 80 Mbyte SSR stores up to 24 h of experiment data. The telemetry transponder provides data transmit (2.6 Mbit/s; 1.25 Gbit storage).

VERIFIED

OSC designed a re-entry vehicle for the NASA Orbital Space Plane programme, cancelled when effort was re-directed toward an Exploration Vehicle for earth orbit and lunar missions (OSC) 0572496

Oremet Titanium

Current Status
Titanium sponge, ingots, mill products and castings. Applications include Shuttle propellant tank ring and elements for the International Space Station including flight hardware and ground test structures. Ground support equipment and stability rigs for propulsion tests.

Precision Rolled Products, Inc (PRP)

Current Status
Founded in 1974 PRP is a subsidiary of Krupp VDM USA, Inc, which is a wholly owned subsidiary of ThyssenKrupp VDM in Germany. The company is an approved superalloy manufacturer and supplier to the gas turbine industry. It is a speciality mill with integrated facilities to hot roll, thermal treat and finish bar products in titanium, titanium alloys, high-T cobalt and nickel-base alloys, and stainless steels for the aerospace industries. Casings for down-lock thrust restraints on ground test structural restraint rigs. Has provided shell cases for module simulators and heavy duty ground test rigs.

Princeton Synergetics, Inc (PSI)

Current Status
PSI specialises in policy and economic issues related to space commercial development, transportation and insurance. Analytical tools include simulation models that evaluate launcher and satellite attributes in terms of telecommunications satellite business financial performance; a computer-aided small satellite design tool aimed at mission planners, satellite designers and technology programme planners; a computer model for comparing advanced space transportation architectures explicitly taking into account uncertainty and risk. PSI has supported the Department of Transportation in its setting of space insurance requirements and has supported space insurance companies in establishing space insurance programmes. Also provides integrated futures projection of new technology. Incorporates international analysis and project evaluation. Has helped evaluate integrated space architecture for government and non-government programmes. PSI framed Colorado's Strategic Plan for Space, an initiative implemented by the Governor of Colorado in 1999 and delivered in September 2000.

Programmed Composites Inc

Current Status
Previously owned by Ketema, provides hot components for aero engines. Programmed Composites are specialists in design/fabrication of precision, dimensionally stable, high-strength, lightweight composite components for space applications, including solar-array substrates,

bus structures, reflectors, tubes, struts and components for telescopic and furlable antennas. PC has developed a high tensile unfurlable antenna design, capable of manufacture to large diameter specifications. This technology has been applied to other requirements for high-strength composite materials and for the development of rigid shell structures that could be applied to modular design for spacecraft sections.

Raytheon Optical Systems Inc (BF Goodrich)

Current Status

Formerly known as Hughes Danbury Optical Systems Inc, Raytheon Optical Systems Inc had been in business for more than 30 years and currently employs more than 700 people, reporting sales of US$165 million in 2001. BF Goodrich completed its purchase of Raytheon Optical Systems in December 2000.

HDOS designs and manufactures precision optics and electro-optical systems for defence and science space applications. Products include ground- and space-based telescopes, science and remote sensing instruments, star trackers, and optical components and systems. It also specialises in surveillance and directed energy systems, with emphasis on visible light sensors, optical systems and multispectral sensors. The company manufactured the Optical Telescope Assembly of the Hubble Space Telescope, including the 2.4 m main and 30 cm secondary mirrors, and the three Fine Guidance Sensors. In 1995, it completed the eight grazing incidence optics for AXAF. The Special Optics Facility, completed in September 1986, supports design, manufacture and complete environmental test capabilities up to the 4 m class, in optics of any geometry.

ROS completed the acquisition of Itek Optical Systems from Litton Industries in February 1996. Itek company designed/developed large, precision lightweight optical systems for space deployment, adaptive optics and optics for very large astronomical instruments. It produced high-resolution film cameras and real-time electro-optical sensors, with 15 mm to 350 cm focal lengths, for reconnaissance and surveillance missions. Equipment was also produced for processing the electro-optical imagery. Itek built cameras for lunar mapping operations in the Apollo programme and NASA's Large Format Camera, which flew aboard Shuttle in 1984 and returned approximately 2,200 large film frames. These capabilities have continued in ROS's main Danbury, Connecticut plant, as well as in facilities located in Lexington, Massachusetts and San Jose, California. ROS has provided new technology development programmes for adaptive optics designed to enhance long life high resolution imaging systems.

R-Cubed Composites Inc

Current Status

R-Cubed was purchased by Composite Optics in 1998, its name thereafter being R-Cubed Composites Inc. The Utah facilities were moved to San Diego in 2001.

Composite structures for satellites and launch vehicles, including Orbcomm, CBERS, Pegasus, Taurus, Clementine, Clark and MicroLab: customers include Hughes, TRW, Space Systems/Loral and Motorola. Also produces support frame assemblies for subsystem elements in satellites. Specialisations permit advanced composite integration structures to be installed in complex load paths.

Rosemount Aerospace Inc

Current Status

Now a division of BFGoodrich, Rosemount Aerospace was established in 1956. Manufactures industrial gas turbine instrumentation, thermocouples, speed and torque sensors, optical flame detectors, pyrometers, pressure sensors for frame and aero-derivative engines. The company designs, develops and produces temperature, pressure and air data sensors for measuring critical flight and engine control system parameters. It has provided development/testing facilities including sub/supersonic wind tunnels, icing and flow tunnels, and extensive environmental facilities. Rosemount sensors have been used in every US manned space programme from Mercury to Gemini and Apollo, including temperature sensors in Apollo's Lunar Surface Experiments Package. Shuttle sensors monitor and control main engine functions such as cryogenic and hydraulic temperature, hot gas temperature, and speed/flow rates. Approximately 12,500 sensors were supplied for onboard and ground support systems. Ground support sensor elements feed critical GSE. In orbit, flow sensors monitor the operation of the Shuttle fuel cells; during entry, two air data probes measure the Orbiter's angle of attack, total temperature and airspeed. The measurements are made from 30 km altitude (and at speeds in excess of M4.0) to touchdown. Flow sensors and meters for ISS applications.

Scaled Composites, LLC

Current Status

Scaled Composites is building what purports to be the world's first feasible, privately funded, vehicle for human space flight. The plan involves a three-person spaceship (SpaceShipOne) carried by a turbojet powered aircraft (White Knight) to an altitude of 50,000 ft (15,240 m) where it separates, ignites its integral rocket motor reaching a speed of 402.25 km/hr coasting to a peak altitude of 100 km (62 miles). During the coast and fall back to earth the spaceship is weightless and converts to a high-drag configuration to allow safe, stable atmospheric entry allowing a 17 min glide from peak deceleration at 80,000 ft (24.4 km). Concept design began in 1996 and preliminary development started in 1999. In 2000 Scaled Composites conducted a study of rocket motor technologies and selected a hybrid combination using nitrous oxide (liquid N2O) and rubber (HTPB) propellants with a motor mounted to the spacecraft by skirt flanges on the oxidiser tank. The nitrous oxide tank consists of a composite liner laid up on to titanium flanges with a graphite overwrap provided by Thiokol. The unified fuel casket and nozzle components use a high-temperature composite insulator with a graphite/epoxy structure laid up on to an ablative nozzle supplied by AAE Aerospace. The ground firing development programme started in November 2002 with a 15 second run by the SpaceDev team. White Knight made its first flight on 1 August 2002. The first captive carry flight took place on 20 May 2003 with SpaceShipOne mated to White Knight. The first glide flight took place on 7 August 2003 when SpaceShipOne was released from an altitude of 47,000 ft (14,325 m). The first powered flight, the eighth flight of SpaceShipOne, took place on 17 December 2003 when it was released from the White Knight carrier aircraft with burnout at Mach 1.2. No first-flight date to space has been announced and the company has said that it has no urgency in achieving interim milestones.

Background

Founded in 1982 by Burt Rutan, Scaled Composites provides air vehicle design, tooling and manufacture utilising specialist composite structures design, analysis and test.

Scaled Composites provides complete composite aerospace vehicle design, manufacture and test services. It builds the fins and wing for Orbital Sciences Corporation's Pegasus launcher, and provided the aeroshell for the DC-X demonstrator. It is currently working on the K-1 vehicle for Kistler Aerospace Corporation and the HMX Roton. SCI has produced new composite structures for replacement elements on new launch vehicles proposed by Kistler and OSC.

Schonstedt Instrument Company

Current Status

Founded in 1953, SIC designs, manufactures and services instrumentation for domestic utilities, construction and exploration. In 1998 the company moved to West Virginia. The company's magnetometers have flown on more than 400 satellites and planetary probes as spacecraft attitude-control sensors and local magnetic field measuring instruments. They were added to the Hubble Space Telescope during its December 1993 in-orbit servicing. Magnetometers for rocket aspect measurement include: SAM-73 tri-axis fluxgate sensor, SAM-72 dual-axis fluxgate sensor and SAM-71 single-axis fluxgate sensor; RAM-5C, S29-S30 single-axis fluxgate sensor. Magnetometers for long duration, deep-space missions where extended life requirements are specified.

Schuller International

Current Status

With a facility also in Canada, Schuller is a specialist in product recycling and specialises in manufacturing fibre glass products from used glass. All Schuller products are manufactured from at least 25 per cent recycled glass.

Acoustic and thermal insulating material for aerospace applications, including Space Shuttle tiles and thermal insulation for rocket motors. Has provided thermal insulation for heating and environmental control systems. It is presently working advanced, repairable, thermal protection systems for space vehicle applications.

Science Applications International Corporation (SAIC)

Current Status

Founded by a small group of scientists in 1969, now a Fortune 500 company, the largest employee-owned research and engineering firm in the USA, SAIC employs more than 40,000 people. SAIC is an employee-owned company specialising in computer systems integration, training-device systems, electronic warfare and countermeasures, systems modelling/simulation, electro-optical and signal processing, military space sciences and technology, aeronautical engineering, astronomy and astrophysics, software development and systems integration. Has conducted systems studies on next generation telescopic concepts. Reported revenue of US$6.1 billion for fiscal year ended 31 January 2002, net income US$19 million. For the fiscal year ended 31 January 2001, SAIC achieved a revenue of US$5.9 billion compared with US$5.5 billion the previous year. 1996 revenue was US$2.4 billion (1995: US$1.9 billion, 1994: US$1.7 billion). SAIC derives about 55 per cent of its revenue from systems integration, research and development and other US government services. SAIC now has 35 companies, subsidiaries and equal partners.

Sedgwick Space Services

Current Status

Now absorbed into Marsh Inc, an operating unit of Marsh & McLennan Companies Inc, Sedgwick provides all aspects of space insurance brokering, specialising in technical, legal, insurance and risk management on behalf of manufacturers, owners and operators. It provides the design and implementation of insurance, self-insurance and captive programmes in respect of build, ground property, pre-launch and in-orbit operations. SSS was one of the early participants in the business of space brokerage and has conducted consultative analyses for operators and users. Sedgwick has performed major roles in restructuring space insurance and in creatively providing new forms of risk assessment and management.

SETI Institute

Current Status
The non-profit institute incorporated in 1984 is making microwave observations looking for extra-terrestrial life. Government support has declined from a financial subsidy that supported extensive frequency searches to an almost total lack of support from federal funds. An innovative approach to garnering public support comes from the SETI at Home project embracing the use of home computers for conducting frequency searches. The SI has established a goal to create a US$100 million sustaining endowment. Supported by philanthropist Paul Allen, who gave an initial grant of US$11.5 million, the Institute has embarked upon a programme to construct an array of 350 small (20 ft) radio telescopes in a 'massively parallel array' as a prelude to the Square Kilometre Array. The Allen array will cost an estimated US$40 million and is scheduled for completion in 2005.

Space Analysis & Research, Inc (SAR)

Current Status
SAR offers satellite and launch database software systems for IBM-type PCs. SPACE 2000 contains all past launches and attempts plus plans through 2012. Includes all 25,000+ catalogued objects detected by US Space Command, orbital element sets for all orbiting payloads and basic orbital data for the approximately 7,500 pieces of debris on orbit. Offers discrimination software for measuring parameters of high orbit objects. SAR has provided orbital debris analysis and derivations of future collision avoidance requirements for trajectory determination. SAR has conducted discriminatory analysis of category selection whereby separate orbital bodies are automatically analysed and categorised.

Space Commerce Corporation (SCC)

Current Status
SCC was formed in 1986 to market Soviet launcher vehicles and space services. It now acts as a consultant to high-technology firms doing business in the Russian Federation and has negotiated commercial deals between science research organisations in Russia and the US. SCC developed blueprints for linking Russian space technology with commercial opportunities in the USA and Europe, concepts taken up by major manufacturers. SSC performed studies on joint ventures before the collapse of the communist ruling party in Russia and has played roles in bringing together engines and scientists from the US and Russia. It conducted analysis of integrated propulsion systems and of mechanisms to adapt variable propulsion chemistries for hybrid propellant combinations.

Space Electronics Inc (SEI)

Current Status
SEI manufactures Rad-Pak™ space-qualified radiation-hardened microelectronic devices for space applications. SEI has developed hardened devices capable of surviving high-intensity environments and hard radiation from natural and artificial sources including EMP from nuclear detonations. 1996 revenues were US$10.6 million. 1995 sales were US$4.0 million (revenue US$3.75 million for US$325,000 profit), with a backlog of approximately US$6 million. Experience includes ISTP, TOMS, Geotail, Mars-96, SeaStar, DMSP and Cassini. Globalstar contracts total US$1.7 million. A new family of microcircuits includes 1M EEPROM (128k×8), 256K EEPROM (32k×8), 1 Mbit SRAM (128k×8), Microprocessor (80386RP + 80186RP) and DRAM (4 Mbit). Devices survive a minimum of 100 krad (Si) in most orbits and have SEU tolerance more than 43 MeV/mg/cm^2 and single event latchup tolerance approximately 120 MeV/mg/cm^2.

Maxwell Technologies Instruments Products, formerly a part of SEI, separated to form its own business as Maxwell Technologies.

Space Industries, Inc (SII)

Current Status
SII was founded in 1982 by Dr Max Faget, the chief engineer responsible for the designs of Mercury, Apollo and Shuttle. At that time it existed to focus development, funding and marketing of a free-flying, man-tended, earth orbiting laboratory capable of carrying a large number of commercial experiments for private companies and government establishments. In March 1996, SII was purchased by GB Tech Inc and is now 'Space Industries Inc, a division of GB Tech Inc'. The company is responsible for the design/fabrication of the Wake Shield Facility, the Meteor capsule and several thermally controlled experiment modules. WSF successfully grew seven thin film substitutes on STS-80, demonstrating its capabilities as an ultra-vacuum research platform. SII has developed the 7 kg, less than US$1 million, Nanosat for a range of applications, including communications and ISS inspection. The 58 × 25 × 36 cm satellite could manoeuvre around ISS for 12 hours using cold gas thrusters and miniature gyroscopes controlled from a joystick terminal. Studies continue and contact with NASA has resulted in revised applications planning. NASA has been keen to continue studies with remotely operated free-flyers based on initial evaluation and analysis from SSI. The company has also proposed a combination of nanosat inspection vehicles and unmanned free-flyers and has conducted evaluation of hybrid constellations of unmanned and man-tended platforms.

Space Machine Advisors, Inc

Current Status
The company provides risk management analysis, insurance and financial services to the space and satellite communications industries. Produces risk analysis assessment and financial estimates for satellite packages, launch vehicle options and ground segment investment. Diversified into general risk analysis and application of space-based technology assessment to earth-bound industries and practices. Adds predictive probability analysis, which provides an assessment of future trends by examining the way historic and actuarial flow affects performance, reliability and failure mode analysis.

Space Policy Institute

Current Status
Established in July 1987, the Institute is located within the Elliott School of International Affairs at the George Washington University and conducts research on space policy issues, organises seminars, symposia/conferences, and offers graduate courses on space policy. It focuses on US space policy issues and competitive and co-operative interactions with other countries. Conferences cover international issues, space-based weapons and space law.

If is an affiliate of the International Space University.

Space Studies Institute (SSI)

Current Status
Formed in 1977 by the late Dr Gerard O'Neill, SSI is a non-profit organisation for conducting and sponsoring research on the use of space resources. Extensive research, think-tank analysis and conferences on alternative US national space programme goals. Focused work on space manufacturing, use of Shuttle External Tank as a low-cost on-orbit facility and energy generation in space. Current projects include the design of solar-power satellites built from lunar materials and mass drivers capable of lifting materials from the moon. Studies connecting this work with the ISS have been conducted. The Institute publishes a bimonthly newsletter. There has been support for SSI objectives for low-cost fast-track human-tended observatories and high earth orbiting laboratories for investigation of space physics phenomena which are unable to attract funds from government led projects but through which research agencies and academic institutions would find value.

Sparta Inc

Current Status
Founded in 1979, Sparta is a systems engineering and advanced technology company. High-technology company focusing on US military space contracts and the application of advanced technology to space-based defence systems. Provides advanced software packages for modelling and simulation exercises. Research and analysis on NMD and ASAT linked technologies with operational simulation and deployment strategies. Sparta has conducted space warfare simulation exercises and performed simulated wargames involving space- and ground-based anti-ballistic missile capabilities. These are continuously updated to reflect new technologies developed by the Department of Defense and contractors.

Specialty Steel & Forge

Current Status
Distributor of high-temperature metals to the aerospace industry, including stainless steel, aluminium, titanium, nickel, copper, brass, glass sealing alloys and tungsten alloys. Hybrid alloys and special steel products for unusual applications. Specialised in high strength, high temperature resistant products. Users include the Space Shuttle and the ISS and high-quality specification users in satellite and spacecraft fabrication. SS&F has developed unique soft forging techniques designed to accommodate high specification requirements.

Spectrum Astro, Inc

Current Status
Founded in 1988, Spectrum Astro provides research, design, development and manufacture of low-cost, advanced technology space systems. 1995 sales were US$7 million and the company had 68 staff. In 2003 the company had 450 employees, sales of US$120 million in 2000 and US$160 million in 2001. Spectrum is constructing an additional facility close to the main centre at Gilbert, Arizona, capable of assembling medium sized satellites. The six product lines are: research and development for advanced space systems; development of rapid prototype advanced space systems; production of multimission spacecraft; design/manufacture of high-reliability space-qualified disk memory systems, space electronics and complete subsystems, and EGSE. Spectrum Astro has completed a detailed conceptual design for the NASA High Energy Solar Spectroscopic Imager for the next primary mission in NASA's Goddard Small Explorer (SMEX) series of space science missions.

Spectrum Astro was awarded a contract in 1991 to design, fabricate, integrate, test and deliver on-orbit the MSTI Miniature Sensor Technology Integration satellites for BMDO/Phillips Laboratory. MSTI 1 was launched successfully in November 1992 and MSTI on 2 May 1994, both providing five months of mission data, meeting all mission objectives. MSTI 3 departed in May 1996. JPL's first New Millennium bus contract was awarded to Spectrum Astro in September 1995. The bus for Deep Space 1 was delivered in February 1997 for launch on the satellite in October 1998. DS1 was switched off in December 2001, two years after completing its design mission. Spectrum designed and manufactured the MightySat II. 1 launched by the Air Force in July 2000. Designed and built Coriolis, a USAF/USN programme, launched January 2003.

The company has developed, space-qualified and flown a low-cost, low-mass 8.7 Gbit Magnetic Disk Data Mass Memory unit for ARPA's Advanced Space Technology Program (see the Recorders & Memories section). Carried by MSTI 3 and STEP M3.

The SA line of multimission small platforms offers standard payload interfaces and high payload mass fraction. The SA-2 first flew for MSTI 2. The SA-1000 would typically carry a small communications payload in GEO. No customers have yet been announced.

Specifications

The specifications below are provided in SA-2/SA-3/SA-1000 order:
Orbit: LEO Sun-oriented/LEO all inclinations/GEO
Schedule: 12/18/24 months
Design life: 3/5/10 years
Dry bus mass: 113/181/295 kg
Payload mass: 181/181/363 kg
Payload power: 190/760/1,200 W
Pointing/control: 0.25°/0.1°/0.1°; 3-axis.

Taber Industries

Current Status

Stainless steel bonded strain gauge pressure transducers covering from 0 to 2,040 atmospheres and from −210 to +190°C. Users include Insat, Intelsat, Arabsat, Tethered Satellite, EOS, Landsat and numerous commercial communication satellites and launch vehicles. TI developed special strain gauge transducers for extreme pressure environments and has a wide range of product suitable for off-nominal mission requirements or specifications.

Teledyne Brown Engineering (TBE)

Current Status

Established in 1953 to support rocket research by the Wernher von Braun team and later to support Huntsville, USA, based technology developments, Brown Engineering merged with Teledyne in 1966. Merged with Allegheny Ludlum in 1996. Teledyne Technologies Inc was formed in 2000.

TBE specialises in mission analysis and planning, hard/software development and payload integration. It is the Systems Engineering and Technical Assistance contractor for the US Army Space & Strategic Defense Command and is the Payload Integration Contractor for Spacelab. Current space-qualified hardware emphasis is on materials-processing systems for Shuttle and Space Station: TBE's high-temperature Crystal Growth Furnace, zeolite CGF and Protein Crystal Growth apparatus flew on Shuttle USML 1/2 in 1992/95.

For the US Army, TBE integrates BMDO experiments and provides technical and programme support for the development of exoatmospheric surveillance sensors, ground-based interceptors, airborne sensors, the Ground-Based Radar and BM/C^3 elements.

Provides space systems training and operations support. In July 2002, TBE received a US$22 million contract from the US Army Space and Missile Defense Command for continued development of Teledyne's Missile Defense System Exerciser. Formed in 2000, Teledyne Technologies is investing research into internet-based information systems and a new method for recycling and destroying propellants and explosives.

Teledyne Electronics and Communications

Current Status

TEC is the consolidation of the former Teledyne Systems Co, Teledyne Lewisburg, Teledyne Electronics and Teledyne Ryan Electronics. It specialises in redundant, fault-tolerant and radiation-hardened electronic systems such as launch computers, on-orbit mission computers (such as Milstar's), navigation systems (inertial, Doppler, GPS), flight-control systems/sensors, telemetry, remote multiplexers, inertial attitude control/pointing, data links, encryption and cryptographic modules, signal conditioning and interface electronics. Supports airborne communication systems, analytical instruments, aviation products and medical devices. TEC is a JPL support contractor for space electronics and has developed software systems for flight control and mission management evaluation. Has 2,500 employees.

Telegenix/Grim Corporation

Current Status

An Inductotherm Industries Company, Telegenix provides electronic devices for a wide range of industries. Specialists in electronic switching systems, technical control equipment, intelligent illuminated graphic displays and electronic contract manufacturing. Product developed for high-speed control mechanisms for fail-safe electronic devices. Microswitching and nano-technology producing control switching systems for robotic devices. Potential applications for long duration spacecraft missions with autonomous self-switching capability and self repair capability.

Textron Inc

Current Status

Founded in 1923 Textron was responsible for the design, development and production of Peacekeeper's re-entry system and expanded its manufacturing facilities for producing the Mk21 Re-entry Vehicle. In 1991, the Electronic Systems business line was formed, including specialised surveillance systems and communications antennas. The Aerostructures Division designs/manufactures structures for spacecraft: it was responsible for the intertank structure of Shuttle's External Tank, and for the nose cones and aft skirts of Titan 4's solid boosters. TSM is a developer/manufacturer of high-strength, lightweight advanced composite and fire-protection materials. Composite materials include boron and carbon fibre, silicon carbide reinforced aluminium, titanium and ceramics, and carbon-carbon for structures, turbine engines and rocket motors. Carbon-carbon throat inserts were built for Trident and MX. The company manufactures 37 tonnes of carbon-carbon annually. TSM developed a silicon carbide-reinforced titanium composite for NASP/X30's skin and internal support.

Textron reported 2001 revenues exceeding US$11 billion and currently employs 49,000 people worldwide. The company owns Bell Helicopter, Cessna Aircraft Company, Lycoming, among others.

Thomas Electronics Inc

Current Status

Founded in 1948 Thomas Electronics furnishes special-purpose CRTs for the aerospace industry, including the Shuttle Orbiter's set of triple displays (these were replaced by flat LCD screens in 1999). TEI also provided CRTs in ground facilities, control stations and text and diagnostic desks. Specialises in photo-imaging, projection systems, air traffic control and aerospace. Has designed and tested new multifunction displays linked to hologram projection for telepresence tests and experiments in distant control mechanisms.

Unisys Defense & Space Systems (UDSS)

Current Status

UDSS, with 1,700 personnel, supplies software and systems engineering, products and services for spacecraft and operations. It manages, modifies and maintains Shuttle ground-based software, which totals approximately 19 million lines of computer code. The software encompasses 15 programming languages, including Fortran, Assembler, PL1 and DB2. Eighteen operating systems are used, along with equipment produced by eight manufacturers, including Unisys, IBM, Perkin-Elmer and SEL. The software runs on almost 300 computers in 13 separate facilities connected through LANs.

UDSS has software and systems engineering responsibilities for Space Station operations, including software development, integration, testing and sustaining engineering, and systems requirements development, design support and performance analysis. It performs sustaining software engineering, testing, analysis and systems support for the Control Center Complex for Shuttle and Space Station at Johnson. At Goddard, it has evaluated the safety, reliability and quality of electronic, mechanical and other spacecraft components since 1965. Manned space activities are performed at the primary Houston facility. In 1995 Unisys Space Systems was awarded the NASA George M. Low Award for support at the Johnson Space Center, Hamilton, Texas. Has recently conducted information processing and sampling criteria for mission support requirements and real-time analytical activities and these have been applied to new requirements for hard data storage concepts with hardware and software integration resulting from that work.

Valcor Engineering Corporation

Current Status

Fuel, hydraulic, cryogenic, pneumatic and special purpose valves for the aerospace industry. Provided base support infrastructure for military airfields and operational bases in CONUS. Supplied to ground processing installations at KSC and VAFB in addition to supplies, contractors and cryogenic production plants. Supplies solenoid latching valves, pressure regulators, relief valves and accumulators designed to work with chemical and electric propulsion.

Veda Incorporated

Current Status

Specialising in engineering and technical services, Veda has 1,320 employees in offices throughout the US. Space activities focus on analysis, engineering and management support: satellite communications; constellation design, analysis, optimisation; TT&C of aerospace systems; satellite communications system engineering and integration; communications network scheduling, simulation, control; spacecraft design; GPS applications and analysis; COTS telemetry systems; launch-vehicle operations analysis and support; economic analysis and market assessment; LEO + MEO satellite systems engineering; legal technical assistance; advanced engineering seminars. Has developed COTS/government support systems integration service allowing customers to link separate specifications in joined-up elements or hardware sets. Conducts briefings and seminars to develop new space applications concepts and evolve radical and innovative ways of achieving space goals. Veda has developed the capability to knit into an integrated whole, several separate elements in space project conceptualisation, enabling planners and mission managers to converge on solutions.

Wear-Cote International Inc

Current Status

Electrolysis coatings for ferrous/non-ferrous metals. The company's electroless nickel is employed in Shuttle main engine ignition and mechanical systems. Developed restartable engine support technologies for NASA engine studies programme.

Now applied to other cryogenic propellant systems and to ground handling equipment. It can provide hybrid metal-to-metal joins and ferrous couplings for propulsion systems.

Wesgo Metals

Current Status

The Metals Division of Wesgo Inc (a division of Morgan Advanced Ceramics, Inc) was formed in 1927 and provides high-purity low vapour pressure brazing alloys. Applications include rocket engines, vacuum tubes and lasers and have been selected for the Titan programme. Also for long life, restartable motors, cryogenic motors and hypergolic propulsion systems. Has provided new methods of brazing hot parts for high heat resistant sections in boilers and high-temperature devices. Morgan Crucible, which owns Morgan Advanced Ceramics, is a holding company for 150 separate companies in 39 countries with sales of over US$1 billion to 160 countries.

Willis Corroon Inspace

Current Status

Operating since the early 1980s and now within the Willis Corroon Group's Aerospace division, Willis Corroon Inspace provides full space insurance services, including launch, in-orbit, product liability and third-party legal liability. Also ground facilities, loss of revenue coverage and protection of customer service warranty in the event of in-orbit failure. A major UK space insurance specialist with large range of international clients. Specialist advisory role for national and international broking consortia.

Wyle Laboratories

Current Status

Founded in 1949 as an engineering consultancy, Wyle Laboratories now serve aerospace, telecommunications, space science, electronics, transportation, nuclear, meteorological and airports industries. Specialists in engineering, testing and support services, small payload integration and small experiment payload engineering. Can integrate several national experiments in an international package for launch or support in a third country. Wyle is setting up life sciences facilities and extending its activities into the field of bio-astronautics. Wyle can provide analytical surveys and has studied a broad range of space related projects and programmes including civil and military activities.

LAUNCHERS, PROPULSION SYSTEMS AND LAUNCH SITE SERVICES

Canada

Bristol Aerospace Limited

Current Status
A member of the Magellan Aerospace Corporation, Bristol operates at three sites in Manitoba. It was acquired by Magellan in 1997. In addition to the supply of basic rocket vehicle hardware, Bristol provides a range of self-contained vehicle/payload support systems, including thrust termination, parachute recovery, payload separation, despin, telemetry systems and fairing deployment. Two launcher types are usually employed: a rail-type typically providing 6 to 8 m of guided rail travel and a 30 m travel tower. In cases where the normal dispersion is unacceptable, the Saab/SES S19 boost guidance system is available; Saab's Spinrac attitude control system provides attitude control for a spinning exoatmospheric stage. Facilities for vibration, temperature and humidity testing, salt fog testing and mechanical testing. More than 800 Black Brant sounding rockets launched to date.

China

Government sponsored organisations

Current Status
China has determined to make a concerted effort to consolidate and expand civil, commercial and military space activities, extending its reach into manned flight, planetary explorations and the establishment of a permanent human presence in orbit. In October 2003 China became the third nation in space flight history to design, develop, build and launch its own manned space vehicle carrying one of its own citizens into space. The China Great Wall Industry Corp is the foreign trade company, under the aegis of the China Aerospace Corp, responsible for launch services marketing, commercial negotiations and contract execution. Historically, National launches are charged to the government at about 60 per cent of the commercial rate; some 30 per cent of GWIC's profits are paid in tax. The China Satellite Launch and TT&C General organisation, with a workforce of more than 20,000, operates the launch sites and is responsible for all TT&C services through to the conclusion of a mission. It operates under the Commission of Science, Technology & Industry for National Defence but for commercial activities it is responsible to GWIC. China's primary space and military launcher manufacturing facility is the China Academy of Launch Vehicle Technology (CALT, formerly the Beijing Wan Yuan Industry Corporation) and near the town of Nan Yuan about 15 km south of the capital. Here, the CZ-2, DF-5 ICBM (a CZ-2 variant) vehicles and the CZ-3 cryogenic upper stages are manufactured in a complex of 13 institutes and seven factories. CALT is the technical co-ordinator for GWIC and provides on-site technical supervision of launcher procedures. The cryogenic engines are built here, but the other stages' engines are built by its Shanxi Liquid Rocket Engine Company. The plant also operates large stands for vibration and thermal testing andhas a large anechoic chamber for antenna development. Some of the launcher production is conducted at the Shanghai Academy of Space Technology (SAST, also known as the SBA Shanghai Bureau of Astronautics) with a 30,000 workforce (8,000 engineering/technical personnel) manning 17 research laboratories and 11 factories: CZ-3 first two stages, CZ-3 cryogenic third-stage attitude control system, and all the launchers' guidance systems. It is responsible for the entire CZ-4/2D vehicles. The Chinese Academy of Space Technology in Beijing is responsible for sounding rockets but is primarily known for the recoverable FSW platforms now offered for μg/Earth resources activities. China is developing a heavy lift launch vehicle equal to the latest Delta 4 and Atlas 5 configurations and has announced development of a relatively low cost solid propellant small satellite launcher to compete with Europe's Vega project.

Space launcher production capacity for the two manufacturing sites is assessed to be 10 to 12 vehicles annually. In June 1989, GWIC and Avibras of Brazil established International Satellite Communications Ltd (Inscom) to market their combined launcher, spacecraft and tracking expertise, but the company fell into abeyance. By the end of 2002, China had launched 97 satellites with a launch success rate of more than 90 per cent.

France

Fortech Airforge

Current Status
Formerly known as L'Air Liquide, with more than 450 plants around the world, Fortech Airforge specialises in the production, handling and utilisation of gases and cryogenic liquids. Its contribution to Ariane 4 includes stage 3's tanks and ground equipment for handling cryogenic liquids. Arianespace signed a FFr400 million joint contract with L'Air Liquide and Aerospatiale in 1989 for 50 Ariane 4 stage 3 tanks in the P9 production lot. The tanks, being delivered 1991–98, are manufactured by the company at its Sassenage facility and then integrated by Aerospatiale at Les Mureaux. The two companies established the Cryospace GIE (55 per cent L'Air Liquide, 45 per cent Aerospatiale) in 1988 to develop/fabricate the main tanks for Ariane 5's principal cryogenic stage. The first tank completed main structural tests in November 1991. L'Air Liquide was contracted in 1990 to build a LH_2 production plant, with an annual 10^6 litre capacity, at Kourou to meet Ariane 4/5 requirements. It entered service in 1991 and now supports all Ariane 5 flights the first of which took place in June 1996. Development programme for research activity at ESA on the next generation of launchers emphasising cryogenic tank and recovery requirements.

Pyrospace

Current Status
The company was established in 1989 by Aerospatiale and SNPE of France and OEA of the US to develop/manufacture pyrotechnic components and systems. For Ariane 4/5, Pyrospace provides stage separation, destruct, strap-on ignition/separation, fairing separation and non-contaminating dual-satellite separation. For satellites, it offers open/close of hydraulic and propellant valves, and solar array and antenna deployment by separation nuts and rod cutters. Can provide actuation controllers for cryogenic and hypergolic propellants and for high-temperature environments. Pyro-cutters for multisegmented large antenna arrays and unfurlable composite antennas and relays for serparation devices. Pyrospace has developed a new generation of rapid fire pyrocutters for high-speed separation requirements which can be applied to rocket stage, satellites or spacecraft.

Snecma Moteurs

Current Status
The principal space business of Snecma Moteurs is the design, development and construction of propulsion systems for launch vehicles and satellites, including Ariane's Viking, HM-7, Vulcain and P230 (partnered with BPD in Europropulsion), motors. In satellite work, SM produces Mage apogee kick motors, hydrazine thrusters for attitude control, bipropellant liquid engines and electric/ion engines. The company also produces solar array drive mechanisms, μg furnaces, surface tension tanks, magnetic bearings, transducers and flow regulators. In 2002, personnel totalled 3,800, with a turnover of FFr7,200 million. Aerospatiale in January 1994 sold its 13.6 per cent interest for FFr178.9 million. SM and Spain's Empresarios Agrupados established Iberespacio in October 1989 to undertake joint development of space propulsion systems. SM and its SNECMA parent established Hyperspace in January 1990 to co-ordinate propulsion activities for hypersonic aircraft and launchers. Snecma has studied air breathing hypersonic propulsion systems and bench tests of components have been performed. Snecma continues to study these economical propulsion concepts for reusable launch vehicles. Other interests include: Arianespace 8 per cent, Europropulsion (50/50 with BPD), G2P 75 per cent (large solid-propulsion systems for military applications), Spot Image 12 per cent, Carbone-Industrie 100 per cent (carbon brakes), TECHLAM 50 per cent (laminated elastomeric components) and S2M 54 per cent (magnetic bearings).

Test/Assembly Facilities: SM maintains a 4,500 m^2 Ariane propulsion assembly hall containing 14 Viking engine assembly stands, five for HM-7 and nine for stage 1 propulsion systems. There is also a 400 m^2 class 100,000 cleanroom for cryogenic engine component assembly. TDF propulsion was installed in an integration building provided with two 300 m^2 class 100,000 rooms. The Vulcain Engine Assembly Building can produce eight engines annually. The 1,220 m^2 class 100,000 cleanroom accommodates 12 assembly cells and 25 class 100 laminar flow booths/hoods for storage before assembly. Also in the building are a 700 m^2 warehouse, a 1,400 m^2 engineering and mechanical area and 1,150 m^2 of office space.

Engine test facilities are concentrated at the 1.5 km^2 Vernon site but some are located at the 24,300 m^2 Villaroche area. For storable propellant motors, there are four component test stands (PF1-3, A48, F22) and one engine stand (PF2). Cryogenic testing is undertaken on three stands: the two of Vernon's PF41 (one with altitude simulation) and one at Villaroche (with altitude simulation). Eleven existing cryogenic component stands were joined in 1988 by PF52 for Ariane 5 Vulcain gas generator and turbopump development. The PF50 stand for complete Vulcain engine tests was inaugurated September 1990 at Vernon. SM inaugurated its 11,000 m^2 production plant for Ariane 5's P230 nozzles in October 1990 at Haillan. A FFr180 million investment, it was fabricating 10 nozzles annually by year 2000. Assembly space is 650 m^2, plus 825 m^2 class 100,000 cleanrooms.

Solid rocket motor development and production is the responsibility of Europropulsion. A new research department into hybrid propellants has been proposed to respond to initiatives from Russian engineers for an international development programme on hybrid rocket motors.

UPDATED

Société Nationale des Poudres et Explosifs (SNPE)

Current Status
The Société Nationale des Poudres et Explosifs (SNPE) Group, Explosives and Propellants division, created as a state-owned company in 1971 and now

with 4,500 personnel and sales exceeding FFr5 billion (US$715 million). About 50 per cent of products are exported to more than 70 countries. Manufactures liquid rocket propellants, with emphasis on UDMH, and develops/produces solid propellants. Along with BPD Difesa e Spazio in the Regulus partnership, the company provides the propellant for Ariane 5's P230 solid boosters using facilities at Kourou constructed by subsidiary SNPE Ingénierie. SNPE maintains facilities at Saint-Médard en Jalles near Bordeaux (solid propellant grains), Toulouse (UDMH, AP), Toulon (pyrotechnics) and Le Bouchet near Paris (HQ and R&D). The Toulouse site manages the production of ammonium perchlorate with a capacity of 6,000 tonnes annually to ensure Ariane 5 supplies. SNPE formed a second partnership with BPD, Société Européenne de perchlorate d'ammonium (EUPERA; SNPE 66 per cent, BPD 34 per cent), to handle production. Toulouse also provides the UDMH for Ariane 4 and the MMH for Ariane 5's upper stage. Subsidiary Pyromeca manufactures detonating cords and flexible linear cutting charges and ultra-fast pyrotechnic valves for Ariane 4 release from the launch pad. Also provides cutting explosives for point or linear separation requirements. Conducts tests on cutting re-charge explosives for shaped detonation and timed explosions. In cryogenic propulsion, SNPE provides the main igniter and turbine starter grain for Ariane 4's HM-7B stage 3 engine and the shaped ducts for the Ariane 5 Vulcain igniter and starter. Experimental tests and chemical analysis seek a more efficient solid propellant mix for greater efficiency in specific impulse and better cost effectiveness.

The propulsion package was designed/produced for the 153/227 kg Eclipse T1/T2 supersonic rocket and autonomous launch ramp in partnership with CAC Systèmes. Applications include meteorological sounding and µg research. T1/T2 capacities are 20/25 kg to 12/72 km. A series of advanced upgrade concepts have been drawn up but no firm commitment has yet been made about development.

UPDATED

STARSEM SA

Current Status

STARSEM is a joint-venture company, set up to market the Soyuz family of launch vehicles on a commercial basis. The partners are EADS (35%), RASA (25%), Samara Space Centre (25%), and Arianespace (15%). The agreement for the creation of STARSEM was signed by the partners on 17 July 1996 in Moscow and on 6 August the statutes of the company were signed in Paris. The company is a French-registered corporation, set up with FFr500,000. The backing for the project is split with Aerospatiale having 35 per cent, Samara and RSA 25 per cent each and Arianespace 15 per cent. Starsem employs more than 40 specialists supporting sales, programmes development and finance. Starsem has achieved high levels of flight utilisation in a shrinking launcher market.

The company markets the Soyuz-U with the added Fregat (fourth) stage and the Soyuz/ST with a 4 m diameter fairing and digital flight control system. The first Starsem flight took place on 9 February 1999 placing four Globalstar satellites in orbit. The tenth flight carried Cluster II on 9 August 2000.

On 18 December 2000 Starsem signed an agreement for the launch of the Metop satellites for Eumetsat, the first in 2005 aboard a Soyuz/ST.

The 1,671st flight of a Soyuz launcher took place on 30 October 2002. STARSEM has opened discussions with international bankers to leverage debt burden on launch agreements through reserved loans against profit, this being deemed a convenient way of easing launch costs.

Germany

Daimler Chrysler Aerospace AG (EADS)

Current Status

Previously DASA, DaimlerChrysler Aerospace AG (DBAA) is the main contractor for Ariane's second stage and the PAL liquid strap-on boosters, and provides the thrust chamber for the HM-7B third-stage engine; it similarly provides the injection head, combustion chamber and expansion bell for Ariane 5's Vulcain first-stage engine, under a contract signed with SEP in April 1989. During 1991–98, DBAA is providing 50 second stages and 120 PALs under a DM1,600 million contract, in addition to 50 HM-7B thrust chambers under a DM52 million award. Another 15 stage 2s and 26 PALs were ordered for 1998–99. The company leads Ariane 5's EPS upper stage, the first development of an Ariane stage outside of France. Fourteen were built between 1997 and 2002. DBAA, as MBB, also built the P3.2 test stand for Vulcain chamber trials. Its other engines include storable bipropellant (4, 10, 400, 27,500 N), hydrazine monopropellant (0.02–350 N) and electric (10–15 and 150–200 mN).

As MBB, the company began development of the world's first 50 kN topping cycle engine in 1958 and began testing at 85 atmosphere chamber pressure in 1963. In 1968, MBB's 180 kN LOX/LH_2 engine attained the existing chamber pressure record of 282 atmosphere, leading to NASA and Rocketdyne acquiring a licence for utilising the technology in the Space Shuttle Main Engine. The current 10 N and 400 N NTO/MMH thrusters are combined to form a Unified Propulsion System (12 × 10 N + 1 × 400 N), employed by the Galileo Jupiter orbiter and the TV-Sat/TDF, Tele-X, DFS Kopernikus, Eutelsat 2, Turksat and Amos telecom satellites. Bipropellant 4 N thrusters are currently under development and qualification.

DBAA's Lampoldshausen and Ottobrunn sites are equipped with test stands for high-pressure thrust chambers up to 1 MN (P3.2 stand for Vulcain) and for gas generators and turbopumps with supply pressures up to 800 atmosphere, stands for LH_2 ramjets, sea level and vacuum stands for 4 N to 50 kN bipropellant and 0.5–500 N monopropellant engines (Trauen site), and a surface tension tank propellant behaviour simulation facility.

In August 1999 DaimlerChrysler Aerospace began studies of potential reusable launch vehicles in a joint initiative with OHB System GmbH and ZARM of the University of Bremen. Integrated as a component of the EADS grouping, the former DaimlerChrysler Aerospace has important capabilities for evaluating post-Ariane launcher concepts as pressure on launch cost becomes acute. The company has provided several optional technology paths applicable to possible future launch vehicle concepts, a new approach rather than looking at specific designs first followed by a pursuit of enabling technology.

MAN Technologie AG

Current Status

Studies with MT acting as ESA's prime contractor include structural components such as apogee boost motor casings, CFRP shells for Rosat, Italsat high-pressure gas tanks, the CFRP structure for Orfeus, Silex structures, and radio astronomy high-precision reflectors. Research and test programme on composite structures for beam deployment systems and semi-rigid structures for large antennas and deployable solar cell arrays.

For Ariane 4, MT Space Systems provides the stage 1 thrust frame, the load-bearing gas generator water coolant tank (fibre composite/aluminium, 30 per cent/150 kg less than conventional design), Viking turbopumps and gas generators, aft/forward booster adaptors, and booster separation and release mechanisms. In December 1996, MT took over the production of the second-stage main tank.

For Ariane 5, MT builds the flow-turned steel booster casings in Augsburg, the main tank's front skirt, the GAT/GAM high-pressure vessels for TVC of Vulcain and booster nozzles, Vulcain's gimbal joint, and the steering mechanism's heat shields. In December 1996, MT took over the production of the main tank bulkheads. MT has production facilities in Augsburg, Munich and Oberpfaffenhofen. MT is also responsible for erection and maintenance services at Ariane launch facilities.

MT played a leading role in the development of Hermes' thermal protection system, drawing on its expertise in ceramic composites, high-temperature materials technology and complex system modelling and testing. MT also developed methods and tooling for manufacturing the hot structural elements. The experience is reflected in its development of Sänger's hypersonic intake ramp model.

MT upgraded and extended its know-how in the fields of development and the production of components incorporated into space transportation systems, in order to qualify itself to be a 'general contractor' for entire systems. The company being commissioned with the overall responsibility for the co-ordination of the configuring of Germany's portion of work to be carried out on the European Crew Transport Vehicle's (CTV) systems represents an initial step towards that goal. MT was also able to secure orders for key components (structures, tanks) of the Automated Transfer Vehicle (ATV) project. MT is also participating in studies of post-Ariane 5 launch vehicles.

MT has delivered almost 800 Viking turbopumps and associated gas generators at a production rate of 60 units annually. The two single-stage centrifugal pumps use eight-blade steel wheels to move 270 kg/s propellant while raising its pressure from 5 to 57 bar. The Curtis velocity-compounded impulse turbine incorporates two rows of rotating blades to drive the shaft from the gas generator's water-cooled exhaust. The generator's spherical chamber is cooled by eight water jets facing the propellant's 40 injection ports. The turbopumps also serve as part of the thrust structure.

India

Alfa-Laval (India) Ltd

Current Status

Incorporated in 1937 the company undertakes precision fabrication of liquid engine components and is contracted to provide elements of the GSLV. Alfa-Laval (India) Ltd was owned by Tetra Laval until a re-structuring in 1993. Designing high-strength, long life, components for GSLV upgrade concept studies leading to an improved reliability value. Alpha-Laval has been asked to look at high-strength structures for advanced launch vehicles more powerful than the GSLV and to suggest appropriate structural design conditions accordingly.

Larson & Toubro Ltd

Current Status

Founded in 1949, Larson & Toubro specialises in precision fabricated and machined components for light engineering and specialised projects. It provides the motor casings in 15CDV6, MDN250 and T16AL4V, fuel tanks and gas bottles for India's PSLV and GSLV satellite launch vehicles. It has also provided specialist machinings for satellite fittings and support brackets and has developed ultra-high strength couplings, bellows and fixtures for high-temperature environments such as upper thrust structure mountings in the vicinity of rocket motors and turbines.

International

Europropulsion

Current Status

Europropulsion was established in 1985 by France's SEP and Italy's BPD to bid for solid motor development contracts within Europe's civil space programme. Its principal activity is Ariane 5's P230 strap-on boosters. The FFr3.8 billion fixed price contract, following two preliminary awards, was signed with CNES in late 1989, covering development/qualification of the motors, including 10 ground tests to mid-1995. Individually, BPD provides the lead for segment and igniter development/integration, SEP contributes the nozzle assembly and insulation liner, and sub-contractors MAN Technologie provide the segment casings, SNPE propellant development and Regulus (the SNPE-BPD partnership) for the two large segment casting operations at Kourou. The small forward segment is completed by BPD in Italy. The EUPERA SNPE/BPD partnership produces the ammonium perchlorate at SNPE's Toulouse site (the AP for 501/503 was provided by Kerr McGee Corporation of the US and for 502 by SNPE; future division remains to be decided). SEP inaugurated its nozzle production plant (UTB: Usine Tuyère Booster) in October 1990 at Haillan. A FFr180 million investment, it will fabricate

10 nozzles annually by 2000. MAN commissioned its production centre in Augsburg during September 1988: this 5,400 m^2 facility includes an 1,800 m^2 air conditioned area for final machining of the clevis/tang connection between the booster segments. The 500 tonne counter-roller flow-forming machine rolls the initial 40 mm thick, 1 m high, 3 m diameter steel cylinders into 8 mm thick 3.5 m high segments before heat treatment produces an ultimate strength of 1,500 N/mm^2.

The first demonstration motor, a reduced scale version loaded with 15 tonnes of propellant, was fired successfully in December 1989. The first full scale P230 was successfully tested 16 February 1993. Seven full scale tests were then planned: B1 with battleship (thick) casings, five (M1-M5) development and two (Q1-Q2) qualification. The last two tested the entire stage. M1, the second firing, was successful 25 June 1993. This was the first in-flight configuration, confirming thrust profile, intersegment seals and nozzle gimballing. M2 was planned for mid-November 1993 but voids were found in the motor's propellant and that motor was shelved (it is expected to be fired after an ageing programme). The third firing, M3, was successful 20 June 1994. M4, with a flight-type actuation system and lighter thermal protection, was successful 30 September 1994. Flight-standard M5 was successful 15 December 1994. Q1 was successful 10 March 1995, Q2 21 July 1995. The first complete Ariane 5 was launched 4 June 1996. Of the 17 Ariane 5 flights completed by 1 September 2003, 15 have been successful for a success rate of 88 percent but all solid motors on all flights have performed as required.

International Launch Services (ILS)

Current Status

After the merger of Lockheed and Martin Marietta in 1995, ILS was created to handle the joint commercial marketing of Proton and Atlas for the Lockheed Khrunichev Energia venture and Lockheed Martin Commercial Launch Services (LMCLS). LKE was formed in early 1993 as a joint venture to market Russia's Proton launch vehicle and associated launch services. LMCLS has been doing business since 1988 as the commercial marketing company for the Atlas launch vehicles. ILS provides integrated marketing of both vehicles, with the concept of mutual back-up.

The ILS services cover spacecraft integration, Proton and Atlas supply, mission management, insurance brokering, launch site support, post-mission support and customer support. Backlog of US$3 billion representing more than 30 launches. Over US$1 billion in contracts signed in 2000 (16 launches). 14 launches in 2000 including first flight of Breeze M upper stage.

By the end of August 2002, ILS had accomplished 52 Atlas launches and 24 commercial Proton launches. Khrunichev has in addition launched 26 Proton variants for the Russian government since the end of 1995. During 2003 ILS launched six times, won contracts for 11 new missions and received nine additional assignments for future US government launches. During 2003, five launches were with various configurations of the Atlas and one with the Proton using a Breeze M upper stage.

International Space Technology, Inc (ISTI)

Current Status

Space Systems/Loral and Russia's OKB Fakel and RIAME Research Institute of Applied Mechanics & Electrodynamics formed the ISTI joint venture in 1992 to market Fakel's SPT-100 0.08 N Hall xenon stationary plasma thrusters using western electronics. They provide a 500 kg mass saving over a conventional system on a 3.5 tonne satellite over a 15-year life. The venture was later joined by SEP and Atlantic Research Corp. ISTI's thrusters became available in 1995. Fakel has built approximately 100 of the devices, which have flown on approximately 50 Meteor polar meteorological satellites since 1972 to provide orbit control. Gals 1, launched January 1994, is carrying a set of eight to provide the first NSSK on a Russian satellite. SEP joined ISTI in 1993, receiving commercialisation rights in Europe in exchange for the other partners to have equivalent rights in North America and Russia on the Mk 2 version that SEP is developing. This will increase SI from 1,650 to 2,200 seconds. Development began in May 1993 at SEP Villaroche with a series of test runs on site at low operating levels. France's 1999 Stentor satellite was to be equipped with them, but technical problems meant the satellite was equipped with only slightly modified versions of the original design instead of the Mk 2 Stentor which was lost in an Ariane 5 launch failure in 2002.

UPDATED

A Russian Proton launch vehicle lifts off from Baikonur with an Inmarsat payload (ILS) 0572503

Regulus

Current Status

Regulus is a partnership between France's SNPE (40 per cent) and Italy's BPD Difesa e Spazio (60 per cent), formed to operate Kourou's Ariane 5 booster propellant facility Usine Propergols en Guyane (UPG). The plant, comprising 40 buildings totalling 26,000 m^2 on a 300 hectare site, was commissioned 24 October 1991. Cost was AU180 million (about FFr1,250 million). Each strap-on contains 237 tonnes of HPBT solid in three sections. Regulus provides the propellant required for the mid and aft units, while the forward element is shipped from BPD. When the facility became fully operational in 1995, its 150 personnel were able to support eight flights a year, requiring 3,300 tonnes of propellant for 32 segments. The first segment was poured November 1991 for the B1 battleship motor, tested 16 February 1993. First flight of full segment booster on Ariane 5 took place 4 June 1996. By the end of 2002 full segment boosters had been launched on 14 Ariane 5 flights. Although Ariane 5 has suffered 3 failures in 14 attempts the boosters have performed close to specification. Regulus continues to research propellant combinations for improved efficiency and more cost-effective production commitments.

Sea Launch Co, LDC

Current Status

The Sea Launch joint venture was created in 1994 for three-stage Zenits to be shipped to the US, assembled and then transported into international waters for near-Equator launches. The consortium comprises Boeing Commercial Space Co (40%), Norway's Kvaerner group, NPO Yuzhnoye and RKK Energia. Hughes Space and Communications Company, placed a firm order December 1995 for 10 launches. With options, the contract could be worth around US$1 billion. In early 1997, there were 13 firm orders from Hughes and five from Loval, with Hughes holding additional options. The first launch of the Zenit-3SL with a demonstration payload took place on 27 March 1999 followed by the first commercial launch on 9 October 1999 with the DIRECTV satellite. The third launch on 12 March 2000 was unsuccessful resulting in the loss of the Hughes HS-601 ICOF-1 satellite. The fourth launch on 28 July 2000 placed the PAS-9 satellite in orbit followed by Thuraya-1 on 20 October 2000. The sixth flight successfully launched on 18 March 2001 carried the XM-2 satellite into GTO followed by XM-1 to GTO on 8 May 2001. The eighth flight took Galaxy 3C into orbit on 15 June 2002 with Thuraya 2 launched on 10 June 2003 and the 10th launch carrying EchoStar IX/Telstar 13 into GTO on 7 August 2003. The 11th flight on 30 September 2003 carried the Galaxy XIII/Horizons-1 satellite into GTO. Kvaerner Rosenberg converted a

30,000 tonne oil drilling rig into the ocean-going launch platform in Stavanger under a US$78 million contract; Kvaerner Govan in Scotland built the 34,000-tonne 200 m Assembly & Command Ship (ACS) for US$93 million.

Israel

Israel Military Industries Ltd (IMI)

Current Status

IMI is a government-owned company comprising an HQ group and four operating groups of 10 divisions and 18 plants. There are 4,250 personnel. The Rocket Systems Division (RSD) of the Systems Group develops/manufactures solid rocket motors, weapon systems and metal/composite material products. For example, it provides the propulsion systems for the Gabriel sea to sea and Shafrir air-to-air missiles. In the space field, it developed/produces the stage 1/2 motors for the Shavit/NEXT launchers. These are HTPB cylindrical grains in filament wound composite cases. Steering options are jet vane, LITVC and flexseal. The motors are marketed as the ASTM family in the US by Atlantic Research Corporation. RSD is deeply involved in both Israel's space programme and the SD10 Arrow. The propulsion systems and the pyrotechnics for these programmes are IMI developed and produced. IMI also produces solid propellant motors for satellite applications. New forms of solid propellant have been reported which can be stored for long periods and controlled by liquid propellant injectors. IMI has worked new high-energy propulsion concepts and tested new and innovative forms of propulsion although details are unknown. It is claimed that IMI is developing propulsion systems for hypersonic scramjet concepts yet to be built.

Rafael Manor

Current Status

Rafael's Propulsion Dept (RPD) specialises in the R&D and production of solid, liquid and air-breathing propulsion systems. It has researched ramjet and scramjet technologies for long range air-to-air missiles. It developed Shavit's AUS-51 solid motor (marketed in the US by Atlantic Research Corp) and Ofeq's hydrazine AOCS. Also used in a new air-to-air missile project. This module provides the means for despin, final orbit positioning, station-keeping and MW unloading. It comprises hydrazine thrusters, tank, latch valves, pressure transducers, filters, thermal control and fill/drain valves. Conceptual development of hypersonic scramjet propulsion for airbreathing missile. Development of suborbital upper stage for cryogenic launcher. Studies continue on supersonic and hypersonic missile propulsion systems.

Rotem Industries Ltd

Current Status

Rotem is the commercial arm of Israel's Nuclear Research Centre-Negev (NRCN). It is the official hydrazine supplier to the Israel Space Agency, IAI, Rafael and other major users. A full range of aniline-free hydrazine-based fuels for thrusters and gas generators is offered: space grade (more than 99 per cent), monopropellant (more than 98.5 per cent), standard (more than 98 per cent), H92, H80, H70 and blends/derivatives.

Rotem also specialises in vacuum/pressure vessels, space simulators (up to 5 m diameter), vacuum chambers (to 10^{-10} atmosphere), cryogenic thermal systems and thermal shrouds for missile and satellite launcher markets. Produces high-stress resistant pressure vessels for high-temperature and high-pressure environments where accelerations and mass loadings are experienced. Developing new protective enclosures for high-velocity boost stages.

Italy

Alenia Aerospazio

Current Status

Alenia Aerospazio is the prime contractor for all projects managed by the Italian Space Agency and has a workforce of 2,800. IRIS (Italian Research Interim Stage) is Italy's most powerful upper stage and the first designed for operations from a dedicated cradle in NASA's Shuttle cargo bay. Alenia Aerospazio, a Finmeccanica company, is prime contractor; BPD is responsible for the IRIS Spinning Stage (ISS) portion of the assembly. The spinning, solid propellant stage is optimised for injecting 900 kg science and small telecom satellites into GTO from the standard Shuttle parking orbit, with a maximum 25 per cent propellant offloading available for 600 kg payloads. IRIS' only scheduled mission, however, was targeted at a lower orbit. The Lageos 2 geodetic satellite, also with Alenia Spazio as prime, was placed in 5,610 × 5,950 km, 52.8° this way in 1992. No further applications have been manifested but it is available for other vehicles. It could also be employed as a planetary kick stage with a launcher such as Atlas Centaur.

The IRIS system is designed for a Shuttle load factor of 0.125 (1/8 of cargo bay volume), with the ASE – essentially a cradle providing structural/thermal protection, power/command links and spinup motors – designed for a minimum of five missions. The ASE interfaces with the Shuttle via four lateral and one keel fittings. The 4.5 m diameter, 2.1 m deep cradle restrains the IRIS Spinning Stage (ISS) with two 'grabber' devices on the Payload Attach Fitting and releases them for the 45 to 100 rpm spinup with redundant electric motors. All the deployment electronics are mounted in the ASE, with the computer responsible for commanding spinup and severing of the base clamp by explosive bolts for four springs to impart 0.4 m/s. An aluminium platform around the motor's dome end supports stage subsystems such as avionics/telemetry systems. Separation switches trigger the stage sequence. 128 channels provide telemetry at 2 kbit/s on 2.2270 GHz through the 5 W transmitter for 50 minutes. BPD's Nutation Control System uses N_2 for controlling coning during coast. All coast operations are commanded by two redundant electrical sequencing systems, finally releasing a yo-yo tumble mass to avoid payload contact. Motor ignition occurs about 45 minutes after deployment some 50 km from Shuttle.

Payload dynamic envelope is a 1,900 mm diameter, 2,455 mm high cylinder. The Payload Attach Fitting 937.62 mm diameter upper flange mates with the payload's fitting with a two-sector clamp band, separated pyrotechnically; four springs provide 0.5 m/s separation rate. Two 37-pin flyaway pins provide electrical interfaces at the clamp band level.

The Shuttle ASE is now stored in Turin.

The IRIS design has been used by Alenia to develop derivative concepts for upper stage applications on future European launch vehicles but to date no configuration has been determined.

Alenia Aerospazio has been central to the design and development of the Automated Transfer Vehicle. Developed versions of the ATV envisage application to lunar and Mars bases.

BPD Difesa e Spazio

Current Status

BPD Difesa e Spazio, wholly part of FiatAvio since January 1996, is involved with propulsion systems, including the study and production of solid, liquid and gaseous propellants, as well as studies on magneto-plasma dynamic systems. This involvement includes the design, development and manufacture of apogee, stage separation and space launcher motors, strap-on boosters, orbital transfer systems and attitude control equipment. FiatAvio has a 50 per cent investment in Europropulsion. 1995's space revenue was projected at Lit280 billion (1994 Lit262 billion; 1993 Lit323 billion). It is Europe's seventh largest space company.

BPD has provided bi/monopropellant and gaseous attitude control systems for satellites such as Italsat 1/2, Lageos 2, SAX and Artemis. It provided the propulsion systems for Sirio and GEOS 1/2, and the motors for the European Mage family of apogee motors for which it is responsible for the propulsion and thermal protection, as well as overall integration/testing. This motor was selected for ESA satellites such as Giotto, Telecom, ECS 2–5, Meteosat, Hipparcos and Lageos 2. BPD also provided the motor for Italy's IRIS upper stage. A large propulsion unit, known as ALFA, was also developed and, as a result of this programme, BPD designed and produced the Ariane 3/4 solid propellant strap-on boosters, and is a partner with SEP in Europropulsion, responsible for developing Ariane 5's P 230 solid boosters. Regulus is a partnership between BPD and France's SNPE formed to operate Kourou's Ariane 5 booster propellant facility. Each strap-on contains 237 tonnes of HPBT solid (68 per cent AP, 18 per cent aluminium, 14 per cent binder) in three sections. Regulus provides the propellant for the mid/aft units, while the forward element's 23 tonnes is shipped from BPD. The facility became fully operational in 1995. BPD has conducted extensive tests on improved lightweight solid propellant booster cases for P230 stages as an unsolicited initiative. New solid motor propellant combinations developed for potential high-efficiency, low-case weight, motors.

FiatAvio SpA

Current Status

FiatAvio entered the space propulsion field in 1985 as an extension of its primary aero gas turbine business. Under SEP contract, it designed, developed and manufactures Vulcain's LOX turbopump. Preliminary development of Vulcain Mk 2's pump, increasing power from 3 to 5 MW, was underway and testing began in 1998. Test facilities accommodate specialised testing of critical cryogenic components. It collaborated with DASA and Techspace Aero on ESA's FESTIP studies, and it is performing R&T studies for advanced cryogenic turbomachinery under the Euro/Russian RECORD program led by SEP. This resulted in further engineering tests in Russia. FiatAvio SpA is also working on propulsion requirements for advanced launch systems under ESA programmes. Previous activities include studies of component development for 30 to 80 kN liquid engines for orbit propulsion modules and launcher upper stage applications. FiatAvio is a stock holder of Intospace and Arianespace, and since January 1996 has held 100 per cent of BPD Difesa e Spazio. Studies with BPD on future propulsion system requirements. Has produced optional development paths for new and more powerful stages to lift a new generation of reusable rocket stages with recoverable motors and complete propulsion systems.

Japan

Daicel Chemical Industries Ltd

Current Status

Daicel makes rocket motor and propellant products. Products include new lubricants and seal gases for component combinations, extreme pressure and temperature-resistant oils and thermal barrier fluids. Employees total more than 300. Develops low slip-flow lubricants for cooling and lubrication.

IHI Company Ltd

Current Status

In space propulsion, IHI provides the cryogenic turbopumps for the H2's stage 1/2 cryogenic engines, in addition to H2's hydrazine Reaction Control System modules. It previously provided the pumps and gas generator for H1's stage 2 engine. ISAS, IHI, KHI and MHI are pursuing development of the ATREX-500 air turboramjet, which could power the flyback booster of a two-stage spaceplane. The hydrogen expander ATREX would take the vehicle to M6/30 km. The 5.4 kN ATREX-500 ¼-scale version, 2.2 m long and fan inlet 30 cm diameter, totalled 1,400 seconds in 40 static sea level runs during 1990–95 at ISAS' Noshiro Testing Centre. Efforts now centre on a 1.4 tonne flying testbed launched from Japan and parachuting into the sea. A jet engine will take the FTB to M0.5 from a horizontal rail for the 370 kg ATREX-500 with 190 kg of LH_2 to attain 25 to 35 km altitude 200 km downrange. The initial metal turbo machinery will achieve M4.5, with M6 targeted for the carbon-carbon composite design. Goals for the composite ATREX are 17.0 to 18.9 kN and 3,150 to 3,300 seconds SL static.

IHI's Rocket Test Centre was established in October 1975 at the Aioi Plant as a cryogenic test centre of IHI's Research Institute. It was incorporated in July 1980 into the Space Development Division for testing the cryogenic engine and bipropellant apogee engines.

IHI developed the 2 kN bipropellant apogee engine under NASDA contract (in co-operation with NAL) for ETS 6 and COMET. IHI is independently developing a 10 kN CUS Cryogenic Upper Stage LOX/LH_2 expander cycle engine for upper stage or orbital transfer vehicle applications. SI 471 seconds, chamber pressure 28 atmospheres, LOX turbopump 43,000 rpm, hydrogen turbopump 85,000 rpm.

UPS provides GEO insertion and attitude/orbit control for 2 tonne-class satellites; it is derived from the LAPS of ETS 6. It carries a single 1.7 kN bipropellant Apogee Kick Engine plus 4 × 50 N + redundant 8 × 0.86 N hydrazine catalytic thrusters. COMETS's flight model was shipped March 1996.

IHI is also contracted to provide the turbopumps for the H2A. IHI has developed high-speed turbomachinery for industrial applications and research in this sector has in turn stimulated improved manufacturing processes for the rocket motor products. IHI is developing new techniques for cutting costs as mandated by the Japanese government decision to rationalise the aerospace industry and to commercialise the H2A launch vehicle.

Iwatsu Electric Company, Ltd

Current Status

Iwatsu performs rocket motor check out and hot firing test, control and telemetry/data acquisition systems. Calibrates thrust and temperature and pressure fluctuations, performs high-speed data capture for combustion instability and thrust variations, performs historic fluctuation profiles for instability predictions and computes variable output rates. The company provided H1's propulsion system console and its engine data acquisition system and H2 data transfer equipment. Has developed new performance monitoring consoles for H2A and new telemetry for performance measurement transmission.

Mitsubishi Heavy Industries Ltd

Current Status

MHI has been given government approval to begin development of an expanded capability H2A with two, instead of one, cryogenic LE-7A motor in the first stage. MHI markets the H2A and and launch costs for the improved version will be about ¥10.5 billion at current prices compared with ¥8.5 billion yen for the standard H2A. However, the improved H2A will lift 6,000 kg to LEO compared with 3,000 kg for the standard launcher and send 4,000 kg to GTO versus 2,000 to 2,500 kg. MHI will provide 25 per cent of the estimated ¥20 billion needed to develop the improved version, the Ministry of Education, Culture, Sports, Science and Technology funding the remaining 75 per cent. The improved capability version is needed to supply the Kibo research laboratory when it is attached to the ISS later in the decade. Mitsubishi provides Japan's primary capability in liquid rocket engines, most recently developing the cryogenic oxygen/hydrogen LE-7A stage 1 and LE-5B stage 2 engines for NASDA's H2A launcher. The company has prime industrial responsibility for H2A (as it did for H2). The 10-year US$108 million LE-5 development programme produced Japan's first operational LOX/LH_2 engine, providing H1's stage 2 with a restart capability. The technology formed the basis for developing the larger and more advanced LE-7 and the LE-5A direct outgrowth. The LE-7A engine was developed for the first stage of the H2A.

MHI also fabricated Rocketdyne MB-3 engines under licence for the H1 and produced the indigenous LE-3 storable propellant engine for the seven N1 rockets. Two advanced smaller engines, the 9.8 kN RE6 and 2.94 kN RE10-300, have been tested for possible applications in future orbit transfer vehicles.

MHI is testing Liquefied Air Cycle Engine (LACE) hardware, in which LH_2 is used to liquefy atmospheric oxygen, using the LE-5 as a demonstrator. Mitsubishi's Tashiro Test Field includes stands for LE-5/5A/5B engine and stage sea level firings, a short-duration LE-7 stand and an H2 stage 1 battleship tower (full-duration firings of flight-type engines and stages are undertaken at Tanegashima's H2 launch complex). MHI completed a 105 m long 30 m wide 20 m high assembly/test facility at its Nagoya plant for the H2 in May 1988, allowing stages to be shipped from Nagoya bay to the launch site. MHI is supplying stage 1 fuel tank and stage 2 LHz tank for Boeing's Delta 3.

MHI developed the LE-5B for the H2A with increased flexibility, chamber expander bleed, and throttle capability. New test facilities and performance monitoring programmes for test rigs in static firing stands. Continues research on reusable cryogenic upper stage configurations.

Nippon Propellant Industry Company Ltd

Current Status

Nippon does R&D and production of anhydrous hydrazine, MMH, UDMH and hydrazine decomposition catalyser for launcher and spacecraft. Monopropellant system and configuration design for long-life storage. Provides propellant combinations for apogee and altitude control motors for satellite and spacecraft. Has developed exotic combinations of propellant for unique propulsion requirements.

Nissan Motor Company Ltd

Current Status

Nissan's Aerospace Division is Japan's primary developer/manufacturer of solid propellant rocket motors and launch vehicles. These include the M-5 satellite launcher, the MT-135P, S-310, S-520 and K-9M sounding rockets, boosters for the H2 and the H2A, and several kick stages. The company is also involved in propulsion studies for Japan's spaceplane and is responsible for the experiment logistics module exposed section of the Japanese Experimental Module contribution to NASA's Space Station.

Manufacture/assembly is undertaken at Nissan's Ogikubo plant (planned to move to Tomioka City); the Taketoyo Test Facility opened in 1980 for vacuum firings of spinning upper stage motors. Taketoyo's Spin Static Firing Test Stand provides a 39 to 78 mbar vacuum for 1.8 m diameter motors of 0 to 147 kN at spin rates up to 150 rpm. The company provided test equipment for the H2 solid booster static stand at NASDA's Tanegashima site. R&D activities are conducted at the Kawagoe plant, including research into liquid bipropellant satellite thruster systems, ram propulsion and gimballing TVC. The company is also a major supplier of pyrotechnic devices for space applications. Has created a unique range of pyrotechnic devices for long term storage and high-precision timing. These are utilised in unlocking rapid sequential phases of spacecraft release and deployment.

Nissan developed the four-segment solid strap-ons for Japan's H2 launcher, with four demonstration firings during 1988–91 at NASDA's Tanegashima Space Centre. First test was conducted April 1988, prototype number 1 was fired June 1989 and prototype number 2 13 December 1989; the qualification model firing was achieved 29 May 1991. The units represent the country's largest and most powerful indigenously developed solids. The first flight with the solid rocket booster took place on 3 February 1994. Seven sets had been launched by the end of 2000. The improved H2's SRBs will be reduced to three segments. Upgraded H2A, uses a monolithic composite case design. Developed manufacturing techniques and assembly methods for low cost production.

NOF Corporation

Current Status

NOF has participated in solid rocket propellant production since 1954 and is now the sole supplier to NASDA/ISAS. Plant for satisfying H2 and H2A booster requirements was installed 1987. The company also manufactures pyrotechnics. These range from small pyro devices to string pyros capable of severing attachment adaptors securing rocket stages and satellite mounting adaptors. NOF has produced propellant combinations for satellite PKM and AKM applications in satellites. Developed hybrid propellant and tri-propellant combinations for special research projects.

Rocket Systems Corporation

Current Status

RSC was incorporated on 5 July 1990 by 75 Japanese aerospace companies, insurance organisations, banks and trading companies. RSC was inspired by an unsuccessful attempt by Japan's government to transfer launch vehicle procurement, manifesting, manufacture and launch to a private organisation. Its original business was to block buy H2 launchers from the suppliers for selling on to NASDA for national missions. With H2's major contractors, RSC has expanded to cover production management and launch preparation activities at Tanegashima, in addition to marketing H2 commercially. RSC also offers the TR-1A suborbital μgravity vehicle. H2 prime contractor Mitsubishi Heavy Industries leads the corporation, with Nissan Motor Co, IHI, NEC, Kawasaki Heavy Industries, Fujitsu Ltd, Japan Aviation Electronics Industry and Toshiba Corp. RSC has concluded launch services agreements with two satellite suppliers: both were in 1996, the first being with Hughes Space and Communications for 10 launches during 2000–05 (plus additional optional launches) and the other is with Space Systems/Loral for 10 launches during 2000–05. Lack of confidence in the H2 and uncertainty regarding the performance and reliability of its successor, the H2A. These plans have been shelved. The Japanese government is presently re-evaluating the possibility of contracting to an industrial consortium. Privatised marketing of the H2A is now in the hands of the Mitsubishi Corporation and that company is presently funding 25 per cent of the development cost of an improved version which will have a heavy lift capacity to LEO and GTO.

Showa Denko K.K.

Current Status
SDK manufactures LH_2 at its Oita Petrochemical Complex for NASDA's H2 launcher, and in 1986 formed a joint company, Pacific Hydrogen Co Ltd. (PHC), with L'Air Liquide SA of France and Teisan Kabushiki Kaisha to create a 7.4 million litre annual production capacity. Hydrogen gas of 96 per cent purity is generated as a by-product at an adjacent ethylene plant and purified to 99.9999 per cent or better by pressure swing and cryogenic absorption techniques. The gas is then liquefied at –253°C by Claude's process by passing it through Freon/LN_2 coolers and into a cold box comprising a multi-stage catalyst bed and coolers. PHC operates 11,000 and 18,000 litre insulated vehicles for shipping. Approved for production of LH2 for the H2A. Also for new upper stage concepts proposed to NASDA by manufacturers. Company has studied the production challenges of disassociation of LH2 using less complex processes that may be required in lunar surface-base extraction from surface rocks and crustal materials. This is also for the relevance of earth-based processes for planetary site applications.

Netherlands

Aerospace Propulsion Products BV

Current Status
APP is owned equally by TNO and Stork. APP, Stork Product Engineering BV and TNO 1986–95 developed Vulcain's starter and igniters. APP is responsible for complete production as a subcontractor to Stork. Following five years of R&D, the feasibility of using Hydrazinium NitroFormate (HNF) in high performance solid propellants (7 per cent greater than current AP/Al/HTPB) was demonstrated in 1995. APP, under ESA contract, is preparing for the industrialisation of HNF propellants. ESA wants to centralise this process to reduce costs. APP's pilot plant produces 5 kg batches. Active in developing long storage life solid propellants for satellites and launcher stages. Has proposed hybrid tri-propellant combustion chemistries for exotic propulsion requirements. Also involved in gaining better scheduling and manufacturing timelines to reduce costs of propellant.

Russian Federation

Eurockot Launch Services GmbH

Current Status
The commercial Rockot launcher, derived from the SS-19 2-stage ICBM, is marketed by Eurockot, the 10-man Khrunichev/Astrium 49/51 per cent joint venture announced 23 March 1995. Eurockot is responsible for all commercial activities, acting as the customer's prime contractor, but Khrunichev develops and produces all of the hardware. Launches are made from a tube on a modified Cosmos pad at Plesetsk, replicating a silo, on stage 1 skids jettisoned pyrotechnically. Launch facilities were commissioned in May 2000.

There were three demonstration launches from Tyuratam, all silo-based. The first two (20 November 1990; 20 December 1991) were suborbital, demonstrating basic vehicle performance and KB Salyut's new Breeze stage 3 (inactive on the first ascent). The first orbital flight was 26 December 1994, into 1,875 × 2,253 km 64.80° carrying the 70 kg RS-15 Radio-Rosto amateur relay. Commercial Rockot carries a large new fairing developed by Khrunichev. Breeze was originally designed to dispense multiple communications satellites into low orbits in the event of nuclear conflict. The SS-19s would have resided in silos awaiting launch on demand. The derived Breeze-M was developed as Proton's new stage 4. It was flown for the first time on a Proton in 2000. Two test flights with the commercial Rockot-km were made in May 2000 and March 2002. The latter included the launch of the NASA Grace satellites. The next launch took place on 20 June 2002 with the launch of two Iridium satellites. On 30 June 2003 the Rockot system successfully carried eight micro- and nano-satellite payloads from the Plesetsk cosmodrome into various orbits through the Multiple Orbit Mission. Eurockot Launch Services has a second launch lined up for 2003, the SERVIS-1 spacecraft belonging to the Japanese USEF foundation to be deployed into a sun-synchronous orbit at 1,000 km altitude.

KB Khimautomatiki

Current Status
The bureau of Semyin A Kosberg (1903–65), formed 1941, began investigating liquid propellant rocket propulsion for aircraft in the early 1950s and later became responsible for some Soviet space launcher upper stages. In 1974 it was reorganised as KB Khimautomatiki. Has carried out studies in association with TsAGI on air-launched satellite carriers and on equatorial launch sites with a new launch system utilising decommissioned rockets and ballistic missiles.

KB Khimmach

Current Status
A bureau headed by Alexei M Isayev was established in 1943 for research into storable liquid propellant rocket engines. Whereas Glushko's GDL concentrated on major launcher engines and Kosberg's bureau on upper stage propulsion systems, Isayev's produced smaller spacecraft engines, including manned, Earth orbit and deep space systems. A number of designations have been adopted but 'KTDU' (Korrektyroushchaya-Tormorznaya Dvigatelnaya Ustanovka, which means "Corrective-Braking Rocket Motor") is most frequent. Information released in early 1991 indicated the bureau is part of NPO Soyuz in Kaliningrad. It is responsible for the 73.58 kN cryogenic engine of the new Proton KM version, originally designed for the L-3M manned lunar mission, and recently sold to India for its GSLV vehicle. A newly-developed DMT-600 600 N NTO/UDMH thruster was referred to in 1994. Khimmach developed all of the engines for the Heavy Cosmos TKS transport craft. These were adapted for application to early versions of the Mir 2 concept, later becoming the ISS Service Module. Additional examples of the Service Module were offered to the ISS partnership, ostensibly on the basis of cutting costs but in reality an attempt to maintain Russian production lines.

Starting in 1971 and at the instigation of Isayev, the bureau started the development of low-thrust engines which could be used for attitude control of spacecraft. A series of eleven engines with thrusts up to 2,206 kN (225 kg) and eight with thrusts up to 49 kN (5 kg) were developed. They are capable of burns lasting from hundredths of a second through to hundreds and thousands of seconds. The company has developed long duration thruster combinations for low thrust operation but these have failed to find applications.

Korolev Design Bureau

Current Status
One of the most famous engineering design organisations in Russia, Korolev's bureau developed Proton's Block D/DM and Molniya's Block L engines. Approximately 250 separate rocket motor designs and 40 patents for innovative technologies. Mikhail Melnikov designed the Proton engine.

N D Kuznetsov Company SSC

Current Status
Nikolai Dmitrievich Kuznetsov (24 June 1911–30 July 1995) established an aircraft engine design and development bureau in April 1946 at Kuibyschev (now Samara). Information finally released in 1989 revealed that the bureau had been assigned the responsibility for the design and development for engines for Korolyov's four-stage N-1 launch vehicle. Previously the bureau had developed engines planned for use on Korolyov's R-9 ICBM, but politics ensured that the missile used other engines.

In July 1993, Aerojet signed a teaming agreement with N D Kuznetsov SSC to market these engines within the United States. Two NK-33 engines were delivered to Aerojet in 1995. One of them was used by Aerojet for five firings which totalled 410 seconds operating time within a 57 to 114 per cent thrust range which followed the proposed Atlas-2AR launch profile. The engines were under consideration to power the first stage of the Atlas-2AR, but they lost out to the Energomash RD-180 engine. By 1994 aircraft engines accounted for 64 per cent of plant production.

The United States Kistler Aerospace Corporation planned to use NK-33 engines on its reusable K-1 launch vehicles: three would be on the first stage and one on the second stage. Kistler teamed with Aerojet anticipating 58 AT26-NK33 engines.

In 1996–97 Aerojet purchased 36 motors of this type with 20 per cent paid on order. The first batch of 12 NK-33 arrived with Aerojet on 30 August 1998. Kistler still bears ambitions to develop an efficient and cost effective launch system but funding is not fully in place and despite NASA's award of a SLI contract it is unlikely Kuznetsov will find an export outlet in this company. The Bush administration is averse to the award of such contracts to non-US companies.

NII Machinostroenye R & DIME

Current Status
Research & Development Institute of Mechanical Engineering was established in 1958 as part of NII Thermal Processes, but became independent in the early 1980s. Skills based on small combustion elements and machined motor components for engines and motors utilising rocket and gas propulsion concepts. Development and production of 0.8–400 N spacecraft thrusters (implying connection with old Isayev bureau), including Soyuz, Progress and Mir. Research and testing for extended life thrusters designed for long-duration missions. Development and small batch production of engines up to 6 kN can be accommodated. Test facilities can handle hydrogen (up to 3 kN), methane and kerosene fuels. The Elektro Earth observation satellite carries 16 DEN-16 electrothermal thrusters. For the thrusters specified below, LTRE means Low Thrust Rocket Engine, or RDMT in Russian (Raketnay Dvigatel Maloi Tyagi).

NII Thermal Processes

Current Status
NII TP performs research in all forms of launcher and space propulsion, including ramjet, electric and nuclear. NII Thermal Processes has developed nuclear shuttle concepts and researched the type of reactor that would be appropriate to interplanetary transportation requirements in a similar role to that envisaged for the cancelled NERVA project. This was developed by NASA and the US Atomic Energy Agency in the 1960s prior to funding being withdrawn during the tenure of the Nixon Presidency. BMDO took delivery in October 1993 of a T-100 1.5 kW Hall-type Stationary Plasma Thruster for evaluation. This was subjected to extensive trials and although it failed to be adopted at the time

the technology reappeared in Russian proposals for a BMD system during the late 1990s. N_2O_4/UDMH thrusters include 10 and 1,000 N models. NIITP developed the Phobos spacecraft's network of hydrazine thrusters for attitude control and minor orbital modifications. 28 (24 × 50 N, 4 × 10 N) were mounted on four spherical tanks. A set of 0.5 N hydrazine microthrusters of an unusual design was also included to avoid the need for separate N_2 cold gas jets; the propellant was burned internally and cooled inside an expansion tank before passing to the thrusters.

NPO Energomash

Current Status

The history of NPO Energomash can be traced back to 15 May 1929 when the Group for the Development of Rocket Engines was organised at the Gas Dynamics Laboratory (GDL) in Leningrad (now St Petersburg). For more than 60 years the organisation repeatedly changed its name and location, but remained at the forefront of rocket engine design. The founder and the permanent head of the GDL as Academician Valentin P Glushko, considered to be the initiator of the Russian rocket engine industry.

In May 1974, the GDL merged with Sergei Korolyov's design bureau, the new organisation being re-named NPO Energiya, with Glushko as its head. Following Glushko's death in January 1989, the former GDL left Energiya to become NPO Energomash, continuing to concentrate on the development of rocket engines: the bureau was headed by Vitali P Radovsky during 1974–1994. The current general director and chief designer is Boris I Katorgin. In 1994, Energomash was transferred to the Russian Space Agency.

NPO Energomash incorporates the design bureau, plant and test facilities, with the capability to test engines with thrusts of up to 10 MN either as components or complete. An agreement was signed with Pratt and Whitney in October 1992 providing exclusive US marketing rights for Energomash's LOX/kerosene and tri-propellant products and technology.

In 1998, Energomash began transforming itself from a state enterprise to an open-stock holding company but with majority shareholding held by the state. Rocket motors under development include RD-161, RD-180, RD-120K, RD-701, RD-704.

Energomesh designed and built the RD-180 which now powers the most powerful variant of the Atlas launch vehicle, the Atlas 5 series. The first flight took place on 21 August 2002.

NPO Saturn

Current Status

The cryogenic engine which has been described in the 1990s as the D-57 would appear to be the 11D57 gimballing engine which was developed during the early 1970s for use on the Block S of the modified N-1 launch vehicle which would have been flown for the dual-launch L-3M manned lunar missions. The bureau was also tasked to develop the 11D54 fixed-chamber cryogenic engine with the same thrust level of 40 tonnes (390 kN). This engine would have been used in a cluster of six to eight on a modified Block V of the N-1 vehicle the terminal stage of the moon launcher. NPO Saturn developed a plan for a series of N-1 derivatives, scaled both up and down from the standard lunar launch vehicle design.

OKB Fakel

Current Status

Fakel specialises in spacecraft attitude control thrusters, ion engines and plasma sources. Research and development began in the 1960s and a considerable range of designs were tested in the early 1970s. Its SPT-70 and SPT-100 (Stationary Plasma Thruster) Hall electric thrusters have flown on approximately 50 Meteor polar meteorological satellites since 1972 to provide orbit control. The numerical designators indicate beam diameter in mm. SPTs employ a DC-gas discharge in an annular chamber, in which a radial magnetic field traps the electrons. These Hall currents ionise the Xe propellant, the ions of which are accelerated inside a quasi-neutral plasma, without grids, and by the discharge voltage itself. The advantages are their rugged simplicity (no grid system or high voltage supply), but they provide lower efficiencies, lower exhaust velocities and increased Xe consumption. These have been developed to power in-orbit platforms but none have been manufactured for operational programmes.

Sweden

Volvo

Current Status

Volvo specialises in the design, development and production of rocket engine and hypersonic propulsion systems for space applications. 1996 turnover SKr237 million; 1995 space business totalled SKr232 million (1994 SKr181 million; 1993 SKr142 million; 1992 SKr124 million; 1991 SKr221 million; 1990 SKr150 million). Space personnel total 150. Current products include the combustion chambers/nozzles for Ariane 4's Viking 4/5 engines, and production of the turbines (subcontracted by FiatAvio for the LOX turbine and by SEP for the LH_2 turbine) and nozzle extension for Ariane 5's cryogenic Vulcain. Volvo is currently developing the new nozzle and Lux turbine for the Vulcain 2, in co-operation with SEP/FIAT. The company also has an agreement to deliver nozzle skirts for the LE-7B engine to be used on the Japanese H2A launcher. This has been modified and tests show a lack of compatibility in the original design. Research activities include combustion, hypersonics and turbopumps and nozzles for storable propellant upper-stage engines and MTU. Also, a small amount of study on hypersonic air-breathing engines. Volvo has conducted research into hybrid propulsion technology to understand the requirements and evaluate existing capabilities to determine key areas unmatched by current systems capabilities. Volvo conducts research into future design concepts to prepare new test programmes.

Ukraine

NPO Yuzhnoye

Current Status

Yuzhnoye produces verniers for some of its own vehicles (Zenit stage 2 and Tsyklon stages 1–3) and the 11D25 main engine for Tsyklon's stage 3. Auxiliary propulsion systems for upper stages. Has exported some components to other countries, it is believed China and India.

United Kingdom

Air Products PLC

Current Status

A wholly owned subsidiary of Air Products and Chemicals of the US the company provides liquid oxygen/hydrogen and other gases, plus cryogenic engineering expertise. Capable of extended liquifaction process. Cryo-engineering for small scale systems and extreme operating environments. Has conducted tests on high-temperature fluid conduits for small spacecraft.

Dowty Aerospace Wolverhampton

Current Status

DAW is a member of the Dowty Aerospace Division of the TI Group, which comprises 60 operations worldwide. The Space Projects department manufacturers diaphragm, capillary and pressurant tanks. Abilities include cryogenic and corrosive propellants storage dewars.

DAW's space manufacturing facilities were opened in 1989 to provide a fabrication capacity for tanks up to 120 × 150 cm long. The 400 m^2 class 100,000 cleanroom, with 150 m^2 test area, includes a computer-controlled inert gas welder for welding titanium tank shells, computer-controlled resistance welder for fabricating internal components, and orbital tube welder for titanium and stainless steel piping. Test facilities include ultra-sonic cleaning, bubble testing, helium leak testing, pressure and expulsion testing, cryogenic proof pressure testing, dye penetrant and radiographic inspection. High-pressure leak tests under extreme temperature.

The capillary tanks are surface tension propellant management device tanks and are under contract for MMS, Hughes HS-601 and BPD.

DAW manufactures diaphragm tanks for mono hydrazine systems. ESA's Infra-red Space Observatory carries them and Orbital Sciences Corp contracted in 1991 for the 96-litre titanium alloy elastomeric diaphragm hydrazine AOCS tanks for SeaStar. In addition, DQW provided 40 titanium alloy centrifuge tanks for Cluster.

DAW is under contract by MMS UK to supply Xe tanks for the UK ion thruster system on ESA's Artemis. Net volume is 30 litres and it can store 44 kg of Xe at 125 atmosphere.

The company is teamed with Vacco Industries to manufacture titanium etched disc filters, latch valves, fill drain valves and non-return valves for the European market. Conducts leak tests and provides diagnostic equipment.

Fairey Microfiltrex Ltd

Current Status

The company designs, develops and manufactures filtration equipment for fuel, hydraulic, gas and oil systems. Has produced multi-system interlink connectors with high heat and pressure conduits. Fairey Microflex has expertise in spacecraft propulsion systems and has supplied the filter element assemblies for the hydrazine thruster reaction control systems of ESA's ISO and Soho satellites. Ariane 4 and 5 also utilise the company's filters. FM have designed long life, high-temperature resistant coupling conduits for corrosive propellants. Filter systems for environmental control systems and heat-pipe systems.

Reaction Engines Ltd

Current Status

Reaction Engines was set up to design the Skylon SSTO spaceplane to deliver up to 12 tonnes into LEO. Reaction Engines conducted supportive investigations in conjunction with Bristol University and AEA Technology Harwell to investigate some of the more technically contentious features. Project funding is sought; Reaction Engines has brought together an interim UK working group before the formation of an industrial consortium, which it is hoped will be expanded to include international partners. Reaction Engines continues to upgrade the conceptual design of the Skylon concept and has maintained contact with Russian and European launcher organisations. Continues to refine and develop the concept of air-breathing propulsion systems and maintains an information and technical contact with specialists in associated fields.

Rotadata Limited

Current Status

Rotadata has an active programme of developing measuring techniques in turbomachinery and aerospace development hardware. Following the design of a number of fully-instrumented full-scale model turbines for the Space Shuttle Main Engine turbopump, the company has developed the following new products, both hardware and software, for applications to aerospace turbomachinery development and monitoring.

Rotodata new product line features a new high-performance pyrometer sensor with a better sensitivity and a high-frequency bandwidth for turbine blade surface temperature mapping at full rotation speed. This technology has applications to the aircraft and power industry. They make turnkey digital and analogue telemetry systems for rotor instrumentation, and high-temperature strain-gauging for turbine blades and discs. In other NDT applications, they make miniature capacitance probes for blade tip clearance and rotor orbit monitoring and an optical smoke density meter for engine emissions. Atomised chemical analysis of ejected, combusted, products permits engine performance evaluation.

The company is also working on a novel CO_2 sensing unit for cabin condition monitoring. Meanwhile, the traditional Rotadata activities in turbomachinery instrumentation, aero-thermal rakes and probes, specialist positioning systems and high quality instrumented test hardware continues to provide a core business from an increasing range of customers.

Rotodata Ltd has developed fire-scale measuring equipment and associated instrumentation for toxic gas bleed in high-pressure environments, a technology applicable to human habitation in space. Also researching the use of installed monitoring systems in turbomachinery and telemetry transmission capabilities where hard-wiring is inappropriate.

United States

Aerojet

Current Status

A segment of GenCorp of Akron, Ohio, Aerojet was founded in 1942 by astronautics pioneer Dr Theodore von Karman and initially developed Jet-Assist Take-Off (JATO) rockets for aircraft. It was the first US company to produce storable bipropellant rockets. The propulsion plant, as Aerojet TechSystems (formerly Liquid Rocket Co), provided liquid engines for Aerobee, Bomarc, Nike, Vanguard, Delta, Atlas Able, Titan 1 and Apollo, and continues to manufacture Titan, Delta and Shuttle engines. Aerojet, Rocketdyne and P&W formed the Space Transportation Propulsion Team in 1990 for work on the STME Space Transportation Main Engine.

Aerojet manufactured the solid motors for MX, Minuteman, Small ICBM, Standard Missile (rolling out the 10,000th motor in June 1989) and Hawk, and successfully bid in partnership as Aerojet Space Boosters (becoming Aerojet ASRM Division) with Lockheed for NASA's Shuttle Advanced Solid Rocket Motor contract. Congress cancelled ASRM in October 1993. Solid propulsion products range in size from 7 to 155 cm in diameter with thrust levels up to 1,361 kN each. Aerojet produces the SRB for the Atlas 5 launch vehicle, up to five of which can be attached to the first stage. Propulsion activities are performed at the 52.6 km^2 Sacramento plant, founded 1951 and capable of testing cryogenic and storable propellant motors up to 6,670 kN. The Mach 8/30 km-altitude hypersonic Hytest facility was added for National AeroSpace Plane engines and components tests of up to 35 seconds, and the company has access to the Hypulse M25 facility at subsidiary General Applied Sciences Laboratories Inc (Ronkonkoma, New York), which it acquired in 1989.

Aerojet teamed with Lyulka in 1993 to improve and market the 400 kN LOX/LH_2 D57 engine, originally developed in the 1960s for the improved N1-L3M manned lunar mission. It is a candidate for an SSTO demonstrator. A teaming agreement was signed July 1993 with Kuznetsov/NPO Trud to use the NK series exN1 engines and their technology in the US market. Aerojet made five tests totalling 410 seconds with an NK-33 in October and November 1995. If it had been selected for a US vehicle as the AJ26-NK33A (it lost to the RD-180 for Lockheed Martin's new Atlas 2AR), a US production line would have been established. Unit cost would have been about US$4 million.

A similar agreement was signed October 1994 with KB Khimautomatiki for the RD0120, principally developing it into a tripropellant engine. Work is being performed under a US$17.2 million NASA Marshall contract.

Aerojet is contracted to provide all propulsion systems for the Kistler K-1 reusable launcher including three NK-33 for the first stage and one NK-43plus two new Lox/alcohol second stage engines.

In November 2000 Aerojet was awarded a US$7.9 million NASA contract to develop technologies for the Next Generation Reusable Launch Vehicle, part of the Space Launch Initiative. In May 2001 Aerojet received a US$10.4 million contract from the US Air Force for development of a non-toxic peroxide Advanced Reusable Rocket Engine for the Space Maneuver Vehicle concept. Also in May 2001, a US$115 million NASA contract for development of reusable rocket engines for the SLI in a joint venture with Pratt & Whitney. Aerojet is also developing the reaction control system engine for the SLI and conducted the first hot-fire test on 12 December 2001. In a teamed effort with Pratt & Whitney, Aerojet developed the COBRA prototype engine, a reusable, cryogenic, booster and second stage motor in the 2,669kN thrust class. The team also developed a cryogenic expander cycle engine known as RLX.

Air Products & Chemicals, Inc

Current Status

Provision of LH_2/LOX propellants and LN_2/LHe for propulsion systems serving NASA and launch vehicle provider. Monitors for purity and density factors to improve chemical identity of propellant hybrids. Annual revenues of US$5.4 billion and operates in 30 countries. Employs 17,200 people. Based in the UK, Air Products Europe offers a similar service for ESA and government contractors.

Alliant Techsystems, Inc

Current Status

Alliant Techsystems acquired Hercules Aerospace Company in March 1995 for US$296 million. Hercules was selected in October 1987 as the propulsion contractor for Titan's Solid Rocket Motor Upgrade programme, potentially worth US$725 million for the development, qualification and production of 15 sets of Titan 4 solid boosters. Lightweight graphite composite materials replaced the steel used in current motor cases, and a high-performance propellant similar to that developed for the Delta 2 strap-on is employed. PQM-1 (Preliminary Qualification Motor) was the first full-scale test, but the motor failed after a few seconds at Edwards AFB 1 April 1991. The added PQM-1 firing was successful 12 June 1992. The QM-2 third test, run at 2.5°C, was successful 21 February 1993. QM-3 at the 41°C upper limit was successful 2 June. Lockheed Martin awarded the company a contract for Titan launch vehicles support which ended in 2003 extended by follow-on contracts.

The company was awarded a McDonnell Douglas contract in 1987 to develop the stretched solid rocket GEM Graphite Epoxy strap-on Motors for Delta 2; they first flew November 1990. The initial contract for 144 motors (16 flight sets) was followed in 1991 by a second for 117 (13 sets) beginning production in 1993, and a third in June 1995 for 144 (16 sets). GEM production rate is six per month. For Delta 2, from 1996, the expansion ratio for the airlit nozzles will be increased from 10.6 to 16.3 by lengthening the nozzle by 30 cm and increasing its diameter by 19.5 cm. The GEMVN version incorporates a vectorable nozzle; the qualification firing was made May 1994. The TVC system is provided by AlliedSignal and the nozzle by BP-Hitco as part of the strategic partnership. Delta 3's nine strap-ons will be lengthened by 1.22 m and their diameters increased to 116.8 cm. Three of the six ground lit motors will carry gimballed nozzles for TVC, and the three airlit will improve performance with extended nozzles. The first of 16 sets ordered June 1995 were delivered late 1997.

Alliant produces Pegasus' Orion solid motors, derived from GEM, and the fairing. Taurus also uses the Orion motors. Hercules also designs/fabricates composite structures: spars, struts and optical benches used in satellites.

Alliant Techsystems employs 11,400 people in aerospace and defence sectors with turnover in 2000 of US$1.7 billion. In 2001 it acquired Thiokol Propulsion from Alcoa. Research into advanced forms of solid propellant motor for upper stage application and apogee kick motion for large communication satellites.

UPDATED

Ardé Inc

Current Status

Ardé was formed in 1952 by engineers from the Rocket Group of M W Kellog. In the space field it specialises in lightweight pressure gas storage vessels and propellant tanks. Programmes include Shuttle/Centaur, Mars Observer, Atlas 2 and GOES helium pressurant tanks, GOES satellite and Shuttle SRB TVC auxiliary power hydrazine tanks, and Shuttle EMU backpack breathing oxygen tank. The product lines include cryoformed 301/304L pressurant and propellant tanks, prestressed composite fibre over-wrapped metal tanks, and ring stabilised and apex rolling diaphragm expulsion devices. Vacuum pressure expulsion and ullage balance control systems, high-pressure diaphragm release vessels and high-temperature resistant gas and fluid containers.

Cryoformed 301/304L Tanks Under-sized 'preform' tanks are conventionally fabricated from austenitic stainless steel and then immersed/filled with LN_2 before pressurisation to a level about equal to its room temperature design burst pressure. The membranes/welds are stressed beyond the 195°C yield and the vessel plastically stretches: 7 per cent in spheres and 15 per cent for cylinders. The temperature causes a transformation to martensite steel in both parent metal/welds with an attendant 5:1 increase in room temperature yield strength. approximately 3,600 flight tanks using this process have been delivered over 25 years. The Shuttle EMU backpack O_2 tank is a 36.3 cm long, 13 cm diameter cylinder pressurised to 71.4 atmospheres; mass 0.9 kg.

PSC Pressurant Tanks Prestressed composite fibre over-wrapped pressurant (PSC) tanks provide pressure vessels up to 50 per cent lighter than equivalent all titanium tanks. Kevlar or graphite fibre is wrapped on a metal liner and immersed in LN_2 as in the cryoforming process to stress both materials.

Ardé Inc has developed micro-vessels capable of storing gaseous and liquid fluids for delivery under high pressure in nano-technology remote operating systems such as those likely to be needed in planetary explorers. Adapted for cryogenic applications and for research into hybrid propulsion systems.

Assured Space Access, Inc

Current Status

Assured Space Access and Russia's AKO Polyot signed an agreement 21 September 1995 to establish the Cosmos USA joint venture for exclusive marketing of Polyot's Cosmos-3M small satellite launcher outside of the Russian Federation and the former Soviet states. In addition to marketing, Assured Space Access serves as the integration services agent between Polyot and commercial customers: support with export licensing and certifications; integration, transport and logistics planning; interface requirements and documentation; provision of cleanrooms, integration area, crew accommodation; satellite integration and launch support.

A request for an exemption from the US Government (USG) policy prohibiting the foreign launch of USG payloads made by NASA was rejected. As a result the Cosmos-3M was ineligible to win the Med-Lite competition. Since informal US government policy was to give the Med-Lite contract to a US launcher provider operating with vehicles designed in the US the bid was unsuccessful.

Assured Space Access has sought to package small launcher technologies for other companies. Lack of a sound market for such a vehicle continues to discourage investment. However, technical evaluation continues at a sustained level effort. This has periodically attracted interest from investors and some concerted efforts have been made by undisclosed clients to obtain the technology behind the launch vehicle.

Atlantic Research Corporation

Current Status

Atlantic Research Corporation (ARC), founded in 1949, is a manufacturer of solid propulsion motors and gas generators, and in 1987 expanded into liquid propulsion with the acquisition of Bell Aerospace Textron. The company's Virginia and Arkansas facilities take solid motor concepts from design through to high-volume production; the company's annual propellant production capacity is 9,000 tonnes. ARC produces propulsion units for missiles such as Stinger, Tomahawk, Peacekeeper and Trident 2, but also undertakes space-related projects: the company teamed with Hercules in December 1987 to bid (unsuccessfully) for NASA's Shuttle Advanced Solid Rocket Motor. Israel's 1.3 m and AUS51 'Marble' motor, flown as Shavit's stage 3, is marketed by ARC/Rafael. ARC markets the ASTM family of Shavit and NEXT launcher stage 1/2 motors in the US under agreement with TAAS Israel Industries Ltd. ARC also manufactures rocket motor components, including igniters, initiators, casings and nozzles using advanced materials and filament winding/braiding techniques. November 1987's acquisition of Bell Aerospace Textron added liquid bipropellant attitude control and apogee satellite propulsion systems, and their associated tanks and valves. ARC's 22 N (5 lbf) bipropellant thruster was developed for Intelsat 6's attitude control system and has demonstrated 400,000 pulses, 32 hours aggregate firing and 748 kg total throughput. This thruster is carried by Optus B, UFO and Intelsat 7. As Bell, ARC/LP developed the Agena propulsion system that flew on approximately 370 missions. The engine remains available for 31.1120 kN thrust. Agena 2000 available as an upgraded version of the classic Agena.

ARC has delivered more than 1,200 thrusters with more than 800 currently operating in orbit. Recently, ARC completed pre-EMD programme for the Agena which made more than 360 successful flights. ARC has successfully developed the LEROS series of bipropellant engines, including the LEROS IC, the highest Isp AKM flown to date.

Boeing Rocketdyne Propulsion and Power

Current Status

Rocketdyne was a division of Rockwell International and in December 1996 was included in the Rockwell merger with Boeing. It is devoted primarily to the design and manufacture of rocket engines – it is the Western world's largest producer, operating from its Canoga Park engineering and 16 km distant 10.5 km^2 Santa Susana test facilities. 1995 revenue was US$880 million; 5,400 personnel. It was established as a separate division of North American Aviation Inc on 8 November 1955 and following its first major propulsion system for the Navaho missile has delivered around 3,500 engines for Jupiter, Redstone, Thor, Delta, Saturn, Atlas and Shuttle. It currently provides the power plants for Atlas and Delta, and continues to test, upgrade and refurbish Shuttle's main engines, still the only large reusable system in service. It designed and fabricated Peacekeeper's stage 4, its first integration of a full stage.

Rocketdyne was selected in December 1987 to develop Space Station's silicon photovoltaic solar arrays, batteries and power management/distribution. 1994 Station revenue was S$282 million; 1993 US$340 million.

Rockwell's Shuttle Operations Co, established in 1985, was renamed the Space Operations Co in 1989 to reflect a broader range of involvement in space activities, including Space Station. A further change of name took place following the Boeing merger with Rockwell in late 1996. RSOC, a Rockwell Space Systems Division subsidiary, manages Shuttle mission planning, flight design, mission data production, flight crew and flight controller training, ground facility engineering and operations support, and direct mission support. NASA's STS Operations Contract (STSOC), worth US$685 million over four years, was awarded to RSOC in September 1985, effective 1 January 1986. A US$2.3 billion extension through December 2000 was exercised March 1991, bringing total contract value to US$4.8 billion. RSOC became responsible for managing 22 Shuttle operations functions previously performed by 17 contractors.

RSOC was selected in 1990 for Johnson's Operations Support Contract, beginning April 1990. The 10-year cost plus award fee contract is valued at about US$814 million. Rocketdyne provides engines for Delta 2, Delta 3, Delta 4, the X-33 and end-to-end electric power systems for the ISS. Conducted solar power satellite studies began by George Woodcock (Boeing) in the 1970s. Adapted integrated electric power systems for combined space vehicle energy and has developed power systems for electric propulsion including Brayton cycle reactors.

Rocketdyne is currently developing the RS-68 large liquid propellant engine for the Delta 4 family of evolved expendable launch vehicles, developing a thrust of 3,314 kN. Teamed with Mitsubishi Heavy Industries to develop the MB-60, a LOX/LH_2 engine producing a thrust of 267 kN. Teamed with Astrium to produce the 55.4 kN thrust RS-72 pump-fed, gas generator cycle, motor.

Boeing Space and Communications Group

Current Status

In September 2001 Boeing Launch Services was formed headed by Will Trafton, former president and general manager of Sea Launch of which Boeing owns 40 per cent. The new BLS markets Delta and Sea Launch vehicles.

Boeing Space and Communications Group includes expendable launch systems, Sea Launch, and Rocketdyne Propulsion and Power. Boeing signed an agreement 1 August 1996 to acquire Rockwell's space and defence businesses, which was renamed Boeing North American Inc, part of Boeing's Space and Communications Systems group. 1995's combined space revenue was approximately US$3.5 billion.

The former Rockwell's Space Systems Division was one of Rockwell's Aerospace and Defense businesses, along with the former Rocketdyne and Autonetics and Missile Systems. Space Systems designs, develops, builds and integrates space vehicles and supporting hardware. As prime contractor for NASA's Shuttle Orbiters, it also maintains their technical integrity and configuration; assists NASA in the integration of Shuttle system elements; and provides support services, including mission planning and simulation, flight design, mission data production, flight crew/controller training, ground facility engineering and operations support, payload integration, launch operations support, flight analysis, direct mission support, Orbiter turnaround assistance, logistics and engineering support.

As part of United Space Alliance Space Systems supports Orbiter logistics operations at KSC, in addition to Orbiter processing and cargo/system integration for NASA Johnson. Also at Houston, the Rockwell Space Operations Company, a subsidiary, provides further Shuttle and advanced programme services. The Florida Operations encompass all work associated with the purchase of Orbiter and Ground Support Equipment (GSE) spares and repair of existing Orbiter and GSE hardware. The NASA Shuttle Logistics Depot is certified to perform Orbiter component overhaul and repair on approximately 3,200 of the Shuttle's 4,000 line replaceable units; 14,864 m^2 was added to the 10,219 m^2 facility in 1990. In October 1991, NASA extended Rockwell's logistics contract to 1994 with a US$453 million award, followed by a US$600 million five-year award in October 1994 (with options totalling US$800 million through 2004).

The Orbiters' forward/aft fuselages, crew compartment, forward RCS, vertical stabiliser and secondary structures were fabricated at SSD's 223,605 m^2 Downey facility. The Orbiters are assembled or modified at the 10,350 m^2 Palmdale, California site, near Edwards AFB, the only facility capable of handling a complete vehicle. All manufacturing was shifted by 1995 from Downey to Palmdale to save NASA US$20 million annually.

McDonnell Douglas now subsumed into Boeing performs Delta marketing/sales and contracting for the new generation series.

Space Systems is also responsible for Sea Launch, the Inertial Upper Stage, and studies on future space transportation systems.

Group has approximately 30,000 employees.

EER Systems Inc

Current Status

A wholly owned susidiary of L-3 Communications Corporation, EER Systems was set up in 1979 and employs 1,500 people. Set up by former astronaut Donald ('Deke') Slayton in 1982, Space Services Inc (SSI) was acquired in December 1990 by EER Systems. SSI designed the Conestoga solid propellant family of orbital launchers capable of handling up to 2.1 tonnes into LEO. It was selected in early 1991 to provide Conestoga launch services for the NASA sponsored Meteor µgravity carrier programme. A preliminary demonstration launch using a single Minuteman 1 second-stage motor was conducted in September 1982. In March 1989, SSI fired its first commercial vehicle: a two-stage Starfire sounding rocket for the University of Alabama under a US$1 million contract carrying a µgravity payload.

SSI's first launcher, Percheron, with a liquid first-stage, exploded on its debut from Matagorda Island off the Texas coast 12 August 1981, and the company switched its efforts to the all solid Conestoga designs. Conestoga 1, using a single 206.8 kN thrust Minuteman 1 second-stage motor, was successfully fired as a demonstration on a 10.5 minute, 313 km altitude suborbital trajectory 9 September 1982 from Matagorda. The current Conestoga orbital versions are based on Thiokol's Castor 4A/4B solids with Star injection stages. The 1679 was selected for BMDO's MSTI-5, although no funding was provided. The debut (1620) October 1995 launch self-destructed at 45 seconds after faulty navigation data produced extensive nozzle movements and depleted the TVC hydraulic fluid.

Development of Conestoga versions has been suspended but the hardware and design test equipment is held pending revitalised interest.

Core services now embrace IT services systems engineering, weapons systems support and aircraft modifications and the company has applied space technology to general aviation tasks and engineering activities.

E'Prime Aerospace Corporation

Current Status

Established in 1987, EPAC is offering its Eagle S-series solid propellant orbital launchers employing stages derived from Peacekeeper motors. Improvements include changing from Kevlar casings to graphite. The primary launch site would be at the Devil's Ash Pit on the UK's Ascension Island (polar missions from Vandenberg AFB), with the vehicle ejected by high-pressure steam from its canister before stage 1 ignition. Eagle and Eaglet, S-I and S-II would be launched from Cape Canaveral. The Eagle and Eagle S-I carry a bipropellant PostBoost Vehicle (PBV) derived from Peacekeeper's stage 4. The S II would carry the USTM (Unified Satellite Transfer Module) or the PASS (Payload Assist SubSystem). USTM is planned as a generic satellite bus containing the essential subsystems and a payload compartment for a customer payload (usually, the electronics and antennas that are integrated into the bus). USTM contains a gimballed bipropellant low-thrust engine for apogee or other orbital manoeuvreing. PASS, attached to USTM and later separated, is a gimballed high-thrust bipropellant engine module for perigee or other orbital manoeuvring. The larger SIII/S-IV would be launched in the future from Cape Canaveral.

The Eagle/PBV combination would deliver 1,361 kg into LEO for US$10 million, the Eagle SI/PBV 2,860 kg LEO for US$20 million and the Eagle SII/PASS 2,062 kg GTO for US$30 million.

The company's debut Loft 1 was launched 17 November 1988 from Canaveral's complex 47 to 4.6 km altitude carrying a 15.5 kg educational payload.

Following an amendment to its Commercialisation Agreement with the USAF, E'Prime was given the right in 1987 to develop MX peacekeeper motors for commercial satellite launchers. Company has not yet received an order for launch services.

In February 2001 company signed agreement with NASA for support of commercial launch programmes enabling it to proceed with programme plan, selection of launch sites, staffing, hardware contractor selection and finalising financial arrangements.

Based on Peacekeeper technology and ICBM design, the proposed Eagle series of launchers provides a wide range of payload capabilities. Peacekeeper performed more than 40 consecutive flights without failure.

Hughes Research Laboratories

Current Status

Hughes has developed an 18 mN Xenon Ion Propulsion Subsystem (XIPS) for NSSK of three-axis or spin-stabilised telecom satellites. It made its debut on Hughes' own Galaxy 3R satellite in October 1995; Astra 1G and ASC 1/2 also used it. XIPS comprises four 13 cm thrusters (two primary + two back-up), power supply (439 W input each thruster off 29 to 34 V DC bus) and a propellant storage/control unit. Total XIPS package mass is 68 kg, for a saving of 400 kg on an HS-601 satellite. 30,000 hours of operating time was accumulated by 2001. In normal NSSK operation, the thrusters will fire for 2 to 3 hours daily.

Xe is fed through a heated (1,000°C) hollow cathode inside the thrust chamber. A ring cusp magnetic field confines the plasma and a three-grid ion optics assembly extracts the thrust beam as 3,125 beamlets. Xe is stored at 74.8 atmospheres and reduced to 0.68 atmospheres for feeding into the cathode.

The 13 cm XIPS was developed from work on a 25 cm diameter 63.5 mN thruster that provided SI 2,800 seconds at 1.3 kW input. It was also operated at 179 mN, 4,000 seconds, 4.5 kW.

Ion propulsion work developed for projected mission scenarios compliant with NASA's new initiative.

ICI Explosives

Current Status

ICI designs, develops and manufactures a wide range of explosively actuated aerospace, ordnance and commercial products, including: actuators/motors, cutters, detonators/explosive leads, gas generators, igniters, switches, electric frangible link assemblies and pyrotechnic actuated valves. Product fitted to spacecraft, satellites, launch vehicles, ground support equipment and test rigs, simulation chambers and destruct systems. Test rigs set up for failure simulation and analysis. Technology developed for these rigs and the subsequent test and failure analysis have been used for accident investigation.

Kaiser Marquardt

Current Status

The Marquardt company was formed in 1944 to undertake R&D leading to the first US subsonic ramjet in 1945. Its principal engineering business continues to be advanced aerospace propulsion and the supply of ram air turbine power systems. The company has been developing and manufacturing bipropellant thrusters since 1959, notably the attitude control thrusters for Apollo spacecraft and the Space Shuttle Orbiter. Precision bipropellant engines are available for 0.004, 0.02, 0.06, 0.11, 0.22, 0.44, 0.89, 4 kN sizes, and requests up to 8 kN can be accommodated. The Apollo R4D thrusters used on the Service Module RCS quads have been developed into propulsion devices for current generation satellites and spacecraft. The monopropellant hydrazine thruster line of UTC's Hamilton Standard Division was acquired in 1993.

In 2000 Kaiser Marquardt bipropellants became part of PRIMEX Space Systems. In 2001 General Dynamics acquired PRIMEX Technologies. Biopropellant systems developed for low-cost small to medium satellites. Marketed on license offer but unsuccessful to date.

Kelly Space and Technology

Current Status

Kelly Space and Technology Inc (KST) is a technology development company which was incorporated in April 1993 under the laws of the State of California. The company's current focus is the development of the 'Eclipse' family of low-cost, reusable, commercial orbital and suborbital space launch vehicles which are intended to make access to space both affordable and routine. KST has focused initial development activities on solving this problem by offering what could become a significantly more operationally efficient launch system at substantially lower cost, with plans to reduce costs even further by incrementally incorporating new technology into improved operational vehicles as it becomes available. The Eclipse launch vehicles are designed to address this need in the near-term by combining the best of available, proven technology to mitigate development risk and improve operational flexibility while defining an implementation plan that minimises non-recurring investment to ensure sound financial performance.

The Eclipse concept utilises a Boeing 747 to tow the Eclipse winged launch vehicle from a conventional runway to the launch altitude of approximately 12 km. At this altitude, the Eclipse vehicle's rocket engine is ignited, the tow line is released and the vehicle climbs to the payload separation altitude of approximately 120 km. Following ejection from the Eclipse vehicle, the upper stages are ignited to deliver the payload to the specified destination while the Eclipse vehicle descends as a glider. For final descent from approximately 9 km, the Eclipse deploys two turboprop engines located aft to support powered approach and landing.

Importantly, KST has mitigated its development risk by employing only existing, proven technology in its launch vehicle configurations. No technological advances or breakthroughs are required.

The Eclipse tow launch concept is currently in a flight demonstration programme being conducted by KST under a United States Air Force Phase II Small Business Innovation Research (SBIR) contract. Under this cost-share contract KST will demonstrate the key features of the Eclipse launch concept using a QF-106 jet powered supersonic aircraft towed behind a C141A aircraft. The USAF has provided the Company with two QF 106 aircraft under the SBIR contract to serve as scaled representatives of the Eclipse launch vehicle. The USAF Flight Test Centre at Edwards AFB is supporting flight test operations by providing a C141A aircraft for use as the tow plane. Flight tests are being supported by and conducted at NASA's Dryden Flight Research Center (DFRC) at Edwards Air Force Base. Flight test activities continue using the QF-106 aircraft throughout the remaining useful life of the aircraft.

The Eclipse family of reusable launch vehicles consists of three basic, sequentially developed, flight vehicle configurations: Eclipse Sprint, Eclipse Express and Eclipse Astroliner. Each vehicle supports the development of the next and is placed into commercial use following testing to service incrementally larger segments of the space launch market. Adopting this approach enables each vehicle to serve initially as a test vehicle and to then generate revenues concurrent with the development of the next vehicle.

The first vehicle in the family, a small launch vehicle named Eclipse Sprint, will be constructed to serve both as a development test bed for the Eclipse Astroliner and as a commercial suborbital launch vehicle to generate early revenue. The Eclipse Sprint vehicle will be used to develop flight operations procedures to ensure safety, evaluate and demonstrate various recovery options, gain operational flight experience with the Eclipse tow launch method, support test activities of the Eclipse Astroliner, and generate early revenues. Eclipse Space Lines Inc, will be established and licensed by KST to commercially provide Eclipse Sprint suborbital launch services. The Eclipse Sprint vehicle, a small-scale rocket powered glider which is representative of the Eclipse Astroliner, will support successive Eclipse development activities and provide suborbital launches to service approximately half of the existing sounding rocket market. Sounding rockets provide payloads with up to 20 minutes of exposure to the space environment supporting microgravity experiments, Earth observation, atmospheric sampling and numerous US Department of Defense applications. The Eclipse Sprint service offers significant price advantages with reliability not provided by any of the expendable rockets currently servicing this market.

The Eclipse Express is a follow-on and enhanced version of the Eclipse Sprint vehicle which incorporates upper stages that could be ejected with a payload to service the balance of the existing suborbital market as well as being capable of launching small satellite payloads (under 90 kg) into low-earth orbit. Development of the Eclipse Express will be initiated as a follow-on to Eclipse Sprint operations in order to escalate the development of the Eclipse Astroliner. The Eclipse Sprint vehicle will serve as the test vehicle to support Eclipse Astroliner development activities. As currently envisioned, Eclipse Express would be developed under the same financing approach and business model as the Eclipse Sprint and is expected to be deployed before flight testing of the Eclipse Astroliner.

The flagship of KST's family of launch vehicles will be the Eclipse Astroliner. This vehicle will be capable of launching payloads of up to 1,600 kg into a 465 km polar orbit or up to 2,300 kg into a 465 km equatorial orbit. Commercial operations are projected to commence around mid-2000. The Eclipse Astroliner recoverable flight vehicle consists of the airframe, propulsion systems, thermal protection systems, avionics packages (including software), attitude control system, ordnance initiation system, and landing gear. The Eclipse Astroliner airframe will be a low to mid-wing delta monoplane with a single vertical stabiliser. The airframe will include an articulated nose door similar to a 747 Freighter or C-5 aircraft for both ground loading of the payload and payload deployment at altitude. Access doors servicing the payload and all other units requiring inspection and maintenance are also incorporated into the airframe.

In 1998, Kelly successfully demonstrated the Eclipse concept under a USAF small business innovative research (SBIR) programme. In September 1999, Kelly received a US$2.1 million NASA contract to perform a long-term transportation architecture study. On 7 January 2000 NASA awarded a risk reduction contract worth US$3.1 million.

In August 2000, NASA awarded Kelly a US$3.1 million contract for risk reduction and analysis of possible shuttle replacement contenders for initial service in 2010.

On 23 January 2001, Kelly Space and Technology Inc and Vought Aircraft Industries Inc announced a teaming agreement to refine 2nd generation RLV architectures and further develop the Eclipse Astroliner concept. Studies have responded to NASA's future launch vehicle studies enshrined within the newly re-launched Space Launch Initiative. The architecture has been redesigned to match potential opportunities.

Kistler Aerospace Corporation

Current Status

Kistler Aerospace Corporation is proposing the development of a two-stage, fully reusable launch vehicle, designated the K-1. The vehicle will use Russian engines designed and built by NPO Trud (previously the Kuznetsov design bureau) for the aborted Soviet manned lunar programme.

The company completed the phase 1 preliminary design in September 1995 and has secured the right of first refusal for all NK33 and NK43 engines which were built for the N-1 manned lunar landing vehicle. Phase 2 development and testing was completed in 1996 and Kistler is currently undertaking phase 3 development and testing. Phase 3 calls for completion of the detailed design of the launch vehicle, the establishment of an advisory group comprising contractors and Kistler supporters, finalisation of the contract team, fund raising, assembly and test of the first K-1 vehicle, conduction of flight tests and the initial commercial operations. Two launch sites are being established at the Woomera range in South Australia.

Each of the two stages of the K-1 vehicle would be recovered using a combination of parachutes and airbags. Ideally, both stages would automatically return to the launch site, the second stage after initially entering orbit with the payload.

When complete a fleet of five K-1 vehicles is proposed, with each vehicle designed for 100 flights. Each vehicle would have a turnaround time of nine days, with each stage being returned to the launch site within 24 hours of launch. The launch vehicle would be assembled horizontally, transported and erected on a launch stand using a wheeled mobile launcher. A Memorandum of Understanding with the State of Nevada to establish a launch site was signed as part of the phase 1 work for the vehicle.

Kistler is forecasting a price of US$1,700 per lb (or about US$770 per kg) to a 28.5°/km LEO. Since 1998, KAC has been re-financing and technical development work has slowed.

NASA has awarded Kistler a US$135 million contract under the Space Launch Initiative for 13 embedded technology evaluations on the first four flights of the K-1.

Redirecting its efforts from a reusable launch vehicle to a low cost conventional launcher, deferring reusability to a later development.

Lockheed Martin Astro Space

Current Status

The USAF Space & Missiles Systems Center began the IABS Integrated Apogee Boost Subsystem programme in 1988 after Atlas 2 was selected for the 10 DSCS 3B military communication satellites. Cost of 10 stages and their launch vehicle integration is about US$200 million; unit cost is about US$15 million. DSCS contractor LMAS was awarded the contract, with the first stage delivered during 1991.

IABS is used to inject DSCS into a near GSO orbit from GTO, permitting the use of a smaller launcher. Payload capacity is 1,180 kg. Spin stabilisation and commands are provided by the payload. For DSCS, the main 60-minute burn is made at apogee 5 into a 0.4° intermediate 'burn short' circular orbit; an 80-second trim at apogee 7 establishes a sub-synchronous drift orbit. The payload separates for its own thrusters to achieve GEO; IABS remains in the drift path. The pair can remain in transfer orbit for up to 12 days for mission flexibility. The peripheral solar array generates 240 W and DSCS's stowed array adds 190 W (there are no batteries; IABS employs DSCS's). IABS has been adapted for several speculative projects which were never funded but which were integrated with existing stages.

IABS has been evaluated for a range of military requirements of a classified nature.

Lockheed Martin Space Launch Systems

Current Status

Now integrated with Astronautics Operations. Lockheed Missiles & Space Co in May 1993 announced plans for a family of all-solid LLV Lockheed Launch Vehicles to deliver payloads ranging from 800 to 3,200 kg to LEO for US$15 million to US$26 million. LMLV 1 employs a Castor 120 stage 1 and Orbus 21D stage 2. LMLV 2 inserts a second Castor 120 between the two, while LMLV 3 would add at least two Castor 4A strap-ons (3/4/6 strap-ons also possible). All versions carry a hydrazine orbit adjust module, which also houses the vehicles' avionics. CDR was conducted April 1994; PDR July 1993. Transfer orbits and Earth escape could be provided by a Thiokol Star stage 4 with TVC. The Castor 4A could also be replaced by the 4AXL version.

The programme moved from LMSC in Sunnyvale to Denver in late 1995, although the vehicles (now named Athena 1 and Athena 2 for LMLV-1 and LMLV-2 respectively) continue temporarily to be built at the Californian site. LMSC and CTA Inc signed a less than US$5 million contract 18 February 1994 for the launch of CTA's 113 kg GEMStar 1 into a 480 km polar orbit from VAFB on the debut LMLV 1. Launch was from an adaptor on one of the solid booster mounts of the Shuttle SLC-6 complex. This flight used an existing Orbus 21 modified to 21D; propellant was offloaded for another mission not flown. The vehicle was destroyed by range safety after 159 seconds, when separate stage 1 TVC and IMU problems left it uncontrollable.

LMSC began investigating the possibilities of converting excess missiles in 1987 (it was responsible for the Polaris, Poseidon and Trident SLBMs) but dropped the approach in 1990 when studies showed the vehicle would not be reliable enough (90 to 95 per cent) for the commercial market. Lockheed Corporation approved the LLV development programme in January 1993. The philosophy is to use existing or imminent hardware, aluminium structures rather than composite (lower cost), higher margins in lieu of acceptance tests and not to push performance at the expense of reliability or lower cost.

Lockheed Martin Space Systems Company Astronautics Operations

Current Status

Lockheed Martin Astronautics is one of the operating elements of Lockheed Martin's Space and Systems Company. It encompasses Titan and Atlas production and operations, as well as Athena 1 and 2 (LMLV-1 and LMLV-2), multiservice launch system, Evolved Expendable Launch Systems and the EELV. Denver is also a centre of satellite production, particularly classified satellites such as Lacrosse and other National Reconnaissance Office and National Security Agency assets.

The Astronautics element of the Martin Marietta Corporation was established in 1956 to develop and produce the Titan 1 ICBM. Headquarters for Astronautics is the Deer Creek facility, 32 km southwest of downtown Denver in the Rocky Mountain foothills. Acquired in 1987, it provides 66,140 m^2 of office space on a 4 km^2 site in Deer Creek Canyon. The main Astronautics facility is located on 20.5 km^2 8 km south of HQ and houses advanced production, space simulation and test facilities. Included in the US$550 million plant are approximately 70 laboratories, as well as computerised engineering design, manufacturing and administration facilities. Astronautics also operates the USAF Titan and Atlas launch facilities at Vandenberg AFB and Cape Canaveral.

NASA programmes include Viking, Magellan, Tethered Satellite and Shuttle's Manned Manoeuvring Unit. It is believed to be prime for the USAF Lacrosse imaging radar spacecraft. The Transfer Orbit Stage was built under contract to Orbital Sciences. A US$30 million five-year contract was awarded in May 1992 to convert Minuteman 2 ICBMs into suborbital launchers for the MultiService Launch Systems Programme. A three-stage orbital version, costing US$4 million to convert, was due to debut in 1997/98. The company acquired a 15 per cent interest in Canada's Radarsat International June 1994 for US$3 million, giving it an exclusive licence for marketing Radarsat data to the US Federal government.

Astronautics Operations employs more than 7,000 people.

A Titan 4 launch vehicle stands ready for flight carrying the Cassini Saturn bound spacecraft (Lockheed Martin)
0572499

Lockheed Martin Space Systems Company Michoud Operations

Current Status

Manned Space Systems was formed in 1973 to design, assemble and test external tanks for NASA's Space Shuttle. It was originally part of Lockheed Martin's Information and Technology Services Sector. Staff total about 2,500. A US$1,797 million contract extension in August 1989 added a further 60 tanks to the 59 already under order. The first was completed in late 1991; the 60th was delivered in 2000. The first tank, the main propulsion test article, was delivered to NASA 7 September 1977. Manned Space Systems is located at NASA's 3.37 km^2 Michoud Assembly Facility, which features one of the world's largest manufacturing plants (174,000 m^2) under one roof and a port with deep water access for barging ETs to Florida. Three ocean-going barges are available for the five-day journeys. Michoud Operations was awarded a US$172.5 million contract to develop the Super Lightweight Tank, using the 2195 Weldalite aluminium lithium alloy instead of 2219 aluminium, saving 3,400 kg. The total saving reached 3,600 kg by fine tuning the foam insulation's spray-on process and optimising other tank structures. First was ET-96 for first Space Station assembly mission. A 12.2 m-long full diameter version, essentially a segment of the hydrogen tank with a LOX tank dome at one end, was delivered to NASA Marshall 1 February 1996 for six months' dynamic and pressure testing to verify the structural design. The alloy is 30 per cent stronger and 5 per cent less dense than the current material and comprises 1 per cent Li, 4 per cent Cu, 0.4 per cent Ag, 0.4 per cent Mg and 94.2 per cent Al. The first SLT was launched by STS-91 in June 1998.

Michoud ordered materials to build 60 new ETs in April 1999. NASA had purchased 119 ETs to this date. Michoud delivered its 100th shuttle ET on 20 August 1999.

In October 2000, Michoud Operations completed negotiations for a further 35 ET elements under a US$1.15 billion contract extending to 30 September 2006. LMSSC developed a new and more efficient method of fabricating and assembling ET segments using a Friction Stir Welding (FSW) technology first developed in the UK. ET manufacturing processes and production finishing techniques have been examined in the investigations into the loss of the NASA Shuttle Columbia on 1 February 2003.

Manned Space Systems studies include advanced and reusable launch vehicles, hybrid propulsion systems, on-orbit use of ETs, and the development of related composite materials technologies. Lightweight pressurant tanks for the propulsion system of Lockheed Martin Astro Space's A2100 satellites are in production.

Microcosm

Current Status

Microcosm is developing a pressure fed LOX/kerosene 22.2 kN engine with a claimed manufacturing cost of less than US$5,000 each, excluding the injector. There are only 38 parts, full duration 200 seconds, ablative film cooled; performance is comparable to solid rockets, although key specifications remain confidential. A 200-second run was successful 20 November 1995. Lack of support from financial backers instigated the demise of the motor. Assisted by USAF Phillips Lab funding (US$3.1 million through September 1996), Microcosm proposes the Scorpius family of vehicles based on the same engine: SR-S sounding rocket (723 kg GLOW, 50 kg to 225 km; US$99,000 recurring cost); SR-1 sounding rocket (400 kg to 190 km; US$295,000); Mini-Lift (100 kg into LEO; US$700,000); Liberty Light Lift (71.7 tonne GLOW, 4 stages, 7 propulsion pods totalling 49 engines, 1 tonne into LEO; US$1.8 million); Exodus Medium Lift (6800 kg LEO; US$8.5 million). Non-recurring development cost of the range through Liberty is less than US$30 million (FY96). Absence of market potential or technical veracity. Further efforts to market the engine aim to exploit assistance from the USAF Phillips Laboratory. Studies performed for unnamed military research institutes with some hardware sets tested on hard stands for unspecified applications giving the propulsion design new applications.

Northrop Grumman Space Technology

Current Status

Known as TRW before it was acquired by Northrop Grumman in December 2002 whereupon it was renamed as Northrop Grumman Space Technology, (NGST) develops and manufactures propulsion components, engines and systems for spacecraft attitude and velocity control, orbit insertion, manoeuvring and adjust, tactical missile steering, low-cost orbit transfer stages and low-cost engines for Earth to orbit systems. Since the company produced the first space operated monopropellant hydrazine system for 1960's Able lunar probe, it has provided approximately 2,300 mono/bipropellant thrusters and reaction control systems. The bipropellant Lunar Module Descent Engine was responsible for landing 12 astronauts on the Moon 1969–72, and Apollo 13's rescue, before it was converted into a fixed thrust model for Delta's second stage; 84 were built for the lunar programme and 75 flew on Delta from 1974. It remains under consideration for future upper stages. A further derivative was to power later versions of NASA's Orbital Manoeuvring Vehicle (cancelled). Propulsion research covers low-thrust monopropellant, bipropellant, colloid, ion, radioisotope, MPD, electrothermal and low-cost engines. NGST's pintle injection concept has been used on several rocket motor designs including the LMDE.

Propulsion test facilities are centred at TRW's 8 km^2 Capistrano Test Site (CTS) 108 km south of Redondo Beach, equipped for thermal vacuum, vibration, shock, cold flow calibration, pressure/leak, functional and electrical, and accurate hot fire performance and life testing. Engine designs burning storable, cryogenic, gelled and high energy bipropellants can be evaluated under sea level and simulated altitude conditions. Continuous operation, steam exhaust pumped altitude chambers can test high-expansion ratio 450–670 N engines at ambient atmospheric pressures equivalent to an altitude of 46 km. The HEPTS High-Energy Propellant Test Stand has four 222 kN test positions, two of which simulate altitude firings. Cold flow and hot fire tests of low-thrust hydrazine, ion/plasma electric and resistojet gas thrusters are undertaken at Redondo Beach.

NGST has developed a simple, low-cost, liquid propellant motor delivering a thrust of 2,890 kN designated TR-106 for the NASA Space Launch Initiative. Also providing the TR-308 liquid apogee motor used by the Chandra x-ray observatory, the TR-312 pending qualification for commercial geostationary applications and the TR-711 gel booster engine.

NGST has been developing electric propulsion since the 1970s and is currently engineering a 200 W Hall motor for precise station-keeping application.

Olin Aerospace Company

Current Status

OAC, previously the Rocket Research Company, specialises in monopropellant hydrazine engines and gas generators for spacecraft and upper stage applications; approximately 9,500 assemblies and 112 propulsion systems have been produced since the first system was qualified in 1964. There are more than 500 personnel at the Redmond and Moses Lake sites. The MR508 hydrazine arcjet was qualified in 1991 and Telstar 4 in 1993 became the first commercial satellite to carry an arcjet for station-keeping. The model is superseded by the MR-509/510. OAC has delivered approximately 250 gas generators for Shuttle's Auxiliary Power Unit. APU developments anticipate more stringent specifications for new requirements from design studies on potential shuttle successors. It has pioneered new integrated propulsion and gas generator systems for reusable and non-reusable successors to Shuttle.

Facilities: OAC's 0.29 km^2 site at the Grant County Airport in Moses Lake houses 2,840 m^2 of solid propellant manufacturing plant, a hazardous device assembly area, storage magazines, a short-range flight test facility, a sea level large liquid engine test complex, ballistic test bays and chemistry laboratory.

Olin Chemicals

Current Status

Olin is the world's largest supplier of hydrazine and hydrazine-based propellants for space propulsion applications, including anhydrous hydrazine, MMH and UDMH. Users include Titan, Delta, Shuttle and most commercial/military spacecraft. In addition, hydrazine and hydrazine blends are used in auxiliary power units such as Shuttle's APU. An AF contract worth US$92 million over five years was awarded January 1995 and has since been extended for a further five years. It is developing refined hydrazine-based propellants for high-performance specifications. It is also developing fast-burn propellant combinations for military spacecraft with a requirement for long duration missions and has conducted tests with prototype designs seeking high efficiency and long duration capabilities.

Orbital Sciences Corporation

Current Status

OSC has joined the team led by Lockheed Martin competing to develop the NASA Orbital Space Plane. Also including Northrop Grumman, the team is competing against Boeing and will capitalise on OSC's work with the DART (Demonstration for Advanced Rendezvous Technology) programme. OSC's role on the OSP will be autonomous rendezvous and docking and with ISS integration. NASA has accelerated the OSP programme by targeting a return ISS crew capability by 2008 and a two-way crew transfer capability by 2012.

Background

OSC was established in 1982 to develop Space Shuttle upper stages, resulting in the solid propellant Transfer Orbit Stage. In June 1988, OSC announced a partnership with motor manufacturer Hercules Aerospace (now Alliant Techsystems) to develop the Pegasus air-launched small orbital delivery system. OSC was contracted in 1989 by DARPA for the Taurus rapid response orbital launcher based on Pegasus. Space Data Corp of Tempe, Arizona became a subsidiary in November 1988, extending OSC's capabilities to a wide range of sounding rockets: it has manufactured/launched approximately 600 suborbital boosters in 35 different configurations, in addition to producing up to 1,000 meteorological rockets annually. The division has participated in approximately 60 Minuteman 1 re-entry tests since 1971. A series of winged air-launched suborbital vehicles was announced September 1992. OSC's L-1011 aircraft can carry three 1- or 2-stage suborbital vehicles derived from the company's other carriers, totalling 11,340 kg with payloads up to 1,130 kg. The OSC/Rockwell team was selected by NASA in March 1995 to develop the semi reusable X34 launcher, the same month the OSC/McDonnell Douglas team was awarded NASA's MedLite launch contract, involving Taurus for the smaller satellites. At this date the X-34 was a two-stage, small, reusable winged launcher, partially reusable, on which NASA would spend about US$70 million. Moving with uncharacteristic haste, NASA selected an Orbital Sciences Corporation (OSC)/Rockwell team to develop the two-stage X-34 aiming for a suborbital flight in 1997 and orbital flights beginning a year later. The team formed American Space Lines as an 'Arianespace' lookalike for commercial development but in early 1996 it pulled out, believing the costs could not produce a viable commercial return. NASA re-contracted with OSC for a single-stage suborbital demonstrator, still designated X-34. Three X-34 test vehicles were scheduled to support flight trials planned to begin in 2000 with the winged vehicle launched from the under-fuselage cradle of a converted L-1011 operated by OSC. Capable of a maximum speed of Mach 8, the X-34 was to have demonstrated all-weather capability and pilotless landings as a technology precursor to fully reusable launch systems of the future. X-34 was scheduled for flight drop tests in 2001–2002. Frustrated by repeated delays and prompted by pressure from the White House, NASA terminated the X-34 programme in February 2001. Active in the development of missile defence systems and

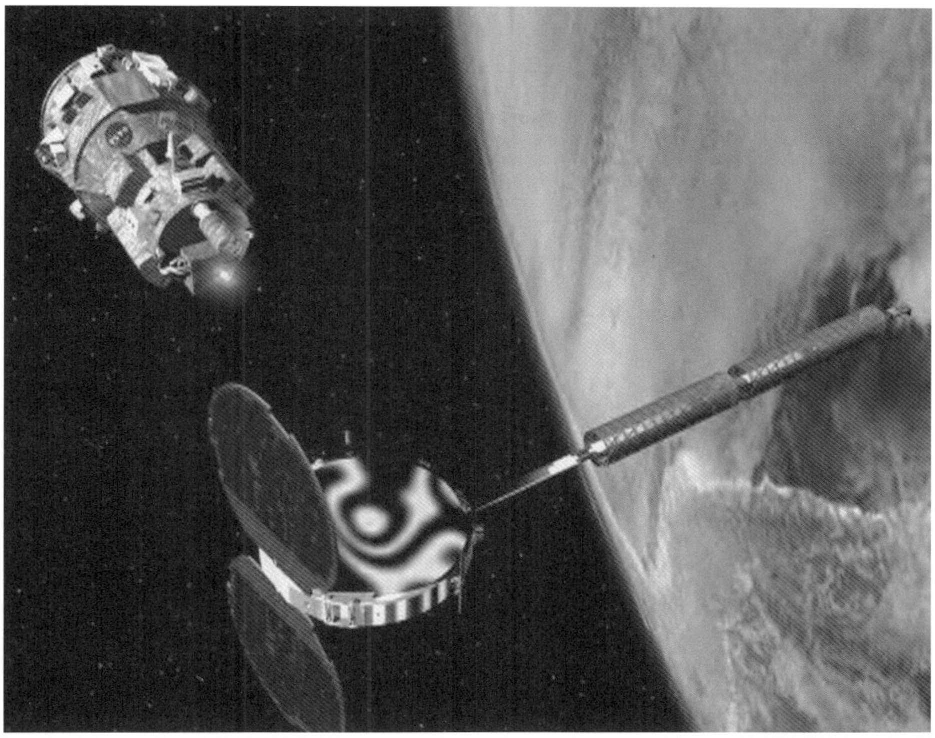

An artist's impression of the DART flight test demonstrator of the space manoeuvring vehicle (OSC) ***NEW***/1047530

its Orbital Boost Vehicle supplying interceptor boosters for 'kill' vehicles. On 24 May 2001 OSC announced that it had been selected for two contracts totalling US$53 million as part of NASA's Space Launch Initiative (SLI). These consisted of a US$47 million contract for the Demonstration of Autonomous Rendezvous Technology (DART) programme where OSC would build, test and launch a space manoeuvring vehicle derived from the Pegasus rocket to demonstrate autonomous rendezvous and close proximity operations. The second contract was for a US$6 million study effort to design and develop the Crew Transfer Vehicle (CTV) concept that the company proposed in 2000 under NASA's Space Transportation Architecture Study (STAS) programme. The DART flight demonstration was scheduled for be flown aboard a Pegasus rocket in 2004 to conduct a series of rendezvous manoeuvres with an orbiting satellite to demonstrate autonomous proximity operations using a modified Pegasus upper stage and avionics system. The space taxi CTV was designed to perform a variety of future NASA missions with significant improvements in cost and safety over the current Space Shuttle. The multifunctional Space Taxi vehicle would serve as an emergency Crew Return Vehicle (CRV) for the ISS, a two-way human space transportation system, a small cargo delivery and return vehicle and a platform for performing satellite servicing or in-orbit construction.

UPDATED

Para-Flite Incorporated

Current Status

Para-Flite develops/manufactures 0.5 to 280 m^2 ram-air gliding parachutes for aerospace recovery systems, including orbital capsules and sounding rocket payloads. Providing a lift-to-drag ratio of three, the parachute carries an Airborne Guidance Unit to process signals from a ground-based transmitter (GPS is an option) to manipulate the control lines and home in on the target. The automatic mode provides 60 m accuracy, complemented by 15 m for manual control. Payloads can be recovered from suborbital rocket and balloon carriers: 100 to 2,500 kg capacity systems are available off-the-shelf. Concepts for launch vehicle stage recovery have been studied but no applications have been adopted. Retarded deceleration techniques tested for soft-landing at ground sites with heavy loads. Technology demonstrated in sub-scale tests and simulation. Para-Flite have also studied specialised applications for retardation in high-dynamic stress environments for planetary exploration and this has been applied to other applications including devices accomodating defence specifications and commercial users.

Pressure Systems, Inc

Current Status

PSI is the world's largest independent manufacturer of titanium pressure vessels for launchers and spacecraft; it now ships over 200 annually, worth about US$25 million in total. More than 4,000 have been delivered, without flight failure, including approximately 625 propellant management device tanks of more than 20 designs and more than 750 elastomeric diaphragm tanks of about 20 designs. PSI acquired Programmed Composites Inc in June 1995, allowing it to offer tanks integrated with satellite core structures, in addition to precision reflectors, solar array substrates and antenna support structures. PSI tanks provide the propellant for most commercial communications satellites, including Hughes HS 601, Loral FS1300 and Lockheed Martin A2100. PSI tanks are also used on Milstar, Cassini, Soho, NEAR, Tirus, Inmarsat, Intelsat, ETS 7, TRMM and Axaf. PSI is completing a US$4 million contract from Lockheed Martin for 75 hydrazine tanks for Iridium satellites. In 1996, PSI delivered the 1000th upper stage high pressure helium tank and the 100th upper stage propellant tank for the Lockheed Martin Atlas Centaur launch vehicle family. Very high-pressure dewars for positive expulsion concepts and new composite materials for high-strength and burst resistant designs marketed to customers according to specified requirements and user demand.

Sandia National Laboratories (SNL)

Current Status

Sandia was established in 1949 and is now operated for the US Dept of Energy by Lockheed Martin's Sandia Corp. The Laboratories are active in defence and energy national security programmes, with principal emphasis on nuclear weapons R&D. Personnel total 7,900, including the associated facilities at Livermore in California. The HQ and main laboratory are located on Kirtland AFB on the southeast edge of Albuquerque. Aerospace interests include flight control/telemetry systems, robotics, materials, and computerised and physical laboratory testing of space debris impact on satellites (the Hypervelocity Launcher gas gun accelerates particles up to 12 km/s). It also develops/operates suborbital launch vehicles for the DoE, the military and other users. Tonopah Test Range at the north end of Nellis AFB, Nevada houses test facilities for small rockets.

Sandia operates the Kauai Test Facility in Hawaii on the Navy's Pacific Missile Range Facility, maintaining its own launch, handling and launch control equipment. For example, the Strypi family of solid propellant rockets has been employed in re-entry research programmes and as targets for orbital sensors. Strypi also launches the Laboratory's SWERVE Sandia Winged Energetic Re-entry Vehicle Experiment as an M14 hypersonic testbed.

The STARS strategic target system has been developed to extend these capabilities, particularly for BMDO demonstrations, and could be modified for modest orbital missions by adding a stage 4. STARS is based on Lockheed's two-stage solid propellant Polaris A3 SLBM motors carrying the new UTC/CSD Orbus 1 solid as stage 3, featuring a flex nozzle for steering. The configuration is capable of throwing 295 kg some 4,255 km downrange at a unit flight cost of about US$5 million. On a typical target mission, STARS is launched from Kauai to impact in the Kwajalein Atoll. First launch was projected for 1991, but the project was stalled by environmental study demands. The first launch was on the MSX satellite launched for the BMDO on 24 April 1996.

Sandia Laboratories conducts research on high-velocity atmospheric entry vehicles and aerothermal designs for hypersonic aerospace vehicles. Also, SAR development for air or space borne applications with high resolution and definition. Radiation hardened microelectronics applications from nuclear weapons research. In the aftermath of the Shuttle Columbia disaster on 1 February 2003, SNL has accelerated studies of possible successors to this reusable transportation system, integrating requirements that originated the Orbital Space Plane concept with its own in-house design configurations for hypersonic vehicles.

Space Vector Corporation

Current Status

Founded in 1969, SVC is a wholly owned subsidiary of Pemco Aviation Group and specialises in the design and manufacture of spacecraft and launcher structures, inertial guidance and control systems, telemetry subsystems and launch support. The Aries suborbital vehicle, based on government furnished Minuteman booster motors, was developed at SVC primarily for experiments of up to 2,270 kg and altitudes up to 480 km. The company serves as booster integrator, performs all design analyses and provides the electronic interfaces, ground support equipment and experiment modules. Approximately 40 Aries-class vehicles have been launched. An orbital version has been proposed, capable of delivering 226 kg into LEO. Aries is now changing from Minuteman stage 2 M56 motors for its first stage, to the more powerful Minuteman 2 stage 2 SR19. This model was used for BMDO's HERA TMD, the GBI development test vehicle and Altair. In addition, SVC has provided attitude control systems to approximately 130 smaller rockets and approximately 600 Miniature Inertial Digital Attitude System (MIDAS) platforms. The strapdown Advanced Guidance System (AGS) is capable of ±0.25° accuracy or better. The Magnetic Control System (MCS) is a gyro-less system that senses Earth's magnetic field. SVC also provides 0.4111 N cold gas thruster packages for sounding rockets and satellites.

Israel's Shavit was offered in 1994 for NASA's ultralight launcher competition (won by Pegasus). Team leader SVC would have integrated and launched the vehicle.

ABIE (Airborne Interceptor Experiment) was a joint exercise flown on 9 July 1996 designed to demonstrate theatre ballistic missile intercept during early boost.

Provides 6 DOF trajectory simulation tools including vehicle heading modes. Utilises Series 3000/Omega NT telemetry ground station.

SSI, Inc

Current Status

Formed in 1993 SSI was a partnership between ITT Industries and California Commercial Spaceport Inc to provide commercial launch services for small to medium polar vehicles from VAFB beginning in 1997. Typical users would be LMLV, Taurus and Delta-Lite. SSI is building a new pad and is leasing the US$250 million mothballed Payload Preparation Room at SLC-6, renovated as the IPF Integrated Processing Facility (GEMstar 1 and STEP M0 has been processed by end 1995). Annual capacity would be 24 launches, using the Integrate-Transfer-Launch (ITL) approach, allowing the use of simple stationary pad structures. The 25-year lease was signed by WCSC/AF for the undeveloped 0.10 km^2 site near SLC6; the annual fee is US$70,000 plus about US$10,000 monthly for operations costs and utilities. The government is providing US$10 million, mostly as AF grant money, and ITT is committing US$30 million. WCSC was formed as a non-profit corporation in May 1992 to pursue commercial opportunities at VAFB. It is designed by the State Legislature as The California Spaceport Authority. Subsidiary CCSI was established in August 1993 as a for-profit company to meet grant requirements for matching funds from space companies. Awarded the first US Commercial Space Launch Site Operator licence in September 1996.

The SLF Spaceport Launch Facility operates a single pad at 34°34′35″ north/120°37′53″ west (92 m altitude) between two SCF Stack & Checkout Facilities on an east-west axis. Most vehicles will depart due south (180° azimuth), so will not overfly the SCFs. 168 to 220° azimuths are available without impacting the local community or off-shore oil rigs; 150° is possible. A vehicle is stacked on an MLP Mobile Launch Platform and transferred launch-ready to the pad on launch day. SCF-E is intended for Delta 2 and Delta-Lite (but can handle Castor 120); SCF-W is optimised for Castor 120-based vehicles. IPF handles booster and payload processing, fairing/storage and payload encapsulation, and houses three launch control rooms in the 12.2 × 42.7 m Launch Control Center (LCC). The processing areas include a 9 × 30 m airlock, three 10.7 × 13.4 m cells and 9 × 45 m high bay, with 47 m high 68-tonne crane.

United Technologies Corporation Pratt & Whitney

Current Status

UTC combined the space propulsion activities of Chemical Systems Division (CSD), USBI and P&W in 1990 within P&W's Government Engines & Space Propulsion unit based at West Palm Beach in Florida. USBI is under contract to NASA Marshall to provide assembly, test and refurbishment of the non-motor segments of Shuttle's solid rocket boosters. NASA extended the contract in 1994 for recovering and refurbishing boosters to September 1997, worth US$1.8 billion. The previous US$1 billion award in January 1989 ran through September 1994. It was extended in 1996 through September 1999. The contracts have been suspended following the loss of Shuttle Columbia on 1 February 2003 pending a determination on utilisation rates at the end of 2004 prior to resumption of Shuttle missions no earlier than the second quarter 2005.

P&W produces the Orbus 21 solid propulsion system for applications such as the first stage of the IUS, PKM for Intelsat 6 and the TOS and provides the Orbus 21D for the Athena series of launchers, the RD-180 and the RL-10 in addition to high-pressure turbopumps for the SSME. Also produces the booster separation motors for the Shuttle SRB, eight being provided for each booster. P&W is developing a series of engines for the NASA Space Launch Initiative in co-operation with Rocketdyne division of Boeing.

CSD's Coyote facility in San Jose includes the world's largest solid rocket vertical test stand: 29 m high, it can handle 26.7 MN thrust motors and was most recently used for qualification firings of the seven segment USAF Titan 4 strap-on motor. Small/medium-sized horizontal and vertical test stands are controlled from a central complex and structural test facilities can apply omni-axial simulated flight loads of up to 4.9 MN to motor cases and attach skirts. CSD's integral rocket/ramjet test complex can simulate booster, transition and ramjet operations in a single sequence under sea level to high altitude conditions up to M6.0. The Coyote site also houses a 5,400-tonne annual capacity solid propellant mixing facility comprising two 2,270 litre, one 2,840 litre and one 1,510 litre mixers.

Universal Propulsion Company, Inc

Current Status

The company makes rocket motors, escape systems, rocket catapults, initiators/gas generators, electro-explosive devices, safe/arm units and pyrotechnic simulators for satellites, spacecraft and aircraft. Has adapted non-aerospace electro-explosive devices to ground test facilities applications. Produces rocket powered deceleration devices for possible space related applications. New and advanced detonator triggers for human-related escape systems in land, sea, air or space-based applications.

Western Electrochemical Company

Current Status

Pacific Engineering & Production satisfied half (18,000 tonnes) of US ammonium perchlorate solid propellant oxidiser requirements until explosions destroyed its plant in May 1988. Users included Delta, Titan and Shuttle boosters; each Shuttle launch consumes 771 tonnes. Launch operations were affected by the dramatic reduction in propellant supply. A new facility operated by Western Electrochemical (WECCO), a wholly owned subsidiary of American Pacific Corp, a Las Vegas based manufacturer of specialty chemicals, was completed during 1989, offering an initial annual production rate of 14,000 tonnes capable of rising to 18,000 tonnes with only minor capital additions. The 33-building complex was constructed at a cost of US$92 million and employs a staff of almost 200 people. The batch process under automated control can produce AP to custom chemical characteristics and physical distributions, as well as standard materials in large homogeneous lots. High purity AP is used in low smoke reduced signature rocket motors. Chemical purification and matrix composition analysis for batch examination and sample analysis. Improvements to production, inspection and verification techniques enhancing manufacture and supply reliability.

MICROGRAVITY FACILITIES, EQUIPMENT AND SERVICES

France

Novespace

Current Status
Novespace was established in July 1986 by CNES and eight banks as a consultancy for commercial utilisation of space technology, focusing on µgravity applications and technology transfer. 1995 revenue was FFr17 million (profit FFr1.3 million); 1994 FFr16.8 million (profit FFr1.6 million), 1993 FFr16 million (profit FFr0.89 million); 1992 FFr16.5 million; 1991 FFr14.5 million; 1990 FFr9.5 million; 1989 FFr 6million. The company offers µgravity campaigns aboard an Airbus A300 aircraft flying parabolic curves. Each week-long campaign costs about US$28,000 for a 150 kg experiment and two people flying for three days. Each day provides 40 parabolas of about 25 seconds µgravity. Testing by CNES began in January 1989 with a Caravelle; Novespace flew more than 40 campaigns by end-1995. The A300 took over in June 1996.

The creation of Space Business International was announced in June 1993 by Novespace, OHB-System GmbH and Space Industries International, Inc to co-operate on space contracts. Novespace developed international microgravity study groups across Europe and capitalised on ESA microgravity experiments aboard Spacelab to stimulate ground research. Novespace is a founder member of the Spacelink Europe consortium, a network of space technology transfer companies spanning 14 European countries. Novespace has specialised in focusing research tasks for µ gravity experiments on European elements of the ISS and in facilities of the ISS operated by ESA astronauts. Commercial diversification from the space-based hardware has directly applied the space-qualified designs to products developed for ground based facilities. Novespace has begun development of new microgravity support technologies with a new generation of experiment support equipment and test and experiment containers.

UPDATED

Germany

DaimlerChrysler Aerospace AG

Current Status
DaimlerChrysler extensive µgravity flight facilities range from the 4 minutes provided by the Mini-Texus sounding rocket vehicle, through the MAUS NASA GAS-based canisters, to SPAS and Eureca. The company manages the Texus and Maxus commercial programmes. Mini-Texus, Texus and MAUS can accommodate small payloads for µgravity levels up to 10^{-5} *g* whereas free-flying SPAS and Eureca offer conditions enhanced by one to two magnitudes. SPAS can be deployed from the Shuttle cargo bay as a free-flyer but its autonomous lifetime is constrained to hours/days by battery supplies. Eureca is based on a double-SPAS structure, but solar arrays permit a typically six month active life followed by a period of dormancy up to three months before retrieval. The larger Maxus rocket was introduced in 1991 in co-operation with Sweden to provide about 14 minutes µgravity. This stimulated several new experimental flight development technologies for short duration missions. The Express capsule flew 15 January 1995 but failed because of a launcher problem. No further missions are planned although Express can be adapted to customer requirements and could be flown on demand. Studies have been conducted on developing high-altitude, sustained zero-*g* capabilities and research has been conducted on adapting ballistic microgravity research flights on short-duration sounding rockets for long-duration orbital flights.

UPDATED

Leybold DurferriT AG

Current Status
Leybold produced high-vacuum furnaces for Spacelab D1/D2; the Isothermal Heating Facility is a multi-purpose system for isothermal and temperature gradient experiments. The company also produced TEM high-vacuum furnaces for dedicated experiments on the Eureca free-flying platform and has proposed similar devices for the ISS. Design evolution has produced several possibilities for furnace experiments on Columbus. Also, tests with equipment adapted from earlier, Russian, microgravity flights on both ballistic and orbital missions. These have resulted in several proposals for larger furnace vessels and for experiment chambers suitable for pressurised and non-pressurised environments. Company has developed a new generation of microgravity experiment control loops.

UPDATED

OHB-System GmbH

Current Status
Orbital- und Hydrotechnologie Bremen-System GmbH, founded in 1958, undertakes system engineering, hardware development/production and project management for space and environmental programmes. 1995 sales were DM41 million. Principal activities are: design/development of mechanical and electronic equipment; µgravity systems, experiment facilities/equipment; manned/unmanned space systems and mission operations, re-entry technology and aerodynamics/aerothermodynamics; small satellites. Its µgravity activities include industrial lead in the Mikroba programme, parabolic flights in chartered aircraft (NASA's KC-135 and CNES' Caravelle), collaboration in the construction and utilisation of ZARM's drop tower at Bremen University (providing the drop capsule), development of the Holop holographic camera for Germany's Spacelab D2 mission and the High Speed Centrifuge for ESA's Anthrorack. The company also developed the electronic control unit for Intospace's Cosima crystal growth facility, first flown aboard a recoverable Chinese platform in August 1988. OHB has continued to develop software controlled electronic processing units for microgravity research on recoverable modules.

OHB manufactures the flight harness for the BPDU Spacelab rack, it developed the video control for IML 2's NIZemi, and it was responsible for the design of ergometry, blood and urine monitoring experiments flown on Germany's Mir-92 mission. On Spacelab D2, OHB provided the Gravitational Biology element of incubators and centrifuges. OHB is MGSE prime contractor for ESA's Polar Platform, it was awarded the Phase A study for Columbus' Anthrolab facility and within the Europayload Consortium, it is working on the Fluid Science Laboratory and Automated Bio Laboratory for Columbus integrated as Europe's contribution to the ISS and due for launch in 2005. OHB-System reconfigured these elements following an ESA decision on the future of European policy on the reduced ISS programme plan reducing microgravity research time.

OHB developed Bremsat for the University of Bremen and its SAFIR-R1 demonstration package was launched November 1994 attached to a Russian Resurs-O1, preceding 1997's SAFIR 2 small satellite to provide global two-way data services. Instrumental in developing joint programme studies for international projects involving ESA and Russian activity in microgravity research.

The creation of Space Business International was announced in June 1993 by OHB, Space Industries International, Inc and Novespace to co-operate on space contracts.

VERIFIED

Italy

Alenia Spazio

Current Status
The company was awarded the 18-month Lit8 billion Phase B work in mid-1990 on a small recoverable capsule for µgravity applications. Work was completed October 1991; the 15-month Phase B2 definitive design phase was authorised January 1993 and concluded April 1994. As Phase C/D requires 32 months, first launch will be no earlier than 1998. Carina (Capsula di Rientro Non Abitata, unmanned re-entry capsule) would be launched into 360 km 2.9° by Vega from San Marco (Pegasus is a possible back-up), carrying more than 160 kg of payload for five days. The 1.067 m diameter, 1.248 m high Apollo-type REM capsule (13° wall slope) permits access up to one day before launch and samples recovery 6 hours after splashdown. Service Module: 1.244 m diameter, 60.95 cm high. Re-entry is initiated typically by a 760-second burn from four 20 N (four back-up) monohydrazine thrusters. The 28.7 kg thermal shielding has 25 mm base and 15 mm wall thickness. Recovery under a 2.1 m conical drogue and then a 13.9 m main parachute, is made following a 7 m/s splashdown. Two flights annually could be made available. The first payload has already been selected by ASI: SCGE Superconductive Crystal Growth Experiment, growth of YBa2Cu3O7-x; MMSFI Marangoni Migration and Solidification Front Interaction to analyse the inclusion, coalescence and migration of bubbles/drops inside liquid and solid matrices; MBI Marangoni Benard Instability in Two Immiscible Liquid Layers.

Since design initiation Vega has been re-designed but the recoverable capsule remains valid for the current configuration. Alenia Spazio has prepared adaptive kits for integrating payload carriage containers for applications in ISS microgravity research.

Specifications
Size: 1.244 m diameter, 1.888 m high
Mass: 599 kg (165 kg payload, 142 kg REM, 292 kg SM)
Microgravity levels: 10^{-5} *g* for 0–100 Hz, 10^{-3} *g* above 100 Hz (re-entry loads less than 9 *g*)
Mission duration: five days
Payload volume: 0.3 m^3
Power supply: eight lithium batteries provide 180 W mean to the payload
Telemetry: 100 kbit/s (6 kbit/s payload) and 2 kbit/s real time; command uplink 2 kbit/s. 40 Mbit mass memory.

VERIFIED

Japan

Japan Microgravity Center (JAMIC)

Current Status
JAMIC was established in March 1989 and began full operation in Kamisunagawa, Hokkaido

in 1991. The total number of drop experiments in the fields of combustion, materials, fluid, science and engineering and life science reached to 4,600 times by March, 2002. JAMIC is the world's largest 'drop-shaft' microgravity research facility providing 10 seconds of microgravity by dropping a rocket-shaped capsule of 7.85 m long, 1.8 m diameter and 5.5 ton in weight toward the bottom of 710 m drop shaft. The drop capsule is a double structure of inner and outer capsules. The outer capsule consists of a thruster module, a payload module and a bus module. The maximum payload weight is 500 kg and its dimensions are 870 mm × 870 mm × 918 mm. Further, through acceleration control by gas thrusters, it is possible to obtain a sustained microgravity condition up to 0.00001 g. The deceleration force exerted on the capsule in braking is less than 8 g. Monitoring of experiments by real-time colour image transmission and analyser is available and it is possible to make immediate feedback for succeeding experiments. JAMIC facility is best suited to experiments for basic research aimed at the development of new technologies, as well as for preliminary tests for experiments on shuttle or space station. Testing new concepts for microgravity experiment integration for space operations and linked analytical evaluations of returned samples.

VERIFIED

Space Technology Corporation

Current Status
STC provides expertise on μgravity materials experiments from experience gained on Spacelab (such as D2), sounding rocket and dropshaft experiments. STC has developed μgravity payload experiments for Columbus, the ESA ISS module. Developed drop-shaft techniques for new microgravity experiments where 1 *g* control testa are used to calibrate the science. STC is planning a programme of materials experiments that can be applied to both Columbus and JEM and these will involve several experiments contributed by industry to integrate government and non-government experiments.

VERIFIED

Netherlands

Bradford Engineering

Current Status
Bradford Engineering is a manufacturer of glovebox systems for space applications. A glovebox provides two levels of environmental containment in the Shuttle and International Space Station, permitting the performance of experiments which might contaminate the living environment. A video link and two-way audio ground link give ground-based scientists the ability to follow the experiments which an astronaut is conducting in orbit. The Bradford glovebox has flown on the USML-1 and 2 Spacelab missions. An improved design, the Middeck Glovebox (MGBX) has flown on the Shuttle middeck and is installed in the Priroda module which is part of the Mir Complex. Glovebox development supports ISS requirements and specific user requirements regarding power pick-up points, telemetry and switching functions. Bradford Engineering has developed concepts for complex storage and power-applied scientific work boxes. These devices can be used to power and control remotely operated expriments and science tasks. Applications include unmanned space platforms, experiment modules, human-tended tasks and remotely operated systems.

Present developments include the Microgravity Science GloveBox (MSG) and the Biological GloveBox (BGB) for ISS, as well as the Universal Transportable GloveBox (UTGB), a flexible low-cost glovebox design which can be carried by individual astronauts.

Bradford is additionally active in equipment such as valves, video systems, cooling loops and software. Many of these components are derived from proven glovebox designs. The Avionics Air Assembly (AAA) air-to-water heat exchanger will provide 500 to 1,500 W cooling power for ISS racks. Control valves from Bradford Engineering were selected by Alenia for use in the water-cooling system of the MPLM logistics module and the Columbus Orbital Facility. This valve is safety critical and therefore required an extensive testing and qualification programme. The company also developed application and control and monitoring software for microgravity science equipment. This was utilised in drop-shaft tests and in short microgravity ballistic flights in addition to orbital operations on European, Russian and US experiments.

CCM designed, developed and built Spacelab biological experiment units that were later modified for sounding rocket application. CCM has also designed a protein crystallisation module in co-operation with Fokker/Stork Comprimo to accommodate 400 samples. Currently preparing experiment cannisters for Columbus, the ESA ISS module.

VERIFIED

Norway

Prototech A/S

Current Status
Prototech provides sounding rocket payload design, development, testing and integration, and is involved in production of μgravity flight facilities. The group's debut in manned space technology came with the design/production of a sensor and electronics box for a Spacelab 1 plasma experiment. It has produced a series of biological cell chambers and, for ESA, a gas supply module and associated ground handling equipment for a μgravity furnace. Prototech has provided experiment cannisters and tailored fitments for experimenters and users and is developing glovebox style enclosures for commercial use aboard ISS or other independent station elements. Prototech has designed several unique experiment installations for science experiments that can be applied to unmanned and manned space habitation systems and which can be applied to air lifted test stands. Prototech developed the first Norwegian fluid physics experiment to be flown on Texus. These have been adopted for multiple applications. Prototech designed a containment system that can be used with any sounding rocket.

VERIFIED

Sweden

Swedish Space Corporation (SSC)

Current Status
SSC offers μgravity flight services within the Maser (Material Science Experiment Rockets) programme, in addition to design/manufacture of specific experiment modules. By using an improved Black Brant 9, 7-minute μgravity can be provided for a 375 kg experiment module. Skylark 7 can provide 6-minute μgravity for 370 kg. Launches are made almost annually. Modules are interchangeable between Maser, Texus and MiniTexus with slight modifications. Maser number 1 flew at Esrange 19 March 1987; number 2: 29 February 1988; number 3: 10 April 1989; number 4: 29 March 1990; number 5: 9 April 1992; number 6: 4 November 1993 (Skylark 7); number 7 (Skylark 7) 3 May 1996, with ESA as the principal customer. Maser 8 was launched in 1997. SSC provides microgravity opportunities to domestic and international project teams.

VERIFIED

United States

Instrumentation Technology Associates, Inc

Current Status
ITA provides commercial Shuttle flight opportunities, fabricates and leases space hardware and offers support services (payload development, integration, consulting and engineering services) for μgravity experiment activities. It provides turnkey leasing of its MDA Automated Lab, Liquids Mixing Apparatus, CMIX Flights and ISEM-G GAS canister modules for Shuttle, sounding rocket and recoverable capsule flights.

Materials Dispersion Apparatus (MDA) is designed for Shuttle, sounding rocket, satellite recovery capsule and Space Station applications for mixing of up to 100 samples of two-three fluids in space at precisely timed intervals. MDA weighs 1.82 kg and occupies less than 2,000 cm^2, operating on 16 VDC. Up to six units can fit into NASA's R/IM Refrigerator/Incubator Module or Commercial R/IM within a single Shuttle mid-deck locker. MDA's first flight was made 29 March 1989 aboard the Consort 1 sounding rocket mission. The second flight ended with launcher failure in November 1989, and during the Consort 3 mission of 17 May 1990, one of two MDAs malfunctioned. The problem was solved and the unit requalified. MDAs flew on the Joust 1 (vehicle failed 18 June 1991) and Consort 4 (16 November 1991) and number 5 (10 September 1992) sounding rockets. An integrated package of four flew on STS-37, 43, 52, 56. MDAs flew on STS-67 in March 1995 and STS-69 August 1995.

The MDA was designed to accommodate a large number of data sets in small volumes. Principal experiments are biotechnology, focusing on protein crystal growth. Experiments in bioprocessing, thin-film membrane casting, zeolite crystal growth, microencapsulation and cell research have also been performed. Two blocks of inert material, each with an equal number of sample test wells in the upper and lower halves, are held together with a sealing mechanism in an aerospace housing. The wells are misaligned at launch, thus separating the fluids to be mixed. After μgravity has been achieved, the blocks are aligned by a motor-cam mechanism. There is an option to mix a third or fourth fluid to fix the process while in μgravity or before re-entry. Each MDA can have up to 100 test cells, and each well can accommodate samples in the 0.125 to 0.5 μlitre range.

The LMA (Liquids Mixing Apparatus) complements MDA by allowing the mixing of larger volumes (up to 6 ml) at variable rates, from laminar to highly turbulent. Each vial can also be loaded individually. A vial comprises a clear plastic tube for the mixing of two or three fluids with the actuation of a plunger. 16 LMA vials flew for the first time as part of CMIX-3 on STS 67 in March 1995. Each was manually operated on that flight but on subsequent missions, beginning 1996, all vials will be actuated simultaneously.

The CMIX (Commercial MDA ITA Experiments) commercial agreement offers six Shuttle flights of 4–6 MDA units in a T-controlled mid-deck locker. Only five were flown. Individual samples or the whole package may be leased. CMIX-1 flew STS-52 October 1992, CMIX-2 April 1993, CMIX-3 on STS-67 in March 1995, CMIX-4 on STS-69 in September 1995; and CMIX-5 on STS-80 in November 1996.

The ISEM-G (ITA Standardised Experiment Module) for NASA's Get-Away Special programme provides an aluminium structure with mounting hard points, support avionics and house-keeping instrumentation for user payloads. A customer is awaited. ISEM-G fits in the 0.07/0.14 m^3 canisters. The standard support equipment in the lower section comprises a power supply, recorder, sequencer, and temperature, pressure and µgravity accelerometer instrumentation. Up to 10 experiments can be accommodated. ISEM-G can also be flown on Black Brant sounding rockets.

ITA developed a standard range of experiment and equipment modules for the ISS, offering customers an interface between their own experiments and the utilities and requirements of the space station experiment racks.

Specifications

The specifications below are given in 0.07/0.14 m^3 GAS version order:
ISEM mass: 13.6/27.2 kg
Experiment mass: 31.7/63.5 kg
Experiment volume: 0.037/0.1 m^3
Power: 1.2 kWh by lead acid cells at 5/16/24 V.

NAVIGATION, SEARCH AND RESCUE FACILITIES AND EQUIPMENT

Canada

BAE Systems

Current Status

Formerly known as Canadian Marconi Corporation (CMC) BAE Systems (Canada) designs and manufactures GPS navigation sensors and receivers and airborne satellite communications antennas. CMC has developed technology to interpolate space and airborne systems. Also, has proposed sensor systems capable of integrating solutions from several different satellite navigation constellations.

The 12-channel CMA-3012 GPS sensor unit is the first designed for certification under anticipated rules that will allow GPS as a sole means of airline navigation. The CMA-3112 is a single card engine derived from the 3012; 116 cm^2, 1.1 kg. The ALLSTAR mobile is a self-contained 12-channel all-in-view GPS receiver packaged in a ruggedised enclosure for applications such as automatic vehicle location. The ALLSTAR base version is designed for differential GPS applications, such as mapping and site positioning.

CAL Corporation

Current Status

CAL corporation is a subsidiary of EMS Technologies Inc and its Space and Technology division is selling antennas, digital command and control systems, RF switching and routing networks and satellite broadband. CAL is a supplier of Sarsat satellite-aided search/rescue ground terminals, providing 75 per cent of the world's total. It received a C$1.4 million contract in late 1989 for an Advanced Technology Local User Terminal (ATLUT) and two Operations Control Consoles for Hong Kong's Kai Tak airport, CAL's 17th for the system. In the first quarter 1991, Japan acquired two ATLUTs co-located at Yokohama, and Indonesia installed one ATLUT in Ambon and one with an Operations Control Console in Jakarta. New Zealand installed an ATLUT-320 at Wellington in autumn 1991. Singapore began operations mid-1992 with a dual ATLUT-320 and three OCC workstations under a C$2 million contract.

France

Informatique Electronique Sécurité Maritime

Current Status

SERPE-IESM offers several versions of its Ka and 406 Cospas/Sarsat beacon. 406 m: pocket-sized Personal Locator Beacon, targeted at pilots or military forces on special assignment. 406 ATP: aircraft, activated by a *g* switch or manually. 406S: hand-held for pleasure crafts, small fishing vessels and life rafts. 406FH: float-free for large fishing and commercial vessels, housed in a case containing the hydrostatic release mechanism which ejects and activates the beacon automatically at 4 m depth. IESM also manufactures Argos transmitters and Local User Terminals, and the Gonio 400 direction-finder to pinpoint and identify Sarsat and Argos beacons. IESM has designed personal locator beacons for time-lapse activation in the event of response failure from the initiating individual. IESM provides search vehicle sector adaptors for position location advancement. The company has also provided longer-life options. New beacons have been developed for radio location finding and for commercial needs including fishing and merchant fleets and these have been given automated controls as optional.

Sextant Avionique

Current Status

Thomson-CSF purchased Aerospatialés 50 per cent interest in ATEV, the parent company of Sextant Avionique, in 1999. It is now a member of the Thales group. The company made its first deliveries in 1992 of its Topstar 200 GPS receiver for civil aircraft. The 3 kg unit combines GPS with Omega. The 10-channel Topstar 100 GPS receiver is available in both P-code (100-P) and C/A-code (100-S) military forms. Europe's first satellite-borne GPS receiver, the eight-channel C/A Topstar 300 is carried by the US HETE satellite under CNES contract. A Topstar 100-P was carried by ESA's ARD Atmospheric Re-entry Demonstrator in 1998. Several GPS space application studies are under way with CNES and ESA. Company is working on a new European navigation system and has presented technical specifications for ratification by the appropriate national authorities across Europe. Company is now developing resources for proposed Galileo system of European navigation satellites and playing a central role in coordinating planning for enhanced receiver facilities integrating multichannel applications for two-way tracking.

Germany

Man Technologie AG

Current Status

MT acted as prime contractors for the ESA NAVSAT Phase B1 study and for the Design and Development of an Experimental Health/Integrity Monitoring Unit for the GPS and GLONASS Satellite Navigation Systems. Company also performed feasibility studies for an autonomous European satellite navigation system later realised as the Galileo programme. These studies provided technical standards recommendations later adopted by the industry. Within ESA's Global Navigation Satellite System (GNSS) 2 Mission Analysis study and the European Geostationary Navigation Overlay System (EGNOS) Project, MT is co-operating with Thomson-CSF (France) in the international teams. In co-operation with 35 Navigation, USA, MT developed and marketed the world's first commercially available integrated GPS/GLONASS receiver, the GNSS200/300. Within the German Luftfahrtforschungsprogramm, MT is developing in co-operation with Litef GmbH, Freiburg, a combined airborne navigation system called GLOGINAV. Various national projects funded by DARA cover new technological developments and different applications. Man Technologie AG has conducted studies supporting the European navigation satellite system and has extended early technical standards assessment into this area. Man Technologie AG employs about 1,000 people on space projects.

Japan

New Japan Radio Company Ltd

Current Status

New Japan Radio Company was formed in 1959 out of the Japan Radio Company and the Raytheon Company. Publicly traded on the stock exchange since 1991. The JLR-6800 GPS Ship Navigator is a single channel multiplexing receiver that can maintain constant track of up to eight satellites. It can be integrated with other systems such as Loran C, and provides fully automatic satellite selection, reception and position fixing under microprocessor control. It displays position data, waypoints, bearing, range and time to go, course deviation, route tracking, clock and stopwatch. JRC is providing its JLR-3300 8-channel land mobile GPS receivers to the car industry. JRC has built a prototype of a wristwatch-size GPS receiver, the Micro GPS-A. The 50 × 56 mm, 28 mm-high unit houses an eight-channel receiver in a case that also acts as the radome for its dielectric patch antenna. A lithium battery provides up to 7 hours continuous use, or five days for receiving every 10 minutes. Another receiver, 54 × 86 mm, 5 mm thick, is designed for insertion in compact portable personal computers with PCMCIA type 2 slots.

JRC's JQE-3A Cospas/Sarsat EPIRB can be provided with two mounting brackets: float-free with hydrostatic release (to satisfy Category 1 requirements), or standard with quick release for manual deployment (Category 2). Activation can be via remote control or sea water contact. The 406 MHz transmission includes a digital message bearing the vessel's identity and nationality, and a separate 121.5 MHz homing signal provides location information for nearby rescue services. A 0.75 candela strobe light flashing 20 to 30 times per minute is included as a visual aid.

Norway

Jotron Electronics AS

Current Status

Jotron and Total Marine Norsk A/S jointly developed the Tron 30S Mk2 Sarsat/Cospas beacon for maritime applications in two basic models. The FB3 version is mounted in a hydrostatic float-free release bracket, while its associated MB3 is designed for attachment to a bulkhead without the automatic release feature. The free versions are released at a preset water depth, typically at 2 to 3 m, and their mass distributions cause them to invert, activating a mercury trigger switch in conjunction with a seawater contact switch. Improvements provide depth and surge protection with impact resistant casing for high-shock and thermal protection. Options include a flashlight and homing transmit on 121.5 MHz. New beacon designs being considered for extreme sea and ocean states with applications on the non-marine market.

Kongsberg Simrad

Current Status

Kongsberg Simrad is part of the Kongsberg Group. Kongsberg is a supplier of precision position and navigation products and services. GPS activities are based on differential applications: DiffStar includes 15 reference stations that provide high-precision services through a range of receivers in the Barents, Norwegian and North Seas. KN is developing integration systems for GPS, Glonnass and Galileo satellite constellations. Each reference station houses a GPS receiver with 12 channels and a real-time computer reading pseudo-range and ephemeris from the receiver. It calculates the difference between the measured and theoretical pseudo-range for transmission to users' GPS systems on 300 kHz/2 MHz bands. Kongsberg Simrad is developing a unique set of navigation transceivers for multiplatform accessibility adding flexibility and new applications.

United Kingdom

Marconi Radar and Defence Systems

Current Status
The division developed the world's first miniature P-code GPS receiver and has established a range of GPS user equipment for military aircraft and marine applications. GPS receivers designed for integration within confined spaces and low mass acceptability. Current products include the five-channel PA9052E Nu airborne (fixed- and rotary-wing aircraft) and the PA9150 naval receiver series. Both provide the full 16 m accuracy and can be fitted with a module to ensure the equipment remains unclassified. A range of antenna subsystems is also available. Complex antenna shapes conform to exterior mould surfaces.

PA9052E Nu GPS Receiver is a five-channel receiver providing either stand-alone operation or it can be incorporated into an aircraft's integrated navigation system. Size/mass are designed for high dynamic and rotary-wing aircraft where space is at a premium. It provides a navigation solution at 10 Hz and a separate navigation filter to estimate INS errors. It supports dual redundant 1553, ARINC 429/575 and PTTI interfaces, but alternative interfaces can be provided. In service with RAF Tornado, Harrier and Nimrod aircraft.

PA9150 Series GPS receivers were designed for the marine environment but employ GPS modules common with the airborne equipment. The built-in Control/Display Unit provides navigation displays, including route planning. Interfaces permit integration into platforms with either Synchro or digital interfaces. The back-up battery allows operations through power outages of up to 15 minutes. The unit also supports remote Control/Display Units. Selected for Royal Navy, Royal Netherlands Navy, Turkish Navy, Royal Norwegian Navy ('Oslo' class frigates) and Royal Australian Navy ANZAC frigates.

Navstar Systems Ltd

Current Status
Nova's RT260M is a Cospas-Sarsat EPIRB. Stable 5 W RF on 406.025 MHz ensures global satellite location; it also transmits at 121. 5MHz for location by aircraft. Frequency transmission is simultaneous and signals are encoded with a vessel's individual call sign or MMSI number to ensure recognition of multiple alerts.

US subsidiary, NavSymm, provides sales and field support. Main product in Mapping Information Database Acquisition System II. Company in UK employs about 40 people. Has participated in development of plans for the European Galileo system.

VERIFIED

Sartech Engineering

Current Status
Lokata produced three versions of a 406 MHz Emergency Position Indicating Radio Beacon (EPIRB) for use on marine craft, designed for compatibility with the Cospas-Sarsat international satellite system. More than 8,000 are in use. The 406H/406MH activate automatically when the vessel sinks by using a hydrostatic release mechanism. The 406M, designed for life rafts, has its own mounting bracket and is activated manually, as is the 406P personal version. All 406 MHz beacons transmit the user's identity to the satellite, which then relays the information and the location of the beacon to within 2 to 5 km to the ground stations. Once alerted, the search and rescue authorities use the 121.5 MHz transmissions from the beacon to home in on the vessel in distress.

Locata Ltd were bought out by Kelvin Hughes (part of the Smiths Group) in 1991 and ceased production in 2001. Sartech Engineering maintain all existing beacons.

Satellite Navigation Group

Current Status
The Satellite Navigation Group (previously UKCSG) was established by RIN in 1988, and represents UK interests on the US Civil GPS Interface Committee. Membership is 265 corporate/individual members. Services include the organisation of meetings and conferences and the publication of a newsletter. It also acts as a UK node for dissemination of material from the US Coast Guard Navigation Information Service. The SNG is assisting with the development of the European Galileo navigation satellite system. The SNG operates in the USA through the US Civil GPS Service Interface Committee (CGSIC) and the International Information Sub-Committee of the CGSIC, reflecting the views of UK civil GPS users to GPS administrators.

United States

3S Navigation

Current Status
3S offers a family of Glonass and integrated Glonass/GPS receivers. Attempts have been made to develop a system compatible with Glonass and Navstar systems and the company has designed a range of receivers designed to operate with Europe's Galileo GPS system.

The R-100 line of receiver products is designed to provide the scientific community with the highest possible precision measurements in the areas of time transfer, Ionospheric measurement and precision navigation using a combined solution of the GPS L1 C/A-code and GLONASS L1/L2 C/A and P- code signals.

The GNSS-300 is a 12-channel, broadband receiver with the capability of dynamically configuring any channel to utilise the GPS L1 C/A-code or GLONASS L1 C/A-code. Applications include precision navigation in harsh environments such as cities and precision survey in remote regions. The GNSS-300 can also be configured as a differential correction reference station producing standard RTCM-SC-104 correction messages for both the GPS and GLONASS, or accepting these same messages for sub-metre positioning accuracy in real time.

3S has researched combined access facilities for navigated equipment linking land, sea and air receivers and a variety of non-space based nav- systems.

Allen Osborne Associates, Inc (AOA)

Current Status
Founded in 1978, AOA have built more than 3,000 GPS systems. It has developed a market for integrated multisatellite receiver systems. TTR series for joint USAF/USN/US Army projects. GPS transfer systems are used by ESA and 33 countries. GPS receivers cover geodesy/survey, precise time/frequency, differential, ionospheric calibration, spaceborne and SA/A-S capability. AOA's TurboStar receiver has been supplied for Orbital Sciences Corp's MicroLab 1 satellite and for Denmark's Ørsted. For MicroLab, tracking of dual-frequency carrier phase and amplitude at 50 Hz from selected satellites are yielding atmospheric temperature and moisture content. TurboStar allows both code (C/A, P) and P-codeless (cross correlation) operation. It automatically acquires and tracks up to eight satellites at power-up without initialisation. AOA introduced Rascal, the first handheld, dual frequency GPS receiver for geodetic survey.

Ashtech, Inc

Current Status
Ashtech is owned by Thales and products are now marketed under the Thales Navigation brand name. Thales also owns Magellan products. Ashtech manufactures and markets 26 types of GPS receiver systems and related subsystems for precision geodetic survey, GIS mapping, and land, marine and aerial navigation. Its major surveying markets include: land boundary and cadastral, including real-time static, rapid-static and kinematic; precise geodetic; hydrographic and bathymetric. The product line includes the all-in-view Z-12 receiver and the less costly Z-12 Field Surveyor, both based upon Ashtech's Z-Tracking technology that mitigates the effects of anti-spoofing. Available in 1996 was a new line of GIS field acquisition tools, featuring Ashtech's Super C/A receiver capable of sub-m accuracy in real time or through the use of Windows-based GIS post-processing software. The company also produces the GG24 receiver, the world's first single-board receiver to integrate the GPS and GLONASS satellite-based navigation technologies. Ashtech is working on a receiver capable of reading data from several separate satellite navigation systems and has developed a multichannel receiver for integrating signals from several separate sources for summed computation.

CAST

Current Status
CAST's GPST GPS evaluation system features include a 10-channel signal generator (L1, L2, C/A, P-code), standard avionics busses, and fully integrated jamming capability with up to eight jammers of six types. Options include dual navigation points (differential or relative), externally generated trajectory, and SA/AS. Can be programmed for multi-mode operation and channel selectivity by incoming signal scanning. Optional stop-start switching capability for multi-access sourcing. New high-performance contact integrators. Capable of developing systems for substantial growth and expansion of separate navigation systems based on multichannel access.

Datum Inc

Current Status
Datum's dual-channel GPS receivers primarily provide time transfer and optional frequency measurement information, accessing to UTC through the GPS master clock system. The 9390-5500 is designed for mounting in a standard 48 cm rack; the 9390-5700 is the airborne/portable version. Timing precision less than 1 μs (100 ns with respect to GPS tracking intervals if reference position accurate to 10 m). Position performance less than 25 m with at least 5 minutes of averaging. Only one visible satellite is required for time/frequency monitoring providing that the receiver's location is known. Otherwise, the sets themselves provide automatic positioning through four Navstars. Initial data entry of geodetic co-ordinates is not required because worldwide geographical positioning is automatically performed by an 'anywhere' algorithm. When interfaced with an external printer, the 9390 generates hardcopy reports detailing satellite position, time/frequency, navigation data and elevation plot. The 9390–52000

was introduced in 1993. Timing accuracy is less than 300 ns under current selective availability. Position determination now less than 10 m. Datum Inc have developed multisystem applications software for dual channel GPS receivers and these have been marketed to national and international customers.

Del Norte Technology, Inc

Current Status
Del Norte's GPS receivers incorporate a PC processor and mass memory to provide logging and interfacing capabilities. The 1009+ ™ 12-channel C/A receiver can operate as a geodetic receiver with post-processing software, as a differential reference receiver or as a differential mobile receiver. Divergent access nodes fitted as standard and multisignal source capability. The 4012 12-channel C/A receiver operates as a differential mobile unit and, like the 1009+, can accept differential corrections in the RTCM104 format. Both units are RTK/OTF capable. Channel selectivity and phase discrimination also available. Multi-access formats available from 2003.

Garmin International, Inc

Current Status
Set up in 1989, Garmin International produced the GPS 100 AVD, the world's first portable or panel-mounted GPS that interfaces with autopilot avionics, in 1991 and the first portable GPS with moving map in 1993. The GPS 10 receiver module was introduced in 1992 for embedding in other products. The GPS 150 was introduced in 1993 and GPS 155 in 1994 for general aviation panel mounting; the GPS 95 AVD is a hand-held aviation receiver with moving map display. Garmin's GPS 100 Euro personal navigator provides 250 waypoints and a 10-route storage capability.

The FCC, in November 1995, approved the Starsys application (initially filed 4 May 1990) to build, launch and operate a system of 24 satellites in 1,000 km 53° orbits to provide two-way positioning and messaging services for remote user terminals at VHF/UHF using Spread Spectrum Multiple Access (SSMA). GE Americom at the same time acquired an 80 per cent interest in Starsys, with North American CLS retaining 20 per cent (Hughes STX sold its 5 per cent share to NACLS). GE Americom continues to pursue its separate and later 24-satellite system. Building and launching two Starsys satellites in late 1997 would have cost GE Americom US$46 million, including one year of operations. The full system would cost US$150 million to US$190 million, depending on the launchers. Alcatel Espace was selected as the satellite manufacturer but the system has been put on hold.

Capacity would be approximately 50,000 messages hourly at each ground station. Position determination is based on the Argos Doppler system plus ranging for 100 m accuracy. The FCC issued an experimental licence in April 1992. The company demonstrated the spread spectrum messaging concept in the allocated frequency bands 1992–93 using a fixed station working with leased capacity on CNES's S80/T satellite. Starsys has entered into a number of Early Entry Programme contracts with companies around the world for demonstration and testing of satellite positioning/messaging using the Argos transponder aboard Tiros.

WARC 92 in March 1992 allocated VHF/UHF frequencies for the Little LEO mobile satellite service; the FCC followed suit domestically in January 1993. The forward link will be FDMA: 149.9–150.05 MHz up and 400.15–401 MHz down. The return link will be SSMA: 148.0–148.905 MHz from terminal to satellite and 137–138 MHz down to main station. The initial core of the system was to have been CNES' TAOS satellites, which the French agency decided in March 1994 not to pursue. At that time the concept was abandoned but resurrected later as a fully international venture that would have potential in developing countries but this too failed to attract backers.

Garmin has facilities in the US, UK and Taiwan.

Honeywell Inc

Current Status
The company, a wholly owned subsidiary of Figgie International Inc, was primarily a military contractor after its founding in 1956. It is now increasingly targeting commercial markets, with its GPS receiver family and the range of bandwidth on demand satellite communication terminals. Interstate introduced the SCS 2400 GPS simulator in 1995, generating 24 independent RF channels of L1/L2 GPS signals that can be routed to two RF ports. The signals simulate ground, air/space vehicles in motion with up to 30 *g* acceleration. Multipath, wide area augmentation, common GPS jammers, GPS ground transmitters, ionosphere, S/A and anti-spoofing effects can be simulated. Interstate is jointly developing with Airport Systems International a differential GPS system for airports to transmit corrections to aircraft on final approach.

The SEM-E size P(Y) and C/A code GPS five-channel continuous-tracking receiver is designed for military environments and is packaged for embedded applications. Dual antennas minimise losses of signal and allow for simultaneous tracking of satellites and pseudo-satellites. A factory-installed satellite almanac minimises acquisition time by eliminating the sky search. An option obtains stabilised fixes in less than 5 seconds.

B2 bombers were fitted in 1996 with 900 kg Mk 84 bombs carrying a Northrop Grumman tail kit incorporating Interstate Electronics' GPS receivers for 6 m CEP accuracy.

Honeywell has conducted simulations with GPS guided field munitions available to land forces and has produced combined logic software for multiple applications.

Leica, Inc

Current Status
Magnavox's commercial navigation satellite division was sold to Leica in 1994. It provides complete turnkey DGPS system solutions, including communication links and application software. Has developed adaptive software for multiconcept systems. Provides GPS and Glonass interoperability.

Litton Aero Products

Current Status
Litton's LTN-2001 is an eight-channel, third-generation C/A GPS receiver, featuring continuous tracking for every satellite in view and advanced integrity monitoring. A contract was signed with Boeing in January 1992 to provide the system for all of Boeing's production aircraft. First hardware was delivered late 1992 for flight testing. The LTN-92 ring laser gyro inertial navigation system also offers GPS operations with the addition of a single- card plug-in module. The LTN-311 Omega/ VLF navigation system can be enhanced with a GPS package. The LTN-101 Flagship GPS inertial system was certified in 1996 as a primary means of navigation; its inertial system detects GPS errors before they can significantly affect navigation. The LTN-700 single channel C/A navigation set was one of the first designed specifically for commercial aviation applications. Developed versions have been used in non-space applications from which advanced technologies have been backed into aerospace applications for higher technical standards. Litton is developing receivers capable of accepting different signals from non-GPS navigation satellite constellations. Integrating software available for dual connections and combined civil and military applications.

Magellan Corporation

Current Status
On 12 July 2001 Thales, a European electronics company, acquired Magellan Corporation and merged it with its subsidiary Thales Navigation SA. Thales employs 600 people worldwide with sales outlets in more than 100 countries.

A pioneer in developing positioning and navigation systems using the United States government's Global Positioning System (GPS), Magellan brought to market the world's first hand-held commercial GPS receiver. Focusing on expanding markets and high-volume production, Magellan's product line offering has swelled from a single GPS receiver to over 30 satellite navigation and communications products and systems. Since 1989, the company has shipped more than 2 million GPS receivers while annual sales have grown from zero to more than US$100 million.

Key worldwide markets include aviation, geosciences, marine, mapping, outdoor recreation, survey, vehicle tracking and navigation, wireless communications and systems integration. Magellan's products are sold throughout the United States and in more than 70 foreign countries. Foreign sales of the company's communications and navigation products account for about 35 per cent of Magellan's total annual sales. New product accommodates requirement for demanding user specifications where receivers are incorporated within existing electronics/avionics.

Incorporated in 1986 as a venture capital startup, Magellan began operations in 1987 with five employees. Today, the company employs more than 350 people at its headquarters in San Dimas, California and assembly facility in Mexicali, Mexico. In 1994, Magellan merged with Orbital Sciences Corporation (NASDAQ: ORBI), a diversified commercial space products company with interests in low-cost launch vehicles, small satellites, space sensors, avionics, satellite tracking systems, Earth observation systems and related services and products. Magellan Systems operates as an independent subsidiary within OSC's Communications and Information Systems Group.

Marine users, including commercial and recreational boaters and sailors, represent a key market for GPS receiver products. The world leader in portable GPS for the marine market, Magellan's product line offers boaters a combination of differential GPS, full graphics plotter capability and portability. Products include the panel-mount NAV 6500™ GPS/chartplotter; the hand-held NAV 6000™ chartplotter; the hand-held GPS 2000 XL™, GPS 3000 XL™, NAV DLX-10™ and Meridian XL™ receivers; the low-cost NAV 1200XL™ panel-mount receiver, the GPS Sensor™ and the Differential Beacon Receiver and DBR-2.

Used in geology, geography, GIS data collection, survey, mapping, oil and gas exploration, forestry, utility management, wildlife studies, public safety, environmental remediation and other professional applications, Magellan's ProMARK X-CM™ with centimetre accuracy offers outstanding 10-channel GPS performance, unlimited data collection and multiple differential GPS approaches. Magellan's professional product line also includes the hand-held ProMARK X, FieldPRO V™, the MBS-2™ base station, and the DBR-2 DGPS reference receiver. Magellan's SkyNav 5000™ panel-mounted GPS receiver and its three portable GPS receivers, the hand-held SkyBlazer XL™ with moving map display, the lower-cost SkyBlazer LT™ and the EC-10X with terrain cartography and aviation database, are targeted to the general aviation market to meet the position, navigation, flight planning and operations needs of pilots around the world. In 1995, Magellan introduced the CNS-12™ integrated two-way digital communications, GPS navigation and automatic dependent surveillance avionics unit.

Magellan developed advanced GPS receiver modules adaptable to a wide range of OEM markets. The 10-channel AIV-10™ is used by systems integrators in applications ranging from vehicle tracking and navigation to avionics instrumentation and marine autopilots.

In 1995, Magellan introduced the world's first hand-held GPS priced under US$200. The GPS 2000 XL™, as well as the GPS 4000 XL™ and Trailblazer XL™ receivers, is designed specifically for the outdoor recreation market, including hunting, hiking, fishing, snowmobiling, camping, skiing, off-road driving and many other activities.

In 1996, Magellan introduced the GSC 100, the world's first hand-held global satellite communicator, offering worldwide two-way, e-mail messaging via the ORBCOMM low-Earth-orbit satellite network. With the Magellan handset, the user can send and receive digital text messages in an e-mail format to and from any Internet e-mail address or other ORBCOMM communicator product anywhere in the world.

Magellan also offers the microCOM-M, the world's smallest, lightest and lowest priced satellite communications telephone for global voice, fax and data communication via the Inmarsat-M satellite system. About the size of a notebook computer, and weighing just 5.5 lb, it will slip comfortably into a briefcase or carry-on.

Micrologic

Current Status

Established in 1994, Micrologic specialises in consumer navigation electronics, using GPS and Loran. The company was acquired in 1995 by Singapore's Vikay Industrial Ltd, a leader in OEM navigation electronics. The Millennium set of GPS products was introduced in late 1996: ML-150 hand-held; ML-250 fixed mount; ML-350 fixed GPS and chartplotter.

The US$1,895 Admiral maritime receiver continuously tracks up to five satellites and offers a 128 × 160-pixel high resolution graphic display. 250 waypoints can be programmed (100 protected), nine routes with 20 waypoints each.

The US$1,195 hand-held Supersport waterproof receiver continuously tracks up to five satellites. 133 map datums are set in the memory and 250 waypoints can be programmed (with 100 protected by a locking system).

Micrologic also produces the Vehicle Tracking System utilising GPS applied to ships and trucks, public bus systems, taxis, banks, police vehicles, security and surveillance systems, service fleets.

Expanded capabilities for integrating civil and military functions by patented Micrologic switching offsets. Has developed a tracking system utilising VTS for GPS into a multiformat system capable of operating with any navigational position-fixing satellite constellation.

UPDATED

Motorola Inc

Current Status

Motorola is one of the 100 largest US companies and a leading world electronics supplier. It introduced the six-channel hand-held GPS receiver in 1992 and the eight-channel Oncore in 1994. The Government Space Systems Operations has developed space-qualified GPS receivers for LEO autonomous satellite navigation and for MEO systems other than the US GPS network. Motorola pioneered a unique signal adaptor for encrypting forward position information fixing and provided military users with an opportunity to code signals. The Viceroy is a low-power 12-channel two-antenna C/A code receiver provided for SeaStar, MOMSNAV, MSTI, Forte and Meteor. Monarch is a 12 to 24 channel one to four antenna C/A, P-Y code receiver capable of anti-spoof/selective availability, radiation hardened to 100 krad (Si). Technologies include ASIC, MMIC and VHSIC. The receiver plays an important role in NASA's Explorer Platform and the Topex mission. Motorola has developed a receiver capable of processing combinations of Glonass and Navstar systems and for integrating future European navigation satellites.

Sokkia Corporation

Current Status

A subsidiary of Japan's Sokkia Co Ltd, the company is a supplier of precision products for surveying, mapping, measurement and positioning. The company has a wide and expanding range of geophysical mapping and geodetic survey equipment and has integrated various data sets within common platforms for computerised integration, enhancing availability to a wide range of customer devices. These include electronic field books and total stations, GPS/GIS systems, software and 3-D positioning systems. Sokkia Corporation has integrated air and satellite-based mapping systems into a common data management system capable of inputting digital and optical information formats. Amalgamated GPS/GIS systems have been developed with other suppliers and these provide full end-to-end coverage. The SGS Sokkia Global Surveying GPS system comprises an eight-channel receiver, antenna and post-processor to determine 3-D positions with static surveying accuracy of 5 mm ±2 PPM and kinematic surveying accurate to 2 cm ±2 PPM. Data are stored on removable 256 kbyte cards, each holding up to 9 hours. Battery allows up to 4 hours continuous use.

Sony Electronics Inc

Current Status

Sony's first GPS product, the four-channel Pyxis IPS-360, receiver was launched in 1992. Designed for outdoor sports enthusiasts including inshore boating, it offers a range of modes in addition to the standard latitude, longitude and altitude display: navigation (distance/direction to destination, with current direction/speed), track (current direction, altitude relative to start, and distance travelled) and mark (store details of 50 waypoints and 50 markpoints in memory). Cost is about US$1,150/ £685. The IPS-760 was introduced in 1993 as the first portable GPS system incorporating Arnav's Aviation Database on removable InfoCards. The eight-channel receiver offers 1,000 waypoint storage. In Japan, the NVX-F10 car navigation system is produced at 12,000 units monthly. The US$2,000 system was introduced June 1993 and comprises a GPS receiver, a CD-ROM, small GPS antenna and 10 cm colour LCD monitor capable of adapting to different input data formats. Sony has developed a lapse-sequence Glonass/Navstar receiver only available in Japan but marketed in the USA under a subsidiary.

Stanford Telecom

Current Status

Stanford Telecom produces GPS satellite signal generators to simulate the exact signal environment experienced by high performance aircraft during manoeuvres. This system is also used for tactical evasion training for out-of-sight target vectoring. It has developed a high-performance GPS receiver embedded in a Doppler Navigator for US Navy and Marine Corps helicopter applications under subcontract to Teledyne Electronics Systems. This has been developed into a field mobile electronics position prediction and evaluation system. Applications evolved from this technology which have been directed to civil and military requirements and subsequent evolutions of the system were applied to an integrated position fixing design.

Tecom Industries, Inc

Current Status

TECOM provides a range of GPS antennas that includes Glonass and space qualification. One version of the 401163 is optimised for L1/L2, and another at L1/Glonass frequencies. The 401163–5 is fully qualified for space applications. The 401226 houses the basic 401200 volute L1 antenna and can house a GPS receiver. The range of military GPS antennas provides gains of more than –2.5 to more than –3.0. Tecom is developing a GPS receiver antenna capable of differentiating between different operating systems and is producing a more durable and survivable structure for operating in high-stress environments. Studies have been conducted on embedded systems for position-fix in mobile equipment for land, sea, air or space forces. Has also developed multisystem integration concepts for diverse operating constellations.

Universal Navigation Corporation

Current Status

The GPS-950 five-channel receiver interfaces with UNS Flight Management and Navigation Management Systems. It is housed in a 1 MCU box weighing 1.50 kg. The UNS-1M NMS, available from January 1994, incorporates a GPS receiver. The company introduced its UNS 764-1 combined two-channel GPS/Omega/VLF aeronautical civil receiver in 1990. Priced at US$38,500, series production is undertaken by Canadian Marconi. UNC offered lightweight units designed for low volume storage and reduced power levels. Also provides ruggedised systems. Multisystem access is also available through modification to existing receivers.

Welnavigate, Inc

Current Status

Welnavigate offers GPS signal test generators; more than 500 have been sold. Current models are Orbiter I/II, GS 100 and GS 500. The Orbiter C/A dynamic simulator provides five channels (up to 16 available) of GPS RF, an 80486 desktop computer, VGA video card and Multisynch Colour Monitor. Version I emulates plane, boat or car movements; II mimics missile or LEO satellite motion. I: US$81,750; II: US$129,000. A constellation of up to 32 satellites can be specified, along with data messages, parity, health, power levels, visibility, ephemeris errors and ionosphere/troposphere effects. Precomputed receiver vehicle motion is supplemented by selectable manoeuvres (turn, climb, accelerate), automatically controlled jerk, and vehicle silhouette masking and signal outage. It tests receiver sensitivity, tracking performance with variable south/north, dynamic tracking capability, antenna and preamp performance, and data demodulation and decoding. Advanced, tailored versions have specially adapted high-stress simulation clearance and have been produced for unspecified customers and operators. Much of the technology thus stimulated and developed has been applied to specialised project work for undisclosed clients.

SATELLITE ACTUATION SYSTEMS AND THERMAL CONTROL DEVICES

Canada

Novatronics Inc

Current Status

Novatronics Inc is engaged in the design and development of custom components and systems for the sensing and actuation of rotary and linear motion. These include brushless DC and stepper motors, linear actuators, brakes and clutches, solenoids, RVDTs and LVDTs, and electronic units to integrate them with other systems. Novatronics is working with Spar on the development and manufacture of motor modules and brakes for the Space Station and Shuttle manipulator systems. Development of highly rugged rotary and linear motion sensor activated motors for remote manipulators capable of autonomous operation necessitating highly reliable operation.

Novatronics has developed multiple line use carriers for automated robot systems working off manned space vehicles. Has combined elements of conventional electric motor technology and miniaturised design concepts to produce small and lightweight systems for lightweight satellites and spacecraft.

Denmark

Alcatel Danmark A/S

Current Status

Alcatel Danmark Space's main product areas are: electronic power conditioners for SSPAs; AC/DC converters for equipment; power distribution units and solid-state power switches; servo amplifiers; magnetics components. It has for almost 25 years supplied custom-designed power electronics for Ariane and satellites such as GEOS, OTS, ECS, Marecs, Viking, Telecom 1/2, Hipparcos, Eureca, Skynet 4, ERS, Soho, Cluster, Spot 4, Inmarsat 3, EMS, Artemis, MSG, Globalstar, Thaicom, MT-SAT, WorldStar, Astra 2B and ICO. Developed conditions for BPTA on large rotating assemblies including rotating SSPA and large satellites.

Advanced SSPA devices for large space structures and rotating artificial-*g* devices.

France

Aerospatiale

Current Status

Aerospatiale has developed a range of rigid solar arrays optimised for 1 to 10 kW outputs. The first generation GSR-1 is employed on Arabsat 1, Italsat 1, Telecom 1, TDF/TV-Sat and Tele-X. The GSR-3 array offers 40 to 45 W/kg and 115 W/m^2 (12-year EOL). The arrays are flying on Spacebus 2000/3000, such as Italsat 2 (5.3 kW) and Turksat (3 kW). The GSR-3 on Spot 4/Helios 1 use ADELE frictionless hinges. Up to seven panels, each a maximum 3 × 6 m, can be accommodated in a variety of 2-D/3-D configurations. Standard or ultra-thin (50 µm) cells can be employed, up to 6 × 8 cm. The panels comprise an aluminium honeycomb sandwich faced by 2-D carbon skins and insulated by a single Kapton sheet. Final panel mass density is 1.050 kg/m^2. The GSR-4 family was qualified in 1996-97. The panels of co-cured aluminium sandwich with 50 µm Kapton reinforced carbon fibre face skins (max density 1 kg/m^2) can host advanced Si and GaAs cells. ILSA is a deployable flexible Kapton array for 4 to 7 kW missions such as Spot 1 to 3 and ERS 1/2. A 55 kW system has been studied for ESA and larger systems with high-power SSPA for station modules and free-flying laboratories. Developing the concept of the unfurlable semi-rigid array, company has designed a 10 to100 kW output array for satellite and ISS applications.

The company operates an 875 m^2 class 100,000 solar array integration area at La Frayère near its major Cannes facility, housing a 35 m^3 vacuum chamber, a Spectrolab flasher for solar cell testing and two deployment jigs.

Alcatel/SAFT

Current Status

Alcatel SAFT's Space Department was established in 1963 and has almost 40 years' experience in providing space-qualified batteries and cells. Gates' aerospace battery element was acquired at end-1993; the manufacturing facilities completed transfer to Poitiers in April 1995. SAFT's principal types are: nickel cadmium, silver zinc, nickel hydrogen and lithium couples. The company provides Ariane's 6/10 Ah primary and rechargeable silver zinc batteries. 250 Ah 38 kg $LiSOCl_2$ lithium thionyl chloride batteries have been developed for Titan's Centaur; first launch was May 1995. SAFT's energy density is up to 500 Wh/kg, with a high rate capability of 1 kW/kg. The large high-rate cell is being produced with capacities covering 5 to 250 Ah. Design developments have extended this to 310 to 400 Ah.

SAFT offers a range of hermetically sealed prismatic Ni/Cd cells/batteries designed specifically for space applications. Capacities cover 7 to 50 Ah, providing energy density of 40 Wh/kg and GEO life expectancy of more than 15 years. The VOS range of cell designs can be accommodated to match mission specifications; SAFT batteries are employed by Meteosat, Spot, Olympus, IRS and GPS; Radarsat carries three 50 Ah batteries. Ni/Cd batteries for launcher and micro-satellite applications: 1 to 6 Ah VRE cylindrical cells. VRE cells have been carried by Ariane from its first flight. VRE derivatives for advanced launcher requirements.

A 23-cell 60 Ah battery was developed in 1979 for Artemis and DRS and high-output batteries are in research and development for high-power satellites.

Cell rated capacity (Ah): 40-100 (VHS-CM series), 35-200 (AN series)

Depth of discharge: 70 per cent for 15 years (1,500 cycles) in GEO, 40% for 5-6 years (30,000 cycles) in LEO

Energy density: 50-60 Wh/kg

Cell design: cylindrical Inconel 718 metal case containing a stack of electrochemically impregnated nickel (positive) electrodes. Two insulated feed throughs are located on one hemispherical end. Minimum burst pressure is 190 atmospheres; maximum pressure is 75 atmospheres when 100 per cent charged.

Battery construction: the modular architecture can accommodate all space platform requirements, allowing extension of the standard configuration with minimised battery height.

Germany

DaimlerChrysler Aerospace AG

Current Status

DASA's Solar Generator Department in Munich (see the General and Launcher/Propulsion sections for company information) was responsible for designing, developing and manufacturing the solar arrays for Intelsat 6 and the Space Systems/Loral FS-1300 satellite communication bus. Intelsat's 10.66 kW is generated at 64 W/kg and could be further improved by replacing the Kevlar face sheets with carbon fibres. Japan's Superbirds, based on SS/L's FS-1300, are the first to use DASA's Mark 4 40 W/kg array, featuring superlight open net substrates with multi-insulation. The motor-controlled deployment mechanism can also be used for retraction. 22 satellites were launched 1990-95 with DASA arrays (Intelsat 6 F1/3/5, Inmarsat 2 F1-4, Superbird A1/B/B1, JCSat 2, DFS 2/3, Eutelsat 2 F1-6, Meteosat 5/6, Italsat 1). A DASA-led consortium under ESA contract in 1994 completed development of 25.1 per cent-efficiency 4 × 6 cm Si cells for use under low-intensity conditions. This is the highest Si efficiency attained without optical concentration devices. The cells are intended for deep space missions such as ESA's Rosetta comet rendezvous. Improvements are being studied to raise efficiency to the 27 to 28 per cent at −150°C (illumination 0.03 solar constant) required by Rosetta. CISE developed 23 per cent efficiency GaAs cells under subcontract to DASA. Developing active power systems for planetary spacecraft.

The former Telefunken Systemtechnik, part of DASA since 1992, provides electrical power systems, including solar cells and complete solar arrays (rollout, flexible foldout, rigid deployable and body-mounted), power conditioning, storage/distribution, in addition to spacecraft power/signal harnesses, static converters/inverters, and data-handling equipment. Electrical switching nodes for multiple equipment outlets and utilities. High power level surge protection for step-down systems.

Friwo Silberkraft

Current Status

Silberkraft designs, develops and manufactures silver zinc, nickel cadmium and lithium batteries for space applications. Lithium primary cells covering 0.35 to 36 Ah (cell voltage 3 V) are available offering shelf lives up to 10 years. Vehicle users include Helios, AMPTE, MAUS, Mikroba, Cosima, SPAS, GAS, Texus, ASTRO-SPAS, Navex, Falke, IBBS. FS has developed energy storage systems for long life dormancy and longevity of operation when activated. These devices apply to secondary payloads carried for long periods before activation. FS is developing limpet batteries capable of being transferred from one payload to another in flight. Also, a new range of high-power lithium batteries with applications for small- and medium-size human space flight support packages and portable modular systems.

Linde AG

Current Status

Refrigerator and cryostat systems for handling oxygen, nitrogen, helium, hydrogen and neon. Specialised systems for long-life storage of cryogenic products, also capable of supplying monitoring equipment ranging from lightweight devices for small satellites or major ground facilities for large volume supply. The company participated in developing the cryogenic cooling system for ESA's Infra-red Space Observatory, and provided the project's cryogenic ground support equipment. It is involved in developing a freezer system for Space Station and lunar surface base systems and has designed such systems for application to Mars surface base requirements. Micro-cryosat development for small voids and volumes. Linde has designed freezer systems for environmental control systems.

TELDIX GmbH

Current Status
TELDIX, a member of Litton Industries Inc (USA), was founded in 1960 as an affiliate of Telefunken GmbH and the Bendix Corporation. In 1973 the Bosch GmbH acquired the BENDIX interest and in 1981 that of AEG-Telefunken. In February 1986 Litton Industries Inc acquired TELDIX GmbH from Bosch. In April 2001 Northrop Grumman Corporation acquired Litton Industries. The company specialises in high-precision devices for satellite attitude-control systems, such as momentum/reaction wheels, solar array drive assemblies and antenna pointing mechanisms. Teldix has developed control mechanisms for integrated reaction and attitude control thrusters applicable to large structures. This has been extended to design and development of stable telerobotic control and attitude stabilisation systems.

TELDIX developed its Solar Array Drive Assembly system for West Germany's DFS Kopernikus but it is suitable for pivoting the solar arrays of any high-power three-axis satellite; it is used on China's DFH-3. It was qualified under the DFS programme for 15-year missions. The modular design has three subassemblies: bearing unit; redundant drive motor with sensors; slipring unit. Control is by redundant electronic units.

Specifications
Electrical transmission (A): 2 × 22.5, 8 × 4, 3 × 1.3, 12 × 0.25
Size: 22.5 cm diameter, 13.0 cm height, less than 5.6 kg
Stepwidth: 0.0375°
Torque: more than 3 Nm hold, more than 0.7 Nm stepping
Power consumption: less than 3.3 W
Lifetime: more than 10 years.

Italy

FIAR SpA

Current Status
FIAR provides spacecraft full power systems (including batteries), regulators and ancillary equipment, as well as designing and supplying GaAs solar arrays. For Argentina's SAC-B, it provided four deployable carbon fibre panels carrying a total of 1,032 30 × 40 cm GaAs cells, generating 205 W EOL at 28 V. For France's Arsene, it provided six panels with a total of 986 2 × 4 cm GaAs cells. Arsene was the first satellite to be powered completely by European GaAs cells. Performance of cell reconstruction has exceeded specifications. New technologies linked to GaAs cell design has been marketed to international partners. FIAR has developed GaAs cell technology for large satellites and high-power requirements and has tested new lightweight cells arrays for long duration use on deep-space systems.

Japan

Sharp Corporation

Current Status
Sharp has provided silicon solar cells for almost 70 satellite programmes since 1971, including Inmarsat, Intelsat and ETS 6 (50 μm). New technology development offers increased power-life usage. Development of extended environment specifications requirements.

Specifications
Cell thickness: 50–200 μm
Cell size: 2 × 4 to 4 × 6 cm
Conversion efficiency: 14.8% (typical) for 100 μm BSFR cells; 17.0% (typical) for 100 μm NRS/BSF cells
Cell type: bare cell, glassed cell and CIC
Silicon diodes
Thickness: 300 μm maximum
Size: 2 × 2 cm
Absolute max rating: V_R 140 V, I_F 3 A, T_{op}–160/150°C.

VERIFIED

Netherlands

Dutch Space BV

Current Status
Dutch Space has been producing solar arrays since 1970, beginning with the ESRO 4 science satellite. Dutch Space specialism is rigid, deployable solar arrays and body-mounted arrays. The Dutch Space solar array program covers the power range of approximately 0.5 to 20 kW. Dutch Space also produces and delivers solar array-related components and assemblies like hold-down and release mechanisms.

In 1973 Dutch Space started the development of the Advanced Rigid Array technology (ARA). The first-generation ARA Mk I, designed for 1 to 4 kW at 25 W/kg after 7 years, has been used in the ECS, MARECS and SKYNET programs. Its successor, ARA Mk II, designed for 2 to 4 kW at 40 W/kg after 10 years with panels up to 7 m^2, was selected for Inmarsat 2, Telecom 2, SAX, SOHO, Hispasat, Artemis and NASA's Chandra; first launch was in 1990 aboard Inmarsat. As a side development, a 5.5 kW retractable ARA version was built for ESA's EURECA platform. It was launched in 1992 and after nine months retrieved by the Space Shuttle. Development of ARA Mk III, the successor of ARA Mk II, began in 1992, offering up to 12 kW after 15 years using Gallium Arsenide solar cells. It can accommodate up to five panels of 8 m^2 each. First launch was on Hotbird in 1996 and further it was selected for WorldStar, XMM, Integral, Sicral, ST-1, Nilesat, Astra, Rosetta, Atlantic Bird, AMOS-2 and AMOS-HP.

For the large LEO platforms Dutch Space-developed the 'Flatpack' array concept, covering a power range of 5 to 12 kW. It's first launch in 2002 was aboard ESA's ENVISAT platform, providing 6.6 kW after 4 years from 1 wing (Silicon) comprising 14 1 × 5 m^2 panels. Flatpack solar arrays were also selected for NASA's EOS-Aqua (launched in 2002) and EOS-Aura platforms.

For ESA's ATV Dutch Space developed and built a special solar array concept compatible with the stringent load requirements due to rendezvous and docking with ISS. This solar array type was also selected and built for ESA's SMART mission.

For the small missions up to 1 kW Dutch Space has developed the SMRT concept, which has been selected for EROS-B.

Switzerland

Cryophysics SA

Current Status
Cryophysics was established in 1967 to provide state-of-the-art cryogenic equipment and instrumentation to European industry and research laboratories. Its activities were broadened in the 1970s to include high-power carbon dioxide lasers. Product lines include helium liquefiers, cryogenic sensors and temperature controllers, super-insulation, closed cycle cryogenic refrigerators and cryogenic storage vessels. Entrance to space project activity was predicated by a requirement from ESA for storage and transfer of small quantities of cryogenic fluids. The Barber-Nichols cryogenic pumps are the industry standard for the efficient transfer of liquid cryogens, including LHe, LN_2 and LH_2. Cryophysics has studied super-insulation for long duration exposure of cryogenic systems in high-temperature environments. It has designed liquifaction systems for small satellites and for miniaturised cryogen storage vessels. Also specialises in cryomagnetics, vacuum pumping systems and accessories for cryogenic systems operation.

ETEL SA

Current Status
Founded in 1974, Études Électromécaniques SA (ETEL) specialises in the development and manufacture of high-performance electric motors and controllers: brushless torque ring motors (PM synchronous), brushless high and very high-speed motors (up to 200,000 rpm), stepper (200 and 400 steps), limited angle motors, reluctant motors, linear motors (voice-coil, moving magnet), linear or rotative geared actuators and EM actuators. All can be delivered frameless or with housing, bearings, position sensors, brake and associated electronics. Examples include all motor brakes, resolvers and associated electronics for ESA's External Robot Arm; brushless motors for driving Telecom 2's optical sensors (1 W consumption, 60 to 600 rpm, stability better than 1:1000 over 10 years); electromagnets for opening ISO's container. Integration of linear motors to electromagnetic doorless containers. New ISS standard international specifications.

The 2,400 m^2 facility, built in 1995, includes a class 100 laminar air flow unit in a laboratory cleanroom equipped for electronic testing. 1996 turnover was more than SFr14 million, with space accounting for about one-third of revenue; personnel total about 100. ETEL Inc (USA) was started in 2001 with ISO 9001 certification and ETEL BV (Netherlands) and ETEL in France were started in 2002.

Reusser AG

Current Status
Space-related activities cover design/fabrication of cooling systems for space applications: water/freon/ammonia pumps of 2 to 800 W input. The company's pump package was used to cool experimental furnaces flown on the Spacelab 1/D1/D2 and Eureca 1 missions. Cooling systems for Columbus laboratory and, from this research, development of advanced cooling insulation and thermal protection systems. Developed cooler facilities for science experiments on Columbus and ISS facilities. Reusser has pioneered several compact furnaces for science experiments in space-related programmes and for ground based facilities where size and mass are critical.

United Kingdom

Control Techniques Dynamics Ltd

Current Status
The company provides high precision electric motors for space applications: solar array deployment drives, instrument positioning drives, shutter drives and antenna deployment drives. Users include Eureca, Inmarsat and ERS. Advanced motor drive design for large solar arrays in 0.1 *g* acceleration environments. Extended acceleration loads accommodated by unique user requirements. Applications extend to Columbus and ISS and the company has developed high speed motors for deep-space applications with long storage and infrequent power-up/power-down cycles with high temperature, pressure and loads requirements.

EADS (Matra Marconi Space)

Current Status
The former British Aerospace Space Systems element designs, develops and manufactures

spacecraft power and active/passive thermal control systems, and offers cryogenic coolers derived from a unit developed for NASA's Upper Atmosphere Research Satellite. The company has experience in rigid, advanced rigid, flexible fold-out and roll-out retractable solar arrays, including the twin 16 m^2 silicon wings for NASA's Hubble Space Telescope. Studies are underway to evaluate the application of InP and GaAs cells in the development of flat and concentrator arrays. BAe developed an 8 kg array regulator introduced aboard France's Telecom 2 series. Using a high-speed switching technique, sections of unregulated solar array are switched in and out of circuit to match demands for regulated power within 1 per cent and with a 98 per cent efficiency. In addition, BAe developed a Solar Array Simulator. Developed versions have been designed for high-rigidity structures available in large or small diameter units. Combined, these have been developed into compressed design concepts applicable to small spacecraft in diverging environments.

BAe's Ni/Cd batteries are all designed/tested for GEO and LEO operational conditions. The electrical designs require that all of the cells used in assembling a flight battery are selected from one cell manufactured lot. Thermal control is achieved using secondary surface mirrors or a radiator, and a (resistive) heater chain. The battery is mounted on the spacecraft using standoffs providing electrical and thermal isolation.

The cell mechanical design requires a hermetically sealed, prismatic, thin-wall case providing a high strength to mass ratio. Cells and divider plates are sandwiched between end plates using tie bars, and the assembly bolted to a baseplate. The cells are electrically connected in series, but each cell case is isolated.

Nickel cadmium batteries

Model 3686 Nickel cadmium battery
Rated capacity: 32 Ah at 32.0 V average, 3.5 A max discharge current
Mass: 30 kg (26 cells)
Size: 336 × 425 × 202 mm
Environmental: tested to 20 *g* at 30 Hz for 60 s, –5/25°C
Model 3600 Nickel cadmium battery
Rated discharge: 35 Ah at 17.0 V average, 15.0 A max discharge current
Mass: 17 kg (16 cells)
Size: 183 × 439 × 128 mm
Environmental: tested to 17 *g* at 30 Hz for 60 s, –5/15°C.

HST solar array

For design and construction of the HST Solar Array, BAe was awarded a £9.5 million contract in June 1988 for a replacement high-power version of the Hubble Space Telescope's solar array. The two wings deliver 5.1 kW BOL, reducing to 4.5 kW after two years, at 34 V when operating at 90°C. The array is qualified for LEO applications over 400–600 km, encountering 30,000 thermal fatigue cycles (–90 to +100°C) during a five year mission. The replacement array was installed December 1993.
Size: 2.6 × 12.0 m each wing deployed, 4.5 × 0.65 × 0.5 m stowed
Mass (2 wings): 87.2 kg four Si cell blankets, 115.1 kg for two Secondary Deployment Mechanisms, 81.8 kg for two Primary Deployment/Solar Array Drive Mechanism and Jettison Adaptors, 10.6 kg for two primary electrical harnesses, 14.4 kg for drive electronics. Total mass for two arrays 322.6 kg.

Solar array simulator

The solar array power simulator comprises up to 38 modules, each with variable control capable of outputting up to 150 V/10 A. Modules are designed as either a complete racked system, with controller and mains distribution functions available or in a stand-alone mode. The master cabinet houses eight array simulator modules and can accommodate three slave cabinets of 10 simulator modules each.

Solar array drive subsystems

MMS manufactures low, medium and high-power solar array drive mechanisms and their associated drive electronics and array Sun sensors. The medium power drive (carried by Inmarsat 2, Hispasat, Orion 1 and Telecom 2) employs indirect drive, with rigidly pre-loaded large diameter ball bearings capable of withstanding all launch and array deployment loads. The high-power version (employed by Olympus) uses two redundant stepper motors driving through a reduction gear. Both operate in open-loop mode and require no launch lock. The drive electronics provide pointing control for two prime and two redundant drive motors, including a differential or common offset pointing facility for solar sailing. The 75 g (52 × 23 × 35 mm) Sun sensor provides 10mV/degree error signals for Sun angles less than 4°. The low power range BAPTA (Bearing And Power Transfer Assembly) permits up to 0.8 kW power transfer and is incorporated on Telecom 1, ECS and ECS derivatives.

Specifications

High power drive mechanism
Drive motors: two stepper motors driving through a 30:1 reduction gear, generating an 11.8 Nm torque per motor at the array flange
Sliprings: 26 or 52 power sliprings each rated at 8.0 A; 50 signal sliprings each rated 0.5 A; 18 pyrotechnic sliprings each rated at 10 A for 100 ms
Friction torque: 3.2 Nm (52 sliprings) to 2.0 Nm (26 sliprings)
Power consumption: 7.5 W for Sun tracking and acquisition
Power transfer: 2.4–10.2 kW per mechanism
Position setting: two orthogonal axes to ±0.05°
Size: 288 mm diameter × 250 long max; 9.8 kg (52 sliprings) to 8.7 kg (26 sliprings)
Drive electronics: 50 VDC, 23 W (worst analytical case), 8.5 kg, 337× 222 × 169 mm
Medium power drive mechanism
Drive motor: one stepper motor with dual stack (one redundant) driving through a 60:1 reduction gear, generating a 15.6 Nm torque at the array flange
Sliprings: 12, 18, 24, 30 or 36 power sliprings each rated at 8.0 A; 30 signal sliprings each rated 0.5 A
Friction torque: 2.4 Nm (36 sliprings) to 1.8 Nm (12 sliprings)
Power consumption: 0.07 W for Sun tracking at 1 rev/day; 6 W max for Sun acquisition at 96 rev/day
Power transfer: 0.8–5.6 kW per mechanism
Position setting: two orthogonal axes to ±0.05°
Size: 230 mm diameter × 265 long max; 7.2 kg (36 sliprings) to 5.1 kg (12 sliprings)
Drive electronics: as high power model.

BAPTA

The BAPTA is a low power mechanism using direct drive DC torque motors operating in closed-loop mode in conjunction with solar array Sun sensors. A pyrotechnic device offloads launch and deployment forces. The control electronics provides pointing control for two units and can operate in Sun acquisition, normal Sun tracking, inhibit and differential or offset pointing modes. Sun tracking error is less than 2°.
Drive motors: two DC brushed torque motors each producing 0.71 Nm at stall
Sliprings: 13 power sliprings rated at 3.5 A each, 19 signal sliprings at 0.1 A
Friction torque: 0.16 Nm
Power consumption: 0.04 W for Sun tracking at 1 rev/day; 0.85 W for Sun acquisition at 1°/s
Power transfer: up to 0.8 kW per mechanism
Size: 171 mm diameter × 267 mm long; 4.2 kg
Control electronics: –50 VDC, +5 VDC and ±15 Vdc; 2.5 W typical, 262 × 171 × 89 mm, 2.2 kg.

Cryogenic cooler

Oxford University and Rutherford Appleton Laboratory developed a Stirling Cycle 80 K cooler for the Improved Stratospheric And Mesospheric Sounder (ISAMS) on NASA's Upper Atmosphere Research Satellite (UARS), and a design variant is incorporated in the Along-Track Scanning Radiometer on ESA's ERS. MMS offers a production version of the cooler for Earth observation, upper atmosphere measurements, and short wavelength/IR/radio sensor applications. The cooler operates on the Stirling Cycle principle and comprises compressor and displacer units with drive/control electronics. The compressor sets up the pressure cycle and heat is shuttled from the cold tip of the displacer by a regenerator operating in a precise phase relationship with the pressure cycle. Long life is attained from the low dynamic stress levels and the absence of contacting sliding surfaces. A two-stage 20 K Stirling Cycle cooler has been developed and qualified under ESA contract. MMS has a separate ESA contract to build another 20 K cooler together with an additional Joule-Thompson expansion stage for achieving down to 4 K.

Of the first commercial batch of six 80 K coolers, one was delivered to NASA's Jet Propulsion Laboratory and two to the US Naval Research Laboratory in January 1990; a fourth was delivered to TRW; Goddard took delivery of the fifth for evaluation in late 1990; the sixth was ESA's pre-qualification cooler. Of the second batch, three have been delivered to Japan and two to Canada; one remains at MMS on life test. Of the third batch, two have been sold to Toshiba (for Japan's ADEOS satellite) and four to Rockwell (for Brilliant Eyes). MMS is now building the Mk2 cooler (increased heat lift and power efficiency), a 50–80 K cooler based on an ESA contract to optimise the 80 K cooler. An initial batch of eight is being built. The first two were ordered at £460,000 by Com Development Atlantic for the MOPITT instrument, to fly on EOS-AM1 in 1998. Other orders are for Envisat 1: three for AATSR and four for MIPAS. The 4K cooler was delivered in mid-1994 to ESA for evaluation.
Cold tip temperature: 80 K standard (60 K possible)
Cooling power at 80 K: 800 mW
Lifetime: 10 years
Size (mm): 200 long, 120 diameter; 3.2 kg compressor 210 long, 78 diameter; 1.08 kg displacer
290 × 205 × 105; 3.9 kg electronics
Power consumption: 40 W (28 VDC, adaptable to other voltages)
Operating temperature tolerance: –20/40°C
Vibration (non-operating): 20–80 Hz increasing at 3 dB/octave; 80–350 Hz level at 0.24 g^2/Hz; 350–2,000 Hz decreasing at 3 dB/octave.
General Specification for 50–80 K cooler
Cold tip temperature: 50–80 K nominal
Cooling power (mW): 800/55 K, 1,100/65 K, 1,750/80 K
Lifetime: 10 years
Size (mm): 200 long, 120 diameter compressor; 3.24 kg typical
230 long, 78 diameter displacer; 1.09 kg typical
Power consumption: 50 W
Operating temperature tolerance: –10/40°C
Vibration: 0.22 N rms back to back compressor; 0.044 N rms back to back displacer
General Specification for 20–50 K cooler
Cold tip temperature: 20–50 K nominal
Cooling power (mW): 60/20 K, 300/30 K
Lifetime: 10 years
Size (mm): 165 long, 140 diameter compressor; 3.5 kg each 210 long, 75 diameter displacer; 1.3 kg
Power consumption: 37 W
Operating temperature tolerance: –10/40°C
Vibration: 0.22 N rms back to back compressor; 0.044 N rms back to back displacer.

VERIFIED

EEV Ltd

Current Status

A subsidiary of GEC plc, EEV manufactures specialist electronic components and subsystems based on solid state and vacuum tube technology.

EEV develops and manufactures advanced high-performance GaAs cells and complete panels. The MGA series of GaAs/GaAs cells offer BOL efficiencies up to 20 per cent AMO (25C) and standard devices up to 16 cm^2. The MGE series of GaAs/Ge cells offer lower mass and BOL efficiencies up to 18.5 per cent AMO (25C), with device sizes up to 36 cm^2. EEV offer a full design, solar cell/panel manufacturing service for panels up to 2 m^2 within a fully integrated facility. Flight proven since 1991 (UoSat 5: Europe's first GaAs array), the panels have a 100 per cent reliability record. Missions include SSTL's UoSat 4/5, Chile's FASats, Pakistan's

Badr-B, UK's STRV 1, Thailand's TMSat. Brazil's SACI-1, Malaysia's MYSAT. Ultrathin (5 μm) GaAs and cascade (23 per cent AMO efficient) cells are under development. Has produced experimental compressed density high-power output GaAs for radar satellite SSPA requirements.

A dedicated CCD facility produces a range of custom and standard Si CCDs tailored for optical, ultra-violet and x-ray applications. Almost any size of linear or 2-D device can be manufactured, subject only to the limitation of the 125 mm Si wafer. EEV is ESA's supplier of choice: Envisat (CCD25-20 for MERIS optical imaging spectroscopy; CCD26-10 for GOMOS ultra-violet/optical spectroscopy) and XMM (CCD15-30 for RGS high resolution x-ray spectroscopy; CCD22-20 for EPIC x-ray imaging spectroscopy). Devices can be supplied to agreed specifications and screened/qualified to the required levels. Standard CCDs have been supplied for micro-satellite missions such as UoSat and FASat. Special user requirements accommodated by purpose-built devices assembled before delivery.

Pilkington Space Technology

Current Status

PST represents the civil business of Pilkington Optronics, a joint venture company owned equally by Pilkington plc and Thomson-CSF. PST specialises in thin space-qualified glass used for solar cell covers and second surface mirrors. Almost 300 satellites, including Landsat, Hubble, GOES, Giotto, Intelsat 5–8, NStar and Cluster carry PST products. Facilities include a custom-designed glass melting furnace, advanced vacuum coating plant, automatic cleaning lines and an accelerated UV machine. The coverglasses and SSMs are manufactured from borosilicate-based glasses carrying a nominal 5 per cent cerium dioxide content to provide radiation stability by preventing the formation of colour centres under proton/electron irradiation. The glasses are available at 0.05 to 0.5 mm thickness and with a variety of coatings. PST has developed thick, high-density, borosilicate glass window panes for high temperature environments. Specialist user requirements accommodated through joint development initiatives.

Saft Nife Limited

Current Status

The company manufacturer of nickel cadmium, silver zinc, lithium and thermal batteries for the space industry. Saft Nife Limited has produced hard, durable, compact battery packs and power units for specialised purposes to government and non-government customers. Exports to the US and European countries. New Ni-cad batteries for long storage requirements. Suppliers to ESA and its contractor base through direct sale or project management groups.

Space Innovations Ltd

Current Status

It produces the NC-X series of Ni/Cd batteries for LEO eclipse operations. Nominal operating voltage 20 V and, operated to 20 per cent depth of discharge, are suitable for LEO missions of up to two years with no reconditioning. Nominal capacities are 4/7/10 Ah, and can be paralleled. 16-cell 4 Ah pack less than 3.8 kg, 7 Ah less than 5.7 kg, 10 Ah less than 85 kg. Pakistan's Badr-B carries a 4 Ah pack. SIL also produces the 30 to 300 W PCS-28R power conditioning system which generates a fully regulated +28 V power bus. Research and development department provides facilities for research module design interface. Has also extrapolated long life battery systems for extended delay activation following long periods of dormancy under variable and changing temperature environments. Developed new specifications for extreme environment survivability standards and has created new range of power storage systems to meet those requirements.

United States

EaglePicher Technologies LLC

Current Status

Now known as EPT, the original company began battery development in 1920 and established the Electronics Division in 1940 producing large numbers of batteries for the aircraft industry. In 1995 sales were about US$800 million. EPT is one of three divisions of EaglePicher, the others being EaglePicher Automotive and EaglePicher Filtration and Minerals. The EPT division accounts for 95 per cent of batteries used on NASA manned missions, 85 per cent in US missiles/weapons and 90 per cent worldwide of satellite Ni/H_2 batteries. Batteries produced for Mercury, Gemini and Apollo spacecraft. Ni/H_2 batteries aboard more than 75 LEO/GEO satellites have demonstrated more than 250 million cell-hour in-space without failure. Satellites include Eutelsat 2, Milstar, Hubble, Telecom 2 and Iridium. 320 Ni/H_2 cells are being supplied to Space Systems/Loral for Space Station. Ground testing has exceeded 110,000 cycles at depths of discharge up to 80 per cent. Production capacity is 3,000 cells annually in the 4,600 m^2 dedicated facility supported by 300 personnel. Some 90 per cent of materials and components involved is produced inhouse. Test facilities can accommodate more than 1,500 cells simultaneously. Test facilities have been made available for combined technology development with NASA research facilities and other government laboratories.

Also available are ready to mount spacecraft power subsystems, including solar arrays, interconnecting cabling, conditioning electronics and mechanical/thermal support structures and enclosures.

The company is developing three Lithium Ion cell sizes: 0.50; 10; 50. Advantages include 125 Wh/kg, conventional rectangular battery packaging, and 3.6 V/cell for lower cost.

Engineered Magnetics, Inc (EM)

Current Status

EM, founded in 1955, designs, develops and manufactures custom power conditioning and conversion equipment for space applications, specialising in switching power converters. Programmes include Landsat, ERBS, GRO, EUVE, Topex, UARS and DMSP. Products cover voltage/power regulators, DC/DC converters, DC current sensors, battery chargers and Ah meters. The electronics assembly area of the 9,300 m^2 plant is certified for ESD; facilities cover solar array simulators, solar simulator (space chamber), vacuum chamber and vibration tables/chambers. Has developed power switching and conversion systems for complex electro-mechanically actuated systems and coupled several different power packages for automated docking and power transfer functions.

Their DC current sensors are a family of non-contact current sensor products for space applications using the saturable reactor principle. Modules cover full scale ranges over 1–250 A unipolar and to 65 A bipolar. Isolated output is linearly proportional to the sensed current. Optional configurations and multiple units can be accommodated. Converters and power regulators for multiple output applications also available. High-specification requirements accommodated through new test facilities funded part by company and part by government funds.

General Atomics

Current Status

General Atomics makes space nuclear power devices, electromagnetic launchers, neutral particle beam devices, robotics, unmanned air vehicles, image intensifiers, power conditioning and advanced survivability materials. Has designed an electromagnetic-magnetically levitated accelerator for propelling buckets of mined material from lunar resource mines to low lunar orbit for collection and cislunar transports. GA is a system contractor for the design, development and testing of thermionic power systems. GA develops active optics and Millimetre wave imaging systems for surveillance. Under a USAF contract, GA is designing, building and testing a prototype system for class 1.1 solid rocket motor propellant disposal using cryogenic washout for propellant removal and supercritical water oxidation for disposal. The only current methods are open burning or detonation. Lightweight miniature power supplies for single and dual microchannel plate 12 to 40 mm image intensifiers. GA has conducted numerous studies for several government agencies supporting research work in support of high-level on orbit power requirements. GA has developed new concepts for nuclear electric power devices for space vehicles and nuclear reactors for space-based propulsion systems and has responded to the new directive introduced by NASA Administrator Sean O'Keefe.

Hamilton Sundstrand

Current Status

Now a part of the Hamilton group owned by United Technologies Sundstrand Aerospace specialises in electrical generating systems, actuation systems/components, and environmental control systems, secondary and emergency power systems, and auxiliary and missile power units (such as Shuttle's APUs). Westinghouse's Electrical Systems Division was acquired by Sundstrand in 1992. Westinghouse in the 1960s entered the space market as the developer/manufacturer of Power Static Inverters for Apollo's Command Module. This led to its selection in the 1970s to develop both the inverter and a family of solid state power controllers (RPCs - Remote Power Controllers) for the Space Shuttle. Each Orbiter employs approximately 600, in three package sizes and six current ratings up to 20 A, working at 28 V DC. Current projects include design/development of candidate load conditioners (converters and inverters) both AC/DC, Remote Power Controllers both AC/DC, transformers and related components for the Space Station testbed programmes.

Sundstrand's Mechanical Systems specialises in the development and manufacture of lubrication systems for space applications; thermal hot gas systems for shutdown/startup in space; and other long life reusable equipment (seven APUs per shipset) for Shuttle Orbiter and SRBs; missile hydraulic power units. Long-life lubrication system research on unique applications for hot environments where highly reliable operating levels are mandatory and for cryogenic systems where long cold-soak periods are experienced.

Sundstrand's Power Systems Division specialises in small gas turbine engines, generator sets, Auxiliary Power Units and pneumatic system components. Has developed integrated power units for future manned platforms and remotely operated laboratory modules.

John Crane Belfab

Current Status

The company designs and manufactures welded metal bellows products for missiles, space vehicles and satellites, including reservoirs, accumulators, volume compensators, temperature/pressure sensors and mechanical shaft seals. It has provided several types of accumulators for Space Station. Accumulators have also been designed for new high pressure requirements and added to sensor packages. New and unique bellows systems have been applied to high-acceleration rocket projectiles. The Urine Water Recovery System version, tested aboard Shuttle, incorporates 20 cm diameter metal bellows with more than 100 individually welded stainless steel diaphragms. Another, 40.6 cm diameter and 61 cm long, welded with Inconel diaphragms, is used in the Ultra Pure Water System. JCB has designed closed-loop water accumulator systems for large manned space facilities applicable to lunar or Mars bases.

Lockheed Martin Missiles & Space Company

Current Status
In a 5,700 m^2 facility (recently expanded from 3,700 m^2 to accommodate Space Station work), using an automated weld process, Lockheed fabricates and tests solar arrays with flexible or rigid substrates. Over the past 27 years, approximately 1,100 m^2 of array surface area has been space qualified. Based on technology that was flight proven on the Shuttle Solar Array Flight Experiment, solar arrays are being built for Space Station.
Cell type: silicon
Cell size: 2 × 4 cm to 8 × 8 cm
Power to area ratio: 120 W/m^2 (rigid)
Power to weight ratio: 63 W/kg (flexible).
Extra large arrays designed for rigidity and shock survivability designed for high-stress environments.

Cryo coolers
The coolers are based on the Reverse Split Stirling Cycle, featuring: compressor and displacer moving components located by low axial and high radial rate suspension springs to maintain alignment within the gas clearance seals; linear motor-driven displacer for precise phase control with the compressor for optimum thermal performance and minimum vibration; helium working gas; low exported vibration (less than 0.18 N from a pair of compressors and 0.004 N for a pair of displacers) by the electronic controller precisely slaving the moving masses together.
Cold tip temperature: 50–120 K
Cold fingers diameters (mm): 7, 10, 12
Cooling power at 65 K: ranging 0.5–4 W
Target life: 90,000 h running, 1,000 on-off operation cycles
Connecting pipe length: 20–600 mm
Environmental tolerance: –50/70°C

VERIFIED

Lockheed Martin Vought Systems

Current Status
The former Loral Vought supplies the reinforced carbon-carbon thermal protection system for Shuttle Orbiter's nose cap, wing leading edge and chin panel. It also supplies the Orbiter's heat rejection radiators and fluid thermal controls. The division is developing two types of large deployed radiators for Space Station. More recently Vought Systems has produced experimental results from an evaluation of thermal radiator and environmental control system radiators for large orbital or planetary surface structures.

Loral Corporation acquired the missile division of LTV Aerospace & Defense in 1992 in a deal worth US$260 million. This ended a lengthy bidding competition for LTV's aircraft and missile businesses, put up for sale as a means of helping to lift LTV Corp from bankruptcy. Lockheed Martin acquired most of Loral's businesses in April 1996. LTV's best-known space involvement was the all-solid Scout orbital launcher. It completed its service in mid-1992 as NASA's smallest launcher and the last departed in 1994 for DoD.

Lockheed Martin is actively supporting the investigation into the Columbia accident where fragility of the wing leading edge RCC is implicated in vulnerability of that system. The findings of the Columbia Accident Investigation Board released in August 2003 exonerated the RCC except where it had been misapplied in the lack of appropriate testing for life fatigue.

Schaeffer Magnetics Inc

Current Status
Schaeffer Magnetics was established in 1967 to produce electric motors, actuators and actuation systems for space applications. Employees total more than 100. The company provided the solar array drive system for the NASA/Fairchild Explorer platform, the Magellan Venus orbiter and Iridium. It also supplied the biaxial drive for Telecom 2's spot beam antenna, the shutter motor and filter wheel assemblies for the Soft X-ray Telescope on Japan's Yohkoh solar observatory, the 8 hp 800 rpm brushless DC motor for the Tethered Satellite's deployment system, the scan mirror assembly for Mars Observer's Pressure Modulator IR Radiometer instrument, the solar array drive and antenna positioning mechanism on Japan's Muses-B, and the biaxial antenna pointing mechanisms (using the type 55 drive) for ADEOS and COMETS. Its 6.35 kg scan/motor encoder will drive the 4.3 kg beryllium mirror of the MODIS instrument for NASA's EOS. The 16-pole two-phase brushless DC motor includes a 14-bit optical encoder with 11 µrad accuracy. Schaeffer is also introducing a family of reaction and momentum wheels and, later, control moment gyros. A class 100 cleanroom became operational in 1990 for assembly, cleaning, inspection and lubrication of RW bearings.

The Hubble Space Telescope carries 57 actuators, for unlatching and opening the main aperture door (type 5 actuator), unlatching and deploying the high-gain antennas, unlatching its solar panels, positioning the secondary mirror, figure controlling its primary mirror, selecting the Fine Guidance System optical filters, selecting its Wide Field Planetary Camera optical filters, and operating its tape recorders. Actuators are carried by Pioneer 10/11 and TDRS' antenna gimbal drives. Actuators also developed for the Free Flying Platform conceived for the NASA space station but cancelled when ESA declined to fund the project.

The Rotary Incremental Actuator can also be supplied with brushless DC motors. These have been adapted to BAPTA designs and are coupled to two-dimensional solutions.

Specifications
Types: TY-1, -2, -3, -5, -7
Output step angles: 0.06, 0.02, 0.01, 0.0075, 0.004°/step
Harmonic drive ratios: 60, 100, 160, 200
Output step rates: 31.25, 9.00, 3.75, 2.25, 0.80°/s
Power: 5, 8, 10, 12, 16 W; units can be wound for any voltage or current
Output capabilities: 1.5, 6.8, 45.2, 108.5, 361.5 Nm; estimated minima, higher capability exists
Powered holding torques: 1.5, 16.9, 45.2, 108.5, 361.5 Nm; adjustable to requirements
Unpowered holding torques: 0.7, 5.7, 19.8, 33.9, 163.8 Nm, adjustable to requirements
Torsional stiffness: 0.34, 0.68, 4.52, 11.29, 42.93 kNm
Shaft load capabilities
axial: 4.9, 7.8, 11.1, 13.3, 53.4 kN
transverse: 4.9, 7.1, 9.3, 11.1, 41.4 kN
moment: 61, 106, 203, 298, 1,628 Nm
Total actuator mass: 0.50, 0.91, 1.50, 2.0, 7.7 kg.

Spectrolab Inc

Current Status
A Hughes Electronics company, Spectrolab in the space field specialises in compound semiconductor and lighting products, solar cells, panels and arrays, and provides solar simulator test equipment. Spectrosun solar simulators provide up to 2,000 solar constants (pulsed) and 200 continuous: X-10 1 kW Xe; X-25 2.5 kW Xe; X-200 for large areas (hexagonal beam 127 cm diameter at 3.6 m distance) at 2 solar constants; the LAPSS (Large Areas Pulsed Solar Simulator) can illuminate and characterise large arrays in 1 ms. Spectrolab also produces the SCATS Solar Cell Automatic Tester Sorter for production line testing of up to 2000 cells/h. When used in a semiautomatic cell test system, the operator task is limited to handling the cell.

The K7700B Silicon Solar cell represents the company's capabilities in space power production. Next generation systems supported by government research facilities for long-life cell technologies.

Specifications
K7700B
Cell thickness: 62 µm
Cell size: up to 8 × 8 cm
Efficiency: 14.6% minimum average
Jsc: 42.5 mA/cm^2; Jmp: 40.2 mA/cm^2
Vmp: 0.490 V;
Voc: 0.600 V
Pmp: 19.7 mW/cm^2
Cff: 0.77
Mass (mg/cm^2, bare): 24
GaAs/Ge
Cell thickness: 175 µm standard; 90 µm available
Cell size: up to 6 × 6 cm
Efficiency: 18.5% minimum average
Jsc: 29.7 mA/cm^2; Jmp: 28.1 mA/cm^2
Vmp: 0.89 V;
Voc: 1.020 V
Pmp: 25.0 mW/cm^2
Cff: 0.82
Mass (mg/cm^2, bare): 100 for 175 µm, 55 for 175 µm
Cell thickness: 175 µm standard, 90 µm available
Cell size: up to 6 × 6 cm
Efficiency: 19% minimum
Jsc: 30 mA/cm^2, Jmp 28.4 mA/cm^2
Vmp: 0.905 V
Voc: 1.020 V
Pmp: 25.7 mW/cm^2
Cff: 0.84
Mass (mg/cm^2): 100 for 175 µm (bare) 55 for 90 µm (bare)
GaAs/GaAs Concentrator Cells
Cell thickness: 200 µm
Cell size: application dependent
Efficiency: 21.5% minimum average
Jsc: 650 mA/cm^2; Jmp: 630 mA/cm^2
Vmp: 0.925 V;
Voc: 1.080 V
Pmp: 582 mW/cm^2
Cff: 0.83
Mass (mg/cm^2, bare): 115

VERIFIED

Tayco Engineering, Inc

Current Status
Tayco develops and manufactures cryogenic Platinum Resistance Temperature Detectors (PRTDs) for space applications. Users include Shuttle, Delta and commercial, civil and military satellites. The model 2921 was originally developed for Delta's propulsion system, designed for a –260/260°C range, 150 ms response in flowing liquid and able to cope with high levels of vibration and mechanical shock. Since improved in response time and durability. Tayco has developed more reliable versions and new engineering derivations which increase the projected component life. It has also developed new and more durable versions of cryo PRTDs.

Tecstar Inc

Current Status
Allied Solar Energy Corp changed its name to Tecstar in February 1996 to reflect its expanding interests. With the March 1996 acquisition of Honeywell Electro Components of Durham, North Carolina, it can offer solar arrays with deployment and motion mechanisms. 1995 revenue was about US$40 million; growth since at about 5 per cent/annum. ASEC began manufacturing silicon solar cells in 1958, with its 5 per cent-efficient cells used to power Vanguard 1. Its cells will help to power Space Station. Key products include space solar cell assemblies, panels and array subsystems of Si (13.8 to 17 per cent efficiency) and GaAs (18.5 to 20 per cent). 21 to 22 per cent multijunction cascade cells are produced for panels and arrays, with development work under USAF contract directed at 24 to 26 per cent. Electromechanical products include solar array drive and solar power transfer assemblies, reaction wheels, and rotary and linear actuators, components and systems. It operates the world's largest cell production facility. New research and development initiatives sponsoring next-generation technologies for high-performance cells.

Previously affiliated with Honeywell Satellite Systems Operation, Durham Electro Components was acquired by Tecstar Inc in March 1996. It designs and manufactures electromechanical and electromagnetic motion control devices and control electronics for spacecraft. These include AC, DC, brushless DC and stepper motors or eddy current dampers, coupled with gearheads, differentials and output shafts for solar array deployment/stowage drive units, solar array drive assemblies, boom/mast deployment drive units, as well as motors, clutches and brakes for the Shuttle RMS. The facility also produces motors, tachometers and resolvers for control moment gyros and reaction wheels. Tecstar designs large solar array configurations for lightweight boom requirements and unique high-tolerance specifications. A space products catalogue is available. Tecstar has developed GaAs SSPA units capable of long term storage in vacuum and wide thermal dispersions before deployment.

TRW Space & Electronics Group

Current Status

Now part of Northrop-Grumman, TRW offers a range of cryocoolers (30 to 300 K), including two innovations: a miniaturised 65K 0.25 W Stirling cooler, and a line of pulse tube coolers. Flexure bearings compressors and passive pulse tube cold heads promise 10-year lives. Life tests of both types are into their third year. Pulse tubes increase reliability and producibility by replacing the Stirling cooler moving displacer piston with a fixed geometry and pretuned orifice to control cold-end mass flow. Helium is the working gas. Efficiencies comparable to Stirling coolers have been demonstrated down to 30 K. Vibration output of the compressor is controlled to very low levels 0 to 1,000 Hz at the fundamental drive frequency (40 to 70 Hz) and its harmonics using either a driven balancer 180° out of phase or an opposed piston. All compression and balancer components are arranged symmetrically along a single axis in a hermetically sealed pressurised metal housing. Cooler mass range 1.4 to 22 kg, with 0.4 to 10 W capacity at 80 K. Maximum input 15 to 250 W.

TRW has designed and delivered four different flight qualified pulse tube cooler designs under AF Phillips Lab contracts: 1 W 35 K; 2 W 60 K; 0.3 W 35 K; miniature (two are cooling IR instruments on NASA's Lewis satellite). The flight cooler was delivered in 1996 for the Multispectral Thermal Imaging satellite. A miniature Stirling cooled payload, including a high temperature superconducting device, was delivered to NRL in 1994 for the HTSSE experiment on 1997's AF Argos satellite. NASA's Earth Observing System AIRS redundant cooler system (1.5 W at 55 K) was completed in 1996.

Research and development initiatives have provided applications in high-intensity pulsed tube coolers. This research has successfully resulted in a new class of cooler systems and pulse-tube coolers capable of extremely long life.

Yardney Technical Products, Inc

Current Status

YTP, established in 1944 as Yardney Electric Corporation serving the US defence industry, is a supplier of batteries for space applications. Since 2001 Yardney Technical Products, which includes Alupower, and Lithion, have been operating as separate divisions of Ener-tek. Current and recent applications include Inertial Upper Stage, Delta, Atlas, Shuttle, Landsat, Space Suit Life Support System, EMU, MMU, PAM, SLSS, Discoverer, Explorer, IMP, GEOS, Telstar, Tiros, Transit, Vanguard, Burner II, Atlas-Centaur, Giotto, IRIS, Tethered Satellite and Cluster. The company has approximately 23,250 m^2 of floor space for the design, development, qualification and manufacture of space-qualified hardware. Facilities include a class 300,000 cleanroom, several Dry Room facilities and a complete Battery Test Laboratory.

YTP has developed and qualified 1–400 Ah 6 to 31 V space batteries. At battery level, energy density yields vary from 60 Wh/kg up to 130 Wh/kg.

Silver cadmium cells are space-qualified silver cadmium cells of 3/5/10/16 Ah and providing 45 to 70 Wh/kg are available as are lithium thionyl chloride batteries of 250 Ah, and providing 200 to 300 Wh/kg at battery level.

YTP is developing prismatic Li-ion cells for 3/4/5/20/100 Ah space batteries. This new technology provides 115 Wh/kg, 250 Wh/l and 1,200 cycles.

YTP has conducted research into 150–170 Wh/kg battery systems for nanosatellites. This technology has been coupled with new and highly compressed storage systems.

SATELLITE BUSSES

United Kingdom

Astrium

Current Status
Astrium and its predecessor companies (British Aerospace Space Systems, Matra Espace and Matra Marconi Space) had been prime on 14 GEO communications programmes by early 1996, totalling 38 satellites. Beginning September 1977, 29 were launched and 26 achieved operational orbit, the remainder suffering from launch failures. Accumulated in-orbit time was 156 per year by January 1996, with 21 satellites still operational. BAe provided ESA's OTS satellite and derived from it three maritime Marecs, the ECS family of five European Communications Satellites (operated by Eutelsat) and the six Skynet 4s for the UK Ministry of Defence and two NATO 4s. BAe/Matra in the Satcom International partnership jointly developed the Eurostar platform from ECS as a medium-class communications satellite. with 2½ times the power of ECS. For example, BAe was prime contractor for four Inmarsat 2s based on Eurostar and Matra similarly built four Telecom 2s and two Hispasats. BAe was also prime contractor for ESA's Olympus technology demonstration satellite.

BAe was appointed prime contractor for ESA's technology demonstration Orbital Test Satellite (OTS), which formed the basis for the European Communications Satellite (ECS). The system busses are represented by the ECS 5 and Skynet 4A-C satellites specifications shown below.
Total ordered: 21 by March 1996
Total launched: 18 by March 1996
Operating: 11 as of January 1996
Future launches: Skynet 4D/4E/4F
Launch mass: 1,172 kg (Skynet 4 1,433 kg)
On-orbit mass: 680/576 kg BOL/EOL ECS (Skynet 4 790 kg BOL)
Deployed dimensions: body 2.1 m high, 1.9 m deep and 1.4 m wide; 13.8 m solar array span (17 m for Skynet 4)
Power: 1,260/1,000 W BOL/EOL derived from twin 24 kg 3-segment AEG BSR silicon wings, each panel 122 cm wide, 173 cm long; two SAFT 24 Ah Ni/Cd batteries (Skynet 4 34 Ah)
In-orbit lifetime: 7 years
Structure: communications module is U-shaped unit sitting on octagonal service module; major structural elements are predominantly aluminium alloy honeycomb with aluminium or CFRP skins. Payload components are mounted on both sides of the upper floor and on the sidewalls. The central thrust cone supporting the 91.4 cm diameter solid AKM is CFRP honeycomb with aluminium support struts
Pointing: 0.07° roll/pitch, 0.35° yaw (Skynet 4); antenna accuracy 0.2° half-cone (ECS), 0.1° (Skynet)
AOCS: momentum biased three-axis control and ±0.1°E-W stationkeeping. Two Teldix 25 Nms fixed MWs (and one 16 Nms bi-directional transverse MW for redundancy) provide pitch/roll from redundant Galileo IRES. Yaw controlled during manoeuvres by rate integrating gyro. 20 redundant 0.7/2 N hydrazine thrusters drawing on four N_2-pressurised tanks totalling 118 kg propellant (Skynet 0.7/2/5/20 N, two tanks carry 71 kg).
Thermal control: heaters plus 2.4 m^2 passive radiators, including secondary surface mirrors and 10 to 15 layer Kapton blankets.

The Eurostar Platform Eurostar was developed 1983-86 directly from the ECS family by BAe/Matra under the Satcom International partnership as a medium-class communications platform originally offered in three models: 1000 (1,500 kg GTO mass), 2000 (2,200 kg) and 3000 (3,000 kg). It has since evolved into a general range with payload capability up to 550 kg/5.5 kW (Eurostar 2000+), increasing to 650 kg/6.5 kW with the 2000+i ion propulsion option. Eurostar 3000 was developed to be available from early 1997 for, initially, at least 850 kg/10 kW. Eurostar 3000 has been selected for Skynet 5 which will involve the first satellite launched in 2006 and the second in 2007.

BAe was awarded a £150 million contract in April 1985 for three Inmarsat 2s (1000), with options for a further six. The fourth was approved in March 1988. Matra was appointed prime contractor in 1988 for three Telecom 2s, in June 1989 for two Hispasats, and in June 1989 for two Locstars (later cancelled). Telecom 2D was ordered December 1993. BAe/Orion Satellite Corp signed a contract 1 September 1989 for the in-orbit delivery of one satellite (2000), marking the first US commercial sale by Europe. Under the turnkey contract, Lockheed Martin provided Atlas launch and insurance services. Eurostar 2000+ was selected in March 1994 for Eutelsat's Hot Bird 2. The first was exercised December 1994 for HB3; HB4 was ordered in July 1995 and Hot Bird 5 in March 1996. Orion Atlantic signed the contract 24 October 1995 for Orion 2 (2000+; 475 kg/5 kW), with an Orion 3 option. Nilesat (2000) was ordered in October 1995 for in-orbit delivery; the contract includes the complete ground segment and parts for a second satellite. Alcatel Espace chose 2000+ in December 1995 for the three WorldSpace satellites. Singapore Telecom selected Eurostar in March 1996 for ST-1. SES selected Eurostar for Astra 2A in June 1996. In January 1997, Intelsat procured a Eurostar for Intelsat K-TV, the first time that Intelsat has selected a European company as prime contractor for one of its satellites.
Total ordered: 23 to February 1997
Total launched: 12 to February 1997 (4 Inmarsat 2, 4 Telecom 2, 2 Hispasat, Orion 1)
Operating: 12 to February 1997
Launch mass: 1,190 kg Inmarsat 2, 2,145 kg Telecom 2, 2,350 kg Orion 1, 2,910 kg Hot Bird 2, 3,200 kg Orion 2
On-orbit mass: 730 kg Inmarsat 2, 1,100 kg Telecom 2, 1,450 kg Orion 1, 1,600 kg Hot Bird 2, 2,000 kg Astra 2B
Deployed dimensions: Inmarsat 2 body 2.557 m high, 1.586 m long, 1.48 m wide, solar wing span 15.23 m; Telecom 2 2.1 × 1.7 × 1.9 m, span 22.02 m; Orion 1 2.1 × 2.5 × 1.7 m, span 21.6 m
Platform reliability: 0.87 after 10 years (0.99 to station acquisition, typically 700 h after launch)
Payload mass: up to at least 550 kg; 650 kg (2000+i); 850 kg (3000). Hispasat payload mass 250 kg, Orion 1 370 kg, Telecom 2 390 kg, Hot Bird 270 kg, Orion 2 475 kg.
Payload power: up to 5,500 W (2000+); 6,500 W (2000+i); 10 kW (3000). Telecom 2 2,600 W, Hispasat 2,250 W, Orion 1 2,350 W, Hot Bird 2 4,200 W, Orion 2 5,000 W, Astra 2B 7,000 W, Intelsat K-TV 7,600 W
In-orbit lifetime: up to 15 years, depending on tank selection
Structure: 1.4/1.6/2.2 m high carbon fibre/aluminium honeycomb thrust cone with shear walls separating the propellant tanks. The payload mounted on north/south panels and Earth-facing floor. East/west faces carry hardpoints for antenna mountings. Structure designed for all launcher load cases.
Power: 1 to 7 kW from twin fold-out wings of 3 to 5 panels each; each panel up to 7.5 m^2 Si or GaAs cells. Eclipse power from 160 Ah Ni/H_2 batteries (Ni/Cd on Inmarsat 2). Single or dual bus systems, regulated at 42.5 V sunlight (50 V option), regulated at battery voltage in eclipse.
Pointing: antenna pointing accuracy typically 0.1° half-cone; bias correction provides greater accuracy
AOCS: automatic momentum-biased three-axis control avoids the need for ground intervention. The fixed MWs provide pitch control, with solar sailing controlling yaw/roll using data from IRES.
Propellants: NTO/MMH helium-pressurised in two sets of twin tanks
Thrusters: Kaiser Marquardt R-4D 490 N liquid apogee motor, two sets of seven DASA platinum-rhodium 10 N thrusters (2000+; sets of six on other models)
Thermal systems: standard heat pipes, heaters and thermostats. Radiator area on north/south faces.

VERIFIED

United States

Lockheed Martin Commercial, Space Systems

Current Status
LMSSC headquartered in Denver, Colorado, comprises four operations units and three joint ventures. Space Systems is part of the Astronautics Operations unit. Other LMSSC units are Management and Data Systems, Michond Operations, and Missile and Space Operations.

The former General Electric Astro Space Division in 1993 became part of Martin Marietta's Space Group and in March 1995 part of Lockheed Martin. GE Astro Space was part of RCA Aerospace & Defense until it was acquired in 1986 by General Electric, which already operated its own space facility at Valley Forge. The East Windsor and Valley Forge production complexes house more than 40 environmental test chambers and 10 class 10,000 integration bays. Workforce is about 5,000. Lockheed Martin announced in June 1995 that both would close after most of the work was transferred to Sunnyvale, California, although existing commercial, Landsat 7 and EOS AM 1 contracts were completed first. Sunnyvale, with four times the area, became responsible for commercial, military and civil projects, including DSCS, GPS, DMSP and NOAA. The new commercial production line built two A2100 satellite communication platforms on spec to demonstrate the capability. Some 550 jobs moved to Sunnyvale, plus 1,100 at a new East Pennsylvania facility, building satellite communication payloads and power elements. The first commercial A2100 were launched for SES Americom in 1999.

Together, the two existing sites were responsible for 220 launched satellites through March 1996, and are working on a backlog of 62. Their record covers Tiros (38 launched over 35 years), DMSP (31 in 29 years), 14 Earth resources/observation (Nimbus 1-7, Landsat 1-6), 14 planned DSCS-3, NASA's Upper Atmosphere Research Satellite, Mars Observer, Dynamics Explorer, Transit, Nova, Polar, Wind, EOS and GPS' Block 2R satellite series. Astro provided the Viking Lander communications subsystems, the TV cameras used by Apollo on the Moon and Shuttle's closed circuit TV system. Valley Forge was selected by the Dept of Energy in mid-1990 to provide the RTGs for the Cassini Saturn mission. The 10 DSCS-3 departing Atlas 2s are delivered into GEO by the company's IABS Integrated Apogee Boost Subsystem, the first of which was delivered in early 1991.

Communications satellites form the largest single space element, beginning in 1958 with RCA providing the payload carried into orbit aboard an Atlas to transmit a taped message from President Eisenhower. The three-axis Satcom platform was introduced in 1975 and now operates with the 3000, 4000 and 5000 series (designating the approximate launch mass in pounds); the 5000 class debuted in 1991, the 7000 class in 1993 and the A2100 in 1996.

The 3000 series is the original member of the Satcom family and is second only to Hughes' HS-376 in its numerical popularity. Satcom 1 introduced the model in December 1975.
Total ordered: 34 through March 1996 (Aurora 1/2, PAS 1, ASC 1/2, Anik B, BSE, BS-2a/b, BS-3a/b, BS-2x, BS-3H, Satcom 1-3R, Satcom 1-4, Satcom C1/C3/C4, GStar 1-4, Spacenet 1-3, Spacenet 3R, Koreasat 1/2, BS-3N)
Total launched: 34 through March 1996
Operating: 18 through March 1996
Launch mass: about 1,290 kg (about 545 kg dry)
On-orbit mass: about 760 kg
Deployed dimensions: 163 × 132 × 99 cm bus and 14.33 m solar array span
Communications section reliability: not available
Communications payload mass: not available
Payload power: not available
Design life: 10 years minimum

Structure: electronic units, batteries, propulsion and AOCS equipment mounted on six honeycomb structural pallets or panels. Transponders and housekeeping components mounted on four panels, two each on the north/south sides. Additional housekeeping equipment mounted on anti-Earth base panel.
Power: twin Si wings of three panels each provide 1,400/1,350 W BOL/EOL supported by 3 × 35 Ah Ni/H_2 batteries.
Pointing: 0.05° roll, 0.05° pitch, 0.10° yaw
AOCS: three-axis by 6,000 rpm MW (pitch) and magnetorquing (roll/yaw) controlled by Earth Sensor Assembly error signals. Propulsion subsystem is conventional 41 blowdown mono hydrazine type: four surface tension tanks (181 kg hydrazine) manifolded to two redundant sets of eight thrusters (six Olin Aerospace MR-103C catalytic and two MR-501 NSSK electro-thermal). GEO insertion by Thiokol Star 30B/C/E solid AKM
Thermal systems: passive and heaters.

The 4000 class was introduced with the launch of General Electric Americom's Satcom Ku-2 in November 1985 by Shuttle. NASA's ACTS is based on the platform, which was also selected in 1990 for Inmarsat's third generation spacecraft (Matra Marconi Space UK is responsible for the payload). The DoD's Navstar GPS Block 2R navigation satellites is derived from the 4000 series; GE was awarded the initial US$93.1 million contract in 1989, including options worth up to US$916 million for 20 to 26 satellites (21 are baselined 1996–2002).
Total ordered: 30 (Satcom Ku-1/2, Astra 1A, ACTS, Inmarsat 3 F1-5, 21 GPS 2R)
Total launched: five through April 1996 (Inmarsat 3 F2-F5, GPS 2R remain)
Operating: five
Launch mass: 1,820 kg
On-orbit mass: 1,050 kg BOL
Deployed dimensions: 170 × 213 × 152 cm body, 19.3 m span across solar arrays, total height 305.8 cm at launch
Communications section reliability: not available
Communications payload mass: not available
Payload power: not available
Design life: 10 years minimum
Structure: equipment is mounted on eight aluminium honeycomb structural pallets or panels; transponders and some housekeeping units mounted on north/south faces; antenna and reflectors, including the primary Kevlar honeycomb reflectors, and Earth sensors on Earth-facing panel.
Power: twin four-panel solar wings using N-on-P Si cells provide 3,576/2,309 W BOL/EOL; 3 × 50 Ah Ni/H_2 batteries provide 100 per cent eclipse protection
Pointing: not available
AOCS: three-axis by MW, magnetorquing and 16 blowdown mono hydrazine thrusters divided into two redundant sets of eight: six 0.89 N MR-103 Olin Aerospace catalytic Reaction Engine Assemblies (REAs) accompanied by two 0.18–0.33 N Electrothermal Hydrazine Thrusters (EHT) for increased performance NSSK, drawing on four tanks. GEO insertion by Thiokol 37F solid AKM.
Thermal systems: passive and heaters

The first 5000 satellite, Anik E2, was launched in 1991. GE provided the bus to Canada's Spar Aerospace for Anik E1/2; the stored Satcom Ku-3/4 are flying as Astra 1B and Intelsat K, respectively.
Total ordered: four (Astra 1B, Intelsat K, Anik E1/2)
Total launched: four
Operating: four
Launch mass: 2,580 kg (Astra 1B)
On-orbit mass: about 1,540 kg BOL
Deployed dimensions: 2.84 × 2.26 × 2.18 m body, 24.3 m span across solar wings, total height 2.56 m at launch (3.30 m deployed)
Communications section reliability: not available
Communications payload mass: not available
Payload power: not available
Design life: 10 years minimum
Structure: equipment is mounted on eight aluminium honeycomb structural pallets or panels; transponders and some housekeeping units mounted on north/south faces; Kevlar honeycomb antenna reflector assembly deployed from west side, Earth sensors on Earth-facing panel.
Power: twin four-panel solar wings using N-on-P Si back surface reflector cells provide 4,850/3,460 W BOL/EOL. During transfer orbit operations, the wings remain stowed against the north/south faces, the outer panels providing housekeeping power. Four 50 Ah Ni/H_2 batteries provide 100 per cent eclipse protection.
Pointing: not available
AOCS: three-axis by MW, magnetorquing and 20 Olin Aerospace blowdown mono hydrazine thrusters, divided into two redundant sets of 10: eight catalytic Reaction Engine Assemblies (REAs) accompanied by two 0.36–0.50 N MR-502 Improved Electrothermal Hydrazine Thrusters (IEHT) for increased performance NSSK. GEO insertion by dual-redundant 500 N Royal Ordnance Leros (used on Astra 1B) or 454 N TRW (used on Anik E & Intelsat K) bipropellant motors. After insertion, the bipropellant manifolds are isolated from the RCS system. Four hydrazine spheres, single N_2O_4 central cylinder.
Thermal systems: passive with heater-augmentation. Transponder panel equipment cooled by ammonia heat pipes and radiators.

AT&T requested bids in 1988 for the largest US domestic communications satellites to date: two hybrid Telstar 4s plus a ground spare. Astro was awarded the contract in July 1989; first launch was December 1993. The 7000 was selected September 1992 by Intelsat for two satellites, worth US$165 million, and in October 1992 for AsiaSat 2 (with option for a second for a potential contract of US$133 million). Intelsat 806 was ordered March 1994 for US$72.3 million, plus US$10 million in-orbit performance incentive.
Total ordered: 12 (Telstar 401/402/402R, Intelsat 801–806, AsiaSat 2, EchoStar 1/2)
Total launched: five through March 1996
Total operating: four (Telstar 402 failed minutes after launch)
Launch mass: 3,375 kg (Telstar 4); payload 420 kg
On-orbit mass: not available
Deployed dimensions: 2.16 × 2.46 × 3.15 m body, 24.3 m solar array span
Communications section reliability: not available
Communications payload mass: not available
Payload power: not available
Design life: 12 years
Structure: as 5000
Power: twin four-panel solar wings totalling 48.6 m^2 using N-on-P Si BSR cells providing 7.2 kW BOL. During transfer orbit operations, the wings remain stowed against the N/S faces, the outer panels providing housekeeping power. Two 50 Ah Ni/H_2 batteries provide 100% eclipse protection.
Pointing: ±0.1° pitch/roll, ±0.25° yaw accuracy; ±6.0° pitch, ±2.0° offset pointing control
AOCS: three-axis by MW, magnetorquing and monopropellant hydrazine thrusters. NSSK provided by two Olin Aerospace 0.20–0.23 N 1.8 kW hydrazine MR-508 arcjets. GEO insertion provided by dual-redundant 500 N Royal Ordnance Leros 1 or 454 N TRW bipropellant motors. After insertion, the bipropellant manifolds are isolated from the RCS system. Four hydrazine spheres, single NTO central cylinder.
Thermal systems: 112–312 W passive with heater-augmentation. Transponder panel equipment cooled by ammonia heat pipes and radiators.

The new A2100 platform is part of GE Americom's fleet, carrying 24 C- + 24 Ku-band transponders. It is expected that the A2100 will supersede the 7000 series as it can carry the same payload for 25 per cent lower total mass and has a shorter production cycle. Lockheed Martin announced 6 July 1995 the contract for two Garuda satellites for the ACeS Asean Cellular Satellite System. It will employ the A2100AX platform, the largest variant in the family.
Mass: 2,585 kg at launch, 1,575 kg on-station BOL.
Total ordered: seven (GE 1/2, Garuda 1/2, DBSC 1/2, ChinaStar 1)
Total launched: 19 through December 2002
Total operating: 19 through December 2002
Design life: 15 years
AOCS: three-axis by MWs, arcjets and hydrazine thrusters (redundant sets)
Power: twin four-panel solar wings, 5,000 W EOL, 2 × 100 Ah Ni/H_2 batteries.

VERIFIED

SATELLITE COMMUNICATIONS FACILITIES AND EQUIPMENT

Belgium

Alcatel Bell NV

Current Status
Bell Telephone is active in the spaceborne and Earth station fields. Since 1987 it has been part of the Alcatel NV group; personnel total 5,700 (less than 100 space). Spaceborne applications include equipment for Intelsat 7, ECS 1–5, Marecs, Giotto, Telecom 2, Spacelab, Envisat, Artemis and AC&S. It provided the communications payload for LLHS. Equipment/subsystems supplied cover satellite electronics, power conditioning systems, Ka-band systems, LNAS. The company also provided the cooler/freezer and the incubator electronics for Biorack. Developed Biorack applications for on-orbit laboratories and space station modules led initially as a potential contributor to private module operators. These have been developed into candidate receptors for unmanned platforms and are proposed as part of an international endeavour to provide extensive biomedical experiment options for international projects.

Complete turnkey capabilities are available for satellite control Earth stations; Alcatel Bell has delivered more than 20 currently operational terminals. Examples are the three S-band terminals installed at Toulouse, Kourou and Pretoria as part of CNES' 'Reseau'. Kiruna's Tele-X and Michelstadt's Meteosat and Hipparcos terminals were also provided by the company. It also provides receiving equipment for Earth observation satellites (for example, Landsat, Spot and ERS). Alcatel Bell is also providing onboard switching control systems and ATM switching for satellites.

Newtec CY NV

Current Status
Newtec CY NV includes research, development and manufacturing of various types of digital satellite modulators and demodulators (IDR, DVB, SDH with QPSK, 8PSK, 16QAM, and so on) for television broadcast and PTT operators. Built-in and stand-alone up/down converters are also available from 50 MHz to 30 GHz. Special break converters designed for channel switching modes adapted to specific customer requirements. The company is also a manufacturer of C-, Ku- and Ka-band Earth stations (Mesh, VSAT-DAMA, and so on). Equipment has been supplied to all major broadcasting stations and PTT administrations and several unique designs have been made available for undisclosed customers, designs which are available for order. Developed multiplexers for special customer requirements and high-stress specifications.

SAIT Systems SA

Current Status
Part of the SAIT RadioHolland Group and specialising in telecom satellite control and Earth stations, including data processing subsystems. The company has provided Intelsat A, IBS/SMS and ECS Earth stations and mobile communications systems. SAIT Systems is system designer for the IRIS (Intercontinental Retrieval of Information via Satellite) satellite system for commercial messaging/localisation. Debut was planned for 1997. SAIT Marine offers Inmarsat A/B/C/M terminals, maritime e-mail systems, GMDSS consoles and IRCS (Integrated Radio Communications System). Also provides remote-control systems for mobile antenna systems.

TSAT manufactures TSAT 2000 which is a low-cost full duplex system for 300–2400 bit/s data transfer via satellite. A common hub station controls remote terminals in a closed star network. Open network solutions for common hubs and centralised duplex routing.

International

Alcatel Qualcomm

Current Status
Qualcomm Inc of the US began development of the OmniTracs mobile communications system in 1985 to provide low-cost two-way satellite communications to mobile units via existing Ku-band transponders. The service is directed at trucking fleets and other mobile vehicles. The EutelTracs European version, operating through commercial Eutelsat transponders, is marketed by the Alcatel Qualcomm joint venture, formed in March 1990 as part of the Alcatel Radio-communication, Space & Defense Group; Qualcomm Inc holds 49 per cent. It is commercialised throughout the coverage zone by a network of 18 national service providers. The QASPR automatic satellite position reporting service provides 200 m accuracy covering the whole of western and eastern Europe, the Mediterranean Basin and the Middle East using two Eutelsat satellites. In fact, accuracy has been 80 m since September 1993 when messaging was transferred from Eutelsat 1-F2 at 1° E to 1-F4 at 25.5° E; positioning remains on 2-F4 at 7° E. As Eutelsat operates at Ku-band, the antenna locks on to the first satellite, rotates to the second (Ranger) and back – the sequence requiring about 10 seconds. The Ranger provides information on the time difference, precise satellite positions and the terrain altitude model stored at the hub. Computed position accuracy depends on this model, timing errors and the satellites' orbital separation. Computer differentials between separate tracker systems and integrated satellite systems. Tracker modules linked to hooked webs facilitated through open network synergies. This allows almost limitless user uptake.

Individual customer dispatch facilities connect via regional centres to the Network Management Facility (9 m antenna) at Rambouillet in France to send or receive messages of up to 1,900 characters to any terminal in their fleets. A message is coded and dispatched via satellite to the prescribed vehicle or group of vehicles. The Facility waits for acknowledgement that it was received correctly; if not, it automatically repeats the procedure according to a system-wide retransmission algorithm. The driver's unit can retain the last 99 messages or 256 lines.

Testing within Europe began in June 1989 through Eutelsat, using the UK's Goonhilly as the hub. Commercial operations began 1 January 1991, offering a system capacity of 55,000 users within the Eutelsat region. The primary market is the road transport industry, with others such as non-civil, tourism and emergency links. Projected market size is about 80,000 for the first 10 years; there were 5,500 active terminals by end-1994 and 10,000 by end-1995 (11,000 delivered for 18 service providers). 10,000 were required to break even. 19,000 are projected by end-1996 for 24 service providers.

The Mobile Communications Terminal (MCT) comprises three units: Outdoor Unit, Communication Unit and Display Unit. The Outdoor Unit contains the antenna assembly and front end electronics. Including electronics, it is 28.95 cm diameter, 17.0 cm high and weighs 3.94 kg. The Communications Unit contains an analogue section, digital electronics and the QASPR receiver, all on four circuit card assemblies. Size: 32.4 × 23.4 × 11.4 cm, 4.07 kg. The Display Unit consists of a 40 character by four line display with keys for pre-programmed user functions.

South Africa

Grinaker Telecom

Current Status
GEL introduced the Thinroute M Inmarsat M package in 1994 for communications from remote areas. Developed as a field mobile system with applications as a high-powered man-portable system or as a miniaturised lightweight unit. It comprises a conventional NERA Inmarsat M satellite phone with a specially developed 2.3 m antenna. The 60 kg modular parabolic reflector is designed for fixed/semi-fixed external applications, connected by a single coaxial cable to the satellite phone inside a building. An upgrade supports 18 simultaneous Inmarsat M voice, fax and data channels. Can be deployed in fixed or wheel or tracked configurations and can be attached to multiple fixed sites. Mobile, fixed or conveyable antennas. Can be truck-mounted or air lifted to user site.

Spain

Alcatel Espacio SA

Current Status
Alcatel Espacio was created in 1988 as an Alcatel Standard Eléctrica subsidiary to specialise in spaceborne communications equipment and subsystems: RF passive equipment (microwave filters and diplexers, Ku-band input multiplexer); RF active (TT&C S-band transponder); digital electronics (antenna pointing system, data handling, baseband). Alcatel Espacio is the Spanish subsidiary of Alcatel Space and is located in Tres Cantos (Madrid) with sales close to €27 million and a sustained average growth better than 14 per cent during the last 5 years. The core product is TT&C S-band Transponder and Input Multiplexer (IMUX) and is supporting development and manufacture of an advanced Interactive Broadband Multimedia Satellite Communication System which will be used on the HISPASAT Amazonas satellite by ILS Proton from Baikonur in mid-2004. Alcatel Espacio is also developing passive microwave equipment (Ku-, X- and Ka-Band Input Multiplexers (IMUX)), filters, diplexers and other passive devices from S- through Ka-Band and microwave assembly units. The company also produces digital processing units (9343 DVB On-Board Processor, digital processing modules, on-board data management units and control electronics equipment). Research and development accounts for more than 6 per cent on sales with products currently operating on more than 50 satellite programmes worldwide including Aces, Arabsat, Globalstar, Hot Bird, Sesat, Worldstar, Spot 5, Helios 3, Ariane 5, Meteosat 2nd Generation, Rosetta, Mars Express, Metop, Galileosat, Scisat, Syracuse III, Stellat, GEA12, Integral, XMN, Envisat, Leostar, Hispasat, Cluster, Eutelsat W24, ATV, HTV, MTSAT, GOCE, Amerhis and Amazonas.

Telefónica Sistemas De Satelites SA

Current Status
A subsidiary of Telefónica de España, TSS was created to provide satellite business telecommunications. As Spanish service provider, it specialises in satellite communication provision,

development, integration, space segment access and provision, installation, operation and maintenance of VSAT networks and business TV as main market segments. Coverage is Spain and Latin America; the merger with Unisource Satellite Services provides access to the whole of Europe. TSS's centre in Madrid operates approximately 24 VSAT networks using Hispasat, Eutelsat and Intelsat. It also offers Inmarsat services. TSS has participated in ESS studies for new communication systems capabilities and for integrated Eutelsat and Inmarsat services. From this extended activity TSS has evolved a strategy for new, unique, system integration for co-locating frequencies on several satellite systems or constellations simultaneously. This has been extended to integrated satellite and terrestrial systems.

Taiwan

Jonsa

Current Status
Jonsa was formed in 1988 and manufactured approximately 90,000 C/Ku-band satellite TV reception dishes monthly in its 8,200 m^2 facility. The 0.35 to 4.2 m offset/prime focus dishes offer ground, wall and polar mounts. Jonsa has also designed dishes and antennas for mobile bases and transportable satellite receivers. These have design specifications allowing acceptability within ITU requirements and are compatible with international requirements. Has continued to provide purpose-built antenna systems for specified customer requirement.

United Kingdom

Andrew Limited

Current Status
Andrew Limited builds Earth station antenna systems, ranging from 1.8 m VSATs to 9.3 m Intelsat B. Adaptability sets permit utilisation with mobile or fixed links. Newly configured customer requirements accommodated.

Andrew Ltd has designed unique custom built structures for a wide range of applications.

COM DEV Europe Ltd

Current Status
The company was established in 1985 as a wholly owned subsidiary of COM DEV Canada to design, develop, manufacture and test multiplexers, filters, receivers, amplifiers, ferrite devices and antenna beam-formers. Project involvement includes GE 1/2, Telecom 2, Artemis, Cassini and Intelsat 7/8. Test facilities include RF benches of up to 40 GHz and thermal/vacuum chambers. Has developed composite beam and antenna band formers for ferrite formers and can provide flexible, compact, folded, structures from composite and hybrid materials for highly demanding specifications. Adapted to provide new waveguide formers.

Kudos Technology Ltd

Current Status
Kudos Technology provides consultancy in space/ground-segment satellite communication systems and supplies ground terminals (including Inmarsat portable terminals) and subsystems for large Earth stations. Has developed lightweight terminal capable of being lifted by helicopter to high elevations and altitudes. Provides end-to-end technology advisory service for operators and manufacturers. KT has designed total end-to-end telecommunications systems for specialised operators.

M/A-Com Limited

Current Status
Microwave amplifiers, switches, phase shifters and microwave semiconductors for space applications. Project-qualified for more than 30 payloads, including Skynet, Italsat, Telecom 2 and Eutelsat. Semiconductors ESA qualified to ESA/SCC 5010. A member of the Tyco Electronics group of companies.

Marconi Communications Ltd (GMCL)

Current Status
GMCL is a Management Company of Marconi Ltd and designs/manufactures systems and equipment for satellite communication ground stations; tropospheric scatter communications; sound broadcasting and TV; line transmission systems; civil, military and naval communications in most frequency bands; mobile radio communications, including fixed/mobile; cellular car phone systems; network management; digital data systems; LF to VHF antennas; and computer software management. All manufacturing, testing, design, development systems engineering and supporting activities are carried out at this site.

GMCL is structured on a divisional basis; the Civil Communications Division specialises in tropospheric scatter radio; specialised supervisory systems for monitoring, control and management of communication networks; satellite stations for Intelsat, Eutelsat, Inmarsat & domsat; Inmarsat ship/land terminals are handled by Marconi Marine.

Marconi entered the space communications business in the early-1960s with the design, manufacture and installation of an Earth station for Cable & Wireless on Ascension Island in connection with the early NASA space programme. Since then, it has been involved in building all classes of satellite Earth stations, ranging from Intelsat A/B C-band and Intelsat C and Eutelsat Ku-band stations to the transportable/portable Earth stations for SNG and portable communications. The product range was expanded in 1990 to include stations for offshore platform deployment and in 1991 to include standard EBU Earth stations. Development of large antenna structures for multiple deployment in clustered networks. The Systems Engineering Dept supported by the GEC-Marconi Research Centre undertakes system studies, planning and link analysis as well as the system design of Earth stations. Marconi's equipment includes modems and frequency conversion equipment for digital communications, allowing the upgrade of existing Earth stations from analogue to cost-effective digital operation.

Marconi Marine

Current Status
Marconi Marine, a division of Marconi Selenia Communications Ltd, a Finnmeccanica Company are providers of a range of satellite communications equipment and systems.

These include Inmarsat A, B, C, mini-M and the new Fleet 77 service. In addition, Marconi Marine are a major provider of Iridium global satellite communications with the Scansat 7701 Iridium Terminal and the Scansat Mini-M.

The range therefore covers all requirements from simple, low-speed messaging via Inmarsat C up to 64 kbit high-speed services and the packet data service available via F77. Products are available to meet the requirements of all classes of marine installation as well as both fixed and mobile (including hand-held) land-based applications.

The services offered include the supply, installation and maintenance of these systems, together with competitively priced airtime and customised billing services.

Satellite television systems are also supplied to provide all classes of vessels with high-quality TV reception, whether in port or at sea.

Marine Technology International Ltd

Current Status
STC International Marine's communications satellites and marine radio business was acquired by Mobile Telesystems Inc of the USA in January 1993 as its marine specialist arm. MTI has specialist product in receivers and can provide integrated packages including antenna systems. Has provided integrated clusters capable of individual frequency tailoring and has conducted analyses of integrated systems modelling based on Navstar GPS, Glonass and the pending Galileo constellation.

Multipoint Communications Ltd

Current Status
Founded in 1982, Multipoint Communications provides a range of earth stations for up-linking and reception of IP, MPEG 2 digital TV, analogue TV, radio and business data for national and international operators. The Jetstream Cadet earth station provides economical and simple installation for smaller satellite uplinks and is available in two sizes to house one rack of equipment in a 1:W or phase combined configuration and can be located close to the antenna for minimal waveguide losses. While the larger earth stations for major broadcasters accommodate multichannel operating systems utilising 15 or more racks of equipment. Produces VSAT, a versatile SCPC or MCPC terminal capable of point to point communications in MESH, STAR and hybrid topologies. Suitable for business data, communication networks, telephone systems and internet IP applications. Research and development programmes for extremely lightweight structures applicable to specified customer requirements. Can produce highly compact furlable large diameter antenna systems for cold and hot climatisation.

Racal Antennas Ltd

Current Status
Earth terminals for all applications, including Inmarsat C and airborne systems. Designs for specific requirements and multi-point systems installations. Purpose built antenna structures for unique applications. Racal has conducted design tests with advanced composite antenna structures with flexible impact resistance designed to survive minor impacts at high velocity.

Racal Avionics Ltd

Current Status
In June 2000, Racal merged with Thomson-CSF, now called Thales. The division is the group's centre of expertise for military satellite communication and Earth observation systems. The leading UK contractor for military ground stations, Siemens Plessey Systems supplies turnkey networks, large anchor stations, network control centres, spacecraft operations facilities, medium/small fixed terminals, medium/small transportable terminals, man-pack terminals and Earth observation ground stations. Studies are underway in connection with Skynet 5, the UK's next generation of military satellites.

Acting as prime contractor and system design authority, it supplied the £40 million FSC 646 satellite anchor station at RAF Oakhanger to the UK MoD as the hub/central control point of the UK Military Satellite Communications System (UKMSCS). The station was handed over in

1986 but in 1990 the company was awarded a further £10 million contract to provide significant expansion. It has now implemented the £60 million expansion and diversification: a second large anchor station, a new Satellite Network Control Centre, improvements to Oakhanger and another existing large ground station and provision of two large transportable terminals with network access points. These facilities are linked by terrestrial bearer systems to provide an integrated survivable ground segment network transparent to the users.

The company supplies full hemispheric coverage ground stations for polar orbiting Earth observation satellites such as ERS, Landsat and Spot. For example, it constructed the UK's 13 m ERS 1/2 station at West Freugh in Scotland. Updated for unspecified user requirement.

The Transportable Rapid Deployment Steerable Mount Antenna (Tradesman) is a 7 m diameter transportable antenna that provides high RF gain to satellite ground terminals requiring a rapid deployment capability. The X-band version, supplied to the UK MoD for the TSC 666 Transportable Satellite Ground Terminals, is designed for high-capacity, multicarrier terminals that handle a variety of communications circuits and are equipped with anti-jam facilities. The S-band version, used for the FSC 671 Satellite Ground Terminal, is for TT&C of satellites on-station or in drift or transfer orbits. A fully transportable TT&C station is also being supplied to NATO. Continuous upgrade programme.

Provost is a self-contained medium TSGT for rapid-deployment SHF satellite communications. It comprises a single 8.5 m trailer package with an electronics cabin and a 4.5 m Cassegrain antenna with two hinged sections on the main reflector folding inwards for transportation. It can be set up by a two-man crew and left unattended for supervision by a remote Monitoring/Control computer.

The X-band Flyaway Terminal is a 1.8 m antenna military terminal packaged à la civil SNG equipment, transported in rugged transit cases for an aircraft hold or small van. It can be deployed by one man in minutes, operated in a stand-alone configuration or for user circuits and networks. Multipoint network collection configuration option.

Satpac is a man-portable terminal that can be carried in a small suitcase or rucksacks. They employ DAMA demand assignment multiple access to minimise usage of satellite transponder power/bandwidth, allowing hundreds to operate within a network. Satpac offers secure voice and data traffic over all current military SHF satellites.

EOSGT can directly acquire data from all current X/S-band observation satellites, capable of tracking down to 500 km altitude and receiving up to 144 Mbit/s. Initially designed for use with ERS, Spot and Landsat, the modular approach allows for future expansion to Radarsat, Envisat and US and Japanese Earth observation systems. A terminal was acquired by the Royal Aerospace Establishment (now part of the Defence Research Agency) as the UK's primary data reception station, at West Freugh. The high gain antenna offers a figure of merit better than 35 dB/K at X-band for high-quality data reception (BER at 1 in 10^{-5} for data recorded from ERS at 105 Mbit/s). Single channel monopulse tracking ensures rapid satellite acquisition near the horizon, and the 7° antenna tilt axis eliminates the keyhole problem with zenith passes.

Dual-band operation is achieved by the use of an S-band feed at the prime focus and an X-band feed at the vertex. The integrated control/monitoring subsystem provides full control from a single console.

The Management of Satellite and Terrestrial Resources & Operations system provides management of complete satellite communication networks. A version has been supplied to the RAF for managing the diversified UKMSCS ground segment. The System Management Facility is the central point of management and enables system, terrestrial network and satellite access using resource and connectivity databases; the Access Monitoring & Policing Subsystem conducts automatic spectrum monitoring by scanning the downlink and comparing the carriers with the SMF access plan; the Satellite Network Expert helps to identify failures and anomalies.

The Siemens Integrated Monitor And Control System, although originally designed for satellite ground stations, can be applied in any installation where control/monitoring of diverse equipment is required from a central location. Based on COTS hard/software, it offers a graphical user interface presenting the operator with mimic diagrams grouping functionally related information. The Windows operating environment allows users to monitor important system data, such as alarms, while simultaneously performing lower level tasks.

Siemens Plessey Systems was recently awarded a contract by the NATO CIS Agency, won under international competition, to supply a Transportable Satellite Ground Terminal (TSGT) for telemetry and command of NATO spacecraft. The terminal design is based on a similar system already successfully in service within the UK SATCOM Anchor Station System, for which the company was also the prime contractor.

Operating in the 2 GHz S-band, the TSGT will provide a secure and reliable RF link to ensure the health and integrity of the spacecraft and to command the satellite in response to changing operational requirements. The electronics equipment is contained in a standard ACE II equipment shelter and the antenna is the established Tradesman design as supplied to the Royal Air Force. Full remote control and status monitoring to allow unmanned operation is provided by SPS's proprietary SIMACS system.

The MATELO project upgrades the UK component of the NATO Maritime Air Telecommunications Organisation facility, which is part of a NATO-wide distributed communications network providing telegraph and voice circuits between UK command centres and NATO aircraft, in support of maritime operations within the Eastern Atlantic region.

The upgrade is being achieved by expanding an existing RAF system, the Strike Command Integrated Communications System (STCICS) through the provision of new terminal and control equipment and by connecting existing MATELO HF radio facilities (suitably modernised) into the expanded STCICS. The resulting system combines the capacity to accommodate a major increase in traffic with improved operational flexibility for command staff.

As prime contractor for MATELO, SPS has managed the whole project, including the civil works. The project team has been responsible for the system design and integration, including interfaces with existing systems (principally STCICS and UNITER). Work has been carried out at nine sites, as far apart as Cornwall and the North of Scotland, with civil works at two of those sites. One of the key requirements has been the maintenance of full operational capabilities during the implementation programme.

For those elements of the system located at RAF St Mawgan and RAF St Eval (both Cornwall), SPS was responsible for the design, supply, installation and commissioning. On both these sites new buildings have been constructed and at RAF St Eval a new antenna system has been erected. At the other sites – RNWS Penhale Sands, RAF Bampton Castle, RAF Kinloss, RAF Chelveston, RAF Milltown, Northwood and Pitreavie Castle – SPS was responsible for design and supply, including the provision of installation design packages.

SPS delivered the first Containerised Relocatable Satellite Terminal (CREST) in the autumn of 1994. CREST was developed by Siemens Plessey Systems to cater for those satellite communication users who need low-cost, high-capacity, high-reliability ground terminals which can be easily 'relocated' from one operating site to another if requirements or priorities change.

Set-up and tear-down takes two or three days, but CREST has the advantage that it can be shipped complete as standard commercial freight. In addition, the use of Commercial-Off-The-Shelf (COTS) equipment throughout means that the costs of initial purchase and through-life support are minimised.

Individual CREST components are selected for their reliability, thus achieving a high system availability. This, combined with a comprehensive graphics-based control and monitor system, based on SIMACS, allows worry-free, unmanned operation.

CREST provides a flexible solution to most communication needs – each system can be configured to match individual customer requirements. Systems are available in C, X and Ku-band configurations and have been delivered with communications capacities ranging from 64 kbit/s to 8 Mbit/s.

The Joint Tactical Information Distribution System (JTIDS) is a secure, jam-resistant, extended line-of-sight RF communications bearer for Link 16, providing integrated digital and voice communications, navigation and identification capabilities. Communications survival is ensured by the use of frequency hopping and spread spectrum techniques.

Siemens Plessey Systems has been executing the Royal Navy JTIDS Ship System (RNJSS) contract since June 1995. The contract is for Full Development and Production, including installation into Type 42 Destroyers and CVSs (Carrier Versatile Support). The order is for 12 complete systems and modification kits for three JTID S terminals bought by MOD under the US Foreign Military Sales programme. These terminals are currently installed in HMS Illustrious and at the Land-Based Test Site situated at Portsdown.

RNJSS consists of two major elements – the JTIDS Class 2 Terminal and the Network Control Initialisation and Data Preparation Sub-System (NCIDPSS) – plus two antennas and audio ancillaries.

The Class 2 Terminal comprises JTIDS equipment, currently in service with the US Navy, adapted to meet Royal Navy requirements. The terminals, modification kits and spares are being procured from Rockwell Collins Avionics and Communications. The first Terminal is due for delivery to Christchurch in March 1997, whereupon it will form the initial building block of the Integration and Test system. The remainder were delivered on a monthly basis until completion in 1998.

The NCIDPSS provides the JTIDS network management facility. Its major function is to initialise, monitor and control the JTIDS Ship System, to manage the JTIDS/Link 16 Network, to prepare initialisation data for the Aircraft Initialisation Peripheral in the CVS and to supply initialisation data parameters to the Data Link Processing System for Link 16 and multi-link operations. The NCIDPSS is a new design and represents the main development item in the RNJSS project. To date the majority of software modules have been completed and integration has commenced. The prototype hardware has completed environmental and EMC testing and is now undergoing informal TEMPEST testing. Inmarsat approved satcom integrator.

United States

Andrew Corporation

Current Status

The parent company of Canada's Andrew Antennas Co Ltd; annual sales for 2001 were US$1.049 billion with net income of US$61.6 million compared with US$1.034 billion for 2000 with net income of US$79.6 million. Can provide innovative solutions and unique systems layouts for customers. Widespread applications serving a range of different user needs from very small operators and antenna requirements to sophisticated programmable search systems.

Antenna Technology Communications Inc

Current Status

Formed in 1979, ATCI manufactures the Simulsat Multibeam Antenna, a semi-parabolic antenna capable of receiving 35+ satellites simultaneously within a 70° view arc at C/Ku-band. Available in 3/5/7 m equivalent sizes. The company also provides new/refurbished 1.8 to 32 m parabolic antennas, satellite head end electronics and maintenance. Developed versions of systems adapted for high-capacity user. Diversification into broadband and general telecom solutions has generated growth and expansion with data, voice, audio-video solution, a significant part of the business. ATC can integrate terrestrial and space-based systems on a specified basis and experience with general telecommunication systems provides access to global networks. ATC have also developed cross frequency integration systems connecting several separate antenna complexes into a single integrated signal-handling centre.

Astroguide International

Current Status

The company was created in 1990 to develop/market three satellite products: the Trax II satellite tracking system, a refrigeration system for cooling focal point electronics of satellite dishes (40°C differential), and the Quad Trax VI motor (designed to eliminate counting errors). Trax was originally designed to track the Gorizont 3 Statsionar 12 satellite in its inclined orbit for customers in the Middle East receiving CNN broadcasts. It continuously measures the signal strength by interfacing with the automatic gain control or tuning meter and moves the antenna to pick up the strongest available signal. Each search is triggered by the drop in signal strength at a user-set value (strength is displayed on a range of 1–250). The IIC is a rack-mounted version and was adapted to facilitate mobile and transportable systems which have wide commercial applications. Search and tracker facilities are integrated in a free motion rocker link to multiple users.

AT&T Tridom

Current Status
The Aegis Network Management System provides real-time event-driven telemetry for navigating, monitoring and troubleshooting large complex satellite-based networks. Aegis can detect, isolate and assess network problems, control growth or configuration changes, manage capacity and control access. The company has provided more than 8,000 of its Clearlink VSAT terminals since 1986. Clearlink products are manufactured at Tridom's HQ and at the AT&T facility in Merrimack Valley, New Jersey. The Clearlink System 400 third generation, typically 1.8 m diameter, features a software-based architecture that can be programmed for a range of satellite transmission formats. It provides complete Ku/C-band data, voice and video. VSATs transmit data to the hub at 64 kbit/s and receive transmissions from the hub at 128/256/512 kbit/s. Outgoing data are transmitted in a first come, first served continuous time division multiplexed fashion. Access to the inbound channels is shared by multiple VSATs using a multiple access technique such as Aloha, fixed frame TDMA, reservation TDMA or fixed assignment. Addition of the Clearlink Voice Link Module provides two full-duplex digital voice channels. FDMA option on selected bandwidth options. New uplink scrubbing facility.

The hub station (3.5, 4.5, 6.1 m diameter) acts as a central satellite access and switching node, providing connections to the Clearlink Network Control Computer and to Host Interfaces that connect host computers to the network. Hubs can be shared or private; AT&T Tridom operates shared hubs in the USA. All hub equipment is fault tolerant, with automatic switchover to prevent interruption in the event of a failure. Also now includes pocket switching survivability for multi-use systems.

Aydin Corporation

Current Status
Telecommunication, telemetry and avionics equipment for satellites, spacecraft, launch vehicles and missiles. Has provided integrated range tracking support for advanced telemetry demands. Aydin produces satellite uplink HPAs, low noise GaAs FET amplifiers, up/down convertors, continuous and burst-type satellite modems, and TDMA systems. Turnkey Earth stations and integrated transmission networks at C/Ku-band have been provided to Belgium, Chile, China and South Korea, in addition to US common carriers. The company's 1000 series GPS amplifier is a linear power UHF unit employed with the navigation satellite tracking system. The SP-100 Microprocessor Based Data Reformatter is used in missiles and standoff weapons to provide GPS tracking information to communications downlinks. Aydin's airborne measurement system is installed in the redesigned Shuttle solid booster applied to all missions since the loss of orbiter Challenger in January 1986. Employees total more than 1,200, including 500 engineers and computer specialists.

Ball Aerospace & Technologies Corporation

Current Status
BATC is part of the Ball Corporation which reported sales of US$3.7 billion for 2000. BATC reported 2000 sales of US$383 million and employs approximately 1,700 people worldwide. The company was founded in 1956 and comprises the Aerospace Division and the Telecommunications Products Division. The latter division produces antennas and video products to customers in the scientific, military and aviation communities. Ball Telecommunications Products has been involved in the aviation industry since the early 1970s when engineers designed the first-of-its-kind airborne antenna for the FAA: this was the precursor to the Airlink satellite communications system used by more than 20 airlines around the world, providing telephone and fax services to airline passengers. Airlink provides high-gain and low-gain capabilities and support software packages. Ball-built antennas that can access the GPS provide navigation information, returning accurate location data to military pilots. Developed advanced information and data transfer capabilities. GPS relay. Ball antennas are also found on many space missions: the Manpack antenna is launched on each Shuttle mission, for example. Flown twice aboard the Shuttle in 1994, the Spaceborne Imaging Radar C returned radar images of the Earth and has performed a third in 2000 on the Shuttle Radar Topographic Mapper. BATC has formed partnerships with Spar Aerospace of Canada, the space division of Japan's Fujitsu Ltd, and Spain's Instituto Nacional de Technica Aerospacial.

California Microwave, Inc

Current Status
Founded in 1968, the company specialises in satellite and wireless communications, including satellite Earth stations. Satellite Transmission Systems, Government Communications Systems, EFData and Sunnyvale products are business units with space-related products including antenna farms and telecommunications buffers for customer specifications. Designed unique two-way profiling relay for high-spec users.

Center for Satellite & Hybrid Communication Network

Current Status
The centre was established in October 1991 as one of NASA's Centers for the Commercial Development of Space affiliated with the University of Maryland, the DoD, Aerospace Corporation, AT&T and Boeing. Its primary goal is development and commercialisation of advanced space-based communications, emphasising hybrid satellite/terrestrial networks. Builds integration networks for new users and integrated telecommunication systems for high-use networks. Specialist areas cover teleconferencing, DBS, HDTV, database access, supercomputing linkage, satellite digital cellular networks, narrowband mobile networks and personal communications. Has developed interactive relay systems for hybrid interconnects and has stimulated research in industry related to technology stemming from new design initiatives, provides a clearinghouse for creative ideas in electronics and telecommunications products.

Collins Air Transport Division (CATD)

Current Status
Formed in 1979, CATD is now a part of Rockwell Avionics and Communications with sales of more than US$500 million annually. CATD's SAT-906 six-channel Satcom system has replaced the SAT-900 as the standard Collins ARINC 741 offering for commercial aircraft communications satellites; the first entered service December 1993. Since then, more than 600 have been delivered. Approximately 130 of the low data rate SAT-900 were delivered; all have been upgraded to the 906. Operating through Inmarsat, SAT-906 comprises three line replaceable units: satellite data unit, high-power amplifier and RF unit. Any channel can support cabin or cockpit voice, fax or data transmission. The modular channel architecture allows the loss of one channel without total system failure. Twin channel redundancy has been made available and can be accommodated through switching and CATD continues to expand development of multisourced systems capable of accepting signals from separate satellite constellations.

Collins Avionics & Communications Division (CACD)

Current Status
CACD introduced its two Inmarsat M satellite phones in 1993: SecSat provides an STU-111 phone to encrypt all voice, fax and data communications; ExecSat operates without this. Several unique variants of this have been produced for some government departments. The STU-111 is the US government standard phone encryption device; several hundred thousand are in use by government agencies worldwide. Each satellite phone is packaged in a 20.3 × 45.7 × 33 cm briefcase; total mass 11.4 kg. The flat antenna is removed and aimed at the satellite; a dial tone is heard in the handset when the satellite is acquired. Used as the basis for satellite occlusion tests and has been developed into a new GAN fielder for omni-directional applications.

Communications & Power Industries, Inc (CPI)

Current Status
CPI, formerly Varian Electron Device Business, designs/manufactures klystrons, travelling-wave tubes, magnetrons, cross-field amplifiers, gyrotrons, mm-wave tubes, power-grid tubes, microwave amplifiers, power supplies and radar components. CPI's Satcom Division is a market leader in high-power uplink amplifiers, with more than 12,000 amplifiers in almost 155 countries. It offers L/C/X/Ku/Ka bands up to 10 kW. Accessory products include computer and remote control, output switching and power-combining networks. Integrated network circuit capabilities for multiple users. CPI has participated in technology development for hardened electronics and for military systems requirements to Mil-Spec standards.

COMSAT RSI

Current Status
Radiation Systems Inc, founded in 1960, was acquired by COMSAT in 1994 and merged with COMSAT Technology Services to create COMSAT RSI, able to provide approximately 80 per cent of complete satellite communication ground systems. Turnkey systems range up to 32 m. Boosted air links extend this range via equipment provided by accessory suppliers. Has since specialised in integrated telecommunications systems linked to satellite relay stations.

Comsearch

Current Status
A member of The Allen Group, Comsearch provides engineering services and software to the telecommunications industry, including mobile, microwave and PCS. It conducted research into multithematic telecoms solutions including switchable packet systems. Comsearch has developed integration hardware for switching between constellations of geosynchronous satellites. Participates in integrated systems analysis and is available for all-up testing of existing equipment.

ComStream

Current Status
ComStream was founded in 1984 by management and engineers from Com Corporation's communications division as a developer/manufacturer of products for digital satellite data networks, from modems and protection switches to fully integrated Earth stations. In December 1992 it became a wholly owned subsidiary of Canada's Spar Aerospace. The company has supplied more than 10,000 receive-only VSATs for use in the Reuters Information Services data-broadcast network worldwide. Has also developed interactive networks for accessing separate media inputs. Technical development of alterntive route path switching gear and special applications circuits for switching routes. Latest product is Spar's VSAT Plus II C- or Ku-band multiservice system for low-traffic-volume locations on private or public networks. Data rate 1 to 15 Mbit/s; for example, four voice-encoding rates up to 64 kbit/s and data transmit/videoconferencing up to T1/E1. It does not require a hub because of its distributed network control architecture. Development of this has produced integrally switched network systems. Others are the DT7000 Earth station (1.2 to 2.4 m antenna system for two-way Ku- or C-band services) and the IntelliCast 501 75 cm to 1.8 m Ku- or C-band rx-only terminal. The DT5000 offers higher-power RF units or C-band. Modems available are: CP101 continuous or burst mode for transmission and burst mode for receiving; CM701 IDR- IBS- SMS-compatible variable-rate modem with electronic selection of data rates in a 4.8 to 2,048 kbit/s range, digitally tunable in 1 Hz steps. Protection switches, high-performance coding and Doppler buffers are also marketed. IntelliCast 501 Data Broadcast Receiver: IntelliCast is an integrated rx-only Earth station operating at C- or Ku-band and data rates over 9.6 to 2,048 kbit/s. Variable symbol rates, programmable in 1 bit/s steps. The terminal comprises an antenna, low-noise block down converter and the receiver indoor unit housing the demodulator, RF converter and power supply. Applications include online document or image retrieval, database transfer and real-time news or financial information networks.

Comtech Antenna Systems Inc

Current Status
A division of Comtech Systems Inc, the company manufactures a range of satellite communications antennas. This includes the 3.0/3.8 m C-band transmit/rx systems and 2.4 × 5.5 m Offsat transportable antennas (designed specifically to meet the FCC specifications for 2° spacing). The 3.8 m inclined-orbit antenna uses a dual-axis polar mount to follow satellites straying beyond GEO. The 2.4 m Ku-band is designed for the VSAT market. The 7.3 m Earth station features a dual-axis polar mount. The 1.8 m C- or Ku-band flyaway terminal can be packed in three cases totalling 150 kg and shipped as air baggage. Integrated into two separately packaged units capable of quick installation for remote use or mobile applications. Developing miniaturised systems for suitcase application to Inmarsat and regional services. Total production has included several thousand troposcatter antenna systems.

Cynetics Corporation

Current Status
Communications systems engineering and small-satellite, off-the-shelf digital and analogue transponders. CC has developed transponder packages for mobile and transportable relocation systems. Has produced systems components and transistors/thyristors for high velocity (high g) accelerations. Miniaturisation has resulted in several compact sets with enhanced performance requirements. The CTL 57 mode L digital transponder was provided for SEDSat 1, 0.83 kg, 3 to 7 W out, 400 to 450 MHz. The CTA 1 mode A linear transponder translates a 60 kHz bandwidth at 146 MHz down to 29 MHz; three-year life, 10 W out, 0.85 kg.

Datron Systems Inc

Current Status
Datron provides satellite communication antenna systems for shipborne, ground-based and tracking requirements; its image-processing division in San Jose was incorporated into Simi Valley in July 1996. It delivered 13 m Multi-Satellite Data-Acquisition Systems to the Swedish Space Corporation and the National Central University of Taiwan and, in 1995, completed the 12.8 m antenna and related complex for the Center for Remote Imaging, Sensing and Processing at the National University of Singapore.

Datron has installed X/S-band remote sensing data-reception terminals in the US, UK, Norway, Sweden, Kenya, Thailand, Taiwan, Indonesia, Australia and South Africa to handle Landsat, Spot, ERS, JERS, MOS and Radarsat imagery, supporting 55 to 105 Mbit/s. The 9.2 m dual-band satellite-tracking antenna system comprises an antenna, series 8400 pedestal, RF feeds, antenna control unit, data receivers and associated patch facilities. The system is typically configured to receive 2.2 to 2.3 GHz (Landsat, Spot, DMSP) and 1.67 to 1.72 GHz (NOAA, Meteosat, GOES, GMS). Adapted for multisystem use and capable of satellite acquisition search. Can be equipped to give SHF access to several satellites simultaneously and upgrades to provide automatic channel selection have been installed.

Datron's Metrack-8 was the first system specifically designed for use with polar-orbiting meteorological satellites; 16 were installed in 12 countries. The Mk II provides simultaneous reception at LH/RHCP over 1.540 to 1.710 GHz. It allows autotrack of NOAA's HRPT downlink, in addition to the Sarsat search/rescue NOAA packages. Orbital parameters are entered for any number of satellites and prediction software determines satellite positions for the daily tracking schedule. The 70 kg positioner is a two-axis elevation over azimuth configuration mounted on a 1.2 m riser base. The company is now focusing on mobile satellite TV reception systems for recreational vehicles, busses and boats.

Decom Systems Inc

Current Status
Mobile/fixed telemetry ground equipment for flight testing and satellite communication signal acquisition and processing. Mobile systems checkout hardware with off-site remote access through secondary antennas. Can track and locate common-frequency availability by searching and identifying available satellites. Complete front-end telemetry acquisition systems, including bit synchronisers, modems and decommutators; high-performance real-time telemetry data processing and information distribution in highly integrated packages. Multi-point communications systems for integrated users.

Deskin Research Group, Inc (DRG)

Current Status
DRG offers ground antennas of up to 5.0 m in diameter, including the 1.8/3.0 m alti/azimuth Isohex antenna, a prime focus antenna comprising seven hexagonal (4.76 mm thick) reinforced glass fibre panels. Pointing is manual. Specifications are provided in C/X/Ku-band order for 1.8; 3.0 m. DRG has designed large antenna structures employing carbon fibre support ribs for folding or unfurlable deployment. Advanced versions carry hybrid materials formed into highly complex shapes with long-term memory facilitating extended periods of compact folding before release into a programmed shape. This obviates the need for deployment fixtures. Developed into long-life, multicycle use for extended operations on satellites or spacecraft.

Dorne & Margolin Inc

Current Status
Design, development of antennas for communications and navigation: GPS, Satcom (UHF, Ku, Inmarsat) telemetry, DF and EW for 1.5 to 18 GHz. Capabilities include base and mobile antennas for wireless markets. New directional adaption for multiple inputs capable of giving operators selected relay satellite terminals. Diversified into recreational uses for navcom systems and for ship and airborne satellite or terrestrial communication systems. Has developed parallel applications for tracking and for vehicle location.

ElectroMagnetic Sciences, Inc (EMS)

Current Status
EMS's 1995 sales were US$129 million (profit US$2.3 million) generated by three operating companies: EMS Technologies Inc; LXE Inc, CAL Corporation. There are about 800 employees. EMS has been involved in more than 60 space projects over three decades. It specialises in space-qualified beam-forming networks for satellite communications antennas, such as those used on NASA's ACTS vehicle. Conformal antennas for beam-shaped selection specified by users and operators. Also, new hybrid materials technology developed for rigid, flexible and re-adaptable structures used in ground and space-based systems. EMS owns 74 per cent of Canada's CAL Corporation.

ESSCO Collins Limited

Current Status
Now a subsidiary of L3 Communications, ESSCO (Electronic Space Systems Corporation) was founded in 1961 to develop ground-based radomes and precision antenna systems. The precision antenna systems are generally used at 30 to 300 GHz for applications in which precision reflector-surface accuracy, pointing and tracking are required, ranging from mm-wave radar to radio astronomy. Reflector sizes range from 6.1 to 26 m diameter and systems are radome-enclosed. The company was responsible for the first large radio telescope in the southern hemisphere, a 13.7 m antenna in Brazil. The radomes include sandwich, metal and dielectric space frame, solid laminate and air supported. More than 3,000 are operational worldwide. Several unique design solutions available for remote or inaccessible regions and difficult locations. Applications include 2-D/3-D radar, air traffic control systems, communications satellites and weather radar. Sizes from 1 m to more than 49 m. ESSCO has supplied radomes for major military and commercial programmes worldwide. ESSCO supplies almost 200 tuned sandwich radomes for FAA programmes including TDWR, ARSR-4 and the Radome Replacement Programme. The company has developed a unique range of small antenna systems for lightweight subsystems requirements on telecommunications and TV broadcast satellites.

ESSCO's Corporate HQ is on a 142,000 m^2 site, including a 610 m test range, 5,580 m^2 of manufacturing space; an Irish facility has 2,800 m^2.

E-Systems Inc (ECI Division)

Current Status
The company was acquired by SSE Telecom in 1996 for US$6.2 million. It specialises in high-speed digital modems and other Earth station products, accounting for more than 25,000 installations by 2002. Network hook-ups for ultra high-speed systems. Company provides a wide range of IT services including network management, application development and technical career placement.

FEI Microwave, Inc

Current Status
FEI Microwave designs/builds spaceborne microwave assemblies, including receivers, frequency converters and solid-state switch matrix products, utilising

qualified MIC/MMIC technology. FEI Microwave has diversified into terrestrial systems support in addition to its satellite related businesses. Development of system interactive networks for flexible user access. It has provided new development techniques for microwave systems on low-mass satellites.

Glocom, Inc

Current Status
Glocom in 1994 introduced the system's third Inmarsat B satellite phone, the GP3000, supporting 16 kbit/s voice, 9.6 kbit/s fax and data services. Requires 160 W in transmit mode. Prices start at about US$24,000. Glocom's Inmarsat M Global Phone 2000t and 2000r models were type-approved September 1993. The 2000t is an 8 kg self-contained briefcase model, while the 6 kg 2000r can remain outdoors for more rugged duty. The user is guided in pointing the built-in patch antenna towards the satellite by software, a signal level indicator and multilingual graphic and synthesised voice instructions. Ports allow addition of optional cordless phone and Gp 3 fax. Developed to radio (wireless) links between phone and transmitters.
Frequencies (GHz): 1.6265 to 1.6605 transmit; 1.5250 to 1.5590 rx
EIRP: 19 or 25 dBW (nominal) per carrier
G/T: –12 dB/K minimum, 14.5 dB gain
Data rate: 6.4 kbit/s full duplex voice with IMBE voice coded; 2.4 kbit/s for fax and data
Power: 96 W transmit; 37 W rx; 90 to 250 VAC or 10 to 16 VDC
Size: 457 × 340 × 127 mm, 8 kg (2000 t).

VERIFIED

Harris Corporation

Current Status
Harris has recently been awarded an advanced architecture study for the GOES-R environmental satellite ground segment programme from the National Oceanic and Atmospheric Administration. It has also been selected by the US Navy for the Advanced EHF Satellite Communications Terminal Prototype programme. GCSD is responsible for a range of satellite communication systems, Earth stations, and transmission and ground terminal products. In addition to a number of next-generation tactical MILSATCOM terminal programs, such as the USAF Ground Multiband Terminal (GMT), the USAF/USMC Lightweight Multiband Satellite Terminal (LMST) Block 5, and the US Navy Multiband Shipboard Satellite Communications Terminal (MSSCT) programs, Harris is also upgrading the Defense Satellite Communications System's (DSCS) worldwide network of strategic AN/GSC-52 SATCOM terminals.

Hughes Electron Dynamics Division

Current Status
Hughes EDD is the only US supplier of space-qualified TWT/TWTAs. More than 8,000 units have been delivered, recording 300 million hours in space. Available designs cover 1.5 to 65 GHz at output power from 10 W to greater than1 kW. EDD's EPCs have accumulated more than 88 million hours operating in orbit. Linearised TWTAs are also available. Research into exceptionally high-powered TWTAs for DBS systems and into high power TWTAs for advanced deep space missions involving satellite data and telecommunications relay around terrestrial planets and Jovian planets with high-intensity magnetospheres.

Hughes Network Systems Inc (HNS)

Current Status
HNS was founded as Digital Communications Corporation in 1971, became COM Telecommunications Inc in 1978 and was acquired by Hughes Aircraft Company in 1987. The company is a leader in the VSAT industry, claiming almost 70 per cent of the world market and delivering more than 150,000 terminals through 2001. 1996 revenue was about US$1.06 billion, of which about half was from communications satellites. A US$30 million contract was signed February 1995 with Argentina's Impsat for 2,000 VSATs and five shared hubs. 6,000 units are being provided to Ford Motor Company and 2,000 to Best Western hotels. A contract was won in 1990 to supply 10,000 1.8 m units for General Motors' Pulsat Network. In 1993, it provided Personal Earth Station and Telephony Earth Station VSATs at 300 locations for the People's Bank of China, working through AsiaSat 1. Shanghai-Hughes Network Systems Ltd (SHNS) was created in 1994 as a joint venture with two major Chinese telecommunications companies, primarily for manufacturing/installing VSATs in China. HNS developed special low-cost USATs for use in China and for China to deploy in conjunction with domestic requirements.

The company also provides satellite trunking, modem and telephony products such as the Model 4100 digital modem, digital SCPC, Telephony Earth Stations and videoconferencing with its InTELEconference product. Subsidiary Hughes Escorts Communications Ltd (HECL) began providing India's first private satellite shared hub service in January 1995, using Insat 2 leased capacity. The UK subsidiary HNS Ltd provides the Inmarsat C 3100 land mobile station.

Integral Systems, Inc

Current Status
Integral Systems Inc offers satellite ground command and control systems. The EPOCH 2000 software can handle as many satellites of different types as required, control ground communications networks and be used in satellite integration and testing. It has exported several versions and has an almost limitless capacity for multisatellite and multifunctional integration. It is used by China's DFH-3 telecommunications satellite system and has been adapted in China for control systems at tracking sites. Ground command and control systems have been developed for hybrid LEO and GEO communication satellite systems. These have now been extended to hybrid connections and switchable packet systems software controlled.

Keiser Engineering, Inc

Current Status
The company was established in 1975 to perform feasibility studies and engineering analyses for satellite telecommunications, including voice, video and data. Included are modulation and multiplexing techniques, (FDMA, TDMA, CDMA), error-correction coding, interference analyses, antenna design and propagation studies. Developed rationales for high-data compression telecommunications techniques interlinked with TDMA and FDMA. Has conducted research on EMP survivability standards and capabilities. Keiser has also developed new applications for satellite interlink through laser or SHF communications frequencies.

LNR Communications Inc

Current Status
LNR develops/manufactures communications satellites equipment, specialising in Earth stations, LNAs, frequency converters and solid-state TWT drivers. LNR's amplifiers are used by the TDRS space/ground segments and by Shuttle. High-power options developed by LNR are interchangeable with standard amplifiers. Can be changed according to customer requirements. The Tramp-T trailerised clamshell Earth station provides C-, X- and/or Ku-band links via a 2.2 m antenna (1.2/1.8 m options). Meets Intelsat IESS 308/309 and DSCS/NATO requirements. The FTSAT Flyaway Triband Satellite Terminal is being produced for the US Army under a US$27 million 1995 contract to provide C/X/Ku-band links via DSCS, NATO, Intelsat, Eutelsat and PanAmSat. The DVF series of digital flyaway stations provides SNG. Options: 1.8/2.4 m diameter, C/Ku-band, 20 to 400 W amplifiers, 220 to 325 kg, digital video encoders for multirate selectable to 3.072 Mbit/s or full broadcast quality at 8.448 Mbit/s.

Lockheed Martin Conic-Terracom

Current Status
Formerly a division of Loral Corporation, supplying satellite transponders, transmitters, receivers and RF power amplifiers for projects such as Radarsat, Rosat, Landsat, STEP, Tiros, Clementine, GGS, Olympus, Lewis and Clark. The CXS-600B 5 W transponder provides TT&C with STDN and DSN ground stations. It demodulates the PCM/PSK telecommand signal and outputs command data and bit timing. For transmission, it receives data from a telemetry encoder and modulates it on an internal subcarrier and/or directly on the S-band downlink. MIL-STD-2000; designed to withstand 10 years in GEO environment. Improvements pending extending survivability to 14 years. Developed long-life, high-power, power amplifiers for 10 to 20 W transponder output.
Size: 11.2 × 18 × 14.7 cm; 3.9 kg max
Power: 34 W at 5 W; 24 W at 3 W
Rx/Transmit: 2.025 to 2.120/2.200 to 2.300 GHz.

VERIFIED

Lockheed Martin Missiles & Space Company, Inc (LMMSC)

Current Status
LMMSC designs, builds and tests antennas and microwave devices for space vehicles. Antennas range from simple omniconical spirals to broadband multimode systems incorporating integrated beam-forming networks and low-noise amplifiers. It has designed, fabricated and delivered more than 45 deployable antenna reflectors in 30 years, the largest being the (9.1 m aperture) reflector for NASA's ATS 6 in the mid-1970s. The Wrap-Rib™ reflector design is space-qualified over 1 to 10 m apertures and 0.15 to 12 GHz. It features low stowed volume, lightweight and proven reliability, never having had a deployment failure. Extended aperture for ultra lightweight carbon and composite antennas capable of deploying from highly compressed, folded packages after extreme environmental conditions over long periods.

LMMSC is also designing, building and testing a line of lightweight phased-array antennas using LMSC T/R modules. Arrays being developed include capacitive-fed, probe-fed and aperture-coupled arrays with steerable, shaped and multiple beams, using printed feed lines with built-in beam shaping. LMMSC has adapted these designs for specified requirements for future DoD satellite designs.

MCL Incorporated

Current Status
MCL was founded in 1961 as a design and manufacturing specialist in coaxial and waveguide components, and now operates with about 100 personnel in a 5,500 m^2 facility. MCL produces satellite communication amplifiers (TWT & Klystron), airborne TWT amplifiers, SNG transportable amplifiers, coaxial cavities and power systems. High power units developed for high-spec customers. The satellite communication amplifiers range from 50 W to approximately 3,000 W CW, covering L- to Ka-band and DBS. More than 5,000 Earth station amplifiers are in service worldwide. Coaxial cavities and power systems range from 10 W to approximately 500 kW over 0.010 to 18 GHz. High density cable for high energy circuits.

Microdyne Corporation

Current Status
Microdyne ATD is involved in the manufacture, design and integration of aerospace telemetry RF acquisition, reception, processing and display. Recent new products include the 700-MR receiver, 1620-PC diversity combiner, WCS-CS cosecant squared tracking antenna system, VXI-2000 multibus Quick Look system and the TSS-2000 Telemetry Signal RF Simulator. It has researched interference related RF jamming. It has also developed high-density RF packages for multiplexed RF systems and processors and has produced filter systems for encrypted signals used in future military reconnaissance satellites.

MITEQ, Inc

Current Status
Specialists in communications technology and front-end systems. Subsystems and components for satellite communication ground segments, including up/down converters, frequency translators, integrated RF front ends, FM modulators and demodulators, RF amplifiers (1 MHz to 50 GHz), signal converters/low-noise front ends to 50 GHz, and oscillators/frequency sources. Interactive space-related and terrestrial systems support at component level. Provides clean rooms, environmental laboratories and test facilities. MITEQ Inc is ISO 9001 certificated. Research into robust converter systems produced new amplifier markets for component applications.

Mobile Telesystems Inc

Current Status
Mobile Telesystems Inc (MTI), leading supplier of Inmarsat communications terminals, introduces a new line of state-of-the-art video transmission products suitable for wireless news gathering, wireless cable TV and distance-learning applications.

MTI's Multichannel Multipoint Distribution System (MMDS) broadcasts TV programmes at microwave frequencies from a central point, or head end, to small receiving antennas on the subscriber's roof. MTI's 8 × 8 in receiver/down converter is the smallest in the industry.

MTI's V-Link is an Audio/Video Microwave Transmission System, designed to relay live TV programmes from the camera, outside broadcast car or studio to relaying systems such as SNG, microwave or cable/fibre systems, or directly transmit TV programmes to remote TV stations or CATV head-end systems for broadcasting or distributing to subscribers. V-Link's compact size and wireless transmission capabilities make it a cost-effective solution to traditional time-consuming installations of heavy coaxial cables.

MTI's CompaLink Radio System is a compact, economical, and high-performance digital microwave radio for point-to-point applications in fast-growing wireless rural telephony, rural telephone corridor networks, remote cellular links, wireless local-loop backbones, video teleconferencing, and telemedicine. Due to the ruggedness feature, the CompaLink has the capability to withstand severe operating conditions in most climates. Has been verified survivable under sub-Arctic conditions for long term dormancy and operation. Also capable of low- or high-pressure operation in hostile environments.

Motorola Space & Systems Technology Group (MSST)

Current Status
The division specialises in the design, development and production of tactical and secure communications equipment and systems. Products include communications equipment for land, sea and air forces and for specialised space related applications. The LST-5D Tactical DAMA Satcom/LOS Transceiver is 5.3 kg, 11.4 cm high, 15.2 cm wide, 24.9 cm deep, and integrates into man-pack, vehicular, airborne and fixed-station units. Embedded DAMA, COMSEC, TRANSEC. Nine programmable transmit/rx pairs. Narrow/wideband dedicated satellite communication (MIL-STD-188-181); 5 kHz DAMA satellite communication (MIL-STD-188-182); 25 kHz DAMA satellite communication (MIL-STD-188-1832). The transceiver is packaged in the LSSC Series UHF Satcom/LOS Terminals, which add a 9 dB fold-out antenna, battery/charger, interconnect wiring, data port, and voice/data interface module. Also has EHF capability. MSST has developed broad frequency options for existing packages and high-density RF terminals.

Raytheon Company

Current Status
The Satellite Terminal Systems Directorate of the Equipment Division develops and manufactures UHF/SHF/EHF airborne, ship- and ground-based communications terminals and peripheral equipment for military applications. Raytheon has developed new terminal equipment for specialised applications and radiation hardened specifications written by users and operators. It was awarded a US$122.6 million contract in November 1994 for the main mission antennas and transceiver modules for Iridium satellite, following a US$94 million award for five prototypes. A family of AF/USN EHF and UHF/EHF terminals has been developed for Milstar and other EHF-capable satellites. Other products include high-power amplifiers, modems and terminals to operate with SHF satellites. Development of radiation hardened amplifiers and terminals. Raytheon has developed hardened subsystems modular packages for small satellites and nanosat design concepts applicable to rapid-replacement concepts involving high-g acceleration launches.

Raytheon E-Systems Richardson Division

Current Status
Formerly Electrospace Systems, Inc, Raytheon E-Systems Richardson designs, integrates, and installs ground, airborne, and ship SATCOM systems. Its AN/USC-54 VME Integrated Communications System (VICS) is the only full-duplex transceiver in the world to receive Demand Assigned Multiple Access (DAMA) certification from the US Defense Information Systems Agency. Its SHF antenna system supports ship links via DSCS, NATO 4, and Skynet. Its modular ground-based LF/VLF solid-state transmitters provide power savings with efficiencies in excess of 85 per cent. Its lightweight 2.4 m 'flyaway' antenna is DISA/DSCS certified in X-band and meets Intelsat E1 certification requirements for Ku-band and Standard G certification for C-band. Its HMMWV-transportable 6.1 m triband SATCOM antenna system provides a rugged, stable platform for mobile communications requirements. Ultra lightweight antenna systems have been developed for specific customers and robust structures built for rugged endurance specifications. All antennas operate with Raytheon E-Systems' state-of-the-art 93C-30 antenna positioning controller, which provides high-speed processing control based on the industry's premier trajectory-estimation algorithm.

Scientific-Atlanta, Inc/ Network Systems Group

Current Status
The company is a leading manufacturer of satellite communication Earth stations, with more than 55,000 systems in more than 140 countries. The product line ranges from telephony and TV Earth stations to encrypted and digital transmission systems, including Intelsat A-Z gateway stations, VSAT and mobile. It also provides the system engineering, installation and training for turnkey systems. SkyRelay is a VSAT network designed specifically for distributed computing environments. The remote VSATs feature Ethernet and Token Ring LAN support, providing bridging/routeing capabilities. The Skylinx.DDS (Digital DAMA System) family of terminals provides full-mesh single-hop connectivity for voice/data at 9.6 to 2,048 kbit/s. For thin-route sites, the µDAMA provides one to two channels. The µMCPC terminal provides point-to-point connectivity. The company was awarded a US$3.7 million contract by China's Ministry of Posts and Telecommunications in September 1995 for an emergency satellite communication network to support relief operations. A new lightweight mobile antenna system was developed out of this research and applied to the needs of releif organisations in developing countries. The initial phase of the Skylinx.DDS network became operational in early 1996 using AsiaSat 2 Ku-band capacity. It employs a 4.5 m master station in Shanghai, working with 3.6 m stations in seven provinces and 37 mobile 1.2/1.8 m VSATs. The service will eventually be expanded to include a 3.6 m station in all 22 provinces, with at least five mobile units per province. Skylinx.DDS is also being used by Teleport-TP to create Russia's largest private VSAT network. The first phase, completed April 1996, links 30 remote sites to the 18 m Intelsat A hub in Moscow, working through Intelsat 704 at 66° E. On completion, there were remote sites by the end of 1999.

The Model 9821 MariStar-M (for maritime users) and Model 9826 TerraStar-M (land users) terminals provide global voice (6.4 kbit/s), fax/data (2.4 kbit/s) Inmarsat M services. MariStar-M comprises an above-deck radome with a 20 kg, 66 cm diameter, stabilised antenna and a small below-deck electronics/communications unit. The Model 9820 MariStar Multi-M delivers 3 to16 channels. The 5.2 kg Model 9805 TerraStar-C terminal and 5.5 kg Model 9801 MariStar-C Satellite Terminal (for maritime users) provide global data/telex Inmarsat C services, including data reporting and automatic polling of terminal position. The optional Model 9803 GPS Receiver (2 kg, C/A code, three-channel, eight-satellite, one-second update rate, 25 m SEP without Selective Availability) provides position determination through the same antenna. The Model 9806 TerraStar-C Portable totals 23 kg in a 356×572×243 mm flight case, including a Ni/Cd pack providing four-hour operation. EIRP 12 dBW min, G/T −23 dB/K at 5° elevation. Is developing new mobile antenna systems for integrated Intelsat and Inmarsat capabilities.

Sea Tel, Inc

Current Status
Sea Tel is the world's largest manufacturer of stabilised antenna platforms, with almost 40,000 delivered since 1979. Antenna sizes cover 0.3 to 3.66 m with accompanying radomes of 4.27 m diameter. Applications include Inmarsat A/B/M, marine digital satellite TV and TVRO systems, C/Ku-band VSAT high-speed data and multichannel voice for oilfield and cruise-ship communications, and spread-spectrum systems. Compact and omni-directional, a new generation of antennas are proposed for a variety of specialised users. Sea Tel offers more than two dozen basic models of which 16 are in volume production. Sea Tel derives more than 80 per cent of its revenue from international sales and maintains facilities in Southampton, UK, with servicing agents all around the world.

SierraCom

Current Status
Formerly Lucas Espec Division and now part of Sierra Networks Inc, SierraCom designs/manufactures VSAT terminals/antennas, SSPAs, microwave components and subsystems for communications satellites. Impact resistant, hardened, antennas applicable to high-level military specifications. Has developed terrestrial link systems for integration with space-based elements. Active in the design and development of shaped antennas for specified requirements.

Space Communications Technology Center

Current Status

SCTC was established in November 1991 as one of NASA's Centers for the Commercial Development of Space. It focuses on developing critical technologies for commercially viable digital transmission systems for video, audio and data by satellite where a large commercial market can be shown to exist. Another focus is in digital signal processing, HDTV signal processing and RF propagation studies for LEO satellite and Ka-band transmission using ACTS. LEO to GEO relay and forward managing relay also possible. SCTC has developed technical strategies for frequency search software and TDMA prediction. It has also designed satellite-to-satellite transmission systems.

SSE Technologies Inc

Current Status

SSE Technologies began in 1936 as an electronics distributor and now manufactures digital modems and C/Ku/X-band transceivers utilised in VSATs, hubs or point-to-point applications. Products include the Ku-band ASAT 1214 (SSPA up to 25 W; TWT 75 to 300 W), C-band ASAT N406 (SSPA 60 W; TWT 75 to 400 W); X-band 5 to 40 W SSPA; up to 300 W TWT. The S-series offers complete RF terminals in one-piece outdoor units for communications satellites in rural and remote areas: C-band ASAT S406 (SSPA up to 25 W); Ku-band ASAT S1214 (SSPA up to 16 W). The C-STAR/K-STAR transceivers incorporate MMIC technology: Ku-band 2 W or 4 W SSPA built-in (external booster for up to 60 W); C-band 2 W or 5 W SSPA built-in (external boost up to 25 W). SSET has developed survivable antennas for folded transport in remote geographic areas used by aid workers from relief organisations.

SSE Telecom is the parent company of SSE Technologies and in 1996 acquired Fairchild Data. It had an 80 per cent interest in the Directsat Corporation (DSAT). Directsat merged with Echostar Satellite Corporation in January 1994 and merged their DBS licences for 119° W. Directsat had been allocated 10 BSS channels at 119° W by the FCC in December 1993. SSE Telecom has an 8 per cent interest in Direct Broadcast Satellite Corporation (DBSC), which has 11 channels at 61.5° W and 11 at 175° W for western USA.

ST Systron Donner Corporation

Current Status

ST Systron Donner, part of Signal Technology, specialises in instrumentation and GaAs FET HPAs. It has provided the first 150 W 1.61 to 1.626 GHz SSPA for Iridium; efficiency is 24.5 per cent. In 1993, it delivered a 100 W C-band unit for US Naval links and in 1994 a 125 W X-band unit. Development of 150 to 200 W SSPA for advanced LEO and GEO applications. Research into concepts involving high-power SSPAs for specialised purposes. Research previously anticipating a new range of very high power transmitters has been terminated.

Stanford Telecommunications Inc

Current Status

Stanford Telecommunications Inc was acquired by ITT Industries on 14 December 1999 and this and other business units are part of ITT Industrie's Defense Electronics & Services segment. The company develops/manufactures Earth terminal and satellite network control systems and VLSI integrated circuits. Major products include: GPS satellite signal generators to simulate the exact signal environment experienced by high-performance aircraft during manoeuvres; anti-jamming satellite-communication equipment for worldwide secure communications; TDMA modems; tracking/ranging receivers for automated remote tracking stations; automated satellite signal-monitoring systems for optimum employment of communication satellites' resources; multipoint communication network systems for real-time, high-speed, subscriber-to-subscriber network controlled communications; and semi-custom high-speed digital signal-processing integrated circuits for forward-error correction, digital synthesis and spread-spectrum applications. Research and development activities in high-speed jamming and frequency scanning.

The company has developed the AN/FSC-111 ICBM SHF Satellite Terminal (ISST) for USAF's satellite communication system. It continues to produce the AFSATCOM AN/FSC-97 Single Channel Transponder Injection System, the uplink transmitting and satellite control system for AFSATCOM payloads now in orbit. A Milstar test terminal was developed and delivered to Lockheed to support the satellite's factory testing; a second terminal undertakes on-orbit tests. TT&C subsystems for USAF's Automated Remote Tracking Stations were delivered under subcontract to Loral Space and Range Systems. They service almost all military satellites, tracking the range, receive telemetry and transmit commands. A Transportable Ground Station was developed under internal R&D funds to incorporate state-of-the-art subsystems. In September 1992, it was awarded a five-month contract with four one-year options totalling US$37.8 million by the US Army for support to the DSCS Operations Control System at Colorado Springs. It provides continued on-site operations, maintenance and training, plus maintenance support at the other DSCS centres around the world. It is designing/building Intelsat's new Communications System Monitor for high-speed measurements on satellite downlink signals, including signal power, carrier frequency and bandwidths.

TECOM Industries, Inc

Current Status

TECOM designs and manufactures antennas and computer-controlled electromechanical positioners for communications, aerospace, navigation, surveillance and command control. These include Inmarsat aeronautical antenna systems, American Mobile Satellite Corporation (AMSC) land mobile antennas, and GPS antennas. New developed version of conventional land mobile antennas adapted into fold-up arrays for specific user requirements. High-gradient routing performance antennas developed for non-civil applications.

The new T-4100 low-gain blade antenna supports aeronautical data services. An intermediate-gain antenna system is being developed from the T-4000 for the new Aero I service of the Inmarsat 3 satellites.

TelSys Inc (TSI)

Current Status

TSI develops and manufactures ground-station communication systems for both traditional telemetry processing and bridging/switching functions to interconnect local/wide-area networks to space-ground communications networks. It has produced hybrid switching systems for advanced high-frequency requirements and has tested the opportunities which may accrue from laser communication systems integrating with terrestrial-based systems. Research and development programmes on new and unusual telecom systems in addition to tailored systems for individual users. Can provide dual service systems for terrestrial and space-based bridging networks.

TIW Systems, Inc

Current Status

TIW Systems is a privately held company founded in 1976. It is a turnkey supplier of antenna systems and Earth stations for communications satellites, tracking/telemetry, radio astronomy and optical astronomy. Its stations are used in the Intelsat and Inmarsat networks, and its antennas by COMSAT, Teleglobe, Telespazio, Indosat, British Telecom, Embratel, GE, Hughes Communications, SES, Lockheed, JPL and NASA. Unspecified customers supplied with unique combinations for unusual customer requirements and specifications.

In 1996, TIW Systems completed the Uplink Antennas for MEASAT Broadcast Network System (MBNS) Direct-To-Home (DTH) digital video and audio system. Antennas were installed for EchoStar's DISH Direct Broadcast services as well as DirecTV Latin American DBS services. Installation and upgrade was completed on the JPL Deep Space Network antenna system. An installation of the Missile Engagement Simulation Area (MESA) for the US Naval Weapons Center was also completed. TIW has developed compact folding arrays for deployment in remote areas and in geographic locations with spare utilities.

Trimble Navigation

Current Status

One of the first companies to design, develop and market GPS equipment. Trimble has introduced the Galaxy combined Inmarsat C/GPS transceiver. Produced ground breaking designs for Navstar GPS. Developed key components for DoD interface and unique transceiver codes for level 1 access. Has developed new product range for signal pickup from several different GPS systems.

W L Pritchard & Co, Inc

Current Status

Named after one of the late founding fathers in satellite telecommunications, Pritchard and Co provide intersystem co-ordination and engineering; economic, competitor and market research for communications satellites, radio, fibre optics and private telecom networks. Full licensing and ITU processing for new telecoms operators or satellite service providers. Advise major telecom operators, agencies and users on configurations of satellite systems. Provides briefings and tutorials on space telecommunications and has conducted several advisory functions to government and satellite operating agencies.

SATELLITE GUIDANCE, NAVIGATION AND CONTROL COMPONENTS

France

Aerospatiale

Current Status
Aerospatiale has developed a family of magnetically suspended MW/RWs, accumulating more than 500,000 hours on Spot 1–4 and ERS 1. A new generation has been qualified for Helios 1, Spot 4 and 5 and Envisat. These have been applied to experimental test programmes. Specifications in Spot 1–3 and Helios 1 order are:
Angular momentum: 15/40 Nms
Torque: 0.2/0.45 Nm
Diameter: 350/345 mm
Mass: 8.5/16 kg
Maximum power: 65/160 W
Speed: 2,450/2,500 rpm
Induced vibrations: 5/0.5 μm
Life: approximately 7/5 years

VERIFIED

SAGEM SA

Current Status
SAGEM, one of Europe's principal inertial systems manufacturers, comprises three divisions: Defence and Security; Terminals and Telecommunications; and Electronics and Industry. Research and Development, prototype design and system integration are concentrated in four sites (Argenteuil, Eragny, Saint-Christophe, Cergy), with four manufacturing centres (Montluçon, Saint-Etienne-du-Rouvray, Fougères and Poitiers). Total floor area of 200,000 m^2 of class 10–10,000 cleanrooms. Principal specialisations for satellites and launchers cover gyros, linear and rotating mechanisms (with associated electronics), bubble memories, optics, microelectronics and navigation/guidance systems. The linear and rotating mechanisms and the mechanical systems used in telescopes and optical instruments, mirror refocusing, optical systems and structures for spectrometers include: Spacelab spectrometer, mirror scanning system for Metop's IASI instrument, Mir MIRAS spectrometer, and Helios refocusing mechanism. Combined linear and rotating systems have been developed for adaptive optics in advanced astronomy telescopes. Adaptive optics have been specified for the Next Generation Telescope and Sagem has conducted primary level research for that programme.

Regys 3S, the newest member of the Regys family of dry-tuned two-axis inertial reference units developed for space applications, it incorporates the Gildas 3 gyro. A fully autonomous unit, it provides error-compensated angular rates in digital form ready for AOCS computer processing. No thermal control is needed as there is built-in compensation for thermal-induced inertial errors. The separate gyro and electronics units are connected by a flexible harness; the electronics units can be stacked. Regys 35 has been selected for the following programmess: ODIN, XMM, Integral, Stentor and Proteus.
Size: 103 × 96 × 95 mm sensor unit; 283 × 164 × 45 mm electronic unit; less than 2.3 kg total
Reliability: 0.99 for a three-axis redundant configuration on a 15-year telecom application
Power requirement: 15 W maximum, at 22–51 V DC
Output interfaces: OBDH/RU, 1553, MACS (optional)
Range (°/s): ±2 fine, ±40 coarse
Resolution: less than 0.1 arcsec
Bandwidth: more than 5 Hz (up to 20 Hz by software adaptation)
Scale factor stability: 10^{-3} BOL; 10^{-2} EOL
g-insensitive drift stability: 0.15° max over 1 h, with ΔT±2IC
Environmental tolerance: –10/50°C; vibration 22 *g* rms 0–2 kHz random, 20 *g* 0–100 Hz 0-peak; radiation 15 year GEO.

The Regys 10 rate gyro package is a digital-output package for sensing satellite angular rates about two orthogonal axes, incorporating a Gildas 1/4 dynamically tuned gyro adapted from an air navigation system, and mounted with a shock absorber in a cradle that can be adjusted for required axis orientation. Regys 10 is used by Spot 4, Helios 1 and Envisat, and the newer version aboard Spot 5 and Helios 2. The computed angular displacement is delivered on request at least every 200 ms. If the measured rate is approximately 2°/s, the OBC can order the gyropackage to operate at quarter speed in security mode and provide coarse data (0.400 arcsec/impulse in contrast to 0.0288 arcsec/impulse in fine mode).
Size: 16 × 16 × 16 cm; 2.5 kg
Power requirement: 21 W null speed, 29 W at 8°/s
MTBF: 200,000 h
Range (°/s): ±0.6 fine, ±2 coarse, ±8 security
Scale factor (arcsec/impulse): 0.0288 fine, 0.0972 coarse, 0.400 security
Scale factor stability: 10^{-4} short-term; 10^{-3} long-term
Environmental tolerance: 0/50°C operation, –20/50°C transient; 40 *g* /4 ms shock: 100 krad radiation

SAGEM began developing a space-qualified brushless DC motor under ESA/CNES contracts in 1975, producing a high resolution unit offering 1,200 steps/rev for application in: stepper mode for solar array drive mechanisms on LEO/GEO spacecraft, synchronous mode for de-spinning onboard structures (as on Giotto), precision positioning mode for mirror pointing (as on Spot) and mini-stepping mode for manipulator articulation. 100 μrad accuracy has been demonstrated for the stepper version, with a 6 per cent maximum variation at 0.01 rpm in the synchronous variant. SAGEM's 18EM brushless DC motor drives the rotating mirror of Sodern's STD 12 Earth horizon sensor aboard Spot.

Sydem is a motorised unfolding and pointing system designed to actuate and precisely position various appendages including large antennae on satellites. Typical use includes also single-shot movement with final locking.

Sydem (Type 42) is qualified for space applications, and represents the top of the Sydem range. A smaller and lighter Sydem (Type 32) supports less demanding applications.

VERIFIED

Sextant Avionique

Current Status
It provides a range of gyroscopes, accelerometers, onboard digital computers and data handling equipment for missiles and space launchers. The company is contracted to provide 87 Quasar ring laser gyro systems for Ariane 4's strapdown inertial system; 72 had been manufactured and 65 launched by end-1995. This comprises an Inertial Measurement Unit of three 33 cm-path length laser gyros and three accelerometers, linked to a Sextant series 7800-based processing unit. The system was qualified by July 1986 and flown aboard the first Ariane 4 in June 1988. The unit is now radiation hardened. Sextant also provides Ariane 5's RLG inertial reference system for Ariane 5+. This features an improved RLG, a 68020/68882 processing unit, radiation hardened memories, new algorithms and reduced power. Updates and improvements are in development. These include improved RLG specifications for strap-down systems.
Mass: 20 kg, size 390 × 230 × 210 mm
Power requirement: 35 W at 28 V
Range: 100°/s, 40 *g* shock
Bias drift: 0.03°/h
Drift at 14 g rms vibration: 0.9°/h
Random drift: 0.005°/h½

VERIFIED

SODERN

Current Status
The Société Anonyme d'Etudes et Réalisations Nucléaires (SODERN) specialises in instrumentation for particle radiation detection, Sun/Earth/star sensors for satellite attitude control systems, and spaceborne rendezvous/docking cameras. Personnel total more than 300. More than 130 SODERN attitude sensors have been operated in space since the first launch of a SODERN sensor in 1974 (Symphonie). Sensor development began with the STR 01 Earth horizon-crossing and STA 01 Earth static sensing devices for the GEO Symphonies, launched 1974/75. All the Earth-sensing devices operate at 14 to 16 μm IR. An improved STA 03 followed on OTS and Marecs, and four STR 03s on each Meteosat. A slightly improved STA 03A is employed by the ECS/Eutelsats. 35 STR 04 horizon-crossing sensors were provided for Intelsat 5 and the self-redundant STA 04 was introduced on Telecom 1 in addition to successor derivatives. The STD 12 two-axis conical scan sensor was developed for Spot and ERS, and has been further developed as the STD 15/16 for operations in all orbits from low altitude to GEO. The SED 04 star tracker was employed by Spacelab 2's Instrument Pointing System. In 1985, CNES awarded SODERN the contract for developing the SED 12 CCD star tracker for the Franco-Soviet Granat X/γ observatory. SODERN also developed the Earth-imaging cameras for Spot 1–4 and Helios, and Envisat's MERIS spectrometer and has developed the Spot 5 and Helios II cameras. New sensor development undertaken by an expanded product research and technology development facility. SODERN provides the capability to integrate star sensors with earth limb scanning sensors in a unique unit slaved to directional target sensors for attitude orientation and control.

SED 04 star/sun tracker
The SED 04 was developed to update the reference gyros of ESA's Instrument Pointing System, first flown aboard Spacelab 2 in 1985.

SED 12 CCD star tracker
The SED 12 was designed and built to provide attitude data to CNES's control system for the French Sigma telescope aboard the Soviet Granat astronomical observatory. The SED 12D model is used by France's Helios 1 reconnaissance satellite and has been adapted for Helios 2, Envisat and Sweden's Odin. The tracker is based on Thomson's TH7861 CCD area array, each element measuring 23 × 23 μm and providing 40 mV/lx sensitivity. Acquisition probability is more than 0.997 for stars of magnitude +8 to –1. Simultaneous tracking of five stars over the 7.5 × 10° rectangular Field Of View (FOV) is possible. From an initial standby mode, the sensor can operate in search mode for up to 30 stars over the whole FOV (requiring minimum detectable magnitude, number of stars to be tracked after search, and apparent star motion), or in tracking mode for a maximum of five local fields (requiring co-ordinates of five stars, expected magnitude and motion). For both modes, the sensor outputs the located stellar co-ordinates, magnitudes, time delay between charge collection and read-out, detector/optical head/electronics temperatures.
Field Of View: 7.5 × 10° rectangular
Accuracy (arcsec): 3 after one observation, 0.5 after filtering/calibration (1σ); 17 on two axes after one observation (3σ), 5 after filtering/calibration
Acquisition time: less than 8 seconds overall field; less than 1.5 seconds in 0.25° square field. Probability of single star acquisition more than 0.997
Maximum star rate: 0.1°/s for magnitude brighter than 9, 1.0°/s for magnitude brighter than 3
Size: 141 mm diameter, 230 mm length/2.5 kg optical head without baffle; 271 × 222 × 217/7.2 kg electronics unit
Input power: 2 W optical unit with cooler off; 2–8 W optical unit with cooler on; 40 W electronics unit typical; 20–50 V DC (PWM 1525)
Temperature range: –20 to +50°C
Memory capacity: 40–64 kbyte ROM; 16–32 kbyte RAM depending on mission
Failure rate: 3,800 × 10^{-9}/h

SED 15 CCD star tracker
The SED 15 CCD Star Tracker is under development for GEO satellites to provide pitch/roll data. One version will track only the Pole Star. 30 arcsec accuracy in 5 × 7° FOV. By adding barycentric signal processing, the accuracy holds for a larger Field of View (FoV), such as 25 × 35°.

STD 15 scanning earth sensor
The STD 15 scanning earth sensor derived from the STD 12 of Spot and ERS 1/2. Its first flight was aboard Telecom 2 in December 1991; 12 are now flying on Telecom 2A/2B, Hispasat 1A/1B and Hot Bird 2. The sensor has been selected by NPO PM (Russia) for a new telecommunications platform. The sensor permits the attitude control subsystem to generate roll/pitch angular deviation data by using the four Earth-space/space-Earth transitions of the two scan traces (generated by the rotating mirror and two fixed mirrors); an optical encoder measures the angular position of each transition with respect to the spacecraft reference axes. Data rate 1.25 Hz in normal mode, 5 Hz in stationkeeping mode.
Field Of View: 1.5° roll/12° pitch; 14° roll/30° pitch in acquisition mode. Instantaneous FOV 1.8 × 1.8°
Accuracy: 0.025° 1 second
Mass: 3.4 kg
Input power: 7.5 W at 22 to 50 V DC
Temperature range: –20- to +50°C operating; –40 to +60°C storage
Vibration tolerance: 20 *g* rms over 20 to 2,000 Hz
Lifetime: 15 years; 0.9 × 10^{-6}/h failure rate.

STD 16 scanning earth sensor
STD 16 scanning earth sensor is Sodern's latest sensor, designed for LEO satellites: Helios 1, Spot 4, Envisat, ADEOS and ETS 7. Operation principle is as STD 15. Nine were supplied to Matra for Spot 4/ Helios and Envisat: six three were supplied 1992–93 to Toshiba for ADEOS and ETS 7.
Field Of View: effective scan for one trace is 152°
Altitude range: 200 to 2,000 km
Accuracy: 0.045° bias, 0.03° noise; 3 seconds
Mass: 3.5 kg; length 386 mm; height 175 mm
Input power: 7.5 W at 22 to 52 V DC
Temperature range: –20 to +50°C; –40 to +65°C storage
Failure rate: 1.3 × 10^{-6}/h
Lifetime: 10 years.

DTA 01 high-resolution imager
Spot 1–3 incorporate Sodern's DTA 01 high-resolution visible push broom imaging unit. Each of the four spectral bands employs four Thomson TH7801A 1728-element CCD arrays (Spot 1's utilised Fairchild's 122 DC detectors). PAN passes directly through the prisms; its spectral range is determined by absorption and interference filters on the CCDs. Green band B1 is reflected off to the side by a dichroic mirror; red band B2 and near-IR B3 are separated by a second mirror. The rigid primary titanium structure holds the beam splitter and the CCD assembly of four butted lines; each line of four CCDs is straight to within 2 µm. A mechanically isolated secondary structure carries the front-end electronics. Resistance heaters help to maintain CCD temperature to within 2°C.
Spectral response (µm, at half-ht): 0.495–0.580 B1; 0.610–0.665 B2; 0.775–0.785 B3, 0.490–0.715 PAN
Number of pixels: 3,000 (1,326 µm) B1-3; 6,000 (1,313 µm) PAN
Resolution (from 800 km): 20 m B1-3; 10 m PAN
Typical peak beam splitter transmittance: 0.8 B1; 0.75 B2; 0.8 B3; 0.85 PAN
Saturation spectral irradiance (W/m²/µm): 0.004 B1; 0.005 B2; 0.003 B3; 0.006 PAN
S/N for max/min illuminance: 700/90 B1; 650/40 B2; 800/50 B3; 500/70 PAN
Response uniformity: 5% B1-3; 3% PAN
Size: 211 mm wide, mounting flange 255 mm diameter; 2.5 kg
Power consumption: 2.7 W
Thermal: 18 to 22°C operating, –20 to +60°C storage.

DTA 03 high-resolution imager
The DTA 03 high-resolution imager, derived from DTA 01 for Spot 4, enhanced by a 1.6 µm SWIR channel and by using the B2 channel (instead of PAN) for 10 m resolution. The SWIR scanline is formed by 3,000 30 × 30 µm InGaAs Thomson detectors in two staggered rows of 1,500 each with a centre to centre pitch of 26 µm. The three B channels use TH7811 detectors, derived from the TH7801. Short-term stability is 0.01°C at 5°C for SWIR and 0.1°C at 20°C for the CCDs.
Spectral response (µm, at half-ht): 0.495–0.580 B1; 0.610–0.665 B2; 0.775–0.785 B3, SWIR 1.54–1.75
Typical peak beam splitter transmittance: 0.8 B1; 0.7 B2; 0.8 B3; 0.7 SWIR
S/N for max/min illuminance: 900/250 B1; 350/60 B2; 1,000/100 B3; 650/3 SWIR
Response uniformity: 4% B1/2; 4% B3; 10% SWIR
Size: 200 × 250 mm, mounting flange 250 mm diameter; 7.4 kg
Power consumption: 10.5 W
Thermal: 18 to 22°C operating, –15 to +50°C storage.

DTA 02/04 very high-resolution imagers
SODERN developed the first prototype of the imager (DTA 02) for the Samro military satellite. When the programme re-started as Helios, the company became responsible for the whole DTA 04 detection chain, including both focal plane and video electronics. An improved new detection chain DTA 06 is being developed for the new-generation Helios II. The programme is classified.

DTA 05 multispectral cameras
The DTA 05 multispectral cameras were developed for Spot 4's Vegetation payload, this imaging subassembly comprises separate focal plane detection units: 1,728-pixel CCDs for 0.49, 0.58, 0.64, 0.83 µm visible and a 1,500-pixel AsGaAs hybrid detector for the 1.6 µm SWIR channel. The front-end electronics are linked to a common video electronic unit, each CCD being read sequentially in series. Each detection unit is mounted on optics (developed by SODERN subsidiary Cerco) with an interposed dedicated interference filter. Optics are telecentric, ±50.5° FOV with registration less than 0.3 pixel regardless of incident angle. Global MTF 0.3 along track. 23 W power consumption.

DTA 10 spectro imager
The DTA 10 spectro imager was developed for Envisat 1's MERIS. Five identical cameras each analyse a 1,200 km swath (300 m resolution) with a spectrometer. Up to 15 bands can be programmed in position and width within the swath. Optics developed by Cerco.
Spectral range: 400 to 1,050 nm
Spectral resolution: 2.5 nm
Radiometric accuracy: less than 2%.

Other imagers
A new imager is being developed for Spot 5. The detection unit is built around a five spectral band beam splitter. Four bands are fitted with a Thomson TH7834 12,000 element array and the fifth band (SWIR) is fitted with a 30,000 element GoAo detector line. The PA channel is designed to achieve a ground resolution of 5 m.

ALICE BOE/BOS
The ALICE BOE/BOS was developed for the ALICE 1 and ALICE 2 instruments and used aboard the Mir Complex for microgravity experiments, the ALICE BOE and ALICE BOS achieve the optical diagnosis of cells which are filled with liquid subjected to critical conditions. Cells are temperature-regulated by high-precision thermostats and are analysed during the physical phase transitions by the optic-electronic devices of BOE and BOS. The ALICE optical capabilities include: microscopic inspection at various magnifications, interferometry, optical transmission, index gradient visualisation and optical diffusion. The total volume is 0.15 m³ and the mass is 55 kg.

POLDER-BOM
The POLDER-BOM was developed for the POLDER instrument aboard Japan's ADEOS 1 and 2 spacecraft, the optical-mechanical sub-assembly POLDER-BOM comprises a wide angle lens, a rotating filter wheel and a CCD focal plane. The optics have a 114° field of view, low distortion and constant illumination over the local plane. The filters and polarisers achieve spectral analysis from violet to the near-IR in different directions of polarisation. Eight spectral channels are available, three of which have polarisers.
Spatial resolution: 244 × 276 pixels
Temperature: 15 to 20°C operating –20 to +60°C non-operating.

SIS 01 very broad band 1-axis seismometer
The SIS 01 very broad band 1-axis seismometer, a vertical-axis seismometer, was developed for the Russian Mars 8 (Mars-96) mission for operations on the Martian surface.
Band pass: 0.03 to 8 Hz
Acceleration resolution: 10^{-8} m/s²
Power consumption: 60 mW
Dimensions: 90 × 90 × 90 mm
Mass: 350 g.

SIS 02 very broad band 3-axis seismometer
The SIS 02 very broad band 3-axis seismometer is an improved-technology development of SIS 01 with three seismic masses on a pyramid corner. Developed for planetary seismology.
Band pass: 0 to 5 Hz
Acceleration resolution: 2 × 10^{-8} m/s²
Power consumption: 1 mW
Dimensions: sphere, diameter 22 cm
Mass: less than 1.6 kg.

VERIFIED

Germany

Daimler Jena-Optronik GmbH

Current Status
DJO (51 per cent DASA; 49 per cent Carl Zeiss) was founded in January 1992 from the Space Research and Development elements of Jenoptik; the Scientific and Technical Equipment Development Division was established in 1980, building on the space research and development team formed in 1973. The company specialises in the design, development and manufacture of: space optical and opto-electronic systems and components; systems, subsystems and components for science observations, measurement, pointing and tracking; precision mechanisms; optical and opto-electronic calibration and test equipment.

All space involvement before German unification in 1990 was with Soviet organisations. For example, the division was responsible for the science instrumentation on the Soyuz 22 manned mission. More recently, its six-channel MKF-6MA multispectral camera and Astro 1 star sensor package (developed 1984–87) were launched to the Mir space station aboard the Kvant 2 module. The improved Astro 1M was contracted for the Spektrum-X/γ astronomy mission but was cancelled because of Russian funding problems. Further development has begun for ESA space station applications.

The Astro 5 autonomous star sensor had been developed for small satellite missions with up to 5 years mission lifetime in LEO with emphasis on minimum weight and cost. Since 1992, DJO has produced Precission Sun Sensors for a range of satellite programmes including SIRIUS, Eutelsat W2, W3, W4 and Arabsat.

TELDIX GmbH

Current Status
TELDIX GmbH was founded in 1960 as a subsidiary of TELEFUNKEN GmbH and the BENDIX Corporation, USA. In 1973, the Bosch GmbH acquired the BENDIX interest and in 1981 also that of AEG-TELEFUNKEN.

In February 1996, Litton Industries Inc, Woodland Hills, California, USA, acquired TELDIX GmbH from Bosch and made TELDIX a 'Center of Excellence' for high-technology products in the fields of computer technology, displays and space products.

In April 2001 Northrop Grumman Corporation, Los Angeles, California, USA acquired Litton Industries. TELDIX now belongs to the Navigation Systems Division within Northrop Grumman's Electronic Systems sector. Space products include high-precision satellite mechanisms, ball-bearing momentum and reaction wheels with integrated of external wheel drive electronics, low-noise magnetic bearing stabilisation wheels for advanced satellites, solar array drive assemblies, antenna pointing mechanisms and Systems for Optical Satellite Communication (SILX).

TELDIX is DIN-EN ISO-9001, AQAP-110, aqap-150, JAR 21 and JAR 145 certified.

For more than 30 years TELDIX has been successful in the field of gyroscopic actuators such as Momentum and Reaction Wheels.

TELDIX Momentum Wheels DRALLRAD® are utilised for the stabilisation of all kinds of satellites. As of June 2002, 411 wheels installed in 182 satellites are in orbit representing nearly 1,350 years of flawless operation.

Range of products include:

High-precision mechanisms for satellites, for example

- Ball bearing momentum and reaction wheels with Wheel Drive Electronics (WDE)
- Low-noise magnetic bearing stabilisation wheels
- Solar Array Drive Assemblies (SADA)
- Antenna pointing mechanisms
- Systems for optical satellite communication SILEX.

Ball bearing momentum and reaction wheels

TELDIX is the sole commercial manufacturer of ball bearing momentum and reaction wheels in Germany and leading manufacturer in Europe.

TELDIX ball bearing momentum and reaction wheels are the ultimate choice for advanced satellite stabilisation.

Reference projects

European satellite programs:

Abrixas, Artemis, Astra-2B, -1K, Beppo-SAX, Demeter, DFS, ECS, ESSAIM, Eurasiasat 1, Europe*Star, EUROSTAR 2000+, EUROSTAR 3000, Eutelsat II, Eutelsat W, GE-1E/Sirius 2, FBM, Hispasat 1A, 1B, 1C, Hot Bird, Inspector, ISO, Italsat, MARECS, MAROTS, Mars Express, Microscope, OTS, Parasol, Picard, Proba, Proteus/Jason, ROSAT, Skynet 4, Spacebus, Stentor, Symphonie A/B, TDF-1, TDF-2, Telecom-1, Telecom II, TELE-X, TUBSAT-B, Turksat 1, TV-SAT and TV-SAT-2.

International satellite programs:

Agila 2, Amos, Apple, Aqua (PM-1), Arabsat II, Arabsat III, AsiaStar, Aura (Chemistry), Beidou 1A, 1B, BSat 2a, 2b, Chandra (CXO), Chinasat 8, 22, DFH3, DFH4, Echostar VI, ETS-V, FEM, GE 5, GOES, Inmarsat II, Insat-1D, Insat-2, Insat-3, Intelsat V, Intelsat VII, IRAS, iSKY (KaStar), KaistSat, KitSat 3, METSAT, MOS-1, MS-T5, MT Sat, Nahuel, NATO IV, Nilestar, N-Star, NSS-6 (K-TV), OmegaSat, Orbcomm, Orion 1, Orion 2, PanAmSat 6, 7, 8, Pioneer, Planet-A, SBIRS Low, Sinosat, Sirius 1–3, Sky-1, Sky-2, ST-1, Step 4, Superbird. Telstar 5, Telstar 6–12, Tempo, Thaicom, Worldstar.

In 1990 TELDIX started the development of an experimental low-cost reaction/momentum wheel with integrated WDE for use in very small satellites. The wheels of this type were delivered in 1993 for the satellites TUBSAT B and C, built by the Technical University of Berlin and were also selected and built for KITSAT-3 (Korea) and a small spacecraft called Inspector built by DASA, Bremen. The more stringent requirements of small commercially used satellites require an adaptation of the experimental design to a highly reliable, medium-life space-qualified reaction wheel as for the ORBCOMM constellation.

Magnetic bearing momentum wheel MWX

The TELDIX MWX, a 5 DOF magnetic bearing momentum wheel means a real breakthrough with regard to vibrations exerted on the satellite, and a quantum step forward in wheel technology towards high-accuracy space missions. The MWX combines the following highly attractive features:

- extremely low vibrational noise due to active vibration suppression feature
- capability to tilt the rotor actively allowing very high pointing control
- capability to damp structural vibrations generated by other sources in the spacecraft
- and, as a general advantage of magnetic bearing wheels: higher speed capability, no lubrication, no stiction and low friction.

Thus the MWX is the most appropriate wheel for:

- scientific missions, μ-gravity
- optical communication links
- space telescopes.

Drive assemblies and pointing mechanisms

As a highly experienced mechanism manufacturer, TELDIX has been working for many years in the field of low-speed mechanisms as well. Within the framework of various contracts, pointing mechanisms for different applications were designed and built:

- Antenna pointing mechanisms
- Fine Pointing and Flip-Flop Mechanisms for SILEX (Semiconductor Intersatellite Laser Link Experiment)
- Pancake-type stepper motors
- Gimbal mechanisms
- Solar Array Drive Assemblies (SADA).

These SADA's are being used on DFS-Kopernikus and the Chinese DFH-3 satellites.

India

ISRO Inertial Systems Unit

Current Status

ISRO's IISU specialises in inertial systems and components for launch vehicles and spacecraft with applications in the GSLV. Development of inertial guidance systems for improved upper stages on GEO launchers. Guidance systems development for advanced orbital re-start capability and multiple stellar navigation updates. Integrating national effort to improve celestial navigation equipment for in-space trajectory adjustment routines through improved accuracy and enhanced reliability for long-duration missions to deep space objectives.

Laboratory for Electro-Optic Systems (LEOS)

Current Status

LEOS provides India's capabilities in star, Earth and Sun sensors, responsible for the units on Aryabata, Bhaskara, Apple, IRS, SROSS and Insat 2. ISAC has also provided tri-axial magnetometers (range ±45,000 γ, accuracy 500 γ) for every Indian satellite and has begun marketing to foreign satellite design teams in Europe and the USA. Discriminatory sensor guidance packages linked to laser ring gyroscope platforms. Former Soviet countries have also expressed interest and orders have been received from some Russian companies. LEOS has marketed these products to several former states of the USSR and has received interest from several agencies and research institutes.

Israel

Elop Electro-Optics Industries Ltd

Current Status

Elop's Remote Sensing Operation is principally concerned with airborne and ground electro-optical observation systems, warning systems and optical processing. Elop has developed key elements of Israel's defence reconnaissance and surveillance system from space and has developed advanced optical systems for its next-generation reconsats. Civil Research and Development activities cover remote sensing, astronomy, Optical sensing instruments for high-resolution telescope systems applicable to ground-, air- and space-based systems. Earth resources monitoring and electro-optical navigation sensors. It developed Israel's TAUVEX UV telescope in co-operation with Tel Aviv University for Russia's Spektrum-X/γ and developed the static horizon sensor and low resolution camera for Technion's TechSat 1. It was also responsible for Ofeq 3's camera. Elop is providing the camera for the German-Israeli David Earth observation satellite and has been involved in developing advanced optical systems for successors designed to operate in near-infra-red.

The compact static horizon sensor, using state of the art Si-monolithic non-cooled line of 32 14–16 μm infra-red detectors, was developed for Israel's TechSat 1 but is suitable for all small LEO satellites. The signals from the four identical optical units are analysed by the integral computer to locate the radiance profile in the horizon's vicinity. This is compared to the memory data and the angle to the horizon in each of four directions (hence the nadir angle) is computed. It can also compute the angle rate of change.

Accuracy: 0.1–0.15° (1 σ)
Field of view: 14.6° total
Size: 110 × 120 × 120 mm; 1.25 kg
Power requirement: less than 1 W; 11 to 14.5 V DC or 5 V DC regulated

The 6.5 kg earth resources monitoring system is a pushbroom linear CCD 0.5 to 0.90 μm visible/near infra-red 300 mm fl f/5 15° FOV camera. It provides 16 to 25 m resolution (40.6 μrad) in a 100 km swath from 400 km aboard a small satellite. It is currently a PAN instrument but it can be converted to XS if required. Resolution can also be improved. Ofeq 3's camera is similar but resolution enhancement is a feature of future Ofeq requirements.

VERIFIED

TAMAM Precision Instrument Industries

Current Status

An element of IAI's Electronics Division, TAMAM specialises in inertial navigations/reference systems, inertial sensors and stabilised platforms. TAMAM is developing long storage moment wheels for large satellites.

TAMAM's MW 1243 momentum wheel is flying on the small HETE satellite (the first was lost on TechSat in Mar 1995). The electronics are accommodated within the wheel case. The balanced flywheel is driven by an ironless armature brushless DC motor. Developed for specialised access capabilities on advanced programmes.

Torque: 20 mN-m maximum
Momentum: 4 Nms maximum
Size: 255 mm diameter, 60 mm height; 2.5 kg
Power consumption: 14 W at 3,500 rpm (20 mN-m); 4 W at 1,700 rpm (10 mN-m)

The MW 2427 is designed for medium and large satellites. It is flying on Ofeq 3.

Torque: 200 mN-m maximum
Momentum: 16 Nms maximum
Size: 355 mm diameter, 95 mm height; 5.5 kg
Power consumption: 3 W at no torque; 60 W at 2,000 rpm (200 mN-m); 13 W at 750 rpm (100 mN-m)

The Coarse Rate Gyro model 2424 provides analogue information of the three mutually orthogonal angular rates sensed by three microminiature rate gyros. It also provides back-up to the main stabilisation system. It has flown on Ofeq 1–3.

Dynamic range: 200°/s X/Y, 600°/s Z
Threshold/resolution: 0.01°/s max X/Y, 0.03°/s max Z
Size: 60 × 70 × 90 mm; 0.580 kg
Power consumption: 50 W max start, 8 W max run

The Inertial Spacecraft Stabilisation Unit is a fully redundant strapdown attitude reference unit for satellite stabilisation and pointing. No mission has yet been claimed. It provides compensated angular incremental pulses and instantaneous digital angular velocity. Accuracy better than 1 mrad in three axes can be achieved when updated by Earth horizon, Sun sensor and magnetometer. The main unit comprises three dual axes thermally modelled dry tuned gyros in orthogonal configuration on four shock absorbers. The electronics unit is separate.

Attitude accuracy: 0.8 mrad
Resolution: 1.6 μrad/pulse
Dynamic range: 2°/s fine, 8°/s coarse
Size: 8 × 13 × 18.9 cm sensor unit, 15 × 14.8 × 22.2 cm electronics unit; 6.6 kg

The model 2425 fluxgate magnetometer is designed for small satellites; it was carried by TechSat 1. It comprises a sensor of coils wound on high-permeability ferromagnetic cores, and an electronic unit with an oscillator that excites the sensors and three control capture loops that process the sensor signals and provide the output analogue signals.

Dynamic range: ±0.6 gauss
Linearity: ±1% of full scale
Scale factor: 10 V/gauss
Size: 27 × 53 × 90 mm; 0.200 kg
Power consumption: 0.9 W maximum.

VERIFIED

Italy

Alenia Difesa

Current Status

The Optics and Space Division (DIOS) was established in 1983, with a staff of 270 by early 2002, to complement the activities of the Military Systems and Environment divisions. It is concerned principally with onboard equipment and instrumentation for spacecraft, including electro-optical sensors for attitude determination (IR Earth, Sun and star sensors), CCD video systems, optical sensors for robotic rendezvous and docking, remote sensing electro-optical instruments (visible and near-infra-red), and mechanisms and components for optical payloads (lightweight mirrors, baffles and optics). DIOS has provided more than 250 attitude sensors for Eureca, Turksat, Soho, SAX, Inmarsat, Telecom, Hipparcos, Olympus, Eutelsat, Italsat, Tethered Satellite, ISO, DFH 3, Cassini, Skynet, Orion, Hispasat, Arabsat 2 and Nahuelsat. The light baffle and aluminium mirror for Giotto's multicolour camera were produced by DIOS, as were the CDD monitoring cameras of Eureca's Protein Crystallisation Facility. Other instruments have been delivered or are under development for astrophysics research and Earth observation projects. UVCS/Soho precision mechanisms for Soho's three-channel UV coronograph spectrometer. JET-X + XMM Optical Monitor ancillary telescopes to detect faint stars in visible/near-UV close to the optical axis. Cassini VIMS camera visible channel on the orbiter; Huygen's Atmospheric Structure Experiment to probe Titan's atmosphere; optical heads for star reference unit. IASI IR atmospheric sounding interferometer for Metop 1 polar satellite for temperature and humidity soundings and trace gases. GOME (Global Ozone Monitoring Experiment) for ERS 2. MIPAS (Michelson Interferometer for Passive Atmospheric Sounding) for Envisat 1. Sounding rocket μg: MITE (Measuring Interfacial Tension Experiment); INEX MAM (Interactive Experiment for Marangoni Migration).

The MSS, Modular Star Sensor, designed for three-axis satellites such as ISO, Soho and SAX, is a two-axis star tracker incorporating a catadioptric head providing attitude control data in addition to the precise pointing of astronomical instrumentation. In the latter role, it also provides magnitude measurements of stars brighter than 9^{mv}. It first flew aboard STS-52 in October 1992 for ESA's In-Orbit Technology Demonstration Programme. With a few modifications (10 × 10° FOV), it was developed for the Cassini spacecraft launched to the planet Saturn in 1997. Advanced navigation systems development for very long duration space missions and complex navigation requirements in deep space where limb surfaces are not available.
Field of view: 4 × 3°
Accuracy: 6 arcsec uncalibrated
Size: 137 × 134 × 227 mm/3.5 kg (sensor head) 137 × 228 × 187 mm/2.5 kg (electronics unit)
Detector: silicon CCD matrix of 288 × 384 pixels
Input power: 10.5 W

The IFPSS, Integrated Focal Plane Star Sensor is a focal plane star tracker developed for ISO but is suitable for operation in any IR astronomical satellite with provision for cooling down to 2 K.
Accuracy: 1 arcsec rms
Size: 70 × 70 × 30 mm/1 kg (optical head at cryogenic temperatures) 200 × 150 × 70 mm/1.5 kg (electronics unit at –30 to +60°C)

The FDSS, Fine Digital Sun Sensor, is a two-axis wide angle static precise Sun sensor for application aboard three-axis satellites such as Eureca, Italsat and TSS.
Number of optical heads: two or more
Field of view: ±45°
Accuracy: 0.02° rms
Size: 170 × 94 × 88 mm/1.0 kg (optical head) 204 × 152 × 121 mm/2.7 kg (electronics)
Input power: 5 W

The CASS, Coarse Analogue Sun Sensor is a two-axis sensor designed for coarse determination of Sun angle aboard three-axis satellites such as Italsat. The sensor head incorporates four redundant Si cells assembled on a truncated pyramid with light baffle.
Field of view: 2π steradians
Accuracy: 0.1° rms
Size: 86 × 86 × 53 mm/0.230 kg

The ESS, Earth Sun Elevation Sensor, incorporates two 15 μm IR telescopes and two visible light solar split sensors to measure Earth/Sun elevation angles for spin-stabilised satellites. Users include ECS, Hipparcos, Giotto and Inmarsat.
Field of view: ±55° Sun
Accuracy: 0.03° both Earth/Sun, rms
Size: 166 × 150 × 127 mm/1.4 kg total package
Input power: 0.7 W

LACES, the Low Altitude Conical Earth Sensor is a two-axis IR (15 μm thermistor bolometer) Earth horizon sensor designed for pitch/roll determination of three-axis LEO satellites. Users include Eureca and TSS; it also flew October 1992 aboard STS-52 with MSS.
Number of sensors: 2 optical heads with conical scanning equipment
Field of view: 360° pitch, 90° roll
Accuracy: 0.1° rms
Size: 170 × 100 × 116 mm/1.55 kg (optical head) 204 × 172 × 107 mm/2.5 kg (electronics)
Input power: 6 W

The IRES, Infra-red Earth Sensor, is a two-axis IR (15 μm thermistor bolometer) Earth horizon sensor designed for pitch/roll determination of three-axis GEO satellites. Users include ECS, Inmarsat, Olympus, Skynet, Eutelsat, DFH-3, Turksat, Italsat, Orion, Arabsat 2 and Nahuelsat.
Field of view: ±11° pitch, ±15° roll
Accuracy: 0.02° rms
Size: 137 × 190 × 158 mm/3 kg integrated unit
Input power: 2.2 W

IRES with new electronics is replacing the original, improving accuracy to less than 0.02°rms and failure rate to 1,700 × 10^{-9}/h, while reducing cost and mass. Qualified for Sinosat and Eutelsat 3.

The CCD onboard camera is a 288 × 384 pixel CCD TV-standard camera (625 lines/50 Hz) is designed for space operations such as rendezvous/docking, robotic vision, ERA teleoperations and routine monitoring. The optics, processing electronics and LED illumination system are integrated in a single housing.
Field of view: 30 × 40°, with 1.8 × 10^{-13} W lower detection limit
Size: 106 × 100 × 95 mm/1.5 kg
Power input: 2 W camera head, 10 W processing electronics and illumination system.

VERIFIED

Japan

Japan Aviation Electronics Industry Ltd

Current Status

The company provides the strapdown RLG-based inertial measurement unit for Japan's H2 launcher, and the strapdown FOG-based inertial sensor package for JAXA's TR-1A suborbital rocket. It also provides liquid propellant gauges for H2 and H2A. Has developed IMU equipment for re-startable upper stages and for complex multimanoeuvre stage restarts. Researched IMU technologies for start-up from isolated guidance and navigation equations.

NEC

Current Status

The ESG-100A/110A/120A Earth sensors generates roll/pitch errors for geosynchronous satellites, employing an infra-red telescope assembly with two pyroelectric detectors and a scanning mirror with an optical encoder of 0.0025° resolution. Users: ETS 6 (1994), EchoStar (1995), Intelsat 7A/8 (1996), PAS (1996), Tempo (1997), COMETS (1997). Type 2 sensors for multi-altitude satellite designs for complex and frequent altitude changes. Is being developed for commercial applications in earth sensor requirements for remote sensing technologies.
Instantaneous field of view: 1.3° diameter
Accuracy: ±0.03° bias error
Output: 8 Hz
Failure rate: 619 × 10^{-9}/h
Size: 122 × 157 × 124 mm; 1.95 kg
Power consumption: 5.0 W

Following the successful flights of the microprocessor-controlled STT-100B star trackers flown on Astro-C (1987) and Solar-A (1991), Muses-B (1996) will carry a new version, the STT-200A utilising a 510 × 1,000-pixel CCD image sensor for satellite attitude reference.
Field of view: 64°-sq (focal length 61.5 mm, effective aperture 48 mm)
Position accuracy: 16 arcsec max (static/random 3 σ)
Sensitivity: better than +6 magnitude
Exposure time: 62.5 ms-2 s (selectable)
Size: optics 24 × 10 cm diameter, 3.5 kg (excluding hood); electronics 9 × 20 × 27 cm, 3.1 kg plus 9 × 26 × 29 cm, 4 kg
Input power: 16 W (27 W max with cooling device).

VERIFIED

Sweden

Celsius Tech Electronics AB

Current Status

The company specialises in optics and electronics. Space activities include a yaw attitude sensor, optical processing, spectrometers and optical subsystems, wide optical scan and pattern recognition systems and near-visible portions of the spectrum. Company has pioneered several unique ways of adapting fixed optical surfaces to new wavelengths for non-visible portions of the spectrum. The company has expanded research and development efforts to increase its vacuum and high-temperature optical surface and coatings mapping activities with new mechanisms for low cost production.

United Kingdom

Marconi Radar & Defence Systems Ltd (Navigation & Electro-Optic Systems Division)

Current Status

NESD manufactures a range of inertial navigation, guidance, control and stabilisation systems for air, land, sea, subsea and space applications. At its Silverknowes and Granton sites in Edinburgh, NESD employs about 1,100 personnel. Of these, approximately 300 are qualified engineers and scientists engaged in the research, design, development and manufacture of inertial systems and components.

For almost 30 years, NESD has become a specialised supplier of gyroscopes, gyro packages and inertial guidance systems for satellites, experimental payloads and launch vehicles. NESD's first space-related equipment were the SF500 and FE 610 Attitude Reference Units for ELDO's Europa launcher, 1966–69. These were followed by the FE 650 system for the four UK national Black Arrow launchers of 1969–71. Both systems incorporated the Type 125 floated rate integrating gyro. This gyro has since been the mainstay of the company's space units. In 1974, NESD was selected to supply Ariane's Inertial Sensing System. A similar system was built under licence in Japan for NASDA's H1 launch vehicle. A dozen 125s have been running continuously since 1976 as part of an open-ended life test programme; there have been no failures and all remain within specifications.

The 125 gyro has also been employed in the series of gyro package sensors flown on Miranda, IRAS, Exosat, Olympus, Rosat and Spacelab's IPS Instrument Pointing System. Closed loop stabilisation and pointing accuracies of less than 1 arcsec and open loop drifts of less than 2 arcsec/h have been achieved. Further packages have been supplied for NStar, Tempo, Ar0temis, ISO, Superbird and Intelsat 7 communications satellites. Named DIRA, this gyro package had received 63 orders by the end of 1996. NESD is now developing a smaller, low-cost system, DIRA Mk 2, which will be used on scientific, communications and observation missions well into the next century. Technical development of new gyro systems for advanced space applications including surface and orbital vehicles.

NESD continues to supply Ariane 4 with strapdown RLG guidance systems using the Type 162 32 cm-path ring laser gyro and has been requested to look at providing a new guidance system for Ariane-5 based on the new Type 164 24-cm ring laser gyro. In addition to the gyro package and launcher GNU, NESD developed the dynamic monitoring accelerometer package for ESA's Cluster mission, with a specified accuracy of 0.05°.

Assembly is conducted in the 490 m^2 category B cleanroom at Silverknowes and in the 52 m^2 class A 100,000 cleanroom; commissioning of all systems takes place in the latter.

Path length: 32 cm
Mass: 0.5 kg
Power: 4 W
Maximum rate: 400°/s
Scale factor: 0.5 arcsec/pulse
Bias drift: 0.004°/h
Random drift: 0.003°/h½
Scale factor stability: 3 PPM

The company also makes the Type 125 Floated Rate Integrating Gyro:

Mass: 0.45 kg
Size: 83 mm long, 53 mm diameter
Scale factor: 42.24°/h/ma
Maximum rate: 5000°/h
Vertical short term drift: 0.0007°/h
Vertical day to day drift: 0.0033°/h

It also makes a standard generic gyro package:

Resolution: 0.5 arcsec (ISO), 0.002 arcsec (IPS)
Maximum rate: 2°/s
Drift stability: 0.0005°/h in zero-g orbit
Scale factor: 0.1% or better
Noise: 0.2 arcsec/s rms (0–2 Hz) or better
Bandwidth: 2 Hz–10 Hz
Power: 5.5 W per channel
Heater power: 1 W at 50°C, 12 W at –20°C

The DMA Package has the following performance:

Accelerometer measuring range: ±20 m *g*
Accelerometer measuring resolution: 0.01 m *g*
Accelerometer scale factor: 222 mV/m *g* nominal
Temperature sensitivity of scale factor: 210 PPM/°C
Bandwidth: 22 Hz
Input power: ±13.9 V less than 0.65 W.

VERIFIED

Sira Electro-Optics Limited

Current Status

Sira specialises in the design and manufacture of space instrumentation, from the study phase through to the manufacture, integration and test of flight hardware. Prinicipal areas of interest are pointing sensors based on CCD and APS technologies, such as star trackers and sensors for Inter-Satellite Links (ISL), visible and infra-red optical systems for remote sensing, and testing of detectors for electro-optical performance and radiation damage. Sira's SILEX inter-satellite laser link was successfully operated for the first time on 20 November 2001 when the Spot-4 earth observation satellite hooked up to the Artemis communications technology research satellite more than 30,000 km away. Using modulated laser beams data was exchanged at 50 Mbps. Aiming accuracy was one microradian. Other areas where Sira has delivered space hardware include the Exobiology and Radiation Assembly (ERA) core payload for Eureca, and the Digital Electronics for the Compton Observatory's Comptel instrument.

Sira supplied the CCD star trackers for the ROSAT and Astro-SPAS satellites and for the ROSAT wide field camera. More recently, an autonomous star tracker has been developed and is flying on the ESA PROBA small satellite mission. Sira is now undertaking development of next-generation miniaturised star sensors using Active Pixel Sensors. Acquisition and tracking sensors supplied for ESA's SILEX optical ISL experiment, flying on SPOT4 and Artemis. Sira is now developing the acquisition sensor and the optical head electronics for the Inter-Satellite Link Front End (ISLFE) programme.

Sira developed four CCD modules for ENVISAT's GOMOS (Global Ozone Monitoring by Occultation of Stars) instrument: the star acquisition and tracking unit, the two spectrometer detector modules, and the fast photometer for scintillation monitoring of the star image. The company is now developing the detection units for the Aladin wind lidar payload for the ESA ADM-Aeolus Mission.

Sira has vonsiderable experience in the design and development of imaging spectrometer systems, including the initial design of the MERIS Medium Resolution Imaging Spectrometer for ESA's ENVISAT and a number of studies on the PRISM (Process Research by an Imaging Space Mission) and LSPIM (Land-Surface Processes and Interaction Mission) sensors. Sira designed and developed the Compact High Resolution Imaging Spectrometer (CHRIS) as the core payload for the PROBA mission which was launched to a sun-synchronous orbit on 22 October 2001. This patented hyperspectral instrument weighs just 14 kg, consumes less than 9 W, yet provides up to 160 fully programmable wavebands with a ground resolution of 17 m. Further imaging spectrometer instruments are also under study.

Sira provided the experimental infra-red telescope for launch on STRV-2. The 200 cm diameter telescope images a 15 km swath of the Earth's surface with a ground-level resolution of 15 to 30 m. It has successfully demonstrated the detection of aircraft and other targets of potential military interest within the operational mission.

Space Innovations Ltd

Current Status

SIL makes a low-cost attitude control system that is based on the SACE-MDS modular electronics unit, which can drive and interface with various combinations of sensors and actuators.

The unit is called the SACE-MDS Attitude Control Electronics. Its baseline performance can support two rod magnetorquers, two two-axis DSS and one three-axis magnetometer. Space Innovations has developed a new method of controlling altitude orientation through electronic controls and hybrid actuators.

Size: 110 × 170 × 195 mm; 2.5 kg
Powerconsumption: lessthan2.5W(magnetorquers off)
Environmental tolerance: –40 to +50°C, more than 25 krad, EMC MIL-STD-461C, 25 *g* sine 25–80 Hz.

Space Innovations makes the SSM-20/2 Star Mapper, a high-performance three-axis star mapper designed for LEO small satellites but with capabilities and radiation tolerance (approximately 20 krad) suitable for GEO missions. 576 × 770-element passively cooled CCD. The FOV guarantees at least three stars brighter than 6.0.

Accuracy: 20 arcsec (3σ)
Field of view: 12 × 16°
Power requirement: 6 W
Size: 100 × 100 × 120 mm 1.5 kg optical head unit; 40 × 120 × 160 mm 2.2 kg processing unit.

In addition it manufactures the DSS-256 Digital Sun Sensor. The sensor head utilises a pair of orthogonal photodiode arrays behind thin slits. It converts the signal into a 10-bit digital word corresponding to the solar aspect angle. DSS is suitable for gravity gradient, three-axis and spin (up to 10 rpm).

Accuracy: 0.5° up to 5 rpm; 1° 5–10 rpm
Field of view: ±60° azimuth, ±60° elevation
Power requirement: less than 0.8 W total
Size: 38 × 51 × 78 mm 0.150 kg sensor head; 18 × 118 × 159 mm 0.180 kg interface card.

One other Earth/Sun sensor is known as the FES-140 Fan Beam Earth Sensor/FSS-140 Sun Sensor. The configuration is typically two FSS and one FES on each of two triplets, with additional FSS at 90° to detect coning.

Attitude reconstruction accuracy: better than ±0.5°
Field of view: 140 × 1°
Range: 250 to 120,000 km orbits, 1–60 rpm spin
Power requirement: less than 0.7 W
Size: FES-140 22 × 44 × 52 mm 0.85 kg; FSS-140 16 × 44 × 52 mm 0.70 kg

The MFM-3L Fluxgate three-axis Magnetometer has a configuration featuring a cylindrical head with three orthogonal and one redundant sensitive axes. It can be supplied with or without boom.

Resolution: 250 nT
Dynamic range: ±50,000 nT low gain, ±2,500 nT high gain
Orbital range: 300 to 10,000 km
Power requirement: less than 1 W total
Size: 110 × 50 mm diameter 0.40 kg (no boom), interface card 18 × 118 × 159 mm 0.180 kg

The MTR-25 Rod Magnetorquer is suitable for any orbit with perigee less than 1,000 km. Larger units are available.

Magnetic dipole moment: 2.2 × 10^{-5} Wbm
Drive current: 0.1 A
Total power dissipation: less than 1.0 W per rod from +5 V
Size: rod 30 × 60 × 270 mm 0.400 kg; interface card 18 × 118 × 159 mm 0.180 kg.

VERIFIED

United States

Adcole Aerospace Products

Current Status

Adcole Corporation's Aerospace Division specialises in spacecraft Sun attitude sensors, including units for spin-stabilised vehicles, two-axis digital sensors, analogue sensors and gating/switching units. The company's devices have seen service aboard more than 450 satellites, including Tiros, DMSP, GPS, Magellan, COBE, Compton Observatory, Radarsat, Inmarsat 3, Meteor, Amos, XTE, Intelsat 7, MOS, Insat, UARS, EUVE, IUE, Landsat, Satcom, Geotail, Mars Observer, ACTS, GOES, Wind, Polar, Topex, Soho, ISO, Milstar and numerous classified US programmes; they were also selected for GPS 2R, TRMM, SAC-B, Cassini and ACE.

The company's popular Sun angle sensor, the Model 17061, was flown aboard the USAF Block 5 DMSP and NOAA Tiros meteorological satellites, consists of a single-axis Sun sensor head and a package of signal processing electronics. The head incorporates two reticles for Sun angle sensing: a binary code reticle for coarse angle determination and a periodic pattern version for fine. AAP has developed a new generation of sun angle sensors for advanced meteorological satellites. Research has provided some technical development toward integrated systems for multifunction satellites.

Max number of sensors: 1
Field of view: ±50°
Accuracy: 0.05°
Size: 109 × 64 × 28 mm/0.322 kg (sensor) 102 × 86 × 51 mm/0.517 kg (electronics)
Input power: +5 V DC at 65 mW; +28 VDC at 0.7 W
Operating range: –5 to +45°C
Acceleration tolerance: 25 *g* (40 *g* shock, 1 ms)

The Model 17140/28520 is a single-axis analogue Sun sensor used for solar array pointing, instrument platform pointing, and initial stabilisation from spinning to three-axis stabilised. Programme involvement includes GOES, Intelsat 5/7, Arabsat, Superbird, Insat, MOS, ETS 5, SCS.

Max number of sensors: 3 (6 for redundant electronics)
Field of view: ±10 to 60° acquisition, ±1/2 to 20° linear range, ±8–60° insensitive axis
Null accuracy: ±0.01 to 1.5°
Size: varies with specific model
Input power: varies with specific model
Operating range: –40 to +60°C
Acceleration: typically 20 *g* rms

The Model 18656 is a high resolution (0.1°) Sun angle spin sensor system. Users include Geotail, Wind, Polar, GPS 2R, Lockheed Martin Satcom series and classified satellites.

Max number of sensors: 2
Field of view: 128° (each sensor)
Accuracy: 0.1° (incidence is less than or equal to 40°); 0.25° (incidence is greater than or equal to 40°)
Size: 66 × 33 × 25 mm/0.109 kg (sensor) 104 × 58 × 94 mm/0.726 kg (electronics)
Input power: +28 V DC at 400 mW
Operating range: –20 to +60°C

The Model 20910: two-axis fine digital sensor system designed originally for NASA's Solar Max and comprising a two-axis sensor head with processing electronics for coarse and fine Sun angle sensing. Users includeTopex, Radarsat, JERS 1, ISO and EUVE.

Max number of sensors: 1
Field of view: 64° each axis
Accuracy: 0.017°
Size: 84 × 109 × 24 mm/0.317 kg (sensor) 198 × 114 × 51 mm/1.044 kg (electronics)
Input power: +28±7 V DC; 2.8 W
Operating range: –10 to +50°C
Acceleration tolerance: 25 *g* (50 to 200 *g* for 200 to 600 Hz shock)

The Model 21730/20850 is a two-axis precision digital sun sensor employed as the primary pointing sensor for Solar Max. This Fine Pointing Sun Sensor (FPSS NASA Standard) operated successfully throughout Solar Max's life (1980–89). The redundant system comprises two sensors co-aligned on a mounting bracket and two electronics units. A similar sensor, offering greater resolution, was supplied for ESA's Soho.

Sensors/electronics: 1
Field of view: ±4° acquisition, ±1/2° linear
Accuracy: 5 arcsec
Resolution: 0.11 arcsec (0.025 arcsec Soho)
Size: 123 × 84 × 36 mm/0.5 kg (sensor) 165 × 90 × 120 mm/1.0 kg (electronics)
Input power: 28 V DC
Operating range: –10 to +50°C
Random vibration: 14.6 *g* rms

The Model 27290/27460 is a high-accuracy wide field of view digital Sun sensor used for yaw measurement. Autonomous yaw update can be achieved by using a Sun crossing pulse generated when the Sun is in the yaw plane. Designed and qualified for Milstar; also flown on SDS and classified programmes. Similar units flown on NOAA Tiros and USAF DMSP.

Sensors/electronics: 1
Field of view: ±50° measurement direction, ±50° insensitive axis
Accuracy: ±0.017°
Size: 168 × 87 × 26 mm/0.8 kg (sensor) 152 × 152 × 62 mm/1.2 kg (electronics)
Input power: 28 V DC
Operating range: –40 to +60°C
Random vibration: 18.6 *g* rms.

VERIFIED

AlliedSignal Aerospace Company (Government Electronic Systems)

Current Status

AlliedSignal Aerospace has merged with Honeywell and its name is absorbed into that of its new business and is itself now known as Honeywell. GES designs, develops and produces gyroscopic equipment, accelerometers, digital computers and other electronic and navigation devices. These include spacecraft pointing and control equipment for Hubble Space Telescope, star scanners such as that for Galileo, and star sensor assemblies designed for spacecraft attitude control using very faint stars. Attitude control equipment using an RW/MW is used aboard satellites such as GPS and various orbiting observatories and technology satellites. It is developing a new generation of angular rate sensors and is conducting research in a life tolerant rate gyro system for long duration storage in adverse temperatures and under varying accelerations. Utilising lasers, fibre optics and solid-state technology, these gyroscopes will be smaller, less expensive and more reliable than their spinning mass counterparts. It has developed a low-cost MW assembly and a three-axis RLG assembly for Iridium. Other efforts involve the application of advanced technologies such as charge coupled devices and Alexandrite lasers.

Engineering and manufacturing facilities and field support offices include the primary facilities in Teterboro, New Jersey and major operations in Boyne City, Michigan; Mishawaka, Indiana; Orlando, Florida; Westlake Village, California. GES also has one subsidiary company, the Cheshire Corporation in Cheshire, Connecticut, a manufacturer of precision rate gyros. Research into a linked altitude sensor and control electronics for development of new position fixing systems. Developed life tolerant systems for new gyro-sensor links.

The Trihex single-fault tolerant flight control and inertial measurement unit debuted aboard Delta 2 in December 1995. Six RL-20 ring laser gyros and six accelerometers are contained in a triply redundant modular architecture. Each lane contains two sensor channels, a 1750A processor and I/O interfaces. Two lanes have a 1553B bus for vehicle data and the third incorporates RS-422 for GSE.

GES builds single- and double-gimbal CMGs covering a stored angular momentum range of 7 to 4,750 Nms and a torque output range of up to 4,600 Nm. Reliability of these units is indicated by accumulated life testing approximately 400,000 h. GES is under contract to build five double-gimbal CMGs for the Space Station.

Recent spacecraft using GES reaction wheels include Insat, DSP and GPS. Over the past three decades, the Division has delivered approximately 600 RWs with a cumulative operating time of approximately 7 million hours in orbit. The small wheels range over 0.4 to 136 Nms with torque over 0.028 to 0.353 Nm. The large MWs range over 814 to 2,440 Nms with 0.177 Nm torque.

GES provided the precision rate sensor system for Hubble's Pointing Control System, capable of measuring spacecraft motion to a resolution of 0.00025 arcsec. This gyroscopic inertial instrument utilises a hydrodynamic gas bearing to achieve spacecraft pointing to 0.007 arcsec and support mission life requirements of 15 years.

Innovations include on-orbit performance optimisation over a wide focus range, and 'apriori' star crossing location and identification to minimise hardware channels.

Galileo's star scanner provides absolute position information to the attitude control system. Designed to operate in Jupiter's intense radiation fields, which comprise primarily high-energy electrons and protons, the optics and photomultiplier tube are configured to minimise fluorescence and Cerenkov effects and the electronic components to minimise ionisation and displacement damage. The scanner can detect the 200 brightest stars at a 20°/s scan rate. The focal plane's V slit modulates the star signal and the clock and cone angles of each star are uniquely determined from the time relationships between each star crossing.

Specifications

Field of view: 10°
Accuracy: 60 arcsec two-axis position (1σ)
Sensitivity: 200 brightest stars
Power consumption: 3 W
Size: 368 × 321 × 251 mm without sunshade, 482 diameter × 975 mm with sunshade; 20 kg
Inputs: commands for power, threshold, channel; 400 Hz AC 28 V DC power
Outputs: star transit pulse; star intensity signal
Operating range: –10 to +50°C.

UPDATED

Ball Aerospace and Technologies Corporation

Current Status

Formerly known as Ball Aerospace Systems Group, Ball Aerospace and Technologies Corporation was created in 1988 to provide electro-optical, cryogenic and tactical subsystems. The hardware includes science instruments such as HST's second-generation ST imaging spectrograph, Near IR camera and advanced camera, plus the Corrective Optics Space Telescope Axial Replacement (COSTAR); multispectral sensors and imaging; attitude sensors; pointing, tracking and electromechanical subsystems; and stored cryogens and closed-cycle refrigerators. Ball provides the Shuttle Orbiter's power reactant storage assembly pairs of cryogenic LOX/LH_2 tanks. It is building three Stratospheric Aerosol and Gas Experiment (SAGE) sensors under a US$20 million contract for NASA. It provided the antenna, star tracker and instrument integration for NASA's SWAS Small Explorer, plus a star tracker for the similar WIRE satellite. BASG has developed specially insulated pressure vessels for long duration missions in hostile environments.

Ball has extensive experience in attitude sensors for rockets, satellites and ground-based systems; more than 2,500 star trackers and scanners have been delivered and flown more than 350 times. Recent programmes include the NASA standard fixed head star tracker flown on Landsat 4/5, HST, UARS and Compton Observatory; the solid-state star scanners flown on Pioneer-Venus, Magellan and the Inertial Upper Stage; the Shuttle Star Tracker, flown on every Orbiter and now adapted as the CT-611 to charge-coupled device (CCD) technology. The Division has completed research, development and engineering for CCD and charge-injection device (CID) sensors for space target acquisition, tracking and vehicle range/attitude determination. The technology applies also to ground-based astronomic position determination. Completed hardware includes the Retroreflector Field Tracker and the Rendezvous/Docking Tracker (two flights each on Columbia and Endeavour), and RME's Coarse Acquisition Sensor.

BASG operates 190 m^2 of production rooms and 95 m^2 of alignment and test areas dedicated to attitude sensors. A high-accuracy test facility is contained in a class 10,000 cleanroom for testing star sensors to accuracies less than 0.03 arcsec. Thermal vacuum test facilities provide 10^{-8} torr and environmental vibration testing is available with up to 133 kN shakers.

Ball makes the CS-203 solid-state star scanner.

Specifications

Field of view: 4.6° elevation, 5.8° azimuth
Sensitivity: +1.6 to –1.4 magnitude at 1° spin
Spin rate: 0.2–1.0°/s
Accuracy: 18 arcsec
Power consumption: 7 W
Size: 16.8 × 18.0 × 24.1 cm (no shade); 5.5 kg

The first production units of Ball's CT-601 High Accuracy Star Tracker are flying on BMDO's MSX and NASA's SWAS Explorer satellites. It will be used by Space Imaging's satellite.

Field of view: 8 × 8°
Accuracy: 3 arcsec (1σ)
Sensitivity: +1 to +6 magnitude
Numbers of stars tracked: five
Power consumption: 12 W, 22–34 V DC
Size: 17.8 cm diameter, 29.5 cm long; 8.2 kg (no shade)
Output: MIL-STD-1553
Environment: –30 to +50°C

The CT-611 Space Shuttle star tracker is the world's only man-rated reusable general purpose attitude sensor. It emulates the original image dissector tube tracker but has 420 parts in contrast to 1,100. NASA has certified it for 100 missions. CCD is 512 × 512 pixels.

Field of view: 10 × 10°
Accuracy: 15 arcsec (1σ)
Sensitivity: –1 to +3.9 magnitude
Numbers of stars tracked: one
Power consumption: 12.5 W, 22–34 V DC
Size: 21.2 × 16.8 × 28.6; 8.26 kg (no shade)
Output: Manchester biphase
Environment: 10 to 60°C

The CT-621 Wide Field Star Tracker FOV is enlarged at the cost of high-end performance. It has the same electrical/mechanical configuration as the CT-601. The wider FOV improves roll determination by allowing greater separation of tracked stars.

Field of view: 20 × 20°
Accuracy: 11 arcsec (1σ)
Sensitivity: +0.1 to +4.5 magnitude
Numbers of stars tracked: five
Power consumption: 12 W, 22–34 V DC
Size: 17.8 cm diameter, 21.6 cm long; 5.4 kg (no shade)
Output: MIL-STD-1553
Environment: –30 to +50°C

The CT-631 Lightweight star tracker repackages the basic CT-600 electronic design with wide field lens into a small unit intended primarily for the miniature satellite market.

Field of view: 20 × 20°
Accuracy: 20 arcsec (1σ)
Sensitivity: +0.1 to +4.5 magnitude
Numbers of stars tracked: five
Power consumption: 12 W, 22–34 V DC
Size: 12.7 cm diameter, 12.7 cm long; 2.3 kg (no shade)
Output: MIL-STD-1553
Environment: –30 to +50°C.

UPDATED

Barnes Engineering Division

Current Status

A division of EDO Corporation, Barnes Engineering was founded in 1952 to develop and produce IR detectors, instruments and systems. It has delivered more than 1,400 mission-critical systems, including approximately 1,150 infra-red Earth sensors and 110 science payloads. Space systems currently in production or operating on-orbit include: scanning Earth sensors, miniature static Earth sensors, GEO static Earth sensors and infra-red detectors. Current programmes include Earth sensors for Hughes (HS-601), Lockheed Martin Astro Space (DMSP, GPS, Landsat, Tiros) and miniature Earth sensors for Lockheed Martin (Iridium), OSC (Orbcomm) and Space Systems/Loral (Globalstar).

Scanning radiometers for atmospheric observation connected to limit scanners for discriminating between sensor levels.

The company's Static GEO Earth Sensor, model 13–405 has been developed for the Boeing HS-601 + CTA's Indostar, combining horizon crossing indicators used for attitude control during the spin-stabilised transfer orbit, with a static sensor suitable for 15-year life. First used with GPS.

Specifications

Accuracy: ±0.1° (3σ)
Reliability: 0.998 5 years, 0.996 10 years, 0.994 15 years
Detector: thermopile
Size: 16.0 cm diameter, 37.3 cm long; 4.72 kg
Power consumption: 3.2 W
Redundancy: full internal, transfer orbit and operational static sensors

The single-axis miniature static earth sensor measures one attitude component of a three-axis satellite with respect to Earth's IR horizon. Multiple sensors can be configured for high accuracy over a limited manoeuvring range. It employs the same large-area thermopile IR detector technology as classic GEO static sensors, but with order of magnitude improvements in mass, power, size and cost. There are two standard versions: CHS coarse horizon sensor; FHS fine horizon sensor. A US$15 million contract was awarded in 1993 by Lockheed for Iridium, in 1994 for approximately US$1 million by OSC for Orbcomm, and in 1995 for US$10.5 million (including the Sun sensor specified below) by DASA for Globalstar.

Accuracy: CHS 1.0° (3σ) over 22° operating range; FHS 0.1° (3σ) over 4° (accuracy 0.08° at optimum point)
Detector: IR thermopile
Size (cm): CHS 4.1 diameter 5.6 cylinder, 0.11 kg; FHS 16.5 diameter 8.1 cylinder, 0.28 kg
Power consumption: CHS less than 30 mW; FHS less than 0.30 W

The analogue sun sensor assembly, model 13–517 assembly provides medium accuracy Sun position sensing for three-axis satellites in any orbit. 13–517 provides dual-axis Sun tracking coverage over a central 128° square region and single axis over 168°. Fully redundant and self-calibration maintains accuracy. A contract was awarded in 1995 by DASA for Globalstar.

Accuracy: 0.3° (3σ EOL)
Reliability (15 years): 0.997 non-redundant, 0.99999 redundant
Size (cm): 13.0 × 6.1 × 4.8; 0.43 kg
Power consumption: 0.3 W satellite conditioned.

VERIFIED

Condor Pacific Industries, Inc

Current Status

On 22 October 2002 BAE Systems announced that it had agreed to the purchase of Condor Pacific Industries for US$58.5 million. Established in 1964, Condor Pacific manufactures rate gyros, tuned gyros, solid-state resonant gyros and the world's smallest multisensor (two gyros and two accelerometers). The T-100 tuned gyro is used in inertial reference units for space applications. CPI has developed integrated gyro systems for advanced space vehicle control mechanisms. Users include Hughes' HS-601 series (Aussat, Galaxy, Astra, MSAT, UFO, DBS, Solidaridad). Rate gyros and pendulous accelerometers for advanced positioning systems uses new linear state gyro software interaction. Specifications are given in gimballed, strapdown and high-rate strapdown order (series 1/2/4):

Specifications

Mass: 100 g for flange mounted
Dynamic range: ±20/±100/±300°/s continuous
Random drift: 0.01 to 0.05/0.02 to 0.05/0.03 to 0.1°/h 1σ
G sensitive drift: 5/5/10°/h/g
Non-G sensitive drift: 20/25/30°/h
Linearity: 0.05/0.05/0.05%
Run power: 2/2/2 W
Typical run up time: 10/10/10 s
Vibration: MIL-STD-810
Operating temp: 0 to 85/–54 to +85/–54 to +85°C
Shock hard mounted: 50/300/300 *g*.

VERIFIED

Delco Systems Operations

Current Status

Delco Systems Operations is part of the Delco Electronics Corporation, a subsidiary of GM Hughes Electronics Corporation, and responsible for the company's space activities. Delco's Magic 3 series of guidance computers is employed in the Delta and Titan family guidance systems and was adapted for the USAF/Boeing Inertial Upper Stage. The M352 Delta Inertial Guidance System (DIGS) computer has been in service since 1975 and was retained for Delta 2 and Delta 3 until the latter is replaced by the Delta 4. Delco's Universal Space Guidance System provided Titan 3 control since the early 1970s and was integrated into the Titan 2 ICBM in the Rivet Hawk vehicle update of 1975. An improved USGS was carried by the Titan 2–4 space launchers, although improvements were introduced for the upgraded Titan 4 series. New derivative versions have been retrofitted into the last of the Titan 4 family due for retirement in 2005.

Delco's Space Inertial Reference Unit is a highly accurate fault-tolerant three-axis inertial reference. The Hemispherical Resonator Gyro (82 cm^3, 0.14 kg) has the highest performance to size ratio of state-of-the-art gyros. The system contains redundant sensors and electronics that enable continued performance to specification after any combination of one gyro, one processor, one power supply or one accelerometer failure. SIRU was supplied for Cassini, NEAR, Clark and a proposed Martin Marietta telecom satellite. Guidance platforms for advanced launch vehicles as part of NASA's Space launch Initiative developed from Delta systems. Delco is researching integrated stabilisation systems for use in high performance, high-g, rocket systems.

Specifications

Size: 20.3 × 9.5 × 30.2 cm; less than 5 kg
Power requirement: 21 W
Bias stability: less than 0.015°/h 3σ
Angle random walk: less than 0.001°/h1/2 3σ
Scale factor stability: 100 PPM 3σ
MTBF: more than 6 million hours

The M362S derivative of the Magic 3 series is utilised as the onboard computer for the USAF/Boeing IUS Space Shuttle and Titan 3/4 upper stage. Each stage carries two computers for redundancy. The memory subsystem offers 66,536 16-bit words of high-speed CMOS RAM, electrically alterable and non-volatile via a self-contained backup battery, producing a central processor computational performance more than 10^6 operations/s. A memory error detection and correction system is used, correcting single-bit errors in real-time without affecting throughput. The unit is sealed in a dry nitrogen atmosphere and passively cooled; a single cover provides access to the 25 modules for individual replacements.

Size: 36.6 × 36.6 × 15.2 cm; 23 kg
Power requirement: 180 W at 28 V DC.

UPDATED

Honeywell Inc

Current Status

Space Systems' reaction wheels have been provided for Landsat 7, Topex, Explorer platform, HST, Milstar, TDRSS, Magellan, Compton Observatory, Globalstar; spin bearing and scan drive assemblies for Galileo, and IMU and star sensors for DMSP. The GG1308 ring laser gyro flew on Clementine in 1994. Advanced ring laser gyro technology developed for new platform requirements.

The Miniature Inertial Measurement Unit is being supplied for the Aerospatiale Spacebus and a variety of other commercial, military and scientific spacecraft. The electronics are hardened to approximately 100 krad.

Specifications

Range: ±375°/s, ±25 *g*
Size: 0.041 m^3/3.45 kg
Power: 33 W
Computer: Honeywell 1750A Generic VISIC
Accelerometers: three Sundstrand QA3000
Gyros: three Honeywell GG1320C RLG
Reliability: more than 173,000 h MTBF
Life: 15 years

Honeywell's Space Systems operation has qualified the Fault Tolerant Inertial Guidance and Navigation System for the next generation of launchers and upper stages. It provides detection and isolation of up to two gyro and accelerometer failures, and improved accuracy using the output of all sensors to compute body rates. Processing and redundancy management is performed by a dual self-checking pair of 1750A radiation-hardened processors arrangement to provide protection against single event upsets and hardware processor failures. With two accelerometer and two gyro failures, the IMU provides 0.999999 reliability. It was adopted by the USAF/Boeing IUS beginning with IUS-27 in 2003.

Size: 0.047 m^3/33 kg
Power: 135 W
Accelerometers: five Sundstrand QA3000
Gyros: five Honeywell GG1320 RLG

Honeywell's IFOG Interferometric Fibre Optic Gyro was provided for NASA's XTE, the first IFOG used on a NASA satellite.

Performance: 1–10°/h low, 0.01–1°/h medium, 0.001–0.01°/h high, 0.0001–0.0001°/h very high

The Laser Inertial Navigation Unit was introduced on Centaur in December 1991.

Size: 50.8 × 30.5 × 22.9 cm; 31.75 kg
Power: 105 W average
Accelerometers: three Sundstrand QA3000
Gyros: three Honeywell GG1342 RLG
Processors: two Performance Series Corporation 1750A 1 MIPS
Reliability: 25,000 hours MTBF ground benign.

Control Moment Gyroscopes from Honeywell's Phoenix operation features high momentum-to-weight ratios, low-torque ripple and emitted vibration, long life and proven performance.

Honeywell's Arizona facility provides six different versions of reaction/momentum wheel assemblies for commercial, military, scientific and special purpose applications ranging from LEO to GEO. Masses range from 3.8 to 19.7 kg: power requirements 14 to 44 V.

VERIFIED

Hughes Danbury Optical Systems

Current Status

The HD-1003 miniature star tracker started deliveries in 1996 for NASA and DoD programmes; earlier models were ordered for Topex/Poseidon. System performance since in-orbit operations began has exceeded specifications. New technology development initiatives have stimulated improvement to existing models and co-operative research projects have linked HDOS to spacecraft design teams. Several new devices linked to the technology developed for the HD-1003 star tracker have been marketed. The all-solid-state HD-1003 employs a frame transfer 512 × 1024 CCD to obtain the required accuracy, sensitivity and stability over the FOV. Lower cost models are available for missions requiring less pointing accuracy; other specifications remain as below.

Specifications

Field of view: 8 × 8° or 20 × 20°
Accuracy: up to 2.5 arcsec rms. Accuracy per star at 6.0_{mv}: 6 arcsec pitch/yaw, 40 arcsec roll
Update rate: 10 Hz
Number of stars tracked: 6 maximum
Size: 38 × 15.7 × 11.2 cm; 3.3 kg (with light shade)
Detector: CCD matrix of 512 × 1,024 pixels
Input power: 11 W
MTBF: 10^6 hours
Environment: –20 to +60°C, 12 *g*.

UPDATED

Humphrey Incorporated

Current Status

The company was founded in 1946 and incorporated in 1951 to design and manufacture precision electromechanical instruments for defence and aerospace applications. HI played a leading role in developing new technologies for the instrument industry. Several innovative design solutions to electromechanical operations, fine-scale positioning devices and drift-rate monitors were developed. Product lines include gyroscopes, rate sensors, pendulous and linear accelerometers, magnetometers and switches. The RT02-0800 series three-axis solid-state rate sensor consists of three single axis rate transducers packaged with electronics in a 6.6 × 5.8 cm unit weighing 0.4 kg. Rates up to 5,000°/s can be accommodated; life is in excess of 10,000 hours. HI has researched integrated systems combining electrical and mechanical technologies for improved compliance. These have been aimed at solving problems with operating techniques.

UPDATED

ITHACO Incorporated

Current Status

ITHACO designs, manufactures and sells attitude determination and control components and systems for spacecraft. In recent years the company has developed adaptive attitude sensor and control systems operating in off-nominal applications such as long duration missions where sensor targets change and convert orientation login from one format to another. Their Earth horizon sensors, electromagnetic torquers (TORQRODS™), reaction and momentum wheels, magnetometers, Sun sensors, and associated electronics have been used on more than 140 US satellites and on many international spacecraft. Specialised test equipment and post-delivery support are also offered to meet customer specific needs. Random test equipment installed for quality control used for new systems development and product upgrade. Test projects have framed the technology for advanced colloidal attitude reference systems and these have been integrated with adaptive sensors for advanced spacecraft specifications.

The T-Scanwheel is a momentum/reaction wheel with an integral high accuracy scanning Earth sensor. The wheel portion provides both angular momentum and control torque, while the sensor portion provides precise two-axis attitude information. The wheel consists of a precision balanced aluminium flywheel driven by an ironless armature brushless DC motor and suspended by a pair of angular contact ball bearings. The T-Scanwheel optics consist of a scan mirror coupled to the flywheel shaft, a parabolic objective mirror, a 14.25 to 15.65 μm spectral filter and an infra-red detector. The scan mirror reflects radiation causing the Field Of View (FOV) of the T-Scanwheel to sweep a 45° half apex angle cone in space. The signals produced by the alternating space-Earth and Earth-space transitions are amplified and sent to a separate electronics box for processing into phase and chord or pitch and roll attitude information. The T-Scanwheel is available in two sizes, A and B, with the following specifications:
Operating speed (rpm): 500–3,000
Angular momentum (Nms): 2.25/9.75 at 3,000 rpm
Available reaction torque (mN-m): more than 13/31
Mass (kg): T-Scanwheel; 3.2/5.8; electronics 2.6
Steady state power: less than 9/15 at 3,000 rpm.

The T-Wheel series of fully capable momentum wheels and bi-directional reaction wheels is built specifically for attitude control of small- and medium-sized spacecraft to provide both reaction torque and angular momentum storage for either three-axis reaction control systems or momentum bias stabilised systems. The T-Wheel is available in three sizes, A, B and E. The Type A and B designs can be augmented with horizon sensor optics to provide both attitude determination and attitude control (see T-Scanwheel). The Type E design provides a significant increase in momentum storage capability and is primarily targeted for medium to large satellites. The following specifications are given for the Type A/B/E wheels, respectively:
Angular momentum (Nms): 4/16.6/50
Available reaction torque (mN-m): more than 20/40/300
Mass (kg): T-Wheel 2.55/5.1/10.5; electronics 0.9/0.9/3.6
Steady state power at max speed (W): less than 9/17/40
Peak power at max speed (W): less than 25/50/280

The Conical Earth Sensor is a scanning horizon sensor used to gather attitude information for Earth orbiting satellites. The most common application utilises one or more CESs to provide pitch and roll attitude data for three-axis stabilised Earth oriented satellites. The CES can also support autonomous navigation and attitude determination for spin-stabilised satellites. The CES can be used from low Earth orbits to above geosynchronous orbits, as well as during transfer orbits. The sensor head is usually mounted with its spin axis located in either the satellite roll/yaw plane or the pitch/yaw plane, while the associated electronics is conveniently mounted inside the satellite. Output signals can be presented in either an analog or a digital format.
Accuracy (deg): less than 0.10° post-processed
Mass (kg): SH 1.1; SE 1.9
Power (W): 8

The ITHACO Torqrod is an electromagnet designed to provide complete momentum management on an Earth orbiting satellite (even at geosynchronous). Typically, one Torqrod is aligned with each axis of the satellite. Torqrods should be mounted away from any instruments sensitive to magnetic fields and should also be separated from one another in order to avoid cross coupling effects. Numerous Torqrod sizes have been developed, with the following range of specifications:
Dipole magnitude (Am^2): 12–790
Length (cm): 16–55.1
Mass (kg): 0.9 to 30
Power (W) at max linear voltage: 0.7 to 8.8

The ITHACO two-axis (IM-102) and three-axis (IM-103 and IM-203) magnetometers produce continuous vector measurements of the Earth's magnetic field as the satellite orbits the Earth. These measurements are used by the satellite's magnetic attitude control system to generate the control commands that stabilise the satellite attitude and/or spin rate. The magnetometer design incorporates both the sense coils and the signal processing electronics into a single box. Both the IM-102 and the IM-103 require regulated input voltages, while the IM-203 incorporates a DC/DC converter and can accommodate an unregulated input voltage. The following specifications are for the IM-102, IM-103, and IM-203 units, respectively:
Field measurement range (gauss): ±0.6/±0.6/±1.0
Mass (g): less than 220/227/500
Volume (cm^3): 165/233/419
Input voltage (V): +5/±15/+28±6.

The Horizon Crossing Indicator is intended for use on a spinning spacecraft for the purpose of providing attitude and timing information. While a single HCI is sufficient for a circular orbit, either two HCIs or the use of a single Steerable HCI is effective for an elliptical orbit. The sensor head contains infra-red optics which respond to the infra-red energy reflected from the Earth as compared to cold space. The sensor electronics converts the detected signals into pulses that are accurately referenced to the Field Of View (FOV) and the Earth's infra-red horizon.
Accuracy (deg): less than 0.10° (post-processed)
Mass (kg): HCI – 0.65; SHCI – 2.2
Power (W): HCI – 0.7; SHCI – 2.5 static/10.0 stepped.

VERIFIED

Kearfott Guidance & Navigation Corporation

Current Status

Kearfott is a subsidiary of the Astronautics Corp of America, specialising in space-qualified inertial guidance and inertial reference systems. Employees total almost 2,000. The company has now produced more than 100,000 inertial devices and more than four million precision components. Principal space hardware covers one-, two- and three-axis attitude control and pointing systems, star trackers, rotary components and cryogenic systems. In its New Jersey facility, Kearfott also produces a line of gimballed and strapdown GNC systems, using both ring lasers and dynamically tuned gyros (DTGs), test equipment and inertial sensors. In North Carolina and Mexico, it designs and manufactures precision rotary components, actuators and electronics. Developing new nano-sized guidance sensors and tracking equipment. This technology underpins a new generation of precision guidance packages which Kerfott have been researching since the early 1990s.

Kearfott initially furnished floated rate integrating gyros to Vela Hotel, OAO, Mariner, Nimbus 2–4, Surveyor, Viking, Skylab and a three-axis rate system (GRA) for classified satellites. These programmes involved approximately 180 gyros, accruing more than 360,000 hours in space without failure.

In the early 1970s, Kearfott developed the first of the Space-qualified Kearfott Inertial reference Unit (SKIRU) three-axis attitude reference systems employing the Gyroflex DTG, which has 6.2 million hours of data on 82 life test units. SKIRU is used on Galileo, DSCS III, Voyager, TDRS, Milstar, XTE, TRMM, EOS AM 1 and AXAF-I, plus classified satellites. 70 SKIRUs have been delivered. Some systems have 45,000 to 110,000 hours in space without failures; total is 1.9 million operating hours. SKIRU has also been developed into a fine pointing system, utilising the same basic components with temperature control and circuit modifications to permit very accurate space pointing over long periods. SKIRU I to V are two-gyro single-string units; the D-II is a three-gyro internally redundant unit that is interchangeable with NASA's DRIRU II Dry Gyro Inertial Reference Unit.

Specifications

Performance: 0.002–0.009°/h, 3σ, 8 h
Size (cm): 17.8 × 22.8 × 9.5 IV; 17.8 × 22.8 × 14.2 IV-H; 17.8 × 22.8 × 15.0 V; 17.8 × 35.5 × 17.8 D-II
Mass (kg): 3.2 IV; 5.0 IV-H; 5.0 V; 12.7 D-II
Power (max quiescent; W): 25 (typical 17) IV; 25 (typical 22.5) IV-H; 26 (typical 17) V; 26 D-II

TARA, the Two-Axis Rate Assembly utilises an IRU comprising a two-axis CONEX MOD I/S DTG. 15,000 CONEX gyros have been built. TARA's modular concept allows two packages to be used for three-axis control, while three would provide complete redundancy. It was developed and qualified by early 1993 under a company-funded programme. Users are TOMS-EP, Lewis, MSTI 2/3, Rocsat 1 and Komsat. The MOD I/S CONEX is also being delivered as part of a three-axis IRU for Hughes' HS-601.
Performance: 0.05°/h
Size (cm): 16.0 × 14.3 × 8.6
Mass: 1.8 kg
Power (max quiescent; W): 11 (typical 7) operating from unregulated DC input.

UPDATED

Litton Applied Technology

Current Status

Applied Technology, a division (ATD) of Litton Industries, was founded in 1959 to develop and produce microwave amplifiers, reconnaissance and surveillance receivers and countermeasure jammers. The Applied Technology Advanced Computer (ATAC) was developed in the 1970s; the company now offers space processor designs with ATAC, 1750A and RISC architectures to optimise radiation hardening and fault tolerance with maximum thermal survivability. ATAC is now performing attitude and articulation control functions for NASA's Galileo Jupiter orbiter; it was also carried by the Magellan Venus orbiter. LAT developed attitude control and off-axis articulation systems for large satellites and some classified programmes. LAT has also started research into highly discriminating sensor coupling for high-accuracy and fine pointing actuators. High fault tolerance components with high impact resistance and radiation hardening have been developed for new performance requirements.

Litton's Guidance and Control Systems division develops and produces navigation and guidance systems for space launchers, missiles and tactical aircraft. Some 1,900 people are employed in the division's Southern California locations near Los Angeles, and another 850 in its main manufacturing facility in Salt Lake City, Utah.

Litton's first space launcher inertial guidance system, the LN-100L, debuted on Lockheed Martin's LMLV in August 1995. A product line derivative of the LN-100 strengthened for the space launch environment, it comprises three Zero-Lock laser gyros, an A-4 triad of miniature accelerometers, sensor electronics, a digital processor and two power supply modules.

Specifications

Performance: 1.48 km/h, 0.76 m/s rms, 0.05° azimuth rms, 0.015° pitch/roll rms
Size: 17.8 × 17.8 × 24.9 cm; 8.4 kg
Power: 27 W running; 27 V DC
Operating ranges: 20 *g* all axes, unlimited attitude, approximately 400°/s three-axes, approximately 1,500°/s^2 three-axes
Outputs: two dual MIL-STD-1553B, three RS-485/RS-422, one bidirectional SDLC, one input-only SDLC
Navigation processor: 32-bit 80960, Ada
Operational service life: 10,000 h or 15 years
Environmental tolerance: MIL-E-5400, sea level to vacuum, –54 to +71°C, vibration (random) 22.6 *g* rms endurance and 8.1 *g* rms performance, vibration (sine) ±5 *g* 5 to 2,000 Hz, shock 21 *g* 40 ms handling and 1,130 *g* pyro, acoustic 140 dB
Zero-Lock laser gyro: 18 cm non-dithered, 0.75 arcsec/pulse resolution (0.003 enhanced), 0.003°/h bias repeatability, 0.2 PPM scale factor stability, less than 0.0015°/h½
A-4 Accelerometer Triad: 25 μ*g* bias repeatability, 120 PPM scale factor, 2 arcsec alignment.

UPDATED

Lockheed Martin Missiles and Space Company

Current Status

LMMSC is the largest producer of US Earth sensors. In 1964, it designed the first sensor for spin-stabilised satellites to use the 15 μm filter (flown 1965). More than 350 sensors for more than 20 different types of spinning spacecraft (0.16 to 180 rpm) have since been delivered. Conceptual work on a sensor for three-axis satellites began in 1968, incorporating an oscillating mirror approach now used on all LMMSC scanning sensors. In 1970, the concept of a single mirror with two detectors was proposed, permitting the simultaneous determination of pitch and roll for GEO satellites. The NOHS and PESA sensors (see below) are current examples.

Lockheed began development of the Wide Angle Sun Sensors (WASS) in 1974. Users include HST. The Intermediate Sun Sensor (ISS) is used for high-accuracy Sun-pointing experiments. This technology is suitable for high specification requirements on classified military satellites which require five pointing angles. Applied also to specialised mission requirements for classified military projects. The technology has been applied to astronomical telescopes.

The Null Operating Horizon Sensor is a scanning mirror Earth sensor for application aboard three-axis stabilised GEO spacecraft, the NOHS provides pitch and roll information. Versions have flown on Intelsat, GOES and DSCS 2. An upgraded version, the VLSIC Earth Sensor, combines the NOHS optical system with new VLSIC electronics for reductions in size, mass and input power.

Specifications

Instantaneous field of view: 1.65° square
Accuracy: ±0.04° (3σ)
Range: ±6.4° pitch, ±2.24° roll
Size: 10.4 × 13.8 × 16.5 cm (11.5 cm VLSIC)
Mass: 1.6 kg (1.36 kg VLSIC)
Input power: 4 W (3 W VLSIC)
Output rate: 4/s
Output signals: serial digital 0.01° LSB
Failure rate: 0.8 × 10^{-6}/h.

The Precision Earth Sensor Assembly, or PESA, is designed to handle a large dynamic range in both vehicle attitude and orbit altitude. Its three detectors compute the Earth's position in parallel to generate a serial-digital output word containing the information from all three, linear to within 0.01° over up to ±16°.

Instantaneous field of view: 1.10° square
Accuracy: ±0.03° (3σ)
Range: ±16°
Size: 11.4 × 17.0 × 22.9 cm/3.37 kg
Input power: 8 W
Output rate: 1/s
Output signals: serial digital 0.01° LSB
Failure rate: 2.5 × 10^{-6}/h

The Dual-Horizon Sensor Assembly contains two identical sensors in a single housing for application on spinning satellites. Fixed level threshold detectors are used for high accuracy in roll at GEO altitude. A bi-level signal proportional to Earth's chordwidth crossed by the field of view is available as an output. Amplifier outputs providing radiometric information are also available and can be externally processed for improved accuracy.

Instantaneous field of view: 1.65° square
Accuracy: ±0.05° (3σ)
Size: 7.6 × 8.3 × 12.1 cm/0.68 kg
Input power: 2.5 W
Output rate: one per revolution
Output signals: 5 V sq wave width of Earth
Failure rate: 0.3 × 10^{-6}/h (× 2).

Wide Angle Sun Sensor is available in several configurations; the detectors may be wired as a single-axis or two-axis control sensor or as a hemispherical sensor. The 1.8 mm Si photo detectors are mounted on a pyramid, sized to accommodate up to three detectors per face for full redundancy, and the pyramid enclosed under a fused silica dome. Each detector output is a cosine function peaking at 620 μA when the Sun is normal to the face. WASS is qualified for a –100 to +85°C thermal operating range and the new design is nuclear hardened.

Accuracy: up to ½° at null
Range: 2π steradian
Size: 6.1 × 5.1 × 4.8 cm; electronics 6.6 × 5.8 × 5.1 cm
Mass: 0.155 kg; electronics 0.132 kg
Input power: none for head, 70 mW electronics
Output: 620 μA peak/detector; 0–5.2 V electronics

VERIFIED

Lucas Control System Products

Current Status

Schaevitz designs, develops and manufactures acceleration, pressure and displacement sensors to military aircraft and space standards. There have been several technical developments with standard position and acceleration sensors which meet both space and non-space customers and the company has achieved a high degree of customer suitability across a wide range of product. The company's linear and angular servo accelerometers include the LSXH, designed specifically for flight control applications such as satellite nutation sensing systems and are employed on both civil and military satellites. LCSP developed tri-axis damper systems for military satellites. These have been developed into similar requirements and specifications for civil satellite projects.

Specifications

Type: linear
Size: 44.4 × 31.8 × 39.6 mm/0.114 kg
Range: ±5 *g*
Scale factor: 1 V/*g* ±1%
Scale factor stability: 500 PPM over 60 days
Bias: 5 mg
Natural frequency: 35 Hz
Linearity: 0.10% over full range
Temperature: –55 to +71°C operating, –65 to +90°C storage
Shock survival: 1,000 *g* for 0.5 ms

UPDATED

Meda, Incorporated

Current Status

Meda offers its TAM 1 and Tam 2 three-axis magnetometer for attitude control systems and science instrumentation. The series meets MIL-STD-97 5G Class S (Class B also available). Meda has also designed magnetometers for science payloads aboard planetary spacecraft and has developed miniaturised packaging for low volume/low weight requirements. Tri-axis magnetometer concepts for advanced planetary and deep space spacecraft. Meda have responded to government contractor interest by developing a new generation of magnetometers.

UPDATED

Servo Corporation of America

Current Status

Founded in 1946, Servo was a pioneer of IR technology, now covering satellite attitude determination, Earth resource measurements, gas analysis, imaging and laser power monitoring. Its bolometers and pyroelectric detectors form the sensing element of horizon sensors for most telecom satellites; they have served on hundreds of satellites with no known failures. Servo also builds its own Earth sensors for three-axis stabilised and spinning satellites. Servo has connected several separate technologies to field a new generation of interlinked stabilisation systems.

Servo's Horizon Sensor and IR Detection Facility comprises research and development laboratories, environmental test facilities, Earth sensor test facility, manufacturing facilities and cleanrooms, including dedicated equipment for all thin film, ceramic thick film and wet chemistry processing of infra-red detection cells, fabrication, testing and environmental screening of infra-red systems and special optics. Specialised equipment includes vacuum evaporator stations, sputtering stations, electron beam equipment, leak detectors, humidity chambers, environmental chambers, spectrophotometers, a full range of electro-optic test equipment, multispindle polishing machines, optical grinder, sand blasters, lens generators, specialised microscopes and a JEOL scanning electron microscope.

The MiDES, Mini Dual Earth Sensor is a state-of-the-art Earth horizon sensor designed to provide pitch and roll information for LEO satellites, this being accomplished by two pairs of pyrotechnic arrays positioned 90° apart in the image plane. Each array is spatially separated into two eight-element columns to provide for Sun/Moon rejection. An algorithm calculates the position of the horizon from the voltage obtained from the pixels which subtend the horizon gradient (typically two) and the pixels that look at space and Earth. The lifetime is improved by there being no motor but a low-power tuning fork like a shutter mechanism. The MiDES can also be adapted for GEP operations. The MiDES first test flight was aboard the NASA/DARA CRISTA SPAS 2 flown as part of the STS-85 mission in August 1997. Development for MEO applications and derivatives for high-orientation demands.

Specifications

Accuracy: 0.06° (3 sigma)
Field of view: ±5.5°
Dimensions: 13.3 × 13.3 × 13.0 cm
Mass: 1 kg
Spectral band: 14–16 micron
Power input: 28 V DC, +7/–6 V DC
Power consumption: less than 800 mW
Signal output: RS-232, -422 or -485 digital data stream of 8-bit words representing 32 pixels voltages plus temp
Environment: –20 to +60°C, 19 g rms, 50 krads

The Model 110 Horizon Crossing Indicator is a state-of-the-art component for use in attitude determination of micro-spinning satellites with short lives. The low-power electronics and pyroelectric detector make it self-powered. It will operate for a minimum three years, with a five-year shelf life.

Spin rate: 1–50 rpm
Mass: 0.120 kg
Power consumption: 100 μW

The Series 13000 Broadband Radiometer Sensor/Telescope is used as the primary sensing element in an Earth resource monitoring system requiring uncooled operation, broad spectral sensitivity and large radiance dynamic range. It was planned to fly as part of the CERES instrument on NASA's EOS AM 1 in 1998 and PM 1 in 2000 and NASA/NASDA's TRMM in 1997. Optional filters provide specific selectivity up to 150 μm.

Spectral response: 0.3–50 μm
Detector: uncooled thermistor bolometer
Time constant: 9 ms maximum

The Servo Wide Angle Earth Sensor (SWAES) is a 120 × 10° field of view Wide Angle Earth Sensor

which has been developed for use on missions which are prohibited from viewing the Earth with the payload sensor. Developed especially for NASA's Wide-Field Infra-Red Explorer (WIRE) mission, the WAES consists of two sensor heads and an electronic control unit. On the WIRE mission the sensor heads are arranged back-to-back 180° apart so that they can monitor one axis of rotation, while a Sun sensor is used for the other axis. This is of extreme importance to the mission because the telescope features a detector which is cooled by a solid hydrogen cryogen. Accidental viewing of the Earth or Sun would cause the cryogen to boil off and terminate the mission.
Accuracy: ±1.5° (exclusive of radiance)
Mass: 1 kg
Power: less than 1 W

VERIFIED

Smiths Aerospace

Current Status

Created out of the merger between Smiths Industries Aerospace and Dowty Group, Smiths Aerospace is a UK company emphasising aerospace electronics, medical systems and specialised industrial products, deriving about half of its revenue from US activities. Generated aerospace 2001 sales in excess of US$1.8 billion and employs 12,000 people. The Group, of which Smith Aerospace is one of four business units, has sales of US$4.5 billion and capitalisation of US$5.6 billion. 1999 aerospace turnover was US$852 million versus US$704 million for 1998, US$622 million for 1996 and US$591 million for 1995. The company provides six Space Shuttle elements: engine interface unit, mission/event timers, FM signal processor, ground command interface logic controller and attitude director indicator. Its fibre optic gyro inertial measurement unit (0.4 kg, 10.7 × 7.1 × 6.6 cm) is used by Boeing's LEAP BMDO demonstration kinetic kill vehicle. System and subsystem upgrades for critical space elements and development of new layout concepts for satellite design engineering.

UPDATED

Systron Donner

Current Status

Formed more than 40 years ago, Systron Donner comprises three divisions, of which the Inertial Division is the most space-orientated. It provides precision inertial systems for space purposes, including inertial angle sensors, accelerometers, gyros, rate sensors and vibration transducers. Equipment has been provided for Shuttle, Landsat, Meteosat, Intelsat, Galaxy and several communications satellites. The division and Rockwell Autonetics in 1994 began producing the tactical-grade DQI digital quartz inertial measurement unit. It is the first solid-state inertial assembly in Rockwell's family of Miniature Integrated GPS and Inertial Navigation System Tactical Systems (MIGITS). Developed integrated IMU and guidance packages linked to altitude control systems.

The Edcliff Division provided pressure transducers for the Shuttle and several NASA science and applications satellites. SD has produced G&N equipment sensors for several classified military projects and has applied this technology to some remote sensing contracts in non-specified countries. Also developing new guidance systems for small and nanostallite projects with requirements for long-duration on-orbit storage and high-*g* tolerant environments. The company has developed its solid-state quartz technology into diverse commercial applications in the automotive and health-care industries.

UPDATED

SATELLITE INFORMATION STORAGE COMPONENTS

France

Alcatel Espace

Current Status
Alcatel Espace and IBM Montpelier Technologies were selected by CNES in February 1995 to provide solid-state recorders for Spot 5 and Helios 2. High duty rate recorders for future satellites in this class have been developed. Capacity of each is nearly 100 Gbit, enough for 200 Spot images. These units formed the basis of a development programme which resulted in new and expanded capabilities fed into the Spot programme. For Helios 2, Alcatel Espace is also responsible for the image telemetry subsystem. Alcatel Espace has successfully penetrated markets in China and India and is providing subsystem elements for communications and remote sensing satellites in those countries.

EADS – Matra Marconi Space

Current Status
MMS has produced mass memories for remote sensing and science satellites and is a lead provider for data storage and retrieval equipment. Multichip module technology enables 1 Gbit/kg capacity. Reed Solomon encoders, decoders and data compression circuits have been built using ASIC technology, allowing further improvement in capacity and bit error rate. High-data rate specifications met with new and improved deadband systems.

Schlumberger Sema

Current Status
Schlumberger Ltd is a global technology services company with central headquarters in Paris, New York and The Hague with a revenue of US$13.7 billion in 2001. The company, which employs more than 80,000 people, was formed in April 2001 when Schlumberger Ltd acquired Sema plc. The telecommunications division is a leader in high-data-rate mass storage magnetic recording. Product lines cover ground station and spaceborne systems, complemented by telemetry capabilities. The DV 6000 series of rotary head digital ground recorders (MIL-STD-2179) include the DV 6410 designed for harsh environments. It has flown on Shuttle for the SIR-C and ATMOS science missions. New and improved digital ground recorders developed from the DV 6410.

Schlumberger's principal spaceborne magnetic tape recorder is the high capacity (50 Mbit/s, 120 Gbit) heavy duty (2,200 h/five year on-orbit) digital recorder for Spot 4/Helios 1. It comprises two modules: a sealed tape transport module (70 × 60 × 41 cm, 95 kg) and an electronics module containing an error detection and correction system (50 × 42 × 20 cm, 26 kg). Power requirement is 175 W record; 260 W replay; 12 W standby. Bit error rate less than 10^{-7} (typically 10^{-8}).

Designed for unattended operation in automatic system environments, the 3801 TT&C digital processor unit is the key communication unit for ground stations operating satellites, from integration, through the positioning phase and in-orbit operation. Contracts have been won for Astra and Nilesat TT&C. The 3301 CCSDS Mux/Demux is a modular stand-alone 3U high chassis housing up to seven Euronorn format VME Cards. It handles CCSDS formats at 3 Mbit/s with Reed Solomon encoders and decoders and/or convolution coders and Viterbi decoders.

Germany

Thales-Heim Systems GmbH

Current Status
In 1996 Racal Recorders Ltd (Thales) in Southampton (UK), acquired the German company Joseph Heim kg in Germany, and as a result the instrumentation activities were renamed Racal-Heim GmbH and were relocated to Germany as shown above. In 2000, Racal plc was acquired by Thomson CSF and the new group name is Thales. Racal-Heim Systems GmbH became Thales Heim Systems GmbH in 2001. The new company manufactures instrumentation recorders for transport, aerospace, science and research and development applications. Research department develops new and untried technologies and makes facilities available for leasing. Storeplex and 'A' series portable recorders offer bandwidth up to 2.56 MHz analogue and 51.2 Mbit/s digital. The rugged 'D' series records both analogue and digital signals (including ARINC and 1553 busses) up to 12 Mbit/s total bandwidth. Storehouse is a modern IRIG recorder compatible with both 2 and 4 MHz standards. Racal-Heim has developed standards for new recording devices optimised for use aboard the ISS and is developing a unique range of high-speed data recorders and information processing and forward store systems for unmanned platforms and for human tended modules.

United Kingdom

Ampex (GB) Ltd

Current Status
Formed in 1944, Ampex was set up by Alexander M Poniatoff in California. Currently the UK office of Ampex Corporation Inc designs and manufactures advanced data recording and processing systems for mobile and fixed installations. High-speed systems capable of surviving harsh and demanding environments. New radiation-hardened systems design emerge from technology development. Users include satellite ground receiving stations such as Helios. The DCRsi transverse digital recording unit is a widely used rotary cassette system, with approximately 1,450 units installed. It stores 50 Gbyte on a single tape cassette, and can sustain 0 to 240 Mbit/s continuous transfer or 0 to 300 Mbit/s burst. Ampex also offers a range of high-capacity (12,800 Gbytes) D-2 technology-based mass storage devices with a maximum data rate of 160 Mbit/s. The DST 914 series provides up to 26.1 terabyte capacity. Ampex has developed scalability to the DST 712 library system allowing multiple DST 712 cabinets to be connected via a simple cartridge pass through mechanism.

Penny and Giles Data Systems Ltd

Current Status
Formed in January 1998, Penny & Giles was an amalgamation of three former individual companies. P&G employs 400 people. A member of Spirent plc (formerly Bowthorpe plc), the company specialises in advanced high-density digital and analogue magnetic tape recording systems for Earth resources satellite ground stations, radio astronomy, pre-launch check-out, meteorology, range telemetry, rocket motor development and general instrumentation data acquisition. P&G also supplies linear and rotary position sensors, solenoids and audio and vision components. Principal space contracts include HDDRs for Landsat, ERS 1/2, IRS 1C, Radarsat, Spot, MOS 1, JERS 1 and Spacelab. Recorders have also been supplied to SEP, DLR and DASA for Ariane Vulcain engine development, to CNES for Ariane 4 engine pre-launch testing and to ISRO for range telemetry and satellite development. Expanded recorder resilience developed to specifications driven by customer requirements.

The Pegasus cassette-based recorder, introduced in 1992, is capable of recording rates approximately 100 Mbit/s on a standard D-1(L) cassette. Used for IRS 1C, ERS 1/2, Landsat, Spot and Radarsat.

Input/output data rate: 100 kbit/s to 108 Mbit/s continuous, variable or burst; 120 Mbit/s burst for 750 ms (repetition not less than 1 s)
Recording duration: 30 minutes at 100 Mbit/s; pro rata other rates
Storage capacity: 22 Gbytes per cassette
Interfaces: serial, parallel

The Sirius laboratory recorder is capable of recording 42 4 MHz channels simultaneously or a single 144 Mbit/s serial digital stream; alternatively, multiple streams of lower rates. Four were provided to China for recording Landsat imagery.

The VLBI is under a user agreement with the US National Radio Astronomy Observatory. Penny & Giles offers a range of VLBI-compatible recorders, reproducers, data acquisition racks and headstacks. VLBI recorders have been selected for radio telescopes in Chile and Poland and at VLBI correlator sites in China, Germany and Netherlands.

VERIFIED

SATELLITE AND PROPULSION SUPPORT HARDWARE

United Kingdom

Dowty Aerospace Wolverhampton

Current Status
Dowty Aerospace Wolverhampton is a member of the Dowty Aerospace Division of the TI Group, which comprises more than 50 operations worldwide. The Space Projects department manufacturers diaphragm, capillary and pressurant tanks.

DAW's space manufacturing facilities were opened in 1989 to provide a fabrication capacity for tanks up to 120 by 150 cm long. The 400 m^2 class 100,000 clean room, with 150 m^2 test area, includes a computer-controlled inert gas welder for welding titanium tank shells, computer-controlled resistance welder for fabricating internal components, and orbital tube welder for titanium and stainless steel piping. Test facilities include ultra-sonic cleaning, bubble testing, helium leak testing, pressure and expulsion testing, cryogenic proof pressure testing, dye penetrant and radiographic inspection.

The capillary tanks are surface tension propellant management devices. Tanks are under contract for MMS, Hughes HS-601 and BPD.

DAW manufactures diaphragm tanks for mono hydrazine systems. ESA's Infra-red Space Observatory carries them and Orbital Sciences Corporation contracted in 1991 for the 96-litre titanium alloy elastomeric diaphragm hydrazine AOCS tanks for SeaStar.

In addition DAW provided 40 titanium alloy centrifuge tanks for Cluster.

DAW is under contract by MMS UK to supply Xe tanks for the UK ion thruster system on ESA's Artemis. Net volume 30 litres; can store 44 kg of Xe at 125 atmospheres.

The company is teamed with Vacco Industries to manufacture titanium etched disc filters, latch valves, fill drain valves and non-return valves for the European market. Cryogenic conduits and piping for propellant supply and replenishment market. High-storage volume requirements met with purpose-built storage systems.

Fairey Microfiltrex Ltd

Current Status
The company designs, develops and manufactures filtration equipment for fuel, hydraulic, gas and oil systems. Fairey Microflex has expertise in spacecraft propulsion systems and has supplied the filter element assemblies for the hydrazine thruster reaction control systems of ESA's ISO and Soho satellites and long-life, high-temperature resistant coupling conduits for corrosive propellants. Has also designed fluid conduits for highly toxic propellants in self-sealing pipes. Registered to ISO 9001.

Firth-Rixson Plc

Current Status
Firth makes open and closed die forgings, rolled rings and superalloys for aerospace applications. Products include steel, nickel and titanium alloy closed die forgings for missile and rocket motor casings. Specialised ports manufactured using exotic combinations of material sand metals. Majority of orders support sounding rocket fabrication. Has the capability to produce high-quality casings for ultra-high temperature applications in propulsion elements or furnaces.

Rotadata Limited

Current Status
Rotadata has an active programme of developing measuring techniques in turbomachinery and aerospace development hardware. Following the design of a number of fully instrumented full-scale model turbines for the Space Shuttle main engine turbopump, the company has developed a number of new products, both hardware and software, for applications to aerospace turbomachinery development and monitoring. Development of turbine applications for the non-aerospace markets has stimulated research in new product for the nano-tech industry.

Rotodata new product line features a new high-performance pyrometer sensor with a better sensitivity and a high-frequency bandwidth for turbine blade surface temperature mapping at full-rotation speed. The company makes turnkey digital and analogue telemetry systems for rotor instrumentation, and high-temperature strain-gauging for turbine blades and discs. In other NDT applications the company manufactures miniature capacitance probes for blade tip clearance and rotor orbit monitoring and an optical smoke density meter for engine emissions.

The company is developing a novel CO_2 sensing unit for cabin condition monitoring. Meanwhile, the traditional Rotadata activities in turbomachinery instrumentation, aero-thermal rakes and probes, specialist positioning systems and high-quality instrumented test hardware continues to provide a core business from an increasing range of customers. ROSCOM was de-merged from Rotadata on 1 May 2003 transferring all personnel and products.

United States

Ardé Inc

Current Status
Ardé was formed in 1952 by engineers from the Rocket Group of M W Kellog. In the space field it specialises in lightweight pressure gas storage vessels and propellant tanks. Programmes include Shuttle/Centaur, Mars Observer, Atlas 2 and GOES helium pressurant tanks, GOES satellite and Shuttle SRB TVC auxiliary power hydrazine tanks, and Shuttle EMU backpack breathing oxygen tank. The product lines include cryoformed 301/304L pressurant and propellant tanks, prestressed composite fibre over-wrapped metal tanks, and ring stabilised and apex rolling diaphragm expulsion devices. Design of unusual tank configuration requirements for void shaping and conformal attachment. Storage vessels with fans and internal heaters for pressure expulsion. Designed unique storage dewars for high-pressure gases in extreme environments and for applications in long-range missions on long-duration flights.

Cryoformed 301/304L Tanks: Under-sized 'preform' tanks are conventionally fabricated from austenitic stainless steel and then immersed and filled with LN_2 before pressurisation to a level about equal to its room temperature design burst pressure. The membranes and welds are stressed beyond the 195°C yield and the vessel plastically stretches: 7 per cent in spheres and 15 per cent for cylinders. The temperature causes a transformation to martensite steel in both parent metal and welds with an attendant 5:1 increase in room temperature yield strength. Approximately 3,000 flight tanks using this process have been delivered over 20 years. The Shuttle EMU backpack O_2 tank is a 36.3 cm long, 13 cm diameter cylinder pressurised to 71.4 atmosphere; mass 0.9 kg.

PSC Pressurant Tanks: Pre-stressed composite fibre over-wrapped pressurant (PSC) tanks provide pressure vessels up to 50 per cent lighter than equivalent all titanium tanks. Kevlar or graphite fibre is wrapped on a metal liner and immersed in LN_2 as in the cryoforming process to stress both materials. Ardé is developing a long life oxygen tank system for a new generation of space suits.

Brunswick Defense

Current Status
Brunswick Defense has three major product lines for space applications: fluid control devices; thin foil extendible booms; fire detection and suppression systems. The Lincoln facility produces composite filament wound pressure vessels for life support systems and control mechanisms. Each Space Shuttle carries 23 He, O_2 and N_2 vessels. Numbers carried can vary according to mission demands. Reduced scale versions can be supplied for supplementary (payload) requirements. Supplementary numbers of vessels can be provided for extended duration missions.

Brunswick Corp acquired Wintec Aerospace in 1972 as specialists in filters and propellant management systems, including high-performance ultra-clean wire mesh filters; large screen propellant management assemblies; rocket engine tubular heat exchangers; Shuttle bay vents; Shuttle shock and vibration mounts; and high-capacity GSE cryo filters. Wintec has a design inventory of approximately 1,500 filter part numbers. The filters are available in all stainless and/or stainless screen with Ti body details, are manufactured in a cleanroom and many are processed to 50 PPM or better. Wintec developed the fabrication design technology of a number of Propellant Management Devices (PMDs), some of which required the joining of stainless screen to larger Ti frames; notably for the Shuttle OMS tank and Peacekeeper missile PMDs. Others include the Harpoon missile, Shuttle RCS tanks, a new satellite design and several developmental units, including a toroidal design.

EG&G Wright Components, Inc (WCI)

Current Status
WCI designs, develops and manufactures solenoid and pressure operated valves, mechanical valves, relief valves and manifolds for aerospace applications, including rocket propulsion systems. Ground servicing equipment support for the launcher and launch site industry. Test facility support. The major space products are hydrazine thruster valves 0.45 to 778 N. WCI has 75 valves on Space Shuttle, ranging from hydraulic valves controlling the landing gear and main engine gimballing, to helium systems for purging oxygen lines. Shuttle's Auxiliary Power Units use hydrazine and relief valves. WCI has begun development of high temperature flexible conduits for high pressure gas systems.

Kaiser Marquardt

Current Status
The Marquardt company was formed in 1944 to undertake research and development leading to the first US subsonic ramjet in 1945. Its principal engineering business continues to be advanced aerospace propulsion and the supply of ram-air turbine power systems. On 14 June 2000 Primex Technologies Inc announced that it had completed its acquisition of the bipropellant rocket motor and related businesses of Kaiser Marquardt, a wholly owned subsidiary of Kaiser Aerospace & Electronics, for US$22.4 million. The company has been developing and manufacturing bipropellant thrusters since 1959, notably the attitude control thrusters for Apollo spacecraft and the Space

Shuttle Orbiter. Precision bipropellant engines are available for 0.004, 0.02, 0.06, 0.11, 0.22, 0.44, 0.89, 4 kN sizes, and requests up to 8 kN can be accommodated. The monopropellant hydrazine thruster line of UTC's Hamilton Standard Division was acquired in 1993 and continues to produce thrusters for military and civil satellites. Bipropellant engineering technology research provides tested product design option for ultra low thrust, low acceleration/extended turn time.

Moog Inc

Current Status

Moog is a worldwide manufacturer of precision fluid control components and systems. With an involvement in space and rocketry dating to 1951, when the US government was looking for a reliable method of controlling anti-aircraft missiles, the Space Controls Division specialises in mono/bipropellant valves, hot/cold gas thrusters, regulators, latch valves and electric propulsion system components. Moog led the transition from hydraulic flight controls such as those on the NASA Shuttle to electromechanical controls designed for the now cancelled X-38 Crew Return Vehicle. The late 1990s saw the evolution from hydraulic thrust vector controls on theTitan 4 to the electromechanical TVC on the Atlas 5 and Delta 4 programmes. It developed the Xe propellant feed system for the SPT-100 plasma thruster. Moog's Model 51172 latch/thruster combination bipropellant torque motor valve is used in the Tempo satellite. It is based on the Model 5289 thruster/thruster valve used on Aerojet's Satprop, for which approximately 180 have been provided. Propulsion components have been provided for Eurostar, Spacebus, Italsat, Cluster, HS601, Hot Bird, Cassini, Artemis and Ariane 5. Cold gas thrust control is provided for Pegasus, Meteor and SAFER. Development of high-performance specification components for Xe feed systems. Fiscal Year 2002 sales were US$107 million with 2003 sales projected as US$98 million. SPD revenues make up about 15 percent of total company sales.

Para-Flite Incorporated

Current Status

Para-Flite develops and manufactures 0.5 to 280 m^2 ram-air gliding parachutes for aerospace recovery systems, including orbital capsules and sounding rocket payloads. Providing a lift-to-drag ratio of 3, the parachute carries an Airborne Guidance Unit to process signals from a ground-based transmitter (GPS is an option) to manipulate the control lines and home in on the target. The automatic mode provides 60 m accuracy, complemented by 15 m for manual control. Payloads can be recovered from suborbital rocket and balloon carriers: 100 to 2500 kg capacity systems are available off-the-shelf. High-velocity retardation systems and heavy (high-mass/volume) recovery equipment. Viewing the potential for safe and affordable launch vehicle boost stage recovery, studies have been carried out on parachute recovery systems with autonomous guidance to a designated landing point.

Parker Bertea Aerospace

Current Status

Parker Bertea Aerospace, part of the Parker Hannifin Corporation, designs, manufactures and services fluid systems, components and related electronic control systems for aerospace applications. Personnel total about 5,000. The space-related products of its divisions are specified below.

Air and Space. This division specialises in the design and manufacture of valves and coupling devices to handle gaseous, liquid and cryogenic propellants, including hydrogen, nitrogen, oxygen, helium, NTO and MMH. Such systems fly on Shuttle, RL10 engine, Peacekeeper and Pegasus. Extended cryogenics containment vessels and dewars.

Parker's high-pressure helium regulator is used for ullage pressure control in a lightweight 680 atmosphere blowdown propellant feed system for missile and spacecraft propulsion systems. The regulator controls ullage pressure within ±10 per cent with a 10:1 variation in inlet pressure and from lockup to rated flow. It is a welded assembly constructed entirely from titanium alloys. The basic regulator design can be used for a variety of fluids and gases. Advanced regulator designs for high-velocity acceleration systems.

The company's cold gas thruster module is designed to provide three-axis control for small launch vehicles; it is flown on Pegasus and has been selected for the Artemis project. The module comprises three cold gas thrusters operated by integral independent solenoid valves. The central thruster generates 111.2 N and the two lateral thrusters 55.6 N when operated with a 136 atmosphere N_2 gas supply. SI of each is 68 seconds. The solenoid valves are pilot operated and have a response time of more than 5 ms at maximum gas inlet pressure.

Pressure Systems, Inc

Current Status

Pressure Systems was incorporated by Doug Juanarena to exploit technology developed by NASA when he was a member of the Langley team which developed a unique sensing system for electronic communication and measuring wind tunnel model pressures. PSI, is the world's largest independent manufacturer of titanium pressure vessels for launchers and spacecraft: it now ships 200/300 annually, worth about US$25 million. More than 4,500 have been delivered, without flight failure, including about 700 propellant management device tanks of approximately 20 designs and more than 750 elastomeric diaphragm tanks of approximately 15 designs. PSI acquired Programmed Composites Inc in June 1995, allowing it to offer tanks integrated with satellite core structures, in addition to precision reflectors, solar array substrates and antenna support structures. PSI tanks provide the propellant for most commercial communications satellites, including Hughes HS 601, Loral FS1300 and Lockheed Martin A2100. PSI tanks are also used on Milstar, Cassini, Soho, NEAR, Tirus, Inmarsat, Intelsat, ETS 7, TRMM and Axaf. PSI is completing a US$4 million contract from Lockheed Martin for 75 hydrazine tanks for Iridium satellites. In 1996, PSI delivered the 1000th upper-stage high-pressure helium tank and the 100th upper-stage propellant tank for the Lockheed Martin Atlas Centaur launch vehicle family. Increased impact resilience for shock environments and stress resistance.

Puroflow Corporation

Current Status

Surface tension devices for spacecraft propellant tanks. Filters and strainers for propellants, pneumatics, hydraulics and cryogenic systems. Users include Shuttle, Titan 3, GOES, Milstar and Eurostar. Sludge and ullage systems and equipment support with positive expulsion properties. An ultra-violet system is used to treat the deluge water on the Shuttle launch pad, a system which Puroflow maintains. New techniques for cleaning and purging toxic flow from line leakage and spillage.

WORLD SPACE CENTRES

Launch sites
Training facilities
Telemetry, tracking and control
Research and development facilities
Space monitoring facilities

LAUNCH SITES

Australia

Cape York

Location: 12.25°S/143°E

Current Status

Environmental studies conducted by the local government agencies continue to mobilise objections to licensing. Nevertheless, intense interest remains because of the regional proximity to the equator for geosynchronous orbital positioning. The Balkanu Cape York Development Corporation has set up a strong regional support infrastructure with extensive business promotion and new telecommunications infrastructure. There are commercial interests throughout Asia that would like to use the facility and there are indications that major space-faring countries would fund a near equatorial launch site at this location.

Background

The Australian government began a feasibility study in October 1986 into the possibility of an international equatorial launch site on the Queensland coast, offering launches over the Pacific. By August 1987, five specialist studies had reported favourably on the project, which would be operated by private enterprise. Two commercial consortia then undertook studies. The Australian Spaceport Group included Martin Marietta and Aussat, and focused on the peninsula's western mining town of Weipa. The group concluded in 1989 that the space market did not require such a facility. However, the Cape York Space Agency submitted a proposal to the Queensland and federal government in October 1989 to develop a 607 km^2 site at Temple Bay, on the eastern side, as the Cape York Space Port.

Capital investment would be about US$470 million. The goal was to operate Ukraine-supplied Zenith launchers using local crews, trained initially by the former Soviets. Full commercial operations would fly five missions annually. Russian satellites would be barred from the site. Facilities would include a 30 × 60 × 10 m processing building and horizontal transporter, and a 50 × 50 m concrete pad with 70 m service tower and support services. The project would require the building of a town to accommodate some 700 people and a runway to handle Boeing 747F and equivalent aircraft.

The government endorsed the proposal in December 1989, subject to a full study of environmental impacts and agreement with local Aboriginal communities. CYSA planned to contract United Technologies' USBI to manage and operate the site (a Memorandum of Understanding was signed June 1989). USBI received a US Technical Assistance Agreement export licence in summer 1990 covering the initial design/development stage. A further licence would have been required for construction. CYSA, owned by Essington Developments Ltd, was wound up in October 1991. Space Transportation Systems Ltd was awarded the mandate to develop the spaceport and Kaiser Engineers Australia was to be project engineers.

1992's INSP report recommended that no further resources should be invested unless financial backers were positively in place by end-1992. STS withdrew in 1992 to concentrate on studying a Proton launch site in Papua New Guinea. The Euro-Pacific Capital Group consortium, based in Sydney, was reported in 1993 to be investigating Cape York's viability. Former Soviet space officials are reported to have reopened discussions about the facility being the basis for the launch site.

Darwin

Location: 12.28°S/130.5°E

Current Status

Renewed interest in small sounding rocket and small satellite launch pads operated from this location. The original parties interested in this initiative have encouraged support from major small-launcher market users. Potential business partners have continued to market the prospective launch location but have as yet not obtained the necessary environmental clearance.

Background

An Australian joint venture in 1993 signed an agreement with AeroAstro of the US for two launches of small communications satellites from Australia. KITComm Pty Ltd, an Australian-owned joint venture between Kennett International Technology Pty Ltd of Queensland, and several private investors, would team with AeroAstro to provide low-cost launches from a new site near Darwin. The commercial facility would be regionally focused to cater for the growing satellite requirements of Asian countries. AeroAstro planned to launch its PA-2 liquid rocket, capable of carrying 230 kg into orbit for US$6 million, from the site. These plans failed to reach fruition. As with other proposals for launch sites elsewhere in Australia, environmental impact concerns have discouraged local authorities from developments of this kind.

Woomera

Location: 31.1°S/136.8°E

Current Status

For more than 10 years the Australian Space Council has been studying whether to establish a new space launch facility first at Woomera and then at Darwin under its five year rolling plan. Its goal is to make Australia a significant launch site for Asian small satellites and space-based services. However, the government has yet formally to endorse the launch programme.

Background

About 430 km north of Adelaide and running eastwards for 2,000 km across the desert, the Woomera range in South Australia was established jointly by the UK and Australia in 1946. Australian hopes that what was originally a test range for ballistic missiles and sounding rockets, and for target practice at pilotless aircraft, would have a future as a satellite launch centre, were not fulfilled. The first Skylark sounding rocket launched from Woomera was flown on 13 February 1957. First Black Knight launched 7 September 1958 and 10 Europa flights between 1964 and 1970.

During the period when the UK's Blue Streak was under development as the main stage of ELDO's Europa rocket, a township of 4,500, with more than 500 houses supplied with power and water across 160 km of desert, was created. However, Woomera was not suitable for equatorial launches and France preferred to develop its own centre at Kourou. Europa research and development flights were conducted between 5 June 1964 and 12 June 1970, the last three were unsuccessful attempts to place a satellite in orbit, before launches switched to the Kourou launch site in French Guiana. After successive cuts in its contributions to the range, the UK announced there would be no more work for Woomera after 1976. NASA also decided to close down its Carnarvon tracking station, which had played a major part in manned and unmanned space flights. The small Wresat test satellite, launched into polar orbit by a US Redstone in November 1967, and the UK's Prospero in October 1971 remain the only two satellites orbited from Woomera. The range reopened in August 1987 for the launch of a West German-funded Skylark sounding rocket to undertake about 5 minutes of observations of Supernova 1987A.

The South Australian government has long wanted to revitalise Woomera as a commercial site for aircraft and aerospace systems testing, but rejected initial proposals in 1990 for relying excessively on the use of the Australian armed forces. Woomera's advantages include polar orbit access, a sparsely inhabited downrange area, an existing infrastructure and largely cloud-free weather.

Woomera has been proposed in the past for redevelopment and refurbishment to handle the Southern Launch Vehicle, capable of delivering 800 kg into LEO. BAe, Auspace and Hawker de Havilland, supported by an A$1.25 million ASO contract, completed preliminary feasibility studies. However, the joint venture was terminated in 1993.

Germany's Express orbital capsule was to be recovered at Woomera 21 January 1995 following ascent from Japan, but was lost soon after launch. NASA Goddard launched six Black Brant 9 sounding rockets October/November 1995 to study the Large Magellanic Cloud. Japan's 760 kg Alflex spaceplane automatic approach/landing experiment was released from a helicopter on a series of tests using the 45 × 2,600 m runway in 1996 to demonstrate GN&C for automatic landings.

Brazil

Alcantara

Location: 2°28′S/44°38′W

Background

The Ministry of Aeronautics runs the larger of Brazil's two launch sites, Alcantara. The other site is Barreira do Inferno. The CLA Alcantara Launch Center was expanded on a 520 km^2 site on the Atlantic coast outside Sao Luis to handle the VLS orbital launcher. CSA also built pads for the Sonda 3 and Sonda 4 sounding rockets, meteorological rockets and other science vehicles (NASA, for example, launched four Nike Orions here in 1994). The position near the equator offers 25 per cent greater initial launch energy due to the earth's rotation compared to Cape Canaveral.

Brazilian government and military officials held a formal opening in February 1990. The fifth and last Sonda 4 was launched from this site on 21 February 1990. Most recent launches took place when two unsuccessful VLS flights were attempted on 2 November 1997 and 11 December 1999. On 22 August 2003, the third VLS launch vehicle exploded while being prepared for flight three days hence when one of the four solid propellant strap-on boosters ignited unexpectedly resulting in the death of 22 people. This explosion, which could be heard 18 km away, completely destroyed the launch pad. Space officials have been developing the launch complex for an international array of launch vehicles including the Ukrainian Cyclone and the Israeli Shavit.

Barreira do Inferno (Barrier of Hell)

Location: 5°55′S/35°10′W

Background

A smaller site than Alcantara near Natal, established by the Minister of Aeronautics in 1964 for Sonda 1, BdI used until the late 1980s for Sonda (the first Sonda 4 departed from here in November 1984) and international vehicles such as Skylark. All launches have now transferred to CLA. It was upgraded in 1979 to provide radar, telemetry and real-time data processing services for Ariane launches from French Guiana. Similar support is provided for sub-orbital and orbital launches from other sites. Sonda 3 flights are made in co-operation with several international organisations including the German DLR, the Max Plank Institute and the US Air Force Geophysic Laboratory. Interest has been shown in using the facilities for missile and weapons development and testing.

Canada

Churchill

Location: 58.46°N/94.1°W

Current Status

Churchill Rocket and Research Centre Committee composed of representatives of the town of Churchill and the government of Manitoba has been established to examine alternative uses for the site. Local authorities are concerned at the limited use of the range and seek mechanisms to broaden its use.

The site had been closed in 1958, a year after it had been opened as Canada's launch site for sounding rockets during the 1957 International Geophysical Year, reopened by the Army in 1959 and transferred to the Air Force in 1962. The National Research Council of Canada took it over in 1966 but it was closed in 1984 due to government cutbacks. In September 1993, a commercial venture to be known as SpacePort Canada was granted an operating lease but the venture folded in 1998 and the range is now owned by the Province of Manitoba.

SpacePort Canada's Orbital Complex number 2, became operational in 1999, and included the following new facilities: launch site, mission control centre, vehicle assembly building, launch vehicle storage building, payload preparation facility and instrumentation sites; supporting infrastructure includes power, base time, and communications; SpacePort operations and data processing systems; emergency services. Hudson Bay Helicopters are available in the area for payload delivery, recovery or logistical support. The Churchill launch site is presently offered for payloads of up to 4,300 kg gross weight.

Background

Akjuit signed a 30-year lease with the Manitoba government on 18 July 1994 for an 80 km^2 site, including the former Churchill Research Range, for the development of SpacePort Canada, a commercial launching facility for sub-orbital and LEO payloads. Akjuit estimates that the first phase of expansion leading to orbital missions will cost C$100 million. The C$550,000 refurbishment of existing sounding rocket facilities was completed in September 1994, this including an upgrade to the universal launcher. The launcher handles vehicles up to 18 m long and with a mass of up to 4,530 kg. The enclosed launch bay is temperature controlled.

In the past, Nike Tomahawk, Black Brant 5B/5C/8/9/10, Taurus Orion and Nike Orion, have all flown from Churchill. The hazardous assembly building, payload assembly area, blockhouse and above-ground tunnels link all operations.

The launch site (Flat Pad) includes a concrete foundation with embedded anchor hooks for installing launch stands for the launching of Russian Start and Start-1 vehicles, plus a movable launch building. The mission control centre will be equipped with separate consoles for the payload owner, launch vehicle provider, senior director, operations director, range system control, mission flight control, range safety control and launch control. GPS information as well as computer simulation and base time will be available at the control centre, which will also act as the hub for the operations and data processing system. A separate room for customer (payload owner) use will also be available. The centre will have a dual-processing capability and thus two launch vehicles can be processed in parallel.

The vehicle assembly building is designed for the horizontal processing of two Start-size launch vehicles in parallel. It has a climate-controlled, 10,000 cleanliness-class assembly room and separate rooms for two users' technical and launch vehicle teams. It includes mechanical, electrical and boiler rooms and an independent back-up power generator, as well as a control room. Two 20 tonne bridge cranes are available for operations.

The payload preparation facility will provide the opportunity for the parallel preparation of up to four satellites up to 150 kg mass under climate-controlled conditions and with a cleanliness class not less than 100,000. The facility will have air lock access, climate-controlled storage for satellites and special storage for satellite propellants. Separate rooms for the customer's team, as well as separate storage for the customer's equipment, will be available.

Instrumentation includes: Churchill site (IS-1), comprising two identical tracking receive systems with S-band antennas and telemetry receive equipment and one UHF command-destruct system; Chesterfield Inlet (IS-2), comprising two identical tracking higher sensitivity receive systems with S-band antennas and telemetry receive equipment and one UHF command-destruct system; P-band telemetry receiving systems will be available for the Russian launch vehicles.

The mission operations and data processing system provides instantaneous impact predictions, flight termination criteria and instrumentation status displays. It supports data types from sources including TT&C instrumentation, facilities, utility systems and range systems. It uses graphics-based software architecture, which permits decisions to be made quickly and accurately. The architecture is based upon the distributed client/server model.

SpacePort Canada will be connected with the North American power network through the special 148 kV line; a back-up standby power plant will be constructed.

Over 40 years the site has hosted more than 3,500 flights with sounding rockets and research projectiles.

China

Hainan/Harbin

Location: 19°N/109.5°E

Background

A sounding rocket site on the west coast of Hainan Island was inaugurated in late 1988 with the launches of four Zhinui 1 ('Weaver Girl') vehicles over Beibuwan Bay. A Zhinui 3 attained 120 km in January 1991. In the previous Five Year Plan, the Chinese began to upgrade the facilities at Hainan to handle 300 km-altitude vehicles. The Harbin site in northeast Manchuria was established in the late 1960s to provide a 3,500 km land range for missile tests, including the CSS- 2 and CSS- 3. Ballistic missile test and performance evaluations has increased in recent years and some buildup has been reported together with upgrades and improvements to tracking and data transmission facilities.

Jiuquan Satellite Launch Centre

Location: 41.10°N/100.30°E

Background

The Chinese built Jiuquan in the late 1950s in the Gobi deserts 1,600 km west of Beijing near the Soviet border, north of Jiuquan city in Kansu province. The first launch took place on 1 September 1960, the first Chinese launch of a Chinese R-2, followed on 5 November 1960 by an indigenous Chinese R-2. The first attempt at an orbited launch on 1 November 1969 was a failure but the first successful launch came on 24 April 1970. Hundreds of missile tests have been reported, including into the Lop Nor nuclear test site. Formerly known in the West as Shuang Cheng Tzu (still used as the US Space Command designator), this was China's first launch site but it is limited to southeastern launches into 57 to 70° orbits to avoid over-flying Russia and Mongolia. Sounding rockets also utilise the site. It consists of two pads 416 m apart, with a shared mobile service tower that can be employed separately for CZ2-/FB-1 and CZ-1 space vehicles. In the early 1980s, the latter added the capability for CZ-2C missions. CZ-2D flew from this pad as well, beginning in 1992.

The site's base area (called Dong Feng, 'East Wind') houses the Huxi Xincun range control centre, power station and residential accommodation. The technical centre, where vehicles are assembled and checked out, is 18 km north. Supporting facilities BL 1 and BL 2 can each integrate two launchers and several satellites simultaneously. BL 1, an enormous 140 m long building has an 18 × 90 m test and assembly hall, where the launch vehicles are delivered on rail from the connected transfer hall. There are 25 test rooms down each side of the main hall, including a class 100,000 satellite assembly and test room. The pads are another 30 km further north.

Jiuquan provides two pads for CZ-1/2 launchers but the site's location limits missions to 57 to 70° 0517071

Within the chamber, 41.6 m towers share a 55 m mobile service tower on 17 m span rails for preparation and delivery of the launch vehicle. The mobile tower's maximum speed is about 15 m/min. It has 15/5 tonne-capacity main/auxiliary cranes. To prepare the satellite, commercial users have a class 100,000 clean room which employs a 6 × 6.5 m 13.1 m high cabin to protect the vehicle's upper section. Temperature can be maintained at 20 ±5°C, with 50 to 70 per cent humidity.

A launch control centre is 200 m from pad 1. Both road and rail systems link the facility to the nearest airport, 94 km to the south. A 270 km line connects to the main railway network so satellites can be delivered for launch from virtually any point in China. Jiuquan will continue to be used primarily for recoverable Earth observation and μgravity missions but, because of its geographical constraints, greater commercial activities are focused on the other two launch bases.

The manned space flight plan that began in 1992 and grew out of the 1996–2000 Five Year Plan saw development of a new launch complex at Jiuquan to support flight operations with the CZ-2F, developed from the CZ-2E first flown in 1990. The first flight of the CZ-2F carrying an unmanned precursor took place on 20 November 1999. Further unmanned test flights of the Shen Zhou spacecraft took place on 9 January 2001, 25 March 2002 and 29 December 2002. It is from this site that the first Chinese astronaut was sent into space, an event that took place on 15 October 2003 when the astronaut was placed in a 343 km circular orbit and returned later the same day.

Taiyuan Satellite Launch Centre

Location: 37.8°N/111.5°E

Background

Taiyuan was inaugurated on 6 September 1988 with the first of the CZ-4A designed to place satellites into polar orbits for remote sensing, meteorological and reconnaissance missions. The second launch was made during September 1990. There is a single pad. It was initially operated for missile testing as an extension of Jiuquan for larger vehicles, including the DF-5. The first photographs of the site were released in 1990. CZ-2C launches began with a load of Iridium satellites. US Space Command still calls the site Wuzhai. The site had supported 25 launches by the end of 2002.

Long March 4 at the Taiyuan launch site used for Sun-synchronous missions (Theo Pirard) 0054384

The China Academy of Launch Vehicle Technology (CALT, launcher manufacturer) provides on-site supervision of launcher procedures. These have recently been updated and some work has been reported in the installation of new tracking and telemetry reception facilities.

Xichang's Mission Command and Control Centre 0517251

Xichang: the main launcher processing facility can handle three vehicles simultaneously. A CZ-2E requires all three tracks because of the strap-ons 0517072

Xichang Satellite Launch Centre

Location: 28.25°N/102.0°E

Background

Xichang was selected from a shortlist of 16 sites, of 81 surveyed, for a more favourable GEO mission base than Jiuquan. Construction work in the canyon site at the foot of Mt Liang Shan, 65 km north of Xichang city, began in 1978 and resulted in the first launch in January 1984. The site hosted its first commercial mission when it launched the Hong Kong AsiaSat 1 communications satellite on a CZ-3 in April 1990, followed by Australia's two Optus Bs in 1992. By the end of 2000, the Xichang launch site had hosted 28 orbital launch attempts.

The local population has traditionally been allowed to live close to the pads. When the maiden CZ-3B crashed on 14 February 1996, into a hillside 1.5 km from the newer complex, the official death toll was six plus 57 injured, and 250 homes were damaged or destroyed. When the CZ-2E of 25 January 1995 exploded at 51 seconds, the debris killed six and injured 23 others in a village 7 km downrange.

As it stands now, the dry season, running from October to May, provides the most suitable time for launch campaigns, but the site enjoys 320 days of sunshine annually. Annual temperature range is −10 to +33°C, because the site sits at 1,839 m above sea level. Xichang's main airport, upgraded to handle B-747 and C-130s, is 50 km south of Xichang city.

The single CZ-3 pad can support up to five missions annually. It has a fixed 77 m high/11-level gantry. In order to set the launch azimuth a technician hand cranks a firing table. The firing room is buried in the hillside 60 m away. Originally, the Chinese stacked vehicles on the pad and mated the payload in the open air but, as part of the CZ-3 commercial marketing, a stage 3/payload class 100,000 clean room was added.

Vehicle stacking requires about two days, followed by 14 days of checkout with a launch rehearsal on the day 15 aiming for the actual launch on the day 20. Stage 1/2 UDMH loading from underground reservoirs begins 16 hours before launch, followed by N_2O_4. Stage 3's cryogenic loading about 5½ hours before launch. 36 days are allowed for launcher and payload checkout.

CZ-2E ready for launch at Xichang. The older pad is in the background 0517250

Xichang launch site used by CZ-2E and CZ-3 rockets for geosynchronous transfer (Theo Pirard) 0054385

The second complex is sited 300 m distant for CZ-2E/3A/3B vehicles with add-on boosters. It was first used on 16 July 1990. The pad is at 28.25° N/ 102.02° E, 1,826 m above sea level. Like the original pad, this one has a fixed 74 m 1,025 tonne tower plus 98 m 4,300 tonne mobile tower. Just before launch the ground crew withdraws the tower 130 m.

Payloads are encapsulated within their fairings in the class 100,000 preparation facilities and delivered to the pad for vehicle mating. Vehicle erection begins 11 weeks before launch, the encapsulated payload is mated at six weeks, launch rehearsal concluded at three weeks and fuelling completed at one week.

The Mission Command and Control Centre (MCCC) lies 7 km southeast of the pads and was upgraded 1994–95. At this site, a communications centre includes an 11 m Intelsat B Earth station. Vehicle/payload processing facilities are 2.2 km from the pads in the technical centre.

Launcher stages are received by rail directly inside a 31 × 14 m transit hall and transferred to an associated 92 × 28 m assembly room capable of processing three complete vehicles. The launchers are assembled, checked out and then broken up for transfer by road for stacking in stages on the pad. The Non-hazardous Operations Building (BS2) houses the main 756 m^2 42 × 18 × 18 m satellite test hall, class 100,000, climate-controlled (20 to 25°C, humidity 35 to 55 per cent) facility. Personnel access it by an airlock with 5.4 m wide 12.5 m high door. The building includes offices, test rooms and storage.

The associated Hazard Operations & Fuelling Building (BS3) is a 324 m^2 spacecraft fuelling and assembly hall, with a class 100,000 clean room. The door is 5.4 m wide 13 m high. Solid kick motors are installed in here and the payload encapsulated in its fairing.

After the stack begins its preparation, the Checkout/Preparation Building (BM) is used to inspect the solid motors. It is a 108 m^2 12 × 9 × 9.5 m hall (door 3.6 m wide 4.2 m high). Next door sits the SPM x-ray Building (BMX), a solid motor test facility with an x-ray hall sized 125 m^2 12.5 × 10 × 15.4 m high (the door is 3.2 m wide 4.5 m high). In the same building the Cold soak chamber, 9.6 m^2 3.3 × 3.0 × 4.0 m, door 3.2 m wide, 3.5 m high, provides 0 to 15°C.

France

Kourou (Centre Spatial Guyanais—CSG)

Location: 5° 14′N; 52° 46′W

Current Status

Launch facilities available for Ariane and Vega launch vehicles. On 15 February 2003, the last

CSG's Vertical Assembly Building can stack one Ariane 4 on a mobile launch platform in a month. The assembled payloads/fairing are integrated on the pad, as are the solid propellant boosters 0517077

Ariane is transported 950 m by rail on its mobile launch platform; the central turntable (with cover off, at right) permits two to pass. The launch control centre (CDL-2) is the low building to the VAB's left in the background (Arianespace) 0517078

Ariane 4 was launched making it the 159th Ariane series flight from Kourou.

Background

Although owned by CNES, the CSG is made available to ESA/Arianespace under a governmental agreement that guarantees access to the ESA-owned Ariane launch facilities. CSG is the most favourable major site for GEO launches, its near equatorial position providing a 15 percent payload advantage over Cape Canaveral for eastward launches. Guiana's coastline permits launches into both equatorial and polar/Sun-synchronous paths, with inclinations up to 100.5° possible. The annual operating cost for the site is FFr1 billion, with France, including its contribution through ESA, paying about 70 per cent. Operating costs have continued to go down since 1991 and now are about 15 per cent less than they were that year. CSG staff total 1,400.

CSG extends more than 30 km along the coast from Kourou town to Sinnamary, with the Ariane facilities about 18 km northwest of Kourou. CSG's Technical Centre, close to Kourou, incorporates spacecraft reception and checkout facilities (able to handle five satellites in overlapping campaigns) and the Jupiter range control centre. The 2,600 m^2 Jupiter 2 handled its first campaign in January 1996 and will be used for all future launches. Its predecessor, Jupiter, had been operational since 1968. Radar and telemetry stations are sited at Montagne des Pères and Montabo, respectively, 10 and 60 km southeast of the Technical Centre, with downrange stations for easterly trajectories in Natal (Brazil), Ascension Island and near Libreville in Gabon.

Detailed study of 14 potential launch sites worldwide resulted in the selection of Kourou and since CSG became operational in April 1968,

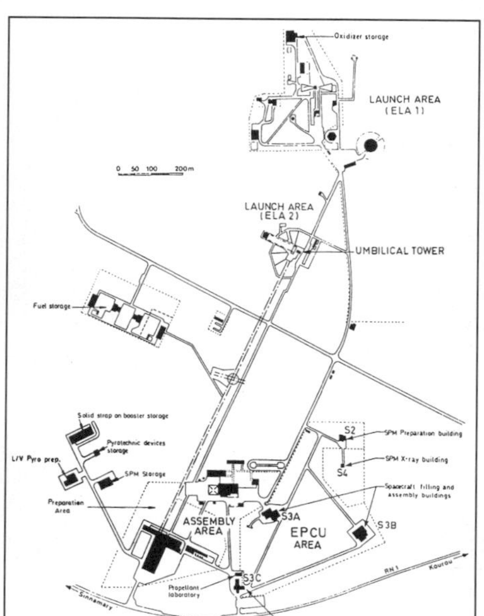

Kourou ELA-2 launch complex layout 0517079

Ariane 5 pad. The 90 m tower holds 1,500 m^3 of water for trench cooling and sound suppression. Four pylons provide lightning protection (ESA) 0517428

more than 700 launches of sounding rockets and orbital vehicles have been made through mid-2003. Eight flights of France's Diamant B were conducted from this site from March 1970 followed in 1971 by the Europa launch vehicle. The site has also witnessed more than 60 balloon flights (the last in 1980).

CSG's meteorological conditions reflect its tropical location. The mean daytime temperature is 26.4°C, with maximums of 34.0°C, and minimums of 18.1°C. The windspeed has an annual mean of 3.3 m/s, a maximum of 23 m/s and the rainfall is 2.9 m annually, 438 h/year over 235 days.

Ariane 4 launch requirements preclude launches if storms exist within 10 km at T–10 minutes or if convective clouds exist at 5,000 to 6,500 m altitude at T-0. These criteria have been used since flight 19.

ESA's Ariane Launch Base comprises two elements: the ELA (Ensemble de Lancement) Ariane launch areas, including facilities for launcher final assembly/checkout, and the EPCU Ensemble de Préparation Charges Utiles for satellite processing.

The major installations are:

ELA-1: Pad 5°14′09″N; 52°46′29″W

This design work for ELA-1 began in 1973 and this pad was responsible for the first Ariane launches from 1979. Responsible for 25 launches under first ESA and then Arianespace auspices, ELA-1 was closed down in December 1989 and handed back to ESA. The 70 m service tower was lowered to the ground in June 1991.

Vehicles were stacked on the pad, limiting campaigns to two month intervals. Platform 8 in the service tower provided a class 100,000 100 m^2 clean room for payload mating/checkout. The 34 m diameter protected CDL-1 (Centre de Lancement) blockhouse was sited 300 m from the pad to handle pre-launch and countdown operations.

Ariane 5 emerges from the BAF final assembly building (ESA) 0517429

MAN tractor pulls the Ariane 5 launch table from the Final Assembly Building (ESA) 0054382

Launcher Integration Building of the Preparation and Launch Control Centre (ESA) 0054386

ELA-2: Pad 5°13′56″N; 52°46′32″W
Design, construction and validation of the ELA-2 complex was conducted between 1981 and 1985 and has been operational since March 1986. Vehicles are stacked up to equipment bay level on one of two mobile platforms in the Vertical Assembly Building, 950 m by rail from the launch stand, permitting monthly departures of Ariane 4s. The CDL-2 control centre is adjacent to the VAB. Payloads are mated on the pad within the 86 m, 5,500 tonne service tower's class 100,000 clean room (PFCU Plateforme Charge Utile, platform 16). ESA/Arianespace considered making ELA-3 compatible with Ariane 4 but instead hardened ELA-2, which CNES guarantees to restore in nine to 10 months in the event of a launch accident.

ELA-3: Pad 5°13′56″N; 52°46′32″W
Construction of the Ariane 5 facilities began in 1988 to allow up to eight departures annually (the current plan is for five), including two a month apart. Each launcher campaign will last 22 days. The new pad cost the GSC FFr5,000 million, with a similar amount invested in upgrading other industrial facilities in Guiana. The simplified pad concept deletes the requirement for large cryogenic arms on the umbilical tower by feeding the propellants from below the mobile launch table. The 600 ha ELA-3 includes the Batiment d'Intégration Lanceur (BIL) launcher integration building where the core is erected on the mobile platform and the boosters added, and the Batiment d'Assemblage Final (BAF) assembly building for completing the vehicle.

To the south is the Batiment d'Intégration Propulseur (BIP) integration hall for the P230 boosters. Solid propellant production and loading for the two large segments is undertaken at the 300 ha Usine Propergols en Guyane (UPG) plant, managed by Regulus, a BPD-SNPE Franco-Italian partnership, and commissioned 24 October 1991 at a cost of FFr1,250 million (ECU180 million). The 40 buildings total 26,000 m^2. UPG made the first casting, unsuccessfully, in November 1991. The forward segment is shipped from BPD in Italy.

The filled segments are taken by road into the 45 m high BIP and stacked as a single P230 on an 80 tonne 8 × 8 m 5 m high rail transporter. From BIP, it can be moved 1 km south to BEAP or to BIL 2.8 km north. P230's seven firing tests were performed 1993–95 on BEAP's 50 m vertical stand over the 200 m long, 60 m deep 35 m wide pit cut into the granite bedrock. No further tests are planned, but BEAP is maintained for possible future development tests and could even be converted to a small satellite launcher pad.

While the P230s are being stacked in BIP, the core is stacked on the 900 tonne 21 × 25 m 7 m high mobile platform in the BIL's 30 × 43 m 58 m high integration hall. The platform can move at a maximum 4 km/h on the 16 m gauge track. At this stage, the platform has only the 30 m integral lower umbilical mast. The BIL is 31 × 127 m overall, including the 26 × 83 m storage hall. This hall is only partly clad, on the rain-prone side, and used to store stage 1/2 in their containers. Stage 1 is rotated to the vertical in the erection hall.

Each P230 is rolled in to the integration hall and attached, remaining on its own pallet throughout the campaign. With the boosters now providing rigidity, the upper stage and vehicle equipment bay can be added. Meanwhile, final spacecraft preparation and encapsulation takes place in the upper portion of the 85 × 52 m 83 m high BAF. The upper payload is mated with the Speltra in the 30 × 30 m class 100,000 encapsulation area and the fairing fitted. The stack is rolled in to BAF's launcher integration hall and the lower payload attached. The upper portion then fits over. The platform's upper mast is added, batteries fitted, the upper stage fuelled under remote control, the hydrazine attitude control system loaded manually and the countdown started. CDL-3, located near BIL, houses two control rooms in separate wings for conducting simultaneous campaigns. The 1½ to 2 hour rollout to the pad 2.8 km to the north begins 8 hours before launch through BAF's 4-section 24 × 62 m door. The pad is designed to be operational again six months after an explosion.

Since 1992 ELA-3 facilities have been used for development and qualification tests on the VULCAIN engine. ELA-3 also comprises a production plant for liquid oxygen. liquid nitrogen and liquid hydrogen.

ELA-4
A fourth site for Ariane 5's successor will not be required for 15 to 20 years but potential locations have been considered because of the expected growth of nearby Kourou town.

Payload Preparation Complex (EPCU, Ensemble de Préparation Charges Utiles)
Payload preparations are conducted in three phases,1) satellite processing/checkout,) hazardous operations, and 3) integration with the launcher on the pad. Buildings S1A and S1B, in the CSG Technical Centre, provide clean room facilities for satellite preparation. S2 and S4, near the ELAs, are designed for solid kick motor preparation and X-ray operations. S3A/3B, also close to the two ELAs, are assigned to payload propellant filling operations, final integration, assembly of the Spelda dual launch carrier, and encapsulation in the Ariane 4 fairing. S3C is used for monitoring the hazardous operations. Ariane 5 payloads continue to use these facilities, although final encapsulation is performed in the BAF.

Ariane 5 emerges from the BAF final assembly building (ESA) 0054383

S1A/B
S1A, the older of the two satellite preparation/checkout buildings, provides a 488 m^2 class 100,000 area accessed by a 4.5 m wide, 10 m long, 10 m high airlock for simultaneous processing of two Ariane 3 class satellites in three bays. S1B was added to the north for handling two Ariane 4 class craft simultaneously. It provides a class 100,000 800 m^2 clean area accommodation with a total area of 300 m^2 for three satellite checkout stations, and air-conditioned 340 m^2 and clean 60 m^2 storage rooms, in addition to offices and meeting rooms for two satellite teams. Payloads pass from this Phase 1 processing to Phase 2 in buildings S3A/B.

S2/4
S2 is the restricted-access Solid Propellant Motor (SPM) preparation building for processing motors and pyrotechnic devices before integration with

Integration of Ariane 5.501 in the launcher integration building 0084630

The main entrance of the centre spatial Guyanais, Kouron, French Guiana 0084626

payloads in buildings S3A/B. Two halls of 97 and 48 m^2 include mechanical test equipment for Mage-type motors and electrical systems for Mage and Star 30/37 safe/arm devices. Building S4 is used for cold soaking and x-ray inspection of SPMs up to 3,000 kg 2 m diameter, with the 75 m^2 x-ray hall including a Linatron 2000 source (5.5/8/10 MeV) linked to an automatic, 32-position rotating table. The 9 m^2, 4 m high cold soak room allows motors to be held at 0 to 15°C.

S3A/B

S3A was designed for handling Ariane 3-class payloads and Syldas but it continues in use for smaller Ariane 4 craft. Hydrazine propellant tank filling, integration of ordnance, balancing and loading into the pad transfer containers for two satellites are undertaken in class 100,000 facilities. These are a 185 m^2 assembly hall, equipped with filling positions over a drain and a Schenck-type E-50S balancing machine, and a 172 m^2 Sylda assembly room. For Ariane 3 launches, the satellites were then encapsulated and transferred to the pad for the fairing to be added. On Ariane 4 campaigns, the payload is transferred into the newer S3B. This includes a 320 m^2 assembly and filling room equipped with a weighing machine (5,000 kg operator ± 1 per cent, resolution 1 kg), and a 400 m^2 encapsulation room. Encapsulation inside Spelda and the fairing is completed before pad transfer. Ariane 5 payloads are transferred to the BAF for their final handling.

India

Balasore Rocket Launching Station

Location: 21°25′N/87°00′E.

Background

Balasore has been set up by ISRO as a meteorological research station located in the north-east supporting sounding rocket flights from facilities adjacent to weather balloon release sites. Balasore has been considered for due south departures into polar paths, enhancing PSLV's Sun-synchronous capability by 500 kg. ISRO currently uses it for RH-200 sounding rockets. A second launch pad was under construction from where the PSLV would gain an additional 550 kg payload capability. Development of this site has been halted pending a decision on funding allocations and programme priorities. The Indian government is reviewing pad operations at all relevant sites in the country with a view to determining the impact of upgrades that may be involved with commercial operations for international operators.

Satish Dhawan Space Centre, SHAR

Location: 13°41′N/80°14′E

Current Status

Completed launch complexes now available for sounding rockets, SLV-3, ASLV and GSLV. Modifications for GSLV, the first of which was launched on 18 April 2001, have provided facilities for the eventual commercialisation of the GSLV although plans for that are not yet funded. The second development GSLV was launched on 8 May 2003 carrying the 1,825 kg GSAT-2 experimental satellite in a GTO.

Background

India's primary orbital launch site is housed at ISRO's Shar Centre on Sriharikota Island 100 km north of Chennai, but the first rocket firings were made from the Thumba Equatorial Rocket Launching Station in 1963. Together with the northerly Balasore Rocket Launching Station, the facilities are operated under the ISRO Range Complex (IREX) headquartered at Shar. Shar launches the full range of Indian vehicles from sub-orbital up to the PSLV but range safety considerations limit Shar to less than 140° launch azimuths.

Shar, encompassing the Sriharikota Island on the east coast of Andhra Pradesh, is ISRO's satellite launching base and additionally provides launch facilities for the full range of Rohini sounding rockets. The Vehicle Assembly, Static Test and Evaluation Complex (VAST, previously STEX) and the Solid Propellant Space Booster Plant (SPROB) are located at Shar for casting and testing PSLV solid motors.

The ASLV orbital launcher was integrated vertically, beginning with motor and subassembly preparations in the Vehicle Integration Building (VIB) and completed on the pad within the 40 m tall Mobile Service Structure. The PSLV launch complex was commissioned in 1990. It has a 3,000 tonne 76.5 m high Mobile Service Tower (MST) which provides the SP-3 payload clean room. The MST is rolled back 180 m for launch. Shar also has a 52 m umbilical tower, an open steel structure next to the launch pedestal. Modifications are under way to allow the complex to handle GSLV. The Launch Control Centre is 5.6 km distant and the CTR Checkout Terminal Room is next to the pad. A second launch pad is nearing completion.

PSLV's Mobile Service Tower provides payload clean room facilities. At bottom is the Checkout Terminal Room for vehicle testing before the more distant Launch Control Centre takes over (ISRO) 0517080

Liquid stage fuelling is done remotely from the LCC. All solid motors are processed in the SMPF Solid Motor Preparation Facility. As a launch campaign nears completion, stage 1's segments are stacked on the pad. Satellite preparation/integration is undertaken in three SP clean room facilities. SP-1 at the LCC has three bays: class 100,000 17 × 15 × 8 m high bay, materials airlock bay and dynamic balancing room. It provides satellite inspection, subsystem tests, RCS leak tests and solar panel deployment tests. SP-2 next to the pad is used for RCS filling and checking compatibility with PSLV's adaptor. SP-3 is part of MST for integrating payloads with the launcher.

The VAST was originally designed to handle motors up to 1 m diameter, 10 m length and 980 kN. ISRO modified it for PSLV for 2.8 m diameter and 4.9 MN levels (first PSLV stage 1 test was made in 1989). The horizontal firing stands provide 1-/6-component facilities. The vibration platform was upgraded from 13.5 tonnes to a 30 tonne dual shaker operating over 5–2,000 Hz with maximum acceleration of 100 *g*. Specimens up to 2 tonnes can be shock tested to 80 *g* by free-fall to simulate motor ignition shocks, and upper stage motors are fired in a 50 km-equivalent high-altitude test facility. The 25 km^2 SPROB was commissioned in March 1977 for SLV-3's motors but it was then augmented to cope with PSLV's 2.8 m diameter 25 tonne segments, requiring a 2.5 tonne vertical mixer and 15 MeV linear accelerator for the radiographic facility in addition to the previous 8 MeV machine.

Approximately 2,300 people were employed at the show centre in 2001.

Thumba Equatorial Rocket Launching Station

Location: 8°32′N/76°52′E.

Background

TERLS supports RH-200/300 sounding rockets. It was the site of India's first sub-orbital launch in 1963 and in 1968 was dedicated to the United Nations. The site has been surveyed for possible expansion based on some interest in using the site for scientific research flights and for development of new solid propellant sounding rockets.

Indonesia

Paneungpeuk

Location: 1.0°S/136°E

Background

Indonesia launches its one- and two-stage solid propellant sounding rockets by the LAPAN national space agency for upper atmospheric research from the Paneungpeuk range some 150 km from Bandung. China considered contracting with Indonesia to build an equatorial launch site nearby for China's CZ-3 during 1987. One possible location under consideration was Biak Island. Such plans have now been abandoned. Indonesia has unsuccessfully sought to attract foreign finance for developing the site as an international research station. These efforts have attracted several potential foreign investors but there is no interest from the Indonesian government in supporting such an initiative.

Israel

Palmachim Air Force Base

Location: 31.52°N/34.45°E

Current Status

Site capable for LK-1 and Shavit launch vehicles with recent upgrade and improvements to facilities and tracking capabilities.

Background

Israel launches its Ofeq satellites orbital shots from an area near or within Israel's Palmachim air force base south of Tel Aviv, and near the town of Yavne. The Israelis inaugurated it as an orbital site with the launch of the first Shavit launcher carrying the Ofeq 1 satellite 19 September 1988. The facilities are classified although they are visible from the coast road. Located at the eastern end of the Mediterranean, the site is restricted to retrograde departures for range safety considerations (although the trajectory still passes over southern Europe). The second orbital launch was made in April 1990 and the third in April 1995 following a launch vehicle failure in April 1994. Ofeq 4 was launched from here in January 1998 but failed to reach orbit. The Ofeq 5 observation technology development satellite was launched on 28 May 2002. Commercial satellites are placed in orbit by launch vehicles outside Israel.

Italy

San Marco

Location: 2.93°S/40.21°E

Background

Italy's offshore orbital launch platform was last employed in March 1988 – its first firing in 12 years. The Italian Vega (previously called San Marco Scout) would use the site, although the vehicle's future is highly uncertain. San Marco is owned and operated by the Italian government. It comprises two platforms in Formosa Bay 4.8 km off the coast of Kenya. The San Marco platform provides the launch pad, while the Santa Rita platform houses blockhouse facilities. The range is maintained in operational condition and the telemetry and tracking station still works under ESA contract. A new MoU was signed with Kenya in March 1995.

The range became operational in 1966 and by May 1976 eight satellites had been launched, four Italian, one UK and three US. The last of nine launches from San Marco was in 1988. The launch platform has 20 steel legs sunk into the sandy seabed, and is linked by 23 cables to the control platform, a modified oil rig, 920 m distant. Explorer 42, the small astronomy satellite launched by an Italian crew on 12 December 1970, was the first US satellite launched by a foreign country. Because of the position of this equatorial site, the Scout put the 195 kg Explorer 43 into orbit, which it could not have done, from a more northerly US location.

In addition to the Scout facilities, there is a ramp for sounding rockets such as US Nike Apache and Nike Tomahawk. The facility has been identified as a potential site for ballistic flight tests with suborbital boost stages designed for orbital space launchers.

Japan

Kagoshima

Location: 31.2°N/131.1°E

Current Status

Major improvements in Japan's launch vehicle programme have influenced the modifications of Kagoshima and Tanegashima now underway. Both launch sites are now operated through the auspices of JAXA which became effective from 1 October 2003.

Background

Established in 1962 ISAS' assigned launch site, Kagoshima, is primarily a sounding rocket facility, but handles a science satellite every one to two years. It was restricted, by government directive, to all-solid vehicles of 1.4 m diameter but approval for the 2.5 m diameter M-5 was made in 1989 to undertake modest deep space missions. The Sakigake and Suisei Halley's Comet probes of 1985–8 provided Japan's first experience of beyond Earth orbit.

Kagoshima's protected pad is used for sounding rockets. This is the ISAS/Nissan S-520 single stage vehicle, capable of delivering 150 kg to 350 km
0517082

Japan's first six satellites were launched from these levelled hilltops facing the Pacific Ocean at Uchinoura on the southern tip of Kyushu Island. By the end of 1999, orbital launchings totalled 23 successful, with 351 launches of all types. Construction of a sounding rocket site began in February 1962, with extensions for science satellite launchings by M rockets completed in 1966. The first Lambda 4 was fired in 1964 but it was another six years before the first successful orbital launch was achieved. Kagoshima's M-3SII all-solid ISAS/Nissan orbital launcher retired in January 1995. The S-310, S-520 and K-9M sounding rockets and the MT-135 meteorological rocket are currently supported. Annual launch rate is typically three to four for the larger sounding rockets.

A satellite processing facility is linked with the M pad and assembly room, and a sounding rocket payload assembly centre with the covered pad. Facilities have been upgraded to support the new M-5 orbital vehicle.

Departures from this site are normally restricted to 45 day January–February/August–September 'launch seasons' because of range safety procedures and the influential fishing industry lobby.

In addition to Usuda's 60 m station, ISAS operates a 20 m dish at Kagoshima built originally for lunar fly-by communications. It provides S-band uplink and S/X-band downlink.

ISAS' seven other principal facilities are the Kagoshima Space Centre, Noshiro Testing Centre (established 1961 on Asanai Beach for solid/liquid engine static testing), Usuda Deep Space Centre (64 m S/X band deep space tracking and TT&C station), Sanriku Balloon Centre (500 kg payloads on 200,000 m^3 helium balloons up to 40 km altitude), the Space Utilisation Research Centre (established 1988 on the new campus), the Centre for Planning & Information Systems (established 1993), and the Centre for Advanced Spacecraft Technology (1995).

By the end of 2002 Japan had completed 24 orbital launches from the Kagoshima facility.

Takesaki

Location: 30°22′25″N/130°57′56″E

Background

The Takesaki Launch Site sits to the southeast of Tanegashima for small sub-orbital solid propellant rockets. It is little used but its pad is activated as required for tests related to other JAXA programmes. Included is the launch of the TR-1A solid propellant sub-orbital vehicle flown from here

NASDA's Takesaki Range Control Centre (top) assumes responsibility for H2 after lift-off, taking over from Yoshinobu's blockhouse (bottom) (NASDA) 0517083

since 1991, LS-C, JCR, TT-500A and TR-1 series. It is proposed to develop this site for testing medium-thrust solid propellant strap-on boosters.

Tanegashima

Location: 30°24'N; 130°58'E.

Current Status

A government review of launch site effectiveness has concluded that Tanegashima can absorb a higher launch rate. Two successful flights of the H2A launch vehicle have demonstrated a 50 per cent reduction in launch costs and Japan is keen to use Tanegashima for commercial flights competing against Ariane, Delta, Proton and Sea Launch/Zenit.

Background

Departures from both sites are normally restricted to 45-day January-February/August-September 'launch seasons' because of range safety procedures and the influential fishing industry lobby (March 1995's H2 launch was a unique exception). The solid strap-ons of JAXA's H1 impacted some 19 to 29 km out, but H2's fall beyond the 190 km range safety limit, provided the potential for year-round firings (although launch failures would still be potentially hazardous). Rocket Systems Corporation is negotiating to increase launch opportunities for commercial H2. NASDA orbital launch facilities are centred on Tanegashima Island, with the Tsukuba Tracking & Control Centre acting as the primary satellite command facility. Subordinate stations are located at Katsuura, Okinawa and Masuda, and facilities on Ogasawara and Christmas Island provide downrange launch tracking. The Kakuda Propulsion Centre is responsible for testing high-performance propulsion systems, but JAXA's major technical facility is the Tsukuba Space Centre. New facilities include a second NDTF Non-Destructive Test Facility for computed tomography examination of SRB segments. Kick motors are still tested in H1's NDTF-1.

JAXA's TNSC orbital launch centre is sited on the southeast tip of 8.6 km^2 Tanegashima Island, 1,000 km southwest of Tokyo. The more northerly Osaki Launch Site includes the H2/J1 pads and liquid engine static test facilities, while the Takesaki Launch Site handles sounding rockets and provides facilities for H2 solid booster static firings and the H2 Range Control Centre.

H2's 150,000 m^2 Yoshinobu complex was built from 1985 to 1992 for launching the larger vehicle from the headland west of J1/H1's site. H2 is integrated in the Vehicle Assembly Building on a mobile platform. A single platform is adequate for two annual launches, but increasing the rate to four would require a second platform. A new pad was be completed in 1999 for the Upgraded H2. LE-7 stage 1 engines are static fired in a dedicated facility off to one side, and the strap-on boosters have the uprated N/H1 solid static site at the Takesaki Launch Site.

The pad at 30°23'45"N; 130°58'22"E handled its last H1 launch in February 1992 and was then modified for the heavier J1, requiring a larger crane, adapted access platforms and elimination of one umbilical tower. J1 stands on a 6 m stool on the launch deck to minimise the changes needed for the Mobile Service Tower. The complex was adapted for H1 by adding cryogenic facilities and a second, taller, umbilical tower to the existing N1/2 complex. The 6.4 m high 12 m wide launch deck was serviced by the two umbilical towers 43 and 49 m high (43 m removed for J1) and the 67 m 2,650 tonne MST, rolled back 100 m before launch.

Rocket Assembly Buildings 1/2 received H1's first two stages for processing before erection on the pad. MB-3 engines and components could be static tested north of the complex. Solids and pyrotechnics are received in the Solid Motor Test Building, with NDTF-1 for ultra-sonic and X-ray inspection of solid motors, and the Spin Test Building used for integration of H1's stage 3 with its spin table. Payloads are processed in the class 100,000 STA-1 in preparation for mating with their upper stage in the Third Stage & Spacecraft Assembly (TSA) building. The assembly was then shipped to the Mobile Service Tower in a protected container for mating with stage 2.

J1 countdowns are still controlled from the Osaki Block House, a reinforced concrete partially submerged structure 170 m from the pad. Control after lift-off, and until the payload separates, is the responsibility of the more distant Takesaki Range Control Centre.

Now satellites are processed in the new class 100,000 3,930 m^2 STA-2 Spacecraft Test & Assembly building and then moved to H1's modified class 100,000 4,010 m^2 SFA Spacecraft & Fairing Assembly building, where propellants are loaded, pyrotechnics and solid motors inserted, and the fairing is added. The complete payload assembly is then transferred to the pad to be attached to the waiting H2.

Orbital launch facilities at NASDA's Tanegashima site. The static test stand at centre right is used for LE-7 firings. H2 is stacked on its mobile platform in the Vehicle Assembly Building at right and moved the 500 m by rail to the pad. At top centre is the J1 pad. Beyond it, the old Osaki Range Control Centre stands unused as it is too close for H2. Launcher and payload processing buildings are at top right (NASDA) 0517427

H2's stages and SRBs are received in the VAB's 32 × 46 m 19.1 m high low bay before hoisting on to the mobile launcher platform in the 32 × 27 m 66 m high high bay in solids/stage 1/stage 2 order. The 800 tonne 18 × 22 m 4 m high ML is powered by electric motors on four dollies at 0 to 0.48 km/h for the 500 m rail transfer to the pad. ML provides LOX/LH_2 to stage 1 and is protected by a water deluge system. The exhaust opening is 4 × 9 m.

The PST Pad Service Tower comprises a fixed tower plus two rotating service structures. The 8 × 15 m 67 m high 1,500 tonne tower provides 10 floors, H2 umbilicals and payload assembly installation. The 2,500 tonne main RSS provides 12 floors, the top two adjustable, and 20 tonne crane. The 800 m^2 launch blockhouse, submerged 4.8 m, is next to the VAB. Control after lift-off, and up to about 30 minutes later when the payload separates, is the responsibility of the distant new Takesaki Range Control Centre, built because the existing Osaki centre is within the launch danger zone.

By the end of 2002 JAXA and the former NASDA had launched 34 orbital shots from Tanegashima.

Norway

Andøya Rocket Range

Location: 69°17'N/16°01'E

Background

The first sounding rocket, a Nike Cajun, was launched from Andøya on 18 August 1962. ARR is now an operational unit within the Norwegian Space Centre, with 27 staff. Andøya is the world's northernmost permanent launch facility for sounding rockets and scientific balloons. There were more than 650 launchings of all types to the end of 2002, primarily to investigate the upper atmosphere in the polar region and to study ionospheric and magnetospheric processes at high latitudes.

The range has been supported since 1972 by ESA through the Special Project Agreement, under which it is maintained by and made available to some ESA states. It is also operated for commercial and bilateral projects. The range has a very large vacant impact area offering users a wide range of possible rocket configurations and launch vehicles without guidance, reducing costs and expanding the user market. The largest launch facility can handle vehicles up to 20,000 kg. Vehicles currently handled include Viper, Dart, Loki, Nike Orion, Orion, Skylark and Black Brant. There are eight pads, including three universal launchers (two with 3 tonne and one with 20 tonne load limits) and a small rail launcher (over-slung type). The new universal launcher 3 (U3) with 20.5 m rail length and 20 tonne safety weight limit became available from late 1993.

A vehicle is prepared horizontally in the protective 36 m long 5.5 m wide 5.2 m high concrete housing and raised shortly before launch. At launch, the launch elevation can be set from 65 to 90° (nominally 85°), the azimuth 260 to 20° (nominally 340°). Four-stage rockets up to 23 m long, 1.2 m diameter, 520 kN thrust (Black Brant 12 class) can be launched into an impact area about 1,900 km downrange.

U3 can also handle the launching of small satellites up to 250 kg into high-inclination orbits. If ESA ever exercised the option this site would be the only satellite pad in Europe. The Polar Satellite Launch Service has passed PDR and is searching for a suitable vehicle. Several US and Russian launchers have been studied, but none satisfies the PSLS cost-performance requirement. PSLS seeks the capacity to put 250 kg into a 750 km orbit. NSC's studies showed that it could capture three orbital launches annually, with a capacity of six. PSLS will be limited to vehicles with propellant totalling less than 40 tonnes because of ground safety considerations.

The 272 m^2 Main Assembly Hall accommodates vehicles up to 15 tonnes.

The User Science Operations Centre is located 300 m northeast of the control centre with two laboratories and a recording room. Payload recovery is performed routinely from the Norwegian Sea using ships or helicopters. New tracking facilities using radar and slant range systems are designed to improve impact prediction to 500 m for sea-recovered capsules.

A downrange telemetry station and a sounding rocket mobile launch facility became operational at Ny-Ålesund (79° N) on Svalbard in the Sababaru Islands during 1998. Climatically, Svalbard is an arctic desert, warmed by the Gulf Stream (mean monthly winter low of –14°C) with 200 to 300 mm annual precipitation. Longyearbyen is the administrative centre, with 1,000 residents and an international-standard airport. The position of the Svalbard Satellite Station (SvalSat) on the Plataberget close to Longyearbyen will allow it to cover every orbit of any polar satellite. Data would be relayed to the mainland via Telenor's terminal at Isfjord Radio.

There is a 20 m VLBI antenna to measure continental drift, and the Norwegian Mapping Authority also operates GPS/Glonass receivers as part of the systems' core network to determine precisely the satellites' orbits. The SvalRak launch facility allows 3 tonne rockets to climb parallel and perpendicular to the magnetic field, as well as directly into the polar cusp and cleft regions.

ARR handles balloons of up to 50,000 m^3. A mobile tracking telemetry unit allows flight times of about 72 h. The balloons drift east in winter and west in summer. An increase in environmental research campaigns (including aircraft, long-duration balloons of up to 600,000 m^3, sounding rockets and ground-based observations) is planned for the updated range. The ALOMAR Arctic Lidar Observatory for Middle Atmosphere Research became operational in July 1994. For the first time, data from several lidars are combined for routine atmospheric observations, including composition, over 10 to 100 km. One lidar is designed for ozone monitoring.

By the end of 1999 Andøya had conducted 698 rocket launches and hosted scientists and engineers from more than 70 institutes and universities.

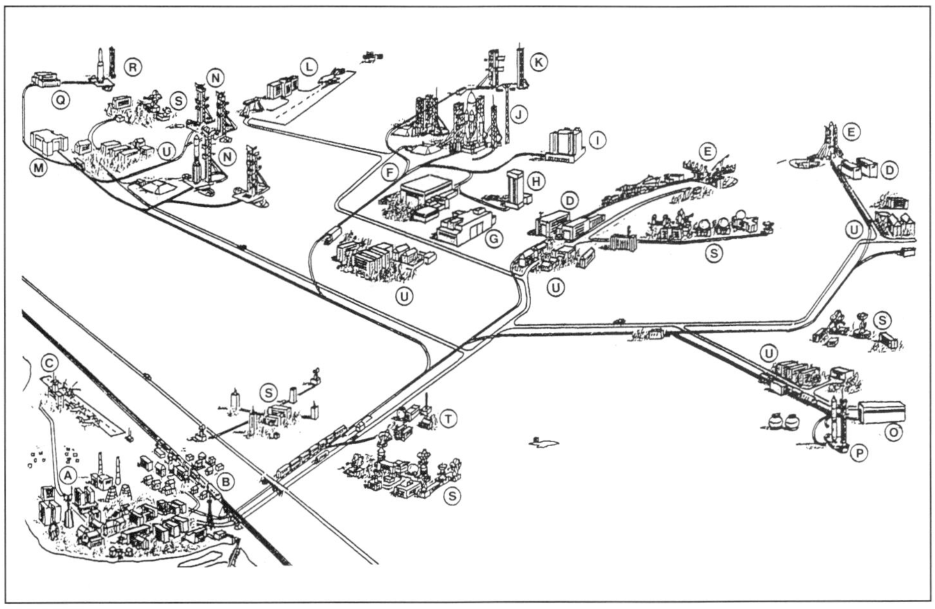

Baikonur Kosmodrome launch and processing facilities. A: Leninsk. B: Tyuratam. C: Krainiy airport. D: Soyuz launcher/payload processing (older is site 2; newer is site 49). E: Soyuz pads; Gagarin pad (often referred to as Pad 1) is site 2; newer is site 31. F: Energia processing building, building 112 site 130 (ex-N1). G: Buran processing building, site 254. H: Energia mechanical test stand. I: Buran/Energia fuelling building. J: Energia/Buran pads (2, site 110; ex-N1). K: Energia pad; complex 113 (but site 250 also referred to). L: Buran Yubileini landing strip (site 251). M: Proton processing building, newer is site 92. N: Proton pads, 81L/R plus 200L/R. O: Zenit processing building, site 42. P: Zenit pads, 45L/45R. Q: Tsyklon processing building. R: Tsyklon pad, 90L/90R. S: launch tracking system. T: KAZ LOX/LN$_2$ plant, site 3. U: technical facilities 0517073

Russian Federation

Baikonur/Tyuratam

Location: 45.6°N/63.4°E

Current Status

A programme of modifications has been delayed by lack of agreement over the responsibilities of user agencies, by differences between Kazakhstan and Russia and by a general lack of funds. Most launches conducted by the complex are for commercial customers or as part of international agreements. The facility has assumed a vital role in processing and launching replacement caretaker crews for the International Space Station since Shuttle flights were halted following the loss of Columbia on 1 February 2003 and are unlikely to resume before the second half of 2004.

Background

All manned flights and planetary missions are launched from Tyuratam. The site is prohibited from using due-east azimuth launches (the most efficient) because the lower stages would impact in China. Lieut Gen Alexei Shumilin, appointed in 1992, commands the facility. Interestingly, he is an officer of the Kazakhstan armed forces, overseeing a facility developed by the Soviet Rocket Forces. VKS personnel still totalled 14,500 by 1995, but all civil launch operations are being transferred to the RKA. Some 700 RKA staff, in blue uniforms, were already in position for the Soyuz-TM 22 launch in September 1995. 300 had transferred from VKS.

The Baikonur Kosmodrome offers nine launch complexes, with 15 pads, to accommodate space and missile activities. It is the only Kosmodrome supporting Proton, Zenit and Energia. Energia's three pads include the original facilities of the ill-fated N1 lunar launcher, which failed in four attempts between 1969 and 1972.

The Kosmodrome was built by a team led by V P Barmin, to house a site for missile and rocket tests which began in the area in the early 1950s. As a historical site, the first pad built launched both Sputnik 1 and Yuri Gagarin. Western media were allowed, in May 1988, to observe the departure of a mapping satellite from this facility. It is now designated as the Gagarin Launching Complex.

A second pad with manned launch capability was completed in 1965 about 30 km from the first. It is now used exclusively for military vehicles.

A Japanese astronomer pinpointed the site's location in 1957 as being near Tyuratam in Kazakhstan, east of the Aral Sea and about 370 km southwest of the real Baikonur. However, the Soviets continued for at least 17 years to give its latitude/longitude as that of the town of Baikonur. Kazakhstan finally named it after Tyuratam in 1992, although locals still refer to it as Baikonur, as does the entrance sign at the main gate.

Not until Soyuz-T 6 in June 1982 was a Western journalist admitted, but US reconnaissance satellites provided detailed information for the military, while Landsat and Spot images brought the sprawl of pads into the public domain in 1974.

Leninsk, 2,100 km from Moscow and founded 5 May 1955, has an official population of 60,000, down from the 100,000 peak of the mid-1980s, although the recent influx of Kazakh peasants probably yields a much higher total. It grew up alongside the launch facilities as the administrative centre of the 1,560 km^2 kosmodrome. 46,011 km^2 is allocated for downrange stage impacts. Cosmonauts are accommodated in a hotel complex with swimming pool and sports centre. RKK Energia earlier maintained a permanent staff of 2,000 at Tyuratam, reduced by 1993 to 800 to 1,000 (mainly for Mir and Proton) because of the loss of Buran and Energia work.

Tyuratam's largest known accident came on 24 October 1960, when at least 165 were killed by the on-pad explosion of a developmental SS-7 ICBM. Although the West had known for years that a major catastrophe had occurred, the USSR did not acknowledge it until 1989 nor give the death toll until the 30th anniversary. Marshall M I Nedelin, Cdr of the Strategic Rocket Forces, was under pressure from Khruschev to hold to schedule when a fault prevented launch. Rather than drain the propellants, Nedelin saved time by ordering engineers to work on the vehicle as it stood. Stage 2 ignited and detonated stage 1. SS-7's designer, Mikhail Yangel, narrowly escaped death. A Moscow TV programme 26 February 1995, claimed that 91 deaths were caused by a 1969 launch failure.

The Minsk agreement of 30 December 1991 declared that Baikonur would provide free access for other states. Although Russia holds up to 90 per cent of space industry capability, Baikonur's loss would have been a major blow. Under the Russian/Kazakh agreement of 25 May 1992, Russia pays 94 per cent of operating costs and Kazakhstan 6 per cent.

Before the USSR's break-up, Tyuratam cost Rb3 billion annually to run. Disagreements over Baikonur have continued to rumble on, with the military complaining of lack of investment. Conscripts from Kazakh battalions have rioted because of the poor conditions. Russian premier Viktor Chernomyrdin signed an MoU December 1993 that Russia would rent some of the 4.5 million ha currently occupied by Baikonur. The MoU also called for gradual demilitarisation; there were 28,000 servicemen based there (it was agreed later that 16,000 would remain for space activities).

Presidents Yeltsin and Nazarbayev agreed 28 March 1994 that Russia has sovereignty and jurisdiction over Tyuratam and Leninsk and their personnel for 20 years (with a further 10-year option) for the equivalent of US$115 million annually. The final agreement was signed 10 December 1994. Leninsk formally returned to Russian jurisdiction 1 August 1995. That US$115 million might not involve actual payments, but is likely credited to the Kazakh national debit to Russia.

The climate can be extreme, +60°C in summer, while suffering violent snowstorms and –40°C frosts in winter.

Soyuz and Progress spacecraft and their launchers are integrated in the MIK building. Following mating, the vehicle is transferred on a rail transporter and erector to the pad, 5 km distant. The cosmonauts suit up in a small room in the rear of MIK and ride to the pad in a bus. Launch operations are controlled from a blockhouse 500 m from the pad; a public viewing stand is only 300 m further out.

Northwest of the Soyuz buildings is the Energia and Buran area, originally built for processing the N1. Energia was integrated horizontally in the 40 m high 274 × 160 m old N1 Building 112 after the core was shipped in by a Mya 201M aircraft. The payload was added and the combination lifted on to the

Tyuratam's Gagarin pad has supported more than 350 orbital launches. Vehicles are moved to the pad horizontally by rail – the transporter is visible at right 0517074

The Cosmonaut Yuri Gagarin was the flagship of the Soviet tracking fleet, sporting pairs of 25 m diameter 'Ship Shell' and 'Ship Bowl' dishes. It was the only ship to carry the 240 tonne larger antenna (Jane's Soviet Intelligence Review/LL van Ginderen) 0517075

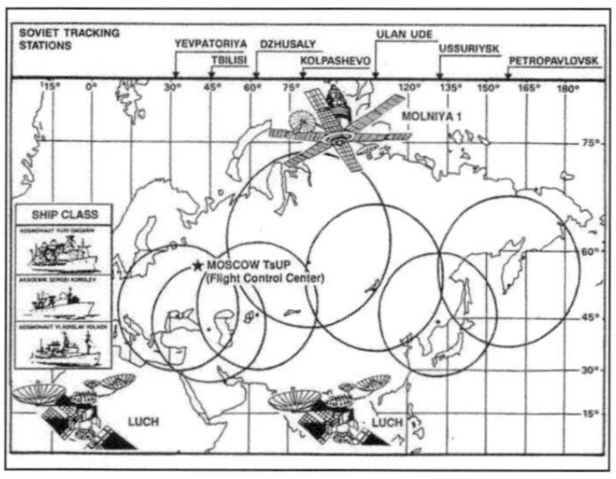

The network for controlling manned missions is augmented by Luch satellites at the central and western SDRN geostationary positions. Molniya provides links from the ships and remote land sites to TsUP (ships now not operational) (Teledyne Brown Engineering) 0517076

transporter/erector. The stack moved to a separate building for Buran fuelling.

In September 1995, a fully integrated Energia and two cores sat in residence. The adjacent Buran building will be upgraded for processing payloads to Western standards. Only the flown Buran remained in residence by September 1995, whereas in October 1994 three other models were also nearby. Since then, Mir's Spektr module became the first to be processed for launch since Buran. Space Station elements are also processed here.

There are two Energia launch complexes 20 km apart: the static test stand (number 113) used for the maiden launch and a pair originally built for N1 capable of handling the Shuttle (number 110). Energia's static stand was purpose-built because there was no existing N1 equivalent – the lunar vehicle was considered too large for static tests.

Some 12 km further northwest is the 4,500 m long 84 m wide Shuttle Yubileini landing strip on a northeast-southwest orientation. It supported jet-powered test and training flights and provided a microwave system for fully automatic landings. Operations were controlled from a nearby building. The runway fell into disrepair after Buran's 1988 landing, but a German company repaved the 52 × 3,200 m central portion in 1995, under contract to LKE, to receive commercial Proton payloads.

To the west of the Shuttle strip is the Proton preparation and launch area. The building (number 92) can handle up to six launchers simultaneously. Rail transfer of the fully assembled vehicle to the pad requires about 7 hours, followed by five days' preparations before launch. Most launches are currently from pad 81L, but will eventually switch to the refurbished 81R. 200R has been reactivated for commercial and Space Station launches (and transferred from VKS to RKA control). 200L is certified for only a handful more launches before it must be refurbished. It will then transfer to RKA. Number 81 is the original complex. Number 200 was built in the 1970s several km to the east with pads 600 m apart. By the time of Granat's launch on 1 December 1989, number 200 had serviced 114 launches.

Zenit's pad 45R was severely damaged 4 October 1990 when a vehicle exploded within seconds of departure; 45L was inaugurated with August 1991's departure. Together, they would be capable of supporting 15 missions annually. 45R was so severely damaged, collapsing the launch table, that it was not repaired until the late 1990s.

By the end of 2002 a total of 1,113 orbited flights had been launched from Baikonur since Sputnik 1 on 4 October 1957.

Kapustin Yar

Location: 48.4°N, 46.5°E

Current Status

Kapustin Yar provides facilities for the United Start company based in the USA but using launch complexes 86 and 107 for Cosmos launches.

Background

Known as the Volgograd Station to its personnel, this was the first rocket development centre, hosting its first launch in 1947. It became a source of renewed interest when it was used in June 1982 for Kosmos 1374, the first test of the subscale spaceplane flights. Its two launches in 1984 included the fourth spaceplane test. Kapustin Yar in later years handled only occasional missions, possibly for radar calibration. There has been only one launch since 1987, leaving the total at 84. It is no longer under VKS control but there are still some missile testing activities. The facility is capable of handling Kosmos sub-orbital launches continue and AKO Polyot offers commercial orbital launches, partly using Kosmos USA processing facilities transported from Plesetsk. Its first orbital launch was Kosmos 1 in 1962. In 1975, a Landsat 2 image revealed publicly that its facilities extend over a 96 × 72 km area northeast of the River Volga and 965 km southeast of Moscow.

During its early years, Kapustin Yar tested captured V2 missiles and conducted sounding rocket experiments carrying dogs and other animals up to 500 km. By 1980, there had been 70 orbital launches, mostly small Kosmos science satellites but, as work was switched to Plesetsk, the annual launch rate fell to an average of only one. The last launch from Kapustin Yar took place in June 1988, a suborbital test of a 1/8th scale model of Buran to a height of 120 km and a pitch-down manoeuvre to drive the orbiter down to a speed of Mach 18.5. By the end of 2002, 84 flights had been launched from Kapustin Yar. Modification and refurbishment plans have been temporarily suspended.

Plesetsk

Location: 62.8°N/40.1°E

Current Status

United Start, based in the USA, has access to launch complexes 131 and 132 for flights of Start1/Start 2, Cosmos and Cyclone 2 launch vehicle booked by commercial customers. Special payload handling facilities have been set up and new and improved acceptance test and checkout capabilities.

Background

Plesetsk was, for a long time, by far the world's busiest spaceport but has now been overtaken by Baikonur. Orbital launches have declined for several years, falling from a steady 62 per year in 1983–85 to 56 in 1986, 47 in 1987, 47 in 1988, 45 in 1989, 48 in 1990, 37 in 1991, 33 in 1992, 26 in 1993, 18 in 1994, 13 in 1995, 10 in 1996, 9 in 1997, 7 in 1998, 4 in 1999 and 5 in 2000. The decline has accelerated as programmes transfer to newer vehicles at Tyuratam. Plesetsk's total record of 1,461 vehicles by the end of 2000 is 378 higher than Baikonur's.

The site's 1,762 km^2 base is located 170 km south of Archangel, which enables communications and spy satellites to be placed in polar (but never retrograde) and highly elliptical orbits. Range safety restrictions now limit most flights to 62.8°, 67.1°, 73–74°, 82–83°.

Plesetsk's existence was established publicly by Geoffrey Perry, of Kettering Grammar School (England), following the first orbital launch, Kosmos 112 on 17 March 1966 from pad 1. It was officially acknowledged only in 1983. Originally intended as an ICBM site, construction of the first of several pads for the R7/A class vehicle began in April 1957, and was completed in December 1959. The pad and ICBM began active duty in January 1960. Three other pads were active by July 1961.

Landsat imagery taken in 1973 showed Plesetsk to be about 100 km long, housing at least four complexes and defended by surface-to-air missiles. Mirny has grown alongside to house staff and their families; population is 40,000.

There are nine operational pads: three for Cosmos at complex 133 (one converted for Rockot), four Soyuz/Molniya (number 16 converted for Russia, number 41 being converted for Rus, 2 at number 43) and two Tsyklon number 32 (at 62°54′26″ north/ 40°47′23″ east). Complex 35 has two Zenit pads and vehicle/spacecraft processing facilities now under construction for polar missions. The new Angara will also use the complex. One pad was retired in 1989 after the launch of the Cosmos 2044 international biosatellite.

Climatic conditions are extreme ranging from –47 to +34°C, with up to 85 cm of snow cover.

The most powerful launcher available at Plesetsk is still the Molniya. Tsyklon is processed horizontally in a six-storey building, where stage 3 is fuelled, and not delivered to one of the two pads 40 km away by train until about T–3 h.

Two major accidents were acknowledged in 1989. Nine technicians and soldiers were killed on 26 June 1973 (SL8 Cosmos) and a further 50 on 18 March 1980 (SL-3 Vostok with Meteor). On both occasions a vehicle exploded during fuelling operations. Collection of the hundreds of strap-ons and first stages discarded over the area started in 1991 in response to complaints from local residents over environmental contamination; compensation was also claimed. The junk is now processed to recover useful materials. By the end of 2002 Plesetsk had launched 1,476 orbited shots since the first on 17 March 1966 making it the most used launch site in the world.

Svobodny

Location: 50°00′–51°40′N/128°00′–128°30′

Current Status

United Start has an agreement to use the Svobodny cosmodrome for commercial customers.

Background

Used as a missile test range off limits to civilians during the Soviet era. Proposals for a new Kosmodrome at the decommissioned missile site at Svobodny, near the Chinese border, surfaced during Russia's negotiations with Kazakhstan over Baikonur's future. They were widely perceived as a negotiating ploy, but President Yeltsin signed a decree on 1 March 1996 approving the Kosmodrome's creation. Start used it first, employing its own transportable launcher. Rockot will follow, employing one of five silos. The largest investment will be a two-pad Angara complex. 60 SS-11s were silo based at the site. The maiden orbital launch for Svobodny took place on 4 March 1997 when Zeya was placed into orbit by a Start-1 flown from a mobile launch platform. A second Start-1 flight on 24 December 1997 carried the Early Bird 1 commercial satellite for EarthWatch Inc. Another Start-1 launched the Israeli satellite EROS A1 from Svobodny in December 2000 followed by a Start-1 carrying Sweden's Odin satellite in February 2001.

South Africa

Overberg Toetsbaan (Overberg Test Range)

Location: 34°35′S/20°19′E

Background

Developed by the Houwteq organisation in the 1980s, OTB is a 43,000 ha site along 70 km of southeast coastline of the Western Cape. Its position near Africa's southernmost tip is ideal for launches into Sun-synchronous, polar and Earth-synchronous orbits down to about 38° inclination. Originally, it was set up to provide launch facilities for ballistic missile tests. The site is a division

OTB's central control room (Denel-OTB) 0517430

of Denel, with a permanent staff of around 25. Employment peaked at 500 in the early 1990s. Construction of the test range and launch facility began in 1983, allowing the first simple tests in 1987. A first stage test of South Africa's planned orbital vehicle was launched in June 1989, followed by stage 1 and 2 in July 1989 and November 1990, leading to the facility's formal qualification by the end of March 1991. RSA cancelled its satellite launch programme here in 1994, but OTB's core capability is maintained through regular use on a wide variety of flight tests. South Africa joined the MTCR as a full member in 1995, allowing OTB to be used by other parties subject to non-proliferation controls. The site is expanding its utilisation of existing facilities for aircraft and weapons systems testing.

The range, close to the villages of Bredasdorp and Waenhuiskrans, is divided into east/west sectors by the De Hoop Nature Reserve, with most of the instrumentation and infrastructure in the western half. An adjacent air force airfield, part of the SAAF Test Flight and Development Centre, includes a 3,000 m runway and can accommodate a Boeing 747 and is available for OTB clients. A good quality highway between Cape Town and Bredasdorp allows the 200 km to be covered by car in 2 hours. OTB's own accommodation caters to 100 people.

The area has a mild Mediterranean climate with a temperature of +5/25°C, 40 cm annual rainfall (mostly May to September) and 80 per cent humidity).

OTB's preparation and launch facilities are:

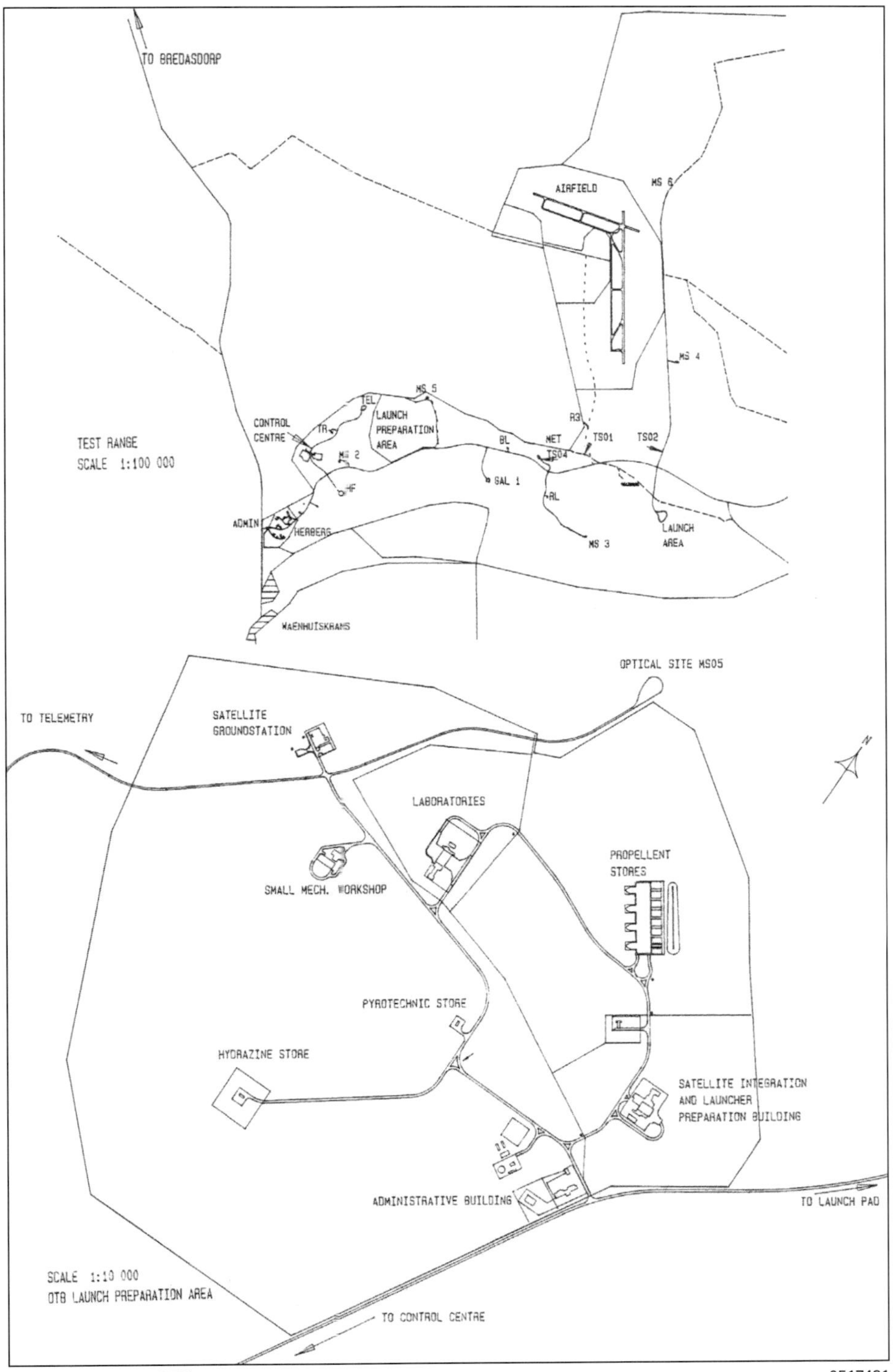

OTB layout (Denel-OTB) 0517431

Launch Preparation Complex

The LPC handles final preparation/checkout and mating of satellite and launcher. The preparation control centre for monitoring this work is in an LPC basement area. LPC comprises:

S1

Spacecraft Launch Preparation for final satellite preparation, mating with the final stages, fairing encapsulation, EM integration with the Satellite Ground Station via S-band. 12 × 17 m 13 m high class 100,000, 5 tonne overhead crane with 11.4 m free hook height. A 3.5 × 5.5 m stabilised column provides satellite alignment. 12 to 20°C clean N_2 at 10 m^3/min provides satellite cooling after encapsulation. EGSE can be housed in S1 or C1/LCC.

S2

Hydrazine Filling is a class 100,000 for fuelling satellite and launcher final stage from adjacent S3 bulk container. Water drenching and neutralisation for spillage. 4 ×10.5 m, 7 m high.

S4

Spacecraft Long-Term Storage, a 4 × 10.5 m, 7 m high class 100,000, for storing fully integrated satellite/final stages for fast reaction launches. At least two 2 × 2 m, 4 m high satellites can be stored simultaneously.

L1

Launch Vehicle Preparation made up of two sections, a 21 m and a 14 m high high bay, both serviced by a 90 tonne crane, 11 m high hook. In high bay, first stages can be mechanically and electrically integrated horizontally on mobile launch platform, and elevated to vertical for system functional testing.

C1

Launch Preparation Control Centre for controlling and monitoring satellite and vehicle preparations, and the first few seconds of flight. The total launch operation is controlled from the Test Range control centre. 5.6 × 10.5 m, 4 m high.

A1

Hydrazine Storage (storage and chemical analysis).

A2

Pyrotechnics Magazine.

A3

System Magazine (solid motor and fully assembled stage storage).

A4

Flammable Liquid Store.

Launch is made up of three main areas:

FIC

Final Integration Centre, the final integration of first stages with final stages/satellite before pad transfer. Fibre optic and RF links with LCC. Cooling N_2 for payload cooling. 20 × 50 m; 10 tonne overhead crane.

Launch Pad

A launch platform is towed by road from FIC. Fibre optic and RF links with LCC. Cooling N_2 for payload cooling. Launch rehearsal verifies system functionality and countdown procedures.

Range Instrumentation

This sits apart from the L1 fixed lateral site in the eastern sector (fixed telemetry station, Doppler receiver, tracking radar) and similar western sites (with two radars).

OTB has numerous measurements sites spread over 20 × 60 km. Sites are linked with the command and control centre by fibre optic and microwave. Some instruments can be moved outside the Range for a 300 km baseline. A fully self-contained telemetry station with stabilised antennas can be deployed on land or onboard ship. All instrumentation is linked to a central computer in the Range Control centre, and can be slaved to the tracked object's state vector, computed by the central computer and based on prioritised pointing and direction information from the instruments and telemetry. Instrumentation includes, three tracking radars, four mobile cinetheodolites, one mobile optical high-speed cameras track mount (with film development laboratory and film/video readers, two telemetry stations, one downrange telemetry station (ship or land), four Doppler (two mobile); Satellite Ground Station (10 m S/X band, 2 m S-band TT&C), Command Destruct System; electronic sky screens.

There is also a fully equipped meteorological station, with rocket sondes that can probe up to 65 km.

VERIFIED

Cape Canaveral's Titan operations. Core stacking is performed in the Vertical Integration Building in the centre, followed by strap-on mating in the next building along the causeway. Further along is the new Solid Motor Assembly and Readiness Facility for processing the strap-ons. Straight on is pad 41, off to the right is pad 40 (USAF) 0517249

Sweden

Esrange

Location: 67°56′N/21°04′E

Background

Esrange is the Swedish Space Corporation's sounding rocket and satellite operations range north of the Arctic Circle and was founded on 14 June 1962 by the European Space Research Organisation, a forerunner of ESA. More than 450 launches were made here between 1966 and 2000. Esrange includes seven permanent launch facilities able to handle most types of sounding rocket (including Terrier, Black Brant and Skylark 12). The seventh became operational in 1991 to handle the large Maxus vehicle. With altitudes of up to 900 km possible, the Swedish German Microgravity Sounding Rocket Programme uses it most frequently to provide ESA members with μgravity conditions in preparation for Spacelab missions. Sweden has been studying the commercial use of Esrange for orbiting small polar satellites.

Esrange has a 75 × 120 km impact area available for land recovery. Payload and control facilities have been upgraded, partly to accommodate the Maxus long-duration sounding rocket, based on Thiokol's Castor 4B and developed jointly by SSC/DASA. The 23 m high Maxus building has a 15 m-high umbilical mast. A 145 m^2 storage building can house two Castors at 20±2°C.

Stratospheric balloons of up to 10^6 m^3 are launched to 40 km and regularly recovered within Sweden, Finland and Russia.

Since 1978, facilities for receiving, processing and disseminating data from remote sensing satellites have been used for Landsat, and have been extended to handle data from France's Spot, Japan's MOS 1 and JERS 1, and Russia's Resurs O1. ESA's station dedicated to ERS is located at Salmijärvi near the range. Esrange is contracted for operation/maintenance. Esrange employs about 150 people.

United States

Cape Canaveral Air Station

Location: 28°26′48.73″N/80°33′54.58″W

Current Status

Operates Atlas, Delta, Pegasus, Taurus and Titan launch vehicles. In early 1998, industry spokesmen and US congressmen called on the United States air force to invest in a new launch infrastructure to bring Cape Canaveral up to the same standards as the US' major space competitors. By historical accident the air force has authority over Cape Canaveral and Vandenberg. A spokesman for the air force said, in March 1998, that the air force would not commit any additional funds to upgrades at this time. That has since been reiterated although State pressure has raised the matter in congress. As of the end of 2002, CCAFS and KSC had hosted 668 launches successfully carrying payloads to orbit. Responding to recommendations of the Columbia Accident Investigation Board, set up to report on the loss of Shuttle Columbia on 1 February 2003, a doubling of tracking camera coverage has been planned for the Air Force Eastern Range and the Kennedy Space Center. NASA and the USAF had 14 film and 12 video systems in place for the launch of Columbia but these will be increased to 23 film and 20 high-definition TV systems with long range trackers increased from 5 to 10 video and from 5 to 10 film systems while mediumn range trackers will be increased from three to seven video and film systems each. Short range tracking equipment will also be increased in addition to the 42 existing fixed cameras at LC39A and B.

Background

The 45th Space Wing (45SW) designation in November 1991 replaced the Eastern Space & Missile Center organisation, established 1 October 1979. Its origins date back to September 1948 when the Banana River Naval Air Station was transferred to the USAF as a base for operations as a joint-service missile range. The Headquarters opened on 10 June 1949 but joint operations proved unwieldy and it was replaced by the air force's Long Range Proving Ground Division on 16 May 1950. On 30 June 1951, in keeping with policy to designate intermediate headquarters as 'centres', the long Range Proving Ground Division became the air force Missile Test Center. It was renamed the air force Eastern Test Range in 1964. A unit of air force Space Command, it is the host organisation for Cape Canaveral AS, Patrick AFB and the two island stations (Ascension and Antigua) of the Eastern Range (ER). The eastern and western launch ranges and bases were managed by air force Systems Commands' Space Systems Division until 1 October 1990. The Wing comprises four major Groups: Operations; Support; Logistics; Medical.

CCAS encompasses active Titan, Atlas and Delta complexes in addition to providing support facilities for the military, NASA and commercial organisations. Military, civilian and contractor personnel at Patrick and the Cape total about 12,000 and of these, up to 4,000 are involved in processing vehicles and satellites. 455 space launches had been made by the end of 1995. No other site has been used for US manned missions.

The annual launch rate is now limited to 25 to 30 because, despite upgrading, the Cape is based on 1950s concepts. The pads are recognised as the choke points, with missions requiring an average of 26 to 40 days processing before launch.

Launch Complex 17 (Delta) and LC 36 (Atlas Centaur), pads, formerly under NASA control, were turned over to the air force following the 1987 announcement of a mixed fleet policy and 1988's revised national space policy endorsing the growth of a commercial ELV industry. The last NASA-managed Atlas Centaur launch took place from LC 36 in September 1989 and of a Delta in November 1989 from Vandenberg. NASA continues to have oversight responsibility for ELVs carrying agency payloads, but now contracts with either USAF or vehicle manufacturer for launch services.

Extensive support is provided for Shuttle launch operations by the military. The USAF Eastern Range, of which the Cape is part, provides weather forecasting, range safety, tracking and preparation of DoD payloads. USAF also operates Titan pad 41 on a 1.44 km^2 patch inside the KSC boundary. KSC operates a Vandenberg Launch Site Resident Office as the interface with USAF for base support of all NASA payloads flying from the west coast.

The Operations Group provides operational control over all aspects of DoD space launches. The 1st Space Launch Squadron (SLS) was activated 1 October 1990 to assume all responsibility for Delta operations. The 3rd SLS activated 3 August 1992 to take over Atlas work. The 5th SLS activated 14 April 1994 to oversee all Titan operations. Each Squadron exercises field operations management/control of all DoD satellites launched from CCAS.

A single storey 11,800 m^2 Range Operations Control Center became operational in March 1995 as the ER control facility under a US$134 million Harris Corporation contract. The goal is to allow flights every 12 hours. Operationally, it serves the ER by processing worldwide radar metric data, generating inter-range vectors to aid in acquisition, calculating state vectors for navigation uplink, displaying telemetry and orbital navigation information, and functioning as LRCC Lead Range Control Center to co-ordinate radar tracking and retrieval. Facilities cover communications, range control and scheduling, data processing, range safety and weather information.

CCAS offers five principal hangars for non-hazardous spacecraft processing. Hangar AO with a 14 × 53.3 m class 100,000 high bay clean room capable of handling two payloads simultaneously. Hangar AE with class 10,000 facilities and the Delta mission director's centre. Hangar AM with a 4.4 × 5 m clean room and solarium, the SPIF Shuttle Payload Integration Facility, and the SAB Satellite Assembly Building. Loading of payload propellants and pyrotechnics can be performed in three facilities, Explosive Safe Area 60A and KSC's Spacecraft Assembly and Encapsulation Facility number 2, both of class 100,000 standard, or KSC's Cargo Hazardous Servicing Facility Complex, built to handle Shuttle-sized elements.

Principal current launch activities are Titan operations at pads 40/41, Delta from Space Launch Complex 17 (A/B) and Atlas Centaur from SLC 36 (A/B). The orbital inclination is limited without doglegging to 57° by range safety considerations. Titan pad 41 is actually north of the CCAS boundary on a 1.44 km^2 site allocated to the air force. The ER's tracking network stretches 16,000 km over the Antigua and Ascension Island stations and into the Indian Ocean, where it meets the Western Range. Mainland Florida stations are maintained at the CCAS, Jonathan Dickinson and Malabar annexes, augmented by the fleet of ARIA advanced range instrumentation aircraft routinely scheduled from the 452nd Test Squadron/ENO Edwards AFB. In addition, ER can use instrumentation operated by NASA on Bermuda and Wallops Island. Another station is available at Argentina in Newfoundland for high inclination launches on northeastern flights of about 37°.

The status of the launch complexes is:

1–4 Inactive. Snark, Bomarc, Bumper, Redstone, Jason, X-17, Polaris launches; 'Fat Albert' tethered aerostat tests and radar testing of air defence equipment to be used at other locations.
3 First Cape launch, Bumper, 24 July 1950.
5–6 Dismantled. First US manned launches: Mercury Redstone 3/4 (Shepard/Grissom suborbital flights, May/July 1961); Jupiter C, Explorer, Pioneer. Part of CCAS air force Space Museum since January 1964.
9/10 Dismantled. Used for Navaho, Minuteman 1–3 and Pershing 1A and later renumbered 31/32 Challenger debris stored here from 1987.
11 Dismantled. Used for Atlas.
12 Dismantled. Used for Atlas and Atlas Agena: ATS, Mariner, OAO, OGO, Pioneer, and Ranger.
13 Atlas and Atlas Agena: Mariner, Lunar Orbiter and DoD missions.
14 Dismantled. Mercury Atlas 6–9 manned orbital flights, starting with John Glenn, February 1962. Demolished in 1976 because of dangerous state from salt corrosion.
15 Dismantled. Titan 1/2.
16 Inactive. Previously Titan 1/2 and Pershing 1/1A/2.
17 Active. Delta Pads A/B (A: 28°26′48.73″ north 80°33′54.58″ west; B: 28°26′44.12″ north 80°33′57.78″ west), providing an annual capacity of 12. Plans had been made to close them down at end-1986, but selection of Delta 2 to launch GPS satellites ensured their survival. One has been modified to accommodate the cryogenic Delta 3 from 1998. Delta stage 1/2 checkout typically takes place over 14-7 weeks before launch in USAF Hangar M and the Delta Mission Checkout Facility. Stage 1/interstage erection occurs five to six weeks before launch, followed immediately by strap-on and stage 2 attachment; simulated launch/flight in second/third week; stage 3 and mated payload during second; final week sees fairing installation and stage 2 propellant loading. GPS satellites and stage 3 are processed at Area 59. Commercial payloads meet the stage 3 assembly for the first time in the Payload Spin Test Facility, 2.8 km west of complex 17, where it is mated/balanced at 0 to 300 rpm.
18 Dismantled. Vanguard, Blue Scout, Thor.
19 Dismantled. Titan 1/2; all 10 Gemini Titan 2 manned flights 1965–66.
20 Dismantled Titan 1/2 pad; modified for Titan 3A testing 1964/65. Renovated to launch Starbird suborbital vehicles as targets for SDI's Starlab Shuttle mission (cancelled). Demonstration Starbird launched 18 December 1990 – the first launch since the 1960s. The unsuccessful Joust 1 materials processing flight for the University of Alabama in Huntsville's Consortium for Materials Development in Space was launched 18 June 1991 by Orbital Sciences Corporation. Successful Aries Red Tigress launches were made by OSC for SDIO 20 August 1991, 14 October 1991, 28 May 1993. Florida Spaceport Authority is modifying the complex for commercial operations from 1997.
21–22 Dismantled. Mace and Bull Goose
25 Inactive. Trident 1, Poseidon and Polaris.
26 Deactivated. Now at now USAF Space Museum. Two pads launched 24 Jupiter IRBMs, two Jupiter bio monkey flights, two Redstone IRBMs, two Jupiter C (Explorer 1/2), and six Juno 2 (Explorer 8/11).
29 Deactivated. Polaris and UK's Chevaline.
31/32 See 9/10.
34 Dismantled. Apollo Saturn 1/1B. Apollo 7 October 1968 was that programme's only (and NASA's last manned) launch from CCAFS.
36 Active. Atlas Centaur. From 1985 only Pad B remained for Atlas launches. Pad A was converted at a cost of US$20 million into a Shuttle Centaur Checkout Facility in time for the 1986 launches of Ulysses and Galileo – delayed by the Shuttle accident. The centre would then have handled the common Centaur stage being developed for NASA/DoD Shuttle operations but abandoned in June 1986. In October 1987, USAF and General Dynamics signed a commercialisation agreement, supplementing an earlier agreement with NASA as owner of the complex, allowing the company to use Complex 36. Both commercial and government-sponsored Atlas Centaur launches are involved, with General Dynamics reimbursing the government for the direct cost of goods and services provided. Pad 36A is used exclusively for USAF Atlas 2 launches, supporting five annually (six under surge conditions). As part of the modifications, all Shuttle Centaur structures were removed from 36A. 36B was modified by the end of 1990 to accommodate all new Atlas 2 types. The last NASA-managed launch was September 1989 departure of AC 68 with the last FltSatCom. Now supports Atlas 2 from both pads but only Complex 36B capable of handling Atlas 2 and Atlas 3 at a rate of 10–12 launches per year. More than 100 Atlas-Centaur have been launched from Complex 36.
37 Pads A/B used for Apollo Saturn 1/1B and Pegasus 1–3. SLC-37 is being modified for Delta 4 launch operations that began in 2002.
40 Titan 4, previously Titan 3C/34D and commercial Titan 3. The Intelsat 6 launch of June 1990 was the pad's 35th. Martin Marietta and engineering contractor Bechtel National Inc then closed it down for a US$–340 million rebuild. It was ready to handle September 1992's Mars Observer launch, after which work was completed for it to accommodate all Titan 4 models. The essentially new complex has a 15-year life. The 5,000 tonne 79.3 m high Mobile Service Tower is encased by 7,000 m^2 of shielding against electromagnetic interference of the vehicle and payload. The major 8 × 37 m vehicle and 5.2 × 26.8 m crane doors are sealed by inflatable bladders. The new 55 m umbilical tower and the rest of the facility can withstand 185 km/h winds. Four 90 m towers with strung cables between were added in 1993 around both pads as lightning protection.

Titan's core is stacked on its launch platform at the Vertical Integration Building. The VIB houses four high bays, with Bay 1 dedicated to commercial Titan operations, Bay 2 to USAF Titan and Bay 3 for Centaur. The new LOCC Launch Operations Control Centre replaced the Launch Control Centre in the VIB in 1993 in a different area. The stack is then moved 1.4 km by rail to the Solid Motor Assembly Building where the strap-ons are stacked and attached before moving the final 4.2 km to pad 40. The SMAB also houses the USAF Spacecraft Processing Integration Facility, a secure equivalent of NASA's Vertical Processing Facility for preparing classified payloads. The new 5,600 m^2 SMARF Solid Motor Assembly and Readiness Facility for simultaneous processing of up to six sets of Titan 4 boosters was completed in 1994. The SMAB is also used for IUS stage assembly/processing. Construction began January 1993 of a separate Centaur Processing Facility to de-couple Centaur pad checkout from other Titan 4 elements.
41 Formerly used for military Titan 3C and NASA Titan 3E Centaur launches such as Viking and Voyager, subsequently modified for Titan 4 (first launch June 1989). Now converted for supporting Atlas 5 class vehicles able to support up to 15 launches per year.
42 Once proposed as a second Titan 4 pad on the same 1.44 km^2 USAF island, as pad 41 within KSC.
43 Deactivated. Used for weather rockets.
45 Deactivated. Used to launch ROLAND mobile missiles
46 USN Trident 2 but being modified by Florida Spaceport Authority under US$4.89 million in Federal grants and US$2 million in industry matching funds to accommodate small orbital vehicles, initially those employing Thiokol's Castor 120 solid motor. Modifications include a mobile access tower and adaptable launch mount. Naval Ordnance Test Unit retains Trident launch capability.
47 Active: sounding rocket launches.

UPDATED

Kennedy Space Center (KSC)

Location: 28°36′33″N/80°36′15″W

Current Status

All Shuttle operations are dictated by the recommendations of the Columbia Accident Investigation Board (CAIB) set up to enquire into the loss of orbiter OV-102 and its seven crew members on 1 February 2003. NASA expects to return the Shuttle vehicles to flight in May or June 2005. Changes to facilities occasioned by the CAIB report include a doubling of tracking camera coverage of launch operations after finding that existing coverage was inadequate for detailed engineering assessments of ascent events. Existing TV and film systems such as the USAF Distant Object Attitude Measurement System will be equipped with high definition TV and linked to a new digital image analysis facility situated locally.

By the end of 2003 the CCAFS/KSC complex had launched a total of 683 orbital shots since the first on 31 January 1958, with the launch of the first US satellite Explorer I.

Background

KSC is NASA's site for processing, launching and landing the Space Shuttle and its payloads, including Space Station components. Located on Merritt Island adjacent to the USAF launch facilities of the Cape Canaveral Air Station, Kennedy was originally built to support the Apollo lunar landing programme of the 1960s. After the last Apollo lunar launch in 1972, Launch Complex 39 supported Skylab in 1973–74, Apollo-Soyuz in 1975 and Shuttle from the late 1970s. Shuttle mission. To the end of 2001 NASA had launched 124 human space flights from KSC. The installation covers 567 km^2. All

Columbia occupies pad 39A in the foreground during STS-35 preparations, while Discovery is on 39B for STS-41. The white Payload Changeout Room in the Rotating Service Structure is used for installing vertically processed payloads on the pad. A total of 82 Shuttle launches had been made by February 1997 (NASA) 0517254

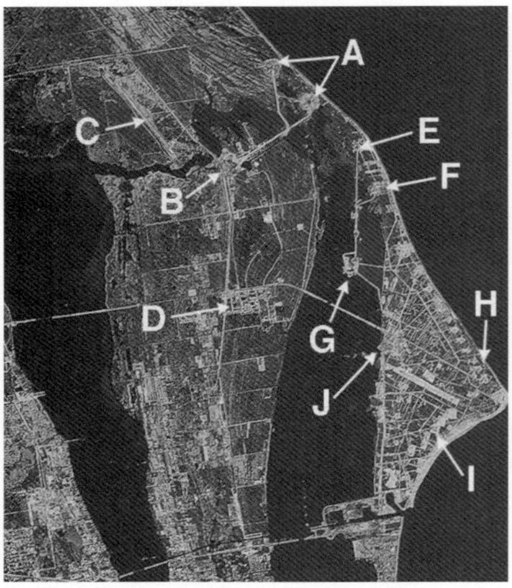

Landsat image of Cape Canaveral. A: Shuttle pads; B: VAB; C: Shuttle runway; D: NASA HQ and industrial area; E: Titan complex 41; F: Titan complex 40; G: Titan Vertical Integration Building; H: Atlas complex 36; I: Delta complex 17; J: CCAS industrial area 0517255

Vehicle Assembly Building and Shuttle landing strip (background). The Launch Control Center is the low building angled at the VAB's base. Note Mobile Launcher Platform 3 at right, awaiting modification from its Apollo configuration (NASA) 0517256

non-operational areas are part of the Merritt Island National Wildlife Refuge. KSC extends some 55 km north to south and 16 km across its widest point.

Land acquisition to build the centre began in 1962, and clearing for the Vehicle Assembly Building began in November 1962. The VAB was structurally completed in mid-1965. The first LC39 launch was Apollo 4 in November 1967. Shuttle processing and launch facilities are in the LC 39 area. About 8 km to the south, office buildings, payload processing facilities (including the new Space Station Processing Facility) and flight crew quarters are in the Industrial Area. NASA also uses some buildings on neighbouring Cape Canaveral AS. The USAF Titan pad 41 lies within a 1.44 km^2 island assigned to USAF inside KSC.

About 14,000 personnel work at Kennedy. 1,925 are civil service employees, but contractors employ the majority. There are four major contracts: Lockheed Martin Space Operations Co was the Shuttle Processing Contractor (SPC) during 1983 to 1996, McDonnell Douglas Aerospace handles payload ground operations, and Rockwell supports logistics related to the spaceplane, including spare parts. EG&G holds the base operations contract.

In October 1996, United States Alliance officially began work on the space flight operations contract – a consolidation of Shuttle processing contracts at KSC.

Today, final Shuttle assembly takes place in the VAB and the stack is then transported either 5,535 m east to pad 39A or 6,828 m northeast to 39B. LC 39's design reversed the fixed launch concept of earlier missions, under which assembly, checkout and launch were conducted at the pad. Kennedy's first director, Dr Kurt Debus, developed the mobile launch concept. The Shuttle processing and launch facilities are:

Vehicle Assembly Building (VAB)

160 m high, 218 m long, 158 m wide, divided into four 160 m high bays and a 64 m low bay, transected by a north to south transfer aisle. One of the world's largest buildings by volume, it covers 32,400 m^2. A maintenance shop for Shuttle main engines and holding areas for the SRB forward assembly and aft skirt is in the low bay. External Tanks are stored and checked out in high bays 2/4, with 4 also used for payload canister handling. high bays 1/3, facing east, are used for assembly of the three main Shuttle elements (two SRBs, ET, Orbiter) on the Mobile Launch Platform. The stack is transported from here to the pad.

Launch Control Center (LCC)

A four-storey building adjacent to the VAB on the east side, housing four firing rooms. Firing rooms 1/3 provide full control of launch and orbiter operations; number 3 was previously secured for classified DoD departures. Number 2 is usually used for software development and testing. Number 4 is only a partial firing room, primarily used as an engineering analysis and support area for launch and checkout operations. Plans are currently under way to turn number 4 into a full-up firing room, with construction beginning in late 1997. Computer and software equipment comprising the LPS Launch Processing System is also housed in the LCC. Designed at Kennedy for

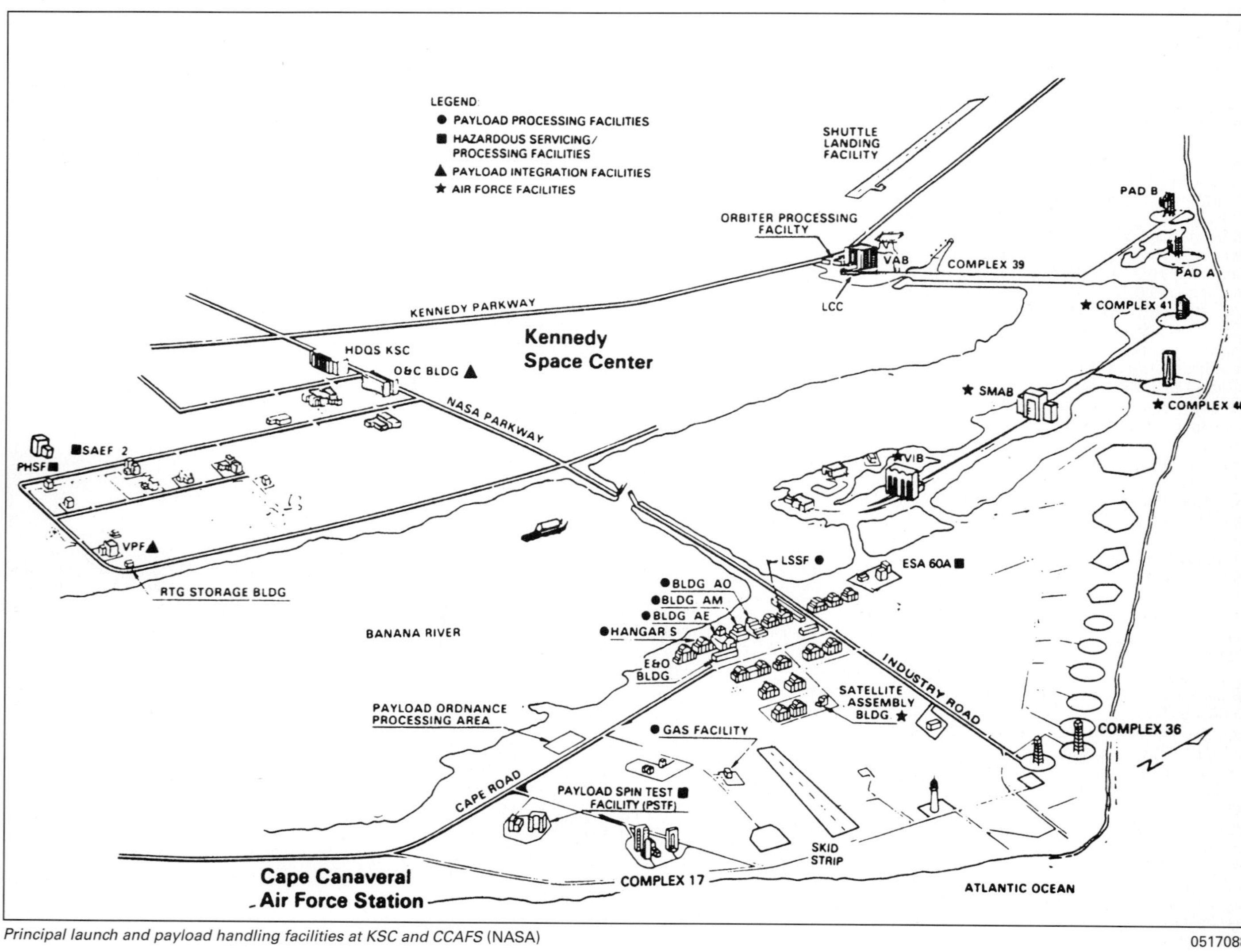

Principal launch and payload handling facilities at KSC and CCAFS (NASA) 0517088

Preserved at the CCAFS historical site, Launch Pad 14 was used to launch John Glenn into orbit in February 1962. (NASA) ***NEW***/1047503

the Shuttle programme, the LPS automatically controls many of the procedures formerly performed mechanically, while also requiring fewer personnel (about 90). Shuttle checkout, countdown and launch are conducted from the LCC using the LPS. A countdown begins typically about three days before scheduled lift-off, at the T-43 h mark and including an average of 28 hours of built-in holds. As soon as the two solids ignite at T-0 control of the mission shifts from Kennedy to Mission Control at Johnson Space Center in Houston. KSC again assumes responsibility after the crew exits (usually within an hour of touchdown) and ground cooling has been established.

Mobile Launcher Platform

Final Shuttle assembly, transport to the pad and launch are conducted on the MLP. A two-storey steel structure, 7.6 m high, 49 m long, 41 m wide, 4,190 tonnes. MLP-3, previously used for Apollo 11, manned Skylab and ASTP, was added to the existing pair in late 1989 and first used by STS-32.

Technicians use heat lamps to dry adhesive during tile fixture at KSC (NASA) ***NEW***/1047501

Crawler-transporter

Two crawlers, each weighing 2,700 tonnes, were in service by early 1967 for the Apollo lunar programme. About 40 m long, 35 m wide and height adjustable to 6 to 8 m, with an operator control cab at each end. There are eight tracks, each with 57 0.9 tonne shoes. Two 2,750 hp diesel engines drive four 1 MW DC generators powering 26 traction motors. Two 750 kW AC generators, driven by two 1,065 hp diesels, provide jacking, steering, lighting and ventilation. Two 150 kW AC generators provide MLP power. Maximum speed is about 3.2 km/h unloaded, 1.5 km/h loaded, consuming about 350 litres of diesel oil per km. It requires between 6 and 8 hours to move Shuttle from the VAB to 'hard down' at the pad. Vehicle apex maintained vertical to within ±10 arcmin. Crawler 2 passed its 1,000 mile (1,609 km) mark, 22 April 1990 carrying Columbia to 39A for STS-35.

Launch Pads

39A/B identical roughly octagonal design. Central structure on hard stand is the 106 m Fixed Service Structure, including 24.4 m lightning mast. FSS has three service arms, the ET gaseous oxygen vent arm (with beanie hood), the Orbiter access arm, the ET hydrogen vent umbilical and intertank access arm. Attached to the FSS is the Rotating Service Structure, which rotates 120° on a track to encase the Orbiter so that its Payload Changeout Room fits flush with the cargo bay for vertical payload installation. All manned flights after Apollo 7 have been launched from complex 39, and all except Apollo 10 during the lunar programme from 39A. The first 24 Shuttles also went from 39A; the fateful Challenger 51L mission was the first to use 39B. During the post-Challenger hiatus, 39B underwent extensive weather protection modifications. Others included safety improvements to the crew's emergency escape system. The Orbiter access arm was covered with solid panels for fire protection and a water spray system was incorporated. Two slidewire baskets were added to the existing five, along with brakes; and improvements were made to the emergency shelter bunker near the end of the slidewires.

Modifications were made to improve clearance between the Shuttle as it lifts off and the ET hydrogen vent umbilical as it retracts. Environmental controls were improved in the Payload Changeout Room. Also added were an SRB joint heater umbilical to keep booster field joints at 24°C, a system to recirculate pad water services to prevent them from freezing, and installation of debris traps in the ground interfaces between the Orbiter and LOX/LH_2 servicing systems on the mobile launcher platform. The Apollo-era elevators were also replaced. Pad B was used for the new series of launches while 39A underwent identical modifications. Pad A returned to service in January 1990 for STS-32. Pad B was again removed from service after STS-40 in June 1991 to undergo US$3.3 million of modifications by Lockheed Space Operations Corportation to prepare it for Endeavour's roll-out in March 1992. Modifications to both Shuttle pads are scheduled periodically. Incidentally, Pad B was the launch complex from where Columbia was launched on 16 January 2003 on its ill-fated mission.

Orbiter Processing Facility

The original OPF is capable of handling two Orbiters in identical high bays 29 m high, 60 m long, 46 m wide. A third bay came online in September 1991 in the converted Orbiter Modification & Refurbishment Facility, north of the OPF. The bulk of Orbiter processing between missions is performed in the OPF. Equipped with movable work platforms that completely envelop an Orbiter, they are equipped to fuel the Auxiliary Power Units and thrusters with hypergolic propellants, in addition to draining them after missions. Horizontal payloads such as Spacelab are installed in the Orbiter within the OPF (vertical payloads are installed on the pad). Before transfer to the VAB for final assembly with the ET/SRBs, the Orbiter is weighed and its centre of gravity located. Conversion of the OMRF into OPF bay 3 cost US$85 million; Discovery was the first to use it, after STS-48 in 1991. Some of the equipment was transferred from the mothballed Shuttle facilities at Vandenberg AFB, including US$40 million of ground support equipment and US$6 million to US$7 million of unique work platforms.

Thermal Protection System Facility

Two storey 4,088 m^2 opposite the original OPF for final manufacture and repair of thermal tiles, and gap fillers.

Rotation/Processing Facility
One of several facilities devoted to SRB processing. Includes four buildings north of the VAB. New and reloaded SRB segments shipped by manufacturer Thiokol in Utah are received here. Recovered casings, refurbished at Hangar AF, are installed on rail cars for shipping to Utah. Inspection, rotation and aft booster build-up occur in the processing building. Completed aft skirt assemblies, brought from the ARF (see below), are integrated here with aft segments. Two nearby surge buildings store up to eight segments each (two flight sets). Final assembly and stacking into complete SRBs is performed in the VAB.

SRB Assembly & Refurbishment Facility
Managed by NASA Marshall for refurbishment and subassembly of inert SRB hardware, including the forward and aft skirt assemblies, frustrum, thrust vector controls, recovery systems and electronic controls takes place in this seven-building complex. The three-storey Manufacturing Building includes an ordnance area for installing separation devices; 24.4 × 61 m high bay; class 100,000 area for building up aft skirt elements; two 13.6 tonne bridge cranes and three of the world's largest overhead gantry industrial robots for painting and spraying on insulation. A hot fire test stand on the southeast corner of the 178,000 m^2 site supports testing of the SRB's hydrazine TVC system. Completed aft skirt assemblies are transported to the RPF for integration with the motor segments.

Shuttle Landing Facility
4.8 km northwest of the VAB on a northwest/southeast alignment. Paved runway is 4,572 m long (305 m overrun at each end), 91 m wide (plus 15 m asphalt shoulders) and 41 cm thick at the centre (38 cm at sides). The 61 cm centre to edge slope facilitates drainage. Post-Challenger, a section 1,067 m long at each end was ground down to create a smoother surface without cross grooves to reduce Orbiter tyre wear. The 6.3 mm grooves provide a skid-resistant surface and rapid water runoff. The centre's surface was also smoothed off (but without reducing the grooves) in September 1994. The reduced tyre wear might allow the current crosswind constraint of 15 kt/h (27.8 km/h) to be increased to 20 kt/h (37.1 km/h), cutting the number of launch scrubs. STS-63 in February 1995 was the first to use it. The strip is considered as two runways: Runway 15 with approach from northwest; Runway 33 from southeast. The TACAN system provides range/bearing measurements from 44.2 km altitude; the Microwave Scanning Beam Landing System provides more precise slant range, azimuth and elevation from 5.5 to 6.1 km. A 149 × 168 m aircraft parking apron is at the southeast end.

On the northeast corner of the ramp is the Mate/De-mate Device that attaches the Orbiter to or lifts it from the Shuttle Carrier Aircraft during ferry operations. There are also movable platforms for access to some Orbiter components. The device can lift 104.3 tonnes and withstand winds up to 200 km/h. A similar unit at NASA's Dryden Flight Research Facility supports Shuttle landings in California. Five missions ended on the Kennedy runway to April 1985 (STS-51D) but future landings were deferred because of tyre, braking and weather problems; STS-38 became the 6th in November 1990. Upgraded main landing gear, carbon brakes, additional nosewheel steering and improved tyres (plus later drag 'chutes), meant that KSC in 1991 was again designated a primary landing site, with equal priority to Edwards depending on weather. Through February 1997, 35 missions landed at KSC.

The Destiny lab for the ISS is made ready for installation in the Shuttle Orbiter payload bay (NASA) ***NEW***/1047502

Logistics Facility
21,351 m^2 main building south of the VAB. Houses some 160,000 Shuttle parts under an automated storage/retrieval system. The separate NASA Shuttle Logistics Depot maintains and repairs 95 per cent of Shuttle's 4,000 Line Replaceable Units and parts.

Operations Support Building
Six-storey building, 27,700 m^2, completed mid-1990, southwest of VAB, housing technical documentation centre, library and photo analysis area.

Shuttle Launch Processing
A KSC-led team prepares a returning Orbiter for turnaround whether it lands at KSC or EAFB. It is returned from Edwards atop the Shuttle Carrier Aircraft (a modified 747) to the Shuttle Landing Facility, where it is removed using the Mate/De-mate Device. Processing for the next flight then begins in the OPF, including ordnance safing, payload removal, payload bay reconfiguration, removal and installation of the three main engines, horizontal payload insertion and any required maintenance. During this work, the SRBs are stacked in the VAB atop the MLP and the ET attached. The Orbiter is then transferred on a transporter, lifted vertically and mated to the ET. Crawler moves the stack to the pad, where connections and hookups are made and checkout begins. Equipment in a room below the pad hard stand provides the links to the Launch Control Centre. If the payload is not installed in the OPF, it is inserted on the pad.

Processing Control Center
The three-storey 9,200 m^2 PCC between the OPF and Operations Support Building, began operations in 1992 for Orbiter testing (three control rooms monitor Shuttle processing in the three OPF high bays), launch team training and LPS maintenance.

Shuttle Payload Processing
Shuttle payloads can be processed horizontally or vertically. Facilities are located in the Industrial Area and across the river on Cape Canaveral AS. Horizontal payloads such as Spacelab are received, assembled and tested in the Operations & Checkout Building, the Industrial Area's largest building. They are then transferred to the OPF for installation in the cargo bay; returned payloads follow a reverse procedure. Astronaut crew quarters are also housed in the O&C. Vertically processed payloads are handled on the Cape in:

Hangar L
Live specimens are received here, also a life sciences research facility.

Hangar S
Satellite reception and storage; plus storage of SCAPE Self-Contained Atmospheric Protective Ensemble suits for ground crews.

Buildings AE/AM/AO
In the CCAFS Industrial Area contain high bay clean rooms for processing deployable satellites. The management team in AE's Mission Directors Center oversees launches of NASA, USAF and commercial payloads on Delta and Atlas. AE includes Launch Vehicle Data Center and Telemetry Facility.

Vertical Processing Facility
Processing of vertical payloads, including upper stages, takes place in the Vertical Processing Facility south of the O&C in the KSC Industrial Area. Contains Cargo Integration Test Equipment (CITE) for verification of orbiter/payload interfaces. 32 m high bay and airlock and single storey support facilities along the sides of the high bay. The entire facility is a class 100,000 clean room.

Spacecraft Assembly & Encapsulation Facility 2
SAEF 2, also in the southern part of the Industrial Area, is used for assembly, testing, encapsulation

The Docking Module is lowered into the Shuttle prior to launch (NASA) ***NEW***/1047518

and sterilisation of spacecraft, particularly large payloads.

Payload Hazardous Servicing Facility
A nearby class 5,000 facility opened in 1986 for large payload processing, integration of solid motors and fuelling operations. The west end of the 21.3 × 33.6 m 28.7 m high service bay has a 6.1 × 12.2 m sloping area for fuelling. PHSF operations are handled from the separate (for safety) Facility Control Building, with two 112 m^2 control rooms.

High Energy Radiographic Facility (HERF)
Operations on Cape Canaveral AS began in early 1990 for non-destructive Linatron testing of kick and other large motors, including those on upper stages. Tests are performed in a 16.5 × 19.2 m high bay with wall/door of reinforced concrete 203 cm thick.

The Canister Rotation Facility
With 670 m^2 high bay was officially activated December 1992 for maintaining Shuttle's two environmentally controlled 19.8 m long payload canisters. Previously, rotation had to be performed in the VAB.

At the Cape, Hangar AF receives recovered SRBs for cleaning and disassembly (some components are reconditioned here) and Building E&O is the headquarters for operations related to ELVs.

Space Station Processing Facility
In order to prepare for the coming Space Station era, NASA has built the Space Station Processing Facility, a three-storey 42,500 m^2 facility east of the O&C which receives, processes and checks out Space Station flight hardware as it arrives at KSC. In fact, SSPF is a generic facility capable of handling any Shuttle payload. The 5,830 m^2 class 100,000 clean room includes a high bay and an intermediate bay, accessed by a 465 m^2 airlock. Eight 13.7 × 27.8 m areas each support a complete Shuttle payload. Full-scale work began in April 1991 and initial occupancy began April 1994. Total cost will be about US$400 million. Personnel total 1,040. The operational readiness date is June 1997, but the first processed payload was STS-74's Mir Docking Module, arriving in June 1995.

UPDATED

Kodiak Launch Complex (KLC)

Background
The Alaska Aerospace Development Corporation (AADC) was established in July 1991 to promote development of the aerospace industry in Alaska. As the spaceport authority for the State of Alaska, it is constructing the commercial US$18 million to US$20 million KLC Kodiak Launch Complex on Narrow Cape, Kodiak Island on 12.55 km^2 of state-owned land, provided free for 30 years. It will be the only non-federally run commercial launch range in the US. Kodiak Island is home to the largest US Coast Guard support centre and has a full complement of ocean support vessels. The Coast Guard Air Station currently consists of several HC-130 aircraft and HH-60 and HH-65 helicopters.

KLC will target launches of up to 3.6 tonne payloads into LEO ±40° polar orbits, taking advantage of the wide launch azimuth. The launch pod can accommodate Castor 120 size vehicles with a flame trench rated up to 544.5 kg thrust. Pod gantry can accommodate large vehicles which is also plumbed for liquid fuelled launchers. AADC completed the design and siting of KLC's facilities in early 1995 and broke ground in April 1996, aiming for an initial operating capability some time before the end of the century. It is still looking for strategic partners. The design represents the state-of-the-art in launch facilities: all-weather, flexible and adaptable to all current small vehicles. Facilities will comprise Launch Control & Management Centre; Payload Processing Facility; Integration & Processing Facility; Spacecraft & Assemblies Transfer Facility. One pad and service structure is planned but the site allows for two additional pads. The KLC, on all weather complex, is located on a 3,100 acre site. The facility covers 27 acres and is divided into four sites: the launch control and management center, the payload processing facility, the integration and processing facility/spacecraft assembly transfer facility and the launch pod and service structure.

The first launch from KLC took place on 5 November 1998 when the US Air Force launched the AIT-1 atmospheric intercepter test rocket. AIT consists of two solid rocket notes: the Thiokol Castor IV with a flexseal as first stage and the M57 as second stage.

KLC can support payloads up to 3,630 kg sent into polar, Molniya or geostationary orbit. It can handle Castor 120 size vehicles and smaller.

Poker Flat Research Range (PFRR)

Location: 65° 07′N, 147° 28′W

Current Status
Launch activity has decreased in recent years but still maintains a robust schedule. Sounding rocket types include Super-Arcos, Nike-Tomahawk, Black Brant, Aries, Nike-Black Brant, Terrier-Malemute, Orion, Nike-Orion, Taurus-Orion, Taurus-Tomohawk, Viper-Dart and Taurus-Nike-Tomahawk. Some modest upgrade and improvement programmes have been implemented.

Background
Set up in the early 1960s by Neil Davis of the local geophysical institute, Poker Flat Research Range (PFRR), primarily a sounding rocket launch facility dedicated to auroral and middle to upper atmospheric research, is 50 km northeast of Fairbanks. Operated by the University of Alaska Geophysical Institute since 1968, it is the world's only university-owned launch range. It is also the only high-latitude and auroral zone rocket launch facility on US soil. PFRR's 5,132 acre central complex lies on 21 km^2, on a site which includes telemetry, optical observatory, five pads, blockhouse, rocket storage/assembly buildings, maintenance and communication facilities, a payload assembly building, radar and the range office. PFRR has been operating under a cooperative agreement between NASA and the Geophysical Insitute since 1979.

With flight zones covering over 67,000 km^2 of northern Alaska, it is the largest land-based launch facility in the world and, as such, permits extensive payload recovery programmes along with established down range instrument sites.

Poker Flat is funded through contracts with DoD, NASA, NOAA and the National Science Foundation. NASA currently funds the US$1.1 million annual operation cost. Five employees work year-round directing design and construction of new facilities, co-ordinating science projects, maintaining the physical plant, obtaining the various waivers, approvals and agreements necessary for each launch, and providing launch services. Additional employees are added during major campaigns.

Although NASA launch campaigns only occur every other year on odd-numbered years, an average of two to 10 major sounding rockets are launched annually. Since construction of the facility began in the Summer of 1968, 270 major rocket experiments were conducted through 1997.

The 5,132 acre launch area contains five launchers rated from 1 to 20 tons, two movable launcher enclosures and separate facilities for payload assembly, rocket motor assembly, motor storage, scientific support, telemetry, radar and data communication (Credit Evelyn Trabant/Poker Flats) 0008102

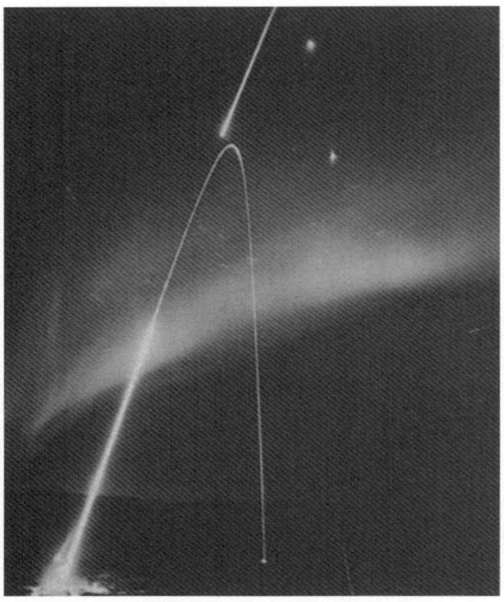

Black Brant 10 launch from Poker Flat March 1984. Two barium releases are visible at 800 and 1,000 km (apogee) (Technology International Corp) 0517089

Nike Tomahawk made the most flights from here, with 64 launches, but the larger and more complex payloads of recent years have led to the use of larger motors and combinations as large as the Aries and Sargent. A Black Brant 12 established a Poker Flat altitude record of 1,422 km on 2 April 1993.

Most missions require simultaneous ground-based data from a worldwide network of stations or are in co-ordination with satellite experiments. The T. Neil Davis Science Operations Center (SOC) houses magnetometers and riometer data displays, all-sky and narrow field auroral TV cameras, meridian scanning photometers and other optical and RF observing instruments for auroral and upper atmospheric research. A large worldwide data display is available at the SOC for launch team use. A major international co-operative effort with the Japanese Communications Research Laboratory is under way. A 256-element imaging riometer is operational and design for large-scale MF radar has begun.

The range operates five launch pads. Pads 1 and 2 each have a 3,400 kg MRL rail launcher. Pad 3 has a 9,000 kg rail launcher. Pad 4 has a 16,000 kg Athena launcher. Pad 5 has a 1,800 kg twin boom launcher and is currently dedicated to the NASA-sponsored Student Rocket Programme, which is affiliated with the Space Grant Programme at the university. A static test stand for cold weather testing of rocket motors is in the design stage in this programme. De-activated Pad 6 was employed for meteorological rockets.

Range telemetry support is provided by two S-band autotrack systems, incorporating 2.44 and 4.9 m dishes, provided by NASA and located in a 15.3 m dome at the top of the hill overlooking the complex. A new telemetry building served one 10 and two 8 m S-band antennas for rocket and satellite telemetry. A NASA C-band radar unit was established at Poker Flat in 1983 and was replaced with a new C-band system in 1995.

Authorisation for overflight, landing and recovery of rockets and payloads on 67,000 km^2 of land, roughly northeast of the range head to the coast of the Beaufort Sea, is granted by the Bureau of Land Management, the US Fish and Wildlife Service, the State of Alaska Dept of Natural Resources (Div of Lands), the Venetie Tribal government, the Traditional Councils of Arctic Village and Venetie and Doyon, Ltd (the Interior Alaska Regional Native Corporation). The Federal Aviation Administration approves rocket flight zones and co-ordinates air space use during launches and the Alaska Department of Transportation and Public Facilities authorises range personnel to stop road traffic on the Steese Highway during launches. In addition to the normal NOTAMS procedure, each launch is co-ordinated with NORAD.

Improvements to launch facilities and research infrastructure are continuing. Major additions completed in 1996 include satellite and rocket telemetry antennas, a refurbished Athena launcher, two launcher enclosures and road improvements. In a separate initiative, the State of Alaska commissioned a feasibility study through the Alaska Industrial Development and Export Authority (AIDEA) on the market potential for commercial launch activities, and established the Alaska Aerospace Development Corporation (AADC) in July 1991 to encourage commercial space applications. AADC won a US$1.1 million federal grant in 1993 to design an all-weather orbital launch facility to handle payloads of up to 3.6 tonnes. The initial site study of environment, construction and safety led AADC to choose a site at Cape Chiniak on Kodiak Island. Orbital launches from there would be directed over the Pacific into 0 to ±40° polar orbits.

By the end of 2002 more than 1,500 meteorological and more than 240 major high-altitude sounding rockets had been launched from the PFRR.

Spaceport Systems International

Location: 34.4°N/120.35°W

Current Status

In September 2002, SSI signed the interface control document for a National Reconnaissance Office mission. NASA's Aqua satellite had been launched from SLC-2 at VAFB by Delta 2.

Background

SSI is a partnership between ITT Federal Services Corporation and California Commercial Spaceport Incorporated to provide commercial launch services for small to medium polar vehicles from VAFB. SSI was the first federally licensed commercial space launch site operator in the USA. The license was issued by the Office of Commercial Space Transportation on 19 September 1996. Typical users would be LMLV, Taurus and Delta-Lite.

The spaceport's Integrated Processing Facility was SLC-6's payload preparation facility. The Shuttle pad is off to the right and the spaceport's to the left (WCSC) 0517252

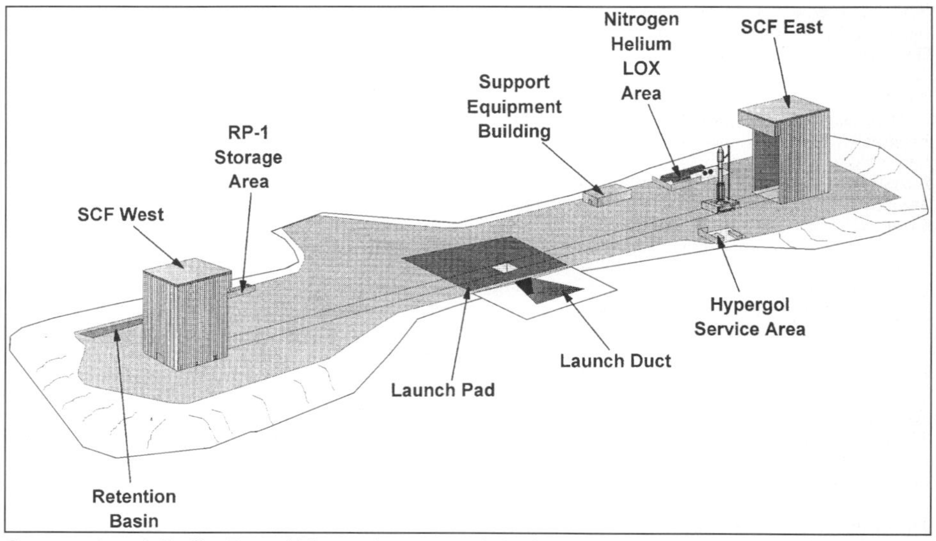

Spaceport Launch Facility. Two vehicles can be processed simultaneously, for launch due south from the single hardstand (WCSC) 0517425

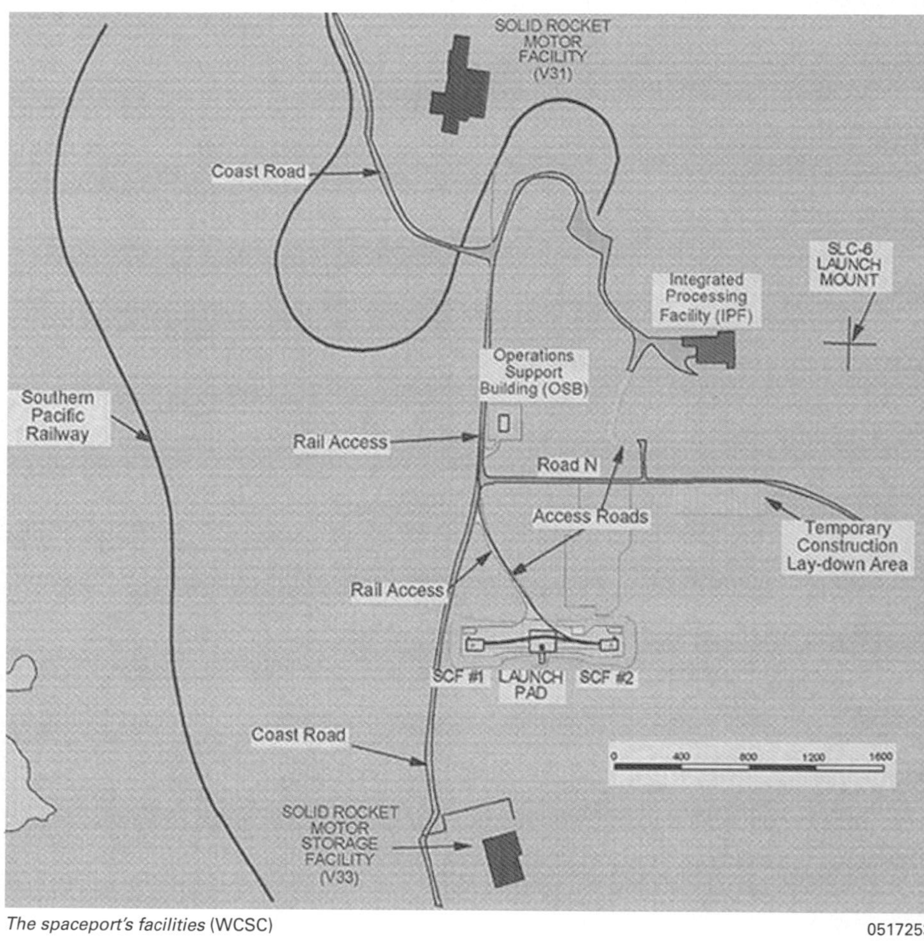

The spaceport's facilities (WCSC) 0517253

SSI is building a new pad and is leasing the US$250 million mothballed Payload Preparation Room at SLC-6, renovated as the IPF Integrated Processing Facility (GEMstar 1 and STEP M0 had been processed by end-1995).

Annual capacity would be 24 launches, using the ITL integrate-transfer-launch approach, allowing the use of simple stationary pad structures. The 25-year lease was signed by WCSC/AF for the undeveloped 0.10 km^2 site near SLC6; the annual fee is US$70,000 plus about US$10,000 monthly for operations costs and utilities. The government is providing US$10 million, mostly as AF grant money, and ITT is committing US$30 million. WCSC was formed as a non-profit corporation in May 1992 to pursue commercial opportunities at VAFB. It is designated by the State Legislature as the California Spaceport Authority. Subsidiary CCSI was established in August 1993 as a for-profit company to meet grant requirements for matching funds from space companies.

The SLF Spaceport Launch Facility operates a single pad at 34°34′35″ north/120°37′53″ west (92 m altitude) between two SCF Stack & Checkout Facilities on an east-west axis. Most vehicles will depart due south (180° azimuth), so will not overfly the SCFs. 168 to 220° azimuths are available without impacting the local community or off-shore oil rigs. 150° is possible.

A vehicle is stacked on an MLP Mobile Launch Platform and transferred launch-ready to the pad on launch day. SCF-E is intended for Delta 2 and Delta-Lite (but can handle Castor 120). SCF-W is optimised for Castor 120-based vehicles. IPF handles booster and payload processing, fairing/storage and payload encapsulation, and houses three launch control rooms in the 12.2 × 42.7 m Launch Control Center (LCC). The processing areas include a 9 × 30 m airlock, three 10.7 × 13.4 m cells and 9 × 45 m high bay, with 47 m high 68 tonne crane.

Vandenberg

Location: 34.4°N/120.35°W

Background

Since 1956 Vandenberg has been responsible for missile and space launches on the west coast and operates the Western Range tracking network extending into the Indian Ocean, where it meets the ER system. Vandenberg provides the US with access to polar orbits using due south launches and was to have provided a base for Shuttle departures on high-inclination missions. Delta missions are supported by the Space Launch Complex (SLC)-2W pad, Titan by SLC-4 and Atlas by SLC-3. Scout retired with its last launch May 1994 from SLC5. The Taurus maiden flight, March 1994, used pad 576E for its transportable launch equipment.

The Western Space & Missile Center in 1991 became the 30th Space Wing. Now elements of air force Space Command, the east and west launch ranges and bases were managed by AF Systems Commands' Space Systems Division. The 30th Space Wing is made up of four groups: 30th Operations Group, 30th Logistics Group, 30th Medical Group and 30th Support Group. 30OG operates the Western Range. Its 2nd Space Launch Squadron is responsible for small/medium launch operations (including Atlas II, Delta 2, Pegasus and Taurus 2) and its 4th SLS for heavy lift vehicles (Titan 2 and Titan 4). The 576th Flight Test Squadron handles ballistic operations.

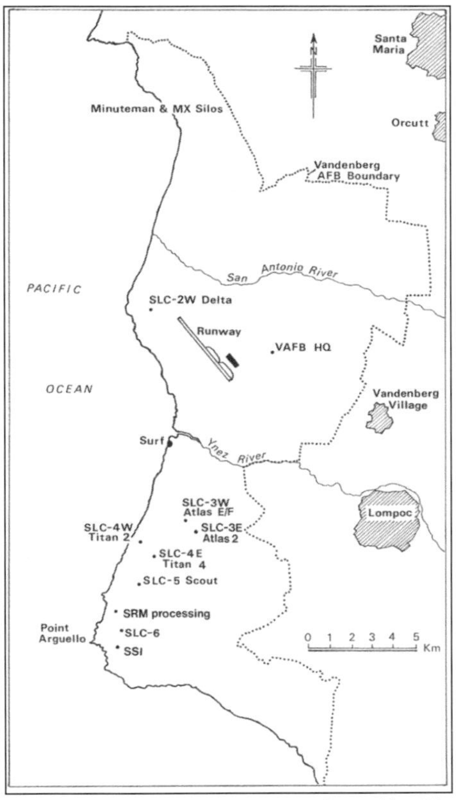

Vandenberg provides access to polar orbits. The northern area is generally employed for missile development and operational test launches; most are de-activated but Minuteman and Peacekeeper ICBM silos and pads remain in use. Apart from the SLC-2W Delta pad, the orbital pads are clustered in the Point Arguello area: SLC-3E/3W Atlas, SLC-4W Titan 2 (previously Titan 3B), SLC-4E Titan 4 NUS (previously for Titan 34D) and SLC-5 Scout (retired 1994). SLC-6 was converted for Shuttle. SLC-7, south of SLC-6, was to have been a second Shuttle pad and was then proposed for Titan 4. Spaceport Systems International's pad is being built at 34.576° north/120.631° west south of SLC-6 for commercial operations starting in 1997
0517426

Vandenberg is also an important base for development and operational ICBM flight testing. Above-ground pads and underground silos support Peacekeeper and Minuteman launches. General Dynamics signed agreements with NASA/USAF by October 1987 allowing up to eight Atlas launches annually from Cape Canaveral, on a cost reimbursable basis. GD has refurbished surplus Atlas ICBMs at Vandenberg and launched them as space vehicles since 1970; the last departed March 1995.

The 399 km^2 base was created along 40 km of coastal scrub, similar to the Canaveral area. The 1,100 buildings house 53 host organisations, 37 associate units and 49 contractor organisations, supporting more than 13,000 military, civil and contractor personnel and service families. From here are launched classified satellites and missiles are test-fired under operational conditions over the Western Range, extending 8,000 km across the Pacific and into the Indian Ocean.

From its southern tip, Point Arguello, spacecraft are launched into polar orbits for both NASA and DoD.

As of the end of December 2003 Vandenberg had launched 567 orbital shots since the first on 28 February 1959.

The major facilities are:

SLC-2

Built in 1958 as part of a group of seven launch installations for the Thor IRBM. Pad 2W completed upgrading in 1995 to accommodate Delta 2 (Radarsat was the first launch). The mobile service tower was raised hydraulically to insert a 3.7 m spacer to handle the longer Delta. Before that, NASA's COBE in November 1989 was its first use since 1984. The second Delta pad, 2E, is no longer operational. 2W can support Delta operations up to the 7920-10 model; it can cope with six launches annually. The 54.3 m gantry supports nine working levels, including a white room enclosing Levels 4–6. Payloads are delivered to the Spacecraft Laboratory in Vandenberg's Building 836, housing assembly areas, laboratory areas, clean rooms, an RF screen room, computer facility and the T/M Doppler station. The 404 m^2 Spacecraft Laboratory 1 includes a 6.1 × 7.5 m curtain clean room providing class 100,000 conditions. A 4.5 tonne bridge crane is used. The 521 m^2 Laboratory 2 includes a 9.2 × 5.1 × 5.1 m class 100,000 clean room and a smaller clean room in the high bay area. Laboratory 3, 284 m^2, accommodates a curtain clean room similar to that in Laboratory 1. Following processing, payloads are transported to the Delta spin test facility some 3 km from SLC-2. The 10.7 × 18.6 × 13.7 m building houses a Trebel dynamic balancing machine and necessary equipment for build-up, alignment and balancing of the upper stage solid motors and spacecraft. Payload propellants are loaded in this facility. Launch operations are conducted from 2W's nearby blockhouse and the Mission Director's Center in Building 840.

SLC-3

SLC-3E became the nation's newest launch complex when it attained initial launch capability in October 1996 following a three year, US$300 million modification project to launch the Atlas II family of launch vehicles. To reach this milestone, Air Force Material Command and Lockheed Martin Astronautics designed and built a state-of-the-art mobile service tower and umbilical tower while transferring countdown operations and blockhouse functions to building 8510 on north VAFB. SLC3E will provide customers with access to sub-synchronous and polar orbits. The 2nd Space Launch Squadron operates SLC-3E. The last launch from SLC-3W was in March 1995. Currently kept in caretaker status by the 2nd Space Launch Squadron. SLC-3W's future missions are yet to be determined.

SLC-4

The two Titan pads, SLC-4E/4W, suffered US$50 to 60 million of damage when a Titan 34D exploded

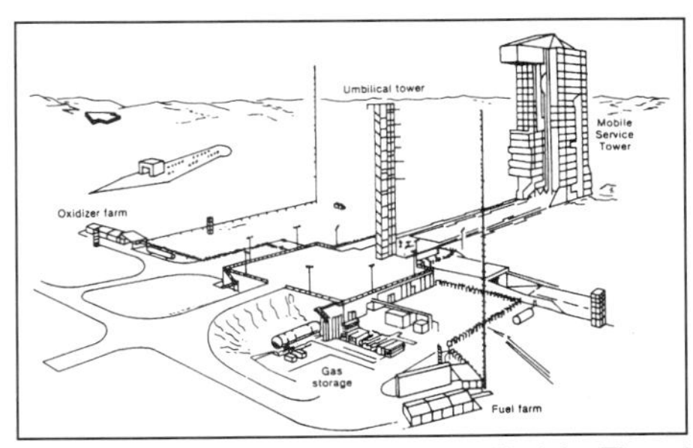

Vandenberg's SLC-4E has been modified for Titan 4 NUS launches (Aerospace Corporation) 0517087

SLC-4 and SLC-3 complexes. From left: SLC-4W (Titan 2), SLC-4E (Titan 4), SLC-3E (being refurbished for Atlas 2 family from 1998) and SLC-3W (Atlas E/F) (USAF)
0517084

SLC-4 and SLC-3 complexes. From left: SLC-4W (Titan 2), SLC-4E (Titan 4), SLC-3E (being refurbished for Atlas 2 family from 1998) and SLC-3W (Atlas E/F) (USAF)
0517085

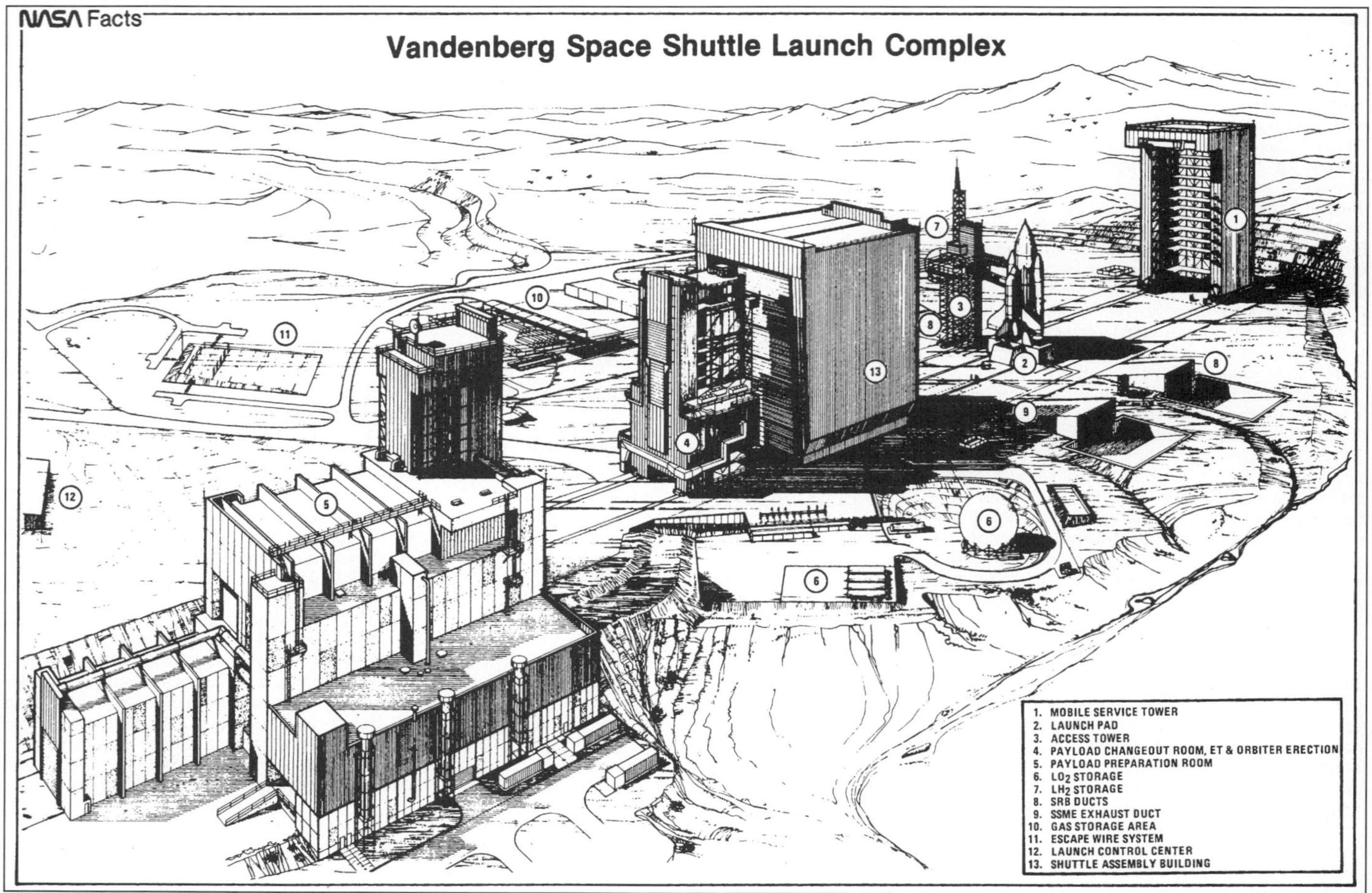

SLC-6 was created in the 1960s for the Manned Orbiting Laboratory, before it was converted for Shuttle operations (shown here). Lockheed Martin's small LMLV 1 uses one of the SRB mounts. The Payload Preparation Room (5) is being leased to Spaceport Systems International (qv) as a processing facility and launch control centre 0517086

8 seconds after lift-off on 18 April 1986. Debris damaged ground support equipment and ignited brush fires up to 1 km away. Following this failure, SLC-4's blockhouse no longer conducts the entire launch: it now takes the countdown from 54 hours to within 8.5 hours, when it is vacated. A remote facility in north Vandenberg takes over. Titan 2 launches continue to be fully handled from the blockhouse's other firing room. The first Titan 2 ICBM refurbished as a space launcher was mounted on the repaired SLC-4W in November 1987 and launched the following Sep. 4W, previously devoted to Titan 3B, is now used exclusively by Titan 2. 4E was refurbished from its Titan 34D configuration at a cost of US$235 million and declared operational for Titan 4/NUS in May 1990; it can handle two missions annually. The first Titan 4, on the pad since June 1990, departed 8 March 1991. Martin Marietta was awarded a US$10 million contract in 1990 to install a sound suppression system on 4E as protection against the solid motors' ignitions. The system dumps water into the exhausts for 2 seconds before and 7 seconds after ignition to reduce the pressure wave.

SLC-6

Selected on the launch site for the Delta 4 to be utilised for launch at azimuth 151–210° and available from 2002. Lockheed Martin's LMLV 1 vehicle was the first launch from the pad, in August 1995. It sits on an adaptor on one of the pad's Shuttle SRB mounts, allowing it to use an existing flame duct. Lockheed Space Operations Corporation was awarded the contract in July 1990 to modify the existing SLC-6 as the TCLC Titan 4 Centaur Launch complex, aiming for a first launch in 1996. However, reduction in projected Titan launches by early 1991 again brought into question the need for converting SLC-6 at all. Like SLC-4E, it would also have been able to handle the Titan 4 NUS version. Projected cost was US$314 million for the integrated collection of structures, systems and equipment required for direct ground processing and launch capability. Design work began 1991, with site modifications planned for 1992–94 and testing for 1995. The conversion was cancelled 1991.

SLC-6 was created in the 1960s for the Manned Orbiting Laboratory. When MOL was cancelled in 1969, the mothballed facility consisted of the pad, mobile service tower, flame duct and launch control centre. In 1975, on the basis that using the existing

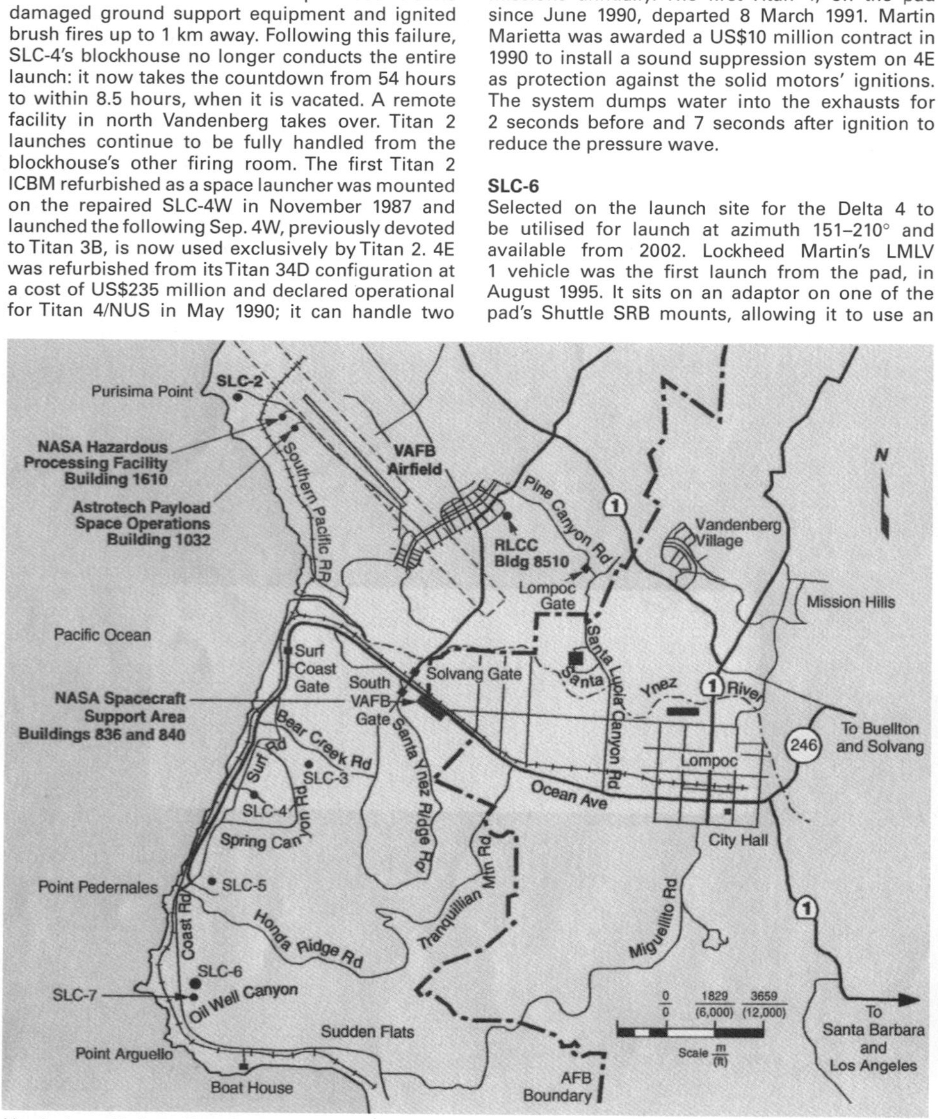

Vandenberg Air Force Base facilities 0132804

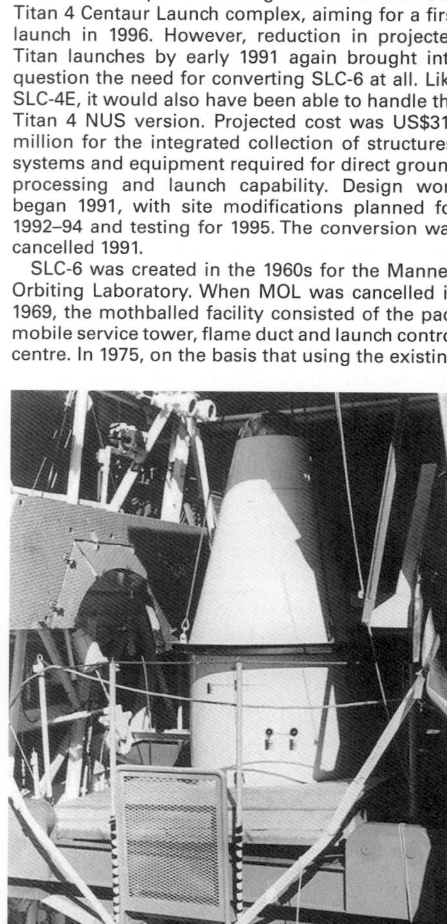

Readied for an early launch from VAFB, a Midas satellite is encapsulated in the launch vehicle shroud (USAF) 1047508

facilities could save US$100 million, it was decided to make it the West Coast Shuttle launch site, and construction began in 1979. Up to 20 flights annually were projected at one stage. US$3.2 billion was expended on expanding the runway to 4,500 m, on building the Orbiter maintenance/checkout facility, and building accommodation for support personnel at north Vandenberg. The funds included modifications to SLC-6 proper, which took in the launch control centre (only 347 m from the pad), payload preparation/changeout facilities, Shuttle Assembly Building (SAB), access tower, launch mount, mobile service tower and three exhaust ducts at south Vandenberg.

Construction delays were compounded when it was found that high winds and low temperatures would not permit the planned assembly of SRB/ET/Orbiter on an unprotected pad. At the end of 1981, a 71 m high movable assembly building to protect the Shuttle cluster and technicians during launch preparations was approved. Lengthening of the existing runway, 25.7 km north of the pad, was completed in 1983. The Orbiter was carried from there to the pad on a 76-wheel transporter, a journey requiring 3 to 4 hours. Shuttle processing contrasted with Kennedy's: instead of the cluster being moved to the pad after assembly, stacking took place on the pad under the protection of three mobile buildings. The first step was for the service tower, using a 200 tonne crane, to mount the SRBs; two cranes would then position the ET between. Then the 4,000 tonne SAB would move in from the opposite direction to mate with the service tower and enclose the assembly area.

With a 125 tonne crane in the roof of the SAB, the Orbiter was to be lifted into position and attached to the SRBs. The third mobile building, an 18 storey Payload Changeout Room, would then move through the SAB and transfer satellites and other payloads to the Orbiter. These payloads would come from another, stationary building, the Payload Preparation Room (this building is being leased to Spaceport Systems International to support its commercial launch services from a new pad south of SLC-6). After fuelling, the three buildings would move back on their tracks for the launch. To protect the pad, 1 million gallons of cooling water would flood into the three exhaust ducts in 23 seconds. At lift-off, control was to pass from Vandenberg to Houston for NASA flights or to the new Support Operations Center at Cape Canaveral AS for DoD missions. The complex was placed in an operational caretaker status in 1986–87 to save US$1 billion following the Challenger accident. The mothballing process was completed by the end of 1989. By end-1986, 1,700 Shuttle contractor employees had been laid off. In December 1986, a General Accounting Office report to Congress estimated a cost of US$200 million in FY89 to maintain SLC-6 in operational status, capable of full reactivation in 18 months. Congress rejected the USAF claim for a new SLC-7 for Titan 4 Centaur and directed instead that SLC-6 should be modified; cost was estimated in 1990 at US$700 million. Before Challenger, subject only to a successful Flight Readiness Firing with Discovery in January 1986, Vandenberg was to provide the Shuttle with a polar orbit launch and landing site for use by NASA and USAF from July 1986. That first DoD flight was to carry the US$500 million Teal Ruby satellite.

UPDATED

Wallops Flight Facility

Location: 37.85° N, 75.46° W

Current Status

WFF manages NASA's Sounding Rocket Program from launch vehicle to payload design and development and data retrieval. NASA maintained the small Wallops centre for Scout vehicles, although only one was launched from 1979 until Scout's 1994 retirement and it is now primarily a sounding rocket base. The maiden Conestoga failed from there in October 1995. As of the end of 1999 a total 25 orbital flights had been launched from this facility. Air launches for Pegasus vehicles are assigned to the drop zone at 37.00° N × 72.00° W.

Background

Established in 1945 under NASA's predecessor, the National Advisory Committee for Aeronautics, Wallops Flight Facility is located on the Virginia Eastern Shore. The first successful launch from Wallops Island took place on 24 August 1945 when a two-stage Tiamat carried the first telemetry system used in a rocket propelled flight research vehicle.

Now administered by NASA Goddard, Wallops Flight Facility became the fifth space port and the third US orbital site when NASA launched Explorer 9 from this island on Virginia's east coast on 16 February 1961. In all, 19 vehicles entered space from this point through the end of 1985, when NASA and the air force largely suspended orbital operations. Although NASA actively maintains the service buildings for sounding rocket tests and other classified experiments, Wallops' space launch history ended with the retirement of the Scout vehicle in 1994. A commercial consortium, EER Systems tried to revive the site with a launch of their Conestoga in October 1995, but the vehicle failed and the programme's future remains uncertain. As a flight control centre, Wallops provided orbital support and control for the B-52/Pegasus launch of Brazil's SCD February 1993. The B-52 took off from Florida.

Wallops served NASA and its predecessor NACA (the National Advisory Committee on Aeronautics) for many years before emerging as a spaceport. NACA launched its first sounding rocket, a Tiamat on 4 July 1945, inaugurating the site as a missile research facility. Since 1945, NACA, NASA, the air force, the navy, SDIO and many foreign governments have launched more than 800 sounding rockets from a pad on the far end of the island. Even today more than 30 launches a year occur at Wallops using vehicles as diverse as Super Areas, Black Brant, Taurus-Tomahawk, Taurus-Orion and Terrier-Malemute.

Wallops also provides support for the University Class Explorer Missions and, recently, the SPARTAN and SPARTAN-LITE programmes which have now been moved from NASA's Goddard Space Flight Centre.

NASA and the air force built extensive facilities on the island over the years and the active maintenance of these facilities reflects a commitment on the part of the US government to maintaining this infrastructure. The buildings, which are still operational, include:

Launch Areas

Launch Area 1 for heavy-lift sub-orbital rockets. Launch Area 2 and associated blockhouse for most sounding rocket launches. Wallops used this pad for the maiden 1945 sounding rocket firing. Launch Area 3 for the Scout orbital launch vehicle. Launch Area 4 for classified payloads and special projects. Launch Area 5 most recently used for US Navy Vandal missile tests. Launch Area Oar for EER System, which built and owns the pad.

Block Houses

Blockhouse 2 to service Launch Area 2. Blockhouse 3, serving all pads.

Workshops and preparation facilities

Scout Assembly Shop which NASA and the air force used to prepare all Scouts in the Scout Assembly Shop. When the Scout was ready the ground crew rolled back the shelter building on rails and raised the launch rail into the firing position.

Assembly Shop 1 for sounding rocket assembly and checkout.

Spin Balance Facility for vertical/horizontal spin testing of payloads and motors up to 13.6 tonnes. A Trebel FVD-3000 machine spins the test object to test and modify its dynamic balance. Typically 32 motors, six satellites and three sounding rocket payloads pass through the test procedure annually, including 80 per cent of AKM and PKM motors for GEO satellites.

Wallops Orbital Tracking Station established by NASA in 1986 to provide tracking support for projects such as IUE, IMP 8, Nimbus 7 and COBE. NASA performs high-speed data transfer to Goddard at Greenbelt through a satellite link located at Wallops. To assist NASA, WFF has installed telemetry receiving and command antennas ranging from 2.4 to 8 m. Wallops' tracking radars include 8.8 m parabolic FPS 6 radar that can track objects out to 60,000 km with a precision of ±3 m and with a line of sight velocity resolution of 9 cm/s. The antenna radiates a peak output 2.5 MW.

NOAA's Command/Data Acquisition Station of nine 7.3 to 26 m antennas to track, command and take data from the NOAA and GOES satellites.

White Sands

Location: 32.23°N/106.28°W

Background

The 8,100 km^2 White Sands is operated by the US Army and was the site of the first major US rocket firings after the Second World War before the move was made to Florida's larger range. The first V-2 flight from US territory took place on 16 January 1946 from this site south of Route 70. Considered as a launch site for the Space Shuttle, but used as an emergency landing strip for STS-3 on 30 March 1982 when bad weather precluded a landing at Edmonds Air Force Base, California. It is also the site of the first atomic explosion; the Trinity Site is on the northern part of the range. White Sands is still a major sounding rocket firing base, however, and supports BMDO flight-testing.

Some 26 km northeast of Las Cruces, White Sands Missile Range has two separate propulsion test sites, designated areas 300 and 400, established for trials of Apollo CSM/LM main engines/thrusters. Area 300 began testing the Service Propulsion System for Apollo's CSM in September 1964, while area 400 accommodated hundreds of Lunar Module ascent/descent engine firings. There are seven test stands, four altitude chambers and three open. The largest vacuum stand is 302, 10 m diameter 17.7 m high; it was employed for Viking's lander thrusters. Stand 405 is an altitude facility for spinning solid motors up to 125 rpm, on cylinders as large as122 cm diameter. It produces 111.2 kN thrust. Stand 401 accommodates cryogenic testing; the others utilise hypergolic propellants. Space Station propulsion module test article firings began November 1992 (The Space Station's redesign no longer requires this element). Materials testing expanded after 1967's Apollo 1 fire, emphasising flammability work.

WSTF also provides precision cleaning facilities, including a class 100 clean room for spacecraft parts. For example, it ensured that Viking's soil sampler and the tools used in the Apollo Lunar Receiving Laboratory at JSC were free of terrestrial contaminants.

White Sands Space Harbor provides the third Shuttle landing site in the US after Edwards and Florida; the third mission landed on the Northrup Strip, consisting of two 11 km long gypsum-sand runways 1,200 m above sea level, in March 1982 after the primary Edwards site became waterlogged.

In 1999 White Sands has a employment inventory of 6,000 people, 5,700 of which were civilian or contractor personnel. To the end of Fiscal Year 1999 (September 30, 1999) White Sands had conducted 42,415 missile firings since 1945.

TRAINING FACILITIES

International

European Astronaut Centre (EAC)

Background
EAC was founded in 1990, to provide expertise for ESA's manned space activities and act as the astronauts' home base. It is responsible for all astronaut-related matters, such as selection, medical surveillance, support and training. Since its foundation, ESA astronauts have flown on STS-42/Soyuz-TM 20 (Ulf Merbold), STS-46/61/75 (Claude Nicollier), STS66 (Jean-Francois Clervoy), Soyuz-TM 22 (Thomas Reiter) and STS-75 (Maurizio Cheli). Following 1992's selection of six new astronaut candidates, EAC organised their basic training leading to their qualification as European astronauts in December 1993. EAC provided payload training to the astronauts and cosmonauts involved in the EUROMIR 94 and 95 missions. On 25 March 1998, the ESA council decided to establish a single European Astronaut Corps and a group of seven astronauts joined the programme in 1998 followed by four more in 1999. The integration process was wound up in June 2000 with an established pool of 16 astronauts (four each for Germany, France and Italy and four for the remaining ESA countries). Six ESA astronauts are currently in active status and training with ESA's partners under EAC co-ordination, preparing for the International Space Station. Delays due to protracted assembly sequence between 1998 and 2002 and the Shuttle Columbia loss in February 2003. The European astronaut pool is in a state of sustained preparation and flight engineering and science assignments pending the return to flight of the NASA Shuttle fleet presently anticipated for the end of 2004. Not until well into 2005 on the third flight resumption flight will the resumption of ISS assembly become clear.

Russian Federation

Zvezdny Gorodok (Star City)

Background
The Gagarin CosmonautTraining Centre (TsPK) was established at Star City 30 km northeast of Moscow in 1960 and trained some 300 in its first 35 years, including back-ups and unflown programmes. RKA/VKS jointly fund and manage TsPK now, assuming control from the air force. The Director and most section heads are veteran cosmonauts. Previous Director Gen Vladimir Shatalov retired in 1991. He succeeded Gen Georgi Beregovoi. Alexei Leonov was head of training until he, too, stepped down in 1991. Star City has a population of 3,500. Only 10 to 20 experienced cosmonauts are active at any one time, plus an equal number of rookies. Of these, six two-man crews (usually comprising one experienced man and one rookie) train for missions. Three organisations are responsible for crew selections: the Russian Space Agency, RKK Energia and the Ministry of Health. The Buran pilots were from VKS. Rookies normally fly five to 10 years after the start of one and a half to two years of basic training. Every cosmonaut has to undergo a week's medical examination every year to retain flight status. They are kept in a pool until assignment to a mission. Added to these pairings are specialists as necessary.

Facilities include classrooms, laboratories, sports complex, centrifuges and thermal chambers. A high fidelity mockup of Mir and its modules provide equipment and procedures training. Two Soyuz-TM and one Soyuz-T simulators are housed in a separate room. OneTM offers full simulation, while the other is a visual docking simulator (another is reportedly in the Mir hall). A 23 m diameter 12 m deep water tank containing spacecraft mockups allows EVA training. Shuttle training facilities are housed in a separate building; a Shuttle simulator is based at the TsAGI Central Aero-Hydrodynamics Institute. The Swedish-built 18 m 3-DOF centrifuge became operational in 1979, providing a maximum 30 *g* and 300 tonne rotating mass. The cabin can simulate different altitudes, temperature and air mixtures. Training requires a cosmonaut holding a dead-man's handle to undergo an initial 3 *g* for 14 seconds, rest for 10 minutes, 5 *g* for 30 seconds, break for one day, 4 *g* for 40 seconds, rest for 10 minutes, then conclude with 6 *g* for 30 seconds. There is also a 7 m centrifuge. A modified Il-76 aircraft is available for parabolic flight training, and is equipped with an airlock to provide EVA rehearsals. It is also available for commercial µgravity flights in a joint venture with Germany's Intospace and other western companies. First demonstration flight was 25 November 1992.

Maintains a physical, psychological and technical examination, selection and training service for 'tourist' astronauts carried on Soyuz ISS ferry flights. The facility has undergone some upgrade to match requirements agreed by the international partners in the ISS programme.

United States

White Sands Space Harbor

Background
Part of WSTF, WS Space Harbor is one of the primary training areas for Shuttle pilots, developing skills at approaches and landings in the Shuttle Training Aircraft. Shuttle pilots are required to complete approximately 1,000 simulated landings, using a modified Grumman Gulfstream STA, prior to a Shuttle mission. NASA selected the site in early 1976 within the US Army's White Sands Missile Range and the original 3 km Northrup gypsum lakebed strip was lengthened to 4.6 km in time for the first Shuttle Training Aircraft flight in August 1976. Altitude above sea level is about 1,200 m. About 10 STA training sorties are made weekly; the aircraft are housed in the NASA hangar at El Paso International Airport, Texas. A second runway was added in 1979 and both lengthened to 10,680 m, with landing aids, to allow WSSH to serve as an alternate Shuttle landing site. Columbia landed there in March 1982 after rain prevented use of the primary Edwards AFB facility. A full set of convoy equipment for safing the Shuttle is kept at WSSH. The runways were laser-levelled and widened to 275 m during the post-Challenger standdown. A duplicate of the trans-Atlantic abort runway at Ben Guerir (Morocco) was added in 1989 for TAL practice. Formalised as a Shuttle contingency landing facility.

TELEMETRY, TRACKING AND COMMAND

Australia

Woomera Tracking and Telemetry Station

Background
The United States Air Force's 21st Space Wing's 5th Space Warning Squadron works with its Australian counterpart at the Woomera Air Station in Australia. The Australian 1st Joint Communications Squadron at the Joint Defence Facility, Nurrangar handles DSP downlinked information for the eastern hemisphere.

In February 1959 a survey team selected a location near the Island Lagoon dry lake bed, 56 km south of the Woomera Rocket Range, as the site for the first deep-space station outside the USA. In 1970 Joint Defence Facility at Nurrungar (JDEN) established as part of the USDSP. Operated by the Australian Department of Supply the 26 metre polar mounted tracking antenna was used for tracking a variety of spacecraft until it ceased operation on 22 December 1972 as part of a consolidation of NASA station facilities.

China

XSSC-Xian Satellite Control Centre

Background
China's launch and satellite TT&C network is centred on the XSSC Xian Satellite Control Centre in Shaanxi province, controlling six fixed, three mobile and three ship (Yuanwang 1–3) stations. Yuanwang 3 was delivered in March 1995. The fixed sites are Weinan (near Xian, and the lead site), Min-Xi (Fujian province), Xiamen, Changchun (Jilin), Karshi (Xinjiang) and Nanning (Guanxi), with the first three providing GEO control. XSSC issues de-orbit commands to FSW recoverable satellites through the two mobile stations, while the third locates the re-entry modules during descent. XSSC's hardware covers three NCI 2780, two VAX 3700 and several VAX-II computers. Improved and updated tracking facilities. New telemetry stations.

Germany

German Space Operations Centre (GSOC)

Background
Operated by DLR, GSOC is responsible for all operational tasks of the national space programme and provided the first manned mission control outside of the US/USSR. It first controlled the German funded Spacelab D1 (1985) and then Mir-92, D2 (1993) and ESA's EuroMir-95 (1995–96). It may operate as the Columbus Orbital Facility Control Centre. GSOC also handles unmanned spacecraft, specifically LEOP. TV-Sat 1's rescue attempt was mounted from here and initial TT&C was provided for the operational TV-Sat and DFS satellites. All Eutelsat 2s were positioned by GSOC, which also supported Giotto through its Weilheim station. Others include AMPTE and Rosat. The centre also supports mobile rocket launches for high altitude investigations. Upgraded for support of Columbus ISS operations, delayed now by the suspension of Shuttle flights and consequent delays in lifting the pressure module to orbit. GSOC has run optional mission control scenarios for different ESA Columbus operations. Control simulations have been performed showing simultaneous operations with occupied and unattended space vehicles.

India

ISTRAC-ISRO Telemetry Tracking and Command Network

Background
Istrac operates a network of TT&C stations for ISRO's launch vehicle and satellite activities. The facility employs 465. Stations are sited at Trivandrum (8 m), Sriharikota (10 m), Carnicobar (8 m, principally for launcher downrange TT&C), Lucknow (10 m) and Bangalore (10 m), which also accommodates the network's Spacecraft Control Facility. Its two 10 m Cassegrain dishes (20.5 dB/K, 2.200–2.300/2.025–2.120 GHz down/up link, 9°/s slew rate, 0.8° beamwidth).

ISTRAC has two VAX 11/985 computers to currently support the IRS remote sensing satellite (imagery is routed via the National Remote Sensing Agency's Hyderabad station) with a back-up VHF command system. Insat is controlled from its own dedicated station at Hassan.

Three satellites can be controlled simultaneously, one undergoing orbital checkout and two operational. SSC has a main control room, mission analysis room, two dedicated control rooms and a data control room, the first two operating until orbital checkout is completed and the satellite is handed over to a dedicated room. The data control room acts as a node for all the network stations and SCC. Added data handling and telemetry processing centre. New data analysis section.

The network also operates an S-band receive-only TT&C station near Port Louis, Mauritius. Optical/laser tracking is performed at the Satellite Tracking & Ranging Station (STARS) established at Kavalur in co-operation with Russia.

International

European Space Operations Centre (ESOC)

Current Status
In addition to its network of ground stations comprising 15 m stations using S-band frequencies for telemetry, command and tracking, ESOC has secured augmentation through a 35 m Deep Space Ground Station at New Norcia in Western Australia with additional coverage in X-band (8 GHz) frequencies.

Background
ESOC, with a permanent staff of 272, and a total staff of about 750, provides control of ESA spacecraft during launch and orbital operations through an operations control centre comprising a main control room (for LEOP), dedicated control rooms and computer facilities. The centre became operational in May 1968 for the control of ESRO 2B.

ESOC controls the Estrack network of seven ground stations: Perth (Australia), Kiruna (Sweden), Malindi (Kenya), Kourou (French Guiana), Redu (Belgium), Villafranca (Spain), Odenwald (Germany). Satellites are assigned to individual stations for operational support. Redu has full autonomy for ECS communications satellite control on behalf of Eutelsat and Villafranca for the ISO mission. The Ground Systems Engineering Department (GSED) is responsible for all engineering functions related to the implementation of the ground systems for operations; all operational launches related to the support of missions entrusted by ESO, comprising spacecraft and ground operations are consolidated in the Mission Operations Department (MOD). ESOC hosts the ESA Operations Control Centre with dedicated rooms: Main Control Room; Flight Dynamics Room; Dedicated Central Room; and the Dedicated Mission Support system.

ESOC is also at the forefront of research into orbital debris, maintaining the DISCOS (Database and Information System Characterising Objects in Space). Created in 1990, DISCOS holds more than 2.3 million elements and 25,000 more are added monthly.

Facilities include:

Redu
The site (50.0° N/5.1° E) of some 19 ha, in a valley outside the town of Redu, houses the station control room, as the network interface with ESOC's OCC; the main equipment room housing the telemetry, telecommand and ranging baseband racks, and the station computers. The ECS control room and computer room; and the test and monitor room. The SHF facilities include four fixed in-orbit test antennas (TMS3A, 3B, 3C, 7) for measuring telecom satellites' performances before handing them over for routine operations, four fixed telemetry (TM1, 2, 3, 6), one ranging (RG1), one satellite multi- services terminal and a radiometer. A new S-band 15 m antenna became operational in mid-1995 to support two Cluster satellites. Staffing is about 50, mostly from the Belgian company CISET, responsible to ESA for maintenance/operations.

Bangalore's Istrac complex. Centre: main control centre; right: TT&C station. The twin 10 m dishes are controlling the IRS remote sensing satellites (ISRO) 0517081

The European Space Operations Centre of the European Space Agency was created in 1967 in order to operate the satellites launched by the Agency. Its expertise encompasses operations for scientific, remote sensing, telecommunications and microgravity satellites. ESOC also serves as the operational computer centre of ESA. Nearly 700 staff including 272 permanent employees, are in charge of operations, software development, ground stations and mission preparation (ESA) 0019685

Villafranca
ESA's Villafranca del Castillo satellite tracking station (40.4° N/3.9° W), Vilspa, occupies a 10,000 m^2 site some 30 km northwest of Madrid, and was inaugurated 12 May 1978 for IUE. The principal building comprises the main equipment room (telemetry, command/ranging baseband racks, the station computers, timing system and network communications interface with ESOC's OCC), the computer room (IUE real-time operations and image processing), IUE control room and observatory, the payload test lab (communications testing), and the Villafranca computing centre (computers and terminals for IUE image de-archiving, data processing/editing). A new operations building accommodates ISO's control centre and science centre; the infra-red observatory is operated from Villafranca. The antenna farm covers: two 15 m and one transportable 5.5 m S-band antennas, one 12 m C-band, one 4 m L-band, one 3 m Ku-band, one VHF uplink and nine crossed yagi arrays. The majority of the more than 70 staff are employees of the Spanish INTA. Vilspa has been providing 16 hours of real-time support to IUE daily since 1 October 1995. For the 18 years before that it was 8 hours by Vilspa and 16 hours by NASA Goddard. The site also provides continuous TT&C services on behalf of ESOC for the Marecs B2 satellite over the Atlantic.

Kourou
The Diane S-band station (5.2° N/52.8° W), 27 km from Kourou and also known as Kourou 93, provides telemetry, tracking/command during LEOP of ESA missions. Odenwald uses a transponder for four-way Meteosat ranging measurements. VITROCISET is responsible for maintenance/operations.

Malindi
The station (2.9° S/40.1° E) is at Italy's San Marco Scout base camp in Formosa Bay. The S-band front end belongs to the University of Rome, while ESA provides the baseband equipment.

Perth
The station (31.8° S/115.8° E) is on the Perth International Telecommunications Centre complex of Telstra Corp, which provides management. The S/X-band antenna supports LEOP of ESA missions and, until mission end in 1993, routine Hipparcos operations.

Kiruna
The 15 m S/X-band antenna (67.9° N/21.0° E) at Salmijärvi primarily supports ERS 1/2 TT&C and data reception under Swedish Space Corp management. The station is remotely controlled from ESOC's ERS Mission Management and Control Centre.

Odenwald
(49.7° N/8.9° E). The main function is the support of two Cluster satellites (the other two to be handled at Redu). A VHF command facility provides Redu's ECS back-up. Thorn Security provides maintenance/operations support.

Japan

Tsukuba Space Centre

Background
Located in Tsukuba Science City, 60 km northeast of Tokyo, TKSC was opened in 1972 and is JAXA's largest technical facility, with seven sections responsible for launcher and satellite R&D, design and testing. The Tsukuba Tracking and Control Centre provides Japan's satellite command node.

The Tracking and Control Centre is the command hub for all Japanese-launched satellites, including those of ISAS and commercial organisations. The Okinawa Tracking and Data Acquisition Station (26°30′ north; 127°54′ east) provide support to the Masuda TDA Station (30°33′ north; 131°01′ east) at Tanegashima, and the Katsuura TDA (35°13′ north; 140°18′ east) south of Tsukuba. NASA, CNES and ESA support is provided as required. The Ogasawara (27°04′ north; 142°13′ east) and Christmas Island (2°03′ north; 157°27′ west) stations provide downrange launch tracking and command relay for Tanegashima's Takesaki Range Control Centre, with Christmas Island employed principally in GEO launches. Added low angle trackers and new weather prediction and monitoring equipment.

Japan's Kibo Space Station contribution will be controlled and supported from the SSIP Space Station Integration & Promotion facility at Tsukuba, completed in 1997. It provides the SEL, Space Experiment Lab, user experiment preparation/testing building, the SST Space Station Test building for JEM testing, training, payload integration and operations simulation, the ATF Astronaut Training Facility for crew operations, WET Weightless Environment Test building (16 m diameter 11 m deep water tank) for procedure verification and crew training. It acts as the SSOF Space Station Operation Facility for JEM operation.

VERIFIED

Norway

Tromsø Satellite Station (TSS)

Background
TSS, established in 1967, is Norway's national receiving station for data from polar satellites, emphasising near-realtime data processing and distribution. Its northerly position provides coverage for 10 out of 14 revs for polar satellites. A fourth antenna, for SAR reception, was added in 1994 to the 10 m L/S/X-band, 2.4 m and 3 m L-band dishes. TSS became a shareholding company in 1995, equally owned by NSC and the Swedish Space Corporation. The number of staff is about 25.

The station's previous main activity was reception and archiving of NOAA data as part of ESA's Earthnet but it completed upgrading in 1991 as the national facility for near-realtime processing and distribution of ERS SAR data. Includes data interpretation service and an analytical tape storage archive. The CESAR fast processor generates a SAR full resolution, a SAR low resolution and a SAR low-resolution calibration image in less than 8 minutes, covering 100 × 100 km. TSS offers the same products from Japan's JERS 1; the 75 × 75 km images are available some 20 minutes after data acquisition. Radarsat was added in 1996. Since summer 1994, TSS has been operating a pre-operational pilot service for near-realtime detection of oil spills using ERS SAR data. NOAA data have been received since 1973; quick look, raw data sets, geo-corrected data sets, SST and weather images are available for near-realtime delivery via Internet or ISDN. Added speed capability for faster processing.

TSS operates a Cospas/Sarsat Local User Terminal in direct contact with the Main Rescue Co-ordination Centre in Bodø. A new LUT was installed in 1993. TSS also provides telemetry support for Andøya launches. Tracking facilities have been upgraded and expanded to include simultaneous tracking and telemetry transmission capabilities.

Russian Federation

Support Ships

Background
As its manned spacecraft are within direct visibility of its territory for only 9 hours daily, the RFAS/USSR traditionally compensated for its lack of land bases with a flotilla of support ships and, increasingly, relay satellites. However, it was noted in early 1992 that Mir mission control was operating without satellite relays because of the cost, restricting contact to 15 hours daily. It also appeared that the ships were returning to port. Indeed, 1993–95 saw no ship activity at all – indicating this element had been retired. Some have been working as cruise ships and making promotional tours. Operation of many of the space surveillance and TT&C sites is now erratic, with facilities dropping out and then returning to service. The relay satellites continue to be used only when necessary because of the large fees – the manned and communications programmes come under different ministries.

The Soviets and their Russian successors put more than 10 ships into service from 1960, operating under the flag of the USSR Academy of Sciences. The flagship, *Kosmonaut Yuri Gagarin* (45,000 tonnes), carried 75 antennas, including two 25 m 'Ship Shell' and two 12 m 'Ship Bowl' dishes. Communications could be maintained simultaneously with two or more satellites, manned craft or orbital stations. The flagship was supported by the *Akademik Sergei Korolev* (21,465 tonnes) and four 9,000 tonne-class Kosmonaut ships: *Pavel Belyayev, Georgi Dobrovolski, Viktor Patsayev* and *Vladislav Volkov*. The *Kosmonaut Vladimir Komarov* was transferred to Leningrad for conversion to an ecological data collection platform. It is claimed that *Komarov, Gagarin* and *Korolev* sit in Odessa, too expensive for Russia to hire from Ukraine. They may even be up for sale, possibly with China interested. The *Morzhovets, Borovichi, Kegostrov* and *Nevel*, former timber carriers, were taken out of service by mid-1989 and sold to India for scrap. The *Marshal Nedelin* was completed in 1983 but was rarely mentioned in relation to the space programme, although she did support Buran's maiden flight in November 1988. Sister ship *Marshal Krylov* was completed in 1989. On 12 April 1988 the keel of the *Akademik Nikolay Pilyugin*, lead ship of yet another large class of space support ship, was laid in Leningrad.

Western navies monitored the fleet's activities because their positioning provides advance warning of manned and other launchings. Molniya relay communications to TsUP. Increased reliance on satellite relay and on US and ESA communications and tracking facilities for human space flight operations now involving International Space Station. No none-ISS flights launched since Mir support.

Russia fully intended to mothball the entire fleet as geosynchronous relays, particularly SDRN, entered full operation but both have suffered from financial cuts. Many of the functions previously carried out by the support ships for human space flight is now covered by the international inventory of tracking capabilities.

UPDATED

TsDUC Russian Deep Space Network

Background

High Earth orbit and deep space missions are controlled by the TsDUC Long Range Space Communications System centred at Yevpatoria in the Ukraine. The Soviet Academy of Sciences and VKS controlled the centre until 1991's independence. It has the responsibility for controlling missions such as Mars, Granat, Interball, Spektrum, Phobos, Astron, Vega and Venera. It took over full control of the Ukraine's Sich 1 11 October 1995 after working in conjunction with VKS. For deep space missions in very high orbits and planetary probes, a special six-station network (Medvezhi Ozera, Fushino, Simeiz, Ulan Ude, Ussurisk and Yevpatoria) of 22 to 70 m radio-telescopes is employed.

This Crimean site was selected by Korolev and Keldysh in 1957 and equipped with eight 16 m antennas mounted on battleship gun turrets. It now operates 70 and 32 m dishes and is complemented by 70 and 32 m antennas at Ussurisk and a 25 m system at Ulan Ude. The 70 m RT-70 radio telescope was being built on the Suffa Plateau, but its status is unclear. This core system also calls on radio telescopes at other sites: 64 m at Medvezhi Ozera (Moscow Power Engineering Institute), 22 m at Simeiz (Crimean Astrophysical Observatory) and 22 m at Pushchino (Lebedev Institute of Physics). Six 32 m telescopes were under construction for VLBI work from the mid-1990s, but have probably been abandoned.

It was claimed in early 1994 that TsDUC would that year become part of the ground system for more 'profitable' satellites such as Molniya, Meteor and Gorizont. This is reflected by its control of Sich 1. CNES has been making payments since 1992 to keep it open for Granat operations. No backup capability for Granat and no backup for Soyuz. All communications and tracking delegated to non TsDUC facilities. System and installation upgrades have been approved although little is known about the scale or the prospective funding but enhanced capabilities linked to military requirements have been acknowledged and observed.

TsUP-Tsentr Upravlenyia Polyotom

Background

To maintain control of the some 150 operational satellites, the Russian Military Space Forces (VKS) operate a far-flung network of about a dozen sites across the former USSR. The HQ of this KIK (komandno-izmeritelnyy kompleks) command and control system is at Golitsyno-2 in the Moscow region. Political problems associated with operating non-Russian stations prompted the first deployment in January 1994 of a mobile station (eight vehicles and 20 officers). Mir operations are controlled primarily using a ground network of seven stations (Dzhusaly, Kolpashevo, Petropavlovsk, Tbilisi, Ulan Ude, Ussurisk and Yevpatoria) co-ordinated by the FCC Flight Control Centre at Kaliningrad outside of Moscow.

The manned and unmanned Flight Control Centre (abbreviated in Russian to TsUP, pronounced 'soop'), 10 km northeast of Moscow, began construction in 1970. 1973's Soyuz 12 was its first manned mission; previous missions were controlled from Yevpatoria. TsUP is manned constantly by 1,350 staff working in three shifts, earning an average salary of Rb230,000 monthly in mid-1995 (about US$50 at the official exchange rate). TsNII Mashinostroenye manages it, where it is located. The main control room was responsible for controlling Mir and its supporting Soyuz and Progress missions. Ten PS-2000 and three Elbrus mainframe computers handled the more demanding missions. There were 287 Mir controllers divided into five shifts (including one reserve), each lasting 24 h. A similar, but independent, control room for Buran was used for the first time in 1988 to support the maiden flight. It was used for Mir support when the main room is undergoing modifications and will be updated as the main Space Station control room.

The centre can handle 10 spacecraft simultaneously, assuming control from Tyuratam at launcher separation. A smaller control room handles Soyuz and Progress flights independent of Mir. Another room accommodates weekly two-way TV broadcasts with cosmonauts' families. The facility is now committed to support of the military satellites and the manned element has been removed. From 2001 a significant increase in operating funds and investment in development work for upgrade and refurbishment.

United States

Army Space Command Ground Stations

Background

Army Space Command (forward) is the operational arm of the US Army Space & Missile Defense Command and is the Army Component of US Space Command. Command of the DSCS System Operations Centres and management of the joint tactical use of DSCS satellites is a primary ARSPACE mission. The Joint Staff assigned ARSPACE the operation of five worldwide DSCSOCs, three worldwide AN/MSQ-114s and three worldwide Regional Space Support Centres. ARSPACE operation of the JTAGS Joint Tactical Ground Stations in Europe and Korea provides theatre commanders with a transportable system for processing/disseminating near-realtime warning of theatre ballistic missile launches.

Using DSP data, JTAGS can process in stereo, improving the sensor's capability. JTAGS ties directly to theatre communications systems to transmit launch point information, impact area prediction, time of flight and positional information. ARSPACE also provides oversight of space surveillance operations at the US Army Kwajalein Atoll. ARSST Army Space Support Teams have deployed with units such as the 10th Mountain Division worldwide, providing weather, terrain analysis, mapping, imagery, communications and early warning information.

Processes signals intelligence from DSCS/Milstar Sources and provides coded forward transmissions to field commanders.

Consolidated Space Test Center (CSTC)

Background

Previously known as the US Air Force Satellite Control Facility, CSTC controls satellite systems such as the Keyhole series of reconnaissance satellites through the worldwide AF Satellite Control Network. It also controls DoD Shuttle payloads, beginning with the STS-4 mission in 1982, routeing secure communications and crew instructions through NASA Johnson and ending when the Department of Defense shifted all satellite payload launches to expendable vehicles. In addition, CSTC operates satellite checkout facilities at Cape Canaveral AS and Vandenberg AFB. Until 1987, it was in sole control of day to day satellite activities, now passed to 50th SW. It was directly responsible to the Space & Missile Test Organisation until SAMTO was absorbed into its parent Space & Missile Systems Center in October 1989.

The operators of the facility, the 50th SW, are a component of Air Force Space Command at Peterson AFB. The wing was originally established 8 July 1985 as the 2nd SW and then redesignated 30 January 1992. The facility at Falcon was built under the air force programme known as the CSOC Consolidated Space Operations Center. Its mission is to command/control *operational* DoD satellites and to manage the worldwide AFSCN AF Satellite Control Network. Previously, the air force Space & Missile Systems Center's Consolidated Space Test Center in California was the prime control facility.

The 50th SW's satellites include DSP, DMSP, GPS, DSCS, NATO 3, NATO 4/Skynet 4, Fltsatcom, IUS, UHF Follow-On and Milstar. The wing has 3,300 military and civil personnel assigned to its 45 units at 16 locations worldwide. It comprises four groups: 50th Operations, 50th Logistics and 50th Support Groups at Falcon AFB, plus the 750th Space Group at Onizuka AS, California. The 50th Operations Group commands/controls the satellites, trains space operations crews and provides operational support and evaluation for managing the satellite operations centres and ground stations. The group's seven Space Operations Squadrons are:

1st SOPS

The 1st SOPS supports operations during launch, early orbit, mission and anomaly phases of GPS and DSP. This includes routinely operating older, more troublesome, satellites and eventually de-orbiting them. Once satellites are checked out, day to day control is turned over to the other SOPS.

2nd SOPS

The 2nd SOPS provides day to day GPS command/control. The Navstar GPS Master Control Station took command of the first GPS operational satellite following launch 14 February 1989. Detachment 1 at Cape Canaveral AFS operates a ground antenna for prelaunch compatibility testing and on-orbit control.

3rd SOPS

The 3rd SOPS is responsible for day to day command/control of 14 communications satellites, including DSCS 3, Fltsatcom and UFO (Fltsatcom transferred to NAVSOC at Pt Mugu on 19 June 1996, to be followed by UFO in 1999). The operators also provide LEOP support, trend analysis and anomaly resolution, which they also do for Milstar (before handing over to 4th SOPS). A total of 20 DSCS, FLTSAT and NATO 3 satellites were transferred 11 July 1991 from CSTC to what were then CSOC.

4th SOPS

The 4th SOPS handles the day to day operations of Milstar, including satellite command/control, communications management and ground segment maintenance.

5th SOPS

Based at Onizuka AS in California the 5th SOPS plans and conducts launch and on-orbit operations for several DoD, allied and commercial missions, including IUS for NASA/DoD, NATO 4/Skynet 4 and DSCS 3. It also supports Shuttle and Titan 4 missions. The operation is responsible for day to day operations of NATO 3 and DSCS 3. Staffed by Lockheed Martin personnel.

6th SOPS

The 6th SOPS located at Offutt AFB in Nebraska for day to day command/control of DMSP (will relocate to Falcon). CSOC assumed DMSP control 1 April 1987. Detachment 1 at Fairchild AFB in Washington is responsible for emergency command/control antenna operations; it will close September 1997 and Vandenberg will provide TT&C station.

7th SOPS

The first reserve unit in AFSPC. It provides launch, early orbit, on-orbit and emergency support for DSP and GPS in conjunction with 1st SOPS. Staffed by experienced operators, many also with civilian satellite jobs.

The 750th Space Group at Onizuka AS is responsible for the daily operation of AFSCN, which AFSPC took over in late 1987. The network comprises nine tracking stations. They are the 23rd SOPS, New Boston AFS (New Hampshire), Detachment 1 Vandenberg AFB (California), Detachment 3 Thule AFB (Greenland), Detachment 4 Mahe Island (Seychelles; planned to close 1997), Detachment 5 Andersen AFB (Guam), Detachment 6 Kaena Point (Hawaii), 22nd SOPS Colorado Tracking Station (Falcon AFB), Detachment 2 Diego Garcia (Chagos Archipelago), OL-AE Oakhanger (UK). Two resource control complexes, the 21st SOPS at Onizuka AS and 22nd SOPS at Falcon AFB, schedule satellite contacts for the tracking stations.

Some 400 contacts of 518 minutes each are scheduled daily for about 90 satellites. Camp Parks at Pleasanton in California is responsible for satellite on-orbit testing. The Transportable Vehicle Checkout Facility at Cape Canaveral AS is responsible for pre-launch compatibility testing.

AFSCN was recently upgraded, automating many operations. Satellites were controlled from the 1960s under the CDS Current Data System from CSTC, using a 26 m roll of graph paper specifying the contact schedule of each ground station over seven days. The computerised CCS Command & Control System took over at 50th SW and CSTC in spring 1992.

VERIFIED

Deep Space Network (DSN)

Current Status

With an unprecedented sequence of closely spaced time slots for a broad range of interplanetary spacecraft the DSN is being rescoped to accommodate US and international planetary missions including Galileo, Cassini, Mars Global Surveyor, Mars Odyssey, Mars Explorer Rover A, Mars Explorer Rover B and several national and international spacecraft including Mars Express and the Beagle 2 lander. Only Beagle 2 failed to achieve its operating site.

Background

The NASA Deep Space Network – or DSN – refers to three antennas located roughly equidistant around the globe that receives downlinked information from interplanetary spacecraft and performs radio and radar astronomy observations. The three active locations for DSN nodes are Goldstone, in California's Mojave Desert; near Madrid, Spain; and near Canberra, Australia. The Goldstone complex is situated on the US Army Fort Irwin Military Reservation about 72 km (45 miles) northeast of Barstow, California. The Spanish complex is located 60 km (37 miles) west of Madrid. The Australian facility is 40 km (25 miles) southwest of Canberra near the Tidbinbilla Nature Reserve. As a two way communications system the DSN is the largest and most sensitive scientific instrument in the world.

DSN draws its heritage from a JPL programme begun in January 1958. JPL, then under contract to the US Army, travelled to Nigeria, Singapore, and California to place mobile tracking and communications dishes in anticipation of the army-launched Explorer 1, the first successful US satellite. NASA, which received an official charter on 1 October 1958, thus consolidated the separately developing space-exploration programmes then under way in the three branches of the US military.

On 3 December 1958, NASA accepted a transfer of JPL from the army, thus inheriting the precursor network. At the same time, NASA turned to JPL to design and carry out the lunar and planetary exploration programmes using remotely controlled spacecraft. As a means to communicate with these new satellites, NASA conceived of a Deep Space Network, initially under a separately managed programme. This new autonomous network would avoid the need for each flight project to reinvent a specialised space communications network. As an autonomous entity the Deep Space Network standardised communications to and from satellites regardless of the builder.

DSN researches and investigates new communications theory and as a result has set a standard for low-noise receivers, tracking, telemetry and command systems, digital signal processing, and deep space navigation.

Each complex has at least four DSN stations equipped with 34 m (111 ft) High Efficiency antenna, 34 m (111 ft) Beam Waveguide antenna, 26 m (85 ft) antenna and 70 m (230 ft) antenna.

All DSN antennas are steerable, high-gain, parabolic reflector antennas. They specifically: acquire spacecraft telemetry; transmit commands to spacecraft; locate spacecraft and measure their flight characteristics such as velocity; perform very-long-baseline interferometry observations; measure variations in radio waves for radio science experiments; gather science data: monitor and control the performance of the network; and the JPL continues to manage the facility for NASA.

Eastern Range

Background

This stretch of ocean and islands from Florida to the South Atlantic had its origins in 1950, when the army and air force jointly established the Joint Long Range Proving Ground. The air force assumed full control in 1951, redesignating it the Air Force Missile Test Center. In October 1952, the Air Research and Development Command needed to extend the Eastern Range's length to 5,000 miles in order to be able to communicate with the new SNARK and NAVAHO missile, over its entire flight path. The army and air force tentatively selected Antigua, St Lucia, Fernando de Noronha and Ascension as new tracking hosts, and the Eisenhower administration signed agreements with the governments of United Kingdom, St Lucia, Brazil and Ascension to secure their co-operation.

By the end of 1954, Cape Canaveral had three tracking stations up and running at the Jupiter Auxiliary Air Force Base and Grand Bahama Island. The air force was also building smaller installations capable of tracking satellites on the islands of Eleuthera, San Salvador, Mayaguana and Grand Turk. That same year the Eisenhower administration entered into agreements with the Dominican Republic and Puerto Rico and had already broken ground on still more tracking stations on those islands. In all, nine tracking stations spread out along 1,000 miles of the Atlantic, entered service before the end of the year. To communicate with this far-flung network, the US set up five separate government contracts, to lay submarine cable. One-by-one the installations entered operation. The last submarine cable linking Florida to Puerto Rico sent its first message in August 1954.

In December 1954, Mayaguana opened for operations followed by Eleuthera, San Salvador and Grand Turk between 18 July 1955, and the end of August 1955. The stations in the Dominican Republic and Puerto Rico became operational on 5 December 1956. The St Lucia site was activated on 5 December 1956, and Antigua and Ascension were ready for operations in October 1957. The Fernando de Noronha station was activated off the coast of Brazil on 18 September 1958.

A coverage gap still opened up between Antigua and Ascension though and the navy provided 12 small telemetry ships downrange in 1957 and 1958. With the ships in place, the air force fired its first 5,000 mile test, a SNARK test flight, on 31 October 1957.

In arguments between themselves and with the civilian government, the US military services finally wrested control of the new space exploration programme from the civilians associated with the International Geophysical Year. In part, the military argued that it alone had the tracking stations to support the mission, whereas civilians did not. In secret memos, the air force cited the ETR as the principal reason it should host all space launches. In fact, the ETR did not cover enough of the globe to be useful to IGY and the new NASA had to place portable antennae around the world. But eventually, after the launch of Sputnik the National Science Foundation had to capitulate and launch its satellites from Cape Canaveral. This ensured the future of Canaveral as America's space port.

All of the Eastern Range stations stared out with single-point radars. These radars came directly from the Second World War vintage SCR-584 radar systems. The so-called MOD I detected winged craft such as the Snark the easiest and so became the most economical solution for winged missile requirements. By the end of 1957, the ranges had optical systems, infra-red tracking equipment, cinetheodolites and ribbon-framed cameras. The CZR-1 ribbon-framed cameras covered the missile during the first 1,000 ft of flight, and cinetheodolites followed the vehicle out to a distance of 20 miles. Wild BC-4 ballistic cameras captured optical data beyond the tracking radars' beamwidths.

Ballistic missiles forced the ETR to upgrade its tracking systems in the 1950s and 1960s. An AZUSA continuous wave tracking system began operation at Cape Canaveral in the mid-1950s and on Grand Bahama in the early 1960s. The AN/FPS-16 radar system went into service at Cape Canaveral, Grand Bahama, San Salvador, Ascension and East Island between 1958 and 1961, followed in the early 1960s, by the MISTRAM (Missile Trajectory Measurement) system at Valkaria, Florida and Eleuthera Island. This new system tracked Minuteman's throughout their flight.

Pershing missiles required an even better system, the UDOP (Ultra-High Frequency Doppler) trajectory measuring system and an AN/MPS-25 radar placed at Grand Bahama's Cay and Carter Cay.

As this system reached out to cover the entire Atlantic, the navy's role diminished. The navy retired its Eastern Range's FS-type telemetry ships in 1960.

Alongside the ground-based systems, the air force routinely supported missile launches with specially instrumented aircraft for communications and telemetry support. The 6550th Operations Squadron preceded important missile tests with the flight of a C-54 aircraft in the 1950s. Eleven C-130s acting in the same role replaced the C-54s in 1961 and 1962. At the dawn of the Apollo programme the ETR introduced the more capable Advanced Range Instrumentation Aircraft (ARIA). Eight C-135A aircraft for the ARIA programme joined the C-130s in January 1966. Each ARIA carried a steerable telemetry antenna.

Half a dozen range stations, no longer had a purpose after the last MERCURY mission in May 1963, and the air force retired them. Mayaguez station shut down in August 1961, and the East Island Annex in 1963. The Dominican Republic reabsorbed the Dominican station on 6 November 1962. Project FRESH LOOK eliminated the MOD II radar network in 1965. San Salvador had a caretaker staff only after March 1965, finally closing for good on 31 January 1970. St Lucia reverted to its original owners in early December 1967. Mayaguana facilities were abandoned on 16 June 1970, except for a still active runway. The Missile Guidance Annex on Fernando de Noronha was returned to Brazil on 14 January 1969.

When the Shuttle Challenger exploded on 28 January 1986, it carried a TDRS satellite meant to speed the replacement of ground installations. Congress directed a national space recovery effort to consider a revitalisation of America's unmanned space vehicle industry. It included an effort to modernise the Eastern Range with a new Range Operations Control Center (ROCC), fibre optic communications, consolidated instrumentation facilities on Antigua and Ascension and radar, telemetry and optics improvements. The USNS Redstone was deactivated on 6 August 1993, but a new range site was completed in Argentina, Newfoundland in June 1993 to support northbound flights of the TITAN IV from Cape Canaveral. Operational testing for the ROCC continued in 1993 and 1994, and the ROCC was declared operational on 1 March 1995. Extensive modifications to Atlas and Delta launch facilities for the new generations of expendable vehicles. As a result of recommendations by the Columbia Accident Investigation Board following the loss of the Shuttle Columbia on 1 February 2003, extensive upgrades to the tracking camera coverage of Shuttle flights have been carried out. The USAF Distant Object Attitude Measurement System equipped with 200 mm and 400 mm tracking lenses is receiving high definition TV linked to a new image analysis facility at KSC.

Space Surveillance Network/NORAD

Current Status

NORAD and US NORTHCOM have participated in detailed analysis regarding the loss of the Shuttle Columbia on 1 February 2003. Several changes are being made to the SSN for supporting NASA human space flight missions.

Background

NORAD is a bi-national US-Canadian command charged with safeguarding the sovereign airspace of the two countries. Through its component commands, US Strategic Command operates a worldwide network of missile warning sensors providing tactical warning and attack assessment of ICBM and SLBM launches posing potential threats against North America, and provides tactical warning of ballistic missile attacks to US Commanders worldwide. The missile warning sensors are also used for space surveillance.

Together with dedicated space surveillance sensors (operated by the 21st Space Wing at Peterson AFB, NAVSPACECOM at Dahlgren and contributing ARSPACE and civil assets), they form the Space Surveillance Network, which provides timely and accurate detection, tracking and identification of space objects and events. The network makes 50,000 to 80,000 daily observations for the command's Space Control Center in

Cheyenne Mountain and maintains a catalogue of more than 10,000 orbiting objects.

Data are relayed via satellite almost instantaneously to four primary locations: the command's Missile Warning Center (Cheyenne Mountain AS), Strategic Command HQ (Offutt AFB, Nebraska), National Military Command Center (Washington DC) and Alternate NMCC (Maryland).

The 21st Space Wing's ground-based sensors until April 1995 comprised BMEWS and SLBM. The two BMEWS radar units are the 12th SWS at Thule AB in Greenland and the 13th SWS at Clear AS in Alaska. SLBM warning system units are the 6th SWS at Cape Cod AS, Massachusetts, 7th SWS at Beale AFB, California. These sensors also provide essential tracking reports on space objects. The 73rd Space Group at Falcon AFB was absorbed 29 April 1995, along with its eight space tracking squadrons. The 11th SWS at Falcon AFB, Colorado, operates the ALERT Attack & Launch Early Reporting to Theater system.

As host wing for the Peterson Complex, the 21st provides complete base support services for Peterson AFB and Cheyenne Mountain AS, plus some support functions for Falcon AFB. Personnel total 4,200 military, 846 air force civilian and 3,350 contractor employees. Formed in October 2002, US Northern Command (NORTHCOM) is charged with protecting the US homeland.

The NORAD/NORTHCOM Command Center in Cheyenne Mountain Operations Center also receives the data.

RESEARCH AND DEVELOPMENT FACILITIES

Belgium

Centre Spatial de Liège (CSL)

Current Status

The EGUNA consortium worked with Regione Wallonie to develop a major holographic laboratory at CSL. The initial phase was implemented in late 2000 but the full-scale development effort slowed due to restrained release of funds.

Background

CSL (IAL Space until April 1992) is an autonomous unit within the University of Liège and, along with Intespace (France) and IABG (Germany), is one of the Co-ordinated European Test Facilities group under ESTEC. The test facilities are based around a 480 m^2 class 10,000 clean room and include three thermal vacuum optical test facilities. The 150 m^3 5 m diameter horizontal 33 tonne Focal 5 chamber provides a 10^{-9} atmosphere vacuum and 10 temperature settings over −268°C to +70°C and optical bench vibration levels of 10^5 *g* on a 350 tonne seismic block. There is also an associated 30 m^2 class 100 laminar flow clean zone. In the same area is the Focal 2 chamber, with identical performance but 21 m^3/2 m diameter. The 1.2 m^3 Focal 1.5 provides a similar environment but with a 1.5 m vertical diameter and 80 m^2 class 10,000 cleanroom and 4 m^2 class 100 laminar flow clean zone. CSL undertakes testing/calibration of camera tubes, image intensifiers, CCDs, photomultipliers and photon counting systems, in addition to the design/manufacture of meteorological instruments.

Canada

Canadian Space Agency – David Florida Laboratory

Background

Operated on government funding and collocated with the Communications Research Center (CRC), CSA's DFL is the national facility for spacecraft AIT; its facilities are available on a fee for service basis. Built in the early 1970s for the Canadian-US Hermes Communications Technology Satellite project, it has since supported Anik C-E, Shuttle remote manipulator, Brasilsat, Olympus, MSAT and Radarsat development, and will play an integral role in the next Anik generation and Canada's contribution to Space Station. Also supported AMSC-1, Indostar and BSAT 2A and 2B.

The current activity centres on the testing of Canada's contribution to the International Space Station programme – the Mobile Servicing System (MSS). In 2001 commenced work on Radarsat 2.

Facilities include:

Clean rooms

Integration and assembly areas comprise three temperature (22±3°C) and humidity (45±5%) controlled clean rooms: Bay 1: 440 m^2; Bay 2: 560 m^2; Bay 3: 1,100 m^2. All three high bays provide class 100,000 conditions, with class 10,000 available when required, and are traversed by travelling bridge cranes with hook heights of 9, 11 and 16 m, respectively. A 6 × 12 m low bay is used for subsystem assembly/checkout. The clean rooms house spin balance and mass property-measuring equipment. The MRC Mk VII-6 Vertical

Radarsat in the CSA's David Florida Laboratory (CSA) ***NEW***/1047515

An artists depiction of Canada's Radarsat 2 in orbit (CSA) ***NEW***/1047516

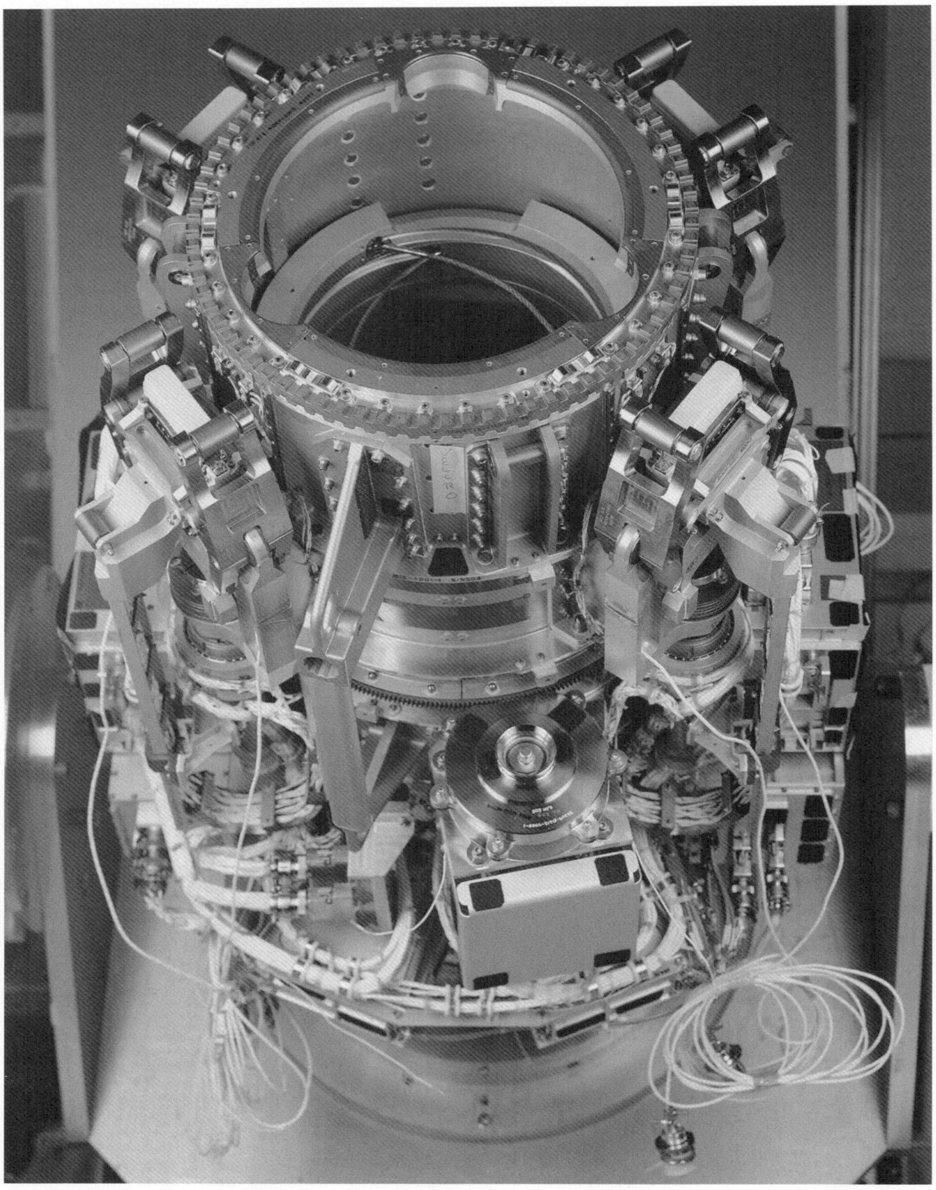

Canadarm 2 end effector developed from the STS RMS but destined for the ISS (CSA) ***NEW***/1047517

The much photographed Richat structure in Mauritania, once thought to have been an impact crater, is now recognised through satellite imagery to be a symmetrical uplift exposed by erosion (NASA) ***NEW***/1047542

Axis Measurement System (VAMS) measures static/dynamic imbalances and moments of inertia for 25 to 2,300 kg payloads spun at up to 200 rpm, and calculates the required balance masses and their locations. A Horizontal Axis Measurement System (HAMS) provides complementary data, for payloads of up to 2,700 kg, 3 m diameter. Both systems can be housed within DFL's 7 × 10 m vacuum chamber.

Thermal Vacuum Test Facility

Four principal chambers of 1, 2.5, 3 and 7 m diameter exist. PC-based data processing system.

7 × 10 m vertical chamber: 1.3×10^{-6}–10^{-7} mbar, 1.1 solar constants/50 per cent shroud area (maximum load 260 kW) using infra-red lamps, 6.7 × 10.7 m test volume within LN_2 shrouds, accessible via 7 m diameter removable top lid and 2.2 and 1.5 m man-doors. Hardpoints permit 27 tonnes mounted on floor, 2,200 kg from lid, or 13,600 kg from 36 hardpoints in six rings on chamber walls.

3 × 9 m vertical chamber: 1.3×10^{-5}–10^{-7} m bar, LN_2 temperature range plus thermal cycling with infra-red lamp system, side- loading 1.8 × 2.1 m door.

2.5×2.5 m horizontal chamber: 1.3×10^{-5}–10^{-7} mbar, LN_2 or ±150°C GN_2 temperature range.

1 × 1 m horizontal chamber: 1.3×10^{-5}–10^{-7} mbar, LN_2 or ±150°C GN_2 temperature range.

Ambient pressure thermal chamber: thermal cycling and/or temp soak, and (modified) vibration or RF testing at controlled temperatures. Can be configured to different volumes, –150 to +149°C, ±1°C.

Vibration Test Facility

UD 4000: 178 kN sinusoidal, 165 kN random, up to 183 cm^2 slip plate, 122 cm diameter head expander. *MB C-150:* 76 kN sinusoidal, 65 kN random, up to 152 cm^2 slip plate, 81 cm diameter head expander.

Ling A395: 27 kN sinusoidal, 23 kN random, up to 61 cm^2 slip plate and 61 cm diameter head expander.

Static Load Test Facility

The new SLT facility, first used in 1992 by the Radarsat bus structural model, incorporates a 12 channel Cyber Fatigue Master 7000 Digital Control System combined with a new data processing/ acquisition system accommodating 200 strain gauge channels and up to 40 LVDTs.

RF Test Facility

Comprises four anechoic chambers, a rooftop range, an EMC facility and PIM (Passive Intermodulation Measurement) facility. The chambers and rooftop turntable provide outdoor ranges for automatic antenna measurements; the three-axis positioners and sources are controlled via a fibre optic link between the 62 m tower and control room. DFL has added a spherical near field antenna measurement facility for performing near field to far field transformations, and is developing an RCS radar cross section measurement capability. DFL is designated by Inmarsat as both an AATH Authorised Antenna Test House, authorised to conduct antenna tests to Inmarsat specifications, as well as a Designated Inmarsat Representative (DIR), providing the DFL representative with the authority to approve antennas on Inmarsat's behalf.

Chamber 1: 24 × 12 × 20 m high for half its area, with 9,000 kg gantry crane access, reflection co-efficient –30 to –55 dB over 0.250–20 GHz. A two-axis positioner has total vertical load of 18,145 kg and 0.05° pointing accuracy, 5 m tower for short (9 and 400 m) outdoor range tests.

Chamber 2: 6.1 × 6.1 × 6.7 m, reflection coefficient –30 to 55 dB over 0.500–20 GHz, magnetic/ electric field shielding 90/100 dB in 200 kHz–15 GHz range. A three-axis positioner provides 0.05° accuracy for 4,536 kg loads, 3 m pyramidal horn with 1.8 × 1.8 m aperture for 6 m indoor range tests.

Chamber 3: 6.1 × 4.3 × 2.9 m, shielded, reflection coefficient better than –40 dB, 2–100 GHz. Azimuth positioner accuracy +1±0.005°. R&D into RCS measurements, near field systems and EHF-bands RF testing.

Rooftop Range: 0.250–18 GHz, 6 × 6 m platform, two-axis positioning of 13,959 kg loads, source inclination 5° for 400 m outdoor range, <1° for 16 km long range to Kingsmere.

EMC Facility: 4.8 × 5.5 × 3.6 m shielded room is lined with high performance anechoic material to provide electric/magnetic field shielding of 100 dB over 200 kHz-15 GHz; 2 × 3 × 2 m Faraday cage; 3 × 2.5 × 2.4 m anteroom. The EMC facility is equipped to test to MIL-STD-461A/B/C/D/462/462D RTCA/DO-160A/B/C. Emission susceptibility up to 40 GHz.

PIM Facility: the passive intermodulation measurement facility evaluates the magnitude of intermodulation products generated by the components and subsystems of satellite transponders. The three DFL-designed systems operate in three separate bands (275–330, 862–873, 1,511–1,578 MHz) and measure down to –146 dBm.

HF Anechoic Chamber: 7.6 × 3.35 × 3.35 m high, reflection co-efficient –30 to –50 dB over 1–40 GHz, two-axis high speed/accuracy, total vertical load 30 kg, <0.005° pointing accuracy. Currently set up for spherical near field measurements and automated gain extrapolation.

To complement and support classical vibration and acoustic testing, DFL is developing modal analysis through both conventional and innovative modal test techniques. A 366 tonne seismic mass has been built as a modal analysis testbed; a new LMS vibration control/modal analysis system capable of supporting swept sine, random or shock control and analysis has been procured.

UPDATED

Denmark

Danish Space Research Institute (DSRI)

Current status
DSRI has focused on the Danish Small Satellite Programme, Solar System Physics, Astrophysics and the preparation of interdisciplinary research reports. DSRI is restructuring for better integration with Danish satellite activities. The Institute facilitates links between Denmark's astrophysical and space sciences research centres and experimental laboratories and with other national resources, the universities and other European institutes.

Background
The Danish Space Research Institute (Dansk Rumforkningsinstitut) is a government-funded research organisation, founded in 1966, under the direct supervision of the Research Directorate of the Ministry of Education & Research. Its principal areas of interest are space astronomy, cosmic rays and space plasma physics. Four wide angle x-ray transient and gamma burst detectors, which form the Watch all-sky monitor, are flying on the RFAS Granat satellite. A single Watch unit was provided to ESA for Eureca's 1992 mission. Two large area x-ray telescopes, four imaging detectors and a Bragg crystal spectrometer (Xspect) are being produced as part of the CIS/Danish Sodart telescope for the Spektrum-X satellite in 1998. Ground support equipment was supplied for the Isophot instrument on ESA's Infra-red Space Observatory in co-operation with Heidelberg's Max Planck Institute of Astronomy, and DSRI is a member of the Hipparcos Northern Data Consortium. The Swedish Viking magnetospheric mission carried a wave experiment built at DSRI; current activities include participation in the Wave and Magnetometer experiments on ESA's Cluster. A series of plasma physics sounding rocket flights in co-operation with Scandinavian countries and NASA is also under way. DSRI is part of Denmark's Ørsted satellite project.

DSRI receives a line item budget voted each year by Parliament as part of the national budget. This includes salary, operating and development costs. Additional support is obtained from the Danish Space Board on a project by project basis. DSRI is headed by a Director, who reports to the Research Directorate on financial and personnel matters and to an external Institute Board for science/programme affairs. Otherwise, the Institute is autonomous. 1996's budget is DKr26.1 million; 1995 DKr19.3 million; 1994 DKr19.4 million; 1993 DKr24 million; 1992 DKr22 million; 1991 DKr22.544 million; 1990 DKr21.844 million; 1989 DKr21.335 million. This supports a fixed staff of 40, including 18 physicists and engineers. Included in the total is funding from the Danish Space Board to support seven engineers, scientists and technicians, and a development programme for thin foil x-ray optics. Facilities include state-of-the-art equipment for determining x-ray optical, scattering and diffraction properties of surfaces, crystals and gratings (transmission/reflection) over a wide range of energies. A high-power rotating anode x-ray tube was installed most recently. Technical personnel were trained/certified by ESA/NASA for flight hardware fabrication; the mechanical workshop is equipped for precision manufacturing of materials such as glass fibre epoxy composites and titanium. CAD/CAM facilities are maintained, electronic circuits can be designed/manufactured and, when required, circuits can be assembled/tested under clean room conditions. At other institutes, DSRI has access to vibration tables, a small thermal vacuum chamber, shock testing equipment and units for determining centre of gravity and moment of inertia.

France

Intespace

Current status
CNES currently holds 35 per cent of the capital with a workforce of 143.

Background
Intespace (Ingéniére Tests en Environnement Spatial) has operated CNES's Toulouse test facilities since 1983 (taking over from Sopemea), incorporating 20,000 m^2 of test halls, laboratories and clean rooms able to handle Ariane 4/5-class 3/4 tonne satellites. 1995 sales were FFr120 million (1994 FFr119 million; profit FFr3.6 million), 1993 FFr108 million, 1992 FFr105 million, 1991 FFr94 million. The facility employs 135 personnel.

CNES/Sopemea hold 35 per cent each, Aerospatiale/Matra Marconi 9 per cent each, Alcatel Espace 3 per cent and others 9 per cent. It is part of the Co-ordinated European Test Facilities group, along with IABG (Germany) and Centre Spatial de Liège (Belgium), organised by ESTEC. Intespace also supervised the installation of new facilities, opened June 1988, at Brazil's INPE, plus those of Taiwan's NSPO and South Korea's KARI. The 1996 principal test programmes included SILEX (LEO and GEO); Hot Bird, MPLM, Telecom 2, Mars-96, Ariane 4 and Ariane 5.

Three 280 m^2 preparation/integration halls support Intespace's own key facilities, each with an access airlock and meeting standard cleanroom requirements. Facilities include:

Thermal chambers
600 m^3 6 m diameter with 1,600 W/m^2 3.8 m diameter solar simulator for objects to 2.5 to 4 m, –170/120°C; 30 m^3 with 1 m diameter solar simulator; also 1, 1.5, 8 m^3 chambers with controlled humidity and –170/120°C temperature range.

Thermal vacuum
SIMMER (Simulateur de l'Environnement) became operational in 1992 (first used by Hispasat 1B) for 5 tonne (vertical), 2.5 (horizontal) subjects. Primary pumps 2.2 × 10^{-2} atmospheres in 2 hours; cryogenic/turbomolecular pumps 4 × 10^{-6} atmospheres in 10 hours. 10 m diameter 13 m long, –173/120°C temperature range, 400 temperature measurements, 200 electrical measurements. Other chambers: 0.25, 3, 20 m^3 volume, temperature –170 to +120°C, vacuum down to 10^{-9} atmospheres.

Other capabilities
Acoustic: 156 dB, 1,100 m^3 chamber
Salt spray chamber: 1.2 m^3 volume

MPLM during acoustic testing at Intespace 0008101

Electrical: 45 + 1,700 m^3 Faraday cages; susceptibility analyses over 20 Hz–40 GHz in 17 × 10 × 8 m anechoic chamber; impulse magnet charger and magnetic measurement facility rated at 0–5,000 gauss
Mechanical: 27, 80 (×2), 150, 170, 300 kN shakers operating over 5–2,000 Hz and driving a 5 × 5 m block under a 13 m ceiling. 10,000 *g* /kg centrifuge; shock simulator (400 *g* for 4 ms); 8 × 10 × 13 m high 1,100 m^3/156 dB and 1 m^3/173 dB acoustic chambers; mass properties (moments of inertia, mass, centre of gravity) for articles up to 2 t in 150 m^2 central hall.
Compact payload test range: the FFr65 million Mistral range entered operation in 1998.

Office National d'Etudes et de Recherches Aérospatiales (ONERA)

Current status
ONERA is conducting research into reusable spaceplane technology and lunar observations while sustaining development programmes for Ariane 5 booster performance improvement and investigation of high-frequency combustion instability in liquid propellant engines. The broad research agenda for ONERA has widened participation in trans-European research into the next generation of post-Ariane 5 launch vehicles.

Background
The Office National d'Etudes et de Recherches Aérospatiales was founded in 1946 as a national aeronautical research centre under the General Delegate for Armament (DGA). 2,200 staff are employed at eight major centres, specialising in aerodynamics, fluid mechanics, flight mechanics, structures, materials, optics, acoustics, electronics, radars, computer sciences, and ground, wind tunnel and flight testing. The Châtillon site houses HQ, operational departments and laboratories, Chalais-Meudon research wind tunnels, Palaiseau specialises in energetics, Lille in flight mechanics and Toulouse accommodates the CERT Center d'Etudes et de Recherches de Toulouse. Large test facilities are operated at two principal centres: Modane-Avrieux and le Fauga-Mauzac. Modane-Avrieux maintains the S1MA continuous sonic wind tunnel (8 m diameter, 14 m long), the S2MA continuous transonic/supersonic (1.8 × 1.75 m), the S3MA transonic/supersonic blowdown (0.8 m high test sections), the S4MA hypersonic (M12.0) and R4.3 transonic/supersonic cascade tunnel. The S4B 15 mbar pressure chamber (2.5 m diameter, 10 m long), is primarily used today for the calibration of nacelles and air-powered engine simulators. Fauga-Mauzac operates the subsonic F1 (pressurised) and F2, and the high enthalpy F4 wind tunnels. A new laboratory was created in 1992 at l'Ecole de l'Air de Salon-de-Provence for common research such as flight mechanics and optronics.

In April 1998, ONERA signed an agreement with CNES creating four core focus areas for launcher research, directly related to Ariane 5 upgrades. A major test campaign began in 1999, testing re-entry shapes for the Mars Sample Return Orbiter which will use aerobraking.

Germany

Industrieanlagen-Betriebsgesellschaft GmbH

Background
IABG was established in 1961 to provide aircraft and spacecraft test capabilities; it is now an element of

the Co-ordinated European Space Test Facilities group, along with Intespace (France), Centre Spatial de Liège (Belgium) and ESTEC. For space, IABG offers thermal vacuum, thermo-mechanical, thermo-environmental, vibration, acoustic, modal, static, magnetic and EMC testing on launcher and satellite components and assemblies. IABG also conducts structural and mechanical tests and electromagnetic compatibility evaluation. It has tested more than 100 satellites and major satellite subsystems; recent activities include Huygens, Cassini, Artemis, ISO, Cluster, Soho, Freja and Ariane 5 stages and major components. IABG employs approximately 1,200 people and has diversified into several discrete business units: Information and Communications; Defence; Aerospace; Environmental and Management Services; Transportation, and Automotive. Facilities include:

Electromagnetic (EMC) Test Facility
The facility comprises a shielded test chamber, fitted with RF absorbers and providing class 100,000 conditions, and EMC measuring equipment. For emission testing, a computer-controlled 10 Hz to 40 GHz spectrum surveillance system, current probes and antennas are employed. For susceptibility measurements, high-power broadband amplifiers provide up to 600 V/m field strength from 10 kHz to 18 GHz. Special pulse generators and current probes allow various susceptibility tests, including electrical tests for power supply simulation.

Usable dimensions: 10.5 m length, 7 m width, and 8 m height.

Access door: 4 m width, 6 m height.

Shielding attenuation: H-field >60 dB for 10 kHz, >100 dB above 1 MHz; E-field and plane waves 100 dB for 1 MHz to 10 GHz.

Internal reflectivity: >20 dB above 100 MHz, >40 dB for 300 MHz to 10 GHz.

Vibration Test Facility
A 300 kN vibration system can operate horizontally or vertically under class 100,000 conditions, powered by four coupled electrodynamic shakers. Specimen response information is fed into a 360-channel digital data acquisition system. Sensors include 500 accelerometers with graded performance ranges. There are also several smaller single-shaker facilities.

Mounting table: 3 × 3 m (80 × 80 mm M10 hole pattern).

Maximum specimen mass: 15,000 kg.

Maximum acceleration: 15 *g* without load, 4.5 *g* with 5.0 t load.

Maximum displacement: ±25 mm.

Frequency ranges: 4–2,000 Hz sine low level, 4–300 Hz sine high level, 10–2,000 Hz random high level.

Modal Test Facilities
The systems provide tuned sinusoidal and non-tuned broadband excitation. The largest is housed in a transportable container; the two smaller ones are easily transported. The software undertakes test control, data acquisition/processing, modal analyses and result presentation. A separate hall is available for testing a larger range of structures, under clean room conditions if needed.

Excitation: sine, random 8 channels.

Vibration exciters: 26, 10–7,000 N.

Measurement: 400 accelerometers, plus force and displacement transducers.

Data acquisition: 882 channels (in 7 blocks) up to 2 kHz; 384 channels (in 8 blocks) up to 4(20) kHz; 16 channels up to 25 kHz

Acoustic Test Facility
Acoustic environments are simulated by electro-pneumatic noise generators that can reproduce overall sound pressure levels and spectra of operational fields. The total acoustic power available is 80 kW AC. The recording system includes 20 microphone channels, 128 accelerometer channels and 24 strain gauge channels. The facility comprises several reverberation chambers, control/computer rooms, air supply systems (5 kg/s) and 9.6 m height × 15.4 × 11.5 m preparation hall. The 800 m^3 chamber is an irregular pentagon 5.7 m high, 13.0 m wide, 10.8 m deep and can create 150 dB levels in class 100,000 conditions. The 206 m^3 chamber, 4.7 m high, 8.0 m wide, 5.5 m deep, can attain a maximum of 162 dB. The Progressive Wave Tube provides up to 170 dB in a 0.8 × 1.2 m test section. A thermoacoustic facility can simulate combined environments such as surface temperatures up to 1,300°C on a specimen plus acoustic excitation up to 158 dB.

Magnetic Field Simulation
The facility consists of a square, triaxial coil system with four coils per axis and with an edge length of 15 m. The test volume measures 10 × 10 × 10 m with free access of 4 × 4 m. It is used for measuring the magnetic cleanliness of objects, recording magnetic moments and eddy current fields, attitude control testing of magnetically stabilised spacecraft, calibrating magnetometers, and magnetising or demagnetising objects to determine permanent, remnant and induced fields. DC field values up to 75,000 nT can be produced, corresponding to Earth's magnetic field level at the poles. The limit for uniform AC fields is 25 kHz. Hall probes, fluxgate magnetometers and search coil magnetometers are available; for precise measurements, proton spin magnetometers and optically- pumped magnetometers are employed.

Thermal-Vacuum Facilities
A 6.8 m diameter chamber (6.2 m diameter test volume) providing <10^{-8} atm vacuum conditions and a 3.6 m diameter or 3 × 4.5 m rectangular solar beam for test objects up to 2.5 tonnes, 4 m diameter/5 m high was introduced in 1983. The solar beam, with ±2° collimation angle, provides up to 1.4 solar constants. The cryogenic shroud creates 100 to 385 K conditions within the 13 m long chamber. Test specimens can be mounted on a two-axis motion simulator, providing up to 10 rpm continuous spin and ±200° attitude excursions and levelling capability.

Other Thermal-Vacuum Chambers
These chambers are used for subsystem and component trials. 3 m chamber: 3.2 m diameter, 3.8 m long shroud, 10^{-8}atm, 100 to 400 K, infra-red radiators; 1.3 m chamber: 1.35 m diameter, 2.2 m high shroud, 10^{-8}atm, 100 to 400 K. Clean room conditions from class 100,000 to 100.

Thermo-Mechanical Vacuum Facility
This is a 1.5 m diameter cylindrical vacuum chamber, with two-zone heating by graphite elements up to 1,600°C, mechanical loading by vacuum-tight feed-through and hot load introduction. Displacement measurement in hot zones relative to fixation points. 600 × 600 × 100 mm maximum sample.

Thermo-Mechanical Facilities
Mechanical loading is applied to shingle-type test samples (up to 350 × 350 mm) by pressure difference in a cylindrical pressure chamber. Seven infra-red modules with total 252 kW_e. Maximum temperature is 1,450°C applied thermal loading. Maximum pressure difference is 500 mbar. Reaction forces, temperature, pressure and displacements can be measured and visualised online. In addition, cover furnaces with resistance heaters up to 1,600°C in air or inert gas atmosphere; metallic and carbon heater up to 2,000°C in vacuum or inert gas available. Samples can be loaded and measured by extensometer and/or displacement by LVDT.

Thermo-Environmental Test Facility
A 1.5 m diameter cylinder of nominal 1.5 m length, extendible by modules. Houses a water-cooled reference system for, for example, displacement measurements. Several heating systems can be installed; maximum sample temperature =1,800°C. Different atmospheres and pressure profiles can be simulated. Mechanical loads can be applied by a hydraulic external loading device.

VERIFIED

India

ISRO Satellite Centre (ISAC)

Background
ISAC is the lead centre for ISRO's satellite development and construction. It grew from the Indian Scientific Satellite Project established in 1972 to build the Aryabhata satellite to a current staff of 2,400 responsible for the design, fabrication and testing of the IRS, SROSS and Insat 2 series. To date ISAC has designed and developed 23 satellites for scientific, communication and remote sensing activities including the new generation Insat 2E, IRS-P4, IRS-P5, IRS-P6, and G-Sat satellites.

ISAC is organised into five main operating groups: Spacecraft Electronics (including responsibility for spacecraft integration and ground testing), Mechanical Systems, Control, Mission, and Technical Support Services. Three independent divisions cover: Quality Assurance, Technical Physics (rocket and satellite science instrumentation), and Programme Planning & Evaluation. The PP&E division provides technical and managerial planning, resource allocation and progress monitoring on each project. Facilities include:

Thermal-Vacuum Chambers
The US$25 million, 9 m diameter 9 m high Large Space Simulation Chamber (LSSC) was commissioned in 1991 for testing Insat 2-class spacecraft. The 7 m diameter auxiliary chamber houses 12 20 kW Xenon lamps (11 used at any one time) to produce a 3.5 m diameter 1.7 kW/m^2 beam for solar simulation. The pumping system achieves 10^{-9} atmosphere in 3 to 4 hours. Two LN_2 tanks each store 1,000 m^3.

4 m diameter chamber, 10^{-8} atmosphere, walls cooled by LN_2. No solar simulation is available. Previously the main test chamber, used for IRS.

2 and 1.2 m diameter chambers for instrument package testing to 10^8 atm/±100°C. A –10/50°C chamber provides thermal cycling testing for instrumentation.

Mechanism Laboratory 40 × 12 × 24 m high, for boom and solar panel deployment testing. A second facility provides horizontal deployment across a water tank.

Motion Simulation Laboratory, 2 m diameter Earth and 25 mm diameter Sun simulators in conjunction with three-axis servotable or three-axis air table for sensor and AOCS testing. SROSS and IRS are the most recent subjects. A moving Earth simulator for GEO/GTO satellite systems has been added, in addition to a thermal chamber for infra-red Earth sensors.

Vibration Facility
Two Ling Dynamics Systems (UK) tables accommodate SROSS and IRS/Insat 2, respectively, with both providing 2.5 cm peak-to-peak sinusoidal (100 *g*) and 30 *g* random.

Antenna and EMC
EMC testing is conducted in a 12 × 40 m anechoic chamber. Satellite integration is conducted in 12 × 12 m class 10,000 (for SROSS-class satellites) or 55 × 14 m class 100,000 (for IRS/Insat 2 class) cleanrooms.

Laboratory for Electro-Optic Sensors
Work under the overall umbrella of ISAT for government and non-government projects.

Space Applications Centre (SAC)

Background
ISRO's prime centre for space applications research, employing 2,200 (including DECU), specialising in communications, satellite remote sensing, meteorology and geodesy. Facilities include mechanical fabrication, electronic fabrication, payload fabrication and environmental tests. The facility also conducts non-space related government test activities and has also been used by commercial companies. The centre is organised into SCA Satellite Communications Area, Remote Sensing Area and Microwave Remote Sensing Programme. The Insat and IRS payloads are developed by SAC. It also supplied extended C-band Earth station equipment to the Insat Master Control Facility and operates the Deltic Earth Station for satellite communication.

Vikram Sarabhai Space Centre (VSSC)

Background
VSSC, located around the village of Thumba near Trivandrum and with a personnel total of 5,600, is ISRO's largest centre. It provides the development base for the country's indigenous launchers: SLV3, ASLV, PSLV, GSLV and sounding rockets such as Rohini, along with the solid motor programme (Trivandrum's Liquid Propulsion Systems Centre operates as a separate entity). A Space Physics

Laboratory carries out research in atmospheric and related space sciences.

Supporting the programmes are specialised R and D groups in: Avionics and Mission Dynamics, Solid Propulsion, Propellants and Chemicals, MMS (Materials and Mechanical Systems), Systems Reliability, Computers and Information Systems, and Programme Planning and Management. Extension facilities are located at Valiamala for the PSLV launcher project, at Vattiyoorkavu for composite development, and an experimental ammonium perchlorate plant at Aluvaye. A space physics laboratory conducts research in atmospheric and related space sciences.

International

European Space Research Institute (ESRIN)

Background

Based in Frascati, Italy, ESRIN is one of four ESA establishments and operates with a permanent staff of 140 and a contractor staff of about 250. Its main focus is Earth observation satellite data handling, including management of the ground segment to acquire, pre-process and archive data, and handling distribution, either directly or via a distributor. ESRIN handles missions for non-ESA clients as well including Landsat, Tiros, MOS, JERS 1 and Spot but the principal emphasis is on ERS and preparations for Envisat launched in 2002.

The ESRIN ERS Central Facility (EECF) maintains constant links with the ERS Mission Management and Control Centre at ESOC. ESOC executes the mission operation plan prepared by the EECF, with ERS ground stations for the scheduling of near-realtime distribution of Fast Delivery products. The Processing and Archiving Facilities handle data product orders, providing users with access to the online worldwide catalogue of ERS data. To this end, ESRIN is responsible for stations in Europe and Canada, ERS PAFs in Italy, Germany, France and UK, and has contracts with worldwide national stations for ERS data acquisition. ESRIN manages more than 30 ground stations that receive data from ESA satellites.

ESRIN also operates the ESA-IRS Information Retrieval Service which provides online access to bibliographic and factual databases covering most fields of science and technology. The ESIS European Space Information System caters for the astronomy and space physics community.

European Space Research and Technology Centre

Current News

ESTEC has reorganised ESA astronaut assignments in the wake of the Columbia accident. It initially agreed to defer flights aboard scheduled Soyuz missions to the ISS in April and October 2003, relieving these spacecraft to carry two-people crews to the ISS for housekeeping duties. Subsequent to the release of the Columbia Accident Investigation Board report on the loss of the Shuttle these flights are deferred to a schedule yet to be determined.

Background

ESTEC is ESA's largest single establishment, with about 1,600 employees. It provides project management for science, communications, earth observation, ügravity and space station programmes, executes the space science programme, performs future satellite programme studies, and undertakes design, development and testing of components and complete space vehicles. Half of the on-site staff belongs to the spacecraft project teams reporting to the Programme Directors at ESA Headquarters; the majority of the remainder belong to the specialised technical divisions based at ESTEC.

Capital investment is about US$55 million for buildings and US$150 million for the technical facilities. The development laboratories include Mechanical Systems, Propulsion, RF Systems, European Space Battery Test Centre, Materials/Processes, Components, Onboard Data Processing, Simulation and Electrical Facilities for Automated and Manned Missions, including a robotics testbed. The Satellite Communications Building supports special telecom services such as videoconferencing, mobile links and other in-orbit testing. The Fuel Cell Test Facility is an annexe to the Space Battery Centre.

Views of ESA's main establishments: HQ, ESTEC, ESOC, ESRIN 0019686

The largest laboratory, the ESTEC Test Centre, operates a wide range of environmental test facilities. Working with Intespace (Toulouse), IABG (Ottobrunn) and Centre Spatial de Liège, ESTEC heads the Co-ordinated European Test Facilities to ensure that national facilities can be used for agency projects. All test areas are air-conditioned (19 to 23°C; 40 to 60 per cent RH), class 100,000.

ESTEC's test facilities include:

Large Space Simulator (LSS)

Europe's largest high-performance solar/vacuum simulator became operational in 1986 and was inaugurated January 1987 for Ariane 4-class payload testing. The first major test was on Alenia's IRIS upper stage test model. The LSS consists of a main 10 m diameter chamber with a removable lid for ease of access and a 5 m side port, with an auxiliary chamber containing the collimating mirror. Payloads can be supported on a vibration-isolated (-10^{-3} *g*) 3.2 × 3.2 m platform, or suspended from the upper volume, or mounted on a two-axis motion simulator. An array of 19 unfiltered 25 kW high-pressure Xenon lamps provides a 6 m diameter collimated beam, with a maximum 1.3 solar constant intensity. The LSS is depressurised using both the test centre's central pumping system and a dedicated high-vacuum system of turbo-molecular pumps and a liquid helium cryo-pump to attain 3×10^{-7} mbar in about 10 hours. Both chambers incorporate stainless steel shrouds operating on LN_2 or GN_2.

Chamber capacity: 10 m diameter × 15 m high, auxiliary chamber 8 to 11.5 m diameter × 14.5 m long. The contractor is BSL.

Vacuum conditions: 3×10^{-7} mbar in 10 hours; repressurised by GN_2 to 100 mbar followed by filtered air over 4–24 hours.

Illumination: 1 solar constant (1,360 W/m^2) by 12 of 19 × 25 kW Xe lamps. 32 kW lamps can be used if needed. 6 m diameter horizontal beam with ±4% uniformity; 7.2 m collimation mirror consists of 121 hexagonal segments, collimation angle 1.9°. Contractor Carl Zeiss.

Shroud temperature: <100K by LN_2 circulation, 150–350K by GN_2. Cool/warm-up period ±2 hours. Shrouds supplied by Leybold-Heraeus.

Motion simulator: spinbox (1–6 rev/min + 1–24 rev/day), turntable (1–24 rev/day). The contractor is SIGRI.

Thermal Facilities

HBF3 thermal vacuum chamber. The chamber is employed for subsystem vacuum temperature cycling in a 3 m diameter usable test volume. The facility is equipped with a demountable infra-red rig for solar panel testing, comprising a LN_2 shroud system with two compartments 3.85 m long, 2 m deep each fitted with 38 500 W infra-red lamps. Maximum temperature range on carbon fibre-backed solar arrays is 170/+90°C. The basic chamber incorporates five shrouds temperature controlled by LN_2 or GN_2. Loading is via removable top lid.

Vacuum conditions: 2–6×10^{-6} mbar in 5 h (0.02 mbar in 1 hour) from atmospheric pressure.

Shroud temperature: <100K by LN_2 circulation, 200–373 K by GN_2. 2 K/min cooling; 3 K/min heating.

Small thermal vacuum chambers: VTC 1.5 usable volume 1.5 m diameter × 2.5 m high, up to 500 kg test subjects, 1.3×10^{6} mbar vacuum, <100 and 125–425 K using LN_2/GN_2. Corona horizontal cylinder for up to 400 kg masses in 1.8 m diameter × 3.2 m long test volume. The LN_2 shroud with two 1.21 m deep, 2.9 m long compartments each equipped with 32 500 W infra-red lamps for solar panels testing, featuring rapid de-pressurisation (down to 1.3×10^{-4} mbar in 7 min), down to 6.6 × 10^{-6} mbar, 170/+110°C achievable on carbon fibre-backed solar arrays (the chamber can also be used for outgassing tests); Ultra-High Vacuum Chamber UHV 0.5 0.120 m^3 test volume in 45 cm diameter × 76 cm high cylinder, attaining <1.3 × 10^{-10} mbar with bake-out facility up to 523 K; Accelerated Thermal Cycling Chamber ATC II for rapid automatic thermal cycling of small lightweight items at atmospheric pressure, T range 93–403 K using dry N_2 in 48 cm high × 50 cm long × 25 cm wide usable volume, typical cycle +100 to −100°C and back in 3 minutes.

Electrodynamic Shakers

Located in the same complex as the Large Space Simulator, the centre is equipped with a 280 kN multishaker and a 70 kN shaker. The larger system incorporates two Ling Dynamics Systems 984 LS vibrators, each capable of 144 kN thrust. The 70 kN system uses one Ling Dynamics 964 LS exciter. All systems generate 52,000 Hz sinusoidal or random vibrations in horizontal and vertical modes. The data acquisition system allows for 250 channels, processed on-line.

HYDRA Hydraulic Shaker the 5.5 m diameter 22 tonne octagonal table (flush with the floor) of the 6-DOF shaker is driven by four actuators vertically and two in each lateral direction. Each actuator has a ±70 mm stroke, max 0.8 m/s piston velocity and 630 kN force rating, providing about 5 *g* for a 5 tonne test payload and 3.5 *g* for a 15 tonne mass in the vertical axis. Frequency range 0.1–100 Hz. In addition to the traditional sine dwell and sine sweep tests, HYDRA can provide transient excitations in 6-DOF. The 6-DOF allows the specimen to be sine tested along the vertical/lateral axes with one single set-up. Adding transients was a major HYDRA objective: multidirectional transients at the launcher/spacecraft interface produces a more realistic structural response. The shaker is supplied by Mannesmann-Rexroth.

Large European Acoustic Facility (LEAF)

The 154.5 dB 1624 m^3 LEAF, the largest in Europe, became operational in 1990. It provides an internal height of 16.4 m, allowing testing of Ariane 4/5 payloads. The noise generation system, employing pressurised N_2, comprises four horns with cut-off frequencies of 25, 35, 80 and 160 Hz. The overall noise level may be increased to 158.5 dB in the future.

Chamber size: 1624 m^3, 9 × 11 × 16.4 m (W × L × H), accessed by 7 × 16.4 m (W × H) door.

Cleanliness level: class 100,000.

Temperature range: 20 ±2°C during test.

Suspension points: 9 of 80 kN capacity, 35 of 15 kN; max crane load 160 kN.

Overall sound pressure level range: 125–154.5 dBL.

Specified octave band pressure levels in empty chamber at 154.5 dBL: 136.5 dBL at 31.5 Hz centre frequency, 141.5 at 63, 147.5 at 125, 150.5 at 250, 147.5 at 500, 144.5 at 1,000, 137.5 at 2,000, 131.5 at 4,000, 125.5 at 8,000.

Field homogeneity in test volume: ±2 dBL.

Control tolerance of sound field: ±1.5 dBL overall.

Noise measurement/data acquisition: 16 microphones, 250 accelerometers, and 50 strain gauges.

Compact Payload Test Range (CPTR)

The CPTR became operational in mid-1992 primarily for measuring the electrical performance characteristics of Ariane 4-class RF-radiating payloads in their operational configurations. It comprises a shielded anechoic chamber with a feed scanner room and control room together with a test preparation area. The PWZ Plane Wave Zone test volume is obtained using two large reflectors in an offset Cassegrain configuration. Payloads of up to 5 tonnes are located at the PWZ centre by means of a positioner providing movement in azimuth, elevation and polar. Test data can be presented in a wide range of formats, including 2-D raxdiation patterns, contour plots and projections in 3-D. System performance parameters can also be measured. By locating feed horns at different positions in the focal plane (in the scanner room), the direction of the electrical boresight can be changed. In this way, the performance of the satellite transmit/rx from stations at different locations is verified. Similarly, Earth coverage contours or in-orbit reconfiguration can be verified. As the chamber is shielded, it can undertake a range of electromagnetic compatibility measurements.

Internal size: 10.9 × 9.6 × 24.5 m.

Frequency range: 1.5–40 GHz.

Plane wave zone: 7 × 5 × 5 m.

PWZ performance: ±0.2 dB amplitude ripple, ±4° phase ripple, <0.4 dB taper, <–40 dB cross polar.

Reflectors: 9.2 × 8.0 m subreflector, 10.2 × 7.6 m main reflector, less than 100 μm peak-to-peak surface accuracy.

Environmental: class 100,000, 20±2°C, 50±10% humidity.

EMC/ESD Facility

The electromagnetic compatibility/electrostatic discharge facility comprises a shielded anechoic chamber and two operating rooms, including a complete range of automated test equipment. The 20 Hz 40 GHz range satisfies all test requirements for science satellites (low magnetic) and communications satellites (high RF power). All emission and susceptibility measurements are fully automated with on-line data reduction narrow/broadband identification. Data output is corrected for probe antenna factors and compensated for resonance effects.

Test area: 6 × 6 × 4.5 m, accessed by 3.5 × 4.5 m door.

Environmental: class 10,000, 20 to 24°C, 610 to 775 mm pressure, 40 to 55 per cent humidity.

VERIFIED

Japan

Tsukuba Space Centre

Background

Located in Tsukuba Science City 60 km northeast of Tokyo, TKSC is JAXA's largest technical facility, with seven sections responsible for launcher and satellite R&D, design and testing. The Tsukuba Tracking and Control Centre provides Japan's satellite command node. Recently upgraded to ensure higher definition fidelity on tracking antenna.

TKSC offers a 25 m high 8.5 m diameter thermal vacuum test chamber, a radiometer test laboratory, ion engine test facility, two anechoic chambers (the larger 22 × 25 × 39 m), vibration platforms for 13.6/7.9 tonne payloads, a 5.5 × 3.5 m surface plate for testing satellite separation and solar array deployment, a spacecraft magnetic test facility, and a four-chamber acoustic facility. A 10,950 m^2 Spacecraft Integration & Test building was completed August 1988 for handling/testing H2-class 2 tonne payloads. 6,000 m^2 of SITE offers class 10,000 conditions.

Facilities include:

Acoustic chamber

9 × 10.5 × 17 m, 1,607 m^3, compressed air generates max 151 dB (31.5 Hz-8 kHz). Concrete walls are 50 cm thick.

Solar simulation chamber

Internally 13 m diameter, 16 m long, <10^{-7} torr, 6 m diameter solar simulator using Xe lamps. 1,150 measuring channels. Evaporated nitrogen is re-liquefied and recycled.

Vibration test

3 × 3 m table to handle 4.5 tonne payloads. 318 kN vertical and 238 kN horizontal, sinusoidal 5 to 100 Hz and random 5 to 200 Hz.

Mass characteristics measurement chamber

One facility for centre of gravity and moment of inertia determination tests payloads up to 6 t/98 kNm^2. Second facility for aligning satellite sensors and other devices; payload up to 5 tonnes, 10 m high; accuracy ±0.5 mm/±20 arcsec.

VERIFIED

United Kingdom

Central Laboratory of the Research Councils

Background

The Central Laboratory was created 1 April 1995 as an independent entity from the merged Daresbury and Rutherford Appleton Labs, reporting to the Office of Science & Technology and exercised by the Department of Trade and Industry. It is operated by the Council for the Central Laboratory of the Research Councils (abbreviated to CCLRC). Its space funding is mostly from PPARC and NERC but some 20 per cent comes from commercial contracts. RAL's Space Science Department is a partner member of the British National Space Centre. RAL is the focus for the UK's space science projects and directs both ESA and bilateral activities, such as the Spectrum-X astronomical satellite collaboration with the Soviets. It hosts 200 personnel. Policy recommendations made to various UK government bodies including DTI and BNSC. CLRC directs external university participation.

Centralised facilities for university research groups include a satellite operations centre, data processing facilities, and the Starlink data analysis network, the British Atmospheric Data Centre and assembly/testing of space payloads. Projects include:

Badr-B

CCD camera for cloud monitoring. Chilton will act as back-up to Lahore ground station. The same camera is used for Morocco.

Cassini

Participation in Saturn orbiter's dust analyser and plasma electron spectrometer, and Titan lander's surface science package.

Cluster

Participated in the development of the four satellites' 16 plasma sensors.

SOHO

Led development of Coronal Diagnostic Spectrometer.

ISO

RAL managed development of the Long Wavelength Spectrometer.

Spektrum-X

Participation in the JET-X grazing incidence multimirror x-ray telescope.

IUE

Managed UK participation until shutdown in 1996.

Rosat

Participated in development of the Wide Field Camera.

Minisat

Built star tracker for Legri gamma telescope.

UARS

Joint developer of the Improved Stratospheric & Mesospheric Sounder, incorporating a Stirling cycle cooler now available from Matra Marconi Space.

ERS 1/2

ATSR Along-Track Scanning Radiometer developed by RAL-led consortium. Advanced ATSR calibration carried out at RAL. The centre provides data processing for all three.

IRAS

Controlled from RAL's Satellite Control Centre, via the 12 m antenna. The IRAS database is available at RAL for the UK astronomical community.

Ginga

Leicester University/RAL provided the Large Area Counter to Japan's X-ray satellite (re-entered 1991).

Solar Max

Major collaboration in the X-ray Polychromator for studying solar flares.

Spacelab 2

CHASE Coronal Helium Abundance Spectrometer Experiment developed by RAL/Mullard Space Science Laboratory.

UKS

RAL and MSSL satellite and sub-satellite.

Yohkoh

Developed Bragg Crystal Spectrometer with MSSL and US Naval Research Laboratory.

Rosetta

Management and a system engineering for sample analysis instrument on the lander and mother spacecraft to visit comet Wirtanen.

EOS

Management of High-Resolution Dynamics Limb Sounder to measure global temperature and atmospheric chemical species.

Polar

RAL provided hardware on two instruments to study Earth's magnetosphere.

Facilities include:

Clean rooms

A 13 × 8 m room containing 4 × 3 m Class 100 tunnel and three class 100 laminar flow benches; a 90 m^2

room housing 7.5 × 3.5 m Class 100 working area in form of vertical laminar flow downdraft units.

Thermal vacuum

5.5 × 3.0 m diameter tank with cryopumps down to 10^{-10} atmosphere −190/100°C, plus 1.7 × 1.0 m diameter with turbomolecular pump down to 10^{-10} atmosphere, −170/100°C. Bakeout Tank: 90 × 60 cm diameter with LN_2 traps on diffusion pumps down to 10^{-10} atmosphere. Ambient temperature to +150°C.

Vibration facility

Electromagnetic vibrator with 90 × 90 cm slip table capable of vibrating a 35 kg mass to ESA/NASA specs in three-axes. Max displacement ±19 mm, max rate 1.78 m/s. The facility has a cryostat to fit on the vibrator for testing 30 kg objects at down to 4 K. A second electromagnetic vibrator can test a 5 kg mass to ESA/NASA specs. Maximum displacement ±22 mm, max rate 1.2 m/s.

VERIFIED

United States

Applied Physics Laboratory of The Johns Hopkins University

Current Status

In January 2003, APL announced that NASA had awarded a contract to formulate, implement and operate multiple spacecraft for the agency's Geospace missions to study the effect of the Sun on earth's magnetic field.

Background

Founded in 1942, APL is a not-for-profit R&D laboratory and independent division of The Johns Hopkins University. It operates primarily under a contract with the US Naval Sea Systems Command. The DOD sponsors some three-quarters of APL's effort. Full-time staff total 2,700, of which 400 work within the Space Department's two branches, named Engineering and Technology and Science and Analysis.

APL is a major centre for satellite development, construction and operation. Through 1996, it had built 57 satellites (45 of them 50–180 kg). The Laboratory gained early experience by putting instruments on captured German V2 rockets after World War II. Based on its analysis of Sputnik 1's Doppler signals, APL conceived and built the Transit satellite navigation system for USN. It built and operated Geosat, provided the altimeters on Geos-C, Geosat, Seasat and Topex, and led the AMPTE project for NASA. Topex also includes APL's laser retro-reflectors and frequency reference unit, the most stable oscillator ever orbited. Earlier satellites include Dodge, Anna, Triad, Traac, Magsat, Nova and three small astronomy satellites, Injun, Hilat, Polar Bear. The HUT Hopkins UV Telescope (a joint project with the Department of Physics & Astronomy) flew on Shuttle's 1990 Astro 1 mission and March 1995's Astro 2. Voyager, Galileo and Ulysses carry particle radiation detectors. Japan's Geotail includes the EPIC Energetic Particles and Ion Composition experiment. Sweden's Freja and NASA's UARS carry APL magnetic field instruments.

Current projects include the NEAR Near-Earth Asteroid Rendezvous craft (launched February 1996) successfully orbiting the asteroid Eros since February 2000 until it was put to the surface in February 2001. Mid-course Space Experiment (MSX, launched April 1996), the ACE Advanced Composition Explorer, the TIMED spacecraft and the Magnetospheric Imaging Instrument for Cassini. Facilities include:

Richard B Kershner Space Integration and Test Facility

The test facility, 7,350 m^2, was dedicated in 1983. It houses seven clean rooms: three class 100,000 (139 + 2 × 93 m^2); two 93 m^2 class 10,000; two 22 m^2 class 100 rooms are suitable for assembling small space instruments; they can also be held sterile for assembling implantable biomedical devices.

The Space Simulation Laboratory

The SSL includes two 2.4 m diameter 3 m high vertical thermal vacuum chambers (−196/+121°C; 10^{-7} torr cryopump and diffusion pump), a 0.91 m diameter 0.91 m long space simulation chamber (−78/+126°C; 10^{-8} torr) and a 1.52 × 1.83 × 2.13 m T/humidity/altitude chamber (−100/+125°C; 35 to 95 per cent humidity; 0 to 61 km altitude), in its 447 m^2 area.

The Vibration Test Laboratory

A facility that offers: two Unholtz-Dickie shakers (T4000: 178 kN, 152 × 152 cm Team bearing-line slip table; T1000: 89 kN, 4.4 cm peak-to-peak displacement, 61 × 61 cm combined slip table assembly) and an MRC mass properties machine (2,270 kg capacity).

The Solar Array Laboratory and Simulator Control Room

A Large Area Pulsed Solar Simulator (LAPSS) characterises and calibrates solar panel and coupon performance using a Xe source. LAPSS can calibrate panels up to 1.83 m across in 1 ms. Smaller laboratories serve the Satellite Reliability Group, where new parts are inspected and radiation tested.

Satellite Communications Facility

Originally developed as a dedicated Transit station, the 623 m^2 facility has been upgraded for S/L/X-band station for a wide range of NASA and DoD satellites. It is used for continuous MSX operations (a 450 m^2 addition houses a new command/control centre and a data-processing centre for MSX). Antennas: 18.3 m parabolic, 2.2 to 2.3 GHz autotrack feed, receive G/T 23 dB/K, simultaneous RH/LHCP. A 5 m parabolic, 2.2 to 2.3 GHz + 1.650 to 1.750 GHz feeds with selectable RH/LHCP, rx G/T 14 dB/K at 2.2 GHz; 10 m parabolic X/S/L-band simultaneous coverage, fully operational SGLS site with 0.2/2 kW transmit, associated 1 m acquisition antenna. Real-time computers steer the dishes and redundant back-up control and configure the 10 m dish.

Time and Frequency Standards Laboratory

Supports APL projects requiring highly stable time (1 pps), frequency (1, 5, 10 MHz) and time code signals (IRIG-A/B). Other activities include measurement of frequency standards stability, design of atomic hydrogen masers, spaceborne oscillators and synthesisers, and superconducting oscillators.

Anechoic Chamber

7.62 × 7.62 m internal, accessed by 1.83 × 2.44 m door, attenuates outside interference by >70 dB over 0.1 to 10 GHz.

VERIFIED

Arnold Engineering Development Center (AEDC)

Background

AEDC operates the largest and most advanced complex of aerospace simulation test facilities in the world, with some 40 units for simulating conditions up to 1,600 km altitude and speeds up to 32,000 km/h. Construction of the J6 Large Rocket Test Facility, costing more than US$200 million, began in FY89 as a national site for handling large solid propellant rocket motors at simulated 30 km altitudes from late 1994. The J4 Rocket Development Test Cell provides vertical orientation for simulated altitudes up to 30.48 km (100,000 ft). The J5 Test Cell is for horizontal orientation tests. AEDC also maintains a wide range of test and development fixtures for manned and unmanned space systems. Regularly conducts failure tests for government agencies including NASA where analyses are sought to rectify failed or suspect rocket motor designs. Also performs engineering tests and mechanical evaluations relating to government programmes and commercial projects where the national interest is paramount.

Los Alamos National Laboratory (LANL)

Background

LANL, founded in 1943, is a multidisciplinary research facility operated by the University of California for the Department of Energy, undertaking research in nuclear and non-nuclear defence programmes, nuclear safeguards/security, biomedical sciences, computational sciences, materials science, environmental cleanup and basic science. Its first director was J Robert Oppenheimer, who helped to found the Laboratory to develop the first atomic weapons. It encompasses 111 km^2 of mesas and canyons in northern New Mexico, has an annual budget of about US$1.5 billion and employs 6,800. It houses a number of interdisciplinary research centres, including the Center for Materials Science, the Advanced Computing Laboratory, the Center for Nonlinear Studies, the Center for Human Genome Studies, a branch of the Institute for Geophysics & Planetary Physics, the Center for International Security Affairs, and the Exploratory Research & Development Center. Continues research in nuclear medical science and routinely conducts simulation science for the nuclear and non-nuclear industries.

Los Alamos continues to have responsibility for monitoring outer space for nuclear explosions, employing its detectors aboard the DSP early warning and lower altitude GPS navigation satellites. DSP also carries LANL's Magnetic Plasma Analyser and Synchronous Orbit Particle Analyser to monitor spacecraft charging and radiation environment. Its Alexis satellite was launched in 1993 to conduct an all-sky survey of the soft x-ray background glow. LANL/Sandia National Laboratory launched their FORTE Fast On-orbit Recording of Transient Events satellite on Pegasus to demonstrate technology to look for pulses from low-technology nuclear explosions.

LANL provided the x-ray detectors for the HETE mission, three instruments on NASA's Polar, four imaging spectrometers on ESA's Cluster, two γ-ray detectors on Mars-96 and a γ-ray spectrometer on Rosetta's comet lander. It is collaborating with Finland's VTT to provide Cassini's Ion Beam Spectrometer.LANL operates the only US facility capable of making Pu-238 RTGs; Cassini has three, Galileo two and Ulysses one. The laboratory provided technical direction for SDI's BEAR (Beam Experiments Aboard a Rocket) flight of July 1989, which fired the first neutral particle beam in space.

Naval Research Laboratory/Office of Naval Research

Background

NRL was established in July 1923 as the Naval Experimental & Research Laboratory. Now an element of ONR, it is the navy's corporate laboratory. Staff total about 3,300, mostly civilian, in Washington DC, Stennis Space Center (Bay St Louis, Missouri), Monterey (California) and Orlando (Florida). NRL's Naval Center for Space Technology (NCST) maintains space technology expertise through two departments: Space Systems Development and Spacecraft Engineering. The Remote Sensing Division (RSD) and the Space Science Division (SSD) come under NRL's Ocean & Atmospheric Science & Technology Directorate.

NRL directed the Vanguard programme and since 1960 has developed more than 85 satellites, most recently the Clementine technology demonstrator lunar/asteroid probe in conjunction with the Lawrence Livermore National Laboratory. It has also flown numerous payloads and experiments, such as the Hercules unit carried by STS-53 in December 1992 to allow astronauts to photograph surface features while automatically recording positions within 2 km. It flew again on STS-56 in April 1993 and on STS-70 in July 1995. Another NRL project, the HTSSE High Temperature Superconducting Space Experiment flew aboard STP's Argos satellite.

The RSD undertakes a broad programme in sensing applications over frequencies from UV to radio. Sensor systems include RAR Real Aperture Radar, scatterometers, lidars, optical/radio interferometers and passive microwave imagers. Its middle atmosphere sensors include the POAM Polar Ozone & Aerosol Monitor on Spot 3, the MAS Millimetre-wave Atmospheric Sounder on Shuttle Atlas missions (three flown), and a suite of water vapour and ozone monitors as part of the Network for Detection of Stratospheric Change. ONR has conducted intensive study of the upper atmosphere in a series of activities involving NASA and the NOAA. This information is useful to

NASA in re-defining characteristics of the extreme stratosphere for engineering information in the OSP.

RSD's facilities include digital image processing, a tactical environmental visualisation centre, an aerosol measurement facility, 25.6/25.9 m radio antennas at Maryland Point Observatory and an optical interferometer at Mount Wilson Observatory for monitoring background environmental emissions at high angular resolution.

SSD maintains facilities for designing, constructing, assembling, calibrating and analysing space experiments, principally upper atmosphere ultra-violet sensing, solar atmosphere spectrometry and celestial radiation over ultra-violet to cosmic rays. For example, it provided the SUSIM Solar ultra-violet Spectral Irradiance Monitor for UARS and Atlas, the RAIDS Remote Atmospheric & Ionospheric Detection System to measure airglow, the OSSE Oriented Scintillation Spectrometer Experiment gamma instrument on NASA's Compton Observatory, solar x-ray measurements using a Bragg crystal on Japan's Yohkoh, and the LASCO Large Angle Spectrometric Coronograph on Soho. SSD provided three of the main instruments (USA, HIRAAS, and GIMI) on Argos.

Phillips Laboratory

Background

An element of the AF Space & Missile Systems Center, the Phillips Laboratory was founded in December 1990 as part of the AF's reorganisation of the 14 existing laboratories and research centres into four 'super laboratories'. Phillips studies military space, missile, directed energy, propulsion and geophysics R&D. It replaced the USAF Space Technology Center (created October 1982) at Kirtland, the USAF Weapons Laboratory (Kirtland), AF Geophysics Lab (Hanscom) and AF Astronautics Lab (Edwards). In fiscal year 1996, the air force provided a budget request of US$510 million; FY95 US$629 million; FY94 US$552 million; FY93 US$756 million. Personnel at Kirtland total 576 civilian and 442 military. Hanscom has 250 civilians and 64 military. Edwards employs 184 civilian and 99 military. In 1997, Phillips Laboratory unveiled a Space Manoeuvre Vehicle developed by The Boeing Company. Studies related to recoverable spaceplanes and ballistic return vehicles explore rapid launch concept for manned or unmanned vehicles. The laboratory provides detailed engineering and programme assessments of prospective technology requirements for new generations of launch vehicles and provides input to the balance between unmanned and human space flight systems.

The laboratory's seven technical directorates are:

Advanced Weapons and Survivability

AWS performs and directs R&D of technologies for directed energy weapons (microwave, laser and particle beam) and related advanced weapon concepts, space systems survivability and EM effects. It measures and models directed energy effects on air and space systems and determines lethality to threat targets, develops hardening countermeasures and transitions survivability technology to product and logistic centres. High-energy density plasma and radiation source research and pulsed power development are performed at the Shiva Star facility, a 10 MJ 3×10^{12} W fast capacitor bank, one of the nation's largest.

Lasers and Imaging

This new optical observatory began operating 1994 as part of the Starfire Optical Range. The 3.5 m telescope, the DoD's largest and 12th in the US, augments the existing 1.5 m telescope and 1 m laser beam director for research into space object tracking and atmospheric compensation techniques. This directorate also operates the AF Maui Optical Station in Hawaii, devoted to satellite and rocket tracking and imaging. Its existing 1.6 m and 2×1.2 m telescopes was to be enhanced with a US$120 million 3.67 m telescope for completion in early 1997 and fully operational in spring 1998. The directorate also operates the NC-135 Argus aircraft as a flying laboratory for space imaging work. The former AFWL handles the directed energy weapon research for BMDO, which included the Alpha hydrogen fluoride chemical laser. It was also responsible for BMDO's Relay Mirror Experiment.

Space and Missiles Technology

SMT focuses on spacecraft structures and controls, power/thermal management, sensors, communications, electronics, satellite control and simulation, orbital dynamics, ballistic missile technologies and reusable launch vehicles. Its Thermionic System Evaluation Test is testing an unfuelled Topaz 2 Russian nuclear reactor. The Space Nuclear Thermal Propulsion Program aimed at developing an advanced nuclear rocket engine with twice the specific impulse of the best current liquid engines, but it was cancelled May 1993 on cost grounds, after expending US$55.5 million FY93 plus US$65 million FY92.

Space Experiments

This branch conducts balloon and sounding rocket payload integration, and designs, integrates and tests whole spacecraft systems. Its MightySat programme will provide low cost and regular access to space for testing new technologies. It is responsible for the TAOS mission of the Space Test Program, and BMDO's HABE programme supports ATP systems for boost phase DEW missile defence.

Geophysics Directorate

The centre for air force research into the near-Earth space environment, the geophysics directorate includes ionospheric, meteorological, optical backgrounds and Earth dynamics measurements in its charter. All research is related to military space applications, such as radar propagation, crustal motion effects on ballistic missiles, and vehicle IR signatures against natural backgrounds. Its projects include MSX.

Propulsion

This directorate undertakes R&D of advanced solid, liquid, hybrid and electric missile and space propulsion on a 170 km^2 site in Edwards AFB's northeast corner. Facilities include test stands accommodating engines up to 45 MN thrust and altitude chambers in which attitude control thrusters can fire up to 7 hours continuously. The National Hover Test Facility to support BMDO was inaugurated in November 1988 by the first hovering trials of a space-based interceptor demonstration model.

Airborne Laser

The ABL programme is developing an airborne demonstrator of a high energy laser for defence against theatre missiles in the boost phase. Two contractor teams: Boeing (lead)/TRW/Lockheed and Rockwell (lead)/Hughes/E-Systems. From the two concept designs produced by the current Phase I, one team will be selected to develop a demonstrator aircraft, culminating in a lethality demonstration against boosting missiles after 2000.

SPACE MONITORING FACILITIES

Russian Federation

The SSS Space Surveillance System

Background

The USSR, during 1969–70, officially established the SKKP System for Monitoring Outer Space using the then recently deployed first generation BMEWS radars, known in the USSR as Dnepr. In 1970, the Soviet satellite catalogue was more than 250, or about 10 per cent of the US catalogue. Today, Russia's SSS Space Surveillance System relies primarily on Dnepr and Darayl-UM radars operating in the VHF range near 150 MHz at eight sites. The Russian sites are Irkutsk, Murmansk and Pechora in Russia, Sevastopol and Uzhgorod in Ukraine, Balkhash in Kazakhstan, Mingechaur in Azerbaijan, and Riga in Latvia (Riga's Darayl-UM was scheduled for 1995 demolition. Dnepr should continue through 1998). Another site at Baranovichi was under construction, but its status is unclear. Operation of many of the space surveillance and TT&C sites is now erratic, with facilities dropping out and then returning to service. Several key installations have been shut down and remaining assets consolidated around primary sites in Russia mentioned above.

C-band and higher frequency radars are beginning to be acknowledged. Observations are also provided to the SSS by two ABMD Anti-Ballistic Missile defence radars in the Moscow region near Sofrino and Chekhov operating at UHF near 400 MHz. This combined network generates some 50,000 observations daily to maintain a catalogue of nearly 5,000 objects, most in LEO. Owing to geographical limitations, Earth satellites at low inclinations are difficult or impossible to track. As a whole, the SSS radar sensors appear to be limited to a range of about 4,000 km.

Also supporting the SSS are more than 20 optical and electro-optical facilities at 14 locations: Irkutsk, Kourovka, Krasnodar, Uzhno-Sakhalinsk and Zvenigorod in Russia, Kiev, Odessa, Simeiz and Uzhgorod in Ukraine, Alma Ata in Kazakhstan, Burokan in Armenia, Abastumani in Georgia, Dushanbe (Hissar) in Tadjikistan, and Ashgabad in Turkmenia. Seven of these sites are supported jointly by SSS and the Russian Academy of Sciences (RAN), while five are run independently by SSS and two by RAN. A wide range of Soviet and German instruments of optical diameters 45–100 cm are employed, including the AZT-8, AZT-14, VAU, SBG and Zeiss-series telescopes. As in the US, these optical and electro-optical facilities perform the majority of deep space observations, especially in the eastern hemisphere GEO arc. RAN previously maintained sites in Bolivia, Chile, Ecuador and Egypt.

In addition to generating positional and ephemeris data, the SSS maintains a catalogue of satellite characteristics such as size and stability. Papers published by RAN scientists reveal an ability to discern the configuration (shape and materials) of high altitude satellites through an inverse photometric process by extended observations at multiple wavelengths, typically U/B/V/R. One of the principal instruments involved in this research is at Simeiz. The 101.6 cm diameter Zeiss-1000 with 2.5 arcmin FOV and about 17.5 limiting magnitude is augmented with a PG-3M two-stage cascade image tube with a 256 × 256 CCD.

United States

Ground-based Electro-Optical Deep Space Surveillance (GEODSS)

Background

A more recent monitoring system than the Space Surveillance Network, GEODSS is also operational. Three sites around the world each house three telescopes on rapid-motion mounts positionable to within 1.5 arcsec. The air force claims they are able to identify basketball-sized objects in orbits between 5,500 to 37,000 km. They supersede the network of Baker-Nunn film cameras in place since 1958 (of which the remaining example at St Margarets, New Brunswick, Canada was closed April 1992). Data from GEODSS's TV image are relayed to SCC's computers. If a satellite is not detected immediately at its expected co-ordinates, a search for it is automatically triggered on-site. The sites, at Socorro (New Mexico), Choe Jong San (South Korea) and Maui (Hawaii), were operational by 1983. A fourth, on Diego Garcia in the Indian Ocean, was completed in February 1987. A fifth was originally planned for Portugal, but cancelled. The Korean site closed in 1993 to save money (the weather made it a poor site). Each site has three observatories: two with a 102 cm 2° FOV Cassegrain to look for faint (to +16th magnitude) slow objects at high altitude and the third with a 38 cm 6° FOV Schmidt to search lower altitudes for faster objects (Diego Garcia has three Cassegrains and no Schmidt). MIT Lincoln Laboratories continue to operate the GEODSS prototype for R&D. The air force is interested in flying a similar system in space, believing an operational system would cost no more than operating an international ground system, nor would cloud cover be a constraint. Some of these activities have been incorporated into intelligence gathering activities located on existing space-based assets. Moreover, consideration is underway about utilisation of optical reconnaissance satellites for this purpose. GEODSS will play a larger role in the optical evaluation of Shuttle orbiter in-space verification following the lack of orbit-to-orbit or ground based imaging of the Shuttle Columbia before its loss during descent to earth on 1 February 2003.

Maui Space Surveillance Site (MSSS)

Background

The USAF's Phillips Laboratory's Lasers & Imaging Directorate operates several facilities for testing advanced imaging and acquisition, pointing and tracking techniques. The Maui Space Surveillance Site at 3,000 m and an altitude of 20.7° north on the summit of Mount Haleakala comprises GEODSS, the AMOS AF Maui Optical Station and the MOTIF Maui OpticalTracking & Identification Facility. MOTIF's paired 1.2 m telescopes became dedicated SSN sensors in October 1979 and provide satellite imaging, tracking and infra-red signature data down to +19th magnitude. AMOS's adjacent 1.6 m telescope with deformable mirror coupled to a 6,940 Å laser radar is primarily for research but also contributes to SSN. Rockwell Power Systems under a five year contract operate MOTIF and AMOS. Under development is the AMOS Daylight Optical Near Infra-red System (ADONIS), which will extend imaging capabilities to 24 hours daily. A major addition to AMOS is the AEOS (Advanced Electro-Optical System): the US$120 million 3.67 m telescope was completed in early 1997 and fully operational by Spring 1998. Upgrade and replacement strategies include high-intensity laser sighting systems supplementing AMOS and ADONIS.

NAVSPASUR

Background

The Space Surveillance Network includes Naval Space Command's electronic fence, known until 1993 as NAVSPASUR. It is a series of 216.980 MHz transmitters and receivers, operational since 1961, forming an electronic fence across the southern US, stretching 4,800 km from Georgia to California, 1,600 km off each coast and 24,100 km into space. Three transmitters are sited at Lake Kickapoo (Texas; primary transmitter, 766.8 kW, array 3,269 m long), Gila River (Arizona; 40.5 kW; 497 m) and Jordan Lake (Alabama; 38.4 kW; 314 m). They work in conjunction with six receivers at San Diego (California), Elephant Butte (New Mexico), Silver Lake (Mississippi), Red River (Arkansas), Tattnall and Hawkinsville (Georgia). The Hawkinsville and Elephant Butte sites operate 25 to 366 m arrays for high altitude satellites; the others use 12 to 122 m systems for lower objects. The data are processed at NAVSPACOM headquarters in Dahlgren (Virginia), which since 1984 has also served as the Alternate Space Control Center. Interlinks with transmitters at Texas, Arizona and Alabama to connect processing structures. Upgrades to enhance discrimination standards.

VERIFIED

The Maui Space Surveillance Site on Mount Haleakala is devoted to satellite and rocket tracking and imaging. MOTIF (top left) uses paired 1.2 m telescopes for satellite imaging, tracking and IR signatures. AMOS's 1.6 m telescope (top right) is coupled to a laser radar. GEODSS is housed in the three identical domes. The two small central domes illuminate targets with lasers for night imaging. The AEOS Advanced Electro-Optical System 3.67 m telescope with adaptive optics will be added to AMOS (Phillips Laboratory) 0516813

Starfire Optical Range

Background

A new optical observatory began operating 1994 at Kirtland AFB in New Mexico as part of the Phillips Laboratory's Starfire Optical Range. The 3.5 m telescope, DoD's largest and 12th largest in the US, joined an existing 1.5 m telescope and 1 m laser director for research into space object tracking and atmospheric compensation techniques. The directorate also operates the NC-135 Argus aircraft as a flying laboratory for space imaging research. The 1.5 m mirror carries 341 actuators 7 mm apart on its 1 mm-thick glass. The system samples atmospheric distortion 1,700 times/s by observing laser scattering off molecules in the lower atmosphere, and the continuously adjusts the deformable mirror. During the 1980s the Starfire Optical Range did much of the pioneering work in developing adaptive optical technology. In 1993, it produced the first important astronomical observation via adaptive optics: ionised hydrogen clouds around Orion nebula stars. Starfire facility routinely tracks objects in space and others descending through the atmosphere. Two technical workers used the re-entering Shuttle Columbia as a test target during the fateful descent on 1 February 2003 but those images were not taken by the facility's main telescope. Recommendations made by the Columbia Accident Investigation Board may result in operational responsibilities for imaging of in-orbit Shuttles to provide additional evaluation of physical integrity.

USSPACECOM Space Surveillance Network

Current News

Assisting NASA with data concerning objects near to or on transecting orbits with the Shuttle Columbia during its period on orbit between 16 January and 1 February 2003, prior to its destructive re-entry. The results of the Columbia Accident Investigation Board review of Shuttle programme processes and flight support has resulted in several key shifts in the way space surveillance systems will be required to support US human space flight operations in the future. These activities will enable additional data sets to be obtained which could support verification of on-orbit Shuttle integrity.

Background

Currently, NORAD, under the control of the US Space Command's Space Control Center and Missile Warning Center inside the hollowed-out Cheyenne Mountain, monitor all space debris and active satellites. The 15 steel buildings house a series of control rooms and offices, cushioned on steel springs inside vast chambers cut into the granite mountain. More than 1,100 personnel (US Army, Navy, Marine Corps, Air Force, Canadian Forces and civil technicians) keep the complex permanently operating. When the site was chosen in the 1950s, it was believed to be impregnable to a nuclear attack.

USSPACECOM plays an important role in monitoring orbital objects in the vicinity of the NASA shuttle.

The centre issues warnings when objects appear likely to pass within 5 km radially, 25 km downtrack or 5 km out of orbital plane of the Shuttle. The Shuttle makes avoidance manoeuvres whenever the values are 2/5/2 km, respectively. Collision avoidance information is increasingly important. Facilities include:

HAX Haystack Auxiliary

The HAX Haystack Auxiliary 16.7 GHz 13 m dish, also known as the NEAR Near-Earth Assessment Radar, became operational in 1993, although it cannot detect low inclination objects. However, it should handle the new 51.6° Station orbit well. The main 36.6 m diameter 400 kW 10 GHz Haystack Long Range Imaging Radar has been used periodically since 1990 to observe LEO debris, capable of seeing objects down to 1 cm at 1,000 km. One discovery is that Russia's retired nuclear-powered Rorsats appear to be leaking liquid metal coolant; some 70,000 droplets were found around 900 to 1,000 km. Six solid spheres (two 15 cm diameter, 5 kg, aluminium; two 10 cm diameter, 4.3 kg, stainless steel and two 5 cm diameter, 0.53 kg, stainless steel) were released from STS-60 in February 1994 to help calibrate radar and optical tracking systems. The primary objective was to fine tune the Haystack Radar. STS-63 in February 1995 deployed a second array. This array had a 5 cm 0.53 kg polished stainless steel, 10 cm 4.24 kg white aluminium and 15 cm 5.0 kg black aluminium spheres, as well as three platinum iridium alloy dipoles. Two of the dipoles were 1.5 g, and 13.3 cm long, the other one was 0.5 g and 4.42 cm long. All were 1.0 mm diameter).

BMEWS

Development of US space surveillance, operated by the US Air Force with help from the UK, Canada and other countries, began in the late 1950s to cover both missile firings and satellite launches. As the USSR developed ICBMs, the West responded with the BMEWS Ballistic Missile Early Warning System. It extended the then existing range of tracking radars from a few hundred km to about 5,000 km. The most probable flight path of a Russian ICBM attack is over the Arctic, so three BMEWS radar stations were built at Thule in northern Greenland, at Clear in Alaska and at Fylingdales, northern England. Thule and Clear had detection radars the size of football fields to pick up a missile as it rises above the horizon. Trackers (Thule and Clear each had 25.6 m dishes; Fylingdales had three but no detection radar).

The BMEWS system has undergone a major update. The first site, Thule, which covers the northern perimeter, was given a new US$80 million phased-array antenna. This is a 50 m high, 122 m long structure, providing two phased-array faces accommodating 2,560 active elements and 1,024 additional elements per face. It began test transmissions in 1987 and was completed in summer 1988. A similar array was constructed at Fylingdales Moor in Yorkshire, England, operated by the RAF. The array replaced the familiar giant 'golf ball' radomes that dominated the landscape of this area of northern England (the radomes were dismantled June 1994). Raytheon completed the new US$200 million system in 1992. Unlike Thule, it has three 25.5 m diameter active faces, providing 360° coverage and improving detection of SLBMs. It can track 800 objects simultaneously. The warning time is about the same as under the old system but radar resolution is much improved. Upgrading the Clear facility is planned but not yet funded. It is often the first to pick up a Plesetsk launch.

Passive Space Surveillance System

Air Force Space Command has recently deployed a new sensor network: the PASS Passive Space Surveillance System using satellite RF emissions to compute angle of arrival observations. The air force deployed tow versions by 1994, the DSTS Deep Space Tracking System and LASS Low Altitude Space Surveillance. A third, CROSS Combined RF-Optical Space Surveillance, became operational August 1991 in San Vito, Italy using an optical telescope and an RF receive antenna but was closed in 1993. The LASS sites in Edzell (Scotland), San Vito (closed 1993) and Osan (Korea), which became operational May 1992, provide RF surveillance of near-Earth satellites. DSTS sites at Feltwell (England), Grifiss (New York) and Misawa (Japan) provide worldwide coverage of geosynchronous satellites.

PAVE PAWS

Detection and warning of SLBMs started in 1968 with an interim system based on eight radar sites located on the Atlantic, Pacific and Gulf coasts. They could detect the launch of a missile from a patrolling submarine and then track its course. It was replaced in 1980 by the USAF/Raytheon Pave Paws, a pair of solid-state 420 to 450 MHz (carriers 422/435/439 MHz) 31 m diameter radar systems on the US east/west coasts. With no mechanical parts to limit the speed of its radar scan, it embodies phased-array technology - thousands of small radar antennas co-ordinated by two large computers - allowing it to detect, track and predict the impact point and numbers of missiles, the technology now extended to BMEWS. Until 1995, four Pave Paws sites provide continuous overlapping coverage from the northeastern US, across the southern periphery and up to the northwest. All four sites also fed display data on the position and velocity of satellites to Space Command. They have achieved 99 per cent reliability. Additional automated data processing equipment and security improvements became operational in 1992. The southeast and southwest sites at Robins AFB in Georgia and Eldorado AS in Texas became operational in 1987 (but were de-activated in 1995), joining the two original operational sites at Cape Cod AS in Massachusetts and Beale AFB in California.

It was the development of Pave Paws that led to the abandonment of Over-The-Horizon Radar (OTHR), which went into operation briefly in March 1968. This was to meet the threat of the fractional orbit bombardment system (FOBS), Soviet missiles launched into orbit via the South Pole, thus avoiding detection by BMEWS. They would travel three quarters of the way around the world before re-entering to strike their target. OTHR was not a conventional radar system, consisting instead of a series of transmitters generating a continuous signal that reflected back and forth repeatedly between the ionosphere and Earth's surface until received at another station several hundred km away. A missile penetrating the ionosphere disturbed the signal and was thus detected. One of these stations operated at Orfordness on England's east coast for several years until it was closed down in 1974. The problem was that the ionosphere was not always suitably positioned – such as at night, when looking north. One solution proposed space mirrors for heating the ionosphere, making it conform to military requirements.

Cobra Dane

Another phased-array radar with greater sensitivity also performed dual missile warning and space surveillance until 1994. The Cobra Dane radar (AN/FPS-108, 1,175 to 1,375 MHz) on Shemya Island (52.7° N/174.1° E) at the western edge of the Aleutian Islands served from 1977 as a primary source of data on Soviet ICBM and SLBM tests impacting near Kamchatka Peninsula. Its location ensured that it was often the first US radar to track Soviet space launches. A major upgrade to the system was completed in 1993; testing was still under way in early 1994 - but it was closed down 1 April 1994. The new Cobra Dane radar could track more than 100 objects simultaneously, identify targets more rapidly, provide more accurate measurements and maintain a significantly larger (12,000 *vs.* 5,000) catalogue. Cost of the Raytheon upgrade was more than US$58 million.

Other Phased-Array Radars

SSN made the first use of the Eglin AFB phased-array radar, consisting of two fixed electronic arrays set in a 10-storey high bank of concrete sloped at 45° to the horizon providing 120 to 240° azimuth coverage. One array, of 5,184 transmitter modules, emits an electronically steered beam that sweeps the area of coverage in ms; the other array, 4,660 modules in a hexagonal arrangement, receives the signal as it bounces back from an orbiting satellite. Florida was chosen for the first of these arrays because most space objects pass within its coverage daily. The Eglin radar system was to be shut down when the Robins Pave Paws site received a power upgrade. However, concerns during 1988 about the Robins upgrade led to an indefinite postponement. Transmitter/receiver upgrades were completed in 1995 at Eglin, where 140 civilian and more than 30 military personnel maintain the facility. Eglin's primary mission was shifted in 1987 to space surveillance. Since the mid-1970s it has provided most of SSN's radar cross-section data. The second oldest phased-array system, the PARCS Perimeter Acquisition Radar Characterisation System, was originally constructed to support the US Safeguard ABM system in North Dakota in the 1970s. Although it is one of the most capable sensors in the Space Surveillance Network, the air force is now considering its closure.

The UHF 26 m radar at Pirinclik in Turkey, operational since 1969, provides early tracking of missiles and space launches from Tyuratam and surveillance of GEO satellites over the eastern hemisphere. C-band radars at Kaena Point (Oahu), Antigua and Ascension Island also provide important early observations: Tyuratam's manned, GEO and deep space missions pass directly over Ascension during the first orbit. Kaena Point sees Tyuratam's high-inclination missions during the first half rev; Antigua tracks most LEO satellites in orbit before Conus sensors.

LAUNCH VEHICLES (ORBITAL AND SUB-ORBITAL)

LAUNCH VEHICLES (ORBITAL AND SUB-ORBITAL)

Brazil

Sonda and VLS

Current Status

On 23 August 2003, the third prototype VLS-1 launcher was destroyed in a pad explosion at the Alacantra launch site in which 21 people were killed. It is believed the first stage prematurely ignited causing the S-43 propulsion system to fire, an event possibly caused by electrical discharge, electromagnetic pulse or a piece of metal in the propellant according to the Brazilian space agency. This launch, designated V3, was carrying the Brazilian Satec-1 and Unosat satellites. VLS-1 and VLS-2 were launched in 1997 and 1999 respectively and both were destroyed by the range safety officer when they failed during ascent.

Background

Sonda 1 was a two-stage vehicle first launched 1964, and flown more than 200 times before retirement. Sonda 2's development then took over in 1966 and has flown about 60 times, offering 44 kg capacity to 80 km. Single stage, 300 mm diameter, 4.1 m long, 360 kg launch mass. Three versions are still employed, principally for technology-proving flights. Sonda 3 development began 1969 to provide a two-stage vehicle capable of delivering 50 kg to 500 km, providing three-axis payload control and sea recovery. The S3 basic version handles 50 to 80 kg to high altitudes using the Sonda 2 as the S20 stage 2. A reduced Sonda two-stage 2 (S23) is carried for S3-M1 missions with 130 to 160 kg payloads destined for lower altitudes. (The M2 version, using the smaller S24, has been used once, 23 August 1979.)

Preliminary studies of a Sonda 4, began in 1974 and produced the decision to launch five prototypes for vehicle qualification and as technology demonstrators for the VLS (Veiculo Lancador de Satelites) satellite launcher, which incorporates clustered Sonda 4 stage 1 motors. Four launches have been made: 21 November 1984, 19 November 1985, 8 October 1987, and 28 April 1989. There have been no flights since the first and last Sonda 4 on 21 February 1990 from the new launch site Centro de Lancamento de Foguetes in Alacantra. Stage 2 failed to separate in 1987's flight. The programme has developed 300M steel for the motor casings, digital control system (although Sonda has so far carried an analogue system), payload fairing ejection and three-axis control (including liquid injection TVC and movable nozzle TVC). MBB provided technical assistance in

A 1/3 VLS model was launched May 1989 to demonstrate strap-on separation (IAE/CTA) 0516861

Specifications

VLS motor

Motor	S-43	S-43TM	S-40TM	S-44
Motor mass (kg)	8,210	8,400	5,340	917
Propellant mass (kg)	7,180	7,180	4,450	810
Burn time (web, s)	58.9	58.9	56.4	67.9
Action time (s)	62.0	62.0	62.0	72.0
Average pressure (atmospheres)	56.2	55.3	57.2	39.5
Average thrust (kN, vacuum)	309.0	327.0	212.5	33.9
Total vacuum imp (MNs)	18.42	19.52	12.00	2.31
SI (s, vacuum)	260.0*	277.0	274.9	281.6
Expansion ratio	12.82	37.68	25.97	66.0
Nozzle exit diameter (mm)	700	1,200	800	602

*sea level SI 230.5 s. S-44 employs Kevlar epoxy casing; others are steel.

developing the LITVC system. #4 demonstrated movable nozzle TVC for the first time on stage 2.

IAE, under the aegis of the Ministry of Aeronautics, has developed the Veiculo Lancador de Satelites. Originally intended to launch the MECB satellites. But the programme ran afoul of the Missile Technology Control Regime signed by the G-7 group of the world's seven most industrialised nations. Many components such as the liquid roll control package planned forward of stage 2 and replaced by solid motors (but is now again liquid), had to be reduced to less capable technology.

The strap-ons and two of the three core stages are derived from Sonda 4's stage 1. The four strap-ons ignite on the pad, employing nozzle flexure for steering. The core stage 1 is nested within the strap-on cluster and ignites at 20 km altitude. A liquid stage 1 is under consideration as a future upgrading. Stage 2 is a shortened version of the S-43 motor. The stage 3 orbit injection motor is newly developed and spin-stabilised with a fixed nozzle. By end-1995, two of four planned strap-on static firings had been made, two of three stage 1, four of five stage 2 and four of six stage 3. The 6.6 tonne 9.52 m 'VS-40' vehicle was launched 2 April 1993 on a 24 minute, 1,248 km altitude flight testing VLS's stages 2/3. A 9.76 tonne 12.0 m VS-43 vehicle using VLS stage 1/3 could deliver 1.2 m diameter 200 to 500 kg suborbital payloads to 1,000–2,000 km.

Brazil had major problems acquiring the inertial guidance platform and the vehicle has yet to be launched. The four flights of the original MECB were planned in 1989 to begin 1992, followed by the second in 1993 and the remaining two during 1994. This never materialised but the first launch attempt for VLS-1 took place on 2 November 1997. It failed to reach orbit. The 50 tonne four-stage all-solid vehicle stands 19 m high, capable of placing 200 kg into a 750 km circular 25° orbit or a 450 km Sun-synchronous orbit.

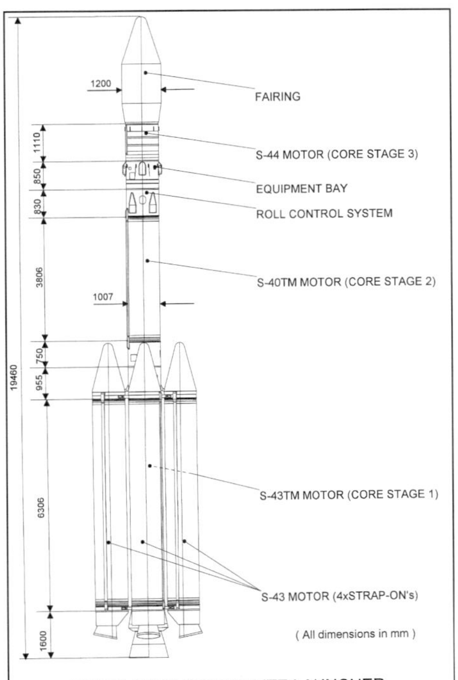

VLS principal features (IAE/CTA) 0003408

The second launch attempt took place on 11 December 1999 and again the VLS failed to reach orbit, the SACI-2 satellite being destroyed in the atmosphere.

Sonda 3/3-M1

First launch: 26 February 1976
Number launched: 28 (2 failures: 31 October/14 November 1983) to end-1996; M1 4 flights, successful
Number of stages: 2 (solid propellant)
Overall length: 8.0 m both versions
Principal diameter: 557 mm
Launch mass: 1,521 kg (M1: 1,527 kg), excluding payload
Typical performance: 60 kg to 600 km (M1: 140 kg to 275 km).

Sonda 3/3-M1 stage 1

Overall length: 373 cm
Principal diameter: 557 mm
Stage mass: 1,205 kg both versions
Propellant mass: 860 kg both versions
Average thrust: 102 kN sea level both versions
Burn time: 24 s.

Sonda 3/3-M1 stage 2

Designation: S20 (M1: S23)
Overall length: 290 cm (M1: 160 cm), excluding payload
Principal diameter: 300 mm both versions
Stage mass: 316 kg (M1: 182 kg)
Propellant mass: 229 kg (M1: 113 kg)
Average thrust: 33 kN vacuum (M1: 18 kN)
Burn time: 15 s (M1: 15 s).

Sonda 4

First launch: 21 November 1984
Number of launches: 4, to end-1995 (8 November 1987 stage separation failure)
Number of stages: 2 (solid propellant)
Overall length: 11.0 m
Principal diameter: 1,008 mm
Launch mass: 6,800 kg, excluding payload
Typical performance: 500 kg to 650 km
Guidance: the inertial platform and control system are housed in the equipment bay atop stage 2. Three-axis control during stage 1 burn is provided by LITVC (2.5° max vector deflection; 81.6 atmospheres injection by N_2; 9 ms valve response time; Three valves per quadrant) for pitch/yaw and thrusters for roll.

Sonda 4 stage 1

Overall length: about 537 cm
Principal diameter: 1,008 mm
Stage mass: 5,670 kg
Propellant mass: 4,220 kg
Average thrust: 203 kN sea level
Burn time: about 60 s

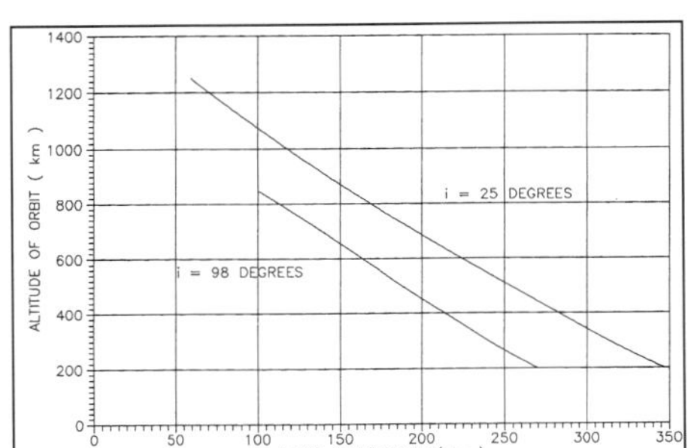

VLS circular orbit performance (IAE/CTA) 0516860

The VS-40 tested VLS stages 2/3 (IAE/CTA) 0517188

End view of Sonda 4 rocket motor propellant grain, for which AGQuimica manufactures the ammonium perchlorate 0516862

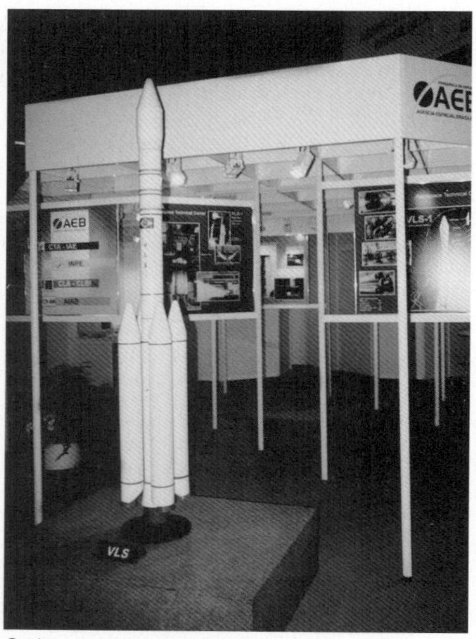

Scale model of the VLS launcher with four solid propellant strap-on booster rockets (Theo Parard) 0054363

Sonda 4 stage 2
Overall length: about 325 cm, excluding payload
Principal diameter: 555 mm
Stage mass: 1,130 kg
Propellant mass: 869 kg
Average thrust: 95 kN vacuum
Burn time: 28 s.

VLS
First launch: 1997 (fail); second launch 11 December 1999 (fail)
Launch site: Alcantara
Principal uses: delivery of small payloads to LEO
Schedule of missions: SCD 2A fourth quarter 1997?; first half 1998 SCD 3; SSR 1 1999; SSR 2 2000
Typical performance: 200 kg into 750 km, 25°
Number of stages: 3 + 4 strap-ons (all solid)
Overall length: 19.46 m
Principal diameter: 1.00 m
Launch mass: 50 t
Guidance: inertial, mounted in equipment bay below stage 3
Launch sequence (min:s):

0.00	strap-ons ignite
0.55	stage 1 ignites
1.07	strap-ons burn out/separate
1.58	stage 1 burnout/separation and stage 2 ignition
2.07	heatshield separates
3.00	stage 2 burnout
3.08	stage 2 separates
7.12	equipment bay separates/stage 3 Ignites
8.23	stage 3 burnout/separate

VLS strap-ons
Designation: S-43
Overall length: 8.92 m
Principal diameter: 1,006.6 mm
Stage mass: 8,550 kg each (7,180 kg propellant + consumables)
Average thrust: 309 kN vacuum each
Burn time: 58.9 s (62 s. action time)
Attitude control: movable nozzles provide up to 3.0° TVC for three-axis control from four motors
Separation: each strap-on is attached by an aft spherical thrust transmission pin and paired forward/aft arm mounts. Following burnout, the arms are severed pyrotechnically and internal gas-pressurised actuators provide separation velocity.

VLS stage 1
Designation: S-43TM (similar to strap-on but nozzle reconfigured optimised for altitude operations)
Overall length: 8.86 m
Principal diameter: 1,006.6 mm
Stage mass: 8,720 kg (7,180 kg. propellant plus consumables)
Average thrust: 327 kN vacuum
Burn time: 59 s (62 s. action time)
Attitude control: nozzle flexure up to 3° provides pitch/yaw TVC; roll managed by solid thrusters.

VLS stage 2
Designation: S-40TM
Overall length: 6.24 m
Principal diameter: 1,006.6 mm
Stage mass: 5,660 kg (4,450 kg propellant plus consumables)
Average thrust: 212.5 kN vacuum
Burn time: 56.4 s (62 s. action time)
Attitude control: nozzle flexure up to 3° provides pitch/yaw TVC; roll managed by liquid thrusters.

VLS stage 3
Designation: S-44
Overall length: 1.75 m
Principal diameter: 1,006.6 mm
Stage mass: 1,025 kg (810 kg propellant plus consumables)
Average thrust: 33.9 kN vacuum
Burn time: 67.9 s (72 s. action time)
Attitude control: spin initiated by solids before separation of equipment bay
Heatshield/payload accommodation: The payload is mounted on its adaptor by a V-clamp and separated by springs following pyrotechnic initiation. The two heatshield aluminium halves provide an internal diameter of 1,180 mm (external 1,200 mm) and cylindrical payload section 1,180 mm high. Mass <150 kg. Pyrotechnic clamps and springs separate the fairing after stage 2 ignition.

VERIFIED

Canada

Black Brant

Current Status
NASA has been the most prolific user of Black Brant sounding rockets and has asked Magellan to evaluate a larger and more powerful variant. There have been some studies conducted by other manufacturers with a view to improving the lift capability.

Background
Bristol Aerospace is a member of the Magellan Aerospace group of companies. Named after a small, stocky goose with black head and neck, the Black Brant sounding rocket is manufactured by the Defence and Space business unit. Bristol Aerospace's solid propellant Black Brant launcher is offered in five basic versions (5, 8, 9, 10, and 12). Payloads of 70 to 850 kg can be delivered to 150 to 1,500 km from conventional boom rails or three to four fin towers, with the BB10 providing 18 minutes of μgravity conditions for 100 kg. >800 vehicles of all types have flown since 1962, of which some 280 (BB6/7) were meteorological rockets. The single stage BB5 forms the basis of all currently available versions, with the addition of upper stages and/or US Army surplus boost motors. Combination with Nike/Terrier boosters, respectively, creates the BB8/9 models. NASA continues to be the major user and the vehicle also flies under the CSA μgravity materials research programme. With the closure of the Churchill Research Range in 1984, the Canadian programme was interrupted but one or two missions are now conducted annually from US/European ranges. The BB9 also forms the basis for EER Systems Space Services Division's Starfire vehicle, first launched March 1989.

In addition to the supply of basic rocket vehicle hardware, Bristol provides a range of self-contained vehicle/payload support systems, including thrust termination, parachute recovery, payload separation, de-spin, telemetry systems and fairing deployment. Two launcher types are usually employed: a rail-type typically providing 6 to 8 m of guided rail travel and a 30 m travel tower. In cases where the normal dispersion is unacceptable, the Saab/SES S19 boost guidance system is available; Saab's Spinrac attitude control system provides attitude control for a spinning exoatmospheric stage.

A 16.0 kg and 22.6 cm high igniter housing module incorporating de-spin and payload separation systems caps the stage 1 motor. The forward end provides a separation interface, comprising a deploying manacle ring with associated pyrotechnically actuated release guns and springs to generate 0.5 to 5 m/s separation. A yo-yo system can accommodate the standard range and profiles of despin requirements. The housing also incorporates the deployment logic for each system, including battery power, sequencing timers and safety altitude switches. The payload can provide the latter functions.

A thrust termination system manufactured by Explosive Technology Inc is provided as a complete assembly with the exception of the range authority-supplied command receiver. Two 'paddles' housing the shaped pyrotechnic charge are attached to the motor's forward end, but all other components, including explosive transfer lines, safe/arm assembly, batteries and command receiver, are mounted in the igniter housing. If activated, two holes 180° apart are cut in the motor but the payload is unaffected and events continue as pre-programmed.

Black Brant has demonstrated reliability of over 98 per cent in more than 800 launches to date.

Specifications
Black Brant 5
Description: The single stage BB5, which forms the basis for all BB variants, is available in 3/4-fin versions, designated the 5B/5C, respectively.
First launch: June 1965.

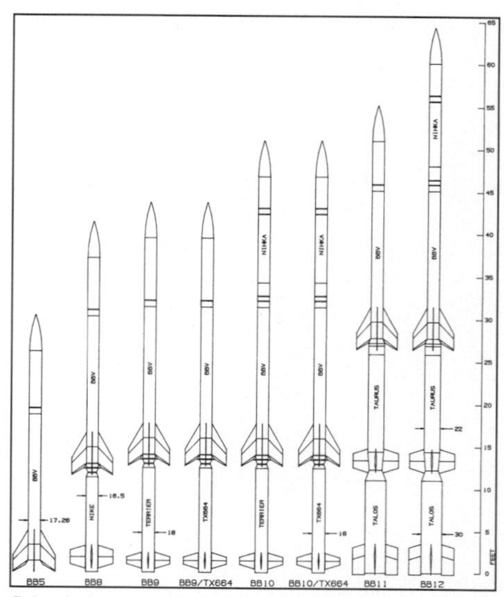

Bristol Aerospace's family of Black Brant rockets. Dimensions are in feet and inches 0517163

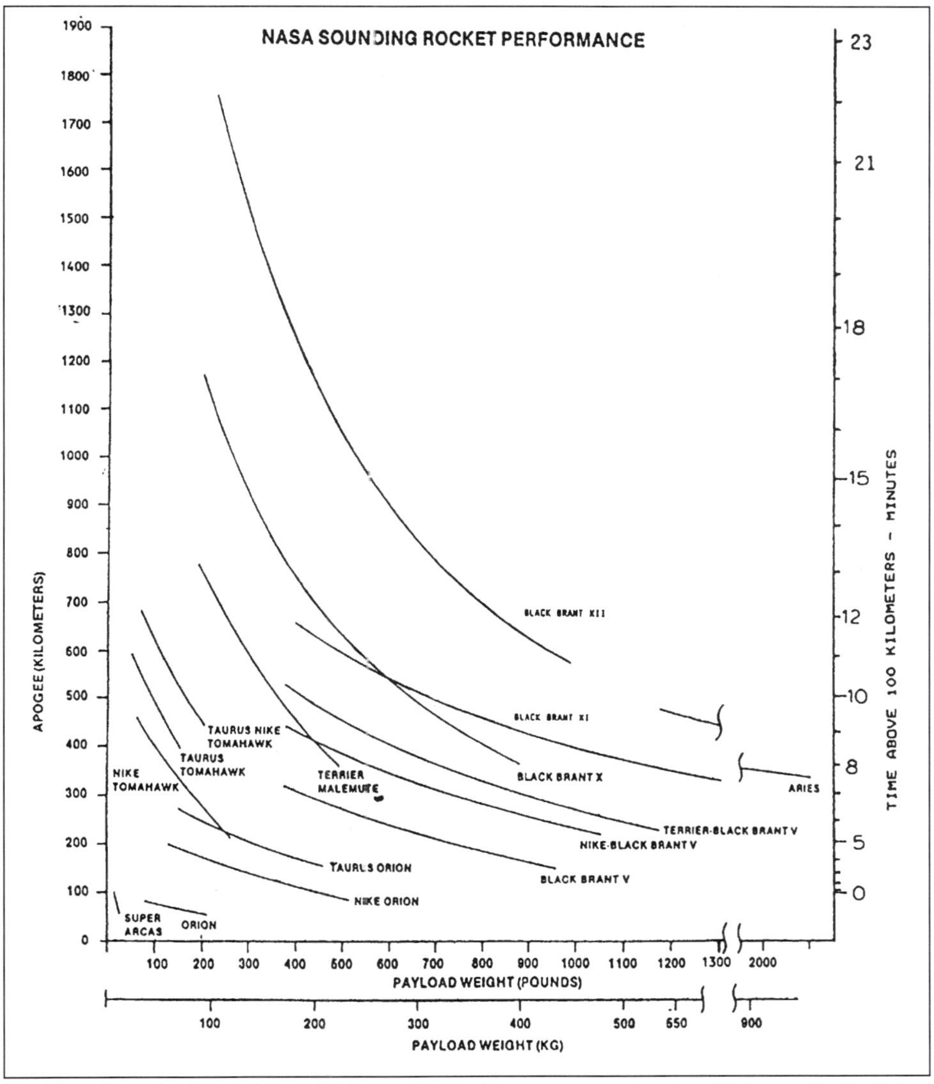

The Black Brant family provides NASA's sounding rocket programme, based at Wallops, with its highest performance vehicles (NASA) 0516864

Number operational launches: 139 to end-1995.
Success rate: 97% to end-1995.
Number of stages: 1 (solid).
Overall length: 5.29 m + payload bay/nosecone.
Principal diameter: 43.8 cm (151.4 cm fin circle diameter).
Launch mass: 1,263 kg + payload/nosecone.
Typical performance: 200 kg to 290 km.
Payload recovery system: two basic (forward/aft-mounted) designs are available. The forward version is housed in a 3:1 ogive stainless steel fairing, the forward section of which is deployed soon after burnout to expose the recovery section heatshield. The aft-mounted recovery system is housed in a cylindrical aluminium bay immediately forward of the payload separation module. Both are self-contained with the necessary deployment logic equipment. The system is automatically armed by the closure of timers and altitude switches on the upleg phase. The heatshield is deployed, beginning the parachute deployment sequence, at about 5 km during re-entry. Both systems are available with either an 8.5 m circular main canopy or a 14 m paraform type. Each recovery system is available in two versions. The forward type comes as a 44.1 kg/1.32 m long unit qualified to 340 kg/320 km apogee or a 50.9 kg/1.38 m long unit qualified to 450 kg/500 km. The aft cylindrical type comes as a 33.2 kg/40 cm long/43.8 cm diameter unit qualified to 340 kg/320 km or a 44.1 kg/52 cm long/44 cm diameter unit qualified to 450 kg/500 km. Two new recovery systems based on the existing forward-mounted hardware have been introduced. The HARS High-Altitude Recovery System extends the recoverable mass/apogee to 725 kg/1,000 km; it flew successfully on the first three Maxus flights, in May 1991 (150 km), November 1992 (718 kg/715 km) and November 1995 (505 kg/706 km). The second system is a water recovery version (WRSA) of the standard 450 kg/500 km system. It uses an integral parachute/flotation system and can accommodate both floating and sinking payloads. First operational flight was made successfully 9 December 1992. Two (successful) flights were made through 1995; no others are yet planned.

Black Brant 5 stage 1
Motor: Bristol Aerospace 26 KS 20000.
Overall length: 5.29 m.
Principal diameter: 43.8 cm.
Stage mass: 1,265 kg (propellant 1,000 kg).
Average thrust: 76.87 kN sea level.
Burn time: 33.0 s.
Payload environment/accommodation: A range of deployable and fixed nose fairings are available in conical (5:1) or ogival (3:1 or 4.25:1) fineness ratios. BB9/10 utilise lightweight 1.5 mm aluminium skins, the others 1.5 mm stainless steel. Deployable fairings are available in split clamshell design with a horizontal (for low spin rates) or vertical (high spin) hinge. Forward-ejecting fairings provide typical separation rates of 5 m/s and a lateral impulse can be provided to prevent vehicle re-contact or ensure a clear field of view for instrumentation.
Peak longitudinal acceleration: 13 *g* /181 kg payload.
Shock: instrument qualification to 100 *g* /11 ms.
Vibration: <10 *g* at any frequency.
Temperature: 50°C maximum.

Black Brant 8
Description: BB8 was developed in 1974 as the first upgrade of the BB5, with the US Army surplus Nike booster providing a 40% apogee performance increase. Otherwise, vehicle configuration is identical.
First launch: December 1975.
Number operational launches: 104 to end-1995.
Success rate: 99% to end-1995.
Number of stages: 2 (solid).
Overall length: 8.995 m + payload/nosecone.
Principal diameter: 43.8 cm.
Launch mass: 1,850 kg + payload.
Typical performance: 200 kg to 400 km.
Payload recovery system: as BB5.

Black Brant 8 stage 1
Motor: Nike M5E1.
Overall length: 370.4 cm.
Principal diameter: 41.9 cm.
Stage mass: 601 kg (propellant 361 kg).
Thrust: 195.6 kN average, sea level.
Burn time: 3.35 s.
Comment: At burnout, the two stages immediately separate by differential drag, augmented by a drag ring at the Nike's forward end. Stage 2 coasts for 5 s before ignition at 8.5 s/2.0 km.

Black Brant 8 stage 2
Configuration: Same as BB5, and the payload accommodation is identical to BB5's. Peak longitudinal acceleration is 10.3 *g* for 435 kg payload. Other payload environment specifications are the same as for BB5.

Black Brant 9
Description: The BB9 adds a US Army surplus Terrier booster to the BB5 to augment apogee performance by 75%. BB9/10 are also available with the Thiokol TX664 booster; 'mod 1' is added to the BB designation. It provides about 30% additional performance than the Terrier but is identical in geometry.
First launch: December 1981.
Number operational launches: 109 (+ 12 mod 1) to end-1995.
Success rate: 99% (+ 92% mod 1) to end-1995.
Number of stages: 2 (solid).
Overall length: 9.56 m + payload/nosecone.
Principal diameter: 43.8 cm (159.6 cm fin circle diameter).
Launch mass: 2,128 kg + payload.
Typical performance: 200 kg to 475 km.
Payload recovery system: as BB5.

Black Brant 9 stage 1
Motor: US Army Terrier Mk 12 mod 1.
Overall length: 4.27 m.
Principal diameter: 45.7 cm.
Stage mass: 878.0 kg (propellant 534 kg).
Thrust: 257.7 kN average, sea level.
Burn time: 4.4 s.
Comment: Staging occurs at burnout using the same interface hardware as BB8. In this case, a 7.5 s coast precedes stage 2 ignition at 11.9 s.

Black Brant 9 stage 2
Configuration: Same as BB5's. Payload accommodation is the same as for BB5. Peak longitudinal acceleration is 13.4 *g* for 431 kg payload. Other payload environment specifications are the same as for BB5.

Black Brant 10
Description: BB10 was introduced in 1981 by adding a new Bristol Aerospace Nihka upper stage to the BB9 configuration. The flight sequence is as for BB9 until stage 2 burnout: the vehicle coasts for 33 s until the stage 2/3 separation mechanism is activated and the finless stage 3 ignites at 82 s/85 km. The payload fairing can be deployed before stage 3 ignition for a 4–8% apogee performance increase, typically at 60 s when dynamic pressure has fallen below 480 N/m^2. The Thiokol TX664 booster replaces the Terrier in the mod 1; see the BB9 for comments.
First launch: 14 August 1981.
Number operational launches: 30 (+ 4 mod 1) to end-1995.
Success rate: 93% (+ 100% mod 1) to end-1995.
Number of stages: 3 (solid).
Overall length: 11.88 m + payload/nosecone.
Principal diameter: 43.8 cm.
Launch mass: 2,560 kg + payload.
Typical performance: 200 kg to 750 km, 400 kg to 400 km.
Payload recovery system: high altitude version of the forward mount system.

Black Brant 10 stage 1
Configuration: Same as BB9 stage 1

Black Brant 10 stage 2
Configuration: Same as for BB5

Black Brant 10 stage 3
Motor: solid propellant Bristol Aerospace Nihka 17 KS 12000, developed specifically for BB10.
Overall length: 1.92 m.
Principal diameter: 43.8 cm.
Stage mass: 402.1 kg (propellant 322 kg).
Thrust: 50.41 kN average, vacuum.
Burn time: 17.8 s.
Attitude control: none
Payload accommodation: Same as for BB5. Peak longitudinal acceleration 22 *g* for 136 kg payload.

Black Brant 12
Description: Bristol Aerospace and NASA Wallops completed studies in 1987 for a vehicle capable of delivering a 140 kg payload to 1,500 km altitude to satisfy plasma physics requirements for investigating ion acceleration over 900–2,000 km. A four-stage vehicle was selected, produced by replacing the BB10's Terrier booster with US Army surplus Talos and Taurus motors. Two demonstration flights were made from Wallops Flight Facility and the first operational flight from Poker Flat (Alaska) in March 1990. On the first demonstration flight in September 1988

the Nihka stage 4 suffered premature loss of thrust. The second vehicle, fired 5 December 1989, qualified the re-design and used a new spin motor to reduce vehicle dispersion. Of the six BB12s flown to end-1995, the one failure was due to Talos explosion at ignition. A BB11 3-stage version, without the Nihka stage 4, is available; its debut launch of 1 February 1990 carried a 547 kg plasma experiment to an altitude of 375 km from Poker Flat.
First launch: 30 September 1988.
Number launched: 2 BB11 + 7 BB12 to end-1995.
Success rate: 100% BB11 + 86% BB12 to end-1995.
Number of stages: 4 (solid).
Overall length: 16.11 m + payload.
Principal diameter: 43.8 cm (upper stages).
Launch mass: 5,253 kg + payload mass.
Typical performance: 140 kg to 1,500 km, 300 kg to 800 km.
Payload recovery system: as BB10

Black Brant 12 stage 1
Motor: US Army Talos.
Overall length: 3.51 m.
Principal diameter: 79.0 cm.
Stage mass: 2,053 kg (propellant 1,285 kg).
Thrust: 489.1 kN average, sea level.
Burn time: 5.42 s.
Separation: drag

Black Brant 12 stage 2
Motor: Taurus.
Overall length: 4.18 m.
Principal diameter: 58.1 cm.
Stage mass: 1361 kg (propellant 755 kg).
Thrust: 490.7 kN average, sea level.
Burn time: 3.30 s.
Separation: spring

Black Brant 12 stage 3/4
Configuration: Same as for BB10 stages 2/3, respectively.

UPDATED

China

Long March

Current Status

The most recent derivative of the Long March Series is the CZ-2F first flown on 19 November 1999. It followed the first flight of the CZ-4B which took place in May 1999. The CZ-2F is a man-rated variant of the CZ-2E, first launched in 1990. The CZ-2F represents the latest development in a programme that began almost a decade before the first flight. Subsequent flights were conducted on 9 January 2001, 25 March 2002, 29 December 2002 and 15 October 2003 when China's first astronaut was successfully launched into orbit at the start of a mission lasting 21 hours 23 minutes. As part of its long range objective China is developing a heavy lift launcher vehicle known as the KT utilising cryogenic high-energy propellants to match capabilities. China is also developing the KT-1, a four-stage solid propellant launch vehicle, designed to compete against US and European launchers for the lightweight and small satellite market.

Background

Rocket propulsion was included as one of the key technologies in the 'Twelve Year Development Plan of Science and Technology' approved in 1958. This plan led to the launches of the first indigenously developed sounding rocket on 19 February 1960, and first successful ballistic missile on 29 June 1964 (failure on the first test 21 March 1962 led to a major redesign). The initial CZ-1 space launcher, derived from the CSS-3 MRBM, utilised nitric acid/UDMH but each subsequent operational and planned stage 1/2 has used N_2O_4 as oxidiser. Development of a LOX/LH_2 cryogenic upper stage for GEO missions began in 1977, leading with its launch as CZ-3 stage 3 in January 1984 to China becoming only the third such user after the US and European Space Agency (ESA). The China Academy of Launch Vehicle Technology (CALT, formerly the Beijing Wan Yuan Industry Corporation) near the town of Nan Yuan 15 km south of the capital develops and builds the cryogenic engines and its Shanxi Liquid Rocket Engine Co the storable engines; SLREC also handles solid motors. The Beijing Rocket Test Centre 50 km southwest of Beijing maintains five major test stands. No 1 handles altitude testing of spacecraft thrusters of up to 490 N, No 2 provides single-engine cryogenic facilities, with No 4 accepting a complete H-8 cryogenic stage, and No 5 is devoted to hydrazine engines. Other test facilities are operated at the launch sites and SLREC probably has stands for the storable engines.

CZ-1/Long March-1
Development of the CZ-1 space launcher from the 2-stage CSS-3 (Chinese designation DF-4, Dong Feng=East Wind) 1,500 km-range IRBM began in October 1965, leading to the launch of China's first satellite in 1970. It was replaced by the more capable CZ-2 following its second space launch but in 1985 it was noted that a model with an uprated third stage would be offered commercially for modest payloads into high inclination LEO orbits. Cost was reported as US$4.1 million, with a 3 year lead time. This CZ-1C variant was subsequently joined by the CZ-1M (Italian IRIS stage 3) and CZ-1D (Chinese solid stage 3).

Specifications

CZ-1 Launcher Family Record
First orbital launch: 24 April 1970
Number launched: 3 (2 orbital)
Launch sites: Jiuquan
Vehicle success rate: 100%

CZ-1 Specifications
Launch site(s) deployed: Jiuquan
Maiden launch: 10 January 1970 (sub-orbital test)
Orbital launch record: 2 launches, 2 successful
Latest launch: 3 March 1971
Number of stages: 3
Overall length: 29.86 m
Diameter: 3.81 m
Launch mass: 81.75 t
Launch thrust: 1,020 kN sea level

CZ-1 stage 1
Engines: 1 × YF-2 liquid bipropellant single start engine
Overall length: 17.835 m
Diameter: 2.25 m
Oxidiser: nitric acid
Fuel: UDMH
Propellant mass: 65.25 t
Dry mass: 4.07 t
Thrust: 1,020 kN sea level
Specific impulse: 2,349 N-s/kg sea level
Burn time: 140 s

CZ-1 stage 2
Engines: 1 × YF-3 liquid bipropellant single start engine
Overall length: 7.486 m
Diameter: 2.25 m
Oxidiser: nitrogen tetroxide
Fuel: UDMH
Propellant mass: 11.21 t
Dry mass: 2.27 t
Thrust: 294.2 kN (vacuum)
Specific impulse: 2,746 N-s/kg (vacuum)
Burn time: 102 s

CZ-1 stage 3
Engines: 1 × GF-02 solid propellant motor
Overall length: 4.565 m
Diameter: 2.25 m
Propellant: solid
Propellant mass: 1.8 t
Dry mass: 0.2 t
Thrust: 111 kN (vacuum)
Specific impulse: 2,472 N-s/kg (vacuum)
Burn time: 40 s

CZ-1 payload shroud
Overall length: 4.63 m
Max diameter: 1.50 m
Mass: 0.27 t

CZ-1C
The Chinese noted in 1985 that a CZ-1 variant with a liquid stage 3 was under development, but it is now not expected to appear. Few specifications were released.
Performance: 400 kg into 70°, 600 km; 400 kg into 400 km Sun-synchronous; 600 kg into 70°, 300 km.

Overall length 33 m; launch mass 86 t.

First stage powered by the YF-2A, four providing a thrust of 1,224 kN. Second stage carried one YF-2A, single engine in third stage.

CZ-1D Specifications
The availability of a CZ-1D variant by 1991 was noted in 1988, exhibiting uprated capability over the original CZ-1 but below that of the CZ-1C. Chinese officials at 1990's Farnborough Air Show noted that the 1D would be ready for a 1991 debut but it would wait for a commercial order - there was no national satellite requiring the vehicle. In Jan 1994 it was again quoted as waiting for a commercial order; cost would be US$10 million. There are some rumours that the launch on 29 May 1995, atributed to a DF-31 missile, was actually a (failed) sub-orbital CZ-1D flight, but the Chinese have not commented on these rumours. The vehicle would carry upgraded electronics and control system (digital and 3-axis platform instead of strapdown systems), derived from the CZ-3/3A.

CZ-1D
Launch site(s) deployed: Xi Chang
Maiden launch: 29 May 1995 (suborbital)
Orbital launch record: no orbital attempts to end of 2002: 3 sub-orbital launches, 1 sub-orbital success, 2 sub-orbital failures
Latest launch: 3 January 2002 (sub-orbital)
Number of stages: 3
Overall length: 28.22 m
Diameter: 3.81 m
Launch mass: 81.075 t
Launch thrust: 1,101.2 kN sea level

CZ-1D stage 1
Engines: 1 × YF-2A liquid bipropellant single start engine
Overall length: 19.735 m
Diameter: 2.25 m
Oxidiser: nitrogen tetroxide
Fuel: UDMH
Propellant mass: 60.86 t
Dry mass: 4.08 t
Thrust: 1,101.2 kN sea level
Specific impulse: 2,378 N-s/kg sea level
Burn time: 132 s

CZ-1D stage 2
Engines: 1 × YF-40 liquid bipropellant single start engine
Overall length: 6.037 m
Diameter: 2.25 m
Oxidiser: nitrogen tetroxide
Fuel: UDMH
Propellant mass: 12.38 t
Dry mass: 1.1 t
Thrust: 100.9 kN (vacuum)
Specific impulse: 2,981.2 N-s/kg (vacuum)
Burn time: 360 s

CZ-1D stage 3
Engines: 1 × FG-36 solid propellant motor
Overall length: 1.696 m
Diameter: 2.054 m
Propellant: solid
Propellant mass: 0.792 t
Dry mass: 0.523 t

CZ-1D payload shroud
Overall length: 4 m
Max diameter: 2.054 m
Mass: 0.25 t

CZ-1M
A CZ-1M variant was first offered in 1987 incorporating the Italian IRIS solid propellant third stage, which would not be available until 1990 at the earliest (first flight made 1992). Performance: 900 kg into 57°, 300 km circular; 830 kg into 70°, 300 km circular; 450 kg into 99°, 903 km Sun-synchronous.

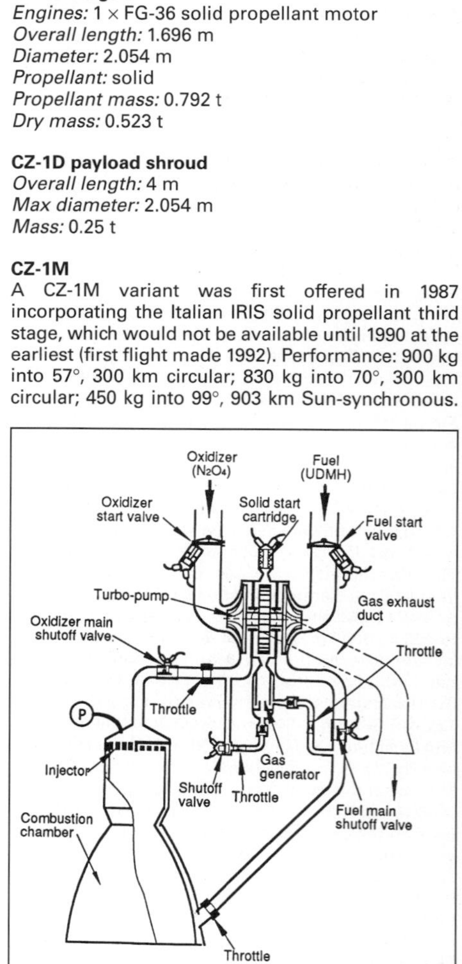

YF-20/21/22 schematic 0516865

CZ-3's H-8 third stage is powered by the 4-chamber cryogenic YF-73 0516867

Overall length 29.6 m; launch mass 85 t. No orbital flights.

CZ-2/Long March-2
Development of a successor to CZ-1 began at BWYIC in 1970 with the parallel development of the CSS-4 (Chinese designation DF-5; first launch September 1971) ICBM and the Shanghai bureau's FB-1 (qv) space launcher. The CZ-2 debut launch in 1974 failed after a few seconds but the CZ-2C variant has since become the most-utilised Chinese launcher. The 2C flew the first recoverable payload mission in November 1975, introducing China as only the third nation with such a capability and permitting reconnaissance and Earth resources imaging. CZ-2 has flown only once and its differences from the 2C are not understood; possibly they are only minor and 2C designates the operational vehicle. It appears that 2C's stage 2 is being stretched for use with a solid upper stage (CPKM) as a GTO launcher. Iridium will also use the stretched version from 1996 for ten launches. The 2E, which introduced a GTO capability into the CA-2 family, includes a stretched stage 1 and four liquid propellant boosters. A stretched 2E is being proposed for the 1996 to 2000 5 year plan to handle a 2-man craft. The CZ-2D appeared unexpectedly in 1992; it is the 2-stage version of CZ-4. Although the designations are no longer used, CZ-2A originally referred to the CZ-4, and CZ-2B became the CZ-3.

CZ-2 Launcher Family Record
First launch: 5 November 1974
Number launcher: 27 (1-CZ-2A, 14-CZ-2C, 3-CZ-2D, 7-CZ-2E, 2-CZ-2F): 2 failure (CZ-2E)
Launch sites: Xichang, Jiuquan
Vehicle success rate: overall 92.6% to end 2001

CZ-2A
Launch site(s) deployed: Jiuquan
Maiden launch: 5 November 1974 (failed to reach orbit)
Orbital launch record: 4 launches, 3 successful, 1 first stage failure
Latest launch: 26 January 1978
Number of stages: 2
Overall length: 31.17 m
Diameter: 3.35 m
Launch mass: 192 t
Full details of this launch vehicle are not known: the CZ-2C was derived from it, and it is believed that the performance details of the CZ-2A and CZ-2C were almost identical.

CZ-2C Specifications
The 2C is offered principally for LEO and Sun-synchronous launches in conjunction with the FSW recoverable μg platform, capable of returning 150 kg payloads to Earth and covered in detail in the Microgravity section. October 1992's FSW launch also carried Sweden's 259 kg Freja piggyback science satellite, at a cost of US$4.3 million. The CPKM version employs a stretch stage 2, as will the 10 Iridium launches planned for 1996 to 2002 (contract signed April 1993). Irdium will include the SD- Smart Dispenser orbit transfer stage to release the two satellites. The main solid and mono liquid thrusters will be controlled by SD's own guidance system.

CZ-2C
Some alternative lengths are given in parentheses.
Launch site(s) deployed: Jiuquan
Maiden launch: 9 September 1982
Orbital launch record: 11 launches, 11 successful
Latest launch: 8 October 1993
Number of stages: 2
Overall length: 31.17 m (33.151 m)
Diameter: 3.35 m
Launch mass: 192 t
Launch thrust: 2,786 kN sea level

CZ-2C stage 1
Designation: L-140
Engines: 4 × YF-20 liquid bipropellant single start engines (overall designator YF-21)
Overall length: 23.72 m
Diameter: 3.35 m
Oxidiser: nitrogen tetroxide in forward tank
Fuel: UDMH in aft tank
Propellant mass: 143 t
Dry mass: 8.6 t
Thrust: 2,786 kN sea level
Specific impulse: 2,540 N-s/kg sea level
Burn time: 130 s

CZ-2C stage 2
Designation: L-35
Engines: 1 × YF-22 liquid bipropellant single start main engine; 1 × YF-23 liquid bipropellant single start vernier engine (overall designator YF-24)
Overall length: 8.706 m
Diameter: 3.35 m
Oxidiser: nitrogen tetroxide in forward tank
Fuel: UDMH in aft tank
Propellant mass: 35 t
Dry mass: 3.2 t
Thrust: 720 kN main engine, 46 kN vernier engine (vacuum)
Specific impulse: 2,834 N-s/kg main engine, 2,762 N-s/kg vernier engine (vacuum)
Burn time: 112 s main engine, 287 s vernier engine

CZ-2C payload shroud
Overall length: 3.144 m (7.125 m)
Max diameter: 2.2 m (3.35 m)

CZ-2C/Smart Dispenser
The CZ-2C/Smart Dispenser took a modified CZ-2C launch vehicle and added a third stage named Smart Dispenser. This vehicle has been used exclusively for the launch of pairs of US Iridium communications satellites (each with a mass of 667 kg).

This launch vehicle variant has not been described in great detail. The Smart Dispenser third stage used solid propellant motors for the orbital manoeuvres, with hydrazine thrusters being used for attitude control. The stage was approximately 3.35 m diameter and 0.8 m long, with a dry mass of 0.25 tonnes.
Launch site(s) deployed: Taiyuan
Maiden launch: 1 September 1997
Orbital launch record: 7 launches, 7 successful
Latest launch: 11 June 1999
Number of stages: 3
Overall length: 39.9 m
Diameter: 3.35 m
Launch mass: 213 t

CZ-2C/CTS
The CZ-2C is a development of the CZ-2C/Smart Dispenser launch vehicle, but no details of the vehicle appear to have been published. The first two stages are believed to be identical with those used on the CZ-2C/Smart Dispenser vehicle and the CTS third stage is believed to be a modified version of the Smart Dispenser. The first launch was of the *Tan Ce* 1 (DoubleStar equatorial satellite) from Xi Chang and the second launch was scheduled for June 2004 from Taiyuan to place *Tan Ce* 2 (DoubleStar polar satellite) into orbit.
Launch site(s) deployed: Taiyuan, Xi Chang
Maiden launch: 29 December 2003
Orbital launch record: 1 launch, 1 successful
Latest launch: 29 December 2003
Number of stages: 3

CZ-2D
Chinese statements during 1989 indicated that a 2D version would remain only as a design study. It would have incorporated a stage 2 stretched by about 2 m and McDonnell Douglas' PAM-D solid propellant spinning upper stage (under an agreement signed in December 1987) to create a modest 1,250 kg GTO capability. In fact, western observers were surprised by the first CZ-2D launch in 1992 carrying the first FSW-2 return craft. It was disclosed in 1993 to be merely the 2-stage version of the CZ-4, possibly intended only for FSW-2 platforms. Development cost was quoted as 40 per cent of one launch; the launch cost is the same as CZ-2C's, thus reducing the cost per unit mass. A perigee kick stage has been suggested; employing a single YF-40, it could deliver 1,850 kg into GTO.
Launch site(s) deployed: Jiuquan
Maiden launch: 9 August 1992
Orbital launch record: 3 launches, 3 successful
Latest launch: 20 October 1996
Number of stages: 2
Overall length: 33.667 m
Diameter: 3.35 m
Launch mass: 236.966 t
Launch thrust: 2,961.6 kN sea level

CZ-2D stage 1
Engines: YF-21B liquid bipropellant single start engine
Overall length: 24.66 m
Diameter: 3.35 m
Oxidiser: nitrogen tetroxide in forward tank
Fuel: UDMH in aft tank
Propellant mass: 183.038 t
Dry mass: 9.82 t
Thrust: 2,961.6 kN sea level
Specific impulse: 2,550 N-s/kg sea level
Burn time: 154.2 s

CZ-2D stage 2
Engines: 1 × YF-22B liquid bipropellant single start main engine; 1 × YF-23B liquid bipropellant single start vernier engine
Overall length: 9.007 m
Diameter: 3.35 m
Oxidiser: nitrogen tetroxide in forward tank
Fuel: UDMH in aft tank
Propellant mass: 34.736 t
Dry mass: 3.122 t
Thrust: 719.8 kN main engine, 46.1 kN vernier engine (vacuum)
Specific impulse: 2,822 N-s/kg main engine, 2,762 N-2/kg vernier engine (vacuum)
Burn time: 113.48 s main engine, 274.42 s vernier engine

CZ-2D payload shroud
No payload shroud was carried on the three launches of FSW-2 satellites.

Uprated CZ-2D
Launch site(s) deployed: Jiuquan
Maiden launch: 3 November 2003
Orbital launch record: 1 launch, 1 successful
Latest launch: 3 November 2003
Number of stages: 2
No details of this launch vehicle have been published by the Chinese. When the debut launch took place it was noted that the second stage was longer compared with the previous CZ-2D vehicles and that the first stage had four fins added to the base.

CZ-2E Launch Record
CZ-2E/PKM:
First launch: 16 July 1990
Number launched: 1 (third stage failure)
Launch site: Xichang
CZ-2E/STAR-63F:
First launch: 12 August 1992
Number launched: 4 (2 failures, possibly of payloads)
Launch site: Xichang
CZ-2E/EPKM:
First launch: 25 November 1995
Number launched: 2 (0 failures)
Launch site: Xichang

CZ-2E Specifications
The 2E is the most powerful of the CZ-2 range, beginning commercial operations in August 1992 with Australia's Optus B1 as the first

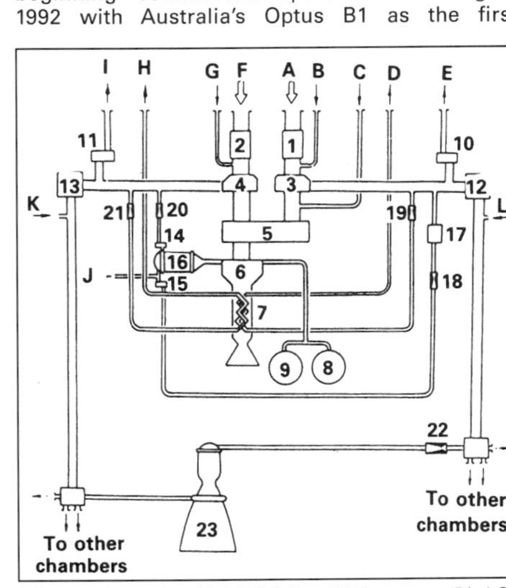

YF-73 schematic. A) LOX inlet, B/C) purge, D) LOX tank pressurisation, E) bleed, F) liquid hydrogen inlet, G) purge, H) LH2 pressurisation, I) bleed, J–L) purge; 1/2) isolation valves, 3) LOX pump, 4) LH2 pump, 5) gearbox, 6) turbine, 7) heat exchanger, 8/9) start tank, 10/11) bleed valve, 12/13) main valve, 14/15) gas generator valve, 16) gas generator, 17) pressure stabiliser, 18–22) venturi, 23) combustion chamber (one of four) 0516868

Lift-off of LM-2E from Xichang launch site (Theo Pirard) 0054375

payload. The agreement was signed with Hughes Communications Inc on 1 November 1988, believed to be at a promotional US$15 million per satellite. A demonstration launch was made 16 July 1990 as part of the contract, carrying a mass dummy + PKM and Pakistan's 50 kg Badr 1 satellite at a charge of US$300,000. It was subsequently reported that stage 2 did not provide the required orientation for dummy release because of a wiring fault; modifications were made. The PKM was intended to boost the dummy to 340 × 25,740 km, 24°, APStar 2 failed to attain orbit in January 1995; high Winter winds were blamed, causing either premature fairing opening or failure of the adapter between APStar and CZ. The failure of Optus-B2 and APStar 2 have been the subject of much speculation and they are discussed in China's national section. 1998's Globalstar launch requires development of a dispenser. 2E might be stretched to carry a 2-person capsule from Jiuquan if the project is accepted as part of the 1996 to 2000 5 year plan. A liquid PKM has been suggested to increase GTO capacity to 4,650 kg, employing two YF-40 engines.

A CZ-2-4L variant was initially described (the 4L indicating four liquid propellant boosters) but the 2E was exhibited at the 1987 Paris Air Show and subsequently confirmed as the new designation. Vehicle characteristics have been contradictory but it appears that the YF-20 engines have been uprated from 696 kN to 740 kN, stage 1 stretched from that of the CZ-2C, and a new stage 2 incorporated. Total launch thrust is 5,923 kN, against the 5,570 kN specified earlier. Solid propellant strap-ons have been mentioned as a possibility for the future. Note that LEO performance is comparable to that of the CZ-3A. A CZ-2E/HO version adding the 3A's large cryogenic H18 upper stage was expected to appear in about 1995 to provide a 4.5 t GTO capability, but information that emerged in 1992 identified CZ-3B as a 3A using 2E's four strap-ons. HEXI Corp's 1.70 m Ø EPKM perigee kick stage was first used on AsiaSat 2 and then EchoStar 1. An early version was carried on the first 2E.

The Chinese are proposing the introduction of a CZ-2E/TS vehicle which will add a new solid-propellant Third Stage (TS) to the basic CZ-2E vehicle. The TS will include control avionics and a multiple payload dispenser which can carry up to twelve satellites similar to GlobalStar. The target orbit is 52°, 900 km circular altitude, to which a payload capability of 5,760 kg is claimed. Launches would be made from Jiuquan (from where the CZ-2E has yet to fly) and payloads could also be flown to polar orbits.

Note: the basic CZ-2E is a two-stage launch vehicle with four strap-on boosters: it is designed to fly with various third stages to allow payload deployments in geosynchronous Earth orbit.
Launch site(s) deployed: Xi Chang
Maiden launch: 16 July 1990
Orbital launch record: with PKM third stage: 1 launch, 1 third stage failure; with STAR-63F third stage: 4 launches, 2 successful, 2 satellite explosions; with *EPKM third stage:* 2 launches, 2 successful
Latest launch: with PKM – 16 July 1990; with STAR-63F – 25 January 1995; with EPKM – 28 December 1995
Number of stages: 3 + 4 strap-on boosters
Overall length: 49.686 m
Diameter: 11.45 m
Launch mass: 462.46 t
Launch thrust: 5,923.2 kN sea level

CZ-2E strap-on booster
Number of strap-ons: 4
Designation: LB-40
Engines: 1 × YF-20B liquid bipropellant single start engine
Overall length: 15.326 m
Diameter: 2.25 m
Oxidiser: nitrogen tetroxide in forward tank
Fuel: UDMH in aft tank
Propellant mass: 37.754 t
Dry mass: 3 t
Thrust: 740.4 kN sea level
Specific impulse: 2,556.2 N-s/kg sea level
Burn time: 127.26 s

CZ-2E stage 1 (core)
Designation: L-180
Engines: 4 × YF-20B liquid bipropellant single start engine (overall designator YF-21B)
Overall length: 28.465 m
Diameter: 3.35 m
Oxidiser: nitrogen tetroxide in forward tank
Fuel: UDMH in aft tank
Propellant mass: 186.28 t
Dry mass: 12.55 t
Thrust: 2,961.6 kN sea level
Specific impulse: 2,556.2 N-s/kg sea level
Burn time: 160.43 s

CZ-2E stage 2
Designation: L-90
Engines: 1 × YF-22B liquid bipropellant single start main engine; 1 × YF-23B liquid bipropellant single start vernier engine (overall designator YF-24B)
Overall length: 14.223 m
Diameter: 3.35 m
Oxidiser: nitrogen tetroxide in forward tank
Fuel: UDMH in aft tank
Propellant mass: 84.759 t
Dry mass: 4.955 t
Thrust: 738.4 kN main engine, 47.1 kN vernier engine (vacuum)
Specific impulse: 2,922.4 N-s/kg main engine, 2834.1 N-s/kg vernier engine (vacuum)
Burn time: 301.18 s main engine, 414.68 s vernier engine

CZ-2E stage 3
Designation: EPKM
Engines: [EPKM]
Overall length: 3.62 m
Diameter: 1.7 m
Propellant: solid
Propellant mass: 5.4 t
Dry mass: 0.685 t
Specific impulse: 2,863.3 N-s/kg (vacuum)
Burn time: 70 s

CZ-2E payload shroud
Overall length: 10.5 m
Max diameter: 4.2 m
Mass: 1.9 t

CZ-2E(A)
During the mid-1980s the Chinese revealed a family of launch vehicles which they planned to introduce, and one-by-one those vehicles have been introduced but with some re-design and under new designators. In early 1998 some details appeared of a variant of the CZ-2E launch vehicle which had four longer strap-on boosters than are used on the 2E and 3B vehicles: the strap-ons on the 2E(A) were almost as long as the first stage core. Each strap-on carried two YF-20B class engines, and thus the 2E(A) was equivalent to the 1980s CZ-2/8L variant which had eight strap-on boosters. Some basic numerical details were given of the 2E(A) in September 1998: launch mass - 695 tonnes, launch thrust – 906 tonnes [=8,884 kN], length – 53.628 m. A payload capability of 14.1 tonnes to a 48 deg, 200 km orbit was quoted.

Some alternative and additional data for the vehicle were quoted in 2000 and are given below. This launch vehicle variant is no longer discussed by the Chinese and with the development of the new CZ-5 family of launch vehicles it seems likely that this variant has now been cancelled.
Launch site(s) deployed: Jiuquan
Number of stages: 2 + 4 long strap-on boosters
Overall length: 50 m
Diameter: 3.35 m core, 2.25 m strap-on booster
Launch mass: 670 t
Launch thrust: 8,886 kN
Payload capability: 11.8 t to 48°, 350 km
Fairing diameter: 4.2 m and 5.2 m

CZ-2F
The CZ-2F has not been described in much detail, although some overall figures have been given. It is the modification of the CZ-2E which is used in the piloted programme to launch *Shen Zhou* spacecraft. The performance has been increased compared with the CZ-2E, with an LEO payload capability of about 9.5 tonnes if the heavy man-related payload shroud and tower are not carried. Avionics and computer systems have been improved compared with the CZ-2E.
Launch site(s) deployed: Jiuquan
Maiden launch: 19 November 1999
Orbital launch record: 5 launches, 5 successful
Latest launch: 15 October 2003
Number of stages: 2 + 4 strap-on boosters
Overall length: 58.34 m (including tower atop the payload shroud)
Diameter: 11.45 m
Launch mass: 479.7 t

CZ-2F payload shroud
Overall length: 15.1 m (including tower)
Max diameter: 3.8 m
Mass: 11.26 t

CZ-3/Long March-3
Development of a LOX/LH_2 stage 3 for uprating the CZ-2 into a GEO-class vehicle began in 1977, following feasibility studies into such a vehicle starting in 1975 (the vehicle was then referred to as CZ-2B). January 1984's CZ-3 debut established China as only the third user of cryogenic propulsion, after the US and European Space Agency (ESA), but on that occasion the stage shut down shortly after re-ignition for GTO injection (which also happened on the 8th mission, December 1991). The second launch 3 months later produced the first Chinese GEO satellite. Departures are made from Xichang, which seems to have been established specifically for GTO/GEO operations. In August 1996 the third stage of the eleventh CZ-3 shut down early, leaving Zhongxing 7 in a lower orbit than intended. CZ-3A doubles the GTO capability of the parent model by stretching stage 1 and incorporating a significantly larger cryogenic stage 3. A 3A-4L variant with four liquid propellant strap-ons as used on the CZ-2E was discussed, but for a long time it appeared that instead the 3A's cryogenic H18 stage would be added to the 2E to provide a 4.5 t GTO capability. It emerged in 1992, however, that the 3A-4L (called the 3B) was intended; Intelsat 708 was lost in the February 1996 debut failure when the inertial platform veered after 2 s.

Long March family (left to right): LM-3B, LM-2E, LM-3, LM-4A, LM-2D, LM-2C and LM-1D 0084624

CZ-3
Launch site(s) deployed: Xi Chang
Maiden launch: 29 January 1984
Orbital launch record: 13 launches, 10 successful, 3 third stage failures
Latest launch: 25 June 2000 (retired)
Number of stages: 3
Overall length: 44.86 m
Diameter: 3.35 m
Launch mass: 204.88 t
Launch thrust: 2,961.6 kN sea level

CZ-3 stage 1
Designation: L-140
Engines: 2 × YF-20B liquid bipropellant single start engine (overall designator YF-21B)
Overall length: 23.49 m
Diameter: 3.35 m
Oxidiser: nitrogen tetroxide in forward tank
Fuel: UDMH in aft tank
Propellant mass: 143.692 t
Dry mass: 9.378 t
Thrust: 2,961.6 kN sea level
Specific impulse: 2,556.2 N-s/kg sea level
Burn time: 121 s

CZ-3 stage 2
Designation: L-35
Engines: 1 × DaFY-24D liquid bipropellant single start main engine; 1 × YF-23F liquid bipropellant single start vernier engine
Overall length: 9.47 m
Diameter: 3.35 m
Oxidiser: nitrogen tetroxide in forward tank
Fuel: UDMH in aft tank
Propellant mass: 35.841 t
Dry mass: 3.599 t
Thrust: 742 kN main engine, 46 kN vernier engine (vacuum)
Specific impulse: 2,922.4 N-s/kg main engine, 2,762 N-s/kg vernier engine (vacuum)
Burn time: 130 s main engine, 135 s vernier engine

CZ-3 third stage
Designation: H-8
Engines: 1 × YF-73 liquid bipropellant multiple start engine
Overall length: 10.634 m
Diameter: 2.25 m
Oxidiser: liquid oxygen in aft tank
Fuel: liquid hydrogen in forward tank
Propellant mass: 8.731 t
Dry mass: 1.965 t
Thrust: 44.43 kN (vacuum)
Specific impulse: 4,119 (vacuum)
Burn time: 729 s (total)

CZ-3 payload shroud
Alternative shroud data are in parentheses.
Overall length: 6.54 m (7.106 m)
Max diameter: 1.7 m (2.6 m)
Mass: 0.45 t (0.603 t)

CZ-3A Specification
Announced in 1988 as available commercially from 1992, Chinese officials noted in late 1990 that its introduction had been delayed until 1994. The CZ-3A incorporates a stretched/uprated stage 1 (also utilised by the CZ-4) and the H-18 cryogenic stage 3, about doubling the basic model's GTO capability.

CZ-3A
Launch site(s) deployed: Xi Chang
Maiden launch: 8 February 1994
Orbital launch record: 8 launches, 8 successful
Latest launch: 14 November 2003
Number of stages: 3
Overall length: 52.52 m
Diameter: 3.35 m
Launch mass: 241 t
Launch thrust: 2,961.6 kN sea level

CZ-3A stage 1
Designation: L-180
Engines: 4 × YF-20B liquid bipropellant single start engines (overall designator YF21B, also called DaFY6-2)
Overall length: 26.972 m
Diameter: 3.35 m
Oxidiser: nitrogen tetroxide in forward tank
Fuel: UDMH in aft tank
Propellant mass: 172.05 t
Dry mass: 11.23 t
Thrust: 2,961.6 kN sea level
Specific impulse: 2556.2 N-s/kg sea level
Burn time: 146 s

CZ-3A stage 2
Designation: L-35
Engines: 1 × DaFY21-1 liquid bipropellant single start main engine; 1 × YF-23B liquid bipropellant single start vernier engine (overall designator is YF-24B)
Overall length: 11.276 m
Diameter: 3.35 m
Oxidiser: nitrogen tetroxide in forward tank
Fuel: UDMH in aft tank
Propellant mass: 30.555 t
Dry mass: 3.561 t
Thrust: 742 kN main engine, 47 vernier engine (vacuum)
Specific impulse: 2,922.4 N-s/kg main engine, 2,834 N-s/kg vernier engine (vacuum)
Burn time: 114 s main engine, 119 s vernier engine

CZ-3A stage 3
Designation: H-18
Engines: 2 × YF-75 liquid bipropellant multiple start engines
Overall length: 12.375 m
Diameter: 3 m
Oxidiser: liquid oxygen in aft tank
Fuel: liquid hydrogen in forward tank
Propellant mass: 18.193 t
Dry mass: 2.742 t
Thrust: 156.9 kN (vacuum)
Specific impulse: 4.315 N-s/kg (vacuum)
Burn time: 469 s

CZ-3A payload shroud
Overall length: 8.887 m
Max diameter: 3.35 m
Mass: 0.5 t

CZ-3B and CZ-3C
It appeared that a 4.5 t GTO capability would be created by adding the 3A's large cryogenic H18 upper stage to 2E. This configuration was referred to as the CZ-2E/HO. Information emerged in 1992 that indicated the 3B would be formed by adding 2E's four stap-ons to the CZ-3A, providing a 4.8 t GTO capacity from 1995. This creates the most powerful vehicle in the Chinese inventory, equivalent to Russia's Proton and the top of Europe's Ariane 4 range. The February 1996 debut failed when the inertial platform veered after 2 s; the fault was determined to be caused by poor workmanship. CZ-3C is identical to the 3B except that only two strap-ons will be utilised.

CZ-3B
Launch site(s) deployed: Xi Chang
Maiden launch: 14 February 1996
Orbital launch record: 5 launches, 4 successful, 1 first stage failure
Latest launch: 18 July 1998
Number of stages: 3 + 4 strap-on boosters
Overall length: 54.838 m
Diameter: 8.45 m
Launch mass: 426 t
Launch thrust: 5,923.2 kN sea level

CZ-3B strap-on booster
Number of strap-ons: 4
Designation: LB-40
Engines: 1 × DaFY5-1 liquid bipropellant single start engine

Preparation for launch of LM-4 at Xixhang launch site 0084634

Overall length: 15.326 m
Diameter: 2.25 m
Oxidiser: nitrogen tetroxide in forward tank
Fuel: UDMH in aft tank
Propellant mass: 37.75 t
Dry mass: 2.8 t
Thrust: 740.4 kN sea level
Specific impulse: 2,556.2 N-s/kg sea level
Burn time: 125 s

CZ-3B stage 1 (core)
Designation: L-180
Engines: 4 × YF-20B liquid bipropellant single start engines (overall designator YF21B, also called DaFY6-2)
Overall length: 23.272 m
Diameter: 3.35 m
Oxidiser: nitrogen tetroxide in forward tank
Fuel: UDMH in aft tank
Propellant mass: 171.8 t
Dry mass: 12.12 t
Thrust: 2,961.6 kN sea level
Specific impulse: 2,556.2 N-s/kg sea level
Burn time: 146 s

CZ-3B stage 2
Designation: L-35
Engines: 1 × DaFY21-1 liquid bipropellant single start main engine; 1 × YF-23B liquid bipropellant single start vernier engine (overall designator is YF-24B)
Overall length: 9.943 m
Diameter: 3.35 m
Oxidiser: nitrogen tetroxide in forward tank
Fuel: UDMH in aft tank
Propellant mass: 45.752 t
Dry mass: 3.848 t
Thrust: 742 kN main engine, 47 vernier engine (vacuum)
Specific impulse: 2,922.4 N-s/kg main engine, 2,834 N-s/kg vernier engine (vacuum)
Burn time: 178 s main engine, 184 s vernier engine

CZ-3B stage 3
Designation: H-18
Engines: 2 × YF-75 liquid bipropellant multiple start engines
Overall length: 12.375 m
Diameter: 3 m
Oxidiser: liquid oxygen in aft tank
Fuel: liquid hydrogen in forward tank
Propellant mass: 18.242 t
Dry mass: 3.062 t
Thrust: 156.9 kN (vacuum)
Specific impulse: 4.315 N-s/kg (vacuum)
Burn time: 478 s

CZ-3B payload shroud
Overall length: 9.561 m
Max diameter: 4 m
Mass: 1.5 t

CZ-3B(A)
Just as the CZ-3B can be considered to be a three-stage variant of the CZ-2E, the CZ-3B(A) can be considered to be a three-stage variant of the CZ-2E(A), adding a third LOX/LH stage to this vehicle.

No performance details appear to have been released for the 3B(A) vehicle.

Since it is unlikely that the CZ-2E(A) development (*qv*) is going ahead, it seems unlikely that the CZ-3B(A) will reach flight status either.

CZ-3C
Launch site(s) deployed: Xi Chang
Maiden launch: (No launches to the end of 2002)
Number of stages: 3 + 2 strap-on boosters
Overall length: 54.838 m
Diameter: 8.45 m
Launch mass: 345 t
Launch thrust: 4,442.4 kN sea level

CZ-3C strap-on booster
Number of strap-ons: 2
Designation: LB-40
Engines: 1 × DaFY5-1 liquid bipropellant single start engine
Overall length: 15.326 m
Diameter: 2.25 m
Oxidiser: nitrogen tetroxide in forward tank
Fuel: UDMH in aft tank
Propellant mass: 37.75 t
Dry mass: 2.8 t
Thrust: 740.4 kN sea level
Specific impulse: 2,556.2 N-s/kg sea level
Burn time: 125 s

CZ-3C stage 1 (core)
Designation: L-180
Engines: 4 × YF-20B liquid bipropellant single start engines (overall designator YF21B, also called DaFY6-2)
Overall length: 23.272 m
Diameter: 3.35 m
Oxidiser: nitrogen tetroxide in forward tank
Fuel: UDMH in aft tank
Propellant mass: 171.8 t
Dry mass: 12.12 t
Thrust: 2,961.6 kN sea level
Specific impulse: 2,556.2 N-s/kg sea level
Burn time: 146 s

CZ-3C stage 2
Designation: L-35
Engines: 1 × DaFY21-1 liquid bipropellant single start main engine; 1 × YF-23B liquid bipropellant single start vernier engine (overall designator is YF-24B)
Overall length: 9.943 m
Diameter: 3.35 m
Oxidiser: nitrogen tetroxide in forward tank
Fuel: UDMH in aft tank
Propellant mass: 45.752 t
Dry mass: 3.848 t
Thrust: 742 kN main engine, 47 vernier engine (vacuum)
Specific impulse: 2,922.4 N-s/kg main engine, 2,834 N-s/kg vernier engine (vacuum)
Burn time: 178 s main engine, 184 s vernier engine

CZ-3C stage 3
Designation: H-18
Engines: 2 × YF-75 liquid bipropellant multiple start engines
Overall length: 12.375 m
Diameter: 3 m
Oxidiser: liquid oxygen in aft tank
Fuel: liquid hydrogen in forward tank
Propellant mass: 18.242 t
Dry mass: 3.062 t
Thrust: 156.9 kN (vacuum)
Specific impulse: 4.315 N-s/kg (vacuum)
Burn time: 478 s

CZ-3C payload shroud
Overall length: 9.561 m
Max diameter: 4 m
Mass: 1.5 t

CZ-4/Long March-4
Built by the Shanghai bureau, the CZ-4 (probably based upon a design designated CZ-2B) utilises the first two stage of the CZ-3A with a new storable propellant upper stage optimised for Sun-synchronous payloads. The first two flights carried FY-1 metsats. It has been claimed that the vehicle was originally designed in the late 1970s as a carrier for GEO telecom satellites, but that its inadequate performance meant that it was cancelled and the CZ-3 with a cryogenic third stage was developed instead. CZ-2D is the 2-stage version. Preliminary details were released in 1994 of the CZ-4B, which appears to employ an uprated stage 3 with a restart capability. A GTO capability was also claimed. The addition of solid strap-on boosters has been mentioned as a possibility. Carrying six 7 m long, 1.4 m Ø, 559 kN average thrust, 235 s SI, 66 s burn time would provide 5,700 kg into 200 × 400 km polar. Eight would yield 6,300 kg.

CZ-4A
Launch site(s) deployed: Taiyuan
Maiden launch: 6 September 1988
Orbital launch record: 2 launches, 2 successful
Latest launch: 3 September 1990
Number of stages: 3
Overall length: 41.901 m
Diameter: 3.35 m
Launch mass: 241.092 t
Launch thrust: 2,942 kN sea level

CZ-4A stage 1
Designation: L-180
Engines: 4 × YF-20B liquid bipropellant single start engines (overall designator YF-21B)
Overall length: 24.65 m
Diameter: 3.35 m
Oxidiser: nitrogen tetroxide in forward tank
Fuel: UDMH in aft tank
Propellant mass: 177.43 t
Dry mass: 9.83 t
Thrust: 2,942 kN sea level
Specific impulse: 2,550 N-s/kg sea level
Burn time: 150.6 s

CZ-4A stage 2
Designation: L-35
Engines: 1 × YF-22 liquid bipropellant single start engine; 1 × YF-23F liquid bipropellant vernier single start engine
Overall length: 10.4 m
Diameter: 3.35 m
Oxidiser: nitrogen tetroxide in forward tank
Fuel: UDMH in aft tank
Propellant mass: 35.5 t
Dry mass: 3.08 t
Thrust: 719.8 kN main engine, 46.1 kN vernier engine (vacuum)
Specific impulse: 2,835 N-s/kg main engine, 2,742 N-s/kg vernier engine (vacuum)
Burn time: 126.8 s main engine, 136.8 s vernier engine

Long March family of launchers: (left to right) LM-3B, LM-2E, LM-3A, LM-2C, LM-4 (Theo Pirard) 0054376

CZ-4A stage 3
Designation: L-14
Engines: 2 × YF-40 liquid bipropellant single start engine
Overall length: 4.954 m
Diameter: 2.9 m
Oxidiser: nitrogen tetroxide in forward tank
Fuel: UDMH in aft tank
Propellant mass: 11.34 t
Dry mass: 1.8 t
Thrust: 98 kN (vacuum)
Specific impulse: 2,942 N-s/kg (vacuum)
Burn time: 329.8 s

CZ-4A payload shroud
Overall length: 4.908 m
Max diameter: 2.9 m
Mass: 0.612 t

CZ-4B
Launch site(s) deployed: Taiyuan
Maiden launch: 10 May 1999
Orbital launch record: 6 launches, 6 successful
Latest launch: 21 October 2003
Number of stages: 3
Overall length: 45.576 m
Diameter: 3.35 m
Launch mass: 248.47 t
Launch thrust: 2,971 kN sea level

CZ-4B stage 1
Designation: L-180
Engines: 4 × YF-20B liquid bipropellant single start engines (overall designator YF-21B)
Overall length: 24.65 m
Diameter: 3.35 m
Oxidiser: nitrogen tetroxide in forward tank
Fuel: UDMH in aft tank
Propellant mass: 183.339 t
Dry mass: 9.998 t
Thrust: 2,971 kN sea level
Specific impulse: 2,550 N-s/kg sea level
Burn time: 154 s

CZ-4B stage 2
Designation: L-35
Engines: 1 × YF-22 liquid bipropellant single start engine; 1 × YF-23F liquid bipropellant vernier single start engine
Overall length: 10.4 m
Diameter: 3.35 m
Oxidiser: nitrogen tetroxide in forward tank
Fuel: UDMH in aft tank
Propellant mass: 35.394 t
Dry mass: 2.932 t
Thrust: 742 kN main engine, 47.1 kN vernier engine (vacuum)
Specific impulse: 2,922.4 N-s/kg main engine, 2,834 N-s/kg vernier engine (vacuum)
Burn time: 126.8 s main engine, 136.8 s vernier engine

CZ-4B stage 3
Designation: L-14
Engines: 2 × YF-40 liquid bipropellant multiple start engine
Overall length: 4.932 m
Diameter: 2.9 m
Oxidiser: nitrogen tetroxide in forward tank
Fuel: UDMH in aft tank
Propellant mass: 12.814 t
Dry mass: 1.727 t
Thrust: 101.03 kN (vacuum)
Specific impulse: 2,971.4 N-s/kg (vacuum)
Burn time: 359.74 seconds

CZ-4B payload shroud
Overall length: 8.483 m
Max diameter: 3.35 m
Mass: 1.35 t

CZ-5 Family
In 1992 a paper presented at the International Astronautical Federation Congress by Xiandong Bao discussed a possible new family of launch vehicles, built around the development of a few standard rocket stages which could be clustered in different ways. The rocket stages would use liquid oxygen/kerosene (for the first time in the Chinese space programme) lower stages and liquid oxygen/ liquid hydrogen upper stages.

The smallest of these vehicles – a two-stage vehicle with the stages in tandem – would be capable of placing ~11 tonnes into a 60°, 300 to 500 km orbit. The most powerful of these vehicles- using six strap-on boosters, each almost identical with the first stage core – would place ~70 tonnes into the same orbit as the smallest vehicle.

In retrospect, it seems likely that the 1992 paper was one of a series of design concepts underway in China with a view to developing a new family of launch vehicles.

As part of the programme to develop LOX/ kerosene engines the Chinese wanted to purchase an RD-170/171 class engine from Russia (the engine used on the strap-on boosters of the Energiya launch vehicle and the first stage of the Zenit-2 launcher), but the Russians refused to sell one of these 7,900 kN thrust engines. Instead the Chinese purchased three RD-120 class LOX/kerosene engines (thrust 835 kN), which are used on the second stage of the Zenit-2 vehicle.

In June 2000 the Chinese announced that they had successfully test-fired their first LOX/kerosene engine at the Fengzhou Test Center. This was described as being "a major milestone in China's rocket engine development programme which paves the way in the new-generation, large thrust, non-pollution space launch vehicle".

A paper presented at the International Astronautical Federation Congress in 2000 briefly mentioned the new LOX/kerosene launch vehicles which were under development, stating that: "A new expendable launch vehicle using non-poisonous propellants is being studied. It is a one-stage launch vehicle with four [strap-on] boosters, which is used to send satellites into LEO. When it is used to launch GTO satellites a second stage will be added."

These new launch vehicles have been identified as being the largest elements in the CZ-5 programme. The vehicles would be about 50–55 m high and have launch masses of about 800 tonnes. The strap-on boosters and the central core stage would use LOX/ kerosene, while the third stage used for GTO – and probably Earth escape – missions would use LOX/LH. The payload capacity to LEO was given as ~23 tonnes and the GTO payload would be ~11 tonnes.

Many more details concerning the CZ-5 plans were revealed by the Chinese in a journal article in February 2003 and at the International Astronautical Federation Congress held in Bremen during September to October 2003. The CZ-5 programme comprises the development of three classes of launch vehicles, based upon core stages having diameters of 5 m, 3.35 m and 2.25 m. Modifications of the 3.35 m and 2.25 m first stage

Standard Rocket Modules for the CZ-5 Launch Vehicles

Stage designator	K2-1	K3-1	H5-1	H5-2
Oxidiser	LOX	LOX	LOX	LOX
Fuel	Kerosene	Kerosene	LH2	LH2
Number of engines	1	2	2	1
Engine thrust, tonnes (sl)	122	122	55	
Engine thrust, tonnes (vac)	137	137	67	16.3
Specific impulse, sec (sl)	300	300	441	
Specific impulse, sec (vac)	335	335	426	438
Diameter, m	2.25	3.35	5.0	5.0
Length, m	25.0	26.3	31.0	
Propellant mass, tonnes	63.0	135.0	158.0	22.9

cores are planned for use as strap-on boosters for the vehicles using core states of 5 m and 3.35 m.

The maiden flight of the CZ-5 vehicle is expected around 2005, although the Chinese have not announced a definitive schedule for the vehicle's development.

The standard first stage modules planned for development are summarised in the accompanying Table.

One Chinese paper from the 2003 IAF Congress gives some slightly different data for the first stages: K2-1 – length 23.277 m, lift-off mass 64 tonnes, K3-1 – length 24.6 m, lift-off mass 144 tonnes and H5-1 – length 29.6 m, lift-off mass 173.5 tonnes. The difference in lengths could be because the shorter figures exclude either the protruding rocket engine chambers or assemblies atop the stage, while the mass differences could be that the lower figures are truly at lift-off, and exclude propellant burned on the launch pad as the engine thrusts are being built up.

H5-2

The H5-2 stage is a modification of the CZ-3B third stage and is flown for CZ-5 missions beyond low Earth orbits: it has not been described in as much detail as the other standard modules detailed in the accompanying Table.

It is believed that the launch pad used at Jiuquan for the CZ-2F/*Shen Zhou* missions will be compatible with the CZ-5 launch vehicle family. For launches to geosynchronous orbit and beyond it would be reasonable for the variants using the H5-2 upper stage to be deployed at the more southerly Xi Chang, while the Chinese have hinted at deployment on Hainan Island for such missions: to 2004 there have been no indications that new launch facilities are being constructed at either Xi Chang or Hainan Island in support of the CZ-5 programme. Some of the smaller variants – based upon the use of the K3-1 and K2-1 stages, not using the H5-1 – could be deployed at Taiyuan, should the Chinese wish to use the vehicles for launches to polar or Sun-synchronous orbits.

At present the priority seems to be on the development of the CZ-5/500 family of launch vehicles – and even then it seems unlikely that all of the variants noted below are currently being funded. If the Chinese were to go ahead with the development of the full CZ-5/500 and CZ-5/335 launch vehicle variants – and even the CZ-5/225 vehicle – then these new vehicles could replace all of the existing launch vehicles in the Chinese inventory.

The designs for the CZ-5 launch vehicles are said to have passed the planning stages and official government funding is expected to be approved in Summer 2004. It is only when this funding is authorised that one may expect to see the start of construction facilities at Jiuquan and Hainan Island. The start dates and speed of such work will give an indication of where the vehicles can be expected to be launched from and the Chinese priorities for launching the vehicles. It is possible that the CZ-5/500 class vehicle can be stacked in the vertical assembly building, constructed at Jiuquan in support of the Project 921 manned programme.

The CZ-5/500 launch vehicle family

The launch vehicles based upon the 5 m diameter core stage represent the CZ-5/500 family and all of the variants described by the Chinese use four strap-on boosters which are based upon the K2-1 and/or K3-1 modules. For missions to low Earth orbit the second stage H5-1 stage (ignited at launch) operates all of the way to orbital injection. When used for missions to geosynchronous orbit or possibly deep space missions the H5-2 stage is added and the H5-1 stage is probably sub-orbital.

In considering lunar applications, it should be noted that both the CZ-5/504/HO and CZ-5/522/HO variants are capable of launching a full *Shen Zhou* spacecraft on a circumlunar mission, even when using a heavy ~11.3 tonnes payload shroud/escape tower assembly.

The payload capabilities which are quoted below – for the 500, as well as the 335 and 225 variants – are those shown in Chinese literature. The Chinese-announced payload capabilities seem to imply unusually high velocity losses to reach orbit, so it is possible that the actual payload capabilities may be higher than these figures.

CZ-5/500/504 variant
Strap-on boosters: 4 × K3-1
Second stage core: 1 × H5-1
Total length: 47.6 m
Launch mass: 779.5 t
Payload capacity: 25 t to LEO

CZ-5/500/504/HO variant
Strap-on boosters: 4 × K3-1
Second stage core: 1 × H5-1
Third stage: 1 × H-2
Total length: 50 m
Launch mass: 789.5 t
Payload capacity: 14 t to GTO 10.6 t to trans-lunar trajectory

CZ-5/500/522 variant
Strap-on boosters: 2 × K2-1 plus 2 × K3-1
Second stage core: 1 × H5-1
Total length: 47.6 m
Launch mass: 614.5 t
Payload capacity: 18 t to LEO

CZ-5/500/522/HO variant
Strap-on boosters: 2 × K2-1 plus 2 × K3-1
Second stage core: 1 × H5-1
Third stage: 1 × H5-2
Total length: 50 m
Launch mass: 636.6 t
Payload capacity: 10 t to GTO 8.1 t to trans-lunar trajectory

CZ-5/500/540 variant
Strap-on boosters: 4 × K2-1
Second stage core: 1 × H5-1
Total length: 42.6 m
Launch mass: 442.5 t
Payload capacity: 10 t to LEO

CZ-5/500/540/HO variant
Strap-on boosters: 4 × K2-1
Second stage core: 1 × H5-1
Third stage: 1 × H5-2
Total length: 47 m
Launch mass: 462 t
Payload capacity: 6 t to GTO 4.4 t to trans-lunar trajectory

The CZ-5/335 launch vehicle family

The launch vehicles based upon the use of a 3.35 m diameter core module are here called the CZ-5/335 family by a logical extension of the Chinese naming of the CZ-5/500 family. The Chinese designate the variants with the third stage by adding 'H' to the basic vehicle designator; this practice is followed here. The vehicle can fly as a basic tandem-staged version or can use either two or four strap-on boosters based upon the K2-1 module.

Two new upper stages are being developed but they have not been described in Chinese literature in any detail. The K3-2 will use LOX/kerosene and carry four engines, each with 15 tonnes (vacuum) thrust: the H3-3 will use LOX/hydrogen and may be a further modification of the CZ-3B third stage.

When trying to model this launch vehicle, two separate classes of K3-2 stage total masses appear: for the 300 and 300H variant the second stage appears to have two engines and a mass of 34–46 tonnes, while the other variants appear to have a second stage carrying four engines and a mass of 91–111 tonnes. Further modelling suggests that the full range of performances shown below can be obtained with the first of these possible K3-2 stages, therefore it is not clear whether the apparent variations in the stage data are due to errors or misunderstandings of the original Chinese data.

The comments in the CZ-5/500 section concerning the velocity losses and performance capabilities also apply to these vehicles - although the implied velocity losses are higher than those seen on the CZ-5/500 launch vehicles.

CZ-5/335/300 variant
Strap-on boosters: None
First stage core: 1 × K3-1

CZ-2F rated for human space flight 0131955

Second stage: 1 × K3-2
Total length: 48 m
Launch mass: 195 t
Payload capacity: 3 t to LEO

CZ-5/335/300H variant
Strap-on boosters: None
First stage core: 1 × K3-1
Second stage: 1 × K3-2
Third stage: 1 × H3-3
Total length: 43 m
Launch mass: 203 t
Payload capacity: 1.8 t to GTO

CZ-5/335/320 variant
Strap-on boosters: 2 × K2-1
First stage core: 1 × K3-1
Second stage: 1 × K3-2
Total length: 54 m
Launch mass: 380 t
Payload capacity: 6 t to LEO

CZ-5/335/320H variant
Strap-on boosters: 2 × K2-1
First stage core: 1 × K3-1
Second stage: 1 × K3-2
Third stage: 1 × H3-3
Total length: 53 m
Launch mass: 390 t
Payload capacity: 4 t to GTO

CZ-5/335/304 variant
Strap-on boosters: 4 × K3-1
Second stage core: 1 × K3-1
Third stage: 1 × K3-2
Payload capacity: 10 t to LEO

CZ-5/335/322 variant
Strap-on boosters: 2 × K2-1 plus 2 × K3-1
Second stage core: 1 × K3-1
Third stage: 1 × K3-2
Payload capacity: Not available

CZ-5/335/340 variant
Strap-on boosters: 4 × K2-1
First stage core: 1 × K3-1
Second stage: 1 × K3-2
Total length: 58 m
Launch mass: 520 t
Payload capacity: 10 t to LEO

CZ-5/335/340H variant
Strap-on boosters: 4 × K2-1
First stage core: 1 × K3-1
Second stage: 1 × K3-2
Third stage: 1 × H3-3
Total length: 55 m
Launch mass: 540 t
Payload capacity: 6 t to GTO

CZ-5/335/300/HO variant
Strap-on boosters: None
Second stage core: 1 × K3-1
Third stage: 1 × K3-2
Fourth stage: 1 × H3-3
Payload capacity: 1.6 t to GTO

CZ-5/335/304/HO variant
Strap-on boosters: 4 × K3-1
Second stage core: 1 × K3-1
Third stage: 1 × K3-2
Fourth stage: 1 × H3-3
Payload capacity: 6 t to GTO

CZ-5/335/322/HO variant
Strap-on boosters: 2 × K2-1 plus 2 × K3-1
Second stage core: 1 × K3-1
Third stage: 1 × K3-2
Fourth stage: 1 × H3-3
Payload capacity: Not available

CZ-5/335/340/HO variant
Strap-on boosters: 4 × K2-1
Second stage core: 1 × K3-1
Third stage: 1 × K3-2
Fourth stage: 1 × H3-3
Payload capacity: Not available

The CZ-5/225 launch vehicle
Most Chinese literature suggests that there will be a single launch vehicle variant based upon the use of the 2.25 m diameter core stage. This will add the K2-2 stage (using a single 15-tonne thrust engine) atop a K2-1 to give the smallest launch vehicle in the CZ-5 family. Here it is designated CZ-5/225/200, following the logic of the CZ-5/500 and CZ-5/335 launch vehicle designators.

One Chinese source – from February 2003 – indicates that a three-stage variant may be planned for launches to Sun-synchronous orbit. This paper calls the launch vehicle defined in the previous paragraph the 225A and the three-stage version the 225B.

The planned third stage for the 225B is not described in any Chinese literature but since the 225A is said to have two engines (one on each of the K2-1 and K2-2 stages) and the 225B three engines, this implies that the 225B third stage also has a single engine. Analysis of the Chinese data suggests that the third stage may have a mass of ~1–2 tonnes, which is extremely low for a liquid-propellant stage. However, this mass is in the same range as the solid-propellant third stages of the CZ-1 and CZ-1D vehicles. It is therefore deemed likely that if the 225B vehicle is developed it will use a solid-propellant third stage, derived from the stage used on the CZ-1D vehicle.

CZ-5/225/200 variant
Strap-on boosters: None
First stage: 1 × K2-1
Second stage: 1 × K2-2
Payload capacity: 2 t to LEO

CZ-5/225/225A variant
Strap-on boosters: None
First stage: 1 × K2-1
Second stage: 1 × K2-2
Total length: 38 m
Launch mass: 95 t
Payload capacity: 1.2 t to LEO

CZ-5/225/225B variant
Strap-on boosters: None
First stage: 1 × K2-1
Second stage: 1 × K2-2
Third stage: 1 × (solid ?)
Total length: 38 m
Launch mass: 95 t
Payload capacity: 0.5 t to SSO

A Giant Launch Vehicle?
The Chinese are clearly thinking in terms of a manned lunar programme at some future time, and the CZ-5 launch vehicle family is only capable of supporting missions to loop around the Moon. It is possible that dual launches with Earth orbit rendezvous (manned spacecraft with lunar orbit propulsion system on one CZ-5/504, a fuelled trans-lunar stage on the second) would allow the Chinese to fly a crew into orbit around the Moon but to complete a manned lunar landing, a much larger vehicle than the planned CZ-5/504 will be required.

It might be possible to add more K3-1 strap-on boosters around the H5-1 core stage, but with eight strap-ons (the maximum which can realistically be mounted around the H5-1 core stage) an LEO payload capability of "only" about 45 tonnes can be accomplished with what would be an extremely inefficient launch vehicle: three launches of such a vehicle would be required for anything other than a "token" manned lunar landing.

The Chinese themselves have probably not started to think about defining a giant post-CZ-5 launch vehicle for the manned lunar mission, and *if* such a programme were to go ahead with a possible manned landing target date around 2020 then the Chinese would not need to start defining such a vehicle until around 2010 with flights starting about five years later.

Whenever a manned lunar landing programme is authorised, it is clear that a Chinese giant launch vehicle in either the Soviet-era N-1 or U.S. Saturn-5 class will be required.

FB-1
Launch site(s) deployed: Jiuquan
Maiden launch: 10 August 1972 (sub-orbital test flight)
Orbital launch record: 8 launches, 4 successful, 4 second stage failures
Latest launch: 19 September 1981
Number of stages: 2
Overall length: 32.57 m
Diameter: 3.35 m
Launch mass: 192.679 t
Launch thrust: 2,746 kN sea level

FB-1 stage 1
Engines: 4 × YF-20 liquid bipropellant single start engines (overall designator YF-21)
Overall length: 23.819 m
Diameter: 3.35 m
Oxidiser: nitrogen tetroxide in forward tank
Fuel: UDMH in aft tank
Propellant mass: 142.73 t
Dry mass: 8.992 t
Thrust: 2,746 kN sea level
Specific impulse: 2,535 N-s/kg sea level
Burn time: 128.09 s

FB-1 stage 2
Engines: 1 × YF-22 liquid bipropellant single start main engine; 1 × YF-23 liquid bipropellant single start vernier engine (overall designator YF-24)
Overall length: 8.607 m
Diameter: 3.35 m
Oxidiser: nitrogen tetroxide in forward tank
Fuel: UDMH in aft tank
Propellant mass: 36.095 t
Dry mass: 3.535 t
Thrust: 716.0 kN main engine; 44.1 kN vernier engine (vacuum)
Specific impulse: 2,815 N-s/kg main engine; 2,736 N-s/kg vernier engine (vacuum)
Burn time: 117.41 s main engine, 323.41 s vernier engine

FB-1 payload shroud
Overall length: 3.144 m
Max diameter: 2.2 m
Mass: 0.242 t

Propulsion
Fewer details are known of Chinese thrusters and solid propellant motors. The CZ-1D stage 3 is a solid of 625 kg propellant and satellites such as FSW incorporate solid kick motors. CZ-2E's 1.7 m diameter EPKM solid stage 3 debuted in 1995. CZ-2C's 1.2 m diameter CPKM is imminent. Sounding rockets with solids are also flying. Solid propellant ICBM/IRBMs are under development. Hydrazine thrusters of unspecified nature provide three-axis control for the H-8 cryogenic stage. However, CZ-2E's stage 2 employs a hydrazine system acknowledged as derived from H-8: 4 × 45 N pitch thrusters, 2 × 45 N yaw + 4 × 11 N roll. The FSW recoverable capsule is also three-axis controlled by cold gas, and the CZ-1D solid stage is spin-stabilised by N_2 thrusters.

China's space launchers draw on a small range of main engines, the majority utilising hypergolic NTO/UDMH although more recently two cryogenic upper stages were added. No LOX/kerosene engines as used extensively elsewhere are known, reflecting the space variants' parallel development alongside storable propellant long-range missiles. The engines carry YF designations, indicating Yei-ti Fa-dong-ji (liquid state engine).

YF-20/21 and YF-20B/21B
The YF-20/21 is the backbone of the Chinese launcher range, clustered in fours (creating the YF-21/21B assembly) for every first stage of the CZ-2 to -4 families and modified into the high altitude YF-22 for the second stages. In the L-140 stage 1s (the 140 indicating 140 tonne propellant loading) it operates at 696 kN thrust, but the stretched L-180 stages and strap-ons of the more powerful vehicles run it at 740 kN. Chinese statements suggest that still higher levels are possible. The LB-40 strap-ons of the CZ-2E/3B/3C each incorporate a single YF-20B.

Development of the cryogenic YF-73 began in 1976 for the restartable H-8 third stage, the first use of LH_2 propulsion technology outside of the US and Europe. Principal use is delivery of payloads into GTO. Firings of flight-type engines began 1981. After completion of acceptance testing, the first flight engine was delivered in the fourth quarter of 1983. By end-1983, about 120 tests totalled more than 30,000 seconds. The four chambers, gimballed ±24° for attitude control, and are serviced by a single 37,000 rpm turbopump. A pair of high pressure nitrogen spheres initiates the two starts. The valves, pumps, gearbox, gas generator and propellant lines carry 10 to 20 mm of polyurethane foam to minimise heat transfer. The engines are manufactured at CALT. There have been two flight failures. The first was January 1984's debut mission when the re-start after coast attained 90 per cent thrust and then rapidly decayed after a few seconds. Cause was attributed to propellant flow fluctuations following the μgravity coast. Similarly, the second (December 1991) suffered a pressure drop 58 seconds into the second burn and shut down after a further 77 seconds. Cause was believed to be helium supply failure to control valves.

Development of the cryogenic restartable YF-75 drew extensively on YF-73 experience to create the H-18 stage, comparable to Centaur D. Its employment in pairs aboard the CZ-3A vehicles provides a GTO capability equivalent to the lower range of Ariane 4. Aboard CZ-3B, it surpasses Ariane 4 and almost equals Proton. The engines are manufactured at CALT.

The solid motor carried by the CZ-1D appears to be related to the GF-02 solid flown on the original CZ-1 vehicle. Work began in 1965 at the Solid Fuel Engine Research Academy under deputy head Yang Nansheng. The first spin firing took place 26 January 1968 but the motor exploded after 30 seconds, causing some fatalities. During 1968 to 1970 a total of 19 tests, including five altitude simulations, was conducted successfully.

The HEXI Corp's indigenous EPKM FG-46 commercial perigee kick stage debuted for AsiaSat 2 in November 1995, providing 3,460 kg into GTO. An early version was carried on the first 2E. Nine test firings included two failures at the nominal 40 rpm spin rate. 31 May 1993, test M1, prototype/design qualification. 24 October 1993, M2, prototype/structural qualification. 31 January 1994, C1, 40 rpm spin (fail). 5 May 1994, C2, altitude simulation. 6 July 1994, C3, 40 rpm spin (fail). 30 October 1994, C5, 40 rpm spin. 8 November 1994, C4, altitude simulation. 16 January 1995, C6, 40 rpm spin. 20 March 1995, Z1, 40 rpm spin/acceptance.

YF-2A
Applications: cluster of four for CZ-1 family stage 1 and CSS-3 IRBM
First flown: June 1964?
Oxidiser: nitric acid
Fuel: UDMH
Propellant flow rate: 115 kg/s?
Cycle: gas generator
Thrust: 275.3 kN sea level
Specific impulse: 241 s sea level
Burn time: about 130 s operational

YF-3
Applications: CZ-1 family stage 2, CSS-3 IRBM
First flown: June 1964?
Oxidiser: nitrogen tetroxide
Fuel: UDMH
Propellant flow rate: 50 kg/s?
Thrust: 294.2 kN vacuum
Specific impulse: 287 s vacuum
Burn time: about 130 s operational

YF-20
Applications: CZ-2C/3 stage 1 for YF-20/21; CZ-2E/3B/3C strap-on for YF-20B; CZ-2D/2E/3A/3B/3C/4 for YF-21B
First flown: 5 November 1974 (first CZ-2 flight)?
Oxidiser: nitrogen tetroxide
Fuel: UDMH
Propellant flow rate: 270 kg/s?
Thrust: 696.2 kN sea level for YF-20/21 of L-140 stage 1 of CZ-2C/3; 740.35 kN, E 12.69, exit area 7,728.6 cm^2 for YF-20B of LB-40 strap-on and YF-20B/21B of L-180 stage 1 of CZ-2E/3A/3B/3C/4.
Specific impulse: 259 s SL YF-20/21; 260.66 s SL YF-20B/21B
Mixture ratio: 2.10 YF-20/21; 2.12 YF-20B/21B
Burn time: 125–160 s operational

YF-22/22B and YF-24/24B
Description: The YF-22/22B is a high-altitude derivative of the YF-20. It is flown singly with four gimballed YF-23/23B verniers, the assemblies designated 24/24B, providing steering on the L-35 stage 2. YF-22B/23B use larger expansion ratios for improved SI
Applications: CZ-2C/3 stage 2 for YF-24; CZ-2D/2E/3A/3B/3C/4 stage 2 for YF-24B
First flown: 5 November 1974 (first CZ-2 launch)?
Oxidiser: nitrogen tetroxide
Fuel: UDMH
Propellant flow rate: 270 kg/s?
Thrust: 719.8 kN vacuum for YF-22, 740.4 kN for YF-22B, but 845.3 kN attainable
Specific impulse: 289 s vacuum YF-22; 298 s vacuum YF-22B
Mixture ratio: 2.181 YF-22 and -22B
Burn time: 300 s CZ-2E; 110 s others

YF-23/23B
Description: Four gimballed YF-23/23Bs provide steering for the fixed single YF-22/22B on the second stages

Applications: as YF-22/22B
First flown: as YF-22/22B
Oxidiser: nitrogen tetroxide
Fuel: UDMH
Propellant flow rate: 16 kg/s for all four chambers with single pump?
Thrust: 46.1 kN YF-23 and 47.1 kN YF-23B total (4 chambers) vacuum
Specific impulse: 281.7 s YF-23 and 289 s YF-23B vacuum
Mixture ratio: 1.57 YF-23/23B
Burn time: up to 300 s. 136 s for CZ-4

YF-40
Description: The engine appears to have been uprated for CZ-4B. Where known, the different specifications are provided in 4A/4B order
Applications: flown paired on CZ-4 stage 3; uprated for 4B?
First flown: 6 September 1988
Mass: each 70/83 kg dry
Oxidiser: nitrogen tetroxide
Fuel: UDMH
Thrust: 98.1/101.03 kN vacuum for pair
Specific impulse: 305/303 s
Mixture ratio: 2.14
Chamber pressure: 45.39 atmospheres
Expansion ratio: 55
Exit diameter: 630 mm
Burn time: 321/412 s (single/single or dual burn)
Nozzle materials: niobium alloy, radiation cooled

YF-73
Application: H-8 third stage of CZ-3
First flown: 29 January 1984 (failed to restart)
Number flown: 9 sets, to end-1995 (2 failures)
Engine cycle: gas generator
Dry mass: 236 kg
Length: 1,438 mm
Maximum diameter: 2,220 mm
Oxidiser: liquid oxygen at 9.086 kg/s
Fuel: liquid hydrogen at 1.817 kg/s
Thrust: 44.147 kN total (4 chambers) vacuum
Specific impulse: 420 s vacuum
Mixture ratio (O/F): 5.0
Chamber pressure: 25.96 atmospheres
Expansion ratio: 40:1
Chamber materials: stainless steel, regeneratively cooled by supercritical hydrogen
Injector: brazed stainless steel, 38 coaxial elements
LOX/LH2 pumps: 37,000 rpm single stage centrifugal pump with helical inducer. LH2 and turbine share common shaft; LOX shaft is driven via reduction gears. Gears cooled by hydrogen.
Burn time: two burns of 500 s + 300 s

YF-75
Application: H-18 stage 3 of CZ-3A/3B/3C
First flight: 8 February 1994
Number flown: 2 sets to end-1995
Oxidiser: liquid oxygen
Fuel: liquid hydrogen
Propellant flow rate: 19 kg/s each chamber?
Thrust: 78.4 kN vacuum each chamber
Specific impulse: 442 s vacuum
Mixture ratio: 5.0
Chamber pressure: 37.15 atmospheres
Expansion ratio: 80:1
Burn time: about 470 s total, dual start

GF-02/CPKM/EPKM/FG-46
First flown: 28 November 1995
Number flown: 2, to end-1995
Mass: 6,001 kg (5,444 kg propellant, 529 kg burnout, 28 kg ablated)
Length: 2.5 m
Diameter: 1,700 mm
Propellant: HTPB
Burn time: 87 s
Thrust: 210 kN max vacuum
Specific impulse: 292 s vacuum
Total impulse: 15.5881 MNs vacuum
Pressure: 6.0 MPa max
Casing materials: glass fibre
Nozzle: carbon-carbon

***Kaitouzhe* Launch Vehicles**
The Chinese are planning the development of a family of solid-propellant launch vehicles based upon the DF-21 missile technology: there have been western reports that the launch vehicles will be derived from the DF-31 missiles, but the launchers have diameters of 1.4 m (like the DF-21), rather than the DF-31 with a diameter of approximately 2 m.

The solid propellant launch vehicle family is named *Kaitouzhe* (KT) which translates as either 'pioneer' or 'explorer'. The smallest variant has flown twice but failed to reach orbit each time. Larger variants are planned using solid propellant strap-on boosters.

The *Kaitouzhe* launch vehicles are being developed by the Aerospace Solid Launch Vehicle Corporation, a subsidiary of the China Aerospace Sciences and Industry Corporation. The KT-1 vehicle is said to be capable of being readied for launch in twelve hours.

The first KT-1 launch attempt has not been confirmed by the Chinese, but took place in September 2002: it is believed that a failure involving the launch vehicle's second stage prevented the orbiting of the 50 kg OlympicSat payload (a 300 km polar orbit was intended).

The second KT-1 launch was announced by the Chinese and took place almost exactly a year after the first launch. This time the Chinese claimed that this was a successful R&D test of the launch vehicle, even though a satellite named PS 2 (mass 40 kg, intended once more for a 300 km polar orbit) had been carried and presumably orbital injection was planned.

KT-1
Launch site(s) deployed: Taiyuan
Maiden launch: 15 September 2002 (failed to reach orbit)
Orbital launch record: 2 launches, 2 failures to reach orbit
Latest launch: 16 September 2003 (failed to reach orbit)
Number of stages: 4
Overall length: 13.6 m
Diameter: 1.4 m
Launch mass: 19.3 t
Payload capacity: 100 kg to 220 km LEO, 50 kg to 400 km SSO

The Chinese have discussed the use of a Russian Illyushin-76 transport aircraft to carry a KT-1 to altitude and complete an air-launch: it is said that this could increase the payload capability to 100 kg to a 400 km SSO.

KT-2
Launch site(s) deployed: Taiyuan?
Maiden launch: 2005 (planned)
Number of stages: 4?
Diameter: 1.7 m (first stage), 1.4 m (upper stages)

No performance data are available for the KT-2 vehicle.

KT-2A
Launch site(s) deployed: Taiyuan?
Maiden launch: 2005 to 2006 (planned)
Number of stages: 2 strap-ons plus 4?
Diameter: 1.4 m (strap-ons) 1.7 m (first stage), 1.4 m (upper stages)

No performance data are available for the KT-2A vehicle.

VERIFIED

India

General History

Current Status
India has achieved its long sought goal of launcher autonomy. The first GSLV was successfully launched on 18 April 2001 followed by a PSLV flight on 12 September 2002 carrying Metsat 1. The second GSLV was successfully launched on 8 May 2003. With its operational debut the GSLV has given India parity with some of the most capable launch systems available in the US and Russia, providing it with a full suite of launch vehicles capable of LEO, Sun-synchronous and geosynchronous orbits. After the detonation of a new nuclear device in June 1998, the western economies imposed sanctions on India, but currently, every part on the Indian space launch vehicles is or can be indigenously produced.

Background
India has undergone a gradual evolution from the all-solid SLV-3 of 1980–83 to today's geo capability. The first all-India launch had the capacity to only place a 42 kg satellite in LEO. The ASLV expanded this to 150 kg for the Augmented SLV largely by adding strap-ons derived from the existing stage 1. ASLV retired in 1994 after four launches. Six of those same boosters fly on the new Polar SLV, allowing India to launch its own 1 tonne-class IRS remote sensing satellites into Sun-synchronous orbits. PSLV has been offered commercially through Antrix Corp since March 1996's flight. Seven PSLV flights have been conducted to date, the last in 2002. The Geosynchronous SLV builds upon the PSLV by replacing the six solid strap-ons with four liquids and substituting a cryogenic stage for the two upper stages. This provides a 2.5 tonne GTO capacity, freeing India from dependence on foreign vehicles for Insat 2-class satellites. PSLV's pad has been modified to accept the new vehicle.

GSLV

Current Status
The first launch of the GSLV took place on 18 April 2001 with the launch of the 1,500 kg GSAT, a scaled down test satellite for future geosynchronous communications satellites. A major upgrade discussed would place it in direct competition with Ariane 5, Atlas 5, Delta 4 and Proton/Angara and could be ready by 2008. ISRO's long-range plan envisages consolidation of the basic launcher with two flights in 2003 to 2004 and another in 2004 to 2005. The second flight of the GSLV took place with the launch of GSLV-D2 on 8 May 2003 from Satish Dhawan Space Centre, Sriharikota. The payload comprised the experimental GSAT-2 weighing 1,825 kg. About 17 min after lift-off the satellite was placed in a 180.4 km × 36,000 km orbit at 19.2° inclination. The two test flights have qualified the GSLV for launching 2 tonne satellites to GTO. ISRO is testing the indigenously developed cryogenic stage which will replace the Russian stage used in the basic version and increase payload capability by 25 per cent.

Background
ISRO is developing an Ariane 4-class Geosynchronous SLV with a 3.4 m diameter fairing, capable of handling 2.5 tonne Insat 2-class satellites. To develop the plans, some 500 configurations were reduced to four candidates during 1986–88, divided equally between solid/liquid cores and both with solid/liquid strap-ons. GSLV replaces PSLV's six solid strap-ons with four liquids powered by Vikas engines (although the initial version will use PSLV strap-ons) and substitute a cryogenic stage for the two upper stages. The vehicle is comparable to Ariane 44L except for the lower performance solid stage 1.

India's orbital launch record

	Date	Vehicle	Payload
1*	10 August 1979	SLV-3	Rohini (35 kg)
2	18 July 1980	SLV-3	Rohini 1 (35 kg)
3	31 May 1981	SLV-3	Rohini 2 (32 kg)
4	17 April 1983	SLV-3	Rohini 3 (41.5 kg)
5*	24 March 1987	ASLV-D1	SROSS 1 (150 kg)
6*	13 July 1988	ASLV-D2	SROSS 2 (150 kg)
7	20 May 1992	ASLV-D3	SROSS C (106 kg)
8*	20 September 1993	PSLV-D1	IRS 1E (846 kg)
9	4 May 1994	ASLV-D4	SROSS C2 (113 kg)
10	15 October 1994	PSLV-D2	IRS P2 (804 kg)
11	21 March 1996	PSLV-D3	IRS P3 (930 kg)
12*	28 September 1997	PSLV-D4	IRS P4 (1,250 kg)
13	26 May 1999	PSLV-D5	Kitsat 3 (107 kg) DLR-Tubsat (45 kg) Oceansat (1,048 kg)
14	18 April 2001	GSLV-1	GSAT-1 (1,520 kg)
15	22 October 2001	PSLV-D6	TES (1,108 kg) Proba (94 kg) Bird (94 kg)
16	12 September 2002	PSLV-C4 Metsat-1 (1,050 kg)	
17	8 May 2003	GSLV-2	GSAT-2 (1,520 kg)

* indicates vehicle failure (SROSS C entered low orbit because stage 4 spun up to 80 rpm instead of 180 rpm). All launches made from Sriharikota.

Ground test version of the Vikas engine to be used on the strap-on boosters for the GSLV; the original Vikas variant flew on the second stage of the PSLV
(Liquid Propulsion Systems Centre) 0003431

The Vikas engine is derived from the SEP Viking 2 motor.

Glavkosmos signed a Rs2,350 million agreement with ISRO in January 1991 to supply India with the cryogenic engine. This KVD-1 76 kN KB Khimmach engine was originally developed to power Russia's N-1 manned booster for the L-3M programme in the 1970s. The contract called for two engines plus the technology transfer required for subsequent Indian production. However, US pressure under the Missile Technology Control Regime forced the Russians to formally terminate the agreement in August 1993. ISRO noted that two years had been lost on the GSLV programme and forced a decision on the development of an indigenous cryogenic engine. Later in the year, the Indian government approved plans for a domestic water-cooled LOX/GH_2 subscale engine which ISRO tested successfully 21 July 1989. Critics of international control regimes point out that the US action to stop the Russian sales actually created another source of such engines, namely, India, and that a resentful India may not be restrained in efforts to sell its technology overseas.

The course of engine development was not easy for India. The contractor LPSC's first attempt at firing a 10 kN LOX/LH_2 engine ended with an explosion at ignition in early July 1993. LH_2 became available in the country only after a US company built a plant at LPSC's Mahendragiri site at end-1992.

Model of the GSLV rocket being developed by ISRO for operational duty early in the next century
(Theo Pirard) 0054377

Programme has been restructured to include basic GSLV with 1,500 kg GTO compatibility, GSLV Mk2 (KVD-1) with 2,500 kg GTO compatibility and GSLV Mk 3 (C-20) with 3,000/3,500 kg GTO capability. Flight dates are uncertain.

The GSLV utilises Insat 2's 119 kg (dry mass) unified bipropellant liquid propulsion system made up of a single 440 N LAM Liquid Apogee Motor and two redundant networks of 8 × 22 N RCS thrusters for final orbital placement. The qualification LAM was tested for a total 9,550 seconds, including a single burn of 3,000 seconds. RCS testing accumulated 30,500 seconds including a single continuous burn of 10,000 seconds. Pulsed firing mode logged >250,000 pulses, each component being tested over three life cycles. The engines, pressure regulator, check valve, fill, drain and vent valve, pyro valve and pressure transducer were developed indigenously; the tanks, gas bottles, filters and latch valves were procured from foreign suppliers. The system carries about 450 welded joints.

The payload fairing is 3.4 m in diameter and 7.8 m in length.

Specifications

GSLV
First launch: 18 April 2001
Launch site: SHAR Centre (Sriharikota)
Principal uses: medium-class GTO
Cost: US$45 million (in 1999 currency values)
Availability: to be announced
Schedule of missions: 7
Performance 5,000 kg to LEO (C-2 version): 2,500 kg into GTO (1,500 kg for initial version)
Number of stages: 3 (1 solid/2 liquid) + 4 liquid strap-ons (6 PSLV solids initially)
Overall length: 50.9 m
Launch mass: 402 t
Guidance: similar to PSLV

GSLV liquid strap-ons
Comment: PSLV's Vikas engine is being adapted for GSLV's strap-ons. A successful 200 s run 24 July 1995 tested the new indigenously-developed silica-phenolic throat, designed to withstand the operational 1,300°C
Designation: L40
Overall length: 19.7 m
Principal diameter: 2.1 m
Dry Mass: 4 × 5,5 t
Total mass: 4 × 45.5 t
Propellant: 4 × 40 t $UDMH/N_2O_4$
Thrust: 4 × 735 kN vacuum average
Chamber pressure: 52.6 bar
Specific impulse: 260 s (vacuum)
Burn time: 158 s
Attitude control: gimballed ±5° for 3-axis

GSLV stage 1
Designation: S125
Other specifications: same as PSLV

GSLV stage 2
Designation: L37.5
Other specifications: same as PSLV

GSLV stage 3
Designation: CS
Overall length: 8.72 m
Principal diameter: 2.80 m
Total mass: 14.60 t
Propellant: 12.50 t LOX/LH_2
Thrust: 76 kN vacuum average (450 s Isp), including two gimballed LOX/GH_2 verniers fed by main chamber pumps
Specific impulse: 76 s (vacuum)
Burn time: 720 s
Attitude control: gimballed, auxiliary and cold gas
Payload fairing: 3.4 m and, 7.8 m long, aluminium

LAM
First flight: July 1992 (Insat 2A)
Dry mass: 4 kg
Length: 570 mm
Maximum diameter: 276 mm
Mounting: fixed
Engine cycle: pressure-fed in regulated mode
Propellants: MMH/N_2O_4 from tanks at 16 atmosphere
Mixture ratio (O/F): 1.65
Thrust: 440 N vacuum nominal
Specific impulse: 310 s vacuum minimum
Expansion ratio: 160
Chamber pressure: 6.9 atmosphere
Injector: coaxial titanium welded to chamber
Combustion chamber: silicide-coated columbium alloy radiatively cooled
Burn time: qualified to 3,000 s continuous; minimum impulse bit not available.

Insat 2 RCS Thruster
First flight: July 1992 (Insat 2A)
Dry mass: 850 g
Length: 249 mm
Maximum diameter: 60 mm
Mounting: fixed
Engine cycle: pressure-fed in blowdown mode
Propellants: MMH/NTO from tanks at 16 atmosphere
Mixture ratio (O/F): 1.65
Thrust: 22 N vacuum nominal
Specific impulse: 285 s vacuum minimum
Expansion ratio: 100
Chamber pressure: 6.9 atmosphere
Throat temperature: 1,200°C
Injector: coaxial titanium welded to chamber
Combustion chamber: silicide-coated columbium alloy radiatively cooled
Burn time: qualified to 1,000 s continuous; minimum impulse bit not available.

PSLV

Current Status

India has committed to moving its payloads from foreign launches to national launch vehicles and the PSLV is scheduled for an average of one flight a year for the foreseeable future. India has no plans to commercialise the PSLV for marketing.

Background

India's Polar Satellite Launch Vehicle (PSLV) represents the first stage in acquiring launcher autonomy for applications satellites. Sized for placing 1 tonne IRS Earth resources satellites in 817 km Sun-synchronous orbits from Sriharikota. The Indians have paid Rs4,155 million, to develop the PSLV through its first two flights and the ground infrastructure. Each launcher itself costs about Rs450 million. The government, in October 1994, approved six further tests known as PSLV C1-C6, (C = continuation) vehicles, after the first three development flights named PSLV D1-D3, (D = development). Some estimates state the total cost for all six flights is Rs4,000 billion. The C version should improve performance to 1,300 kg. IRS-1D (1,200 kg) will use the PSLV.

PSLV employs an unusual combination of liquid and solid stages, using the first Indian liquid systems for stages 2 to 4 and clustering six ASLV strap-ons around the solid first stage. Stage 2's single Vikas engine is based on SEP's Viking from Ariane. In order to prepare for the eventual programme, the Indian Space Research Organisation began long-duration testing on the engine in January 1988 at LPSC's Mahendragiri facility.

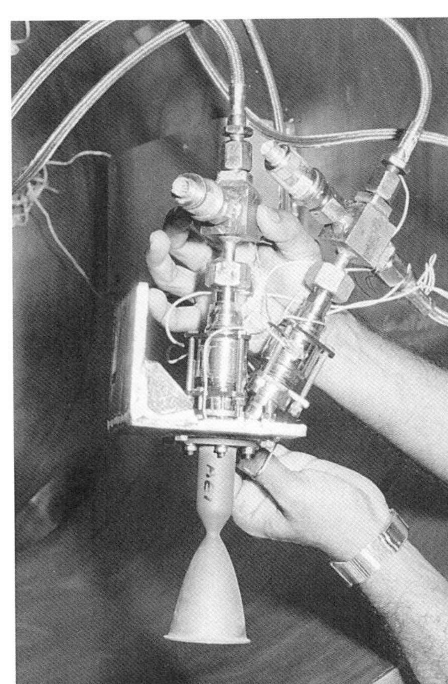

The 22 N RCS thruster carried on INSAT 2
(Liquid Propulsion Systems Centre) 0003444

The 440 N LAM carried on INSAT
(Liquid Propulsion Systems Centre) 0003445

In practice, Sriharikota has a launch constraint on azimuths of less than 140°, forcing PSLV to execute a 55° yaw manoeuvre after 100 seconds and stage 1 separation to reach a Sun-synchronous path. Without this constraint the PSLV could launch 1.6 tonne into Sun-synchronous orbit.

PSLV-D1 performed well but an uncorrected pitch problem after stage 2 separation left it at only 340 km altitude after stage 3's burn instead of the planned 414 km. Stage 4 had insufficient reserve to attain a useful orbit. Investigators later found stage 2 engine nulling occurring about 3.7 seconds before stage 3 ignition and as a consequence, the stages recontacted because two small solid retros failed, producing a pitch disturbance that exceeded the preset limit. PSLV could have removed it but a software error in the control loop prevented action. D2, with modified software, was successful, followed by D3 in March 1996. C1 launched in September 1997 placed the IRS satellite in an eccentric orbit instead of the planned circular orbit due to a fourth stage malfunction. C2 placed three satellites in Sun-synchronous orbit in May 1999. C3 also carried three satellites to orbit in October 2001. C4 was launched in September 2002 carrying Metsat to a GTO followed by C5 on 17 October 2003 carrying IRS-P6 (Resourcesat-1) to a sun-synchronous orbit.

The liquid propellant engine, two of which are carried on the fourth stage of the PSLV
(Liquid Propulsion Systems Centre) 0003446

Specifications

PSLV

First launch: 20 September 1993
Number launched: 8 through end of October 2003
Launch sites: SHAR Centre (Sriharikota)
Principal uses: 1 t-class remote sensing payloads into Sun-synchronous orbit, 3 t-class into LEO
Vehicle success rate: 87.5 per cent
Cost: Rs450 million each D1-D3, Rs650 million C1-C6
Availability: launches projected every year
Performance: Sun-synchronous: 1,000 kg into 900 km, 99.1° from SHAR (1.6 t without range safety constraints)
GTO: 450 kg
LEO: 3,000 kg into 400 km
Number of stages: 4 (2 solid/2 liquid) plus 6 solid strap-ons
Overall length: 44.2 m
Fairing diameter: 3.2 m (5.1 m base circle with strap-ons)
Launch mass: 283 t
Guidance: Inertial Guidance System located in vehicle equipment bay surrounding stage 4 base. Redundant strapdown inertial navigation system (Resins, incorporating three dry-tuned gyros + four servo accelerometers) feeds the navigation processor, which produces navigation data every 500 ms for the guidance & control processor to issue steering commands at similar intervals. Open loop guidance is employed during stage 1 burn before switching to closed loop from stage 2 onwards. Target 3Σ injection accuracy for Sun-synchronous payloads ±35 km altitude, ±0.2 in c.
Launch sequence (typical IRS mission, min/s):

00:00	stage 1 ignite
00:00	2 solids ignite
00:30	4 solids ignite at 3 km
00:54	2 solids burnout at 17 km
01:13	2 solids sep at 24 km
01:19	4 solids burnout at 36 km
01:30	4 solids sep at 38 km
01:45	stage 2 ullage motors ignition, stage 1
–	sep + stage 1 retro motors ignition
01:45	stage 2 ignition at 48 km
02:32	fairing sep at 105 km
02:37	closed loop guidance begins
04:21	stage 2 shutdown, sep, retro
–	ignition, stage 3 ignition at 232 km
05:33	stage 3 burnout at 350 km
05:07	stage 3 sep at 405 km
10:17	stage 4 ignition at 700 km
17:06	stage 4 shutdown at 817 km

PSLV solid strap-ons

Designation: PSOM (PSLV strap-on motors).
Comment: The segmented motors are almost identical to the ASLV strap-ons. Two are ignited on the pad, followed 30 s later by the remaining four, with separation occurring at 73/90 s respectively. Nozzles are canted 9°. One ground-lit and one air-lit strap-on include TVC secondary injection to augment roll control.
Overall length: 11 m
Principal diameter: 1.0 m
Propellant: 8,628 kg of HTPB solid
Average thrust: about 440 kN vacuum each, 662 kN max
Burn time: 47 s
Attitude control: exhaust secondary injection
Separation: pyrotechnic nuts and springs

PSLV stage 1

Designation: PS1
Overall length: 20.3 m
Principal diameter: 2.8 m
Total mass: 156.0 t
Propellant: 129.0 t of HTPB solid in five maraging steel segments, each 3.4 m long and cast individually (centre three are interchangeable); segments held by 144 pins in tongue/groove joint.
Thrust: max 4,600 kN vacuum, 3,500 kN thrust at launch
Burn time: 103 s
Attitude control: pitch/yaw provided by secondary injection TVC of strontium perchlorate from two 70 cm diameter 1,650 litre steel tanks through quadrants of six valves from a total of 24 into stage 1 nozzle (35% of distance from throat to exit). Each quadrant can generate a 200 kN side force; valve flow rate is 12 litres maximum. Roll control provided by two continuous-burn swivelling 6.4 kN vacuum NTO/MMH thrusters below SITVC tanks.
Separation: flexible linear shaped charge and eight 48.5 kN mean thrust 1,414 × 209 mm diameter retros, each burning 26.8 kg propellant in 1.3 s

PSLV stage 2

Comment: The integrated system was first tested at Mahendragiri 21 March 1990. Final qualification test October 1992 for 150 s. Four 15.3 kN solid motors burn for 5.5 s at stage 1 separation to provide ullage for stage 2.
Designation: PS2/L37.5
Engine: Vikas open gas generator cycle engine (based on SEP Viking), gimballed for pitch/yaw control, single start, film + radiatively cooled, pumps driven by single 9,400 rpm shaft powered by gas generator.
Overall length: 11.5 m
Principal diameter: 2.8 m
Dry mass: 5.3 t
Propellant: 37.5 t of UDMH/NTO
Stage thrust: 725 kN vacuum
Burn time: 149 s
Attitude control: Vikas gimballing for pitch/yaw; roll control provided by two on/off 300 N nozzles drawing hot gas from Vikas gas generator
Separation: at 249 s/243 km, with four 22.6 kN 1.6 s retros providing separation from third stage

PSLV provides a 1 t Sun-synchronous capacity (ISRO)
0516904

PSLV stage 3

Comment: Stage 3 is suspended inside a 2.8 m diameter shroud from the fourth stage skirt
Designation: PS3
Overall length: 354.1 cm
Principal diameter: 2.0 m
Propellant: 7,260 kg of HTPB in Kevlar casing
Thrust: max 386 kN vacuum
Burn time: 73.1 s (78.3 s action time)
Attitude control: pitch/yaw by ±2° nozzle flexure (qualified to ±3°) controlled by two actuators; roll control from the 50 N roll thrusters on stage 4.
Separation: ball release and springs

PSLV stage 4

Comment: The integrated stage was first tested at Mahendragiri on 27 September 1989, beginning a programme including two full duration trials and a demonstration of re-start capability.
Designation: PS4/L2
Overall length: 2.65 m
Principal diameter: 1.335 m
Dry mass: 920 kg
Propellant: 2 t MMH/NTO
Average thrust: 14 kN
Burn time: 425 s (D2: 397 s)
Attitude control: six 50 N thrusters in two blocks drawing propellants from stage 4's tanks provide 3-axis control during coast period; main engine gimballing employed during burn.
Separation: Merman clamp and springs

PSLV stage 4 engines

Description: two ISRO 7.5 kN pressure-fed bipropellant engines, gimballed up to ±3° in orthogonal axes for 3-axis control, 8.37 atmosphere chamber pressure, 60:1 expansion ratio. The titanium alloy tanks are pressurised at 19.2 atmospheres by helium stored at 300 atmospheres in five titanium spheres. Tank incorporates surface tension-type propellant acquisition system to ensure supply under adverse acceleration conditions.
First flight: 20 September 1993
Dry mass: 28 kg
Length: 1.1 m
Maximum diameter: 0.63 m
Mounting: gimballed up to ±3° in two orthogonal planes by a closed loop servo control system to provide stage pitch, yaw and roll control during thrust phase.
Engine cycle: pressure-fed

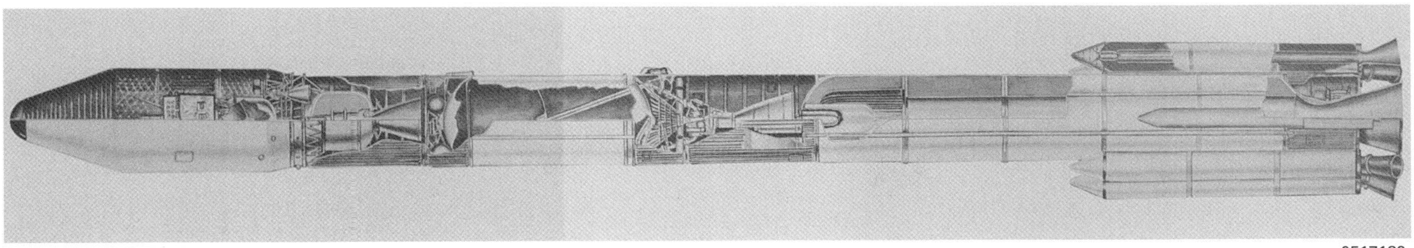

0517189

Propellants: MMH/NTO from tanks at 19.2 atmosphere
Mixture ratio (O/F): 1.4
Thrust: each 7,500 N vacuum nominal
Specific impulse: 308 s vacuum minimum
Expansion ratio: 60
Chamber pressure: 8.37 atmosphere
Injector: AISI 304 stainless steel; 45 sets of triple elements
Combustion chamber: AISI 304 stainless steel film cooled along single helical grooving
Nozzle: silicide-coated columbium alloy radiatively cooled
Burn time: 425 s mission nominal
Qualification time: 530 s (continuous)

PSLV L37.5 Vikas engine
Comment: India's first large liquid propellant rocket engine is derived from Viking technology acquired from SEP. The engine achieved its first full-duration 150 s firing in January 1988 on the principal test stand at the Liquid Propulsion Test Facility, Mahendragiri, and completed the first integrated stage test 21 March 1990 (qualification completed October 1992). A 686 kN version will be used for GSLV's strap-ons. A successful 200 s run 24 July 1995 tested the new indigenously-developed silica-phenolic throat, designed to withstand the operational 1,300°C.
First flown: 20 September 1993
Number flown: 4 to May 1999
Dry mass: 876 kg
Length: 3.509 m
Mounting: gimballed for pitch/yaw control
Engine cycle: gas generator (9400 rpm turbine)
Propellants: UDMH/NTO
Mixture ratio (O/F): 1.86
Thrust: 72 kN vacuum (686 kN for GSLV)
Specific impulse: 295 s vacuum
Time to full thrust: 2.4s
Expansion ratio: 31 (13.88 C-SU strap-on)
Chamber pressure: 51.9 atmospheres
Injector: 216 like on like doublets set in six rows for each propellant in light alloy annular injector
Chamber cooling: UDMH film supplied through additional channels on the injector's lower section
Igniter: hypergolic
Nozzle materials: cobalt alloy with SEPHEN (phenolic resin/silica fibre) throat
Burn time: 160 s nominal mission

Payload fairing/accommodation
Hindustan Aeronautics Ltd. builds 3.2 m diameter aluminium fairing, 8.3 m long. The 4.5 m long isogrid central cylindrical section is topped by a nose with a 70 cm radius cap/20° sloping wall; the boat-tail narrows along a 14° slope to 2.8 m diameter. A Merman clamp, explosive bolt cutters and zip cord, effects separation. 937 mm payload adapter using 12.2 kN Merman clamp; 4 springs provide 80 cm/s max separation velocity.
Acceleration: 6.0 *g* max, during stage 3
Acoustic: payloads qualified to 146 dB overall for two min, acceptance testing to 142 dB for one min

PSLV stage 1 motor
PSLV's first stage is the most powerful model produced by ISRO; it was successfully static fired for the first time 21 October 1989. The second was successfully fired 23 March 1991.
Application: PSLV stage 1
First flight: 20 September 1993
Number flown: 6 to end 2001
Mass (kg): 145,000
Length (cm): 2,034.5
Diameter (cm): 280.4
Propellant: type: HTPB, segment-cast with three central segments interchangeable
shape: forward segment deep 10-fin to burn 22 t loading in 20 s to generate high launch thrust; remainder tubular with average 120 cm diameter bore.
mass fraction: 0.890 (129 t loading)
Burn time (s): 100
Thrust (kN, vacuum): 3,463 mean, 4,600 max at 17 s
Specific impulse (s, vacuum): 265
Total impulse (kNs, vacuum): 326,000
Pressure (atmosphere): 58 max at 17 s
Nozzle: throat diameter (cm): 80
length (cm): 2,034.8
exit diameter (cm): 227
expansion ratio: 8
materials: carbon phenolic throat insert, tape-wound carbon phenolic on divergent cone changing to silica phenolic below TVC inlets.
Casing materials: five segments of M250 maraging steel, 3.4 m long. Joints tongue/groove with two O-rings and 144 pins. Insulated by nitrile rubber, with silicon potting compound on joints.
Igniter type: pyrogen head-end 30 cm diameter, 125 cm long containing Pedpro 2661 HTPB and its own 2 kg-charge igniter

PSLV stage 3 motor
Comment: This is ISRO's most advanced solid propellant motor, designed for a 0.915 mass fraction and the first to adopt a submerged flex nozzle design for TVC. Off-loading is also possible. Test #5 made 29 July 1991, #6 17 October 1991. Tests 8, 9 and 10 were for qualification.
Application: PSLV stage 3 (PS3)
First flight: 20 September 1993
Number flown: 6 to end 2001
Mass (kg): 7,975
Length (cm): 244.20 (casing 208.35)
Diameter (cm): 198.8 (202.5 skirt)
Propellant: type: HTPB casebound, 86% solid
shape: slotted
mass fraction: 0.910 (7,260 kg standard loading)
Burn time (s): 73.1; 78.1 action time
Thrust (kN, vacuum): 243.2 mean, 386 max
Specific impulse (s, vacuum): 293
Total impulse (kNs, vacuum): 20,850
Pressure (atmosphere): 57.4 max
Nozzle: the 19.6 per cent submerged contoured nozzle is provided with a ±3° (±2° flight) conical flex bearing. The elastomer, based on natural rubber of nominal 3 mm thickness is bonded with steel reinforcements of 4.5 mm thick run. Each seal consists of six reinforcements and seven elastomer pads. The seal positioned beyond the throat housing is driven by two 11.8 kN electromechanical actuators 90° apart; a thermal boot protects the bearing.
throat diameter (cm): 19.4
length (cm): 180.2
mass (kg): 280
exit diameter (cm): 141.4
expansion ratio: 52.3
materials: graphite throat insert, carbon and silica phenolic composites for nozzle.
Casing materials: Kevlar filament-wound, EPDM insulation and aluminium forgings for skirt and end-opening fittings; total mass 340 kg. Nozzle opening 750 mm diameter, 200 m head end.
Igniter type: pyrogen head-end

PSLV strap-on motor
Comment: The same basic motor, derived from the SLV-3's stage 1, was employed for ASLV's stage 1 and strap-on, and adopted as PSLV's initial strap-on. The strap-ons are essentially identical to the ASLV stage 1 version, with the principal exception of canted nozzles. For PSLV, the secondary injection TVC systems are deleted on one ground-lit and three air-lit strap-ons. The specifications below are for ASLV's stage 1 model; differences for the strap-ons are noted where applicable.

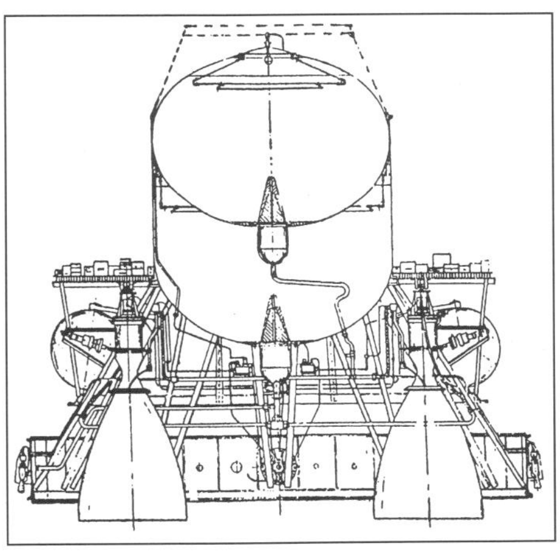

PSLV's stage 4 incorporates two 7.5 kN engines and, for coast attitude control, six 50 N thrusters. Surrounding the tank is PSLV's vehicle equipment bay. On top is IRS 1E (ISRO) 0516906

PSLV stage 2: ISRO developed the Vikas engine from Ariane's Viking to power the first large Indian liquid stage. The toroidal water coolant tank is visible inside the base skirt, with solid ullage and retro motors on the exterior (ISRO) 0516905

Application: ASLV stage 1 + strap-ons, PSLV strap-ons
First flown: 24 March 1987, ASLV-D1 launch
Number flown: 24 PSLV strap-ons, 4 ASLV stage 1, 8 ASLV strap-ons to May 1999
Mass: 10.4 t
Length (cm): 998.99
Diameter (cm): 100
Propellant: type: AP/18%Al/HTPB case-bonded
shape: star
mass fraction: 0.853 (8.9 t loading; 8.63 t for strap-on)
Burn time (s): 46 action time (49.95 action strap-on)
Ignition delay: about 200 ms strap-on
Thrust (kN, vacuum): 502.6 mean, 643 max (strap-on: 440 mean, 580 max)
Specific impulse (s, vacuum): 259 (252 strap-on)
Total impulse (kNs, vacuum): 22,600 (21,450 strap-on)
Pressure (atmosphere): 43.5 max (40.2 max strap-on)
Nozzle: secondary injection ports for TVC; 9°-canted nozzles
expansion ratio: 6.7
materials: graphite throat, carbon phenolic composite for convergent section, tape-wound carbon and then silica phenolic composites for divergent.
Casing materials: 3.5 mm thick low carbon steel with nitrile rubber insulation. Three-segment casing with tongue/groove joints.
Igniter type: pyrogen head-end

Rohini Sounding Rockets (RSRs)

Background

India's space programme began in November 1963 with the launch of a two-stage US Nike-Apache sounding rocket from Thumba, near Trivandrum. Production of indigenous Rohini Sounding Rockets (RSRs) at VSSC created the core of expertise for an orbital launcher programme, beginning with the 75 mm diameter RH-75. Four Rohini models remain in service, launching at an average of one a week for meteorology (RH-200), middle atmosphere (RH-300) and ionosphere (RH-560) investigations from SHAR, Thumba and Balasore. The Indians favour the RH-200s, launching about 40 during 1987. These rockets carry radar chaff out of Balasore to test and calibrate Indian Meteorological Department equipment.

The RSR programme is sited at the Vikram Sarabhai Space Center, which includes a Rocket Propellant Plant and Rocket Fabrication Facility.

Several organisations in India have begun an extensive marketing campaign to offer Rohini Sounding Rockets in Europe and the United States.

The Rohini 300 provides access to the middle atmosphere for Indian scientists (ISRO) 0516907

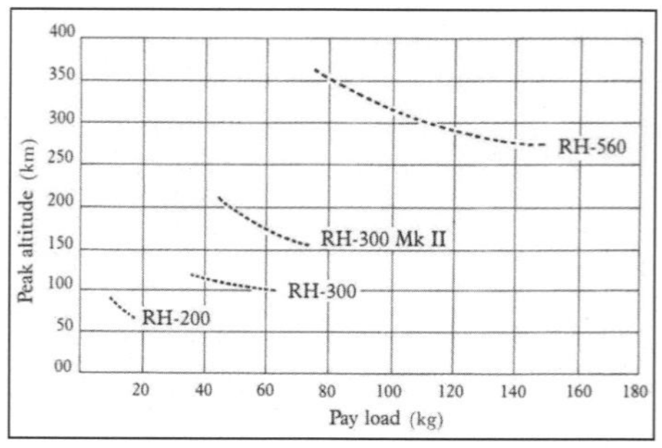

Rohini sounding rocket performance curves. The improved RH-560 will deliver 100 kg to 500 km 0516908

Specifications

Rohini 200
Applications: meteorological sonde up to 80km
Number launched: 0 in 1995, 18 in 1994, six in 1993
Number of stages: 2 (solid)
Overall length: 3.60 m
Principal diameter: 207 mm
Launch mass: 108 kg
Typical performance: 10 kg to 80 km

RH-200 stage 1
Motor: solid from VSSC
Overall length: 1.48 m
Principal diameter: 0.21 m
Average thrust: 16.89 kN sea level
Burn time: 5.7 s
Attitude control system: spin at max 270 rpm

RH-200 stage 2
Motor: RH-125 solid from VSSC
Overall length: 2.12 m
Principal diameter: 0.12 m
Average thrust: 10.49 kN sea level
Burn time: 2.9 s action time

Rohini 300
Applications: middle atmosphere probing above 100 km
Number launched: 8 to May 1999
Number of stages: 1 (solid)
Overall length: 4.14–6.84 m depending on payload
Principal diameter: 305 mm
Launch mass: 370 kg
Typical performance: 50 kg to 140 km

RH-300 stage 1
Motor: solid from VSSC
Overall length: 2.844 m
Principal diameter: 305 mm
Propellant mass: 240 kg
Average thrust: 37.94 kN sea level
Burn time: 14.5 s

Rohini 300 Mk II
Comment: The RH-300 Mk II was introduced during 1987 for investigations of the lower ionosphere, beyond the RH-300's capabilities. Major modifications included stretching motor length 1.2 m, reducing casing thickness by 0.4 mm from 2.0 mm, and adopting case-bonded HTPB propellant for high volumetric loading and burning characteristics resembling those of a two stage system. The resulting propellant mass increase was 100 kg.
Applications: upper middle atmosphere, lower ionosphere probing.
First launch: 8 June 1987
Number launched: 6 to end-1995
Number of stages: 1 (solid)
Overall length: 5.39–5.89 m depending on payload.
Principal diameter: 304.7 mm
Launch mass: 510 kg
Typical performance: 60 kg to 160 km
Spin rate: 6 rps

RH-300 Mk II stage 1
Motor: solid from VSSC
Overall length: 3.4 m
Principal diameter: 304.7 mm
Propellant mass: 330 kg of HTPB propellant
Average thrust: 39.14 kN sea level
Burn time: 20.2 s

Rohini 560
Comment: The RH-560 is India's largest sounding rocket, although a 560/300 MkII is under development to deliver 100 kg to 500 km. A static test of an RH-560 with case-bonded HTPB was successful in 1995, and flew in 1995.
Applications: ionospheric and astronomical payloads
Number of stages: 2 (solid)
Number launched: 33 to end-1995 (1995:1; 1994: 0; 1993: 1)
Overall length: 8.376–9.176 m depending on payload
Principal diameter: 561 mm
Launch mass: 1,350 kg
Typical performance: 100 kg to 350 km
Spin rate: 6 rps

RH-560 stage 1
Motor: solid from VSSC
Overall length: 3.29 m
Principal diameter: 0.56 m
Average thrust: 75.97 kN sea level
Burn time: 19.3 s

RH-560 stage 2
Motor: solid from VSSC (based on RH-300)
Overall length: 3.78 m
Principal diameter: 0.30 m
Average thrust: 37.94 kN sea level
Burn time: 14.5 s action time

RH-560 Mk II
Overall length: 9.55 m
Number of stages: 2
Lift-off weight: 1,635 kg
Diameter: 56.1 cm
Payload: 100 kg
Spin rate: 6 rps

International

Ariane

Current Status

Success with Ariane 5 has precipitated a decision to retire the Ariane 4 family of satellite launchers, the last flight taking place on 15 February 2003 when the last Intelsat 9 series satellite was launched from Kourou, French Guiana. In all 116, Ariane 4s were launched since June 1988, of which the last 74 were successfully carrying 118 satellites in dual-launch configuration pioneered by this vehicle. In 2002, Arianespace booked 11 launch contracts for Ariane 5. The first Ariane 5 EC-A launched on December 2002 failed to deliver its two commercial satellites to orbit. The 116th and last Ariane 4 launch took place on 15 February 2003.

Background

France designed and optimised Ariane as a direct ascent GTO launcher, beginning with the Ariane 1 variant in 1979. ESA and CNES, with Aerospatiale as the industrial architect, then further developed the Ariane 1 to 4 range of orbital launchers. Arianespace now procures, markets and launches the vehicles. The company plans to phase out the six Ariane 4 versions but ESA's Ariane 5 development programme has been quite successful. It was approved in 1987, to produce a vehicle offering 60 per cent additional GTO capacity at significant cost reduction and able to carry a manned spaceplane. Following the development flight programme, Ariane 5 will be assigned to Arianespace for commercial launches, beginning in late 1998. The cost to customers for Ariane 5 is 20 per cent less expensive than a launch using the most powerful Ariane 4, except that Ariane 5 provides a free relaunch in the event of failure.

Kourou can launch up to eight annual departures from the dedicated ELA-3 pad, but only five are now projected. Arianespace made its first order for 14 commercial vehicles on 12 June 1995 for launch as: 503–505 in 1997, 506–509 in 1998, 510–515 in 1999. Following the loss of Ariane 501 this was revised to 503–506 in 1998, 507–511 in 1999 and 512–518 in 2000.

Ariane 3 introduced the Sylda (Système de Lancement Double Ariane) for dual launches, encapsulating the lower satellite and mounting the forward payload on its front face. Ariane 4 introduced British Aerospace's (now Matra Marconi Space) larger aluminium honeycomb Spelda (Structure Porteuse pour Lancements Doubles Ariane) as part of the fairing assembly. The lower 3.97 m diameter, 2.=78 m high cylinder (Short Spelda) is attached to the VEB by 180 bolts and topped by either an 80 cm high cylinder, with a 1 m high frustum on top presenting a 1,920 mm diameter interface to the upper payload. Short Spelda masses 408 kg and extends the fairing by 280 cm; it provides a 34 m^3 payload envelope and can carry a 2,800 kg forward load. Separation is achieved by pyrotechnic line charge around the

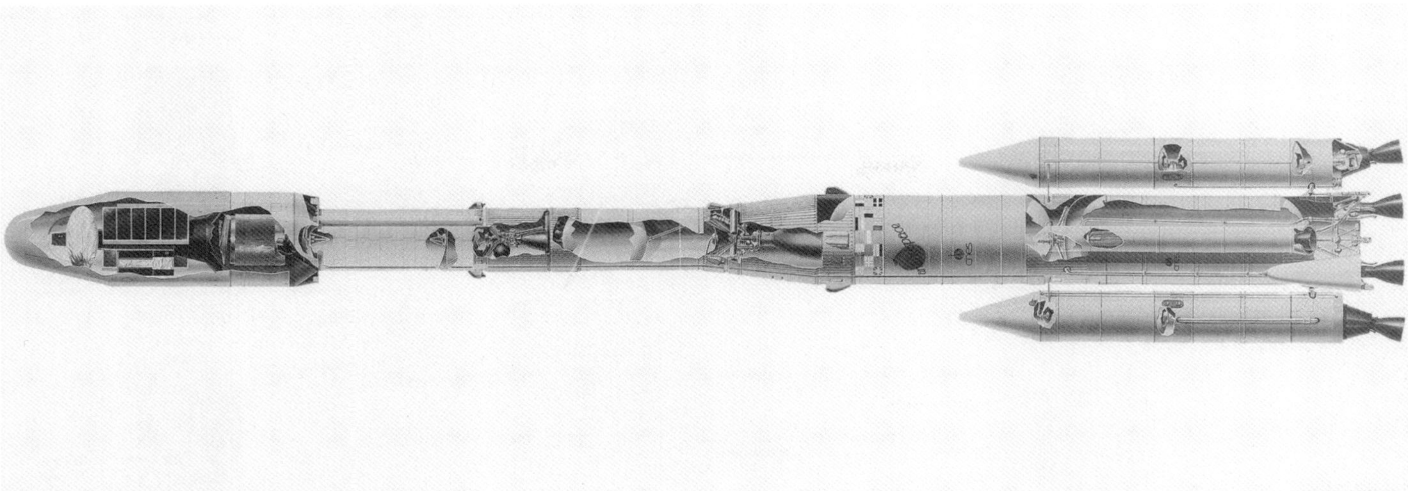

The 44LP is Ariane 4's most powerful version 0517346

circumference, and six springs. The 458 kg Long Spelda adds a further 1 m cylindrical section, provides a 42 m^3 payload envelope and can carry a 1,900 kg upper load. Arianespace began offering an SDS Spelda Dedicated Satellite service in 1990, available from 1992 (Insat 2B in July 1993 was the first; using the 3,080 mm-long Mini-Spelda to carry 24 m^3 payloads of 400 to 1,100 kg). This 337 kg unit can handle a 3,200 kg forward load. The 350 kg Mini-Spelda+300 version became available in 1996, stretching Mini-Spelda's cylinder by 300 mm for 26 m^3 payloads. It can handle a 3,200 kg forward load.

The development of a larger lift Ariane variant to become the standard vehicle through the mid-1990s, was formally approved by ESA's Council in January 1982, and management responsibility assigned to CNES. Arianespace completed technical and scheduling negotiations in 1988 with contractors for a single purchase of 50 Ariane 4 vehicles (increased in March 1995 to 55, and in February 1996 to 75) as the 'P9' final batch, for first launch mid-1992. The contracts were signed 15 February 1989, totalling about FFr18 billion; the large order yielded a 20 per cent saving in production costs. 1996's purchase of another 10 is worth about FFr5 billion. P9 introduced a stretched stage 3 (H10+) on V50 in April 1992, carrying an additional 340 kg of propellants and increasing burn time by 30 seconds, to accommodate satellite growth. The modification and weight-saving measures improved GTO capability by 130 kg, increasing 44L's to 4,460 kg. The H10-3 modification, introduced on V70 in December 1994, increased that to 4,720 kg. Other weight-saving, improved engine calibration and new trajectory optimisation, yielded 4,901 kg from V93.

Percentage interests in the Ariane 4 development programme were: France 62.60, Germany 16.28, Italy 5.87, UK 4.64, Belgium 4.05, Spain 1.77, Switzerland 1.55, Sweden 1.15, Netherlands 1.76, Denmark 0.17, and Ireland 0.07. Development programme cost was €650 million, including launcher development (€340 million), ELA-2 construction (€210 million) and V22 demonstration flight (€100 million). Ariane 4 was designed

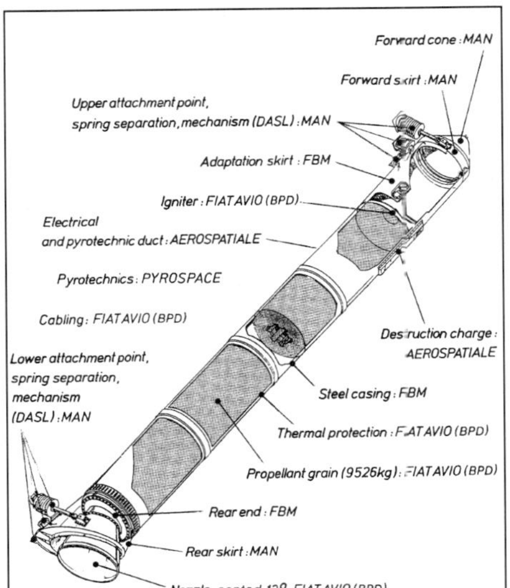

Ariane 4's PAP solid propellant strap-ons are stretched versions of those carried by Ariane 3: 126 should be used operationally within the Ariane 4 programme (Arianespace) 0003432

primarily to handle Intelsat 6 and to cater for IUS-class payloads. Mixing pairs of solid and/or liquid strap-on boosters creates six variants; in order of increasing performance, they are:

Ariane 40	no strap-ons
Ariane 42P	two solids
Ariane 44P	four solids
Ariane 42L	two liquids
Ariane 44LP	two solids/two liquids
Ariane 44L	four liquids

While the Ariane 2/3 launch vehicle variants were based upon the proven technology of Ariane 1 and Ariane 4 was similarly based upon the experience and hardware of the earlier versions, Ariane 5 marked the introduction of a totally new launch vehicle design and associated technology.

Ariane 5 is optimised for dual GTO launches, but will not be as attractive for polar missions, as there will be fewer Earth observation satellites that can be paired with a primary payload. Ariane 4 can be better matched with single polar payloads. Arianespace projects that, for GTO, 3 to 3.5 tonne satellites will dominate in coming years. Ariane 5's current 5.9 tonne dual capacity will be upgraded by 2003 to allow easier mixing of payloads.

Three principal changes achieved the 7.4 tonne GTO capacity. The dual payload carrier will be reduced by 400 kg, translating into a direct 400 kg GTO increase. The booster casings will be welded instead of bolted, yielding a 1.8 tonne saving for a 150 kg GTO increase. The principal increase (800 kg), and the most costly, accrues from improving the core stage. Vulcain's mixture ratio will be increased 5.3 to 6.2 by increasing the LOX flow rate by 18 per cent, accommodated by lowering the core's common bulkhead by 70 cm to raise propellant mass from 155 to 172 tonne. Vulcain thrust will be raised to 1,350 kN vacuum, achieved by a 10 per cent wider throat (expansion ratio from 45 to 61.7), 10 per cent chamber pressure increase,

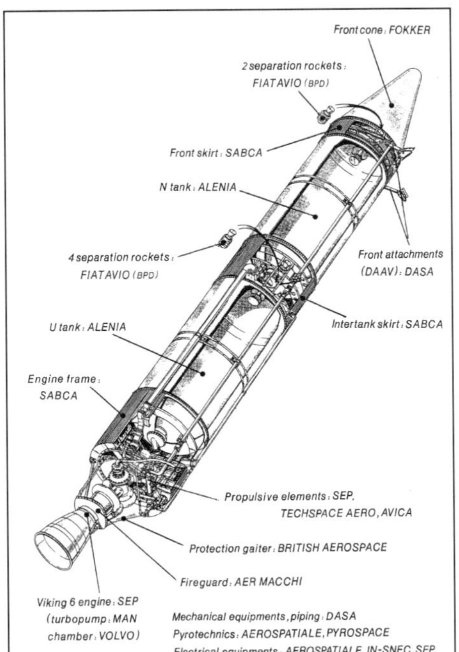

Ariane 4's PAL liquid boosters are of comparable size and performance to stage 2: 194 should be used operationally within the Ariane-4 programme (Arianespace) 0003406

the higher mixture ratio, and extending the nozzle, its lower part using a simple film-cooled sheet. The launcher's structures are sized to accommodate such upgrades. Should the €167 million be unblocked, stage 2 improvements would add another 400 kg GTO. The propellant load would be increased to 15 tonne and thrust to 35 kN.

Other options studied include using two current Vulcains and stretching the core to accommodate 210 tonnes of propellants. This would increase LEO capacity by 3.5 tonnes, but hardly improve GTO performance because staging would no longer be optimised. Other studied enhancements include replacing EPS with a cryogenic upper stage (H10) and adding a short segment to the booster to create a P270 strap-on. A heavy lift version for 35 tonnes into a trans-lunar path would employ four P230s, a 7.5 m diameter core carrying 620 tonne propellants for five Vulcain MkIIs, and a cryogenic upper stage with 80 tonne propellants and a single restartable Vulcain derivative.

On 6 June 1996, the first Ariane 5 (Ariane 501, Volume 88) was launched from Kourou. About 40 seconds after the start of the flight sequence and at an altitude of 3.7 km the launch vehicle veered off its flight path, broke up and exploded with the total loss of the four Cluster scientific satellites. Within a few days an independent inquiry board was established to determine the cause of the launch failure. They investigated whether the qualification and acceptance tests were appropriate in relation to the failure, and recommended corrective action to remove the cause of the 'anomaly', and other possible weaknesses of the system found to be at fault. The report was published on 19 July 1996.

The countdown on 6 June had been trouble-free, with the exception of a hold at HO–7 minutes because of weather problems (HO is the time of the command for the main cryogenic engine ignition). When the conditions improved the launch sequence resumed at 12 hours 33 minutes 59 seconds UT, with ignition of the first stage core and two solid booster engines being nominal. The vehicle performed a nominal flight until HO+36 seconds when there was a failure of the back-up inertial reference system, immediately followed by a failure of the active inertial reference system. The nozzles of the two solid-propellant boosters swivelled to their extreme positions. An instant later the Vulcain engine's nozzle did the same, causing the launch vehicle to abruptly veer off-course. This triggered (correctly) the self-destruction of the launch vehicle by the rupture of links between the strap-on boosters and the core stage. Thus, the origin of the failure was quickly narrowed down to the flight control system and specifically the inertial reference system, which had ceased operating simultaneously at HO+36.7 seconds.

The inquiry board's report found that at HO+22 seconds, variations of 10 Hz frequency started to appear in the hydraulic pressure of the actuators which commanded the nozzle of the main engine. Although this was considered to be significant and was not fully explained in the report, it was not considered to be relevant to the failure of the launch.

At HO+36.7 seconds the computer within the back-up inertial reference system – which was operating on stand-by for guidance and attitude control – became inoperative, this being caused by an internal variable related to the horizontal velocity of the launch vehicle exceeding a pre-programmed limit. Approximately 0.05 seconds later the active inertial reference system (identical to the back-up system in terms of hardware and

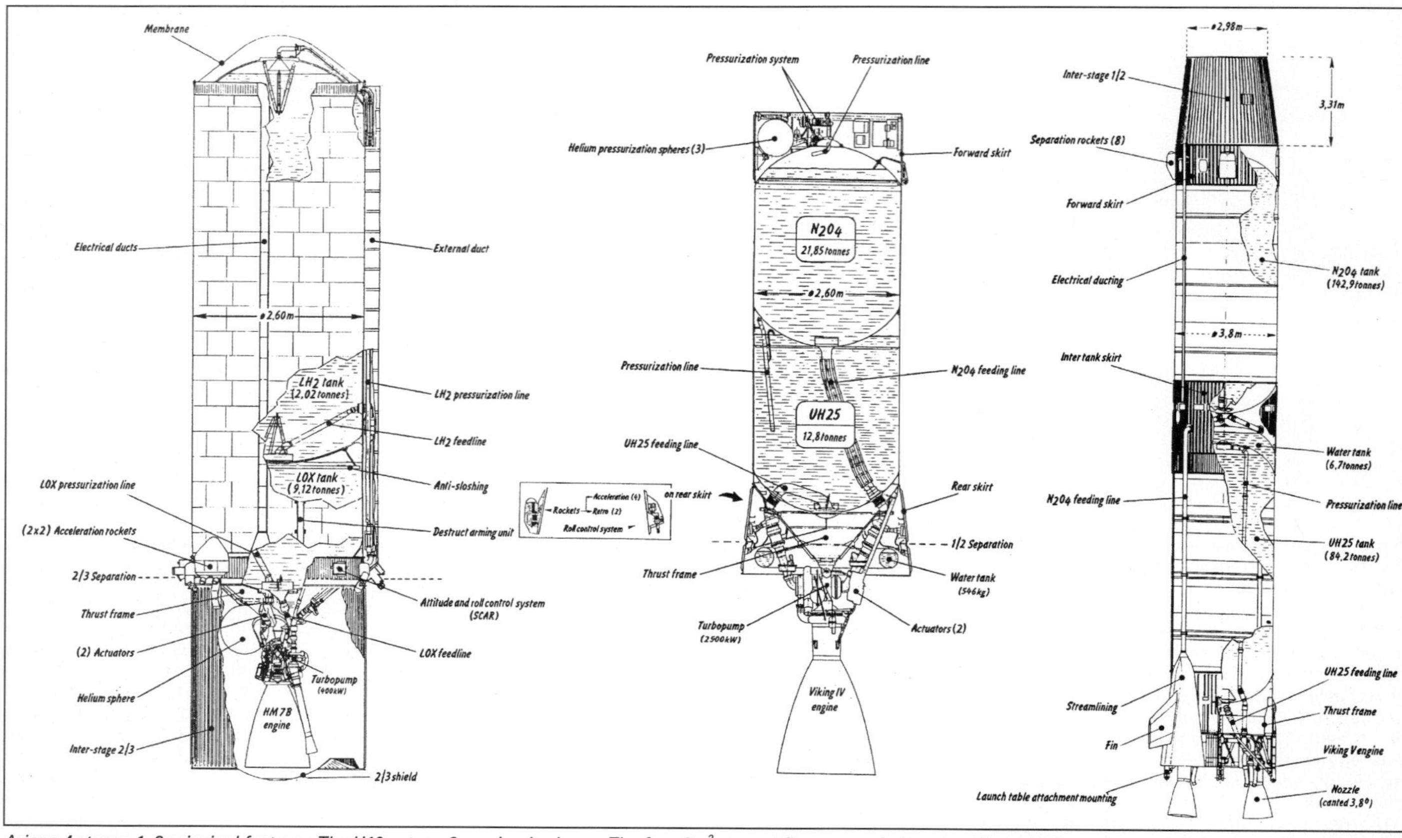

Ariane 4 stages 1–3 principal features. The H10+ stage 3 version is shown. The four 2m² stage 1 fins are carried on the 42L and 44L versions only (Arianespace) 0516852

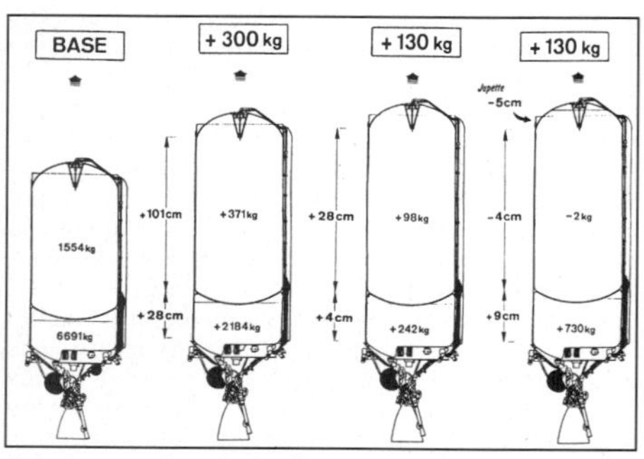

Ariane stage 3 evolution (Arianespace) 0516853

software) failed for the same reason. With neither system operating, the correct guidance and attitude information could no longer be obtained. The fate of the mission thus became inevitable.

As a result of its failure, the active inertial guidance reference system transmitted what was essentially diagnostic information to the launch vehicle's main computer, where it was interpreted as actual flight data and used for flight control calculations. On this basis, the main computer commanded the booster nozzles, and slightly later, the core's main engine nozzle, to make a major correction for what was an imaginary deviation in the vehicle's attitude. The rapid change of attitude caused the launch vehicle to disintegrate at HO+39 seconds due to excessive aerodynamic forces. The vehicle's destruction was automatically initiated at an altitude of about 4 km and at a distance of 1 km from the launch pad.

The Ariane 5 inertial reference system was essentially common to a system which is used for Ariane 4. The part of the software which caused the interruption in the inertial system computers is used before launch to align the inertial reference system and for Ariane 4 is also used to permit a rapid re-alignment of the system in the case of a late hold in the countdown. This re-alignment function does not serve any useful purpose when included in the Ariane 5 software, but it was retained for commonality reasons, and as on Ariane 4 flights it would be allowed to operate for approximately 40 seconds after launch.

During the design of the inertial reference system used for both Ariane 4 and Ariane 5, Arianespace made a decision not to protect the inertial system computer from being made inoperative by an excessive value of the variable related to the horizontal velocity. This projection was provided for several other variables of the alignment software. When this decision was made it had not been realised which values this variable might assume when the alignment software was allowed to operate after the launch of an Ariane 5.

For Ariane 4 flights the horizontal-velocity variable cannot reach (with adequate operating margins) the level which is pre-set in the software. However, Ariane 5 has a high initial acceleration and a trajectory which leads to a high horizontal velocity component which is five times the value for Ariane 4: when the limiting value of the variable was reached during the Ariane 501 launch the inertial system computers ceased operations.

Since it was designed for Ariane 4 launches, the specifications for the inertial reference system and the tests performed at an equipment level did not include simulated Ariane 5 trajectory data. Thus the re-alignment function was not tested under Ariane 5 flight conditions. As a result the design error was not discovered until after the loss of Ariane 501. Post-flight simulations were carried out on a computer running the inertial reference system and with a simulated environment which included the actual trajectory data from Ariane 501. These simulations faithfully reproduced the chain of events leading to the failures of the inertial reference systems.

As a result of its deliberations, the inquiry board made a series of recommendations which included switching-off the inertial reference system immediately after launch: more generally, no software should be run in-flight unless it is specifically required. All the flight software should be reviewed to ensure that all assumptions which were made when the code was written, are validated against the restrictions of the actual launch vehicle and also verification should be made of any internal or communication variables in the software.

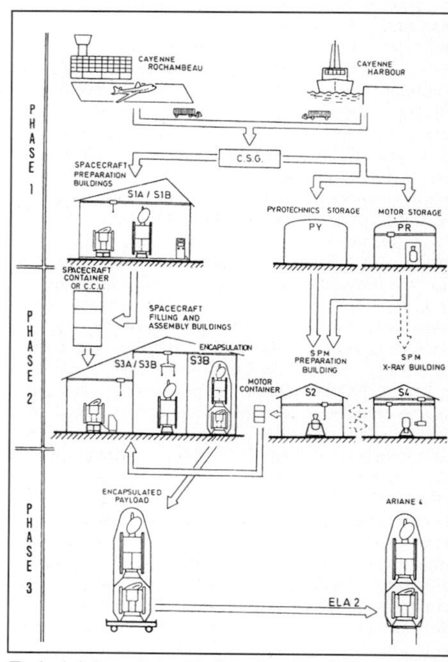

Typical Ariane 4 payload processing flow at Kourou 0516854

The second launch of Ariane 5 carried MAQSAT H and MAQSAT B, instrumented platforms to measure the launch and ascent environment. The two were successfully placed in orbit on 31 October 1997. The third flight (Ariane 503) successfully carried the Atmospheric Re-entry Demonstrator (ARD) to a ballistic re-entry and MAQSAT 3 (a mockup of the Eutelsat W communications satellite) to GTO on 21 October 1998. The fourth Ariane flight (Ariane 504) took place on 10 December 1999 when the

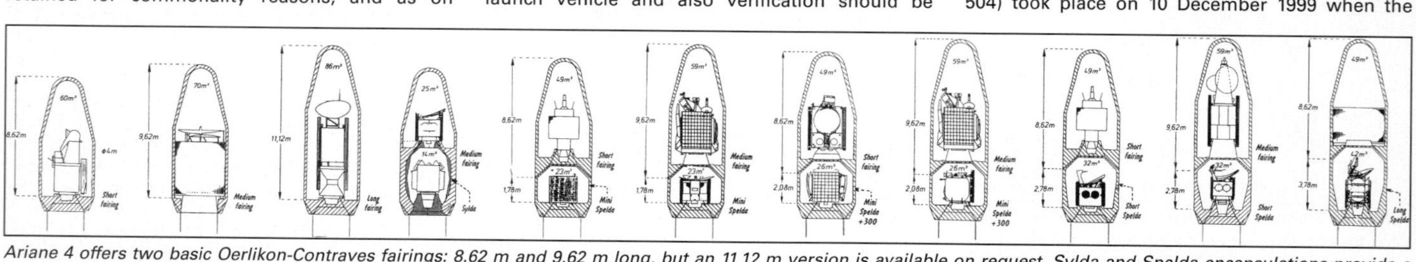

Ariane 4 offers two basic Oerlikon-Contraves fairings: 8.62 m and 9.62 m long, but an 11.12 m version is available on request. Sylda and Spelda encapsulations provide a dual-launch capability. Mini-Spelda debuted in 1993. The Mini-Spelda+300 version became available in 1996, stretching Mini-Spelda by 300 mm (Arianespace) 0517345

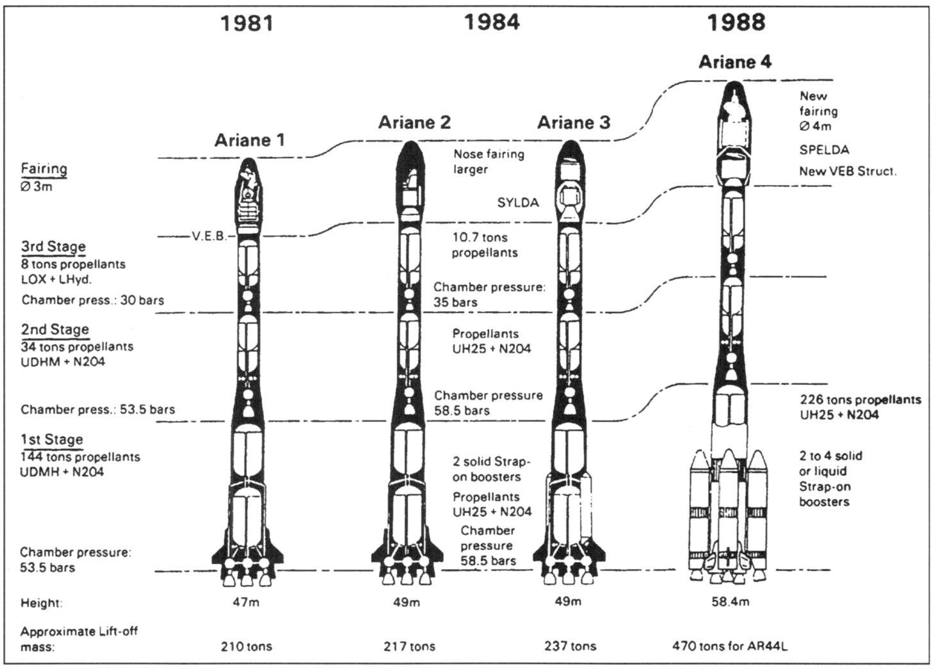

124 Ariane 1–4s have been ordered in 10 batches: 4 development (Ar 1 1979–81), 6 P1 (Ar 1 1982–84), 4 P2 (1 x Ar 1+3 x Ar 2/3 1984–85), 5 P3 (Ar 2/3 1985–87), 7 P4 (Ar 2/3 1987–88), 3 P5 (Ar 4 1988–89), 6 P6 (2 x Ar 2/3+4 x Ar 4 1989), 6 P7 (Ar 4 1989–90), 8 P8 (Ar 4 1990–91), 75 P9 (Ar 4 1992–2000). Totals are 11 Ar 1, 17 Ar 2/3, 96 Ar 4 0516855

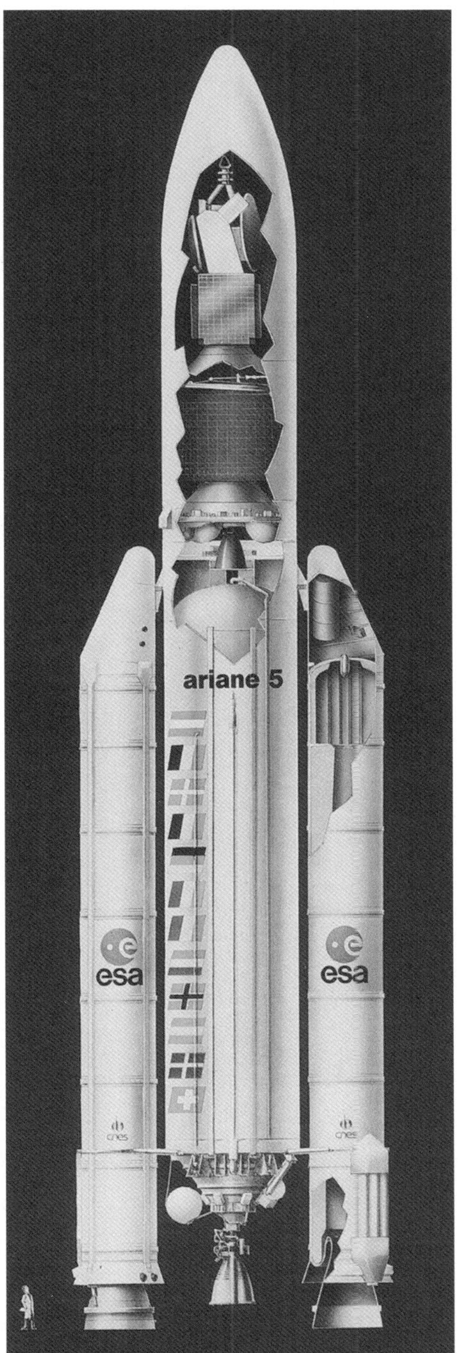

Ariane 5 in dual-launch Speltra configuration (ESA) 0517175

XMM X-ray Multi-Mirror space observatory was placed in an elliptical orbit.

The fifth Ariane 5 launch carried AsiaStar and Insat 3B into orbit on 21 March 1999.

The sixth Ariane 5 flight took place on 15 September 2000 with the successful launch of Astra 2G and GE7.

The seventh Ariane 5 flight on 16 November 2000 carrying PAS 1R, Amsat P3D, STRV 1C and 1D.

The eighth flight on 20 December 2000 carried the LDREX Ariane stage, Astra 2D, GE-8/Aurora 3.

The ninth flight on 8 March 2001 lifted Eurobird and BSAT-2A to geostationary orbit.

The 10th Ariane 5 on 12 July 2001 suffered an upper-stage malfunction dumping Artemis in the wrong orbit, along with BSAT-2B.

The 11th flight on 1 March 2002 lifted ESA's Envisat to orbit.

The 12th Ariane 5 on 5 July 2002 put N-Star and Space Stellat 5 in orbit.

The 13th Ariane 5 on 28 August 2002 put Atlantic Bird and MSG-1 into orbit.

The 14th flight on 11 December 2002 carrying the first modified upper-stage in the EC-A configuration failed and left Hot Bird 7 and Stentor stranded.

The 15th flight on 9 April 2003 lifted Insat 3A and Galaxy 12 to orbit.

The 16th flight on 11 June 2003 put Australia's Optus 1/Defence 1 and Japan's BSAT-2c into orbit.

The 17th flight on 27 September 2003 carried Europe's E-bird, India's Insat 3E and ESA's SMART-1 lunar mission.

The 18th flight on 2 March 2004 put ESA's Rosetta science satellite in orbit.

The 19th flight on 18 July 2004 placed in orbit the 5.9 tonne Anik F2, the heaviest single payload to GTO to date.

The costs for building Ariane are divided among France 46.2 per cent, Germany 22.0 per cent, Italy 15.0 per cent, Belgium 6.0 per cent, Spain 3.0 per cent, Netherlands 2.1 per cent, Sweden 2.0 per cent, Switzerland 2.0 per cent, Norway 0.6 per cent, Austria 0.4 per cent, Denmark 0.2 per cent, Ireland 0.2 per cent. Following go-ahead in January 1985 for the Ariane 5 Preparatory Programme, Arianespace received AU718 million (AU780 million 1995 terms) to proceed. Then October 1995's ESA Ministerial conference approved €1,026.2 million over 1995–2003 for the Ariane 5E upgrade. Of this, €167 million is blocked, intended to increase stage 2 propellant load to 15 tonnes and its thrust to 35 kN. The same Ministerial meeting also approved €351.5 million 1996 to 2000 for ARTA Ariane Research & Technology Accompaniment, to maintain the vehicle's overall reliability, plus €335.7 million 1996 to 2000 to maintain ESA's launch and production facilities.

ESA is developing an orbital transfer vehicle to fit atop the Ariane 5 for Space Station operations. Arianespace is awaiting advances in the Space Station programme before expending funds to man rate the Ariane 5. During its flight programme the core stage had 0.99 reliability, an order of magnitude greater than for Ariane 3/4. The unmanned vehicle's overall reliability target is 0.985. During 1988, Aerospatiale/BAe and MBB/Matra teams undertook preliminary studies of an Ariane 5 Transfer

Ariane 5's Vehicle Equipment Bay (MMS) 0517337

Vehicle, then aiming for a 1996 initial operating capability, based on enhanced versions of Ariane's EPS upper stage, Vehicle Equipment Bay and hardware from other projects as much as possible to minimise development costs. The baseline ATV is intended to be expendable; the cost effectiveness of a reusable version has been shown to be less attractive.

ESA projects one or two flights annually with about 16 tonnes of cargo in logistics carriers. For Space Station missions, ATV would execute an automatic approach on a collision avoidance trajectory. Safety and contamination regulations may demand that only cold-gas thrusters are used for the final approach and grappling by Canada's Shuttle-type Remote Manipulator System. Dwell time at the Station could be up to six months.

The Aerospatiale/MMS (ex-British Aerospace) proposal considerably modifies Ariane's EPS/VEB into a compact 3.936 m diameter (4.57 m diameter envelope), 2.47 m high stage of 2,998 kg dry mass (plus 2,983 kg propellant). The propulsion module incorporates gimballed twin 500 N bipropellant thrusters drawing on 2,400 kg MMH/NTO in four 108.5 cm diameter He-pressurised tanks, supported and backed up by four 4 × 250 N hydrazine thruster clusters. Proximity operations switch to four clusters of 20 N nitrogen jets.

The MBB/Matra studies show an ATV encompassing the VEB and elements of the EPS, maintaining the basic Ariane 5 upper stage configuration as much as possible. The VEB guides Ariane during ascent and, after ATV separation, becomes an integral part of ATV. Dry mass is 3,903 kg with 3,086 kg propellants. The 5.4 m diameter ATV is powered by a single 27.3 kN bipropellant gimballed engine, supported by 4 × 4 350 N hydrazine thrusters and 4 × 4 10 N cold-gas jets for proximity operations.

ESA began the AU37.7 million (1994 rate) Future European Space Transportation (FESTEP) Investigation Programme in February 1994 to define which reusable launchers are relevant to Europe and to develop the required technology. Contributions are 33.5 per cent Germany, 15 per cent Italy, 6 per cent Belgium, 5 per cent Spain, 4 per cent Netherlands, 3.35 per cent Sweden, 2.3 per cent Austria and 1.6 per cent Norway, totalling 70.75 per cent. DASA led the system studies (AU13 million), beginning October 1994. Five technology studies (AU8 million) were placed in first half 1995 on aerothermodynamics, materials, structures, propulsion and heat management. Proposals for FESTIP's next slices, covering larger scale work and experimental vehicles, were discussed in 1996. The continuation of the programme is under request. Concepts which were defined included single-and two-stage launch vehicles, essentially using rocket propulsion and being either partially or fully reusable. The design work is managed by an integrated system concept located at

Ariane 5's L9 EPS stage (ESA) 0517176

DASA Ottobrunn. Technologies being investigated are in the fields of aero-thermodynamics, materials, structures, propulsion and heat management.

For either concept, manned safety rules require ATV to back off automatically if it suffers major failures, providing fail-operational/fail-safe capacity.

Ariane family statistics
First launch: 24 December 1979
Number launched: 159 to end of September 2003
Launch sites: Kourou ELA-1 (Ariane 1 to 3; site closed December 1989) and ELA-2 (Ariane 2 to 4); ELA-3 completed 1995 for Ariane 5
Vehicle career success rate: 94.7% through end of September 2003
Success rate, past 50 launches: 98% to the end of September 2003
(these figures exclude Ariane 501)

Specifications

Ariane 4
First launch: 15 June 1988
Number launched: 116 at programme termination 15 February 2003
Principal uses: delivery of 2,100–4,720 kg payloads into GTO for apogee kick motor insertion into GEO and using the H-10+ third stage

Ariane launch record

	Launch	Type	Payload		Launch	Type	Payload
L01	24 December 1979	1	CAT (monitoring)	V80	17 November 1995	44P/3+	ISO
L02*	23 May 1980	1	CAT, Firewheel, Amsat 3A	V81	6 December 1995	44L/3+	Telecom 2C/Insat 2C
L03	19 June 1981	1	CAT, Apple, Meteosat 2	V82	12 January 1996	44L/3+	PAS 3R, Measat 1
L04	20 December 1981	1	Marecs A	V83	5 February 1996	44P/3+	NStar b
L5*	10 September 1982	1	Marecs B, Sirio 2	V84	14 March 1996	44LP/3+	Intelsat 707
L6	16 June 1983	1	ECS 1, Oscar 10	V85	20 April 1996	42P/3+	MSAT 1
L7	19 October 1983	1	Intelsat 507	V86	16 May 1996	44L/3+	Palapa C2, Amos 1
V8	5 March 1984	1	Intelsat 508	V88*	4 June 1996	5	Cluster
V9	22 May 1984	1	Spacenet 1	V87	13 June 1996	44P/3+	Intelsat 708
V10	4 August 1984	3	ECS 2, Telecom 1A	V89	9 July 1996	44L/3+	Arabsat 2A, Turksat 2C
V11	10 November 1984	3	Spacenet 2, Marecs B2	V90	8 August 1996	44L/3+	Italsat 2, Telecom 2D
V12	8 February 1985	3	Arabsat 1A, Brasilsat 1	V91	11 September 1996	42P/3+	Echostar 2
V13	8 May 1985	3	GStar 1, Telecom 1B	V92	13 November 1996	44L/3+	Arabsat 2A, Measat 2
V14	2 July 1985	1	Giotto	V93	30 January 1997	44L/3+	GE 2, Nahuel 1A
V15*	12 September 1985	3	Spacenet 3, ECS 3	V94	1 March 1997	44P/3+	Intelsat 801
V16	22 February 1986	1	Spot 1, Viking	V95	16 April 1997	44L/3+	Thaicom 3, B-Sat 1A
V17	28 March 1986	3	GStar 2, Brasilsat 2	V97	3 June 1997	44L/4+	Immarsat-3 F4, Insat 2D
V18*	31 May 1986	2	Intelsat 514	V96	25 June 1997	44P/4+	Intelsat 802
V19	16 September 1987	3	Aussat/Optus A3, ECS 4	V98	8 August 1997	44P/4+	PAS-6
V20	21 November 1987	2	TV-Sat 1	V99	2 September 1997	44L/2+P/2+	Hot Bird-3, Meteosat-7
V21	11 March 1988	3	Spacenet 3R, Telecom 1C	V100	24 September 1997	42L/2+	Intelsat 803
V23	17 May 1988	2	Intelsat 513	V101	30 October 1997	5	Magsat H, Magsat B
V22	15 June 1988	44LP	Meteosat 3, PAS 1, Oscar 13	V102	12 November 1997	44L/4+	Sirius-2, Cakrawarta-1
V24	21 July 1988	3	ECS 5, Insat 1C	V103	2 December 1997	44P/4+	JCSAT-5, Equator-S
V25	8 September 1988	3	GStar 3, SBS 5	V104	22 December 1997	44L/2+	Intelsat 804
V26	28 October 1988	2	TDF 1	V105	4 February 1998	44L/4+P/2+	Brasilsat B3, Inmarsat-3 F5
V27	11 December 1988	44LP	Skynet 4B, Astra 1A	V106	27 February 1998	42P/2+	Hot Bird-4
V28	27 January 1989	2	Intelsat 515	V107	24 March 1998	40	Spot 4
V29	6 March 1989	44LP	JCSat-1, Meteosat 4	V108	29 April 1998	44P/4+	Nilesat 101, BSAT-1B
V30	2 April 1989	2	Tele-X	V109	25 August 1998	44P	ST-1
V31	5 June 1989	44L	Superbird 1, DFS 1	V110	16 September 1998	44LP	PAS-7
V32	11 July 1989	3	Olympus 1	V111	5 October 1998	44L/4+	Eutelsat W2, Sirius 3
V33	8 August 1989	44LP	Hipparcos, TV-Sat 2	V112	21 October 1998	5	Maqsat 3
V34	27 October 1989	44L	Intelsat 602	V113	28 October 1998	44L/4+	GE 5, AfriStar
V35	22 January 1990	40	Spot 2, UoSAT 3 & 4,	V114	6 December 1998	42L/2+	SATMEX 5
			Microsats 1–4	V115	22 December 1998	42L/2+	PAS-6B
V36*	22 February 1990	44L	Superbird B, BS-2X	V116	26 February 1999	44L/4+	Arabsat 3A, Skynet 4E
V37	24 July 1990	44L	TDF 2, DFS 2	V117	2 April 1999	42P/2+	Insat 2E
V38	30 August 1990	44LP	Eutelsat 2 F1, Skynet 4C	V118	12 August 1999	42P	Telkom 1
V39	12 October 1990	44L	SBS 6, Galaxy 6	V120	4 September 1999	42P	Koreasat 3
V40	20 November 1990	42P	Satcom C1, GStar 4	V121	25 September 1999	44LP	Telstar 7
V41	15 January 1991	44L	Italsat 1, Eutelsat 2 F2	V122	19 October 1999	44LP	Orion 2
V42	2 March 1991	44LP	Astra 1B, Meteosat 5	V123	13 November 1999	44LP	GE-4
V43	4 April 1991	44P	Anik E2	V124	3 December 1999	40	Helios-B+ Clementine
V44	17 July 1991	40	ERS 1, UoSAT 5, SARA,	V119	10 December 1999	5	XMM
			Orbcomm-X, Tubsat A	V125	21 December 1999	44L	Galaxy X1
V45	14 August 1991	44L	Intelsat 605	V126	24 January 2000	42L	Galaxy XR
V46	26 September 1991	44P	Anik E1	V127	17 February 2000	44LP	Superbard-4
V47	29 October 1991	44L	Intelsat 601	V128	21 March 2000	5	AsiaStar + Insat 3B
V48	16 December 1991	44L	Inmarsat 2 F3, Telecom 2A	V129	19 April 2000	42L	Galaxy IVR
V49	26 February 1992	44L	Arabsat 1C, Superbird B1	V131	17 August 2000	44LP	Brasilsat B4/Nilesat 102
V50	15 April 1992	44L+	Telecom 2B, Inmarsat 2 F4	V132	6 September 2000	44P	Entelsat W1
V51	9 July 1992	44L	Insat 2A, Eutelsat 2 F4	V130	14 September 2000	5G	Astra 2G, GE7
V52	10 August 1992	42P	Topex-Poseidon,	V131	6 October 2000	42L	N-Sat-110
			S80/T, Kitsat 1	V132	29 October 2000	44LP	EuropeStar 1
V53	10 September 1992	44LP+	Hispasat 1A, Satcom C3	V133	16 November 2000	5G	PAS-1R/Amsat P-3D/STRV
V54[1]	28 October 1992	42P+	Galaxy 7				1C/STRV 1D
V55	1 December 1992	42P+	Superbird A1	V134	22 November 2000	44L	Arik F1
V56	12 May 1993	42L	Astra 1C, Arsene	V135	20 December 2000	5G	Astra 2D/GE 8/LDREX
V57[1]	25 June 1993	42P+	Galaxy 4	V136	10 January 2001	44P	Turksat 2A
V58	22 July 1993	44L+	Hispasat 1B, Insat 2B	V137	7 February 2001	44L	Sicral/Skynet 4A
V59	26 September 1993	40	Spot 3, Stella, Kitsat 2, Itamsat	V138	8 March 2001	5G	Eurobird/BSat 2a
			PoSAT 1, Healthsat 2, Eyesat	V139	9 June 2001	44L	Intelsat 901
V60	22 October 1993	44LP	Intelsat 701	V140	12 July 2001	5G	Artemis/BSat 2G
V61	20 November 1993	44LP+	Solidaridad 1, Meteosat 6	V141	30 August 2001	44L	Intelsat 902
V62	18 December 1993	44L+	DBS 1, Thaicom 1	V142	25 September 2001	44P	Atlantic Bird 2
V63*	24 January 1994	44LP+	Turksat 1, Eutelsat 2 F5	V143	27 November 2001	44LP	DirecTV4S
V64	17 June 1994	44LP+	Intelsat 702, STRV 1A/1B	V144	23 January 2002	44L	Insat 3C
V65	8 July 1994	44L+	PAS 2, BS 3N	V145	23 February 2002	44L	Intelsat 904
V66	10 August 1994	44LP+	Brasilsat B1, Turksat 1B	V146	1 March 2002	5	Envisat 1
V67	9 September 1994	42L+	Telstar 402	V147	29 March 2002	44L	JCSat 8/Astra 3A
V68	8 October 1994	44L+	Solidaridad 2, Thaicom 2	V148	16 April 2002	44LP	NSS-7
V69*	1 November 1994	42P+	Astra 1D	V149	4 May 2002	42P	SPOT-5
V70*	1 December 1994	42P/3+	PAS 3	V150	5 June 2002	44L	Intelsat 905
V71	28 March 1995	44LP+	Hot Bird 1, Brasilsat B2	V151	5 July 2002	5	STELLAT5/N-Star 3
V72	21 April 1995	40+	ERS 2	V152	28 August 2002	5	Atlantic Bird/MSG-1
V73	17 May 1995	44LP/3+	Intelsat 706	V153	6 September 2002	44L	Intelsat 906
V74[1]	10 June 1995	42P/3+	DBS 3	V154*	11 December 2002	5	Hot Bird-7/Stentor
V75	7 July 1995	40/3+	Helios 1A, Cerise, UPM-Sat 1	V155	20 December 2002	44L	NSS-6
V76	3 August 1995	42L/3+	PAS 4	V156	15 February 2003	44L	Intelsat 907
V77	29 August 1995	44P/3+	NStar a	V157	9 April 2003	5G	Insat 3A/Galaxy 12
V78	24 September 1995	42L/3+	Telstar 402R	V158	11 June 2003	5G	Optus/Defence C1/BSAT-2c
V79	19 October 1995	42L/3+	Astra 1E	V159	27 September 2003	5G	Insat 3E/E-BIRD/SMART-1

*indicates launcher failure (L02: stage 1 Viking combustion instability. L5: HM7 turbopump lubrication. V15/V18: stage 3 igniter. V36: stage 1 engine cooling. V63: stage 3 LOX pump bearing. V70: blockage in LOX stream to stage 3 gas generator). Arianespace assumed operational responsibility from flight ('vol') 9, following ESA's L01–04 development and L5–8 promotional phases. Notes: + indicates use of H10+ stage 3, +3 = H10+3 stage 3; 1: use of PVA Perigee Velocity Augmentation.

Ariane 42P launches the Hot Bird 4 satellite in February 1998 (ESA) 0054364

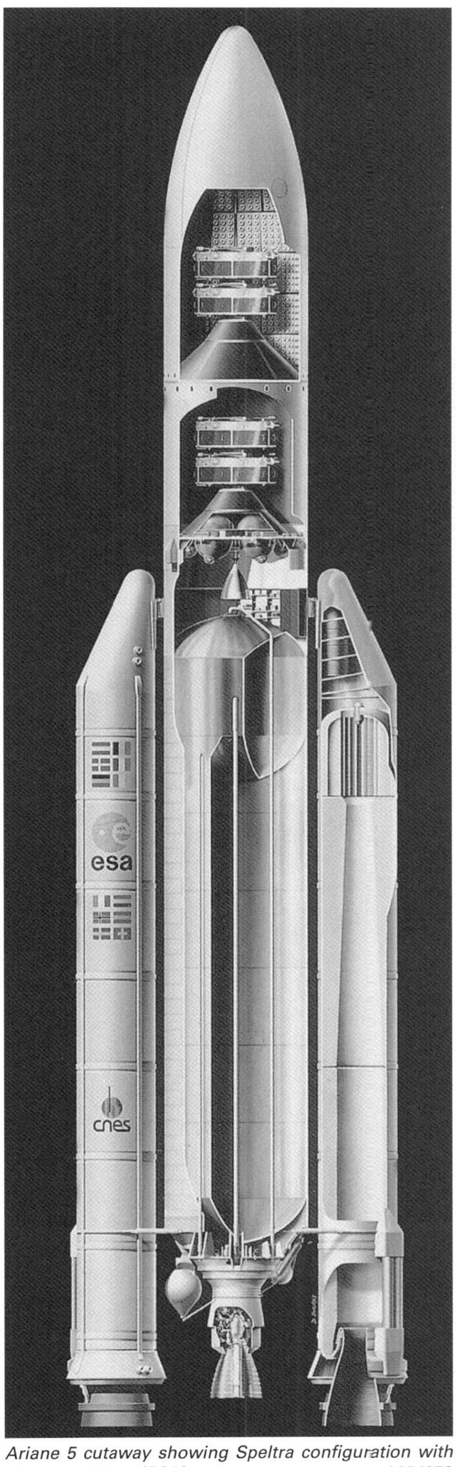

Ariane 5 cutaway showing Speltra configuration with cluster satellites (ESA) 0054378

Launch sites: Kourou ELA-2
Vehicle career success rate: 96.9% through end September 2000
Success rate, past 50 launches: 100% through end 2001
Cost: US$40–50 million for 1,200–1,600 kg (A40), US$55–65 million for 2,000–2,500 kg (A42P), US$65–80 million for 2,500–3,000 kg (A44P), US$90–110 million for >3,000 kg (A42L and above), US$1 million for ASAP platform, US$19.5–26 million for 400–1,100 kg SDS Spelda Dedicated Satellites using the Mini-Spelda. (European governments typically pay a 15–20% surcharge, but ESA in October 1995 agreed to drop the requirement.) Users are separately offered a FFr2.5 million package of services, including payload transport from Cayenne airport to Kourou and the use of the integration facilities. The payment plan typically calls for a down payment on contract signature of 10–15% of the total price, followed by quarterly instalments of 10% and four payments of 5% in the last eight months before launch. Contract signature is preferred 36 months before launch. A reflight is guaranteed (paid for by insurance) within nine months of request in the event of vehicle failure. If a satellite fails after <27 months in orbit a new launch can be made within 9–12 months of request. Third party liability is covered by a mutual non-recourse clause and insurance subscribed by Arianespace for damages up to FFr400 million (the French government covers higher claims).
Availability: 10 launches per year.
Performance:
dedicated 7° GTO: 1,900 kg (A40; H10+ 2,020; H10-3 2,105), 2,600 (A42P; H10+ 2,740; H10-3 2,930), 3,000 (A44P; H10+ 3,290; H10-3 3,465), 3,200 (A42L; H10+ 3,350; H10-3 3,480), 3,700 (A44LP; H10+ 4,030; H10-3 4,220), 4,200 (A44L; H10+ 4,460; H10-3 4,720).
Sun-synch: 2,740 (only used by A40).
Number of stages: 3 plus up to 4 strap-ons.
Overall length: 56.35–60.13 m depending on fairing/payloads (minimum length uses short fairing; maximum uses medium fairing plus short Spelda).
Principal diameter: 380.0 cm.
Launch thrust (kN): 2,720 (A40), 3,944 (A42P), 4,060 (A42L), 5,140 (A44P), 5,270 (A44LP), 5,400 (A44L).
Ignition mass (H10+/10-3): 245 t (A40), 324 t (A42P), 356 t (A44P), 363 t (A42L), 421 t (A44LP), and 484 t (A44L).
Guidance: Matra Marconi Space inertial in stage 3 Vehicle Equipment Bay, with Sextant laser gyro providing attitude data (GEC-Marconi gyro back-up) from stage 1 separation. Saab Ericsson Space provide the computers. Some 600 parameters are telemetered on 2.1 GHz. Ariane follows

The first Ariane 5 lifts off ELA-3 in May 1996 (ESA) 0054379

Ariane 501 on its roll-out from the Launcher Integration Building showing the enclosed aft area of the first stage (ESA) 0054380

pre-programmed attitude profile during first stage burn. Typical GTO standard deviation values are 50 km apogee, 1 km perigee, 0.018° inclination

Ariane 4 solid strap-ons
Stretched versions of Ariane 3's PAP (Propulseur d'Appoint Poudre) strap-ons are carried by the 42P/44P/44LP variants, integrated on the pad. Nozzle canted 12°. The solid pair ignites at launch, 4.2 s after the main engines. Two spring pairs (66 kN forward, 59 kN aft) ensure 6°/s rotation away from base at 5 m/s separation rate. Separation in pairs 89 s after launch (42P), 70 + 70.3 s (PAP 1/3 + 2/4, 44P) and 67 s (44LP).
Contractor: Aerospatiale (stage), Fiat Avio (BPD (propulsion), MAN (mechanical equipment)
Length: 1,204.8 cm
Principal diameter: 107.1 cm
Mass at ignition: 12,660 kg each
Propellant mass: each 9,500 kg of CTPB 1613
Average thrust: 650 kN each, sea level (625 kN at H0)
Burn time: 236 s
Casing material: AISI 4130 steel, 5 mm thick

The first two of four cluster satellites attached to the Ariane 5 adapter 0084635

Ariane 4 liquid strap-ons
Pairs of PAL (Propulseur d'Appoint Liquide) strap-ons, comparable in size/performance to stage 2, are carried by 42L/44LP/44L (the last two pairs). Each comprises a fixed Viking 6 engine drawing on propellants from identical 2.1 mm-thick stainless steel tanks (weighing 830 kg each), with water coolant provided via stage 1's intertank skirt. The PALs are pyrotechnically released at 149.1 s/37.5 km after burnout at 143.6 s and separated by six solid BPD motors identical to the stage 1 retrorockets.
Contractor: DASA (prime, integration), Alenia (main tanks), Sabca (thrust frame, intertank, forward skirts), Fokker (nose cone).
Engine: single SEP Viking 6 canted at fixed 10°
Length: 1,900 cm
Principal diameter: 221.6 cm (internal 217.4 cm)
Dry mass: 4,540 kg
Oxidiser: nitrogen tetroxide, in forward tank pressurised at 4.45 atmospheres.
Fuel: UH25 (UDMH+25% hydrazine hydrate), in aft tank pressurised at 4.45 atmospheres.
Propellant mass: 39,000 kg (A44L/44LP), 37,000 kg (A42L).
Average thrust: 739 kN each vacuum (670 kN at launch).
Burn time: 140 s (igniting with main engines, 3.4 s before launch)

Four cluster satellites carried on the first Ariane 5 flight on 4 June 1996 0084627

Ariane 4 stage 1
The Ariane 3 stage was stretched 6.7 m to increase propellant capacity from 144 t to 227 t and the four Viking 5 engines qualified to 300 s to accommodate the burn increase from 138 to 205 s. Fins are carried only on the 42L/44L variants. The two 10.09 mm long, 3,800 mm diameter identical 15CDV6 steel tanks are separated by a 2,688 mm high interstage and an 6,736 l water tank 730 mm high with a GFRP forward dome and sharing the UH25 tank's forward bulkhead. The water coolant is also routed to the PAL liquid strap-ons, when carried. The stage's 1,500 mm high forward skirt supports the eight Fiat Avio/BPD separation retrorockets. Stage 1 + 4 PALs (i.e. 44L configuration) provide about 3,040 m/s.
Designation: L220
Contractor: Aerospatiale (prime): Man (thrust frame, water tank), Fokker (interstage), CASA (inter tank, forward skirt), SABCA (streamlining, fins)
Engine: four gimballed SEP Viking 5, canted at fixed 3.8°. For each, a 35 atmosphere gas generator drives a single 10,000 rpm turbopump to deliver propellants and water coolant in addition to maintaining tank pressures at 5 atmospheres. UH25/NTO delivery pressures are 67/69 atmospheres.
Overall length: 28.388 m (including 3.31 m interstage 1/2)
Principal diameter: 380 cm
Dry mass: 17,560 kg
Oxidiser: nitrogen tetroxide, in forward tank
Fuel: UH25 (UDMH + 25% hydrazine hydrate) in aft tank
Propellant mass: 167.5 t (A40), 217.2 t (A42P), 201 t (A42L), 227.1 t (A44P/44LP/44L).
Stage thrust: 2,718 kN sea level
Burn time (s): 150 (A40), 196 (A42P), 181 (A42L), 205 (A44P/44LP/44L)
Attitude control: engine gimballing
Separation: stage 1/2 separation is commanded by the VEB when stage 1 thrust decays to 50%. The

Four cluster satellites after release of the fairing and separation between the first group of two cluster with the second group inside spectra 0084628

four stage 2 acceleration solids fire, followed 2 s later by the interstage explosive cord + stage 1's eight Fiat Avio/BPD solid retros. Stage 2's Viking 4 ignites after a further 1 s, its solids burn out 4 s later and are jettisoned with a 4 s delay.

Ariane 4 stage 2

Essentially the same as Ariane 3's stage 2 but structurally stiffened. The aluminium alloy propellant tanks form a cylinder, height 6,515 mm, with hemispherical bulkheads, divided into two vessels by a common hemispherical bulkhead with its concave face forward. The NTO feed line passes through the lower tank. The thrust frame includes a 188 mm high cylindrical section and a 1,350 mm high cone bearing the gimbal unit mounting flange. The frame connects to the stage 1/2 interstage conical skirt via the aft conical skirt, height 1,570 mm. This aft skirt incorporates the Viking 4 engine's toroidal water coolant tank (overall diameter 2,240 mm, tube diameter 340 mm). The 1,245 mm high forward skirt connects to the carbon fibre stage 2/3 interstage and houses three 300 atmosphere helium spheres for 3.5 atmosphere tank pressurisation.

Designation: L33

Contractor: DASA (prime, integration), DASA/Dornier (tanks), Zeppelin (water tank), Aljo (thrust frame, skirts).

Engine: single SEP Viking 4 (chamber pressure 58.5 atmospheres), gimballed for pitch/yaw control

Overall length: 1,161 cm

Principal diameter: 260.0 cm

Oxidiser: nitrogen tetroxide

Fuel: UH25 (UDMH + 25% hydrazine hydrate)

Propellant mass: 34.6 t

Dry mass: 3.4 t

Average thrust: 794 kN vacuum

Burn time: 125 s

Attitude control system: pitch/yaw by engine gimballing; roll by two sets of 50 N thrusters mounted on rear skirt fed from Viking's gas generator. Guidance begins about 10 s after stage 1 separation (stage 1 executes pre-programmed profile).

Separation: at about 344 s, 147.4 km, 5,379 m/s, assisted by two solid deceleration rockets; sequence begins when stage 2 has added required ΔV.

Ariane 4 stage 3

Essentially the same as Ariane 3's stage 3 to provide direct burn into GTO (although sub-GTO first used on V54), but structurally stiffened. The associated aft carbon fibre interstage skirt, developed by Fokker Space, is 53 kg lighter than its aluminium predecessor, adding 12.5 kg to GTO capability. The cryogenic propellants are housed in 7020 aluminium alloy tanks separated by a phenolic honeycomb bulkhead under vacuum; external surfaces are treated with thermal insulation. The tank assembly is extended by a 451 mm high forward skirt connecting to the equipment bay, and at the rear by a 288 mm high skirt that provides a mounting for the thrust frame, consisting of a 1,184 mm high cone and 499 mm high cylindrical section. This cylinder carries the acceleration rockets and pyrotechnic cutting system for stage 2/3 separation. The stage 2/3 skirt, 2,730 mm high, is sealed at its lower end to minimise heat exchange. The P9 Ariane production batch introduced a stretched stage 3 (H10+) on V50 in 1992, carrying an additional 340 kg of propellants (242 kg LOX + 98 kg LH_2) in stretched tanks (4 cm LOX + 28 cm LH_2) and increasing burn time by 30 s, to accommodate satellite growth. Dry mass was reduced by 30 kg by reducing wall thickness to 1.5 mm, 1.4 mm with internal integral longitudinal stiffeners. The modification and vehicle weight-saving measures improved GTO capability by 200 kg (stage 3 contributing 130 kg), increasing 44L's to 4,460 kg. The last of the old tanks flew on V60 in October 1993. PVA Perigee Velocity Augmentation was used for the first time on V54. Stage 3 established a sub-GTO with 27,720 km apogee, allowing the payload itself to establish GTO. The H10-3 modification was introduced on V70 in December 1994, improving 44L capacity to 4,720 kg. This is achieved by reducing the thermally stratified unused LH_2, increasing the flow rate and moving the internal bulkhead slightly (see diagram) to carry more LOX (increased mixture ratio). These measures together make another 730 kg LOX available. The safety margin of unused propellants was previously 180–200 kg and once was 250 kg. The performance of the turbopump exhaust heater was improved to maintain the same margin on the He pressurisation system. V93 introduced an increase to 4,900 kg GTO by further weight saving, better engine calibration and new trajectory optimisation. Specifications are given in H10/H10+/H10-3 order.

Designation: H10/H10+/H10-3

Contractor: Aerospatiale (prime, integration), Air Liquide (tanks), Fokker (thrust frame, interstage 2/3).

Engine: single cryogenic open-cycle SEP HM-7B, gimballed for pitch/yaw control. The LOX/LH2 gas generator drives a turbopump including two 60,000 rpm LH2/13,000 rpm LOX shafts. LOX is routed directly to the injectors; LH2 is heated to 150 K as it cools the engine nozzle. This gas also pressurises the fuel tank; before ignition both tanks are pressurised by helium from a single 90K/200 atmosphere sphere. Chamber pressure is 36.2 atmospheres (37.2 atmospheres H10-3).

Overall length: 1,073.4/1,105.4/1,105.4 cm

Principal diameter: 260.0 cm

Oxidiser: liquid oxygen

Fuel: liquid hydrogen

Propellant mass: 10.80/11.14/11.86 t

Dry mass: 1.2 t/1.24/1.24 t, excluding interstage 2/3

Mixture ratio: 4.76/4.66/5.04

Average thrust: 63/63.2/64.8 kN vacuum

Burn time: 720/750/780 s, until desired GTO has been achieved in single direct ascent burn beginning about 347 s after launch.

Attitude control system: the Roll & Attitude Control System (SCAR) maintains roll control during main engine burn (engine gimballing provides pitch/yaw) and three-axis following cut-off. SCAR consists of two sets of diametrically opposed triple thrusters fed with GH2; total impulse >15 kNs. Two aft-facing nozzles depressurise the LOX tank after cut-off and are sequenced as part of the attitude control system and collision avoidance manoeuvre following payload separation. Spin stabilisation of about 10 rpm is provided for payload deployment.

Ariane 4 Vehicle Equipment Bay

Matra Marconi Space's 520 kg 1,037 mm high VEB performs guidance (Sextant or GEC-Marconi laser gyros), data processing, sequencing, telemetry, tracking and destruction function, as on all Ariane variants, but was completely reconfigured for Ariane 4. A digital flight control system replaced Ariane 2/3's analogue equipment. There are four principal elements fabricated from aluminium honeycomb with carbon fibre facing (provided by CASA). They are the internal cone, providing 1,920 mm diameter interface to the payload, the external 1,000 mm high cone mating to either the 3,936 mm diameter fairing or Spelda at its forward face, a horizontal annular plate carrying the electronics, and 12 removable carbon fibre panels enclosing the equipment compartment while providing simple access on the pad.

Payload fairing/carriers

Three basic lengths of Oerlikon-Contraves (prime contractor) 25 mm-thick aluminium alloy honeycomb two-piece fairings are available. They are 8.6 m (type 01, 740 kg, 60 m^3), 9.6 m (type 02, 800 kg, 70 m^3) and 11.1 m (type 03, 86 m^3, available

Solid Booster structures manufactured by SABCA of Belgium for Ariane 5 0084631

Rear skirt of Solid Booster for Ariane 5 0084632

Hydraulic servo-activators and mechanical structures for Ariane 5 0084633

on special request). Ariane 4 is the first fairing to provide a standard 4 m diameter, with 3,650 mm payload envelope. Pyrotechnic cord and piston separate the fairing halves when the heating flux has reduced to 1,135 W/m^2, at about 285 s/115.3 km during stage 2's burn. The base clamp band is released first. Integration is the responsibility of SF Emmen. The following standard payload adaptors are available (numerical designation indicates diameter in mm at satellite interface):

937 truncated cone providing four springs for 0.5 m/s separation. Eight versions: 937/937A, 1.5 t, 48 kg carbon fibre; 937B, 2.3 t, 60 kg carbon fibre; 937C, 1.4 t, 54 kg metallic; 937V, 1.8 t, 75 kg carbon fibre; 937VB, 2.3 t, 80 kg carbon fibre, 937VD, 2.4 t, 88 kg, metallic; 937D, 3.2 t, 90 kg, metallic.

1194 in four versions: 1194, 2.6 t; 1194A, 3.5 t, 45 kg carbon fibre; 1194B, 2.8 t, 60 kg metallic truncated cone; 1194V, 3.5 t, 72 kg carbon fibre.

1666A, 2.8 t, 50 kg aluminium alloy truncated cone, 4–8 actuators provide 0.5 m/s separation.

1920, adapts 1666A for HS-601 satellites. Three versions, 400 mm height (22 kg), 500 mm height (25 kg), 550 mm heightt (26 kg). The prime contractors for the payload adapters are Aerospatiale (937/937A, 937C, 937D, 937VD, 1194A, 1194B, 1920), CASA (937B, 937V, 937VB, 1194V, 1666A) and Matra Marconi Space (1194).

Acceleration load: 4.0 *g* at end stage 1 burn, 4.1 *g* at end stage 2, 1.6 *g* end stage 3 (for A40); 7.3 *g* rms max vibration.

Acoustic load: 142 dB integrated at launch + transonic

Thermal load: max 500 W/m^2 radiated by fairing/Spelda

The 60 kg ASAP Ariane Structure for Auxiliary Payloads developed by Matra Marconi Space, was introduced in January 1990 to carry up to six small (<50 kg) piggyback payloads. (Up to a total of 200 kg on a 40 mm-thick aluminium honeycomb circular plate with an outer diameter 290 cm, and an inner 210 cm extending into the base of the payload fairing volume below the primary satellite). ASAP is designed for missions with a single primary payload, usually found on polar launches. Each payload and its separation system are contained within a 350 × 350 × 450 mm volume. First flight was V35.

Vehicle/payload processing

The Operations Directorate is responsible for the vehicle stages from the time they leave the factory until launch. Launch campaigns now require typically 27 working days and some 400 people for the vehicle, payloads and site. New procedures introduced in 1992 (with the P9 batch) reduced the number of launch operations from 1,200 to 700. Kourou's ELA-2 complex is designed to accommodate up to 10 Ariane 4 departures annually. Stages are shipped down the Seine from Aerospatiale typically nine weeks before launch from Les Mureaux to Le Havre, then to Cayenne (Kourou's Pariacabo harbour from 1995) for road transfer to CSG. Propellants (except LOX/LH$_2$) are also shipped in. Erection of the three stages and liquid boosters in the assembly building requires about three weeks before transfer to the pad, two weeks before launch and a 38 h countdown distributed over two days. The solid strap-ons are added on the pad. Payloads are received and undergo the first processing phase in building S1, while kick motors are handled in S4 (x-ray) and S2. Spin balancing and hydrazine filling are conducted in S3A/B, completed by encapsulation and transfer to ELA-2 typically six days before launch, for mating the following day. Stages 1/2 + PALs are fuelled the day before launch and final countdown begins 16 hours 40 minutes before ignition (times are referenced to ignition; lift-off occurs 4.4 seconds later). Service tower removal at 5:25 before launch, stage 3 filling start at 3:35, apogee kick motor armings at 50/45 minutes, and finally initiation of the automatic launch sequence 6 minutes before ignition. Ariane switches to internal batteries at 60 seconds and the inertial platform is released at 9 seconds. The two computers in the Ariane Launch Center confirm that all liquid propellant engines are burning satisfactorily 4.1 seconds after ignition, and then fire the solids and release the hold-down arms. If any hold in the synchronised sequences is called up to 5 seconds before ignition, Ariane automatically resets to its 6 minute configuration.

Ariane 5

Launch sites: Kourou, French Guiana

First launch: 4 June 1996

Number of launches: 19 through end July 2004 (3 failed to deliver payloads to advertised orbit)

Success rate: 84% through end July 2004

Performance: 6,800 kg single or 5,970 kg double payload version into 7° GTO. 10,000 kg into 800 km

PanAmSat PAS-1R was launched by Ariane 5 (Boeing) ***NEW***/1047468

98.6° Sun-synchronous. 18 t into 70 × 300 km, 51.6° (Space Station transfer). 4,450 kg dedicated into lunar transfer orbit, shared GTO mission could deliver 400 kg lunar orbiter + 200 kg-payload lunar lander.
Number of stages: 2 + 2 strap-ons
Overall length: 45.71–51.37 m
Principal diameter: 5.40 m^2
Launch mass: 746 t
Guidance: provided by the Vehicle Equipment Bay encircling stage 2. Accuracy (1 Σ): 26 km semi-major axis, 0.02° inclination GTO; 4 km, 0.04° 800 km 98.6° Sun-synchronous

Ariane 5 boosters
Designation: P230 (EAP Etage Accélérator à Poudre)
Contractor: Aerospatiale (stage integrator), Europropulsion (motors), Parachutostroenye (recovery parachutes).
Overall length: 31.16 m
Principal diameter: 3.049 m

Ariane 5 was selected as the launch vehicle for the Geosynchronous Spacecraft Life Extension System for launches beginning in 2005 ***NEW***/1047498

Propellant type: PCA + HTPB + aluminium solid
Propellant mass: 237.7 t each
Empty mass: 40 t
Thrust: 5,250 kN sea level (5,904 kN vacuum) at launch
Burn time: 132 s

The boosters from two launches annually will be recovered from the Atlantic 450 km downrange following separation at about 55 km altitude for inspection to check design/manufacturing margins. Six chutes (auxiliary, three drogues, secondary, main) for 27 m/s impact. Originally sized to carry 170 t of propellant each, the boosters were upgraded to cope with satellite growth (including Hermes) and a demand for uprated GTO performance over initial requirements. First static test 16 February 1993; last two of the seven used flight-type motors (#6 first quarter was successful 10 March 1995; number 7 second quarter 21 July 1995). See the Europropulsion entry for motor details.

Ariane 5 stage 1 (core)
The core stage delivers the upper composite into 50 × 1,300 km, 7° on GTO missions (which ARD will employ for its re-entry mission).
Designation: H155 (EPC Etage Principal Cryotechnique)
Contractors: Aerospatiale (stage integrator), SEP (engine), Fokker (thrust frame), MAN (forward skirt), Cryospace (tanks), SABCA (engine actuation system).
Engine: single cryogenic gas generator cycle SEP Vulcain, began stage-level qualification firings 17 November 1994 using a battleship stage (concluded 27 January 1995); flight-type stage testing May 1995 – January 1996. Thrust structure 1.6 t, 3.6 m ht, 5.40 m diameter. 501 carried Vulcain M14, 502 M16, 503 M17 (M15 used for tests).
Overall length: 30.7 m
Principal diameter: 5.40 m
Oxidiser: 130.60 t max of liquid oxygen in 120 m^3 forward tank, pressurised to 3.5 atmosphere by helium (140 kg in 830 mm diameter 400 atmosphere spheres)
Fuel: 25.60 t max of LH2 in 390 m^3 aft tank, pressurised to 2.5 atmosphere by GH2 from Vulcain
Dry mass: 12.6 t; launch mass 167.5 t
Average thrust: 1,145 kN in vacuum, 900 kN at launch
Burn time: 580 s
Attitude control system: Vulcain gimballed ±6° for pitch/yaw

Stage 1 will decay naturally from its low perigee. The core, originally sized to carry 120 t of cryogens, was upgraded because of satellite growth (including Hermes') and a demand for uprated GTO performance. PDR was conducted March 1990. As the aluminium 2219 5 t tank is <2 mm thick. The tank is first riveted to the forward skirt from which it is suspended. Then it is placed on the thrust frame. A 2 cm layer of expanded polyurethane insulates the tank. 18.5 cm diameter lines carry the propellants to the engine. On the launch pad the two boosters at the level of the forward skirt support the vehicle, and the EPC is suspended. The thrust from the boosters is transferred to the vehicle via the skirt.

Ariane 5 stage 2 (upper stage)
An upgrading of stage 2 from L5 to L7 and then L9.7 EPS was necessary to optimise the configuration and accommodate project payload growth. Reliability goal is 0.997. A design requirement calls for the stage to accommodate 60-day standbys after propellant loading (precluding cryogenics). PDR was conducted October 1989, CDR February 1995. Stage safing at mission's end will prevent explosive creation of debris. First complete stage testing began September 1994; the first 1,075 s long duration burn was successful 5 October 1994. Qualification using a flight-type stage at Lampoldshausen was conducted 18 July – 20 September 1995 with two 1,100 s and nine 10–100 s tests simulating the various loading, pressurisation and burnout configurations.
Designation: L9.7 (EPS Etage à Propergols Stockables).
Contractor: DASA (stage integrator, propulsion); sub-contracts to CASA (structure), Zeppelin (tanks).
Engine: single DASA 29 kN Aestus storable propellant, open cycle, pressure-fed, reignitable, gimballed ±6° for pitch/yaw control.
Overall length: 3.3 m
Principal diameter: 3.936 m
Propellants: 6.5 t N_2O_4 + 3.2 t MMH in four 1.410 m diameter tanks at 20 atmosphere, two 0.3 m^3 helium pressurant 830 mm diameter spheres (400 bar).
Dry mass: 1,190 kg
Thrust: 29 kN vacuum
Burn time: 1,100 s.

Ariane 5 VEB

Matra's Vehicle Equipment Bay commands missions from its position encircling stage 2 and, unlike its predecessor Ariane 4 VEB, incorporates an attitude control system, providing roll control during stage 1/2 burns and three-axis control after burnout. The 5.4 m diameter, 1.56 m high, 1,400 kg unit carries 70 kg of hydrazine.

Payload fairing/carriers

Contraves provides two standard 5.40 m diameter fairings to accommodate 4.57 m diameter payloads: 12.70 m/2.3 t short and 17 m long units. The ogive fairing jettisons at about 191s/106 km during stage 1 burn when aerodynamic heating has reduced to 1,135 W/m^2. Dornier's 880 kg Speltra (Structure Porteuse Externe de Lancements Triples Ariane), the equivalent of Ariane 4's Spelda, increases the length of the upper composite by 5.66 m and permits the launch of double payloads. Dornier was awarded the contract in early 1990; the first Speltra was delivered first quarter 1995. Standard payload adaptors are available with interface diameter 937; 1,194; 1,666; 2,624 mm.

Acceleration load: 4.2 *g* at end P230 burn, 3.4 *g* at end EPC burn, 0.4 *g* at end EPS burn; lateral acceleration does not exceed 0.25 *g*.
Acoustic load: 142 dB at launch/transonic
Thermal load: max 1,000 W/m^2 radiated by fairing + VEB

Ariane 5 vehicle processing

Kourou's ELA-3 and associated processing areas were constructed as dedicated Ariane 5 facilities to permit up to eight launches annually. Unlike Ariane 4, the payload assembly is integrated with the vehicle before they are transported to the pad only 8 h before launch, in order to minimise pad operations (similar to CIS launch operations). A launch campaign covers 21 days; the payload is mated six days before launch. The simplified pad concept deletes the requirement for large cryogenic arms on the umbilical tower by feeding the propellants from below the mobile launch table. It also reduces vulnerability to pad and launch accidents. There are four principal buildings in the preparation zone. The BIP Batiment d'Intégration Propulseur integration hall for the P230s to be assembled/checked out, the BIL Batiment d'Intégration Lanceur launcher integration building where the core is erected on the mobile platform and the boosters added, the BAF Batiment d'Assemblage Final assembly building where the payload composite is assembled and erected, the stage 2 tanks filled and the final electrical checkout conducted, and the Launch Center (CDL-3) for launch operations with two vehicles simultaneously. The pad was also used for stage 1 static firings. LOX/LH$_2$ production and mixing/casting P230's two large segments are performed on-site.

Ariane 5 P230

First flight: 4 June 1996 (February 1993 first static firing, July 1995 end of ground qualification).
Mass: 268.7 t (237.7 t propellant): 106.9 t aft, 107.4 t centre, 23.4 t forward; nozzle 6 t; internal insulation 4.3 t; casing 20 t; igniter 0.3 t; 0.4 t miscellaneous).
Length: 26.774 m (stage 31.16 m)
Diameter: 3.049 m
Propellant: type: HTPB 68% ammonium perchlorate, 18% aluminium, 14% PBHT liner.
 shape: main section cylindrical, forward segment 15 rectangular slots mass fraction: 0.884 for 237.7 t loading.
Burn time: 132 s
Thrust: 5250 kN sea level at launch (5904 kN vacuum); 4931 kN average vacuum, 6637 kN max vacuum.
Specific impulse: 271 s vacuum
Total impulse: better than 630 MNs
Maximum pressure: 59.9 atmosphere
Nozzle: mass: 6 t
 throat diameter: 90.0 cm
 length: 3.420 m (2.22 m exposed)
 exit diameter: 2.826 m (initial expansion ratio 9.6)
 materials: carbon-carbon throat; nozzle flexbearing provides 6° deflection (6.6° max). Employs two actuators 90° apart with ±145 mm strokes. Actuator system mass 150 kg, hydraulic 1100 l, 345/270 atmosphere initial/end pressure.
Casing materials: 8 mm-thick D6AC steel heat-treated at 14,800 atmosphere ultimate strength; thermal insulation EPDM/Silica (GSM55) or Kevlar (EG2). Six 3.351 m-long segments are assembled into two motor sections of three segments each with two factory joints and 1.005 m-long aft dome. Tipped by 3.398 m-long forward section.
Igniter type: pyrogen head-end (25 kg composite propellant).

UPDATED

Ariane 5 Plus

Current Status

ESA expects the Ariane 5 Plus launcher to be available by 2005 and is to be achieved sequentially through Ariane 5 Versatile, Ariane 5 EC-A and Ariane 5 EC-B. The first EC-A variant was launched on 11 December 2002 but a malfunction in the propulsion system prevented the satellite from reaching orbit. A mishap report indicated that tube cracks in the active cooling system prevented the Vulcain 2 engine from overheating. The grounding of Ariane 5 postponed the launch of Rosetta beyond its acceptable launch window.

Background

Initiated by the European Space Agency in June 1998 and authorised for full scale development in May 1999, Ariane 5 Plus is a development which will allow the launcher to meet major growth in satellite weight and allow multiple satellite launches to similar orbits as well as lifting space station modules on future missions. This is a response to the expanding demand for constellations of satellites delivered through a common trajectory. Management of the Ariane 5 Plus programme is handles by the Ariane Development Department jointly operated b CNES and Arianespace. Ariane 5 Plus is a continuation of the Ariane 5 Evolution effort and involves improvements to the launcher's lower composite comprising main cryogenic stage and solid booster stage.

Improvements focus on structural weight reduction, major modifications to the solid boosters and the main cryogenic stage as well as a thrust increase the Vulcain main engine to 1,353 kN in vacuum. Introduced in paralled with derivatives of the new Ariane 5 EPS upper stage Ariane 5 Plus will increase payload capability from 7,300 kg on a dual GTO launch to 11,000 kg in dual GTO launch with Speltra and 12,000 kg in a single payload mission to GTO. Ariane 5 Plus has been developed to meet a wide range of payload requirements, including dual satellite launches to GTO, high-perigee GTO missions, direct geostationary insertion, interplanetary launches, deployment of satellite constellations in medium-earth orbit and launches to low-earth and sun-synchxronous orbit. The manoeuvrability of the new upper stages will enable the launch of combined missions with different payload deploy orbits.

Modifications to the Ariane 5 lower composite include shifting the location of the common bulkhead between the two propellant tanks to increase the quantity of LO2 that can be loaded. The liquid oxygen/hydrogen ration will also be increased for the new Vulcain 2 engine producing 1,353 kN thrust, 20 per cent more than the basic Vulcain. The solid booster cases bolted joints will be replaced with welded casing joints resulting in a weight reduction and an additional 2.4 t of solid propellant loaded in the booster's upper case segment will increase maximum thrust from 6,718 kN to 7,002 kN each. The combination of new lower composite and EC-A cryogenic upper stage configuration was ready in 2002 while the lower composite with EC-B was scheduled for 2005 or 2006. Known as Ariane 5 Versatile, the ultimate evolved version will increase the GTO capability to 7,300–8,000 kg through improvements to the existing upper stage.

Cryogenic Upper Stage

Two new cryogenic upper stage configurations are at the heart of Ariane 5 Plus, the ESC-A and the ESC-B. They derive from the improvements being made to the basis EPS upper stage to produce Ariane 5 Versatile, which has a payload capability of 7,400 kg in dual payload missions to GTO with Speltra and 8,000 kg in a single-satellite launch to GTO and GTO+ (high perigee orbit).

EC-A: Powered by the same 63.8 kN HM-7B engine used in the Ariane 4 third stage and which is designed for a single in-flight re-ignition. ESC-A will carry 14 t of L02/LH2 propellant and allow Ariane 5 Plus to place 9,500 kg into GTO on a dual payload mission and 10,500 kg into GTO while carrying a single large payload.

EC-B: Powered by a new 152 kN Vinci expander-cycle engine with 25 t of L02/LH2 propellant and a capability for multiple in-flight re-start. Payload capability will allow dual GTO launch of 11,500 kg with Speltra 12,000 kg in a single payload GTO mission.

Rockot

Background

Eurockot markets the commercial Rockot launcher, derived from the SS-19 two-stage ICBM, a 10-man Khrunichev/DASA joint venture announced 23 March 1995. Eurockot is responsible for all commercial activities, acting as the customer's prime contractor, but Khrunichev develops and produces all of the hardware. It is a joint venture of EADS Space Transportation (51 per cent) and Khrunichev Space Centre (49 per cent). The first revenue launch was planned for the fourth quarter of 1998 from Plesetsk, using a modified Kosmos pad now designated exclusively for Rockot. Launches are from a tube, replicating a silo.

To prepare, Eurockot has performed three demonstration launches from Tyuratam, all silo-based. The first two (20 November 1990 and 20 December 1991) were suborbital, demonstrating basic vehicle performance and KB Salyut's new Breeze stage 3 (inactive on the first ascent). The first orbital flight was 26 December 1994 from Baikonur, into 1,875 × 2,253 km 64.80° carrying the 70 kg RS-15 Radio-Rosto amateur relay. The second orbital flight on 16 May 2000 carried Simsat-1 and Simsat-2 from complex LC133 at the Plesetsk site to 86.4° orbit at approximately 550 km. The third orbital flight launched from Plesetsk on 17 March 2002 lifted the NASA/German Gravity Recovery And Climate Experiment (GRACE) mission satellites into low earth orbit. Commercial Rockot carries a large new fairing developed by Khrunichev. The fourth flight conducted on 20 June 2003 lifted two Iridium satellites to orbit and the fifth, on 30 June 2003 with the launch of eight micro-satellites (Mimosa, DTUat, MOST, Cute-1, QuakeSat, AAU-Cubesat, Can X-1, Cubesat-XI and Monitor-E GVM). The sixth flight on 30 October 2003 lifted Japan's SEVIS-1 satellite into sun-synchronous orbit from Plesetsk cosmodrome.

The standard Rockot has a 1,850 kg payload capability to a circular 370 km orbit launched from Plesetsk to an orbital inclination of 63°. It can place 1,340 kg in an 800 km, 90° circular orbit or 1,530 kg in an 370 km, 90° orbit.

Development of the silo-launched missile began in 1965 and it was first deployed operationally in 1974. The latest Mod 3 entered service in 1979. At the height of the Cold War, the Soviets deployed a peak 360 SS-18s in silos by 1982. Some would be retired under the START 2 treaty. In 1991, Russia and other countries had 300 silos at four operational sites, Kozelsk (60 silos), Tatischevo (110), Khemilnitsky (90, Ukraine) and Pervomaysk (40, Ukraine). In December 1994, the operational stockpile had dwindled to 290 missiles and Khrunichev had 15 at its factory available for conversion as of October 1995.

Military secrecy still surrounds this space launcher. It is known that stage 1 carries four engines, but it is not known whether the RD-0233 or the RD-0234 carries the tank pressurant system: there will be one engine of the type which carries the pressurant system and three of the other engine.

Soviet nuclear planners originally built the Breeze to dispense multiple communications satellites into low orbits in the event of nuclear conflict. The RS-18s would have resided in silos awaiting launch on demand. A derivative, the Breeze-M is being developed as Proton's new stage 4.

Breeze's restart capability, plus 7 hour flight life, allows payloads to be dispensed into different orbits. A single ignition is used for orbits up to 400 km. Breeze performs collision avoidance and can also de-orbit. During coast, it holds three-axis (1 to 10° accuracy) in any selected attitude within ±120° of Sun axis. It can provide 5 rev/minute about the longitudinal axis to the payload on release. Three

Rockot will depart from Plesetsk using a tube to replicate a silo launch
(Theo Pirard/Space Information Centre) 0517332

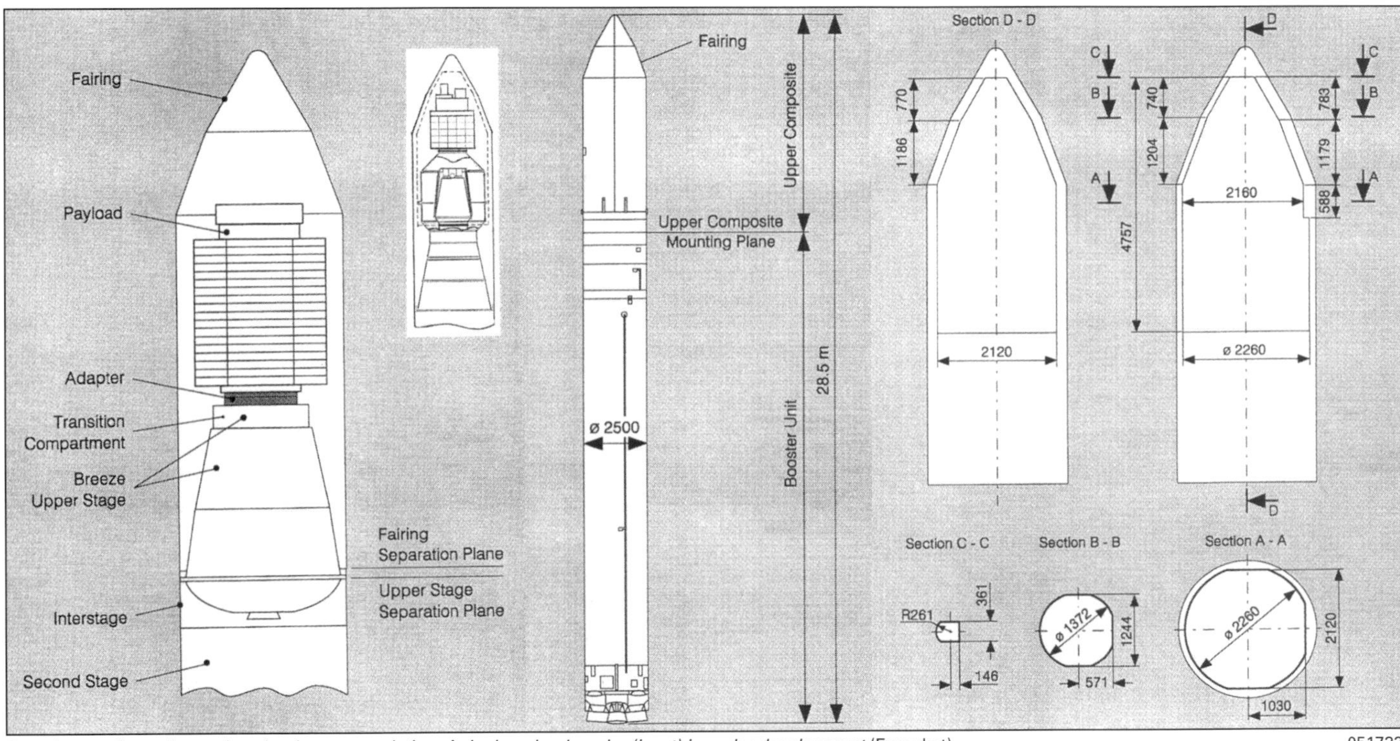

Eurockot configuration and payload accommodation. A dual-payload carrier (inset) is under development (Eurockot) 0517331

burns are required for Sun-synchronous missions: 1, to attain 200 × 700 km, 93°; 2, circularise at 700 km, 63°; 3, plane change to 700 km, 98°.

No designators are available for the main or vernier engine systems used on the Breeze (Briz) third stage on Rockot. It has been indicated that the main engine has flown on Phobos, VEGA, Venera and Mars spacecraft, suggesting that it is (or derived from) the Isayev bureau's (now Khimmash) KTDU-425. As a comparison with the third stage main engine data the KTDU-425A thrust is variable within the range 9.86 to 18.89 kN and the specific impulse within the range 293 to 315 seconds.

Eurockot is developing an 80 kg double launch system (DOLASY). Payload integration and checkout is performed in a 180 m^2 class 100,000 facility at 18 to 25°C and 30 to 70 per cent humidity. After encapsulation in the adjacent 90 m^2 clean room, air conditioning maintains the payload at 10 to 25°C, 15 to 60 per cent humidity, class 100,000. Air conditioning is interrupted only during two periods, each of 1 hour max: mounting on the pad and final assembly of the 'silo' container. There is no access after payload encapsulation; the fairing is not radio transparent. Customer accommodation is provided in Mirny, some 42 km from the launch site, during campaigns. The airport is 35 km from the complex. A campaign requires 30 days: stage 1/2 assembly/testing L-30 to L-22; Breeze testing/fuelling L-21 to L-13; satellite fuelling L-13/12; upper composite integration L-11 to L-8; stage 1/2 transfer to pad L-7; transfer upper composite to pad L-6; launch preparations L-5 to L-1; fuelling/launch L-).

Specifications

Rockot

First orbital launch: 26 December 1994
Number launched: 6, through end of October 2003 (plus 2 sub-orbital)
Launch sites: Baikonur, Plesetsk
Principal uses: small-medium payloads (up to about 1,900 kg) into intermediate orbits
Vehicle success rate: 100% through end October 2003.
Past 25 launches; last 25 RS-18 100% successful
Missile career: 142 successes from 145 launches
Future commercial missions: none announced
Availability: 5/yr (up to 12/yr) on 20-month cycle between contract signature and launch
Cost: on application to Eurorockot
Insurance: information not available
Number of stages: 3 (liquid)
Overall length: 28.496 m
Principal diameter: 2.500 m
Launch mass: 107 t
Guidance: inertial, carried in compartment attached to front end of stage 3, provides three-axis control until payload separation commanded. Accuracy for 300 km 82° orbit ±3 km altitude, ±0.03° inclination; 700 km 63° orbit ±14 km altitude, ±0.05° inc
Launch sequence (min:s) for 700 km, 63° (some values have been estimated)

0.00	launch
0.52	Q_{max}, 605 m/s, 11 km altitude
2.01	stage 1 shutdown & separation (stage 2 verniers provide hot sep)
2.27	stage 2 main ignition
3.06	fairing separation, 3.74 km/s, 119 km
4.44	stage 2 main shutdown
5.04	stage 2 verniers shutdown and separation
08.40	Breeze ignition 1,
–	max altitude (257 km)
14.06	Breeze shutdown 1
–	orbit 200 × 700 km
63.50	Breeze ignition 2 (circularisation)
64.26	Breeze shutdown 2
–	orbit 700 × 700 km

Rockot stage 1

Bureau: Khrunichev/KB Salyut
Engines: RD-0233 and RD-0234
Overall length: 17.2 m
Diameter: 2.5 m
Oxidiser: nitrogen tetroxide
Fuel: UDMH
Dry mass: not available

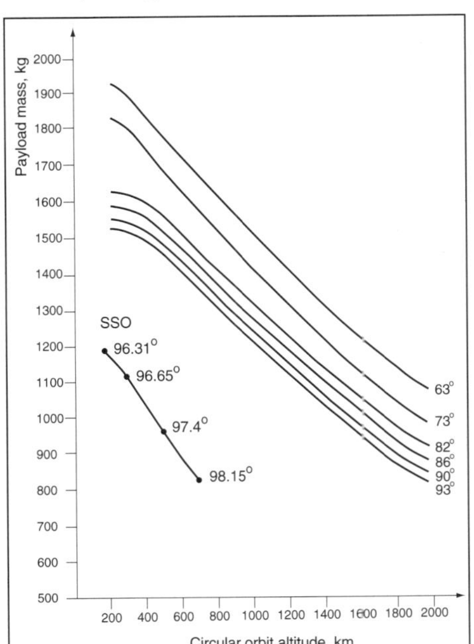

Rockot circular orbit performance from Plesetsk. Orbits up to 400 km are achieved by a single stage 3 burn; >400 km require Breeze re-ignition at apogee for circularisation. 63–93° inclinations are allowable without plane changes during launch (Khrunichev/DASA) 0519179

Rockot's new stage 3 was first displayed at 1995's Paris Air Show (Ted Hooton) 0517192

Rockot is launched by ejecting it from a tube simulating the silo from which the SS-18 is usually fired (Theo Pirard) 0054371

Propellant mass: 81 t
Total thrust: 2,030 kN (vacuum), 1,834 kN (s1)
Specific impulse: 310 s (vacuum), 285 s (s1)
Burn time: 121 s
Comment: Once the first stage engines have ignited they cannot be shut down: there are no brackets to hold the launch vehicle onto its launch table and no engine parameter verification before launch. The first stage carries four solid-propellant retro-rockets for separation from the second stage. The propellant tanks have a common bulkhead and tank pressurisation is achieved by means of hot gas from one of the main engines.

Rockot stage 2
Bureau: Khrunichev/KB Salyut
Engines: RD-0235 (main engine), RD-0236 (verniers)
Overall length: 3.9 m
Diameter: 2.5 m
Oxidiser: nitrogen tetroxide
Fuel: UDMH
Dry mass: not available
Propellant mass: 15 t
Total thrust: RD-0235 – 235 kN (vacuum), RD-0236 – 14.7 kN (vacuum)
Specific impulse: RD-0235 – 320 s (vacuum)
Burn time: RD-0235 – 158 s, RD-0236 – 183 s
Comment: Separation of the first and second stages is a 'hot' one since the RD-0236 vernier engines are ignited just before separation: the exhaust gases are diverted using special hatches within the first stage.

Rockot stage 3 (Breeze)
Bureau: Khrunichev/KB Salyut
Engines: KTDU-425? (main engine) ? (verniers)
Overall length: 3.38 m
Diameter: 2.28 m (maximum)
Oxidiser: nitrogen tetroxide
Fuel: UDMH
Dry mass: = ~3.1 t
Propellant mass: 3.5 t
Total thrust: main engine – 20 kN (vacuum), verniers –3.9 kN (vacuum)
Specific impulse: main engine – 325.5 s (vacuum), verniers – 275 s (vacuum)
Burn time: main engine – 564 total (max six burns)

In addition, the third stage carries 12 attitude-control thrusters: each has a thrust of 0.13 kN and a specific impulse of 270 s.

Payload processing/accommodation
The 8.5 m long upper composite includes the fairing, Breeze and interstage. The 7.9 m long, 2.5 m diameter cylindrical/biconic two-piece composite fairing, developed specifically for Rockot, provides a payload envelope of 4.757 m long 2.260 m diameter. Bolted payload adaptor is 591 or 937 mm diameter. The separation system comprises pyrolocks (four for 937 mm, three for 591 mm), three springs, guide pins (eight for 937 mm, six for 591 mm) and two separation sensors. The upper ring of each adaptor remains bolted to the payload: 15 kg for 937 mm, 12 kg 591 mm. After the pyrolocks are cut by pyrotechnics, the springs impart a minimum 0.5 m/s. A Western standard system using a 937 mm Marman clamp band is also available. The adaptors carry up to eight 50-pin payload electrical connectors, relaying up to 20 A total 28 V DC during launch preparation. 100 lines allow payload checkout until launch. During launch, Rockot's batteries in the transition section between Breeze and the adapter can provide up to 5 A totalling 15 Ah over 7 hours. Up to 50 lines can be provided for Rockot's computer to issue payload commands. 32 payload telemetry channels are available.
Acceleration load: longitudinal static 7.2 *g* at end of stage 1 burn, 3.0 *g* at end of stage 2 burn, 1.6 *g* max at end of Breeze burn (depends on payload mass); vibration 7.5 *g* rms long/lat.
Acoustic load: 142 dB max integrated level during lift-off and transonic
Thermal load: <500 W/m^2 at any point of the fairing; <1,135 W/m^2 after fairing jettison.

Iraq

Tamouz 1

Background
The Tamouz 1 vehicle launched from the Al Anbar facility 80 km west of Baghdad on 5 December 1989 was claimed by the Iraqis as a satellite launcher test. Only the first of the three stages was live, reportedly attaining 12 km altitude. A range of 2,000 km was claimed for the Al Aabed (The Worshipper) missile version, apparently with a 750 kg payload. On 22 March 1990 project originator Gerard Bull was assassinated, ending the Tamouz project. Plans revealed in 1990 to build a 1 m-bore 'supergun' under Project Babylon by the Brussels-based Space Research Corp included the firing of small payloads into low Earth orbit. In April 2000 renewed work on the Tamouz 1 was reported to have started in the second quarter of 1999. This work continued during 2000 and 2001 but slowed in 2002. No evidence for its continuation as a potential satellite launcher has been found by coalition forces in Iraq.

Specifications
Tamouz 1
First launch: 5 December 1989
Number launched: 2 to end 2001 (two Al Aabed missile versions were apparently launched previously).
Launch site: Al Anbar
Performance: about 150 kg into LEO
Number of stages: 3 (liquid)
Overall length: 24.4 m
Launch mass: about 48 t
Guidance: inertial

Tamouz 1 stage 1
Comment: The stage appears to be a cluster of five units derived from the Al Abbas version of the SS-1 Scud missile.
Length: about 11 m
Principal diameter: about 2.3 m
Propellants: liquid
Stage thrust: about 686 kN sea level

Tamouz 1 stage 2
Comment: The stage appears to be a single unit derived from the Al Abbas version of the SS-1 Scud missile
Length: about 9 m
Principal diameter: 90 cm
Propellants: liquid

Tamouz 1 stage 3
Comment: Little is known of the third stage, but length is about 3.5 m and it is believed to utilise liquid propellants

Israel

Shavit/NEXT

Background
The Shavit ('Comet') all-solid orbital launcher, built by Israel Aircraft Industries, is believed to be derived from the two-stage Jericho 2 missile, first fired into the Mediterranean in 1986. A full range firing of 1,400 km is believed to have taken place from South Africa's Overberg site near Cape Town in June 1989, followed on 14 September 1989 by a 1,300 km test into the Mediterranean. The Jericho 1 500 km-range missile is reported to have been developed in the 1960s with French assistance, based upon the Dassault MD-600. The Israelis launched the second Shavit carrying Ofeq 2 on 3 April 1990. Ofeq 3 appeared 5 April 1995 and Ofeq 4 failed in 1998. It has been claimed there was a failed launch between Ofeq 2/3. Ofeq 5 was launched by Shavit on 28 May 2002.

Launches took place from an area within the Palmachim Air Force Base south of Tel Aviv, Israel, near the town of Yavne, inaugurated with the launch of Ofeq 1 on 19 September 1988. A second satellite was successfully launched on 3 April 1990 but a third launch on 15 September 1994 failed and was never publicly acknowledged. The fourth launch on 5 April 1995 carried the Ofeq 3 satellite into orbit and was followed by Ofeq 4 on 22 January 1998 which failed to reach orbit.

Delta Research Inc of Huntsville, Alabama joined forces with IAI to propose Shavit in 1990 for NASA's Meteor payload carrier. It was then offered in 1994 for NASA's ultra-light launcher competition (won by Pegasus). Team leader Space Vector would have integrated/launched the vehicle. Atlantic Research Corp would have loaded the stage 1/2 casings provided by IAI. Manufacturer Rafael/ARC already markets stage 3's AUS-51 motor. TRW in 1995 approached the Shavit manufacturer for the launch of its TSX 5. Israel is pressing for exemption from the White House rule that US vehicles must be used for government payloads from US soil.

IAI is working on the NEXT commercial version of Shavit, stretching stages 1/2 and adding a bipropellant stage 4. Development depends on commercial commitments. Unconfirmed reports suggest that a 4,800 km-range Jericho 3 is under development, in common with NEXT.

Specifications
Shavit
First launch: 19 September 1988 (Ofeq 1 test satellite)
Number launched: 6 through end of 2002
Launch sites: Palmachim Air Force Base
Vehicle success rate: 60%
Principal uses: small payloads into LEO
Performance: 156 kg into 248 × 1,170 km, 142.9° orbit
Number of stages: three solid propellant
Overall length: 17.7 m
Principal diameter: 1.352 m
Launch mass: 22–23 t
Guidance: inertial, mounted on stage 2. Redundant; MIL-STD-1553B bus. Flight angle commanded as function of time (zeroed for stage 1 separation)

Shavit stage 1
Motor: TAAS Israel Industries Ltd. PBAN solid with filament wound graphite epoxy casing. SI 240 s sea level, expansion ratio 9

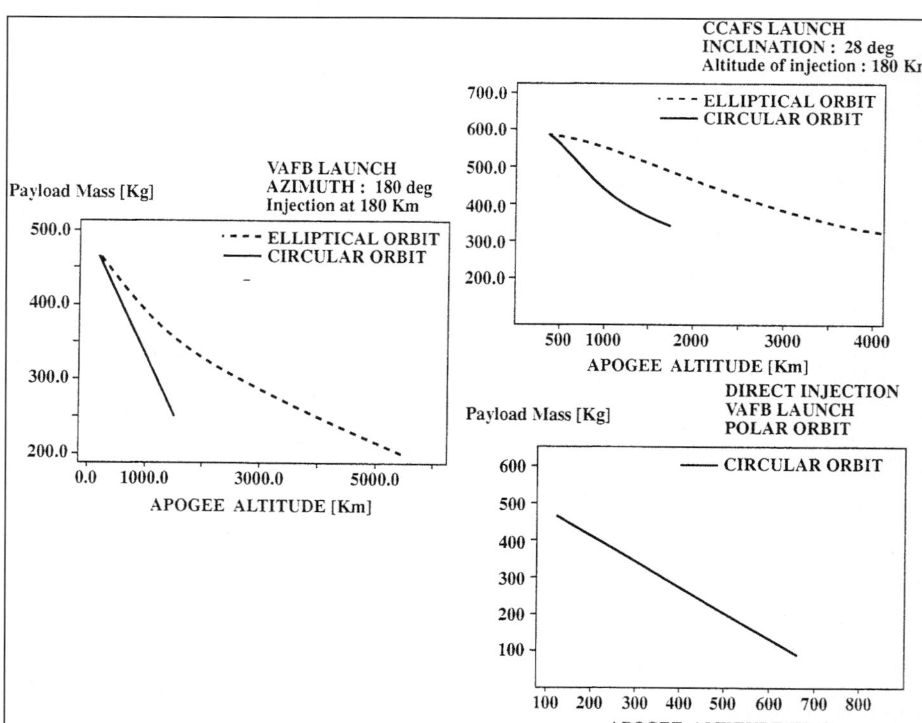

NEXT payload performance (IAI) 0517166

Overall length: 5.25 m
Principal diameter: 1.352 m
Stage mass: unavailable (9.1 t propellant)
Average thrust: 610 kN vacuum
Burn time: 43 s
Attitude control: four jet vanes (jettisoned after vertical phase) plus four air vanes
Staging: pyrotechnic separation of external V-clamp holding the two bulkheads with trapezoid teeth

Shavit stage 2
Motor: similar to stage 1 but with 23.4 expansion ratio for altitude performance
Overall length: 5.676 m
Principal diameter: 1.352 m
Stage mass: unavailable (9.1 t propellant)
Average thrust: 564 kN vacuum
Burn time: 52 s
Attitude control: pitch/yaw by LITVC modules. Four 0.8 kg modules each injects 0.35 kg/s (at 51.0 atmosphere) strontium perchlorate through a single orifice; max operating pressure 85.0 atmosphere)
Staging: as stage 1, hot separation

Shavit stage 3
Motor: Rafael AUS-51 'Marble'
Overall length: about 2.1 m
Principal diameter: 1.3 m
Stage mass: about 2 t (1,895 kg propellant)
Average thrust: 55
Burn time: 92.5 s
Attitude control: spin-stabilised, spun up/down by Rafael's ST-200N thrusters. NEXT's AUS-51 will employ LITVC. Rafael's ACT-25N thrusters provide attitude/nutation control

NEXT
Comment: IAI is working on the NEXT commercial version of Shavit, stretching stages 1/2 and adding a bipropellant stage 4. Development depends on commercial commitments. All propulsion systems were expected to be qualified by end-1995 but no test flight appeared.
First launch: 2000?
Principal uses: small payloads into LEO
Performance: see diagrams
Number of stages: three solid-propellant plus one liquid

The Shavit satellite launch vehicle flanked on either side by missiles with propulsion systems developed by Israel Military Industries (Israel Military Industries) 0003371

Overall length: about 21 m
Principal diameter: 1.352 m
Launch mass: about 31 t
Guidance: gimballed platform on stage 2, succeeded by GPS receiver on stage 4. MIL-STD-1553B bus

NEXT stage 1
Motor: as Shavit, but lengthened about 20%. A flexseal nozzle has been developed that might be employed by future NEXT.
Overall length: about 6.8 m
Principal diameter: 1.352 m
Stage mass: 14.3 t (12.87 t propellant)
Average thrust: 637 kN vacuum
Burn time: 51 s
Attitude control: as Shavit
Staging: as Shavit

NEXT stage 2
Motor: as Shavit, but lengthened about 20%
Overall length: about 7.3 m
Principal diameter: 1.352 m
Stage mass: 14.4 t (12.87 t propellant)
Average thrust: 705 kN vacuum
Burn time: 60 s?
Attitude control: as Shavit
Staging: as stage 1, hot separation

NEXT stage 3
Same as Shavit, but three-axis attitude control by LITVC modules as used on stage 2

NEXT stage 4
Designation: BPM Bipropellant Module
Engine: IAI bipropellant (see Propulsion section for specifications)
Overall length: unavailable
Principal diameter: 1.35 m
Dry mass: 60 kg; structure is composite
Propellants: 120 kg MMH/N_2O_4
Stage thrust: 3,000 N
Burn time: 120+30 s 2-phase circularisation
Attitude control: two clusters of three thrusters for three-axis control; 6 × 20 N or 6 × 150 N depending on mission requirements
Separation: unavailable

Italy

Vega

Current Status
Vega is an approved ESA project and a first flight in 2006 is scheduled following several years delay. In December 2000 European governments agreed to spend US$301.5 million on development of the Vega launcher and US$109.9 million on a parallel advanced technologies development programme called P-80. Italy has agreed to finance Vega development and 45 per cent of the P-80 effort. France declined to participate in Vega but agreed to fund 35 per cent of P-80, which will directly assist the Ariane 5 programme since it adopts the Vega first stage as first stage strap-ons. In February 2003 the ELV Consortium, Vega's prime contractor, signed a US$200 million contract with ESA for final stages of development leading to manufacture and testing. The ELV Consortium is owned by FiatAvio (70 per cent) and the Italian Space Agency (30 per cent). Together, Italian interests account for 65 per cent of the total project, other contributors being France, Belgium, Spain, the Netherlands, Switzerland and Sweden.

Background
The Italian government accepted proposals from industrial concerns in 1988 to upgrade Scouts, with plans to launch them from San Marco. Among the various proposals was one from BPD to double the Scout's base performance. Italian government approval followed, in July 1991, of the country's 1990–94 space plan which allowed ASI to award BPD the Lit5,120 billion industrialisation contract on 3 December 1991. ASI's plan included little US involvement by making all stages indigenously

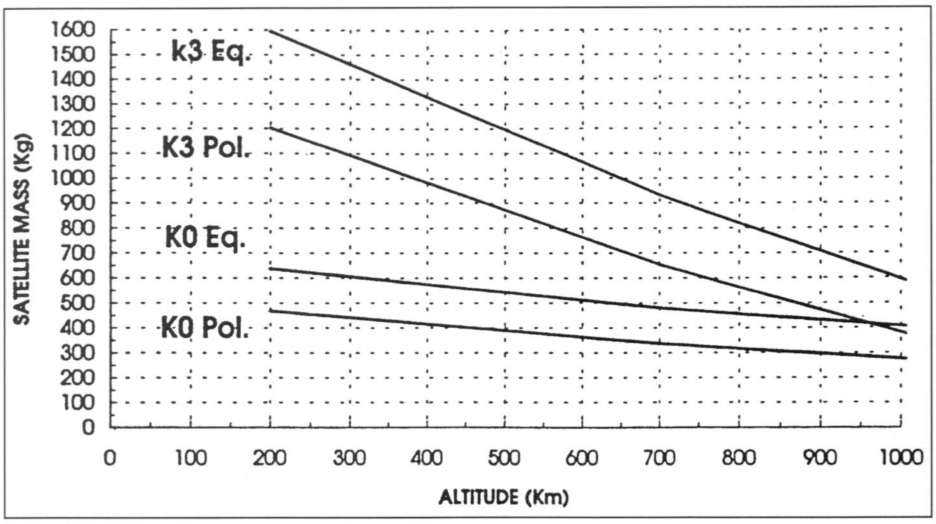

Vega K0/K3 polar and equatorial orbit performance (BPD) 0517333

produced. Versions without, or with two or four strap-ons might allow the Italians to modularise the design for economic efficiency.

BPD's offered its Zefiro (Zephyr) motor, developed from the company's Ariane experience, with thrust vector control. It conducted four successful static tests during 1991, but the first flight test ended by detonation from range safety 10 seconds into the planned 40 second burn on 18 March 1992 from the Sardinian military range. Ground qualification continued until final completion in mid-1995 and two qualification flights at the end of the year.

A design rework in 1994 dropped the strap-on approach and increased Zefiro's diameter from 1.35 to 1.90 m. The two base designs are now called Vega K0 and Vega K3. A K0 minus stage 1 could substitute for Capricornio (150 kg, US$9 million) and an enlarged K4 could become the European Small Launcher (1,100 kg, US$18 million).

Italy's 1990–94 five-year plan approved development of the San Marco Scout launcher for operations from Italy's equatorial platform off the Kenyan coast. Payloads were to include recoverable μgravity carriers such as the Carina capsule. The Italians expected to make their first launch in 1995. However, the University of Rome proposed a separate San Marco Scout programme closely based on the US Scout, prompting legal wrangling. BPD is continuing Vega development with internal funding plus Ministry of Research funding for Zefiro development. Should government support be provided, BPD will press on with its programme to the qualification flight.

In the meantime, Italy's San Marco site has become too expensive for the Italian government to maintain, and Vega's principal sites would be Kourou and Vandenberg AFB using transportable equipment to support campaigns of 7 to 15 days. Discussions are under way for commonality with Spain's Capricornio and with Europe's ESL at a total development cost of US$200 million. The vehicle family must be commercially viable, addressing the 200 to 600 kg market for 200 to 1,000 km orbits. The configurations are not optimised but yield lower development and recurring costs.

In May 1999, the European Space Agency approved US$340 million (€317 million) for development but only after successful completion of additional technical, financial and economic assessments in September 1999. Support for this launcher has grown in the last year and a first flight is scheduled for 2006.

Specifications

Vega/San Marcos
First launch: 2006
Launch sites: ELA-1 pad Kourou, French Guiana
Principal uses: 300–700 kg payloads into LEO
Performance: 1,450 kg into LEO; 800 kg into 550; 1000 kg into 700 km PEO; see diagram
Cost: 18.5 M ero
Number of stages: 4 K0; 3 K3
Overall length: 27 m
Maximum diameter: 3 m
Launch mass: 128 t
Guidance: inertial above stage 3. Accuracy ±20 km altitude, ±0.05° inc. Stage 1/2 TVC for pitch/yaw, stage 3 ACS for roll control during stage 1/2 and three-axis during coast and stage 3 burn; stage 4 ACS for three-axis stage 4, orbit trim and payload release. MIL-STD-1773 optical bus.

Vega stage 1 (SRM)
Motor: Europropulsion P85 with TVC
Overall length: 10.7 m
Diameter (max): 3 m
Stage mass: 85 t
Average thrust: 3,000 kN
Burn time: 101 s
Specific impulse: 279 s

Vega stage 2 (Zefiro)
Motor: Fiat Avio P16 with TVC
Overall length: 5–6 m
Diameter: 1.9 m
Stage prop mass: 16 t
Average thrust: 910 kN
Burn time: 70 s
Specific impulse: 286 s

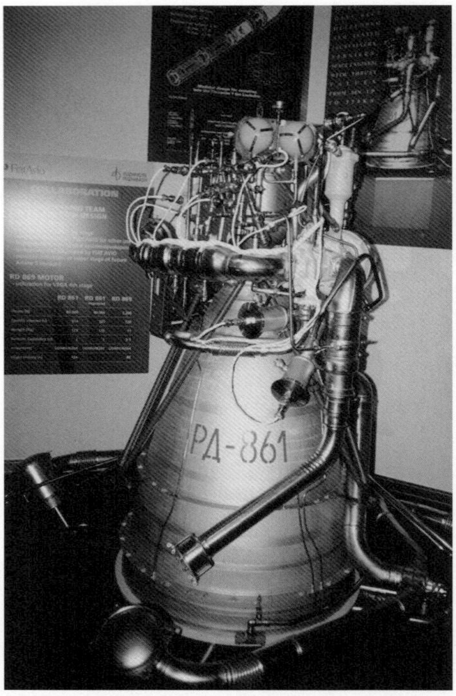

RD-861 motor proposed by Fiat Avio to equip second stage of the Vega Launch vehicle (Theo Pirard) 0054372

Vega stage 3
Motor: Europropulsion P7 with TVC
Overall length: 2.9 m
Diameter: 1.9 m
Stage prop mass: 7 t
Average thrust: 250 kN
Burn time: 106 s
Specific impulse: 106 s

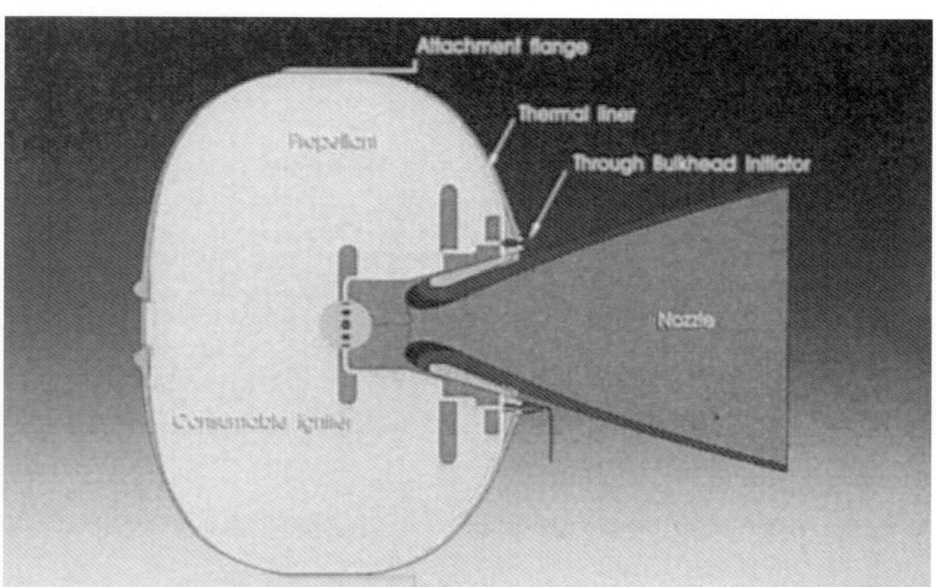

BPD's End Burning Motor is an improved version of the IRIS motor (ESA) 0516909

Japan

H1/H2/H2A

Current Status

Abandoning the expensive and costly to operate H2, the first five H2A have been successfully launched from Tanegashima, the most recent on 28 March 2003. In a decision reached at Cabinet level, the Japanese government has approved the development of an advanced capability H2A which has yet to receive a formal designation. The improved launcher will have two cryogenic LE-7A motors in the first stage, which will retain the same length but increase in diameter from 4 to 5 m, adopt a standard configuration of four solid propellant strap-on boosters versus two but retain the same second stage with its single LE-5B upper stage. The advanced capability H2A will be capable of lifting 6,000 kg to LEO compared to 3,000 kg for the standard H2A, and send 4,000 kg to GTO versus 2,000 to 2,500 kg for the basic launcher. The improved version will have the lifting capacity of the Atlas 2AS and Delta 4 class launchers and allow greater growth through the possible adoption of liquid propellant strap-on boosters. The driving imperative to expand the lift capability of the H2A has been the need to supply the Kibo pressure module which will be attached to the International Space Station and to deliver cargo pallets which will dock automatically and relieve the NASA Shuttle of some elements of re-supply.

Background

Japan's Space Activities Commission gave the go-ahead, in February 1984, for the development of a large all-Japanese H2 capable of placing 2 tonnes in GEO and a small spaceplane in LEO. Work had already started on the current H1 in 1979 as an interim step. N1/2's McDonnell Douglas stage 1 with Rocketdyne's MB-3 engine transferred directly to the programme, but the cryogenic stage 2, inertial guidance and solid upper stages were all-Japanese. Two demonstration missions flew 1986/87, with seven operational departures (five GEO) following. H2 completed the transition by adopting indigenous solid strap-ons and large cryogenic first stage, creating a capability comparable to that of Ariane 4 and Titan 3 but with a significantly lower launch mass because of the more

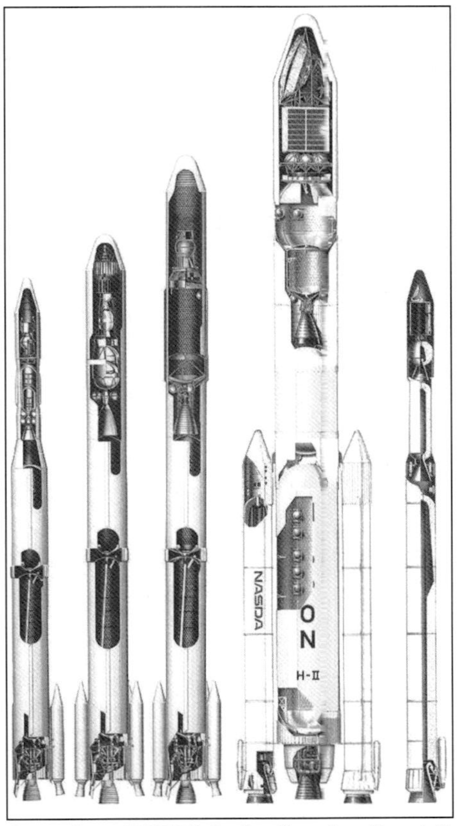

Comparison of NASDA's family of launchers (from left): N1, N2, H1 (essentially the N2 with improved upper stages), H2 and J1 (NASDA) 0517177

Fourth H2 launch in August 1996 carrying ADEOS and JAS-2 into orbit (NASDA) 0003372

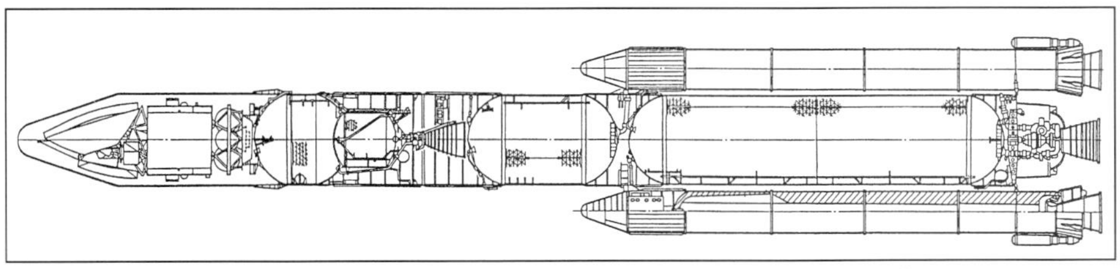

H2 launch vehicle, shown with ETS 6 payload in a 4S single fairing. H2 can add two small strap-ons to stage 1 (NASDA) 0516910

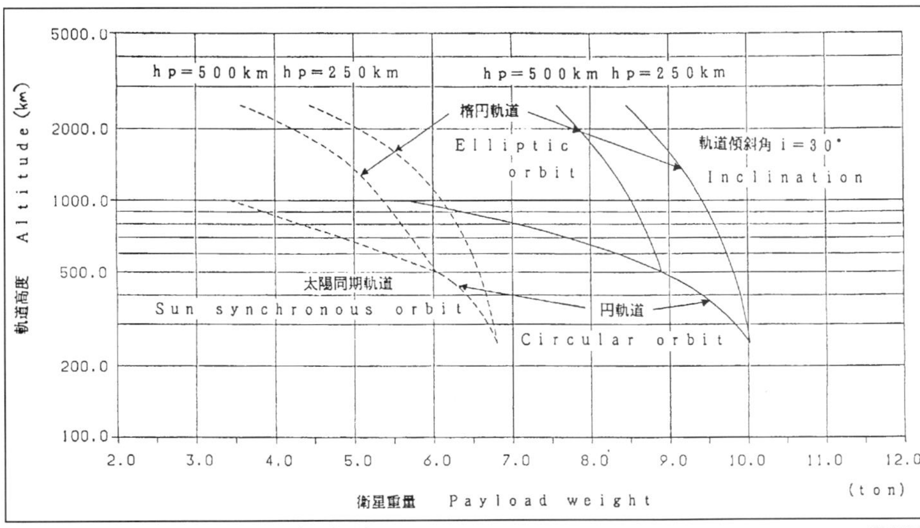

H2 performance from Tanegashima (NASDA) 0516911

advanced stages. Development of the Improved H2 and the Upgraded H2 (H2A) is under way.

The N1/2 launchers provided the initial expertise but they depended on McDonnell Douglas Delta stages built under a 1969 licence that prohibited orbiting of third party payloads without US approval. The interim H1 introduced Mitsubishi's cryogenic stage 2 and all-Japanese inertial guidance; H2 completed the transition in 1994 by adopting a large cryogenic first stage and solid strap-ons.

To enhance the cost efficiency of the H2, the Improved H2 and Upgraded H2 (H2A) programmes are under way. The Improved H2 will debut in 1999 with a production cost reduced by about 30 per cent. The H2A development began FY96 and should debut about 2000, giving a launch vehicle family with a payload capability of up to 15 tonnes/3 tonnes (LEO/GEO) for the H2A. A growth potential of up to 20 tonnes/4 tonnes (LEO/GEO) is under consideration in the design.

These increased capabilities will allow it to fly logistics missions to the International Space Station and carry the HOPE spaceplane.

Introduction of the H2 in 1994 provided Japan with its first large indigenous launcher, capable of placing 4-tonne payloads in GTO, and dispatching deep space probes. Freed from US licensing restrictions, it is offered commercially but cost renders it a difficult sell. Unit cost is ¥19 billion, but the Improved H2 is under development for debut in 2000, principally to reduce cost by 30 per cent. It will provide *direct* GEO injection. The H2A plans to debut, as a family capable of delivering up to 4 tonnes GEO and 20 tonnes LEO. Unit cost will be about ¥8.5 billion. That last capability will allow it to fly Space Station logistics missions and carry the HOPE spaceplane.

An initial batch of six H2s was built, the first two considered to be prototypes.

The development of the Improved H2 began in 1995, principally to reduce unit cost and increase reliability to >0.96. GTO capacity will be 4 tonnes. First flight will carry MTSAT in 1999. Stage 1 will be little changed. Payloads will be attached to the adaptor before reaching the pad. Stage 2 will be extensively reworked, principally separating the main tanks and using the improved LE-5A engine. The current common LOX/LH_2 bulkhead is expensive and thermal leakage limits coast duration. Improved H2 will use tanks separated by a truss structure, reducing cost and allowing a 5 hour long coast. LH_2 tank 4.0 m diameter, 43.9 m^3, 2.3 atmosphere; LOX tank 3.2 m diameter, 12.7 m^3, 2.75 atmosphere. Propellant mass 16.6 tonnes, thrust increased to 137.3 kN. Electromechanical actuators replace hydraulic TVC. The ambient He bottles will be changed from titanium to composite. Captive Firing Test of the first Flight Model Stage 2 came in 1998.

Development of the H2A began FY96 to debut 2000 as a family which eventually will cover the launch capability from the existing H2 to 15 tonnes to LEO, 3 tonnes to GEO, with a growth potential up to 20 tonnes to LEO, 4 tonnes to GEO being considered in the design. Unit cost will be less than ¥8.5 billion for the 2-tonne GEO configuration. These increased launching capabilities will allow it to fly Space Station logistics missions using the HTV H2 Transfer Vehicle and to carry the HOPE spaceplane. Improved H2's SRBs will be replaced by monolithic composite designs. Liquid boosters, identical to stage 1 with two LE-7A engines on each, will provide LEO/GTO/GEO capabilities of 15/6/3 tonnes (one booster) and 20/8/4 tonnes (two boosters) respectively. Other improvements include propellant cross-feeds, an intelligent electrical system, automatic checkout functions and simplified launch operations. The new pad was completed in 1999. Hughes Space and Communications and Space Systems/Loral have each reached agreements (in 1996) with RSC for 10 launches each during the period 2000–2005.

Nissan developed the four-segment solid strap-ons for Japan's H2 launcher, with four demonstration firings during 1988–91 at NASDA's Tanegashima Space Center. The first test was conducted April 1988, prototype number 1 was fired June 1989 and prototype number 2 13 December 1989, with the qualification model firing following on 29 May 1991. The units represent the country's largest and most powerful indigenously-developed solids. The Improved H2's SRBs were reduced to three segments, debuting in 1999. Following a launch failure on 15 November 1999 the H2 programme was terminated in January of the H2A. The upgraded H2A, from 2001, will use a monolithic composite case design.

Cost to first launch was estimated at ¥200 billion (then about US$800 million), with 23 per cent allocated to LE-7 development and 20 per cent for launch site construction. A 1993 estimate was ¥270 billion, including facilities and the first two launches. ¥35.5 billion was allocated in FY89 for

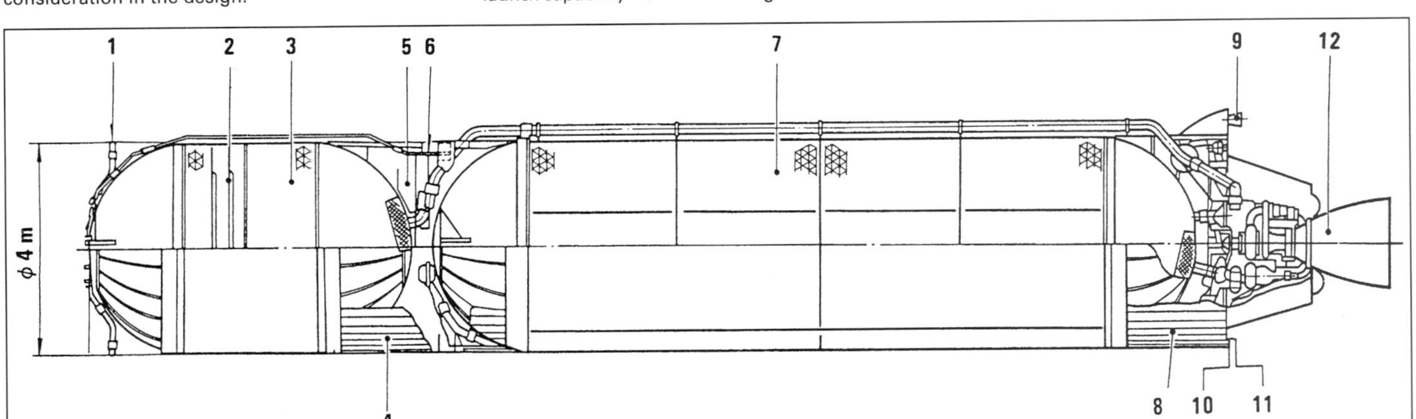

H2's stage 1 is powered by a single LE-7 LOX/LH_2 engine. Original stage length was lengthened by 2 m to 28 m as a result of detailed design studies. 1) oxygen gas vent, 2) baffles, 3) LOX tank, 4) centre body section, 5) electronics, 6) H_2 gas vent, 7 LH_2 tank, 8) engine section, 9) auxiliary engine, 10) umbilical connector, 11) LOX/LH_2 fill port, 12) LE-7 (NASDA) 0516912

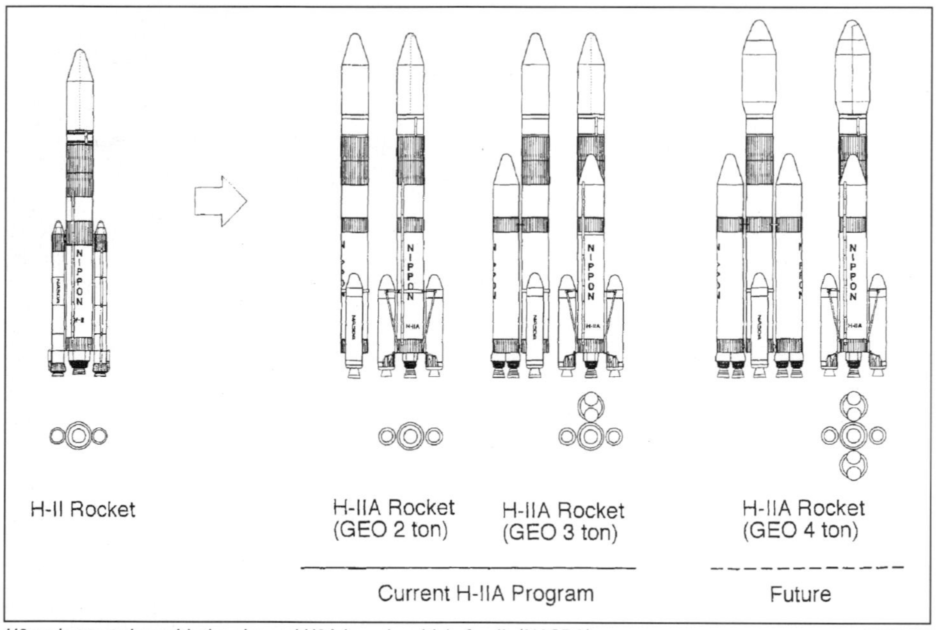

H2 rocket together with the planned H2A launch vehicle family (NASDA) 0003447

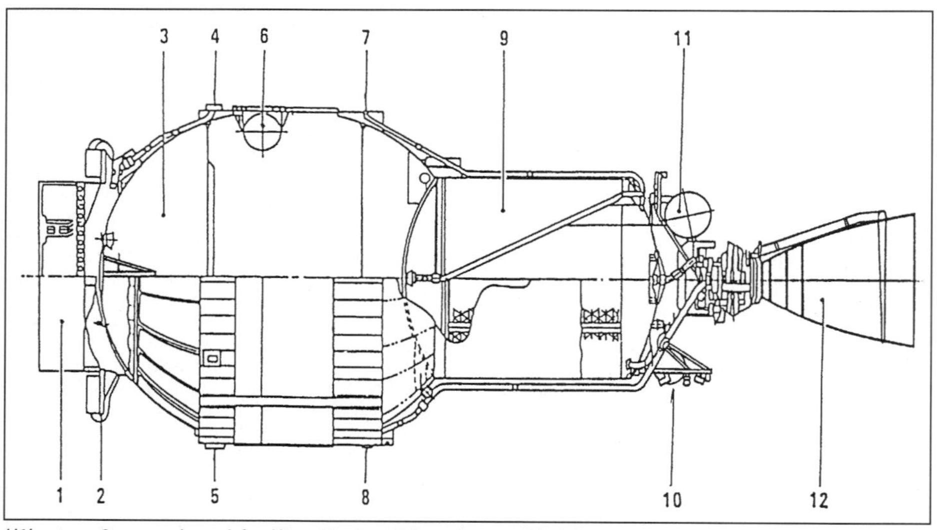

H1's stage 2 was enlarged for H2, with the LE-5 engine modified for increased thrust and improved re-start characteristics. 1) payload attach fitting, 2) guidance section, 3) LH_2 tank, 4) H_2 gas vent port, 5) umbilical connector, 6) cryogenic helium bottle, 7) LOX fill port, 8) LH_2 fill port, 9) LOX tank, 10) RCS thrusters, 11) ambient helium bottle, 12) LE-5A engine 0516913

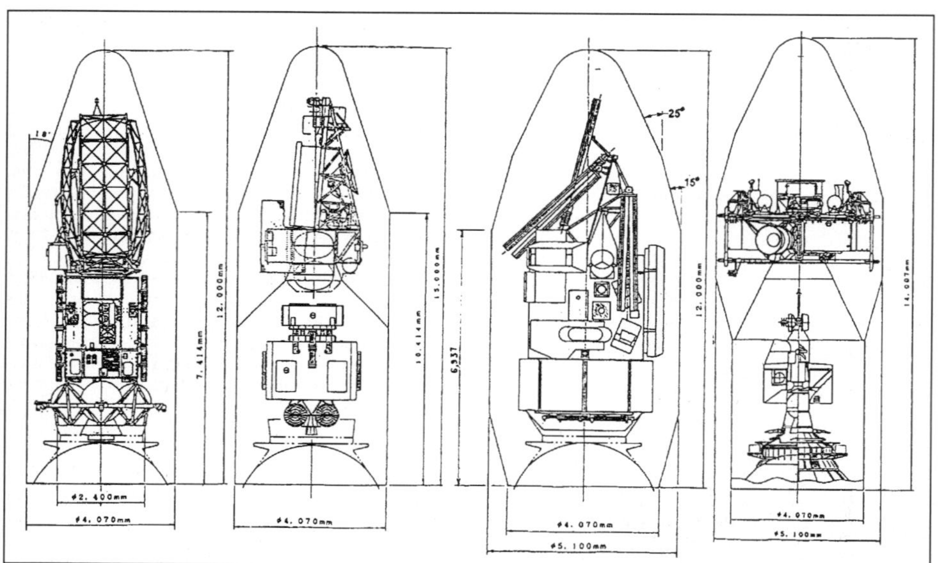

H2 payload accommodation (scales differ). From left: type 4S (single 4.1 m), 4/4D (dual 4.1 m), 5S (single 5.1 m), 5/4D (5.1 + 4.1 m). The 5.1 m fairing handles 4.6 m diameter payloads. 5/4D first flew on mission 3 (NASDA) 0516914

vehicle development, followed by ¥37.80 billion, plus ¥10 billion supplemental for the previous year, in FY90, ¥32.407 billion FY91, ¥28.13 billion FY92, ¥16.93 billion FY93, ¥1.656 billion FY94 request. LE-7 development had cost about ¥50 billion by the beginning of 1990. In 1989–90, the Japanese deferred the first launch into early 1993. Because of continuing, the country made a decision in July 1990 that LE-7's target performance would be reduced to 92 per cent, a level already almost achieved in tests. Overall H2 capability would not be affected, however, because SRB and stage 2 performance exceeded target. Further LE-7 problems in 1992 deferred first launch to early 1994.

Tanegashima can handle four flights annually, but acceleration to six would require a second pad.

The H2 has now been retired. All future launches will be with the H2A, the first of which was launched on 29 August 2001.

Specifications

H2/H2A

First launch: 3 February 1994
Number launched: 11 to end of 2002
Industrial integrator: Mitsubishi Heavy Industries, Nagoya
Launch sites: Yoshinobu Launch Complex, Tanegashima
Principal uses: delivery of 4 t-class payloads into GTO
Vehicle success rate: 85.7% for 7 launches
Cost: ¥19 billion; Improved H2 goal is ¥14 billion
Availability: fishing/range safety requirements currently constrain launches to 45-day periods in January/February and August/September each year. With second mobile platform, plan to have 4/year capability starting 1999.
Performance:
LEO: 10,000 kg into 250 km circular 30° orbit
GTO: 4,000 kg
GEO: 2,000 kg
Sun-synchronous: 4,300 kg into 800 km
Escape: 2.8 t Moon, 2 t Venus, 1.7–2.2 t Mars (for 3,650 m/s transfer injection), 700 kg Mercury, 400 kg Jupiter
Number of stages: 2+2 strap-ons
Overall length: 50 m with 4S fairing; 51.1 m with 5/4D fairing
Principal diameter: 4.0 m
Launch mass: 260 t plus payload for 4S fairing; 278.9 t plus payload for 5/4D fairing and SSBs
Guidance: strapped-down inertial mounted at top of stage 2. NEC inertial guidance computer, 14.2 kg, 45 W consumption, mission reliability >0.999, 32 kword main memory capacity (16-bit words), qualified to 12.8 *g* rms random vibration. Three Japan Aviation Electronics laser gyros and accelerometers.
Launch sequence (min:s) for GTO single fairing/LEO+GTO dual fairing + SSBs:

–00:06/–00:06	stage 1 ignition
00:00/00:00	SRBs ignite/lift-off
/00:10	SSBs ignite
/01:16	SSBs burnout
/01:30	SSBs separation
01:34/01:34	SRBs burnout
01:39/01:37	SRBs jettison
03:46/03:39	fairing/upper fairing jettison
05:46/05:47	stage 1 shutdown
05:56/05:56	stage 1 separation
06:02/06:02	stage 2 ignition 1
12:45/12:20	stage 2 shutdown
/13:16	LEO payload separation
/16:11	lower fairing separation
24:44/25:03	stage 2 ignition 2
28:01/27:02	stage 2 shutdown
28:21/27:52	GTO payload separation

N, H and J series orbital launch history

	Date	Vehicle	Payload
1	9 September 1975	N1	ETS-1
2	29 February 1976	N1	Ume-1
3	23 February 1977	N1	ETS-2
4	16 February 1978	N1	Ume-2
5*	6 February 1979	N1	Ayame-1
6	22 February 1980	N1	Ayame-2
7	11 February 1981	N2	ETS-4
8	11 August 1981	N2	GMS-2
9	3 September 1982	N1	ETS-3
10	4 February 1983	N2	CS-2a
11	6 August 1983	N2	CS-2b
12	23 January 1984	N2	BS-2a
13	3 August 1984	N2	GMS-3
14	12 February 1986	N2	BS-2b
15	13 August 1986	H1	EGS/Fuji-1
16	19 February 1987	N2	MOS-1
17	27 August 1987	H1	ETS-5
18	19 February 1988	H1	CS-3a
19	16 September 1988	H1	CS-3b
20	5 September 1989	H1	GMS-4
21	7 February 1990	H1	MOS-1b/ Fuji-2/DEBUT
22	28 August 1990	H1	BS-3a
23	25 August 1991	H1	BS-3b
24	11 February 1992	H1	JERS-1
25	3 February 1994	H2	OREX/VEP
26	28 August 1994	H2	ETS-6
27	18 March 1995	H2	SFU, GMS-5
28[1]	11 February 1996	J1[1]	Hyflex
29	17 August 1996	H2	ADEOS
30	27 November 1997	H2	ETS-8, TRMM
31	21 February 1998	H2	Kakehashi
32	15 November 1999	H2*	MTSAT
33	29 August 2001	H2A	LRE, VEP-2
34	4 February 2002	H2A	MDS-1, DASH
35	10 September 2002	H2A	DRTS, USERS
36	14 December 2002	H2A	Fedsat, WEOS, Micro-Labsat

* indicates vehicle failure. All launches made from Tanegashima. *N/H Breakdown* (launch total for each type, followed by number of failures; through May 1996): N1 7(1); N2 8(0); H1 9(0); H2 3(0); J1 1(0).
[1]J1 #1 was suborbital.

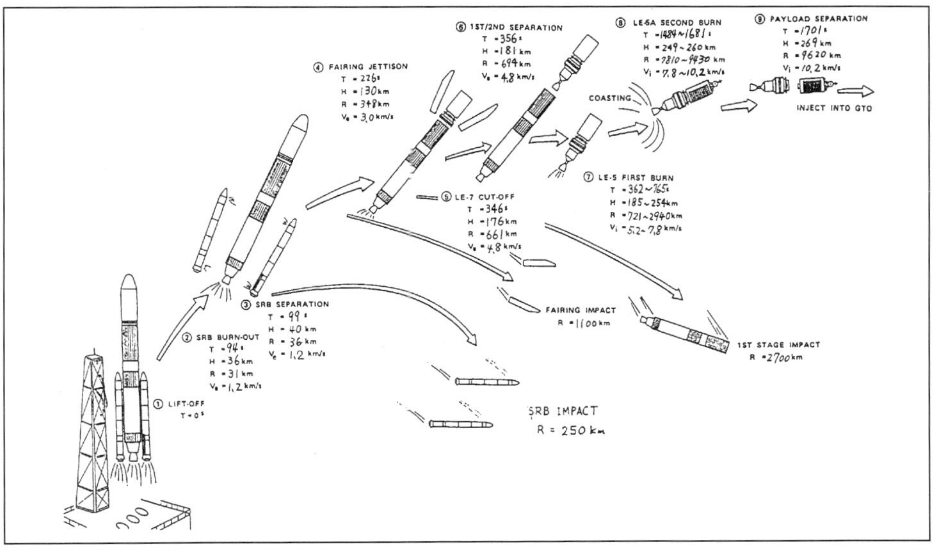

Typical H2 GTO flight sequence (NASDA) 0516915

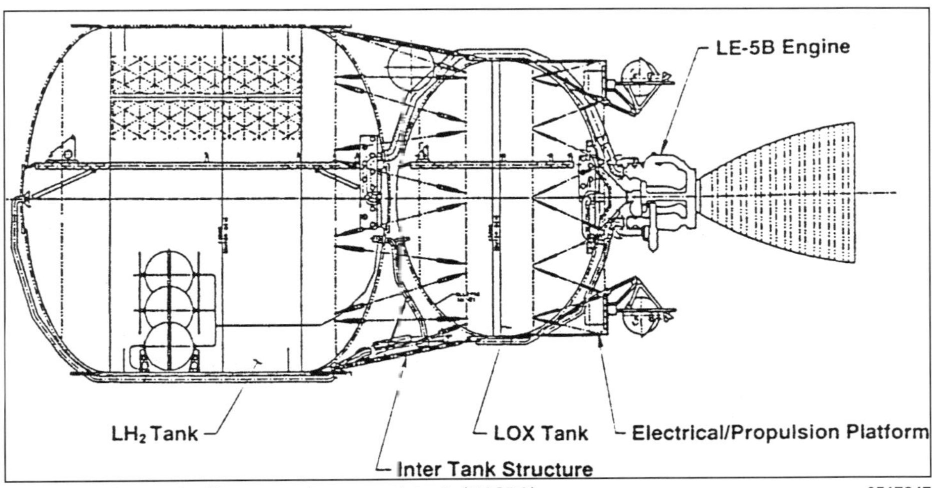

Improved H2's stage 2 will separate the cryogenic tanks (NASDA) 0517347

H2A will have new monolithic composite strap-on boosters and greatly improved payload capability (Theo Pirard) 0054367

H2 SRB solid strap-ons
Comment: The first full-scale static test was undertaken at Tanegashima 15 April 1988 (including 5° nozzle deflection), followed by prototype model (PM) firings June and 13 December 1989 at Tanegashima. The single qualification model (QM) run was conducted 29 May 1991.The first, successful, booster separation test was conducted in July 1988.
Motor contractor: Nissan Motor Co
Length: 23.362 m (19.26 m motor)
Principal diameter: 1.81 m
Mass at ignition: 70.4 t each
Propellant: 59.15 t each of 14% HTPB/ 68%AP/ 18%Al, in four-segment NT-150 steel case held by 108 bolts at each joint.
Average thrust: 1,560 kN each average sea level
Burn time: 94 s
Steering: TVC by ±5° deflection of flexible nozzle joint provides vehicle three-axis control

H2 SSB solid strap-ons
Comment: The twin Solid Sub-Boosters are carried by stage 1 only when required by mission performance. One flight so far: #3.
Motor contractor: Nissan Motor Co
Length: 9.053 m
Principal diameter: 1.125 m
Mass at ignition: 10.5 t each
Propellant: 8.4 t each of 14% HTPB/ 68%AP/18%Al in steel case.
Average thrust: 34.5 kN each average sea level (SI 267 s vacuum)
Burn time: 66 s
Steering: none, fixed nozzle canted 10° towards H2's centre of gravity.

H2 stage 1
Comment: The stage 1 Captive Firing Test series began 23 February 1993 with a 10 s run on the Tanegashima pad. The full 350 s was achieved 15 June 1993. The last run, 60 s, was performed 8 July 1993. Battleship tests were held at Mitsubishi's Tashiro site October 1990 to May 1992. The bracketed values below indicate goals, reduced during 1990 because of LE-7 development problems.
Contractor: Mitsubishi Heavy Industries
Engine: single Mitsubishi LE-7 staged combustion cycle, single start, liquid oxygen/hydrogen, regenerative cooling, 54:1 (60) expansion ratio, 6.0:1 mixture ratio, gimballed for pitch/yaw control.
Overall length: 28.0 m
Principal diameter: 4.0 m
Total stage mass: 98.1 t.
Oxidiser: 73.6 t LOX in forward tank; both tanks of semi-monocoque 2219 aluminium alloy, external polyisocyanurate foam thermal insulation. Pressurised at 4.26 atmospheres by helium from spheres immersed in LH_2 tank.
Fuel: 12.6 t LH_2 in aft tank, pressurised at 3.39 atmospheres by GH_2 tapped off from LE-7 nozzle extension.
Stage thrust: 843.4 kN (911.96 kN) sea level; 1,078 kN (1,180 kN) vacuum.
Burn time: 346 s
Attitude control system: two 1,500 N base auxiliary engines bleed GH_2 from LE-7 for roll control following SRB separation

H2 stage 2
Comment: An enlarged variant of H1's stage 2. It originally selected the same LE-5 engine, but NASDA decided to increase thrust from 102.96 kN to 121.5 kN (vacuum SI increase of 3 s to 452 s, chamber pressure from 37 to 39 atmospheres) by switching from the gas generator cycle to the hydrogen bleed cycle, tapping gaseous fuel from the nozzle skirt to drive the turbines. The design also provides improved re-start characteristics. LE-5A prototype firings concluded May 1988 after achieving a total run time of 2,800 s. Stage 2 battleship firing tests were held January to March 1990 at Mitsubishi's Tashiro test site, followed by flight-type stage testing August to October 1991. The Improved H2 will significantly redesign stage 2.
Contractor: Mitsubishi Heavy Industries
Engine: single Mitsubishi LE-5A hydrogen bleed cycle, dual start, LOX/LH_2, regenerative cooling, 130:1 expansion ratio, 5.0:1 mixture ratio, gimballed 3.5° square for pitch/yaw control.
Overall length: 10.6 m
Principal diameter: 4.0 m
Total stage mass: 19.7 t
Oxidiser: 13.9 t of liquid oxygen in aft tank
Fuel: 2.8 t of liquid hydrogen in forward tank
Stage thrust: 121.5 kN vacuum, but 5% idle mode provides precise orbit control + de-orbit capability.
Burn time: 609 s max; two burns (403 s to establish LEO parking orbit + 197 s for 28.5° GTO) for typical GEO missions.
Attitude control system: LE-5A gimballing provides pitch/yaw control during burn; two IHI hydrazine

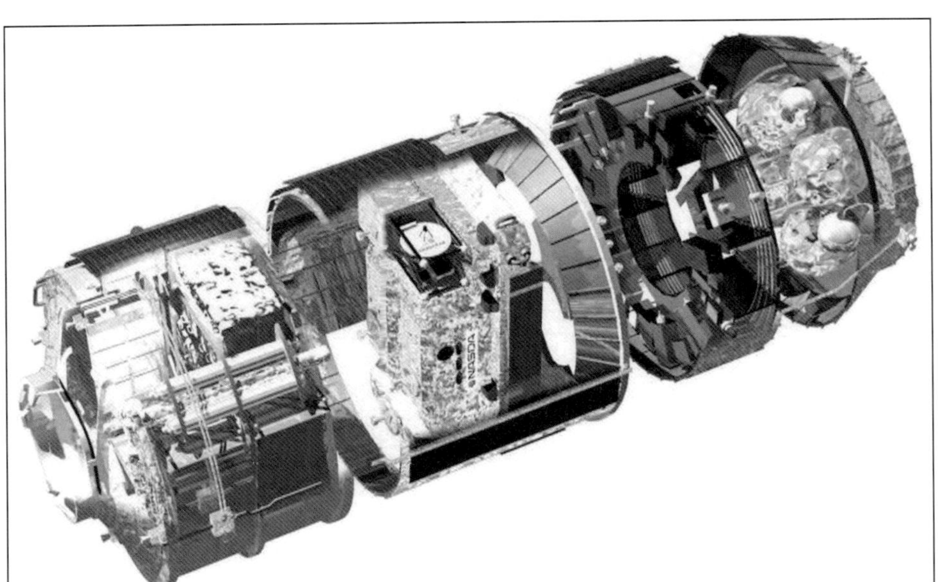

Planned as a transfer stage for the H2A launch vehicle this module would be used to access the ISS (JAXA)
NEW/1047527

The evolutionary development of the H2, the H2A, seen here in model form (right) alongside its predecessor
0084629

thruster modules (each 4 × 50 N + 2 × 18 N) provide roll control during burn and three-axis during coast.

Payload fairing/attach fitting

Comment: Kawasaki's standard 4.1 m diameter, 12.0 m high 1.4 t aluminium honeycomb core/skin 4S fairing provides a 3.7 m diameter envelope for single payloads. The 4/4D is stretched to 15.0 m for dual main payloads, with an 11 m aluminium upper section and 4 m CFRP lower portion; it will be used for ETS 7/TRMM. A 5.1 × 12.0 m version (5S) offers accommodation for Shuttle-class (4.6 m diameter) payloads; ADEOS used it. Flight #3 was the first to carry the '5/4D' 14.1 m-long dual fairing, with 5.1 m diameter 9.6 m-long aluminium upper portion and 4.1 m diameter 4.5 m-long CFRP lower. A Mild Detonating Fuse effects separation of the clamshell halves and springs when aerodynamic heating has fallen to 1,135 W/m^2 during stage 1 burn.

Acoustic loads: 141 dB peak during lift-off and transonic flight

Vibration: sinusoidal 2 *g* longitudinal, 1 *g* lateral

Acceleration: peak 4 *g* longitudinal at stage 1 shutdown, design requirement for 5 *g*

Thermal environment: 500 W/m^2 radiation from inner wall

Pad environmental control: class 100,000, 15–25°C air conditioning.

UPDATED

HOPE (H2 Orbiting Plane)

Current Status

Phase 1 of the High Speed Flight Demonstration phase using 1/10th scale HOPE-X design vehicles was conducted during October and November 2002 and the phase 2 test took place in July 2003. Under the new reformatted Japan Aerospace Exploration Agency (JAXA), the Institute of Space Technology and Aeronautics is managing tests eventually leading to the HOPE-X orbital spaceplane.

Background

NASDA initiated studies of the HOPE unmanned reusable spaceplane in 1986, intending to provide servicing missions from 1998 to Japan's Experimental Module attached to the Space Station. Both HOPE and the Space Station have been considerably delayed. From their work, the Space Activities Commission in mid-1993 released a report recommending development of a winged reusable vehicle using H2. However, given the problems with the Shuttle and the Soviet equivalent, the Commission recommended that Japan hold back on a full commitment to the programme until after demonstration of critical technologies.

Japan seemed especially interested in understanding hypersonic re-entry and building a guidance and control system so the spaceplane could have autonomous operation and not require the same expensive overhead associated with training a crew. These technologies were actively pursued through three experimental vehicles. The OREX re-entry vehicle carried by H2 number 1 in February 1994 tested thermal protection and navigation. J1 number 1 in February 1996 carried the Hyflex hypersonic flight experiment spaceplane model on a suborbital path. The Alflex automatic approach/landing experiment was released and landed successfully 13 times from a helicopter July-August 1996 to demonstrate automatic GN&C beginning at 1,500 m altitude and 2,700 m out from Australia's 2,600 m Woomera runway.

The 760 kg 6.11 m length, 3.79 m span Alflex is a 37 per cent scale version of HOPE's FY92 04C double delta design. It includes an IMU, differential GPS, radio altimeter and microwave landing system. Fuji Heavy Industries built two models. The full scale HOPE-X demonstrator, approved FY96 and started Phase C, HOPE-X was planned for launch by a single-stage H2A to a 120 to 200 km orbit, completing one circuit of the Earth. HOPE-X will re-enter, glide and land automatically within 2 hours. HOPE-XA was planned for a brief LEO mission. FY97 funding was ¥11 billion (NASDA 10.5 plus NAL 0.5), FY96 was ¥9.5 billion (8.5 + 1.0). HOPE will be modified from HOPE-X and increase the ISS cargo capacity from 3 to 5 tonnes utilising standard 500 kg, 1 × 1 × 1.8 m ISS racks. Earth-return capacity would be 5 tonnes (for wing loading of 250 kg/m^2).

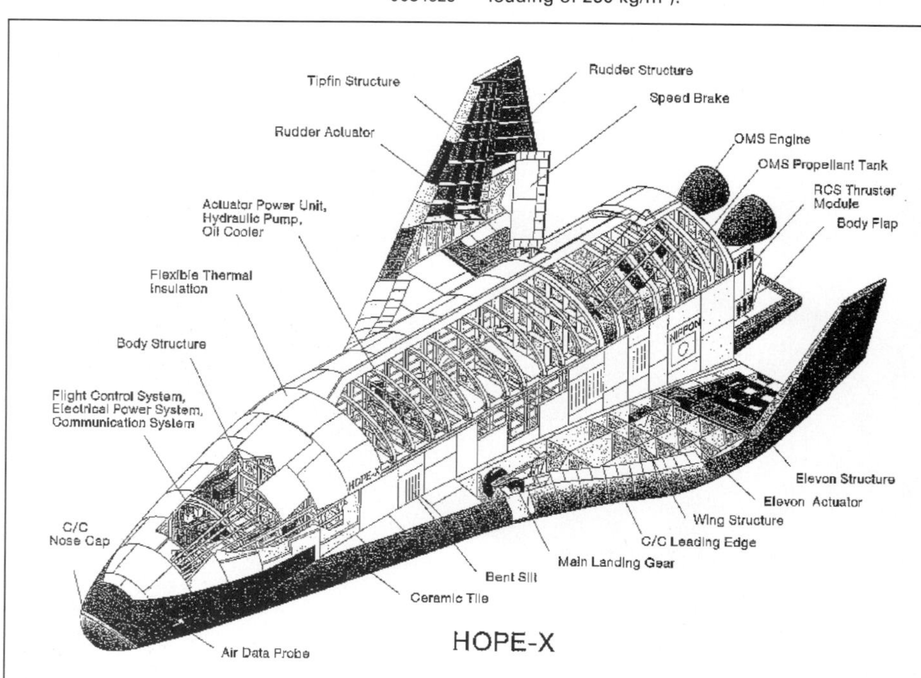

HOPE was designed to deliver 3 tonnes of cargo to Space Station from about 2010 (NASDA)
0003454

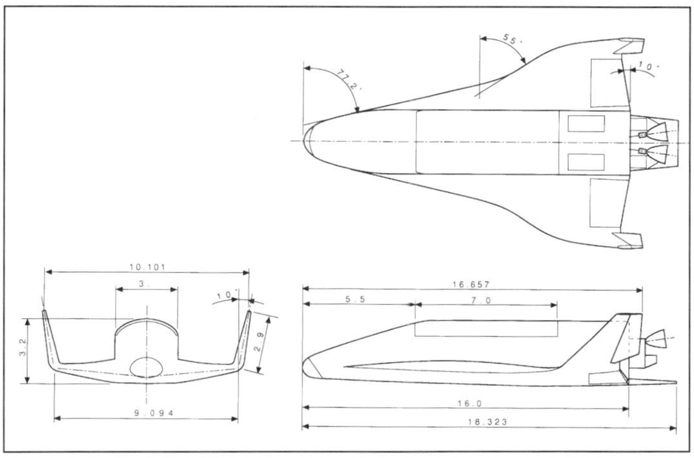

HOPE-X's double delta design provides lower mass and better controllability in all flight regimes (NASDA) 0003452

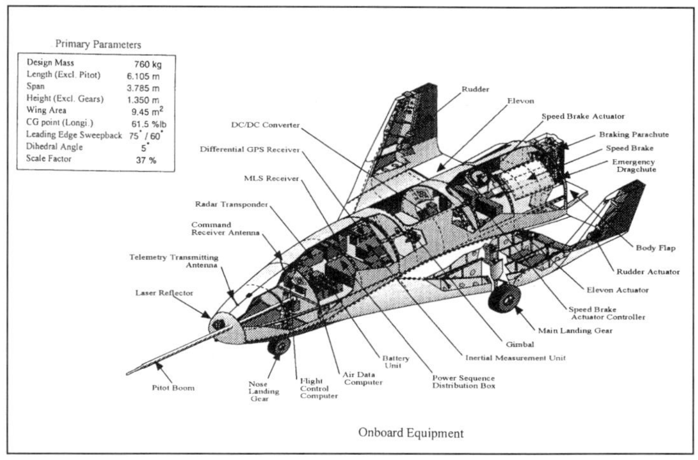

Alflex in 1996 demonstrated HOPE's landing sequence (NASDA/NAL) 0003450

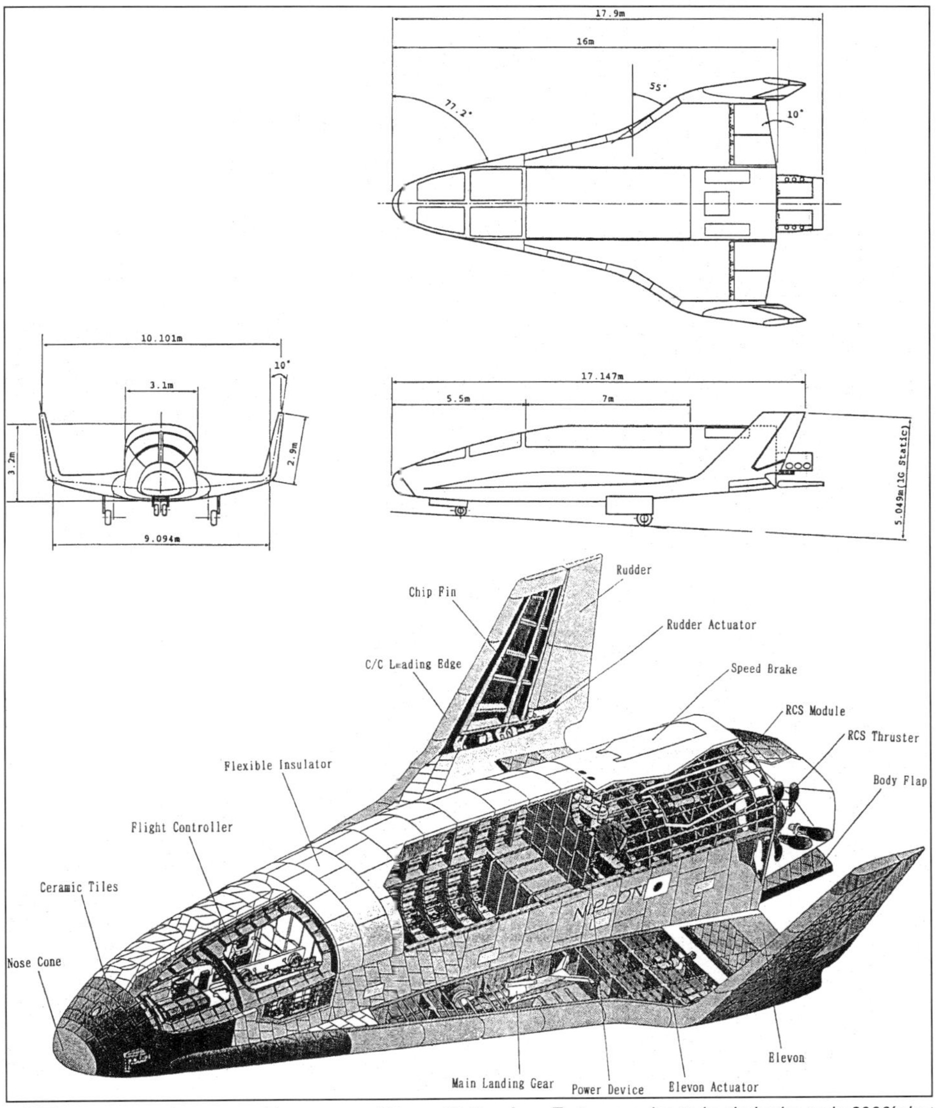

HOPE was proposed as a reusable unmanned Space Station ferry. Tests were due to begin in the early 2000's but have been delayed by budgetary constraints. (NASDA) 0517353

Alflex in 1996 demonstrated HOPE's landing sequence (NASDA/NAL) 0003449

HOPE's own propulsion system circularises its path at Space Station altitude and then the craft performs an automatic rendezvous/docking after two days. After exchanging cargoes and undocking, it would re-enter within one day and execute an automatic 300 km/h touchdown on a 3,000 m runway. Crossrange capability is projected at 1,500 km, with maximum time on-orbit about 100 h. In addition to Station and platform servicing, HOPE would act as a Japanese technology demonstrator for future spaceplane development. Principal contractors are Mitsubishi, Kawasaki and Fuji. Citing financial and technical difficulties and uncertainties with the H2 launch vehicle, NASDA stopped work on HOPE-X in August 2000 after spending US$286 million. Technology studies continued and subscale tests were approved as demonstrators under the High Speed Flight Demonstration (HSFD) programme managed by NASDA and Japan's National Aerospace Laboratory (NAL). The two-phase flight demonstration programme began in October 2002 and involved two vehicles, one of which was powered. The HSFD programme began in April 1999 as a flight verification effort for the Hope spaceplane and the two experimental demonstrators were rolled out from the Fuji Utsunomiya plant in April 2002. Phase 1 flights are conducted from Christmas Island in the Pacific Ocean and involve the powered vehicle fitted with a 4.4 kN thrust Teledyne Continental Motors J69-T406 turbojet. The vehicle takes off and lands on a 1,800 m runway and the flight follows a pre-set course of 50 km during which a maximum altitude of 8,000 m and a maximum speed of Mach 0.7 at 6,000 m is attained. The powered air vehicle is a refined development of the original Hope and carries two outward-canted vertical stabilisers and an air intake located beneath the mid-fuselage. It is 3.34 m long and has a delta wing span of 2.96 m with a wing area of 4.4 m (space wing) and a maximum weight of

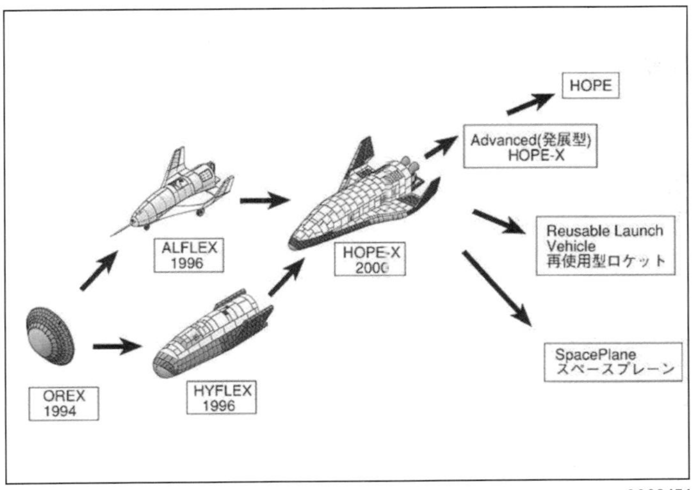

HOPE families 0003451

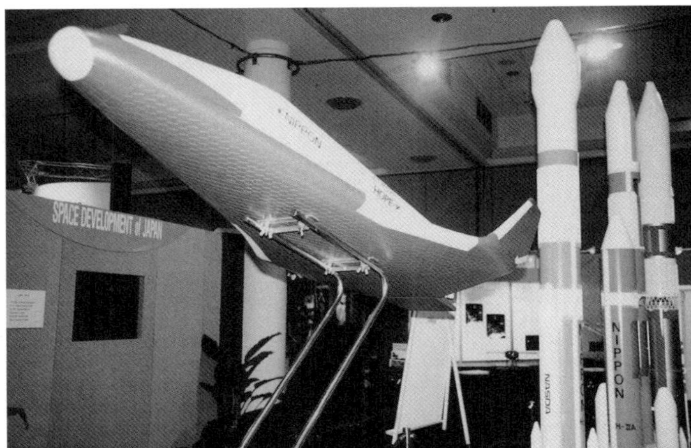

Scale model of HOPE spaceplane shows double-delta wing planform and simulated thermal tiles (Theo Pirard) 0054365

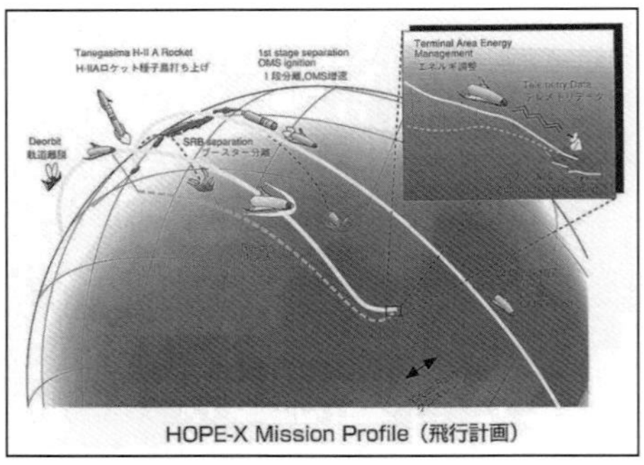

HOPE-X mission profile 0003453

735 kg. Phase 2 of tests was scheduled for 2003 using a smaller unmanned air-dropped vehicle for transonic flight studies. The French space agency CNES is helping NASDA and the NAL develop in return for shared data. These tests took place from the Esrange balloon test site in Kiruna, Sweden. This smaller unpowered air vehicle has a 2.42 m wingspan and a maximum weight of 500 kg. Dropped from a maximum altitude of 30 km, it will achieve a target speed of Mach 1.2, flare for 10 to 12 seconds and deploy a parachute prior to touchdown, impact cushioned by airbags. HSFD tests are now seen as a bridge between the OREX/Hyflex/Alflex tests and a Hope but rather between those tests and a two-stage launcher with upper stage fly-back capability leading to a SSTO and return concept.

Specifications

HOPE-X

Length (m) × width (m) × height (m): Approx. 16 × 10 × 5
Total weight (kg): Approx. 14,000 (at launch)/9900 (at re-entry)
Launcher: Single Stage type H-11A Rocket (Improved)
Landing: Automatic landing on 1800 m class runway
Shape: Delta wing with tip fins
Structure: Al alloy and carbon/polyimide composite
Thermal protection system: carbon/carbon composite, ceramic tile and flexible thermal insulator
Propulsion system: Orbital Manoeuvring System (OMS) and Reaction Control System (RCS)
Attitude control method: Aerodynamic control surface and RCS.

UPDATED

J1

Current Status

The former NASDA decided to retire the J1 in 2001 and replace it with the J-2 (J-1A) in March 2002.

J1 debut was successful in 1996 (NASDA) 0517352

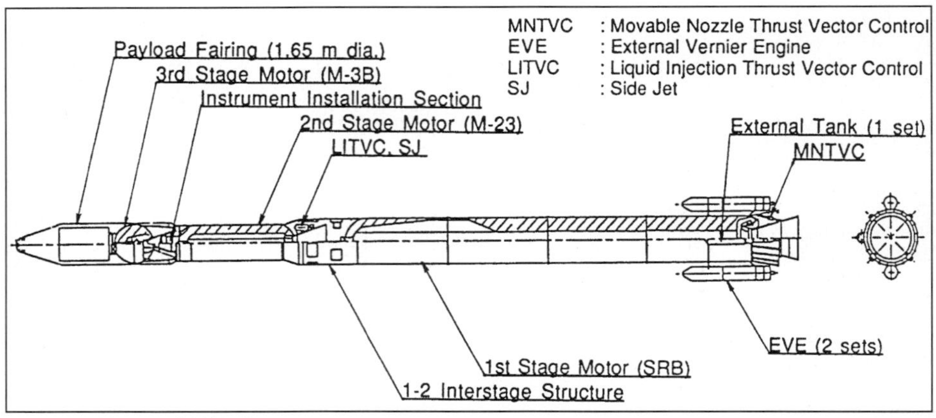

J1 principal features (NASDA) 0517184

Funds for this vehicle have been cut and a change in Japan's launcher development strategy has been introduced. The reconfiguration of Japan's space and aeronautics research agencies has eliminated the J launcher.

Background

NASDA has, unusually, developed the all-solid J1, capable of delivering 900 kg into LEO. Stage 1 employs H2's strap-on, while stages 2 and 3 use M-3SII's second and third stages. The debut two-stage suborbital test in February 1996 was successful, but J1's manifest remains sparse - only one other launch is scheduled.

The Space Activities Commission in July 1991 approved development of the ¥11.30 billion all-solid J1 intermediate launcher. ¥434 million was authorised for FY92, ¥3,215 million FY93, ¥2,909 million FY94, ¥4,767 million FY95, ¥0 FY96. NASDA sought minimum development cost and risk by adapting existing hardware. Tanegashima's old H1 pad is utilised. Two strap-ons derived from the TR-1 would increase LEO capacity to about 1 tonne; using two H2 SRBs would yield 2 tonnes.

Specifications

J1

First launch: 11 February 1996 (2-stage suborbital)
Number launched: 1 by end of 2001
Launch site: Tanegashima
Principal uses: delivery of intermediate payloads to LEO
Vehicle success rate: 100 per cent
Cost: about ¥5 billion, excluding development cost (not planned commercially).
Availability: potentially two annually. Fishing/range safety requirements constrict launches to January-February/August-September annually.
Performance: 870 kg into 250 km 30° circular
Number of stages: 3
Overall length: 33.110 m
Principal diameter: 1.81 m
Launch mass: 87,700 kg plus payload
Guidance: inertial/radio, largely derived from M-3SII. The B2CNE stage 2 control electronics unit is mounted on stage 2's forward end, and includes the RIG Rate Integrating Gyro on the SFAP Spin Free Analytic Platform for pitch/yaw/roll data. Ground commands are received by B2CNE should the Nogi Radar Station on the Tanegashima Islands 115 km south of Kyushu detect deviation from the programmed path. B2CNE routes radio commands and attitude error signals to stage 1's B1CNE, which also uses a stage 1 rate gyro package.

J1 stage 1

Motor contractor: Nissan Motor Co
Length: 20.992 m
Principal diameter: 1.8094 m
Mass at ignition: 70.9 t

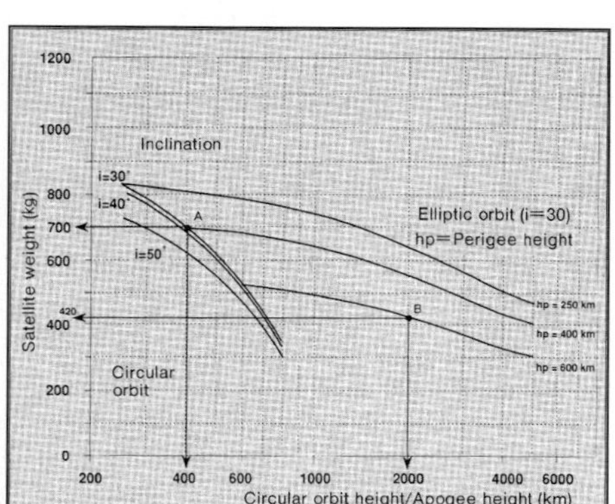

J1 payload capability from Tanegashima 0517194

Propellant: 59.15 t of 14% HTPB/ 68%AP/18%Al, in 4-segment NT-150 steel case held by 108 bolts at each joint
Average thrust: 1,560 kN average sea level
Burn time: 89 s
Steering: TVC by ±5° deflection of flexible nozzle joint provides vehicle pitch/yaw control. Two aft External Vernier Engine pods ignite 5 s before launch to provide roll control during burn, then three-axis during coast (previously planned solid motors on fins for roll control, as on M-3SII). EVE 4.177 m long, 56 cm diameter, carrying 110 kg N_2O_4/hydrazine. 3.43–2.45 kN vacuum blowdown (from hydraulic accumulator), gimballed ±45° pitch/roll, ±11° yaw.

J1 stage 2

Designation: M-23
Contractor: Nissan Motor Co
Overall length: 6.3 m
Principal diameter: 140.7 cm
Mass at ignition: 12.7 t (excluding fairing)
Propellant: 10.4 t of solid seven-star grain HTPB composite in steel case
Average thrust: 524 kN vacuum
Burn time: 55 s
Attitude control system: pitch/yaw control during burn maintained by LITVC, injecting 40 litres of $NaClO_4$ solution into nozzle at eight points. Roll control during burn and three-axis control in preparation for stage 3 spin-up/separation provided by four clusters of 4 × 150 N Side Jet (SJ) thrusters at stage 2's base drawing on 40 litres hydrazine

J1 stage 3

Designation: M-3B
Contractor: Nissan Motor Co
Overall length: 2.7 m
Principal diameter: 149.5 cm
Mass at ignition: 3.59 t
Propellant: 3.28 t of seven-star HTPB composite solid in welded titanium alloy casing
Average thrust: 132 kN vacuum
Burn time: 71 s
Attitude control system: upper stages spin induced by stage 2 attitude control system before separation

Payload fairing

The 500 kg glass fibre honeycomb 2 cm-thick sandwich clamshell fairing, as used by the M-3SII, is pyrotechnically separated after stage 2 burnout. Nissan's 6.86 m long, 1.65 m diameter structure, which also covers stage 3, provides a payload envelope of about 3.0 m long, 1.4 m diameter. Peak vehicle acceleration is 7 *g* longitudinal (2 *g* at launch).

L and M series orbital launch history

	Date	Vehicle	Payload
1*	26 September 1966	L-4S	Test satellite
2*	20 December 1966	L-4S	Test satellite
3*	13 April 1967	L-4S	Test satellite
4*	22 September 1969	L-4S	Test satellite
5	11 February 1970	L-4S	Ohsumi
6*	25 September 1970	M-4S	Science sat 1
7	16 February 1971	M-4S	Tansei 1
8	28 September 1971	M-4S	Shinsei
9	19 August 1972	M-4S	Denpa
10	16 February 1974	M-3C	Tansei 2
11	24 February 1975	M-3C	Taiyo
12*	4 February 1976	M-3C	Corsa A
13	19 February 1977	M-3H	Tansei 3
14	4 February 1978	M-3H	Kyokko
15	16 September 1978	M-3H	Jikiken
16	21 February 1979	M-3C	Hakucho
17	17 February 1980	M-3S	Tansei 4
18	21 February 1981	M-3S	Hinotori
19	20 February 1983	M-3S	Tenma

	Date	Vehicle	Payload
20	14 February 1984	M-3S	Ohzora
21	8 January 1985	M-3SII	Sakigake
22	18 August 1985	M-3SII	Suisei
23	5 February 1987	M-3SII	Ginga
24	21 February 1989	M-3SII	Akebono
25	24 January 1990	M-3SII	Hiten/ Hagomoro
26	30 August 1991	M-3SII	Yohkoh
27	20 February 1993	M-3SII	Asuka
28*	15 January 1995	M-3SII	Express
29	12 February 1997	M-5	Hanika
30	3 July 1998	M-5	Nozomi
31	10 February 2000	M-5	Astro E

* indicates vehicle failure. All launches made from Kagoshima. *M-3S Breakdown* Launch total for each type, followed by number of failures. M-3S: 4(0); M-3SII: 8(1); M-5 0(0).

Preparing for the first M-5 (ISAS) 0517351

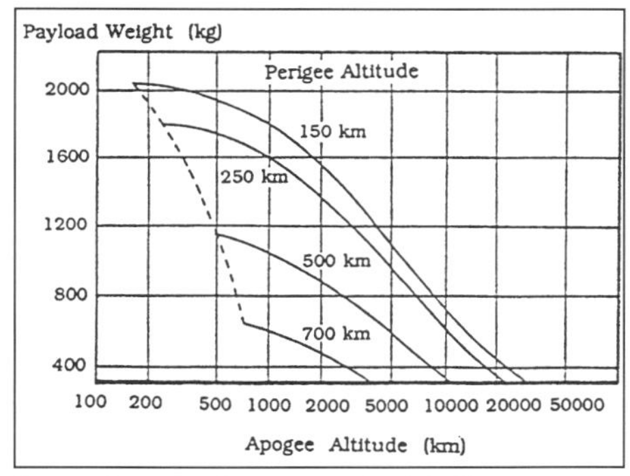

M-5 performance from Kagoshima (ISAS) 0516917

M Launch Vehicles

Current Status

ISAS has flown three M-5 launches of which one has failed. Launch costs are approximately US$36 million. ISAS has an ambitious programme of science missions in earth orbit and schedules as ongoing sequence of flights.

Background

ISAS has maintained its own stable of orbital and suborbital rockets for science missions since the late 1950s. All solid-fuelled, these have been integrated by ISAS with Nissan Motor as prime industrial contractor. The largest sounding rocket, L-3H, was augmented by a kick stage to create the L-4S orbital vehicle as a technology demonstrator for the M-4S launcher, flying Japan's first satellite in 1970 after four failures. The M-3SII model was responsible for launching two Halley's comet probes in 1985 - the first deep spacecraft dispatched by all-solid rockets. It retired January 1995 after its eighth (and first unsuccessful) mission.

The division of launcher responsibilities between ISAS/NASDA limited the scientific agency to rockets with main diameters no greater than 1.4 m, but studies were initiated in 1987 of the 2.5 m diameter M-5 design that will almost triple LEO capacity and permit a wider range of deep space missions. Development was approved during 1989, then aiming for first launch in 1995.

Studies were undertaken during 1987–88 of a large successor to the M-3SII, capable of placing 2.2 tonnes in LEO and dispatching 520 kg to the Moon. Development cost was estimated at US$133 million, with a US$36 million unit cost. ISAS estimated in 1990 a unit mission cost of ¥5.85 billion.

After the feasibility studies showed promise for the rocket, the Japanese government approved the M-5 in 1989. A September 1994 test launch of an ST-735-2 vehicle succeeded in demonstrating stage 2's 'fire-in-the-hole' separation technique. Then ISAS/Nissan began studies in 1990 of an air-launched three-stage winged version, capable

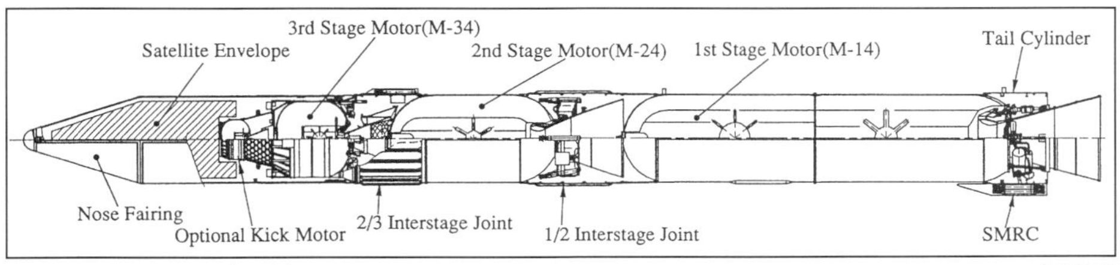

The new M-5 launcher (ISAS) 0516916

Models of the M5 and J1 launch vehicles (Theo Pirard) 0054366

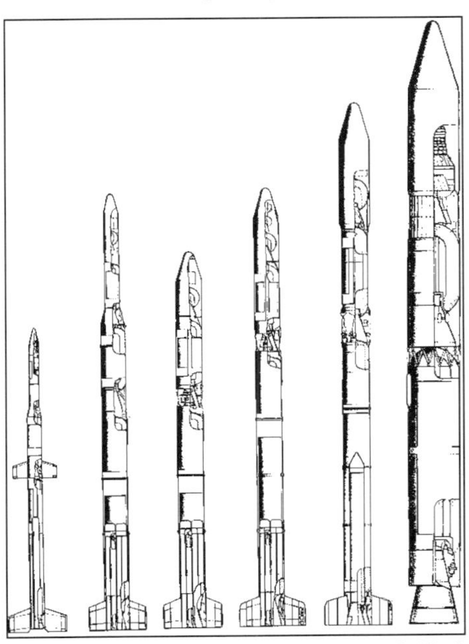

ISAS orbital launchers (from left): L-4S, M-4S, M-3C, M-3S, M-3SII, M-5 0517178

of delivering 1,270 kg into a 250 km circular orbit at a cost of ¥360,000/kg. The 17 m-long 51.85 tonne vehicle would be released from the back of a 747-200 at an altitude of 10 km and speed of 200 m/s. This vehicle derives its stage 1 from M-5's stage 2 M-24, its stage 2 from M-5's stage 3 M-34, and its stage 3 from M-3SII's M-3B stage 3. The advantages would be a 2.5 per cent payload ratio (1.5 per cent for M-5) and relaxation of the fishing season launch constraints. The first M5 was launched on 12 February 1997, carrying the first satellite dedicated to Very-Long Baseline Interferometry, followed by the second on 3 July 1998, which launched the first Japanese Mars orbiter scheduled to arrive in early 2004 but which was lost prior to encounter, and the third on 10 February 2000. In May 2003 an M5 launched the Hayabusa engineering spacecraft on a sample return mission to an earth-crossing asteroid.

Specifications

M-5

Launch site: Kagoshima
Performance: 250 km, 31°: 1,800 kg
500 × 500 km, 31°: 1,200 kg
Sun-synchronous: 800 kg
GTO: 800 kg
Lunar probe: 520 kg
Mars/Venus, C3 9.0 km^2/s^2: 450 kg
Number of stages: 3 plus optional stage 4
Overall length: 31.2 m
Principal diameter: 2.50 m
Launch mass: 128.4 t (excluding payload)
Guidance: active radio guidance with inertial using fibre optic gyros and microprocessor as back-up on early flights. Inertial system may be used later. Other electronics, such as telemetry, ranging and command systems transferred from M-3SII, are also in a bay at top of stage 3.

M-5 stage 1

Comment: Prototype stage 1 was successfully fired 21 June 1994. A second firing, using a thinner nozzle, was successful June 1995.
Designation: M-14
Contractor: Nissan Motor Co
Overall length: 13.65 m
Principal diameter: 2.50 m
Mass at ignition: 82.5 t
Propellant: 71.7 t solid propellant in HT-230M maraging steel case
Average thrust: 4,119 kN sea level
Burn time: 47.1 s (effective)
Attitude control system: movable nozzle, actuated by gas generator

M-5 stage 2

Comment: Stage 1 separates at 70 s and stage 2 ignites simultaneously (fire-in-the-hole) to reduce gravity losses. The lattice structure of the 1,762 kg 3.84 m high interstage vents the exhaust and the three 120° supporting panels are released by FLSC Flexible Linear Shaped Charges and pushed out by springs from top. Stage 2 incorporates an extending nozzle, increasing expansion ratio from 27.9 to 37.1 and Isp vacuum from 288 s.
Designation: M-24
Contractor: Nissan Motor Co
Overall length: 6.71 m
Principal diameter: 2.51 m
Mass at ignition: 34.3 t
Propellant: 31.1 t solid in HT-230M maraging steel case
Average thrust: 1,206 kN vacuum
Burn time: 70.2 s (effective)
Attitude control system: liquid injection TVC

M-5 stage 3

Comment: 904 kg 2.45 m high aluminium interstage. Extending nozzle saves 300 kg and 1 m length.
Designation: M-34
Contractor: Nissan Motor Co
Overall length: 3.55 m
Principal diameter: 2.2 m
Mass at ignition: 11.1 t
Propellant: 10.1 t solid in carbon fibre 12-layer filament wound case. Expansion ratio 96.0, Isp 301 s vacuum.
Average thrust: 289 kN vacuum
Burn time: 103.5 s (effective)
Attitude control system: movable nozzle (electrical actuation)

Payload accommodation

715 kg clamshell fairing of CFRP skins on honeycomb sandwich, length about 9 m, 2.50 m diameter. Outer surface covered by cork for thermal protection

UPDATED

N1/2

Current Status

The last N1 was launched in September 1982. Of the seven launches flown two were failures. All eight N2 launches flown between 1981 and 1987 were successful.

Background

The former NASDA began developing the Delta-based N1 orbital launcher in 1970 under US license, flying seven missions 1975–1982. The 130 kg GEO capability was increased to 350 kg for the eight N2s launched between February 1981 and February 1987. Considerable technical support from (then) McDonnell Douglas company at N1 inception stimulated modifications which were later retro-fitted into the US Delta launch vehicles.

The former ISAS and NASDA continue the operation of two distinct launcher types. ISAS operated under constraint by a 1970 ruling limiting their missions to science objectives, using only solid propellant vehicles, while NASDA provided the national capability for launching applications satellites aboard liquid vehicles of increasing capabilities. Until 1990, ISAS was limited to making vehicles with base diameters no greater than 1.4 m but development of the 2.5 m diameter M-5 was approved in 1990. NASDA, on the other hand had the charter to build an indigenous vehicle for 2-tonne GEO applications, with a view to commercialisation. Now both organisations have merged into the new JAXA those limitations on launcher assignment have subsumed under a more urgent national need for rationalisation.

S-520 Sounding Rocket

Background

Nissan's S-520 is the largest of JAXA's range of sounding rockets, capable of delivering 250 kg to 400 km. In addition to space science missions, it is employed for recoverable µgravity experiments. The single solid propellant motor, with steel casing, is supported by four aluminium honeycomb tail fins with niobium/titanium alloy leading edges. The SS-520 is derived from the S-520 but with a more powerful second-stage and was first launched in January 1998. The SS-520-2 was launched 4 December 2000 and reached an altitude of 1,040 km. In its former guise as the ISAS, JAXA continues to conduct analysis of improvements to the S-520 for national and internationally cooperative activities.

Specifications

S-520

First launch: 18 January 1980 (first recoverable flight 5 September 1981)
Number launched: 24 to end-April 2000 (20 Kagoshima, 4 Andøya)
Success rate: 100% to end-April 1999
Launch site: Kagoshima (Japan), Andøya (Norway)
Performance: 150 kg to 350 km, using 82° launch angle
Number of stages: 1 (solid)
Overall length: 931.0 cm
Principal diameter: 524 mm
Launch mass: 2,285 kg
Guidance: fin spin stabilisation, but de-spin system included.

S-520 stage 1

Contractor: Nissan Motor Co
Overall length: 6,053 mm
Principal diameter: 524 mm (179 cm across four fins)
Thrust: average 143.17 kN vacuum
Burn time: 28.7 s effective

Payload accommodation

GFRP clamshell ogive nose, providing 494 mm payload base diameter, narrowing to 156 mm after 1,252 mm length. The support section below houses telemetry, timing, ignition relay, radar transponder and power supply equipment. Immediately below are the attitude control and recovery sections.

Russian Federation

Angara

Current Status

Severe financial restraints slowed development and has delayed inaugural flights probably beyond 2006. Funding is restricted to some engineering tests, including the Energomash test firing of the RD-191 engine for the first stage.

Background

Khrunichev/KB Salyut conceived the Angara series to replace Proton by early in the 21st century. Khrunichev's design won out in a competition by RKA over a concept forwarded by Energia-M and others.

It promised to significantly improve Proton performance but relied on existing or imminent technology, drawing on Zenit for stage 1, Energia for stage 2 and the Breeze-M and cryogenic stages already under development for Proton KM. It was to have employed Zenit's Plesetsk pad, freeing Russia of Proton's reliance on Kazakhstan. A fly-back stage 1 was possible, as was replacing the stage with an Antonov An-225 to deliver 6 tonnes into 200 km 51° orbit for US$1,500/kg.

By the second half of the 1990s the configuration had changed to one with more conventional design layouts and a series of launchers capable of meeting a broad range of user needs. The Angara core would comprise a single chamber RD-191M, a derivative of the four-chamber RD-170 used on Zenit and the two-chamber RD-180 used with Atlas. Using its own funds, Khrunichev designed the Baikal reusable fly-back booster concept for Angara as an alternative to the conventional layout but no government support has been forthcoming.

Specifications (early concept)

Angara

First launch: Unknown
Launch sites: Plesetsk Zenit complex 35
Contractors: Khrunichev/KB Salyut (prime), NPO AP (GN&C), NPO Energomash (main engines)
Principal uses: geostationary, escape and large LEO
Performance (from Plesetsk): 26.0 t into 200 km 63° two-stage; 3.5 t GEO using Breeze-M; 4.5 t GEO using Cryogenic. See diagram for further information.
Number of stages: two for LEO; three for higher orbits
Overall length: 55.5 m (5.0 m diameter fairing); 52.4 m (4.35 m diameter fairing).
Principal diameter: 4.10 m core
Launch mass (with payload): about 640 t
Launch thrust: 7,622 kN sea level
Guidance: unknown

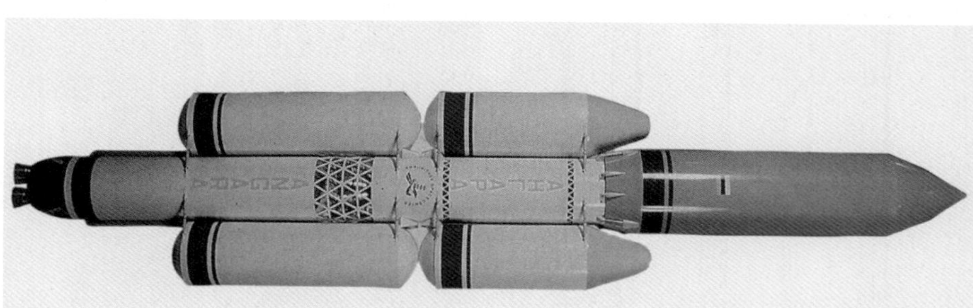

An early concept for Angara employed an unusual suspended tank configuration for both main stages. Stage 1's outer tanks carry LOX, stage 2's LH_2 (Khrunichev)
0517326

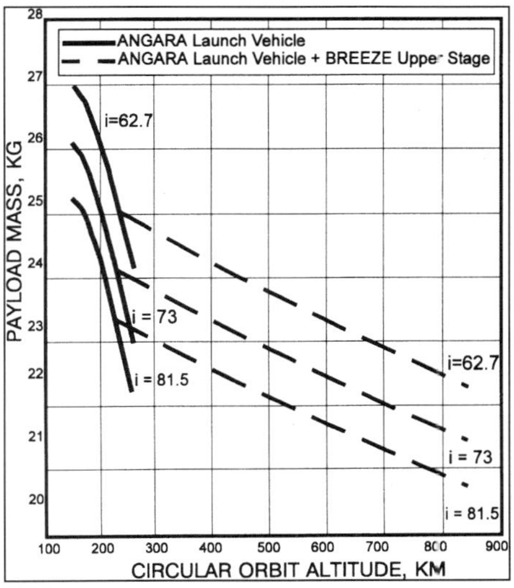

Angara performance from Plesetsk for the 2-stage (solid lines) and Breeze-M (dashed) versions (Khrunichev) 0517327

Three proposed configurations for the Angara launcher series (Theo Pirard) 0054361

Angara stage 1
Engines: RD-174 from NPO Energomash, four gimballed chambers.
Overall length: 14.5 m
Diameter: 4.10 m core, 3.90 m suspended tanks
Thrust: 7,622 kN sea level
Burn time: not available
Dry mass: not available
Oxidiser: LOX in two suspended tanks
Fuel: kerosene in core

Angara stage 2
Engines: NPO Energomash RD-0120 single fixed reignitable chamber with verniers
Overall length: 9.5 m
Diameter: 4.10 m core, 3.90 m suspended tanks
Thrust: not available
Burn time: not available
Dry mass: not available
Oxidiser: LOX in central core
Fuel: LH_2 in two suspended tanks

Angara stage 3
Configuration: Angara will employ the same Breeze-M and cryogenic stage 3 as Proton KM's stage 4. See Proton's entry for further details.
Payload accommodation: two fairings are planned: 5.00 m diameter, 22.50 m long and 4.35 m diameter, 19.42 m long.

Dnepr-1 (SS-18K)

Background

When NPO Yuzhnoye started to market this launch vehicle – based upon the R-36M2 missile – they took the NATO designator 'SS-18' and added a 'K' to indicate a space application. It remained the SS-18K until 1996 when the name Dnepr started to be applied to the vehicle. Yuzhnoye offered the booster for the cost of US$20 million to US$40 million in mid-1992. Tsyklon's S5M stage 3 or Lavotchkin's Fregat stage could be adapted for orbital injection/circularisation.

Development of the silo-launched missile began in 1964 and it was first deployed in 1974. Different modifications carried single or 10 warheads. Yuzhnoye is particularly interested in launching µgravity payloads of up to 700 kg, providing up to 2,000 hours of 10^{-6} *g* µgravity on orbital missions and 2 hours on ballistic flights.

Members of the design and launch team claimed the missile achieved a 97 per cent success rate on 157 launch attempts between 1973 and April. Production ended in 1991 under the first phase of the START arms control agreement. By 2003 174 rounds were destroyed despite START-2 not being ratified and 154 missiles were available to be converted into satellite launchers. US Defense Secretary William Perry, confirmed in May 1996, that China had been seeking to obtain SS-18 technology from Russia and possibly Ukraine.

The first Dnepr was launched successfully from a silo at Baikonur on 21 April 1999, carrying the UK satellite UoSAT-12. The second Dnepr carried five satellites into orbit on 26 September 2000. These comprised Tinngsat 1, MegSat 1, SandiSat 1A, SandiSat 1B and UniSat. The third lifted six satellites to orbit on 20 December 2002. A fourth flight in late 2004 or 2005 will lift the TransOrbital Trailblazer lunar spacecraft capable of imaging using redundant hardware left over from Russian and American moon missions.

Sources originally reported that the RS-36M2 missile used the RD-251 engine on its first stage, but it is now known that this is the designator applied to the first stage of Tsyklon. The first is the RD-264 with a thrust of 4,522 kN. The second stage is a VMZ RD-0228 with a thrust of 760.4 kN and the third stage a dual mode liquid RD-864 with a thrust of 18.9 kN/7.9 kN from the GDL/Energomash.

The payload occupies the volume previously allocated to the warhead re-entry vehicle/s. The 5.25 m long 3.0 m diameter fairing provides a 21.0 m^3 payload envelope; 8.25 m long fairing is possible. Maximum static launch acceleration is 7.5 *g* during powered flight; 1.3 *g* lateral. Advanced versions are in development capable of placing 3,000 kg in sun-synchronous orbit.

Specifications

Dniepr
First launch: 21 April 1999
Launch sites: Tyuratam silo (10)
Principal uses: delivery of medium-class payloads into LEO
Typical performance: 4,500 kg into 200 km circular at 65°; 4,100 kg into 200 km circular at 90°; 2,900 kg to a 200 km, 87.3° orbit; 3,700 kg into 200 × 1,000 km, 65°; 3,500 kg into 200 × 1,000 km, 90°. Injection accuracy: ±3.5 km altitude, ±2.5 s period, ±2.0 arcmin inc.
Number of stages: three (liquid)
Overall length: 34.30 m (37.3 m with extended payload fairing)
Principal diameter: 3.00 m
Launch mass: about 210 t
Guidance: computer-controlled inertial

UPDATED

Kosmos 3M

Current Status

Kosmos 3M is marketed commercially by PO Polyot and Rosvoorouzhenie, launch rates have been declining as demand reduces on world traffic levels. Only four were launched in 2002. Production was halted in 1995 but approximately 15 vehicles remain available for operational launches.

Background

For many years the smallest vehicle in the operational inventory, Kosmos originated as the R14 (11K65)/SS-5 Skean medium range missile, developed by the Yangel bureau (now NPO Yuzhnoye). NPO-PM made it a space launcher, because of Yuzhnoye's heavy workload. In its first incarnation the Soviets called it Kosmos and then the current Kosmos 3M (11K65M). It is now the responsibility of AKO Polyot in Omsk. It remains as the only vehicle to be fired from all three Soviet sites. Its debut space launch in August 1964 orbited the first Soviet triple satellite payload and it is now employed principally for placing small military craft into medium Earth orbits. About half of all launches have performed suborbital hypersonic and orbital entry tests. The Russians continue suborbital flights from Kapustin Yar. Orbital launch frequency has decreased as new generations of large satellites have transferred to Tsyklon. The precipitous decline in the launch rate will only get worse as the VKS Military Space Forces switch to Rockot. At this point, commercial success for the launcher is critical for Polyot's survival. Kosmos was one of only two bidding for NASA's 1995 Med-Lite launch contract (awarded to McDonnell Douglas/Orbital Sciences Corp).

Kosmos released its first commercial payloads during the 24 January 1995 launch. They were Sweden's 28 kg Astrid 1 science satellite and Final Analysis Incorporated's 114 kg store/forward Faisat 1. FAI planned a second satellite in 1996 as a prelude to 26 data messaging satellites. Polyot planned to provide the launch in return for being the Russian service provider, but in April 1997, the satellite was not carried by the planned launch vehicle, apparently because paperwork was incomplete. The 5 September 1996 flight had carried EUNAMSAT 2, a Mexican satellite which replaced an earlier launch in 1995, where the satellites were lost during the failure of the first Start mission. Germany's DLR-Tubsat and the similar Tubsat being built for Morocco may also fly on Kosmos. The German government is expected to buy a launch, through OHB-Systems, for its Abrixas astronomy satellite. The Belgium-led IRIS is a likely payload as well. Polyot and Assured Space Access Inc of the US signed an agreement 21 September 1995 to set up the Kosmos USA joint venture for the global marketing of Kosmos 3 outside of the former Soviet Union. UK's Plowshare Technology Ltd acts as the European representative. Commercial contacts are now made via Assured Space Access, Inc, Arlington, VA, USA, although pre-existing agreements with Final Analysis and OHB-System are still being executed directly by Polyot.

The 11K55 vehicle was designed with state approval in the 1980s by Yuzhnoye/Polyot to replace Kosmos and Tsyklon, but development was prevented by the Soviet Union's collapse. Stage 1 would have employed RD-120.

The improved Vzlet version is under study, increasing performance by 15 to 20 per cent, increasing the precision of the final orbit to ±2.5 km by including Glonass receivers in the updated guidance system, and adding the capability of dispensing multiple payloads into different orbits. Stage 2 would be burned to depletion and de-orbited. Funding problems may delay the upgrade or cancel it altogether. Polyot does not have commercial or state support for the work but Germany may become involved. The Kosmos USA agreement does not include Vzlet.

Specifications

Kosmos 3M
First orbital launch: 18 August 1964 (Kosmos 38–40)
Number launched: 430 attained orbit by end 2002. 744 launches in total 1970–99, including 345 suborbital re-entry tests.

Kosmos is still Russia's smallest operational orbital launcher. One of stage 2's external tanks can be seen at right 0517360

Kosmos launchers stockpiled in Omsk. The vehicle stands on the four protruding feet, which also house graphite jet vanes for steering; the vanes are installed on the pad (Plowshare Technology Ltd) 0517186

Launch sites: Plesetsk, Kapustin Yar (KY currently suborbital only, but orbital available).
Principal uses: small military store/dump communications, navigation and unknown minor missions at medium altitudes; Interkosmos science satellites; commercial.
Vehicle success rate: 94.5 per cent orbital attempts; 23 orbital failures: 23 October 1964, 16 November 1966, 26 June 1967, 27 September 1967 (pad explosion), 4 June and 15 June 1968, 27 December 1969, 27 June 1970, 23 December 70, 22 July 71, 17 October 72, 25 May 73, 3 June 75, 19 December 75, 29 November 77, 20 December 78 (Kosmos 1064 attained orbit but stage 2 failed to circularise), 4 March 82, 18 June 82 (Kosmos 1380 attained orbit but stage 2 failed to circularise), 30 August 82, 24 November 82, 25 January 83, 23 October 85, 25 June 91, 6 October 95 (stage 2 sticking fuel valve left Kosmos 2321 in low orbit, although still claimed as usable). Nine were killed 26 June 1973 at Plesetsk when Kosmos exploded during fuelling operations.
Cost: about US$10 million for complete capacity; US$6,500/kg for secondary payloads (Sweden was charged US$5,000/kg for Astrid).
Availability: main payloads can be accommodated with three months' notice. Capacity of Polyot and facilities is 30 launches annually (four to six military launches planned for 1996).
Performance: 1,400 kg into 180 km circular, but principally used for about 1 t payloads into 800–1,500 km circular orbits. Accuracy into 1,000 km ±40 km perigee/apogee, +0.04/–0.08° inc. 1,280 kg into 400 km circular at 66°; 1,210 kg into 400 km circular at 74°; 1,140 kg into 400 km circular at 83° and 1,070 kg into 200 × 1,500 km elliptical orbit at 51°.
Number of stages: 2
Overall length: 32.4 m
Principal diameter: 2.40 m
Launch mass: 109 t
Launch thrust: 1,726 kN
Guidance: inertial, on stage 2 forward end, by Khartron

Kosmos 3M stage 1
Engines: RD-216 assembly (four fixed chambers) from NPO Energomash with storable propellants. Steering by graphite vane inserted into each exhaust (installed at end of pad preparations).
Overall length: 22.5 m
Diameter: 2.40 m
Thrust: 1,726 kN
Burn time: about 130 s
Oxidiser: N_2O_4 in NO_2
Fuel: UDMH (aft tank)
Propellant mass: 82.0 t; dry mass 5.30 t
Separation: cold separation, assisted by solid braking rockets

Kosmos 3M stage 2
Configuration: Stage 2 adds two side tanks for missions into 1,000–1,500 km circular orbits, using four smaller thrusters for the climb to the required altitude. The combined UDMH/N_2O_4 external tanks run the length of stage 2. They are not carried on the suborbital version, which also ignites stage 2 only once.
Engine: KB Khimmach 11D49: pump-fed fixed single chamber + four steering thrusters (used for steering, orbit trims, manoeuvres for multiple deployments, climb to high orbits).
Overall length: 4.205 m (6.585 m from nozzle to payload mount).
Diameter: 2.40 m (3.2 m across external tanks)
Thrust: 157.3 kN main chamber plus 4 × 1.4–1.8 kN steering/adjust thrusters (each with a pressure-fed 99 N thruster for three-axis coast control and ullage).
Propellants: as stage 1, carried in spherical tanks
Burn time: 325–335 s burn 1 plus 2–8 s burn with 2 main engine
Propellant mass: 18.7 t; dry mass 1,434 kg. Fuel in aft tank.

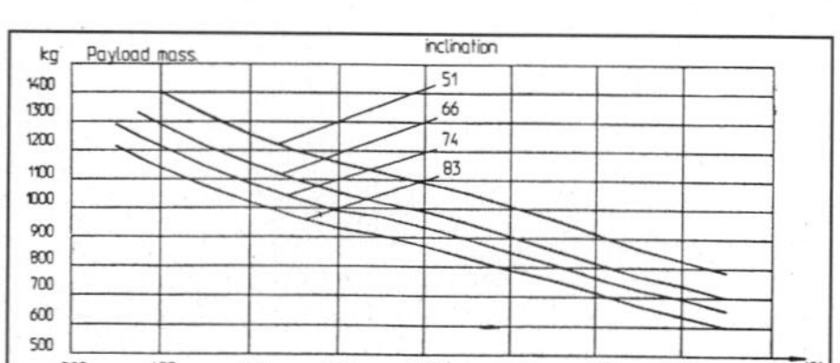

Kosmos 3M circular orbit performance from Plesetsk. The curves indicate different inclinations. Precision into a 1,000 km circular path is ±40 km; ±80 km for 1,600 km 0516873

Fairing/payload environment
The 2.40 m diameter 5.720 m long 40° apex coni-cylindrical fairing provides 12.58 m^3 payload volume, with four access doors in the 1.809 m high cylindrical section and seven in the conical. The vehicle is integrated horizontally and the payload attached in the same building. Access provides two 3 × 3 m class 100,000 cleanrooms in the building for satellite integration/checkout, in addition to the building's own 28 m-long satellite area. Launcher preparation up to pad arrival requires 105 men over 34–36 h; pad activities 120–135 men for 8–10 h; launch is by a crew of 20–25. Fuelling is completed 4 h before launch and air conditioning to the payload fairing ends at T-1 h. Launches are made within –40/50°C and 20 m/s wind surface conditions. Kosmos provides 29.5±2 V DC to the payload. The fairing halves separate mechanically at about 75 km. Payload is separated by four pyropushers within 20 s of shutdown; stage 2 ignites a small lateral solid 1.25 s after separation for collision avoidance. Payload angular disturbance at separation is ±3.0°/s pitch/yaw + ±0.5°/s roll for 1,000 kg.
Mechanical loads: 1.0–6.5 *g* longitudinal, <1.3 *g* lateral
Acoustic load: 140 dB rms max
Thermal: 280°C conical and 180°C cylindrical internal wall maximum temperatures

Molniya-M (8K78M)

Background

Molniya is the most powerful of the Sapwood-based space launchers. It was employed until 1972 for planetary missions, but is now principally used for placing Molniya communications and early warning satellites into highly eccentric Earth orbital paths. The Soviets and their Russian successors have not yet used it for a GEO mission, but the first Sun-synchronous mission came in December 1995.

The Molniya-M is employed primarily for Cosmos early warning and Molniya communications satellites 0516871

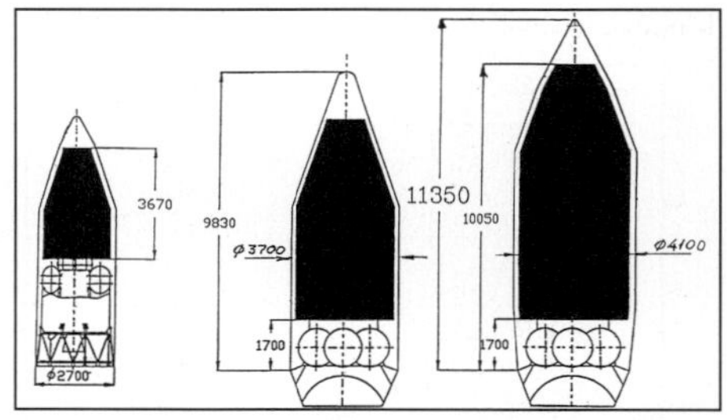

The current Molniya payload envelope and Block L configuration are shown at left. To the right are NPO Lavotchkin's additional proposals for the Rus upgrade: replacing Block L with Fregat and using Proton's larger fairings 0516872

Russians have been rumoured to be upgrading the new derivative, the Rus, but have not yet released figures on its performance.

The 8K78 Molniya first flew successfully in February 1961 carrying the Venera 1 Venus probe. The original 8K78 model employed the RD-0107 engine to power stage 2. This was replaced by the RD-0110 for the current 8K78M, first flown 19 February 1964 (failed to reach orbit). Before that an earlier variant attempted to put two unsuccessful Mars probes on the way to the Red Planet on 10 and 14 October 1960. In general, the Molniya itself is a capable launcher, but the Block L family of escape ('e') stages has proved to be one of the more unreliable, occasionally stranding payloads in the LEO parking orbit. The Kosmos 2084 early warning mission of June 1990 is the most recent example. NPO Lavotchkin proposes replacing Block L with its more powerful and flexible Fregat stage, based on its Phobos Mars stage.

The Soviets used various design bureaus as competing agencies and the Molniya 8K78 variant had at least one rival. The Korolyov bureau proposed an 8K77 launch vehicle using a Block Zh second stage instead of the Block I. A version designated 8K76 was also discussed, this being the 8K77 without the Block L class final stage.

Commercial customers, such as India's IRS 1C use the Molniya as well. The Indian launch occurred in December 1995, accompanied by BMDO's Skipper. Molniya's first Sun-synchronous mission uniquely used Block L for the orbital injection, the previous stages all being suborbital. In mid-1996, Starem, a new joint venture company, opened in France to market the Molniya-M launch vehicle on a commercial basis.

Molniya and Molniya-M annual launch rates

1960: 2	1967: 6	1974: 7
1961: 2	1968: 6	1975: 12
1962: 6	1969: 4	1976: 11
1963: 6	1970: 7	1977: 10
1964: 8	1971: 3	1978: 9
1965: 12	1972: 11	1979: 7
1966: 10	1973: 10	1980: 12

Subsequent missions are covered in greater detail in the table overleaf.

Specifications
Molniya/Molniya-M
First launch: 10 October 1960 (Mars probe)
Number launched: 311 to end-2002, 292 attaining LEO and 276 successfully departing LEO
Launch sites: Tyuratam, Plesetsk
Principal uses: Kosmos early warning, Molniya communications and science satellites, all in highly eccentric Earth orbits
Vehicle success rate: 89% to end-1998 (includes nine failures to LEO + 24 upper stage failures in LEO)
Commercial launches: IRS 1C/Skipper December 1995, Victor 1 (Argentina) August 1996
Cost: about US$15 million for commercial launches. India paid Rs500 million for IRS 1C
Performance: 7,000 kg to 200 km, 51.6° orbit; 1,800 kg to 820 km, 99° orbit; 1,600 kg to 400 × 40,000 km, 63–65° orbit; 900 kg to 400 × 200,000 km, 65° orbit
Sun-synchronous, 800 km: 1,500 kg
Prognoz-type (400 × 200,000 km, 65°): 900 kg
Lunar delivery: 1,640 kg
Venus delivery: 1,180 kg
Mars delivery: 950 kg
Comment: The corresponding version of Rus could deliver 2,100 kg into 200 × 36,000 km, 51.6° orbit from Tyuratam. Replacing Block L by Fregat would increase that to 2,700 kg.
Number of stages: 3 + 4 strap-ons
Overall length: 43.44 m

Molniya-M Launches 1981–1999

	Date	Site	Payload
1	9 January 1981	Plesetsk	Molniya 3–14
2	30 January 1981	Plesetsk	Molniya 1–49
3	19 February 1981	Plesetsk	C1247 early warning
4	24 March 1981	Plesetsk	Molniya 3–15
5	31 March 1981	Plesetsk	C1261 early warning
6	9 June 1981	Plesetsk	Molniya 3–16
7	19 June 1981	Plesetsk	C1278 early warning
8	24 June 1981	Plesetsk	Molniya 1–50
9	4 August 1981	Plesetsk	C1285 early warning
10*	11 September 1981	Plesetsk	Molniya failure
11	17 October 1981	Plesetsk	Molniya 3–17
12	31 October 1981	Plesetsk	C1317 early warning
13	17 November 1981	Plesetsk	Molniya 1–51
14	23 December 1981	Tyuratam	Molniya 1–52
15	26 February 1982	Plesetsk	Molniya 1–53
16	3 March 1982	Plesetsk	C1341 early warning
17	24 March 1982	Plesetsk	Molniya 3–18
18	7 April 1982	Plesetsk	C1348 early warning
19	20 May 1982	Plesetsk	C1367 early warning
20	28 May 1982	Plesetsk	Molniya 1–54
21	25 June 1982	Plesetsk	C1382 early warning
22	21 July 1982	Tyuratam	Molniya 1–55
23	27 August 1982	Plesetsk	Molniya 3–19
24	22 September 1982	Plesetsk	C1409 early warning
25*	8 December 1982	Tyuratam	Molniya failure
26	11 March 1983	Plesetsk	Molniya 3–20
27	16 March 1983	Plesetsk	Molniya 1–56
28	2 April 1983	Tyuratam	Molniya 1–57
29	25 April 1983	Plesetsk	C1456 early warning
30	1 July 1983	Tyuratam	Prognoz 9 science
31	8 July 1983	Plesetsk	C1481 early warning
32	19 July 1983	Tyuratam	Molniya 1–58
33	30 August 1983	Plesetsk	Molniya 3–21
34	23 November 1983	Plesetsk	Molniya 1–59
35	21 December 1983	Plesetsk	Molniya 3–22
36	28 December 1983	Plesetsk	C1518 early warning
37	6 May 1984	Plesetsk	C1541 early warning
38	16 March 1984	Plesetsk	Molniya 1–60
39	4 April 1984	Plesetsk	C1547 early warning
40	6 June 1984	Plesetsk	C1569 early warning
41	3 July 1984	Plesetsk	C1581 early warning
42	2 August 1984	Plesetsk	C1586 early warning
43	10 August 1984	Plesetsk	Molniya 1–61
44	24 August 1984	Plesetsk	Molniya 1–62
45	7 September 1984	Plesetsk	C1596 early warning
46	4 October 1984	Plesetsk	C1604 early warning
47	14 December 1984	Plesetsk	Molniya 1–63
48	16 January 1985	Plesetsk	Molniya 3–23
49	26 April 1985	Tyuratam	Prognoz 10
50	29 May 1985	Plesetsk	Molniya 3–24
51	11 June 1985	Plesetsk	C1658 early warning
52	18 June 1985	Plesetsk	C1661 early warning
53	17 July 1985	Plesetsk	Molniya 3–25
54	12 August 1985	Plesetsk	C1675 early warning
55	22 August 1985	Plesetsk	Molniya 1–64
56	24 September 1985	Plesetsk	C1684 early warning
57	30 September 1985	Plesetsk	C1687 early warning
58	3 October 1985	Plesetsk	Molniya 3–26
59	22 October 1985	Plesetsk	C1698 early warning
60	23 October 1985	Tyuratam	Molniya 1–65
61	28 October 1985	Plesetsk	Molniya 1–66
62	9 November 1985	Plesetsk	C1701 early warning
63	24 December 1985	Plesetsk	Molniya 3–27
64	1 February 1986	Plesetsk	C1729 early warning
65	18 April 1986	Plesetsk	Molniya 3–28
66	19 June 1986	Plesetsk	Molniya 3–29
67	5 July 1986	Plesetsk	C1761 early warning
68	30 July 1986	Plesetsk	Molniya 1–67
69	28 August 1986	Plesetsk	C1774 early warning
70	5 September 1986	Plesetsk	Molniya 1–68
71*	3 October 1986	Plesetsk	early warning
72	15 October 1986	Plesetsk	C1785 early warning
73	20 October 1986	Plesetsk	Molniya 3–30
74	15 November 1986	Plesetsk	Molniya 1–69
75	20 November 1986	Plesetsk	C1793 early warning
76	12 December 1986	Plesetsk	C1806 early warning
77	26 December 1986	Plesetsk	Molniya 1–70
78	22 January 1987	Plesetsk	Molniya 3–31
79	4 June 1987	Plesetsk	C1849 early warning
80	12 June 1987	Plesetsk	C1851 early warning
81	21 December 1987	Plesetsk	C1903 early warning
82	26 February 1988	Plesetsk	C1922 early warning
83	11 March 1988	Tyuratam	Molniya 1–71
84	17 March 1988	Plesetsk	Molniya 1–72
85	26 May 1988	Plesetsk	Molniya 3–32
86	12 August 1988	Plesetsk	Molniya 1–73
87	30 August 1988	Plesetsk	C1966 early warning
88	29 September 1988	Plesetsk	Molniya 3–33
89	3 October 1988	Plesetsk	C1974 early warning
90	25 October 1988	Plesetsk	C1977 early warning
91	22 December 1988	Plesetsk	Molniya 3–34
92	28 December 1988	Plesetsk	Molniya 1–74
93	14 February 1989	Plesetsk	C2001 early warning
94	15 February 1989	Tyuratam	Molniya 1–75
95	8 June 1989	Plesetsk	Molniya 3–35
96	27 September 1989	Plesetsk	Molniya 1–76
97	23 November 1989	Plesetsk	C2050 early warning
98	28 November 1989	Plesetsk	Molniya 3–36
99	23 January 1990	Plesetsk	Molniya 3–37
100	27 March 1990	Plesetsk	C2063 early warning
101	26 April 1990	Plesetsk	Molniya 1–77
102	28 April 1990	Plesetsk	C2076 early warning
103	13 June 1990	Plesetsk	Molniya 3–38
104*	21 June 1990	Plesetsk	C2084 early warning
105	25 July 1990	Plesetsk	C2087 early warning
106	10 August 1990	Plesetsk	Molniya 1–78
107	28 August 1990	Plesetsk	C2097 early warning
108	20 September 1990	Plesetsk	Molniya 3–39
109	20 November 1990	Plesetsk	C2105 early warning
110	23 November 1990	Plesetsk	Molniya 1–79
111	15 February 1991	Plesetsk	Molniya 1–80
112	22 March 1991	Plesetsk	Molniya 3–40
113	18 June 1991	Plesetsk	Molniya 1–81
114	1 August 1991	Plesetsk	Molniya 1–82
115	17 September 1991	Plesetsk	Molniya 3–41
116	24 January 1992	Plesetsk	C2176 early warning
117	4 March 1992	Plesetsk	Molniya 1–83
118	8 July 1992	Plesetsk	C2196 early warning
119	6 August 1992	Plesetsk	Molniya 1–84
120	14 October 1992	Plesetsk	Molniya 3–42
121	21 October 1992	Plesetsk	C2217 early warning
122	25 November 1992	Plesetsk	C2222 early warning
123	2 December 1992	Plesetsk	Molniya 3–43
124	13 January 1993	Plesetsk	Molniya 1–85
125	26 January 1993	Plesetsk	C2232 early warning
126	6 April 1993	Plesetsk	C2241 early warning
127	21 April 1993	Plesetsk	Molniya 3–44
128	26 May 1993	Plesetsk	Molniya 1–86
129	4 August 1993	Plesetsk	Molniya 3–45
130	10 August 1993	Plesetsk	C2261 early warning
131	22 December 1993	Plesetsk	Molniya 1–87
132	5 August 1994	Plesetsk	C2286 early warning
133	23 August 1994	Plesetsk	Molniya 3–46
134	14 December 1994	Plesetsk	Molniya 1–88
135	24 May 1995	Plesetsk	C2312 early warning
136	2 August 1995	Plesetsk	Interball 1 science
137	9 August 1995	Plesetsk	Molniya 3–47
138	28 December 1995	Tyuratam	IRS 1C/Skipper
139	14 August 1996	Plesetsk	Molniya 1–89
140	29 August 1996	Plesetsk	Victor 1 (Argentina) plus Maglen 5 (Czech Republic) plus Interball 2 science
141	24 October 1996	Plesetsk	Molniya 3–48
142	9 April 1997	Plesetsk	C2340 OKO earlywarning
143	14 May 1997	Plesetsk	C2342 OKO earlywarning
144	24 September 1997	Plesetsk	Molniya-1 90
145	7 May 1998	Plesetsk	C2351 OKO earlywarning
146	1 July 1998	Plesetsk	Molniya-3 49
147	28 September 1998	Plesetsk	Molniya-1 91
148	8 July 1999	Plesetsk	Molniya-3 50
149	27 December 1999	Plesetsk	Cosmos 2368
150	20 July 2001	Plesetsk	Molniya-3 51
151	25 October 2001	Plesetsk	Molniya-3 52
152	1 April 2002	Plesetsk	Cosmos 2388
153	24 December 2002	Plesetsk	Cosmos 2393
154	2 April 2003	Plesetsk	Molniya-1T

* indicates launcher failure (all those shown attained orbit but Block L failed from LEO).

Principal diameter: as SL-4
Launch thrust: as SL-4
Launch mass: 306 t (305.481 t for Interbal 1)

Molniya-M stage 3
Description: The 329 kg solid propellant ullage motor package (BOZ Blok Obespecheniya Zapuska: launching provision block) fires to settle the stage's propellants for ignition and is then jettisoned. It has been speculated that this stage might have been scaled up to create the Block D for N1 and Proton. OkO launches employ the Block 2BL version; Molniya satellites Block ML 2BL-SM2 was referred to for Interball 1. There are believed to be others.
Designation: 11D33, Block L family
Engine: Korolev bureau
Stage length: 2.64 m.
Stage diameter: 2.41 m (suspended inside shroud)
Oxidiser: liquid oxygen (2.66 t for Interball 1)
Fuel: kerosene (1.11 t for Interball 1)
Propellant mass: 3.77 t for Interball 1 (2BL dry mass 897 kg); excludes ullage package.
Thrust: 67 kN vacuum
Burn time: about 200 s
Attitude control: unknown
Fairing: The payload and Block L are protected by an 825 kg 7.90 m long 2.70 m diameter fairing, providing a payload envelope of 3.670 m long 2.65 m diameter.

Proton-K

Current Status

The Khrunichev Enterprise embarked on the development of a new derivative of the highly successful Proton family, the Proton-M which, like Proton-K, has the optional Breeze fourth stage.

Proton-M has the capacity to place a payload of 5,500 kg into geosynchronous transfer orbit or inject 2,920 kg directly into geostationary orbit. The first Proton-M/Breeze launch successfully took place on 6 June 2000.

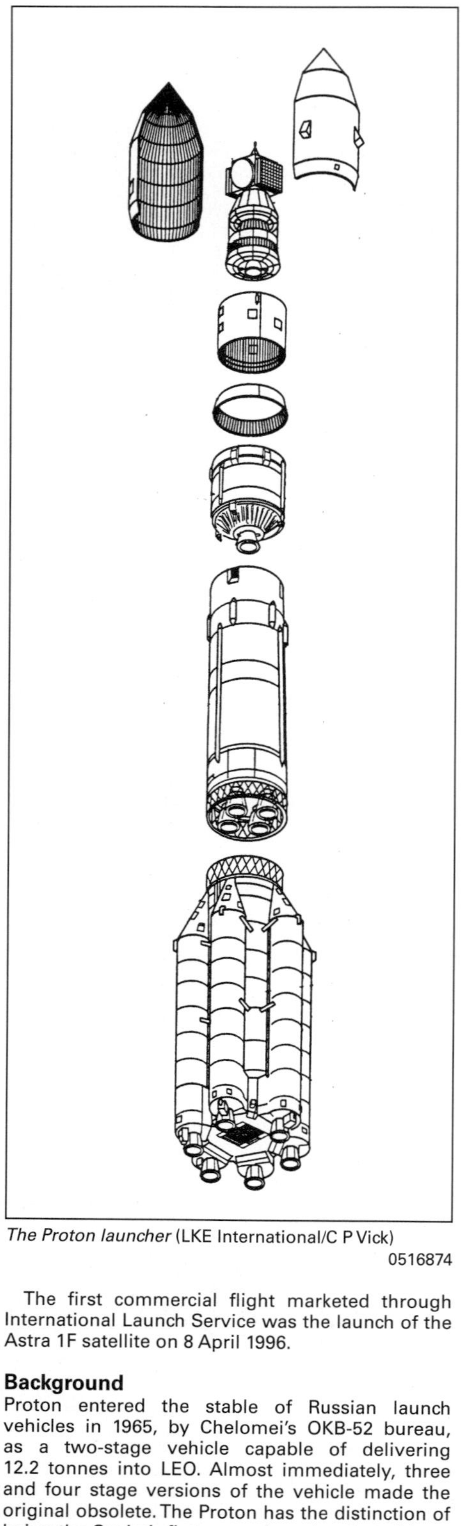

The Proton launcher (LKE International/C P Vick)
0516874

The first commercial flight marketed through International Launch Service was the launch of the Astra 1F satellite on 8 April 1996.

Background

Proton entered the stable of Russian launch vehicles in 1965, by Chelomei's OKB-52 bureau, as a two-stage vehicle capable of delivering 12.2 tonnes into LEO. Almost immediately, three and four stage versions of the vehicle made the original obsolete. The Proton has the distinction of being the Soviet's first space launcher that did not begin as a military ballistic missile. Although the design originated in 1962 as the UR500 ICBM, the military need for that calibre of ICBM diminished as international targeting doctrines changed and the Soviet dropped plans for the missile. A two-stage test version left over from the early missile programme served as the testbed for the new space launch programme.

Chelomei used the original Proton project in his bid to become the father of the Soviet Moon programme. He aimed to use it to launch a one-man circumlunar flight using the original two-stage launcher, thus stealing some of Korolev's fame. This programme, called LK-1, received official approval in early 1965, but on 25 December 1965 Korolev wrested control back from Chelomei, adding it to his own manned landing programme. Slippage of the whole lunar programme prompted the addition of N1's Block D stage to Proton. This was the UR500K-L1 project and was publicly designated Zond. L1 was cancelled in early 1969 and the effort transferred to the Almaz and Salyut space station projects. KB Salyut divided off from Chelomei's bureau following the designer's death in late 1984.

KB Salyut (the design authority) and Khrunichev (the builder) in recent years promoted Proton separately, disagreeing over rights to the vehicle. President Yeltsin issued a decree 7 June 1993 merging them as the 'Khrunichev State Research & Production Space Center', which Lockheed joined as a minority shareholder to form Lockheed-Khrunichev-Energia (LKE) International in early 1993. Because of proliferation concerns, the US and Russia reached a draft accord in May 1993 limiting total sales of the Proton to eight GEO/GTO commercial launches through 2000 (further limited to two in any 12 months). Although not specifically a Proton agreement, in practice it covers principally that vehicle's GEO version. Part of the deal required prices within 7.5 per cent of the western equivalents, indicating the US concerns had more to do with launcher sales than arms control. The agreement excludes Proton's earlier Inmarsat contract. LEO and other non-GEO/GTO launches are dealt with on a case-by-case basis. A more relaxed agreement was signed 30 January 1996, allowing 16 launches (plus another four should market conditions allow) within 15 per cent of western prices.

Confusingly, the Russians refer to both three- and four-stage versions of this launch vehicle as the Proton-K. This is done because the basic launch vehicle is considered to be the Khrunichev three-stage Proton-K and if required one of Energiya's Block D or Block DM variant fourth stages can be added for missions beyond low Earth orbit.

The Proton's original fourth stage started life as the fifth stage of the N-1 manned lunar vehicle developed in the late 1960s. Two basic types of fourth stage are flown on the Proton-K: the Block D family and the Block DM family.

Launches using the Block D family employ the Block D to perform initial orbital injection, the first three stages of the vehicle being sub-orbital. The Block D then performs a second burn to launch the spacecraft out of a low parking orbit. The Block D family does not carry its own command and control system and it relies upon the payload to relay commands for engine firing, separation, and so on. These upper stages have been used exclusively for deep space missions. Different versions have appeared over the years: the original Block D (11S824), lunar and planetary missions during 1967–1983, the Block D-1 (11S824M) for Astron, Venera 11–16, VEGA 1–2 and Granat, and the Block D-2 (11S824F) used for Phobos 1–2 and Mars 8 (=Mars-96). Actual differences between these upper stages have not been described in Russian literature.

Proton's first three stages are assembled by Khrunichev's factory close to Moscow. The production rate declined to eight annually in 1989 from the previous 12. The factory can handle 16 per year and requires orders for six to remain a viable commercial entity. Only the propellant tanks are welded at Khrunichev; the vehicle itself is assembled at Baikonur over 1½ months as components arrive at the main 300 × 400 m, 20 m high assembly hall. The Block D/DM stage 4 is managed and built by RKK Energia for integration with the vehicle at Baikonur. This stage was designed by Korolev's OKB for the N1 lunar launcher.

Until the Vega missions of December 1984, western observers had never seen the launcher in any detail. Then it became apparent that six cylinders, each with one engine, were clustered around a central propellant reservoir. It had been believed previously that these were strap-ons separated at altitude for core engines to continue burning. The unique configuration was adopted to allow rail transport to Baikonur, where the outer tanks could be attached. Stage 1's descent is now controlled to minimise the pollution problem in Kazakhstan. Impact dispersion is limited for a team to clean up the area.

In the past, vehicles up to seven years old have been launched.

Proton's future employment for GEO payloads and planetary missions requires enhancements. The Russian Space Agency announced in late 1993 that it is funding improvements to both Proton and Soyuz. Proton's stage 1 engines will be upgraded (as they were in 1986 and possibly in 1991), Zenit's avionics will be adopted, a 5 m diameter payload fairing will be added, and the interstages strengthened to cope with the heavier loads. These improvements will provide 22.3 tonnes LEO and almost 3-tonne GEO capacities. The coming generations of larger payloads will be beyond this GEO capability. However, the Proton KM (M means 'modernised') is planned for about 2000 to handle 3.2 tonne in GEO using Khrunichev's own Breeze-M (adapted from Rockot), and a new cryogenic stage 4 post-2001 will provide 4.2 tonnes GEO. A new stage 2 has also been mooted, powered by a single RD-0120 LOX/LH_2 engine. Addition of NPO Lavotchkin's Fregat as a fifth stage under the existing fairing has been studied to increase Proton K GEO capacity to 3.3 tonnes. Fregat has flown three as the injection module for 1988's Phobos missions and on Mars-96.

The Proton launcher (LKE International/C P Vick)
0516875

The possibility of flying non-Soviet payloads first arose with an offer to ESA for launching the Marecs maritime satellite, followed by a more vigourous campaign in 1983 to launch the Inmarsat 2 generation. Commercial promotion developed during 1986–87 but, despite the lower rates and assurances that customer representatives can accompany payloads at all times to avoid technology transfer problems, no customers were found. The Soviets claimed in 1990 that firm contracts had been signed in late 1987 for two US telecom satellite launches 1989–90, but the deals were blocked by the refusal of US export licences. It was reported in mid-1990 that discussions were under way with the Brazilian government for commercial operations from the Alcantara site. Zenit replaced Proton in plans for Australia's Cape York spaceport. Australia's Space Transportation Systems Ltd has studied a commercial spaceport based on Proton from the Papua New Guinea island of Manus or Emirau. Requiring US$900 million, the near-equatorial location would improve GEO performance to 4.6 tonnes without KM. Lockheed Martin/ILS completed a study in July 1996 of an alternative launch site for Proton. Candidates included Florida, Brazil and Australia.

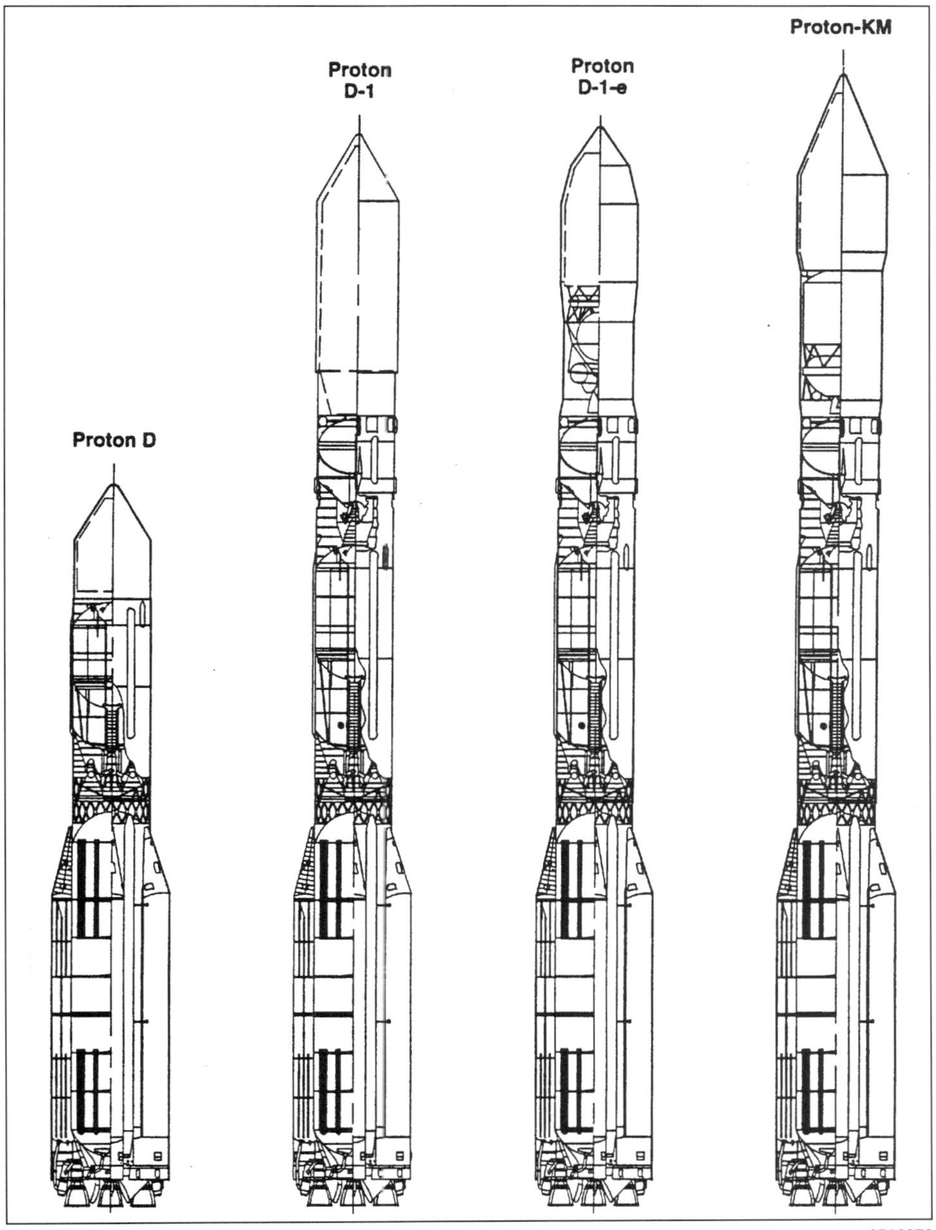

The Proton family (LKE International) 0516876

Proton stage 1 production at Khrunichev's Moscow factory 0517365

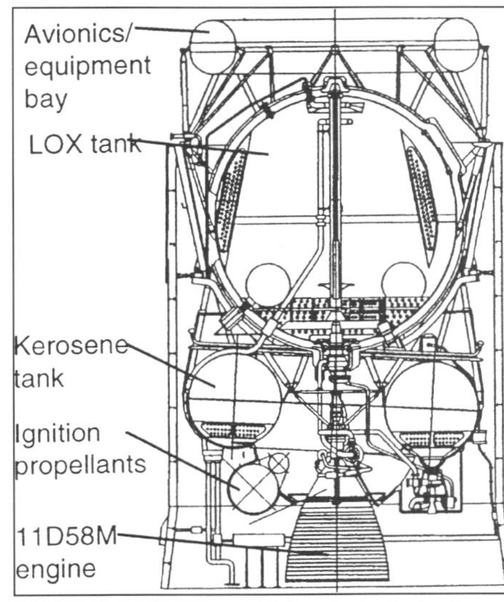

Block DM carries the vehicle's avionics in its forward torus (Boeing) 0517366

The first marketing success came in November 1992 when Inmarsat selected Proton and KB Salyut for a US$36 million contract to launch an Inmarsat 3; the contract was signed 27 April 1993. (Inmarsat has noted its total cost will be closer to US$50 million.) The success built on the Bush administration's June 1992 approval for one Proton launch of a US-built satellite. Motorola, in February 1993, announced that it had signed a contract (subject to US government approval) with Khrunichev Enterprises for three Proton launches of seven satellites each. In return, Motorola expected Khrunichev to invest US$40 million in Iridium Hughes Space & Communications Co, and in February 1995, signed a general launch services agreement for an unspecified number. Proton is baselined for the 2001 launch of ESA's Integral satellite. Launch would be provided in return for observing time with the γ-ray observatory.

The Block DM family of upper stages have been used for launching GLONASS payloads. The Soviets used the original Block DM (11S86) for launches to geosynchronous orbit during 1974–1990 until its last flight for Gorizont 20. The Block DM-2 (11S861) put all GLONASS satellites into position, delivered Kosmos 1700 and the majority of GEO missions starting with Kosmos 1961 (in 1988), and put the Elint launches of Kosmos 1603/1656 into position. The Block DM-2M (11S861-01) has seen limited use in the launches of geosynchronous communications satellites. The commercial launch of Astra-1F introduced a new version to the Block DM family: the Block DM-3.

The three-stage Proton-K places the fourth stage and payload assembly into a low parking orbit and then the fourth stage ignites for the first time in LEO to enter a transfer orbit and then a second time to circularise the orbit with an orbital inclination of around 47 to 48°. At the first pass through apogee the SOZ motors would fire to settle the remaining propellant in the main tanks and they would separate as the main engine ignited to circularise the orbit at geosynchronous altitude and reduce the inclination to 1.5° or less. The Block DM-3 performed the first burn to enter a geosynchronous transfer orbit but retaining the original 51.6° inclination: at the apogee pass the second burn changed the orbit to 11,970 to 35,940 km, but inclined at 6.95°. In addition, no SOZ motors appear on radar screens after the satellite reaches its final position suggesting that they might have remained attached to the Block DM-3.

All of the Block D and DM variants discard two ullage motors as they start their final main engine burns. The ullage motors (designated SOZ by the Russians) fire briefly just before the final burn of the fourth-stage engine to settle the propellants in the rocket stages tanks.

There have been two propellant versions used within the Block DM (and presumably Block D) flight programme. Originally, the Soviets used a standard LOX/kerosene combination, but starting in the 1980s - possibly with the introduction of the Block DM-2 stage – the higher-performance LOX/SYNTIN combination was introduced. It seems likely that the later Block D variants also used the new propellant combination. In September 1996, the Russians halted any further production of SYNTIN. While the Russians have said that they have sufficient SYNTIN to meet

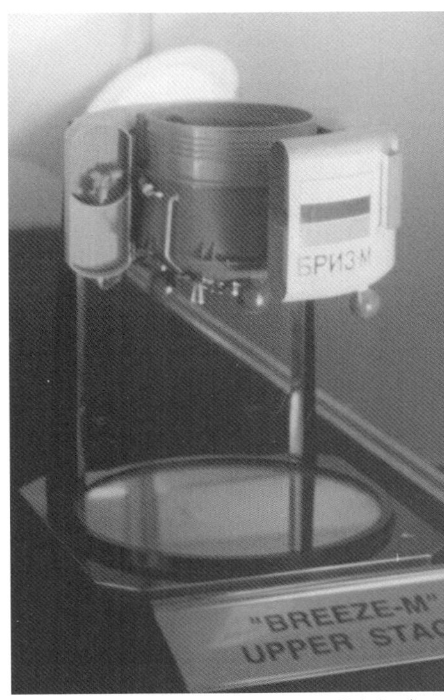

Proton's Breeze-M stage 4 (Khrunichev/KB Salyut) 0517381

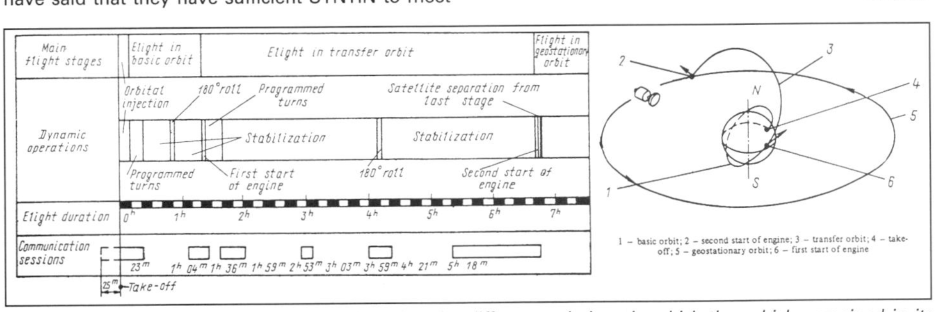

Standard Proton GEO flight profile. 1988 introduced a different technique in which the vehicle remained in its parking orbit for a further 6 hours 0516877

Proton launch history (all launches made from Tyuratam)

	Date	Stages	Payload
1	16 July 1965	2	Proton 1
2	2 November 1965	2	Proton 2
3*	24 March 1966	2	Proton
4	6 July 1966	2	Proton 3
5	10 March 1967	4	Kosmos 146/L1-P
6	8 April 1967	4	Kosmos 154/L1-P
7*	28 September 1967	4	Zond lunar
8*	22 November 1967	4	Zond lunar
9	2 March 1968	4	Zond 4 lunar
10*	23 April 1968	4	Zond lunar
11	14 September 1968	4	Zond 5 lunar
12	10 November 1968	4	Zond 6 lunar
13	16 November 1968	3	Proton 4
14*	20 January 1969	4	Zond lunar
15*	19 February 1969	4	Luna-1969A/Lunokhod
16*	27 March 1969	4	Mars-1969A
17*	2 April 1969	4	Mars-1969B
18*	14 June 1969	4	Luna-1969B
19	13 July 1969	4	Luna 15
20	7 August 1969	4	Zond 7 lunar
21*	23 September 1969	4	C300 (lunar?)
22*	22 October 1969	4	C305 (lunar?)
23*	16 November 1969	4	Kosmos/L1-E
24*	6 February 1970	4	Luna 1970A
25	12 September 1970	4	Luna 16
26	20 October 1970	4	Zond 8
27	10 November 1970	4	Luna 17
28	2 December 1970	4	Kosmos 382/L1-E
29	19 April 1971	3	Salyut 1
30*	10 May 1971	4	C419 (Mars)
31	19 May 1971	4	Mars 2
32	28 May 1971	4	Mars 3
33	2 September 1971	4	Luna 18
34	28 September 1971	4	Luna 19
35	14 February 1972	4	Luna 20
36*	29 July 1972	3	Salyut (DOS 2)
37	8 January 1973	4	Luna 21
38	3 April 1973	3	Salyut 2
39	11 May 1973	3	C557/Salyut
40	21 July 1973	4	Mars 4
41	25 July 1973	4	Mars 5
42	5 August 1973	4	Mars 6
43	9 August 1973	4	Mars 7
44	26 March 1974	4	Kosmos 637
45	29 May 1974	4	Luna 22
46	24 June 1974	3	Salyut 3
47	29 July 1974	4	Molniya-1S
48	28 October 1974	4	Luna 23
49	26 December 1974	3	Salyut 4
50	6 June 1975	4	Venera 9
51	14 June 1975	4	Venera 10
52	8 October 1975	4	Kosmos 775
53*	16 October 1975	4	Luna-1975A
54	22 December 1975	4	Raduga 1
55	22 June 1976	3	Salyut 5
56	9 August 1976	4	Luna 24
57	11 September 1976	4	Raduga 2
58	26 October 1976	4	Ekran 1
59	15 December 1976	3	Kosmos 881–882
60	17 July 1977	3	Kosmos 929
61	23 July 1977	4	Raduga 3
62*	4 August 1977	3	Dual Kosmos
63	20 September 1977	4	Ekran 2
64	29 September 1977	3	Salyut 6
65	30 March 1978	3	Kosmos 997–998
66*	27 May 1978	4	Ekran
67	18 July 1978	4	Raduga 4
68*	17 August 1978	4	Ekran
69	9 September 1978	4	Venera 11
70	14 September 1978	4	Venera 12
71*	17 October 1978	4	Ekran
72*	19 December 1978	4	Gorizont 1
73*	21 February 1979	4	Ekran 3
74	25 April 1979	4	Raduga 5
75	22 May 1979	3	Kosmos 1100–01
76	5 July 1979	4	Gorizont 2
77	3 October 1979	4	Ekran 4
78	28 December 1979	4	Gorizont 3
79	2 February 1980	4	Raduga 6
80	14 June 1980	4	Gorizont 4
81	15 July 1980	4	Ekran 5
82	5 October 1980	4	Raduga 7
83	26 December 1980	4	Ekran 6
84	18 March 1981	4	Raduga 8
85	25 April 1981	3	Kosmos 1267
86	26 June 1981	4	Ekran 7
87	30 July 1981	4	Raduga 9
88	9 October 1981	4	Raduga 10
89	30 October 1981	4	Venera 13
90	4 November 1981	4	Venera 14
91	5 February 1982	4	Ekran 8
92	15 March 1982	4	Gorizont 5
93	19 April 1982	3	Salyut 7
94	17 May 1982	4	Kosmos 1366
95*	23 July 1982	4	Ekran
96	16 September 1982	4	Ekran 9
97	12 October 1982	4	Kosmos 1413–15
98	20 October 1982	4	Gorizont 6
99	26 November 1982	4	Raduga 11
100*	24 December 1982	4	Raduga
101	2 March 1983	3	Kosmos 1443
102	12 March 1983	4	Ekran 10
103	23 March 1983	4	Astron 1
104	8 April 1983	4	Raduga 12
105	2 June 1983	4	Venera 15
106	6 June 1983	4	Venera 16
107	1 July 1983	4	Gorizont 7
108	10 August 1983	4	Kosmos 1490–92
109	25 August 1983	4	Raduga 13
110	29 September 1983	4	Ekran 11
111	30 November 1983	4	Gorizont 8
112	29 December 1983	4	Kosmos 1519–21
113	15 February 1984	4	Raduga 14
114	2 March 1984	4	Kosmos 1540
115	16 March 1984	4	Ekran 12
116	29 March 1984	4	Kosmos 1546
117	22 April 1984	4	Gorizont 9
118	19 May 1984	4	Kosmos 1554–56
119	22 June 1984	4	Raduga 15
120	1 August 1984	4	Gorizont 10
121	24 August 1984	4	Ekran 13
122	4 September 1984	4	Kosmos 1593–95
123	28 September 1984	4	C1603/Tselina 2
124	15 December 1984	4	Vega 1
125	21 December 1984	4	Vega 2
126	18 January 1985	4	Gorizont 11
127	21 February 1985	4	Kosmos 1629
128	22 March 1985	4	Ekran 14
129	17 May 1985	4	Kosmos 1650–52
130	30 May 1985	4	C1656/Tselina 2
131	8 August 1985	4	Raduga 16
132	27 September 1985	3	Kosmos 1686
133	25 October 1985	4	Kosmos 1700
134	15 November 1985	4	Raduga 17
135	24 December 1985	4	Kosmos 1710–12
136	17 January 1986	4	Raduga 18
137	19 February 1986	3	Mir
138	4 April 1986	4	Kosmos 1738
139	24 May 1986	4	Ekran 15
140	10 June 1986	4	Gorizont 12
141	16 September 1986	4	Kosmos 1778–80
142	25 October 1986	4	Raduga 19
143	18 November 1986	4	Gorizont 13
144*	29 December 1986	3	Almaz
145*	30 January 1987	4	Kosmos 1817
146	19 March 1987	4	Raduga 20
147	31 March 1987	3	Kvant 1
148*	24 April 1987	4	Kosmos 1838–40
149	11 May 1987	4	Gorizont 14
150	25 July 1987	3	Kosmos 1870
151	3 September 1987	4	Ekran 16
152	16 September 1987	4	Kosmos 1883–85
153	1 October 1987	4	Kosmos 1888
154	28 October 1987	4	Kosmos 1894
155	26 November 1987	4	Kosmos 1897
156	10 December 1987	4	Raduga 21
157	27 December 1987	4	Ekran 17
158*	18 January 1988	4	Gorizont
159*	17 February 1988	4	Kosmos 1917–19
160	31 March 1988	4	Gorizont 15
161	26 April 1988	4	Kosmos 1914
162	6 May 1988	4	Ekran 18
163	21 May 1988	4	Kosmos 1946–48
164	7 July 1988	4	Phobos 1
165	12 July 1988	4	Phobos 2
166	1 August 1988	4	Kosmos 1961
167	18 August 1988	4	Gorizont 16
168	16 September 1988	4	Kosmos 1970–72
169	20 October 1988	4	Raduga 22
170	10 December 1988	4	Ekran 19
171	10 January 1989	4	Kosmos 1987–89
172	26 January 1989	4	Gorizont 17
173	14 April 1989	4	Raduga 23
174	31 May 1989	4	Kosmos 2022–24
175	21 June 1989	4	Raduga-1 1 (24)
176	5 July 1989	4	Gorizont 18
177	28 September 1989	4	Gorizont 19
178	26 November 1989	3	Kvant 2
179	1 December 1989	4	Granat
180	15 December 1989	4	Raduga 25
181	27 December 1989	4	Kosmos 2054
182	15 February 1990	4	Raduga 26
183	19 May 1990	4	Kosmos 2079–81
184	31 May 1990	3	Kristall
185	20 June 1990	4	Gorizont 20
186	18 July 1990	4	Kosmos 2085
187*	9 August 1990	4	Ekran
188	3 November 1990	4	Gorizont 21
189	23 November 1990	4	Gorizont 22
190	8 December 1990	4	Kosmos 2109–11
191	20 December 1990	4	Raduga 27
192	27 December 1990	4	Raduga-1 2 (28)
193	14 February 1991	4	Kosmos 2133
194	28 February 1991	4	Raduga 29
195	31 March 1991	3	Almaz 1
196	4 April 1991	4	Kosmos 2139–41
197	1 July 1991	4	Gorizont 23
198	13 September 1991	4	Kosmos 2155
199	23 October 1991	4	Gorizont 24
200	22 November 1991	4	Kosmos 2172

	Date	Stages	Payload
201	19 December 1991	4	Raduga 30
202	29 January 1992	4	Kosmos 2177–79
203	2 April 1992	4	Gorizont 25
204	14 July 1992	4	Gorizont 26
205	30 July 1992	4	Kosmos 2204–06
206	10 September 1992	4	Kosmos 2209
207	30 October 1992	4	Ekran 20
208	27 November 1992	4	Gorizont 27
209	17 December 1992	4	Kosmos 2224
210	17 February 1993	4	Kosmos 2234–36
211	25 March 1993	4	Raduga 31
212*	27 May 1993	4	Gorizont
213	30 September 1993	4	Raduga 32
214	28 October 1993	4	Gorizont 28
215	18 November 1993	4	Gorizont 29
216	20 January 1994	4	Gals 1
217	5 February 1994	4	Raduga-1 3 (33)
218	18 February 1994	4	Raduga 34
219	11 April 1994	4	Kosmos 2275–77
220	20 May 1994	4	Gorizont 30
221	6 July 1994	4	Kosmos 2282
222	11 August 1994	4	Kosmos 2287–89
223	21 September 1994	4	Kosmos 2291
224	13 October 1994	4	Ekspress 1
225	31 October 1994	4	Elektro 1
226	20 November 1994	4	Kosmos 2294–96
227	16 December 1994	4	Luch 1
228	28 December 1994	4	Raduga 35
229	7 March 1995	4	Kosmos 2307–09
230	20 May 1995	3	Spektr
231	24 July 1995	4	Kosmos 2316–18
232	30 August 1995	4	Kosmos 2319
233	11 October 1995	4	Luch-1 1
234	17 November 1995	4	Gals 2
235	14 December 1995	4	Kosmos 2323–25
236	25 January 1996	4	Gorizont 31
237*	19 February 1996	4	Raduga 36
238	8 April 1996	4	Astra 1F
239	23 April 1996	3	Priroda
240	25 May 1996	4	Gorizont 32
241	6 September 1996	4	Inmarsat-3 2
242	26 September 1996	4	Ekspress 2
243	16 November 1996	4	Mars 8 (Mars-96)
244	24 May 1997	4	Telstar 5
245	6 June 1997	4	Kosmos 2344
246	18 June 1997	4	Iridium (×7)
247	14 August 1997	4	Kosmos 2345
248	28 August 1997	4	PAS-5
249	14 September 1997	4	Iridium (×7)
250	12 November 1997	4	Kupon 1
251	2 December 1997	4	Astra 1G
252	24 December 1997	4	AsiaSat 3
253	7 April 1998	4	Iridium (×7)
254	29 April 1998	4	Kosmos 2350
255	7 May 1998	4	EchoStar 4
256	30 August 1998	4	Astra 2A
257	4 November 1998	4	PAS-8
258	20 November 1998	3	ISS (Zarya)
259	30 December 1998	4	Kosmos 2362, 2363, 2364
260	15 February 1999	4	Telstar 6
261	28 February 1999	4	Raduga 1–4
262	21 March 1999	4	AsiaSat 3S
263	20 May 1999	4	Nimiq 1
264	18 June 1999	4	Astra 1H
265*	5 July 1999	4	Raduga
266	26 September 1999	4	LM-1
267*	27 October 1999	4	Ekspress-A1
268	12 February 2000	4	Garnda 1
269	6 June 2000	4	Gorizont
270	24 June 2000	4	Ekspress A-3
271	30 June 2000	4	Sirius 1
272	4 July 2000	4	Comsat
273	12 July 2000	3	ISS (Zvezda)
274	28 August 2000	4	Globus
275	5 September 2000	4	Sirius SR-2
276	1 October 2000	4	GEIA
277	13 October 2000	4	Glonass
278	21 October 2000	4	GE6
279	30 November 2000	4	Sirius SR-3
280	7 April 2001	4	Ekran-M
281	15 May 2001	4	PAS 10
282	16 June 2001	4	Astra 2C
283	24 August 2001	4	Kosmos 2379
284	6 October 2001	4	Raduga 1
285	1 December 2001	4	Glonass (×3)
286	30 March 2002	4	Intelsat 903
287	7 May 2002	4	DirecTV 5
288	10 June 2002	4	Express 4A
289	25 July 2002	4	Cosmos 2392
290	22 August 2002	4	Echostar 8
291	17 October 2002	4	INTERGRAL
292*	25 November 2002	4	Astra 1K
293	25 December 2002	4	Glonass (×3)
294	29 December 2002	4	Nimiq 2
295	24 April 2003	4	Cosmos 2397
296	6 June 2003	4	Americom 9

* indicates launcher failure. February 1996 stage 4 failed to reignite for GSO injection. The review board concluded that the triethyl aluminium ignition hypergol had not reached the gas generator or main chamber because of reduced pressure from a leaking joint in the pipe from the TEA bottle. The fitting nut had a broken lockwire, probably from poor installation, allowing it to back off during burn 1. Corrective actions include adding a second lockwire. May 1993 failure due to stage 2/3 engine burn-throughs caused by contaminated UDMH. 1990 stage 2 engine failure believed to be due to a worker leaving a rag in the vehicle. Stage 3 failure January 1988. Proton breakdown launch total for each type, followed by number of failures in ( ): two-stages 4(1), four-stages 235(30), three-stages 27(3): these give success rates of 75%, 87% and 89% respectively. Four stage failures include Block D missions where the failure was with the payload (for example Mars 8). Two-stages 4(1)/75%, four-stages 235(30)/87.3%, three-stages 27(3)/88.9%.

their commercial needs with the Proton-K, it remains to be seen for how long the domestic programme will continue before the Block DM family have to revert to using LOX/kerosene. This problem will be alleviated when the Proton-KM starts to fly with its LOX/hydrogen fourth stage and when the Briz-M replaces the existing upper stage. The first attempt at launching a Proton Breeze-M took place on 5 July 1999 but the second stage exploded in flight

The significantly uprated Proton KM (M = modernised), designated 8K82KM began operations on 6 June 2000 (see launch history), replacing Energia's Block D with Khrunichev's Breeze-M for a 3.2 tonne GEO capacity. The second of this type was launched on 7 April 2001. It had earlier been expected that a cryogenic stage 4 would appear by 1995 for 4.5 tonne GEO, combined with improved stage 3 performance. That is now projected for post-2000, providing 4.2 tonne GEO, 6.4/7.9 tonnes 7/28° GTO, 7.9 tonne Earth escape. The cryogenic stage has been under development since 1988/89 using the KVD-1 engine designed for the N1-L3M manned lunar lander project. India's GSLV will also use this engine and the same stage is planned for Angara.

KB Salyut's Breeze-M, and its 50-hour active life, coupled with multiple restarts, will allow Proton to dispense satellites into different orbits. Its compactness leaves 98 m^3 payload volume under the largest current fairing and 150 m^3 with the new hammerhead design. Breeze-M uniquely adopts a core surrounded by a jettisonable neo-toroidal module of NTO/UDMH tanks with a common bulkhead. The core is based on Rockot's flight-proven Breeze stage, with the fixed main chamber and verniers to provide ullage and three-axis coast during coast and burn. Breeze's core houses Proton's GN&C equipment.

Forward section of a completed three-stage Proton at the Khrunichev factory (Theo Pirard) 0054370

Proton with Block DM3 launches the Astra 1F satellite on 9 April 1996 (Theo Pirard) 0054381

Elements of the Proton-launched vehicle in the Khrunisher production facility 0084618

Typical GEO/GTO missions will jettison the external tank, but it can remain attached for missions to 15,000 to 25,000 km orbits. Breeze-M can fly with the core only for LEO missions.

Comparison of Proton-K and Proton-M four-stage vehicles

	Proton-K	Proton-M
Payload fairing:		
Usable diameter	3.9 m	3.9 m
Usable length	7.3 m	10.0 m
Fairing mass	2,000 kg	2,300 kg
Payload capabilities:		
LEO (three-stage version)	20,100 kg	22,000 kg
GSO (four-stage version)	2,100 kg	3,000 kg
Length	60 m	58 m
Launch mass	688,000 kg	691,790 kg
Stage 1:		
Length	21.18 m	21.18 m
Diameter*	7.4 m	7.4 m
Dry mass	31,000 kg	30,600 kg
Propellant mass	419,410 kg	419,410 kg
Total thrust (vacuum)	10,000 kN	10,497 kN
Specific impulse	316 s	316 s
Stage 2:		
Length	17.05 m	17.05 m
Diameter	4.10 m	4.10 m
Dry mass	11,700 kg	11,400 kg
Propellant mass	156,113 kg	156,113 kg
Total thrust (vacuum)	2,325 kN	2,325 kN
Specific impulse	326.5 s	326.5 s
Stage 3:		
Length	4.11 m	4.11 m
Diameter	4.10 m	4.10 m
Dry mass	4,185 kg	3,700 kg
Propellant mass	46,562 kg	46,562 kg
Total thrust (vacuum)	613 kN	613 kN
Specific impulse	326.5 s	326.5 s
Stage 4:	Block DM-3	Briz-M
Length	6.28 m	2.61 m
Diameter	3.7 m	4.0 m
Dry mass**	3,130 kg	2,370 kg
Propellant mass	15,050 kg	19,800 kg
Total thrust (vacuum)	88 kN	19.6 kN
Specific impulse	361.0 s	325.5 s

These figures have been extracted from an ILS brochure issued in 1997.

* Diameter measured across a pair of external propellant tanks and the central core propellant tank.

** For the Block DM-3 this figure includes the fourth stage shroud discarded in LEO and the two ullage motors. The vehicles use nitrogen tetroxide throughout other than the Block DM-3 which uses LOX/ SYNTIN. Some of the data quoted by ILS differs from those available from other Russian literature: for example it is known that some domestic Proton-K LEO payloads have been in excess of 20,600 kg mass and some GSO payloads are around 2,500 kg.

Specifications

Proton launcher

First launch: 16 July 1965 (two-stage version)
Number launched: 305 to 3 August 2004
Launch sites: Baikonur

Proton K (four stages)

First launch: 10 March 1967
Number launched: 272 to 4 August 2004
Launch sites: Baikonur
Principal uses: geostationary, escape and Glonass navigation satellite missions

Vehicle success rate: 88.07% to end 2001. *Past 25 launches:* 92.0% to end-2001. *Past 50 launches:* 94.0% to end-2001
Availability: two-year lead time
Cost: ILS/LKE declined to provide current commercial costs. US$56 million quoted in 1992 for dedicated (but Inmarsat's 1992 contract is for about US$36 million). About US$40 million quoted October 1994, with Rb14–15 billion for internal flights. Quoted 1989: US$35 million for dedicated GEO (US$28 million GTO), US$12 million dual launch, US$8 million triple.
Insurance: arranged through Ingosstrakh, 12% premium for reflight; commercial broker quotes available through LKE for specific missions.
Performance:
GTO (48°): 4,800 kg
LEO: not used for LEO missions, but vehicle 20 t in 200 km parking orbit before GTO injection
GEO: 2,600 kg using Block DM-2M. 2,400 kg using Block DM and Syntin. 3,200 kg using Breeze-M.
Sun-synchronous: 2,800 kg (not yet flown)
Lunar delivery: 5,700 kg
Venus delivery: 5,300 kg
Mars delivery: 4,600 kg

Core element of the Proton launcher 0084619

Elements of Proton launcher with strap-on boosters attached to core stage in dulame 0084620

Strap-on propulsion elements of Proton core stage 0084621

Breeze-M restartable upper stage adapter for Proton KM 0084623

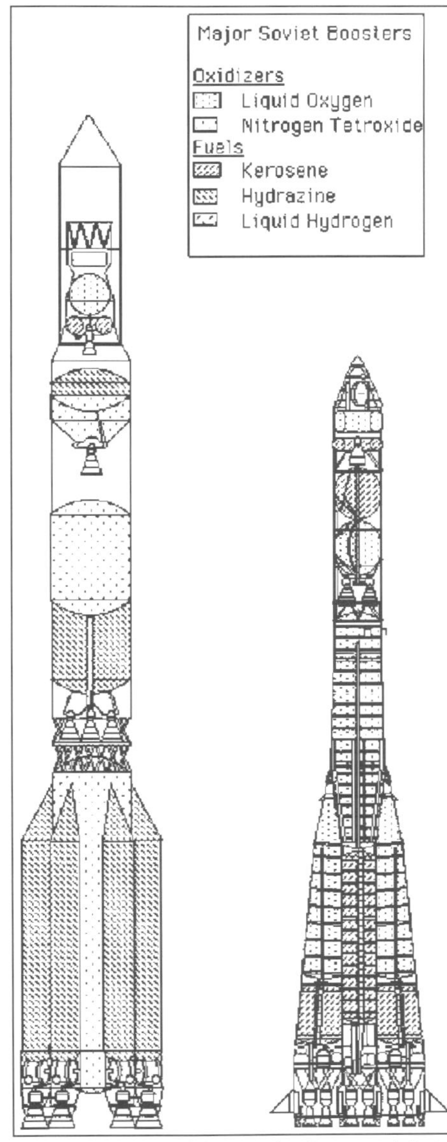

Comparative internal layouts for Soyuz and Proton launchers 0131960

Proton prior to Granat launch 0131959

Number of stages: 4
Overall length: SL-12 57.07 m, SL-13 57.76 m (42.340 m without stage 4/payload)
Principal diameter: 7.400 m stage 1, 4.100 m upper stages
Launch mass: 690 t
Guidance: inertial, carried in stage 4, provides three-axis control by means unknown until payload separation commanded. GEO injection accuracy ±1°, ±20 min orbital period. Analogue control system carried by Block DM or by payload for Block D. GN&C systems provided by NPO AP. Stage 1 conversion to closed loop guidance will add 500 kg LEO by improving efficiency and reducing propellant reserve.

Proton K stage 1
Bureau: Khrunichev/KB Salyut
Engines: single storable liquid non-restartable NPO Energomash RD-253 carried in six cylindrical UDMH side modules around central oxidiser tank.
Overall length: 21.1 m
Principal diameter: 7.4 m overall, 1.6 m cylinders around 4.1 m core
Oxidiser: nitrogen tetroxide in core tank
Fuel: UDMH in side tanks
Propellant mass: 420 t; 31.0 t dry
Stage thrust: 8.8 MN at sea level, 9.8 MN at altitude
Burn time: 130 s

Proton K stage 2
Description: The original stage 2 was about 8.5–9 m long, 6 t dry, 80 t propellant. It was then stretched for the later versions. Chelomei's one-man LK-1 circumlunar mission would have apparently employed the original short version.

Proton launch vehicle on its pad at the Baikonur cosmodrome (ILS) ***NEW***/1047536

Bureau: Khrunichev/KB Salyut
Engines: gimballed liquid chambers, developed by KB Khimautomatiki. Three are RD-0210 and one is RD-0211 version (with gas generator).
Overall length: 14.56 m
Principal diameter: 4.15 m
Oxidiser: nitrogen tetroxide
Fuel: UDMH
Stage thrust: 2,376 kN vacuum
Burn time: up to 300 s
Propellant mass: 156 t, 11.7 t dry

Proton K stage 3
Comment: Stage 3 injects the payload and stage 4 into a 200 km circular, 51.6° parking orbit about 10 min after launch.
Bureau: Khrunichev/KB Salyut
Engine: single fixed RD-0212 version of RD-0210 with four gimballed verniers for steering
Overall length: 6.52 m
Principal diameter: 4.15 m
Oxidiser: nitrogen tetroxide
Fuel: UDMH
Stage thrust: 593.6 kN main chamber plus 4 × 7.875 kN verniers
Burn time: about 250 s
Propellant mass: 46.6 t; 4,185 kg dry

Proton K stage 4 Block DM
Designation: Block D (11S824, first flown 1967), Block 'D-1' [?] (11S824M, 1978), Block D-2 (11S824F, 1988), Block DM (11S86, 1974), Block DM-2 (11S861, 1982), Block DM-2M (11S861-01, 1994), Block DM-3 (?, 1996): possibly a Block DM-4 in 1997.
Bureau: RKK Energia; control system by NPO AP
Engine: 58/58M restartable single chamber developed by Korolev bureau for D/DM.
Overall length: 5.366/6.218 m D/DM-2
Principal diameter: 3.70 m
Stage mass: 17.3 t D/18.46 t DM-2; dry mass 2.5 t D/3.37 t DM-2 (dry masses include the 800 kg casing + SOZ).
Oxidiser: 10,610 kg liquid oxygen in forward sphere
Fuel: 4,330 kg kerosene or syntin in aft torus
Stage thrust (vacuum): 85 kN kerosene; 83.5 kN Syntin.
Burn time: about 600 s total over two burns
Comment: Stage 4/Block D/DM-2 ignites about 80 min after stage 3 separation, making a maximum 450 s burn for injection into a 36,000 km, 48° GTO. About 400 min after launch, it ignites for a maximum of 230 s at first apogee to slot the payload directly into GEO – no satellite apogee boost motor is required. (DM injected Astra 1F into 12,100 × 36,000 km, 7.0° GTO). D/DM-2 carries two 110–120 kg (56 kg dry) 60 × 100 cm NTO/UDMH 'SOZ' thruster packages for three-axis control during coast and to settle the stage's propellants after coasting. Each houses five thrusters: 2 × 22 N pitch/roll, 1 × 44 N yaw and 2 × 11 N ullage. The ullage motors ignite about 300 s before main stage burn starts and are jettisoned 5 s after. About 12 of these packages have broken up years after launch. Vehicle three-axis control is lost at the instant of shutdown and the stage control system orders payload spring separation (at 1.5±0.3 m/s) within 15 s. By this time stage tumbling is typically 2°/s. Combined with separation disturbances, the payload's attitude control system has to cancel 2.5°/s rotation about the longitudinal axis and 3°/s about lateral axes. All Soviet satellites were inserted into GEO close to 90° E for 14 years, sometimes requiring up to several weeks to drift to their operating positions, but in April 1988 Kosmos 1940 was held in its parking orbit for an additional 6 h and then slotted directly into 15° W. This pattern followed with Kosmos 2155 in 1991, Kosmos 2209 (20° W) in 1992 and Kosmos 2282 (25° W) in 1994.
Fairing: The spacecraft is protected until 351 s after launch at an altitude of 150 km. Class 100,000 air conditioning on the pad through a 45 cm^2 hatch provides 10–25°C (30°C possible for up to three days), 30–60% humidity inside the fairing until it is disconnected 2 h before launch. There is also a 4–5 h interruption while the vehicle is erected on the pad. Power and communications links are terminated 90 min before launch; telemetry contact is not available until stage 4 of the flight profile. Surface wind limit 15–18 m/s, depending on fairing type.
Acceleration load: about 4 *g* longitudinal ±1 *g* lateral at 5.7 Hz during ascent, 1.2 ±0.8 *g* longitudinal at 10–15 Hz/±3.0 *g* lateral at 5–7 Hz at lift-off; 1.25 *g* ±4.75 *g* longitudinal at 10–15 Hz/±1.5 *g* lateral at 5–7 Hz during separation.
Acoustic load: 144 dB rms
Temperature: 130°C sidewall max after 200 s

Payload processing
The customer is responsible for ensuring that a payload is protected against –40 to 40°C conditions for the air transfer to Baikonur and the subsequent 100 km truck journey. A commercial payload undergoes

Restartable Breeze-M upper stage for Proton launcher 0084622

NASA's Mars Reconnaissance Orbiter scheduled for launch in 2005 will conduct the most detailed reconnaissance of the planet Mars since missions to the Red Planet began 40 years ago (NASA) ***NEW***/1047538

initial processing in the 270 m^2, class 100,000 20 ± 3°C Satellite Preparation Room 119, with the adjacent 100 m^2 class 10,000 room 120 for electrical tests. It is then transferred into a separate 100 m^2 20 ± 5°C facility for hydrazine and nitrogen loading. The 500 m^2 General Assembly Room number 100 at 15–30°C, 40 to 60% humidity is then used for the horizontal attachment to the spacecraft adaptor, to stage 4 and fairing enclosure. Subsystems testing and storage rooms are also available. Stage 4 is then filled with propellants in a separate area. The old Buran building at site 254 is expected to be converted for processing. The assembly then transfers to the Proton building 70 km distant for vehicle integration; the building can handle up to six launchers simultaneously. The fully assembled Proton then undertakes a 7 h journey to the pad, where five days are normally allocated to preparations before launch. All payload checkout procedures must be completed 8 h before launch; contact is severed at T-90 min. Two complexes several km apart and each with two pads are available, although not all pads are in service simultaneously. Most launches are currently from pad 81L, but will switch to the refurbished 81R in 1996. 200R has been reactivated for commercial and Space Station launches (it has also been transferred from VKS to RKA control); 200L is certified for only a handful more launches before it must be refurbished.

Proton K (three-stages)
First launch: 16 November 1968 (Proton 4 satellite)
Number launched: 28 to end-1999
Principal uses: large LEO spacecraft, principally space station modules.
Cost: LKE declined to provide current commercial costs: US$20 million quoted 1987
Vehicle success rate: 89% to end-1996
Performance: 20–21 t LEO (51.6°), other missions not flown.
Number of stages: 3
Overall length: 44.3 m + payload

Proton KM Breeze-M stage 4
Designation: Breeze-M
Bureau: Khrunichev/KB Salyut
Engine: single fixed restartable pump-fed engine from KB Khimmach, plus three-axis control provided by verniers

Proton KM cryogenic stage 4
Designation: KVRB Kislorodno-Vodourodny Razgonny Block (oxygen/hydrogen upper stage).
Bureau: Khrunichev (stage), KB Khimmach (engine).
Engine: single chamber cryogenic KVD-1M (with two verniers) capable of five burns and 7.5 h coast.
Overall length: 8.6 m
Principal diameter: 4.0 m
Oxidiser: liquid oxygen
Fuel: liquid hydrogen
Dry mass: 3.4 t
Propellant mass: up to 19 t
Stage thrust: 73.58 kN vacuum (69.66 kN main chamber plus 2 × 1.96 kN LOX/GH2 two-axis verniers)
Burn time: 450 s total. First burn starts at 650 s.

UPDATED

Soviet Era Heavy Lift Launch Vehicles

Background

In the early 1960s, there were three contenders for the launch vehicle to place men on the moon: Korolyov's OKB-1 proposed the N-1, Chelomei's OKB-52 proposed the UR-700 and Yangel's OKB-586 proposed the R-56.

The Yangel R-56 proposal formed part of a plan to develop three related missiles and space launch vehicles. The first, the R-36, was developed both as a missile and as the Tsyklon launch vehicle family. The second, the R-46 would have been capable of launching a 50 MT warhead, but it was not developed as either a weapon or a launch vehicle, and the third, the R-56 could have been the basis of a manned lunar programme Considering just sub-orbital (missile?) trajectories, the R-36 could launch a payload of 5 tonnes, the R-46 15 tonnes and the R-56 50 tonnes.

R-56's launch mass would have been about 1,400 tonnes and two different payload capabilities have been quoted, namely, 30 tonnes to an unspecified orbit or 40 tonnes to a 200 km 'polar' orbit. These figures indicate that any manned lunar landing programme based upon this vehicle would

NK-33 engine from the N-1 heavy lift vehicle programme procured by Aerojet (Theo Pirard) 0054373

have required multiple launches and rendezvous and docking of several lunar spacecraft components in Earth orbit. It is therefore little surprise that the R-56 was the first of the lunar launch vehicles to be cancelled, this taking place on 19 June 1964.

Originally, the N-1, the second contender would have been just one of a series of launch vehicles with common stages. The N-11 was planned using Blocks B, V and G: this would have had a launch mass of 700 tonnes and could have placed 20 tonnes into a 300 km orbit. The N-111 would have used the N-1 Blocks V and G plus the second stage of the R-9A missile: this would have had a launch mass of 200 tonnes and could have put 5 tonnes to a 300 km orbit. However, these plans were not developed and only the N-1 reached flight status.

The first N-1 was launched 21 February 1969. On a thrust of 45,000 kN it ascended to a height of approximately 27 km when all 30 first stage engines shut down after a fire broke out when vibration ruptured pressurisation and fuel lines. The wreckage struck the ground 50 km away from the launch site.

The second N-1 was launched 3 July 1969 but it fell over only 200 m above the launch pad and was destroyed. The third N-1 was launched on 27 June 1971 but lost control and broke up 48 seconds later, hitting the ground 20 km away. The fourth and final N-1 flight took place on 23 November 1972 but at 107 seconds the first stage blew up. The configuration of the planned N-1/L-3 stack was for the first three stages of the N-1 (Blocks A, V and G) to be used to place the fourth stage (Block G) and lunar complex into a low Earth orbit. The Block G would perform trans-lunar injection and be discarded. The Block D (also flown as the fourth stage of the Proton-K launch vehicle) would perform lunar orbit injection and then the main descent burn to the lunar surface with the lunar lander (LK), while the lunar Soyuz (LOK) command craft would remain in lunar orbit. The four stages of the N-1 and the Block D would all have used LOX/kerosene propellants.

Korolyov looked towards future versions of the N-1 launch vehicle, using liquid hydrogen as a fuel rather than kerosene. New LOX/LH Blocks S and R would replace the developed Blocks G and D. In support of these plans, OKB-2 (Isayev, now KM Khimmash) developed the 11D56 engine with a thrust of 7.5 tonnes for the Block R and OKB-165 (Lyulka, now NPO Saturn) developed the 11D54 (fixed chamber) and 11D57 (gimballing chamber), both with a thrust of 40 tonnes. It was planned that six to eight 11D54 engines would be used on a modified Block V of the N-1, while a single 11D57 would be used on the Block S.

The Block R which would perform lunar orbit injection and descent to the Moon would have a dry mass of 4.3 tonnes and would carry up to 18.7 tonne of propellant: the diameter would have been 4.1 m and the length 8.7 m.

The first test-firing of the 11D56 closed-cycle engine was carried out in June 1967: at around the same time Kuznetsov's OKB-276 was conducting development of the NK-15V LOX/hydrogen engine with a thrust of 200 tonnes, this being planned for the second stage of the improved N-1 launch vehicle. This new design would make the N-1 appear more like the United States Saturn-5. The US vehicle used LOX/kerosene (RP-1) in the first stage and LOX/hydrogen in the second and third stages. The new N-1, on the other hand, would have had LOX/kerosene in the first stage and LOX/hydrogen in its second, third and fourth stages (the 'fifth' stage, whether Block D or Block R, is considered to be part of the payload rather than the launch vehicle).

In May 1970, the Politburo decided to develop a single new stage, designated Block Sr, which would combine the functions of the Block S and Block R. The Block Sr would have used 11D56M engines having a diameter of 5.2 m, a length of 16.5 m, a dry mass 11.5 tonnes and propellant mass 66.4 tonnes.

The Block Sr would have been used on the dual-launch N-1/L-3M manned lunar missions, but the N-1 programme was cancelled in 1974 when Valentin Glushko became head of the combined Korolyov bureau and GDL (NPO Energiya). Despite this, five Block R stages were built and testing of the technology continued through 1977: the live firing of a Block R came on 12 October 1976.

Finally, while the Korolyov bureau was refining the designs of the N-1, OKB-52 was working on its UR-700 launch vehicle – again seen as part of a family of vehicles which would have included the UR-200 missile (cancelled in 1964) and the UR-500 missile. The UR-500 eventually turned into the Proton-K launch vehicle. The overall design of the UR-700 was finally revealed in a paper presented at the 1996 IAF Congress, although there are still many unknown details concerning its staging technique, and so on.

Chelomei's UR-700 would have had a launch mass of 4,823 tonnes and would have placed 151 tonnes into a 200 km orbit: the payload injected to a trans-lunar orbit would have been 50 tonnes. The design of the UR-700 appears to be unique. To reach Earth orbit three pairs of strap-on boosters (first stage) would have surrounded a two-stage central core (stages two and three). Each of the two core stages would have comprised three independent sets of propellant tanks and engines. The strap-on boosters would have each carried a single RD-270 engine, as would each second-stage module: the third-stage modules would each have carried an altitude-ignition version of the RD-253 engine used on the first stage of the Proton. The requirement for railway transportation meant that each strap-on or core unit was limited to a maximum diameter of 4.1 m.

The unusual aspect of the launch vehicle is that each of the strap-on boosters had three propellant tanks, and for each pair the upper propellant tanks were of different sizes. It would appear that each pair of strap-on boosters carried either an extra tank of nitrogen tetroxide or UDMH atop the tanks used by that strap-on, and this extra propellant load would be fed into the adjacent main central core to 'top-up' the fuel being used in flight. With all nine RD-270 engines firing at launch (six on the first stage strap-ons, three on the second-stage core), the total thrust generated would have been approximately 56,450 kN (9 × 6,272 kN or 5,755 tonnes) sea level, 60,420 kN (9 × 6,713 kN, 6,160 tonnes). The second-stage core would have burned for longer than the strap-ons and by the time that the fuel transfer from each pair of strap-ons to the adjacent second-stage core unit was complete, the strap-ons would be ready to separate after burning of their dedicated propellant.

The third-stage engines would have each had a thrust of 1,716 kN (175 tonnes) and would have burned through to Earth orbit injection. A sketch showing the LK-700 manned lunar landing complex attached to its trans-lunar stage was published in the December 1991 issue of the now-defunct Znanye journal. That article suggested that the stage might be based upon the original short second stage of the Proton vehicle (used on the two-stage variant in 1965–1966).

The high-thrust RD-270 engines were not developed in time to save the UR-700 lunar programme, which was finally passed over in favour of Korolyov's N-1/L-3 programme in 1966. However, some of the engines were subsequently built and tested.

This was not the end of the UR-700 proposal. Although he had lost the chance to fly men to the Moon, Chelomei proposed an even more powerful version of the UR-700, the UR-700M, for a manned Mars expedition. The problems with the RD-270 mean that this engine had to be dropped from the project and in its place each first-stage strap-on and second-stage core module would now have a cluster of four RD-253 engines. By 1968, these engines were flying on the Proton launch vehicle's first stage with a reasonable reliability. This would have meant 36 engines firing at launch – more than the 30 engines on the Block A of the much-criticised N-1 vehicle. Also, a fourth stage using the GDL's RD-410 nuclear motor would have been added, presumably to perform the heliocentric orbit injection from an Earth parking orbit. The UR-700M could have placed 240 tonnes into Earth orbit. When Chelomei was making this proposal – presumably in parallel with Vasili Mishin putting forward the Korolyov bureau's plans for a manned Mars mission based upon the uprated N-1 launch vehicle – the politicians thought that such a programme was too expensive and they had no interest in supporting it. As a result all work on the UR-700 class launch vehicle was stopped around early 1970.

When Glushko took over as head of the new NPO Energia he brought with him plans for another family of launch vehicles which would give the Soviet Union both a launcher for a space shuttle orbiter (Buran) and launch vehicles with different orbital capabilities. The vehicles would all be based around a central core having a different number of strap-on boosters. The launch vehicles which Glushko proposed were Groza (thunderstorm) with two strap-ons, which would become Energia-M, and Grom (thunder) with four strap-ons eventually turning into Energia, and Vulkan (volcano).

Of these three, Vulkan was the real monster. It would have eight strap-on boosters surrounding a modified Energia central core and a second core stage mounted on top. It was intended for a manned lunar mission.

The KVD-1 engine used for the N1-L3M manned lunar landing programme has been used as the basis for the Breeze M stage introduced in 2000 as eventual replacement for the DM stage on the Proton K launcher. It also forms the basis for the Angara launcher and for Indias GSLV.

Specifications

Payload mass: 170 t (LEO), 28 t (GEO)
Maximum diameter: 16.5 m (over strap-ons)
Length: 88 m
Strap-on engines:
8 × RD-170: 8 × 7,255 kN = 58,040 kN (sl)
8 × 7,905 kN = 63,240 kN (vacuum)
Core engines:
4 × RD-0120: 4 × 1,865 kN = 7,460 kN (vacuum)
Specific impulse: 351 s (s1), 452 s (vacuum)
1 × 11D57M: 1 × 410 kN 410 kN (vacuum)
Specific impulse: 460 s (vacuum)

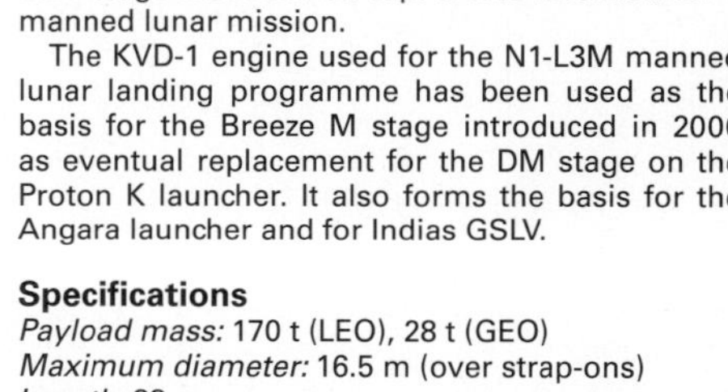

N-1 derivative family (Mark Wade) 0131957

(The original data are given as tonnes for thrusts: in the conversions the results are rounded to the nearest 5 kN. The payload mass includes the mass of the dry Block V.)

The oft-rumoured six strap-on version of Energia does not appear to have been part of Glushko's original plans, and it is unclear at what point it appeared as a design study. Development of the Vulkan launch vehicle ended soon after the proposal, although into the early 1990s the Russians were still talking about an eight strap-on variant of Energia as a distant concept. The Energia-M and standard four-strap-on Energia were cancelled in 1993 due to a lack of funding, although should funding become available once more it is possible that the vehicle might be revived.

RD-704 operating data

Operating Mode	**Mode 1**	**Mode 2**
Fuel	kerosene/liquid hydrogen	liquid hydrogen
Oxidiser	LOX	LOX
Thrust: vacuum	1,960 kN	781 kN
sea level	1,720 kN	n/a
Specific impulse: vacuum	407 (401) s	452 (450) s
sea level	356 (351) s	n/a
Mixture ratio (O/F)	4.38	6.0
Chamber pressure	290 atmosphere	120 atmosphere
Area ration	74	
Mass	2,417 kg	
Single bell diameter	1.8 m	
Length	3.8 m	

n/a – not applicable. For the specific impulse values the nominal value is quoted, followed by a worst case value in parentheses.

Soviet Launch Vehicles

Background

Historically, the general Soviet philosophy was to cluster numbers of relatively modest, low chamber pressure open cycle engines, mass-produced with few modifications in their decades-long careers. Most are assigned RD (Raketnay Dvigatel, rocket motor) designations. NPO Energomash is the major centre for rocket engines; KB Khimautomatiki, was identified in 1990 as responsible for Energia's core engines, and emerged from the Kosberg bureau, which concentrated on upper stage propulsion, as does KB Khimmach of the old Isayev bureau. NII Machinostroenye undertakes more basic research but does offer spacecraft thrusters. The Kuznetsov (late NPO Trud, now RD Kuznetsov Co SSC) aircraft engine bureau developed engines for the abandoned N1 lunar launcher for Korolev. Korolev's bureau itself developed Proton's

Rocket engines used on selected RFAS missiles and launch vehicles

Vehicle	**Designator**	**Stage 0**	**Stage 1**	**Stage 2**	**Stage 3**	**Stage 4**
1) Selected missiles (development related to early launch vehicles)						
R-9	8K75 (original) (developed)		4 × NK-9 1 × RD-111	1 × NK-19 1 × RD-0106		
R-9M	8K77 (not developed)		4 × NK-9	4 × SI-5400		
R-16	8K64		1 × RD-218	1 × RD-219		
GR-1	8K713		4 × NK-9	1 × NK-19	1 × 8D726	
UR-200	8K81		? × RD-0203 ? × RD-0204	1 × RD-0206 1 × RD-0207		
2) Existing and retired launch vehicles						
Sputnik	8K71PS	4 × RD-107	1 × RD-108			
Sputnik	8A91	4 × RD-107	1 × RD-108			
Vostok-L	8K72	4 × RD-107	1 × RD-108	1 × RD-0105		
Vostok	8K72K	4 × RD-107	1 × RD-108	1 × RD-0109		
Vostok-2	8A92	4 × RD-107	1 × RD-108	1 × RD-0109		
Meteor	8A92M	4 × RD-107	1 × RD-108	1 × RD-0109		
Voskhod	11A57	4 × RD-107	1 × RD-108	1 × RD-0107		
Soyuz	11A511	4 × RD-107	1 × RD-108	1 × RD-0108		
Soyuz-L	11A511L	4 × RD-107	1 × RD-108	1 × RD-0108		
Soyuz-M	11A511M	4 × RD-107	1 × RD-108	1 × RD-0108		
Soyuz-U	11A511U	4 × RD-107	1 × RD-108	1 × RD-0110		
Soyuz-U2	11A511U2	4 × RD-107	1 × RD-118	1 × RD-0110		
–	11A59	4 × RD-107	1 × RD-108	1 × RD-0110		
Soyuz-2		4 × RD-120K ?	1 × RD-120K ?	1 × RD-0124		
Molniya	8K78	4 × RD-107	1 × RD-108	1 × RD-0107	1 × SI-5400	
Molniya-M	8K78M	4 × RD-107	1 × RD-108	1 × RD-0110	1 × 11D33	
Kosmos	63S1		1 × RD-214	1 × RD-119		
Kosmos-2	63S1M		1 × RD-214	1 × RD-119		
Kosmos-2M	11K63		1 × RD-214	1 × RD-119		
Kosmos-1	65S1		1 × RD-216	1 × 11D49 ?		
Kosmos-3	11K65		1 × RD-216	1 × 11D49 ?		
Kosmos-3M	11K65M		1 × RD-216	1 × 11D49		
Proton	8K82		6 × RD-253	3 × RD-0210 1 × RD-0211		
–	11A510	4 × RD-107	1 × RD-108			
R-36-0	8K69		1 × RD-218	1 × RD-219		
Tsyklon-2	11K67		1 × RD-251	1 × RD-252		
Tsyklon-M	11K69		1 × RD-251	1 × RD-252		
Proton-K	8K82M		6 × RD-253	3 × RD-0210 1 × RD-0211	1 × RD-0213 1 × RD-0214	1 × RD-54
Proton-K	8K82M		6 × RD-253	3 × RD-0210 1 × RD-0211	1 × RD-0213 1 × RD-0214	
Tsyklon	11K68		1 × RD-251	1 × RD-252	1 × RD-861	
N-1	11A52		30 × NK-33	8 × NK-43	4 × NK-39	1 × NK-31
Zenit-2	11K77		1 × RD-171	1 × RD-120		
Energiya	11K25	4 × RD-170	1 × RD-0120			
Rockot			? × RD-0233 ? × RD-0234	1 × RD-0235 1 × RD-0236		
3) Future launch vehicles						
Angara			1 × RD-174	1 × RD-0120		
Dnepr			4 × RD-264	1 × RD-0228 1 × RD-0257	1 × RD – 864	
Riksha			6 × RD-169	1 × RD-185		
Shtil			? × RD-0244 ? × RD-0245	? ?		

Some launch vehicles use clusters of slightly different engines on the same stage. For example, the Proton family uses three RD-0210 and one RD-0211 on the second stage, the only difference being that the RD-0211 provides the tank pressurisation: the assembly is called the RD-0210. Vernier engines usually have separate designators and these are indicated where known: in such cases there is one main engine and one set of verniers. In the case of the Proton-K third stage it is known that the RD-0213/RD-0214 assembly is designated RD-0212.

Block D/DM and Molniya's Block L engines. The Lyulka aircraft engine bureau (now part of NPO Saturn) developed the LOX/LH_2 D-57 to replace the lunar N1 launcher's stage 3 engines.

The history of NPO Energomash can be traced back to 15 May 1929 when the Group for the Development of Rocket Engines was organised at the Gas Dynamics Laboratory (GDL) in Leningrad (now St Petersburg). For more than 60 years the organisation repeatedly changed its name and location, but remained at the forefront of rocket engine design. The founder and the permanent head of the GDL was academician Valentin P Glushko, considered to be the initiator of the Russian rocket engine industry.

In May 1974, the GDL merged with Sergei Korolyov's design bureau, the new organisation being re-named NPO Energia, with Glushko as its head. Following Glushko's death in January 1989, the former GDL left Energia to become NPO Energomash, continuing to concentrate on the development of rocket engines: the bureau was headed by Vitali P Radovsky during 1974–1994. The current general director and chief designer is Boris I Katorgin.

NPO Energomash incorporates the design bureau, plant and test facilities, with the capability to test engines with thrusts of up to 10 MN either as components or complete. An agreement was signed with Pratt and Whitney in October 1992 providing exclusive US marketing rights for Energomash's LOX/kerosene and tri-propellant products and technology. Lockheed Martin selected the RD-180 to power the Atlas 3 and Atlas 5. The first launch of the Atlas 3 with a dual chamber RD-180 took place on 24 May 2000.

1996 released the designations of many previously-unidentified engines used on missiles and launch vehicles, and these are included under the appropriate design bureaus. For easy cross-referencing between a launch vehicle and the propulsion systems, please refer to the table which accompanies this entry.

Currently the central specialised design bureau Samara is responsible for Molniya M and Soyuz-2 which, from 2002, has the increased payload volume facilitated by a 4.11 m fairing. KB Polyot is responsible for Kosmos-3M, Khrunichev is responsible for Proton K and Rockot, the Makayev Design Bureau handles Shtil and Volna, STC Koonplex is responsible for Start, Yuzhnoye State Design Office produces Cyclone, Dnepr and Zenit 2 and NPO Yuzhnoye produces Zenit 3SL.

Soyuz FG

Background

Developed as an upgrade variant of the Soyuz U launcher with approximately 5 per cent improved performance. An evolution of the Soyuz ST (Soyuz/Ikar) proposed as an uprated variant for commercial customers. FG first launched 20 May 2001 carrying Progress M1-6 to the ISS followed by second launch on 26 November 2001 carrying Progress MI-7. A third launch on 21 March 2002 carried the Progress M1-8 to the ISS while a fourth carried the Progress M1-9 on 25 September 2002. The fifth launch took place on 26 April 2003 on the ISS-6S mission followed by the ISS-11P mission on 8 June 2003.

UPDATED

Soyuz M

Background

Conceived as the first major upgrade of Soyuz launcher using Zenit second stage enquiries (RD-120) with higher efficiency. Soyuz M was abandoned in favour of Soyuz FG with less ambitious upgrade although a large number of potential applications had been identified. If adopted this would have been the first major upgrade to the R-7 family of launchers in 40 years and would have permitted launches from Russia's Plesetsk site with the same payload mass as unmodified Soyuz launchers from Baikonur in Kazakstan. This has the attraction of eliminating payment fees to another country. The Soyuz M was abandoned in favour of the less ambitious Soyuz St and Soyuz FG upgrades.

Soyuz-U

Background

The Vostok, Soyuz and Molniya family of launch vehicles all derived from Sergei Korolyov's R-7 (8K71) ICBM. Called SS-6/Sapwood in the West: the vehicle was first launched on 15 May 1957. In the 8K71PS version – little modified from the missile – it launched the first two Sputniks, while the further modified 8A91 was used for the Sputnik 3 launch. The Vostok and different Soyuz variants have carried all Soviet and Russian manned missions. They are now the responsibility of the Central specialised Design Bureau in Samara. CSDB was founded in 1959 as a division of Korolyov's OKB-1 Moscow design bureau to provide design support and launch vehicle manufacturing at the Samara 'Progress' factory where the vehicles are now built.

The launch vehicles that evolved from the R-7 missile are all built around the standard 'core plus four strap-ons' concept. Soviet era commentators designate the strap-ons as Block B, V, G and D while the core is Block A. Vostok added a stage designated Block E atop the core.

As the Soyuz family matured, the Soviets considered two variants to carry on the design. The first, the 8K72 incorporated the Kosberg RD-0105 engine into the Block E. The second, the 8K73 would have used Glushko's RD-0109 on that stage. The 8K72 eventually won out and the Soviets launched the initial Luna missions during 1958–1960 with it. Some Russian literature has called this the Vostok-L. Later, an improved Block E returned the RD-0109 engine to the first stage to give the 8K72K vehicle, used for the manned Vostok flights. Further improvements gave the Vostok-2 8A92, used within the photo reconnaissance programme and a final variant was called either Vostok-2M or Meteor (since it was used for launching Elint and Meteor satellites), the 8A92M.

The next variant to appear introduced two new upper stages: the Block I replaced the Block E and final Block L stage was added: that in turn spawned the Molniya launch vehicle. The Molniya carried the RD-0107 in its Block I, and this vehicle without the Block L was flown as the Voskhod launch vehicle. Western observers erroneously grouped the Voskhod launches with the Soyuz launch vehicles

SL-4/A-2 launch of a manned Soyuz-TM spacecraft. Icing reveals liquid oxygen tank locations 0516869

since they were very similar vehicles and flew the same types of missions. Although the Molniya-M launch vehicle had switched to an improved Block I engine compared with the original Molniya, there is no evidence that a similar switch took place with the Voskhod vehicle's Block I.

The true Soyuz launch vehicle, 11A511, came along in 1966, when the Soviet Union launched Kosmos 133 (the first Soyuz spacecraft launch). This was an improved version of the Voskhod vehicle, carrying the RD-0108 engine on the Block I. There were two minor variants of the Soyuz which flew a few missions, namely the Soyuz-L (three test flights of the lunar module and the Soyuz-M (eight flights of photo reconnaissance satellites).

The Soyuz-U, 11A511U, replaced the older engine design with the RD-0110 Block I engine in May 1973 and put Kosmos 559 in orbit. In a design sense the U directly replaced and retired the Voskhod vehicle for good. The Voskhod saw its last mission in June 1976. Starting with launches in the ASTP programme and then the more general Soyuz programme which gave the world its first dedicated resupply vessel, the unmanned Soyuz 20, the Soyuz-U became the workhorse for manned and unmanned missions through to 1985. With the recent retirement of the Soyuz-U2 vehicle, Soyuz U has resumed its former role as the backbone of the Russian manned missions.

By the end of September 2003 almost 1,700 R-7 and Soyuz vehicles had been launched since the first flight on 15 May 1957 for a reliability level in excess of 97.5 per cent.

Voskhod and Soyuz variants orbital launches annual launch rates

1963: 1	1977: 37	1991: 24
1964: 5	1978: 45	1992: 24
1965: 13	1979: 45	1993: 17
1966: 15	1980: 45	1994: 15
1967: 22	1981: 42	1995: 12
1968: 34	1982: 43	1996: 2
1969: 37	1983: 43	1997: 10
1970: 31	1984: 44	1998: 8
1971: 33	1985: 40	1999: 12
1972: 31	1986: 36	2000: 13
1973: 40	1987: 43	2001: 9
1974: 36	1988: 42	2002: 6
1975: 41	1989: 38	
1976: 38	1990: 30	

Soyuz-U2 launch record summary

Programme	Flight Period	Number of launches
5th Gen P-R	1982–1995	20
6th Gen P-R	1989–1993	5
Soyuz-TM	1986–1995	22
Progress-M	1989–1994	19

Gen = generation, P-R = photo-reconnaissance

Specifications

Soyuz-U

Launch sites: Tyuratam, Plesetsk

Principal uses: photo-reconnaissance, manned Soyuz, Progress, remote sensing, Bion, other science; all LEO

Vehicle success rate: 96.8% at end of programme in 1995

Commercial launches: 0

Soyuz Fregat rocket stages in production (RSA) ***NEW***/1047484

RFAS/Soviet launchers are transported hor'zontally to the pad by rail and erected for firing. Manned SL-4 shrouds carry four panels deployed on aborts to decelerate the escaping Soyuz craft 0516870

Soyuz U on pod with erector and propellant trucks 0131961

Cost: US$15 million quoted 1989 (during June 1988 Bulgarian cosmonaut mission to Mir space station it was reported that Soyuz-U launcher hardware cost was R 2 million; in November 1992 it was quoted at R 56 million, in March 1995 at R 12 billion), in October 1995 at R 15 billion.
Performance: 6,855 kg into 220 km, 51.6°
Number of stages: 2 plus 4 strap-ons
Overall length (m): 45.22 unmanned (with 10.14 m-long fairing)
Principal diameter: 10.3 m across fins
Launch mass: 309.7 tonnes
Launch thrust (kN, vacuum): 5,932
Guidance: inertial. NPO AP is responsible for the GN&C systems

Soyuz-U strap-ons
Number of strap-ons: 4
Designations: individually Blocks B, V, G, D
Engines: RD-107 from NPO Energomash, each with four fixed chambers and two gimballed verniers
Length: 19.8 m
Diameter: 2.68 m at base
U: 1,013 SL; 1,239 vacuum
U2: 1,030 SL; 1,242 vacuum
Burn time: about 120 s
Oxidiser: 27.8 t liquid oxygen
Fuel: 11.8 t kerosene
Propellant mass: 39.6 t each strap-on; dry mass 3.770 t

Soyuz-U core stage 1
The strap-ons separate at 120 s, but the hammerhead core continues for a further 190 s
Designation: Block A
Engines: single RD-108 from NPO Energomash, based on the RD-107 but with four gimballed verniers for steering
Overall length: 28.75 m
Diameter: 2.95 m maximum at strap-on attach points on LOX tank; 2.15 m lower kerosene tank
Thrust (kN, vacuum): 978 U
Burn time: 310 s
Oxidiser: liquid oxygen
Fuel: kerosene
Propellant mass: about 94 1/2 t; dry mass 6.8 t

Soyuz-U stage 2
Designation: Block I
Engines: RD-0110 from KB Khimautomatiki, incorporating four fixed chambers and four gimballed verniers for steering
Overall length: 8.1 m
Diameter: 2.66 m
Oxidiser: liquid oxygen
Fuel: kerosene
Propellant mass: about 23 t; 2 t dry mass (excluding payload adaptors, and so on)
Thrust: 298 kN vacuum
Burn time: about 245 s
Fairing: A 4.5 t fairing protects the payload until 152–171 s after launch (77–93 km), depending on ascent profile. The final stage's control system commands shutdown at orbit injection about 9 min after launch, and 3.3 s later orders payload release for spring separation at about 1 m/s. Attitude control is not available following shutdown and stage rotation can be up to 3.5°/s at payload separation. The stage executes a collision avoidance manoeuvre 0.7 s after the separation command. Unmanned payloads employ the 10.14 m-long two-piece fairings, consisting of a 6.72 m-long cylindrical section and providing a 9 m long 2.65 m and payload envelope.
Acceleration loads: 4.5 *g* longitudinal, 1.5 *g* lateral; payloads rated for 10 *g* longitudinal, 2 *g* lateral for 10 min
Acoustic: 144 dB for 60 s, payloads rated for 148 dB for 120 s
Vibration: 0.5 *g* 10–40 Hz actual
Thermal environment: fairing inner wall reaches maximum 57°C before separation at about 170 s

Vehicle/payload processing
On the pad, air conditioning provides a 15–25°C environment under the fairing until services are withdrawn 40 minutes before launch, with a 0–30°C variation around the payload possible. All payload checkout operations must be completed 40 minutes before launch, with ground test systems disconnected by 30 minutes later, and no payload telemetry is available to avoid interference with the launcher until the final stage's telemetry is terminated 35 seconds after separation.

Fregat
Comment: NPO Lavotchkin's stage has flown three times on Proton as the injection module for 1988's Phobos missions and for Mars-96
Engine: NPOL single chamber; up to 20 ignitions. Four clusters of 22.2 N hydrazine thrusters (SI 225 s) provides attitude control/ullage
Overall length: 1.500 m
Principal diameter: 3.350 m
Stage mass: up to 6,530 kg (up to 5,350 kg main engine propellant plus 85 kg hydrazine for RCS)
Oxidiser: N2O4 in two neo-spherical tanks
Fuel: UDMH in two neo-spherical tanks
Stage thrust (vacuum): 19.6 kN main (SI 327 s)
Burn time: above values indicate 6.1 kg/s propellant flow rate, leading to 877 s total burn time

Summary Vostok, Soyuz and Molniya launch record summary

Launch vehicle		First launch	Final launch	Successes	Failures	Total
Vostok-L	8K72	23 September 1958*	26 April 1962	7	7	14
Vostok	8K72K	28 July 1960*	16 June 1963	10	2	12
Vostok-2	8A92	1 June 1962*	12 May 1967	45	5	50
Molniya	8A92M	28 February 1967	29 August 1991	87	2	89
		Totals	149	16	165	
(No name)	11A59	1 November 1963	20 July1966	2	0	2
Voskhod	11A57	16 November 1963	29 June 1976	293	13	306
(No name)	11A510	27 November 1965	20 July 1966	2	0	2
Soyuz	11A511	28 November 1966	24 May 1975	28	2	30
Soyuz-L	1A511L	24 November 1970	12 August 1971	3	0	3
Soyuz-M	11A511M	27 December 1971	31 March 1976	8	0	8
Soyuz-U	11A511U	18 May 1973	(operational)	634	20	682
Soyuz-U2	11A511U2	28 December 1982	3 September 1995	66	0	66
		Totals	1,036	35	1,077	

The 11A59 launches were the core and strap-ons plus the Polyot satellites as upper stages: the 11A510 used a Tsyklon upper stage as a technology test. Final launches were: Voskhod – June 1976, Soyuz-L – August 1971, Soyuz-M – March 1976 and Soyuz-U2 – September 1995. The number of Soyuz-U launches is through to the end of March 1997.
*These maiden flights were launch vehicle failures.

Soyuz-U2

Background

Soyuz-U2 was an uprated version of the Soyuz-U, flying with improved control systems. The Soviets, by introducing LOX/SYNTIN, liquid oxygen and synthetic kerosene, on the central core stage, rather than LOX/kerosene improved the Soyuz performance by a factor of 25 per cent. There are reports that an RD-118 engine replaced the RD-108 second stage of the Soyuz-U and other vehicles within the family.

On 5 September 1996, ITAR-TASS announced that because of the halt in the production of Syntin fuel, for budgetary reasons the Soyuz-U2 had been retired from use. As a result, launches of Soyuz-TM spacecraft to Mir cannot reach the altitude at which Mir operated and it flew around 10 km lower to compensate, so that a three-person crew could still reach the station using the Soyuz-U launch vehicle.

The final launch of the venerable vehicle put Kosmos 2320 in orbit on September 1995. There appear to have been 66 orbital launches of the

Soyuz-U2 and Russian literature states that there were no launch failures.

Recent interest, reported to have stimulated further development in the production of Soyuz-U2 class launchers, is not confirmed. Some work has been done on derivative evolutions of RD-118 based proposal, first conceived as an evolutionary step.

Soyuz-2

Background

Coxmmercially marketed by Starsem based in France, Soyuz-2 replaces the existing Soyuz-U launch vehicle. As a commercial vehicle Russian literature first referred to it as the 'Rus' and occasionally as Soyuz-M, a name which would cause confusion with the Soyuz-M variant flown in the 1970s.

The new launch vehicle retains the concept of the existing and well-proven Soyuz design by keeping the four strap-on boosters surrounding a central core with an orbital stage on top. The strap-on booster and central core will both use engines derived from the RD-120 which has flown on the second stage of the Zenit-2 launch vehicle.

A 1991 paper from NPO Energomash described the development of different variants of the RD-120 to be used on the first stages of launch vehicles. While the literature is sketchy and possible unreliable, it is possible that the RD-120.01-2 variant forms the basis of the strap-on boosters engine, while the RD-120.01-1 might be used on the core. Other references state specifically that the RD-120K engine is used on the Soyuz-2. A further range of optional upper stage engines continues to be evaluated.

A new payload fairing with a diameter of 3.3 m will be introduced to complement the existing 3.0 m diameter fairings used on Progress-M and Soyuz-TM missions. The payload capability to LEO will be around 8.2 tonnes.

The first commercial Soyuz-2 was launched on 16 July 2000 carrying satellites Samba and Salsa of the ESA Cluster series.

Soyuz/Ikar

Background

In mid 1996, a new joint venture company called STARSEM offered to market a modified booster based on the Soyuz-U design. Based in France, the company added an Ikar fourth stage. The Ikar stage is described as 'an orbital module which has already been used successfully for more than 30 missions'. The Ikar upper stage is derived from the Yantar reconnaissance satellite propulsion module. The Ikar stage is capable of performing many small orbital corrections or a smaller number of larger ones, limited in total magnitude by the propellant supply.

For the Soyuz-U/Ikar launch vehicle the mechanical interfaces (which are not specified), together with equipment such as the infra-red Earth sensor had to be redesigned, necessitating a new flight control programme. In this new design, controllers keep the Ikar stage under ground control during its operations. A computer generates commands for the manoeuvres on the ground, transmits them to the stage and receives updates on the attitude and position.

The addition of Ikar to the Soyuz-U permit a much greater flexibility to the variety of orbits available to commercial users. Historically, the Soyuz class of launch vehicles simply placed payloads into low Earth orbits (perigee 170 to 200 km, apogee typically 300 to 400 km, rarely higher), but Ikar will bring higher orbits within range. If used like the Molniya-M fourth stage with a single burn then Ikar can reach higher-apogee orbits than can the Soyuz-U. With its restart capability Ikar can also circularise the orbit. STARSEM literature states that 2.5 tonnes can be placed in a 51.8°, 1,400 km orbit or 2.35° to a 90°, 1,400 km orbit.

The Soyuz/Ikar made its first flight on 9 February 1999. Total of five flights, the last on 22 November 1999, all of which carried four Globalstar satellites each. The official designation of the launch vehicle is the Soyuz 11A511U/Ikar optimised for sun-synchronous orbital payloads.

RD-0124M developed as the upper stage engine for the Soyuz-2 (Theo Pirard) 0054369

Space Clipper (SS-24)

Current Status

The proponents of Space Clipper have reopened the plausibility of air-launched medium class satellite launchers with proposals for a commercial initiative supported by US or European interests. Interest has heightened during the last few months of 2003 but nothing has been funded and no specific research programme has been approved. Nevertheless, there is considerable interest in the next generation of launch vehicles and Space Clipper is being re-examined for cost analysis and budget projections.

Background

In 1991, NPO Yuzhnoye proposed conversion of the SS-24/RS-22 Scalpel MIRV ICBM into a medium-class LEO vehicle air-launched from an Antonov An-124. By 1994, sources estimated the entire programme would cost the Russians US$220 million for system development and US$11 million per launch. Operationally the Russians can fly the vehicle to the customer's payload rather than vice versa. This neatly sidestepped objections of US export control restrictions regarding shipping US technology satellites into the former Soviet Union. Since the original proposal this concern has diminished however and the company is looking for western investment to develop a canister-based service. User cost would be about US$23 million to US$25 million. The Talisman launcher, using SS-24 technology, is being jointly studied with Dassault Aviation.

SS-24 development began in 1971 as an answer to the US MX. The ballistic missile achieved the first test flight October 1982 from Plesetsk. Deployment of the rail mobile version began in 1987, followed by the silo model two years later. It was announced in August 1990 that missile production would end in 1991, and the Russians would retire all 92 silo/rail-based missiles under the START 2 treaty.

To make the Space Clipper a viable space launch vehicle designers must add a new control system, stage 4, fairing and attitude control thruster module. Stage 1 yaw/pitch control is provided by nozzle flexure and stage 1/2 roll control is from fairing fins. The new thruster module, probably using elements of the SS-18/24 NTO/UDMH units, would provide the remaining control. The 105 tonne vehicle would be ejected from its launch canister by a cold gas system before ignition. Stage 4 could be liquid or solid. A liquid stage would be more suited to circular LEO orbits, delivering 2,500 kg into 200 km circular (accuracy 10 km altitude, 0.05° inclination). A solid stage is preferable for elliptical paths: 1,000 kg into 150 × 36,000 km (accuracy ±1,000 to 2,500 km, ±0.2 to 0.6°).

Even more remarkable than flying the missile to the payload, the Space Clipper can be launched directly from its transport vehicle. The launcher is rear-loaded and checked aboard the An-124 before flying to the customer's airfield. Once at the delivery site a customer can use the onboard clean room by loading the payload through the front hatch. Then the Antonov flies to the release area, while the launcher's control/guidance system is initialised, and the rear doors are opened. At an altitude of 10 to 11 km and speed of 800 to 850 km/h, a pilot parachute extracts the vehicle, which ignites some 3 to 5 seconds later.

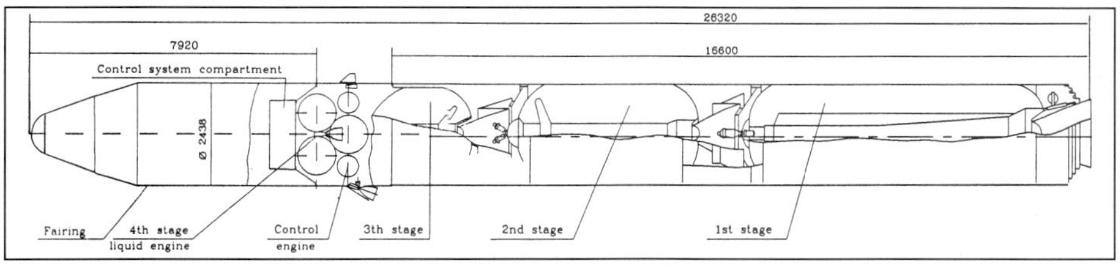

Principal features of the proposed SS-24 canister-launched vehicle. Shown is the version with a liquid stage 4. The solid stage 4 would sit forward of the avionics compartment. The attitude control thruster module (labelled 'control engine') would be retained. Dimensions in mm (Commercial Space Technologies) 0517190

Start installed in its launch canister before transfer to its launch stand
(Theo Pirard/Space Information Centre) 0517334

Specifications

Space Clipper

Typical performance: 570 kg (1234) or 330 kg (2234) into 400 × 40,000 km 65°; 820 kg (1234), 530 kg (2234) or 160 kg (0234) into 150 × 20,000 km 63.5°; 450 kg (1234), 200 kg (2234) or 60 kg (0234) on escape trajectories.
Number of stages: 3 or 4 (solid)
Overall length: 21.00 m (1234/1235), 20.50 m (2235), 14.30 m (0235)
Principal diameter: 2.40 m
Thrust (kN, vacuum): 1: 2,059; 2: 1,049; 3: 206; 4: 59; 5: 20
Launch mass: 64 t (1234/1235), 52 t (2234/2235), 32 t (0234/0235)
Guidance: computer-controlled inertial. Control can be provided for 60 min post-boost. Injection accuracy: ±3.5 km altitude, ±2 s period, ±2 arcmin inc.
Fairing/payload accommodation: the two-piece aluminium alloy fairing separates at about 100 km altitude. Two versions available: payload envelope 2.140 m diameter, 3.590 m long for 1235/2235/0235 and 2.150 m diameter, 4.790 m long for 1234/2234/0234. Maximum static acceleration 10 g during powered flight. 25 ± 5°C maintained within the fairing during air transport.

Start-1/Start

Current Status

No launches in 2002. The last Start-1 launch carried the ODIN satellite to orbit on 20 February 2001.

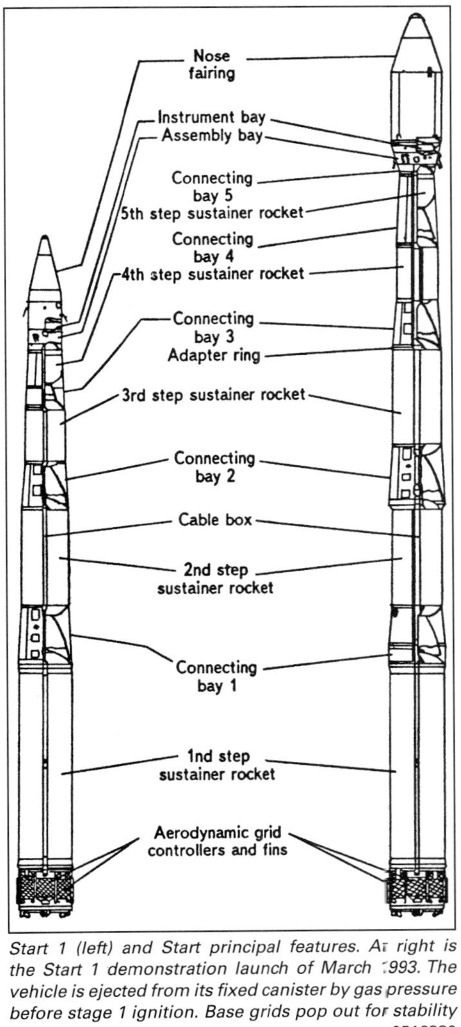

Start 1 (left) and Start principal features. At right is the Start 1 demonstration launch of March 1993. The vehicle is ejected from its fixed canister by gas pressure before stage 1 ignition. Base grids pop out for stability 0516880

Background

The first demonstration launch of the four-stage Start 1 was made from Plesetsk 25 March 1993, delivering a 260 kg communications payload into 695 × 966 km, 75.8°. The second Start-1 launch trailed along on 4 March 1997, inaugurating the Svobodny launch site. The third launch, on 5 December 2000, successfully launched EROS AI, an Israeli earth observation satellite. Its sister version, the first five-stage Start was lost 28 March 1995. Stage 4 shut down 12 seconds early and stage 5 did not ignite. Israel's TechSat, Mexico's Unamsat and STC Complex's own test payload were lost in the failure. Start-1 is a converted three-stage solid propellant road mobile SS-25 Sickle (RS-12M Topol) ICBM with an added solid stage 4 which apparently repeats stage 2. A Start launch costs about US$6 million, although costs of up to US$10 million have also been claimed. 1995's launch was free to the users.

Some elements from the related two-stage SS-20 IRBM are included. Both missiles came from the Nadiradze bureau and built at the Votkinsk plant, Udmurt. Unusually, the project is performed under the aegis of the Strategic Rocket Forces. STC Complex provides customer contact. The SS-25 began development in 1971 and had its first test flights in 1982 (the first, 27 October 1982, was unsuccessful). Testing completed with the launch of 23 December 1987. It is capable of delivering a single 1 tonne 550 kT nuclear warhead over 10,500 km. It entered service in 1985 and in 1991 there were 288 missiles deployed at nine sites. The current road mobile form is expected to be replaced by the Topol-M silo-based version. It is heavier and has a larger-diameter stage 1; first launch was 28 March 1995. It will be deployed in former SS-18 silos.

Start 1 (left) and Start principal features. At right is the Start 1 demonstration launch of March 1993. The vehicle is ejected from its fixed canister by gas pressure before stage 1 ignition. Base grids pop out for stability 0516881

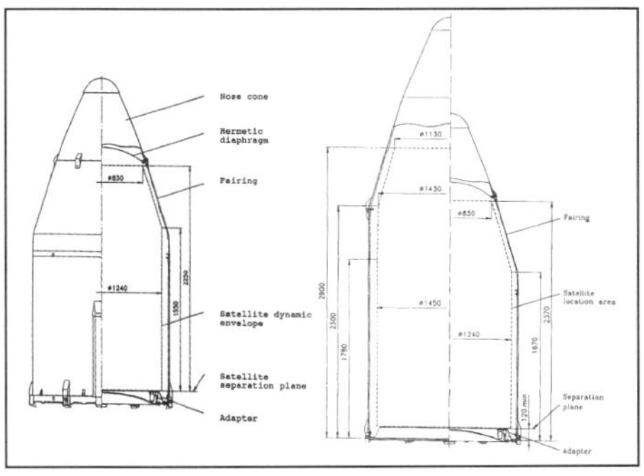

Start-1 (left) and Start payload envelopes
(STC Complex) 0517335

All stages burn out fully (about 1 minute each). The 10 to 20 second coast after stage 1 burnout reduces dynamic pressure so that stage 2 is controllable after separation. The base aerodynamic fins and exhaust vanes provide control during stage 1 coast. N_2 jets in stage 5's aft section (stage 4 on Start-1) are used for roll control during stage 1 and three-axis during subsequent burns. There are no coast periods for the intermediate stages, but one of several 100 seconds before last stage firing, while the N_2 jets provide control.

EarthWatch planned to launch its EarlyBird Earth observation satellite on Start in 1996 but the flight was delayed. South Africa's Houwteq, in 1994, selected it for the Greensat, before the project was cancelled. Another proposal is for Russia's own Courier system of LEO relay satellites in a joint venture with NPO ELAS but this has been abandoned.

Specifications

Start 1/Start

First launch: 25 March 1993/28 March 1995
Number of launches: 4/1, through end of 2002
Success rate: 20%, through end of 2002
Number of stages: 4/5 (plus each can carry 70/80 kg post-boost stage)
Launch mass: 47/60 t
Length: 22.7/28.8 m
Maximum diameter: 1.8/1.8 m
Guidance: inertial; avionics in sealed compartment atop last stage
Performance: with post-boost stages 420/645 kg into 300 × 300 km, 90°; 300/530 kg into 500 × 500 km, 90°; 110/275 kg into 1,000 × 1,000 km, 90°. Accuracy 5 km apogee ±3′ inclination with post-boost stage (solid propellant gas generator); 60–140 km apogee uncertainty without. Performance values increased by 70/80 kg without post-boost. Start-1's performance is reduced by 50 kg if larger fairing used.
Payload accommodation: Launcher is assembled horizontally in an 18 × 18.5 60 m class 100,000 building. Payload attachment requires 3 h, fairing/canister installation 5 h, tests 10 h and final launch operations 0.5 h. The campaign requires two to three days from when the payload is ready. The facility, canister and launch site maintain the payload at 10.5–26°C, 25–70% humidity. The payload can be accessed until 2–3 h before launch. The vehicle is transported horizontally to the 5 × 8 m launch box and raised. Start provides payload power and issues commands. Telemetry transmit: 203.27, 219.70, 75.67 MHz. On release, payload motion does not exceed 2°/s pitch/yaw + 1.5°/s roll with post-boost stage; 4°/s + 3.5°/s without.
Acceleration (g): 2.8/2.3 ejection, 0.15/3.3 stage 1, 6.5/4.4 stage 2, 6.5/6.1 stage 3, 9.0/6.3 stage 4, –/8.0 stage 5.
Acoustic load: ≤135 dB over 50–4,000 Hz

Shtil

Current Status
No further launches have been recorded since the Tubsat satellites were carried into orbit on 7 July 1998.

Background
The Makeyev bureau offers satellite launch services using its converted SLBMs. The bureau was founded by V P Makeyev (1925–1985) in 1955 as an offshoot of Korolev's bureau and developed the RSM-25 Zyb (SS-N-6 Serb), RSM-40 Vysota (SS-N-8 Sawfly), RSM-50 Volna (SS-N-18 Stingray), RSM-52 Rif (SS-N-20 Sturgeon) and RSM-54 Shtil (SS-N-23 Skiff) SLBMs. A Volna was launched 7 June 1995 from the 'Kalmar' Delta 4 boat submerged in the Barents Sea on a 20 minute suborbital μgravity flight carrying a 120 kg experiment from Germany's ZARM (electrically-driven thermal convection in a fluid shell between two concentric spheres). Recovery was 5,600 km downrange on Kamchatka. There were apparently three SLBM demonstrations 1991–93. A Shtil 2N orbital demonstration launch was reportedly planned for 1996 carrying a 100 kg Izmiran satellite, but the first known launch did not take place until 7 July 1998 when TUBSAT N and TUBSAT N1 were launched from a submerged submarine in the Barents Sea using the Shtil 3N. No launches of Shtil have been recorded since the inaugural flight. Other variantsavailable include the Shtil 2/2N, the Shtil 3A and the Shtil 3N.

Surf
Based on the SS-N-21 104 tonne Shtil satellite launcher attaches a Rif solid propellant 2.4 m diameter stage 1 to two stages of the 1.90 m diameter liquid propellant Shtil ('Calm Sea') with new stages 4/5. Length is 33.9/34.8/34.8 m depending on the 4/8/12 m^3 fairing. Water-based launches are planned but land ascents are possible using the Antonov An-225 as transporter. Up to three vehicles could be assembled and transported on an *Ivan Rogov*-class ship and released aftwards to float vertically in the sea for launch. Commercial launches would cost <US$11 million. Capacities into circular orbits are (kg, 0°/90° inclination): 2,400/1,840 200 km; 2,200/1,650 400 km; 2,020/1,500 600 km; 1,850/1,360 800 km; 1,700/1,230 1,000 km; 1,570/1,110 1,200 km; 1,460/1,000 1,400 km; 1,360/910 1,600 km; 1,270/830 1,800 km; 1,200/770 2,000 km; 170/-6,500 km.

Shtil
The individual vehicles are also offered commercially. The Shtil is available in three versions: –1N 430 kg into 200 km, 185 kg into 700 km, 40 tonne launch mass, 1.5 m^3 payload volume using original warhead vehicle; –2N 265 kg into 200 km, 2 m^3 payload volume; –3N 670 kg into 200 km, 410 kg into 700 km, 46 tonne launch mass, 3.6 m^3 payload volume. Aerolaunched, the Shtil 3A can send a 950 kg payload to 200 km orbit or 620 kg to a 400 km orbit. SLBM testing began in 1983 and Shtil entered service in 1986; there were 112 aboard Delta 4 boats in 1991. Inertial guidance plus stellar reference update; 8,300 km range.

Zyb
The Zyb is offered with two types of re-entry capsules for suborbital microgravity missions: 450 kg Sprint capsule (100 kg payload, original warhead vehicle) and 650 kg Efir (65 to 80 kg payload). A Zyb/Sprint test launch was made December 1991 and a Zyb/Efir December 1992. Development of the two-stage liquid propellant Zyb began in 1958 and the 2,400 km-range missile entered service in 1967. A 3,000 km-range version began flight tests October 1972. Launch mass 14,200 kg, length 9.65 m, diameter 1.5 m, inertial guidance. 600 Zybs were built and 192 were in service aboard Yankee 1 boats in 1991. A total of 419 will become available for use as they were removed from service from the mid-1990s.

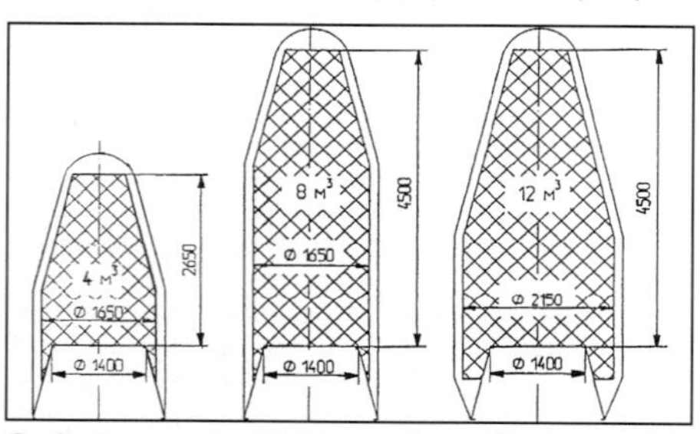

Surf's three payload fairing options; all are watertight (Glavcosmos) 0517325

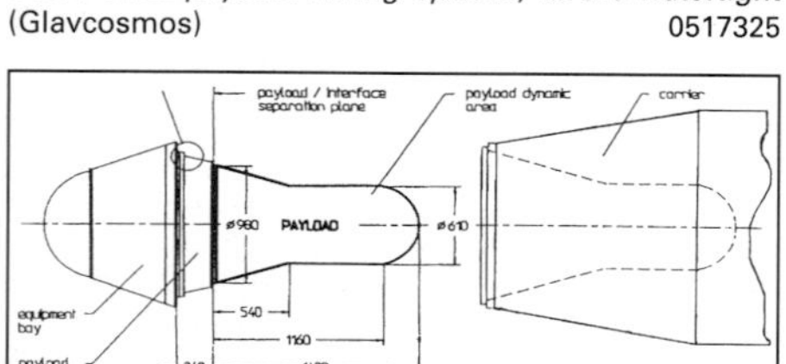

Vysota payload accommodation. The payload hangs below the nose in the space previously occupied by the warhead re-entry vehicle 0516882

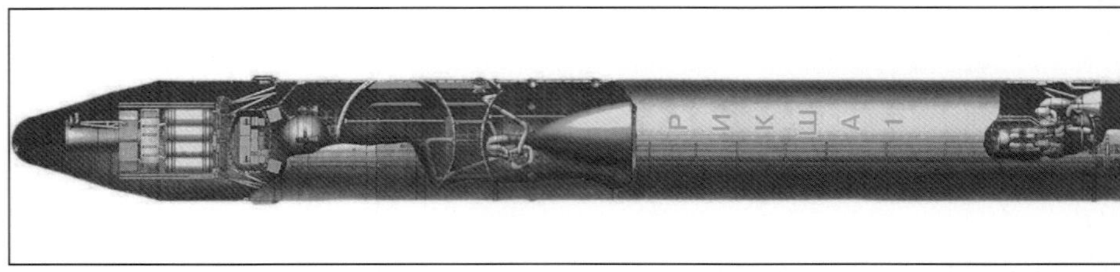

Riksha would be powered by liquid natural gas 0517349

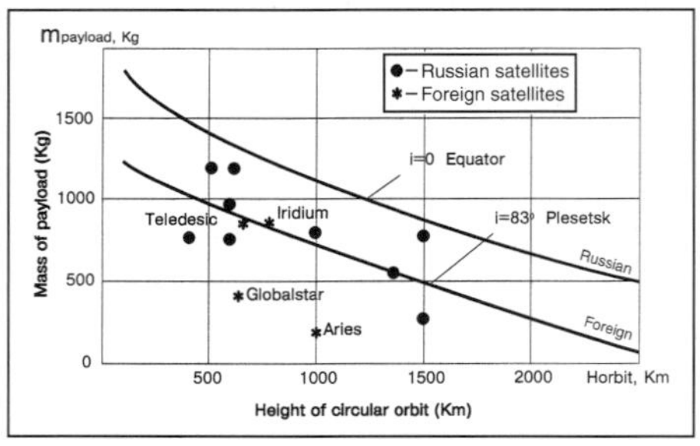

Riksha payload performance 0517350

Vysota
The two-stage liquid propellant Vysota ('Altitude') began development in 1961 and entered service in 1971; there were 244 operational missiles in 1991 plus a further 177 in storage. It seems likely that Vysota and its boats will be decommissioned to meet the requirements for reductions in SLBM numbers under the START 1/2 agreements. In December 1994, it is believed that two Delta1 boats had been taken out of service, reducing the number of operational missiles to about 220. Typical space performance would be 130 kg into 200 to 250 km circular at 0 to 25°; 130 kg into 170 to 220 × 1,000 to 1,500 km 0 to 25°. Length 14.2 m, diameter 1.80 m, launch mass 33 tonnes, burn time about 230 seconds. The payload occupies the volume previously allocated to the warhead re-entry vehicle, hanging inside the carrier below the nose equipment bay. Separation of the forward assembly is achieved by residual air pressure in the carrier. Temperature is maintained at 15 to 30°C; 200 W is available to the payload. Maximum static launch acceleration is 10.6 *g* during powered flight; 13.5 *g* for 0.1 second at stage separation.

Volna
Based on the R-29RL three-stage liquid propellant Volna ('Wave') began development in 1968; there were 224 missiles in service in 1991 aboard Delta 3 boats and a further 100 in storage (no reduction in operational numbers had been reported as of December 1994). Typical orbital performance is as Vysota but with a larger volume. Length 15.6 m, diameter 1.80 m, launch mass 35.3 tonnes.

Rikscha
The new Riksha is proposed, requiring US135 million and four years to develop, and capable of delivering 300 to 1,700 kg into orbit for US$10 to 11 million. New Energomash's LOX/methane engines would power the 64 tonne 24.5 m 2.4 m diameter two-stage vehicle. Six RD-169s would be clustered for stage 1, and a single RD-185 altitude version would power stage 2 (see Propulsion for further details). NPO Energomash, but now comes under the KompoMach consortium with Makeyev as lead initiated the project.

Tsyklon-2

Current Status
No launches since 21 December 2001.

Background
In the early 1960s, the Yangel design bureau (now Yuzhnoye) proposed a family of launch vehicles to be called Tsyklon. These would be based upon the bureau's R-16 and R-36 missiles and topped by a common C5M third stage for orbital missions. A government decree of 24 August 1965 ordered Yangel to develop the R-36 into the 11K67 to launch naval intelligence satellites and the Tsyklon 2 replaced the 11K67 from 1969. The R-16 had been approved for development as an ICBM in December 1956 and basic design was completed by November 1957. The first launch took place on 2 February 1961 following a launch pad disaster on 24 October, which killed almost 100 people. The Tsyklon-1 was the R-16 missile with the C5M added (launch mass 145.4 tonnes, payload to a polar 1,000 km orbit = ~700 kg). The Tsyklon-2 would simply be a modified two-stage R-36 missile (launch mass 179 tonnes, 1.5 tonnes to a polar 200 km orbit). And the Tsyklon-3 would be the Tsyklon-2 with the S5M stage added (launch mass 185.5 tonnes, 2.5 tonnes to polar 1,000 km orbit). Of these three options, the Soviets backed away from the Tsyklon-1 and never developed it.

The Soviets began their first series of test launches of the R-36 missile in either late 1965 or early 1966. Reports differ because the vehicles involved in various launch failures have never been positively identified. For years, rumours of a R-36-0 Fractional Orbit Bombardment System (FOBS) test on 16 December 1965 and 5 February 1966 have circulated. On 16 March 1966, an R-36-0 exploded on the launch pad, followed by a possible FOBS test on 20 May 1966. Sorting out the payload for these vehicles is difficult because, by their very nature, the FOBS launches had the dummy warheads recalled after slightly less than a single orbit. Since the difference between a successful partial orbit and a failed full orbit is unclear to observers the exact purpose of the launch in some cases depends on the interpretation of the observer. Since the FOBS programme was looked upon as an extension of the ICBM programme the Soviets might have considered that flights which entered orbit but were recalled after less than one orbit were not 'true' satellites to be announced publicly and registered via the United Nations.

Two unannounced orbital launches on 17 September and 2 November 1966 are particularly curious. On each occasion the dummy FOBS exploded shortly after orbital injection, and since

Tsyklon is based on the SS-9 Scarp ICBM, with an unusual six-chamber first stage configuration 0516878

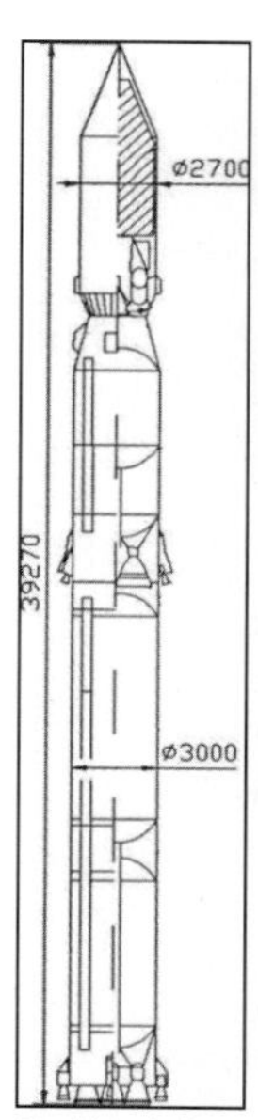

Tsyklon is based on Tsyklon M 0517328/0517329

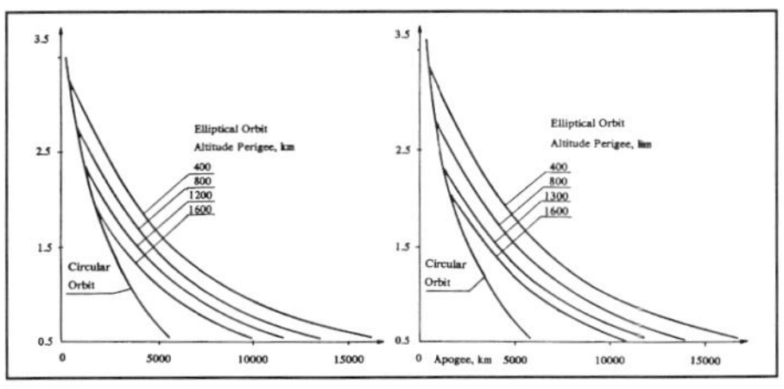

Tsyklon performance (t) into 82.5° (left) and 73.5° (right) 0517330

there was operating or intact payload in orbit, the Soviets might not have thought that they should register the launches. In reality, the debris was tracked in the West and the launches appeared in all of the western launch catalogues – public and otherwise. This seems to have brought a change of heart within the Soviet bureaucracy, since the next launch in the series - Kosmos 139 on 25 January 1967 - was announced by TASS.

On 22 March 1967, a FOBS failed to reach orbit, and then a series of generally successful tests of the system took place during May 1967 to August 1971. Kosmos 160 (17 May 1967) seems to have been a payload failure since it apparently decayed from orbit rather than being de-orbited. Following the August 1971 launch the FOBS system was considered operational, although the system was subsequently dismantled under arms reductions treaties.

The Russians seem to be in two minds concerning the R-36-0 launches since the flights were of little-modified R-36 missiles and were part of the ICBM programme: therefore sometimes they are considered to be part of the two-stage Tsyklon launch programme and at other times they are not.

Launches of the dedicated space version of the R-36, the Tsyklon-2 vehicle started in October 1967, but this variant had a short period of flights and was replaced by the launch vehicle referred to as both Tsyklon-2A and Tsyklon-M. For the now-retired RORSAT launches the payload used its own propulsion system to reach orbit, the two Tsyklon rocket stages being sub-orbital. While the spacecraft propulsion systems have failed either on their way to orbit or in orbit, the actual Tsyklon-2A launch vehicle has never failed in flight.

Summary R-36-0 and two-stage Tsyklon launch record

Launch vehicle		First Launch	Final Launch	Successes	Failures	Total
R-36-0	8K69	16? December 1965**	August 1971	22	2	24
Tsyklon-2	11K67	27 October 1967	1 November 1968**	7	0	7
Tsyklon-M*	11K69	6 August 1969	(operational)	100	0/5	100

* Also sometimes called the Tsyklon-2A. ** Launch vehicle failure. Tsyklon-2 and Tsyklon-M failures are shown as X/Y where X is the number of launch vehicle failures and Y the number of payload propulsion system failures.

Specifications

Tsyklon 2

First launch: 25 January 1969
Number launched: 104 to end of 2001 (102 attaining orbit)
Launch sites: Plesetsk
Vehicle success rate: 98.05 to December 2002
Cost: US$12 million in 1994
Number of stages: 3
Performance: 2,800 kg to 200 km orbit at 52.0°
Overall length: 39.7 m
Principal diameter: 3.00 m
Launch mass: 182,000 kg
Launch thrust: 2,336 kN SL

Tsyklon 2 stage 1

Engines: Tsyklon 2 uses the RD-251 with six chambers and four gimballed NPO Yuzhnoye verniers.
Overall length: 18.9 m
Diameter: 3.00 m
Thrust: 2,366 kN SL (2,648 kN vacuum); verniers 296 kN SL (334 kN vacuum)
Mass: 122,300 kg; 6,400 kg empty
Burn time: 120 s
Isp: 301 s

Tsyklon 2 stage 2

Engines: Tsyklon 2 stage 2 employs the RD-252 engine produced by NPO Energomash
Overall length: 9.4 m
Diameter: 3.00 m
Thrust: 940.5 kN vacuum
Mass: 49,300 kg gross; 3,700 kg empty
Burn time: 160 s
Isp: 317 s

Tsyklon 2 stage 3

Engines: NPO Yuzhnoye RD-861 single restart embedded in toroidal tanks. Three-axis control during burns by eight fixed nozzles (6 x 20 N + 2 x 98 N) fed by main engine gas generator. Three-axis coast control by 10 thrusters of separate system.
Overall length: 2.5 m
Diameter: 2.0 m
Mass: 3,200 kg gross; 400 kg empty
Thrust: 78.0 kN
Burn time: 112 s
Isp: 317 s

Tsyklon 3

Current Status

No launches since 28 December 2001.

Background

As part of a plan to reduce the number of military space launch vehicles, a government decree of 2 January 1970 authorised development of what would become the Tsyklon 3. Based on the two stage 8K68 (SS-9 Mod 2), the Tsyklon 3 would adopt the S5M stage – a development of the R-36-0

Tsyklon-2 (M) launches

Launch	Date	Payload
1	25 January 1969	Kosmos US-A
2	6 August 1969	Kosmos 291
3	23 December 1969	Kosmos 316
4	3 October 1970	Kosmos 367
5	20 October 1970	Kosmos 373
6	23 October 1970	Kosmos 374
7	30 October 1970	Kosmos 375
8	25 February 1971	Kosmos 397
9	1 April 1971	Kosmos 402
10	4 April 1971	Kosmos 404
11	3 December 1971	Kosmos 462
12	25 December 1971	Kosmos 469
13	21 August 1972	Kosmos 516
14	25 April 1973	Kosmos US-A
15	27 December 1973	Kosmos 626
16	15 May 1974	Kosmos 651
17	17 May 1974	Kosmos 654
18	24 December 1974	Kosmos 699
19	2 April 1975	Kosmos 723
20	7 April 1975	Kosmos 724
21	29 October 1975	Kosmos 777
22	12 December 1975	Kosmos 785
23	16 February 1976	Kosmos 804
24	13 April 1976	Kosmos 814
25	2 July 1976	Kosmos 838
26	21 July 1976	Kosmos 843
27	17 October 1976	Kosmos 860
28	21 October 1976	Kosmos 861
29	26 November 1976	Kosmos 868
30	27 December 1976	Kosmos 886
31	23 May 1977	Kosmos 910
32	17 June 1977	Kosmos 918
33	24 August 1977	Kosmos 937
34	16 September 1977	Kosmos 952
35	18 September 1977	Kosmos 954
36	26 October 1977	Kosmos 961
37	21 December 1977	Kosmos 970
38	19 May 1978	Kosmos 1009
39	18 April 1979	Kosmos 1094
40	25 April 1979	Kosmos 1096
41	14 March 1980	Kosmos 1167
42	18 April 1980	Kosmos 1174
43	29 April 1980	Kosmos 1176
44	4 November 1980	Kosmos 1220
45	2 February 1981	Kosmos 1243
46	5 March 1981	Kosmos 1249
47	14 March 1981	Kosmos 1258
48	20 March 1981	Kosmos 1260
49	21 April 1981	Kosmos 1266
50	4 August 1981	Kosmos 1286
51	24 August 1981	Kosmos 1299
52	14 September 1981	Kosmos 1306
53	11 February 1982	Kosmos 1337
54	29 April 1982	Kosmos 1355
55	14 May 1982	Kosmos 1365
56	1 June 1982	Kosmos 1372
57	18 June 1982	Kosmos 1379
58	30 august 1982	Kosmos 1402
59	4 September 1982	Kosmos 1405
60	2 October 1982	Kosmos 1412
61	7 May 1983	Kosmos 1461
62	29 October 1983	Kosmos 1507
63	30 May 1984	Kosmos 1567
64	29 June 1984	Kosmos 1579
65	7 August 1984	Kosmos 1588
66	31 October 1984	Kosmos 1607
67	23 January 1985	Kosmos 1625
68	18 April 1985	Kosmos 1646
69	1 August 1985	Kosmos 1670
70	23 August 1985	Kosmos 1677
71	19 September 1985	Kosmos 1682
72	27 February 1986	Kosmos 1735
73	21 March 1986	Kosmos 1736
74	25 March 1986	Kosmos 1737
75	4 August 1986	Kosmos 1769
76	20 August 1986	Kosmos 1771
77	1 February 1987	Kosmos 1818
78	8 April 1987	Kosmos 1834
79	18 June 1987	Kosmos 1860
80	10 July 1987	Kosmos 1867
81	10 October 1987	Kosmos 1890
82	12 December 1987	Kosmos 1900
83	14 March 1988	Kosmos 1932
84	28 May 1988	Kosmos 1949
85	18 November 1988	Kosmos 1979
86	24 April 1989	Kosmos 2033
87	27 September 1989	Kosmos 2046
88	24 November 1989	Kosmos 2051
89	14 March 1990	Kosmos 2060
90	23 August 1990	Kosmos 2096
91	14 November 1990	Kosmos 2103
92	4 December 1990	Kosmos 2107
93	18 January 1991	Kosmos 2122
94	30 March 1993	Kosmos 2238
95	28 April 1993	Kosmos 2244
96	7 July 1993	Kosmos 2258
97	17 September 1993	Kosmos 2264
98	2 November 1994	Kosmos 2293
99	8 June 1995	Kosmos 2313
100	20 December 1995	Kosmos 2326
101	11 December 1996	Kosmos 2335
102	9 December 1997	Kosmos 2347
103	26 December 1999	Kosmos 2367
104	21 December 2001	Kosmos 2383

All launches from Tyuratam; all of ocean surveillance satellites. Failures 16 March 1966, 18 September 1966 (in orbit), 2 November 1966 (in orbit), 22 March 1967, 25 January 1969, 25 April 1973.

Tsyklon 3 launches

Launch	Date	Payload	Launch	Date	Payload
1	24 June 1977	C921	61	18 August 1987	Meteor 2–16
2	24 September 1977	C956	62	7 September 1987	C1875-80
3	27 December 1977	C972	63	20 October 1987	C1892 electronic intelligence
4	28 June 1978	C1025 electronic intelligence	64	6 January 1988	C1908 electronic intelligence
5	26 October 1978	C1045/RS 1/2	65	15 January 1988	C1909-14 communication satellites
6	12 February 1979	C1076 ocean			
7	23 January 1980	C1151 ocean	66	30 January 1988	Meteor 2–17
8*	23 January 1981	GEO-IK	67	15 March 1988	C1933 electronic intelligence
9	24 August 1981	C1300 ocean	68	30 May 1988	C1950 GEO-IK
10	21 September 1981	Aureole 3	69	14 June 1988	C1953 electronic intelligence
11	30 September 1981	C1312 GEO-IK	70	5 July 1988	Okean 1 ocean
12	3 December 1981	C1328 ocean	71	26 July 1988	Meteor 3–2
13	25 March 1982	Meteor 2–8	72	11 October 1988	C1975 electronic intelligence
14	10 June 1982	C1378 ocean	73	23 December 1988	C1985
15	16 September 1982	C1408 ocean	74	10 February 1989	C1994-99
16	24 September 1982	C1410 GEO-IK	75	28 February 1989	Meteor 2–18
17	23 April 1983	C1455 electronic intelligence	76*	9 June 1989	Okean
18	23 June 1983	C1470 electronic intelligence	77	28 August 1989	C2037 GEO-IK
19	28 September 1983	C1500 ocean	78	14 September 1989	C2038-43
20	24 November 1983	C1510 GEO-IK	79	28 September 1989	Interkosmos 24
21	15 December 1983	C1515 electronic intelligence	80	24 October 1989	Meteor 3–03
22	8 February 1984	C1536 electronic intelligence	81	26 December 1989	C2053
23	15 March 1984	C1544 electronic intelligence	82	30 January 1990	C2058 electronic intelligence
24	5 July 1984	Meteor 2–11	83	28 February 1990	Okean 2 ocean
25	8 August 1984	C1589 GEO-IK	84	27 June 1990	Meteor 2–19
26	28 September 1984	C1602 ocean	85	30 July 1990	C2088 GEO-IK
27	18 October 1984	C1606 electronic intelligence	86	8 August 1990	C2090-95
28*	27 November 1984	C1612/Meteor 3	87	28 September 1990	Meteor 2–20
29	15 January 1985	C1617-22	88	28 November 1990	C2106 electronic intelligence
30	24 January 1985	C1626 electronic intelligence	89	22 December 1990	C2114-19
31	6 February 1985	Meteor 2–12	90	24 April 1991	Meteor 3–4
31	5 March 1985	C1633 electronic intelligence	91	16 May 1991	C2143-48
33	14 June 1985	C1660 GEO-IK	92	4 June 1991	Okean 3
34	8 July 1985	C1666 electronic intelligence	93	13 June 1991	C2151 electronic intelligence
35	8 August 1985	C1674 electronic intelligence	94	15 August 1991	Meteor 3–5
36	9 October 1985	C1690-95	95	28 September 1991	C2157-62
37	24 October 1985	Meteor 3–1	96	12 November 1991	C2165-70
38	22 November 1985	C1703 electronic intelligence	97	18 December 1991	Interkosmos 3/Magion 3
39	12 December 1985	C1707 electronic intelligence	98	13 July 1992	C2197-2202
40	26 December 1985	Meteor 2–13	99	20 October 1992	C2211-16
41	17 January 1986	C1726 electronic intelligence	100	24 November 1992	C2221 electronic intelligence
42	11 February 1986	C1732 GEO-IK	101	22 December 1992	C2226 GEO-IK
43	19 February 1986	C1733 electronic intelligence	102	25 December 1992	C2228 electronic intelligence
44	15 May 1986	C1743 electronic intelligence	103	16 April 1993	C2242 electronic intelligence
45	27 May 1986	Meteor 2–14	104	11 May 1993	C2245-50
46	12 June 1986	C1758 electronic intelligence	105	4 June 1993	C2252-57
47	28 July 1986	C1766 ocean	106	31 August 1993	Meteor 2–21/Temisat
48	30 September 1986	C1782 electronic intelligence	107	25 January 1994	Meteor 3–6/Tubsat
49*	15 October 1986	Sextuplet communication satellites	108	12 February 1994	C2268-2273
			109	2 March 1994	Coronas-I
50	2 December 1986	C1803 GEO-IK	110*	25 May 1994	electronic intelligence?
51	10 December 1986	C1805 electronic intelligence	111	11 October 1994	Okean 4
52	18 December 1986	C1809 science	112	29 November 1994	GEO-IK 1
53	5 January 1987	Meteor 2–15	113	26 December 1994	C2299-2304
54	14 January 1987	C1812 electronic intelligence	114	31 August 1995	Sich 1/FASat-Alfa
55	20 February 1987	C1823 GEO-IK	115	19 February 1996	C2328-30/Gonets
56	3 March 1987	C1825 electronic intelligence	116	14 February 1997	C2337-9/Gonets
57	13 March 1987	C1827-32	117	15 June 1998	C2352-7/Strela-3
58	27 April 1987	C1842 electronic intelligence	118*	27 December 2000	Gonets/Strela-3
59	1 July 1987	C1862 electronic intelligence	119	31 July 2001	Karonas-F
60	16 July 1987	C1869 ocean	120	28 December 2001	Gonets/Strela-3

* indicates vehicle failure. All launches from Plesetsk.

deorbit stage – and be capable of lifting payloads up to 4 tonnes in mass to low earth orbit. Following the disintegration of the USSR Russia decided to cancel the programme and the Ukrainian government did not take over the production of this launcher. In 120 flights up to the last on 28 December 2001 the Tsyklon 3 had eight failures, a reliability rate of 93.3 per cent.

The three-stage variant of Tsyklon – referred to as either Tsyklon-3 or simply Tsyklon, was part of the original 'Tsyklon family' proposed in the early 1960s. It took over from the Meteor launch vehicle (a modification of the Vostok launcher) to accomplish the launches of Meteor and Elint satellites. Two Meteor launches carried paying piggyback payloads, Italy's Temisat (August 1993) and Germany's Tubsat B (January 1994). The UK's Surrey Satellite Technology Ltd used Tsyklon in August 1995 for Chile's 50 kg FASat-Alfa and was charged a fee of only US$400,000. FASat-Bravo will also use Tsyklon. However, the launch rate is falling because of funding limitations and the loss of suppliers for the ageing avionics. Tsyklon production appears to have halted, but the Russians have hinted that some of Russia's R-36 strategic stockpile may be converted into Tsyklon's.

Tsyklon is the three-stage version and was first offered as the commercial Tsyklon launcher in October 1987 at Sputnik's 30th anniversary forum in Moscow. It supports a range of medium military and civil missions but its single most numerous payloads are electronic intelligence spacecraft. It is launched only from Plesetsk, but Tyuratam's Tsyklon 2M facilities should be compatible.

Specifications

Tsyklon 3

First launch: 24 June 1977 (Kosmos 921)
Number launched: 120 to end of 2002 (116 attaining orbit)
Launch sites: Plesetsk
Principal uses: electronic intelligence, Meteor, oceanographic, geodetic, remote sensing
Vehicle success rate: 96.6% to December 2000
Cost: US$10 million quoted in 1989
Performance: see diagrams
Number of stages: 3
Overall length: 39.270 m
Principal diameter: 3.000 m
Launch mass: 185.5 t plus
Launch thrust: 2,364 kN SL
Guidance: inertial; accuracy into 1,500 km orbit ±25 km altitude, 12 s period, 5 arcmin inclination; 600 km orbit ±15 km, 5 s, 3 arcmin

Tsyklon 3 stage 1

Engines: Tsyklon-M uses the six-chamber RD-251 engine from NPO Energomash and four gimballed NPO Yuzhnoye verniers: there are some reports that the three-stage Tsyklon uses a modification of this assembly which is designated RD-261.
Overall length: 19.38 m
Diameter: 3.000 m
Thrust: 2,364 kN SL (2,643 kN vacuum); verniers 296 kN SL (334 kN vacuum)
Mass: 127.0 t at launch; 8,300 kg dry
Burn time: 120 s (280 s quoted for combined stage 1/2 burn times)
Oxidiser: nitrogen tetroxide; 84,860 kg consumed. 60.15 m^3 tank pressurised to 2.5 atmospheres
Fuel: UDMH; 33,810 kg consumed. 44 m^3 tank pressurised to 2.1 atmospheres
Propellant mass: 120.7 t

Tsyklon 3 stage 2

Engines: Tsyklon-M uses the two-chamber RD-252 engine from NPO Energomash: there are reports that the three-stage Tsyklon uses a modification of this assembly which is designated RD-262
Overall length: 10.9 m
Diameter: 3.000 m
Thrust: 940.5 kN vacuum
Mass: 53.3 t at launch; 4,800 kg dry
Burn time: 160 s
Oxidiser: nitrogen tetroxide; 34,820 kg consumed. 25.00 m^3 tank pressurised to 3.7 atmospheres
Fuel: UDMH; 13,640 kg consumed. 18.14 m^3 tank pressurised to 2.2 atmospheres
Propellant mass: 49.8 t

Tsyklon 3 stage 3

Designation: S5M
Engines: NPO Yuzhnoye RD-861 single restart embedded in toroidal tanks. Three-axis control during burns by eight fixed nozzles (6×20 N + 2×98 N) fed by main engine gas generator. Three-axis coast control by 10 thrusters of separate system.
Overall length: 2.58 m
Principal diameter: 2.25 m
Thrust: 78.90 main kN vacuum
Mass: 4.63 t at launch; 1,407 kg dry
Burn time: 125 s in two burns (dual burn used for orbits >250 km)
Oxidiser: nitrogen tetroxide; 2,030 kg consumed. 1.6 m^3 tank pressurised to 7.5 atmospheres

Fuel: UDMH; 970 kg consumed. 1.5 m^3 tank pressurised to 4.2 atmospheres
Propellant mass: 3.2 t
Fairing
Length: 10.0 m.
Diameter: 2.700 m; payload volume 19 m^3 (15 m^3 in 2.4 m diameter cylindrical section).
The fairing can be installed up to 5 h before launch. Peak loads: 4.5/10 *g* stage 1/2 longitudinal, 1.5 *g* lateral

Vehicle/Payload Processing
Plesetsk vehicles are processed horizontally in a six-storey building, where stage 3 is fuelled, and not delivered to one of the two pads 40 km distant until about T-3 h by train. Processing takes 38 hours. Once at the pad, all operations are performed remotely. Propellant loading through the base requires about 1 hour from T-75 min. The complex can handle three launches monthly. A launch can be made 72 hours after request if Tsyklon is specially stored and the payload/s ready.

Zenit-2/Zenit-3

Current Status
No flights since the multiple satellite launch of 10 December 2001.

Background
When first proposed, Zenit was planned to be one member of a family of launch vehicles based upon the commonality of rocket stages. The 11K55 and 11K66 vehicles would have been used for small satellite launches, the 11K77 (Zenit-2 itself) for medium satellites and the 11K37 was offered as a variable-configuration assembly for the launching of heavy payloads within the range 3 to 6 tonnes.

The family of vehicles would all have used the same first stage as a building block. Presumably the 11K55 and 11K66 vehicles would have used second stages (different second stages for the different designators) which were smaller than the existing Zenit-2 second stage. The 11K37 would have used clusters of two, three or four Zenit first stages to give the proposed range of payload performances. Of these options only the 11K77 vehicle was actually built, and this was the first new Soviet launcher developed since the unsuccessful N-1 of 1969–1972. It provides a payload capability midway between those of Proton and Soyuz.

Zenit-2/3 and the Energia Zenit-1 strap-on are the responsibility of NPO Yuzhnoye in Dniepropetrovsk, and its chief designer Dr Yuri A Smetanin. In the cavernous integration hall at Yuzhnoyes, workers can assemble about 20 first stages simultaneously. Titanium forging and tank production is performed on-site. The Soviets only reluctantly released information about the programme because of its military pedigree even as the country tried to market the launcher commercially. The first brief glimpse came in Energia footage released in 1988, and a relatively detailed description appeared during the second 'Progress in Space Transportation' conference held in Bonn, Germany, May 1989 to support commercial marketing. Further information, particularly on the three-stage version, was provided at the Montreux commercial space conference in February 1990.

The Soviets claimed a manned capability (stage 1 is man-rated as Energia's strap-on), for the booster and stated a plan for carrying a two-man spaceplane called Uragan (Hurricane) to intercept US Shuttle missions out of Vandenberg. The project was cancelled in 1987 after west coast Shuttle launches were abandoned in the wake of *Challenger's* loss. Zenit was also originally intended to replace Soyuz manned launches.

A new cryogenic stage 3 was designed for Zenit 3 but discarded in favour of the existing Proton Block DM (which will be used by Sea Launch's Zenit 3SL version). It seems that Yuzhnoye has continued working on its own stage 3, although its status is unclear. The Zenit M3 would provide 900 kg GEO and 5.16 tonne GTO capacity with the 23.25 kN NTO/UDMH stage 3. The single chamber (SI

Zenit 2 roll out to one of the two Tyuratam pads
0517362

325 seconds, 1,600 seconds total burn), gimballed for pitch/yaw control, would be embedded in a single 2.72 m diameter spherical tank, with 10 × 29 N verniers for roll control, three-axis coast control and ullage. The Zenit M4 would add a 2.6 tonne NTO/UDMH apogee stage for 1.7-tonne GEO capacity from Baikonur. An air-launched version was described in 1994 for 1998 first test: released from an Antonov 225, it could handle 9-tonne LEO and 1-tonne GEO. Yuzhnoye began consideration of a Heavy Zenit in early 1989. Capable of delivering 25 tonnes to LEO, it was intended to supplant Proton by adopting the Proton approach of clustering Zenit stage 1's around a central tank. The Zenit's first stage appeared in parallel with Energia's strap-ons (designated Zenit-1), although Yuzhnoye's original proposal was based on the SS-18 ICBM.

The Ukrainian Space Agency said in September 1994 that an agreement had been reached with Russia to use Zenit for Space Station resupply (a 1996 Progress launch to Mir was expected to demonstrate the capability but it did not take place). At the same time, the Russian military indicated a requirement for two to three vehicles annually through 2000. Zenit's

Zenit-2 flight history

Launch	Date	Payload	Mass kg	Inclination degrees	Altitude km	Comments
1985	13 April	Kosmos	3,250 ?	Launch failure	Elint test: no details of failure	
	21 June	Kosmos	3,250 ?	64.41	197–340	Elint test: only debris reached orbit: vernier engine on second stage exploded 511 s after launch
	22 October	Kosmos 1697	3,250 ?	70.99	850–854	Elint
	28 December	Kosmos 1714	3,250 ?	70.99	163–853	Elint: payload shroud did not separate: second stage engine exploded preventing orbital circularisation
1986	30 July	Kosmos 1767	10,000 ?	64.88	196–207	Purpose unknown
	22 October	Kosmos 1786	500 ?	64.88	190–2,564	Atmospheric density sphere
1987	14 February	Kosmos 1820	10,000 ?	64.83	180–252	Purpose unknown
	18 March	Kosmos 1833	3,250 ?	71.01	849–852	Elint
	13 May	Kosmos 1844	3,250 ?	71.01	849–853	Elint
	1 August	Kosmos 1871	11,000 ?	97.03	179–199	First Zenit Sun-synchronous attempt
	28 August	Kosmos 1873	11,000 ?	64.84	177–255	Purpose unknown
1988	15 May	Kosmos 1943	3,250 ?	71.01	849–851	Elint
	23 November	Kosmos 1980	3,250 ?	71.00	849–854	Elint
1990	22 May	Kosmos 2082	3,250 ?	71.00	849–855	Elint
	4 October	Kosmos	3,250 ?	Launch failure	Elint: first stage exploded	
1991	30 August	Kosmos	3,250 ?	Launch failure	Elint: second stage failure	
1992	5 February	Kosmos	3,250 ?	Launch failure	Elint: second stage failure	
	17 November	Kosmos 2219	3,250 ?	71.01	849–855	Elint
	25 December	Kosmos 2227	3,250 ?	71.02	849–854	Elint: second stage exploded in orbit after satellite deployment
1993	26 March	Kosmos 2237	3,250 ?	71.02	849–853	Elint: second stage exploded in orbit after satellite deployment
	16 September	Kosmos 2263	3,250 ?	71.00	849–855	Elint
1994	23 April	Kosmos 2278	3,250 ?	71.01	849–855	Elint
	26 August	Kosmos 2290	10,600 ?	64.81	212–293	Photo-reconnaissance
	4 November	Resurs-01 1	1,900 ?	98.05	661–663	Remote sensing
	24 November	Kosmos 2297	3,250 ?	71.00	849–854	Elint
1995	31 October	Kosmos 2322	3,250 ?	71.02	849–852	Elint
1996	4 September	Kosmos 2333	3,250 ?	71.01	849–852	Elint
1997	20 May	Kosmos	3,250 ?	Launch failure	Elint; first stage exploded 48 s after launch	
1998	10 July	Resurs-02/01	2,800?	98.79	817–818	remote sensing
		Fasat-Bravo	50?	98.79	817–818	Chilean Air Force
		Tinsat	50?	98.79	816–818	Thai remote sensing
		Techsat 1B	50?	98.79	817–818	remote sensing
		Westpac	24?	98.79	817–818	Geodetic
		Safir 2	60	98.79	815–819	Relay
	28 July	Kosmos 2360	3,250?	71.02	848–854	Elint
	9 September	Globalstar×12	12×450	Launch failure	ComSatellite failure	
1999	17 July	Okean 5	6,300	98.06	661–663	Oceanographic
2000	3 February	Cosmos 2369	6,000	71.0	849–860	Elint
	25 September	Cosmos 2372	12,000	67.8	343-211	Surveillance
2001	10 December	Meteor 3 m	2,500	99.7	1,015-996	Weather
		Badr B	70	99.7	1,015-996	Surveillance
		Kompas	80	99.7	1015-996	Seismology
		Maroc Tubsat	45	99.7	1,015-996	Technology
		Reflector	8	99.7	1,015-996	Reflector

The 1985 launch vehicle failures are not acknowledged as such in Russian/Ukrainian marketing literature, although when the May 1997 launch failure was reported, the ITAR-TASS announcement confirmed that this had been the seventh failure of the launch vehicle in 28 flights. In turn this confirmed that the 1985 missions were indeed failures.

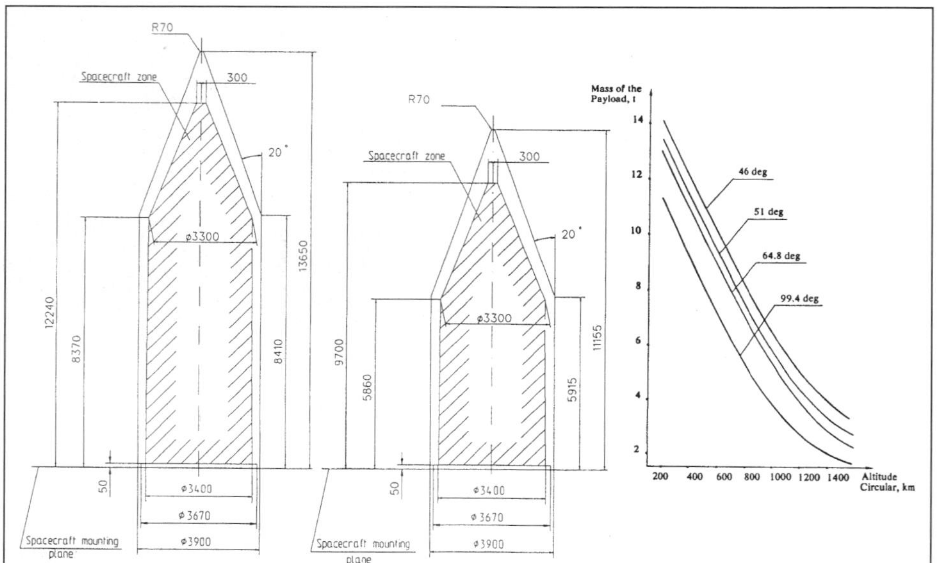

Zenit 2 fairings and performance 0517363

first commercial success came with its May 1995 selection by Space Systems/Loral for three launches to carry 36 Globalstar satellites. Zenit's survival thus appears assured.

The Soviets used the vehicle in the early programme to launch Tselina-2 heavy electronic intelligence satellites.

The Ukrainian literature issued in support of marketing the Zenit-2 has claimed that all four 1985 launches succeeded, although in reality three of the four were failures. At the time Western observers thought that the launches in April and June 1985 were simply sub-orbital test flights, and it was decided by the Ukrainians to continue this story. In reality both had been orbital attempts with Tselina-2 test satellites, confirmation coming from Russian sources in 1995. Additionally, Kosmos 1714 in December 1985 was stranded in an unplanned eccentric orbit following the failure of the payload shroud to separate and the explosion of a second stage engine.

By implication, these launches were finally confirmed as having been failures by ITAR-TASS when announcing the launch failure on 20 May 1997. The news agency stated that this had been the seventh launch vehicle failure in the Zenit-2's history – a figure which can only be reconciled if the three 1985 launches were also failures.

Specifications

Zenit 2

First launch: 13 April 1985
Number launched: 36 to end of June 2004

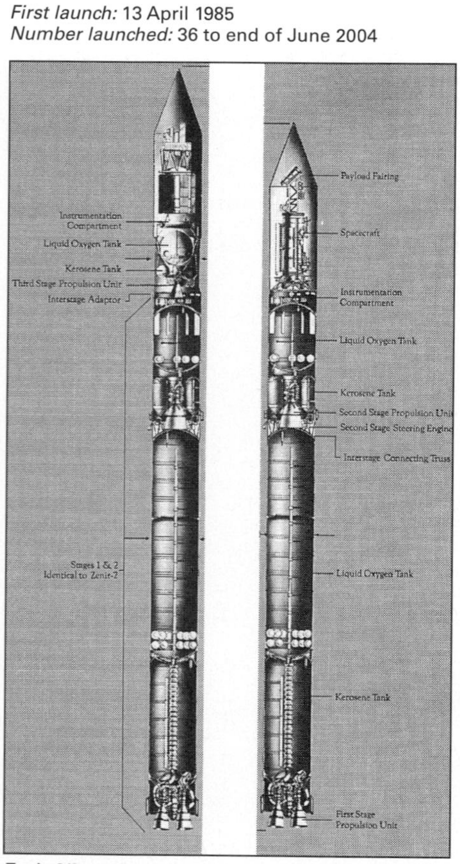

Zenit 2/3-stage versions. Zenit 2 is depicted with an Okean ocean radar satellite 0516879

Launch sites: Tyuratam (two complex 45 pads, capable of supporting 15/year; pad 45R severely damaged October 1990 and still not in service). Plesetsk complex 35 was operational about 1998.
Contractors: NPO Yuzhnoye (prime), NPO AP (GN&C), NPO Elektropribor (avionics), PO Komunar (inertial guidance package), NPO Energomash (main engines).
Principal uses: Tselina 2 electronic intelligence, Resurs-O, Kuban; potential: commercial, Advanced Okean, super-Progress, manned to LEO.
Cost: Rb23 billion quoted October 1995
Vehicle success rate: 73% to end of 2000
Performance: Zenit 2 13,740 kg into 200 km 51.6°, 11,380 kg into 200 km 99°, both from Tyuratam; 15.7 t into 200 km 12° for equatorial site. Injection accuracy: ±3.5 km altitude, ±2 arcmin inclination, ±2.5 s period. Zenit 3 capabilities from Tyuratam: 5,370 kg GTO, 1,600 kg GEO
Number of stages: 2 (3 for Zenit 3)
Overall length: 57.00 m (61.40 m) including 13.65 m payload fairing; 11.155 m fairing available
Principal diameter: 3.90 m
Launch mass: 449 t (466 t) without payload
Launch thrust: 7,259 kN (7,622 kN) sea level
Guidance: inertial. 1 Mbit/s telemetry provides information on 1,000 parameters during launch. Receives navigation update just before launch via laser optical receiver package mounted on side; package is then blown free.
Comment: Zenit 2 stage 1 Stage 1 and SL-17 Energia's strap-ons were developed in parallel. Both stages are aluminium with integrally machined stiffeners.
Engines: RD-171 from NPO Energomash, four gimballed chambers. RD-172 will be used on Zenit 3 5% thrust increase).
Overall length: 32.94 m
Diameter: 3.90 m
Thrust: 7,911 kN vacuum, 7,259 kN sea level
Burn time: about 144 s (146 s Resurs-O1)
Dry mass: 28,080 kg
Propellants: 318,800 kg of liquid oxygen (210.55 m^3 forward tank) and kerosene (107.31 m^3 aft)
Separation: four retro solids at aft end Zenit 2 stage 2
Engines: NPO Energomash RD-120 single fixed reignitable chamber with four NPO Yuzhnoye verniers
Overall length: 11.50 m
Diameter: 3.90 m
Thrust: 833.5 kN main + total 78.4 kN verniers, vacuum
Burn time: about 300 s total, verniers can burn for up to 1,100 s (Resurs-O1: 285 s main, 499 s verniers)
Dry mass: 8,300 kg
Propellants: 80,600 kg of liquid oxygen (53.43 m^3 forward tank) and kerosene (27.87 m^3 aft torus)
Separation: four retro solids at aft end

Vehicle/payload processing

Zenit is assembled horizontally, with the payload integrated on stage 2 before stage 1/2 mating. Assembly of the vehicle alone requires 80 h, increasing to 116 h with the payload. The vehicle can be stored for a year. Transfer to pad (45L/45R) is by rail; erection/launch processing is highly automated, requiring 21–80 h between initial integration and launch. Umbilicals retract 12 min before launch, returning to their protected positions. At 4 min, the train returns to the integration building. 4 m^3/s of water is sprayed into the flame pit beginning at 15 s to reduce acoustic loads and temperature. The pad is automatically reconfigured within 5 h for the next launch. Zenit stages 1/2 separate at 150 s (Resurs-O1: 145 s) and the payload shroud during stage 2 burn between 160–1,240 s (Resurs-O1: 160 s). Payload volume is 90 m^3 for the 13.65 m long shroud, reduced to 43–50 m^3 for three-stage missions. An 11.155 m shroud is also available. Payload separation is achieved at 450–1,250 s, depending on the mission (Resurs-O1: 640 s). Acoustic load 140 dB.

UPDATED

Zenit-3SL

Background

Boeing initiated the Sea Launch joint venture in 1994 for three-stage Zenits to be shipped to the US, assembled and then transported into international waters for near-Equator launches. The consortium comprises Boeing Commercial Space Co, Norway's Kvaerner group, NPO Yuzhnoye and RKK Energia. Hughes Space & Communications Co, placed a firm order December 1995 for 10 launches beginning in 1998. With options, the contract could be worth around US$1 billion. In early 1997, there were 13 firm orders from Hughes and five from Loval, with Hughes holding additional options. Kvaerner Rosenberg is converting a 28,000 tonne oil drilling rig into the ocean-going launch platform in Stavanger under a US$78 million contract; Kvaerner Govan in Scotland is building the 30,000 tonne 200 m support/control ACS Assembly & Command Ship for US$93 million. The first Zenit-3 was successfully launched on 28 March 1999 carrying DEMOSAT, a dummy manufactured by Boeing to simulate the mass of a Hughes HS-702 satellite. The second launch successfully placed the Direc TV 1R HS 601 satellite in orbit on 9 October 1999. The third flight, on 12 March 2000, carrying the HS-601 ICOF-1 satellite was a failure. The fourth flight, on 28 July 2000, carrying PAS-9 and the fifth flight, on 21 October 2000, carrying Thuraya-1 were successful. The sixth flight, launched 18 March 2001, carried the XM-2 Rock satellite into orbit followed by the seventh flight on 8 May 2001 with XM-1 Roll satellite. The eighth flight on 15 June 2002 placed Galaxy 3C in orbit and was followed by the ninth flight on 10 June 2003 (Thuraya-2), the tenth flight on 7 August 2003 (EchoStar IX/Telstar 13) and the eleventh launch on 30 September 2003 (Galaxy XIII/Horizons-1). The twelfth launch on 11 January 2004 took Skynet Brazil 1 (Telstar 14) to orbit, the thirteenth launch on 4 May 2004 carried DirecTV7S to orbit and the fourteenth launch on 29 June 2004 carried Apstar 5.

Specifications

Zenit 3SL

First launch: 27 March 1999
Number launched: 14
Launch sites: typically 0°North/152°West, South of Hawaii.
Contractors: NPO Yuzhnoye (stage 1/2), RKK Energia (Block DM, vehicle assembly, range services), Kvaerner (vessels, sea-going operations), Boeing Commercial Space Co (project manager,

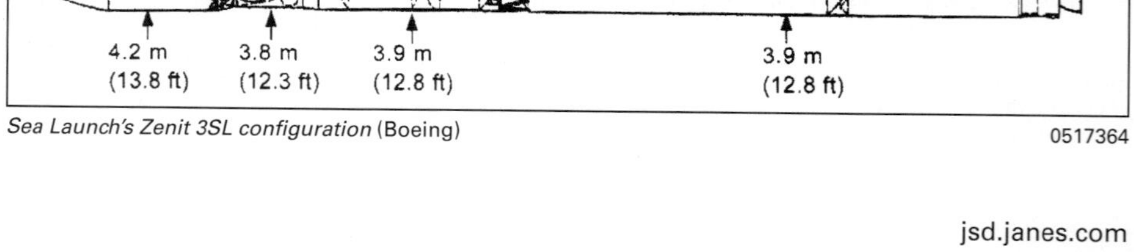

Sea Launch's Zenit 3SL configuration (Boeing) 0517364

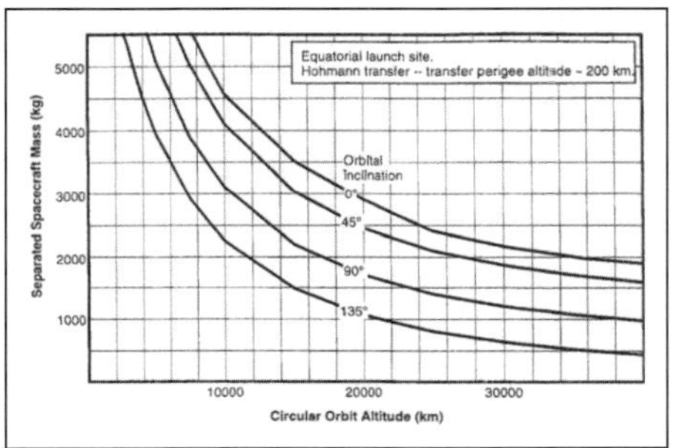

Zenit 3SL capability to circular orbit (Boeing) 0517367

mission operations, spacecraft processing, Home Port, fairing, adaptor).
Principal uses: commercial GEO launches
Cost: US$90–100 million
Availability: 18 months after contract signature (12 months for repeat satellites).
Vehicle success rate: 100% (of 1)
Performance: 5.25 t GTO, 2.0 t GEO. GTO accuracy: 5/40 perigee/apogee altitude, 0.20° inclination.
Number of stages: 3
Overall length: 59.8 m with 11.390 m fairing
Principal diameter: 3.90 m
Launch mass: 470 t with 5t payload
Launch thrust: 7,370 kN sea level
Guidance: as Zenit 2
Launch sequence (typical GTO mission, min:s):

00:00	stage 1 ignite/lift-off
00:08	begin pitch over
01:04	max Q (5,300 kgf/m^2)
01:48	max acceleration (3.8 *g*)
01:49	begin throttle to 75%
02:09	begin throttle to 50%
02:21	stage 2 vernier ignite
02:23	stage 1 shutdown
02:26	stage 1 separation
02:31	stage 2 main engine ignite
03:37	fairing jettison (117 km, <1,135 W/m^2)
07:09	stage 2 begin throttle to 85%
07:29	stage 2 main engine shutdown
08:44	stage 2 vernier shutdown/separation
08:49	stage 3 casing jettison
08:54	stage 3 ignition 1
12:46	stage 3 shutdown 1 (LEO coast)
42:46	stage 3 ignition 2
49:02	stage 3 shutdown 2/GTO insertion
49:17	payload separation

Zenit 3SL stage 1 As Zenit 2
Zenit 3SL stage 2 As Zenit 2
Zenit 3SL stage 3
Engine: 11D58M restartable single chamber developed by Korolev bureau.
Overall length: 6.2 m
Diameter: 3.70 m
Thrust: 85 kN vacuum
Burn time: about 600 s total over two burns
Dry mass: 2.4 t + 800 kg casing + 200 kg residual propellants
Propellants: 14.6 t liquid oxygen in forward sphere and kerosene in aft torus

Vehicle/payload processing
The vehicle stages and payload are delivered to Sea Launch's Home Port in Long Beach, California. The stages are loaded into the ACS Assembly & Command Ship via the aft ramp for assembling on the main deck (nine days). ACS can handle three vehicles in parallel. Block DM is handled in the bow, where its kerosene is loaded (LOX is loaded on-pad). Meanwhile, the payload is processed and encapsulated in Boeing's two-piece composite fairing in land facilities. The bolted or Marmon clamp adaptors are 937, 1194 and 1666 types. The 3,000 m^2 Payload Processing Facility includes two 15 × 20 m class 100,000 cells for fuelling and adding ordnance. The encapsulated unit is loaded on to the ACS for integration with Zenit (three days). The completed vehicle is then loaded into the Launch Platform's horizontal hangar (two days), where it remains for the 11-day journey to the launch site. There, the LP semi-submerges 22.5 m and Zenit is erected for launch. Fuelling begins at T-3 hours. Launch operations are controlled from the Launch Control Center aboard ACS, 5 km distant. Water deluge into the flame bucket protects the LP and reduces acoustic effects. ACS provides tracking and telemetry for 410 s, then the Selena-M Russian tracking ship until LEO parking and then Russian Altair satellites (Russian ground stations also available). Launch can occur with waves up to 2.5 m high.
Acoustic loads: 142 dB overall
Maximum internal fairing T: 65°C

UPDATED

Spain

Capricornio

Current Status
Development ceased due to lack of commercial interest.

Background
INTA's began its own programme to develop an orbital launcher by building upon its own small INTA 100 and 300 sounding rockets. Development cost US$32 million. The agency calls the result of its evolutionary programme the Capricornio's. Its first and second stages were expected to fly as the 9.2 m Argos sounding rocket in 2002. A lack of support brought cancellation of the project.

Specifications
Capricornio
First launch: originally planned 1998
Launch sites: Isla de El Hierro, Canary Islands
Principal uses: 60–140 kg payloads into LEO up to 600 km
Cost: unavailable
Number of stages: 3 (solid)
Overall length: 18.25 m
Principal diameter: 1.0 m
Launch mass: 15,035 kg
Guidance: inertial above stage 3. Accuracy ±50 km altitude, ±1° inclination.

Capricornio stage 1
Motor: Thiokol Castor 4B
Overall length: 8.990 m
Principal diameter: 1.019 m
Stage mass: 11,674 kg (9,974 kg propellant)
Average thrust: 429 kN average (26.142 MNs total impulse)
Burn time: 61 s
Attitude control: ±6° provides pitch/yaw TVC; roll by stage 3 ACS.

Capricornio stage 2
Designation: Deneb-F
Overall length: 3.000 m
Principal diameter: 0.830 m
Stage mass: 2,549 kg (motor 2,436 kg; 2,124 kg propellant).

Capricornio's launch pad will be built on the Isla de El Hierro (Theo Pirard/Space Information Centre) 0517358

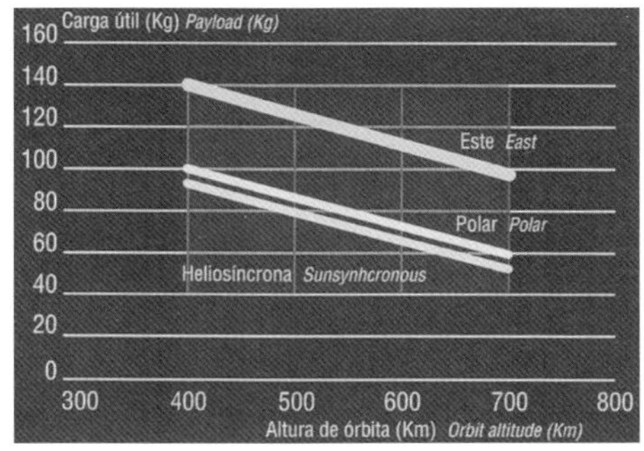

Capricornio performance from the Canary Islands (27° north) (INTA) 0517359

Average thrust: 167.9 kN average (5.976 MNs total impulse)
Burn time: 35.6 s
Attitude control: nozzle pitch/yaw TVC; roll by stage 3 ACS?

Capricornio stage 3
Designation: Mizar-B
Overall length: 2.104 m
Principal diameter: 0.830 m
Stage mass: 717 kg (motor 686 kg, propellant 607 kg)
Thrust: 50.29 kN average (1.700 MNs total impulse)
Burn time: 33.8 s
Attitude control: unknown.

Heatshield/payload accommodation
0.55 m^3 in a 25 kg 1.722 m-long 83 cm diameter fairing, able to accommodate two satellites.

United Kingdom

HOTOL/SKYLON

Current Status
Interim HOTOL project initiated by Dr Bob Parkinson in 1991, essentially a scaled back version of HOTOL utilising Russian propulsion systems in the form of four rocket motors (RD-0120) and with a length of 36.5 m, span of 21.6 m and fuselage diameter of 10.0 m. Gross mass 250,000 kg, empty mass of 33,100 kg. Informal studies and re-design activities continues in the absence of government or private/corporate support.

Background
The unmanned Hotol Horizontal Take-Off and Landing concept emerged from British Aerospace in September 1984, with the goal of reducing launch costs of LEO payloads by 80 per cent (700 to 1,000 ECU/kg) of current Shuttle levels. BAe and the UK government put up £3 million for a two-year proof-of-concept study. It concluded in September 1987, and the UK government declined to fund the submitted three year feasibility study proposal, believing Hotol development cost would be about £4.5 billion. Rolls-Royce exercised its option in April 1988 to buy designer Alan Bond's patent on the RB-545 dual-mode engine but halted work after government funding ran out. Some basic information was released during 1988 and the patent was declassified during 1991, allowing a detailed description to be published in 1993. Bond's Sabre engine design has not been patented in order to prevent its classification. It differs by driving the air compressors from a separate power loop rather than using the hydrogen stream directly.

The Hotol SSTO fully reusable concept relied upon a dual mode engine, using atmospheric oxygen and onboard LH$_2$ for take-off, and onboard LOX/LH$_2$ as it left the atmosphere. The air is cooled by two heat exchangers to just above its condensation point using LH$_2$ and then compressed to 150 atmospheres (reaching 750 K) to feed into the rocket chamber,

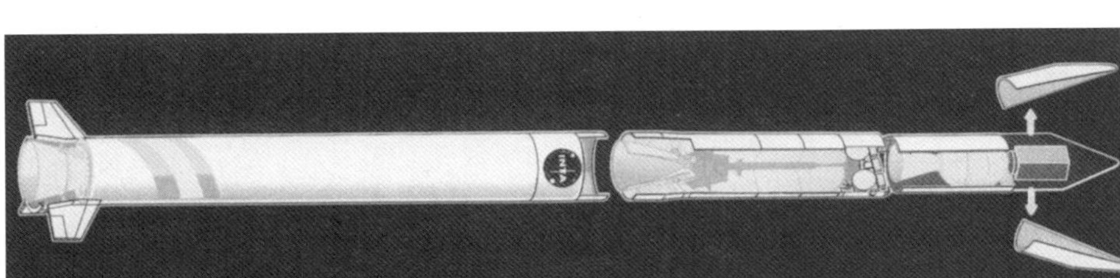

Capricornio will employ one US and two Spanish solid motors (INTA) 0517357

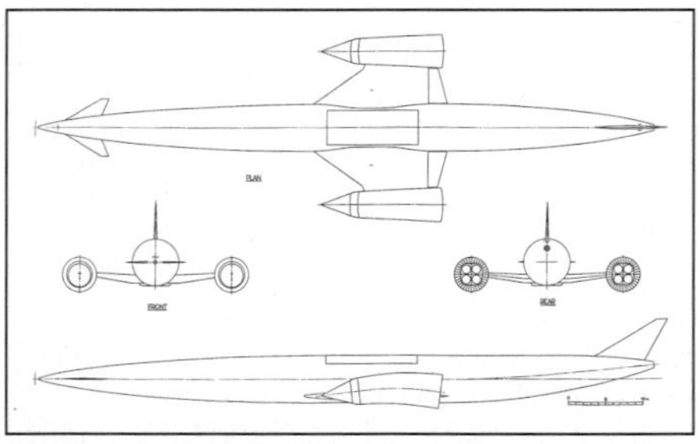

Skylon configuration C1 (Reaction Engines) 0517322

where it burns at 100 atmospheres and 2,700 K with hydrogen to generate 340 kN thrust. Air-breathing continues up to M5.5/26 km, for about 10 minutes, when the engine would switch to pure rocket mode using onboard oxygen, providing 735 kN thrust from 3,400 K and 150 atmosphere chamber conditions (SI 459 s). The engine also carries a small ramjet to utilise the excess air gathered at low speeds (M0.5 to 4.5). The 250 tonne vehicle could deliver 7 tonnes into LEO. BAe studied an Interim Hotol with CIS partners, powered by four RD-0120 engines and released from the back of an An-225. The full Hotol and interim Hotol entries are on pp246–247 of 1994–95's edition.

Building on Bond's design, Reaction Engines is designing the Skylon SSTO spaceplane to deliver up to 12 tonnes into LEO. Like Hotol, Skylon would be unpiloted (although eventually capable of carrying passengers) and it would take off and land horizontally using conventional runways on its own undercarriage. The airframe is designed for 200 flights. Aircraft-type refuelling and checkout procedures would minimise turnaround time. Two twin-chambered SABRE (Synergetic Air-Breathing and Rocket Engine) engines are mounted on each wingtip in a circular nacelle. These dual-mode engines would employ an air-breathing hydrogen-fuelled mode from a standing start before switching to the closed cycle LOX/LH_2 rocket mode at about M5/26 km. Skylon is designed for maximum abort capability, flyable with one nacelle shut down. Projected launch cost, including amortised purchase cost, insurance, fuel, maintenance, range activity and spaceport handling charges, is estimated at about US$10 million.

Reaction Engines is conducting supportive investigations in conjunction with Bristol University and AEA Technology Harwell to investigate some of the more technically contentious features. Project funding is sought; Reaction Engines has brought together an interim UK working group before the formation of an industrial consortium, which will eventually be expanded to include international partners.

In rocket mode, each Sabre engine driving the vehicle provides 729.2 kN thrust at 459 seconds SI and 145 bar chamber pressure. The air entry temperature into the engine in air-breathing mode is up to 1,000°C at high Mach numbers, so it is cooled before compression by passing it through a closed-cycle helium precooler and using the cryogenic hydrogen fuel flow as the heat sink. After compression, the airflow is divided between the main chamber and the fuel-rich preburner. The preburner's hot gas, en route to the main chamber, passes through a heat exchanger where it raises the helium's temperature to drive the compressor. The helium is then cooled by the hydrogen heat exchanger: the warmed hydrogen drives the hydrogen pump and the helium's circulator; the cold helium passes to the precooler to cool the incoming air, completing the cycle. Owing to the thermodynamics, the fuel flow for this type of engine is greater than stoichiometric but the installed SI is still three to six times better than a pure rocket engine. The airflow is drawn into the engine via an axi-symmetric intake which, at supersonic speeds, contains one oblique and one normal shock wave. To minimise spill drag, the centrebody translates to maintain the shock on the lip and finally closes altogether to seal the nacelle for re-entry. The intake operates at maximum capture during air-breathing, but as the core engine operates at an almost constant air mass flow the excess flow at moderate Mach numbers is diverted down a spill duct where it is heated by combustion with part of the hydrogen flow and expelled through a separate nozzle.

The fuselage and wing load-bearing structure is made from CFRP; the 2014-T6 aluminium propellant tank is suspended within, free to move under thermal and pressurisation displacements. The aeroshell is made from hot-pressed SiC fibre reinforced glass-ceramic and carries only aerodynamic pressure loads, transmitted to the fuselage structure through flexible suspension points. The 0.5 mm aeroshell is corrugated for stiffness. It is free to move under thermal expansion during the latter stages of the aerodynamic ascent and re-entry. It reaches a maximum 855 K during ascent, and the trajectory is controlled during re-entry via active feedback of measured skin temperatures to keep it down to 1,100 K.

On-orbit the main tanks are vented and allowed to warm to ambient. AOCS is provided by the LOX/LH_2 OMS/RCS system, which can remain operational for seven days. The RCS employs gaseous propellants, which are also supplied to the fuel cells and APU turbines.

Specifications

Skylon

Fuselage length: 83 m
Wing span: 25 m
Mass: 275 t gross take-off, 220 t propellant (ascent: 150.2 t LOX + 66.8 t LH_2; 2.36 t OMS/RCS), 55 t maximum landing mass
Performance: 12 t into 300 km 0°; 10.5 t 460 km 0°; 9.5 t 460 km 28.5° from equatorial site
Payload bay: 4.6 m diameter, 12.319 m long (compatible with ISO standard containers).

Skylark sounding rocket

Background

The Royal Aircraft Establishment originally developed Skylark for the International Geophysical Year of 1957–58, and consequently it is therefore the longest-running current space programme in the World. It proved so successful that, following the first launch on 13 February 1957, its launch total for all models was 437 by the end of 2001. The first parachute recovery occurred in 1961. In 1964, the 100 total was passed. The 200 mark was achieved five years later, followed by 300 in 1972 as the launch rate began to moderate. Skylark 7 was introduced in late 1974 with launch 335 and Skylark 12 in mid-1976.

MMS, the prime industrial contractor, offers Skylark launchers in three versions (5/7/12 variants) although other motor combinations are available on request, such as the four Skylark 2s flown 1988–89. The launch rate is maintained primarily by Germany's Texus programme and Sweden's Maser with the Skylark 7 from Kiruna, attaining apogees of about 250 km with 370 kg payloads to create μgravity conditions for 6 minutes.

The Skylark 7 consists of a Raven XI solid boosted by a Goldfinch IIA; the 12 model adds a Cuckoo IV upper stage for lofting a 100 kg payload above 1,000 km.

A cast interstage adaptor is bolted to the front end of the Goldfinch steel casing and interfaces with the Raven motor spacer bay by dowel pins to permit clean drag separation. This adaptor houses a spin-up system of six Imp-18 motors armed by a lanyard at the 'instant of move' and triggered by microswitches leaving the launcher rail. A transient roll rate of about 250°/s is achieved in 0.5 seconds.

Skylark 5 is the smallest in the range, comprising a single Raven XI capable of delivering a 250 kg payload to 200 km. A Skylark 17 model was proposed unsuccessfully in 1988 for ESA's long duration sounding rocket programme, with the

Skylark launches from 1981

No	Type	Date	Site	Payload/Customer
383	7	April 1981	K	Texus 3B/DFVLR
384	7	May 1981	K	Texus 4/DFVLR
385	7	December 1981	K	Energy/DFVLR
386	7	April 1982	K	Texus 5/DFVLR
387	7	May 1982	K	Texus 6/DFVLR
388	12	November 1982	A	Caesar/DFVLR
389	7	March 1983	K	Texus 7/DFVLR
390	7	April 1983	K	Texus 8/DFVLR
391	6	November 1983	K	Mapwine/DFVLR
392	7	March 1984	K	Texus 9/DFVLR
393	7	April 1984	K	Texus 10/DFVLR
394	12	March 1985	N	Interzodiac 1/DFVLR
395	7	April 1985	K	Texus 11/DFVLR
396	7	April 1985	K	Texus 12/DFVLR
397	7	April 1986	K	Texus 13/DFVLR
398	7	April 1986	K	Texus 14/DFVLR
399	7	April 1987	K	Texus 14B/DFVLR
400	7*	April 1987	K	Texus 15/DFVLR
401	7	August 1987	W	Super Nova/DFVLR
402	7*	November 1987	K	Texus 16/DFVLR
403	7	May 1988	K	Test flight/BAe
404	12	September 1988	N	Interzodiac 2/DFVLR
405	7	November 1988	K	Texus 19/MBB-ERNO
406	2	November 1988	A	Rose 1/DFVLR
407	7	November 1988	K	Texus 20/MBB-ERNO
408	2	December 1988	A	Rose 2/DFVLR
409	2	January 1989	K	Rose 3/DFVLR
410	2	February 1989	K	Rose 4/DFVLR
411	7	April 1989	K	Texus 21/MBB-ERNO
412	7	April 1989	K	Texus 22/MBB-ERNO
413	7	November 1989	K	Texus 23/MBB-ERNO
414	7	December 1989	K	Texus 24/MBB-ERNO
415	6	March 1990	K	SISSI 1/DLR
416	7	May 1990	K	Texus 25/MBB-ERNO
417	7	May 1990	K	Texus 26/MBB-ERNO
418	6	July 1990	K	SISSI 2/DLR
419	7	August 1990	K	SISSI 3/DLR
420	7	November 1990	K	Texus 27/MBB-ERNO
421	12	November 1990	K	Maxus test/SSC & MBB
422	7	March 1991	K	SISSI 4/DLR
423	7	November 1991	K	Texus 28/MBB-ERNO
424	7	November 1992	K	Texus 29/DASA-ERNO
425	7	May 1993	K	Texus 30/DASA-ERNO
426	7	November 1993	K	Maser 6/SSC
427	7	November 1993	K	Texus 31/DASA-ERNO
428	7	May 1994	K	Texus 32/DASA-ERNO
429	7	November 1994	K	Texus 33/DASA-ERNO
430	7	March 1996	K	Texus 34/DASA-ERNO
431	7	May 1996	K	Maser 7/SSC
432	7	November 1996	K	Texus 35/DASA-ERNO
433	7	February 1998	K	Texus 36/DASA
434	7	May 1999	K	Maser 7/SSC
435	7	March 2000	K	Texus 37/DASA
436	7	April 2000	K	Texus 38/DASA
437	7	May 2001	K	Texus 39/DASA

* indicates vehicle failure. Texus 14 (April 86) was rated as a partial success because the payload failed to achieve μg conditions. Texus 7 (March 83) suffered a nozzle burn-through and was rated as partially successful. Texus 17/18 flew aboard Black Brant. Skylark 2 comprised Raven VIII plus Cuckoo I for the four Rose (Rocket Scatterometer Experiment) firings of 1988/89. SISSI = Spectroscopic Infra-red Structure Signatures Investigation. Skylark 6 was essentially identical to Skylark 7. Key: A=Andoya, K=Kiruna, N=Natal, W=Woomera.

The Skylark 12 is the most powerful vehicle in the range, capable of delivering 100 kg to 1,000 km 0516920

Stonechat-Mage 2 combination designed to provide 15 minutes of μgravity conditions for a 350 kg payload.

Skylark sounding rocket propulsion has found application in other derivative sounding rockets manufactured in the USA.

UK	266
ESRO	82
Germany	72
NASA	4
Sweden	6
UK/Argentina	2

Skylark launch sites are located in Argentina, Australia, Brazil, Italy, Norway, Spain, Sweden and the UK.

Specifications

Skylark 7
First launch: November 1974
Number launched: 57 to end-1995
Success rate: 96.5% to end-1995
Success rate for last 25 launches: 92% to end-1995
Number of stages: 2 (solid)
Overall length: 15.3 m + payload bay/nosecone as required (≤5.5 m).
Principal diameter: 43.8 cm (1.96 m fin circle diameter)
Launch mass: 1,700 kg plus payload/nosecone
Typical performance: 100 kg to 460 km; 300 kg to 270 km; 400 kg to 210 km (250 kg to 210 km for Skylark 5) for standard launches at 5° from vertical.
Payload recovery system: a standard 36.1 cm-long Skylark recovery parachute bay is fitted at the rear of the payload. A pair of barometric switches initiates a pyrotechnic sequence at about 4 km altitude to deploy a stabilising drogue, followed 5 s later by the 7 m diameter main canopy for a 13 m/s final descent rate. Internally mounted equipment is generally recovered undamaged and can be reflown after refurbishment.

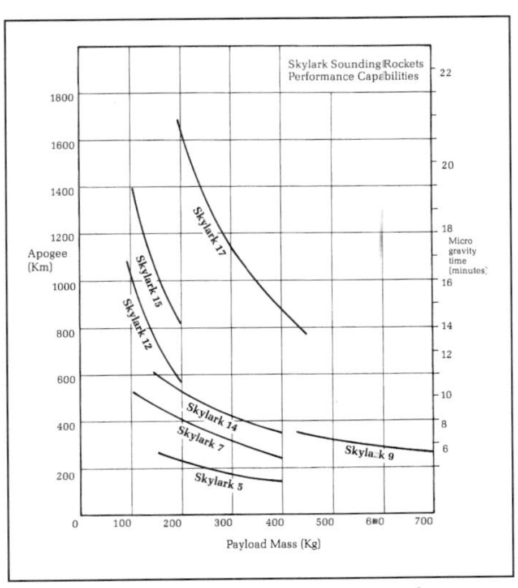

Skylark family apogee and microgravity performances, for 88° launch angle. Skylark 5, 7 and 12 are available, the rest remain as studies (Matra Marconi Space) 0516921

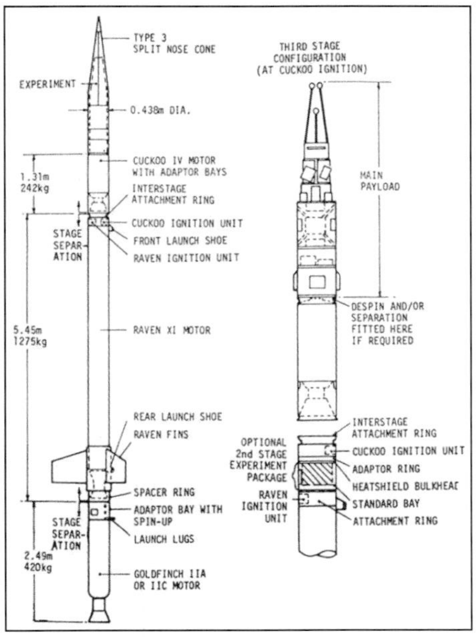

Typical Skylark 12 configuration (MMS) 0516922

Skylark 7 stage 1
Motor: Goldfinch IIA (Royal Ordnance)
Overall length: 249 cm
Principal diameter: 43.8 cm
Total mass: 420 kg (propellant 305±5 kg)
Peak thrust, sea level: 191 kN (total impulse 701 kNs)
Burn time: 3.7 s

Skylark 7 stage 2 (stage 1 for Skylark 5)
Motor: Raven XI (Royal Ordnance)
Overall length: 545 cm
Principal diameter: 43.8 cm
Total mass: 1,280 kg (propellant 990 kg)
Peak thrust, sea level: 105 kN (total impulse 2,502 kNs)
Burn time: 39 s
Comment: A three-blade fin unit is attached to the bottom end with the fin cant angle of 0.40° providing roll control. For single rail launcher systems, the rear launch shoe is positioned on the fin barrel, with the second on the motor head end ring. A 23 cm-long spacer bay is bolted to the barrel's rear face to protect the nozzle during Goldfinch separation. The self-contained Raven Ignition Unit includes time delay switches, inertia switches, redundant batteries and arming sockets. An acceleration-triggered time-delay unit at 6 or 12 s initiates the igniter sequence after launch (the latter for lower payload masses to avoid excessive heating). A yo-yo despin system is carried above this unit to cancel the fin-induced roll for attitude stabilisation to begin. A manacle ring and three separation plungers are fired by a gas pressure source at a programmed time to release a 300 kg payload at 2 m/s.

Payload environment/accommodation
Peak longitudinal acceleration: 10.5 *g* for 300 kg, 13.5 *g* for 100 kg
Shock: 20 *g* for 15 ms
Vibration: the complete payload assembly is tested at 5.2 *g* rms over 20–2,000 Hz for 2 min (longitudinal), and 5.6 *g* for lateral.
Temperature: 157°C max skin for parallel bay; 253°C for nosecone
Comment: Standard magnesium alloy body sections are available in 8.3, 20.3, 30.5, 40.6, 50.8 cm lengths clamped together by quick release manacle rings. Hatches/access holes can be cut in the skin where required providing adequate reinforcement is added. Located at each body section junction is either a sealing ring or a bulkhead for mounting experiments. The whole or part of the payload can be sealed at ground pressure by fitting rubber O-rings in the body junctions; venting is also possible. Generally, the experiments are housed in the forward end, with the user-provided telemetry and housekeeping instruments, tracking aids and attitude determination systems occupying the remainder of the volume. A range of stainless steel/glass fibre nosecones is available: split versions released pneumatically are employed when the payload protrudes into the nose space, forward-ejecting (3 m/s) for non-intrusive payloads, and fixed for payloads with no exposure requirements (such as μgravity payloads). Glass fibre cones are available for experiments with RF or magnetometer elements.

Skylark 12
Skylark 12 is basically a Skylark 7 with a Cuckoo IV third stage motor, almost doubling the peak apogee performance. It is designed for a third-stage payload range of 100–200 kg, which includes everything forward of the Cuckoo motor adaptor ring. The first two operational launches occurred from Andøya in November and December 1976. Since then, its primary employment has been for science payloads requiring high altitudes.
First launch: November 1976, from Andøya, Norway.
Number launched: 15 to end-1995
Vehicle success rate: 100% to end-1995
Number of stages: 3 (solid)
Overall length: 9.25 m + payload/nosecone
Principal diameter: 43.8 cm (1.96 m diameter fin circle).
Launch mass: 1,937 kg + payload/nosecone
Performance: 100 kg to 1,030 km, 200 kg to 575 km.
Payload recovery system: as Skylark 7
Typical flight sequence (time in s):

0	Goldfinch ignition
0.5	Spin-up
6	Raven ignition
45	Raven all-burnt
60	ejection of nose-cone and experiment hatches, under control of payload timer
70	stage 3 separation, ignition at 70.4 s
85.4	Cuckoo all-burnt
96	earliest time for payload separation, boom deployment and so on

Skylark 12 stage 1
as Skylark 7

Skylark 12 stage 2
as Skylark 7, but with fins canted 0.5° for roll and the availability of a small experiment bay atop the Raven (see payload accommodation below).

Skylark 12 stage 3
Motor: Cuckoo IV (Royal Ordnance)
Overall length: 131 cm
Principal diameter: 43.8 cm
Mass: 242 kg (186 kg propellant)
Peak thrust, vacuum: 44 kN (total impulse 452 kNs)
Burn time: 15 s at 20°C
Comment: Magnesium alloy adaptor bays are bolted to each end of the motor and provide the standard manacle feature; the short 78 mm forward ring can accommodate a de-spin or payload separation system described in the Skylark 7 entry. At 70 s, the Cuckoo ignition unit at Raven's forward end actuates two pistons that push the third stage away at 0.2 m/s; 87 mm separation triggers the firing pulse from the Raven along a flying lead.

Payload environment/accommodation
Peak longitudinal acceleration: 40 *g* for 100 kg, 30 *g* for 150 kg
Peak lateral acceleration: 1.5 *g*
Shock: 20 *g* for 15 ms
Vibration: the complete payload assembly is tested for 2 min at 5.2 *g* rms over 20–2,000 Hz longitudinal, 5.6 *g* for lateral.
Spin: 230 rpm final
Temperature: 283°C on nosecone, 182°C on side wall for 150 kg payload.
Comment: A 30 cm-long, 40 kg experiment bay can be carried at the Raven's forward end, with a solid bulkhead as a heatshield against the Cuckoo efflux. The magnesium body sections for the main section are described in the Skylark 7 entry. The recommended nosecone for Skylark 12 is the stainless steel type 3 split version since it has to be ejected without inducing significant pitch rate before Cuckoo burn. Both halves incorporate a sealed pneumatic system with the gas provided by a Conax cylinder fitted with a pyrotechnic valve; the half cones are held by a tip latch and an aft manacle body ring.

United States

ATHENA (LMLV)

Current Status
Athena was added to the NASA Launch Services contract calling for a capability of launching 1,500 kg or more to a 200 km circular, due east, orbit.

Background
LMSC began investigating the possibilities of converting excess missiles in 1987 (it was responsible for the Polaris, Poseidon and Trident SLBMs)

LMLV 1 debut, August 1995 0517377

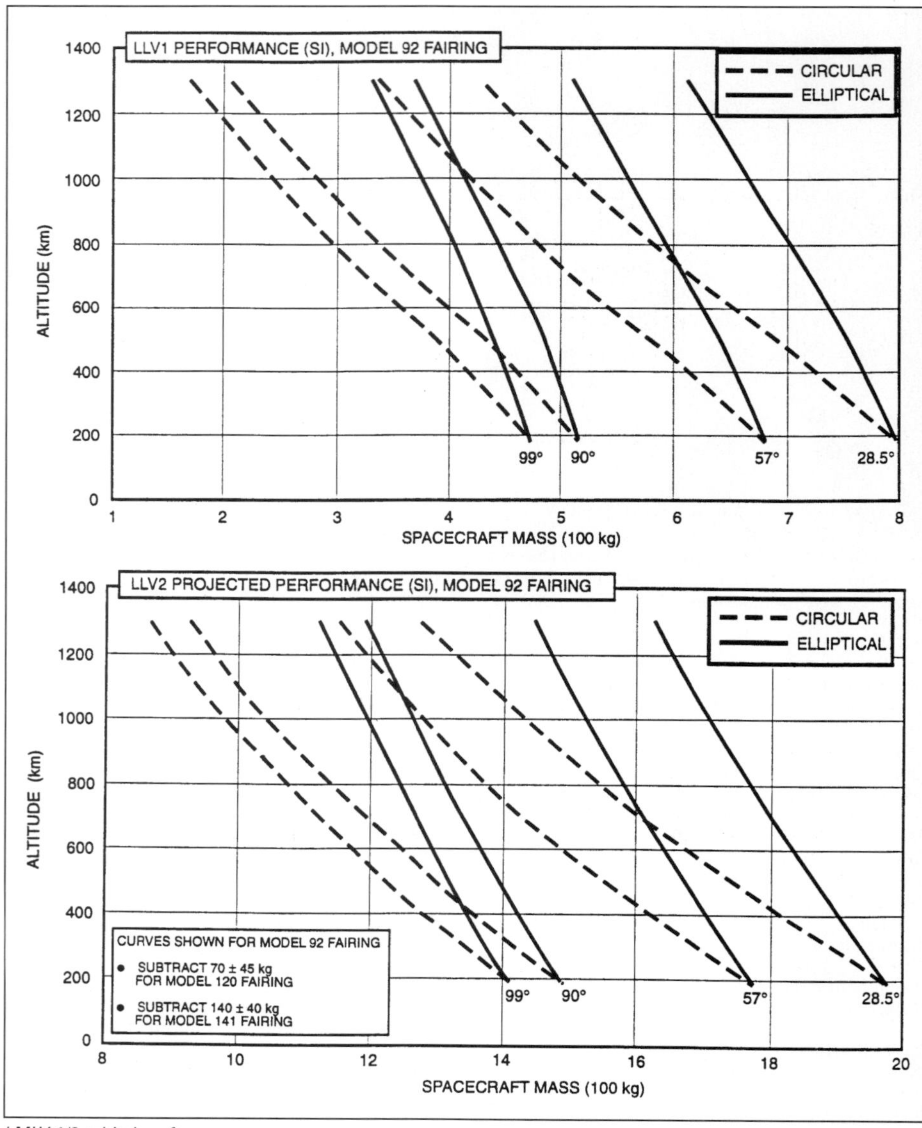

LMLV 1/2 orbital performance 0517182

but dropped the approach in 1990 when studies showed the vehicle would not be reliable enough (90 to 95 per cent) for the commercial market. Lockheed Corp approved the LLV development programme in January 1993. The philosophy is to use existing or imminent hardware, aluminium structures rather than composite (lower cost), higher margins in lieu of acceptance tests and not to push performance at the expense of reliability or lower cost.

Lockheed Launch Vehicles should deliver 800 to 3,200 kg to LEO for US$15 million to US$26 million. LMLV 1 employs a Castor 120 stage 1 and Orbus 21D stage 2. LMLV 2 inserts a second Castor 120 between the two, while LMLV 3 would add at least two Castor 4A strap-ons (3/4/6 strap-ons also possible). All versions carry a hydrazine orbit adjust module, which also houses the vehicles' avionics. Transfer orbits and Earth escape could be provided by a Thiokol Star stage 4 with TVC. The Castor 4A could also be replaced by the 4AXL version.

The programme moved from LMSC in Sunnyvale to Denver in late 1995, although the vehicles (now named LMLV) continue temporarily to be built at the Californian site. LMSC and CTA Inc signed a roughly US$5 million contract 18 February 1994 for the launch of CTA's 113 kg GEMStar 1 into a 480 km polar orbit from VAFB on the debut LMLV 1, 15 August 1995. Launch occurred from an adaptor on one of the solid booster mounts of the Shuttle SLC-6 complex. This flight used an existing Orbus 21 modified to 21D. The vehicle was destroyed by range safety after 159 seconds, when separate stage 1 TVC and IMU problems left it uncontrollable. Second launch carried the satellite Lewis into orbit on 23 August 1997. Before the launch of the first LMLV-2, the vehicle programme was renamed Athena. Athena 2 made its maiden flight with the launch of Lunar Prospector on 7 January 1998. A third Athena 1 launch carried the ROCSAT satellite for Taiwan into orbit on 27 January 1999. The second Athena 2 failed when the payload shroud failed to separate four minutes after launch on 27 April 1999. The third Athena 2 was successfully launched on 24 September 1999 carrying the Ikonos imaging satellite. The fourth Athena 1 launched on 30 September 2001 was the first orbited launch from Kodiak Island, Alaska. It carried the 90 kg Starshine 3, the 16 kg Sapphire, the 10 kg PCSat and the 68 kg Picosat.

Specifications

LMLV 1 Athena 1

First launch: 15 August 1995 from SLC6 Vandenberg AFB
Number launched: 4, to end 2001
Launch sites: SLC-6 VAFB; planned LC46 Cape Canaveral.
Principal applications: small payloads to LEO
Vehicle success rate: 75%
Availability: current production capacity can support 12 flights annually (increase to 24 if required). Typical mission flown 18 months after receiving order
Cost: US$16 million (excluding range + insurance)
Number of stages: 2
Overall length: 18.9 m with 2.34 m Model 92 fairing
Principal diameter: 236.3 cm
Launch mass: 66,345 kg
Launch thrust: 1,780 kN
Avionics/guidance: flight electronics, batteries, telemetry, inertial navigation unit and attitude control carried in OAM. Autopilot is hosted in Lockheed RISC 3000-based avionics processor and uses a Litton LN-100L IMU with software modified for LMLV to provide incremental velocity and angle data to Lockheed's navigation software. Strapdown

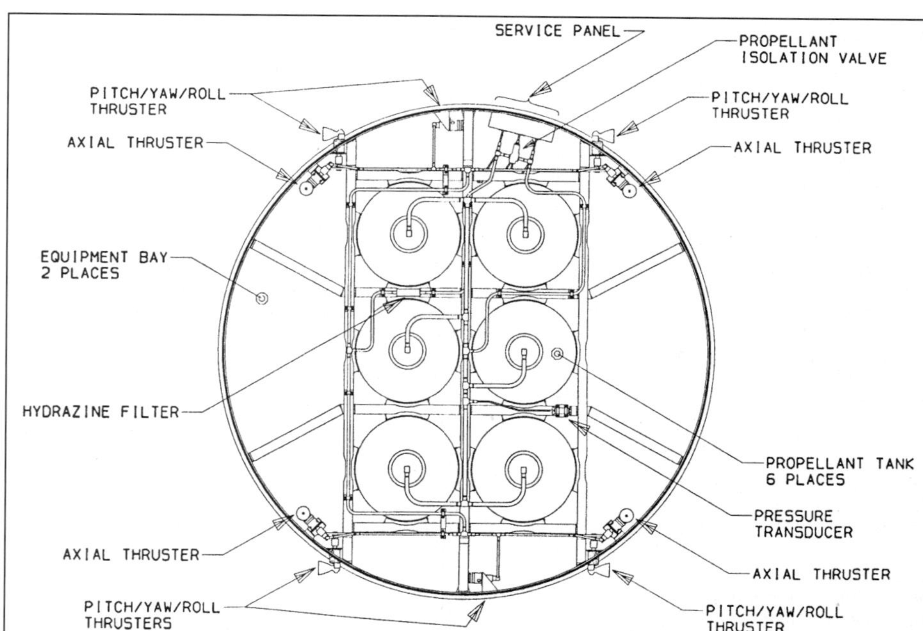

OAM configuration (Lockheed Martin) 0517378

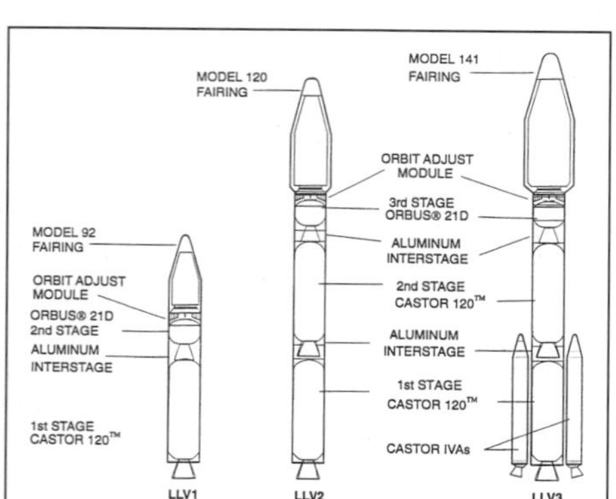

LMLV family. The market requirements for an LMLV 3 are under evaluation. The LMLV 3 strap-ons would not separate from stage 1 0517181

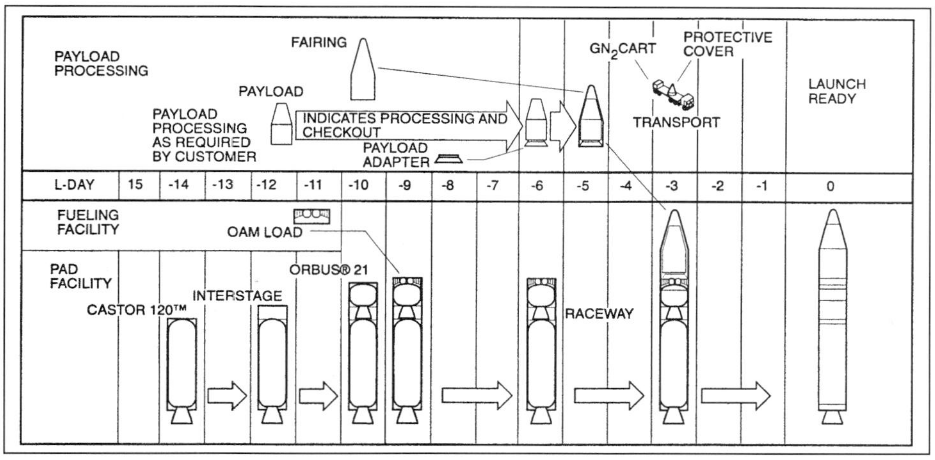

LMLV 1 processing flow (Lockheed Martin) 0517379

IMU uses three non-dithered Litton Zero-lock 18 cm path length RLGs and three A-4 accelerometers. Avionics performance is monitored by the avionics processor and formatted for downlink. Castor 120 provides ±5.5° movable nozzle TVC pitch/yaw control (enabled at T-7 s but restricted until pad cleared). Orbus 21D provides ±4° movable nozzle TVC pitch/yaw control. OAM provides roll control during burns, three-axis during coast and orbit adjust. Accuracy: 3Σ ±5.4 km at 1,100 km circular altitude, 0.06° inc.

Launch sequence (s) into 463 km 28.5°:

00.0	stage 1 ignite
85.9	stage 1 burnout 57.1 km/3.33 km/s
132.0	fairing separation 109.1 km/3.18 km/s
137.0	stage 1 sep/stage 2 ignition 113.9 km/3.16 km/s
291.0	stage 2 burnout 208.6 km/7.54 km/s
517.0	OAM ignition 329.6 km/7.39 km/s
1,885.0	OAM burnout 463 km/7.64 km/s

LMLV 1 stage 1
Motor: Thiokol Castor 120
Stage length: 9.02 m
Diameter: 236.3 cm
Thrust: 1,780 kN at launch, 1,554.8 kN average vacuum, 2.00 MN peak vacuum at 20 s
Burn time: 85.9 s
Mass at ignition: 52,852 kg
Propellant mass: 48,809 kg
Separation: hot separation; linear shaped charge separates the interstage's aluminium skin. Interstage is 184.7 cm high

LMLV 1 stage 2
Comment: The first launch employed an Orbus previously assigned to another mission and offloaded.
Motor: Orbus 21D employing carbon phenolic nozzle instead of previous carbon/carbon because of US$700,000 lower cost
Stage length: 317 cm
Diameter: 233.7 cm
Thrust: 192.4 kN average vacuum, 267.8 kN peak
Burn time: 154 s
Mass at ignition: 10,644 kg
Propellant mass: 9,779 kg
Separation: clean Zip tube fractures the OAM separation ring.

LMLV 1 OAM orbit adjust module
Olin Aerospace integrates the ACS into OAM for each vehicle. OAM can carry 2, 4 or 6 hydrazine aluminium tanks for 118/236/354 kg loading (total 607/653/818 kg). LMLV 1/2 typically carry 236 kg (354 kg on flight 1). 10 Olin MR-107 blowdown thrusters: 4 × 222 N axial for orbit trims + 6 × 111 N for three-axis control. Tanks pressurised to 31.3 atmospheres by N_2. Total impulse 254.4 kNs. Elastomeric bladders. Activated 1 s after stage 1 ignition. After payload release, the axial thrusters fire for separation and to deplete the tanks.

Fairing and payload accommodation
The Model 92 233.7 cm diameter 792 kg two-piece clamshell fairing provides a 10.6 m^3 198.4 cm diameter × 429 cm long payload envelope. LMLV 2/3 can also employ: Model 120 305 cm diameter, providing 29.5 m^3 274.3 × 710.1 cm envelope; Model 141 358 cm diameter, providing 56.3 m^3 327.7 × 872 cm. A Zip tube cuts the base and springs separate the halves. 74 copper conductors + two fibre optic cables are available for payload links via the T-0 umbilical. 10 analogue + 10 discrete monitors, plus five signal conditioned continuity loops and five separation indicators, are standard via telemetry. The avionics computer can issue eight discrete payload commands during ascent.

Acceleration load: 8 *g* longitudinal (stage 1/2), ±2.5 *g* lateral.
Acoustic load: 133.5 dB overall
Maximum internal fairing temp: 107°C
Recommended acceptance vibration: unavailable.

Vehicle and payload processing
LMLV 1 assembly/checkout requires about 14 days by a crew of 2,025 (LMLV 2 18 days; LMLV 3 25 days). OAM is checked out at the factory and delivered directly to the hydrazine loading facility. It arrives at the pad on T-9 days. The payload is mounted to its adaptor (mating to LMLV's 167.6 cm diameter bolt circle) and the fairing added before transfer to the pad three-days before launch. Launch control is provided from a 12 m long van 2,700 m from the pad connected via two-way fibre optics in the T-0 flyaway umbilical.

LMLV 2 Athena 2
First launch: 7 January 1998
Number launched: 3, to end of 2001 (of which 1 failure)
Launch sites: as LMLV 1
Performance: see diagram. Also, 450 kg into lunar transfer.
Cost: US$21 million
Number of stages: 3
Overall length: 28.2 m with 2.34 m fairing; 30.2 m with 2.95 m fairing; 31.7 m with 3.58 m fairing.
Launch mass: 120,150 kg.

LMLV 2 stage 1
Same as LMLV 1.

LMLV 2 stage 2
Same as LMLV 1 stage 1 but expansion ratio increased from 17 to 24 to improve altitude performance.
Thrust: 1,606 kN average vacuum
Mass at ignition: 53,020 kg.

Fairing and payload accommodation
Acceleration load: 8 *g* longitudinal stage 1/2, ±2.5 *g* lateral.
Acoustic load: to be determined
Maximum internal fairing temp: to be determined
Recommended acceptance vibration: unavailable.

UPDATED

Atlas

Current Status
Marketed by ILS, the Atlas family has been consolidated around the 2A, 2AS, 3 and 5 series and with the adoption of the Energomash RD-180 propulsion system for Atlas 3 and 5 launchers has received major payload capability expansion. The basic Atlas 2 was retired in March 1998. Atlas 5 made its first flight on 21 August 2002 carrying the Hot Bird 6 satellite for Eutelsat. Actual apogee was 45,856 km versus a planned 45,845 km with an actual perigee of 314.9 km versus 314.8 km. Actual inclination was 17.827° versus a planned 17.829°. The second flight took place on 14 May 2003 (401 configuration) with the launch of HellaSat from SLC-41. The third flight (521 configuration) took place on 17 July 2003 with the launch of Rainbow 1 from SLC-41 at the CCAFS, Florida.

Background
Atlas was developed as the first US ICBM, becoming operational in September 1959. A total of 159 were at one time deployed across the US before the last was withdrawn from service in 1965. As they were deactivated, they were refurbished as space launchers. More than 500 in the Atlas family have been launched. They are responsible for orbiting the first US astronaut and dispatching the first successful probes to Mars, Venus, Mercury and Jupiter. Of the 140 Atlas E/F types originally stored at Norton AFB, the last departed in March 1995. A new build Atlas J was studied to complement the refurbished Titan 2, providing a 3,400 kg capacity (with spacecraft kick motor) into polar orbit. Studies were also made of using the Atlas 2 family for Space Station resupply.

The vehicle's future depends on the Atlas Centaur combination, first launched in 1962. In January 1987, it lost out to the Delta as the USAF Medium Launch Vehicle for the GPS Navstar satellites. General Dynamics in 1987 committed US$100 million of company funding to 18 new commercial Atlas Centaur vehicles providing a 2,335 kg GTO capability. The company later increased the plan to US$400 million/62 vehicles. Eutelsat was the first to select this Atlas 1, for one of its second generation satellites in January 1990 (it actually flew on the first of the Atlas 2 family, December 1991). In October 1987, NOAA booked three GOES at a fixed price of about US$200 million for delivery into GTO and a free reflight in the event of a launcher failure. The last Atlas 1 departed in 1997.

Atlas became the Air Force's choice in May 1988 as the USAF MLV-2 to handle DSCS3, beginning in 1992. 62 vehicles are being built to support projected launch requirements through 2000. Another eight Atlas 2AS were added in 1995 for launch through 2002. Martin reportedly charges the USAF US$45 million per launch. Atlas 2 achieved the required DSCS 3 capability by stretching the booster and upper stage tanks by 2.75 m and 92 cm, respectively, uprating the Atlas booster powerplant from 1,668 kN to 1,841 kN and incorporating new lightweight and low cost avionics. Lockheed Martin's IABS Integrated Apogee Boost Subsystem provides insertion of DSCS into near GEO. This new Atlas 2 (the previous 18 Atlas Centaurs were retrospectively renamed Atlas 1) provides 2,812 kg GTO capability. The last Atlas 2 was launched on 16 March 1998.

The MLV2 success permitted GD to offer a derived Atlas 2A commercial version (3,045 kg GTO) from 1992. The 2A is an enhanced version of the 2 model for commercial and government applications. The major improvement is the incorporation of P&W's RL10A-4 engines on the Centaur stage. A 508 mm long all welded columbium (niobium) extension skirt is deployed following Centaur separation and provides a 6.5 second SI increase to 449.5 second vacuum (thrust 185 kN total vacuum). There is no thrust compartment length increase required and no additional mass. The Atlas 1/2 avionics is further updated by replacing the sequence control unit and the servo inverter unit with a remote-control unit, providing 128 channels of solid state switching for vehicle/spacecraft sequencing instead of the current electromechanical relays.

GD began offering a 2AS version in 1993, uprated with the inclusion of four solid propellant jettisonable Castor 4A boosters. Two are ignited on the pad and jettisoned at 88 seconds after their 54 second burn. The second pair ignite at 57.5 seconds for jettison at 114 seconds. The Atlas structure required strengthening to accommodate the boosters. 22 are being built (eight were added in 1995 as insurance against 2AR problems). Intelsat's three launches are contracted at US$93 million to US$98 million each.

The 2AS was conceived to bid for the Intelsat 7 contract; the original intention was to propose the 2A but Intelsat's mass growth dictated the addition of strap-ons. Intelsat awarded a US$9.5 million contract in 1991 to pay for enhancing Atlas 2AS capacity into GTO by 121 kg, the additional satellite propellant providing a 16 year life. Atlas is not able to handle the heavier Intelsat 7A. The Block 1 upgrade became available in 1994 for 2AS/2A, able to improve GTO capacity by 8 per cent. In addition to software improvements, the two air-lit Castor 4As on the 2AS replace their 11° canted nozzles with 7°, the Centaur's RL10A-4 is upgraded to the 10A41 with thrust increase per engine of 6.67 to 99.19 kN. First flight was January 1995 for Intelsat 704. The Block 2 upgrade is planned to handle larger payloads such as EOS AM 1 into intermediate LEOs. The fairing will be stretched 90 cm, and the payload adaptor and Centaur equipment module strengthened.

Lockheed Martin, on 7 November 1995, announced development of the 2AR (R means re-engined) to replace all previous models and, potentially, as the basis of a common core family, for handling medium to heavy payloads. The radical redesign will

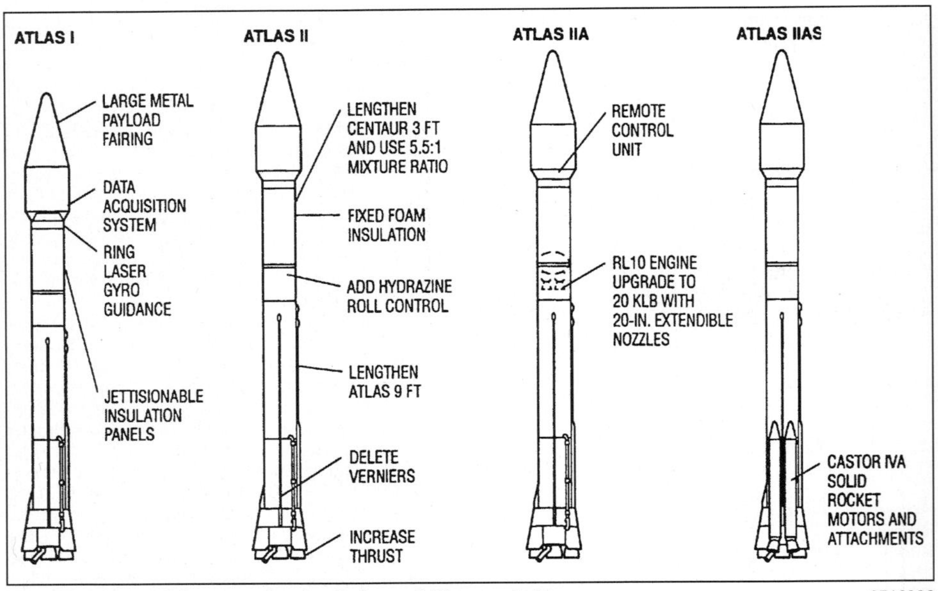

Final Atlas assembly is now undertaken in Denver's Waterton facility 0516926

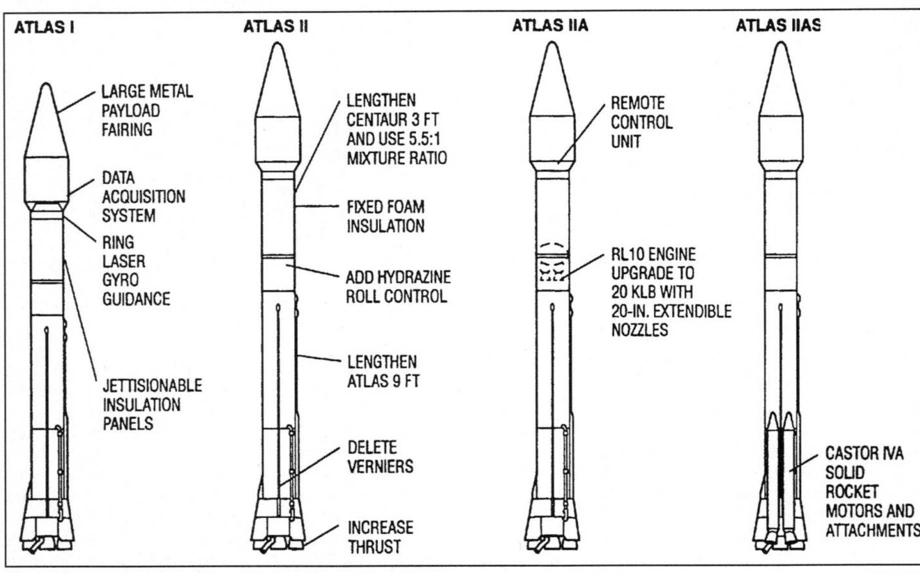

The Atlas Centaur 1 and 2 series (Lockheed Martin) 0516928

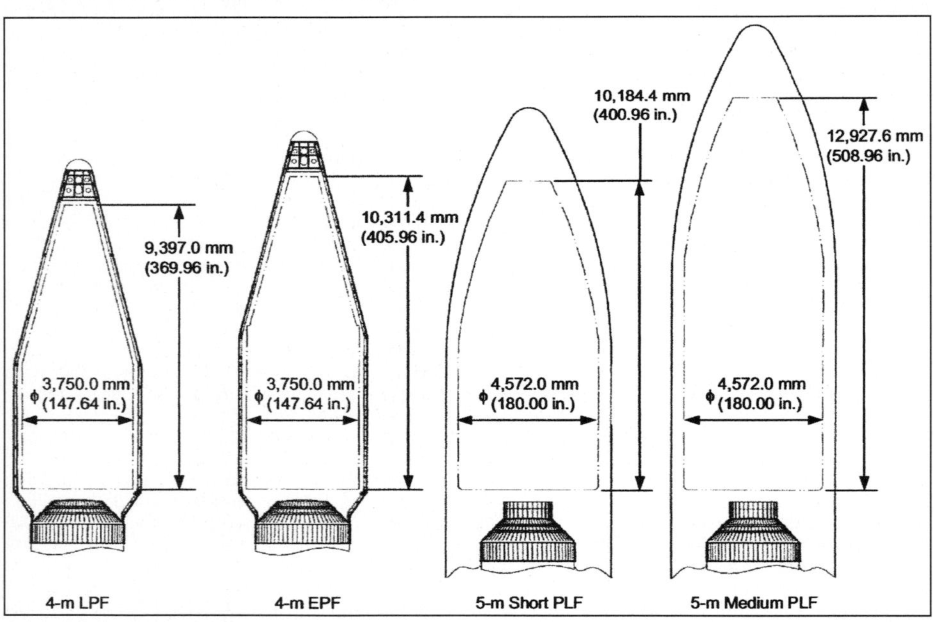

Atlas static payload envelope 0131862

reduce cost by more than 20 per cent, principally by adopting Russia's RD180 main engine and a single engine Centaur. It will carry 3,810 kg GTO.

Initially, the company will build 18, using Russian built engines. US built engines will then become available (US law requires US engines for government payloads). The two-engine Centaur may be offered for high inclination or multiple satellite LEO missions.

Systems integration and final assembly of the Atlas boosters and Centaur was transferred to Denver in 1995 when GD sold its space division to Martin Marietta for US$208.5 million. Denver's existing Waterton facility accommodates seven vehicles, with four being actively worked on at any one time. A new class 100,000 3,720 m^2 Final Assembly Building now handles Centaur. Tanks continue fabrication at the Naval In Service Engineering West facility (formerly Air Force Plant 19) in San Diego (26 annual capacity) and fairings, thrust structures, adaptors and other components continue to be provided from a company facility in Harlingen, Texas.

Lockheed Martin can launch up to four vehicles annually (five under surge conditions) from each of Canaveral's two complex 36 pads. 36A is used for USAF launches but open slots are available for commercial flights. 36B is dedicated to commercial

NPO Energomash RD-180 engine adopted by Lockheed Martin for the 3 and V series (Theo Pirard) 0054362

operations. Work is under way on converting VAFB's SLC-3E to handle the Atlas 2 family.

On 8 April 1998, Lockheed Martin announced changes to the designation system whereby Atlas 2AR would become Atlas 3A and a variant known as the 2ARC would become the Atlas 3B. The Atlas 3B incorporates two RL-10A cryogenic engines to power a stretched version of Centaur and extends the lift capability to 4,500 kg to GTO, an increase of more than 770 kg over the Atlas 2AS, the most powerful launch vehicle in the Atlas family presently flying. Atlas 3A made its first flight on 24 May 2000 followed by the first Atlas 3B on 21 February 2002. In October 1998, Lockheed was awarded contracts to complete development of a new family of launchers under the USAF Evolved Expendable Launch Vehicle (EELV) programme. The air force envisions that these vehicles will eventually replace current Atlas series vehicles through utilisation of the new Common Core Booster (CCB) stage and reduce launch costs by up to 25 per cent. The CBC is essentially the Atlas 3A first stage, 3.8 m in diameter and 27 m tall with a single Russian built RD-180 engine and equipped with fixtures for solid rocket boosters that could be attached to increase lifting capacity.

On 29 July 1998, an RD-180 was test fired in a prototype Atlas 3A for the first time for 10 seconds at the NASA Marshall Space Flight Center Advanced Engine Test Facility. This was the first time a Russian rocket motor had been fired at a US government test facility. The advanced Engine Test Facility had previously been used to test Shuttle Orbiter main engines and the Saturn V F-1 first-stage engines before that. The RD-180 delivers 3,825.28 kN giving the Atlas 3 series more than twice the lift-off thrust of the original Atlas space launch vehicle and at the date of this test nine development engines had been successfully tested for more than 10,000 seconds. In flight, the RD-180 will operate for about 186 seconds in providing first-stage energy.

On 2 February 1999, Lockheed Martin named this new family Atlas 5, the baseline intermediate class vehicle being able to lift payloads weighing more than 5,000 kg to GTO. The common Centaur upper stage has the performance option of one or two RL10 engines and the new Centaur has been flight proven on Atlas 3. By carrying up to five Aerojet solid rocket boosters attached to the CCB, each producing 1,110 kN thrust, the most powerful commercial variant of Atlas 5 will be able to carry payloads weighing up to 8,200 kg to GTO. Atlas 5 is available in two classes, optional variants of SRBs and upper stage carrying a 4 m diameter payload fairing (400 series) or a 5 m diameter fairing for large satellites (500 series). Further sub-division indicates number of SRB by the second digit (0, 1, 2 or 3) and number of engines in Centaur stage (1 or 2). For instance on Atlas 5 421 indicates it to carry a 5 m fairing, two SRBs and a single Centaur engine. Developed for the USAF EELV requirement, the Atlas 5 Heavy configuration will comprise three CCB stages placed side by side with the common Centaur upper stage placed atop the centre CCB and encapsulated in a 5.4 m diameter Contraves composite payload fairing. The Atlas 5 Heavy will be capable of placing a payload in excess of 13,000 kg to GTO or directly injecting a payload of 6,500 kg into geostationary orbit. As of April 2000, Lockheed Martin has a backlog of 31 launch orders for Atlas 2A, 2AS, 3A and 5.

Using the Common Core Booster developed for the Atlas 5 series, a Heavy Lift Vehicle is offered capable of carrying 12,650 kg to GTO or 25,000 kg to low earth orbit. Although they are essentially the same, the three CCB elements which provide thrust for lift off are designated as the primary CCB in the centre supporting the second stage flanked by two liquid rocket boosters. The second stage is the single-engine RL10A-42 carried by Atlas 3 and 5 versions with the twin-engine version as an option.

Specifications

Atlas 1

First launch: 25 July 1990 (AC-69 CRRES)
Number launched: 11 to 1997 when the last in the series was launched
Launch sites: Canaveral pad 36B
Principal uses: medium-class telecom and meteorological satellite payloads into GTO.
Vehicle success rate: 82%
Performance
LEO (185 × 185 km, 28.5°): 5,900 kg medium fairing/5,700 kg large.
GTO (167 × 35,788 km, 27.0°): 2,375 kg medium fairing/2,255 kg large; 3Σ errors of 106 km apogee, 2.4 km perigee and 0.02° inc
Earth escape: 1,520 kg medium fairing, 1,400 kg large
Number of stages: 2 1/2 (booster engines burn in parallel)
Overall length: 42.0 m with medium fairing, 43.9 m with large
Principal diameter: 3.05 m
Launch mass: 163,900 kg with medium fairing, 164,290 kg with large.
Launch thrust: 1,953 kN sea level
Guidance: Honeywell's Inertial Navigation Unit mounted on Centaur's forward equipment module performs the inertial guidance and attitude control computations for both Atlas and Centaur. Some initial Atlas 1s retained the existing Honeywell inertial unit and Teledyne flight control computer but subsequent vehicles incorporate a Honeywell ring laser gyro INU + Gulton digital data acquisition unit, saving 36 kg and enhancing reliability.

Atlas stage 1

Comment: Atlas uniquely incorporates two booster engines fired in parallel with the central sustainer until the base section is jettisoned by the release of pneumatically-actuated latches at about T+156 s/ 5.5 *g* longitudinal acceleration. The propellant tanks are thin-walled (0.361.04 mm), fully monocoque stainless steel separated by an ellipsoidal bulkhead. Structural integrity is maintained in flight by the pressurisation system and on the ground by either internal pressure (N_2 at 0.330.68 atmosphere) or application of mechanical stretch. The tank design is driven primarily by the 'max α Q condition', the angle of attack due to jet stream winds near the trajectory's maximum dynamic pressure. Atlas separation is ensured by eight solid propellant retros around the base firing angled at 40° to the vertical to prevent spacecraft contamination. An external Atlas pod houses range safety, propellant utilisation, pneumatics and instrumentation.
Engines: Rocketdyne MA-5 propulsion system of two booster, one sustainer and two vernier single start liquid bipropellant engines. Each booster yields 839.5 kN sea level, the sustainer 269 kN sea level, and the verniers 2,975 N each. The engines are gimballed hydraulically to provide three-axis control during Atlas burn.
Overall length: 22.16 m
Principal diameter: 3.05 m
Propellant: 137.53 t LOX (forward tank) + RP1
Burn time: 156 s boosters, 266 s sustainer

Interstage adaptor

Mass: 477 kg
Dimensions: 3.96 m long, 3.05 m diameter
Comment: The ISA supports Centaur until separation at about 268 s is effected by a flexible linear shaped charge around the forward circumference. Construction is an aluminium skin/stringer and frame.

Fairing/payload accommodation

Two fairing designs are available for spacecraft protection during ascent: 4.19 m diameter 12.22 m long 2,005 kg, or 3.30 m diameter 10.36 m long 1,375 kg mass. Usable diameters are 365.0 cm and 292.1 cm, respectively. Both employ an aluminium skin/stringer/frame structure and non-contaminating pyro separation bolts for jettison in halves at about 205 s before sustainer engine cut-off when the heating rate has reduced to 1,135W/m^2. Four 762 mm^2 doors in the aft section provide payload access.

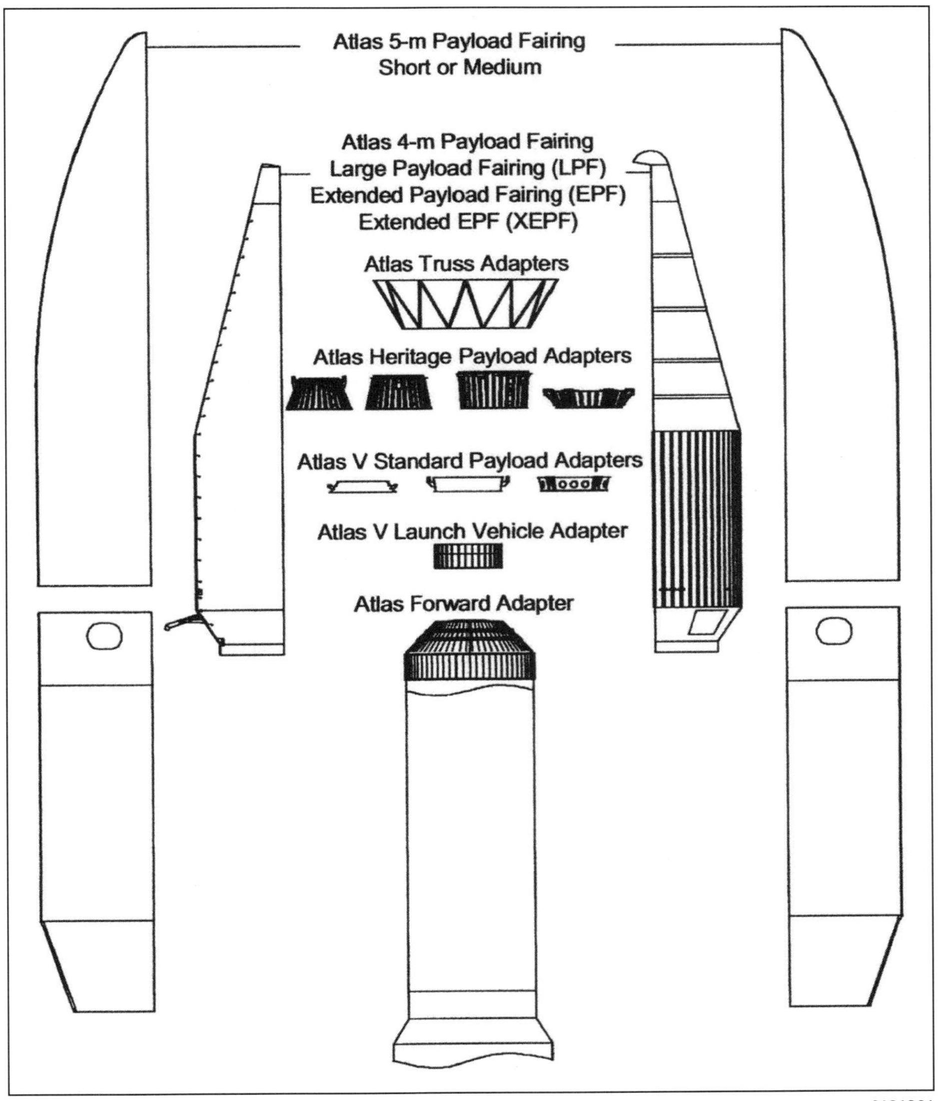

Atlas payload interface options 0131861

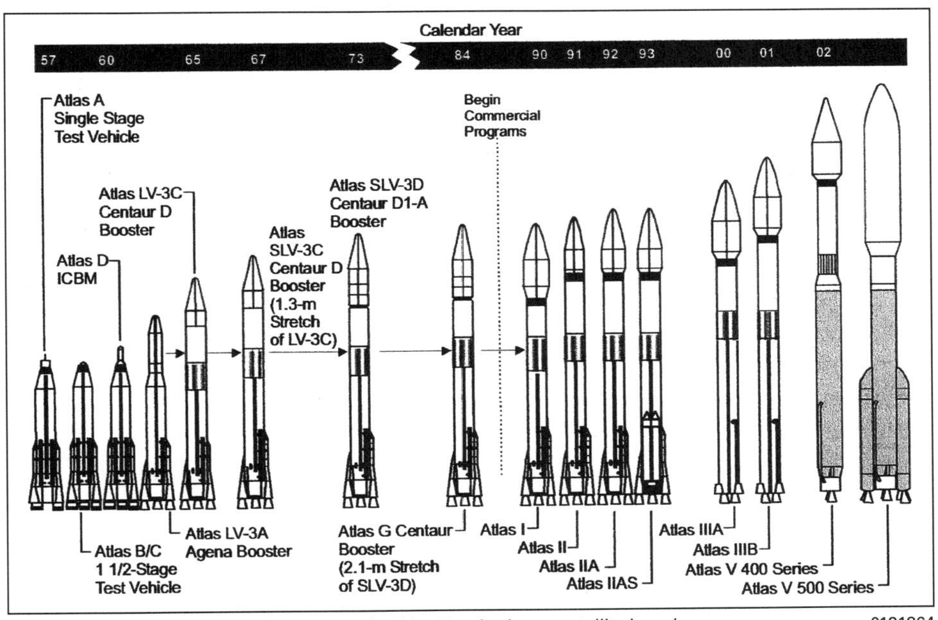

Atlas evolution from 1 CBM in the 1950s to the Atlas 5 series heavy satellite launcher 0131864

The highest steady state longitudinal acceleration occurs at Atlas booster engine cut-off. It is typically 5.5 *g* but can be reduced with some loss of performance. Laterally, the highest steady state acceleration is 0.4 *g* caused by the vehicle reacting to winds. During the assembled payload/ fairing transport to the pad for vehicle mating, air conditioning can provide a 15–25°C environment around the spacecraft. On the pad, air with a 2°C dew point is used until 2 h before launch, when a switch to N_2 is made. Gas is provided to class 10,000 (5,000 available) with 1,029°C inlet temperature. *Acceleration load:* 5.5 *g* maximum longitudinal, 0.4 *g* lateral *Acoustic:* max 138.9 dB overall.

Atlas provides five basic payload adaptors compatible with the Ariane 937B (Atlas type A), 937A (A1), 1194A (B, B1) and 1666A (D, Hughes HS 601) interfaces, and two (C1, C2) providing bolted interfaces for satellite adapters or mission peculiar requirements. The numerical designators indicating diameter in millimetres. Separation is effected by springs following V-clamp release commanded by Centaur's guidance system. The upper stage's sequence control unit can also issue up to 10 control commands to the spacecraft.

Payload Processing

Cape payload processing facilities, including those of Astrotech Space Operations in nearby Titusville, are described under the McDonnell Douglas Delta entry. Each pad at complex 36 includes a 64 m Mobile Service Tower, 50 m Umbilical Tower and a Blockhouse for launch operations. The Atlas is erected on the pad typically 71 days before launch, followed by Centaur (69 days), its insulation panels and forward equipment module (45 days).

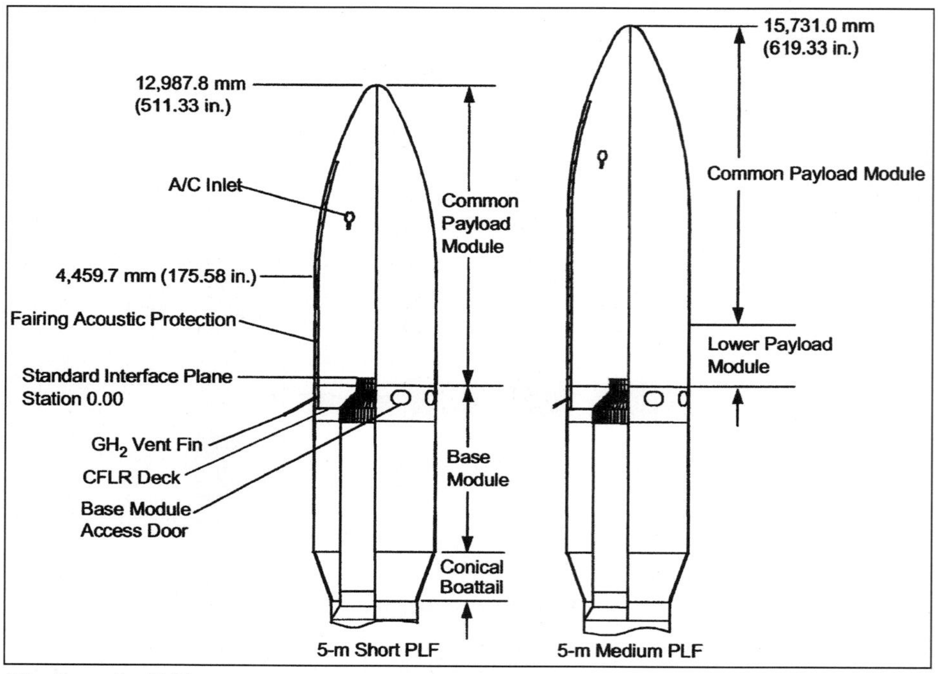

Atlas 5 m payload fairings

0131863

Payload separation tests and terminal countdown demonstration are held 40 days and 29 days before launch, respectively. The spacecraft is encapsulated at one of the remote safe facilities and the assembly installed atop Centaur at 13–8 days. Atlas' RP1 fuel is loaded at 3 days. A countdown typically requires 910 h, including two 30 min and 5 min built-in holds. Centaur's LOX is loaded at T-75 min, followed by Atlas' LOX at 55 min and Centaur's LH_2 at 43 min. The booster and its upper stage switch to internal power at 4 min and 2 min, respectively.

Atlas 2/2A/2AS
First launch: 7 December 1991
Number launched: 59 to end 2002
Launch sites: Canaveral pads 36A/B, VAFB SLC 3E from 1998 for access to Sun-synchronous, polar and 63.4° orbits for military and Earth observation satellites.
Principal uses: delivery of DSCS-3 satellites into GTO; future: high inclination missions.
Vehicle success rate: 100% to end 2001
Schedule of further missions:
Performance:
LEO (185 × 185 km, 27.0° Canaveral): 6,780 kg medium fairing, 6,580 1kg large
LEO (185 × 185 km, 90.0° VAFB): 5,510 kg large
GTO (160 × 35,788 km, 28.5°): 2,950 kg medium fairing, 2,810 kg large
Earth escape: 2,000 kg large fairing
Number of stages: Same as Atlas 1
Overall length: 46.8 m with medium fairing; 47.4 m with large fairing
Principal diameter: 3.05 m
Launch mass: 187,170 kg with medium fairing, 187,560 kg with large
Launch thrust: 2,159 kN sea level
Guidance: see Atlas 1

Atlas 2 stage 1
Comment: To complement the higher performance MA5A engines, stage 1 tanks were stretched 1,702 mm (LOX) plus 1,041 mm (RP-1), adding 2,585 kNs total impulse.
Engines: Rocketdyne MA-5A single start liquid bipropellant consisting of two booster engines (1,841 kN total SL) and one sustainer (269 kN sea level). The sustainer remains unchanged from that of Atlas 1 but the booster portion now uses the Delta RS-27 engine for an SI increase from 259.1 s to 263.1 s sea level. The side mounted verniers have been deleted; hydrazine thruster modules on the interstage adaptor have replaced their roll control function.
Overall length: 24.9 m

Comparison of current and future launch vehicle variants

	Atlas 2A	Atlas 2AS	Atlas 3A	Atlas 3B
Usable diameter	3.65 m	3.65 m	3.65m	3.65 m
Usable length	9.39 m	9.39 m	10.3 m	10.3 m
Fairing mass	2,076 kg	2,076 kg	2,215 kg	2,215 kg
Payload capabilities:				
LEO at 28.5°	7,316 kg	8,618 kg	8,641 kg	9,095 kg
LEO at 90°	6,192 kg	7,212 kg	7,121 kg	7,398 kg
GTO	3,066 kg	3,719 kg	4,037 kg	4,264 kg
Overall vehicle:				
Length	47.5 m	47.5 m	52.8 m	52.8 m
Launch mass	185,427 kg	233,750 kg	214,535 kg	238,044 kg
Stage 1: strap-on boosters		2 pairs		1 pair
Motor		Castor 4A		Castor 4A
Length		9.9 m		9.9 m
Diameter		1.0 m		1.0 m
Dry mass ground ignited		3,101 kg		3,101 kg
Dry mass altitude ignited		3,224 kg		3,244 kg
Propellant mass		10,200 kg		10,200 kg
Thrust (sl)		433 kN		433 kN
Specific impulse (sl)		229 s		229 s
Stage 1: Atlas				
Motor	MA-5A	MA-5A	RD180	RD180
Length	24.9 m	24.9 m	29.0 m	29.0 m
Diameter	3.05 m	3.05 m	3.05 m	3.05 m
Dry mass	7,151 kg	8,856 kg	11,613 kg	11,613 kg
Propellant mass	156,260 kg	156,260 kg	183,530 kg	183,530 kg
Booster thrust (sl)	1,843 kN	1,843 kN	3,826 kN	3,826 kN
Sustainer thrust (sl)	269 kN	269 kN		
Booster specific impulse (sl)	261.8 s	261.8 s	311.3 s	311.3 s
Sustainer specific impulse (sl)	220.4 s	220.4 s		
Stage 2: Centaur				
Motor	2 × RL-10A-41	2 × RL-10A41	1 × RL-10A41	1 × RL-10A 42
Length	10.1 m	10.1 m	10.1 m	10.1 m
Diameter	3.05 m	3.05 m	3.05 m	3.05m
Dry mass	1,735 kg	1,735 kg	1,680 kg	1,680 kg
Propellant mass	17,060 kg	17,060 kg	16,946 kg	16,946 kg
Total thrust (vacuum)	198.4 kN	198.4 kN	99.2 kN	99.2 kN
Specific impulse (sl)	450.5 s	450.5 s	450.5 s	450.5 s

Principal diameter: 3.05 m
Propellant: 156,260 kg of LOX/RP1
Stage thrust: 2,159 kN sea level
Burn time: 169 s boosters, 277 s sustainer

Interstage Assembly
The ISA is similar to that on Atlas 1 but two hydrazine thruster modules similar to those on Centaur provide roll control.

Atlas 2 stage 2
Comment: Centaur's LOX/LH_2 tanks are stretched 338/577 mm to increase propellant loading by 2,990 kg. Foam panels bonded to the tanks. replace the four jettisonable insulation panels of Atlas 1.
Engines: as Atlas 1 but oxidiser/fuel ratio increased from 5.0 to 5.5
Overall length: 10.06 m
Principal diameter: 3.05 m
Propellant: 16,780 kg LOX (aft) + LH_2
Stage thrust: 146.8 kN vacuum
Burn time: typically as Atlas 1

Fairing/payload accommodation
Same as Atlas 1.

Atlas 2A
First launch: 10 June 1992 (Intelsat K)
Number launched: 10 to end-1996
Vehicle success rate: 100%
Performance
LEO (185 × 185 km, 28.5° Canaveral): 7,316 kg large.
LEO (185 × 185 km, 90.0° VAFB): 6,190 kg large.
GTO (167 × 35,788 km, 27.0°): 3,040 kg medium, 2,900 kg large; Block 1 upgrade: 3,180 kg medium fairing, 3,066 kg large.
Earth escape: 2,160 kg large fairing

Atlas 2AS
First launch: 15 December 1993
Number launched: 24 to end 2001
Vehicle success rate: 100% to end 2001
Performance
LEO (185 × 185 km, 28.5° Canaveral): 8,610 kg large fairing.
LEO (185×185 km, 90.0° VAFB): 7,210 kg large fairing.
GTO (167 × 35,788 km, 27.0°): 3,730 kg medium, 3.620 kg large; Block 1 enhancement: 3,810 kg medium, 3,700 kg large.
Earth escape: 2,680 kg large fairing

Castor 4A strap-ons
Motor contractor: Thiokol, Huntsville division. TX7803 ground-ignited (quads 1 + 3), TX7804 air-ignited (quads 2 + 4) to minimise umbilical tower plume impingement.
Length: 11.28 m
Diameter: 1,016 mm
Average thrust: each 433.6 kN sea level
Burn time: 53 s
Mass at ignition: 11,649 kg (ground), 11,743 kg (air).
Burnout mass: 1,452 kg (ground), 1,529 kg (air).
Propellant: 10,200 kg of HTPB solid.

Atlas 3A
First launch: 24 May 2000
Launch sites: Canaveral pad 36B, Vandenberg SLC3E.
Principal uses: medium-class telecom and meteorological satellite payloads into GTO.
Performance
LEO (185 × 185 km, 28.5°); 8,641 kg.
GTO (167 × 35,788 km, 27.0°): 3,810 kg (4,060 kg with two Castor 4Bs); 3Σ errors unavailable.
Availability: typically four launches/year per pad. Launch typically 24–30 months following contract.
Cost: about US$45 million into GTO
Number of stages: 2
Guidance: Honeywell's Inertial Navigation Unit mounted on Centaur's forward equipment module performs the inertial guidance and attitude control computations for both Atlas/Centaur.
Launch sequence (s) into GTO 27.0°:

00.0	stage 1 ignite (75 per cent thrust)
04.8	stage 1 throttle up to 85 per cent, 30 m
12.9	start pitch programme, 290 m
24.4	ramp to zero angle of attack
32.4	zero angle attack, 2.44 km
44.2	pre-max Q throttle down to 61 per cent
67.6	post-max Q throttle up to 85 per cent
147.9	max 5.0 *g*, begin throttle down
184.9	stage1cut-off/separation(49percent throttle)
199.9	Centaur ignition 1
209.9	jettison payload fairing
760.2	Centaur cut-off 1
1,449.9	Centaur ignition 2
1,625.3	Centaur cut-off 2

Atlas 3A stage 1
Comment: The two-chamber throttleable RD-180 eliminates the booster section and provides a more benign environment. Max thrust 85%; 49%

Atlas launch history, from 1980

	Launch	Atlas/Centaur	Site	Payload
1980				
453	17 January	5029D/3220	36A	Fltsatcom 3. AC 49
454	9 February	35F	3E	GPS Navstar 5
455	3 March	67F	3W	NOSS 3
456	26 April	34F	3E	GPS Navstar 6
457*	29 May	19F	3W	NOAAB
458	30 October	5037D/3228	36A	Fltsatcom 4. AC 57
459	6 December	5034D/3225	36B	Intelsat 502. AC 54
460*	8 December	68E	3W	NOSS 4
1981				
461	21 February	5023D/3213	36A	Comstar 1D. AC 42
462	23 May	5036D/3227	36B	Intelsat 501. AC56
463	23 June	87F	3W	NOAA 7
464	6 August	5039D/3230	36A	Fltsatcom 5. AC59
465	15 December	5035D/3226	36B	Intelsat 503. AC55
466*	18 December	76E	3E	GPS Navstar7
1982				
467	4 March	5038D/3229	36A	Intelsat 504. AC 58
468	28 September	5040D/3231	36B	Intelsat 505. AC 60
469	20 December	60E	3W	DMSP F6
1983				
470	9 February	6001H	3E	NOSS 5
471	28 March	73E	3W	NOAA 8
472	19 May	5041D/3232	36A	Intelsat 506. AC 61
473	10 June	6002H	3E	NOSS 6
474	14 July	75E	3W	GPS Navstar 8
475	17 November	58E	3W	DMSP F7
1984				
476	5 February	6003H	3E	NOSS 7
477*	9 June	5042G/3233	36B	Intelsat 509. AC 62
478	13 June	42E	3W	GPSNavstar 9
479	8 September	14E	3W	GPS Navstar 10
480	12 December	39E	3W	NOAA9
1985				
481	12 March	41E	3W	Geosat
482	22 March	5043G/3234	36B	Intelsat 510. AC 63
483	29 June	5044G/3236	36B	Intelsat 511. AC 64
484	28 September	5045G/3237	36B	Intelsat 512. AC 65
485	8 October	55E	3W	GPS Navstar 11
1986				
486	9 February	6004H	3E	NOSS 8
487	17 September	52E	3W	NOAA 10
488	4 December	5046G/3238	36B	Fltsatcom 7. AC66
1987				
489*	26 March	5048G/3239	36B	Fltsatcom 6. AC67
490	15 May	6005H	3E	NOSS 9
491	19 June	59E	3W	DMSP F8
1988				
492	2 February	54E	3W	DMSP F9
493	24 September	63E	3W	NOAA 11
1989				
494	25 September	5047G/3240	36B	Fltsatcom 8. AC 68
1990				
495	11 April	28E	3W	Stacksat
496	25 July	5049/3241	36B	CRRES. AC69; A1
497	1 December	61E	3W	DMSP F10
1991				
498*	18 April	5050/3242	36B	BS-3H. AC70; A1
499	14 May	50E	3W	NOAA12
500	28 November	53E	3W	DMSP F11
501	7 December	8102/202	36B	Eutelsat 2 F3. AC102; A2
1992				
502	11 February	8101/203	36B	DSCS 3B14. AC101; A2
503	14 March	5052/3244	36B	Galaxy 5. AC 72; A1
504	10 June	8105/205	36B	Intelsat K. AC105; A2A
505	2 July	8103/201	36A	DSCS 3B12. AC-103; A2
506*	22 August	5051/3243	36B	Galaxy 1R. AC71; A1
1993				
507*	25 March	5054/3246	36B	UFO 1. AC 74; A1
508	19 July	8104/204	36A	DSCS 3B9. AC 104; A2
509	9 August	34E	3W	NOAA 13
510	3 September	5055/3247	36B	UFO 2. AC75; A1
511	28 November	8106/206	36A	DSCS 3B10. AC106; A2
512	16 December	8201/601	36B	Telstar 401. AC108; A2AS
1994				
513	13 April	5053/3245	36B	GOES8. AC 73; A1
514	24 June	5056/3248	36B	UFO 3. AC 76; A1
515	3 August	8107/207	36A	DBS 2. AC-107; A2A
516	29 August	20E	3W	DMSP F12
517	6 October	8202/602	36B	Intelsat 703. AC-111; A2AS
518	29 November	8109/208	36A	Orion 1. AC 110; A2A
519	30 December	11E	3W	NOAA14
1995				
520	10 January	8203/603	36B	Intelsat 704. AC113; A2AS
521	29 January	8110/210	36A	UFO 4. AC112; A2
522	22 March	8204/604	36B	Intelsat 705. AC 115; A2AS
523	24 March	45E	3W	DMSP F13
524	7 April	8111/211	36A	AMSC 1. AC 114; A2A
525	23 May	5057/3249	36B	GOES 9. AC77; A1
526	31 May	8112/212	36A	UFO 5. AC 116; A2
527	31 July		36B	DSCS 3B-7. AC-118; A2A
528	29 August	8205/605	36B	JCSat 3. AC 117; A2AS
529	22 October	8114/214	36A	UFO 6. AC119; A2
530	2 December	8206/606	36B	Soho.AC-121; A2AS
531	15 December	8115/215	36A	Galaxy 3R. AC 120; A2A
1996				
532	1 February		36B	Palapa C1. AC126; A2AS
533	3 April		36A	Inmarsat 3F1. AC 122; A2A
534	30 April		36B	SAX. AC-78; A1
535	25 July		36B	UFO7. AC125; A2
536	8 September		36B	GE 1, AC123, Atlas 2A
537	21 November		36B	Hot Bird 2, AC 124, Atlas 2A
538	18 December		36A	INMARSAT-3 F3, AC-129, Atlas 2A
1997				
539	17 February		36B	JCSAT 4, AC-127, Atlas 2AS
540	8 March		36A	Tempo 2, AC 128, Atlas 2A
541	25 April		36B	GOES 10, AC-79, Atlas1 (final Atlas1 launch)
542	28 July		36B	Superbird C
543	4 September		36A	GE 3
544	4 October		36B	EchoStar 3
545	25 October		36A	USA 134
546	8 December		36B	Galaxy 8I
1998				
547	29 January		36A	USA 137
548	28 February		36B	Intelsat 806
549	16 March		36A	UFO 8
550	18 June		36A	Intelsat 805
551	9 October		36B	Hot Bird 5
552	20 October		36A	UFO 9
1999				
553	16 February		36A	JC-SAT 5
554	12 April		36A	Eutelsat
555	23 September		36A	Echastar 5
556	23 November		36A	UF010
557	18 December		SLC3E	Terra
2000				
558	21 January		36A	DSC5388
559	3 February		36B	Hispasat 1C
560	3 May		36A	GOES L
561	24 May		36B	Eutelsat W4
562	30 June		SLC36	TDRS 8
563	14 July		36B	Echostar VI
564	20 October		36A	DSCS3BII
565	6 December		36A	NRO
2001				
566	19 June		36B	ICO
567	23 July		36A	GOES M
568	8 September		36B	MLV-10
569	11 October		36A	MLV-12
2002				
570	21 February		SLC36	EchoStar 7
571	8 March		36A	TDRS-I
572	21 August		SLC36	Hot Bird 6
573	18 September		SLC36	Hispasat 1D
574	4 December			
2003				
575	12 April		SLC36	Asiasat
576	14 May		SLC41	HellaSat
577	17 July		SLC41	Rainbow 1

* denotes launch failure (June 1984 due to Centaur leak after abnormal separation shock; March 1987 due to lightning strike; April 1991 and August 1992 due to Centaur engine failures; March 1993 due to Atlas booster low thrust, payload attained inadequate low orbit). 36A/B = Canaveral launch; 3E/W = Vandenberg. AC is Atlas Centaur designator (Atlas 2 series began at #101). Launch number at left includes all Atlas vehicles flown. A1= Atlas 1; A2 = Atlas 2; A2A = Atlas 2A; A2AS = Atlas 2AS; A2AR = Atlas 2AR

at cut-off. The LOX tank is extended by 2.60 m to accommodate the change in mixture ratio.
Engines: NPO Energomash/P&W RD180 two chamber engine throttleable 40–105%, nominal 3,827 kN SL/4,152 kN vacuum. The chambers are gimballed hydraulically to provide three-axis control during Atlas burn.

Overall length: 28.91 m
Principal diameter: 3.05 m
Inert mass: 13,725 kg
Propellant: 183,200 kg LO_2 + RP1 (aft)
Stage thrust: 3,827 kN (sea level); 4,152 kN (vacuum)
Burn time: 184 s

Interstage adapter
Comment: The 4.42 m long, 3.05 m diameter ISA supports Centaur until separation at 185 s is effected by a flexible linear shaped charge around the forward circumference. Construction is an aluminium skin/stringer and frame.

Atlas 3A stage 2
Comment: Essentially the same Centaur stage as that used by the Atlas 2AS but with a single-engine RL10A-41.
Overall length: 10.06 m
Principal diameter: 3.05 m
Propellant: 16,930 kg LH_2 + LOX(aft)
Stage thrust: 99.2 kN
Isp: 450.5 s
Burn time: ~651 s

Fairing/payload accommodation
Comment: 3A will adopt the Block 2 fairing upgrade, to be introduced on 2AS. Other specifications as for earlier models.
Acceleration load: 5.0 *g* maximum longitudinal, 0.4 *g* lateral.
Acoustic: unavailable
Max Q: 427 N/cm^2

Atlas 3B
First launch: 21 February 2002
Launch site: Canaveral pad 36B, Vandenberg SLC3E
Performance
LEO (185 × 185 km, 28.5° Canaveral): 10,718 kg (extended fairing), 10,759 kg (large fairing)
GTO (167 × 35,786 km, 27.0°): 4,119 kg (single-engine Centaur), 4,400 kg (large fairing) or 4,477 kg (extended fairing) both with twin-engine Centaur.
Number of stages: Two
Overall length: 54.5 m
Diameter: 3.05 m
Launch mass: 225,392 kg
Launch thrust: 3,827 kN sea level
Guidance: From upper stage

Atlas 3B stage 1
Comment: The two-chamber throtteable RD-180 identified for the Atlas 3A is the same as that used for the Atlas 3B and all stage parameters are the same.

Atlas 3B stage 2
Comment: The Centaur stage can be fitted with single- or double-engine configuration, each designated RL10A4-2. In either case the extended length Centaur is standard with an additional 180 kg inert mass for the double engine version.
Engines: One or two Pratt & Whitney RL10A-42 restartable engines
Overall length: 11.74 m
Principal diameter: 3.05 m
Propellant: 20,830 kg LH_2 + LOX (aft)
Stage thrust: 99.2 kN (single), 198.4 kN (twin)
Burn time: ~907 s (single); 354 s (twin)
Isp: 450.5 s

Atlas 5 400 series
First launch: 21 August 2002
Launch sites: CCAFS LC-41, Vandenberg SLC3W
Principal users: heavy and medium class telecom and meteorological satellites into GTO
Performance
LEO (185 × 185 km, 28.5°): 12,500 kg (no solids)
GTO (167 × 35,786 km, 27°): 4,950 kg, no solids; 5,950 kg one solid; 6,830 kg two solids; 7,640 kg three solids.
Number of stages: 2
Guidance: Honeywell Inertial Navigation Unit mounted on Centaur forward equipment module
Launch sequences into GTO 27°(sec):

00.0	Liftoff
236	Core engine cutoff
241	Core jettison
251	Centaur main engine start
259	Payload fairing jettison
937	Centaur main engine cutoff
1,162	Start turn to Centaur second start
1,502	Centaur second start
1,720	Centaur second cutoff
1,730	Start turn to separation attitude
1,945	Spacecraft separation
2,045	Start turn to CCAM attitude
4,070	Centaur end of mission

Atlas 5 400 stage 1 Common Core Booster (CCB)
Comment: The CCB is a new, bigger, core stage using the RD-180 for propulsion and facilitating options for up to three solid rocket boosters, or none.
Engines: One dual-chamber Energomash RD-180 booster engine with eight solid propellant retro rockets for separation.
Overall length: 32.46 m
Principal diameter: 3.81 m
Inert mass: 20,743 kg
Propellant: 284,089 kg LOX + RP-1 (aft)
Stage thrust: 3,827 kN (sea level); 4,152 kN (vacuum)
Burn time: 236 s

Atlas 5 400 Solid Rocket Boosters
Comment: Atlas 6 can carry up to five SRBs with the 5 m diameter large fairing but the 400 series with the 4 m diameter fairing can carry up to three SRBs. All SRBs are ground lit.
Overall length: 19.5 m
Principal diameter: 1.55 m
Fuelled mass: 46,494 kg
Thrust: 1,361 kN (each)
Impulse: 275 s
Nozzle cant: 3°
Burn time: 99 s

Atlas 5 Centaur Interstage Adaptor
Length: 3.13 m
Diameter: 3.05 m
Mass: 374 kg
Structure: Aluminium lithium skin stringer and frame

Atlas 5 CCB Conical Interstage Adaptor
Length: 1.65 m
Diameter: 3.81 m (bottom); 3.05 m (top)
Mass: 420 kg
Structure: Graphite epoxy composite with aluminium ring frames

Atlas 5 400 Stage 2
Comment: Centaur for Atlas 5 is identical to that developed for the Atlas 3 series but with extended high-performance motor nozzle.
Length: 12.68 m (with extendable nozzle)
Principal diameter: 3.05 m
Engines: One or two Pratt & Whitney RL10A-42 restartable
Propellant: 20,830 kg LH_2 + LOX
Stage thrust: 99.2 kN (single); 198.4 kN (twin)
Isp: 450.5 s
Burn time: ~904 s

Atlas 5 500 series
First launch: 17 July 2003
Launch sites: CCAFS LC-41, Vandenberg SLC3W
Performance:
LEO (185 × 185 km, 28.5°): 10,300 kg, no solids; 12,590 kg, one solid; 15,080 kg, two solid; 17,250 kg, three solid; 18,955 kg, four solid; 20,520 kg, five solid.
GTO (167 × 35,786 km, 27°): 3,970 kg, no solid; 5,270 kg, one solid; 6,285 kg, two solid; 7,200 kg, three solid; 7,980 kg, four solid; 8,670 kg, five solid.
GSO (35,786 km circ, O°): 2,680 kg, two solid; 3,190 kg, three solid; 3,540 kg, four solid; 3,810 kg, five solid.

Atlas 5 HLV
Launch sites: CCAFS LC-41 Vandenberg SLC 3W
Principal users: Heavy and medium class telecom and meteorological satellites into GTO or GSO
Performance:
LEO (185 × 185 km, 28.5°): 25,000 kg
GTO (167 × 37,786 km, 27°): 12,650 kg
GSO (37,786 km circ, 0°): 6,350 kg

Atlas 5 HLV Stage 1
Comment: CCB as developed for Atlas 3 and 5.

Atlas 5 Liquid Rocket Booster
Comment: Uniquely developed from the CCB differing only in structural attachments. All parameters except length and mass are the same.
Overall length: 35.99 m
Inert mass: 21,902 kg

Atlas 5 Centaur Interstage Adapter
Length: 3.53 m
Diameter: 3.81 m
Mass: 1,096 kg

Atlas 5 Cylindrical Interstage Adapter
Length: 0.6 m
Diameter: 3.81 m
Mass: 278 kg

Atlas 5 HLV Stage 2
Comment: Centaur C3 as Atlas 3 and 5.

UPDATED

Centaur Upper Stage

Current Status
Supporting evolution of the Atlas series of launch vehicles, performance and operating capabilities have been improved utilising upgraded engine specifications and extended burn times. Last Centaur launched atop a Titan 4 was sent into space on 9 September 2003.

Background
GD began development of the world's first LOX/LH_2 upper stage in the late 1950s. Through 1995 it was launched 120 (including nine R&D) times with seven operational flight failures. Seven out of 97 operational Atlas Centaur launches failed, but none of the 14 operational Titan Centaurs did. The stage continues in service with the Atlas 1/2 vehicles.
A Centaur G prime derivative debuted February 1994 on Titan 4. NASA purchased the Centaur G and the G prime for applications aboard the Shuttle by increasing LH_2 tank diameter to 4.32 m. The longer G prime version can accommodate almost 50 per cent greater propellant loading than

Each Centaur upper stage requires paired RL10s, designated C-1 and C-2 (Lockheed Martin) 0516972

Final Centaur assembly is now undertaken in a new building in Denver. Two Titan 4-type stages are at left and two Atlas-type at right. The tanks are either pressurised or mechanically tensioned to prevent the thin-wall structures from collapsing (Lockheed Martin)
0516927

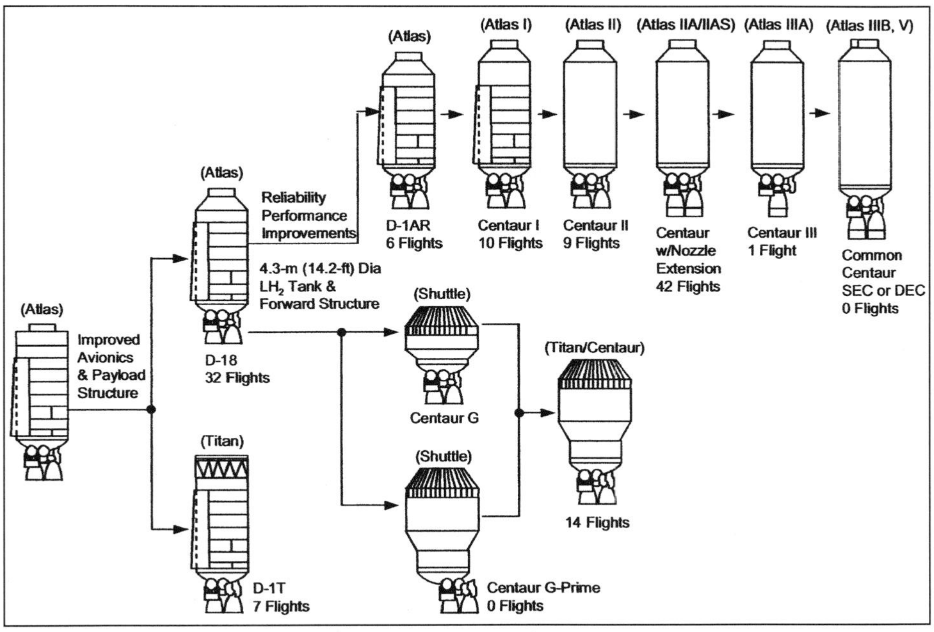

Centaur evolution over 40 years since 1962 when the stage first flew 0131860

the Atlas 1 version. The USAF's G model was 3 m shorter to provide maximum payload length within the cargo bay but could still deliver 4,500 kg to GEO.

By late 1985, there were seven orders, four G variants for classified payloads and three of the longer G prime for Galileo, Ulysses and Magellan. The G version development cost US$280 million, with NASA adding a further US$88 million for modifications to the *Challenger* and *Atlantis* Shuttle orbiters, and US$80 million for changes to the Shuttle launch pad. Among the modifications, the air force changed the old Atlas Centaur pad 36A for the new version of the upper stage. Orbiter modifications included dump valves to permit jettison of 20 tonnes of cryogenic propellants in a few seconds in an abort.

Centaur G and G prime were cancelled in June 1986 as Shuttle stages because of increased safety concerns over carrying a liquid vehicle inside a manned spaceplane. Development and cancellation costs totalled an estimated US$1 billion. Those built were scrapped.

Centaur G prime was then adapted for employment with Titan 4, creating the West's most powerful launcher combination and providing the air force with the required 4.5 tonne GEO capability (5.7 tonnes with upgraded Titan boosters). The first Centaur for Titan was delivered to the Cape in December 1990 under a US$1.3 billion contract with Martin Marietta that calls for 15 such stages over five years. This model will continue to use the RL10A33A engine. A single engine version employing the RL10C is under consideration to increase capacity (545 kg more into GSO) and improve reliability and reduce cost; such an Atlas stage would cost about US$105 million to develop.

The upper stage Atlas version carries LOX/LH_2 separated by a double wall vacuum insulated intermediate bulkhead. External insulation on Atlas 1 is provided by glass fibre honeycomb panels jettisoned 25 seconds after booster cut-off. Like Atlas, Centaur is constructed of thin-wall fully monocoque stainless steel, stabilised only by helium pressurisation. An advanced avionics programme for both commercial and Titan Centaurs has replaced the Teledyne digital computer and Honeywell's three-axis inertial platform with a Honeywell INU incorporating RLGs and state of the art processors.

Specifications

Centaur

Engines: one or two P&W RL 10A-42 cryogenic multiple start engines of 194.8 kN total vacuum thrust gimballed for three-axis control; 4 × 27 N (Olin Aerospace MR-106) and 8 × 40 N hydrazine thrusters provide three-axis control during coast, spin-up for payload separation and collision avoidance manoeuvre.
Overall length: 11.74 m (Atlas 3) or 12.68 m (Atlas 5)
Principal diameter: 3.05 m
Propellant: 20,830 kg LOX (aft) + LH_2
Stage thrust: 194.8 kN vacuum
Comment: The single engine configuration, pneumatic system self-test capability and replacing the hydraulic engine gimbal actuators with an electromechanical system reduces cost and increases reliability (reduces part count by 4,200).
Burn time: two burns of 560.3 s + 175.4 s.

Conestoga

Background

Managed by the late Donald K (Deke) Slayton, one of the original Mercury astronauts. SSI designed the Conestoga solid propellant family of orbital launchers capable of handling up to 2.1 tonnes into LEO. It was selected in early 1991 to provide Conestoga launch services for the NASA sponsored Meteor μgravity carrier programme. A preliminary demonstration launch using a single Minuteman 1 second stage motor was conducted in September 1982. In March 1989, SSI fired its first commercial vehicle, a two-stage Starfire sounding rocket for the University of Alabama under a US$1 million contract carrying a μg payload.

SSI's first launcher, Percheron, with a liquid first stage, exploded on its debut from Matagorda Island off the Texas coast 12 August 1981, and the company switched its efforts to the all solid Conestoga designs. Conestoga 1, using a single 206.8 kN thrust Minuteman 1 second stage motor, was successfully fired as a demonstration on a 10.5 min, 313 km altitude suborbital trajectory 9 September 1982 from Matagorda. The current Conestoga orbital versions are based on Thiokol's Castor 4A/4B solids with Star injection stages. The 1679 was selected for BMDO's MSTI-5, although no funding was provided. The debut (1620) October 1995 launch self-destructed at 45 seconds after faulty navigation data produced extensive nozzle movements and depleted the TVC hydraulic fluid. A lack of customer support has failed the programme and Conestoga is a casualty of diminished launcher prospects in a static market.

Specifications

Conestoga 1620

First launch: 23 October 1995 from Wallops Island
Number launched: 1, to March 2002
Launch sites: Wallops Island
Principal applications: small payloads to LEO
Vehicle success rate: 0%.
Schedule of future missions: none known
Performance: 945 kg into 463 km, 40°
Availability: 20 month baseline
Cost: about US$18 million (65% is cost of solids)
Number of stages: 4
Overall length: 15.2 m with 4.88 m long fairing
Principal diameter: 1.2 m
Launch mass: 87,407 kg
Avionics/guidance: all configurations use TVC for three-axis control through the stage 1/2 phases. The six motor cluster uses only four motors for control authority; the other two have fixed (noncanted) nozzles. TVC provides pitch/yaw control during core burn; RCS system in Payload Attach Fitting provides roll during stage 3 and three-axis during stage 4 coast/burn. The H-MACS Hydrazine Manoeuvring & Attitude Control System can be carried on top of Star for three-axis control and orbit trim. The avionics module is housed in the interstage structure above the core. The Mars guidance system (two tuned rotor gyros + three accelerometers; Guidance Dynamics Corp) interfaces with the 1553 avionics bus to provide all TVC, sequencing, staging and prelaunch monitoring/control. Injection accuracy: ± 37 km/±0.20° 3Σ (±18.5 km/±0.05° 3Σ with H MACS)
Performance:
Conestoga 1229 295 kg to 180 km, 90°; 363 kg to 180 km, 38°. Two stage 1 Castors strapped to stage 2 core Castor. Injection by Star 48V stage 3.
Conestoga 1620 1,000 kg to 180 km, 90°; 1,180 kg to 180 km, 38°. Six Castors (4 stage 1 + 2 stage 2) clustered around stage 3 core Castor. Injection by Star 48V stage 4.
Conestoga 1679 1,250 kg to 180 km, 90°; 1,500 kg to 180 km, 38°. As 1620 but Star 63 stage 4.
Launch sequence (s) into 463 km 40°:

00.0	stage 1 ignite
56.4	stage 1 4A sep 16.35 km
59.4	stage 2 ignite 18.01 km
66.4	stage 1 4B sep 22.03 km
119.6	stage 3 ignite 64.58 km
125.6	stage 2 sep 70.93 km
147.6	fairing sep 96.80 km
185.6	stage 3 sep 157.87 km
475.7	stage 4 ignite 460.3 km
562.0	stage 4 burnout 463.0 km
742.0	payload sep 463.0 km

Conestoga 1620 stage 0

Motor: two Castor 4A + two Castor 4B (Thiokol solids) 4Bs provide 6° blowdown TVC
Stage length: 9.14 m
Thrust: 1,884 kN avg
Burn time: 52/61 s Castor 4A/4B
Mass at ignition: 47,137 kg
Propellant mass: 40,164 kg.

Conestoga 1620 stage 1

Motor: two Castor 4B (as stage 1), 6° blowdown TVC
Stage length: 9.14 m
Thrust: 869.8 kN average vacuum, 1,049.7 kN peak
Burn time: 61 s
Mass at ignition: 23,764 kg
Propellant mass: 19,949 kg.

Conestoga 1620 stage 2

Motor: one Castor 4B, 6° blowdown TVC; as stage 1 but with increased expansion ratio (12.97) for altitude operation.
Stage length: 9.24 m
Stage diameter: 101.6 cm
Thrust: 434.9 kN average vacuum, 524.9 kN peak.
Burn time: 61 s
Mass at ignition: 11,914 kg
Propellant mass: 9,974 kg.

Conestoga 1620 stage 3

Motor: Thiokol Star 48V, with 4° pitch/yaw TVC (Meteor cold gas thrusters provide roll).
Stage length: 2.03 m
Stage diameter: 124.5 cm
Thrust: 66 kN average vacuum
Burn time: 52 s
Mass at ignition: unavailable
Propellant mass: 9,974 kg.

Payload accommodation

1.829 m diameter composite fairing provides lengths of 4.27 6.56 m. Separation system derives from Delta technology.
Acceleration: 11 *g* longitudinal
Acoustic: 128.5 dB for acceptance tests.

DC-XA Delta Clipper

Current Status

In the wake of the loss of Shuttle Columbia on 1 February 2003, technologies developed for Delta Clipper are being re-examined regarding thermal protection concepts. Delta Clipper continues to be quoted as a concept for private enterprise or initiative for future reusable launcher requirements but no investment has been attracted in the wake of a lack of interest from NASA or the DoD.

Background

McDonnell Douglas developed the DC-X vehicle for BMDO's Single Stage Rocket Technology programme, making a total of five flights at White Sands Missile Range by mid-1994. All flights in this first series ended in failure. Three further flights costing US$5 million to US$6 million were successful in May-July 1995 to complete the series.

McDonnell Douglas Space Systems was awarded BMDO's 24-month US$58.9 million Phase 2 contract 16 August 1991 for the design of an

SSTO to provide a launch and rapid replacement capability for Brilliant Eyes and Brilliant Pebbles. The experimental DC-X 1/3 scale 11.9 m high, 3.7 m base diameter version (7.2 tonnes dry, 18.88 tonnes at launch, 1.5 tonne LH_2, 7.3 tonne LOX) of their Delta Clipper was demonstrated in summer 1993 at White Sands. Four P&W RL10A5 provide propulsion oxygen/hydrogen engines derived from the RL10A-3 proven by long service on the Centaur upper stage. P&W's US$12 million contract called for four engines plus spares for a fifth. Each 143 kg engine, incorporating a conical nozzle for low altitude operations, can be throttled 30 100 per cent (19.5–65.7 kN) and gimballed ±8°. Four Aerojet gaseous O_2/H_2 1.78 kN thrusters provide RCS.

Phase 1 evaluated three basic launch/recovery concepts: vertical take-off/landing; vertical take-off/horizontal landing; horizontal take-off/landing. By eliminating the dead weight of wings, vertical/vertical is the lightest and most cost effective design. Even without wings, cross-range would be more than 2,970 km. Very high controllability from four body flaps, GPS navigation and powered landings (at 15 to 20 per cent thrust) would enable it to operate in rain and high winds that would force a diversion for Space Shuttle. Flight control would be totally autonomous and handled by onboard computers. The crew module would be removed for unmanned missions. Airline type operations were planned:

The first ascent (18 August 1993) was a hop to 80 m,demonstrating throttling but remaining vertical throughout. The second (11 September) attained 100 m and the third (30 September) 370 m. Two flights were planned 20/23 October to attain 1.3 km plus various manoeuvres and demonstrate a three-day turnaround, but an engine problem delayed the first flight and funding ended 22 October.

NASA, in early 1994, provided US$1 million to prepare DC-XA for further tests, and DoD in May 1994 released another US$5.1 million for five flights over June to July 1994. Flight 4 attained 790 m on a 135.9 second venture 20 June 1994, for the first time with a full propellant load. Flight 5, intended to last for 136 seconds on 27 June 1994, was aborted after an explosion at ignition blew a 1 × 4 m hole in the composite aeroshell. Controller Pete Conrad triggered the automatic abort sequence at 17 seconds and DC- X landed safely at 78 seconds. The cause was identified as vehicle-vented GH_2 collecting in a service trench.

Flight 6, the first of the new series, came 16 May 1995: 124 seconds, 1,300 m altitude, 15° angle of attack, landed 340 m downrange. Firsts included GPS data for navigation and differential throttling for primary attitude control. Number 7 on 12 June, 132 seconds, 1,900 m altitude, 070° AOA, 590 m downrange. The gaseous attitude thrusters were used for the first time. Number 8: 7 July, 124 seconds, 2,500 m altitude. Pitched to 10° to show that an SSTO could enter the atmosphere nose first and then flip over for landing. Faulty radar altitude readings produced a heavy landing, cracking the aeroshell and cancelling flight 9.

As a result of White House policy decisions on the future route for US launchers, DC-X was transferred to NASA after flight 8.

Following BMDO's Phase 2, the full size 38.7 m-high 458 tonne (36 tonne dry) DC-Y prototype was proposed. Payload cost for the operational version would be around US$250 2,500/kg, depending on flight rates. The goal was for 4,500 kg into 740 km polar and 10,000 kg LEO.

McDonnell Douglas predicted that less than 100 people could maintain all operational and ground support needs for a fleet of six, each with a three-day turnaround. Cargo containers would be loaded à la commercial airfreight into a 4.57 × 4.57 × 6.7 m bay. A major factor in reducing ground processing is the lack of Shuttle type thermal tiles; a maximum 1,430°C around the carbon/carbon nose is predicted, some 440°C below maximum protective capability. The flaps and the lower temperature nose areas would be protected by a carbon/silicon carbide composite. The underbody and other medium temperature areas would carry Inconel 718 metal multiple layers. The aeroshell would be a graphite epoxy/aluminium honeycomb sandwich supported by a P75/P100 graphite/epoxy truss. The LOX tank would be an aluminium/copper/lithium alloy (Weldalite 049); LH_2 woven P75 graphite cloth and toughened epoxy. DC-Y would carry eight bell nozzle engines, four with extending nozzles. They would fire at 85 per cent for the 1.3 *g* launch, reserving 100 per cent for engine-out situations and 110 per cent in more extreme cases. The goal was for 200 flights between tear-down inspection. Currently there are no plans to conduct further flights.

Delta Series

Current Status

Pinnacle of a development success story almost 45 years old, the latest and most powerful Delta launch vehicle derivative successfully carried the Eutelsat W5 satellite into orbit when the first Delta 4 lifted off from LC-37B on 20 November 2002. Almost 30 Delta 4 flights are booked through 2007. Built for the NASA Saturn 1 and 1B rockets in the 1960s, LC-37 was extensively modified for Delta 4 flights.

Since 1960, through to 1 September 2003, the overall Delta record has achieved a 94.7 per cent success factor, with 285 out of 301 flights.

Background

In 1959, NASA's Goddard Space Flight Center contracted the Douglas Aircraft Company to develop an interim space launcher from the company's Thor IRBM and Vanguard upper stages programmes. Douglas had been working on the Thor since 1954 for the USAF. NASA envisioned a vehicle capable of putting about 45 kg into GTO. As the space race matured, Delta's growth potential and the need for a medium launcher for communications, meteorological and science satellites paced a continuous set of improvements to the original vehicle. The first Delta launch took place on 13 May 1960 but a second stage malfunction prevented the vehicle from reaching orbit. On the second launch, on 12 August 1960, the Delta successfully launched NASA's Echo passive communications satellite. The basic Delta was used between 1960 and 1962 successfully launching Tiros weather satellites, scientific research satellites and the first UK satellite Ariel 1. In October 1962 the Delta A was introduced with the uprated MB-3 first stage motor and a payload capability of 320 kg with the Delta B incorporating increased tankage and a modified second stage motor increasing payload capacity to 375 kg.

Strap-ons were first added in 1964, and in 1965 the booster propellant capacity was increased, a larger fairing was introduced and both upper stages were upgraded. The number of strap-ons was increased to six in 1970 and to nine in 1972, at the same time as Aerojet's Titan Transtage engine was adopted for stage 2. Further improvements were made in 1973 with the introduction of Rocketdyne's RS-27 stage 1 engine and in 1975 with the introduction of Thiokol's Castor 4 strap-ons to replace the Castor 2s. These latter two configurations, when using TRW's TR201 engine, derived from the Apollo Lunar Module's descent engine, were designated the Delta 2914 and 3914, respectively. First employed on Delta in 1974, a total 69 TR201 engines between this date and the last in 1988.

Under the four-digit system of type designation, the first digit indicated the type of first stage engine and solid rocket motors, the second digit indicates the number of solid rocket motors, the third digit indicates the type of second stage and the fourth digit indicates type of third stage.

UPDATED

Delta-2

Current Status

Delta 2 continues to serve customers until the Delta 4 replaces it. Launch rates have gone from six in 2000 to seven in 2001 and three in 2002. Four Delta 2 launches had taken place in the first eight months of 2003. Reliability for Delta 2 has been 99.1 per cent since inception in 1989, or 100 per cent in the last 50 launches.

Background

NASA ordered Delta's Huntington Beach production line to close down in 1984 but Shuttle's failure in January 1986 resulted in its reactivation eight months later to produce three further vehicles (#182 to 184) under a US$130 million contract. Delta's subsequent emergence as a commercial vehicle resulted from its selection as USAF's Medium Launch Vehicle (MLV) in January 1987 under an initial US$316.5 million contract. The basic buy of seven eventually expanded to 20 when the air force exercised two options in February 1988 (seven vehicles) and December 1988 (six vehicles), totalling US$680 million, to launch the network of GPS Navstars offloaded from Shuttle.

Delta became associated with civil science again when the US$140.6 million contract for the Geotail, Wind and Polar science satellites was signed with NASA in December 1990. It included VAFB pad modifications and pre-priced options for 12 further vehicles to mid-1998. The previous 3920/PAM models could handle 1,284 kg to GTO, but the Delta 2 6925 raised that capacity to 1,447 kg, and the 7925 to 1,841 kg.

The most recent NASA/USAF ELV reimbursement agreement of October 1987 sets the cost to NASA for a Delta at US$25.6 million cost (FY89 rate), and in 1990 the USAF agreed to pay US$37 million plus US$4 million launch services.

USAF Delta 2 requirements called for a switch to the more capable 7925 following their first nine launches to cope with the heavier GPS Navstar Block 2A models. The 7925 similarly incorporates a 2.9 m diameter fairing and 3.6 m longer first stage but adds graphite epoxy strap-ons and uprated stage 1 thrust. 6925 made its last flight with Geotail in 1992. 7925 was selected April 1993 for USAF's MLV3 contract. 25 are required (+11 options) to launch the GPS Block 2R replenishment architecture from 1997–2002.

Launches are required within 60 days of request. GPS performance is improved to 2,141 kg and GTO to 1,882 kg, principally by increasing the expansion ratios of the three air-lit strap-ons. MLV-3 employs the triply redundant RIFCA Redundant Inertial Flight Control Assembly and the improved FTS Flight Termination. The fairing is modified so that the satellite, stage 3 and fairing can be transported to the pad as a single unit. The 7920 series differs from the 6920 principally with the adoption of the 12:1 expansion ratio RS27A (also known as RS-2701C) stage 1 engine and more powerful solid strap-ons. The 7925A additional option (not exercised) is to enlarge the strap-ons to 116.8 cm diameter. This would improve GPS performance to 2,254 kg and GTO to 1,987 kg.

Delta 7320, 7326, 7420, 7425, 7426 series

MDA/OSC teamed to bid for NASA's Med-Lite contract, providing a performance in the current gap below Delta at half the cost (US$25 million). The team was selected March 1995, and the contract awarded February 1996 for five firm launches from 1998. Two vehicles were to have satisfied the requirements: MDA's DeltaLite and OSC's Taurus, but MDA now has no plans to develop Delta Lite. That would have comprised Castor 120 stages 1/2, the current Delta 2 stage 2 as stage 3 and a Star 37 kick motor. This combination would deliver 1,930 kg into 185 km, 28.7°; 1,463 kg 185 km, 90°; 650 kg GTO; 331 kg C^3 10 km^2/s^2. With two Castor 4B strap-ons, these capacities would be 2,540/1,968/837/440 kg, respectively. Taurus equivalent capacities would be: 1,770/1,435/445/230 kg and 2,180/1,750/640/275 kg. Instead, the Delta 7325 will be used at a special discount. Equivalent capacity is: 2,865/2,095/1,004/594 kg. It was reported in February 1996 that this vehicle, although double the cost, will replace Delta-Lite and the excess capacity be offered for other payloads.

Delta 7925 GPS launch from Canaveral's pad 17A
0517179

Delta launches (from 1978)

No	Payload	Date	Pad	Type
138	IUE	26 January 1978	17A	2914
139	Landsat 3/Oscar	5 March 1978	2W	2910
140	BSE	7 April 1978	17B	2914
141	OTS 2	11 May 1978	17A	3914
142	GOES 3	16 June 1978	17B	2914
143	Geos 2	14 July 1978	17A	2914
144	ISEE3	12 August 1978	17B	2914
145	Nimbus 7	24 October 1978	2W	2910
146	NATO 3C	19 November 1978	17B	2914
147	Anik 4	16 December 1978	17A	3914
148	Scatha	30 January 1979	17B	2914
149	Westar 3	10 August 1979	17A	2914
150	RCA Satcom 3	7 December 1979	17A	3914
151	Solar Max	14 February 1980	17A	3910
152	GOES 4	9 September 1980	17A	3914
153	SBS 1	15 November 1980	17A	3910/PAM
154	GOES 5	22 May 1981	17A	3914
155	Dyn Exp 1/2	3 August 1981	2W	3913
156	SBS 2	24 September 1981	17A	3910/PAM
157	SME + Uosat 1	6 October 1981	2W	2310
158	RCA Satcom 3R	20 November 1981	17A	3910/PAM
159	RCA Satcom 4	16 January 1982	17A	3910/PAM
160	Westar 4	26 February 1982	17A	3910/PAM
161	Insat 1A	10 April 1982	17A	3910/PAM
162	Westar 5	9 June 1982	17A	3910/PAM
163	Landsat 4	16 July 1982	2W	3920
164	Anik D1	26 August 1982	17B	3920/PAM
165	RCA Satcom 5	28 October 1982	17B	3924
166	IRAS	26 January 1983	2W	3910
167	RCA Satcom 6	11 April 1983	17B	3924
168	GOES 6	28 April 1983	17A	3914
169	Exosat	26 May 1983	2W	3914
170	Galaxy 1	28 June 1983	17B	3920/PAM
171	Telstar 3A	28 July 1983	17A	3920/PAM
172	RCA Satcom 7	8 September 1983	17B	3924
173	Galaxy 2	22 September 1983	17A	3920/PAM
174	Landsat 5 + UoSat 2	1 March 1984	2W	3920
175	AMPTE	16 August 1984	17A	3924
176	Galaxy 3	21 September 1984	17B	3920/PAM
177	NATO 3D	13 November 1984	17A	3914
178*	GOES G	3 May 1986	17A	3914
180	DoD (SDI)	5 September 1986	17B	3920
179	GOES 7	26 February 1987	17A	3924
182	Palapa B2P	20 March 1987	17B	3920/PAM
181	DoD (SDI)	8 February 1988	17B	3910
184	Navstar 21	14 February 1989	17A	6925
183	Delta Star	24 March 1989	17B	3920
185	Navstar 22	10 June 1989	17A	6925
186	Navstar 23	18 August 1989	17A	6925
187	Marcopolo 1	27 August 1989	17B	4925
188	Navstar 24	21 October 1989	17A	6925
189	COBE	18 November 1989	2W	5920
190	Navstar 25	11 December 1989	17B	6925
191	Navstar 26	24 January 1990	17A	6925
192	RME/LACE	14 February 1990	17B	69208
193	Navstar 27	26 March 1990	17A	6925
194	Palapa B2R	13 April 1990	17B	69258
195	Rosat	1 June 1990	17A	692010
196	Insat 1D	12 June 1990	17B	4925
197	Navstar 2–8	2 August 1990	17A	6925
198	Marcopolo 2	18 August 1990	17B	6925
199	Navstar 29	1 October 1990	17A	6925
200	Inmarsat 2 F1	30 October 1990	17B	6925
201	Navstar 210	26 November 1990	17A	7925
202	NATO 4A	8 January 1991	17B	7925
203	Inmarsat 2 F2	8 March 1991	17B	6925
204	ASC 2	13 April 1991	17B	7925
205	Aurora 2	29 May 1991	17B	7925
206	Navstar 211 +Losat	4 July 1991	17A	7925
207	Navstar 2–12	23 February 1992	17B	7925
208	Navstar 2 13	10 April 1992	17B	7925
209	Palapa B4	14 May 1992	17B	7925 8
210	EUVE	7 June 1992	17A	6920-10
211	Navstar 2 14	7 July 1992	17B	7925
212	Geotail	24 July 1992	17A	6925
213	Satcom C4	31 August 1992	17B	7925
214	Navstar 2 15	9 September 1992	17A	7925
215	DFS 3	12 October 1992	17B	7925
216	Navstar 2–16	22 November 1992	17A	7925

No	Payload	Date	Pad	Type
217	Navstar 2 17	18 December 1992	17B	7925
218	Navstar 218	3 February 1993	17A	7925
219	Navstar 2-19/SEDS	30 March 1993	17A	7925
220	Navstar 220	12 May 1993	17A	7925
221	Navstar 2–21	26 June 1993	17A	7925
222	Navstar 222	30 August 1993	17B	7925
223	Navstar 2 23	26 October 1993	17B	7925
224	NATO 4B	8 December 1993	17A	7925
225	Galaxy 1RR	19 February 1994	17B	79258
226	Navstar 224/SEDS	10 March 1994	17A	7925
227	Wind	1 November 1994	17B	792510
228[1]	Koreasat 1	5 August 1995	17B	7925
229	Radarsat	4 November 1995	2W	792010
230	XTE	30 December 1995	17A	792010
231	Koreasat 2	14 January 1996	17B	7925
232	NEAR	17 February 1996	17	7925
233	Polar	24 February 1996	2W	7925 10
234	Navstar 225	28 March 1996	17	7925
235	MSX	24 April 1996	2W	7920 10
236	Galaxy 9	24 May 1996	17B	7925
237	Navstar 226	16 July 1996	17	7925
238	Navstar 227	12 September 1996	17	7925
239	Mars Global Surveyor	7 November 1996	17	7925
240	Mars Pathfinder	4 December 1996	17	7925
241*	Navstar 2R1	17 January 1997	17	7925
242	Iridium 4 8	5 May 1997	17	7920
243	Thor 2A	20 May 1997	17	7925
244	Iridium (×5)	9 July 1997	17	7920
245	Navstar 29	23 July 1997	17	7295
246	Iridium (×5)	21 August 1997	17	7920
247	ACE	25 August 1997	17	7920
248	Iridium (×5)	26 September 1997	17	7920
249	Navstar 38	6 November 1997	17	7925
250	Iridium (×5)	9 November 1997	17	7920
251	Iridium (×5)	20 December 1997	17	7920
252	Skynet 4D	10 January 1998	17	7925
253	Globalstar (×4)	14 February 1998	17	7420
254	Iridium (×5)	18 February 1998	17	7920
255	Iridium (×5)	30 March 1998	17	7920
256	Globalstar (×4)	24 April 1998	17	7420
257	Iridium (×5)	17 May 1998	17	7920
258	Thor 3	15 June 1998	17	7925
259*	Galaxy 10	27 August 1998	17	Delta 3
260	Iridium (×5)	8 September 1998	17	7920
261	Deep Space-1/ SEDSAT	24 October 1998	17	7326
262	Iridium (×5)	6 November 1998	17	7920
263	Bonum 1	23 November 1998	17	7925
264	Mars Climate Orbiter	11 December 1998	17	7424
265	Mars Polar Lander	27 January 1998	17	7425
266	Stardust	7 February 1999	17	7426
267	Argos/Orsted/Sunsat	23 February 1999	17	7920
268	Landsat 7	15 April 1999	17	7920
269*	Orion 3	4 May 1999	17	8930
270	Globalstar 3	10 June 1999	17	7925
271	FUSE	24 June 1999	17	7925
272	Globalstar 4	10 July 1999	17	7925
273	Globalstar 5	25 July 1999	17	7925
274	Globalstar 6	17 August 1999	17	7925
275	GPS	7 October 1999	17	7925
276	Globalstar 7	8 February 2000	17B	7925
277	IMAGE	25 March 2000	SLC2W	7925
278	GPS	11 May 2000	17A	7925
279	GPS	16 July 2000	17A	7925
280	DMF3	23 August 2000	17B	8930
281	GPS	10 November 2000	17A	7925
282	F01	21 November 2000	SLC2W	7320
283	GPS	30 January 2001	17A	7925
284	Mars Odyssey	7 April 2001	17A	7925
285	Geolite	18 May 2001	17B	7925
286	MAP	30 June 2001	17B	7425
287	Genesis	8 August 2001	17A	7326
288	QuickBird	18 October 2001	SLC2W	7320
289	Jason 1/ Timed	7 December 2001	SLC2W	7920-10
290	Iridium (×5)	11 February 2002	SLC2W	7920
291	Aqua	4 May 2002	SLC2W	7920-10L
292	Contour	3 July 2002	17	7425
293	Eutelsat	20 November 2002	37B	4M

* indicates launcher failure. Note 1: Koreasat 1's Delta did not achieve GTO because number 6 GEM failed to separate; the satellite's own station-keeping propellant was required to reach final GEO. The pad designator indicates launch site; 17A/B = Cape Canaveral; 2E/W = WTR. The full list to the end of 1977 is included on p 276 of the 1990/91 edition.

Delta 4925/5920

Based on the 3920, these models were produced as interim models to help ease the RS27 first stage engine shortage until the full Delta 2 manufacturing capacity became available. The 4925 was essentially a 3920/PAM but with uprated Castor 4A strap-ons (as used on Delta 2) compensating for an old 765.06 kN Thor IRBM Rocketdyne MB-3 Block 3 stage 1 engine from storage at Norton AFB, California. Orbital performance was thus similar to that of the 3920/PAM. The 5920, flown only with NASA's Cosmic Background Explorer in November 1989, was a 3920 (that is retaining the RS 27) but with Castor 4A strap-ons providing 1,406/3,849 kg GTO/LEO capacities. Only two 4925s were flown, carrying BSB 1 and Insat 1D, because only two MB3 engines were available, built in 1978 from USAF spares. As the engine required RJ-1 fuel, no longer produced in the US, it had to be imported from Japan, where it remained in use aboard NASDA's H1 vehicle.

Company sources claim a production capacity of 12 to 18 per year is possible from their Huntington Beach (manufacture) and Pueblo, Colorado (assembly; 250 work force) facilities, although the rate is 12 initially. Eight sales appear to be the average at least for the near future.

Delta 2 had problems with the launch of Koreasat 1: on this flight one of the GEM solid propellant boosters failed to separate. Orbit was reached, but it was lower than planned and the satellite had to use its own propellant to reach geosynchronous orbit thus reducing its operational life.

The first complete loss of a Delta-2 came on 17 January 1997 when a 7925 was destroyed seconds after launch, this being due to a malfunction

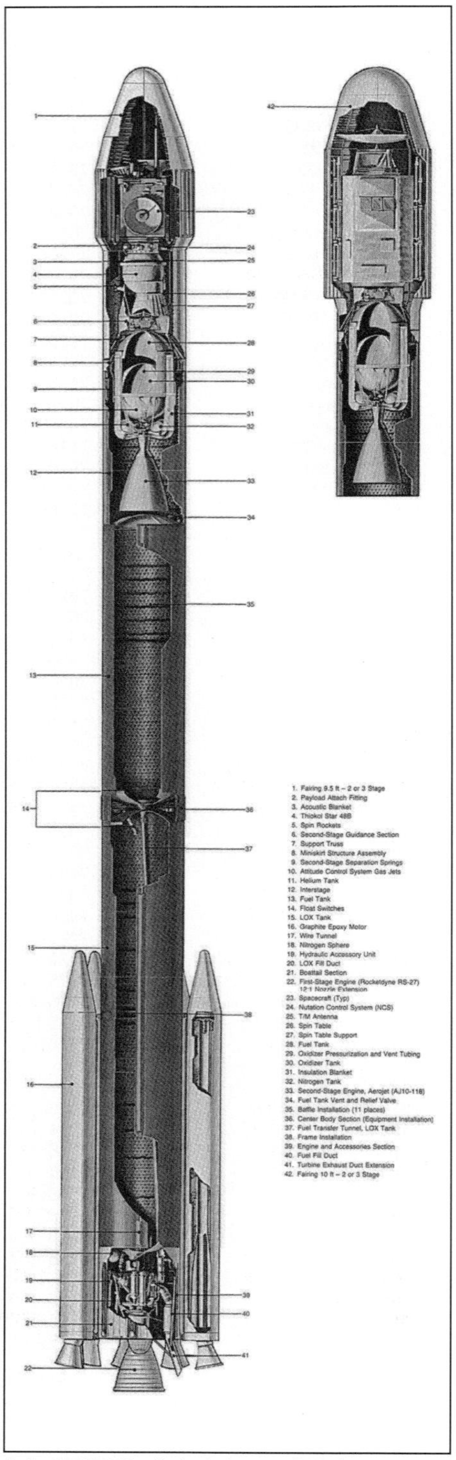

Delta 7925/7920 principal elements (MDA) 0516931

Hoisting Delta stage 2 at pad 17A (NASA) 0517180

of one of the strap-on boosters. The payload – the first Navstar2R satellite was destroyed.

Specifications

Delta 2/Delta 7925/7920

First launch: 14 February 1989
Number launched: 109 to 1 September 2003
Launch sites: Canaveral pads 17A/B, VAFB SLC2W.
Principal uses: launches of GPS Navstar Block 2A; medium class telecom satellites into GTO.
Vehicle success rate: 99.1%
Performance:
GTO: 1,842 kg (1,882 kg from December 1995)
GPS: 2,090 kg (2,141 kg from December 1995)
LEO: 5,039 kg (28.7°) 7920 two-stage
Sun-synchronous: 3,175 kg, 830 km 7920 VAFB
Molniya-type (63.4°, 370 × 40,094 km): 1,275 kg
Lunar delivery: 1,240 kg
Availability: typically 30 months from contract signature to launch.
Number of stages: 3 plus 9 strap-ons (7920 excludes stage 3)
Overall length: 38.41 m
Principal diameter: 2.44 m
Launch mass: 231,870 kg
Launch thrust: 3,110 kN (six solids fire at launch)
Guidance: the digital inertial guidance system is mounted inside a cylinder at stage 2's forward end, controlling the vehicle during stage 1/2 and commanding spinup/separation of stage 3 in addition to triggering its fuze based sequencing system. The system is a strapped down all-inertial unit incorporating a Delco guidance computer and a Delta Redundant Inertial Measurement System (DRIMS) containing three gyros, four accelerometers and conditioning electronics. The computer also issues preprogrammed sequence commands and provides attitude control. GTO perigee altitude error is typically <5.6 km + 0.6° inclination. Typical 3Σ errors for 2-stage missions into 1851,850 km circular orbits are 2 km/0.05°. The triply redundant

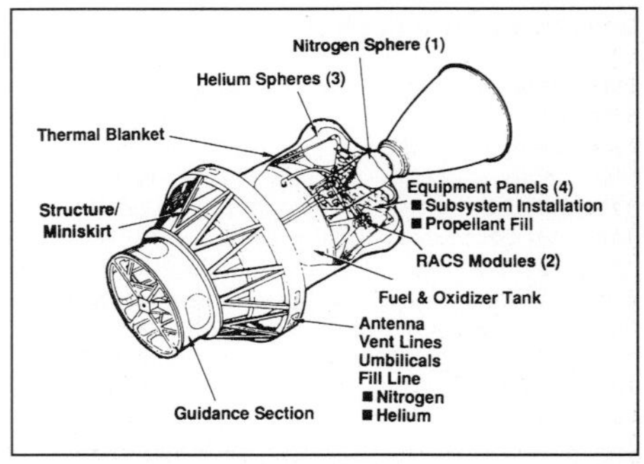

Delta second stage features (MDA) 0516932

RIFCA Redundant Inertial Flight Control Assembly replaced DRIMS from the XTE launch of December 1995. The single fault tolerant RIFCA comprises six RL 20 AlliedSignal ring laser gyros and six QA3000 Sundstrand accelerometers in a triply redundant modular architecture. Each lane contains two sensor channels, a 1750A processor and I/O interfaces. Two lanes have a 1553B bus, for vehicle data, and the third incorporates RS422 for GSE. There are two separate power sources. The data are continuously processed, updating the required course throughout ascent.

Alliant GEM solid strap-ons

Comment: Hercules' Aerospace Group (now Alliant Techsystems) of Magna, Utah was contracted in February 1987 to provide 144 (16 flight sets) higher performance, lightweight Graphite Epoxy Motors for Delta 7925. A second contract was awarded 1991 for a further 117 (13 sets), to begin production 1993. A third award, for 16, was made June 1995. For MLV3, the expansion ratio for the air-lit nozzles is increased from 10.6 to 16.3 by lengthening the nozzle by 30 cm and increasing diameter by 19.5 cm (debut was on XTE December 1995).

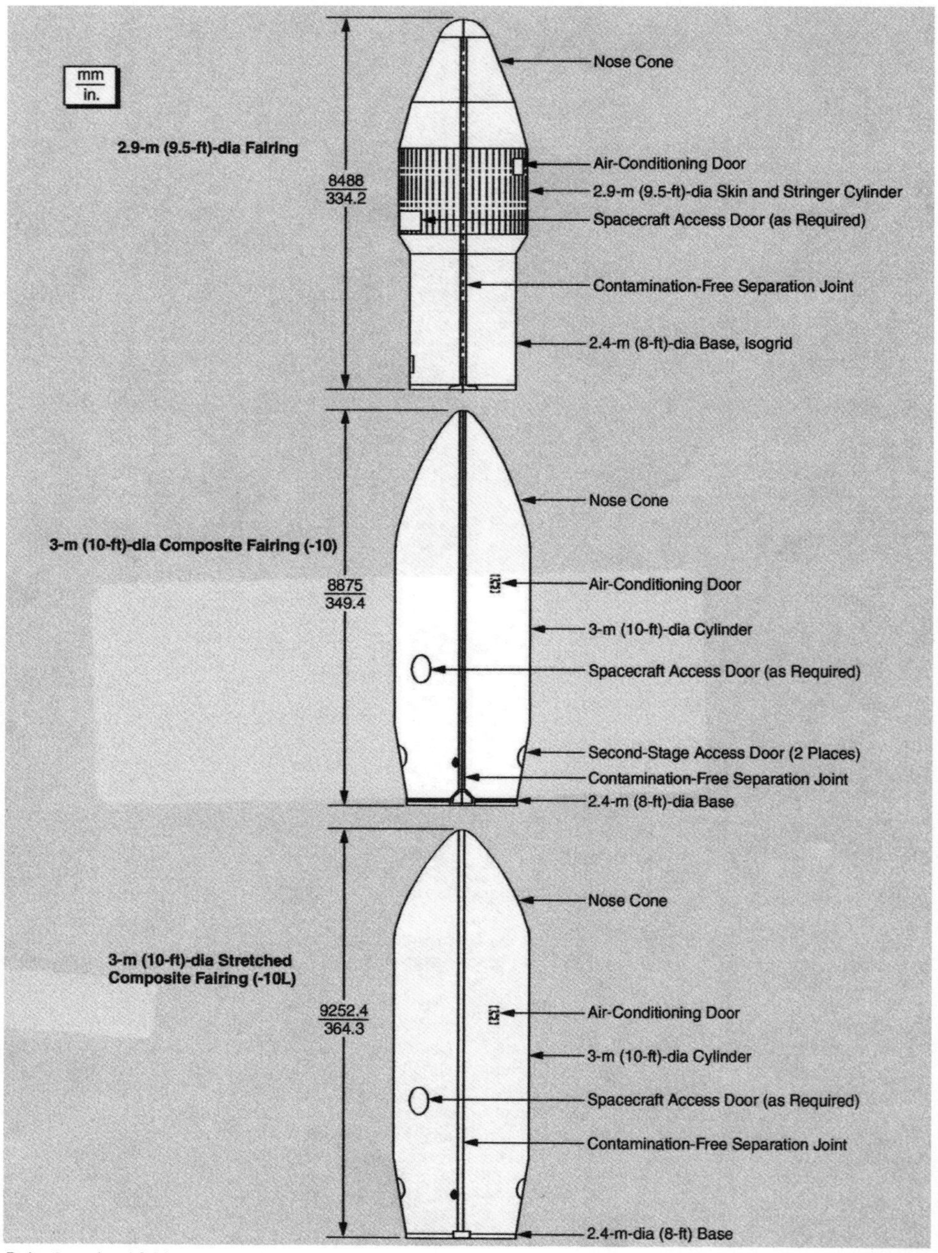

Delta 2 payload fairings 0131859

Length: 12.96 m
Diameter: 101.6 cm
Average thrust: each 446.02 kN SL; 499.18 kN vacuum
Burn time: 63.3 s
Mass at ignition: 13,232 kg each for six ignited on pad; 13,061 kg each for three air-lit
Propellant mass: 11,765 kg each
Propellant type: solid
Burn sequence: six GEMs ignite at launch, followed by the remaining three 2.5 s after the others burnout at 63 s. The original six are spring separated (under command of a sequencer between stage 1's tanks) in two symmetrical sets at 66/67 s; the final three burnout at 129 s and separate 3 s later.

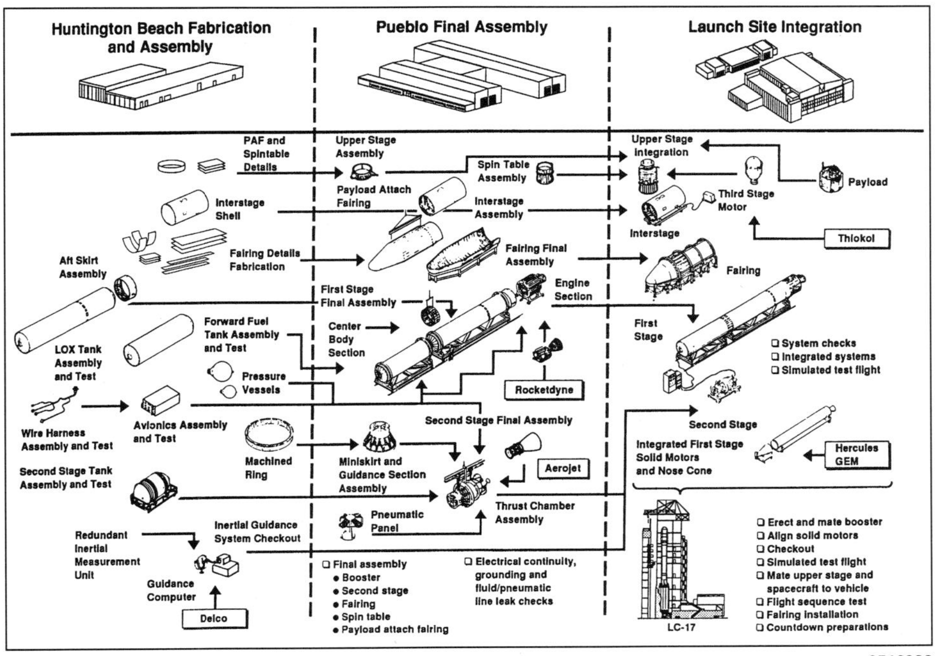

Delta 2 production flow (MDA) 0516933

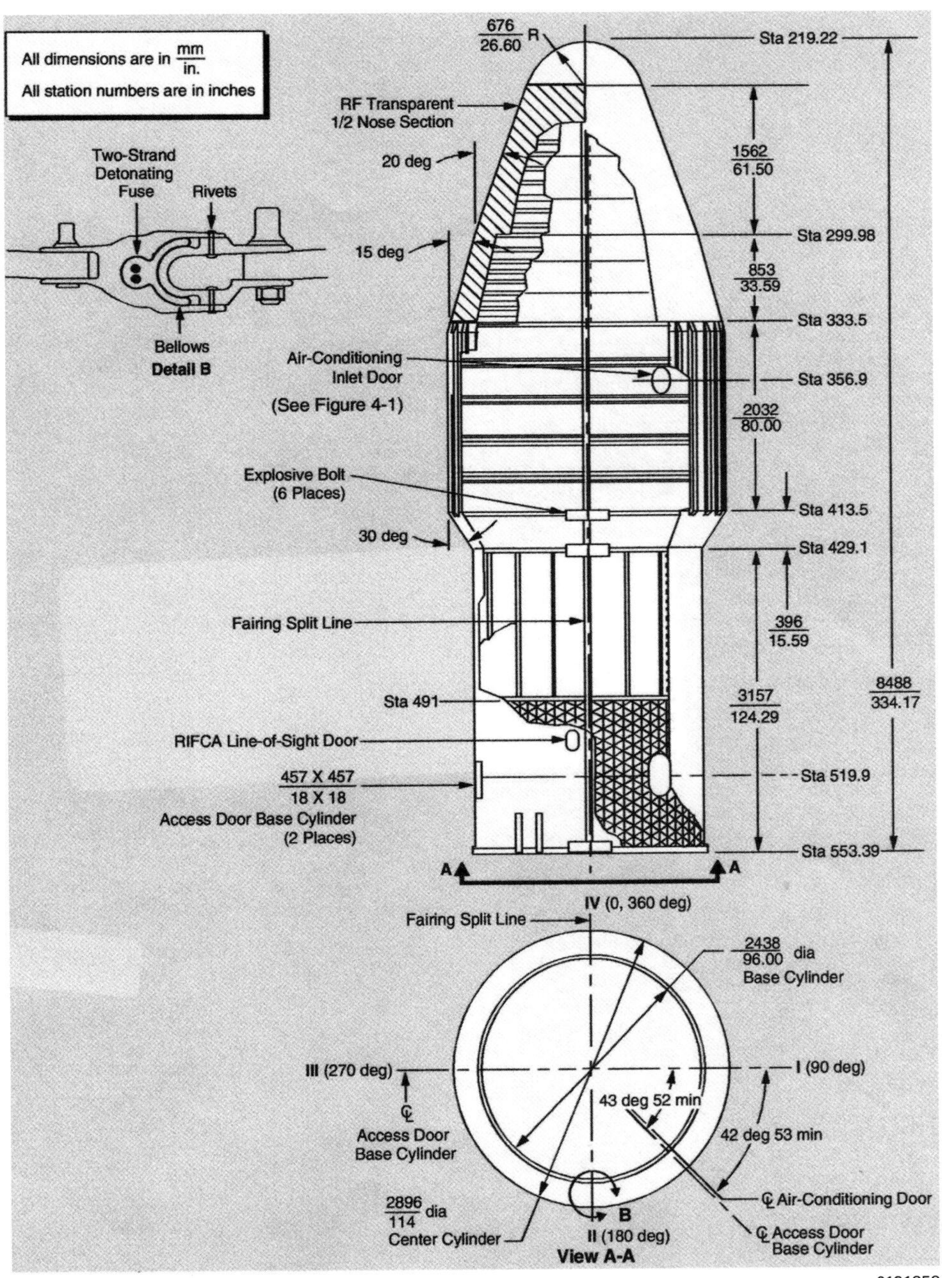

Delta 2 2.9 m payload fairing 0131858

Delta 7920/7925 stage 1
Comment: Incorporating uprated strap-ons, which provide the majority of the launch thrust, permitted increasing RS 27A main engine expansion ratio from 8:1 for improved altitude performance.
Engines: Rocketdyne 12:1 expansion ratio RS 27 single start liquid bipropellant of 890 kN sea level thrust (1,084.8 kN vacuum), hydraulically gimballed for pitch/yaw control; two Rocketdyne 4.45 kN LR101-NA11 verniers provide roll control.
Overall length: 26.047 m
Diameter: 2.44 m
Stage mass: 101,718 kg
Oxidiser: liquid oxygen
Fuel: RP-1
Propellant quantity (usable): 96,033 kg
Oxidiser tank length: 13.00 m, extended 2.23 m over 3920 version
Fuel tank length: 8.38 m, extended 1.43 m over 3920 version
Stage thrust: 890 kN SL + two 4.45 kN verniers
Burn time: 260.5 s
Comment: The two isogrid tanks are separated by a 77 cm centre body section that houses control electronics, ordnance sequencing and telemetry equipment. The package receives commands from RIFCA and drives servo amplifiers for engine gimbal; DC batteries in the centre section provide power. A rate gyro was added forward of this section to ensure adequate stability margins with the extended tanks and larger fairing.

Interstage
The 4.72 m-long isogrid interstage extends from the top of stage 1 to stage 2's miniskirt, encircling most of the upper stage until explosive bolt detonation 8 s after stage 1 burnout. Six spring driven separation rods at the forward end then separate the stages and stage 2 ignition occurs 5 s later.

Delta 7920/7925 stage 2
Engine: pressure-fed restartable 65:1 Aerojet AJ10 118K, developed for the USAF Improved Transstage Injector programme
Overall length: 5.89 m.
Diameter: 1.70 m, suspended by 2.44 m diameter mini skirt and support truss.
Stage mass: 6,930 kg.
Oxidiser: nitrogen tetroxide.
Fuel: Aerozine-50.
Propellant mass (usable): 6,006 kg.
Thrust: 43.37 kN vacuum.
Burn time: 432 s (restartable).
Attitude control system: the nitrogen cold gas jet Redundant Attitude Control System (RACS) provides three-axis control during coast periods and roll control during powered flight; hydraulic gimballing of the main engine provides pitch/yaw control under the command of RIFCA at the forward end. Three helium spheres provide tank pressurisation. Four aft isogrid panels provide equipment attach points and tank fill ports; the gas spheres are filled through the miniskirt on pad.
Burn time: On a typical 7925 GTO mission, stage 2 will burn for 343 s to enter a low circular orbit, coast under control of RACS for 620 s and then burn near the equator for a final 70 s. For the two-stage 7920, the burns are 395/25 s, respectively.

Delta 7925 stage 3
Designation: PAM-D
Motor: Thiokol Star 48B solid
Stage length: 2.04 m
Diameter: 1.24 m
Propellant: 1,756–2,025 kg solid, depending on mission requirements.
Thrust: 66.4 kN vacuum
Burn time: 87.1 s
Mass: 2,141 kg before ignition; 132 kg after burnout.
Comment: Delta 2's third stage is derived from components and concepts used on the PAM-D and the USAF SGS-2 upper stage. The Star 48B is base supported by a spin table that mates to the top of the second stage guidance section. Spin rockets fire typically 50 s after stage 2 burnout to create a stabilising spin of 30,110 rpm for stage 3 firing. Four or eight spin rockets can be used in 334/823/934 N thrust versions, firing for 1s. Stage 2's RIFCA triggers the stage 3 fuze based sequencing system, which produces spin-up, pyrotechnic/spring separation 2 s later and ignition after a further 38 s. A Nutation Control System derived from SGS2 and mounted above the PAM suppresses coning during motor burn and post-burn phases; a single axis rate gyro senses the motion and fires a counteracting hydrazine 178 N thruster system (2.7 kg propellant loading). Stage 3 can also house a 3.6 kg C-band radar transponder tracking system to aid early orbit determination; if integrated with the telemetry unit it adds 2.3 kg. The latter returns

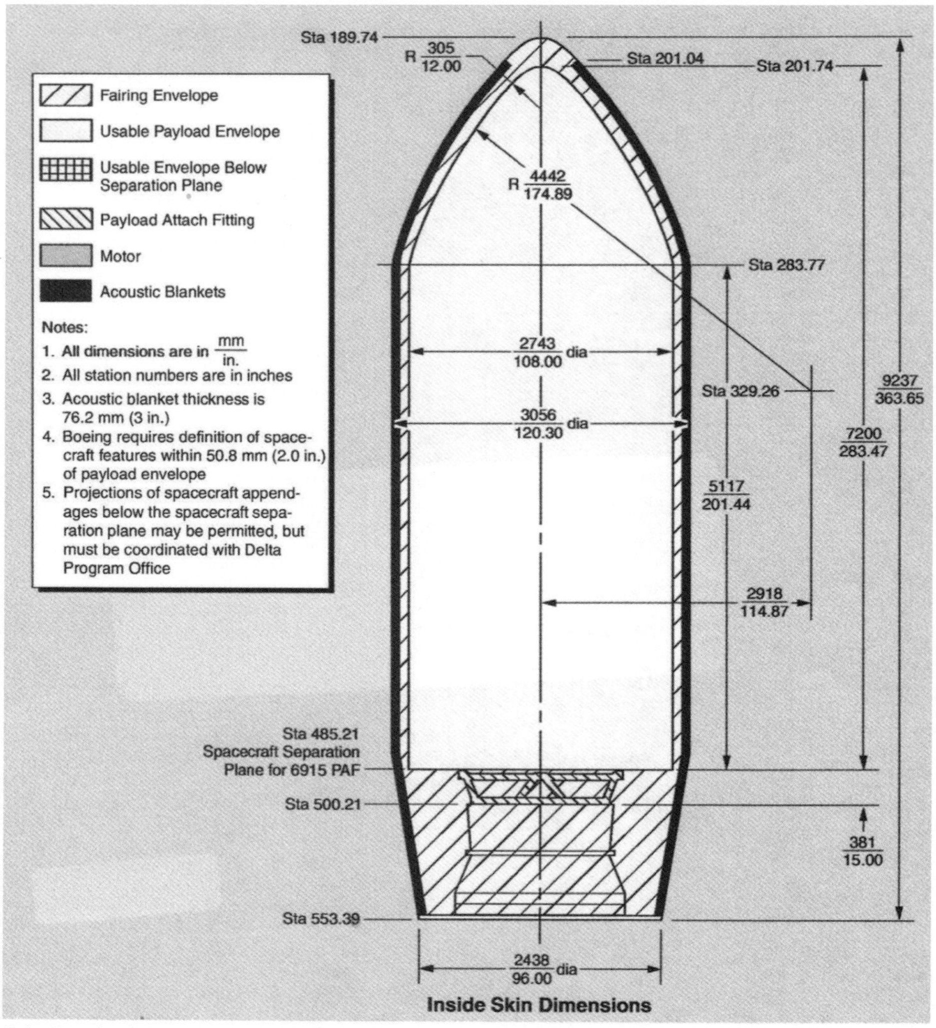

Delta 2 payload static envelope, 3 m diameter stretched composite fairing, two-stage configuration (6915 PAP)
0131857

spacecraft environment data such as vibration, acceleration transients, shock, velocity increment, indication of separation and temperatures.

Three stage missions adopt the standard 3712 Payload Attach Fitting at stage 3's forward end, capable of holding 3,084 kg. The spacecraft is secured by means of a two-piece V-block clamp assembly, held by two instrumented studs. Two ordnance cutters sever the studs for four springs to effect a 0.612.4 m/s separation rate. A yo-yo weight tumble system despins and imparts a coning motion to the expended motor 2 s after separation to modify its momentum vector and prevent recontact with the spacecraft. On Delta 6920 missions, a 6019 Payload Attach Fitting was used designed originally for NASA's Multimission Modular Spacecraft and capable of holding 2,177 kg with three explosive nuts and a secondary latching system. Stage 2's nitrogen gas thrusters ensure a separation speed of 31 cm/s.

Fairing

The spacecraft is protected by an 848.8 cm long, 839 kg aluminium fairing 243.8 cm diameter at the base and 289.6 cm at the widest portion. Essentially it consists of a Titan 3C section atop the previous Delta fairing cylinder. The halves are pyrotechnically separated during second stage burning at a height of about 111 km where aerodynamic heating has reduced to 1,135 W/m^2. However, 243.8/304.8 cm diameter versions are available; in these cases, 8/10 are added to the normal Delta 4-digit vehicle configuration designations, respectively. A composite 304.8 cm diameter 8.89 m long fairing has been developed Iridium. Air conditioning on the pad can maintain 14 26.7°C, max humidity 50%. Four standard fairing doors provide limited access on the pad and others can be provided. Payload closeout and fairing installation occurs typically three days before launch. An umbilical to stage 2 provides 64 hardwire payload links via the fairing.
Acceleration load: 6.25 *g* maximum at end of stage 1 burn for heaviest GTO payloads; 4.24 *g* at stage 3 burnout; 1.31 *g* at launch.
Acoustic load: maximum about 130 dB at launch/transonic; payloads subjected to 140 dB for 60 s for flight acceptance test.
Temperature: 33°C max on fairing inner wall after 300 s.

Vehicle/Payload processing

Delta stage 1/2 checkout typically takes place over 13 to 7 weeks before launch in Hangar M of the Cape Canaveral Air Station (CCAS). Stage 1/interstage erection followed by strap-on attachment occurs in the sixth week before launch, with stage 2 mating in the fifth, checkout over the fourth to second week, stage 3 with attached spacecraft mating seven days before, followed by ordnance installations. The final week sees fairing installation, mission rehearsal and stage 2 propellant loading. Launch is typically scheduled for a Thursday. MDA provides transport/handling for payloads between local airports and processing facilities. CCAS offers three principal hangars for non-hazardous spacecraft processing. Hangar AO has a 14 × 53.3 m class 100,000 high bay clean room capable of handling two payloads simultaneously. Hangar AE has class 1,000 facilities and the Delta launch control centre, and Hangar AM with a 4.4 × 5 m clean room and solarium. Loading of payload propellants and pyrotechnics can be performed in three facilities: Explosive Safe Area 60A and KSC's Spacecraft Assembly& Encapsulation Facility #2, both of class 100,000 standard, or KSC's Cargo Hazardous Servicing Facility Complex, built to handle Shuttle-sized elements. The spacecraft then meets the stage 3 assembly for the first time in the Payload Spin Test Facility 2.8 km west of complex 17, where it is mated and balanced at 300 rpm. The combination is then ready for transport to the pad in MDA's 6.1 m high, 3.05 m diameter N_2 purged handling can.

Alternatively, Astrotech Space Operations offers equivalent commercial processing facilities at its Titusville site. Building 1 provides three 18.3 × 12.2 m class 100,000 high bays for non-hazardous processing, building 2 handles the hazardous processing of PAM, payload fuelling and ordnance loading, mating and spin balancing.

Delta-3

Current Status

The first Delta-3 launch, on 26 August 1998 and the second launch on 5 May 1999 were both failures. The third launch, on 23 August 2000, was a partial success, placing a dummy payload in a simulated GTO, albeit much lower than planned due to fuel temperature and atmospheric conditions. Always intended as a transitional programme between Delta 2 series and the Delta 4 family of expendable launchers, Delta 3 has been retired.

Background

Delta-3 provides a low Earth orbit payload capability of 8,290 kg and a geosynchronous transfer orbit capability of up to 3,810 kg, more than twice the payload capability of the Delta 2. To meet the specification Delta-3 has a larger composite fairing to house bigger payloads, larger and more powerful strap-on solid rocket motors, a new cryogenically propelled single engine upper stage.

Both Delta-2 and Delta-3 use the same first-stage engine, as well as similar avionics systems, launch operations infrastructure and liquid oxygen tanks to minimise cost and maximise flexibility. The first stage is powered by a Rocketdyne RS 27A main engine and two Rocketdyne vernier engines to control roll during main engine burn and attitude control between main engine cut-off and second-stage separation. The diameter of the booster fuel tank was increased from Delta-2 to reduce length and improve control margins.

Nine Alliant Techsystems-built strap-on solid rocket motors supplement the first stage: these are directly evolved from the Delta-2's proven motors, but provide 25 per cent more thrust. Three of the nine are equipped with thrust vector control to further improve vehicle manoeuvrability and control. The second stage carries a Pratt & Whitney RL10B-2 engine which is derived from the RL10 engine in use for more than three decades aboard the Centaur stage used on Atlas and Titan launch vehicles. The RL10B-2 also incorporates a larger exit cone for increased specific impulse and payload capability: it relies heavily on the flight-proven RL10 engine components, with operability improvements for increased reliability.

Mitsubishi Heavy Industries supplied 16 sets of stage 1 fuel and stage 2 LH2 tanks, delivery starting in 1996. These will only be used for commercial launches. The Delta-3 will use the SLC 17B at Cape Canaveral. Workers will modify the white room platform locations (part of the 10-storey mobile service tower used to protect the vehicle during processing), add a ground liquid hydrogen propellant tank and increase the overhead crane capability for Delta-3. The solid rocket motor storage area, Delta checkout facility, and high-pressure test facility at Cape Canaveral will also be modified to assemble and check out the Delta 3. Vehicle control and launch operations will be conducted from the operations building.

Delta 3 growth capability was not possible due to the dynamic characteristics of the re-structured first stage and the extended hammerhead configuration.

Specifications

Delta 3

First launched: 21 August 1998
Number launched: 3
Launch sites: LC-17 CCAFS
Principal users: Heavy satellite operators from GSO
Performance
GTO: (185 × 35,786 km): 3,810 kg
LEO: (185 km circ) 8,292 kg
Overall length: 39.1 m

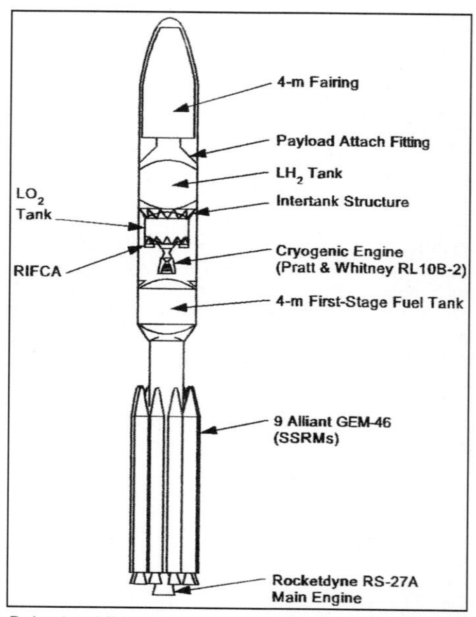

Delta 3 exhibits shorter first stage and more powerful SRMs
0131856

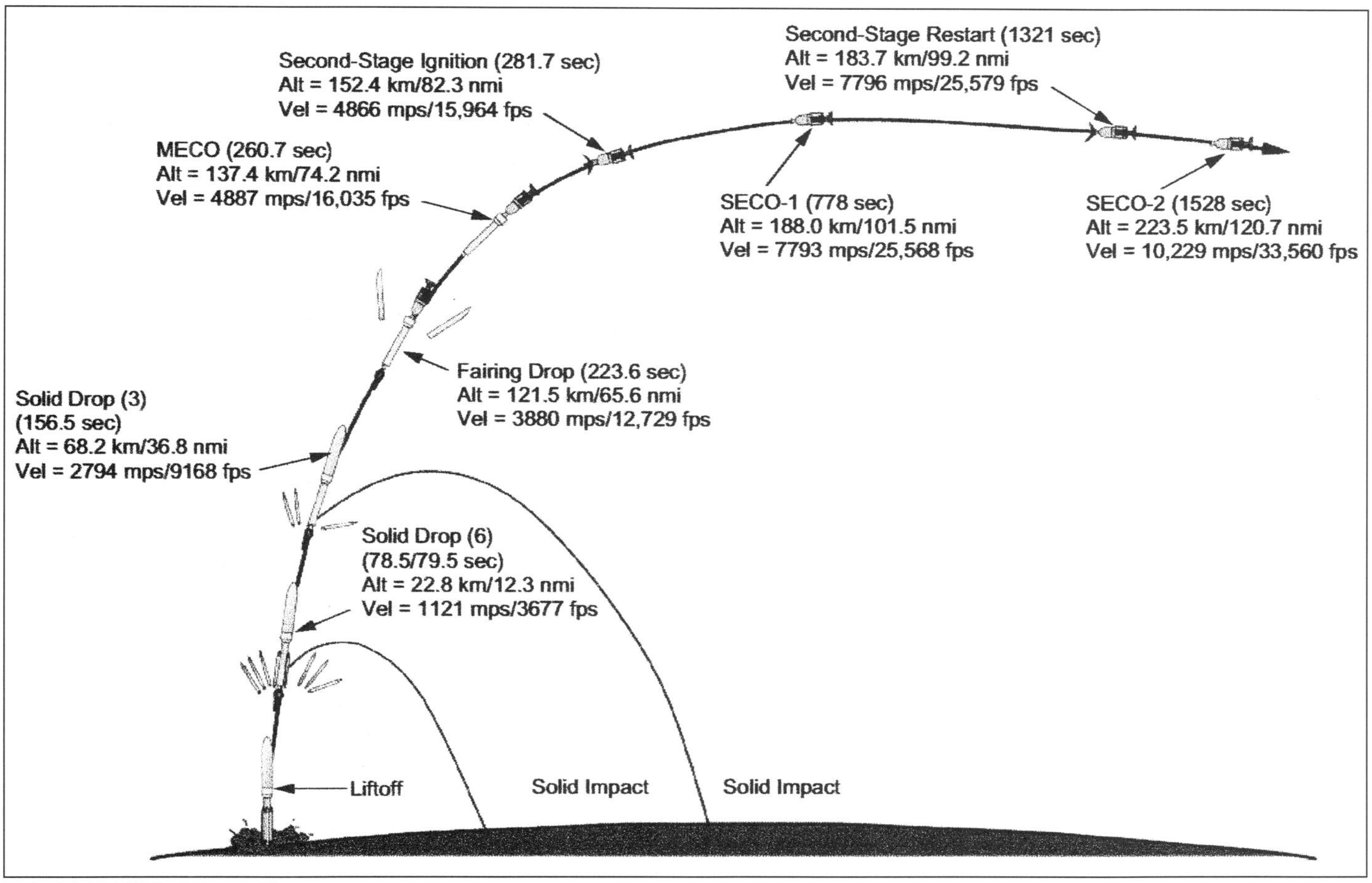

Typical Delta 3 GTO mission profile 0131855

Launch mass: 301,540 kg
Launch thrust: 4,890 kN (6 SRM ignite for launch)
Number of stages: 2 plus 9 strap-on SRM

Delta 3 Stage 1
Comment: Although the Delta 3 first stage adopts the same RS-27A main engine and verniers of the Delta 2, the first stage structure is radically different in that the nominal diameter of 2.44 m is retained only for the LOX tank, the RP-1 fuel tank being shorter by 6 m and greater in diameter by about 1.6 m, creating a shorter and fatter upper first stage section.
Overall length: 20 m
Diameter: 2.44 m aft, 4.0 m forward
Stage mass: 104,377 kg
Oxidiser: LOX
Fuel: RP-1
Propellant quantity (useable): 102,095 kg
Burn time: 260.7 s

Delta 3 solid strap-ons
Comment: The nine Alliant graphite epoxy GEM-46 strap-on solid rocket motors (SSRM) augment first stage performance and are an evolution of the GEM motors on Delta 2. Three of the six ground-ignited SSRMs have thrust vector control to increase control authority.
Overall length: 14.7 m
Diameter: 1.2 m
Stage mass: 19,327 kg
Empty mass: 2,282 kg
Thrust (vac): 628 kN
Burn time: 273 s
Specific impulse: 273 s

Delta 3 Stage 2
Comment: The second stage accommodates an upgraded cryogenic Pratt & Whitney RL10B-2 incorporating an extendible exit cone for increased Isp and payload capability. The basic engine and turbopump are unchanged relative to the RL-10. The engine gimbal system uses electro-mechanical actuators that increase reliability while reducing both cost and weight.
Overall length: 8.8 m
Diameter: 2.4 m in 4 m shroud
Empty mass: 2,476 kg
Gross mass: 19,300 kg
Propellant: LOX/LH_2
Thrust: 109.9 kN
Burn time: 703 s in a typical GTO mission via an inertial burn of 496 s followed by a second burn of 207 s.

Delta 3 Stage 3
Comment: Depending on mission needs a third stage is employed to increase capability and consists of a Star 48B solid rocket motor, a payload attach fitting with rotation control system and a spin table containing small rockets for spin-up of the third stage and satellite. This stage has not been used in any of the three Delta 3 launched to dated.

Delta 4

Current Status
The first Delta 4 was successfully launched from the converted ex-Saturn 1 launch complex 37B at CCAFS on 20 November 2002 carrying the Eutelsat W5 telecommunications satellite. It was followed by the second Delta 4 on 10 March 2003 lifting the DSCS 3-A3 military communications satellite and the third Delta 4 on 29 August 2003 carrying DSCS 3-B6.

Background
The Delta 4 series of medium-weight and heavyweight launchers will become the mainstay of the Delta family and provide a wide range of optional payload capabilities for military and commercial users. In a development cycle that began in 1995, the US Air Force began a programme leading to procurement of a series of satellite launchers in the Evolved Expendable Launch Vehicle (EELV) programme. During the first phase four competitors completed a 15-month contract to validate low-cost concepts and in December 1996 two contractors were selected to participate in the second phase known as the Pre-Engineering, Manufacturing and Development (Pre-EMD) phase under a fixed-price US$60 million contract for each company. In November 1997, the US Air Force announced that it would introduce competition into the EELV programme by sustaining a dual-source procurement strategy.

On 16 October 1998, the US Air Force announced that it was to buy 19 Boeing Delta 4 launchers for the EELV programme at a cost of US$1.38 billion. The Delta 4 family includes five launch vehicles with payload capabilities ranging from 4,173 to 13,154 kg: medium, heavy and three variants of the medium vehicle collectively known as the medium-plus variants. All vehicles will use the Common Booster Core (CBC) powered by the cryogenic liquid hydrogen/liquid oxygen fuelled Boeing (Rocketdyne) RS-68 engine delivering 2,900 kN thrust. The RS-68 is designed for low-cost

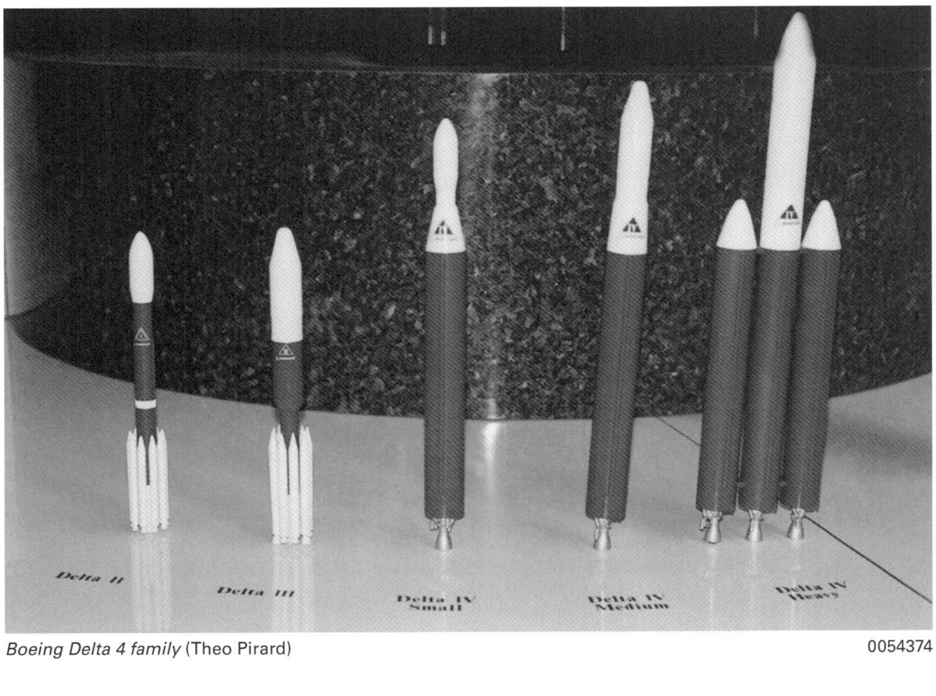

Boeing Delta 4 family (Theo Pirard) 0054374

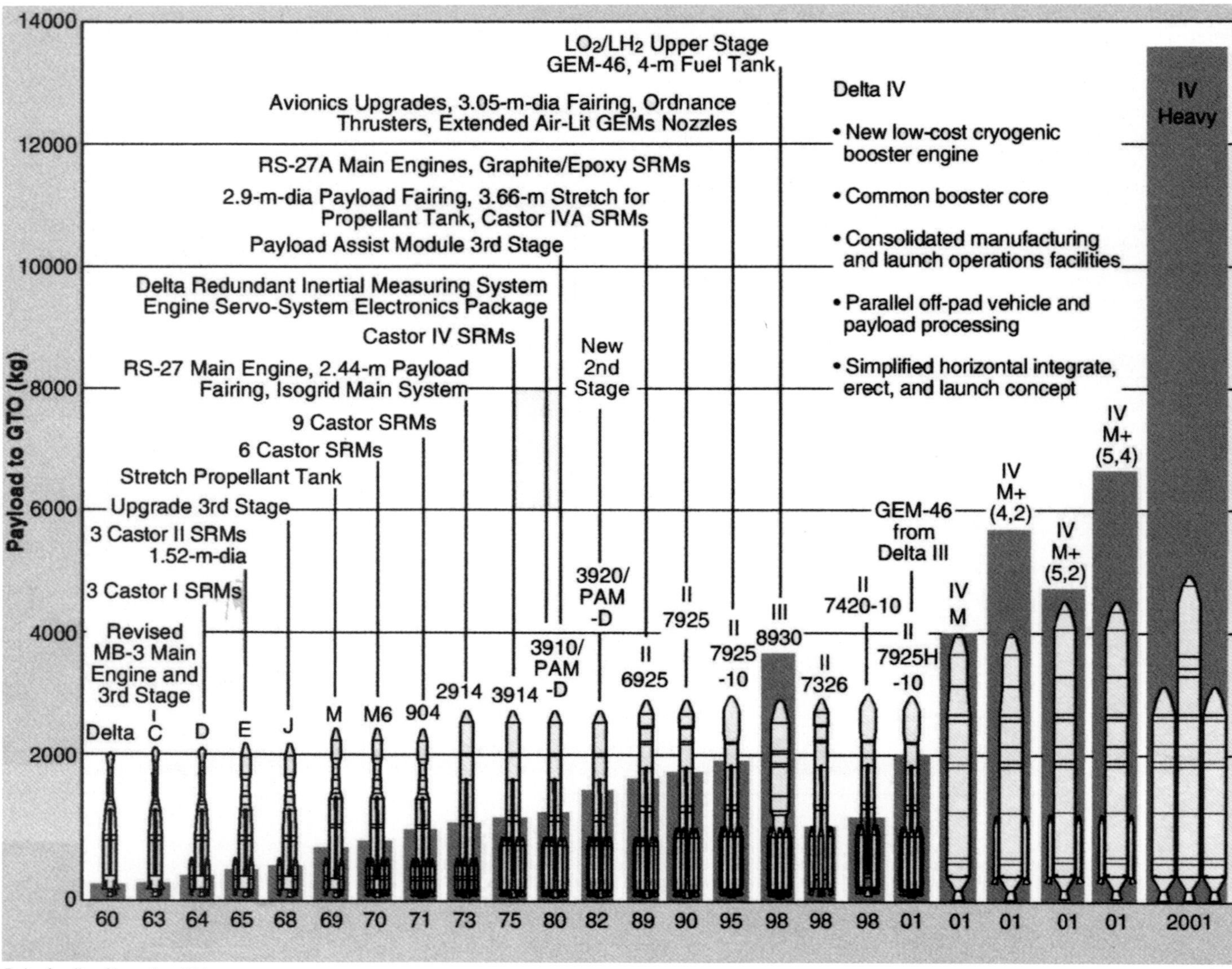

Delta family of launch vehicles

0131854

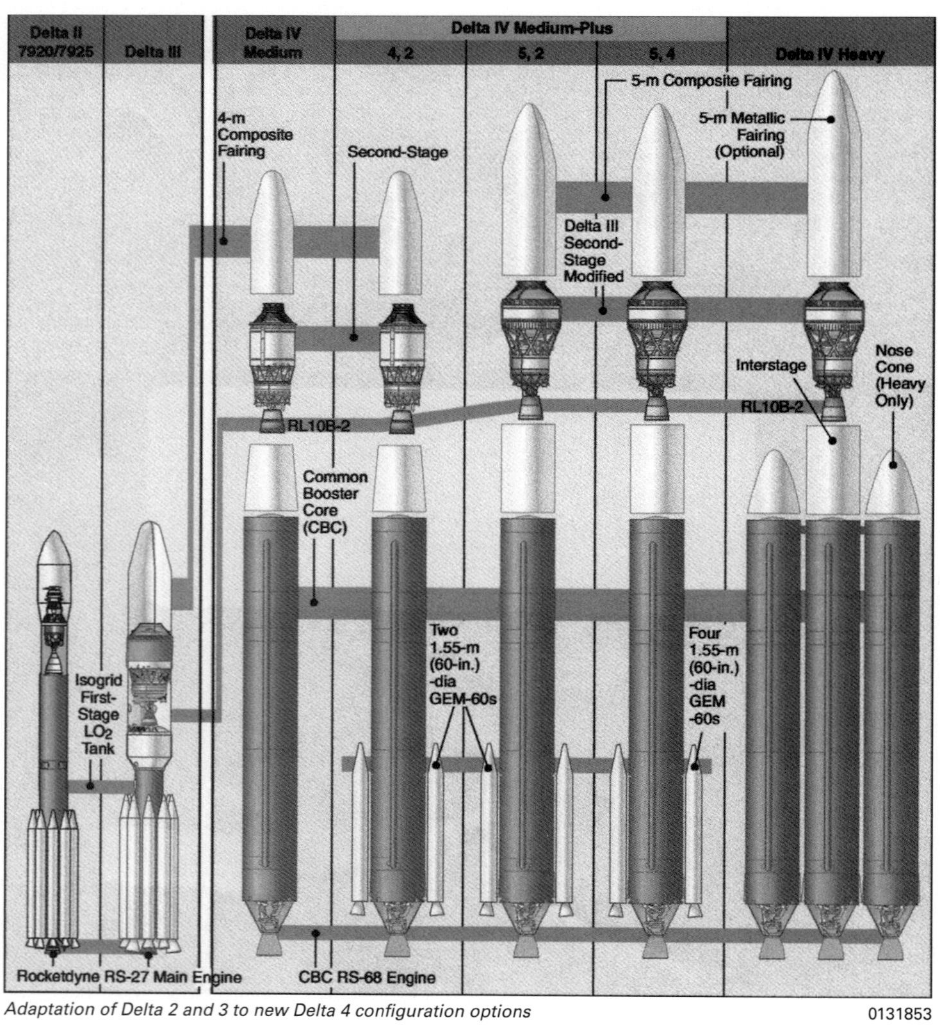

Adaptation of Delta 2 and 3 to new Delta 4 configuration options

0131853

production and has 80 per cent fewer moving parts than the Shuttle Main Engine. Modified Delta 2 and Delta 3 upper stages will be added to the CBC to complete each vehicle.

There are three variants of the Delta 4 in five configurations: Delta 4-M (Medium); Delta 4-M+ (Medium Plus); Delta 4-H (Heavy). The Delta 4-M carries the cryogenic second stage of the Delta 3 and that launcher's composite 4 m diameter fairing to provide a 4,200 kg GTO capability. Medium-Plus members also available for commercial operations include the 4,2 with two solid rocket motors, a 4 m payload fairing and a GTO capability of 5,845 kg. The 5,2 also has two solid rocket boosters but a 5 m fairing currently manufactured by Boeing for the Titan 4 and a GTO capability of 4,640 kg. The 5,4 has the same payload shroud but four solid propellant rockets and can lift 6,565 kg to GTO. The Delta 4-H links three of the CBCs together in parallel and carries an enlarged Delta second stage for increased propellant. With a 5 m diameter fairing it can lift 13,130 kg to GTO.

Boeing selected Decatur, Alabama, as the home of the Delta 4 CBC manufacturing programme and construction of the 139,350 m^2 facility began in November 1997. The CBC cores are 37.92 m in length with a diameter of 4.85 m – roughly the size of a Boeing wide-body airliner fuselage. Completed CBCs will be moved to docks on the Tennessee River using a specially designed transporter, loaded onto a cargo vessel and delivered to launch sites at Cape Canaveral Air Force Station (CCAFS), Florida, and Vandenberg Air Force Base, California. Employment at the Decatur facility began with 100 people in 1998, grew to almost 600 by the end of 1999, increasing to 2,000 by 2004. The RS-68 engines are manufactured at the Boeing (Rocketdyne) Canoga Park facility. Start-up for production commenced early 1999 with the first Delta 4 leaving the facility in 2001. First launch occurred in November 2002.

In July 1998, Boeing unveiled plans for a major expansion of launch capabilities at SLC-37, CCAFS, to accommodate Delta 4. A new launch pad, mobile service tower, fixed umbilical tower, 6,967 m^2 horizontal integration facility as well as support and test facilities will be provided. This former Saturn I launch facility was topped off in March

2000. It had previously been used to launch eight Saturn Is between 1964 and 1968. In addition, Boeing is modifying SLC-6 at Vandenberg AFB to handle west coast launches.

Specifications

Delta 4

First launch: 20 November 2002
Reliability: 100%
Launch sites: LC-37 CCAFS, SLC-6 VAFB
Performance:
GTO (185 × 35,786 km, 28.5°):
3,960 kg Delta 4-M
5,735 kg Delta 4-M+ (4.2)
4,586 kg Delta 4-M+ (5.2)
6,470 kg Delta 4-M+ (5.4)
10,819 kg Delta 4-H
LEO:
11,700 kg (4.2)
10,300 kg (5.2)
13,600 kg (5.4)
25,800 kg (Heavy)
Overall length:
Delta 4-M: 63.0 m
Delta 4-M+ 4.2: 66.0 m
Delta 4-M+ 5.2: 77.2 m
Delta 4-M+ 5.4: 77.2 m
Delta 4-H: 77.2 m
Diameter: 5.1 m
Launch mass:
Delta 4-M: 249,500 kg
Delta 4-M+ 4.2: 292,732 kg
Delta 4-M+ 5.2: 292,732 kg
Delta 4-M+ 5.4: 404,600 kg
Delta 4-H: 733,400 kg
Lift-off thrust:
Delta 4-M: 2,893 kN
Delta 4-M+ 4.2: 4,182 kN
Delta 4-M+ 5.2: 4,148 kN
Delta 4-M+ 5.4: 5,865 kN
Delta 4-H: 8,666 kN
Number of stages
Delta 4-M: 2
Delta 4-M+ 4.2: 2 + 2 SRM
Delta 4-M+ 5.2: 2 + 2 SRM
Delta 4-M+ 5.4: 2 + 4 SRM
Delta 4-H: 1 × triple CBC + 1

Delta 4 stage 1 (CBC)

Comment: The first stage common to all Delta 4 variants is the Common Core Booster (CBC) with a single Boeing (Rocketdyne) RS-68 cryogenic motor throttleable between 60 and 100 per cent thrust. Stage elements include an engine section incorporating thrust mount, an aft LH_2 tank, centrebody section and forward LOX tank. Delta 4-M and 4-M+4.2 variants have tapered interstage sections while the remainder have constant-diameter sections. The two CBC elements operating as liquid propellant boosters on the Delta 4-H carry nosecone elements.
Overall length: 40.8 m
Diameter: 5.1 m
Empty mass: 26,700 kg
Gross mass: 226,400 kg
Oxidiser: LOX
Fuel: LH_2
Isp: 420 s
Launch thrust: 3,313 kN
Burn time: 249 s

Delta 4 solid strap-ons

Comment: Derived from the smaller GEM-46 used on Delta 3, the Alliant Techsystems graphite epoxy motors GEM-60. Separation is accomplished by initiating ordnance thrusters that provide radial thrust to jettison expended SRMs away from the first stage. Two GEM-60s are carried by the Delta 4-M+ 4.2 and 5.2 and four are fitted to the 5.4 variant.
Overall length:
Diameter: 1.55 m
Gross mass: 33,798 kg
Empty mass: 3,849 kg
Thrust (vac): 826.3 kN
Isp: 275 s
Burn time: 243 s

Delta 4 stage 2

Comment: Two second-stage configurations are offered: a 4 m diameter version used on Delta 4-M and 4-M+ (4.2) and a 5 m diameter version on the Delta 4-M+ (5.2), 4-M+ (5.4) and 4-H. Both use the flight-proven RL-10 family. The 4 m diameter stage is modified from that used on Delta 3 but with total propellant capacity increased and burn time. The 5 m diameter stage is based on the 4 m version with the LOX tank increased in length by 0.5 m and total propellant load and burn times increased. Propellants are managed by directing hydrogen boil-off through aft-facing thrusters to provide settling thrust and by the use of attitude control system as required. Minimum duration is nominally 2.3 h but may be increased to over 7 h by adding hydrazine bottles and batteries.
Overall length: 12 m
Diameter: 4 or 5.1 m
Gross mass: 24,170 kg or 30,710 kg
Empty mass: 2,850 kg or 3,490 kg
Thrust (vac): 110 kN
Isp: 462 s
Oxidiser: LOX
Fuel: LH_2
Motor: RL10B-2
Burn time: 850 or 1,125 s

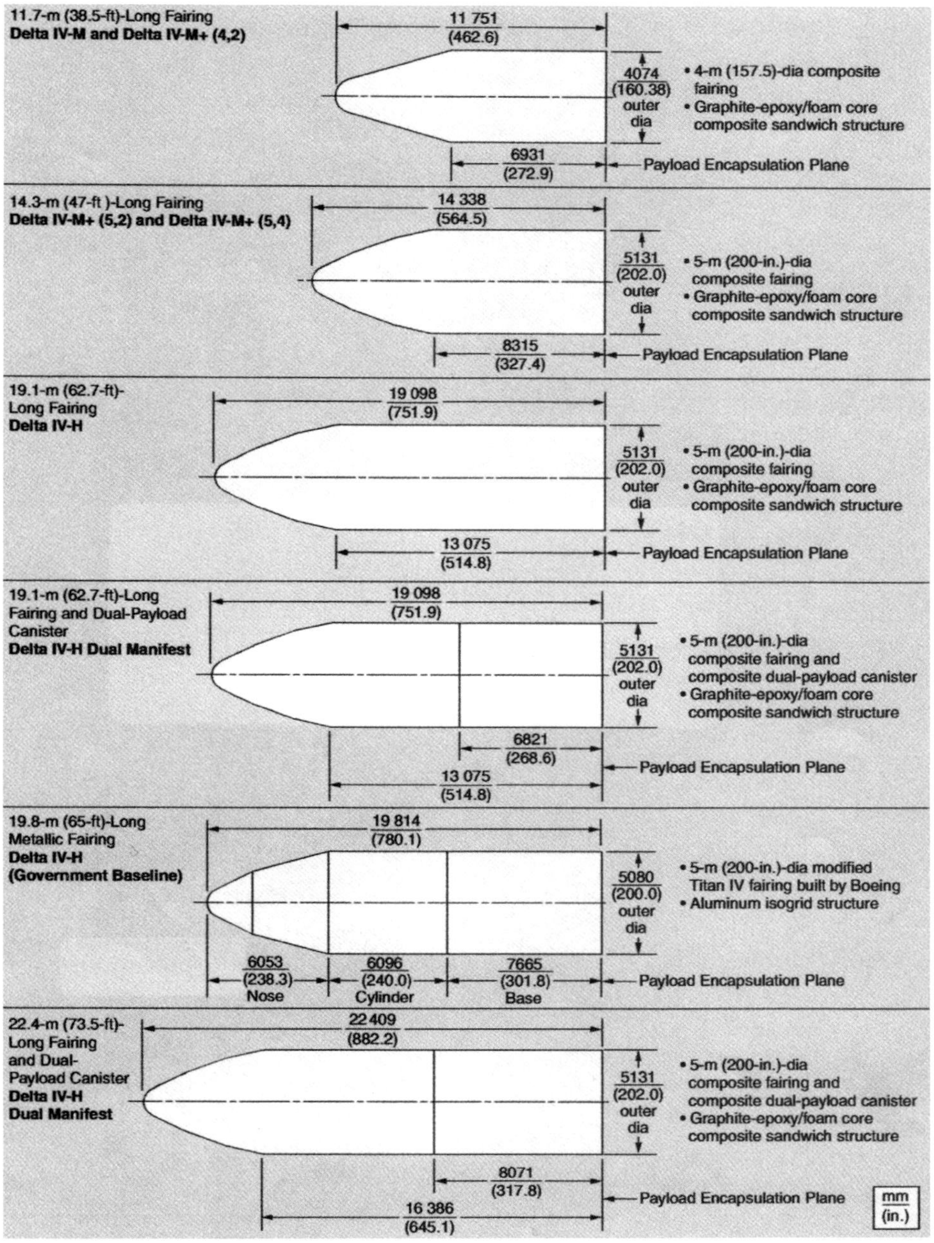

Delta 4 fairing configuration 0131852

Delta 4 Payload Attach Fittings

The PAF provides the mechanised interface between the payload and the launch vehicle. The Delta 4 launch system offers a selection of standard and modifiable PAFs to accommodate a variety of payload requirements. The customer had the option to provide the payload separation system and mate directly to a PAF provided by Boeing; or Boeing can supply the entire separation system. Payload separation systems typically incorporated on the PAF include clampband systems or explosive attach-bolt systems.

Boeing has extensive experience designing and building satellite dispensing systems for multiple satellite launches. These dispensers have a high success rate.

Payload Fairings (PLF)

The fairings protect the payload once the payload is encapsulated through boost flight. The Delta IV launch system offers PLFs for different launch vehicle configurations.

The 4 m fairing is a stretched Delta III 4 m composite bisector design. The 5 m composite fairing for single-manifest missions is also based on that of Detla III and comes in two standard lengths: 14.3 and 19.1 m. The dual-manifest fairing consists of two sections, a 5 m composite bisector fairing and a lower 5 m composite Dual-Payload Canister (DPC), and is available in two lengths: 19.1 and 22.4 m.

The 5 m metallic trisector fairing (the baseline for government programs) is a modified version of the flight-proven Titan IV aluminium isogrid fairing designed and manufactured by Boeing.

All PLFs are configured for off-pad payload encapsulation to enhance payload safety and security, and to minimise on-pad time. Interior acoustic blankets as well as flight-proven contamination-free separation joints are incorporated into the fairing design for payload protection. Mission-specific fairing modifications can be made as required by the customer. These include access doors, additional acoustic blankets, and Radio Frequency (RF) windows.

Dual-manifest capability

The Delta IV launch system offers dual-manifest capability using the heavy configuration. This dual-manifest system provides payload autonomy similar to a dedicated launch, but at a significant cost reduction compared to a dedicated launch. The dual-manifest approach has the capability of launching two spacecraft totalling up to 10,700 kg to a standard 27° GTO orbit using a 5 m composite fairing that is 22.4 m long.

The 5 m dia by 19.1 m long dual-manifest system consists of a 12.3 m long composite fairing and a 6.8 m long DPC. The 5 m dia by 22.4 m long dual-manifest system consists of a 14.3 m long composite bisector fairing and an 8.1 m long DPC. Using standard PAFs, both payloads are mounted within two independent payload bays that are similar in volume and vented separately. Separate fairing access doors of standard size are provided in the cylindrical section of each payload bay. As with the single-manifest mission, existing acoustic blankets are provided from the aft end of the fairing to the nose cone.

A contamination-free separation system that runs along the full length of the fairing is detonated to

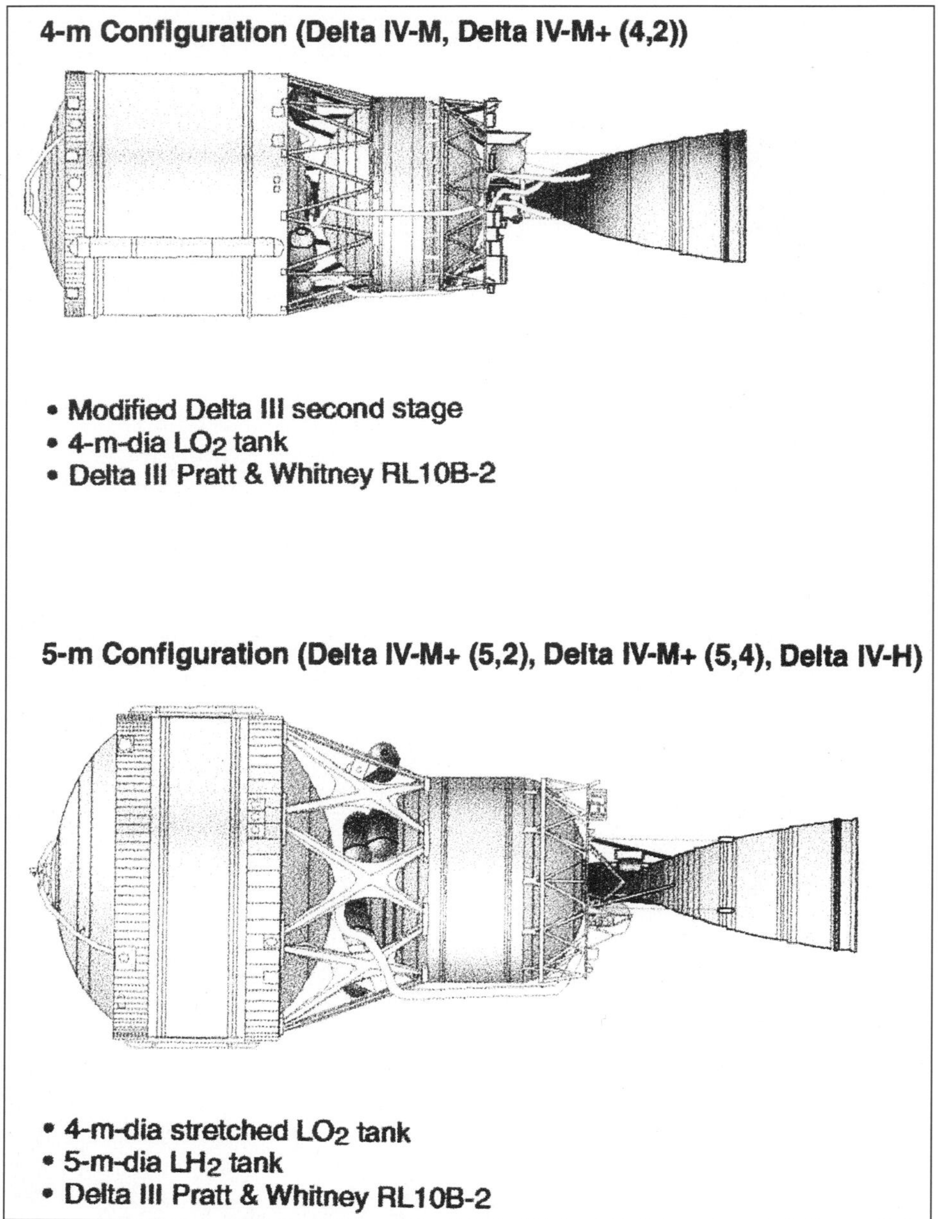

Delta 4 second stage configurations 0131851

separate the fairing into halves, exposing the upper payload. For the lower bay, a circumferential Sure-Sep system (patented by Boeing) with spring actuators is used to deploy the DPC over the payload.

Avionics and flight software

The Delta IV launch system uses a modified Delta III avionics system with a fully fault-tolerant avionics suite, including a Redundant Inertial Flight Control Assembly (RIFCA) and automated launch operations processing using an advanced launch control system.

The RIFCA, supplied by AlliedSignal, uses six RL20 ring laser gyros and six QA3000 accelerometers to provide redundant three-axis attitude and velocity data. In addition to RIFCA, both the first- and second-stage avionics include interface and control electronics to support vehicle control and sequencing, a power and control box to support power distribution, and an ordnance box to issue ordnance commands. A Pulse Code Modulation (PCM) telemetry (T/M) system delivers real-time launch vehicle data directly to ground stations or relays through the Tracking and Data Relay Satellite System (TDRSS). If ground coverage is not available, instrumented aircraft of TDRSS may be available, in co-ordination with NASA, to provide flexibility with telemetry coverage.

The flight software comprises a standard flight program and a mission-constants database specifically-designed to meet each customer's mission sequence requirements. Mission requirements are implemented by configuring the mission-constants database, which is designed to fly the mission trajectory and to separate the satellite at the proper attitude and time. The mission-constants database is validated during the hardware/software functional validation tests and the system integration tests. The final software validation test is accomplished during a full-length simulated flight test at the launch site.

The RIFCA contains the control logic that processes rate and accelerometer data to form the proportional and discrete control output commands needed to drive the control actuators and hydrazine control thrusters.

Position and velocity data are explicitly computed to derive guidance steering commands. Early in flight, a load-relief mode turns the vehicle into the wind to reduce angle of attack, structural loads, and control effort. After dynamic pressure decay, the guidance system corrects trajectory dispersions caused by winds and vehicle performance variations, and directs the vehicle to the nominal end-of-stage orbit. Payload separation in the desired transfer orbit is accomplished by applying time adjustments to the nominal engine start/stop sequence, in addition to the required guidance steering commands.

Delta 4 sequence of events for GTO mission

Time (sec)	Altitude (km)	Accel (g)	Event
Delta 4-M:			
0	0	1.15	Liftoff
83	9.4	1.6	Mach 1.0
94	12.9	1.7	Max q
246	99.7	6.0	Main engine cutoff
251	105	0.0	Stage 1/2 separation
268	122	0.37	Stage 2 ignition
278	131	0.41	Jettison fairing
854	190	0.83	Stage 2 cutoff (SECO-1)
4,101	172	0.84	Stage 2 ignition
4,336	227	1.44	Stage 2 cutoff (SECO-2)
Delta 4-M+ (5.2)			
0	0	1.33	Liftoff
49	7.4	1.64	Mach 1.0
65	12.7	1.92	Max q
100	28	1.76	Jettison two GEM-60
246	117	5.17	Main engine cutoff
251	123	0.00	Stage 1/2 separation
268	141	0.31	Stage 2 ignition
278	151	0.33	Jettison fairing
981	184	0.70	Stage 2 cutoff (SECO-1)
4,089	168	0.70	Stage 2 ignition
4,371	242	1.20	Stage 2 cutoff (SECO-2)
Delta 4-H			
0	0	1.21	Liftoff
50	3.5	1.47	Start core throttledown
56	4.5	1.28	Core throttledown
86	11.1	1.44	Mach 1.0
86	11.3	1.44	Max g
238	69	4.20	Start SRM throttledown
244	72	3.01	SRMs at 60%
249	74	3.13	SRM cutoff
251	75	1.79	SRM jettison
252	76	1.79	Start core throttleup
257	81	2.75	Core at 100%
327	116	4.40	Main engine cutoff
332	119	0.00	Stage 1/2 separation
349	129	0.23	Stage 2 ignition
358	134	0.25	Jettison fairing
873	198	0.35	Stage 2 cutoff (SECO-1)
3,872	152	0.35	Stage 2 ignition
4,431	401	0.61	Stage 2 cutoff (SECO-2)

IUS

Current Status

The pace of IUS launch scheduling has slowed significantly and will continue to decline as Delta 4 takes over some of the missions hitherto assigned to Titan/IUS flights.

Background

Boeing's two-stage all-solid Inertial Upper Stage is designed to deliver payloads of more than 2 tonnes directly into GEO from LEO following launch by either Shuttle or expendable Titan. The USAF Space & Missile Systems Center IUS Office in Los Angeles maintains overall responsibility for development, procurement and operations, allocating vehicles to NASA Marshall for civil use. Most of the manufacturing is conducted at Boeing's Kent Space Center south of Seattle, but the motors themselves come from United Technologies/Chemical Systems Division of California.

NASA and the USAF conceived of the Interim Upper Stage as a stop-gap solution while they awaited the introduction of a reusable orbital transfer vehicle. Boeing's built their concept around their successful dual-stage solid propellant Burner 2, but growth of mass requirements into GEO dictated a more powerful vehicle. The company was awarded the development contract in August 1976 following a 2½ year competition, leading to the final qualification in 1981. The first, US$700 million, contract called for eight stages, with a US$277.3 million option for a further six to be delivered by April 1985 subsequently exercised. A third batch (vehicles 15 to 19) was contracted at

TDRS using an IUS for geo injection, departing from Shuttle (NASA) 0517342

Inertial Upper Stage flight record

	Launched	Vehicle	Payload
1	30 October 1982	T 34D	DSCS II-15/III1
2*	4 April 1983	STS6	TDRS 1 communication satellites
3	24 January 1985	STS 51C	Magnum electronic intelligence
4	3 October 1985	STS 51J	DSCS III 2/3
5*	28 January 1986	STS51L	TDRS 2 communication satellites
6	29 September 1988	STS26	TDRS 3comms
7	13 March 1989	STS29	TDRS 4 communication satellites
8	4 May 1989	STS 30	Magellan probe
9	14 June 1989	T4-1	DSP early warning
10	4 September 1989	T 34D	DSCS II 16/III-4
11	18 October 1989	STS34	Galileo probe
12	23 November 1989	STS33	Magnum electronic intelligence
13	6 October 1990	STS41	Ulysses probe
14	13 November 1990	T4-3	DSP early warning
15	2 August 1991	STS43	TDRS 5 communication satellites
16	24 November 1991	STS44	DSP16 early warning
17	13 January 1993	STS54	TDRS 6 communication satellites
18	22 December 1994	T411	DSP 17 early warning
19	13 July 1995	STS70	TDRS 7 communication satellites
20	24 February 1997	T4-20	DSP 18 early warning
21	9 April 1999	T4B	DSP 19 early warning
22	23 July 1999	STS 93	Chandra observatory
23	8 May 2000	T4B	DSP 20 early warning
124	6 Aug 2001	T4B	DSP21 early warning

T = Titan. STS = Shuttle mission. *stage 2's TVC system failed on second mission but payload separated and attained final GEO with its own thrusters; TDRS 2 and its IUS were lost in the Shuttle Challenger explosion. DSP-19 lost in failure of SRM2 to separate from SRM1 following successful Titan mainstage launch. Originally, Boeing did not acknowledge all of the IUS missions, but the WWW site for the IUS put together by Boeing now lists all missions. STS-38, in November 1990, has been claimed as a Magnum flight but a second IUS only 2 days after DSP's is unlikely.

a 32 per cent lower unit cost. By September 1984, the USAF programme was estimated to have cost US$1.7 billion.

IUS, qualified to Mil-Std 1540A standards, was designed as the most reliable upper stage ever built (98 per cent). Centaur's 4,540 kg GEO performance delivered special payloads for an increasing number of demanding military and NASA planetary missions from Shuttle's cargo bay, but the June 1986 decision to cancel the highly energetic liquid Centaur stage from the manned spaceplane on safety grounds ensured further IUS employment. The Magellan, Ulysses and Galileo deep space probes were consequently switched to Shuttle/IUS deployment. The stage, now renamed the Inertial Upper Stage, is also offered for the Titan 4 and commercial Titan 3 vehicles.

Boeing completed studies in 1993 on an IUS upgrade that could be available 1999. The second stage replaced the solid with storable liquid propellants and improved avionics to reduce mass and cost. Previous studies considered increasing GEO capacity to 3,765 kg by using Aerojet's Transtar engine on a liquid stage 2. Other possibilities included adding a third stage (a PAM was for Ulysses) and enlarging stage 2.

Originally, the unit cost was US$60 million to US$70 million, depending on mission unique requirements, within production lots of threes. The air force ordered 23 IUSs in the initial buy and the last was delivered in mid1991. A follow-on contract delivered one in September 1994 to carry TDRS 7 in 1995. The first serial buy also included two for USAF Titan 4, to be delivered in 1997, but they were cancelled. NASA, in May 1994, decided to order another under a US$25 million contract to carry its AXAF astronomy satellite in 1998. Shuttle/IUS flight turnaround is eight months.

Specifications

Inertial Upper Stage

First flight: 30 October 1982
Number flown: 24 to end 2001
Vehicles: Shuttle, Titan 34D, Titan 4; Titan 3
Success rate: 91.7% through end 2001 (TDRS1 payload on failed stage 2 was able to attain operating orbit)
Number of stages: 2 solid.
Overall length: 5.17 m
Principal diameter: 2.9 m
Mass (excluding payload): 14,760 kg (Shuttle version), 13,100 kg (Titan 34D)
Performance: (motors can be offloaded 0–50%, vehicle sized for up to 3,600 kg payloads, but 7,200 kg possible with modifications)
GEO: 2,268 kg (Shuttle), 1,817 kg (Titan 34D), 2,364 kg (Titan 4 with CSD SRM strap-ons), 2,860 kg (Titan 4B with Hercules SRMU strap-ons)
GTO: 4,944 kg (using payload for apogee kick into GEO)
Polar (96°, 24 h): 1,860 kg from 57° Cape Canaveral orbit
Molniya (925 × 39,450 km, 63°, 12 h): 3,645 kg (T4 SRM via 185 km, 55°), 3,866 kg (T4 SRM via 185 km, 57°), 4,189 kg (T4 SRM via 185 km, 63°), 4,933 kg (T4 SRMU via 185 km, 55°), 5,253 kg (T4B SRMU via 185 km, 57°).
Escape: 820 kg for C^3 of 60 km^2/s^2, 3,550 kg for 7.0 km^2/s^2.

IUS Motors

Description: IUS incorporates two UTC Chemical Systems Div solid propellant motors. On a typical GEO mission, the larger SRM1 performs GTO injection, followed by SRM2 insertion into GEO. The motors are designed from a common approach, with Kevlar filament wound cases, 3-D carbon/carbon integral throats and 2-D carbon-carbon exit cone nozzles. Thrust vector pitch/yaw control is provided by the nozzles moving on a Techroll joint pressurised by silicon oil under the action of electromechanical actuators. SRM1 can gimbal 4°, SRM2 7°. The Techroll joint failed on the second mission near the end of the SRM2 burn but the payload was able to separate and attain GEO with its own thrusters. SRM2 can be flown without its Extendible Exit Cone (EEC), resulting in a reduced specific impulse.

IUS SRM1 stage 1

Motor length: 3.52 m
Motor/stage diameter: 2.34 m
Propellant loading: 4,853–9,709 kg
Thrust: 185.1 kN average vacuum
Specific impulse: 295.5 s vacuum
Burn time: 152 s 100% loading

IUS SRM2 stage 2

Motor length: 2.08 m without nozzle extension
Motor diameter: 1.61 m
Propellant loading: 1,361–2,749 kg
Thrust: 78.41 kN average vacuum
Specific impulse: 289.1 s (without EEC), 303.5 s (with EEC), vacuum
Burn time: 103.35 s 100% loading

IUS interstage and ESS

Description: An aluminium skin stringer interstage connects the SRM1 to the Equipment Support System (ESS), which houses SRM2, most of the avionics and the payload interfaces (3.05 m diameter payload mounting ring and electrical connectors). The eight longerons on the aluminium skin stringer ESS assembly provide primary load paths and the spacecraft attach points.

Reaction control subsystem

Description: Six modules each with two Kaiser Marquardt 133 N hydrazine thrusters provide roll control during main motor burns, three-axis control during coast periods, vernier orbit corrections after SRM2 burnout, and collision avoidance manoeuvres after payload separation. The configuration provides dual-nozzle redundancy and there are no forward facing thrusters to avoid payload contamination. Two tanks hold a total of 109 kg of hydrazine but one and three tank versions are possible. RCS burns to depletion after payload release in order to remove IUS as far as possible from GEO.

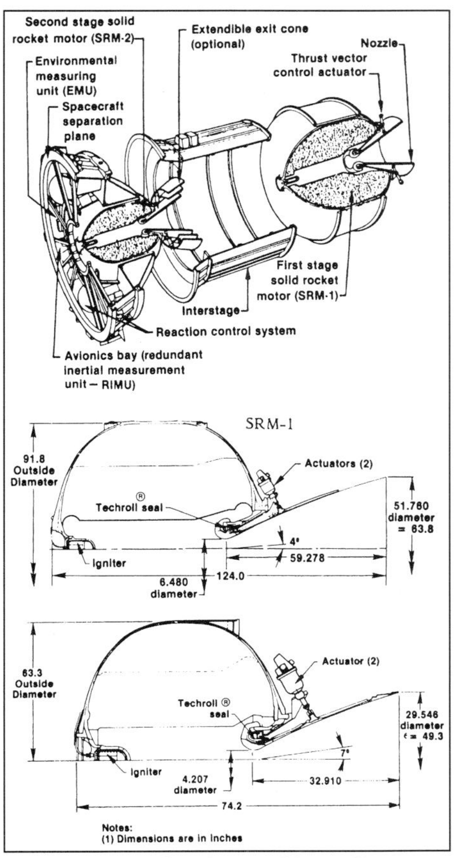

The Inertial Upper Stage's two solid propellant motors (UTC Chemical Systems Division) 0516924

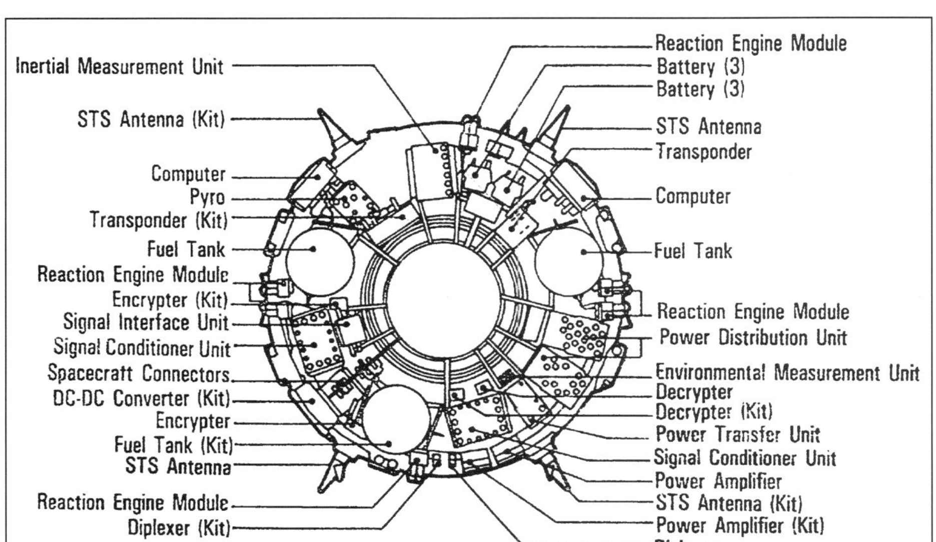

IUS Equipment Support Section in Shuttle configuration (NASA) 0516925

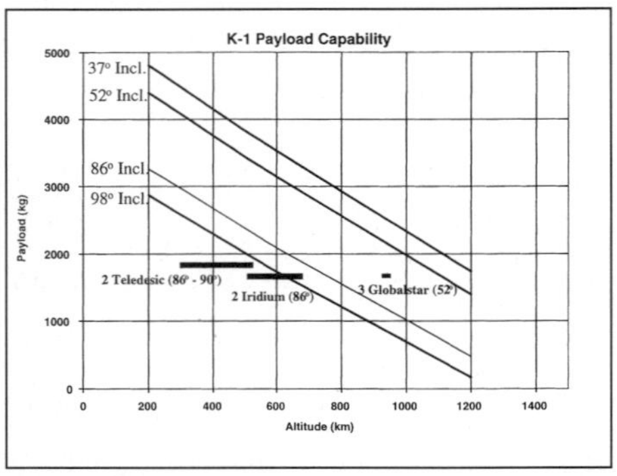

K-1 payload capacity 0003456

Avionics

Description: IUS is completely autonomous once separated from its launcher. Avionics are redundant throughout; the A side is normally in control but the B side performs the same functions and issues the same commands in parallel and takes over should A fail. This redundancy saved three missions through TDRS 6, on IUS1/6/7. IUS1 lost its A computer memory before SRM2 ignition because of a cosmic ray upset, an IUS6 thruster stuck on, and IUS7's SRM2 failed to arm. On IUS13 (TDRS 6), side B's power was lost. The avionics consist of five subsystems: TT&C, Guidance & Navigation, Data Management, TVC, and Electrical Power. Telemetry is provided at 16/64 kbit/s rates, allowing 1/4 kbit/s, respectively, from the payload. Eight primary and eight back-up discrete commands can be sent to the payload. The Guidance & Navigation Subsystem (GNS) provides measurements of angular rates and linear accelerations to the two Delco M362S 65,536, 16 bit word computers of the Data Management Subsystem from a redundant strapdown inertial measurement unit (RIMU) and a star scanner. The RIMU incorporates five Hamilton Standard RI1010 rate integrating gyros and five Kearfott 2401 accelerometers. IUS uses an explicit guidance algorithm (gamma guidance) to generate thrust steering commands, SRM ignition time and RCS commands. The vehicle is oriented to a thrust attitude before each burn based on normal performance of the remaining propulsion stages. Stage 1 carries four batteries: two 100 Ah + one 140 Ah for the avionics, plus one 100 Ah for the spacecraft. Five 13 Ah (one for the payload) provide power for stage 2 operations. All are AgZn 28±4 V batteries. The current Delco computers and Hamilton Standard RIMU may be replaced by a new Honeywell fault tolerant system.

IUS/payload deployment

Description: On Shuttle missions, the IUS/Payload is spring released from the ASE tilted at 58°. The RCS does not activate until 10 min later at a separation of about 60 m. SRM1 ignition does not occur for a further 50 min until the Orbiter is some 18 km distant. The times in the following schedule are from Orbiter/IUS deployment (min:s):

10:00	Activate RCS three-axis control
34:00	Star scanner attitude update
47:10	manoeuvre to SRM1 burn attitude; data to 64 kbit/s
59:48	SRM1 ignition
61:39	RCS corrects SRM1 errors
66:39	RCS manoeuvre to coast attitude, with roll for thermal control
370:55	switch to stage 2 batteries
371:12	separate stage 1, extend SRM2 nozzle 7 s later
371:47	manoeuvre to SRM2 burn attitude; data to 64 kbit/s
373:47	SRM2 burn
379:14	RCS corrects SRM2 errors
389:14	Separate payload
394:56	Collision avoidance manoeuvre by IUS

IUS/payload processing

The IUS, without motors, is fabricated and tested at the Boeing Space Center and shipped to the Cape, where CSD's motors are added. Processing/checkout are performed in the east low bay of the USAF Solid Motor Assembly Building (SMAB) at Cape Canaveral, followed by spacecraft mating in the Shuttle Payload Integration Facility (SPIF) in the SMAB west bay. For NASA missions, integration is performed in the Vertical Processing Facility. In both cases, the assembly is transferred to pad 39 in the MultiMission Support Equipment canister. For Titan missions, the IUS is processed as for the Shuttle, but the payload is mated on the pad following Titan/IUS integration.

Orion and Orion Derivatives (Sounding Rockets)

Current Status

Several proposed initiatives for development of an improved Orion sounding rocket matching reaction from the user community.

Background

The Orion is a single-stage unguided rocket, incorporating a surplus US Army motor, optimised for 39 to 68 kg payloads. An Orion version with 20 per cent increased total impulse was tested successfully 5 April 1994 as the second stage on a Terrier. The test included a GPS receiver, attitude/rate gyros and a new water recovery system. This design has been packaged into a much more powerful derivative with a variety of stage configurations.

Specifications

Orion

First launch: May 1974
Number launched: 33 to end-1996
Success rate: 100%
Number of stages: 1 (solid)
Overall length: 534 cm
Principal diameter: 35.6 cm
Launch mass: 422.3 kg plus payload
Typical performance: 39 kg/88 km, 68 kg/71 km
Payload recovery system: payload can be separated if required.
Payload environment: separable clamshell nose accommodates standard 35.6 cm diameter, 183–254 cm length.
Guidance: unguided, three fins provide 4 rev/s roll stabilisation at burnout.

Orion Derivative

Terrier-(Improved) Orion

Comment: A new two-stage vehicle employing surplus military motors. Stage 2 employs an enhanced version of the basic Orion. Performance is similar to Taurus Orion, but the environment is more benign, allowing bulbous (42.8 cm diameter) payloads.
First launch: 5 April 1994
Number launched: 3 to end-1996
Success rate: 100%
Number of stages: 2 (solid)
Overall length: 10.6 m (6.6 m without payload)
Principal diameter: 45.7 cm
Launch mass: 1,316 kg plus payload
Typical performance: 290 kg to 118 km
Payload recovery system: IRMA Ignition and Recovery Modular Assembly
Payload environment: 35.6 cm or 42.8 cm diameter payloads with various deployable nosecones. 19 *g* peak for 290 kg payload.
Guidance: unguided, fin spin-stabilised. Spin motors. Burnout roll rate about 6.5 Hz.

Terrier-(Improved) Orion stage 1

Motor: US Army Aerojet solid, surplus
Overall length: 268.1 cm
Principal diameter: 35.6 cm
Stage mass: 422.3 kg
Thrust: 7.74 kN average sea level
Burn time: 32 s.

Terrier-(Improved) Orion stage 2

Motor: improved XM22E8 Orion
Overall length: 268 cm
Principal diameter: 35.56 cm
Stage mass: 438 kg (293 kg propellant)
Thrust: unavailable
Burn time: 25.4 s (T+15 s ignition typical).

Nike Orion

Comment: A boosted version of the basic Orion vehicle, extending payload performance to >200 kg
First launch: September 1975
Number launched: 103 to end-1996
Success rate: 97%
Number of stages: 2 (solid)
Overall length: 880.9 cm
Launch mass: 1,021.8 kg plus payload
Typical performance: 68 kg/190 km, 204 kg/90 km
Payload recovery system: Wallops' 35.56 cm (14 inch) recovery system
Payload environment: as for Orion
Guidance: unguided, fin spin stabilisation.

Nike Orion stage 1

Specifications as for Nike Tomahawk.

Nike Orion stage 2

Specifications as for the Orion vehicle, but the motor ignites 9 s after lift-off and the four fins are canted to generate 4 rev/s at burnout.

Nike Tomahawk

First launch: March 1965
Number launched: 230 to end-1996
Success rate: 97%
Number of stages: 2 (solid)
Overall length: 7.3 m
Principal diameter: 22.9 cm
Launch mass: 885 kg
Typical performance: 45 kg/370 km, 113 kg/220 km
Payload recovery system: non-standard
Payload environment: normally 22.9 cm diameter under 3:1 tangent ogive nose; a 30 cm diameter version is available but requires the larger fins. Payload length 183–305 cm. Clamshell nose is separable; payload can be separated if required. De-spin module is standard.
Guidance: unguided, fin spin-stabilised.

Nike Tomahawk stage 1

Motor: Nike M5-E1 solid (US Army Allegheny Ballistics Laboratory)
Overall length: 345 cm
Principal diameter: 41.9 cm
Stage mass: 597 kg (342 kg propellant)
Thrust: 190.29 kN average sea level, total impulse 651.8 kNs at sea level
Burn time: 3.2 s
Comment: An interstage adaptor is bolted to Nike's forward face, its conical portion slip fitting into stage 2's nozzle, allowing drag separation. Four fins, each 0.23 m^2, are canted to generate 2 rev/s at Nike burnout.

Nike Tomahawk stage 2

Motor: Thiokol TE-M-416 solid
Overall length: 361 cm
Principal diameter: 22.9 cm
Stage mass: motor 245 kg (175 kg propellant)
Thrust: 48.9 kN average sea level, total impulse 416.6 kNs sea level
Burn time: 9 s, ignited 12 s after launch
Comment: Fin and shroud assembly mass 22 kg. Four fins, each 0.13 m^2, are canted to generate 6 rev/s at burnout. 0.17 m^2 fins are available for increased aerodynamic stability.

Taurus Orion

First launch: July 1977
Number launched: 53 to end-1996
Success rate: 96%
Number of stages: 2 (solid)
Overall length: 986.8 cm
Principal diameter: 57.8 cm
Launch mass: 1,800 kg plus payload
Typical performance: 68 kg/260 km, 227 kg/140 km
Payload recovery system: Wallops' 35.56 cm (14 inch) recovery system)
Payload environment: standard payload 35.6 cm diameter, length 183–380 cm. Separable clamshell nose; payload can be separated.
Guidance: unguided, four-stage 10.45 m^2 Ajax fins are canted to generate 2 rev/s spin at burnout.

Taurus Orion stage 1

Motor: US Army surplus solid (Allegheny Ballistics Laboratory)
Overall length: 4.5 m
Principal diameter: 57.8 cm
Stage mass: 1,363 kg
Thrust: 457 kN average sea level
Burn time: 3.5 s.

Taurus Orion stage 2

Specifications as for Nike Orion.

Taurus Tomahawk

First launch: October 1978
Number launched: 14 to end-1995
Success rate: 100%
Number of stages: 2 (solid)
Overall length: 821 cm
Principal diameter: 22.75 cm
Launch mass: 1,621 kg plus payload
Typical performance: 27 kg/590 km, 59 kg/490 km
Payload recovery system: non-standard
Payload environment: as Nike Tomahawk
Guidance: unguided, fin spin stabilisation.

Taurus Tomahawk stage 1

Specifications as for Taurus Orion stage 1.

Taurus Tomahawk stage 2

Specifications as for Nike Tomahawk stage 2, but ignition is 18 s after launch.

Taurus Nike Tomahawk
A boosted version of the Nike Tomahawk. Stage 2 hardware includes an interstage locking device that holds the Nike/Tomahawk motors together during the coast before Nike ignition. Standard stage 3 hardware includes a Tomahawk Firing De-spin Module.
First launch: September 1983
Number launched: 15 to end-1996
Success rate: 100%
Number of stages: 3 (solid)
Overall length: 11.64 m
Principal diameter: 22.86 cm (Tomahawk)
Launch mass: 2,229.9 kg plus payload
Typical performance: 32 kg/700 km, 125 kg/400 km
Payload recovery system: non standard
Payload environment: typically 22.9 cm diameter, length 137 366 cm under 3:1 ogive cone. 30 cm diameter is possible but special analysis is required because of high dynamic load factors.
Guidance: unguided, fin spin stabilisation.

Taurus stage 1
Specifications as for Taurus Orion stage 1 but fins canted for 1 rev/s spin at stage 1 burnout.

Nike stage 2
Specifications as for Nike Tomahawk stage 1 but ignition is made 11 s after launch.

Tomahawk stage 3
Specifications as for Nike Tomahawk stage 2 but ignition after 18 s and action time is 11.5 s.

PA-X Sounding Rocket

Background
AeroAstro's PacAstro subsidiary is developing the PA-X suborbital rocket for USAF Phillips Laboratory, primarily as a target for ballistic missile defence testing. PA-X is also available commercially for US$1 million. On 4 April 1996 AeroAstro successfully completed a 50 second firing of the 53.4 kN rocket motor designed for PA-X. It was based on designs done at Nasa's Lewis Research Center. Funding for the US$1,020 million development cost was under negotiation at the beginning of 1996. The future PA-1 (suborbital) and PA-2 (orbital) vehicles would employ Russia's RD-120M (stage 1) and 11D58M (stage 2).

Specifications
PA-X
Launch sites: compatible with Vandenberg AFB, White Sands Missile Range, NASA Wallops
Principal applications: small suborbital payloads
Performance: 45 kg to 64 km at 90° elevation
Cost: US$1 million
Overall length: 9.1 m
Principal diameter: 91 cm
Launch mass: 2,400 kg
Thrust: 62.3 kN SL/68.9 kN vacuum
Engine: PacAstro's PA-X engine. Fixed, liquid propellant, 10.2 atmosphere chamber pressure, He pressure-fed stainless steel injector, graphite over-wrapped chamber, ablative cooling, restartable, throttleable.
Fuel: 435 kg RP-1 kerosene
Oxidiser: 1,010 kg liquid oxygen
Burn time: 54 s
Payload accommodation: conical envelope 122 cm high, 91 cm diameter at bottom, 61 cm diameter at top. 15 *g* max acceleration. Payload integration and final assembly performed on unfuelled rocket.

PA-Y
First launch: planned for fourth quarter 1998 from site TBD
Launch sites: compatible with Vandenberg AFB, White Sands Missile Range, NASA Wallops
Principal applications: small orbital payloads
Performance: 20 kg into 100 × 280 km
Cost: US$3–4 million
Overall length: 11.1 m
Principal diameter: 1.22 m
Launch mass: 3,035 kg (800 kg dry)
Launch thrust: 72.1 kN SL/83.4 kN vacuum
Guidance: inertial

PA-Y stage 1
Engine: PaCAstro's enhanced PAX
Overall length: 7.1 m
Principal diameter: 1.22 m
Fuel: 580 kg RP-1 kerosene
Oxidiser: 1,345 kg liquid oxygen
Dry mass: 725 kg
Thrust: 72.1 kN SL/83.4 kN vacuum
Burn time: 72.5 s

PA-Y stage 2
Engine: PacAstro's PA-Y-2 engine. Restartable and throttleable.
Overall length: 4.0 m
Principal diameter: 0.91 m
Fuel: 128 kg RP-1 kerosene
Oxidiser: 298 kg liquid oxygen
Dry mass: 73 kg
Thrust: 14.7 kN vacuum
Burn time: 94 s.

PA-Y stage 3 options
Comment: PA-Y delivers the payload into an elliptical orbit, but AeroAstro's miniature cold gas package or the user's own kick stage can circularise. 54 m/s delta-V is required at 280 km.

Payload accommodation
Conical envelope 61 cm high, 61 cm diameter at bottom, 46 cm diameter at top. 15 *g* max acceleration. Payload integration and final assembly performed on unfuelled rocket.

UPDATED

Payload Assist Module (PAM) (Upper Stage)

Background
As the Delta family grew and evolved, NASA sought ways to put ever larger payloads into orbit. One chief means was the introduction of the new PAM, or Payload Assist Module, a spinning upper stage based around Thiokol's Star 48 motor. The Delta 3920/PAM configuration first flew in 1982 and incorporated a larger capacity Delta stage 2 with an Aerojet AJ10 118K engine to create a 1,284 kg GTO capacity with the PAM.

MDA's Payload Assist Module was first conceived to promote the transition of payloads from the company's Delta launcher to the Shuttle. It would later serve as an interchangeable option for commercial satellite customers when both Delta and the NASA Space Transportation System was available for non-government payloads. It could fly as a conventional upper stage on the expendable, or be released into LEO with its payload from a dedicated cradle in the Orbiter. The loss of Shuttle Challenger on 28 January 1986 and resultant offloading of commercial payloads meant that PAM's application was exclusively with Delta; it is not now considered as a separate stage. 18 PAM-Ds and two PAM-D2s flew from Shuttle, one PAM-D2 from Titan 3 and the one-off PAM-S from Shuttle with Ulysses solar probe. The company also provided the SGS-2 stage for initial Navstar GPS launches, and the Payload Assist Stage for SDI Delta launches. PAM-D2 was developed for shuttle and Titan applications. For a while early in its application confidence with the PAM was blunted by failures in March 1984 when two motors blew their carbon phenolic cones and left two satellites stranded in orbit from two sequential deployments on the same Shuttle mission. A review board identified problems with identifying imperfections in the composite cones and stringent inspection procedures prevented recurrence.

Specifications
PAM-D (TE-M-711-18)
Motor: Star 48
Gross mass: 2,141 kg
Empty mass: 232 kg
Thrust (vac): 67.1 kN
Isp: 292 s
Burntime: 88 s
Length: 2.0 m
Diameter: 1.2 m

PAM-S (TE-M-711-17)
Motor: Star 48
Gross mass: 2,182 kg
Empty mass: 220 kg
Thrust (vac): 66.7 kN
Isp: 270 s
Burn time: 80 s
Length: 2 m
Diameter: 1.2 m

PAM-D2
Motor: Star 63
Gross mass: 3,697 kg
Empty mass: 431 kg
Thrust (vac): 107 kN
Isp: 282 s
Burn time: 120 s
Length: 1.8 m
Diameter: 1.6 m

Pegasus

Current Status
Launch rates for Pegasus have progressed from five in 1997 to six in 1998, three in 1999, two in 2000 and one in 2002. To date, there have been four launches of Pegasus XL from CCAFS, Florida, in 2003, the first on 25 January (SORCE), the second on 28 April GALEX), the third on 26 June (OrbView-3) and the fourth on 12 August (SCISAT-1).

Background
OSC was established in 1982 to develop Space Shuttle upper stages, resulting in the solid propellant Transfer Orbit Stage. In June 1988, OSC announced a partnership with motor manufacturer Hercules Aerospace (now Alliant Techsystems) to develop the Pegasus air-launched small orbital delivery system. The 28-month development schedule was then already under way as a private venture, aiming for a first launch in July 1989 carrying a DARPA Glomar data relay and NASA Pegsat chemical release satellites.

Pegasus is capable of placing a 201 kg payload into a 460 km polar orbit (289 kg equatorial) following aircraft release at 11.9 km/Mach 0.8. Six launches (at 12.2 km) utilised the USAF-owned B-52 NASA-operated aircraft employed for X-15 hypersonic drops in the 1960s. A Lockheed L1011 Tristar, named 'Stargazer', was acquired by OSC from Air Canada in 1992 on a 10-year lease as the belly carrier beginning with STEP M1 in 1994. It is now leased from the owner, ITT Financial. Stargazer can handle 35.5 tonne vehicles and operate from most major runways.

The base vehicle consists of three graphite epoxy composite case Alliant solid motors of conservative design, a fixed delta planform composite wing, an aft skirt assembly including three composite control fins, an avionics section forward of the stage 3 motor, and a two-piece composite payload fairing. Some 94 per cent of the structural mass is graphite epoxy, 5 per cent aluminium, 1 per cent titanium. Air-launching at 11.9 km altitude reduces ΔV requirements by 10 to 15 per cent.

Work began on the 18 tonne, 15 m long, 6.7 m wingspan Pegasus in 1987 with US$40 to 50 million private funding divided equally between the partners. Commercial launch costs were estimated at US$10 to 12 million, requiring 16 to 18 missions in 2 to 2½ years to recoup costs. DARPA contracted for six launches (responsibility from number 3 was transferred April 1993 to the AF Space & Missile Systems Center), paying US$6 million for the first and US$7.3 million each for numbers 2/3. The increased cost reflects inclusion of the liquid HAPS Hydrazine Auxiliary Propulsion System stage 4, added to cope with increased mass of flight number 2's Microsats and to improve injection accuracy. This variant, known as Pegasus XL, lengthened version improves performance into LEO by about 10 per cent.

NASA Goddard selected Pegasus in March 1991 to provide Small Expendable Launch Vehicle Services for its Small Explorer science satellite programme. The total contract, worth some US$80 million, covers seven firm and three optional launches from 1993. It was selected April 1991 for one firm (late 1992) and 39 optional launches under the

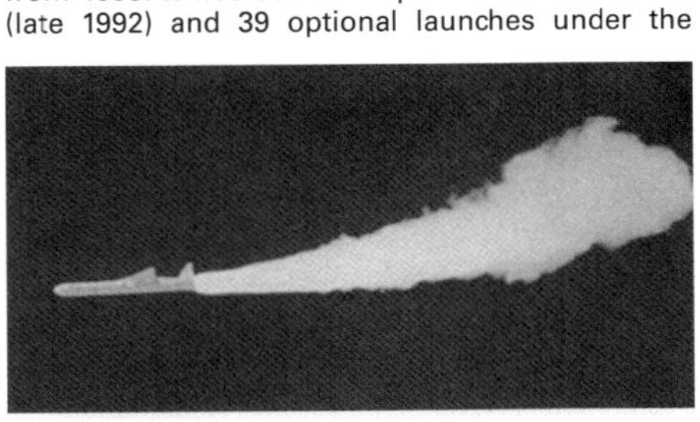

Pegasus stage 1 ignites 5 seconds after release (OSC)
0516934

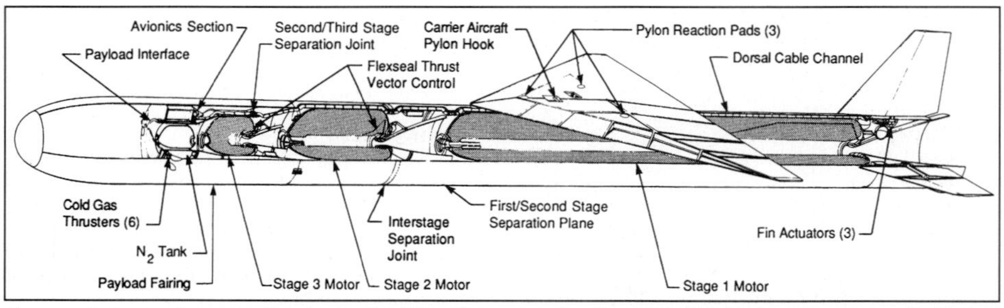

Pegasus features all-composite wing and solid rocket motor structures. The HAPS stage 4 replaces the N_2 tank shown here (OSC/Alliant)
0516935

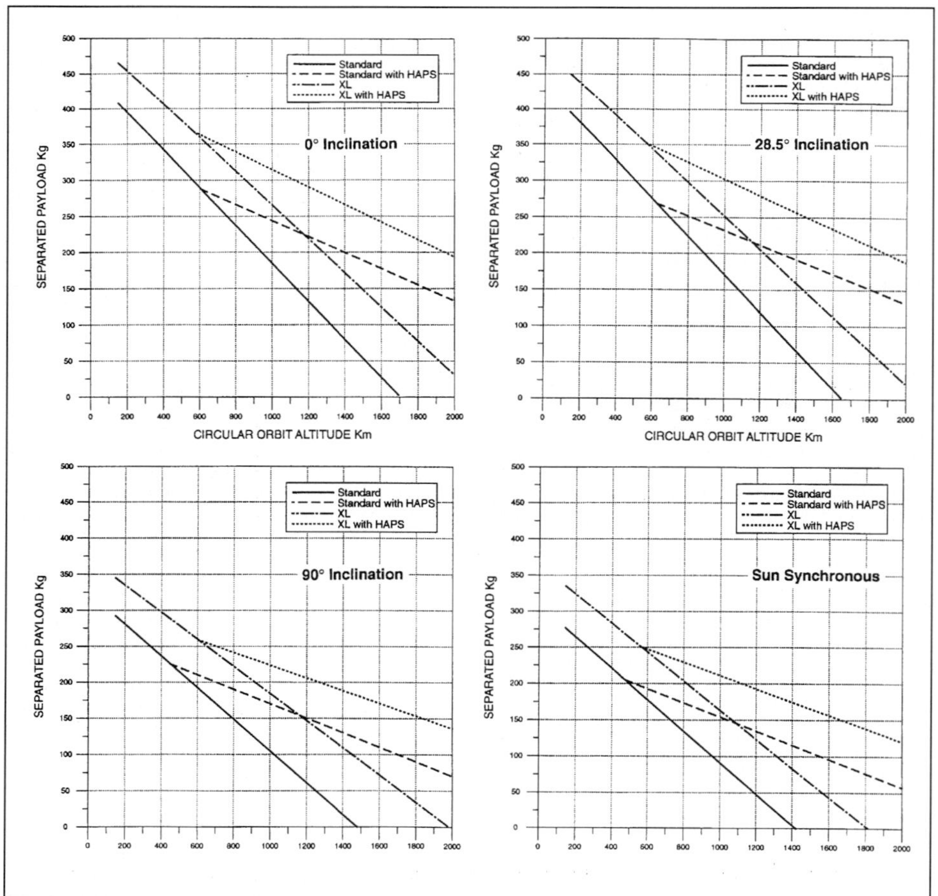

Pegasus payload performance (OSC)
0516936

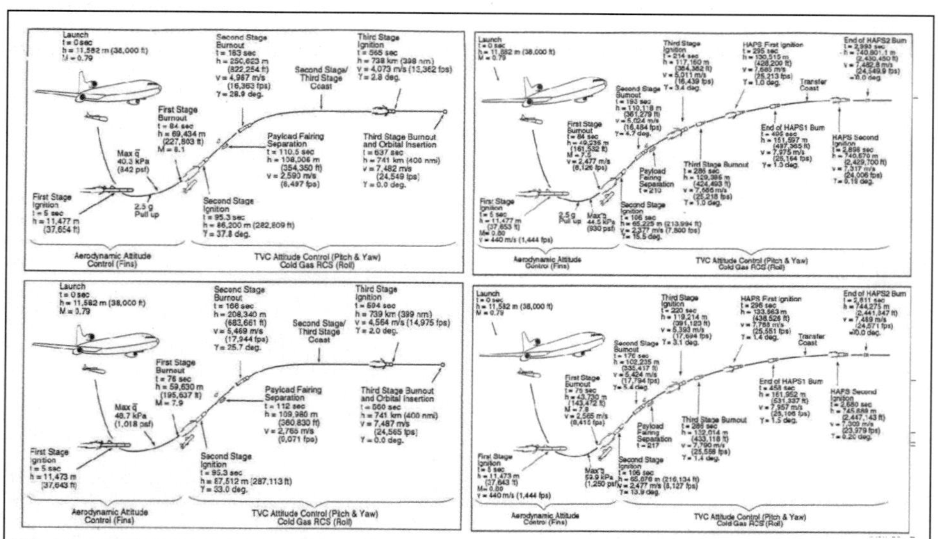

Pegasus mission profiles into 741 km circular polar. From top: standard Pegasus, 162 kg payload; Pegasus XL, 219 kg; standard + HAPS, 190 kg; XL + HAPS, 244 kg (OSC)
0516937

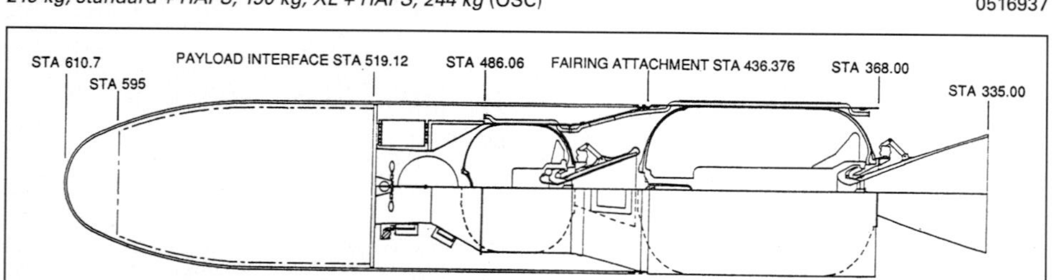

Pegasus payload accommodation. Station positions (standard Pegasus) are given in inches. Length is reduced 10 cm with HAPS (OSC/Alliant)
0516938

AFSLV Air Force Small Launch Vehicle programme, bringing orders to 15 firm and about 45 optional. BMDO, in July 1992, awarded a US$14.7 million contract for one launch (plus nine options) for its MSTI 3 satellite. NASA selected Pegasus as its Ultralight launcher in November 1994, with two firm (USRA in 1997) and eight optional launches. Spain's INTA selected it in October 1994 for Minisat 01, which will be the first west European orbital launch (from Spain's Torrejon AB). Contract value is US$10.2 million.

Specifications

Pegasus

First launch: 5 April 1990
Number launched: 35 through end September 2003 (10 Pegasus; 25 Pegasus XL)
Launch sites: now hosted by L-1011 operating out of Vandenberg AFB, NASA Wallops and Cape Canaveral (Spain's Minisat will be from Torrejon AB near Madrid)
Principal applications: small payloads into LEO
Vehicle success rate: 91.5% for Pegasus; 88% for Pegasus XL
Performance: 3-stage typically 200 kg (XL: 279 kg) into 463 km polar, 288 kg (XL: 382 kg) into 463 km equatorial (payload fraction 2.2%). For 740 km polar orbit, accuracy is ±0.2° inclination, ±85 km 3Σ deviation from circular; with HAPS: ±0.1°/±19 km. 165 kg GTO capacity projected with XL + stage 4
Availability: 1 flight/45 days projected
Cost: US$53,000/kg or US$12 million for 227 kg payload
Number of stages: 3
Overall length: 15.512 m; XL: 16.8 m
Principal diameter: 127.1 cm (wingspan 670.6 cm)
Launch mass: 18,518 kg; XL: 22,583 kg
Guidance: autonomous autopilot forward of stage 3, with Oettle & Reicher's 4.46 kg 68430-based flight computer (previously AlTech's 8.2 kg 5E 102 MS 68020 based) containing mission specific mission data load (MDL). Inertial/attitude data are provided from 5.8 kg Litton LR81 Inertial Measurement Unit (IMU); GN&C performance is monitored by the autopilot processor and processor data are formatted within the flight computer for direct downlink via the 56 kbit/s S-band vehicle telemetry system. Flight number 2 demonstrated a Trimble GPS receiver for navigation. It is now standard; IMU is retained for attitude data. XL carries upgraded avionics for increased payload mass/dynamic volume.

Pegasus stage 1

Stage 1 incorporates a Scaled Composites Inc graphite composite wing with 45° sweptback leading edge and 670.6 cm span. Subsonic L/D is 4.0. The airfoil is a double wedge with 2.5 cm radius leading edge. The wing thickness is truncated to 20.3 cm, with upper/lower parallel surfaces facilitating attachment to the motor case wing saddle. Three active aft fins provide control during stage 1 burn as part of an aluminium assembly attached to the motor's skirt extension. Control is augmented at end of stage 1 burn by firing small fin-mounted motors.
Motor: Alliant Techsystems Orion 50S, stretched by 137 cm for XL. All three motors use IM7/55A graphite epoxy composite cases with aramid filled EPDM rubber internal insulator and integral skirts. Each nozzle consists of a carbon phenolic exit cone with 3D carbon/carbon integral throat/entry. Propellant is HTPB with 88% solids, grain designed for low burn rates. The igniter is mounted on the forward dome. A flight termination charge is mounted on the aft dome of stage 1 and forward dome of stage 2 for range and aircraft safety requirements. Stage 1 employs a fixed nozzle.
Stage length: 9.39 m; XL: 10.76 m
Diameter: 127.1 cm
Thrust: 486.696 kN average vacuum, 583.773 kN max vacuum attained about mid-way through burn (total impulse 35.05 MNs); XL: 619.9 kN average vacuum.
Burn time: 72.3 s; XL: 64.3 s
Mass at ignition: 14,020 kg; XL: 17,157 kg
Propellant mass: 12,152 kg; XL: 15,051 kg
Separation: the forward skirt, which also serves as interstage adaptor, incorporates two linear-shaped charges.
Burn sequence: vehicle is released from L 1011, activating sequencer and autopilot. At 2 s, the arming sequence begins for stage 1 arming/ignition at 3/5 s, respectively. Burnout occurs 76 84 s.

Pegasus stage 2

Motor: Alliant Techsystems Orion 50 (50XL stretched 45.0 cm). See stage 1 for comments. Stage 2 employs a silicon elastomer flexseal nozzle and Parker Bertea Aerospace electromechanical actuators for TVC

Pegasus XL with the Celistis payload launched 21 April 1997 (Theo Pirard) 0054368

Pegasus expendable launcher attached to Lockheed Martin L1011 0131958

Pegasus launches

	Date	Type	Carrier/Base	Payloads
1	5 April 1990	std	B-52/DFRC	192 kg Pegsat + 68 kg SECS
2	17 July 1991	HAPS	B52/DFRC	7 × 21.8 kg Microsats
3	9 February 1993	std	B 52/KSC	115 kg SCD 1 + 14.5 kg CDS
4	25 April 1993	std	B52/DFRC	109 kg Alexis
5	19 May 1994	HAPS	B52/DFRC	180 kg STEP M2
6*	27 June 1994	XL	L 1011/VAFB	348 kg STEP M1
7	3 August 1994	std	B52/DFRC	261 kg APEX
8	3 April 1995	std	L1011/VAFB	2 × 40 kg Orbcomm 1/2 + 68 kg MicroLab 1
9*	22 June 1995	XL	L1011/VAFB	268 kg STEP M3
10	9 March 1996	XL	L 1011/VAFB	113 kg REX 2
11	17 May 1996	std	L1011/VAFB	211 kg MSTI 3
12	2 July 1996	XL	L-1011/VAFB	294 kg TOMSEP
13	21 August 1996	XL	L1011/VAFB	180 kg FAST
14*	4 November 1996	XL	L-1011/WI	2200 kg SAC B+NETE
15	21 April 1997	XL	L1011/C1	209 kg MINISAT 1/Celestis
16	1 August 1997	XL	L-1011/VAFB	309 kg Seastar
17	29 August 1997	XL	L-1011/VAFB	215 kg FORTE
18	22 October 1997	XL	L-1011/WI	395 kg STEP M4
19	23 December 1997	XL	L-1011/WI	8x43 kg Orbcomm
20	26 February 1998	XL	L-1011/VAFB	132 kg SNOE; 45 kg Teledisc T-1
21	2 April 1998	XL	L-1011/VAFB	250 kg TRACE
22	2 August 1998	HAPS	L-1011/WI	8x43 kg Orbcomm
23	23 September 1998	HAPS	L-1011/WI	8x43 kg Orbcomm
24	23 October 1998	Std	L-1011/K5C	540 kg STEX
25	6 December 1998	XL	L-1011/VAFB	283 kg SWAS
26	5 March 1999	XL	L-1011/VAFB	270 kg WIRE
27	18 May 1999	HAPS	L-1011/VAFB	123 kg TERRIERS; 45 kg MuBLcom
28	4 December 1999	XL	L-1011/VAFB	7 × 43 kg Ovbcomm
29	7 June 2000	XL	L-1011/VAFB	TSX-5
30	9 October 2000	XL	L-1011/VAFB	HETE-2
31	5 February 2002	XL	L-1011/KSC	HESSI
32	25 January 2003	XL	L-1011/KSC	SORCE
33	28 April 2003	XL	L-1011/KSC	GALEX
34	26 June 2003	XL	L-1011/KSC	OrbView-3
35	12 August 2003	XL	L-1011/KSC	SCISAT-1

* launcher failure. DFRC = Dryden Flight Research Center; KSC = Kennedy Space Center; VAFB = Vandenberg Air Force Base; WI = Wallops Island. 1996 Minisat launch is planned from Spain's Torrejon Air Base, near Madrid. Number 6 failure: the review board concluded the guidance software was unable to cope with unpredicted aerodynamic responses on the first XL vehicle. The design was then tested in wind tunnels to develop an aerodynamic model, rather than previous reliance only on computer modelling. Number 9 failure: stage 1/2 interstage remained attached during stage 2 burn (one of three separation skids improperly installed) satellite failed to separate from Pegasus final stage. Flight 14 from Wallops Island, 15 from Canary Islands.

Stage length: 383.7 cm; XL: 428.7 cm
Diameter: 127.1 cm
Thrust: 113.874 kN (160.136 kN) average vacuum, 131.667 kN (195.266 kN) max vacuum. Total impulse 8.67 MNs (11.22 MNs)
Burn time: 75.9 s; XL: 70.1 s
Mass at ignition: 3,370 kg; XL: 4,314 kg
Propellant mass: 3,024 kg; XL: 3,919 kg
Attitude control: motor nozzle flexing provides pitch/yaw control during powered flight; roll control and three-axis control during coast provided by six N_2 thrusters in two Parker Bertea Aerospace modules on stage 3.

Pegasus stage 3
Motor: Alliant Techsystems Orion 38. See stage 1/2 for comments. Stage 3 also employs flexing for pitch/yaw control during burn. Igniter is toroidal. Head-end grain design maximises propellant density. XL uses the same stage 3 as the standard Pegasus.
Stage length: 208 cm
Diameter: 96.5 cm
Thrust: 34.314 kN average vacuum, 38.686 kN max vacuum. Total impulse 2.18 MNs.
Burn time: 63.8 s
Mass at ignition: 893 kg
Propellant mass: 771 kg
Attitude control: motor nozzle flexing provides pitch/yaw control during powered flight; roll control and three-axis control during coast provided by six N_2 thrusters in two Parker Bertea Aerospace modules.

Pegasus stage 4
The HAPS Hydrazine Auxiliary Propulsion System is an optional stage carried to improve injection accuracy and performance by about 36 kg for 720 km, 82°. It was ordered by DARPA to cope with the Microsat mass growth (launch number 2, July 1991). It occupies the payload adaptor interior, replacing the N_2 tank of the 3-stage version. The attitude jets are supplied instead from two high-pressure (395 atmosphere) N_2 tanks on the adaptor's outer surface. On typical missions, the first burn establishes an elliptical orbit, followed by circularisation.
Thrusters: three fixed blowdown monopropellant Olin Aerospace MR-107s
Propellant: 72.6 kg of hydrazine stored in central sphere
Stage thrust: 3 × 222 N vacuum
Burn time: typically 2 burns of 131 and 110 s
Attitude control: nitrogen jets and differential firing of main thrusters.

Fairing and Payload Accommodation
Alliant's 127 kg 2-piece composite fairing, which also covers stage 3 and the avionics packages (and stage 4 when carried), is jettisoned during stage 2 burn. The payload may remain attached to stage 3 or be separated by clamp, following pointing to ±2° accuracy or spin up. The payload can be provided with 60 V DC (up to 140 W) from the carrier aircraft until release, and up to 16 discrete commands can be issued by Pegasus' computer. Telemetry at 3 kbit/s from the vehicle's total 56 kbit/s downlink is available to the payload. On the ground, air conditioning maintains 21 ± 5°C.
Payload volume: 1.17 m diameter, 2.13 m long for 3 stage; 1.76 or 1.79 m long when HAPS carried
Acceleration load: 8 *g* longitudinal max at end of stage 1 burn (lighter payloads can exceed 8 *g* during stage 3 burn).

Vehicle and Payload Processing
Vehicle and payload processing are performed in two 15 × 36 m bays of the OSC Vehicle Assembly Building at VAFB, controlled at 21 ± 5°C and humidity of 40 ± 10%. The stages and other subassemblies arrive typically 4 months before launch. Payload mating at T12 days is in a 3.66 × 7.32 m class 100,000 soft-walled clean room. The N_2 ACS tank is then filled and the fairing attached. All processing is undertaken horizontally. Pegasus is attached to the carrier aircraft at T–1 day. Takeoff is 1 hour before release.

Space Transportation System

Current Status
Shuttle orbiter 102, Columbia, was destroyed during re-entry at the end of STS-107 on 1 February 2003. Chaired by retired Admiral Hal Gehman, the Columbia Accident Investigation

Board was established to investigate the cause of the accident. It was given a broader remit than that given to the Rogers Commission, set up to investigate the loss of Challenger in January 1986, in that it was required to report on the management infrastructure of NASA and review the wider issues of human space flight. The Board comprised 13 members, approximately 120 Board investigators and several thousand NASA and support personnel. In its report, issued late August 2003, it presented a critique of the entire Shuttle programme and drew conclusions which informed 29 recommendations, 15 of which were mandatory prior to flight resumption. Key elements were the development of tools and procedures to repair on orbit defective thermal protection identified through imagery from ground, in space and by cameras manipulated to view the undersurfaces of the vehicle. It required the establishment of a Technical Engineering Authority to verify launch readiness, oversee and co-ordinate requests for waivers and to decide on what constitutes an anomalous event. NASA has tentatively placed a date of late September 2004 as the earliest it will return the Shuttle to flight but this date is contingent upon several critical developments to meet the mandatory requirements from the CAIB.

Background

NASA's Space Transportation System relies for its propulsion on four elements: two Solid Rocket Boosters (SRB), the Orbiter main engines (SSME), an External Tank (ET), and small manoeuvring engines on the Shuttle Orbiter (OMS). At an altitude of about 43 km, after some 2 minutes, the boosters separate for recovery and refurbishment. The Orbiter continues under SSME power until about 8 minutes 50 seconds after launch, when the ET separates for destructive re-entry. One or two Orbital Manoeuvring System firings establish operational orbit.

Specifications

Space Shuttle Launch Operations

First launch: 12 April 1981 from Kennedy Space Center complex 39A
Number launches: 113 to the STS-107 mission January 2003
Launch sites: KSC pads 39A/39B
Vehicle success rate: 98.23% to February 2003
Principal uses: US manned capability to beyond 2015 (four reusable Orbiter fleet), 25 t payload delivery to LEO, satellite retrieval/in situ repair, short duration science platform, Space Station assembly/servicing.
Availability: typically 4–5 flights/year manifested.

Performance

LEO (204 km, 28.45°)
21,140 kg OV-102, 24,950 kg OV-103/104/105; each additional 1 km altitude reduces capacity by 25 kg, each crew member beyond 5 persons reduces capacity by 230 kg into basic orbit.
LEO (204 km, 57°)
14,800 kg OV-102, 18,600 kg OV-103/104/105 (57° is usually the highest inclination flown from Florida, but dog-legging provides access to higher angles)
LEO (204 km, 98° Sun-synchronous VAFB)
13,426 kg OV-103/104/105, assuming use of Advanced Solid Rocket Motors. Neither VAFB nor ASRM will be used.
Molniya (925 × 39,450 km, 63°)
3,563 kg using IUS via 222 km, 57° parking orbit.
Crossrange: 2,040 km max; 1,020–1,300 km used operationally by NASA.
Number of stages: 1 plus 2 strap-ons plus payload injection stages
Overall length: 56.14 m (Orbiter 37.24 m)
Launch mass: about 2,040 t
Landing mass: 104,328 kg max, defined by abort requirements
Launch thrust: 34,677 kN (three SSMEs firing at 104%/1,754 kN each, two SRBs at 14,680 kN each; SSMEs throttle down to 65% by 35 s and up to 104% by 65 s).

Solid Rocket Boosters (SRB)

Description: Thiokol was selected as Shuttle SRM prime in 1974. NASA announced July 1991 a US$2.6 billion Buy 3 contract extension to Thiokol for 67 further flight pairs through 1999/STS105, plus eight ground test motors. Another extension is expected for launches through 2012. The SRM is cast in four segments in Utah and then transported by rail to Kennedy Space Center, Florida, where they are stacked vertically in pairs either side of the External Tank on the mobile launch platform in the Vehicle Assembly Building. The boosters provide the majority of launch thrust, with TVC provided by a flexible bearing nozzle driven by two hydraulic actuators. After burnout at about 2 minutes, the boosters are separated pyrotechnically and fall into the Atlantic for recovery. The motors are cleaned, disassembled and returned to Utah for refurbishment and reloading. The segments are designed for 20 flights; nine is currently the maximum. The configuration underwent extensive reworking and testing following Challenger's loss, and the new booster was redesignated RSRM (Redesigned Solid Rocket Motor, changed by NASA in 1995 to Reusable SRM).

A High-Performance Motor (HPM) design for an improved SRM incorporates lightweight steel cylindrical case segments, a higher expansion nozzle, and altered motor thrust characteristics to increase Shuttle payloads by >1,540 kg. This design was static tested in 1982 and went into service on STS-8 in August 1983. Graphite epoxy composite cylindrical SRM case segments were also developed to increase Shuttle capacity by 2 t; the configuration was shelved following Challenger's loss. Thiokol was subcontracted to Aerojet to develop/produce the nozzle for the Advanced SRM that was to enter Shuttle service in about 1998. ASRM was cancelled in 1993.
Mass: 569,636 kg.
Dimensions: Length 45.46 m, diameter 3.7 m
Propellant configuration: 504 t of ammonium perchlorate/aluminium powder/iron oxide/binding agent in four segments with 11 point star, with max thrust at launch, reducing by one-third after 50 s (maximum dynamic pressure period).

Propellant Data

Type: TP-H1148 HB Polymer, 16% Al, 70% AP shape: 11-point star forward; double taper cylinder two centre segments; triple taper cylinder aft.
Mass fraction: 0.883 (501,746 kg)
Burn time (s): 123.7 action, 111.4 web
Thrust (vacuum): 11.52 MN average, 14.77 MN max
Specific impulse: 268.0 s vacuum
Total impulse: 296.3 MNs vacuum
Pressure (atmosphere): 62.1 max (MEOP 69.1); 45.0 avg
Throat diameter: 136.8 cm
Length: 454.0 cm
Exit diameter: 380.1 cm (expansion ratio 7.72)
Materials: D6AC steel for fixed housing, throat and forward exit cone housings; aluminium for nose inlet and aft exit cone housing
Casing materials: D6AC steel, 1.165 cm minimum thickness
Igniter: forward end internally mounted. 60.7 kg of TP H1178 case bonded 40-point star.
Control: Nozzle gimbals 8° for steering
Design lifetime: 20 missions (current max 9)
Thrust: 29.36 MN at sea level (70% of total lift-off thrust)
Comment: SRB separation occurs at about 124 s, 45 km altitude (triggered when pressure falls to 3.4 atmospheres), landing speed <100 km/h under 3 × 41 m diameter parachutes.
Cost: US$64 million/pair (1983), US$25 million per reflight.

Space Shuttle Orbiter-Main Engines (SSME)

Description: Three LOX/LH2 staged combustion Rocketdyne SSMEs fed by External Tank
Burn time: typically 520 s
Design life: 55 missions or 7½ h at 1,668 kN sea level (specific impulse 363 s)/2,090 kN in vacuo (SI 455 s).
Performance: Typically throttled over 65 104% rated power setting (109% was to have been used for Vandenberg polar missions, but is now available for transatlantic aborts). Constant 104% setting, extending lifetime, would have been possible with Advanced Solid Rocket Boosters providing required thrust profile (cancelled 1993). Pump and other improvements will allow 106% for Space Station assembly.
Control: The engines are individually gimballed hydraulically by two APU powered actuators at up to ±10.5° pitch and ±8.5° yaw; each engine was originally controlled in 20 ms steps by a 97 kg top-mounted package housing two redundant Honeywell 360 16.4 kbit-memory computers. The controllers are commanded by the Orbiter's General Purpose Computers, or by the crew in emergencies. Block 2 Honeywell controllers are now installed (introduced on OV105 and retrofitted to the others), providing four times the memory using the first space-qualified 68000 series microprocessors. They also allowed elimination of a separate 13.6 kg vibration monitoring box.
Pump performance: The incoming oxidiser is boosted by the first turbopump from 7 to 29 atmospheres before the High-Pressure Oxidiser Turbopump increases it to 500 atmospheres in two stages. This pump has suffered durability problems, with some engines requiring overhaul after three missions (33 min). Pratt & Whitney, one of the original SSME bidders, was awarded a US$188 million contract, as part of a 10-year US$1 billion Phase II SSME improvement programme beginning in 1983. In 1986 to develop a new power head containing liquid oxygen and hydrogen high-pressure turbopumps and a single pre-burner chamber (the Rocketdyne design incorporates a pre-burner for each). The goal is for 7½ h firing (55 missions) between major overhauls, but problems increased development cost to US$504 million. NASA thus halted fuel pump work from December 1991 in order to focus on the more critical oxygen unit (it restarted work May 1994 for STS86 September 1997 debut once the LOX pump was on schedule for 1995 debut). First flight was aboard STS70 in July 1995 with one pump. As that was successful, all three engines were equipped for STS 77 in May 1996.
Comment: Some of the oxygen turbopump output is routed into a pre-burner where it is mixed with fuel and burned hydrogen-rich at 760°C. The hydrogen half of the system likewise employs a low-pressure pump (2 to 18.7 atmosphere), high-pressure pump (18.7 to 442 atmosphere) and a pre-burner. Some GH_2 is directed from the high-pressure unit as a coolant through the 390 main chamber wall channels and the 1,080 tubes composing the 2.87 m long, 2.38 m max diameter nozzle. The products of the two pre-burners are fed into the main chamber with additional oxygen through 600 elements in 13 annular rows in a 56 cm diameter injector plate and burned at 3,300°C/204 atmospheres (100% thrust level) in a 6:1 oxygen to hydrogen ratio. Some 770 kg LH_2 + 1,680 kg LOX are trapped in the system at cut-off, producing an 18 cm aft shift in the Orbiter's centre of gravity. Sporadic venting during orbital flight would affect guidance and possibly contaminate the payloads, and the hydrogen could combine during re-entry with atmospheric oxygen to form a potentially explosive mixture. Instead, after ET separation, the LOX is dumped through the SSME nozzles and the LH_2 is released through the starboard T0 umbilical overboard fill/drain valves.

External Tank (ET)

Dimensions: Length 46.88 m (LH2 tank 29.47 m, LOX tank 16.64 m), diameter 8.40 m.
Dry mass: 35.43 t for early version, reduced to 29.93 t in later models, 26.33 t for aluminium lithium alloy Super LightWeight version, loaded typically 760 t.
Flight pressure: 3.0 atmosphere LH2, 1.43 atmosphere LOX.
Contractor: Lockheed Martin Manned Space Systems at Michoud, Louisiana
Unit cost: US$30 million (US$33.5 million for AlLi version).
Capacity: 104,308 kg LH2 at 253°C in 1,514 m^3 aft, 625,850 kg LOX at 183°C in 559 m^3 forward aluminium alloy semi-monocoque tanks.
Supply lines: Orbiter's SSMEs, by two 43.2 cm outlets, discarded after about 530 s, 110 km altitude.
Insulation: 25 mm-thick polyurethene foam insulation. Super LightWeight version first flight ET-96 December 1997.

Space Shuttle overall external dimensions

Wing span	23.79 m (78 ft 0.7 in)
Orbiter length	37.24 m (122 ft 2 in)
Orbiter height	17.27 m (56 ft 8 in)
Length overall	56.14 m (184 ft 2.4 in)
Length of external tank	46.88 m (153 ft 9.6 in)
Length of boosters	45.46 m (149 ft 2.0 in)
Height overall	23.35 m (76 ft 1.6 in)

Masses

Shuttle stack	2,040,000 kg (4,500,000 lb)
OV-102 inert	82,288 kg (181,445 lb)
OV-103 inert	78,448 kg (172,978 lb)
OV-104 inert	78,687 kg (173,505 lb)
OV105 inert	79,135 kg (174,493 lb)
Max landing	104,328 kg (230,000 lb)
External tank (full)	750,975 kg (1,665,600 lb)
Boosters (2), each	589,670 kg (1,300,000 lb)

Thrust

Total, at lift-off	34,677 kN (7,781,000 lb)
Orbiter, SSME (3), each	1754 kN (393,800 lb)
Boosters (2), each	14,680 kN (3,300,000 lb)

Space Shuttle Orbiter-OMS/RCS

Description: The Orbital Manoeuvring System and aft Reaction Control System engines are housed in two 6.45 m long 8.6 t McDonnell Douglas pods mounted on the aft fuselage either side of the tail. The 1.8 t composite (skin panels and frames) and 2,124 aluminium (main structural members) pods are each attached by 11 bolts accessible through 24 doors. Each carries 4.1 t MMH plus a 6.7 t NTO

in 2.45 × 1.25 m titanium tanks for OMS, although this supply can also be made available to RCS. Cross lines provide redundancy and balancing for re-entry.
Performance: 26.7 kN/313 s SI vacuum engine
Mass: 118 kg dry mass
Dimensions: 1.96 m long
Operating pressure: 8.5 atmospheres chamber pressure, helium pressure fed
Control: Gimballed ±6% pitch and ±7° yaw
Design lifetime: 10 years/100 mission/1,000 starts/ 15 h firing duration
Comment: Earlier missions required two OMS burns to attain operational orbit but a direct ascent technique first demonstrated on mission 41C is now employed, omitting the OMS-1 burn following ET separation and relying on the OMS-2 burn at apogee about 45 minutes after launch. The OMS engines are normally fired in pairs for a minimum of 2 s, but for fine orbital adjustments of 1.8 m/s or less one can be gimballed through the Orbiter's centre of mass for solo firing. Some 80–95 m/s capacity is reserved from the about 305 m/s total ΔV available for the 120–170 s re-entry burn. Aerojet was contracted in 1984 to add turbo pumps for increased chamber pressure and SI; no fitting schedule has been announced. 12 OMS engines were originally produced under US$140 million contracts, including four spares stored at KSC. A further pair was ordered for Challenger's replacement.

The pods also house part of the Orbiter's RCS of 44 hypergolic thrusters providing attitude control and fine orbit tuning. Each pod incorporates 12 Kaiser Marquardt 3,870 N R40 and two 106 N R1E radiatively and fuel film cooled engines drawing on 675 kg NTO plus 422 kg MMH in titanium spheres. The larger motors, operating with 10.3 atmosphere chamber pressures, are grouped into three up, three down, two aft and four side, rated for 50,000 individual firings totalling 330 minutes in 1,150 s bursts or 80 ms (min) pulses. The vernier pairs, modified from USAF Manned Orbiting Laboratory motors, are capable of 500,000 ignitions over 35 h lifetimes of 125 s bursts and 80 ms (min) pulses. The system can also take propellants from OMS tanks if necessary.

The forward RCS module is removable for maintenance. It houses identical motors and 99 cm titanium spheres but the nozzles are machined flush with the thermal tiles. The 14 R40s are grouped into three forward, three up, two each side and two angled 45° down each side for roll control, as are the twin solo verniers. Propellants are loaded on the pad via servicing panels protected by thermal tiles (forward RCS) or through the T0 panels (OMS pods). Post-mission, the propellants are drained in the Orbiter Processing Facility.

Starfire Sounding Rocket

Current Status

Starfire 1 first launched 29 March 1989 was basically a commercial version of the Terrier Black Brant and was the first commercial space flight operated under the US Department of Transportation's licensing authority. Several marketing initiatives aimed at recruiting user interest from small operators has failed to attract custom.

Background

SSI's first commercial launch was successfully undertaken 29 March 1989 using a Magellan Black Brant 9 boosted by a Thiokol Mark 70 solid motor. SSI calls the configuration 'Starfire'. This first launch carried a 295 kg package under a US$1 million contract from the Consortium for Materials Development in Space at the University of Alabama. Operations were conducted from White Sands' complex 36 under US Navy control. Vehicle guidance is provided by Saab's S19 package, incorporating a Space Vector Corporation gyro platform.

The second flight failed 15 November 1989 because a capacitor broke loose in the guidance system. Number 3 succeeded on 17 May 1990 with a repeat payload from number 2 and Consort 4, launched 16 November 1991, also succeeded. The company was awarded a contract worth up to US$9.8 million for up to seven further Consorts through 1996. Consort 5 flew 10 September 1992 but stage 2's burn ended 3 s early and μg data were not obtained. Consort 6 flew successfully 19 February 1993 with seven experiments, attaining 301 km. The US$2 million funding was provided from the insurance following C5's partial failure. Conquest 1, the first of a new series, flew 3 April 1996 carrying eight experiments.

SSI offers larger Starfire models for longer duration μg missions, but none has flown. Starfire 2 is a one-stage Castor 4B capable of providing 15 minutes μgravity time for 454 kg; Starfire 3 adds two Castor 4B strap-ons to create 20 minutes μgravity for 900 kg payloads.

Specifications

Dimensions: Length: 15.25 m, 46 cm diameter
Mass: 2,234 kg
Performance: Up to 8 min μ*g* can be provided for 454 kg payloads.

Super Arcas Sounding Rocket

Current Status

Primarily used for meteorological research and plasma physics. Can carry payloads of 3.5 to 8.2 kg. Super Arcas has been adapted to carry several separate dispensers in a carry-tray which can eject instruments at sequential periods.

Background

Super Arcas has been used since 1965 to carry meteorological sensors up to around 100 km. A high-altitude decelerator slows and stabilises the payload for making measurements in the mesosphere; it also acts as part of the recovery system.

Specifications

First launch: January 1965
Number launched: 251 to end 2002
Success rate: 97%
Number of stages: 1 (solid)
Overall length: 268.4 cm
Principal diameter: 11.4 cm
Launch mass: 37.6 kg
Typical performance: 4.5 kg to 92 km (86° launch)
Payload recovery system: high-altitude radar reflector decelerator; payload section separates 148 s after launch
Payload environment: 7 *g* peak acceleration and 25 revs/s at burnout. Payload diameter 11.4 cm, typical length 100 cm, accommodated under tangent ogive nosecone
Guidance: unguided, fin spin stabilised.

Super Arcas stage 1
Overall length: 192.5 cm
Principal diameter: 11.4 cm
Stage mass: 37.7 kg
Thrust: 1,445 N average sea level
Burn time: 40.2 s.

Taurus

Current Status

Taurus has successfully completed 5 out of 6 launches but no flights have taken place since September 2001. The first flight on 13 March 1994 carried the STEP and Darpasat payloads into orbit and was followed by the second flight on 10 February 1998 carrying the GFO and Orbcomm satellites. The third flight was launched on 3 October 1998 with the STEX experiment for the National Reconnaissance Office. The fourth flight, on 20 December 1999 lifted the Kompsat and Acrimsat satellites into orbit followed by the fifth flight on 12 March 2000 launched the MTI satellite. The sixth flight, on 21 September 2001, was a failure and the Orbview-4 QuickTOMS payload was destroyed.

Background

DARPA contracted OSC in 1989 to build the Taurus rapid response orbital launcher using the company's Pegasus as a baseline. OSC had already taken on Space Data Corporation of Tempe, Arizona in November 1988. SDC had launched a wide range of sounding rockets, including more than 600 suborbital boosters in 35 different configurations. The division has participated in more than 60 Minuteman 1 re-entry tests since 1971, and so it is not surprising that the joint corporation announced that it would make a series of winged air-launched suborbital vehicles using some Minuteman technology.

According to the contract, Taurus had to be a self-contained road transportable system capable of rapid response from austere sites. The vehicle and all its support equipment was required to be transported up to 1,600 km before the launch site was established and the vehicle integrated/tested within five days. After this setup, the vehicle can wait for the payload to arrive, from when launch takes a further 72 hours.

The original SSLV requirement was for a 450 kg payload into a 740 km polar orbit, increased to 815 kg in 1991 as a result of a DARPA-funded modification to add a vectorable nozzle to stage 1. The vehicle's three upper stages are derived from Pegasus, supported for the first launch (and any subsequent USAF mission) by a Thiokol Peacekeeper stage 1.

After that successful debut, the commercial Taurus uses Thiokol's Castor 120, based on Peacekeeper stage 1 technology, and a 120 per cent larger fairing. Uprating options are the Taurus XL (same stretched motors as Pegasus XL) and Taurus XLS (two strap-ons from 1998), increasing the 1,400 kg LEO capacity from Canaveral to 1,550 kg (XL) and 1,950 kg (XLS). XLS would carry two Thiokol Castor 4B motors. For transfer orbit or deep space missions, stage 3 can be replaced with a spin-stabilised Thiokol Star 37 for perigee kick, or followed by Star 37. The OSC/McDonnell Douglas team was awarded NASA's MedLite launch contract in March 1995. Taurus XL/XLS, along with optional upper stages, will be used for the smaller payloads in the MedLite spectrum. After five successes.

Specifications

Taurus
First launch: 13 March 1994 from VAFB
Number launched: 6 to end of 2002 (including 1 failure)
Launch sites: number 1 from VAFB 576E; potential Canaveral complex 46 and VAFB SSI; transportability allows launch from almost any range
Principal applications: medium LEO, small GTO or Earth escape payloads
Vehicle success rate: 83.3%
Performance: typically 800 kg into 695 km polar, 1,300 kg into 400 km at 28.5°. For 645 km polar orbit, accuracy is projected at ±0.2° inclination, ±37 km deviation from circular. See performance diagrams. 470 kg GTO and 320 kg lunar transfer with Star 37 stage 3.
Availability: current production capability can support six flights annually. Typical mission cycle is 18 months after receiving order.
Cost: US$14,000/kg standard; US$13,000/kg XL/XLS
Number of stages: 4 (optional 3/5 stage)
Overall length: 26.72 m launch 1; 27.56 m standard; 29.12 m XL/XLS
Principal diameter: 236.3 cm stage 1 (233.7 cm for launch 1)
Launch mass: 72,576 kg (68,040 kg for launch 1)
Guidance: autonomous autopilot hosted in Oettle & Reicher RCOM06OSC 68020-based flight computer contains the mission specific mission data load (MDL). Inertial/attitude data provided from Litton's LR-81 Inertial Measurement Unit (IMU) aided by position data from Trimble GPS receiver. The autopilot processor monitors GN&C performance and processor data are formatted within the flight computer for direct downlink via the vehicle telemetry system.

Taurus Strap-ons
Comment: Optional paired strap-ons augment XLS performance. They mount to reinforcing rings on the Castor 120 forward/aft ends using machined ball fittings mated to receptacles.
Motor: Thiokol Castor 4B. Welded 4130 steel case with TI R300 internal insulator. Nozzle features carbon phenolic ablative exit cone with 4130 steel structural shell. Nozzle mounted in a laminated rubber/glass fibre flexseal for ±6° vectoring with AlliedSignal hydraulic cold gas blowdown TVAs. Igniter mounted in forward dome. Redundant conical shaped charges on forward dome allow flight termination.
Stage length: 8.98 m
Diameter: 101.9 cm
Thrust: each 429 kN average vacuum (total impulse 26.1 MN)
Burn time: 60.9 s
Mass at ignition: 15,001 kg each
Propellant mass: 9,990 kg each
Separation: clamp bands retaining ball fittings are pyrotechnically released, with separation force provided by hot gas pushoff thrusters
Flight sequence: TVA enabled after stage 0. Internal count sequencer performs nozzle sweep then commands simultaneous ignition of stage 0/strap-ons. Strap-on nozzles vectored to augment core control authority and provide roll control.

Orbital's Taurus rocket on the launch pad for GFO/ORBCOMM mission 9 February 1998 at VAFB 0024769

Taurus stage 0
Comment: The first vehicle (and any later USAF missions) utilised a Peacekeeper stage 1. Commercial launches adopt Thiokol's Castor 120 motor. The Peacekeeper specifications are given first, followed by those related to Castor 120.
Motor: Thiokol TU-903 solid, filament wound Kevlar epoxy casing structure of 23 helical layers and 55 hoop plies. Attachment to these skirts provided by aluminium rings. TU903 employs HTPB propellant and a silica-filled EPDM internal insulator. All stages' motors employ 88% solids, with grain designed for low-burn rates. Protection from aerodynamic heating and static charge build-up provided by an external rubber insulating layer impregnated with graphite. Ignition by pyrogen igniter on forward dome. Each stage nozzle consists of a carbon phenolic exit cone with 3-D carbon/carbon integral throat/entry. Stage 0 uses a silicone elastomer flexseal nozzle with hydraulic actuators. Hydraulic power is provided by a hot gas-driven turbopump. Redundant flight termination charges are mounted along the casing.
Stage length: 8.48 m
Diameter: 233.7 cm
Thrust: 2,204.5 kN average vacuum, 2,583.7 kN max vacuum.
Burn time: 56 s
Mass at ignition: 48,960 kg
Propellant mass: 44,660 kg
Separation: aft skirt extension incorporates redundant MDF to sever connection with launch stand. Aluminium interstage uses linear-shaped charge to release stage 1. Interstage vent covers released when bolt cutters separate forward and aft clamp bands.
Flight sequence: stage 0 TVA enabled at T2 s. Internal count sequencer performs nozzle sweep then commands simultaneous release from stand and ignition. Flies preprogrammed pitch profile to limit aerodynamic loading; roll not controlled. Vent panels released just before burnout, relieving pressure for stage 1 ignition (keyed to predetermined thrust tail-off; staging occurs 0.9 s later).
Motor: Thiokol Castor 120 solid. Employs Toray 100068 graphite epoxy wound composite case with an aramid-filled EPDM rubber internal insulator. External protection from aerodynamic heating also provided by EPDM rubber layer. The nozzle features a carbon phenolic ablative exit cone with a graphite epoxy structural over-wrap and a 3-D carbon/carbon integral throat/entry. Nozzle mounted in a laminated rubber/glass fibre flexseal for ±5° vectoring. The 88% solids HTPB propellant is machined for fine tailoring of thrust profile. Igniter mounted in the forward dome. Redundant linear shaped charge (LSC) in external raceway allows flight termination.
Stage length: 7.72 m
Diameter: 236.3 cm
Thrust: 1,703 kN average vacuum, 1,942 kN max vacuum attained after 29 s (total impulse 133.9 MN).
Burn time: 83 seconds
Mass at ignition: 53,070 kg
Propellant mass: 48,988 kg
Separation: bolted connection to launch stand removed several hours before launch. Aluminium interstage LSC releases stage 1. Interstage vent covers released when bolt cutters separate forward/aft clamp bands
Flight sequence: stage 0 TVA enabled at T-30 s. Internal terminal count sequencer performs nozzle sweep then commands ignition. Remaining sequence similar to TU 903.

Taurus stage 1
Motor: Alliant Techsystems Orion 50S (50XL stretched by 137 cm for XL/XLS). All three Alliant motors use IM7/55A graphite epoxy wound composite cases with aramid filled EPDM rubber internal insulator and integral skirts. Flight termination charges mounted on the aft domes of stage 1/2 motors for range safety requirements. Stage 1 employs silicon elastomer flexseal nozzle with AlliedSignal hydraulic cold gas blowdown TVAs. Changes from Pegasus include removal of the wing saddle, thickening of the aft skirt and skirt doublers, revision of the aft attach ring and aft skirt for a tapered unit that mates to the interstage, and a vectored nozzle (4°) to replace fins for vehicle control.
Stage length: 763.8 cm (943.9 cm)
Diameter: 127.1 cm
Thrust: 471.511 kN (627.199 kN) average vacuum, 556.028 kN (720.612 kN) max vacuum attained about mid-way through burn. Total impulse 34.10 MNs (43.58 MNs).
Burn time: 72.4 s (69.3 s)
Mass at ignition: 13,270 kg (16,230 kg)
Propellant mass: 12,143 kg (15,072 kg)
Separation: the forward skirt, which also serves as interstage adaptor, incorporates LSC
Flight sequence: stage 1 TVA enabled just before stage 0 burnout. Following release of the interstage vent panels, stage 1 ignites, followed immediately by the LSC firing in the 0/1 interstage. Stage 1 also follows a preprogrammed pitch profile. Nozzle vectoring provides pitch/yaw control during powered flight. Burnout occurs typically 154 s (151 s), followed by staging at 158.2 s (150 s). Springs push the stack away, providing clearance for stage 2 ignition.

Taurus stage 2
Motor: Alliant Techsystems Orion 50 (50XL stretched 45.0 cm for XL/XLS). Changes from Pegasus are reinforcement of the forward/aft skirts and attach rings. Stage 2 also employs silicon elastomer flexseal nozzle and Parker Bertea Aerospace electromechanical actuators for TVC. A graphite epoxy facesheet/syntatic foam core sandwich structure boat-tail transitions from the stage 2 diameter to the 160 cm diameter fairing. A self-contained non-contaminating frangible separation system on the boat-tail's forward end allows stage 2 to be jettisoned. Four springs provide separation for the upper stage/payload.
Stage length: 265.7 cm (310.6 cm)
Diameter: 127.1 cm
Thrust: 113.874 kN (160.136 kN) average vacuum, 131.667 kN (195.266 kN) max vacuum. Total impulse 8.67 MNs (11.22 MNs).
Burn time: 75.9 s (70.1 s)
Mass at ignition: 3,370 kg (4,314 kg)
Propellant mass: 3,024 kg (3,919 kg)
Separation: frangible joint at boat-tail forward end allows separation of avionics skirt/stage 3/payload
Flight sequence: stage 2 ignites at 165 s as the remainder of the 1/2 interstage is jettisoned. Payload fairing jettisoned at 168 s. Stage 2 burnout at 237 s. Roll control during burn and three-axis control during coast by six N_2 thrusters in two Parker Bertea Aerospace modules on avionics section.
Attitude control: as stage 1.

Taurus stage 3
Motor: Alliant Techsystems Orion 38 (also for XL/XLS). Stage 3 also employs vectored nozzle for pitch/yaw control during burn. Head end grain with toroidal igniter. Unlike Pegasus, the motor is not a load-carrying member but is cantilevered from its forward attach ring. The annular avionics shelf encircles the motor to mount most of the vehicle's electronics in addition to the N_2 attitude control system. This arrangement allows replacement of the motor with a spinning upper stage, such as Star 37, for perigee kick.
Stage length: 133.9 cm
Diameter: 96.5 cm
Thrust: 34.314 kN average vacuum, 38.686 kN max vacuum. Total impulse 2.18 MNs.
Burn time: 63.8 s
Mass at ignition: 893 kg
Propellant mass: 771 kg
Flight sequence: stage 3 ignites at apogee for orbit circularisation, using nozzle flexing for pitch/yaw control and the N_2 jet system for roll control and three-axis control during coast. Following injection, stage 3 manoeuvres to prescribed payload deployment attitude and provides a separation command to the payload, typically 30 s after burnout.

Fairing and payload accommodation
Two bisector fairings are available: 1.6 and 2.3 m diameter external. Fairing shells are fabricated by Courtaulds Structural Composites, Inc (1.6 m) and R Cubed (2.3 m) from graphite epoxy facesheets and aluminium honeycomb core. An OSC proprietary frangible joint separation system secures the fairing halves at the base and along the longitudinal seams until jettison during stage 2 burn. A pneumatic thruster system ejects the fairing, which is guided by fall-away hinges. Class 10,000 is maintained inside the fairing at all times, beginning from encapsulation until launch. On the ground, air conditioning maintains 15.6+4/7°C and 2065% humidity. The payload can be provided with 28 V DC (up to 140 W) until T0. 90 passthrough circuits allow payload commanding and monitoring before launch. The Taurus flight computer can issue up to five discrete payload commands and can monitor five discrete talkbacks. RS-422/485 serial data interfaces can also issue commands and accept payload telemetry in the vehicle data stream. Four pyro commands can be issued. Data for 2.6 m fairing in ().
Payload volume: 1.37 m (4.06 m) diameter × 2.8 m (5.84 m) long
Acceleration load: 6.9 *g* longitudinal max at end of stage 1 burn
Acoustic load: 133.5 dB OASPL (XLS: 138 dB OASPL).

Vehicle and Payload Processing
Stages are pre-integrated and tested at a Stage Integration Facility. For early missions, this is the Missile Assembly Building at VAFB. Stage and launch support equipment are transported to the launch site as late as 8 days before launch. The LSV launch support van is positioned outside the caution/hazard corridor and connected to the pad via a fibre optic bundle. Portable generators can power the system if facility power is not available. Once the launch stand is bolted to the concrete pad, stage 0 is mounted and scaffolding erected up to its top. In parallel, the upper stages are processed in an adjacent integration tent. Jumper harnessing between the mated upper stages and the erected stage 0 allows end-to-end integrated system tests and mission simulations.

In parallel with (or before) vehicle integration, the payload is processed and encapsulated in an appropriate facility. OSC provides the payload cone, mounted on a GSE handling adaptor, for spacecraft integration/test. The fairing is then mated to the cone, encapsulating the cargo element. The environmental control system is connected and the cargo element moved to the launch. This can be as late as 72 h before launch. At the launch site, the cargo element is electrically and mechanically mated to the upper stages. This stack is then lifted and mated to stage 0. Final systems tests are completed, followed by alignment of the IMU. Vehicle closeout and final countdown starts at T7 h, controlled from the LSV. Dedicated consoles are provided for payload personnel, range safety personnel and the customer launch director as well as the OSC personnel controlling the launch.

Launch log

No	Date	Payload
1	13 March 1994	DARPASAT/STEP-0
2	10 February 1998	ORBCOMM G1/G2/ CELESTIS/GFO
3	3 October 1998	STEX/ATEX
4	21 December 1999	ACRIMSAT/CELESTIS/ KOMPSAT
5	12 March 2000	MTI
6	21 September 2001	QUICKTOMS/ORBVIEW 4/CELESTIS/SBD

Terrier Malemute Sounding Rocket

Current Status
Terrier Malemute has proven popular for plasma physics and is capable of lifting a 90 kg payload to an apogee of 700 km or a 227 kg payload to 400 km. Also selected for interactive plasma/magnetosphere experiments and measurements. NASA has proposed some improvements to the Terrier Malemute design for enhanced stability in the post-boost phase but these have not been funded although some additional upgrades have been carried out.

Background
NASA's sounding rocket programme is based at the Wallops Flight Facility in Virginia using vehicles from commercial sources or developed by the agency. Some 25 launches are made each year from a range of sites, including Wallops itself, Poker Flat and White Sands. NASA currently employs 15 sounding rocket types (Aries use ended 1993), including several utilising Bristol Aerospace's Black Brant 5 motors. Extensive use is made of military surplus motors and all are unguided. Saab's S19 Boost Guidance System can be added to the Nike Black Brant and Black Brant 5/9/10. The gyro platform controls aerodynamic canards, decreasing impact dispersion by a factor of 5 to 10, allowing higher apogees and launches in higher winds.

This combination is particularly suited to lower mass payloads; performance reduces appreciably as the weight increases. Bulbous payloads can be accommodated but high-dynamic pressures result in high-aerodynamic heating rates and vehicle structural loads. In general, it is employed for relatively lightweight plasma physics experiments and rail launched.

Specifications
Terrier-Malemute
First launch: June 1975
Number launched: 34 to end 2002
Success rate: 92%
Number of stages: 2 (solid)
Overall length: 7.2 m
Principal diameter: 40.6 cm
Launch mass: 1,478 kg plus payload
Typical performance: 91 kg to 670 km
Payload recovery system: custom built
Payload environment: 26 *g* peak acceleration for 90 kg payload, accommodated typically in 40.6 cm 3:1 ogive nose
Guidance: unguided, spin stabilised by four 0.22 m^2 fins.

Terrier-Malemute stage 1
Motor: US Navy Terrier Mk12 mod 1 (Hercules Powder Co and Allegheny Ballistics Laboratory)
Overall length: 4.27 m
Principal diameter: 45.7 cm
Stage mass: 878 kg (534 kg propellant)
Thrust: 257.7 kN average sea level
Burn time: 4.4 s.

Terrier-Malemute stage 2
Motor: Thiokol Malemute TU-758
Overall length: motor 3,302 mm
Principal diameter: 40.6 cm
Stage mass: 600 kg (499 kg propellant)
Thrust: 64 kN average SL (total impulse 1,223 kNs SL)
Burn time: 20 s.

Titan 2

Current Status
Managed by Lockheed Martin Space Systems Company, Astronautics Operations. A total of 24 launches, including 12 Gemini-Titan flights, by the end of September 2003. The last flight of a USAF Titan 2, the 25th, is scheduled to carry DMSP F-16 to orbit in the last quarter 2003.

Background
Martin Marietta began development of the Titan 2 LGM-25C ICBM in 1960, achieving the first of 85 ICBM-related launches in March 1962 and beginning deployment in 1963 in three wings of 18 missiles each at Little Rock AFB (Arkansas), Davis Montham AFB (Arizona) and McConnell AFB (Kansas). The missile carried 5 MT thermonuclear warheads. Over the years, Martin Marietta had delivered 141 vehicles to USAF: 36 Research and Development N types and the rest operational B versions. Production attained a peak rate of 12 per month during the mid 1960s.

The missile was declared operational after 32 launches, but a further 49 test/training launches were made from Vandenberg until the last in June 1976. Twelve flew in the manned Gemini programme of 1964–1966, 10 carrying two-man crews. ICBM deactivation began July 1982 and was completed with the removal of the last missile from its silo at Little Rock 23 June 1987 after its last duty cycle 5 May.

The Titan 2G is the base version of the Space Launch Vehicle. 2B and 2S models were also planned. 2S competed unsuccessfully in 1993 for USAF's MLV3 medium-launch vehicle contract. It was originally planned for BMDO's MSX mission, which switched to Delta. A 2L version was also considered, adding two liquid strap-ons to handle 8,165 kg into 185 × 185 km 28.5° from the Cape.

Martin Marietta received a US$45.2 million letter contract in January 1986 to proceed with the final system design review of the Titan 2 as a space launcher. The US government signed a US$483.7 million contract in October 1986 to refurbish eight ICBMs, with a later exercised option for five more (a 14th was added in November 1987), bringing the total contract value to US$638.7 million to September 1995. The work included modifying the stage 2 forward structure to accommodate McDonnell Douglas' 3.05 m Titan 34D-type payload fairing, refurbishing the Aerojet engines, upgrading the Delco inertial guidance systems (to Titan 4 standard), developing command, destruct and telemetry systems, and modifying Vandenberg's Titan 3B SLC4W pad to accept the smaller vehicle. The first refurbished Titan, designated as the Titan 23G Space Launch Vehicle, was rolled out at the company's Denver plant 3 August 1987 scheduled initially for April 1988 launch. The launch actually occurred the following September.

This first vehicle employed the B56 stage 1 and B98's stage 2 (Titan 2 was designed to allow stage interchanging). B106 was used for the second mission.

Martin Marietta finished refurbishing the first 14 in December 1990, each vehicle requiring typically 12 months' work. The remainder of the 55 were stored at Norton AFB, California but they began October 1993 moving to an army storage facility near Pueblo, Colorado when Norton shut its doors as part of the Base Closings Act. There are no plans for either Canaveral launches or commercial marketing. The most recent USAF/NASA ELV reimbursement agreement of October 1987 provides for space agency use at a basic US$24.6 million cost (FY89 rate). Launch cost to USAF was reported in 1990 at US$33 million. NOAA was charged US$36.5 million by USAF for 1993's Landsat 6 launch vehicle and services.

The Titan 2 Basic is similar to the 2G configuration but with minimal modification to the ICBM. It incorporates new wiring and avionics, and offers the option of an attitude control system. Instead of the McDonnell Douglas fairing, it uses the existing re-entry vehicle fairing.

The Titan 2 Solid Thrust Augmented version enhances the 2G's performance by adding 28 Castor 4A strap-ons. This requires a stage 1 extension to provide attach points. Alliant's higher performance GEM graphite epoxy motors (now flying on Delta 2) may also be used to increase payload capability.

Specifications
Titan 2 SLV
First orbital launch: 8 April 1964 (Gemini Titan 1)
Number of orbital launches: 24 to end September 2003, plus 3 suborbital
Launch sites: Canaveral pad 19 (deactivated), Vandenberg SLC 4W
Principal uses: currently small and medium payloads into low polar orbits; 10 manned Gemini launches 1965/66
Vehicle orbital success rate, career: 100% to end September 2003
Performance: 2,177 kg into 185 km polar, 2,360 kg into 185 km 63.5°; with Star 37 kick stage 3,028 kg into 546 km Sun-synchronous polar; 3,175 kg into 185 km, 28.6° from Canaveral; 1,043 kg into GTO with SSPS upper stage. In addition, Castor 4A strap-ons can be added (Titan 2S) to increase LEO polar orbit capacity to 3.7 t, depending on inclination (3,540 kg 185 km 99°); performance to 1,100 km circular orbits >1.9 t. 2L: 8,165 kg 185 km 28.5°

Titan 2 space launch history

	Date	Site	Payload
1*	8 April 1964	ETR 19	Gemini 1
2	19 January 1965	ETR 19	Gemini 2
3	23 March 1965	ETR 19	Gemini 3
4	3 June 1965	ETR 19	Gemini 4
5	21 August 1965	ETR 19	Gemini 5
6	4 December 1965	ETR 19	Gemini 7
7	15 December 1965	ETR 19	Gemini 6
8	16 March 1965	ETR 19	Gemini 8
9	3 June 1965	ETR 19	Gemini 9
10	18 July 1965	ETR 19	Gemini 10
11	12 September 1965	ETR 19	Gemini 11
12	11 November 1966	ETR 19	Gemini 12
13	5 September1988	SLC4W	electronic intelligence
14	5 September 1989	SLC4W	electronic intelligence
15	25 April 1992	SLC4W	electronic intelligence
16*	5 October 1993	SLC4W	Landsat 6
17	25 January 1994	SLC4W	Clementine1
18	4 April 1997	SLC4W	DMSPSD2 14
19	13 May 1998	SLC4W	NOAA-15
20	20 June 1999	ETR 19	Quickscat
21	12 December 1999	SLC4W	DMSP5D3 15
22	21 September 2000	SLC4W	NOAA-16
23	24 June 2002	SLC4W	NOAA-17
24	6 January 2003	SLC4W	Coriolis

* Gemini 1 was a sub-orbital mission. Landsat 6 mission failed but Titan's suborbital portion was successful.

Number of stages: 2 plus optional kick stages
Overall length: 31.4 m
Principal diameter: 3.05 m
Launch mass: 153.7 t plus payload
Launch thrust: 1,913 kN
Guidance: Delco Electronics Titan 4 digital inertial mounted on stage 2
Launch sequence (min:s):

–00:05	go inertial
00:02	stage 1 ignition signal
00:09	roll to flight azimuth
00:19	end roll
00:30	begin zero lift
01:50	end zero lift
02:37.7	stage 1/2 hot separation (65 km, 2,704 m/s)
03:31	payload fairing separation (119 km, 3,393 m/s)
05:32.8	stage 2 shutdown (189 km, 7,785 m/s)
05:49	enable attitude control system
06:18	converge to commanded attitude, begin velocity trim
06:40	end velocity trim

Comment: An Enhanced Attitude Control System, carrying 726 kg hydrazine, became available in 1993 to reach higher orbits and/or maintain control for longer.

Titan 2G stage 1
Engines: refurbished gimballed paired Aerojet LR87AJ5 single-start hypergolic.
Overall length: 21.4 m
Principal diameter: 3.05 m
Stage mass: 4,220 kg dry
Propellant: 118.3 t N_2O_4/Aerozine 50
Stage thrust: 1,913 kN sea level
Burn time: 158 s (154 s Landsat 6)

Titan 2G stage 2
Engines: refurbished Aerojet LR91 AJ5 single-start hypergolic single chamber.
Overall length: 12.2 m
Principal diameter: 3.05 m
Stage mass: 2,860 kg dry
Propellant: 28.44 t N_2O_4/Aerozine 50
Stage thrust: 445 kN vacuum
Burn time: 175 s (182 s Landsat 6)

Fairing
McDonnell Douglas supplies the 3.05 m diameter 6.1–9.2 m Titan 34D-type aluminium skin/stringer payload fairing, providing a 2.83 m diameter, 9.1 m long envelope. A 15.3 m long fairing is feasible. Martin Marietta completed development in 1993 of an aluminium honeycomb/graphite epoxy composite skirt and payload adaptor that will yield an 87 kg payload increase.

Titan 2S
Motor contractor: Thiokol, Huntsville division
Length: 11.16 m
Diameter: 1.01 m
Average thrust: 434 kN sea level
Burn time: 53.1 s
Mass at ignition: 11.6 t
Propellant: TP-H8299 HTPB polymer, 20 per cent aluminium
Propellant mass: 10.1 t

UPDATED

Titan 3/34D

Current Status
Titan 3 launch vehicle has now been retired, having completed 156 flights between 1964 and 1992 with a success rate of 91.6 per cent. The commercial Titan 3 made four flights between 1990 and 1992 of which the first was a failure. Titan 2 has also been retired leaving only the Titan 4 operational.

Background
Martin Marietta announced at the UK's Farnborough Air Show in September 1986 the availability from spring 1989 of a commercial version of its Titan 34D USAF launcher. Federal Express made the first reservation that month. Commercial Titan is essentially a 34D with a stretched stage 2 and 4 m diameter fairing to accommodate single or dual Shuttle-class payloads with perigee kick motors to LEO. Vehicles are handled on the same production line as Titan 4, allowing low-volume commercial operations.

Commercial Titan Inc was established in May 1987, but absorbed into Space Launch Systems in summer 1990. In order to launch the missile commercially, Martin concluded an agreement to utilise Canaveral pad 40 with the Air Force in August 1987. GE Astro Space made a 15 satellite block reservation in January 1988 to offer in orbit delivery of their satellites. GE and Martin later mutually terminated the deal.

Emphasis is now placed on flying single large payloads. The contract for launching Mars Observer in 1992 reportedly brought in US$156.6 million, including US$20 million for mission specific modifications. In order to make the vehicle available to a wider commercial base, Martin Marietta once considered making a restartable stage 2 and adopting Titan 4's fairing and uprated strap-ons for Space Station support and assembly.

A Titan 3L version was mooted in 1991, employing two liquid strap-ons to handle 9.1 tonnes LEO at about 70 per cent of the current Titan 3 cost. The Titan 3 family has now been retired, its place taken by Titan 4, larger versions of Atlas and by the new generation of more powerful Atlas 3 and 5 and Delta 4 launcher which are much more flexible and cost effective.

Titan 3 family record
First launch: 1 September 1964 (18 June 1965 for strap-ons)
Number launched: 156 to end of programme in 1992
Launch sites: Canaveral pads 40/41, Vandenberg SLC4E/4W
Vehicle success rate, career: 91.0%
Vehicle success rate, last 25 launches: 84% to end 1999 (four failures in 10 years before last launch)

Specifications
Commercial Titan 3
First launch: 1 January 1990
Number launched: 4 to end-2001
Vehicle success rate: 75.0% for commercial flights, 91.6% for series
Launch sites: Canaveral pad 40
Principal uses: delivery of Shuttle class payloads to LEO for perigee kick motor insertion into GTO.
Cost: US$110 million for a dedicated flight; US$100,000 reservation deposit required; reflight insurance offered at 10 per cent premium.
Availability: 33 months after contract go-ahead; potentially 2/3 flights/year.
Schedule of future missions: none

Performance
LEO (148 × 259 km, 28.6°): 14,334 kg dual payload, 14,742 kg single payload.
GTO 26.4°: 1,279 kg with PAM D; 1,851 kg with PAM-D2, dual carrier; 4,944 kg with IUS; 4,990 kg with TOS, single carrier
GEO: not applicable
Space Station transfer orbit: 11,700 kg with no upper stage.
Sun-synchronous: not available
Lunar delivery: 3,400 kg with TOS
Venus delivery (Type II): 2,600 kg with TOS
Mars delivery (Type II): 2,430–2,600 kg with TOS
Number of stages: 2 plus 2 strap-ons
Overall length: 47.3 m dual carrier, 44.06 m single carrier
Principal diameter: 3.05 m
Launch mass: 680 t
Guidance: Delco Systems Operations inertial, incorporated in stage 2 with Olin Aerospace's 178 N MR 107 hydrazine attitude control thrusters. 3Σ injection accuracies: apogee 259 ± 11.1 km, perigee 148 + 1.9/–0 km, inclination 113.3 ± 3.0°
Launch sequence (min:s)

00:00	solids ignition
00:54	max dynamic pressure
01:48	stage 1 ignition
01:56	solids jettison
04:29	stage 2 ignition
04:30	stage 1 separation
04:40	fairing jettison
08:14	stage 2 shutdown
08:30	park orbit insertion
08:40	trim burn
60:30	payload1 deploy attitude
63:30	payload 1 spin-up
67:16	payload 1 deploy
77:00	avoidance manoeuvre
112:16	payload 1 GTO burn
148:00	eject payload carrier
151:00	payload 2-deploy attitude
152:00	payload 2 spin-up
155:50	payload 2 deploy
188:00	avoidance manoeuvre and burn to depletion
200:50	payload 2 GTO burn.

Titan 3 solid strap-ons
Motor contractor: United Technologies' Chemical Systems Division, San Jose, California. CSD was awarded a $100 million contract in January 1988 for commercial Titan motor supply. Each motor comprises 5½ segments, each joint secured by 237 hand-placed clevis pins. The base segment extension bears the weight of the whole vehicle on the pad.
Length: 27.57 m
Principal diameter: 3.11 m
Mass at ignition: 250.387 t each
Propellant: UTP-30001 B solid
Propellant mass: 210.63 t
Average thrust: 6,227 kN each, vacuum
Burn time: 113.7 s
Steering: secondary injection through 24 nozzles into exhaust of N_2O_4 from 3,630 kg capacity tank (8.5 m long, 1.1 m diameter), although typically less than half consumed. Injection adds about 17.8 kN thrust to each motor.
Separation: jettison at 116 s by eight 20 kN solid motors firing for 1 s.

Titan 3 stage 1
Contractor: Lockheed Martin, Denver
Engines: paired Aerojet gimballed LR87AJ11
Overall length: 24.0 m
Principal diameter: 3.05 m
Oxidiser: 71.7 t nitrogen tetroxide in forward tank, loaded 3 days before launch.
Fuel: 38.0 t Aerozine-50 in aft tank, loaded 2 days before launch.
Stage thrust: 2,340 kN vacuum
Burn time: about 160 s, igniting at 118 s
Attitude control: gimballed main engines for pitch/roll; forward boat-tail carries four external modules, each with three nozzles (pitch, roll, yaw), providing three-axis control. 102 kg hydrazine, thrust 80–133 N. The thrusters also provide ΔV corrections and payload spin-up before deployment.
Separation: at 270 s.

Titan 3 stage 2
Comment: Stage 2 employs a single burn but a restart capability has been considered for enhancing payload

Titan 3 launch history

	Date	Type	Site	Payload
1*	1 September 1964	3A	ETR 20	test; stage 3 failure
2	10 December 1964	3A	ETR 20	test
3	11 February 1965	3A	ETR 20	LES 1
4	6 May 1965	3A	ETR 20	LES 2
5	18 June 1965	3C	ETR 40	test
6*	15 October 1965	3C	ETR 40	OV2 1; stage 3 failure
7	21 December 1965	3C	ETR 41	LES 3/4, OV2-3
8	16 June 1966	3C	ETR 41	IDCSP 1–7, GGTS1
9	29 July 1966	3B	SLC4W	KH 8 close-look
10*	26 August 1966	3C	ETR 41	IDCSP
11	28 September 1966	3B	SLC4W	KH 8 close-look
12	3 November 1966	3C	ETR 40	OV43, 4-1R, 1-6, 4 1T
13	14 December 1966	3B	SLC4W	KH 8 close-look
14	18 January 1967	3C	ETR 41	IDCSP 8 15
15	24 February 1967	3B	SLC4W	KH 8 close-look
16*	26 April 1967	3B	SLC4W	KH 8 close-look
17	28 April 1967	3C	ETR 41	Vela Hotel 7/8, OV5-3/5 4
18	20 June 1967	3B	SLC4W	KH 8 close-look
19	1 July 1967	3C	ETR 41	IDCSP 1618, LES 5, Dodge
20	16 August 1967	3B	SLC4W	KH 8 close-look
21	19 September 1967	3B	SLC4W	KH 8 close-look
22	25 October 1967	3B	SLC4W	KH 8 close-look
23	5 December 1967	3B	SLC4W	KH 8 close-look
24	18 January 1968	3B	SLC4W	KH 8 close-look
25	13 March 1968	3B	SLC4W	KH 8 close-look
26	17 April 1968	3B	SLC4W	KH 8 close-look
27	5 June 1968	3B	SLC4W	KH 8 close-look
28	13 June 1968	3C	ETR 41	IDCSP 19–26
29	6 Aug 1968	3B	SLC4W	KH 8 close-look
30	10 September 1968	3B	SLC4W	KH 8 close-look
31	26 September 1968	3C	ETR 41	LES 6, OV25, 5-2, 54
32	6 November 1968	3B	SLC4W	KH 8 close-look
33	4 December 1968	3B	SLC4W	KH 8 close-look
34	22 January 1969	3B	SLC4W	KH 8 close-look
35	9 February 1969	3C	ETR 41	Tacsat1
36	4 March 1969	3B	SLC4W	KH 8 close-look
37	15 April 1969	3B	SLC4W	KH 8 close-look
38	23 May 1969	3C	ETR 41	Vela Hotel 9/10, OV5
39	3 June 1969	3B	SLC4W	KH 8 close-look
40	22 August 1969	3B	SLC4W	KH 8 close-look
41	24 October 1969	3B	SLC4W	KH 8 close-look
42	14 January 1970	3B	SLC4W	KH 8 close-look
43	8 April 1970	3C	ETR 40	Vela Hotel 11/12
44	15 April 1970	3B	SLC4W	KH 8 close-look
45	25 June 1970	3B	SLC4W	KH 8 close-look
46	18 August 1970	3B	SLC4W	KH 8 close-look
47	23 October 1970	3B	SLC4W	KH 8 close-look
48	6 November 1970	3C	ETR 40	DSP1 early warning
49	21 January 1971	3B	SLC4W	KH 8 close-look
50	21 March 1971	34B	SLC4W	Jumpseat
51	22 April 1971	3B	SLC4W	KH 8 close-look
52	5 May 1971	3C	ETR 40	DSP2 early warning
53	15 June 1971	3D	SLC4E	KH 9 Big Bird #1
54	12 August 1971	3B	SLC4W	KH 8 close-look
55	23 October 1971	3B	SLC4W	KH 8 close-look
56	3 November 1971	3C	ETR 40	DSCS 1/2
57	20 January 1972	3D	SLC4E	KH 9 Big Bird #2
58*	16 February 1972	34B	SLC4W	Jumpseat
59	1 March 1972	3C	ETR 40	DSP-3 early warning
60	17 March 1972	3B	SLC4W	KH 8 close-look
61*	20 May 1972	3B	SLC4W	KH 8 close-look
62	7 July 1972	3D	SLC4E	KH 9 Big Bird #3
63	1 September 1972	3B	SLC4W	KH 8 close-look
64	10 October 1972	3D	SLC4E	KH 9 Big Bird #4
65	21 December 1972	3B	SLC4W	KH 8 close-look
66	9 March 1973	3D	SLC4E	KH 9 BigBird #5
67	16 May 1973	3B	SLC4W	KH 8 close-look
68	12 June 1973	3C	ETR 40	DSP-4 early warning
69*	26 June 1973	3B	SLC4W	KH 8 close-look
70	13 July 1973	3D	SLC4E	KH 9 Big Bird #6
71	21 August 1973	34B	SLC4W	Jumpseat
72	27 September 1973	3B	SLC4W	KH 8 close-look
73	10 November 1973	3D	SLC4E	KH 9 Big Bird #7
74	13 December 1973	3C	ETR 40	DSCS II3/4
75*	11 February 1974	3E	ETR 41	test/Viking mass Centaur failure
76	13 February 1974	3B	SLC4W	KH 8 close-look
77	10 April 1974	3D	SLC4E	KH 9 Big Bird #8
78	30 May 1974	3C	ETR 40	ATS 6
79	6 June 1974	3B	SLC4W	KH 8 close-look
80	14 August 1974	3B	SLC4W	KH 8 close-look
81	29 October 1974	3D	SLC4E	KH 9 Big Bird #9
82	10 December 1974	3E	ETR 41	Helios 1
83	10 March 1975	34B	SLC4W	Jumpseat
84	18 April 1975	3B	SLC4W	KH 8 close-look
85*	20 May 1975	3C	ETR 40	DSCS II
86	8 June 1975	3D	SLC4E	KH 9 Big Bird #10
87	20 August 1975	3E	ETR 41	Viking1
88	9 September 1975	3E	ETR 41	Viking 2
89	9 October 1975	3B	SLC4W	KH 8 close-look
90	4 December 1975	3D	SLC4E	KH 9 Big Bird #11
91	14 December 1975	3C	ETR 40	DSP5 early warning
92	15 January 1976	3E	ETR 41	Helios 2
93	15 March 1976	3C	ETR 40	LES 8/9, Solrad 11A/B
94	22 March 1976	3B	SLC4W	KH 8 close-look
95	2 June 1976	34B	SLC4W	SDS
96	26 June 1976	3C	ETR 40	DSP6 early warning
97	8 July 1976	3D	SLC 4E	KH 9 Big Bird #12
98	6 August 1976	34B	SLC4W	SDS
99	15 September 1976	3B	SLC4W	KH8 close-look
100	19 December 1976	3D	SLC4E	KH 11 #1
101	6 February 1977	3C	ETR 40	DSP7 early warning
102	13 March 1977	3B	SLC4W	KH 8 close-look
103	12 May 1977	3C	ETR 40	DSCS II 7/8
104	27 June 1977	3D	SLC4E	KH 9 Big Bird #13
105	20 August 1977	3E	ETR 41	Voyager 2
106	5 September 1977	3E	ETR 41	Voyager1
107	23 September 1977	3B	SLC4W	KH 8 close-look
108	25 February 1978	34B	SLC4W	Jumpseat
109	16 March 1978	3D	SLC4E	KH 9 Big Bird #14
110*	25 March 1978	3C	ETR 40	DSCS II 9/10
111	10 June 1978	3C	ETR 40	Chalet electronic intelligence
112	14 June 1978	3D	SLC4E	KH 11 #2
113	5 August 1978	34B	SLC4W	SDS
114	14 December 1978	3C	ETR 40	DSCS II 11/12
115	16 March 1979	3D	SLC4E	KH 9 Big Bird #15
116	28 May 1979	3B	SLC4W	KH 8 close-look
117	10 June 1979	3C	ETR 40	DSP8 early warning
118	1 October 1979	3C	ETR 40	Chalet electronic intelligence
119	21 November 1979	3C	ETR 40	DSCS II 13/14
120	7 February 1980	3D	SLC4E	KH 11 #3
121	18 June 1980	3D	SLC4E	KH9 Big Bird #16
122	13 December 1980	34B	SLC4W	Jumpseat
123	28 February 1981	3B	SLC4W	KH 8 close-look?
124	16 March 1981	3C	ETR 40	DSP9 early warning
125	24 April 1981	34B	SLC4W	SDS
126	3 September 1981	3D	SLC4E	KH 11 #4
127	31 October 1981	3C	ETR 40	Chalet electronic intelligence
128	21 January 1982	3B	SLC4W	radar test satellite?
129	6 March 1982	3C	ETR 40	DSP-10 early warning
130	11 May 1982	3D	SLC4E	KH 9 Big Bird #17
131	30 October 1982	34D	ETR 40	DSCS II15 & III1/IUS
132	17 November 1982	3D	SLC4E	KH 11 #5
133	15 April 1983	3B	SLC4W	KH 8 close-look
134	20 June 1983	34D	SLC4E	KH 9 Big Bird #18
135	31 July 1983	34B	SLC4W	SDS or Jumpseat
136	31 January 1984	34D	ETR 40	Chalet?/Trans-stage
137	14 April 1984	34D	ETR 40	DSP 11/Trans-stage
138	17 April 1984	3B	SLC4W	KH 8 close-look
139	25 June 1984	34D	SLC4E	KH 9 Big Bird #19
140	28 August 1984	34B	SLC4W	SDS
141	4 December 1984	34D	SLC4E	KH 11 #6
142	22 December 1984	34D	ETR 40	DSP 12/Trans-stage
143	8 February 1985	34B	SLC4W	SDS or Jumpseat
144*	28 August 1985	34D	SLC4E	KH 11 #7
145*	18 April 1986	34D	SLC4E	KH 9 Big Bird #20
146	11 February 1987	34B	SLC4W	SDS or Jumpseat
147	26 October 1987	34D	SLC4E	KH 11 #8
148	29 November 1987	34D	ETR 40	DSP 13/Transtage
149*	2 September 1988	34D	ETR 40	Chalet (Vortex)/ Trans-stage failure
150	6 November 1988	34D	SLC4E	KH 11?
151	10 May 1989	34D	ETR 40	Chalet (Vortex)/ Trans-stage
152	4 September 1989	34D	ETR 40	DSCS II-16 & III 4/IUS
153	1 January 1990	3	ETR 40	Skynet4A/JCSat 2
154*	14 March 1990	3	ETR 40	Intelsat 603
155	23 June 1990	3	ETR 40	Intelsat 604
156	25 September 1992	3	ETR 40	Mars Observer/TOS

ETR indicates Canaveral launch; SLC Vandenberg.
* indicates launcher or upper stage failure.

capability to circularised orbits. For example, restart would raise the 11,700 kg injection capacity into transfer orbit for 370 km circular 28.5° by 2,270 kg.
Contractor: Lockheed Martin, Denver
Engines: gimballed Aerojet LR91AJ 11, igniting at 269 s.
Overall length: 9.85 m
Principal diameter: 3.05 m
Oxidiser: 18.3 t nitrogen tetroxide in forward tank, loaded 3 days before launch.
Fuel: 10.3 t Aerozine-50 in aft tank, loaded 2 days before launch.
Thrust: 467.04 kN vacuum
Burn time: 225 s (269–494 s after launch)
Separation: stage 2 can provide spin of up to 2 rpm in addition to optional spin table performance; payload spring separated at 0.6 m/s.

Payload fairing/carrier

Contraves AG of Zurich, Switzerland was awarded the contract in 1987 for the 10.4 m long, 3.95 m internal diameter aluminium honeycomb/graphite epoxy fairing, based on their Ariane 4 design. The standard fairing sits atop the Payload Carrier in dual or single form; Germany's Dornier was awarded an initial US$18 million contract in October 1987 (in competition with Aerospatiale's Ariane Sylda and British Aerospace's Ariane Spelda) for one test, two single and two dual carriers. In the single form, it accommodates the payload and Contraves' fairing, creating a payload unit 12.69 m long. With a dual carrier, the fairing is separated in halves at 280 s during stage 2 burn when atmospheric heating has reduced to 1,135 W/m^2. The forward payload is deployed 67 minutes after launch, the 5.59 m high aft carrier cover is jettisoned forward as a unit at 148 minutes for payload number

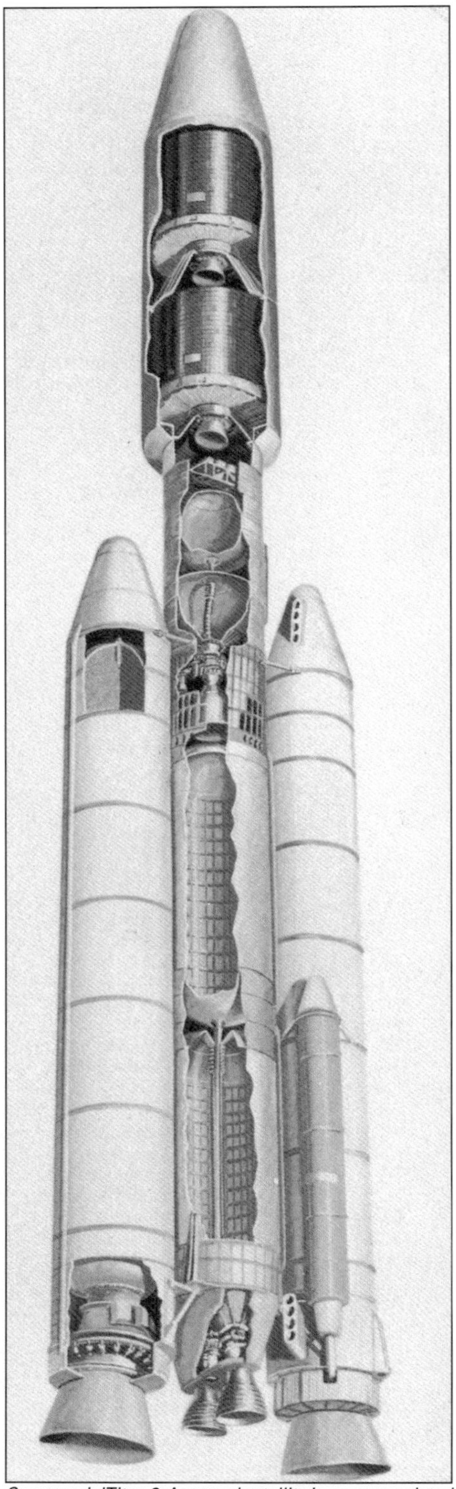

Commercial Titan 3. A second satellite is accommodated in the carrier below the first 0516929

2 deployment 7:50 minutes later. Standard Titan 34D 3.05 m diameter fairings are also available, as are spin tables providing up to 70 rpm.
Acoustic loads: 142 dB peak after 10 s and during transonic flight (54 s)
Vibration: typically 4.2 *g* rms for 60 s
Acceleration, longitudinal: 2.8 *g* strap-on burn peak, 4 *g* stage 1 peak, 2.5 *g* stage 2 peak. Payload design requirement for 6 *g*.
Thermal environment: fairing internal wall surface 100°C peak by 280 s.
Pad payload umbilical: 171 hardwire links to the two payloads are available through a pad umbilical at the top of stage 2, below the aft carrier.
Pad environmental control: 9.4 37°C, 30–50% humidity, class 10,000 filtration air conditioning provided inside fairing.

Vehicle/payload processing
Titan's core is stacked on its launch platform at launch complex 40/41's Vertical Integration Building. The VIB houses four high bays: Bay 1 is dedicated to commercial Titan operations, Bay 2 to USAF Titan and Bay 3 for Centaur. VIB's Launch Control Center was replaced in 1993 by the new Launch Operations Control Center elsewhere. The stack is then moved 1.4 km by rail to the SMAB Solid Motor Assembly Building where the strap-ons are stacked and attached before moving the final 4.2 km to pad 40/41. The new SMARF Solid Motor Assembly & Readiness Facility was added in 1994 to prepare Titan 4's strap-ons.

Payload processing facilities, including those of Astrotech Space Operations in nearby Titusville.

Titan 3 breakdown
Launch total for each type, followed by number of failures in ()/followed by success rate. Upper stage failures included.

3 family: 156(14)/91.0%	*3B:* 54(3)/94.4%
3A: 4(1)/75%	*3C:* 36(4)/88.9%
34B: 14(1)/92.9%	*34D:* 15(3)/80%
3D: 22(0)/100%	*Comm 3:* 4(1)/75%
3E: 7(1)/85.7%	

Titan 4

Current Status
Managed and manufactured by Lockheed Martin Space Systems Company, Astronautics Operations. Two Titan 4 were launched in 2003, the most recent on 9 September being the last with a Centaur upper stage. Only three more launches of this vehicle are planned: two for 2004 from CCAFS and one for 2005 from VAFB.

Background
The US Air Force began a programme in March 1984 for a Complementary Expendable Launch Vehicle (CELV) because of Shuttle's failure to provide a 14,515 kg (32,000 lb) capability into polar orbit for the new generation of surveillance platforms and as a hedge against operating difficulties with the manned craft. Martin Marietta's standard Titan 34D launcher of KH11s was upgraded to a 34D7 version by stretching the 5½ segment strap-ons to seven segments, stretching stages 1/2, incorporating a 5.09 m fairing and adding, when required at all, either an IUS or Centaur upper stage (NUS = No Upper Stage).

The company was awarded a US$2.100 billion contract in February 1985 for 10 of these Titan 4s, extended in August 1986 to 23 by a further US$1.970 billion contract. The awards included US$250 million to Aerojet TechSystems for 23 stage 1/2 engine sets (now US$350 million for 41 sets). A US$1.600 billion contract for a further 18, reducing unit cost from US$150 million to US$89 million, was announced in November 1989, with an option for eight more, bringing the total to 41 by September 1995 under Martin Marietta's US$7.8 billion contract. USAF Space Systems Division planned to increase the buy through 1995 to 55 (and 78 through 1997), comprising 21 for USAF payloads and 34 for other users such as the CIA. 30 would have been launched from the Cape.

A 1990 GAO report noted that cost overruns totalled US$298 million, with development problems delaying Initial Launch Capability from October 1988 to February 1989, and the Centaur version's ILC by 14 months to spring 1991. USAF's 1990 projection was for 90–100 launches through FY2000, but by early 1991 programme cutbacks had reduced the plan to 61 launches. Martin Marietta's contracted 41 vehicles were due to be completed in 1995 but that schedule has been stretched for the 41st launch in 2004.

The reduced production rate of five and a half years from 1991, instead of the planned 10, proved too high and vehicles are being stockpiled. A USAF order reducing the rate to three a year was made in mid-1993 (a US$216 million contract increase was awarded December 1994 to maintain production capability during the stretch-out). The purchase of another five is under discussion, the last departing in 2007. The company receives a US$7 million fixed incentive for each successful launch and loses US$45 million for each failure.

As the liquid Centaur upper stage was deleted from Shuttle scheduling on safety grounds in June 1986, NASA requires Titan 4 Centaur for its most demanding deep space missions. The cost of the Cassini mission in 1997 however, stunned NASA and means that one civil use will probably stand as a unique instance.

Space Station assembly or resupply missions could also employ the vehicle, but none has been scheduled. Martin Marietta claimed a Titan 4 cost of US$200 million for NUS and US$250 million for Centaur (1993 rates). In 1996 the figures were reported as US$250 million and US$350 million, respectively.

Air force planners awarded Martin Marietta US$64.3 million contract in February 1990 for modification design work to allow CCAS pad 40 to handle Titan 4 in addition to Titan 3; the conversion was completed in 1992. The new 5,600 m^2 SMARF Solid Motor Assembly & Readiness Facility for simultaneous processing of up to six sets of Titan 4 boosters began full operations in August 1994. The SMAB is also used for IUS stage assembly/processing.

Construction began January 1993 of a separate Centaur Processing Facility to decouple Centaur pad checkout from other Titan 4 elements. Vandenberg pad SLC4E, declared operational for Titan 4 in May 1990, was refurbished at a cost of US$235 million. Congress rejected the USAF claim for a new SLC 7 and directed instead that the SLC-6 Shuttle pad should be modified. Cost was estimated in 1990 at US$700 million. USAF announced in September 1990 that work would begin in 1991 for a 1996 first launch. By early 1991, however, the reduction in projected launch rates brought into question the need for SLC6; it was cancelled that year. Titan 4 and IUS authority was transferred from the Systems Command to the Space Command in 1994: 5th Space Launch Squadron for Canaveral operations 4th SLS for VAFB.

The most powerful Titan 4 version, 401 (Centaur), debuted in February 1994 with Milstar. The second departed May 1994, although the original vehicle was stacked on the pad June 1991, with launch planned for mid-1992. However, it was de-stacked after corrosion and water were found in the joints June 1992 (the SRMs are certified for 1 year on the pad). New SRMs and core were built up but, largely because of delays caused by investigations into two Centaur failures on Atlas launches, a third vehicle had to be assembled.

Specifications
Titan 4
First launch: 14 June 1989
Number launched: 36 to end 2003
Vehicle success rate: 83%
Launch sites: Canaveral pad 41 for LEO/GEO missions (plus pad 40 from 1992), Vandenberg SLC4E for NUS version from 1991 (capacity 2/yr). Canaveral pad 41 was used in the 1970s for Titan 3E Centaur D1T launches of deep space probes. A new pad 42 complex was at one stage under consideration for Cape Canaveral.
Principal uses: large military payloads into Sun synchronous orbit from Vandenberg; large military payloads such as electronic intelligence, DSP and Milstar into GEO from Canaveral, BMDO tests in lower orbits (no upper stage), NASA deep space launches.
Availability: no commercialisation plans

Performance
LEO (28.6°): 17,770 kg NUS plus SRM 21,900 kg NUS + SRMU
GEO: 4,545 kg with Centaur plus SRM. 5,773 kg with Centaur plus SRMU: 2,364 kg with IUS plus SRM: 2,860 kg with IUS plus SRMU.
Sun-synchronous (VAFB): 14,090 kg NUS plus SRM

Molniya (12 h, 925 × 39,450 km, 63°) from CC
Deep space: 4,545 kg at 20.5 km^2/s^2 (SRM)
Number of stages: (2 or 3) plus 2 strap-ons
Overall length: 63.14 m for Centaur, 53.99 m for IUS, 53.99/57.04 m for NUS.
Principal diameter: 3.05 m.

Launch mass
Centaur: 868,644 kg (SRM); 939,301 kg (SRMU).
IUS: 910,018 kg (SRM); 924,515 kg (SRMU).
NUS: 906,937 kg (SRM); 910,018 kg (SRMU).
Guidance: GM's Delco Systems Operations digital inertial. 4B's standardised avionics from vehicle #24 will use Honeywell system: 3 Honeywell GG 1342 RLGs, 3 Sundstrand QA 3000 accelerometers, 1750A 1 Mips processors, 25,000 h MTBF, 30 kg, 0.03 m^3, 110 W. 38 kg 250 Ah SAFT lithium thionyl chloride battery.

Titan 4 strap-ons (stage 0)
Comment: Titan 4 is initially employing United Technologies Chemical Systems Division's steel casing strap-ons. Lack of capacity to satisfy at least eight launches per year and the requirements of some missions for higher performance resulted in a US$725 million contract (cost in July 1993 reported at US$1.2 billion) for 15 booster sets being awarded to Hercules Aerospace (now Alliant Techsystems) in October 1987.

The 3-segment filament-wound graphite composite casings and the high-performance HTPB propellant similar to that used in Peacekeeper's stage 1 will increase Titan 4 capacity by typically 27%. Development firings were due to begin March 1990 but were delayed by bonding problems with the second segment, cast December 1989. The new motors were to have debuted on launch 11, in spring 1992, and have taken over from #17. As insurance, USAF authorised long-lead procurement of seven further CSD sets.

On 7 September 1990, a 136 t SRMU practice segment ignited in an accident at the USAF Astronautics Laboratory in California. The unit, incorporating an aft skirt/nozzle, was being lifted from a test stand when its crane overbalanced, killing a nearby worker. The segment was ignited by friction sliding into the flame trench. The first full-scale SRMU firing was planned for the same stand later that year in the schedule to qualify the booster by October 1991. The PQM1 (Preliminary Qualification Motor) exploded after a few seconds 1 April 1991 because of unpredicted internal aerodynamic choking. The added PQM1's firing was delayed to 12 June 1992, largely because the test stand had to be rebuilt. The QM-2 third test, run at the 2.5°C lower limit, was successful 21 February 1993. QM3 at the 41°C upper limit was successful 2 June. The final QM4, 12 September 1993, would have allowed flight introduction in 1995, although the initial requirement is for K24 fourth quarter 1996. The first production flight set was delivered first quarter 1994.

CSD's seven segment boosters are based on the model developed for the USAF Manned Orbiting Laboratory Titan 3M launcher, cancelled 1969. Two qualification firings only, therefore, were required for Titan 4, accomplished 21 December 1987 and 14 February 1988. Martin Marietta was awarded a US$10 million contract in 1990 to install a sound suppression system on 4E as protection against the solid motors' ignitions. The system dumps water into the exhausts to reduce the pressure wave. The company also received a US$5.3 million contract for upgrading motor storage facilities by April 1991 to handle the SRMU. Titan 4 #7 exploded after 101 seconds when the casing of SRM1's segment 3 burned through. The cause was a slit in the grain from repairs to de-bonded restrictor material on top of the segment. This material retards burning on top. It was expected that compression would close the slit in flight, but tests showed ignition pressure opened a route to the wall.

Contractor: UTC CSD for heavyweight steel casing (D6AC) SRM versions; Alliant for SRMU lightweight filament wound composites.
Length: 34.43 m (CSD), 34.25 m (Alliant)
Diameter: 3.11 m (CSD), 3.20 m (Alliant)
Mass at ignition: 342.8 t (CSD), 349.6 t (Alliant)
Propellant: PBAN (CSD), HTPB (Alliant)
Propellant mass: 295.5 t (CSD), 344.4 t (Alliant)
Average thrust: 7.117 MN each sea level (CSD), 7.562 MN each vacuum (Alliant)
Burn time: 126.5 s (CSD), 145 s (Alliant).

Titan 4 stage 1
Comment: Aerojet was contracted in December 1990 for a modified stage 1 nozzle; it became available in 1994. It required an increased expansion ratio for ignition at greater altitude when Alliant's higher-performance strap-ons are carried. The nozzle exit closure is enlarged and improved to provide better protection during early flight, and a thermal protection clamshell is fitted around the combustion chamber. Projected LEO capacity increase is 136 kg; future modifications under consideration could add a further 318 kg.
Contractor: Lockheed Martin

Canaveral pad 40 preparations for launching the first Titan 4 Centaur (Lockheed Martin) 0516930

Engines: paired gimballed Aerojet LR87AJ11A single start pressure fed. First Titan 4 full duration 200 s test firing 20 July 1987.
Overall length: 26.38 m
Principal diameter: 3.05 m
Oxidiser: 111 t of nitrogen tetroxide
Fuel: 59 t of Aerozine-50
Stage thrust: 2,433.8 kN vacuum
Burn time: 186 s (ignites at 116 s)
Separation: stage 2 hot separation

Titan 4 stage 2
Contractor: Lockheed Martin
Engine: gimballed Aerojet LR91 AJ11A single start liquid propellant pressure fed
Overall length: 9.94 m
Principal diameter: 3.05 m
Oxidiser: 24.5 t of nitrogen tetroxide
Fuel: 13.9 t of Aerozine-50
Stage thrust: 472.0 kN vacuum
Burn time: 240 s

Titan 4 stage 3
Comment: Centaur and IUS are the principal upper stage choices but a range of spinning solids is possible. Centaur performs three burns to: achieve low Earth parking orbit, establish GTO and inject into GEO. The two-stage IUS is released in LEO and performs a stage 1 burn for GTO, followed by the stage 2 injection into GEO.
Designation: Centaur or IUS (designated Titan 401 when Centaur carried; 402 for IUS)
Contractor: Lockheed Martin (Centaur), Boeing (IUS)
Type: liquid (Centaur), solid (IUS)
Engine: P&W RL10-3-3A (Centaur), CSD Orbus (IUS)
Overall length: 8.94 m (Centaur), 5.00 m (IUS)
Principal diameter: 4.51 m (Centaur), 2.896 m (IUS)
Centaur propellants: 23 t of liquid oxygen/hydrogen
IUS propellant: 9,818 kg stage 1, 2,722 kg stage 2
Stage thrust: 146.8 kN vacuum (Centaur), 202.8 kN stage 1/82.3 kN stage 2 (IUS)
Mass at ignition: 16.2 t (IUS), 26 t (Centaur)
Burn time
IUS: 152 s for stage 1 GTO insertion, 289 s for stage 2 GEO insertion
Centaur: 617 s total
Attitude control: 3-axis both Centaur/IUS

Titan 4 fairings
Contractor: McDonnell Douglas
Construction: isogrid aluminium 6061
Length: 17.08 m NUS, 26.23 m Centaur, 17.06 m IUS (15.25, 20.1, 23.2 m versions also available). 23.2 m version used for first time by Milstar DFS-1.
Diameter: 5.09 m
Mass: 4,033 kg NUS, 6,073 kg Centaur, 4,026 kg IUS
Separation: low-explosive detonating fuse along seams and 12 explosive bolts divides fairing into three sections in 0.2 s

Titan Family Development History
Titan 1 (HGM-25A) Development began in October 1955 as a back-up to Atlas ICBM. First of 67 launches January 1959, entered service 1962 with 54 missiles in six squadrons; deactivated 1965. Launch mass 99.79 t, length 29.87 m with Avco Mk4 warhead, stage 1: 3.05 m diameter, stage 2: 2.44 m diameter, propellants LOX/RP-1, engines: two Aerojet LR87AJ1s, stage 1 (1,334 kN), single LR87AJ-1, stage 2 (356 kN). The only IRBM/ICBM developed 1950s/early1960s not adapted as space launcher
Titan 2 144 built (36 R&D, 108 operational), first launch 1962, last of 54 deployed ICBMs deactivated May 1987. Modified as space launcher with last launch in 2003
Titan 3A 4 launched 1964–5 as demonstrations of Titan 3C's core. Essentially a Titan 2 with added Trans-stage upper stage. Length 37.8 m. Retired.
Titan 3B 54 launched, all with KH8 close look reconnaissance satellites from VAFB's SLC-4W, 1966–1985. Essentially Titan 2 with added Agena upper stage. A further 14 were launched as the **Titan 34B** (also known as Titan 3B Ascent Agena) 1971–87 from VAFB with stretched stage 1 and 3.05 m diameter payload fairing, carrying the highly classified SDS communications or Jumpseat electronic intelligence into Molniya-type orbits with apogees over the Soviet Union. Retired.
Titan 3C 36 launched 1966–82 from Florida carrying Vela and DSP early warning, DSCS communications, GEO electronic intelligence and unclassified satellites, including NASA's ATS 6. Titan 3A as core stage with five segment solid strap-ons, 25.9 m long, 3.05 m diameter each generating 5.34 MN. Stage 1's engines ignited at altitude for greater efficiency, producing 2.09 MN. Retired.
Titan 3D 22 launched from VAFB 1971–83 with KH9 Big Bird and KH11 reconnaissance satellites. Essentially west coast version of the 3C but without Trans-stage. **Titan 34D** introduced in 1982, incorporating 34B's stretched stage 1, either Transtage or IUS as upper stage, and adding a half segment to each strapon. Number 15/last 34D launched September 1989. Retired.
Titan 3E Seven launched 1974–77 from Canaveral with Helios, Voyager and Viking. Titan 3D modified for NASA deep space missions: Centaur cryogenic upper stage and 4.27 m diameter fairing. Retired.
Titan 3M Seven-segment strap-on version of the 3C, intended as the launcher of the USAF Manned Orbiting Laboratory but development ended when MOL was cancelled in 1969. Titan 4 eventually introduced the seven-segment variant. Retired.

Titan 4 launch history

	Date	Site	Stage	Type	Payload
1	14 June 1989	ETR 41	IUS	402	DSP14
2	8 June 1990	ETR 41	–	405	ocean surveillance?
3	13 November 1990	ETR 41	IUS	402	DSP 15
4	8 March 1991	SLC4E	–	403	Lacrosse?
5	8 November 1991	SLC4E		403	ocean surveillance?
6	28 November 1992	SLC4E		404	Adv KH 11?
7*	2 August 1993	SLC4E		403	ocean surveillance?
8	7 February 1994	ETR 40	Centaur	401	Milstar DFS 1
9	3 May 1994	ETR 41	Centaur	401	Adv Jumpseat?
10	27 August 1994	ETR 41	Centaur	401	Adv Vortex?
11	22 December 1994	ETR 40	IUS	402	DSP 17
12	14 May 1995	ETR 40	Centaur	401	Adv Orion?
13	10 July 1995	ETR 41	Centaur	401	Adv Jumpseat?
14	6 November 1995	ETR 40	Centaur	401	Milstar DFS 2
15	5 December 1995	SLC4E		403	?
16	24 April 1996	ETR 41	Centaur	401	Adv Orion?
17	12 May 1996	SLC4E		?	?
18	3 July 1996	ETR 40	–	405	ocean surveillance
19	20 December 1996	SLC4E		404?	Advanced KH-11?
20	24 February 1997	ETR 40	IUS	402	DSP18 (Titan 4B)
21	15 October 1997	ETR 40	Centaur	401	Cassini/Huygens
22	24 October 1997	SLC4E		403	Lacrosse/Vega
23	8 November 1997	ETR 41	Centaur	401	Trumpet (Elint)
24	9 May 1998	ETR 40	Centaur	401	Advanced Orion (Elint)
25*	12 August 1998	ETR 41	Centaur	401	Mercury (Vortex)
26*	9 April 1999	ETR 41	IUS	402	DSP-19
27*	30 April 1999	ETR 40	Centaur	401	Milstar 2
28*	22 May 1999	SLC4E	–	405	Ocean surveillance
29	8 May 2000	ERT40	IUS	402	DSP 20
30	17 August 2000	SLC4E	–	405	NRO Orynx
31	27 February 2001	SLC-40	Centaur	401	Milstar
32	6 August 2001	SLC-40	IUS	402	DSP
33	5 October 2001	SLC-4E		405	Keyhole
34	16 January 2002		Centaur	401	Milstar 2

* vehicle failure (SRM 1 burn through on #7 caused vehicle to explode after 101 s). 401 = Centaur, 402 = IUS, 403 = NUS from VAFB, 404 = NUS plus 15.25 m fairing, 405 = NUS from Canaveral.

Titan 3 commercial Detailed below; first launch 1 January 1990. Now retired.
Titan 4 In service but due for retirement after last launch in 2005. Stretched stages 1/2, seven-segment strap-ons, IUS, Centaur or no upper stage. First launch June 1989. For the standardised **Titan 4B**, strap-ons replaced from end 1996 by the Solid Rocket Motor Upgrade (SRMU) of larger three-segment strap-ons employing filament wound graphite epoxy cases.

X-33 VentureStar

Current Status

Technologies developed for the thermal protection system are being re-examined in the light of a suspected problem on the Shuttle Columbia on 1 February 2003.

At the behest of US Defense Secretary Donald Rumsfeld, the X-33 project was examined in Spring and Summer 2001 as a possible candidate for a deep strike system capable of hitting targets anywhere on earth within 60 minutes. The concept is retained but the X-33 has been dropped as a candidate.

US government support for this programme was cancelled on 1 March 2001 when NASA responded to budget charges demanded by the newly inaugurated Bush administration. Due to insurmountable technical problems NASA decided against including X-33 in the Space Launch initiative, a programme designed to develop technologies for concepts capable of carrying payloads into space and crew members to the International Space Station. NASA has spent US$912 million on the X-33. On 6 February 2001, tandem aerospike-engines were successfully tested at NASA's John C. Stennis Space Center for 1.1 s with no observed anomalies. On 3 November 1999, a composite fuel tank structurally failed during tests at NASA's Marshall Space Flight Center. By September 2000 NASA and Lockheed Martin had agreed a strategy for X-33 development which envisaged the use of aluminium fuel tanks, a revised payment schedule and a deferred first flight date of 2003. Additional funds needed for this work were to be competed through the Space Launch Initiative.

Background

The Reusable Launch Vehicle (RLV) programme originated from a NASA study in 1993, which addressed the United States' future launch vehicle requirements. The Access to Space Study concluded that the most beneficial option would be the development and deployment of a fully reusable launch vehicle based upon pure rocket propulsion.

VentureStar 0003457

The Phase 1 studies during 1995/1996 called for concept verification which was government-led and funded, and industry executed. For Phase 2 from 1996 to 1999 – Lockheed Martin was chosen to develop the X33 flight demonstrator which would be a scaled version of what could become an operational launch vehicle, the full-scale Venture Star. Phase 3, which is 2000 and beyond will see the development and operations of a RLV which will be industry led and investor financed.

The Lockheed Martin consortium which won the contract for the development of the Venture Star vehicles is working in collaboration with AlliedSignal (systems and avionics), Boeing (linear aerospike engine and RCS), Rohr (thermal protection system) and Sverdrup (launch facilities).

The first sub-orbital flight was planned for late 2003 at the earliest and the vehicle will be capable of flying to a maximum altitude of 80 km and with speeds of up to Mach 10: launch and landing will be from Edwards Air Force Base. NASA has budgeted US$941 million for the project, with Lockheed Martin investing more than US$300 million of its own money. If the X-33 meets with NASA's approval then Lockheed Martin and its partners will have to decide whether it is economically feasible to develop the estimated US$5 billion fully Reusable Launch Vehicle.

The full-scale VentureStar – should it be developed – would be capable of matching the Shuttle's current capability to LEO, carry about 11 tonnes of payload to the International Space Station or carry a payload of 5.9 tonnes to geosynchronous transfer orbit with a suitable upper stage.

Comparison of Space Shuttle and VentureStar

	Space Shuttle	VentureStar X-33	VentureStar Full Scale
Launch mass, tonnes	2,050	124	990
Length, m	56	20.5	38.8
Width, m	24	20.8	39.1
Payload to 28.5°, 185 km orbit, tonnes	23	–	>23
Payload bay size, m	4.6 × 18.3	1.5 × 3.1	4.6 × 13.7
Propulsion	3 × LOX/LH2 + 2 SRBs	3 × LOX/LH2 + 2 SRBs	7 × LOX/LH2

X-34

Current Status

Responding to changes stipulated by the incoming Bush administration on 1 March 2001, NASA cancelled its funding for the X-34, saying that "the benefits to be derived from continuing the programme did not justify the cost". Elements of the X-34 design have been proposed for the new Space Launch Initiative but there is little likelihood that major technologies from the X-34 will have application. Engineering analyses indicate several technology areas developed for X-34 have applications to the NASA Space Launch Initiative.

Background

The Space Transportation Policy, signed by President Clinton 5 August 1994, gave the DoD responsibility for upgrading existing medium and heavy ELVs. Clinton allowed NASA to take the lead after 2000. To respond, NASA, in January 1995, invited industry to submit competing designs for two reusable launch vehicles. The first, the X-33, was to have tested the feasibility of replacing Shuttle and the larger ELVs by 2012 with the Reusable Launch Vehicle.

On paper, the 54-tonne 27 m long X-34 should handle satellites of about 1.1 tonnes. Orbital Sciences Corporation and Rockwell were selected March 1995, agreeing to invest US$100 million in addition to NASA's US$70 million through FY99. The two-stage liquid propellant vehicle was to be air-launched from NASA's Boeing 747 and the reusable stage 1 returned to a conventional runway after releasing stage 2 at 100 km altitude and M16. NASA wanted to keep mission costs below about US$3 million and the vehicle would be proposed for its next mission by a 15-man crew in less than three weeks. In the first development of the plane, NASA wanted to use existing engines. Russia's RD 120 was selected in favour of Rocketdyne's RS27. Later advanced, more efficient units that may be supplemented by air breathing engines to increase payload without significantly increasing operating costs would replace these engines.

NASA wanted the X-34 also to demonstrate RLV technologies, including advanced thermal protection, all composite structures, reusable cryogenic tanks and state-of-the-art avionics. Moving with uncharacteristic haste NASA selected an Orbital Sciences Corporation (OSC)/Rockwell team to develop the two-stage X-34, aiming for a suborbital flight in 1997 with orbital flights beginning a year later. The team formed American Space Lines as an 'Arianespace' lookalike for commercial development of the small launcher market but in early 1996 the team pulled out believing the costs could not produce a viable commercial return. NASA re-contracted with OSC for a single-stage suborbital demonstrator, still designated X-34. Three X-34 test vehicles were scheduled to support flight trials originally planned to begin in 2000 with the winged vehicle launched from the under-fuselage cradle of a converted L-1011 operated by OSC. Capable of a maximum speed of Mach 8 the X-34 was to have demonstrated all-weather, pilotless landings as a technology precursor to fully reusable launch systems of the future.

UPDATED

LAUNCH VEHICLE PROPULSION

LAUNCH VEHICLE PROPULSION

China

Propulsion

China's space launchers draw on a small range of main engines, the majority utilising hypergolic NTO/UDMH although more recently two cryogenic upper stages were added. No LOX/kerosene engines as used extensively elsewhere are known, reflecting the space variants' parallel development alongside storable propellant long range missiles.

Rocket propulsion was included as one of the key technologies in the 'Twelve Year Development Plan of Science and Technology' approved in 1958, leading to the launches of the first indigenously developed sounding rocket on 19 February 1960 and first successful ballistic missile on 29 June 1964 (failure on the first test 21 March 1962 led to a major redesign). The initial CZ-1 space launcher, derived from the CSS-3 MRBM, utilised nitric acid/UDMH but each subsequent operational and planned stage 1/2 has used N_2O_4 as oxidiser. Development of a LOX/LH_2 cryogenic upper stage for GEO missions began in 1977, leading with its launch as CZ-3 stage 3 in January 1984 to China becoming only the third such user after the US and European Space Agency (ESA). The China Academy of Launch Vehicle Technology (CALT, formerly the Beijing Wan Yuan Industry Corporation) near the town of Nan Yuan 15 km south of the capital develops/builds the cryogenic engines and its Shanxi Liquid Rocket Engine Co the storable engines; SLREC also handles solid motors. The Beijing Rocket Test Centre 50 km southwest of Beijing maintains five major test stands: No 1 handles altitude testing of spacecraft thrusters of up to 490 N, No 2 provides single engine cryogenic facilities, with No 4 accepting a complete H-8 cryogenic stage, and No 5 is devoted to hydrazine engines. Other test facilities are operated at the launch sites and SLREC probably has stands for the storable engines.

Fewer details are known of Chinese thrusters and solid propellant motors. The CZ-1D stage 3 is a solid of 625 kg propellant and satellites such as FSW incorporate solid kick motors. CZ-2E's 1.7 m dia EPKM solid stage 3 debuted in 1995. CZ-2C's 1.2 m dia CPKM is imminent. Sounding rockets with solids are also flying. Solid propellant ICBM/IRBMs are under development. Hydrazine thrusters of unspecified nature provide 3-axis control for the H-8 cryogenic stage. However, CZ-2E's stage 2 employs a hydrazine system acknowledged as derived from H-8: 4×45 N pitch thrusters, 2×45 N yaw + 4×11 N roll. The FSW recoverable capsule is also 3-axis controlled by cold gas, and the CZ-1D solid stage is spin-stabilised by N_2 thrusters.

The engines carry YF designations, indicating Yei-ti Fa-dong-ji (liquid state engine).

YF-2A
Applications: cluster of four for CZ-1 family stage 1 and CSS-3 IRBM
First flown: June 1964?
Oxidiser: nitric acid
Fuel: UDMH
Propellant flow rate: 115 kg/s?
Cycle: gas generator
Thrust: 275.3 kN sea level
Specific impulse: 241 s sea level
Burn time: about 130 s operational

YF-3
Applications: CZ-1 family stage 2, CSS-3 IRBM
First flown: June 1964?
Oxidiser: nitrogen tetroxide
Fuel: UDMH
Propellant flow rate: 50 kg/s?
Thrust: 294.2 kN vacuum
Specific impulse: 287 s vacuum
Burn time: about 130 s operational

YF-20/21 and YF-20B/21B
The YF-20/21 is the backbone of the Chinese launcher range, clustered in fours (creating the YF-21/21B assembly) for every first stage of the CZ-2 to -4 families and modified into the high altitude YF-22 for the second stages. In the L-140 stage 1s (the 140 indicating 140 t propellant

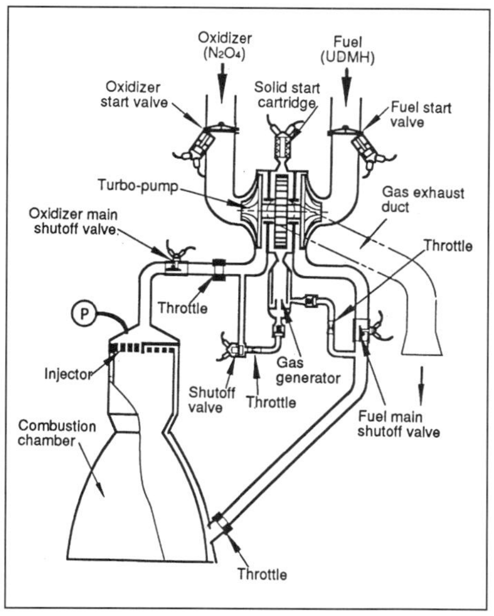

YF-20/21/22 schematic 0516865

loading) it operates at 696 kN thrust, but the stretched L-180 stages and strap-ons of the more powerful vehicles run it at 740 kN. Chinese statements suggest that still higher levels are possible. The LB-40 strap-ons of the CZ-2E/3B/3C each incorporate a single YF-20B.
Applications: CZ-2C/3 stage 1 for YF-20/21; CZ-2E/3B/3C strap-on for YF-20B; CZ-2D/2E/3A/3B/3C/4 for YF-21B
First flown: 5 November 1974 (first CZ-2 flight)?
Oxidiser: nitrogen tetroxide
Fuel: UDMH
Propellant flow rate: 270 kg/s?
Thrust: 696.2 kN sea level for YF-20/21 of L-140 stage 1 of CZ-2C/3; 740.35 kN, E 12.69, exit area 7,728.6 cm^2 for YF-20B of LB-40 strap-on and YF-20B/21B of L-180 stage 1 of CZ-2E/3A/3B/3C/4
Specific impulse: 259 s SL YF-20/21; 260.66 s SL YF-20B/21B
Mixture ratio: 2.10 YF-20/21; 2.12 YF-20B/21B
Burn time: 125–160 s operational

YF-22/22B and YF-24/24B
The YF-22/22B is a high altitude derivative of the YF-20. It is flown singly with four gimballed YF-23/23B verniers, the assemblies designated 24/24B, providing steering on the L-35 stage 2. YF-22B/23B use larger expansion ratios for improved SI.
Applications: CZ-2C/3 stage 2 for YF-24; CZ-2D/2E/3A/3B/3C/4 stage 2 for YF-24B
First flown: 5 November 1974 (first CZ-2 launch)?
Oxidiser: nitrogen tetroxide
Fuel: UDMH
Propellant flow rate: 270 kg/s?
Thrust: 719.8 kN vacuum for YF-22, 740.4 kN for YF-22B, but 845.3 kN attainable
Specific impulse: 289 s vacuum YF-22; 298 s vacuum YF-22B
Mixture ratio: 2.181 YF-22 & -22B
Burn time: 300 s CZ-2E; 110 s others

YF-23/23B
Four gimballed YF-23/23Bs provide steering for the fixed single YF-22/22B on the second stages listed above.
Applications: as YF-22/22B

CZ-3's H-8 third stage is powered by the 4-chamber cryogenic YF-73 0516866

CZ-3's H-8 third stage is powered by the 4-chamber cryogenic YF-73 0516867

First flown: as YF-22/22B
Oxidiser: nitrogen tetroxide
Fuel: UDMH
Propellant flow rate: 16 kg/s for all four chambers with single pump?
Thrust: 46.1 kN YF-23 and 47.1 kN YF-23B total (4 chambers) vacuum
Specific impulse: 281.7 s YF-23 and 289 s YF-23B vacuum
Mixture ratio: 1.57 YF-23/23B
Burn time: up to 300 s. 136 s for CZ-4

YF-40
The engine appears to have been uprated for CZ-4B. Where known, the different specifications are provided in 4A/4B order.
Applications: flown paired on CZ-4 stage 3; uprated for 4B?
First flown: 6 September 1988
Mass: each 70/83 kg dry
Oxidiser: nitrogen tetroxide
Fuel: UDMH
Thrust: 98.1/101.03 kN vacuum for pair
Specific impulse: 305/303 s
Mixture ratio: 2.14
Chamber pressure: 45.39 atm
Expansion ratio: 55
Exit diameter: 630 mm
Burn time: 321/412 s (single/single or dual burn)
Nozzle materials: niobium alloy, radiation cooled

YF-73
Development of the cryogenic YF-73 began in 1976 for the restartable H-8 third stage, the first use of LH_2 propulsion technology outside of the US and Europe. Principal use is delivery of payloads into GTO. Firings of flight-type engines began in 1981. After completion of acceptance testing, the first flight engine was delivered in the fourth quarter of 1983. By end-1983, about 120 tests

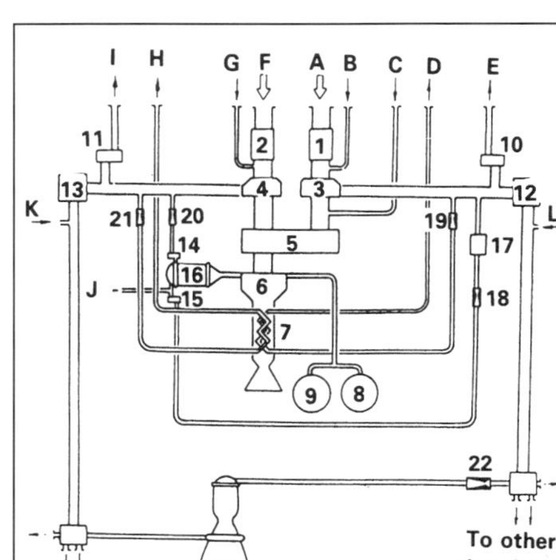

YF-73 schematic. A LOX inlet, B/C purge, D LOX tank pressurisation, E bleed, F liquid hydrogen inlet, G purge, H LH_2 pressurisation, I bleed, J-L purge; 1/2 isolation valves, 3 LOX pump, 4 LH_2 pump, 5 gearbox, 6 turbine, 7 heat exchanger, 8/9 start tank, 10/11 bleed valve, 12/13 main valve, 14/15 gas generator valve, 16 gas generator, 17 pressure stabiliser, 18–22 venturi, 23 combustion chamber (one of four) 0516868

totalled >30,000 s. The four chambers, gimballed ±24° for attitude control, are serviced by a single 37,000 rpm turbopump. The two starts are initiated by a pair of high pressure nitrogen spheres. The valves, pumps, gearbox, gas generator and propellant lines carry 10–20 mm of polyurethane foam to minimise heat transfer. The engines are manufactured at CALT. There have been two flight failures. The first was January 1984's debut mission when the re-start after coast attained 90 per cent thrust and then rapidly decayed after a few seconds. Cause was attributed to propellant flow fluctuations following the μg coast. Similarly, the second (December 1991) suffered a pressure drop 58 seconds into the second burn and shut down after a further 77 seconds. Cause was believed to be helium supply failure to control valves.

Application: H-8 third stage of CZ-3
First flown: 29 January 1984 (failed to restart)
Number flown: 9 sets, to end-1995 (2 failures)
Engine cycle: gas generator
Dry mass: 236 kg
Length: 1,438 mm
Maximum diameter: 2,220 mm
Oxidiser: liquid oxygen at 9.086 kg/s
Fuel: liquid hydrogen at 1.817 kg/s
Thrust: 44.147 kN total (4 chambers) vacuum
Specific impulse: 420 s vacuum
Mixture ratio (O/F): 5.0
Chamber pressure: 25.96 atmospheres
Expansion ratio: 40:1
Chamber materials: stainless steel, regeneratively cooled by supercritical hydrogen
Injector: brazed stainless steel, 38 coaxial elements
LOX/LH_2 pumps: 37,000 rpm single stage centrifugal pump with helical inducer. LH_2 and turbine share common shaft; LOX shaft is driven via reduction gears. Gears cooled by hydrogen
Burn time: two burns of 500 s + 300 s

YF-75

Development of the cryogenic restartable YF-75 drew extensively on YF-73 experience to create the H-18 stage, comparable to Centaur D. Its employment in pairs aboard the CZ-3A vehicles provides a GTO capability equivalent to the lower range of Ariane 4. Aboard CZ-3B, it surpasses Ariane 4 and almost equals Proton. The engines are manufactured at CALT.

Application: H-18 stage 3 of CZ-3A/3B/3C
First flight: 8 February 1994
Number flown: 2 sets to end-1995
Oxidiser: liquid oxygen
Fuel: liquid hydrogen
Propellant flow rate: 19 kg/s each chamber?
Thrust: 78.4 kN vacuum each chamber
Specific impulse: 442 s vacuum
Mixture ratio: 5.0
Chamber pressure: 37.15 atm
Expansion ratio: 80:1
Burn time: about 470 s total, dual start.

GF-02/CPKM/EPKM/FG-46

The solid motor carried by the CZ-1D appears to be related to the GF-02 solid flown on the original CZ-1 vehicle. Work began in 1965 at the Solid Fuel Engine Research Academy under deputy head Yang Nansheng. The first spin firing took place 26 January 1968 but the motor exploded after 30 seconds, causing some fatalities. During 1968 to 1970 a total of 19 tests, including five altitude simulations, was conducted successfully.

The HEXI Corp's indigenous EPKM FG-46 commercial perigee kick stage debuted for AsiaSat 2 in November 1995, providing 3,460 kg into GTO. An early version was carried on the first 2E. Nine test firings included two failures at the nominal 40 rpm spin rate: 31 May 1993, test M1, prototype/design qualification; 24 October 1993, M2, prototype/structural qualification; 31 January 1994, C1, 40 rpm spin (fail); 5 May 1994, C2, altitude simulation; 6 July 1994, C3, 40 rpm spin (fail); 30 October 1994, C5, 40 rpm spin; 8 November 1994, C4, altitude simulation; 16 January 1995, C6, 40 rpm spin; 20 March 1995, Z1, 40 rpm spin/acceptance.

The 1.2 m dia CPKM is imminent on the CZ-2C.

EPKM specifications

First flown: 29 May 1995
Number flown: 2, to end-1995
Mass: 6,001 kg (5,444 kg propellant, 529 kg burnout, 28 kg ablated)
Length: 2.5 m
Diameter: 1,700 mm
Propellant: HTPB
Burn time: 87 s
Thrust: 210 kN max vacuum
Specific impulse: 292 s vacuum

Total impulse: 15.5881 MNs vacuum
Pressure: 6.0 MPa max
Casing materials: fibre glass
Nozzle: carbon-carbon

VERIFIED

France

Société Européenne de Propulsion

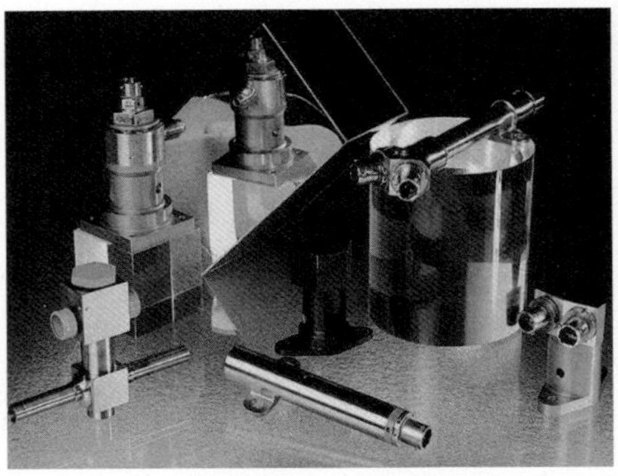

Pyrospace pyromechanisms for space applications. It provides more than 230 for each Ariane 4 0517172

Current Status

On 31 October 1997 Snecma acquired 100 per cent of SEP which then became the Rocket Engine Division. In an effort to increase payload carrying capacity of Ariane 5, SEP is developing in association with ESA and CNES the Vinci cryogenic upper stage. Designed for ECA-B, it has a thrust of 180 kN giving the launch vehicle a lift capability of 12,000 kg to GTO. It is expected to be available for service from 2006.

Background

The principal space business of SEP, a subsidiary of SNECMA, is the design, development and construction of propulsion systems for launch vehicles and satellites, including Ariane's Viking, HM-7, Vulcain and P230 (partnered with BPD in Europropulsion, qv) motors. In satellite work, SEP produces Mage apogee kick motors, hydrazine thrusters for attitude control, bipropellant liquid engines and electric/ion engines. The company also produces solar array drive mechanisms, μg furnaces, surface tension tanks, magnetic bearings, transducers and flow regulators. In 1995, personnel totalled 3,500, with a turnover of Fr5,400 million generating a Fr150 million profit. 1994: 3,550/4,800/133.5 million profit; 1993: 3,500/4,250/110; 1992: 3,600/4,355/140; 1991: 3,800/4,540/87.2; 1990: 4,050/4,447/103, respectively. Aerospatiale in Jan 1994 sold its 13.6 per cent interest for FFr178.9 million. SEP and Spain's Empresarios Agrupados established Iberespacio (qv) in Oct 1989 to undertake joint development of space propulsion systems. SEP and its SNECMA parent established Hyperspace in Jan 1990 to co-ordinate propulsion activities for hypersonic aircraft and launchers. Other interests include: Arianespace 8 per cent, Europropulsion (50/50 with BPD), G2P 75 per cent (large solid propulsion systems for military applications), Spot Image 12 per cent, Carbone-Industrie 100 per cent (carbon brakes), TECHLAM 50 per cent (laminated elastomeric components) and S2M 54 per cent (magnetic bearings).

Test/Assembly Facilities: SEP maintains a 4,500 m^2 Ariane propulsion assembly hall containing 14 Viking engine assembly stands, five for HM-7 and nine for stage 1 propulsion systems. There is also a 400 m^2 class 100,000 cleanroom for cryogenic engine component assembly. TDF propulsion was installed in an integration building provided with two 300 m^2 class 100,000 rooms. The Vulcain Engine Assembly Building can produce eight engines annually. The 1,220 m^2 class 100,000 cleanroom accommodates 12 assembly cells and 25 class 100 laminar flow booths/hoods for storage before assembly. Also in the building are a 700 m^2 warehouse, a 1,400 m^2 engineering & mechanical area and 1,150 m^2 of office space.

Engine test facilities are concentrated at the 1.5k m^2 Vernon site but some are located at the 24,300 m^2 Villaroche area. For storable propellant motors, there are four component test stands (PF1-3, A48, F22) and one engine stand (PF2). Cryogenic testing is undertaken on three stands: the two of Vernon's PF41 (one with altitude simulation) and one at Villaroche (with altitude simulation). Eleven existing cryogenic component stands were joined in 1988 by PF52 for Ariane 5 Vulcain gas generator and turbopump development. The PF50 stand for complete Vulcain engine tests was inaugurated Sep 1990 at Vernon. SEP inaugurated its 11,000 m^2 production plant for Ariane 5's P230 nozzles in Oct 1990 at Haillan. A FFr180 million investment, it fabricated10 nozzles annually by year 2000. Assembly space is 650 m^2, plus 825 m^2 class 100,000 cleanrooms.

SPT 100 Mk 2

SEP joined the International Space Technology Inc partnership (qv) in 1993. ISTI is marketing OKB Fakel's SPT-100 0.08 N Hall xenon stationary plasma thruster using western electronics. France's Stentor satellite will carry a slightly modified version. SEP began development in May 1993 of a Mk 2 version at Villaroche to increase SI from 1,600 s to 2,200 s. See the ISTI and OKB Fakel entries for further details.

Ariane Viking Engines

SEP provides four first stage, one second stage and the required number of liquid strap-on engines for the Ariane launcher (the cryogenic HM-7B is specified separately below). India's Vikas engine is derived from Viking technology. The design of the Viking engine was based on the M40 400 kN feasibility demonstrator. The M55, first fired 8 Apr 1971, was renamed Viking in the tradition of using names beginning with 'V' (for Vernon). At that time it was intended to power ELDO's Europa III, cancelled in 1972. Full development of Viking 2 began 1973 for Ariane; replacement of the conical nozzle with a bell form in 1976 created the Viking 5 and increased sea level thrust from 588.5 kN to 623.7 kN. Three qualification firings were successful in 1979. Viking 3 was a ground test version of the Viking 4 altitude variant of Viking 5. By end-1994, Viking engines had logged 110,400 s of operation, including 68,100 s in flight and 1,261 start-up sequences; a total of 680 units had been fabricated. Almost 800 will have been produced by the end of its career when Ariane 4 is phased out. The initial Ariane 1 vehicle was powered by Viking 5A (611 kN thrust each) and Viking 4A (721 kN); Ariane 2/3 switched to UH25 fuel (75 per cent UDMH/25 per cent hydrazine hydrate) from UDMH and carried the more powerful Viking 5B (752 kN) and Viking 4B (805 kN) models. The current Ariane 4 launcher retains the second stage Viking 4B, but its first stage Viking 5C's firing time is extended by 72 s and up to four Viking 6 engines, derived from Viking 5, can be carried as strap-ons.

Viking 5C & 6

Applications: Ariane 4 first stage propulsion (four Viking 5C), Ariane 4 PAL liquid strap-on propulsion (single Viking 6, specifications as for Viking 5C but with different layout for PAL accommodation)
First flown: 15 June 1988 on Ariane V22-401
Number flown: 400 × Viking 5C, 200 × Viking 6, to end-2000
Dry mass: 826 kg
Length: 287.3 cm
Maximum diameter: 99.0 cm
Mounting: gimballed for pitch/yaw/roll control
Engine cycle: open gas generator
Oxidiser: NTO at 173.3 kg/s (Viking 6 173.7 kg/s)
Fuel: UH25 at 101.9 kg/s (Viking 6 101.5 kg/s)
Mixture ratio (O/F): 1.70 (Viking 6 1.71)
Turbopump: 10,000 rpm, 2,500 kW power rating. Single-shaft turbopump driven by gas generator consuming 1.2 kg/s propellants. A third pump adds water to limit turbine operating T. A main flow control unit maintains thrust at a pre-determined reference level by adjusting the relative flows of the three liquids entering the gas generator. Another control loop adjusts the thrust chamber pressure by controlling gas production and regulating the turbopump
Thrust: 752 kN vacuum, 678 kN sea level
Specific impulse: 278.4 s vacuum
Expansion ratio: 10.5
Nozzle materials: cobalt alloy with SEPHEN (phenolic resin/silica fibre) throat
Combustion chamber:
pressure: 58 atm
temperature: about 3,000°C
injector: 216 like on like doublets set in six rows for each propellant in light alloy annular injector
cooling: UDMH propellant film supplied through additional channels on the injector's lower section
ignition: hypergolic
Burn time: 209 s Viking 5B, 143 s Viking 6

Viking 4B specifications

Ariane's second stage is powered by a single derivative of the Viking 5 engine, optimised for operation at altitude. Gimbal mounting provides pitch/yaw control and roll control is exercised by auxiliary jets feeding from the main engine's gas generator.

Applications: Ariane 2–4 stage 2 engine
First flown: 4 Aug 1984 on Ariane V10
Number flown: 123 to end-2000
Dry mass: 826 kg
Length: 350.9 cm
Maximum diameter: 170.0 cm
Mounting: gimballed for pitch/yaw control
Oxidiser: nitrogen tetroxide at 175.0 kg/s
Fuel: UH25 at 103.0 kg/s
Mixture ratio (O/F): 1.70
Turbopump: as Viking 5B/6
Thrust: 805 kN vacuum
Specific impulse: 295.5 s
Expansion ratio: 30.8:1
Chamber pressure: 58.5 atm
Burn time: 125 s

Ariane HM-7B engine

Ariane 4's stage 3 cryogenic engine was developed by SEP from experience with the 40 kN HM-4, tested 1962–69. The HM-7 model was introduced on Ariane 1 in 1979 but subsequent launcher versions have utilised the HM-7B, offering 265 s longer firing duration. Recent enhancements to Ariane's stage 3 have required engine modifications: values are given below for the H10, H10+ (first flown V50, 1992) and H10-3 (debut V70, Dec 1994) stages. Ariane V15 (Sep 1985) and V18 (May 1986) failed because of stage 3 engine ignition problems. The pyrotechnic igniter was redesigned to expel its hot gases at 45° instead of the previous vertically downwards parallel to the propellant flow. SEP conducted extensive ignition trials with the new design, concluding with nine firings of one engine totalling 3,130 s, including three long-duration runs of about 900 s. Ariane V63 (Jan 1994) was lost because of abnormal heating of the immersed LOX pump bearing. Insufficient cooling and aggravating factors resulted in failure. The solution was installation of a purge line in the bearing cavity to enhance cooling and helium purging, and fitting the bearing with a self-lubricating MoS_2 coating to decrease its sensitivity to possible aggravating factors. V70 failed because stage 3's gas generator was starved of LOX. The engine stabilised at 70 per cent and shut down 40 s early. As a result, a filter was added to the oxygen line upstream of the generator to prevent blockage in future and more stringent contamination prevention measures introduced. By end-1995, HM-7B had made 898 firings totalling 180,225 s (48,240 s in-flight). Firing trials of an HM-7B equipped with a Novoltex-reinforced SEPCarbinox 25 kg composite nozzle were made in 1989 as part of a technology programme. The nozzle does not require active cooling to cope with the 1,800°C, allowing more efficient use of the liquid hydrogen. An **HM-7S** featuring a deployable nozzle has also been considered for a future Ariane 5 upper stage: 62.2 kN, SI 447 s, 4.56 mixture ratio.

Applications: Ariane stage 3 propulsion system
First flown: 4 Aug 1984 on Ariane V10
Number flown: 121 HM-7/7B to end-2000, including five failures (11 HM-7 flown, one failure)
Dry mass: 155 kg
Length: 201.3 cm
Maximum diameter: 99.2 cm
Mounting: gimballed for pitch/yaw control
Oxidiser: liquid oxygen
Fuel: liquid hydrogen
Mixture ratio (O/F): 4.76:1 (gas generator 0.87) for H10 stage, 4.66 H10+, 5.04 H10–3
Turbopump: 60,500 rpm, 380 kW single turbine powered by gas generator requiring 0.25 kg/s propellants. Single-stage hydrogen pump raises pressure from 3 atm to 55 atm; single-stage oxygen pump 2.5–50 atm. Gas generator exhaust temperature 800–900K
Thrust (vac): 62.3 kN H10/H10+, 63.8 kN H10–3
SI (s, vac): 446 H10, 446.1 H10+, 445 H10–3
Expansion ratio: 83.1:1
Cooling method: hydrogen is heated to 150K during cooling of combustion chamber wall along 128 longitudinal channels for injection in gaseous form. The nozzle cooling system employs 242 helical tubes; 0.15 kg/s of hydrogen is dumped overboard
Combustion chamber:
pressure: 35.5 atm
ignition: pyrotechnic, comprising Isolite 1431 Rauline grain, with redundant initiators, providing two flame jets angled at 45° to propellant flow
materials: stainless steel
Burn time (s): 735 H10, 760 H10+, 790 H10–3

Vulcain

CNES initiated studies in 1978 for launcher requirements following the Ariane 4 series, while SEP conducted preliminary studies of a 500 kN hydrogen/oxygen engine. During 1979–81, a 900 kN HM-60 engine emerged to power Ariane 5's H-60 second stage employing the gas generator cycle instead of staged combustion because of cost considerations (development cost saving was estimated at about 30 per cent for a 10–15 s performance loss). Subsequently, HM-60 was assigned to Ariane's core stage and uprated to >1 MN (vacuum). Target reliability is 0.9946; programme cost about ECU1 billion. Contractor responsibilities are: SEP prime, gas generator and LH_2 turbopump; DASA thrust chamber/nozzle; Volvo Flygmotor nozzle (DASA subcontract) and pump turbines; FiatAvio LOX turbopump; Techspace Aero and Microtechnica fluid control valves; MAN gimbals.

The preparatory programme ran October 1984–87, followed by development 1988–94. Subsystem testing began during 1987. The first engine, the short-nozzle M1, was installed on Vernon's PF50 60 m vertical stand 4 April 1990 and performed its first ignition test 5 July. M1 achieved 15 runs, including two at full thrust for 8 s. PF50 allows one test per week. DLR's identical P5 stand at Lampoldshausen received the short-nozzle M2 10 August. Engine M3 took over on PF50 in early 1991 for flight duration 600 s tests: its first attempt on 29 May was halted after 104 s and on 5 June after 185 s as a result of T sensor problems. M3 achieved the first 605 s full duration full-rating run 13 June 1991. This was the 42nd Vulcain test and M3's 13th, bringing it to 1153 s. On 4 July it ran for 886 s until the stand's propellants were exhausted. A full nozzle was added to M1 (redesignated M1R) for trials on P5, the first 65 s run 2 July 1991 leading to the full 605 s later that month. It was then replaced by new engine M4. By end-August, the three engines totalled 3,048 s. This first test phase, running April 1990 to end-1992, required one engine to achieve 6,000 s over 20 runs; attained were 6,819 s in 28 starts. By end-June 1992, 97 tests on the six engines had accumulated 11,329 s, attaining a 1,275 kN max thrust and running for 793 s at the 1,120 kN flight thrust. Seven development engines had been tested by end-1992. Through July 1993, 145 tests on nine engines (two revalidated) totalled 29,000 s. 26 were >580 s and 13 >780 s. The longest, 960 s, was achieved by M8 (which completed 20 tests totalling 9,700 s through July 1993). By April 1994, 180 tests on 13 engines had achieved 44,730 s; M8 totalled 10,000 s. M6-R1 suffered a serious LOX turbopump fire 11 April 1994 as a result of impeller friction with the casing (this was an old design; the new version avoids the problem). The stiffer nozzle to prevent buckling was introduced in early 1994. By end-1994, 207 tests on 18 engines had achieved 56,920 s. By end-1995, 275 tests on 18 engines had achieved 85,837 s. Pad firings at Kourou began 17 November 1994 using M9 on a battleship stage 1, ending 27 January 1995. Qualification firings with a flight-type stage were completed 6 January 1996 (the first 680 s firing was achieved 16 June 1995). Ariane 5's debut, flight 501, employed engine M14; 502 M16; 503 M17.

Other facilities include the PF52 stand at Vernon (capable of running both turbopumps and gas generator together or independently for 100 s), P5.9 LOX turbopump and gas generator stand at Ottobrun (DASA), and the P3.2 thrust chamber stand at Lampoldhausen. The specifications below are typical of a GTO mission.

Applications: Ariane 5 core stage propulsion
First flight: 4 June 1996
Number flown: 8 (to end-2000)
Dry mass: 1,475 kg
Length: 300 cm
Maximum diameter: 176.0 cm
Mounting: gimballed ±6° for pitch/yaw control
Engine cycle: open gas generator
Propellants: LOX at 225.6 kg/s into chamber, LH_2 at 36.4 kg/s into chamber; total propellant flow 271.5 kg/s includes 9 kg/s for gas generator, attitude control thrusters, nozzle dump cooling and pressurisation
Mixture ratio: 6.2:1 in chamber, 0.94:1 gas generator. Valves are pneumatically operated and provide only two mixture ratio settings
Turbopumps: separate fuel/oxidiser turbopumps. Hydrogen pressure raised to 160 atm by 34,500 rpm 11.9 MW (range 30,340–39,330 rpm, 7.7–16.9 MW) 240 kg 2-stage centrifugal pump; oxygen to 150 atm by 130 kg 13,700 rpm 3.3 MW (range 11,300–14,880 rpm, 1.8–4.3 MW) 1-stage pump
Thrust: 1,145 kN vacuum, 900 kN sea level
Specific impulse: 431.2 s vacuum
Expansion ratio: 45:1
Combustion chamber:
pressure: 110 atm
cooling: regenerative hydrogen through 360 longitudinal channels in copper-silver-zirconium alloy liner
materials: stainless steel
injector: Inconel 718 516-element co-axial
ignition: pyrotechnic. Start-up and shutdown sequences are controlled by microprocessor-based system incorporated into the engine
Nozzle:
diameter: 176.0 cm (surface area 7.6 m^2)
mass: 170 kg
cooling: hydrogen circulated through helical tubes and dumped into exhaust through 228 ports. Capacity about 23 MW. Nozzle is composed of these 456 Inconel 600 helical 4 × 4 mm square tubes (wall thickness 0.4 mm) joined by 2 km of welding. Additional hatbands were added around the lower nozzle following tests, to cope with startup/shutdown transients
Burn time: 580 s (design life: 6,000 s + 20 starts)

Vulcain MkII

Commercial payloads require the upgrading of Ariane 5 performance by 15–20 per cent, a programme approved by ESA in Oct 1995. The Vulcain MkII preparatory phase ran January 1995 to March 1996 for full development to begin 1996 and qualification in 2002. 120 tests totalling 56,000 s are planned from mid-1998 to end-2001 on seven engines. Vulcain's mixture ratio will be increased to 6.2 by increasing the LOX flow rate by 18 per cent, accommodated by lowering the core's common bulkhead by 70 cm to raise propellant mass from 155 t to 172 t. The other requirement is to improve Vulcain thrust to 1,350 kN vac. This will be achieved by a 10 per cent wider throat (expansion ratio from 45 to 61.7), 5.5 per cent chamber pressure increase, the higher mixture ratio, and extending the nozzle, its lower part using a simple sheet film-cooled by the injected turbine exhaust. Dry mass will increase to 1,850 kg, mainly because of the larger nozzle. The launcher's structures are sized to accommodate such upgrades. The 185 kg LOX pump will increase its flow by 23 per cent, pressure by 19 per cent, power to 5.4 MW and outlet T to 875K. The LOX dome will be redesigned to cope with the increased flow. The thrust chamber will use a 7.1 mixture ratio, 306 kg/s flow and 116 atm pressure. 1.7 kg/s for dump cooling.

Liquid AOCS Thrusters

SEP has developed two sizes of monopropellant hydrazine thrusters for spacecraft attitude and orbit control subsystems. The 3 N and 15 N class thrusters are based on the catalytic decomposition of hydrazine using CNESRO catalyst and are/have been used by D5A, Spot, Geos, Exosat, ERS, Helios and Envisat. The specifications below are in 3/15 N order.

Dry mass: 0.320/0.355 kg
Length: 107.7/145.1 mm
Mounting: fixed
Thrust: 3.5/15.6 N vacuum at max pressure
Inlet pressure range: both 22–5.5 atm
Specific impulse: 230/232 s vacuum
Start temperature: both 180°C
Number of hot pulses: 380,000/116,000
Number of starts: 9,000/13,500
Total on time: 40/17 h
Total impulse: 520/615 kNs
Valve power supply: both 27 V and 6 W
Heater power supply: 5.2/6 W at 34 V

SEP has developed NTO/UDMH thrusters for military applications which form the basis for space versions suited to Ariane Transfer Vehicle (8 kN + 200 N) and satellite AOCS (20 N) applications.

Mage Solid Propellant Motors

SEP is the industrial lead for ESA's range of three solid propellant apogee kick motors. Mage 1, first launched 1981 on the third Ariane vehicle, is designed for GEO injection of 400–500 kg class satellites. Its 1S derivative carries a greater propellant loading and is intended for spin-stabilised platforms. Mage 2 can accommodate propellant loadings of up to 490 kg and is used by 500–680 kg payloads. As lead, SEP is responsible for 55 per cent of the programme, BPD for 35 per cent (propellant, thermal protection, test firings), and Germany's MAN for 10 per cent (structure/casing). See pp 242 1995–96 for full details.

Composite Motors

SEP's composite activities include development of carbon-ceramic structures for the Hermes spaceplane and SEPCarbinox and CeraSEP (carbon Novoltex/silicon carbide) composite liquid rocket engine nozzles. Trials of an Ariane HM-7B with a composite nozzle were successfully performed in 1989. Endurance trials were performed on a range of NTO/MMH CeraSEP thrusters into 1989: 5 N (50 h),

200 N (3 h) and 6.2 kN (925 s). Nozzle temperatures of 1,700K were routinely maintained. SEP has been working since 1988 on an 8 kN pressure-fed engine featuring an advanced injector and CeraSEP chamber/bell. Propellants are fed through 509 holes: 89 pentads (central fuel + four surrounding oxidixer holes) + 32 doublets. The pentads intersect for efficient propellant mixing, while the doublets spray a fuel-rich film against the chamber wall. Specifications are:

Applications: launcher upper stages and space vehicle manoeuvring
Dry mass: 20 kg (excluding actuators)
Length: 1.5 m
Maximum diameter: 150 cm
Mounting: gimballed for pitch/yaw control
Oxidiser: NTO at 1.557 kg/s
Fuel: MMH at 0.943 kg/s
Propellant feed pressure: 21.8 atm
Mixture ratio (O/F): 1.65:1
Thrust: 8 kN vacuum
Specific impulse: 320 s vacuum
Expansion ratio: 150:1
Cooling method: spray film and radiative
Combustion chamber:
pressure: 14.8 atm
ignition: hypergolic
materials: C/SiC ceramic matrix
Burn time: 1,000 s total.

A single Vulcain cryogenic engine (left) powers Ariane 5's first stage. Vulcain MkII (right) is being developed to improve Ariane 5 performance (SEP) 0517354

VERIFIED

SNPE

Le Bouchet Research Center
9 rue Lavoisier, BP2, F-91710 Vert le Petit.
Tel: +33 1 64 99 12 34
Fax: +33 1 64 99 13 55
Web:http://www.snpe.fr
Chairman/CEO: Jean Faure
Advanced Technologies & Propulsion: Jacques Cardin
Space Propulsion Programs: David Quancard
Space Propulsion Programs
BP57, F-33166 Saint-Médard en Jalles Cedex.
Tel: +33 5 56 70 50 16
Fax: +33 5 56 70 50 52
Pyromeca
Chemin Charles Battezzati, Quartier Lagoubran,
BP2148, F-83063 Toulon.
Tel: +33 4 94 22 86 86
Fax: +33 4 94 22 86 99

The Société Nationale des Poudres et Explosifs SNPE Group, created as a state-owned company in 1971 and now with 5,000 personnel and FFr4.4 billion 1994 turnover (FFr100 million profit), manufactures liquid rocket propellants, with emphasis on UDMH, and develops/produces solid propellants. Along with BPD Difesa e Spazio in the Regulus partnership (see the international propulsion section), the company provides the propellant for Ariane 5's P230 solid boosters using facilities at Kourou constructed by subsidiary SNPE Ingénierie. SNPE maintains facilities at Saint-Médard en Jalles near Bordeaux (solid propellant grains), Toulouse (UDMH, AP), Toulon (pyrotechnics) and Le Bouchet near Paris (HQ and R&D). The Toulouse site manages the production of ammonium perchlorate with a capacity of 6,000 t annually to ensure Ariane 5 supplies. SNPE formed a second partnership with BPD, Société Européenne de perchlorate d'ammonium (EUPERA; SNPE 66 per cent, BPD 34 per cent), to handle

746 Viking engines of all types have been produced through 1995 (SEP) 0516898

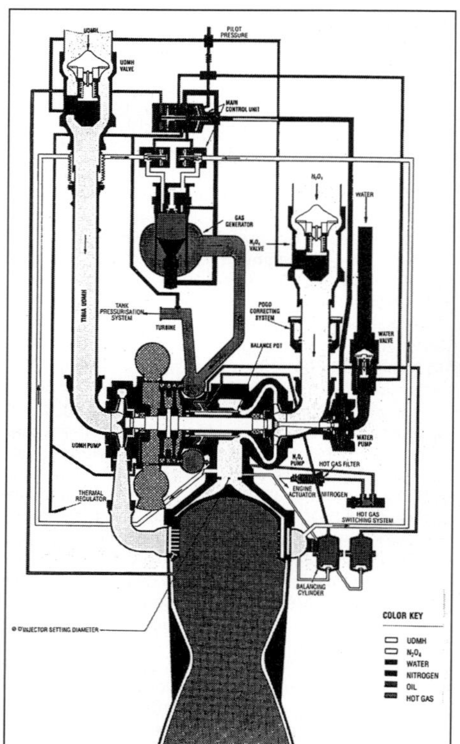

Viking 5 is clustered to power Ariane's first stage 0516899

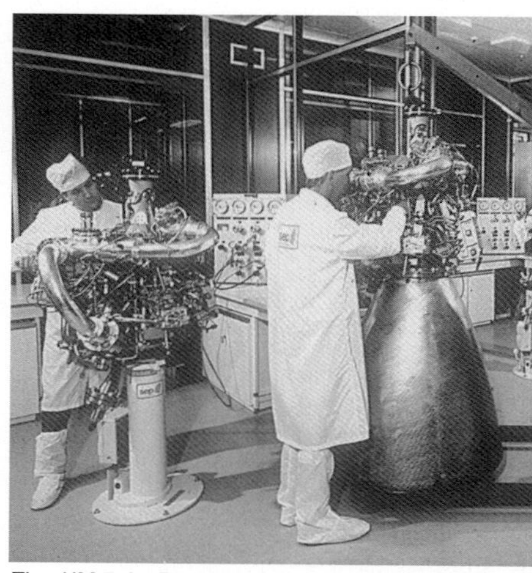

The HM-7 is Europe's first liquid hydrogen/oxygen engine. The 63.8 kN version debuted in Dec 1994. Future developments could include a composite nozzle (SEP) 0517185

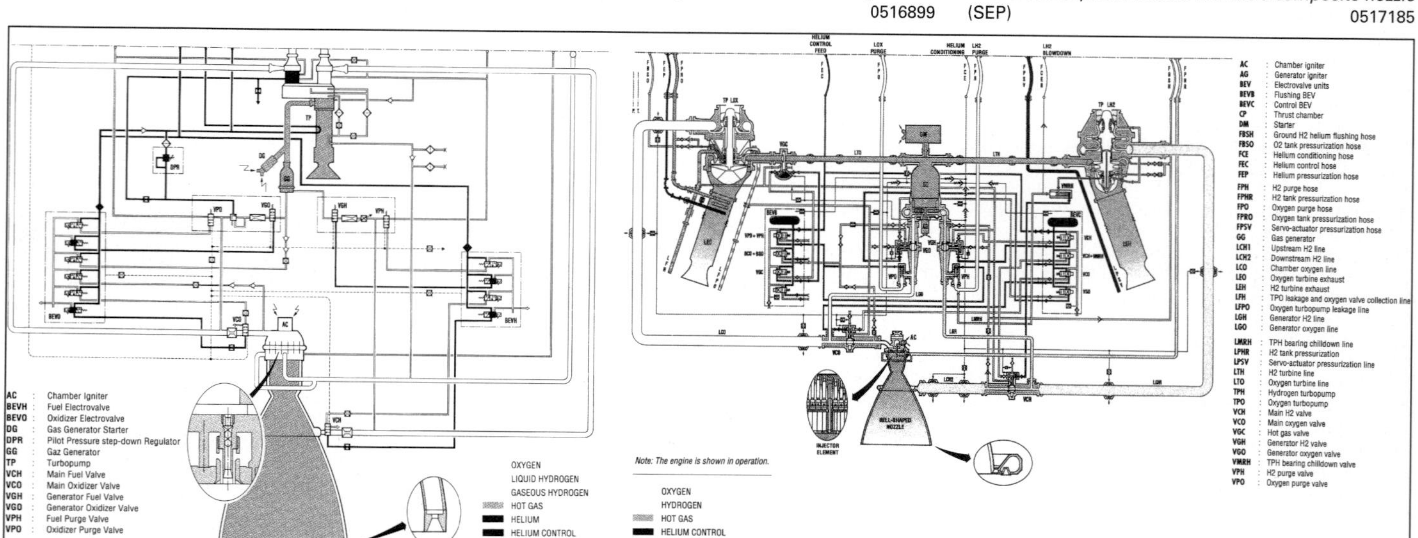

Left: HM-7B engine flow diagram. Hydrogen is tapped off for the attitude control thrusters. Right: Vulcain propellant flow (SEP) 0517191

A Vulcain 2 engine in a test and evaluation stand (SEP) 0572505

production. Toulouse also provides the UDMH for Ariane 4 and the MMH for Ariane 5's upper stage. Subsidiary Pyromeca manufactures detonating cords and flexible linear cutting charges and ultra-fast pyrotechnic valves for Ariane 4 release from the launch pad. See also the Pyrospace entry. In cryogenic propulsion, SNPE provides the main igniter and turbine starter grain for Ariane 4's HM-7B stage 3 engine and the shaped ducts for the Ariane 5 Vulcain igniter and starter.

The propulsion package was designed/produced for the 153/227 kg Eclipse T1/T2 supersonic rocket and autonomous launch ramp in partnership with CAC Systèmes. Applications include meteorological sounding and µg research. T1/T2 capacities are 20/25 kg to 12/72 km.

Created a small research division to develop combination propellants, tri-propellants and cryogenic systems applicable to the new ESA launch vehicle futures study. More associated with optional upper stages.

STARSEM SA

33 rue Fernand Forest
F-95150 Suresenes.
Tel: +33 1 46 25 03 60
Fax: +33 1 45 06 05 78

President: Francois Calaque (Managing Director of Aerospatiale's Space Business Unit)
Director-General: Viktor Kuznetsov (Russian Space Agency)
Commercial Operations: Jean-Charles Vincent (Arianespace)

STARSEM is a joint-venture company, set up to market the Soyuz family of launch vehicles on a commercial basis. The partners are Aerospatiale and Arianespace in France, the Russian Space Agency (RSA) and the Samara Space Center (comprising the TsSKB design and engineering bureau and the Progress production facilities) which produces the launch vehicles. The agreement for the creation of STARSEM was signed by the partners on 17 July 1996 in Moscow and on 6 August the statutes of the company were signed in Paris. The company is a French-registered corporation, set up with FFr500,000. The backing for the project is split with EADS having 35 per cent, Samara and RSA 25 per cent each and Arianespace 15 per cent. The Starsem Baikonur facilities include Site 112 and Site 31. Site 112 is the location of the assembly and integration facility for the former Energia launch vehicle and houses the dedicated clean rooms and provision for the customer payload and mating with the Fregat upper stage. Site 31 includes the launch area, assembly and integration facilities and administrative buildings.

The company proposes to market three basic launch vehicles: the current Soyuz-U, the current Molniya-M and a new Soyuz-U/Ikar version which adds a new upper stage to the existing Soyuz-U.

The Soyuz ST is being introduced in 2003, adding increased payload volume and a digital control system.

To date 10 Soyuz flights have been managed by Starsem, the first on 9 February 1999 and the most recent on 9 August 2000.

Germany

Daimler-Benz Aerospace AG

Previously DASA, Daimler-Benz Aerospace AG (DBAA) is the main contractor for Ariane's second stage and the PAL liquid strap-on boosters, and provides the thrust chamber for the HM-7B third stage engine; it similarly provides the injection

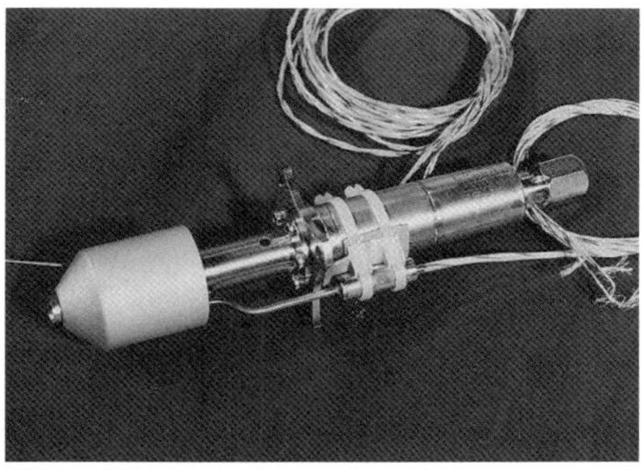

CHT 1.0 N hydrazine thruster for Globalstar 0517374

head, combustion chamber and expansion bell for Ariane 5's Vulcain first stage engine, under a contract signed with SEP in Apr 1989. During 1991–98, DBAA is providing 50 second stages and 120 PALs under a DM1,600 million contract, in addition to 50 HM-7B thrust chambers under a DM52 million award. Another 15 stage 2s and 26 PALs have been ordered for 1998–99. The company leads Ariane 5's EPS upper stage, the first development of an Ariane stage outside of France. 14 will be built 1997–2002. DBAA, as MBB, also built the P3.2 test stand for Vulcain chamber trials. Ariane engine specifications are included in France's SEP company entry. Its other engines include storable bipropellant (4, 10, 400, 27,500 N), hydrazine monopropellant (0.02–350 N) and electric (10–15 and 150–200 mN).

As MBB, the company began development of the world's first 50 kN topping cycle engine in 1958 and began testing at 85 atm chamber pressure in 1963. In 1968, MBB's 180 kN LOX/LH_2 engine attained the existing chamber pressure record of 282 atm, leading to NASA and Rocketdyne acquiring a licence for utilising the technology in the Space Shuttle Main Engine. The current 10 N and 400 N NTO/MMH thrusters are combined to form a Unified Propulsion System (12×10 N + 1×400 N), employed by the Galileo Jupiter orbiter and the TV-Sat/TDF, Tele-X, DFS Kopernikus, Eutelsat 2, Turksat and Amos telecom satellites. Bipropellant 4 N thrusters are currently under development and qualification.

Facilities: DBAA's Lampoldshausen and Ottobrunn sites are equipped with test stands for high pressure thrust chambers up to 1 MN (P3.2 stand for Vulcain) and for gas generators and turbopumps with supply pressures up to 800 atm, stands for LH_2 ramjets, sea level and vacuum stands for 4 N to 50 kN bipropellant and 0.5–500 N monopropellant engines (Trauen site), and a surface tension tank propellant behaviour simulation facility.

Monopropellant Hydrazine Thrusters

DBAA has developed a 0.5–2,500 N range of monopropellant catalytic hydrazine thrusters (CHTs), together with complete mono propulsion systems. It has produced >550 units for satellites such as Skynet 4, ECS, Telecom, Giotto, Ulysses and Hipparcos.

CHT 0.5 (–1 straight nozzle, –2 90°)

Applications: OTS 2, ECS, Telecom 1, Skynet 4, NATO 4
First flight: May 1978 aboard OTS 2
Number produced: 160 units
Dry mass: 0.19 kg
Length: 113.6 mm
Mounting: fixed
Engine cycle: decomposition of hydrazine over heated catalyst bed of Shell 405 ABSG or KC-12 GA
Thrust: 0.2–0.75 N vacuum (0.5 N nominal)
Specific impulse: 216–227 s vacuum
Chamber pressure: 5.5–22 atm
Power requirement: 5 W valve, 2.5 W heater

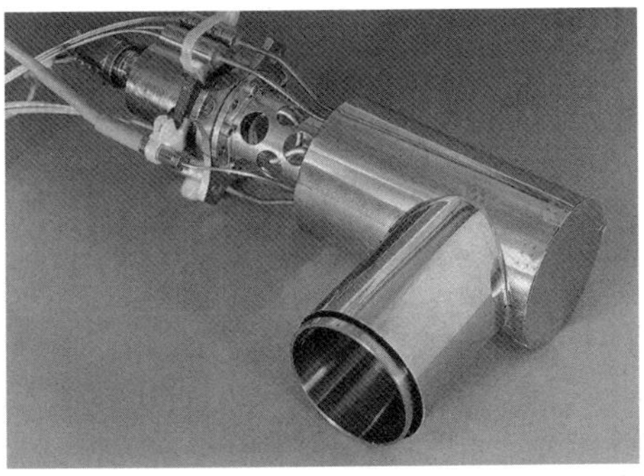

DASA's 20 N hydrazine thruster is available in straight and 90°-canted nozzle configurations (DASA) 0517375

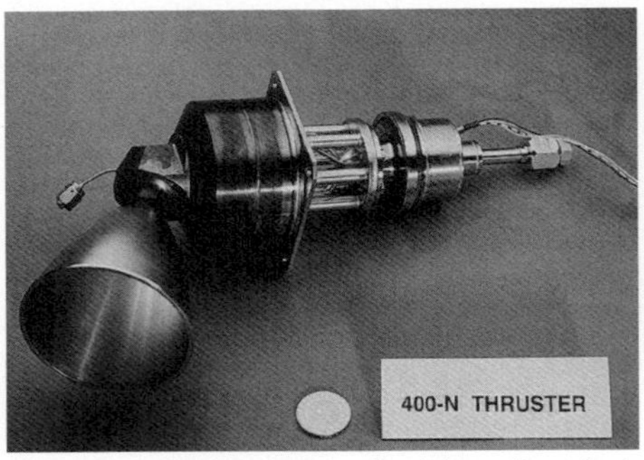

CHT 400 hydrazine thruster for Ariane 5's attitude control system (DASA) 0517376

Burn time: qualified for 25.2 s single burn, 143 h total
Minimum impulse bit: 0.005–0.015 Ns

CHT 2.0 (–1 straight nozzle, –2 90°)
Applications: OTS 2, ECS, Telecom 1, Giotto, Ulysses, Skynet 4, Meteosat MOP, ISO
First flight: May 1978 aboard OTS 2
Number produced: 211 units
Dry mass: 0.200 kg
Length: 123.9 mm
Mounting: fixed
Engine cycle: see CHT 0.5
Thrust: 0.6–2.0 N vacuum
Specific impulse: 214–227 s vacuum
Chamber pressure: 5.5–22 atm
Power requirement: 5.0 W valve, 1.0 W heater
Burn time: qualified for 36 s single burn, 22.5 h total
Minimum impulse bit: 0.015–0.036 Ns

CHT 5.0
Applications: Skynet 4, NATO 4, Hipparcos
First flight: Dec 1988 aboard Skynet 4B
Number produced: 69 units
Dry mass: 0.220 kg
Length: 129.3 mm
Mounting: fixed
Engine cycle: see CHT 0.5
Thrust: 1.85–6.0 N (5.0 N nominal) vacuum
Specific impulse: 216–228 s vacuum
Chamber pressure: 5.5–22 atm
Power requirement: 5.0 W valve, 3.37 W heater

DASA has developed a second generation 400 N MON/MMH bipropellant thruster 0517341

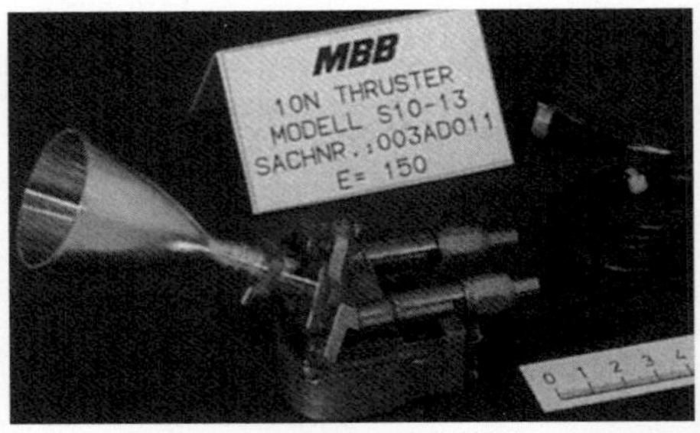

DASA's 10 N bipropellant thruster (DASA) 0516900

Burn time: qualified for 3.6 s single burn, 12.5 h total
Minimum impulse bit: 0.038–0.096 Ns

CHT 10.0
Applications: Meteosat MOP series, SAX
First flight: Mar 1989 aboard Meteosat 4
Number produced: 51 units
Dry mass: 0.240 kg
Length: 142.34 mm
Mounting: fixed
Engine cycle: see CHT 0.5
Thrust: 3.0–10.0 N (10.0 nominal) vacuum
Specific impulse: 220–230 s vacuum
Chamber pressure: 5–22 atm
Power requirement: 5.0 W valve, 2.4 W heater
Burn time: qualified for 1.5 s single burn, 3.4 h total
Minimum impulse bit: 0.070–0.190 Ns

CHT 20.0
Applications: Eureca retrievable platform; –1 version utilises constant pressure, –2 is blow down version
First flight: July 1992 on Eureca 1
Number produced: 46 units
Dry mass: 0.360 kg
Length: 195.74 mm
Mounting: fixed
Engine cycle: see CHT 0.5
Thrust: 7.2–24.0 N (20 N nominal) vacuum
Specific impulse: 222–235 N vacuum
Chamber pressure: 5–22 atm
Power requirement: 13.0 W valve, ≤6.0 W heater
Burn time: qualified to 3.6 s single burn, 4 h total
Minimum impulse bit: 0.165–0.370 Ns

CHT 400
Applications: Ariane 5 attitude control system
First flight: June 1996
Number produced: 53 units to end-1995
Dry mass: 1.8 kg
Length: 307 mm
Mounting: fixed
Engine cycle: see CHT 0.5
Thrust: 110–450N (350N nominal) vacuum
Specific impulse: 218–235 s vacuum
Chamber pressure: 5.5–22 atm
Valve power: 70 W
Minimum impulse bit: 5 Ns

Bipropellant (MON/MMH) Thrusters
DBAA has developed a 4–27,500 N range of bipropellant thrusters, together with fill systems and fill/drain valves. 600 units have been produced for satellites such as Galileo, TDF, Telecom 2, Hispasat, Turksat, Amos and Cluster.

Aestus thrust chamber assembly (DASA) 0516901

DASA's propulsion system for telecommunications satellites incorporates a single 400 N and 14 10 N bipropellant thrusters (DASA) 0516902

Aestus Ariane L9 Engine
DBAA is responsible for development of Ariane 5's L9 EPS upper stage, the first development of an Ariane stage outside of France. Development firings were completed in late 1994, totalling almost 11,000 s on six engines in >800 tests, including a single 1,380 s run. EPS achieved its first nominal duration firing, 1,075 s, 5 Oct 1994. Engine qualification firings started Dec 1994 for completion Mar 1995. Pre-development work began in Jan 1995 on uprating Aestus to 35 kN, including increasing the number of injector elements to about 160, for potential future Ariane 5 upgrades.
First flight: June 1996 (8 to end-2000)
Dry mass: 111 kg
Length: 2,195 mm
Max diameter: 1,263 mm
Mounting method: gimballed ±6° by electro-mechanical actuators
Engine cycle: pump-fed (feed pressure 18.0 atm)
Oxidiser: nitrogen tetroxide at 5.89 kg/s
Fuel: monomethyl hydrazine at 2.87 kg/s
Mixture ratio: 2.05:1
Thrust: 27.5 kN vacuum
Specific impulse: 324 s vacuum (potential 331 s)
Expansion ratio: 83.3:1
Combustion chamber:
pressure: 10.42 atm
cooling: regeneratively (fuel) cooled nozzle to 10:1 area ratio
materials: stainless steel inner liner with milled cooling passages and electrolytically deposited nickel layer for outer closure
injector: 132-element coaxial arranged in spirals. Fuel enters injector from the chamber cooling channels. Oxidiser is distributed to elements via manifold at rear
Nozzle:
cooling: radiation cooled extension below 10:1 area ratio.

S400 Engine
The 400 N bipropellant thruster is incorporated in the DFS, TV-Sat, Tele-X, TDF, Eutelsat 2, Amsat, Turksat, Amos 1, Cluster and Artemis propulsion systems and provides the Galileo Jupiter probe's primary propulsion capability. See also the S10 specifications. DASA has developed a second generation engine under ESA contract to increased SI to 317 s. Performance optimisation is underway for formal qualification by end-1996. Modifications: the pilot operated pneumatic valves were replaced with Moog 53–193 single seat dual coil valves; the combustion chamber was completely redesigned (regenerative cooling loop eliminated, chamber pressure increased, throat section recontoured and made from platinum/rhodium alloy); reduced throat diameter increased expansion ratio to 220 without changing engine length or exit diameter. It has been tested over sea level and vacuum conditions over 1.2–1.8 mixture ratio and 110–145 g/s flow rates. Pulse mode demonstration tests were performed with on-times of 100, 250, 500 and 800 ms and off-times of 1,000 and 100 ms. Specifications are provided in brackets.
Dry mass: 2.8 kg (3.120 kg)
Length: 531 mm (498 mm)
Max diameter: 248 mm (248 mm)
Mounting method: fixed
Engine cycle: pressure fed, at 18 atm (inlet presure 13.8 atm)
Oxidiser: MON-1 (NTO + 1%NO)
Fuel: monomethyl hydrazine
Mixture ratio: 1.64:1 (1.65:1)
Thrust: 400 N (404 N, flow rate 130 g/s)
Specific impulse: 308 s (317 s)

Expansion ratio: 102/150 (220)
Expansion bell: stainless steel (stainless steel and platinum-rhodium for higher heat loads)
Chamber pressure: 6.9 atm (9.87 atm)
Chamber materials: stainless steel using MMH regenerative cooling (stainless steel upper and platinum-rhodium lower radiative cooling)
Injector: stainless steel swirl injector, double cone

S10 Thruster
This 10 N (nominal) bipropellant thruster is teamed with the 400 N thruster (see above) in the Galileo, DFS, TV-Sat, TDF, Tele-X, Eutelsat 2, Turksat, Amos 1 and Cluster propulsion systems. >100 were also provided for Inmarsat 2, Telecom 2 and Hispasat. A second generation version, without regenerative cooling, has been qualified; its different specifications are noted in (). Cluster, Amos 1 and Artemis are users.
Dry mass: 0.350 kg
Length: 138 mm
Diameter: 67 mm
Mounting method: fixed
Engine cycle: pressure fed
Oxidiser: MON
Fuel: monomethyl hydrazine
Mixture ratio: 1.55–1.72 over thrust range
Thrust: 8–11.4 N vacumm
Specific impulse: 287 s (292 s)
Expansion ratio: 90 (150)
Chamber pressure: 18–12 atm over thrust range

S4 Thruster
The S4 will be utilised by the next generation of telecom satellites.
Dry mass: 0.270 kg
Length: 105 mm
Diameter: 29 mm
Mounting method: fixed
Engine cycle: pressure fed, at 5–25 atm
Oxidiser: MON
Fuel: 4 monomethyl hydrazine
Mixture ratio: 1.5–1.85 over thrust range
Thrust: 1.77–5.62 N
Specific impulse: 290 s
Expansion ratio: 190

S3K Engine
S3K is a technology programme that could form the basis for a future orbital transfer vehicle.
Dry mass: 14.5 kg
Length: 1,030 mm
Max diameter: 530 mm
Mounting method: fixed
Engine cycle: pressure fed
Oxidiser: MON 3
Fuel: monomethyl hydrazine
Mixture ratio: 1.6–2.1
Thrust: 3,500 N
Specific impulse: 352 s
Expansion ratio: 125
Chamber pressure: 9–12 atm

VERIFIED

Institut für Raumfahrtsysteme

Universität Stuttgart
Pfaffenwaldring 31, D-70550 Stuttgart.
Tel: (+49 711) 685 23 78
Fax: (+49 711) 685 35 96
Web: http://www.irs.uni-stuttgart.de

Senior staff
Director: Prof Dr-Ing habil Monika Auweter-Kurtz
Deputy: Dr Ulrich Schöttle

IRS's Artus 1.5 kW thermal arcjet suspended from a pendulum thrust balance. Nozzle temperatures reach 1,600K under steady state firing 0516903

The Space Systems Inst (IRS) was founded in 1970 and plasma thruster development has always been a principal interest. Facilities allow MW-class testing at 0.5–2 g/s propellant flow rates (argon) under selected 10^{-6}–10^{-3} atm conditions. 0.5–1 MW MPD stationary thrusters of different geometries have been developed, mainly under AFOSR grants. Based on this experience, IRS has designed & built a range of thermal arcjet thrusters. Current arcjet projects include: Atos 600–700 W ammonia, Artus 2 1–2 kW NSSK hydrazine (with DASA; DARA contract); 10 kW & 100 kW devices. Atos is a simplified version of Artus 2 for flight on Amsat amateur satellite.

Four plasma wind tunnels are used to investigate re-entry vehicle thermal protection materials and to validate aerothermodynamic CFD codes. Two have MPD plasma generators (0.10–1 MW) and are especially suited for high specific enthalpy (up to 150 MJ/kg) and low total pressure conditions. One is equipped with a thermal plasma generator for higher total pressure conditions and one is equipped with a thermal plasma generator for higher total pressure and low specific enthalpy areas. The fourth plasma wind tunnel has an inductive plasma generator which is used for catalycity investigations simulating entry into planetary atmospheres with reactive gases.

The Mission and System Analysis division of IRS addresses the development and application of numerical simulation tools employing analysis, CAD and optimisation methods for flight performance prediction and post-flight evaluation, for subsystem development (for example propulsion, thermal protection, guidance and control, flight experiments), and for preliminary design of future space transportation systems (launch vehicle, orbit transfer and reentry systems). In order to determine mission requirements, the interaction between system properties, flight controls and trajectories, mechanical and aerothermodynamic flight loads and vehicle performances are investigated together with asociated system design needs. Advanced plasma-phase, radiation and plasma-surface-interaction models are developed within the aerothermodynamics division. They are implemented into Navier-Stokes Solvers which are used in numerous research contracts for the numerical simulation of nonequilibrium flows in plasma wind tunnels and around reentry vehicles for interplanetary and LEO missions.

Artus 2 typical operation specifications
Thrust: 0.188 N
Power level: 1.5 kW at 18 A
Mass flow: 35 mg/s simulated hydrazine (2:1 H_2:N_2)
Thrust efficiency: 33%
Specific impulse: 537 s

VERIFIED

Man Technologie AG

Address
PO Box 10 22 35
D-86012 Augsburg
Germany
Tel: (+49 82) 15 05 01
Fax: (+49 82) 15 05 10 00
e-mail: SpaceTransport@mt.man.de

Infrastructure Systems
P.O. Box 34 80
D-55024 Mainz
Germany
Tel: (+49 61) 31 21 55-0
Fax: (+49 61) 31 21 55-20 2

Senior Staff
Executive Board Members:
Wolfgang Brunn (Chairman)
Walter Köppel
Carl-Ferdinand Kolbow
Representative of the Board for External Relations:
Alex Gerhard
Aerospace Sales, Marketing, Projects, Engineering:
Walter Köppel
Aerospace Production: Horst Choinowski
Business Administration, Personnel: Carl-Ferdinand Kolbow
Public Relations: Ingrid Wagner
Space projects represent >50% of MT's turnover: system studies, development and manufacture of advanced metallic and composite lightweight structures and mechanisms, thermal protection units, mobile satcoms, satnav and system monitoring (see the Navigation entry and the System Electronics Division entry in the Satcom section). Studies with MT acting as ESA's prime contractor include the definition of the AROC recoverable μg capsule, the NAVSAT phase B1 system study, and the definition

Ground test version of the Vikas engine to be used on the strap-on boosters for the GSLV: the original Vikas variant flew on the second stage of the PSLV (Liquid Propulsion Systems Centre) 0003431

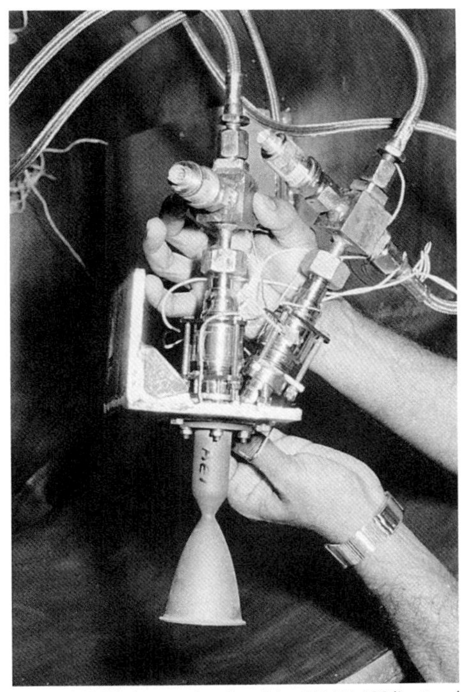

The 22 N RCS thruster (top) and the 440 N LAM (bottom) carried on INSAT 2 (Liquid Propulsion Systems Centre) 0003444

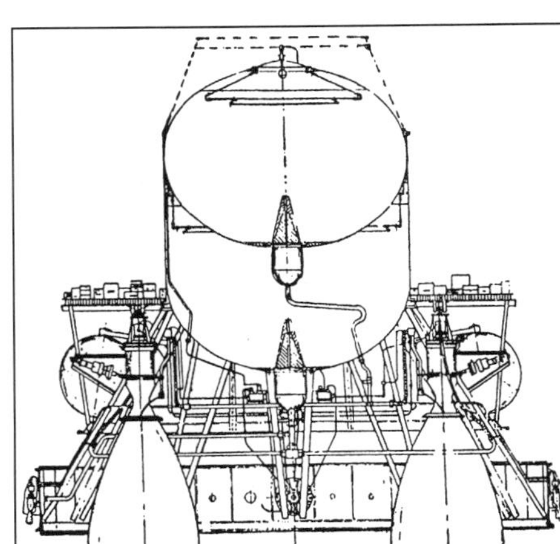

PSLV's stage 4 incorporates two 7.5 kN engines and, for coast attitude control, six 50 N thrusters. Surrounding the tank is PSLV's vehicle equipment bay. On top is IRS 1E (ISRO) 0516906

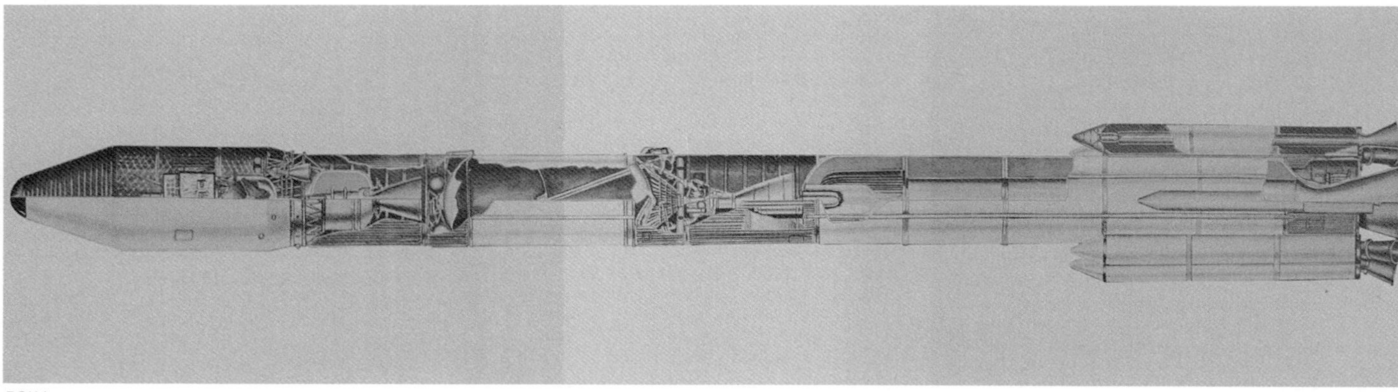

PSLV 0517189

PSLV stage 2: ISRO developed the Vikas engine from Ariane's Viking to power the first large Indian liquid stage. The toroidal water coolant tank is visible inside the base skirt, with solid ullage and retro motors on the exterior (ISRO) 0516905

PSLV provides a 1 t Sun-synchronous capacity (ISRO) 0516904

and development of an experimental health monitoring unit for GPS and Glonass. Structural components include apogee boost motor casings (see below), CFRP shells for Rosat, Italsat high pressure gas tanks, the CFRP structure for Orfeus, Silex structures, and radio astronomy high precision reflectors. New structures research and engineering development for high stress resistant carbon composite and hybrid materials applications.

For Ariane 4, MT Space Systems provides the stage 1 thrust frame, the load-bearing gas generator water coolant tank (fibre composite/aluminium, 30%/150 kg less than conventional design), Viking turbopumps and gas generators (see below), aft/forward booster adaptors, and booster separation and release mechanisms. In Dec 1996 MT took over the production of the second stage main tank.

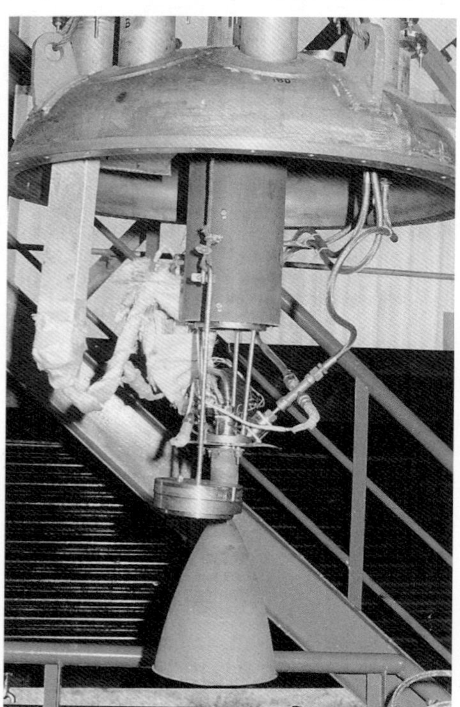

The 22 N RCS thruster (top) and the 440 N LAM (bottom) carried on INSAT 2 (Liquid Propulsion Systems Centre) 0003445

MT has production facilities in Augsburg, Munich and Oberpfaffenhofen.

For Ariane 5, MT has developed and produced the SRM booster cases, aluminium front skirt and fibre composites, high-pressure vessels, main and upper cryogenic stage bulkheads, thermal protection shields for main and boost stages, and storable second-stage propellant tanks. For ESA, MT is investigating reusable launch vehicles and related R&D. For CNES, MT designed AROC, the Ariane Recoverable Orbital Carrier, MT is also conducting in-house studies for a liquid fly-back reusable booster.

MT has delivered >850 Viking turbopumps and associated gas generators at a production rate of 60 units annually. The two single-stage centrifugal pumps use 8-blade steel wheels to move 270 kg/s propellant while raising its pressure from 5 to 57 bar. The Curtis velocity-compounded impulse turbine incorporates two rows of rotating blades to drive the shaft from the gas generator's water-cooled exhaust. The generator's spherical chamber is cooled by eight water jets facing the propellants' 40 injection ports. The turbopumps also serve as part of the thrust structure. Key characteristics are:

The liquid propellant engine, two of which are carried on the fourth stage of the PSLV (LPSC) 0003446

Viking engine	4B	5C	6
Propellant flow (kg/s)	278.1	276.4	276.4
Turbine speed (rpm)	10,000	10,000	10,000
Turbine power (kW)	2,500	2,500	2,500
Turbopump mass (kg)	331	325	326
Gas generator mass (kg)	28	28	28
Gas flow (kg/s)	6	7	7

MT's apogee boost motor cases consist of a pressure reservoir precision-wound from high strength aramide fibres, with two different pole openings for the nozzle and the cylindrical transition structure for spacecraft mounting. Key casing characteristics are:

	Mage 1	Mage 1S	Mage 2	IRIS
Mass (kg)	9	11.4	13.7	53
Volume (l)	212	260	309	985
Length (mm)	628	728	845	1,020
Diameter (mm)	766	766	766	1,310

India

Propulsion

Current Status

The first development flight of the GSLV, GSLV-D1, was successfully completed on 18 April 2001. The second development flight, GSLV-D2, was launched successfully on 8 May 2003 carrying the GSAT-2 satellite. Developed with assistance from Russia, the liquid propellant strap-ons are the L37.5 Vikas motors used in the PSLV.

Background

India's original orbital launchers utilised solid propellant main motors, drawing on liquid systems only for interstage control systems. The PSLV, however, incorporates two major liquid engines. ISRO's Liquid Propulsion Systems Centre with headquarters in Trivandrum and units in Bangalore and Mahendriagiri, is responsible for R&D of all Indian liquid motors, from satellite attitude control thrusters to the 725 kN Vikas. Solid motors are developed by the Vikram Sarabhai Space Centre at Trivandrum and the SHAR Centre.

Air-breathing technology for the ABR 200(I) engine was flight tested 14/19 Oct 1992 on sounding rockets launched from Sriharikota.

Liquid Propulsion Systems Centre
Valiamala, Trivandrum 695547.
Tel: +91 471 572257
Telex: 0435–396 LPSCIN
Fax: +91 471 462686
Director: Mr G Madhavan Nair
LPSC is responsible for the development of launcher liquid and cryogenic propulsion stages and auxiliary propulsion for launch vehicles and spacecraft. It is growing in importance as main liquid engines are introduced to India's orbital launchers and with the increasing size of indigenous satellites. Current main projects cover the 720 kN Vikas PSLV stage 2 engine, PSLV's stage 4 dual 7.5 kN system, Insat 2's 440 N Liquid Apogee Motor and unified network of 22 N ACS thrusters, and the 680 kN Vikas GSLV strap-on. Seven Russian 76 kN cryogenic engines are being provided for the initial GSLVs beginning 1998, but India's own engine will complete development in 1999.

LPSC's test facilities are sited southeast of Trivandrum at Mahendragiri. Vikas' Principal Test Stand (PTS) was commissioned during 1987 and used in Jan 1988 for the engine's first full-duration 150 s firing. Altitude facilities are also available for PSLV's 7.5 kN motor, Insat 2's LAM and smaller thrusters. Cryogenic engine and stage facilities will be commissioned in 1997.

Insat 2 Propulsion System
Insat 2's 119 kg (dry mass) unified bipropellant liquid propulsion system comprises a single 440 N LAM Liquid Apogee Motor and two redundant networks of 8 × 22 N RCS thrusters. The qualification LAM was tested for a total 9,550 s, including a single burn of 3,000 s. RCS testing accumulated 30,500 s including a single continuous burn of 10,000 s. Pulsed firing mode logged <250,000 pulses, each component being tested over three life cycles. The engines, pressure regulator, check valve, fill, drain & vent valve, pyro valve and pressure transducer were developed indigenously; the tanks, gas bottles, filters and latch valves were procured from foreign suppliers. The system carries about 450 welded joints.

LAM
First flight: July 1992 (Insat 2A)
Dry mass: 4 kg
Length: 570 mm
Maximum diameter: 276 mm
Mounting: fixed
Engine cycle: pressure-fed in regulated mode
Propellants: MMH/NTO from tanks at 16 atm
Mixture ratio (O/F): 1.65
Thrust: 440 N vacuum nominal
Specific impulse: 310 s vacuum minimum
Expansion ratio: 160
Chamber pressure: 6.9 atm
Injector: coaxial titanium welded to chamber
Combustion chamber: silicide-coated columbium alloy radiatively cooled
Burn time: qualified to 3,000 s continuous; minimum impulse bit not available

Insat 2 RCS Thruster
First flight: July 1992 (Insat 2A)
Dry mass: 850 g
Length: 249 mm
Maximum diameter: 60 mm
Mounting: fixed
Engine cycle: pressure-fed in blowdown mode
Propellants: MMH/NTO from tanks at 16 atm
Mixture ratio (O/F): 1.65
Thrust: 22 N vacuum nominal
Specific impulse: 285 s vacuum minimum
Expansion ratio: 100
Chamber pressure: 6.9 atm
Throat temperature: 1,200°C
Injector: coaxial titanium welded to chamber
Combustion chamber: silicide-coated columbium alloy radiatively cooled
Burn time: qualified to 1,000 s continuous; minimum impulse bit not available

PSLV stage 4 engines
PSLV's final stage carries two identical gimballed liquid engines to provide guided injection.
First flight: 20 Sep 1993
Dry mass: 28 kg
Length: 1.1 m
Maximum diameter: 0.63 m
Mounting: gimballed up to ±3° in two orthogonal planes by a closed loop servo control system to provide stage pitch, yaw and roll control during thrust phase
Engine cycle: pressure-fed
Propellants: MMH/NTO from tanks at 19.2 atm
Mixture ratio (O/F): 1.4
Thrust: each 7,500 N vacuum nominal
Specific impulse: 308 s vacuum minimum
Expansion ratio: 60
Chamber pressure: 8.37 atm
Injector: AISI 304 stainless steel; 45 sets of triple elements
Combustion chamber: AISI 304 stainless steel film cooled along single helical grooving
Nozzle: silicide-coated columbium alloy radiatively cooled
Burn time: 425 s mission nominal
Qualification time: 530 s (continuous)

PSLV L37.5 Vikas engine
India's first large liquid propellant rocket engine is derived from Viking technology acquired from SEP. The engine achieved its first full-duration 150 s firing in Jan 1988 on the Principal Test Stand at the Liquid Propulsion Test Facility, Mahendragiri, and completed the first integrated stage test 21 Mar 1990 (qualification completed Oct 1992). A 686 kN version will be used for GSLV's strap-ons. A successful 200 s run 24 July 1995 tested the new indigenously-developed silica-phenolic throat, designed to withstand the operational 1,300°C.
First flown: 20 Sep 1993
Number flown: 3 to end-1996
Dry mass: 876 kg
Length: 3.509 m
Mounting: gimballed for pitch/yaw control
Engine cycle: gas generator (9400 rpm turbine)
Propellants: UDMH/NTO
Mixture ratio (O/F): 1.86
Thrust: 72 kN vacuum (686 kN for GSLV)
Specific impulse: 295 s vacuum
Time to full thrust: 2.4 s
Expansion ratio: 31 (13.88 C-SU strap-on)
Chamber pressure: 51.9 atm
Injector: 216 like on like doublets set in six rows for each propellant in light alloy annular injector
Chamber cooling: UDMH film supplied through additional channels on the injector's lower section
Igniter: hypergolic
Nozzle materials: cobalt alloy with SEPHEN (phenolic resin/silica fibre) throat
Burn time: 160 s nominal mission

Solid Propulsion Group
Vikram Sarabhai Space Centre
Trivandrum 695022.
Tel: +91 471 562 444/562 555
Telex: 0435-201, 0884-202
Deputy Director: R Nagappa
India's first, 75 mm dia, solid motor was produced in 1967, followed by a 125–560 mm range for sounding rocket applications and, in the early 1970s, motors for the SLV-3 satellite launcher. Work on the most powerful so far began in 1984: the 2.8 m dia 3,500 kN thrust solid powers PSLV's first stage. Although the Solid Propulsion Group is an element of VSSC, the primary Solid Propellant Space Booster Plant (SPROB) is sited at SHAR Centre on Sriharikota Island, along with the Vehicle Assembly Static Test & Evaluation Complex (VAST) for solids. At VSSC the motor cases are produced in the Mechanical Engineering Facility and the propellant binders in the Centre's Propellant Fuel Complex.

PSLV stage 1 motor
PSLV's first stage is the most powerful model produced by ISRO; it was successfully static fired for the first time 21 Oct 1989. The second was succesfully fired 23 Mar 1991.
Application: PSLV stage 1
First flight: 20 Sep 1993
Number flown: 8 to end October 2003
Mass (kg): 145,000
Length (cm): 2,034.5
Diameter (cm): 280.4
Propellant
type: HTPB, segment-cast with three central segments interchangeable
shape: forward segment deep 10-fin to burn 22 t loading in 20 s to generate high launch thrust; remainder tubular with avg 120 cm dia bore
mass fraction: 0.890 (129 t loading)
Burn time (s): 100
Thrust (kN, vacuum): 3,463 mean, 4,600 max at 17 s
Specific impulse (s, vacuum): 265
Total impulse (kNs, vacuum): 326,000
Pressure (atm): 58 max at 17 s
Nozzle
throat diameter (cm): 80
length (cm): 2,034.8
exit diameter (cm): 227
expansion ratio: 8
materials: carbon phenolic throat insert, tape-wound carbon phenolic on divergent cone changing to silica phenolic below TVC inlets
Casing materials: five segments of M250 maraging steel, 3.4 m long. Joints tongue/groove with two O-rings and 144 pins. Insulated by nitrile rubber, with silicon potting compound on joints
Igniter type: pyrogen head-end 30 cm dia, 125 cm long containing Pedpro 2661 HTPB and its own 2 kg-charge igniter

PSLV stage 3 motor
This is ISRO's most advanced solid propellant motor, designed for a 0.915 mass fraction and the first to adopt a submerged flex nozzle design for TVC. Off-loading is also possible. Test #5 made 29 July 1991, #6 17 Oct 1991. Tests 8, 9 and 10 were for qualification.
Application: PSLV stage 3 (PS3)
First flight: 20 Sep 1993
Number flown: 8 to end October 2003
Mass (kg): 7,975
Length (cm): 244.20 (casing 208.35)
Diameter (cm): 198.8 (202.5 skirt)
Propellant
type: HTPB casebound, 86% solid
shape: slotted
mass fraction: 0.910 (7,260 kg standard loading)
Burn time (s): 73.1; 78.1 action time
Thrust (kN, vacuum): 243.2 mean, 386 max
Specific impulse (s, vacuum): 293
Total impulse (kNs, vacuum): 20,850
Pressure (atm): 57.4 max
Nozzle: the 19.6% submerged contoured nozzle is provided with a ±3° (±2° flight) conical flex bearing. The elastomer, based on natural rubber of nominal 3 mm thickness is bonded with steel reinforcements of 4.5 mm thick run. Each seal consists of six reinforcements and seven elastomer pads. The seal positioned beyond the throat housing is driven by two 11.8 kN electromechanical actuators 90° apart; a thermal boot protects the bearing
throat diameter (cm): 19.4
length (cm): 180.2
mass (kg): 280
exit diameter (cm): 141.4
expansion ratio: 52.3
materials: graphite throat insert, carbon and silica phenolic composites for nozzle
Casing materials: Kevlar filament-wound, EPDM insulation and aluminium forgings for skirt and end-opening fittings; total mass 340 kg. Nozzle opening 750 mm dia, 200 m head end
Igniter type: pyrogen head-end

PSLV strap-on motor
The same basic motor, derived from the SLV-3's stage 1, was employed for ASLV's stage 1 and strap-on, and adopted as PSLV's initial strap-on. The strap-ons are essentially identical to the ASLV stage 1 version, with the principal exception of canted nozzles. For PSLV, the secondary injection TVC systems are deleted on one ground-lit and three air-lit strap-ons. The specifications below are for ASLV's stage 1 model; differences for the strap-ons are noted where applicable.
Application: ASLV stage 1 + strap-ons, PSLV strap-ons
First flown: 24 Mar 1987, ASLV-D1 launch
Number flown: 30 PSLV strap-ons, 4 ASLV stage 1, 8 ASLV strap-ons to end-2000
Mass: 10.4 t
Length (cm): 998.99
Diameter (cm): 100
Propellant
type: AP/18%Al/HTPB case-bonded
shape: star
mass fraction: 0.853 (8.9 t loading; 8.63 t for strap-on)
Burn time (s): 46 action time (49.95 action strap-on)
Ignition delay: about 200 ms strap-on
Thrust (kN, vacuum): 502.6 mean, 643 max (strap-on: 440 mean, 580 max)
Specific impulse (s, vacuum): 259 (252 strap-on)
Total impulse (kNs, vacuum): 22,600 (21,450 strap-on)
Pressure (atm): 43.5 max (40.2 max strap-on)
Nozzle: secondary injection ports for TVC; 9°-canted nozzles
expansion ratio: 6.7
materials: graphite throat, carbon phenolic composite for convergent section, tape-wound carbon and then silica phenolic composites for divergent

Casing materials: 3.5 mm thick low carbon steel with nitrile rubber insulation. Three-segment casing with tongue/groove joints
Igniter type: pyrogen head-end

VERIFIED

International

Advanced Technology Engine (ATE)

Royal Ordnance (UK), FiatAvio (Italy), SEP (France) and Volvo (Sweden) have been undertaking ESA studies since 1986 on an advanced 20 kN European engine for early next century. Development through ground qualification would require 8–9 yr at a cost of 140 MAU ($190 million); recurring cost is estimated at 1.7 MAU. ATE's outline specification was set at a minimum specific impulse of 345 s vac, 10 restarts/mission and an operational firing time/mission of 2 h, its 4 h qualification life thus permitting its incorporation in re-usable stages. Four principal mission models were considered: re-usable orbital transfer vehicle, Ariane 5 upper stage (10% improvement by replacing DASA's 27.5 kN Aestus), deep space missions (specifically comet sample return) and planetary missions, emphasising Mars and Earth-return capabilities.

ESA selected a staged combustion cycle, resulting in a 74 kg (metallic option) or 58 kg (ceramic option), 1.7 m high, turbopump-fed 9 MPa chamber engine using MON_3/MHH propellants. The metallic chamber would be regeneratively cooled, while the ceramic version would be film cooled. Such an engine would provide a 787 kg increase in GTO injection capacity over current 309–316 s SI engines. Used with an Orbital Propulsion Module, ATE yields a <800 kg improvement over the 27.5 kN engine for LEO-GEO or LEO-GTO missions.

In the metallic chamber option, most of the MON_3 flows at supercritical pressure through small channels to cool the chamber and nozzle up to a 1:35 expansion ratio. Part of the heated MON_3 flows directly to the injector, but most is routed to the pre-burner. Most of the MMH flows to the combustion chamber, where some acts as a film coolant.
Dry mass: 74.2 kg metallic, 57.9 kg SEP ceramic
Length: 1,710 mm metallic, 1,720 mm ceramic
Maximum diameter: 764 mm
Oxidiser: NTO + 3% NO (MON_3)
Fuel: MMH
Propellant feed method: 162 kW 76,400 rpm turbopump (Fiat version), 0.4 MPa inlet, 18.1 MPa MMH pump exit, 19.5 MPa NTO pump exit
Thrust: 20 kN vacuum
Specific impulse: 351 metallic, 347 ceramic)
Mixture ratio (O/F): 1.86:1
Expansion ratio: 400 (exit diameter 760 mm)
Expansion bell: high temperature stainless steel (Volvo) or C/SiC Ceramic (SEP)
Chamber pressure: 9 MPa
Chamber materials: Narloy-Z or similar copper alloy, carrying 122 NTO coolant channels or SiC/SiC film cooled. The cool outside of the metallic chamber liner is closed by an electro-deposited metal transition layer and lightweight fibre overwrap
Injector: metallic option employs central pintle or swirl with annular fuel sheet and fuel cooling; ceramic option employs a classical impingement injector with fuel film cooling.

Electric Propulsion
ESA's Artemis mission (see ESA's Communications section entry) will carry an ion propulsion subsystem, conceived and sized to support the full 10 yr of NSSK required for commercial telecommunications missions. It will carry north/south face packages, each with a 15 mN RIT-10 German and an 18 mN UK-10 British thruster. An engineering model of the RIT-10 thruster has already demonstrated more than 2,000 h of operations on the ground.

FEEP microthruster vacuum chamber firing at ESTEC
0517340

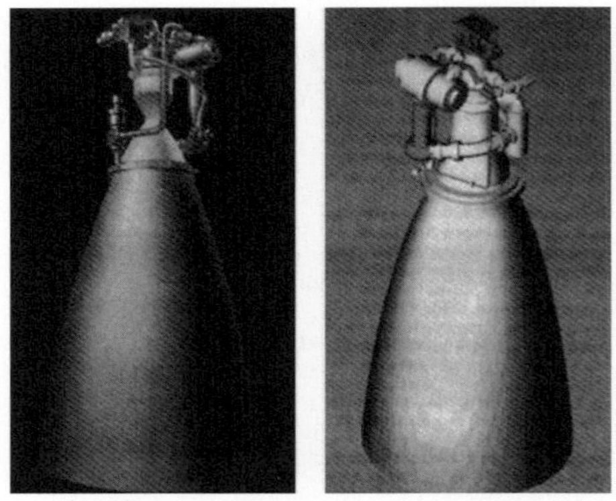

ATE would provide Europe with an advanced 20 kN reusable engine. Shown left is the Royal Ordnance version, at right the FiatAvio version (ESA/ESTEC)
0516857

FEEP The Field Emission Electric Propulsion (FEEP) thruster offers the unique characteristics of extremely low thrust levels (down to 1 mN) and pulsing and any desired frequency. This makes it the only system suitable for ultra-fine AOCS and drag compensation. Several science missions have chosen it as the only propulsion system capable of fulfilling their ultra-fine pointing requirements. In addition, the compact design and high specific impulse of the FEEP system make it a good candidate for future commercial missions like small satellites and constellations in low Earth orbit. A <7 kV electric field ionises liquid caesium and provides acceleration. ESTEC and CentroSpazio (Italy) are responsible for the testing and development. Current plans foresee a FEEP flight test in 1998 aboard a Shuttle Get Away Special canister.
Mass of thruster module: 450 kg
Mass of PCU: 2,500 g
Thrust: 1 to 1,000 mN
Thrust resolution: 0.1 mN
PCU input power: 9–60 W
Specific impulse: 8,000 s

LMIS A space-proven Liquid Metal Ion Source (LMIS) developed by the Austrian Research Centre, Seibersdorf (ARC) and flown on many science missions for charge control, is being investigated by ESA for micro-propulsion applications within the range 1–10 mN. The device uses liquid indium as propellant, flowing on a tungsten needle which is then ionised and accelerated at the needle top by field emission.

ESA-XX Primary electric propulsion, particularly for high delta-V planetary missions, requires thrust levels of 200–300 mN. ESA is supporting the development of the ESA-XX xenon thruster, combining TIT's radio frequency electrodeless discharge, the simple and reliable grid system of the 25 cm 200 mN UK-25 (see the UK propulsion section) and Italy's high power electronics. The thruster was studied jointly by Germany and Russia for interplanetary missions. The self-sustaining RF plasma is ignited by electrons from the neutraliser(s). The ion beam is extracted, accelerated and focused by an inward-dished 9,100-hole grid system, comprising the holder grid (+1.5 to +2.0 kV), the accelerator grid (–0.5 kV) and the (normally) grounded decelerator grid. Prototype tests at Giessen University during 1993–1994 mapped the motor over the 50–200 mN and up to 2.0 kV beam voltage. This prototype is being followed by an advanced breadboard-EM for further parametric testing.
Diameter: 25 cm beam exit; 39 cm thruster
Length: 39 cm
Mass: 9 kg (target mass), 12.6 kg (prototype)
Thrust: 200 mN nominal
Exhaust velocity: 54 km/s (at 2 kV)
Specific impulse: 5,365 s at 200 mN, 2.0 kV
Acceleration voltage: 2.0 kV at <100 mN

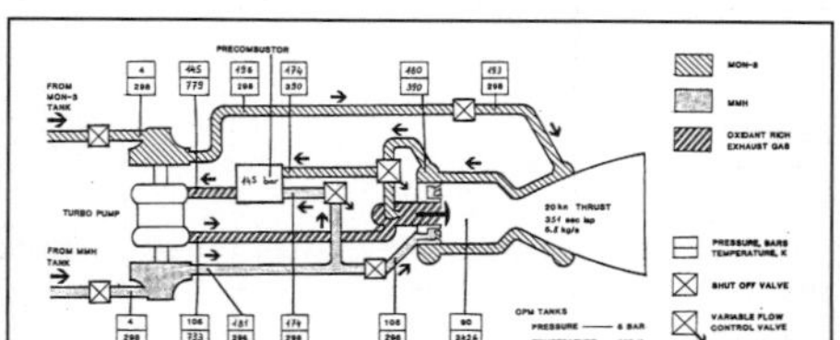

ATE's pre-combustion cycle. Values refer to Royal Ordnance's turbopump (ESA/ESTEC)
0516858

Discharge power: 650 W at 200 mN
Power input: 6.2 kW at 200 mN
Specific power consumption: 31 kW at 200 mN
Specific propellant consumption: 19 mg/Ns
Total efficiency: 83% maximum
Lifetime: 15,000 h (target)

VERIFIED

DASA/ERNO and SSC

Astrium and the Swedish Space Corporation jointly developed the Maxus suborbital vehicle to extend the μg time provided by the Skylark-based Texus and the Terrier Black Brant-based Maser rockets. Thiokol's Castor solid propellant motor was selected following 1987–88 studies. Already qualified as a Delta strap-on booster, it can handle 420 kg useful payload to 850 km altitude, providing >14 min μg. Sweden's Guidance Control System ensures the rocket and payload descend within Esrange with <7 km 1 Maxus 1A launch preparations at Esrange (SSC) gs dispersion. Laboratories offering basic equipment are provided at the site to assist experimenters' final flight preparations. At least one flight was planned annually, but ESA budget limitations now restrict them to one every 1½–2 yr. A launch campaign typically requires 3 weeks in Apr or Nov/Dec, balancing daylight and the need for frozen lakes to avoid water recovery. User cost is about $10,000/kg. DASA/ERNO has invested about DM20 million (principally in 10 Castors) and SSC about SKr60 million.

The debut launch on 8 May 1991 attained only 200 km altitude instead of the 800 km planned. The flight was smooth to 37 s/21 km but then began pitching; at 46 s the all-ESA payload separated and was recovered. It was later determined that electrical cables of the nozzle's TVC system had burned through; thermal protection was added for future vehicles. ESA paid about DM17 million for the flight and was guaranteed a reflight in the event of failure; insurance was 3%. The second flight, Maxus 1B was fully successful 8 Nov 1992, providing 12 min 32 s μg time to the seven experiments. Five were controlled by telescience, including one via ESA's Olympus satellite. The third flight, Maxus 2 was successful 28 Nov 1995, again with an ESA payload (505 kg), attaining 706 km to provide min μg to eight experiments. The fourth flight, Maxus 3, took place on 24 November 1998, followed by the fifth flight, Maxus 4, on 29 April 2001. Maxus 5 is due for launch in 2004.
First launch: 8 May 1991
Number launched: 3
Success rate: 67%
Launch site: Esrange, Sweden
Number of stages: one (solid)
Overall length: 14.877 m min, 15.460 m max
Principal diameter: 1.018 m
Launch mass: 11,800 kg + 420 kg payload + 300 kg service systems
Typical performance: 720 kg total to 850 km, providing 14 min of useful microgravity time
Payload recovery system: flat spin deceleration, subsonic parachute descent (10 m/s landing), helicopter land retrieval

Maxus stage 1
Motor: Thiokol Castor 4B (±6° nozzle TVC)
Overall length: 9.61 m
Principal diameter: 1.018 m
Thrust: 450 kN average
Burn time: 63 s
The motor is topped by an adapter, motor telemetry module and Saab Ericsson's Guidance Control System.

Maxus 1A launch preparations at Esrange (SSC) 0516856

Payload Environment/Accommodation

Total length/mass of all payload modules must be <350 cm/420 kg; 6–8 can be carried. Modules are available on a rental basis, 640 mm dia, 5 mm aluminium wall (0.5 mm zirconium oxide thermal coating), 640 mm Radax mechanical joints. Access to experiments is possible up to 20 min before launch; the payload is returned to the launch site within an hour of landing. The module below the payload section provides telemetry (2.2–2.3 GHz S-band) + telecommand (450 MHz P-band) to all experiments via a standardised interface. If required, TV signals are transmitted at S-band in PAL B/G using pre-emphasis according to CCIR 405/625. After separation, the cold gas fibre optic gyro Attitude & Rate Control System in the service module above the payload section performs a despin manoeuvre before the μg phase begins at 90 s. If roll rate <1.5°/s or the cone angle <45° (for the RF downlink) during the μg phase, ARCS activates in low thrust mode to minimise experiment disturbance below 2×10^{-4} *g*. At T-891 s, a pyrotechnic valve opens to dump the residual gas through two roll nozzles to create a <60 rpm spin for thermal distribution during re-entry.

UPDATED

Europropulsion

Europropulsion was established in 1985 by France's SEP and Italy's BPD to bid for solid motor development contracts within Europe's civil space programme. Its principal activity is Ariane 5's P230 strap-on boosters. The FFr3.8 billion fixed price contract, following two preliminary awards, was signed with CNES in late 1989, covering dev/qual of the motors, including 10 ground tests to mid-1995. Individually, BPD provides the lead for segment and igniter dev/integration, SEP contributes the nozzle assembly and insulation liner, and sub-contractors MAN Technologie provide the segment casings, SNPE propellant development and Regulus (the SNPE-BPD partnership) for the two large segment casting operations at Kourou. The small forward segment is completed by BPD in Italy. The EUPERA SNPE/BPD partnership produces the ammonium perchlorate at SNPE's Toulouse site (the AP for 501/503 was provided by Kerr McGee Corp of the US and for 502 by SNPE; future division remains to be decided). SEP inaugurated its nozzle production plant (UTB: Usine Tuyère Booster) in Oct 1990 at Haillan. A FFr180 million investment, it will fabricate 10 nozzles annually by year 2000. MAN commissioned its production centre in Augsburg during Sep 1988: this 5,400 m^2 facility includes an 1,800 m^2 air conditioned area for final machining of the clevis/tang connection between the booster segments. The 500 t counter-roller flow-forming machine rolls the initial 40 mm thick, 1 m high, 3 m dia steel cylinders into 8 mm thick 3.5 m high segments before heat treatment produces an ultimate strength of 1,500 N/mm^2.

The first demonstration motor, a reduced scale version loaded with 15 t of propellant, was fired successfully in Dec 1989. The first full scale P230 was successfully tested 16 Feb 1993. Seven full scale tests were then planned: B1 with battleship (thick) casings, five (M1-M5) development and two (Q1-Q2) qualification. The last two tested the entire stage. M1, the second firing, was successful 25

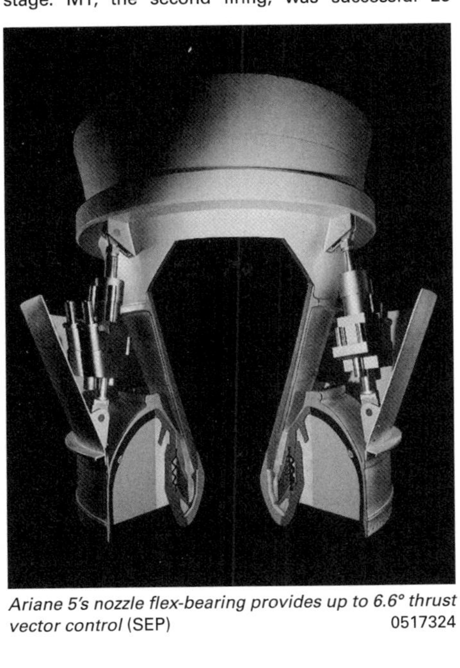

Ariane 5's nozzle flex-bearing provides up to 6.6° thrust vector control (SEP) 0517324

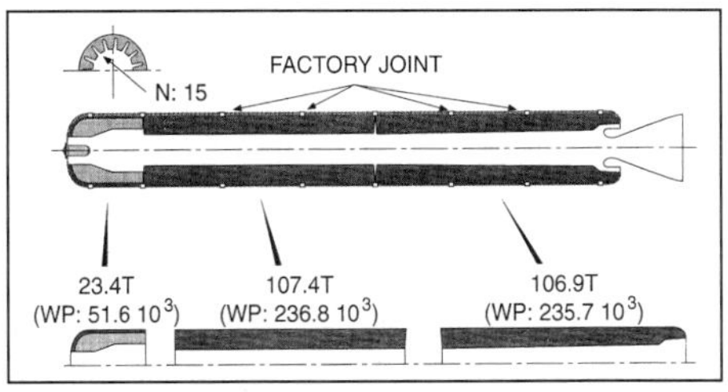

Ariane 5's strap-on booster was developed by the Europropulsion partnership. The masses shown are the segment propellant loadings 0517323

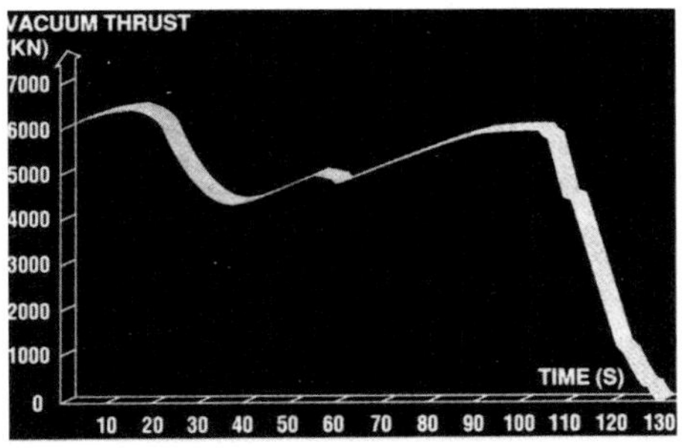

P230 typical thrust curve (Europropulsion) 0516859

June 1993. This was the first in flight configuration, confirming thrust profile, intersegment seals and nozzle gimballing. M2 was planned for mid-Nov 1993 but voids were found in the motor's propellant and that motor was shelved (it is expected to be fired after an ageing programme). The third firing, M3, was successful 20 June 1994. M4, with a flight-type actuation system and lighter thermal protection, was successful 30 Sep 1994. Flight-standard M5 was successful 15 Dec 1994. Q1 was successful 10 Mar 1995, Q2 21 July 1995. Europropulsion supports booster assembly and integration at the Booster Integration Building, Kourou and co-ordinates EADS Space Transportation activities with the facilities and work cells needed for stage integration.

Prime contractors include SNECMA, MAN Technologie and Fiat Avio for structures and mechanisms, SNECMA and Fiat Avio for propulsion and aerothermodynamics and subcontractors Arianespace and CNES.

Ariane 5 P230 specifications

First flight: 4 June 1996 (Feb 1993 first static firing, July 1995 end of ground qualification)
Mass: 268.7 t (237.7 t propellant): 106.9 t aft, 107.4 t centre, 23.4 t forward; nozzle 6 t; internal insulation 4.3 t; casing 20 t; igniter 0.3 t; 0.4 t miscellaneous)
Length: 26.774 m (stage 31.16 m)
Diameter: 3.049 m
Propellant:
type: HTPB 68% ammonium perchlorate, 18% aluminium, 14% PBHT liner
shape: main section cylindrical, forward segment 15 rectangular slots
mass fraction: 0.884 for 237.7 t loading
Burn time: 132 s
Thrust: 5250 kN sea level at launch (5904 kN vac); 4931 kN avg vac, 6637 kN max vac
Specific impulse: 271 s vacuum
Total impulse: better than 630 MNs
Maximum pressure: 59.9 atm
Nozzle:
mass: 6 t
throat diameter: 90.0 cm
length: 3.420 m (2.22 m exposed)
exit diameter: 2.826 m (initial expansion ratio 9.6)
materials: carbon-carbon throat; nozzle flexbearing provides 6° deflection (6.6° max). Employs two actuators 90° apart with ±145 mm strokes. Actuator system mass 150 kg, hydraulic 1100 l, 345/270 atm initial/end pressure
Casing materials: 8 mm-thick D6AC steel heat-treated at 14,800 atm ultimate strength; thermal insulation EPDM/Silica (GSM55) or Kevlar (EG2). Six 3.351 m-long segments are assembled into two motor sections of three segments each with two factory joints and 1.005 m-long aft dome. Tipped by 3.398 m-long forward section
Igniter type: pyrogen head-end (25 kg composite propellant).

VERIFIED

Regulus

Regulus is a partnership between France's SNPE (40%) and Italy's BPD Difesa e Spazio (60%), formed to operate Kourou's Ariane 5 booster propellant facility (UPG Usine Propergols en Guyane). The plant, comprising 40 buildings totalling 26,000 m^2 on a 300 hectare site, was commissioned 24 Oct 1991. Cost was 180MAU (about FFr1,250 million). Each strap-on contains 237 t of HPBT solid in three sections. Regulus provides the propellant required for the mid & aft units, while the forward element is shipped from BPD. When the facility became fully operational in 1995, its 150 personnel were able to support eight flights a year, requiring 3,300 t of propellant for 32 segments. The first segment was poured Nov 1991 for the B1 battleship motor, tested 16 Feb 1993. Regulus has been instrumental in developing increased safety and enhanced manufacturing techniques which have increased the reliability levels and decreased the probability of failure.

Japan

IHI COMPANY LTD

Current Status

Space Development Division

As a major Japanese aerospace company, Ishikawajima-Harima Heavy Industries' main details are to be found in the Space Industry section. In space propulsion, IHI provides the cryogenic turbopumps for the H2A's stage 1/2 cryogenic engines, in addition to H2's hydrazine Reaction Control System modules. It previously provided the pumps and gas generator for H1's stage 2 engine. ISAS, IHI, KHI and MHI are pursuing development of the ATREX-500 air turboramjet, which could power the flyback booster of a 2-stage spaceplane. The hydrogen expander ATREX would take the vehicle to M6/30 km. The 5.4 kN ATREX-500 ¼-scale version, 2.2 m long and fan inlet 30 cm dia, totalled 1,400 s in 40 static sea level runs during 1990–95 at ISAS' Noshiro Testing Centre. Efforts now centre on a 1.4 t flying test bed launched from Japan and parachuting into the sea. A jet engine will take the FTB to M0.5 from a horizontal rail for the 370 kg ATREX-500 with 190 kg of LH_2 to attain 25–35 km altitude 200 km downrange. The initial metal turbo machinery will achieve M4.5, with M6 targeted for the carbon-carbon composite design. Goals for the composite ATREX are 17.0–18.9 kN and 3,150–3,300 s SL static. IHI continues to study air-breathing propulsion concepts and has refocused its studies toward civil aviation applications as the plausibility of an air-breathing space launch systems becomes less likely. Since the changes to Japanese planning predicated by a decision to continue development of the HOPE spaceplane IHI has begun preliminary definition of a new conventional propulsion system which could be used in the first stage of a successor to the H2A.

IHI's Rocket Test Centre was established in Oct 1975 at the Aioi Plant as a cryogenic test centre of IHI's Research Institute. It was incorporated in

IHI's LE-7 turbopumps: the liquid hydrogen unit is at bottom, oxygen at top 0517171

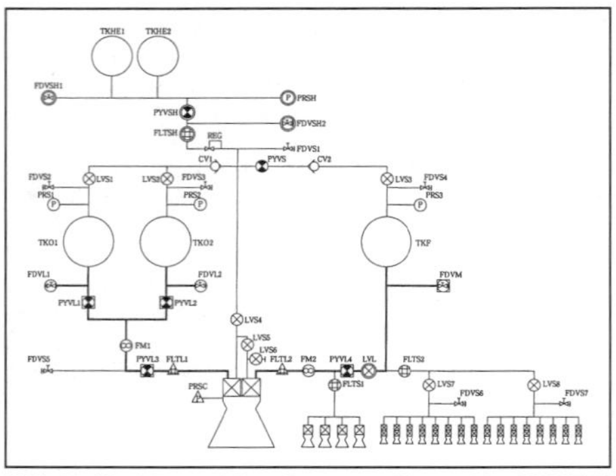

IHI's Unified Propulsion System will provide COMETS with apogee kick and attitude/orbit control 0517336

July 1980 into the Space Development Division for testing the cryogenic engine and bipropellant apogee engines described below.

Spacecraft Propulsion

IHI developed the 2 kN bipropellant apogee engine under NASDA contract (in co-operation with NAL) for ETS 6 and COMETS (see below). ETS 6's LAPS Liquid Apogee Propulsion Systems was specified in detail p258 1995–96. IHI is independently developing a 10 kN CUS Cryogenic Upper Stage LOX/LH_2 expander cycle engine for upper stage or orbital transfer vehicle applications. SI 471 s, chamber pressure 28 atm, LOX turbopump 43,000 rpm, hydrogen turbopump 85,000 rpm.

Unified Propulsion System

UPS provides GEO insertion and attitude/orbit control for 2 t-class satellites; it is derived from the LAPS of ETS 6. It carries a single 1.7 kN bipropellant Apogee Kick Engine plus 4 × 50 N + redundant 8 × 0.86 N hydrazine catalytic thrusters. COMETS's flight model was shipped Mar 1996.

Dry mass: UPS total 163 kg, AKE 15.7 kg (including valves)

AKE length: 102.5 cm

AKE maximum diameter: 58.4 cm

Engine cycle: regulated pressure-fed

Oxidiser: nitrogen tetroxide at 0.255 kg/s

Fuel: hydrazine at 0.284 kg/s

Propellant feed: GHe 17 atm in two 82 cm dia NTO tanks + 130 cm dia N_2H_4 tank; titanium alloy. 223 atm GHe in two 48 cm dia CFRP spheres

Thrust: 1,700 N AKE

Specific impulse: 321.4 s AKE

Time to 90% thrust: 1 s max

Mixture ratio (O/F): 0.9

Expansion ratio: 240 (exit dia 584 mm)

Expansion bell:
- length: 83.8 cm
- material: columbium (niobium)/silicide coating
- cooling method: radiative

Combustion chamber:
- pressure: 8.3 atm
- temperature: 1,770K max wall
- stability: ±5%
- materials: as expansion bell
- cooling method: fuel film + radiative
- ignition method: hypergolic

Duty cycle: AKE 3,500 s total, 10 burns max (typically two required); 50 N thruster 40,000 pulses min total; 1 N thruster 270,000 pulses min

LE-7/5/5A turbopumps

IHI developed H1's cryogenic turbopumps in co-operation with NASDA/NAL. For LE-7, the LOX/LH_2 pumps provide flow rates of 35.7/211 kg/s at discharge pressures of 266/254 atm working at 42,000 rpm. For LE-5A, the pumps provide 17.4/51.6 l/s at discharge pressures of 56/64 atm working at 16,500/50,800 rpm. In the LE-5, the pumps were driven by combustion gases from the gas generator: LOX/LH_2 burned at 840K/25.4 atm and a mixture ratio of 0.85 to produce an exhaust flow rate of 0.43 kg/s.

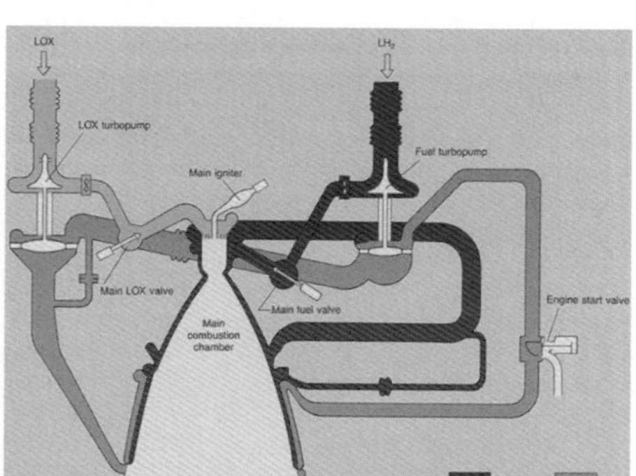

LE-5A nozzle hydrogen bleed cycle 0517348

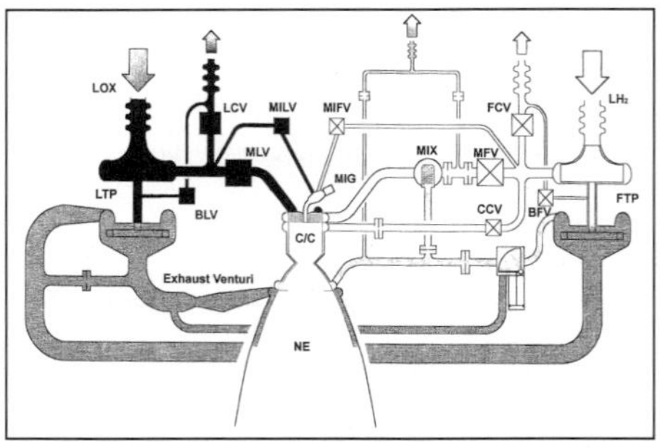

LE-5B's improvements include switching to a chamber hydrogen bleed cycle (IHI) 0003455

Reaction Control Systems

IHI developed hydrazine RCS systems for the ETS 3–6, MOS, CS-3 and JERS satellites and the H1/H2 second stages. It developed an EHT thruster for NASDA (see below) and a magnetoplasmadynamic (MPD) thruster for ISAS' Space Flyer Unit. H2 roll control during burn phases, 3-axis control during coast periods and ullage acceleration is provided by two IHI thruster modules at stage 2's base, either side of the LE-5A engine. Each houses six anhydrous hydrazine thrusters (4 × 50 + 2 × 18 N) pressure-fed at 20 atm. For direct ascent trajectories, without LE-5A re-start, a 15 kg dry mass version carries 6 kg of hydrazine to provide 10.67 kNs total impulse. On re-start missions, the 20 kg (dry mass) module houses 27 kg of propellant for a 52.27 kNs total impulse. ETS 6's blowdown RCS system comprises 16 × 1 N (0.95 N ± 10%) thrusters + two EHTs (see below). Dry mass 29.6 kg; propellant (hydrazine) 118 kg; SI 221-192 s; inlet pressure 24.6–7.0 kg/cm^2, 463,000 pulses demonstrated.

Electrothermal Hydrazine Thrusters

IHI/Toshiba developed an experimental 0.2–0.5N EHT under NASDA contract. Two sets are carried by ETS 6 for NSSK trials; they were tested successfuly over 4.2 h.

Dry mass: 2.0 kg (heater control unit adds 2.85 kg)

Length: 13 cm

Mounting: fixed

Engine cycle: decomposition of hydrazine over a heated catalyst bed. The 600°C gases then pass to the heat exchanger chamber where they are pre-heated by channeling along the wall (which also provides wall cooling) and then heated by a 16.5 A double coil tungsten filament to 1,100°C (low power) or 1,600°C (high power)

Thrust: 0.2–0.5 N vacuum

Specific impulse: goal is 290 s vacuum. 170°C without heating, 240 s low level heating and 280 s high level heating demonstrated. Thermal efficiency 80% demonstrated; 85% goal

Inlet pressure: 6.8–23.8 atm

Power requirement: 300 W low (1,400°C heater), 500 W high (1,900°C heater); 9.8 W valve (plus 4.6 W valve heater), 3.0 W catalyst bed heater, total power 1,132 W under maximum operating conditions

Total impulse: >10^5 Ns demonstrated, 2.7 × 10^5 Ns goal.

VERIFIED

MITSUBISHI HEAVY INDUSTRIES LTD Space Systems Department

Current Status

MHI Ltd has engineered a complete re-work of the H-2 into the H2A with about 50 per cent of the launch price of its predecessor. Commercial opportunities are integral to how the vehicle has been rescoped and further commercialisation has been expressed as part of new government policy following the integration of NASDA and ISAS into JAXA.

Background

Mitsubishi provides Japan's primary capability in liquid rocket engines, most recently developing the cryogenic oxygen/hydrogen LE-7A tage 1 and LE-5B stage 2 engines for NASDA's H2A launcher. The company has prime industrial responsibility for H2A (as it did for H2). The 10 yr $108 million LE-5 development programme produced Japan's first operational LOX/LH_2 engine, providing H1's stage 2 with a restart capability. The technology formed the basis for developing the larger and more advanced LE-7 and the LE-5A direct outgrowth. Nine LE-5s flew, successfully, 1986–92 (see p271 of the 1991–92 edition for detailed information). MHI also fabricated Rocketdyne MB-3 engines under licence for the H1 and produced the indigenous LE-3 storable propellant engine for the seven N1 rockets. Two advanced smaller engines, the 9.8 kN RE6 and 2.94 kN RE10-300, have been tested for possible applications in future orbit transfer vehicles. MHI is testing LACE liquefied air cycle engine hardware, in which LH_2 is used to liquefy atmospheric oxygen, using the LE-5 as a demonstrator. Mitsubishi's Tashiro Test Field includes stands for LE-5/5A/5B engine and stage sea level firings, a short-duration LE-7 stand and an H2 stage 1 battleship tower (full-duration firings of flight-type engines and stages are undertaken at Tanegashima's H2 launch complex). MHI completed a 105 m long 30 m wide 20 m high assembly/test facility at its Nagoya plant for the H2 in May 1988, allowing stages to be shipped from Nagoya bay to the launch site. MHI is supplying stage 1 fuel tank and stage 2 LHz tank for Boeing's Delta 3.

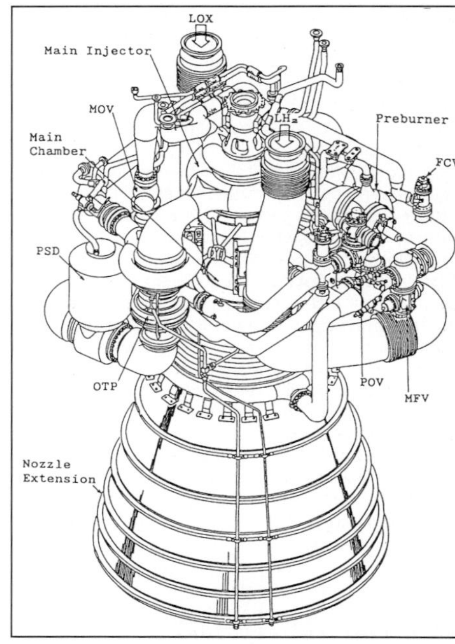

The LE-7 is the first Japanese engine to employ the staged combustion cycle (NASDA) 0516918

LE-5A

H1's LE-5 was initially intended for direct application aboard H2's stage 2 but NASDA decided at the beginning of the detailed design phase in 1987 to uprate performance and improve re-startability by increasing chamber pressure and throat diameter, and by switching from gas generator to hydrogen bleed cycle, respectively. IHI's turbopumps are

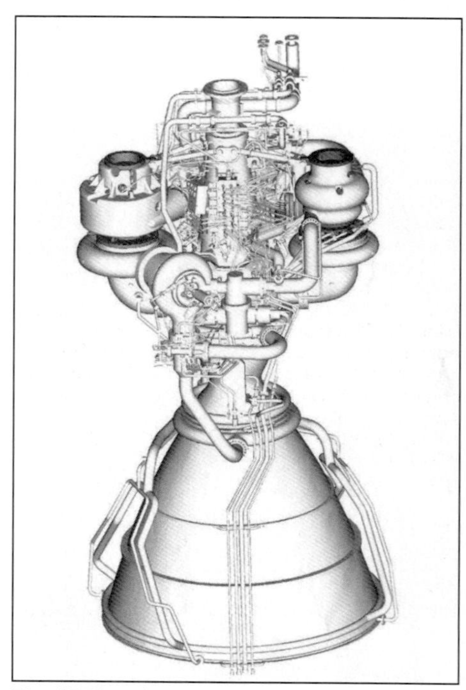

The LE-7A reduces cost and increases reliability compared to the LE-7 0525666

powered by tapping GH_2 from the nozzle coolant loop, reducing complexity and mass.
Application: H2 stage 2 main engine
First flight: 3 Feb 1994
Number flown: 11 to end of 2002
Dry mass: 244 kg
Length: 2,668 mm
Maximum diameter: 1,625 mm
Engine cycle: nozzle hydrogen bleed
Oxidiser: liquid oxygen at 22.79 kg/s total
Fuel: liquid hydrogen at 4.57 kg/s total
Mixture ratio: 5.0:1 (20 for 5% idle)
LOX turbopump: contractor IHI, 17,370 rpm rotor speed for pump discharge pressure of 60.3 atm
LH_2 turbopump: contractor IHI, 52,240 rpm rotor speed for pump discharge pressure of 68.1 atm
Thrust: 121.5 kN vacuum; 5% idle mode provides precise orbit control and stage de-orbit
Time to full thrust: 9 s
Specific impulse: 452.9 s (>200 s for 5% idle)
Expansion ratio: 130:1
Chamber pressure: 40.6 atm
Cooling method: regenerative hydrogen circulation through tubular walls
Burn time: 400 + 210 s dual burn

LE-5B

Improved H2 in 1999 introduced the LE-5B, reducing cost and increasing flexibility, particularly restarts for deploying multiple satellites into different LEOs and direct injection into GEO by a third burn at apogee. The engine cycle will be changed from a nozzle expander bleed to a chamber expander bleed; the heat exchanger function is thus integrated into the chamber cooling function. This reduces nozzle complexity – the skirt can use dump cooling with a simple sheet metal structure. Chamber cooling will be changed from the current brazed tube walls to machined channels for better heat absorption. The injector elements will be reduced from 208 to 127. The engine can operate in throttled condition with throttling valves added. Feasibility test (8 firings, 236.9 s total duration) was conducted in 1995. Phase 1 development test (13 firings, 930.7 s total duration) was conducted in 1996. Phase 2 test (qualification) was conducted in 1997.
Dry mass: 285 kg
Length: 2,625 mm
Maximum diameter: 1,713 mm
Engine cycle: chamber hydrogen bleed
Oxidizer: liquid oxygen at 26.07/15.63 kg/s total
Fuel: liquid hydrogen at 5.22/3.13 kg/s total
Mixture ratio: 5.0:1
LOX turbopump: 17,630/12,210 rpm rotor speed for pump discharge pressure of 55.9/29.2 atm
LH_2 turbopump: 52,110/42,360 rpm rotor speed for pump discharge pressure of 71.3/49.9 atm
Thrust: 137.3/82.4 (throttled) kN vacuum; 5% idle mode provides precise orbit control and stage de-orbit
Specific impulse: 447/448 s
Expansion ratio: 110:1
Chamber pressure: 36.5/22.1 atm

LE-7

Development of Japan's largest and most advanced engine was authorised in 1985 simultaneously with that of its host vehicle, the H2. Like Rocketdyne's Space Shuttle Main Engine, it employs staged combustion in which the propellants are partially burned in the pre-burner, routed to drive the turbopumps and then combined with more oxygen in the main combustion chamber. Complexity is reduced, however, by the lower main chamber pressure and the requirement for only a single start. LE-7 development covered 21 engines, 282 firings and 15,639 s, including 20 mission duty cycles. EG307 flew on H2#1.

Because of the development delays, it was decided during the CDR in July 1990 that LE-7's target performance would be reduced to about 90%, a level already achieved in tests. Overall H2 capability would not be affected, however, because SRB and stage 2 performance had exceeded target. NASDA's projected development cost increased from ¥80 billion to ¥100 billion. The specifications provided below reflect the revised performance; the original specifications are retained in parentheses.
Application: H2 first stage
First flight: 3 Feb 1994
Number flown: 11 to end of 2002
Dry mass: 1,714 kg (1,560 kg); vacuum thrust: mass 64
Length: 340 cm (350 cm)
Maximum diameter: 180 cm (190 cm)
Mounting method: gimballed to provide ±7.5° pitch/yaw control, hydraulic pump powered by auxiliary turbine mounted on vehicle
Engine cycle: closed, staged combustion
Oxidiser: liquid oxygen at 18.7 kg/s in pre-burner and 211.1 kg/s in main chamber
Fuel: liquid hydrogen at 29.4 kg/s in pre-burner and 35.2 kg/s in main chamber (resultant fuel-rich gas then burned in main chamber with added oxygen)
Mixture ratio: 0.64:1 in pre-burner; 6.0:1 main chamber, 5.92 overall
LOX turbopump: contractor IHI, 18,100 rpm (20,000 rpm) single-stage centrifugal pump with pre-burner pump; 177.6 atm (210 atm)/263.2 atm (323 atm) main/pre-burner discharge pressures, respectively
LH_2 turbopump: contractor IHI, 42,200 rpm (46,000 rpm) two-stage centrifugal pump; 274.8 atm (315 atm) discharge pressure
Thrust: 84.5 kN (910 kN) SL, 1,079 kN (1,180 kN) vac
Specific impulse: 446.5 s (449 s) vacuum
Expansion ratio: 52 (60); exit diameter 173.7 cm
Main combustion chamber: 452-hole coaxial main injector, 130 atm (145 atm) combustion pressure
Pre-burner: 237-hole coaxial injector, with damp-cooled combustion wall; 214 atm (237 atm) pressure, 537°C (698°C) temperature
Cooling method: regenerative hydrogen; LH_2 at 44 K in both channel-structured chamber and tubular nozzle extension
Expansion bell length: 216 cm
Burn time: 350 s single start

LE-7A

The H2A will debut the LE-7A, reducing cost and increasing robustness. For the main injector assembly: reduce and relocate the welding lines, cast the LOX dome, reduce the number of injector elements, remove the hydrogen heat exchanger, reduce the injector dia from 1,400 mm to 1,160 mm. The main chamber throat area will be enlarged by up to 10%. For the pre-burner the number of elements is reduced and the LoX dome weld assembly is replaced by a single-cast part. The turbopumps will use more precise cast housings. The design will allow longer operation at current thrust. Specifications are given below in that order:
Length: 3.4 m
Max diameter: 2 m
Chamber mixture ratio: 6.0
LOX turbopump: 18,300 rpm
LH_2 turbopump: 42,100 rpm
Thrust: 1,074 kN vac
Specific impulse: 430 s vacuum
Chamber pressure: 125 atm
Pre-burner temperature: 443°C
Burn time: 400 s single start

RE6/RE10

The RE6 cryogenic expander bleed cycle engine has been tested for potential orbit transfer vehicle and launcher stage 3 applications. The lower-thrust RE10 was investigated as a high performance space-storable apogee engine for satellite main propulsion and orbiter manoeuvring systems. Turbopump, pre-burner and main combustion chamber tests were completed since development began in 1986; sea level engine system trials followed. Work ended 1992. See p293 of the 1992–93 edition for details.

VERIFIED

Nissan Motor Company Ltd

Nissan's Aerospace Div is Japan's primary developer/manufacturer of solid propellant rocket motors and launch vehicles. These include the M-5 satellite launcher, the MT-135P, S-310, S-520 and K-9M sounding rockets, boosters for the H2, and several kick stages. The company is also involved in propulsion studies for Japan's spaceplane and is responsible for the experiment logistics module

Nissan developed the TR-1A rocket, providing some 6 min µg (NASDA) 0003448

exposed section of the Japanese Experimental Module contribution to NASA's Space Station. Since the rescoping of the H2 programme and a greater emphasis on commercialisation, redesign of the spaceplane concept toward a more fiscally economic posture.

Manufacture/assembly is undertaken at Nissan's Ogikubo plant (planned to move to Tomioka City); the Taketoyo Test Facility opened in 1980 for vacuum firings of spinning upper stage motors. Taketoyo's Spin Static Firing Test Stand provides a 39–78 mbar vacuum for 1.8 m dia motors of 0–147 kN at spin rates up to 150 rpm. The company provided test equipment for the H2 solid booster static stand at NASDA's Tanegashima site. R&D activities are conducted at the Kawagoe plant, including research into liquid bipropellant satellite thruster systems, ram propulsion and gimballing TVC. The company is also a major supplier of pyrotechnic devices for space applications.

H2 Solid Rocket Boosters

Nissan developed the 4-segment solid strap-ons for Japan's H2 launcher, with four demonstration firings during 1988–91 at NASDA's Tanegashima Space Centre. First test was conducted Apr 1988, prototype #1 was fired June 1989 & prototype #2 13 Dec 1989; the qualification model firing was achieved 29 May 1991. The units represent the country's largest and most powerful indigenously-developed solids. The Improved H2's SRBs will be reduced to three segments, debuting in 1999. Upgraded H2A, from 2001, will use a monolithic composite case design.
First flight: 3 Feb 1994
Number flown: 11 sets, through end-2002
Mass: 70.4 t
Length: 2,336 cm total booster, 1,926 cm for motor without nose section
Principal diameter: 1,809 cm
Propellant:
  type: 14HTPB-68AP-18Al
  mass: 59.15 t in 4 segments
Burn time: 94 s
Thrust: 1,560 kN, sea level
Specific impulse: 273 s vac
Chamber pressure: 46 atm average

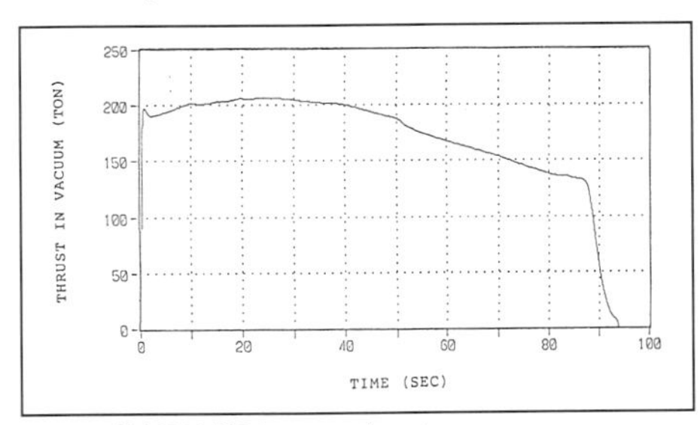

Nissan/NASDA H2 booster thrust curve (thrust given in tonnes) 0516919

Nozzle: throat 53.4 cm dia, expansion ratio 10; nozzle gimballed ±5°
Case materials: four 4.0 m-long segments of 4.7 mm-thick NT-150 low carbon steel

S-520 rocket motor
Nissan produces a range of sounding rockets for ISAS, NASDA, the Japan Meteorological Agency and the National Inst of Polar Research. One of the largest is the S-520 single stage sounding rocket for ISAS (qv). Motor specifications are given below.
First flown: 18 Jan 1980
Length: 6,053 mm
Principal diameter: 524 mm
Propellant:
  type: polybutadiene composite
  mass: 1,610 kg
Burn time: 28.7 s effective, 35.7 s action
Thrust: 143.168 kN avg vac
Specific impulse: 265 s vacuum
Total impulse: 4,109 kNs
Chamber pressure: 55 atm maximum
Nozzle:
  throat diameter: 161 mm
  expansion ratio: 7.0
  materials: CFRP/GFRP conical
Case materials: high tension steel
Igniter type: pyrogen head end, SAD ignition with CFRP case.

VERIFIED

Russian Federation

PROPULSION

Historically, the general Soviet philosophy has been to cluster numbers of relatively modest, low chamber pressure open cycle engines, mass-produced with few modifications in their decades-long careers. Most are assigned RD (Raketnay Dvigatel, rocket motor) designations. NPO Energomash is the major centre for rocket engines; KB Khimautomatiki, was identified in 1990 as responsible for Energia's core engines, and emerged from the **Kosberg** bureau, which concentrated on upper stage propulsion, as does KB Khimmach of the old Isayev bureau. NII Machinostroenye undertakes more basic research but does offer spacecraft thrusters. The Kuznetsov (late NPO Trud, now RD Kuznetsov Co SSC) aircraft engine bureau developed engines for the abandoned N1 lunar launcher for Korolev. Korolev's bureau itself developed Proton's Block D/DM and Molniya's Block L engines. The Lyulka aircraft engine bureau (now part of NPO Saturn) developed the LOX/LH_2 D-57 to replace the lunar N1 launcher's stage 3 engines.

In 1996 the designations of many previously-unidentified engines used on missiles and launch vehicles were released, and these are included under the appropriate design bureaus. To permit easy cross-referencing between a launch vehicle and the propulsion systems the following table has been prepared.

Some launch vehicles use clusters of slightly different engines on the same stage. For example, the Proton family uses three RD-0210 and one RD-0211 on the second stage, the only difference being that the RD-0211 provides the tank pressurisation: the assembly is called the RD-0210. Vernier engines usually have separate designators and these are indicated when known: in such cases there is one main engine and one set of verniers. In the case of the Proton-K third stage it is known that the RD-0213/RD-0214 assembly is designated RD-0212.

N/A – not applicable. For the specific impulse values the nominal value is quoted, followed by a worst case value in parentheses.

NPO ENERGOMASH
The history of NPO Energomash can be traced back to 15 May 1929 when the Group for the Development of Rocket Engines was organised at the Gas Dynamics Laboratory (GDL) in Leningrad (now St Petersburg). For more than sixty years the organisation repeatedly changed its name and location, but remained at the forefront of rocket engine design. The founder and the permanent head of the GDL as Academician Valentin P Glushko, considered to be the initiator of the Russian rocket engine industry.

In May 1974 the GDL merged with Sergei Korolyov's design bureau, the new organisation being re-named NPO Energiya, with Glushko as its head. Following Glushko's death in Jan 1989, the former GDL left Energiya to become NPO Energomash, continuing to concentrate on the development of rocket engines: the bureau was headed by Vitali P Radovsky during 1974–1994. The current general director and chief designer is Boris I Katorgin.

NPO Energomash incorporates the design bureau, plant and test facilities, with the capability to test engines with thrusts of up to 10 MN either as components or complete. An agreement was signed with Pratt and Whitney in Oct 1992 providing exclusive US marketing rights for Energomash's LOX/kerosene and tri-propellant products and technology.

RD-100
Developed directly from the V-2/A-4 engine, this was the first large Soviet engine employing a turbopump feed system, powered by catalytically decomposed hydrogen peroxide. Tested May 1948. Development period 1947 to 1953.
Applications: R-1 missile and research rockets
Dry mass: 885 kg
Oxidiser: liquid oxygen
Fuel: 75% ethyl alcohol solution
Feed method: turbopump driven by hydrogen peroxide gas generator
Thrust: 267 kN sea level, 307 kN vacuum
Specific impulse: 199 s sea level, 232.5 s vacuum
Chamber pressure: 16 atm
Burn time: about 65 s

RD-101
Direct development from the RD-100 and used by the R-2 missile and V-2A research rocket. Solid catalyst in gas generator replaced hydrogen peroxide decomposition. Development period 1947 to 1953.
Thrust: 363 kN sea level, 404 kN vacuum
Specific impulse: 210 s sea level, 237 s vacuum
Chamber pressure: 21 atm
Burn time: 85 s

RD-103 (8D52)
Applications: developed 1947 to 1953 for the R-5M missile and V-5V research rocket.
Dry mass: 870 kg
Oxidiser: liquid oxygen
Fuel: 92% ethyl alcohol solution
Thrust: 432 kN sea level, 500 kN vacuum
Specific impulse: 216 s sea level, 243 s vacuum
Chamber pressure: 24 atm
Burn time: about 120 s

RD-105
This single-chamber engine was proposed for use on the strap-on boosters for the R-7 early in the missile's development. There were design difficulties during the development of the engine due to burning instabilities in the combustion chamber. It was also realised that the engine would not be powerful enough for the R-7, which also went through major design changes.

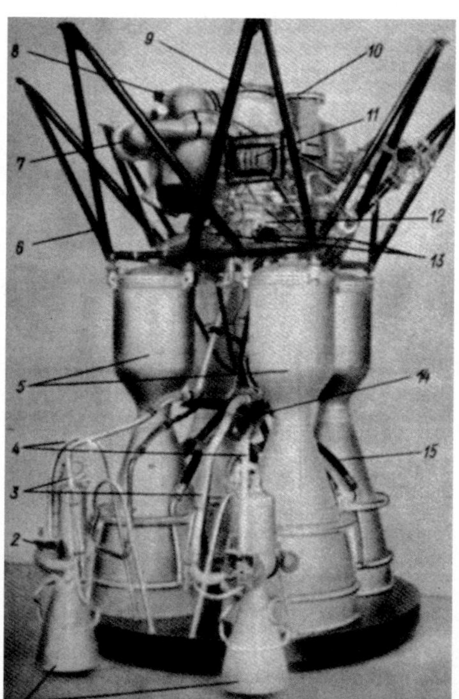

RD-107: the world's most utilised rocket engine, with four main chambers and two steering verniers 0516883

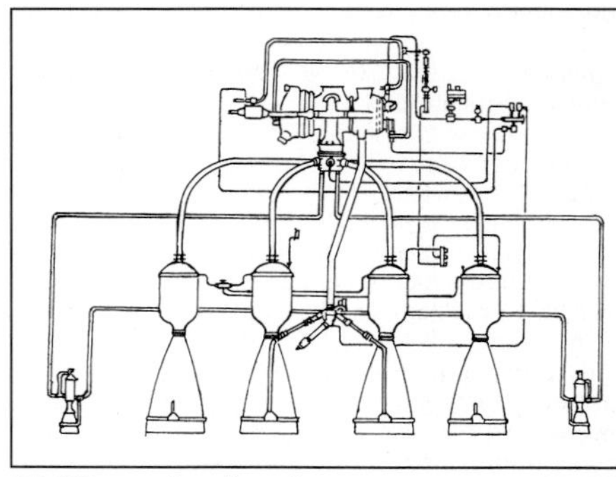

RD-107 propellant flow diagram 0516884

Application: Early concept R-7 strap on
Oxidiser: liquid oxygen
Fuel: kerosene
Thrust: 540 kN sea level

RD-106
This single-chamber engine was proposed for use on the central core stage for the R-7 early in the missile's development. There were design difficulties during the development of the engine due to burning instabilities in the combustion chamber. It was also realised that the engine would not be powerful enough for the R-7, which also went through major design changes.
Application: Early concept R-7 central core
Oxidiser: liquid oxygen
Fuel: kerosene
Thrust: 520 kN sea level, 645 kN vacuum

RD-107 (8D74)
The world's most used rocket engine. Developed 1954 to 1957. Those now flying on Soyuz-U are designated 11D511 and those on Molniya-M 11D728. Each of the four strap-ons incorporates a single RD-107 of four fixed combustion chambers, a single turbopump and two ±45° gimballed verniers. These engines and their RD-108 derivative of the launcher's sustainer core were responsible for launching the first satellite, man in space, lunar probes and still account for every manned flight. Developed over 1954–57 by Glushko's GDL, the original versions were uprated after the first Sputnik launches. The RD-107/108 engines are built by the PO Samara Frunze Engine Building factory.
Applications: R-7 ICBM, Vostok/Soyuz/Molniya space launcher family
First flown: May 1957
Dry mass: 1,155 kg
Engine cycle: open
Oxidiser: liquid oxygen at 226 kg/s
Fuel: kerosene at 91 kg/s
Feed method: this engine introduced a more efficient turbopump to feed four main chambers, consuming four times the peroxide flow of the RD-100 but yielding ten times the power. The single 8,300 rpm shaft carries a double-sided shrouded impellor for LOX, a single-sided impellor for kerosene, a small centrifugal pump for the two vernier chambers and for feeding LN_2 to the evaporator in the turbine exhaust, and a small pump to feed the H_2O_2 to the gas generator (previously done by compressed air). Turbine power is 3,800 kW. The kerosene is routed from the nozzle base through the cooling channels before reaching the injectors at 210°C
Thrust: 1,000 kN vacuum, 821 kN sea level (verniers about 38 kN each); uprated from original 976 kN vacuum
Specific impulse: 257 s sea level, 314 s vacuum; uprated from original 308 s vacuum
Chamber pressure: 60 atm, gas temperature 3,520°C. For maximum thermal conductivity the inner wall was originally copper, later replaced by a brass-chromium alloy. The outer wall is alloy steel, the two shells joined by both soldering/welding in an inert atmosphere, the joints being on ribs milled on the outer face of the inner shell or to a corrugated plate between the shells. The heads are made from two flat plates welded with concentric tubular rings in which the propellants are pre-mixed before injection through the inner plate. The 337 injector holes are arranged in 10 rings, the outer ones spraying on to the inner wall to hold T at 380°C
Chamber internal diameter: 430 mm, throat 166 mm
Burn time: 120 s, but 140 s possible.

RD-108 (8D75)
Developed in parallel during 1954–57 with the RD-107 but with reduced thrust and chamber pressure to permit longer operating time (330 s instead of 120 s) as the core engine for the Vostok and related family of launchers. As only one engine is carried, the number of verniers was increased

Rocket Engines Used on Selected CIS Missiles and Launch Vehicles

Vehicle	Designator	Stage 0	Stage 1	Stage 2	Stage 3	Stage 4
1. Selected Missiles (development related to early launch vehicles)						
R-9	8K75 (original)		4 × NK-9	1 × NK-19		
	(developed)		1 × RD-111	1 × RD-0106		
R-9M	8K77 (not developed)		4 × NK-9	4 × SI-5400		
R-16	8K64		1 × RD-218	1 × RD-219		
GR-1	8K713		4 × NK-9	1 × NK-19	1 × 8D726	
UR-200	8K81		3 × RD-0203	1 × RD-0206		
			3 × RD-0204	1 × RD-0207		
2. Existing and Retired Launch Vehicles						
Sputnik	8K71PS	4 × RD-107	1 × RD-108			
Sputnik	8A91	4 × RD-107	1 × RD-108			
Vostok-L	8K72	4 × RD-107	1 × RD-108	1 × RD-0105		
Vostok	8K72K	4 × RD-107	1 × RD-108	1 × RD-0109		
Vostok-2	8A92	4 × RD-107	1 × RD-108	1 × RD-0109		
Meteor	8A92M	4 × RD-107	1 × RD-108	1 × RD-0109		
Voskhod	11A57	4 × RD-107	1 × RD-108	1 × RD-0107		
Soyuz	11A511	4 × RD-107	1 × RD-108	1 × RD-0108		
Soyuz-L	11A511L	4 × RD-107	1 × RD-108	1 × RD-0108		
Soyuz-M	11A511M	4 × RD-107	1 × RD-108	1 × RD-0108		
Soyuz-U	11A511U	4 × RD-107	1 × RD-108	1 × RD-0110		
Soyuz-U2	11A511U2	4 × RD-107	1 × RD-118	1 × RD-0110		
–	11A59	4 × RD-107	1 × RD-108	1 × RD-0110		
Soyuz-2		4 × RD-120K	1 × RD-120K ?	1 × RD-0124		
Molniya	8K78	4 × RD-107	1 × RD-108	1 × RD-0107	1 × SI-5400	
Molniya-M	8K78M	4 × RD-107	1 × RD-108	1 × RD-0110	1 × 11D33	
Cosmos	63S1		1 × RD-214	1 × RD-119		
Cosmos-2	63S1M		1 × RD-214	1 × RD-119		
Cosmos-2M	11K63		1 × RD-214	1 × RD-119		
Cosmos-1	65S3		2 × RD-216	1 × RD219		
Cosmos-3	11K65		2 × RD-216	1 × RD219		
Cosmos-3M	11K65M		2 × RD-216	1 × RD219		
Proton	8K82		6 × RD-253	3 × RD-0210		
			1 × RD-0211			
–	11A510	4 × RD-107	1 × RD-108			
R-36-0	8K69		1 × RD-218	1 × RD-219		
Tsyklon-2	11K67		1 × RD-251	1 × RD-252		
Tsyklon-M	11K69		1 × RD-251	1 × RD-252		
Proton-K	8K82M		6 × RD-253	3 × RD-0210	1 × RD-0213	1 × RD-54
				1 × RD-0211	1 × RD-0214	
Proton-K	8K82M		6 × RD-253	3 × RD-0210	1 × RD-0213	
				1 × RD-0211	1 × RD-0214	
Tsyklon	11K68		1 × RD-251	1 × RD-252	1 × RD-861	
N-1	11A52		30 × NK-33	8 × NK-43	4 × NK-39	1 × NK-31
Zenit-2	11K77		1 × RD-171	1 × RD-120		
Energiya	11K25	4 × RD-170	1 × RD-0120			
Rockot			? × RD-0233	1 × RD-0235		
			? × RD-0234	1 × RD-0236		
3. New Launch Vehicles						
Angara			1 × RD-174	1 × RD-0120		
Dniepr			?	1 × RD-0256		
				1 × RD-0257		
Riksha			6 × RD-169	1 × RD-185		
Shtil			? × RD-0244	?		
			? × RD-0245	?		

to four for steering. The current Soyuz-U model is 11D512, for Molniya 11D727.
Applications: R-7 ICBM, Vostok/Soyuz/Molniya launchers
First flown: May 1957
Dry mass: 1,250 kg
Engine cycle: open

RD-108: core engine with four steering verniers, flown with RD-107 strap-on stage engines 0516885

Oxidiser: liquid oxygen at 52 kg/s each main chamber
Fuel: kerosene at 21 kg/s each main chamber
Feed method: as RD-107
Thrust: 745 kN sea level, 941 kN vacuum; uprated from original 912 kN vacuum
Specific impulse: 248 s sea level, 315 s vacuum; uprated from original 308 s vacuum. More recent uprating to 323 s is possible
Chamber pressure: 52 atm
Burn time: up to 330

RD-109
Little is known about this engine. Developed 1958 to 1960. It was proposed as the orbital stage engine for the original Vostok launch vehicle, but the RD-0105 was chosen in its place. Presumably it was a single-chamber engine, possibly with verniers.
Application: Proposed for early Vostok Block E.
Oxidiser: liquid oxygen
Fuel: UDMH
Thrust: =~100 kN

RD-111
This engine replaced the rival NK-9 as the first stage of Korolev's R-9 missile, first flight 9 Apr 1961 (failure) and entering service 21 July 1965.
Applications: first stage R-9 missile
Dry mass: 1,480 kg
Engine cycle: open
Oxidiser: liquid oxygen
Fuel: kerosene
Feed method: turbopump, feeding four gimballed chambers
Thrust: 1,408 kN sea level, 1,628 kN vacuum
Specific impulse: 270 s sea level, 311 s vacuum
Chamber pressure: 80 atm

RD-119 (8D710)
Developed 1958–1963 for the Cosmos/Cosmos-2 launcher second stage, the Soviet Union's smallest satellite vehicle and retired in 1977. The single fixed chamber is constructed principally of titanium and

GDL's RD-111 four-chamber engine which was used on the first stage of the R-9 missile (William C Ingalls) 0003410

unusually burns UDMH as fuel. Steering is effected by routing the turbopump exhaust through four large and four small fixed verniers.
Application: Cosmos satellite launcher stage 2
Dry mass: 168 kg
Engine cycle: open (turbine exhaust utilised for steering)
Oxidiser: liquid oxygen
Fuel: UDMH
Feed method: turbopump
Thrust: 105 kN vacuum
Specific impulse: 347 s vacuum
Chamber pressure: 80 atm
Burn time: 260 s

RD-120 (11D123)
The single fixed chamber is supported by four NPO Yuzhnoye 'RD-8' (actually the true designator might be in the 'RD-800' series) verniers for steering. Each engine is test-fired at the factory for the intended flight duration. Reliability level is 99.5%. The design minimises the upper section for packaging; on Zenit it protrudes from a toroidal tank. An uprated version, the RD-120.03, has been mooted; its characteristics are included below in (). Two RD-120.01 versions are being considered for stage 1 applications: thrust 715.8/784.5 kN sea level (784.5/853.1 kN vac); SI 302/304 s sea level (329/330 s vac); chamber pressure 160.6/174.2 atm; length 2,430 mm; 1,400 mm dia; dry mass 1,065/1,079 kg. (It has been claimed the 120.03 and 120.01 versions *have* been built and tested.) The second 120.01 version is now designated the RD-120M, with 4.96 expansion ratio, 2.6 ± 10% mixture ratio, –40/+3.5% throttling and ±6° 2-axis gimballing. It was selected for the OSC/Rockwell X-34 before the companies pulled out. It has been suggested for the Rus Soyuz upgrade.
An RD-120 production engine was tested Oct 1995 by Pratt & Whitney in West Palm Beach, Florida, believed to be the first of an in-production and in-service engine outside of Russia. The three <5 s runs showed the engine was compatible with US ground systems; further testing requires a sea level nozzle and a customer.
Application: Zenit stage 2
Dry mass: 1,125 kg dry, 1,285 kg with propellant; (1,125 kg)
Length: 3,872 mm (3,872 mm)
Maximum diameter: 1,954 mm (1,954 mm)
Engine cycle: closed gas generator
Oxidiser: liquid oxygen at 175.4 kg/s

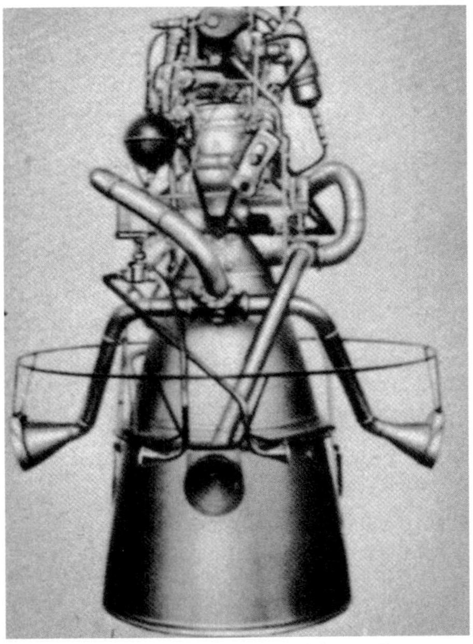

RD-119: single main chamber and eight verniers 0516886

The RD-120 powers Zenit's stage 2
(Commercial Space Technologies) 0516887

Fuel: kerosene at 64.46 kg/s
Feed method: 19,000 rpm turbopump; gas generator 462°C
Thrust: 833 kN vacuum (throttling –8.5/+3.5%); (882.5 kN)
Specific impulse: 350 s vacuum; (353 s)
Expansion ratio: 106:1
Chamber pressure: 160.6 atm (nozzle 0.126 atm); (170.3 atm)
Burn time: 200–315 s

RD-161
Soyuz-2 may use the single chamber engine. The engine gimbals ±6° in two planes and can ignite up to 17 times in flight. It also provides high pressure LOX for attitude control verniers. The specifications are given below in order of with/without the uncooled nozzle extension.
Application: launcher upper stage
Dry mass: 141/119 kg
Length: 2,205/1,700 mm
Maximum diameter: 1,020/780 mm
Engine cycle: closed gas generator
Oxidiser: liquid oxygen
Fuel: kerosene
Mixture ratio (O/F): 2.6
Feed method: turbopump
Thrust: 19.9/19.6 kN vacuum
Specific impulse: 365/360 s vacuum
Expansion ratio: 19.25/18.75
Chamber pressure: 116.1 atm
Burn time: 900 s maximum

RD-169/RD-185
The RD-169 is proposed to power KB Machinostroenye's new Rikscha small launcher. Six would be clustered for stage 1 (gimballing 6°), followed by the RD-185 altitude version for stage 2 (gimballing 3°). Specifications are given below in RD-169/185 order.
Dry mass: 215/445 kg
Length: 1.5/3.0 m
Maximum diameter: 0.5/1.5 m
Engine cycle: closed gas generator

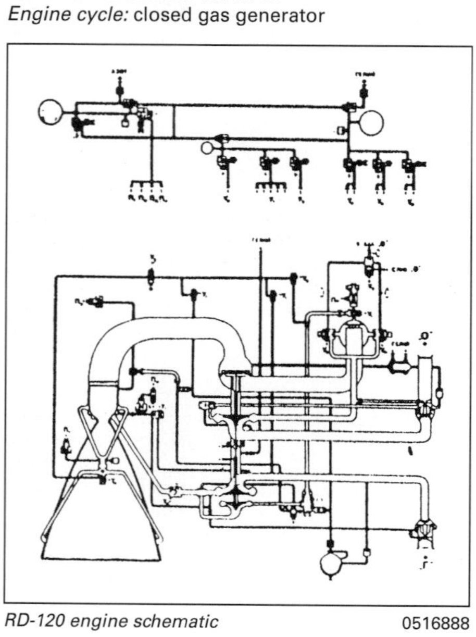

RD-120 engine schematic 0516888

The RD-170/171 was publicly displayed for the first time at 1989's Paris Salon 0517372

Oxidiser: liquid oxygen
Fuel: methane
Mixture ratio (O/F): 3.4/3.4
Feed method: turbopump
Thrust: 147/- kN SL; 167/179 kN vac
Specific impulse: 307/- s SL; 349/376 s vac
Expansion ratio: not available
Chamber pressure: not available
Burn time: not available

RD-170/171/172/191
With the Energia core engine, the most recent Soviet rocket engine development and the highest known chamber pressure. It is also the highest thrust liquid propellant rocket engine ever flown, surpassing Rocketdyne's single chamber F-1. Development began in 1976. The 4-chamber engine (paired chambers) flies in the first stage of the Zenit 2 medium lift launcher, which is also modified as Energia's reusable strap-ons. It can make 10 flights although there is no evidence that it has been re-used. There are only minor differences between Energia's RD-170 (11D520) and Zenit's RD-171 (11D521); for example, 1-axis gimballing for Zenit and 2-axis for Energia. Engines were manufactured from each batch by Polyot in Omsk; one was taken through three life cycles while the remainder underwent a single acceptance test. Polyot production for both RD-170/171 has halted and the plant dismantled, but Energomash itself can build up to 20 annually. The 190 MW 14,500 rpm turbopump of each chamber pair is driven by pre-burners (see diagram). The design permits smooth transition to full thrust over 2 s. At burnout, thrust drops to 70% over 30 s, to 50% over the next 2 s, where it holds for 10 s, and finally shuts down in 0.5 s.

Development of the improved RD-172 for Zenit 2 was expected to achieve certification in 1996 but there has been no word on this recently. A 5% thrust increase is provided by a chamber pressure of 255.4 atm. The similar RD-174 version is expected to be available in 1996. The single-chamber RD-191, utilising 70% of RD-170 elements, would provide 2,059 kN.
Applications: Zenit stage 1, Energia strap-ons
First flown: 13 Apr 1985
Dry mass: 8,755 kg
Length: 3.56 m
Maximum diameter: 399 cm
Engine cycle: staged combustion

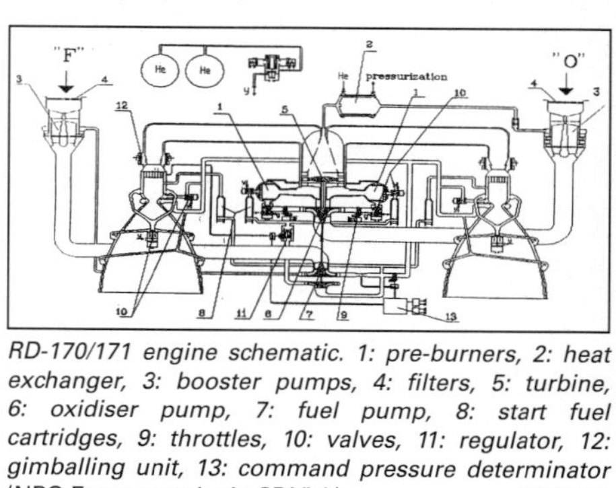

RD-170/171 engine schematic. 1: pre-burners, 2: heat exchanger, 3: booster pumps, 4: filters, 5: turbine, 6: oxidiser pump, 7: fuel pump, 8: start fuel cartridges, 9: throttles, 10: valves, 11: regulator, 12: gimballing unit, 13: command pressure determinator
(NPO Energomash via CP Vick) 0516889

Oxidiser: liquid oxygen at 432 kg/s
Fuel: kerosene at 166.2 kg/s
Thrust: 7,903.77 kN vacuum, 7,256.56 kN SL (but can be throttled back to 4,413 kN vac) for Energia, 7,911 kN vac and 7,259 kN SL for Zenit 2/3
Specific impulse: 308 s SL, 336 s vacuum for Energia; 337 s for Zenit-2 vacuum
Expansion ratio: 36.8
Chamber pressure: 242 atm; 250 atm maximum
Chamber mass: 480 kg
Burn time: 140–150 s

RD-180
NPO Energomash developed the RD-170's 2-chamber version in co-operation with Pratt & Whitney for Lockheed Martin's Atlas 2 (qv) and 5 series. As only the main turbopump and boost pumps require development (all other components are interchangeable with the RD-170), relatively easy adaptation facilitated an early transition to the US launch vehicle. The first 18 vehicles will use Russian-built engines. It was the baseline engine for Lockheed Martin's proposal for the USAF Evolved Expendable Launch Vehicle (EELV) competition.
Applications: launcher stage 1
First flown: 24 May 2000
Dry mass: 4,164 kg
Length: 3.56 m
Maximum diameter: not available
Engine cycle: staged combustion
Oxidiser: liquid oxygen
Fuel: kerosene
Thrust: 4,164 kN vacuum, 3,826 kN sea level, throttleable 40–105%. Atlas 2AR will use it at max 85% (75% at launch)
Specific impulse: 304 s sea level, 337 s vacuum
Expansion ratio: 36.8
Chamber pressure: 242 atm; 250 atm maximum
Chamber mass: 480 kg
Burn time: 185 s on Atlas 2AR

RD-190
The RD-190 is a single chamber derivative of the RD-170 engine. No application for the engine is known other than a potential first stage of a launch vehicle. The engine can be throttled within the range 40–100%.
Dry mass: 2,517 kg
Length: 3.7 m
Maximum diameter: 1.8 m
Engine cycle: stages combustion
Oxidiser: liquid oxygen
Fuel: staged combustion
Thrust: 1,900 kN sea level, 2,075 kN vacuum
Specific impulse: 311 s sea level, 337 s vacuum

RD-211
Little is known of this engine, which completed development in 1963 with specific impulse of about 314 s. It has been suggested to be an upper stage engine.

RD-214
The first Soviet engine to be identified which uses storable nitric acid as oxidiser for a rapid response

The RD-180 will power Lockheed Martin's Atlas 2AR
(P&W) 0516890

The RD-216 assembly powers Polyot's Cosmos 3M stage 1 (AKO Polyot) 0517187

time from the R-12 IRBM, the RD-214 carried four fixed chambers and no verniers – steering was effected by refractory vanes protruding into the exhaust.
Applications: Cosmos satellite launcher stage 1, R-12 IRBM stage 1
First orbital flight: Mar 1962
Number flown: 144 on orbital missions
Dry mass: 645 kg
Engine cycle: open
Oxidiser: nitric acid
Fuel: kerosene
Feed method: single turbopump driven by hydrogen peroxide gas generator feeding four fixed chambers
Thrust: 635 kN sea level, 730 kN vacuum
Specific impulse: 225 s sea level, 259 s vacuum
Chamber pressure: 45 atm
Chamber: 480 mm dia internal, 176 mm throat dia
Burn time: 140 s

RD-216
The first Soviet engine to utilise the storable nitric acid/UDMH combination as the first stage propulsion system for the R-14, subsequently adapted for the Cosmos 3 launcher and, until Start, the smallest satellite vehicle in the inventory. RD-216/11D614 comprises four (two pairs) combustion chambers. The engines are built by the Ust-Katavski Machinebuilding Plant at Ust-Katav in the Urals.
Applications: Cosmos 3/3M satellite launcher stage 1, R-14 IRBM stage 1
First orbital flight: Aug 1964
Dry mass: 662 kg
Engine cycle: open
Oxidiser: nitric acid/27% N_2O_4
Fuel: UDMH
Feed method: single turbopump feeds each chamber pair
Thrust (4 chambers): 1,469 kN sea level, 1,728 kN vacuum
Specific impulse: 248 s SL, 291.3 s vac
Chamber pressure: 75 atm
Burn time: 170 s (130 s on Cosmos 3M)

RD-218 (8D712)
The RD-218 engine was originally reported to have been the first stage of the R-36 missile and the

RD-218 six-chamber engine which was used on the first stage of the R-16 missile (William C Ingalls) 0003411

The RD-219 incorporates two thrust chambers fed by a single turbopump 0516891

Tsyklon launch vehicles. It is now known that the RD-218 was used on the R-16 missile's first stage. It comprised three engine assemblies, each with two combustion chambers and a turbo-pump assembly. The development period is given as 1958–1961.
Application: R-16 missile first stage
First flown: 1961
Dry mass: 1,960 kg
Length: 2.188 m
Maximum diameter: 2.789 m
Engine cycle: closed
Oxidiser: nitric acid
Fuel: UDMH
Thrust: 2,220 kN sea level, 2,609 kN vacuum
Specific impulse: 246 s sea level, 289 s vacuum
Chamber pressure: 73.5 atm

RD-219 (8D713)
Two-chamber, high-altitude engine developed during 1958–1961 for the R-16 missile second stage (although originally it was reported that it was used on the R-36 missile and Tsyklon launch vehicles).
Application: R-16 second stage
First flown: 1961
Dry mass: 665 kg
Length: 2.030 m
Maximum diameter: 2.162 m
Engine cycle: open
Oxidiser: nitric acid
Fuel: UDMH
Feed method: single turbopump feeding two chambers
Thrust: 883 kN vacuum
Specific impulse: 287.5 s vacuum
Chamber pressure: 75 atm
Burn time: 125 s

RD-250
No details are available concerning this engine, other than it was an early version of what became the RD-253, the engine used on the first stage of the Proton-K. Presumably it had a single thrust chamber.
Oxidiser: nitrogen tetroxide?
Fuel: UDMH

The RD-251 six-chamber engine which was used on the first stage of the R-36 missile and the two-stage Tsyklon-M launch vehicle. The engine was developed in parallel with the RD-218 used on the related R-16 missile (William C Ingalls) 0003412

RD-251
The RD-251 is used on the first stage of the R-36 missile and the Tsyklon-2 and Tsyklon-M launch vehicles. It was developed 1961 to 1965 in parallel with the RD-218 used on the R-16 missile. Like the RD-218, it comprises three independent two-chamber engines clustered together and operating as a single unit. The development period is given as 1961–1965.
Application: R-36, Tsyklon first stage
First flown: 1965
Dry mass: 1,729 kg
Length: 1.762 m
Maximum diameter: 2.520 m
Engine cycle: open
Oxidiser: nitrogen tetroxide
Fuel: UDMH
Feed method: turbopump
Thrust: 2,364 kN sea level, 2,643 kN vacuum
Specific impulse: 269 s sea level, 301 s vacuum
Chamber pressure: 85 atm

RD-252
Like the RD-251 being developed between 1961 and 1966 in parallel with the RD-218, the RD-252 for the second stage of the R-36 missile and Tsyklon launch vehicle was developed in parallel with the RD-219. The engine has two combustion chambers and was developed during the period 1961–1966.
Application: R-36, Tsyklon second stage
First flown: 1965
Dry mass: 715 kg
Length: 2.186 m
Maximum diameter: 2.590 m
Engine cycle: open
Oxidiser: nitrogen tetroxide
Fuel: UDMH
Feed method: single turbopump
Thrust: 940.5 kN vacuum
Specific impulse: 318 s vacuum
Chamber pressure: 91 atm

RD-253 (11D43)
The first known Soviet closed cycle and N_2O_4/UDMH engine, the gimballed single chamber provides the first stage power for Proton's six clustered external tanks. The engine was displayed at the 1987 Paris Air Show. A modified configuration with improved performance (SI increased by 30 s) was introduced on Proton in 1986. The original version incorporated a high-mounted pump but the new version is side-mounted. The engines are built by the PPO Motorostroitel (=Engine Builder; see their National section entry).
Application: Proton stage 1
First flown: 16 July 1965
Dry mass: 1,280 kg
Length: 2.72 m
Maximum diameter: 1.50 m
Engine cycle: closed
Oxidiser: nitrogen tetroxide
Fuel: UDMH
Propellant flow rate: about 528 kg/s; mixture ratio 2.67
Feed method: 18.7 MW turbopump driven by pre-burner gas
Thrust: 1,635 kN vacuum (1,474 kN sea level), uprated from 1,635 kN (1,460 kN sea level). Profile is 40% full thrust in 0.1 s, held at 40% for 2 s, then to 100% in 0.1 s
Specific impulse: 310 s vacuum, 279 s sea level
Chamber pressure: 150 atm
Burn time: 130 s

RD-261 and RD-262
There have been some reports that while the RD-251 and RD-252 engines are used on the first and second stages (respectively) of the Tsyklon-M vehicle, modifications of these engines designated RD-261 and RD-262 (respectively) are carried on the first and second stages of the three-stage Tsyklon. Russian confirmation of these reports has not been found.

RD-270 (8D420)
Although apparently not tested beyond component level, the RD-270 was designed for Chelomei's UR700 lunar vehicle and then rejected for Korolev's N1. It would have been the largest Soviet storable propellant engine, operating with a chamber pressure greater than RD-170's.
Application: launcher stage 1
Dry mass: 4,770 kg (5,603 kg wet)
Length: 4.850 m
Maximum diameter: 3.300 m

RD-253 employs a side-mounted turbopump (Andrew Wilson) 0517373

Engine cycle: gas-gas closed staged
Oxidiser: nitrogen tetroxide
Fuel: UDMH
Propellant flow rate: unknown
Feed method: turbopump
Thrust: 6,713 kN vacuum, 6,272 kN sea level
Specific impulse: 322 s vacuum, 301 s sea level
Chamber pressure: 266 atm
Burn time: unknown

RD-301
Developed 1969–75, the RD-301 was intended as Proton's stage 3 engine but flight tests were cancelled when it was decided to be too toxic. It was the first known example of experimentation with high SI exotic propellants, although details on the earlier RD-502 subsequently emerged.
Mass: 183 kg
Length: 1.88 m
Maximum diameter: 98 cm
Engine cycle: open
Oxidiser: fluorine
Fuel: ammonia
Mixture ratio (O/F): 2.7:1
Feed method: 1,265 kW turbopump driving 2-stage oxidiser & single-stage fuel pumps and a turbine
Thrust: 98.1 kN vacuum
Specific impulse: 393 s vacuum
Expansion ratio: 108.7
Chamber pressure: 120 atm
Burn time: 750 s

RD-502
Developed 1960–66, the RD-502 was intended as Proton's stage 3 engine but work was halted when it was decided, like RD-301 before it, to be too toxic.

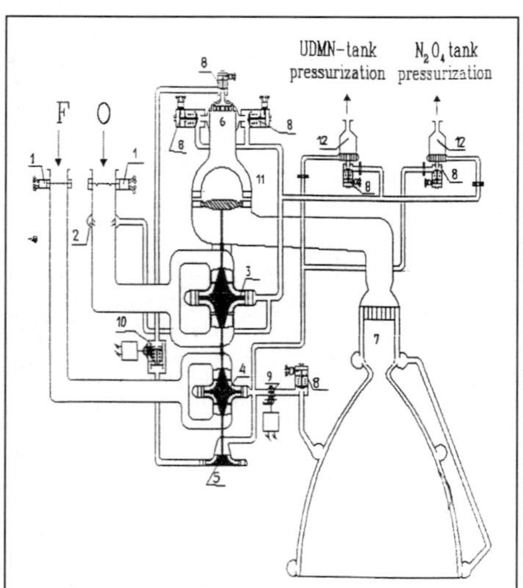

RD-253 engine schematic. 1: valves, 2: ejector, 3: oxidiser pump, 4: main fuel pump, 5: pre-burner pump, 6: preburner, 7: combustion chamber, 8: shutoff valves, 9: fuel throttle, 10: proportional flow regulator, 11: turbine, 12: tank pressurisation generators (NPO Energomash) 0516892

Mass: 132 kg (140 kg wet)
Length: 2.510 m
Maximum diameter: 1.180 m
Engine cycle: closed staged
Oxidiser: hydrogen peroxide
Fuel: pentaborane
Mixture ratio (O/F): unknown
Feed method: turbopump
Thrust: 98.06 kN vacuum
Specific impulse: 380 s vacuum
Expansion ratio: unknown
Chamber pressure: 150 atm
Burn time: unknown

RD-701
NPO Energomash has been working on this tripropellant RD-170 derivative engine since 1989 to propel a small 22 t shuttle-type vehicle designed by NPO Molniya for release from an An-225 at altitude. For the first 2½ min, it would burn principally kerosene and oxygen (mode 1), using LH_2 for cooling from the separate tank of 205 t of propellants. It would then switch (mode 2) to the less dense LH_2 at altitude for the final 6–8 min to orbit. The required 5 yr engine development would cost <$500 million; international partners are sought. Each engine would comprise two thrust chambers gimballed in 2-axes. Each chamber would employ two turbopumps: one channeling LOX/kerosene into a pre-burner where they are burned oxygen-rich for the product to burn with hydrogen in the main chamber. US marketing is being provided by P&W. NASA awarded P&W a $5.4 million contract in Aug 1994 towards converting the RD-701 into a tripropellant engine for a US reusable launcher. The RD-704 single chamber derivative is also proposed for US applications.
Mass: 3,990 kg assembly, 1,840 kg each dry
Length: 5.4 m extension down, 3.8 m extension up
Diameter: 2.4 m
Engine cycle: closed staged combustion
Oxidiser: liquid oxygen
Fuel: mode 1: hydrogen/kerosene; mode 2: hydrogen
Mixture ratio (O/F; mass): mode 1 5.27:1 kerosene and 13.2:1 hydrogen; mode 2 6.1:1
Flow rates (each chamber, kg/s): mode 1 388.4 LOX, 29.5 LH_2/73.7 kerosene; mode 2 148.5 LOX/ 24.7 LH_2
Thrust (each chamber, vac): 1,960 kN mode 1, 785 kN mode 2 (throttle 40–100%)
Specific impulse (s): mode 1 415 vac, 330 SL; mode 2 460 vac
Chamber pressure: mode 1 290 atm; mode 2 122 atm
Expansion ratio: 170 extension down, 70 extension up

KELDYSH RESEARCH CENTER
Some details of engines designed by the Keldysh Research Center are available, although none of the engines described here is believed to have actually flown.

Hybrid Rocket Engines
The Keldysh Research Center is undertaking work on the development of hybrid rocket engines which could be used on lightweight launch vehicles or upper stages of launch vehicles. The advantages of hybrid engines are seen as: simplicity of design and reduction of payload cost by 30%, friendly to the environment during production and when in service, manufacture and service safety, fire and explosion safety preventing the failure of launch vehicles, high reliability, a multiple burn capability while throttling over a wide thrust range and the use of proven designs, advanced technologies and production equipment which reduce the cost of engine production by a factor of 1.5–2 compared with existing engines.

The principles of the design of hybrid engines have been developed at the Center and a model and demonstration prototype with a thrust of 29 kN (Keldysh literature states 3 t) have been developed. The design performances of hybrid engines have been proved by actual performances and the documentation for a modular hybrid engine giving thrusts of 177 kN (18 t) and 294 kN (30 t).
Oxidiser: oxygen, hydrogen peroxide
Fuel: rubber
Specific impulse: 3,140–3,335 kN (first stage), 3,335–3,725 kN (second stage)
Chamber pressure: 20–60 atm
Burn time: 400 s (max)

Ion Thruster for Small Space Vehicles
An ion thruster has been developed for the attitude control of small satellites intended for either remote sensing or communications. Performance data are as follows:
Power input: 50–150 W

The tripropellant dual mode RD-701 could power a spaceplane (William C Ingalls) 0003413

Thrust: 2–5 mN
Specific impulse: 3,000–3,500 s

RD-704 Operating Data

Operating Mode	**Mode 1**	**Mode 2**
Fuel	kerosene/liquid hydrogen	liquid hydrogen
Oxidiser	LOX	LOX
Thrust: vac	1,960 kN	781 kN
sea level	1,720 kN	(N/A)
Specific impulse: vac	407 (401) s	452 (450) s
sea level	356 (351) s	(N/A)
Mixture ratio (O/F)	4.38	6.0
Chamber pressure	290 atm	120 atm
Area ration	74	
Mass	2,417 kg	
Single bell diameter	1.8 m	
Length	3.8 m	

Liquefied Air Cycle Engine (LACE)
LACE is an advanced hybrid engine being studied for an aerospace plane. Advantages of the design are given as: 2.5–4.5 fold improvement of performance compared with liquid propellant engines, specific thrust in the atmosphere is 1.5–2 times higher compared to other air-breathing engines, use of a single-type engine as the propulsion system, maximum possibility of final development using the existing test models and ease of engine integration into spacecraft. When developed using LACE in aerospace planes would

RD-0109 engine used on the Vostok 8K72K launch vehicle's orbital stage: it was this variant which was used for the six manned launches in the Vostok programme (KB Khimavtomatiki) 0003414

RD-0110 engine used on the Soyuz-U and Molniya-M launch vehicles' orbital stages (KB Khimavtomatiki) 0003415

permit a payload of 10 t to be launched with a single-stage take-off mass of 340 t: compared with a rocket-based single stage launch vehicle the launch mass would be reduced by a factor of 1.5–2 and the cost of launching each kilogramme into orbit would be reduced by a factor of 15–20 compared with an expendable launch vehicle.

The Keldysh Research Center states that concepts and configurations for a LACE have been developed and experimental investigations of the main stages of the operation of the engines have been studied experimentally using sub-scale engines. A sub-scale demonstration LACE is being assembled at the Center's test facility and technical proposals for the LACE and the aerospace plane are being prepared with other specialist Russian design bureaus. The investigations are being undertaken within the framework of a joint programme between the Russian Space Agency and the European Space Agency.

A few numbers have been released concerning LACE. Two modes of operation are noted, with a specific impulse within the range 1,300–2,000 s being attained. The engine would operate in two modes: using liquefied air for Mach 0–6 and in rocket mode for Mach 6–25. In the short term a structural specific mass of 60 kg/t of thrust should be obtainable, reducing to 35 kg/t of thrust in the longer term.

'Rocket Engine for the 21st Century'

A new liquid engine is being studied as the basis for the development of a new generation of 21st century launch systems. The requirements for the launch system are: increased reliability and safety in operation, everything should be reusable and major savings in the cost of putting spacecraft into orbit. The Keldysh Research Center is proposing a new family of liquid propellant rocket engines which would use ecologically-friendly propellants. It is hoped that these proposed engines would provide: reduction of costs as well as ease of development, testing and manufacture of the engine, high energy and mass characteristics, minimum delays between inflight servicing and reuse.

Model prototypes of these engines have undergone successful test firings. The most attractive propellants for the programme appear to be oxygen/methane for the first stage and oxygen/hydrogen for the second stage. The engines could be developed by 2004 (one assumes that finances are the limiting factor) and the Center Claims that there is nothing like their proposed engines being developed elsewhere. However, there are no further details available concerning the engines.

The RD-0120 is Russia's first operational cryogenic engine (KB Khimavtomatiki) 0003370

T-100 Electric Thruster

The T-100 electric thruster is described by the Keldysh Research Center as being 'a superb generator of a high-velocity directed flow of xenon plasma', and applications in both the space programme and new technologies within Russian industry are foreseen. The thruster was developed as an actuator for spacecraft manoeuvring systems for orbital transfers, injection of a spacecraft into a planned orbit, attitude control and orbital corrections. The T-100 is described as being highly efficient with a low-level of radiation, a long life and a short time for preparation. It is unclear whether this thruster has actually been flown on spacecraft.

Dry mass: <32 kg
Dimensions: 140 × 135 × 230 mm
Propellant: high-purity xenon
Thrust: 80 mN
Specific impulse: 1,600–1,800 s (literature says 16,000–18,000 m/s)
Efficiency: 50% (minimum)
Propellant flow rate: 5 mg/s
Supply voltage: 300 V AC
Input power: 1.4 kW
Preparation time for operation: <1 s
Burn time: 8,000 h
Start/shutdown cycles: 6,000

KB KHIMAUTOMATIKI

The bureau of Semyin A Kosberg (1903–65), formed 1941 as OKB-296, began investigating liquid propellant rocket propulsion for aircraft in the early 1950s and later became responsible for some Soviet space launcher upper stages. In 1974 it was reorganised as KB Khimautomatiki. In 1994 it was transferred to the Russian Space Agency.

RD-0105

Also called the RO-5, this engine was the original one used on the Vostok launch vehicle's Block E. The engine assembly comprised a single main combustion chamber plus four verniers. It has not been used since the failed Luna launches in Apr 1960.

Application: Original Vostok Block E
First flown: 1958
Dry mass: =~125 kg
Feed method: pump
Thrust: 49 kN vacuum
Specific impulse: 316 s vacuum
Chamber pressure: 44 atm
Burn time: 480 s

RD-0106

The RD-0106 was the first engine in the family which would be used on the Block I orbital stages of the Molniya, Soyuz and Voskhod families of launch vehicles. The RD-0106 itself was used on the second stage of the R-9A missile. Assuming that it was like the later variants, it had four main combustion chambers plus four vernier chambers.

Application: R-9A missile
First flown: 1961
Oxidiser: liquid oxygen
Fuel: kerosene

RD-0107/8D715

The RD-0107 was the first development from the RD-0106, and it was used on the Block I of the Molniya and then Voskhod launch vehicles. The engine has also been designated the RO-9. The engine had four main combustion chambers and four vernier chambers.

Application: Molniya, Voskhod Block I
First flown: 1960
Dry mass: 409 kg
Oxidiser: liquid oxygen
Fuel: kerosene
Thrust: 294 kN vacuum
Specific impulse: 326 s vacuum
Chamber pressure: 67 atm
Burn time: 250 s

RD-0124 engine planned to fly on the orbital stage of the new Soyuz-2 ('Rus') launch vehicle: as such, this engine is a replacement for the RD-0110 (KB Khimavtomatiki) 0003416

RD-0108

The RD-0108 was used on the Block I of the original Soyuz, Soyuz-L and Soyuz-M launch vehicles. The performance details are probably similar to those of the RD-0110 which replaced it. Like the earlier versions of the same basic engine, it carried four main combustion chambers and four vernier chambers.

Application: Soyuz, Soyuz-L, Soyuz-M Block I
First flown: 1966
Oxidiser: liquid oxygen
Fuel: kerosene

RD-0109/8D719

This engine replaced the RD-0105 on the orbital stage of the Vostok launch vehicle: it has also been referred to by the RO-7 and RD-448 designators. The designator 11D719 has also been used rather than the 8D719, and it is unclear which one is correct. The engine has a single main combustion chamber plus four vernier chambers.

Application: Vostok Block E
First flown: 1960
Dry mass: 121 kg
Length: 1.535 m
Maximum diameter: 2.1 m?
Oxidiser: liquid oxygen
Fuel: kerosene
Feed method: pump
Thrust: 54.8 kN vacuum
Specific impulse: 324 s vacuum
Chamber pressure: 49 atm
Burn time: 430 s

RD-0110/11D55

The RD-0110 replaced the RD-0107 on the Block I of the Molniya launch vehicle and the RD-0108 on the same stage of original group of Soyuz launch vehicles. The engine has four main combustion chambers plus four vernier chambers.

Application: Molniya-M and Soyuz-U Block I
First flown: 1964
Dry mass: 408 kg
Length: 1.575 m
Maximum diameter: 2.240 m
Oxidiser: liquid oxygen
Fuel: kerosene
Thrust: 298 kN vacuum
Specific impulse: 330 s vacuum
Chamber pressure: 69 atm
Burn time: 240 s

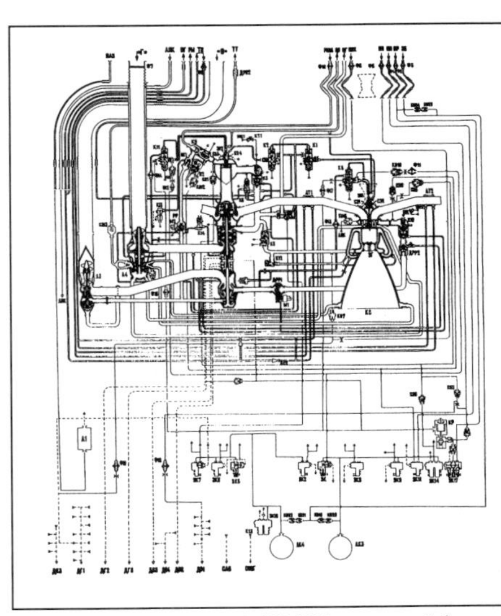

RD-0120 engine schematic (KB Khimautomatiki) 0516893

Khimmach's 11D49 engine power's Polyot's Cosmos 3M stage 2 (KB Khimmash) 0003373

RD-0120/RO-200/11D122
Energia's four single chamber core engines is the first operational Soviet cryogenic system, following space introduction by the US (1962), ESA (1979), China (1984) and Japan (1986). Only the US Shuttle had previously flown a cryogenic first stage. The engine was first displayed in Moscow in Sep 1990. Each chamber gimbals ±11° in two axes for steering. >800 tests exceeding 170,000 s in total had been performed to end-1995 since runs began in 1979. A version capable of 25 flights was proposed for the joint Interim Hotol vehicle (see the UK Launchers section). An agreement was signed Oct 1994 with Aerojet of the US to codevelop and market derivatives in the US; it could be developed into a tripropellant engine for a future reusable vehicle. Aerojet's work is being performed under a $17.2 million NASA Marshall contract. Europe's RECORD (Russia-Europe Cooperation on Rocket Engine Demonstration) project, initiated in 1993 by SEP and its Vulcain industrial partners, began tests at NII Khimmach, Russia's largest test site, in Aug 1995 to help validate modelling for future engines. The ECU20.5 million cost over 3 yr is covered by: 13 European Commission; 6 ESA; 1.5 industry.
Application: Energia core, 4 engines
First flown: 15 May 1987
Number flown: 8 to end-2000
Dry mass: 3,449 kg
Length: 4,549 mm
Maximum diameter: 2,420 mm
Engine cycle: closed
Oxidiser: liquid oxygen at 376.7 kg/s
Fuel: LH_2 at 62.78 kg/s
Mixture ratio (O/F): 6.0:1
Feed method: 33,385 rpm dual-stage turbopump on each chamber. A single pre-burner burns fuel-rich at 527°C/408 atm to drive the single-shaft high pressure turbopump. Some of the pre-burner gas drives the oxygen low pressure pump; the fuel low pressure pump is driven by GH_2 from the main chamber cooling loop. Main LH_2 discharge pressure 448 atm; LOX stage 1 342 atm; LOX stage 2 571 atm
Thrust: 1,961 kN vacuum, 1,451 kN sea level; throttle range 25–114%
Expansion ratio: 85.7:1 (throat dia 261 mm, exit dia 2,420 mm)
Specific impulse: 455 s vacuum
Chamber pressure: 215.7 atm
Burn time: about 480 s (600 s max operational)

The KVD-1 cryogenic engine will fly on India's GSLV 0003369

RD-0124
This engine is planned to replace the RD-0110 when the new Soyuz-2 launch vehicle is introduced.
Application: Soyuz-2 Block I
First flown: 1997–1998?
Dry mass: 450 kg
Length: 1.575 m
Maximum diameter: 2.273 m
Oxidiser: liquid oxygen
Fuel: kerosene
Thrust: 294 kN
Specific impulse: 359 s
Chamber pressure: 157 atm

RD-0202
RD-0202 is the designator applied to the cluster of four engines on the first stage of the UR-200 missile: it possibly comprises three RD-0203 engines and one RD-0204 engine.

RD-0203/RD-0204
These two engines were used on the first stage of the UR-200 missile. The stage carried four single-chamber engines, and it was thought that one of these carried the tank pressurisation system while the other three did not (hence the two different engine designators for the stage). However, according the KB Khimautomatiki neither of the engines carries a tank pressurisation system, and therefore the difference between them is not known. The Proton-K launch vehicle's RD-0210 and RD-0211 are believed to be altitude-ignition variants of these two engines. The development period is given as 1961–1964.
Application: UR-200 missile first stage
First flown: 1964
Engine cycle: closed
Oxidiser: nitrogen tetroxide
Fuel: UDMH

RD-0205
This designator is applied to the combined RD-0206 main engine and RD-0207 vernier engines on the UR-200 missile's second stage.

RD-0206/RD-0207
The RD-0206 is the main engine carried on the second stage of the UR-200 missile, with the RD-0207 being the vernier engine on the same stage. It is possible that the RD-0206 is a modification for altitude ignition of the RD-0204 and RD-0205 engines. The RD-0206 carries a tank pressurisation system. The development period is given as 1961–1964.
Application: UR-200 second stage
First flown: 1964
Engine cycle: RD-0206 – closed, RD-0207 – open
Oxidiser: nitrogen tetroxide
Fuel: UDMH

RD-0210/8D411
Three RD-0210 engines are carried on the second stage of the Proton-K launch vehicle. They are single-chamber engines.
Application: Proton-K second stage
First flown: 1965
Engine cycle: closed (oxidiser pre-burner gas routed to main chamber after driving turbine)
Oxidiser: nitrogen tetroxide
Fuel: UDMH
Thrust: 594 kN
Specific impulse: 327.4 s
Chamber pressure: 148 atm
Burn time: 210 s

RD-0211/8D412
A single RD-0211 engine is carried on the second stage of the Proton-K launch vehicle. They are single-chamber engines and carry the tank pressurisation system. The performance data are as for the RD-0210.

RD-0212/8D49
This designator is applied to the RD-0213/RD-0214 engine combination which is used on the third stage of the Proton-K.

RD-0213
The RD-213 is the main engine on the Proton-K third stage, and is a modification of the RD-0210/RD-0211 engines used on the launch vehicle's second stage.

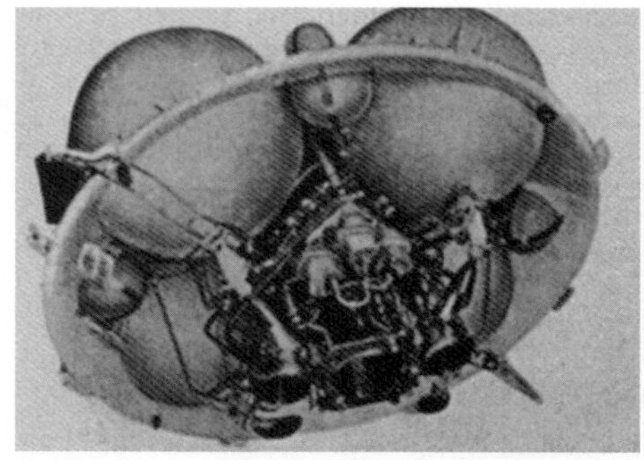

Soyuz manned spacecraft KTDU-35 propulsion system consists of a single primary central chamber and two reserve chambers. The four smaller associated chambers are part of the separate hydrogen peroxide attitude control network 0516894

Application: Proton-K third stage main engine
First flown: 1967
Engine cycle: closed
Oxidiser: nitrogen tetroxide
Fuel: UDMH
Thrust: 593.6 kN
Specific impulse: 325.3 s
Chamber pressure: 148 atm
Burn time: 240 s

RD-0214
The RD-0214 is the four-chamber vernier engine assembly carried on the third stage of Proton-K. It carries the tank pressurisation system for the stage.
Application: Proton-K third stage vernier engine
First flown: 1967
Engine cycle: open
Oxidiser: nitrogen tetroxide
Fuel: UDMH
Thrust: 31.5 kN
Burn time: 240 s

RD-0216
Three RD-0216 engines were carried on the first stage of the UR-100 missile. It was developed during 1963–1966. It carries a single main combustion chamber.
Application: UR-100 missile first stage
First flown: 1965
Engine cycle: closed
Oxidiser: nitrogen tetroxide
Fuel: UDMH

RD-0217
A single RD-0216 was used on the first stage of the UR-100 missile and it carried the stage's pressurisation system. It was developed during 1963–1966. It carries a single main combustion chamber.
Application: UR-100 missile first stage
First flown: 1965
Engine cycle: closed
Oxidiser: nitrogen tetroxide
Fuel: UDMH

The 11D58M (left) powers Proton's stage 4; its 17D12 modification was used on Buran (Andrew Wilson) 0517320

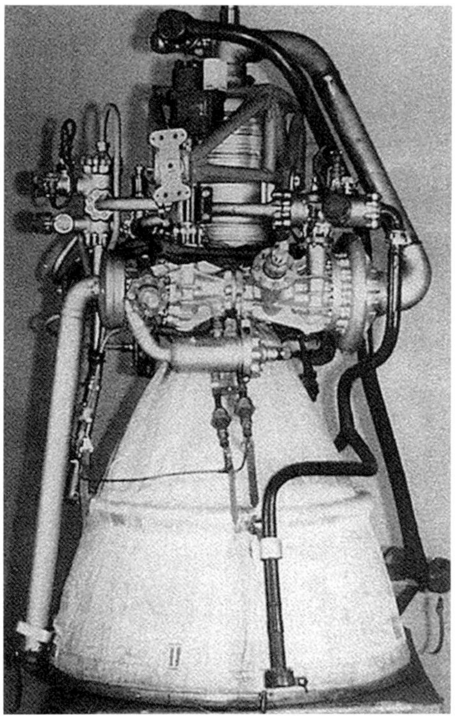

Molniya's Block L engine 0006583

RD-0225

Developed as the main manoeuvring engine of the military Almaz manned space station. A 3.9 kN Khimmach engine is used by the Heavy Cosmos modules.
Applications: paired as Almaz orbital manoeuvring engine
First flown: 3 Apr 1971, Almaz 1
Engine cycle: pressure fed
Oxidiser: nitrogen tetroxide
Fuel: UDMH
Thrust: 3.9 kN vac
Specific impulse: 291 s vac
Chamber pressure: 8.8 atm
Burn time: ?

RD-0229

A single RD-0229 is used as the main engine on the second stage of the R-36M missile, and it carries the tank pressurisation system for the stage. The development period is given as 1967–1974.
Application: R-36M missile second stage main engine
Engine cycle: closed
Oxidiser: nitrogen tetroxide
Fuel: UDMH

RD-0230

The RD-0230 is the four-chamber vernier engine system for the R-36M missile. The development period is given as 1967–1974.
Application: R-36M missile second stage vernier engine
Engine cycle: closed
Oxidiser: nitrogen tetroxide
Fuel: UDMH

RD-0231

This engine was developed during the period 1968–1970 and was used on a spacecraft designated 'Granit': however it is not known to which particular spacecraft this name refers – it is certainly not the Granat astrophysical observatory.

Lyulka's only rocket engine is the LOX/LH_2 D-57, developed for the improved N1 manned lunar launcher (SEP) 0516895

Application: 'Granit'
Engine cycle: closed
Oxidiser: nitrogen tetroxide
Fuel: UDMH

RD-0233

First stage engine for the UR-100N missile and the Rockot launch vehicle. It carries the tank pressurisation system for the stage, and therefore one of these engines is used on the stage. It was developed during 1969–1974.
Application: UR-100N missile first stage, Rockot first stage
Engine cycle: closed
Oxidiser: nitrogen tetroxide
Fuel: UDMH
Specific impulse: 291 s
Chamber pressure: 20.5 atm

RD-0234

Three of these single-chamber engines are used on the first stage of the UR-100N missile and Rockot launch vehicle: they are almost identical with the RD-0233, except that they lack the tank pressurisation system. They were developed during 1969–1974.
Application: UR-100N missile first stage, Rockot first stage
Engine cycle: closed
Oxidiser: nitrogen tetroxide

RD-0235

The RD-0235 is the main engine for the second stage of the UR-100N missile and Rockot launch vehicle: it is a single-chamber engine, developed during 1969–1974.
Application: UR-100N missile and Rockot second stage main engine
Engine cycle: closed
Oxidiser: nitrogen tetroxide
Fuel: UDMH

RD-0236

The RD-0236 is the four-chamber vernier engine system for the UR-100N and Rockot second stages. It carries the pressurisation system for the propellant tanks. Development period was 1969–1974.
Application: UR-100N missile and Rockot second stage vernier engine
Engine cycle: open
Oxidiser: nitrogen tetroxide
Fuel: UDMH
Specific impulse: 293 s
Chamber pressure: 17.5 atm

RD-0237

The small RD-0237 is carried on the post-boost third stage of the UR-100N missile. It was developed during 1969–1974.
Application: UR-100N third stage
Engine cycle: open
Oxidiser: nitrogen tetroxide
Fuel: UDMH

RD-0244

First stage engine for the RSM-54 submarine-launched ballistic missile, having a single combustion chamber. It carried the propellant tank pressurisation system for the stage. The development period is given as 1977–1985.
Application: RSM-54 SLBM first stage main engine
Engine cycle: closed
Oxidiser: nitrogen tetroxide
Fuel: UDMH
Specific impulse: 301 s
Chamber pressure: 27.5 atm

RD-0245

Four-chamber vernier engine system for the first stage of the RSM-54 SLBM. The development period is given as 1977–1985.
Application: RSM-54 SLBM first stage main engine
Engine cycle: open
Oxidiser: nitrogen tetroxide
Fuel: UDMH

RD-0256

This engine replaced the RD-0229 on the R-36M to give the R-36M2 and it carries the pressurisation system for the propellant tanks. The development period is given as 1983–1987.
Application: R-36M2 missile second stage
Engine cycle: closed
Oxidiser: nitrogen tetroxide
Fuel: UDMH

RD-0257

The four-chamber vernier engine system for the R-36M2 missile, replacing the RD-0230 of the

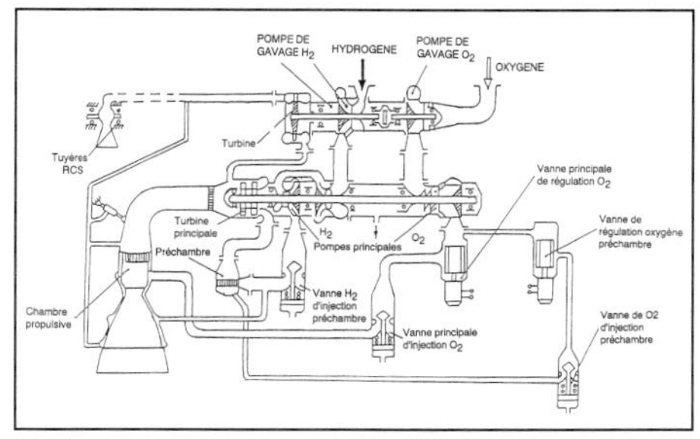

D-57 propellant flow schematic (SEP) 0516896

earlier R-36M. The development period is given as 1983–1987.
Application: R-36M2 missile second stage vernier engine
Engine cycle: open
Oxidiser: nitrogen tetroxide
Fuel: UDMH

RD-0410

Development of the 35 kN nuclear thermal design began in 1965 as a prototype for manned Mars vehicles. Specific impulse of the 1.6 m dia, 3.5 m high liquid hydrogen motor was 910 s.

KB KHIMMASH

A bureau headed by Alexei M Isayev was established in 1943 for research into storable liquid propellant rocket engines. Whereas Glushko's GDL concentrated on major launcher engines and Kosberg's bureau on upper stage propulsion systems, Isayev's produced smaller *spacecra* ft engines, including manned, Earth orbit and deep space systems. A number of designations have been adopted but 'KTDU' (Korrektyroushchaya-Tormorznaya Dvigatelnaya Ustanovka = Corrective-Braking Rocket Motor) is most frequent, with the others variations on this theme. Information released in early 1991 indicated the bureau is part of NPO Soyuz in Kaliningrad. It is responsible for the 73.58 kN cryogenic engine of the new Proton KM version, originally designed for the L-3M manned lunar mission, and recently sold to India for its GSLV vehicle. A newly-developed DMT-600 600 N NTO/UDMH thruster was referred to in 1994. Khimmach developed all of the engines for the Heavy Cosmos TKS transport craft.

Starting in 1971 and at the instigation of Isayev, the bureau started the development of low-thrust engines which could be used for attitude control of spacecraft. A series eleven engines with thrusts up to 2,206 kN (225 kg) and eight with thrusts up to 49 kN (5 kg) were developed. They are capable of burns lasting from hundredths of a second through to hundreds and thousands of seconds.

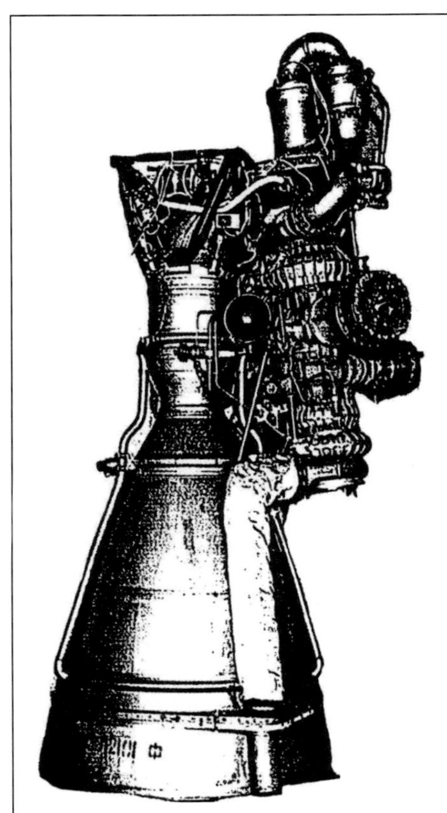

A cluster of 30 of Kuznetsov's 1,510 kN engines powered the lunar N1 stage 1 (NPO Trud) 0516897

Aerojet offers modified NK-33s for future US launch vehicles (Aerojet) 0517361

11D49
It was confirmed in 1994 by the *Directory* that the bureau is responsible for the Cosmos 3M stage 2 engine. The development period was 1965–1971.
Application: Cosmos 3M stage 2
First flown: May 1967
Dry mass: 225 kg
Length: unknown
Maximum diameter: unknown
Engine cycle: pump-fed
Oxidiser: nitrogen tetroxide
Fuel: UDMH
Mixture ratio: 2.33/0.35
Thrust: 157.3 kN main + 4 × 1.4–1.8 kN steering thrusters (each with a 25 N pressure-fed thruster for 3-axis coast + ullage)
Expansion ratio: main 103.4; steering 12.0
Specific impulse: 303 s vac (176 s steering)
Chamber pressure: 102 atm
Burn time: main chamber about 350 s operationally

DOK-10
Application: Spacecraft thruster
Mass: 0.6 kg
Propellant: hydrazine
Thrust: 10 kN
Specific impulse: 229 s
Chamber pressure: 1.5 atm
Burn durations: 0.05–600 s
Total burn time: 1,500 s
Number of firings: 4,000

DOK-50
Application: Spacecraft thruster
Mass: 1.1 kg
Propellant: hydrazine
Thrust: 50 N
Specific impulse: 229 s
Chamber pressure: 1.5 atm
Burn durations: 0.05–600 s
Total burn time: 1,500 s
Number of firings: 4,000

RD-0213 main engine and RD-0214 vernier system used on the third stage of the Proton-K launch vehicle. (KB Khimavtomatiki) 0003419

DOT-5
Application: Spacecraft thruster
Mass: 0.9 kg
Propellant: hydrazine
Thrust: 5 N
Specific impulse: 230 s
Chamber pressure: 1.2 atm
Burn durations: 0.05–8,000 s
Total burn time: 120,000 s
Number of firings: 55,000

DOT-25
Application: Spacecraft thruster
Mass: 1.3 kg
Propellant: hydrazine
Thrust: 25 kN
Specific impulse: 235 s
Chamber pressure: 1.2 atm
Burn durations: 0.05–6,000 s
Total burn time: 25,000 s
Number of firings: 6,000

DST-25
Application: Spacecraft thruster
Mass: 0.9 kg
Oxidiser: nitrogen tetroxide
Fuel: UDMH
Mixture ratio (O/F): 1.85
Thrust: 25 N
Specific impulse: 285 s
Chamber pressure: 1.5 atm
Burn durations: 0.03–4,000 s
Total burn time: 25,000 s
Number of firings: 300,000

DST-100
Application: Spacecraft thruster
Mass: 1.1 kg
Oxidiser: nitrogen tetroxide
Fuel: UDMH
Mixture ratio (O/F): 1.85
Thrust: 100 N
Specific impulse: 276 s
Chamber pressure: 2.5 atm
Burn durations: 0.05–300 s
Total burn time: 10,000 s
Number of firings: 10,000

DST-100A
Application: Spacecraft thruster
Mass: 1.5 kg
Oxidiser: nitrogen tetroxide
Fuel: UDMH
Mixture ratio (O/F): 1.85
Thrust: 100 N
Specific impulse: 304 s
Chamber pressure: 1.6 atm
Burn durations: 0.05–4,000 s
Total burn time: 50,000 s
Number of firings: 450,000

DST-200
Application: Spacecraft thruster
Mass: 1.3 kg
Oxidiser: nitrogen tetroxide
Fuel: UDMH
Mixture ratio (O/F): 1.85
Thrust: 200 N
Specific impulse: 280 s
Chamber pressure: 2.5 atm
Burn durations: 0.03–300 s
Total burn time: 5,000 s
Number of firings: 10,000

DST-200A
Application: Spacecraft thruster
Mass: 1.7 kg
Oxidiser: nitrogen tetroxide
Fuel: UDMH
Mixture ratio (O/F): 1.85
Thrust: 200 kN
Specific impulse: 300 s
Chamber pressure: 1.6 atm
Burn durations: 0.05–4,000 s
Total burn time: 10,000 s
Number of firings: 100,000

DST-600
This engine can use either UDMH or MMH as the fuel: hence two sets of mixture ratios and specific impulses are given.
Application: Spacecraft thruster
Mass: 4.2 kg

RD-0216 engine on the first stage of the UR-100 missile (KB Khimavtomatiki) 0003418

Oxidiser: nitrogen tetroxide
Fuel: UDMH/MMH
Mixture ratio (O/F): 1.85/1.65
Thrust: 25 N
Specific impulse: 301 s/306 s
Chamber pressure: 1.6 atm
Burn durations: 0.05–1,500 s
Total burn time: 5,000 s
Number of firings: 6,000

KDU-414
This engine was used on the first-generation of planetary missions which were launched using the Molniya and Molniya-M vehicles. The final use for a deep space mission was for Venera 8 and the failed Cosmos 472 Venus probe in 1972, but it is still used on Molniya-1 and probably Molniya-3 communications satellites. It has a single fixed combustion chamber.
Applications: Mars 1, Zond 1–3, Venera 2–8, Molniya 1
First flight: Nov 1962 (Mars 1, but earlier, failed, missions likely)
Oxidiser: nitric acid
Fuel: UDMH
Feed method: pressure fed
Thrust: 1.96 kN vacuum
Specific impulse: 271 s vacuum
Chamber pressure: 12 atm
Burn time: 40 s in at least two burns

KRD-61
Apparently designed specifically for the 2.708 km/s V final leg of lunar sample return missions, achieving three successes 1970–1974. Single fixed main chamber with four large verniers providing

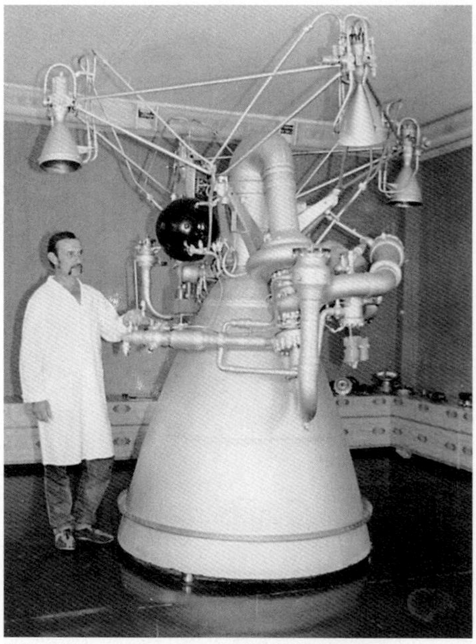

RD-0229 main engine and RD-0230 vernier engine assembly used on the original versions of the R-36M missile. This assembly was placed by the RD-0256/RD-0257 on the later R-36M2 missiles. (KB Khimavtomatiki) 0003417

RD-0233 engine on the first stage of the UR-100N missile and the derived Rockot launch vehicle (KB Khimavtomatiki) 0003420

pitch/yaw control and four small verniers for roll. Single ignition.
Applications: third generation lunar sample return missions (Luna 16/20/24 successful)
First flight: 14 Apr 1969 (failed Luna attempt)
Oxidiser: nitrous oxide
Fuel: UDMH
Feed method: turbo pump
Thrust: 18.8 kN vacuum
Specific impulse: 313 s vacuum
Chamber pressure: 95 atm
Burn time: 53 s

KRD-79
No details are known about this engine other than it is an engine used on Mir core module: it is not known whether it is the main engine or one of the attitude control thrusters.

KRD-442
No details are known about this engine other than it is an engine used on Mir core module: it is not known whether it is the main engine or one of the attitude control thrusters.

KTDU-5A
Retromotor for semi-soft landing or braking into orbit of second generation Luna craft. Single fixed chamber and four peripheral verniers fired for about 43 s to cancel the 2.6 km/s direct arrival speed from Earth. The first successful lunar landing (Luna 9, Feb 1966) and lunar orbiter (Luna 10, Apr 1966) were made possible by this engine.
Applications: spacecraft single burn retro-motor
First flight: lunar landing attempt of Jan 1963?
Oxidiser: nitric acid
Fuel: amine-based
Feed method: pressure fed

RD-0235 main engine and RD-0236 vernier engine assembly used on the UR-100N missile and Rockot satellite launcher. (KB Khimavtomatiki) 0003421

Thrust: 45.5 kN vacuum
Specific impulse: 278 s vacuum
Chamber pressure: 64 atm
Burn time: 43 s in two burns

KTDU-35/-53/-66
Principal application of the KTDU-35 system was the orbital manoeuvring and de-orbit system of the original manned Soyuz (1–40) spacecraft. A single multi-start chamber was fed with propellants by a turbopump. Retrofire required a burn of typically 150 s. >4,000 ground/space ignition tests were conducted to qualify the engine. There were also twin nozzles either side providing a single start 4.03 kN backup capability; this was called on during Soyuz 33 when the primary system failed. The **KDTU-53** was essentially the same system used for unmanned Zond lunar round trips but without the backup chambers and with reduced propellant supply. It was further modified as the Salyut 1 space station **KDTU-66** orbital manoeuvring system.
Applications: primary Soyuz/Zond propulsion, Salyut 1 space station orbital adjustment
First flown: Nov 1966 (Soyuz unmanned test)
Engine cycle: open
Oxidiser: nitric acid
Fuel: UDMH
Feed method: turbopump
Thrust: 4.09 kN vacuum primary single nozzle; 4.03 kN vacuum backup twin nozzles
Specific impulse: 281 s vacuum prime, 270 s vacuum backup
Chamber pressure: 40 atm primary
Burn time: up to 1,000 s in multiple burns
Few details are known of the Soyuz-T propulsion system and almost nothing of the current Soyuz-TM model. A switch was made to hypergolic UDMH/N_2O_4 and the attitude control thrusters were integrated with the larger system, providing a second de-orbit backup capability. Thrust of the primary chamber is 3.1 kN vacuum. A space station version was introduced on Salyut 6 for AOCS, with two primary chambers. The current Mir station appears to incorporate a similar system, with a thrust of 5.9 kN.

KTDU-80
No details are known about this engine other than it is the main propulsion system for the Soyuz-TM and Progress-M spacecraft.

KTDU-417
Introduced for the soft-landing stage of the third generation Luna craft (starting in 1969). In this role the engine included a single 11-start main chamber for the greater part of the descent, followed by twin low-thrust chambers for the final approach and landing. Both were variable thrust. The S5.92 engine (qv) appears to be a modernised version of the KTDU-417. (This engine has also been described in Russian literature using the designator KRD-417: it is unclear which is the correct designator.)
Applications: lunar landing and planetary propulsion stage?
First flight: 14 Apr 1969 (failed Luna)
Oxidiser: nitric acid
Fuel: UDMH
Feed method: pressure fed
Thrust: 7.35–18.92 kN vacuum variable prime, 2.06–3.43 kN vacuum variable secondary
Specific impulse: 308–314 s vacuum primary, 249–254 s vacuum secondary
Burn time: 650 s primary, 30 s secondary

KTDU-425A
The KTDU-425 was used as the main propulsion system for the Mars 2/3 missions in 1971: presumably it was also used for the failed Mars probe Cosmos 419 (an intended Mars orbiter) and the two Mars probes launched and lost in 1969. In 1973 it was replaced by the KTDU-425A, first launched on Mars 4. It is the later version which has been described in Russian literature. A single chamber, pump-fed and gimballed.
Applications: Mars 4–7, Venera 9–16, Vega 1–2, Phobos 1–2
First flight: July 1973 (Mars 4)
Oxidiser: nitrous oxide
Fuel: amine-based
Feed method: pump
Thrust: 9.86–18.89 kN vacuum variable
Specific impulse: 293–315 s vacuum
Chamber pressure: 95 atm
Burn time: 560 s in up to seven burns

KVD-1/(11)D56
This cryogenic engine was originally developed for the Block R of the abandoned N-1 launch vehicle which would have flown the dual launch L-3M manned lunar missions in the 1970s: the N-1 programme was cancelled in 1974. When the

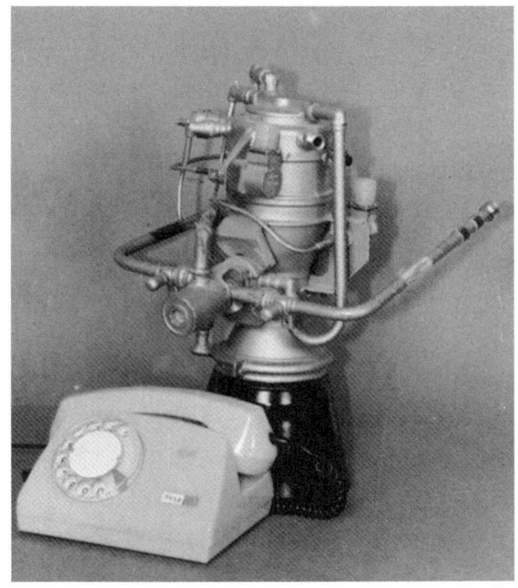

RD-0237 engine used on the post-boost stage of the UR-100N missile. (KB Khimavtomatiki) 0003428

engine was finally revealed in the 1990s it was identified as the KVD-1 or D-56, but clearly it is the same engine as the 11D56 developed in the 1970s. The engine is now to be used on the Proton-M launch vehicle's planned cryogenic fourth stage and it has been sold to India to be used as an upper stage engine on the GSLV.
Applications: upper stages (Proton, Angara, GSLV, possibly Zenit)
First flight: planned 1998 on India's GSLV
Dry mass: 292 kg
Length: 2.146 m
Diameter: 1.28 m
Engine cycle: closed staged combustion
Oxidiser: liquid oxygen
Fuel: liquid hydrogen
Feed method: turbopump
Thrust: 69.6 kN vacuum; 73.58 kN with two gimballed LOX/GH_2 verniers fed by main chamber pumps; 76 kN quoted for GSLV
Specific impulse: 463 s vacuum
Expansion ratio: unknown
Chamber pressure: 58.1 atm
Burn time: 24,000 s in 6 starts (450 s for Proton)

R2.2000
This engine appears to be a replacement for the KTDU-425A (qv). It can operate in either low or high thrust modes, and thus corresponding low/high thrust mode parameters are shown for some data.
Applications: Mars 8 and Phobos spacecraft main engines
First flown: 1988
Dry mass: 74 kg
Length: 1.025 m
Maximum diameter: 0.920 m
Oxidiser: nitrogen tetroxide
Fuel: UDMH
Mixture ratio (O/F): 2.0/1.95

RD-0244 single-chamber main engine and the RD-0245 four-chamber vernier engine assembly, used on the first stage of the RSM-54 submarine-launched ballistic missile. (KB Khimavtomatiki) 0003427

RD-0256 main engine and RD-0257 vernier engine assembly which replaced the RD-0229/RD-0230 assembly for the second stage of the R-36M missile (KB Khimavtomatiki) 0003426

Thrust: 13.725 kN/19.613 kN
Specific impulse: 316 s/325 s
Burn time: 1,000 s (max 25 burns)

S5.92
This engine has not been described in great detail, but could be a modification of the KTDU-417 (qv). It has been stated that it was used on the Phobos 1 and Phobos 2 missions for launching from Earth orbit, course corrections and entry into Mars orbit, and therefore it is clearly the engine used on the Fregat module – also carried by the ill-fated Mars 8 (Mars-96) mission in 1996. It is also possible that this engine (or a modification) is used on the Breeze (Briz) third stage of the Rockot launch vehicle.
Applications: Phobos 1/2, Mars 8 main propulsion modules, Fregat, possibly Breeze on Rockot
First flown: 1988
Dry mass: not known
Length: not known
Maximum diameter: not known
Oxidiser: nitrogen tetroxide
Fuel: UDMH
Mixture ratio (O/F): not known
Feed method: not known
Thrust: 20 kN
Specific impulse: 327 s
Expansion ratio: not known
Chamber pressure: not known
Burn time: 2,000 s (max of 50 burns)

KOROLEV DESIGN BUREAU
Korolev's bureau developed Proton's Block D/DM and Molniya's Block L engines. Mikhail Melnikov designed the Proton engine.

RD-58M/11D58M: Proton stage 4
The engine provides a direct GEO and planetary trajectory injection capability. The stage was originally developed for Korolev's N1 lunar vehicle and then transferred to Proton. There are two versions flying: the standard kerosene version and one using 'sintin' for higher Isp. Sintin is known only as a synthetic hydrocarbon-based fuel. The same engine might power Zenit's stage 3 and possibly Energia's Retro & Correction Stage. Buran utilised two **17D** 12 sintin versions as its main orbital manoeuvring engines. The configuration was changed only slightly (see photograph); the performance and overall dimensions are as given here for the sintin 11D58M.
Applications: Proton stage 4, Zenit stage 3?, Buran OMS
First flown: 1967
Dry mass: 303 kg
Length: 2.268 m
Diameter: 1.167 m
Oxidiser: liquid oxygen
Fuel: kerosene or sintin
Thrust (vac): 85 kN kerosene; 83.5 kN sintin
Specific impulse (s, vac): 352 kerosene, 362 sintin
Chamber pressure: 76.4 atm kerosene, 78.4 sintin
Burn time: 600 s in multiple burns

S5.92 engine used as the main propulsion system on the Fregat module used on the Phobos and Mars 8 missions: the same engine might be used on the Breeze third stage of the Rockot launch vehicle. (KB Khimmash) 0003429

C1–5400/11D33: Molniya final stage
Little is known of the propulsion system on the Molniya Block L final stage but it is a fixed single chamber. It was also used as the injection stage for deep space missions until Proton was introduced.
Applications: Molniya stage 3 Block L
First flown: Oct 1960? (as planetary probe escape stage)
Oxidiser: liquid oxygen
Fuel: kerosene
Thrust: 67 kN vacuum
Specific impulse: 340 s vacuum
Burn time: about 200 s

N D KUZNETSOV COMPANY SSC
Nikolai Dmitrievich Kuznetsov (24 June 1911–30 July 1995) established an aircraft engine design and development bureau during the 1940s in Kuibyschev (now Samara). Information finally released in 1989 revealed that the bureau had been assigned the responsibility for the design and development for engines for Korolyov's four-stage N-1 launch vehicle. Previously the bureau had developed engines planned for use on Korolyov's R-9 missile, but politics ensured that the missile used other engines.

In July 1993 Aerojet signed a teaming agreement with N D Kuznetsov SSC to market these engines within the United States. Two NK-33 engines were delivered to Aerojet in 1995. One of them was used by Aerojet for five firings which totalled 410 s operating time within a 57–114% thrust range which followed the proposed Atlas-2AR launch profile. The engines were under consideration to power the first stage of the Atlas-2AR, but they lost out to the Energomash RD-180 engine.

The United States Kistler Aerospace Corporation is planning to use NK-33 engines on its reusable K-1 launch vehicles: three would be on the first stage and one on the second stage. See the Kistler entry in the United States Launcher section for further details.

NK-9 and NK-19
These two engines were proposed for use on the Korolyov bureau's R-9 missile: the first stage would have used four NK-9 engines and the second stage a single NK-19. After political considerations became dominant, the missile used the GDL's RD-111 on the first stage and Kosberg's RD-0106 on the second stage. The N-1's NK-39 and NK-31 engines were developed from the NK-19.

NK-31
Application: N-1 Block G fourth stage (1 engine)
First flown: 20 Feb 1969
Dry mass: 722 kg
Wet mass: 798 kg
Maximum diameter: 1.8 m
Engine cycle: closed-loop (similar to NK-33)
Oxidiser: liquid oxygen
Fuel: kerosene
Feed method: single-shaft turbopump, 26,150 rpm
Thrust: 402 kN vacuum
Specific impulse: 353 s vacuum
Burn time: 1,140 s (530 s for N-1)

NK-33
It is expected that Aerojet modifications will add about 136 kg to the masses quoted below.
Applications: N-1 Block A first stage (30 engines) Kistler K-1 first and second stages
First testbench firing: 12 Feb 1969
First flown: 20 Feb 1969
Dry mass: 1,236 kg
Wet mass: 1,393 kg
Maximum diameter: 1.5 m
Engine cycle: closed-loop with staged kerosene combustion in oxygen
Oxidiser: liquid oxygen
Fuel: kerosene
Propellant flow rate: 500 kg/s
Mixture ratio (O/F): 2.547
Feed method: two-shaft turbopump, 17,300 rpm
Thrust: 1,510 kN sea level

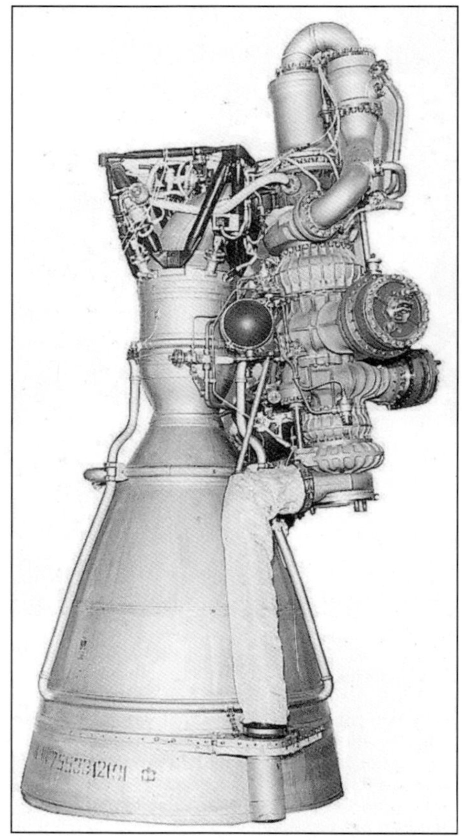

N D Kuznetsov with his NK-33 engine intended for the first stage of the N-1 launch vehicle (bottom) and a full length view of the NK-33 (top): thirty of these engines would have been clustered on the first stage of the N-1. It is now planned to use these engines on the Kistler K-1 launch vehicle. (N D Kuznetsov Company SSC) 0003425

Specific impulse: 331 s vacuum 297 s sea level
Burn time: 600 s (120–150 s for N-1)

NK-43
Application: N-1 Block B second stage (8 engines)
First testbench firing: 29 July 1970
First flown: 20 Feb 1969
Dry mass: 1,359 kg
Wet mass: 1,542 kg
Maximum diameter: 2.5 m
Engine cycle: closed-loop with staged kerosene combustion in oxygen
Oxidiser: liquid oxygen
Fuel: kerosene
Propellant flow rate: 500 kg/s
Mixture ratio (O/F): 2.547
Feed method: two-shaft turbopump, 17,300 rpm
Thrust: 1,757 kN vacuum
Specific impulse: 346 s vac

NK-39
Application: N-1 Block V third stage (4 engines)
First flown: 20 Feb 1969

N D Kuznetsov with his NK-33 engine intended for the first stage of the N-1 launch vehicle (bottom) and a full length view of the NK-33 (top): thirty of these engines would have been clustered on the first stage of the N-1. It is now planned to use these engines on the Kistler K-1 launch vehicle. (N D Kuznetsov Company SSC) 0003430

The NK-43 engine was a modification of the NK-33, used on the second stage of the N-1 launch vehicle. (N D Kuznetsov Company SSC) 0003424

Dry mass: 584 kg
Wet mass: 635 kg
Maximum diameter: 1.8 m
Engine cycle: closed-loop (similar to NK-33)
Oxidiser: liquid oxygen
Fuel: kerosene
Feed method: single-shaft turbopump, 26,150 rpm
Thrust: 402 kN vacuum
Specific impulse: 353 s vac
Burn time: 1,140 s (380 s for N-1)

NII MACHINOSTROENYE

R&DIME Research & Development Institute of Mechanical Engineering was established in 1958 as part of NII Thermal Processes, but became independent in the early 1980s. Development and production of 0.8–400 N spacecraft thrusters (implying connection with old Isayev bureau), including Soyuz, Progress and Mir. Development and small batch production of engines up to 6 kN can be accommodated. Test facilities can handle hydrogen (up to 3 kN), methane and kerosene fuels. The Elektro Earth observation satellite carries 16 DEN-16 electrothermal thrusters. For the thrusters specified below, LTRE = Low Thrust Rocket Engine, or RDMT in Russian (Raketnay Dvigatel Maloi Tyagi).

LTRE-400

This 400 N man-rated engine is employed in clusters aboard Mir's Kvant, Kristall, Spektr and Priroda modules. It was developed for the manned Almaz station and the Heavy Cosmos transport craft. It was test-fired by Aerojet in the US in 1994 using MMH, attaining 293 s SI. A redesigned version is estimated to be capable of 310–320 s SI.
Dry mass: 2.5 kg
Length: not available
Max diameter: 15 cm exit dia
Oxidiser: NTO
Fuel: UDMH
Propellant feed method: pressure
Inlet pressure: not available
Valves: 2 A at 27 V (20–34 V supply)
Thrust: 400 N vacuum
Specific impulse: 255 s vacuum nominal (after 30 s)
Response time: 30 ms to 90% thrust (30 ms decay to 10%)
Mixture ratio (O/F): 1.9:1 (fuel rich); 30–40% of NTO flow used for film cooling. 1,038°C limit; 93°C nominal throat
Expansion ratio: 56:1
Expansion bell: stainless steel, NTO film-cooled
Chamber pressure: 6.5 atm
Chamber materials: stainless steel, NTO film-cooled

LTRE-12

This 12 N thruster is employed in clusters aboard Mir's Kvant, Kristall, Spektr and Priroda modules.
Dry mass: 0.55 kg
Oxidiser: NTO
Fuel: UDMH
Propellant feed method: pressure
Valves: 0.2 A at 27 V (20–34 V supply)
Thrust: 12 N vacuum
Specific impulse: 279 s vacuum nominal
Response time: 15 ms to 90% thrust (10 ms decay to 10%)

LTRE-50

Application: unmanned satellite
Dry mass: 1.0 kg
Oxidiser: NTO
Fuel: UDMH
Propellant feed method: pressure
Valves: 1 A at 27 V (20–34 V supply)
Thrust: 50 N vacuum
Specific impulse: 255 s vacuum nominal
Response time: 30 ms to 90% thrust (30 ms decay to 10%)

LTRE-100/135

This man-rated engine is employed in clusters aboard Mir, Soyuz-T/TM, Progress and unmanned satellites in 98 N or 137 N versions.
Dry mass: 1.2 kg
Oxidiser: NTO
Fuel: UDMH
Propellant feed method: pressure
Valves: 1 A at 27 V (20–34 V supply)
Thrust: 98 N or 137 N vacuum
Specific impulse: 260 s vacuum nominal
Response time: 30 ms to 90% thrust (30 ms decay to 10%)

LTRE-200

Application: Almaz module
Dry mass: 2.0 kg
Oxidiser: NTO
Fuel: UDMH
Propellant feed method: pressure
Valves: 2 A at 27 V (20–34 V supply)
Thrust: 200 N vacuum
Specific impulse: 255 s vacuum nominal
Response time: 30 ms to 90% thrust (30 ms decay to 10%)

LTRE-200K

Application: Buran AOCS
Dry mass: 5.5 kg
Oxidiser: GOX
Fuel: kerosene
Valves: 2.5 A at 27 V (20–34 V supply)
Thrust: 200 N vacuum
Specific impulse: 265 s vacuum nominal
Response time: 50 ms to 90% thrust (60 ms decay to 10%)

NII THERMAL PROCESSES

NII TP performs research in all forms of launcher and space propulsion, including ramjet, electric and nuclear. BMDO took delivery in Oct 1993 of a T-100 1.5 kW Hall-type Stationary Plasma Thruster for evaluation. N_2O_4/UDMH thrusters include 10 N and 1,000 N models. NII TP developed the Phobos spacecraft's network of hydrazine thrusters for attitude control and minor orbital modifications. 28 (24 × 50 N, 4 × 10 N) were mounted on four spherical tanks. A set of 0.5 N hydrazine microthrusters of an unusual design was also included to avoid the need for separate N_2 cold gas jets; the propellant was burned internally and cooled inside an expansion tank before passing to the thrusters.

NPO SATURN/LYULKA

The cryogenic engine which has been described in the 1990s as the D-57 would appear to be the 11D57 gimballing engine which was developed during the early 1970s for use on the Block S of the modified N-1 launch vehicle which would have been flown for the dual-launch L-3M manned lunar missions. The bureau was also tasked to develop the 11D54 fixed-chamber cryogenic engine with the same thrust level of 40 t (390 kN). This engine would have been used in a cluster of 6–8 on a modified Block V of the N-1 vehicle.

It has been claimed that the D-57 (ie, 11D57) is the only rocket engine developed by the bureau, suggesting that the 11D54 either had development problems and was abandoned or that the N-1/L-3M programme was simply scrapped before the engine's development was completed.

Development of the engine now called D-57 is said to have started in 1960 (which is too early for the N-1/L-3M programme: perhaps it was designed in the hope of finding a later application?) and work ended in 1977. When production of the engine ended in 1975 – a year after the L-3M programme was terminated – 105 of the engines had been built with more than 53,000 s of testing having been completed in 470 runs, including a single burn of 5,400 s and 22 starts. Only four complete unused engines remain, two used engines (one with the 2-position nozzle), 33 thrust chambers/skirts and other spares. Aerojet teamed with Lyulka in 1993 to improve and market the engine. NASA has studied employing it on an SSTO demonstrator; for example, 1xNK-33 + 2xD-57.
Application: N1-L3M stages 3/4
First flown:
Dry mass: 840 kg
Length: 3,660 mm
Maximum diameter: 1,860 mm
Mounting: ±2.5° 2-axis gimbal (powered by H_2)
Engine cycle: staged combustion
Oxidiser: liquid oxygen
Fuel: liquid hydrogen
Mixture ratio (O/F): 5.8:1
Feed method: main/boost ungeared pump turbines on separate single shafts are powered by 540°C fuel-rich exhaust; pump outputs 254 atm LOX and 170 atm LH_2. Most of exhaust is routed into main chamber; some provides roll control. GOX injection from chamber LOX cooling
Thrust: 400 kN vacuum; throttle down to 10%
Expansion ratio: 143:1; also version 2-position nozzle providing 170:1 and 88:1 (throat dia not available, exit dia 1,860 mm)
Specific impulse: 456.5 s vacuum (461 for 170:1 nozzle)
Thrust chamber
  materials: non-tubular stainless steel
  cooling: hydrogen
Combustion chamber
  pressure: 109 atm (100% thrust)
  temperature: not available
  materials: milled copper liner
  cooling: LOX cooled, providing GOX for injection
  ignition: not available
Burn time: 800 s

NPO YUZHNOYE

Yuzhnoye produces verniers for some of its own vehicles (Zenit stage 2 and Tsyklon stages 1–3) and the 11D25 main engine for Tsyklon's stage 3.

RD-800 Series Engine

The engine's four nozzles gimbal ±33° for vehicle steering.
First flown: Apr 1985 (Zenit 2 stage 2)
Cycle: closed
Dry mass: 400 kg (418 kg wet)
Length: 1,000 mm
Maximum diameter: 430 mm
Oxidiser: oxygen
Fuel: kerosene
Thrust: 78.4 kN total four nozzles
Mixture ratio: 2.40:1
Specific impulse: 342.0 s vac
Expansion ratio: 103.7:1 main
Chamber pressure: 75.5 atm main
Burn time: 3,300 s (1,100 s nominal)

RD-861

Each engine includes 6 × 20 N + 2 × 98 N fixed thrusters for steering during the fixed main chamber's burn.

Four NK-39 engines were carried on the third stage of the N-1 which would have performed Earth orbit injection. (N D Kuznetsov Company SSC) 0003422

The NK-31 is a modification of the NK-39 and one was carried on the N-1 fourth stage which would have performed trans-lunar injection
(N D Kuznetsov Company SSC) 0003423

Application: Tsyklon third stage
First flown: June 1977
Cycle: open, with pump turbine exhaust routed to steering thrusters
Dry mass: 123 kg (138 kg wet)
Length: 1,555 mm
Maximum diameter: 795 mm
Oxidiser: nitrogen tetroxide
Fuel: UDMH
Thrust: 78.90 kN main vacuum
Mixture ratio: 2.1:1 main, 1.6:1 steering
Specific impulse: 317 s main vac; 230 s steering vac
Expansion ratio: 111.2:1 main
Chamber pressure: 90.5 atm main
Burn time: 118 s

OKB FAKEL

Fakel specialises in spacecraft attitude control thrusters, ion engines and plasma sources. Its SPT-70 and SPT-100 (Stationary Plasma Thruster) Hall electric thrusters have flown on >50 Meteor polar metsats since 1972 to provide orbit control. The numerical designators indicate beam dia in mm. SPTs employ a DC-gas discharge in an annular chamber, in which a radial magnetic field traps the electrons. These Hall currents ionise the Xe propellant, the ions of which are accelerated inside a quasi-neutral plasma, without grids, and by the discharge voltage itself. The advantages are their rugged simplicity (no grid system or high voltage supply), but they provide lower efficiencies, lower exhaust velocities and increased Xe consumption than ion thrusters such as the UK-10.

The 0.08 N SPT-100 has been sold to Space Systems/Loral for commercial applications. Thrust range 0.02–0.2 N (nominally 84 mN), SI 1,600 s, 3.2 kg, 1,350 W, 4,000 h life, 48% thrust efficiency. The International Space Technology, Inc (qv) joint venture with SS/L, SEP, ARC and Moscow's RIAME Research Inst of Applied Mechanics & Electrodynamics holds exclusive marketing and distribution rights outside of Russia. ISTI now offers complete propulsion systems. Gals 1, launched Jan 1994, is carrying a set of eight to provide the first Russian NSSK in GEO. SEP is developing a Mk 2 version that will increase SI to 2,200 s. Development began in May 1993 at SEP Villaroche. RIAME/Fakel are investigating SPT thrusters up to 90 cm dia and power levels up to 100 kW for primary propulsion. Krypton is a possible propellant.

The SPT-70 offers: thrust 40 mN, SI 1,600 s, 1.5 kg, 640 W, 9,000 h life, 50% thrust efficiency. An improved SPT-50 is under development for the small satellite market, drawing 300 W. Fakel has developed the SPT-140, SPT-200 and SPT-290 for primary propulsion. The specifications below are given in 140/200/290 order:
Diameter: 4/20/29 cm beam exit; 18/26/37 cm thruster
Length: 10/20/23 cm
Mass: 6/12/18 kg
Thrust: 200/500/1,000 mN for 200–600 V discharge voltage
Specific impulse: 1,530–3,160 s
Acceleration voltage: 0.2–0.6 kV discharge
Discharge power: 3–6/6–12/12–25 kW
Power input: 4.5/8.0/18.0 kW
Specific power consumption: 22.5/16.0/18.0 kW/N
Specific propellant consumption: 52.0/51.2/50.8 mg/Ns
Total efficiency: –/50–71/50–70%
Lifetime: 6,000/8,000/12,000 h

Plasma Thrusters

In addition to OKB Fakel's SPT-100 (see above), the US Ballistic Missile Defense Organization is evaluating several other Russian plasma thrusters. For example, BMDO took delivery in Oct 1993 of NII TP's T-100 1.5 kW Hall-type SPT. See the BMDO entry in the Military section for further details.

VERIFIED

United Kingdom

British Aerospace Defence Co Ltd

Royal Ordnance RMD provides the principal UK capability in propellants and rocket motors, including double base/composite solids and mono/bipropellant liquid engines. Solid rocket performances cover 0.008–600 s burning time, 0.012–425 kg charge mass and 0.8–200 kN thrust.

Liquid Propellant Engines

The 500 N Leros liquid apogee engine represents the first phase of RO's expansion into civil space propulsion, with the long-term objective of developing a comprehensive range of mono and bipropellant engines. Leros 1 was the first to be qualified; Leros 1b, 2, 20 and 20H have been developed since. RO is improving performance by using advanced materials in thrust chambers for prolonged operation at elevated temperatures.

Olin Aerospace's MR 103C 0.5 N mono thruster is marketed in Europe under licence by RO. It is being supplied to MMS for Skynet 4. RO has a long history of work with catalytic thrusters for attitude control applications or stage attitude control systems. For example, in 1974 RO was appointed as Technical Authority for Chevaline's Hydrazine Actuation System.

Leros 1

RO began development of Leros in 1986 to satisfy the orbital manoeuvring requirements of large telecom satellites using dual mode propulsion, which provides a high performance bipropellant MON/hydrazine liquid apogee engine combined with mono hydrazine thrusters for attitude control and stationkeeping. The design approach was based on proven design features, optimisation of chamber length for smooth combustion over a wide operating envelope, a long combustion chamber to reduce thermal soakback and to ensure a cool running injector assembly and stable film conditions, conventional fabrication and assembly, and in-process injector screening in a slave chamber for acceptable performance.

Leros 1 design verification testing was completed in spring 1989, with the qualification programme at Boeing Tulalip for Martin Marietta Astro Space demonstrating by mid-1990 that the engine can handle a 3 h burn (double the worse-case mission

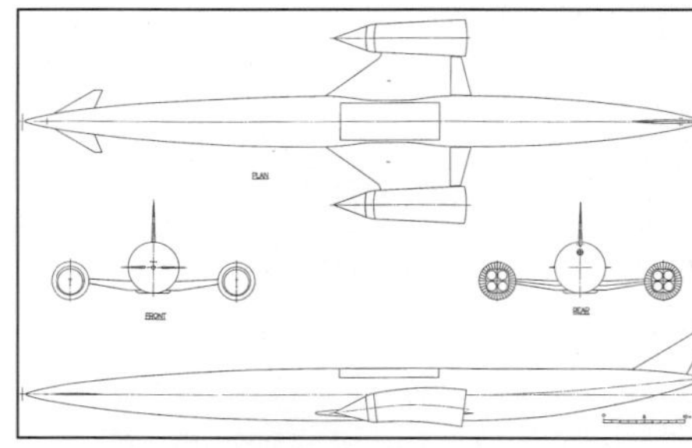

Skylon configuration C1 (Reaction Engines) 0517322

duration identified). Tests covered oxidant depletion, bubble ingestion, multiple hot restarts, hard vacuum ignition and helium-saturated propellants. The 140 starts accumulated 23,000 s and accounted for 3.7 t propellant throughput. Lockheed Martin Astro Space has purchased 29 Leros 1 engines for use in telecom platforms. Leros 1 became the world's first flight-proven dual mode LAE when a pair placed Astra 1B into GEO in Mar 1991. Other users are Telstar 4, AsiaSat 2, Echostar and Intelsat 8/8A.
Applications: dual mode satellite propulsion systems
Dry mass: 4.2 kg
Length: 610 mm
Max diameter: 288 mm (nozzle)
Oxidiser: MON3
Fuel: hydrazine
Propellant feed method: pressure, by helium
Inlet pressure: 15 atm
Valves: Moog model 53–177 torque motor valve, dual coil single seat configuration, based on the model 53–135 unit used on Rocketdyne's Peacekeeper attitude control bipropellant thrusters. >1,500 produced
Thrust: 500 ± 25 N vacuum
Specific impulse: 314 s vacuum nominal
Response time: <10 ms
Mixture ratio (O/F): 0.8
Expansion ratio: 150:1 with 8.5° exit plane half angle
Expansion bell: disilicide-coated niobium C103, radiatively cooled
Chamber pressure: 7.0 atm
Chamber combustion stability: ±12% above 100 Hz, ±3% below 100 Hz
Chamber materials: disilicide-coated niobium C103, T≤1,360°C
Injector: 6Al/4V titanium alloy, jet impingement type using eight unlike doublets for the core around which are 16 fuel film coolant orifices. A backing plate encloses the injector elements and provides the welded interface with the engine mounting flange and mounting points for the propellant control valves
Duty cycle: unlimited, but qualified to 3 h

Leros 1b

This improved version delivers 318 s Isp. Design/construction is identical in all respects to Leros 1 with the exception of the propellant valve, which has been replaced by two Moog solenoid valves (model 53–200). Leros 1b was qualified in 1995 for Lockheed Martin and flight engines have been developed.

Leros 1c

Under development for a major US company using high T materials for the thrust chamber. Target vacuum SI is 325 s.

Leros 2

The Leros 2 apogee engine is designed for use in unified MON/MMH bipropellant systems. It differs from Leros 1 only in its injector configuration and combustion chamber geometry. An alternative valve option is available. The engine has completed development and is being used as a springboard for the development of the higher performance

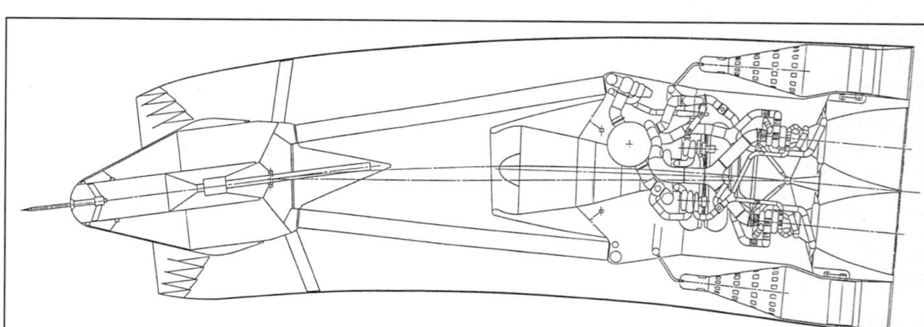

Sabre nacelle vertical cross section. Each nacelle carries two Sabre engines (Reaction Engines) 0517321

(320 s Isp) **Leros 2A** version under ESA contract employing an advanced high temperature material chamber capable of withstanding prolonged operation at up to 2,000°C. Design verification testing is planned for 3Q 1997. **Leros 2B** features conventional columbium/disilicide technology for the thrust chamber and the 300:1 expansion cone. Demonstrated vacuum SI >317 s.
Dry mass: 3.40 kg
Inlet pressure: 16.0 atm
Thrust: 556 N
Mixture ratio (O/F): 1.65
Specific impulse: 312 s

Leros 20
This second generation bipropellant attitude control thruster was developed for MMH propulsion systems.
Dry mass: 0.73 kg
Length: 202 mm
Max diameter: 60 mm (nozzle)
Oxidiser: MON3 at 4.8 gm/s
Fuel: MMH at 2.9 gm/s
Propellant feed method: pressure, by helium
Inlet pressure: 10–20 atm
Valves: Moog single or dual seat
Thrust: 22 N nominal vacuum
Specific impulse: 294 s vacuum nominal
Mixture ratio (O/F): 1.65
Expansion ratio: 180:1
Expansion bell materials: niobium C103 or titanium
Chamber pressure: 7.5 atm
Chamber materials: niobium C103, T≤1,400°C
Injector: multiple impinging jet element with three unlike doublets and six fuel film cooling orifices
Duty cycle: unlimited
Minimum impulse bit: 25 mNs
The **Leros 20H** MON/hydrazine engine is aimed at dual mode systems under ESA contract: SI 300 s, mixture ratio 0.7–0.8, minimum impulse bit 15 mNs. Performance characterisation testing is scheduled for 3Q 1996.

Test and Production Facilities
The company most recently commissioned two major high altitude test facilities for its liquid thrusters. Westcott's new facility handles attitude control thrusters, while the Boeing site at Marysville (Washington state) is used for the apogee motors. Westcott's unit provides:
Pumping system: two stage ejector, 52 km at 10 gm/s (airflow) and 0.57 mbar for Leros 20
Vessel volume: 3.5 m^3
Propellant feed system: 30 l oxidant tank, 20 l fuel tank, 35 bar working pressure, 4–45°C thermal conditioning, helium saturation
Thrust measurement: single axis, high response, piezo electric transducer
Instrumentation: 64 channels, 120 kHz, analogue recorders, Agema thermal imaging system, Cyclops IR thermometer, CCTV high speed ciné.

The Boeing facility provides:
Pumping system: 5-stage steam ejectors with condensors and engine diffuser
Performance: 55 km at 195 gm/s (airflow) and 0.22 mbar for Leros 1
Vessel volume: 285 m^3
Propellant feed system: 450 l oxidant tank, 450 l fuel tank, 20 bar working pressure, ambient to 30°C thermal conditioning
Thrust measurement: six component, multi-axis fast response
Instrumentation: 50 channels, 1 kHz, analogue recorders, Agema thermal imaging system, Cyclops IR thermometer, CCTV high speed ciné.
Westcott provides other firing sites for mono/bi system testing at SL/altitude. The altitude site utilises three interconnected chambers totalling 600 m^3; two positive displacement pumps reduce the pressure in a single chamber to about 4 mbar, and to 6 mbar with all three connected. The 60 l hydrazine expulsion rig can supply propellant at up to 45 atm. The sea level facilities accommodate all storable propellants and thrusts up to 100 kN.
Westcott also offers:
Climatic chambers six chambers covering −55/90°C; seven cold chambers for long-term storage testing down to −60°C, with additional hot air and hot water long-term chambers
Vibration two tables, one covering 5–2,000 Hz, 144 kN sine thrust (±25.6 mm amplitude) and the other 10–3,000 Hz, 27 kN sine thrust (±12.7 mm amplitude)
Shock 1,000 kg test items subjected to 250 *g*
Centrifuge two centrifuges for testing solid propellant motors under varying *g* conditions. One provides up to 100 *g* for up to 33.4 kN motors on a 3 m arm, the second 60 *g* for 800 kN motors on a 5 m arm
Dynamic balancing up to 1,000 kg items on 56–300 rpm 1 m dia table; minimum measurable out-of-balance 0.3 kg-mm at 270–300rpm.
Westcott's **Liquid Motor Processing Facility** was established to process propellant and gas storage assemblies for Chevaline's post boost propulsion system and has handled >1,000 tanks and gas storage assemblies. The LMPF processes, fills, tests and performs final closure welds on aluminium and stainless steel tanks, with class 100,000 clean rooms. Propellants in use include IRFNA, MAF 1 and 4 and hydrazine.

VERIFIED

UK Ion Propulsion

The UK-25 xenon ion thruster was performance-mapped in Culham's vacuum facility (UKAEA Culham) 0516923

UK ion thruster development began in the 1960s with a 10 mN 10 cm mercury Kaufman thruster at RAE/Culham. The 10 cm work was re-activated in 1985 using Xe because of projected satcom growth in the 1990s. UK-10 in its qualified form provides a thrust of typically 10–25 mN but can be smoothly throttled from 0.2 to 30 mN. Culham has also performed extensive trials on the UK-25 25 cm version, capable of delivering a throttlable 50–300 mN. ESA's Artemis will carry two 18 mN UK-10 versions for NSSK and elements of the UK-25 are being adapted for the ESA-XX (qv) interplanetary thruster. A thruster of intermediate size, in the 50 mN category is also under development by the DRA (ex RAE).

DRA provides: programme technical lead, thruster design/manufacture, testing, electronics and hollow cathode development. Culham: thruster testing (including diagnostic life tests), plasma physics, ion beam extraction modelling, UK-25 development/testing. Matra Marconi Space (MMS): development of the UK-10 system for Artemis, including the power conditioning and control equipment and the propellant supply/monitoring equipment and the manufacture and qualification of the flight hardware for Artemis. An element of AEA Technology, Culham houses Europe's largest Xe propellant ion thruster test chamber: 5.8 m long, 1.3 m dia, equipped with helium cryopumps for handling the inert gas. During 1996 MMS commissioned a new test facility, capable of handling two 25 mN thrusters simultaneously.

UK-10 ion thruster development in the UK began in the late 1960s at RAE/Culham, concentrating on a 10 cm dia thruster with a nominal 10 mN thrust and mercury propellant. The T4A engineering model achieved very high efficiency, stable operation and a long operational lifetime. The propellant was changed to Xe in 1985 and new versions of the T5 flight model thruster have been manufactured/tested in a collaboration between DRA, Culham and Matra Marconi Space. 0.2–70 mN thrust has been demonstrated; qualification testing at 18 mN is underway at MMS and DRA, specifically for operational NSSK application on ESA's Artemis. This will be extended to 25 mN during 1997. A major experimental programme was completed in 1995 at the Aerospace Corp in Los Angeles, using USAF funding and a thruster supplied by DRA. This concentrated on a detailed characterisation of the ion beam. A lower level effort continues at Aerospace funded by DRA, MMS and the USAF. The thruster comprises a cylindrical discharge chamber closed at one end by a soft iron backplate and with a set of closely-spaced grids at the other. A magnetic field is generated by six equispaced peripheral solenoids. The field lines inside the chamber link an inner cylindrical soft iron pole and a larger diameter outer pole. Propellant gas is introduced through the axial hollow cathode and a bypass distributor on the backplate. A DC discharge is set up between the cathode and the cylindrical anode. This ionises the gas, the efficiency being enhanced by the magnetic field and the correct design of the inner pole/baffle disc arrangement. The positive ions are extracted and accelerated by a high electric field between the grids, attaining typically 30–60 km/s. A triple grid design (see diagram) is used to minimise damage caused by the impact of charge-exchange ions. The ion beam's charge is neutralised by electrons from an external cathode, and the whole thruster is surrounded by an earthed screen to prevent those electrons from reaching other parts of the device. MMS UK is responsible for the electronics and propellant feed systems, and commercial exploitation of the fully qualified operational system. Philips Components Ltd previously manufactured the hollow cathode and neutraliser; the work has now been transferred to DRA. DRA/Culham in 1994 produced a low cost version of the complete system for experimental applications. This was accepted for flight on Johns Hopkins Univ's NEPSTP Nuclear Electric Propulsion Space Test Program, since cancelled, which was intended to test the Russian Topaz 2 nuclear reactor. Other flight opportunities are being sought.

DRA's UK-10 thruster under test at the Aerospace Corp. Detailed characterisation of the ion beam was completed in 1995 (The Aerospace Corp) 0003458

ESA's Artemis carries two 18 mN versions. Another application may be ESA's Gravity Explorer Mission; DRA/Dornier studied the possibility in 1995/6 with encouraging results. DRA's STRV-1A (see UK National), launched in June 1994, included an experiment to allow the hollow cathode assembly to demonstrate spacecraft electrostatic discharging. Although the cathode assembly operated correctly, charging was not observed. The nominal operating parameters given below are in 18/25 mN order.
Thrust: 18/25 mN
Beam voltage: 1,100/1,100 V
Accelerator voltage: −250/−250 V
Decelerator voltage: −50/−50 V
Beam current: 0.329/0.457 A
Exhaust velocity: 40,233/40,233 m/s
Specific impulse: 3,084/3,131 s
Discharge power: 80/106 W
Beam power: 362/503 W
Power/thrust ratio: 26.4/25.7 W/mN
Electrical efficiency: 76/78%
Total efficiency: 57/60%
UK-25 design of a laboratory model 25 cm thruster began in early 1986 using scaling laws formulated during the UK-10 programme. Culham testing began Oct 1986; nominal thrust was 200 mN (highest obtained was 316 mN). Following the successful conclusion of the lab model work, an engineering model was manufactured at the beginning of 1989 and completed a comprehensive test/evaluation programme in 1992. Included were studies of life-limiting factors. Investigation of other

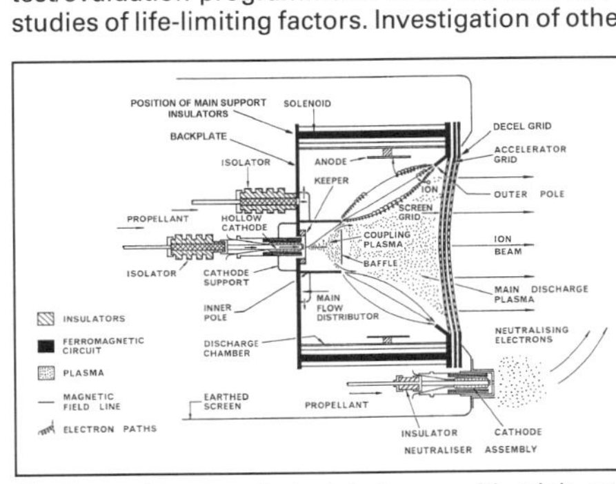

Kaufman thruster principal features, with triple grid (DRA) 0517338

propellants was completed in 1995 at Southampton Univ (although an ideal inert propellant, xenon is costly). Development of a power conditioning system was undertaken at Birmingham Univ and high current hollow cathodes were developed by DRA/Philips Components. Further work on the ESA-XX hybrid design, using features from this thruster and the German RIT-35, is underway with ESA funding. The first prototype of this thruster, intended for interplanetary missions, has been successfully tested at the Univ of Giessen in Germany.

VERIFIED

United States

AEROJET

Current Status

Aerojet has been awarded a contract by the Air Force Research Laboratory for computational capability improvements in advanced technology upper stage engines. The three-phase programme begins with the development of a conceptual engine design followed by a computational tools phase and concluding with a hardware and computational tools validation phase. The contract was awarded on 2 February 2004 and will run for nine months.

Background

A segment of GenCorp of Akron, Ohio, Aerojet was founded in 1942 by astronautics pioneer Dr Theodore von Karman and initially developed JATO jet-assist take-off rockets for aircraft. It was the first US company to produce storable bipropellant rockets. The propulsion plant, as Aerojet TechSystems (formerly Liquid Rocket Co), provided liquid engines for Aerobee, Bomarc, Nike, Vanguard, Delta, Atlas Able, Titan 1 and Apollo, and continues to manufacture Titan, Delta and Shuttle engines (see individual entries below). Aerojet, Rocketdyne and P&W formed the Space Transportation Propulsion Team in 1990 for work on the STME Space Transportation Main Engine.

Aerojet manufactured the solid motors for MX, Minuteman, Small ICBM, Standard Missile (rolling out the 10,000th motor in June 1989) and Hawk, and successfully bid in partnership as Aerojet Space Boosters (becoming Aerojet ASRM Division) with Lockheed for NASA's Shuttle Advanced Solid Rocket Motor contract. Congress cancelled ASRM in Oct 1993. Propulsion activities are performed at the 52.6 km^2 Sacramento plant, founded 1951 and capable of testing cryogenic and storable propellant motors up to 6,670 kN. The Mach 8/30 km-altitude

Aerojet's paired LR-87 AJ11 engines power the Titan 3/4 first stages 0517369

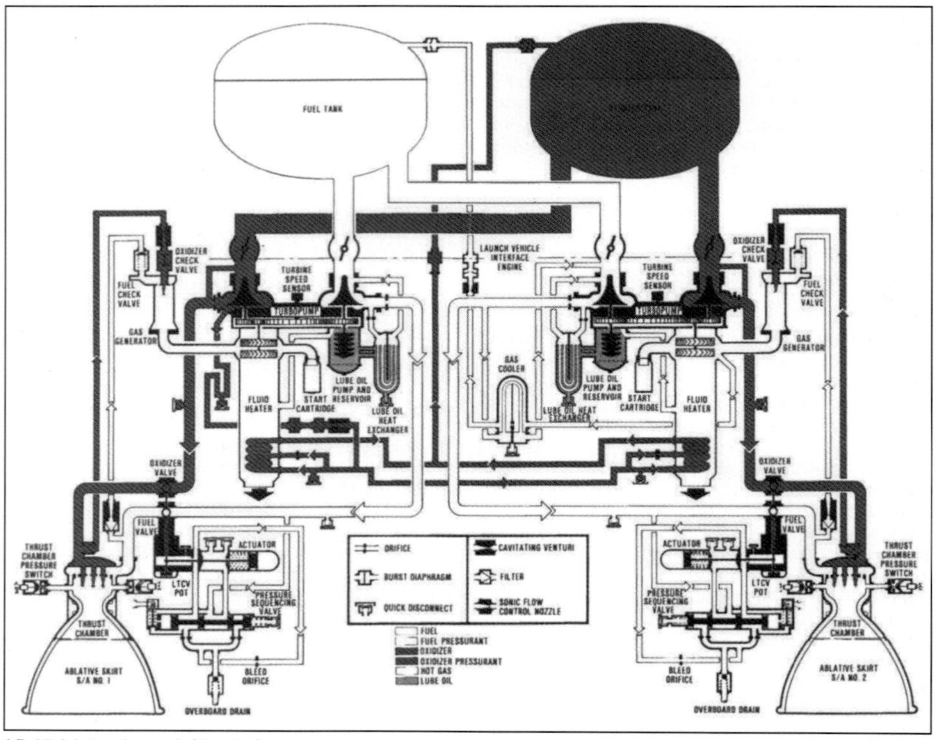

LR-87-AJ-11 schematic (Aerojet) 0517368

hypersonic Hytest facility was added for National AeroSpace Plane engines and components tests of up to 35 s, and the company has access to the Hypulse M25 facility at subsidiary General Applied Sciences Laboratories Inc (Ronkonkoma, NY), which it acquired in 1989.

Aerojet teamed with Lyulka in 1993 to improve and market the 400 kN LOX/LH_2 D57 engine, originally developed in the 1960s for the improved N1-L3M manned lunar mission. It is a candidate for an SSTO demonstrator. A teaming agreement was signed July 1993 with Kuznetsov/NPO Trud (qv) to use the NK series exN1 engines and their technology in the US market. Aerojet made five tests totalling 410s with an NK-33 in Oct Nov 1995. If it had been selected for a US vehicle as the AJ26-NK33A (it lost to the RD-180 for Lockheed Martin's new Atlas 2AR and 5 launch vehicles), a US production line would have been established. Unit cost would have been about $4 million. See the CIS Propulsion section for further information. A similar agreement was signed Oct 1994 with KB Khimautomatiki for the RD0120, principally developing it into a tripropellant engine. Work is being performed under a $17.2 million NASA Marshall contract.

Aerojet presently employs about 500 people.

AJ10-118K

This pressure-fed engine, optimised for altitude operation, has flown as the stage 2 propulsion system for McDonnell Douglas' Delta and the related Japanese N vehicle, in addition to flying paired (as the AJ10138) on Titan 3's Transtage upper stage. Contracts continue through 2000, building about 10 annually.

Applications: Delta stage 2
First flown: Aug 1982, Delta 164
Dry mass: 124.7 kg
Length: 269 cm
Mounting: fixed
Engine cycle: pressure-fed
Oxidiser: nitrogen tetroxide at 9.1 kg/s
Fuel: Aerozine-50 at 4.76 kg/s
Mixture ratio: 1.9:1
Thrust: 43.38 kN vacuum
Specific impulse: 320.5 s vacuum
Expansion ratio: 65:1
Combustion chamber pressure: 8.84 atm
Cooling method: ablative chamber, radiative skirt
Burn time: qualified up to 500 s (unlimited starts)

Orbital Maneuvering System (OMS)

NASA's Shuttle Orbiter carries two OMS pods, each housing a single Aerojet OMS engine for orbit insertion, manoeuvring and reentry initiation. The engines are capable of 100 missions and 500 starts in space. 12 were originally delivered for the Shuttle fleet (including four reserves), with two more added following *Challenger's* loss (#102/112) in 1986 (these were completed as #115/116 in summer 1989 and first flown on STS50 in June 1992). Aerojet was awarded a $3.6 million 38 month contract in Nov 1986 to continue development of an uprated OMS featuring increased chamber pressure through the addition of a pumpfeed system. Successful testing beyond 600 s was completed in autumn 1990 at NASA's White Sands altitude simulation facility. OMS is derived from Aerojet's Apollo Service Propulsion System, designed with high reliability for manned lunar missions. The first OMS prototype and demonstration tests were completed 19723 and the company was awarded the full development contract in 1974. The thrust chamber was successfully tested for the first time in 1976 and the first prototype development engine was delivered to NASA in Feb 1977 for extensive White Sands testing to begin the following Aug. A further engine, used to qualify the system as part of the Orbiter's pod, was delivered in Jan 1979 and fired 270 times that year, accumulating 10,817 s. The first two production engines, for *Columbia,* were delivered Mar 1979, and testing of the final qualification engine, delivered Oct 1979, was completed in June 1980. In the first 24 Shuttle missions, nine of the 12 engines were carried and fired 254 times for a total of 21,875 s. In the 27 missions following *Challenger* to end 1992, eight engines were used for 377 firings totalling 39,995 s without anomaly.

Applications: Space Shuttle orbit/de orbit insertion, circularisation
First flown: 12 Apr 1981, Orbiter
Columbia Number flown: 14, to end-1995 (103/104 have never flown)
Dry mass: 118 kg
Length: 195.6 cm
Maximum diameter: 116.8 cm
Mounting: gimballed ±7° yaw/±6° pitch by two electromechanical actuators for thrust vector control
Engine cycle: pressure-fed (improvement programme underway for pumpfeeding)
Oxidiser: 6,743 kg nitrogen tetroxide in each pod (pods can be cross linked)
Fuel: 4,087 kg of monomethyl hydrazine in each pod (pods can be cross linked)
Mixture ratio: 1.65:1
Thrust: 26.7 kN vacuum
Specific impulse: 316 s vacuum
Expansion ratio: 55:1
Combustion chamber pressure: 8.62 atm
Cooling method: fuel regenerative for chamber, radiative for nozzle
Burn time: qualified for 500 starts, 15 h/100 mission life, longest firing 1,250 s, de orbit burn typically 150–250 s.

Titan Space Launcher Engines

Aerojet has developed and manufactured the engines for four generations of USAF/NASA Titan ICBMs and space launchers since 1958, attaining a peak annual production rate of 270. Each Titan core carries a dualchamber engine in the first stage and a scaleddown single-chamber version to power stage 2. The company has refurbished 14 sets of Titan 2 engines for orbital launcher applications, and is manufacturing new 11family engines for Titan 4. The Titan 4 contracts, worth some $350 million, call for 41 USAF and four NASA engine sets.

AJ-5 Titan 2 engines
Aerojet began testing the first stage 1/2 experimental engines in 1960 and completed production of the 474 chambers by end1967. The ICBMs were overhauled 1974–82 and the company subsequently refurbished 14 sets for the space launcher version. The Gemini Titan 2 manned version flown 1964–66 utilised manrated 7-type engines.
Designation: Aerojet LR-87-AJ-5
Configuration: twin fixed motors with individual turbopump assemblies
Applications: Titan 2 stage 1
First flown: 1962 ICBM; Sep 1988 orbital
Dry mass: 1,266 kg for full assembly
Length: 2.3 m
Maximum diameter: 1.1 m
Engine cycle: gas generator
Propellants: hypergolic nitrogen tetroxide and Aerozine50, delivered at 754 kg/s
Mixture ratio: 1.93:1
Thrust: 1,913 kN sea level
Specific impulse: 259 s sea level
Expansion ratio: 8:1
Combustion chamber pressure: 53.7 atm
Burn time: about 165 s
Designation: Aerojet LR-91-AJ-5
Configuration: scaled-down version of stage 1 engine, featuring fixed single chamber
Applications: Titan 2 stage 2
First flown: as stage 1 engine above
Dry mass: 472 kg
Length: 2.80 m
Maximum diameter: 1.68 m
Engine cycle: gas generator
Propellants: as stage 1, at 146.5 kg/s
Thrust: 444.8 kN vacuum
Specific impulse: 312 s vacuum
Expansion ratio: 49.2:1
Combustion chamber pressure: 56.2 atm
Burn time: about 185 s

LR-87-AJ-11
Titan 4's first stage (and Titan 3's if it flies again) are powered by paired AJ-11 engines. The 9 model was first tested in 1963 and 476 were produced before being replaced by the 11 in late 1968, first tested earlier that year. Together with its stage 2 derivative, it is the only US engine to be operated on storable N_2O_4/A-50 *and* tested on LOX/RP1. Ablative skirts with 12:1 and 15:1 expansion ratios are available, producing 254 s (sea level) and 302 s (vacuum) SI. USAF awarded a contract to Martin Marietta in 1990 for a modified Titan 4B nozzle. This requires an increased expansion ratio for ignition at greater altitude when Hercules' higher performance strap ons are carried.
Applications: Titan 3/4 stage 1
First flown: 1968 Titan 3, 1989 Titan 4
Dry mass: 1,874 kg (paired), 758 kg (single)
Length: 3.84 m to top of thrust structure, 3.23 m to top of turbopump assembly
Maximum diameter: 1.6 m
Mounting: gimballed pair
Engine cycle: gas generator
Oxidiser: nitrogen tetroxide at 513 kg/s

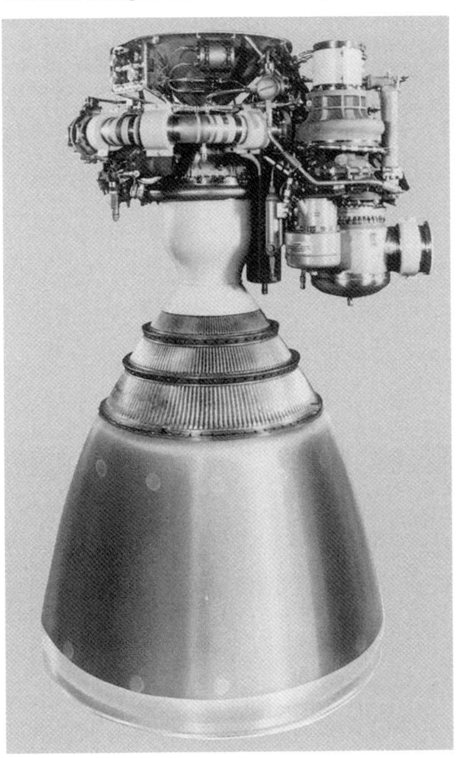

Aerojet's LR-91-AJ-11 Titan 3/4 stage 2 engine 0517371

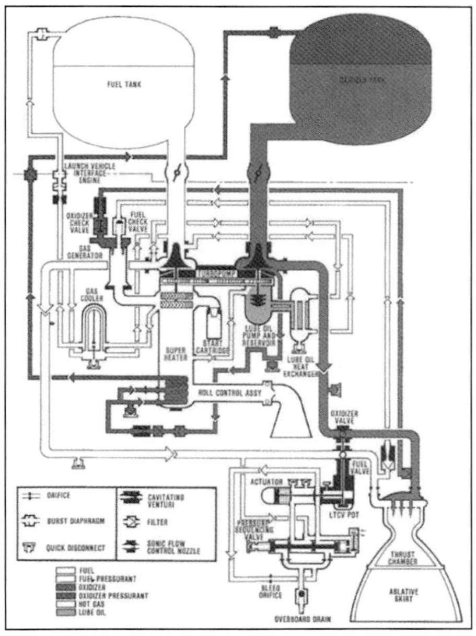

LR-91-AJ-11 schematic (Aerojet) 0517370

Fuel: Aerozine-50 at 268 kg/s
Mixture ratio (O/F): 1.91:1
Thrust: 2,340 kN vacuum paired
Specific impulse: 301 s vacuum
Expansion ratio: 15:1
Combustion chamber pressure: 55 atm
Cooling method: fuel regenerative and ablative skirt
Burn time: about 200 s

LR-91-AJ-11
Development of the current Titan 3/4 stage 2 engines followed that of the stage 1 version. It has flown without failure on every operational Titan 3 mission.
Applications: Titan 3/4 stage 2
First flown: late 1968 Titan 3, 1989 Titan 4
Dry mass: 589 kg
Length: 281 cm
Maximum diameter: 163 cm (skirt outer diameter)
Mounting: gimballed, turbine exhaust utilised for roll control
Engine cycle: gas generator
Oxidiser: nitrogen tetroxide at 97.0 kg/s
Fuel: Aerozine-50 at 54.7 kg/s
Mixture ratio: 1.86:1
Thrust: 467 kN vacuum
Specific impulse: 316 s vacuum
Expansion ratio: 49.2:1
Combustion chamber pressure: 58.5 atm
Cooling method: fuel regenerative thrust chamber, with separate ablative skirt
Burn time: about 247 s

Transtar
Transtar was developed as an upper stage engine using injector, chamber and nozzle derived from OMS. However, the propellants are pump-fed, increasing chamber pressure and SI, and permitting use of lowpressure lightweight tankage.
Applications: upper stage
First flight: not flown
Dry mass: 67 kg
Length: 127 cm
Mounting: gimballed ±10° by two electromechanical actuators
Propellants: N_2O_4/MMH
Mixture ratio: 1.8:1
Thrust: 16.680 kN vacuum
Specific impulse: 328 s vacuum
Expansion ratio: 132:1
Combustion chamber pressure: 23.8 atm
Cooling method: fuel regenerative for chamber, radiative for extension
Burn time: not available, 15 starts

Aerojet Satellite Engines
Aerojet also produces 2, 21, 62 and 445 N bipropellant (NTO/MMH) thrusters for satellite orbitadjust and attitude control systems. Qualification of a 21 N thruster is being completed for a USAF satellite. Principal characteristics are:

Thrust (N, vac)	*2.00*	*21.35*	*62*	*445*
Specific impulse (s, vac)	265	285	287	309
Dry mass (kg)	0.27	0.57	1.13	1.86
Expansion ratio	150	150	75	150
Mixture ratio (O/F)	1.65	1.60	1.65	1.65

VERIFIED

ALLIANT TECHSYSTEMS, INC (ATK)

Current status
ATK has consolidated its space propulsion capabilities with the acquisition of Hercules Aerospace Company in 1995, Thiokol Propulsion in 2001 and Composite Optics in 2003, a specialist composite space structures company which ATK has applied to innovative design concepts.

Background
Alliant Techsystems acquired Hercules Aerospace Co in Mar 1995 for $296 million. Hercules' rocket motor, weapons and ordnance businesses totalled $660 million revenue in 1993, for a $105 million operating profit. Hercules was selected in Oct 1987 as the propulsion contractor for Titan's Solid Rocket Motor Upgrade programme, potentially worth $725 million for the development, qualification and production of 15 sets of Titan 4 solid boosters. Lightweight graphite composite materials replaced the steel used in current motor cases, and a high performance propellant similar to that developed for the Delta 2 strapon is employed. PQM-1 (Preliminary Qualification Motor) was the first full scale test, but the motor failed after a few seconds at Edwards AFB 1 Apr 1991. The added PQM1' firing was successful 12 June 1992. The QM2 third test, run at 2.5°C, was successful 21 Feb 1993. QM-3 at the 41°C upper limit was successful 2 June. Delivery of the first flight set was made 1Q 1994. Lockheed Martin awarded the company a contract for Titan launch vehicles support which extended to 2003 and was followed by contract extensions.

The company was awarded a McDonnell Douglas contract in 1987 to develop the stretched solid rocket GEM graphite epoxy strapon motors for Delta 2; they first flew Nov 1990. The initial contract for 144 motors (16 flight sets) was followed in 1991 by a second for 117 (13 sets) beginning production in 1993, and a third in June 1995 for 144 (16 sets). GEM production rate is 6/month. For Delta 2, from 1996, the expansion ratio for the airlit nozzles will be increased from 10.6 to 16.3 by lengthening the nozzle by 30 cm and increasing its dia by 19.5 cm. The GEMVN version incorporates a vectorable nozzle; the qualification firing was made May 1994. The TVC system is provided by AlliedSignal and the nozzle by BP-Hitco as part of the strategic partnership. Delta 3's nine strapons will be lengthened by 1.22 m and their diameters increased to 116.8 cm. Three of the six groundlit motors will carry gimballed nozzles for TVC, and the three airlit will improve performance with extended nozzles. The first of 16 sets ordered June 1995 were delivered by early 1998.

Alliant produces Pegasus' Orion solid motors, derived from GEM, and the fairing. Taurus also uses the Orion motors. Hercules also designs/fabricates composite structures: spars, struts and optical benches used in satellites.

UPDATED

ATK Thiokol Propulsion

Thiokol was acquired by ATK. Established in 1929, Thiokol Chemical Corp produced and marketed the first synthetic rubber manufactured in the US. In 1943, the discovery by Thiokol of liquid polymer, a new type of synthetic rubber, paved the way for the development of the case bonded principle of rocket power plant design. The company's polysulphide liquid polymer proved to be the catalyst for the first mass production of efficient solid propellant rocket motors, as well as for the development of *large* solid propellant motors. Thiokol Corp merged with Morton Norwich Products in Sep 1982, but separated off in July 1989. It reported $52.3 million net income in FY95 on sales of $956.8 million ending June 1995; FY94 60.3, FY93 63.8, FY92 63.0, FY91 53.4. Personnel total >7,500.

Thiokol's Space Operations produces the Space Shuttle Solid Rocket Motor (SRM). FY95 revenues were $467.4 million for an operating income of $60.4 million (FY94 500.8/51.5; FY93 519.3/61.5; FY92 555.6/58.5). Utah DLV Operations (until 1995 Strategic Operations) in a joint venture with Hercules Inc produces stage 1/2 of the USN Trident II D-5 missile. It provided the complete stage 1, ignition systems and flight termination

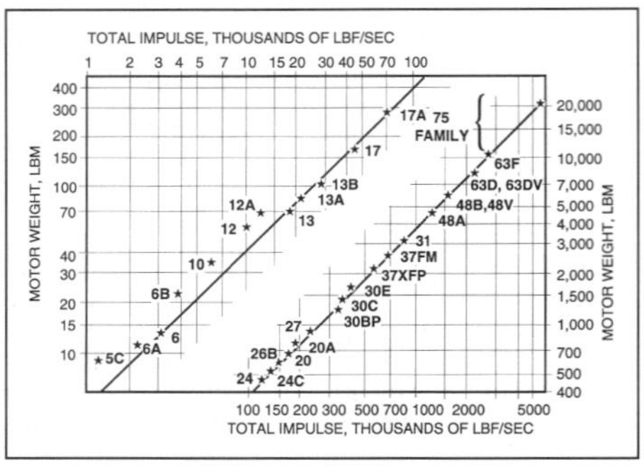

Star family performance. Star 26B and 20A have been deleted (Thiokol) 0517173

systems for USAF's Peacekeeper. It also produces the Castor 120 and Castor 4 motors, flares and the HARM and Standard Missiles. Elkton produces gas generators, multiple pulse motors and the Star series of spacecraft kick and launcher stage motors. Most Castor motors, long used as Delta strapon boosters and more recently on Atlas 2AS, were handled by Huntsville. The Castor family has flown on >1,800 launches since 1959. Huntsville also produced Sidewinder, Hellfire and Maverick motors. The facility closed in 1996. Huck produces fasteners for Space Shuttle.

A 10 yr licensing agreement with the UK's Royal Ordnance Rocket Motors Div was announced in Nov 1989 that allows Thiokol to manufacture solid propellant and motor casings developed by RO. These include smokeless flameless elastomer modified cast double base propellant and steel strip laminate cases.

ATK Thiokol is producing the Titan Solid Rocket Motor Upgrade providing 25 per cent increase in performance. Thiokol is also researching new derivative solid propulsion systems for

Castor 4 (TX526)

The TX526-0 motor, combined with four Recruit strapon motors, was the booster for the Athena H launch vehicle. The TX526-2 (with an 11° canted nozzle) and the TX526 3 (7° nozzle) were used as strap-on boosters for Delta, each of which had nine motors. A total of 342 flew on 38 Delta 3000 series missions (#118 to #183); 243 of the 244 ignited on the pad were successful.

First flown: 1971
Number flown: 360 to end-1995 (0 in 1995; 0 in 1994; 0 in 1993; 99.72% success)
Mass: 10,534 kg
Length: 9,072 mm
Diameter: 1,019 mm
Propellant:
type: TP-H8038 PBAA Polymer, 14% aluminium
shape: radially slotted centre perforate
mass fraction: 0.880 (9,265kg)
Burn time: 53.9 sec at 27°C
Thrust: 428kN vac at 27°C
Specific impulse: 228.4 s
Total impulse: 23,748 kNs vac at 27°C
Pressure (atm): 45.4 max; 37.0 avg
Nozzle:
throat diameter: 29.54 cm
length: varies – SL/altitude/canted available
exit diameter: 81.66 cm
materials: 4130 steel
Casing materials: 4130 steel

Thiokol's new Castor 120 fills the gap between the Castor 4 and large segmented motors 0516955

Castor 4A (TX780)

Castor 4A began development in the early 1980s under a NASA Goddard programme to improve Delta performance by 11% by switching to HTPB propellant. A development motor was fired 25 Feb 1983 and a qualification motor 31 Aug 1983 before the programme was halted when it appeared that Delta would be terminated. Development restarted when Delta 2 was selected as the USAF GPS launcher; three further qualification firings followed. Each vehicle carried nine strap-ons (later Deltas switched to Hercules' motors). The TX780-0 with an 11° canted nozzle is used as the ground ignited booster; the TX7801 (7° cant) and TX7802 (11°) are ignited at altitude. The TX780-3/-4 were developed as the ground and altitude ignited boosters, respectively, for Atlas 2AS (first flight Dec 1993; six, successful, by end1995). The TX7805 (straight nozzle) powered OSC's Prospector suborbital vehicle and two flew on Conestoga in Oct 1995. **4H** (114.3 cm dia) and **5** (127.0 cm dia) versions have been studied. See the separate **4AXL** entry below. The Huntsville plant closed in 1996; Castor 4 production transferred to Utah DLV Operations.

First flown: 14 Feb 1989, Delta 184
Number flown: 211 to end-1995 (100% success)
Mass: 11,578 kg (propellant 10,121 kg) ground ignited
Length: 8,087/9,199 mm loaded case with/without nozzle
Diameter: 1,016 mm
Propellant:
type: TP-H8299 HTPB Polymer (20% aluminium)
shape: cylindrical, 76% web fraction; transitions to four longitudinal slots. Tailored for ignition/burnout thrusts to be similar to Castor 4
mass fraction: 0.874 for standard loading
Burn time: 53.1 s web
Thrust (kN, sea level, 21°C): 476.661 max, 436.736 avg
Specific impulse: 237.6 s sea level 21°C
Total impulse: 23,846 kNs sea level 21°C
Pressure (atm): 46.60 avg, 49.86 max sea level 21°C
Nozzle:
throat diameter: 278.6 mm
materials: carbon-phenolic
Casing liner: standard HT polymer
Igniter type: pyrogen, head-end

Castor 4AXL

The 4AXL version, extended by 2.44 m, was first tested May 1992. Its 30% performance increase would improve performance of vehicles such as Atlas and Conestoga (Martin Marietta included it as part of its unsuccessful bid for the USAF MLV3 bid, won by Delta in 1993). QM-1 was successfully tested Nov 1992 and QM2 completed qualification 3 Feb 1993. Thiokol has invested about $12 million in the programme. 4AXL was to have been combined with 4B's TVC system to create the 4BXL stage 1 motor for CTA's ORBEX small launcher.

Mass: 14,851 kg (propellant 13,128 kg for ground ignited strap-on)
Length: 12,279 mm with nosecone adapter, 13,711 mm with nosecone
Diameter: 1,018.5 mm
Propellant:
type: TP-H8299 HTPB Polymer, 20% Al, 68% AP
shape: forward cylindrical perforate, seven aft longitudinal slots
mass fraction: 0.884 ground ignited strap on
Burn time: 60.1 s
Thrust (kN, vac): 599.81 avg, 700.5 max
Specific impulse: 269.2 s vacuum
Total impulse: 34.679 MNs vacuum
Pressure (atm): 41.70 avg, 54.42 MEOP
Nozzle:
throat diameter: 318.8 mm
length: 1,216.8 mm
exit diameter: 937.3 mm
materials: 4130 steel, graphite phenolic throat insert, carbon phenolic exit cone
Casing material: AISI 4130 steel 0.28 mm thick
Igniter type: TX544 (>500 units flown successfully) forward internal pyrogen, 2.45 kg TPH8027 HTPB propellant, cartridge loaded

Castor 4B (TX859)

The 4B, which incorporates TVC in the series for the first time, was developed for ESA's Maxus μg carrier and first flown in 1991. Ten were ordered. Five flew on Conestoga's maiden launch, Oct 1995. It is now used by the US Army Theater Missile Defense programme, planned to debut in mid1996. CTA's cancelled ORBEX was to use the **4BXL** version, which would combine the 4AXL motor with 4B's TVC system (±6°). Thiokol may still qualify it.

First flown: 8 May 1991
Number flown: 8 to end-1995 (100% success: 3 Maxus + 8 Conestoga)
Mass: 11,482 kg

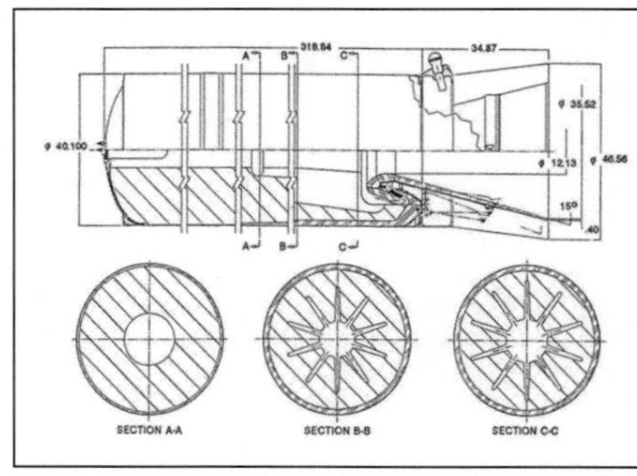

Castor 4B first flew in 1991 on the suborbital Maxus launcher. Dimensions are in inches 0516956

Length: 8,984 mm
Diameter: 1,019 mm
Propellant:
type: TP-H8229 HTPB Polymer, 20% aluminium
shape: radially slotted centre perforate
mass fraction: 0.868 (9,969 kg)
Burn time: 60.9 s at 25°C
Thrust: 429 kN vac at 25°C
Specific impulse: 267.4 s vac
Total impulse: 26,141 kNs vac at 25°C
Pressure (atm): 50.6 max; 30.7 avg
Nozzle:
throat diameter: 30.84 cm
length: varies – SL/altitude/canted available
exit diameter: 90.2 cm
materials: 4130 steel
Casing materials: 4130 steel

Castor 120

The motor is similar in size to the Peacekeeper missile stage 1 motor, filling the gap between Castor 4A and the large segmented motors, but using proven technology throughout. Primary goals are >0.999 reliability and 50% cost reduction from Peacekeeper. It represents the upper limit for a monolithic motor that can be handled and transported by commercial methods with few restrictions. Unit cost would be about $4.0 million for volume production, rising to $5.5 million for one annually. Thiokol has invested >$50 million in its development. Its first application was as stage 1 of Lockheed Martin's LMLV small launcher in Aug 1995, but failed because of TVC cable overheating. OSC's Taurus satellite launcher will use it beginning with the second vehicle. Three flights are planned for 1996 (two LMLV + one Taurus); six motors are in production for 199798 flights. A potential application is as a strap-on, such as for Delta. It can be used as a stage 1 or 2 or strapon because both the forward/aft ends can take tension or compression forces. ±5.0° TVC is provided by movable nozzle, but the cold gas blowdown TVC system (provided by AlliedSignal Equipment Systems) can be deleted. The grain can be tailored to reduce thrust during max-q, for high initial thrust or for a regressive thrust to reduce acceleration.

Fabrication of the first 120 (DM1201) was started in late 1990; first static firing was made successfully 22 April 1992. Second test (DM1202), at the 1°C lower limit, was successful 4 March 1993.

First flight: Aug 1995
Number flown: 1 (0% success)
Mass: 53,118 kg (propellant 49,032 kg)
Length: 9,017 mm
Diameter: 2,363 mm
Propellant:
type: HTPB
shape: three aft conical slots, one forward slot
mass fraction: 0.923 (49,032 kg)
Burn time: 81 s
Thrust: 1,650 kN vac average
Specific impulse: 280 s vacuum
Total impulse: 133,440 kNs vac
Pressure (atm): 101 max; 78 avg
Nozzle:
throat diameter: 368 mm
length: 1,448 mm

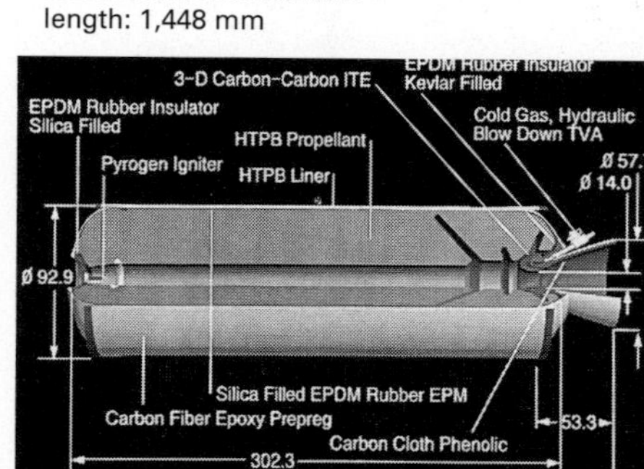

Castor 120. Dimensions are in inches 0516957

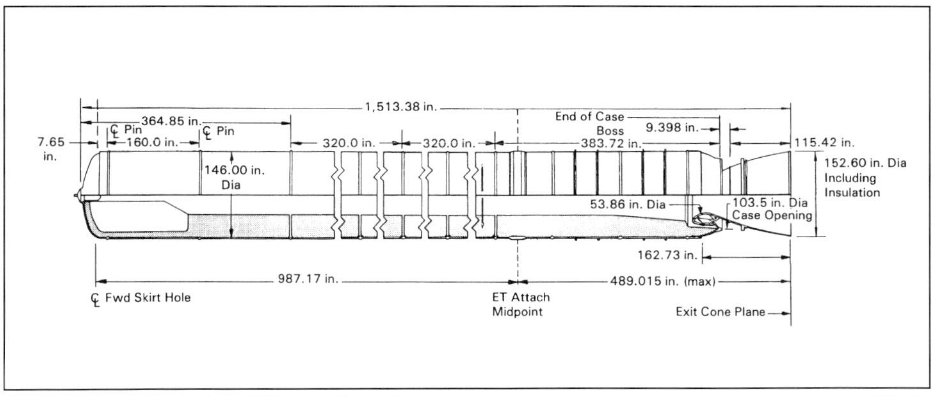

Thiokol's Shuttle Reusable Solid Rocket Motor design 0517339

exit diameter: 1,600 mm
materials: glass epoxy/polyisoprene flex bearing; 3D carbon carbon throat; carbon phenolic exit cone; aluminium stationary shell; steel throat support
Casing materials: carbon epoxy, lined with silica filled EPDM rubber

Space Shuttle Solid Rocket Motor
Thiokol was selected as Shuttle SRM prime in 1974. NASA announced July 1991 a $2.6 billion Buy 3 contract extension to Thiokol for 67 further flight pairs through 1999/STS105, plus eight ground test motors. Another extension is expected for launches through 2012. The SRM is cast in four segments in Utah and then transported by rail to Kennedy Space Center, Florida, where they are stacked vertically in pairs either side of the External Tank on the mobile launch platform in the Vehicle Assembly Building. The boosters provide the majority of launch thrust, with TVC provided by a flexible bearing nozzle driven by two hydraulic actuators. After burnout at about 2 min, the boosters are separated pyrotechnically and fall into the Atlantic for recovery. The motors are cleaned, disassembled and returned to Utah for refurbishment and reloading. The segments are designed for 20 flights; nine is currently the maximum. The configuration underwent extensive reworking and testing following *Challenger's* loss, and the new booster was redesignated RSRM (Redesigned Solid Rocket Motor, changed by NASA in 1995 to Reusable SRM). Further information is provided within the US National entry. A High Performance Motor (HPM) design for an improved SRM incorporates lightweight steel cylindrical case segments, a higher expansion nozzle, and altered motor thrust characteristics increased Shuttle payloads by >1,540 kg. This design was static tested in 1982 and went into service on STS-8 in Aug 1983. Graphite epoxy composite cylindrical SRM case segments were also developed to increase Shuttle capacity by 2 t; the configuration was shelved following *Challenger's* loss. Thiokol was subcontracted to Aerojet to develop/produce the nozzle for the Advanced SRM that was to enter Shuttle service in about 1998. ASRM was cancelled in 1993.

Star 48 provides the propulsion for McDonnell Douglas' Payload Assist Module (MDSSC) 0516958

Reusable Solid Rocket Motor Mass: 569,636 kg
Length: 38.47 m
Diameter: 3,708 mm
Propellant:
type: TP-H1148 HB Polymer, 16% Al, 70% AP
shape: 11-point star forward; double taper cylinder two centre segments; triple taper cylinder aft
mass fraction: 0.883 (501,746 kg)
Burn time (s): 123.7 action, 111.4 web
Thrust (vac): 11.52 MN average, 14.77 MN max
Specific impulse: 268.0 s vac
Total impulse: 296.3 MNs vac
Pressure (atm): 62.1 max (MEOP 69.1); 45.0 avg
Nozzle:
throat diameter: 136.8 cm
length: 454.0 cm
exit diameter: 380.1 cm (expansion ratio 7.72)
materials: D6AC steel for fixed housing, throat and forward exit cone housings; aluminium for nose inlet and aft exit cone housing
Casing materials: D6AC steel, 1.165 cm minimum thickness
Igniter: forward end internally mounted. 60.7 kg of TP H1178 case bonded 40 point star

Star 5CB (TE-M-344-16)
Reduced aluminium HTPB propellant (2.1 kg) minimises contamination when used as Titan 4 stage separation motor. 121 mm main dia, 341 mm length, 4.5 kg total mass, 5.5 kNs total impulse, 2.77 s action time. Adapted for Mars Pathfinder.

Star 6B (TE-M-790)
Space kick motor, principal dia 166 mm, length 403 mm, total mass 10.26 kg (propellant 6.1 kg). Delivers 2.6 kN thrust vac over 6.65 s action time; total impulse 16.4 kNs. Offloading provides impulse tailoring to 7.6 kNs. Sweden's Freja scientific satellite used a 6B to circularise its orbit in Oct 1992.

Star 13/13B (TE-M-458/763)
Braking motor used in NASA's Anchored Interplanetary Monitoring Platform programme. 31 kg charge of AP/Al urethane contained in a spherical case of 6Al-4V titanium, with graphite/vitreous silica phenolic nozzle. Loaded mass 35.65 kg; average thrust 3.8 kN for burn time of 21.8 s. Sweden's Freja scientific satellite used it Oct 1992 to raise apogee. The 13A, with 33 kg propellant, is used to deorbit Meteor's return capsule. The 13B was developed by adding a 56 mm cylindrical section to the spherical Star 13. Propellant mass 40.86 kg; first flown 1984.

Star 17/17A (TE-M-479/521)
The TE-M-479 is a 442 mm spherical motor developed for NASA's Radio Astronomy Explorer as an AKM. 687 mm long; mass 78.8 kg (propellant 69.7 kg), titanium case. Star 17A, 444 mm dia, 980 mm long, developed by adding a 175 mm straight section to the spherical 17. 17A mass 126 kg (propellant 112 kg). Circularised orbit for Skynet 1, NATO 1 and IMPH/J.

Star 27 (TE-M-616)
694 mm dia, 1303 mm long AKM, 338 kg charge of AP/HTPB/Al in 6Al 4V titanium case and graphite/carbon phenolic nozzle. Mass 365.7 kg; mass fraction 0.925 (propellant mass can be reduced up to 25%). Average thrust 27 kN vac for 34.5 s burn, SI 289.5 s. AKM for Canada's CTS, Japan's GMS/BS, and several USAF GPS and NOAA GOES satellites. GMS 4 in 1989 employed the TEM 6165, propellant mass 348 kg, Isp 287.8 s vac. A TVC nozzle was demonstrated on a Star 27 in 1987.

Star 30BP/C/E (TE-M-70020/18/19)
TE-M-700-20 (STAR-30BP) 762 mm dia 1,506 mm long AKM. 505 kg charge of AP/HTPB/Al in 6Al4V titanium case; nozzle with carbon/carbon throat insert; aft end igniter. Mass 542.8 kg (mass fraction 0.951); propellant mass can be reduced 20%. Burn 54 s, vacuum total impulse 1,460 kNs, effective Isp 292.0 s. Used on Thaicom. TEM700 18 (Star 30C) has shorter exit cone + 126.2 mm longer case for total 1,493 mm length. Mass 626.9 kg (propellant 585 kg max). Vacuum total impulse 1647 kNs, effective Isp 284.6 s. Used on Lockheed Martin series 3000 telecom satellites.
TE-M700-19 (Star 30E) has longer case and expansion cone than Star 30C: 1,683 mm motor length and 668 kg. Vacuum total impulse 1,780 kNs, effective Isp 289.2 s. CTA planned to use it for stage 4 of the ORBEX small orbital launcher. Used by Thaicom and Koreasat (628 kg propellant, mass fraction 0.940).

Star 31 (TE- M-762 Antares 3)
762 mm dia, 2,873 mm long motor carries 1,287 kg of AP/HTPB/Al in a filament wound Kevlar case and a carbon/phenolic nozzle with a carbon/carbon throat insert. 1,393 kg including external insulation. Average thrust 80 kN over 46 s burn. Qualified as an improved Scout stage 3; first flight late 1979, last 1994.

Star 37FM (TE-M-783)
Developed originally as AKM for USN's Fleetsatcom. Offloading propellant weight by 20% is a standard option. 37FM will be employed by the GPS 2R navsat series. Clementine 1 used it Feb 1994 to enter a lunar transfer orbit.
Mass: 1,148 kg (propellant 1,066 kg, qualified for offloading to 1,024 kg)
Length: 1,676 mm loaded case with nozzle
Diameter: 933 mm
Propellant:
type: AP/HTPB/Al
mass fraction: 0.929 for standard loading
Burn/action time: 62.7/64.1 s
Thrust (kN, sea level, 21°C): 54.80 max, 47.26 avg
Specific impulse: 289.8 s sea level 21°C
Total impulse: 30,512 kNs sea level 21°C
Pressure (atm): 36.73 avg, 43.67 max sea level 21°C
Nozzle:
throat diameter: unavailable
materials: the semi-submerged, contour nozzle has a carbon phenolic expansion cone and carbon carbon throat insert
Casing material: 6Al- 4V titanium
Casing liner: unavailable
Igniter type: aft end, initiated from a remotely located Model 2134 safe/arm device

Star 37XFP (TE-M-714-16/-17)
The GPS 2A navsats use the -16 version for orbit insertion. Propellant offloads to 20% are also produced. It has flown from Atlas Centaur, Ariane, Delta and Shuttle.
Mass: 955.7 kg (propellant 884 kg)
Length: 1,515 mm loaded case with nozzle
Diameter: 933 mm
Propellant:
type: AP/HTPB/Al
shape: unavailable
mass fraction: 0.925 for standard loading
Burn/action time: 66/67 s
Thrust (kN, sea level, 21°C): 42.48 max, 37.72 avg
Specific impulse: 290.0 s sea level 13°C
Total impulse: 2,537 kNs vac 13°C
Pressure (atm): 35.58 avg, 39.18 max vac 13°C
Nozzle:
throat diameter: unavailable
materials: the semi-submerged nozzle has a carbon-carbon throat insert
Casing material: 6Al-4V titanium
Casing liner: unavailable
Igniter type: aft end, initiated remotely

Star 48B (TE-M-711)
TE-M-711-17 1,245 mm dia, 2,032 mm long PKM with 2,011 kg charge of AP/HTPB/Al in 6Al-4V titanium case; semi submerged nozzle. Aft end igniter initiated from a remotely located safe/arm device. 2,137 kg (mass fraction 0.941). Average vac thrust 66 kN over 84 s burn; total impulse 5,647 kNs, effective Isp 286 s. Employed in McDonnell Douglas's PAM for satellite launches from Shuttle and Delta. The TEM71118 has a 203 mm longer exit cone for use in launches from Delta, mass 2,142 kg, delivers effective Isp 292.1 s, total impulse 5,799 kNs.

Star 48V (TE-M-940-1)
Developed for Conestoga's upper stage, providing a ±4° capability at 30°/s using the same loaded case as the 48B. First test was made June 1992; the second and final firing was made 2 Mar 1993. Debut Oct 1995.

A Star 63DV was fired Jan 1994 with the TVC system developed for Star 48V 0517174

Star 63D/F (TU-936/TE-M-963-2)
Perigee kick for the McDonnell Douglas PAM D2 system. 3,502 kg (propellant 3,254 kg). 1,600 mm dia, 1,779 mm long, SI 283 s. 63F is an extended version, overall length 2,685 mm, 4,266 kg propellant, SI 298 s. 63F provided Optus B's perigee kick from China's Long March. A Star 63DV was fired 26 Jan 1994 with the TVC system developed for Star 48V.

Star 75 (TE-M-775, 105-KS45, 220)
A tested demonstration motor as a first step in developing PKMs in the 4,0807,940 kg propellant range. It features a slotted, centre perforated grain in a graphite epoxy filament wound case, a semi submerged nozzle with carbon phenolic exit cone, and a consumable wafertype igniter. Propellant mass fraction 0.9299, burn rate 0.56 cm/s (at 34.0 atm, 15.5°C), length 259.0 cm, 190.5 cm dia, burn/action time 105/107 s, max thrust 242.83 kN, SI 289.6 s, total impulse 21.330 MNs.

VERIFIED

Atlantic Research Corp

Current Status
ARC, founded in 1949, is a manufacturer of solid propulsion motors and gas generators, and in 1987 expanded into liquid propulsion with the acquisition of Bell Aerospace Textron. The company's Virginia and Arkansas facilities take solid motor concepts from design through to high volume production; the company's annual propellant production capacity is 9,000 t. ARC produces propulsion units for missiles such as Stinger, Tomahawk, Peacekeeper and Trident 2, but also undertakes space-related projects: the company teamed with Hercules in Dec 1987 to bid (unsuccessfully) for NASA's Shuttle Advanced Solid Rocket Motor. ASRM was subsequently cancelled by NASA. Israel's 1.3 m and AUS51 'Marble' motor, flown as Shavit's stage 3, is marketed by ARC/Rafael. ARC markets the ASTM family of Shavit and NEXT launcher stage 1/2 motors in the US under agreement with TAAS Israel Industries Ltd.

ARC also manufactures rocket motor components, including igniters, initiators, casings and nozzles using advanced materials and filament winding/braiding techniques. Nov 1987's acquisition of Bell Aerospace Textron added liquid bipropellant attitude control and apogee satellite propulsion systems, and their associated tanks & valves. ARC's 22 N (5 lbf) bipropellant thruster was developed for Intelsat 6's attitude control system and has demonstrated 400,000 pulses, 32 h aggregate firing and 748 kg total throughput. This thruster is carried by Optus B, UFO and Intelsat 7. As Bell, ARC/LP developed the Agena propulsion system that flew on >370 missions. The engine remains available for 31.1120 kN thrust.

ARC is developing the Agena 2000, providing the thrusters for the X-37 and is a partner of International Space Technology, Inc.

International Space Technology's plasma thruster is flight proven on Russian Meteor satellites (Space Systems/Loral) 0517183

Boeing Space Division

Current Status
Now a part of Boeing, Rocketdyne is designing and developing the RS-83, a hydrogen fuelled, fuel rich staged combustion cycle engine. Also advancing the design maturity of a highly reusable oxidizer – rich staged combustion hydrocarbon engine. These propulsion systems are funded in part by contracts under the NASA Space Launch Initiative. Upper stage hydrogen peroxide for upper stage SLI applications.

Background
Rocketdyne was a division of Rockwell International and in Dec 1996 was included in the Rockwell merger with Boeing. It is devoted primarily to the design and manufacture of rocket engines - it is the western world's largest producer, operating from its Canoga Park engineering and 16 km distant 10.5 km^2 Santa Susana test facilities. 1995 revenue was $880 million; 5,400 personnel. It was established as a separate division of North American Aviation Inc on 8 Nov 1955 and following its first major propulsion system for the Navaho missile has delivered around 3,500 engines for Jupiter, Redstone, Thor, Delta, Saturn, Atlas and Shuttle. It currently provides the powerplants for Atlas and Delta, and continues to test, upgrade and refurbish Shuttle's main engines, still the only large reusable systems in service. It designed and fabricated Peacekeeper's stage 4, its first integration of a full stage.

MA-5 Atlas Engine System
The MA-5 propulsion system comprises a YLR89NA7 dual chamber liquid propellant booster engine, a YLR105-NA-7 single chamber liquid sustainer and two YLR101-NA-15 verniers to control vehicle roll and provide final velocity and directional control following sustainer burnout. The single-start main engines are fed from the same propellant tanks but Atlas uniquely separates the two outer boosters some 172 s into flight, leaving the altitude configured sustainer to provide propulsion to nearorbital velocity. This system was introduced into spacelauncher Atlas versions in 1961 while its direct predecessor continued in service with Atlas E/F ICBMs, subsequently refurbished as orbital launchers and last flown in 1995.
Applications: Atlas 1 propulsion system
First flown: 18 Nov 1961, Atlas 117D
Number flown: 239 MA-5 (to end-2000), 152 MA 3, 118 MA2
Dry mass: 1,423 kg for both boosters, 470 kg sustainer
Length: 340 cm booster, 269 cm sustainer (with aspirator)
Max diameter: 119 cm booster, 117 cm sustainer
Mounting: all thrust chambers are gimballed
Engine cycle: gas generator
Oxidiser: liquid oxygen at 458 kg/s for booster, 89 kg/s for sustainer
Fuel: RP-1 hydrocarbon at 203 kg/s for booster, 39 kg/s for sustainer
Mixture ratio: 2.25:1 booster, 2.27:1 sustainer
Oxidiser turbopump: each 1,678 kW, 6,732 rpm, 68 atm discharge pressure for boosters, 846 kW, 10,568 rpm, 73 atm discharge pressure for sustainer
Fuel turbopump: each 1,116 kW, 68 atm discharge pressure for boosters; 508 kW, 73 atm discharge pressure for sustainer
Thrust: 1,882 kN for booster pair vacuum/1,681 kN SL; 374 kN sustainer vacuum/269 kN SL; verniers are 2,975 N each
Specific impulse (s): 292 vac/259 SL for boosters; 309 vac/220 SL for sustainer
Time to full thrust: 2.0 s
Expansion ratio: 8:1 boosters, 25:1 sustainer
Thrust chamber:
length: 249 cm both booster/sustainer
materials: nickel 200
cooling method: regenerative by two passes of fuel through 292 tubes in booster chambers, 240 tubes in sustainer
Combustion chamber:
pressure: 44 atm booster/50 atm sustainer, at injector end
temperature: 3,316°C both booster/sustainer
materials: nickel 200
cooling: as thrust chamber
ignition: hypergolic fluid cartridge enclosed in burst diaphragms
Burn time: 167 s max booster, 368 s max sustainer
Verniers: each 24.1 kg mass, 2.24/2.97 kN SL/vac thrust, 172/231 s SL/vac SI, 1.8 mixture ratio, 5.66 expansion ratio (exit dia 9.65 cm), 391 s burn time.

MA-5A/RS-56
Atlas Engine SystemThe MA-5 system was uprated for Atlas 2 by replacing the booster engines, now designated RS 56BA, with Rocketdyne's RS 27 (qv), providing a sea level SI increase of 4 s. The side mounted verniers were deleted; their roll control and final adjustment functions were assumed by thruster modules on the vehicle's interstage. Sustainer engine (RS56SA) specifications are as for the MA-5 except that total oxidiser/fuel flow rates are 90/38 kg/s, respectively; only booster engine specifications are given below. MA5B and MA-5C have been studied for future Atlas versions, but lost out to Russia's RD-180 for Atlas 2AR.
Applications: Atlas 2 primary propulsion system
First flown: 7 Dec 1991
Number flown: 50 to end-2000
Dry mass: 1,610 kg
Length: 343 cm
Maximum diameter: 119 cm
Mounting: all thrust chambers are gimballed
Engine cycle: gas generator
Oxidiser: liquid oxygen pump-fed at 505 kg/s
Fuel: RP-1 hydrocarbon pump-fed at 224 kg/s

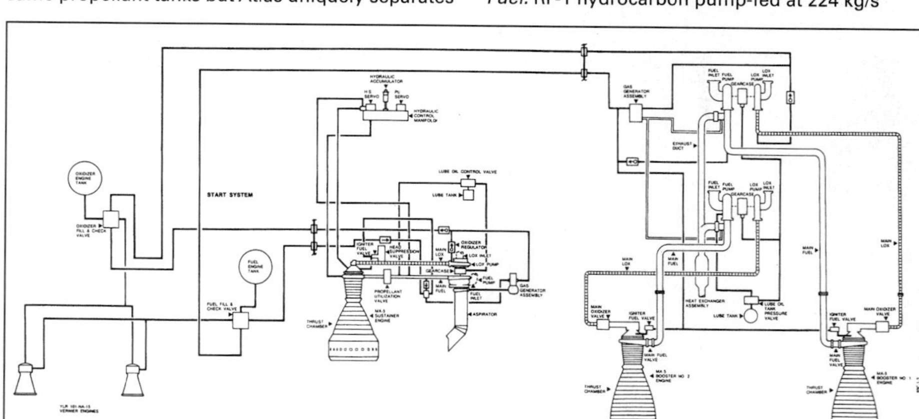

Atlas MA-5 propulsion system schematic (Rocketdyne) 0516945

RocketdyneYLR105-NA-7 sustainer on Atlas Centaur 68. Note the booster unit separation rail (MM) 0516946

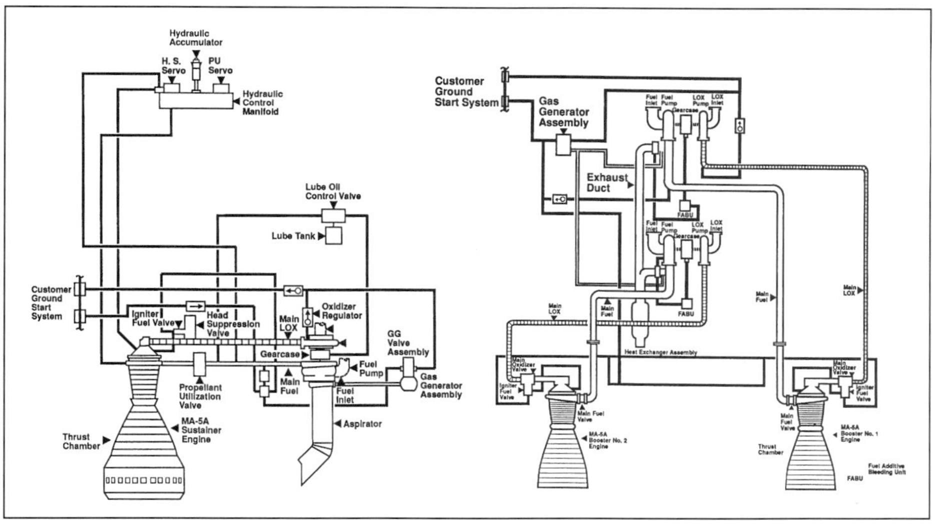

Atlas 2 MA-5A propulsion schematic. Note that Atlas 1/MA5's integrated thrusters have been deleted 0516948

Mixture ratio: 2.25:1
Oxidiser turbopump: each 1,903 kW, 6,730 rpm, 70 atm discharge pressure
Fuel turbopump: 1,362 kW, 75 atm discharge pressure
Thrust: total 2,100 kN vacuum, 1,890 kN SL
Specific impulse (s): 295 vac, 263 SL
Time to full thrust: 2.0 s
Expansion ratio: 8:1
Thrust chamber:
length: 249 cm
materials: 347 CRES austenitic stainless steel
cooling method: regenerative, two passes of fuel through 292 chamber tubes
Combustion chamber:
pressure: 48 atm at injector end
temperature: 3,316°C
materials: 347 CRES austenitic stainless steel
cooling: as thrust chamber
ignition: hypergolic fluid chamber enclosed in burst diaphragms
Burn time: 167 s max flight duration.

RS-27/RS2701A, B, RS-27A/RS2701C

The RS-27 powerplant comprises an RS2701A/B main engine and twin LR101-NA 11 verniers. It was introduced in 1974 on McDonnell Douglas' Delta 2000 series launcher, replacing the MB3, completed its Delta service on the 6000 model in 1992, and continues in service as part of Atlas' MA-5A powerplant. The uprated RS-27A version featuring a 12:1 expansion ratio for improved altitude performance is employed by the 7000 series; its specifications are given below, where different, in parentheses. The engine is a hybrid design utilising the turbopump, turbine, gas generator, valves and thrust chamber of Rocketdyne's H1 Saturn engine, and the control system, start system and component packaging arrangement of the MB3.
Applications: Delta 6000 series (RS27A: 7000 series)
First flown: 18 Jan 1974 on Delta 100/Skynet 2A (RS 27A debut 1990)
Dry mass: 1,027 kg (1,091 kg)
Length: 363 cm from top of thrust ring (378 cm)
Maximum diameter: 170 cm envelope
Mounting: gimbal-mounted for pitch/yaw control + gimballed verniers for roll control

The MA-5A propulsion system for Martin Marietta's Atlas 2 incorporates RS-27 engines from Delta's powerplant as the boosters 0516947

Engine cycle: gas generator
Oxidiser: liquid oxygen at 250 kg/s
Fuel: RP-1 hydrocarbon at 111 kg/s
Mixture ratio: 2.245:1
Oxidiser turbopump: 1,900 kW, 6,784 rpm (7,085 rpm at altitude), 70 atm discharge pressure
Fuel turbopump: 1,289 kW, 70 atm discharge pressure
Thrust: 971 kN SL/1032 kN vac (890 kN SL/ 1,054.2 kN vac)
Specific impulse (S): 264 SL/295 vac (255 SL/ 302 vac)
Expansion ratio: 8:1 (12:1)
Thrust chamber:
length: 219 cm (234 cm)
materials: 347 CRES austenitic stainless steel
cooling: regenerative, two passes of fuel through 292 tubes
Combustion chamber:
pressure: 48 atm at injector end
temperature: 3,315°C
materials: 347 CRES austenitic stainless steel
cooling: as thrust chamber
ignition: hypergolic fluid cartridge enclosed in burst diaphragms
Burn time: 274 s
Verniers: each LR101-NA-11 21.8 kg mass, 4.63/ 5.30 kN SL/vac thrust, 209/246 SL/vac SI, 1.8 mixture ratio, 5.6 expansion ratio (9.8 cm exit dia), 283 burn time.

RS-68

Rocketdyne is working on the 2.89 MN engine for the Delta 4 EELV proposal. Larger than SSME, it would be simpler and cheaper, derived from SSME, STME and J2. Subscale injector tests at NASA Marshall in late 1995 demonstrated >98 per cent combustion efficiency.

Space Shuttle Main Engine (SSME)

Rocketdyne was selected by NASA in July 1971 to develop Shuttle's high pressure propulsion system. Three oxygen/hydrogen staged combustion SSMEs burn typically for 520 s to propel the vehicle slightly short of orbital velocity. Typically throttled over 65104 per cent rated power setting, with 109% in abeyance (originally intended for Vandenberg polar missions). Some Space Station assembly missions might require a 106 per cent setting, allowed by the improvements outlined later. Slightly smaller than Rocketdyne's F1 engine used in the Saturn 5 lunar launcher, it was designed as the first large reusable engine, capable of 55 missions or 7½ h firing time between major overhauls. Flight certification was achieved in Dec 1980, after two certification cycles had been completed on each of two engines. Each cycle required a minimum of 13 tests and 5,000 s, including simulation of nominal and abort mission profiles. 40 tests totalling 12,750 s were made on one engine, and 34 tests, totalling 10,650 s, on the other. Another cycle in the Full Power Level certification programme was completed Apr 1983, requiring four test series of 5,000 s each involving engine power levels of 104/105/109/111 per cent. Each flight engine undergoes three tests at NASA's Stennis: 0.5 s ignition, 300 s calibration and 550 s mission duty cycle. The first ascent including 104% power was made by *Challenger* in Apr 1983 for STS-6. The engines are individually gimballed hydraulically by two APUpowered actuators at up to ±10.5° pitch and ±8.5° yaw; each engine is controlled in 20 ms steps by a 97 kg top mounted package housing two redundant Honeywell 360 16.4 kbit memory computers. The controllers are commanded by the Orbiter's General Purpose Computers, or manually in emergencies. Block 2 Honeywell controllers are installed on OV105 (and are being retrofitted to the other Orbiters), using the first spacequalified 68000 series microprocessors. They also allow elimination of a separate 13.6 kg vibration monitoring box and add 20 engine parameter monitoring channels to the current 120. The first hot firing with a Block 2 controller was made in Apr 1990. The last of the 30 ordered was delivered Apr 1993.

The incoming oxidiser (propellants are channelled into the Orbiter from the External Tank via 43 cm dia pipes and routed to individual engines by 30.5 cm dia versions) is boosted by the first turbopump from 7 atm to 29 atm before the High Pressure Oxidiser Turbopump increases it to 500 atm in two stages. This pump has suffered

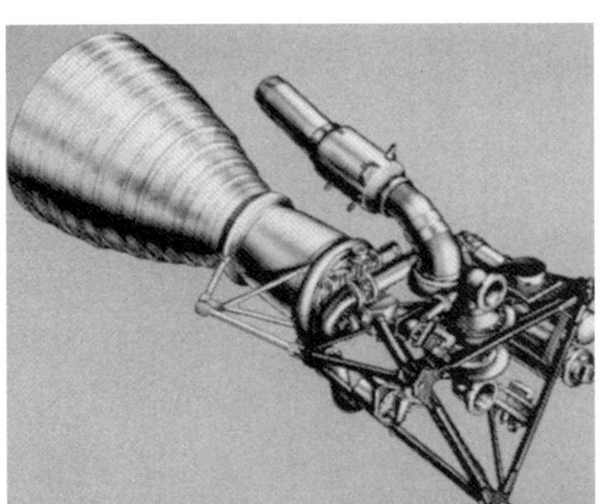

Rocketdyne's RS-27 was in service with Delta 1974 92; it continues aboard Atlas 2 0516949

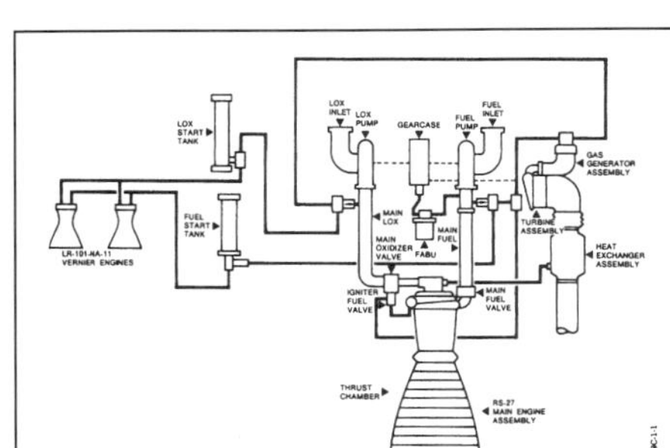

RS-27 engine schematic (Rocketdyne) 0516950

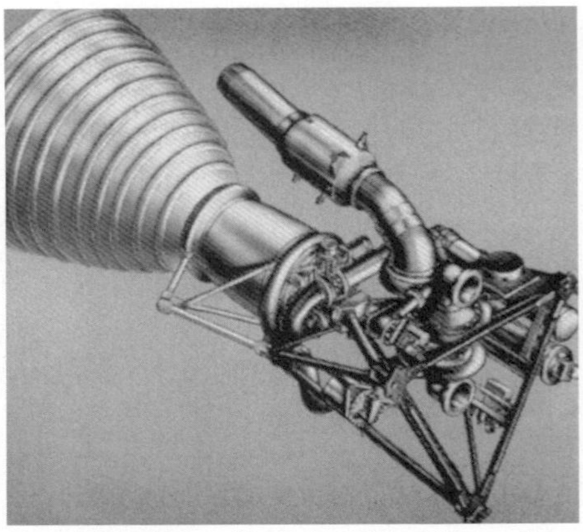

The RS-27A, designed for Delta 7000 vehicles, replaces the 8:1 expansion bell with a 12:1 for improved altitude performance (Rocketdyne) 0516951

durability problems, with some engines requiring overhaul after three missions (33 min). Pratt & Whitney was awarded a $188 million contract in Aug 1986, as part of NASA's 10 yr $1 billion Phase II SSME improvement programme that began in 1983, to develop a new powerhead containing two LOX and LH_2 high pressure turbopumps and a single preburner chamber (the current design incorporates a preburner for each). The goal is 7½ h firing (55 missions) between major overhauls. Problems increased turbopump development cost to $504 million, so NASA halted fuel pump work from Dec 1991 in order to focus on the more critical oxygen unit for the Block 1 upgrade. It restarted work May 1994 for 1997 debut as part of the Block 2 upgrade (see below) once the LOX pump was on schedule for its July 1995 flight debut. The Block 1 upgrade includes a single coil heat exchanger (eliminating welds) and changing the 3-duct hot gas manifold to a 2-duct design, which allows the fuel pump to run at lower T/P. A single Block 1 upgrade engine (#2036) flew flawlessly on STS 70 in July 1995. All three engines were upgraded for STS 77 in May 1996. The pump is now certified for 10 flights between detailed inspections. Advanced fabrication techniques have produced a more reliable engine, with the 157 kg increase offset by a 1 s SI improvement. Testing of the Block 2 upgrade began at NASA Stennis on #0521, once Block 1 certification was completed on #0422/0423 26 May 1995 (0423 ran for a final 761 s; it was later converted to 0523 Block 2). Block 2 includes P&W's fuel pump and a 10 per cent dia larger throat combustion chamber that will reduce wear, reducing pump discharge pressure by 4.56% and pump operating T by 712 per cent. The $90 million large throat contract covers development and eight flight units. All three engines on STS86 in Sep 1997 will debut Block 2. STS-85 will be the last mission to fly with an old engine; Block 1 or 2 will be used on all subsequent missions.

Some of the oxygen pump output is routed into a preburner where it is mixed with fuel and burned hydrogen rich at 760°C. The hydrogen half of the system likewise employs a low pressure pump (2 to 18.7 atm), high pressure pump (18.7 to 442 atm) and a pre burner. The preburner products are

SSME static firing at Rocketdyne's Santa Susana Field Laboratory 0516952

SSME delivered in 1977 for NASA's three engine Main Propulsion Test Article. The engine carries a 35:1 expansion ratio bell for ground firings (Rocketdyne) 0516953

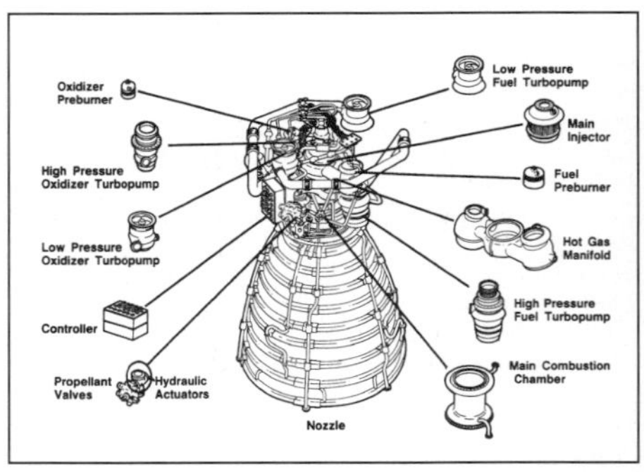

SSME principal features and propellant flow scheme (Rocketdyne) 0517380

fed into the main chamber with additional oxygen through 600 elements in 13 annular rows in the 56cm dia injector plate.

In the 18 months before the resumption of Shuttle missions following *Challenger's* loss in Jan 1986, SSME testing at NASA Stennis accumulated the equivalent of 40 launches. By July 1990, an aggressive test campaign covered 293 equivalent ascents, all elements apart from the turbopumps exceeding 55 missions. The programme, including flights, attained 500,132 s 15 Nov 1991 in >1,900 firings, the equivalent of 320 ascents. By end 1993, SSME attained 601,303 s in >2,100 firings. 1993's record was 37,806 s in 88 tests. 1994's was 49,181 s in 104 tests. Jan 1995 totalled 7,296 s in 14 tests, including 12 on the Block 1 certification engines (0422/0433). Rocketdyne's contract received a 4 yr $819.8 million extension in Aug 1990. SSME unit cost is up to $40 million. Including test failures, total reliability is 0.9991 in >500,000 s. Production is expected to continue to at least 2010, although no more than the 36 ordered are planned. Because of the reduced Shuttle mission rate, NASA is aiming for only 30 reuses of each engine.

Applications: Space Shuttle Main Engine
First flown: 12 Apr 1981, Orbiter
Columbia Number produced: 36 planned
Number flown: 56 to end-2002
Dry mass: 3,177 kg (3,334 kg Block 1)
Length: 4,242 mm
Maximum diameter: 2,388 mm
Mounting: gimballed by hydraulic actuators ±10.5° pitch/yaw
Engine cycle: staged combustion
Oxidiser: liquid oxygen at 408 kg/s
Fuel: liquid hydrogen at 68.0 kg/s
Mixture ratio (O/F): 6:1
Oxidiser turbopump: 261 kg, 17,900 kW, 28,500 rpm, discharge pressure 292.5 atm
Fuel turbopump: 351.1 kg, 46,230 kW, discharge pressure 415 atm
Thrust: 1,668/2,091 kN SL/vac at 100%, throttlable 65 109% in 1% increments. Launch is made at 104%
Specific impulse (s): 452.9 vac (Block 1 goal 453.3)
Time to full thrust: 4.2 s
Expansion ratio: 77.5:1
Thrust chamber:
materials: stainless steels
cooling: GH_2 is directed from the high pressure fuel pump through the 1,080 tubes comprising the chamber wall
Combustion chamber:
pressure: 204 atm (100% thrust)

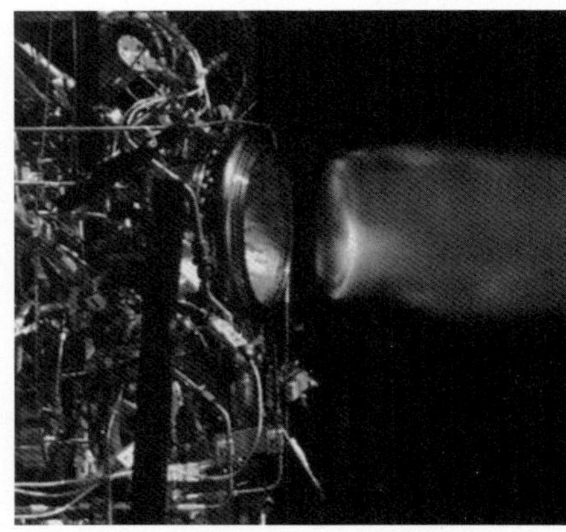

XLR-132 flightweight engine test (Rocketdyne) 0516954

temperature: 3,300°C (100% thrust)
materials: NARloy-Z copper alloy and stainless steels
cooling: GH_2 is directed from the high pressure fuel pump through the 390 main chamber wall channels
ignition: augmented spark igniters
Burn time: 520 s typical, 761 s max.

XLR-132

Rocketdyne has completed the advanced technology development phase of the pump fed high performance upper stage engine. Its performance, mass and extended space storability make it a candidate for perigee/apogee stages, transfer vehicles and lunar and Mars missions. 10-start, 4,000 s demonstration testing was performed on the injector, turbopump, combustion chamber and gas generator. Sea level and simulated altitude testing has been completed in both the breadboard (1991) and flightweight engine programmes (1992), including a 500+ s run in Feb 1992. Current efforts include design studies of engines with thrusts to 111.2 kN, adjustments of the expansion nozzle configuration, and a heavy emphasis on design refinements and producibility to achieve demanding cost goals.

Applications: kick stages, deep space, Space Transfer Vehicle
Dry mass: 54 kg
Length: about 120 cm
Maximum diameter: about 60 cm
Engine cycle: gas generator
Oxidiser: nitrogen tetroxide at 3.3 kg/s
Fuel: monomethyl hydrazine at 1.68 kg/s
Thrust: 16.68 kN vacuum
Specific impulse (s): 340 vacuum minimum
Expansion ratio: 400:1
Thrust chamber:
materials: columbium
cooling: radiative
Combustion chamber:
pressure: 102 atm at injector end
cooling: regenerative by oxidizer
ignition: hypergolic
Burn time: 4,000 s in 10 starts.

VERIFIED

Hughes Research Laboratories

Set up in 1948, HRL space research effort has placed research emphasis on electric propulsion systems or solar system spacecraft on planetary or science missions as an expansion of ion propulsion research for satellite station-keeping. HRL is jointly owned by The Boeing Company, General Motors Corporation and the Raytheon Company. The primary research facilities comprise the Communications & Photonics Laboratory, the Information Sciences Laboratory, the Microelectronics Laboratory and the Sensors and Materials Laboratory.

Hughes has developed an 18 mN xenon ion propulsion subsystem (XIPS) for NSSK of 3 axis or spin stabilised telecom satellites. It will debut on Hughes' own Galaxy 3R satellite in Oct 1995; Astra 1G and ASC 1/2 will use it. XIPS comprises four 13 cm thrusters (two primary + two backup), power supply (439 W input each thruster off 29–34 VDC bus) and a propellant storage/control unit. Total XIPS package mass is 68 kg, for a saving of 400 kg on an HS-601 satellite. 12,000 h of operating time was accumulated by early 1995. In normal NSSK operation, the thrusters will fire for 2–3 h daily.

Xe is fed through a heated (1,000°C) hollow cathode inside the thrust chamber. A ringcusp magnetic field confines the plasma and a 3grid ion optics assembly extracts the thrust beam as 3,125 beamlets. Xe is stored at 74.8 atm and reduced to 0.68 atm for feeding into the cathode.

The 13 cm XIPS was developed from work on a 25 cm dia 63.5 mN thruster that provided SI 2,800 s at 1.3 kW input. It was also operated at 179 mN, 4,000 s, 4.5 kW. The 25 cm version will be used by the HS702's improved XIPS.

Thrust: 17.8 mN
Beam voltage: 751 V
Accelerator voltage: –300 V
Beam current: 0.405 A
Accelerator current: 0.97 mA
Exhaust velocity: 30 km/s
Specific impulse: 2,585 s
Beam power: 304 W
Power/thrust ratio: 24.6 W/mN
Electrical efficiency: 88–90%
Total efficiency: 51.3%
Mass: 5.0 kg each thruster (x4); 6.8 kg each power processor (x4), 2.0 kg each Xe tank (x2)

VERIFIED

International Space Technology, Inc

Space Systems/Loral and Russia's OKB Fakel and RIAME Research Inst of Applied Mechanics & Electrodynamics formed the ISTI joint venture in 1992 to market Fakel's SPT-100 0.08 N Hall xenon stationary plasma thrusters using western electronics. They provide a 500 kg mass saving over a conventional system on a 3.5 t satellite over a 15 yr life. The venture was later joined by SEP and Atlantic Research Corp. ISTI's thrusters became available in 1995. Fakel has built >100 of the devices, which have flown on >50 Meteor polar metsats since 1972 to provide orbit control. Gals 1, launched Jan 1994, is carrying a set of eight to provide the first NSSK on a Russian satellite. Attempts have been made to develop export markets for the Fakel thrusters. SEP joined ISTI in 1993, receiving commercialisation rights in Europe in exchange for the other partners to have equivalent rights in N America and Russia on the Mk 2 version that SEP is developing. This will increase SI from 1,650 s to 2,200 s. Development began in May 1993 at SEP Villaroche. France's 1999 Stentor satellite was to be equipped with them, but technical problems mean the satellite is equipped with only slightly modified versions. See the Fakel entry at the end of the CIS propulsion section for other SPT information.

IST Inc.
SPT 100 specifications
Thrust: 83 mN
Specific impulse: 1,600 s
Efficiency: 0.48
Electric power: 1,350 W
Mass flow rate: 5.3 mg/s
Size: 12.5 × 15 22 cm; 3.5 kg
Design cycles: 4,000
Design total impulse: 1 MNs (>2 MNs demonstrated)

VERIFIED

Kaiser Marquardt

On 4 May 2000 Primex Technologies, Inc., announced that it had acquired the bipropellant and rocket motor business of Kaiser Marquardt, hitherto a wholly owned subsidiary of Kaiser Aerospace & Electronics, for US$22.5 million. The Marquardt company was formed in 1944 to undertake R&D leading to the first US subsonic ramjet in 1945. Its principal engineering business continues to be advanced aerospace propulsion and the supply of ram air turbine power systems. The company has been developing and manufacturing bipropellant thrusters since 1959, notably the attitude control thrusters for Apollo spacecraft and the Space Shuttle Orbiter. Precision bipropellant engines are available for 0.004, 0.02, 0.06, 0.11, 0.22, 0.44, 0.89, 4 kN sizes, and requests up to 8 kN can be accommodated. KM has performed a modest upgrade to the R-40 under a technology study contract seeking cost effective solutions to future launch vehicles. The monopropellant hydrazine thruster line of UTC's Hamilton Standard Div was acquired in 1993; this is described separately at the end of this entry.

Kaiser Marquardt R-1E

This 110 N hypergolic thruster was qualified for the USAF Manned Orbiting Lab and then produced as the Space Shuttle Orbiter's vernier attitude control/orbit adjust thruster. Six are employed in conjunction with 38 3,870 N R-40 thrusters.
Applications: Shuttle Orbiter vernier, satellite orbit adjust
First flown: Apr 1981 on Shuttle
Columbia Dry mass: 3.7 kg
Length: 279 mm, 312 mm satellite version
Maximum diameter: 15.2 cm satellite version, 14.0 cm Shuttle version
Mounting: fixed
Engine cycle: pressure-fed at 15 atm (range 6.8 27.2 atm)
Oxidiser: nitrogen tetroxide at 0.0256 kg/s
Fuel: monomethyl hydrazine at 0.0354 kg/s
Mixture ratio (O/F): 1.65 (range 1.0 2.7)
Thrust: 110 N (range 67–155.7 N)
Specific impulse: 280 s vacuum
Expansion ratio: 100:1 standard
Combustion chamber: single chamber, min expansion ratio 26 with orthogonal and scarfed nozzles. Made of silicide-coated columbium (niobium) and insulated for buried installation. Started by electrical signal to on/off solenoid valve. Single doublet injector with hypergolic ignition
Burn time: min 82,000 s demonstrated, 0.89 Ns min impulse bit

Kaiser Marquardt R-4D

Development of the R-4D began in 1962 as an attitude control thruster for Apollo's Service/Lunar Modules, each employing four quadruple clusters; about 800 were built. The thruster is currently employed in the R4D 10 version by the US Navy's Leasat, R4D11 by Insat 1, Arabsat 1, HS376W, HS393, Milstar, Intelsat 6, Italsat, Olympus and Eurostar. Versions are also used on GOES, Superbird, HS601, Italsat, IABS, Mars Observer and Cassini.
Applications: apogee/perigee manoeuvres, orbit adjust, attitude control
First flown: Feb 1966 (Apollo 201)
Number produced: >1,000
Dry mass: 3.63 kg

SPT 100 MkII under test at SEP 0517193

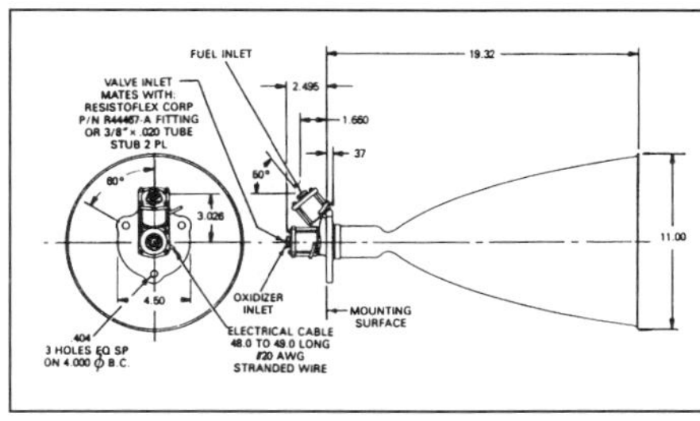

Marquardt's R-4D bipropellant thruster was initially developed for the Apollo manned spacecraft but is now incorporated in satellites for orbital manoeuvres. Dimensions are in inches. The photograph shows the R-4D-11-300 model, which has a 38 cm dia nozzle for a 300:1 expansion ratio 0516939

Length: 55.41 cm
Maximum diameter: 27.94 cm (38 cm 11300 model)
Mounting: fixed
Engine cycle: pressure-fed
Propellants: hypergolic N_2O_4 and MMH or hydrazine, pressurefed at 0.056 kg/s fuel and 0.091 kg/s oxidiser at 15.2 atm
Mixture ratio: 1.65 (1.02.4 range)
Thrust: 490 N vacuum standard
Specific impulse: 311.5 s (164:1) and 315.5 s (300:1) vac
Expansion ratio: 164:1 and 300:1
Chamber pressure: 6.84 atm
Combustion chamber: single chamber/nozzle of welded coated columbium (niobium), radiatively and film cooled, controlled by on/off solenoid, multiple doublet injector with hypergolic ignition
Burn time: 45,000 s total demonstrated (including 10,000 s continuous, limited by test facility); 2.67 Ns min impulse bit

Kaiser Marquardt R-6C

The 6C is a derivative of an engine developed by Marquardt for the Advent communications satellite in the early 1960s; it was qualified in 1981 for Insat 1, Arabsat 1, Olympus and HS 393 satellites. Olympus, for example, employed a network of 22 for attitude control. Milstar also incorporates it. The 22 N thrust is reduced in the R-6C2.2 version to 10 N.
Applications: satellite AOCS
Dry mass: 0.66 kg
Length: 251.6 mm
Mounting: fixed
Engine cycle: pressure-fed
Oxidiser: N_2O_4 at 0.005 kg/s at 1517 atm
Fuel: MMH at 0.003 kg/s at 1517 atm
Mixture ratio (O/F): 1.6 ± 0.1
Thrust: 22 N (range 6.2–32.9 N). R6C 2.2 version operates at 10 N nominal
Specific impulse: 290 s vacuum at 22 N thrust
Exit diameter: 55.9 mm
Expansion ratio: 100:1
Combustion chamber: single chamber of silicide coated columbium (niobium). Pressure 6.84 atm. Started by electrical signal to on/off solenoid valve; multiple doublet injector with hypergolic ignition
Burn time: unlimited

Kaiser Marquardt R-40A

The R-40 was developed for Shuttle Orbiter orbit control: the vehicle carries 38 in long scarf, short scarf or no scarf configurations, depending on location. It has also flown on classified satellites.
Applications: spacecraft orbital manoeuvring, perigee kick engine
First flown: Apr 1981 on Shuttle
Columbia Dry mass: 10.25 kg
Length: 55.4 cm short-scarf version, 103.9 cm longscarf, 88.9 cm for satellite version with 100:1 expansion ratio
Maximum diameter: 30.9 cm longscarf version, 51.8 cm satellite version of 100:1 expansion ratio
Mounting: fixed
Engine cycle: pressure-fed
Oxidiser: N_2O_4 at 0.838 kg/s at 16.4 atm
Fuel: MMH at 0.526 kg/s at 16.4 atm
Mixture ratio (O/F): 1.6
Thrust: 3,870 N (range 3,114 5,338 N)
Specific impulse: 281 s (306 s for 120 expansion ratio)
Expansion ratio: 20 (40–150 available for satellite version)
Exit diameter: 267 mm (328–635 mm for satellite version)
Chamber pressure: 10.5 atm
Combustion chamber: single chamber of silicide coated columbium (niobium) with welded on orthogonal and scarfed nozzle extension in same material. Internal film cooling. Exterior insulated for buried installation. Started by electrical signal to on/off solenoid valve. Multiple doublet injector with hypergolic ignition

The Marquardt R-6C 22 N bipropellant thruster is utilised by the Milstar military satellites 0516940

Burn time: max 500 s for single burn, qualified to 15,319 s life; 289 Ns min impulse bit

Kaiser Marquardt R-42

The R-42 was developed for a classified programme. It will be used for Space Station reboost.
Applications: satellite orbit adjust
First flown: classified
Dry mass: <0.45 kg (5.2 kg with series redundant valves)
Length: 787 mm
Maximum diameter: 406 mm
Mounting: fixed
Engine cycle: pressure-fed at 15.0 atm (range 6.8 28.9 atm)
Oxidiser: nitrogen tetroxide
Fuel: monomethyl hydrazine
Mixture ratio (O/F): 1.65 (range 1.0 2.2)
Thrust: 890 N (range 578–1,200 N)
Specific impulse: 303 s vacuum
Expansion ratio: 160:1
Combustion chamber: C-103 silicide coated columbium (niobium), radiation and fuel film cooling. Started by electrical signal to on/off solenoid valve. Single doublet injector with hypergolic ignition
Burn time: >22,000 s, 15.6 Ns min impulse bit, 20.0 MNs max

Kaiser Marquardt R-53

Applications: satellite AOCS
First flown: unflown, but development completed
Dry mass: 0.41 kg
Length: 165 mm
Maximum diameter: 51 mm
Mounting: fixed
Engine cycle: pressure-fed at range 6.8 20.4 atm
Oxidiser: nitrogen tetroxide
Fuel: monomethyl hydrazine
Mixture ratio (O/F): 1.65 (range 1.0 2.2)
Thrust: 9 N (range 4.0–12.9 N)
Specific impulse: 295 s vacuum
Expansion ratio: 300:1
Combustion chamber: C-103 silicide coated columbium (niobium), radiation and fuel film cooling. Started by electrical signal to on/off solenoid valve. Single doublet injector with hypergolic ignition
Burn time: 70,000 s demonstrated, 0.03 Ns min impulse bit

Monopropellant Thrusters

The hydrazine thruster line of UTC's Hamilton Standard Div was acquired in 1993. Hamilton Standard was the world's largest supplier of catalytic hydrazine thrusters and propulsion systems; it delivered >80 monopropellant systems.

Five of the Space Shuttle Orbiter reaction control system engines. The smallest is Marquardt's R-1E; the other four are R40A models in long scarf and short scarf versions (NASA) 0516941

It provided the complete propulsion system for the USAF DSCS 3 milcomsat and the reaction control modules for the IUS upper stage. The IUS carries six modules, each with two 133 N hydrazine thrusters, to provide roll control during main motor burns, 3 axis control during coast periods, vernier orbit corrections after stage 2 burnout, and collision avoidance manoeuvres following payload separation. The configuration provides dual-nozzle redundancy and there are no forwardfacing thrusters, in order to avoid payload contamination. Two tanks hold a total of 109 kg of hydrazine but 1 and 3 tank versions are available. The low thrust (0.8926.7 N) engine range includes the KMHS 10 (0.89 N), KMHS 176 (2.22 N), KMHS 1712 (4.45 N), KMHS 16 (22.2 N) and KMHS 392 (22.2 N). The higher power motors are used for precession, despin and {d}V manoeuvres; the 0.89 N versions are employed for wheel desaturation, N S/EW stationkeeping and precise attitude control. The 22.2 N thrusters, first flown 1967, were the first space qualified catalytic engines. The KMHS 39 is capable of either one million cycles or >20 h continuous operation without thrust degradation. The range of mid thrust (44.5556 N) engines includes the KMHS 22-5 (53.4 N), KMHS 22-2 (91.6 N), KMHS 22-16 (133.4 N), KMHS 2217 (177.9 N) and KMHS 20-4 (556 N). These are used generally for major {d} V manoeuvres and atmospheric drag make-up. The KMHS 20-4 is capable of operating for >4 h in total, delivering 4.45 MNs impulse and consuming >2 t of hydrazine.

KMHS Model 10

Applications: satellite AOCS
First flight: 1971. Topex, IUE, Apple, BS2
Number produced: 7,200
Dry mass: 0.33 kg
Length: 146 mm
Maximum diameter: 31 mm
Engine cycle: decomposition of hydrazine over heated catalyst bed
Thrust: 0.89 N vac (0.45–1.42 N range)
Specific impulse: 226 s vac
Inlet pressure: 15.0 atm (range 6.0 27.2 atm)
Burn time: 0.01 Ns min impulse bit, 111.2 kNs max
Total pulses: >45,000 cold start, >350,000 hot

KMHS Model 16/39

Applications: satellite AOCS
First flight: 1971. Skynet 4, Clementine, COBE, CRRES, Topex, TRMM
Number produced: 7,200
Dry mass: 0.51 kg
Length: 254 mm
Maximum diameter: 32 mm
Engine cycle: decomposition of hydrazine over heated catalyst bed
Thrust: 22.2 N vac (7.1–25.4 N range)
Specific impulse: 235 s vac (220 s at 5.1 atm inlet)
Inlet pressure: 3.4–27.2 atm
Burn time: 0.31 Ns min impulse bit, >501 kNs max
Total pulses: >60,000 and 227 kg throughput Model 16; $>10^6$/520 kg Model 39

KMHS Model 17

The thruster was originally designed and qualified for 1.78 N at 14.3 atm inlet pressure.
Applications: satellite AOCS
First flight: 1970. DSCS 3, SeaStar, Geosat
Number produced: 7,200
Dry mass: 0.38 kg
Length: 203 mm
Maximum diameter: 32 mm
Engine cycle: decomposition of hydrazine over heated catalyst bed
Thrust: 4.45 N vac (1.56–5.34 N range)
Specific impulse: 230 s vac (220 s at 14.3 atm inlet)
Inlet pressure: 5.1–29.3 atm
Burn time: 0.089 Ns min impulse bit, 311.4 kNs max
Total pulses: $>10^6$ and 141 kg throughput

KMHS Model 22

Applications: satellite AOCS
First flight: 1980. IUS, TOS, ASAT, TRS, AXAF
Number produced: >200
Dry mass: 1.02 kg
Length: 305 mm
Maximum diameter: 84 mm
Engine cycle: decomposition of hydrazine over heated catalyst bed
Thrust: 88.9 N vac (22–200 N range)
Specific impulse: 230 s vac (225 s at 6.8 atm inlet)
Inlet pressure: 5.1–30.6 atm
Burn time: 0.89 Ns min impulse bit, >3,496.6 kNs max
Throughput: 771 kg demonstrated

VERIFIED

Northrop Grumman Space Technology (TRW Space and Electronics Group)

On 12 December 2002 Northrop Grumman completed its acquisition of TRW Inc.

TRW develops and manufactures propulsion components, engines and systems for spacecraft attitude and velocity control, orbit insertion, manoeuvring and adjust, tactical missile steering, low cost orbit transfer stages and low cost engines for Earth to orbit systems. Since the company produced the first spaceoperated monopropellant hydrazine system for 1960's Able lunar probe, it has provided >2,000 mono/bipropellant thrusters and reaction control systems. The bipropellant Lunar Module Descent Engine was responsible for landing 12 astronauts on the Moon 19691972, and Apollo 13's rescue, before it was converted into a fixed thrust model for Delta's second stage; 84 were built for the lunar programme and 75 flew on Delta from 1974. It remains under consideration for future upper stages. A further derivative was to power later versions of NASA's Orbital Maneuvering Vehicle (cancelled). Propulsion research covers low thrust monopropellant, bipropellant, colloid, ion, radioisotope, MPD, electrothermal and low cost engines.

Propulsion test facilities are centred at TRW's 8 km^2 Capistrano Test Site (CTS) 108 km south of Redondo Beach, equipped for thermal vacuum, vibration, shock, cold flow calibration, pressure/leak, functional and electrical, and accurate hot fire performance and life testing. Engine designs burning storable, cryogenic, gelled and high energy bipropellants can be evaluated under sea level and simulated altitude conditions. Continuous operation, steam exhaustpumped altitude chambers can test high expansion ratio 450–670 N engines at pressures equivalent to 46 km. The HEPTS High Energy Propellant Test Stand has four 222 kN test positions, two of which simulate altitude firings. Cold flow and hot fire tests of low thrust hydrazine, ion/plasma electric and resistojet gas thrusters are undertaken at Redondo Beach.

ESEX Arcjet

TRW has developed a 30 k We 2 N ammonia arcjet under contract to the AF Phillips Lab for test aboard the Argos satellite in 1997. Objectives of this ESEX electric space experiment are to measure plume deposition, EM interference, thermal radiation and acceleration. The mission calls for 10 firings of 15 min each at 26k We avg power.
Dry mass: unavailable
Fuel: ammonia at 0.25 gm/s steady state
Specific impulse: 800 s
Design life: 1,500 h and 400 on/off cycles

Secondary Combustion Augmented Thruster

SCAT was qualified in 1995 as a high performance bipropellant thruster for stationkeeping, TVC, attitude control and orbit insertion/change. It uses a gas/gas mixing and combustion process that provides a wide operational range of mixture ratio and thrust for both regulated and deep blowdown systems. It features a standard catalyst decomposition chamber, a low cost regeneratively cooled thrust chamber (no refractory material or oxidationresistant coatings) and operates near 100% efficiency in the bipropellant mode. It can operate in either mono hydrazine or bipropellant mode and can readily be designed as a bilevel thruster for TVC. Flight unit production is underway (SCAT will be carried on Odyssey), and advanced designs are in test for higher performance orbit change/insertion applications.
Dry mass: 1.8 kg
Length: 30.5 cm
Maximum diameter: 7.9 cm (125:1 nozzle)
Mounting: fixed (four threaded standoffs)
Engine cycle: pressure-fed 8–24 atm. Decomposition of N_2H_4 over Shell 405 catalyst bed mixed with vaporised N_2O_4 (gasgas mixing)
Oxidiser: N_2O_4
Fuel: N_2H_4
O/F ratio: 0.9 to 2.0:1
Thrust: 20–67 N vacuum
Specific impulse: 287–326 s vacuum
Expansion ratio: 125:1
Chamber pressure: 6.8 atm
Ignition: hypergolic
Burn time: >150,000 s total, >7,000 s single burn demonstrated. >1,362 kg total throughput, >600 cycle life.

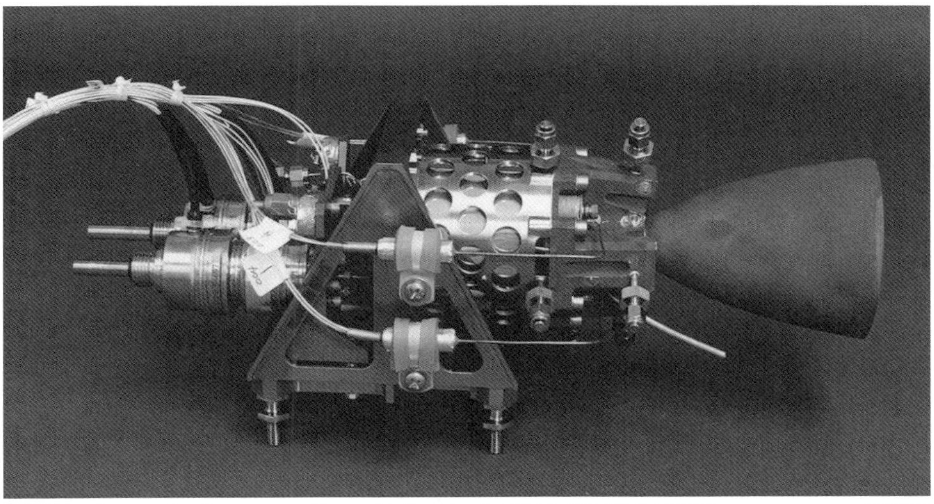

SCAT's gas/gas combustion process allows wide thrust and mixture ratio ranges (TRW) 0517343

Dual Mode-Liquid Apogee Engine

DM-LAE was developed for reduced propulsion system complexity and lighter mass, in which the hydrazine supply is used for both the apogee engine (with N_2O_4) and the mono attitude control thrusters. The coaxial pintle injector is tuned for an optimum combination of performance and operating temperature margin. This basic injector concept began with the design of TRW's Apollo throttleable Lunar Module Descent Engine. Two DMLAEs were successfully used to inject each of the Canadian Anik E2/E1 telecom satellites into GEO in Apr/Sep 1991. Intelsat K followed in 1992. No further deliveries have been made but an improved version (SI 321 s and 328 s) qualified in 1996. A 330 s model will fly on TRW's Odyssey satellites and the 321 s version is qualified for AXAF-I.

First flight: Apr 1991 (Anik E2)
Number flown: 6 to end-1995
Dry mass: 4.8 kg
Length: 56.05 cm (68.40 cm total)
Maximum diameter: 30.15 cm
Mounting: fixed (3 lugs at 120°)
Engine cycle: pressure-fed (15 atm)
Oxidiser: MON 3; flow rate 0.076 kg/s
Fuel: high purity hydrazine, at 0.071 kg/s
O/F ratio: 1.070:1
Thrust: 454 N vacuum
Specific impulse: 314.5 s vac (320 s + 328 s versions expected to be qualified in 1996)
Expansion ratio: 204:1
Cooling method: radiation
Chamber pressure: 7.0 atm
Ignition: hypergolic, initiated by 28 V electrical signal to on/off bipropellant torque motor valve
Burn time: 24,000 s total, 3,600 s max single burn.

OMV Variable Thrust Engine (VTE)

This deep throttling engine is derived from TRW's manrated Apollo LMDE. Development began 1986; qualification was due to be completed in 1991 but NASA's Orbital Manoeuvring Vehicle was cancelled in 1990. It completed preliminary development and is ready to enter prequalification. It was recently considered as a lander engine for precursor planetary probes. A set of four was to provide low acceleration primary propulsion for OMV, using low thrust ramp up starts to minimise dynamic interaction and propellant sloshing. Thrust modulation of the four provides vehicle TVC.

TRW's 454 N DM-LAE dual mode liquid apogee engine is being improved for Odyssey 0516959

Applications: OMV primary propulsion; rendezvous, interception and inspection missions
First flight:
Dry mass: 6.8 kg
Length: 70.05 cm
Maximum diameter: 27.13 cm
Mounting: fixed
Engine cycle: pressure fed (tanks at 19.7 atm)
Oxidiser: N_2O_4; 0.119 kg/s at max thrust
Fuel: MMH; 0.0745 kg/s at max thrust
O/F ratio: 1.64
Thrust: 57.8–578 N vac (10:1 throttle range, 110 N/s throttle rate)
Specific impulse: 280–308 s vacuum
Expansion ratio: 125:1
Cooling method: radiative columbium (niobium) alloy chamber, fuel film barrier; designed to operate buried with full radiation heat shield
Chamber pressure: 0.75–6.8 atm (inlet pressure 19.7 atm)
Ignition: hypergolic
Burn time: 28 h total. 2,400 s max single burn at max thrust, 7,200 s at min thrust; 30 s min single burn at both max/min thrusts

Delta TR-201

The TR-201 is a fixed thrust version of the LMDE for Delta's stage 2. Multi start operation is adjustable up to 55.6 kN and propellant throughput up to 7,711 kg, and the engine can be adapted to optional expansion ratio nozzles. The engine made 75 flights during its Delta career over 197488. It is under consideration for future upper stages (at one time for PacAstro's PA2). Upgraded versions are under development for advanced high performance LOX/LH_2 upper stages. It has been demonstrated at NASA Lewis using LOX/LH_2 at 97% performance efficiency at 89 kN. It is under contract for McDonnell Douglas's EELV proposal, scaled up to 2.89 MN.

First flown: Jan 1974, Delta 100
Number flown: 75
Dry mass: 136 kg
Length: 227.1 cm
Maximum diameter: 148 cm
Mounting: gimbal attachment above injector
Engine cycle: pressure fed (15.5 atm reservoir)
Oxidiser: 50/50 N_2O_4/UDMH at 8.92 kg/s

TRW's primary OMV bipropellant thruster, based on Apollo's descent engine, was designed to provide a 10:1 throttling range 0516960

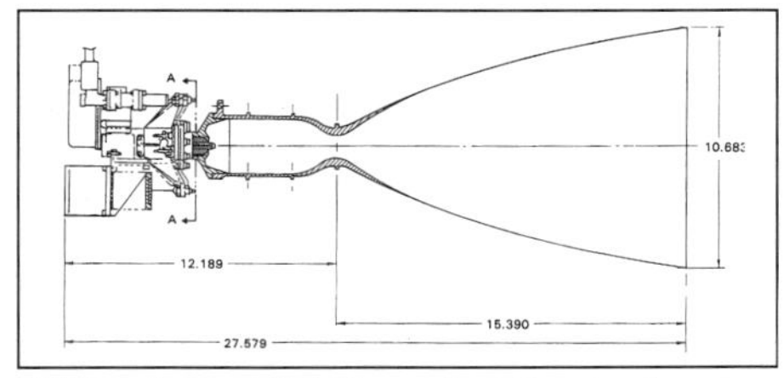

TRW's primary OMV bipropellant thruster, based on Apollo's descent engine, was designed to provide a 10:1 throttling range 0516961

Fuel: monomethyl hydrazine at 5.62 kg/s
O/F ratio: 1.60
Thrust: 44.035 kN vac
Specific impulse: 303 s vacuum
Expansion ratio: 43:1
Cooling method: quartz phenolic chamber ablation and columbium (niobium) nozzle radiation
Chamber pressure: 7.1 atm
Ignition: hypergolic, started by 28 V electrical signal to on/off solenoid valves
Burn time: 500 s for total of 5 starts; 10 350 s single burn

Monopropellant Hydrazine Thrusters

TRW produces a family of six RCS thrusters ranging over 0.44 222 N and has provided hydrazine RCS systems for DSP, DSCS 2, Mariner, Pioneer, Fleetsatcom, HEAO, ISEE, TDRS and Compton Observatory.

MRE-0.1

Fleetsatcom's RCS incorporates TRW's MRE 0.1 + MRE-1 to manage spin/despin, precession, attitude, positioning, stationkeeping and MW unloading. The pressure-fed system includes sets of redundant thrusters and a complement of latching isolation valves for system redundancy management. Each system comprises 16 MRE1 + four MRE 0.1 thrusters in dual modules, supplied from 122 kg of hydrazine in two 56 cm dia pressure spheres. System dry mass is 30 kg.

Applications: long duration missions requiring continuous precision pointing
First flight: 1978, Fleetsatcom 1
Number produced: 32
Length: 8.9 cm
Engine cycle: decomposition of hydrazine over heated Shell 405 catalyst bed
Thrust: 0.3–0.9 N vac
Specific impulse: 205–220 s vacuum
Total impulse: 71.2 kNs vac from 34 kg throughput
Inlet pressure: 24–79.6 atm
Burn time: 28,800 s maximum single burn
Minimum firing bit: 0.02 s
Total pulses: >370,000
Random vibration: 21 g rms

MRE-1

MRE-1 is NASA's standard 5 N hydrazine thruster and has passed six different qualification programmes, the latest as a low cost version for the STEP/TOMS lightsat class. In addition to its deep space application

TR-201: Delta's fixed thrust version of the Lunar Module Descent Engine. Upgraded versions are being developed for LOX/LH_2 upper stages 0516962

TRW's MRE-1 is NASA's standard 5 N thruster 0516963

aboard Pioneer 10/11, it was responsible for saving the TDRS 1 mission by boosting the 2 t satellite up to its operational GEO, consuming 363 kg during 45 h of cumulative firing. TDRS's system employs redundant groups of 12 thrusters, with associated latching isolation valves, drawing on 680 kg of hydrazine in two 102 cm dia tanks; dry mass is 86.5 kg.
Applications: satellite attitude and velocity control; Pioneer 10/11, Fltsatcom, HEAO, TDRSS, DoD
First flight: Mar 1972, Pioneer 10
Number produced: >1,500
Mass: 0.82 kg dual thruster module
Length: 15 cm
Engine cycle: decomposition of hydrazine over heated Shell 405 catalyst bed
Thrust: 5 N nominal, range 0.9–5 N vac
Specific impulse: 210–220 s vac steady state, 185217 s vac pulsed
Total impulse: ≥1,156 kNs vac for 544 kg hydrazine throughput
Inlet pressure: 5.4–38 atm
Burn time: 11,200 s max single burn, >90 h demonstrated total firing time
Minimum firing bit: 0.02 s
Total pulses: 458,000
Random vibration: 21 *g* rms

MRE-4
Development began in 1967 for Intelsat 3. It was responsible for re-routing NASA's ISEE 3 in 1985 to the first comet intercept. The longest demonstrated firing time is almost 2 h. Nozzle angle is selected to fit the geometry of the host spacecraft.
Applications: spacecraft attitude/velocity control; Intelsat 3, DSCS 2, ISEE 3, DSP, Atmospheric Explorer
First flight: Dec 1968, Intelsat 3F2
Number produced: >300
Mass: 0.41 kg
Length: 20 cm
Engine cycle: decomposition of hydrazine over heated Shell 405 catalyst bed
Thrust: 2.2–18.2 N vac
Specific impulse: 212–230 s vac steady state, 208 222 s vac pulsed
Total impulse: 512 kNs vac; demonstrated propellant throughput >227 kg
Inlet pressure: 3.4–272 atm
Burn time: 7,080 s maximum single burn
Minimum firing bit: 0.025 s
Total pulses: 507,000
Random vibration: 26.8 *g* rms

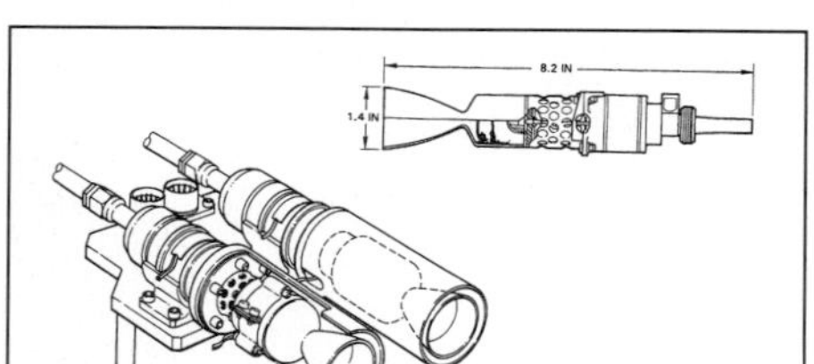

TRW's MRE-5 is used by NASA's Compton Observatory for attitude control 0516964

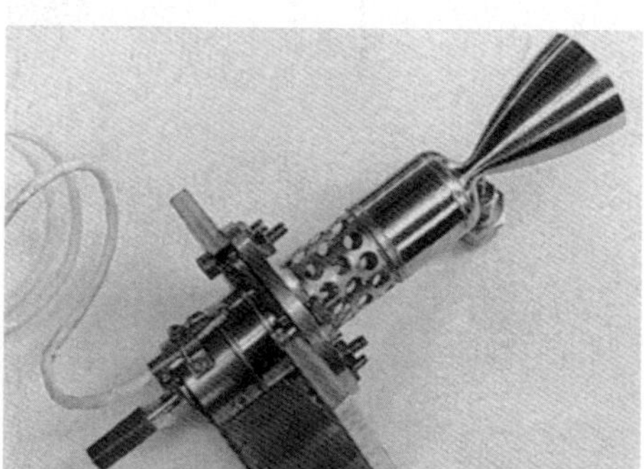

TRW's MRE-5 is used by NASA's Compton Observatory for attitude control 0516965

MRE-5/Compton Observatory
Development began in 1975 for USAF programmes; 11 flights units were delivered in 1984 for attitude control of NASA's Compton Observatory. Operational design life is 2 yr, with 450 kg hydrazine throughput and capable of 120 cold starts from <32°C. Carrying 1,920 kg of hydrazine in four tanks and weighing 281 kg dry, it is the largest mono system built, incorporating four dual thruster modules (each a primary + redundant 22 N thruster) and one orbit adjust module of four 445 N thrusters. The MRE15 (see below) is a scaledup version for OMV.
Applications: missions requiring duty cycle flexibility, high reliability and long life in steady state/pulse modes
First flight: Apr 1991, Compton Observatory
Number produced: 10 development units, 11 flight units
Mass: 1.13 kg dual thruster module
Length: 21 cm
Engine cycle: decomposition of hydrazine over heated Shell 405 catalyst bed
Thrust: 12–24.5 N vac
Specific impulse: 232–240 s vac steady state
Inlet pressure: 6.8–31.6 atm
Burn time: 0.050 s to several hours
Minimum firing bit: 0.025 s
Total cycles: 50,000
Random vibration: 21 *g* rms

MRE-15/OMV
The MRE-15 completed development and design verification (pre qualification) as a scaled up version of the MRE-5 for use on the NASA/TRW Orbital Maneuvering Vehicle (the programme was cancelled in 1990). 28 were to be clustered in sets of seven in each of four RCS orbital replacement units. Propellant throughput of each thruster was 450 kg of hydrazine. TRW was also to supply the MMH/N_2O_4 propulsion module engines (see below) and the 24 × 22 N nitrogen thrusters. No other applications are yet planned.
Applications:
Mass: 1.13 kg
Length: 26.9 cm
Engine cycle: decomposition of hydrazine over heated Shell 405 catalyst bed
Thrust: 44.5–89 N vac
Specific impulse: 225 s vac steady state
Inlet pressure: 12–27 atm
Burn time: 0.020 s to steady state
Total cycles: 100,000

Positive Expulsion Devices
>200 of TRW's elastomeric bladders and diaphragms have flown in hydrazine systems. Large opening diaphragms and smallopening bladders have been tested under low/moderate accelerations for resistance to slosh/vibration and longterm compatibility with rocket and ramjet propellants. Oxidiser compatible diaphragms up to 80 cm dia have been fabricated and expulsions demonstrated over 40/370°C.

Olin Aerospace Co

Current Status
Primex Technologies was acquired by General Dynamics in 2001 and in 2002 Aerojet acquired GD Space Propulsion Systems. In 1985, Rocket Research Corporation was acquired by the Olin Corporation and Olin Aerospace was born. Primex Technologies spun off from Olin in 1997 and Primex Aerospace was created. Specialises in monopropellant hydrazine engines and gas generators for spacecraft and upper stage applications; >9,300 assemblies and 100 propulsion systems have been produced since the first system was qualified in 1964. There are 600 personnel at the Redmond and Moses Lake sites. The MR508 hydrazine arcjet was qualified in 1991 and Telstar 4 in 1993 became the first commercial satellite to carry an arcjet for stationkeeping; see p295 1995–96 for specifications. The model is superseded by the MR-509/510. OAC has delivered >250 gas generators for Shuttle's Auxiliary Power Unit. Olin is working on designs for high temperature rocket motors and on microthruster applications with extended life.
Facilities: OAC's 0.29 km^2 site at the Grant County Airport in Moses Lake houses 2,840 m^2 of solid propellant manufacturing plant, a hazardous device assembly area, storage magazines, a shortrange flight test facility, a sea level large liquid engine test complex, ballistic test bays and chemistry lab.

MR-50 22.2N
Applications: attitude control for SMS, Viking, Meteosat, GOES, Voyager, GPS, Intelsat 5, Scatha, Lockheed Martin 5000, Delta Star, Magellan, Wind/Polar, EOS AM 1
First flown: SMS metsat (May 1974)
Flown: 695 (748 produced) to end 1995
Dry mass (g): 680
Length (cm): 18.3
Max diameter (cm): 6.6
Mounting method: bolted three places
Propellant: hydrazine at 4.67 17.33 g/s (Shell 405 catalytic decomposition)
Feed method: GN_2 or GHe at 4.832.7 atm, through 29 W solenoid dual seat valve
Thrust (N): 9.79–38.7 (22.2 nominal)
Specific impulse (s, vacuum): 215 228
Time to full thrust (ms): 150 to 90% PC
Nozzle area ratio: 40:1
Nozzle length (cm): 3.95
Nozzle cooling method: radiative
Combustion chamber:
pressure (atm): 2.9 11.3
temperature (°C): 800
cooling method: radiative
Duty cycle (s): 0.020 min to 5,400 max; 471,000 pulses
Total impulse (kNs): 459.4

MR-103 0.89N
Applications: attitude control thruster for Voyager, GPS, Intelsat 5, Lockheed Martin 3000, 4000, 5000 & 7000, Mars Observer, ACTS, Magellan, Cassini, Iridium
First flown: ATS 6 (May 1974)
Flown: 1,260 (2,705 produced) to end 1995
Dry mass (g): 332
Length (cm): 14.7
Max diameter (cm): 3.43
Mounting method: bolted three places
Propellant: hydrazine at 0.091 0.499 g/s (Shell 405 catalytic decomposition)
Feed method: GN_2 at 4.8–28.6 atm through 9W solenoid dual seat valve
Thrust (N): 0.19–1.12 (0.89 nominal)
Specific impulse (s, vacuum): 206227
Time to full thrust (ms): 150 to 90% PC
Nozzle area ratio: 100:1
Nozzle length (cm): 0.98
Nozzle cooling method: radiative
Combustion chamber:
pressure (atm): 4.35–23.9 atm
temperature (°C): 800
cooling method: radiative
Duty cycle (s): 0.008 min to unlimited max burn; 750,000 pulses
Total impulse (kNs): 158.46

MR-104 445N
Applications: attitude control and {d} V corrections, Voyager, Magellan, DMSP, Tiros N, Landsat
First flown: Voyager 2 (Aug 1977)
Flown: 68 (101 produced) to end 1995
Dry mass (g): 1,860
Length (cm): 46.0
Max diameter (cm): 15.2
Mounting method: bolted three places
Propellant: hydrazine at 91-290 g/s (Shell 405/LCH202 catalytic decomposition)
Feed method: GN_2 or GHe at 6.8 28.6 atm through Wright Components 30 W single seat solenoid valve
Thrust (N): 205–572 (445 nominal)
Specific impulse (s, vacuum): 228239
Time to full thrust (ms): 50 to 90% PC
Expansion ratio: 53:1 (dia 15.2 cm)
Nozzle length (cm): 17.8
Nozzle cooling method: radiative

Magellan thruster module: 2 × 445 N MR 104; 3 × 0.89 N MR 103; 1 × 22.2 N MR-50 (OAC) 0516943

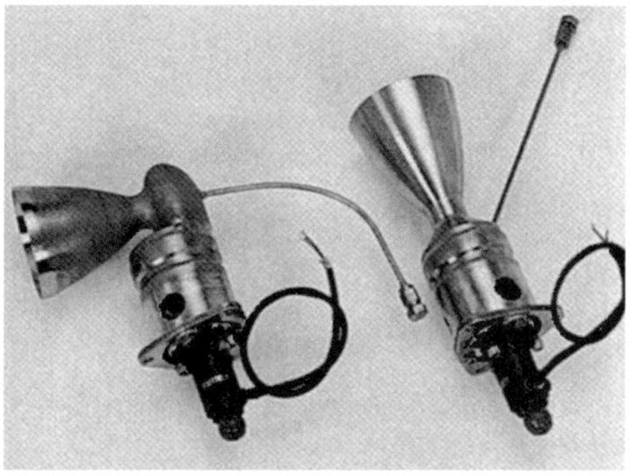

The MR-107 provides Delta, Titan and Atlas attitude control (Olin Aerospace Co) 0516944

Combustion chamber:
pressure (atm): 3.8–19.0
temperature (°C): 800
cooling method: radiative
Duty cycle (s): 0.022–2,000 single firing; 2,654 cumulative; 1,742 pulses
Total impulse (kNs): 693.9

MR-106 27N
Applications: spacecraft and upper stage attitude control and {d} V corrections, PAM A/S, Radarsat, GPS Block 2R, HAS/Peace Courage, Titan Centaur, Atlas Centaur, A2100, Lunar Prospector, NEAR
First flown: HAS/Peace Courage
Flown: 396 (2,381 produced) end 1995
Dry mass (g): 476
Length (cm): 17.8
Max diameter (cm): 6.4
Mounting method: bolted three places
Propellant: hydrazine at 4.0811.79 g/s (LCH 227/202 catalytic decomposition)
Feed method: GN_2 or GHe at 6.8 30.6 atm through Wright Components 27 W solenoid valve
Thrust (N): 8.9–26.7 (26.7 nominal)
Specific impulse (s, vacuum): 218–232
Time to full thrust (ms): 200 to 90% PC
Expansion ratio: 61:1
Nozzle length (cm): 4.72
Nozzle cooling method: radiative
Combustion chamber:
pressure (atm): 6.12–10.8
temperature (°C): 800
cooling method: radiative
Duty cycle (s): 0.016 min to 2,000 max; 12,397 pulses
Total impulse (kNs): 167

MR-107 178N
Applications: spacecraft and upper stage attitude control and {d} V corrections, Delta 2, Titan 2, Commercial Titan, PAM D, Small ICBM, HAS/Peace Courage, Atlas roll control module, STEP, Pegasus, LMLV/OAM
First flown: HAS/Peace Courage
Flown: 328 (1,898 produced) to end 1995
Dry mass (g): 885
Length (cm): 21.8
Max diameter (cm): 6.6
Mounting method: bolted three places

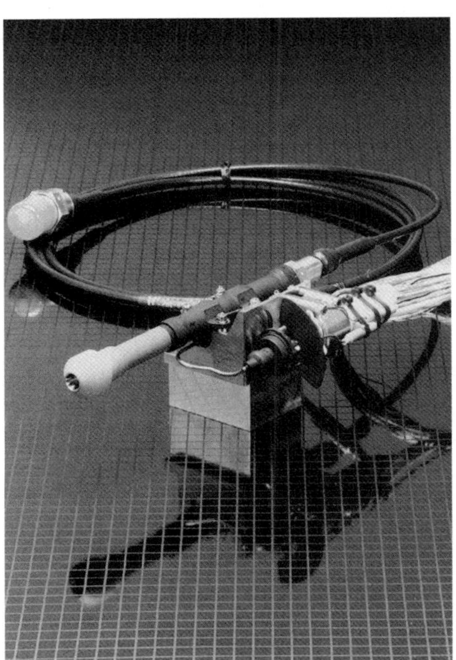

OAC's MR-509 arcjet was introduced on AsiaSat 2 0517356

Propellant: hydrazine at 24-113 g/s (LCH 207/LCH 202 catalytic decomposition)
Feed method: GN_2 or GHe at 6.8 34 atm through 50 W solenoid valve
Thrust (N): 51.2–257.9 (178 nominal)
Specific impulse (s, vacuum): 217 236
Time to full thrust (ms): 200 to 90% PC
Expansion ratio: 21:1
Nozzle length (cm): 4.72
Nozzle cooling method: radiative
Combustion chamber:
pressure (atm): 1.9–9.5
temperature (°C): 800
cooling method: radiative
Duty cycle (s): 0.016 min to 2,137 max; 7,005 pulses
Total impulse (kNs): 332.3

MR-111 2.2-4.4N
Applications: attitude control, Intelsat 5, ERBS, ACTS, Radarsat, Lockheed Martin 4000, 5000, 7000, Wind/Polar, Landsat, Mars Observer, Radarsat, ACE, Skynet 4, MGS, EOS AM 1, MSTI 2/3, Mars Pathfinder, Mars 98, Clementine 1
First flown: Intelsat 5 (Dec 1980)
Flown: 410 (627 produced) to end 1995
Dry mass (g): 345
Length (cm): 16.7
Max diameter (cm): 35.6
Mounting method: bolted three places
Propellant: hydrazine at 0.19 2.4 g/s (Shell 405 catalytic decomposition)
Feed method: GN_2 or He at 4.08 27.2 atm through Wright Components 9 W dual seat solenoid valve
Thrust (N): 0.44–5.34 (2.2–4.4 nominal)
Specific impulse (s, vacuum): 213 229
Time to full thrust (ms): 150 to 90% PC
Expansion ratio: 200:1
Nozzle length (cm): 2.18
Nozzle cooling method: radiative
Combustion chamber:
pressure (atm): 3.1–13.9
temperature (°C): 800
cooling method: radiative
Duty cycle: 0.020 s min to 15 h max; 420,000 pulses
Total impulse (kNs): 260.21

MR-501
Electrothermal Hydrazine Thruster
Applications: communications satellite NSSK, Lockheed Martin 4000
First flown: Satcom 1R (Apr 1983)
Flown: 72 (92 produced) to end 1995
Dry mass (g): 816
Length (cm): 19.4
Max diameter (cm): 8.84
Mounting method: bolted four places
Propellant: hydrazine at 0.059 0.127 g/s over Shell 405 catalyst and 350510 W electric heater
Feed method: GN_2 or GHe at 6.8 23.8 atm through Wright Components 9 W dual seat solenoid valve
Thrust (N): 0.18–0.33
Specific impulse (s, vacuum): 280 304
Time to full thrust (ms): not applicable
Expansion ratio: 100:1
Nozzle length (cm): 0.879
Nozzle cooling method: radiative
Combustion chamber:
pressure (atm): 3.4–6.5
temperature (°C): 800
cooling method: radiative
Duty cycle: 300 s min to 1.7 h max; 500,000 pulses
Total impulse (kNs): 311.4

MR-502
Applications: communications satellite NSSK, Lockheed Martin 5000
First flown: Astra 1B (Mar 1991)
Flown: 16 (32 produced) to end 1995
Dry mass (g): 848
Length (cm): 19.4
Max diameter (cm): 8.84
Mounting method: bolted four places
Propellant: hydrazine at 0.119 0.167 g/s over Shell 405 catalyst and 610839.6 W electric heater
Feed method: GN_2 or GHe at 21.8 23.8 atm through Wright Components 9 W dual seat solenoid valve
Thrust (N): 0.360.50
Specific impulse (s, vacuum): 280304
Time to full thrust (ms): not applicable
Expansion ratio: 100:1
Nozzle length (cm): 0.879
Nozzle cooling method: radiative
Combustion chamber:
pressure (atm): 3.4–6.5
temperature (°C): 800
cooling method: radiative
Duty cycle: 300 s min to 2.0 h max; 430,000 pulses
Total impulse (kNs): 525

MR-509 Low Power Hydrazine Arcjet
Applications: communications satellite NSSK, Lockheed Martin 7000
First flight: AsiaSat 2 Nov 1995
Flown: 8 (36 produced) to end 1995
Dry mass (g): 1,338
Length (cm): 24.4
Max diameter (cm): 9.27
Mounting method: bolted four places
Propellant: hydrazine at 0.068 0.077 g/s over Shell 405 catalyst and 1.8 kW (PCU input) electric arc
Feed method: GN_2 or GHe at 14.920.4 atm through Wright Components 9 W dual seat solenoid valve
Thrust (N): 0.21–0.25
Specific impulse (s, vacuum): 511
Time to full thrust (ms): not applicable
Expansion ratio: 100:1
Nozzle length (cm): 0.879
Nozzle cooling method: radiative
Combustion chamber:
pressure (atm): 4.1–4.7
temperature (°C): 800
cooling method: radiative
Duty cycle: 5 min minimum to 65 h max
Total impulse (kNs): 883

MR-510 Low Power Hydrazine Arcjet
Applications: communications satellite NSSK, Lockheed Martin A2100
First flight: planned GE 1 1996
Flown: 0 (44 produced) to end 1995
Dry mass (g): 1,338
Length (cm): 24.4
Max diameter (cm): 9.27
Mounting method: bolted four places
Propellant: hydrazine at 0.068 0.077 g/s over Shell 405 catalyst and 2.2 kW (PCU input) electric arc
Feed method: GN_2 or GHe at 14.9 20.4 atm through Wright Components 9 W dual seat solenoid valve
Thrust (N): 0.210.25
Specific impulse (s, vacuum): 586
Time to full thrust (ms): not applicable
Expansion ratio: 100:1
Nozzle length (cm): 0.879
Nozzle cooling method: radiative
Combustion chamber:
pressure (atm): 4.1–4.7
temperature (°C): 800
cooling method: radiative
Duty cycle: 5 min minimum to 65 h max
Total impulse (kNs): 812

VERIFIED

Parker Aerospace

Air/Space Division
18321 Jamboree Road, Irvine, California 92715.
Tel: +1 714 833-3000
Fax: +1 714 851-3341

Senior staff
General Manager: Jim Sabin

Control Systems Division
14300 Alton Parkway, Irvine, California 92718.
Tel: +1 714 833-3000
Telex: 678427
Fax: +1 714 586-8456

Senior staff
VP/General Manager: Robert Barker
VP Marketing: Jim Ryder
1425 West 2675 North, Ogden, Utah 84404.
Tel: +1 801 782-3100
Telex: 388424
Fax: +1 801 786-3045

Parker Bertea Aerospace produces the nitrogen thruster packages for Pegasus 0516942

Through to Feb 1997 Pressure Systems delivered 60 titanium hydrazine tanks for the Iridium satellites. (PSI) 0517165

Senior staff

Plant Manager: Mike Romito

Parker Aerospace, part of the Parker Hannifin Corp, designs, manufactures and services fluid systems, components and related electronic control systems for aerospace applications. Personnel total 5,300. The space related products of its divisions are specified below.

Air/Space This division specialises in the design/manufacture of valves and coupling devices to handle gaseous, liquid and cryogenic propellants, including hydrogen, nitrogen, oxygen, helium, NTO and MMH. Such systems fly on Shuttle, RL10 engine, Peacekeeper and Pegasus.

High Pressure Helium Regulator

Applications: ullage pressure control in a lightweight 680 atm blowdown propellant feed systems for missile and spacecraft propulsion systems. The regulator controls ullage pressure within ±10% with a 10:1 variation in inlet pressure and from lockup to rated flow. It is a welded assembly constructed entirely from titanium alloys. The basic regulator design can be used for a variety of fluids/gases.

Inlet pressure: 680–102 atm

Regulated pressure: 87 atm ±10%

Flow rate: 0.16 kg/s helium min at 102 atm inlet

Inlet/outlet ports: 6.35 mm dia × 0.89 mm wall tube stubs

Startup transient: 95.8 atm max regulated pressure at 680 atm inlet pressure

Mass: 0.077 kg

Cold Gas Thruster Module

This module is designed to provide 3-axis control for small launch vehicles; it is flown on Pegasus. The module comprises three cold gas thrusters operated by integral independent solenoid valves. The central thruster generates 111.2 N and the two lateral thrusters 55.6 N when operated with a 136 atm N_2 gas supply. SI of each is 68 s. The solenoid valves are pilot operated and have a response time of <5 ms at max gas inlet pressure.

Operating pressure range: 136–13.6 atm

Thrust at 136 atm: 4.5–111.2 N + 8.9 55.6 N

Voltage: 28 ± 4 VDC

Response time: 8 ms max

Mass: 0.086 kg per thruster

Size: 25.4 mm dia × 63.5 mm per thruster.

Control Systems The division, with 1,700 employees, designs/manufactures flight control actuators and associated electronics for space applications that provide system functions of proportional control, performance enhancement and availability enhancement utilising available hydromechanical, electrohydraulic, electromechanical and electro-optic technologies. CSD provides the stage 2 3axis fin control actuator system for Pegasus and the stage 2/3TVC actuator system.

VERIFIED

Pressure Systems, Inc

PSI, is the world's largest independent manufacturer of titanium pressure vessels for launchers and spacecraft: it now ships 200 to 300 annually, worth about US$25 million. >3,800 have been delivered, without flight failure, including >600 propellant management device tanks of >20 designs and >700 elastomeric diaphragm tanks of >15 designs. PSI acquired Programmed Composites Inc in June 1995, allowing it to offer tanks integrated with satellite core structures, in addition to precision reflectors, solar array substrates and antenna support structures. PSI tanks provide the propellant for most commercial communications satellites, including Hughes HS 601, Loral FS1300 and Lockheed Martin A2100. PSI tanks are also used on Milstar, Cassini, Soho, NEAR, Tirus, Inmarsat, Intelsat, ETS 7, TRMM and Axaf. PSI is completing a $4 million contract from Lockheed Martin for 75 hydrazine tanks for Iridium satellites. In 1996 PSI delivered the 1000th upper stage high pressure helium tank and the 100th upper stage propellant tank for the Lockheed Martin AtlasCentaur launch vehicle family. PS developing new high-pressure propellant vessels for exotic propellants sought for NASA's SLI. This has now been adapted for several non-government launch vehicle programmes and propellant vessels originally designed for specific SLI applications have been offered as alternatives to existing vessels on current launch vehicles.

Puroflow Corp

Surface tension devices for spacecraft propellant tanks. A wide range of non-aerospace applications and patented fluids management systems. Filters/strainers for propellants, pneumatics, hydraulics and cryogenic systems. Users include Shuttle, Titan 3, GOES, Milstar and Eurostar. Developed new methods of pad cleanup operations and special fluids handling including toxic propellants and washdown procedures. A UV system is used to treat the deluge water on the Shuttle launch pad. See also the Decca Valves entry in the Space Industry section.

VERIFIED

Quantic Industries, Inc

Quantic designs, develops, manufactures and tests ordnance devices (primarily electrically initiated), including pyrotechnic initiators, bolts, valves and cutters. The PAM upper stage and Delta launch vehicle carry the company's explosive bolt cutters. It also offers a line of lasers and laser ordnance. Development of high-intensity laser communications technologies for DoD and NASA research projects. Company has diversified into niched products linking pyrotechnic technologies with laser applications.

Facilities: manufacturing facilities cover propellant mixing, powder blending, glassto-metal sealing, protective finishing, inert assembly, welding/soldering and automated explosive loading. The 0.42 km^2 site houses 26 buildings, providing 5,600 m^2 under roof for administration, engineering, manufacturing, inspection and testing, in addition to inert storage and explosive magazines.

United Technologies Corporation (Solid Propulsion)

Current Status

United Technologies markets the Orbus 21D as a derivative of the Orbus 21 for commercial applications. It is used as the upper stage for the Athena family of launch vehicles. The Orbus 21S has the same basic thrust as the Orbus-21. A full variation firing took place in June 1994.

Background

P&W SPO is responsible for UTC's space propulsion activities. Liquid engine programmes are pursued at the West Palm Beach facilities (see the separate entry), while solid programmes are run by CSD in San Jose. CSD specialises in the design, development and production of solid propellant, hybrid and ramjet propulsion systems for space, strategic, tactical and launch booster systems. Its Orbus upper stage family includes Orbus 6/6E/21 (IUS), Orbus 7S (JCSat), Orbus 21S (Intelsat 6), Orbus 21 (TOS), Orbus 21D (LMLV launcher) and Orbus 1 (STARS and Starbird upper stage). The numerical designator corresponds to propellant mass to the nearest 1,000 lb, S denotes spin stabilisation and E the use of an extending exit nozzle (EEC) system. CSD also manufactures the segmented SRMs for Titan 4 and the Space Shuttle SRB separation motors. It was responsible for the Algol stage 1 motor on Scout, which retired in 1994.

Orbus 21D was qualified as LLV's stage 2 motor by its 30 June 1994 firing (CSD) 0517167

CSD's Coyote facility in San Jose includes the world's largest solid rocket vertical test stand: 29 m high, it can handle 26.7 MN thrust motors and was most recently used for qualification firings of the 7 segment USAF Titan 4 strap-on motor. Small/medium-sized horizontal and vertical test stands are controlled from a central complex and structural test facilities can apply omniaxial simulated flight loads of up to 4.9 MN to motor cases and attach skirts. CSD's integral rocket/ramjet test complex can simulate booster, transition and ramjet operations in a single sequence under sea level to high altitude conditions up to M6. The Coyote site also houses a 5,400 t annual capacity solid propellant mixing facility comprising two 2,270 l, one 2,840 l and one 1,510 l mixers.

Orbus

The main Orbus range derives from the two motors developed for the Inertial Upper Stage programme. The US Air Force awarded the Boeing/CSD team a contract in Oct 1976 to develop the IUS for Shuttle and Titan 34D. The stage completed its validation phase 19 Dec 1977 after a successful test firing of the Orbus 21 at the Arnold Engineering Development Center. Key components, such as the advanced flightweight nozzle, proved their survivability at a simulated altitude of 33 km. Full scale development of the IUS propulsion system began 1 Mar 1978 under an award from Boeing Aerospace. The 2 stage IUS incorporates the large Orbus 21 as the first stage SRM-1, topped by the Orbus 6E with EEC extendable exit cone as SRM-2. In May 1978 CSD was contracted by Boeing to develop the TVC electromechanical actuator system with redundant drive motors. The following Dec, CSD was contracted to develop the actuators for raising the stage and attached payload in the Shuttle cargo bay prior to deployment.

The first offloaded Orbus 21, at minimum 50% propellant loading, was fired 13 Aug 1980. By Jan 1983, no failure had occurred in 30 static firings despite the combination of such advanced features as Kevlar fibre case, carbon/carbon integrated throat/entrance, the world's first flightworthy EEC, Techroll seal and lightweight TVA. The first IUS flight placed two DSCS milcom satellites in GEO; the first Shuttle application came aboard STS-6 in Apr

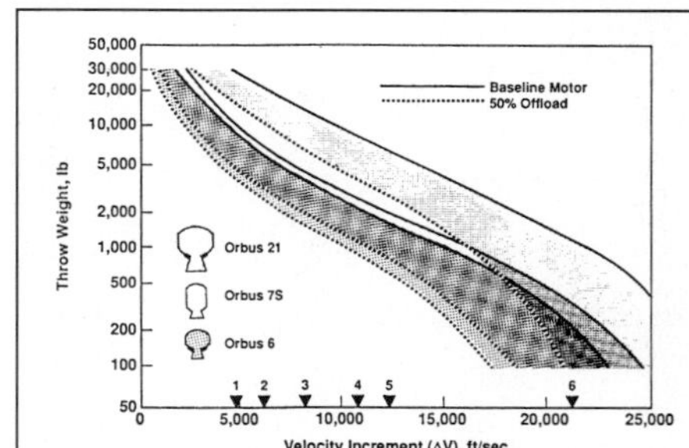

Baseline Orbus motor performances. Numbered triangles indicate 1 apogee burn for circularising 185 km perigee GTO, 2 apogee burn for circularising 185 km perigee GTO with 28.5° plane change, 3 185 km perigee burn for GTO, 4 parabolic escape from 185 km circular orbit, 5 Mars flyby from 185 km circular orbit, 6 Jupiter flyby from 185 km circular orbit (CSD) 0516967

Orbus 21 is the most powerful solid developed for in space applications. Shown is Intelsat 6'2 21S version
0517168

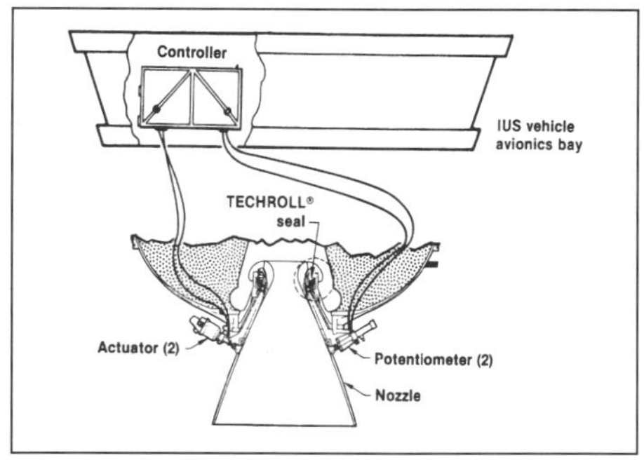

Orbus 21/6E provide steering for the 3axis Inertial Upper Stage through the Techroll nozzle joint and orthogonal actuators. Orbus 21 produces a maximum deflection of 4°, with 7° for the upper stage Orbus 6E
0516966

1983 with NASA's TDRS 1 data relay satellite. SRM1 fired successfully on that occasion, but SRM 2's Techroll joint overheated and failed, leaving the vehicle in an incorrect orbit. See the IUS entry for flight history. On a typical GEO mission, SRM1 performs GTO injection, followed by SRM2 insertion after a 6 h coast into GEO. Variations in mission requirements are accommodated by propellant offloading of up to 50%; SRM-2 can also fly without EEC, resulting in reduced specific impulse, or deployed before launch. Orbus 21 was also adapted as the Intelsat 6 perigee motor (21S spinning model) and for Orbital Science Corp's TOS upper stage which has been used for NASA's ACTS and Mars Observer. All the 'S' models (6, 7, 21) fly with fixed nozzles. The 21D version was test fired 30 June 1994 to qualify it as the upper stage of Lockheed's LMLV. Differences are a carbon phenolic exit cone in place of the previous 2D carbon-carbon (this adds 136 kg inert mass but is $700k cheaper), 4D throat replacing the 3-D throat, and state of the art TVC actuators (digital *vs* analogue). Specifications provided below are for the IUS and LMLV versions (21D specifications given only where different).

Orbus 21
First flown: 30 Oct 1982 on Titan 34D IUS
Number flown: 23 on IUS, 2 TOS, 6 Intelsat, 1 LMLV, through Feb 97
Applications: IUS stage 1 motor, TOS motor, Intelsat 6 perigee boost motor, LMLV stage 2
Mass: 10,398 kg fully loaded (10,619 kg 21D); 639 kg burnout (780 kg 21D)
Length: 315.0 cm (317.2 cm 21D)
Maximum diameter: 233.7 cm

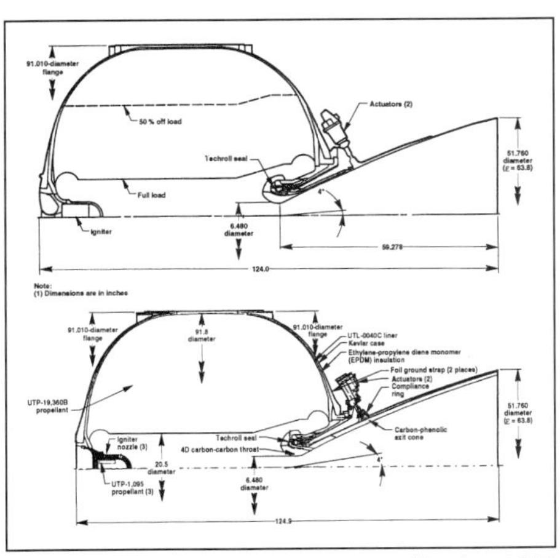

Orbus 21 cutaway principal features: 21 ((top) and 21D (bottom) (CSD)) 0517169

Propellant:
type: 86% solids HTPB UTP 19360A
shape: based on 52 cm dia bore
mass fraction: 0.935 fully loaded (9,709 kg; 0.941 for 21S model without TVC system), 0.47 max 50% offloading (4,853 kg). 21D 0.921 (9779 kg), 0.457 max offloading (4,583 kg)
Burn time: 154 s (146 s 21D)
Thrust (vacuum): 195.7 kN average, 267.77 kN peak (192.4 kN avg 21D)
Total impulse (MNs, vacuum): 28.38 (28.14 21D)
Specific impulse (s, vacuum): 295.5 (293.5 21D)
Pressure (atm): 57.9 atm max
Nozzle:
throat diameter (cm): 164.6 mm
length (cm): 150.56 cm
exit diameter (cm): 1,315 mm (63.9 expansion ratio)
materials: single-piece integral throat and entrance of 3D carbon-carbon (4D for 21D); exit cone 2D carbon-carbon (carbon phenolic with graphite/epoxy structural overwrap for 21D)
Casing materials: 353.8 kg, Kevlar/epoxy with silica filled EPDM
Igniter type: pyrogen head-end, 16.3 kg with 3 nozzles, UTP 1,095 propellant
Thrust vector control: ±4°TVC (up to 20°/s) provided by batterypowered orthogonal electromechanical actuators and Techroll movable nozzle joint pressurised by silicon oil; system mass 22.4 kg

Orbus 6/6E
First flown: 30 Oct 1982 on Titan 34D
Number flown: 23, on IUS, to end-1999
Applications: IUS stage 2 motor
Mass: 3,018/2,749 kg fully loaded with/without EEC
Length: 198 cm with EEC stowed, 320 cm EEC deployed
Maximum diameter: 160.0 cm
Propellant:
type: 86% solids HTPB UTP 19360A
shape: cylindrical

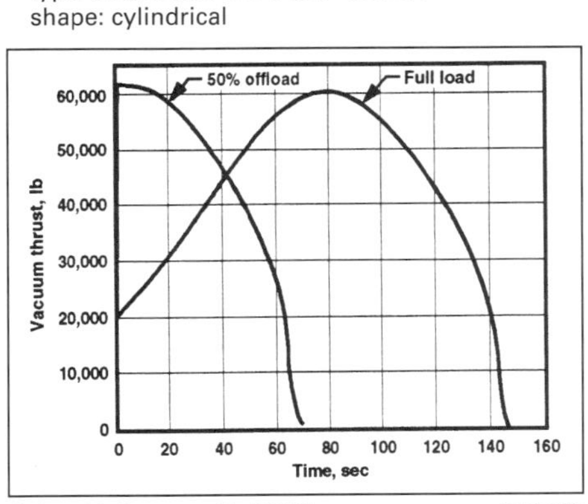

Orbus 21 thrust profile (CSD) 0517170

mass fraction: 0.902/0.914 with/without EEC fully loaded (0.921 without TVC system), 0.45 with max 50% offloading
Burn time: 103 s fully loaded
Thrust (kN, vacuum): 80.95 average, 113.87 peak
Total impulse: 8.14 MNs
Specific impulse (s, vac): 303.5/289.1 with/without EEC
Pressure (atm): 57.1 atm max
Nozzle:
throat diameter (cm): 106.9 mm
length (cm): 122/83.6 with/without EEC
exit diameter (mm): 1,438.0 (181.1 expansion ratio) with EEC; exit dia 750.3 (47.3 expansion ratio) without EEC
materials: single-piece integral throat and entrance of 3D carbon-carbon; exit cone/s 2D carbon-carbon
Casing materials: 90.7 kg Kevlar/epoxy with silica filled EPDM
Igniter type: 9.5 kg pyrogen head end, three nozzle, UTP1095 propellant
Thrust vector control: as Orbus 21 but 7°

Orbus 7S specifications
The Orbus 7S was qualified as the PKM for Hughes' HS393 family of communications satellites. Derived from CSD's Minuteman 3 stage 3 motor, with the thrust termination and vector control systems removed, it could be stretched mid cylinder and incorporate a lengthened nozzle. An Orbus 7 version is available incorporating a liquid injection TVC system.
First flown: 30 Aug 1984 as part of USN/Hughes Leasat 2 milcomsat
Number flown: 8 to end-2000
Applications: PKM, launcher upper stage
Mass: 3,547 kg fully loaded, 196 kg burnout
Length: 227.3 cm
Maximum diameter: 132 cm
Propellant
type: 86% solids CTPB ANB 3066; forward end finocyl
shape: stepped cylinder
mass fraction: 0.935
Burn time: 60.1 s
Thrust in vacuum: 153.467 kN average, 191.274 kN max
Total impulse: 12.5 MNs
Specific impulse (s): 281.9
Pressure (atm): 35.3 average, 43.5 max
Nozzle:
throat diameter: 174.7 mm
length: 90 cm
exit diameter: 84.76 cm (expansion ratio 23.55)
materials: tungsten throat insert, carbon phenolic exit cone, total mass 63.2 kg
Casing materials: 93.94 kg S901 fibreglass/epoxy with silicafilled Buna-N
Igniter type: 4.54 kg head-end pyrogen, 6 nozzle

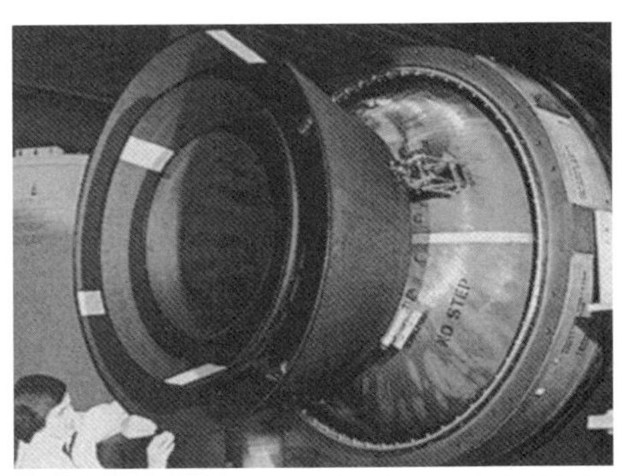

Orbus 6E incorporates an extending nozzle for Shuttle cargo bay compactness, increasing IUS GEO capacity by 100 kg. The motor can fly without the 122 cm extension (CSD) 0516968

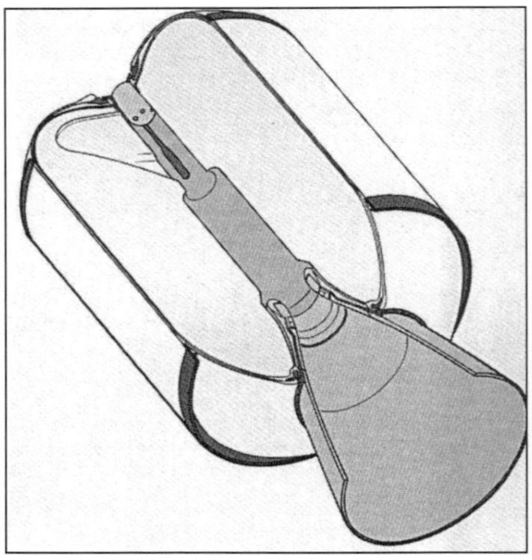

Orbus 7S derives from Minuteman 3's stage 3 motor 0516969

Orbus 1
The smallest motor in the series was developed for the upper stages of the Starbird and STARS target vehicles of the US Army Strategic Defense Command. The motor is a complete stage including a flexseal movable nozzle, electromechanical actuators, TVC control electronics and thermal battery, flight termination system and cables. The Orbus 1S spin version is also available.
First flown: 17 Dec 1990, Starbird stage 3/4
Number flown: 3 to end 1995
Applications: upper stage and space motor
Mass: 470.4 kg fully loaded, 53.2 kg burnout
Length: 124.9 cm
Maximum diameter: 69.2 cm
Propellant:
type: 90% solids HTPB
shape: head-end web
mass fraction: 0.88 (including TVC/TVA)
Burn time: 39 s
Thrust in vacuum: 30.38 kN average, 36.25 kN max
Total impulse: 1.19 MNs
Specific impulse (s): 293.3
Pressure (atm): 58.4 average, 66.7 max
Nozzle:
throat diameter: 55.1 mm
length: 525.7 cm
exit diameter: 375.4 mm (expansion ratio 46.4)
materials: carbon/carbon throat, carbon phenolic exit cone
Casing materials: T-40 graphite epoxy with silica filled EPDM insulation
Igniter type: aft-end toroidal pyrogen

Shuttle Booster Separation Motors
NASA selected CSD in late 1975 to develop and produce solid propellant Shuttle booster separation motors (BSMs) for the first six development missions. Each strapon fires a cluster of four motors at the forward and aft ends to separate from the vehicle's External Tank and begin the recovery sequence. The motor was accepted by the space agency in Sep 1978 and the last of 104 originally ordered was shipped in early 1980. CSD also provides separation motors for the Titan launcher strapons (see below). >1,500 have been produced.
First flown: 12 Apr 1981 on Shuttle mission 1
Number flown: 16 per Shuttle mission (1,664 by end April 2000 on 104 missions)
Mass: 73.0 kg
Length: 838 mm
Maximum diameter: 327 mm
Propellant:
type: 86% solids HTPB (UTP 19048)
shape: 16-point star
mass fraction: 0.472 (34.5 kg propellant)

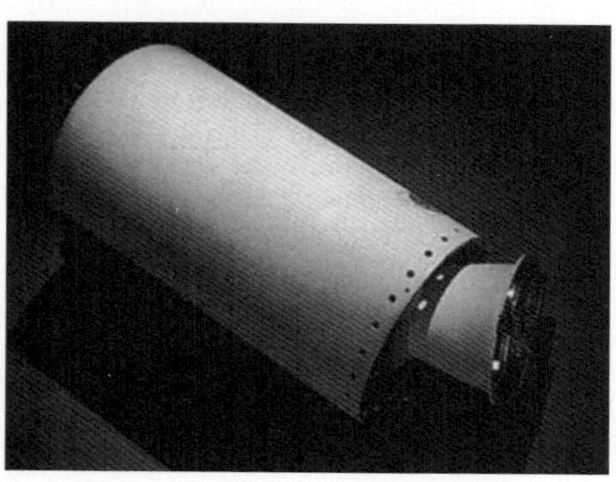

Each Shuttle strap on carries eight CSD separation motors to initiate the recovery sequence 0516970

CSD tested this first 7 segment motor in Apr 1969 before the Manned Orbiting Laboratory programme was cancelled. The same motor is now utilised by Titan 4 0516971

Burn time: 0.8 s
Thrust (kN, vac): 82.6 average, 129.5 max
Total impulse: 78.1 kNs
Nozzle: 20° canted; graphite throat insert with steel exit cone
Casing materials: aluminium with silica filled EPDM insulation
Igniter type: each cluster of four is ignited by firing redundant NSD (NASA Standard Detonators) pressure cartridges into redundant confined detonating fuse manifolds.

Titan segmented boosters
CSD test fired the first large segmented solid propellant motor on 15 Dec 1960 and in May 1962 was contracted to provide Titan 3's strapon motors. This 120-inch programme (referring to the motor diameter) achieved its first full scale static firing 20 July 1963; 13 others followed by its conclusion on 28 Apr 1965 shortly before Titan 3C's debut. This booster comprised five segments with pinned clevis joints, fixed canted nozzles, LITVC system, aft skirt, nosecone, separation system and destruct systems. Production ended in 1979; the last of 65 missions was in Nov 1982.

A '5 1/2 segment' version was created by adding a top segment 60% of normal 320 cm length for Titan 34D; the 15th/last departed in Sep 1989. Lockheed Martin's commercial Titan 3 launcher is derived from this vehicle; further information is provided under that company's entry. Separation of insulation from the steel casing because of a manufacturing/processing anomaly was determined to be the cause of the Titan 34D failure 18 Apr 1986; during the revalidation phase heater strips were also added to the segment joints to avoid *Challenger* type failures.

A 7-segment version was selected for the USAF Manned Orbiting Laboratory programme and first fired 26 Apr 1969. Three other full-scale trials followed successfully in 1970, but MOL had already been cancelled. The design was selected for the Titan 4 launcher in the 1980s, requiring only two qualification firings, because of the previous development work, on 21 Dec 1987 and 14 Feb 1988. CSD's steel-casing motors are utilised on Titan 4 launches; later missions will employ Alliant Techsystem's lightweight filament wound graphite composite casing motors, although development problems prompted procurement in early 1990 of long lead items for seven further sets of CSD's design. 20 of the 26 ordered flight sets had been delivered by end 1995. Titan strapons provide steering by injection of N_2O_4 into the exhaust from a 3,630 kg capacity tank; less than half is normally consumed. The injection adds typically 17.8 kN thrust to each motor. CSD also provides the eight separation motors per booster, mounted in clusters of four fore/aft. Each 1.5 m long 15 cm dia 38.5 kg motor contains 25 kg of solid propellant and provides 20 kN thrust in the 1 s burn time.

Titan solid rocket motor specifications
The specifications given below apply to the Titan 3 5 1/2 segment SRMs; where Titan 4's 7 segment characteristics differ they are included in parentheses.
First flown: Oct 1982 on Titan 34D, Jan 1990 on commercial Titan 3 (Titan 47 segment 1989)
Number flown: 15 sets on Titan 34D + four sets on commercial Titan 3 to end 1995 (32 sets on Titan 4 by end-2000)
Mass (t): 238 (316.6)
Length (m): 27.57 (34.43), each segment 328 cm
Maximum diameter (cm): 311 (311)
Propellant:
type: 84% solids PBAN (UTP 3001B)
shape: forward segment has fins; others are tubular, tapering from 98.3 cm thick to 85.6 cm aft for desirable tailoff characteristics
massfraction:0.885/210.6tpropellant(0.846/268.1t Titan 4 propellant)
Burn time (s): 113.7 (119.5)
Thrust (kN): 6,227 (7,117) vacuum average
Specific impulse: 265.2 (272.0) s vacuum
Total impulse (vacuum): 545 MNs (706.8 MNs)
Pressure (atm): 58 max (joints designed for 72.5/1.25 times MEOP)
Casing materials: D6AC high strength steel <12.7 mm thick. Segments are stacked with clevis joints incorporating 240 pins and single O ring. Segments are interchangeable. Insulation provided by silica-filled Buna-N rubber.

VERIFIED

United Technologies Pratt & Whitney (Liquid Propulsion)

Government Engines & Space Propulsion PO Box 109600, West Palm Beach, Florida 33410-9600.
Tel: +1 407 796-6796
Fax: +1 407 796-7258

Senior staff
President, P&W GESP: John A Balaguer
Public Relations: Patrick W Louden

Address P&W Space Propulsion Operations
PO Box 49028, San Jose, California 951619028.
Tel: +1 408 779-9121
Fax: +1 408 778-4599

Senior staff
Executive VP/GM: Douglas A North
VP Liquid Space Programs: Joseph P Zimonis
RL10 Program Director: Peter G Scharf
SSME/ATD Program Manager: John L Price
P&W Space Propulsion is responsible for the company's space propulsion activities. Liquid engine programmes are pursued at the West Palm Beach facilities, while solid rocket programmes are run by the Chemical Systems Div in San Jose (see the separate UTC CSD entry). USBI provides Shuttle booster services; see the separate entry. UTC combined the space propulsion activities of CSD, USBI and P&W in 1990 within GESP. P&W has supplied RL10 cryogenic engines for the Centaur upper stage since the early 1960s. In 1986 the company was awarded a NASA contract to develop alternate turbopumps for the Space Shuttle main engine. An agreement was signed with NPO Energomash (qv) in Oct 1992 providing exclusive US marketing rights on the Russian company's products and technology. An RD-120 production engine was tested Oct 1995 in West Palm Beach,

Pratt & Whitney's RL10 was the first operational liquid oxygen/hydrogen engine and will continue in service next century. Shown is the RL10A-4 (P&W) 0517344

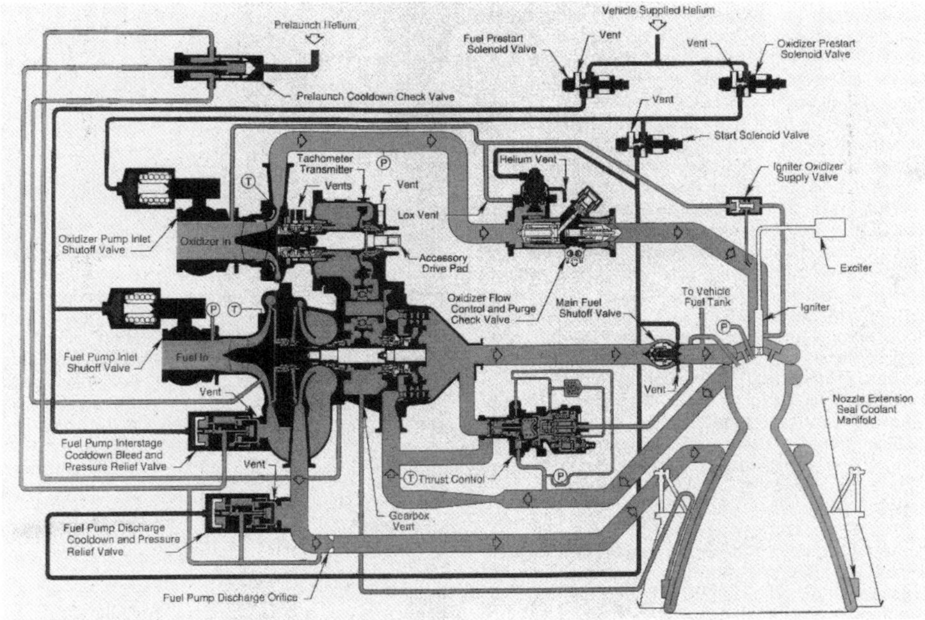

The RL10B-2 will power Delta 3's cryogenic stage (P&W) 0516973

believed to be the first test of an inproduction and inservice engine outside of Russia. Modified versions are proposed for several US vehicles, including the X34 and Kistler's K1. The three <5 s runs showed the engine was compatible with US ground systems; further testing requires a sea level nozzle and a customer. A major activity continues to adapt variants of the 4 chamber RD170/171. The RD180 2-chamber derivative has been selected by Lockheed Martin for the Atlas 2AR; it is being developed by NPO Energomash and provided through a P&W/Energomash joint venture. The first 18 vehicles use Russian built engines, followed by P&W built motors (commercial vehicles can employ Russian engines, but US government payloads must use US engines by law).

Test facilities: at peak, the Florida site operated seven RL10 test stands, including one for Centaur-type dual engine firings. Stand E6 has been refurbished for the new production RL10s and stand E8 accommodates Space Shuttle ATD turbopump work.

P&W has partnered with Russia's Energomash to develop the RD-180 (which see) for Atlas applications. A derivative of the RD-170 it was first fired in the US in Julyy 1998. P&W-Aerojet Propulsion Associates has developed a high-energy propulsion system for duty as a high-thrust, reusable booster engine incorporating low-cost, low-risk, LH_2/LOX single burner, fuel-rich staged combustion cycle known as the Cobra.

P&W RL10

The company's RL10, initially fired in July 1959 as the world's first LOX/LH_2 engine and first delivered to NASA in Aug 1960, continued in paired service for the Centaur upper stage to the new century and could be developed further to produce 156 kN for a future Space Transfer Vehicle. RL10 design work began in Oct 1958, leading to the first Centaur ground test firing in Mar 1962 and the stage's flight debut in Nov 1963. Oct 1966 marked the first completely successful in-space restart of a hydrogen engine. Fluorine and methane have also been used separately in test firings. The current A33A version completed qualification tests in Nov 1981 and first flew on AC62 in June 1984; its development was supported by Intelsat to handle their satellites. By end 1990, 178 RL10s of all types had flown in space, accumulating 290 firings totalling <21 h without a single failure. The first failure occurred 18 Apr 1991 in the AC70 launch of Japan's BS-3H satellite, although the blame was fixed on pre-launch gantry procedures and not on RL10's operation. General Dynamics' review board decided that turbopump foreign object damage, possibly combined with water/ice contamination, was a likely cause. The Feb 1992 Atlas 2 launch included the 300th space firing. AC71 on 22 Aug 1992 produced the second failure: icing in the hydrogen pump was identified as the cause for both AC71/70. The engines were chilled before launch to 256°C by liquid helium from a pad umbilical. It appears that a spring-operated check valve stuck open when the umbilical pulled free, allowing air to be sucked in where the nitrogen could condense and freeze. Subsequent vehicles carry electrically operated valves and are not as chilled so that air cannot freeze even if it does penetrate. Before this second failure, RL10 had achieved 190 flights and 312 firings. By end 1995, 246 RL10s of all types had flown in space, accumulating 436 firings totalling >29.3 h. In addition to Centaur applications, clusters of six RL10A3 powered Saturn 1's S4 stage on six missions during 1964/65. The original RL10A-1 model generated 66.72 kN vacuum thrust with 412 s SI and 20.4 atm chamber pressure. The current RL10 is available in the A-3-3A version offering 5:1 oxidiser: fuel mixture ratio for the Atlas Centaur combination (Atlas 1) and the USAF Titan 4 Centaur. Atlas 2 employs a 5.5:1 mixture ratio for 442.3 s SI and 73.39 kN thrust; Atlas 2A required a significant enhancement to 92.5 kN. Further information on this RL10A-4 model is provided below. Titan 4 will continue to use the A3-3A, but if its Centaur converts to a single engine then A4 could be adopted. Further upratings are under consideration, particularly for improving ELV performance and space transfer vehicles. The 93.4 kN RL10B-2 will emphasise performance improvement by increasing SI to 466.5 s; expansion ratio 285:1. Its selection for Delta 3's stage 2 was announced in June 1995; first delivery is due mid 1997. The RL10C-1, providing 156 kN using a 147:1 expansion ratio, has been studied by NASA for a single engine Centaur. P&W expects to produce 1624 RL10s annually to end 1999 for the Atlas, Titan 4 and Delta 3 programmes. 26 were manufactured in 1995 (1994: 14; 1993: 7; 1992: 12; 1991: 13; 1990: 18).

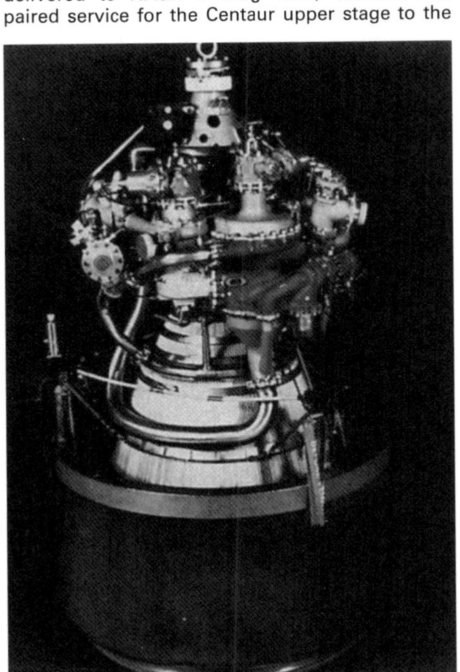
RL10A-4's 508 mm skirt extension improves specific impulse by 6.5 s for the Atlas 2A (P&W) 0516974

P&W was awarded a $12 million contract by McDonnell Douglas Aerospace to provide four RL10A 5 for BMDO's DC-X SSRT single stage rocket technology suborbital demonstrator. Each 143 kg engine, incorporating a 10° conical nozzle for low-altitude operations, can be throttled 30 100 per cent (19.565.7 kN) and gimballed ±8.5°. The settings correspond to 380.5/373 s SI, respectively, and 9.66/32.3 atm chamber pressures. Length 106.9 cm, expansion ratio 4.28:1.

RL10A-3-3A

Applications: Centaur stage of Atlas & Titan
First flown: Nov 1963 (3-3A first flight was June 1984)
Number produced: 380 of all types and including ground test engines to end-1995
Number flown: 291 of all types to end-2000 (two inflight failures caused by vehicle preflight procedures and not attributed to engine operations)
Dry mass: 138 kg
Length: 178 cm
Maximum diameter: 102 cm
Mounting: ±4° gimbal for pitch/yaw control
Engine cycle: expander
Oxidiser: liquid oxygen at 14.0 kg/s
Fuel: liquid hydrogen at 2.79 kg/s
Mixture ratio (O/F): 5.0:1
Oxidiser pump: 11.3 kg mass, 88 kW power, 13,100 rpm rotor speed producing 45.6 atm discharge pressure
Fuel pump: 34 kg, 32,800 rpm rotor speed, 76.2 atm discharge pressure
Thrust: 73,390 N vacuum
Specific impulse: 444.4 s vacuum
Time to full thrust: typically 2.15 s
Nozzle area ratio: 61:1
Expansion bell:
length: 117 cm
diameter: 102 cm (expansion ratio 61)
materials: AISI 347 stainless steel
cooling method: regenerative by hydrogen
Combustion chamber:
pressure: 32.2 atm
temperature: 3,340°C
stability: ±0.5%
materials: AISI 347 stainless steel
cooling: regenerative by hydrogen
ignition: spark igniter. Qualified for 20 starts (seven made on Helios 2 mission)
Burn time: about 600 s required on Titan 4 Centaur, engine qualified to 4,000 s.

RL10A-4

The commercial Atlas 2A/2AS versions incorporate RL10s improved to 92.5 kN thrust each. Incorporation of the extending nozzle contributes a 6.5 s SI gain without additional thrust compartment length increase. The all welded columbium (niobium) 508 mm long skirt is electromechanically deployed following Centaur separation from the Atlas booster to create an 84:1 expansion ratio. The A-4 was flight certified in May 1991, followed by delivery in June 1991 of the first of 74 ordered by General Dynamics. By end-1992, ground testing accumulated 41,946 s in 275 firings; first flight was made successfully June 1992. The A41 version is available to improve Atlas 2A/2AS performance. First flight was AC113 10 Jan 1995 to cope with the larger Intelsat 7 series; second AC-115 22 Mar 1995. Thrust is increased by 6.67 kN and SI by 2 s by a modified injector improving propellant mixing in the thrust chamber. The tangential swirl oxidiser injector was developed in the 1960s for the USAF XLR129 engine. The RL10E version was developed for the single engine Centaur of Atlas 2AR, redesignated Atlas 3A on 8 April 1998 and first flown on 24 May 2000. It replaces the hydraulic and pneumatic valves with electromagnetic actuators, uses EMAs for gimballing and employs a solid state controller.

Specifications are given below for the RL10A4 where they differ from the A3 3A (A-4-1 values are given in parentheses).
First flown: June 1992 (Jan 1995)
Number flown: 41 by end-2000
Dry mass: 168 kg
Length: 178 cm stored, 229 cm deployed
Maximum diameter: 117 cm
Oxidiser: LOX at 17.78 kg/s (19.10); initial P 2.9 atm
Fuel: LH_2 at 3.22 kg/s (3.45); initial P 1.95 atm
Mixture ratio (O/F): 5.5:1
Oxidiser pump: 14,300 rpm (14,950) rotor speed producing 56.5 atm (57.6) discharge pressure
Fuel pump: 35,800 rpm (37,400) rotor speed, 92.5 atm (94.6) discharge pressure
Thrust: 92,500 N (99,190) vacuum
Specific impulse: 448.9 s (451.0) vacuum
Nozzle area ratio: 84:1
Expansion bell:
length: 119.4 cm primary + 51 cm extension
diameter: 102 cm primary; 117 cm extension (expansion ratio 84)
Combustion chamber:
pressure: 39.46 atm (41.5)
temperature: unavailable
stability: better than ±0.5%
Burn time: qualified to 3,000 s in multiple starts

RL10B-2
Delta 3 employed the B-2 for its cryogenic stage 2. Development began in mid 1995, and both certification and first production delivery were completed in 1997. The B-2 features a very high area ratio nozzle to produce the highest SI of any operational chemical engine. SEP is developing the 100 kg 3 m high carbon-carbon extending nozzle. Specifications are given below where they differ from the A-3-3A.

Applications: Delta 3 stage 2
Dry mass: 259 ± 5.5 kg
Length: 415.3 cm extended
Maximum diameter: 222.3 cm
Oxidiser: liquid oxygen at 19.9 kg/s
Fuel: liquid hydrogen at 3.3 kg/s
Mixture ratio (O/F): 6.0:1
Oxidiser pump: 14,676 rpm rotor speed, 55.6 atm discharge pressure
Fuel pump: 36,690 rpm rotor speed, 96.7 atm discharge pressure
Thrust: 105,645 ± 1,955 N vacuum
Specific impulse: 466.5 s vacuum
Nozzle area ratio: 285:1
Expansion bell:
length: 349 cm extended
diameter: 222.3 cm maximum
Combustion chamber:
pressure: 42.2 atm
stability: better than ±0.5%
Burn time: 3,500 s in multiple starts

Shuttle Main Engine Turbopumps
NASA awarded P&W a $188 million contract, subsequently increased by $20 million, in Dec 1986 for new Shuttle SSME high pressure turbopumps as part of the engine's Block 1/2 upgrades. Alternative Turbopump Development (ATD) design was completed in May 1988 for component testing of bearings, seals and turbine blades to begin. Hardware procurement for the initial pumps also began in 1988. Development problems, including casting the LOX main housing in three pieces to eliminate the old welded configuration, coupled with a rotor vibration and bearing wear problem, increased programme cost to $504 million. Testing started in early 1990 at P&W's reactivated and modernised facilities in West Palm Beach. The fuel pump attained rated power by year's end and the first development pump set was delivered in 1991 for engine testing at NASA Stennis. NASA halted fuel pump work from Dec 1991 in order to focus on the more critical oxygen unit. It restarted work May 1994 once the LOX pump was on schedule for its 1995 flight debut. LOX pump certification was completed 15 Mar 1995, testing two pumps >10,000 s each in >20 tests. Block 1 first flight was successful STS70 in July 1995 with one pump. All three engines were equipped for STS-77 in May 1996. The LOX pump completed 235 firings by end 1995 at Stennis, totalling 110,000 s with flight representative hardware. It will be fully retrofitted into the fleet by 1998. The fuel pump will fly as part of the Block 2 upgrade, debuting on STS86 in Sep 1997 and completing fleet retrofit by 2000.

A separate contract provides for 15 production LOX units in 1995-98 and 15 fuel 1997–2000. The pumps are not intended to increase engine performance but instead offer improved durability of 7½ h firing, or 55 missions, at 109% rated thrust. The design eliminates the 469 welds in Rocketdyne's fuel pump and all but seven of 300 in Rocketdyne's oxidiser pump, and houses fewer rotating parts. Stiffer rotors with robust single row ball and roller bearings are aimed at meeting the long life requirements. The fuel pump delivers 55,650 l/min at up to 455 atm; 36,300 rpm, 54.66 kW. The LOX side provides 28,000 l/min at 520 atm; 24,000 rpm, 19.46 kW.

Cobra
A product of Pratt & Whitney and Aerojet teaming to provide a booster motor for second generation reusable launch vehicle (2GRLV) requirements.
Applications: 2GRLV booster stage
Thrust: 889,600–4,448.000 N
Dry weight: 3,629 kg
Specific Impulse: 455 sec
Cycle: staged combustion
Mixture ratio: 5.5:1–6.5:1
Shutdown reliability: 0.9995
Catastrophic reliability: 0.999995
Mission life: >100 missions
Time between overhauls: ≥50 missions
Time between overhauls: >50 missions
Scheduled maintenance per flight: <100 man-hours
Turnaround between flights: <16 hr

VERIFIED

United Technologies USBI Co

UTC combined the space propulsion activities of CSD, USBI and P&W in 1990 within P&W's Government Engines & Space Propulsion unit based at West Palm Beach in Florida. USBI is under contract to NASA Marshall to provide assembly, test and refurbishment of the nonmotor segments of Shuttle's solid rocket boosters. NASA extended the contract in 1994 for recovering and refurbishing boosters to Sep 1997, worth $1.8 billion. The previous $1 billion award in Jan 1989 ran through Sep 1994. It was extended in 1996 through Sep 1999.

Western Electrochemical's 400 μm grain ammonium perchlorate 0517164

Western Electrochemical Co

Western Electrochemical is a wholly owned subsidiary of American Pacific Corp, a Las Vegas based manufacturer of speciality chemicals. Pacific Engineering & Production satisfied half (18 kt) of US ammonium perchlorate solid propellant oxidiser requirements until explosions destroyed its plant in May 1988. Users included Delta, Titan and Shuttle boosters; each Shuttle launch consumes 771 t. Ammonium perchlorate (NH_4ClO_4) is the oxidiser component of choice in solid rocket motors employed by NASA and the US Department of Defense. A new facility operated by Western Electrochemical (WECCO) was completed during 1989 on a 115 acre site in Iron County, Utah, offering an initial annual production rate of 14 kt capable of rising to 18 kt with only minor capital additions. The 33 building complex was constructed at a cost of $92 million and employs a staff of 175. The batch process under automated control can produce AP to custom chemical characteristics and physical distributions, as well as standard materials in large homogeneous lots. High purity AP is used in low smoke reduced signature rocket motors.

CIVIL AND COMMERCIAL SATELLITES

Communications
Data Relay
Earth Observation and Meteorological
Engineering Test
Geodetic
Microgravity and Materials Science
Scientific
Search and Rescue

COMMUNICATIONS

Argentina

NAHUEL Series

Current Status

With a projected lifetime of 12 years, the Nahuel 1A communications satellite was launched by Ariane 44LP on 30 January 1997, to a geostationary position at 288° E.

Nahuel 2 is scheduled for launch in 2004 providing broadcast and business services, DTH, internet access, telecommunications and VSAT nets.

Background

In 1991, Argentina issued a RfP for a domestic/regional satellite communication system in late 1991. Argentina wanted to operate two Nahuel ('Jaguar' or 'Tiger') satellites at one of two sets of locations 59° W, 71.8° W, 75.8° W, 80° W, 85° W. The country planned to pay about US$250–300 million for the capability and generate US$60–70 million revenue annually. Bolivia, Brazil, Chile and Uruguay all planned to benefit from the system. To establish a framework, the Argentine government organised the Comision Nacional de Telecomunicaciones on 16 December 1992 to negotiate with a consortium headed by DASA, and including Aerospatiale, Alcatel Espace and Alenia Spazio. The agency signed a contract 27 May 1993 for a term of 24 years and five orbital slots, to serve the southern part of the continent for the first time with high performance Ku-band links. NahuelSat SA created by the three European negotiation partners has US$100 million in equity to own/operate the system. The consortium plans to hold about half the shares.

Argentina operates the system as a concession and began service in 1997 following the successful launch of Nahuel 1A. Nahuel provides the equivalent of 36 TV programmes (180 with digital compression), 18,000 duplex telephone calls or 9,000 VSAT interactive connections. Connections require 1–2 m dishes.

Before Nahuel 1A orbited, Argentina used Telesat Canada's Anik C1/C2 satellites (as Nahuel A/B) to provide a Ku-band interim service. Telesat Canada sold the satellites in January 1993 for C$37 million over 3 years to the Paracom SA consortium. Telesat maintained a 10 per cent interest to secure participation in marketing services. Nahuel 2 will carry 24 C-band and 48 Ku-band transponders and operate from 81° W.

Specifications

Nahuel 1A

Launch: 30 January 1997 by Ariane 4 from Kourou, French Guiana
Location: 71.8° W geostationary
Design life: 12 years
Contractors: Aerospatiale (prime), Alcatel Espace (payload), DASA (AOCS, solar arrays)
Transponders: 18 55 W TWTA 14/12 GHz Ku-band South American beams, bandwidths 54 MHz, EIRP 50 dBW, orthogonal linear polarisation
Principal applications: TV distribution, telephony, VSAT data and business services

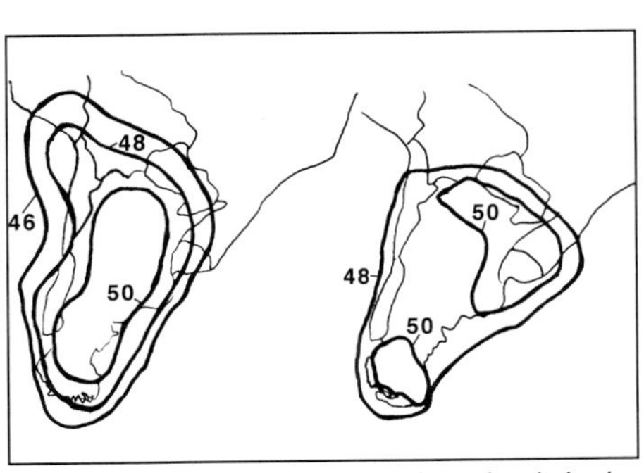

Nahuel EIRP (dBW) contours: horizontal polarisation (right) and vertical (left) 0517213

Configuration: Spacebus 2000 platform, 3-axis control, 22.4 m span across solar panels, single dual reflector multifeed antenna. GEO location maintained within ±0.05°
Mass: 1,820 kg at launch; about 1,100 kg on-station BOL
AOCS: provided by DASA's 22 × 10 N + 400 N MMH/MON bipropellant system
Power system: twin solar wings providing 2.9 kW equinox EOL.

Canada

Anik A Series

Background

Anik initiated commercial communications over Canada's sprawling 9.22 million km^2 land mass. The US launched the initial series beginning with Anik A1, 9 November 1972 by Delta, and followed by Anik A2, on 20 April 1973, Anik A3, 7 May 1975. All went up on Delta launch vehicles from Cape Canaveral. Canada retired A1 at 114° W, in July 1982; Anik A2 at 109° W, in October 1982; Anik A3 at 119° W capable of handling 10 colour TV channels or 9,600 phone circuits on 21 November 1984. The satellite linked 37 earth receive stations. Telesat Canada pioneered domestic and regional satellite operations and served as a consulting engineering source for other national satellite systems elsewhere.

Anik B Series

Background

Canada upgraded the Anik A specification to the Anik B, the world's first dual-band (C/Ku-band) satellite, which followed up CTS with experiments in education, telemedicine, conferencing and Inuit broadcasting. Essentially an RCA Astro Satcom bus, Anik B took on the bulk of the C-band East-West traffic from the retired Anik As. It weighed 474 kg on-station and went into orbit on 16 December 1978 aboard a Delta from Cape Canaveral with a design life of seven years. A wide range of new satellite services were introduced and in 1979 televised sessions from Canada's House of Commons were circulated across the nation. In October 1980 the national edition of The Globe and Mail was launched when it was transmitted for the first time through a satellite channel. Canada retired it in December 1986.

Anik C Series

Background

Canada purchased Aniks C1-3 from Hughes under a 1978 contract worth C$53.6 million plus C$13.7 million incentive awards. C3 operated in the Canadian orbital slot 117.5° W as the prime satellite until May 1989, with C1 held in orbital storage. The satellite took advantage of pre-1986 Shuttle tariffs. Service switched to C1 because the introduction of Mexico's Morelos required the orbital slot 1° away. C1 remained at 107.3° W, where E2 joined it in 1991. It continued to provide Ku-band services, while E2 employed C-band.

Between 1983 and May 1984, GTE Spacenet leased Anik C2, which in turn leased capacity to United Satellite Communications Corporation to provide DBS service to northern US states. The relatively low-power beam required receiving antennas in the 1.2–1.8 m range, a standard that did not succeed commercially. C2/C3 began inclined orbital operations 20 August/20 July 1991, respectively, to extend their lives. C3 still sees service as a backup to fibre optic providers for the Canadian telephone companies.

C1, with only 4.5 years of station-keeping propellant, began to move from 109.25° W on 29 March 1993 to take up service for Argentina. It arrived at its new 71.8° W slot on 15 April 1993 and began service 1 June 1993. C2 likewise began the move from 109.15° W 5 April 1993 and arrived at its new 75.8° W slot 18 June 1993.

Specifications

Launched: C3 11 November 1982, C2 18 June 1983, C1 13 April 1985, all by Shuttle
Locations: C1: 63° W (August 2000); C2: 114.9° W (December 1997); C3: 114.9° W GEO (inclination 3.68° prior to deorbit manoeuvre June 1997)
Design life: 10 years
Contractors: Hughes Space & Communications
Transponders: 16 15 W TWTA 14.003–14.497/11.703–12.197 GHz up/down Ku-band in four overlapping spot beams covering most of populated Canadian landmass, 54 MHz bandwidth, EIRP 46.5 dBW min, linear polarisation
Principal applications: 32 TV channels (1.2 m home antennas) or 21,504 simplex voice links, business communications
Configuration: Hughes HS-376 spin-stabilised bus
Mass: 562.5 kg in orbit (including 99 kg hydrazine)
TT&C: primary from Allan Park (Ontario) with full backup from Edmonton (Alberta) controlled from HQ
Power: 1,135 W from body array, K7 cells; three Ni/Cd batteries.

Anik D/Telstar 301 Series

Background

The fourth Anik series Anik E inaugurated the true commercialisation of Canadian services. Canada initially stored D2 in orbit, to benefit from advantageous pre-1986 Shuttle tariffs. It operated at 110.9° W until its services were transferred to nearby E1 in 1991. Then GE Americom bought it and moved it to 82° W in December 1991 as cover for Satcom 4, designating it Satcom 4R. GE paid C$18 million. Telesat Canada bought it back in April 1993 and put it in an inclined orbit to conserve NSSK fuel. Arabsat (who called it Arabsat 1D) bought it for C$5.5 million, to provide mostly video services from 19 to 20° E beginning in August 1993. Telesat still provided control from Telesat HQ, via a full TT&C facility in Tunisia. D2 left service and exited above GEO in January 1995 as it came to the end of its propellant. D1 left GEO 16 December 1991 after being replaced by E.

Following the Anik E control problems from January 1994, Telesat acquired Telstar 301 from AT&T on 28 March 1994. It arrived at its new 107.1° W slot on 4 April 1994 from AT&T's 96° W to restore services until E2 began commercial services again 1 August 1994. Telesat sold Telstar to Arabsat on 7 June 1994 and began moving it on 3 August to 20° E. It arrived February 1995 and began service as Arabsat 1DR. Anik D1 is currently at 104.5° W, Anik D2 at 110.5° W.

Specifications

Launched: D1 26 August 1982 by NASA Delta 3920; D2 9 November 1984 on Shuttle mission 14
Design life: 10 years
Contractors: Spar Aerospace prime, technology licensed from Hughes. Canadian content >50%
Transponders: 24 11 W TWTA 6/4 GHz up/down C-band national and north US beams, 36 MHz bandwidth, EIRP 36 dBW min (33 dBW US coverage), orthogonal polarisation

Anik E series satellites inaugurated service in 1991 (Telesat Canada) ***NEW***/1047519

Launched 17 July 2004, Anik F2 was built on a Boeing 702 bus ***NEW***/1047520

Principal applications: voice/data services, broadcast distribution (24 TV channels or 21,600 simplex voice circuits)
Configuration: based on Hughes HS-376 spin-stabilised bus
Mass: 633 kg in orbit
TT&C: as Anik C
Power: 1,000 W from body array.

UPDATED

Anik E Series

Background

Telesat contracted with Spar Aerospace in November 1986 to build two hybrid Anik Es at costs of C$85.9 million and C$68.1 million, respectively, excluding performance incentive payments. Marketed directly as improved broadcast and business services satellites. The Anik E series were Lockheed Martin AS5000 model satellite bus configurations. The first in this series was launched on 2 March 1991 as the Astra 1B owned and operated by SES, the second in a network of four satellites. The combination of its C/Ku-band service requirements within a single system has provided considerable economies; each E replaces one C and one D model. Technical problems during 1989, requiring design changes in the liquid apogee engines and the C-band solid-state power amplifiers, caused delivery and launch delays. E2's service got started 3 months late after its C-band antenna refused to deploy, probably because of thermal snagging. It was finally released 3 July after an unprecedented series of manoeuvres.

E2 provided broadcast services over east/west beams, E1 carried only national beams. E1 suffered a catastrophic failure of a momentum wheel assembly and E2 lost both, probably as a result of electrostatic discharges built up during 10 days of geomagnetic disturbances. E1 returned to service after 7 hours using MWA 2. E2's traffic was switched to E1 and Galaxy 6, and then to Telstar 301. Engineers concentrated on installing software to use thrusters for attitude control, which would reduce E2's life by about 3 months. Full control returned 20 July 1994 and commercial services resumed 1 August. E2 is not now operating.

Specifications

Launched: E2 4 April 1991 by Ariane 44P (entered service 1 September 1991, collocated with C1); E1 26 September 1991 by Ariane 44P
Locations: E2 111.1° W, E1 118.7° W GEO
Design life: 13.5 years (E2 12 years projected)
Contractors: Spar Aerospace (prime, bus from GE Astro-Space); David Florida Lab (final integration/testing)
Transponders: 24 (plus 6 backup, 5-for-4) 11.5 W SSPA 6/4 GHz up/down 36 MHz beams. All-Canada beam, with extended coverage to all of continental USA/Alaska with horizontal polarisation, and to northern states with vertical polarisation. Edge EIRP 35.5 dBW, 37.5 dBW for specified cities. 4-for-2 receiver redundancy

16 (plus 2 backup) 50 W TWTA 14.003–14.497/11.703–12.197 GHz up/down 54 MHz Ku-band beams: National (E1 carries only National), East (vertical polarisation) and West (horizontal). Seven E2 West channels can be switched to the East or National beams; two East-beam channels can also be switched to the National beam. Edge of coverage EIRPs: 44.0 dBW (regional beams), 40.5 dBW (National beam) increasing to 49-44 dBW for specified cities
Configuration: 3-axis GE Satcom 5000 bus. Two large deployable antennas, each generating horizontally/vertically polarised beams. C-band antenna steerable by ground command. Employs GE SCOTS integrated liquid propulsion systems a 454 N TRW bipropellant apogee injection motor sharing hydrazine tanks with mono AOCS thrusters. The two redundant sets of thrusters comprise 12 catalytic versions for attitude control, east/west station-keeping to within ±0.05°, spin-up/nutation control, and four high performance Electrothermal Hydrazine Thrusters for NSSK
Mass: 2,932 kg at launch, 1,781 kg on-station BOL, 1,335 kg (dry)
TT&C: as Anik C
Power: 3,888 W EOL, 4 × 56 Ah Ni/H_2 batteries.

UPDATED

Anik F Series

Background

On 27 November 1997, Telesat Canada announced that it had obtained government approval for two orbital slots to launch Anik F series satellites to replace current Anik E satellites from 2000. Telesat

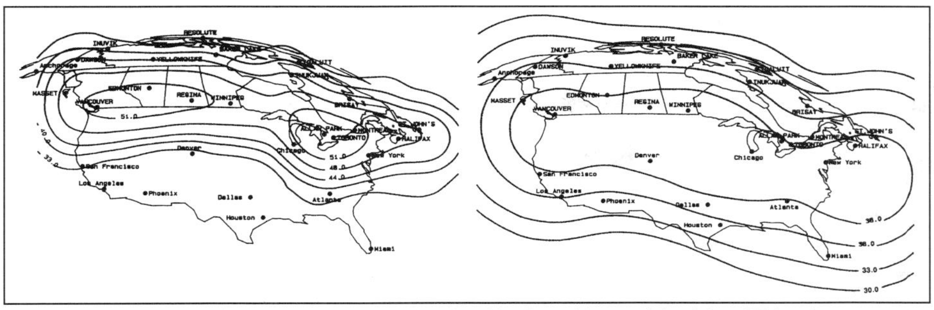

Typical Anik E2 EIRP (dBW) contours. Left: Ku-band national horizontal beam; right: C-band (Telesat) 0517394

Anik E series satellites used the Lockheed Martin AS5000 satellite bus (Lockheed Martin) ***NEW***/1047519

selected Hughes 702 series as the platform for its Anik F series, each of which would carry 84 active transponders to provide telecommunications service for North and South America. The next generation Anik F series adopts the Hughes HS702 bus with a payload of 36 C-band and 48 Ku-band transponders. Anik F1 weighs 4,600 kg and was launched on 21 November 2000 by Ariane 44L. Anik F2 was ordered from Boeing Satellite Services in April 2000. It has 114 transponders consisting of 50 Ka-band, 40 Ku-band and 24 C-band. BSS-702 weighs 5,910 kg at launch, scheduled for 2004 on an Ariane 5, and is designed for 15 kW EOL.

Specifications

Launched: F1 21 November 2000 by Ariane 44LP
Location: 107.3° W

Anik F series are an on-going family of high-powered satellites, F2 being built by Boeing for communication services throughout North America (Boeing) ***NEW***/1047520

Design life: 15 years
Contractor: Boeing Satellite Systems
Transponders: 36 (8 redundant) 40 W TWTA C-band for North and South America via two radio antennas, 48 (10 redundant) 115 W TWTA Ku-band for North and South America. One west and one east antenna for North America Ku-band and C-band coverage, respectively
Configuration: Boeing 702 bus 3-axis stabilised. Two 2.2 m reflectors for South American Ku-band and C-band coverage. One west and one east reflector, each 2.4 m, for North American Ku-band and C-band coverage. One 433 N liquid propellant apogee motor, 10 × 43.3 – 26 N × IPS station-keeping thrusters. Stowed size 4 × 2.1 × 3.4 m; deployed size 40.4 m across solar arrays and 9 m across antennas.
Mass: 4,710 kg at launch, 3,015 on-orbit BOL
Power system: Two solar wings each carrying 5 panels of dual-junction GAaes solar cells, 17.5 kW BOL and 15 kW EOL with 56-cell NiH batteries.

UPDATED

MSAT (Hockeypuck)

Current status

MSAT access is provided via Infosat, providing business and government clients for office WAN down to individual users.

Background

Telesat Mobile Inc began operations in 1988 with the mandate to construct and operate MSAT, Canada's first commercial, mobile-communications system. Telesat Canada (80 per cent) and Japan's ITOCHU Corporation held the only shares. Telesat Mobile Incorporated announced in the spring of 1993 that it had been unable to secure additional financing in the face of cost increases. Telesat Canada was unable to continue funding under the circumstances. Accordingly, on 5 April 1993, the company filed a 'Notice of Intention to Make a Proposal' pursuant to the Bankruptcy

MSAT will provide mobile communications over North America (Spar Aerospace) 0517099

and Insolvency Act, in order to provide the time to pursue the possibilities of a plan for reorganisation.

Canada based TMI launched its initial service offerings in spring 1990 through leased capacity on an Inmarsat satellite. Known as KITs (Keep In Touch), these provided full two-way digital messaging, automatic vehicle location and fleet management services for Canadian users throughout North America. The services ceased as a result of the reorganisation. The North American mobile satellite system (MSAT), began providing the US and Canada with services in 1995. TMI and American Mobile Satellite Corporation signed contracts with Spar Aerospace in 1990 to build their respective satellites for the initial system. MSAT-I operates at 106.5°W with MSAT-2 (AMSC-1) at 101°W. MSAT-2 was launched April 1995 followed by the Canadian built MSAT-1 in April 1996. Each satellite has capacity to support up to 2,000 simultaneous radio channels.

Specifications

MSAT 1

Launch: 20 April 1996 by Ariane 4 from Kourou, French Guiana
Location: 106.5° W geostationary
Design life: 12 years
Contractors: Hughes Space & Communications Co (management, bus) and Spar Aerospace (payload, integration, testing)
Transponders: 16 (plus 4 backup in 10-for-8 open ring) 35W SSPA 1.6315–1.6465/1.530–1.545 GHz up/down and 1.6465–1.6605/1.545–1.559 GHz up/down in 2 × 8 matrix feeding 6 elliptical spot beams (4 North America, a Caribbean beam for Puerto Rico, Virgin Islands and Mexico, and Alaska/Hawaii. Aggregate edge of coverage EIRP 56.5–57.3 dBW, circular polarisation. Bandwidth capacity for 2,000 5 kHz channels. Each '35 W SSPA' comprises four 20 W SSPAs in parallel but operated at 35 W for required 16 dB noise/power ratio

1 (plus 2 backup) 110 W TWTA 13.000–13.15 and 13.20–13.250/10.750–10.950 GHz up/down Ku-band 36 dBW feeder/return link in single North American beam
Principal applications: mobile radio, telephone, data, aeronautical, maritime, safety/distress, position location, wide area paging. High quality voice, packet and circuit switched data
Configuration: Hughes' HS-601 3-axis bus, 2.3 m box-shaped body, 3.63 m diameter stowed, 4.44 m stowed ht, twin solar arrays spanning 20.96 m, 18.9 m across deployed antennas. Two 4.9 × 6.7 m 20 kg graphite mesh reflectors deployed from east/west faces provide separate L-band transmission/receive, respectively, illuminated by 23 cup dipoles. Single 76 cm Ku-band shaped reflector for North America coverage.
Mass: 1,710 kg on-station BOL, 2,855 kg at launch (1,330 kg dry)
AOCS: onboard control processor for control of momentum wheel assembly. Unified 12 × 22 + 1 × 490 N bipropellant thrusters (NTO/MMH in four tanks).
Antenna pointing: 0.125° accuracy using Earth sensors
TT&C: two facilities for redundancy and security: Satellite Operations Centre (satellite control) and Network Operations Centre (network control and channel access on demand)
Power system: twin 9 m long 3-panel solar arrays producing 3.6 kW BOL, 3 kW EOL from K43/4 large

MSAT 1's two 4.9 × 6.7 m graphite mesh antennas weigh only 20 kg each (Communications Research Centre) 0517387

area silicon cells; 160 Ah Ni/H_2 battery provides 100 per cent eclipse protection
Thermal control: dissipation by north/south radiators, with heat transported by heat pipes.

China

Apstar Series

Background

The Chinese government, which owns 75 per cent of The Asia Pacific Telecommunications Satellite Company launches and maintains the APstar satellites. APT plans to deliver a three satellite system for the Asia-Pacific region. The consortium signed a contact with Hughes Communications International Incorporated on May 25 1992 to build the APStar 1, and the Macau ground station for US$135 million. Hughes increased the price to US$200 million when APT included the hybrid HS-601 APStar 2 option, primarily for television broadcasting. Apstar also asked Hughes to extend the coverage region to Europe, India and Russia. Apstar 1 was launched 21 July 1994 and was operated until 1999 from 138° E until 1999. Apstar changed its plans when Apstar 2 suffered a launch failure on 25 January 1995 and instead substituted the APStar 1A HS-376 in March 1995. Apstar chose the smaller design because Hughes promised it could be ready in 11 months, rather than the HS-601's 18 months. Apstar 3 (1A) was launched by CZ-3 from Xichang launch site, a near identical twin to Apstar 1, on 3 July 1996.

APStar 1 was joined in 1996 by the near-identical 1A (Hughes) 0517202

The company originally placed APStar 1 at 131° E for operational service, but Rimsat & Japan (CS-3a) complained of signal interference and consequently APT moved Apstar 1 to 138° E. It presently resides at 138.04° E. APStar 4 (2R) was launched on 16 October 1997. APStar 5 will be launched in 2003 at 130° E as replacement for APStar 1. Based on a FS1300 bus it will have 38 60 W C-band transponders and 16 141 W Ku-band transponders. APStar 5 will be launched by LM-3B. In August 1999 Loral Space and Communications Ltd leased the satellite as a replacement for Orion 3.

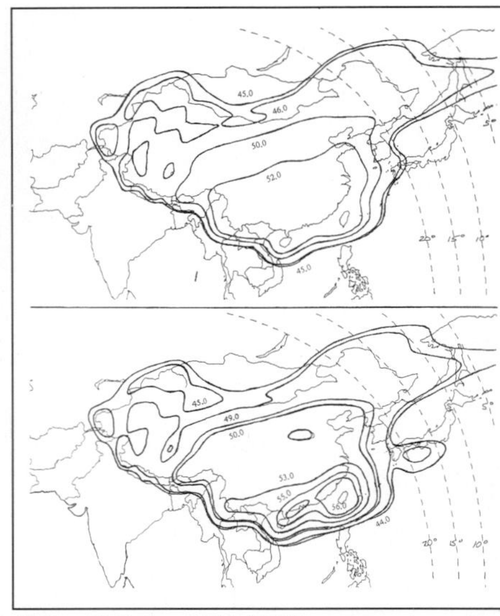

APStar 2R Ku-band EIRP (dBW) coverage; beam 1 top, beam 2 bottom (APT) 0517384

Specifications

APStar 1

Launch: 21 July 1994 by CZ-3 from Xichang, China, People's Republic
Location: 138° E GEO. Initially located at 137° E for testing and was then to move to 131° E for operational service, but complaints from Rimsat & Japan (CS-3a) of signal interference with their nearby satellites resulted in positioning at 138° E, reportedly leased from Tonga
Design life: 12 years
Contractors: Hughes Space & Communications Co (contract with Hughes Communications International, Inc)
Transponders: 24 (plus six backup, 30-for-24) 16 W TWTA 5.850–6.425/3.625–4.200 GHz up/down C-band regional beam, 20 × 36 + 4 × 72 MHz bandwidth, EIRP (dBW) >35 China, 36.0 Hong Kong, 35.9 Philippines, 36.1 Thailand, 36.4 Korea & 35.1 Singapore, orthogonal polarisation
Principal applications: regional telecommunications; VSAT, data & video
Configuration: Hughes' HS-376 platform, cylindrical body 2.16 m diameter total height 6.59 m
Mass: 1,383 kg at launch, 557 kg dry
AOCS: spin-stabilised at 50 rpm by hydrazine thrusters. ±0.05° E/west and north/south

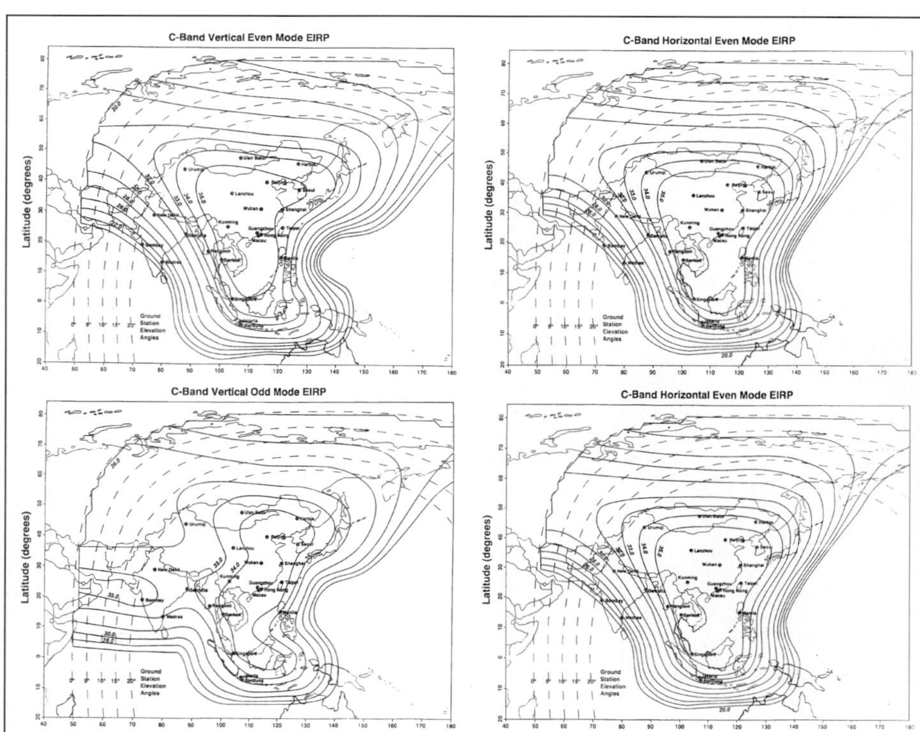

APStar 1A EIRP contours (dBW). Odd-numbered transponders are stretched for Indian coverage (APT) 0517383

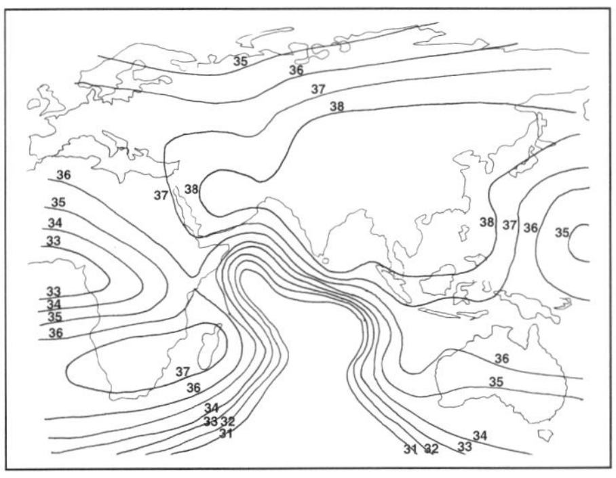

APStar 2R C-band EIRP (dBW) coverage (APT) 0517385

station-keeping; antenna pointing accuracy 0.021° pitch & roll & 0.199° yaw
TT&C: through Macau station. Command uplink at 6,422 MHz; telemetry down at 4,198.125 and 4,199.625 MHz.
Power system: solar array mounted on cylindrical body and extension provide 1,070/970 W at solstice/equinox EOL.

APStar 3
Launch: 3 July 1996 by CZ-3 from Xichang, China, People's Republic
Location: planned for 76.5° E GEO
Transponders: as APStar 1 but additional feedhorns extend coverage to all India (see EIRP contour diagram) Other specifications as for APStar 1.

APStar 4 (2R)
Launch: 16 October 1997 by CZ-3B from Xichang, China, People's Republic
Location: planned 76.5° E GEO
Design life: 15 years minimum
Contractors: Space Systems/Loral (prime)
Transponders: 28 (plus 12 backup, 36-for-28) 60 W TWTA 5.850–6.425/3.625–4.200 GHz up/down C-band regional beam, EIRP 38 dBW main region (see map), 27 × 36 MHz + 1 × 30 MHz bandwidth, linear orthogonal polarisation
16 (plus 8 backup, 24-for-16) 110 W TWTA 14.00–14.50/12.25–12.75 GHz up/down Ku-band China beams 1&2, EIRP see map, 15 × 54 MHz + 1 × 36 MHz bandwidth, linear orthogonal polarisation. Beam 1 has eight even-number channels; beam 2 eight odd number channels; the four horizontal polarisationbeam 2 channels can switch to beam 1
Principal applications: China (Ku) and regional services
Configuration: 3-axis stabilised SS/L FS-1300 2.41 × 2.58 × 2.20 m box-shaped bus with twin solar wings on North/South sides, 27.3 m span
Mass: 3,700 kg at launch, 1,415 kg dry
AOCS: 3-axis bias momentum system; single Marquardt R-4D 490 N apogee thruster and two sets of six 22 N RCS thrusters, common NTO/MMH propellant system
Antenna pointing: 0.13°
TT&C: as APStar 1/1A
Power system: twin 4-panel GaAs wings providing 8,125 W BOL. Ni/H_2 battery provides 100 per cent eclipse protection.

AsiaSat Series

Current Status

The Asiasat 3 satellite, a fully functional HS601HP, ended up in an incorrect orbit when the fourth stage of the Proton booster, carrying the satellite, failed. The final orbit, achieved after launch on 25 December 1997, resembled a GTO rather than a GEO orbit. On 10 April 1998 Hughes began a series of orbital manoeuvres resulting in a translunar fly-by in an attempt to reposition the satellite in its correct orbit.

The manoeuvre consisted of a series of satellite thruster firings by 7 May 1998 converting the elliptical orbit of 350 km by 36,000 km into a circumlunar trajectory. This is the first commercial use of the moon and the first lunar mission attempted by a non-government entity.

When owners Asia Satellite Telecommunications Co Ltd of Hong Kong filed an insurance claim of $215 million, Hughes Global Services, Inc obtained title for the salvage operation and renamed the satellite HGS-1. HGS would share any profits with etc insurers.

The first encounter on 13 May carried the satellite within 6,200 km of the lunar far side before looping back to Earth. Engineers adjusted the Earth encounter to effect a second fly-by of the Moon at a distance of 34,000 km on 6 June. Thrusters were fired at the second Earth fly-by to place HGS-1 in a 46 h orbit of 36,000 km by 82,000 km. Subsequent thruster firings settled it into a geostationary orbit at 36,000 km.

AsiaSat 3 is planned for October 1997 (Hughes) 0517402

AsiaSat 3SA was launched by Proton K on 21 March 1999.

AsiaSat 4 was scheduled for launch in 2003 by Atlas to a position at 122° E as a replacement for AsiaSat 1. Satellite assembly at Boeing Satellite Systems officially began in September 2000. Based on an HS601HP bus, AsiaSat 4 will carry 20 transponders supporting Ku-band BSS services for DTH services to Hong Kong and 28 C-band transponders serving Asia, the Middle East, the CIS and Australasia in the 6/4 GHz band.

AsiaSat 1

AsiaSat 1 was manoeuvred off-station over 105.5° E and relocated to 122° E during June–August 1999. The satellite was manoeuvred away from this location at the end of January 2003, and was still drifting in mid-March 2003.

AsiaSat G

In order to fill a gap in its communications requirements, AsiaSat leased the Russian Gorizont 30 satellite which originally had been leased to RIMSAT. During December 1997 Gorizont 30 was relocated to 122° E where it was operated for nearly two years as 'AsiaSat G'. The satellite was relocated to 142° E in June 1999 when AsiaSat's agreement to use the satellite terminated.

Background

Lockheed Martin Astro Space, in competition with Hughes and Matra Marconi signed a contract for a hybrid AsiaSat 2 satellite on 26 October 1992. The contract had a potential worth of US$133 million

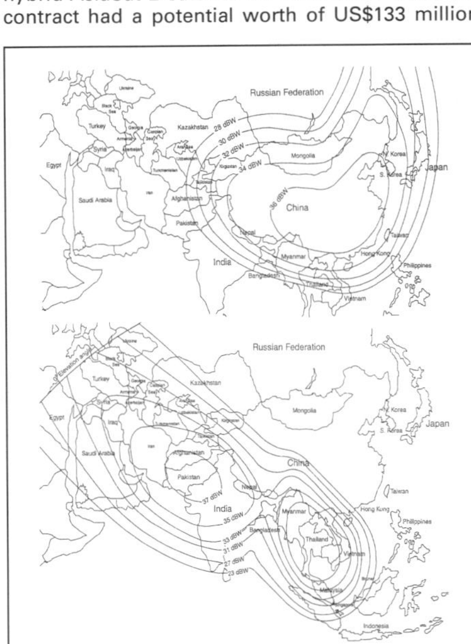

AsiaSat 1 provides two beams. The northern beam covers China, while the southern footprint is optimised for Thailand and Pakistan (AsiaSat) 0517203

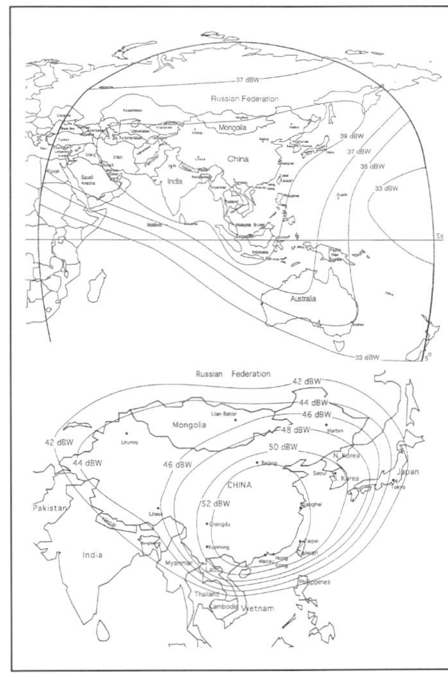

AsiaSat 2 dBW contours: C-band (top) and Ku-band (bottom) (AsiaSat) 0517204

if China exercised all of the options including a second satellite purchase. China paid COMSAT to give additional technical support.

The CZ-2E launch vehicle received the nod to put the satellites in space, but China withheld a launch decision until 1995 to allow time to determine the cause of the Telstar 402 (which used the same bus) and CZ- 2E failures to be identified. The first CZ-2E/EPKM launch vehicle combination built by the Chinese put AsiaSat 2 in space on 28 November 1995. It commenced commercial operations in January 1996.

With success assured, the consortium backing AsiaSat took out a US$220 million loan on 16 February 1996 to finance AsiaSat 3, and then signeda contract with Hughes Space & Communications International for the manufacture and on-orbit delivery of an HS-601 HP. On 22 August 1996, a US$250 million loan facility entered effect with 12 international banks called upon to refinance the existing loans for AsiaSat 1, AsiaSat 2 and AsiaSat 3 as well as to support future capital expenditure.

AsiaSat 3 carries 28 C-band and 16 Ku-band high-powered linearised transponders. The C-band footprint can cover Asia, the Middle East, the CIS and Australasia. There are also two separate footprints providing high-powered Ku-band coverage. In addition, a steerable Ku-band beam provides added market flexibility.

AsiaSat 4 has a launch delivery schedule of late 2001 when AsiaSat 1 reaches the end of its

After the failure of the geosynchronous placement motor on AsiaSat 3, manufacturer Hughes Space and Communications Company chose to salvage the satellite by repositioning it with a lunar gravity assist manoeuvre and onboard station-keeping propellant (Hughes Space and Communications Company) 0022572

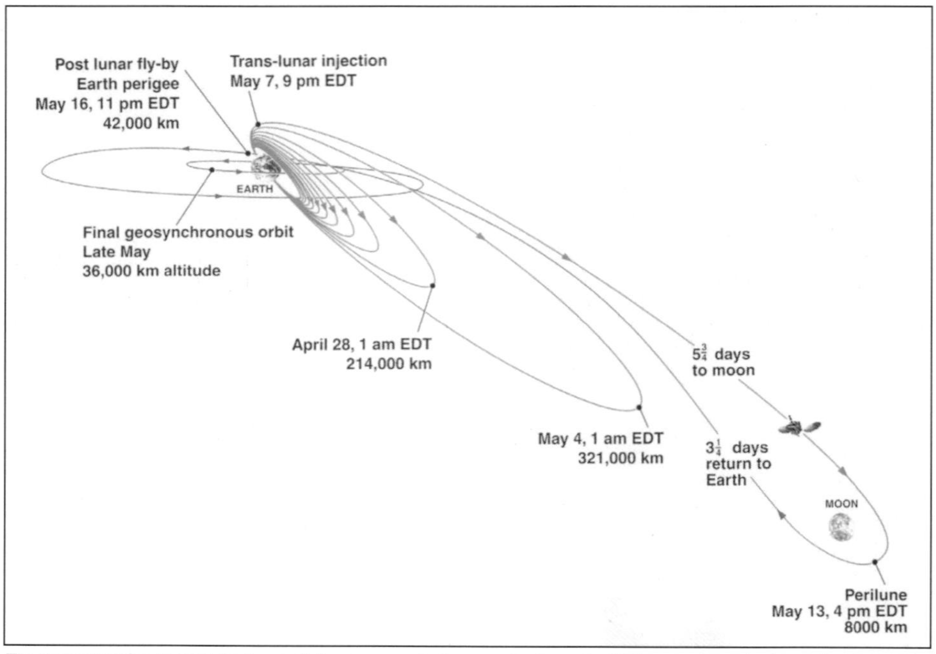

The trajectory for Asiasat 3's Lunar flyby rescue mission required nearly 45 days and used a large fraction of the satellite's station-keeping propellant. All dates shown on the diagram occurred in 1998 (Hughes Space and Communications) 0022575

planned life. Asia Satellite Telecommunications selected Hughes to build a replacement for the AsiaSat 3 after its launch was unsuccessful on 25 December 1997. The AsiaSat 3S is an exact replica of AsiaSat 3 and was launched by Proton on 22 March 1999 toward its operating slot at 105.5° east. In November 1998, Hughes was contracted to build a backup, designated AsiaSat 3SB until its predecessor had been successfully launched subsequent to which the order has been converted to a future AsiaSat requirement.

AsiaSat's ultimate holding company went public in June 1996. Both the Hong Kong and New York Stock Exchange list it with 31 per cent of the company's shares held by the public. The remaining shares reside equally with Cable and Wireless PLC, China International Trust and Investment Corporation (CITIC) and Hutchinson Whampoa.

Specifications

AsiaSat 1

Launch: 7 April 1990 by CZ-3 from Xichang (launched initially as Westar 6 by STS-41B, 3 February 1984; retrieved by STS 51-A in November 1984)
Location: 105.5° E geostationary
Design life: 9–10 years
Contractor: Hughes Space & Communications Co
Transponders: 24 (plus six back-up) 8.2 W TWTA 5.925–6.425/3.700–4.200 GHz up/down C-band, 12 each in two beams (see map), 36 MHz bandwidth, China 32–36 dBW, Thailand 35–37 dBW, Pakistan 35–37 dBW, Korea 35 dBW, Japan 30–34 dBW, Singapore 32 dBW, linear polarisation (12 V+12 H)
Principal applications: TV, radio, teleconferencing and VSAT links. TV users include Star TV, China, Mongolia, Myanmar, Pakistan. Each transponder can handle 3,360 telephone channels via a 7 m station. TVRO diameter 2.6 m for 36 dBW
Configuration: Hughes' HS-376 platform, cylindrical body 2.16 m diameter, total height 6.59 m
Mass: 1,280 kg at launch, 654 kg on-orbit BOL
AOCS: spin-stabilised at 50 rpm by hydrazine thrusters
Power system: solar array mounted on cylindrical body and extension provide 900 W at BOL, about 700 W at EOL
TT&C: Stanley Earth Station, Hong Kong.

AsiaSat 2

Launch: 28 November 1995 by CZ-2E from Xichang
Location: 100.5° E GEO (arrived 14 December 1995, entered service mid-January 1996)
Design life: 13 years
Contractor: Lockheed Martin Astro Space
Transponders: 24 (plus 8 backup, 16-for-12) 55 W TWTA 5.845–6.425/3.620–4.200 GHz up/down C-band, 20 × 36 + 4 × 72 MHz bandwidth, 40 dBW EIRP max, linear orthogonal polarisation
9 (plus three backup, 12-for-9) 115 W TWTA 14.000–14.300/12.200–12.500 GHz up/down Ku-band, 54 MHz bandwidth, 53 dBW EIRP max, covering Hong Kong, China, Taiwan, Japan, Korea, linear orthogonal polarisation
Principal applications: TV and radio distribution, telecoms, VSAT
Configuration: series 7000 platform, 3-axis box-shaped bus with twin solar arrays spanning 23 m
Mass: 3,351 kg at launch
AOCS: nominal pointing accuracy ±0.1° pitch/roll, ±0.25° yaw. 1.8 kW hydrazine arcjets provide NSSK, China's EPKM perigee kick motor, Liquid Apogee Engine provides GEO insertion, Reaction Engine Assemblies for attitude control
Power system: twin 4-panel wings totalling 48.6 m^2, providing 4,780 W EOL, supported by Ni/H_2 battery
TT&C: Stanley Earth Station, Hong Kong.

AsiaSat 3

Launch: 25 December 1997 by Proton-K from Baikonur, Kazakhstan. Replacement AsiaSat 3S launched 21 March 1999.
Location: 105.5° E GEO (AsiaSat 35)
Design life: 15 years
Contractor: Hughes Space & Communications
Transponders: 28 (plus 6 back-up, 34-for-28) 55 W TWTA 5.845–6.425/3.620–4.200 GHz up/down C-band, 36 MHz bandwidth, 41 dBW EIRP maximum, coverage as AsiaSat 2, linear orthogonal polarisation
16 (plus 4 backup, 20-for-16) 138 W TWTA 14.000–14.500/12.25/12.75 GHz up/down Ku-band, 54 MHz bandwidth, 55 dBW EIRP max, three beams (North Asia, East Asia, steerable), linear orthogonal polarisation
Principal applications: TV distribution, telecoms, VSAT
Configuration: improved HS-601 HP (High Power) platform, featuring GaAs solar arrays, enlarged payload and lightweight contoured reflectors
Mass: 3,480 kg at launch, 2,560 kg on-station BOL
AOCS: Bipropellant system.
Power system: twin solar wings of four panels providing 8.7 kW BOL (7.8 kW total required); 350 Ah, Ni/H_2 battery provides 100% eclipse protection
TT&C: Stanley Earth Station, Hong Kong.

AsiaSat 3SA

Launch: 21 March 1999 by Proton K from Baikonur as replacement for AsiaSat 3 launched 25 December 1997
Design life: 15 years
Location: 105° E
Contractor: Hughes Space and Communications
Transponders: (as AsiaSat 3) plus two fixed 140 W ku-band
Mass: 3,465 kg at launch, 2,530 kg BOL, 1,770 kg dry.

UPDATED

With fully extended solar array panels, AsiaSat nears eclise (Boeing) ***NEW***/1047526

Lunar Programme

Current Status

Following the launch of China's first manned flight in October 2003 the country has become more specific on long term goals and projects. A lunar programme is asserted but there is little evidence of this being anything other than a determination to send unmanned spacecraft to the lunar surface. China is working toward a lunar orbit mission but the date of launch is uncertain. Human flight to earth's nearest celestial neighbour is probably beyond its capabilities for at least another 15 years.

Background

In January 1995, the Xinhua news agency announced that China had started preparations to launch a spacecraft to the Moon in about 2000. Professor Jiang Jingshan, the director of the space centre under the Chinese Academy of Sciences, was quoted as saying that China had started preparations for a lunar exploration programme, and that a scientific payload would be launched as the first phase of a lunar programme.

Clearly, no such launches were conducted up to 2000 – or indeed up to the present, and one might wonder whether the reported comments were a result of the preliminary study programme rather than a fully funded programme.

The existing Chinese launch vehicles, CZ-3A and CZ-3B, can be used for a lunar programme and two launch approaches can be considered. One would parallel the launch of the European Giotto spacecraft on top of an Ariane-1 in 1985: in this instance the Ariane-1 simply performed its standard launch profile to geosynchronous transfer orbit and then the spacecraft itself performed the actual manoeuvre into a trans-lunar orbit. The advantage of this approach is that the comparatively large dry mass of the launch vehicle's final stage does not have to be accelerated all of the way to escape velocity and above.

The second approach is, historically, the normal one for deep space launches, where the final stage of the launch vehicle accelerates itself and the spacecraft to a deep space or escape trajectory but, of course, this means that the dead mass of the rocket stage is replacing what could be useful spacecraft payload.

Some preliminary calculations can be performed to show the approximate masses of spacecraft which various Chinese launchers could propel to the Moon (TLI), and these are shown below together with the GTO payload capabilities for comparison. The CZ-3 in particular is greatly affected by having to accelerate its third stage to the trans-lunar trajectory.

CZ-3	GTO	1,400 kg
	TLI	725 kg
CZ-3A	GTO	2,320 kg
	TLI	1,415 kg
CZ-3B	GTO	4,850 kg
	TLI	3,300 kg

Some Chinese comments, made in early 2003, might indicate that the CZ-3 vehicle is no longer operational.

When the CZ-5 vehicle becomes available, the most powerful variant should be capable of launching more than 10 tonnes into a trans-lunar trajectory (see separate entries detailing the many CZ-5 variants).

Project Chang'e

In March 2003 China announced Project *Chang'e* to launch unmanned spacecraft to the Moon. The name refers to a young fairy who flies to the Moon in Chinese legends. (The name in Chinese is two characters '*Chang*' and '*e*' written as a single word: the apostrophe is introduced in the English transliteration so that the name is not confused with the word 'change'.)

Project *Chang'e* has not received government funding but, according to Ouyang Ziyuan, described as the chief scientist of China's lunar programme, the first launch could come approximately two and a half years after the official go-ahead being given.

Luan Enjie, the director of China's National Space Administration (CNSA), has said that the first phase of unmanned lunar exploration should be completed by 2010.

After years of preparation, China's unmanned lunar programme is now seen as having three phases: orbiting the Moon, landing on the Moon and returning samples of lunar rock to Earth.

Chang'e – Phase 1

The authorisation and funding of the first phase of the *Chang'e* programme was announced on January 7, 2004 by a spokesman from the Chinese Commission of Science, Technology and Industry for National Defense (CSTIND). The funding for the programme was said to be 1.4 billion yuan (US$170 million). The Chinese have talked about *a* lunar satellite, but it would be reasonable for a series of launches (2–3 spacecraft ?) to be planned.

On 27 February 2004 Luan Enjie was announced as the 'chief commander' of the *Chang'e* programme. The chief scientist has been identified as Ouyang Ziyuan, a member of the Chinese Academy of Sciences.

The maiden launch is expected in late 2006 and the spacecraft will be launched atop a CZ-3A vehicle. Launch will initially be to a geosynchronous transfer orbit, then the spacecraft itself will perform three perigee manoeuvres which will raise apogee from a ~12 hour period GEO to a 24-hour, then 48 hour period orbit and finally to a trans-lunar orbit.

The transit time to the Moon will probably be about 3–4 days and the spacecraft will initially enter a~12 hour period orbit around the Moon, with periselene of 200–300 km over the lunar far side. The orbit will be lowered to initially one with an orbital period of 3.5 hours and then to a circular one.

The *Chang'e* lunar orbiter will be based upon the existing DFH-3A satellite bus (described elsewhere in *Jane's Space Directory*) with suitable modifications for operating in the lunar environment.

The objectives of the Phase 1 programme will be to obtain three-dimensional images of the lunar surface, remotely analyse the lunar surface composition, probe the depth of the lunar soil and investigate the environment between the Earth and Moon.

The Director General of the experiments, to be carried by the lunar orbiter, has been identified as Sun Huixian of the Center for Space Science and Applied Research (part of the Chinese Academy of Sciences). Sun has displayed two sensors to be carried aboard the lunar orbiter: a gamma-ray spectrometer and an X-ray spectrometer.

Development of the experiments was started in July 2003 and the first testing started on 15 December 2003.

Chang'e – Phase 2

The second phase of the *Chang'e* programme calls for the landing of spacecraft on the Moon. Although these flights have not been funded, it is expected that the first landing will be accomplished by about 2010.

The Chinese have discussed the Phase 2 programme in terms of landing roving vehicles on the lunar surface, but it would be surprising if the first launch in the programme carried a rover: it is more likely to be a static spacecraft, although it will probably use the same descent stage as a rover. The Chinese have not specified whether they plan to have the spacecraft enter lunar orbit and then descend to the lunar surface (like the Soviet-era third generation Ye-8 Luna spacecraft) or make direct descents to the lunar surface from a trans-lunar orbit.

Articles in *Aerospace China* have reviewed the use of small rovers rather like the United States' Sojourner (mass 10.6 kg) which flew to Mars aboard the 990 kg Mars Pathfinder spacecraft or the Spirit/Opportunity rovers (mass 174 kg) aboard the 1,062 kg Mars Exploration Rover spacecraft. It is therefore likely that something on this scale is being considered by the Chinese, rather than the large 756 kg Lunokhod 1 rover that the Soviet Union flew to the Moon aboard Luna 17 in 1970.

The launch vehicle for the Phase 2 mission has not been specified, but it would be reasonable for such a mission to be supported by the CZ-3B vehicle. One Chinese report has indicated that one launcher will be the "CZ-2F with an added stage", which is a passable description of the CZ-3B.

Chang'e – Phase 3

The third and final phase of the *Chang'e* programme calls for the automatic return of lunar samples to the Earth for analysis, therefore it could be the equivalent of the Soviet-era Ye-8-5 third-generation Luna sample-return missions. The Soviet spacecraft (like Luna 16, September 1970) had masses of ~5.4 tonnes, and required launches aboard the Proton-K vehicle with the Blok-D fourth stage added.

It is unlikely that the Phase 3 mission can be accomplished using the CZ-3B launch vehicle, but by the time that missions are ready to begin the new-generation CZ-5 class of launch vehicles will be available, and such a mission could possibly be supported by the more capable members of the CZ-5/335 family.

Perhaps the Chinese will accomplish a mission discussed by but not flown by the Soviet Union: using a roving vehicle to gather lunar samples from different sites over a period of some months and then delivering them to a separately-flown Earth-return spacecraft.

Manned lunar exploration

While the Chinese are interested in a manned lunar programme, they have said that there is nothing planned at present. This subject is discussed in more detail in the Chinese 'Piloted Space programmes – Future Manned Programmes' entry.

UPDATED

STTW (Chinasat) Series

Current Status

Chinasat 5 launched as Spacenet 1 by Ariane 1 on 23 May 1984 currently operated by China Telecom and Broadcasting Satellite. This was the first non-military satellite in the Chinasat series.

Background

MBB/CGWIC signed a DM51.2 million contract on 14 July 1987, effective from 8 March 1988, to develop China's communications and broadcast infrastructure and some satellite subsystems. The Chinese Space Agency CAST acts as prime contractor for both space/ground segments. MBB supported the system definition work and built the satellite antennas and solar array mechanical elements.

The result of this collaboration, the DFH-2A/Chinasat-1 flew successfully three times between 1988 and 1990. Each satellite carries four instruments (electron, proton and solar X-ray detectors and potentiometer) to monitor the GEO radiation environment, the satellite's surface charging and solar activity.

DASA and China Aerospace Corp signed a MoU 16 November 1993 to co-operate in building and launching the next generation satellites. These entities formed the EurasSpace joint venture 8 July 1994, with an initial staff of 12, to help produce China's domestic satellites and to bid for international contracts.

When it was completed, 80 per cent of China's first 3-axis GEO satellite came for indigenous sources. US sanctions did create delays. The satellites incorporate Teldix GmbH's SADA solar array drive assemblies. Officine Galileo provided the IRES IR Earth sensor to determine pitch/roll in GEO. M/A-COM Inc had contracted to supply a transponder but President Bush imposed a ban on export of US components for Chinese spacecraft in the summer of 1991. TIW Systems of the US received a contract in 1993 for the TT&C complex in Beijing.

Specifications

STTW-T1

Launch: 29 January 1984 by first CZ-3 from Xichang, initially into 451 × 309 km parking orbit then after 12 revs boosted to 359 × 6,479 km, 36°. First attempt at placing experimental DFH-1 telecom satellite at 125° E STW-1 but the stage 3 failed to re-ignite for GTO insertion. Satellite kick motor was fired as a demonstration.

STTW-T2

Launch: 8 April 1984 by CZ-3 from Xichang to 125° E GEO (STW-1 position). 420 kg on-station BOL, 915 kg at launch; design life 3 years. Provided two 8 W TWTA 6/4 GHz up/down 20 dBW global beams using horn antenna working to 10 m ground stations. The DFH-1 configuration is similar to STTW 2, but was 3.1 m high and employed a de-spun horn antenna. Operational life apparently ended 1987 when allowed to drift off-station.

STTW-1

Launched: 1 February 1986 by CZ-3 from Xichang to 103° E STW-2 position. DFH-2 design; specifications similar to STTW-2. Broadcast to 8 July 1989; corrections made only occasionally during 1990.

STTW-2/Chinasat-1A

Launched: 7 March 1988 by CZ-3 from Xichang
Location: 87.5° E geostationary (initially 88° E)
Design life: 4 years
Contractors: CAST Chinese Academy of Space Technology
Transponders: four 10 W TWTA 6.050–6.425/3.825–4.200 GHz up/down C-band regional beams using parabolic reflector providing 36 dBW centre, 32 dBW beam edge, 36 MHz bandwidth. Payload can handle 5 TV + 3,000 phone calls simultaneously
Configuration: 2.1 m diameter, 1.6 m height spin-stabilised cylindrical bus, including single de-spun parabolic antenna. Total height 3.68 m. 77.6 kg station-keeping propellant. Antenna pointing accuracy 0.6° N/South, 0.42° E/West
Mass: 441 kg on-station BOL, 1,024 kg at launch
TT&C: provided through XSCC Xian Satellite Control Centre
Power system: 351 W provided by 20,000-cell Si (11% efficiency) array on cylindrical exterior, supported by 15 Ah Ni/Cd batteries
Apogee kick motor: indigenous solid propellant

STTW-3/Chinasat-1B

Launched: 22 December 1988 by CZ-3 from Xichang
Location: 110.5° E geostationary (initially 111° E)
Other specifications as for STTW-2.

STTW-4/Chinasat-1C

Launched: 4 February 1990 by CZ-3 from Xichang
Location: 98° E geostationary
Other specifications are the same as for STTW-2.

STTW-5

Launched: 28 December 1991 by CZ-3 from Xichang but short stage 3 GTO burn left it in 219 × 2,451 km, 31.1°. Its kick motor was fired 29 December to attain 205 × 35,087 km, 31.6°. 4th/last flight of DFH-2A/Chinasat-I design. Possibly intended as STTW-2 replacement. Satellite was a ground spare upgraded to flight status.

Zhongxing Series

Specifications

Zhongxing 5 /Spacenet 1

Launched: 23 May 1984 to 102° W by Ariane, was owned/operated by GTE at 120° W until it was sold in December 1992 to China and moved to 115.5° E by June 1993. CNPTAC China National Postal & Telecommunications Appliances Corp bought it for use by Chinasat for about US$30 million. See p326 1995–96 for specifications. It exhausted its propellant in second quarter 1996. Deorbited mid-December 1999. When DFH-3 1 failed to replace it, Chinasat 7 was ordered from Hughes.

Zhongxing 6/DFH-3/Chinasat-III
Launched: 29 November 1994, abandoned in geosynchronous drift orbit.

Zhongxing 7/ChinaStar 7
Launched: 18 August 1996, left stranded in a 27.2°, 200–17,230 km orbit when the second burn of the C2-3 third stage ended prematurely
Mass: 1,200 kg with propellant, Hughes HS376 satellite bus.

Zhongxing 8
Launched: 11 May 1997 using C2-3A from Xichang
Location: 115.3° E
Design life: >8 years
Transponders: six 16 W TWTA 6.2–6.4/4.0–4.2 GHz up/down C-band regional beams. EIRP >37 dBW (allowing 3 m TVRO antennas)
18 8 W SSPA 6.2–6.4/4.0–4.2 GHz up/down C-band regional beams. 35 dBW for telephony (8,000 duplex calls)
Configuration: 3-axis 2.2 × 1.72 × 2.0 m bus, total height 5.71 m with single 2 m diameter reflector deployed, 18.1 m solar array span. Antenna pointing 0.15° pitch/roll, 0.50° yaw. North South/East West station-keeping 0.10°. Liquid bipropellant motor for GEO insertion
Mass: 1,145 kg on-station BOL, 2,230 kg at launch
TT&C: provided through Beijing complex
Power system: two 3-panel solar wings provide 2,000 W BOL/1,700 W EOL.

France

STENTOR Satellite

Background
France approved the Stentor (Satellite de Télécommunications pour Expérimenter les Nouvelles Technologies en Orbite) on 4 October 1994 (simultaneously with Spot 5) to develop and demonstrate a telecommunications platform, aimed at helping French industry compete with the US and Japan. The French government expected to pay FFr1.8 billion for the satellite, plus FFr800 million for launch, operations and ground segment. France cut this by FFr520 million in 1996. CNES heads the project committee, which includes France Telecom and DGA, in close collaboration with two co-contractors including Alcatel and Astrium. Stentor was launched by Ariane 5 on 11 December 2002 but destroyed in a launch vehicle propulsion failure after liftoff.

Specifications
Launch: 5 July 2002 by Ariane 5 from Kourou, French Guiana
Design life: 9 years
Contractors: Three co-primes: Aerospatiale (platform prime, solar arrays, AOCS), Alcatel Espace (payload prime), Astrium (power supply, onboard data management, attitude control, plasma experiments)
Transponders: 1 (plus 1 back-up) 110 W linearised TWTA Ku-band Western Europe beam, 55 dBW EIRP. Phased array antenna with 48 transmission elements (100 W total RF output), reconfigurable coverage and beam hopping
1 L-band providing four European spots, 42 dBW. Ku-band receive
1 20 W TWTA 20.7 GHz Ka-band beacon, 45 dBW France, 44 dBW French Guiana
Principal users: new Ku-band services, L-band air traffic management, Ka-band military link
Configuration: Spacebus platform with Eurostar electronics. NSSK by SEP's modified Russian SPT-100 xenon plasma thrusters (Mk 2 design was to have been employed but will not be ready in time); conventional propellants will provide 3 years life as backup. Deployable radiator
Mass: 2,000 kg at launch
TT&C: via the Toulouse Space Centre working through Ku-band antennas at Aussaguel-Issus near Toulouse
Power system: 2,500 W after 2 years and 1,000 W for the payload from GaAs/Ge cells and lithium carbon batteries.

TELECOM Series

Current Status
Post-Telecom 2 series analysis has yet to reach decisions regarding successors. Probability of leasing from Inmarsat for satellite maritime services.

Background
France authorised France Telecom to build the Telecom satellite system in 1979 to link domestic and overseas territories with television, telephone, digital data and military links. To meet these needs, the quasi-public company immediately ordered the Telecom 1 satellite and then later ordered three Telecom 2s in December 1987 from Matra/Alcatel. The Telecom 2 has three times the power of the Telecom 1. Each carries a Syracuse (Système de Radio Communications Utilisant on Satellite) package for secure Ministry of Defence links. The X-band transponders work through three Earth stations: in Brest (8 m dish), Paris (8 and 18 m dishes and the tracking, telemetry and command) and France-Sud (2, 18 m dishes, including back-up tracking, telemetry and command).

Syracuse 2 expands access to 40–90 cm antennas, permitting 75/200 bit/s telegraph; 2–4/16 kbit/s telephone; 75/2,400/16,000 bit/s data routeing and a secure backup in case of damage to the primary communication centre.

France Telecom may rely in future on leased Eutelsat coverage under privatisation. It is a member of the Satellite Aircom consortium providing aeronautical communications services through the Inmarsat system. Telecom 2A, 2B, 2C and 2D were still operating in February 2002.

Specifications
Telecom 1A
Launched: 4 August 1984, was raised above GEO in late 1992 after being replaced at 8° W by 2A. 1B, launched 8 May 1985, lost both attitude control systems 15 January 1988. 1C, launched 11 March 1988, retired from 3° E in February 1996 after being replaced by 2C.

Telecom 2A
Launched: 16 December 1991 by Ariane 44L from Kourou, French Guiana. In-orbit acceptance tests completed 15 March 1992 at 3° E; moved April to 8° W
Location: 8° W geostationary
Design life: 10.25 years minimum
Contractors: Telecom 2A: Matra (platform, integration), Alcatel Espace (payload), Sodern (Earth sensors), Galileo (Sun sensors), SEP/MMS (propellant tanks), Sagem (rate gyros), MBB (thrusters), Fokker (solar arrays), Eagle Picher (batteries), Teldix (reaction wheels), Thomson (TWTs)
Transponders: 10 (plus 4 back-up; 6-for-4 + 8-for-6) 8.5 W min SSPA 5.925–6.425/3.700–4.200 GHz up/down C-band Atlantic regional beams, 6 × 50 MHz + 4 × 92 MHz bandwidths, circular polarisation. Semi-global coverage at 32.5 dBW EIRP on channels C1-C4: mainland France, Reunion, Antilles/Guiana, St Pierre/Miquelon. Spot coverage: Antilles/Guiana (39.0 dBW; C5A, C6A, C7A), mainland France (42.4 dBW; C5B, C6B, C7B), St Pierre/Miquelon

Telecom 2C in Matra Marconi Space's Toulouse test facilities 0517403

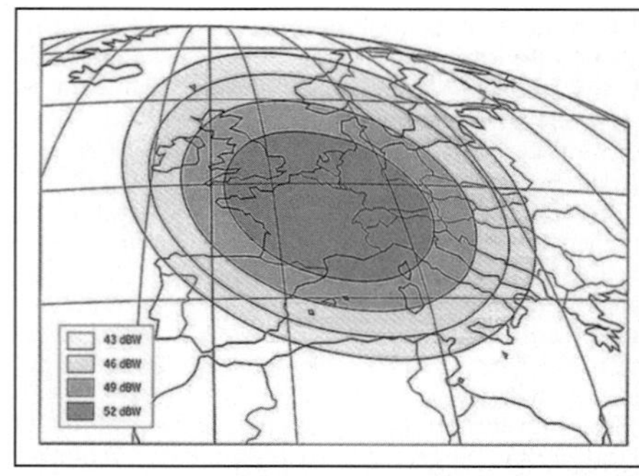

Telecom 2 Ku-band beam coverage 0517198

(39.0 dBW; C5B, C6A, C7A); channels C6A/C7A can be reconfigured onboard between Antilles/Guiana and St Pierre/Miquelon, C5B is divided

11 (plus 4 backup in 15-for-11) 55 W min TWTA 14.00–14.250/12.50–12.75 GHz up/down Ku-band European regional beams, 36 MHz bandwidth, 52.4 dBW min EIRP central zone, 49.4 dBW min EIRP surrounding zone, vertical polarisation. This payload also transmits a 12 GHz beacon signal for ground station pointing

3 (plus 3 backup; two groups, 3-for-2 + 3-for-1) 20 W min TWTA 7.900–8.395/7.250–7.745 GHz up/down X-band beams, circular polarisation, channel X1 60 MHz bandwidth central European beam EIRP 44.0 dBW, X4 40 MHz steerable spot EIRP 31.3 dBW, X5 80 MHz global 28.5 dBW

2 (plus 1 backup, 3-for-2) 40 W min TWTA 7.900–8.395/7.250–7.745 GHz up/down X-band beams, circular polarisation, channels X2/X3 40 MHz bandwidth global EIRP 31.3 dBW
Principal users: domestic/overseas telephony, data, radio, TV, military
Configuration: Matra/BAe Eurostar 2000 bus. Four military antennas: LHCP global horn transmitter, RHCP global horn receiver, 2.2 m diameter offset paraboloid 1.65 m fl for European coverage, and a steerable offset paraboloid spot reflector
Mass: 2,275 kg at launch, 1,380 kg on-station BOL, 1,124 kg dry mass (400 kg payload)
Pointing: antenna pointing accuracy typically 0.13°
AOCS: 3-axis, 12 × 10 N DASA hydrazine thrusters + Marquardt R-4D 490 N bipropellant apogee motor in unified bipropellant propulsion system
TT&C: via the Toulouse Space Centre working through antennas at Bercenay-en-Othe near Troyes and Aussaguel-Issus near Toulouse
Power system: twin 4-panel solar arrays, 2.050 m wide, spanning 22.02 m provide 3,770 W equinox EOL with full eclipse battery protection (78 Ah Ni/H_2). 2,600 W payload power requirement.

Telecom 2B
Launched: 15 April 1992 by Ariane 44L from Kourou, French Guiana
Location: 5° W from mid-1992 after commissioning at 3° E, now 24.98° EOther specifications as for Telecom 2A.

Telecom 2C
Launched: 6 December 1995 by Ariane 4 from Kourou, French Guiana. Replaced 1C February 1996 after testing at 1° E
Location: 5.05° E (replacing 1C)
Mass: 2,283 kg at launch, 1,360 kg on-station BOL, 1,120 kg dry mass (400 kg payload)
Other specifications as for Telecom 2A.

Telecom 2D
Launched: 8 August 1996 by Ariane 4 from Kourou, French Guiana
Location: 8° W (collocated with 2A)
The 11 Ku-band transponders use 14.25–14.50/11.45–11.70 GHz up/down. Other specifications as for Telecom 2A.

TELEDIFFUSION DE FRANCE Series

Current Status
TDF-1 launched 28 October 1988 has been retired and TDF-2, launched 24 July 1990, is still operating.

Background
TDF, created in 1975 as a limited company with the French government as majority shareholder (and a France Telecom subsidiary since 1988), operates the TDF television repeater satellites.

The Eurosatellite consortium was awarded the space segment contract in 1984, and the

TDF 1 during final flight readiness review. The solar arrays have yet to be attached (Aerospatiale) 0517409

company built the original satellites basing TDFs design on Aerospatiale's Spacebus 300 platform. Aerospatiale acted as prime for TDF 1 and 2 and backup for TV-Sat 1 and 2. The French made the decision to build TDF 2 in December 1984. By 1986 the programme verged on cancellation because of developing competition from medium-power systems such as Astra, but by then the l.5 billion francs of the 3.5 billion francs allocated to TDF had been spent. The French government ruled that it should proceed despite the financial difficulties in order to promote D2MAC as the standard for high definition television in the 1990s.

The TDF's under construction underwent minor modifications after TV-Sat 1 failed because one solar array remained stuck in November 1987 and thruster overheating occurred during orbital manoeuvres. At the same time, France decided there would be no TDF 3 and that the future system would lease Eutelsat space. By 1992, only about 35,000 locations had TDF dishes. France cut its annual transponder leasing cost from 80 million francs to 35 million francs. Both TDFs suffered from malfunctions that limited their usefulness. TDF 1's channel 1 transponder failed in August 1989 because of arcing promoted by leaking thruster propellant and TDF shut down channel 17 in September 1990 after 5 months of current variations. TDF 2's two Thomson TWTAs (channels 1/13) failed simultaneously in October 1990, when safety mechanisms irreversibly shut them down.

Specifications

TDF 1

Launched: 28 October 1988 by Ariane V26 from Kourou, French Guiana

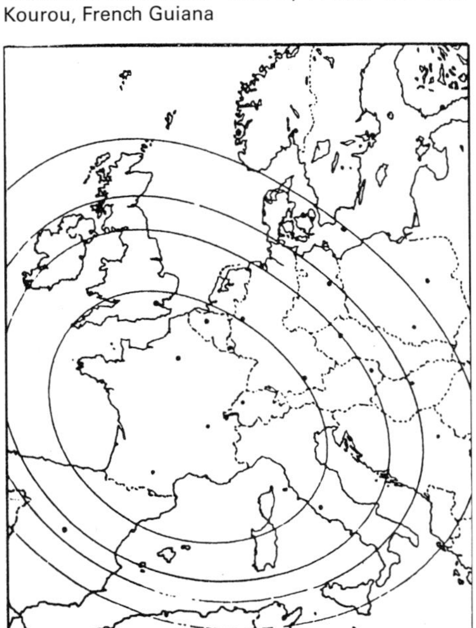

TDF works to 45 cm 70 home receivers in the inner France-centred ellipse 0516998

Location: During August-September 1996 TDF 1 was relocated from 19° W to 21–22° W, and then on 4 October the satellite was boosted out of the geosynchronous orbit band, to retirement
Design life: 8 years (sufficient AOCS propellant after orbital positioning for 9 years operations)
Contractors: Eurosatellite (Aerospatiale/MBB prime, Alcatel Espace payload), AEG (three 260 W TWTAs), Thomson Tube Electroniques (three 230 W TWTAs)
Transponders: 5 (plus 1 backup only for channel 9) 230 W min TWTA 17.3–17.7/11.714–12.045 GHz up/down Ku-band elliptical beams providing 64 dBW min EIRP coverage of France, 27 MHz bandwidth, RHCP
Principal applications: direct TV/radio broadcast channels 1/5/9/13/17 and HDTV demonstrations
Configuration: Spacebus 300 platform, body 1.65 × 2.4 × 7.1 m box-shaped, solar array span 19.3 m, 2 m diameter circular receive antenna, 2.4 × 0.9 m elliptical transmission antenna (9-horn feed), both carbon fibre
Mass: 2,136 kg at launch, 1,318 kg on-orbit BOL
AOCS: 3-axis control by 14 × 10 N MBB thrusters of unified bipropellant system incorporating 400 N apogee kick motor
Antenna pointing: ±0.01° driven by RF sensor detecting beacon signal from control centre
TT&C: orbital positioning, station-keeping and performance monitoring by CNES from Toulouse; programme feeds allocated to TDF's Bercenay-sur-Othe station in the Champagne region
Power system: twin 4-panel solar arrays (43,000 cells) providing 4.3/3.3 kW BOL/EOL (8 years).

TDF 2

Launched: 24 July 1990 by Ariane V37 from Kourou, French Guiana
Location: 19° W geostationary
Mass: 2,096 kg at launch, 1,255 kg on-station BOL, 1,040 kg dry mass
Other specifications as for TDF 1.

Germany

DFS-Kopernikus Series

Background

Beginning as an arm of the West German government, Deutsche Telekom became a stock corporation on 1 January 1995 and began public quotations on the stock market during that year. DT commissioned three satellites in December 1983 at DM81.5 million each, plus launch costs and ground segment equipment. DT reportedly paid 250 million marks for its first DFS Ariane launches. In contrast, the 1990 DFS 3 launch contract awarded to McDonnell Douglas cost 75 million marks.

DFS-Kopernikus, which stands for Deutscher Fernmeldesatellit, employs Ku/Ka-bands for television and cable distribution, telephony, voice, facsimile, data and VSAT applications. A ground network of 32 Ku-band and two Ka-band stations handles 7,500 simultaneous telephone conversations and seven television/stereo radio programmes. Licensed by DT, the DFS operate through Raistang and service the former East

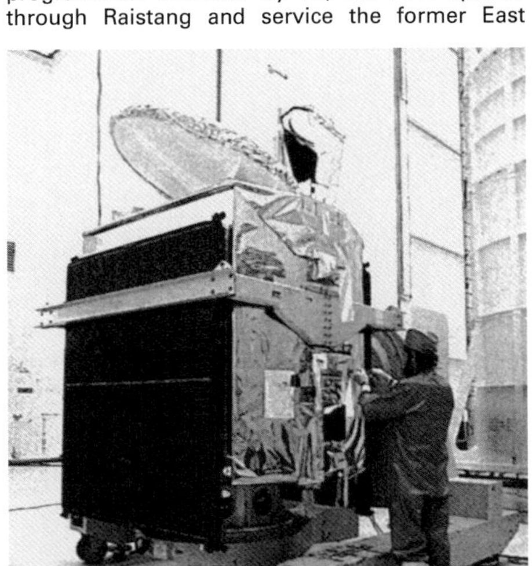

DFS-Kopernikus provides public and business telecom services for Germany; a Ka-band transponder was included for propagation experiments and, ultimately, assignment to commercial operations (DASA) 0516732

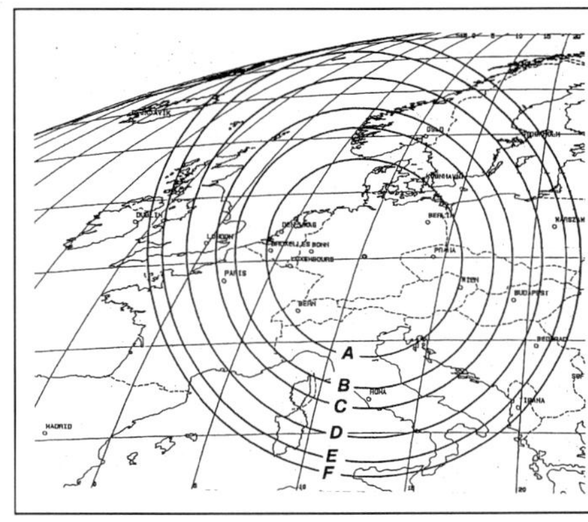

DFS 3 footprint. Antenna diameters (11.5/12.6 GHz): A 1/0.6 m, B 1.3/0.8 m, C 1.5/0.85 m, D 2.2/1.4 m, E 3.35/2.2 m & F 4.35/3.0 m 0516999

Germany as well. At the time of reunification there were four million telephones in East Germany, compared with 40 million in West Germany.

In early 1993, DFS 3 replaced the original DFS 1, which entered service over 23.5° E on 5 June 1989.

DFS satellites are no longer operating.

Specifications

DFS 1

Launched: 5 June 1989 by Ariane and positioned at 23.5° E
Mass: the 1,416 kg prototype model
Current Position: 90.12° E from 33° E (1993–1995) and 23° E (1989–1993)

DFS 2

Launched: 24 July 1990 aboard Ariane V37 from Kourou, French Guiana
Location: 32.3° E from 28° E (1990–1999)
Design life: 10 years (Ariane accuracy left sufficient propellant for 11 years); 67% reliability after 7 years
Contractors: GESAT (MBB, ANT, Dornier, Siemens, SEL), ANT (TWTs), AEG (TWTs)
Transponders: Seven (plus three back-ups) 20 W TWTA 14.25–14.5/12.5–12.75 GHz up/down Ku-band providing 49 dBW min EIRP coverage (54.2 dBW max) of Germany, 44 MHz bandwidth, orthogonal polarisation. Each 1 TV or 450 telephone channels

Three (plus two back-ups) 20 W TWTA 14.0–14.25/11.45–11.7 GHz up/down Ku-band providing 49 dBW min EIRP coverage (53.7 dBW max) of Germany, 90 MHz bandwidth, orthogonal polarisation. Each two TV or 900 telephone channels

One (plus one back-up) 20 W TWTA 29.5–29.6/19.7–19.8 GHz up/down experimental Ka-band providing 48 dBW min EIRP (51.9 dBW max), 90 MHz bandwidth, linear polarisation
Principal applications: national telecom systems, TV/cable distribution, VSAT and business applications, TV-Sat backup. The narrow bandwidths can handle digital data at 60 Mbit/s, the wideband and Ka-band 140 Mbit/s. DFS 3 dedicated to TV/radio broadcasts; DFS 2 to other services
Configuration: ECS/Spacebus 100 platform, span 15.40 m, height 4.15 m
Mass: 1,418 kg at launch, 850 kg on-station BOL, 645 kg dry
AOCS: 3-axis by 14 × 10 N MBB thrusters plus associated 400 N engine for GEO insertion; GEO position maintained to within 0.07°; antenna pointing accuracy 0.16°
TT&C: initial orbital operations/checkout by German Space Operations Centre at Oberpfaffenhofen (15 m antenna), operational control via DT's Usingen Earth Station (18/9.5/4.5 m)
Power system: twin solar wings provide 1,700 W BOL reducing to 1,580 W EOL; two 35 Ah Ni/Cd batteries provide full eclipse protection.

DFS 3

Launch: 12 October 1992 aboard Delta 7925 from Cape Canaveral
Location: 23.45° E from 23° E (1993–1999) and 31° E (1992)
Mass: 1,411 kg at launch
Other specifications as for DFS 2.

TV-Sat Series

Background

Originally, Deutsche Telekom planned to use TV-Sat to provide Europe's first direct broadcast services, offering four television channels through 35–45 cm dishes, collocated with France's equivalent TDF satellites.

An Ariane launched TV-Sat 1 on 20 November 1987, but one solar panel failed to deploy 25 minutes after launch and could not be shaken free with thrusters firings. DT abandoned the satellite in February 1988; and the insurers paid a US$51 million insurance claim. Initially, DT allocated three channels to private broadcasters, including SAT 1 and RTL, with the fourth used by the ARD association of public broadcasters from 6 pm-1 am for TV and the remaining time for 16 digital sound broadcasting channels. DT have conducted telecommunications experiments with TV-Sat 2, leading to improved operating procedures.

DT modified TV-Sat 2/TDF to change the thrusters because of overheating discovered on TV-Sat 1. TV-Sat 2 was launched on 8 August 1989 and was moved from its 19.2° W slot in February 1995 for leasing Norway's Telenor at 0.6° W in March 1995 for the rest of its life. TV-Sat 2 was retired in early September 1999 and deorbited. TV-Sat-1 has been retired.

India

Insat 1 Series

Background

Indian plans for the Insat 1 system originally envisaged two multipurpose satellites, each providing two high-power television broadcast and twelve national coverage transponders, in addition to meteorological services. These two original satellites both ran into difficulties. Orbit inaccuracies forced Insat controllers to use propellant faster than anticipated reducing the satellite's useful life to just 4 months. Then it took Ford and ISRO engineers at India's Hassan Satellite Control Centre several months to deploy the second satellite's solar array. By the end of 1991, 11 transponder channels on Insat 1D, supplemented by two leased on an Indian Ocean Intelsat satellite operated in the Insat system. The Insat system has 140 point-to-point routes with 133 Earth stations of various types.

India purchased a replacement satellite for the lost Insat 1A but a power system failure from a solar array isolation diode short soon removed half of its capacity. However, the meteorological services, powered through either bus, remained unaffected. Then on 22 November 1989 Insat 1C lost Earth lock and India had to abandon it.

India procured an Insat 1B with expanded battery and propellant capacities and christened it 1D. During payload encapsulation the ground crew damaged the satellite and the repairs delayed the launch. It was repaired at a cost of US$10 million. It also sustained US$150,000 worth of damage during the October 1989 California Earthquake.

The failed 1A/1C reportedly produced insurance payouts of about US$70 million each. The long-lived 1B, launched 30 August 1983 by Shuttle, retired in 1991. It served as a spare from 17 July 1991 and from 1992 was used only for experiments. 2B replaced it at 74° E in 1992 and it retired from 93.5° E when 2B arrived August 1993. NSSK had not been provided since August 1989. Insat 1D has now been retired from operations leaving Insat 2DT (formerly Arabsat 1C) at 55 ° E, Insat 2E at 83 ° E and Insat 3B at 83 ° E in service.

Specifications

Insat 1D

Launched: 12 June 1990 by Delta 6925 from complex 17, Cape Canaveral. Assumed primary role 17 July 1990 from 1B

Insat is controlled through Hassan's Master Control Facility, with the two 14 m antennae joined by two 11 m dishes for Insat 2 (DOS) 0517000

Location: 83° E geostationary
Mass: 1,292 kg at launch
Design life: 7 years
Contractors: Ford Aerospace (prime)
Transponders: 12 (plus one back-up, for numbers 11/12) 4.5 W TWTA 5.935–6.425/3.710–4.200 GHz up/down C-band all-India beam, 36 MHz bandwidth, 32 dBW min EOL EIRP over primary coverage area, linear polarisation

Two (plus one backup) 50 W TWTA 5.855–5.935/2.555–2.635 GHz up/down S-band all-India beam, 42 dBW EOL EIRP over primary coverage area linear polarisation

Principal applications: domestic communications on C-band, one S-band channel dedicated to direct TV broadcast and other to five low-level carriers for services such as radio programme distribution, disaster warning, and so on. The Insats also relay 402.75 MHz transmissions from 100 hydrological, meteorological and oceanographic data collection platforms

Spacecraft: 2.18 × 1.42 × 1.55 m box-shaped 3-axis stabilised bus with asymmetrical 5-panel 11.5 m^2 solar wing providing 1,185 W BOL (930 W EOL) to two 12 Ah Ni/Cd batteries and creating total span of 19.4 m. The asymmetrical solar array configuration provides an unobstructed view into space for the radiative cooler of the VHRR; a solar sail is used for balance. A magnetorquer with current coil around the body provides fine control; a unified bipropellant system is used for orbit transfer, station-keeping and attitude maintenance

Mass: 1,152 kg at launch, 650 kg in-orbit BOL (550 kg dry).

Insat 2 Series

Current Status

Insat 2A, Insat 2B, Insat 2C and Insat 2D have all been retired or are no longer operating. Launched on 3 April 1999, Insat 2E is still operational at 55 ° E.

Background

The Department of Space forged ahead with the Insat 2 indigenous satellite following approval of funding in April 1985. Initially, the DOS sized the Insat 2 to fit aboard the Space Shuttle, but the continued launch delays associated with the Challenger accident forced a redesign of the satellite so it could fit on the Ariane, and the first satellite appeared on station in 1992.

India is now close to accomplishing its original goal of complete space communication autonomy with the introduction of the Insat 2 communications payload. India had the four first-generation Insat 1 satellites built in the United States, but the advanced Insat 2 family is now indigenously produced. India's Department of Space (DOS), Department of Telecommunications, Meteorological Department, All India Radio and Doordarshan all participate in the Insat system. Overall co-ordination resides in the inter-ministerial Insat Co-ordination Committee (ICC). The DOS establishes and operates the space segment, with the entire system generating revenues of about 4 billion rupees annually.

The construction of the Insat 2 Test Spacecraft began with approval in April 1985. India expected to launch the satellites by Ariane in 1992/3, followed by three operational satellites beginning 1994/5 by its own vehicles. The design was sized initially for Shuttle/PAM-D2 deployment. Because of increasing demand, India added to the number

Insat 2: the first Indian-built new generation of multipurpose GEO satellites appeared in 1992. India will be providing transponders to Intelsat from 1998. Shown is the Insat 2C/2D configuration, which does not carry the Earth imager and thus has two solar wings (ISRO) 0517137

Insat 2C replaces the meteorological payload with Ku-band and mobile services (DOS) 0517391

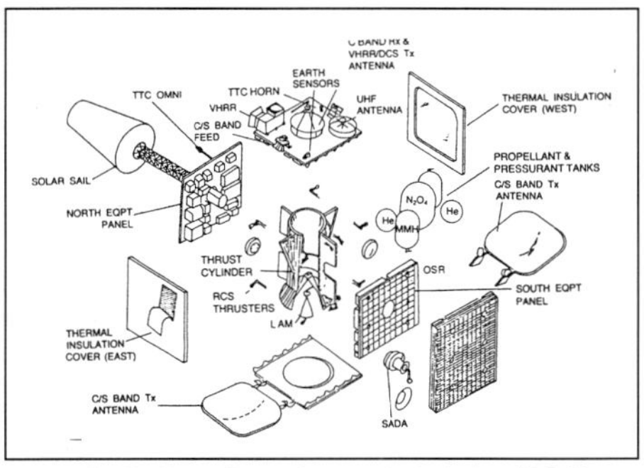

Insat 2A/2B/2E configuration exploded view (DOS) 0517001

of C-band transponders bringing the total to 18 (six at extended C-band) and a 406 MHz transponder adds a distress beacon detection capability. 2C/D had three Ku-band TWTAs and increased C-band power to improve TV coverage and business and remote area services.

A ground station under the direction of the Insat Master Control Facility (MCF) at Hassan Kamataka handles all station-keeping duties through two Satellite Control Earth Stations. One has a 14 m fully steerable antenna and the other a 7.5 m limited steering dish. The facility includes one additional 14 m fully steerable antenna and an Insat 1 Satellite Control Centre (SCC) with associated tracking, telemetry and command equipment, on-orbit checkout equipment, computer facilities and auxiliary power services. The Indian government added an Insat 2 satellites control centre at the Master Control Facility MCF at Hassan, Karnataka for the next generation satellite.

Insat 2D-R has been inducted into service with the acquisition and re-naming of Arabsat 1C.

Specifications

Insat 2A

Launch: 9 July 1992 on Ariane 4 V51 from Kourou, French Guiana (entered service 6 August 1992)
Location: 73.9° E GEO
Design life: 7 years
Contractors: ISRO and Indian companies; HAL Bangalore (structural elements), HAL Hyderabad (C-band SSPAs)
Transponders: 12 (plus one back-up, for numbers 11/12) 4.5 W SSPA 5.930–6.410/3.705–4.185 GHz and six (three back-up; 3-for-2) 6.735–6.975/4.510–4.750 GHz up/down C-band and extended C-band, respectively, all-India beam, 36 MHz bandwidth, 32 dBW EOL EIRP for 16 transponders, 34 dBW for two

Two (plus one back-up, 3-for-2) 50 W TWTA 5.850–5.930/2.550–2.630 GHz up/down S-band all-India BSS beam, 36 MHz bandwidth, 42 dBW EOL EIRP

Single 402.75 MHz Data Collection System transponder

Single 406 MHz Cospas/Sarsat search/rescue package

Principal applications: as for Insat 1
Spacecraft: 193 × 164 × 170 cm box-shaped 3-axis bus around 930 mm diameter central aluminium cylinder with asymmetrical 5-panel solar wing providing 1,400 W at BOL (1,180 W EOL) through 2 × 18 Ah Ni/Cd batteries and creating 23 m span. Reflectors of 1.77 m diameter for S/C-band services located on East/West faces. A 90 cm body-mounted antenna on Earth face performs the C-band receive and DCS/VHRR/S&R data transmit functions
Mass: 1,906 kg at launch, 1,162 kg BOL (911 kg dry); communications payload 179 kg, meteorological 56 kg

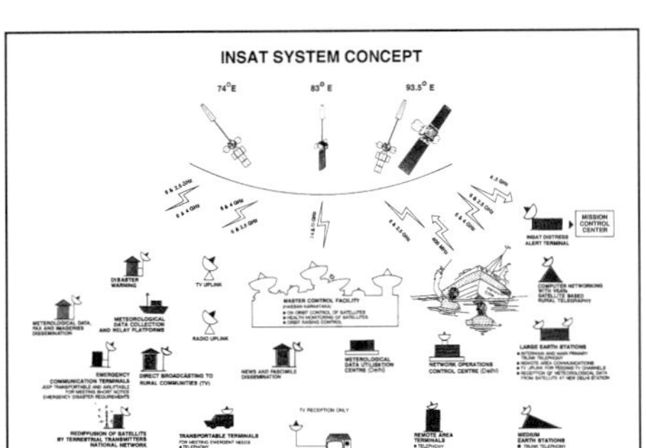

Insat 2 system concept (DOS) 0517218

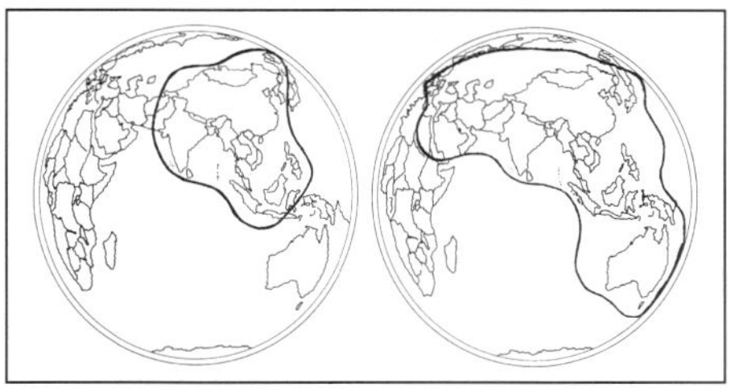

Insat 2E coverage: zonal beam (left) and wide beam (DOS) 0517219

AOCS: unified liquid bipropellant 440 N LAM Liquid Apogee Motor and two redundant networks of 8 × 22 N RCS thrusters. Propellant capacity is 1,015 kg. N_2O_4 carried in 520 l central cylinder inside thrust structure, MMH in two 260 l cylinders, pressurised at 16.1 atmospheres by helium stored at 232 atmospheres in two bottles. Latch valves isolate LAM after apogee operations.

Insat 2B

Launched: 22 July 1993 on Ariane 4 V58 from Kourou, French Guiana. Arrived on-station 6 August, declared operational 10 August
Location: 93.5° E GEO
Mass: 1,931 kg at launch.

Insat 2C

Launch: 6 December 1995 on Ariane 4 from Kourou, French Guiana. LAM number 1 firing 01.28–02.32 GMT 8 December raised GTO to 14,420 × 35,790 km, 1.8°; LAM 2 08.32–09.02 GMT 9 December 31, 894 × 35,786 km, 0.5°; LAM 3 186 s starting 05.20 GMT 11 December, established GEO. The solar arrays and antennas deployed 12 December
Location: 93.5° E GEO (collocated with 2B)
Transponders: 12 (2 × 50 W TWTA + 7 × 10 W SSPA + 3 × 4 W SSPA; 3-for-2 channel 11/12) 5.930–6.410/3.705–4.185 GHz up/down C-band all-India beam, 36 MHz bandwidth, 36+36+32 dBW edge EIRP

6 (2 × 10 W + 4 × 4 W SSPAs, 3-for-2), 6.735–6.975/4.510–4.750 GHz up/down extended C-band all-India beam, 36 MHz bandwidth, 35+32 dBW edge EIRP

3 (plus one back-up, 4-for-3) 20 W TWTA 14.252–14.493/11.452–11.693 GHz up/down Ku-band all-India beam, 1 × 77 MHz + 2 × 72 MHz bandwidth, 41 dBW edge EIRP

Two (plus one back-up, 3-for-2) 50 W TWTA 5.850–5.930/2.550–2.630 GHz up/down S-band all-India BSS beam, 36 MHz bandwidth, 42 dBW EOL edge EIRP

One 50 W TWTA 2.671–2.690/2.501–2.519 GHz up/down S-band all-India MSS beam, 18 MHz bandwidth, 35 dBW edge EIRP

One (plus one back-up) 4 W SSPA 6.452–6.468/3.681–3.699 GHz up/down C-band all-India MSS link, 18 MHz bandwidth, 30 dBW edge EIRP

Spacecraft: box-shaped 3-axis bus around 896 mm diameter central aluminium cylinder 1.685 m high. Anti-Earth panel carries most of propulsion elements. Twin 3-panel (each panel 1.8 × 2.15 m) solar wings spanning 14.6 m, totalling 23 m^2, providing 1,620 W EOL (summer solstice; 1,100 W payload required), 2 × 24 Ah Ni/Cd batteries. 1.77 m diameter deployed parabolic reflectors on East/West faces for C-band FSS & S-band BSS transmission; 90 cm Earth face S-band MSS transmission/receive; 70 cm C-band receive; 60 cm Earth face Ku-band transmission/receive; 15 cm Ku-band global horn
Mass: 2,050 kg at launch (747 kg bus, 233 kg payload, 1,070 kg propellant)
AOCS: unified liquid bipropellant system as 2A/2B. Antenna pointing better than ±0.2° pitch/roll, ±0.4° yaw. 3-axis body stabilised by MW/RWs, momentum dumping by four 22 N thrusters on East/West faces.

Insat 2D

Launch: 3 June 1997 on Ariane 4 from Kourou, French Guiana
Location: 74° E geostationary (collocated with 2A)
Mass: 2,050 kg
Transponders: as 2C, but orthogonal polarisation
Other specifications as Insat 2C.

Insat 2E

Launch: 2 April 1999 by Ariane 42P from Kourou
Location: 55° E geostationary
Design life: 12–14 years
Transponders: 16–20 36 MHz up/down C-band Asia & Australia to Western Europe beams

Single 402.75 MHz Data Collection System transponder

Single 406 MHz Cospas/Sarsat search & rescue package

Principal applications: as for Insat 1, plus Intelsat services
Spacecraft: box-shaped 3-axis bus central aluminium cylinder with asymmetrical 4-panel GaAs solar wing on south side providing 2,000 W EOL through two 50 Ah Ni/H_2 batteries and creating 27 m span. C-band transmission reflectors of 1.7/2.0 m diameter on East/West faces; C-band receive antenna on Earth face
Mass: 2,500 kg at launch, 1,100 kg dry
AOCS: unified liquid bipropellant 440 N LAM Liquid Apogee Motor and two redundant networks of 8 × 22 N RCS thrustersOther specifications as Insat 2A/2B.

Insat 2DT (Arabsat 1C)

Launch: 26 February 1992 by Ariane 4 from Kourou, French Guiana
Design life: 7 years
(See entry Arabsat 1C for all other data).

Indonesia

Indostar Series

Current Status

Plans were laid for Indostar 2 to be replaced at 118.9° E, Indostar 3 at 107.5° E, Indostar 4 at 119.1° E. Indonesia's Indostar (Cakrawarta-1) developed battery charger problems in November 1997 and the operators silenced one of the satellite's five transponder during both the spring and autumn eclipse periods in order to compensate. Nevertheless, the system has achieved success with the first satellite as the world's first commercial communication satellite system to use S-band transmission frequencies.

Background

Indostar provides the world's first DBS dedicated to radio/TV for a single nation. The first of four planned satellites provides 5 S-band transponders for digital TV, using 8:1 compression for 49 channels. The second was to have added L-band for CD-quality radio. The US$100 million contract for the first satellite and its launch was signed with CTA International 8 December 1993. Funding is provided by 40 per cent private consortium, 30 per cent PT Amcol Graha Ltd. (a Jakarta electronics company) and 30 per cent PT Bimantra Citra. The Indostar 1 DBS system costs US$271 million. PT MediaCitra Indostar is the satellite company of the Indovision group. Indovision is a multichannel pay television company. National TV cannot be received by 32 per cent (US$57 million) and radio by 20 per cent (US$36 million). The commercial Indostar system is received via US$100 hand-held radios, US$100 analogue receivers and US$500 digital

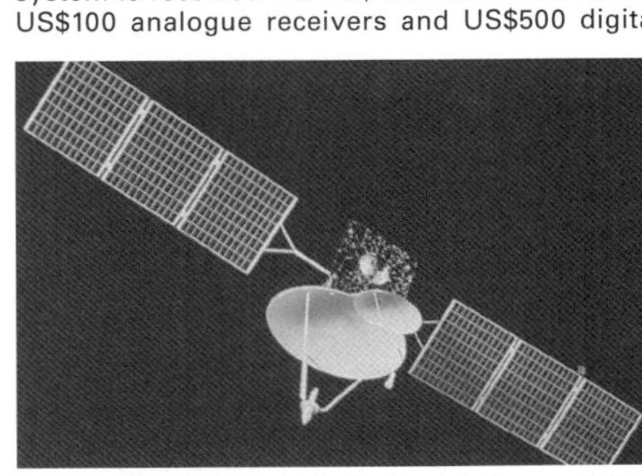

Indostar 1 was the world's first geostationary DTH TV/radio lightsat (CTA) 0517216

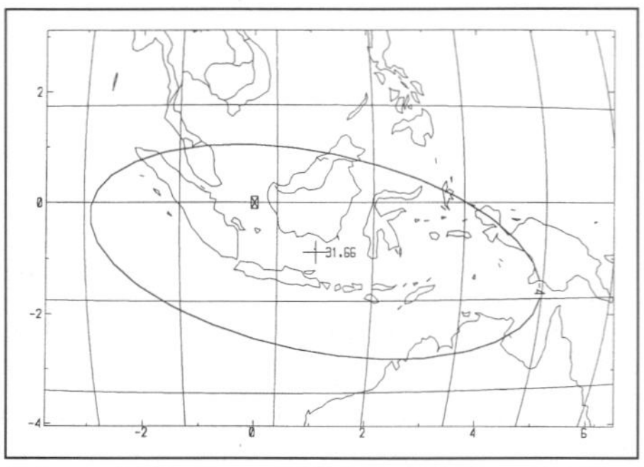

Indostar 1 TV coverage. Indostar 2 will add L-band transponders for CD-quality radio (CTA) 0517404

satellite decoders with satellite dishes less than 1 m diameter. It is designed to guarantee quality reception within Indonesia, whatever the weather conditions.

Specifications

Indostar 1

Launch: 12 November 1997 on Ariane 44L from Kourou, French Guiana
Location: number 1 107.7° E
Design life: 10 years (12 years propellant)
Contractors: CTA International
Transponders: 5 (plus 4 back-up, 5-for-4) 70 W TWTA 8.120–8.270/2.520–2.670 GHz up/down S-band domestic TV 3.6 × 8.35° beams (8:1 compression for 40 channels), 24 MHz bandwidth, EIRP 48–50 dBW min Indonesia, linear orthogonal polarisation
From Indostar 2: 2 TWTA 8.067–8.092/1.467–1.492 GHz up/down L-band domestic radio beams, 300 kHz bandwidth, EIRP 48–50 dBW min Indonesia, linear orthogonal polarisation
Principal applications: direct domestic TV/radio
Configuration: CTA's Star lightsat, bus 165 × 185 × 150 cm. AKM Star 30C
Mass: 1,400 kg at launch; 430 kg on-station BOL
TT&C: through Master Control Station at Jakarta; Backup Control Station at Surabaya
Power system: 1,700 W EOL; two SAFT 52 Ah Ni/H$_2$ batteries.

Palapa Series

Current Status

In December 2000 Palapa B2R was moved from 108° E to 42.5° E inclined by January 2001.

Background

Indonesia witnessed the commercially successful domestic satellite systems in Canada (Anik) and US (Westar) for large territorial communication and determined that its 6,064 inhabited islands arcing across 5,000 kms along the equator would benefit from the same type of system. The country built 40 large Earth stations, some 200, 5 m telephony stations and more than 1,000 TVRO stations (television receive only). Indonesia decided to begin its satellite programme with the Palapa ('Fruit of labour'), but did not announce the plans until after negotiations with Hughes had been completed by Perumtel (now PT Telkom). The Palapa became fully operational in August 1976 after the launch of Palapa AI by a Delta on 8 July 1976. Indonesia had a completely functioning network by March 1977 after Palapa 2 completed orbit tests.

Palapa C1 expands coverage to southeast Asia, China and Australasia (Hughes Space & Communications) 0517405

The satellites ended their operational lives in June 1985 and January 1988 respectively but in 1979 Indonesia had contracted with Hughes to build a follow-on system for two second-generation satellites at a cost of US$74.5 million. NASA threw in two Shuttle launches of at a cut-rate price of US$18 million. The B1 series Palapa proved a success operating from its June 1983 Shuttle launch continuously until retirement in 1990. PT Pasifik Satelit Nusantara purchased it and wrung three more years of life out of it.

Indonesia originally launched the B2 from the Space Shuttle in February 1984 but the PAM-D GTO insertion stage failed stranding it in a low earth orbit. Indonesia used the US$75 million insurance payment to purchase a replacement of the lost Palapa B1P (P means pengganti or 'replacement' in Indonesian). Shuttle mission 51A recovered it and the insurers sold it in November 1984 to Sattel Technologies Incorporated of Van Nuys, California for about US$18 million. Sattel planned to launch the refurbished satellite by Delta 2 in 1990 and then resell it to Indonesia once in-orbit. Relaunched 13 April 1990 by Delta 6925. PT Telkom retired the Palapa B1 at 118° east and sold it to PSN in 1991 to provide Pacific Rim services. Still operated by PT Telkom, PSN's major shareholder, the company has allowed the inclination to drift in order to extend its life by three years.

Indonesia set up PT Satelit Palapa Indonesia (PT Satelindo) in January 1993 to finance and operate the US$240 million Palapa C programme. Indonesia plans to increase the five telephones per 1,000 population to 200 by 2019 using follow-on systems to the Palapa. The third-generation Palapa C appeared in February 1996, expanding coverage to China, India and Japan. Indonesia created PT Satelindo to finance and operate the new generation. In May 1995 the German company T-Mobil, a subsidiary of Deutsche Telekom MobilNet GmbH, purchased 25 per cent of Satelindo's shares. Satelindo plans to develop the Palapa D series but plans have been delayed due to economic difficulties and schedules are undetermined.

Specifications

Palapa B2/B2R

Launched: 6 February 1984, retired 12 November 1984, refurbished and relaunched 13 April 1990 by Delta 6925 from Cape Canaveral
Location: 108° E GEO (originally at 118° E it replaced Palapa B1 in 1990). Replaced by C2 in 1996
Mass: 1,240 kg at launch, 652 kg on-station BOL
Other specifications as for Palapa B2P.

Palapa B4

Launched: 14 May 1992 by Delta 7925 from Cape Canaveral
Location: 118° E geostationary (to be replaced by C3)
Mass: 692 kg on station BOL
Other specifications as for Palapa B2P.

Palapa B2P/Palapa Pacific 2/Agila 1

Launched: 20 March 1987 by Delta from Cape Canaveral
Location: 144° E geostationary from August 1996 (replaced at 113° E 1996 by C1)
Design life: 8 years
Contractors: Hughes Space & Communications Co
Transponders: 24 (plus 6 back-up) 10 W TWTA 5.925–6.425/3.72–4.16 GHz up/down C-band regional beam, 36 MHz bandwidth, EIRP 34 dBW min Indonesia, 32 dBW min ASEAN countries (Philippines, Singapore, Thailand, Malaysia), linear orthogonal polarisation
Principal applications: telecoms, regional/domestic TV

The second-generation Palapa satellites will remain in service into the early 2000s (Hughes Space & Communications) 0517002

Based on the powerful Hughes HS-601 bus, Palapa-C series was introduced in 1996 (Hughes Space & Communications) 0054340

Configuration: Hughes' cylindrical HS-376 bus, 2.16 m diameter, 6.83 m deployed height, spin-stabilised, position/spin by hydrazine thrusters, GEO injection by solid AKM
Mass: 1,200 kg at launch; 628 kg on-station BOL
Pointing: 0.05° accuracy
TT&C: through PT Telkom's Master Control Station at Cibinong, near Jakarta
Power system: 1,062 W BOL/830 W EOL from K7 solar cells on cylindrical body and skirt.

Palapa C1
Launched: 1 February 1996 by Ariane 4 from Kourou, French Guiana
Location: 113° E GEO (replacing B2P). Initially: in March 1996 manoeuvred off-station and re-located over 112° E again manoeuvred off-station August 1996 and the following month re-located over 150° E
Lifetime: 12 years full performance; 14 years basic performance
Transponders: 24 (plus 6 back-up) 21.5 W SSPA 5.925–6.425/3.700–4.200 GHz up/down C-band regional beam, 36 MHz bandwidth, EIRP 37 dBW min Indonesia, ASEAN countries (Philippines, Singapore, Thailand, Malaysia), Vietnam, Brunei, Papua New Guinea, New Zealand/Eastern Australia, linear orthogonal polarisation

6 (plus 2 back-up) 26 W SSPA 6.425–6.665/3.400–3.640 GHz up/down extended C-band beams for PT Pasifik Satelit Nusantara, 36 MHz bandwidth, EIRP 37 dBW min Pakistan, India, Burma, China, People's Republic, Korea, Taiwan, Japan, linear orthogonal polarisation

4 (plus 2 back-up, 6-for-4) 130 W TWTA 13.750–14.490/10.950–11.690 GHz up/down Ku-band regional beams, 72 MHz bandwidth, EIRP 50 dBW min Indonesia, Japan, Korea, Eastern China, Thailand, Vietnam, Singapore, linear orthogonal polarisation. A battery supply problem means this payload cannot operate while C1 is in eclipse.
Principal applications: telecoms, regional/domestic TV
Configuration: Hughes HS-601 platform, 3-axis, solar array span 21.0 m (satellite within 2.7 × 3.1 × 3.5 m envelope at launch). Payload mounted on North/South panels and Earth-facing floor; East/West faces carry hard points for two 216 cm diameter C-band antennas, fixed antenna mountings on Earth-facing wall for 178 cm diameter extended C-band + 152 cm diameter Ku-band; all antennas octagonal dual-gridded with single feed horns
AOCS: 3-axis, unified 13 ARC 22 N and one Marquardt 490 N bipropellant thrusters incorporating onboard control processor, Sun/Earth sensors, and two 61 Nms 2-axis gimbaled momentum bias wheels. The NTO/MMH is contained in four spheres. Spin-stabilised in transfer orbit
Mass: 2,989 kg at launch, 1,775 kg on-station BOL
TT&C: as Palapa B
Power system: twin solar wings of three 2.16 × 2.54 m panels carrying K43/4large area Si cells on Kevlar substrate provide 3,730 W; EOL 3,400 W. Ni/H_2 batteries provide 100% eclipse protection.

Palapa C2
Launch: 16 May 1996 by Ariane 4 from Kourou, French Guiana
Location: 123° E GEO (replacing B2R).
Initially: during June 1996 relocated to ~112° E
Mass: 2,989 kg at launch, 1,803 kg on-station BOL, 1,669 kg dry
Other specifications as for Palapa C1.

International

Arabsat 1 Series

Current Status
Arabsat 1C has been transferred to operations with India and has been redesignated as Insat 2D-R.

Background
In May 1981, Aerospatiale and Ford Aerospace received a contract worth US$134.4 million for a regional satellite communication network for the Gulf region. The contract, from the newly formed Arabsat Organisation, asked for development and fabrication of three satellites (one a ground spare) for launch aboard either Ariane or Shuttle. In November 1982, Arabsat and NASA signed an agreement for a Shuttle/PAM launch for US$20.1 million (plus US$6 million for the PAM-D perigee kick stage); an earlier Ariane launch had been negotiated for US$23 million.

The Arabsat 1 system cost about US$500 million in 1986 dollars. The Arab Fund for Economic & Social Development made a loan of US$53 million to the 11 members Arabsat Organisation in 1985 to build Earth stations, in addition to US$123 million it had loaned earlier to 13 members for modernisation of communications systems. Then Arabsat paid Japan's NEC to build the two TT&C stations, while Aerospatiale operated the stations and trained Arabsat personnel over a 2 year period.

At the outset a lack of funding hampered the Arabsat programme. By the beginning of 1989, outside sources estimated the total debt at US$60 million with only 40 per cent utilisation of its total capacity on one active satellite. A drop in oil revenues throughout the Arab world precipitated a debt-reduction plan and gave incentive to Arabsat to increase revenues. The review indicated that the 50 W S-band semi-DBS links were not operational, and some 25 per cent of the utilisation shortfall arose because Egypt did not participate due to its suspension in 1979 from the Arab Telecommunications Union resulting from its dealings with Israel. When Egypt rejoined in April 1989 the debt crisis ended.

Arabsat launched Arabsat 1A on 8 February 1985. One solar panel failed to open during the GTO orbital placement phase but thruster firings successfully freed the unit. About one month later, power to two AOCS gyros failed and consequently NSSK at 19° E had to be exercised manually from Saudi Arabia's Dirabh TT&C station. Arabsat declared the satellite, although still operational an in-orbit spare and claimed US$75 million insurance based on failure to meet specified operational quality. The company allowed the satellite to be used from an inclined orbit to add three years to its seven year lifetime by saving station-keeping propellant. These operations began at the end of 1990, but the fuel ran out September 1991 and the satellite began drifting. Arabsat declared it lost 30 March 1992, and deactivated it 31 July 1992.

Arabsat 1B went up 18 June 1985 by Shuttle to 26.2° E and it began inclined operations in late 1991 to extend useful life by 3 years but, again, there was insufficient fuel and Arabsat took it out of service in October 1992 and deactivated 3 April 1993. Arabsat representative Prince Sultan Salman Al-Saud accompanied the second satellite into orbit.

Aerospatiale/Ford completed the ground spare, 1C, in 1985 and held it in storage at Aerospatiale's facility in Cannes. Intelsat negotiated purchasing it during 1989 for Pacific operations and Arabsat withheld payment of the last 30 per cent of the cost. Repayments began again in June 1990 and China's GWIC received a US$25 million launch contract in January 1990 for departure November 1991. Then Arabsat applied for both US and French export licences because some 60 per cent of the components are US-derived. In March 1991, the US claimed that the licences for the Chinese launch were not complete, Arabsat cancelled the launch for this reason and the Persian Gulf war. Then Arabsat renegotiated the launch with Arianespace instead on 15 May 1991.

Specifications
Arabsat 1C/Insat 2D-R
Launched: 26 February 1992 by Ariane 4 from Kourou, French Guiana
Location: 30.5° E GEO
Design life: 7 years
Contractors: Aerospatiale (prime; structure, thermal control, solar arrays, integration/testing), Ford Aerospace (now SS/L; communications payload, attitude control, propulsion, power conditioning/storage, telemetry, command & ranging), Alenia (antenna reflectors, TT&C transponders), TNT/AEG (solar array electrical network), MBB (propulsion & attitude control integration)
Transponders: 25 8.5 W TWTA 5.925–6.425/3.70–4.20 GHz up/down C-band regional beams, 33 MHz bandwidth, EIRP 31 dBW min, circular polarisation. Two (plus one back-up) paralleled 50 W TWTA 5.925–6.425/2.540–2.655 GHz up/down S-band downlink for semi-DBS broadcasts (1.5 m dishes), 33 MHz bandwidth, 41 dBW min EIRP, linear polarisation
Principal applications: regional TV, telephony, data, fax relay. Each C-band can handle 1,466 SCPC telephone channels or one TV
Configuration: Aerospatiale/MBB Spacebus 100 (now classed as 1000) 3-axis stabilised 2.26 × 1.64 × 1.49 m box-shaped bus with 20.7 m-span twin solar arrays
Mass: 1,310 kg launch, 785 kg on-station BOL, 600 kg dry
TT&C: through Saudi Arabia's Dirabh station, plus back-up station at Eddikhila near Tunis
Power system: 1,400 W BOL/1300W EOL provided by twin silicon solar arrays.

Arabsat 1D-R (ex-Telstar 301)
Launched: 29 July 1983 as Telstar 301 by Delta from Cape Canaveral
Location: 20° E GEO, February 1995–September 1996
Design life: 10 years
Contractors: Arabsat 1D-R: Hughes Space & Communications Co
Transponders: 18 5.5 W SSPAs 5.925–6.425/3.700–4.200 GHz up/down C-band regional & spot beams, 36 MHz bandwidth, 34 dBW min EIRP, linear polarisation. 12 5.5 W TWTAs, other specs as for SSPAs
Principal applications: regional telephony, data, TV broadcast
Configuration: Hughes' HS-376 spin-stabilised bus, 2.2 m diameter, 6.8 m high deployed
Mass: 3,423 kg with PAM-D GTO stage attached, 625 kg on station BOL
TT&C: from Telesat Canada HQ via Tunis station
Power: 800 W BOL/670 W EOL provided by 15,588 silicon cells on cylindrical exterior
Apogee kick motor: Thiokol solid propellant.

Arabsat 2 Series

Current Status
To satisfy a growing demand, and to guarantee services, Arabsat launched Arabsat 2B in July 1996. Previously, Arabsat intended to hold the satellite as a ground spare, dispatching it at the latest in 1999 before 1C is retired. Arabsat 3A was launched in February 1999 to provide direct TV broadcasting to North Africa and the Near East. Arabsat 2A C-band with one channel each for Bahrain, Egypt, Mauritius, Oman, Sudan and the UAE, and three for Saudi Arabia. Ku-band provides three channels for Saudi Arabia and one each for Jordan, Kuwait, Qatar and Syria.

Background
Arabsat requested bids for a successor system to Arabsat 1 in early 1990 and allowed a 15 October 1990 deadline. The Iraqi invasion of Kuwait forced a delay. The final Arabsat 2 RFP went to bidders in 1991 projecting a new launch date. In October of 1992, Arabsat tentatively chose Hughes Space & Communications Company to build the new satellites on a US$258 million bid for two HS-601 satellites. British Aerospace and Aerospatiale also submitted proposals. However, Hughes and BAe reportedly refused to agree to some of Arabsat's terms and the organisation instead placed the order for US$257.9 million for Aerospatiale's Spacebus 3000 platform on 17 April 1993. The contract included first launch on Ariane 4.

Specifications
Arabsat 2A
Launched: 9 July 1996 by Ariane 4 from Kourou, French Guiana
Location: planned 26° E geostationary
Design life: 12 years (16 years propellant)
Contractors: Arabsat 2A: Aerospatiale (prime; structure, thermal control, solar arrays, integration/testing), Alcatel Espace (communication payload), Detecon (contract monitoring on behalf of Arabsat), DASA (propulsion/attitude control integration), Alenia (TT&C)
Transponders: 14 (plus 6 back-up: 10-for-7) 15 W SSPA 5.925–6.425/3.70–4.20 GHz up/down C-band

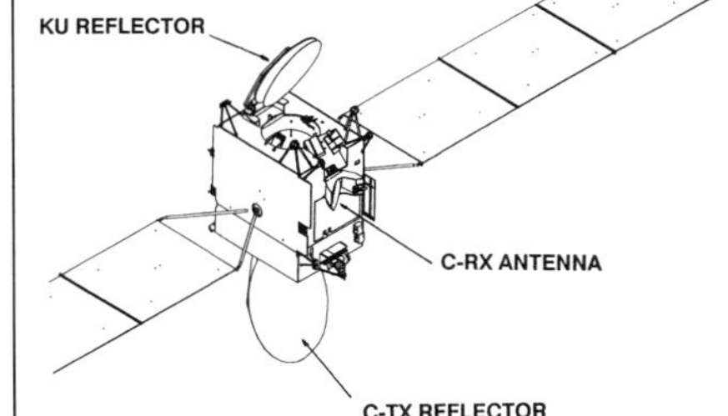

The Arabsat 2 generation became operational in 1996 (Aerospatiale) 0517197

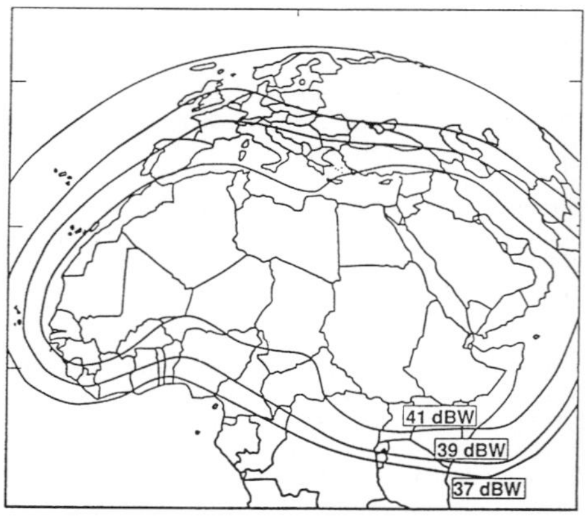

Arabsat 2 high-power C-band EIRP contours (Aerospatiale) 0516975

regional beams, 12 × 36 + 2 × 54 MHz bandwidths, EIRP 35 dBW min, circular polarisation for telecom services.

8 (plus 4 back-up, 12-for-8) 57 W TWTA 5.925–6.425/3.70–4.20 GHz up/down C-band regional beams, 36 MHz bandwidth, EIRP 41 dBW min, circular polarisation for semi-DBS via 0.8–1.0 cm dishes

12 (plus 5 back-up, 17-for-12) 95 W TWTA 13.75–14.0/12.5–12.7 GHz up/down Ku-band regional beams, 8 × 36 + 4 × 30 MHz bandwidths, 47 dBW min EIRP, linear polarisation

Principal applications: regional TV, telephony, data, fax relay

Configuration: Aerospatiale/DASA Spacebus 3000 3-axis stabilised 1.8 × 2.6 × 2.3 m box-shaped bus (7 m deployed height) with 26.3 m-span twin solar wings. C-band: 2.1 × 2.2 m transmission west panel dish + 60 cm receive Earth face dish; Ku-band: single 1.6 m dual gridded dish on east panel. C-band repeater assembly on N panel; Ku-band on S panel

Mass: 2,617 kg launch, 1,570 kg on-station BOL, 1,108 kg dry, 260 kg payload

TT&C: as Arabsat 1

Power system: 5,074 W EOL provided by twin Si solar wings totalling 46 m^2. Two 52 Ah Ni/H_2 batteries.

Arabsat 2B

Launch: 13 November 1996 by Ariane 4 from Kourou, French Guiana

Location: 30.5° E geostationary.

Arabsat 3 Series

Current Status

Alcatel Space Industries was contracted to provide the first of two direct TV broadcast satellites based on the SB-3000B bus. Arabsat 3A was launched on 26 February 1999 for the Arab Satellite Communications Organisation based in Saudi Arabia. On 7 December 2001, eight transponders failed, possibly as a result of a solar array power system. It was declared a total loss and Panamsat 5 was co-located to assist. A contractor for Arabsat 3B has yet to be determined. Arabsat 3B is planned for location at 30.5° E but launch plans have yet to be determined.

Specifications

Arabsat 3A

Launched: 26 February 1999 by Ariane 44L

Location: 26° East co-located with Arabsat 2A

Design type: 15 year

Transponders: 20 active in Ku-band BSS

Configuration: 3.83 × 3.35 × 2.26 m; solar array span 29 m; onboard power 6,400 W

Mass: 2,708 kg at launch; 1,646 kg BOL; 1,200 kg dry.

Artemis Satellite

Current Status

A launch failure prevented Artemis from achieving GTO but onboard propulsion moved it into position, finally achieving a location at 21.5° E on 31 January 2003. The rescue operation began shortly after launch on 12 July 2001 when the Ariane 5 Aestus engine failed to reach full thrust and cut off early stranding the satellite in a 592 × 17,528 km orbit compared to the planned 858 × 35,853 km ellipse. In the first rescue operation utilising electric propulsion the four German xenon thrusters were used to gradually shift the satellite to its planned position in orbit.

Artemis includes large reflectors to provide L-band mobile beams (left reflector) and S/Ka-band inter-satellite data relay (right reflector). Their feeder link antennas are centrally mounted: Ku-band for L-band (top right) and Ka-band (top left) for S/Ka/laser. The Silex laser system is the box unit at bottom right (Alenia) 0516976

Background

The ESA Council approved three new experimental payloads in June 1990 and folded them into the ARTEMIS research satellite. The new experiments ESA wants to investigate are the Silex (Semiconductor Laser Intersatellite Link Experiment), SKDR data relay as a DRS precursor, and an L-band mobile channel to demonstrate services for European land vehicles in addition to backing up EMS. Technology experiments. Originally, ESA planned a launch aboard an Ariane 502 in April 1996 but technical problems delayed the launch date to late 1997.

Estimated to cost ECU665 million at 1992 rates, the cost of the Artemis satellite has swollen to about ECU920 million. The cost overrun has prompted NASDA to offer a free H2 launch in return for making Japan a full partner. Italy provided about 45 per cent of the original funding. ESA signed an ECU446 million contract with Alenia Spazio on 4 October 1993. Phase B2 began July 1989. The ultimate objective of the Artemis satellite is to act as a relay satellite between low-earth orbit satellites and ground stations and to serve the planned EGNOS European navigation satellite system by 2010.

Specifications

Launch: 12 July 2001 by Ariane 5 from French Guiana.

Location: 16.4° E, then 59° E when DRS appears

Design life: 10 years minimum

Mass: 3,100 kg at launch, 1,500 kg dry

Technicians at Estec check out the Artemis satellite bus (Theo Pirard) 0054353

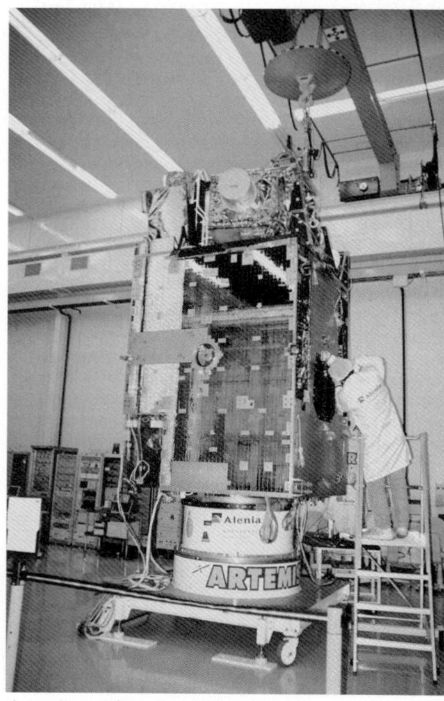

Artemis and Silex (top) (Theo Pirard) 0054355

Contractors: Alenia Spazio (prime), Matra Espace (Silex prime), BPD/DASA (unified propulsion), DASA (S400 + S10MkII thrusters for AOCS), SAFT (Ni/H_2 batteries), CASA (structure), FIAR (power conditioning/distribution), Galileo (attitude sensors)

Transponders: three (plus one back-up, 4-for-3) feeder link for Silex and SKDR, 27.5–30/18.1–20.2 GHz up/down, EIRP 43 dBW, G/T 0 dB/K, 234 MHz

DRS could add S-band phased arrays, for multiple access, to Artemis's Silex and SKDR payloads. The 48-element receive array is deployed at bottom; the 24-element transmit array is mounted on the Earth face (Alenia) 0516977

bandwidth, linear vertical polarisation Silex. Data rate 50 Mbit/s, bit error rate $<10^{-6}$. Two optical terminals are being built: the GEO element for Artemis and the LEO unit on Spot 4. Under an agreement with CNES, a pre-operational Silex service will be provided to Spot 4 after the initial experimentation phase. Each terminal employs a 25 cm diameter optical telescope mounted on a coarse pointing mechanism; total 140 kg; pointing accuracy 0.2 µrad (random), 0.8 µrad (static). Optical power source: 8,300 Å GaAlAs semiconductor laser diode with a peak output of 160 mW (60 mW continuous), and a beamwidth of 0.0004°
Receiver: SI-APD silicon avalanche photodiode followed by a low noise trans-impedance amplifier; 1.5 nW useful received power. Sira's CCD acquisition/tracking sensors direct Teldix's fine pointing mechanism of orthogonal mirrors. Phase B concluded October 1990; C/D started mid-1991. A 1 m telescope at the Teide Observatory on Tenerife in the Canary Islands will be used for experiments with Artemis.

Comstar/SBS Series

Current Status

SBS-1 was retired in June 1990, SBS-2 was retired in September 1996 and was de-orbited, SBS-3 was retired in June 1995 and de-orbited, SBS-4 is still operating, SBS-5 was retired in March 2000 and was replaced by Galaxy 10R, SBS-6 is still operating as of late February 2004. It is now used primarily for occasional access for coverage of special events and for breaking news.

Background

Comsat established the Comsat General Corporation subsidiary in 1973 to act as the satellite operator of its growing inventory of orbiting hardware. The subsidiary now manages the five satellites of the Marisat, Comstar and SBS 2 networks. Another subsidiary, Comsat Video Enterprises owns and operates a satellite-based network distributing entertainment and video conference services to the US lodging industry.

Comsat General's 4-satellite Comstar system was the first to be integrated into the US national telephone network, providing voice, data and TV services for the whole of the US, Alaska, Hawaii and Puerto Rico. Hughes received COMSAT's US$65 million contract in September 1973. Comsat immediately leased all of the capacity to AT&T for the 7-year satellite operational lives at US$1.3 million monthly for each. Comstar D1, launched 13 May 1976, took its slot at 76° W next to D2 as their capacity declined. D3 replaced it 29 June 1978. Each satellite provided 18,000 duplex voice channels. COMSAT announced that it planned to move D4 over China in 1993; AsiaSat had an option on D4 but it expired 31 July 1993. D2, launched 22 July 1976, drifting January 1994.

Comsat also established a new Satellite Business Systems (SBS) subsidiary in December 1975 with ownership held at 41.3 per cent. IBM holds an equal 41.3 per cent share and Aetna Life & Casualty 17.4 per cent. In April 1984, after the successful launches of three satellites, Comsat sold its interest to its partners. The Federal Communications Commission approved of transfer of SBS to MCI Communications Corporation in November 1985. Comsat then announced the purchase of SBS 1 and 2 from MCI in March 1987.

Then, after Comsat successfully launched three SBS satellites through April 1984, Comsat sold its interest to its partners. Part of the partnership arrangement of commercial owners entailed a transfer of SBS to MCI Communications Corporation, approved by the FCC in November 1985. But SBS 4 and subsequent satellites remained majority partner IBM's wholly-owned Satellite Transponder Leasing Corporation (STLC) subsidiary. SBS 4 began operations in October 1984 to be joined in late 1988 by the last of the series' HS-376-class satellites, SBS-5. The more powerful HS-393-class SBS 6 had a Shuttle launch date of September 1988 when the *Challenger's* loss occurred in January 1986. It eventually departed October 1990 aboard Ariane.

STLC services provided by SBS 4 and 5 included network TV, wideband TDMA (multiple voice and high speed data) and corporate network VSATs for data networks and video broadcasting. HCI reached agreement with IBM for purchase of the SBS system in July 1989.

SBS 6 is the only HS-393 satellite of the SBS series, offering triple the capacity of its predecessors. Galaxy 4 replaced it at 99° W in November 1993. HCI moved SBS 6 to 95° W in March 1994 until Galaxy 3R's December 1995 launch. With the Galaxy in place HCI then moved the SBS satellite to 74° W. Galaxy 7 replaced SBS 4 at 91° W in January 1993. SBS 4 then moved in February 1993 to 77° W pending a decision on its future. NBC bought its entire capacity in September 1993, and began using it on a full-time basis by September 1995. It is now operating in an inclined orbit to extend its life beyond 2000.

Specifications

Comstar D4

Launched: 21 February 1981 by Atlas Centaur from Cape Canaveral complex 36
Location: 76° W geostationary, inclined 8.6°
Design life: 10 years
Contractors: Hughes Space & Communications
Transponders: 24.5/5.5 W TWTA 5.925–6.425/3.70–4.20 GHz up/down C-band CONUS / Hawaii / Alaska / Puerto Rico beams, edge EIRP 33 dBW, 34 MHz bandwidth, orthogonal polarisation. 19.0/28.6 GHz beacons for propagation research
Principal users: national voice, data, TV
Configuration: similar to Intelsat 4A. Spin-stabilised (56 rpm) cylindrical satellite with despun communication platform incorporating two offset parabolic 127 × 178 cm reflectors each handling 12 channels. Stationkeeping provided by hydrazine thrusters
Mass: 1,516 kg at launch, 911 kg on-station BOL
TT&C: Comsat's Washington facility working through Southbury, CT and Santa Paula, CA stations
Power: 570 W BOL from 17,000 silicon cells on cylindrical surface; two Ni/Cd batteries
Apogee kick motor: Thiokol solid propellant.

SBS 2

Launched: 24 September 1981 by Delta/PAM from Cape Canaveral
Location: 71° W GEO, inclination 6.7°; 97° W May 1992 to October 1993; 71° W June 1994 to August 1996
Design life: 7 years
Transponders: 10 (plus 4 back-up) 20 W min TWTA 14.025–14.466/11.725–12.166 GHz up/down Ku-band CONUS beams, 43 MHz bandwidth, EIRP 43.7 dBW edge CONUS
Principal users: domestic business voice, video, data, and VSAT services
Configuration: based on Hughes' HS 376 spin-stabilised bus, 2.16 m diameter, 6.6 m height with solar array drum extended (2.8 m stowed). Single antenna comprising two offset grid reflectors with horizontal/vertical feed system on despun platform
Mass: 546 kg on-station BOL
Attitude control: position + 50 rpm spin control maintained by four Hughes 22.2 N hydrazine thrusters
TT&C: through Comsat's Clarkburg, MD facility
Power system: >16,000 Si body cells provide 900 W EOL in conjunction with two Ni/Cd batteries
Apogee kick motor: Thiokol Star 30 solid.

SBS 4

Launched: 30 August 1984 by Shuttle STS-41D/PAM-D from Cape Canaveral
Location: 77° W GEO, inclination 2.4°; 91° W May to October 1992; 77° W from October 1993
Design life: 10 years minimum GEO
Mass: 1,117 kg at launch, 571 kg on-station BOL
Other specifications as for SBS 5 except that SBS 4 cannot double-up its transponders for 40 W output and does not incorporate the four 110 MHz units.

SBS 5

Launched: 8 September 1988 by Ariane V25 from Kourou, French Guiana
Location: 123° W geostationary (to be replaced by Galaxy 10)
Design life: 10 years minimum
Transponders: 10 (plus 4 back-up) 43 MHz and 4 (plus 2 back-up) 110 MHz bandwidth 20 W min TWTA 14.0–14.46/11.72–12.16 GHz up/down Ku-band CONUS plus Alaska and Hawaii spot beams, EIRP 39 dBW edge CONUS (45–46 dBW E/W concentrations), orthogonal polarisation. Four of the 43 MHz and all four of the 110 MHz transponders can be paralleled to provide 40 W output, principally for VSAT services
Principal users: domestic business voice, video, data, and VSAT services
Configuration: based on Hughes' HS-376 spin-stabilised bus, 2.16 m diameter, 6.6 m height with solar array drum extended (2.8 m stowed). Single antenna comprising two offset grid reflectors with horizontal/vertical feed system on despun platform
Mass: 1,241 kg at launch, 725 kg on-station BOL
AOCS: 50 rpm spin control and station-keeping maintained by four Hughes 22.2 N hydrazine thrusters
TT&C: through the Castle Rock, Colorado station
Power system: >16,000 Si cells on cylindrical surface provide 1,200/1,150 W BOL/EOL in conjunction with two Ni/Cd batteries
Apogee kick motor: Thiokol Star 30 solid.

SBS 6

Launched: 12 October 1990 by Ariane 4 from Kourou, French Guiana. Four apogee firings 13/14 October for GEO insertion
Location: 77° W geostationary; May 1992 to October 1995 at 95° W; from November 1995 at 74° W
Design life: 10 years (hydrazine for 15.6 years remaining after Ariane launch)
Transponders: 19 (plus 11 back-up; 30-for-19 ring redundancy) 41 W min TWTA 14.00–14.49/11.700–12.19 GHz up/down Ku-band CONUS plus switchable Hawaii/Alaska spot beams, 43 MHz bandwidth, CONUS EIRP 49–53 dBW, orthogonal polarisation
Configuration: Hughes HS-393 spin-stabilised bus 3.66 m diameter, 3.6 m stowed height/10.0 m deployed, payload and single 2.4 m diameter antenna mounted on despun platform
Mass: 2,478 kg at launch, 1,514 kg on-station BOL, 1,138 kg dry
AOCS: spin-stabilisation maintained by six 22 N thrusters, GEO insertion and major manoeuvres performed by two Marquardt R-4D 489 N thrusters, both systems drawing on 1,200 kg of NTO/MMH in four tanks each
Antenna pointing: 2-axis RF beacon tracking maintains 0.06°-accuracy pointing, backed up by Earth sensors
TT&C: Same as SBS 5
Power system: 2,300 W equinox BOL/2,200 W solstice EOL provided by Si solar array on cylindrical body (K7 cells) and extension skirt (K43/4 cells) in conjunction with two 48 Ah Ni/H_2 batteries
Thermal control: quartz mirror radiator band encircling upper body provides primary heat rejection path.

ECS Series

Background

The first-generation European Communications Satellite system OTS-1 was launched on 14 September 1977 but was lost when the Delta 3914 exploded. OTS-2 was successfully launched on 12 May 1978 and placed at 10° E. Launched on 12 October 1983, ECS-1 (Eutelsat 1-F1) operated initially from 13° E. ESC-1 (Eutelsat 1-F1) operated initially from 13° E, ECS-2 followed in 1984, ECS-3 was lost in an Ariane launch failure, ECS-4 was launched in 1987 followed by ECS-5 in July 1988. The second generation dropped the 'ECS' prefix and were known as Eutelsat-2. ESA declared it fully operational in October 1984 and handed ECS-2 over to Eutelsat at that time. ESA then embarked on a programme to procure and launch five ECS for Eutelsat, because by then, the demand for channels had led to three more satellites being ordered. ESA handed over the last of these satellites in-orbit in 1988.

At 1,175 kg launch mass, the ECS satellites are about twice the size of OTS, and in addition to telephone, telex and business traffic the system transfers TV between member countries of the European Broadcasting Union. Under a 10 year agreement, ESA provided the first generation space segment for Eutelsat, which became the owner and manager of each satellite following in-orbit testing. Able to carry 12,000 phone circuits plus two TV channels, they came from the production line of the Mesh consortium, led by British Aerospace.

ESA intended to launch ECS 1/2, by Deltas but eventually transferred them to Ariane. It ordered the first ECS in 1978 at a cost of US$114 million plus Ariane launches at US$26.4 million each. Ariane failures and Marecs electrostatic problems delayed the launches. The ECS 1/2 development phase probably cost ESA about 220 MAU, and ECS 3–5 cost 221 MAU in 1984 rate currencies. Last of the OTS/ECS platform series, Skynet 4F was launched on 7 February 2001 and was first positioned at 6° E. The satellite had a mass of 1,489 kg and a dry mass of 830 kg, more than twice the dry mass of the OTS bus.

Eutelsat 1 Series

Background

ESA designed, commissioned and financed the first generation of the European Communications Satellite (ECS) and transferred them to Eutelsat, once checked-out in orbit, for commercial exploitation. The international organisation launched the first satellite, known initially as ECS 1 until it began operations as Eutelsat 1-F1, in 1983. ECS 5 (Eutelsat 1-F5) completed the first network in July 1988. Each Eutelsat 1 operates up to 10 Ku-band transponders, simultaneously providing up to five beams.

ESA handed over Eutelsat 1 at 10° E on 12 October 1983 primarily for TV distribution services. Eutelsat moved it to 16° E after the launch of F5 in July 1988. After 2-F3 appeared in December 1991, Eutelsat allowed the satellite to begin a slow drift to 25.5° E. In August 1993, Eutelsat then moved it to 48° E, where it arrived 14 December, to provide CIS services and retain connectivity to the West. Eutelsat moved it again in March 1996 to 36° E, where a quick motor firing boosted it out of the geosynchronous orbit band. Eutelsat 1 lost part of the solar array capacity after 15 months through short circuits across the Kapton insulating layer between the cells and their grounded panel structure. But, in all, the satellite logged 640,410 channel-hours by 30 June 1993. Eutelsat 1 was retired out of service in December 1996, more than 13 years after launch.

Eutelsat 1 F2 added the Satellite Multiservice System to provide companies with access to a European network of dedicated links for high speed data transmission, video conferencing, facsimile and remote printing. It handled 2.4–2,000 kbit/s data. ECS 2 and successors had expected lifetimes of only 4.9 years NSSK because of heavier payload and the Eutelsat retired 1-F2 from service at 1° E 13 September 1993 by raising it 396 km above GEO. In all, it provided 634,969 channel-hours in its life. ECS 3 was destroyed in an Ariane launch vehicle failure on 13 September 1985.

ECS 4 and 5 cost US$37 million. ECS 4 became the property of Eutelsat 30 October 1987 after employing the new Arianespace launch-to-separation insurance coverage, on the open market for final deployment. Initially positioned at 10° E, then 13° E for TV distribution, Eutelsat moved it in late 1990 after 2-F1's commissioning to 7° E for fulltime leases, SMS traffic and occasional use, and then in 1992 to 25.5° E before most of 2000 at 21.5° E before a shift to 33.5° E and retirement in November 2002. It had provided 480,461 channel-hours by 30 June 1993. F5 completed the first-generation system of four operational satellites in 1988. It operated initially at 10° E and moved in 1991 to 21.5° E to carry telephony and occasional use traffic. In 1999 it moved to 12.5° W, then to 14.8° W and finally to 4° E from where it was deorbited in 2000.

Eutelsat tariffs cover the satellite transponder alone and do not include charges for uplinks and other services on the ground. A lease for an Eutelsat 1 transponder costs a user ECU2.8 million annually for a pre-emptible narrowband transponder when a three-year contract is signed. Eutelsat also offers a non-pre-emptible service over three years, which costs ECU3.8 million annually, and ECU3.4 million if a user selects a five-year option for at least five transponders. Eutelsat revises the transponder lease costs based on the consumer price index for all European Union members (if the result is a tariff increase, the Signatory has the right to terminate the agreement).

Specifications

Eutelsat 1-F1 (ECS 1)

Launched: 16 June 1983 by Ariane L6 from Kourou, French Guiana

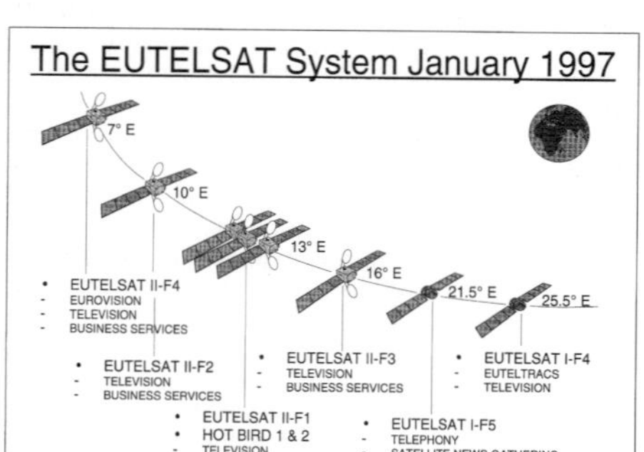

The Eutelsat System in January 1997 0005564

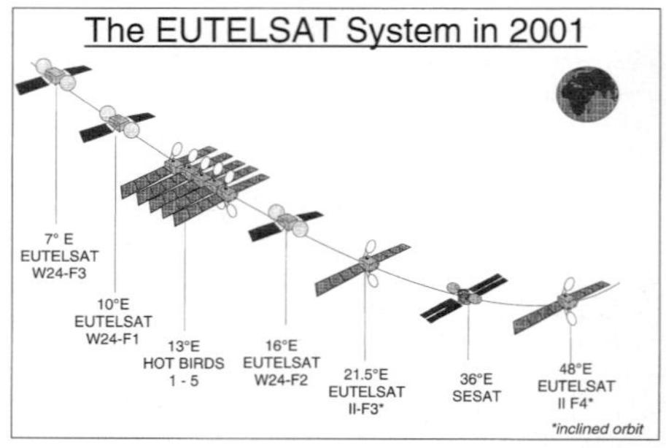

The Eutelsat System in 2001 0005565

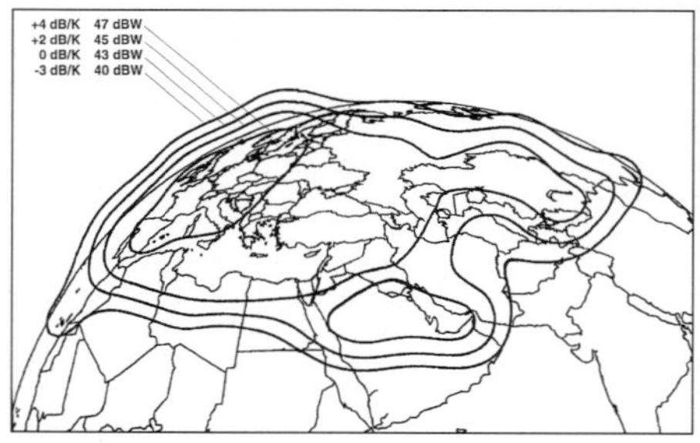

Sesat coverage from 48° east (Eutelsat) 0517397

Location: 36° E geostationary, inclination increased to 5.2° by end-1995; retired in December 1996
Design life: 9 years, with 74 per cent reliability for 16 channels
Contractors: Satcom International (MMS)
Transponders: ten (plus two back-up) 20 W TWTA 14.00014.500/10.950–11.200 + 11.450–11.700 GHz up/down Ku-band three spot beams (telephony, spots SA Atlantic, SE east, SW west) and one Eurobeam (TV), 72 MHz bandwidth, spot EIRP 40.8–46.3 dBW, Eurobeam EIRP 34.8–41 dBW, linear orthogonal polarisation
Principal applications: leases
Configuration: ECS/OTS platform, 3-axis controlled
Mass: 1,185 kg at launch, including 475 kg Mage AKM and 107 kg hydrazine AOCS propellant (2.5 kg remained August 1995), 680 kg on-station BOL
TT&C: launch and on-orbit TT&C performed by ESOC at Darmstadt, Germany; operational TT&C exercised from ESOC's station at Redu, Belgium through 13.5 m antenna
Power system: twin 5.2 × 1.3 m wings spanning 13.8 m providing 1 kW BOL, 900 W (solstice) EOL, supported by 2 × 24 Ah Ni/Cd batteries.

Eutelsat 1-F2 (ECS 2)

Launched: 4 August 1984 by Ariane V10, given to Eutelsat 12 October 1984
Transponders: Same as Eutelsat 1-F1 plus two 20 W TWTA on 12.5 12.75 GHz Ku-band downlink offering 39.8–43.5 dBW EIRP Satellite Multiservices System business links.

Eutelsat 1-F3 (ECS 3)

Destroyed in the Ariane's 12 September 1985 launch failure.

Eutelsat 1-F4 (ECS 4)

Launched: 16 September 1987 by Ariane V19 from Kourou, French Guiana
Location: 25.5° E geostationary, inclination 2.0° (last NSSK made April 1993)
AOCS: 116 kg hydrazine loaded; 14 kg remained as of August 1995
Other specifications as for Eutelsat 1-F2.

Eutelsat 1-F5 (ECS 5)

Launched: 21 July 1988 by Ariane V24 from Kourou, French Guiana
Location: 21.5° E GEO, inclination 1.5°
Mass: 1,185 kg at launch, about 700 kg on-station BOL. 122 kg hydrazine at launch; 22 kg as of August 1995 (mean usage 1.23 kg monthly)
Other specifications as for Eutelsat 1-F2.

Eutelsat 2 Series

Eutelsat itself manages the follow-on Eutelsat 2 programme that began with the 2-F1 satellite launched in August 1990. The cost of the six satellites plus launch and insurance ran about ECU1,094 million at 1994 rates. Each F2 offers 16 Ku-band transponders and improved on-board redundancy. When it built its economic model for the second generation, Eutelsat assumed all the traffic carried by the first generation would

The Eutelsat 2 series is based on Aerospatiale's Spacebus 2000 platform. Each carries 16 channels working through the East and West 1.6 m {70} reflectors 0516978

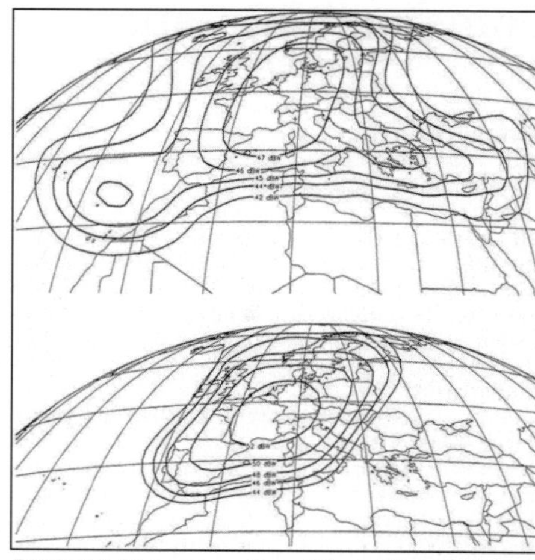

Eutelsat 2-F1 coverage from 13° east. Top: Widebeam; bottom: Superbeam 0516980

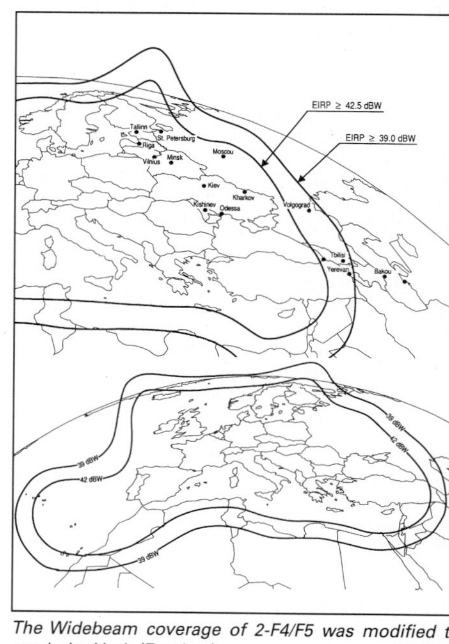

The Widebeam coverage of 2-F4/F5 was modified to reach the Urals (Eutelsat) 0516981

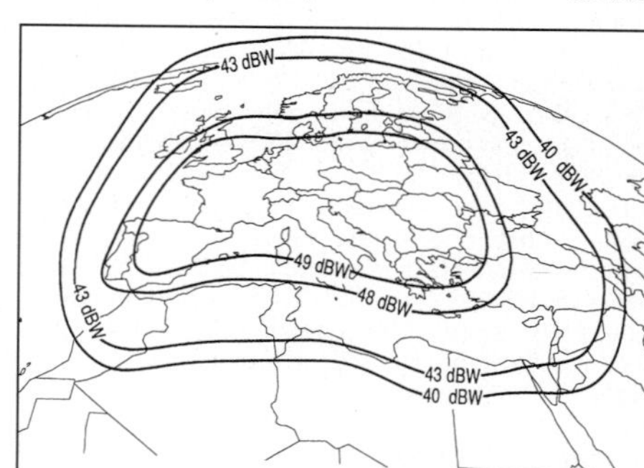

Instead of being divided into Widebeam/Superbeam coverage, 2-F6/Hot Bird 1 offers increased power European beams stretching to the Urals. Its new frequencies and collocation with 2-F1 at 13° east provide up to 39 TV channels from the one slot (Eutelsat) 0516982

Eutelsat 2-F4 integration at Aerospatiale's Cannes facility 0516979

Eutelsat satellite for East-West Communications (Theo Pirard) 0054354

Eutelsats' purpose-built Paris Headquarters 0084764

continue and included sufficient capacity for expansion of services.

The second generation offers 16 simultaneously-operational transponders. 2-F1 occupied the prime 13° E slot since 1990 to guarantee services without the need for users to re-direct their antennas. It was moved to 36° E, replacing TDF 2 in 1999 and then to 48° E until 2002 when it was moved to 70.5° E and, in January 2003, on to 76° E. Eutelsat contracted with Aerospatiale for three follow-on satellites in May 1986 at the price of ECU225 million. They took out an option on the fourth in June 1987, and planned to launch the new series at intervals of 6 months. Then due to increased demand Eutelsat added a fifth to the overall purchase in March 1989 at a cost of ECU52.8 million, and a sixth asking for delivery during 1990.

2-F6 features with higher power TWTAs in the expansion band frequencies to provide 49 dBW over Europe and to reach to the Urals. Collocation with 2-F1 at 13° E provides up to 39 TV channels from the one slot without receiver modification. Direct reception is possible with 70 cm antennas in all of central/western Europe, and with slightly larger antennas for cable and community reception as far east as Turkey and Ukraine. Commercial services began 29 April 1995. Built by Matra Marconi, the Eutelsat 2-F7/-8/-9/-10 series were launched between November 1996 and October 1998 providing digital and analogue television and multimedia series. The first in a series of widebeam telecommunications satellite built by Alcatel Eutelsat W-F2 was launched by Ariane 44L on 5 October 1998 to 16° east.

Logica built a new TCR system, for Eutelsat 2 including the Satellite Control Centre in Paris and TCR equipment centres in Rambouillet (France) and Sintra (Portugal). Demand exploded when Eastern Europe left the Soviet Union in 1990. Eutelsat met the demand for services to eastern Europe, by stretching 2-F4 and 2-F5's widebeams to the Urals and modifying 2-F6 (Hot Bird 1) by adding radically new and more powerful TWTAs.

Specifications

Eutelsat 2-F1

Launch: 30 August 1990 by Ariane V38 44LP from Kourou, French Guiana

Location: 13° E geostationary; 36° E in 1999, 48° E in 1999, 70.5° E in 2002, 76° E in 2003

Design life: 9 years, reliability 0.74 for 16 channels and 7 years

Contractors: Aerospatiale (prime), Marconi Space Systems (payload), MBB (20% share: AOCS, solar arrays, TT&C equipment) Alcatel Espace, CASA, Aeritalia, ETCA, ERA

Transponders: 16 (plus 8 back-up; 12-for-8 ring redundancy) 50 W TWTA 14.0–14.5/10.95–11.20 + 11.45–11.7 + 12.50 12.75 GHz up/down Ku-band regional Superbeam and Widebeam, bandwidths 72 MHz (seven transponders) and 36 MHz (nine), EIRP 44–52 dBW Superbeam over western Europe + 44–47 dBW pan-European Widebeam, linear orthogonal polarisation

Principal applications: TV/radio distribution to cable networks and domestic satellite systems

Configuration: derived from Spacebus 2000 platform, 3-axis control, 22.4 m span across solar wings, two 1.60 m diameter dual reflector multifeed antennas, one providing transmission/receive, the other transmission only, switchable between Widebeam/Superbeam as required. GEO location maintained within ±0.05°

Mass: 1,878 kg at launch, propellant for 10 years operation; 1,123/866 kg on-station BOL/EOL

AOCS: provided by MBB's 22 × 10 N + 1 × 400 N MMH/MON bipropellant system

TT&C: initially through GSOC German Space Operations Centre at Oberpfaffenhofen, Germany, then via Eutelsat's TCR system, comprising the Satellite Control Centre in Paris and TCR equipment centres in Rambouillet (France) and Sintra (Portugal; also equipped as the remote SCC back-up facility)

Power system: twin solar wings 3,500/3,000 W BOL/EOL.

Eutelsat 2-F2

Launch: 15 January 1991 by Ariane V41 from Kourou, French Guiana

Location: 10° E geostationary; 21.5° W in 1999; 48° E in 2002

Principal applications: TV/radio distribution, domestic telephony, SMS business traffic

Other specifications as for Eutelsat 2-F1.

Eutelsat 2-F3

Launch: 7 December 1991 by first Atlas 2 (AC-102) from Cape Canaveral

Location: 16° E geostationary; 36° E in 1999; 21.5° E in 2000

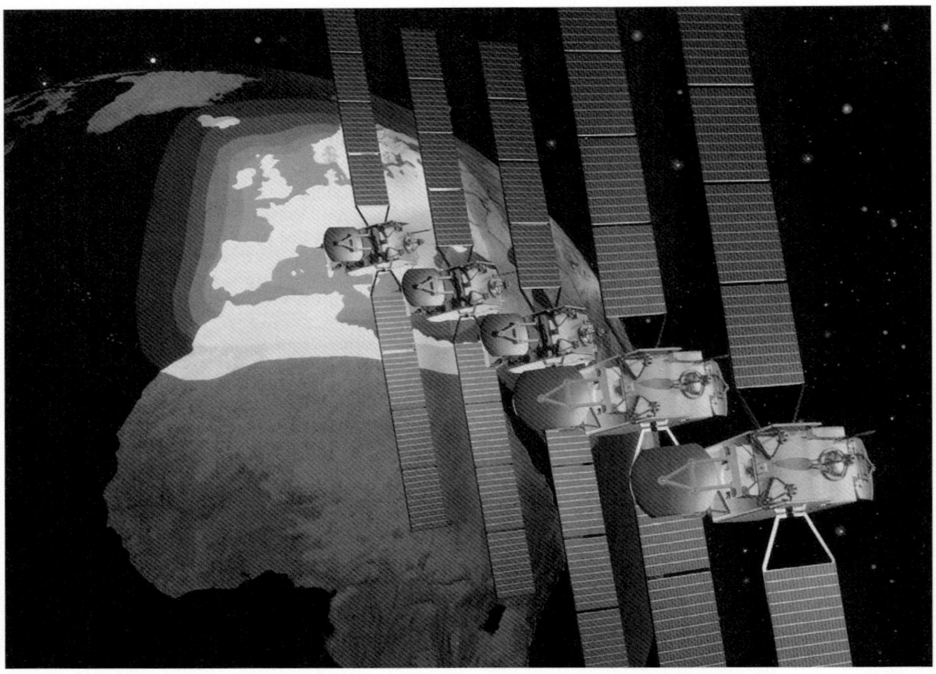

Eutelsat coverage with (from left): Hot Bird 1, Hot Bird 2, Hot Bird 3, Hot Bird 4 and Eutelsat 2F1 0084765

A mock-up of the Eutelsat W satellite bus was test flown on the third Ariane 5G flight on 21 October 1998 (Arianespace) **NEW**/1047472

Principal applications: TV/radio distribution, national/international telecom services, SMS business traffic
Other specifications as for Eutelsat 2-F1.

Eutelsat 2-F4
Launch: 9 July 1992 by Ariane 4 from Kourou, French Guiana
Location: 7° E geostationary; 10° E in 1999; 28.5° E in 2000; 25.5°E in 2002; 12.5° W in 2002
Other specifications as for Eutelsat 2-F1 but Widebeam modified for coverage as far as Urals, providing >42.5 dBW to Moscow and >39.0 dBW to Volgograd.

Eutelsat 2-F5
Lost in Ariane's 24 January 1994 failure. Planned for 36° E, with specifications as 2-F4 (Widebeam modified for coverage to Urals). Launch/satellite fully insured for ECU180 million.

Eutelsat 2-F6/Hot Bird 1
Launch: 28 March 1995 by Ariane 4 from Kourou, French Guiana
Location: 13° E geostationary (collocated with 2-F1)
Design life: 11 years
Transponders: 16 (plus 8 back-up; 12-for-8 ring redundancy) 70 W TWTA 12.895–13.25/11.20–11.55 GHz up/down Ku-band European beams, 36 MHz bandwidths, EIRP 40–49 dBW, linear orthogonal polarisation
Configuration: one transmit 1.60 m diameter dual reflector multifeed antenna and one elliptical receive antenna
Mass: 1,798 kg at launch, 1,078 kg on-station BOL, 840 kg dry
Other specifications as for Eutelsat 2-F1.

Eutelsat 2-F7/Hot Bird 2
Launch: 21 November 1996 by Atlas 2A from CCAFS
Location: 13° east geostationary (colocated with 2-F6)
Design life: 15 years
Transponders: 20 × 115 WTWTA in Ku-band (11.7–12.5 GHz)
Configuration: two 2.3 m transmit antennas for widebeam coverage from Kazakhstan to Persian gulf and 'super beam' coverage for Europe and North Africa.
Mass: 2,923 kg at launch; approx 1,700 kg BOL
Bus: 3.65 × 2.30 × 3.99; span 27.9 m; onboard power 5,500 W (EOL).

Eutelsat 2-F8/Hot Bird 3
Launch: 2 September 1997 by Ariane 44LP
Location: 13° east geostationary
Lifetime: 12 years
Configuration: as Eutelsat 2-F7/Hot Bird 2
Mass: 2,915 kg at launch; 1,715 kg BOL; 1,310 kg dry.

Eutelsat 2-F9/Hot Bird 4
Launch: 27 February 1998 by Ariane 42P
Location: 13° east geostationary
Lifetime: 12 years
Mass: 2,885 kg at launch; 1,770 kg BOL; 1,310 kg dry.

Eutelsat W-F2
Launch: 5 October 1998 by Ariane 44L
Location: 16° east
Transponders: 24 × Ku-band (72 MHz and 36 MHz)
Configuration: 1 fixed widebeam over Europe, North Africa and the Middle East (40–49 dBW), 1 steerable beam positioned over Mauritius and Reunion Island (42–53 dBW)
Mass: 2,950 kg at launch, 1,810 kg BOL; 1,375 kg dry
Bus: 4.6 × 2.5 × 1.8 m; span 29 m; onboard power 5,900 W (EOL).

Eutelsat 2 F10/Hot Bird 5
Launch: 9 October 1998 by Atlas 2A
Location: 13° east

Eutelsat W-F3
Launch: 12 April 1999 by Atlas 2AS
Location: 7° east

Sesat
Launch: 18 April 2000 by Proton K
Location: 36° east

Eutelsat W-F4
Launch: 24 May 2000 by Atlas 3
Location: 36° east

Eutelsat W-F1
Launch: 6 June 2000 by Ariane 4
Location: 10° east

Eutelsat W1R
Launch: 6 September 2000 by Ariane 44P
Location: 10° east

Eurobird 1
Launch: 8 March 2001 by Ariane 5
Location: 28.5° east

Atlantic Bird 2
Launch: 25 September 2001 by Ariane 44P
Location: 8° west

Atlantic Bird 3
Launch: 5 July 2002
Location: 5° west

Hot Bird 6
Launch: 21 August 2002 by Atlas 5
Location: 13° east

Atlantic Bird 1
Launch: 28 August 2002 by Ariane 5
Location: 12.5° west

Eutelsat W5
Launch: 20 November 2002 by Delta 4
Location: 70.5° east

e-Bird
Launch: 27 September 2003
Location: 33° east

W3A
Launch: 16 March 2004
Location: 7° east

UPDATED

Inmarsat 1 (Marecs) Series

Background

In 1975, the Inter-Government Maritime Consultative Organisation (IMCO) began discussing the need for a global satellite system serving shipping in the same way that Intelsat links countries. This body followed up with an agreement to establish Inmarsat, with headquarters in London in 1979. Several countries contributed to the operating costs on inauguration day including the US 23.37 (per cent), USSR 14.09, UK 9.89, Norway 7.88, and Japan 7.00.

With an initial London-based staff of 65 (now totalling 500), and with 37 (participating now from more than 80) countries, Inmarsat took over responsibility for maritime links from the US-run Marisat on 1 February 1982. By March 1987, Inmarsat membership had grown to 48 countries. More than 5,000 of the world's 74,000 ships greater than 100 tonnes, plus oil rigs and other maritime platforms, have each been equipped with terminals costing about US$25,000 and providing direct access to high-quality telephone, telex and other services. The Inmarsat system includes dial phones and an emergency button giving priority to distress and other urgent messages. Despite the Soviets' large initial stake, only 10 of their ships had been equipped with receiving stations, which either the US or Japan built. In comparison the US had equipped 240 US ships by inauguration day.

Initially, Inmarsat leased Comsat's 10 telephone circuit capacity aboard each of the existing Marisats, stationed over the Atlantic at 15° W, the Indian Ocean at 73° E and the Pacific at 176.5° E. Intelsats 5-F5 to 5-F9 leased maritime packages to Inmarsat as well, each providing 30 telephone circuits to ships and providing connections to Inmarsat shore stations through Intelsat's normal frequencies. But this still left spotty service. ESA built upon its ECS experience to propose and develop three Marecs satellites, a maritime version of ECS, produced by the Mesh consortium.

ESA's Marecs A was launched on 20 December 1981, offering an additional 40 telephone circuits over the already saturated Atlantic route at 26° W. Inmarsat leased the Marecs A for an initial five-year period for a total of US$65 million. Marecs A was moved to 22.8° E before retirement in August 1996. On 1 May 1982 after problems had been overcome with MARECS A telemetry and command systems, a build up of electrostatic charge occurred on the satellite bus. ESA traced the problem and began modifications to Marecs B and ECS 1, to fix the situation. These modifications delayed the Marecs B1 launch by 6 months and the satellite foundered in the 10 September 1982 launch failure of Ariane L-05 due to a stage 3 malfunction. ESA declared its cost as US$37.5 million plus US$26.4 million for the launch. ESA paid 160.5 MAU (1983 rate) to complete Marecs A and 107 MAU (1984 rate) for Marecs B.

Marecs B2 remains operational with Inmarsat's system
0517206

In March 1983, the USSR brought an Inmarsat station into use covering both Atlantic and Indian Oceans from Odessa, and the service assisted Soviet space tracking ships. Another major ground station followed at Nakhodka with Soviet electronics and Japanese aerials in 1985, to cover the Pacific/Indian oceans. When the new coast Earth station at Jeddah (Saudi Arabia) opened in January 1987, Inmarsat stations had reached 17, which then grew to 20 by October 1987.

BAe Dynamics, as prime contractor for Marecs, received the go ahead for Marecs B2 in October 1979. An Ariane launched it on 10 November 1984 into 55° W from January 1985 until February 1997 when it was moved to 14.9° W until January 2002 and a shift to 26° W where it was deorbited. It was to have been used for communications with Antarctica before the terrorist attacks of September 2001 and a shift in funding. (Intelsat was covering the Indian Ocean). With Marecs A providing 50 per cent more capacity over the Atlantic than its specified capability, and Marecs B2 underemployed over the Pacific, ESA decided to switch the two satellites in the early months of 1986 to better match their respective capacities. By March 1988, B2 functioned as the prime Atlantic link, with its predecessor relegated to back-up status over the Pacific.

By mid-1985, Inmarsat covered the Atlantic with Marecs A at 26° W (Intelsat MCS as a spare at 18.5° W), the Indian Ocean with Intelsat MCS at 63° E (Intelsat MCS as spare at 60° E), and the Pacific with Marecs B2 at 177.5° E (Marisat F3 as spare at 176.5° E). ESA leased the A/B2 to Inmarsat initially for 5 years for a total of US$65 million.

The system underwent its most fundamental rearrangement in autumn 1990 since becoming operational in 1982, in order to support its expansion into land and aeronautical services. On 29 September 1990, it split into two regions, and Marecs B2 moved west from 26° W to 55.5° W during 5 October – 4 November under the control of ESOC at Darmstadt via the Spanish Villafranca station. The move also closed the coverage gap over parts of North America and the Pacific. This new crowding of GEO slots required Inmarsat to certify precise frequency co-ordination to avoid interference with the four satellites it passed en route – three Intelsats at 27.5/34.5/53° W and a PanAmSat at 45° W. Today the satellite remains active at Inmarsat's Atlantic East 15° W slot.

ESA removed Marecs A from Inmarsat Pacific (178° E) service in mid-1991 after high solar activity degraded its array output and then moved it to 22.5° E (inclined 7.7°) where it served as a testbed for mobile communications experiments. Another section of the solar array stopped functioning in July 1996 and ESA abandoned operations at that time. On 21 August 1996, ESA commanded a series of manoeuvres to boost the satellite above the geosynchronous orbit band to retirement.

Inmarsat dropped the lease on Marecs B2 at the end of 1996. ESA still owns the satellite, and from 1 July 1997 and later, began renting it to Telespazio for back-up of the EMS system on Eutelsat-2 F-2.

Marisat and Marecs continued service even after the Inmarsat 2s became operational. Marisat F1, for instance, provided Inmarsat's first polar links in 1993 from its inclined orbit, and the US National Science Foundation's South Pole Station transmitted data for routing through Comsat's Southbury station.

Any ship can use the Inmarsat 1 service, irrespective of its country, or status as a shareholder. To receive the signal, a ship must have the necessary 1 m diameter parabolic antenna inside a protective radome above deck. Panama, with 130 ships on inauguration day, exemplifies the typical non-shareholding user. Inmarsat adjusts shareholdings periodically according to national use of the system.

On 15 April 1999 the privatisation of Inmarsat was concluded, the first intergovernmental organisation to become a limited liability company. The new organisation is known as Inmarsat Holdings Ltd.

Inmarsat plans to initiate operations with Inmarsat 4 satellites in 2004. One will be stationed over the Atlantic Ocean at 64° east and a second stationed over the Indian Ocean at 54° west, for interoperability with Inmarsat 3 series satellite.

As of April 2000, Inmarsat satellites were deployed to cover four regions as follows: Atlantic Ocean Region-West (AOR-W) Inmarsat 3 F-4 at 54° W and Inmarsat 3 F-2 at 15.5° W. (Inmarsat 2 F-2 as a spare at 98° W, Inmarsat 3 F-2 as a spare at 15.5°, Inmarsat 3 F-5 as a spare at 25° E and Inmarsat 3 F-4 as a spare at 54° W). Indian Ocean Region (IOR) Inmarsat 3 F-1 at 64° E (Inmarsat 2 F-3 as a spare at 65° E). Pacific Ocean Region (POR) Inmarsat 3 F-3 at 178° E (Inmarsat 2 F-1 as a spare at 179° E).

Inmarsat 2 Series

Current Status

Inmarsat currently operates all four Inmarsat 2 class satellites. All satellites survived the solar coronal mass ejection in October 2003 almost completely unaffected by the unprecedented level of activity.

Background

Inmarsat requested bids for its first dedicated satellites in August 1983 and the consortium awarded a contract to British Aerospace in March 1985 for three satellites based on the Eurostar 1000 platform, with options for a further six. The Inmarsat board authorised a fourth satellite in March 1988. The overall programme cost about US$675 million, excluding cost of financing and TT&C capital/operating costs. Each satellite provides 125 voice channels forward, and 250 return, five times the previous capacity.

Inmarsat finalised an agreement in October 1988 to sell and lease back its first three satellites from the North Sea Marine Leasing Co (a partnership of the leasing arms of four major UK banks: NatWest, Barclays, Lloyds, Midland (now HSBC)) for 10 years. The consortium insured each satellite for up to US$160 million (including an element of self insurance) to cover launch through the first year's operation.

These satellites mark the first time Inmarsat manage its own satellites from the HQ's US$10 million Satellite Control Centre, working through ground stations in Pennant Pt (Canada), Lake Cowichan (Canada), Fucino (Italy) and Beijing (China).

Each Inmarsat 2 has a launch mass of 1,310 kg (Ariane) and 1,385 kg (Delta).

Specifications

Inmarsat 2-F1
Launch: 23.16 GMT 30 October 1990 by Delta 6925 from Cape Canaveral pad 17B. Three apogee burns 1–4 November provided GEO injection at 44° E;

Inmarsat 2-F1 EMC test preparations 0516983

arrived 30° E 9 November, where payload activated 13 November for testing. The final move began 26 November, arriving/entering service 8 December, a week earlier than planned. It became a spareat the same location when 3-F1 became operational
Location: 64.4° E GEO over Indian Ocean, relocated to 179° E November 96 currently deployed as a spare
Design life: 10 years minimum; estimated BOL sufficient hydrazine for 14 years
Contractors: BAe (prime; assembly, integration/test; thermal; power subsystems), Matra (AOCS), Fokker (solar arrays), Lockheed (propellant tanks), Marquardt (liquid apogee thruster), Hughes (payload, repeaters and, with Spar, antennas)
Transponders: four (plus two back-up) 45 W TWTA 1.621.64/1.53–1.54 GHz L-band global beams, 5 + 22.5 kHz bandwidths, 39 dBW min EIRP, circular polarisation, to provide 250 voice circuits shore to ship through the 43 cup-dipole radiating elements

one (plus one back-up) 10 W TWTA 6.42 6.44/3.60–3.62 GHz C-band global beams, 5+22.5 kHz bandwidths, 24 dBW min EIRP, circular polarisation, providing 125 voice circuits ship to shore through seven cup-dipole radiating elements. Each ring is separately excited for beam shaping, providing higher power at edge of coverage
Principal applications: maritime, air and land mobile
Configuration: based on box-shaped Eurostar 1000 bus, 2.557 m high, 1.586 m deep, 1.48 m wide, centred on 1.8 m high thrust cone. Payload and solar arrays mounted on North/South panels and Earth-facing floor. The payload uses separate receive/transmission antennas for both L/C band, all four cup-dipole phased arrays mounted on the Earth-facing wall
Mass: 1,385 kg, 824 kg on-station BOL; 130 kg payload mass, 624 kg dry
AOCS: unified redundant 3-axis control and apogee motor thrusters drawing on 110 kg MMH + 130 kg NTO; Kaiser Marquardt R-4D 490 N AKM and two sets of six MBB 10 N thrusters. Roll/yaw control provided by solar sailing with solar array flaps. Single MW with redundancy
Antenna pointing: typically 0.1° half-cone
TT&C: through Inmarsat's London HQ on C-band (control assumed 10 November 1990 from CNES Toulouse facility)
Power systems: 1,200 W BOL/1,142 W EOL provided by twin Si wings, each comprising three 122 cm wide, 173 cm long panels and spanning 15.23 m. Regulated 42.5 V voltage. Eclipse protection provided by two Ni/Cd batteries.

Inmarsat 2-F2
Launch: 8 March 1991 by Delta 6925 from Cape Canaveral pad 17B. Entered service 13 April
Location: 15.5° W GEO over Atlantic Ocean until September 1996, located at 55° W from January 1997 until July 1999 when deployed as a spare at 98° W
Other specifications as for F1.

Inmarsat 2-F3
Launch: 16 December 1991 by Ariane 4 from Kourou. Entered service 19 January 1992
Location: 178.0° E GEO over Pacific Ocean until March 1997, relocated to 65° E and in 2001 when it was moved to 142° W
Mass: 1,310 kg launch
Other specifications as for F1.

Inmarsat 2-F4
Launch: 15 April 1992 by Ariane 4 from Kourou. Entered service 31 May 1992
Location: 54° W GEO over Atlantic Ocean (tested over 32° W); 17° W October 1997 to May 1999; 109° E from July 1999
Other specifications as for F1.

Inmarsat 3 Series

Background

Inmarsat issued a request for a proposal on 2 October 1989 for its third generation of at least three satellites which it then planned to launch in the mid-1990s. The consortium wanted to build a network of global, high-power multiple spot beams for mobile personal communications services, and L-band to L-band connections for direct mobile-to-mobile communications which it expected to be valuable in emergency and search/rescue operations. Responses to the RfP came in during February 1990 from British Aerospace leading a TRW/NEC/Matra team; Martin Marietta Astro Space/MMS UK; Aerospatiale/Alcatel with Ford Aerospace/Mitsubishi; Hughes; and the Indian Space Research Organisation. Inmarsat let the US$320 million contract for four satellites, with options for up to five more, to GE Technical Services Co on behalf of GE Astro Space/Marconi Space on 1 February 1991.

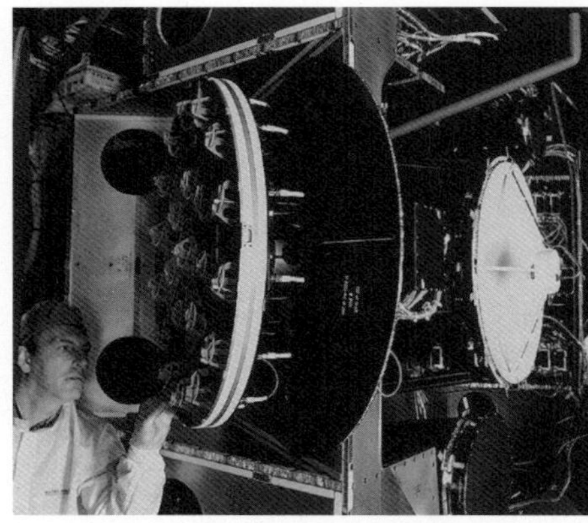

Inmarsat 3 antenna feed assembly. At right is the navigation antenna (MMS) 0517222

The satellites can reallocate power and bandwidths between the five spots and a single global beam to meet changing traffic requirements. Each satellite provides a capacity 10 times that of Inmarsat 2 and can transmit overlay signals almost identical to those of GPS/Glonass to provide three civil navigation functions: improvement of accuracy, compensation for coverage gaps; and continuously updated, independently monitored integrity monitoring. These functions identify faulty satellites, which is essential for air navigation safety.

Inmarsat has tentatively allocated its transponders to Comsat for the Pacific and West Atlantic satellites, France Telecom/Comsat for East Atlantic and to Deutsche Telekom for Indian Ocean service. Inmarsat had 1 million terminals in operation by 2000.

Specifications

Inmarsat 3-F1
Launch: 3 April 1996 by Atlas 2A from Cape Canaveral
Location: Initially 28° E for testing, moved to operational location of 64° E in May 1996
Design life: 13 years
Contractors: Lockheed Martin Astro Space (prime), MMS UK (communications payload), Olin Aerospace (hydrazine thrusters)
Transponders: 490 W SSPA matrix 1.63–1.66/1.53–1.56 GHz up/down L-band global + five spot beams, 34 MHz bandwidth, EIRP 48 dBW total (switchable between global/spot in any proportion of total EIRP), circular polarisation

two (plus two back-up) 12 W SSPA 6.43 6.45/3.60–3.63 GHz up/down C-band global beams (feeder for L-band), 29 MHz bandwidth, EIRP 27 dBW in each of two circular polarisations

one (plus one back-up) 20 W SSPA 1,575.42 MHz GPS L1 L-band global beam, 2.2 MHz bandwidth, EIRP 27.5 dBW (translated from C-band uplink). GPS/Glonass overlay signals compensating for coverage gaps and enhancing their accuracy. Also, continuously updated information on these systems' integrity broadcast to identify faulty satellites. A C-to-C communication channel used by the uplink station for more precise signal tracking (less susceptible to ionosphere disturbances than L-band); channel can also be used for precise time and frequency applications
Principal applications: maritime, air and land mobile
Configuration: based on Astro Space boxed-shaped 4000 bus, body 2.1 × 1.8 × 1.7 m, 2.5 m high, 3.2 m radial envelope, 16.7 m span, centred on

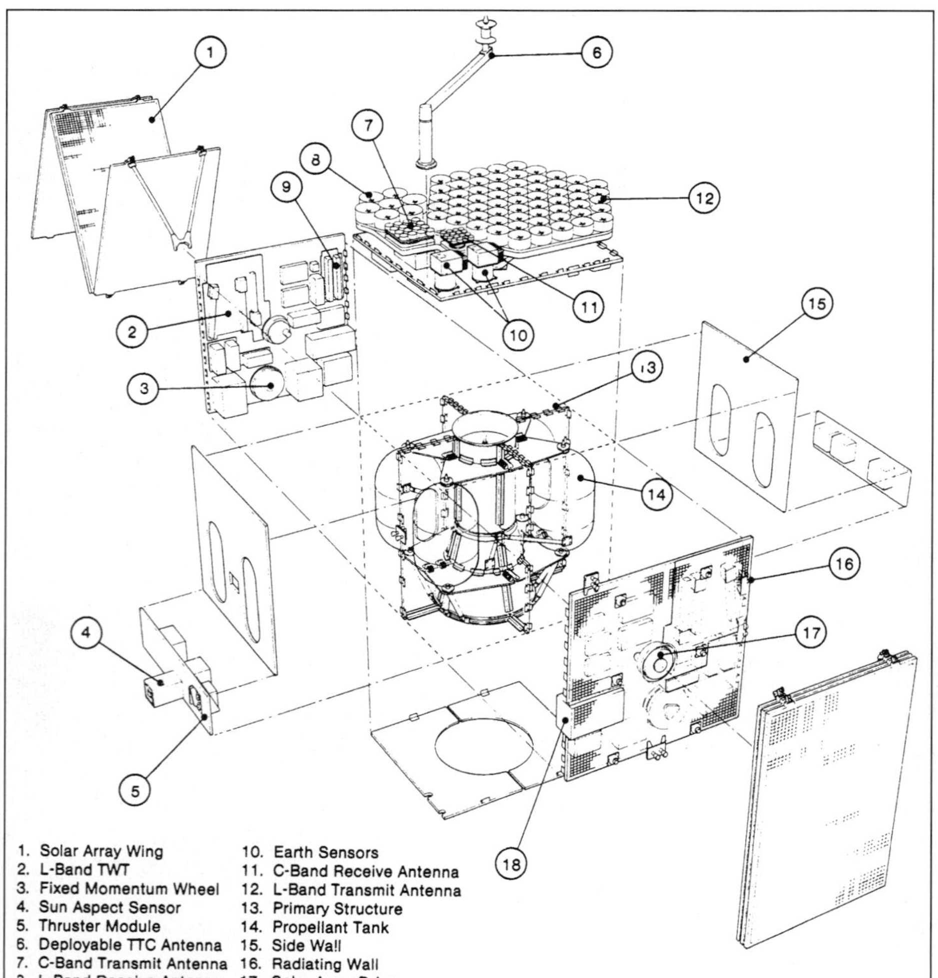

1. Solar Array Wing
2. L-Band TWT
3. Fixed Momentum Wheel
4. Sun Aspect Sensor
5. Thruster Module
6. Deployable TTC Antenna
7. C-Band Transmit Antenna
8. L-Band Receive Antenna
9. C-Band TWT
10. Earth Sensors
11. C-Band Receive Antenna
12. L-Band Transmit Antenna
13. Primary Structure
14. Propellant Tank
15. Side Wall
16. Radiating Wall
17. Solar Array Drive
18. Battery

Inmarsat 2 principal features (Inmarsat) 0516984

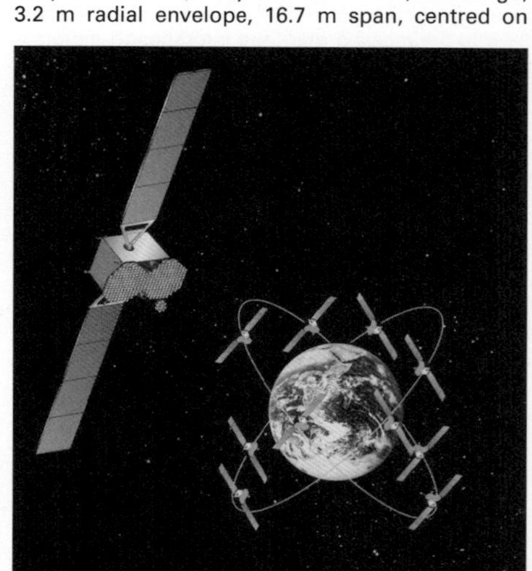

Mobile communication services provided by Inmarsat P satellites utilising the Hughes HS-601 bus (Hughes Space & Communications) 0054339

COMMUNICATIONS PAYLOAD
1 L-band receiver reflector panel
2 Reflector panel support structure
3 Hinged reflector support strut
4 Strut hinge
5 Launch position shear ties
6 L-band receiver feed assembly with thermal sunshield
7 Gold film multi-layer insulation, applied to exterior of satellite except for North and South radiant panels
8 West feed panel with return combiner
9 East feed panel
10 Feed assembly mounting
11 Feed motor drive electronics
12 L-band output network
13 L-band transmitter feed assembly
14 L-band transmitter reflector
15 C-band receiver antenna
16 C-band transmitter antenna
17 Navigation antenna
18 Graphite/epoxy antenna support structure
19 Pilot tone injection system [PTIS]

NORTH TRANSPONDER PANEL [Z+ Wall]
20 Wall structure [aluminium face panel with honeycomb core]
21 C-band receiver [CMR]
22 Navigation solid state power amplifiers [NAV HPA] [2]
23 L-band solid state power amplifiers [SSPA] [33]
24 Output switches
25 Forward beamforming matrix [BFMJ
26 Power supply electronics [PSE]
27 Power junction assembly [PJA]
28 Return IF processor [RIFP]
29 Forward IF procesor [FIFP]
30 Remote terminal unit [RTU]
31 Frequency generator [FGEN]

TELEMETRY TRACKING AND COMMAND SYSTEM
46 TTC Omni-antenna
47 Omni-antenna hinged mounting [ODS]
48 Telecommand lOgic processor [TCP]
49 Central logic extender [CLX]
50 Redundant telemetry module [RTM]
51 Pyrotechnic relay assembly 1 [PRA 1]
52 Pyrotechnic relay assembly 2 [PRA 2]

ATTITUDE AND ORBITAL CONTROL SUB-SYSTEM [AOCS]
53 Roll control torquer
54 Yaw control torquer
55 Roll rate measuring assembly
56 Yaw rate measuring assembly
57 Pitch rate measuring assembly
58 Attitude system processor [ASP]
59 Momentum wheel assembly 1 [MWA 1]
60 Momentum wheel assembly 2 [MWA 2]
61 Horizon sensing assembly [HSA]
62 Earth sensor assembly 1 [ESA 1]
63 Earth sensor assembly 2 [ESA 2]
64 Sun sensor detector 1 [SSD 1]
65 Sun sensor detector 2 [SSD 2]

REACTION CONTROL SUB-SYSTEM
66 Catalytic rocket engine assemblies [REA'S] [12], attitude control
67 Improved performance electrothermal hydrazine thrusters [IMPEHT's] [4], North - South control
68 IMPEHT power electronics
69 Hydrazine propellant tanks [4] with helium pressurant
70 Propellant tank aluminized Kapton multi-layer insulation
71 Hydrazine feed lines to thrusters
72 Propellant fill and drain valves
73 Tank sump, propellant management device [PMD]
74 Tank support struts

BUS MODULE
75 Cylindrical core structure
76 Radial bulkheads/shearwalls
77 Launch vehicle interface/separation ring
78 STAR 37XFP solid propellant apogee kick motor [AKM]
79 Carbon phenolic exhaust cone
80 AKM mounting/thrust ring frame
81 Plume shield

SOLAR ARRAY
82 Single-sided solar panels, inner - mid and outer
83 Outer solar panel provides trickle charge for batteries prior to full deployment
84 6cm x 4cm solar cells
85 Interpanel hinges
86 Hinged berrylium Support boom
87 Solar array boom hinges
88 Solar array launch Position latch
89 Solar array drive motor [SAD]

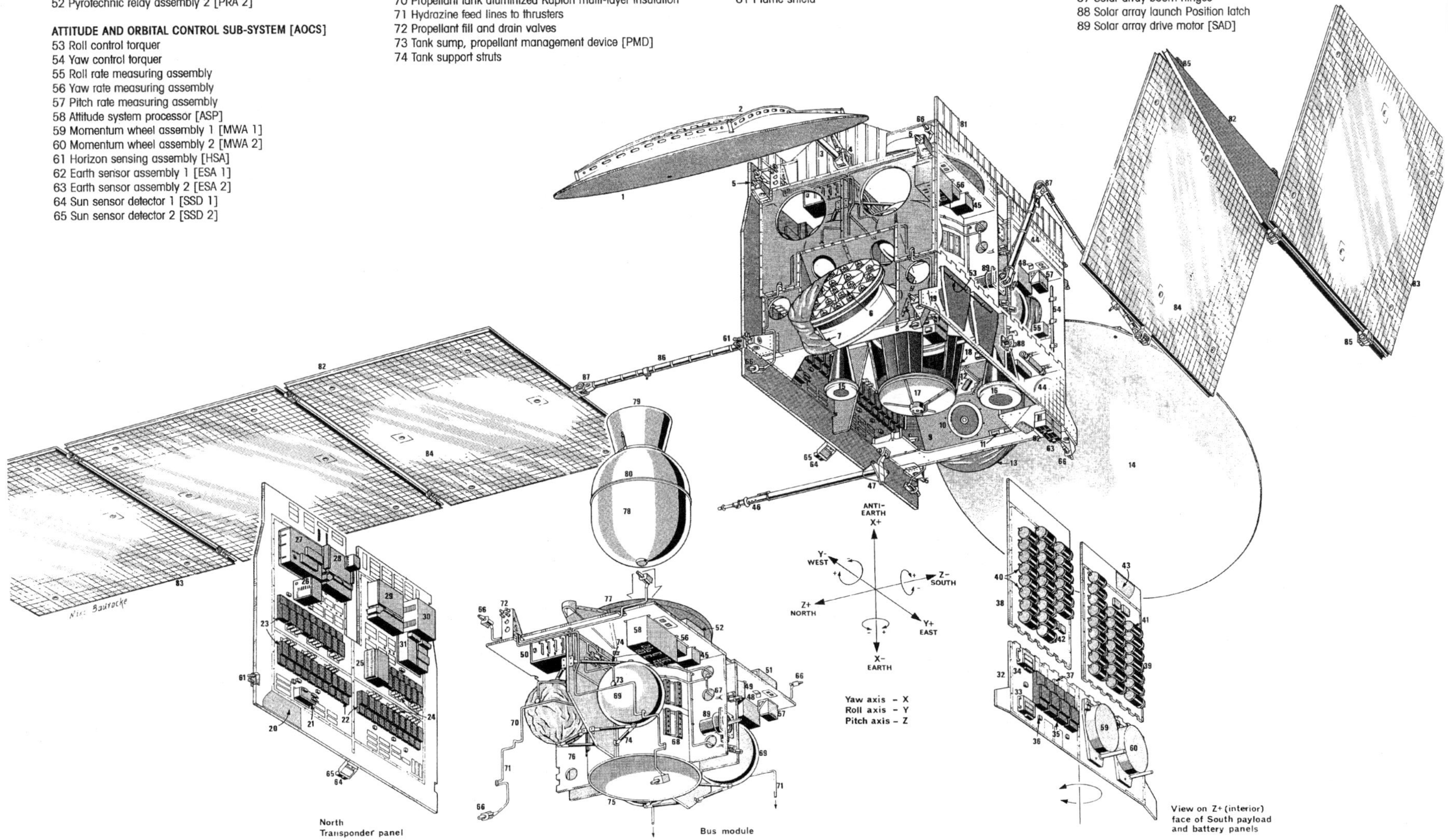

SOUTH PAYLOAD AND BATTERY PANELS [Z- wall]
32 Aluminium honeycomb wall structure
33 No 1 beacon transmitter [CBX 1]
34 No 2 beacon transmitter [CBX 2]
35 C-band power amplifiers [CHPA]
36 Power amplifier output switch
37 Power amplifier input switch
3e Battery panels [2]
39 Nickel-Hydrogen batteries [46]
40 Low emmisivity gold plated battery end caps
41 High emmisivity black painted battery centre sections
42 Battery pressure monitor [BPM]
43 Aluminium honeycomb wall structure
44 North and South panels optical solar reflector external radiant mirror surfaces
45 Battery control voltage monitor [BVCM]

Inmarsat 3 principal features (Inmarsat/Mike Badrocke) 0517406

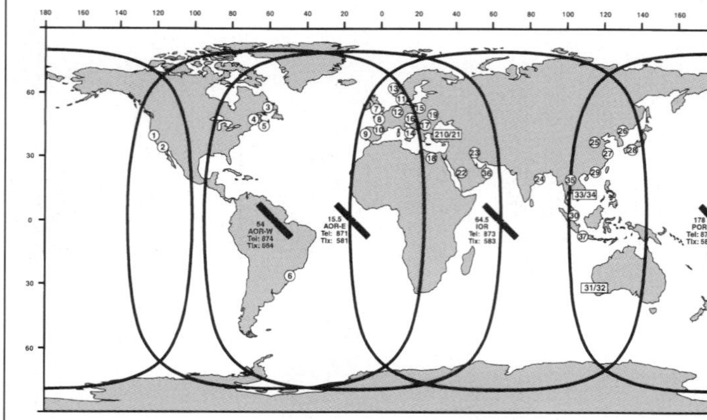

System Availability	Inmarsat-A	Inmarsat-C	Inmarsat-M/B	Inmarsat-Aero
1. Niles Canyon	●			●
2. Santa Paula	●	●	●	●
3. Laurentides			●	●
4. Southbury	●	●	●	●
5. Staten Island	●			
6. Tangua	●	●		
7. Goonhilly	●	●	●	●
8. Pleumeur Bodou	●			
9. Sintra		●		
10. Aussaguel		●	●	●
11. Blaavand		●		
12. Burum	●	●	●	
13. Fucino	●	●	●	
14. Eik	●	●	●	●
15. Psary	●			
16. Raisting	●	●	●	
17. Thermopylae	●	●	●	
18. Maadi	●			
19. Odessa	●			
20. Anatolia	●			
21. Ata	●	●		
22. Jeddah	●			
23. Boumehen	●	●		
24. Arvi	●	●	●	
25. Beijing	●	●	●	
26. Nakhodka	●			
27. Kumsan	●	●		
28. Yamaguchi	●	●	●	●
29. Hong Kong	●		●	
30. Singapore	●	●	●	●
31. Perth	●	●	●	●
32. Gnanagara	●			
33. Comsat Eurasia			●	
34. Kuantan			●	
35. Nonthaburi		●	●	
36. Towi Al Saman			●	
37. Jafilahur			●	

Inmarsat system stations and their capabilities 0517398

thrust cone. Payload and solar wings mounted on N/S panels. L band receive/transmission reflectors mounted on E/W panels and fed by cup-shaped radiating elements
Mass: 2,068 kg at launch, 1,100 kg on-station BOL, 190 kg payload, 860 kg dry
AOCS: 3-axis momentum bias stabilisation with MWs and magnetorquers; pivoted MW provides antennas pointing for up to 2.7° inc. Mono hydrazine thrusters, 283 kg capacity, 12 Reaction Engine Assemblies + four Improved Electrothermal Hydrazine Thrusters. GEO insertion by Thiokol Star 37XFP or 37FM
Power system: 2.8 kW EOL provided by twin Si wings (six panels total 30.5 m^2) spanning 16.7 m. Eclipse protection by 2 × 62 Ah Ni/H_2 batteries.

Inmarsat 3-F2
Launch: 6 September 1996 on Proton from Tyuratam
Location: 15.5° W Atlantic Ocean
Other specifications as for F1.

Inmarsat 3-F3
Launch: 18 December 1996 on Atlas-2A from Kourou
Location: Initially tested over 158° E, then relocated to operational 178° E 25 January 1997
Other specifications as for F1.

Inmarsat 3-F4
Launch: 3 June 1997 on Ariane-4 from Kourou
Location: 54° W in service from 26 July 1997
Other specifications as for F1.

Inmarsat 3-F5
Launch: 4 February 1998 on Ariane 4 from Kourou
Location: 24.5° E and currently deployed as a spare
Other specifications as for F1.

Intelsat 1 (Early Bird)

Background
Conceived as a means of transmitting news and information about the Apollo lunar landings and inspired as a global telecommunications service, the Intelsat system was conceived in late 1963 and early 1964. Manufactured by Hughes Space and Communications. Design life 1½ years; 50 mHz bandwidth; 240 voice circuits.

Description: 70 cm diameter, 60 cm height, 68 kg at launch, 38.5 kg after apogee motor burn. Capacity 240 telephone circuits (triple the trans-Atlantic capacity of the time) or one TV channel. The single Early Bird was launched 6 April 1965 and became operational over the Atlantic at 325° E on 28 June 1965. The world's first commercial communications satellite, it made scheduled trans-oceanic TV possible for the first time, its antenna covering North America/Europe with a 6 W transponder, 50 MHz bandwidth. Design life 18 months, but operated for 3½ years; reactivated in 1984 in celebration of Intelsat's 20th anniversary.

VERIFIED

Intelsat 2 Series

Background
Satellite manufactured by Hughes Space and Communications as part of the expansion designed to provide global coverage of US events such as the (then) upcoming lunar landings in the Apollo Programme. Design life three years; 130 mHz bandwidth; 240 voice circuits.

Description: Extended coverage to two-thirds of the world, though only the last two achieved the planned three-year life. 162 kg at launch, 86 kg after apogee burn. F1, 26 October 1966, failed. F2, launched 11 January 1967 was operational for 2 years over Pacific. F3, launched 22 March 1967, was operational 3½ years over Atlantic. F4, launched 27 September 1967 was operational 3½ years over Pacific.

VERIFIED

Intelsat 3 Series

Background
Manufactured by TRW with a five-year design life, and breaking the pattern of Intelsat satellites built by Hughes, each providing 1,500 voice circuits with 300 mHz bandwidth.

Description: Capacity increased to 1,200 circuits or four TV channels, with 5 years design life (not achieved). The satellite weighed 293 kg at launch and 151 kg after apogee burn. F1, 18 September 1968, failed. F2, launched 18 December 1968, was operational 1½ years over Atlantic. F3, launched 5 February 1969, was partially operational 7 years over Indian Ocean. F4, launched 21 May 1969, was operational 3 years over Pacific. F5, 25 July 1969, failed. F6, launched 14 January 1970, was operational 2 years over Atlantic. F7, launched 22 April 1970, was operational 16 years over Atlantic. F8 failed. Starting with this series, satellites were boosted above GEO if possible at end of life.

VERIFIED

Intelsat 4 Series

Background
Manufactured by Hughes Space and Communications in a reversal of contractor set when TRW won the contract to build the third series the capacity of the basic Intelsat 4 series was a quantum step forward in telecommunications capability with the payload section increased to 4,000 circuits and two TV channels, with 7 years design life. 1,414 kg, reducing to 730 kg on station. Two TV channels and 500 mHz bandwidth. F2, launched 25 January 1971, was operational 12 years over Atlantic. Raised above GEO and deactivated 1983. F3, 19 December 1971, primary Atlantic 3 years, operational >12 years, placed in orbital storage 1983. F4, 22 January 1972 (first commercial satellite assembled outside US, by British Aircraft Corp at Bristol, UK; provided TV link between US/China for President Nixon's visit to Peking), operational over Pacific >7 years, raised above GEO orbit and deactivated 1983. F5, 13 June 1972, operational over Indian Ocean >7 years, raised above GEO and deactivated 1983. F7, 23 August 1973, major-path satellite Atlantic 2 years, operational >6 years, retired and placed in orbital storage 1983. F8, 21 November 1974, operational Pacific from December 1974, in use for international and domestic lease 1983. F6, 20 February 1975, failed to achieve transfer orbit. F1, 22 May 1975 after 4 years storage, operational Indian Ocean 3 years, moved to Atlantic as major-path satellite 1979, domestic lease service 1983.

VERIFIED

Intelsat 4A Series

Background
Description: Manufactured by Hughes Space and Communications using the standard Intelsat 4 satellite as the basis for a major upgrading of the payload sector. Capacity increased to 6,000 circuits plus two TV channels, with seven years life and 800 mHz bandwidth. Significant improvements to the main bus section of the satellite. Intelsat 4A weighed 1,515 kg at launch and 825 kg after apogee motor burn. F1, launched 25 September 1975, was the primary Atlantic satellite from February 1976. F2, 29 January 1976, major-path Atlantic 2 years from 1976, then leased. F4, launched 26 May 1977, leased for Atlantic until December 1977, then major-path satellite. The last fourth-generation satellite to be removed from service, 17 August 1989. F5, 29 September 1977, launch failure. F3, 6 January 1978, leased over Indian Ocean until 1979, then primary satellite.

VERIFIED

Intelsat 5/5A Series

Background
Manufactured by Ford Aerospace. Intelsat began its new quasi-public communication service by purchasing the basic Intelsat 5 satellite model (501–509) which it enhanced by the addition of three cross-polarised spot beams at 6/4 GHz for domestic leased services. The last three of the six 5As operate at both 14/11 GHz and 14/12 GHz to facilitate Intelsat Business Services. The organisation ordered nine satellites in total at US$300 million. Launches included: 501 (Atlas Centaur) 23 May 1981, retired February 1997; 502 (Atlas Centaur) 6 December 1980, retired April 1998; 503 (Atlas Centaur) 15 December 1981, retired January 1998; 504 (Atlas Centaur) 4 March 1982, retired November 1995; 505 (Atlas Centaur) 28 September 1982, retired August 1999; 506 (Atlas Centaur) 19 May 1983, retired July 1998; 507 (Ariane) 18 October 1983, retired August 1996; 508 (Ariane) 4 March 1984 retired; 509 (Atlas Centaur) 9 June 1984 (launcher failure); 510 (Atlas Centaur) 22 March 1985, retired July 1999; 511 (Atlas Centaur) 29 June 1985, currently at 29.5° W since 1998; 512 (Atlas Centaur) 28 September 1985, retired July 1998; 513 (Ariane) 17 May 1988, currently at 177° W; 514 (Ariane) 30 May 1986 (launcher failure); 515 (Ariane) 27 January 1989, retired November 2002.

Each satellite has three-axis stabilisation rather than spin stability. They carry 12,250 circuits plus two TV channels, with 7 year life. On station each weighed 1,012 kg on station and had 19.9 m solar wing span. 505–509 carried maritime packages for lease to Inmarsat which carried the equivalent of 30 phone calls in L-band (1.5 to 1.6 GHz).

Intelsat ordered six 5As from Ford Aerospace in January 1981 and contracted four launches to Atlas Centaur and two to Ariane. By the end of the production run Intelsat increased capacity to 15,000 2-way circuits, plus two TV channels and made improvements to the basic Intelsat 5 package through three additional cross polarised spot beams at 6/4 GHz for domestic leased services. The last two Series 5 satellites operate both in the 14/11 and 14/12 GHz bands to provide the Intelsat Business Service.

As of September 2003 all 13 Intelsat 5/5A satellites successfully launched had been retired from service.

Specifications
Design life: 7–9 years
Contractors: Ford Aerospace (prime), Aerospatiale (modular structure), Thomson-CSF (11 GHz TWTs), MBB (largest subcontractor, at 10 per cent; AOCS, solar array)
Transponders: 32 (plus 12 back-up) 8.5 W TWTAs 5.9 6.4/3.7–4.2 GHz C-band hemispherical zonal and spot beams, 36/72 MHz bandwidth, EIRP 29 dBW (hemisphere), 29 dBW (zone), 32.5 dBW (spot), dual polarisation
6 (plus 4 back-up) 10 W TWTAs 14.0 14.5/10.9–11.7 GHz Ku-band spot beams, 77 MHz bandwidth, 41–44 dBW EIRP, dual polarisation satellites 505–509 carry maritime packages for lease to Inmarsat, each capable of carrying the equivalent of 30 telephone circuits. L-band*:* 1.6365–1.644/1.535–1.5425 GHz up/down, 33 dBWEIRP.C-band*:*6.4175–6.425/4.1925–4.200GHz up/down, EIRP 21.0 dBW
Principal applications: Intelsat services
Configuration: 3-axis box-shaped bus with tower and 15.6 m span solar wings; total height 6.4 m. Panels provide 1,300/1,271 W BOL/EOL. Thiokol solid AKM
Mass: 1,998 kg at launch, 1,188 kg on-station BOL
TT&C: through SCC at Intelsat HQ.

VERIFIED

Intelsat 6 Series

Background
Industry proposals for a new communication satellite with three times the Intelsat 5 capacity, went out to industry in 1980. Many international consortia bid on the request led by Hughes and Ford. Intelsat closed the competition in April 1982 and awarded a US$700 million contract to Hughes for five satellites, with first launch in 1986. The contract called for an option on another six, which would bring the contract to US$1,300 million.

Hughes proposed a spin-stabilised hybrid that offers 33,000 2-way phone circuits and four TV channels. Originally Hughes designed the satellite for launch by Shuttle or Ariane 4; and the first (602) flew aboard an Ariane on 27 October 1989 and entered full service 7 April 1990.

In full service, the five satellites ordered provide the equivalent capacity of up to 120,000 simultaneous 2-way telephone circuits using digital modulation techniques and at least three TV channels utilising 6/4 and 14/11 GHz communication links. Intelsat became the first commercial satellite to employ SS/TDMA Satellite Switched/Time Division Multiple Access, allowing beam interconnections according to traffic requirements. The satellites reuse the C-band frequencies six times through two hemispherical beams and four zone beams with dual circular polarisation. The K-band frequencies undergo reuse twice by spatially isolated east west spots using linear orthogonal polarisation. All provide higher K-band power to promote the use of small Earth stations.

Intelsat launched the first (602) on 27 October 1989 on an Ariane and it entered service 7 April 1990. Titan 3 stranded the second (603) in LEO following launch on 14 March 1990. Shuttle flight STS-49 carried a propulsion module to the stranded satellite and re-boosted it to a correct orbit May 1992, providing a 10 year life; still operating from 19.95° W. 604 went to space on a Titan 23 June 1990; still operating from

Intelsat 6 communications platform (Hughes Space & Communications) 0516987

157° E, followed by 605 (14 August 1991 by Ariane), operating from 33° E; 601 (29 October 1991 by Ariane) still operating from 64.25° E. 604's success allowed Intelsat to move six satellites in the second half of 1990, followed by a further six during 1991, in two batches after 605/601 appeared.

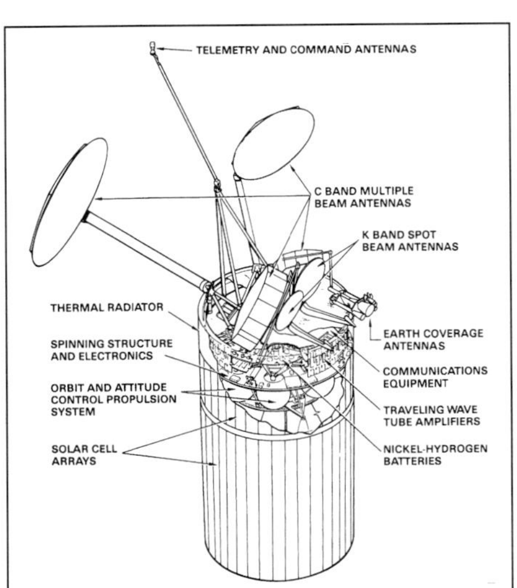

Principal Intelsat 6 features (Intelsat) 0516986

Specifications
Design life: 13 years min 602/3; 15 years 605
Contractors: Hughes Space & Communications Co (prime), British Aerospace (Ku/C-band reflectors, power electronics, Ariane adapter), Pilkington (solar cell covers), Alcatel Espace (Ku/C-band receivers, output multiplexer), Thomson-CSF (Ku-band TWTs), Alenia Spazio (spot antenna, telemetry transmitters/receivers, remote/central telemetry units), AEG (solar cells), MBB/ERNO (solar panels, contract worth about DM50 million), NEC (Ku-band receivers, SSPAs, master oscillators)
Transponders: 8 (plus 3 back-up) 20 W and 2 (plus 1 back-up) 40 W TWTA 14.0–14.5/10.95–11.2 + 11.45–11.70 GHz Ku-band steerable spots, 72 MHz bandwidth, EIRP 45 dBW min, dual polarisation
6 (plus 2 back-up) 16 W TWTA 5.85 6.42/3.62–4.20 GHz C-band hemispherical and zone beams, 72 MHz bandwidth, EIRP 30 dBW min, dual polarisation
32 (plus 10 back-up) 10 W SSPA 5.85 6.42/3.62–4.20 GHz C-band hemispherical and zonal beams, 72 MHz bandwidth, EIRP 30 dBW min, dual polarisation
Principal applications: Intelsat services
Configuration: Hughes' HS-393 platform spin-stabilised, cylindrical body 3.63 m diameter, deployed height 11.84 m (height aboard Ariane 4 6.4 m, height with PKM 8.06 m). Each 3.2 m (30 kg carbon, Kevlar/boron fibre composite) + 2.0 m C-band antenna provides two fixed beams for hemispherical coverage and four isolated beams for zone coverage (these can be reconfigured in orbit to match the requirements of the three ocean regions). Steerable 1.1 m + 1.0 m diameter Ku-band reflectors provide spot coverage
Mass: 4,600 kg at launch (includes adapter), 2,546 kg on station BOL, 1,896 kg dry
Attitude control: 30 rpm spin-stabilised; thrusters fired automatically when attitude excursions exceed pre-set limits
Antenna pointing: ±0.05° maintained in beacon tracking mode, ±0.10° with Earth/Sun sensors
TT&C: through Intelsat's SCC HQ station
Power system: drum-mounted solar arrays provide 2.6 kW BOL/2.252 kW EOL; K7 cells on skirt, K4¾ on fixed body. Two Ni/H_2 batteries provide eclipse protection
Propulsion: perigee kick from Titan 3 or Shuttle provided by 8.76 t UTC/CSD Orbus 6S solid propellant stage. ACS unified with apogee kick*:* MMH/NTO supply four 22 N radial thrusters for E-W station-keeping + spinup/down, two 22 N axial for NSSK + attitude control, and two 490 N for apogee kick + reorientation manoeuvres.

VERIFIED

Intelsat 7 Series

Background
The Intelsat Board requested proposals in October 1987 for fixed price bids for 2–3 satellites, with an additional bid for a fourth, for a new generation of Pacific region satellite. Intelsat expected to put the satellites in service in late 1992 or early 1993. Previous satellites had met Atlantic region

The first Intelsat 7 entered service in 1994 (Intelsat) 0516985

requirements; which differed from Pacific service. Space Systems/Loral received the contract in June 1988. Matra served as a back-up source. On 9 September 1988, Intelsat awarded a US$394.28 million contract for five series 7 satellites with the first launches in mid-1992 and early 1993.

The satellites, combining an Intelsat 5-type payload with Ford's FS-1300 bus, offer increased power and coverage for smaller ground stations and flexibility through three independently steerable Ku-band spots and a steerable C-band spot. Intelsat decided in December 1989 to increase the steerability and Ku-band power of F3 and its successors. The satellite operates inverted at 66° E/1° W, simplifying antenna design. General Dynamics' Atlas 2AS has launched three, charging Intelsat US$75 million each. ArianeSpace's Ariane 4 launched two at US$86.5 million each.

Intelsat awarded General Dynamics a US$9.5 million contract in 1991 to pay for enhancing Atlas 2AS capacity into GTO by 121 kg to give the new satellite additional propellant providing a 16 year life. Intelsat decided in March 1993 to order a sixth satellite (709) and completed the order September 1993.

Specifications
Launch schedule: 701 22 October 1993 (Ariane 44LP; entered service 15 January 1994) operating from 180° E, 702 17 June 1994 (Ariane 44LP), operating from 55° E, 703 6 October 1994 (Atlas 2AS) now retired, 704 10 January 1995 (Atlas 2AS; entered service 25 February) now retired, 705 (Atlas 2AS) 22 March 1995 (Atlas 2AS) operating from 50° W since May 2002, 709 15 June 1996 (Ariane 44P) operating from 85° E (Ku-band only), 706 (Ariane) 17 May 1995 tested at 56° W prior to problems, now operating from 307° E; 707 (Ariane) 14 March 1996 now operating from 359° E; 708 (Ariane) 14 February 1996, lost in launcher failure.
Design life: 65% transponder operability after 10.9 years; sufficient propellant for 15 years
Contractors: Space Systems/Loral (prime), Alcatel Espace (payload), NEC (20/30 W SSPAs), MELCO (10/16 W SSPAs), AEG (TWTs)
Transponders: 26 SSPA (10, 16, 20, 30 W) in 7-for-5 rings 5.8–6.5/3.6–4.2 GHz up/down C-band global (26–29 dBW), hemispherical + zonal (33 dBW) and spot (33–36 dBW), 36/72 MHz bandwidth, dual polarisation

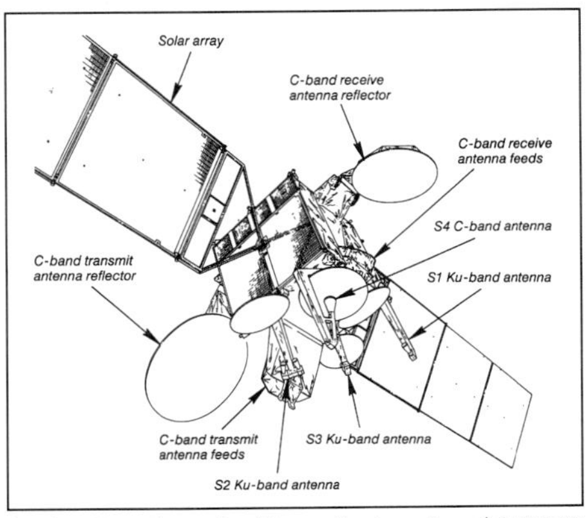

Intelsat 7 configuration (Space Systems/Loral) 0516988

5 × 35 W (7-for-5) + 5 × 50 W (8- for-5) TWTA 14.0–14.5/11.0–11.2 + 11.5–11.7 GHz Ku-band spot beams, 44–47.5 dBW EIRP, 6 × 72 MHz + 4 × 112 MHz (allows 155 Mbit/s) bandwidth, dual polarisation
Configuration: box-shaped SS/L FS-1300 bus with Intelsat 5 type communication platform with three steerable Ku-band dishes and one C band, in addition to fixed C-band receive (1.57 m diameter) + transmission (2.44 m diameter)
Mass: 1,450 kg dry mass, 3,650 kg GTO (for Ariane, with 2,160 kg propellant load), typically 1,800 kg on-station BOL
AOCS: 3-axis (from immediately after launcher separation). 12 × 22 N N_2O_4/MMH, 490 N thruster permits Perigee Velocity Augmentation during Atlas orbital insertion, adding 2.1 years to life. Propellant loading 2,160 kg for 92% fill
TT&C: through Intelsat's SCC HQ station
Power system: twin 3-panel solar wings, spanning 21.843 m, providing 3.9 kW after 10.9 years. Supported by four Ni/H_2 batteries totalling 85.5 Ah.

Intelsat 7A Series

Background

Intelsat authorised the purchase of two 7A satellites in December 1990 at a total fixed price cost of US$202.4 million (33.3 per cent in-orbit performance incentive) to replace the 5As over the Atlantic during 1995 and 1996. Almost identical to the 7 model, they offer increased Ku-band capacity and higher power global transponders. The increased mass requires a more powerful Ariane booster. Each new satellite costs about US$85 million each. Intelsat 706 was tested at 56° W and it became operational on July 1995, though there were technical problems with the power system. Intelsat ordered a third 7A (708) from Space Systems/Loral in September 1992 for US$140 million to keep up with demand. China supplied a US$60 million CZ-3B launcher, but 708 was lost in the February 1996 failure of the Long March.

Specifications

Launch schedule: 706 17 May 1995, 708 14 February 1996 (launch failure), 707 14 March 1996
Locations: 706 307° E, 707 359° E (708 was planned for 50° W)
Transponders: C-band global EIRP 29.0 dBW, C-band spot 36.1 dBW
Ku-band 35 W + 50 W TWTAs replaced by 49 W + 73 W, respectively, and can be paralleled for 49.5 dBW EIRP. Four wideband added for total of 6 × 72 MHz + 8 × 112 MHz
Configuration: body stretched about 45 cm for additional Ku transponders, increased radiator area and stretched propellant tanks
Mass: 1,748 kg dry mass, 4,180 kg GTO (for Ariane 44LP/H10- 3, with 2,820 kg 100% propellant load)
Power system: twin 4-panel solar wings with improved cells, providing 5.3 kW after 10.9 years. Supported by Ni/H_2 batteries totalling 120 Ah.

Intelsat 8 Series

Background

Intelsat issued a RfP in December 1991 for three new satellites, with options for a further four if required. Instead of the full purchase, it allocated funds for a third 7A from SS/L, while Lockheed Martin's 7000 platform received the contract for the two new generation 8 satellites at US$165 million. Generation 8 satellites feature a design to meet the growing Pacific region need, using 6-fold C-band and 2-fold expanded C-band frequency reuse.

Intelsat selected Ariane 4 boosters to launch the satellites and announced the decision in December 1992, then it decided to exercise options for three further satellites, including the first 8A (805) landmass satellite. The contract provides the satellites at a favourable price as part of the settlement with Martin Marietta Astro Space over the loss of Intelsat 603.

Specifications

Launch schedule: 801 (Ariane 44LP) 1 March 1997, damaged in on-orbit testing, currently at 31.5° W; 802 (Ariane 44LP) 25 June 1997 tested at 170° E, located at 174° E; 803 (Ariane 44LP) 23 September 1997, transferred to New Skies Satellites N.V. on 30 November 1998, moved from 27.5° W to 21.5° W in 1998, to 19.8° W in 2002 and to 177° W to replace 513 (now called NSS-5); 804 (Ariane 42L) 22 December 1997 operated from 44° E 1997 to 2002, then to 64.15° E in 2002 and 176° E in November 2002; Atlas 11AS; 805 (Atlas 11AS) 18 June 1998 operating at 304.5° E.

The first Intelsat 8 was launched in 1997 (Intelsat)
0005566

Design life: 10 years (13–19 years propellant)
Contractors: Lockheed Martin Astro Space
Transponders (801–804): 18 27–38 W variable power SSPA C band, providing 6 global (EIRP 29.0 dBW) + 12 hemispherical (EIRP 34.5–36.0 dBW) bandwidths 34, 36, 41, 72, 77 MHz
20 10–20 W variable power SSPA C-band zonal beams, EIRP 34.5–36.0 dBW, bandwidths 34, 36, 41, 72, 77 MHz
6 43 W TWTA in two Ku-band steerable spots (2.8° circular + 1.9 × 4.3°), EIRP 44/47 dBW inner/outer, 602 MHz bandwidth (as 34, 72, 77, 112 MHz)
Principal applications: Intelsat services
Configuration: 7000 series platform. 3-axis box-shaped 216 × 246 × 315 cm bus with central cylinder
Mass: 3,245 kg at launch, 1,538 kg dry
AOCS: spin-stabilised in transfer orbit; 3-axis on-station using roll/yaw magnetorquers in normal mode, hydrazine thrusters for station-keeping. Redundant 1.8 kW hydrazine arcjets provide NSSK. Pointing: ±0.1° pitch/roll, ±0.25° yaw accuracy; ±6.0° pitch + ±2.0° offset pointing control
TT&C: through Intelsat's SCC HQ station
Power system: twin 4-panel wings (totalling 48.6 m^2) providing 4,800 W BOL. 2 × 45 Ah Ni/H_2 batteries provide 100% eclipse protection.

Intelsat 8A Series

Background

Intelsat, in September 1993, exercised its option on the generation 8 contract and purchased the first 8A (805) landmass satellite. It provided expanded C/Ku-band coverage for Asia-Pacific users, with a minimum 36 dBW EIRP C-band and fully-manoeuvreable Ku beams to cover all of China, Korea and most of Japan. Intelsat ordered 806 in March 1994 for US$72.3 million, plus US$10 million in-orbit performance incentives, to cover the Americas.

Specifications

Launch: 806 28 February 1998 (Atlas-2AS), 805 18 June 1998 (Atlas2AS)
Locations: 805 operating from 304.5° E, 806 tested at 55.5° W until April 1998 when moved to 40.5° W and transferred to NSS N.V. in November 1998, now retired.
Transponders: 28 60 W TWTA C-band broad hemispheric beams, 805 EIRP 37.2/36 dBW inner/outer (806 37.5 dBW). 3 130 W TWTA Ku-band steerable shaped beams, 805 EIRP 49/42 dBW inner/outer (806 48 dBW)
Mass: 3,570 kg at launch, 1,595 kg dry
Other specifications as for Intelsat 8.

Intelsat 9 Series

Background

In December 1996, Intelsat management chose to procure two follow-on spacecraft (FOS-11) from Space Systems/Loral, which it wanted to operate over the Indian Ocean as replacement for Intelsat 6 satellites. Re-named Intelsat 9, the programme has now been expanded to include seven satellites.

Digital voice and data services will be provided in both C-band and Ku-band. Intelsat is investing approximately $1 billion in the first few primary satellites with a fifth under contract for back-up or reassignment.

Key parameters include 32 × 72 MHz and 12 × 36 MHz C-band transponders with uplink frequency of 5.850 to 6.425 GHz and downlink frequency of 3.625 to 4.200 GHz. Typical e.i.r.p range is 43.0–37.0 dBW for the hemi-beam, 44.8–37.0 dBW for the zone beam and 34.0–31.0 dBW for the global beam. In addition, the satellites each have 8 × 72 MHz and up to 8 × 36 MHz Ku-band transponders. Uplink frequencies are in the range 14.00–14.50 GHz with downlink frequencies of 10.95–11.20 GHz and 11.45–11.70 GHz. Typical e.i.r.p. ranges are 52.0–47.0 dBW for Spot 1 and 52.0–47.0 dBW for Spot 2. In all, Intelsat 9 has 60 transponders. Carrying the target satellite capacity in the Intelsat system, Intelsat 9 will have the flexability and power to provide DTH services to 60 cm antennas.

Spacecraft delivery and launch periods*:*

Spacecraft	Launch Vehicle	Launch	Orbit Location
Intelsat 901	Ariane	9 June 2001	67.5° W until September 2001, then 53° W, then 18° W in November 2001
Intelsat 902	Ariane	30 August 2001	62° E
Intelsat 903	Ariane	30 March 2002	325.5° E
Intelsat 904	Ariane	23 February 2002	Tested at 60° E
Intelsat 905	Ariane	5 June 2002	Tested at 25.5° W
Intelsat 906	Ariane	6 September 2002	Operating at 64° E
Intelsat 907	Ariane	15 February 2003	Operating at 332.5° E

Intelsat K Satellite

Background

In 1989, Intelsat purchased the partially completed GE Satcom K4, with extensive payload modifications (coverage, frequencies and switching matrix), to guard against potential transponder shortages. It is Intelsat's first all-Ku satellite and the first the organisation bought from an outside source. It operates in collocation with Intelsat 512 (C-band) over the Atlantic. The satellite is a modified Satcom bought to fill urgent needs and should have been designated Saturn K4. It cost Intelsat an estimated US$80 million for procurement and modifications, plus US$10–15 million for on-time delivery and in-orbit performance incentives. Intelsat K was transferred to NSS N.V. in August 1998 and was retired out of service in August 2002.

Specifications

Launch: 10 June 1992 by Atlas 2A from Cape Canaveral. Commercial service began 20 August 1992
Location: 21.5° W geostationary but now retired from service
Design life: 10 years (10–12 years station-keeping propellant)
Contractors: GE Astro Space Division (through GE Technical Services Co)
Transponders: 16 (2 rings, 11-for-8) 62.5 W TWTA 14.00 14.50/11.45–11.70 (all beams), 11.70–11.95 (North/South America), 12.50 12.75 (Europe) GHz up/down Ku-band beams providing 47 dBW edge (50 dBW centre) over North America + Europe and 45 dBW edge over South America (42.7 dBW North/South America in split mode), 54 MHz bandwidth, orthogonal linear polarisation. In normal mode, North America/Europe covered by 8 transponders each. Eight North America downlinks can be switched between North America/South America downlinks, or operate in a split North America/South America mode. Four of these can also be switched to Europe (making 12 in all). South America coverage is a satellite transmit-only service
Principal applications: primarily video (27 MHz) and business services. The high EIRP/gain are designed for small Earth stations (<1 m for receive only). Italian/German TV/radio began DTH to 66–88 cm dishes in the Americas in 1992
Configuration: GE 5000 series platform with a 3-axis box-shaped bus with central cylinder. 90 × 80 cm fixed reflector dish and tower on Earth face for South America coverage and deployable 2.15 m dual gridded reflector on west face for North America and Western Europe coverage
Mass: 2,836 kg at launch, 1,547 kg BOL on-station, 1,219 kg dry
AOCS: spin-stabilised at 10 rpm in transfer orbit; 3-axis on-station using roll/yaw magnetorquers in normal mode, hydrazine thrusters for

Intelsat K is collocated at 21.5° west with Intelsat 512 0516989

station-keeping. Two 445 N dual mode liquid apogee motors for GEO insertion using NTO/MMH. Redundant branches of eight catalytic hydrazine thrusters for AOCS (each 1 × 4.45 N, 5 × 0.89 N, 2 × 22.2 N) plus redundant 2 × 0.44 N EHT for station-keeping
TT&C: through Intelsat's SCC HQ station
Power system: two 15.6 m span solar wings of four panels each (totalling 34.70 m^2) provides 4.8 kW BOL/3.5 kW EOL (10 years). 4 × 50 Ah Ni/H_2 batteries provide 100% eclipse protection and EHT firing support.

Marisat Series

Background

The United States government decreed in early 1963 that it would form a global commercial satellite communication service in conjunction with other countries. Shares in the new company became available in February 1963 as the public offering of the Communications Satellite Corporation (Comsat). Comsat General introduced pre-Inmarsat maritime service through its three Marisats, positioned over the Atlantic, Pacific and Indian oceans during 1976, and then leased them to Inmarsat when that organisation became active in February 1982. By 1980, Marisat had 300 commercial users, including cargo ships, oil tankers and offshore platforms. The US and UK navies both used the system and Comsat reserved three UHF channels for US government use. Despite their five year design lives, all three Marisats continued into 1995 as Inmarsat spares over the ocean regions. The US National Science Foundation's South Pole Station transmits data for routing through Comsat Southbury station.

Specifications

Marisat 1
Launched: 19 February 1976 by Delta from Canaveral
Location: 105° W geostationary from launch to 1976, then to 106° W before retirement in April 1997.
Design life: 5 years
Contractors: Hughes Space & Communications
Transponders: single L-band 1.6385–1.642/1.537–1.541 GHz up/down, EIRP 27 dBW with a 4 MHz bandwidth. Single C-band 6.420–6.424/4.195–4.199 GHz up/down, EIRP 18.8 dBW, 4 MHz bandwidth. Three UHF channels

Three Marisats provided a pre-Inmarsat maritime service 0517018

Principal users: spare in Inmarsat's maritime system
Configuration: spin-stabilised cylindrical platform with despun antenna section; height 231 cm. Solar array of 7,000 Si cells on cylindrical surface provided 330 W BOL
Mass: 655 kg at launch, 362 kg on-station BOL
TT&C: Comsat's Washington facility working through the Southbury, Connecticut and Santa Paula, California stations.

Marisat 2
Launched: 10 June 1976 by Delta from Canaveral
Location: 73° E geostationary through March 1995, then to 72.5° E inclined 11.4° prior to retirement in September 1996.
Other specifications as for Marisat 1.

Marisat 3
Launched: 14 October 1976 by Delta from Canaveral
Location: 15° W from launch through October 1976, then to 178° W in March 1995, then to 72° E in October 1996 and to 33.9° W inclined in August 2000 available for Antarctic telecommunications traffic.
Other specifications as for Marisat 1.

Olympus Satellite

Background

Olympus 1 introduced the world's largest civil communications platform when ESA launched it in 1989. An Ariane put the Olympus in orbit on 12 July 1989 and positioned it at 19° W. During the ESA experimental phase it suffered its most serious in-orbit problems when ground controllers lost control in May 1991 due to a stabilisation problem, which caused electrical drop-outs. As it drifted over the Pacific and Atlantic Oceans, it was acquired by radar and the Darmstadt control station resumed authority over it. A complex recovery programme returned it to 19° W after 77 days, on 13 August 1991. But the rescue drew heavily on propellant reserves, and ESA began to lose hope that it would maintain the projected nominal 5-year mission, in early 1994. When control ceased again on 11/12 August 1993, due to a meteor impact from the Perseid shower the few remaining kg of station-keeping propellant proved insufficient for recovery, resumption of services and end of life de-orbiting. ESA decided instead to lower Olympus from GEO and ended the mission formally on 30 August 1993. The satellite nonetheless demonstrated new market applications and helped to establish requirements for the projected Data Relay Satellite (DRS).

Beginning development in 1979 and known until 1983 as L-Sat, the 2,612 kg Olympus, covered direct and semi-direct TV/radio broadcasting, inter-city telephone routing, business communications and mm-wave communications. It carried in total four payloads: a 2 channel 230 W DBS 18/12 GHz a 4-channel 12/14 GHz Specialised Services Payload; the 20/30 GHz advanced experiments; and a 12/20/30 GHz beacon package for propagation studies.

The CCTS established a group of Experimenters for Olympus assisted by Eutelsat, which approved 22 experiments by March 1990 for using the DBS, 20/30 GHz and Specialised Services payloads. Technical institutes across Europe took measurements on the propagation payload and co-ordinated their results. On board the Olympus the IOC (In-Orbit Communications) experiment operated the first Ka-band inter-satellite datalinks, working with Eureca 1 between August 1992 and 24 June 1993. More than 100 organisations in 12 countries used the facility.

Outside auditors estimated the Olympus cost ECU700 million at 1988 rates for the complete programme, including two spacecraft, launcher, insurance, ground segment investments and 5 years of operations. Austria 0.9 per cent, Belgium 3.2 per cent, Canada 11.0 per cent, Denmark 1.2 per cent, Italy 31.5 per cent, Netherlands 10.4 per cent, Spain 2.7 per cent, UK 39.1 per cent all participated in the programme.

OTS Satellite

Background

ESA used its initial communications satellite programme, the OTS Orbital Test Satellite technology demonstration mission, to provide the basis for developing the medium-class ECS European Communications Satellite and its Marecs maritime derivative.

The ESA board approved OTS in 1971, but the programme began inauspiciously. OTS 1, intended to be ESA's first communications satellite, disappeared in the 14 September 1977 Delta 3914 explosion, which occurred 54 seconds after lift-off from Cape Canaveral. ESA, had already suffered losses and delays as a result of a series of US launch failures, and to protect itself, it insured the launch for a total of US$29 million. The identical OTS 2 back-up successfully went into space aboard a Delta on 12 May 1978 at 10° E GEO. Its six antennas covered Western Europe, the Middle East, North Africa and Iceland.

Hawker Siddeley (later part of British Aerospace) built the 865 kg (444 kg on-station), 6-sided satellite, and acted as prime contractor for the 10 Mesh countries. The satellite became among the first to operate in the 11/14 GHz Ku-band. This OTS satellite provided four wideband channels able to accommodate up to 7,200 telephone circuits; alternatively, each channel could handle one or two TV transmissions.

After 1 year of non-revenue experimental use, covering telephone, data transmission and TV (including transmissions between Europe, Egypt and Morocco), ESA announced that the commercial potential of OTS had been proved and released the satellite to Eutelsat. Under a joint management structure with ESA, Eutelsat scheduled and controlled all communications traffic through OTS for itself and its customers.

Eutelsat ceased using OTS at the end of 1983. During 1984, ESA conducted a series of end of service tests, involving some risky manoeuvres. The space agency put the satellite in an intentional flat spin and then recovered it, the first time a 3-axis GEO craft had been so recovered. ESA also established a method of station-keeping by solar sailing allowing future spacecraft to tighten control specifications from ±0.1° to within ±0.035° without consuming significantly more propellant.

Solar sailing found an operational application for more than 2 years during normal mode control. In late 1984, OTS then entered hibernation for final studies of long term subsystem degradation. In May 1988, it broadcast celebrations marking its 10th anniversary. The last of its eight TWTAs failed at the end of 1990. ESOC commanded multiple burns during 2/3 January 1991 to raise its orbit 320 km above GEO, leaving it to complete a westward circuit every 90 days.

The European Communications Satellite system became operational when OTS-2 was handed over to Eutelsat in October 1984 and three satellites under the ECS designation were ordered, each about twice the weight of the OTS types.

Israel

Amos Series

Current Status

Israel Aircraft Industries (IAI) plans to launch the Amos 2/CERES Central European Regional Satellite on 16 December 2003 aboard the Russian

The completed Amos 1 at IAI/MBT 0517395

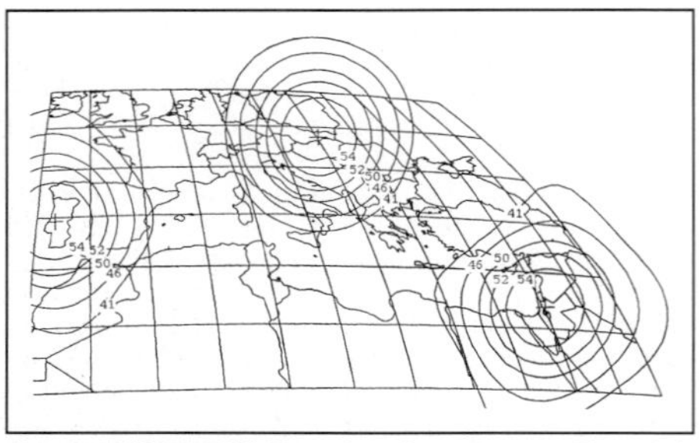

Amos 1 EIRP (dBW) footprints from 4° west. Boresight EIRP is 56 dBW (IAI/MBT) 0517196

Soyuz launcher from Baikonur. For the project to proceed, the Hungarian government agreed to lease two transponders, on 30 April 1995. The satellite will be owned by Magyarsat in Hungary, a 50/50 joint venture by IAI and Antenna Hungaria, and collocated with Amos 1 with eight or sixteen active 72 MHz transponders (plus two back-ups). The Israel beam will provide Amos 1 back-up (there will be no Portugal beam).

Background

Israel Aircraft Industries began the Amos programme in 1990, with the full go-ahead formally approved in January 1992, to develop, build, operate and market a medium-class GEO satellite, principally for sub-regional TV broadcasting. The satellite is an acronym for Affordable Modular Optimised Satellite. IAI retains ownership of Amos 1. It created a commercial entity named Spacecom Ltd in 1993 to sell capacity. Spacecom is in equal partnership with IAI, Gilat Communication Engineering, General Satellite Services Co and Mer Services Group Ltd. Industry sources project the cost through launch, plus operating costs over 10 years at US$250 million. Israeli banks provided US$100 million, guaranteed by the government and the government supports the programme at US$15 million per year. The Israeli government initiated the project by guaranteeing it would use three transponders over the 10 years life and thereby replacing the three previously leased from Intelsat. Transponder cost is about US$3 million annually non-pre-emptible. Israel hoped that Arab users would be signed up before launch since the majority of the 250 million people within the 700,000 km^2 Middle East beam are Arabic speakers. Amos 1 still provides DBS service providers, cable TV companies and TV broadcasters with channels and HBO uses Amos 1 to deliver programmes to Hungary, Poland, Slovakia, Romania and the Czech Republic. It is also the "YES" digital DBS service provider to Israel.

Specifications

Amos 1

Launch: 16 May 1996 by Ariane 4 from Kourou, French Guiana
Location: 4° W (Israel also holds registrations at 39° E/1.5° E)
Design life: 10 years
Contractors: IAI (MBT; prime, systems engineering, mission life operations, structure, AOCS, TT&C, data handling, thermal), Alcatel (Ku-band payload), DASA/Dornier (power), DASA/MBB (thrusters), ANT (transponders), CDI (OBCs), Gates (Ni/H2 battery, integrated by SS/L), Teldix (momentum wheels)
Transponders: 7 (plus 2 back-up in 7-for-2 ring) 36 W TWTA 14.0045–14.4955/10.9545–11.1935 & 11.4565–11.6955 GHz Ku-band up/down in two of three beams using single reflector: Israel beam can take 3–6 transponders, EIRP 55 dBW max, 14.0 dB/K max, –99 to –74 dBW/m^2; European beam (centred on Hungary) of 1–4 transponders, one switchable to Portugal. 72 MHz bandwidth, orthogonal polarisation
Applications: projected 70–80% TV distribution; DTH to 50–60 cm dishes; VSAT video, voice and data (including interactive learning) HDTV
Configuration: 1.21 × 1.670 × 1.930 m bus, solar array span 10.550 m. Single 1.73 m diameter deployed Ku-band antenna on west face. 3-axis stabilisation by MWs (no gyros); wheel unloading/station-keeping by redundant sets of 7 × 10 N thrusters. 400 N thruster accomplishes GEO insertion. Control by two 1,750A OBCs. Internal structure + East/West faces are composite; North/South are aluminium honeycomb. Ku-band horn for TT&C
Mass: 961 kg at launch, 479 kg dry, 580 kg on-station BOL
TT&C: from MBT's Yehud complex, Ku-band via two 9 m antennas. Same control room as Ofeq programme
Power system: 1.23 kW EOL from twin Si solar wings supported by 25-cell 50 Ah Ni/H$_2$ battery.

Italy

Italsat Series

Background

Italsat 1 is a Lit545 billion pre-operational programme to investigate the use of Ka-band in domestic voice/datalinks. It features onboard baseband-switched TDMA and makes available up to 12,000 telephone channels. ASI managed the demonstrations over the first two years, before transfer to the PTT. Telespazio provided in-orbit control. The contract included six Earth stations.

Italsat is 77 per cent Italian (Alenia 52 per cent) and the ground segment 100 per cent (Alenia 75 per cent). Alenia served as both prime contractor and systems integrator for a Lit160 billion contract awarded in 1990 for a second satellite. Italsat F2 adds the European Mobile Services (EMS) payload for ESA. Telespazio plans to build about 30 ground stations for the 2-satellite system. Italsat's switching capability means that it can transfer capacity to areas hit by disasters and to holiday regions during the summer. ESA's Artemis is based on the Italsat platform.

Italsat 1 was taken out of service in January 2001 and deorbited. Italsat 2 service commenced 31 October 1996. In early 2001 it was announced that a partnership with Eutelsat was formed to begin Ka-band services prior to the launch of Hot Bird 6. Out of propellant, it was taken out of service in July 2002.

Specifications

Italsat 1

Launched: 15 January 1991 by Ariane 44L from Kourou, French Guiana; spacecraft commissioned March 1991; payload testing completed end-September 1991
Location: 13.2° E geostationary
Design life: 5 years minimum for Ka-band demonstrations (Ariane launch accuracy could allow 7 years life)
Contractors: Alenia Spazio (prime, bus, payload, structure, thermal control), BPD (propulsion), Matra (AOCS), Aerospatiale (solar panels), SS/L (batteries), Fiat (power control/distribution), and Galileo (attitude sensors)
Transponders: 3 (plus 2 back-up) 20 W TWTA 29.5–30/19.7–20.2 GHz up/down Ka-band 1.55 × 1.06° national beams (Global Beam payload), 36 MHz bandwidth, EIRP 46.2 dBW min, 25 Mbit/s data rate, linear polarisation. Primary application business point-to-point

6 (plus 4 back-up) 20W TWTA 27.7–29.5/18.5–20.5 GHz up/down Ka-band Multibeam payload, 120 MHz bandwidth, EIRP 57.0 dBW min, 147.5 Mbit/s data rate, linear polarisation. TWTAs can be paired to burn through precipitation. Each of two non-steerable 2 m antennas generates three 0.59° spots (360 km diameter), four covering mainland + one each Sicily/Sardinia. 32 ms TDMA bursts handled by the synchronous baseband switching matrix. Capacity 12,000 duplex voice at 32 kbit/s; also analogue video at 2 Mbit/s and digital TV to 70 Mbit/s. Up to 16 ground stations can operate within a beam

1 (plus 1 back-up) 1 W 40/50 GHz Propagation beacon, providing 27 dBW EIRP over 3 × 3° European coverage
Principal applications: demonstration of Ka-band and onboard baseband switching for domestic TV, telephony and datalinks
Configuration: Alenia Spazio's platform, body 3.46 m high, solar array 21 m span, 3-axis stabilisation, Eurostar-type AOCS provided by 16 × 22 N + one Marquardt R-4D 490 N (for apogee insertion and major manoeuvres) thrusters drawing on 915 kg MMH/NTO, two 2 m diameter reflectors for six spot beams and single 64 × 88 cm elliptical reflector for national beam
Mass: 1,865 kg at launch, 890 kg on-station BOL, 950 kg dry mass, 915 kg propellant, 290 kg payload
Antenna pointing: ±0.03° for spot beams, ±0.15° for national beam; maintained by RF auto-tracking
TT&C: through Telespazio's Fucino Earth station. System performance is investigated through stations at CSELT in Turin, CNR's Centre for Satellite Telecommunications Studies in Spino d'Adda, and at the PTT's Higher Institute in Rome. Two beacon stations for antenna pointing control at Cagliari (Sardinia) and Courmayeur (foot of Mt Blanc). Three propagation beacon stations at Milan, Rome and Turin.
Power system: twin solar wings providing 1,600/1,400 W BOL/EOL, in conjunction with two 30 Ah Ni/H$_2$ batteries.

Italsat 2

Launched: 8 August 1996 by Ariane from Kourou, French Guiana
Location: 16.4° E
Design life: 8 years minimum
Contractors: as Italsat 1, plus Alenia Spazio is prime for EMS payload
Transponders: Same as Italsat 1, except the propagation beacon is replaced by ESA's 47 kg EMS European Mobile Services payload:

4 (2-for-1 redundancy) 23 W SSPA 1.6315–1.6355, 1.6415–1.6455, 1.6565–1.6605/1.530–1.534, 1.540–1.544, 1.555–1.559 GHz up/down L-band Europe/North Africa/Turkey EMS receive/forward beams, 4 MHz bandwidths (divided into 1 MHz segments), EIRP 41.5 dBW min (42.5 dBW over 95%), RHCP (switchable to LHCP), uses 2 m reflectors of Ka-band Multibeam

1 (2-for-1 redundancy) 4.5 W SSPA 14.236–14.240, 14.246–14.250, 14.241–14.245/12.736–12.740, 12.746–12.750, 12.741–12.745 GHz up/down Ku-band Western Europe EMS feeder/return beams, 4 MHz bandwidths (divided into 1 MHz segments), EIRP 32.5 dBW min, linear polarisation
Principal applications: as Italsat 1 plus mobile service demonstrations
Mass: 1,990 kg at launch, 1,200 kg on-station BOL, 1,025 kg dry mass, 975 kg propellant, 299 kg payload
Power system: as Italsat 1, except 1,880 W BOL
Other specifications as for Italsat 1.

Temisat Satellite

Background

Telespazio invested US$31 million in the Telespazio Micro Satellite (Temisat) project to demonstrate a commercial service for pulling data from environmental platforms over eight 2.4 kbyte/s UHF channels. Germany's Kayser-Threde built the 32 kg 35 cm^3 Temisat and a spare under a US$10 million contract that included a Russian piggyback launch from a Tsyklon 3 vehicle on 31 August 1993. It was carried on a Meteor 2–21 satellite and released to orbit at 15:20 UT, 10 hours and 40 minutes after launch. Telespazio designed the satellite to last for three to five years but it failed unexpectedly after a little over a year. Telespazio marketed 1,000 transmitters and 50 Data Collection Centre terminals to Italian customers. In addition to receiving data, the DCCs could also send commands and data to their platforms.

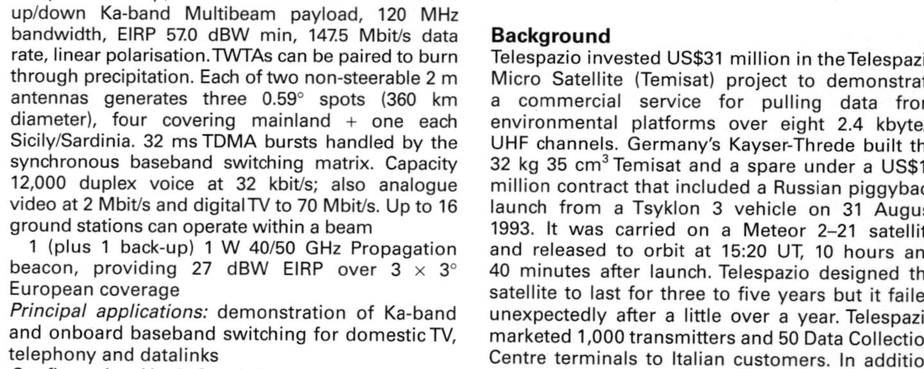

Italsat 1 was launched in 1991 to demonstrate Ka-band applications (Alenia Spazio) 0516734

Japan

BS Series

Current Status

Japan took the decision to de-orbit the BS-3A and the Ministry of Posts and Telecommunications began the process beginning 30 April 1998. The satellite's remaining fuel boosted the orbit by 340 km up and out of its geostationary position. Bsat 1A replaced BS-3A as prime in August 1997. Bsat 1B was launched in April 1998 and began commercial operations in August 1998. Bsat 2A was launched on 26 April 2001 followed by Bsat 2B on 12 July 2001. A launch vehicle failure prevented Ariane placing it in the correct orbit and a contract for Bsat 2C was awarded to Orbital Sciences Corporation in October. The Bsat 2C was launched on 11 June 2003.

Background

Until 1989, Japan owned and operated all of its communications satellites through the Telecommunication Advancement Organisation of Japan (TAO). Organisationally the National Space Development Agency (NASDA) procured and launched (or obtained launch services for) the BS and CS satellites and then handed them over to TAO for operational use.

When it launched the second generation Broadcast Satellites (BS), also known as Yuri, Japan became the first country to provide direct broadcast services to home receivers. Japan investigated the concept for two years with BS-1, launched 8 April 1978 by Delta from Cape Canaveral. The total programme cost the Japanese government 28.8 billion yen. Six years later, Japan launched the first operational satellite, BS-2A, on 23 January 1984 by N-2 from Tanegashima. Two of its three 100W transponders failed within 3 months and Japanese DBS service then relied on the sole BS-2B. The Japanese removed the BS-2A from GEO in April 1989.

BS-2B brought the system to full capacity in 1986, but NHK purchased a 3000-class GE Astro Space satellite carrying three 200 W transponders, to maintain the system until the BS-3 series became operational in 1991. The total system cost the Japanese US$24.48 million for the satellite, about US$49 million for the launch and insurance.

BS-2X disappeared in the Ariane failure of 22 February 1990 and NHK initially indicated they would not seek a replacement craft because of BS-3's imminent appearance. Nonetheless, after a discussion with GE Astro Space, NHK purchased a second satellite for launch in January 1991. NHK had awarded GE the US$88 million contract (mostly paid for by the BS-2X insurance) to deliver the replacement BS-3H satellite in orbit. The replacement BS-3H was then itself lost in its April 1991 launch. BS-3A, launched 28 August 1990, replaced BS-2B but it operates on marginal power. 3B completed the system in 1991. Again faced with growing demand NHK procured BS-3N in late 1992 (launched July 1994) to guarantee services.

BSAT-1A began services, including high-definition TV broadcasting, in 1997 (Hughes Space & Communications International Inc) 0517209

BS-3N provides three active 120 W DBS transponders (Martin Marietta Astro Space) 0517195

Japan carried on the tradition it pioneered with BS-2's DBS services, on the current generation BS-3 Yuri. NHK announced in late 1991 that it intended to procure a BS-3N for launch spring 1994. The proportion of indigenous systems and components on the BS-3s has reached 83 per cent. The programme cost Japan about 78.8 billion yen. As it prepared to launch the first in the series NHK reportedly self insured BS-3A. In-orbit testing revealed that one of the four solar array power segments failed, leaving the satellite with only marginal power supply.

Specifications

BS 3A/Yuri 3A

Launch: 28 August 1990 by H1 from Tanegashima
Location: 110° E geostationary (arrived end-September 1990; AKM fired 30 August); collocated with BS-3b
Design life: 7 years (specified EOL reliability >0.8 bus, >0.9 broadcast transponder, >0.8 wideband transponder)
Contractors: NEC (prime, communications payload) with GE Astro Space (bus), Nissan (solid AKM)
Transponders: 3 (plus 3 back-up) 120 W TWTA 14.069–14.219/11.766–11.919 GHz up/down Ku-band domestic DBS beams, 27 MHz bandwidth, EIRP 55 dBW for mainland, circular polarisation. Channel allocations (up/down GHz, centre frequencies) number 3: 14.06584/11.76584; number 7 14.14256/11.84256; number 11 14.21928/11.91928

1 (no back-up) 20 W TWTA 14.37/12.64 GHz up/down Ku-band for 60 MHz wideband relay
Principal applications: DBS TV services: two channels (number 7 + 11) NHK (3rd channel, number 3, not used because of power limitation)
Configuration: GE 3000 series, 3-axis stabilised box-shaped 1.3 × 1.6 × 1.6 m bus with twin 3-panel 1.5 × 5 m solar wings and single 11 kg 80 × 170 cm offset parabolic elliptical reflector. In-orbit span 15 m, height 3.2 m
Mass: 1,115 kg at launch, 550 kg on-station BOL
AOCS: bias momentum 3-axis, antenna beam pointing accuracy ±0.1°, ±0.1° E/West + North/South. 4 × EHT + 12 × catalytic REA thrusters 0.39–0.91 N drawing on four hydrazine (170 kg total) surface tension tanks. Pitch control using MWs, roll/yaw using magnetorquers, active nutation damping. Solid AKM
TT&C: through TAO Kimitsu Satellite Control Centre; NHK programming uplinked through 8 m Tokyo antenna
Power system: 1,443/1,093 W BOL/EOL to be provided by twin solar wings (11,515 high voltage + 4,183 low voltage N-on-P 2 × 4 cm BSR) on North/South walls, but one of the four supply segments is not contributing. Supported by 2 × 17 Ah Ni/Cd batteries.

BS 3B/Yuri 3B

Launch: 25 August 1991 by H1 from Tanegashima
Location: 110° E GEO (collocated with BS-3a)
Design life: 7 years, as BS-3a, but insertion precision left 9 years propellant
Transponders: as BS-3a but using channels 5/9/15 (up/down GHz, centre frequencies): 14.1042/11.8042; 14.18092/11.88092; 14.2960/11.0660
Applications: one channel is providing JSB's DBS; one has been demonstrating HDTV 8 h daily since 25 November 1991 to some 400 sets; SDAB digital audio broadcasting is also a user
Other specifications as for BS-3a.

BS-3N/Yuri 3N

Launch: 8 July 1994 by Ariane 4 from Kourou, French Guiana
Location: 109.85° E geostationary, as 3a/b back-up
Transponders: 120 W TWTA, as BS-3a
Applications: as BS-3a/3b
Mass: 1,210 kg at launch, 699 kg on-station BOL, 575 kg dry
Configuration: as BS-3a except 50 cm circular antenna added for 17 GHz uplink
Power system: as BS-3a except Ni/H_2 batteries carried
Other specifications as for BS-3a.

BSAT-1A

Launched: 16 April 1997 by Ariane 4 from Kourou, French Guiana
Location: Initially over 121° E, but planned for 110° E (collocated with 1b)
Design life: 10 years minimum (12.5 years propellant)
Contractors: Hughes Space & Communications
Transponders: 4 (plus 4 back-up, 8-for-4) 106 W TWTAs 17.3–17.8/11.7–12.2 GHz up/down Ku-band domestic beams, 27 MHz bandwidth, >55 dBW EIRP mainland (as BS-3), RHCP
Principal applications: DTH, including high-definition TV
Configuration: Hughes' HS-376spin-stabilised, 7.97 m high deployed (3.15 m stowed), 2.17 m diameter. Despun single reflector and communications shelf
Mass: 1,250 kg at launch, 720 kg on-station BOL
Pointing: <0.1° accuracy by RF beacon
AOCS: 50 rpm spin-stabilisation and station-keeping by four 22.2 N hydrazine thrusters. GEO insertion provided by solid Thiokol Star 30BP AKM
TT&C: by B-SAT
Power system: K43/4solar cells mounted on cylindrical exterior and deployed drum providing 1,200 W BOL. Two Advanced Ni/Cd batteries provide 100 per cent eclipse protection.

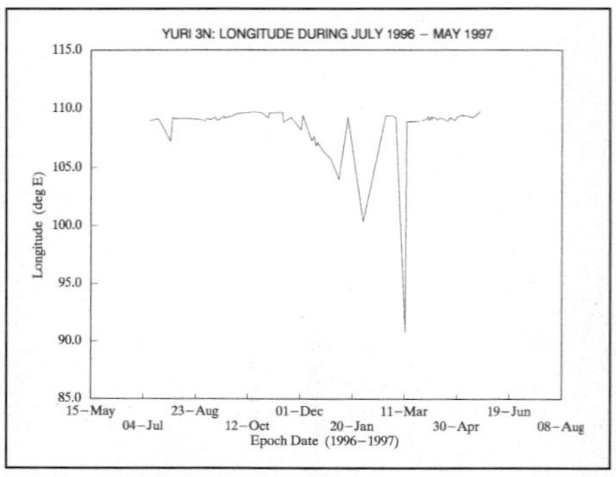

Although operating primarily over 110° east, data from USSPACECOM suggests that from September to December 1996 the Yuri 3N satellite drifted to about 105° east. Some of the USSPACECOM data might be in error (eg, single sets of data for 100° east or 90° east), but this drift appears to have been real. During 1997 the satellite has been shown over 110° east by USSPACECOM 0005567

BSAT-1B

Launch: 29 April 1998 by Ariane 44P from Kourou, French Guiana
Location: planned for 110° E (collocated with 1a)
Other specifications as for BSAT 1a.

BSAT-2A

Launch: 8 March 2001 by Ariane 5 from Kourou, French Guiana
Location: 110° E, geostationary
Design life: 11.5 years
Contractor: Orbital Sciences Corporation
Configuration: OSC Star™ platform, 3-axis stabilised box-shaped 3.8 × 2.8 m bus with twin solar arrays spanning 15.7 m
Mass: 1,317 kg at launch
Transponders: 4 of 8 × 130 W Ku-band conduction-cooled TWTAs. Ku-band 17.3–17.6 GHz/11.7–12.0 GHz receive/transmit
Propulsion: Solid propellant apogee motor, redundant liquid monopropellant systems for AOCS.
Application: DTH for Broadcasting Satellite System Corporation digital programming.

BSAT-2B

Launch: 12 July 2001 by Ariane 5 from Kourou, French Guiana
Location: Planned for 110° E, but a launch vehicle failure prevented the satellite reaching appropriate orbit.

BSAT-2C

Launch: 11 June 2003 by Ariane 5 from Kourou, French Guiana
Location:

CS-3/JCSat Series

Current Status

Successor to the ageing JCSat 2 launched on 1 January 1990, the JCSat-8 (JCSat 2A) satellite was carried to a GTO by Ariane 44L on 29 March 2002.

Background

Two Communications Satellite-3 (CS-3), also known as Sakura, superseded CS-2A/B and provide domestic Ka/C-band services, with both C-band area beams and seven of the 10 primary Ka-band spot beams dedicated to NTT operations. Japan estimates the programme cost 66.3 billion yen.

Japan created the JSB Japan Satellite Broadcasting Incorporated in 1984 by the Ministry of Posts & Telecommunications to commercially develop and launch communication satellites. The government ownership of JSB incited US pressure to open more of Japan's satellite purchases to foreign competition. Japan did relent and bids for the 'BS-4' generation resulted in the selection of Hughes Space & Communications as the prime contractor to build the satellites. Until then The Japan Broadcasting Corporation (NHK: Nippon Hoso Kyokai) held a large stake in broadcast through BS-3 transponders.

The hybrid JCSat 3 provides Ku/C-band services to the Pacific Rim and Asia (Hughes Space & Communications) 0517382

Then in 1989, the Diet released all restrictions on commercial systems development and several projects resulted. Nippon Telegraph and Telephone provided communications services for the CS series through CS-3, but NTT turned to another manufacturer, Space Systems/Loral (SS/L), to build its own two NStars which it launched in 1995. Satellite Japan Corporation (Sajac) planned to launch its own Sajac Hughes HS-601 satellites in 1994 until financial problems forced it to merge with JCSat in 1993, creating Japan Satellite Systems (JSAT); owner and operator of the JCSat series of satellites. JCSat 3 began commercial services in November 1995 and JSAT ordered JCSat 4 in February 1996.

With no further Japanese government restriction on commercial venture, Broadcasting Satellite System Corporation (B-SAT) incorporated itself on 13 April 1993 as a consortium of NHK, WOWOW, five other broadcasters and eight banks to procure and operate the 'BS-4' generation of broadcasting satellites for launches which took place in 1997 and 1998.

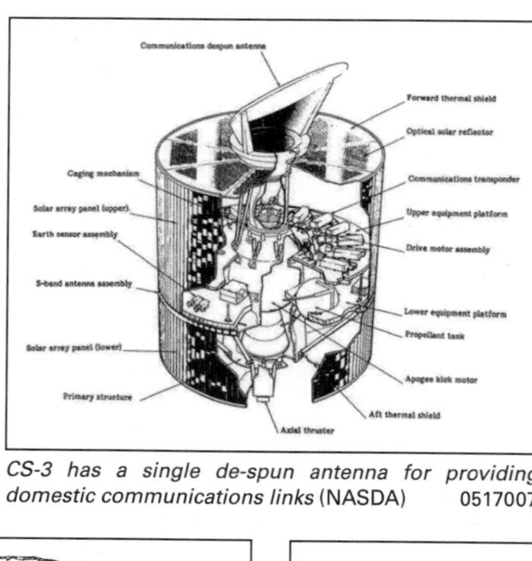

CS-3 has a single de-spun antenna for providing domestic communications links (NASDA) 0517007

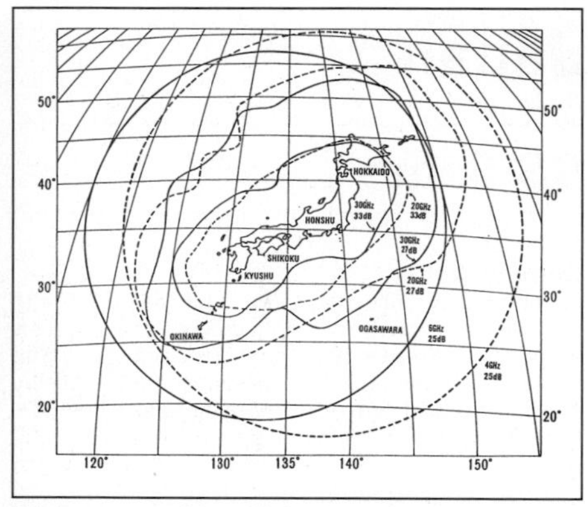

CS-3 coverage: C-band links the remoter communities, the higher-capacity Ka-band the main islands. The solid lines indicate CS-3a coverage from 132° east, the broken CS-3b from 136° east (NASDA) 0517008

About 65 per cent of the 64 transponders on the JCSat's satellites are now leased; and the company claims there are 55,000 subscribers for five television channels. In addition to network and cable television distribution, Japan Satellite Communications Network Corporation (JSNet) employs JCSat to provide VSAT business data services. Three transponders are used for the video services and two carry 12 stereo channels of high quality Digital Audio Broadcast programming. JCSat charges 300–650 million yen annually to lease a transponder.

Specifications

CS-3A / Sakura 3A

Launched: 19 February 1988 by H1 from Tanegashima Satellite operated over 132° E until late September 1996: it was then boosted to a retirement orbit
Design life: 7 years
Contractors: Mitsubishi Electric Corp (prime and antenna), NEC (transponders)
Transponders: 2 (plus 1 back-up) 4 W min TWTA 5.925–6.425/3.7–4.2 GHz up/down C-band global beams, 180 MHz bandwidth, output power ≥36.0 dBm, 25 dB min gain, circular polarisation
10 (plus 5 back-up) 4.2 W min TWTA 27.5–29.25/17.75–19.45 GHz up/down Ka-band spot beams, 100 MHz bandwidth, output power ≤36.2 dBm, 33 dB min gain to main islands, 27 dB min to Okinawa islands, circular polarisation, plus 19.450 GHz beacon
Principal applications/users: NTT (7 Ka + 2 C; routine and emergency telephone, TV and datalinks for major and remote islands), National Police Agency (1 Ka), Construction Ministry (0.4 Ka), other users <0.4 Ka-band transponders
Configuration: 218.4 cm diameter, 242.9 cm high cylindrical bus, and overall height 356 cm including single despun antenna. Monocoque shell supporting two horizontal equipment platforms
Mass: 550 kg BOL on-station, 1,099 kg at launch
AOCS: spin-stabilised at 90 ± 9 rpm, 0.12° BOL spin axis adjustment, ±0.26° measuring accuracy of Earth width from GEO, four 20 N thrusters (2 radial, 2 axial; continuous mzxode or 90 ms pulses) drawing on 118.5 kg hydrazine in three spheres for attitude control and ±0.05° E/West-North/South station-keeping. GEO insertion by Thiokol Star 30B solid
Antenna pointing: ≤0.2° half cone angle accuracy, ±0.03° stepping angle by despin assembly
TT&C: through TAO's Kimitsu Satellite Control Centre; 1–3 W DSN-compatible, near-isotropic S-band transponder up/down-converted to C-band; 125 bit/s 253-command command detector/decoder
Power system: 2-segment GaAs cylindrical solar array (36,621 P-on-N GaAs primary power;

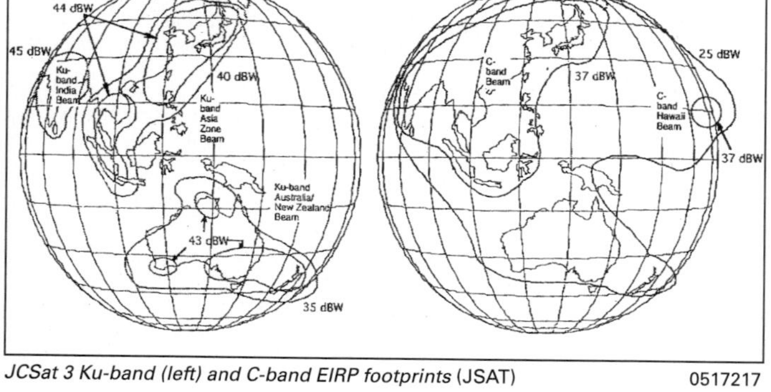

JCSat 3 Ku-band (left) and C-band EIRP footprints (JSAT) 0517217

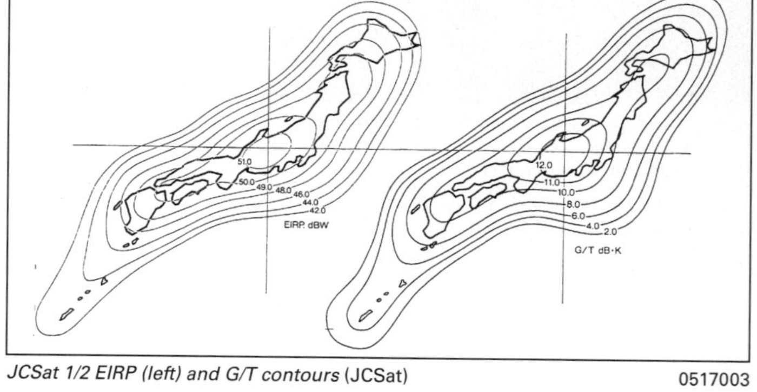

JCSat 1/2 EIRP (left) and G/T contours (JCSat) 0517003

With nearly 55,740 sq metres of manufacturing space the Hughes Integrated Satellite Facility accommodates a 1,783 cu metre thermal vacuum chamber in which here seen is JCSat-6 (Hughes Space & Communications) 0054335

396 N-on-P Si cells for battery charging), 833/750 W autumnal equinox/summer solstice EOL, two 35 Ah Ni/Cd batteries (55% discharge depth, 616 charge cycles in 7 years), main bus voltage 29.4 ±0.2 VDC
Thermal control: passive, augmented by heaters. Primary heat emitted through solar array and end shields.

CS-3B/Sakura 3B
Launched: 16 September 1988 by H1 from Tanegashima
Location: Initially over 136° E: during April 1997 relocated to 153–154° E
Other specifications as for CS-3a.

JCSat 1
Launched: 6 March 1989 by Ariane 44LP V29 from Kourou, French Guiana. It formally entered service 1 May 1989
Location: 150° E geostationary
Design life: 10 years (launch accuracy left sufficient propellant for 14 years operation)
Transponders: 32 (plus 8 back-up, 5-for-4) 20 W TWTA 14.0–14.5/12.25–12.75 GHz up/down Ku-band domestic Japanese beams, 27 MHz bandwidth, EIRP 50 dBW min (51.8 dBW peak) over most densely populated regions, orthogonal linear polarisation

Undergoing tests in the Hughes Space Environment Chamber, a JCSat Satellite has its extendable solar cell drum retracted (Hughes Space & Communications) 0054338

Principal applications: TV/cable distribution, telephone, business services, and satellite news gathering. Each transponder can carry one high-quality TV channel, a 45 Mbit/s data stream or 250 full duplex telephone circuits
Configuration: Hughes' HS-393 cylindrical platform, 3.66 m diameter, 3.6 m height stowed at launch, 10.0 m height deployed, payload and single 2.4 m diameter antenna, of two orthogonally offset grid reflectors, mounted on despun platform
Mass: 2,280 kg at launch, 1,376 kg on-station BOL, 1,097 kg dry
AOCS: station-keeping + 30 rpm spin-stabilisation maintained by six 22 N thrusters (two axial, four radial); GEO insertion and major manoeuvres performed by two Marquardt R-4D 489 N thrusters, both systems drawing on 1,200 kg NTO/MMH in four tanks each
Antenna pointing: 2-axis RF beacon tracking maintains <0.06°-accuracy pointing, backed up by Earth sensors
TT&C: though the company's Yokohama station with 11 m antenna for transfer orbit operations and two 5.5 m for on-station TT&C. A back-up facility is maintained in Gunma Prefecture
Power system: 2,240 W equinox BOL/1,808 W solstice EOL provided by Si solar array mounted on cylindrical body (K7 cells) and extension skirt (K43/4cells); 2 × 38.4 Ah Ni/H$_2$ batteries

Engineers check out the payload section of a Hughes JCSat (Hughes Space & Communications) 0054337

A Hughes HS-393 derived from the Intelsat UI bus (Hughes Space & Communications) 0054336

Thermal control: quartz mirror radiator band encircling upper body provides primary heat rejection path.

JCSat 2
Launch: 1 January 1990 by Titan 3 from Canaveral
Location: 154° E geostationary
Principal applications: as JCSat 1 plus DTH TV + DAB
Other specifications as for JCSat 1, but Titan launch required perigee kick stage derived from Minuteman stage 3.

JCSat3
Launch: 29 August 1995 by Atlas 2AS from Cape Canaveral
Location: 128° E geostationary
Design life: 12 years
Transponders: 28 (plus 6 back-up, 17-for-14) 60 W TWTA 13.75–15.0/12.25–12.75 GHz up/down Ku-band beams: Japan (>55 dBW mainland, >51 dBW Okinawa, >46 dBW other), India (45 dBW), Asia zonal (>45 dBW), Australia/New Zealand (>46 dBW).12 × 36 MHz + 16 × 27 MHz bandwidth (27 MHz switchable to 54 MHz), orthogonal linear polarisation
12 (plus 4 back-up, 16-for-12) 34 W SSPA 6.22–6.49/3.93–4.2 GHz up/down C-band zonal beam covering Russia, Korea, Japan, China, India, South Asia, Indonesia, Hawaii, Australia, New Zealand. 36 MHz bandwidth, orthogonal linear polarisation
Principal applications: as JCSat 1/2
Configuration: Hughes' HS-601 3-axis bus, 26.1 m solar array span
Mass: 3,100 kg at launch, 1,820 kg on-station BOL
Power system: twin solar wings providing 5.2 kW BOL, supported by 200 Ah Ni/H$_2$ battery.

JCSat 4
Launch: 17 February 1997 by Atlas 2AS from Cape Canaveral
Location: Initially over 150° E, but planned 124° E geostationary
Transponders: 16 60 W TWTA 13.75–14.5/12.25–12.75 GHz up/down Ku-band beams as JCSAT 3 except four 27 MHz transponders can be switched to 120 W/54 MHz
4 (plus 4 back-up, 8-for-4) 90 W TWTA 13.75–14.5/12.25–12.75 GHz up/down Ku-band beams: Japan (>55 dBW mainland, >53 dBW Okinawa), 36 MHz bandwidth, orthogonal linear polarisation.
C-band as JCSat 3
Mass: 3,100 kg at launch, 1,820 kg on-station BOL
Other specifications as for JCSat 3.

JCSAT 8 is checked out prior to delivery 0132824

Launched December 1997 JCSAT 5 provided Ku-band services from 150° east 0132826

JCSat 5
Launch: 2 December 1997 using an Ariane-4 from Kourou
Location: 150° E
Contractors: Hughes Space & Communications Co
Bus: Hughes HS-601
Transponders: Ku-band, 60 W/90 W, 16 × 27 MHz and 16 × 36 MHz (4 switchable to 2 × 72 MHz)
Service area: Japan, Asia and Hawaii.

JCSat 6
Launch: 16 February 1999 aboard an Atlas 2 AS
Location: 124° E
Transponders: Ku-band, 70 W, 32 × 27 MHz
Service area: Japan.

JCSat 8 (JCSat 2A)
Launched: 29 March 2002 by Ariane 44L from Kouron
Location: 154° E
Contractor: Boeing Satellite Systems
Bus: HS-601
Transponder: Ku-band 16 × 120 W TWTA; C-band 16 × 34 W SSPA
Mass: 2,460 kg at launch, 1,500 kg BOL.

UPDATED

CS-4/NStar Series

Background
The CS-4 series dates back to 1984 when two Super CS satellites were envisaged to replace CS-3 with an enlarged capacity of 60–90 multibeam transponders on a 2 tonne platform tailored around the H2 launcher. ¥2.57 billion was allocated to NASDA in FY90 to continue CS-4 studies. In 1989, MPT approved development of two satellites for launches in 1994/96. Transponders working at S/L-band would provide operational mobile services. The project would be closed to foreign bidders because it was designated as R&D. However, although NASDA was to provide 25 per cent of the funding, the domestic telephone monopoly NTT would fund the remainder in return for use of the operational transponders. US pressure resulted in early 1990 with CS-4 being transformed into two distinct projects: NASDA's 25 per cent was applied to the COMETS satellite and NTT acquired a satellite system on the open market.

NStar appeared in August 1995 0517211

NTT called for bids 20 June 1991, closing the competition 5 September 1991. SS/L received the contract December 1991, against GE and Hughes competition, for the NStar satellites to be launched 1995 to replace CS-3. The on-orbit delivery contract was worth some US$600 million; the launch contract signing with Arianespace was announced 10 April 1992. Each carries 26 active transponders: six Ka multibeam, five Ka-shaped beam, eight Ku-shaped beam, five C-shaped beam and one S plus its C feeder for mobile links. NStar A was damaged en route to the launch site and the launch was subsequently delayed. After launch in August 1995, it was delivered to the customer in October 1995.

Specifications
CS-4A/ NStar A
Launch: 28 August 1995 by Ariane 4 from Kourou, French Guiana
Location: 132° E (replacing Sakura 3A)
Design life: 10 years minimum; sufficient propellant for 15 years
Contractors: Space Systems/Loral (prime), MELCO, NEC, Thomson Tubes Electroniques (TH 3781 12 GHz; TH 3919 20 GHz TWTs), Hughes Electron Dynamics Div, Aerospatiale (reflectors)
Transponders: 5 (plus 2 back-up) 15 W min TWTA 27.5–29.25/17.75–19.45 GHz up/down Ka-band domestic beams, 100 MHz bandwidth, EIRP 44 dBW, circular polarisation

6 (plus 3 back-up) 30 W min TWTA 27.5–29.25/17.75–19.45 GHz up/down Ka-band multibeams, 200 MHz bandwidth, EIRP 51 dBW, circular polarisation, for 1.56 Mbit/s data transmission or broadband ISDN, with onboard switching capacity

8 (plus 3 back-up) 55 W TWTA 13.84–14.56/12.19–12.91 GHz up/down Ku-band domestic beams, EIRP 53 dBW, 54 MHz bandwidth, linear polarised

5 (plus 2 back-up) TWTA 5.925–6.425/3.7–4.2 GHz up/down C-band domestic beams, EIRP 38 dBW Tokyo, 72 MHz bandwidth, circular polarised; plus 6th TWTA as feeder link for S-band (see below)

1 (plus 1 back-up) 250 W TWTA 2.655–2.690/2.500–2.535 GHz S-band domestic beam, EIRP 52 dBW, 15 MHz bandwidth; one C-band feeder link (as other C-band transponders; one of Ku-band payload serves as feeder backup). Principally for maritime links, replacing 200 terrestrial stations along Japan's coast
Principal applications/users: continues CS-3 services, mobile communications, and business services
Configuration: box-shaped FS-1300 bus, two 2.6 × 3.25 m C/S + 2.6 × 4.5 m Ku/C deployable + one 2.1 m reflectors
Mass: 3,410 kg launch, 1,617 kg dry, 2,057 kg on-station BOL
AOCS: 3-axis (from immediately after launcher separation). 12 × 22 N N_2O_4/MMH; 490 N thruster permits Perigee Velocity Augmentation. Propellant loading 2,160 kg for 92 per cent fill
TT&C: through TAO Kimitsu Satellite Control Centre; transponders controlled through NTT's Inuishi and Tokyo stations
Power system: twin 4-panel solar wings, spanning 27 m. Supported by Ni/H_2 batteries 5,200 W BOL

CS-4B/NStar B
Launch: 5 February 1996 by Ariane 4 from Kourou, French Guiana
Location: 136° E (replacing Sakura 3B)
Mass: 3,420 kg launch, 1,618 kg dry, 2,062 kg on-station BOL
Other specifications as for Nstar A.

CS-4C/Nstar C
Launch: 5 July 2002 by Ariane 4 from Kourou, French Guiana
Mass: 800 kg dry
Dimensions: 3.3 × 1.9 × 1.5 m (12.6 m span deployed)
Stabilisation: 3-axis
DC Power: 2,600 W BOL
Lifetime: 15 years

ETS Series

Current Status
The Communications and Broadcasting Engineering Test Satellite (COMETS) failed to achieve the correct orbit after lift-off on 21 February 1998. Japanese investigators traced the failure to a premature shutdown of the H2's, LE-5A second-stage engine. NASDA immediately announced a plan to rescue the satellite and the orbit out of passage through the van Allen belts where it was stranded. In a sequence of seven orbital manoeuvres the satellite was placed in a temporary orbit from which several experiments were conducted. The routine operational phase of the mission ended 31 January 1999 and the extended mission was terminated on 6 August 1999.

Background
NASDA still conducts communication experiments on its own set of Engineering Test Satellites. NASDA launched four between 1975 and 1982, before privatising its space network. NASDA demonstrated the fruits of its research in August 1987 when it launched the ETS 5, the first Japanese 3-axis stabilised satellite in geosynchronous orbit and the first to use an indigenous apogee kick motor. It also carried an experimental mobile communications payload. It was deorbited in October 1997.

COMETS under integration at NEC (NEC) 0517134

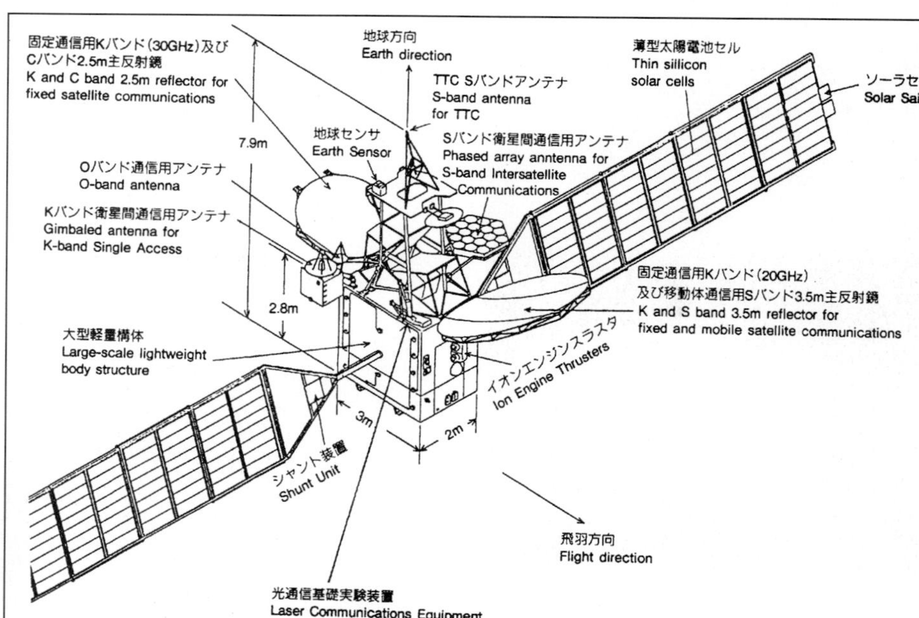

Principal ETS 6 features (NASDA) 0517004

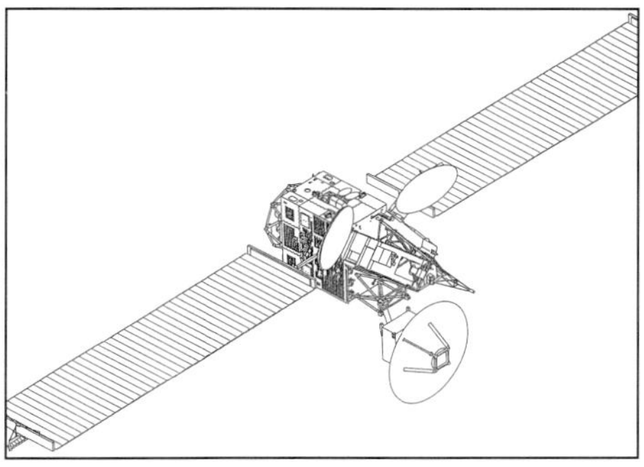

COMETS will demonstrate technology for future data relay, mobile communications and DBS. The largest antenna provides inter-orbit links 0517400

ETS 6 appeared in 1994 incorporating inter-satellite data relay and mobile communications packages. NASDA had intended to demonstrate a new ion propulsion engine to maintain north-south station-keeping on this two-tonne platform. The launch vehicle left it short of geosynchronous orbit however after launch in August 1994. The first of three planned firings by the new ETS 2 kN LAPS (Liquid Apogee Propulsion System) ended when controllers found the engine attained only 10 per cent of full thrust, probably because a fuel line valve had failed to open. The tanks ran out of oxidiser supply on the fifth attempt and ground controllers separated the LAPS on 31 August by command via NASA Goldstone, leaving ETS 6 in 7,796 × 38,707 km, 13.1° inclination orbit. However, in early 1996 bi-directionaly laser communications between JPL and the satellite were carried out.

NASDA planned to launch the COMETS Communications & Broadcasting ETS in 1997 to develop new technologies, preceding the operational DRTS Data Relay & Test Satellites in 2000. The OICETS Optical Inter-orbit Communications Engineering Test Satellite should follow in 1998 to test laser links with ESA's Artemis. ETS 8 in 2002 will demonstrate a single-hop cellular phone system and a CD-quality radio broadcast service.

The Ministry of Posts & Telecommunications established the Partners (Pan-Pacific Regional Telecommunications Network Research Satellite) project in 1992, using Kiku 5 to transmit educational teleconferences (64 kbit/s) and material to less developed Asia and Pacific sites, such as Fiji and Papua New Guinea.

Specifications

ETS-5/Kiku 5

Launched: 27 August 1987 by H1 from Tanegashima
Location: 150° E GEO, 3.7° inclination
Design life: 1½ years (hydrazine depletion projected imminent)
Contractors: Mitsubishi Electric, Nissan (AKM)
Transponders: Two 30 W SSPA 1.6445–1.6472/1.543–1.545 GHz up/down L-band beams for 23.2 dBW EIRP Experimental Mobile Satellite System experiment covering North Pacific (Japan-Alaska) and the Philippines/Indonesia/Australia regions for voice/datalinks with small maritime and aircraft terminals

One 6 W SSPA 5.95–5.97/5.22–5.24 GHz up/down C-band beam for NASDA link
Principal applications: spacecraft technology demonstration but experimental mobile communications package included
Configuration: 3-axis box-shaped bus with twin solar wings, GEO insertion by solid AKM
Mass: 1,081 kg at launch, 550 kg on-station BOL
TT&C: through NASDA's Tsukuba station.

OICETS will demonstrate laser inter-satellite links 0517005

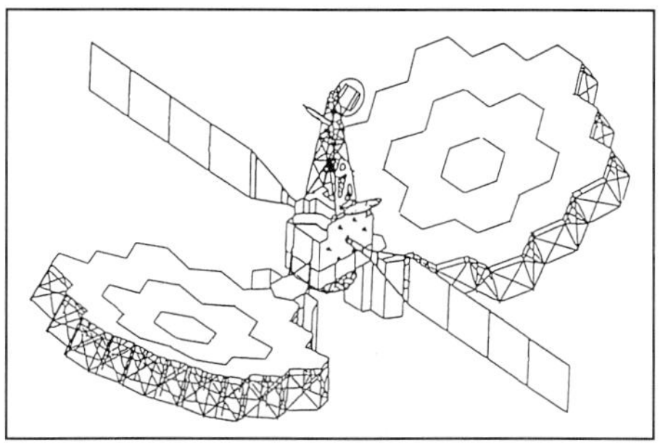

NASDA's ETS 8 will demonstrate single-hop phone links and CD-quality radio broadcasting 0517401

ETS8

Launch: January/February 2002 by H2 from Tanegashima
Locations: 135° E, 129° E or 146° E GEO
Design life: 3 years
Contractors: to be determined
Transponders: S-band for mobile-mobile connections

BSS mobile broadcasting of CD-quality radio
Configuration: to be determined, but with two 10 m-class deployable antenna
AOCS: to be determined
Power: to be determined.

COMETS

Launch: 21 February 1998 by H2 from Tanegashima
Location: planned 121° E geostationary
Design life: 3 years
Contractors: NEC (system integrator)
Transponders: S-band 2.0–2.1/2.2–2.3 GHz forward/ return inter-orbit link, working through 3.6 m antenna (steerable ±10°). Redundant 50 W 20 MHz SSPA/3 W 10 MHz SSPA forward/return, max EIRP 45.0 dBW. Developed by NASDA

Ka-band 23/26 GHz forward/return inter-orbit link (working through same 3.6 m antenna). Redundant 3 W 30 MHz SSPA/30 W 150 MHz TWTA forward/return, EIRP >56.0 dBW. Developed by NASDA

Ka-band 30/20 GHz + O-band 50/40 GHz for high quality mobile communication via 2.0 m feeder link antenna. Developed by Communication Research Lab

2 (plus 1 back-up, 2-for-1) Ka-band 21 GHz 200 W TWTA 120 MHz high quality DBS, covering Japan in two beams (Tokyo area + Kyushu Island) via 2.3 m antenna. Developed by CRL/NASDA
Technical Engineering Data Acquisition: TEDA payload will monitor the space environment and its effect on the spacecraft with Single Event Upset Monitor, Dosimeter, Magnetometer, Potential Monitor, Solar Cell Monitor
Principal applications: development of technology for inter-orbit data relay and tracking services, mobile communications and high definition DBS
Configuration: 2 t-class, similar to ETS 6. Solar array span 33.2 m, bus 1.96 × 2.96 × 2.8 m
Mass: 3.9 t at launch, 2 t on-station BOL
AOCS: highly accurate ACS (<±0.05° pitch/roll, <±0.15° yaw) required to control large antenna pointing. MELCO's 20–30 mN ion thrusters (35 kg Xe fuel) paired on East/West panels, provide NSSK. GEO insertion and attitude/orbit control by UPS Unified Propulsion System derived from ETS 6's LAPS: single 1.7 kN bipropellant Apogee Kick Engine plus 4 × 50 N and redundant 8 × 0.86 N hydrazine catalytic thrusters
Power: twin solar wings 2.93 × 15.6 m (rolled up for apogee motor firings) provide 4.7 kW EOL; Ni/H_2 battery provides 100 per cent eclipse protection.

OICETS

Launch: J1 from Tanegashima
Orbit: planned 550 km circular, 35°
Design life: 1 year mission planned
Contractors: NEC (system integrator)
Transponders: 2.04425/2.22000 GHz S-band forward/return inter-orbit links (COMETS, Artemis); EIRP 4.4 dBW, G/T –30.6 dB/K

2.04425/2.22000 GHz S-band up/down link, EIRP 12.3 dBW, G/T –47.5 dB/K

Laser communications demonstration: 8,010 Å beacon, 8,190 Å receive (2 Mbit/s), 8,470 Å transmit (50 Mbit/s) 200 mW (CW) AlGaAs diode; 26 cm diameter telescope
Principal applications: demonstration of optical inter-satellite communications
Configuration: 3-axis stabilised box-shaped 78 × 110 × 150 cm bus, 9.36 m span across twin solar wings, 2.5 m high from base to top of optical assembly. Coarse pointing assembly provides 0–120° elevation, ±190° azimuth range
Mass: about 550 kg in orbit
Power system: twin 3-panel solar wings spanning 9.36 m, 1.75 m wide; 13 Ah NiMH battery.

Superbird Series

Current Status

In September 2001 Boeing Satellite Systems was awarded a contract from Space Communications Corp of Japan for a 601 series platform to fill the specification for Superbird 6. It was launched in 2003. Superbird 6 will carry 23 active Ku-band and four active Ka-band transponders and occupy a location at 158° E longitude providing business telecommunications services using a Japan beam for both Ku-band and Ka-band services with a steerable spot beam for additional Ka-band services.

Background

A Japanese consortium planned to launch two Superbird satellites built by Space Systems/Loral and based on the FS-1300 platform. The Japanese stated the programme cost ¥70 billion (US$636 million), covering two satellites, launch services and insurance at premiums of about 20 per cent. An Ariane launched the Superbird A on 5 June 1989 to 158° E. Most of the satellite's station-keeping oxidiser bled off in December 1990 and commercial operations ended. Some customers transferred to the rival JCSat. Superbird claimed a reported US$170 million for the satellite. Superbird B, delayed from a December 1989 launch because of transponder problems, disappeared in the Ariane 4 accident of 22 February 1990. Insurance covered US$94.3 million. SS/L delivered the replacement Superbird B1 satellite within 24 months. Hughes received the order for Superbird C to expand coverage to the Asia Pacific region, including HDTV and 150 Mbit/s data. The satellite cost about 35 billion yen (US$412 million) with launch insurance. On 6 April 1998 HSCI received an order for a fourth satellite, a HS60HP to be designated Superbird-4, launch was successfully carried out on 17 February 2000 by Ariane 4.

Specifications

Superbird B1

Launched: 26 February 1992 by Ariane 4 from Kourou, French Guiana; entered service 6 April 1992
Location: 162° E geostationary
Design life: 10 years minimum
Contractors: SS/L (prime), DASA (solar array), Hughes Electron Dynamics Div (EPCs, TWTA), AEG (TWTs), Teldix (BAPTA), Gates (battery cells)
Transponders: 23 (plus 8 back-up, in 3 × 8-for-6 + 1 × 7-for-5) 50 W TWTA 14.00–14.50/12.25–12.75 GHz up/down Ku-band national Japanese beams, min EIRP 47.5 dBW, 36 MHz bandwidth, linear polarised (11 horizontal, 12 vertical)

3 (1 reserved as back-up) 29 W TWTA 28.4–28.8/18.615–18.715, 18.735–18.835 + 18.855–18.955 GHz up/down Ka-band national Japanese beam, min EIRP 43 dBW, 100 MHz bandwidth, LHCP up, RHCP down)
Principal applications: national TV and cable distribution, telephone, business services, and government
Configuration: 3-axis stabilised SS/L FS-1300 2.41 × 2.58 × 2.20 m box-shaped bus with twin solar wings on north/south sides, 20.3 m span; 2.1 m diameter Ku-band dual gridded reflector on West face, 2.0 m Ka-band reflector on East face
Mass: 2,560 kg at launch, 1,532 kg on-station BOL, 1,224 kg dry

Superbird C provides business and TV services to Japan, south/east Asia and Hawaii. The steerable spot could extend coverage to Australia/New Zealand (Hughes Space & Communications) 0517223

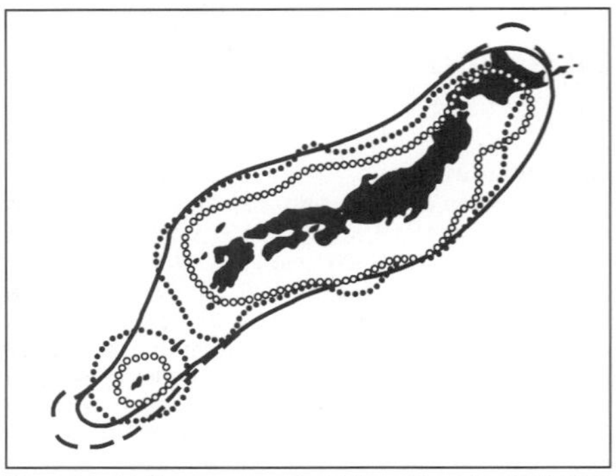

Superbird A/B coverage. Solid line: Ku 47.5 dBW EIRP; dashed: 4 dB/K. Dotted line: Ka 43 dBW; open dotted: 3 dB/K 0517006

From its location at 158° east, Superbird 6 will serve the business community in Japan 0132825

AOCS: 3-axis bias momentum system; single Marquardt R-4D 490 N apogee thruster and two sets of 6 × 22 N AOCS thrusters, common NTO/MMH propellant system
Antenna pointing: 0.09° azimuth/0.08° elevation Ku-band; 0.09° azimuth/0.07° elevation Ka-band
TT&C: 14 GHz command frequency, 12 GHz telemetry frequency
Power system: twin 3-panel wings providing 3,984 W BOL summer solstice, 4,460 W BOL autumnal equinox, 3,479 W EOL summer solstice, 3,778 W EOL autumnal equinox. 42 ±0.5 V bus (sunlight), 27.0–42.5 V bus (eclipse). Two 85.5 Ah Ni/H_2 batteries.

Superbird A1
Launch: 1 December 1992 by Ariane 4 from Kourou, French Guiana. Entered service 20 February 1993
Location: 158° E geostationary
Design life: 10 years (propellant for 13 years)
Mass: 2,780 kg at launch, 1,665 kg on-station BOL, 1,255 kg dry (propellant increase over B1 allowed by uprated Ariane stage 3)
Other specifications are the same as for Superbird B1.

Superbird C
Launch: 28 July 1997 on Atlas 2AS from Cape Canaveral
Design life: 13 years
Contractors: Hughes Space & Communications (prime), MELCO (steerable spot)
Transponders: 24 90 W TWTA Ku-band beams, orthogonal linear polarisation
Principal applications: as Superbird B + Asia-Pacific services
Configuration: Hughes' HS-601 3-axis bus, 26.1 m solar array span. Two 216 cm shaped-beam antennas plus

Superbird 4 (B2)
Launch: 17 February 2000 on Ariane 4 from Kourou French Givana.
Design life: 13 years
Contractor: HSCI (now Boeing Satellite Systems)
Transponders: 23 82 W TWTA Ku-band beams, six 50 W TWTA Ku-band
Configuration: Hughes' 601 HP 3-axis bus, 26.1 m span, 7 m width across antennas (4 × 2.7 × 3.6 m stored). Three primary antennas: One 1.9 × 2.5 m ku-band dual-gridded, one 1 m Ku-band steerable

Mass: 4,060 kg launch 2,460 kg BOL
AOCS: 1 × 490 N liquid apogee motor; 4 × 10 N N-S bi-propellant and 4 × 22 N E-W bi-propellant station-keeping thrusters
Power system: Twin solar wings each with 3 panels of silicon cells and one panel GaAs + 31-cell NiH, 200 Amp batteries; 5.5 kW BOL.

UPDATED

Korea, South

Koreasat (Mugunghua) Series

Background
Korea Telekom, formerly owned by the South Korean government, asked satellite builders to supply a US$260 million Koreasat (Mugunghwa after the national flower) domestic satellite telecommunication system in March 1991. Four teams headed by GE Astro, Space Systems/Loral, Hughes and British Aerospace responded and KT issued a US$145 million contract for two satellites and the ground control system to GE Astro in December 1991, supported by Matra Marconi Space UK (payload), Goldstar Information & Communications of Korea, and Korean Air. Hughes Network Systems of the US provided a hub and network of 70–80 cm VSAT dishes.

At launch, Koreasat 1 had to expend its own onboard propellant to achieve orbit, because the Delta launch vehicle fired incorrectly for geostationary transfer orbit when a strap-on failed to jettison. Koreasat 1 used about half of its station-keeping propellant to achieve final geosynchronous orbit after the apogee kick motor fired. The satellite launch insurer withheld payment of US$104 million because the satellite remains operational although lifetime has been reduced from 10 to 4½ years. Beam now focused on Hungary.

Koreasat 5 based on a Spacebus-4000 from Alcatel Space will carry 36 transponders in Ku-band, C-band and SHF (military) band delivering advanced broadband multimedia and digital television transmission services.

Specifications
Mugunghua 1 (Koreasat 1)
Launched: 5 August 1995 by Delta 7925 from Cape Canaveral complex 17
Location: 116° E geostationary (with number 2) until July 2000, then at 45° E (Europestar B) and to 47.5° E in October 2000
Design life: 10 years (+2 years on-orbit storage), but Delta shortfall expected to limit to <5 years
Contractors: Lockheed Martin Astro Space (prime), MMS UK (payload), and Hughes Electron Dynamics Div (TWTAs)
Transponders: 12 (plus 4 back-up, 16-for-12) 14 W TWTA 14.00–14.50/12.25–12.75 GHz up/down Ku-band S Korean FSS 0.86° beam, 36 MHz bandwidth, EIRP peak/edge 52.5/50.4 dBW, orthogonal polarisation
3 (plus 3 back-up, 6-for-3) 120 W TWTA 14.50–14.80/11.70–12.00 GHz up/down Ku-band South Korean DBS 0.86 × 1.06° beams, 27 MHz bandwidth (4 channels per transponder), EIRP peak/edge 62/59.4 dBW, LHCP
Principal applications: domestic telecommunications, VSAT, data, video and DBS (HDTV from 1998)
Configuration: LMAS 3000 3-axis 163 × 142 ×174 cm box-shaped platform with 15.45 m solar array span and 1.52 × 1.83 m diameter reflector. Electronic units, batteries, propulsion and AOCS equipment are mounted on six honeycomb structural pallets or panels. Transponders and housekeeping components are mounted on four panels, two each on North/South sides. Additional housekeeping equipment is mounted on anti-Earth base panel
Mass: 1,459 kg at launch, 833 kg on-station BOL, 641 kg dry, 144 kg payload
AOCS: 3-axis by 6,000 rpm MW (pitch) and magnetorquing (roll/yaw) controlled by Earth Sensor Assembly error signals. Propulsion subsystem is a conventional blow-down monopropellant hydrazine *type:* four surface tension 50 cm diameter tanks (max capacity 50.3 kg each, initial 26.2 atmosphere; max blow-down 4:1) are manifolded to two redundant sets of eight Olin Aerospace thrusters (six catalytic 0.89 N MR-103C + two NSSK electro-thermal 0.4 N MR-501); 187.5 kg loaded. GEO insertion provided by Thiokol Star 30E solid AKM
TT&C: Yong-In primary, Taejon back-up; 11 m for TT&C, testing and manoeuvres, 6.4 m for on-station TT&C
Power system: twin Si arrays of three panels each provide 1,686/1,533 W BOL/EOL (10 years) supported by 2 × 42.5 Ah Ni/H_2 batteries.

Mugunghua 2 (Koreasat 2)
Launch: 14 January 1996 by Delta 7925
Location: 116° E GEO (with number 1), then to 113° E in June 2000 until present
Other specifications as for Mugunghua 1.

Mugunghua 3 (Koreasat 3)
Launch: 4 September 1999
Location: 116° E
Design life: 15 years
Contractor: Lockheed Martin A2100 platform and Spar Aerospace
Transponders: 6 × 120 W Ku-band DBS, 27 mHz bandwidth; 24 (plus 8 back-up) 45 W Ku-band FSS, 36 mHz bandwidth; 3 (plus 2 back-up) 70 W Ka-band, 200 mHz bandwidth
Mass: 2,790 kg at launch; 1,322 kg dry.

Luxembourg

Astra Series

Current Status
Astra currently operates through 12 on-orbit satellites (1A, 1B, 1C, 1E, 1F, 1G, 1H, 2A, 2B, 2C, 2D and 3A). Luxembourg has filed for eight further 10.70–12.75 GHz slots: 24.2, 26.2, 28.2, 31.5, 35.5, 37.5, 41.2, 43.2° E. SES ordered its 10th Boeing (Hughes) satellite, Astra 3A, in August 2000, launched on 28 March 2002 to a GSO at 23.5° E. A Boeing 376 HP, it joined 12 other satellites in the SES Astra fleet. Astra 2D launched December 2000 was the first spin-stabilised satellite in the series.

Background
The Grand Duchy of Luxembourg permitted SES to be incorporated as a private company in March 1985 for the purpose of establishing a medium-power satellite system for television distribution. SES now operates under a franchise extending to 2010 from the Duchy, which also holds a 20 per cent interest in the company. The company contracted with RCA-Astro Electronics in October 1986 for a single series 4000 satellite. The resulting Astra 1 launched on 11 December 1988 and became Europe's first private satellite when it began operational programming on 1 February 1989, after its orbital trials.

The success of this first trial encouraged SES to expand and it acquired Satcom K3, previously owned by Crimson Satellite Associates. This 5000-class satellite had almost reached the end of the assembly line and the manufacturer needed to

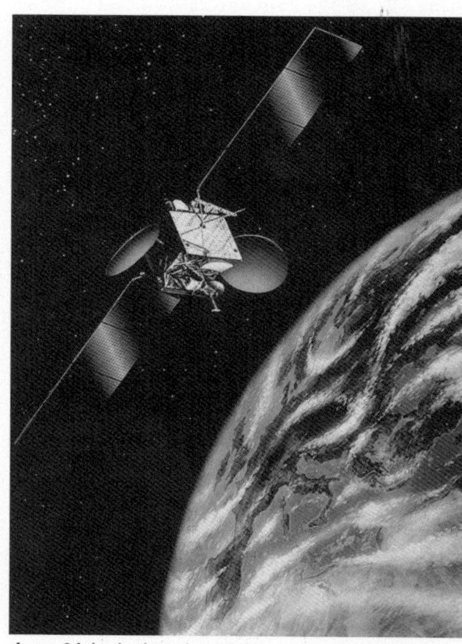

Astra 2A is deployed at 28.2° east, using ion thrusters for NSSK (Hughes Space & Communications) 0517407

Astra 1C-1H employ Hughes' HS-601 platform. Shown is Astra 1G/1H, featuring the new lightweight contoured antenna and GaAs solar wings 0517201

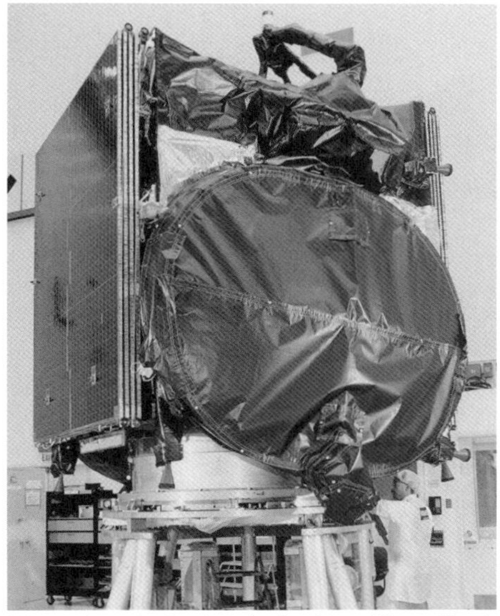

Wrapped ready for launch the Astra 1F satellite is shipped to Russia for a flight on a Proton launcher (Hughes Space & Communications) 0054346

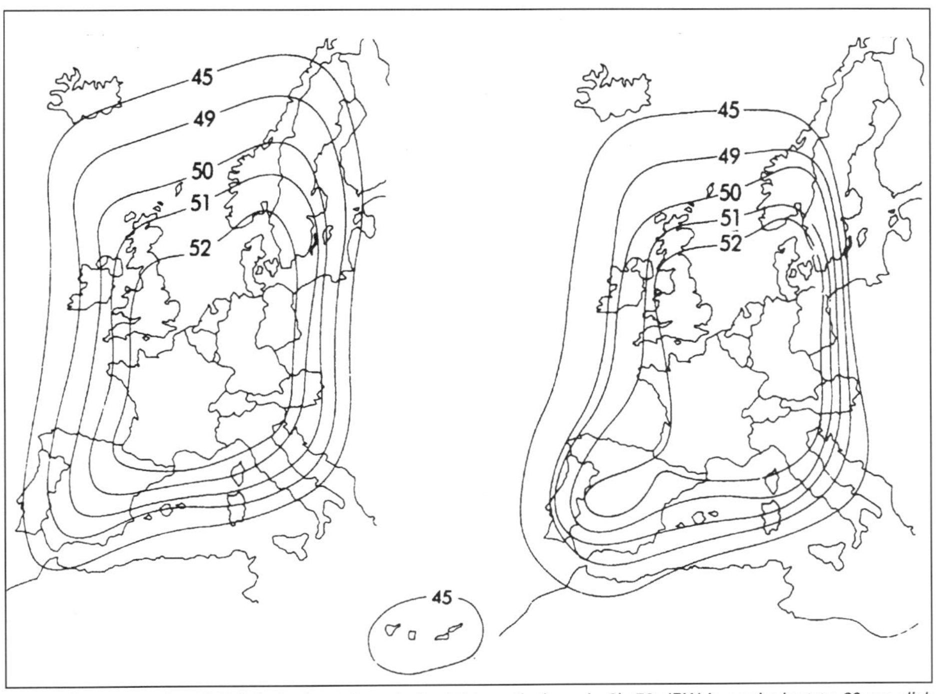

Astra 1B coverage in dBW (left: horizontal mode 2; right: vertical mode 2). 52 dBW is equivalent to 60 cm dish reception; 50 dBW to 85 cm. Astra 1A's lower power provides a smaller central region. Beginning with 1E, the higher power ellipse is stretched eastwards, providing the Moscow region with 75 cm dish coverage (SES) 0517009

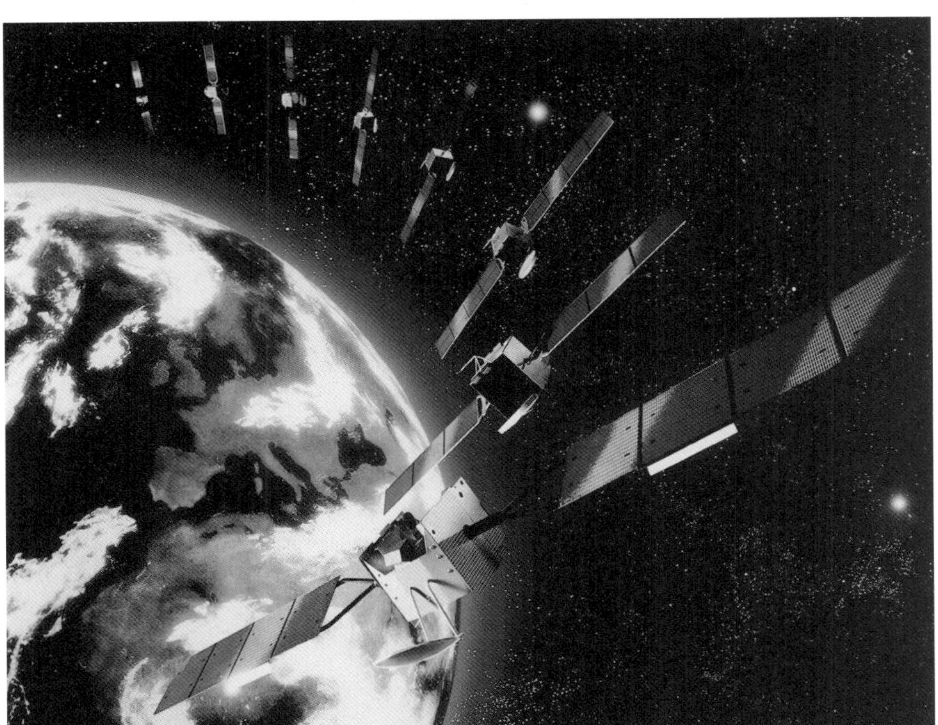

Up to eight satellites are employed in the Astra constellation 0084768

make only modest changes to its transmission characteristics to make it acceptable for SES. Rather than using its more powerful tubes to increase EIRP, the 52 dBW extended coverage instead. When it arrived at geosynchronous orbit it doubled the number of channels to 32.

SES continues to build on its transmission infrastructure. It awarded Hughes a US$428 million contract in December 1990 for two further satellites, based on the HS-601 platform. These new satellites, located within a 140 km cube of space, act primarily as back-up in the event of the primary satellite's failure. Astra 1D backs up 1B and 1C. It can also switch to BSS frequencies that can be employed for high definition television. The Duchy approved the purchase of Astra 1E in 1992 to bring digital programming to Europe. SES borrowed €220 million in February 1993 to finance it, bringing total SES capital investment to €1 billion. Approval for Astra 1F followed in 1993.

Astra 1F approved in 1993 provides DTH service from 19.2° E and the enhanced 1G, ordered in January 1995 and the identical 1H in July 1995 back up 1E-1G. SES used an €175 million loan contracted September 1995 to purchase the satellites, which it then supplemented with an additional €360 million loan in July 1996. The first two satellites of the second generation, replacing the first and expanding into other business areas from the new 28.2° E slot were ordered in June 1996.

SES Astra currently transmits more than 1,000 analogue and digital TV and radio channels across 176 transponders to more than 91 million households.

Specifications

Astra 1A

Launched: 11 December 1988 by Ariane 4 V27 from Kourou, French Guiana. To be replaced by Astra 2A in 1999

Location: 19.2° E geostationary (above Zaire)

Design life: 10 years min (projected 12.1 years)

Contractors: GE Astro Space (prime), COM DEV (transponder input-output multiplexers, filters), Parsons (structure), Composite Optics (reflectors, tower), Lockheed Space Systems (Earth sensors), Adcole (Sun sensors)

Transponders: 16 (plus six back-up, 11-for-8) 45 W TWTA 14.25–14.5/11.20–11.45 GHz up/down Ku-band European beams, 26 MHz bandwidth, EIRP 50 dBW min (52 dBW in central region), orthogonal linear polarisation

Principal applications: TV/radio distribution (DTH, cable, SMATV) to Europe

Configuration: GE 4000 platform, 170 × 213 × 152 cm body, 19.3 m solar array span, total height 305.8 cm at launch, 3-axis stabilisation. Equipment mounted on eight aluminium honeycomb structural pallets or panels; transponders and some housekeeping units mounted on North/South faces; antenna/reflectors, including the primary Kevlar honeycomb reflectors, and Earth sensors on Earth-facing panel

Mass: 1,817 kg at launch, 1,043 kg on-station BOL

AOCS: 3-axis by 6,000 rpm MW, magnetorquing and 16 blow-down mono-hydrazine thrusters. Thrusters divided into two redundant sets of eight: six catalytic Reaction Engine Assemblies (REAs) accompanied by two Electro-thermal Hydrazine Thrusters (EHT) for increased performance NSSK, drawing on four tanks. Apogee insertion provided by Thiokol's 808 kg Star 37XFP solid

TT&C: through SES' Betzdorf control centre 25 km east of Luxembourg. TT&C routed through 11 m dish and programme uplinking via 9.5 m antenna. Each of the Satellite Control Facility's Hewlett-Packard A900s can process six telemetry streams and command two satellites simultaneously. A 4.5 m transportable Earth station provides backup TT&C

Power system: twin 4-panel solar arrays using N-on-P Si cells provide 2,790/2,309 W BOL/EOL; 3 × 50 Ah Ni/H_2 batteries provide full eclipse protection.

Astra 1B

Launch: 2 March 1991 by Ariane 44L V42 from Kourou, French Guiana. Broadcasting began 15 April

Location: 19.2° E geostationary (above Zaire)

Design life: 10 years min (projected 14 years)

Contractors: GE Astro Space (prime), AEG (TWT), Rocket Research (RCS thrusters), Royal Ordnance (liquid apogee engines), TRW Pressure Systems (tanks), COM DEV (transponder input/output multiplexers, filters, diplexers, switches), Lockheed Space Systems (Earth/horizon sensor assemblies), Program Components (antenna reflectors), Parsons of California (structure), Adcole (Sun sensors), Northrop (rate measuring assembly), Composite Optics (solar array substrates)

Transponders: 16 (plus six back-up, quasi 22-for-16) 60 W TWTA 14.00–14.25/11.450–11.700 GHz

The location for the Astra ground central facilities at Betzdorf, Luxembourg 0084767

up/down Ku-band European beams in 250 MHz band adjacent to Astra 1A, 26 MHz bandwidth, EIRP 50 dBW min (52 dBW in central region), orthogonal linear polarisation
Principal applications: TV/radio distribution (DTH, cable, SMATV) to Europe and the Canary Islands (45 dBW)
Configuration: GE 5000 platform, 2.84 × 2.2 × 2.18 m body, 24.3 m solar array span, total height 2.56 m at launch (3.30 m on-orbit), 3-axis stabilisation. Equipment mounted on eight aluminium honeycomb structural pallets or panels; transponders and some housekeeping units mounted on North/South faces; 213 cm antenna reflector assembly deployed from west side; Earth sensors on Earth-facing panel
Mass: 2,580 kg at launch, 1,537 kg on-station BOL
Attitude control: 3-axis by 6,000 rpm MW, magnetorquing and 20 Olin blow-down mono hydrazine thrusters divided into two redundant sets of 10: eight catalytic REAs accompanied by two Improved EHTs for increased performance NSSK. GEO insertion provided by dual redundant Royal Ordnance Leros 1 500 N bipropellant motors. Dual mode system drawing on four hydrazine spheres + single N_2O_4 central cylinder
TT&C: as Astra 1A
Power system: twin 4-panel solar arrays using N-on-P BSR Si cells providing 4,850/3,700 W BOL/EOL; 4 × 50 Ah Ni/H_2 batteries provide full eclipse protection.

Astra 1C
Launch: 12 May 1993 by Ariane 4 V56 from Kourou, French Guiana. Broadcasting began 1 July
Location: 19.2° E geostationary (above Zaire)
Design life: 15 years min
Contractors: Hughes Space & Communications Co
Transponders: 18 (plus six back-up) 63 W TWTA Ku-band European beams, 2 at 10.90–10.95 GHz +16 at 10.95–11.20 or (to replace Astra 1A) +16 at 11.20–11.45 GHz down, 26 MHz bandwidth, EIRP 50 dBW min (52 dBW in central region), orthogonal linear polarisation
Principal applications: as 1B
Configuration: HS-601 platform, 3-axis, solar array span 21 m (satellite within 229 × 254 × 254 cm envelope at launch). Payload mounted on North/South panels and Earth-facing floor; East/West faces carry hard-points for two dual-gridded antennas, fixed antenna mountings on Earth-facing wall
Mass: 2,790 kg at launch, 1,700 kg on-station BOL, 1,180 kg dry
AOCS: 3-axis, unified 12 ARC 22 N + one Marquardt 490 N bipropellant thrusters incorporating onboard control processor, Sun/Earth sensors + 2 × 61 Nms 2-axis gimbaled momentum bias wheels. 1,658 kg NTO/MMH is contained in four spheres. Spin-stabilised in transfer orbit.
TT&C: as Astra 1A
Power system: twin solar wings of three 2.16 × 2.54 m panels carrying K43/4 large area Si cells on Kevlar substrate to satisfy 3.3 kW requirement. Ni/H_2 batteries provide 100 per cent eclipse protection.

Astra 1K was launched in 2001 on a Proton launcher 0084766

Astra 1D
Launch: 1 November 1994 by Ariane 4 from Kourou, French Guiana. Broadcasting began 1 January 1995
Location: 28.2° E
Transponders: 18 (plus six back-up) 63 W TWTA Ku-band European beams, switchable as 16 at 10.70–10.95 or 10.95–11.20 (1C back-up) or 11.45–11.70 GHz (1B back-up), 26 MHz bandwidth or 18 at BSS 11.70–12.07 GHz (1E back-up) 33 MHz bandwidth, EIRP 50 dBW min (52 dBW in central region), orthogonal polarisation
Mass: 2,924 kg at launch, 1,700 kg on-station BOL, 1,250 kg dry
Other specifications as for Astra 1C.

Astra 1E
Launch: 19 October 1995 by Ariane 4 from Kourou, French Guiana
Location: 19.2° E (above Zaire)
Transponders: 18 (plus six back-up) 85 W TWTA Ku-band European beams, switchable as 18 at BSS 11.70–12.10 GHz 33 MHz bandwidth, or 16 at 10.70–10.95 (Astra 1D back-up) or 10.95–11.20 (1C back-up) or 11.45–11.70 GHz (1B back-up), 26 MHz bandwidth, EIRP 50 dBW min (52 dBW in central region), orthogonal polarisation
Mass: 3,010 kg at launch, 1,803 kg on-station BOL, 1,343 kg dry
Power system: twin solar wings (spanning 26 m) of four 2.16 × 2.54 m panels carrying K43/4 large area Si cells on Kevlar substrate providing 4.7 kW.
Other specifications as for Astra 1C.

Astra 1F
Launch: 8 April 1996 on Proton-K from Tyuratam, Kazakhstan
Location: 19.2° E (above Zaire)
Transponders: 22 (plus two back-up) 82 W TWTA 12.10–12.50 GHz Ku-band European beams (switchable to 18 at 11.70–12.10 GHz as 1E back-up, or 16 at 11.2–11.45 GHz as 1A back-up), 33 MHz bandwidth, EIRP 50 dBW min (52 dBW in central region), orthogonal polarisation
Mass: 3,010 kg at launch, 1,900 kg BOL
Other specifications as for Astra 1E.

Astra 1G
Launch: 2 December 1997 on Proton-K from Baikonur, Kazakhstan
Location: 19.2° E (above Zaire)
Design life: 15 years min
Transponders: 32 100 W TWTA 12.50–12.75 GHz Ku-band European beams (switchable to 18 at 11.70–12.10 GHz as 1E back-up, or 22 at 12.10–12.50 GHz

Astra 3A during final tests at Boeing Satellite Systems 0132805

A Boeing 601 HP, Astra 2C was launched by Proton from Baikonur on 16 June 2001 0132809

as 1F back-up), 26/33 MHz FSS/BSS bandwidths, EIRP 52 dBW min, orthogonal polarisation
Configuration: improved HS-601 HP (High Power) platform, featuring GaAs solar arrays, enlarged payload and lightweight contoured reflectors
Mass: 3,300 kg at launch, 2,485 kg on-station BOL
AOCS: Hughes' 68 kg XIPS xenon ion propulsion system will provide NSSK: two (+ two back-up) 18 mN 13 cm thrusters
Power system: twin solar wings of four panels providing 8 kW BOL (7,075 W total required) and 100% eclipse protection
Other specifications as for Astra 1E.

Astra 1H
Launch: 18 June 1999 on Proton from Baikonur
Location: 19.2° E (above Zaire)
Transponders: Same as Astra 1G, except two 29.50–30.00/18.80–19.30 GHz up/down Ka-band beams will introduce interactive services
Other specifications as for Astra 1G.

Astra 1K
Launch: December 2001 on Proton from Baikonur
Location: 19.2° east
Design life: 13 years
Transponders: 52 Ku-band 2 Ka-band for first 5 years; 46 Ku-band and 2 Ka-band thereafter.
TWTA power output: Ku-band 105 watts/Ka-band 66 watts.
EIRP: Ku-band: 44.2–52.2 dBW over UK/Ireland 46.1–53.5 dBW over continental Europe.
EIRP: Ka-band: 55 dBW over Betzdorf
Configuration: Alcatel Space Spacebus 3000 B3S. Power consumption 12.4 kW.
Mass: 5,250 kg at launch.

Astra 2A
Launch: 30 August 1998 on a Proton-K from Baikonur
Location: 28.2° E
Design life: 15 years min
Other specifications as for Astra 1G.

BSS technicians pack the Astra 2D HS376 HP for shipment prior to launch 0132808

Astra 2B
Launch: 14 September 2000 on an Ariane 44P.
Location: 28.2° E
Design life: 14 years min
Transponders: 30 108 W TWTA 11.70–12.75 GHz Ku-band European beams (16 switchable to steerable beam), EIRP 52 dBW min, orthogonal polarisation for first 5 years, 28 thereafter
Configuration: MMS Eurostar 2000+i
Mass: 3,300 kg at launch
Power system: twin solar wings providing >6.4 kW BOL; 100% eclipse protection.

Astra 2C
Launch: 16 June 2001 by Proton
Location: 28.2° east
Design life: 15 years
Transponders: 32 98.5 W TWTA Ku-band European beams (8 spare at BOL)
Configuration: Boeing 601 HP platform featuring GaAs solar arrays on two wings each with one panel dual-junction and three panels single-junction
Mass: 3,643 kg at launch: 2,494 kg BOL
AOCS: Hughes 490 N liquid apogee motor; 12 × 18 mN XIPS xenon ion propulsion system for NSSK
Power system: twin solar arrays of 4 panels providing 8.5 kW BOL, 6.7 kW EOL
Dimensions: length (solar arrays) 26 m; width, 10 m. Stowed 5.5 m × 3.3 m × 3.3 m.

Astra 2D
Launch: 19 December 2000 by Ariane 5
Location: 28.2° east
Design life: 12 years
Transponders: 16 × 39 W TWTA for DTH
Configuration: Spin stabilised Boeing (HS) 376 HP with 18 TWTA, 2 spare, featuring two telescoping cylindrical panels
Mass: 1,445 kg at launch; 824 kg BOL
AOCS: Thiokol Star 30C solid apogee motor; 2 × 22.2 N N-S bipropellant; 2 × 22.2 N E-W bipropellant
Power system: Two telescoping cylindrical solar panels with large area GaAs solar cells and 16-cell, 141AhNiH batteries. Provides 1.6 kW BOL; 1.4 kW EOL
Antennas: 2.03 m diameter shaped reflector precision pointing antenna accuracy >0.10°. Comprises two shaped surfaces, one for vertical linear polarisation for communication and one for horizontal linear polarisation for communications and tracking
Dimensions: Stored height 3.15 m, width 2.17 m; deployed, height 7.97 m, width, 2.17 m.

Astra 3A
Launch 2 March 2002 by Ariane 44L
Location: 23.5° east
Design life: 10 years
Transponders: 20 × 30 W, 52 dBW. 24 Ku-band channel; 11.45–11.70 GHz on 12 channels, 12.50–12.75 GHz on 12 channels
Configuration: Spin stabilised Boeing (HS) 376 HP with 24 TWTA featuring two telescoping cylindrical solar panels
Mass: 1.5 kg BOL; 908 kG EOL
AOCS: Thiokol Star 30C solid apogee motor; 2 × 22.2 N N-S bipropellant; 2 × 22.2 N E-W bipropellant
Power system: as Astra 2D
Antennas: as Astra 2D
Dimensions: as Astra 2D.

UPDATED

Malaysia

MEASAT Series

Current Status
Binariang Satellite Systems plans to launch MEASAT 3 in 2004 carrying 24 C-band transponders

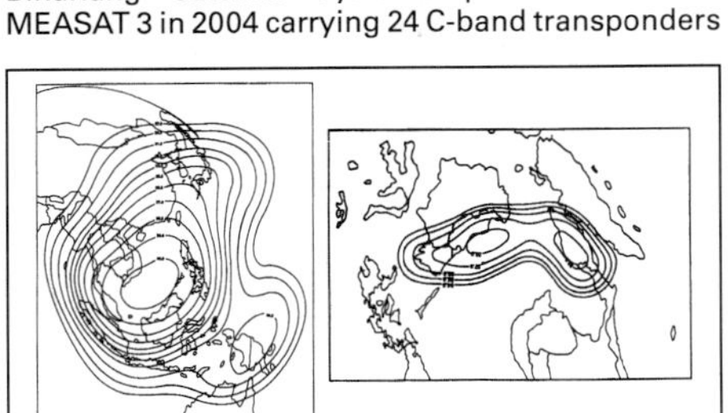

MEASAT coverage from 91.5° east. C-band (left) central EIRP is 40.0 dBW (33 for secondary peak); Ku-bands are both 58.0 dBW. The contours reduce in 1 dBW steps 0517393

Hughes HS 376 MEASAT built for Binariang, the first of its type to carry GaAs solar cells 0132811

with 36 mHz bandwidth each. It will be collocated with MEASAT 1 at 91.5° E with switching capabilities between Malaysia, India and China with the global footprint covering Asia, Middle East, eastern Europe and eastern and central Africa. Measat 4 would carry C-band coverage to areas from the Middle East across Asia to all areas south of Kazakhstan and Mongolia and to a longitude as far east as Japan and Papua New Guinea and Australia but not New Zealand. Measat 5 will cover Africa with C-band and Ku-band states.

Background
Malaysia which had spent US$30 million annually on Intelsat and Palapa leases made a national decision to take over responsibility for its own communications network in 1990. It uses Binariang, a private company to purchase satellites and launch services. Binariang signed a memorandum of understanding with Hughes in November 1991 to buy two HS-376 satellites to provide domestic and regional services. Considering Hughes production schedule at the time, Malaysia scheduled the first MEASAT (Malaysian-East Asian Satellite) launch for 1994 aboard Ariane. But Malaysia did not sign a final contract with Hughes Communications International Incorporated until 17 May 1994 following a 4-month international competition. The agreement specified that Malaysia would launch one satellite by late 1995. The contract also kept an option open for a second satellite, which the Malaysian government authorised in January 1995, as well as the construction of a ground control station and training for the Malaysian operators. This HS-376 is the first to carry GaAs solar cells, a lightweight shaped antenna (improves gain and eliminates multiple feedhorns) and a bipropellant station-keeping propulsion system.

Specifications
MEASAT 1
Launched: 12 January 1996 by Ariane 4 from Kourou, French Guiana
Location: 91.5° E geostationary
Design life: 12 years minimum
Contractors: Hughes Space & Communications (prime)
Transponders: 12 (plus three back-up) 12 W SSPA 5.925–6.425/3.7–4.2 GHz up/down C-band regional beam, EIRP 35 dBW edge (40 dBW centre), 40 MHz bandwidth (36 MHz usable), orthogonal linear polarisation
5 (plus one back-up) 112 W TWTA 14.0 – 14.25 + 13.75 – 14.0/10.95 – 11.2 + 12.2 – 12.5 GHz up/down Ku-band domestic beam (India and Philippines selectable), EIRP 55 dBW edge (58 dBW centre), 54 MHz bandwidth, orthogonal linear polarisation
Principal users: Ku-band for domestic direct to user services, including DTH TV via 50 cm dishes to Malaysia, Philippines and India; C-band for regional telecom services
Configuration: Hughes' HS-376 spin-stabilised bus, 2.16 m diameter, and 6.6 m height with solar array drum extended (2.8 m stowed). Single-shaped antenna on despun platform
Mass: 1,450 kg at launch, 839 kg on-station BOL, 626 kg dry
AOCS: 50 rpm spin control maintained by four (two radial and two axial) 22.2 N NTO/MMH bipropellant thrusters. GEO insertion provided by Thiokol Star 30 solid
TT&C: Satellite Control Facility on Langkawi Island
Power system: GaAs cells on cylindrical surface provide 1,700 W (1,200 W BOL to payload); 16-cell 142 Ah Ni/H_2 battery.

MEASAT 2
Launch: 13 November 1996 on Ariane 4 from Kourou, French Guiana
Location: initially over 148° E
Transponders: 3 (plus one back-up) 12 W SSPA 5.925–6.425/3.7–4.2 GHz up/down C-band regional beam, EIRP 35 dBW edge (40 dBW centre), 72 MHz bandwidth, orthogonal linear polarisation
6 65 WTWTA Ku-band domestic, 54 MHz bandwidth, orthogonal linear polarisation
6 100 W TWTA Ku-band domestic, 54 MHz bandwidth, orthogonal linear polarisation
Other specifications as for MEASAT 1.

Mexico

Morelos Series

Current Status
Hughes provided the new Morelos 3 (Satmex 5) satellite, initially called Morelos-2-FOS, launched by Ariane 42L on 6 December 1998. Satmex 5 replaced Morelos 2 and carried the XIPS ion engine station-keeping system. Satmex 6 currently under discussion but no programme or launch plan announced.

Background
Despite its participation in international communications bodies, Mexico decided in 1982 to develop an independent communications system. During that year, SCT ordered two Hughes satellites to expand domestic television and other services available to the entire population of the country. Before that time, telecommunications only served the 18 million people living in and around Mexico City.

The Mexican government's Departamento Especial of the DGT Direccion General deTelecomunicaciones for Mexico's telecommunications systems gave authority to SCT, Secretaria de Comunicaciones y Transportes (SCT) to develop and service a communications network. Then in 1990 the Mexican government created Telecomm, an independent non-stock company owned entirely by the federal authority, with SCT operating as a regulating body.

Mexico concluded a US$92 million contract with Hughes which included the control centre at Iztapalapa and the Mexican government selected the name Morelos for its satellites to honour the revolutionary Jose Maria de Morelos y Pavon a hero of the Mexican revolution. The Shuttle launched the first Morelos on 18 June 1985 to 113.5° west. Solidaridad 1 retired

An HS-601 HP, SATMEX 5 relies on Zenon propulsion for North/South station-keeping
(Hughes Space & Communications) 0054329

As replacement for Morelos 2, SATMEX 5 carries 10 × the wattage payload power and 48 C-band and Ku-band transponders
(Hughes Space & Communications) 0054328

the Morelos in March 1995 when it began service. NASA allowed Mexican payload specialist Rodolfo Neri Vela, to accompany Morelos 2 when it was deployed. The satellite entered a two-year geosynchronous drift orbit because Hughes had not completed the ground segment to control when it was launched. It entered service May 1989 (requiring Telesat Canada to move Anik C3 from 117.5° west to avoid interference) and it should continue operations into 2001. Morelos is now operated by GE Americom under the designation K2.

The specifications for a new Morelos 3 satellite took more than a year to develop, but the bidding process started on 18 March 1996. Five companies responded initially, but only three presented proposals at the end of May.

Specifications

Morelos 2 (K2)

Launched: 27 November 1985 by Shuttle mission 61B
Location: 116.5° W geostationary
Design life: 10 years
Contractors: Hughes Space & Communications Co
Transponders: 12 (plus 4 back-up, 14-for-12) 7 W TWTA 5.925–6.425/3.70–4.20 GHz up/down C-band domestic beams, 36 MHz bandwidth, EIRP 36 dBW min, orthogonal polarisation

6 (plus 2 back-up, 8-for-6) 10.5 W TWTA 5.925–6.425/3.70–4.20 GHz up/down C-band domestic beams, 72 MHz bandwidth, EIRP 39 dBW min, orthogonal polarisation

4 (plus 2 back-up, 6-for-4) 20 W TWTA 14.00–14.50/11.70–12.20 GHz up/down Ku-band domestic beams, 108 MHz bandwidth, EIRP 44 dBW min, circular polarisation

Principal applications: domestic TV distribution, telephony and data services
Configuration: Hughes HS-376 cylindrical spin-stabilised bus, 2.16 m diameter 6.60 m high, single 1.8 m diameter dual reflector for C-band and Ku-band transmission, planar array for Ku-band receive (first Hughes civil satellite with planar array); AOCS provided by hydrazine thruster system, GEO insertion by Thiokol solid AKM, GTO insertion by McDonnell Douglas PAM-D
Mass: 1,140 kg launch mass (including 133 kg hydrazine), 645 kg on-station BOL
TT&C: through the Iztapalapa Earth station 10 km South East of Mexico City. Full tracking antenna + two 11 m antennas
Power system: solar array of K-7 cells mounted on cylindrical body and skirt extension provides 950 W BOL/800 W EOL; two Ni/Cd batteries.

Morelos 3 (Satmex 5)

Launch date: 6 December 1998 by Ariane 42L
Location: 243.2° east
Design life: 15 years
Mass: 4,144 kg at launch; 2,267 kg BOL
Power: 10.9 kW beginning of life, 9.2 kW end of life
Transponders: 24 × C-band for continental coverage: bandwidth 36 MHz, TWT 36 W, EIRP 38.0 dBW 24 × Ku-band for regional and continental coverage: bandwidth 36 MHz, TWT 110 and 132.5 W, EIRP–Ku-1 49.0 dBW (EOC), Ku-2 – 46.0 dBW (EOC).
Propulsion: 490 N liquid apogee motor; station keeping by 8 × 10 N bipropellant and 4 × 22 N bipropellant thrusters
Power system: 2 × 4 panel wings of dual junction GaAs solar cells providing 10 kW BOL, 8.8 kW EOL
Dimensions: stowed: height 5.5 m width 3.5 m × 2.6 m; deployed, length 26 m, width (antennas) 9.4 m.

Solidaridad Series

Current Status

On 28 April 1999 Solidaridad 1 suffered a malfunction and services were moved to Satmex 5 (Morelos 3) before normal service resumed a day later. On 27 August 2000 the satellite control processor failed and after 65 attempts Solidaridad 1 was switched off and declared a total loss.

Background

In March 1991, Hughes emerged as the leader in the contractor race to provide Mexico with two Solidaridad satellites. Hughes won with its HS-601 platform. Mexico signed the US$183.47 million contract 8 May in Mexico City, calling for first delivery in 28 months. In addition to continuing C/Ku-band services, Solidaridad has an L-band payload providing Mexico with its first domestic mobile links in the Americas. Arianespace signed the dual launch contract 24 September 1991, against Atlas 2A and Long March competition.

Specifications

Solidaridad 1

Launch: 20 November 1993 on Ariane 4 V61 from Kourou, French Guiana
Location: 109.2° W geostationary
Design life: 14 years
Contractors: Hughes Space & Communications C
Transponders: 18 (plus 6 back-up) 10–16 W SSPA 5.925–6.425/3.70–4.20 GHz up/down C-band beams serving Region 1 (Mexico, South West US, Northern Central America), R2 (R1 + Caribbean, Central America, Northern South America), R3 (most of remaining South America except Brazil), 12 × 36 MHz + 6 × 72 MHz bandwidth, R1/R2/R3 EIRP 37.5/37/37 dBW min for 36 MHz, 40.5 dBW min for 72 MHz, orthogonal polarisation. All 72 MHz transmission to R1 only, 4 × 36 MHz switchable between R1/R2, 4 × 36 MHz between R1/R3 and 4 × 36 MHz between R2/R3. R1/R2 feeds use W reflector, R3 E

16 (plus 4 back-up) 42.5 W TWTA 14.00–14.50/11.70–12.20 GHz up/down Ku-band beams serving Region 4 (as R1) and R5 (major US cities and small parts of Canada and Cuba), 54 MHz bandwidth, R4/R5 EIRP 47.0/46.4dBW min, orthogonal polarisation. All transmission/receive for R4 but two receive (and one of those two transmission) switchable to R5

One channel using four paralleled 21 W 1.6295–1.6605/1.528–1.559 GHz up/down L-band SSPAs (plus 2 back-up) for mobile links throughout Mexico and up to 370 km offshore, 17 MHz selectable out of available 34 MHz bandwidth, EIRP 45 dBW min, RHCP. 42.5 W 14.248–14.265/11.9515–11.9885 GHz up/down Ku-band feeder link provides 46.4 dBW min, 17 MHz bandwidth

Principal applications: domestic TV distribution, telephony, data and mobile services; international services
Configuration: Hughes HS-601 platform, 3-axis, solar array span 21.0 m (satellite within 2.7 × 3.1 × 3.5 m envelope at launch). Payload mounted on North/South panels and Earth-facing floor; East/West faces carry hard-points for two dual-gridded antennas, fixed antenna mountings on Earth-facing wall: 1.8 × 2.4 m C-band on West side; 1.8 × 2.4 m Ku-band on East side; 26 L-band cup dipoles on Earth face
AOCS: 3-axis, unified 13 ARC 22 N and one Kaiser Marquardt 490 N bipropellant thrusters incorporating onboard control processor, Sun/Earth sensors and two 61 Nms 2-axis gimbaled momentum bias wheels. The NTO/MMH is contained in four spheres. Spin stabilised in transfer orbit.
Mass: 2,776 kg at launch, 1,671 kg on-station BOL, 1,291 kg dry
TT&C: as Morelos
Power system: twin solar wings of three 2.16 × 2.54 m panels carrying K41/4 large area Si cells on

Solidaridad 2 assembly, with the 26-element L-band array prominent (Hughes Space & Communications) 0517214

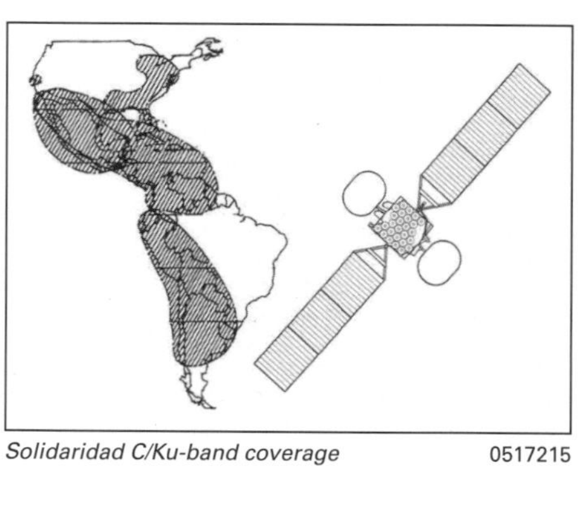

Solidaridad C/Ku-band coverage 0517215

Part of the communications payload for Solidarad 2
(Hughes Space & Communications) 0054330

Solidarad 1 undergoes vibration tests to simulate launch vehicle dynamics (Hughes Space & Communications) 0054331

Kevlar substrate to satisfy 3.37 kW requirement. 160 Ah Ni/H_2 battery provides 100 per cent eclipse protection.

Solidaridad 2
Launch: 8 October 1994 by Ariane 4 from Kourou, French Guiana
Location: 113.5° W geostationary
Mass: 2,776 kg at launch, 1,640 kg on-station BOL
Other specifications: as for Solidaridad 1.

Norway

Thor Series

Current Status
On 18 May 1997, Norway announced that Hughes Space & Communications would build Thor 3. The satellite is a high-power HS-376. The agreement contains an option for an additional satellite. Selection of a contractor to produce Thor 4 is underway.

Background
Thor 1 began life as the Marcopolo 2 satellite, part of the UK's British Satellite Broadcasting company. BSB merged into rival Sky Television in November 1990, in preparation to moving all programming to Sky's Astra carrier. In the meantime, Hughes received a follow-up contract for US$300 million contract in 1987 for the on-orbit delivery of two HS-376 satellites to provide five TV channels employing the D-MAC transmission standard. The contract represented Hughes' first sales of the HS-376 models configured for DBS, using paired 55 W TWTAs to achieve the required power levels. Norway bought the satellite for US$42 million.

Specifications
Thor 1 (BSB 2)
Launched: 00.42 GMT 18 August 1990 by Delta 6925 from Cape Canaveral. Satellite separated at 13.10 GMT, followed by AKM firing at 16.17 GMT 20 August for positioning at 50° W for a month of in-orbit testing. Ownership transferred to BSB in mid-December 1990 at 31° W collocated with Marcopolo 1. Transferred over 15 September to 29 October 1992 to 1° W under Norwegian ownership

Hughes' HS-376 bus was uprated to meet DTH power requirements. Thor 2A is similar 0517010

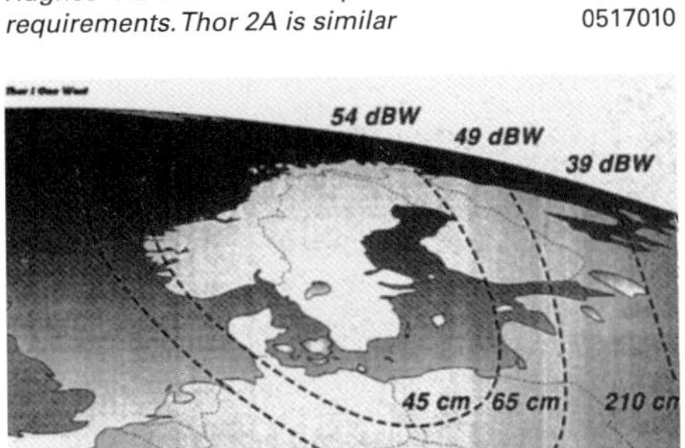

Thor 1 coverage from 1° west 0517011

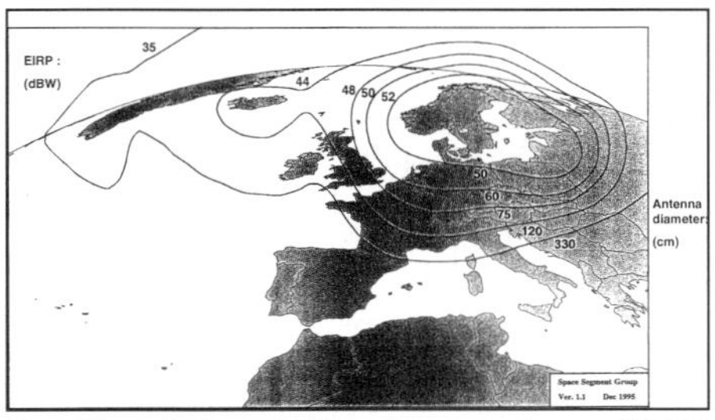

Thor 2A preliminary EIRP and antenna sizes 0517396

A future Thor 3 0005568

Location: 0.8° W geostationary until December 2002 when moved to 7.3° W and retired
Design life: 10 years minimum
Contractors: Hughes Space & Communications Co
Transponders: six paired 55 W TWTAs 17.385–17.992/11.7715–12.1054 GHz up/down Ku-band Nordic beam (BSB channels 4, 8, 12, 16, 20), 27 MHz bandwidth, 59 dBW min EIRP (63 dBW max), RHCP
Principal applications: DTH via 45 cm dishes
Configuration: Hughes' HS-376 spin-stabilised, 7.2 m high deployed (2.7 m stowed), 2.16 m diameter. Despun single reflector and communications shelf
Mass: 1,250 kg at launch, 660 kg on-station BOL
Pointing: 0.05° accuracy by RF beacon
AOCS: station-keeping and spin-stabilised at 50 rpm by four 22.2 N hydrazine thrusters; GEO insertion by solid Thiokol Star 30B AKM
TT&C: via Nittedal station
Power system: K7 solar cells mounted on cylindrical exterior and deployed drum providing 1,024 W BOL (drum extended 25 cm over normal 376). Two Ni/Cd batteries provide 100% eclipse protection.

Thor 2
Launch: 20 May 1997 by Delta 7925 from Cape Canaveral
Location: 0.8° W geostationary
Design life: 10–12 years minimum
Transponders: 15 40 W TWTAs 17/11.2–11.45 GHz up/down Ku-band Nordic beam, 26 MHz bandwidth, 52 dBW min EIRP
Principal applications: DTH via 45 cm dishes
Configuration: Hughes' HS-376HP spin-stabilised, 7.2 m high deployed (2.7 m stowed), 2.16 m diameter. Despun single reflector and communications shelf
Mass: unavailable
AOCS: station-keeping and spin-stabilised at 50 rpm by four 22.2 N hydrazine thrusters; GEO insertion by solid Thiokol Star 30B AKM
TT&C: via Nittedal station
Power system: GaAs solar cells mounted on cylindrical exterior and deployed drum providing 1,200 W BOL for payload. Batteries provide 100 per cent eclipse protection.

Thor 3
Launch: 11 June 1998 by Delta 2
Location: ~1° W
Design life: >10–12 years
Transponders: 14 × 47 W Ku-band.

Russian Federation

Ekran (Screen) Series

Background
Conceived in the late 1960s, Ekran was the Soviet Union's first geosynchronous satellite and was originally to have been powered by a 5 kW nuclear reactor.

Ekran 1, launched on 26 October 1976, brought direct broadcast services to the world for the first time. By March 1996, 20 had been placed into Statsionar-T positions near 99° E over the Indian Ocean. The longevity of the collocated Ekran-M 2, 3, 4 required no further launches 1989–91; M 5 appeared October 1992.

Cosmos 1817 in January 1987 introduced the improved Ekran-M, perhaps explaining the reduction in launch rate. Ekran-M4 in orbit since December 1988 performed a small end-of-life manoeuvre in early October 1996 and drifted off-station. This left Ekran M5, launched October 1992, as the sole operating Ekran satellite.

The Ekran satellites, each carry a single operational 200 W transponder. NPO PM has been controlling satellites itself instead of relying on the military since January 1994 through its Persei subsidiary at tracking, telemetry and command stations in Krasnoyarsk and Gauss Khroustalni. To service the system, the Soviet Union and its successor states built more than 5,000 receivers and now this network broadcasts television to some 20 million viewers in Siberia and the far north east. Since no other signals interfere with the Ekran in these remote regions the satellite can broadcast directly at UHF (0–7 Ghz).

Specifications
Launched: total 26 between 1976 and 1992 previously typically twice annually by 4-stage Proton-K from Tyuratam of which 3 were improved Ekran-M series. Most recent in the series launched 7 April 2001 after a nine-year gap
Locations: 99° E geostationary
Design life: >3 years (previously 1 year)
Contractors: NPO PM (spacecraft), NPO Radio (payload)
Transponders: one (plus one back-up) 200 W Klystron 6.188 6.212/0.702–0.726 GHz up/down beam covering Siberian region, 24 MHz bandwidth, EIRP 55.5 dBW, circular polarisation
Principal users: direct broadcast TV
Configuration: similar to Gorizont, but planar array carries 96 helical transmit aerials. Solar wings are configured to avoid antenna shadowing
Mass: 1,970 kg (320 kg payload)
AOCS: 3-axis; east-west station-keeping to 0.5° by liquid thrusters
Power: Ekran-M's solar wings were enlarged to provide 1.8 kW.

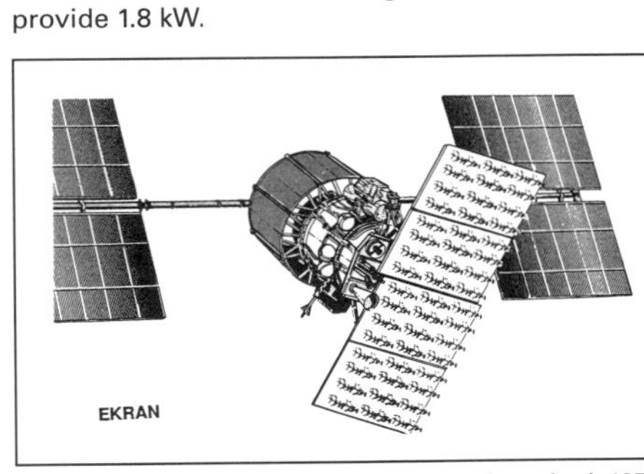

Ekran direct broadcast satellite, first launched 1976. Its 200 W transmitter provides TV coverage to remote Russian territory. Ekran-M has added a smaller panel to each of Ekran's four to boost power to 1.8 kW 0516991

Elekon Series

Background
The GEO-IK Musson geodetic satellite launched 29 November 1994 by Tsyklon 3 from Plesetsk into 1,480 × 1,527 km, 73.61° includes an Elekon payload which Russia used to demonstrate location and data relay for cargo transport services. This was the last of 14 GEO-IK satellites launched since January 1981. NPO PM and Germany's Elbe Space & Technology Dresden GmbH, entered into a joint agreement to operate the global store/forward system consisting of L/S/C-band payloads on seven Tsikada navigation satellites.

Express Series

Current Status

On 11 June 2001 Express 2 suffered an emergency deactivation and services were transferred to Express A3. Express A1R was launched 10 June 2002 by a Proton K/DM-2M but the parking orbit was off-nominal corrected by the second burn of the DM-21 upper stage.

Background

Gorizont's successor, called the Express, appeared in 1994. InterKosmos operated and marketed the satellite service through its consortium of NPO PM, NIIKP, NPO Radio and the Vostok Bank (along with Gals). Most of the Express payload directly duplicates the Gorizont, although it deletes the L-band Volna service (transferred to Marathon). Express is the second RFAS/Soviet satellite, after Gals, to provide NSSK. Intelsat agreed in March 1993 to lease up to three satellites beginning in mid-1994, but the agreement lapsed in 1994 because InformKosmos required payment in advance. Intelsat only agreed to begin leasing once each satellite was on-station.

Rimsat Ltd, an Asia-Pacific telecom company based in Fort Wayne, Indiana, planned to lease four Express satellites from InformKosmos for the Asia-Pacific region, with first launch in October 1994, but payments reportedly stopped in April 1994. Following restoration of the Express programme the first Express A satellite was destroyed in the launch failure of a Proton, 27 October 1999. Express 6A was launched on 12 March 2000 by Proton and redesignated A2 on orbit. It was followed by 3A on 24 June 2000.

Specifications

Express 1–2 satellite

Launch: number 1 13 October 1994 by Proton-K from Tyuratam; located over 70° E before moving to 14° W in February 1995 for Intersputnik. Express 2 launched 26 September 1996, located over 80° E and operated by Intersputnik
Locations: initially 14° W and 80° E
Design life: 5–7 years
Contractors: NPO PM (spacecraft), NIIKP (payload)
Transponders: one (plus one back-up) 60–80 W TWTA 6.000/3.675 GHz up/down C-band beam, 40 MHz bandwidth, transponder number 6, switchable to A1 global (EIRP 35.6/32.6 dBW max/min) or A2 steerable (7.3° N, 5° S, 5.5° E/West) 4.6° spot (EIRP 46.1/43.0 dBW max min), RHCP, for single channel TV distribution to 2.5 m Moskva antenna

Nine 10.5 W TWTA 6.050–6.450/3.725 4.125 GHz up/down C-band global, hemispheric, zonal and spot beams, 36 MHz bandwidth, numbers 14/16 jointly switchable to A2 steerable (7.3° N, 5° S, 5.5° E/west) 4.6° spot (EIRP 38.0/35.0 dBW max/min) or A4 steerable 7.5° N-S 4.6 × 10.6° elliptical (EIRP 34.7/31.7 dBW max/min), numbers 8/10 jointly A4 steerable 7.5° N-S 4.6 × 10.6° elliptical (EIRP 39.2/36.2 dBW max/min) or A3 fixed 14.6° quasi-global (EIRP 33.4/30.4 dBW max/min), numbers 7/9/11 jointly A5 fixed 14.6° quasi-global (EIRP 28.9/25.9 dBW max/min) or A6 north-south steerable 4.6 × 10.6° elliptical (EIRP 34.7/31.7 dBW max/min), numbers 15/17 as 7/9/11, RHCP transmission.

Two 15 W TWTAs centred at 14.325/11.525 and 14.425/11.625 GHz up/down Ku-band beams, numbers 12/20 working through fully steerable (north-south and east-west) 4.6° spot antenna,

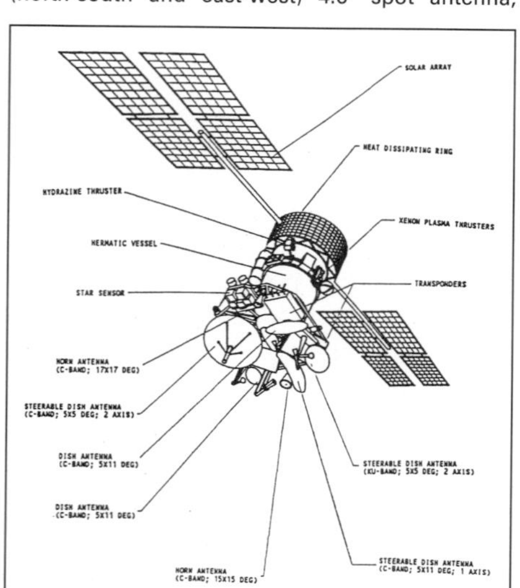

Express is replacing Gorizont (Intersputnik) 0517389

Express M2000, the most powerful communications satellite designed by Russian telecommunications engineers (Theo Pirard) 0054358

36 MHz bandwidth, EIRP 39.2/36.6 dBW maximum/minimum, orthogonal polarisation (vertical receive and horizontal transmission) for Luch service
Principal users: TV distribution via Moskva system for Russia, telecommunications services
Configuration: 3-axis stabilised, based on pressurised 2 m diameter central pressurised cylinder (maintained at 0–40°C). Seven antennas provide zonal, hemispheric and global beams. Power comes from twin four-panel solar wings spanning 21.0 m, cylindrical base also carries solar cells; 2,400 W EOL. AOCS is provided by liquid hydrazine thrusters (25.5 kg hydrazine); north-south/east-west station-keeping to ±0.2° by eight OKB Fakel SPT-100 plasma thrusters (56 kg Xe loaded). ±0.1° pointing accuracy. Express can operate autonomously for 30 days
Mass: about 2,500 kg at launch (payload 430 kg).

Express-6A

Launch: 12 March 2000 by Proton
Locations: initially 40° E, 103° E, 140° E now 80° E
Design life: 10 years
Transponders: 8 50 W TWTA 14.000–14.250/10.950 11.200 GHz up/down Ku-band 3 × 3° beams, 72 MHz bandwidth, EIRP 46.9 dBW, for 1,024 Mbit/s data services to 1.5–2.5 m dishes

12 (plus 4 back-up, 3 × 4-for-3 + 1 × 2-for-1) 32 W TWTA 14.340–14.500/11.540–11.700 GHz up/down Ku-band 3 × 3° beams, 2 × 36 MHz bandwidth, EIRP 44.9 dBW, for TV distribution and telecom services
Principal users: TV distribution, telecommunications services
Configuration: 3-axis stabilised, based on pressurised 2 m diameter central cylinder. 4 transmission/receive antennas. Twin 6-panel solar wings spanning 26.6 m providing 6 kW (Ni/H_2 batteries) and the base also carries solar cells. AOCS by hydrazine thrusters; north-south/east-west station-keeping to ±0.1°, by Xe ion thrusters (120 kg Xe)
Mass: 2,570 kg at launch.

Express-3A

Launch: 24 June 2000
Location: 11° W
Specification: as Express A2.
Mass: 2,600 kg.

Express-A1R

Launch: 10 June 2002
Location: 40° W
Specification: Built by NPO PM and Alcatel
Mass: 2,600 kg.

Gals/Gelikon Series

Background

Three Gals satellites, the first of which appeared January 1994 were built to replace the Ekran. Gals 1 was the first Russian civilian controlled space applications project. Gals introduced North-South station-keeping to Russian satellite operations. Russia has updated the satellite to be competitive with the West by building the Gals-R version. Each R derivative carries four or six (reports vary) active

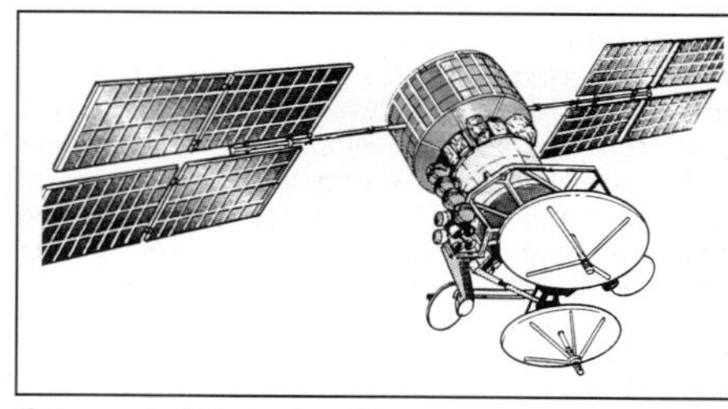
Gals was built to replace Ekran (Teledyne Brown Engineering) 0517390

transponders. Gorizont and Ekran satellites were to be replaced by the Gelikon series conceived in the early 1990s by the Applied Mechanics NPO. Satellites were to be located at 23° east, 44° east, 74° east, 110° east, and 140° east at the five major broadcast time zones. Together the Russians call these three systems the STV-12. No further launches as of March 2002.

Specifications

Gals series

Launched: number 1 20 January 1994 and number 2 17 November 1995 by Proton from Tyuratam
Locations: number 1 initially 44° E GEO and then 70° E by 13 June 1994. During May–June 1996 number 1 relocated to 35° E, number 2 relocating to this longitude by early July 1996
Design life: 7 years
Contractors: NPO PM (spacecraft), NPO Radio (payload)
Transponders: Two (plus 1-for-1 back-up) 90 W + one (plus 1 for-1 back-up) 45 W TWTA 17.3–18.1/11.72748–12.47550 GHz up/down working through antennas A1 0.9 × 1.2° and A2 1.25 × 2.5°, 27 MHz bandwidth, peak EIRP 57 dBW (56.0 dBW central region), circular polarisation. Both antennas are north-south and east-west steerable and A2 can rotate ±30°
Principal users: direct broadcast TV to 90 cm home (50 cm in 56.0 dBW region) and 1.5–2.5 m professional
Configuration: similar to Express. 21.0 m-span solar array provides 2.40 kW EOL. 100 per cent eclipse protection comes from Ni/H_2 batteries. All four main antennas (A1/A2 transmission + A4/A5 receive) are carbon fibre
Mass: 2,500 kg on-station (375-kg payload)
AOCS: 3-axis; ast-West station-keeping to 0.5° by hydrazine thrusters, NSSK by eight OKB Fakel SPT-100 plasma thrusters
TT&C: C-band via horn antenna. 12 m ground stations at four sites, headed by NPO PM's Krasnoyarsk. Gals allows 30 days' autonomous operations.

Gelikon series

Launch: by Proton from Tyuratam
Locations: planned 23, 44, 74, 110, 140° E
Design life: 12 years
Transponders: 12 TWTA 17.3–18.1/11.7–12.5 GHz up/down, peak EIRP 55 dBW, circular polarisation
Principal users: direct broadcast TV to 60–90 cm home and 1.5–2.5 m professional
Mass: 2,500 kg on-station
AOCS: 3-axis; north-south/east-west station-keeping to 0.1°.

Gonets-D1 Constellation

Current Status

Gonets D1-1 to D1-3 were launched by Tsyklon 3 on 18 February 1996 followed by Gonets D1-4 to D1-6 by Tsyklon 3 on 14 February 1997. The system was never deployed to its planned constellation of 12 satellites in two orbital planes of six and commercial application failed through lack of funding. Only two sets of three satellites have been successfully launched to date.

Background

The civil Gonets system of small communications satellites come directly from the equivalent military Strela-3 communications satellite programme. Russia flew two Gonets-D demonstrator missions named Cosmos 2199 and Cosmos 2201 as part of the sextet launched on 13 July 1992. The two sextet missions launched on 19 February 1996 and 14 February 1997 each carried three of the military Strela-3 satellites and three Gonets-D1 test satellites known as Gonets D1-3 and D4–6, respectively.

The Gonets- D1 (Gonets means 'messenger') are cylindrical (covered in solar cells), 0.8 m diameter and 1.6 m long, plus an extended boom for gravity stabilisation and cone-antennae: the mass of each satellite is 225–230 kg. AKO Polyot builds the satellites for NPO PM. The Gonets-R satellite was to have been equipped with satellite intralinks with 45 satellites in five planes of nine each operating at L-band and Ku-band. Each satellite would have had the capacity of 15 ground-space and three space-ground channels with 8 Mb of on-board storage and a data transmission rate of 1.2–64 kbps.

The Smolsat consortium, including NPO PM, NPO Precision Instruments, Selkhoz Bank and Moscow's Soyuzmedinform Program Management (connected with the Ministry of Health), planned two systems. The first was to provide e-mail store/forward and real-time (within the 3,000 km diameter footprint) bulk data relay. The second, Gonets-R, was to offer mobile phone links.

Network Services International in New York claimed to be the western representative for Gonets and the Russian Space Agency provides state support. The first 36-satellite system was projected to be operational by end-1996, employing six satellites in each of six 82.5° planes at 1,300–1,500 km altitude for 4.8/9.4/64 kbit/s links. 312–315/387–390 MHz up/down 10 W link. Each 230 kg satellite would handle 1.85 Mbit daily, plus 250 Gbit for the 1.6/1.5 GHz up/down L-band, EIRP 10 dBW. This plan did not materialise and after a series of test launches to validate hardware the first trio of satellites was launched on 14 February 1997. A second trio was launched 27 December 2000 but failed to reach orbit. The third trio launched to date were placed in orbit on 28 December 2001.

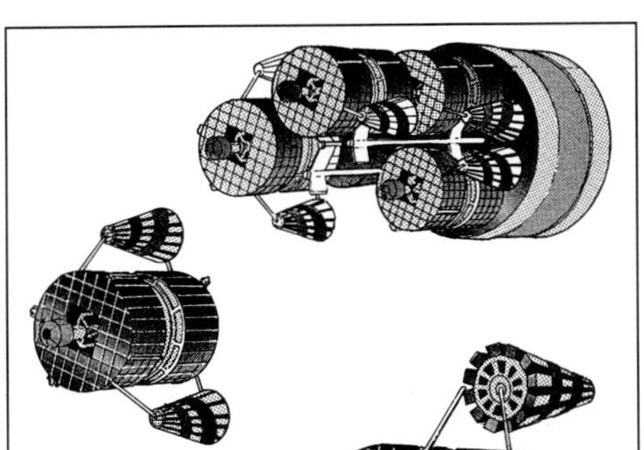

Derivatives of the military sextet store and forward satellites are offered commercially
(Teledyne Brown Engineering) 0516993

UPDATED

Gorizont (Horizon) Series

Background

The Gorizont system was inspired by the 1980 Olympic Games from Russia under design article number 11F662 built by the MOM agency. The geostationary satellite series was later made a part of the YeSSS Unified Satellite Communication System. Gorizont 1 appeared 19 December 1978, although it achieved only 22,553 × 49,023 km, 11.3°. Despite its non-GEO path, it provided some service and was joined by three others for coverage of 1980's Moscow Olympic Games. A single 40 W transponder, similar to its Molniya 3 equivalent, relays TV to more than 1,000 2.5 m diameter dishes in the Moskva system. Gorizont now carries the US/Russia hotline previously supplied by Molniya 3 (Intelsat has also carried it since 1974). The full Gorizont capacity was offered by Glavcosmos at about US$40 million, including launch. Intersputnik provides most of its services but Express will eventually replace them.

Gorizont 22 appeared 23 November 1990, and was positioned at 40° E, the Statsionar 12 location. It was the first of three satellites for the Russian Federation's Ministry of Communications & Information Technology, which paid Rb100 million to the Ministry of Defence for launch services. The satellite provided TV services to 21,000 settlements in European Russia, while the second, Gorizont 23 launched 1 July 1991 to 103° E Statsionar 21, added 3,500 in the Urals and Siberia. The third appeared 2 April 1992 as Gorizont 25, also over 103° E (number 23 apparently failed mid-1992). Gorizont 20 carried a Mayak transmitter, possibly to demonstrate mobile L-band communications. What should have become the original Gorizont 28 was lost in the Proton launch failure of 27 May 1993. It was to have been the first occupant of the Statsionar 16 145° E slot, from where transponders numbers 9–11 were to have been marketed by SovCan Star and Vista Satellite.

Russia and the former Soviet Union launched a total of 32 named Gorizont satellites by end of 1995 into Statsionar and, now, Tonga slots. Russia said January 1996's Gorizont 31 would complete the series, Gorizont 32 appeared in May 1996. Gorizont 29 sits at 130° E (a Tonga slot), and is leased from InformKosmos as Rimsat 1 by Rimsat Ltd. Rimsat initially leased 'Tongasat 1' (Gorizont 17) at 134° E from July 1993, but when the satellite moved January 1995, the lease ended. Gorizont 30 (Rimsat 2) was launched 20 May 1994 towards 142.5° E.

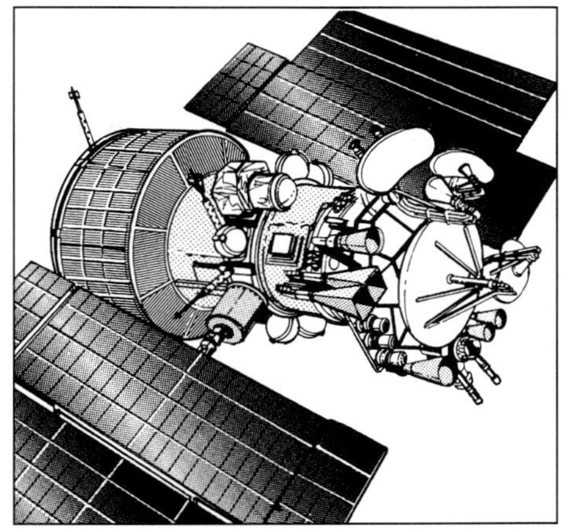

The Gorizont multipurpose communications satellites are available commercially 0516992

Specifications

Launched: until 1993 typically twice annually, by Proton-K from Tyuratam
Locations: 40/53/80/90/96.5/103/140° E and 14/11° W geosynchronous (plus Rimsat/Tonga slots), inclined 1.5° BOL
Design life: 3 years
Contractors: NPO PM (spacecraft), NIIKP (payload)
Transponders: One (plus one back-up) 65 W TWTA 6.000/3.675 GHz up/down C-band spot or global beam for transponder number 6, 40 MHz bandwidth, switchable between A3 4.6° spot EIRP 46.1/43.0 dBW max/min or A1 global 35.6/32.6 dBW maximum/minimum, RHCP, for TV distribution to 2.5 m Moskva antenna. The six C-band (also see below) together comprise the Statsionar service.

Five 15 W TWTA 6.050–6.250/3.725– 3.925 GHz up/down C-band global and hemispheric beams, 34 MHz bandwidth, numbers 8/10 on A4 9 × 18° EIRP 31.0/28.0 dBW max/min, numbers 7/9/11 switchable between A6 global 28.5/25.5 dBW max/min and A8 6 × 12° EIRP 34.0 dBW max; RHCP. The six C-band (also see above) together comprise the Statsionar service.

One (plus one backup) 15 W TWTA 14.325/11.525 GHz up/down Ku-band beams, 36 MHz bandwidth, number 12 on A9 4.6° spot EIRP 39.2/36.0 dBW max/min 4.6° Luch service

One (plus one back-up) 1.6377 1.6386/1.5362–1.5371 GHz up/down L-band; transponder cross- strapped with a 15 W C-band transponder for link with Earth station, for Volna maritime service, EIRP 40 dBW
Principal users: TV distribution via Moskva system, telecommunications services, maritime and aeronautical mobile

Gorizont satellites (as of March 2002)

Launch Date	Satellite	Location	Location Name
19 December 1978	Gorizont 1	–	Failed to orbit
6 July 1979	Gorizont 2	14° E	Statsionar-4
28 December 1979	Gorizont 3	53° E	
14 June 1980	Gorizont 4	14° W	
15 March 1982	Gorizont 5	53° E	
20 October 1982	Gorizont 6	90° E	
1 July 1983	Gorizont 7	14° W	
30 November 1983	Gorizont 8	90° E	
22 April 1984	Gorizont 9	53° E	
2 August 1984	Gorizont 10	80° E	
18 January 1985	Gorizont 11	140° E	
10 June 1986	Gorizont 12	345° E	
18 November 1986	Gorizont 13	90° E	
11 May 1987	Gorizont 14	140° E	
18 January 1988	Gorizont 25L	–	Failed to orbit
31 March 1988	Gorizont 15	–	
18 August 1988	Gorizont 16	80° E	
26 January 1989	Gorizont 17	53° E	
6 July 1989	Gorizont 18	140° E	
28 September 1989	Gorizont 19	97.5° E	
20 June 1990	Gorizont 20	25° E	
3 November 1990	Gorizont 21	145° E	Statsionar-16
23 November 1990	Gorizont 22	140° E	Statsionar-7
2 July 1991	Gorizont 23	103° E	
23 October 1991	Gorizont 24	80° E	Statsionar-13
2 April 1992	Gorizont 25	103° E	Statsionar-21
14 July 1992	Gorizont 26	349° E	
27 November 1992	Gorizont 27	96.3° E	Statsionar-14
28 October 1993	Gorizont 28	90° E	Statsionar-6
18 November 1993	Gorizont 29	161° E	
20 May 1994	Gorizont 30	142.5° E	
25 January 1996	Gorizont 31	40.5° E	Statsionar-12
25 May 1996	Gorizont 32	53° E	Statsionar-5
6 June 2000	Gorizont 452	–	–

Three Gorizonts are located at positions which have not been registered to the RFAS: Gorizont 19 is close to the Statsionar-2 location (35° E) and Gorizont 20 close to Statsionar-19 (23° E), but there is no RFAS registration close to the Gorizont 29 location.

Configuration: 3-axis stabilised, based on pressurised 2 m diameter central cylinder, about 5 m long. 11 antennas provide zonal, hemispheric and global beams. Twin solar wings, of two major/two minor panels each, spanning 9.460 m generate 1,280 W; cylindrical base also carries solar cells. Liquid bipropellant thrusters provide AOCS. East-West station-keeping to within 0.5° is exercised but north/south excursions are uncontrolled: 2° is possible after 3 years
Mass: about 2,120 kg originally, now about 2,500 kg.

Informatr Satellite

Background

The 600 kg Informatr 1 Koskon prototype was launched by Kosmos C on 29 January 1991 and placed in a 955 × 1,005 km, 82.9° orbit. AKO Polyot derived this satellite from a military store/forward design and the Elas Scientific Production Association provided the core payload. The press release described it as being launched for the Ministry of Geology to relay data collection platform data, provide communications for survey parties and to support disaster relief operations. The real purpose of the satellite was to serve as a prototype for the proposed Koskon Global Space Communication System which was to have consisted of 32–45 satellites launched between 1997 and 2001. Five mobile ground stations worked through the satellite, four in Krasnoyarsk and one in Archangelsk.

Informatr also carried the RS-14 amateur transponders for the Soviet Amsat. Two transponders provided: 435.102–435.022 MHz up and 145.852–145.933 MHz down, and 435.123–435.043 MHz up and 145.866–145.946 MHz down. Part of this Radio-M1 amateur payload constituted a second version of Amsat Germany's Rudak digital packet communications unit (designated Oscar 21). Amateurs continued to use the satellite until it was deactivated in October 1994 by its VKS Military Space Forces controllers at Golitsyno-2.

Kupon Satellite

Current Status

Originally mooted as third generation GKKRS (Global Space Command and Communications System) it was the first communications satellite for the Lavotchkin bureau but the project collapsed with the breakup of the Soviet Union. In its place the Kupon satellite was redesigned to support the Russian Federation Central Bank.

Plans to build a system of three Kupon satellites, featuring independently steerable spot beams and full inter-transponder switchability fell through and only one was launched. Satellite is the first to support the Russian banking system and the first commercial satellite sold by Lavochkin.

Background

Lavotchkin proposed the Bankir project in 1990 to provide messaging services on 400–800 MHz UHF through 6–8 LEO satellites. Housed on a platform derived from Lavotchkin's Phobos craft, the payload would have been developed by NPO ELAS. The projected plan was for four Kupon satellites at locations above the eastern Atlantic, the eastern and western Indian Ocean. Up to 10,000 transmissions of messages up to 400 characters could have been handled daily. The major user would be the Bankir network providing Russian banks with a 40,000-terminal VSAT network with a centralised control centre (the service began in 1993 using existing Potok satellites). Bankir cost US$100 million to establish, and requires 20 per cent of Kupon's capacity.

Seen at NPO Lavotchkin, the launch configuration shows the radiator at left, solar wing at right and the phased array at top 0517388

The company, GIS, became the commercial backer of Kupon in 1991 with shareholders NPO Lavotchkin; NPO ELAS; SovFinTrade, Financial & Trading Co; Pragma Electronics and a consortium of Russian banks. It appears that Russia's Central Bank took control in 1994 after the joint venture failed to raise sufficient capital for the full system.

At the beginning of 1994, the Central Bank of Russia supposedly guaranteed financing of the first three satellites, but failure of the joint venture in 1994 to raise sufficient capital reportedly left only one satellite in prospect. A February 1993 agreement with the Manila based Communication & Broadcast Managers, Inc to set up the Galaxy Satellite & Telecommunications fell through and plans for follow-on satellites have languished since then.

Specifications

Launches: 12 November 1997 on a Proton-K from Baikonur
Locations: 55° E in 1997–1998, 86.25° E in September 2001
Design life: 6 years minimum (8 years expected)
Contractors: NPO Lavotchkin (bus); NPO ELAS (payload)
Transponders: 16 14.020–14.500/10.960–11.200 and 11.460–11.700 GHz up/down Ku-band spots. Each of four transmission blocks comprises 64 SSPAs and can shape up to four independent spot beams. Each spot is independently steerable ±8.5° and adjustable on either axis from 2 × 2° to 3.5 × 3.5° in 0.5° steps. Each repeater block comprises 6 × 36 MHz 'transponders': four for normal traffic (one per beam) and two either for switching traffic between any beam on any block or as additional transponders in the block. Total power of each block is 20 W and redundancy is provided by the multiple SSPAs (10 per cent failure would provide –0.8 dB loss in overall EIRP). EIRP is adjustable on each block 35 to 50 dBW; equalised power for all beams (2°) would be 39 dBW linearised EIRP (equivalent to 49 dBW saturated), SFD –80 dBW/m^2 to –95 dBW/m^2, orthogonal linear polarisation
Principal user: Bankir VSAT network, international diameter domestic VSAT, DBS, cable/TV broadcast, SNG, videoconferencing, point to point
Configuration: 3-axis platform. 4 × transmission and 4 × receive phased array antennas on Earth face, 3-panel solar wings on North/South faces, radiators on East/West faces, electronics housed in central pressurised cylinder
Mass: 2.5 tonnes BOL on-station
TT&C: control centre and ground station both outside of Moscow
Power system: twin 3-panel solar wings.

UPDATED

Molniya (Lightning) Constellation

Current Status

Eight Molniya 1 and four Molniya 3 satellites still operate in planes spaced at 45° intervals. Russia

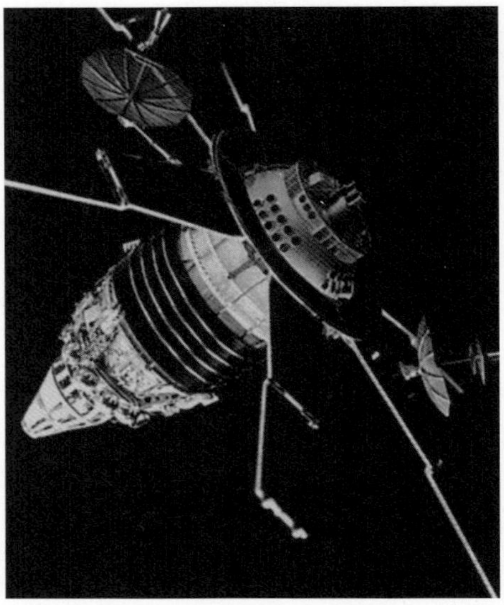

Each Molniya 1 carries at least one 40 W X-band transponder for government and military links 0516994

has formally ended the Molniya 2 programme. Russia continues to launch Molniyas, but not at the same rate as their original introduction in April 1965. Russia launched only four Molniya types between August 1996 and the end of 1999. No Molniya were launched in 2000 but two Molniya 3 series were launched in 2001 and one Molniya-1 was launched in April 2003. The most recent launch carried a Molniya 3 into orbit on 19 June 2003.

Background

Sergei Korolev's design bureau first conceived of Molniya in 1962 as a method to cover the USSR's northern expanses from highly inclined orbits. Since boosters had not yet attained the ability to place a satellite in geosynchronous orbit, when the USSR launched the first Molniya it put it in a sub-geosynchronous orbit that became a unique signature of the satellite. With a usual perigee over the Southern Hemisphere, each satellite lingers around apogee over the northern territories for about 8 hours each day. Their unusual orbits also avoid the severe launch penalty imposed on geosynchronous satellites from latitudes at Tyuratam and Plesetsk. By coincidence, the 63.4° Molniya inclination also nulls orbital perturbations created by the oblateness of the Earth and so the satellites require much less station-keeping propellant than most other highly elliptical orbits. On the other hand, the orbit passes through the Van Allen radiation belt, which imposes a radiation dosage five to six times the amount experienced in geosynchronous orbit. Hardening to mitigate this radiation dose makes the satellites about 20 per cent heavier than their non-hardened counterparts.

The USSR inaugurated communication service with the first Molniya 1 series satellite, launched 23 April 1965. This Molniya carried the first operational television and telephony links between Moscow and Vladivostok before decaying 16 August 1979. Molniya 1–3, launched almost exactly one year later, added Earth imaging to complement the Meteor series, and exchanged colour television signals with France. Three satellites in a constellation with the same orbital parameters provide 24-hour coverage. Generally however, the USSR placed eight satellites in planes 45° apart to ensure un-interruptible service. The original network of Molniya 1 satellites probably carried military/government links over one transponder. An improved Molniya-1T appeared in June 1970 equipped with the "Beta" transmitter and became operational in 1972. The satellite was part of the Korund system facilitating strategic communications and telephony with military units

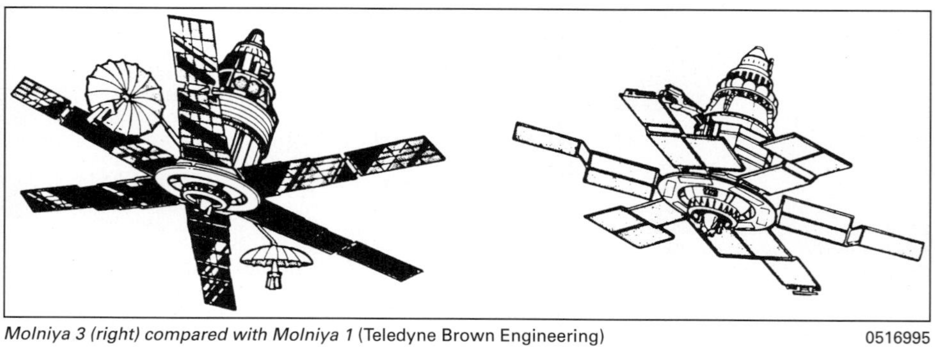

Molniya 3 (right) compared with Molniya 1 (Teledyne Brown Engineering) 0516995

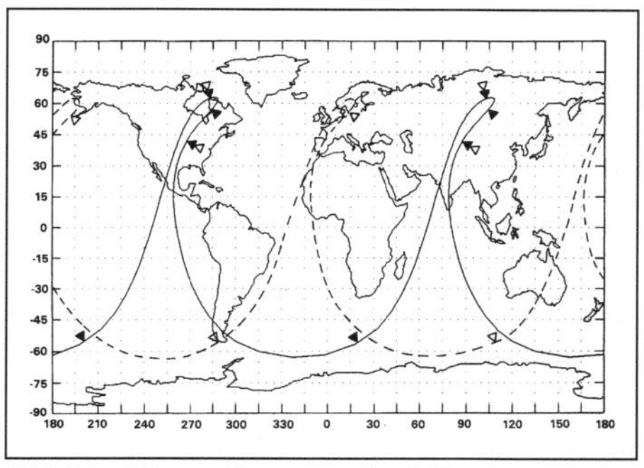

Molniya 1/3 satellites usually follow a common groundtrack taking them over the heart of the Russian landmass. Since 1983, some Molniya 3s have flown with groundtracks (dotted line) between the original constellations (Teledyne Brown Engineering) 0516996

in Siberia and the Far East. The complete system included eight satellites on orbit continuously. The 16-constellation network were divided into two types and four separate groups. Eight Molniya-1T types were operated in two constellations of four with both consisting of four orbital planes spaced 90° apart with the ascending node of one displaced 90° to the other. Beginning with Molniya 1–75 all Molniya-1T types are launched from Plesetsk by Molniya launcher. The last Molniya-1T was launched in September 1998 followed by Molniya 1–92 in April 2003.

The Molniya 1 appears to be employed primarily for military and government links through at least a single 40 W X-band transponder. Some early models incorporated a camera to add synoptic Earth views to low-altitude Meteor imagery. Molniya 1–20, launched 4 April 1972 from Plesetsk carried France's piggyback research SRET 1. Molniya 1–30 similarly released SRET 2 on 5 June 1975. Molniya 1S, different from the standard series acted as a geosynchronous demonstration.

Beginning with Molniya 1–11 in 1969, the number above the horizontal at any one time was increased from two to three by reducing the equatorial spacing to 90°. Following the introduction of the Molniya 2 series in 1971 and Molniya 3 in 1974, they were positioned in groups of three employing different frequencies. As traffic increased, the Soviets had to decrease the spacing between Molniya 1s to 45°.

The Molniya constellation achieved great success at bringing communications to remote parts of the USSR. Starting with Molniya 1–11 in 1969, the USSR decreased the orbital stationing to 90° instead of 120° in order to keep at least two satellites above the horizon at all times. The Molniya 2 constellation followed in 1971, and the Molniya 3 constellation in 1974. But the newly increased traffic demands caused the USSR to rely more on geosynchronous satellites and the last Molniya 2 began its service in 1977.

The first Molniya 2, a more capable series employing higher frequencies orbited on 24 November 1971. The 17th and last satellite appeared on 11 February 1977.

Molniya 3 first appeared in November 1974 and was used to create the Orbita communications system for northern regions. It was derived from the Molniya 2M, development of which began in

A late Molniya 3 0132812

1922. From early 1976, Molniya 3 provided one of the two independent satellite-based systems (the other via Intelsat) to maintain the Washington-Moscow hotline. Presidents Nixon and Brezhnev agreed to the hotline concept during the SALT negotiations of 1971. To ensure contact, the United States agreed to continually track the entire Molniya constellation from the US station at Fort Dietrick (Frederick, Maryland). Dietrick tested the circuits hourly. (Traffic from the hotline has now shifted mostly to the Gorizont.)

Until 1984, all Molniyas had a common ground track roughly bisecting Asia/North America. In late 1984, Molniya 3–21 manoeuvred to re-establish its ground track over the Pacific and the extreme eastern edge of the Atlantic. During 1985, Molniyas 3–25, 3–26 and 3–27 followed suit. These new satellites had a planar spacing of 90°, but the ascending nodes occurred mid-way between the other Molniya 3s. Western observers speculate that this new constellation might be connected with maritime communications.

The Molniya 3, NPO-PM's newest class of satellites offers three C-band transponders for general telecommunication services, including television distribution over the 100 or more ground stations of the Orbita 2 network. 12 m ground stations in the CIS and affiliated countries such as Cuba and Mongolia.

157 named satellites have been launched in the three series. Molniya 1–85 was at first described as a Molniya-1T, although the Russians subsequently indicated that Molniya-1T satellites had been operating for many years before this. In the same source the Russians referred to Molniya 1–88 as the 48th 1T, but they gave no indication of the difference with the standard design.

Specifications

Molniya 1

Launched: typically one a year by Molniya-M from Plesetsk (very rarely Tyuratam)

Orbit: typically 400 × 40,000 km, 63°; placed in eight orbital planes separated by 45°

Design life: 2 years

Contractors: NPO PM

Transponders: At least one 40 W 1.0/0.8 GHz up/down X-band; possibly total of three transponders

Principal users: military/government

Configuration: pressurised cylinder with domed ends, 1.6 m diameter, 3.4 m long, accommodates electronics and provides mounting for two boom-mounted 90 cm diameter dishes. Six windmill solar wings generate 500–700 W. AOCS provided by liquid bipropellant engine assembly at other end; 3-axis control

Mass: about 1,600 kg.

Molniya 3

Launched: typically now one or two per year on average

Orbit: typically 400 × 40,000 km, 63°; placed in eight orbital planes at 45° intervals

Design life: 4 years?

Transponders: 2 (plus 1 back-up) 30 W TWTA 5.9756.225/3.650–3.900 GHz up/down C-band global beams, 50 MHz bandwidth, 35 dBW edge EIRP, circular polarisation

Principal users: domestic TV/telecommunications

Configuration: pressurised cylinder with domed ends, 1.6 m diameter, 4.2 m long, accommodates electronics and provides mounting for six 3-panel windmill solar wings spanning 8.13 m and generating 1.2 kW (950 W payload requirement). AOCS provided by 1.96 kN liquid bipropellant engine assembly at other end; 3-axis control

Mass: 1,750 kg (payload 220 kg).

UPDATED

Spain

Hispasat Series

Current Status

The increase in demand on Hispasat's capacity due to digital television and the liberalisation of satellite services has led Hispasat SA to purchase of a third and fourth satellite, Hispasat 1C before the first generation satellites require replacement. Built by Alcatel Hispasat 1C is located over 30° W (with 1A and 1B) and provides improved performance and capacity over Europe and the Americas. Hispasat 1D was launched 18 September 2002 and is collocated with Hispasat 1C. Hispasat 1C carries 24 Ku-band transponders and was launched on 3 February 2000. Hispasat 1D will have 30 Ku-band transponders and is contracted to be launched by ILS aboard an Atlas 2A5 in the third quarter of 2002 for DTH services to Europe, the Iberian peninsular and the Americas. Hispasat 1D is built around an Alcatel Space Industries Spacebus 3000. Hispasat 1E is being designed for Ka-band multi-media services. Hispasat 1A and 1B services are expected to retire by the end of 2003.

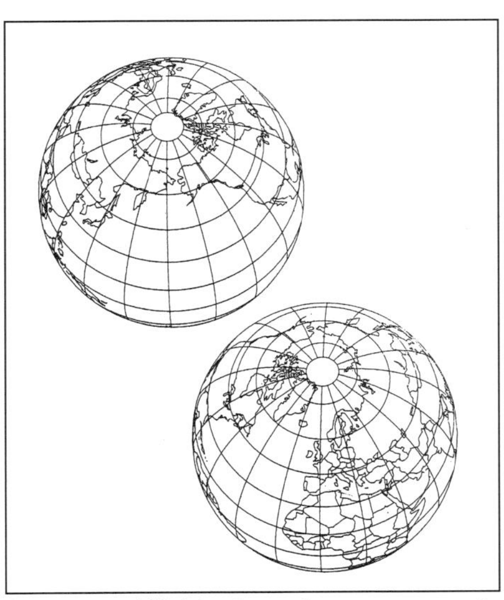

One group of Molniya 3s reaches apogee over the Pacific and East Atlantic, possibly for maritime links. Shown is the vantage point on each of the two daily apogees (Teledyne Brown Engineering) 0516997

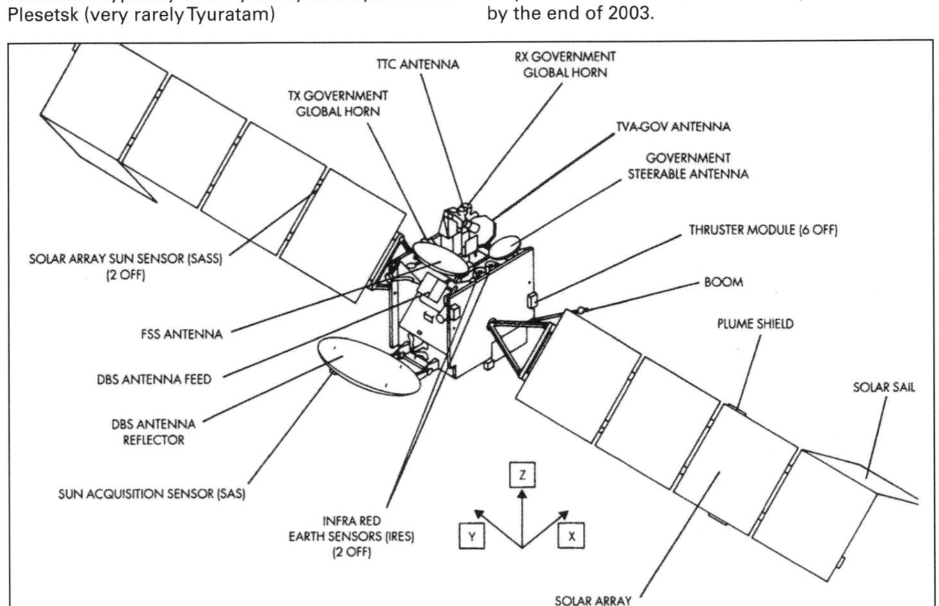

Hispasat configuration: the large dish provides DBS services to 60 cm receivers within Spain 0517012

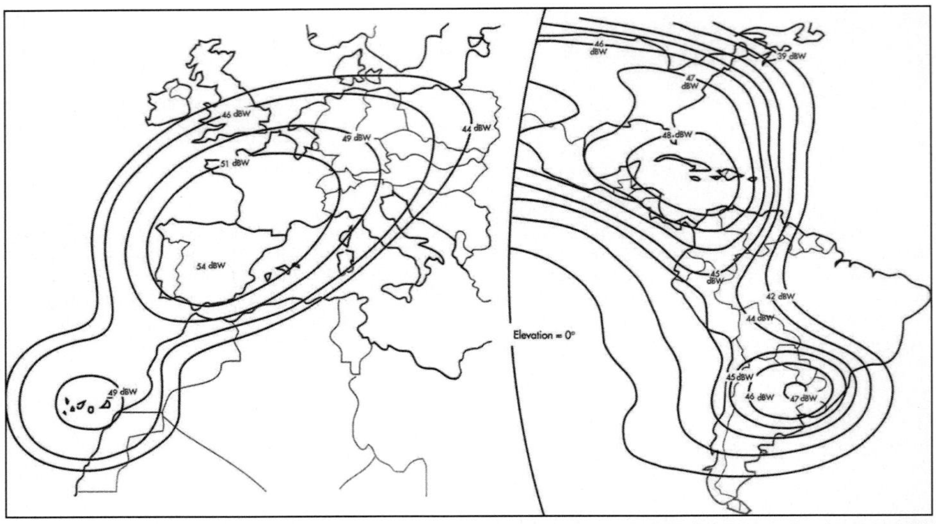

Hispasat is the first European satellite to address American countries. Left: FSS (DBS is similar but 5–6 dBW higher); right: broadcast to America. Not shown is the classified footprint of the military payload 0517392

Tele-X is based on the Spacebus 300 platform (Aerospatiale) 0517013

Background

The Spanish government originally considered building its own satellite network in 1985, but then rejected the idea. Parliament resurrected again in July 1988 to address the communications needs for the Barcelona Olympic Games and the Columbus quincentenary celebrations. After approval, The Secretaria General de Comunicaciones issued tenders for a two-satellite domestic broadcasting and communications system, plus ground segment. The programme received government approval in April 1989 for two satellites, and a ground spare, to provide three to five Spanish direct broadcast service channels, two North and South America television distribution channels, 8 to 16 television broadcast and communications channels and two government/military X-band channels.

Spain gave the satellite construction contract to the Satcom International partnership of Matra/British Aerospace in June 1989 for Pts19,898 million. The design reflected closely Matra's Telecom 2 because Spain sought to have the satellite in space by 1992. Spain chose an Ariane 4 instead of an Atlas 2 to launch the satellite despite the lower US bid because of the presumed greater stability of the peseta against the franc.

Hispasat 1A became operational in January 1993 despite the fact the 2.2 m direct broadcast satellite antenna pointed slightly too far north, probably because of thermal distortion. The addition of 1B created a total of five operational DBS channels. The Spanish military used the Hispasat for the first time in March 1993 when Loral Western Development Labs installed an SCT-10 anchor terminal in Madrid to provide 12 voice/ data circuits to a transportable unit in Bosnia.

Specifications

Hispasat 1A

Launch: 10 September 1992 by Ariane 4 from Kourou, French Guiana
Location: 30° W geostationary
Design life: 10 years min, 12 years possible
Contractors: Satcom International (MMS France prime), MMS UK (payload), Thomson Tubes Electroniques (TWTs), ANT (EPCs)
Transponders: three (plus two back-ups shared with TVA) 110 W TWTA 17.314–17.648/12.136–12.470 GHz up/down Ku-band supporting two DBS beams over Spain and the Canary Islands working to 30–60 cm dishes, 27 MHz bandwidth, 55–58 dBW Spain (54–56 dBW Canary Islands) EIRP

One (plus two back-ups shared with DBS) 110 W TWTA 14.249/12.078 GHz (12.015 GHz on 1B) Ku-band supporting TVA Spanish-speaking TV distribution service to North/South America through 1.5–2.5 m stations, 36 MHz bandwidth, 41–47 dBW EIRP

Eight (plus four back-up) 55 W TWTA 14.00–14.50/ 11.450–11.700 + 12.50–12.750 GHz up/down Ku-band FSS European/Canary Islands beams supporting TV broadcast + telecom services, 4 × 36, 2 × 54, 2 × 72 MHz bandwidth, 49.5–52 dBW

Three (plus one back-up) 40 W TWTA 7–8 GHz X-band Spanish beams for government/military applications, 20–40 MHz bandwidth, EIRP 42 dBW
Principal applications/users: DBS, TV distribution, datalinks, point-to-point applications, rural telephony, and defence. FSS controlled by Telefonica, Retevision and Spanish Post Office; military/government by INTA
Configuration: Matra/BAe Eurostar 2000 bus
Mass: 2,194 kg at launch, 1,325 kg on-station BOL, 1,013 kg dry, 280 kg communications payload
Attitude control: 3-axis employing inertial/Earth sensors in conjunction with hydrazine thrusters
TT&C: identical to Telecom 2 Satellite Control Centre Three dedicated Ku-band + single S-band antennas at the Arganda facility, near Madrid
Power system: twin 4-panel solar wings, 2.050 m wide, spanning 22 m, providing 3,237 W at equinox BOL and full eclipse protection with Ni/H_2 batteries. Communications payload requirement 2.50 kW
AOCS: unified 3-axis control and apogee GEO insertion motor using NTO/MMH; two sets of 10 N thrusters and single 490 N motor.

Hispasat 1B

Launch: 22 July 1993 by Ariane 4 V58 from Kourou, French Guiana
Location: 30° W (collocated with Hispasat 1A)
Transponders: Same as 1A but two FSS channels (1 × 54 + 1 × 72 MHz) have switchable uplink to relay TV (TVR) from the Americas to Spain
Mass: 2,210 kg at launch, 1,330 kg on-station BOL, 1,052 kg dry
Other specifications are the same as for Hispasat 1A.

Hispasat 1C

Launch: 3 February 2000 by Atlas 2 AS from CCAFS
Location: 30° W (collocated with Hispasat 1B)

Hispasat 1D

Launch: 18 September 2002 by Atlas 2 AS
Location: 30° W co-located with Hispasat 1A, 1B and 1C

Sweden

Tele-X/Sirius Series

Current Status

The Swedish Space Corporation signed a contract for a single Sirius 2 large satellite to replace Tele-X on 4 July 1995 with Aerospatiale. The contract included a duplicate payload for Sirius 3 but SSC subsequently went to Hughes Space & Communications (now Boeing Satellite Systems) for this satellite placed under contract in May 1997 and launched in October 1998.

Background

The Nordic countries of Denmark, Finland, Iceland, Norway and Sweden began studying the feasibility of a joint regional satellite system in the mid-1970s. From the beginning, these countries sought television direct broadcasting via satellite for their far northern regions. In 1977, the World Administrative Radio Conference on broadcast satellite services (WARC) allocated eight channels to the Nordic group as an entity (excluding Iceland), plus another 17 to the individual countries (including five for Iceland). What became the Tele-X project assumed its current form in 1982, when the Nordic countries sought to provide both direct broadcast satellite and business communications services within Scandinavia, and also as a means of stimulating the space engineering capabilities of these countries. The 'X' represented Experimental, to acknowledge the developmental nature of this indigenous satellite.

By then, Denmark and Iceland had withdrawn from the project, though Denmark had earlier expressed interest in a beam covering southeast Greenland. Sweden and Norway agreed to continue in a formal agreement signed in 1983, giving them 85 per cent and 15 per cent interest, respectively, in NSAB (Nordiska Satellite AB). Three per cent of Sweden's share of cost/work in the project (but not in NSAB) was allocated to Finland later in a separate bilateral agreement.

Taking advantage of the Franco-German TDF/ TV-Sat programme, initiated in 1980, the Swedish Space Corporation (SSC) ordered a similar satellite based on the Spacebus 300 platform. Nordiska Satellite AB (NSAB) planned to be proprietor of the system, while SSC planned to act as the procurement agent and executive agency. Shortly before launch, Norway withdrew from NSAB, trading its 15 per cent share for a lifetime lease of one channel. NSAB now constitutes a 50/50 partnership between SSC and Teracom, a major distributor of television/ radio programming. SSC became wholly responsible for operating the system.

The Sirius system eventually cost Sweden SKr1.25 billion at fixed 1982 rates, or US$280 million. All three direct broadcast satellite television channels are occupied and the Swedes have sought other capacity. NSAB bought the UK's Marcopolo 1 DBS satellite from National Transcommunications Ltd in 1993 for a reported £2.1 million to add five channels. NTL had acquired the satellite from BSkyB. It was renamed Sirius and moved from 3.1° west in January 1994 to begin services at the end of February from 5° east. Sirius 2 is now operational on behalf of GE Americom and is re-designated GE-1E.

Specifications

Tele-X

Launched: 2 April 1989 by Ariane V30 from Kourou
Location: 5° E GSO 1989–1998, 138.62° E September 2001
Design life: 6 to 8 years
Contractors: Aerospatiale/Eurosatellite were joint prime contractors. Aerospatiale also handled satellite AIT, thermal control, transmit reflectors/ antenna subsystems, and built the solar arrays. Scandinavian industry received 30 per cent of the programme contract value: Saab was associate prime, and was involved in structure, TT&C, antenna integration and OBC work. Ericsson was responsible for the communications payload,

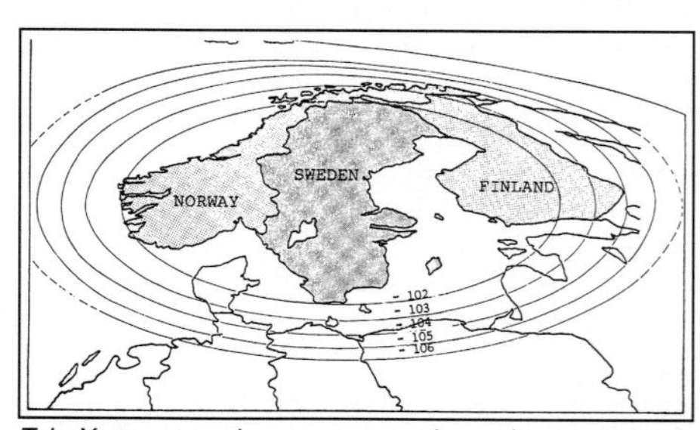

Tele-X transponder coverage; the values represent power density in dBW/m² 0517014

Launched in 1998, Sirius 3 was collocated with Sirius 2 providing Ku-band services to the Nordic countries 0132813

receive antennas & beam-tracking receiver. Other Eurosatellite members involved were AEG (solar cells, TWTs, power subsystems), ANT (TWTs), Alcatel Espace (communication module integration, linear amplifiers, other components), ETCA (power system components) & MBB (AOCS). Ground segment contracts went to Elektrisk Bureau (Norway) and Nokia Research Centre, Teleste, Valmet, VTT Technology (Finland)
Transponders: Two (plus one back-up) 220 W (230 to 260 W planned) TWTA 17.7–18.1/11.7715–12.489 GHz up/down Ku-band Nordic TV DBS beams, 27 MHz bandwidth, EIRP 61 dBW max, LH circular polarised

Two (plus one back-up) 220 W (230–260 W planned) TWTA 14–14.25/12.5–12.75 GHz up/down Ku-band Nordic data/video services, 40 & 86 MHz bandwidths, data rates of 0.64/2/34 Mbit/s, EIRP 61 dBW max at beam edge
Configuration: Aerospatiale/MBB Spacebus 300 3-axis satellite with 2.4 × 1.65 × 2.4 m box-shaped bus. Unified liquid bipropellant propulsion system for GEO insertion + AOCS, in conjunction with MWs, IR Earth sensors. RF antenna pointing system. The same deployable receive/transmission antennas are used for each payload, with 0.165°/0.075° pointing accuracies, respectively. Body pointing accuracy is ±0.3°
Mass: 2,130 kg at launch, 1,277 kg BOL on-station
TT&C: Ku-band operational (S-band during transfer phase), via SSC Earth station at Kiruna, Sweden
Power: 3,200 W EOL, generated by twin 4-panel solar wings with 19 m span. One 18 Ah Ni/Cd battery.

Sirius 1 (Marcopolo 1)
Launched: 27 August 1989 by Delta 4925 from Cape Canaveral pad 17B
Location: 5.2° E geostationary
Design life: 10 years
Contractors: Hughes Space & Communications
Transponders: six paired 55 W TWTAs 17.385–17.992/11.7715–12.1054 GHz up/down Ku-band beam (BSS channels 4, 8, 12, 16, 20), 27 MHz bandwidth, 59 dBW min EIRP (63 dBW max), RHCP
Principal applications: TV direct to home
Configuration: Hughes' HS-376 spin-stabilised, 7.2 m high deployed (2.7 m stowed), 2.16 m diameter. Despun single reflector and communications shelf
Mass: 1,250 kg at launch, about 660 kg on-station BOL
Pointing: 0.05° accuracy by RF beacon
Attitude control: spin stabilised at 50 rpm by four 22.2 N hydrazine thrusters; GEO insertion provided by solid Thiokol Star 30B AKM
Power system: K7 solar cells mounted on cylindrical exterior and deployed drum providing 1,024 W BOL (drum extended 25 cm over normal Hughes 376 configuration). Two Ni/Cd batteries provide 100 per cent eclipse protection.

Sirius 2 (GE-1E)
Launch: 12 November 1997 by Ariane 4 from Kourou, French Guiana
Location: 5° E geostationary 1997–1999, 4.81° E September 2001
Design life: >12 years (15 years propellant)
Contractors: Aerospatiale (prime), Space Systems/Loral (repeater), DASA (AOCS, solar arrays), Saab Ericsson Space (antennas, data handling), Alenia (TCR-RF)
Transponders: 13 57 W TWTA 17.30–18.10/11.70–12.50 GHz up/down Ku-band Scandinavian BSS beams, 33 MHz bandwidth, EIRP 55.5 dBW edge, linear vertical polarised

13.85 W TWTA 17.30–18.10/11.70–12.50 GHz up/down Ku-band European BSS steerable beams, 33 MHz bandwidth, EIRP 49.7 dBW edge, linear horizontal polarised

6.85 W TWTA 14.00–14.25/12.50–12.75 GHz up/down Ku-band Northern/Central European FSS beams, 36 MHz bandwidth, linear polarised
Principal applications: DTH, TV distribution, and data services
Configuration: Spacebus 3000B 3-axis with 1.8 × 2.3 × 2.86 m box-shaped bus. Two 1.8 m fixed dual gridded antennas for FSS and Scandinavian BSS coverage, plus steerable antenna for BSS European beam
Mass: 2,920 kg at launch; 1,760 kg BOL; 1,250 kg dry
Pointing: standard body pointing accuracy 0.1–0.15° following in-orbit calibration. Fine pointing by RF sensors and antenna pointing mechanisms (0.05°)
AOCS: unified bipropellant (NTO/MMH) system of single 400 N and redundant networks of 7 × 10 N thrusters provide GEO injection + 3-axis control. Two titanium propellant tanks, titanium/Kevlar pressurisation tanks. Two 2-axis IRES IR Earth Sensors, two Sun sensors (4 heads each, 60° FOV), two RIGA gyro packages, two 52–68 Nms fixed momentum pitch wheels controlled by digital programmable control electronics. Normal stabilisation is based on pitch by wheel, roll/yaw by Earth sensor using roll/yaw thrusters. Station-keeping by 10 N thrusters; yaw reference by Sun sensors or gyros
TT&C: via SSC Earth station at Kiruna, Sweden
Power system: 5,845 W EOL from twin 4-panel wings spanning 26.3 m. Two 62 Ah Ni/H_2 batteries provide 100% eclipse protection.

Sirius 3
Launch: 5 October 1998 by Ariane 44L from Kourou, French Guiana
Location: 28° E 1998–1999, 5° E September 2001
Design life: 12 years
Contractor: Hughes Space & Communications HS-376
Transponders: 14 Ku-band 11.7–12.5 GHz
Configuration: body diameter 2.16 m, height extended from 3.32 m at launch to 7.76 m in orbit.
Mass: 1,465 kg at launch; 815 kg BOL; 630 kg dry
Power: 1,400 watt.

Thailand

Thaicom Series

Current Status
Thaicom 4 will have C-band, Ku-band and Ka-band transponders.

Thaicom 2 anechoic chamber antenna performance tests (Hughes Space & Communications) 0517207

Thaicom 1, ANHS-376L gets an antenna beam checkout in an anechoic chamber
(Hughes Space and Communications) 0054345

Background
Thailand requested that satellite offerers provide them with quotes for a new satellite broadcasting system in 1987 to reach its entire population of 55 million persons. The announcement predated Thailand's eventual entry into Inmarsat in January 1995. The country expected to offer a 30 year concession to lease transponders to government and private sector users, with excess capacity offered to neighbouring countries.

Shinawatra Computer and Communications Group (SC&C), the largest group of Thai companies operating telecommunications and broadcasting service concessions, won an eight year monopoly in 1991 to supply satellite capacity to domestic users. The founder and major shareholder of Shinawatra is Dr Taksin Shinawatra. SCC began formal negotiations June 1991 with Hughes, and signed a contract on 8 October 1991 for about US$100 million to buy two satellites, ground equipment and training.

As a lighter version of Hughes' HS-376 bus, Thaicom was the second to come within Ariane's SDS (Spelda Dedicated Satellite) 1 tonne launch limit. Shinawatra signed a letter of intent on 1 November 1993 to negotiate with Hughes for a third satellite. The total cost for Thaicom 3 on station was estimated at US$240 million.

Specifications
Thaicom 1
Launch: 18 December 1993 by Ariane 44L from Kourou, French Guiana. Handed over by Hughes to Nonthaburi (Bangkok) station 31 December; entered full service February 1994
Location: 78.5° E GEO (collocated with number 2); moved to 120° E in August 1997 after the launch of Thaicom 3
Design life: 13.5 years minimum, 15 years goal

Technician checks the feedhorn array for Thaicom 1
(Hughes Space & Communications) 0054344

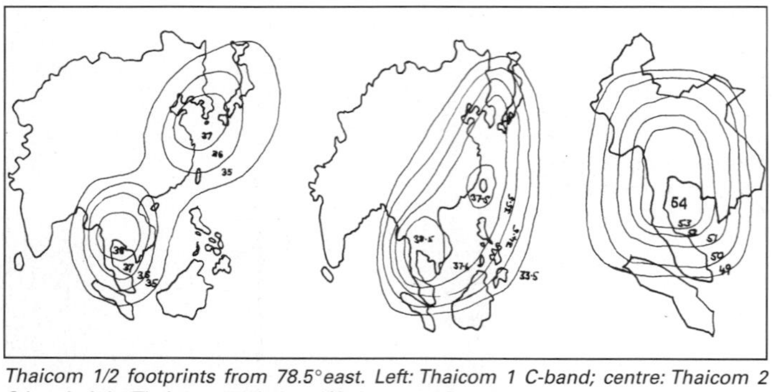

Thaicom 1/2 footprints from 78.5°east. Left: Thaicom 1 C-band; centre: Thaicom 2 C-band; right: Thaicom 1/2 Ku-band (Shinawatra Satellite Public Co Ltd) 0517208

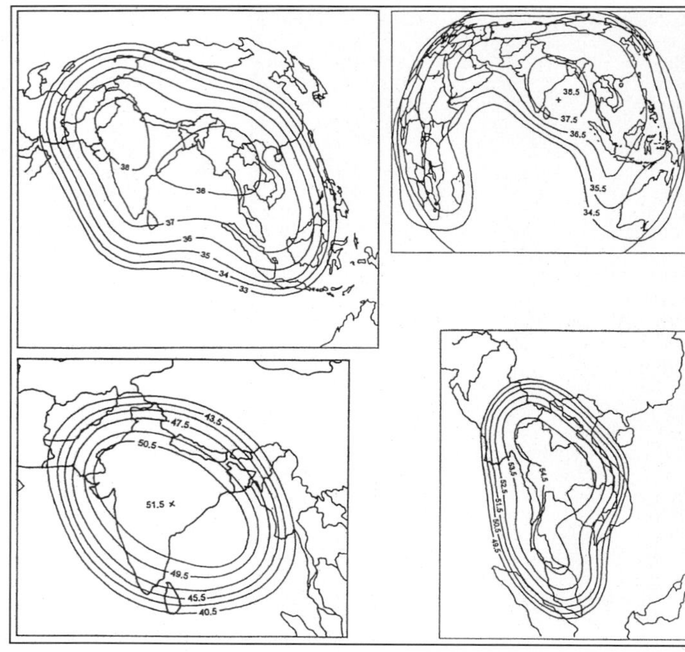

Thaicom 3 footprints (dBW). Top left: C-band India/Indochina; top right: C-band hemispherical; bottom left: Ku-band steerable (India); bottom right: Ku-band Thailand (Shinawatra Satellite Public Co Ltd) 0517386

Contractors: Hughes Space & Communications (prime), Telespace Ltd. (technical support)
Transponders: 10 (plus 2 back-up, 12-for-10) 11 W SSPA 5.925–6.425/3.700–4.200 GHz up/down C-band Thailand, Indochina, North Pacific rim beam, EIRP 35 dBW edge (37.5 dBW over Thailand), 36 MHz bandwidth, orthogonal linear polarisation
2 (plus 1 back-up, 3-for-2) 47 W TWTA 14.3159–14.4951/12.5679–12.7471 GHz up/down Ku-band Thailand beam, EIRP 51 dBW edge (54 dBW centre), 54 MHz bandwidth, orthogonal linear polarisation
Principal users: domestic phone, TV, cable TV, voice, video, data, VSAT services
Configuration: lightweight version of Hughes' HS-376 spin-stabilised bus, 2.16 m diameter, 6.7 m height with solar array drum extended (2.6 m stowed). Single antenna comprising two offset grid reflectors with horizontal/vertical feed system on despun platform
Mass: 1,080 kg at launch, 629 kg on-station BOL, 436 kg dry
AOCS: 55 rpm spin control/position maintained by four Hughes 22.2 N hydrazine thrusters. GEO insertion provided by Thiokol Star 30BP solid
TT&C: through the Nonthaburi station, 15 km north of Bangkok
Power system: Si cells on cylindrical surface provide 801/671 W BOL/EOL (summer solstice) in conjunction with 51.6 Ah Ni/H_2 battery.

Thaicom 2
Launch: 8 October 1994 by Ariane 44L from Kourou, French Guiana
Location: 78.5° E GEO (collocated with number 1)
Other specifications as for Thaicom 1, but see footprint diagram.

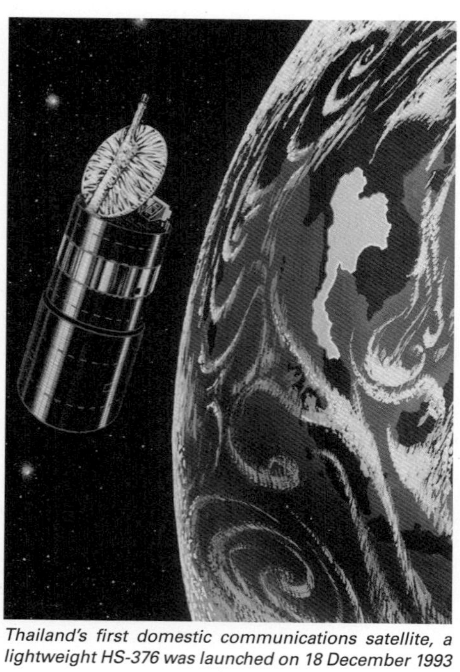

Thailand's first domestic communications satellite, a lightweight HS-376 was launched on 18 December 1993 0054343

Thaicom 3
Launch: 16 April 1997 on Ariane 44LP from Kourou, French Guiana
Location: 78.5° E GEO (co-located with number 2)
Design life: 14 years (16 years propellant)
Contractors: Aerospatiale (prime; structure, thermal control, solar arrays, integration, testing), Alcatel Espace (communication payload), DASA (AOCS integration)
Transponders: 18 (plus 6 back-up; 24-for-18) 19 W SSPA 5.925–6.725/3.405–3.700 GHz up/down C-band India/Indochina regional beams, 36 MHz bandwidths, EIRP 37 dBW min, orthogonal linear polarisation
6 (plus 3 back-up; 9-for-6) 42 W SSPA 6.425–6.725/3.405–3.700 GHz up/down C-band hemispheric beams, 36 MHz bandwidths, EIRP 34.5 dBW min (38.5 maximum), orthogonal linear polarisation
14 (plus 4 back-up, 18-for-14) 97 W TWTA 14.000–14.500/12.200–12.750 GHz up/down Ku-band domestic DTH beams and one steerable beam (max 7 36 MHz transponders), 2 × 54 MHz + 12 × 36 MHz bandwidth, EIRP 54 dBW min Thailand and 45 dBW India (51.5 maximum), orthogonal linear polarisation
Principal applications: Ku-band principally for DTH
Configuration: Aerospatiale/DASA Spacebus 3000 3-axis stabilised 1.8 × 2.6 × 2.3 m box-shaped bus (7 m deployed height) with 25 m-span twin solar wings
Mass: 2,650 kg launch, 2,450 kg on-station BOL, 1,160 kg dry
TT&C: as Thaicom 1/2
Power system: 5,300 W EOL provided by twin Si solar wings totalling 46 m^2. Two 57 Ah Ni/H_2 batteries.

LStar 1
Launch: Ariane 4 from Kourou, French Guiana
Location: planned 116° E GEO (collocated with number 2)
Design life: 14 years minimum
Contractors: Space Systems/Loral (prime), Telesat Canada (consulting, management services)
Transponders: 32 (48-for-32 ring redundancy) 113 W TWTA 17.32748–17.76862/11.72748–12.16862 GHz up/down Ku-band (paired TWT for high power mode, single TWT for medium power) in three beams: 55.5 dBW min West (India, Pakistan, Nepal, Sri Lanka); 56.8 dBW min Central (Thailand, Laos, Cambodia, Vietnam, Singapore, Western Malaysia); 53.8 dBW min East (Eastern China, Hong Kong, Taiwan and Korea). 33 MHz bandwidth, orthogonal linear polarisation
Principal users: DTH TV, VSAT services
Configuration: 3-axis FS-1300 bus, two 4-panel GaAs solar wings (plus Ni/H_2), Gregorian antenna for each beam. 12 × 22 N NTO/MMH thrusters and Marquardt R-4D 490 N for GEO insertion
Mass: 3,500 kg at launch, 2,132 kg on-station BOL, 1,480 kg dry
TT&C: through the Vientiane, Laos station; Perth back-up.

LStar 2
Launch: Uncertain
Location: planned 116° E GEO (collocated with number 1)
Other specifications as for LStar 1.

Turkey

Turksat Series

Current status
Türk Telekom has set up a joint venture with Alcatel Space for the purchase of a second generation Turksat 2A satellite. It is expected to carry 32 high-powered transponders with 20 × 33 mHz fixed beam in BSS band and 12 × 36 mHz transponders with two steerable beams in FSS. Turksat 2A will operate from 42° E.

Background
Turkey chose to build a completely turnkey satellite operation in order to reach Turkish speaking people in the Central Asian land mass. The government released a request for proposals in June 1989 and received offers from Aerospatiale Hughes/Selenia Spazio, BAe-Matra/ANT and Aerospatiale/MBB by the deadline of December 1989. Aerospatiale signed the US$315 million prime contract, 21 December 1990 for turnkey delivery of two satellites, associated ground stations, insurance, financing and training. It is the first complete system ever delivered on a turnkey basis by a European team. Each satellite carries 16 active Ku-band transponders providing television/radio distribution (also serving the large Turkish populations in Germany/Austria and CIS), data and voice services.

Turkey's first satellite disappeared in the Ariane failure of 24 January 1994. In the contract, Turkey required Aerospatiale to provide on-orbit delivery, and the company had taken full insurance, so it provided a replacement in 1996.

Turkish Telecom and Aerospatiale formed the Eurasiasat joint venture in August 1996 before ordering Turksat 2A, which became known as Eurasiasat 1 following launch on 10 January 2001.

Specifications
Turksat 1B
Launch: 10 August 1994 by Ariane 4 from Kourou, French Guiana
Location: 42° E GEO initially, relocated to 31° E during October 1996.
Design life: 10–13 years (10 years contracted)

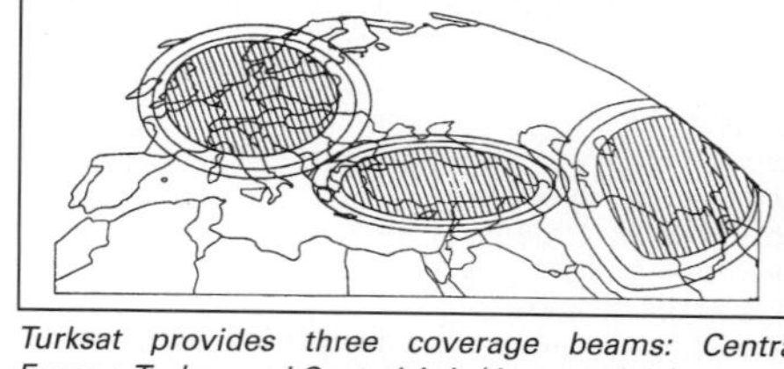

Turksat provides three coverage beams: Central Europe, Turkey and Central Asia (Aerospatiale) 0517015

Contractors: Aerospatiale (prime), Alcatel Espace (payload), DASA/MBB (AOCS, solar arrays), Thomson Tubes Electroniques (TWTAs), Eagle Picher (battery), SEP (BAPTA), Contraves (structure), Sextant Avionique (power control, distribution unit), ETCA (power conditioning), Teletas (TV receive antennas)
Transponders: 16 (plus 8 back-up; 12-for-8 ring redundancy) 55 W TWTA 14.000–14.500/10.950–11.700 GHz up/down Ku-band Central European, Turkey, Central Asia, bandwidths 6 × 72 MHz + 10 × 36 MHz, EIRP 51 dBW Turkish, 48 dBW Central European, 45 dBW Central Asia, linear polarisation. Up to four channels can be switched to Central Europe and up to two to Asia
Principal applications: TV/radio direct (12 TV + 20 radio to 50 cm antennas), telephony, VSAT data and business services, mobile links
Configuration: Spacebus 2000 platform, 3-axis control, 22.4 m span across solar panels, single dual reflector multifeed antenna. GEO location maintained within ±0.05°
Mass: 1,783 kg at launch; 1,078 kg on-station BOL; dry 827 kg
AOCS: provided by DASA/MBB's 22 × 10 N + 400 N MMH/MON bipropellant system
TT&C: Ankara (prime). Istanbul (back-up)
Power system: twin solar wings providing 2.9 kW equinox EOL.

Turksat 1C
Launch: 9 July 1996 by Ariane 4 from Kourou, French Guiana
Location: initially over 31° E for testing in September 1996 then 42° E 1996–1999
Mass: 1,743 kg at launch; 1,078 kg on-station BOL; dry 789 kg.

United States

ACTS Satellite

Background

NASA Lewis developed and is responsible for the Advanced Communications Technology Satellite which it is using to demonstrate new technologies required for multiple antenna beam communications and satellite switched operation in the 30/20 Ghz Ka-band. More than 100 industry, government and university organisations use ACTS in about 80 different application, technology and propagation experiments. Most applications trials are oriented towards services with commercial potential: business networks, ISDN, education, DoD tactical reconnaissance imagery, broadcast video, and aeronautical links.

NASA Lewis contracted with RCA Astro-Electronics for US$260 million to build the ACTS in 1984 with the intention of launching the satellite in 1989. Hughes Communications Galaxy Incorporated/Galaxy challenged the government for using public funds to build the ACTS on the grounds that it was planning a similar project using its own corporate funds. The US government put ACTS on hold at the time, but then Congress

ACTS attached to its Transfer Orbit Stage deployed from STS-51 in September 1993 0132814

directed the space agency to revive the satellite. Within six months the project was again shelved and re-approved – a process repeated several times.

Congress ultimately determined that with the imminent congestion of the spectrum and GEO arc, US leadership in this might erode with respect to Europe and Japan and that NASA needed to reinvigourate its experimental communication work. NASA Lewis resurrected the programme in February 1988, but Congress still demanded that costs should be capped at US$475 million. NASA argued successfully that this ceiling compromised its technology goals, and a limit of US$499 million was agreed to instead.

Motorola is incorporating ACTS-type technology into Iridium and Hughes's Spaceway proposed 48-spot Ka-band video phone satellite system. The US Army used ACTS to provide secure videoconferencing links with Ft Bragg, NC from Haiti following September 1994's occupation.

Specifications

Launch: 12 September 1993 by Shuttle-TOS, ejected 21.13 GMT, TOS fired 21.58 GMT to enter GTO
Location: 100° W geostationary 1993–1999; 105° W in 2000–2001 drifting 0.007° W/day
Design life: minimum goal was 2 years; current station-keeping propellant projection shows operations through May 1998 (longer for inclined orbit)
Contractors: Lockheed Martin Astro Space (prime, system design, bus, payload, integration, flight operations), Motorola (baseband processor), Watkins Johnson (TWTs), Electromagnetic Sciences (beamforming networks), OSC (Transfer Orbit Stage), Comsat Labs (ground system, network operations), Harris (experimenter Earth station), BBN Inc (high data rate experimenter Earth station)
Transponders: 4 Ka-band 19.2–20.2 GHz transmitters and 29.0–30.0 GHz receivers with 46 W TWTAs give 4-for-3 redundancy with stationary beams and 4-for-2 redundancy with the hopping beams. Beam-forming network of 45 feed horns covers about 20% of CONUS area, and provides two 0.3° hopping spotbeams with opposite polarisation, three 0.3° fixed beams and one 1.0° mechanically-steerable beam. EIRP 60 dBW fixed, 59 dBW hopping spots, 53 dBW steerable. Baseband processor stores/switches 64 kbit/s circuits, providing demand-assigned multiple access between all spotbeams. Total throughput 220 Mbit/s. FDM-TDMA access on uplink at 27.5/110 Mbit/s burst rates, TDM only on downlink at 110 Mbit/s. Automatic rain-fade compensation. High speed programmable 3 × 3 switch matrix provides

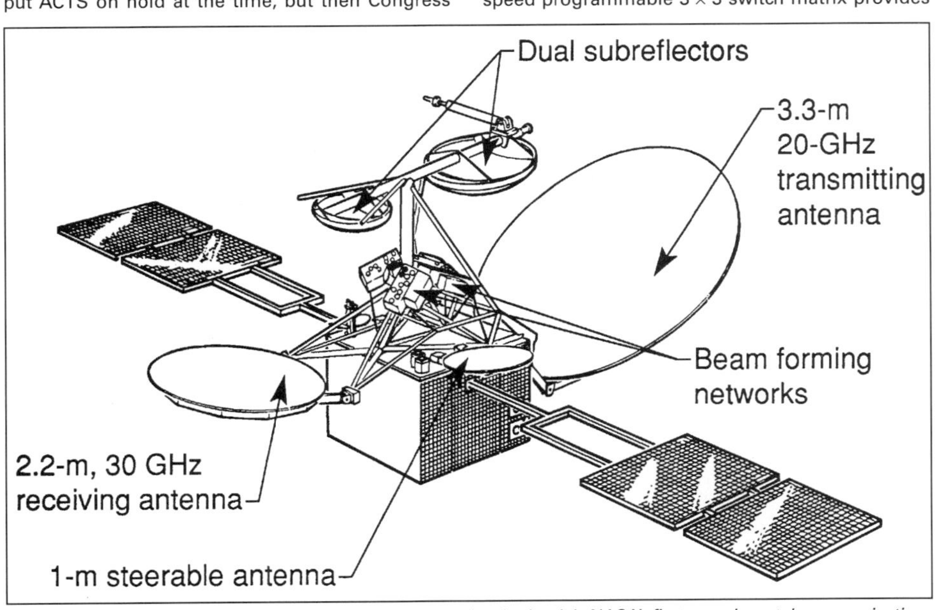

ACTS is demonstrating new Ka-band communications technologies. It is NASA's first experimental communications satellite since 1973 0517026

Final stages of ACTS checkout prior to launch 0132815

4 input + 4 output channels of 900 MHz bandwidth (one for redundancy). 20/30 GHz propagation beacons (stability 1.5 dB over 2 years)
Configuration: modified series 4000 3-axis bus. Two multi-beam offset Cassegrain antennas: receive antennas 2.2 m + 1 m (steerable) diameter; transmission 3.3 m + 1 m (steerable) diameter. Bus 203 cm long, 213 cm wide, 190.5 cm deep; solar array span 14.3 m. Blowdown hydrazine system carrying 263 kg propellant in four tanks supplying 16 thrusters (two redundant half systems) of 0.9/2.2/4.4 N; station-keeping ±0.05°. 3-axis stabilised via Earth/Sun sensors + MW; 0.025° pitch/roll, 0.25° yaw; offset pointing control ±6° pitch ±2° roll
Mass: 2,540 kg at launch, 1,474 kg BOL on-station
TT&C: LMAS's station in E Windsor, NJ working through NASA Lewis ACTS Master Ground Station
Power system: 12.53 m^2 solar wings providing 1,770 W BOL and 1,418 W after 4 years: 2 × 19 Ah Ni/Cd batteries power only essential functions during eclipses.

AMSC Series

Background

In May 1989, the FCC licensed AMSC to provide voice and data Skycell communications to mobile users on land (Conus, Hawaii, Alaska, Puerto Rico and Virgin Islands), up to 320 km out at sea and in the air. AMSC agreed with Canada's TMI Communications to operate compatible systems and use each other's satellite as spares. The companies jointly issued a request on 7 July 1989 for one satellite each to provide mobile communications for North America from 106.5° W (TMI) and 101° W (AMSC) GEO. The Hughes/Spar Aerospace team won the competition in December 1990, with Hughes acting as prime for AMSC and Spar for TMI. Each satellite cost about US$100 million. The two companies claim that their unique joint operation reduced start-up costs by about US$200 million, and ensure a high degree of mutual backup and protection. Users may tap either satellite from anywhere over North America with continuous coverage. Each satellite provides up to 2,000 simultaneous voice channels through L-band spot beams. The satellite is the twin of MSAT 1. While testing, a transmission was sent which caused components in the communications payload to overheat. This damaged one of eight hybrid matrix ports covering the eastern and central United States. Occassional shutdown of 2 SSPAs.

Westinghouse Electronic Systems Group manufactures the satellite phones for both systems. Mitsubishi Electric Corp also manufactures phones for the AMSC system. Combined manufacturing capacity reached 10,000 monthly by mid-1996. AMSC began voice service in January 1996, using US$3,000 transportable phones. The full range of Skycell voice products became available in 1996. The products included land mobile phones, US$1,800–US$2,300, aeronautical phones US$15,000; maritime phones US$4,000–US$6,000.

AMSC subsidiary American Mobile Radio Corp filed an FCC application 15 December 1992 to provide 55 radio broadcast channels (11 CD-quality) from two satellites at 99° W and 103° W. Hughes Space & Communications were to build the US$528 million system's satellites, although no contract has yet been signed.

Specifications

AMSC 1
Launch: 7 April 1995 by Atlas 2A from complex 36A, Cape Canaveral
Location: 101° W GEO (62/139° W also authorised)
Design life: 12 years

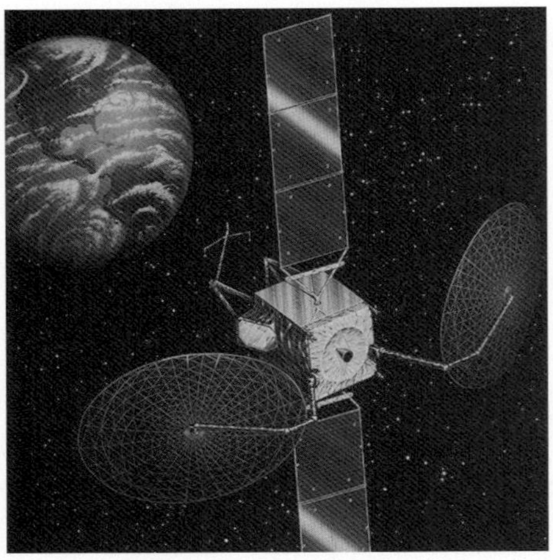

AMSC provides North American mobile services
(Hughes Space & Communications) 0517210

Contractors: Hughes Space & Communications Co HS601 (management/bus) and Spar Aerospace (payload)
Transponders: Forward (Ku/L) 13.023–13.232/1.530–1.559 GHz up/down, 16 35 W 55 PA (4 back-ups in 10-for-8 redundancy scheme) in 2 × 8 matrix feeding 6 elliptical spot beams (4 North America, a Caribbean beam for Puerto Rico, Virgin Islands and Mexico, and Alaska/Hawaii). Aggregate edge of coverage EIRP 56.5–57.3 dBW, circular polarisation. Bandwidth capacity for up to 2,000 5 kHz channels. Each '35 W SSPA' comprises four 20 W SSPAs in parallel but operated at 35 W for required 16 dB noise/power ratio

Reverse (L/Ku) 1.6315–1.6605/10.783–10.932 GHz up/down one (plus two back-up) 110 W TWTA 13.000–13.250/10.750–10.950 GHz up/down Ku-band 36 dBW feeder/return link in single North American beam
Principal applications: mobile radio, telephone, data (gateway to feeder 13.2405–13.2455/10.9405–10.9455 up/down Ku-Ku in single North American beam) aeronautical, maritime, safety and distress, position location, wide area paging. High quality voice, packet and circuit switched data
Configuration: Hughes' HS-601 3-axis bus, 2.3 m box-shaped body, 3.63 m stowed diameter, 4.44 m stowed height, twin solar wings spanning 20.96 m, 18.9 m across deployed graphite mesh antennas. Two Hughes 4.9 × 6.7 m 20 kg Springback mesh reflectors deployed from east/west faces provide separate L-band transmit/receive, respectively, illuminated by 23-cup dipoles. Single 76 cm Ku-band shaped reflector for North America coverage
Mass: 1,650 kg on-station BOL, 2,720 kg at launch (1,270 kg dry)
AOCS: onboard control processor for control of momentum wheel assembly. Unified 12 × 22 N + 1 × 490 N bipropellant thrusters (N_2O_4/MMH in four tanks)
Antenna pointing: 0.125° accuracy using Earth sensors
TT&C: Hughes Communications controls AMSC 1 under a five year contract from its OCC Operations Control Centre in El Segundo via its Spring Creek, New York ground station. AMSC's own Network Operations Centre in Reston, Virginia provides network control/channel access on demand
Power system: twin 9 m long 3-panel solar wings producing 3.6 kW BOL, 3 kW EOL from K4-3/4 large area Si cells; 160 Ah Ni/H_2 battery provides 100 per cent eclipse protection
Thermal control: dissipation by north/south radiators, with heat transported by heat pipes.

DirecTV

Background
Hughes spin off DirecTV in 1992 after Sky Cable collapsed to provide a domestic US DTH service using Ku-band transmissions from the three DBS satellites launched in 1993, 1994 and 1995, renamed DirecTV-1, -2 and -3, respectively. DirecTV-IR was launched in 1999 as an on-orbit spare. DirecTV 4S was launched on 27 November 2001. DirecTV-5 was based on a FS-1300 platform Space Systems/Loral. It was originally planned for launch on a Proton in October 2000, then moved to February 2001, 21 May 2001 and 19 October 2001 as the launch shifted to an Atlas 2AS before finally being launched on 7 May 2002 on a Proton.

Specifications
DirecTV-1
Launched: 18 December 1993 by Ariane 4 from Kourou, French Guiana. HSC handed over control of number 1 to HCI 6 January 1994; broadcasting began 17 June
Location: 101.2° W geostationary
Design life: 12–15 years
Contractors: Hughes Space & Communications Co, Hughes Aircraft Company (prime)
Transponders: 16 (plus eight back-up; 12-for-8) 120 W TWTAs (can be paired into 8 × 240W) 17.3–17.84/12.2–12.74 GHz up/down Ku-band CONUS beams, 27 MHz bandwidth, 48–53 dBW EIRP, 40 Mbit/s, R/LHCP for DBS 1, L/RHCP DBS 2. BSS channels 1–32. Each shaped reflector is fed by a single feed horn; the shape concentrates power on areas such as the SE affected by rain fade
Principal applications: direct broadcast to 45 cm diameter antennas
Configuration: Hughes' HS-601 3-axis bus, 26.2 m solar array span, two 2.14 m transmission reflectors, 102 cm receive reflector, 36 cm beam tracking antenna. Each antenna array is mounted at only one point to minimise thermal distortion to shaped profile
Mass: 2,860 kg at launch, 1,270 kg dry mass, 1,725 kg on-station BOL
TT&C: from HCI's Operations Control Center in El Segundo, CA, via Castle Rock, CO + Spring Creek, NY
Power system: twin solar arrays providing 7,000 W BOL. 100 per cent eclipse protection by Ni/H_2 battery.

DirecTV-2
Launched: 3 August 1994 on Atlas 2A from Cape Canaveral
Location: 100.8° W GEO, about 300 km from number 1
Other specifications as for DBS 1.

DirecTV-3
Launched: 10 June 1995 by Ariane 4 from Kourou, French Guiana
Location: 100.8° W GEO, collocated with DBS 2
Mass: 2,934 kg at launch, 1,259 kg dry mass, 1,707 kg on-station BOL
Other specifications as for DBS 1.

DirecTV-IR
Launched: 9 October 1999 by Zenit 3SL from SeaLaunch
Location: 101° W GSO
Contractor: Boeing Satellite Systems
16 (+4 spare) × 200 W MPA (dual 100 W TWTA) Ku-band (12.2–12.7 GHz) with 50–55 dBW radiated power. Two 272 cm shaped Gregorian reflectors for transmit and one 137 cm shaped single-surface reflector
Configuration: BSS (Hughes) 601HP 3-axis bus, 26 m solar array span, 7 m width across antennas.
Stowed size: H 4 m, W 2.7 × 3.6 m
Mass: 3,446 kg at launch; 2,304 kg at BOL
Power system: Twin solar arrays 8.7 kW BOL, 7.7 kW EOL
AOCS: 1 × 490 N liquid apogee motor; 1 × 17 mN N-S ion propulsion system; 4 × 10N E-W bipropellant thruster; 4 × 22N N-S bipropellant backup thrusters.

DirecTV-4S, a Boeing (HS) 601HP prior to final shipment for launch in November 2001 0132823

DirecTV-4S
Launched: 26 November 2001 by Ariane 4 from Kouron Fr Guiana
Location: 101° W
Contractor: Boeing Satellite Systems
Communications: spot beam payload: 38 × 30 W, 45 W, 65 W, 80 W TWTA to varying combinations.
National beam payload: 2 × 280 W TWTA. Supplemental beam payload: 2 × 240 W TWTA; 6 × 120 W TWTA.
Configuration: as DirecTV-IR
Mass: 4,260 kg at launch; 2,800 kg at BOL
Power system: as DirecTV-IR
AOCS: as DirecTV-IR.

DirecTV-5
Launched: 7 May 2002 on a Proton 8K82K/11S861-01 from Baikonur
Location: 129° W
Contractor: Loral/Space Systems
Configuration: Loral/Space Systems FS-1300 with Tempo 1 payload
Mass: 3,640 kg at launch.

EchoStar Series

Current Status
Following a launch gap after EchoStar 6, the next satellite was built up on the LM-700 bus originated by the Iridium system satellite design and launched on the first Atlas 3B. The EchoStar 8 satellite was an FS-1300 class bus, resuming affiliation to that platform used previously for EchoStar 5 and 6.

Background
When EchoStar Satellite Corporation announced its intentions to enter the direct-to-home broadcast market it became the second large corporation to do so, setting up a competitive market for TV services. EchoStar signed a contract with Martin Marietta Astro Space in October 1992 to build one 7000-series satellite, plus options on six others. EchoStar plans to deliver 100 channels via 45 cm dishes. DirectSat merged with EchoStar in January 1994 and has FCC approval to merge their direct broadcast satellite licences for 119° west. DirectSat had been allocated 10 BSS channels at 119° west by the FCC. EchoStar made the US$52.3 million winning bid at the FCC's 26 January 1996 auction for the 24-channel slot at 148° west.

Specifications
EchoStar 1
Launch: 28 December 1995 by CZ-2E from Xichang; entered service 3 March 1996
Location: 119° W geostationary
Design life: 12 years' propellant carried
Contractors: Lockheed Martin Astro Space
Transponders: 16 130 W TWTAs 14/12.2–12.7 GHz up/down Ku-band CONUS beams (BSS channels, odd numbers 1–21), 24 MHz bandwidth, EIRP CONUS 51 dBW edge, circular polarisation
Principal applications: DTH TV
Configuration: GE 7000 series 3-axis platform
Mass: 3,288 kg at launch
AOCS: station-keeping ±0.5° lat/long. NSSK by 1.8 kW hydrazine arcjets
TT&C: by AT&T Skynet Satellite Services from operations facility at Hawley, Pennsylvania.

EchoStar 1 provides US direct to home TV broadcasts
(Lockheed Martin Astro Space) 0517019

EchoStar 2
Launch: 11 September 1996 by Ariane 4 from Kourou, French Guiana
Location: 119° W geostationary
Mass: 2,865 kg at launch
Other specifications as for EchoStar 1.

EchoStar 3
Launch: 4 October 1997 by Atlas 2AS from Cape Canaveral
Location: 298° E geostationary
Other specifications as for EchoStar 1.

EchoStar 4
Launch: 7 May 1998 on Proton-K from Baikonur, Kazakhstan
Location: 212° E geostationary
Local programming spot beams may complement the full CONUS coverage of EchoStar 1–3.

EchoStar 5
Launch: 23 September 1999 by Atlas 2AS from Cape Canaveral
Location: 260° E
Mass: 3,500 kg at launch

EchoStar 6
Launch: 14 July 2000 by Atlas 2AS
Location: 119° W
Mass: 3,700 kg at launch

EchoStar 7
Launch: 21 February 2002 by Atlas 3B
Location: 119° Wt
Mass: 690 kg at launch

EchoStar 8
Launch: 22 August 2002 by Proton-K
Location: 110° W
Mass: 4,660 kg at launch

Faisat Satellite

Current Status

Launched by Cosmos 3C, on 23 September 1997, to a 967 × 1,021 km, 82.9°, 105 minute, orbit, Faisat 2V experienced problems generating sufficient power while in Earth's shadow. Final Analysis Incorporated planned to develop a new software package to mitigate the problem.

Background

Final Analysis Incorporated entered business as a commercial telecommunication provider after modifying a bus originally built for a classified mission for data store/forward designed by the DARPA. The company received a Department of Commerce export licence 25 March 1994 for the first US commercial launch aboard a Russian vehicle. FAI launched its 114 kg, Faisat 1, as a piggyback payload aboard AKO Polyot's SL-8 Cosmos on 24 January 1995 into 968 × 1,021 km, 82 to 93° inclination orbit. The 46 cm diameter 89 cm high bus is gravity gradient stabilised by a 3 m boom with the assistance of magnetorquers. Faisat is demonstrating data store/forward from interrogated remote terminals in preparation for a 26-satellite constellation. Information is unlinked at 140 MHz/9.6 kbits/s and downlinked at 400 MHz. The company controls the satellite from its Logan, Utah station. Polyot may launch the full constellation in exchange for acting as the system's CIS service provider. FAI is targeting applications such as utility company meter reading, two-way messaging and glacier research. It carries the VITA secondary payload which enables two-way email through gateways around the world. The payload is based on a Loral RS6000 Command and Data Handling processor with 16 MB of storage.

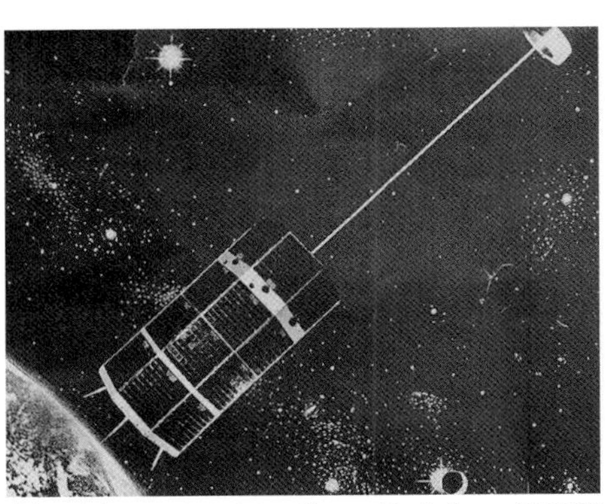

Faisat 1 demonstrated store/forward techniques 0517212

Globalstar Constellation

Curren+t Status

Globalstar has achieved commercial operating capability with its 48-satellite network. Full 48-satellite coverage was obtained at the end of 1999. On 8 February 2000, the full constellation of 52 satellites was achieved with the launch of four spare satellites on a Delta 2 from CCAFS. Four satellites operate as spares. The company currently offers services across 100 countries as well as from most territorial waters and several mid-ocean regions.

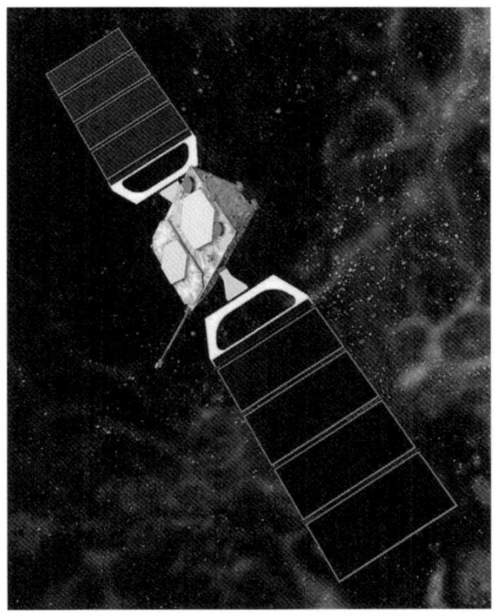

The first Globalstar launch took place in 1998 aboard a Delta 2 (Globalstar) 0005569

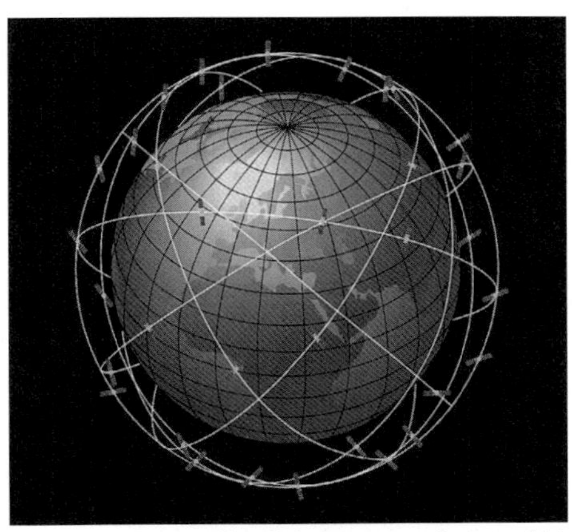

The planned 48-satellite Globalstar constellation, which was completed in 1999 (Globalstar) 0005570

An engineer makes final adjustments to a Globalstar satellite (Theo Pirard) 0054356

Final assembly of the Globalstar satellites (Theo Pirard) 0054357

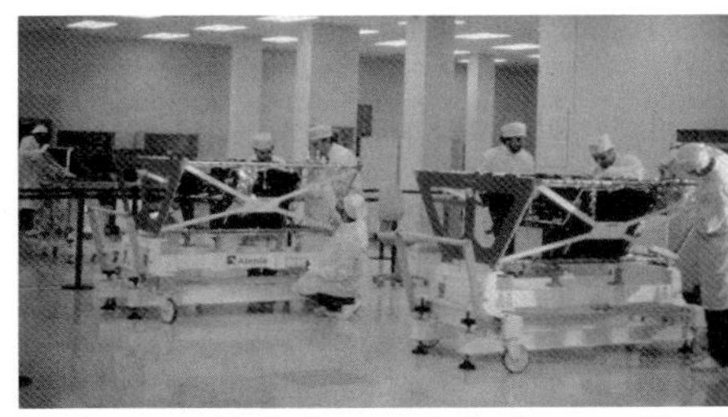

Engineers at Alenia Aerospazio begin assembly of a Globalstar satellite bus 0084770

Specifications

Launch: 1–4 on Delta 7420 (14 February 1998); 5–8 on Delta 7420, 24 April 1998; 9–12 on Soyuz-Ikar, 9 February 1999; 13–16 on Soyuz-Ikar, 15 March 1999; 17–20 on Soyuz-Ikar, 15 April 1999; 21–24 on Delta 7420, 10 June 1999; 25–28 on Delta 7420, 9 July 1999; 29–32 on Delta 7420, 25 July 1999; 33–36 on Delta 7420, 17 August 1999; 37–40 on Soyuz-Ikar, 22 September 1999; 41–44 on Soyuz-Ikar, 18 October 1999; 45–48 on Soyuz-Ikar, 22 November 1999 49–52 on Soyuz-Ikar, 8 February 2000
Orbits: 1,414 km, 52°; 6 active satellites in 8 planes (plus 1 spare in each plane)
Design life: 7½ years (EOL de-orbit)
Contractors: SS/L (prime: satellite and ground/space operations), Loral (operations control centre, TT&C stations), Alcatel Espace (payload modules), Alenia Spazio (payload/platform integration), DASA (AOCS, solar array), Qualcomm (gateways, operations control centres, handsets)
Principal applications: 4.8 kbit/s voice/fax, 9.6 kbit/s data, and position location (300 m)
Transponders: 1.610–1.6265/2.4835–2.5000 GHz user up/down, 5.091–5.250/6.700–7.075 GHz gateway up/down. Overall EIRP 39 dBW
Configuration: 3-axis 0.60 × 1.50 × 1.60 m box-shaped bus. Attitude/orbit determination by onboard GPS receivers using IBM RISC 6000 processors, reducing ground operations and eliminating Earth sensors. 1,200 W from twin solar wings spanning 12 m. AOCS by hydrazine thrusters
Mass: 450 kg at launch, 400 kg dry
TT&C: two SOCC Satellite Operation Control Centres.

UPDATED

Healthsat Satellite

Background

SatelLife is a non-profit organisation that began by the initiative of the 1985 Nobel Peace Prize-winning International Physicians for the Prevention of Nuclear War. The project seeks to reduce the isolation of the world's health workers, particularly in developing countries, and link them with information resources. Its HealthNet programme (which began in five East African medical schools) employs the Healthsat 2 digital store/forward micro-satellite to provide e-mail links with Internet. SatelLife originally owned 90 per cent of UOSAT 3 (Healthsat 1), as it does Healthsat 2. UoSAT 3 was one of six secondary packages launched with SPOT 3 by Ariane 40 on 22 January 1990. UoSAT-builders Surrey Satellite Technology Ltd provided the first dedicated satellite, Healthsat 2, under a US$1.7 million contract, including launch. HealthNet connects about 4,000 users in 25 developing countries with Internet. The user needs only a small radio, antennas and a computer. Messages are stored in 32 Mbyte memory for later downloading by ground stations.

Specifications

Healthsat 2

Launch: 26 September 1993 as Ariane ASAP passenger

Orbit: 793 × 805 km, 98.68°

Communications: store/forward packet communication. Messages stored in 32 Mbyte memory for later downloading by ground stations. 3-channel VHF receiver (two 9.6 kbit/s communication, one command), redundant UHF transmitters (9.6 or 38.4 kbit/s) switchable to 2/12.5 W RF output, 4-element turnstile VHF antenna, one monopole UHF (400 MHz band) antenna

Configuration: based on SST's UoSAT 5 design. 600 × 345 × 345 mm box-shaped body with solar panels on four sides, 6 m boom deployed from top face for gravity gradient stabilisation. Each housekeeping system or payload is housed in a standard module box, mass-produced by a computer-controlled process. The boxes are then stacked to form the main spacecraft structure. Payload is a dual 'hot' redundant OBC supporting both housekeeping and the store/forward function: two 80C186 processors (+80C188 back-up) and 32 Mbyte message memory

Total mass: 47.5 kg

Attitude control: gravity gradient Earth-pointing by 3 kg tip mass on 6 m boom deployed from top face, magnetorquing by six 150-turn aluminium wire coils (one on each spacecraft face); Sun/Earth horizon sensors and 3-axis fluxgate magnetometer (±2 nT res over ±64 μT) provide attitude information

TT&C: SSTL provides TT&C services. Gateway stations at St. John's, Newfoundland and Boston, Massachusetts

Power supply: four GaAs panels (each 34 W) totalling 1,344 cells provide 20 W (orbit average processed); 6 Ah Ni/Cd batteries.

Final integration of HealthSat 2 0517126

Iridium Constellation

Current Status

In December 1998, a CZ-2C/5D launch from T'ai Yuan completed the 72 satellite (66 prime, 6 spares) Iridium constellation. Iridium planned to begin initial operations as scheduled in the early autumn. Ground observers repeatedly noticed the satellites in the sun's glare near evening and in the early morning. In 1999 Iridium LLC filed for Chapter 11 bankruptcy protection in the face of low receiver sales, an inability to secure protected revenue streams and unwillingness of banks to extend further credit. On 20 November 2000, newly formed Iridium Satellite LLC received approval from a New York bankruptcy court to buy the satellites for US$25 million. Subsequently, the Defense Information Systems Agency confirmed a two year contract to Iridium Satellite LLC for US$72 million with options for extending satellite utilisation through 2008 at a value of US$252 million. Operated by Boeing Space Systems, the network will give unlimited airtime to 20,000 US government workers who currently use the system.

Background

Motorola unveiled its Iridium mobile communications system proposal in June 1990 after 2½ years of internal planning. Iridium plans a system of 66 small LEO satellites by end-1998 providing global pocket mobile digital 4.8 kbit/s telephone services via onboard switching and inter-satellite links for US$3/min to the user. Iridium plans to sell handsets for about US$3,000. The company planned to spend about US$3.45 billion for establishing the operational system of 66 satellites and 14 in-orbit spares by 1998, followed by another US$2.8 billion over the next 5 years for operations/maintenance. For an economic break even Iridium needs to sign up 800,000 users within 5 years. Motorola business plans project 1.824 million by 2001, broken down as 42 per cent pager, 42 per cent portable phone and 7 per cent mobile. After 10 years, in 2006, it projects 3.224 million.

Motorola submitted its proposal to the FCC 3 December 1990 and the Commission granted an experimental licence in August 1992 for five satellites. A full licence followed on 31 January 1995. Lockheed Space Systems Division (now Lockheed Martin) received a US$700 million contract in August 1993 for 120 Iridium spacecraft to be built through the year 2003. Lockheed builds the satellite bus in Nashua, New Hampshire at Lockheed Martin Sanders and then integrates it with Motorola's communications module in Motorola's final assembly facility in Chandler, Arizona. Chandler's production line can handle five satellites simultaneously and takes 21 days from start to finish of a satellite.

Lockheed drew on its Agena upper stage experience to build the satellites power and control systems. Volume production allows use of GaAs solar arrays. Motorola said in February 1993 that it planned to use the Khrunichev Enterprises Proton booster to put the first three packages of seven

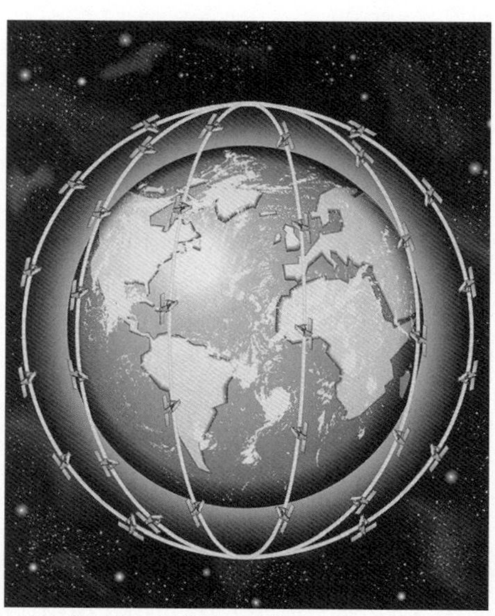

Iridium's network of 66 satellites will provide global pocket mobile telephone services via the L/S-band phased array antennas. Iridium is the 77th element, reflecting the original constellation size 0517224

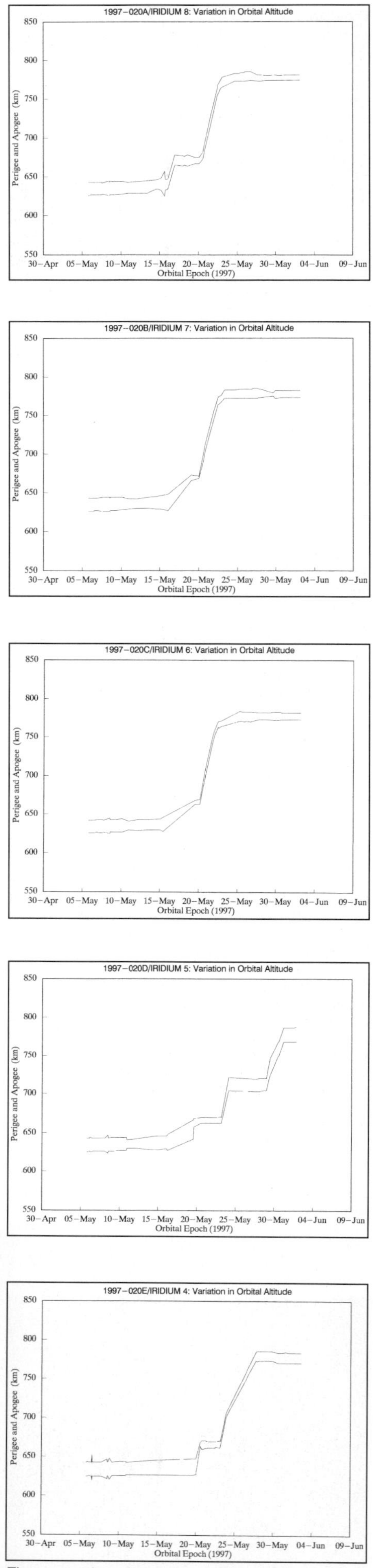

These graphs show the manoeuvres completed during the first month in orbit by the five Iridium satellites which comprised the first launch in the programme. It will be noted that Iridiums 6–8 started their manoeuvres around 16 May and completed them within 10 days. Iridiums 4–5 started their manoeuvres around 20 May with Iridium 4 reaching its planned altitude around 27 May and Iridium 5 in early June (Phillip S Clark) 0005571/0005572/0005573/0005574/0005575

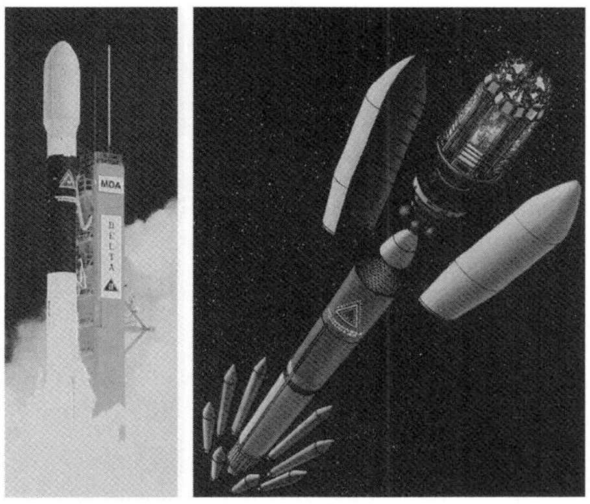

Delta is launching Iridium in clusters of five (McDonnell Douglas Aerospace) 0517225

satellites in orbit. Khrunichev invested US$40 million in return. Then Motorola made a similar agreement with GWIC April 1993 for 10 launches of two satellites each between 1996–2002 on the CZ-2C. Eight Delta 2 launches with five satellites each, will establish the initial constellation in 1998. As replacements are required, Motorola will use Delta 7420 launches between 1998 and July 2000. To avoid on-orbit debris and pollution, Iridium will de-orbit at the end of its useful lifetime.

Individual satellites broadcast to 48 L-band cells 670 km across using FDMA/TDMA. Calls are routed to the Earth gateways and among the space segment at Ka-band. The user's 0.7 kg 0.6 W handset automatically selects a terrestrial cellular network if one is available and cheaper. Each device has a GPS receiver to maintain a log of each handset's last known position. Each device relays its position to a visible satellite for passing on to the nearest gateway. Each gateway has three terminals dispersed sufficiently to avoid weather problems affecting links.

Specifications

Launches: 5 × Delta 2, 5 May 1997; 7 × Proton K4, 18 June 1997; 5 × Delta 2, 9 July 1997; 5 × Delta 2, 21 August 1997; 7 × Proton K4, 14 September 1997; 5 × Delta 2, 26 September 1997, 5 × Delta 2, 9 November 1997; 5 × Delta 2, 20 December 1997; 5 × Delta 2, 18 February 1998; 2 × CZ-2C/SD, 25 March 1998; 5 × Delta 2, 30 March 1998; 7 × Proton K4, 7 April 1998; 2 × CZ-2C/SD, 2 May 1998; 5 × Delta 2, 17 May 1998; 2 × CZ-2C/SD, 19 August 1998; 5 × Delta 2, 8 September 1998; 5 × Delta 2, 6 November 1998; 2 × CZ-2C/SD, 19 December 1998; 2 × CZ-2C/SD, 11 June 1999; 5 × Delta 2, 11 February 2002; 1 × Rockot, 20 June 2002
Orbits: 780 km circular, 86.4°; 11 active satellites in 6 planes (plus 1 spare in each plane)
Design life: 5–8 years
Contractors: Motorola (payload, integration), Lockheed Space Systems Division (bus), Raytheon (phased array antennas), COM DEV (inter-satellite and feeder link antennas), Olin Aerospace (thrusters), Barnes Engineering Division (miniature infra-red Earth horizon sensors), Tecstar Inc (solar cells), Ithaco, Johnson Controls, Schaeffer Magnetics, Pressure Systems Inc (hydrazine tanks)
Principal applications: voice, fax and data global handheld services
Transponders: three phased array antennas each provide 16 user up/down L/S-band beams from 106 elements at 1.610–1.6265/2.4835–2.500 GHz
Single phased array antenna provides inter-satellite up/down Ka-band beam
Single dish antenna provides feeder link up/down Ka-band beam
Configuration: 3-axis triangular bus, 1 m diameter, accommodating twin solar wings, body-mounted Ka-band inter-satellite antenna, and lower Earth-pointing section with three angled 86 × 188 cm 30 kg L/S-band antennas, each with 106 radiating elements. 115 kg hydrazine (sufficient for >10 years) in single 53.5 cm diameter 71.1 cm height titanium tank
Mass: 690 kg at launch
TT&C: 10 worldwide System Control Segment terminals, controlled by master site at Landsdowne, Virginia (back-up at Nuova Telespazio, Rome, Italy). Motorola's Chandler facility will initially control the first 40 satellites
Power: two deployed 5-panel wings of GaAs cells; 50 Ah battery.

Loral Skynet

Current Status

On 15 July 2003 Loral Space & Communications filed for reorganisation under Chapter 11 of US Bankruptcy law. It also announced that it had reached a definitive agreement to sell six of its satellites to Intelsat Ltd for US$1.1 billion in cash. The agreement included the sale of Telstars 4, 5, 6 and 7 currently in orbit and Telstars 13 and 18 yet to be launched. Henceforth, the Loral Skynet fleet will consist of the existing Telstar 10, 11 and 12 satellites and Telstar 18/Apstar V and Telstar 14/Estrela do Sul due for imminent launch. On 11 November 1999, Loral Space & Communications announced that Loral Orion satellites had been renamed to reflect their inclusion in the fleet of satellites managed by Loral Skynet. Thus, Orion 1 became Telstar 11 and Orion 2 became Telstar 12 and Apstar 11R became Telstar 10. Telstar 18 is due for launch in 2004, two years after the introduction of Telstar 13.

Background

AT&T built and operated the world's first commercial satellites, Telstar 1 and 2 in 1962 and 1963. It revived the designation two decades later to commemorate the achievement when the FCC authorised it to proceed with the Telstar 300 series to replace the Comstar satellites.

The first of the series Telstar 301 went into space on 12 July 1982. Following its Anik E2 control problems beginning in January 1994, Telesat Canada decided to acquire 301 from AT&T on 28 March 1994. After purchase, Telesat Canada moved it to its new 107.1 degree west slot on 4 April 1994 to restore services until Anik E2 began commercial operations. Then Telesat Canada sold the Telestar to Arabsat on 7 June 1994 and began moving it to 20° east where it arrived in February 1995 and was designated Arabsat 1DR.

The Iridium LLC filled all of the orbital slots for its constellation in early 1998. Barring any further loss of on-orbit spares, the company planned to offer services beginning in September 1998 (Iridium LLC) 0022577

In October 1987, AT&T further requested approval to build a hybrid Telstar 4 offering 48 Ku-/C-band transponders. Space Systems/Loral contracted with AT&T to build the satellite for Skynet and possibly exercise three more satellite options for on-orbit delivery. AT&T planned to launch the Telstar 5 on a Proton. The company paid about $489 million for three satellites. The first of the series orbited in September 1994 by Ariane 4. 10 minutes after separation ground controllers lost contact with the satellite. An investigation concluded that the firing of a pyrovalve to open a propellant line had ignited some hydrazine. Insurance providers paid out a reported $187 million. AT&T quickly substituted 403 as 402R.

A surge in electromagnetic radiation following a major event on the sun's surface on 6 January 1997 resulted in the electronics on Telstar 401 short-circuiting four days later, causing the loss of the satellite. As a result the aging Telstar 303 was relocated to the Telstar 401 longitude as a replacement.

Space Systems/Loral built Telstar 5,6,7,8 and 9, plus a ground spare, for launches between 1997 and 2001. The new satellites target specific markets, whereas the 400s are more flexible. Telstar 6 replaces Aurora 2 for Alaska services from January 1999. Loral Skynet's VoiceSpan system would use 12 satellites in seven GEO slots. As of the end of 1999, Telstar 5, 6 and 7 had been orbited successfully by Proton K and Ariane launchers.

Specifications

Telstar 302 (3C)
Launch: 30 August 1984 by Shuttle STS-41D
Location: 85° W geostationary; inclined orbit operations began 1 January 1995
Design life: 10 years
Contractors: Hughes Space & Communications
Transponders: 18 5.5 W SSPAs 5.925–6.425/3.700–4.200 GHz up/down C-band CONUS + spot beams, 36 MHz bandwidth, 34 dBW min EIRP, linear polarisation
12 5.5 W TWTAs, other specs as for SSPAs
Principal applications: domestic telephony, data and TV broadcast, including AT&T's range of Skynet business services
Configuration: Hughes' HS-376 spin-stabilised bus, 2.2 m diameter, 6.8 m high deployed
Mass: 3,423 kg with PAM-D GTO stage attached, 625 kg on-station BOL
TT&C: through AT&T's Satellite Management & Control Facility in Hawley, Pennsylvania
Power: 800 W BOL/670 W EOL provided by 15,588 silicon cells on cylindrical exterior
Apogee kick motor: Thiokol solid propellant.

Telstar 303 (3D)
Launch: 17 June 1985 by Shuttle STS-51G
Location: 123° W geostationary, inclined operations began June 1995. Following the loss of Telstar 401 in early 1997, Telstar 303 was manoeuvred off-station from 123° W in January and relocated to 89° W the following month
Other specifications as for Telstar 302.

Telstar 401 (4A)
Launch: 16 December 1993 by Atlas 2AS from Cape Canaveral
Location: 97° W GEO (replacing Telstar 301). Arrived 89° W 24 December 1993 for testing. Services began formally 1 February 1994. Satellite control was lost in January 1997 after damage caused by a solar electromagnetic storm
Design life: 12 years' propellant carried
Contractors: Lockheed Martin Astro Space
Transponders: 24 11–21 W (variable, plus four back-up in 7-for-6) SSPAs 5.925–6.425/3.7–4.2 GHz up/down C-band CONUS and spot beams, 36 MHz bandwidth, EIRP CONUS 37.8/35.0 dBW (in 21/11W order), Alaska 33.8/31.0 dBW, Hawaii 35.8/33.0 dBW, Puerto Rico 33.8/31.0 dBW, linear orthogonal polarisation
24 60–120 W (12 variable 60–120 W, plus 12 back-up in 18-for-12) TWTAs 14/12 GHz up/down Ku-band CONUS and spot beams, 54 MHz bandwidth (eight can be reconfigured to 2 × 27 MHz), EIRP CONUS 44.3/47.1 dBW (in 60/120W order), Alaska 40.3/43.1 dBW, Hawaii 44.3/47.1 dBW, Puerto Rico 44.3/47.1 dBW, linear polarisation
Principal applications: TV broadcast, data, distance learning
Configuration: MMAS 7000 series 3-axis platform, 246 × 216 × 315 cm bus, 24 m solar array span
Mass: 3,375 kg at launch
AOCS: 3-axis control by momentum bias, thrusters + magnetorquers; NSSK by 1.8 kW hydrazine arcjets. Pointing accuracy ±0.1° pitch/roll, ±0.25° yaw; offset pointing control ±6.0° pitch + ±2.0° roll. Station-keeping ±0.05° lat/long (inclined orbit ops to 5°). GEO injection by dual mode liquid apogee engines

TT&C: as Telstar 3
Power: 7.2/5 kW BOL/EOL provided by Si cells on twin 4-panel solar wings, totalling 48.6 m^2. 100 per cent eclipse protection by 2 × 50 Ah Ni/Cd batteries.

Telstar 4 (402R/4B)
Launch: 24 September 1995 by Ariane 4 from Kourou, French Guiana. Services began 20 November 1995
Location: 89° W geostationary
Design life: 13 years
Mass: 3,410 kg launch, 2,097 kg on-station BOL, 1,578 kg dry
Other specifications are the same as Telstar 401.

Telstar 5
Launch: 24 May 1997 by Proton K from Baikonur
Location: 263° E
Design life: 12 years
Mass: 3,650 kg at launch; 1,400 kg dry
Specification: 24 × C-band 36 MHz at 20 W; 4 × Ku-band 54 MHz at 100 W; 24 × Ku-band 27 MHz at 100 W. Frequency band 4/6 GHz and 12/14 GHz.

Telstar 6
Launch: 15 February 1999 by Proton-K from Baikonur
Location: 267° E
Mass: 3,650 kg at launch; 1,400 kg dry
Specification: 24 × C-band 36 MHz at 20 W; 4 × Ku-band 54 MHz at 100 W; 24 × Ku-band 27 MHz at 100 W. Frequency band 4/6 GHz and 12/14 GHz.

Telstar 7
Launch: 25 September 1999 by Ariane 44LP from Kourou, French Guiana
Location: 231° E
Mass: 3,790 kg at launch; 1,537 kg dry
Specification: 24 × C-band 36 MHz at 37 W; 24 × Ku-band 36 MHz at 100 W. Frequency band 4/6 GHz and 12/14 GHz.

Telstar 8
Launch: planned for 2004
Location: planned: 89° W
Specification: 2 × C-band 72 MHZ at 100 W; 20 × C-band 36 MHz at 37 W; 36 × Ku-band 36 MHz at 130 W; 24 Ka-band 66 MHz at 105 W.

Telstar 10/Apstar-2R
Launch: 16 October 1997
Location: 76.5° E
Mass: 3,747 kg at launch; 1,415 kg dry.

Telstar 11
Launch: 29 November 1994
Location: 37.5° W
Mass: 2,358 kg at launch.

Telstar 12/Orion 2
Launch: 19 October 1999
Location: 15° W
Mass: 3,795 kg at launch; 1,564 kg dry.

Telstar 13
Launch: planned for 2003
Location: 121° W planned
Specification: 24 × C-band at 36 MHz.

ORBCOMM GLOBAL Constellation

Current Status

On 24 February 2004 Orbcomm announced it had completed a US$26 million equity financing involving SES Global, OHB Technology, Ridgewood Capital and Northwood Ventures. Currently 35 satellites in orbit: 31 at a 45° inclination, two at 70° inclination and two at 108° inclination. Data transmitted in VHF frequency range. As of April

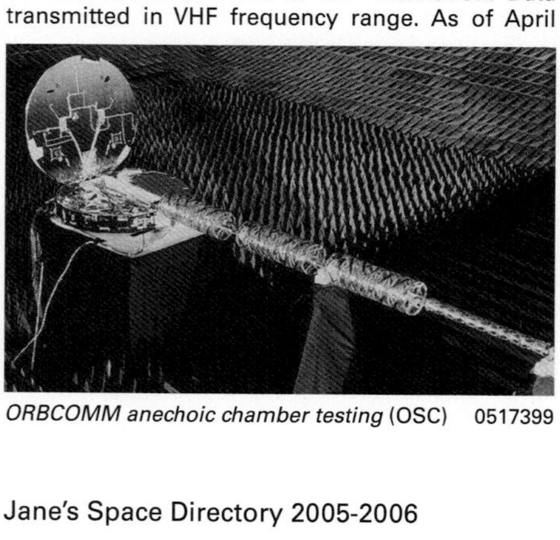

ORBCOMM anechoic chamber testing (OSC) 0517399

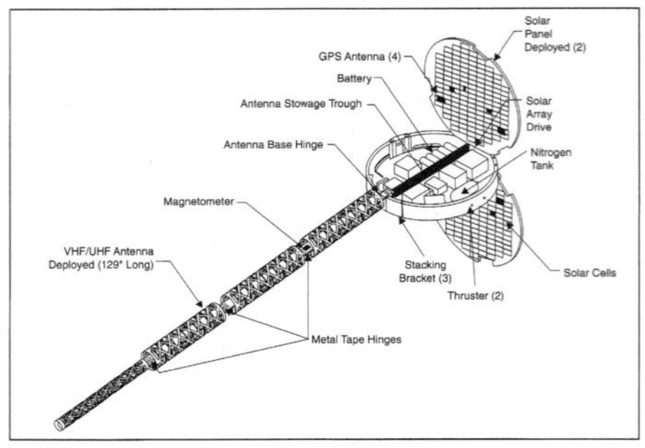

ORBCOMM's design is based on OSC's MicroStar bus 0517027

2002, 35 satellites in LEO (31 at 45° inclination, 2 at 70° inclination and 2 at 108° inclination). In March 2002 Volvo Trucks North America announced that ORBCOMM would provide two-way communication, tracking and monitoring of all its trucks in North America.

Background

OSC's Orbital Communications Corporation subsidiary, created at the beginning of 1990, filed an FCC application 28 February 1990 for a constellation of LEO satellites. OSC planned to provide for two-way communications and geolocation services with low-cost alphanumeric data communications and position determination (375 m) for emergency assistance, data acquisition and messaging services. Circuit design has optimised the system for shorter messages of 6–250 characters. In order to finance the system, OSC and Canada's Teleglobe Inc signed an agreement 26 July 1993. Teleglobe and OSC will jointly operate the system. Teleglobe provided US$10 million of the US$55 million cost of the two first phase satellites. For the second phase, which requires 34 more satellites (including eight ground spare), Teleglobe has increased its equity share in ORBCOMM Global LP to 50 per cent. Teleglobe contributes US$85 million to OSC's US$75 million.

OSC launched the 16.7 kg ORBCOMM-X as an experimental forerunner by Ariane into a polar orbit 17 July 1991. Virginia's Center for Innovative Technology, co-developed the original design and invested US$250,000 as part of its VaStar satellite R&D programme. The state's universities and institutions planned to use it for collecting environmental data from remote platforms. OSC expected ORBCOMM-X to survive for 18 months. OSC's Boulder station made initial contact some 14 hours after launch and verified basic operations, but was unable to establish subsequent links. It abandoned attempts after several weeks. OSC has traced the failure to a software glitch in a power control circuit.

When finished, the ORBCOMM constellation will have 28 43 kg satellites by 1998 at 775 km altitude; 24 at 45° inclination (eight in each of three planes) and two at 70° for high latitude coverage. OSC will conduct four launches aboard its own Pegasus and one on Taurus, plus a set of eight ground spares intended for Pegasus. The first two satellites, in polar orbit, appeared successfully 3 April 1995. Both soon developed receiver problems. ORBCOMM 2's solved its problem by a self-activated reset on 13 May, and ground commands solved ORBCOMM 1's by freeing the receiver 18 May. Apparently battery drain triggered a computer reset.

ORBCOMM 2 relayed the system's inaugural message 25 May to the Arcade gateway from a production prototype Panasonic personal communicator at Dulles airport near Washington, DC. A message from the Elisra EL-2000 Data Communicator near the Dead Sea in Israel routed through the system to Dulles on 25 July 1995 via the Georgia gateway.

ORBCOMM seeks to serve the US market at first, where it foresees a total of 10 million to 20 million users, beginning with 150,000 the first full year of constellation operations and rising to 5.2 million after seven years. Emergency services may provide 75 to 85 per cent of the market. ORBCOMM plans to market three classes of terminals, a 5 W power, signalling and data transmit only, a US$150 position-fixing and a US$400 alphanumeric messaging terminal. The first type requires a US$30–50 annual service charge, with the others supported by US$35 monthly plus usage-linked charges.

Each satellite carries 17 data processors and seven antennas, allowing the relay of 50,000 messages hourly in each direction, or more than five million two-way daily messages for the 28-satellite system. Four regional gateways will cover the US from Oscilla, Georgia, Arcade, New York, St Johns, Arizona and Washington State.

Specifications

Launch: 3 April 1995 (numbers 1–2 Pegasus); 8 × Pegasus-XL, 23 December 1997; 2 × Taurus, 10 February 1998; 8 × Pegasus-XL + HAPS, 2 August 1998; 8 × Pegasus-XL + HAPS, 23 September 1998; 7 × Pegasus XL, 4 December 1999
Orbits: numbers 1–4 736 × 749 km 70.0°; numbers 5–28 775 km, 45°
Design life: 4–6 years
Transponders: 4,800/9,600 bit/s downlink uses 320 kHz in the 137.0–138.0 MHz VHF band (10 channels) and 50 kHz at 400.1 MHz UHF (1 channel, for GPS time data only). EIRP 16.5 dBW at 137 MHz and 19.5 dBW at 400 MHz. For 2,400 bit/s uplinking, a Dynamic Channel Activity Assignment System directs terminal users to a block of 24 out of a possible 760 channels at 148.00–149.90 MHz depending on instantaneous usage by other systems
Principal applications: two-way packet data communications
Mass: 43 kg at launch
Configuration: MicroStar design, allowing eight to be stacked on single Pegasus. Stowed size 104 cm diameter × 16.5 cm. Deployed length 432 cm, width 224 cm (across solar arrays). ±5° nadir pointed by active magnetorquers and gravity gradient; attitude sensing by Earth sensors + magnetometer. 328 cm VHF/UHF antenna deployed by four hinges. Orbital spacing provided by N_2 thrusters
TT&C: via Dulles Satellite Control Centre; 57.6 kbit/s VHF
Power system: twin circular Sun-tracking solar arrays hinge out from main body providing 230 W BOL; 14 V system; 5V regulated bus for electronics. Supported by Ni/H_2 battery.

ORBCOMM workers prepare a stack of eight LEO satellites that will form part of the ORBCOMM constellation. When complete, the constellation will have 28 of these 43 kg satellites circling the globe at 775 km altitude. Orbital Sciences launched its two latest satellites, the sixth launch in a series of ten, on 10 February 1998, by Taurus rocket (ORBCOMM Global, L.P.) 0022576

Orion (Orion Atlantic LP) Series

Current Status

Launched 29 November 1994, Orion 1 is now operating as Telstar 11 and Orion 2, launched 19 October 1999, Telstar 12. Telstar series satellites are operated by Loral Skynet.

Background

BAe and Orion Satellite Corp signed a US$360 million contract on 1 September 1989 for the in-orbit delivery of two satellites. It marked the first US commercial sale by a European company. Construction began 7 January 1992. Under the turnkey contract, Martin Marietta Commercial Launch Services provided launch and insurance services. The FCC granted final authority to proceed 28 June 1991. Orion cancelled the second satellite, planned for launch early 1996 by Atlas 2A to 47° W, in April 1993, but BAe's remaining contract had a value of US$227 million. Orion 1 is the World's first all-solid-state Ku-band satellite. An agreement was reached in October 1995 with Telecomm Mexico to exchange capacity on their respective satellites, providing Orion with access to Mexico, Central and South America.

Orion ordered the all-Ku Orion 2 on 24 October 1995 for 1998 in-orbit delivery. It will eventually pay US$265 million, including launch and insurance. The Eurostar 2000+ design resembles Orion 1, but Orion 2's higher power transponders allow additional services such as DTH. The coverage overlaps with Orion 1's in the US and Europe. Onboard switching will allow beam connectivity reconfiguration according to customer demand. The contract includes an option for the similar Orion 3, which would be owned and operated by ONS' wholly owned Orion Asia Pacific subsidiary. Total cost, including ground segment, launch, insurance and operations, has been estimated at US$275 million. Orion 1 and Orion 2 are now operated by Loral Skynet and have been renamed Telstar 11 and Telstar 12, respectively.

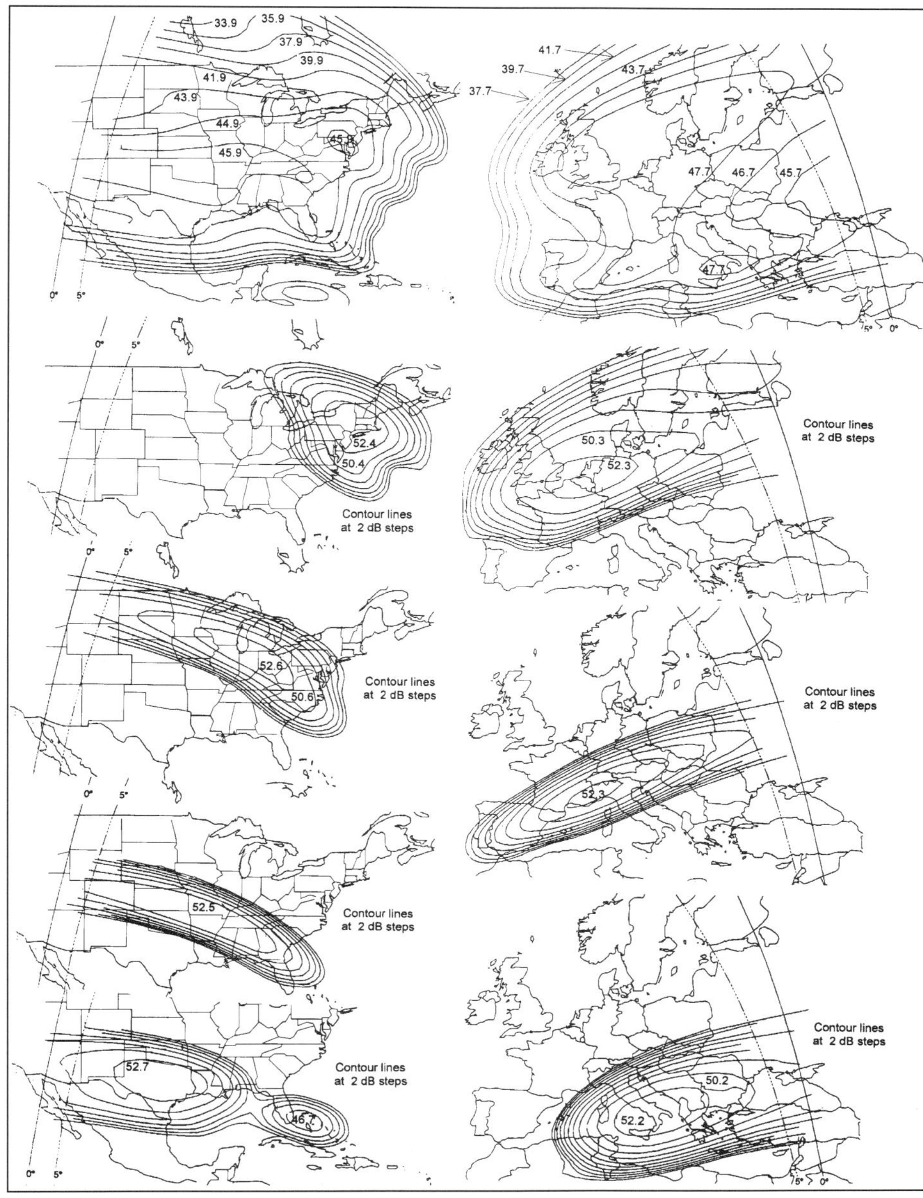

Orion 1 broad and spot beam EIRP (dBW) coverage from 37.5° west. At left from top: North America broad, northeast, north, central and south. At right from top: European broad, north, central and south (Orion Atlantic LP) 0517200

Specifications

Orion 1 (Telstar 11)

Launch: 29 November 1994 aboard Atlas 2A from Cape Canaveral

Location: 37.5° W geostationary (arrived 25 December 1994, MMS handed over control 16.00 GMT 22 January 1995 after completing checkout)

Design life: 12 years

Contractors: BAe Space Systems (prime, now part of Matra Marconi Space), Martin Marietta (launch and insurance services), Matra (platform electronics), NEC (communications payload), Lockheed (propellant tanks), Kaiser Marquardt (apogee motor), MBB (10 N thrusters), Fokker (solar arrays), Spectrolab (solar cells), Eagle Picher (Ni/H_2 batteries), COM DEV (output multiplexers) and Telesat Canada (LEOP services)

Transponders: 34 (plus 8 back-up: 5-for-4 54 MHz and 7-for-6 36 MHz) 15 W SSPAs 14.0–14.5/11.45–12.2 and 12.5–12.75 GHz up/down Ku-band Europe and North America spots & regional beams, bandwidth 28 × 54 and 6 × 36 MHz, EIRP 50 dBW (54 dBW max) spots and 45 dBW (48 dBW max) regional. Beam connectivity via switch matrix provides not only trans-Atlantic services but also regional & domestic coverage for Europe and US. 17 transponders allocated to North America (11.7–12.2 GHz down) and 17 to Europe (11.45–11.7 and 12.5–12.75 GHz down). Spots: 3 Europe (North, South and Central) & 4 North America (Northeast, North, South and Central). Regional: 1 North America and 1 Europe

Orion 1 began providing trans-Atlantic services in January 1995 (Orion Atlantic LP) 0517199

Principal applications: optimised for VSAT applications (1.2 m diameter 2-way; 85 cm diameter receive only). Flexible interconnectivity between high powered spot beams provide for both trans-Atlantic and regional/domestic networking

Mass: 2,361 kg at launch, 1,140 kg dry, payload 330 kg

Configuration: MMS Eurostar 2000 bus. Deployed size 2.1 × 1.7 × 1.9 m; span 22.0 m. Two 2.3 m antennas on east/west faces

AOCS: 490 N R-4D for apogee raising, 12 × 10N MBB for station-keeping and wheel unloading; NTO/MMH in 4 tanks

TT&C: 15 m fully steerable & two 9 m antennas at Mt Jackson, Virginia

Power system: twin 4-segment Si solar wings providing 3.5 kW EOL, supported by two 78 Ah Ni/H_2 batteries. Payload power is 2,200 W.

Orion 2 (Telstar 12)

Launch: 19 October 1999

Location: planned 12° W geostationary

Design life: 13 years

Contractors: Matra Marconi Space (prime), Kaiser Marquardt (apogee motor), DASA (10 N thrusters)

Transponders: 32 TWTA 14.0–14.5/11.45–12.2 & 12.5–12.75 GHz up/down Ku-band Europe & North America spots & regional beams, 54 MHz bandwidth. Beam connectivity via switch matrix provides not only trans-Atlantic services but also regional & domestic coverage for Eastern US & Europe

Principal applications: as Orion 1, plus DTH

Mass: 3,200 kg at launch

Configuration: MMS Eurostar 2000+ bus. Two antennas on East/West faces

AOCS: 490 N R-4D for apogee raising, 12 × 10 N DASA for station-keeping and wheel unloading; NTO/MMH in 4 tanks
TT&C: as Orion 1
Power system: twin 4-segment Si solar wings providing 7 kW EOL, supported by two 160 Ah Ni/H_2 batteries.

Orion 3
Launch: 5 May 1999 aboard Delta 3 which failed due to a premature stage 3 shutdown.
Location: projected 139° E GEO (applied for through Marshall Islands)
Design life: 13 years
Transponders: 32 TWTA 14.0–14.5/11.45–12.2 & 12.5–12.75 GHz up/down Ku-band Asia-Pacific spots & regional beams, 54 MHz bandwidth
Principal applications: as Orion 2
Mass: 3,200 kg at launch
Configuration: as Orion 2
AOCS: as Orion 2
Power system: as Orion 2.

UPDATED

PanAmSat Series

Current Status
In 2001 PanAmSat exercised an option and ordered three satellites from Orbital Sciences Corporation based on its Orbital Star 2 lightweight bus. The first, Galaxy 12 was launched on 8 April 2003. Galaxy 13 was launched 1 October 2003 initiating the premier High Definition TV service to North America.

Background
Hughes Electronics Corporation established Hughes Communications Incorporated, or HCI as a wholly owned subsidiary to operate the world's largest fleet of privately-owned commercial communications satellites, known as the Galaxy. HCI first began satellite operations with the launches of its three initial Galaxy satellites during 1983–84. The system provides point-to-point and point-to-multipoint C-band services for users in the continental US, Mexico, Alaska, Hawaii and the Caribbean Basin. Then in 1993, HCI established DirecTV, Inc to operate a domestic US DBS system using three Ku-band HS-601s. DirecTV became an independent Hughes Electronics subsidiary in November 1993.

From the beginning, HCI intended to make cable TV distribution a priority and dedicated Galaxy flights 1RR/5/7 to this application. Galaxy 3R/4/6 provided business users and audio subscribers with a continental United States service. HCI launched Galaxy 1 on 28 June 1983 and replaced it in March 1994 with the higher power, 16 W Galaxy 1RR. The first planned replacement, Galaxy 1R disappeared in a 22 August 1992 launch failure. Galaxy 2 followed on 22 September 1983 and retired from 74° W in May 1994 after replacement by Galaxy 6. Galaxy 3, launched 21 September

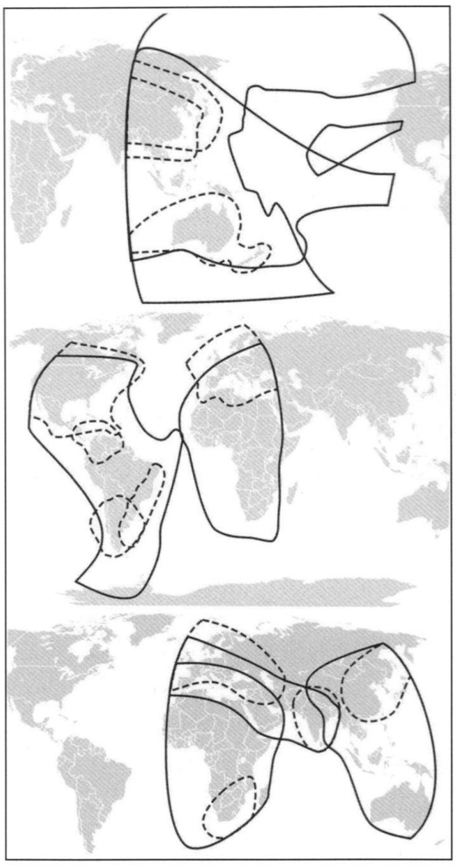

PAS 2–4 footprints (from top). Ku-band is dashed line; C-band is solid 0517017

1984, left its assigned spot at 93.5° W on 30 September 1995 when 3R replaced it. Then HCI acquired the Westar satellite system and used its geosynchronous locations for its own spacecraft, Galaxy 6/4/5, respectively. This expanded the Galaxy fleet to six satellites.

Galaxies 1RR/5 are stretched HS-376s, originally built as the ground spare for Galaxy 4/7. Their 24 Ku-band transponders provide an initial DTH Latin America service via 60 cm dishes. Although its orbital slot is licensed only for US domestic services, HCI successfully argued that it is only a temporary measure as the Ku-band coverage will be switched back to the US when the dedicated Galaxy 8I appears to provide the full service.

The old C-band Westar 4 slot (temporarily serviced by Galaxy 6) and Ku-band SBS 6 slot at 99° W has been occupied since November 1993 by the hybrid Galaxy 4. HCI decided in late 1989 to expand the Galaxy series with Galaxy 4, at 91° W. Ariane launched it into a sub-GTO (apogee 27,670 km) parking orbit and the Galaxy made two perigee burns to attain GTO. The perigee burns optimised the climb to GEO sparing station-keeping propellant and lengthening the satellite lifetime by about 1 year.

Replacement for PanAmSat 3 lost in an Ariane failure in December 1994, PAS-3RR is made ready for a January 1996 launch (Hughes Space & Communications) 0054342

Galaxy 7 joined in 1997 and HCI scheduled Galaxy 8I (I indicates international) for a late 1997 launch to provide dedicated Latin America DTH, replacing the initial service from Galaxy 3R.

Hughes Electronics spun off DirecTV in 1992 after Sky Cable collapsed as a corporate vehicle to operate a domestic US direct-to-home television system using Ku-band transmission from a HS-601 satellite. When it produced its own DBS 1 satellite it became the 100th Hughes-designed telecommunications satellite to be launched. Hughes built a 3,720 m^2 facility for programming uplink facility at Hughes Communications Incorporated's Castle Rock (Colorado) station. The new facility uses 13 m antennas, capable of delivering 216 simultaneous channels of television. DirecTV has received an FCC licence to collocate up to four HS-601 satellites over 101° W with a total of 27 transponders.

Hughes sold 5 of the 16 transponders aboard the first satellite to the United States Satellite Broadcasting Incorporated (USSB), a subsidiary of Hubbard Broadcasting for more than US$100 million. DirecTV owns all 16 channels on DBS 2. The two satellites deliver a combined 175 channels of television. Each transponder can provide four video channels using digital compression, or eight films because of the slower motion. In order to turn a profit Hughes estimates the venture requires three million subscribers. It projects 10 million by 2000. One million homes had subscribed by November 1995.

HCI's Galaxy, SBS and Leasat satellites are controlled from the Operations Control Center (OCC) located at its corporate headquarters.

In May 1997 PanAmSat Corporation and the Galaxy Satellite Series business element of Hughes Communications Inc, merged under the existing name of PanAmSat Corporation.

PanAmSat operates the first global geosynchronous private sector telecommunications system, with satellites over the Pacific, Atlantic and Indian Ocean regions. The company selected Hughes in 1991 to provide four of its HS-601 models for PAS 2–4. PanAmSat paid Hughes about US$300 million for all of the satellites. The fourth satellite remained on the ground as a ground spare. Each bus carries cross-strapped C-/Ku-band 54 MHz transponders optimised for TV programmers and corporate telecommunications. Ariane received the contract to launch the entire series in June 1992, in return for which Credit Lyonnais arranged French export loans covering 85 per cent of the US$240 million launch costs. Then Televisa SA of Mexico stepped up to purchase 50 per cent of PanAmSat's equity, to finance the US$700 million expansion in December 1992, for US$200 million in cash and US$440 million raised from a bond sale.

With financing assured, PanAmSat launched its first satellite PAS 1, on the maiden flight of the Ariane 4 booster. ArianeSpace sold the spot to PanAmSat for the reduced fee of US$8 million. Insurance coverage cost PanAmSat another US$40 million at a 25 per cent premium. PanAmSat after proceeding with its plans, remained prevented by the FCC from connecting with public switched networks. So the service relies on TV programming of US and Mexican origin, and business data traffic. PAS 2 appeared in July 1994 to serve the

PanAmSat is the first private sector operator of a global system (Hughes Space & Communications) 0517016

Another utilisation of the HS-601HP, this PanAmSat has 24 high-power transponders in C-band and Ku-band (Hughes Space & Communications) 0054341

Asia-Pacific region, and by November 1995 it had 17 full-time programmers. Broadcast sales that year made up 67 per cent of revenue.

The 2,985 kg PAS 3, intended for 43° W to provide DTH services to Latin America disappeared in Ariane's 1 December 1994 failure. Insurance providers paid an estimated US$214 million at 17.5 per cent. PanAmSat outfitted its ground spare which became PAS 3R, and began the Latin America service in early 1996.

The company announced plans for further satellites and ordered PAS 5 from Hughes Space & Communications in March 1995 and PAS 6 from Space Systems/Loral in August 1994 to expand the DTH Latin America capacity to more than 120 digital channels per customer. PanAmSat raised US$262 million in April 1995 in a preferred stock offering to finance PAS 5/6. SS/L took an order for PAS 7/8 in April 1996, and PanAmSat filed the necessary paperwork with the FCC in September 1995 to ensure a slot for the PAS 10/11 Atlantic Ka-band satellites.

On 16 May 1997 PanAmSat Corporation and Hughes Communications Inc (HCI) announced that a merger had been completed between PanAmSat and Galaxy Satellite Services, a business element of HCI.

On 1 May 1998 Hughes Electronics Corporation increased its equity in PanAmSat to 81 per cent.

After Panamsat 6 suffered electrical failures limiting use on some transponders, Panamsat announced in March 1998 that Hughes would build a replacement satellite named PAS-6B which was launched 21 December 1998.

Specifications

Galaxy 1R
Launch: 19 February 1994 by Delta 7925 from Cape Canaveral
Location: 133° W geostationary (replacing number 1)
Mass: 1,393 kg at launch, 798 kg on-station BOL
Coverage: North America, Caribbean
Other specifications as for Galaxy 5.

Galaxy 3
Launched: 21 September 1984 by Delta 3920/PAM from Cape Canaveral
Location: drifting
Design life: 10 years, sufficient AOCS propellant for 9 years now predicted eol 2006
Transponders: 24 (plus six back-up; 5-for-4 redundancy) 9W TWTA 5.925–6.425/3.700–4.200 GHz up/down C-band CONUS/Alaska/Hawaii beams, 36 MHz bandwidth, EIRP 34 dBW over CONUS, orthogonal polarisation
Principal applications/users: cable TV distribution
Configuration: Hughes' HS-376 platform, spin-stabilised at 50 rpm, drum-shaped 2.16 m diameter, 6.6 m deployed height (2.8 m stowed) single 1.83 m diameter reflector (vertical/horizontal polarisation) and communications shelf despun
Mass: 1,200 kg at launch, 654 kg on-station BOL
AOCS: station-keeping and spin-stabilised at 50 rpm by four hydrazine thrusters (136 kg propellant); GEO insertion provided by solid propellant Thiokol Star 30 apogee kick motor
TT&C: stations at Spring Creek (Brooklyn) in New York and Fillmore in California relay signals to the Operations Control Center in El Segundo, California. Users are served by Hughes Communications Network Operations (HCNO), collocated with the OCC
Power system: solar cells mounted on 47 panels on cylindrical surface providing 19.7 mW/cm^2, 990 W BOL, 900 W EOL; two Ni/Cd batteries provide eclipse protection.

Galaxy 3C
Launched: 15 June 2002 by Zenit (Sea Launch)
Location: 95° W longitude GSO
Design life: 15 years
Configuration: BSS-702 bus; two solar cell wings each with six panels of improved triple-junction GaAs cells producing 15 kW EOL; 56-cell NiH battery with 328-Amp/hr life; single 490 N liquid apogee motor; 4 × 0.165 N XIPS; 4 × 0.165 N transfer orbit; 4 × 0.08 N attitude control.
Mass: 4,860 kg launch; 2,873 at BOL
Size: length 48 m across extended solar arrays; width 9.4 m across antennas; height 7.3 m to tip of antennas; stowed 7.3 m × 3.2 m × 3.3 m
Transponders: 24 (6 spare) × 36 MHz (34 W SSPA) C-band; 8 (2 spare) × 54 MHz (240 W each configurable to 16 active/4 spare 120 W TWTA), 13 (3 spare) × 27 MHz (127 W TWTA), and 24 (4 spare) × 24 MHz (120 W LTWTA) in Ku-band.
Antennas: one 216 cm, one 140 cm, one 114 cm dual gridded shaped reflectors and two 284.5 cm single surface reflectors.

Galaxy 3R
Launched: 15 December 1995 by Atlas 2A from CCAFS
Location: 95° W (moving to new position on Galaxy 3C launch)
Coverage: North America, Caribbean
Design life: 12 years

Galaxy 4
Launched: 25 June 1993 by Ariane 42P+ V57 from Kourou, French Guiana
Location: 99° W geostationary (replacing Galaxy 6/SBS 6)
Design life: 15 years, but with sufficient propellant for 13.5 years operations
Transponders: 24 (15-for-12 redundancy) 16 W SSPA 5.925–6.425/3.700–4.200 GHz up/down C-band CONUS/Hawaii/Alaska beams, 36 MHz bandwidth, EIRP 37 dBW CONUS, 30 dBW Alaska/Hawaii, orthogonal polarisation
8 54 MHz + 16 27 MHz (15-for-12 redundancy) 50 W TWTA 14.0–14.46/11.72–12.16 GHz up/down Ku-band CONUS and four of the 54 MHz switchable to Alaska, Hawaii, Virgin Islands and Puerto Rico, EIRP 45 dBW CONUS, 41 dBW Alaska/Hawaii, orthogonal polarisation. Two of the 54 MHz transponders can be cross-strapped Ku-to-C/C-to-Ku
Principal applications: network TV, radio, VSAT, business video and data
Configuration: Hughes' HS-601 3-axis platform
Mass: 2,980 kg at launch, 1,692 kg BOL, 1,320 kg EOL
AOCS: provided by unified 12 × 22 N + 1 × 490 N bipropellant thrusters incorporating onboard control processor, Earth/Sun sensors, MWs. 1,658 kg NTO/MMH in four tanks. Solar sailing techniques reduce conventional δV requirements
TT&C: stations at Spring Creek (Brooklyn) in New York and Fillmore in California relay signals to the Operations Control Center in Long Beach, California. Users are served by Hughes Communications Network Operations (HCNO), collocated with the OCC
Power system: twin solar wings spanning 24.7 m of four 2.16 × 2.54 m panels carrying K4¾ large area Si cells generate up to 6 kW (4.7 kW EOL). Ni/H_2 batteries provide 100 per cent eclipse protection.

Galaxy 4R
Launched: 18 April 2000 by Ariane 4 from Kourou, French Guiana
Location: 99° W
Coverage: North America
Design life: 15 years

Galaxy 5
Launch: 00.06 GMT 14 March 1992 by Atlas 1 from Cape Canaveral
Location: 125° W geostationary (replacing Westar 5)
Design life: 10 years (predicted eol 2005)
Transponders: 24 (plus 6 back-up; 30-for-24 ring redundancy) 16 W TWTA 5.925–6.425/3.700–4.200 GHz up/down C-band CONUS/Alaska/Hawaii beams, 24/36 MHz bandwidth, EIRP 40/37.5 dBW peak/min over CONUS, orthogonal polarisation
Configuration: stretched HS-376, 3.10/7.49 m height stowed/deployed
Coverage: North America, Caribbean
Mass: 1,412 kg at launch, 788 kg dry

Galaxy 6/Westar 6S
Launch: 12 October 1990 by Ariane 4 from Kourou, French Guiana. GEO insertion 17 October, solar drum and antennas deployed 19 October
Location: in August 1993 it was moved from 99° W to 103° W temporarily; in late 1991 it was positioned at 99° W to replace Westar 4 until Galaxy 4 appeared; previously at 91° W (replacing Westar 3); it moved from 103° W to 74° W in May 1994 to replace the retiring Galaxy 2
Design life: 10 years (hydrazine for 13.9 years remaining after Ariane launch) predicted eol 2002
Transponders: 24 (plus 6 back-up, in 5-for-4 ring) 10.2 W TWTA 5.925–6.425/3.700–4.200 GHz up/down C-band CONUS, Hawaii, central America, Caribbean beams, 36 MHz bandwidth, EIRP 35–39 dBW CONUS, 28–34 dBW Alaska, 27–30 dBW Hawaii, Caribbean/central America 30–32 dBW, Puerto Rico 28–31 dBW, orthogonal polarisation
Principal applications: system C-band back-up and occasional video services
Configuration: HS-376 bus, solar arrays and attitude control tanks enhanced
Mass: 708 kg on-orbit BOL, 1,212 kg at launch, 509 kg dry mass
Power system: as HS-376 but improved to 869 W EOL.

Galaxy 7
Launch: 28 October 1992 by Ariane 4 from Kourou, French Guiana
Location: 91° W GEO (replacing SBS 4)
Design life: 15 years but with sufficient propellant for 13.5 years operations
Mass: 2,968 kg at launch, 1,680 kg BOL, 1,303 kg dry
Other specifications are as Galaxy 4.

Galaxy 8 (IIIR)
Launch: 15 December 1995 by Atlas 2AS from CCAFS
Location: 95° W
Design life: 12 years
Transponders: 24 × 16 W C-band for cable and broadcast TV; 24 × 63 W Ku-band for data, voice and video
Other specifications as Galaxy 4.

Galaxy 8I
Launch: 8 December 1997 by Atlas 2AS from CCAFS
Location: 280° east
Design life: 12 years but predicted eol late 2002
Dedicated Latin American service: (I for International) replacing Galaxy 8 (IIIR).

PanAmSat's Atlanta, Georgia, US, Operations Centre 0084759

Hughes technicians package an HS 601 for PanAmSat 0084769

A 601HP, PAS 10 series Asia, Africa, the Middle East and Europe 0132819

Galaxy 9
Launch: 24 May 1996 by Delta 7925 from Cape Canaveral
Location: 123° W initially (collocated with SBS 5), then 127° W geostationary
Design life: predicted eol 2008
Transponders: 24 (plus 6 back-up; 30-for-24 ring redundancy) 16W TWTA 5.925–6.425/3.700–4.200 GHz up/down C-band CONUS/Alaska/Hawaii beams, 24/36 MHz bandwidth, EIRP 40/37.5 dBW peak/min over CONUS, orthogonal polarisation
Configuration: stretched HS-376, 3.10/7.49 m height stowed/deployed
Mass: 1,397 kg at launch
Other specifications as for Galaxy 5.

Galaxy 10
Launch: 27 August 1998 on Delta 3 from Cape Canaveral, launch vehicle failure resulted in destruction of the satellite
Location: planned for 237° E geostationary (replacing SBS 5 and Galaxy 9)
Design life: 15 years
Transponders: Same as Galaxy 4 except power increased for 24 × 20 W C-band + 24 × 50 W Ku-band
Configuration: HS-702 platform. GaAs solar arrays span 27 m, providing 10 kW BOL. Ni/H_2 batteries provide 100 per cent eclipse protection. AOCS by bipropellant (1,700 kg NTO/MMH) thrusters incorporating onboard control processor, Earth sensors and MWs. Hughes' XIPS xenon ion propulsion subsystem for NSSK: four 25 cm thrusters (2 primary and 2 backup) consume 5 kg/years
Other specifications as for Galaxy 4.
Comments: The contract for this first HS-702 model includes the Galaxy 11 ground spare, which served initially as the ground spare but was launched.

Galaxy 10R
Launched: 24 January 2000 by Ariane 4 from Kourou, French Guiana
Location: 123° W

Final alignments in the Mass Properties Laboratories for PAS 10, a 601HP series satellite 0132821

Galaxy 3C on its way to a planned launch in May 2002 0132822

Design life: 15 years
Transponder: Same as Galaxy 10
Configuration: HS601HP
Coverage: North America.

Galaxy 11
Launch: 22 December 1999 on Ariane 44L (V125) from Kourou, French Guiana.
Location: 99° W initially then 91° W operationally.
Transponders: 24 × 20 W C-band (6 spare); 24 × 75 W Ku-band (6 spare); 16 × 140 W Ku-band (4 spare).
Configuration: HS-702 platform. GaAs solar arrays span 31 m provide 10.4 kW EOL; 42 cell Ni/H_2 batteries for eclipse protection. Liquid apogee motor (490 W); AOCS by bi-propellent thrusters (4 × 22 N; 4 × 10 W); Hughes Xenon is a propulsion subsystem for NSSK (4 × 0.08 W for transfer orbit; 4 × 0.08 N for altitude control.
Antennas: 2 × 2.4 m/2 × 1.8 m dual-gridded shaped reflections.
Size: Stowed: h 6.2 m × w 3.8 m by 3.3 m; in orbit h 31 m, w 9.0 m; mass 4,488 kg/2,775 kg launch/BOL
Coverage: North America; Brazil.

Galaxy 12
Launch: 8 April 2003 on Ariane 5 from Kourou, Fr Guiana
Location: 74° W geostationary
Design life: 15 years
Configuration: Orbital Star 2 (Orbital Sciences Corporation) platform; power 2.64 kW EOL; tri-axis, zero momentum stabilisation
Transponders: two groups of 16:12 LTWTA (37 W) in C-band
Dimensions: 1.5 m × 1.9 m × 3.3 m stowed
Mass: 1,700 kg (launch)
Coverage: CONUS, Alaska and Hawaii.

Galaxy 13/Horizons 1
Launch: 30 September 2003 on Zenit from Sea Launch platform
Location: 127° W geostationary
Design life: 15 years
Transponders: 24 × 108 W Ku-band (+8 spare); 24 × 40 W C-band (+8 spare)
Configuration: BSS-601HP platform; 2 × four-panel GaAs dual junction solar cell arrays 9.9 kW BOL, 8.6 kW EOL plus 30-cell NiH batteries; one 490 N liquid apogee motor; AOSC by 4 × 13 cm (0.018 N) North-South primary (xenon backup) thrusters; 12 × 10 N thrusters (4 × north-south backup, 4 × east-west, 4 × aft).
Antennas: two shaped Gregorian transmit/receive each 42 cm antennas; two dual gridded surface shaped transmit/receive antennas each 20 cm.
Size: length 26.2 m across solar arrays; width 7 m across antennas; stowed 5.7 m × 2.7 m
Mass: 4,060 kg at launch; 2,630 kg BOL
Coverage: USA and Asia.

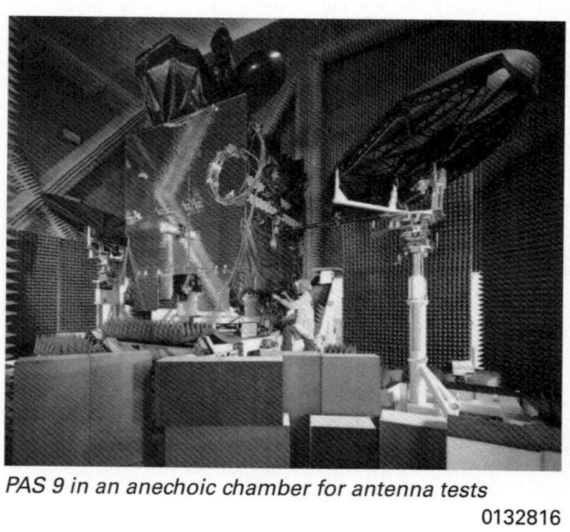

PAS 9 in an anechoic chamber for antenna tests 0132816

PAS 1
Launched: 15 June 1988 by Ariane 401 from Kourou, French Guiana
Location: 45° W geostationary
Design life: 11 years (sufficient propellant for 13¼ years)
Contractors: Lockheed Martin Astro Space
Transponders: 12 (plus 2 back-up) 8.5 W 36 MHz bandwidth SSPA and six (plus 1 back-up) 16.2 W 72 MHz bandwidth TWTA 5.92–6.42/3.7–4.2 GHz up/down C-band global and spot beams for US/Latin America and domestic Latin America; 33–35 dBW EIRP for 36 MHz, 37–40 dBW for 72 MHz, linear orthogonal polarisation
Six (plus 1 back-up) 16.2 W TWTA 14–14.5/11.5–12.2 GHz up/down Ku-band spot beams, 72 MHz bandwidth, 43–4 5 dBW EIRP
Principal applications: TV, voice, data, telex, facsimile
Configuration: Lockheed Martin Astro Space 3000 bus, 1.63 × 1.32 × 1.9 m box-shaped bus, 15.8 m across solar wings. 3-axis stabilised by hydrazine RCS + MW assemblies. GEO injection by solid AKM
Mass: 1,220 kg at launch, 708 kg on-station BOL
TT&C: through Astro Space's Satellite Operations Control Centre at East Windsor, New Jersey. Station-keeping to within 0.1° longitude
Power system: two identical solar wings providing 1,400 W BOL/1,200 W EOL.

PAS 1R
Launch: 15 November 2000 by Ariane 5 from Kourou, French Guiana
Location: 45° W GEO
Design life: 15 years minimum
Contractor: Boeing Space Systems (prime)
Transponders: 12 (plus 3 spare) 34 W SSPA, C-band; 24 (plus 6 spare) 55 W TWTA C-band; 24 (plus 6 spare) 125 W TWTA ku-band; 12 (plus 4 spare) 140 W TWTA ku-band
Configuration: Boeing Space Systems 702 bus, 3-axis, colour array span 40.8 m, antenna width 8.2 m. Stowed dimensions 6.2 m × 3.8 m × 3.4 m. Antennas: 2 × 37 cm and 2 × 27.5 cm dual-gridded shaped reflectors
Mass: 4,792 kg at launch; 3,059 kg bol
AOCS: 3-axis, 1 × 490N liquid apogee motor; 4 × 0.165 N transfer orbit XIPS; 4 × 0.08 N attitude control XIPS
Power system: twin solar arrays each with 5 panels of dual-junction gallium – arsenide cells; 54-cell NiH, 328-Ahr, battery for 100% eclipse protection. System provides 14.3 kW EOL
Coverage: Americas, Caribbean, Europe, Africa.

PAS 2
Launch: 8 July 1994 by Ariane 4 from Kourou, French Guiana
Location: 169° E geostationary
Design life: 15 years min (Ariane accuracy left propellant sufficient for 16½ years) predicted eol 2008
Contractors: Hughes Space & Communications Co (prime)
Transponders: 16 (plus 4 back-up) 30 W SSPA two C-band 5.925–6.425/3.70–4.20 GHz up/down Pacific beams (Australasia, Northeast Asia, China; Western US/Hawaii), 54 + 64 MHz bandwidth, EIRP 33–38 dBW for ≥2.4m Earth stations, orthogonal linear polarisation. 8 transponders can be cross-strapped C-Ku
16 (plus 4 back-up in 20-for-16) 63 W TWTA three 14.00–14.50/12.250–12.750 GHz up/down Ku-band Pacific beams (Australia/New Zealand; Northeast Asia/China; China/Japan), 54 + 64 MHz bandwidth, EIRP 44–51 dBW for ≥90 cm Earth stations, orthogonal linear polarisation. 8 transponders can be cross-strapped Ku-C
Principal applications: TV distribution and business communications
Configuration: HS-601 platform, 3-axis, solar array span 26.2 m (satellite within 2.7 × 3.8 × 4 m envelope at launch). Payload mounted on north/south panels and Earth-facing floor; east/west faces carry two dual-gridded rectangular antennas, Earth-facing wall carries two dual-gridded dishes
Mass: 2,920 kg at launch, 1,802 kg on-station BOL, 1,291 kg dry
AOCS: 3-axis, unified 12 ARC 22 N and one Marquardt 490 N bipropellant thrusters incorporating onboard control processor, Sun and Barnes Earth sensors and two 61 Nms 2-axis gimbaled momentum bias

With a total of 72 transponders, PAS IR launched in 1997 was the second of the powerful 702 series platforms
0132817

wheels. NTO/MMH in four spheres. Spin stabilised in transfer orbit
TT&C: controlled by Hughes Communications Inc from its OCC Operations Control Center in El Segundo, California via its Fillmore, California station
Power system: twin solar wings (spanning 26.2 m) of four 2.16 × 2.54 m panels carrying K43/4 large area Si cells on Kevlar substrate provide 4.225 kW EOL. 100% eclipse protection by Ni/H_2.
Coverage: Asia-Pacific

PAS 3
Launch: 12 January 1996 by Ariane 4 from Kourou, French Guiana
Location: 43° W geostationary
Transponders: Same as PAS 2 but two C-band beams cover Americas and Africa/Europe and five Ku-band US, Canada, Europe and South America
TT&C: controlled by Hughes Communications Inc from its OCC Operations Control Center in El Segundo, California via its Spring Creek, New York station
Mass: 2,918 kg at launch, 1,753 kg on-station BOL, 1,318 kg dry
Other specifications as for PAS 2
Coverage: Americas, Caribbean, Europe, Africa.

PAS 4
Launch: 3 August 1995 by Ariane 4 from Kourou, French Guiana
Location: 72° east geostationary
Design life: predicted eol 2011
Transponders: C-band as PAS 2/3 but three beams cover Africa; Middle East/India; Asia
24 (plus 4 back-up) 60 W TWTA four 14.00–14.50/12.250–12.750 GHz up/down Ku-band Pacific beams (southern Africa; Northeast Asia; Europe/Eastern Russia; Middle East/India), 16 × 27 + 8 × 54/64 MHz bandwidth, EIRP 44–53 dBW for ≥90 cm Earth stations, orthogonal linear polarisation. 8 transponders can be cross-strapped Ku-C
Mass: 3,043 kg at launch, 1,868 kg on-station BOL
Other specifications as for PAS 2
Coverage: Asia, Africa, Middle East, Europe.

PAS 5
Launch: 28 August 1997 by Proton-K from Tyuratam, Kazakhstan
Location: 58° W geostationary
Design life: 15 years min predicted eol 2012
Transponders: 24 (plus 4 back-up in 28-for-24) 50 W TWTA two C-band 5.925–6.425/3.70–4.20 GHz up/down North/South American beams, 36 MHz bandwidth, orthogonal linear polarisation
24 (plus 4 back-up in 28-for-24) 6 × 60 W + 18 × 110 W TWTA 14.00–14.50/12.250–12.750 GHz up/down Ku-band North/South American beams, 36 MHz bandwidth
Principal applications: TV distribution (including Latin America DTH) and business communications
Configuration: improved HS-601 HP (High Power) platform, featuring GaAs solar arrays, enlarged payload and lightweight contoured reflectors
Mass: 3,720 kg at launch
AOCS: Hughes' 68 kg XIPS xenon ion propulsion system will provide NSSK: two (+2 back-up) 18 mN 13 cm thrusters
TT&C: controlled by Hughes Communications Inc from its OCC Operations Control Center in El Segundo, California via its Fillmore, California station
Power system: twin solar wings of four panels providing 8 kW BOL (7,075 W total required); 100 per cent eclipse protection.

PAS 6
Launch: 8 August 1997 on Ariane 44LP from Kourou, French Guiana
Location: 316.5° east geostationary (collocated with PAS 3R)
Design life: 15 years minimum
Contractors: Space Systems/Loral (prime)
Transponders: 36 (plus 12 back-up, in 12-for-9) 100 W TWTA Ku-band South American beams, 36 MHz bandwidth, linear orthogonal polarisation
Principal applications: DTH Latin America
Configuration: 3-axis stabilised SS/L FS-1300 2.41 × 2.58 × 2.20 m box-shaped bus with twin solar wings on North/South sides, 27.3 m span; 2.4 m diameter Ku-band reflectors on East/West faces plus 1.2 m receive on Earth face
Mass: about 3,420 kg at launch, 1,285 kg dry
AOCS: 3-axis bias momentum system; single Marquardt R-4D 490 N apogee thruster and two sets of six 22 N RCS thrusters, common NTO/MMH propellant system
Antenna pointing: 0.09° azimuth/0.08° elevation Ku-band
TT&C: under development
Power system: twin 4-panel GaAs wings providing about 8 kW BOL. Ni/H_2 battery provides 100 per cent eclipse protection
Coverage: South America.

PAS 6B
Launch: 21 December 1998 by Ariane 4 from Kourou, French Guiana
Location: 43° W GEO
Design life: 15 years
Contractor: Boeing Space System
Transponders: 32 × 36 mHz, 105–140 W TWTA
Configuration: Boeing 601 HP
Coverage: South America.

PAS 7
Launch: 16 September 1998 on Ariane 44LP from Kourou, French Guiana
Location: 68.5° E geostationary (collocated with PAS 4)
Design life: 15 years minimum
Transponders: 14 (plus 4 back-up in 9-for-7) 50 W TWTA 6.425–6.725/3.40–3.70 GHz up/down C-band beams, 36 MHz bandwidth, maximum EIRP 39 dBW, orthogonal linear polarisation
30 (plus 10 back-up in 40-for-30) 100 W TWTA 13.75–14.25/10.95–11.70 GHz up/down Ku-band, 36 MHz bandwidth, max EIRP 53 dBW
Principal applications: telecom + DTH
Configuration: 3-axis stabilised SS/L FS-1300 2.41 × 2.58 × 2.20 m box-shaped bus with twin solar wings on N/south sides, 27.3 m span; 1.8 m diameter C/Ku-band reflectors on east/W faces plus 1.2 m receive on Earth face
Mass: about 3,838 kg at launch, 2,118 kg BOL, 1,466 kg dry
AOCS: 3-axis bias momentum system; single Marquardt R-4D 490 N apogee thruster and two sets of six 22 N RCS thrusters, common NTO/MMH propellant system

An engineer attaches a cable to the Galaxy 10R, an HS601HP, prior to thermal testing 0132818

Antenna pointing: 0.09° azimuth/0.08° elevation Ku-band
TT&C: location to be determined
Power system: twin 4-panel GaAs wings providing about 9 kW BOL. Ni/H_2 battery provides 100 per cent eclipse protection
Coverage: Asia, Africa, Middle East, Europe.

PAS 8
Launch: 4 November 1998 on Proton-K from Tyuratam, Kazakhstan
Location: 166° E geostationary
Design life: 15 years minimum
Transponders: 24 (plus 4 back-up in 28-for-24) 50 W TWTA 5.925–6.425/3.70–4.20 GHz up/down C-band beams, 36 MHz bandwidth, orthogonal linear polarisation
24 (plus 8 back-up in 32-for-24) 100 W TWTA 13.75–14.00/12.25–12.75 GHz up/down Ku-band, 36 MHz bandwidth
Principal applications: telecom + DTH
Configuration: 3-axis stabilised SS/L FS-1300 2.41 × 2.58 × 2.20 m box-shaped bus with twin solar wings on N/south sides, 27.3 m span; 2.4 m diameter C/Ku-band reflectors on East/West faces plus 1.2 m receive on Earth face
Mass: about 3,800 kg at launch, 2,100 kg BOL; 1,466 kg dry
AOCS: 3-axis bias momentum system; single Marquardt R-4D 490 N apogee thruster and two sets of six 22 N RCS thrusters, common NTO/MMH propellant system
Antenna pointing: 0.09° azimuth/0.08° elevation Ku-band
TT&C: location to be determined
Power system: twin 4-panel GaAs wings providing about 9 kW BOL. Ni/H_2 battery provides 100 per cent eclipse protection
Coverage: Asia-Pacific.

PAS 9
Launch: 28 July 2000 by Sea Launch
Location: 58° W GEO replacing PAS-5
Design life: 15 years
Contractor: Boeing Space Systems (formerly HSC)
Transponders: 24 × 55 W C-band TWTA; 24 × 108 W TWTA ku-band
Principal applications: C-band and Ku-band services for the Americas, the Caribbean and Wern Europe plus DTH series for Mexico in Ku-band
Configuration: 3-axis stabilised 601 HP bus with twin solar arrays spanning 26 m. Antennas span 7 m.
Stowed size: L 4 × 2.7 × 3.6 m
Mass: 3,659 kg at launch; 2,389 kg at BOL
AOCS: 1 × 445 N liquid apogee engine; 1 × 13 cm, 0.017 N N-S primary xenon in thruster; 4 × 10 N N-S backup bi-propellant thrusters; 4 × 10 N E-W bi-propellant thrusters; 4 × 10 N aft bi-propellant thruster
Power system: twin 4-panel dual-junction gallium arsenide solar cell arrays, 30-cell NiH, 350 Ahr, battery.
Total power levels: 9.9 kW BOL, 8.9 kW EOL
Coverage: Americas, Caribbean, Europe.

PAS 10
Launch: 14 May 2001 by Proton
Location: 68.5° E
Design life: 15 years
Contractor: Boeing Space Systems (formerly HSC)

Vibration tests on Galaxy 4R a HS 601HP launched in 2000 0132820

Transponders: 24 (+8 spare) × 55 W C-band TWTA; 24 (+8 spare) × 98 W Ku-band.
Principal applications: DTH, internet, video and data broadcasting services
Configuration: 3-axis stabilised 601 HP with twin solar arrays spanning 26 m. Antennas span 7 m
Stowed size: H 4 m, W 2.7 × 3.6 m
Mass: 3,739 kg launch, 2,510 kg BOL
AOCS: 1 × 490 N liquid apogee motor; 1 × 13 cm, 0.017 N N-S primary xenon ion thrusters; 4 × 10 N N-S lockup bipropellant thrusters; 4 × 10 N E-W bipropellant thrusters; 4 × 10 N aft bipropellant thrusters
Power system: twin 4 panel dual-junction GaAs solar cell arrays, 30 cell NiH 350 A/h battery, 9.6 kW BOL, 8.5 kW EOL
Coverage: Asia, Africa, Middle East, Europe.

SBS-6
Launch: 12 October 1990 by Ariane from Kourou, Fr Guiana
Location: 74° W longitude.
Design life: 10 years
Configuration: BSS-393 platform; solar cells on telescoping cylindrical sections telescoped for launch and deployed on orbit delivering ~2 kW and with Ni-H batteries for eclipse coverage; 2 × 22.2 N axial and 4 × 22 N radial AOCS thrusters; 2 × 490 N thrusters for apogee manoeuvres and longitudinal transfer.
Transponders: 30 × 43 MHz (41 W) TWTA (inc 11 spares) Ku-band.
Mass: 2,445 kg at launch, 1,481 kg BOL.

UPDATED

SES AMERICOM

Current Status

On 12 November 2001 GE Americom was combined with SES Astra to form a new company SES Global, SA. This new company has a fleet of 29 satellites which carry new names as identified in this entry. AMC-15 is scheduled for launch by Proton M in third quarter 2004 featuring 24 Ku-band transponders operating from 105° W.

Background

RCA spun off its wholly-owned subsidiary RCA Americom in 1976 to own and operate the parent company's domestic satellite system known as Satcom. Following GE's 1987 acquisition of RCA, a new corporate entity GE Americom took over the company as part of GE's Communications & Services Group. In the days when it was still RCA, the company launched the original Satcom 1, on 13 December 1975 and began the Satcom network.

Americom's GE-1 hybrid
(Lockheed Martin Astro Space) 0517020

Satcom 2, launched 26 March 1976, continued the series until Satcom 2R replaced it. Satcom 2R subsequently retired from service 28 February 1995 from 72° W. Satcom 3 foundered following a bad firing of its apogee kick motor 7 December 1979; but 3R quickly occupied its slot. Launched 20 November 1981, 3R retired from 131° W in 1991. Satcom 4 and 5 completed the original C-band system in 1982. Then Alascom bought Satcom 5 and renamed it Aurora 1, working until 1991.

Satcom 5 was the first C-band spacecraft to employ solid state power amplifiers in its 24 active transponders. At the time, AT&T Alascom existed as a subsidiary of Pacific Telecom, until AT&T acquired it in 1995. Aurora 1 transferred hands in the acquisition. The Aurora carries domestic telephone, the Alaskan Television Network and emergency messages. As part of the new structure, Alascom bought the satellite for US$88 million and leased back four transponders to RCA.

Aurora served mostly the remote reaches of Alaska, until its customers had to be moved to Satcom C-1 in late 1990, when Aurora suffered service interruptions because of attitude control problems. Eventually Aurora's aging systems forced its retirement in 1991. The satellite now sits in a 40° inclination orbit at 105° west. GE Aerospace launched Aurora 2 (Satcom C5) in 1991 to replace Aurora 1. The new satellite incorporates 24 active 11 W C-band solid-state power amplifiers. GE Americom continues to provide operational control.

Satcom C-1, launched November 1990, began the next generation of communication satellites. Three replacement 3000-class C-band satellites appeared 1990–92. Satcom C-5 replaced C in May 1991. C-4 launched August 1992 to 135° W took traffic from Satcom 4R at 82° W. C-3 appeared in September 1992 to replace the retiring 1R at 131° W.

Americom's Satcom 4, launched 16 January 1982, failed in October 1991, but Telesat Canada's Anik D2 launched November 1984, leased by Americom under a 1990 agreement, stepped in as a temporary measure (designated Satcom 4R) until C-4 appeared and arrived at 82° W in December 1991.

Then the Satcom K generation introduced by GE Americom established a high-power Ku-band data capacity. The Shuttle launched both Satcom K-1 and K-2 in 1985 and 1986, respectively. Crimson Satellite Associates, a GE Americom/Home Box Office partnership planned to operate Satcoms K-3 and K-4. The company later dissolved, but before that SES purchased K-3 in 1989 for use as Astra 1B. Intelsat bought K-4 the same year to become Intelsat K.

The new GE Americom ordered its first two hybrid satellites from Lockheed Martin Astro Space for debut in 1996, to supply more capacity. Half the payload of Sweden's 1997 Sirius 2 was bought in mid-1995 for the 16 transponders to provide pan-European services.

As a transition, Americom bought all of Spacenet 2's transponders from GTE Spacenet. The WARC originally authorised 85° W for GE-1 (later Gstar-1), where Alaska and Hawaii coverage would have been poor. Americom and Spacenet signed an agreement in 1993 to share the 103° W slot, GStar 1's current position. GE-2, originally slated to remain on the ground as a spare became a firm order when GE's demand grew beyond expectations.

GStar 2 and 4 swapped positions in 1992. Gstar 2 resides next to Hughes' Galaxy 5 for home dish customers. It has combination C-/Ku-antennas and can receive either without repositioning. GStar 2 remained in GEO until June 1995, when GE began operating it in an inclined orbit. After Gstar 3's 55 second GEO insertion burn by the Thiokol Star 30B AKM placed it in a 1.5° inclusive elliptical orbit below GEO, Americom wrote off the satellite as lost. The failure occurred because of a static imbalance in the hydrazine tanks following incorrect ground loading. GStar 3 became a declared US$77 million insurance write-off including US$5 million for Geostar's navigation package.

Controllers, cycling the NSSK thrusters, managed to nurse Gstar 3 all the way to GEO on 15 November 1989 with sufficient propellant for a 7 year life. GTE, the owner, requested FCC permission to operate the satellite from the more accessible 93° W, leaving the planned 125° W to GStar 4. Because GTE wanted to conserve station-keeping propellant, it allowed the satellite to drift in inclination. This inclination required the development and manufacture of a VSAT/TVRO tracking antenna.

In early 1998 GE decided to redirect the footprint of the GE-1A and GE-2A satellites to provide better coverage of China, Taiwan and northeast Asia.

Specifications

Aurora 2 (Satcom C-5)
Launched: 29 May 1991 by Delta 7925 from Cape Canaveral
Location: 139° W geostationary (now 137° west)

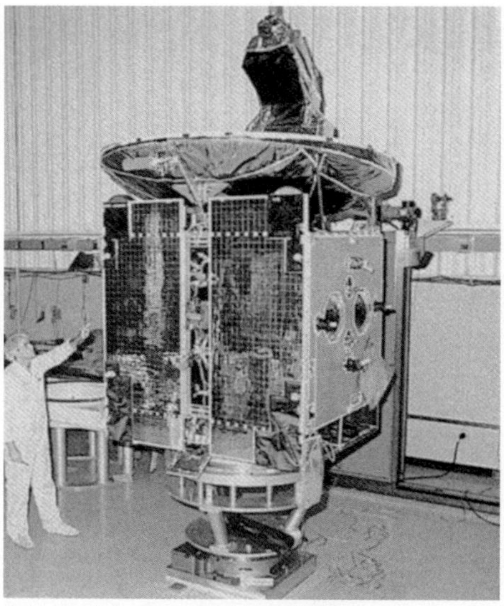

Satcom C-1 appeared in 1990 as the first of GE Americom's C-band replacements (MM Astro Space) 0517021

Design life: 12 years
Contractors: GE Astro Space Division
Transponders: 24 (plus 8 back-up in 8-for-6 configuration) 11 W SSPA 5.925–6.425/3.70–4.20 GHz up/down C-band CONUS/Alaska/Puerto Rico/Virgin Islands beams, 36 MHz bandwidth, edge EIRP 34–39 dBW Alaska, 34 dBW CONUS, 30 dBW Puerto Rico, 28 dBW Virgin Islands, orthogonal linear polarisation
Principal users: telephony, TV networking, data transmit
Configuration: GE Astro Space 3000-series platform. 3-axis 163 × 132- × 99-cm box-shaped platform with 14.33 m solar wing span. Electronics units, batteries, propulsion and AOCS equipment are mounted on six aluminium honeycomb structural pallets or panels. Transponders and housekeeping elements are mounted on four panels, two each on the North/South sides. Additional housekeeping equipment is mounted on the anti-Earth panel
Mass: 1,338 kg launch, 729 kg on-station BOL
AOCS: 3-axis by 6,000 rpm MW (pitch) and magnetorquing (roll/yaw) controlled by Earth Sensor Assembly error signals. About 180 kg hydrazine for 12 Reaction Engine Assemblies + two electrothermal thrusters. Station-keeping accuracy is ±0.1°. GEO insertion by a Thiokol Star 30B solid AKM
Pointing accuracy: 0.1° (±6.0° pitch offset pointing control)
TT&C: through GE Americom's stations at Vernon Valley, NJ + South Mountain, California
Power system: twin Si wings of three panels each, totalling 12.9 m^2, two 40 Ah Ni/H_2 batteries.

GE satellites
Launch: Americom 1 (GE-1) GE-1, 8 September 1996 by Atlas-2A from Cape Canaveral; Americom 2 (GE-2), 30 January 1997 by Ariane-4 from Kourou, French Guiana; Americom 3 (GE-3) 4 September 1997 by Atlas 2AS from Cape Canaveral; Sirius 2 (GE-1E) 12 November 1997; Americom 4 (GE-4), 13 November 1999 by Ariane 44LP from Kourou, French Guiana; Americom 5 (GE-5), 28 October 1998 by Ariane 44L from Kourou, French Guiana; Americom 6 (GE-6), 21 October 2000 by Proton-K from Baikonur; Americom 7 (GE-7) on 14 September 2000; AAP-1 (GE-1A) on 1 October 2000 Americom 8 (GE-8), 19 December 2000.

Deployment of Satcom K-2 from Shuttle Columbia's cargo bay, November 1985. The attached PAM-D2 kick motor provided GTO insertion 0517022

Location: Americom 1 103° W, Americom 2 85° W, Americom 3 87° W, Americom 4 101° W, Americom 5 79° W, Americom 6 72° W, Americom 7 137° W, Americom 8 139° W, APP-1 108.2° E, Sirius 2 4.8° E
Design life: 15 years
Contractors: Lockheed Martin Astro Space (prime)
Transponders: 24 (plus 8 back-up; 16-for-12) 12–18 W variable SSPA 5.925–6.425/3.7–4.2 GHz up/down C-band CONUS/Alaska/Hawaii beams, 36 MHz bandwidth, edge EIRP 37 dBW CONUS, 29 dBW Alaska, 32 dBW Hawaii, 33 dBW Caribbean, linear orthogonal polarisation

24 (plus 12 back-up; 18-for-12) 60 W TWTA 14.0–14.47/11.73–12.17 GHz up/down Ku-band CONUS + Hawaii beams, 36 MHz bandwidth, edge EIRP 45 dBW CONUS, 42 dBW Hawaii beam, orthogonal polarisation
Principal users: cable, broadcast, education, government and business
Configuration: A2100A platform, 3-axis stabilisation
Mass: 2,585 kg at launch, 1,575 kg on-station BOL,
AOCS: 3-axis by MW, magnetorquing, arcjets and hydrazine thrusters. Arcjets and thrusters divided into two redundant sets
TT&C: as Satcom C-1
Power system: twin 4-panel solar wings providing 5,000 W EOL; 2 × 100 Ah Ni/H_2 batteries.

Gstar 4
Launched: 20 November 1990 by Ariane 42P from Kourou, French Guiana
Location: 105° W GEO (swapped 1992 with GStar 2)
Design life: 10 years, but station-keeping propellant sufficient for >13 years projected
Transponders: Same as GStar 1 but no spot beams and additional 27 W transponder replaces one 20 W unit
Principal applications: one-third leased, one-third data/voice services, one-third SNG + business TV
Configuration: as GStar 1 but lighter structure as not intended for Shuttle launch
Mass: 1,295 kg at launch, 741 kg on-station BOL (543 kg dry; 181 kg hydrazine)
Power: as GStar 1.

Satcom C-1
Launched: 20 November 1990 by Ariane 4 from Kourou, French Guiana
Location: 179° W geostationary, providing in-orbit protection for C-band fleet (polarising switching capability allows operations from other locations to restore interruptions in multipoint services)
Design life: 12 years minimum
Contractors: Martin Marietta Astro Space
Transponders: 24 (plus 4 back-up; 7-for-6) 9.6 W SSPA 5.925–6.425/3.7–4.2 GHz up/down C-band CONUS/Alaska/Hawaii beams, 36 MHz bandwidth, edge EIRP 33 dBW CONUS, 32 dBW Alaska, 29 dBW Hawaii, linear orthogonal polarisation (odds horizontal down)
Principal users: cable programming, NBC TV and in-orbit protection
Configuration: RCA Advanced Satcom series platform. 3-axis 163 × 132 × 99 cm box-shaped platform with 14.33 m solar array span. Electronics units, batteries, propulsion, AOCS equipment mounted on six aluminium honeycomb structural pallets or panels. Transponders and housekeeping elements mounted on four panels, two each on North/South sides. Additional housekeeping equipment is mounted on the anti-Earth panel
Mass: 1,169 kg at launch, 682 kg on-station BOL
AOCS: 3-axis by 6,000 rpm MW (pitch) and magnetorquing (roll/yaw) controlled by Earth Sensor Assembly error signals. Blowdown mono hydrazine propulsion: four surface tension tanks manifolded to two sets of seven redundant thrusters. GEO insertion by a Thiokol Star 30B solid AKM
TT&C: through the Astro Satellite Operations Control Center at East Windsor, New Jersey for launch and transfer orbit phases; Americom's own stations at Vernon Valley New Jersey, South Mountain California, Goddard, Maryland, Woodbine, Maryland and Grand Junction, California for operational control
Power system: twin Si wings of three panels each provide about 1,500 W BOL/1,230 W EOL, supported by 3 × 24 Ah Ni/Cd batteries for eclipse operation.

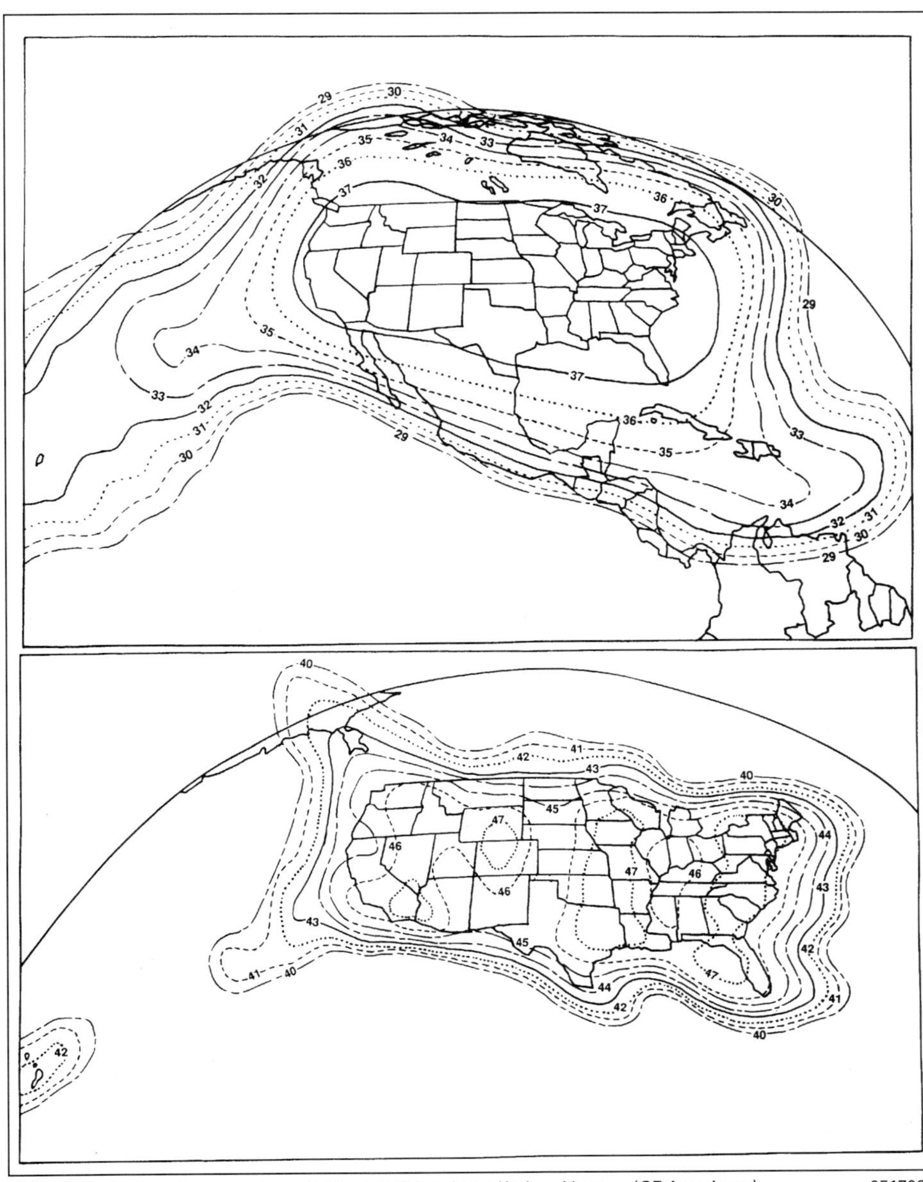

GE-1 EIRP (dBW) performance from 103° west. C-band top; Ku-band bottom (GE Americom) 0517023

Satcom C-4
Launch: 31 August 1992 by Delta 7925 from Cape Canaveral
Location: 135° W geostationary
Transponders: 24 (plus 8 back-up; 8-for-6) 17.5 W TWTA 5.925–6.425/3.7–4.2 GHz up/down C-band CONUS/Alaska/Hawaii beams, 36 MHz bandwidth, edge EIRP 39 dBW CONUS (40 dBW peak), 27 dBW Alaska, 31 dBW Hawaii, linear orthogonal polarisation (evens horizontal down)
Principal users: cable programming
Mass: 1,402 kg at launch, 791 kg on-station BOL
Power: 1,950 W BOL; 2 × 50 Ah Ni/Cd batteries
Other specifications as for Satcom C-1.

Satcom C-3
Launch: 10 September 1992 by Ariane 4 from Kourou, French Guiana
Location: 131° W geostationary
Mass: 1,375 kg at launch, 789 kg on-station BOL
Other specifications as for Satcom C-4.

Satcom K-2
Launched: 27 November 1985 by Shuttle as Morelos 2
Location: 85° W geostationary
Design life: 10 years
Transponders: 16 (plus 4 back-up) 47 W TWTA 14.0–14.47/11.73–12.17 GHz up/down Ku-band CONUS + east/west US beams, 54 MHz bandwidth, edge EIRP 43 dBW CONUS, 48 dBW West beam, orthogonal polarisation
Principal users: PrimeStar Inc, business video, occasional video and data networks
Configuration: MMAS 4000 series platform, 170 × 213 × 152 cm body, 19.3 m span across solar arrays, total height 305.8 cm at launch, 3-axis stabilisation. Equipment mounted on eight aluminium honeycomb structural pallets or panels; transponders and some housekeeping units on north/south faces; antenna/reflectors, including primary Kevlar honeycomb reflectors, and Earth sensors, on Earth-facing panel
Mass: 1,926 kg at launch, 996 kg on-station BOL, about 780 kg EOL
AOCS: 3-axis by MW, magnetorquing and 16 hydrazine thrusters, divided into two redundant sets of eight, drawing on four spherical tanks. GEO injection by Thiokol solid; GTO injection by PAM-D2 solid
TT&C: as Satcom C-1
Power system: twin 4-panel Si solar wings providing 3.7 kW BOL, 2,490 W EOL; 3 × 50 Ah Ni/H_2 batteries.

Spacenet 4
Launch: 13 April 1991 by Delta 2 from Cape Canaveral
Location: 172° W geostationary
Design life: 10 years
Transponders: 12 (7-for-6 redundancy) 8.5 W min SSPA 5.925–6.425/3.7–4.2 GHz up/down C-band US beams, 36 MHz bandwidth, EIRP 34–37 dBW continental US, 31–33 dBW Alaska, 28–30 dBW Hawaii, 27–29 dBW Puerto Rico

6 (4-for-3 redundancy) 16.2 W min TWTA 5.925–6.425/3.7–4.2 GHz up/down C-band US beams, 72 MHz bandwidth, EIRP 37–40 dBW continental US, 32–35 dBW Alaska, 31–32 dBW Hawaii, 30–31 dBW Puerto Rico

4 16.2 W min TWTA 14.0–14.5/11.7–12.2 GHz up/down Ku-band US beams, 72 MHz bandwidth, EIRP 38–43 dBW continental US

2 30.0 W min TWTA 14.0–14.5/11.7–12.2 GHz up/down Ku-band US beams, 72 MHz bandwidth, EIRP 38–43 dBW continental US
Principal applications: private C-band data networks
Configuration: LMAS 3000 3-axis satellite. Hybrid C/Ku-band payload with dual antennas
Mass: 728 kg BOL on-station
TT&C: as Spacenet C-1
Power: 1,212 W EOL; 100 per cent eclipse protection.

515 (Interlat 5)
Launch: 27 January 1989
Specification: as Intelsat V
Location: 37.7° W.

TDRS-5
Launch: 2 August 1991
Specification: see TDRS entry
Location: 174.3° W.

TDRS-6
Launch: 13 January 1993
Specification: see TDRS entry
Location: 47° W.

Western Union (Westar) Series

Background

Western Union owned and operated the original Westar franchise and thus became the first US domestic satellite system. The Westar 4 satellite doubled the number of transponder channels available when it introduced its spin-stabilised platform with an orthogonal polarising reflector. The system relayed voice, data, video, facsimile and other services to the continental US, Hawaii, Alaska, Puerto Rico, and the Virgin Islands.

The first Westar satellite is based on Hughes' HS-333 design. Western Union inaugurated the series beginning with Westar's 1 launch 13 April 1974 to 99° W followed closely by Westar 2 on 10 October 1974 by Delta from Canaveral to 125° W. Each had a mass of 572 kg at launch, 291 kg in-orbit. Westar 3, launched 10 August 1979 by Delta 2914, operated at 91° W as the system's last HS-333 until it retired 26 January 1990. Westar 4 introduced the larger HS-376 design. Launched 26 February 1982 by Delta 3910, it operated at 99° W until retirement 5 November 1991.

Western Union's role as a satellite operator ended in February 1988 when Hughes Communication Incorporated bought all outstanding interest in Westar. HCI took over operating Westar 3–5 satellites and ownership of Westar 6S.

Westar 1 ended operational life April 1983. Westar 3, launched 10 August 1979 by Delta 2914, operated at 91° west as the system's last HS-333 until it retired in January 1990.

Westar 4 introduced the larger HS-376 design, almost tripling power from the solar array by adding an extending skirt and doubling the number of transponders. Launched on 26 February 1982 by Delta 3910, it operated at 99° west until HCI retired it in November 1991. Galaxy 6 provided temporary replacement until Galaxy 4 could be launched in 1993. Western Union launched Westar 5 8 June 1982 by Delta 3910 and then retired the satellite from 122.5° west in March 1992 after replacement by Galaxy 5. Westar 6 went into space on 4 February 1984 aboard STS-41B but the PAM-D upper stage malfunctioned and stranded the satellite in a useless 307 × 1,128 km orbit. To ensure that it did not re-enter, pending decisions about its future, Western Union fired its apogee kick motor 12 May 1984 to raise it to 1,060 × 1,666 km. The Shuttle subsequently retrieved it after its hydrazine motors lowered the orbit for rendezvous Shuttle mission 51A in November 1984.

The underwriters, represented by Merrett Syndicates of London paid NASA US$2.75 million for the retrieval, plus US$5 million to Hughes Aircraft for the necessary Shuttle equipment. Pan Am Pacific Satellite Corporation paid US$20 million for the satellite in mid-1986 to provide services to dispersed nations over the Pacific. It eventually became AsiaSat 1. Westar replaced the satellite with Westar 6S, using the US$110 million insurance, offering greater capacity but had not launched it at the time of purchase by HCI.

DATA RELAY

Brazil

SCD Series Satellite

Background
The Brazilian government launched SCD-1 on 9 February 1993 by Pegasus into 725 × 790 km, 24.97° orbit. Little equipment degradation was observed after one year of life. The 115 kg provided real-time relay of data from environmental data collection platforms using the Argos system frequency (401.65 MHz) and 401.62 MHz, which alone will handle 200 platforms. 65 DCPs were in place at end-1995; another 250 will be added during 1996, mainly for monitoring hydrographic basins and for flood control. The 1.70 m high octagonal prism carries University of Sao Paulo solar cells on the eight outer and top faces to generate 60 W average in sunlight, supported by an 8 Ah Ni/Cd battery. S-band/UHF antennas are mounted on the top and bottom.

SCD-1 monitors the condition of the Amazon River basin and the surrounding rain forests. It also monitors CO_2 and ozone levels, mainly produced from burning vegetation.

The Brazilians spun SCD at release to 120 rpm and thereafter maintained simple attitude control by a magnetorquer. The main requirement for satellite survivability is to hold the bottom face away from the Sun for heat dissipation. The satellite contains 160 telemetry channels.

SCD-1 was ready for flight in 1991, but had to wait until the Brazilian government could select a launcher. OSC's Pegasus won the US$11.5 million contract in October 1991 and it was signed 20 August 1992. The SCD-1 satellite cost US$20 million; 73 per cent of the funding was expended in Brazil.

SCD-2A was launched on 2 November 1997 by indigenously built VLS launcher but a failure during ascent destroyed the vehicle before it reached orbit. SCD-2 was launched by Pegasus 23 October 1998. The 115 kg satellite was placed in a 742 × 768 km, 25°, orbit.

France

Eole Satellite

Background
France launched the 80 kg Eole on 16 August 1971 by Scout from Wallops into 651 × 835 km, 50.2°, 99.7 min orbit. It weighed 84 kg and the country planned to use it to test a concept for data relay of the satellite interrogation of 500 meteorological balloons drifting in the atmosphere at 12,000 m. The satellite was also known as CAS 1 (Co-operative Application Satellite) and FR 2 (the second French experimental data relay satellite for meteorological purposes) and was the first launch by NASA under a co-operative agreement with CNES. The octagonally shaped satellite measured 0.71 m across the opposing corners and was 0.58 m long with eight rectangular shaped solar panels deployed 45° from the upper octagonal structure providing 20 W power. The satellite also had 15 rechargeable silver-cadmium batteries. Constant earth orientation was maintained by a deployable 10.06 m long gravity boom with attitude to within 9° of local vertical. It returned data for five months and the last balloon ceased transmitting in January 1974. It officially retired in July 1974 after a period during which the satellite tracked and received data from ocean buoys, icebergs and ships.

Germany

SAFIR Series

Background
Germany plans to use the SAFIR (Satellite for Information Relay) as a two-way data collection/distribution system using six small satellites in 640 to 800 km 62 to 98° orbits.

Germany launched the 38.5 kg SAFIR R1 core demonstration package 4 November 1994 attached to Resurs-O1 N3 into 551 × 663 km, 98.05°. The satellites launch was delayed because the lifetime of the previous Resurs-O1's went on longer than expected and deferred departure of its replacement. The satellite body houses the SAFIR electronics module, software and TM/TC box; Resurs provides the power. OHB-System operates the hub station in Bremen. SAFIR interrogates fleet stations and stores the data in its 10 Mbit solid-state memory for later transfer and/or multi-user access. In the operational system, 1.2 kg macro-stations would provide 300 bit/s and 2.4 kbit/s of data transmission.

Co-funded by DARA and OHB-System, the SAFIR 2 which draws on Bremsat technology, was ejected from Russia's Resurs-O1 N4 Earth observation satellite on 10 July 1998 for a DM200,000 fee. It was placed in a 815 × 818 km, 98.8°, 101 min orbit after launch by Zenit. Transmission rates of 300, 600, 1,200, 2,400 and 4,800 Band over two telemetry systems.

Specifications
Size: 45 × 45 × 50 cm; 60 kg
Power: 25 W orbit average, 50 W peak, from Si cells on five faces; 4 Ah Ni/Cd and 4 Ni/H_2 batteries
Stabilisation: ±5° nadir pointing accuracy using gravity gradient boom and magnetorquer, determination by magnetometer and Sun/star sensors
Data rate: 300 and 2,400 bit/s
Frequencies: users, UHF up/down; TT and C VHF up/down
Configuration: two modular compartments (lower 30 cm high; upper 20 cm high). The bottom plate carries launcher interface and two communications units. The middle plate houses the batteries, power supply unit, computer and gravity gradient boom interface. Controlled by modified Bremsat 1 OBC, based on Inmos T800 transputer with 10 Mbit mass memory. TT&C on 2 m band.

International

Data Relay Satellite

Background
In order to reduce some of the costs for building and maintaining a satellite ground station network, ESA considered a TDRS-like space-based relay satellite network and proposed to run the project in a series of phases. The first, (Phase A1), ran from the middle of 1987 to early 1988 with the work emphasising system design and ground and spaceborne system configuration. This phase confirmed that a DRS using largely available technology could be developed and deployed by 1996. Phase A2 defined the satellite configuration, and Phase B1, lasting from mid-1989 to Spring 1990 developed the designs for the selected satellite and ground segment facilities down to the subsystem and initial equipment levels.

The ESA Council then approved Phase 1 in June 1990. At their Granada meeting in November 1992, ESA also approved Phase 2 full development for 1999 launch but the project has been under review since early 1995, now that ESA has dropped Hermes and the Columbus free-flyer will not require it. Columbus free-flyer has now been cancelled in the light of ISS and launcher problems. One other possible concept now under investigation by ESA is a more cost-effective two-satellite relay system with Artemis.

Specifications
Location: planned 44° W GEO
Transponders: Silex/SKDR, except feeder link has four active transponders plus two back-ups. DRS adds an advanced service at S-band using phased arrays for multiple access with 24/48 transmission/receive elements (Artemis can handle only one user satellite at a time) and provides more data channels than on Artemis. However, a second SKDR antenna might replace this S-band phased array
Ground system: Fucino (operational control centre), Darmstadt (mission control centre), Villafranca (S-band).
Other specifications as for Artemis.

Russian Federation

Luch/Altair/SDRN/ SSRD Series

Current Status
Employed positions are 40, 45, 53, 85, 90, 96.5,103 and 140° E; 11, 14, 25 and 170° W.

Background
The Soviet Union gave notice in 1981 to the International Frequency Registration Board that it intended to establish a Satellite Data Relay Network (SDRN), similar to NASA's TDRSS. The mission requirements stemmed from a military specification but incorporated a strong civilian applications directive. Several changes in operational requirements were implemented in order to accommodate new requirements from the Ministry of Defence. Soviet officials planned to place satellites at 95° E, 16° W and 160° W between 1983 and 1985. The system would eliminate the Soviet problem of communicating with its spacecraft from the geographically limited Soviet territory. The system would feature an Eastern SDRN for communications with Salyuts and other low-orbiting spacecraft, and using frequencies of 10.82, 11.32, 13.7 and 13.52 GHz for downlink, and 14.62 and 15.05 GHz for uplink.

The first satellite in the series Kosmos 1700, launched 25 October 1985, placed a Luch at 95° E (CSDRN), relayed extensive communications with the new Mir space station in 1986, and subsequently with Soyuz-TM. In October 1986, the satellite began drifting off station. A replacement Kosmos 1897 appeared 26 November 1987. It moved to 12° E in July to August 1988 to support the Buran Shuttle mission, and then returned to 95° E in February 1989 but by the end of 1991 it had drifted to 77° east and it was abandoned.

The WSDRN position at 34.4° E has been occupied by Kosmos 2054 since mid-January 1990 (launched 27 December 1989). With both positions operational, the expanded Mir coverage reduced the requirements for Soviet communications ships. The first named Luch appeared 16 December 1994 for positioning at 95° E, while Kosmos 2054 exceeded its planned operating lifetime. Launched 11 October 1995 Luch 1 was placed at 77° east. At the end of March 1997, Luch-1 left station and in early May re-located itself close to the Kosmos 2054 longitude.

Excess Luch capacity is being leased on a commercial basis through NPO PM's Mercury subsidiary, particularly for TV in the US and Argentina. Today, Luch (Beam) is both a service and a system. Some Gorizont and Raduga satellites carry Luch transponders, usually as only one of the spacecraft's payloads. Luch has been the basis for an attempt to revitalise the system and transfer it to a new SDRN-based service.

Specifications

Locations: 16° W, 77° E, 95° E, 160° W GEO
Design life: 5 years
Contractors: NPO PM (spacecraft), NPO Radio (payload)
Transponders: TWTA 15.05/13.5 GHz up/down Ku-band 0.5° steerable, EIRP 56.2 dBW
TWTA 14.62/11 GHz up/down Ku-band 1.0° steerable, EIRP 42.3 dBWTV/34.2 dBW voice
TWTA 900/700 MHz up/down UHF 5.0° steerable, EIRP 40.7 dBW
Configuration: 16.0 m-span solar array provides 1.8 kW
Mass: 2,400 kg on-station
AOCS: 3-axis; east-west station-keeping to 0.5° by liquid thrusters.

United States

TDRS Constellation

Current Status

The first in a series of three Advanced TDRS satellites (H, I and J) was successfully launched by Atlas 2A on 30 June 2000. The second followed on 8 March 2002 but a loss of pressure in one of four onboard fuel tanks four days later prevented it from reaching GSO and it remained in GTO while contingency operations to raise it to geosynchronous orbit were developed. On 30 June 2000, Boeing reported that five of the 18 communication services on TDRS-H were performing below specification due to a specific material used in the fabrication of the phased array antenna, an anomaly corrected for TDRS I and J. Based on the Boeing Space Systems, (formerly HASC) HS 601, TDRS H uses dual telemetry and command antennas, advanced RF electronics and, uniquely in the TDRS series, on-board beam forming for the multiple access system with Ku- and Ka-band services.

Pre-Phase A studies of an advanced TDRS satellite, were conducted 1981–82 and followed by Phase A preliminary analysis between September 1987 and April 1989. Phase B definition began in August 1990 and were completed in July 1991 but a programme redirection in October 1992 preceded a contract award in February 1995 to (then) Hughes Space and Communications. Under a US$481.6 million contract BSS will provide three Advanced TDRS derived from the HS 601 bus augmenting existing S-band and Ku-band frequencies with Ka-band capability.

A Preliminary Design Review, in July 1996, preceded the Critical Design Review in June 1997. TDRS-H (TDRS 8 on orbit) was followed by TDRS-I on 8 March 2002 and TDRS-J, the last in the series, on 4 December 2002.

Background

TDRSS provides NASA with communications coverage of satellites and manned vehicles below 3,000 km, permitting links during 85 per cent of each orbit and allowing NASA to close much of its extensive ground system. NASA planned for TDRSS to be the backbone of the Space Shuttle communication system, and provide 80 per cent of the voice datalinks instead of the 15 per cent coverage previously available via ground stations.

When it first proposed the system in December 1979, NASA projected a cost of US$796 million, but by late 1985 costs had exploded to about US$3 billion. Part of the cost overrun resulted form NASA's realisation in 1978, when the satellites were already designed and partially constructed, that they might suffer radio frequency interference from Soviet radars in Eastern Europe. NASA immediately made US$70 million available to correct the problems and harden the satellites. Then NASA entertained a USAF proposal in 1981 to add a Batson 11 security system to protect against Soviet electronic interference. The air force offered the system to NASA for US$500 million for the six satellites. In the end, the air force dropped the plan, but the bureaucratic discussion delayed the programme for several months and further added to the costs.

In the first independent satellite launch from the Shuttle, the Boeing IUS failed and the satellite limped to its correct orbit six months later after 30 separate burns of its control thrusters. At that point the system was two years behind schedule and Spacelab 1, originally designed to use both TDRS 1 and 2 for full exploitation, went into space with only the replacement TDRS 1 available. Because of the delays, NASA had to reprogramme the use of data relay for the Landsat Thematic Mapper images and ironically denied the air force the use of the system until full service could be established.

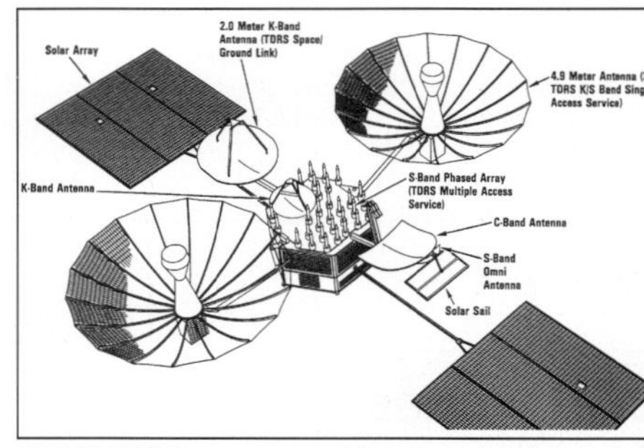

NASA's TDRS system provides communications links for LEO spacecraft over more than 85 per cent of each orbit. The K-band antenna is not used; that and the C-band payload are deleted on TDRS 7 (TRW) 0517025

Up and running, TDRS 1 operations led ground controllers to conclude a fundamental flaw in the satellite design caused several timing-oriented problems, which would have been compounded by multiple satellite operations.

Besides technical problems, TDRS suffered from financing problems as well. NASA attempted to spread the cost over 10 years of operation by leasing the satellite to itself instead of paying the purchase price. It entered into its first large-scale joint venture in the form of a leasing contract with Western Union's SpaceCom. Funding problems, including interest charges totalling US$600 million, created another string of delays. TRW originally contracted to build six satellites; two for lease by NASA, one for SpaceCom for commercial Westar users, the fourth as a joint on-orbit spare for NASA/SpaceCom, and two as flight-ready spares. Two months before the TDRS 1 launch, NASA/SpaceCom agreed on cancellation of the commercial service, making it a 100 per cent NASA system, though still owned by SpaceCom. Though increasing NASA's costs by US$216 million, this decision extended TDRSS' life in the relay role to 15 years, allowing numerous ground stations to be closed.

When the Shuttle resumed operations after the Challenger accident, NASA allotted two out of the first three missions to TDRS. TDRS 1 continued operations despite an additional loss of capacity following an antenna failure on 28 November 1986. NASA planned to replace TDRS 2, lost in the Challenger accident, at a cost of about US$250 million compared with the original cost of US$100 million. In the meantime, NASA kept six ground stations open at Guam, Hawaii, Santiago, Ascension Island, Dakar and Yarragadee, which were to have been closed after TDRS 2 became operational. These stations cost NASA US$35 million per month.

With much less demand for satellite communications during the Shuttle stand down, an international team used TDRS-1 in July and August 1986 for radio astronomy observations. In the first use of a synchronous satellite in such a role, the Jet Propulsion Laboratory linked TDRS-1 with ground radio telescopes in Australia and Japan in a Very Long Baseline Interferometry (VLBI) technique to accurately locate quasar sources.

Despite deployment difficulties with their 4.9 m antennas, TDRS 3 and 4 brought the system to full operational status in September 1989, when NASA formally accepted TDRS 4. One of TDRS 3's two 4.9 m steerable antennas become locked in

Launched by Atlas 2A on 8 March 2002, TDRS-I is the second of three data relay satellites built by BSS for NASA 0132807

Second of three advanced TDRS satellites built by Boeing Space Systems for NASA's Goddard Space Flight Centre 0132806

LHCP (Left Hand Circular Polarisation) in January 1990, preventing RHCP reception from the Shuttle's RHCP datalinks. NASA decided to shunt TDRS 3 to 174° west, augmenting it with TDRS 1 at 171° west to cover for the lost Ku-band capacity.

NASA stopped using the Shuttle as a launch vehicle, when it decided in 1991 to transfer such payloads to expendable vehicles, rather than using the costly and fragile Shuttle.

The appearance of TDRS 6 in 1993 created for the first time a full system of two operational satellites and a stored spare. The military pays some US$100 million annually for use of the system. NASA's operating costs are about US$65 million annually. The satellites work through the White Sands Complex in New Mexico, reporting to Goddard's Network Control Center (NCC). The station can handle up to 2,400 satellite passes each day at data rates up to 300 Mbit/s.

Congress halted funding for an Advanced TDRS successor in 1993. But NASA still awarded a US$481.6 million fixed price contract to Hughes for three modestly enhanced satellites. Congress has yet to approve the funding.

TDRS 7 was launched in July 1995 as a spare, the last of the first-generation type built by TRW plans.

In February 1995, NASA selected Hughes Space and Communications to build three Tracking and Data Relay Satellites (TDRS-H, = -I and J) for launch from 2000. They were each to be functionally equivalent to number 7 but adding Ka-band, to help avoid interference from commercial satellites and to offer higher data rates. Based on HSC's HS-601 bus, each will have two 4.58 m Springback steerable single access antennas. These upgrade satellites will provide simultaneous S-/Ku-band or S-/Ka-band operations: 6/0.30 Mbit/s receive/transmit S-band, 300/25 Mbit/s receive/transmission Ku-/Ka-band. An S-band phased-array will receive from five spacecraft.

Specifications

TDRS 1

Launched: 4 April 1983 by Shuttle STS-6/IUS
Location: 49° W GEO (inclination 8.7° January 1996). Replaced as TDRS-E by TDRS 4 in 1989 and moved to 79° W as system spare; in May 1990 transferred to 171° W as cover for lost TDRS 3 Ku-band capacity. Arrived 85° E 7 February 1994 (after number 6 appeared) to provide Compton Observatory link through Australia's Tidbinbilla for relay to White Sands via Intelsat; moved off-station May 1995, stopped 139° W August to December 1995 for Antarctica science operations, before arriving at 49° W June 1996
Design life: 10 years minimum, still operating in September 2000
Contractors: TRW Space & Communications (spacecraft), Harris Electronics Systems Group (ground/space antennas)
Transponders: 2 (plus 4 back-up) 26 W TWTA 2.030–2.113/2.219–2.290 GHz forward/return S-band 2.08° steerable ±22.5° E -west/±31° N-S beams working via 4.9 m antenna, 20/12 MHz forward/return bandwidth, ≤0.300/12 Mbit/s forward/return, EIRP ≥43.6 dBW min, LH/RHCP selectable

2 (plus 2 back-up) 1.5 W TWTA 13.775/15.0034 GHz forward/return Ku-band 0.28° steerable ±22.5° E-W/±31° N-S beams working via 4.9 m antenna, 50/225 MHz forward/return bandwidth, ≤25/12 Mbit/s forward/return, EIRP ≥46.5 dBW min, LH/RHCP selectable

30 3.5 W SSPA phased-array for accessing up to 20 satellites simultaneously in 13° beam. 30 elements receive: 2.2875 GHz, 5 MHz bandwidth, LHCP, ≤50 kbit/s. 8 elements (plus 4 back-up) also transmission: 2.1064 GHz, 6 MHz bandwidth, LHCP, ≤10 kbit/s, 4° beam, EIRP ≥34.0 dBW

2 (plus 4 back-up) 25 W TWTA 14.6–15.23/13.4–14.05 GHz up/down Ku-band 0.70° beams linking White Sands with TDRS, 650 MHz bandwidth, ≤300 Mbit/s, EIRP ≥50.8 dBW min, orthogonal linear polarisation

12 5 W TWTA 5.925–6.425/3.700–4.200 GHz C-band up/down beams, 36 MHz bandwidth, EIRP 32 dBW, linear polarisation
Principal users: NASA (including orbiting spacecraft), US government agencies and commercial C-band leasing
Configuration: hexagonal bus/payload modules, 17.4 m span across deployed solar wings
Antennas: two 24.3 kg steerable ±22.5° E-W/±31° N-S 4.9 m S/Ku-band single access molybdenum wire mesh dishes (that is, each accesses one target spacecraft at a time), one 2 m 15.0 kg Ku-band ground link antenna, a 30-element S-band receive phased array for accessing up to 20 satellites and 12 of those elements for S-band transmit, 1 m 5.0 kg K-band dish (not used), 1.2 × 1.6 m 13.6 kg C-band, S-band omni TT&C antenna
Mass: 2,200 kg at launch, 2,120 kg on-station BOL
AOCS: 3-axis, maintained by Earth sensors, RWs + 24 × 4.45 N thrusters (385 kg hydrazine)
TT&C: NASA's White Sands Complex
Power system: two 380 × 380 cm solar wings provide 2.4 kW BOL, 1.85 kW after 10 years (1.7 kW required), supported by 3 × 40Ah Ni/Cd batteries.

TDRS 3

Launched: 29 September 1988 by Shuttle STS-26/IUS
Location: 85° E geostationary and still operating in September 2000
Other specifications as for TDRS 1.

TDRS 4

Launch: 13 March 1989 by Shuttle STS-29/IUS
Location: 41° W GEO TDRS-E (replaced number 1)
Other specifications as for TDRS 1.
Comments: In addition to the NASA mission TDRS 4 – as well as TDRS 5 and TDRS 6 – are providing commercial C-band services.

TDRS 5

Launch: 2 August 1991 by Shuttle STS-43/IUS
Location: 174° W GEO TDRS-W (replaced number 3) and still operating in September 2000
Other specifications as for TDRS 4.

TDRS 6

Launch: 13 January 1993 by Shuttle STS-54/IUS
Location: 46° W geostationary (from November 1993) and still operating in September 2000
Other specifications as for TDRS 4.

TDRS 7

Launch: 13 July 1995 by Shuttle STS-70/IUS
Location: 150° W as reserve until May 1996: the satellite was then re-located to 172° W at the end of June 1996 and was still operating in September 2000
Mass: 2,225 kg at launch
Comments: The K-band antenna and C-band packages were deleted (the C-band antenna was replaced by a larger solar sail); SSPAs replaced the S/Ku-band TWTAs of the single-access system.
Other specifications as for TDRS 1.

TDRS 8

Launch: 30 June 2000 by Atlas 2A
Location: 150° W GEO
Mass: 1,600 kg (dry)
Comment: First TDRS built by Hughes, consists of HS-601 bus features S-band phased array antenna and two Ku-. Ka-band reflectors 4.6 m in diameter
Design life: 11 years
Configuration: HS-601 bus with 2 × 4.5 m diameter flexible graphite mesh antenna reflectors folded for launch and deployed on orbit for simultaneous transmit/receive at either S-band or combined Ku-band and Ka-band. Receive data at 300 mbits/s at Ku-band or Ka-band and 6 mbits/s at S-band. Transmit rates at 25 mbits/s for Ku-band and Ka-band or 300 kbits/s. Stored size 3.4 × 8.4 m; solar arrays deployed 21 m length, antennas 13 m across.
Mass: Launch 3,180 kg; BOL 1,781 kg; propellant 1,671 kg
Power system: 2 × solar arrays, 2.3 kW at BOL.

TDRS 9

Launch: 8 March 2002 by Atlas 2A
Location: (planned)
Specification: as TDRS 8

TDRS 10

Launch: 4 December 2002 by Atlas 2A.

UPDATED

Advanced TDRS satellite seen following assembly and prior to pre-delivery tests (Boeing) ***NEW***/1047465

EARTH OBSERVATION AND METEOROLOGICAL

Australia

ARIES 1 Satellite

Current Status
After several delays which saw the launch date slip from 2000 to 2002, the Australian government expected to launch Aries in 2004 but as of March 2003 that programme has been cancelled.

Background
In October 1996, Australia announced it would build a remote-sensing satellite, with the first launch planned for 1999. The Australian Resource Information and Environment Satellites (ARIES) project is a joint programme between CSIRO, Auspace Ltd, the Australian Centre for Remote Sensing (ACRES) and has additional support from Earth Resource Mapping PTA Ltd, Geoimage PTA Ltd and Technical and Field Surveys PTA. Ltd. The initial feasibility study was planned to take six months and cost in excess of A$1.2 million, with the Australian government sponsoring the study with a grant of A$300,000 and 10 international industry and agency sponsors have made commitments of A$75,000 each. The ARIES-1 board appointed the Macquarie Bank as financial advisers and believe that the financial package of A$70–100 million required for the launch of an operational satellite and the associated receiving and processing facilities can be achieved.

The design of the satellite is not finalised, but it is expected to have a mass of around 400 kg and would operate in a near-polar 450 km orbit.

Canada

RADARSAT

Background
After a successful launch aboard a McDonnell Douglas Delta II from Vandenberg Launch Facility in California on 4 November 1995, RADARSAT began full operations 28 March 1996.

Designed, constructed, launched, and operated by the Canadian Space Agency (CSA), RADARSAT is the world's most powerful commercial radar remote sensing satellite totally dedicated to operational applications. During its design life of five years, RADARSAT C-Band Synthetic Aperture Radar (SAR) will gather data for numerous global applications including ice surveillance, crop monitoring, disaster assessment, forest applications including clearcut monitoring, mapping, geological exploration, and coastal zone and ocean monitoring. RADARSAT's unique and state-of-the-art characteristics enable RADARSAT International, the Canadian company tasked with international marketing and distribution of RADARSAT data, to customise products to meet specific client requirements. Seven beam modes of 10 to 100 m resolution and 50 to 500 km swath widths combined with 25 beam positions ranging from 10 to 60° incidence angles mean a wide range of products are available for use in environmental monitoring and natural resource mannagement applications worldwide.

In the ScanSAR Wide beam mode, the whole of Canada can be covered every 72 hours and the Arctic daily. Imagery can be provided to high priority users electronically, such as the Canadian Ice Centre to navigate ships through the high Arctic, and within 4 hours of data acquisition. The RADARSAT system is designed to operate with no backlog, so that all imagery received within a 24 hour period can be processed and distributed (or stored) within 24 hours. For most of its life, RADARSAT will fly with the SAR antenna 'looking' to the right, providing coverage over the Arctic region. However, the spacecraft will be yawed 180° twice in its lifetime to switch the beam to the left for Antarctica mapping, during maximum and then minimum sea ice cover. The RADARSAT mission control centre is in Saint-Hubert, Quebec, Canada (near Montreal), and the principal data processing centre, operated by RADARSAT International, is in Gatineau, Quebec (near Ottawa, Ontario).

Total cost of the RADARSAT-1 programme reached C$642 million, with the Federal government contributing C$520 million, which included the C$75 million launch, the provinces C$59 million, and the private sector, namely RADARSAT International, C$63 million. NASA provided the launch in exchange for US research access to 15 per cent of RADARSAT SAR on-time and US private sector participation in distribution. The projected operations costs of RADARSAT over five years is expected to be recovered from RADARSAT International by the Canadian Space Agency in the form of 20 per cent royalty payments on the estimated C$250 million revenue. US agencies have approximately 15 per cent of observing time. In September 1999 the primary pitch momentum wheel suffered from excess friction and control was shifted to a back-up wheel. This suffered similar problems and was taken off-line on 27 November 2002 leaving the satellite safe but tumbling. Service restored in December.

The next satellite in the series, RADARSAT-2, has been contracted to McDonald Dettwiler and is scheduled for launch late in 2005.

Specifications
RADARSAT 1 Satellite
Launch date: 4 November 1995 by Delta 2
Contractors: Spar Aerospace (prime), Ball Aerospace Systems Division (bus), CAL Corp. (slotted waveguide panels), Telesat, COM DEV (low-power electronics), SED (satellite operation, special test and ground equipment), MacDonald Dettwiler (mission control and ground processor), Odetics (two digital tape recorders), Dornier/AEG (TWT), SAFT (batteries), Astro Aerospace Corp. (SAR antenna deployment mechanism)
Resolution: variable over 7.6–100 m
Mass: 2,749 kg (dry bus, solar array 1,315 kg, payload 1,366 kg, hydrazine 68 kg), plus 55 kg adapter and 352 kg margin
Orbit: 783 × 787 km at 98.6° Sun-synchronous, 18.00 ascending node, 24 day repeat cycle to provide

Beam modes

Fine	36–48° incidence angle, 8 m Resolution:	*50 km swath width, 1 × 1 look	5 beam positions
Standard	20–49° incidence angle, 25 m Resolution:	100 km swath width, 1 × 4 looks	7 beam positions
Wide	20–49° incidence angle, 30 m Resolution:	150 km swath width, 1 × 4 looks	3 beam positions
ScanSAR Narrow	20–46° incidence angle, 50 m Resolution:	300 km swath width, 2 × 2 looks	2 beam positions
ScanSAR Wide	20–49° incidence angle, 100 m Resolution:	500 km swath width, 2 × 4 looks	1 beam position
Extended Low	10–23° incidence angle, 35 m Resolution:	170 km swath width, 1 × 4 looks	1 beam position
Extended High	49–59° incidence angle, 25 m Resolution:	75 km swath width, 1 × 4 looks	6 beam positions

* Tape recorded images may vary in size.

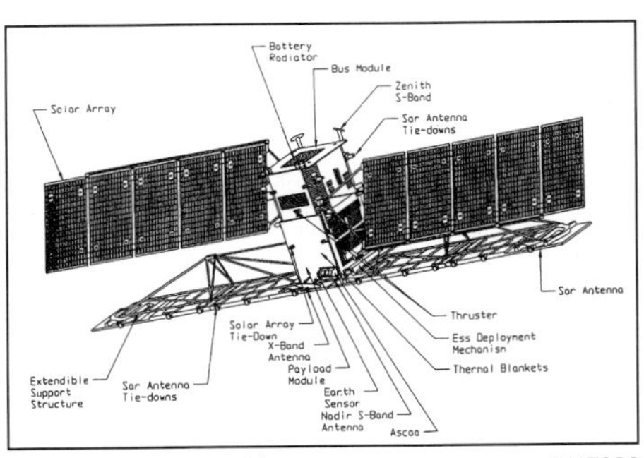

RADARSAT principal features 0517036

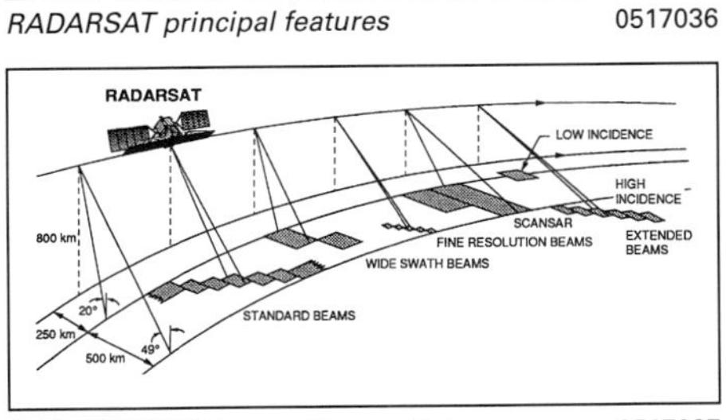

RADARSAT SAR operating modes 0517037

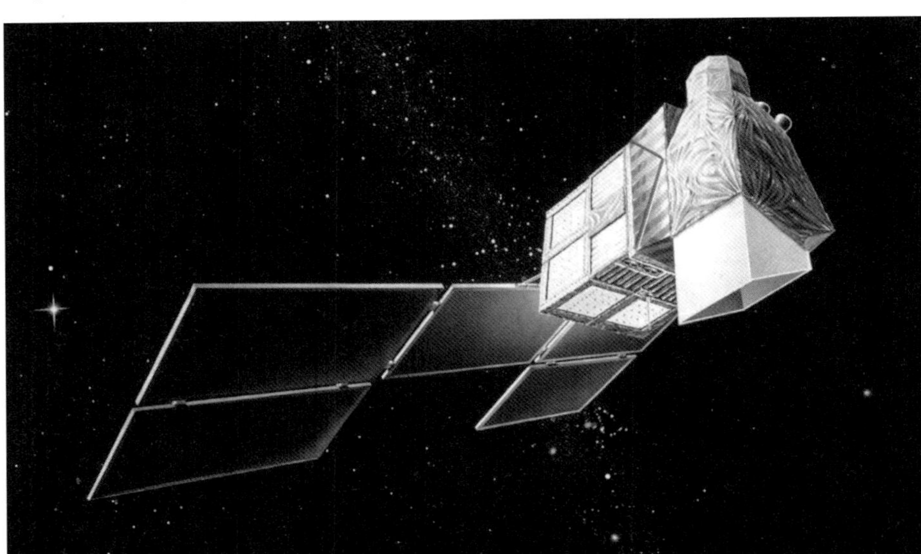

In one application the Canadians use Radarsat images to prepare topographical and relief maps of inaccessible regions. In this view, Radarsat International merged a Radarsat-1, 25 m resolution image with both Landsat and Spot to produce a land use profile of Nagano, Japan prior to the Winter Olympic Games. Normally, Nagano is cloud shrouded and surrounded by mountains making conventional photographic mapping difficult (Radarsat International) 0022564

optimum revisit opportunities (Arctic regions daily; 49–70° N every 3 days). 14 revisits daily (28 min SAR on-time/orbit), (100.7 min/orbit)
Spacecraft: 3-axis controlled to 0.1° in all axes by reaction wheels, magnetometers, magnetorquers, Earth and Sun sensors. Hydrazine thrusters to maintain orbit altitude within at least 10 km to ensure track repeat. Downlinking on two X-band (8.2 15–8.400 GHz) channels (one for real time, other recorder dumping) at 85–100 Mbit/s to Prince Albert (Saskatchewan) and Gatineau (Quebec) stations in Canada, plus NASA's Fairbanks station in Alaska. Additional international network stations currently receiving RADARSAT data include Tromso Satellite Station (Norway), West Freugh (UK), and Singapore. Rigid Spar array of two 5-panel wings provide 3 kW, supported by three 48 Ah SAFT Ni/Cd batteries. Two tape recorders store 10 min of data each at 85 Mbit/s SAR antenna 1.5 × 15 m
Sensors: 5.3 GHz C-band SAR, HH polarisation, 50–500 km selectable swath width, 10–60° selectable incidence angle. The 300 W (average power; 5 kW peak) TWT is slightly modified from ERS-1. RF bandwidth 11.6, 17.3 or 30.0 MHz, transmit pulse length 42.0 μs, pulse repetition frequency 1,270–1,390 Hz
Processing Levels: Path Image, Path Image Plus, Map Image, Precision Map Image, Signal Data, Single Look Complex
Derived products: Mosaics, Digital Elevation Models.

UPDATED

China

Feng Yun (FY)

Background

China launched the first NOAA-compatible FY-1 polar-orbiting meteorological satellite in 1988 by the first CZ-4 from the new Taiyuan polar site. The Chinese began developing the satellite in the late 1970s. Then the country planned to build the FY-2 family and introduce it in April 1994 for GEO observations. The Chinese announced they would launch the FY-2A in April 1994, but the satellite was destroyed in a fire and explosion during final checkout in test hall number 2 at Xichang on 2 April. One worker died later and the fire injured 31 more.

The Shanghai Institute of Satellite Engineering builds the satellites with funds and specification requirements from the State Meteorology Administration. FY-1A, with one year design life, operated for only 39 days but yielded high-quality imagery. Its far Infra-Red (IR) channel, however, failed to produce useful information because water vapour contaminated the detectors and radiation cooler, a fault rectified on 1B. The heavier FY-1B appeared a year later, accompanied by two balloons for upper atmosphere density measurements. It lost attitude control in February 1991 because of radiation upsets, but was recovered after a 50 day effort. The recovery came in time for it to monitor severe floods and help forecast the weather for the Asian Games. However, radiation damage broke control again. Chinese government officials noted in January 1994 that contact was being maintained but data were unusable. The Chinese used the FY-1A/1B as test satellites, preceding the operational 1C/1D in 1997/98. To ensure that the satellites can survive the radiation hazards of space, the Chinese have tested subassemblies of equipment at their Lop Nor nuclear site.

The first FY-3 second-generation polar satellite will carry a 10-channel scanning radiometer, infra-red and microwave radiometers, space environment monitor and an Earth radiation budget scanning radiometer. The 10-channel scanning radiometer will cover: 0.433–0.453, 0.49–0.51, 0.54–0.56, 0.58–0.68, 0.745–0.785, 0.725–1.1, 1.58–1.64, 3.55–3.95, 10.3–11.3, 11.5–12.5 μm. Later satellites in the series may add solar backscatter UV radiometer, microwave imager, microwave humidity detector, medium resolution imaging spectrometer, altimeter and scatterometer. FY-3 series of 10 satellites projected for launch 2003–2011.

Various sources estimated in 1994 that the FY-2 programme cost ¥300 million (US$35 million).

Specifications

Feng Yun-1 Series (Sun-synchronous orbit)
FY-1A
Launched: 6 September 1988 by CZ-4A from Taiyuan (design life one year but failed after 39 days)
Mass: 757 kg at launch
Orbit: 881 × 904 km, 99.13° (afternoon orbit)
Power: 774 W generated BOL.

FY-1B
Launched: 3 September 1990 by CZ-4A from Taiyuan
Mass: 881 kg at launch
Contractors: Shanghai Institute of Satellite Engineering (platform), Shanghai Institute of Technical Physics (radiometers)
Resolution: 1.1 km visible high, 4 km infra-red
Orbit: 886 × 900 km, 98.9° (morning orbit), providing seven passes per day over Chinese territory and real-time image coverage of 55–145° E/0–65° N in conjunction with three stations able to receive data within 2,600 km
Spacecraft: body 1.4 × 1.4 × 1.4 m, height 1.76 m. Solar array span 8.6 m. 750 W 42.5 VDC BOL (design goal) power provided by 14,000 2 × 4 cm + 2 × 6 cm Si cells covering 6.8 m^2, supported by two 48 Ah Ni/Cd batteries. 3-axis attitude control by N_2 gas thrusters, RWs (magnetorquer unloading), gyros, infra-red horizon sensors. High-resolution image transmissions made at 6.654 kbit/s in real time on 1.6955/1.7045 GHz, APT on 137.795/137.035 MHz; housekeeping telemetry on 180.0 MHz. Global data stored on tape recorder
Sensors: Two 95 kg 360 rpm scanning 5-channel Very High-Resolution Scanning Radiometers (VHRSR) operating at (in μm): 0.48–0.53 (ocean colour, land), 0.53–0.58 (ocean colour, land), 0.58–0.68 (daytime cloud, land), 0.725–1.1 (daytime cloud, water, snow, ice, vegetation), 10.5–12.5 (diurnal cloud, land/sea T). 4–24 MeV/n solar/cosmic ray composition monitor.

FY-1C
FY-1C was the first satellite to be launched from the *Feng Yun*-1 batch 2 series, the operational satellites. It was launched with the *Shi Jian* 5 satellite riding piggy-back on the maiden flight of the CZ-4B vehicle.
Launch: 10 May 1999 using a CZ-4B from Taiyuan
Orbit: Sun-synchronous: 98.8°, 849–868 km
Lifetime: nominal mission 2 years: still operating in March 2003
Contractor: Shanghai Academy of Spaceflight Technology
Configuration: hexahedral body, 1.42 × 1.42 × 1.2 m plus two vanes of solar panels, giving a total satellite span of 10.556 m
Mass: 958 kg
Instruments: 10-channel radiometer, operating in 0.433–0.453, 0.49–0.51, 0.54–0.56, 0.58–0.68, 0.745–0.785, 0.725–1.1, 1.58–1.64, 3.55–3.95, 10.3–11.3 and 11.5–12.5 μm

FY-1D
FY-1D, the second batch 2 satellite in the programme, was launched with the *Hai Yang* 1 satellite riding piggy-back. Its launch had been delayed because of FY-1C.
Launch: 15 May 2002 using a CZ-4B from Taiyuan
Lifetime: nominal mission 2 years: still operating in March 2004
Orbit: Sun-synchronous: 98.8°, 851–873 km
Other specifications as FY-1C.

Feng Yun-2 Series (geosynchronous orbit)
FY-2A
Launch: 10 June 1997 by CZ-3 from Xichang
Mass: 1,200–1,250 kg at launch, 600–700 kg on-station
Contractor: as FY-1
Resolution: 1.25–1.41 km visible, 5–5.76 km infra-red, 5–5.76 km water vapour
Orbit: geosynchronous over 105° E. China has stated that a de-spin problem developed on 3 March 2000 and the satellite ceased operations. On 19 April the satellite was manoeuvred off-station over 105° E and five days later the longitude was restabilised over 85–86° E (this was not announced by China). Since that time the satellite has been conducting regular station-keeping manoeuvres, but the status of the meteorological instruments on board is not known
Spacecraft: similar configuration to Meteosat/GMS. Spin-stabilised with despun antenna platform. Control system derived from DFH 2. 2.1 m diameter, 4.5 m height, 280 W generated by drum-mounted solar cells. GEO insertion by solid AKM (separated after insertion)
Sensors: 0.55–0.75 visible, 10.50–12.50 IR, 5.70–7.10 μm water vapour scanning radiometer. Payload includes data collection and image relay systems
TT&C: China's three S/VHF-band stations at Beijing (116°16′ east/40°03′ north), Guangzhou (113°20′ east/23°09′ north) and Urumqi (87°34′ east/43°52′ north) receive real-time high resolution and APT imagery from NOAA and FY polar satellites for relay to the data processing/archiving facility at Beijing's National Satellite Meteorological Centre, part of the State Meteorological Administration. The local station utilises a microwave link to the Centre but the two remote sites communicate via GEO communications satellites through ground stations 10 km distant. The stations receive data when satellites are within 2,600 km, the spread providing pan-China coverage of 55–145° E/0–65° N. Data processing is accomplished by three IBM 4381 model 3s in conjunction with two IBM 7350 image processors and two IBM 5080 graphic processors.

FY-2B
Launch: 25 Jun 2000 using a CZ-3 from Xi Chang
Mass: 958 kg on station
Contractors: as FY-1.

FY-2C
Three FY-2 batch 2 satellites are planned for launch, FY-2C, -2D and -2E. The first of these satellites was expected to be launched in 2003, but it is now scheduled for launch in late 2004 because FY-2B is continuing to operate. The main difference between the batch 1 and batch 2 satellites is that the latter carry radiometers operating in five, rather than three, channels.

Since the CZ-3 launch vehicle has been retired from operations the Chinese have confirmed that FY-2C will be the first in the series to be launched atop the CZ-3A vehicle.

Feng Yun-3 Series (Sun-synchronous orbit)
China has announced plans for the development of an FY-3 series of satellites which will replace the existing FY-1 Sun-synchronous orbit satellites. Originally planned for introduction in 2005, Chinese comments in March 2004 indicated that the maiden launch is scheduled for 2006.
Launch: starting in 2006, using a CZ-4B from Taiyuan
Orbit: Sun-synchronous: 98.728°, 836.4 km
Lifetime: nominal mission 2 years
Contractor: Shanghai Academy of Spaceflight Technology
Configuration: at launch – 4.4 × 2 × 2 m, with solar panel vanes and instrumentation deployed in orbit – 4.46 × 10 (solar panel span) × 3.79 m
Mass: 2,400 kg
Power: average at EOL – 1.1 kW
Instruments: medium-resolution imaging spectrograph, infra-red spectrometer, scan

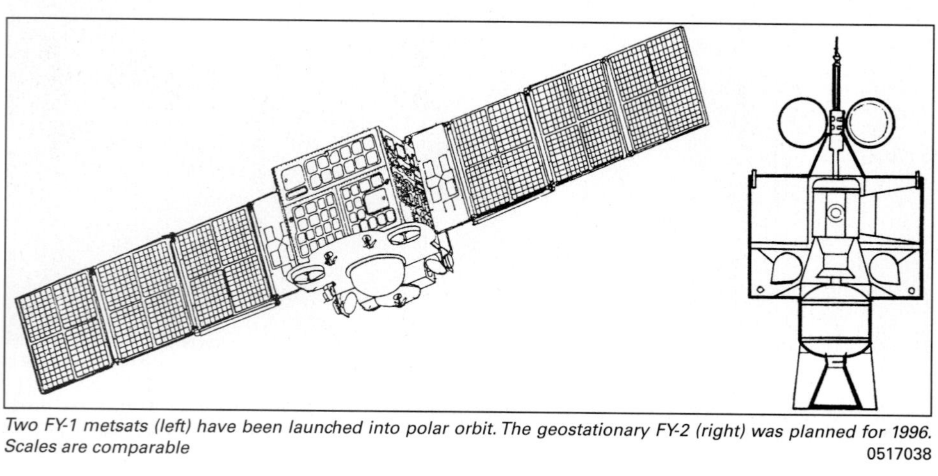

Two FY-1 metsats (left) have been launched into polar orbit. The geostationary FY-2 (right) was planned for 1996. Scales are comparable
0517038

radiometer, microwave radiometer, microwave imager, ultra-violet ozone sounder, Earth-radiation budget instrument and space environment monitor for global, all-weather, multi-spectrum, three-dimension quantitative detection
Data transmission: real time will use L-band BPSK modulation, data rate 4.2 Mbps, and X-band QPSK modulation, data rate 18.2 Mbps: stored data will use X-band QPSK modulation, data rate 93 Mbps.

Feng Yun-4 Series (geosynchronous orbit)
The planned FY-4 series of satellites has not been described in any detail, but it is planned to replace the existing FY-2 meteorological satellites with the FY-4 series, possibly starting around 2006. The launch vehicle will probably be the CZ-3A from Xi Chang.

***Hai Yang* Series (Sun-synchronous orbit)**
The *Hai Yang* ('ocean') oceanographic satellite uses the CAST-968 satellite platform which was test flown as the *Shi Jian* 5 satellite in 1999 (*qv*). The first satellite was launched piggy-back with FY-1D. It is planned that a series of five satellites will be launched, flown at two-year intervals.

HY 1
Launch: 15 May 2002 using a CZ-4B from Taiyuan
Orbit: initial orbit was 98.8, 851–872 km: during the first two weeks in orbit the satellite manoeuvred to a Sun-synchronous 98.8°, 792–795 km orbit
Lifetime: nominal mission 2 years: still operating in March 2004
Contractor: Shanghai Academy of Spaceflight Technology
Configuration: box-shaped body, approximately 1 m each side, plus two vanes of solar panels, giving a total span of approximately 4 m
Mass: 360 kg (some reports quoted 328 kg)
Instruments: 10-channel infra-red ocean scanner and four-band CCD camera

SMMS: Small MultiMission Satellite
China has described a series of small satellites which will form part of the China Environment and Disaster Monitoring Small Satellite constellation. The original project was proposed by China, Pakistan and Thailand in November 1992: the current stockholders are China, Iran and Thailand, with other Asia-Pacific countries said to be welcome to join the group.

Details of the satellite platform are not known, but it would be reasonable if the CAST-968 platform were to form the basis of the satellite (see entries for *Shi Jian* 5 and *Hai Yang* 1).
Launch: first launch late 2004 using a CZ-4B from Taiyuan
Orbit: Sun-synchronous at 650 km
Lifetime: nominal mission >3 years
Contractor: Chinese Academy of Space Technology
Mass: 350–400 kg?
Instruments: Hyper-Spectrum Imager (HSI) and wide-swath CCD camera (WSCCD)

Zi Yuan/CBERS
An agreement was signed in November 1993 between China and Brazil's National Institute for Space Research (INPE). This called for the development and launch of two remote sensing satellites: the Chinese name is *Zi Yuan*-1 meaning 'resource' (*Zi Yuan*-2 is a reconnaissance satellite, discussed in the *Jian Bing* entry) while the programme is called China-Brazil Earth Resources Satellite (CBERS).

The primary imaging system was developed by the Xian Institute of Radio Technology: this would have characteristics similar to the French SPOT, using linear CCDs. Brazil provides the wide field imager: INPE is responsible for the development of the data collection system, structure, power supply, TT&C, electrical ground support equipment, integration/testing (30%) and management (30%). The first satellite was manufactured by CAST and INPE was responsible for the manufacture, integration and testing of the second satellite. Chinese project manager is Chen Yiyuan, Brazilian project manager is Carlos Santana.

Zi Yuan-1 1 (CBERS-1)
Launch: 14 October 1999 using a CZ-4B from Taiyuan
Orbit: Initially 98.54°, 733–746 km, standard operational orbit Sun-synchronous 773–773 km: repeated ground track every 373 circuits (26 days): repeat cycle also permits a 3-day repeat with +32° side-looking permitted
Resolution: 20–260 m visible, 80 m IR (160 m in TIR)
Mass: 1,450 kg at launch
Lifetime: Nominal lifetime was 2 years, but actually operated until 13 August 2003: on this date the satellite's orbit was raised to a 774–782 km retirement orbit.
Spacecraft: 1.8 × 2.0 × 2.2 m body, 6.3 × 2.6 m solar panel providing 1,100 W. 3-axis control provided by RWs plus 1 N thrusters. 20 N thrusters used for orbit maintenance (a small orbit-raising manoeuvre was completed at least once every 6–8 weeks). 53 Mbit/s experimental high-density tape recorder. Image data transmission was at 2 × 53 Mbit/s (each using 20 W TWTA) plus 6.13 MBit/s (8 W SSPA) on 8.103–8.321 GHz. Argos-compatible data collection system can handle up to 500 platforms: 401.65 + 401.62 MHz up and 462.6 + 2,267 MHz down.
Sensors: 5-channel linear CCD providing visible/PAN coverage: 0.51–0.73, 0.45–0.52, 0.52–0.59, 0.63–0.69 and 0.77–0.89 μm. 120 km swath. 2 × 53 Mbit/s IR/thermal XS scanner covers 0.50–1.10, 1.55–1.75, 2.08–2.35 and 10.40–12.50 μm, swath 120 km. 6.13 Mbit/s CCD wide field imager covers two visible channels: 0.63–0.69 and 0.77–0.89 μm, swath 900 km. 1.1 Mbit/s Space environment Monitor also carried.

Zi Yuan-1 2 (CBERS-2)
Launch: 21 October 2003 using a CZ-4B from Taiyuan
Orbit: Initially 98.54°, 731–750 km, standard operational orbit Sun-synchronous 773–773 km
Mass: 1,550 kg
Lifetime: Nominal ~2 years, but based upon the Zi Yuan-1 1 experience the satellite could operate for twice this period.
Other specifications as Zi Yuan-1 1 (CBERS-1).

Zi Yuan-3 (?)/CBERS-3 and CBERS-4
At the end of September 2000 it was announced that China and Brazil had signed an agreement for the development of two second-generation (thus *Zi Yuan*-3 ?) remote sensing satellites, CBERS-3 and CBERS-4. These two satellites would carry improved CCDs which will permit a ground resolution of 5 m. No launch schedule has been announced, other than in October 2003 it was said that both satellites should be in orbit by 2010.

China and Brazil have also undertaken studies for the launch of satellites based upon the CBERS platform to geosynchronous orbit for meteorological/remote sensing or communications missions. No further details are available concerning these studies.

UPDATED

France

Spot

Current Status
The Spot 4 earth observation satellite was successfully launched by an Ariane 40 launcher from the European spaceport in Kourou, French Guiana, on 24 March 1998 at 2.46 am French time. Spot 4's HRVIR instruments possess an additional spectral band in the Short-Wave Infra-Red (SWIR). The SWIR band offers better discrimination between different types of crops and plant cover. With Spot 1 and Spot 2 still in service, the enhanced capability of Spot 4, provides commercial stereographic images at the 5 m resolution level. Spot 5 was launched on 4 May 2002 by Ariane 42P. Today, Spot Image holds a 60 per cent share of the market, supported by a worldwide commercial network of three subsidiaries (United States, Australia and Singapore), nearly 90 distributors and 23 receiving stations.

Currently Spot 2 and Spot 4 are operational with Spot 1 in reserve.

Background
CNES began the Spot programme in 1978 and subsequently established Spot Image to market the data. Spot Image had a start-up capitalisation of US$41.1 million, allocated by public institutions and private companies in Belgium, France and Sweden. Initially the company had several partners. The original percentages were: (current values shown in parentheses), CNES 39.0 (35.32), Institute Geographique National 10.0 (11.27), Matra 8.8 (28.01), Aeropart (5.40), Bureau de Recherches Geologiques et Minieres 7.4 (1.00), Alcatel Espace (3.33), Crédit Lyonnais (1.19), Banque Nationale de Paris (1.19), Société Générale (1.19), Paribas (0.60), Swedish Space Corp 6.0 (5.94), Nuova Telespazio

Spot 3 launch preparation (Arianespace) 0517121

(2.32) and Belgium 4.0 (3.27). Spot Image has three subsidiaries, the Spot Image Corp in the US, Spot Imaging Services in Australia and Spot Asia in Singapore.

As the satellite operator, CNES is at the centre of the organisational structure, directly responsible from Toulouse for orbit maintenance, payload programming, reception and pre-processing. Data downlinks to two SRIS (Station de Reception des Images Spatiales) stations at Aussaguel near Toulouse (SRIS-T) and one at Esrange, near Kiruna, Sweden. These receive both real-time and recorded data as Spot passes over the north polar region, Europe and North Africa within 2,500 km range, in addition to the stored images from other regions. Together, they have a reception capacity of 500,000 images annually. Each SRIS is associated with a pre-processing/archiving centre or CAP (Centre d'Archivage et de Prétraitement), with the equivalent of 700 scenes archived every 24 hours at Toulouse.

Spot Image has responsibility for the Direct Receiving Stations (SDRS) around the world, which receive real-time imagery only. SDRS are located at Prince Albert and Gatineau (Canada), Maspalomas, Spain, station managed by ESA), Cuiaba (Brazil), Lad Krabang (Thailand), Hatoyama (Japan), Islamabad (Pakistan, commissioned 1989), Hartebeesthoek (South Africa, 1989), Riyadh (Saudi Arabia, 1989), Alice Springs (Australia, 1990), Tel Aviv (Israel, 1 February 1991), Cotopaxi (Ecuador, July 1992), Taipeh (Chung-Li University, Taiwan, July 1993), Pare-Pare (Indonesia, October 1993), Fucino (Italy, July 1995) and Singapore (September 1995); the USAF Eagle Vision transportable system began operating in June 1994 (another 20 will

Spot 5 is being developed in parallel with Helios 2 (CNES/D Ducros) 0517120

Spot 5 will provide routine 5 m imaging (CNES/Spot Image/ISTAR) 0517232

In this view of the Spot 4 satellite undergoing testing in the Interspace anechoic chamber, the vegetation camera appears as a cylinder in the horizon position, under the twin imaging cameras (Matra Marconi) 0038492

be ordered). South Korea was added in 1996, and Russia's Resurs-O station at Obninsk will be upgraded for Spot 2 imagery during the two warmest months.

CNES categorises processing operations into several levels of sophistication. Level 1 provides basic radiometric and geometric corrections and it does not involve ground control points or satellite attitude restitution data. Level 1A downlinks essentially raw data, apart from the normalisation of CCD detector response in each spectral band. It is useful for stereoplotting and basic radiometric studies. Level 1B gives full radiometric (desmearing) and limited geometric corrections with the pre-processing level for photo interpretation and thematic analysis. Users may also attain stereoscopic pairs at this level. Level 2 gives advanced rectifications according to a given cartographic projection and Level 2A corresponds to Level 2 precision processing but can be implemented without use of map ground control points. The Spot introduced Level 1AP in 1990, which is optimised for photogrammetric applications using analytical stereoplotters. The commerical suppliers provide products in standard Computer-Compatible Tapes, photographic form and CD-ROM.

Spot's 4 millionth scene was archived 19 July 1995. 1995's revenue from data/product sales and receiving station fees totalled FFr207.5 million (US$41.5 million); 1994 220; 1993 178, profit 0.031; 1992 218, profit 1.38; 1991 204; 1990 165; 1989 123; 1988 90; 1987 55 and 1986 15. 1993's decrease reflected lost sales because of Spot 2 recorder problems before Spot 3 became available. Asia/Pacific (largely Japan) accounted for 25.3 per cent of 1995's revenue; Europe 33 per cent, North America 25 per cent, Middle East 6.7 per cent, Latin America 3.8 per cent, Africa 6.3 per cent. Accumulated sales exceeded FFr1.4 billion (US$273 million) by end-1995 (70 per cent from outside Europe); the government had invested FFr8 billion. Europe provided 32 per cent revenue in the first 10 years, followed by North America 22 per cent and Asia/Pacific 20 per cent. The enhanced Spot 4 design was approved in 1989 for launch about 1994, but actually launched in 1998.

Spot 1

Spot 1 was placed in orbital storage 15 January 1991, but with its health monitored at least every 6 revolutions so that it could resume operational service and meet all image quality specifications within two weeks. It became operational again in March 1992 to help Spot 2 meet increasing demand through to late October 1992 during the northern hemisphere's growing season. It provided real-time transmissions (recorders are inoperative; some 60,000 to 80,000 images were received) for Europe/Middle East, while Spot 2 worked in recording mode (and normally for the rest of the world). The two were phased to optimise stereopair acquisition above 41° latitude. It was revived similarly for April-July 1993, at a total cost of FFr7.7 million, requiring a third control centre, as the second is now used for Spot 3. The European Commission this time contributed €350,000, in addition to paying for each scene used for its agricultural monitoring work. Spot 1 was again retired after Spot 3 entered service in late 1993, but can still be called upon at a few days' notice. Spot 1 and 2 carry two identical push-broom CCD High Resolution Visible (HRV) imagers, pointed for a total swath width of 117 km with a 3 km overlap. Site revisit period is 26 days but up to 27° off-nadir viewing (creating total swath width of 950 km) via a tilting mirror permits a single area to be imaged on seven successive passes at equatorial latitudes and 11 passes at 45°. Image width is 60 km for nadir viewing, stretching to 80 km at maximum mirror tilt. The capability also permits stereo imaging. Through 1995, Spot 1's cameras totalled 2,927 h, returning 1.8 million images.

Spot 1 was returned to active duty on 13 January 1997 to fill a gap caused by the loss of Spot 3. Although the recorders have failed it can operate when in contact with a ground station.

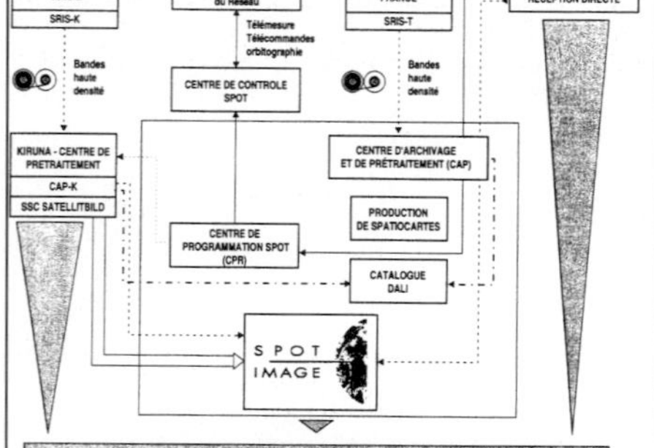

Spot control and data reception network 0517413

Spot 2

The satellite returned its first multispectral image, from HRV 2, at 10.18 GMT 23 January 1990, 32 hours after attaining orbit. It arrived in the desired orbit 180° from Spot 1 on 30 January, having expended 12 kg hydrazine. The image quality review of 5 March concluded it to be of commercial quality, and Spot 2 was declared ready for operational exploitation from 21 March 1990. 1.9 million images were returned by end-1995. One tape recorder failed after one year. A second tape recorder in February 1993 began suffering from recording-head wear, leaving spots on imagery. The prime image telemetry channel has failed but the spacecraft is otherwise in perfect order. Spot 3 replaced it as prime satellite November 1993. Spot 2 requires FFr12 million in annual sales to cover operations costs; contracts from the European Commission allowed simultaneous Spot 2/3 operations through 1994–95.

Spot 3

The first image was transmitted within 33 hours, showing the Strait of Bonifacio separating Corsica/Sardinia. Spot 3 was declared operational 29 November 1993. It expended only 17 kg hydrazine attaining its final orbit, 42 and 208° ahead of Spot 1/2, respectively. It is exhibiting slightly better gain control and dark current characteristics than its predecessors. 686,028 images had been returned by end-1995. Spot 3 cost FFr1.4 billion, including FFr650 million for launch. Spot 3 unexpectedly ceased operating on 14 November 1996 when altitude control was lost after gyroscope problems.

Spot 4 was fast-tracked to launch and Spot 1 was reactivated.

Spot 4

The French government approved Phase C/D in July 1989 for the expanded Spot 4. It employs the same augmented multimission bus as the Helios 1 reconnaissance satellite, adds a mid-infra-red band to the HRV (now HRVIR) by replacing Sodern's DTA 01 detection unit with the DTA 03 model, incorporating Thomson's 7811 CCDs. Sodern's DTA 05 detection unit gives images at 1 km resolution for monitoring crops, natural vegetation and geosphere/biosphere interactions. The bands are equivalent to HRVIR plus a blue channel for atmospheric corrections. The European Union in January 1994 agreed to contribute half of VGT development costs; Sweden, Belgium (10 per cent) and Italy are also contributing separately. The total VGT programme is expected to cost ECU109.7 million.

Spot 4's Vegetation is an independent payload, having its own recording, transmission and management systems, working to its own ground segment for payload programming, image reception/processing and product distribution. Local users can acquire 1 km resolution imagery broadcast in real time at 1.704 GHz L-band 1 Mbit/s on their existing metsat receivers. The global data, recorded on the 2.25 Gbit SSR, are downlinked at 3.4 Mbit/s at 8.153 GHz X-band using Spot 4's main antenna. The CPV Centre de Programmation de VGT in Toulouse draws up the daily plan for uplinking by the main Spot 4 control centre. The SRIV Station Reception des Images VGT at Esrange takes 4/5 data dumps daily totalling about 8 Gbit. The CTIV

The Spot 4 IR instrument receives a final vibration checkout before launch (Matra Marconi) 0038493

Spot 5 will provide routine 5 m imaging (Theo Pirard) 0054347

Centre de Traitement des Images VGT in Mol (Belgium) archives/processes the data for a daily synthesis, increasing progressively to a 10-day global synthesis. It can deliver dedicated products within three days if ordered in advance. The QIV Centre Qualité Image VGT in Toulouse is in charge of instrument calibration and image quality.

Spot 5

Definition of Spot 4's successor began in 1993, following an extensive user market survey in 1991–92. 5 m resolution is necessary to address new markets such as civil engineering and an along-track 5 m stereo capability was to generate simultaneous stereo pairs, whereas its predecessors are limited to separate passes. Resolution is legally barred from being less than 5 m. The goal is to provide simple generation of 1/50,000–1/100,000 standard maps (1/25,000 in some cases) and digital elevation models. The government approved the FFr6.3 billion programme 4 October 1994 for the first of two launches in 2002 (but available late 1999), employing a bus jointly developed for Helios 2. Sweden and Belgium (4 per cent) are again contributing; Belgium is separately paying 10 per cent towards VGT. The programme was reduced in early 1996 to FFr4.4 billion, cutting Spot 5B and losing the simultaneous stereo imaging by dropping one camera and nadir pointing the remaining two. Spot 5 was launched in May 2002.

Specifications

Spot 1

Launched: 22 February 1986 by Ariane 1 from Kourou, French Guiana (2 year design life; initially retired 31 December 1990)
Mass: 1,830 kg at launch
Contractors: Matra (prime, imaging system), Aerospatiale (structure, thermal control, solar array), Sodern (optical sensors)
Resolution: 20 m multispectral, 10 m panchromatic
Orbit: 824 × 829 km (822 km after 3 years) in 98.7° Sun-synchronous, crossing equator at about 10.30 local time, with 26 day repeat cycle (369 revs). Altitude corrections executed every 2 months and inclination adjustments annually.
Spacecraft: 3.5 m high, 2 m^2, consisting of Multimission Platform (PFM) and forward Spot payload module. 2 year design life. 15.6 m solar wing provided 1,382/1,326 W BOL/EOL (equinox values; 36 V) with 3 × 24 Ah Ni/Cd batteries. Panel degradation after 3 year and 18,000 thermal cycles was 4 per cent instead of the projected 21 per cent. 3-axis platform pointing accuracy of 0.1° (0.15° specified) maintained by three MWs unloaded by two magnetic coils. System includes digital Earth sensors, 2 × 2° FOV. orbit control by 8 × 15.6 N hydrazine thrusters (two 75 kg tanks, 130 kg remained after 3 years). Each of the two HRV imagers returns data at 25 Mbit/s on 8.025–8.400 GHz real-time using 20 W Thomson-CSF TWTA
TT&C: Spot Mission Centre is located at Toulouse. 8.307 GHz beacon multiplexed with the transmitted signal to facilitate ground antenna pointing. Communications subsystem mass 240 kg; power consumption 170 W for direct tx, 270 W for recorder playback. The 70 kg Odetics recorder could hold 22 min of data. Recorder number 1 (a qualification model flown to save money) failed in Sepember 1986, number 2 was shut down after problems appeared July 1990, and the prime telemetry TWT failed August 1989. 3.7 kg 2.025–2.120 GHz S-band transponder provides housekeeping telemetry (2,048 bit/s) + TC (2 kbit/s)
Sensors: High Resolution Visible (HRV) Spot was the first satellite to incorporate push-broom sensors, avoiding the need for a mechanical scanning mirror. On Spot 1, each spectral band utilises Fairchild's 122 DC detectors; Spot 2/3 employ four Thomson-CSF TH 7801A 1,728-element CCD arrays, combined in each HRV's Sodern DTA 01 detection unit. The spectral ranges (in μm) of the bands are: Band 1 0.50–0.59, Band 2 0.61–0.68, Band 3 0.79–0.89; panchromatic 0.51–0.73
Passenger: radar transponder for calibration of Kourou's ground radar and operator training.

Spot 2

Launch: 01.35.27 GMT 22 January 1990 by Ariane 40 from Kourou, French Guiana; declared operational 21 March 1990
Mass: 1,870 kg at launch
Resolution: same as Spot 1
Orbit: 819 × 822 km, 98.7°, initially 180° from Spot 1, now 166° behind Spot 1
Spacecraft: same as Spot 1 but with 18.5 kg 20 W DORIS (Doppler Orbitography and Radiopositioning Integrated by Satellite) unit for accurate orbit determination in preparation for 1992's joint Franco- US Topex mission and hydrazine loading 160 kg
Sensors: Same as Spot 1.

Spot 3

Resolution: same as Spot 2
Launch: 26 September 1993 by Ariane V59 from Kourou, French Guiana
Mass: 1,907 kg launch; dry 1,749 kg; 158 kg hydrazine
Orbit: 816 × 818 km, 98.68°
Spacecraft: as Spot 2, but TWTs and recorders modified slightly for longer lives
Sensors: Same as Spot 2, plus 25 kg POAM II Polar Ozone & Aerosol Measurement instrument carried for USAF Space Test Program.

Spot 4

Launch: 24 March 1998 by Ariane 40 from Kourou, French Guiana
Mass: 2,755 kg launch; dry 2,600 kg: payload 1,400 kg
Orbit: 832 km, 98.7°, 101 min period
Contractors: MMS France (prime, assembly, integration, test, onboard data management, AOCS, inertial units, Sun/star sensors), Aerospatiale (VGT, mechanical subsystem, thermal control, solar array), Alenia (structural elements, wheels/magnetorquer drive electronics), Alcatel Espace (transponder), CASA (cabling harness), CRISA (software, EGSE), Sextant (housekeeping unit, bus couplers, decoding/reconfiguration unit), ETCA (electrical power supply), Laben (wheels/magnetorquers drive electronics), Saab (central communications unit), SAFT (batteries), SAGEM (gyroscopes), SEP (propulsion), Sodern (Earth sensors, star sensors, optical head), Schlumberger (tape recorders)
Resolution: HRVIR 20 m multispectral/10 m monospectral; VGT 1 km/4 km
Spacecraft: based on Spot 1–3 design but with additional compartment between platform and payload to accommodate the new tape recorders and passenger experiments. Design life is 5 years (0.8 reliability after four years). New rigid 74 kg 5-panel 1.90 × 2.60 m Aerospatiale GSR-3 solar array will provide 2.5 kW BOL/equinox and 2.2 kW EOL (5 years)/Summer solstice, feeding four SAFT 40 Ah Ni/Cd batteries. 3-axis control to <0.03°, employing three RWs and two magnetorquers, Sodern STD-16 Earth horizon infra-red sensors, Sun sensors and Sagem's Regys 10 gyro orbit and yaw adjustment by four pairs of 15.6 N thrusters; roll/pitch by four pairs of 3.5 N thrusters supplied from four 75 kg hydrazine tanks. Control system based on radiation hardened microprocessor: Central Communication Unit (that is, main computer) incorporates a Fairchild 9450 microprocessor, providing 128,000 words redundant RAM. Telemetry rate at 4,096 bit/s; up 2.025–2.120, down 2.200–2.290 GHz. Two 121 kg 50 Mbit/s 120 Gbit Schlumberger redundant tape recorders store up to 40 min imagery each (175 W required in record mode, 260 W replay, 12 W standby; 70,000 on/off cycle specification). Imagery can be encrypted. VGT has its own recorder and telemetry system: 2.25 Gbit SSR (3 Mbit/s replay at 8.153 GHz); real-time at 1.704 GHz. Spot 4/Helios 1 controlled from the new Fresnel building at Toulouse
Sensors: HRVs paired as in Spot 1–3 are upgraded to HRVIRs (High-Resolution Visible Infra-Red) by the addition of a mid-infra-red band for vegetation water registration; independent and simultaneous observing schedules possible. The new wide-angle VGT is optimised for monitoring of vegetation, crops and geosphere/biosphere interactions

HRVIR Paired instruments with spectral ranges (in μm): Band B1 0.50–0.59; Band B2 0.61–0.68; Band B3 0.79–0.89; SWIR 1.580–1.750; yielding 20 m resolution in spectral mode. In monospectral mode, covering 0.61–0.68, resolution is 10 m (the previous panchromatic band is replaced by sampling the B2 band at 10 m or 20 m resolution). Data rate 25 Mbit/s. Viewing characteristics as Spot 1–3. See Sodern's entry in the Space Industry GN&C section for detailed imager information

VGT bands as HRVIR B2, B3, SWIR but with B1 replaced by B0 0.43–0.47 μm for atmospheric corrections. Swath 2,200 km (101° FOV), allowing daily global coverage
Passenger payloads: DORIS (Doppler Orbitography and Radiopositioning Integrated by Satellite), improved for precise (1 m) onboard real-time autonomous orbit determination and distribution of timing pulse; 10 kg, 14 W

Vega Radar Transponder for calibration of Kourou's ground radars and operator training

PASTEC (Passager Technologique), a group of experiments for studying the orbital environment

PASTEL (Passager Spot de Telecommunication Laser), demonstration of optical 50 Mbit/s datalinks via GEO satellites as part of ESA's Artemis SILEX programme. ESA's Artemis satellite will be used to relay HRVIR images

POAM III Polar Ozone & Aerosol Measurement instrument carried for USAF Space Test Program.

Spot 4 immediately prior to encapsulation on top of the Ariane launcher at Kourou 0084763

Spot 5

Contractors: MMS France (prime, assembly, integration, test, onboard data management, AOCS, inertial units, Sun/star sensors), Aerospatiale

(VGT, mechanical subsystem, thermal control, solar array), ETCA (electrical power supply), Saab (central communications unit), Alcatel Espace/IBM (solid-state recorders)
Resolution: HRG 5 m PAN, 10 m XS, 20 m SWIR; VGT 1 km/4 km
Launch: 4 May 2002 by Ariane 5 from Kourou, French Guiana
Mass: 2,600 kg launch
Orbit: 825 × 826 km, 98.8°, 101.4 min period
Spacecraft: based on Spot 4/Helios 1 design. Design life 5–7 years. 3-axis control. Two 100 Gbit SSRs. 150 Mbit/s real time/storage
Sensors: HRG three 270 kg High-Resolution Geometry cameras operating independently and simultaneously were planned, but there will now be two nadir-pointing (losing the simultaneous stereo pairs). The PAN band is as Spot 1–3 but at 2.5–5 m resolution; the visible/NIR B1/B2/B3 are as Spot 4, but at 10 m resolution; the 20 m SWIR band for vegetation monitoring is as Spot 4. For three cameras, they would be sighted at nadir, 19.2° fore, 19.2° aft. 60 km nadir swath + ±27° cross-track tilting (as on Spot 1–4) provides swath widths up to 160 km for two cameras operating in conjunction, and 240 km for three. Three operating modes: multi-spectral, 60–240 km swath, 10 m B1/B2/B3 + 20 m SWIR; high resolution, 60–240 km swath, 5 m PAN + 10 m B1/B2/B3; stereo, 60–80 km swath, 5 m PAN + 10 m B3 VGT as Spot 4
Passenger payloads: DORIS as Spot 4, PASTEC as Spot 4.

UPDATED

India

Insat

Current Status

Insat 3B was launched 21 March 2000 by Ariane 5 and is the first of five successors of the Insat 2 series. Insat 3C was launched by Ariane 4 on 23 January 2002 and was followed by Insat 3A launched by Ariane 5 on 10 April 2003. The solar array and two antennas were deployed on 15 April and the satellite was placed in a tri-axis stabilisation mode a day later followed by deployment of the solar sail/boom on the north side. The two meteorological cameras were tested over two days beginning 18 April. Insat 3A reached its operating longitude at 93.5 °E on 23 April from where the transponders were tested. Insat 3B is presently at 83 °E with Insat 3C at 74 °E.

Background

The Indian National Satellite System uniquely provides geostationary platforms for simultaneous Indian domestic communications and Earth observation functions (Japan's 1999 MT-Sat adopts a similar approach). The first-generation Insat 1 satellites were all US-built, but the advanced Insat 2 family is indigenously produced. 2C/2D replaces the imaging, DCP and DRT payloads with an enhanced communications element. 2E is more like 2A/B, but with expanded imaging capability.

The Insat system is a joint venture of the Department of Space (DOS), the India Meteorological Department (IMD), the Department of Telecommunications (DOT) and All India Radio. The Insat office at Bangalore is responsible for controlling the system through the Master Control Facility (MCF) at Hassan. Meteorological imagery and data are received at the Delhi Earth Station and relayed by microwave to the key Meteorological Data Utilisation Centre (MDUC) at the India Meteorological Department (IMD) headquarters in New Delhi for processing and analysis. 11 VHRR images (each three visible and one infra-red) are acquired each day. 1D returned more than 24,500, 2A more than 11,900 and 2B more than 9,100 images by end of 1995. The processed material is relayed to 22 Secondary Data Utilisation Centres (SDUCs), initially by land lines but converted by 1989 for retransmission through Insat. Information from Data Collection Platforms is also processed at the MDUC, through Insat's Data Relay Transponder.

Insat 1B retired August 1993 after returning more than 36,000 images. 1C failed in November 1989. 1D assumed the prime role in July 1990 from 1B. The five-satellite Insat 2 system began in 1992 with 2A, which took over the prime imaging role from 1D 6 October 1992. Insat 1A was launched 10 April 1982 and positioned at 74° E but it was abandoned 4 September 1982 following exhaustion of its attitude control fuel during eclipse. 1B was launched by Shuttle in August 1983 but it was not until mid-September that Ford and ISRO engineers at India's Hassan Satellite Control Centre succeeded in deploying its solar array. Full operation was achieved in October 1983 and the satellite was retired in August 1993. 1C was launched 21 July 1988 by Ariane V28. A power system failure (from a solar array isolation diode short) soon removed half of the communications capacity but the meteorological services, powered through either bus, remained unaffected from 93.5° E. However, the satellite lost Earth lock 22 November 1989 and was abandoned. 1C procurement cost was reported at Rs800 million, with Ariane launch cost adding Rs740 million. The failed 1A/1C reportedly produced insurance payouts of about US$70 million each.

Insat 1D was damaged by a crane hook at the launch site and then while being repaired it suffered further damages. Insat 2A was retired in April 2002, Insat 2B in November 2000, Insat 2D in October 1997.

Specifications

Insat 1D
Contractors: Ford Aerospace (prime)
Launched: 12 June 1990 by Delta 6925 from complex 17, Cape Canaveral. Entered service 17 July 1990
Orbit: 74° E geostationary October 1999 to May 2002 then drifting between 68° and 72° E
Resolution: 2.75 km visible, 11 km infra-red
Spacecraft: A 2.18 × 1.42 × 1.55 m box-shaped, 3-axis stabilised bus with asymmetrical 5-panel 11.5 m^2 solar wing providing 1,185 W BOL (930 W end) and creating total span of 19.4 m. The asymmetrical solar array configuration provides an unobstructed view into space for the radiative cooler of the VHRR and a solar sail is used for balance. A magnetorquer with current coil around the body provides fine control; a unified bipropellant system is used for orbit transfer, station-keeping and attitude maintenance

Insat 2C solar array deployment test (Theo Pirard) 0054348

Sensors: 2-channel Very High-Resolution Radiometer (VHRR) providing 0.55–0.75 μm visible + 10.5–12.5 μm infra-red images of full Earth disc every 30 minutes. The Insats also relay 402.75 MHz transmissions from up to 100 hydrological, meteorological and oceanographic data collection platforms (system dormant until Insat 2 appeared because of Insat 1 failures).

Insat 2A
Launch: 9 July 1992 by Ariane 4 V51 from Kourou, French Guiana (design life 7 years)
Mass: 1,162 kg BOL, 911 kg dry
Orbit: 74° E geostationary July 1992 to January 2001, then 48° E from April 2001
Contractors: ISRO and Indian companies; ISRO Space Application Centre (VHRR)
Resolution: 2 km visible, 8 km infra-red
Spacecraft: box-shaped 3-axis controlled bus based around 930 mm diameter central cylinder with asymmetrical 5-panel solar wing providing 1,400 W BOL (1,180 W EOL) and creating 23 m span. Integral liquid propellant (NTO/MMH) apogee boost motor and 16 associated attitude control thrusters
Sensors: VHRR similar to Insat 1's but with improved resolution; redundant Data Relay Transponder for Data Collection Platforms. VHRR detectors are redundant linear arrays of four Si photodiodes (visible) and redundant HgCdTe detectors at 105,000 (using passive radiative cooler) for thermal-IR. VHRR operates in 3 modes: full frame 20 × 20° full Earth scanned in 33 min; normal mode 14° N-S 20° E-W scan 23 min; sector scan 4.5° N-S 20° E-W scan 7 min for rapid repetitive coverage during severe weather conditions such as cyclones.

Insat 2B
Launch: 22 July 1993 by Ariane 4 from Kourou, French Guiana. Returned first image 21 July; declared operational 10 August 1993
Orbit: 93.5° E geostationary.

Insat 2D
Launch: 3 June 1997 by Ariane 44L from Kourou, Fr. Guiana
Mass: 2,079 kg
Location: 74.1° east, suffered electrical problems and declared dead 5 October 1997.

Insat 2E
Launch: 2 April 1999 by Ariane 42P from Kourou
Location: 83° E geostationary
Mass: 2,500 kg at launch
Sensors: 2E configured like 2A/2B, but with 5.7–7.1 μm water vapour channel added to VHRR and a new 3-band CCD camera providing 1 km resolution 0.63–0.69 μm vis, 0.77–0.86 μm NIR, 1.55–1.70 μm SWIR.

Insat 3B
Launch: 21 March 2000 by Ariane 5 from Kourou, French Guiana.

Insat 3C
Launch: 23 January 2002 by Ariane 4 from Kourou, French Guiana.

Insat 3A
Launch: 10 April 2003 by Ariane 5 from Kourou, French Guiana.

IRS Series

Background

India has built its first domestic dedicated Earth resources satellite programme and named it the Indian Remote Sensing satellite system as an element of the National Natural Resource Management System. Between 1988 and 2000 India will launch 12 polar orbit satellites. These satellites put India in a position to satisfy its own applications needs, but also to dominate a significant portion of the global commercial market.

The country launched the first in the series, the Rs650 million IRS 1A by a Soviet Vostok booster in March 1988, the first commercial satellite launched by the organisation Glavkosmos. Then signed a Rs220 million contract for the identical IRS 1B in November 1988 for an August 1991 departure. IRS 1C's December 1995 launch was brought about with a contract signed on 18 January 1991. IRS 1C was the first Russian launch into a retrograde orbit using the Molniya M four stage launcher. IRS 1D launched 29 September 1997 was to have replaced IRS 1C but it was left in a 300 × 823 km orbit by the PSLV launcher and remained in a elliptical sun-synchronous orbit.

IRS 1C undergoes dynamic balancing. The large drum-shaped PAN camera is at top right; WiFS and MOS at top left (IRS) 0517417

1C/1D offer improved resolution, stereo viewing and more frequent revisits, making them the world's most advanced civil remote sensing satellites. The 846 kg IRS 1E, 1A's refurbished engineering model, was lost on PSLV-D1 20 September 1993 carrying Germany's MEOSS Monocular Electro-Optical Stereo Scanner (as flown on the failed SROSS 2) and India's LISS 1 camera and CO_2-band Earth radiance monitor. The demonstration IRS P2 launched on PSLV number 2 in October 1994 carries a LISS 2 camera. IRS P3 on PSLV in March 1996 carries a German scanner and an improved wide-field imager. IRS P4 launched in 1997 performs ocean reconnaissance. P5 in 1998 will have a 10 m resolution vegetation camera. IRS P6 in 1999 will provide 2.5 m cartographic stereo images (improved to 1 m in 2002 by the Cartosat 2 follow-on). India has devoted the IRS P7 in 2000 to ocean and fishing applications and IRS P9 in 2001 will address environmental and atmospheric studies.

EOSAT and ISRO signed an agreement 21 October 1993 for the US company to distribute IRS data globally. EOSAT began receiving IRS imagery at its Norman, Oklahoma station on a daily basis in June 1994. EOSAT and the Indian government signed an agreement 2 February 1995 for exclusive worldwide marketing rights for IRS data. EOSAT expects that, within five years, the majority of its new imagery will be from IRS.

All three IRAS 1A imaging systems remained operational from 7 April 1988 until ISRO retired 1A 17 March 1995 from routine service. It remains on standby to supplement the other satellites. The IRAS returned more than 500,000 scenes by the end of 1993.

IRS P2 after being declared operational on 7 November 1994 demonstrated the 1C/1D bus. During the first year, it returned more than 60,000 images of India, in addition to EOSAT imagery received at two US stations. IRAS 1C/1D offers improved spatial and spectral resolution, onboard recording, stereo viewing capability and more frequent revisits. The VNIR resolution is about 20 m, the SWIR about 70 m and the PAN 10 m. SWIR provides data on water stress and pest infestation. The WiFS Wide Field Sensor with 180 m resolution monitors vegetation. All three cameras were returned their first images, 5 January 1996. Outside observers estimate the launch cost India Rs500 million.

IRAS P3 applies its Earth imagers to four major applications: ocean chlorophyll analysis, vegetation assessment, snow studies and geological mapping

LISS 1 (Linear Imaging Self-Scanning) CCD camera system (ISRO) 0517048

for identifying prospective mineral sites. ISRO activated the WiFS at 04.40 GMT 22 March 1997, during orbit number 15 and the MOS on 23 March during orbit 29.

India conducts TT&C functions at ISRO's Bangalore station, with imaging downlinked to Shadnager near Hyderabad for distribution by the National Remote Sensing Agency (NRSA). Cost of each LISS 1 (1:250,000) and LISS 2 (1:125,000) photo product is Rs3,000.

Specifications

IRS 1A

Launched: 17 March 1988 by Russian SL-3 Vostok from Tyuratam
Mass: 975 kg in-orbit at beginning of 3 year life
Contractors: ISRO Satellite Centre (prime), Hindustan Aeronautics (structure), and ISRO Space Applications Centre (imaging system)
Resolution: 72.5 m LISS 1, 36.25 m LISS 2
Orbit: 867 × 913 km (aiming for 904 km circular) 99.03° Sun-synchronous, crossing equator on descending node at 10.25 local time (allowed to drift to 10.10 after 2 years), with 22 day repeat cycle (307 revs) and 2,872 km ground track separation at equator
Spacecraft: box-shaped 1.6 × 1.56 × 1.1 m bus with two Sun-tracking solar wings totalling 8.58 m^2, providing 709 W EOL. Two Ni/Cd 40 Ah batteries provide eclipse power. 3-axis control with a 0.3° pitch/roll + 0.5° yaw pointing accuracy provided by zero-momentum RW system utilising Earth/Sun/star sensors + gyros; 1 N hydrazine thrusters (80 kg loaded) for AOCS + momentum dumping. Real time LISS 2A/B data downlinked at 10.4 Mbit/s each to the 10 m dish at Shadnager on 20 W X-band + LISS 1 data on 5 W S-band at 5.2 Mbit/s. No onboard recorder. Satellite control exercised from Bangalore, with Lucknow station providing backup
Sensors: Linear Imaging Self-Scanning (LISS). Three push-broom CCD units operate in four bands compatible with Landsat TM and Spot HRV. Band ranges (µm) and applications: Band 1 0.45–0.52 (coastal environment, soil/vegetation); Band 2 0.52–0.59 (vegetation vigour, rock/soil discrimination, turbidity, bathymetry); Band 3 0.62–0.68 (chlorophyll absorption, plant species); Band 4 0.77–0.86 (delineation of water features, land forms).

LISS 1. Four 2,048-element linear CCD imagers with spectral filters; total unit mass 38.5 kg. Focal length 162.2 cm, FOV 9.4°, generating resolution of 72.5 m over 148 km swath, framing LISS 2 image pairs. Data transmitted on S-band at 5.2 Mbit/s

LISS 2A/B. Eight 2,048 element linear CCD imagers with spectral filters. Focal length 324.4 mm, FOV 4.7° each, 36.25 m resolution over 74 km swath. The two 80.5 kg units are positioned either side of LISS 1 and view either side of the ground track with a 3 km lateral overlap; two pairs thus cover a single LISS 1 frame. Data transmitted on X-band at 10.4 Mbit/s each.

IRS 1B

Launched: 29 August 1991 by Russian SL-3 Vostok from Tyuratam. Declared operational 16 September 1991
Orbit: 857 × 919 km, 99.25° Sun-synchronous
Other specifications as for IRS 1A.

IRS 1E

Launched: 20 September 1993 by PSLV from Sriharikota but destroyed in ascent failure.

IRS 1C

Launch: 28 December 1995 by Russian SL-6 Molniya
Mass: 1,250 kg in-orbit at beginning of 3 year life
Contractors: ISRO Satellite Centre (prime), Hindustan Aeronautics (structure), ISRO Space Applications Centre (imaging system), Thomson Tubes Electroniques (40 W X-band TWTA), ISRO Inertial Systems Unit for momentum/reaction wheels
Resolution: 23.5/70.5 m LISS 3, 5.8 m PAN, 188 m WiFS
Orbit: 817 km 98.6° Sun-synchronous, crossing equator on descending node at 10.30 local time, with 24-day repeat cycle (341 revs) for LISS 3 and 5 days for PAN/WiFS; 2,820 km ground track separation at equator
Spacecraft: box-shaped 1.6 × 1.56 × 1.1 m bus with two Sun-tracking solar wings (each three 1.1 × 1.46 m panels) totalling 9.6 m^2, providing 830 W EOL. Two Ni/Cd 21 Ah batteries provide eclipse power. 3-axis control with a 0.15° pitch/roll + 0.2° yaw pointing accuracy provided by a zero-momentum RW system (four 5 Nm wheels) utilising Earth/Sun/star sensors and gyros. Propulsion system employs 1/11 N hydrazine thrusters for orbit control, attitude manoeuvres and momentum dumping. Lockheed 62 Gbit tape recorder
Communications: satellite control exercised from Bangalore, with Lucknow station providing back-up. Recorded data downlinked on the 40 W 8 GHz X-band link
Sensors: three cameras all utilise push-broom CCD units, continuing and expanding the IRS 1A/B imagers.

LISS 3. Similar to LISS 1/2 but replaces one visible band with Short Wave IR (SWIR). 85 W power operating. Three 6,000-element linear visible CCD imagers with spectral filters for bands 2,3 and 4. 23.5 m resolution, 142 km swath width. Data rate 35.70 Mbit/s. One 2,100-linear SWIR InGaAs CCD imager cooled to −10°C for band 5 (1.55–1.70 µm). 70 m resolution, 148 km swath width. Data rate 2.02 Mbit/s.

PAN. Single-band panchromatic (0.50–0.75 µm), 5.8 m resolution, three linear CCDs with 23.9 km swath widths combined for 70 km width. Three off-axis mirrors with focal plane splitting by isosceles prism. Swath steering of ±398 km by Payload Steering Mechanism with ±26° steerability (entire camera rotates, yielding better S/N + MTF). 0.2° PSM step yields 2.57 km nadir. 55 W power operating. Data rate 84.903 Mbit/s

WiFS. Wide Field Sensor similar to LISS 1 of IRS 1A/B. Four 2,048-element linear CCDs, dual band (0.62–0.68, 0.77–0.86 µm), focal length 56.420 mm, 188 m resolution, 774 km swath width (yielding 5-day repeat cycle), FOV ±27° (±13.5° each imager), data rate 40.43 Mbit/s.

IRS 1D

Launch: 28 September 1997 by PSLV which suffered fourth stage malfunctions and left ID in an eccentric orbit of 308 × 822 km rather than the planned circular orbit at 820 km
Other specifications as for IRS 1C.

IRS 2A

Launch: planned for 2000 on PSLV from Sriharikota
Sensors: LISS 3. Same as IRS 1C

LISS 4. 3-band, 5–10 m resolution, steerable WiFS. Same as IRS P5.

IRS 3

Launch: planned for 2005
Sensors: tbd
Data: Demonstrated by Space Imaging EOSAT

IRS P2

Launched: 15 October 1994 on PSLV-D2, from Sriharikota
Mass: 804 kg at launch
Orbit: 817 km 98.7° Sun-synchronous, crossing equator on descending node at 10.40, with 24-day repeat cycle (341 revs). The first IRS with a frozen perigee to minimise scale variations
Spacecraft: as 1C, with 6.42 m^2 4-panel wing providing 510 W (with two 21 Ah Ni/Cd batteries). Data rate 2 × 10.4 Mbit/s
Sensors: LISS 2. Single camera of 1A/1B design except each band carries staggered CCDs for one unit to provide same coverage as two LISS 2 previously. 131 km swath width.

IRS P3

Launch: 21 March 1996 by PSLV from Sriharikota
Mass: 922 kg at launch
Orbit: 819 × 821 km, 98.8° Sun-synchronous, crossing equator on descending node at 10.30 local time, with 22 day repeat cycle
Sensors: WiFS. Same as 1C's but with added 1.5–1.7 µm SWIR for determination of vegetation dynamics. In conjunction with 1C, provides 2–3 day coverage of India.

MOS. German (DLR) Modular Opto-electronic Scanner optimised for oceanography. 18-channel VNIR, 248 km swath. MOS-A: 0.7567, 0.7606, 0.7635, 0.7664 µm, resolution 2,520 × 2,520 m; MOS-B: 0.408, 0.443, 0.445 0.485, 0.520, 0.615, 0.650, 0.685, 0.750, 0.815, 0.870, 1.010 µm, resolution 720 × 580 m; MOS-C: 1.600 µm, resolution 720 × 1,000 m.

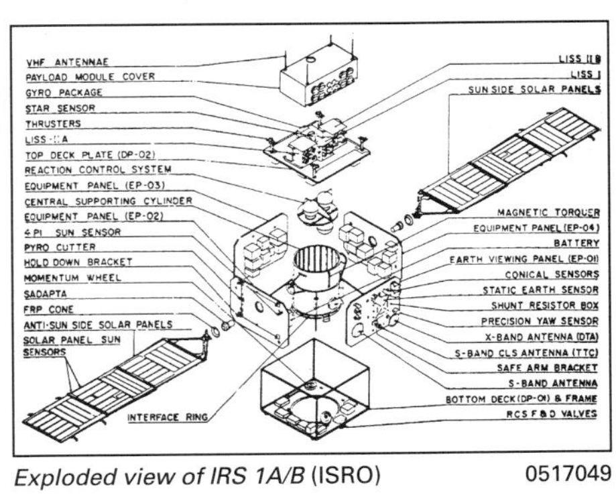

Exploded view of IRS 1A/B (ISRO) 0517049

XAP. X-ray astronomy payload for time variability and spectra of sources and detection of x-ray transients. Three Pointed-mode Proportional Counters 2–20 keV. XSM x-ray Sky Monitor 3–6 keV pinhole camera (PSPC). The Earth imagers have no onboard storage; real-time imagery is transmitted to one Indian and two German stations. XSM operates continuously, but PPC requires P3 to operate in sky-pointing mode, precluding Earth observations.

IRS P4 (Oceansat 1)
Launch: 26 May 1999 on PSLV C1 from Sriharikota
Orbit: planned 720 km circular; 98.28° inclination; Period 99.31 min
Sensors: OCM. 8-band Ocean Colour Monitor. Band 1: 0.402–0.422, 2: 0.433–0.453, 3: 0.490–0.510, 4: 0.500–0.520, 5: 0.555–0.575, 6: 0.655–0.675, 7: 0.745–0.785, 8: 0.845–0.885. IFOV 250 m VNIR, 500 m SWIR. +20/0/−20° along track steering, 17.35 Mbit/s, 75 kg.

MSMR (Multi-frequency Scanning Microwave Radiometer) 6.6/10.6/18/21 GHz, 3 dB beam widths 4.2/2.6/1.6/1.4°, <1.0K T resolution, 120/75/45/40 km resolution over 1,500 km swath, 86 cm diameter antenna. 5 kbit/s, 76 W, 6 kg

Size: 1.8 m × 1.98 m × 2.57 m; length 11.67 m fully deployed
Mass: 1,050 kg
AOC: 3-axis body-stabilised using reaction wheels, magnetic torques and hydrajure thrusters
Power: 9.6 sq m solar array generating 750 W; two 21 Ah Ni-Cd batteries
Life: 5 years.

IRS P5 (CartoSat)
Launch: planned on PSLV from Sriharikota 2003–2004
Sensors: 3-band XS vegetation camera, 10 m resolution over 40 km swath, across-track steering. Will provide multiple-crop discrimination, species-level discrimination, and so on

WiFS. Improved Wide Field Sensor, 125 m resolution over 750 km swath. Improves frequency/accuracy for assessing mono-crop areas.

IRS P6 (ResourceSat 1)
Launch: 17 October 2003 on PSLV C5 from Sriharikota
Sensors: Linear-Imaging Self-Scanner (LISS-4) operating in three spectral bands in the visible and near infrared region (VNIR) with 5.8 m spatial resolution and steerable up to +/–26° across track to obtain stereoscopic imagery and achieve five-day revisit capability; medium resolution LISS-3 operating in three spectral bands in VNIR and one in Short Wave Infrared (SWIR) band with 23.5 m spatial resolution; an Advanced Wide Field Sensor (AWiFS) operating in three spectral bands in VNIR and one band in SWIR with 56 m spatial resolution.

IRS P7/Oceansat
Launch: undetermined
Sensors: Ku-band scatterometer; Ku-band altimeter: Microwave radiometers: Thermal-IR radiometer.

IRS P8/Atmos
Launch: undetermined.

International

Civil Earth Observation Satellites

Satellite	Launch	Orbit	Range
International			
ERS 1	17 July 1991	780 km, 98.5°	radar/infra-red
ERS 2	21 April 1995	780 km, 98.5°	radar/infra-red
Meteosat 5	2 March 1991	0° GEO	visible/infra-red
Meteosat 6	20 November 1993	10° W GEO	visible/infra-red
Meteosat 7	2 September 1997	0° W GEO	visible/infra-red
Kitsat 3	26 May 1999	720 km, 98.4°	visible
Envisat 1	1 March 2002	790 km, 98.6°	visible/infra-red
MSG 1	28 August 2002	0° GEO	visible/infra-red
Canada			
Radarsat 1	4 November 1995	792 km, 98.6°	radar
China, People's Republic			
FY (Fung Yen) 2	10 June 1997	105° E GEO	visible/infra-red
FY (Fung Yen) 1C	10 May 1999	Sun-synch	visible/infra-red
FY (Fung Yen) 2	25 June 2000	GEO	visible/infra-red
21 Yuan 2	1 September 2000	495 km, 97.4°	visible/infra-red
Haiyang 1	15 May 2002	860 km, 98.8°	visible/infra-red
JB-3 2	27 October 2002	780 km, 97.4°	visible/infra-red
China/Brazil			
CBERS 1	Third quarter 1999	778 km, 98.5°	visible/thermal infra-red
France			
Spot 1[3]	22 February 1986	825 km, 98.7°	visible/near infra-red
Spot 2	22 January 1990	820 km, 98.7°	visible/near infra-red
Spot 3	26 September 1993	820 km, 98.7°	visible/near infra-red
Spot 4	24 March 1998	820 km, 98.7°	visible/infra-red
Spot 5	4 May 2002	825 km, 98.8°	visible/infra-red
India			
Insat 1D	12 June 1990	83° E GEO	visible/infra-red
IRS 1A[4]	17 March 1988	900 km, 99.0°	visible/near infra-red
IRS 1B	29 August 1991	900 km, 99.3°	visible/near infra-red
Insat 2A	9 July 1992	74° E GEO	visible/infra-red
Insat 2B	22 July 1992	93.5° E GEO	visible/infra-red
IRS P2	15 October 1994	817 km, 98.6°	visible/near infra-red
IRS 1C	28 December 1995	817 km, 98.6°	visible/near infra-red
IRS P3	21 March 1996	820 km, 98.8°	visible/near infra-red
IRS 1-D	28 September 1998	300 × 823 km, 98.6°	visible
Prota	22 October 2001	553 × 677 km, 97.9°	infra-red
Bird 2	22 October 2001	557 × 581 km, 97.7°	infra-red
Japan			
GMS 4	6 September 1989	140° E GEO	visible/infra-red
JERS 1	11 February 92	568 km, 97.7°	visible/infra-red/radar
GMS 5	18 March 1995	140° E GEO	visible/infra-red
MT-Sat	August 1999	140° E GEO	visible/infra-red
Adeos 2	14 December 2002	800 km, 98.7°	visible/infra-red
RFAS[2]			
Meteor 2–21	31 August 1993	950 km, 82.6°	visible/infra-red
Meteor 3–06	25 January 1994	1,200 km, 82.6°	visible/infra-red
Meteor 3M-2	August 1998	925 km, 82.5°	visible/infra-red
Mir[1]	20 February 1986	375 km, 51.6°	See note 1
Okean 4	11 October 1994	650 km, 82.6°	visible/infra-red/μ/ radar
Sich 1	31 August 1995	670 km, 98°	visible/infra-red/μ/ radar
Elektro 1	31 October 1994	76° E GEO	visible/infra-red
Resurs-O1 N3	4 November 1994	662 km, 98°	visible/infra-red
Resurs F1-M	10 November 1997		
Resurs-O1 N4	10 July 1998	835 km, 98.8°	visible/infra-red
Resurs-O21	10 July 1998	818 km, 98.8°	visible/infra-red radar
Resurs FM1-2	9 September 1999	225 km 92.3°	visible
United States			
GOES 7	26 February 1987	96° W GEO	visible/infra-red
Landsat 4	16 July 1982	705 km, 98.2°	visible/infra-red
Landsat 5	1 March 1984	705 km, 98.2°	visible/infra-red
NOAA 9	12 December 1984	855 km, 98.9°	visible/infra-red
NOAA 10	17 September 1986	815 km, 98.8°	visible/infra-red
NOAA 12	14 May 1991	816 km, 98.7°	visible/infra-red
GOES 8	13 April 1994	75° W GEO	visible/infra-red
NOAA 14	30 December 1994	855 km, 98.9°	visible/infra-red
Microlab 1 (Orbview 1)	3 April 1995	740 km, 70°	visible/infra-red
GOES 9	23 May 1995	135° W GEO	visible/infra-red
NOAA 15 (K)	13 May 1998	855 km, 98.9°	visible/infra-red
SeaStar (Orbview-2)	1 August 1997	705 km, 98.2°	visible/near infra-red
Lewis	23 August 1997	failed in post launch	
Early Bird	24 December 1997	failed in post launch	
Landsat 7	15 April 1999	705 km, 98.2°	visible/infra-red
Ikonos 2	25 September 1999	680 km 98.2°	visible
Terra	18 December 1999	665 km 98.2°	visible/infra-red
MTI	12 March 2000	600 km 97.4°	visible/infra-red
GOES-L	3 May 2000	106° west GEO	visible/infra-red
NOAA-L	21 September 2000	860 km 98.8°	visible/infra-red
Quick Bird 1	20 November 2000	Failed	visible/infra-red
Earth Observing 1	21 November 2000	702 km 98.2°	visible/infra-red
SAC-C	21 November 2000	685 km 98.3°	visible/infra-red
GOES 12	23 July 2001	30° W GEO	visible/ infra-red
QUICKTOMS	21 September 2001	329 km 51.6°	ozone mapper
Quick Bird 2	18 October 2001	465 km, 97.2°	visible/ infra-red
Jason	7 December 2001	1,319 × 1,333 km, 66°	radiometer
Timed	7 December 2001	627 × 629 km, 74°	visible/ infra-red
Israel			
EROS A1	5 December 2000	500 km 97.3°	visible/infra-red
Malaysia			
TiungSat 1	26 September 2000	650 km 64.6°	visible/infra-red

Notes: 1: manned space station carrying array of cameras, spectrometers and modules. 2: Resurs-F3 film-return mission in 1996. 3: Spot 1 retired 31 December 1990 but revived March-October 1992 and April-July 1993 to return real-time imagery; still held in standby mode. 4: IRS 1A retired from routine service 17 March 1995. Italicised text indicates satellites yet to be launched.

UPDATED

ERS

Current Status

ERS-1 mission operations were concluded 18 March 2000. The operational phase was divided into seven phases: orbit acquisition, commissioning phase (5.5 months), first-ice phase (28 December 1991 to 31 March 1992), experimental roll-tilt-mode campaign (4–14 April 1992), multidisciplinary phase (14 April 1992–6 December 1993), second ice phase (1 January-7 April 1994), geodetic phase (April 1994-March 1995).

Background

ESA built the ERS for global measurements of sea wind/waves, ocean and ice monitoring, coastal studies and a small amount of land imagery. Following ERS 1 approval in October 1981, West Germany, as 24 per cent contributor, provided the instrumentation. The Phase C/D contract, worth about DM840 million, was signed with the Dornier group 29 October 1986 for development, construction, launch preparation and the associated ground station. Under a January 1986 ESA/NASA agreement, direct readout of ERS 1 SAR data at the Fairbanks, Alaska, station is permitted, and NASA scatterometer and radar imagery are exchanged for other ERS data of interest. The exchange enhances NASA/ESA-supported polar ice research, and complements NASA experimental activities with the Topex/Poseidon (see International) and Shuttle Imaging Radar-C. By permitting NASA direct data readout from ERS, ESA reciprocates earlier NASA provisions allowing European data readout from Seasat and Nimbus 7.

The European Remote Sensing satellite's three primary all-weather instruments provides systematic, repetitive global coverage of ocean, coastal zones and polar ice caps, monitoring wave height/wavelengths, wind speed/direction, precise altitude, ice parameters, sea surface T, cloud top T, cloud cover and atmospheric water vapour content. The Active Microwave Instrument (AMI) can operate as a wind scatterometer or Synthetic Aperture Radar (SAR, with wave scatterometer mode), the Along-Track Scanning Radiometer & Microwave Sounder (ATSR-M) provides the most accurate sea surface T data to date, and the Radar Altimeter (RA) measures large-scale ocean/ice topography and wave heights. The follow-on ERS 2 was approved in June 1990 to provide continuity until Envisat 1 appears in 1999 (see the POEM entry at end). ERS 1/2 operated simultaneously 16 August 1995 to mid-May 1996, the first time that two identical civil SARs worked in tandem. The orbits were carefully phased to provide 1-day revisits, allowing the collection of interferometric SAR image pairs and improving temporal sampling. Although still working perfectly, ERS 1 funding required it to be put on standby from the end of May 1996. ERS-1 was formally retired in 1999 when it ran out of propellant. ERS-2 is expected to remain operational for several more years. Extending ERS 2 operations to 2000 is under discussion.

ERS 1 SAR image of flooding in Northern Europe in January 1995 (ESA) 0517243

ESA's Council announced in June 1990 that subscriptions were sufficient to start the ERS-2 programme. Cost is 371 MAU (1989 rate). Dornier was awarded the DM480 million prime contract (DM360 million went to non-German subcontractors). Similar instruments and mission to ERS 1 provided coverage until Envisat 1 became available. ERS 2's first SAR image, covering Italy's Campania region, was acquired 2 May 1995. The first ATSR 2 image was acquired 5 May 1995, covering Britain/Ireland. AMI's wind scatterometer mode did not begin operations until 22 November 1995 because of an anomaly. ATSR stopped normal operations 22 December 1995 because of a problem with its scan mirror. The SAR, RA and Microwave Sounder operated nominally since September 1995; ATSR, PRARE and GOME since January 1996.

Matra of France (18.31 per cent) supplied a Spot-type bus. The UK (13.34 per cent) provided the AMI lead contractor (Marconi). Other contributors to the 13-nation project include Italy (10.61 per cent), Canada (9.1 per cent), Netherlands (5 per cent), Sweden (3.9 per cent), Belgium (3.72 per cent), Spain (2 per cent), Denmark (1.99 per cent), Switzerland (1.7 per cent), Norway (1.5 per cent). ERS 1 development phase was estimated at 584 MAU at 1984 rates; ESA projected cost to completion was 728 MAU at 1989 rates.

The ERS 1 payload was delivered to Matra at the end of February 1990 for integration with the platform. Following launch, the solar array began deployment within 1 minute of separation, followed by the SAR antenna after 75 minutes and the wind scatterometer three-part antenna after 4 hours. All the instruments were activated in the first two weeks and the first SAR image (of Spitzbergen) was returned to Kiruna 27 July 1991. The operational phase began January 1992. Apart from PRARE's failure, the only significant in-flight anomalies have been some interruption to AMI operations during 1994 caused by recoverable problems in the transmitter, a sudden gain loss in the SAR telemetry downlink requiring a switch to the redundant unit, and failure of ATSR's 3.7 μm channel. All instrument and satellite performances meet or exceed expectations. >100,000 SAR images were returned in the first year.

ERS 2 final preparations. Prominent is the ATSR (curved apertures) (DASA) 0517414

Specifications

ERS 1

Launch: 17 July 1991 by Ariane 4 from Kourou, French Guiana
Mass: 2,384 kg in-orbit BOL (1,100 kg payload, 317.6 kg hydrazine)
Contractors: Dornier (prime), Matra (bus), Marconi Space Systems (AMI), Alenia Spazio (RA), Laben (payload data handling), Contraves (payload module structure), Fokker (payload module thermal control, harness, integration), Aerospatiale (solar array, laser reflectors), British Aerospace (ATSR), Insitut für Navigation Universität Stuttgart (PRARE)
Resolution: 30 m SAR (100 km swath), 1 km/0.5K ATSR-M, 3 cm Radar Altimeter, 2 m/s wind scatterometer, 5–10 cm PRARE satellite range (failed)
Orbit: controlled within 765–825 km, 98.5° Sun-synchronous for required track repeat period, crossing equator southbound at 10.30 am local time. 3-day (43 rev) repeat cycle for 3-month commissioning phase, then 35-day cycle alternating with 3-day cycle periods in early 1992 and 1994. From April 1994, a 168 day repeat 'geodetic' orbit has been flown
Spacecraft: 11.8 m high, 11.7 m span on-orbit. Based on Matra's 3-axis Spot bus. Payload electronics housed in compartment mounted on platform. Two 2.4 × 5.8 m solar wings (total of 22,260 cells) to provide 2.2 kW after 2 years, supported by 4 × 24 Ah SAFT Ni/Cd batteries. ESOC Mission Management & Control Centre in Darmstadt controls via Kiruna
Sensors: Active Microwave Instrument (AMI) operates in the SAR mode to produce C-band Earth imagery and separately in the AMI-wind mode as a 3-beam C-band scatterometer. Synthetic Aperture Radar: spatial resolution 30 m; frequency 5.3 GHz (C-band), VV polarisation, pulse length 37.1 μs, data rate 105 Mbit/s; swath width 100 km, with

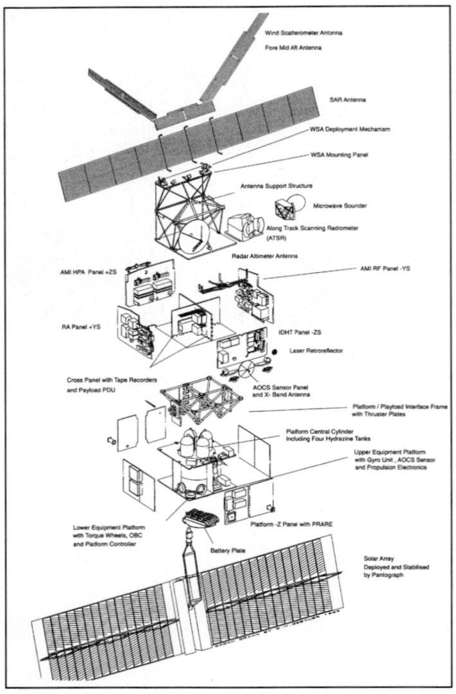

ERS: ESA's ocean and ice monitoring platform 0517028

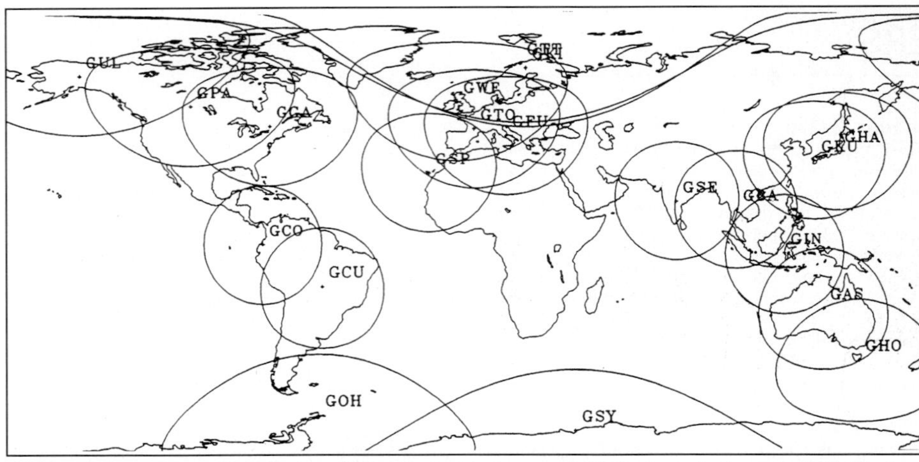

ERS receiving stations. Alice Springs (GAS, Australia), Aussaguel (GTO, France), Bangkok (GBA, Thailand), Cotopaxi (GCO, Ecuador), Cuiaba (GCU, Brazil), Fairbanks (GUL, US), Fucino (GFU, Italy), Gatineau (GCA, Canada), Hatoyama (GHA, Japan), Hobart (GHO, Australia), Hyderabad (GSE, India), Kiruna (GKI, Sweden), Kumamoto (GKU, Japan), Maspalomas (GSP, Spain), O'Higgins (GOH, Antarctica/Germany), Pare-Pare (GIN, Indonesia), Prince Albert (GPA, Canada), Syowa (GSY, Antarctica/Japan), Tromso (GTT, Norway) and West Freugh (GWF, UK) (Eurimage) 0517030

23° incident angle at mid-swath (up to 35° possible via experimental 'roll-tilt' attitude control system mode). Wave Scatterometer: the programmable wave mode operates every 200–300 km providing a 6×6 km image for extraction of information on wave length and direction. Wave length accuracy of ±25% over 100–1,000 m, direction accuracy ±20°. AMI Wind Scatterometer: three antennas (mid 0.35 × 2.3 m, fore/aft 0.25× 3.6 m) providing fore/mid/aft beams sweep a 500 km swath in 50 km cells at 5.3 GHz, allowing the radar returns to be analysed for surface wind vectors: 4–24 m/s + 0–360° with ±20° accuracy. Incident angle varies over 27–58°. Spatial resolution: 50 km.

Radar Altimeter (RA). The nadir-viewing 13.8 GHz Ku-band 1.3° beamwidth 1.2 m dia antenna altimeter measures, in Ocean Mode (330 MHz bandwidth), wind speed (2 m/s accuracy from measuring backscattering coefficient to 0.5 dB), wave heights of 1–20 m with 0.5 m accuracy over 1.6–2.0 km footprints, and determine altitude to about 5 cm (<1 cm resolution). Ice Mode (82.5 MHz bandwidth) operates with a coarser resolution to determine ice sheet topography, ice type and sea/ice boundaries.

Along-Track Scanning Radiometer & Microwave Sounder (ATSR-M). An experimental passive instrument comprising an advanced four-channel infra-red radiometer (UK-Australia) and a two-channel nadir-viewing microwave sounder (France) for T and water vapour measurement, respectively. ATSR-M IR Radiometer: res 0.5K over 50 × 50 km square, 1 km spatial res; wavelengths 1.6/3.7/11/12 μm; swath width 500 km. Microwave Sounder: passive nadir-viewing radiometer operating at 23.8/36.5 GHz measuring the vertical column water vapour content within a 22 km footprint, providing corrective data for ATSR sea surface T + RA measurements.

Precise Range And Range Rate Experiment (PRARE) for orbit determination (ranging accuracy 3–7 cm) using 8.489 GHz X-band 1 W signals transmitted to a network of 60 cm 2 W mobile ground transponders, with 2.248 GHz 1 W S-band transmissions permitting ionospheric corrections. PRARE failed after 3 weeks in orbit, concluded to be destructive proton-induced memory latch-up over the South Atlantic Anomaly. Russia's Meteor 3–06 launched January 1994 is successfully operating a PRARE provided by Kayser-Threde under DARA contract.

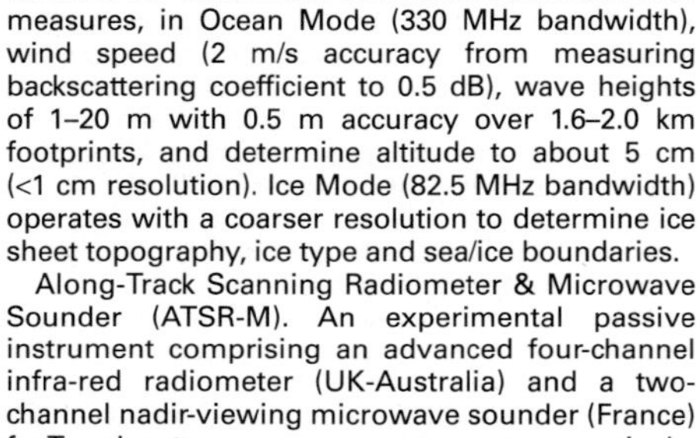

ERS orbital configuration (ESA) 0517029

Laser Retroreflector permits precise range/orbit determination, although less frequently, and RA calibration.

ERS 2

Launch: 21 April 1995 by Ariane 4 from Kourou, French Guiana
Mass: 2,516 kg at launch, 2,516 kg on-station BOL, 2,110 kg dry
Contractors: as ERS 1 except Officine Galileo (GOME), Schrack (microwave radiometer)
Orbit: similar to ERS 1, phased for 1 or 8 day revisits in turn. 771 × 797 km 98.55° at injection
Spacecraft: Saab developed a new computer memory due to ERS 1 design obsolescence
Sensors: Active Microwave Instrument as ERS 1

Radar Altimeter as ERS 1

Global Ozone Monitoring Experiment (GOME) scanning near-UV/visible spectrometer measuring backscattered Earth radiance in 3,500 channels over four bands, 240- 295, 290–405, 400–605, 590–790 nm (2–4 Å res), to determine ozone/trace gases in tropo/stratosphere. Scan angle and integration variable within ±2–31°/0.1–3,000 s. Internal, Sun, Moon calibration. 52 kg, 30 W operating; 50 × 60 × 70 cm. Each band focused on 1,024-pixel photodiode array cooled to −40±1°C by Peltier coolers. Essentially a scaled-down version of SCIAMACHY

ATSR augmented by 0.55, 0.67, 0.78 μm visible channels to improve monitoring of land applications (vegetation moisture, state, species). Also improvements to mechanical and electronics design.

PRARE redesign of control processor and memory with radiation-tolerant components plus redundant unit, to avoid ERS 1-type loss
Ground segment: ERS are controlled from ESOC at Darmstadt, Germany with ESA ground receiving stations at Salmijärvi, near Kiruna (Sweden, the primary station with 15 m antenna; also used for TT&C), Fucino (Italy), Gatineau (Canada), Maspalomas (Spain), plus national stations at Fairbanks (Alaska), Prince Albert (Canada), West Freugh (UK), Alice Springs (Australia). SAR's 105 Mbit/s image data are returned in real time only, available only when the wave/wind modes are inactive (other data are recorded onboard, thus providing global coverage). ESA's ESRIN ERS Central Facility (EECF) facility at Frascati, Italy is the data management centre and prepares the mission operation plan for ESOC, with processing/archiving facilities at Brest (France), Farnborough (UK), DLR Oberpfaffenhofen (Germany) and Matera (Italy). Some products, such as from the wind scatterometer, are available within 3 hours of observation. Dornier was awarded a DM5 million contract in April 1989 for 20 to 30 mobile PRARE ground stations. Some were deployed in 1994 in support of Meteor 3 and the remainder in 1995.

POEM series

To solidify its civil remote sensing plans ESA developed the Polar Orbit Earth Observation Mission Programme, envisaged as two series (Envisat/Metop) of polar platforms providing data to a wide user spectrum. POEM's bus derives from the ESA designed modular platform built as part of the Columbus programme until it was unified in 1994 with Envisat 1. POEM-1 Phase 1 was approved at 1991's Munich Ministerial Council meeting and the 1,174.5 MAU Ph 2 at Spain in 1992. Envisat 1, the largest and most expensive satellite ever built by Europe, will be launched in 1999 for environmental research, providing continuity for ERS. 40 MAU preliminary work on Metop 1 was approved; the full 760 MAU (1994 rates) was reduced in early 1996 by 45 per cent when it became clear it could not be met. The cost was cutting the programme from 15 years to 10 years and deleting the microwave imager. Developed in co-operation with EUMETSAT for 2001 launch, its climatic research instruments will be accompanied by operational meteorological units. The PPF prime contract with MMS is worth ECU502 million (1988 rate; ECU675 million 1995). Ph C/D contract was signed 24 July 1995 with MMS, including Envisat 1 payload integration. DASA has 30 per cent.

France provides 24 per cent of the funding, UK 19 per cent, Germany 18 per cent, Italy 11 per cent; 5 per cent remains uncovered. The Phase C/D contract, worth ECU675 million (1995 rates; ECU502 million 1988) was signed with MMS 24 July 1995 for Polar Platform development and Envisat 1 integration. MMS Bristol is responsible for platform development and Envisat integration, and MMS Toulouse for the service module. ASAR, RA-2, MERIS, MIPAS, GOMOS, MWR and LRR will be provided by ESA; AATSR (UK/Australia), SCIAMACHY (Germany/Netherlands) and DORIS (France) will be funded nationally. The ECU810 million contract for payload development and system engineering was signed 17 July 1996 with Dornier.

Metop 1 is being jointly developed by ESA and EUMETSAT in the frame of a single space segment as the prototype of a series of operational satellites addressing operational meteorology and climate monitoring missions of EMETSAT. The single space segment, comprises Metop 1, 2 and 3 satellites and instruments like ASCAT, GOME and GRAS. The procurement will be done by a single space segment team led by an ESA project manager assisted by an EUMETSAT project manager. The other instruments will be provided by EUMETSAT, NOAA and the 'Centre National d'Etudes Spatiales' (CNES). EUMETSAT is responsible for launcher procurement, ground segment, satellite operations and data distribution. In addition to complementing the meteorological payload, the

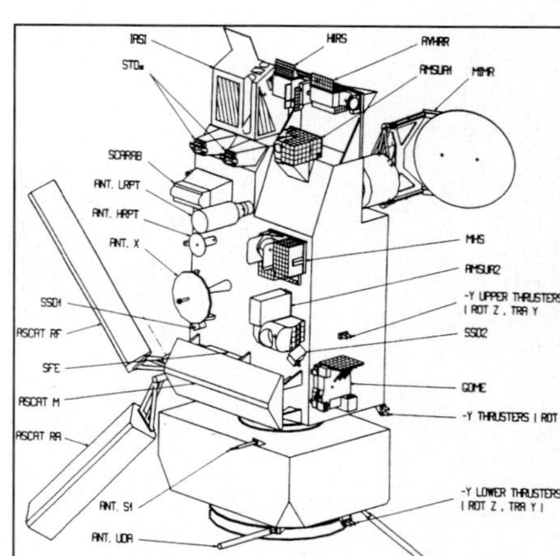

Metop will provide operational meteorology and climate monitoring. Scarab and MIMR have been deleted (ESA) 0517228

The ERS-1 carriesan ESA SAR, and a NASA scatterometer 0038495

climate instruments will establish a global ocean, ice and ozone monitoring system. Phase A started October 1993 for completion 1994; Phase B began July 1996, EUMETSAT's cost is projected at ECU 1.5 billion (1994 rate) for two satellite launches, and 14 years of operations: ESA contributes financially, partially to Metop-1, the development of ASCAT, GOME and GRAS. Metop-2 as planned for 2007 but it is to be available for launch within 18 months of the launch of Metop-1. Metop-3 is envisaged for 2011.

Envisat 1
Launch: 1 March 2002 by Ariane 5 from Kourou, French Guiana
Design life: 5 years
Mass: about 8 t; 1,900 kg payload
Contractors: Dornier (mission prime, MIPAS), MMS (platform, service module, integration, GOMOS, ASAR, AATSR), Alenia Spazio (RA, MWR), Fokker Space (solar array, SCIAMACHY), CASA (service/payload module structures, ASAR panels), Aerospatiale (MERIS)
Orbit: planned 800 km 98.55° morning orbit, crossing equator 10.00 am, 35 day repeat cycle
Spacecraft: 3 axis 10 m long, 3 m wide, comprising the payload module + service module. Single 5 × 14 m 14-panel solar array generates 6.7 kW after 4 years (1.9/4.1 kW average/peak to payload); 8 × 40 Ah Ni/Cd batteries. Data handling: one 100 Mbit/s channel (for ASAR) + 10 × 32 Mbit/s for others. 4 × 30 Gbit tape recorders; record 5 Mbit/s, playback 50 Mbit/s. FOCC Flight Operations Control Centre at ESOC (via Kiruna S-band or DRS); PDCC Payload Data Control Centre at ESRIN. X-band real time and recorded data direct and Ka-band via DRS (steerable 90 cm antenna on 2 m mast); each two channels 50/100 Mbit/s. Hydrazine thrusters (300 kg loading) for attitude control; RWs + magnetorquers for fine control (0.1° 3Σ)
Sensors: Advanced Synthetic Aperture Radar (ASAR) improved version of ERS SAR. 5.331 GHz, 1.3 × 10 m array of 20 66.4 × 99.5 cm radiating panels (each 16 rows of 24 microstrip patches). Imaging mode HH or VV, 29 × 30 m/2.5 dB resolution, 7 swaths 100–56 km at 20–45°, 96.3 Mbit/s, 1,200 W DC power consumption; alternating polarisation mode HH + VV, 29 × 30 m/3.5 dB, 7 swaths 100–56 km at 20–45°, 96.3 Mbit/s, 1,200 W; wide swath mode HH or VV, 150 × 150 m/2.5 dB, >400 km 5 subswaths at 17–42°, 96.8 Mbit/s, 1,200 W; global monitoring mode HH or VV, 1,000 × 1,000 m/1.5 dB, >400 km 5 subswaths at 17–42°, 0.9 Mbit/s (allowing onboard storage), 750 W; wave mode HH or VV, 30 m/2.0 dB, two vignettes of 5 × 5 km every 100 km in any swath at 20–45°, 0.9 Mbit/s, 520 W. Up to 30 min of high resolution imagery each orbit.

Radar Altimeter (RA-2) fully redundant nadir pointing pulse limited radar using single 1.2 m dish at 13.575 & 3.3 GHz. Derived from ERS RA but improved performance without significant increase in mass and power consumption. 3.3 GHz channel added to correct for ionosphere propagation effects. Fixed pulse repetition frequencies of 1,800/450 Hz are respectively used by the two channels. RA-2 onboard autonomous selection of transmitted bandwidth (CW/20/30/320 MHz) avoids dedicated operational mode commanded from ground and makes possible continuous operation over ocean, ice, land and their boundaries. Altitude accuracy after ionospheric correction improved to <4.5 cm for Significant Wave Height up to 8 m. Operation supported by MWR/LRR/DORIS measurements.

Medium Resolution Imaging Spectrometer (MERIS) is the first programmable imaging spectrometer. VNIR (400–1,050 nm), 250 m/12.5 nm resolution (adjustable as required), swath width 1,130 km at 800 km altitude. Aimed at water quality measurements, such as phytoplankton content, depth and bottom-type classification and monitoring of extended pollution. Secondary goals: atmospheric monitoring and land surfaces processes.

Michelson Interferometer for Passive Atmospheric Sounding (MIPAS) is a Fourier transform spectrometer observing mid-IR 4.15–14.6 μm limb emissions with high spectral resolution (<0.03 cm^{-1}), allowing day/night measurement of trace gases (including the complete nitrogen- oxygen family and several CFCs) in stratosphere and cloud-free troposphere. Global coverage, including poles.

Global Ozone Monitor by Occultation of Stars (GOMOS). Two spectrometers observe setting stars over UV to near-infra-red for 50 m vertical resolution and 0.1% annual variation sensitivity of ozone/related gases. Occultation method is self-calibrating and avoids the long term instrumental drift problems of current sensors. Spectrometer A: 2,500–6,750Å, res 0.3 nm/pixel; B: 9,260–9,520 Å (H_{2O}) + 7,560–7,730 Å (O_2), res 0.05 nm/pixel. The Fast Photometer Detection Module provides 1 kHz 2-band scintillation monitoring of the star image.

Advanced Along-Track Scanning Radiometer (AATSR). Performance almost identical to ATSR 2 on ERS 2. Provided by UK Department of Environment + Australia.

MicroWave Radiometer (MWR) is a nadir-viewing 23.8/36.5 GHz Dicke radiometer with 3 dB IF bandwidth 600 MHz. ERS's MWR design modified mainly in the mechanical layout and antenna configuration. Radiometric stability <0.5K over 1 year. Periodic onboard calibration by switching receiver input between two references: a horn pointing at the cold sky and a hot radiator at ambient. MWR determines tropospheric columnary water vapour content by measuring the radiation received from Earth's surface to correct RA-2 altitude measurements.

Scanning Imaging Absorption Spectrometer for Atmospheric Chartography (SCIAMACHY) 2,400–23,800 Å grating spectrometer (limb/nadir viewing) for detrimental trace gas measurement in troposphere/stratosphere. resolution 2.4 Å UV and 2.2–14.8 Å vis/infra-red. Swath 1,000 km in nadir mode.

DORIS as Spot 3/4 + Topex/Poseidon for precise (1 m) onboard real time autonomous orbit determination and distribution of timing pulse; 10 kg, 14 W.

LRR Laser Retro Reflector for precise orbit determination

Ground segment: Comprises two major elements: FOS Flight Operations Segment and PDS Payload Data Segment. FOS employs the FOCC Flight Operations Control Centre at ESOC to provide mission and operations planning, and command/control via the primary S-band TT&C station at Kiruna-Salmijärvi. The PDS is composed of the PDCC Payload Data Control Centre at ESRIN (controlling and monitoring all services offered by PDS facilities); two Payload Data Handling Stations at Kiruna-Salmijärvi (X-band) and at ESRIN (via DRS) for acquiring global and regional data and providing near-realtime and FD services; the PAC Processing and Archiving Centres (offline services). The PDS will offer a comprehensive set of user services, with online access to catalogue, browse and ordering services. ESA's products will range from raw reformatted data (level 0) to geophysical products (level 2) and images (from browse to high resolution). PDS development has been awarded to a consortium led by Thomson-CSF.

ESA plans to gather environmental data with the Envisat using eight major instruments to perform atmospheric ozone and turbulence monitoring 0038494

Alcatel Bell is developing the Ka-band on-board communicators system for the Envisat platform 0084762

Metop 1
Launch: planned for 2005 by Ariane 5 from Kourou, French Guiana
Design life: 5 years
Mass: 4,244 kg, payload 975 kg
Contractors: Matra Marconi Space/Dornier
Orbit: 830 km Sun-synchronous, 09.30 descending node
Spacecraft: reduced Envisat platform, PPF/Spot avionics, 10-panel solar array feeding 5 × 40 Ah Ni/Cd batteries. Eight paired 15 N hydrazine thrusters drawing on four tanks, fine pointing (±0.15°) by three 40 Nms RWs
Sensors: Advanced Visible/Infra-Red High-Resolution Radiometer (AVHRR/3). Day/night imaging in 6 bands over 0.68–12.50 μm, sea surface T, ice, snow and vegetation cover. 31.3 kg, 29 W, 822.0 kbit/s.

High-Resolution Infra-red Temperature Sounder (HIRS/3). 20 bands over 0.69–14.95 μmm, vertical T profile. Moisture content, cloud height, surface albedo. 33.1 kg, 24 W, 2.88 kbit/s.

Advanced Microwave Sounder Unit (AMSU-A1/A2). Two modules covering 23.8 MHz-89 GHz: A1 13 channels + A2 2 channels for T/humidity atmospheric profiles. A1: 53.3 kg, 88.3 W, 2.080 kbit/s; A2: 47.4 kg, 37.25 W, 1,120 bit/s.

Microwave Humidity Sounder (MHS). 90–190 GHz 5-channel self-calibrating for humidity profiling. 66 kg, 100 W, 3.95 kbit/s.

Argos. DCS 2 data collection/location. 47 kg (inc antennas), 27.5 W, 2.56 kbit/s.

IR Atmospheric Sounding Interferometer (IASI). 3.5–15.5 μm Fourier Transform spectrometer. 148 kg, 196 W, 1.5 Mbit/s.

Advanced Wind Scatterometer (ASCAT). 6.255 GHz C-band radar for ocean surface wind vectors, 213 kg, 250 W, 45 kbit/s.

Global Ozone Monitoring Expt (GOME). As ERS 2. 59 kg, 44 W, 50.0 kbit/s for Metop 1, 2.

OMI-Ims considered for Metop-3.

Global Navigation Satellite System Receiver for Atmospheric Sounding (GRAS) Spaceborne GPS/GLONASS receiver for Earth science applications with dual frequency measurement from GPS and GLONASS carrier phase measurements to mm-level precision good quality (<1 m) code phase measurements.

VERIFIED

Meteosat Series

Background

Meteosat constitutes ESA's contribution to the international World Weather Watch of the Global Atmospheric Research Programme (GARP) which began with Meteosat 1, launched 23 November 1977 by Delta from Cape Canaveral. Although fully operational for only two years of its projected three year planned life, it gathered data until it exhausted its supply of hydrazine propellant in October 1985. During this lifetime, the satellite played a key role in the Global Weather Experiment, a major exercise involving almost all of the 147 member nations of the World Meteorological Organisation. GWE had almost ended when, a day after its second anniversary, Meteosat 1 suffered an onboard radiometer failure. Despite efforts spread over some months, its imaging capacity never reached its full programme again, although its data collection function remained unimpaired. By then, however, ESA had gathered more than 40,000 images and submitted an outline proposal for the five-satellite Meteosat programme for 1984–1994.

Engineers modified M2 after M1's radiometer problems and repaired damage during ground tests that were designed to assess its resistance to the additional vibration involved in an Ariane launch instead of the Delta. The launch occurred on 19 June 1981. Its primary imaging mission successfully started, but ESA ground controllers could not activate the DCP system. Despite the problems the two satellites together provided a full service until the end of 1984, when M1 began to malfunction and its hydrazine approached exhaustion. NOAA then loaned ESA the partially operational GOES 4, moving it to 10° W by June 1985, as a return for M2's help following a GOES failure over the Atlantic. Data collection by GOES 4 continued from 43° W into 1988, with M2 providing the images, until M3 (P2) assumed the relay role. Although M2 had a three year design life, the quality of its imagery continued to be excellent into 1991. As part of the effort to extend its life into 1989 (when M4 became available), ESA performed a small inclination manoeuvre at end of 1986 to hibernate it at 10° W. ESA planned to raise it above GEO, using the 2 kg hydrazine remaining, when M5 was commissioned in 1991. The satellite returned a commemorative photo on its 10th anniversary image 19 June 1991. In all, the satellite returned some 284,000 images during its career. Its retirement move began 2 December 1991, when its thrusters fired for 8 minutes to raise altitude by 140 km. During the third burn sequence two days later, the hydrazine became exhausted and the orbit only increased by 334 km instead of the planned 700 km. The transmitter was shut down 6 December 1991.

A three-channel Meteosat image is returned every 30 min: visible (left), infra-red (centre) and water vapour (EUMETSAT) 0008086/0008087/0008088

ESA built M3/P2 as a qualification model for the pre-operational programme and launched it 15 June 1988 as a back-up to the ageing M2 until the delayed MOP series could begin. Financial constraints dictated that the 12 year old spacecraft could not be upgraded to full flight standards, but its primary role was only to provide the DCP service unavailable from M2. Although its electronically despun antenna experienced 3 dB fluctuations, it operated as prime satellite from August 1988 until M4 appeared. ESA considered loaning M3 to NOAA at 50° W to compensate for the GOES 6 failure of January 1989. Both organisations approved the move June 1989 and completed the transfer from 3° W to 50° W on 4 November 1989. It returned to assume the primary role in January 1990 after the appearance of M4's imaging problems. M4 again assumed the prime role in April 1990 after M3 lost dissemination channel 1. M3 moved to 3° W as standby; after Meteosat 5's appearance it began the move again 29 April 1991 to 50° W, arriving 19 July, again to substitute for GOES. Service began 1 August 1991.

Limited visibility to earth stations limited a move further west for Meteosat. After a new relay station came online February 1993 at Wallops Island (Virginia), this problem disappeared and Meteosat began to move again 27 January 1993 to 75° W, arriving 24 February, to provide coverage of the whole continental US. The Wallops station was operated remotely by ESOC. All imagery received there returned to Germany via PanAmSat, then after processing returned to Wallops, where it was uplinked to Meteosat for broadcast to users. M3 moved to 70° W by 14 March 1995, from where it was boosted 21 November 1995 into 36,725 × 36,762 km, 2.82°, using the last 5.1 kg hydrazine.

MOP–1/M4 was launched 6 March 1989 and began service as the prime satellite 19 June 1989. But the satellite began to generate erroneous grey levels in October 1989 and image rectification became impossible. A switch to the second Synchronisation and Imaging Chain (SIC-2) in November 1989 resumed nominal imaging, but it was replaced as prime at 0° by M3 24 January 1990 to allow detailed failure analysis and an intensive test programme. The cause was found to be a malfunction of the DC/DC converters supplying the SICs. All future satellites were modified. Procedures were initiated to switch periodically from one SIC to the other and the image processing software was improved to detect/correct the erroneous grey levels (if not too numerous). Following the failure of an M3 dissemination channel, M4 returned as prime in April 1990. It was then replaced as prime by M5 in February 1994 and held at 8° W until 10 May 1995, when it moved to 8° E before being boosted 9 November 1995 into 36,619 × 36,777 km, 1.50°.

M5 has remained the operational satellite, but the on-board fuel reserves for inclination control have been consumed and thus inclination is increasing at about 0.8°/year. Due to the high inclination and the impact upon user station reception, the operational mission was switched to M6 in February 1997: M6 was relocated from 10° W to 0° in early 1997.

When the EUMETSAT Convention came into force on 19 June 1986, the organisation assumed overall and financial control of the MOP Meteosat Operational Programme beginning on 12 January 1987. A special Council controls EUMETSAT Earth observation programmes and it represents the National Meteorological Services of 17 European states. Each European country makes a contribution to the programmes based on GNP (per cent): Austria 2.23, Belgium 2.70, Denmark 1.76, Finland 1.84, France 16.78, Germany 22.29, Greece 0.96, Ireland 0.54, Italy 15.46, Netherlands 4.03, Norway 1.47, Portugal 0.86, Spain 6.96, Sweden 3.20, Switzerland 3.33, Turkey 1.50 and UK 14.09.

In order to make a smooth transition from a research and technology programme to a fully operational system, ESA planned to develop its earth imaging resources in three phases: the Meteosat Operational Program (MOP), the Meteosat Transition Program (MTP) and the Meteosat Second Generation Program (MSG). Each phase uses distinct satellites and ground procedures. Meteosat 5 serves as the geostationary system's primary satellite, but utilises Meteosat 6 as a standby. Two older satellite's Meteosat 3/4 predated this pair but Eumetsat de-orbited both in 1995. Meteosat 7, the MTP Meteosat Transition Programme satellite, may join the others at geosynchronous in the second half of 1997 to maintain a continuous service until the MSG (Meteosat Second Generation) appears in 2000. Encryption of high resolution image data began 1 September 1995. On 1 December 1995, Eumetsat inaugurated a new facility dedicated to control the satellites at its MTP HQ of satellite launch and operations Primary Ground Station in Fucino, Italy. Before then, ESA provided these functions.

The Meteosat Operational Program (MOP) began 23 November 1983 under ESA's auspices and ended 30 November 1995. ESA transferred overall responsibility to EUMETSAT in January 1987 but in-orbit control remained with ESA. The Meteosat Operational Program receives funds from the Eumetsat member states in contributions of (per cent): Belgium 4.4, Denmark 0.58, Finland 0.35, France 25.60, Germany 26.39, Greece 0.30, Ireland 0.11, Italy 12.00, Netherlands 3.00, Norway 0.50, Portugal 0.30, Spain 5.24, Sweden 0.93, Switzerland 3.03, Turkey 0.50, UK 16.76 (0.01 per cent not covered). Various sources estimate the total cost for the three satellite series in 1982 terms at ECU378 million (revised to ECU721 million, or US$901 million, in 1995 terms).

Aerospatiale received an ECU139.1 million contract in May 1984 for three flight model MOP satellites and one spare. MOP had three primary missions: Earth imaging, dissemination of image and other meteorological data, and data collection/distribution, with two secondary objectives of meteorological processing and data archiving/retrieval. These MOP/Meteosat satellites transmit real-time image data (1,686.833 MHz) with up to 66 channels of DCP data (1,675.181–1675.381 MHz) at

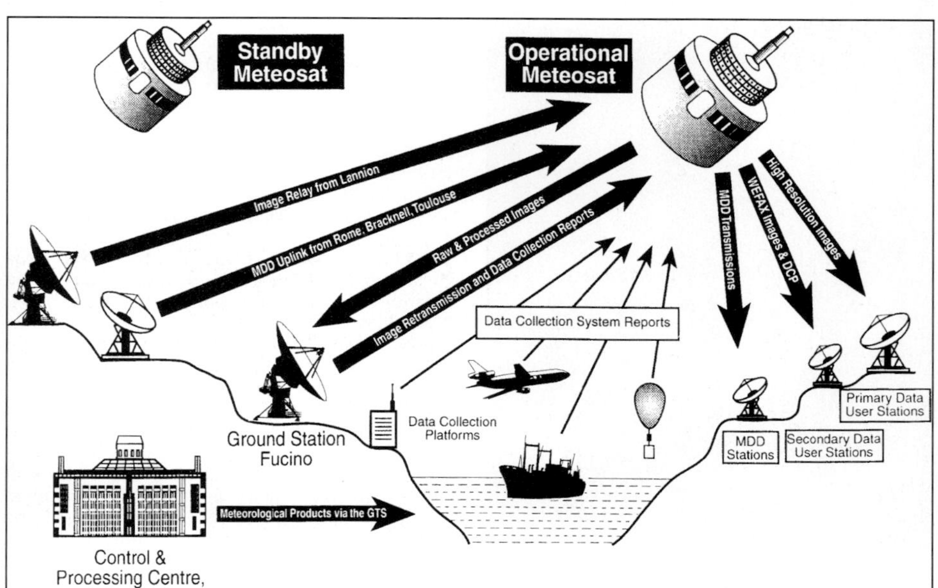

Meteosat system overview (EUMETSAT) 0517419

Meteosat integration at Aerospatiale 0517031

333 kbit/s to EUMETSAT's Primary Ground Station in Fucino for relay to the Mission Control Centre in Darmstadt.

Eumetsat expected to put MOP 2 in service as the prime satellite in Summer 1991, replacing Meteosat 4. It received the first image on 3 April during commissioning at 4° W. However, an imaging anomaly required Eumetsat to make modifications to ground processing software. Then the satellite entered full service as prime in February 1994. ESA formally transferred the satellite to EUMETSAT 14 January 1992.

Then MOP 3 followed MOP 2 as part of the programme after it transmitted its first image (visible) on 29 November 1993. An onboard anomaly in the infra-red/water vapour imagery was detected in further testing, caused by an apparent movement of the cold optics within the radiometer. Additional ground processing software corrects the anomaly sufficiently to allow normal extraction of image products.

Darmstadt processes MOP images into a range of formats covering Europe only, or covering most of the full Earth disc with up to 24 frames of data. Final processed images include lat/long grids and coastlines which are added by transmission up to Meteosat (channel 1 2,101.5 MHz; channel 2 2,105.0 MHz) and relayed to users on two channels. Channel 1 (1,691.0 MHz) operates on conventional analogue WEFAX, and is compatible with other GEO metsats and NOAA's Automatic Picture Transmission service. This information is available to simpler Secondary Data User Stations. Channel 2 (1,694.5 MHz) provides high-resolution digital transmissions to Primary Data User Stations. Some of NOAA's GOES images are also available via Meteosat using the Lannion, northwest France relay station. Three other channels (1,695.725, 1,695.756, 1,695.787 MHz) primarily relay digitised facsimile and selected conventional met observations to Africa, as the Meteorological Data Dissemination (MDD) service.

In 1996, DCP messages received by Meteosat totalled 3,311,644, allowing a distribution of more than 1.1 million images to meteorological services around the world. The MARF Meteorological

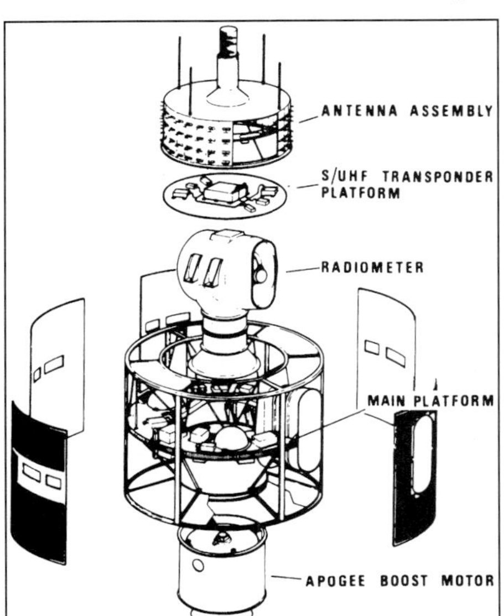

Meteosat principal features 0517032

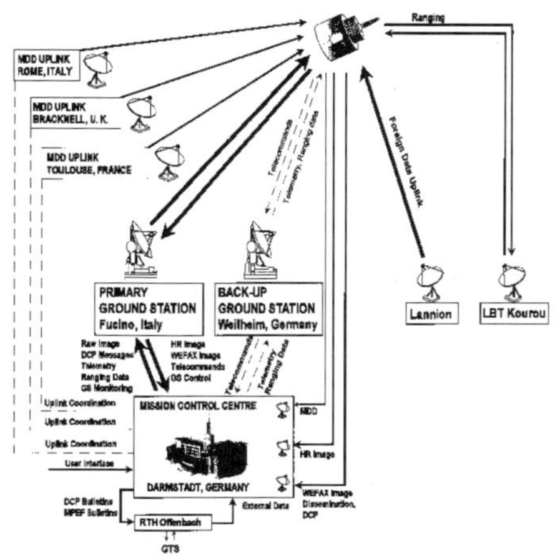

EUMETSAT ground segment from December 1995 (EUMETSAT) 0517033

Archive and Retrieval Facility holds, as the single repository, all Meteosat image data and derived products acquired since 1978. Live data continuously enters the archive and Eumetsat copies historic data for later retrieval. All of the data are recorded on 6.6 Gbyte optical disks, with online access through the MARF's catalogue. During MOP, ESOC gathered some 40,000 tapes' worth of data, dating back to Meteosat 1.

The Meteosat Transition Programme began in 1991 when Eumetsat planned to orbit one Meteosat follow-on satellite, identical to MOP, to maintain a continuous service until MSG appears. The organisation signed FFr630 million contract in June 1993 with Aerospatiale to cover one satellite plus components for a second. ESA manages satellite procurement on EUMETSAT's behalf, but EUMETSAT has responsibility for the ground segment, launch and operations.

A new ground segment was developed by EUMETSAT to take full control of all satellites after 1 December 1995. It includes the MCC Mission Control Centre in Darmstadt and Primary/Secondary Ground Stations, replacing the system previously operated from ESOC. The PGS Primary Ground Station is at Fucino, Italy and includes facilities for operational support of two satellites. A high-speed link connecting PGS/MCC transmits data, telemetry and satellite commands. The back-up station is at Weilheim, Germany.

EUMETSAT budget (ECUmillion)

	1994	**1995**	**1996**	**1997**
MOP	30.5	27.3	–	–
MTP	59.4	55.4	31.0	34.916
MSG	42.0	57.2	86.6	132.507
EPS polar	13.8	11.3	10.8	63.657
General budget	15.0	15.6	22.4	12.738
Total	**160.7**	**166.8**	**150.8**	**243.818**

1993 131.916; 1992 91.720; 1991 64.847; 1990 50.841; 1989 42.8; 1988 59.25; 1987 62.763. Staff total 141 by end-1996.

The Meteosat Second Generation (MSG) Phase C/D plan began in July 1995. Eumetsat planned to launch its first satellite under this programme in

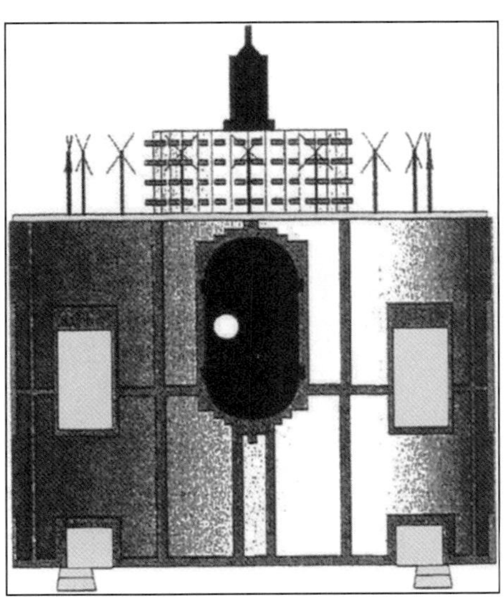

Meteosat Second Generation will retain the spin-stabilisation of the smaller current satellites 0517418

The first satellite in the second generation of meteosats, (MSG-I), is shown here in Aerospatiale's facilities in Cannes during instrument integration (Aerospatiale) 0038496

the fourth quarter of 2000. This did not actually take place until August 2002. Eventually the organisation plans to launch three satellites, the second after 18 months and the third as needed. Arianespace signed an Ariane 5 launch contract for all three, worth more than ECU200 million on 27 July 1995. ESA maintains an active role in the programme as the developer of the first, prototype satellite. In addition to a financial contribution to that satellite, EUMETSAT will fund the three launchers, the next two satellites, the procurement of which has been entrusted to ESA, develop the ground facilities and provide operations until 2012.

MSG's mission is to provide basic multispectral imagery. Improved spatial, temporal and spectral resolution, for weather forecasting. It must also provide high resolution imagery. With AVHRR-type spatial resolution in visible band, for mesoscale convective monitoring over Europe. The information provided includes air mass analysis for Water/CO_2 absorption and meteorological product extraction for improved wind, and temperature data. Eumetsat intends to emphasise support of climate and environment monitoring, continuity of data collection and dissemination and the relay of search and rescue distress signals and observation of the global Earth radiation budget.

Metop, the Meteorology Operational Polar satellite, provides the European component of a joint NOAA/EUMETSAT polar system that aims to provide continuous meteorological global observation from Sun-synchronous LEO. NOAA will launch the morning and afternoon (local time) orbits and EUMETSAT will eventually become responsible for the morning element.

Metop 1, after it is launched in 2005, will be the first in a series of operational satellites: Metop 2 and Metop 3 follows in 2019. The preparatory programme, which prepared the proposals for the full EPS programme of ECU1.5 billion (1994 rates), began in January 1994.

Eumetsat implemented a technical means of controlling access to Meteosat HRI High-Resolution Image data on 4 September 1995. Now the satellite encrypts the data normally received by PDUSs, while the analogue (WEFAX) transmissions remains unencrypted. Over the past decade, a rapid increase in value-added commercial activity by the private sector has developed, based on meteorological forecasts and data provided by meteorological services. Simultaneously, government satellite organisations have brought increasing pressure on space resources to recover their costs by charging for services. The encryption of data began an effort to register all users of the high-resolution Meteosat data. Regular test transmission of encrypted data began in 1994. Decryption requires an MKU Meteosat Key Unit from EUMETSAT and some adaptation to the PDUS for MKU interfacing. Some exceptions exist. For example, HRI images at 00/06/12/18 GMT remain un-encrypted, and the

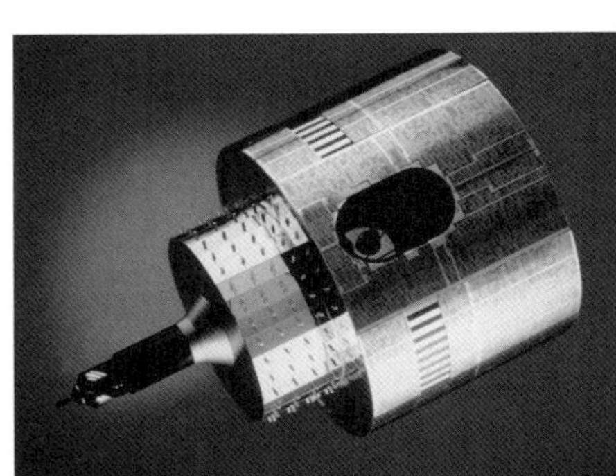

Meteosat contributes ESA's contributors to the World Weather Watch 0084760

One of the first images returned from the Meteosat 7 weather satellite 0084761

3 hourly encrypted data are free to all National Meteorological Services. In general, the NMS of countries with a GNP per capita less than US$2,000 have free access to hourly and half-hourly data for internal use; wealthier countries pay according to their GNPs. Educational and science programmes have free access to all data. At the beginning of 1996, there were more than 300 PDUS and almost 2,000 SDUS.

Specifications

MOP-2/Meteosat 5

Launched: 2 March 1991 by Ariane V42 from Kourou, French Guiana (lifetime 5 years)
Mass: 681 kg at launch, 316 kg after firing of Mage apogee kick motor (including 39 kg hydrazine thruster propellant, with 2 kg conserved for removal from GEO at EOL)
Contractors: Aerospatiale (prime), MMS (radio meter, AOCS), Bosch Telecom (transponders), Alenia (TT&C), ETCA (power supply), MBB (structure, thermal)
Resolution: 2.5 km visible, 5 km water vapour/TIR
Orbit: 0° W geostationary
Spacecraft: 2.1 m diameter, 3.195 m high stepped cylindrical body with solar cells on six main body panels providing 300 W at beginning of 5 year life (200 W EOL). Spin-stabilised at 100 rpm around main axis aligned almost parallel to Earth's axis, with spin regulated by two small hydrazine thrusters. Two pairs of larger thrusters provide spin axis precession/inclination control and east-west station-keeping, respectively. East-west drifting of ±1° about 0° longitude is permitted for the prime satellite. Attitude information provided by pairs of Earth horizon + Sun-slit sensors. Meteosat's smaller cylinder carries radiating dipole antenna elements activated sequentially (that is, electronically despun) for 333 kbit/s S-band (1.670–2.300 GHz) image transmissions and TT&C operations (telecommand 2.0980 GHz; telemetry 1.675928 GHz); the protruding antenna provides toroidal-pattern S-band and low UHF links. Meteosat's other pole houses the radiometer's passive cooler open to space; the radiometer telescope itself scans along the satellite's equator
Sensors: Single Imaging Radiometer operating in three visible-infra-red bands and providing a full-disc Earth image in 25 min, followed by a 5 min retrace/stabilisation period. Each infra-red image comprises 2,500 lines (each of 2,500 pixels) but the two simultaneous visible detectors provide 5,000-line images, corresponding to a resolution of 2.5 km. All four detectors are redundant. The two visible Si photodiode detectors cover 0.4–0.9 μm, the thermal-infra-red HgCdTe detector 10.5–12.5 μm, and the water-vapour/infra-red HgCdTe 5.7–7.1 μm, continuing the pre-operational Meteosat data
Optics: Ritchey-Chrétien 40 cm primary aperture, 365 cm fl telescope mounted on two flexible plate pivots attached to a step motor by a high precision jack screw via a gear box. The telescope is stepped 0.125 mrad every 100 rpm satellite rotation so that Earth's surface is scanned at 5 km intervals south to north, covering a total angular distance of 18°.

MOP-3/Meteosat 6

Launched: 20 November 1993 by Ariane V61 from Kourou, French Guiana (lifetime 5 years)
Mass: 704 kg at launch, 316 kg on-station BOL
Orbit: 10° W geostationary as standby
Other specifications are the same as Meteosat 5.

MOP-4/Meteosat 7

Launched: 2 September 1997 by Ariane 44LP from Kourou (French Guiana)
Mass: 703 kg at launch, 321 kg onstation BOL
Orbit: 349° east initially, operated at 0°
Other specifications are the same as Meteosat 5.

Meteosat Second Generation

Launch: Meteosat 8 (MSG 1) 28 August 2002 by Ariane 5 from Kourou, French Guiana
Mass: about 1,750 kg at launch
Orbit: 0° geostationary
Contractors: The satellite is developed by an industrial consortium led by Aerospatiale, with MMS responsible for the SEVIRI instrument
Resolution: 1 km visible, 3 km infra-red
Spacecraft: the satellites will continue 100 rpm spin stabilisation, though considerably larger: 3.2 m diameter, 3.7 m high. Unified bipropellant system. 750 W BOL from Si panels. 7 year life. The communications package's three channels will downlink 3.2 Mbit/s raw images, housekeeping and relayed DCP data to the primary ground station, and retransmit processed imagery and other data to users in high-/low-rate data schemes (up to 1 Mbit/s without compression). MSG's data volume will be an order of a magnitude greater than the current system. Accomodation of the Search & Rescue package has been approved by EUMETSAT
Sensors: SEVIRI Spinning Enhanced Visible/IR Imager, scanning Earth's disc every 15 min in 12 channels. The 0.50–0.90 μm HRV High-Resolution Visible channel will provide 1 km resolution. The other 11 will return images of max 3,750 lines of 3,750 samples, corresponding to 3 km resolution visible 0.56–0.71, 0.74–0.88, 1.50–1.78; window 3.40–4.20, 8.3–9.1, 9.8–11.8, 11.0–13.00; water vapour 5.35–7.15, 6.85–7.85; ozone 9.38–9.94; CO_2 12.40–14.40 μm. Typical radiometric resolution 0.25 K window channels, 0.75 k water vapour and 1.50 k ozone/1.80 k CO_2.

GERB Global Earth Radiation Budget instrument has been approved by EUMETSAT, provided by a UK-led consortium: Natural Environment Research Council; Rutherford Appleton Lab (technical management, data processing, archiving, distribution); Imperial College of Science, Technology and Medicine (PI Prof John Harries,

One of the early images from MGS-1 declared operational as Meteosat 8 in February 2004 0572904

science, calibration); Royal Meteorological Institute of Belgium (optics unit). 89% of the cost has been subscribed; EUMETSAT and the European Commission are expected to fund the remainder. Together with SEVIRI, GERB will enable study of water vapour and cloud forcing feedback, two of the most important (and poorly understood) feedback processes in climate prediction. It will measure Earth's long-wave and short-wave radiation every 15 min to an absolute accuracy of about 1 W/m^2. Bands: 0.35–4.0, 0.35–30 μm (4.0–30 μm by subtraction), 48 km nadir pixel, absolute accuracy 1% SW/0.5% LW, S/N 1,500 SW/LW, dynamic range 0–450/0–130 Wm^2 SW/LW, <30 kg, <70 kbit/s, <35/40 W avg/peak.

Small Multimission Spacecraft (SMMS)

With assistance from Thailand, China and Iran are leading an international effort in the region to produce a Small Multimission Spacecraft (SMMS) capable of providing an autonomous civil remote sensing and imaging capability. China views the co-operative venture as a platform on which further international space ventures can be built and sees this programme as an investment, creating new Asian markets for its launch services. Pakistan is co-operating with China on regional remote sensing activities and the SMMS is intended to accommodate the needs of that country also. China wants the satellite bus to provide a common platform for other users in the region. Several marketing initiatives have already been made.

The satellite will weigh about 470 kg and is scheduled for launch in 2005 to 2007 by a Chinese launcher. It will be equipped with a low-resolution Charge-Coupled Device (CCD) and will operate from a circular, sun-synchronous, 650 km orbit.

Topex/Poseidon

Current Status

In late 1998Topex/Poseidon was used to experiment with an autonomous orbit control mode. A software programme was uploaded to allow the satellite to control its orbit. In May 1999 the satellite operated with the backup altimeter.

Background

NASA/JPL's combined its Topex (The Ocean Topography Experiment) with CNES' similar Poseidon payload in FY87 to provide long-term observation of global ocean circulation and surface topography. The satellite carries two Ku-band altimeters to determine sea height to within 2 to 5 cm. NASA's dual frequency (C/Ku) altimeter built on experience from NASA Goddard and the Applied Physics Lab at The Johns Hopkins University of 1978's Seasat instrument. The CNES Ku-band device is a new design employing solid-state technology.

Topex was originally planned as a US$270 million NASA mission, targeted for 1989 launch, as

Topex/Poseidon: built by NASA and launched by France, carrying sensors from both 0517231

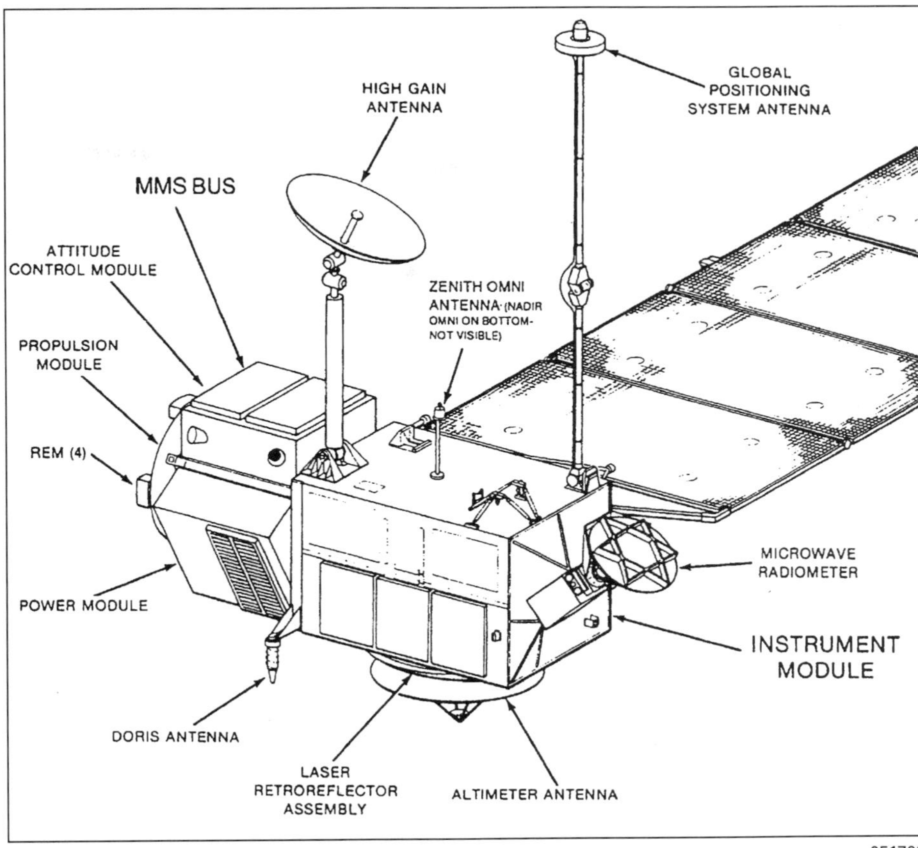

Principal Topex/Poseidon features 0517034

a successor to Seasat. In 1985, NASA merged Topex with Poseidon. In April 1984, JPL awarded an eight month, US$1 million satellite definition contract to Fairchild, RCA Astro-Electronics and Rockwell based on their respective Solar Max, Tiros/DMSP and Navstar platforms. But Congress refused to grant new money in FY86 for the FY87 budget. NASA then issued requests for a proposal to the same three companies, in July 1986 with the intention of selecting a single satellite contractor in December. Fairchild won in January 1987, basing their design on the company's successful Solar Max multimission. Topex became the first complete spacecraft built by Fairchild. The total satellite cost ended 42 per cent over the US$121 million initial budget, largely due to NASA imposing more stringent requirements.

NASA projected total costs including launch and tracking at US$485 million in 1985 dollars and US$125 million for CNES. France's independently estimated its costs in 1991 to be about FFr900 million.

Topex/Poseidon flies as part of the World Ocean Circulation Experiment (WOCE), a major oceanographic field programme under the auspices of the World Climate Research Program. WCRP combines satellite-based data with traditional observations to generate global 3-D ocean current structure model. The operational orbit is phased to overfly a NASA calibration site on the Harvest oil platform near Pt Conception (California) and a CNES calibration site on Lampione Rock near Lampedusa in the Mediterranean (the sites provide independent measures of satellite height and sea level). A three year extended mission began in August 1995 after the US$30 million annual operations costs were reduced to less than US$20 million.

NASA anticipated at first that the inaccuracies in orbital determination would reduce certainty relative to the geoid to 13.4 cm. But the orbit restituted to less than 5 cm (radial) by the DORIS system, orbital determination is no longer the dominant item in the error budget; sea surface height accuracy has proved to be less than 5 cm rms.

Specifications

Topex/Poseidon

Launch: 10 August 1992 by Ariane 42P from Kourou. Operational mission began late February 1993

Mass: 2,380 kg in-orbit BOL

Contractors: JPL (project management, Topex payload), Fairchild Space Co (bus), Alcatel Espace (Poseidon altimeter), Honeywell Satellite Systems Div (antenna pointing system, RW assemblies), Dassault Electronique (DORIS receiver)

Altitude precision: 2–3 cm Topex, 2–5 cm Poseidon

Orbit: 1,331 × 1,332 km, 66.05°, repeating within 1 km every 9.9 days (127 revs). This is a 'frozen' orbit, holding a fixed eccentricity and argument of perigee to minimise manoeuvres to maintain the ground track. The selected altitude was a trade-off between gravity anomalies and atmosphere introducing unacceptable perturbations versus increased altimeter power for higher resolution in a higher orbit

Spacecraft: based on NASA/Fairchild standard 3-axis Multi-Mission Spacecraft (MMS) bus designed for 3 to 5 year lifetime and 3.38 kW BOL solar power (2.14 kW after 5 years). 5.5 m length, 11.5 m span, 6.6 m high. Communications via TDRS. Power supply supported by 3 × 50 Ah batteries. Attitude control (nadir pointing) by RWs + magnetorquers; attitude determination by Earth/Sun sensors, star trackers, magnetometers, 3-axis gyros. Orbit control by 1/22 N hydrazine thrusters providing total ΔV 176 m/s (sufficient for 12 years)

Sensors: Topex/Poseidon carries two altimetry systems, a microwave radiometer to correct for atmospheric effects, and three tracking systems:

Topex (US) Radar altimeter: operating with a prime channel at 13.6 GHz (Ku-band) and a secondary at 5.3 GHz (C-band), the two frequencies permitting corrections for ionospheric effects. The US/French altimeters share one 1.5 m dia dish, providing a 3 dB beamwidth of 2.7/1.1° at C/Ku. US radiated power 2/9 W C/Ku. Topex mass 206 kg, consumed power 237 W. Microwave radiometer: 18/21/37 GHz to correct altimetry data for tropospheric water vapour effects. Orbit determination: operationally by TDRS, precision (science) by laser reflectors (13 cm radial accuracy) and experimental NASA/Motorola GPS Demonstration Receiver operating at 1.2276/1.5754 GHz, using a new technique of GPS differential tracking for accuracy of a few cm.

Poseidon (France) Radar altimeter: solid-state instrument operating on a time-share basis with US version through same 1.5 m dia dish at single Ku-band centred on 13.65 GHz, bandwidth 330 MHz, pulse duration 100 μs, pulse repetition frequency 1,700 Hz. Radiated power 5 W, power consumption 49 W. An 80C86 microprocessor controls the instrument. Microwave unit size 36.5 × 28.8 × 23.2 cm; 13 kg. Processing unit size 35.5 × 25.5 × 20.3 cm; 12 kg. Orbit determination: Doppler ranging by DORIS payload operating on 2,036/401 MHz uplink (10 cm radial accuracy).

TRMM

Current Status

The co-operative Tropical Rainfall Measuring Mission, launched 27 November 1997, entered orbit at 367 × 385 km and 35° inclination. Descending to

TRMM will provide the first comprehensive precipitation data on a global scale. It is the largest spacecraft ever built in-house by NASA Goddard (NASDA) 0517035

350 km, on board propulsion was used to raise the orbit to 402 km in August 2001.

Background

TRMM plans to investigate the interactions of water in all three of its physical phases to determine the contribution it makes to the Earth's process. Water substantially alters incoming and outgoing radiation and affects global air motions and heat fluxes through condensation and evaporation. Moreover, the latent heat released by tropical rain provides 75 per cent of the energy the atmosphere receives, thus playing a key role in driving global atmospheric circulation. Yet scientists know the amount of tropical rainfall to within only a factor of two over the oceans and little better, if any, over the arid continents and jungles. As a result, five of the leading climate models differ widely in their estimates of rainfall under identical boundary conditions - they differ by factors of two to three for the mean annual rain in the equatorial belt and even more widely for rainfall on a regional scale.

TRMM will provide the first comprehensive precipitation data on a global scale. The rainfall package comprises the PR (the first weather radar to fly in space), the multichannel TMI microwave radiometer and VIRS. TMI improves rain remote sensing of its SSM/I Special Sensing Microwave/Imager predecessor, carried by the military DMSP metsats, by adding a 10 GHz channel. TRMM's data system will operate in collaboration with EOSDIS, which will perform its first collection/archiving of very large space/ground validation data with this mission. CERES and LIS, funded by EOS (LIS was deleted from EOS in 1992), were added later to improve investigation of global change. CERES will measure upwelling cloud radiation, helping to identify how clouds warm/cool the planet. LIS will clarify the relation between cloud electrification and rain processes, and why it apparently differs over oceans and land masses.

TRMM will provide monitoring of tropical rainfall over three years, yielding monthly averages over 5×5° (500 × 500 km) cells. The US provided the spacecraft and four instruments (project management at NASA Goddard as an Earth Probe) and Japan launched the satellite aboard an H-2 rocket from Tanegashima. In addition, Japan added one instrument to the payload. NASA approved TRMM in 1990 following a 1986–88 Phase A evaluation programme. NASA put forward FY96 funding of US$23.1 million. (Previous years: FY95 48.1; FY94 63.1; FY93 51.5). NASDA for its part put up FY96 ¥3,630 million. (Previous years: FY95 4,964; FY94 2,537.5; FY93 919.1; FY92 874.) Total costs for the project are estimated at approximately US$650 million. NASA has proposed a similar follow-on mission is being studied, flying at 55° to cover all the continental US.

CERES will help to identify how clouds affect Earth's temperature (TRW) 0517416

Specifications

TRMM Satellite

Launch: 27 November, 1997 by H2 from Tanegashima, Japan

Mass: 3,820 kg (725 kg hydrazine to maintain altitude to ±1.25 km)

Orbit: planned 350 km circular, 35°

Spacecraft: 3-axis (0.2° knowledge) by Earth sensor (primary attitude reference), coarse/digital Sun sensors, gyros, four Ithaco 75 Nms RWs, magnetorquers, hydrazine thrusters, 1.1 kW min from twin GaAs solar wings + 2 × 50 Ah Ni/Cd batteries, S-band via TDRSS: data 32 kbit/s real-time; 2.048 Mbit/s playback. Instrument data rate 166.4 kbit/s. Two 264 Mbyte SSRs.

Sensors: PR Precipitation Radar (NASDA, Japan; Toshiba prime). 0.7 mm/h measurable rainfall. 13.796/13.802 GHz, pulse width 1.67 μs, pulse repetition frequency 2.776 Gz, peak power 578 W, beam width 0.71°, cross track scan angle ±17°, 93.5 kbit/s, 128 slotted waveguide antennas (using SSPAs) in 2.3 × 2.3 m planar array, mass 470 kg, power 250 W.

VIRS Visible Infra-Red Scanner (NASA). Channels (μm): 1 0.63, 2 1.6, 3 3.75, 4 10.8, 5 12. IFOV ±0.210° at 350 km, ±45° cross track. 2 km res at 10 kbit/s/channel. Mass 49 kg, power 36 W.

TMI TRMM Microwave Imager (NASA). 9 channels (GHz): 10.65, 19.35, 21.3, 37, 85.5, mass 53 kg, power 43 W, data rate 8.5 kbit/s

CERES Cloud/Earth Radiant Energy System (NASA EOS; TRW). Channels (μm): 1 0.3–3.5, 2 8.0–12.0, 3 0.3–>50, scan angle ±42.5° in 6.6 s cycle, 8 kbit/s, IFOV 25 km nadir, mass 45 kg, power 45 W.

LIS Lightning Imaging Sensor (NASA MSFC in-house). 0.7774 μm, scan angle ±42.5° cross track, res 5 km nadir, 0.5 kbit/s (6 kbit/s at full background), mass 20 kg, power 24 W.

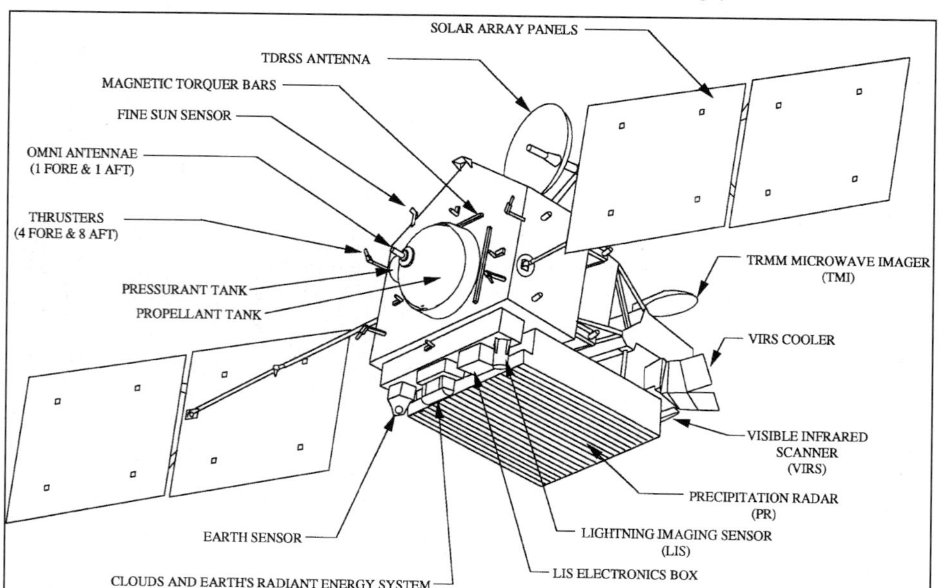

TRMM principal features (NASA) 0517230

Japan

ADEOS/MIDORI Series

Current Status

Thruster problems affected Adeos 1 and prevented it reaching the proper orbit without the 20 N thrusters. On 24 September 1996 the satellite shut down all non-essential systems. In late 1996 the OCTS instrument had a blocked mirror.

Background

NASDA's 1996 Advanced Earth Observation Satellite (renamed Midori, 'Green', and once in orbit) is a modular platform upon which the ADEOS 2 was launched in December 2002. As part of its science and remote sensing research effort NASDA set aside ¥542 million for ADEOS 1 in FY88 to initiate the programme. Phase C/D began August 1990, following sensor selection in August 1989. Following a critical design review in 1991, integration and testing of the Engineering Model and the Structural Thermal Model began 1992.

MELCO, Nippon Electric and Toshiba submitted bids by the 14 April 1990 deadline in time for NASDA to announce the prime contractor in the summer. The Japanese government issued a shared contract in the aftermath of the CS-4 programme cancellation. NASA offered its Scatterometer to provide accurate measurements of ocean surface winds following NASA's successful demonstration of the technique by Seasat during 1978. ADEOS also carries NASA's TOMS Total Ozone Mapping Spectrometer, extending observations from Nimbus 7 and Russian Meteor 3, and 1996's Earth Probe TOMS. The satellite failed in orbit 30 June 1997 due to problems with the solar panel system used to generate electrical power.

The ¥276 million request to start ADEOS 2 was denied for FY92/FY93, but approved by the SAC in 1993 for FY94. Japan allocated ¥7.667 billion FY95, ¥9.896 billion FY96. NASA is developing SeaWinds (similar to NSCAT but with a mechanically scanning 1 m diameter antenna) to avoid a break in scatterometer data until its own EOS appears 2002.

Specifications

ADEOS 1

Launch: 17 August 1996 by H2 from Tanegashima

Mass: 3,560 kg in-orbit BOL (1,300 kg instruments)

Contractors: Mitsubishi Electric, Nippon Electric and Toshiba sharing development contract; Odetics (DDS-6000EC tape recorders), Sodern (STD-16 Earth sensors) BAe Space Systems (IMG cryogenic coolers)

Resolution: from 8 m visible to 700 m thermal IR

Orbit: planned 797 km 98.59° Sun-synchronous, repeating every 41 days (585 revs), descending node 10.30 local time

Spacecraft: modular, providing minimum 3 year lifetime. 3-axis control (within 0.3° + 0.003°/s per axis) provided by zero momentum strapdown system derived from ETS-6; four RWs, two magnetorquers, inertial reference unit, 4 × 20 N + two sets 8 × 1 N RCS hydrazine thrusters supplied from three 55 cm diameter spheres. Single 3 × 24 m GaAs flexible paddle generates 4.5 kW EOL, supported by 5 × 35 Ah Ni/Cd batteries (max depth of discharge 20%). OBC stores 7-day batches of commands; 500 bit/s command uplink and 4,096 bit/s housekeeping down on 2.22 GHz S-band. High-rate data (AVNIR, OCTS, POLDER, IMG, and ILAS) stored at 3/6/60 Mbit/s (playback 60 Mbit/s) on three tape recorders. Low rate (NSCAT, TOMS, and TEDA) at 4,096 bit/s (playback 517 kbit/s) on two recorders (backed up by single housekeeping recorder, record 4,096 bit/s, and playback 33 kbit/s). Data transmitted on three 8.150–8.350 GHz X-band links: X1/X2 each 40 W SSPA 60 Mbit/s; X3 5 W SSPA 6 Mbit/s. ADEOS will also demonstrate IOL links, up to 1126 Mbit/s through COMETS, using 2.2875 GHz S-band and 25.850 GHz Ka-band. Low resolution OCTS real-time data transmitted on 465.00 MHz at 23 kbit/s to local users such as fishing vessels

Sensors: Advanced Visible/Near-Infra-red Radiometer (AVNIR). NASDA core instrument, prime contractor MELCO. Three visible and single near-infra-red band complemented by PAN band. 230 kg, 300 W, 60 Mbit/s. Resolutions: 16 m XS + 8 m PAN for 80 km swath. Si CCD array, electronic scanning. 5.7° FOV, can point 40° either side of ground track. Bands (μm): 0.42–0.50, 0.52–0.60, 0.61–0.69, 0.76–0.89; 0.52–0.69 PAN.

Ocean Colour & Temperature Scanner (OCTS). NASDA core instrument, NEC prime contractor. Three thermal-infra-red, one middle-infra-red, two

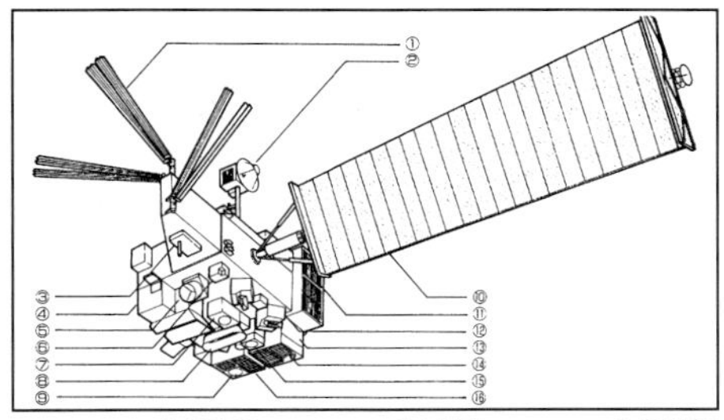

ADEOS instruments and principal features. 1 NASA Scatterometer, 2 DRTS antenna, 3 UHF antenna, 4 OCTS, 5 RIS, 6 NASA TOMS, 7 IMG, 8 AVNIR, 9 X-band antenna, 10 solar array, 11 mission data processing unit, 12 ILAS, 13 control & data handling unit, 14 S-band antenna, 15 POLDER, 16 direct transmit unit (NASDA) 0517051/0517052

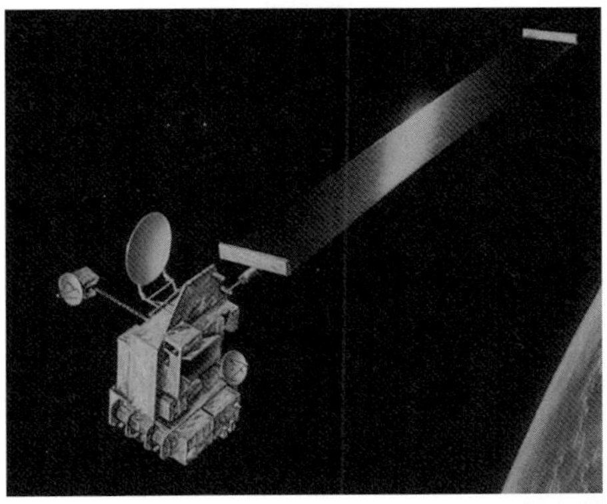

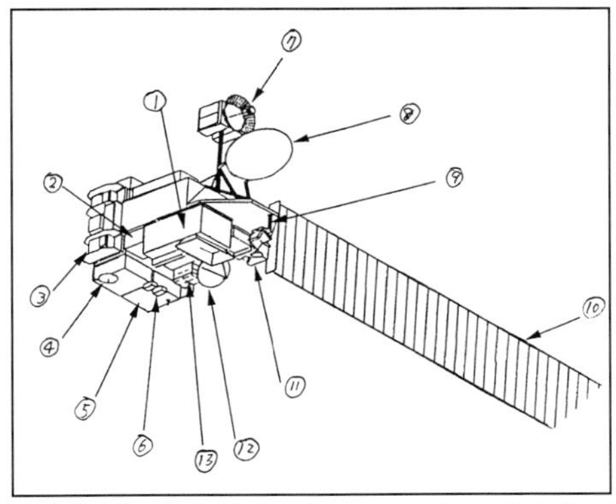

ADEOS 2 instruments and principal features. 1 Global Imager, 2 SeaWinds, 3 mission data processing unit, 4 direct transmit unit, 5 communications/data handling unit, 6 Earth sensor, 7 inter-orbit communications antenna, 8 AMSR scanning antenna, 9 UHF antenna, 10 solar array, 11 ILAS 2, 12 SeaWinds 1 m antenna, 13 POLDER (NASDA) 0517238/0517239

near-infra-red and six visible channels. 360 kg, 315 W, 3 Mbit/s. Rotating scan mirror can tilt 20° forward/back along ground track to minimise Sun glitter. Resolution 700 m, swath 1,400 km. Bands (μm): 0.402–0.422, 0.433–0.453, 0.479–0.500, 0.511–0.529, 0.555–0.575, 0.660–0.680, 0.745–0.785, 0.845–0.885, 3.55–3.88, 8.25–8.80, 10.3–11.4, 11.4–12.7.

Scatterometer (NSCAT). NASA/JPL active microwave radar for near-surface wind speed/direction over ocean accurate to 2 m/s + 20°. 13.995 GHz. 300 kg, 275 W, 3,178 bit/s. Spatial resolution 25 m. Radar cross section measured at 3 azimuth angles in both 600 km swaths.

Total Ozone, Mapping Spectrometer (TOMS). NASA Goddard mapping UV polychromator for total O_3 and SO_2 monitoring. 35 kg, 25 W, 700 bit/s. 3° resolution, swath 2,795 km. Wavelengths (μm): 0.3086, 0.3125, 0.3175, 0.3312, 0.3398, 0.3600.

Polarisation/Directionality of Earth's Reflectance (POLDER). CNES Toulouse. CCD detector behind rotating wheel of filters/polarisers. 33 kg, 40 W, 900 kbit/s. Swath 1,440×1,920 km. Wavelengths (μm): 0.443, 0.670, 0.865 at three polarisations; 0.443, 0.490, 0.565, 0.763, 0.765, 0.910.

Interferometric Monitor for Greenhouse Gases (IMG). MITI (Toshiba prime). Michelson interferometer for observation of CO_2, CH_4, NO and other greenhouse gases. 150 kg, 150 W, 600 kbit/s. 8 km resolution on 20 km swath. 3.3–4.3, 4.0–5.0, 5.0–14.0 μm.

Improved Limb Atmospheric Spectrometer (ILAS). Environment Agency. Trace constituents over 10–60 km (2 km resolution) at high latitudes measured by separate packages observing sunrise/sets. Each houses 12 cm diameter telescope with 44 pyroelectric detectors for 6.21–11.77 μm and 3 cm telescope with photodiode array for 0.753–0.784 μm. 130 kg, 80 W, 500 kbit/s.

Retroreflector In Space (RIS). Environment Agency. 50 cm diameter corner cube laser reflector to monitor O_3, fluorocarbon, CO and other constituents using 0.3–14 μm ground laser. 45 kg.

Technical Engineering Data Acquisition (TEDA) payload monitors the space environment and its effect on the spacecraft with Heavy Ion Telescope, Single Event Upset Monitor, Dosimeter, Contamination Monitor, Potential Monitor.

ADEOS 2
Launch: 14 December 2002 by H2A from Tanegashima
Mass: 3.5 t at launch, 1.2 t payload
Orbit: 803 km 98.6° Sun-synchronous
Spacecraft: modular, providing minimum 3 years (5 year goal) lifetime. 3-axis control, single Si solar paddle providing 5.0 kW (1.2 kW required for instruments)
Sensors: Advanced Microwave Scanning Radiometer (AMSR). NASDA. Advanced version of MOS 1 MSR. Water vapour content, sea surface T+ wind, precipitation, sea ice. 8 bands (6.9, 10.65, 18.7, 23.8, 36.6, 50.3, 52.8, 89.0 GHz), 50/50/25/25/15/10/10/5 km + 0.3–2K resolution on 1,600 km swath.

Global Imager (GLI). NASDA. Improved ADEOS 1 OCTS, 36-band visible to thermal-IR passive sensor for water colour, sea T, vegetation index, clouds, aerosols. 1 km/250 m resolution over 1,600 km swath.

SeaWinds. NASA/JPL. Modified NSCAT at 13.402 GHz for sea wind speed/direction. Resolution 2 m/s speed, 20° direction, 50 km spatial.

POLDER. CNES Toulouse. Same as ADEOS 1

Improved Limb Atmospheric Spectrometer 2 (ILAS 2). JEA. As ILAS, with resolution improved to 1 km.

Data Collection System (DCS): Argos compatible (adding platform command capability). CNES/NASDA.

GMS Series

Current Status

GMS-1 was shut down on 30 June 1989, GMS-2 was retired 20 November 1987, GMS-3 was put out of service in July 1995, GMS-4 ceased operating in March 2000 and GMS-5 ceased operational activity when station-keeping activity was suspended in October 2001 due to low fuel levels.

Background

The Geostationary Meteorological Satellite (GMS) system provides coverage of the Pacific region from Hawaii to India with a single primary satellite positioned over 140° E. It is Japan's contribution to the Global Observing System of the World Weather Watch. In addition to transmitting a full-disc visible and infra-red Earth image every 30 minutes, the satellite relays meteorological data from ships, aircraft, buoys and other weather stations, and relays ground-processed facsimile images to users.

Based on Hughes Aircraft's GOES design, the series began in 1977 with a Delta launch from Florida but the current satellites are launched from Japan. NASDA procures and launches the satellites according to requirements and funding from the Japan Meteorological Agency (JMA).

GMS-4 became operational at 140° E on 14 December 1989 following in-orbit testing at 160° E. When GMS-4 became operational NASDA moved GMS-3 to 120° E. Then GMS-5 subsequently replaced GMS-4 at the prime position in mid-1995.

GMS-5 acted as the last satellite in the series which the Ministry of Transport expects to replace with the MT-Sat in 1999. On GMS-5, the single infra-red channel of its predecessors is divided into two to improve temperature measurement accuracy, and NASDA added a third for atmospheric water vapour measurements, but the SEM detectors carried aboard GMS 1–4 were deleted. GMS-5 also carries an experimental payload to relay Sarsat/Cospas emergency signals. The Japanese government allocated ¥979 million in FY91, ¥1.138 billion FY92, ¥6.0585 billion FY93, ¥4.5068 billion FY94, ¥93 million FY95.

NASDA controls GMS satellites through the end of their in-orbit checkout phases from its Tsukuba Space Centre facilities and associated tracking stations. Operational control is exercised by JMA's Meteorological Satellite Centre, although station-keeping manoeuvres are commanded on JMA's behalf by NASDA (2.28/2.10 GHz up/down). Raw imagery, data collection platform signals and spacecraft telemetry are received on two 18 m Cassegrain dishes by JMA's Command & Data Acquisition Station at Hatoyama 56 km northwest of Tokyo. They are then relayed by 7.5 GHz microwave link to the Data Processing Centre (DPC). The DPC maintains four mainframe computers: two M-1600/2 (64 Mbyte each) for spacecraft control/operations, and three M-1600/10 (128 Mbyte each) for image processing and subsequent extraction of parameters such as sea T and cloud top height. The smaller machines can assume some of the 1600's processing duties in the event of failure. The Stretched-VISSR and WEFAX data are passed to the CDAS for uplinking and broadcast to users: Medium Data Utilisation Stations (MDUS) can receive S-VISSR but the Smaller Data Utilisation Stations (SDUS) receive the lower-resolution (7 to 8.5 km) WEFAX. Ranging data obtained from CDAS and two Turn-Around Ranging Stations (TARS) on Ishigaki Island (Japan) and Crib Point (Australia) are used for orbit determination.

Specifications

GMS-1 (Himawari 1)
Launch: 14 July 1977 by McDonnell Douglas Delta from Cape Canaveral. 670 kg at launch, 315 kg on-station. Operational from April 1978; replaced by number 2 21 December 1981 and moved to 160° E. At 140° E in 1984, 160° E 1984–1989, last known at 10.47° W (September 2001). Shut down 30 June 1989.

GMS-2 (Himawari 2)
Launch: 10 August 1981 by N2 from Tanegashima. Design life 5 years, mission life 3 years. At Operational life ended 27 September 1984; shut down 20 November 1987. At 140° E 1981–1984, 145° E 1984–1985, 120° E 1985–1988, last known at 33.93° E.

GMS-3 (Himawari 3)
Launch: 3 August 1984 by N2 from Tanegashima. 682 kg at launch, 304 kg on-station BOL. Design life 5 years, replaced as GMS prime satellite at 140° E 14 December 1989 and moved to 120° E (1989–1995) last known at 147° E. Possibility of moving it to contribute to the US GOES system, then with only a single operational satellite, was considered in 1989. Removed from GEO and shut down 23 June 1995 after GMS 5 became prime satellite for the series.

GMS-4 (Himawari 4)
Launched: 6 September 1989 by H1 from Tanegashima (design life 5 years)
Mass: 725 kg at launch, 325 kg on-station BOL
Orbit: 120° E GEO from July 1995; 140° E December 1989 to June 1995, last known at 127° E.

GMS 5 provides Pacific region coverage from above 140° east via a scanning radiometer
(Hughes Space & Communications) 0517235

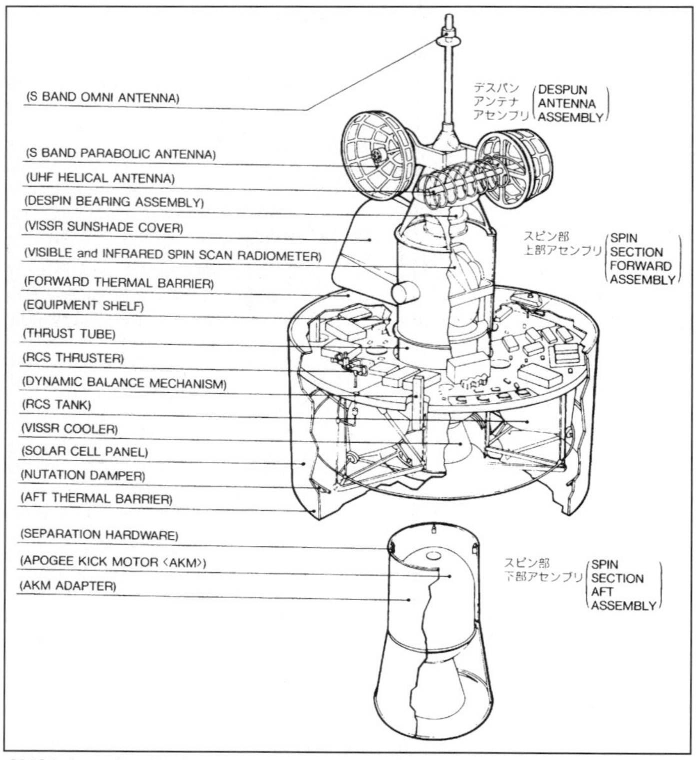

GMS is based on Hughes' spin-stabilised GOES design (NEC Corp) 0517050

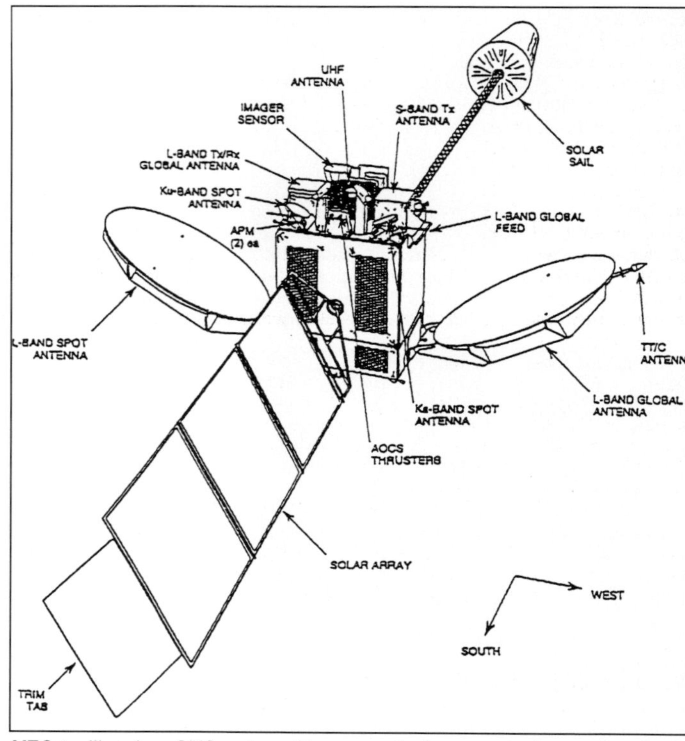

MT-Sat will replace GMS and provide aeronautical communications and navigation services (JRANSA) 0517420

Spacecraft: Design based on Hughes' US GOES. Reliability is greater than 0.5 after 5 years. 214.6 cm diameter, 345.1 cm high cylinder (after AKM separation, 444.1 cm before) with solar cells on main body generating 300 W BOL (265 W Summer solstice EOL). The cells comprise 4,588,000–7 2.2 × 6.2 cm cells for power generation, and 740 2 × 2 cm cells for charging the two 5.5 Ah Ni/Cd batteries (50 per cent max discharge). 1.7 GHz relays spacecraft imagery at 14 Mbit/s and SEM/telemetry at 250 bit/s PCM and IRIG-B FM-PM to CDAS main control centre. CDAS uplinks commands and processed imagery on 2.03 GHz for broadcast to users on 1.687–1.691 GHz; UHF (402/468 MHz up/down) for data collection platform relay
Sensors: Visible/Infra-red Spin Scan Radiometer (VISSR) with redundant detectors operating in one visible (0.5–0.75 μm) + one infra-red (10.5–12.5 μm) bands. Space Environment Monitor of five Si detectors for routine coverage of solar proton (1–500 MeV), electron (up to 2 MeV) and alpha particle (8–390 MeV) radiation levels in GEO.
Other specifications are the same as for GMS-5.

GMS-5 (Himawari 5)
Launched: 18 March 1995 by H2 from Tanegashima (design life 5 years)
Mass: 746 kg at launch, 345 kg on-station BOL
Contractors: Hughes Aircraft Co (spacecraft prime), NEC Corp (Japanese prime), Hughes Santa Barbara Research Center (VISSR)
Resolution: 1.25 km visible, 5 km infra-red
Orbit: 140° E geostationary by 15 June 1995; tested at 160° E (arrived 30 March) moved to 140° E (1995–1999).
Spacecraft: HS-378 design based on Hughes' US GOES. Reliability is >0.5 after 5 years. 214.6 cm diameter, 345.1 cm high cylinder (after AKM separation, 444 cm before) with solar cells on main body generating 300W BOL (291 W summer solstice EOL, 22 W margin). Two 4.8 Ah Ni/Cd batteries. Spin-stabilised at 100 ±1 rpm around main axis, accurate to 0.5°, aligned parallel to Earth's axis with spin regulated by hydrazine thrusters. Two axial + four radial 4.45 N thrusters draw on a 40.8 kg supply of hydrazine. Attitude information provided by Earth horizon + Sun sensors, providing east-west antenna pointing accuracy of 0.39° (Sun sensor)/0.56° (Earth). GEO injection by Thiokol Star 27 solid AKM (providing 1.8 km/s δV) and then separated. GEO position held to within ±1.0Γ north-south/±0.5Γ E-W. UHF helical and S-band omni plus parabolic antenna carried on de-spun platform. 1.7 GHz relays spacecraft imagery/telemetry to CDAS main control centre at 14 Mbit/s. CDAS uplinks commands (omni) and processed imagery on 2.026–22.035 GHz for broadcast to users on 1.682–1.698 GHz. UHF (402/469 MHz up/down) for data collection platform relay
Sensors: Visible/Infra-Red Spin Scan Radiometer (VISSR) with redundant detectors operating in one visible (0.5–0.75 μm) + three IR (10.5–11.5, 11.5–12.5, 6.5–7.5 μm) bands, providing full-disc Earth image in 25 minutes followed by 5 minutes retrace/stabilisation. The visible channel comprises four Si photodiodes (+4 back-up), covering 1.1 km, scanning simultaneously. GMS spin provides east-west scanning, motor-actuated scanning mirror covers north-south in 2,500 steps. The HgCdTe pairs of redundant infra-red detectors are radiation cooled to 90K.

JERS Satellite

Current Status

Problems with SAR antenna deployment but connected, only 4 kW instead of 12 kW electrical power and a malfunction in altitude control system took it out of service on 12 October 1998. Decayed out of orbit 3 December 2001.

Background

The Japan Earth Resources Satellite is the second domestic remote sensing satellite and the first to operate at all-weather radar wavelengths. It carries a Synthetic Aperture Radar accompanied by an optical sensor package. Japan began a study of this new ERS in 1980, and started preliminary design work on the new satellite in April 1986. The critical design review concluded in July 1989, allowing NASD to proceed from a prototype model to complete assembly during first half of 1990. NASDA funded the spacecraft and MITI the instruments.

Japan's Earth Observation Centre at Hatoyama manages all aspects of the mission control and acts as the primary processing, archiving and distribution centre. NASDA's Tsukuba provides tracking and spacecraft control stations (including an 11 m S-band transportable unit at Sweden's Kiruna). Only EOC and the University of Alaska's Fairbanks station (under a 1988 NASA agreement) receive stored data, while other stations handle only real time. Other stations are EOC, Kumamoto (Japan), Bangkok, Showa Base (Antarctica), Fucino, Kiruna, Maspalomas, Tromsoe, Gatineau, Prince Albert, Beijing and Fairbanks.

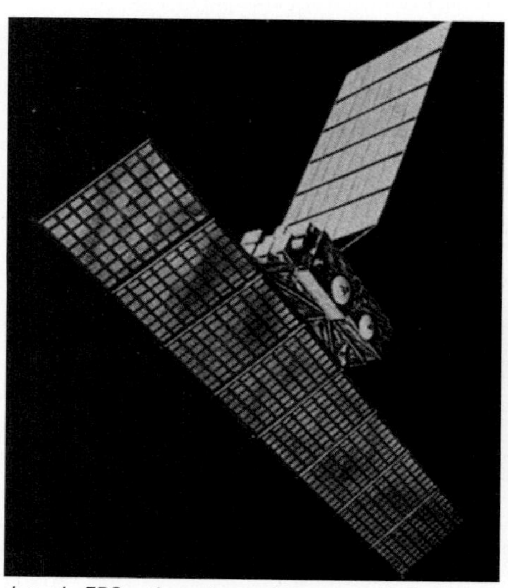

Japan's ERS: the primary instrument is the 18 m-resolution Synthetic Aperture Radar (NASDA) 0517053

The first OPS image was recorded 12 March 1992 over the Sea of Okhotsk. Unfortunately, the two wings of the SAR antenna could not be deployed until 8 April so the first SAR image could not be returned until 21 April 1992. Full operations began formally 1 June 1992. By March 1995, JERS had transmitted more than 140,000 SAR and 90,000 optical scenes.

Japan did not release imagery to the general user community until 1993 because of striping problems on some radar and radiometer images and blurring on the SWIR radiometer. The SAR striping probably developed due to ground radar interference over Japan. VNIR is marred by horizontal striping, which Japanese ground stations now remove by data processing algorithms. SWIR's vertical striping was expected to remain, but the instrument was lost anyway 3 December 1993 because of a cooling fault.

Specifications

JERS 1 (Fuyo 1)
Launch: 11 February 1992 by two-stage H1
Mass: 1,340 kg in-orbit at beginning of 2 year life; propellant mass 115 kg, instrument payload 497 kg
Contractors: Mitsubishi Electric (prime) and Nippon Electric, Toshiba, IHI, instruments from JAROS Japanese Resources Observation Systems Organisation, Odetics (DDS-6000EC tape recorder)
Resolution: 18 m SAR, 18 m OPS in 75 km swathes
Orbit: 567 × 569 km 97.7° Sun-synchronous, crossing equator at 10.30–11am local time, with 44-day repeat cycle (low orbit requires weekly adjustment)
Spacecraft: 3-axis 0.93 × 1.83 × 3.16 m box-like bus of CFRP struts/frames supporting nine aluminium honeycomb panels. Single 56.4 kg 7.03 × 3.46 m CFRP Solar Array Paddle (SAP) of six panels providing 2,053 W EOL from 22,344 2 × 4 cm

JERS 3-D reconstruction of Mount Fuji from the high-resolution stereoscopic imagery acquired by NEC's optical sensor. The colour original clearly shows the effect of land development on the natural vegetation. JERS is the first satellite to provide continuous stereo imaging (NEC) 0517229

Si cells and 4 × 30 Ah Ni/Cd batteries. 11.9 × 2.5 m SAR planar antenna. Zero momentum attitude control system: four skewed 20 Nm RWs and magnetorquers, supported by three gyros, Earth/Sun sensors, provide 0.3° accuracy (design 0.11° roll, 0.18° pitch, 0.10° yaw). Redundant RCS: 22 × 1 N catalytic thrusters drawing on 115 kg hydrazine. Heat rejection via 3.6 m^2 radiator area (21 per cent total surface); 573 W/rev generated when SAR/OPS active. 72 Gbit data recorder holds up to 20 min imaging; 16 Mbit bubble memory for up to 102 minutes telemetry. Transmission at 64 Mbit/s on 8.15 + 8.35 GHz X-band, 50 MHz bandwidth, via 20 W TWTA

Sensors: L-band SAR, 1.275 GHz 1.3 kW peak power, 35 μs pulses, 18 m resolution, 75 km scan width, 64 Mbit/s data rate.

OPS. Optical Sensor covering visible-near infra-red at 18 m resolution (compared to 50 m resolution of MESSR on MOS 1). Seven bands (μm): B1 0.52–0.60, B2 0.63–0.69, B3/B4 0.76–0.86, B5 1.60–1.71, B6 2.01–2.12, B7 2.13–2.25, B8 2.27–2.40. The VNIR Visible/Near-IR Radiometer covers bands 1–4, and 5–8 by the SWIR Short Wavelength IR Radiometer, its detectors cooled to 80K (SWIR bands lost December 1993). Both employ 4,096-element CCDs. Nadir-viewing B3 and forward-looking (15.3°) B4 for continuous stereoscopy.

MOS Series

Current Status

MOS 1A was retired several years ago and MOS 1B was taken out of service on 19 May 1996.

Background

The Marine Observation Satellite was Japan's first domestic Earth resources satellite. NASDA launched and operated it to monitor atmospheric water vapour, ocean currents, sea surface temperature, ice floe distribution and chlorophyll content, in addition to supporting land applications and acting as a relay for Data Collection Platforms. Originally NASDA planned to operate the satellite for just two years, with MOS 1B extending coverage following its launch in February 1990. The two could be operated simultaneously and independently.

Japan's Space Activities Commission approved MOS in 1978 as an element of the space development policy in order to acquire expertise for operational systems in the 1990s. NASDA completed MOS 1A's conceptual design and preliminary systems in 1978 and 1979, respectively, followed by the basic design freeze in July 1981. The last N2 booster carried the satellite into orbit in February 1987 but on 28 July 1987 transmissions ceased owing, most likely, to a failure of a pointing sensor. A back-up system activated successfully and the satellite resumed normal operations 4 August 1987.

Data from the Multi-Spectrum Electronic Self-Scanning Radiometer (MESSR), Visible/Thermal Infra-red Radiometer (VTIR) and Microwave Scanning Radiometer linked in real time to a network of seven ground stations worldwide. The primary station was the Earth Observation Centre at Hatoyama north of Tokyo. The other stations were operated by Tokai University (Kumamoto, Japan), National Institute of Polar Research (Showa Base, Antarctica), National Research Council of Thailand (Bangkok), CSIRO (Alice Springs, Australia) and Canada Centre for Remote Sensing (Prince Albert and Gatineau, Canada). RESTEC provides data to commercial users.

MOS 1A ended data operations in April 1995 and was shut down 29 November 1995. The second MOS was fabricated from 1a's prototype model, incorporating minor modifications gained from operational experience. It was shut down 25 April 1996 after battery failure.

Specifications

MOS 1A (Momo 1)

Launched: 18 February 1987 by N2 from Tanegashima
Mass: 738 kg in-orbit at beginning of life
Orbit: 909 km circular in 99.1°.

MOS 1B (Momo 1B)

Launched: 7 February 1990 by H1 from Tanegashima
Contractors: NEC Corp prime/MESSR sensor, MELCO attitude control system + MSR sensor, Fujitsu Corp VTIR sensor, Toshiba Corp solar array, IHI Corp attitude control thruster system
Resolution: 50 m MESSR, 0.9 km VTIR, 2.7 km VTIR thermal, 23 km (31 GHz), 32 km (23.8 GHz) MSR
Orbit: 912 × 920 km, 99.14° Sun-synchronous, crossing equator at 10.00–11.00 local time, with 17-day repeat cycle
Spacecraft: 1.26 × 1.48 × 2.4 m high box-shaped satellite of aluminium honeycomb construction with single 2.0 m wide, 5.28 m solar wing of three 1.51 m-wide panels. 3-axis controlled by MWs + 4 × 1 N hydrazine thrusters. TT&C performed by Tsukuba Space Centre on 2 GHz, with MESSR/VTIR data returned in real time on 8 GHz and MSR data on 2 GHz. The Data Collection System Transponder relaying data from ground platforms on a random access basis operates in the 400 MHz band, transmitting to EOC at 1.7 GHz

MOS 1a was launched in February 1987, followed three years later by MOS 1b (NASDA) 0517055

Sensors: Multispectral Electronic Self-Scanning Radiometer (MESSR). MOS carries two MESSRs, each covering two wavelengths with 2,048-element CCDs having a total swath width of 200 km. The data rate of each 70 kg, 40 W instrument is 8 Mbit/s. Spectral ranges (in μm) for the four bands are: Band 1 0.51–0.59; Band 2 0.61–0.69; Band 3 0.72–0.80; Band 4 0.80–1.1. Applications include water turbidity of coastal areas, red tide, ice distribution, water resources, land use, vegetation, snow cover distribution, volcanic ash, geological structure.

Visible/Thermal Infra-red Radiometer (VTIR). A mechanical scanning radiometer with Ritchey-Chretien optics; the mirror scan rate is 7.3 rev/s. Visible band sensing is performed by Si photodiodes, with HgCdTe elements used for the three IR-bands. Data rate of the 25 kg, 35 W instrument is 800 kbit/s. Swath width is 1,500 km. Spectral ranges (in μm) with associated applications are: Band 1 0.5–0.7 (snow cover, ice, cloud); Band 2 6–7 (upper atmosphere vapour, cirrus clouds); Band 3 10.5–11.5 (snow cover, ice, cloud, ocean currents, sea surface T); Band 4 11.5–12.5 (applications as Band 3). Swath width 1,500 km.

Microwave Scanning Radiometer (MSR). Dicke-type radiometer operating at 23.8 GHz (resolution 32 km) + 31.4 GHz (23 km resolution) with the offset parabolic reflector rotating at 18 rpm providing the scanning action. Data rate of the 54 kg, 60 W instrument is 2 kbit/s; swath width 317 km. Spectral ranges (in GHz) with applications are: Band 1 23.8 ± 0.2 (water vapour, rainfall area, snowfall) and Band 2 31.4 ± 0.25 (ice, cloud water, snowfall). MSR is no longer operational.

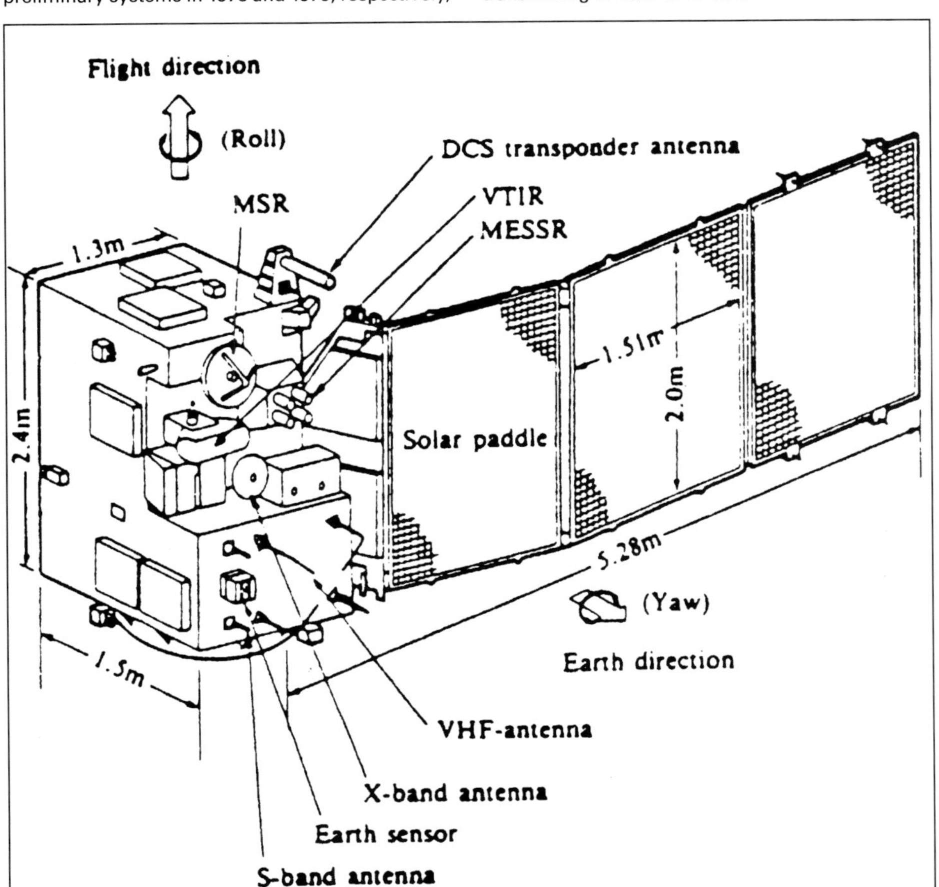

Principal features of Japan's MOS remote-sensing satellite 0517054

Russian Federation

Almaz

Background

The Soviet Union introduced a new class of Almaz ('Diamond') remote sensing satellites in 1987 with the launch of the 18½ tonne Kosmos 1870 on 25 July following a launch attempt failure with the first spacecraft on 29 November 1986. With this test satellite the country intended to demonstrate the satellite's potential, in particular, techniques for oil pollution and ice monitoring. It carried a synthetic aperture radar which provided 30 m resolution all-weather coverage from the slotted waveguide antenna. The first satellite in the series, Almaz 1 appeared March 1991 featuring several improvements over older generation Earth monitoring platforms. The new SAR doubled resolution to less than 15 m and it transmitted data digitally through satellite links (C1870 employed

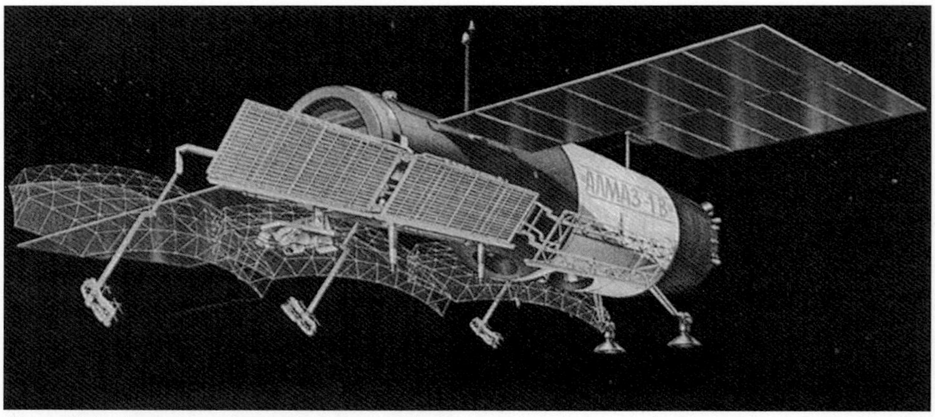

Almaz 1B is planned to carry a wider range of instrumentation. The two articulated white panels at far right are phased array antennas for data relay through GEO satellites
(NPO Machinostroenye, courtesy Space magazine) 0517236

optical processing). After launch the port antenna did not fully deploy, making the production of stereo pairs between 40° N and south difficult.

The Soviet Union originally designed Almaz as a military reconnaissance craft and reportedly intended to launch it in 1981. It was developed by Vladimir Chelomei's bureau, which was also responsible for Proton, the military Salyuts and the Heavy Kosmos modules, and predecessor to NPO Machinostroenye. The Almaz derives directly from the manned military Almaz (flown under the Salyut designation). A disagreement between chief designer Chelomei and Minister of Defence Dmitri Ustinov resulted in cancellation of the programme three days before the planned July 1981 launch and all work halted the following December. Both men died in December 1984 before the programme resumed.

Kosmos 1870 launched 25 July 1987 into 237 × 249 km, 71.9°, preceded the Almaz as a prototype. It required corrections every 12 days to combat orbital decay. In September 1988, ground controllers raised the height to 270 km and corrections switched to a 24-day cycle. The platform was commanded to re-enter 29 July 1989 after an apparently successful demonstration mission. A TV interview with Machinostroenye officials in October 1991 noted that the 1981 Almaz was left at Tyuratam after the cancellation, untouched because it still carried destruct charges, until work restarted in January 1985.

As the Soviet Union broke up in late 1990, Space Commerce Corporation of the US sought exclusive western marketing rights to the data under a joint venture agreement with GlavKosmos and spacecraft manufacturer NPO Machinostroenye. These entities formed the Almaz Corporation to handle the marketing (NPO Machinostroenye had its own Almaz foreign trade company for marketing).

To prepare an Almaz 1 image, SSC had data transmitted via two articulated phased array panels to a Luch GEO satellite for relay to the Data Processing & Customer Support Centre in Moscow, which could handle 100 scenes daily. Imagery cost US$1,600 per 40 × 40 km scene on magnetic tape or floppy disk. Hughes STX signed a further agreement with NPO Machinostroenye for the distribution, processing and analysis of Almaz SAR data. Products included mosaics, annotated mapsheets, custom colourisations, enhancements, interpretation packages and georeferencing. Government support for the international marketing of Almaz data approved partial de-classification of technical specifications.

In December 1992, these companies halved the cost of a 40 × 40 km scene to US$800. By end of 1991, some 60 orders had been received, insufficient to break even, since the company had targeted a goal of US$2 million in sales in 1992. By the time of Almaz's re-entry, only a few hundred images had been sold. Hughes STX had sold about 175 for US$250,000 revenue. In addition to the commercial marketing, primary users could also receive data directly from Almaz using 2 m dishes, although no open source primary users ever appeared. After launch, one of the two SAR antennas failed to deploy fully rendering that side ineffective. Returned images of 10 to 15 m resolution through 17 October 1992. The radiometer provided images of 10 to 30 km resolution over a 600 km swath. Its propulsion system completed 760,000 firings during its 18 month service life.

Russian officials claimed the Almaz 1 programme cost Rb300–400 million. At launch, its supporters sought a minimum 30-month operational life, but increased solar activity required frequent orbit raising, depleting propellant and requiring a controlled re-entry over the Pacific 17 October 1992. Financial difficulties prevented the launch of a second Almaz immediately. The bus for the second satellite is 60 per cent complete and is stored at Khrunichev awaiting funding. A new entity, the joint US-Russian SAR Corp (Sokol Almaz Radar) seeks financing, equivalent to US$126 million. SAR is 49 per cent owned by the US Sokol Group Inc and 47 per cent by Machinostroenye.

Specifications

Almaz 1

Launched: 31 March 1991 by Proton from Tyuratam
Mass: 18.55 tonne on-orbit (4 tonne instrument payload)
Resolution: 15–30 m
Orbit: typically 275 km, 72.7°, revisit 1–3 days. orbit adjusted typically every 24 days
Spacecraft: derived from Mir/Salyut core, about 15 m long, 4.15 m max diameter, two solar arrays total 86 m^2 (average power 2.4 kW), providing up to 10 kW for 20 minutes from the batteries. 3-axis control. Interior pressurised by N_2 and regulated to 5–35 ±1°C. Attitude stability provided by 6 rpm 3 m diameter RW
Sensors: 3.1 GHz synthetic aperture radar using two 1.5 ×15 m slotted waveguide antennas, horizontal polarisation, 190 W peak/ 80 W average pulse power, 0.07/0.1 ms pulses, 3,000 Hz repetition, 3–5 dB pixel accuracy, 30 km beam width in two 350 km swaths. Recorded in 20–240 km lengths. 2-channel radiometer (8 mm, 5 cm), 600 km swath, 0.3°C accuracy, spatial resolution 10–30 km. Carried backup 30 m res radar of C1870 type. Instrumentation for magnetic field mapping was also considered.

Almaz 1B

Launch: uncertain
Mass: 18.55 t on-orbit (4.5 t instrument payload)
Resolution: 5–7, 15–40, 15–60 m (SAR)
Orbit: 350–400 km, 73°
Spacecraft: as Almaz 1. Users could receive data directly at 8.2 GHz: at 122.8 Mbit/s for main stations or 0.960/3 Mbit/s for small/mobile stations
Sensors: SAR-10 (based on Almaz 1 SAR). 3.125 GHz, 320 km swath, 25–51° off-nadir viewing. Starboard antenna provides two modes: observing 15–40 m res, VV, 100–150 km swath; intermediate 15 m res, V/HV, H/HV, 60–100 km swath. Port provides high-resolution 5–7 m res, HH, 25–50 km swath.

SAR-70 (carried on starboard side). 0.429 GHz, 15–60 km res, H/VH, V/VH, 100–150 km swath, 25–46° off-nadir viewing.

4-channel optical stereo imager, 0.5–0.8 μm, 2.5–4 m res, 70 m swath.

Two MSU-E scanners, channels 0.5–0.6, 0.6–0.7, 0.8–0.9 μm, 10 m res, 24 km swath (paired for 48 km).

MSU-SK visible/infra-red scanning radiometer. Two visible channels 0.53–0.59, 0.61–0.69 μm 80 m res; two near-infra-red 0.7–0.8, 0.9–1.0 μm 100 m res; one infra-red 10.4- 12.6 μm 300 m res. 300 km swath.

Sea surface radiometer, 0.1K res. 11 channels 0.4–12.5 μm, 600 m nadir res, 2 × 1,100 km swath.

Balkan 2 lidar to measure upper cloud altitude, 5,320 Å, 40 arcsec beamwidth, 1 s intervals, 3- 10 m vertical res, ±10° off-nadir swath.

Goms/Elektro

Current Status

Elektro 1 became operational on 1 June 1996 and broadcast on 1,691 MHz from a position at 76.5° E until September 1998 when its formal operational period expired and it was taken out of service and left to drift.

Elektro 1 at Baikonur during final launch processing
(Theo Pirard) 0517105

Background

Russia's first GEO meteorological satellite finally appeared in 1994 after years of delay. Plans for a Geostationary Orbit Meteorological Satellite system were announced in 1975 as part of the Global Atmospheric Research Programme, for which Europe, the US and Japan have long provided satellites. The delayed programme was acknowledged during 1988 as still active, with the possibility of the first launch during 1989. VNII Elektromekaniki leads development, in addition to its Meteor and Resurs-O activities. In late 1989, Chief Designer Yuri Trifonov noted the first launch was scheduled for 1991. Elektro 1 was completed that year but there were further delays. A ground station was built in Tashkent, Uzbekistan but the USSR's break-up in 1991 required a new one outside of Moscow. Elektro 1 was reportedly then firmly scheduled for July1993 but was postponed because of May 1993's Proton failure. Elektro is controlled by the VKS Military Space Forces for NPO Planeta, part of Roskomgidromet, the Federal Service for Hydrometeorology & Environmental Monitoring. Data are relayed to NPO Planeta and other Roskomgidromet centres. Slots are reserved at 14° W, 18° W, 166° E and 76° E and 40 to 50° inclination satellites have been proposed for improved northern coverage.

Elektro 1 experienced initial control problems. The Earth sensor never operated and controllers instead provided orientation by ground monitoring the broadcast system's signal strength (it has also been claimed that instead, the coarse Sun sensor was used to aim the solar panels and the resulting power output indicated attitude). Elektro was injected into GEO at 90° E and was planned to reach 76° E 20 November but it overshot 3° because it could not be oriented properly while the plasma thrusters fired to unload the reaction wheels. The unplanned attitudes also caused thermal problems. Brought under initial control, it arrived at 76° E 6 December 1994. It was then planned to use the main parabolic antenna for more precise control for the pole star sensor to lock on; Elektro could then operate normally. Only infra-red images are returned; the visible channel has never worked. GOMS was also intended to carry a military telecom payload, but it was deleted when the military withdrew. This probably accounts for the two dummy parabolic antennas.

Specifications

Elektro 1 Satellite

Launch: 31 October 1994 by Proton SL-12
Mass: Elektro 1 2,580 kg, payload 700 kg
Resolution: 1.5 km visible, 6.5 km infra-red
Orbit: 76° E geostationary
Spacecraft: 3-axis controlled to 2 arcmin roll/pitch, 5 arcmin yaw. 16 DEN-16 electrothermal thrusters from NII Machinostroenye. Payload includes environmental platform data collection system (100 bit/s; 401–403/1,697 or 7,482 MHz up/down). Design life 2–3 years. 1,500 W at 28 V DC from 14.70 m-span 30 m^2 solar wings. OBC provides autonomous operation for 18 days

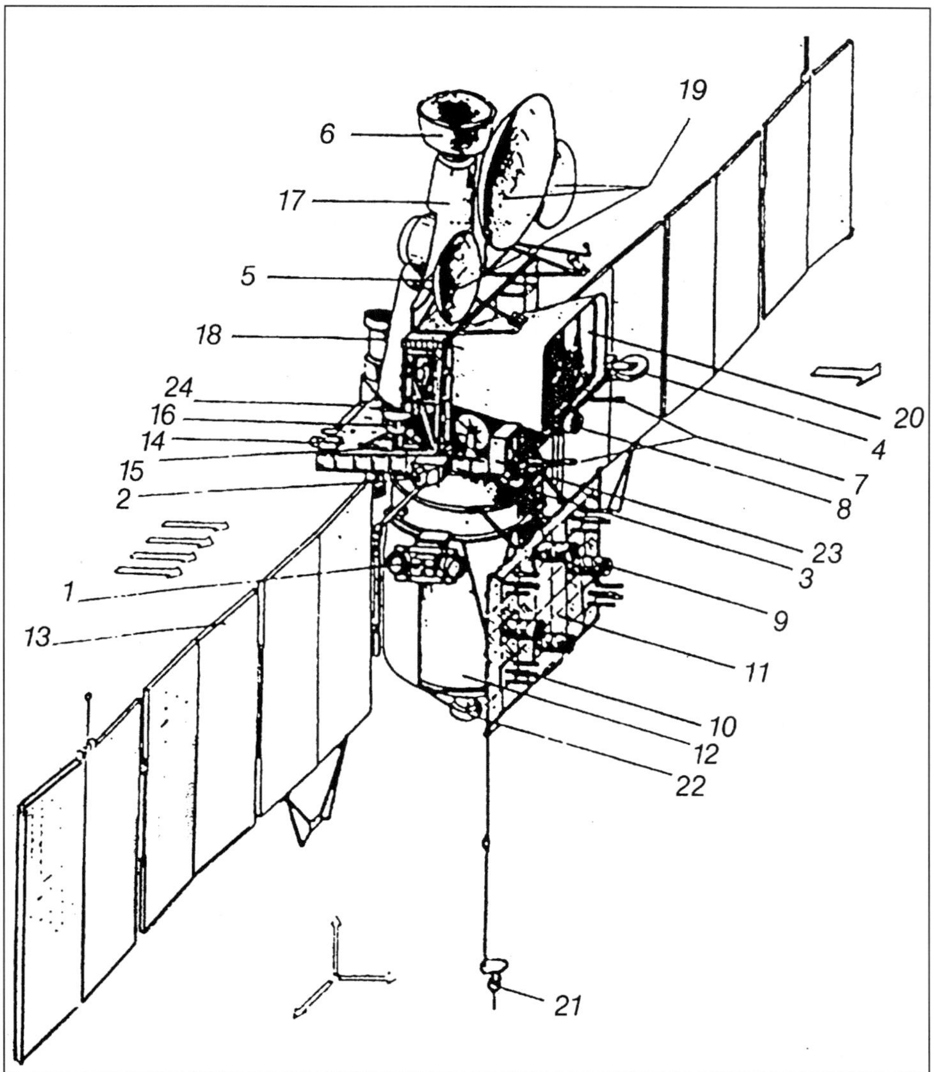

Elektro's key features. 1 arcjet thruster, 2 Sun sensor, 3 instrument platform, 4 reaction wheel, 5 2-axis antenna drive, 6 imaging system radiator cooler, 7 antennas, 8 nadir sensor (failed), 9 antenna for data collection platforms, 10 broadcast antennas, 11 antenna platform, 12 pressurised body, 13 solar array, 14 solar UV sensor, 15 coarse Sun sensor, 16 x-ray sensor, 17 imaging system, 18 polar star tracker, 19 main antennas (outer two are apparently dummies), 20 imaging system shade, 21 magnetometer, 22 thermal screen, 23 proton/electron sensor, 24 low-energy particle spectrometer 0517039

Sensors: Visible/infra-red TV system operating at 0.46–0.7 µm (8,000 lines) and 10.5–12.5 µm (1,400 lines), but visible inoperable. Elektro 2 will add 6–7 µm channel for water vapour. Global image returned every 1 h at 2.56 Mbit/s on 1.685 or 7.465 GHz to Moscow, Novosibirsk and Khabarovsk. WEFAX-standard imagery received at 2.115 or 8.195 GHz for broadcast at 1.691 GHz. Monitored are 0.2–2.5 MeV electrons, 0.2–500 keV protons, 2.0–120 MeVα particles, 2–10 E solar X-rays, 100–1,300 E solar UV and magnetic fields.

Meteor

Current Status

Several Meteor satellites are still operational, including 3–05, still in use and Meteor 2–21, the last of the second generation series in a non sun-synchronous orbit currently with early morning and mid to late afternoon passes. Meteor 3–05 is in a slightly higher orbit than Meteor 2–21. Meteor 3–06 operates still but the onboard French Scarat payload for radiation budget failed in April 1995.

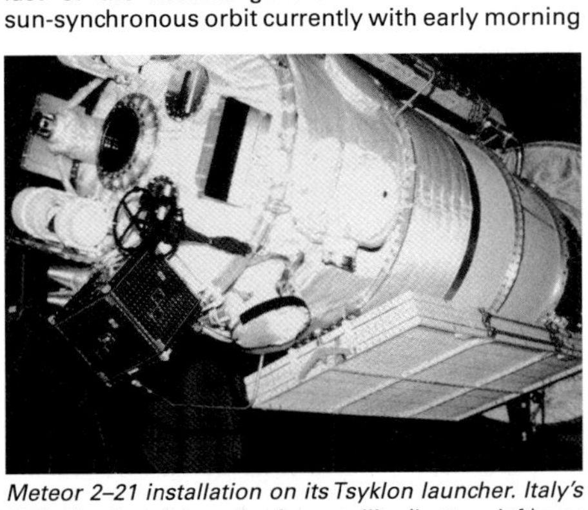

Meteor 2–21 installation on its Tsyklon launcher. Italy's 32 kg Temisat data collection satellite (bottom left) was released by Meteor in orbit (Kayser-Threde GmbH) 0517040

Background

Similar to the US NOAA series, Meteor has launched more than 50 spacecraft over three generations. After three years of testing under the Kosmos label, the first Meteor 1 appeared in 1969, followed by the more capable Meteor 2 in July 1975 and Meteor 3 in November 1984. The Soviet Union retired the last of the 31 Meteor 1s in June 1984. The SL-14 Tsyklon began to replace SL-3 in 1982 as the launch vehicle for putting the Meteors in orbit. Meteor 1/2 typically flew at 800 to 900 km, but when the Soviet Union switched to the SL-14 the satellites increased their altitude to 950 km for Meteor 2 and to 1,200 km for Meteor 3. This provided a more comprehensive coverage. The new generation 3M may fly at 925 km. Four Meteor 1s entered into Sun-synchronous orbits from Tyuratam during 1977–81 as testbeds for the Resurs-O programme. All other Meteors have departed from Plesetsk since 1967. Meteor 3–05 broke new ground in 1991 when it carried NASA's TOMS.

In conjunction with the normal launch some Meteors have released third party satellites. Italy's Temisat entered space on a Meteor 2 and Germany's Tubsat B on a Meteor 3. That Meteor 3 included France's Scarab Earth radiation budget instrument and DARA's PRARE unit.

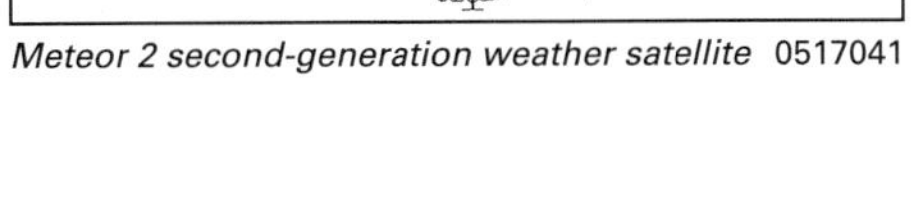

Meteor 2 second-generation weather satellite 0517041

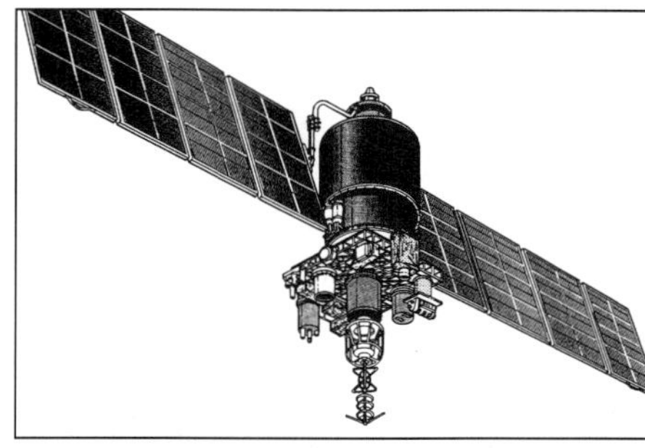

Meteor 3 can accommodate experimental instruments on its 1.6 × 1.8 m truss
(Teledyne Brown Engineering) 0517042

The Russian Federation planned to continue the successful series, but funding problems have hampered progress. The new generation Meteor 3M is based on as much of the old platform as possible.

Meteors provide a daily weather review for more than two-thirds of the globe, on clouds, ice cover, atmospheric radiation, weather fronts and jet streams. Kosmos 122, launched 25 June 1966 from Tyuratam, 589 × 643 km, 65°, later became the first dedicated meteorlogical satellite identified by the Soviets as meteorological. President de Gaulle of France witnessed the launch, the first Westerner to visit a Soviet launch site. By 1978, according to Moscow Radio, Meteors had transmitted one million images. These satellites cut the sailing time of Soviet ships because the satellites indicated the best courses to avoid storms, winds and ice. This maritime contribution alone was officially estimated to be worth more than R1 million. Meteors also monitor clear-air turbulence so that the airliner crews can be warned of potential hazards. In 1980, Meteor satellite made their first hail warnings, a prediction of particular value in Soviet Georgia, with its 600,000 hectares of vineyards.

Meteor 1

Launches, beginning with Meteor 1 on 26 March 1969 from Plesetsk (644 × 713 km, 81°), began at the rate of three to four annually so that Meteor 2 and 3 were always operational and providing a continuous survey of atmospheric conditions from pole to pole in a swath up to 1,500 km wide. The information dumped to ground stations in passes lasting only a few minutes. Three sites received these transmissions, the USSR Hydrometeorological Service in Moscow, Novosibirsk (Siberia) and Khabarovsk (Pacific Coast). It took about 1 hour 30 minutes for the Soviets to process the data and transmit it to users. From Meteor 1–10, launched on 30 December 1971, the Meteor 1s used an 81° orbit situated at about 890 km. This satellite carried the first Soviet meteorological automatic picture transmission compatible with Western receivers. The bus had a small Fakel plasma thruster for orbital adjustments.

From Meteor 1–28, launched 29 June 1977 (re-entered 28 August 1993), orbits were retrograde, Sun-synchronous and circular, at 630 km, 98°. This was the first of four Meteor 1s flown as testbeds for Resurs-O, which became operational in 1988 with Kosmos 1939. The last Meteor 1 was 1–31, launched 10 July 1981.

Meteor 1–31 also known as Priroda ('Nature') went into space 10 July 1981 aboard a SL-3 from Tyuratam into 630 km, 98° Sun-synchronous orbit.

The Meteor 1 had a main sensor consisting of a 32-channel scanning radiometer and 33-channel microwave radiometer for surface observations. The Bulgaria 1300 programme provided some of the equipment. At the end of 1983, this equipment was being used to determine snow depths on all Soviet agricultural regions to assist in planning the spring work season and forecasting the 1984 harvest. Transmissions ceased probably in early 1983.

Meteor 2

Introduced experimentally, while the Meteor 1 programme continued, Meteor 2–01 (11 July 1975), established the new operational orbits of 950 km, 82.5° orbits. The satellites have longer life, and include scanning radiometers. Various combinations of orbital plane have been tried, usually with two to three satellites giving passes at 6 to 8 h intervals. Since 1981, the Soviets maintained two or three in operation at any one time, although the Russians let the two remaining satellites fall silent in Summer 1992. August 1993's new launch remains the sole satellite. Automatic picture transmissions for visible imagery are switched off when, for the satellite, the Earth is in shadow.

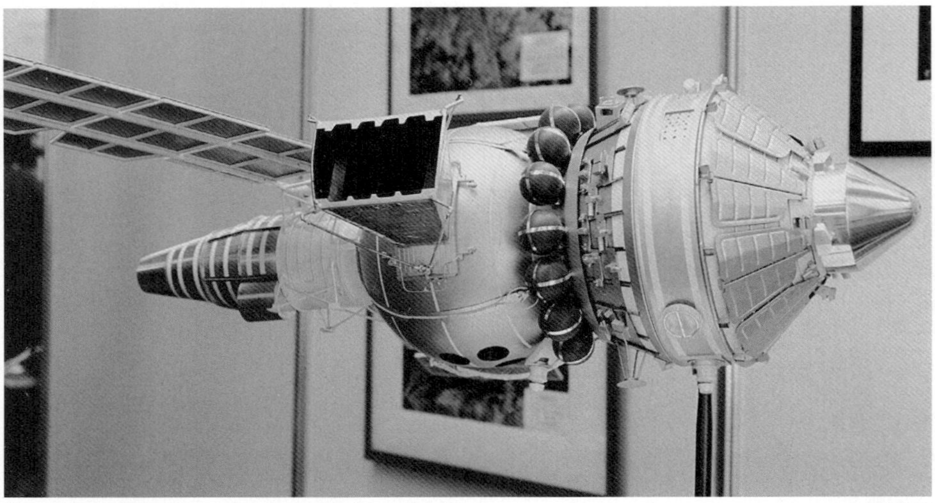

Russia's new Meteor 3 m weather satellite (Theo Pirard) 0054350

Meteor 2–19/2–20 apparently replaced 2–16/2–18, respectively, but all five of 16–20 still operated at the beginning of 1991, probably explaining why there were no 1991 departures. 2–19/2–20 fell silent summer 1992.

Meteor 3

The first official Meteor 3 launch occurred on 24 October 1985 from Plesetsk using SL-14. Although employing the same 82.6° inclination as the SL-14 Meteor 2s, Meteor 3–01 was inserted into a much higher 1,227 × 1,251 km. In retrospect, the launch of Kosmos 1612 on 27 November 1984 (135 × 1,230 km) appears to have been a precursor Meteor 3. This launch failed when Tsyklon's upper stage did not re-ignite. Meteor 3–01 incorporated plasma thrusters, which were used to lower its orbit to a mean altitude of 1,200 km in November-December 1985. In late June 1986, the Soviets lowered the orbit by another 4 km. No new 3s were launched in 1986 or 1987, but 3–02 appeared 26 July 1988, followed by 3–03 24 October 1989. Both satellites operated in early 1991, but were replaced during the year by 3–04 (24 April) and 3–05 (15 August). 3–05 carried NASA Goddard's Total Ozone Mapping Spectrometer to extend Nimbus 7 observations, which discovered Antarctica's ozone hole in 1987. The instrument failed December 1994, well beyond its two year design life. Data went via NASA's Wallops Flight Facility to the Russian Obninsk station.

Meteor 3–06 carries France's Scarab Earth radiation balance instrument. The Scarab (scanner for radiation budget) project began in 1987, made up of three 40 kg units built at FFr65 million. The second scanner is scheduled aboard the Resurs-O1 N4 and may fly on 3M-1. Data from Scarab go first to Obninsk and Medvezhi Ozera for relay to NPO Planeta in Moscow and then to CNES Toulouse. The instrument monitors four bands: visible 0.5–0.7, solar 0.2–4, total 0.2–50, infra-red window 10.5–12.5 μm. The US NOAA and ERBS satellites correlate the data.

The All-Russian Research Institute for Electromechanics (VNII Elektromekaniki), designed and built the Meteor/Resurs-O satellites. Meteor 3 and Resurs-O employ the same platform.

Specifications

Meteor 2/3

Launched: typically annually by Tsyklon from Plesetsk: 3M on Zenit 2

Mass: Meteor 2 1,300 kg on-orbit; Meteor 3 2,215 kg on-orbit, payload 500–700 kg; Meteor 3M 2,500 kg on-orbit, payload 900 kg

Resolution: 0.7–2 km visible, 3–42 km infra-red

Orbit: 950 km 82.5° Meteor 2, 1,200 km 82.5° Meteor 3, 925 km 82.5° Meteor 3M (3M-1 will satisfy SAGE requirements with: 1,020 km, 99.5°, 09.15 ascending node equatorial crossing)

Spacecraft: 3-axis stabilisation (0.5° accuracy; 3M will be improved), 4.2 m long 1.4 m diameter cylindrical body with twin 10 m span 2 m-wide solar wings providing 500 W to the payload (3M: 1 kW). Design life 2 years (3M, 3 years). Meteor 3 adjusts orbit by Fakel SPT plasma thrusters (these thrusters are now offered commercially; see the International Space Technology joint venture entry in the Launchers section). 3M will use GPS/Glonass for precise orbit determination. Transmissions directed at three primary sites (Moscow, Novosibirsk in Siberia and Khabarovsk on the Pacific coast; Tashkent was lost with independence; new sites were planned for Murmansk and Petropavlovsk-Kamchatski) in conjunction with >80 smaller ground stations. Internationally-compatible APT transmissions are made on 137–138 MHz to 15,000 CIS terminals; there is also a 10 W analogue store/forward mode at 466.5±0.125 MHz (3M will use 1.69–1.71 GHz digital). Data are processed within 90 minutes by the Hydrometeorological Centre

Sensors: Meteor 2. 0.5–0.7 μm 2,100 km swath medium resolution (2 km), 0.5–0.7 μm 2,400 km swath high resolution (1 km), 8–12 μm global scanning infra-red radiometer (2,600 km swath, 8 km resolution), and 11/10- 18/70 μm scanning infra-red radiometer (1,000 km swath, 30 km resolution). Two satellites can cover 80 per cent of the Earth's surface in 6 h.

Meteor 3. 0.5–0.8 μm 3,100 km swath 0.7–1.4 km resolution, 10.5–12.5 μm infra-red radiometer 3,100 km swath 3 km resolution (1°C resolution), 9.4–19.68 μm 10-channel infra-red spectrometer 1,000 km swath 42 km resolution, 0.25–0.380 μm 8-channel ultra-violet backscatter spectrometer 200 km swath 3–5 km altitude resolution, Ozon-M 0.25-0.29, 0.37–0.39, 0.6–0.64, 0.99–1.03 μm 4-channel UV ozone monitor 2 km altitude resolution, plus 0.17–600 MeV electron/proton detectors. 1991's Meteor 3–05 carried NASA's TOMS; second was planned for 3M-2 in 2000. Meteor 3–06 carries France's Scarab Earth radiation balance instrument (see below for details), DARA's PRARE unit (see ERS 1 for background), and released Germany's Tubsat B.

Resurs-F

Background

In the mid-1970s, the Russians started to fly recoverable remote sensing satellites within the Kosmos programme, these satellites being based upon the contemporary Zenit-4 class of photo-reconnaissance satellites. Over the years the majority of the launches were to orbits with inclinations of 81.2–82.5° and they were initially announced as simply undertaking remote sensing work 'for the national economy' or undertaking work for the 'Priroda Centre'. The first of these groups of recoverable remote sensing satellites was Kosmos 742, a rather isolated mission because it has been linked with four satellites in the Gektor-Priroda programme which were launched in 1979. A larger group called Fram debuted in September

The nature and schedule of KB Photon's new Resurs-Spektr satellites are uncertain. They appear to be derived from the fifth-generation digital photo-reconnaissance satellites. Spektr-V would provide 3 to 5 m visible stereo resolution, complemented by microwave sensors. Spektr-R (shown here) would provide radar imagery
(Theo Pirard/Space Information Centre) 0517423

Meteor 3 weather satellite first launched in 1983 0054352

Resurs-F2 final flight preparations 0517234

Resurs-F missions		**Launch**	**Orbit (km)**	**Days**
(1986 onwards)[1]				
C1746	F1	28 May 1986	259 × 273, 82.3°	14
C1757	F3	11 June 1986	159 × 391, 82.3°	14
C1762	F1	10 July 1986	259 × 273, 82.6°	14
C1768	F1	2 August 1986	259 × 273, 82.6°	14
C1789	F1	31 October 1986	322 × 342, 82.6°	14
C1846	F1	21 May 1987	323 × 342, 82.4°	14
–[3]	F1	18 June 1987	–	–
C1882	F1	15 September 1987	259 × 275, 82.3°	21
C1906[2]	F2	26 December 1987	257 × 277, 82.6°	–
C1920	F1	18 February 1988	330 × 334, 82.6°	20
C1951	F1	31 May 1988	259 × 275, 82.3°	14
C1956	F3	23 June 88	333 × 368, 82.4°	14
C1957	F1	7 July 1988	260 × 275, 82.6°	14
–[3]	F1	21 July 1988	–	–
C1965	F2	23 August 1988	257 × 277, 82.3°	30
C1968	F1	9 September 1988	260 × 275, 82.3°	14
C1990	F2	12 January 1989	256 × 269, 82.6°	30
C2000	F3	10 February 1989	343 × 395, 82.4°	20
Res 1	F1	25 May 1989	255 × 272, 82.3°	23
Res 2	F1	27 June 1989	260 × 274, 82.6°	14
C2029	F3	5 July 1989	338 × 362, 82.4°	14
Res 3	F1	18 July 1989	260 × 275, 82.6°	21
Res 4	F2	15 August 1989	259 × 268, 82.3°	30
Res 5	F1	6 September 1989	262 × 273, 82.3°	16
Res 6	F1	29 May 1990	259 × 273, 83.3°	16
Res 7	F2	17 July 1990	261 × 269, 82.3°	30
Res 8	F1	16 August 1990	259 × 271, 82.3°	16
Res 9	F1	7 September 1990	261 × 275, 82.6°	14
Res 10[4]	F2	21 May 1991	229 × 237, 82.3°	30
Res 11	F1	28 June 1991	259 × 273, 82.3°	23
Res 12	F1	23 July 1991	258 × 272, 82.3°	16
Res 13[4]	F2	21 August 1991	231 × 234, 82.3°	30
Res 14[4]	F2	29 April 1992	233 × 245, 82.1°	30
Res 15[4]	F1	23 June 1992	229 × 233, 82.3°	16
Res 16[4]	F1	19 August 1992	221 × 239, 82.6°	16
Res 17[4]	F2	21 May 1993	235 × 245, 82.6°	30
Res 18[4]	F1	25 June 1993	223 × 241, 82.6°	17
Res 19[4]	F1	24 August 1993	234 × 239, 82.6°	17
Res 20[4]	F2	26 September 1995	231 × 235, 82.3°	30
Res 21	FIM	18 November 1997	238 × 270, 89.6°	25
Res 021	F1	10 July 1998	817 × 818 98.7°	
Res 22	FIM	28 September 1999	222 × 230, 82.3°	24

Notes: 1: all launches made by Soyuz from Plesetsk; first Resurs-F mission was C1127 5 September 1979 (28 missions performed pre-1986). 2: detonated 31 January 1988 following spacecraft malfunction. 3: 1987 failure not identified until 1995; 1988 failure not identified until 1993; no further details. 4: new low-altitude regime to increase resolution (but Res 20 uniquely then rose to the old standard 256 × 280 km 22 October 1995). Res 20 image coverage was 20.7 million km^2.

1975 with the launch of Kosmos 771 and launches in this series continued through until 1984 (Kosmos 1681 was the last Fram satellite).

Resurs-F employs the Vostok-based photo-reconnaissance spacecraft, with more than 800 missions to its credit, flying for the Priroda centre of Roskartografia (the Federal Service of Geodesy & Cartography). Resurs-F began flying with C1127 in September 1979. The older F1 type flew missions of typically two to three weeks (although active for only up to 14 days) with Kate-200 and KFA-1000 cameras. F2 undertook four-week missions using the MK-4 camera system, although lifetimes of up to 45 days are possible. F3 missions were identified in 1993, flying occasionally with two KFA-3000s for 2 m resolution. The film camera systems are returned in the 2.2 m diameter spherical descent capsules, which are reused an average of three times. The camera systems are also refurbished and reflown. F2's longer duration missions (see table) are supported by solar panels supplementing the standard battery power supply system. F3's appearance is identical externally to F1. The improved F1M had been planned to debut in 1994 but no Resurs-F missions flew that year because of the lack of launch vehicles. The satellite is completed but still lacks launch funding. F1M extends the active period to 19 days (+6 days inactive) and carries three KFA-3000s. F1M can also fly down to 190 km for improved resolution. The F2M was planned for 1997, carrying an improved film for higher resolution. Launch actually took place in 1999.

Resurs-F1 recovery. The two KFA-1000 cameras sit above three Kate-200s. The camera section cover lies on the ground, with the Kate viewpoints just visible (Resurs-F WorldMap) 0517064

Resurs-F3 recovery, showing the paired KFA-3000 cameras. Above the cameras are fans to maintain convection for cooling (Resurs-F WorldMap) 0517065

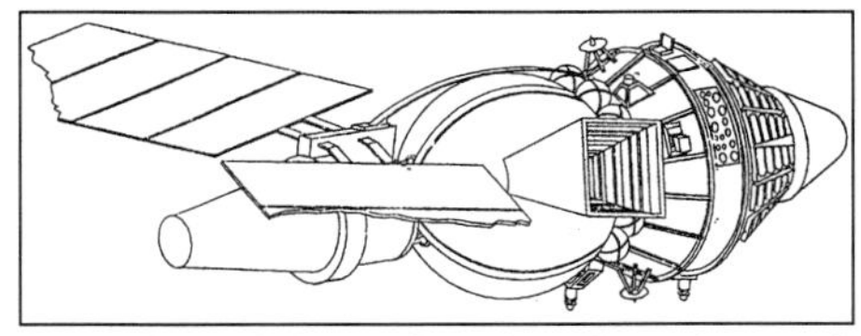

Resurs-F2 in flight configuration. The solar panels allow longer missions than the battery-powered F1/F3 0517046

JEBCO Information Services (UK) and Priroda, the primary non-military remote-sensing acquisition and processing organisation in Russia created the Resurs-F WorldMap consortium in 1993. Priroda's archives contain two million XS/PAN scenes at 2 to 20 m resolution covering most of the globe. The consortium accepts requests for targets for future flights, and would like to reduce lead time. Usually 40 to 60 per cent of the full mission capability is specifically targeted, while the remainder of the film is acquired over targets of opportunity. 12 hour operating plans are uplinked twice daily. F1/F1M/F3 can power down for 11/11/6 days to wait for more suitable target conditions.

It has been reported that Resurs-F will be replaced eventually by KB Photon's Nika-K, although recent reports of funding problems make delays likely. It will provide more propellant, more film, improved orientation and 45 day flights. It is apparently derived from the Generation 6 Kuban photo-reconnnaissance satellite that is also being modified as the Nika-T μg craft to replace Photon.

Kosmos 1906, which was deliberately destroyed in orbit because of a control malfunction, was acknowledged at the time to be the first of a new type of craft carrying multispectral cameras, and later identified as the first F2. The first named Resurs-F satellite was launched 25 May 1989, accompanied by two small Pion passive upper atmosphere density satellites. The F2/MK-4 system completed its test flight programme in 1989, complementing the Kate/KFA cameras of Resurs-F1. Small μg payloads are also flown aboard Resurs: Intospace's Cosima 2 was the first, in September 1989, followed by number 3 on 29 May 1990 and the first Casimir unit in September 1990. August 1992's F1 carried the US Naval Research Laboratory's Be-7 Induced Radiation Experiment (aluminium foil on an extending arm) and released two further Pions. It was the first CIS/USSR satellite to carry a DoD experiment, under Space Test Program sponsorship. The arm failed to extend; there was an identical failure in mid-1993.

Launched by Zenit 2 on 4 November 1994 the first Resurs 01 carried two multispectral scanners, one using a CCD system composed of two images with 45 m ground resolution covering a 90 km swath, the other using a conical scanning device with a 600 km swath. The next Resurs 01 was launched on 10 July 1998.

Specifications

Resolution: 5–8 m visible for MK-4, 5 m visible KFA-1000, 2 m visible KFA-3000
Launched: typically 3–5 years, all by SL-4 Soyuz from Plesetsk, until funding slowed rate
Mass: about 6.3 t on-orbit, 2.3 t descent sphere
Orbit: typically 260–275 km, 82.5° for F1/2 majority (high 330–365 km for Kosmos 1283/1284/1472). Res 10 introduced new low of 230–235 km in 1991 for improved resolution. F3 now 275 km. 14 days active and 11 days quiescent for F1/F3, F1M 19/6 days; 30 days F2 (all active, none quiescent); initial orbit typically 180 × 30 km, then raised. F1M expected to employ 190 km operationally for improved resolution. F3 planned for 180 km to provide 1.6 m resolution
Spacecraft: Vostok-based
Sensors: Three Kate-200 and two KFA-1000 (F1); one Kate-200 + three KFA-1000 (F1M); MK-4 (F2); two KFA-3000 (F3) film cameras

Stellar camera (SA-33 for F1 and SA-3R for F2) provides pointing reconstruction.

UPDATED

Resurs-O

Current Status

The Resurs-01 launched 4 November 1994 had a 21 day cycle and a downlink capability of 7.68 mb/sec.

Background

The Resurs-O1 system, the multispectral digital equivalent of the US Landsat, became operational

Resurs-O1 N3 ready for launch (OHB-System) 0517233

in 1988 with the launch of the series' fifth, Kosmos 1939. It had been preceded by the flight of Kosmos 1689 on 3 October 1985. The Russians expected to replace Kosmos 1939 in June 1993, but the satellite's longevity deferred Resurs-O1 N3 until November 1994. The launch marked the first civil orbit for the Zenit booster. Resurs-01 N4 was launched on 10 July 1998 and in addition to its remote sensing equipment the satellite also carried the Belgian LLMS communications for the IRIS system.

The All-Russian Research Institute for Electromechanics (VNII Elektromekaniki) builds the Resurs-O satellites and the Meteor 3 and Resurs-O1 employ the same platform modified for operations at different altitudes and offering three to five year lifetimes. The history of the two platforms has been intermixed throughout their lifetimes. Kosmos 1939 (20 April 1988) followed Meteor 1–30 (deactivated 18 June 1988 after eight years' observations), Meteor 1–31, Kosmos 1484 and Kosmos 1689. All of these satellites used the same basic design.

Resurs control commands come from the VKS Military Space Forces using its Moscow centre. The imagery product from the satellites goes directly through Moscow, Novosibirsk and Khabarovsk. Commercial imagery for Western purchasers passes through Sweden's SSC Satellitbild, by an arrangement begun in 1995. The first SSC imagery came from MSU-SK downlinked from the November 1994 Resurs-O1. Sweden interest targeted agricultural and coastal water images.

SSC deals with the Sovzond consortium of VNII Elektromekaniki, NPO Planeta and the NIIKP Institute of Space Device Engineering (which builds the MSU sensors). Direct mode images began 3 July 1995, and recorded images arrived by the end of 1995. SSC and its partners had archived 1,000 images by the end of 1995. Archived images cost ECU1,100 path-oriented or ECU1,800 map-oriented.

VNIIEM has developed the USSP Universal Small Space Platform for release from Resurs-O. The 60 kg satellite employs a gravity gradient boom, reaction wheel and magnetorquers for attitude control. The bus is a hexagonal prism.

Specifications

Resurs-O1 N3 & N4

Launch: N3 on 4 November 1994 and N4 on 10 July 1998 both by Zenit from Tyuratam

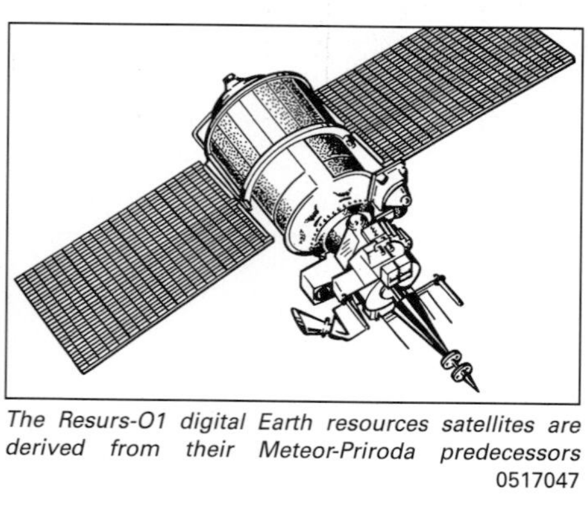

The Resurs-O1 digital Earth resources satellites are derived from their Meteor-Priroda predecessors 0517047

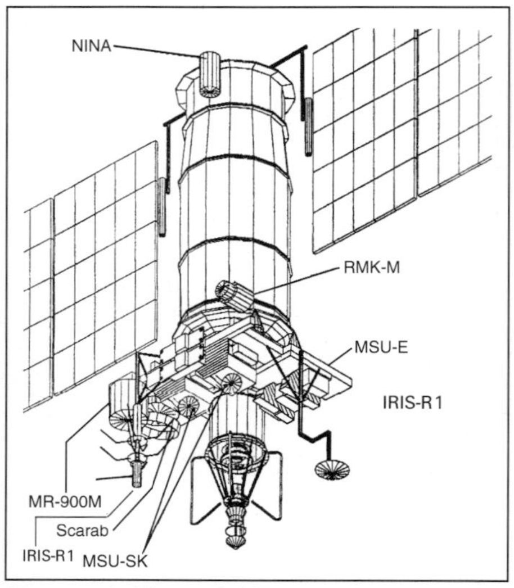

The upgraded Resurs-O1 N4 0517421

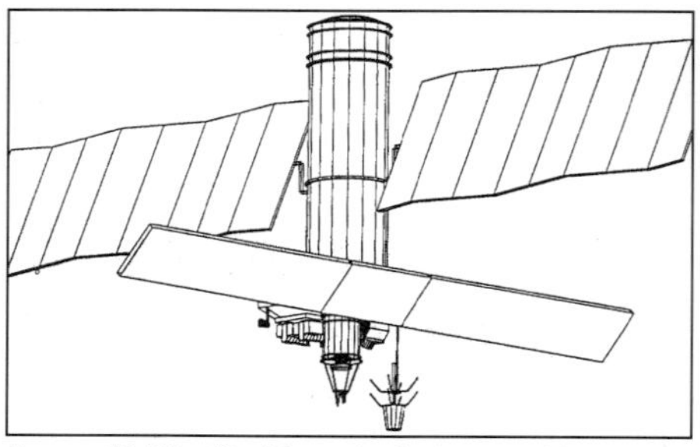

Resurs-O1 N5 will add a radar for ice patrol imagery to aid shipping (VNIIEM) 0517422

Mass: 1,950 kg (550 kg payload)
Dimensions 6 m × 1.4 m
Resolution: 45 m visible, 170 m infra-red, 600 m thermal infra-red
Orbit: 663 × 691 km, 98.05° Sun-synchronous (16 day repeat)
Spacecraft: based on Meteor 3 platform (qv); 3-axis stabilised, about 5 m high, 10 m across solar panels. Real-time data rate 7.68 Mbit/s on 8.192 GHz. Power 500 W average (1.2 kW peak)
Sensors: Multiple multispectral package covering visible to thermal infra-red

MSU-SK conical scanner 0.5–0.6, 0.6–0.7, 0.7- 0.8, 0.8–1.1, 10.4–12.6 µm offers 600 km swath width, 170 m resolution for first four bands and 600 m for band 5.

MSU-E push-broom CCD imagers cover 0.5–0.6, 0.6–0.7, 0.8- 0.9 µm, providing 45 m resolution over 90 km total swath width of both instruments. 1,000-pixel linear CCD.

Ukraine

Okean-O1/Okean-O/ Sich

Background

The Okean-O1 all-weather radar, ice and oceanographic satellite system, became operational with the launch of Okean 1 by Tsyklon 3 on 5 July 1988. Kosmos 1076/1151 flew as precursor satellites, but 1983's Kosmos 1500 prototype was the first to incorporate 1.5 km resolution side-looking radar utilising a 12 m deployable antenna. Data return through the three primary Meteor ground stations. Ice forecasts are uplinked four times weekly to vessels via the Ekran geostationary satellites. Kosmos 1500 in December 1983 helped free 70 Soviet ships trapped in heavy Arctic ice near Wrangel Island. NPO Yuzhnoye in Dniepropetrovsk, Ukraine, built and designed the Okean and the VKS Military Space Forces operate the system for NPO Planeta. Possibly the last of the 1.9 tonne Okean-O1 designs departed in August 1995 on Tsyklon as a joint Russian/Ukrainian project, designated Sich 1. The Ukraine took over full control from 11 October 1995 using Yevpatoria. The first Okean 02 series satellite was launched by Zenit on 17 July 1999 as a joint venture between the RSA and the Ukrainian National Space Agency. Major failures plagued

The first 6.3 tonne Okean-O is expected in 1997 (Theo Pirard) 0517415

operations and it was silent between December 1999 and May 2000.

Sich is the Ukraine space agency's main satellite programme, and the Okean-O platform will be used for the Ukraine-only Sich 2, carrying a large SAR. Sich 3 will carry more scanning radiometers.

As with so many former Soviet programmes, financial constraints have significantly reduced these plans.

Specifications

Okean-O1

Resolution: 1.5 km radar, 0.35–1.5 km visible/infra-red, 6–15 km radiometer
Design life: six months
Mass: 1,950 kg (550–650 kg payload)
Orbit: typically 630–660 km, 82.5°
Spacecraft: stepped cylindrical pressurised bus 3 m high, maximum 1.4 m diameter, 3-axis stabilisation (base nadir pointing, accuracy 0.75–1.0°, determination to 2 arc/min), aided by gravity gradient boom. Twin solar arrays spanning 4.82 m (each an irregular 2 m high, 1.6 m wide); power consumption 1.1 kW operating, 110–270 W daily average. Sensors and single 11.8 m radar antenna mounted on four 2.9 m panels. APT transmissions on 137.4 MHz. Condor 2 digital data collection from small automatic Condor 1 stations at 500 bit/s on 1533.4 MHz, relay at 2 kbit/s on 460.03 MHz). No orbit or yaw control
Sensors: RLS-BO side-looking radar operating at 3.15 cm and 450 km swath width to left of ground track. Power limits operations to 10 min sessions.

MSU-S visible/near-infra-red scanning radiometer (channels 0.5–0.7, 0.7–0.9 µm) providing 350 m resolution over 1,380 km swath width, nadir pointing.

MSU-M multispectral visible/near-infra-red scanner, 0.5–0.6, 0.6–0.7, 0.7–0.8, 0.8–1.1 µm, 1.5 km resolution over 1,930 km swath, nadir pointing.

RM-08 8 mm-wavelength scanning radiometer, 550 km swath, to left of track. C1500/1602/1766 carried the MSU-K circular scanning radiometer, 0.5- 0.9 µm single band, 900 km swath, 500 m resolution, 6.5 kg.

The Okean-O1 ocean radar satellite was first displayed at 1985's Paris Salon. The SLR antenna is seen diagonally at bottom left (Graham Turnill) 0517043

Okean-O1 radar programme launches[1]

	Launch	Orbit	End Ops
E1/C1076[2]	12 February 1979	593 × 620 km, 82.5°	31 March 1980
E2/C1151[2]	23 January 1980	613 × 640 km, 82.5°	13 October 1981
OE1/C1500	28 September 1983	643 × 679 km, 82.6°	16 July 1986
OE2/C1602	28 September 1984	629 ×664 km, 82.5°	5 December 1986
O1/C1766	28 July 1986	631 × 662 km, 82.5°	24 October 1988
O2/C1869[3]	16 July 1987	634 × 667 km, 82.5°	3 May 1989
O3/Okean 1	5 July 1988	635 × 666 km, 82.5°	14 June 1990
O4[4]	9 September 1989	–	–
O5/Okean 2	28 February 1990	639 × 666 km, 82.5°	18 July 1991
O6/Okean 3[5]	4 June 1991	634 × 666 km, 82.5°	4 January 1994
O7/Okean 4	11 October 1994	632 × 666 km, 82.6°	
O8/Sich 1[6]	31 August 1995	632 × 669km, 82.5°	
09/Okean 2–1	17 July 1999	633 × 661 km, 98.1°	

Notes: 1) all launches made by SL-14 Tsyklon 3 from Plesetsk; 2) experimental, did not carry SLR; 3) radar antenna failed to deploy but other sensors operational; 4)stage 3 failure; 5)plane 90° from number 2's; 6) Chile's 50 kg FASat-Alfa failed to separate.

Okean-O
Resolution: 1.5 km radar, 25–200 m visible, 100–600 m infra-red, 16–130 km radiometer
Design life: 3 years (1 year Okean-O number 1)
Mass: up to 6.5 t (up to 2 t payload); Okean-0 number 1 6.3/1.55 t
Orbit: planned 670 km, 98°
Spacecraft: total length 10.6 m, 1.8 m diameter 6 m pressurised bus, aluminium-magnesium alloy structure. 3-axis stabilisation (base nadir pointing, accuracy 5 arcmin, determination to 2 arc/min). orbit adjust/momentum dumping provided by base propulsion system: 10 × 30 N UDMH/NTO thrusters. Attitude control by RWs and magnetorquers. Single articulated solar wing; power consumption 5 kW operating, 1.5 W daily average (Okean-O number 1 3.5/ 0.80 kW). Sensors and radar antenna mounted on sides/base of bus and forward instrument platform. Condor 2 digital data collection and transmission system (receive at 500 bit/s on 1.5334 GHz; transmit at 2 kbit/s on 460.03 MHz). Data transmitted at 15.36–122.88 Mbit/s (Okean-O number 1 up to 64 kbit/s) on 8.2 GHz; APT transmitted on 137.4 MHz. 19 Gbit memory (Okean-O number 1 5.5 Gbit)
Sensors: RLS-BO side-looking radar operating at 3.15 cm and 450 km swath width to left of ground track

MSU-M multispectral visible/near-infra-red scanner operating at 0.5–0.6, 0.6–0.7, 0.7–0.8, 0.8–1.1 μm providing 1–1.7 km resolution over 1,930 km swath.

MSU-A 3-channel 0.5–0.9 μm scanner providing 30 m resolution over ±300 km swath.

MSU-SK visible/infra-red conical scanning radiometer. Two visible channels 0.53–0. 59/ 0.61–0.69 μm 200 m resolution; two near-infra-red 0.7–0.8/0.9–1.0 μm 200 m resolution; one infra-red 10.4- 12.6 μm 600 m resolution. 600 km swath.

MSU-V conical scanner. Eight channels 0.45–0.52 (50 m resolution), 0.52–0.62 (50 m), 0.62–0.74 (50 m), 0.76–0.9 (50 m), 0.9–1.1 (50 m), 1.55–1.75 (100 m resolution), 2.1–2.35 (250 m), 10.0–12.0 (100 m resolution) μm. 180 km swath, 1.28–5.12 Mbit/s. Ice/snow imaging, land use.

Trasser-0 spectroradiometer. 62 channels over 0.430–0.800 μm, <1 n mile resolution.

Land Delta 2 scanning radiometer, 7.0/13.0/22.5/ 36.5 GHz, 2,600 km swath.

R600 microwave radiometer. 6 cm/4.9 GHz, two polarisations, 130 km resolution, 1 kbit/s.

R225 microwave radiometer. 2.25 cm/13.3 GHz, two polarisations, 130 km resolution, 2 kbit/s.

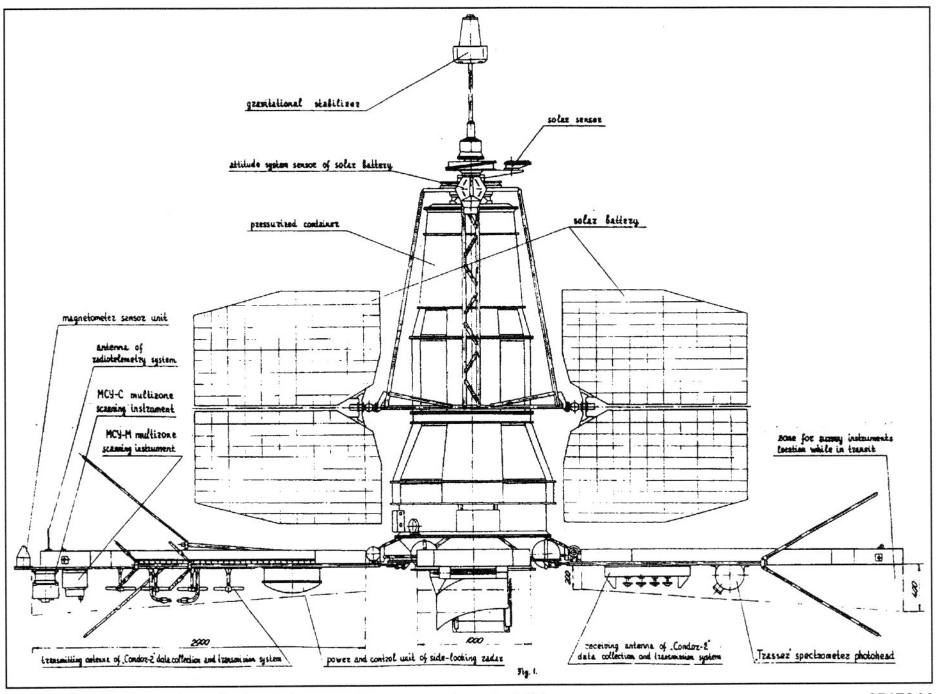

Okean-O1 configuration; the large base radar antenna is not visible 0517044

Sich 2
Resolution: 10/50 m SAR, 1.5 km SLR, 30 m visible
Launch: undetermined
Design life: 3 years
Mass: up to 6.5 t (up to 2 t payload)
Orbit: planned 670 km, 98°
Spacecraft: see Okean-O
Sensors: RLBO double side-looking radar (with scatterometer mode), 12 m antenna, 3.2 cm wavelength, 700 km swath each side ground track.

2.4 × 15 m SAR, 23 cm wavelength, 85–122 km swath provides 50 m resolution, 36–57 km swath 10 m resolution.

MSU 3-channel visible/near-infra-red scanner, 0.5–1.0 μm, 48 km scan with 700 km for 30 m resolution.

Sich 3
Resolution: 3–10 m visible, 20 m infra-red, 3 km microwave
Launch: undetermined
Design life: 3 years
Mass: up to 6.5 t (up to 2 t payload)
Orbit: planned 670 km, 98°
Spacecraft: see Okean-O

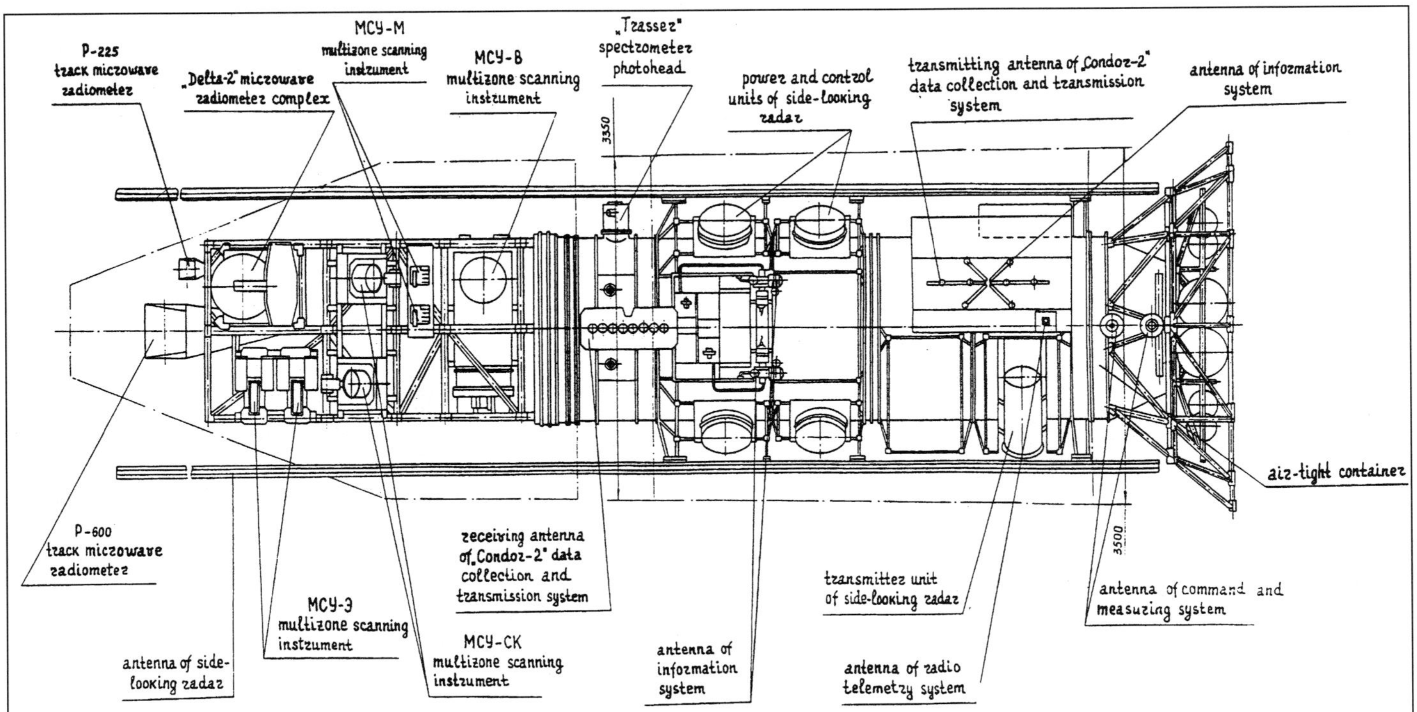

Earth-pointing face of 6.3 tonne Okean-O with potential sensor payload. The SLR antennas run along both sides (only the lower one will be carried); the propulsion system is at far right. The dashed line indicates the Zenit payload envelope 0517045

Sensors: 11-channel spectroradiometer, 0.4–2.4 μm, 10–40 m res, 160–200 km swath within 600 km width, 32–64 Mbit/s. Ice/snow, land use.

3–5 m high resolution 0.4–0.8 μm camera, 30–50 km scan within 150 km width.

3-channel IR spectroradiometer, 0.3–13 μm, 20–60 m res, 50–150 km swath within 600 km width, 16–32 Mbit/s.

10–26-channel scanning radiometer, 5–120 GHz, 3–4 km res, 1,000 km swath, 100 kbit/s.

United States

Landsat Series

Current Status

The USGS, which now has responsibility for operational programming and archiving, reports a better than 6-sigma performance for 2002 with 366 days without operator error, the second year that this has happened, on operations with Landsat 7.

Background

The United States floated the first proposals for a civil remote sensing satellite in the late 1960s following Earth observations from the manned Gemini and Apollo. Serious study began when the Department of Interior (DOI) in 1968 requested funding for an EROS Earth Resources Observation Satellite. Congress denied DOI the funding, because it felt the Department was not authorised to undertake space projects. Nevertheless DOI built a co-operative programme with NASA and funds appeared in 1969.

General Electric built Landsat 1–3. They were enhanced versions of the Nimbus weather and research satellites, originally known as the Earth Resources Technology Satellites (ERTS). NASA planned to launch two in consecutive years, 1972–73, as an experiment in systematically surveying the Earth's surface to study the health of its crops and the potential use and development of its land and oceans. 300 investigators in the 50 participating countries made use of the dramatic and vivid Landsat 1 images. Nations quickly realised these images were directly relevant to the management of the world's food, energy and environment and NASA laid plans for Landsat 2.

Before the Landsat 1 launch, NASA decreed that all ERTS data, including 9,000 weekly images, should be unclassified and made available to the public. The satellite downlinked the pictures through NASA ground stations to the US Department of Interior's EROS Data Center in Sioux Falls, South Dakota where they could be bought for prices starting at US$1.25. Unexpectedly, news agencies and private individuals studied detailed pictures of Soviet- and Chinese-launch centres and missile sites. The immediate success of the project caused the processing and distribution systems to be swamped with requests. As a result, a number of countries subsequently installed their own receiving stations, including South Africa, Canada, Brazil, Italy, Sweden, Japan, Thailand, India, Australia, Argentina, Zaire, Iran and China. They initially paid construction costs of US$4 to 7 million, US$1 to 2 million annual operational costs and an annual service fee of US$200,000.

The first three Landsats carried combinations of Return Beam Vidicon (RBV) and Multi-Spectral Scanner (MSS) imagers. Technical problems resulted in data from the RBVs of Landsat half being seldom used. For Landsat 3, even though RBV resolution was improved from 80 m to 40 m,

Landsat 5 image including an oil slick off Saudi Arabia (EOSAT) 0517241

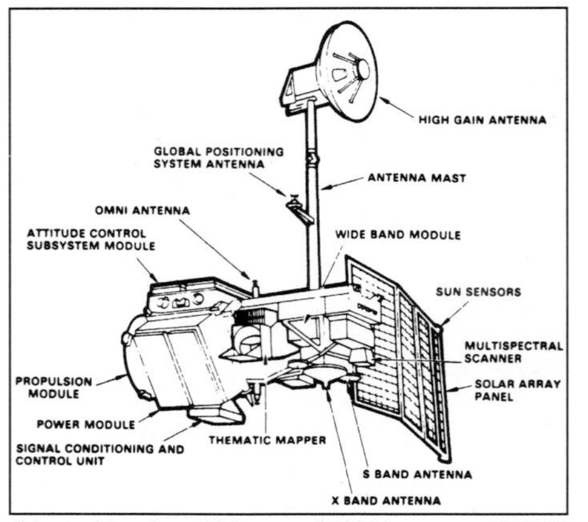

Principal Landsat 4/5 features (NASA) 0517056

the spectrally and radiometrically superior MSS was still preferred. RBVs were deleted from later spacecraft.

Despite its success, by 1982 the long-term Landsat programme had run into both technical and financial troubles. Landsat 2 and 3 had surpassed their design lives and NASA expected them to fail at any time. GE, whose prime contract for Landsat D/D-Prime (as Landsat 4 and its back-up were known before launch) had trouble selling its US$505 million contract in Congress. Hughes also seemed to be behind in the development of the advanced Thematic Mapper.

Finally, NASA managed to solve the technical and bureaucratic problems and got Landsat 4 into space on 16 July 1982. Today, Landsat 4 retains little operational capability. It was already suffering from multiple malfunctions by February 1983. First, the power cabling from the solar panel broke up under the day/night thermal cycling, leaving the satellite with only 50 per cent power capacity. Then the TM X-band link failed after transmitting 6,000 images. NOAA, faced with a monthly revenue loss of US$600,000, because of these failures, requested the US$20.4 million launch of the back-up Landsat D-prime, undertaken in March 1984. NASA studied a Landsat rescue mission to retrieve the US$60 million satellite in June 1986 using the second Vandenberg Shuttle mission. The end to Shuttle preparations at Vandenberg shelved the idea. Instead, EOSAT maintained full operations with the MSS and resumed TM transmissions via the TDRS satellite link in October 1987 after reprogramming the satellite to reactivate the back-up Ku-band.

L5 was built as L4's identical back-up (then modified to prevent L4-type failures) and launched when the prime vehicle suffered failure of its redundant power and transmitter systems. Capability began declining in 1992, as expected, but engineering tests indicate that transmission direct to ground stations is expected to continue through 1997.

The first three Landsats cost NASA a total of US$251 million, including US$149 million for the spacecraft and US$14 million for the launchers. By July 1981, it had become clear that the series would represent a total US investment of US$1 billion. The Reagan Administration decided at that stage that the remote sensing programme must either end, or be absorbed by the private sector. COMSAT, Hughes and several other companies expressed interest subject to government guarantees of a minimum level of data purchase.

The Landsat 4/5 system is operated by the Earth Observation Satellite Co (EOSAT), a partnership formed between Hughes Aircraft (subsequently purchased by GM) and RCA Corporation in May 1984. Under the Landsat Remote Sensing Commercialization Act of 1984, the Department of Commerce (DoC) awarded EOSAT a contract to operate the system for 10 years and to build two new satellites, plus ground systems. It granted EOSAT exclusive marketing and distribution rights to existing Landsat data until July 1994 and to new data for 10 years after collection. The government agreed to provide US$250 million for spacecraft development and to pay for the launches. In 1986, however, DoC eliminated funding for Landsat commercialisation from its FY87 budget request, beginning a series of battles with Congress ultimately accepting a scaled-back commercialisation plan. EOSAT/DoC, on 31 March 1988, signed modifications to the 1985 contract to bring it into line with the compromise reached between Congress and the Reagan Administration, Congress having finally released US$62.5 million for Landsat 6 programme start.

President Bush signed the Land Remote Sensing Policy Act of 1992 on 28 October 1992. This repealed 1984's Act, and transferred the satellites to the joint NASA-DoD Landsat Programme Management from NOAA/DoC. It also provided authorisation for the funding of Landsat 7 as a functional equivalent of Landsat 6. In response to public criticism of data release policy especially the close involvement of the military in the programme, the Act called for negotiations between LPM/EOSAT to modify data policy for Landsat 4 to 6 to reduce data price for US government agencies and their affiliated users.

Following Landsat 6's loss, EOSAT and NASA reached a preliminary agreement in April 1994 giving the company a further five years of rights to archived imagery and exclusive rights to new Landsat 4/5 scenes (and for five years after the satellites' deaths). In return, EOSAT would charge government users reduced rates: US$3,500 per scene for the rest of 1994, and US$2,500 from 1995, instead of the commercial US$4,400 rate. Once bought, these scenes could be copied for other government users at no extra cost. However, the DoC, in July 1994, decided to open the contract for bidding; EOSAT's contract was extended through 30 June 1995 while the matter was pursued; EOSAT was successful.

The 1992 Landsat Act also outlined a proposed Landsat 7 data policy to stimulate the use of data through sales at the cost of fulfilling user requests by the government and/or the competitive marketing of unenhanced data. Landsat 7 will be operated by NOAA and the imagery handled by the US Geological Survey. EOSAT will thus be left as any other value-added company. Landsat 6's launch loss in October 1993 was a severe blow to the system and the US must continue to rely on the ageing Landsat 4/5. There is a high chance that both will fail before Landsat 7 appears in 1998.

The National Space Council established a joint DoD/NASA Landsat 7 programme in early 1992. The NSC designated the air force for the development of the spacecraft, instruments and launch services, and NASA for the ground system. Because of severe budgetary pressures and coupled with Landsat 6's loss, the National Science and Technology Council reassessed the purpose of Landsat 7. Congress appropriated US$84 million for DoD in FY93 for the procurement of L7, allowing launch then planned for 1997. In October 1992, the USAF awarded the US$398 million contract to Martin Marietta Astro Space. DoD agreed to build Landsat 7 in the expectation that it would receive 5 m imagery from the new US$207 million High-Resolution Multispectral Stereo Imager but NASA continued to resist funding the high costs for the required ground system. The Clinton Administration decided, in February 1994, that NASA should develop Landsat 7 for 1998 launch.

This produced Presidential Decision Directive/NSTC-3, dated 5 May 1994, changing it to a joint NASA, NOAA and US Geological Survey programme, dropping the military component. The Landsat 7 programme aims to maintain data continuity and minimise any gap as well as make L7 data available to all users at the cost of fulfilling requests.

NOAA is responsible for operations and data capture, processing, archiving and distribution. USGS will be NOAA's operational arm for the data element, and will continue the operation and maintenance of the National Satellite Land Remote Sensing Data Archive for the long-term preservation of Landsat and other satellite land remote sensing data.

Landsat 5 checkout; Fairchild's MMS section is at bottom, with the instrument section and folded TDRS antenna above (EOSAT) 0517057

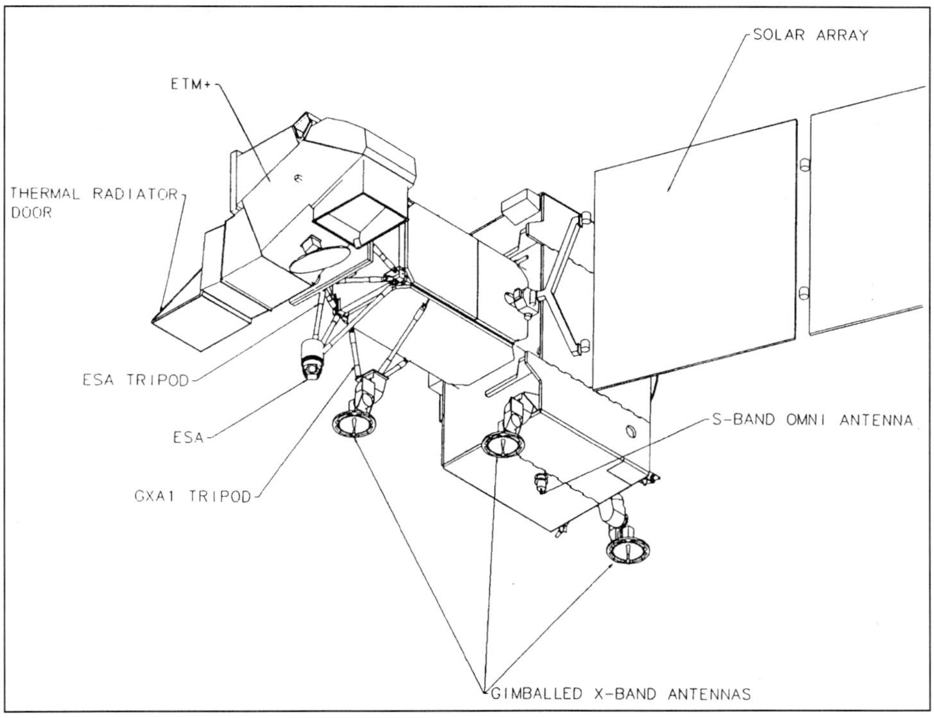

Landsat 7 principal features. ESA is the Earth Sensor Assembly. Velocity vector is to the left (NASA) 0517240

The new Newcastle ground station site at Norman, Oklahoma was developed in late 1991 for integrated TT&C and data reception of Landsat 6-era operations, adding 158 m^2 to an existing 279 m^2 building. The mid-continent location allows real-time reception of data covering the contiguous 48 states. The station was formally dedicated 8 May 1992, discontinuing the datalink through NASA Goddard. The data are then transferred to EOSAT's Lanham headquarters for processing and distribution as CCTs (6,250/1,600 bpi), 8 mm Exabyte tapes and photo products. Swedish Space Corporation's Kiruna site provides back-up TT&C. Control is maintained from HQ's LSOC Landsat Spacecraft Operations Center via Newcastle or Kiruna.

More than 2 million digital images are archived at EOSAT in addition to those held in countries with Landsat ground stations operating under annual US$600,000 fees plus royalty payments: Australia, Brazil, Canada, China, Ecuador, India, Japan, Pakistan, Saudi Arabia, South Africa and Thailand, plus two stations for Europe's Eurimage network.

Specifications

Landsat 1

Launched 23 July 1972 by Delta from Vandenberg. 891 kg; 901 × 920 km, 99°. Operations ended 6 January 1978. By 30 March 1973, when faults occurred in the tape recorder, the North American continent had been covered 10 times; the total of >34,000 images included all the world's major landmasses at least once. The recorder fault meant that images could no longer be stored for transmission, though when the satellite was over one of NASA's three ground receiving stations in California, Alaska and Maryland it was still possible to transmit real-time imagery of the entire North American continent. By the time L1 had to be retired in 1978, it had returned 300,000 images and demonstrated the potential of remote sensing

Sensors: Return Beam Vidicon (RBV). Landsat 1/2 each carried three 80 m instruments sensitive to Band 1 0.475 to 0.575 μm (blue-green), Band 2 0.580 to 0.680 μm (yellow-red); and Band 3 0.690 to 0.830 μm (red-near-infra-red). Simultaneous views were provided of 185 km^2 areas. Landsat 3 carried two single-band 40 m RBVs providing adjacent 98 km^2 images, covering 0.505 to 0.750 μm. TM replaced the RBV on later satellites.

Multi-Spectral Scanner (MSS). All three included 80 m resolution four-band sensors, with L3 adding a 5th, thermal 120 m band. Band 4 0.50–0.60 (green), Band 5 0.60 to 0.70 (red), Band 6 0.70 to 0.80 (red-near infra-red), Band 7 0.80 to 1.10 μm (near-infra-red). L3 only: Band 8 10.40 to 12.50 (thermal-infra-red). Coverage was identical to RBVs and subsequent Landsat MSS.

Landsat 2

Launched 22 January 1975 by Delta from Vandenberg. 816 kg; 907 × 918 km, 99°. Operations ended 25 February 1982. This Landsat incorporated a 14 kg digital computer with 4,096-word memory able to handle 55 separate commands from ground stations and carry out routine operations for up to 24 h – a vital function as it was out of tracking station range for 80 per cent of the time. L2 was intended to supplement and later replace L1, and the two were able to work together until 1978, providing repeat coverage every 9 instead of 18 days. Retired on its fifth birthday in 1980 because of attitude control problems, L2 was revived six months later by Goddard engineers, who developed a magnetic compensation system to provide stabilisation. Though its tape recorders had failed, it still provided real-time imagery in 1981, supplementing the limited activities of L3

Sensors: Same as Landsat 1.

Landsat 3

Launched 5 March 1978 by Delta from Vandenberg. 960 kg; 900 × 918 km, 99°. Operations switched to standby 31 March 1983. Further sensor refinements included the addition of a thermal band to the MSS, able to detect T differences in vegetation, bodies of water and urban areas by day/night; and an improved RBV offering a 50 per cent resolution increase and enabling areas as small as half an acre to be identified. The three year Large Area Crop Inventory Experiment (LACIE), concluded in November 1978, showed that Landsat's multispectral scanners revealed the Soviet Union's wheat crop to be 91.4 million tonnes, less than 1 per cent below the official Soviet figure. There was less success with predictions of wheat yields in Canada and the US, where the long, narrow fields can easily be confused with adjoining spring-planted crops. Improvements in L4's instruments were intended to remedy this, making it impossible for countries such as the Soviet Union to repeat past successes in buying up world wheat surpluses cheaply when anticipating a poor crop of its own. As wheat is the world's most important grain crop, Australia, China, Brazil and Argentina were also included in LACIE. As a follow-on, it was decided to monitor wheat, corn, rice, barley, sorghum, soya bean and sunflower crops in the Soviet Union, China, Brazil, Mexico, Australia, Canada and US. In January 1981, after two years of operations, L3's MSS malfunctioned but the RBV continued to complement L2's operational MSS

Sensors: Same as Landsat 1.

Landsat 4

Launched: 16 July 1982 by Delta 3920 from VAFB
Mass: 1,941 kg at beginning of life
Contractors: General Electric (now Lockheed Martin Astro Space) prime, Fairchild for Multimission Spacecraft, Hughes Santa Barbara Research Center for sensors
Resolution: 30 m Thematic Mapper (120 m thermal-infra-red), 80 m Multi-Spectral Scanner
Orbit: 705 km circular in 98.2° Sun-synchronous, crossing equator at 09.45 local time, with 16 day repeat, 2,760 km ground-track separation at equator between revolutions, 172 km between tracks in completed coverage
Spacecraft: about 4 m long and is based upon the NASA/Fairchild Multimission Modular Spacecraft (MMS), as used initially by Solar Max, attached to the instrument/solar panel/antenna forward section. The MMS is essentially a triangular polyhedron with a propulsion module on one end and three 1.2 × 1.2 × 0.3 m box-like side modules handling power, AOCS and data/communication links. 3-axis control with pointing accuracy of 0.01° provided by MWs in conjunction with an inertial reference system using attitude updates from two star trackers. The aft propulsion module houses 11 N + 0.5 N hydrazine thrusters for maintaining the 16-day repeat cycle and MW unloading. Real-time TM data transmitted at 84.9 Mbit/s on X-band and MSS data at 15 Mbit/s on S-band; an S/Ku-band dish on a 4 m mast accesses the TDRS network. No data recorders. The single 4-panel solar wing was designed to provide 990/814 W BOL/EOL; this failing power system prompted the launch of Landsat 5
Sensors: Thematic Mapper (TM). Landsat 4 was the first to incorporate the TM, with its greater number of spectral bands and enhanced resolution. The 245 kg 1.1 × 0.7 × 2.0 m TM draws 345 W. The spectral ranges (μm) and physical characteristics measured by the bands are: Band 1 0.45–0.52 (chlorophyll and carotenoid for soil/vegetation and deciduous/coniferous differentiation), Band 2 0.52–0.60 (green reflectance by healthy vegetation), Band 3 0.63–0.69 (chlorophyll absorption for plant species differentiation), Band 4 0.76–0.90 (near-infra-red reflectance of healthy vegetation for biomass surveys), Band 5 1.55–1.75 (vegetation moisture and snow/cloud reflectance differences), Band 6 10.40–12.50 (thermal mapping), Band 7 2.08–2.35 (vegetation moisture and hydroxyl ions in soils). Band 6 exhibits 120 m resolution; the remainder are 30 m. Each uses an oscillating mirror to scan the Earth's surface in the cross-track direction as Landsat's motion provides along-track coverage. The 170 km long 185 km wide images have a 14 per cent side overlap at the equator, increasing with latitude.

Multi-Spectral Scanner. The Landsat 4/5 MSS are similar to those flown on Landsat 1–3, with the

Landsat 7 is almost identical to Landsat 6 (shown here) (Lockheed Martin) 0517242

four bands offering 80 m resolution and the same coverage as the TM. The 58 kg 35 × 40 × 90 cm MSS draws 81 W. Spectral ranges (μm): Band 1 0.50–0.60 (green); Band 2 0.60–0.70 (red); Band 3 0.70–0.80 (near-infra-red); Band 4 0.80–1.10 (near-infra-red). Bands 1–3 in the Ritchey-Chrétien telescope system utilise six photomultiplier tubes; band 4 uses four silicon photodiodes. Data quantised to 6 bits (64 digital counts); transmission rate (real time only) 15 Mbit/s
Optics: The prime focal plane assembly of the Ritchey-Chrétien telescope with its 40 × 53 cm scan mirror contains four sets of 16 Si photodiodes for the VNIR bands 1–4. The cooled focal plane assembly contains two arrays of 16 InSb photodiodes for bands 5/7, and four HgCdTe detectors for thermal band 6. The TM also offers improved radiometric sensitivity over the MSS, permitting the data to be quantised to 8 bits (256 digital counts), thus providing a four-fold increase in the grey scale. Data rate (real-time only) 84.9 Mbit/s.

Landsat 5
Contractors: Same as Landsat 4
Resolution: Same as Landsat 4
Launched: 1 March 1984 by Delta 3910 from VAFB
Mass: 1,941 kg BOL (249 kg hydrazine, 150 kg remained May 1995)
Orbit: Same as Landsat 4
Spacecraft: Same as Landsat 4, but modified to prevent repetition of solar array power cabling and X-band transmitter failures
Sensors: Same as Landsat 4.

Landsat 6
Launched 5 October 1993 by Titan 2 from VAFB. It appears not to have attained orbit, although Titan apparently performed as expected. The converted ICBM released it with a 724.7 km (target 724.5 km) apogee and 98.0107° (target 98.006) inclination. L6's Thiokol Star 37XFP solid apogee motor was to have ignited at 819 seconds to establish the near-final 704 km 98.2° circular orbit. However, Landsat's attitude control thrusters were useless because their hydrazine tank had ruptured, leaving it without control during the kick motor's burn and leading to it burning up over the Pacific. The cause was difficult to identify as Landsat telemetry was switched off to avoid interference with Titan's stage 2 transmitter. L6 was almost identical to L7, except that ETM's thermal-IR band provided only 120 m resolution. Launch mass was about 2,750 kg, with 1,740 kg on-orbit BOL. Design life five years. EOSAT signed L6's rescoped development contract with the US Department of Commerce 31 March 1988, Congress having released US$62.5 million for programme start. Total cost was US$268.9 million, including US$36.5 million for launch, US$220 million for the space/ground segments, with US$10.8 million contributed by EOSAT. L6 was to continue its predecessors' TM products, while adding a co-registered 15 m PAN band to the US$92 million 390 kg ETM. See p400 of the 1994/95 edition of *Jane's Space Directory* for the full L6 entry.

Landsat 7
Contractors: Lockheed Martin Astro Space (satellite, flight systems integration), Hughes Santa Barbara Research Center (ETM+)
Resolution: ETM+ 30 m (6 bands), 15 m panchromatic, 60 m thermal-infra-red
Launch: 15 April 1999 by Delta 27920 from VAFB
Mass: About 2,170 kg at launch, about 2,020 kg on-orbit BOL
Orbit: 705 km circular in 98.2° Sun-synchronous, descending node at 10.00±0.15 local time, with 16-day repeat cycle (233 revolutions) and 172 km ground track separation at equator
Spacecraft: About 4.3 m long and based on Tiros/DMSP meteorological satellite bus but with extensive subsystem modifications for Landsat-unique requirements and to improve mission reliability. Sensor mounted on the forward module. Three steerable high-gain X-band antennas operating at 8.0825, 8.2125 or 8.2434 GHz. Each provides two 75 Mbit/s streams (I/Q) of ETM+ data, with bands 1–6 on one channel and 7–8 on other. 375 Gbit SSR can store about 100 full ETM+ scenes. Stored and real-time data acquired at USGS EROS Data Center in Sioux Falls. Real-time data available to international stations on X-band. Command uplink 2.1064 GHz; telemetry downlink 2.2875 GHz. 1,550 W generated by a 4-panel single solar wing driven at orbital rate to follow the Sun. Eclipse power by 2 × 50 Ah Ni/H_2 batteries. 3-axis attitude control by RWs and magnetorquers, accuracy 180 arcsec; attitude information from star sensors, gyros, Earth sensors and magnetometer. Ground track held within 5 km by 12 × 4.45 N hydrazine thrusters on lower module
Sensors: Enhanced Thematic Mapper + ETM+ is a nadir-viewing cross-track scanning 8-band multispectral radiometer, developed by Hughes SBRC. It provides continuity of TM data with Landsat 4/5 and adds a 15 m PAN band, improved resolution (60 m) for thermal, and improved radiometric accuracy (5 per cent). All other bands remain at 30 m. ETM+ can image >525 scenes daily, with a 185 km swath width and 170 km along-track length. Spectral ranges (μm): Band 1 0.45–0.52, Band 2 0.52–0.60; Band 3 0.63–0.69; Band 4 0.76–0.90; Band 5 1.55–1.75; Band 6 10.4–12.5 (thermal mapping); Band 7 2.08–2.35; Band 8 (PAN) 0.50–0.90. Imagery can be collected in low- or high-gain modes; high gain doubles the sensitivity.

Lewis and Clark Satellites

Background

NASA's Small Spacecraft Technology Initiative plans to demonstrate new techniques, to speed the development and reduce the costs of next-generation satellites. The Office let two contracts on 8 June 1994 for Earth observation satellites with unprecedented sensor technology to be developed and delivered on-orbit within 24 months. From final announcement to contract signing consumed 70 days instead of the usual six months to one year.

TRW received US$59 million to build Lewis – the first hyper-spectral imaging system. Lewis' 384-channel HSI Hyper Spectral Imager has commercial applications in forestry, agriculture, water, land-use management and environmental monitoring. The high spectral resolution can distinguish between tree varieties, for example, and their states of health.

CTA's US$49 million Clark offers high resolution with stereo imaging. A key new capability of both is cloud editing, which 'remembers' cloud locations for later replacement with clear data. Both carry additional instruments for space physics and for global atmospheric pollution dynamics information. The high resolution locates utility pipelines and cables, and help town planners with, for example, assessments of transport needs and construction sites. The same imager will fly on EarthWatch's commercial satellite, which CTA Space Systems is building. CTA at one stage proposed to buy Clark back from NASA after one year of operations in order to use it as an EarthWatch back-up.

The Lewis and Clark have introduced key spacecraft technologies which include GPS attitude determination, a fibre optic data bus, CPV common pressure vessel, Ni/H_2 battery, composite structures, miniature star trackers, 32-bit processors, 2 Gbit solid-state recorders and multijunction GaAs solar arrays. After launch in August 1997, Lewis was found to be spinning at an unacceptable 2 rev/min and as this could not be stopped the satellite was inoperable. Subsequently, Clark was cancelled. Total cost to NASA was US$64.8 million with an additional US$6.2 million spent for 12 months storage and maintenance due to a delayed launch.

Specifications

Lewis
Launch: 23 August 1997 by Lockheed Martin LMLV 1 from VAFB SLC-6
Design life: 5 years
Mass: 385.6 kg at launch
Contractors: TRW Space & Electronics Group, Harris Corp, AlliedSignal Aerospace, Gulton, Ithaco, Hughes Danbury, Jackson & Tull, NASA Goddard, NASA Lewis, NASA Stennis, NASA Langley satellite built in only 24 months from the contract being awarded
Resolution: 5 m panchromatic, 30 m multispectral
Orbit: planned 523 km circular, 97.4° Sun-synchronous
Spacecraft: bus 149.86 cm diameter hexagonal prism, 203.2 cm long. Zero momentum bias 3-axis stabilised using magnetically suspended RWs, GPS attitude determination + 20° FOV star tracker (6 star capability). Wheel unloading + orbit adjust by four pairs of 4.45 N hydrazine thrusters. 740 W from dual deployed solar wings, including multijunction GaAs and Amorphous Si. R3000 processor, fibre optic bus, 4 Gbit SSR, S-band STDN-compatible
Sensors/payload: HSI Hyper Spectral Imager, 384 0.01 μm channels over 0.4–2.5 μm, 30 m XS resolution, 5 m PAN.
LEISA Linear Etalon Imaging Spectral Array, 256 channels, 300 m resolution, map reflectance spectra of surface and atmospheric features.
UCB Ultraviolet Cosmic Background, 580-950 Å grating spectrometer, map astrophysical component of EUV cosmic background
Components: TRW pulse tube cryocooler.

Clark
Launch: originally planned for November 1996 by Lockheed Martin LMLV 1 from VAFB SLC-6 but cancelled along with satellite
Design life: 3 years
Mass: 278 kg at launch, including 94 kg payload and 12 kg hydrazine
Contractors: CTA Space Systems, Lockheed Martin Astronautics, Signal Corp, Futron, Odetics, WJ Schafer, CH2M Hill, NASA Goddard, NASA Langley, NASA Stennis, MIT, Stanford University
Resolution: 3 m panchromatic, 15 m multispectral
Orbit: planned 476 km circular, 97.3° Sun-synchronous, 11.15 descending node
Spacecraft: 73.9 × 83.8 × 61.1 cm bus; 163.7 cm high imager on 61.1 cm high bus. 3-axis zero momentum stabilisation (pitch bias back-up) by three 0.025 Nm RWs + three 17 Am^2 magnetorquers (0.15° attitude determination by GPS, 0.1° roll/pitch Earth horizon sensors, 5 arcsec star tracker, 0.015°/h gyro). Orbit correction/drag makeup by two 0.89 N thrusters. 584 W from cells on two wings totalling 3.09 m^2; 0.14 mm-thick GaAs cells on primary panels, CIS and multijunction cells on secondary (reflectors increase output), supported by two 15 Ah CPV Ni/H_2 batteries. RHC3000 processor with 8 Mbit × 40 bit DRAM memory, fibre optic bus, 16 Gbit 160 Mbit/s Odetics SSR. Imagery transmitted on X-band.

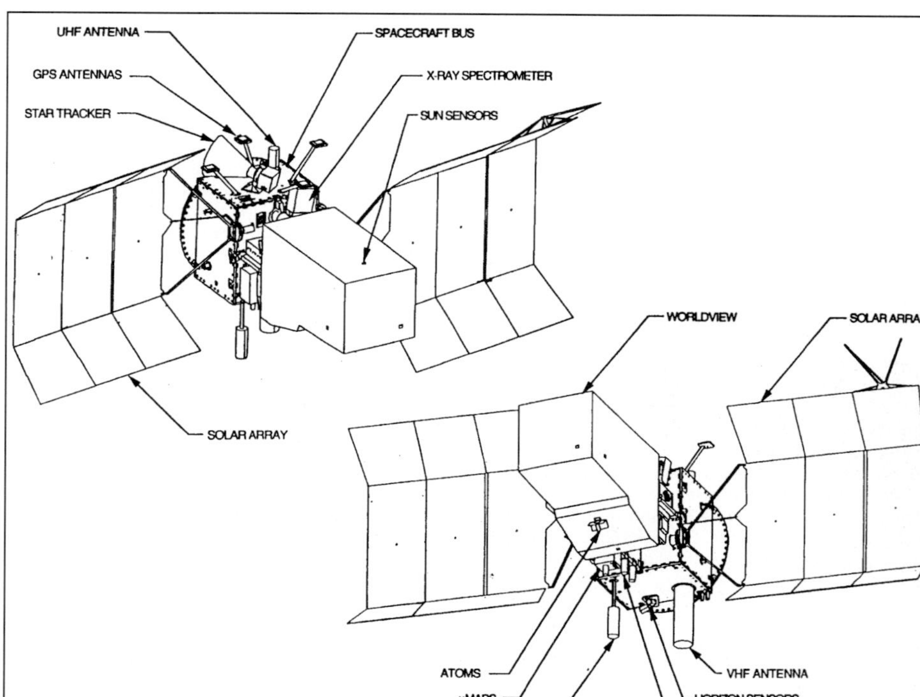

CTA's Clark carries a 3 m-resolution imager. WorldView is now named EarthWatch (CTA) 0517237

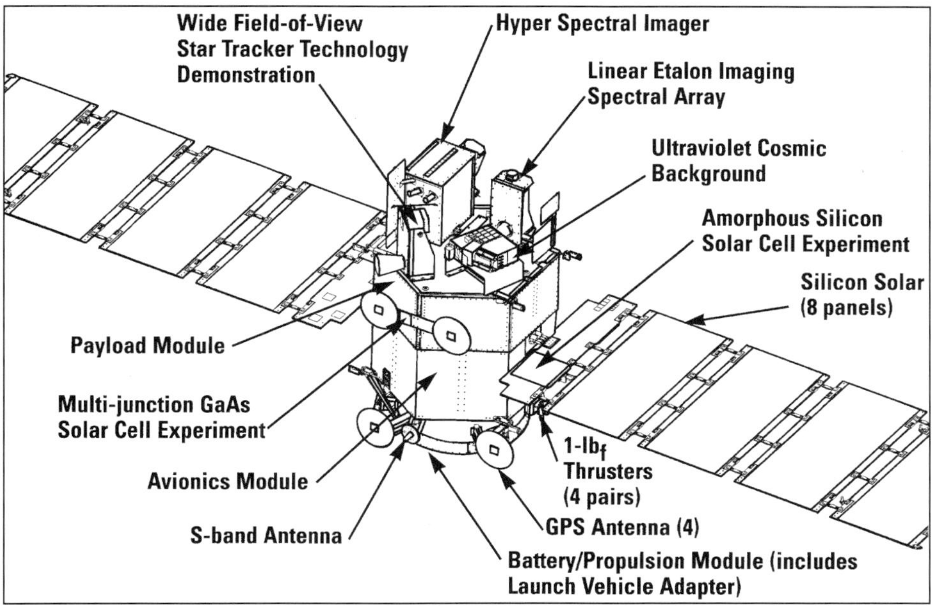

TRW's Lewis carries a 384-channel imager 0517412

Redundant 10 W UHF 2.4/19.2 kbit/s up/down; UHF for TT&C, VHF for store/forward. Shape memory devices replace pyrotechnics

Sensors/payload: PAN. 3 m plus 15 m XS imagery. Step-stare for 3 m collects 6 × 6 km images over 30 × 30 km by gimballed steering mirror. Mirror allows ±30° along/cross track, allowing stereo pairs with 5 m height resolution. Images transmitted at 160 Mbit/s X-band.

μMAPS Measurement of Air Pollution from Satellites is a miniaturised version of the NASA Langley MAPS gas filter radiometer flown on Shuttle. Measures CO mixing ratios in three altitude regions; experimentally uses N_2O capability to estimate image sensor cloud coverage.

Atmospheric Tomography. Proof of concept demonstration of 2-D mapping of atmospheric constituent gases by using ground laser (at air force Maui AMOS station) and satellite corner reflector to map CO using the 4.7 μm absorption band.

XRS X-ray spectrometer, 2–20 keV by eight 8 × 8 mm Si Avalanche PhotoDiodes (APD) and 10–80 keV by 12 8 × 8 mm CdZnTe detectors. Primary goal is to test these new semiconductor detectors, all operating at or near room temperature. Secondary is to determine spectra of solar flares and cosmic γ bursts. The ms timing of γ bursts will help to locate their origin by timing with similar measurements from other satellites. NASA Goddard.

MicroLab 1 Satellite

OSC/NASA Marshall in August 1993 signed an all-inclusive contract for the provision of atmospheric research data from NASA's payload aboard OSC's MicroLab 1. The Optical Transient Detector is studying the spatial/temporal distribution of global lightning to better understand and predict major atmospheric storm systems and climate changes. It is a precursor to the LIS Lightning Imaging Sensor on 1997's NASA/Japanese TRMM mission. 500 images/s are recorded and the real-time event processor handles 10 million pixels/s to extract

MicroLab 1 attached to the Pegasus final stage. The optical sensor is covered by the box at top. OSC's first two Orbcomm satellites were then stacked forward of MicroLab (OSC) 0517244

MicroLab 1 is returning lightning data (OSC) 0517245

lightning signals from the bright background. Microlab data are downlinked twice daily to OSC's tracking station in Fairmont, West Virginia. The data are analysed in NASA Marshall's Global Hydrology & Climate Center and archived/distributed by the EOSDIS Earth Observing System Data Information System. The OTD is the world's first space-based sensor capable of detecting and locating lightning events in the daytime as well as during the nighttime with high detection efficiency.

One useful finding of the satellite has resulted from observations of a 17 April 1995 storm in Oklahoma. Results suggested that increased intracloud lightning presages tornadoes, a possible method to perform tornado predictions and provide population alerts from space.

MicroLab's GPS meteorological experiment uses an onboard eight-channel TurboStar GPS receiver (donated by Allen Osborne Associates) daily recording 500 occultations of GPS signals to map atmospheric temperature and water vapour content. The experiment is managed by the UCAR (University Corporation for Atmospheric Research) for the National Science Foundation, which paid some US$500,000 to OSC. The total price paid to OSC for the two year mission is more than US$7 million. MicroLab is based on OSC's MicroStar small standard platform, developed for the Orbcomm satellite messaging system.

OSC retained ownership of Microlab 1 with NASA renting space for experiments and their operations.

Specifications

Mass: 68 kg
Launch: 3 April 1995
Dimensions: 1 × 0.3 m
Orbit: 733 × 749 km, 70.0°
Design life: 2 years

NOAA Series (TIROS/TOS/ITOS/ NOAA/GOES)

Current Status

NOAA currently operates 17 meteorological satellites in three separate constellations: GOES, POES and DMSP. The future NPOESS constellation will merge the two polar orbiting series into a single programme. NOAA launched NOAA-K on 13 May 1998.

The GOES-West/East satellites, at 135° and 75° W GEO, provide synoptic visible/infra-red imaging and infra-red/thermal sounding for atmospheric temperature profiles. GOES 7 remains in orbit, left as the last representative of the earlier technology, after a bulb failed in GOES 6's scanning encoding system January 1989. The replacement new-generation GOES 8 did not occupy the 75° W slot until February 1995; GOES 9 took over at 135° W in January 1996.

Background

The United States has launched 35 civil Tiros/north polar meteorological satellites (and suffered three launch failures) in three distinct programmes since the beginning of the space age. NASA named the first series, launched during 1960–65, Tiros. Then it followed the Tiros programme with nine more satellites over the 1966–1969 period. It designated these the Tiros Operational Satellite (TOS) and sometimes called them the ESSA (Environmental Science Services Administration) satellites after the organisation that managed the programme during that period. NOAA introduced the second-generation Improved TOS (ITOS) in 1970. In order to enhance the ITOS satellites, NOAA built a prototype Synchronous Meteorological Satellites (SMS) in 1974, to provide synaptic coverage of short-term and other weather features from GEO. These geosynchronous satellites set the stage for a permanent weather station in geosynchronous and acted as the testbed and forerunners of GOES, which now provides continuous coverage of North and South America and its oceans. NOAA polar spacecraft distribute their data to Earth stations in 120 nations as they pass overhead, and thousands of schools, private individuals and others receive NOAA imagery.

TIROS

The Television & Infra-red Observation Satellite programme began as a joint NASA/Department of Defence project to develop meteorological satellites. Tiros 1 was launched in April 1960, and orbited at 692 × 740 km. The satellite demonstrated the value of an orbital viewpoint for meteorological applications. During its 78 day battery life, it returned 22,952 cloud cover images. The original Tiros satellites were basically cylindrical 18-sided prisms, 1.07 m in diameter and 0.55 m high. Solar cells covered the sides and top, with apertures for two TV cameras on opposite sides of the spacecraft body. Each camera captured 16 images a revolution at 128 second intervals. The image system passed its information to two tape recorders capable of storing up to 48 scenes when ground stations were out of range. Tiros 1's 119 kg increased to 138 kg over the course of design changes to Tiros 9/10.

By the time Tiros 10 had been launched, NASA had already begun to phase in the more advanced Nimbus and ESSA satellites. The first eight Tiros', all operating in similar orbits, transmitted several hundred thousand images, together with information on the Earth's heat budget. Tiros 9, marked the first time that NASA attempted to reach polar orbit from Cape Canaveral. It used a series of three Delta dogleg manoeuvres to reach the 82° inclination angle, but a stage 2 overburn meant that the orbit, instead of being 644 km circular, was 700 × 2,578 km.

Tiros 10 transmitted the first photomosaic of the entire world's cloud cover in 450 high-quality

GOES 9 began operations at 135° west in January 1996 (Space Systems/Loral) 0517424

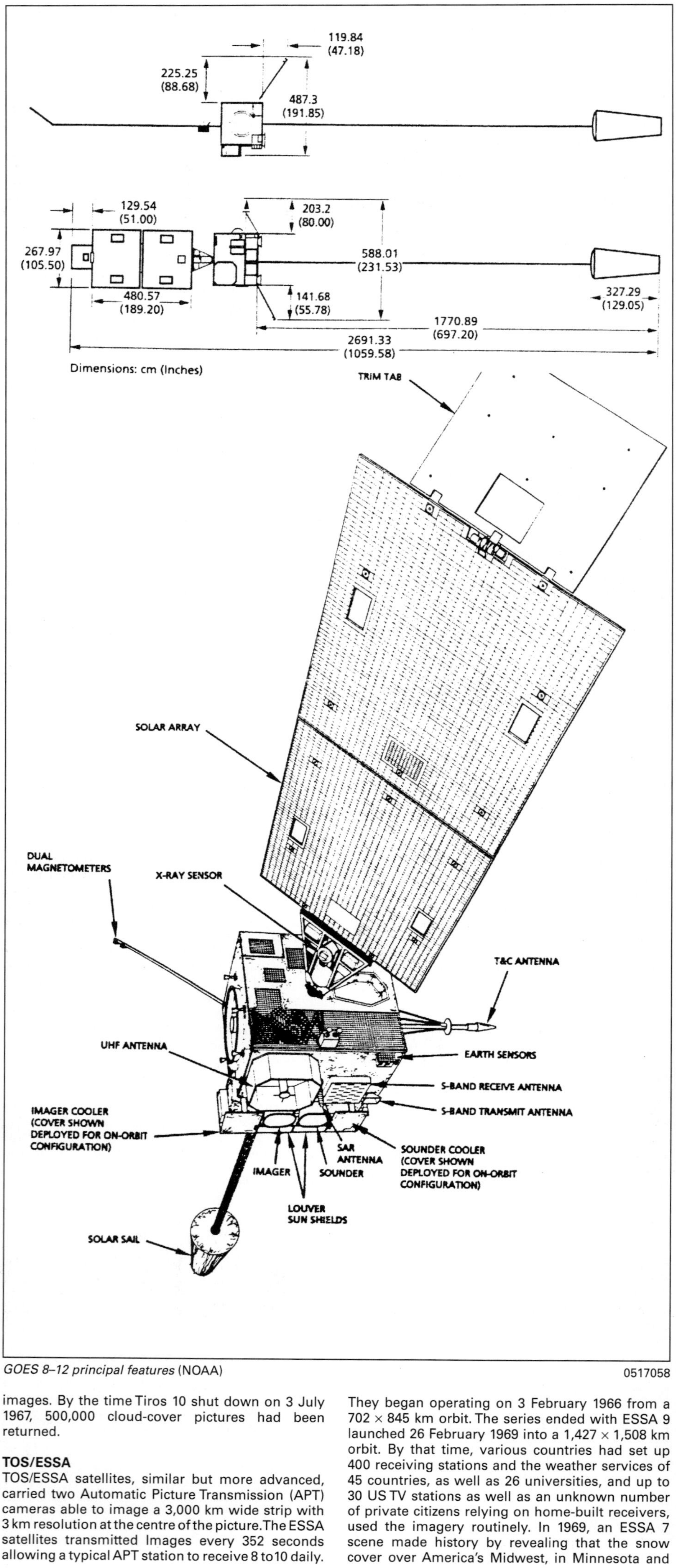

GOES 8–12 principal features (NOAA) 0517058

images. By the time Tiros 10 shut down on 3 July 1967, 500,000 cloud-cover pictures had been returned.

TOS/ESSA

TOS/ESSA satellites, similar but more advanced, carried two Automatic Picture Transmission (APT) cameras able to image a 3,000 km wide strip with 3 km resolution at the centre of the picture. The ESSA satellites transmitted Images every 352 seconds allowing a typical APT station to receive 8 to10 daily. They began operating on 3 February 1966 from a 702 × 845 km orbit. The series ended with ESSA 9 launched 26 February 1969 into a 1,427 × 1,508 km orbit. By that time, various countries had set up 400 receiving stations and the weather services of 45 countries, as well as 26 universities, and up to 30 US TV stations as well as an unknown number of private citizens relying on home-built receivers, used the imagery routinely. In 1969, an ESSA 7 scene made history by revealing that the snow cover over America's Midwest, in Minnesota and the Dakotas, was three times thicker than normal. Measurements showed that it was equivalent to 15 to 25 cm of water covering thousands of square miles. A disaster area was declared before the event, and when the floods came much had been done to control the situation.

ITOS

The Improved Tiros Operational System (ITOS) added IR sensors to permit night observations. NOAA superseded ESSA during the period that the agency launched the satellites and so they were designated ITOS before launch and once in orbit given a new series of NOAA designations. Thus NOAA 1, the second-generation prototype, and launched 11 December 1970 from Vandenberg into a 1,429 × 1,472 km, 101° inclination orbit began the series. ITOS-B since it failed to achieve orbit never received a NOAA designation. And the NOAA designation resumed with NOAA 3/4 launched in November 1973 and 1974 respectively.

NOAA

The US National Oceanic and Atmospheric Administration (NOAA) operates the polar-orbiting NOAA and geostationary GOES environmental systems of two satellites each, procured for NOAA by NASA through its Goddard operations centre. Launched into near-polar Sun-synchronous orbits, each NOAA can view almost all of Earth's surface twice every 24 hours. In addition to returning weather imagery, they monitor atmospheric temperature and humidity, sea surface temperature to 1.6°C accuracy, snow/ice cover, and total ozone content. near-Earth proton/electron fluxes atmospheric aerosols and radiation fluxes, and relay Argos surface platform environmental data and Cospas/Sarsat distress signals. NOAA designs its satellites for a nominal two year lifetime.

NOAA satellites carry two basic sensors (AVHRR/TOVS), and afternoon satellites such as NOAA 11/14 include the Solar Backscatter UV Spectral Radiometer, Mod 2 (SBUV/2, non-scanning 1,600–4,000 Å) for determining ozone levels to 1 per cent absolute accuracy. It also carries a Cospas/Sarsat distress relay, Argos Data Collection & Location System transponder, and a Space Environment Monitor.

NOAA 8 was the first of RCA's Advanced Tiros-N (ATN) satellites operating as part of the international Cospas-Sarsat search and rescue system. In June 1984 it began tumbling following failure of the master timing system, and the refusal of the back-up system to activate. All attempts to save the US$30 million satellite failed, and it had to be abandoned, presumed dead, leaving NASA with no search/rescue capability.

NOAA 9 introduced an ozone mapper being flown on subsequent satellites with near-noon equator crossings. NOAA launched the morning orbit satellite counterpart NOAA 10 in September 1986 as NOAA-G, which it replaced, by NOAA-D/12 in 1991. NOAA 9, after a series of weather and technical delays, restored both weather and rescue services from its afternoon orbit. It also carried

NOAA 14 became the prime satellite in 1995 (Lockheed Martin Astro Space) 0517246

Some NOAA polar satellites also operate as part of the Cospas/Sarsat international search and rescue system, relaying distress calls from ships and aircraft (Lockheed Martin Astro Space) 0517060

NASA's Earth Radiation Budget Experiment scanner and non-scanner instruments. The morning satellite NOAA-10 replaced the degraded NOAA 6, which was brought back into use following the failure of NOAA 8. NOAA 10's Sarsat equipment restored the ability of the US to contribute to the international Cospas/Sarsat search/rescue system. Within six days of becoming operational, it had picked up its first distress signal, which led to the rescue of four Canadians whose aircraft had crashed in a remote area of Ontario.

One of the three attitude control gyros failed on NOAA 11 in August 1989 and the single back-up unit was brought on line. Software for operating with only two gyros was uplinked 8 August 1990 a month before the expected second failure. Software has been developed for operations with no gyros. AVHRR failed 13 September 1994 after operating since 8 November 1988. Placed on standby 1995 after being replaced by NOAA 14. The satellite cost US$53.5 million.

NOAA intended to use NOAA 13 as the prime afternoon satellite. However, failure of the power link from the solar panel lost contact 21 August during checkout. The September 1994 review board report identified a short-circuit in the battery charge assembly that prevented the solar arrays from charging the batteries. The short was probably caused by a 31 mm screw penetrating insulation and contacting a radiator plate. Inspection of NOAA-J (NOAA 14) showed 10 of the 12 screws to be in danger of causing the same fault.

NOAA plans to continue launches, on average, at 16-month intervals for the foreseeable future. It launched NOAA 13 (I) 9 August 1993 to replace NOAA 11 but this new satellite suffered a power failure within days. As a replacement NOAA converted NOAA-J, originally planned to replace NOAA 12, into an afternoon role and launched it in December 1994. In the coming decade NOAA plans to drop its morning operational satellites and turn meteorological monitoring into a co-operative effort with Europe's Metop.

President Clinton, in May 1994, approved a long awaited plan to combine NOAA's polar system with the military DMSP, halving the number of operational satellites to two. NOAA had planned a new series of Tiros-Next satellites (designated O–O) for 2005, 2008 and 2011. The first joint military/civilian satellites may appear in 2007, controlled by NOAA from its Suitland, Maryland centre. To prepare, Suitland will take over DMSP control beginning in 1998, and DoD will phase out its own station retaining only one as back-up. The plan projects a US$300 million saving through 1999,

NOAA 7 transmitted this image of Hurricane Lili over the north Atlantic on 20 December 1984, only the third recorded hurricane in that region (the others were in 1887 and 1954) 0517061

by eliminating duplications and then at least US$1 billion over the next 15 years.

NASA placed NOAA-11 on standby 1995 after being replaced by NOAA 14, crossing the equator in the afternoon. The Centre de Meteorologie Spatiale in Lannion, France can relay stored/real-time data. Afternoon satellites such as NOAA 11/ 14 include the Solar Backscatter UV Spectral Radiometer, Mod 2 (SBUV/2, non-scanning 1,6004,000 A) for determining ozone levels to 1 per cent absolute accuracy.

NOAA planned to place NOAA 14 in the morning orbit as a replacement for NOAA 12, but then the loss of NOAA 13 forced the agency to reconfigure the satellite for NOAA 11's afternoon role. This required replacing the Naval Research Lab's RAIDS Remote Atmosphere Ionospheric Detection System with SBUV. RAIDS is a precursor to two instruments planned for DMSP 5D3 that will improve prediction of ionospheric disruptions to communications and Over-the-Horizon radar systems. It probably will never fly since the next generation of satellites cannot carry it.

NASA built a new fourth-generation prototype for NOAA and named it the NOAA-N, and launched it into an 850 × 866 km orbit at 102° inclination. NOAA 6 joined it on 27 June 1979. Developed, like the earlier generations, by RCA, these were the start of a series based on the Block 5D bus developed for the USAF DMSP. NOAA-B failed to achieve the correct orbit, but by the time NOAA 7 was in orbit (launched 23 June 1981), NASA claimed another success for the programme. Fishermen in California, Oregon, Washington and Alaska noted improved catches as a result of following the NOAA sea surface temperature charts. One 60-vessel towing and transportation company announced fuel savings of 20 to 40 per cent as a result of choosing routes with the help of NOAA's stream and loop current information. NOAA 7 cost US$15 million plus US$7.5 million launch. NOAA-D was the last Tiros-N, launched out of sequence in 1991 as NOAA 12.

Beginning with NOAA-M, the US will fly only afternoon satellites; and Europe will be responsible for the morning orbit making NOAA-K the last US morning satellite. This satellite includes a channel for humidity profiles, delineating sea ice and open water, and monitoring snow thickness and soil moisture.

SMS/GOES

While the NOAA satellites added a significant improvement to the weather gathering capabilities of meteorologists, NOAA sought a continuous scene generator, which required a move to geosynchronous orbit. NASA built two prototype Synchronous Meteorological Satellites (SMS) to answer this need which it later released to NOAA as part of the Geostationary Operational Environmental Satellite system. NASA remains responsible for spacecraft procurement. SMS/GEOS satellites have a Visible/Infra-red Spin Scan Radiometer (VISSR) to differentiate between water land and clouds. By GOES 4 the more advanced VAS added data on atmospheric temperature and water vapour content. NASA stationed SMS 1 initially over the east Atlantic, and then moved it to 75° west over Bogota, to give the first day/night storm watch and alert capability, with full disc pictures of the Western hemisphere every 30 minutes.

As part of the Global Atmospheric Research Program (GARP), SMS-1 provided the first continuous coverage of a major hurricane, designated Carmen, in September 1974. SMS 2 joined the first satellites when NASA stationed it over 115° west, just east of Hawaii, completing total coverage for the western hemisphere. NOAA assigned it the tasks of watching California's forest areas to give warning within 90 minutes of fire outbreaks.

During the Global Weather Experiment of 1977–78, NOAA positioned GOES 1 over the Indian Ocean at 60° east, with ESA's ground station in Spain processing data. The satellite later moved to 116° west and 135° west.

GOES 2 performed as expected until a VISSR primary encoder failed 26 January 1979 and the back-up encoder failed 18 December 1978, ending this satellite's imaging function after returning 20,591 full disc images.

On GOES 3, the VISSR problem continued to plague the GOES system and this satellite the replacement for GOES 1 operated in a degraded mode from 14 September 1979 until the encoder finally failed 6 May 1981 after returning 36,190 full disc images. After its failure, NASA moved the satellite to 176° west from 130° west in June 1990 for use by Pacific island nations as a communication satellites through the Peacesat Pan-Pacific Educational and Cultural Experiments. But perhaps the most successful of all of the meteorological satellites, in 764 operational days, the GOES 4 provided warning of many Pacific storms, transmitting 28,500 scenes.

NASA intended to use the GOES 6 as the east coast satellite, but following GOES 5 failure, the Agency moved it to 98° west to cover both US coasts; then to 108° west to monitor winter storms in the Pacific off the US northwest coast. For almost three years following the failure of GOES 5, GOES 6 had to do the work of two satellites by being moved with the seasons between 98 and 108° west.

At the beginning of 1991, NOAA planned that it would follow the mid-1992 launch of GOES-I, eight months later by GOES-J and by the 1996 debut of GOES-K. However, instrument technical problems surfaced during the year, including the use of incorrect wiring, prompted the delay of GOES-I to at least December 1993. At one time during the correction process, NOAA considered launching at end of 1992 with below specification instruments. Congress set aside an additional US$110 million to ensure the programme continued, if required. The Department of Commerce decided instead to press for a full-performance GOES-I while buying time with Europe's Meteosat.

GOES 7 began at the GOES position, but later moved west following the GOES 6 failure, to 108° west for the winter season, then to 98° west in the Spring for observing hurricanes over the Atlantic and returning further west in the autumn. This GOES carries the Space Environment Monitor, a Data Collection System. To improve station-keeping lifetime, NASA included a small conical sail on a 17 m boom, on GOES 8 to balance solar radiation torque and reduce station-keeping propellant use; a trim tab on solar wing provides fine control.

GOES-L/M will take the series into the next century, with launches tentatively set for 2000/2004. NOAA began work on the series beyond GOES-M during 1988, holding meetings in early 1989 to define requirements. These were passed on to NASA in June 1989 for Goddard Space Flight Center to undertake Phase A, completed during 1990.

NOAA revealed in February 1995, in its FY96 budget request, that it intends to order three

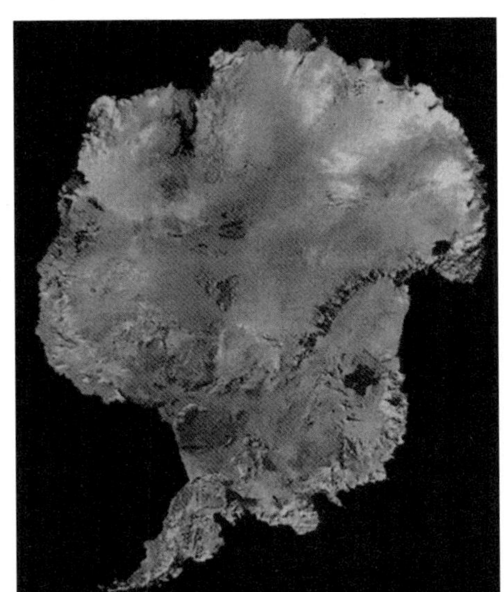

Mosaic of 28 NOAA AVHRR images covering Antarctica produced in 1988 in a joint venture between NOAA and the UK's National Remote Sensing Centre (NRSC) 0517062

GOES 8's first visible engineering test image, 9 May 1994 (NOAA) 0517247

more similar satellites, plus an option on a fourth, before beginning Phase A of the GOES-R next generation in 1998 for first launch around 2008. FY96's request includes funding to begin procurement of GOES-N/O. The contract will be awarded in late 1997 (for 2002 first delivery) after the competition is opened by end-1996. It is expected that sounder resolution will be improved. Adding star trackers will improve location identification to 1 km from the current 4 to 6 km, improving forecasting.

Studies may show that this payload should be divided between several smaller carriers. Similar to the previous polar orbiters and built by Lockheed Martin Astro Space, this new generation carries different instruments. Three were initially ordered at US$160 million (excluding instruments and launches); N/N-prime were added in 1993 (NASA made the US$151 million award January 1995) to allow more time to prepare the next series. Beginning with NOAA-M, the US will fly only afternoon satellites; Europe will be responsible for the morning orbit. DSB has been eliminated, although sounder data are available through HRPT broadcasts. An eight-channel ocean colour instrument for VNIR sensing was considered.

Specifications

Tiros 1

Launched: 1 April 1960
Orbit: 692 × 740 km

TOS

Launched: as ESSA 1 on 3 February 1966 (702 × 845 km) and ended with ESSA 9 (launched 26 February 1969, 1,427 × 1,508 km)

NOAA 1

Launched: 11 December 1970 from Vandenberg
Orbit: 1,429 × 1,472 km, 101° 114.8 min
Mass: 307 kg

NOAA 2 (ITOS-B)

Launched: 21 October 1971
Comment: This launch was never allocated an NOAA designation because it failed to achieve a satisfactory orbit

NOAA 3

Launched: 6 November 1973
Orbit: 1,499 × 1,508 km, 102.2°, 116.1 min
Mass: 340 kg

NOAA 4

Launched: 15 November 1975
Orbit: 1,442 × 1,482 km, 102°, 114.9 min

NOAA 5

Mass: 340 kg
Launched: 29 July 1976.
Orbit: 1,504 × 1,520 km, 102.2°, 116.3 min

Tiros N/NOAA-A

Launched: 13 October 1978 (first of the series known as Tiros N – the NASA prototype)
Mass: 734 kg
Orbit: 850 × 866 km 102°
Comment: The complementary NOAA 6 joined the satellite on 27 June 1979. Developed, like the earlier generations, by RCA, these were the start of a series based on the Block 5D bus developed for the USAF DMSP.

NOAA 6

Launched: 27 June 1979
Orbit: 785 × 800 km, 98.7°, 100.7 min
Out of service: 31 March 1987

NOAA 7

Launched: 23 June 1981
Orbit: 828 × 847 km, 98.9°, 101.7 min
Out of service: June 1986 due to power failure

NOAA-B

Launched: 29 May 1980
Orbit: 264 × 1,445 km, 92.2°, 102.1 min
Comment: Failed to achieve the correct orbit, but by the time NOAA 7 was in orbit (launched 23 June 1981) it was possible to claim that fishermen in California, Oregon, Washington and Alaska were improving their catches by using the sea surface temperature charts. One 60-vessel towing and transportation company announced fuel savings of 20–40 per cent as a result of choosing routes with the help of NOAA's stream and loop current information. Decayed 3 May 1981.

NOAA 8/NOAA-E

Launched: 28 March 1983 by Atlas from Vandenberg
Mass: 1,712 kg
Orbit: 833 km circular at 98.3°
Comment: First of RCA's Advanced Tiros-N (ATN) operating as part of the international Cospas-Sarsat search/rescue system. High-resolution instruments measured both surface and vertical temperature; a UK stratospheric sounding unit monitored upper atmosphere temperature and the French Argos system relayed data from balloons, buoys and remote weather stations. Attitude control problems began once in orbit. In June 1984, the satellite began tumbling following failure of the master timing system, and the refusal of the back-up system to activate. All attempts to save the US$30 million satellite failed, and it had to be abandoned, presumed dead, leaving NASA with no search/rescue capability. NOAA 6, placed on standby following a malfunction of its AVHRR, was reactivated to fulfil NOAA 8's other functions. Then, in June 1985, NOAA 8's back-up attitude control system unexpectedly came back online and its use was fully recovered. However, an overcharged battery burst in December 1985, some contact was restored, but NNOAA finally shut it down on 8 January 1986.

NOAA 9

Launched: 12 December 1984 by Atlas from Vandenberg
Comment: After a series of weather and technical delays, restored both weather and rescue services from its afternoon orbit. Carried NASA's Earth Radiation Budget Experiment scanner and non-scanner instruments (scanner failed 20 January 1987, non-scanner remains operational); see the ERBS satellite entry in the main national US section. Also the first to carry the SBUV/2 ozone mapping instrument. Satellite cost was US$43.5 million plus US$11.4 million launch. Placed on standby following the launch of NOAA 11. Its MSU channel 2 failed 8 March 1987, number 3 7 May 1987, prompting launch of NOAA 11. Collection of SBUV and ERBE non-scanner data continues. S&R function decommissioned March 1995. AVHRR remains usable, although 3.7 µm channel out of specification. The power system is marginal due to array shunt failures and battery problems.

NOAA 10

Launched: 17 September 1986 by Atlas E from Vandenberg after 16 delays (planned for August 1985)
Orbit: Morning 808 × 826 km, 98.8°
Comment: This satellite replaced the degraded NOAA 6, which was brought back into use following the failure of NOAA 8. NOAA 10's Sarsat equipment restored the ability of the US to contribute to the international Cospas/Sarsat search/rescue system. Within six days of becoming operational, it had picked up its first distress signal, which led to the rescue of four Canadians whose aircraft had crashed in a remote area of Ontario
Sensors: Same as NOAA 11 but does not carry SBUV/2 and incorporated Earth Radiation Budget Experiment scanner and non-scanner for NASA Langley. ERBE scanner failed 22 May 1989 (non-scanner solar monitor shutter stuck open, but continues to provide usable data). The Sarsat processor receiver failed 8 September 1988. Replaced by NOAA 12 in May 1991. NOAA 10 remains on standby, but has some degraded performance in the AVHRR, SEM and power systems.

NOAA 11

Launched: 24 September 1988 by Atlas E from Vandenberg AFB; two year design lifetime
Mass: about 1,700 kg BOL
Cost: US$53.5 million
Contractors: GE Astro Space (prime), ITT (HIRS, AVHRR), NASA/JPL (microwave sounding unit)
Resolution: 1.1 km AVHRR, 20 km HIRS/2, 147 km SSU, 105 km MSU
Orbit: 849 × 865 km, 98.9° Sun-synchronous, crossing equator in afternoon (originally 13.40; drifted to 16.02 by May 1993) on ascending node with daily repeat cycle
Spacecraft: Astro Space's Advanced Tiros-N platform, 1.80 m wide, 4.18 m long with single 2.37 × 4.91 m 8-panel 11.6 m^2 solar wing deployed, providing 1,500/1,400 W BOL/EOL; 3 × 26.5 Ah Ni/Cd batteries. Injection provided by Thiokol Star 37S solid. ±0.2° 3-axis control by RWs using magnetic unloading; 8 GN_2 thrusters (4.23 kg supply) provide control during solid burn and until wheels take over (also auxiliary wheel unloading); orbit adjust by 4 hydrazine (28.4 kg supply). Payload 386 kg. Real-time transmissions made for Automatic Picture Transmission (APT, 4 km resolution, 137.5 or 137.62 MHz), High Resolution Picture Transmission (HRPT, full resolution, 1.698 or 1.707 GHz) services and DBS Direct Broadcast Sounder (full resolution sounder data on 1.698, 1.7025 or 1.707 GHz). Spacecraft command/control at 148.56 MHz provided by NOAA's CDA stations in Wallops Island, Virginia and Fairbanks, Alaska; Satellite Operations Control Center is in Suitland, Maryland. The Centre de Meteoroloogie Spatiale in Lannion, France can relay stored/real-time data
Sensors: Advanced Very High-Resolution Radiometer (AVHRR). The instrument operates in five channels (µm)*:* 0.58–0.68 (Si), 0.725–1.0 (Si), 3.55–3.93 (InSb), 10.3–11.3 (HgCdTe) and 11.4–12.4 (HgCdTe), primarily for assisting weather forecasting, snow/ice monitoring and sea surface temperature, but also applied to marine oil pollution mapping, volcanic eruption monitoring, assessment of vegetation vigour on international scales and estimating atmospheric aerosols for climate monitoring. 1.3 mrad IFOV. 20 cm diameter Cassegrain telescope, beryllium scan mirror, ±55° scan angle.

Tiros Operational Vertical Sounder (TOVS) is a three-instrument system consisting of a four-channel Microwave Sounding Unit (MSU) for tropospheric temperature soundings in cloudy regions, a three-channel Stratospheric Sounding Unit (SSU) for 25–50 km stratospheric temperature probing, and the 20-channel High-Resolution Infra-red Sounder (HIRS/2) for vertical temperature profile, water vapour and total ozone content to 40 km. JPL's MSU, 50.30, 53.74, 54.96, 57.95 GHz, 220 MHz bandwidth, ±47.4° scan width from nadir, two scanning reflector antennas (9.5° steps through 360°), 0–350K dynamic range, 0.3K NEδT. Matra Marconi Space's SSU, pressure modulated CO_2 cell in each optical path for 14.926, 14.934, 14.940 µm, 10° IFOV, 0.25 NEδT at 273K, no collecting optics (5 cm aperture), ±40° scan width from nadir. ITT's HIRS/2 channels 1–12 6.72–14.95 µm HgCdTe, numbers 13–19 3.76–4.57 µm InSb, number 20 0.69 µm Si. 24 mrad IFOV, ±49.5° scan width from nadir, 15 cm diameter Cassegrain telescope, 1.8° step scanner covers 56 steps then retraces.

Space Environment Monitor records radiation levels to determine the energy deposited by solar particles in the upper atmosphere and to provide a solar storm warnings. TED cylindrical electrostatic analyzer and spiraltron: 0.3-20 keV protons/electrons in 11 bands. MEPED solid state telescopes and omnidetectors: 30–20 keV protons, 11 bands; >30->300 keV electrons, 3 bands; >6 MeV ions; >16, >36, >80 MeV omniprotons.

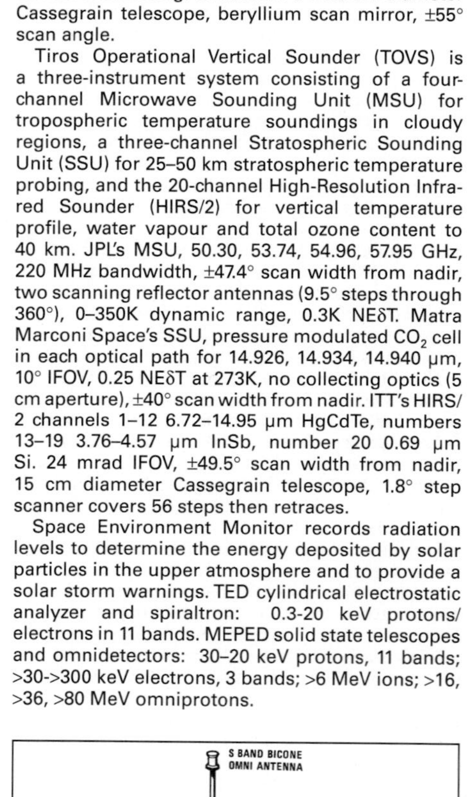

Principal GOES 7 features (Hughes) 0517059

NOAA 12 (NOAA-D)
Launched: 15.52 GMT 14 May 1991 by Atlas 50E from Vandenberg AFB
Orbit: 807 × 826 km, 98.7° Sun-synchronous in morning orbit. Replaced NOAA 10 as the prime morning satellite
Sensors: Same as NOAA 11 but does not carry SBUV/2, the S&R payloads or the SSU instruments. MSU has degraded because of gain changes. Other specifications as for NOAA 11
Comment: The last Tiros-N. First known as NOAA-D, but launched out of sequence in 1991 as NOAA 12.

NOAA 13
Launched: 9 August 1993 by Atlas 34E from Vandenberg AFB to replace NOAA 11
Orbit: Afternoon satellite
Comment: Failure of the power link from the solar panel lost contact 21 August during checkout. The September 1994 review board report identified a short-circuit in the battery charge assembly that prevented the solar arrays charging the batteries. A 31 mm screw penetrating insulation and contacting a radiator plate probably caused the short. Inspection of NOAA-J (NOAA 14) showed 10 of the 12 screws to be in danger of causing the same fault. The same design had flown on 16 NOAA and DMSP satellites
Sensors: Same as NOAA 11 plus the experimental MAXIE and EHIC sensors, sponsored by the Office of Naval Research. The Magnetospheric Atmospheric X-ray Imaging Experiment, provided by Lockheed, assisted by the Aerospace Corp and Norway's University of Bergen, was to map the intensities and energy spectra of X-rays generated by electrons in the upper atmosphere, and the associated auroral and substorm imaging. Good data were returned until the power failure. The Energetic Heavy Ion Composition Experiment, provided by the University of Chicago and Canada's NAC HIA, was to measure the chemical and isotopic composition of energetic particles between hydrogen and nickel over 0.5–200 MeV/nucleon.

NOAA 14 (NOAA-J)
Launch: 30 December 1994 by Atlas 11E from Vandenberg AFB. Control transferred from NASA to NOAA 3 January 1995
Mass: 1,712 kg at launch, 1,030 kg BOL on-orbit
Orbit: 848 × 863 km, 98.9° afternoon Sun-synchronous to replace NOAA 11.
SARP failed, SBUV/2 failed, DTR 4A/B inoperable February 1995.

NOAA 15 (NOAA-K, POES 1)
Launch: 13 May 1998 by Titan 2, first of five POES satellites with improved imaging and sounding capabilities. Imaging instrument failed July 2000
Orbit: 833 km, 98.7°, 101 min
Mass: 2,234 kg at launch, 1,454 kg BOL on-orbit
Power system: as NOAA 11 but 11-panel 17 m^2 solar wing
Sensors: AVHRR/3. Same as NOAA 11 but 6th time-shared channel (3A: 1.58–1.64 μm InGaAs) improves snow/cloud discrimination and changes in several channels improve calculation of worldwide vegetation levels and estimation solar and terrestrial radiation levels previously monitored by the two ERBEs. 33 kg, 28.5 W, ITT Aerospace/Communications Division.
HIRS/2 20-channel High-Resolution IR Sounder for vertical T profile, water vapour and total ozone content to 40 km. Channels 1–12 6.52–14.95 μm HgCdTe, numbers 13–19 3.76–4.57 μm InSb, number 20 0.69 μm Si. 24 mrad IFOV, ±49.5° scan width from nadir, 15 cm diameter Cassegrain telescope, 1.8° step scanner covers 56 steps then retrace. 34 kg, 24 W, ITT Aerospace/Communications Division.
AMSU-A 15-channel advanced microwave sounding unit for atmospheric T soundings from surface to 45 km at 23.8, 31.4, 50.3–57.3 and 89 GHz. 50 km nadir resolution, 2,200 km swath, 2.2 kbit/s.
AMSU-B 5-channel on NOAA-K/L/M for humidity profiles, delineating sea ice and open water, and monitoring snow thickness and soil moisture. 89, 157 and 183 GHz, 15 km nadir resolution, 2,200 km swath, 4 kbit/s. UK Met Office instrument; Matra Marconi Space was awarded a £9.5 million contract in March 1989.
MHS 5-channel microwave humidity sounder on NOAA-N/N-prime for humidity profiles up to 42 km. 89, 150 and 183 GHz, 15.4 km nadir resolution, 2,348 km swath, 3.95 kbit/s, 66 kg, 100 W. EUMETSAT (also on Metop). Argos system capacity is quadrupled, able to handle eight messages simultaneously at 2,560 bit/s over 80 kHz (currently 24 kHz) bandwidth
Cospas-Sarsat S&R. Same as NOAA 11
SEM. Same as NOAA 11
SBUV/2. Same as NOAA 11; not on NOAA-L.

NOAA 16 (NOAA-L)
Launched: 21 September 2000 by Titan 2
Orbit: 870 km, 98.7°, 102 min. APT system failed in September 2001

NOAA 17 (NOAA-M)
Launched: 24 June 2002 by Titan 2
Orbit: 833 km, 98.7°, 101.3 min

SMS 1
Launched: 17 May 1974 by Delta from Cape Canaveral
Mass: 627 kg
Orbit: Initially over the East Atlantic, and then at 75° W over Bogota, 132° W in 1980
Comment: This satellite provided the first day/night stormwatch, with full disc pictures of the western hemisphere every 30 min. As part of the Global Atmospheric Research Programme (GARP), it provided the first continuous coverage of a major hurricane, designated Carmen, in September 1974. Out of service 24 January 1981.

SMS 2
Launched: 6 February 1975
Orbit: 125 ° W, east of Hawaii; 75 ° W in 1980
Comment: Positioned so that it and SMS 1 could cover the western hemisphere. One of its tasks was to keep watch on California's forest areas to give warning within 90 min of fire outbreaks. Out of service 5 August 1982.

GOES 1
Launched: 16 October 1975 by Delta from Cape Canaveral. 293 kg
Mass: 293 kg
Orbit: During the Global Weather Experiment of 1977–78, it was positioned over the Indian Ocean at 60° E, with ESA's ground station in Spain processing data. Later moved to 116° W and 135° W
Comment: Following GOES 4 failure in November 1982, it was functioning only in the visible band, but it was reactivated to transmit images via GOES 4. Retired 7 March 1985.

GOES 2
Launched: 16 June 1977
Orbit: 105° W. At beginning of 1992, positioned over 60° W (±1.0°), inclined at 9.4° for relay of Mode AAA (Stretched VAS) from GOES 7, relay of East WEFAX and for East Data Collection Platform Interrogation (time-code). These functions were principally for western European and African users, who could not see GOES 7. It was moved in 1992 to 135° W to replace GOES 6 supporting all non-imaging functions of the west spacecraft (WEFAX, DCS, and SEM)
Comment: VISSR primary encoder failed 26 January 1979 (redundant encoder failed 18 December 1978), ending its imaging function after returning 20,591 full disc images and 6,838 partial. Its functions were replaced by GOES 7 in January 1995 and it retired out of service in May 2001.

GOES 3
Launched: 16 June 1978, replacing GOES 1
Orbit: Moved to 176° W from 130° W in June 1990. It was held at 176° W, inclined at 10.6°, but it has been drifting since 1995 although still usable
Comment: The move to a new location occurred under an agreement with NASA and DoC's National Telecommunications & Information Administration for use by Pacific island nations through the Peacesat Pan-Pacific Educational & Cultural Experiments by Satellite project to re-establishthecommunicationsserviceinaugurated in 1971 by ATS 6. This provides 2-way voice/data services to small Earth stations at more than 100 sites in 22 countries. Peacesat is controlled for NOAA by NASA from the Kokee Park Geophysical Observatory on Kauai, Hawaii; NTIA co-ordinates user operations.

GOES 4
Launched: 9 September 1980
Orbit: 135° W
Comment: In 764 operational days provided warning of many Pacific storms, transmitting 28,500 scenes. Imaging system failed in November 1982, after which it was used as a standby transponder relaying data. In May 1985, it was loaned to ESA as temporary substitute for Meteosat 1 and was moved 4° daily to place it over the Atlantic at 43° west by June. This was in return for Meteosat 2's help to NOAA in 1984 following the GOES 5 failure. As GOES 4 was incompatible with ESA's ground equipment, NOAA operated GOES 4 for ESA from Suitland, Maryland and Wallops, Virginia. It was to be boosted out of GEO at the end of the loan. In August 1985 however, unidentified UHF interference at 401.9 MHz rendered its data collection capability useless as it was being moved. It was therefore returned to 43° W. It was boosted above GEO on 22 November 1988 and deactivated.

GOES 5
Launched: 22 May 1981
Orbit: 75° W. Hydrazine was depleted November 1989 and it began drifting westwards. It will drift to about 145° W, then return east towards its original starting point of 60° W. This pattern will continue, with outer limits decreasing, until it settles around 105° W
Comment: Intended to monitor US east coast and tropical storm formations in the west Atlantic Ocean. VAS failed 29 July 1984 due to two tungsten filament lamp failures; then used to relay data from GOES 6 and then GOES 7. Relay of GOES 7 imagery to Europe was transferred to GOES 2 in May 1990. On 18 July 1990 all other functions ended and it was deactivated.

GOES 6
Launched: 28 April 1983
Orbit: 135° W, moved to 98° W to cover both US coasts; then to 108° W to monitor winter storms in the Pacific off US northwest coast. Following the failure of GOES 5, GOES 6 had to do the work of two satellites by being moved with the seasons between 98/108° W. Into 1992, it was at 135.6° W (inclined 3.6°) out of service 21 January 1989 when VAS imager failed
Comment: Intended to cover western US. With one of its four encoder lamps burned out, there was concern that the other three might fail; otherwise there was sufficient propellant to keep it operational until late 1988 or early 1989. Meanwhile, it could provide coverage for the main US continent, but could not provide hurricane coverage for some Atlantic islands, the Caribbean and areas west or south of Hawaii. The last lamp failed 21 January 1989 and GOES 6 was reduced to acting as a data relay, supporting all non-imaging functions of the west spacecraft (WEFAX, DCS, and SEM). It was low on hydrazine and each station-keeping manoeuvre could have depleted the supply. This happened in 1992 and it began slowly drifting, following a pattern similar to GOES 5's. GOES 2 took over its relay role, but return of SEM data continues.

GOES 7
Launched: 26 February 1987 by Delta
Mass: 835 kg at launch, 399 kg on-station BOL
Orbit: Initially GOES-East satellite, nominally at 75° W GEO. Following GOES 6 failure, it was moved to 108° W for the Winter season, then to 98° W in the spring for observing hurricanes over the Atlantic, returning further west in the Autumn. At 112° W from 1992. Arrived 135° W 27 January 1995 after GOES 8 began operations; replaced 11 January 1996 at 135° W by GOES 9. Started moving 26 January to 96° W storage position (arriving 10 July). The inclination is being allowed to increase (3.0° by April 1996, increasing at 0.9°/year)
Contractor: Hughes Aircraft Co (prime)
Resolution: 900 m visible, 6.9/13.8 km infra-red
Spacecraft: 2.16 m diameter 3.53 m high cylindrical spin-stabilised satellite; body-mounted solar array provides 450/330 W BOL/EOL. Two Ni/Cd batteries provide TT&C eclipse protection. Injected into GEO by solid AKM
Sensors: Visible Infra-red Spin-Scan Radiometric Atmospheric Sounder (VAS). Scans the full Earth disc in 1,820 successive steps by combination of spacecraft spin and stepping optics. VAS carries single visible channel (eight photomultiplier tube detectors in push-broom array) providing 900 m resolution, and six infra-red channels: two 6.7–14.7 μm with 6.9 km resolution primarily for imaging, two 3.7–7.3 μm and two 7.7–14.7 μm. The satellite also carries a Space Environment Monitor, a Data Collection System and was the first GEO satellite equipped for Cospas/Sarsat trials (see GOES-I to -M). DCS signals are relayed to a Command & Data Acquisition station, which performs error checks and routes them to NOAA offices at the World Weather Building in Camp Springs, Maryland for distribution to users. WEFAX broadcasts are made in the 10 minute intervals between VAS readouts.

GOES 8
Launched: 13 April 1994 by Atlas 1 from Cape Canaveral; lifetime five years (propellant sufficient for 8 years). Formally transferred from NASA to NOAA 26 October 1994; declared fully operational 9 June 1995. One of two momentum wheels lost on 9 January 1997. Now functioning on remaining momentum wheel and reaction wheel for altitude control

Mass: 2,105 kg at launch, 1,140 kg on-station BOL
Orbit: arrived 75° W GEO (GOES-East) 27 February 1995 after testing at 90° W (departed 20 January 1995)
Contractor: Space Systems/Loral (prime), ITT Aerospace/Communications Division (sensors)
Resolution: 1 km visible, 4/8 km infra-red
Spacecraft: Space Systems/Loral's 3-axis stabilisation box-shaped platform, with single two-panel Si solar wing on south side and associated solar sail on north, allowing passive north-facing Imager/Sounder coolers to view cold space. Conical sail on 17 m boom balances solar radiation torque; a trim tab on solar wing provides fine control. Total span 26.91 m. Attitude control by two 51 Nms MWs, one 2.1 Nms yaw RW + two magnetorquers; during orbital manoeuvres by two sets of 6 × 22 N MMH/NTO thrusters. GEO insertion provided by three firings of 490 N Kaiser Marquardt R-4D. 431 kg MMH + 694 kg NTO each in single 103 cm diameter titanium sphere, pressurised by helium. Payload pointing accuracy: ±9.1 µrad roll; ±9.4 µrad pitch; ±73.3 µrad yaw (in 90 min). 268 × 481 cm solar wing can provide 1,057 W EOL Summer solstice; supported by 2 × 12 Ah Ni/Cd batteries; 42.0 V DC (30.0 V DC min in eclipse). Command/control provided via NOAA's Command & Data Acquisition station on Wallops Island, Virginia; the Satellite Operations Control Center is in Suitland, Maryland. Command transmitted 2.03420 GHz. Data are processed at the CDA and uplinked for broadcast
Sensors/payload: Imager. Imaging radiometer sweeping 8 km high swath along east-west/west-east path at 20°/s via servo-driven 2- axis gimballed mirror and 31.1 cm diameter aperture Cassegrain telescope. 120 kg, 119 W daily average. Channel 1: 0.55–0.75 µm, 8 Si detectors, nadir instantaneous FOV 1 km each, cloud cover; 2: 3.80–4.00 µm, 4 InSb detectors, nadir IFOV 4 km each, night cloud cover; 3: 6.50–7.00 µm, 2 HgCdTe detectors, nadir IFOV 8 km each, water vapour; 4: 10.20–11.20 µm, 4 HgCdTe detectors, nadir IFOV 4 km each, surface temperature; 5: 11.50–12.50 µm, 4 HgCdTe detectors, nadir IFOV 4 km each, sea surface temperature and water vapour. Full Earth scan in <26 minutes; 3,000 × 3,000 km in 3 minutes; 1,000 × 1,000 km in 41 seconds. Data rate 2.621 kbit/s. Imaging stability 42 µrad for noon ±8 h, 70 µrad for midnight ±4 h. Passive cooler maintains infra-red detectors at 94/101K Winter/Summer. Full Disk Mode repeatedly scans the full disc every 30minutes,exceptonceevery6hourswhenitperforms an extended northern hemisphere scan to allow 10 minutes for housekeeping work. Routine Mode is a repeated 3 h sequence. It begins with a 30 minute full disc scan, followed by a 30 minute sequence (performed five times) of a northern hemisphere scan, Conus scan and southern hemisphere scan. The last southern scan is omitted every 6 hours for 10 minutes housekeeping. Star looks (for precise attitude registration) and blackbody calibrations occur every 30 minutes. Rapid Scan Mode is similar but it emphasises Conus scans.

Sounder. 19-channel radiometer for atmospheric vertical temperature and moisture profiles, surface/cloud top temperature, and ozone distribution. 127 kg, 106W daily average, 31.1 cm aperture. Optics and scan patterns similar to Imager. Wheel carries 18 filters in three concentric rings, one for each infra-red detector group. Longwave channels: 1–7, 14.71–12.02 µm, 4 HgCdTe detectors, and circular IFOV 242 µrad. Midwave channels: 8–12, 11.03–6.51 µm, 4 HgCdTe detectors, circular IFOV 242 µrad. Shortwave channels 13–18, 4.57–3.74 µm, 4 InSb detectors, circular IFOV 242 µrad; visible channel 19, 0.70 µm, 4 Si detectors, circular IFOV 242 µrad. Data rate 40 kbit/s. Scans in 10 km steps east-west/west-east, moving 40 km north-south at end of scan line. Each step 100/200/400 m/s, controlled by filter wheel.

Space Environment Monitor. Of four instruments for GEO particle radiation and solar X-ray output. EPS Energetic Particles Sensor: 0.8–500 MeV protons, 3.2–400 MeV α, 0.55–4.0 MeV electrons. HEPAD High Energy Proton & Alpha Detector: 370–970 MeV protons, 640->850 MeV α. XRS x-ray *Sensor:* 0.5–3.0 and 1.0–8.0 A real-time solar flux; mounted on solar wing yoke. Magnetometers: redundant 3-axis fluxgate magnetometers on 3 m boom, ±1,000 nT (±4 nT accuracy). GOES-M and successors will include a Solar X-ray Imager to aid prediction of geomagnetic storms. SEM data downlinked on 1.694000 GHz to Environmental Research Lab, Boulder, Colorado.

Cospas/Sarsat Distress Relay (406 MHz beacon relay only, no location service). Downlink at 1.544500 GHz.

WEFAX transponder. Processed imagery is returned to GOES at 2.03300 GHz for broadcast to users at 1.6910 GHz at 11 W to users

Data Collection System: Up to 233 100 bit/s signals received from DCP Data Collection Platforms on 401.9 or 402.2 MHz. Downlinked at 1.69450 or 1.69480 GHz to CDA and users. DCP interrogated by request uplinked to GOES from CDA on 2034.9000 or 2034.9125 MHz; DCP interrogated by GOES at 468.8250 or 468.8375 MHz. >12,000 DCPs active in 1995 (increases 15% yearly).

GOES 9
Launched: 23 May 1995 by Atlas 1 from Cape Canaveral. NOAA assumed control from NASA 31 October 1995. Scheduled for 155° W in mid-2003 to bolster Japanese meteorology support after GMS-5 end of life. Monthly rental US$127,500
Orbit: GEO at 135° W (GOES-West) from 11 January 1996, replacing GOES 7 after testing at 90° W. First (visible) image received 12 June 1995. Other specifications are the same as for GOES 8.

GOES 10
Launched: 25 April 1997 from Cape Canaveral using an Atlas-1 (final Atlas-1 launch). Solar array rotation problem. Satellite inverted for optimum solar energy
Orbit: GEO over 106° W (early June 1997)
Other specifications as for GOES 8.

GOES 11
Launched: 3 May 2000 by Atlas 2A from LC-36A, CCAFS
Orbit: GEO over 106° W

GOES 12
Launched: 23 July 2001 by Atlas 2A
Orbit: GEO at 75° W from where enters service April 2003 replacing GOES 8

OrbView Series

Current Status
Orbimage launched the SeaStar on 1 August 1997 from Vandenberg AFB, California using the Pegasus XL booster. Launched 21 September 2001, OrbView 4 was lost when the Taurus 2110 encountered problems with its T6 second stage and final velocity at cutoff was far too low to reach a sustainable orbit after all succeeding stages fired correctly. The satellite burned up in the earth's atmosphere northeast of Madagascar before reaching orbit.

Background
OrbView 1 (MicroLab 1)
Orbital Sciences Corp's Eyeglass International, was formed in 1994, to build and operate the Eyeglass satellite system and to market imagery products and services. Orbimage, in 1995, decided instead to pursue the smaller, lower cost OrbView satellite, derived from its MicroLab design. The cost of deploying one satellite and establishing the ground system has been estimated at more than US$100 million. OSC's MacDonald Dettwiler Associates is managing the ground segment requirements. Saudi Arabia's EIRAD Co-signed June 1994 to become a major investor (operating a receive station).

The United States' Department of Commerce approved the project 5 June 1995. EIRAD decided to take a 20 per cent stake in Orbimage since US regulations impose a 25 per cent limit.

OrbView 2 (SeaStar)
NASA Goddard selected OSC in March 1991 for a US$43.5 million contract to deliver multispectral ocean colour data to NASA investigators for

OSC's SeaStar will provide Earth colour data 0517063

The 1-metre resolution remote sensing satellite from Orbimage (Theo Pirard) 0054351

five years as part of the agency's Earth Probes programme. This is the first time that the US government has purchased global environmental data from a privately-designed and operated remote sensing satellite. SeaStar, based on OSC's PegaStar bus, carries the SeaWiFS Sea-viewing Wide Field Sensor of Hughes' Santa Barbara Research Center. The instrument, originally planned for Landsat 6, is a next-generation design of Nimbus 7's Coastal Zone Colour Scanner.

SeaWiFS measures ocean surface-level productivity of phytoplankton and chlorophyll for ocean dynamics and marine life research. It will contribute to understanding the global carbon cycle and its effect on global warming. Orbimage retains exclusive worldwide rights to imagery for commercial and operational uses. It expects to find markets in ocean fishing, coastal monitoring, land management and the military. SeaStar was originally designed for ocean colour but SeaWiFS gains were changed during development to add higher land radiance dynamic range. It will provide a more environmentally stable Vegetation Index than NOAA's AVHRR, which is inaccurate under hazy atmospheric conditions because of its single visible and near-infra-red channels. Orbimage will sell to both end-users and value added resellers at a price point of US$500 per image set.

A delay to mid-1994 was announced August 1993 for instrument modifications to cure a problem with stray light. The onboard computer then required redesign (the original computer from OSC's APEX was found to have insufficient throughput to process SeaWiFS data) and 1994's Pegasus XL failure resulted in further delays.

Specifications
OrbView 1 (MicroLab-1)
Launch: 3 April 1995 by Pegasus XL
Resolution: 1/2 m PAN (4/8 km swath), 4 m XS (8 km swath)
Lifetime: 3 years
Mass: 146 kg
Orbit: planned 460 km polar, 10.30 equatorial crossing, revisit <3 days
Spacecraft: 96.5 cm diameter, about 60 cm ht disc. Although there will be some onboard storage, most imagery will be downlinked in real time
Sensors: PAN. Bands: (µm): 0.50–0.90; XS 0.45–0.52, 0.52–0.60, 0.63–0.69, 0.76–0.90.

OrbView 2 (SeaStar)
Launch: 1 August 1997 by Pegasus XL
Mass: 309 kg on-orbit (SeaWiFS 50 kg)
Orbit: 705 km, 98.2° equator crossing at noon, 99 min (16 day cycle)
Resolution: 1.13 km Local Area Coverage and 4.5 km Global Area Coverage. L-band transmits encrypted continuous LAC data when SeaStar is visible to standard HRPT stations equipped with OSC's licensed decoder. GAC data are stored and downlinked every 12 h on S-band to central location
Spacecraft: 213 cm high, 112 cm diameter OSC PegaStar bus. 5-year design life; 10 year goal.

Redundant 3-axis momentum biased ACS (0.5°; 1.23 mrad knowledge) of momentum wheels, magnetorquers, horizon sensors, Sun sensors, and 2-axis magnetometers. Redundant GPS receiver provides precise real-time position data to within 100 m. 200 W BOL orbit average from four 55.9 × 152.4 cm rigid solar panels supported by 10 Ah Ni/H_2 battery. TT&C and 0.665/2.0 Mbit/s data via L/S-band downlinks; 1.25 Gbit recorder. Four 22.2 N hydrazine thrusters to raise/maintain mission orbit
Sensors: SeaWiFS. The main telescope rotates 360° about a pivot axis to scan each scene, thereby avoiding the use of a scan mirror and its associated polarisation effects. Specular Sun reflection is avoided by tilting the telescope in the plane perpendicular to the scan plane to one of three positions: +20°, 0° or −20°. The continuous 360° scan permits reference sources to be viewed during the non-scene portion, in addition to including a deep space scan for zero referencing before the scan begins. The spread of observing bands addresses both chlorophyll and pigment absorption values, as well as water optical properties and suspended sediment measurements. Spectral band centres, (in μm, with bandwidth and application in brackets): Band 1 0.412 (0.020, gelbstoffe), Band 2 0.443 (0.20, chlorophyll), Band 3 0.490 (0.020, pigment), Band 4 0.510 (0.020, chlorophyll), Band 5 0.555 (0.20, suspended sediments), Band 6 0.670 (0.020, atmospheric aerosols), Band 7 0.765 (0.40; atmospheric aerosols), Band 8 0.865 (0.044, atmospheric aerosols). The bands for atmospheric aerosols provide atmospheric corrections.

OrbView 4
Launch: 21 September 2001 by Taurus but satellite lost due to launcher failure.

Space Radar Lab

Current Status
SRT and SAR data have been interpolated and converged with declassified military data for an increasingly broad range of applications including archaeo-geologic studies.

Background
NASA flew an 11-day SRTM Shuttle Radar Topography Mission in February 2000 to help the defence Mapping Agency create a global digital elevation map with 16 m height accuracy at 30 m horizontal intervals. SRTM used SRL's SIR-C antenna working interferometrically with another on a 60 m mast. SRL 1 flew aboard STS-59 in April 1994 and SRL 2 in September to October 1994 aboard STS-68. A third mission was, until 1993, planned for mid-1996. JPL is NASA's lead centre. A Spacelab pallet accommodated JPL's L/C-band SIR-C (Spaceborne Imaging Radar-C) and the German/Italian X-SAR X-band synthetic aperture radar. Novel mission features are multi-frequency radar imaging observations, dual polarisation (L/C) and 15 to 60° variable incidence angles. SRL returned imagery at up to 10 m resolution. JPL/Ball Communication Systems Division provided SIR-C's antenna systems. X-SAR's industrial team was headed by Dornier/Alenia Spazio; the 12 × 0.4 m planar slotted waveguide antenna comprises three leaves on a CFRP structure. Alenia Spazio was responsible for the RF subsystems, antenna electrical design and the mission operations ground segment. X-SAR represents the most extensive Earth observation co-operation yet between Germany/Italy.

SRL 1 returned imagery for 133 h covering 25 per cent of the globe's land surface, and demonstrated its effectiveness for the monitoring of environmental parameters such as vegetation type, coverage, moisture distribution and energy transfer mechanisms. Individually, SIR-C covered 99.67 million km^2 in 90 h and X-SAR 47.89 million km^2 in 43 h. Some imagery was returned in real-time. The crew took >14,000 supporting photographs and 19 'supersites' were monitored intensively by ground teams. SRL 2's trajectory was carefully managed during its last four days to repeat within 800 m of some targets so that SRL interferometery would generate high-resolution topgraphical maps. Combining the data with SRL 1's revealed movements since April 1994 with cm accuracy. If used by a free-flying satellite, the technique would allow the minute movements of glaciers, volcanoes and Earthquake zones to be monitored routinely.

SRL also demonstrated an ocean wave data processor from Johns Hopkins' Applied Physics Laboratory. The data were relayed through TDRS and then processed in near-real-time at NASA's SAR Processor Facility at JPL.

Specifications
X-SAR
Frequency/polarisation: 9.6 GHz (X-band)/VV
Resolution: less than or equal to 30 m azimuth, 10–20 m slant
Swath width: 15–45 km (off-nadir angle 20–55°)
Peak transmit power: 3.2 kW; pulse length 40 μs; max DC power 1,060 W
Beamwidth (3 dB): 5.7° elevation, 0.4° azimuth
Antenna gain: 43 dB
Mass: 210 kg
Data rate: 45 Mbit/s.

SIR-C
Frequency/polarisation: 1.25/5.3 GHz (L/C-band); HH, HV, VH, VV
Resolution: ≤30 m azimuth, 10–20 m slant
Swath width: 70 km
Data rate: 50–100 Mbit/s.

TOMS-EP Satellite

Current Status
In December 1998 satellite was hit by a high-energy particle which disrupted onboard computer and put satellite in safe mode. Stabilisation process exhumed fuel and TRW developed a control method using magnetic torque.

Background
The Total Ozone Mapping Spectrometer Earth Probe (TOMS-EP) extends the work of Nimbus 7's similar instrument in monitoring atmospheric ozone levels. It helps establish the continuity in

TOMS-EP is mapping global ozone levels (TRW) 0517226

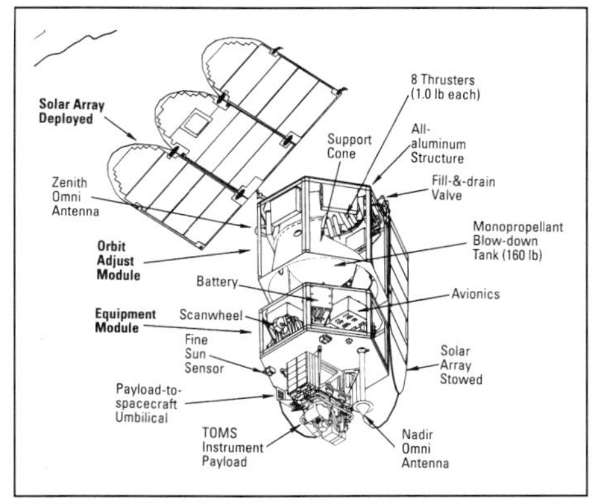

TOMS-EP principal features (TRW) 0517227

the gathered data so a long-term picture of the Earth's ozone layer can be formed. NASA originally selected TOMS in 1989 as the fourth Small Explorer but then dropped its funding before reviving it as the first Earth Probe in NASA's Mission to Planet Earth.

NASA Goddard released a request for spacecraft proposals in late 1990; and the TRW's Eagle bus won out over its competitors for the June 1991 contract of US$29.3 million (now US$57 million). Mission cost, including launch but excluding operations has amounted to about US$80 million for NASA. TRW built the spacecraft and integrated the instrument.

In operation, the satellite stores up to 24 h of data stored on 16 Mbyte solid-state recorder and downlinks it via S-band transponder at 202.5, 50.6 or 1 kbit/s. The Total Ozone Mapping Spectrometer is based on 3,086–3,600 A Fastie-Ebert monochrometer, for 1 per cent/decade ozone trend determination in six wavelength bands.

Specifications
Launch: 2 July 1996 by Pegasus XL from Lockheed L1011 out of Vandenberg AFB
Mass: 294 kg (including 54 kg hydrazine)
Orbit: planned 500 km circular 97.4° Sun-synchronous at 11.00–11.30 ascending node
Contractors: TRW (bus), Orbital Sciences Corp (TOMS instrument)
Resolution: 50 × 50 km
Lifetime: minimum 2 years, goal 3 years
Spacecraft: TRW Eagle platform, two compartments, hexagonal prism of three equipment decks. Bus 99 cm diameter, 168 cm long without TOMS; stowed/deployed 112/388 cm max width, 178/243 cm long with TOMS. Two 3-panel dual-sided solar wings of composite material over honeycomb core providing 6.25 m^2 total cell area, mounted at 45° cant to body, for 233 W EOL (β = 20°). 200 μm Si cells protected by 150 μm cover glass. 9 Ah Super Ni/Cd battery provides eclipse power. 3-axis control to 0.5° pitch/roll (0.25° knowledge, 3 Σ) pitch/roll, <1.0°/0.25° 3 Σ yaw by momentum bias RW system utilising Earth/Sun sensors + gyros. 4 N dual-seat hydrazine thrusters for orbit insertion, trim and station-keeping. Up to 24 h TOMS data stored on 16 Mbyte solid-state recorder and downlinked via 2 W S-band transponder to DSN & Wallops stations at 202.5, 50.6 or 1 kbit/s. Satellite designed for 24 h autonomous operations. Combined Mission Operations Center and Science Operations Center at NASA Goddard
Sensors: Total Ozone Mapping Spectrometer. Based on 3,086–3,600 Å Fastie-Ebert monochrometer, for 1%/decade ozone trend determination in six wavelength bands. 3.0 × 3.0° FOV crosstrack scanning capability allows daily high spatial resolution (50 × 50 km) global O_3 maps. SO_2 monitoring also undertaken. Sensor mass is 32 kg, 1,525 cm^3, 25 W average power consumption (including heaters); onboard solar and electronic calibration capabilities, plus reflectance calibration mode not previously flown. Microprocessor-based electronics provide flexibility in control and data formatting; data rate up to 736 bit/s.

ENGINEERING TEST

France

Proteus Satellite

Current Status
France developed the Proteus satellite as a minisatellite system designed to be adaptable and reconfigurable. This was so that it could be used for scientific missions or applications with each mission using one or two Proteus satellites according to needs. Successor to Topex-Poseidon, Jason 1 is the first application of the Proteus satellite to launch. It was placed in orbit by Delta 7925 on 7 December 2001. CNES planned the TPFO Topex-Poseidon Follow-On mission for launch in 1999 as the first of the 200 to 500 kg Proteus (Platforme Reconfigurable pour l'Observation, les Télécommunications et les Usages Scientifiques) multipurpose LEO platforms. France has shown its determination to make the programme cost effective by establishing a capital cap on the programme of FFr100 million for the platform and payload accommodation. Contractors studied the programme goals in the Phase A investigation, which ran from November 1994 to April 1995. Then CNES selected Aerospatiale as the Phase B prime contractor in May 1996. A final contract calls for one satellite plus one option. NASA plans to launch the TPFO. Candidates for subsequent missions are Tropiques (interaction between radiation, aerosols, cloud, water and atmospheric circulation in the tropics), Vagsat (conical scanning radar) and Irsute (biosphere 40 m resolution thermal-infra-red).

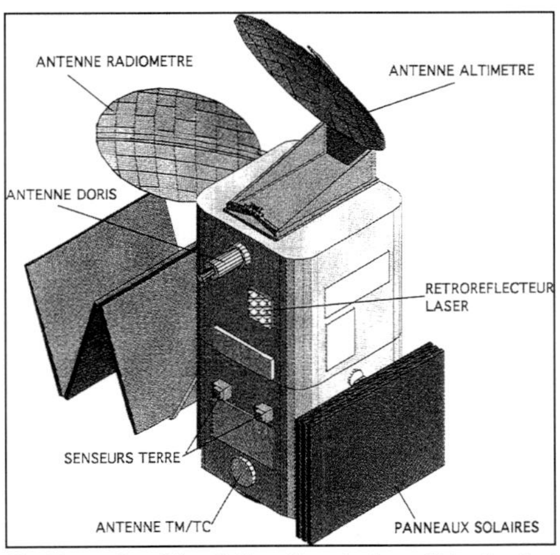

The Topex-Poseidon Follow-On mission will be the first to use the Proteus platform (CNES) 0517262

SRETS Series

Background
At the invitation of the Soviet government, France became one of the first Western governments to participate in Soviet space research. After a meeting with the Soviet premier, Charles de Gaulle approved a small satellite programme to investigate the effects of the space environment on operating satellites. The SRET programme, which resulted, sought to measure the decay in solar array performance due to radiation exposure. The French chose to launch their satellite in conjunction with, and in the same orbit as, a Soviet Molniya satellite. The Molniya semi-synchronous orbit carries a satellite through the van Allen radiation belts twice in one twelve-hour circuit of the Earth. By measuring the decay after each pass through the belts, and knowing the radiation exposure levels, the French could predict the exact decay any satellite would experience in space under different circumstances.

In the second experiment, SRET-2, the French tested a passive cryogenic radiative system for cooling that it expected eventually to incorporate in its Meteosat bus, then being planned for a launch in conjunction with ESA.

Specifications
SRET 1
Launched: 4 April 1972 piggyback with Molniya 1-20 on SL-6 from Plesetsk
Mass: 15.4 kg
Orbit: 460 × 39,248 km
Re-entered: 26 February 1974.

SRET 2
Launched: 5 June 1975 piggyback with Molniya 1-30 on SL-6 from Plesetsk
Mass: 29.6 kg
Orbit: 513 × 40,825 km, 62.8°.

Germany

MIRCA Satellite

Current Status
DARA initiated a joint project between Kayser-Threde and DASA in 1992 using Jena-Optronik to develop and fly the 1 m diameter MIRCA Micro Re-entry Capsule for re-entry investigations. KT and Russia's Central specialised Design Bureau signed a contract in November 1994 to launch MIRCA attached to a Photon in September 1996. To date, the launch has not occurred.

MIRCA will be mounted on the forward face of Photon's own sphere, 12° below the centreline until it is ejected at 0.8-1.6 m/s separation after retrofire. KT's design uniquely uses the heatshield as the main load-carrying structure, comprising hemispheres bolted at the equator. The shield consists of a 1 mm carbon fibre composite core, 20/30 mm ablator for the upper/lower hemisphere and a 1.35 mm C/SiC outer layer. The heatshield has integrated temperature and pressure sensors and a pyrometer in the lower shield. Given the nose tip shape and other re-entry conditions, the Germans predict the shield experiences a 2,100°C stagnation point.

The mass budget is 150.0 kg total (71.0 kg heatshield, 16.7 kg mechanical support structure, 16.6 kg service system, 6.6 kg battery, 16.8 kg experiments, 14.3 kg recovery system, 8.0 kg balance mass) in a payload volume of 110 litres. The satellite will have 20 Ah, 28 V DC from $LiMnO_2$ cells of power which can be expanded to 50 Ah. The planned instrumentation includes HEATIN Heatshield Instrumentation*:* temperature, pressure and heat flux distribution within two meridian planes 90° apart. Also temperature at different depths in material and pressure caused by off-gassing. RAFLEX RArefied FLow EXperiment*:* pressure at stagnation point and 60° behind. PYREX Pyrometer Experiment*:* touchless temperature measurement of C/SiC layer behind shield at stagnation point. Upon completion, the experiment will also space qualify this mini-pyrometer. In flight the system has a command data, transputer-based, containing 16 Mbytes on EPROMs for control/data handling.

A Photon command at 105 seconds before separation activates the system. Three-axis body rates up to ±100°/s and ±64 *g* can be recorded. The payload will be recovered by pushing the cover clear with two 1,700 N solids burning for 80 ms. This action pulls out a conical ribbon chute at about 7 km, which is also activated redundantly by barometric switch and timer after peak deceleration. A reefed cross chute is released 25 seconds later, fully open 10 seconds later. Shortwave beacon (antennas on chute lines) is active for 10 minutes during descent; and a VHF transmitter and strobe remains activated on the ground for 48 hours.

A follow-on tether mission was studied, beginning in 1991 but cancelled by DARA in 1994. Project Rapunzel (Rope Attached Piggyback Unit Zooming on Environment Data at Low Cost) would have reeled out MIRCA on a 50 km tether from Photon to de-orbit when the line was cut. GPS receivers in each module would have provided relative position information. Before then, the first Rapunzel test was planned for a Resurs-F mission carrying a 45 kg 50 cm re-entry sphere developed by the Samara State Aerospace University. Later, Rapunzel studies concentrated on a MIRCA tethered to a Progress-M as it left Mir in 1997. The concept has been explored for a range of mission opportunities associated with a diverse set of re-entry trajectories and flight profiles. Renewed interest in unmanned test activities in materials processing and bioprocessing research has stimulated further studies of unmanned recovery capsules based on this technology.

India

Apple (Ariane Passenger Payload Experiment) Satellite

Background
Apple was India's first indigenous GEO test communications satellite. The French launched the satellite to give ISRO experience in attaining its indigenous satellite goals before the completion of its own GSLV space launch booster. After launch, it relayed TV programmes and educational teleconferences via a C-band transponder despite a jammed solar panel. ISRO de-activated Apple on 19 September 1983.

Specifications
Launched: 19 June 1981 by Ariane from Kourou
Mass: 670 kg
Orbit: 102°E GEO.

Israel

Ofeq Series

Background
The original Ofeq ('Horizon') satellite was a demonstration testbed developed by IAI/MBT for ISA. It is controlled from MBT's ground station. The satellite itself resembles an irregular octagonal prism of conventional aluminium construction, measuring 2.3 m high, and 1.2m at the base, shrinking to a top diameter of 0.7 m. It weighs 156 kg, broken down by: structure 33 kg, power system

Shavit's Ofeq 3 launch (IAI) 0517283

Ofeq 3 is almost identical to South Africa's cancelled Greensat. Rafael's hydrazine propulsion module is visible at top (IAI) 0516733

58 kg, computer 7 kg and communications package 12 kg. The thermal control weighs 5 kg, the wiring 9 kg and the instrumentation 32 kg. To maintain control, the Israelis spin-stabilised the satellite at the rate of 400°/s. Sixteen solar panels provide 246 W through a 7 Ah Ni/Cd battery. The satellite consumes 53 W on average. It transmits 2.5 kbit/s S-band from 128 kbyte memory.

The Israelis refuse to disclose the full payload, but do say it included magnetometers and housekeeping monitoring. Attitude sensing results from the output of a three-axis rate gyro, magnetometer and Sun sensor. During its short lifetime, the only malfunction was in telemetry memory when the computer switched to back-up. Originally, the Israelis expected the satellite to last only a few weeks, but it did not re-enter until 14 January 1989 because of higher than expected orbit.

Several months later and almost identical to Ofeq 1, Ofeq 2 carried an improved computer, telemetry system back-up, thermal protection, gyros, and magnetometer. It could also execute some commands, whereas Ofeq 1 followed a pre-set program. The second satellite had a mass budget of: structure 34 kg, power system 59 kg, computer 7 kg, communications system 14 kg, thermal control 5 kg, wiring 10 kg and instrumentation/balance masses 31 kg. The orbit differed from Ofeq 1's to provide new perspective on radiation belts and to test Shavit's injection precision. It re-entered 9 July 1990 with an orbital life 40 days longer than expected because of reduced solar activity.

Some observers have speculated that the Israelis attempted another launch shortly after the launch of Ofeq 2 but that it failed.

By the time the Israelis were ready to proceed to their second-generation system, they had built a prototype 36 kg satellite. They described it in 1994 as a demonstration satellite with no science or other payload, but post-launch reports suggested a 2 m resolution Earth imager, supported by the orbit providing suitable lighting conditions over the region. Elop, in 1995, confirmed there was a camera similar to its ERMS Earth Resources Monitoring System.

The second-generation Ofeq has power provided by deployed twin three-panel solar wings totalling 3.6 m^2. The Sun sensors were developed by the Netherlands' TNO Institute of Applied Physics. Rafael provided the blowdown Hydrazine Propulsion Module for de-spin from 65 rpm, final orbit positioning, station-keeping and MW unloading. The thrusters are mounted on the top platform supported by eight composite struts, above the single 420 mm diameter positive expulsion sphere carrying 30 kg of hydrazine pressurised at 22 atmospheres. There are two independent branches of six LT-5N 5 N plus one HT-25N 25 N thrusters (HT-25N is principally for final positioning and orbit control). This propulsion system, with two tanks, appears to have been planned for South Africa's Greensat. Photographs show them to be identical.

Specifications

Ofeq 1
Launched: 09.32 GMT 19 September 1988 by Shavit from Palmachim Air Force Base south of Tel Aviv
Orbit: 248 × 1,170 km, 142.86°, 98.79' retrograde.

Ofeq 2
Launched: 12.02 GMT 3 April 1990 by Shavit
Orbit: 208 × 1,584 km, 143.23°
Mass: 160 kg.

Ofeq 3
Launch: 11.16 GMT 5 April 1995
Orbit: 247 × 732 km, 143.4° and raised itself at first apogee to 369 × 730 km, 143.4°. The goal was about 500 km circular but no further changes were made
Mass: 225 kg at launch
Design life: 2 years
AOCS: 3-axis control system providing 0.1° astronomical-class pointing accuracy
Data rate: 15 kbit/s with a command rate of 5 kbit/s.

Ofeq 4
Failed in launch attempt on 22 January 1998 due to a problem with the second stage of the Shavit launcher.

Ofeq 5
Launch: 14: 55 UT 28 May 2002
Orbit: 309 × 770 Km, period 105 min, 83°. Military reconnaissance satellite launched retrograde from Palmahim Beach AFB on Mediterranean coast.

TechSat (Gurwin) Satellite

Current Status

TechSat 1 was lost in the start launcher failure but TechSat 2 was successfully launched by Zenit in July 1998.

Background

A collaboration between the Faculty of Aerospace Engineering at Technion Institute (under the direction of Professor Chaim Eshed), ISA, Israeli industry and Amsat-IL resulted in the design of a simple, low-cost, low-power platform. The consortium completed the Flight Model in October 1994. At launch, TechSat was lost in Start's launch failure. Techsat 1B, a near duplicate, is currently being funded. It will include the Netherlands' TNO-TPD wide field star sensor.

Technion proposes a 250–300 kg second-generation TechSat as a multipurpose platform, created by adding an AOCS unit and GPS navigation to the basic TechSat 1. The 100 kg AOCS module would carry Rafael 1 N liquid propellant and N_2 thrusters. TechSat 2A would carry a high-resolution Earth observation camera, gimballed on command and relaying imagery via a MMIC Ku-band phased-array antenna. Techsat 2B will reportedly be used for data relay and 2C for data and voice communication. The Israeli government has declined all requests for technical information on TechSat 1B.

Specifications

TechSat 1A
Launch: 28 March 1995 on Russian Start from Plesetsk
Orbit: planned orbit was 670 km 75.4°
Contractors/institutes: Technion (management, integration, ground station, ozone sensor, X-ray detector), MLM (solar panels), Rafael (antennas, structure), Tadiran (transceivers), TAMAM (wheel, magnetometer, magnetorquers), El-Op (horizon sensor, CCD camera), Elbit (OBC), Elisra (image compression system), IAI-MBT (battery, power control unit)
Mass: 52 kg
Configuration: box-shaped 460 mm high, 430 mm across (six Al plates, 2.5 mm thick). Earth-pointed face carries antennas, horizon sensor, and camera. 80C186EC-based CPU with two 512 kbyte independent RAM memories. Controlled from Technion station; command reception at 145.910/1,269.85 MHz
Power: solar cells on four faces (1,000 2 × 4 cm Si, 17 W average, 9 W required for housekeeping, Ni/Cd battery)
Attitude control: 3-axis Earth-pointing control (5° coarse, 0.1–0.5° fine) by 2.5 kg 3,600 rpm 4 Nms MW 3 × 1 Am^2 magnetorquers, 1 kg 1 W static horizon sensor (0.05° accuracy), 0.22 kg 0.8 W ±0.6 gauss 3-axis magnetometer
Payload: Amateur digital store/forward 1/3 W transponder. Transmit at 435.225 plus back-up 435.325 MHz band. Receive at 145.850, 145.890, 145.930, 1,269.700, 1,269.800, 1,269.900 MHz at 1,200/9,600 bit/s.

Elisra Earth-observing 350 × 570 pixel CCD camera. Compressed image data broadcast using mm-wave link experiment (also on UHF).

Technion ultra-violet CsTe spectroradiometer to monitor Earth's albedo at 0.24–0.30 µm (0.01 µm/2 mrad resolution), 15° FOV. The spectral profile is a function of the vertical ozone profile. Goals are investigating the long-term stratospheric impact of the Kuwaiti oil fires and the extent of the region's ozone depletion.

Technion X-ray detectors (up to 200 keV). Principal purpose is to prove that CdTe crystals can be used for X-ray imaging.

TechSat 2 (1B)
Launch: 10 July 1998 by Zenit
Mass: 48 kg
Configuration: 48 × 48 × 48 cm box
Stabilisation: 3-axis
Power: 4 solar arrays; 20 W DC power
Design life: 3 years

Japan

DEBUT (Orizuru) Satellite

Background

The Japanese National Aerospace Laboratory's Deployable Boom and Umbrella Test repeatedly deployed a 15 cm diameter 1.2 m coilable lattice boom 34 times and a 24-panel 89.8 cm diameter umbrella 52 times during the 10 day life of the satellite. A 95 Ah lithium battery powered it. The behaviour mirrored the expectations the Japanese published prior to the test.

Researchers claim that the umbrella device could be utilised in future µg processing capsules released from space stations, using atmospheric drag to manoeuvre back to their orbital facility following 10^{-7} *g* operations.

Debut was one of three satellites launched by HI, other payloads being MOS1B and JAS1B.

Specifications

Launched: 7 February 1990, piggyback with MOS-1B by HI
Mass: 50.3 kg
Orbit: 917 × 1,753 km, 99.05°, 112.2 min

ETS (Japanese Communications Engineering Test) Series

Current Status

The Communications and Broadcasting Engineering Test Satellite (COMETS) failed to achieve the correct orbit after lift-off on 21 February 1998. Japanese investigators traced the failure to a premature shutdown of the H2's, LE-5A second-stage engine. NASDA immediately announced a plan to rescue the satellite and change the orbit to prevent passage through the van Allen belt where it was stranded. Engineers commanded the satellite to a contingency operating orbit through a sequence of seven manoeuvres from where it conducted routine experiments until 31 January 1999, extended to August that year by a series of additional test activities. The H2 programme was subsequently cancelled, all launches going to the H2A.

Background

NASDA still conducts communication experiments on its own set of Engineering Test Satellites. It launched four between 1975 and 1982, before privatising its space network. NASDA demonstrated the fruits of its research in August 1987 when it launched the ETS 5, the first Japanese three-axis stabilised satellite in geosynchronous orbit and the first to use an indigenous apogee kick motor.

ETS 6 appeared in 1994 incorporating intersatellite data relay and mobile communications packages. NASDA had intended to demonstrate a new ion propulsion engine to maintain north-south station-keeping on this two-ton platform. The launch vehicle left it short of geosynchronous orbit however, after launch in August 1994. The first of three planned firings by the new ETS 2 kN LAPS (Liquid Apogee Propulsion System) ended when controllers found the engine attained only 10 per cent of full thrust, probably because a fuel line valve had failed to open. The tanks ran out of oxidiser supply on the fifth attempt and ground controllers separated the LAPS on August 31 by command via NASA Goldstone, leaving ETS 6 in 7,796 × 38,707 km, 13.1° inclination orbit.

NASDA planned to launch the COMETS Communications & Broadcasting ETS in 1997 to develop new technologies, preceding the operational DRTS Data Relay & Test Satellites in 2000. The OICETS Optical Inter-orbit Communications Engineering Test Satellite should follow in 1998 to test laser links with ESA's Artemis. ETS 8 in 2002 will demonstrate a single-hop cellular phone system and a CD-quality radio broadcast service.

The Ministry of Posts & Telecommunications established the Partners (Pan-Pacific Regional Telecommunications Network Research Satellite) project in 1992, using Kiku 5 to transmit educational teleconferences (64 kbit/s) and material to less developed Asia and Pacific sites, such as Fiji and Papua New Guinea.

The Japanese will use ETS-7 for automatic and tele-controlled rendezvous and docking with a small target satellite. A GPS, laser, radar and optical sensor will simultaneously guide the satellite to its final rendezvous and docking. It will also demonstrate the capabilities of the 20 kg Orbital Replacement Unit exchange using a 2 m 6 DoF telerobotic arm. MITI's separate Advanced Robot Hand will include NAL's truss-building experiment, Atomic Oxygen Monitor. The work began on this project in FY91 when the Japanese expended £85 million; (FY92 £1,046, FY93 £2,709, FY94 £8,743, FY95 £8,659, FY96 £9,817).

Specifications

ETS-1 (Kiku, 'Chrysanthemum')
Launched: 9 September 1975 by N1 from Tanegashima
Mass: 83 kg
Orbit: 980 × 1,098 km, 47°
Comment: Test of launcher, tracking and antenna extension. Shut down 28 April 1982.

ETS-2 (Kiku 2)
Launched: 23 February 1977 by N1 from Tanegashima
Mass: 130 kg
Comment: First Japanese GEO satellite, manoeuvred to 130° E. Shut down 14 December 1990.

ETS-3 (Kiku 4)
Launched: 3 September 1982 by the last N1 from Tanegashima
Orbit: 968 × 1,229 km, 45°
Mass: 385 kg
Comment: Designed for 1 year of testing 3-axis control, deployable solar paddles and ion thrusters.

ETS-4 (Kiku 3)
Launched: 11 February 1981 by N2 from Tanegashima
Mass: 638 kg
Orbit: 220 × 35,082 km, 28.5°
Comment: Shut down 24 December 1984.

ETS-5 (Kiku 5)
Launched: 27 August 1987 on 3-stage test flight by H1 from Tanegashima
Mass: 1,096 kg (550 kg on station)
Orbit: 150° east, GEO
Comment: Japan's first 3-axis stabilised satellite. L-band mobile communications package. Out of service October 1997 and deorbited

ETS-6
Launched: 28 August 1994 by H2 from Tanegashima
Mass: 3,800 kg at launch; 2,000 kg BOL
Comment: The first 2 t GEO satellite (3.8 t at launch) built and launched by Japan. The satellite provided a demonstration of ion propulsion, mobile communications and other technologies for future projects such as a Data Relay and Test Satellite system. Included the first laser communications demonstration. However, propulsion problem left it in low orbit, allowing only a truncated programme, completed 12 January 1996. Successful intersatellite communication with UARS and bi-directional contact by laser between JPL and ETS-6

ETS-7
Launch: With TRMM on 28 November 1997
Orbit: 550 km, 35°
Mass: 2,370 kg chaser, 403 kg target
Lifetime: 1.5 years. Out of service December 1999

ETS-8
Launch: 2003–2004 by H2A from Tanegashima
Locations: 135° E, 129° E or 146° E GEO
Design life: 3 years
Contractors: to be determined
Transponders: S-band for mobile-mobile connections
BSS mobile broadcasting of CD-quality radio
Configuration: to be determined, but with two 10 m-class deployable antenna
AOCS: to be determined
Power: to be determined.

COMETS
Launch: 21 February 1998 by H2 from Tanegashima
Location: planned 121° E geostationary
Design life: 3 years
Contractors: NEC (system integrator)
Transponders: S-band 2.0–2.1/2.2–2.3 GHz forward/return inter-orbit link, working through 3.6 m antenna (steerable ±10°). Redundant 50 W 20 MHz SSPA/3 W 10 MHz SSPA forward/return, maximum EIRP 45.0 dBW. Developed by NASDA.

Ka-band 23/26 GHz forward/return inter-orbit link (working through same 3.6 m antenna). Redundant 3 W 30 MHz SSPA/30 W 150 MHz TWTA forward/return, EIRP greater than 56.0 dBW. Developed by NASDA.

Ka-band 30/20 GHz + O-band 50/40 GHz for high-quality mobile communication via 2.0 m feeder link antenna. Developed by Communication Research Lab.

2 (plus 1 back-up, 2-for-1) Ka-band 21 GHz 200 W TWTA 120 MHz high-quality DBS, covering Japan in two beams (Tokyo area + Kyushu Island) via 2.3 m antenna. Developed by CRL/NASDA
Technical Engineering Data Acquisition: TEDA payload will monitor the space environment and its effect on the spacecraft with Single Event Upset Monitor, Dosimeter, Magnetometer, Potential Monitor, Solar Cell Monitor
Principal applications: development of technology for inter-orbit data relay and tracking services, mobile communications and high-definition DBS
Configuration: 2 t class, similar to ETS 6. Solar array span 33.2 m, bus 1.96 × 2.96 × 2.8 m
Mass: 3.9 t at launch, 2 t on-station BOL
AOCS: highly accurate ACS (<±0.05° pitch/roll, <±0.15° yaw) required to control large antenna pointing. MELCO's 20–30 mN ion thrusters (35 kg Xe fuel) paired on E/W panels, provide NSSK. GEO insertion and attitude/orbit control by UPS Unified Propulsion System derived from ETS 6's LAPS: single 1.7 kN bipropellant Apogee Kick Engine plus 4 × 50 N and redundant 8 × 0.86 N hydrazine catalytic thrusters
Power: twin solar wings 2.93 × 15.6 m (rolled up for apogee motor firings) provide 4.7 kW EOL; Ni/H_2 battery provides 100% eclipse protection.

OICETS
Launch: August/September 1998 by J1 from Tanegashima
Orbit: planned 550 km circular, 35°
Design life: 1 year mission planned
Contractors: NEC (system integrator)
Transponders: 2.04425/2.22000 GHz S-band forward/return inter-orbit links (COMETS, Artemis); EIRP 4.4 dBW, G/T -30.6 dB/K.
2.04425/2.22000 GHz S-band up/down link, EIRP 12.3 dBW, G/T -47.5 dB/K.
Laser communications demonstration: 8,010 Å beacon, 8,190 Å receive (2 Mbit/s), 8,470 Å transmit (50 Mbit/s) 200 mW (CW) AlGaAs diode; 26 cm diameter telescope
Principal applications: demonstration of optical inter-satellite communications
Configuration: 3-axis stabilised box-shaped 78 × 110 × 150 cm bus, 9.36 m span across twin solar wings, 2.5 m high from base to top of optical assembly. Coarse pointing assembly provides 0–120° elevation, ±190° azimuth range
Mass: about 550 kg in orbit
Power system: twin 3-panel solar wings spanning 9.36 m, 1.75 m wide; 13 Ah NiMH battery.

Japanese Launch Monitoring Test Series

Specifications

Ohsumi
Launched: 11 February 1970 by Lambda 4S-5 from Kagoshima
Mass: 24 kg
Orbit: 323 × 2,416 km, 31°, 113.1 min
Comment: Test satellite was 4th stage of Lambda launcher and the first Japanese satellite, launched from Kagoshima. Transmitted 17 h

Tansei ('Light Blue')
Launched: 16 February 1971 by M-4S-2 from Kagoshima
Mass: 63 kg
Orbit: 986 × 1,106 km, 29.7°, 106.1 min
Comment: Demonstrated launcher capability. Operated for full 10 days. Named after Tokyo University's school colour ('light blue').

Tansei 2
Launched: 16 February 1974 by M-3C-1 from Kagoshima
Mass: 56 kg
Orbit: 290 × 3,240 km, 31°, 121.7 min
Comment: Tested M-3C performance and satellite attitude control. Decayed 22 January 1983.

Tansei 3
Launched: 19 February 1977 by M-3H-1 from Kagoshima
Mass: 130 kg
Orbit: 790 × 3,810 km, 66°, 134.1 min
Comment: Tested launcher performance and satellite attitude control. Shut down March 1977.

Tansei 4
Launched: 17 February 1980 by M-3S-1 from Kagoshima
Orbit: 517 × 608 km, 38.7°, 95.9 min
Comment: Tested M-3S performance and future science satellite subsystems. Decayed 13 May 1983.

VEP (Myojo) Vehicle Evaluation Payload
Launched: 3 February 1994 on H2 number 1
Mass: 2.4 tonnes
Orbit: 449 × 36,261 km 28.6° GTO 641.95 min
Payload: Box-shaped 3 × 2 × 0.8 m bus above ETS-6 type apogee propulsion system. The tanks carried 1,470 kg water for practice tests of pressurisation and depressurisation. Battery-powered from 4, 35 Ah Ni/Cd batteries
Comment: Payload was used to monitor H2 and range performance. The satellite had no attitude control but attitude and position knowledge was derived within less than 15° from solar cell sensors. The satellite transmitted real-time acceleration, vibration and distortion data. Satellite operated for 100 h/10 orbits.

Japanese Spaceplane Test Series

Current Status

Financial constraints and revised programme objectives have reduced spaceplane design goals to requirements similar to those posed for NASA's X-38 CRV. Japan is keen to reinvigorate its spaceplane programme in the wake of the Shuttle Columbia disaster and the increasing commitment of China to humanspace flight and has begun an agency-wide analysis to examine options for possible development.

Background

NASDA, NAL and ISAS designed and are jointly researching technologies for aerospace vehicles. To review Japanese standing in the space transportation market, STA's Spaceplane Committee reported in 1987 that development of a manned space transportation system might be important but strongly recommended that unmanned cargo transports, based on existing technology, should be separated from manned craft, which require greater safety and reliability. As with all major policy decisions, the Space Advisory Committee reviewed the advice, and in mid-1993, released a report recommending development of a winged reusable vehicle launched from the H2.

A fully reusable spaceplane would take off and land horizontally, powered by air-breathing engines such as ramjets, Liquefied Air Cycle Engines (LACE) and scramjets. NASDA initiated Phase A studies of this HOPE unmanned spaceplane in 1986. Japanese government and industry spent some US$150 million in FY82-90 on air-breathing aerospace plane activities, and projected US$3.4 billion FY90-98 if HOPE, the prototype vehicle, moved off the drawing boards. The National Aerospace Laboratory is working with a 350 tonne baseline vehicle, capable of carrying a crew of four into a 500 km orbit. Landing mass would be 101 tonnes. Preliminary HOPE designs have six ramjets, six scramjets and three rocket engines, working with 80 tonnes liquid hydrogen (slush hydrogen is favoured) and 164 tonnes of liquid oxygen. Some economists claim the plane will cost Japan more than £20 billion. The first, uncooled, engine ran for 6 seconds in May 1994.

The OREX re-entry vehicle carried by H2 number 1 in February 1994 tested thermal protection and navigation as a preliminary step towards spaceplane re-entry. Then ALFLEX tested automatic approach and landing techniques by releasing test bodies from a helicopter in a test series. Japan conducted the series during 1996 at Woomera in Australia to demonstrate GN&C for the proposed HOPE Spaceplane's automatic landing. The first J1 in February 1996 carried the HYFLEX suborbital hypersonic flight experiment spaceplane model. The full-scale HOPE-X demonstrator was planned for launch by a single-stage H2A in 2000 on a suborbital flight, but that has since been deleted.

Thus far, Japan has expended £10.5 (NASDA) and £0.5 (NAL) billion in FY97 to permit Phase C to begin.

The HYFLEX was Japan's first hypersonic lifting body experience. HYFLEX's slender shape and flight profile was dictated by J1's fairing and test requirements. At the conclusion of the test, the vehicle should have had a parachute splashdown 1,300 km downrange for recovery 250 km northeast of the Ogasawara Islands. The chute deployed at 694 s and the flotation bag inflated, but HYFLEX sank in 6 km deep water after apparently breaking free of the bag. The thermal protection system could not be examined directly, but 12 of the 14 test objectives and the demonstration of the critical technologies for HOPE were achieved.

Following the successful experiments of OREX, HYFLEX and AFLEX in 1997, the SAC made the decision to proceed to Phase C. In 1997, STA proposed a pre-study of fully reusable launch vehicles and a spaceplane. NASDA started Phase C of HOPE-X in 1997 as a NASDA/NAL joint project, intending to provide servicing missions in the early 2000s to Japan's Experimental Module attached to the Space Station.

In parallel, ISAS is working on the HIMES HIghly Manoeuvrable Experimental Space unmanned suborbital delta-wing boost-glide vehicle, again as a technology demonstrator. The organisation plans to conduct a launch of the 14 tonne, 9.33 m wingspan 13.6 m long HIMES on a reusable LOX/LH_2 sounding rocket. ISAS began low-speed gliding tests from helicopters in 1986. In October 1987, it experimented with a 2 m long, 1.5 m wingspan 100 kg model released from 1 km and 2 km altitude over the Sea of Japan. ISAS further attempted a Mach 4 re-entry gliding test, but the 185 kg 2 m model and its solid rocket were lost 21 September 1988 at sea because of carrier balloon problems after ascent from Kagoshima.

ISAS and IHI are also pursuing development of the ATREX-500 air turbo-ramjet, which could power the flyback booster of a 2-stage spaceplane. The hydrogen expander ATREX would accelerate the vehicle to Mach 6 and carry it up to 30 km. ISAS used the 5.4 kN ATREX-500 1/4-scale version, 2.2 m long, to conduct 40 static sea-level runs during 1990–95 at ISAS' Noshiro Testing Centre. Efforts now centre on a 1.4 t flying test-bed launched from Japan and parachuting into the sea. A jet engine will take the FTB to Mach 0.5 from a horizontal rail for the 370 kg ATREX-500 with 190 kg of LH_2 to attain 25–35 km altitude 200 km downrange. The initial metal turbo machinery will achieve Mach 4.5, with Mach 6 targeted for the carbon-carbon composite design.

Specifications

OREX (Ryusei, 'Shooting Star') Orbital Re-entry Experiment
Launched: 3 February 1994 on H2 number 1
Cost: US$20 million
Vehicle*:* 3.4 m diameter 1.46 m high paraboloid vehicle re-entered at HOPE angle/speed
Mass: 865 kg launch, 761 kg re-entry.
Comment: Testing thermal protection and GPS navigation for HOPE spaceplane. Splashed down 460 km south of Christmas Island after 133 minutes 2 seconds. The recovery was not planned.

HYFLEX
Launched: 11 February 1996 with the HYFLEX hypersonic flight experiment model of HOPE by 2-stage J1
Mass: 1,040 kg
Test Description: 238 second flight at 110 km 3.9 km/s on a suborbital path to enter the atmosphere at Mach 14.4
Vehicle: Length 4.40 m, span 1.358 m, height 1.04 m, and planform area 4.27 m^2. Aluminium alloy primary structure, protected by ceramic tile underbody, flexible blankets on the upper body and carbon-carbon nose cap and elevons. Pitch and roll control during Mach 14.4 glide is performed by pair of elevons with six 40 N GN_2 thrusters for 3-axis exo- and endo-atmosphere control. Strapdown RLG and a C-band transponder for radar tracking perform autonomous navigation. T+P measured around body. 49° angle of attach for entry, 30–35° for later descent.

UPDATED

UME (Apricot) Series

NASDA, under the direction of the Communications Research Laboratory (Ministry of Posts and Telecommunications) launched UME-1 to act as an ionosphere sounding satellite. NASDA used the fourth N-2 launch vehicle to put the satellite into orbit. The Space Agency expected to monitor radio waves in the ionosphere and use the results to forecast transmission conditions for short-wave communications. About one month after it was launched UME-1 failed. The satellite used spin-stabilisation for control, and NASDA hoped it would have a lifetime with a targeted residual rate of over 70 per cent of one and a half years. It decayed from orbit and impacted the Earth on 23 February 1983.

NASDA expected to use Ume-2 as an improved back-up for UME-1 (ISS-1) and pressed it into service shortly after the loss of UME-1. Launched on 16 February 1978 it was placed in a 980 × 1,220 km orbit. It performed well during its originally scheduled mission, and Japan continued to gather data from the satellite after its official lifetime had ended.

Specifications

UME-1 (ISS 1)
Launched: 29 February 1976 by N1 from Tanegashima
Mass: 135 kg
Orbit: 994 × 1,013 km, 70°, 105 min
Comment: This ionospheric sounder experienced a power failure after 1 month but the back-up UME-2 compensated.

UME-2 (ISS 2)
Launched: 16 February 1978 by N1 from Tanegashima
Mass: 141 kg
Orbit: 980 × 1,220 km, 69°, 107.2 min
Comment: The Japanese planned to use this UME-1 back-up satellite to study ionospheric radio propagation. It was shut down 23 February 1983.

Russian Federation

Soviet Spaceplane Test Series

Current Status

Interest has been expressed by the Russian Space Agency in developing a reusable crew return vehicle in co-operation with ESA or an industry consortium. In the aftermath of the cancellation of the X-38, the decision to proceed with an Orbital Space Plane (OSP), the grounding of the Shuttle and the reliance on Russian Soyuz and Progress vehicles to nurse the ISS through this difficult period, renewed efforts to engage European countries in a joint endeavour have been proposed. Developments throughout 2003 have moved the ISS partners away from anticipating co-operation on a X-38 successor as NASA moves forward with its Orbital Space Plane project. This may be little more than a semi-ballistic capsule capable initially of serving as an ISS lifeboat and later as a crew transport vehicle carried into orbit by a commercial launcher derivative such as Delta 4 or Atlas 5 leaving no room for European or Russian ventures.

Recovery of Cosmos 1445 in March 1983. Contemporary speculation suggested it was a one-third model of a manned military vehicle
(Australian Department of Defence) 0516712

Pre-flight view of the BOR-4 shuttle testbed; illustration released 1989 0516713

Background

Currently all practical research into spaceplanes has been shelved due to severe financial constraints but the history of Russian/Soviet spaceplane is important for the possible development of winged recovery vehicles through international co-operation which may take place in the future.

The Soviet Union began investigating the potential of small reusable aerodynamic spaceplanes for short duration manned missions to low Earth orbit in the early 1960s. Efforts became serious with the inauguration of the Mikoyan Design Bureau's 50/50 Spiral project, headed by Gleb Lozino-Lozinsky, in 1965. Mikoyan proposed a small reusable spaceplane carried aloft by a hypersonic aircraft. In the original design, the plane had a total 140 tonne take-off mass and carried one man on the 10.3 tonne 8 m long, 7.4 m span spaceplane. The carrier dropped the plane at high altitude and around M5.5–6.0.

Early tests produced the BOR 2/3 one-third and one-half scale model prototypes which the Soviets tested in wind tunnels and on suborbital ballistic trials using the Cosmos-3. The first BOR launch occurred 15 July 1969 on Cosmos-3. It used a wooden BOR 1 model that burned up at 60–70 km. The Politburo cancelled Spiral in 1969, following this test, but then revived it in 1972, after NASA proposed the Space Shuttle.

Mikoyan continued with a full-scale spaceplane model, constructed for runway trials in May 1976 and equipped with a turbojet. Aviard Fastovets piloted the first test. Igor Volk numbered among the first BOR astronaut corps (although he never got to fly the vehicle), but Volk made the first Buran approach and landing tests and was expected to command the first manned orbital mission. A Tu-95 conducted the first drop test 27 October 1977, followed by a further five piloted tests by Petr Ostapenko in 1978. At that point, the Soviets terminated Spiral altogether. The test vehicle is now displayed at the Monino museum near Moscow.

Chelomei's bureau began designing a new spaceplane in 1975. It was a 25 tonne vehicle, which would be able to carry 4–5 tonnes into LEO atop Proton. A half-size mockup was built and the bureau presented its design to a government commission in 1980, claiming that development would require four years. However, the commission suspended work and accelerated Energia-Buran work.

The Soviet Union had begun the Buran Space Shuttle effort beginning in the 1970s, and it encompassed some of the same spaceplane technology work as a testbed for the other programme. As the technology matured, the Soviets conducted six suborbital flights of the BOR-5 one-eighth scale Buran between 4 July 1983 and 24 June 1988. Each test used an SL-8 Cosmos-3M launcher from Kapustin Yar. These tests verified the overall configuration.

Simultaneously, the Soviets also built a separate (military) spaceplane. Sources refer to this programme as the BOR-4 and indicate the Soviet Union adopted it to counteract the military applications of the US Space Shuttle. The 3.4 m long, 2.6 m span, 1.5 tonnes BOR-4 lifting body bore a strong resemblance to the 1960s' US DynaSoar. Subscale tests flew four orbital missions between 1982–84 of more than 2 h each, apparently using the same vehicle each time. (On 5 December 1980 and 20 October 1987, the Soviets also conducted suborbital tests.)

The Mikoyan full-scale spaceplane model was flown from a Tu-95 in 1977/78. Photograph released 1989
0516714

BOR-4 orbital test flights began in 1982 with Kosmos 1374, followed by the identical Kosmos 1445 (apparently the same vehicle), in March 1983. Both spaceplanes de-orbited and descended by parachute. Orion aircraft of the Royal Australian Air Force monitored these tests. When the Australians released the photos, the US Air Force made an unsuccessful attempt to classify them.

In the third test, a Soviet North Atlantic tracking ship monitored the retrofire of Kosmos 1517 (the test's international designation) instead of the usual tracking conducted from the main Soviet tracking station in the Crimea. A Soviet announcement reported a controlled descent and splashdown in the Black Sea. The smaller body of water indicated increased confidence and provided greater security from Western observation. The fourth subscale spaceplane test, Kosmos 1614, also re-entered into the Black Sea. By then it was believed to be one-third the actual size of the manned vehicle being developed. In 1987, the US Department of Defence suggested that full-scale production of the spaceplane might be already under way.

The Soviets reiterated their claim, in 1989, that this military craft, using the SL-16 Zenit medium-lift booster, was not under development. Western Intelligence services, nonetheless, suspected that a three-man 10 tonne design for quick-reaction military missions was being pursued. It was further claimed, in early 1991, that there was a manned programme, with cosmonauts selected in 1980 and 1983, and that it was ready to fly on Zenit in 1987. It later emerged that the 'Uragan' MiG craft, equipped with a kinetic weapon, was intended to intercept the US Shuttle when it operated out of Vandenberg. The Soviets cancelled this programme immediately when the US dropped the California site as a Shuttle spaceport. Programmatic inertia, however, meant that two Zenit launches in August 1987, of Cosmos 1871/1873, flew full-scale Uragan mockups.

NPO Molniya advertised BOR-4 in 1990 as available for foreign commercial flights to test re-entry materials and other hypersonic systems. A BOR-6 exists, designed to investigate radio transmission during re-entry, but remains grounded because of financial constraints.

The Soviets disclosed proposals in 1989 for several new spaceplane concepts involving air launches from the super-heavy lift An-225 or a dedicated hypersonic aircraft. Departing from the Antonov, a spaceplane accommodating two cosmonauts and 5 to 7 tonne payload could attain LEO, while the hypersonic platform approach could increase the payload to 25 tonnes.

Specifications

Bor-4 Test 1
Launched: 3 June 1982 by Kosmos-3M from Kapustin Yar
Orbit: 158 × 204 km, 50.7°
Comment: Recovered after 109 min 560 km south of the Cocos Islands in the Indian Ocean by a 7-ship Soviet task force.

Bor-4 Test 2
Launched: 15 March 1983 by Kosmos-3M from Kapustin Yar
Orbit: 158 × 204 km, 50.7°
Comment: Recovered 556 km south of the Cocos.

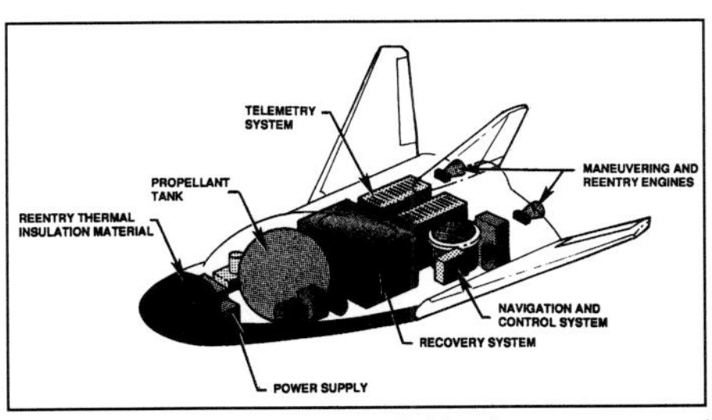

BOR-4 was claimed as available for re-entry and hypersonic trials (Teledyne Brown Engineering)
0516715

Bor-4 Test 3
Launched: 27 December 1983 by Kosmos-3M from Kapustin Yar
Orbit: 158 × 204 km, 50.7°
Comment: Recovered in the Black Sea.

Bor-4 Test 4
Launched: 19 December 1984 by Zenit from Tyuratam
Orbit: 158 × 204 km, 50.7°
Comment: Recovered in the Black Sea.

Spain

Intasat Satellite

Background

Spain, through its INTA space and science agency, developed the Intasat to act as an ionospheric sounding beacon with a two-year operational life. Spain planned to use the results of the test to develop information for later communications satellites. The spacecraft body was a 12-sided polyhedron.

Specifications

Intasat 1
Launched: 15 November 1974 piggyback with NOAA 4/Oscar 7 by Delta from Vandenberg AFB
Orbit: 1,442 × 1,462 km, 115°, 114.8 min
Mass: 24.5 kg

Sweden

Viking/Astrid-Freja Engineering Test Series

Background

The Swedish government funded spaceborne scientific investigations primarily to gain experience in system development/management. It intended to follow up Sweden's sounding rocket studies of the interaction of the solar wind and Earth's magnetosphere. These tests would expand the investigations to 2 Earth Radii (RE) to gain insight into the behaviour of the Aurora Borealis. Since Sweden did not plan to develop an indigenous launcher capability, the government sought a satellite that could ride aboard existing launch vehicles dedicated to other missions. After an initial design phase, the Viking vehicle resulted from the Swedish effort.

Sweden funded its effort through the Swedish Space Corporation. SSC awarded a prime contract to Saab-Scania in September 1980 to begin work on a small precursor satellite. The resulting small satellite platform became known as Mesa. It was based on the USAF Small Scientific Satellites programme and NASA's Atmospheric Explorers. Given the space available on the Ariane launcher at the time of the first design review, the satellite met the objective of fitting below the main payload on Ariane launches. Swedish technical organisations and universities provided the first payload consisting of electric/magnetic field experiments, particle and magnetospheric wave experiments, and two Canadian UV imaging cameras.

As the satellite neared its launch date of 22 February 1986, the Swedish government officially named it the Viking. All of the experiments on the satellite performed well, until it ceased operations as planned 12 May 1987 after 444 days (goal: 8 months). The only reason the satellite finally shut down was due to a degrading power supply. It cost the Swedes SKr238 million (1991 rate) for the satellite and its launch.

By September 1986, the Canadian imaging cameras had returned more than 20,000 images showing the complete auroral oval and previously unobtainable sequences of the full cycle of auroral activity. The cameras were recording sequential shots at 20 second intervals and global views every 80 seconds, for real-time transmission to Kiruna. They were then relayed by tape or telemetry to Canada's Centre for Space Science and the University of Calgary.

SSC followed the programme with the Freja-C ('Compact') platform, which is a scaled-down version of the original Freja. Astrid became the first application for the Freja bus. In keeping with a tradition started by the British with their satellite names drawn from Shakespeare, the Swedes named the satellite after author Astrid Lindgren and the instruments after her characters. SSC states the first Astrid launch cost SKr9 million, including launch. Three instruments successfully monitored magnetosphere neutral particles and electrons and imaged aurora in ultra-violet, but their power supply failed 1 March 1995. Ground controllers had the last Astrid contact on 27 September 1995.

The Freja version, which SSC now offers to clients, is 45 × 45 × 39 cm and weighs 20 kg. The bus can carry 10 kg payload and it has 0.3 mm aluminium facesheets glued to aluminium honeycomb core. SSC promises attitude determination within 2° by analogue Sun sensors, coupled with two ±60 μT two-axis fluxgate magnetometers. A 12 Am^2 magnetorquer gives precession control, along with a 7 Am^2 for spin. In operation, the bus spins up and down by 10–20 Ns HTPB solids space-qualified motors. Power comes from deployable Si solar panels that generate 80 W BOL. Users communicate with the satellite by a 5 W downlink, 512 kbit/s, 2.20–2.30 GHz S-band antenna.

Specifications

Viking
Launched: 22 February 1986 piggyback with Spot 1 by Ariane V16 from Kourou
Mass: 286 kg (535 kg launch)
Orbit: 811 × 13,536 km, 98.8°, 261.6 min
Size*:* 1.9 m diameter, height 50 cm

Astrid 1
Launched: 24 January 1995 by Cosmos 3 m from Plesetsk
Orbit: 965 × 1,026 km, 82.93°, 105 min
Comment: The first application of the Freja-C generic micro platform
Mass: 28 kg
Dimensions: 0.45 × 0.45 × 0.45 m
Power: 4 solar arrays; 28 V DC
Stabilisation: spin-stabilised
Design life: 3 months

Astrid 2
Contractors: prime Swedish Space Corporation on behalf of Swedish National Space Board
Mission life: 1 year
Mass: 30 kg
Orbit: planned 979 × 1,1013 km, 105.04°, 105 min
Cost: budgeted at SKr10 million, including development of S-band ground station
Mission: electric and magnetic field polar phenomena investigations, electron density studies and a continuation of Astrid 1's neutral/charged particle imaging (secondary)
Spacecraft: Astrid 1 mechanical design slightly modified for increased solar cell area to provide more than 55 W peak and deployable booms. Dimensions with stowed solar panels*:* 40 × 45 × 95 cm. Payload power 15 W continuously. Sun-pointing spin axis is under automatic attitude control. A star imager will probably be test flown giving accurate attitude knowledge. 128 kbit/s 5 W S-band downlink to a 1.8 m ground station antenna.

VERIFIED

United Kingdom

R Series/ X-Shakespearean Test Series

Background

The United Kingdom began its own satellite programme in the early 1960s shortly after the United States and the Soviet Union had orbited satellites. Principally the UK wanted to build an indigenous European launch and satellite support service with the same capabilities as the superpowers' programmes. As a first demonstration platform, it built the R series beginning with the R-0 which it planned to launch from its co-operative sounding rocket range in

Woomera Australia. In the first test of the R series, the UK planned to demonstrate stage 1/2 firings and a dummy stage 3 separation. The launch vehicle had to be destroyed at 64 seconds because it began tumbling, but R2 was successful as a demonstration of stage 1 and 2 engine firings, and a test spin up of the satellite. The second stage, as well as the payload, impacted in the Indian Ocean according to plan after 15 minutes, about 3,050 km northwest of Woomera. In the next test, R3, the British intended to use a live third stage and inject a satellite directly into orbit.

Originally intended as part of the Black Arrow launch vehicle programme, the British Parliament cancelled the X-series programme in July 1971. The controversy centred on whether the Europeans had the economic resources to build a launcher that was competitive in cost with the United States. Despite the cancellation, because some of the satellites had been built, the RAE continued with its launch programme. The X-4 (renamed Miranda in orbit, in the Shakespearean tradition of the UK series), was launched 9 March 1974 by Scout from Vandenberg. Despite minor control problems, good data were obtained. Total cost was more than US$8 million, plus US$2.35 million launch costs.

In anticipation of the project success, test officials named the mission the X-2 development payload, including a 13 kg Orba gold-plated drag sphere. As the test progressed, the second-stage shut down 15 seconds too early. In the wake of this failure, the British continued to fund the programme while Birmingham University prepared a new science payload to count micrometeoroids. The satellite had a one year design life. As the final preparations for the launch got under way the British government stopped the Black Arrow programme, but decided to let the one last test occur. The Royal Aircraft Establishment named it the Prospero.

Following a successful launch on 28 October 1971, it made 812 passes over the Woomera test range during a total of 4,960 orbits in the first year. The tape recorder replayed 697 times. It finally failed 24 May 1973 after 730 replays. RAE Lasham contacted the satellite annually until 1989, when the site's VHF system closed down. However, various stations reported continued reception of the 137.56 MHz beacon as late as February 1993. Unfortunately, Prospero's success did not come soon enough to save the Black Arrow programme, which was cancelled July 1971. The programme had sufficient funds available to continue with the X-4 Miranda satellite, which humiliatingly was launched by NASA Scout. Its intended launcher, R-4, is displayed at the Science Museum, London. The series of satellite projects cancelled along with this programme was one element in an industry-wide reappraisal regarding UK participation in global space endeavours. The UK government of the day led by Edward Heath chose to cancel independent participation in space projects, opting to rely on the United States for leadership in scientific space research.

Specifications

R-0

Launched: 27 June 1969 from Woomera, Australia on the Black Arrow.

R-1

Launched: 4 March 1970.

R-2

Launched: 2 September 1970.

R-3 (X-3)

Launched: 28 October 1971 from Woomera, Australia, by UK Black Arrow launcher
Mass: 66 kg
Description: 109 cm diameter
Comment: X-3 Prospero technology demonstration satellite
Orbit: 557 × 1,598 km, 82°.

X-4

Launched: 9 March 1974 by Scout
Mass: 93 kg
Orbit: 703 × 912 km, 98.81°
Comment: UK-built by Hawker Siddeley for DTI, its principal goal was to demonstrate 3-axis ACS to replace the spin-stabilisation of previous UK satellites.

STRV Series

Current Status

Two weeks after launch on 16 November 2000, STRV 1c and 1d suffered catastrophic failures when an electrical continuous drive signal severed receivers from the power supply.

Final preparations before STRV 1B was attached to Ariane's carrier (DRA Space Department) 0517103

Background

The DERA Space Department and JPL designed and built the STRV 1A/1B satellites. They are intended to test new technologies such as solar cells, microelectronics, electrostatic charging alleviation, structural materials and onboard data handling in a harsh environment. The Molniya-type orbit causes the satellites to traverse the van Allen belts four times daily over a year. This provides the approximate equivalent of a 10 year GEO radiation dose. STRV also provided advances in the use of carbon-PEEK composites, radiation-resistant microelectronics (SOS processors) and advanced systems architectures.

After launch, ground controllers lost contact with 1A temporarily during August 1994 because of a power system problem. The anomaly initially prevented full operation of the Xe plasma neutraliser, because of its high power requirement, but battery reconditioning overcame the problem and the tests of plasma discharges progressed satisfactorily. All other experiments have operated routinely but, on 1B, a hardware problem with the solar cell experiment has prevented operation in its primary mode.

With the nominal mission completed in June 1995, the accumulated radiation dose was several times lower than models had predicted. Combined with the high quality of data, this has allowed mission extension, with successful transfer of the day-to-day orbital maintenance of both satellites in September 1996 to the University of Colorado, under NASA funding. In the experiment, the satellite has not seen any transient system anomalies uniquely correlated with the radiation environment. Surprisingly dynamic behaviour in the van Allen belts, particularly the electron belt, has been observed, however.

The project definition phase of follow-on STRV 1C/1D satellites was completed in December 1996 with the paired GTO launch in November 2000 as Ariane-5 auxiliary passengers (see Current status).

Specifications

STRV 1A

Launch: 17 June 1994 Ariane V64 with Intelsat 702 from Kourou, French Guiana
Orbit: GSO at 38°W in 1994, 1°W in 1994–1996, 177° E in 1996–1999, 176° E in September 2001
Configuration: 40 × 50 × 50 cm box carrying no deployed structures. First space use of carbon-PEEK composite panels for main structural elements.
Mass: 50 kg 1A, payload 12 kg; 53 kg 1B payload 14 kg
AOCS: 5 rpm (provided by Ariane), axis 90 ±5° from Sun line (for power/thermal control). Sun/Earth sensors. Three systems, one nutation damper and magnetorquer (to control precession and spin axis during perigee) perform attitude control. Each has two separate windings of aluminium wire (for redundancy). Three Xe thrusters exist on 1A only, two for spin up and one for precession in case of magnetorquer failure. The system uses cold gas from electrostatic discharge experiment
Power: GaAs cells on X/Y faces to provide 30.5/32.5W 1A/1B BOL (normal to Sun), supported by 64 Ah Ni/Cd battery. Only 10 per cent array loss by May 1996
Communications: redundant 2 W transmit at 2.200–2.300 GHz at 1 kbit/s, 2.025–2.120 GHz receive at 125 bit/s. ESA/CCSDS packet telemetry/telecommand (custom chips) and ADA software. First space use of UK MIL-STD 1750/MAS 281 SOS 16-bit SOS µp set, 96/128 k word 1A/1B memories. Control room at DRA Farnborough working through 12 m antenna at DRA Lasham (passes up to 10 h). NASA's Deep Space Network routinely complements Lasham. Each satellite downlinks 60 kbyte experiment data daily
Payload: Charge alleviation package. Xe plasma emitted to discharge electrostatic buildup. First space test of the hollow cathode from the UK-10 ion thruster. Monitored by electrostatic charge detector (a new type of solid-state opto-electronic device to measure surface charge) and cold ion detector.

Atomic oxygen measurement package. Erosion rates of eight coatings (including carbon, Kevlar, Mylar, Kapton, FEP Teflon and SiO_2) by atomic oxygen around perigee. Measured by monitoring increasing resistance of a coated silver strip as it oxidises.

CREDO Cosmic Ray Effects and Dosimeter. Cosmic ray dosage at 5 minute intervals on 3 cm^2 300 µm diodes; total dose of all ionising radiation from RadFET dosimeters spread around STRV.

Battery Recharging Experiment. Temperature fluctuations to detect end of charge more accurately.

STRV 1B

Launch: 17 June 1994
Orbit: 284 × 35,899 km, 7.04°, 10.58 h GTO
Configuration: 40 × 50 × 50 cm box carrying no deployed structures. First space use of carbon-PEEK composite panels for main structural elements
Mass: 53 kg 1B payload 14 kg
Other Specifications: Same as STRV 1A
Payload: SCTE Solar Cell Technology Experiment: New types of cells cover glasses, coatings and manufacturing techniques to determine radiation resistance. Includes Si, high-efficiency Si, amorphous Si, GaAs, GaAs/Ge, ultrathin GaAs, InP, CIS and CdTe.

Vibration suppression: Piezo-electric elements to suppress vibration of Stirling cycle cooler cold finger.

RADMON: Total radiation dose plus proton belt profiles, 4 kbit SRAM devices, results compared with CREDO.

HIP Heterojunction Internal Photo-emission: Novel Infra-Red (IR) detector studying radiation effects by measuring electrical characteristics.

Neural Net. Radiation fault tolerance of analogue neural network VLSI devices.

REM Radiation Environment Monitor (ESA). Trapped electrons, protons and some heavy ions for CREDO comparison. During an astronaut EVA an identical REM was installed on Mir's exterior 9 September 1994.

STRV 1C/SRTV 1D

Launch: 16 November 2000 Ariane 5
Orbit: 615 × 39,269 km, 6.4°, 11 hr 48 min

UoSat Series

Current Status

A new SSTL 100kg enhanced microsatellite based on the heritage of the 50kg microsatellite has been launched to demonstrate new technologies for Enhanced Small Satellite users.

Background

Satellites built by University of Surrey (Surrey Satellite Technology Ltd). UoSAT-1 launched 6 October 1981 to a 372 × 374 km, 97.6° orbit, retired October 1989. UoSAT-2 launched 1 March 1984 by Thor-Delta from VAFB. UoSAT-3 launched 22 January 1990 by Ariane, retired late 1996. UoSAT-4 launched 22 January 1990 but retired 23 January 1990. UoSAT-5 was launched 17 July 1991 by Ariane.

UoSAT-12 was launched by Dnepr on 21 April 1999 from Baikonur. Surrey University built UoSat-12 as a modular multipurpose bus which will provide S-band communications and enhanced OBDH for 180 kg payloads. The first payload carries RWs, a 40 m resolution Earth camera and propulsion system for orbit control. The bus is known as Bus Alpha and it weighs 23 kg and measures 600×600×500 mm, accommodates four body-mounted GaAs solar panels providing 38W average power. A full 3-axis

Surrey University's series of successful 50 kg microsatellites will be joined in 1997 by the 300 kg UoSAT 12 minisat (Theo Pirard/Space Information Centre) 0517291

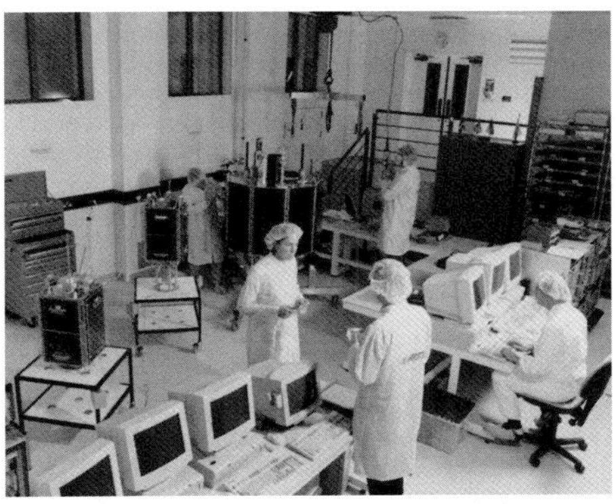

Surrey Space Centre's new mini satellite UoSat-12 was launched on 21 April 1999 on a Dnepr (SSC) 0054359

SSTI built UoSat-12 to demonstrate advanced leo communications and Earth observation payloads (SSTL) 0054360

altitude control system allows payload pointing accuracy better than 0.2°.

Bristol University emphasised simplification by making the satellite transmit-only and stripping the bus of attitude or orbit control. The 2 kg single main experiment can be varied for each satellite. On the first satellite a new type of battery cell will be tested. The other experiments examine GaAs solar cell degradation by exposing 14 experimental cells on the X dimensional face panel, radiation environment measurement, and monitoring the satellite's nutation and spin behaviour. The SSTL enhanced satellite can carry 5 CCD remote-sensing cameras providing 3-band multispectral imaging with 50 metres ground resolution and panchromatic imaging with 20 metres ground resolution.

United States

ASUSat Satellite

Current News

ASUSat was launched 27 January 2000 by Minotaur to a 750 km, 100°, 100.4 min orbit. The satellite failed shortly after launch when the solar panels failed to charge the battery.

Background

ASUSat 1 is the result of a challenge made to Arizona State University students in October 1993 by Orbital Sciences Corp's co-founder Scott Webster, to design and build a 4.5 kg nanosat to perform meaningful science. Total project cost is US$200,000 maximum and has involved 150 students in the Aerospace Research Center. The contractors are Arizona State University which is also building the control station, OSC which will donate the launch services on a Pegasus XL, Marshall Electronics building the cameras, Trimble Navigation providing a GPS receiver, Applied Solar Energy Corporation (GaAs cells), SAFT (batteries), Intel (80C188 processors).

The satellite has an expected design life of 2 years and it weighs only 4.5 kg. To support its primary objective of earth imaging it will be launched into a planned 550 km circular 97.6° Sun-synchronous orbit. In configuration it consists of a 14-sided prism, 24 cm high (dynamic envelope 26 cm height 31 cm diameter set by Pegasus avionics section), and has a structural weight of 1.1 kg. To reduce weight the student project uses carbon fibre in 954-2A epoxy resin, with a 0.8 mm wall. A gravity gradient boom and gravity anchor, which is a ball free to rotate in a fluid-filled spherical casing, and weighted to assume local vertical, provides attitude control by viscous drag to damp satellite. The orbit position determination comes from a commercial GPS receiver and spacecraft attitude (±10° error) from diodes (one Sun, three Earth) on each of 14 sides, supplemented by two Sun diodes on each end face.

Four VVL-1070 CMOS cameras (one side pointing, three Earth reference) provide 2 km/pixel resolution. Each orbit the satellite photographs up to eight 25-kbyte images as power permits. The Intel 80C188EC CPU has a 2 kbyte PROM, 256 kbyte EPROM and 1 Mbyte RAM. A GaAs solar array (30 18.5 per cent efficiency 2 × 2 cm cells on each of 14 sides plus 90 on top) powers the satellite along with two six-pack Ni/Cd switched batteries which provide 8.5–10 W. The satellite has an allocated amateur voice repeater with J-mode frequencies at 436.5 MHz for transmission, and 145.820 for receive. It is controlled from ASU on 2 m band.

ATS Series

Background

The Applications Technology Satellite series was conceived to demonstrate techniques for communications, meteorological and navigation satellites from geosynchronous altitude.

ATS 1 began operations in 1967 by providing the first transmissions of full-Earth, cloud-cover images from GEO, and the first real-time TV from the Pacific with coverage of Apollo 4's splashdown. In 1967, it provided emergency communications during Alaska's floods. In 1968, it demonstrated two-way links with commercial airliners, and in 1971 linked US/Soviet scientists during an atmospheric, sea and ice conditions experiment in the Bering Sea. Under the Peacesat project, ATS 1 transmitted educational, health, research, technology and community services to 23 autonomous terminals in Hawaii, Cook Islands, the Mariana and Caroline Islands, West and American Samoa, the Marshall Islands, Melanesia, New Zealand and Australia.

ATS 3 antenna pattern test at Hughes Aircraft in 1967. ATS 3 operations continue to this day 0516738

ATS-2 was launched on 6 April 1967 but left in a useless orbit due to an upper stage failure.

In 1973, ATS 1 pioneered transmission of electrocardiographs from Hawaii to New Zealand, and from Alaska to the University of Washington. During the flight of the Apollo 11 lunar landing in July 1969, ATS-1 served as the primary communications link between President Nixon and the White House as he watched the recovery operation in the Pacific. Then, during 1971–78, it carried 12-nation medical conferences over the Peacesat network. NASA announced in March 1985 that it had failed to respond to commands to correct its eastward drift from its 162° E Pacific position over the Gilbert Islands, and would drift out of a useful position within six months.

ATS 3 covers most of the Atlantic, East Pacific and US as far as Hawaii. It is supported equally by NASA/NSF and controlled from ATSOCC at the University of Miami as the node for data and telephone networks. Several Antarctica stations use ATS 3 as their primary link for voice, data and message traffic with the rest of the world. South Pole Antarctica uses it for about 5 h daily, while Palmer Station uses it 3 h and Seal Island 1 to 2 h. It also supports datalinks with research ships between 24° and 180° W and Arctic Ocean to Antarctica, and is still used about 5 hours a day by Peacesat for voice circuits to many Pacific basin islands.

The most notable of many ATS 3 achievements was the transmission of the first colour Earth imagery from space; it also obtained many cloud images and monitored severe storms. On 21 November 1967, it provided the first link over the Atlantic during a Pan American flight. ATS 3 then supported the Apollo Moon landings. In a series of maritime experiments covering ship location and ship-to-shore communications, it demonstrated that satellites could bring major improvements to the management of shipping fleets, presaging the Inmarsat system. At the end of 1974, ATS 3, as part of an NOAA programme, predicted and measured rainfall in remote areas by the comparison of satellite images of the brightness and size of clouds with radar density scans. During the 1980 Mount St Helens eruption, it provided links through a ground vehicle at the disaster site. In 1983, stationed at 105° W, it was being used for remote site communications by users of a two-suitcase terminal. Within 24 h of 1985's Mexican earthquake, ATS 3 had been brought into use to provide critical support for the international relief effort.

ATS 6 used its 9 m diameter high-power antenna to relay demonstration TV to small receivers, including a year of educational programming over India. Other ATS 6 achievements included the first aircraft-to-ship relay via satellite; the first direct flight control of an aircraft by an ocean air traffic controller using a satellite; and the first direction of a search/rescue operation, albeit simulated, by satellite.

Specifications

ATS 1

Launched: 7 December 1966 by Atlas Agena D from Cape Canaveral
Mass: 352 kg
Size: 1.42 m diameter, 1.34 m high
Power: 23,870 solar cells
Orbit: In GEO over the Pacific, 149° W 1968–1982; 170° E 1982–1985; now 167° E
Lifetime: 3 years planned life. Inoperable since 2 September 1969

ATS-2
Launched: 6 April 1967 by Atlas-Agena D but partial launch failure stranded the satellite in a wrong orbit.

ATS 3
Launched: 6 November 1967 by Atlas Agena D from Cape Canaveral
Mass: 365 kg after apogee motor burnout
Orbit: Initially at 86° W GEO over the Atlantic, it continues transmission from 105.4° W inclined at 14.9° providing communications with Antarctica.

ATS-4
Launched: 10 August 1968 by Atlas-Centaur but partial launch failure stranded the satellite in a wrong orbit.

ATS-5
Launched: 12 August 1969 by Atlas-Centaur to GSO. Malfunction of nutation damper. Out of service March 1984.

ATS 6
Launched: 30 May 1974
Mass: 1,402 kg.
Orbit: GSO at 95° W, decayed 30 June 1979.

LDEF Satellite

Background

NASA Langley's 9,705 kg Long Duration Exposure Facility (LDEF) was deployed by Shuttle STS-41-C 7 April 1984 and retrieved by shuttle STS-32 on 12 January 1990, very much later than the 12-months it had been intended to leave it in orbit. LDEF's 57 experiments with 10,000 samples covered materials, coatings and thermal systems, power and propulsion, science, electronics and optics. The satellite was a passively, free-flying unit with no central power or data systems or communications and command links. It was placed in a gravity gradient attitude upon release. A database by Langley's LDEF Material Special Investigation Group is maintained to be the most extensive open collection of information on materials behaviour in the space environment. It is being expanded to include material from other sources and is publicly available via Internet.

LDEF's emptied structure was placed in protected storage while the possibilities for further missions were clarified. It is now displayed at Kennedy Space Center.

SEDS Satellite (SEDSSat)

Background

SEDS (Small Expendable Deployer System) began in 1983 as a Small Business Innovative Research programme and, through April 1997, the programme had flown two tests at a cost of US$10 million. Both tests were incorporated into separate flights on Delta 2 launch vehicles as secondary payloads on USAF Navstar GPS flights. A Delta 2 launched 30 March 1993 carried NASA Marshall's 41 kg SEDS 1 to observe the dynamics of a long tether system. It unreeled the 23 kg instrumented end mass on a 0.75 mm diameter polyethylene tether to 20 km in 77 minutes, allowing it to swing like a pendulum for 16 minutes before being detached to burn up. Observations corroborated modelling of tether dynamics. It decayed on 30 March 1993.

The 41 kg SEDS 2 flew 10 March 1994 on a Delta 2 and this time remained attached to stage 2 on its 19.7 km tether. The 25.8 kg 20 × 33 × 41 cm payload included Langley's accelerometer, magnetometer and tensionometer to monitor its motion. Despite its slim diameter, the tether was visible to naked-eye ground observers. SEDS 2 decayed 8 May 1994.

GEODETIC

France

Stella Satellite

Background
Stella is almost identical to 1975's Starlette geodetic laser ranging satellite. In appearance it is a 35.74 kg icosahedron core of depleted U_{-238} alloy, surrounded by 20 AlMg alloy spherical segments each carrying three reflectors of 3.29 cm diameter entry pupil. The satellite spins at 5 rpm imparted by Ariane stage 3. CNES built the satellite as an in-house effort and CEA (the atomic energy authority) at Bruyères-le-Châtel provided the structure and Aerospatiale the reflectors. The satellite was released piggyback by Spot 3's Ariane. It is designed to be tracked by laser with a 1 cm accuracy. Stella continues to be used by a range of international agencies from many countries around the world in addition to the geodetic institutes and relevant university groups in France.

Specifications
Launched: 26 September 1993
Orbit: 718 × 805 km, 98.68°, 100.8 min
Mass: 47.978 kg
Size: 24.000 ±0.002 cm diameter.

Stella joined the network of laser geodetic satellites in 1993 (CNES) 0516731

Germany

GFZ Series

Background
The GeoForschungsZentrum (GFZ, National Centre of Earth Sciences) in Potsdam signed a contract 31 May 1994 with Kayser-Threde to build a laser-reflecting geodetic satellite. Various government sources estimate the project cost DM1 million. GFZ's orbit is the lowest of any laser geodetic satellite yet, improving the modelling of Earth's gravitational field. The National Research Institute for Geology operates a laser station in Potsdam and a joint facility in Santiago de Cuba with the Cuban Centro Nacional de Investigaciones Sismologicas. NII KP provided the 60 laser reflectors. Some 25 observatories around the world execute measurements by laser reflectors. Launched on 9 April 1995, it decayed back down through the earth's atmosphere on 23 June 1999. Some efforts were made to continue the programme by funding a successor capable of sustained operations but no support was forthcoming.

GFZ 1 was released from Mir in April 1995 (Kayser-Threde) 0517141

Specifications
Launched: 9 April 1995 aboard Progress-M 27 to Russia's Mir and ejected from the station 19.12 GMT 19 April
Mass: 20 kg
Size: 21.5 cm diameter
Orbit: 384 × 394 km 51.65°, 92.3 min

Japan

EGS (AJISAI) Satellite

NASDA and Kawasaki Heavy Industries developed the Experimental Geodetic Satellite, (EGS or Ajisai, which means Hydrangea). EGS is a simple 685 kg, 2.15 m sphere carrying 318 mirrors and

EGS, carrying 120 laser reflector assemblies and 318 mirrors, is a passive geodetic satellite. The time-lapse exposure was recorded at Tsukuba Space Centre soon after launch (NASDA) 0517136

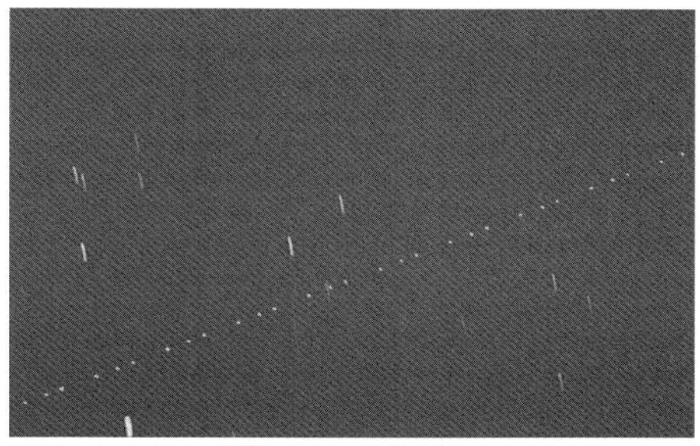

EGS, carrying 120 laser reflector assemblies and 318 mirrors, is a passive geodetic satellite. The time-lapse exposure was recorded at Tsukuba Space Centre soon after launch (NASDA) 0517135

120 laser reflector assemblies (1,436 corner cube reflectors) for precise tracking to provide Earth crustal movement and other data. Stage 2 also carried NAL's 295 kg MABES Magnetic Bearing Flywheel Experimental Satellite (although it was not designed for release) to test a two-axis actively controlled flywheel. A 2-D CCD star sensor and a linear CCD Sun sensor measure the attitude. The battery provided a three day life for the satellite.

Specifications
Launched: 13 August 1986 on inaugural flight of NI
Orbit: 1,483 × 1,497 km, 49.98, 115.7 min
Mass: 685 kg
Diameter: 2.15 m

United States

LAGEOS Series

Background
NASA designed the Laser Geodynamics Satellite as a tool to aid geodetic research, including earthquake prediction. It was NASA's first effort devoted wholly to laser ranging. With a high orbit and resulting long life, LAGEOS 1 acts as a permanent reference point to measure distances between ground stations. From this information scientists derive knowledge of the Earth's crustal motion, Earth's rotational variations, solid Earth and ocean tides and movement of the polar axis. By paying particular attention to earthquake areas such as the San Andreas Fault in California, LAGEOS may ultimately provide information on impending earthquakes. Station co-ordinates are now determined to within 1 to 3 cm and relative motions of 5 to10 mm per year can be measured. NASA Goddard, through its Laboratory for Terrestrial Physics manages the programme.

LAGEOS allows researchers to determine position by bouncing LASER light off of its 426 3.8 cm diameter corner cube reflectors. The four germanium and 422 fused silica reflectors return laser beams to their source, regardless of the incident angle. To be effective, the satellite had to accommodate a large number of reflectors but at the same time be small enough to minimise solar radiation pressure and atmospheric drag effects. By the same arguments, NASA built the satellite from a combination of aluminium and brass. A pure aluminium construction would have been too light and brass too heavy, but their combination yielded a sufficiently high mass to surface area ratio to meet the objectives. NASA had to consider the interaction between the satellite and Earth's magnetic field in its selection of construction materials as well.

The LAGEOS carries a symbolic mission in addition to its scientific ones. Each end of the bolt connecting the hemispheres carries a copy of a

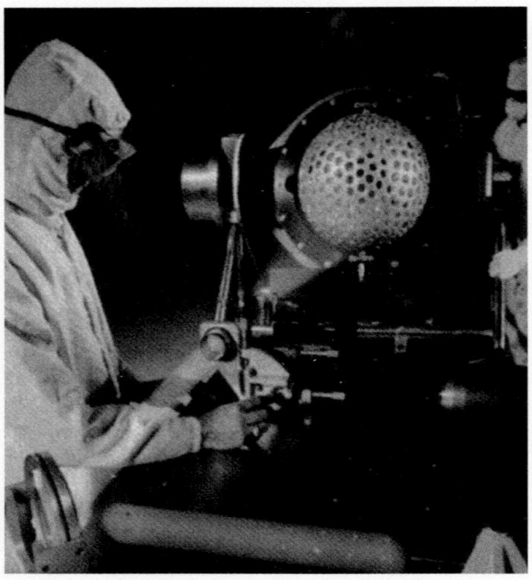

Lageos carries 426 laser reflectors embedded in its aluminium shell (NASA) 0516749

message prepared by Dr Carl Sagan. It includes three maps of Earth's surface. One shows the period 225 million years ago when, it is believed, the landmasses were one super-continent (sometimes called 'Pangaea'). The second shows the position of the continents as they are now, and the third their estimated position 8.4 million years from now when the satellite, so solid it will survive re-entry, falls back to Earth.

LAGEOS 1's first four years were devoted to determining its precise orbit and establishing a global network of 14 permanent Satellite Laser Ranging (SLR) stations supported by mobile truck-mounted SLR systems. By accurately measuring the round trip time for a laser pulse, the position of the laser system could be determined to within about 10 cm. LAGEOS 2 improved this accuracy by a factor of 2.

As part of its Geodynamics Program, NASA in 1979 initiated the international Crustal Dynamics Project (CDP) to measure contemporary tectonic plate velocities, regional crustal deformation and various Earth orientation and rotation parameters. CDP was superseded in 1992 by the DOSE Dynamics Of Solid Earth programme. During the 1980s, 36 permanent SLR stations were operated by 18 countries (US, Mexico, France, UK, Germany, Italy, Netherlands, Poland, Czechoslovakia, Austria, Australia, USSR, Egypt, Israel, China, Japan, Chile, Peru), making repeated measurements with LAGEOS and providing their data to the CDP. NASA currently has 10 stations. Three sites (Texas, Hawaii and France) also range to reflectors on the Moon placed by Apollo 11/14/15 and the unmanned Lunokhod 1/2.

Typical velocities between tectonic plates vary over 1 to 20 cm per year, and the contemporary plate motions measured by space geodetic techniques are in good general agreement with the long-term average motion indicated by the geological record. Models derived from LAGEOS provided the first evidence of gravity field temporal variations, believed to result from continuing crustal relaxation following the last ice age.

Originally due for Shuttle launch, with the IUS upper stage in 1987, Italy's identical LAGEOS 2 was released by STS-52 23 October 1992. The altitude is the same as LAGEOS 1's but the 52° inclination improves coverage of the seismically active Mediterranean Basin and should enable investigators to identify the source of irregularities noted in LAGEOS 1's orbit. Some 16 mobile laser ranging facilities supplement the 10 fixed stations around the world that track both satellites. A complementary LAGEOS 3 is under consideration.

Specifications

LAGEOS 1

Launched: 4 May 1976 by Delta 2913 from Vandenberg
Orbit: 5,858 × 5,958 km 109.8°, 225.4 min
Mass: 405 kg
Lifetime: Operational life of 50 years. Orbital life of 8 million years
Size: 60 cm diameter, 407 kg
Description: Two aluminium hemispherical shells, totalling 117 kg, surrounding a 31.76 cm diameter 27.5 cm long cylindrical brass core
Payload: 426 3.8 cm diameter laser reflectors for ground stations to make precise distance measurements

LAGEOS 2

Released: From the Shuttle STS-52 on 23 October 1992
Orbit: 5,610 × 5,950 km, 52.84°, 222.5 min
Other specifications: Same as LAGEOS 1.

MICROGRAVITY AND MATERIALS SCIENCE

China

FSW μGravity (Jianbing) Series

Current Status

China refers to its recoverable orbital payloads as Jianbing (Progress). China became the third nation after the US and the USSR to recover an object it placed in space when the FSW-0 1 test reconnaissance platform returned to Earth in December 1975.

Development of China's recoverable system began in 1966, but the first eight launches, all made by CZ-2C from Jiuquan into 57 70° orbits carried a mix of Earth resources and reconnaissance imaging payloads. But then China used the same configuration for a μg payload the first time in August 1987. The payload itself consisted of a 15 kg demonstration package from France's Matra. China followed this flight with the FSW-1 design later in 1987. This design had the same external configuration but a completely reworked internal cargo space to improve capacity, duration and control. The FSW-2 debuted in August 1992 aboard the maiden CZ-2D, adding a pressurized equipment module and again improving the avionics. FSW-1 has retired. As of May 1999, the last FSW launch had been that of FSW-2 3 on 20 October 1996. A failed launch on 20 April 1997 was reported by Chinese news sources. FSW-3 1 was launched from the Xichang Space Centre on 3 November 2003 and the recovery capsule returned to earth on 18 December 2003.

Background

Overall, FSW-1 was 3.144 m high and 2.20 m across the base with an additional 200 kg of mass over the FSW-0's 1,800 kg. Chinese engineers deactivated the attitude control in orbit. The 2.4–3.1 tonne FSW-2 model can return up to 350 kg on dedicated missions. It comprises essentially an FSW-1 with added 1.50 m-high pressurized base equipment module in an overall length of 4.644 m and a diameter of 2.200 m. The later version improves on FSW-1's attitude control and avionics by using a triply redundant computer. It can also adjust its orbit using a hydrazine mono system. The Chinese noted in late 1989 that they plan to build a new spacecraft employing a lifting configuration for improved re-entry loads and landing accuracy.

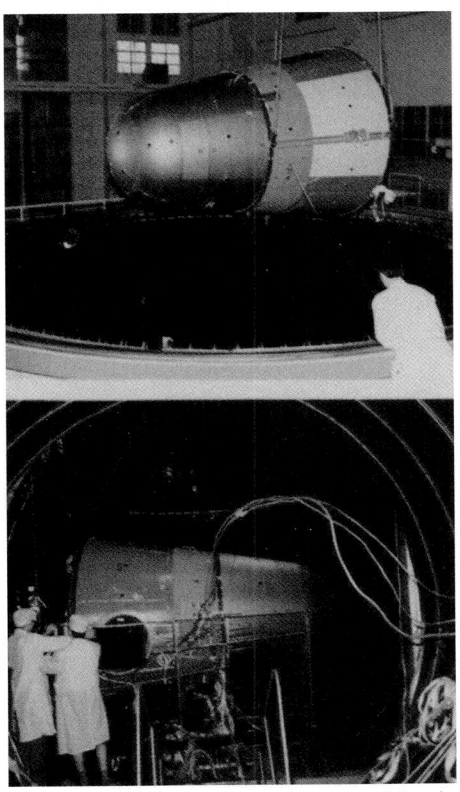

Fifteen FSW capsules had been recovered by the beginning of 1996 0517104

Chinese recoverable missions

	Launch	Days	Payload
1	5 November 1974	Failed to orbit	
1	26 November 1975	3	FSW-0 1 imaging
2	7 December 1976	3	FSW-0 2 imaging
3	26 January 1978	3	FSW-0 3 imaging
4	9 September 1982	5	FSW-0 4 imaging
5	19 August 1983	5	FSW-0 5 imaging
6	12 September 1984	5	FSW-0 6 imaging
7	21 October 1985	5	FSW-0 7 imaging
8	6 October 1986	5	FSW-0 8 imaging
9	5 August 1987	5	FSW-0 9 μg/imaging
10	9 September 1987	8	FSW-1 1 μg/imaging
11	5 August 1988	8	FSW-1 2 μg/imaging
12	5 October 1990	8	FSW-1 3 μg/imaging
13	9 August 1992	16	FSW-2 1 μg/imaging
14	6 October 1992	7	FSW-1 4 μg/imaging/Freja
15*	8 October 1993	–	FSW-1 5 μg/imaging
16	3 July 1994	15	FSW-2 2 μg/imaging
17	20 October 1996	15	FSW-2 3 μg/imaging
–	20 April 1997	Failed to orbit	FSW-2

* indicates failure. All FSW-1 launched by CZ-2C (FSW-2 by CZ-2D) from Jiuquan. 100 per cent recovery rate until October 1993. The CZ-2A failed test launch of 5 November 1974 is believed to have carried an FSW. The long gap after 1978's test was used to increase orbital operations to 5 days, reduce structural mass and improve control. FSW-2 in 1992 was the first to manoeuvre: its adjustments every three days for the first nine days probably indicate Earth imaging; the manoeuvring system then failed and it was recovered 1 day late. First known FSW recovery failure was October 1993. Capsule was to be recovered 16 October but apparent attitude problem resulted in retro raising the orbit; decayed 12 March 1996 over South Atlantic.

The satellite initiates its re-entry over China by an unspecified solid-propellant motor housed in a section immediately below the capsule. The Ian Satellite Control Center at Ian City in Shane province transmits the de-orbit command through two mobile stations, while the third located the re-entry modules during descent. To deorbit, the satellite must generate 650 m/s if it uses the nose forward re-entry position, but only 150 m/s required for against-orbit returns. The higher velocity puts the satellite immediately down inside China. The dramatic descent results from China's lack of space partners to spread a tracking network outside of Hina's borders. Consequently the re-entry must followed immediately over China after the deorbit signal is sent. This manoeuvre produces a 20 *g* re-entry load, working against the goal of a microgravity environment.

The Chinese built their heat shielding from oak wood using a blunt planform to slow the capsule. Then a parachute system takes over to reduce the speed to a terminal velocity of 10 to 15 km/h. FSW's can even survive a water landing since the landing site is a central area of the Sichuan Basin, where there are numerous rivers and lakes. A transponder and two beacons identify the position of the returned craft and aid recovery operations.

Some 275 μg experiments were returned on the six flights between 1987 and 1992, including guinea pigs in October 1990 and several GaAs, HgCdTe and protein crystal growth experiments. The Matra mission landed in central Sichuan Province in southwest China and the Chinese returned the sealed package to the owners 2 days after it was retrieved. August 1988's mission carried Intospace's 20 kg Cosima package of 104 protein crystal growth ampoules. All performance characteristics of the mission followed the flight plan, but large numbers of crystals became damaged during re-entry and landing. October 1992's mission included Japan's 710°C semiconductor furnace.

Chinese sources quote a price of US$45,000–50,000/kg for the small packages, and say the dedicated mission costs are negotiable. China can launch 2 to 3 missions a year, on a 12 month lead time for small packages and 24 months for dedicated missions.

Specifications

Mass: FSW-1 payload: recoverable 20 kg in two 10 kg, 200 × 200 × 250 mm packages; non-recoverable 50 kg in discarded unit. On dedicated missions total of 380 kg permissible, with 150–180 kg recoverable

FSW-2 payload: recoverable 150 kg in 200 × 200 × 250 mm volume; non-recoverable 200 kg in discarded unit. On dedicated missions total of 750 kg possible, with 310–400 kg recoverable

Microgravity level: 10^{-4}–10^{-5} *g*, duration unknown. Micro-accelerometer measurements during 1994's FSW-2 revealed 4–28 μ*g* during quiet periods; atmospheric drag reached 7 μ*g* during daytime perigee

Flight duration: 7–10 days for FSW-1; 15–18 days for FSW-2

Power supply: 1,300 Ah 27 V silver zinc batteries. CAST claims that all power requirements can be satisfied

Ambient internal conditions: ≤50°C, 10^{-3}–10^{6} torr

Experiment control and telemetry: unspecified but multiple programmed and ground commands available. Telemetry can be transmitted real-time or stored for subsequent dumping according to user requirements. Self-contained experiments yield lower costs

Attitude control: 3-axis but assumed de-activated during μg operations. Employs redundant digital computer (double redundancy on FSW-1; triply on FSW-2), inertial measurement unit, Sun/Earth sensors and cold gas thrusters. FSW-2 carries three-1 DOF fluid floated rate integrating gyros, two IR conical scanning Earth sensors, 120 × 128° FOV digital Sun sensors. Control accuracy: FSW-0 1° pitch/roll, 2.2° yaw; FSW-1 0.7°, 1°; FSW-2 0.5°,

The FSW-1 descent capsule, mass about 1.2 tonnes, incorporates oak for re-entry protection (CAST) 0517066

0.7°, angular rate to within ±0.02°/s for fine Earth pointing, 3-axes to ±3°, to within ±0.3°/s for maneuver mode
Acceleration levels: payloads must be tested for 11 *g* longitudinal acceleration, 2 *g* lateral during launch, with 20 *g* for re-entry
Vibration/shock: payloads must be tested to 5–10 *g* sine wave and 70 *g*/6–10 ms shocks for launch; and 2.5 *g* sine wave, 70 *g*/6–8 ms for descent. Landing speed is typically 14 m/s
Acoustic loads: payloads must be tested to 150 dB for 120 s for launch.

Germany

Express Satellite

Background

Germany used Japan's last M-3SII booster to launch its microgravity research satellite the Express from Kagoshima on 15 January 1995 to a 110 × 250 km, 33° orbit. The payload consisted of the 765 kg Express 1, including its 405 kg reusable capsule with 128 kg of µg/technology experiments. Japan provided 80 per cent of the US$93.5 million funding. Originally Germany wanted to recover the satellite on the ground at Woomera, Australia after 5.5 days, but a launch vehicle problem resulted in an approximate orbit of 110 × 250 km, 31° instead of the planned 210 × 398 km, resulting in re-entry after 2.5 orbits. Incorrect orbit caused by second stage altitude control oscillations exhausting altitude control propellant. When the satellite lost contact with the ground after 2 orbits, controllers believed it had been lost in the Pacific. Instead it descended safely in a remote region near Gushiago in northeast Ghana. DARA did not respond to local press reports of an unknown capsule with Russian labeling until November 1995. During this period the satellite lay untouched at Tamale airport until it was returned to Germany on 20 February 1996.

SPAS Series

Background

Developed as a commercial venture under BMFT sponsorship, SPAS has a modular experiment pod built with a carbon fibre tubular structure connected by titanium elements. Subsystems and experiment modules sit within a standard 0.7 × 0.7 × 0.7 m cube assembled according to mission requirements. Experimenters may configure the basic structure in a variety of ways within a total spacecraft budget of 1,500 to 3,000 kg. The original SPAS-01 version was 1,500 kg, with 900 kg dedicated to payload. The 3,600 kg Astro-SPAS version contained the four approved Orfeus/Crista missions holding telescopes up to 3.9 m long, 1.2 m diameter with improved attitude control system for a few arcsec pointing accuracy.

Shuttle Mission STS-7 released the Shuttle Pallet Satellite in June 1983 to make the platform the first retrievable satellite. NASA reflew the SPAS aboard STS-41B in February 1984 and STS-39 in April 1991, SPAS flew again in September 1993 by Shuttle mission STS-51 and in November 1994 on STS-66. After this successful series of tests SDIO (now BMDO) bought the SPAS from ESA for US$16 million. SDIO wanted to fly it as SPAS 3 on STS-64 in 1994, but Congress cut funding in FY94. On 7 August 1997 Shuttle Discovery launched the CRISTA payload with Astro-Spas experiments. CRISTA-SPAS-2 was the second flight on a Shuttle and this was the fourth co-operative venture between DARA and NASA. CRISTA-SPAS was deployed 7 August and recaptured 16 August. Unlike Eureca, SPAS cannot be left in space to be picked up by a subsequent Shuttle. It is therefore constrained to Shuttle-duration missions. Several attempts have been made to reinvigorate the programme by attracting commercial interest. Few companies have shown interest.

Specifications

Mission: 6 (1983–1997)
Mass: 1,500 kg total, 900 kg payload
Microgravity levels: same as Eureca
Flight duration: typically 4–5 days
Power supply: 28 VDC direct to Shuttle, two AgZn batteries of 70 Ah each in free-flight
Telemetry/telecommand: two S-band links to orbiter (2.0603 GHz transmit, 2.2375 GHz receive)
Attitude control: 3-axis rate integrating gyro package, 12 N_2 cold gas 1 N thrusters drawing on four 10 l tanks
Thermal control: passive
Data processing/control: DASA's Modular Digital Universal System
Size: 4.8 m wide, 4.4 m high and 1.5 m deep.

International

Eureca Satellite

Background

Eureca or, the European Retrievable Carrier is a Spacelab follow-on program initiated by ESA in April 1982. It is the first spacecraft designed specifically as a µg carrier and extends the capabilities of SPAS. Eureca 1 and its 15 experiments entered orbit on 31 July 1992 and was two days later released from the STS-46. It attained a 508 km circular orbit after a week and began full operations 18 August. Eureca carried the first inter-orbit communications conducted with ESA's Olympus. After a highly successful mission, STS-57 recovered it in June 1993. Eureca is designed for five missions, each of six to nine months duration spread over a 10 year period. At about 4.4 tonne total mass, it incorporates a hydrazine propulsion system to take it to its 28.5°, 510 km operational altitude and then return it to 315 km for Shuttle recovery. The carrier is stored in Bremen, its future uncertain. ESA's October 1995 ministerial meeting rejected further funding and a Daimler-led commercial consortium took over responsibility for a commercial flight between 1998 and 2000. Interest in possible re-use of this carrier has waned and Eureca is not now being considered for additional flights.

Specifications

Payload mass: 1,000 kg in 8.5 m^3
Microgravity level: worst-case 10^{-5} *g* <1 Hz, 10^{-3} *g* >100 Hz; 5×10^{-7} *g* if solar activity/air drag at minimum and payload close to centre of gravity
Availability: <1.5 year turnaround
Flight duration: 6 months operational, 3 months dormant
Power supply: 1 kW average available payload, 1.5 kW peak (solar array output 5 kW); 28 VDC
Thermal control: liquid freon loop provides 1 kW heat rejection; multilayer insulation
Data management: 256 kbit/s high, 2 kbit/s low, 128 Mbit memory capacity (1.5 kbit/s) for average payload; S-band downlink
Attitude pointing accuracy: 1° at 3σ
Orbit: typically 525 km, 28.5°.

Japan

Space Flyer Unit

Background

Japan launched the SFU-1 free-flying orbital platform on its third manufactured H2 on 18 March 1995. Then in a co-operative effort with the United States of America, Japan retrieved it on 13 January 1996 by the Space Shuttle STS-72. The H2 used three burns to achieve the 486 km 28.46° circular operational orbit by 23 March 1995. Japan planned to lower the orbit to 310 to 318 km for retrieval but thruster failures left it in 463 × 478 km. After rendezvous, NASDA's astronaut Koichi Wakata performed the retrieval. As he worked to bring the SUF back aboard the Shuttle, the solar wings would not latch down after retracting and had to be pyrotechnically jettisoned before the satellite bus and payloads could be safely installed for return to earth.

ISAS managed the project, in collaboration with NASDA and the USEF Institute for Unmanned Space Experiment Free Flyer. SFU-1 had a microgravity quality of 10^{-5} *g*. In 1991, ISAS sources indicated the project cost about US$430 million or £60 billion. ISAS wanted to fly at least five missions and anticipated integration with ISS participation but this project was not sustained and abandoned.

Specifications

Contractors: MELCO (integrator)
Mass: 4.0 t launch, payload 1 t, 3.56 t retrieval
Orbit: 322 × 336 km 28.45°
Configuration: based on a 4.46 m diameter, 2.8 m high octagonal bus of aluminium alloy truss structure supporting six 150 kg box-like payload modules. Carries one keel and four longeron trunnions for Shuttle interface
AOCS: 3-axis control of 1° accuracy; Sun-pointing maintained. GPS and IMU provide 50 m, 0.1 m/s accuracy. RCS by two redundant sets of 6 × 3 N and 2 × 23 N N_2H_4 thrusters, total capacity 100 kg; orbit adjust by two redundant sets of 4 × 23 N N_2H_4, total 650 kg capacity. Attitude control maintained by RWs and 3 N thrusters
TT&C: ISAS Sagamihara Operations Center working via Kagoshima Space Center. S-band 2.0844/2.2636 GHz up/down; housekeeping down at 1 kbit/s; data down at 16 kbit/s packet and 128 kbit/s playback of 80 Mbit and 4 Mbit bubble memory recorders. Includes still b/w TV
Power: two 2.40 m wide retractable Solar Array Paddles spanning 24.42 m deployed generate 2.7 kW BOL, including 850 W for the experiments. Each carries 28,000 2 × 4 cm Si cells. Supported by 4 × 19 Ah Ni/Cd batteries. SAPs retracted before Shuttle grappling, but latching failure required both to be pyrotechnically jettisoned.

Russian Federation

Nika-T Series

Background

KB Photon is developing the enhanced Nika-T for µg missions although there are funding difficulties. It is possibly derived from the Kuban Generation 6/7 photo reconnaissance satellite, which will also be modified as the Nika-K to replace Resurs-F and as Nika-B to replace Bion. Capable of carrying more than 2.5 tonnes of payload, 1,200 kg of it recoverable, on missions of up to 120 days, Nika-T will be launched by Zenit 2. Small satellites of up to 100 kg 0.1 m^3 can also be deployed. Nika retains the 2.214 m diameter Vostok descent sphere adds non-recoverable but pressurized forward/aft payload modules, two large solar arrays, a 16.5 m^2 radiator and an updated service/control module. Capsule access is provided by 100 and 170 cm diameter hatches. The Zona 8 and Konstanta 4 furnaces are under development for Nika. The concept has been raised several times as a precursor micro-gravity test platform capable of supporting the individual research experiments of several score users. International clients are sought.

Specifications

Flight duration: 90–120 days
Spacecraft mass: about 9 t
Length: 9.27 m
Max diameter: 2.72 m
Orbit: 300–400 km circular Sun-synchronous or 200–300 × 300–500 km elliptical, both at 62.8–70.4°
Payload mass/volume: 900–1,200 kg recoverable in 4.5 m^3 capsule, 1,500–2,600 kg in and around non-recoverable forward pressure container and service module.
Power supply: 4.5 kW daily average at 23–32 V provided to payload from 6 kW generated by two deployed solar panels (totalling 72 m^2).
Ambient internal conditions: 0.46–1.5 atmosphere, 15–25°C (increasing to 30°C for high power operations) in recovery sphere
Experiment control: up to 80 commands transmitted in a single pass, up to 75 commands provided from the onboard computer (maximum of 70 in a day)
Telemetry: 50 channels totalling 3,000 measurements/s for recovery module, 280 channels providing 20,000 measurements/s for non-recoverable forward module (which incorporates a 60 Mbit memory). Depending on data rate, 108 min or 432 min of data can be dumped in 5.1 min pass
AOCS: 3-axis to within 1.5°.

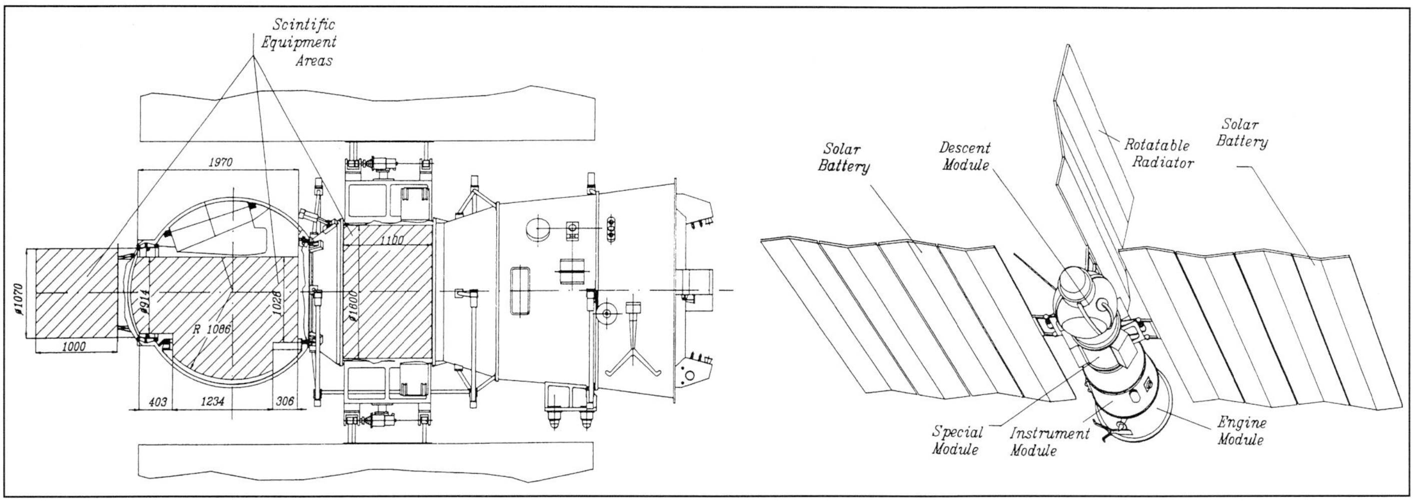

The Nika spacecraft is planned for μg missions of up to 120 days. The hatched areas designate pressurised payload accommodation 0517068

Photon Series

Current Status

The European Space Agency has signed an agreement for Photon missions M2 and M3 each to carry 660 kg of science payloads. The agreement binds the Russian Space Agency, ESA and two Russian partner companies (KBOM and TsKB-Progress) for three years from 2003–2006. The first Photon M1 was lost during a launch failure on 15 October 2002. A modified vehicle, the M series Photon had a mass of 6,425 kg and carried science experiments for ESA. Scheduled to fly in 2005 and 2006, the two missions will help reinstate ESA science experiments lost when the Shuttle Columbia was destroyed during re-entry on 1 February 2003.

Background

The Soviets began marketing space facilities and services to the international μg community in 1987, notably at that year's Paris Salon. The USSR first offered temporary positions aboard the Mir/Kvant orbital complex, with equipment delivered by the Soyuz manned or Progress unmanned ferries, and accompanied by a specialist if required. The Russians have offered the Photon capsule, essentially the mature Vostok craft, for dedicated bookings on 14 to 30 day missions. It is built by the Central Specialised Design Bureau in Samara. The spacecraft consists of a spherical descent module, a service module with solid propellant deorbit motor, and a separate battery pack. Russian and foreign researchers also place small μg payloads (typically 33 kg, 60 × 40 × 25 cm) aboard the Resurs-F film camera Earth observation satellites.

In the following list the first three launches are not given Photon numbers but some sources designate them as Photon 1–3 inclusive, thereby adding three numbers to the following.

Photon is based on the well-proven Vostok spacecraft, introduced in 1960 for the manned programme. In retrospect, it is clear that Photon-type materials processing missions were conducted by Kosmos 1645 (April 1985) and Kosmos 1744 (May 1986), followed by the acknowledged 14 day Kosmos 1841 of April 1987. The last's three primary payloads were the Zona 1 and Splav 2 electric furnaces and the Kashtan electrophoresis unit, all offered commercially.

The first named Photon materials processing spacecraft flew a 14-day mission during April 1988. Photon 2 followed on a 14-day mission beginning 27 April 1989. Its 405 kg payload included a CNES/Matra experiment investigating fluid behaviour. The Russians claimed the total mission cost was R6.5 million. Photon 3 in October 1991 carried CNES' Crocodile (Croissance de Cristaux Organiques par Diffusion Liquide dans l'Espace) equipment to grow NPP organic crystals for non-linear optics applications. The four cells achieved 12 days' crystal growth. CNES' 15 kg Sedex (Synthèse Enzymatique de Dextrane) flew on Photon 4 in October 1991 after the agency signed a 10 year agreement in June of 1991 with Glavcosmos and KB Photon for future Photon and Nika-T flights. Exercising the contract, CNES flew the Gezon experiment aboard Photon 6 in 1994 using Zona 4 to investigate the effects of magnetic fields on the growth of germanium crystals. Ibis flew on Photon 7 in February 1995 with a biology payload under a FFr8 million contract. It will refly on Photon 8.

Eight German experiments flew on Photon 4–6. Photon 5 included four German crystal growth experiments using Zona 4 (continuing the CdTe growth on Photon 4). Then ESA conducted the first Biopan experiment, which it mounted externally to expose micro-organisms to free space (although this test flight carried only technology experiments). DARA used the Photon for a C/SiC test sample at the capsule's stagnation point to determine if this non-ablative material had heatshield applications. Kayser-Threde developed the Biopan biological experiment carrier for the exterior of Photon and Resurs descent capsules under ESA contract. ESA approved two further flights; first was Photon 6 carrying six biology experiments and another C/SiC sample. The next was Photon 8, which carried Germany's 1 m diameter 150 kg MIRCA re-entry capsule, and ESA's Biobox.

Photon 7 performed seven more German experiments under a DM1 million contract. The experiment tested five samples of AlSi for directional solidification in a Zona 4M and two glass interdiffusion experiments in a Konstanta 2M. ESA's Biobox 2 investigated cells responsible for bone mineralisation, flying under an ECU700,000 contract. Unfortunately, Photon 7 was destroyed after a safe landing when it was dropped by its recovery helicopter (Photon 8 is its reflight).

Commercially, the mission of September 1989 carried Intospace's Cosima 2 package (Cosima stands for Crystallisation of Organic Substances in Microgravity for Applied Research), followed by Cosima 3 May 1990 and the first 33 kg Casimir unit September 1990. Germany's Kayser Threde signed an agreement with Glavcosmos in December 1987 for options on Photon and Resurs missions, acting as marketing agent. Four missions flew between 1989 and 1991, carrying Intospace's Cosima 2–4, Casimir and LZZ experiments.

A Soyuz booster launches Photons from Plesetsk into 220–250 km perigee, 300–400 km apogee, 62.8° orbits for missions of 14 to 30 days. So far only a two week capability has been demonstrated. The mass is 6.2 tonnes. Russian trackers determine the spacecraft position to within 17.57 km along orbital track, 0.63 km transverse and 0.95 km in altitude. The 2.21 m diameter return sphere masses around 2.4 tonnes and is probably constructed from aluminium; thermal protection was originally provided by a ≥5 cm thick layer of asbestos-type material embedded in a resin but the current technique is unknown. During the Photon 2 mission, it was acknowledged for the first time that a Soviet spacecraft was being reused: the descent sphere had already flown on C1841 and Photon 1. An attitude control system incorporating N2 jets and Earth horizon sensors align Photon in preparation for the rear section's retroburn of around 90 m/s over Africa, some 8,000 km downrange. Predecessor Vostok utilised a 15.83 kN TDU-1 liquid-propellant retro; the current liquid engine is unknown. The rear section provides attitude control although marketing specifications indicate this is not available for Photon, possibly because of the need for conservation of limited resources for the re-entry sequence following a lengthy mission. Pyrotechnics release the four tension restraining straps and an offset centre of gravity generates the only attitude control during entry at peak 16 *g*, creating a tendency to wallow that was noted by the early cosmonauts. The three

Photon integration with its carrier 0517248

Photon microgravity missions

	Launch	**Orbit**	**Life (days)**
C1645	16 April 1985	215 × 378 km,	62.8° 13
C1744	21 May 1986	218 × 371 km,	62.8° 14
C1841[1]	24 April 1987	218 × 381 km,	62.8° 14
Photon 1[1]	14 April 1988	217 × 375 km,	62.8° 14
Photon 2[1]	26 April 1989	216 × 380 km,	62.8° 15
Photon 3	11 April 1990	225 × 398 km,	62.8° 16
Photon 4	4 October 1991	216 × 396 km,	62.8° 16
Photon 5	8 October 1992	221 × 359 km,	62.8° 16
Photon 6	14 June 1994	221 × 364 km,	62.8° 17 ½
Photon 7[2]	16 February 1995	220 × 369 km,	62.8° 14 ½
Photon 8[3]	9 October 1997	218 × 375 km,	90.45° 13 ¾
Photon 9	7 September 1999	217 × 384 km,	62.8° 16.4

Notes: 1) reused descent capsule; subsequent reuse unknown. 2) capsule dropped by helicopter after safe landing; most of experiments destroyed or unusable. The Photon 1 listed here was the first named Photon launch and its successors were announced as 2–5. However, what should have been Photon 6 was announced as Photon 9 as a result of the three Cosmos predecessors being retrospectively referred to as Photon 1–3. The original designation system is employed here. 3) Photon 8 appears to be largely funded by foreign payloads: Biopan, Biobox, MIRCA, Crocodile 2, Ibis and Cosima 5.

The Photon materials processing return sphere was first displayed at June 1987's Paris Salon (Graham Turnill) 0517067

main 27 m diameter parachutes are deployed at 2½ km (following 4.8 m diameter main drogue), leading to a 7.5–8.5 m/s landing, cushioned by solid retros in the shrouds. Reserve chutes are 5.2 m diameter.

Specifications

User cost: ECU20,000/kg quoted in 1995
Availability: 12–18 months on signature of contract
Payload mass/volume: up to 500 kg in a volume of 4.7 m^3 internally; volume consists of two stacked cylinders 1.8 m dia/0.7 m high + 0.6 m dia/0.4 m high. There is also accommodation externally for =~0.06 m^3 non-recoverable
Microgravity level/duration: less than 10^{-5} *g* for unknown duration
Power supply: 400 W average daily; 700 W max for 90 min/day; 27 ± 4 VDC; 25 A steady, 50 A for 200 ms. Battery supply accommodated in front-mounted unit ejected before re-entry; known as the Nauka module on science missions, 1.8 m diameter, 47 cm deep cylindrical section capped by domed end and mounted on four legs
Processing units available: Splav, Zona and Konstanta electric furnaces, Kashtan electrophoresis unit. Specifications provided under separate entries. If these are to be employed, information on material composition, behaviour and safety, experiment sequencing, T gradients and cooling rates is required. If the equipment is user-provided, dimensions, energy requirements, sequencing and telemetry requirement information must be provided
Ambient internal conditions: 0–40°C; 0.46–1.5 atmosphere; 0.055 rad/day radiation background plus 50 rad/flight from solar flares. During testing, the inert spacecraft and contents are subjected to a pressure differential of ≤0.5 atmosphere by an air-5 per cent helium mixture in a 10^{-9} atmosphere chamber for 10 hours. Re-entry could generate internal temperatures up to 60°C; internal pressure drops to 0.053 atmosphere during parachute descent. External payloads must be able to withstand –150/120°C, 10^{-12} atm; 51.0 rad/day radiation background, 37 krad/flight from solar flares
Storage conditions: 5 to 30°C, 30 to 80 per cent relative humidity for 2½ years (maximum 35°C, 90 per cent humidity for 60 days/year). Prior to installation: –40/50°C, 20 to 80 per cent for 3 months at the launch site
Experiment control: 25 single ground commands, four preprogrammed commands from Photon's control system
Telemetry: 48 channels available. T measured (3.5 per cent error of measurement range) every 10.56 s with real-time transmission, or 42.24 s if stored; data transmission limited to 4 min/pass. The payload can command start/stop of the spacecraft's data storage system
Attitude control: none available (activated for retrofire sequence)
Payload mockups: customers must provide unit mockups for spacecraft fitting, accurate to within 0.5 kg, 5 mm for centre of masses and identical attachment points and means
Acceleration levels: 10 *g* for 600 s during launch; 0.1 *g* for 1,000 s during manoeuvres in-orbit, 0.1 *g*/1,000 s and 8 *g*/100 s during entry. Five 40 *g*/5–10 ms shocks for parachute deployment sequence, ended by a single 40 *g* /40–50 ms shock on landing
Vibration: unit resonance should occur ≥25 Hz. Payloads are tested for 9 *g*/5–10 ms shocks at ≤2 Hz, and 1 *g* vibration over 2–10/10–20 Hz for 2 min each, and 0.5 *g* for 4 min over 10–20 Hz. Acoustic loads are 140 dB for internal equipment and 150 dB for external, both for 120 s over 20–4,000 Hz.

United States

COMET (Meteor) Satellite

Background

As the Reagan Administration moved in the 1980s toward more commercial applications for space research, NASA chose to fund projects in other organisations. The Space Agency's Office of Advanced Concepts & Technology, explored the possibility of building a commercially viable re-entry module known as the COMET (Commercial Experiment Transporter) with both the Department of Transportation and commercial sources. NASA wanted the recoverable satellite to provide university and industry consortia with easy access to space. Plans were laid for a reusable, production-line carrier capable of multiple reuse. In this way it was hoped to stimulate potential customers for Space Station research attracting corporate participation. The concept was later marketed to private companies in an effort to encourage commercial use of orbital carriers for packaged science and manufacturing space industry projects.

Originally, COMET should have debuted in the summer 1994 carrying 11 experiments from Wallops Island into 460 km 40° for 30 days, but as the programme approached this deadline NASA chose in May 1994 to refuse additional funding unless the programme had a solid commercial basis. Nonetheless, NASA signed the launch contract 28 March 1995 with system owner EER Systems, who renamed it Meteor (Multiple Experiment Transporter to Earth Orbit & Return). The new flight plan called for the return capsule to carry four experiments, with eight more on the bus to continue operations for 1 to 2 years.

COMET's (Meteor's) mission ended after 45 seconds on 23 October 1995 when the maiden Conestoga that should have carried it into space self-destructed. NASA spent US$73 million on the programme. EER still takes bookings for Meteor 2, but few customers have shown an interest. A prospective customer has to pay about US$30 million to ensure a dedicated mission.

The CSTAR Center for Space Transportation & Applied Research at The University of Tennessee Space Institute headed the original three-flight programme. Westinghouse Electric Corporation provided the service module and systems engineering, and EER Systems supplied the Conestoga launcher. The launcher contract alone came to about US$45 million. Space Industries built the recovery capsule for about US$15 million, including orbital operations and payload integration.

After NASA lost interest, EER assumed responsibility for the programme in late 1994. Because NASA's FY94 budget only allocated US$14.5 million, the entire system still needed about US$6 million in other money to get to the launch pad. Commercial customers provided the remainder, obligating NASA to release the US$14.5 million by 28 March 1995. For its money, NASA received half of the capacity of Meteor.

Originally, NASA wanted to fly three COMETs, but the budget squeeze required the Agency to modify the three into a single mission costing US$65.8 million. The first budget projected that the three separate missions would cost more than US$158 million.

At mission's end, the RM spins up to 73 rpm, separates and fires its Thiokol Star 13A solid retro motor for re-entry. The drogue and main parachutes are complemented by an airbag inflating just before landing in the Atlantic. Westinghouse hoped to land the first test at Utah Test & Training Range, with subsequent missions ending at White Sands.

The concept and engineering design have been adapted to potential users interested in ISS participation but no commitments have been reported.

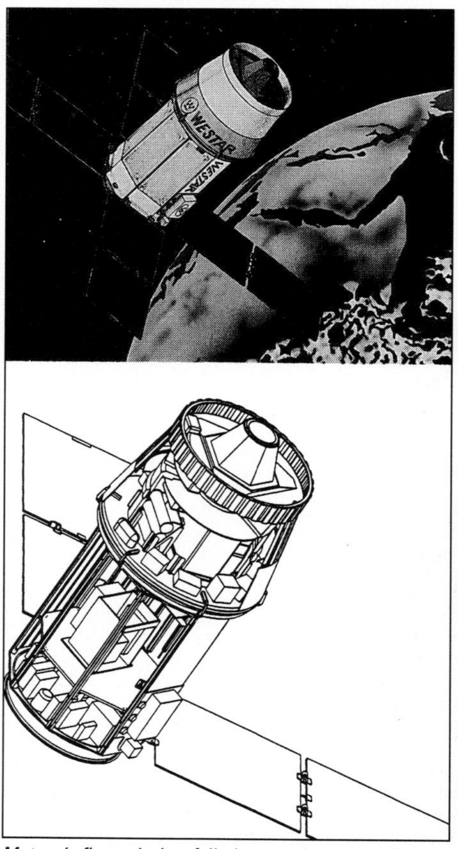

Meteor's first mission failed to reach orbit in October 1995 0517069

Specifications

Launcher: EER Systems' Conestoga 1620
Payload mass/volume: 136 kg/0.227 m^3 recoverable; 68 kg/0.426 m^3 non-recoverable
Payload access: samples can be loaded until 6 hours before launch
Microgravity level: better than 10^{-5} *g* (10^2 *g* during manoeuvres)
Mission duration: 30–100 days recoverable and 180 or more days non-recoverable
Power supply: 350 W continuous, 400 W peak for 200 h, 28±4 VDC, both recoverable/non-recoverable on-orbit from four 2-panel Si wings and Ni/Cd batteries (recovery capsule carries battery for descent, 5 W for 4 hours)
Ambient internal conditions: payload compartment pressurised to 1 atmosphere with dry air in recoverable capsule; unpressurised non-recoverable. Payload mounting plate maintained at 40 ± 2.8°C
Communications: 9.6 kbit/s command uplink (2.075 GHz), 250 kbit/s down (2.315 GHz). 24 Mbit storage. Both event-driven and time-tagged experiment commands can be stored for autonomous execution
Pointing accuracy: roll ±1°, pitch/yaw ±0.1°. Provided by SM cold gas thrusters + RWs
Acceleration levels: <12 *g* ascent, 8 *g* re-entry, 6 *g* parachute, <10 *g* landing.

Wake Shield Facility Series

Background

The Space Vacuum Epitaxy Center at the University of Houston built the Wake Shield Facility (WSF) under a grant from the NASA Office of Advanced Concepts & Technology and Space Industries Inc

SVEC's Wake Shield Facility was released in September 1995 0517070

built the satellite. NASA sought a free-flying satellite it could deploy and retrieve from the Shuttle in order to conduct super vacuum and microgravity studies. The Shuttle launched it the first time in February 1994 (STS-60).

WSF flies in the wake that the Shuttle creates in the solar wind and residual atmosphere to minimise the drag forces that this small amount of material produces in other satellites at that altitude. The wake also establishes a higher vacuum of about 10^{-14} torr known as an ultra vacuum. NASA hoped to investigate the commercial possibilities of producing new electronic, magnetic and superconducting thin-film materials and devices.

It expected to use the first 52-hour mission to grow seven 6 to 7 µm-thick GaAs wafers. However, a series of technical problems forced the crew to hold WSF on the Shuttle arm resulting in a vacuum of about 10^{-10} torr (about the best achievable on Earth).

WSF-02 did fly free during STS-69 in September 1995, achieving its primary objectives in five runs. WSF-03 flew aboard STS-80 and in free-flight in November 1996, growing 100 wafers. The WSF-3 flight achieved an extended duration of 180 hours. By NASA decision, the WSF-04 requires full commercial operation and industrial funding has not been forthcoming. Attempts were made to interest ISS partners and to draw in joint venture participation and fund sharing but no commitments were made. To date no further progress has been made to get the device into orbit.

Specifications

Size: 3.66 m diameter, 1.81 m high, 1,700 kg deployed
Power provision: 50 kWh by three AgZn batteries (WSF 01/02); 1 kW solar array (WSF-03/04)
Orbit: 305 km, 28.5°
AOCS: two MWs, three magnetorquers, conical Earth sensor, 3-axis magnetometer
Vacuum quality: 10^{-14} torr
Microgravity quality: 10^{-6} *g* for 45 hours
Payload: MBE/CBE equipment, mass spectrometers, retarding potential analyser, total pressure gauges, thermocouples, accelerometers and materials exposure experiments.

UPDATED

SCIENTIFIC

Argentina

MuSat (Victor) Satellite

Background

Known as MuSat as it was built, the Argentines changed the name to Victor 1 when the Russians launched it on 29 August 1996 by Molniya-M. The Centro de Investigaciones Aplicades developed it as a technology demonstration satellite. The launch occurred aboard a Russian Molniya-M from Plesetsk. Victor weighed 32 kg and entered a 62.8° orbit at 235–1,170 km. The space craft itself measured 33×33 cm and was 43 cm high. The small experimental payload was developed and supplied by the Institute Universitario Aeronautica de Cordoba in Argentina. The launch cost US$200,000 and the mission ended when the satellite decayed on 12 November 1999.

Argentina sought to develop space platform technology based upon low-cost engineering, to provide scientists and engineers with fast access to space for experiments. They expected the first experiments to include radio communications between Earth stations at a low cost, the demonstration of two CCD cameras to undertake colour photographs of Earth using low resolution (2 km/pixel) and high resolution (0.7 km/pixel) focal plane arrays. The image, sized 700×500 pixels, is immediately transmitted to Earth. The science agency also wanted to stimulate youth interest in space activities by promotion of participation by schools and universities. MuSat teams have suggested continued activity with the International Space Station.

SAC Series

Background

Argentina emphasises applications and space science, in its modest space programme. Until May 1991, the air force executed all space research projects. In that year, the CNIE National Commission for Space Research was transformed into the civil CONAE National Commission for Space Activities, under direct supervision of the country's President. CONAE's proposed its first strategic programme covering 1995–2006 and President Menem approved it 28 November 1994. CONAR projected an expenditure of US$701 million from 1995–2006, although this includes significant contributions from unspecified third parties. The Argentine government reviews the plan every two years.

CONAE and its predecessor have worked on the SAC Satelite de Aplicaciones Cientificas (Scientific Applications Satellite) since 1981. Their research culminated in the SAC programme, first proposed in 1988, in response to NASA's request for Scout Small Explorer payloads. SAC-A would have been placed in a 500 km, 97.4° orbit to monitor solar output and its terrestrial effects during a solar maximum. Its instruments included a Fourier Imaging x-ray Spectrometer (FIXS), soft x-ray Bragg Crystal Locator (BCL), Hard x-ray Spectrometer (HXRS), Broad Band Photometer (BBP) array and a charged particle detector to monitor passage through the van Allen belts.

NASA ranked the SAC proposal in the top five, but it only had funds for the first three. Nonetheless, NASA officials maintained enough enthusiasm for the SAC to conduct a joint six-month study ending April 1991 on future co-operative missions. Then Argentina and NASA signed an MoU in August 1991, agreeing that Argentina would provide the spacecraft, the solar HXRS instrument, and the particle detector and mission operations. NASA provided the launch, other science instruments and back-up tracking coverage. Italy later agreed to provide one instrument (by IFSI) and the solar array (by FIAR). INVAP built the SAC-B as prime contractor INVAP. INVAP principal research prior to this was nuclear studies, in San Carlos de Bariloche.

When the satellite was ready, Brazil's INPE space agency performed all system qualification testing in its Sao Paulo facilities in return for participation in the science. INPE completed testing in 1995 and placed SAC in storage as it waited for the Pegasus programme to restart. Argentina's space segment cost is estimated at US$9 million and NASA's overall cost at US$13.4 million (including US$10.8 million for the Pegasus launch).

A Pegasus launched SAC-B on 4 November 1996, but it and the accompanying HETE (United States) satellite remained attached to the third stage of the Pegasus launch vehicle. NASA attempted to deploy its solar panels and return some experimental data. Although the solar panels emerged, the assembly tumbled, preventing the batteries from charging. The satellite decayed on 25 October 1999.

Argentina wants to launch SAC-D remote sensing, space physics and astrophysics satellite in 2004.

SAC-C is an international co-operative mission between NASA, the Argentine Commission on Space Activities, the French Space Agency, the Brazilian Space Agency, the Danish Space Agency and the Italian Space Agency and includes the Multispectral Medium Resolution Scanner, magnetometer and GPS receiver.

Specifications

SAC-B

Launch: 4 November 1996 by Pegasus-XL from Wallops Island
Mass: 182.7 kg (bus 134.7 kg; experiments 48.0 kg)
Orbit: 550 km circular, 38°, 95.6 min
Spacecraft: 72 cm^3 80 cm high. 3-axis control $\pm 5°$ of Sun-pointing axis and 2° over 2 days X-axis for CUBIC; two 3-axis magnetometers plus two 2.0 Nms Ithaco RWs. Four 60.5×78.5 cm solar panels (224 cm span) using 1,032 30×41 mm GaAs cells provide 22 1/204 W BOL/EOL average (49 W payload requirement) at 28 VDC; 2×10 Ah Ni/Cd batteries. Redundant 2 Mbit memories, downlink 2.25550 GHz S-band at 100 kbit/s to CONAE's San Miguel centre
Payload: HXRS High Energy x-ray Spectrometer
IAFE, 20–320 keV solar x-ray and γ bursts, 7.6 kg
CUBIC Cosmic Unresolved Background Instrument with CCDs, The Penn State University, 0.1–10 keV, diffuse X-ray background, 30.0 kg.
GXRE Goddard x-ray Experiment, NASA Goddard, 4.8 kg, comprising: SoXS Soft X-ray Spectrometer (2–20 keV) plus GRaBS Gamma Ray Burst Spectrometer (>30 keV).
ISENA Imaging Particle Spectrometer for Energetic Neutral Atoms, Italy (IFSI Instituto di Fisica dello Spazio Interplanetario, CNR, Frascati), 5–200 keV, 5.6 kg.

SAC-A

Launch: 14 December 1998 by Shuttle on STS-88. Satellite decayed 25 October 1999
Orbit: 378×395 km, 51.6°, 92.3 min
Mass: 60 kg

SAC-C

Launch: 21 November 2000 by Delta 2
Orbit: 671×707, 98.3°, 98.6 min.

Australia

Wresat Satellite

Background

Australia joined the ranks of spacefaring nations on 29 November 1967, when it launched the Wresat, a test payload developed by the Weapons Research Establishment out of the Woomera range. The satellite weighed 45 kg and entered a 106×777 km, 83°, 99.3 min orbit. A modified US Redstone rocket, remaining from a re-entry test programme launched Australia's first satellite. The modifications included a solid second- and third-stage. Wresat transmitted data for five days and re-entered after 42 days. It carried several science payloads including particle radiation detectors, x-ray counter, ozone sensor, magnetometer, Lyman alpha telescope and solar aspect sensor to study effects of solar radiation on upper atmosphere.

Brazil

SACI 1

Background

SACI 1 was launched on 14 October 1999 riding on a Long March vehicle. It was placed in a 750 km sun-synchronous orbit. Contact with the satellite was lost after one week. This is an independent effort for Brazil and not part of the MECB. The satellite is a $400 \times 400 \times 650$ mm box-shaped body with four deployed solar panels and sun-pointing spin stabilisation. The entire package only costs about US$4.6 million, including US$800,000 for the experiments. The Brazilian Academy of Sciences selected four experiments for the project. They are: (1) ORCAS Solar/Anomalous Cosmic Rays Observation in the Magnetosphere, (2) the Plasma Bubbles study; development and decay over Brazil, to understand the strong influence of bubbles and their associated plasma turbulence on trans-ionosphere communications, space geodesy, radar remote sensing, (3) the Airglow Photometer measuring oxygen OI 557.7 nm, OI 630.0 nm and OH (8,3) and terrestrial airglow, particularly the equatorial ionospheric anomaly and the South Atlantic Anomaly, and (4) the Geomagnetic Experiment. The latter gives simultaneous measurements of the geomagnetic field.

A second satellite, SACI 2, was launched by the second VLS for a total cost of US$10 million. The launcher failed to reach orbit.

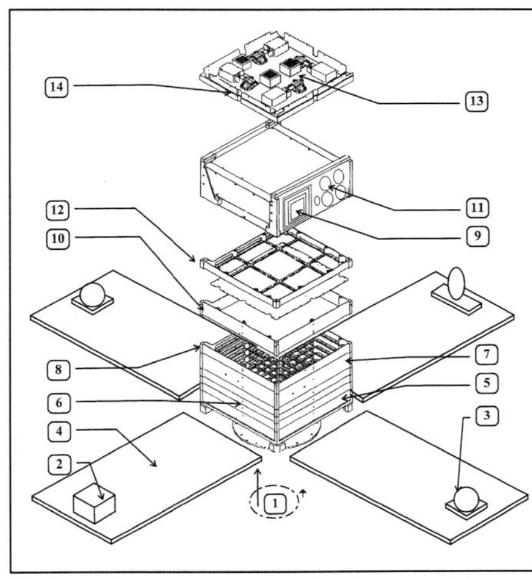

SACI 1 principal features. 1 Sun-facing solar panels, 2/12 magnetometer, 3 plasma sensor, 4 solar panel, 5 communications box, 6 OBC, 7 battery/power box, 8 plasma bubble expt, 9/10 cosmic ray detector, 11 airglow photometer, 13 antenna + open devices, 14 anti-Sun plate (INPE) 0517270

Canada

Alouette/ISIS

Background

Canada's first satellite, the 145 kg Alouette 1, was launched 29 September 1962 by Thor Agena B from Vandenberg into $997 \times 1,026$ km 80° orbit. Canada sought a joint agreement with NASA in order to fly their ionosphere sounder in order to explore the effects of the auroral regions on telecommunications. Canada had identified telecommunications to its vast northern regions as an application of spaceflight early in the space age. Alouette 1 was deactivated on 30 September 1972. They built two satellites and when Alouette 1 performed as expected, they launched the 147 kg back-up Alouette 2, on 29 November 1965, into a $505 \times 2,987$ km 79.8° orbit, as the first of three satellites in the newly created joint Canadian-NASA

International Satellites for Ionospheric Studies (ISIS) programme. Alouette 2 ceased operating in July 1975 after an operating period of 115 months.

After the first data from these satellites became widely available and showed some promise in predicting communication and navigation disruptions due to ionospheric turbulence, NASA and the CSA developed the ISIS which was first launched 30 January 1969 by Delta into 578 × 3,526 km, 88° orbit. The satellite weighed 241 kg. Both agencies agreed to extend the programme by ISIS 2, launched on 1 April 1971 by Delta into 1,358 × 1,429 km, 88° orbit. ISIS 2 weighed slightly more than its predecessor at 264 kg.

Chile

FASat Series

Background

Chile conducts its space affairs under the auspices of the CAE Committee for Space Matters, which it created in 1980, as part of the Ministry of Defence. The MOD's Space Division of the Fuerza Aérea de Chile (FACH, Chilean Air Force) launched the 50 kg FASat-Alfa as a passenger payload, in 1995, into polar orbit aboard a Ukrainian/Russian Tsyklon. Chile purchased from the UK's Surrey Satellite Technology Ltd, in May 1994, the satellite, ground station and on-the-job training for eight FACH engineers (launch cost was additional). A Chilean team built the actual satellite at Surrey working alongside SSTL staff. The programme helped students to gain Masters' Degrees on Surrey University's satellite engineering course. The satellite carried 5 g of Chilean soil on board and was to have transmitted regular information on UV radiation and the ozone layer over the Antarctic.

Unfortunately, at launch, a pyrotechnic problem on a locking system built by the UK left the FASat-Alfa attached to the Russian Sich. FACH thus signed a £1.3 million contract with SSTL for the identical FASat-Bravo for 1997 launch. Educational activities using the satellite will help to create space awareness in the country. FASat-B was launched as a menage relay for the Chilean Air Force by a Zenit 2 from Baikonur on 10 July 1998. FASat-B was retired from service when the batteries ran out in June 2001. Two others to form a constellation providing LEO communications services will follow it. Work is also under way on the FASat-Gamma mini-satellite for remote sensing from 2003. The expertise derived from these projects is expected to lead to a GEO satellite around 2010.

Specifications

FASat-Alfa/Bravo

Launch: Alfa 31 August 1995 piggyback on Sich 1 by Tsyklon from Plesetsk; Bravo 10 July 1998 piggyback on Zenit 2 from Baikonur

FASat-Alfa Flight Model; Bravo is identical. One of the Earth face's four blocks carried 5 g of Chilean soil; the others housed Sun sensors (2) and the 'Earth Underneath Detector'. At bottom is the tip mass of the compacted 6 m boom (FACH) 0517272

Orbit: Alfa 632 × 669 km, 82.5° attached to Sich 1; Bravo 817 × 818 km, 98.8°

Configuration: based on SSTL's UoSat 5 design. 600 × 345 × 345 mm box-shaped body with solar panels on four sides. 6 m boom deployed from top face for gravity gradient stabilisation. Each housekeeping subsystem is housed in a standard module box mass-produced by computer-controlled process (most were manufactured by the ENAER Chilean aerospace company)

Mass: 50 kg on-orbit; 53 kg including launcher fittings

Power supply: four GaAs panels (each 36 W) providing 20 W orbital average processed power; 6 Ah Ni/Cd battery supports eclipse operations

OBCs: OBC386 based on an 80C386EX with 387, 800 kbyte Error Detected/Corrected program memory and 16 Mbyte EDAC mass memory complemented by standard SSTL OBC186-Enhanced with 512k EDAC program and 16 Mbyte file store. The transputer system based on two transputers associated with the imaging payloads provides for high-performance data processing; the 256 Mbyte solid-state memory accommodates large onboard storage

Attitude control: Earth pointing maintained by combination of passive gravity gradient and active magnetorquing. 1-axis RW used for yaw rate control in conjunction with imaging payloads. Sun sensors in azimuth and elevation angle plus 3-axis fluxgate magnetometers (±2 nT resolution over ±64 nT range) provide attitude information. Magnetorquing provided by coils in each axis

Communications: data rates of 9.6, 38.4 or 76.8 kbit/s on 400 MHz using redundant UHF synthesised transmitters 2–10 W variable. 150 MHz VHF up (two communications + one command). 4-element VHF + UHF antenna arrays

TT&C: from ECM-Santiago control station at the Los Cerrillos Air Force Base

Payload: OLME Ozone Layer Monitoring Experiment. Two UV CCD cameras plus four UV photodiodes to monitor the ozone layer in southern Chile by measuring backscattered solar UV at 3,000 Å.

DTE Data Transfer Experiment. Provides enhanced store and forward communications through two synthesised VHF receivers under control of the dual Digital Signal Processing (DSP) computers. They interface with both up/down links for DTE receiver control, up/down modulation experiments and speech processing.

EIS Earth Imaging System. Two monochromatic 578 × 576 pixel CCD cameras provide 100 m/1 km resolution. Images recorded on schedule controlled by the transputer data processing equipment. The transputers recover the images and provide onboard processing before passing to the OBC file system for distribution on the downlink.

GPS navigation experiment. GPS receiver controlled by the transputer data processing experiment. Keplerian elements are generated onboard to assist in image targeting and for downlink distribution. Bravo's receiver, developed by SSTL under ESA contract, will also provide attitude data.

SSDRE Solid State Data Recorder Experiment. The unit was developed by Sanders (Lockheed Martin) in collaboration with the FASat-Alfa project, providing 256 Mbyte storage via the CAN bus.

China

Shi Jian Series

Background

The majority of Chinese satellite launches in the domestic programme have been dedicated to practical uses rather than propaganda type missions. While there have been some science missions flown, these have been few and far between. Most domestic launches have been for either remote sensing (including meteorological) or communications.

Launched 3 March 1971 as the second satellite launched by China and placed in a 268 × 1,830 km, 69.9°, 106.1 min orbit, Shi Jian 1 embodied the first true science payload, acting in a role of exploration and inquiry. It carried a magnetometer, cosmic ray and solar x-ray detectors. Apparently, the Chinese used the same bus for most of their mission applications and the Shi Jian 1 duplicated the DFH-1 configuration. It flew about a year after the successful launch of China's first satellite. The Chinese obviously learned from their earlier experience. This time the satellite flew solar cells for power and controlled the temperature of the platform by using louvres on the satellite. A second attempt to launch a satellite in this class on 28 July 1979 failed. Then the Chinese launched Shi Jian 2, their first triple payload, made up of SJ-2, 2A and 2B launched on 19 September 1981 the triple payload occupied slightly differing orbits close to 234 × 1,600 km. 2B apparently studied atmospheric density since the satellite, a balloon attached to a heavy mass - quickly decayed from orbit.

Shi Jian Series

Shi Jian 3 is a designator used for an abandoned early-1980s meteorological satellite programme, and was never used for an orbital mission. The following satellites have been launched since 1990.

Shi Jian 4

Shi Jian 4 was launched aboard the maiden flight of the CZ-3A, mounted above a second satellite identified as *Kua Fu* 1, named in honour of a character in Chinese mythology who chases the Sun. *Kua Fu* 1 was a drum-shaped satellite, approximately 2.5 m base diameter and 2.2 m high, mass 1,342 kg. No details of the *Kua Fu* 1 programme are known to have been discussed in Chinese literature.

Specifications

Launch: 8 February 1994 using a CZ-3A from Xi Chang

Orbit: 28.5°, 209–36, 118 km

Contractor: Shanghai Institute of Satellite Engineering

Configuration: drum, approximately 2 m diameter, 1 m tall

Mass: 396 kg

Instruments: satellite was devoted to solar research and it carried: 5-channel, 0.5–4 MeV high-energy electron spectrometer, electrostatic analyser (0.1–40 KeV protons/electrons), high-energy proton (3–400 MeV) and heavy ion (4–25 MeV) detector, charging meter registering –6.4/+1 kV at the top and bottom of the satellite and –70/300 V around the waist of the satellite and stationary and dynamic single-event monitors

Power system: solar panels cover the waist of the satellite

Shi Jian5

Shi Jian 5 was launched piggy-back on the maiden flight of the CZ-4B, the primary payload being the third FY-1 meteorological satellite (*qv*). Satellite was the test flight of the new CAST 968 satellite platform.

Launch: 10 May 1999 using a CZ-4B from Taiyuan

Orbit: Sun-synchronous: 98.8°, 849–868 km

Contractor: CAST, Shanghai Institute of Satellite Engineering

Configuration: box-shaped body, approximately 1 m along each side plus two three-segment vanes of solar panels

Mass: 298 kg

Instruments: Satellite carried instruments to monitor the particle precipitation and the Earth's radiation belt particles over the South Atlantic Anomaly: single-event upset monitor, single-event shield effect tester, single-event upset tester, single-event synthetic tester, single-event latch-up tester, high-energy proton and heavy ion detector, high-energy electron detector, radiation dosimeter

Lifetime: on 10 August 1999, China announced that the satellite had successfully completed its planned three months of operations.

SVOM: Space MultiBand Variable Objects Monitor

The SVOM satellite was described in a Chinese paper at the 2002 IAF Congress. The primary research area for the satellite includes the detection and measurements of stellar vibrations, zero-delay X-ray/optical synchronous observation of rapidly varying objects such as gamma and X-radiation bursts and the search for planets. The satellite will probably be launched piggy-back with a meteorological/remote sensing satellite – possibly the first FY-3.

Launch: late-2005 using a CZ-4B from Taiyuan

Orbit: Sun-synchronous: 98.6°, 800 km

Lifetime: nominal mission 2 years, designed for 3 years

Contractor: Tsinghua University

Configuration: box-shaped body, 0.8 × 0.8 × 0.9 m, plus four vanes of solar panels

Mass: 104 kg

Instruments: High-Precision Photometric Measurement (HPPM) telescope, aperture 0.3 m, effective field-of-view 21 arc-min, able to simultaneously record data in the UV, green and

red bands: high-resolution wide field-of-view X-ray telescope using a silicon drift semiconductor X-ray detector: Optical Patrol (OP) telescope, aperture 50 mm, focus f/0.8, imaging sensor a CCD chip with 1,024 × 1,024 pixels
Communications: 9.6 kps uplink, 2 Mbps downlink
Power: 4 folding solar panels with silicon cells: Ni Cd battery packs will deliver 144 W towards the end of operations

Czech Republic

Magion Series

Background

When it still formed part of the Soviet bloc, Czechoslovakia belonged to Interkosmos. Both the Czech and Slovak republics moved away from that organisation after they became separate powers on 1 January 1993. The Czech Republic continued on as a significant contributor to international space science missions.

Within its former borders, Czechoslovakia designed and built a series of small magneto/ionospheric (hence the name) research satellites. They came from a small group in the Upper Atmosphere Department of the Institute of Atmospheric Physics (until April 1994 the Ionosphere Department of the Geophysical Institute).

Interkosmos 18 released Magion 1, the first in the series and the first Czech satellite, shortly after launch by Kosmos C on 24 October 1978 into a 404 × 762 km, 83°, 96.4 min orbit. The satellite only weighed 14.5 kg, and it was 370 mm in diameter. The Institute planned to study low frequency fields between the pair of slowly diverging satellites, but a power malfunction degraded data acquisition. Even in its reduced capacity Magion 1 remained operational until entry 11 September 1981.

Then Magion 2, launched with Interkosmos 24 by Tsyklon 3 on 28 September 1989 into 500 × 2,492 km, 82.6°, 115.3 min orbit continued the series by conducting propagation studies of VLF waves (from the parent) and their effects on the van Allen belts. The parent and the small sub-satellite separated by spring action at 0.2 m/s upon command from a ground station near Prague on 3 October while Interkosmos was nadir-pointing. Soon after release, the temperature fell to –40°C, which compromised the deployment performance since the satellite design only called for –20°C. Only one boom and one solar panel deployed, generating insufficient power for telemetry. Some telemetry did arrive on 21 December when the orbit had shifted into sunlight. Deployment commands retransmitted from the ground deployed all six booms and four solar panels on 24 December. The science programme began 21 January 1990, but power levels remained marginal, probably because of solar array degradation from intense solar activity. The last telemetry contact occurred 20 November 1990.

Magion 2 had a 26-face polyhedron, 560 mm diameter and weighed 52.18 kg. Within that weight budget a 9 kg AOCS package stabilised

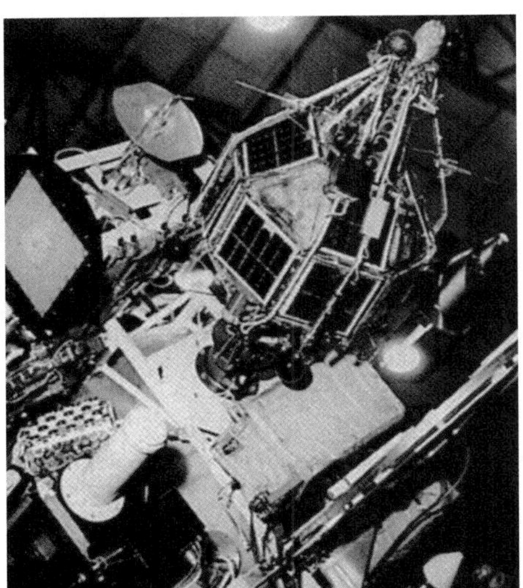

Magion 4/5 appear almost identical to Magion 2 0516730

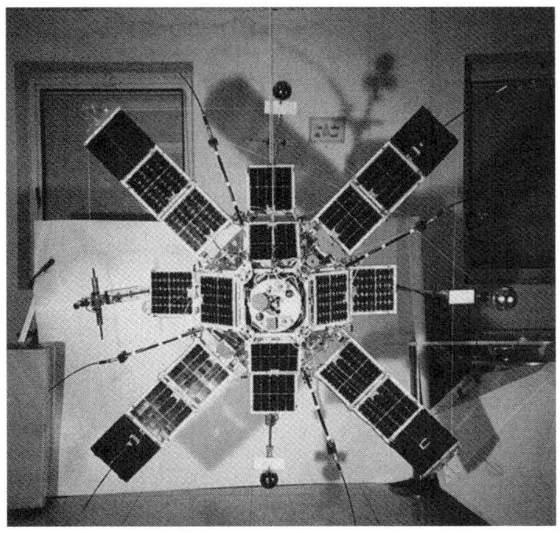

Magion 4 Flight Model viewed from the solar direction (Dr Pavel Triska) 0517289

attitude control by a 30 Am^2 permanent magnet, using attitude information from magnetometer and Sun sensors. All of the science experiments used the 12 W of power from Si solar panels, supported by two Ni/Cd batteries (4 Ah each). Originally, ground controllers expected to adjust the orbit for varying separation from parent, by 150 atmospheres compressed air through 0.2 N thruster (total impulse 300 Ns). A cable failure prevented this from following the design. Telemetry arrived over a transmit carrier of 137/400 MHz, at the rate of 40.96 kbit/s max, using a 4 Mbit memory, 256 command combinations, 196 housekeeping parameters.

To perform its science investigations, Magion 2 carried a 3-axis fluxgate magnetometer, resolution 2–16 nT (Romania, Czech), a wave experiment with search coil magnetometers operating between 10 Hz-40 kHz, double probe electric component sensors 0.1 Hz/120 kHz, frequency analyser 1–120 kHz, filter banks 20 Hz/15 kHz (Czech, Bulgaria), a radiospectrometer 0.01–10 MHz (Poland, Czech). The cold plasma data came from a Langmuir probe and RF probe (East Germany, Czech), and an ion trap (Bulgaria, Czech). The hot plasma experiments examined 0.01–20 keV electrons/protons at 16 energy levels (Bulgaria, USSR, Czech), and a 0.20–1 MeV solid-state detectors at 8 energy levels (Czech). In the visible/near visible wavebands Bulgaria provided a 6,300/5,777 Å photometer and 17 3 cm^2 laser reflectors on lower surface.

Magion 3, launched by Tsyklon 3 18 December 1991 with Interkosmos 25 into an initial 438 × 3,071 km, 82.56°, 121.3 min orbit monitored the propagation of electron beams and plasma injected by the parent. Magion separated at 02.03 GMT 28 December, drifting from its parent because of differential drag. It used its small thruster 9 January 1992 for the first time to close in again by 30 January. Similar manoeuvres followed on 6 May and 11–26 June 1992. The last approach was within 100 m with relative velocity almost zero. The sub-satellite operated with no problems until 20 August 1992 when it suddenly suffered a power shortage, probably caused by radiation damage to the battery charge system after a period of enhanced solar activity. Final telemetry downlinked on 9 September 1992. Magion 3 and its payload were almost identical to Magion 2.

Interball 1 provided the parent craft for Magion 4, launched by Molniya M on 2 August 1995, into a 793 × 191,900 km 62.9°, 5,454 min orbit to study mechanisms transporting solar wind energy into the magnetosphere. It weighed in at 58.7 kg. The design resembled Magion 2 except the onboard memory increased to 32 Mbit, and the design group added a 1.5 GHz band for additional telemetry. Its control improved too, to 3°/s spin-stabilised control around a Sun-pointing axis (within 15°). Magion 4 carried a science payload consisting of a 3-axis fluxgate magnetometer (0.015–8 nT resolution over ±128–65,535 nT), 3-axis search coils for 10–1,000 Hz ULF/ELF waves and 10 or 30 Hz ULF waves, a 1-axis search coil for 0.50–40 kHz VLF waves, and one for 0.04–20 Hz and 0.02–400 kHz electric fields and waves. An x-ray photometer searches for 1.5–500 keV solar x-rays.

Magion 5 went into orbit following launch by Molniya M with the Interball 2 on 29 August 1996. It consisted of an auroral probe on a 750 × 20,000 km 63° orbit. Like Magion 4, the research craft studied the mechanisms transporting solar wind energy into magnetosphere, in conjunction with Magion 4, and it weighed 62 kg. It had a cold plasma detector in place of ion flux and solar x-ray instruments. Problems with the solar panel deployment ended after its release.

France

Cerise Satellite

Background

Helios 1A's Ariane carried the 50 kg Cerise (Caractérisation de l'Environnement Radioélectrique par un Instrument Spatial Embarqué) to characterise the Earth's radio environment in research that could lead to a 'Zenon' electronic intelligence satellite in the next decade. Cerise was launched by Ariane V75 on 7 July 1995 to a 666 × 675 km, 98.2°, 98.2 min orbit. Alcatel Espace acted as Cerise prime contractor for the DGA, using the University of Surrey's UoSAT bus. Cerise should have lasted 2½ years. France sought industry proposals in 1993 for 1998's Clementine Cerise follow-on to investigate different frequencies regimes in the ionosphere. Cerise became the first proven casualty of space debris when it collided with a piece of debris from the Ariane-1 third stage, which had orbited SPOT 1 in 1986. The collision with the debris 1986-019RF was at 09.48 GMT on 21 July 1996 with a relative velocity of 14.8 km/sec and the upper 3.1 m portion of the Cerise gravity gradient boom was broken off. Control of Cerise was regained following the collision.

Cerise is contributing to the development of an electronic intelligence satellite (SSTL) 0517268

French Science Satellite Programme Series

Background

France created the CNES in 1962 with a staff of 27 to execute the national space programme. Since April 1993 it has been under the regulating authority of three ministries (Defence; Industry, Posts, Telecommunications & Foreign Trade; Higher Education & Research). A President heads CNES and a Director General sets the science programme agenda. Today, an 18-member Conseil d'Administration, with six members elected by agency employees, consult and advise the direction of space priorities.

In the beginning, France sought to separate its military and civilian space programmes, but in the interest of efficiency the country consolidated the two in 1993, in co-ordination with the Ministry of Defence. Now CNES has 2,488 personnel. Over the past 42 years, CNES has conducted a vigorous launch programme and has one of Europe's largest space science research programmes for both national and international (ESA) applications.

Specifications (Historical)

French launches

A-1 (Asterix)

Launched: 26 November 1965 by Diamant from Hammaguir and retired out of service the same day (absence of transmission) due to damaged antenna during launch
Mass: 41.7 kg

Orbit: 528 × 1,758 km, 34°
Comment: First French satellite. Transmitted data for 2 days.

D-1A (Diapason)
Launched: 17 February 1966 by Diamant from Hammaguir and was retired on 23 January 1972 when it stopped transmitting
Mass: 20 kg
Orbit: 504 × 2,753 km, 34°
Comment: Third French satellite and the second national launch. Geodetic research.

D-1C (Diademe 1)
Launched: 8 February 1967 by Diamant from Hammaguir and stopped transmitting on 2 January 1970
Mass: 22.6 kg
Orbit: 580 × 1,340 km, 40°
Comment: Geodesy; operational in spite of low apogee.

D-1D (Diademe 2)
Launched: 15 February 1967 by Diamant from Hammaguir but stopped transmitting 5 April 1967
Mass: 22.6 kg
Orbit: 592 × 1,886 km, 39°
Comment: Laser/Doppler ranging data to 5 April.

WikaMika
Launched: 10 March 1970 by Diamant B from Kourou
Orbit: 307 × 1,700/1,746 km, 5°
Comment: West German Wika was first foreign payload orbited by France. It provided 30 days of geocorona and upper atmosphere investigations. The satellite re-entered 5 October 1978. Mika monitored Diamant performance. Its re-entry occurred 9 September 1974.

Peole
Launched: 12 December 1970 by Diamant B from Kourou
Mass: 69.7 kg
Orbit: 580 × 747 km, 15°
Comment: Test flight to qualify Eole meteorology data relay satellite, launched by Scout B from US Wallops 16 August 1971 in collaboration with NASA to track hundreds of 4 m meteorology balloons. Out of service 23 March 1972. Re-entry occurred 16 June 1980.

D-2A (Tournesol)
Launched: 15 April 1971 by Diamant B from Kourou
Mass: 90 kg
Orbit: 456 × 703 km, 46°
Comment: Five instruments studied solar and UV radiation, including hydrogen content of geocorona. First French satellite to utilise active stabilisation (thrusters). Out of service 22 July 1973. Re-entered 28 January 1980.

D-2A (Polaire)
Launched: 5 December 1971 by Diamant B from Kourou
Mass: 97 kg
Comment: Tournesol's back-up; intended for higher inclination (75°) for complementary and simultaneous measurements, but launcher stage 2 failed.

D-5A/5B (Castor/Pollux)
Launched: 21 May 1973 by Diamant BP-4 from Kourou
Comment: Castor/Pollux fell into Atlantic when the Diamant failed to produce sufficient thrust.

Starlette
Launched: 6 February 1975 by Diamant B from Kourou
Mass: 48 kg
Orbit*:* 804 × 1,138 km, 50°, 104.2 min
Comment: 24 cm sphere carrying 60 laser reflectors for geodetic studies. Observations continue. Similar 1993 Stella satellite added polar coverage.

D-5A (Pollux)
Launched: 17 May 1975 by Diamant BP-4 from Kourou
Orbit: 270 × 1,270 km, 30°
Mass: 36 kg
Comment: Intended to space-qualify SEP's 1.5–4 N thruster (7 kg hydrazine). Performed 11 sequences totalling 3,000 pulses of 0.125–100 s. Decayed 5 August 1975.

D-5B (Castor)
Launched: 17 May 1975 by Diamant BP-4 from Kourou
Orbit: 270 × 1,270 km, 30°
Mass: 76 kg
Comment: 800 mm diameter, 26 faces carrying 26 laser reflectors and solar cells to generate 20 W. Intended to space-qualify Onera's 10^{-5}–10^{-9} *g* microaccelerometer, which was also used for air density studies over 300–800 km (in conjunction with laser tracking) and registering micrometeoroid impacts. Re-entered 18 February 1979.

D-2B (Aura)
Launched: 27 September 1975 by Diamant BP-4 from Kourou
Mass: 110 kg
Orbit: 499 × 723 km, 97°
Comment: Solar/stellar UV observations. Out of service 28 December 1976. Re-entered 30 September 1982.

European launches
SARA
Launched: 17 July 1991
Orbit: 770 × 777 km, 98.54°, 100 min
Mass: 26.6 kg
Comment: 34 cm^3, intended to be the first French science amateur satellite. Total project cost FFr3 million, including FFr1.2 million launch fee. Planned to observe Jupiter's 2–15 MHz emissions but electronics failure prevented useful data, although SARA is otherwise in good condition. Switched to sleep mode in May 1992.

S80/T
Launched: 10 August 1992 by Ariane V52
Orbit: 1,302 × 1,329 km, 66.1°, 111.9 min
Mass: 50 kg
Comment: Demonstrated technologies for proposed S80 TAOS 2-way messaging/localisation system for mobile users. MMS France's 1991 FFr25 million contract from CNES included a mobile station for tests. The S80/T employs Surrey University's subcontracted UoSAT platform and UK control station. Communications are tested through stations in Toulouse and Lanham. The 7 kg payload includes Dassault Electronique repeater, up/down 148–149.9/137–138 MHz, 2–3 year life. Stabilised by gravity gradient and magnetorquer. Released from ASAP carrier with Korea's Kitsat. 1993 activities included multipath observations and single event upsets. 1994–96 work includes analysis of the 148–149.9 MHz band in several world regions and platform technology experiments. CNES planned to contract for four or five satellites in 1994, but in March 1994 decided not to pursue the project. TAOS would have formed the core of the Starsys system. The full system would have been operational by 1996/7, with users paying FFr1,500 for a hand-held terminal, a FFr500 annual fee and FFr5 per message.

USSR launches
Aureole 1
Launched: 27 December 1971 Kosmos C from Plesetsk
Mass: 300 kg
Orbit: 410 × 2,500 km, 74°, 108.8 min
Comment: First Soviet-French satellite, launched under international 'Arcad' (Arctic auroral density) to investigate aurora borealis and ionosphere. Instruments provided by Moscow's Space Research Institute and France's Toulouse Centre for Study of Space Radiation.

Aureole 2
Launched: 26 December 1973 by Kosmos C from Plesetsk
Mass: 400 kg
Orbit: 400 × 1,975 km, 74°, 103.3 min

Aureole 3
Launched: 21 September 1981 by Tsyklon 3 from Plesetsk
Mass: 1,000 kg
Orbit: 380 × 1,920 km, 82.6°, 105.8 min
Comment: Third Arcad and the first Soviet satellite to be integrated at least partially in the West. It carried four French instruments and France also supplied the onboard programming and telemetry unit. Both countries received data directly.

SRET 1
Launched: 4 April 1972 piggyback with Molniya 1–20 on Molniya from Plesetsk
Mass: 15.4 kg
Orbit: 460 × 39,248 km, 65.6°, 735 min
Comment: Meant to study radiation effects on solar cells. Re-entered 26 February 1974.

SRET 2
Launched: 5 June 1975 piggyback with Molniya 1–30 on Molniya from Plesetsk
Mass: 29.6 kg
Orbit: 513 × 40,825 km, 62.8°, 737.8 min
Comment: Tested Meteosat cooling technology. Decayed 26 February 1974.

Signe 3
Launched: 17 June 1977 by SL-8 from Kapustin Yar
Mass: 102 kg
Orbit: 459 × 519 km, 50.6°
Comment: Monitored cosmic γ-rays for 733 days. Re-entry occurred 20 June 1979.

United States launches
FR-1
Launched: 6 December 1965 by Scout from Vandenberg
Orbit: 717 × 728 km, 75.9°, 98.8 min
Comment: Meant to study ionosphere and VLF wave propagation by monitoring 16/25 kHz signals from St Assise (France) and Balboa (Panama) stations. 62 kg. Stopped transmitting 28 February 1969.

Eole
Launched: 16 August 1971 by Scout from Wallops
Orbit: 664 × 870 km, 50.1°, 99.7 min
Mass: 84 kg
Comment: Data relay satellite interrogating 500 met balloons drifting in atmosphere at 12,000 m; returned data for 5 months. Out of service July 1974.

Symphonie 1/2
Launched: 19 December 1974/27 August 1975 by Delta from Cape Canaveral into GEO at 11.5° W
Mass: 237 kg on-station
Comment: Franco-German telecom project, each providing two TV channels or 300 duplex phone. Aerospatiale built number 1 and MBB built number 2. Direct result was TV-Sat/TDF project. Out of service 19 February 1983.

VERIFIED

Germany

Astro-SPAS (ORFEUS and CRISTA) Series

Background

MBB began development of the SPAS carrier in 1986 (in preference to Dornier's Robus) into a free-flying astronomical platform. In total, the DARA/NASA agreement calls for four co-operative science missions, with DARA providing the satellite, NASA the Shuttle launch, deployment/retrieval services and the two parties sharing the science instruments. Germany set aside DM115 million, for the first mission falling to DM40 million for the second as the carrier was reused. Germany's costs for the next two are projected at DM20 million each. NASA is providing the Shuttle free of charge, in return for access to data and inclusion of US experiments.

ORFEUS, the Orbiting and Retrievable Far/Extreme UV Spectrometer measuring radiation between 400–1,280 Å flew on STS-51 in September 1993 and was released at 14.06 GMT 13 September, retrieved 11.50 GMT 19 September. Science contributions came from the University of Tübingen, Sternwarte Heidelberg, University of Berkeley and Princeton University (IMPAS).

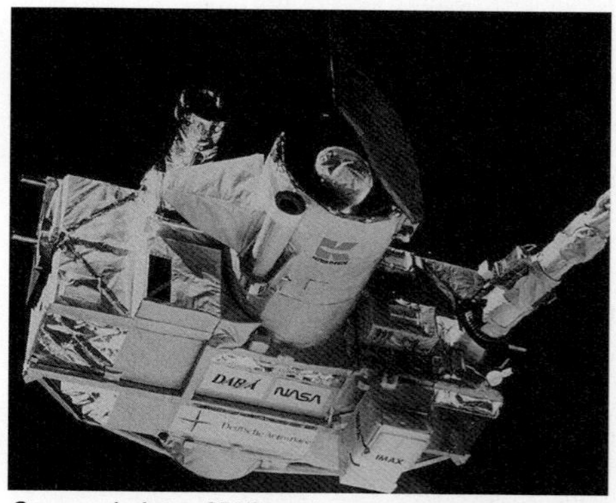

Germany's Astro-SPAS carries astronomical payloads: here it accommodates the Orfeus ultraviolet telescope. This 1993 configuration will be repeated in late 1996
(NASA) 0517118

ORFEUS' telescope was fabricated by Kayser-Threde in Germany; France's REOSC provided the 1 m f/2.5 mirror. The separate 950–1,150 Å IMAPS Interstellar Medium Absorption Profile Spectrograph added to the observations of hot galactic objects and the interstellar medium at high spectral resolution (240,000). Other payloads were DLR's Surface Effective Sample Monitor and Canada's IMAX camera. STS-80's mission will compare combined GPS/star tracker attitude determination against the gyro system.

CRISTA, the Cryogenic Infrared Spectrometers/Telescopes for the Atmosphere, developed by the University of Wuppertal, flew on the STS-66. It was released 12.50 GMT 4 November, and retrieved 13.05 GMT 12 November 1994. It carried as a Secondary instrument the US Naval Research Lab's MAHRSI Middle Atmosphere High Resolution Spectrograph Investigation for nitric oxide and OH chemistry measurements. CRISTA's primary goal is the first study of global atmospheric variability using limb sounding. 51,000 vertical scans were acquired during STS-66, measuring 15 trace gases every second. As this requires precise SPAS position information, it carries a GPS receiver qualified by the Orfeus mission.

The SPAS platform was sought for US Department of Defense tests but the sale was not carried out. Commercial customers were sought prior to an offering for international co-operative initiatives, efforts which were unsuccessful.

Bremsat

Background

ZARM (University of Bremen), in co-operation with OHB-System and HTG Hypersonic Technology Goettingen, launched Bremsat aboard the Shuttle STS-60 flight launched 3 February 1994. It was released 9 February into 340 × 357 km, 57.0°, 91.4 min orbit. The 12-sided 63 kg (30 kg science payload), 480 mm diameter 520 mm high aluminium prism completed two experiments over 48 h before ejection from its GAS canister. The first experiment tested thermal conductivity of a fluid (4 kg) and measured Shuttle's residual acceleration by a 3-axis micro-accelerometer (accurate to 10^{-6} *g*). After release, the satellite concentrated on measuring the near-Earth dust/micrometeoroid population to investigate the growth of manmade pollution. The same type of 0.8 kg detector flew aboard Japan's Hiten satellite.

This orbital phase also measured atomic oxygen concentrations. Once Bremsat had decayed to 180 km, the atmospheric force on a solar panel was measured to determine the accommodation coefficients of solar cells. For the entry phase, the MW was ejected to remove all satellite momentum. Beginning at 120 km (remaining life about 40 minutes before destruction at 60 km), temperature and pressure data were collected. Of particular interest was the transition from free molecular flow to continuum flow.

Bremsat re-entered 12 February 1995; the last data were collected at 16.48 GMT at 110 km altitude, 15 minutes before destruction.

Bremsat 2 has been proposed to DARA. Using a modified Bremsat 1 bus, it would carry seven experiments. In-Orbit: materials science; two technical (GPS orbit determination and new antenna design); active radiator thermal control tile; high resolution measurement of Earth's magnetic field. The main experiment is re-entry, using a 2,375 mm diameter parashield and recovery.

German Science Satellite Programme Series

Background

German space projects have always formed part of wider European and international activities – the country has never pursued an independent programme. When the Bundestadt created DARA, the German Space Agency, created in 1989, it put it in full charge of implementing the space programme, except for the hypersonic technology programme, which remains embedded within the BMFT Ministry for Research & Technology's aeronautics activities.

Despite the lack of a focused national programme, the Federal government has been supporting space research and technology since 1962. The early development programmes were part of the projects funded and approved under the European Space Research organisation (ESRO) prior to the establishment of the European Space Agency. In February 1976 the government decided that there would be no further national or even bilateral research satellite projects such as Symphonie and two-thirds of the 1976–79 space budget of DM2.2 billion would be allocated to ESA.

Specifications

Azur
Launched: 8 November 1969 by Scout from Vandenberg
Mass: 71 kg
Orbit: 387 × 3,150 km, 103°, 110.4 min
Comment: Seven instruments monitored radiation belts and solar particle flux in co-operation with the US to augment data from US Explorer and OGO.

Dial
Launched: 10 March 1970 by Diamant B from Kourou
Mass: 63 kg
Orbit: 301 × 1,631 km
Comment: France's first foreign launch, to study the hydrogen geocorona and ionosphere but excessive vibration during launch made data evaluation difficult. Re-entered 5 October 1988.

Aeros 1
Launched: 16 December 1972 by Scout from Vandenberg
Mass: 127 kg
Orbit: 223 × 867 km, 97°, 95.6 min
Comment: Five experiments (four German, one US); dipped into upper atmosphere 3,844 times for density measurements while monitoring solar UV output. Re-entered 22 August 1973.

Aeros 2
Launched: 16 July 1974 by Scout from Vandenberg
Mass: 127 kg
Orbit: 224 × 869 km, 97°, 95.6 min
Comment: Extended upper atmosphere research. Re-entered 25 September 1975.

Helios 1/2
Launched: 10 December 1974/15 January 1976 by Titan Centaur in collaboration with NASA
Mass: 370 kg
Comment: Solar probes passed within record 48/45 million km of Sun.

Firewheel
Comment: Designed by Max Planck Institute, with ESA/NASA support. The parent craft and all four sub-satellites were lost in Ariane L-02 failure of 23 May 1980. Magnetospheric investigations planned with injections of barium and lithium ions. The AMPTE programme continued the research.

AMPTE CCE
Launched: 16 August 1984 by Delta from Cape Canaveral
Mass: 695 kg
Orbit: 1,121 × 49,671, 4.8°, 939.5 min.
Comment: 3-satellite Active Magnetospheric Particle Tracer Explorer mission. The German Ion Release Module, developed by Max Planck Institute, made seven lithium and barium releases for detection by US/UK satellites. Failed 12 July 1989.

MIRCA

Background

DARA, in 1991, initiated a joint project between Kayser-Threde and DASA Jena-Optronik to develop and fly the 1 m diameter MIRCA Micro Re-entry Capsule for re-entry investigations. KT and Russia's Central Specialised Design Bureau signed the contract November 1994 for launching MIRCA attached to Photon in September 1996, however, the Russian launch was delayed and the satellite was not launched until 9 October 1997 when a Soyuz-U carried the capsule into orbit with the Foton 8 (No 11). It separated from the Foton re-entry module after retro-fire on 23 October 1997 and landed 110 km south-east of Orsk. MIRCA was mounted on the forward face of Photon's own sphere, 12° below the centreline until it is ejected at 0.8–1.6 m/s separation after retrofire. Unusually, the heatshield is also the main load carrying structure, comprising hemispheres bolted at the equator. The shield consists of a 1 mm carbon

Germany's MIRCA re-entry capsule will be released from a Russian Photon craft. The central cylinder is the parachute container (Kayser-Threde) 0517264

fibre composite core, 20/30 mm ablator for the upper/lower hemisphere and a 1.35 mm C/SiC outer layer. Temperature and pressure sensors and a pyrometer are integrated in the lower shield.

A follow-on tether mission was studied, beginning in 1991, but cancelled by DARA in 1994. Project Rapunzel (Rope Attached Piggy-back Unit Zooming on Environment Data at Low Cost) would have reeled out MIRCA on a 50 km tether from Photon to de-orbit when the line was cut. GPS receivers in each module would have provided relative position information. Before then, the first Rapunzel test was planned for a Resurs-F mission carrying a 45 kg 50 cm re-entry sphere developed by the Samara State Aerospace University. Later Rapunzel studies concentrated on a MIRCA tethered to a Progress-M as it left Mir in 1997.

Specifications

Size: 1,000 mm-diameter sphere
Mass: 150.0 kg total (71.0 heatshield, 16.7 mechanical support structure, 16.6 service system, 6.6 battery, 16.8 experiments, 14.3 recovery system, 8.0 balance mass)
Payload volume: 110 litres
Payload power: 20 Ah 28 V DC from $LiMnO_2$ cells; can be expanded to 50 Ah
Command/data: transputer-based for control/data handling. System activated by Photon command 105 s before separation. 16 Mbyte on EEPROMs. 3-axis body rates up to ±100°/s and ±64 *g* recorded
Thermal: –20/70°C internal range specified; heater activated below –20°C
Recovery system: cover pushed clear by two 1,700 N solids burning for 80 ms will pull out conical ribbon chute at about 7 km, activated redundantly by barometric switch and timer after peak deceleration. Reefed cross chute released 25 s later, fully open 10 s later. Shortwave beacon (antennas on chute lines) active for 10 min descent; VHF transmit plus strobe activated on ground for 48 h
Payload: HEATIN Heatshield Instrumentation: T, P and heat flux distribution within two meridian planes 90° apart. Also T at different depths in material and P caused by off-gassing.
RAFLEX Rarefied Flow Experiment: P at stagnation point and 60° behind.

Germany's MIRCA re-entry capsule will be released from a Russian Photon craft. The central cylinder is the parachute container (Kayser-Threde) 0517265

PYREX Pyrometer Experiment: touchless temperature measurement of C/SiC layer behind shield at stagnation point. Will also qualify this mini-pyrometer for other space applications.

TubSat

Background

The Technical University of Berlin's first satellite was released into polar orbit from ERS 1's Ariane launch on 17 July 1991. The 35 kg, 38 cm^3 carries an L-band store/forward payload to demonstrate electronic mail applications, GaAs test cells and a 3-axis star sensor.

Following the successful first launch, TUB put the 40 kg TubSat B technology demonstrator into orbit, 25 January 1994, on Russia's Meteor 3-06 launched by Tsyklon 3. Ground controllers lost contact after 5 March. The 35 kg DLR-TubSat is under construction for 8 m-resolution Earth imaging following a possible launch on Russia's SL-8 Cosmos.

TubSat-N was launched on 7 July 1998 by Shtil-1 and decayed 22 April 2002. Partnered with TubSat N, TubSat N 1 decayed 21 October 2000. TulSat-C was launched on 26 May 1999 by India's PSLV.

India

Indian Science Satellite Programme Series

Specifications

Indian launches

Rohini 1B

Launched: 18 July 1980 by SLV-3 from Sriharikota
Mass: 35 kg
Orbit: 305 × 919 km, 44.7°, 97 min
Comment: India's first successful launch, following the failure of Rohini 1A on 10 August 1979. Re-entered 20 May 1981.

Rohini 2

Launched: 31 May 1981 by SLV-3 from Sriharikota
Mass: 38 kg
Orbit: 186 × 418 km instead of planned 296 × 834 km, 46°, leading to re-entry on the day of launch

Rohini 3

Launched: 17 April 1983 by SLV-3 from Sriharikota
Mass: 41.5 kg
Orbit: 371 × 861 km, 46.6°, 92.7 min
Comment: Carried two cameras and L-band beacon, and returned more than 5,000 Earth images before de-activated 24 September 1984. Re-entered 19 April 1990.

USSR/RFAS launches

Aryabhata

Launched: 19 April 1975 by Kosmos 3 from Kapustin Yar
Mass: 360 kg
Orbit: 563 × 619 km, 50.7°, 92.6 min
Comment: Named after the 5th-century astronomer/mathematician. Designed for 6 months' x-ray astronomy. Control was to have been handed over to Indian scientists on day 3, but the Indian-designed transformer failed and only a few days' useful data were obtained. Decayed 10 February 1992.

Bhaskara 1

Launched: 7 June 1979 from Kapustin Yar by Kosmos 3
Mass: 442 kg
Orbit: 519 × 541 km, 50.6°, 92.5 min
Comment: Named after 7th-century Indian astronomer. With two TV cameras and microwave radiometers, it cost Rs65 million and was to spend 1 year studying India's resources. Useful ocean/land data received but the cameras malfunctioned. Nonetheless housekeeping telemetry was received until re-entry 17 February 1989.

Bhaskara 2

Launched: 20 November 1981 from Kapustin Yar by Kosmos 3
Mass: 444 kg
Orbit: 541 × 557 km, 50.7°, 92 min
Comment: Declared operational after receipt of more than 300 TV images of Indian sub-continent. Housekeeping telemetry was still being received 1991; re-entered 30 November 1991.

European launches

APPLE (Ariane Passenger Pay Load Experiment)

Launched: 19 June 1981 by Ariane L03 from Kourou
Mass: 670 kg
Orbit: 102° E GEO
Comment: India's first indigenous GEO test communications satellite. Indian TV programmes and educational teleconferences were relayed despite a jammed solar panel. De-activated September 1983.

SROSS Series

Background

India took its second step towards an autonomous space programme with the first attempted launch of the Stretched Rohini on 24 March 1987. SROSS 1 disappeared in the vehicle failure of first ASLV. It represented the first in the series of low-cost 150 kg satellites designed to carry scientific and remote-sensing payloads of 15–35 kg on both three-axis and spin-stabilised missions. ISRO postponed the first attempt at launching SROSS 1, from late 1985 to early 1987 because of ASLV stage 1 problems. The Space Agency wanted to use the satellite carrying a battery of inertial measurement instruments to monitor launcher performance. SROSS 1 had an octagonal body and was powered by solar cells on each face and deployed panels, and also carried two retro-reflectors for laser tracking and ISRO's 30–1,000 keV Gamma-ray Burst (GRB) Experiment.

SROSS 2 launched 13 July 1988, also disappeared in a launch failure, this time the second ASLV. The 150 kg, 85 cm diameter, 110 cm high satellite, with eight deployable solar wings feeding 100 W through a 12 Ah Ni/Cd battery, carried two instruments. West Germany provided the first instrument, a Monocular Electro Optical Stereo Scanner (MEOSS), a prototype for later earth resources instruments. The second instrument was ISRO's own 20–3,000 keV Gamma-ray Burst Experiment.

SROSS 3 (C-1) had the honour to be the first fully successful launch in the series when it was sent aloft on 20 May 1992 by ASLV-D3 from Sriharikota to a 251 × 436 km, 46°, 91.4 min orbit. Lighter than its predecessors, at 106 kg, it carried the National Physics Laboratory's Retarding Potential Analyser for low-latitude ionosphere/thermosphere investigations and ISRO's 20–3,000 keV Gamma-ray Burst Experiment (two CsI(Na) scintillation detectors; 37/76 mm diameter, 12.5 mm thick).

During flight, the Stage 4 spun up to only 80 rpm instead of 180 rpm, yielding low perigee and a short life, but the satellite still met mission specification before re-entry 14 July 1992.

On 4 May 1994, the SROSS C2 took off aboard ASLV-D4 from Sriharikota into a 433 × 619 km, 46.0°, 95.1 min orbit. India planned a four-year design life for the payload, which was identical to SROSS C. Ground controllers activated the RPA on 9 May. The GRB flown on C2 was an improved version of the one flown on the first C satellite with enhanced on-board memory and a better measurement of background scatter after each burst event. These led to the discovery of 12 candidate events detected up to 15 February 1995 out of 993 triggers. The C2 satellite is one of the Interplanetary Networks satellites.

This SROSS adjusted its orbit by lowering to an intended 429 × 628 km using RCS over 1–7 July 1994, giving India the potential to develop multiple deployment capabilities. The bus had 1.3 kg of fuel remaining of 5 kg fuel load at the end of the satellite lifetime.

International

European Space Agency Science Series

Background

ESA's contribution to the Inter-Agency Solar Terrestrial Physics programme forms the first cornerstone of the agency's Horizon 2000 space science plan. ESA's cost to completion was to be ECU761 million (1995 rate). The four Cluster satellites were lost in Ariane 5's failed debut in June 1996. Following the loss of the Cluster satellites, consideration was initially given to flying a ground spare satellite as a solo mission, the name being given as Phoenix. However, in April 1997, it was announced that this ground spare together with three other newly-built satellites would be launched as the Cluster-2 mission in 2000. Two launches of Russian Soyuz launch vehicles – each with an added upper stage – would each place a pair of Cluster-2 satellites into orbit. The first two, FM6 and FM7, were launched on 16 July 2000 by Soyuz 11A511U followed by FM5 and FM8 on 9 August 2000 by another Soyuz of the same type.

The X-ray Multi-Mirror (XMM) observatory forms the second, High-Throughput X-Ray Astrophysics, cornerstone of ESA's Horizon 2000 space science programme. ESA's Science Program Committee approved the Science Management Plan and Schedule on 21 June 1988, with a launch in 1999, to provide astrophysicists with a wide-band, high sensitivity instrument for up to 10 years of data gathering. Two competing mirror technologies were tested: the baseline monolithic nickel approach of Italy's Media Lario company and epoxy replication on CFRP carriers at Zeiss. The Mirror Development Model using four of these shells underwent X-ray testing in May-August 1993 with breadboard instruments. CFRP did not meet requirements because of moisture sensitivity; the nickel approach was chosen end-1993. This safer route means that XMM's mass has increased from 3,300 to 3,900 kg at launch, requiring a switch from Ariane 4 to Ariane 5. One nickel mirror module weighs 340 kg, instead of 137 kg for CFRP.

ESA Science Satellites 1967–1989

ESRO-2A: 29 May 1967, 74 kg, solar/cosmic rays, launch fail
ESRO-2B (Iris): 17 May 1968, 164 kg, as 2A but to a 334 × 1,085 km, 97.2°, 99 min orbit. Decayed 8 May 1971
ESRO-1A (Aurorae): 3 October 1968, 84 kg, ionosphere/auroras to a 258 × 1,538 km, 93.8°, 103 min orbit. Decayed 26 June 1970
HEOS 1: 5 December 1968, 108 kg, interplanetary magnetic fields and solar particles 20,020 × 202,780 km, 60.5°, 6,704.3 min on a Delta.
ESRO-1B (Boreas): 1 October 1969, 85 kg, as 1A, to a 298 × 378 km, 85°, 91.3 min orbit on a Scout launcher. Decayed 23 November 1969
HEOS 2: 31 January 1972, 117 kg, as number 1 to a 405 × 240,164 km, 89.9°, 7,477.1 min on a Delta. Decayed 2 August 1974
TD-1A: 12 March 1972, 472 kg, UV sky survey. 523 × 550 km, 97.6°, 95.4 min by Delta. Decayed 9 January 1980
ESRO-4: 22 November 1972, 130 kg, ionosphere/solar particles. 245 × 1,173 km, 91.1°, 99 min orbit by Scout. Decayed 15 April 1974
COS-B: 9 August 1975, 275 kg, γ-ray astronomy to a 442 × 99,002 km, 90.3°, 2,03.9 min orbit by Delta. Decayed.
GEOS 1: 20 April 1977, 573 kg, Earth's magnetosphere
ISEE 2: 22 October 1977, 166 kg, magnetosphere
IUE: 26 January 1978, 445 kg, UV spectroscopy
GEOS 2: 14 July 1978, 573 kg, as number 1
Exosat: 26 May 1983, 510 kg, x-ray astronomy
Giotto: 2 July 1985, 960 kg, comet fly-bys
Hipparcos: 8 August 1989, 1,140 kg, astrometric survey.
Rosat: 1 June 1990, 2,426 kg. West German EUV/X-ray telescope for astrometric survey.
ISO: 17 November 1995, 2,500 kg, IR astrophysics.
SOHO: 2 December 1995, 1,850 kg, solar and heliospheric observatory
Cluster 1, 2, 3 & 4: 4 June 1996, 1,200 kg each, magnetospheric studies. Launch failure.
XMM Newton: 10 December 1999, 3,764 kg, X-ray astronomy
Cluster 2 Salsa and Samba: 16 July 2000. 1,200 kg each, magnetospheric studies
Cluster 2 Rumba and Tango: 9 August 2000. 1,200 kg each, magnetospheric studies.
Integral: 17 October 2002, 4,100 kg, Gamma ray astronomy
SMART-1: 27 September 2003, 370 kg, space science and solar particles studies
Rosetta: 2 March 2004, 3,065 kg, cometary sampler

UPDATED

HIPPARCOS (ESA SCIENCE SERIES) Satellite

Background

As its name implies, the High Precision Parallax Collecting Satellite (HIPPARCOS) launched 8 August 1989 by Ariane from Kourou, undertook the first space-based stellar position/parallax survey.

Scientific investigations continued successfully, despite the failure of the launch craft's apogee motor. Stranded in GTO as it was and passing through the van Allen belts twice each orbit, the high radiation environment took its toll, the satellite stopped transmitting in June 1993 and ESA terminated operations early on 15 August 1993.

HIPPARCOS solved the previous errors on positions, parallaxes and proper motions by taking measurements typically below the 0.002 arcsec target at ninth magnitude for 120,000 stars. The Tycho element provided data on more than one million objects. Final HIPPARCOS results offer parameters 10–100 times more accurate than those from ground observations. It was completed in December 1995 and released to participating scientists in March 1996.

ESA made the HIPPARCOS and Tycho catalogues publicly available in March 1997. Project cost to completion about ECU400 million (1990 rate).

Infra-red Space Observatory (ESA SCIENCE SERIES) Satellite

Current Status
Building on the outstanding success of the ISO, scientists are planning its successor, the Herschel observatory scheduled for launch around 2007. The work begun by ISO will be sustained through the new generation of much bigger observatories.

Background
The Infra-red Space Observatory (ISO) operated from November 1995 to 8 April 1998 and was abandoned on 16 may 1998, almost a year longer than expected and made nearly 30,000 scientific observations. Helium consumption was smaller than planned enabling the satellite to operate over a period of 28 months.

The cryogenically-cooled Infra-red Space Observatory observed 2.5–200 µm IR radiation as a follow-up to the pioneering all-sky survey undertaken by IRAS (8–120 µm) in 1983. ISO has a sensitivity up to 1,500 times greater and ESA operates it as an observatory, studying specific targets, with about two-thirds of its observing time available to the general astronomical community. Among other objects, ISO studies selected galaxies and star formations within relatively cool clouds of gas and dust (typically 15–300,000). Within the solar system, it observes asteroids, comets and the atmospheres of the giant planets and moon Titan.

Estimated cost-to-completion to ESA of the specified 18-month in-orbit phase, excluding science instruments, is ECU750 million (1995 rate). Separately funded international consortia developed the four science instruments (delivered 1993). ESA/Aerospatiale signed the FFr1, 500 million prime contract 17 September 1990. Two major technical problems in 1992 delayed launch by at least eight months to early 1995 and added 28.6 MAU to the cost. The first problem was leakage from the cryostat's liquid helium valves. The second problem was contamination of the primary mirror's gold coating. ISO's spare mirror became the primary after being coated and the flight telescope was delivered February 1993.

The flight model instruments and telescope were integrated into the FM cryostat by end-1993. Isocam returned its first image (galaxy M51) 28 November; Isophot observed star Gamma Draconis 29 November; LWS observed the S106 region 30 November.

The payload consists of a Ritchey-Chretien telescope, with an effective 60 cm aperture and providing a 20 arcmin unvignetted FOV. It is suspended in the cryostat. Four prime focus instruments are mounted below the optical support, each occupying an 80° segment and receiving a three arcmin unvignetted FOV, providing photometric and imaging capabilities over 2.5–200 µm and medium/high resolution spectroscopy over 2.5–180 µm. In general, only one instrument is operated at a time, although the camera can acquire extra data when one of the others is in use. Also, the long wavelength channel of the photometer is active when the satellite is being slewed, yielding a partial 200 µm sky survey (IRAS operated <120 µm).

Specifications
Launch: 19 November 1995 on Ariane 44P from Kourou, French Guiana
Contractors: Aerospatiale (prime), DASA (cryostat payload module), Fokker (attitude control subsystem)
Mass: 2,498 kg at launch, 2,418 kg BOL, 1,515 kg dry
Orbit: 1,038 × 70,578 km, 5.2°, 24 h; initial 500 × 71,850 km, 5.25°. ISO Control Centre is at Villafranca, Spain, allowing about 13 h of useful observing time over each 24 h rev (no onboard recorder). NASA's Goldstone extends observations to almost 17 h daily (in return, NASA has 0.5 h observing time daily)
Spacecraft: 5.3 m high, 2.3 m wide, 3-axis controlled to within a few arcsec. The cylindrical payload module carries a conical sunshade and two star trackers; the service module provides basic spacecraft functions. Attitude control for stable observations of up to 10 h incorporates Earth/Sun sensors, star trackers, four rate integrating gyros, four skewed RWs and two branches of 8 × 2 N hydrazine thrusters. Real-time downlink (no onboard recorders) 33 kbit/s, 24 kbit/s of which is devoted to science data. The payload module is essentially a large cryostat. Inside the vacuum vessel is a toroidal tank holding 2,250 litres of superfluid helium at mission start to provide at least 18 months' observations (two years is now projected). Some of the IR detectors are cooled to 2K by direct connection to the tank; the boil-off gas cools other units. This is first routed through the optical support structure, where it cools the telescope and science instruments to about 3K; it then passes along the baffles and radiation shields before being vented to space. A small toroidal auxiliary tank near the cryostat's neck containing about 60 litres of normal liquid helium provided cooling on the pad of the last 100 h before launch. On 5 September 1996 a measurement of the remaining mass of liquid helium suggested that the operational lifetime of ISO should be around two years rather than eighteen months which was the design lifetime. This would mean that the satellite should operate through to the end of 1997.
Payload: Isocam 2.5–17 µm camera/polarimeter, 1.5/3/6/12 arcsec resolutions, two channels each with 32 × 32 element arrays. PI: Catherine Cesarsky, CEN-Saclay, France.

Isophot: 2.5–200 µm imaging photopolarimeter, operating in three modes as 30–200 µm far-IR camera, 2.5–12 µm spectrophotometer and 3–110 µm multi-band multi-aperture photopolarimeter. PI: Dietrich Lemke, MPI für Astronomie, Germany.

SWS: 2.5–45 µm Short Wavelength Spectrometer incorporating two gratings and two Fabry-Perot interferometers, 10 × 20 arcsec and 20 × 30 arcsec res. PI: Thijs de Graauw, SRON, Netherlands.

LWS: 45–180 µm Long Wavelength Spectrometer incorporating grating and two Fabry-Perot interferometers, 1.65 arcmin resolution. PI: Peter Clegg, Queen Mary & Westfield College, UK.

IASTP Series

Background
The Inter-Agency Solar Terrestrial Physics programme is a 12-satellite project, involving NASA, ESA, ISAS and Russia's Institute of Space Research (IKI). IASTP plans and executes detailed investigations of the Sun, Earth's space environment and Sun-Earth interaction. The origins of the Solar-Terrestrial science programme trace back to a decision to investigate new subjects made by the Inter-Agency Consultative Group (IACG) following the highly successful Comet Halley collaboration. IASTP, started in 1977, with planned completion in 2001, involves more than 100 universities, research labs and major contractors in 16 countries.

NASA budgeted US$18.6 million in FY88 to start development of Wind/Polar (rising to 64.4 FY89, 57.6 FY90, 96.6 FY91, 75.3 FY92, 72.6 FY93, 21.6 FY94, 40.0 FY95, 0 FY96) and provide instruments for Geotail, Soho and Cluster. Russia's KONUS γ-burst detector was delivered 1992 to fly aboard Wind, the first CIS instrument on a US satellite bus. GE Astro Space built the Polar/Wind. Total NASA development cost for GGS, Geotail, Soho and Cluster is estimated at US$732 million. Wind/Polar design/construction difficulties resulted in launch delays, leaving Geotail as the first in orbit following launch on 24 July 1992. Wind was launched by Ariane 412 on 1 November 1994 with Polar by Delta 7925 on 24 February 1996 performing simultaneous, co-ordinated measurements of key regions of earth's near space environment.

Cluster (IASTP SERIES) Satellites

Background
Cluster had four identical spacecraft scheduled on board the first Ariane 5, for field and plasma measurements of the magnetosphere. Cluster planned to map the 3-D extent and dynamic behaviour of small scale structures (from a few hundred km to a few thousand km) in the Earth's plasma. They would observe how solar particles interact with Earth's magnetic field and study the interactions of solar-terrestrial magnetic/electrical fields by direct measurement. Their loss on Ariane 5's debut was a major scientific blow. Only a month later at the SPC meeting held in London on 2–3 July 1996, it was agreed that a Cluster-2/Phoenix mission should be readied as soon as possible, with a final decision coming at the November 1996 meeting.

Studies to determine the right replacement included launching the ground spare which was shelved because the Cluster concept required a group of satellites in orbit, not just one. The next most plausible solution called for ESA to build three new satellites to accompany the ground spare, thus duplicating the planned 1996 mission, or developing a set of four new smaller satellites. At the November 1996 meeting, the ESA Board chose the option of flying the existing ground spare with three new identical satellites, on condition that the cost could be capped at 210 MAU. In April 1997, ESA announced that it had chosen to launch the Cluster-2 mission (as it is now called) using two Russian Soyuz-U launch vehicles, each carrying two Cluster-2 satellites. Satellites FM6 and FM7 were launched by Soyuz 11A511U on 16 July 2000 followed by FM5 and FM8 by a similar launcher on 9 August 2000.

Since the Soyuz-U launch vehicle is designed for launches to low Earth orbit only, the Russians may need to add a new upper stage to launch the two pairs of satellites into their highly eccentric orbits. Whether this will be the Soyuz-U/Ikar launch vehicle being marketed by STARSEM or another Soyuz-U variant has still to be clarified.

Specifications
Contractors: Dornier (prime), Alcatel (TT&C), British Aerospace (AOCS, RCS), Contraves (structure), FIAR (power), Fokker (thermal control), Laben (OBDH), DASA/MBB (solar array, thrusters), Saab (harness, separation system), Sener (booms), Odetics (tape recorder), Cubic (HPA)
Launch: 4 June 1996 by Ariane 501 from Kourou and 9 August 2000 (FM 5, FM 8)/16 July 2000 (FM 6, FM 7) by Soyuz
Mass: FM 1 1,183/531 kg launch/dry; FM 2 1,169/518 kg; FM 3 1,171/520 kg; FM 4 1,184/533 kg. 650 kg propellant, 72 kg payload
Orbit: initial 10° GTO, rising to planned 25,513 × 140,318 km, 90°, 66 h in five manoeuvres (numbers 1–3 each 240 m/s; number 4 630 m/s; number 5 370 m/s for plane change). Tuned so that the four satellites would fly in tetrahedral formation, passing in/out of Earth's magnetic field, crossing cusp, magnetopause, bow shock, magnetotail and other features. Separations adjusted between 200 km in cusp and 18,000 km in plasma sheet. FM 6/FM 7 to a 16, 869 × 121,098 km, 65°, orbit; FM 5/FM 8 to a 19,000 × 119,000 km, 65°, orbit.
Lifetime: 2 years nominal
Spacecraft: 2.9 m diameter 1.3 m high cylinders, with conductive surfaces, spin stabilised at 15 rpm. Magnetic field experiments mounted on two 5 m radial booms; two pairs of 100 m tip to

The four Clusters were launched in two double stacks (ESA) 0517302

An artists view of the Cluster constellation in earth orbit (ESA) 0572507

Cluster satellites in a test cell (ESA) 0572506

tip wire antennas for electric field measurements. Solar array to provide 224 W, including 47 W for the payload. Science data rate 16.9 kbit/s normal operation and 105 kbit/s burst mode; payload data outside of ground coverage stored in 1 Gbit tape recorder or 2.25 Gbit SSR for subsequent dumping. Telemetry downlink 2–262 kbit/s. DASA's S400 400 N MON/MMH thruster to establish the operational orbits, including the inclination increase in five firings from Ariane's 10° GTO, using 500 kg of propellant. RCS by eight DASA S10 MkII 10 N MON/MMH thrusters. Attitude determination <0.25° by star mapper and Sun sensor. Data from four satellites synchronised via highly stable onboard clock and time stamping at ground stations. Operated from ESOC via Odenwald (Germany) and Redu (Belgium). Science operations co-ordinated through a Joint Science Operations Centre in the UK; data distributed via Cluster Science Data System, using ESANET to transfer it to centres in Austria, France, Germany, Hungary, Scandinavia, UK, China and US

Payload: STAFF Spatio-Temporal Analysis of Field Fluctuations (AC magnetometer; France).

EFW Electric Fields & Waves (utilising long wire antennas; Sweden). WHISPER Waves of High Frequency and Sounder for Probing of Density by Relaxation (active density measurement of solar plasma; France).

WBD Wide Band Data (high frequency electric fields of several 100 kHz; US), DWP Digital Wave Processor (controls STAFF, EFW, WHISPER, WBD wave consortium experiments; UK).

FGM Fluxgate Magnetometer (DC magnetometer; UK).

EDI Electron Drift Instrument (measurement of electric field by firing electron beam in circular path for many tens of km around satellite, detector on other side picks up return beam; FRG).

CIS Cluster Ion Spectrometry (composition/dynamics of slowest ions; France).

PEACE Plasma Electron/Current Analyser (distribution, direction, flow and energy distribution of low/medium-energy electrons; UK).

RAPID Research with Active Particle Imaging Detectors (essentially a pinhole camera to measure high energy electrons/ions; FRG).

ASPOC Active Spacecraft Potential Control (removal of satellite excessive charge by emitting indium ions; Austria).

Geotail (IASTP SERIES) Satellite

Background

Geotail investigates the energy mechanisms in the magnetosphere tail and the physical processes in other important regions, such as the dayside magnetopause and bow shock. Scientists are interested in the geomagnetic field lines on the dayside, in a hemisphere of some 10 Earth radii, because they are compressed by the solar wind. The nightside lines are correspondingly stretched into a tail extending beyond 2,000R_E. A substantial fraction of the solar wind energy imparted to the magnetosphere is stored in those stretched lines, and enhanced activity arises in the inner magnetosphere and polar ionosphere when this energy is released.

NEC developed the Geotail under ISAS direction. EPIC, CPI, the Multichannel Analyser and the Inboard Magnetometer on the Magnetic Field Measurement Investigation are funded by NASA Goddard.

Specifications

Launch: 24 July 1992 by Delta 6925 from Cape Canaveral

Japan's Geotail was the first IASTP craft (ISAS) 0516798

Geotail seen atop its kick stage motor on the Delta launch pad at Cape Canaveral (NASA) NEW/1047490

Mass: 1,009 kg (360 kg propellant; 105 kg science payload)
Orbit: for distant tail Sun-synchronous Moon orbit plane double lunar swing-by, max apogee 200R_E; for near tail 7.5° ecliptic inclination, 8 × 30R_E. Orbit 41,363 × 58,542 km, 22.4°, 4,750.6 min. Time spent in Earth's shadow was 1.7 h max, including penumbra. In the first part of the mission, lasting 1.75 years, its apogees were held in the magnetotail (on Earth's night side) by repeated paired lunar swing-bys. After 4½ Earth orbits, the first lunar swing-by was 13,657 km at 15.34 GMT 8 September 1992, producing a 136R_E apogee in the magnetotail 18 days later. Farthest apogee was 199R_E on 25 April 1994. Four δV manoeuvres totalling 154.2 m/s in November 1994 reduced apogee to 50R_E and then six more totalling 161.5 m/s in February 1995 reduced it to 30R_E to study the neutral sheet current. No further δV manoeuvres are planned; it will remain in that orbit to support other IASTP missions
Life: 1.75 years in the distant tail and 1.45 years in the near tail (nominal mission completed July 1996)
Spacecraft: 20 rpm spin-stabilised cylinder 2.2 m diameter 1.6 m height with mechanically despun antennas. 340 W EOL provided by Si N-on-P BSFR cells + 3 × 19 Ah Ni/Cd batteries. Data at 16.384, 65.536 or 131.072 kbit/s real-time or playback (two 450 Mbit Odetics tape recorders) to NASA DSN, Usuda, Kagoshima (Japan). Operation/control from SOCC Sagamihara Operation Control Centre
Payload: Electric Field (EFD). Comprising the Spherical Probe, Wire Antenna and Electron Boomerang.
Magnetic Field (MGF). Fluxgate and Search Coil.
Plasma (LEP). Made up of Ion/Electron Energy Analyser, Solar Wind Ion Analyzer and Ion Mass Spectrometer.
Plasma (CPI). Consisting of the Hot Plasma Analyser, Solar Wind Ion Analyser and Ion Composition Analyser.
Energetic Particles (HEP). Constituted by the Low Energy Detector, Burst Detector and Isotope Telescopes.
Energetic Particles (EPIC). Composed of the Supra-Thermal Ion-Composition Spectrometer and the Ion Composition Subsystem.
The Plasma Waves (PWI). Made up of the Multichannel Analyser, the Sweep Frequency Analyser and the Wave Form Capture (both E/B fields).

UPDATED

Polar (IASTP SERIES) Satellite

Background

Polar monitors the ionosphere's role in substorm phenomena and in overall magnetospheric energy balance. It measures plasma energy input through the dayside cusp, and determine the characteristics of ionospheric plasma outflow and energised plasma inflow to the atmosphere. Like the Wind, it will also study characteristics of the auroral plasma acceleration regions, and provide multispectral auroral images of the footprint of the magnetospheric energy disposition into the ionosphere/upper atmosphere.

Lockheed Martin Astro Space was Polar and Wind prime contractor 0516799

Lockheed Martin Astro Space built the Polar satellite. It carries 11 instruments all with high data compression capabilities including experiments utilising the telemetry transmission at 2265.0 MHz (realtime 55.6 kbps, playlock 512 kbps) and command at 2085.688 MHz.

Specifications

Launch: 24 February 1996 by Delta 7925–10 from VAFB
Mass: 1,300 kg (301 kg propellant, 226 kg science payload, 999 kg dry)
Orbit: 5,141 × 50,605 km, 85.9°
Life: 3 years nominal, extended mission under study
Spacecraft: 2.7 m diameter 2.5 m height cylinder (stowed envelope) with conductive surfaces, 10 rpm spin-stabilised with a 180° precession manoeuvre every 6 months. Despun platform for imagers inertially stabilised to 0.2°. Two deployable 6 m lanyard booms for magnetic field instruments, two deployable 5.5 m Z-axis booms and four radial wire antenna (two 50 m, two 80 m) for electric field measurements. 14.9 m^2 solar array provides 511 W BOL, including 186 W for payload; 3 × 26.5 Ah Ni/Cd batteries. Four Olin MR-111 2.2 N spin, four MR-111 4.4 N spin axis precession + four MR-50 22 N δV thrusters. Telemetry received by DSN over four 45 min nominal daily contacts. Science data 55.5 kbit/s real-time and up to 512 kbit/s playback. Payload data outside of groundcoverage stored on 1.3 Gbit digital tape recorder. Mission operations/data processing from GSFC
Payload: MFE Magnetic Fields Experiment (University of California at LA).
EFI Electric Fields Instrument (University of California, Berkeley).
PWI Plasma Wave Instrument (University of Iowa).
TIDE/PSI Thermal Ion Dynamics Experiment/ Plasma Source Instrument (NASA Marshall).
HYDRA Fast Plasma Analyser Experiment (University of Iowa).
TIMAS Toroidal Imaging Mass-Angle Spectrograph (Lockheed Palo Alto Research Lab, California).
CEPPAD/SEPS Comprehensive Energetic Particle Pitch Angle Distribution/Source Loss Cone Energetic Particle Spectrometer (Aerospace Corp, California).
CAMMICE Charge & Mass Magnetospheric Ion Composition Experiment (Boston University).
PIXIE Polar Ionospheric X-ray Imaging Experiment (Lockheed Palo Alto Research Lab, California).
VIS Visible Imaging System (University of Iowa).
UVI Ultraviolet Imager (University of Washington).

SOHO (IASTP SERIES) Satellite

Current Status

During June 1998, ground controllers at the NASA Goddard facility lost contact with the SOHO satellite. Using the Deep Space Network, Goddard engineers managed to locate the satellite one month later. Apparently operator fault turned the satellite antenna away from ground stations. In March 2003, SOHO continues to operate well and transmit good data, its life extended from a projected 6 years to 20 years as a result of highly accurate launch vehicle insertion parameters.

Soho final flight processing at Canaveral (Theo Pirard/Space Information Centre) 0517301

Background

Until its loss, the Solar Heliospheric Observatory made continuous observations of the solar surface, corona and wind, from a vantage point in the L1 Lagrangian libration point, about 1.5 million km out. ESA and the international scientific community seeks the data to investigate physical processes that form and heat the corona, maintain it and give rise to the expanding solar wind, and to investigate internal structure by helioseismology and solar irradiance variations.

Matra Marconi Space France is the prime contractor, with help from BAe (AOCS), MMS UK (payload module, propulsion subsystem), Alenia (structure, harness), CASA (thermal control), Saab Ericsson (communications, data handling, software), FIAR (power), SAFT (batteries), Odetics (tape recorder), Cubic (HPA).

Specifications

Launch: 2 December 1995 by Atlas 2AS from Cape Canaveral. First solar image recorded 19 December 1995 and commissioning formally completed 16 April 1996
Mass: 1,864 kg at launch (655 kg payload)
Orbit: circling L1 libration point since 14 February 1996, 1.5 million km out on Earth-Sun line
Lifetime: 2 years nominal, consumables sized for 6 years (accurate launch left 20 years of propellants)
Spacecraft: 3.6 m height, 3 m diameter, 3-axis stabilised Sun-pointing with 10 arcsec accuracy and a pointing stability of 1 arcsec per 15 min (10 arcsec per 6 months). Instruments accommodated in payload module, with a high degree of decoupling from the service module. Solar panels generate about 1,400 W (payload requires 440 W). S-band telemetry received by NASA's Deep Space Network over 3 × 1.3 h + 1 × 8 h periods each day. Continuous science data stream 40 kbit/s, increased to 246 kbit/s when MDI solar oscillation imaging instrument is operating in high rate mode; 1 Gbit tape recorder + 2 Gbit SSR. EOF Experiment Operations Facility at NASA Goddard co-ordinate/plan science operations. Its main tasks are to organise the payload's real-time operation and to control the imaging/spectrometric instruments during the daily 8 h contacts. MDI's solar oscillations imaging data requires a specialised data facility at University of Stanford, Palo Alto
Payload: SUMER Solar Ultraviolet Emitted Radiation (FRG)
CDS Coronal Diagnostic Spectrometer (UK)
EIT Extreme UV Imaging Telescope (France)
UVCS UV Chronograph Spectrometer (US)
LASCO White Light/Spectrometric Chronograph (US)
SWAN Solar Wind Anisotropies (France)
CELIAS Charge, Element/Isotope Analysis (FRG)
COSTEP Suprathermal/Energetic Particle Analyser (FRG)
ERNE Energetic Particle Analyser (Finland; COSTEP/ERNE are combined in the single CEPAC instrument)
GOLF Global Oscillations at Low Frequencies (France)
VIRGO Variability of Solar Irradiance (Switzerland)
MDI Michelson Doppler Imager (US).

Wind (IASTP SERIES) Satellite

Background

The Wind was built to investigate sources, acceleration mechanisms and propagation processes of energetic particles and the solar wind. IATSP also wants it to provide complete plasma, energetic particle and magnetic field input data for magnetospheric and ionospheric studies, determine the magnetospheric energy output to interplanetary space in the upstream region, and provide baseline ecliptic plane observations for heliospheric studies.

Lockheed Martin Astro Space built the satellite. Programme cost US$173 million.

Specifications

Launch: 1 November 1994 by Delta 7925–10 from Cape Canaveral
Mass: 1,250 kg (300 kg propellant, 198 kg science payload, 752 kg dry)
Orbit: lunar swing-by (first 27 December 1994) to $250R_E$ in first 2 years; L1 halo (3×10^5 km) from November 1996
Life: 3 years nominal, extended mission plans under study
Spacecraft: 2.6 m diameter 2.5 m height cylinder with conductive surfaces, spin-stabilised at 20 rpm. Two deployable 12 m lanyard booms for magnetic field instruments, two deployable 5.28 m Z-axis booms plus four wire antennas (two 7.5 m, two 50 m) for electric field measurements. 11.5 m^2 solar array provides 472 W BOL, including 144 W for payload; 3 × 26.5 Ah Ni/Cd batteries. Four Olin MR-111 2.2 N spin plus eight MR-50 22 N δV thrusters. Telemetry received by DSN over 2 h nominal daily contact. Science data rates 5.65/11.3 kbit/s real-time and up to 128 kbit/s playback. Payload data outside of ground coverage stored on 1.3 Gbit digital tape recorder. Mission operations/data processing from NASA Goddard. Telemetry 2275.5 MHz (realtime 5.56 kbps, playlock 32/64 kbps). Command 2094.9 MHz (realtime 250 kbps).
Payload: MFI Magnetic Fields Investigation (GSFC)
WAVES Radio & Plasma Wave Experiment (Observatoire de Meudon, France)
SWE Solar Wind Experiment (GSFC)
EPACT Energetic Particles Acceleration, Composition, Transport (GSFC)
TGRS Transient Gamma Ray Spectrometer (GSFC)
SMS SWICS/Mass STICS (Solar Wind Mass/Suprathermal Ion Composition Studies) (University of Maryland)
3D PLASMA Energetic Particles & Three Dimensional Plasma Analyser (University of California, Berkeley)
KONUS Gamma Ray Burst Investigation (Ioffe Physical Technical Institute, St Petersburg).

Wind is installed on Delta (NASA) 0517299

Ibiza/Impact Satellite

Background

The International Investigation BIsatellitaire des Zones Aurorales (Impact: Investigation of Magnetospheric Particle Acceleration and Turbulence) dual satellite mission will pick up where Explorers S3-3 and DE-1, Viking, Akebono and Freja ended their investigations. The former satellite's coverage revealed that high-time resolution using two co-ordinated satellites is essential for a deep understanding of particle acceleration phenomena in the ionosphere.

The steering group, set up in January 1996, issued an Invitation to propose experiments in late March 1996 with a 30 September 1996 due date. Cost to completion is estimated at FFr230 million. 75 per cent will be spread between Germany (DARA, with Max Planck Institute for Aeronomy the major participant), France (CNES) and Sweden (Swedish National Space Board). Canada, UK and US may be contributors.

Once in space, the satellite will investigate, with unprecedented spatial and temporal resolution, the auroral acceleration region at 1,000–10,000 km altitude, where electrons/ions are accelerated by energy from the magnetosphere. Twin satellites will cross the same magnetic field lines simultaneously, but tens to hundreds of km apart.

Several delays due to changes in scientific specifications for payloads.

Specifications

Launch: planned for 2004/2005 aboard a vehicle to be selected, but an Ariane polar mission is a strong candidate
Design life: 2 years, covering the Northern hemisphere twice + Southern hemisphere once
Mass: each 320 kg at launch
Orbit: about 3,000 × 10,000 km, 97.2°. Dual launch with Earth observation satellite into initial 800 km, 98.7° is assumed, then joint solid motor establishes required orbit. Natural rotation of the line of apsides will fully rotate the apogee over 16 months, allowing coverage of the Northern, then Southern, then Northern hemispheres
Spacecraft: 15 rpm spin-stabilised perpendicular to orbital plane (180° switch every 8 months). 650 kbit/s max S-band downlink to 9 m antenna at Kiruna (Antarctica station required when apogee over Southern hemisphere); 1.6 Gbit science data/rev (1 Mbit/s max, typically 100 kbit/s), 3 Gbit memory. Two 50 m wire booms. 140 W required. Onboard GPS orbit determination. 75 m/s δV capacity
Candidate payload: fluxgate magnetometer (2.4 kg, 2.5 W, 8 kbit/s), search coil magnetometer (2.5 kg, 1 W, 8 kbit/s), electric field/waves (9 kg, 22 W, 250 kbit/s), electron drift (7.2 kg, 6.6 W, 2 kbit/s), electron analysers (6.8 kg, 9 W, 250 kbit/s joint for particles analysers+correlator), ion analysers (9.0 kg, 14 W), correlator/FPA (2.0 kg, 3 W), cold plasma (5 kg, 8 W, 30 kbit/s), potential control device (1.5 kg, 4 W, 3 kbit/s), intersatellite propagation experiment (2 kg, 2 W, 10 kbit/s), UV imager (one satellite only; 8.5 kg, 7 W, 120 kbit/s), electron gun (one satellite only; 5 kg, 10 W); two 50 m wire booms (14 kg), radial booms + HF antenna (9 kg).

Interkosmos Series

Current Status

Interkosmos, formally the Council for International Co-operation in the Studies and Uses of Outer Space, built collaborative science projects among its 10 member states and with other nations, until it closed in 1994. The projects continued under IKI, IZMIRAN and other institutes. Current activities include the Aktivny, APEX, Interball and Relikt science satellite projects. After it closed, funds for joint Interkosmos projects evaporated. Russia offered its technology and launchers free of charge and each country bore its own costs. The programme embraced small science satellites, medium to launch series and human space flight participation with national cosmonauts from Interkosmos countries. No new missions have recently been announced and existing projects continue to suffer delays.

Background

The first Interkosmos launch, Cosmos 261 in 1968 from Plesetsk, studied air density and polar auroras. Each member financed its own contributions, with the CIS/USSR supplying the basic satellite and the launch costs. Scientific results became the common property of all member nations.

RFAS/USSR-led International Earth-orbit Science Missions

Project	Launch date	Spacecraft	Orbit	Objective
Aktivny-IK	September 1989	AUOS parent	Low Earth Orbit	Study the effects of ULF waves on the magnetosphere;
		+ Magion sub-sat		Interkosmos project; parent still operational
Granat	December 1989	Astron-type	Prognoz-type	Study of gamma/x-ray radiation with France/Denmark; still operational
Gamma	July 1990	Progress-type	Low Earth Orbit	Study of gamma/x-ray radiation in co-operation with France
APEX	December 1991	AUOS parent	Low Earth orbit	Electron and plasma probing of magnetosphere; Interkosmos
		+ Magion sub-sat		project; parent still operational
Coronas-I	March 1994	AUOS-SM	Low Earth Orbit	Helioseismology and solar activity; Interkosmos project, still operational
Interball	August 1995	Prognoz-M2	Prognoz + intermediate	Study of magnetosphere and plasmasphere in 14-nation
	August 1996	+ Magion sub-sats	Molniya-type	co-operation; one Prognoz/ Magion pair in each orbit type
Priroda	April 1996	Mir module	Low Earth Orbit	Mir remote sensing module, including France's Alissa radar
Bion 11	December 1996	Vostok-type	Low Earth Orbit	Life sciences research on living organisms, including 2 monkeys; NASA involvement

The first programme ran from 1967 to 1980, with Czechoslovakia a noted contributor to every unmanned mission. This culminated in the debut of the first Magion satellite on Interkosmos 18, the first non-Soviet satellite in the series.

Interkosmos' second programme, lasting from 1980 to 1985, saw a move towards Earth surface studies, with the ocean dynamics investigations of Interkosmos 20/21. By October 1984, when the 17th Interkosmos conference was held in Berlin, 11 high-altitude Vertikal research rockets and 22 Interkosmos satellites had been launched, the last in August 1981.

By the time for the third programme in 1985, the programme plan stated an economic justification based on the results already achieved, but made no mention of any new launches. However, as a member of the Inter-Agency Consultative Group, Interkosmos participated in the Vega missions and the international collaboration on the 1985/86 studies of Halley's Comet. It also made major contributions to the 1988/89's Phobos mission. In April 1985, The Soviets renamed Prognoz 10 (Intershock), one of the solar radiation series, Interkosmos 23. Aktivny, APEX and Interball subsequently appeared as major projects in October 1987. Interkosmos announced proposals for two Coronas observatories that would monitor the Sun during its period of maximum activity in the early 1990s. The SAS aeronomy satellites were described in 1989 (but later cancelled) and agreement was near on the Regatta contribution (now postponed indefinitely) to the Inter-Agency Solar-Terrestrial Physics (IASTP) programme.

With the exception of Interkosmos 6, a Vostok-type vehicle, and Interkosmos 15, an AUOS Automatic Unified Orbital Station prototype, Interkosmos 1–16 all were light research satellites. Weighing 320–375 kg, they were based around a standard ellipsoidal pressurised shell. Czech scientists made contributions to all these satellites, and provided the orientation system for the more advanced 422 kg AUOS, introduced in operational form with Interkosmos 17. Features included onboard data processing and provision for a sub-satellite. APEX is flying the last AUOS-Z model (Z means Seemly, or Earth, denoting 3-axis pointing about a z-axis pointing vertically from the centre of the earth). Coronas-I introduced the AUOS-SM in 1994; the Prognoz-M2 debuted on Interball in 1995.

Aktivny-IK/ Interkosmos 24

Background

Conceived as first in a series of three Active Satellites and as the first subsatellite in a series of international scientific projects Active, Apex and Interball. After a two-year delay, Aktivny was launched in 1989 under the Interkosmos programme, releasing Czechoslovakia's Magion 2 to monitor the magnetospheric propagation of VLF waves and their effects on the van Allen belts. The parent craft was also designated Interkosmos 24. It was the first dual-craft active probing of the deep magnetosphere. Both Magion 2 and the Interkosmos 24 craft suffered initial problems. First, the large ORU-20 antenna failed to deploy as expected and Magion could not begin its science programme until 21 January 1990. Then, the failure of the small satellite's thruster prevented control of the separation from its parent. The Interkosmos received its last Magion contact on 20 November 1990. The main satellite is operable but is not used regularly, although some data is still used in support of other missions.

The development of Aktivny cost the Soviets a reported Rb4.8 million. Its payload consisted of a VLF generator, that was a 5 kW 9.6±0.2 kHz transmitter, radiating via 20 m diameter ORA-20 loop antenna. The antenna unfurled from a reel of 1 mm-thick aluminium (deployment problem meant that in-space inductivity was 25 per cent of expected). The Plasmas generator: modulated plasma environment around antenna. A neutral xenon injector pumped xenon around satellite for monitoring of ionisation of neutral gas flow and its propagation effects. A Low-frequency Wave System (LWS) was made up of a 0.30–22 kHz receiver. The Plasma radio spectrometer was a 0.10–10 MHz receiver. Proton/Electron Spectrometer (PES) monitored in 30–700 keV electron plus 40–700 keV proton flux. The Soft Electron Detector (SED) measured 0.01–10 keV electron flux. KM-6 and Langmuir probed plasma electron/ion temperature and number densities. The NAM-5: measured plasma mass composition.

The Institute of Terrestrial Magnetism, Ionosphere & Radio Wave Propagation (Izmiran) support the Interkosmos project, with IKI leading the effort, with contributions from Bulgaria, Former East Germany, Cuba, Hungary, Poland, Czechoslovakia. Soviet (now Russian), Czech, Cuban, US, Japanese, Canadian and Brazilian ground stations provided complementary observations.

Specifications

Launch: 28 September 1989 by Tsyklon 3 from Plesetsk
Mass: about 450 kg parent craft, 51.2 kg sub-sat
Orbit: 500 × 2,493 km, 82.6°, 115.3 min permitting injection at auroral latitudes. Sub-sat separation was planned to vary from 100 m to 100 km but Magion thruster failed
Spacecraft: Interkosmos AUOS 3-based parent craft.

APEX/Interkosmos

Background

Second in a series of three, the first series of Interkosmos subsatellite launches. The Soviet Union launched APEX (Active Plasma Experiment) under the Interkosmos programme, releasing Czechoslovakia's Magion 3 sub-satellite to monitor the propagation of electron beams and plasma injected into the magnetosphere by the parent craft, as their separation varied from tens of m to 1,000–2,000 km. It was the first CIS/Soviet mission of this type, but closely complemented Aktivny's investigations. Simultaneous ground, balloon and sounding rocket observations allowed the spacecraft release and separation to be measured through a variety of paths. It was the last flight of this AUOS-Z model. The AUOS-SM design was introduced in 1994 on the Coronas mission.

APEX launch processing at Plesetsk on the SL-14 Tsyklon. Magion 3 is at far left. The Aktivny mission configuration was identical (Dr Pavel Triska) 0517149

Magion separated and began operating on 28 December 1991. It drifted from its parent because of differential drag. A small thruster fired onboard on 9 January 1992 for the first time to bring the two craft back together again by 30 January. Similar manoeuvres in May and 11–26 June 1992 accomplished the same objective. The last approach was within 100 m with relative velocity almost zero.

About 200 plasma injections lasting 4–20 minutes performed the same task that normally would have required an estimated 200 sounding rocket flights. Magion detected all of these events. The small sub satellite operated with no problems until 20 August 1992 when it suddenly suffered a power shortage, probably caused by radiation damage to the battery charge system after a period of enhanced solar activity. A ground station received the last telemetry 9 September 1992.

Specifications

Launch: 18 December 1991 by Tsyklon 3 from Plesetsk
Mass: 1,000 kg parent satellite; sub-satellite 52 kg
Orbit: initially 440 × 3,083 km, 82.5°, 121.3 min. Separation varied from <100 m to 1,000–2,000 km, using Magion's single 0.2 N compressed air thruster
Spacecraft: Interkosmos AUOS-Z based parent craft; 56 cm diameter sub-sat almost identical to Magion 2. AUOS was introduced on the Interkosmos 15 mission.

Interball/Interkosmos

Background

Interball is a 20-nation collaborative project to investigate the interaction of solar wind with Earth's magnetosphere and the movement of solar wind's particles through the auroral oval. It is the third in a series (Active, Apex and Interball) of Interkosmos science instrument platforms, the first Interkosmos subsatellite projects. Interkosmos launches Interball satellites into tow types of orbits using the new Prognoz-M2 design. The first orbital type monitors magnetotail processes in conjunction with its S2-T sub-sat, and the second set operating at a higher inclination observe auroral phenomena with its S2-A companion, a Magion sub-satellite provided by the Czech Institute of Atmospheric Physics' Upper Atmosphere Department S2-Y sub-satellites have the same Canadian UV imager first flown aboard Sweden's Viking auroral research satellite in 1986.

All four Interball craft carry fields and particle instrumentation. Sweden provided its Promix-3 ion composition spectrometers. France's CESR supplied two types of particles detector, one for the lower craft only to study auroral acceleration mechanisms. LPCE contributed to electric and magnetic field monitoring on the lower pair, and CRPE contributed an ion mass spectrometer. The auroral probe carries the University of Calgary's paired UV cameras, built by Canadian Astronautics. The independent cameras cover different UV bands and each comprise a photocathode, multi-channel plate, fibre optic transfer bundle and two dimensional CCD. Images, taken via an f/1 inverse Cassegrain wide field telescope, are motion-compensated by the radiation hardened read-out electronics and then processed before transmission. The imager provides sequential shots at 20 second intervals and global views in 80 seconds.

Specifications

Institutions/contractors: 13 nations collaborating with RFAS (Austria, Bulgaria, Canada, Cuba, Czech Rep, ESA, Finland, France, former GDR, Greece, Hungary, Poland, Romania, Sweden), Canadian Astronautics Ltd was hardware contractor for University of Calgary's ultra-violet auroral imager

Launches: tail pair 2 August 1995 (Magion 4); auroral pair 29 August 1996 (Magion 5)
Mass: Interball 1 (tail parent) 1,270 kg, Magion 4 sub-sat 58 kg; Interball 2 (auroral parent) 1,400 kg, Magion 5 sub-sat 62 kg
Orbit: tail probe 793 × 191,900 km, 62.9°; auroral probe planned 770–19,200 km, 63° (passing 5,000–15,000 km over the auroral oval). Magion thruster varies separation 10–10,000 km from parent
Spacecraft: 0.5 rpm spin-stabilised (about Sun axis, accurate to 1°) Prognoz-M2 parent craft incorporating pressurised (0.8–1.2 atmospheres, 20°C) quasi-cylindrical equipment bay and four tapered solar panels totalling 8.34 m^2 cell area; 36 Ah Ni/H_2 battery (tail probe also has 135 Ah Li battery for eclipses). ACS by 0.7 N N_2 thrusters (eight tanks holding 8 l/18.6 kg LN_2. Data stored in 30 Mbyte memory, dumped at 16 kbit/s (65 kbit/s
Tail probe: 270 kg science payload. Nine plasma instruments; three energetic particles (0.01–150 MeV) and solar x-rays; 3-axis magnetometer and wave analysers; 0.1–2 MHz radio detector
Auroral probe: 400 kg science payload. Seven plasma instruments (four hot plasma 0.02–25 keV, three thermal 0.1–100 eV), ion emitter for spacecraft potential control; 3-axis magnetometer; wave analysers and 0.1–2 MHz auroral radio detector; energetic particle detectors; paired auroral oval imagers.

Oscar

Background

The Orbiting Satellites Carrying Amateur Radio (Oscar) series of small satellites was initiated for radio amateurs to experience satellite tracking and participate in radio propagation experiments. The World Administrative Radio Conference allocated frequencies for the Amateur Satellite Service, including 29 MHz (10 m), 145 MHz (2 m), 435 MHz (70 cm), 1,270 MHz (24 cm) and 2,400 MHz (13 cm).

Transmitting low-powered signals, initially battery-operated and offering short lives, the satellites have become increasingly sophisticated. They have served school science groups, provided emergency communications for disaster relief, acted as technology demonstrators and transmitted Earth imagery. The UK's UoSAT series alone has involved hundreds of schools and thousands of groups worldwide in activities using simple antennas, receivers and personal computers. The work ranges from telemetry transmission only, through amateur radio communications, to advanced experiments such as testing indium phosphide solar cells and transputers, and monitoring radiation effects on electronics – satellite engineering for a fraction of the time and cost of the more advanced conventional satellites. Somewhat confusingly, US military Transit satellites also carry Oscar designations.

Israel's TechSat, launched 28 March 1995 on a Russian Start, included a store/forward transponder operating at amateur frequencies, but the launch failed. Mexico's Unamsat was lost at the same time. Italy's Itamsat and Portugal's PoSAT include amateur services. Similarly, SA Amsat is providing a voice and digital package to South Africa's Sunsat. Arizona State University's ASUSat will provide a J-mode voice repeater. Chile's Amsat-CE is planning a Microsat-type satellite providing a digital voice transponder. Malaysia's 1997 microsatellite includes amateur services.

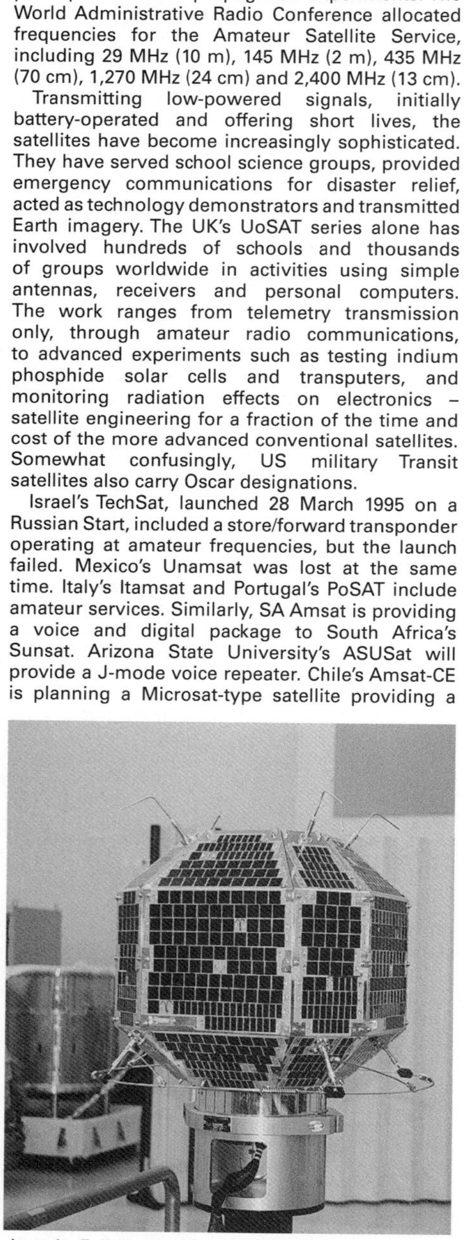

Japan's Fuji 3 provides worldwide amateur radio communications 0517300

Oscar 13 was launched by Ariane 4 in June 1988 and incorporated a liquid thruster to attain a high-inclination elliptical orbit (Amsat-DL) 0516802

Amsat-P3DL has built the 500 kg Phase 3-D satellite for operation in a 4,000 × 47,000 km, 63.4° Molniya-type orbit following launch by Ariane 502. Project cost is US$3.5 million, including DM1.3 million Ariane contract. It is the first to cover 10/24 GHz. Three-axis control (three MWs with magnetic bearings) point the antennas at Earth, along with two color CCD cameras (narrow/wide-angle) provided by Amsat-Japan. 12 GPS antennas allow it to derive its own Kepler elements. The CEDEX experiment monitors particle radiation. The bus is hexagonal prism; 67.50 cm height, 2.23 m diameter. Solar array provides 870 W BOL. 100 mN 600–700 W ATOS ammonia plasma station-keeping thruster from the University of Stuttgart's Institute of Space Systems. It is released into a 500 × 35,000 km, 10° GTO, where a 400 N burn will raise apogee to 47,000 km. A major burn will then lift perigee to 4,000 km while cranking inclination to 60°. Total 400 N burn duration is 20 minutes. Finally, ATOS will create the required 63.4° over two years. Some 250 kg of 3D's initial mass is propellant.

Analogue uplink is (MHz): 21.210–21.250, 145.840–145.990, 435.550–435.800, 1,269.250–1,269.500, 1,268.325–1,268.575, 2,400.350–2,400.600, 2,446.450–2,446.700, 5,668.550–5,668.800. Analogue downlink: 145.805–145.955, 435.475–435.725, 2,400.225–2,400.475, 10,451.025–10,451.275, 24,048.025–24,048.275. Digital uplink: 145.800–145.840, 435.300–435.550, 1,269.000–1,269.250, 1,268.075–1,268.325, 2,400.100–2,400.350, 2,446.200–2,446.450, 5,668.300–5,668.550. Digital downlink: 29.325–29.335, 145.955–145.990, 435.900–436.200, 2,400.650–2,400.950, 10,451.450–10,451.750, 24,048.450–24,048.750. The Rudak digital packet unit allows up to 256 kbit/s.

Specifications

Oscar 1 was launched 12 December 1961 with Discoverer 36 by Thor Agena. A group of enthusiasts in California persuaded the USAF to replace ballast on the Agena upper stage with the 4.5 kg Oscar package. 570 amateurs in 28 countries reported receiving its simple 'Hi-hi' morse code signals on the 2 m band before it re-entered after 312 revolutions. It led to the creation of Amsat in 1969. Out of service (January 1962, deorbited 31 January 1962).

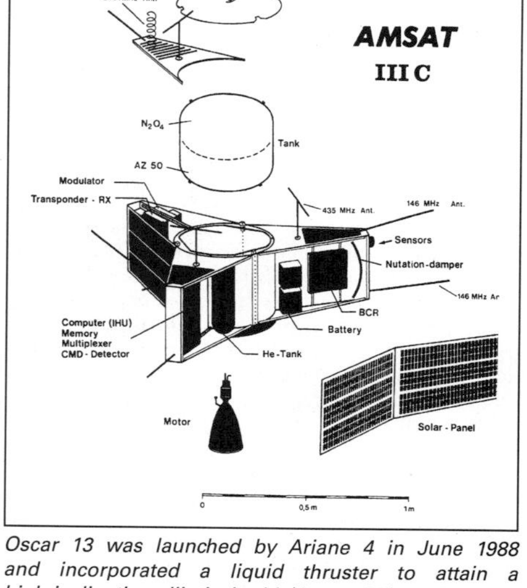

Oscar 13 was launched by Ariane 4 in June 1988 and incorporated a liquid thruster to attain a high-inclination elliptical orbit (Amsat-DL) 0516803

Oscar 2 launched on 2 June 1962 and Oscar 3 on 9 March 1965, were DoD piggybacks. Oscar 3 was the first true amateur satellite, relaying voice contacts at 145 MHz. Decayed 21 June 1962.

Oscar 3 was launched 9 March 1965 by SLV-2/Agena D.

Mass: 18 kg. 2.5 W DC power. Orbit 874 × 902 km, 70.1°, 102.7 min.

Oscar 4 was launched 21 December 1965 by Titan 3C provided the first US-Soviet amateur link. Orbit 162 × 33,549 km, 26.8°, 587.5 min. Decayed 12 April 1976.

Oscar 5 was launched 23 January 1970 with ITOS 1. 18 kg. 1,435 × 1,481 km, 102°, 115 min. Built by Melbourne University, Australia. Battery-powered; transmitted telemetry for 46 days. The university compiled tracking reports from hundreds of stations in 27 countries.

Oscar 6 was launched 15 October 1972 with NOAA 2. 16 kg. 1,450 × 1,459 km, 101.7°, 114.9 min. First Phase 2 satellite. Solar panels powering Ni/Cd battery permitted operations until mid-1977. Provided store/forward for morse and teletype; up 145.90–146 MHz, down 29.450–29.550 MHz at 1 W. Out of service due to battery failure 21 June 1977.

Oscar 7 was launched 15 November 1974 with NOAA 4. 29 kg. 1,444 × 1,462 km, 101.7°, 114.8 min. Phase 2 similar to Oscar 6, built by German, Canadian and Australian radio hams under direction of US Amsat. Two 2 W repeaters provided store/forward morse & teletype. Out of service due to battery failure 1980.

Oscar 8 was launched 5 March 1978 with Landsat 3. 27 kg; 33 cm high, 38 × 38 cm. 903 × 917 km, 99°, 103 min. Last Phase 2 providing similar store/forward service (145.90 up, 29.40 MHz 1–2 W down) of number 7 but also educational applications. Japan transponder 145.90–146.00 up, 435.10–435.20 MHz 2 W down (telemetry 435.095 MHz). Design life 3 years; served to mid-1983. Out of service due to battery failure 24 June 1983.

Oscar 9 (UoSAT 1) was launched 6 October 1981 with SME. Re-entered 13 October 1989. Decayed 13 October 1989.

Oscar 10 First successful Phase 3 satellite and the first Oscar with on-board propulsion, built by Amsat-DL, launched 16 June 1983 by Ariane L6. The first was lost on the failed Ariane L02 of 23 May 1980. Oscar 10 was struck twice by stage 3 soon after separation, damaging an antenna. Its 400 N liquid thruster also fired 80 seconds too long, so that the planned 1,500 × 35,800 km, 55° 11 hours orbit ended as 3,952 × 35,510 km at 25.9°. 140 kg at launch, 90 kg on-station mass, spin-stabilised and three-point star structure (see Oscar 13). It was to have provided emergency communications for disaster relief, transmissions of education programmes direct to classrooms, and two-way long-distance communications for 17 hours daily but its computer failed, through particle radiation, although its transponders are usable. See Oscar 13 for transponder details. It is operational only periodically, depending on the Sun's angle.

Oscar 11 (UoSAT 2) was launched 1 March 1984 with Landsat 5. 60 kg; Sun-synchronous at 651 × 669 km 97.8°, 98 min orbit. Built by Surrey University, UK. With three year design life, it remains operational providing Earth imaging and store/forward. The DCE Digital Communications Experiment pioneered store/forward mailbox links. The Digitalker Speech Synthesiser is used for school demonstrations (in 1988 it supported a Canadian/USSR expedition to the North Pole). Down: 145.825/435.025MHz 1,200/4,800bit/s AFSK-FM.

Oscar 12 (Fuji 1, JAS 1) was launched 13 August 1986 piggyback with EGP into 1,448 km, 50°orbit 50 kg, including 11 kg dummy mass. Japan Amateur Satellite was 26-faced polyhedron, 40 × 40 × 47 cm, powered by 6.9 W/14 V average from 979 2 × 2 cm Si cells and 6 Ah Ni/Cd battery. The satellite carries two transponders: 145.9–146.0/435.8–435.9 MHz up/down analogue (1 W transmit), and 145.85–145.91/435.91 MHz up/down digital (1 W transmit). Developed by Japan Amateur Radio League, with system design/integration by NEC. Number 1 used Flight Model 2; FM 1 was launched 7 February 1990 as Oscar 20. Number 1 was taken out of service November 1989 because of battery failure.

Oscar 13 was launched 15 June 1988 by first Ariane 4. Built by Amsat-DL and similar to Oscar 10. Released into GTO and fired 400 N liquid bipropellant thruster twice (51 seconds 22 June; 341 seconds 6 July) to raise orbit to 2,545×36,265 km, 57.7°. Re-entry came 6 December 1996 resulting from orbital perturbations lowering perigee. 140 kg initial mass decreased to 90 kg at beginning of six-year operational life. 24,000 solar cells on three-point star structure provide 40 W (25 W after five years). Payload included the Rudak digital packet communications unit and a microprocessor for managing/monitoring the satellite's operation.

UoSAT 3/4 and the four Microsats integrated on their Ariane carrier. Webersat is at the rear, slightly taller than the other Microsats. DOVE is at far right (University of Surrey) 0516804

Spot 3's Ariane carried six passengers on its ASAP platform. Clockwise, from far left: Itamsat, Kitsat 2, Stella (apex), PoSAT 1, Eyesat and HealthSat 2 0517307

Three transponders: 50 W 435.423–435.573/145.825–145.975 MHz up/down mode B; 50 W 1,269.641–1,269.351/435.715–436.005 MHz mode L, 144.423–144.473/435.990–435.940 MHz up/down mode J, Rudak 400 bit/s digital 1,269.710/435.677 MHz up/down; 1 W 435.603–435.639/2,400.711–2,400.747 MHz up/down mode S. Rudak failed to work; a second version was launched 29 January 1991 as the Oscar 21 part of the Soviet Informator 1 RS-144 package. Decayed 6 December 1996.

Oscar 14/15 (UoSAT 3/4) was launched 22 January 1990 by Ariane 4 into 789 × 804 km/790 × 805 km, 98.7° orbit. 45kg. First use of Ariane ASAP rack to orbit six payloads simultaneously (see also Oscars 16–19). Number 3 carries store/forward digital transponder and cosmic particle detector; number 4 included CCD Earth imaging camera, Transputer data processing experiment and solar cell technology experiment. UoSAT 3 operations continue but contact with number 4 was lost after 30hours. UoSAT 5 reflew the payload plus an enhanced Packet Communications experiment. All non-amateur links were switched to UoSAT 3 in early 1992 and all amateur services moved to UoSAT 5. UoSat 3 out of service late 1996 due to onboard computer failure. UoSat 4 out of service 23 January 1990.

Oscar 16–19 (Microsat 1–4) was launched 22 January 1990 jointly with numbers 14/15. All four developed jointly by Amsat-NA and Weber State University's Center for AeroSpace Technology. Based on 228 mm cubic bus powered by solar cells (Webersat's height extended to 31.7 cm). Orbits about 787 × 805 km, 98.7°. Numbers 1/4 (10/12 kg) are Pacsat and Lusat packet radio satellites for Amsat-NA and Amsat-Argentina, respectively, number 2 (12 kg) relays messages using a synthesised voice on 145.825 MHz at 4 W for Project DOVE (Digital Orbiting Voice Encoder) for Amsat-Brazil, and number 3, Webersat, (12.25 kg) carries a colour Canon CCD camera (400 × 700 pixel) for Weber State University, Utah, returning 2–3 km resolution Earth imagery. Webersat features visual light spectrometer, flux gate magnetometer, video flash digitiser, horizon sensor, 3.175 × 15.24 cm micrometeoroid impact detector (from Brighton High School in Sandy, Utah), L-band (1.265 GHz) AM receiver for amateur video store/forward, amateur radio packet store/forward, and sensors for TT, voltages and currents throughout the satellite. Solar panels plus Ni/Cd batteries provide 6 W orbit average. All four satellites operational.

Oscar 20 (Fuji 2, JAS 1b) was launched 7 February 1990, piggyback with MOS-1b. 50 kg; 912 × 1,746 km, 99.1° orbit. Identical to JAS 1. Operational but failure of degrading solar array is imminent. JAS 2 replacement.

Oscar 21, Amsat-DL's Rudak II package was launched 29 January 1991 aboard the Soviet Informator satellite in co-operation with Russia's Amsat-U-Orbita. Operations continued until the satellite was deactivated in September 1994 by its VKS Military Space Forces controllers.

Oscar 22 (UoSAT 5) was launched 17 July 1991 on ERS 1's Ariane V44. 47.5kg. Reflight of UoSAT 4 payload plus enhanced Packet Communications Transponder for US SatelLife medical information services in developing countries; it is also used by the US National Science Foundation for Antarctic links. The satellite is both amateur/non-amateur to enable use within the Amateur Satellite Service while supporting the payloads (which paid for the project). Operations continue; all non-amateur links were switched to UoSAT 3 in early 1992 and all amateur services moved to UoSAT 5.

Oscar 23 (Kitsat 1) was launched 10 August 1992 with Topex on Ariane into 1,301 × 1,402 km circular, 66.1° orbit.

Oscar 24 (Arsene) was launched 12 May 1993 by Ariane into 17,699 × 37,094 km, 1.10° orbit. 150.6 kg; 96 kg on-station. Ariane Radio amateur Satellite pour l'ENseignement de l'Espace conceived by CNES radio amateurs. Hexagonal prism, height 618 mm, diameter 785 mm; first satellite completely powered by European GaAs array. After three months of amateur operations, falling power levels forced Arsene into dormancy; the failed command system prevented revival.

Amsat-DL's 500 kg Phase 3-D satellite will raise itself into a Molniya-type path (Amsat-DL) 0517142

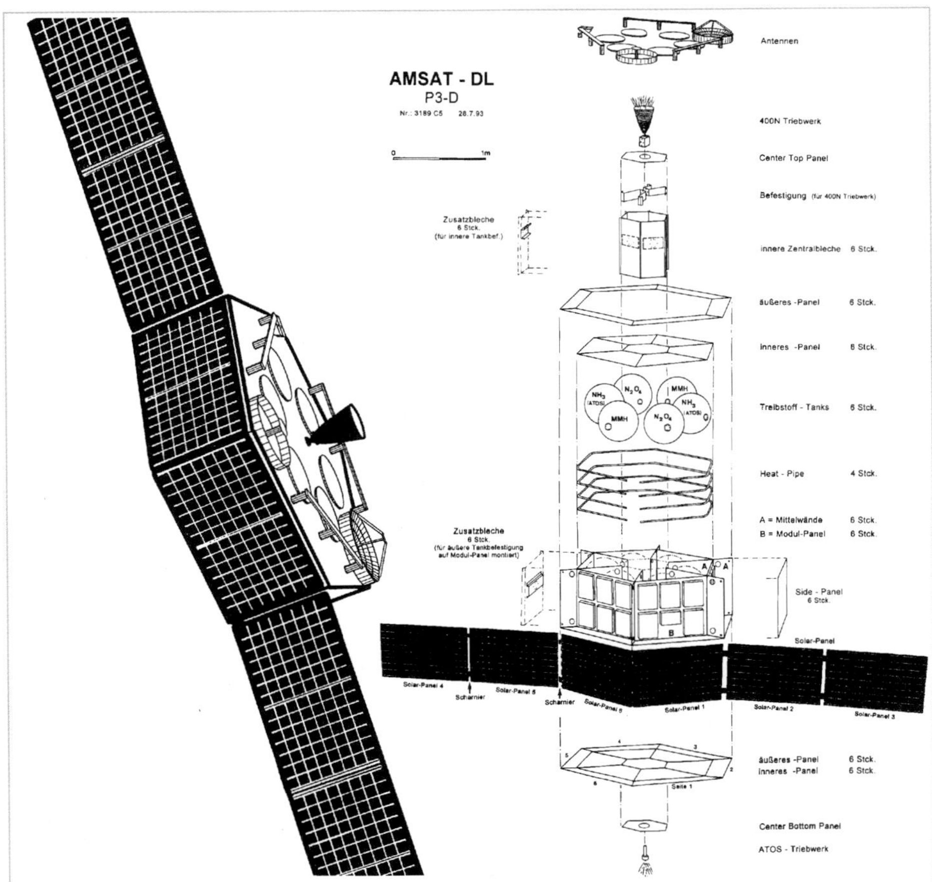

Principal features of Amsat-DL's Phase 3-D satellite (Amsat-DL) 0516801

Oscar 25 (Kitsat 2) was launched 26 September 1993 with Spot 3 by Ariane into 795 × 805 km 98.7° orbit. Released from Ariane's 4th ASAP platform.

Oscar 26 (Itamsat) was launched 26 September 1993 with Spot 3 by Ariane into 796 × 803 km 98.7° orbit. Italy's first amateur satellite, built by Amsat-Italy from Amsat-NA's Microsat design, for store/forward digital communications (started 24 October). 9.7 kg, 23 cm cube carrying no deployables. Five aluminium frame modules bolted together. 264 2 × 4 cm GaAs cells provide 8–10 W orbit average, supported by 6.5 Ah Ni/Cd battery. Passive magnetic stabilisation with solar spin (1.7 rpm); magnets maintain alignment with local geomagnetic field. CMOS V-40 microprocessor, 10 Mbyte RAM, 256 k RAM EDAC memory. 435.867 MHz (primary, 1,200 baud PSK) + 435.825 MHz (secondary, 300–9,600 baud AFSK) downlink, output of each controllable up to 4 W; omni four-element turnstile antenna. 145.875, 145.900, 145.925, 145.950 MHz (300–9,600 baud) user uplink, omni whip antenna; command channel 1 in 144–146 MHz band.

Oscar 27 (Eyesat 1) was launched 26 September 1993 with Spot 3 by Ariane into 791 × 806 km 98.6° orbit. Built by Interferometrics of US to demonstrate commercial store/forward digital communications. 10.5 kg, 23 cm cube carrying no deployables. Eyesat includes the AO-27 experimental amateur payload, built by the AMRAD Amateur Research & Development Corp of Vienna (Virginia), supporting digital communications at 300–9,600 baud. One UHF transmit (436.7975 MHz), output controllable to 0.5 or 2.5 W; omni whip antenna. two-channel VHF rx (145.85 MHz), omni whip antenna. The receiver's audio output can be switched directly to the transmitter output as a single analogue channel.

Oscar 28 (PoSAT 1) was launched 26 September 1993 with Spot 3 by Ariane into 794 × 806 km 98.7° orbit. Built in co-operation with Surrey Satellite Technology Ltd; similar to UoSAT 5.

Fuji 3 (JAS 2) was launched 17 August 1996 piggyback with ADEOS into 98.6°, 800–1,325 km orbit to replace JAS 1b. Design is basically as JAS 1/1b but added are attitude control, 9,600 bit/s packet system and Digitalker for hand-held receivers with simple aerials. 50 kg, 26-faced polyhedron, 44 × 44 × 47 cm, powered by 700 2 × 2 cm and 748 1 × 2 cm GaAs cells and 6 Ah Ni/Cd battery. Spin axis held at 90° to orbit plane by magnetorquer, Sun sensor and geomagnetic aspectometer. Transponders: 145.9–146.0/435.8–435.9 MHz 1 W up/down analogue voice and Morse; 145.85–145.91/435.91 MHz 1 W digital 1,200 or 9,600 bit/s; Digitalker (activated occasionally by ground control) provides 435.91 MHz 1 W 25 second message for FM receivers. Developed by Japan Amateur Radio League, system design/integration by NEC.

Oscar (ASUSat 1). Arizona State University's 4.5 kg student-built satellite will provide a J-mode voice repeater at 145.820/436.5 MHz up/down. See the separate ASUSat entry in the US National section for further information.

ROSAT

Current Status

The Rosat mission commenced with a six-month all-sky PSPC survey and was followed by a series of pointed observations that was to be sustained throughout the life of the mission. On 11 September 1994 the PSPC was shut down to conserve remaining detector gas. This gas was thought to have been used up during 1997 in a series of pointings to complete the all-sky survey coverage but additional PSPC observations were conducted in 1998 and early 1999. Rosat was finally shut down on 12 February 1999.

ROSAT's principal instruments are housed within the body; the UK Wide Field Camera at top right was a later addition (Dornier) 0516805

Background

Roentgensatellit (ROSAT) carries the largest imaging x-ray telescope flown at the time of launch. It is a joint German/UK/NASA project and follows the Uhuru, HEAO and Exosat projects. After a six week calibration and verification phase immediately after launch on 1 June 1990, on 30 July 1990 it began its six months all-sky survey in the course of which it constructed the first all-sky x-ray map at soft x-ray energies obtained with an imaging telescope. These images of the entire sky provided the first high-resolution map of the diffuse but structured x-ray background emission in addition to containing more than 80,000 point-like x-ray sources. A simultaneously all-sky survey at XUV wavelengths with the Wide Field Camera instrument resulted in 500 XUV sources. Since February 1991, more than 6,000 pointed observations have been carried out with the x-ray telescope's focal plane cameras (the PSPC and the HRI) adding more than another 70,000 sources.

The ROSAT has paid in science dividend with the detection of soft x-ray emission from SN 1987A (in 1991), the detection of x-ray emission from SN 1993J only a few days after an optical outburst, the discovery of x-ray emission from the Moon, from Comet Hyakutake and other comets, the detection of x-ray emission from the impact of comet Shoemaker-Levy on Jupiter, the x-ray emission from neutron stars embedded in supernova remnants as well as isolated thermally cooling neutron stars and reheated neutron stars. X-ray shadows produced by optically thick interstellar matter have been discovered high-lighting the structure of the interstellar medium.

In the Andromeda Nebula M31, 560 discrete sources have been resolved for the first time, using ROSAT's camera. Now scientists know that the clusters of galaxies from the merging of subclusters appears a common phenomenon and the radial surface brightness profile of clusters of galaxies has revealed the distribution of hot and dark matter.

X-ray emission from quasars could be traced to a redshift of z = 4.33. The ultra-deep surveys have resolved at least 60 per cent of the so-called extragalactic x-ray background emission as a superposition of faint point sources. The ROSAT sources include many thousands of previously unknown or not studied objects including stars, supernova remnants, galaxies, clusters of galaxies and Active Galactic Nuclei (AGN), including Seyfert galaxies and quasars. More than 2,400 publications in scientific journals have been the harvest of ROSAT's first six years of operation.

The mission is controlled from the German Space Operations Centre at Oberpfaffenhofen via the Weilheim ground station. NASA provided launch services and ground station support from its Deep Space Network. The only consumable on board, the PSPC gas, was exhausted in June 1994, two years after the nominal mission lifetime. Since then x-ray observations have been carried out with the ROSAT HRI only. One of the three star trackers on board failed in late 1990, but the two remaining star trackers are fully sufficient for attitude control and attitude reconstruction (required for the science data). Two of the four gyros have failed, the Y-gyro in May 1991, and the X-gyro in November 1993. After each failure, the onboard software was successfully reconfigured. After the Y-gyro failure, nominal operations could be resumed in November 1991 with a reduced pointing capability being available between May 1991 and November 1991.

In the period between November 1993 and February 1994 there was also only a reduced pointing capability available. The X-gyro failure in November 1993 necessitated the use of magnetometer and Sun sensor data for target acquisition. Since then new targets can be acquired only on the Sunlit side of the ROSAT orbit which led to an overall efficiency loss of about 20 per cent compared to the nominal system performance but otherwise satisfactory observations.

The BMFT German Federal Ministry for Research & Technology provided DM160 million funding. NASA's total ROSAT expenditure is $106 million under a 1982 agreement providing for the launch and HRI, first flown aboard HEAO 2 in 1978. Observing time for the mission's pointed phase is divided 50 per cent US, 38 per cent Germany and 12 per cent UK.

Specifications

Launch: 21.48 GMT 1 June 1990 by Delta 6920–10 from Cape Canaveral complex 17

Contractors: Dornier (prime, telescope structure, thermal control, power supply, data handling, ground support equipment), MPE Max Planck Institut for Extraterrestrial Physics (science management, PSPC, German data analysis centre), SAO Smithsonian Astrophysical Observatory (HRI), MBB (40%, covering AOCS, data transmission system, satellite structure), Carl Zeiss (mirror assembly), Telefunken Systemtechnik (solar arrays), Sira Ltd. (star trackers)

Mass: 2,426 kg at launch, including 1,555 kg payload

Orbit: 580 × 584 km, 53.0°, 95.6 min

Spacecraft: box-shaped bus, 2.40 × 2.15 × 4.50 m at launch, 2.30 × 4.70 × 8.90 m with solar array deployed on-orbit. Array provides 1 kW EOL to satisfy 905 W total consumption, including 340 W for charging 24 Ah battery. 3-axis stabilisation by four RWs and three 0.014N m magnetic coils; attitude measurement by two CCD star trackers (WFC also carries its own), four rate integrating gyros, three coarse Sun sensors and 3-axis fluxgate magnetometer. Real-time data transmitted at 8 kbit/s, stored at 1 Mbit/s; max data volume 700 Mbit per 21 h. Uplink on 2.096/2.771 GHz; downlink on 2.2765 GHz. Controlled from GSOC via Weilheim (Bavaria); German science centre is at MPE Garching

Payload: an 83.5 cm aperture 50 cm long telescope of four nested Wolter I grazing incidence mirrors (gold coated Zerodur), range 6–100 Å, rms roughness 3 Å, focal length 2,400 mm, 5 arcsec on-axis resolution feeds three focal plane instruments mounted on a carousel for operation as required: two 0.1–2 keV Position Sensitive Proportional Counters (MPE) and a High Resolution Imager (3 arcsec; SAO), providing source positional accuracy to within 10 arcsec during the observatory phase. The PSPCs undertook the 6-month survey, scanning the sky in 2° wide great circles perpendicular to the Earth-Sun line.

The UK's 135 kg Wide Field Camera was subsequently added to perform the first all-sky survey at 60–300 Å EUV, concurrent with the x-ray survey. The three nested Wolter Schwartzschild I gold coated aluminium grazing incidence mirrors provide a 475 cm^2 collecting area for a 3.5 arcmin resolution (1 arcmin on-axis). Max aperture 576 mm, focal length 525 mm, 5° FOV, drawing 78 W when fully operating. Focal plane instruments: two microchannel plate detectors. Sponsored by the UK's Science & Engineering Research Council, built by a consortium led by Leicester University.

SAX

Background

The 'Satellite Astronomia raggi-X' (SAX) Italian-Dutch bilateral project is undertaking 0.1–300 keV x-ray observations at moderate angular resolution (1 arcmin) with emphasis on spectral/time variability measurements of x-ray binaries, pulsars, transients, supernovae remnants, stellar coronae, active galactic nuclei and normal galaxies. 80 per cent of observing time is devoted to a set core programme, with the remainder available for opportunity and guest observations.

Shuttle uncertainties slowed the development time of SAX and made its developers decide to make it ELV-compatible, delaying completion of the phase into mid-1988. The consortium selected an Atlas launch in 1988; the contract was signed 31 August 1990 in Rome.

ASI, in December 1992, quoted SAX's budget as Lit408 billion; the cost was reported in January 1994 at Lit614 billion, including launch. The Operations Control Centre and a Science Data Centre are sited at Fucino, Italy. Data are returned via a Kenyan station (required because of the low inclination), with Kourou as back-up. The NIVR Netherlands Agency for Aerospace Programmes

The Italian-Dutch SAX is making x-ray observations 0516806

is responsible for Dutch involvement. SRON Space Research Organisation Netherlands contributed two X-ray Wide Field Cameras and Fokker Space provided the AOCS and solar arrays. The bilateral MoU, worth NLG65 million to Dutch companies, was signed 29 June 1990.

Italian instruments are a concentrator/ spectrometer, 0.1–10 keV 1 arcmin resolution (three concentrators constitute the 1–10 keV Medium Energy Concentrator Spectrometer, a 0.10–10 keV Low Energy Concentrator Spectrometer (provided by ESA's Space Science Department), a High Pressure Gas Scintillation Proportional Counter, 3–120 keV, and a Phoswich Detector System 15–300 keV (also operating as 100–600 keV γ-burst monitor).

Two SRON Wide Field Cameras viewing in opposite directions (each 42 kg, 14.5 W 1.8–30 keV, 5 arcmin resolution over 20 × 20′1 FOV; coded mask 70 cm from 25 × 25 cm multiwire proportional counter filled with 2.1 atmosphere Xe). As the NFI Narrow Field Instrument group studies a source, the 3-axis SAX rotates about its pointing axis for the perpendicular WFCs to scan the sky for simultaneous observations.

By December 2001 all three primary gyros and the three spares had failed. Subsequently now uses the three star trackers and orientates satellite through the reaction wheels.

Specifications

Launch: 30 April 1996 by Atlas 1 from Cape Canaveral
Contractors: Alenia Spazio (satellite, system, assembly, integration/test; structure, thermal control, harness subsystems), Alenia Spazio/Fokker (AOCS, TT&C), Fokker (solar arrays), FIAR (power distribution), Laben (OBDH, NFI Narrow Field Instruments), BPD (RCS), Teldix (four RWs), Nuova Telespazio (ground segment)
Mass: 1,400 kg at launch (490 kg payload)
Orbit: 583 × 603 km, 3.96°, 97 min
Spacecraft: 3.623 m high, 2.718 m diameter at launch, 26-month min life (up to 4 year projected); 3-axis stabilised, 1.5/0.5 arcmin accuracy/stability. Power (2.42 kW EOL) from two fixed deployable wings of three 2.58 × 1.15 m panels (spanning 8.97 m) each plus a single body-mounted panel; supported by 2 × 30 Ah Ni/Cd batteries (650 W mean over eclipses of up to 37 min). Data production 50 kbit/s mean, 100 kbit/s max, 300 Mbit per orbit. Downlinked at 2.2455 GHz S-band from 450 Mbit redundant tape recorder memory at 917 kbit/s (high) or 16 kbit/s (low) or 131 kbit/s real-time over Malindi (Kenya) during 5–8 min contact period each pass. 2.0675 GHz uplink at 2 kbit/s. RCS by 12 × 10 N thrusters. Attitude control by four RWs, three magnetorquers, gyro package, magnetometer, three star trackers, four Sun acquisition sensors (requirement 1.5 arcmin error for 10^5 s observation).

Tethered Satellite System (TSS)

Background

The Tethered Satellite System (TSS), designed to continue Italy's atmospheric studies, was a joint project with NASA. Alenia built the satellite to

Final processing at Cape Canaveral for installing the Tethered Satellite aboard Shuttle (Alenia Spazio) 0006595

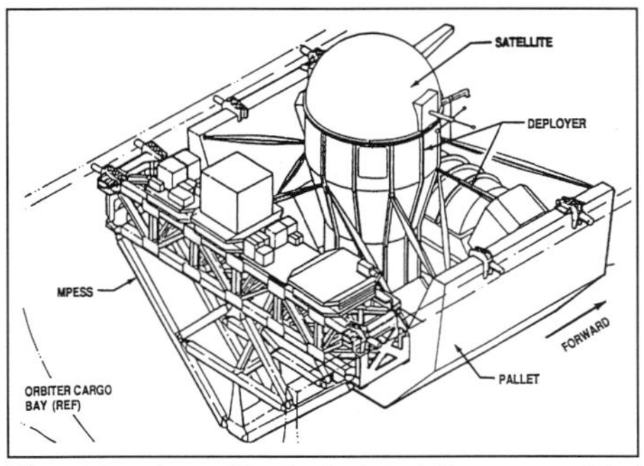

The tethered satellite is deployed from a modified Spacelab pallet 0516810

trail out behind the Shuttle. The two tests cost ASI/ NASA $404 million.

A joint US/Italian project, agreed to in 1984, the Tethered Satellite System comprised an ASI/Alenia Spazio aluminium spherical satellite capable of being reeled upwards or downwards from the Shuttle cargo bay on a 2.5 mm diameter Kevlar tether up to 100 km long. Martin Marietta developed the deployment system for the STS-46 first mission in 1992, involving a 20 km upward deployment, with an insulated copper wire incorporated in the tether, to study electrodynamic and other phenomena. Downward deployment with the help of N_2 thrusters on the satellite could 'troll' the Earth's atmosphere at heights of 130–150 km - normally inaccessible because of drag – for upper atmosphere research at 7.65 km/s.

TSS 1 attained only 256 m (insufficient for significant science returns) before a protruding bolt fouled the U.S. developed winding mechanism. The first TSS deployment should have allowed about 20 km of tether to be played out. The Italians made a second attempt with the replacement satellite TSS 1R in February 1996 on STS-75. This second try succeeded until the tether snapped at the 19.7 km mark. Up to 4 kW of power should have flowed along the core down to the Shuttle, but the cable generated only 40 V at 15 mA during the STS-75.

NASA/ASI are jointly defining requirements for TSS follow-on missions, though none is funded. An atmospheric verification mission would possibly employ the TSS deployer and a NASA reusable satellite. An all-up TSS atmospheric science mission would lower Italy's satellite and its 44 kg science payload to around 130 km altitude on a 100 km-long non-conducting heat resistant tether. The configuration would be essentially as TSS 1/1R but a simple aerodynamic surface and thermal protection would add 22 kg. Battery power supply would be 2 kWh; data rate 54 kbit/s. Consideration was given to modifying TSS 1's structural model as a non-recoverable TSS 2. A TSS 3 might extend TSS 1's investigations but be based at the Space Station.

These plans, being reliant on NASA's Shuttle for implementation, have been suspended pending decisions about the resumption of Shuttle flights and the future of the ISS. Some work has extended the potential application to unmanned platforms and to further iterations of advanced satellite concepts.

Specifications

Contractors: Alenia Spazio (satellite), Lockheed Martin (deployer system, contract value US$68 million)
Mass: satellite 518 kg
Orbit: up to 20.7 km above Shuttle for electrodynamics TSS 1/1R mission; up to 100 km below for atmospheric mission
Spacecraft: 1.6 m diameter spherical shell of conventional Al construction. Allowable payload mass 70 kg, volume 0.43 m^3. Power provided by 2 batteries accessible through four equatorial doors. Attitude control exercised by 12 N_2 thrusters: 4 × 1 N thrust along line direction to aid deployment and maintain tension, 4 × 2.5 N at 90° intervals along equator for in-plane and perpendicular-to-plane thrusting, and 4 × 2.5 N yaw jets in two clusters 180° apart on equator. Attitude information by Earth sensors and four Sun sensors. TT&C via S-band link: 16 kbit/s data rate, 2kbit/s command rate. TSS 1/1R tether 5-layer: Nomex core, copper conductor, insulation, Kevlar strength member and outer Nomex braid; outer diameter 2.5 mm, mass 8.2 kg/km, capable of carrying 1 A current, 10 kV, resistance 0.12 ohm/m. Satellite housed in deployer support structure mounted on enhanced Spacelab pallet, deployed by electric reel mechanism from extended 12 m boom.

Israel

Gurwin

Background

The Technion Institute of Technology proposed building three or four small satellites for continuous early warning coverage. The Gurwin 1 TechSat civil satellite may have been its first attempt to provide technology demonstrations of future imaging systems. The satellite came about as a result of a collaboration between the Faculty of Aerospace Engineering at Technion Institute (under the direction of Professor Chaim Eshed), ISA, industry and Amateur satellite-IL to develop a simple, low-cost, low-power platform. The Flight Model was completed October 1994, but the first TechSat was lost in Start's launch failure. The TechSat 1B near-duplicate is being funded. It will include the Netherlands' TNO-TPD wide field star sensor. Gurwin was launched on the first successful Start launcher. The preceding Start launch attempt suffered a fifth stage failure.

The TechSat 2 is apparently a 250–300 kg second-generation TechSat proposed by Technion as a multipurpose platform, created by adding an AOCS unit and GPS navigation to the basic TechSat 1. The 100 kg AOCS module would carry Rafael 1 N liquid propellant and N_2 thrusters. TechSat 2A would carry a high-resolution Earth observation camera, gimballed on command and relaying imagery via an MMIC Ku-band phased array antenna.

Specifications

Launch: 28 March 1995 on Russian Start from Plesetsk; planned orbit was 670 km 75.4°; Gurwin TechSat 1B was launched on 10 July 1998 by Zenit 2
Mass: 52 kg
Configuration: box-shaped 460 mm height, 430 mm across (six Al plates, 2.5 mm thick). Earth-pointed face carries antennas, horizon sensor, and camera. 80C186EC-based CPU with two 512 kbyte independent RAM memories. Controlled from Technion station; command reception at 145.910/ 1,269.85 MHz
Power: solar cells on four faces (1,000 2 × 4 cm Si, 17 W average, 9 W required for housekeeping, Ni/Cd battery)
Attitude control: 3-axis Earth-pointing control (5° coarse, 0.1–0.5° fine) by 2.5 kg 3,600 rpm 4 Nms MW 3 × 1 Am^2 magnetorquers, 1 kg 1 W static horizon sensor (0.05° accuracy), 0.22 kg 0.8 W ±0.6 gauss 3-axis magnetometer (see GN&C Space Industry entries for detailed specifications)
Payload: Elisra Earth-observing 350 × 570 pixel CCD camera, compressed image data broadcast using millimetre-wave link experiment (also on UHF).

Technion UV CsTe spectroradiometer to monitor Earth's albedo at 0.24–0.30 μm (0.01 μm/2 mrad resolution); 15° FOV. The spectral profile is a function of the vertical ozone profile. Goals are investigating the long-term stratospheric impact of the Kuwaiti oil fires and the extent of the region's ozone depletion.

Technion x-ray detectors (up to 200 keV). Principal purpose is to prove that CdTe crystals can be used for x-ray imaging.

Italy

San Marco Series

Background

Italy became a force in international space science with the introduction of the San Marco, a small but effective University of Rome programme built around Italy's international platform in the Indian Ocean off San Marco, Kenya. University of Rome scientists began their investigations in 1964 and continued the programme through 1974 with a small revival in the mid-1980s. The primary purpose of the San Marco satellites were atmospheric density tests.

Specifications

San Marco 1
Launched: 15 December 1964 by Scout from Wallops Island

Mass: 24 kg
Orbit: 194 × 697 km, 38°, 95.1 min
Decay date: 13 September 1965.

San Marco 2
Launched: 26 April 1967 by Scout from San Marco
Mass: 129 kg
Orbit: 185 × 211 km, 3°, 94.2 min
Comment: First launch from the platform and first in a series to measure atmospheric density: two concentric spheres connected by flexible instrumented arms measured drag
Decay date: 14 October 1967.

San Marco 3
Launched: 24 April 1971 by Scout from San Marco
Mass: 164 kg
Orbit: 222 × 718 km, 3°, 93.8 min
Decay date: 29 November 1971.

San Marco 4
Launched: 18 February 1974 by Scout from San Marco
Mass: 164 kg
Orbit: 231 × 910 km, 3°, 96.1 min
Comment: With Explorer 51, sampled the auroral zone and SM4 over the equator. The satellite detected variations in atmospheric density due to differences in solar heating. Transmitted 806 days
Decay date: 4 May 1976.

San Marco 5
Launched: 25 March 1988 by Scout from San Marco
Mass: 237 kg
Orbit: 262 × 619 km, 2.89°, 93.5 min
Decay date: 6 December 1988
Comment: Drag on 1 m sphere measured from spring arms supporting external 3 kg 6 mm-thick shell. Covered 500 km to re-entry at 141 km. Also monitored charged particles, neutral winds, electric fields, UV solar fluxes and UV airglow.

Sirio Series

Background
Italy began its first investigations of experimental space communication when it launched the Sirio 1 satellite on 25 August 1977 by Delta from Cape Canaveral. The satellite, weighing 398 kg and 220 kg on-station at 15° W GEO, carried two 9 W transponders operating at 12/18 GHz. As an all-Italian project, the ground control operations centred at Fucino. Italy prepared for an eventual communications system for the entire country by investigating SHF propagation in rain, snow and fog. When the experiments were complete, Italy turned the results over to ESA's to build the Sirio 2 follow-on, using Sirio1's back-up bus. Italy provided 74 per cent of funding for Sirio 2 which was lost in Ariane L5 failure September 1982. The 420 kg satellite was designed to investigate meteorological impediments to high-frequency communications. Following the 1983 space co-operation agreement with China, Sirio 1 was moved to 65° E to assist China in development of 18 GHz GEO communications.

Netherlands

ANS Satellite

Background
The first Dutch satellite was the ANS (Astronomische Nederlandse Satelliet) astronomical platform launched in 1974, followed by significant participation in IRAS and a 10 per cent interest in the SAX satellite. The ANS was launched 30 August 1974 by Scout from Vandenberg. It weighed 129 kg and entered a 258 × 1,173 km, 98°, 99.1 min orbit. A Dutch industrial consortium designed and built the Netherlands' first satellite. Scientists in the Netherlands and other investigators wanted a near-polar, Sun-synchronous circular orbit of 500 km for studies of the UV spectra of young, hot stars and soft/hard x-rays from cosmic sources. Due to a guidance failure the satellite was placed in an elliptical orbit causing initial problems with background radiation. Contact with the ground took place every 12 hours.

In appearance, the ANS sits 1.12 m high and has a depth of 70 cm. The solar array spans 1.5 m (totalling 0.95 m^2, 2,000 cells). It has a controls system with 3-axis 1 arcmin pointing accuracy by RWs/magnetorquers. A computer could hold data from a 12-hour observation programme. The science package had three instruments: 5-band UV photometer 150–350 nm (Groningen); soft x-ray detectors 2–4, 4–12, 27–35, 44–55 Å (Utrecht); 2–40 keV hard x-ray telescope (MIT, US).

ANS re-entered 14 June 1977.

Pakistan

BADR (Full Moon) Series

Background
Pakistan has plans to develop an independent remote-sensing, science and communications space presence. The country currently uses Landsat images from a ground station it installed as part of a cooperative international agreement. To pursue its goal of space science, the Pakistanis built the BADR-1 in conjunction with the Chinese. BADR-1 passed over the country about five times a day to give ground controllers experience in handling spacecraft operations and preparing for an eventual store/forward communications capability.

After a successful launch on 16 July 1990 by CZ-2E from Xichang, the satellite entered into a 205 × 983 km, 28.49°, 96.7 min orbit. This demonstrator satellite weighed 52 kg, and had a 483 mm diameter 26-sided polyhedron. It carried a UHF/VHF store/forward payload. Chinese officials quoted the launch cost to Pakistan as US$300,000. The orbit provided four or five usable passes daily for its controllers, each lasting as much as 20 minutes, instead of the two pairs separated by at least 10 h available from a polar orbit. Suparco's Karachi campus provided TT&C, supported by the Sparc station at Lahore. The sites also operated two low-cost tracking and receiving stations. Unfortunately, contact was lost after 35 days, seriously affecting the communications experiments. BADR was taken out of service on 21 August 1990 and re-entered 8 December 1990 after 145 days.

The Pakistanis BADR-2 launched by Zenit on 10 December 2001 into a 986 ×1,015 sun-synchronous orbit at 99.7° with a period of 105 min. The new satellite weighs 50 kg, with a design life of one year. Gravity gradient stabilisation using 4 kg tip mass on a 6 m boom will maintain the transponder orientation. BADR-2 carries a CCD camera for cloud monitoring, provided by the UK's RAL and University of Leicester. Other experiments are SAFE Store/Forward Experiment (Suparco), RadFET Compact Dosimeter (SIL/ESA) and Battery End of Charge Detector (SIL/ESA). S-band/UHF will provide 128/9.6 kbit/s data rates, respectively. RAL's Chilton will act as Lahore's back-up. The UK's Satellites International Ltd is providing the entire attitude control system consisting of one MFM-3L fluxgate magnetometer, two MTR-25 magnetorquers, two DSS-256 Digital Sun Sensors (each a pair) and attitude control electronics. It is also providing the S-band transceiver, 4 Ah Ni/Cd battery, OBDH and power conditioning system. The UK's EEV Ltd is providing the GaAs solar panels. CCD camera: 0.840–0.890 μm, 2.5 kg, 7 W, 17° FOV, 72 mm fl, f/6, EEV 576 × 770-pixel CCD for cloud monitoring with 310 m resolution (from 1,000 km). The ATSR Along Track Scanning Radiometers on ESA's ERS 1/2 measure sea surface T but have to reject pixels covering clouds. However, pixels contaminated by clouds below their 1 km resolution can still be accepted, biasing the readings. BADR's camera will thus quantify the sub-1 km clouds and may demonstrate the need for such monitors on future ATSRs. The camera is carried free of charge for RAL in return for Suparco using it for land imaging.

The UK's cloud camera will fly on Badr 2 (Rutherford Appleton Lab) 0517293

Portugal

PoSat Series

Background
The first Portuguese satellite, PoSat 1, was launched in 1993, built by a Portuguese team within a technology transfer programme between the UK and Portugal. The primary objective is to stimulate a national space industry and generate a nucleus of engineers with first-hand expertise for possible future satellite projects. The PoSat consortium is led by INETI (National Institute of Engineering & Industrial Technology; see below) and includes EFACEC, Alcatel Portugal, OGMA Oficinas Gerais de Material Aeronautico, CPRM Companhia Portuguesa de Rádio Marconi, IST Instituto Superior Técnico, UBI Universidade de Beira Interior and CEDINTEC. Overall cost was about £5 million, funded by the Ministry of Industry through the PEDIP programme, plus each consortium member supported its own manpower. The programme included on the job training for 11 engineers at Surrey Satellite Technology Ltd. in the UK, who also provided the ground station in Portugal.

Specifications
PoSat 1
Launched: 26 September 1993 as an Ariane ASAP passenger
Orbit: 794 × 806 km, 98.68°, 100.8 min
Mass: 47.5 kg. Based on SSTL's UoSAT 5 design
Configuration: Same as UoSAT 5. Platform built by SSTL, 352 × 356 × 670 mm gravity gradient stabilisation, design life 5 years
Power supply: four GaAs panels (each 34 W) totalling 1,344 cells providing 20 W (orbit average processed); 6 Ah Ni/Cd battery supporting eclipse operations
OBC: 80C186 processor with 512k memory and associated 16 Mbyte Ramdisk bulk data store. The payload transputers provide high performance data processing and attitude control
Attitude control: Earth-pointing maintained by combination of gravity gradient stabilisation and active magnetorquing; Sun/Earth horizon sensors and 3-axis fluxgate magnetometer with ±2 nT resolution over ±64 μT provide attitude information. Gravity gradient supplied by 3 kg tip mass on 6 m boom deployed from top face, magnetorquing by six 150-turn Al wire coils (one on each PoSAT facet)
TT&C: control from CPRM's site at Sintra, near Lisbon
Communications: data rate of 9.6 or 38.4 kbit/s, 3-channel VHF receiver (2 communications, 1 command), redundant UHF transmit (four bands) switchable to 2/10 W RF output, 4-element turnstile VHF antenna, one monopole UHF antenna. 148.585 + 148.285 MHz up; 429.950 + 429.450 MHz down (switchable to amateur service on 435.250 + 435.275 MHz down and 145.925 + 145.975 MHz up)
Payload: EIS Earth Imaging System. Two monochromatic CCD 578 × 576-pixel cameras, 2 km/200 m resolution. Image read from each by TDPE Transputer Data Processing Experiment and, after processing, passed to OBC for storage on 16 Mbyte Ramdisk.

DSPE Digital Signal Processing Experiment. Two Texas Instruments C25 + C30 TMS320 processors for relaying speech.

SAF Store/Forward Packet Communications. Stored on OBC's Ramdisk for downloading by other stations.

GPS Navigation Experiment. Based on Trimble TANSII receiver to provide position, velocity and onboard time reference. Allows PoSat to generate its own orbital element sets and ground stations with GPS receiver to experiment with real-time differential applications.

CRE Cosmic Ray Experiment particle radiation detector to monitor environment and effects on electronics.

SS Star Sensor. Patterned after EIS camera but with stellar optics, analysed byTDPE and data passed to OBC. Allows 1 arcsec attitude measurement.

Russian Federation

Coronas I/F

Background

These are the first Russian/Ukraine observatories dedicated to helioseismology (solar oscillations) and identification of coronal/chromospheric acceleration mechanisms. They serve additionally as solar system 'weather' forecasters. Coronas stands for Comprehensive Orbital near-Earth Observations of the Active Sun.

The satellites are placed in Prognoz-type highly elliptical orbits at 500 kms. They monitor solar events such as flares to evaluate their effects on near-Earth space. This facility will be particularly useful for warning crews aboard permanent space stations of potentially hazardous solar activity. The Soviet Union first revealed the missions in 1987 and stated it planned to begin the launches in 1990 and through to 1992.

The Institute of Terrestrial Magnetism, Ionosphere & Radio Wave Propagation (Izmiran, =I), and by the Lebedev Physical Institute (=F) manage the project. German and Polish scientists also collaborate.

Coronas I marks the first flight of NPO Yuzhnoye's AUOS-SM (S=Sun, M=modified) Sun-pointing bus, derived from the AUOS design last employed by Apex. It can accommodate a science payload of 700 kg, providing daily-averaged 300 W of power, and point at the Sun with 10 arcmin accuracy. Electronics are housed in the bus pressurised at 0.77–1.35 atmosphere with nitrogen at 0–40°C.

Each Coronas carries solar-observing and near-Earth particle radiation instrumentation for detection of solar events and resulting terrestrial interaction. The solar instruments include x-ray telescope (10–300 Å), coronograph, heliometer (10–30 keV), Thomson polarimeter, x-ray and visible photometers, 0.03–30 MHz radio detector. The particle radiation monitoring includes 0.1–100 MeV gamma-rays, a monitor for 30 MeV or less neutrons, one for 1 MeV or less protons/nuclei and 0.05 MeV or greater electrons. Yuzhnoye is investigating power production techniques with four 64 × 53 × 3 cm panels carrying different solar concentrators and cell types, each panel contributing a maximum 50 W.

Specifications

Coronas satellite series

Launch: Coronas I 2 March 1994 by Tsyklon 3 from Plesetsk. Satellite decayed 4 March 2001; Coronas F planned 1997–1998 but subsequently cancelled
Orbit: Coronas I 488 × 527 km, 82.49°, 94.7 min
Design life: 5 years
Mass: 2,160 kg (700 kg payload)
TT&C: provided by VKS centre at Golitsyno-2; science data returned to Izmiran's Troitsk station

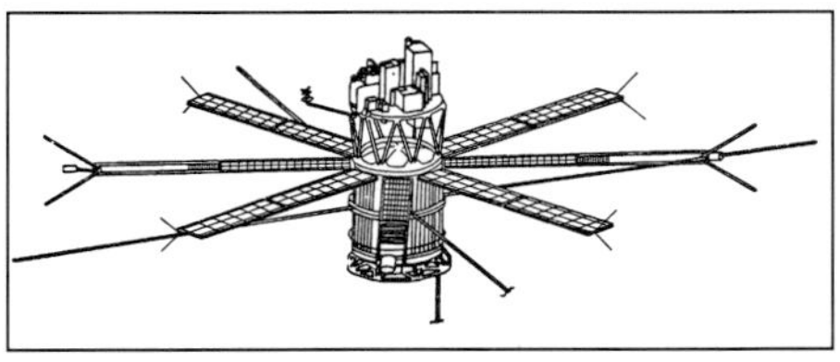

Coronas I began solar observations in 1994. The arrays are also testing sample solar panels 0516791

Bion/Biokosmos

Current Status

Bion 12, the last in the long-running series of CIS/Soviet biological satellites has yet to be launched. It had been planned to fly two Rhesus monkeys on the mission, but this was thrown into doubt when the United States announced in April 1997 that they were pulling out of the monkey research following pressure from animal rights groups – pressure which increased following the loss of one of the monkeys launched on the Bion 11 mission. If monkeys are not flown on Bion 12, then other animals such as rodents and insects, which have flown in previous Bion/Kosmos flights, can be carried. However, Illyin has noted that experiments and equipment planned for the mission would have to be re-designed at the cost of a significant mission delay.

Bion missions employ a Vostok-derived spacecraft, carrying specimens in the recoverable sphere. The two monkey housings are visible through the hatch. The same basic vehicle is offered commercially as the Photon microgravity facility (qv) (Graham Turnill) 0516787

Background

The Bion programme is operated by the Institute of Biomedical Problems in Moscow and is headed by Professor Yevgeni Illyin. Illyin had undergone preliminary cosmonaut training in the mid-1960s as the science officer for a Voskhod mission conducting biomedical research. The Biokosmos programme utilises a modification of the Vostok spacecraft similar to Photon microgravity capsule.

Ten international biosatellites had been launched by the CIS/USSR by end 1995 in the Biokosmos series, with the US participating in the last eight. US interest began after the collapse of the US's own Biosatellite programme in 1969. CIS/USSR and US scientists have used more than 50 biological specimens in this series, ranging from viruses to mammals, and including monkeys, rabbits, guinea pigs, rats, mice, tortoises, higher plants, fish, birds and insects. At least nine countries have participated, with ESA contributing for the first time to Kosmos 1887.

Bion capsules enter space aboard Soyuz-U from Plesetsk into about 220 × 400 km at 62.8° or 82.3°. The capsules weigh about 6 tonnes, with 700 kg internal cargo and 200 kg external. Zonds 5/6/8 and Soyuz 20 also carried biological specimens such as turtles, flies, plants and seeds. The two most recent Bion missions were Kosmos 2229, launched 29 December 1992 by Soyuz-U from Plesetsk into 226 × 397 km, 62.8° orbit, again carrying two Rhesus monkeys (Krosh/Ivasha), and Bion 11.

Bion 10 preparations. The silvery unit in the rear of the compartment is ESA's Biobox (ESA) 0516788

The Russians recovered Bion 10 on 10 January 1993 at 04.16 GMT in a forest some 100 km north of Karaganda, 800 km from the pre-mission site near Kustanai and two days early because of high internal temperatures. 7 of 15 salamanders died from the heat, one monkey went without food for two to three days and two of the four external pans failed to close for re-entry, damaging the experiments. NASA provided some flight hardware for the primate experiments in addition to participating in several primate experiments. NASA also conducted bone and muscle biopsies, neurovestibular pre/post-flight measurements, thermoregulation and immunology experiments.

ESA's 40 kg Biobox debuted with six experiments, investigating bone weakening in μg and 1 *g* centrifuge conditions. ESA hoped this and later flights would settle the question of bone sensitivity to gravity. The incubator's temperature is controlled to 4–37°C and the conditions recorded on a 1 Mbyte memory card. Biobox can also be used for other experiments such as protein crystallisation, plant metabolism and genetics.

ESA/IBMP shared five other experiments: three μg flights for fruit flies and unicellular algae, and two radiation exposure flights for plants and seeds and to test dosimetry. The algae experiment was planned for the last days so was truncated by the early return.

Bion 11 was launched from Plesetsk, 24 December 1996 using a Soyuz-U vehicle. The Russians used the Bion name openly for the first time on this flight – previous flights had always carried a Kosmos number. The spacecraft entered a 62.8°, 217–379 km orbit. On board were two Macaque monkeys (Lapik and Multik), together with newts, snails, drosophila flies and other insects and bacteria. Both Russian and United States researchers carried out experiments. The United States had paid about US$15 million towards the flight.

The spacecraft landed 130 km north of Kustani 7 January 1997. The following day Multik died of a heart attack during medical tests under general anaesthetic, resulting in overseas protests about the flight from animal rights sympathisers. Yevgeni Illyin said that the monkey's death had nothing to do with the two weeks just spent in orbit: while coming around from a general anaesthetic Multik's heart simply stopped, while Lapik who had undergone the same medical procedure came through without any trouble.

Elektron

Background

The Soviet Union became interested in the van Allen belts, principally because its proposed series of communications satellites, the Molniyas, had to traverse the belts twice a day exposing them to a potentially lethal dose of ionised particles. The Russians were also concerned about the levels of radiation which would be encountered by moon-bound cosmonauts and design work began in July 1960. They developed a civilian satellite research programme known as the Elektron to investigate the environment of the van Allen belts.

Elektron 1 and 2 went into space on the same booster, an SL-3 Vostok, launched on 30 January 1964 from Tyuratam. The two satellites weighed 350 and 445 kg respectively, the greater weight on the second being used for propellant to put the satellite in a more energetic orbit than its sibling. Elektron 1 had a 404 × 6,577 km, 60.8°, 162.6 min orbit and Elektron 2 a 9,470 × 58,952 km, 60.8°, 1,356.4 min orbit. This launch marked the first Soviet dual launch. The Soviets announced that the flight would gather information on radiation for protection of manned spacecraft. By the time, however, the needs for manned shielding for one or two brief passages through the van Allen belts was fairly well established and long term exposure to the belts had a higher investigative priority. Elektron 2 decayed 22 April 1997.

Elektron 3 and 4 on similar missions left Earth on 11 July 1964 by SL-3 Vostok from Tyuratam. Both had similar masses, orbits and missions as Elektron 1 and 2. The scientific value of two simultaneous measurements accrued from data obtained from both the inner and the outer radiation belts. Elektron 4 decayed 12 October 1983. During the height of the Cold War, many analysts assigned an early warning role to this mission, but no proof has ever been revealed.

Granat ('Garnet')

Background

Russia's answer to the Hubble Space Telescope, the gamma and x-ray observatory Granat continues operations. It was the second flight of the Astron astrophysical platform provided by the Lavochkin bureau. This was the last flight built around the Venera planetary spacecraft; the Spektrum bus, designed specifically for astronomical missions is superseding it. Its main mission was to conduct observations of 3–2,000 keV γ/x-ray sources, including imaging in 3–500 keV, source pinpointing, recording temporal/spectral characteristics of burst-type and transient sources, monitoring of bright, permanent x-ray sources, and investigation of the afterglow of γ bursts in the soft x-ray region

Granat was launched on 1 December 1989 from Baikonur by Proton-K/D-1 launch vehicle and observations began in January 1990 following in-orbit checkout, which included instrument testing beginning 11 December. Russian planners wanted a mission duration of about 18 months and delayed the launch 11/2 years while they checked out payload development problems. In the first nine months of operations it completed more than 140 observing sessions, some lasting up to 46 h. By early 1992, about 400 sessions typically of 24 h had been logged. 1992 provided 163 sessions and 143 in 1993.

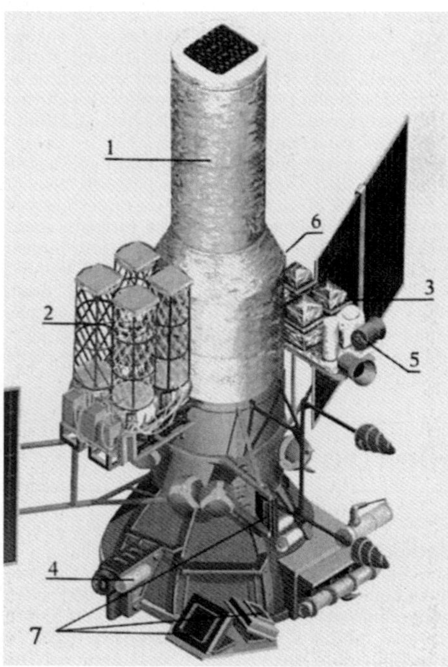

Granat: last of the Venera series of spacecraft. 1 Sigma (France), main gamma telescope; 2 ART-P (USSR), imaging x-ray telescope; 3 ART-S (USSR), x-ray detector; 4 Phebus (France), gamma/x-ray burst detector; 5 Konus (USSR), gamma/x-ray burst detector; 6 Podsolnukh (USSR), detection of burst sources; 7 Watch (Denmark), transient and burst detection (CNES) 0516792

A truly international collaboration of space programmes contributed to building the Granat. CNES provided the primary 1 tonne Sigma (Système d'Imagerie Gamma Masque Aléatoire, or random mask imaging gamma-ray system) telescope at a cost of FFr150 million. NASA and the Russian Space Agency collaborated on observations in September 1991, collating data from the Granat and NASA's Compton Observatory. As its attitude control propellant approached exhaustion, Granat switched on 1 October 1994 from pointing mode to a 30°/min slow yaw, providing its instruments with an all-sky scan every six months. This provided 5,500 h of observations until 9 September 1995, mapping the Galaxy's diffuse gamma emission. For the rest of September 1995, Granat returned to its original pointing mode for 110 h of observations in the hope (unrealised) of confirming pulsars in the galactic centre. It began slow yawing again by month's end, before further galactic centre pointing in May 1996.

Granat marks several major findings among its accomplishments, among them, correlating the diffuse gamma emission with the interstellar gas column density. CNES reported at end-1992 that Granat had identified eight black hole candidates. Granat's engineering model was sold at Sotheby's December 1993 Russian space auction in New York.

Specifications

Granat satellite

Contractors/Institutions: NPO Lavotchin/Babakin Engineering Research Centre (spacecraft), IKI/ CNES Toulouse (science lead), CESR of Toulouse/ CEA Service d'Astrophysique near Paris (Sigma telescope), Sodern (SED 12 star trackers), Denmark (Watch X-ray detector), Bulgarian Space Research Institute (Podsolnukh), Sagem (Sigma's magnetic bubble memory), Sextant Avionique (Sigma's 8 kg onboard management unit, 8 Mbit 5.5 kg buffer memory, 90 W 8.5 kg power converter)

Launch: 1 December 1989 by Proton from Tyuratam. Out of service December 1998, decayed 25 May 1999.

Mass: 4,402 kg, payload 2,146 kg (including 1 t Sigma)

Orbit: Prognoz-type 4-day, 1,760 × 202,480 km, 51.86°, 5,928 min, permitting 56 h observation sessions above radiation belts beginning 60,000 km

Spacecraft: based on Venera/Astron bus, with 2.35 m diameter pressurised base torus housing electronics. 6.5 m high. 3-axis controlled, capable of 20 arcmin pointing accuracy over 24 h. Soviet star sensors complemented by CNES' SED 12 trackers for Sigma pointing. Real-time data transmissions at 3,072 bit/s or 65,546 bit/s playback from 150 Mbit memory. Power (400 W allocated to science instruments) derived from twin double-segment solar wings spanning 8.5 m

Payload: Sigma Mass: 1,050 kg, 3.5 m high, 1.2 m base diameter topped by 49 × 53-element 1.5 cm-thick tungsten coded mask permitting 16 arcmin resolution imaging with 57 cm diameter 1.25 cm thick NaI (Tl) crystal over 0.5–2 MeV. The detector lies 2.5 m below the mask with effective area 800 cm^2. Full FOV 10 × 10°; 16 arcmin resolution available over 4.2 × 4.2°. Pointing information provided by two bore-sighted Sodern CCD SED 12 star trackers; one provides visible imaging of γ bursts identified by the main instrument. Sigma is also the most sensitive γ burst detector aboard Granat and can additionally use its 19,200 cm^2 CsI (Tl) anti-coincidence detectors for 0.10–7.5 MeV monitoring. Sigma's 28 kg 128 Mbit Sagem magnetic bubble mass memory unit also used for data processing and as a telemetry buffer for data downlinks.

ART-P: four identical modules with 41 × 43-element coded masks for 5 arcmin resolution imaging over 3–100 keV. Effective area 4 × 600 cm^2, time resolution 1 ms, 4 × 64 kbit storage.

ART-S: two pairs of 3–150 keV non-imaging detectors, each pair alternately observing source/ background via a rocking collimator. FOV 2 × 2° FWHM, time resolution 100 μs, effective area 4 × 600 cm^2, 4 × 64 kbit memory.

Konus: detection of 0.20–4 MeV bursts with seven NaI (Tl) detectors 20 cm dia/5 cm thick. Location accuracy 1–5°, 40-channel spectrum, time resolution 1/64–1/256 s.

Phebus: detection of 0.10–100 MeV bursts with six bismuth germanate crystals. FOV 4TT, 1/128 s time resolution, energies measured over 102 bands over 0.10–10.0 MeV and 14 bands over 10.0–100 MeV. Instrument provided by France.

Watch: detection of 5–120 keV x-ray transient sources and γ bursts with four Danish 45 cm^2 all-sky 4TT detectors. Accuracy of burst localisation about 0.1°. Each detector is a mosaic of NaI (Tl) + CsI (Tl) scintillators 2 mm thick. The rotation modulation collimator permits real-time localisation of rapidly varying x-ray sources.

Podsolnukh: two 2–25 keV proportional counters mounted on a rotational platform steered from Konus data. Instrument provided by Bulgaria.

Pion

Background

In conjunction with a Resurs-F film return mission, the Soviet Union released pairs of 45–47 kg, 33 cm diameter Pion spheres for upper atmosphere studies. The photo mapping satellites released the spheres, launched from Resurs F01 on 25 May 1989 and from Resurs F03 on 18 July 1989, a few days into each mission. Built by the student design bureau at the Korolev Aviation Institute in Samara, they carried only simple radio transmitters to aid orbit determination at a time of enhanced solar activity. Pion 1 and 2 began their mission 9 June 1989, and re-entered 23 and 24 July. Pion 1 has been described as a 45 kg 'glass sphere' with an internal corner reflector. Pion 2 was a 45 kg sphere with a 'magnesium aluminium alloy surface'. The second two separated from the Resurs-F carrier vehicle on 7 August and decayed 18 and 19 September 1989. Pion 3 matched Pion 2, but with a mass of 47 kg, and Pion 4 matched Pion 1, but with a mass 47 kg.

The Soviets felt the experiment was successful enough to release another pair, Pion-Hermes from a Resurs-F on 1 and 2 September 1992. The Resurs was launched on 19 August. KB Photon (responsible for Resurs-F) built the new satellites under contract to the Russian-US Hermes company, to provide upper atmosphere data. The Russian Defence Ministry invited US Space Command to provide joint tracking, but the US military snubbed the request. Earlier reports described them as 78 kg solid metal spheres. PH-1 decayed 25 September and PH-2 on 24 September 1992. Several attempts were made to re-open the programme but limited resources prevented continuation of the programme.

Prognoz ('Forecast')

Background

The Prognoz series specialises in solar radiation and its interaction with Earth's magnetosphere. Spheroidal, with four cruciform solar panels and weighing 900–1,000 kg, they enter highly eccentric orbits ranging from 500 × 200,000 km to 400 × 720,000 km at 65°. SL-6 Molniya vehicles from Tyuratam launch them.

Prognoz orbits provide comparative measurements from near-Earth space and the upper atmosphere to out beyond the magnetosphere. France, Sweden, Czechoslovakia, Hungary and Poland have provided instruments and conducted experiments in parallel observations undertaken from ground stations.

The Soviets placed the satellites at opposite orbital extremes during their lifetimes. Prognoz 1, 14 April 1972, was at its apogee while Prognoz 2, 29 June 1972, was at perigee. Several launches came in rapid succession. Prognoz 3 15 February 1973, Prognoz 4 22 December 1975, Prognoz 5 25 November 1976, Prognoz 6 22 September 1977, Prognoz 7 30 October 1978 and Prognoz 8, 25 December 1980. All tested instruments to be used by Vega to study VLF plasma waves. Prognoz 9, introducing a second generation of satellite, lifted off 1 July 1983, weighing 930 kg, and adopting a 380 × 720,000 km orbit (taking it beyond Moon). It carried equipment from Czechoslovakia and France.

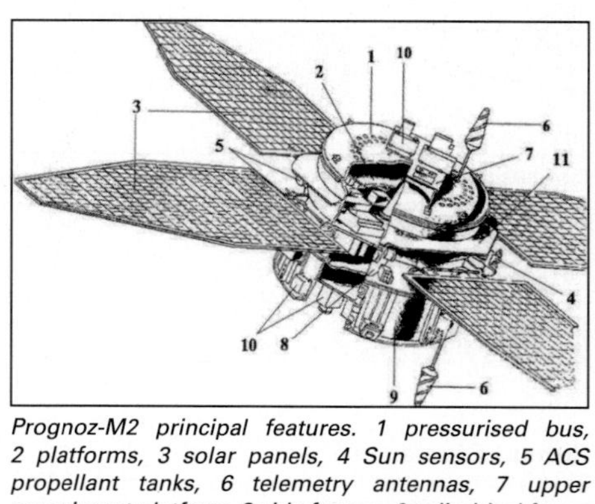

Prognoz-M2 principal features. 1 pressurised bus, 2 platforms, 3 solar panels, 4 Sun sensors, 5 ACS propellant tanks, 6 telemetry antennas, 7 upper experiment platform, 8 side frames, 9 cylindrical frame (Magion released from here), 10 science payload, 11 thermal shield 0517290

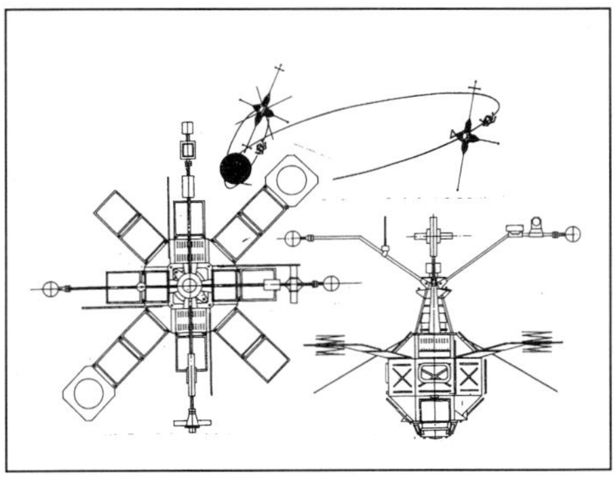

Magion subsatellite configuration and Interball orbital arrangement 0516800

Prognoz 10/Intercosmos 23, launched 26 April 1985, weighed an identical 930 kg and was placed in a 421 × 200,320 km orbit. It was developed under Project Intershock, with Czechoslovakia, to study the structure of shock waves generated from solar wind plasma striking the magnetosphere. Czech scientists developed the computer required to record hundreds of measurements in the few seconds, about half way between Earth/Moon, during which the satellite crossed the front of the near-Earth shock wave. Soviet scientists developed the main instrumentation for measuring plasma waves and particles.

The Soviets began the Interball programme with a first launch of a 1,200 kg Prognoz-M2 (Prognoz 11) satellite on 2 August 1995 to a highly elliptical 505 × 193,064 km, 63.8°, 5,472 min orbit by Molniya launch vehicle. The second followed on Prognoz 12 on 29 August 1996 when a Molniya launcher placed it in a 782 × 19,140 km, 62.8° orbit. There have been no further launches although Argentina has raised the prospect of co-operation on a resumption of this programme.

Relikt/Libris

Background

Relikt 1 part of Prognoz 9's payload launched on 1 July 1983 to search for variations in the Universe's microwave background at 8 mm/37 GHz and placed in a 380 × 720,000 km, 65.5°, 26.7 day orbit. Two horn antennas, one anti-solar and its perpendicular twin scanned the entire celestial sphere every six months as Prognoz rotated at 1–2 rpm. Throughout their lifetimes, they had detected no unevenness at the 0.5mK/5.8° level at mission's end in February 1984. The Soviets subsequently revealed, however, that thermal radiation from the Earth and the Moon had disturbed the instruments.

Relikt 2 (apparently also called Libris) is thus planned for a 180-day halo orbit around the L2 libration point beyond the Moon to avoid this interference. Four radiometers with angular resolution 7, 3 and 1.5° will scan at 22, 35, 59 and 72 GHz. Relikt will also carry four plasma instruments to monitor geotail processes. As the Prognoz spacecraft has limited δV propulsion capacity, the halo orbit lies outside the core geotail, compromising the plasma detectors.

Investigators are interested primarily in large-scale (more than 5°) inhomogeneous regions in the Big Bang and medium-scale (0.5–5°) variations. Relikt was the third mission of the Prognoz-M2 platform. Further missions were planned, although Relikt 2 was never launched.

Libris/Relikt 2 was designed to orbit the L2 libration point to scan the Universe's microwave background 0516697

UPDATED

Spektrum-R/ Radioastron

Current Status

Considerable delay has postponed the originally planned launch date of 2002 due to depleted resources including funds and personnel.

Background

The Astro Space Centre of the Lebedev Physics Institute is the lead centre; Professor Nikolai Kardashev is project manager. High resolution Very Long Baseline Interferometry (VLBI) radio mapping of active galaxies, black holes, neutron stars and other exotic objects will be undertaken in conjunction with 70 m dishes at Yevpatoria (the control centre), Usserisk and a new antenna in Uzbekistan on the Suffa plateau. The Australia Telescope National Facility will play a critical role in the southern hemisphere observations of both Radioastron and Japan's VSOP. Tracking accuracy required is 5–15 cm range and 0.3 mm/s range-rate.

The US announced in March 1990 that it would participate in data processing, provide four data recorders on extended loan, and support tracking operations with its Deep Space Network (doubling coverage to 90 per cent). Total cost to the US was estimated at $20 million over five to six years. Finland is providing the 22 GHz ground/space receivers (designed/manufactured by Juhana Ylinen Electronics Co). BAe Australia built the 1.66 GHz L-band receivers and amplifiers. India made the 0.327 GHz unit and Europe the 4.83 GHz.

IKI issued a report in September 1987, outlining a 20 year plan of space-based radio astronomy missions. What had been believed to be a single Radioastron launch of 1990 was revealed to be a series of up to six craft, the first three of which would be based on Spektrum. Radioastron-cm comprised two satellites for cm-wavelength observations beginning in 1994/5, followed by at least one Radioastron-mm after 1996. The ultimate three-craft network of 30 m-dish Radioastron-KK beyond 2001 in GEO, 27-day and L2 halo orbits appeared to require a follow-on generation to Spektrum - payload mass was described as 20 tonnes.

Spektrum-UV was originally planned for launch around 2002 and it is intended as an ultra-violet observatory, which the Russians see as complementing the United States' Hubble Space Telescope. The observatory is to be built around a primary mirror (designated T-170) with a diameter of 1.7 m.

Specifications

Spektrum RG

Planned launch: undetermined

Contractors/Institutes: Lavotchkin/Babakin Engineering Research Centre (bus), IKI (payload; PI. Driving force behind project is Academician R Sunyaev, Head of Dept of High Energy Astrophysics).

Mass: 6.0 t at launch, science payload 2,750 kg

Orbit: planned 500 × 200,000 km, 51.6–65°, 96 h, launch by Proton-K from Tyuratam with apogee over northern hemisphere

Spacecraft: sized for accommodation within Proton's 3 m diameter × 5 m payload envelope. Central pressurised cylinder about 2 m diameter houses most electronic systems. 3-axis control provided by Sun/star/Earth sensors, RWs and microthrusters yielding 2–2.5 arcmin pointing accuracy (30–40 arcsec stability) for up to 30 h and 0.0001°/s angular stability. Twin Sun-following 3-panel solar wings provide 500 W to the science payload at 27±1 VDC. Data transmitted at 16–65 kbit/s or 18–128 Mbit/s through articulated 90 cm parabolic dish. Daily total 2–2.5 Gbyte for 4–5 h as each instrument in turn dumps its data load. Thermal control provided by radiator below cylindrical equipment bay. Design life 2 years

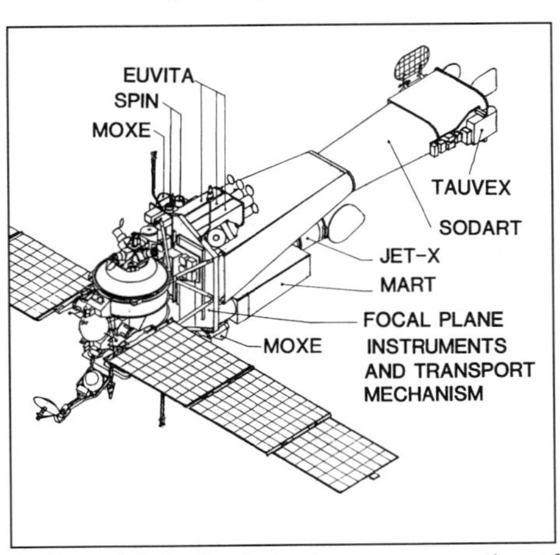

Spektrum-RG is the first of a new generation of astronomical observatories. The assembly at top right is the Sodart telescope, folded out 180° from its launch position (Danish Space Research Institute) 0516725

Radioastron's 10 m dish will operate in conjunction with ground telescopes and Japan's VSOP satellite for VLBI studies 0516726

Payload: SODART two multimirror (145 shells, 60 cm outer diameter) grazing incidence telescopes packages, one with Bragg crystal, each with focal plane detectors selected by a sliding mechanism. Telescope focal length 8.0 m, 1° FOV. Instrumentation covers 0.1–20 keV, 2 arcmin resolution, 30 arcmin FOV. For launch, SODART's upper section, housing the mirror modules, star/optical monitors and Bragg crystal, folded against the lower unit. Deployed by electric motor through 180°. SODART total mass 1,570 kg. Telescope A, with Bragg crystal: two soft x-ray XSPECT position-sensitive proportional counters (Denmark/Finland) for 0.5–3 keV and 2–25 keV, 2 arcmin resolution. Sealed position-sensitive gas proportional counter (CIS), 1.5–15 keV, 2 arcmin resolution. Two SXRP stellar x-ray polarimeters (USA, Italy) 0.26–20 keV, with 0.3 arcmin resolution are included. Telescope B carries two XSPECT counters as Telescope A, a solid state Si(Li) spectrometer (Finland/CIS) radiatively cooled to 100K, 00.3–20 keV, 3 arcmin resolution. Part of SODART is the 33.5 kg TAUVEX (Israel), providing simultaneous UV photometry via six filters covering 1,400–3,000 Å. It also provides offline aspect solutions for the x-ray experiment and is part of Spektrum's acquisition/pointing system to improve pointing to tens of arcsec.

JET-X 551 kg grazing incidence multimirror dual 3.5 m fl telescope with UK, Italian and Russian participation, 0.1–10 keV, 10–30 arcsec resolution, 30 × 40 arcmin FOV, optical star tracker provides pointing, optical monitor images faint sources to magnitude +23 for correlation with x-ray imagery.

MART coded mask x-ray telescope (Italy, CIS), 4–100 keV, 6 arcmin resolution, 8 × 8° FOV.

Euvita six EUVE telescopes (CIS, Switzerland) 0.031–0.12 keV, 10 arcsec resolution, 2 × 2° FOV, 10 cm^2 effective area.

Spin two γ-burst and scintillation spectrometers (CIS) 0.010–10 MeV, 30 arcmin resolution, 4π steradian FOV.

Spektrum-ultra-violet may carry a 1.7 m telescope 0516727

MOXE four coded mask x-ray burst monitors (USA) 3–12 keV, 15 × 600 arcmin resolution, 4π steradian FOV.

Spektrum-R Radioastron satellite
Contractors/institutes: Lavotchkin/Babakin Engineering Research Centre (spacecraft), Astro Space Centre (payload)
Mass: 5.22 t, payload 2.5 t
Orbit: 2,000 × 80,000 km, 51.5°, 24 h
Spacecraft: as Spektrum platform description but payload reduced at 1.5 t. Pointing accuracy required 60 arcsec
Payload: 10 m parabolic radio telescope (27 panels around 3 m central disc, surface accuracy rms. 0.5 mm) operating at 0.327, 1.66, 4.83, 22.2 GHz. Resolution 2.1, 0.4, 0.14, 0.031 marcsec, respectively.

Spektrum-UV satellite
Contractors/Institutes: Lavotchkin/Babakin Engineering Research Centre (spacecraft), Crimean Astrophysical Observatory (science payload)
Mass: 5.87 t, payload 2.5 t
Orbit: 51.5°, 500–300,000 km, period – 7 days.

Spektrum Series

Background

Soviet scientists introduced a new class of astronomical observatory platform design to supersede the Venera-based Astron/Granat type. Until recently, sources indicated that the Spektrum would appear in late 1995, but RKA financing problems have delayed the series and the first launch date has still not arrived. The Spektrum class bus provides 128 Mbit/s data rates, 30–40 arcsec pointing stability and 500 W power for a 2.7 tonne payload. While these features seemed impressive, in fact, it became apparent in 1993 that Lavotchkin's bus is already in service for the Prognoz GEO early warning satellites, the first of which flew as Kosmos 1940 in April 1988. Considerable effort has been expended on garnering international support to spread resources and funding among several investing research groups.

The first design in the new series, Spektrum-RG, includes extensive international x-ray instrumentation. The second derivative, Spektrum-R accommodates the Radioastron VLBI payload due to fly in 2002, followed by the Spektrum-UV observatory in 2004. The cryogenically cooled Spektrum-IR for sub-millimetre observations appears to have been cancelled. Despite the turmoil, sufficient funding is at least allowing the projects to survive.

As a follow-up to 1989's Granat x-ray observatory, the Spektrum-RG will cost Russia an estimated Rb3 billion (1992 Roubles). Russia spent Rb500 million through 1992. When the Spektrum series was first described in 1987 the cost per mission was estimated at Rb480 million. Denmark, Italy, UK and the US are contributing major x-ray and UV telescopes and instruments. The UK proposed to fly the Euvita UV instrument in October 1986, but UK participation was subsequently cancelled and a consortium led by Leicester University and RAL, and including Birmingham University and the Mullard Space Science Lab, together with Italian, Russian and German groups, was formed to develop Jet-X. The planned Dutch/UK TTM coded mask telescope was subsequently replaced by the Soviet's own Art-SP (now also cancelled). US MOXE is slated to replace the UK's Fourpi (4π) all-sky burst monitor. Some instruments were cancelled in 1991 when the scan platform was dropped because of mass problems (which continued).

Spektrum is planned to fly in a highly eccentric orbit reaching above Earth's radiation belts which permits protracted observations of x-ray sources and monitoring for γ-burst events. Up to 30,000 x-ray sources are expected to be mapped, including 3,000 active galactic nuclei.

Tenuous references to a subsequent complementary Spektrum-gamma mission in a 400 km orbit were made but no firm details have emerged and the Spektrum-X payload was extended to include a γ-ray capability. Several additional claims that a successor programme is in development have not been substantiated but Russian scientists insist that they are in discussion with fellow colleagues in former Soviet states for such a project.

VERIFIED

Znamya (Banner) Reflector Satellite

Background

In 1984, the Moscow Institute of Aviation proposed an experimental 200 kg satellite reflector with a 110 m^2 working area capable of reflecting sunlight to illuminate 9 km^2 regions in different time zones at night. Operational reflectors, the Institute thought, would provide seven times as much light as a full moon over areas of 10 km diameter, and should be able to prolong daylight for several hours in cities like Moscow. The first designs used a metallised polymer film. An early systems study of the scheme said the cost would be repaid in four to five years by electricity savings. The reflected light could be switched from one group of towns to another at short notice, providing a useful facility when night work was needed. When they launched Energia, in May 1987, Soviet officials linked it to this idea and future projects were designed to provide both illumination and power to the Earth from satellites. Radio Moscow reported in April 1989 that work on the Institute's experimental satellite had been halted, partly because of concerns over ecological effects of illuminating large areas of the Earth and its atmosphere.

The first flight test of a solar reflector was achieved 4 February 1993 when Progress-M 15 undocked from Mir and deployed a rotating 20 m 5 μm thick aluminised sail, directing sunlight to the ground. Rotation maintained the shape since there was no framework. The Znamya (Banner) Reflector remained visible after it was jettisoned from M15's nose after four orbital revs. The reflector weighed 40 kg, including the 4.2 kg sail. RKK Energia proposed a 200 m full-scale sail for late 1993 or early 1994 to travel beyond the Moon as part of the international race to celebrate Christopher Columbus' 1492 voyage. RKK Energia also proposed using it to illuminate the Earth's night side, followed by a plan to launch four or five reflectors by 1995. Funding has not been forthcoming. Concept has been revitalised as part of a drive to involve east European research centres, universities and institutes in a revitalised effort to develop unique space technologies. To date some limited progress has been made in encouraging independent research and limited development of a deep-space explorer concept utilising the solar sailing principle for very long duration flights. The Znamya work has been used in this effort.

South Africa

Sunsat

Current Status

Sunsat, a micro-satellite built by post-graduate engineering students at the University of Stellenbosch. South Africa's first satellite.

Background

Sunsat was announced in June 1991, to promote development of industrial and educational space engineering. It is part of a programme to establish an internationally recognised satellite engineering activity at Stellenbosch, started with the Grinaker company's sponsorship of a Chair of Satellite Systems. The control centre will be sited at the university over the five-year life. Stellenbosch already has an Oscar ground station and a 4 m antenna. Students are supported by major electronics companies: Altech and Siemens of Johannesburg, Grinaker Electronics of Pretoria, Plessey SA of Cape Town, Analysis Management & Services of Stellenbosch, and First National Bank of Johannesburg. Cost is estimated at US$800,000. NASA provides launch free in exchange for the GPS receiver and laser reflector payloads. Sunsat was originally planned for Ariane's Helios 1A launch but the US$400,000 cost was prohibitive. Instead, it replaces the balance mass for Denmark's Ørsted on Delta's Argos launch. Breadboard prototypes were completed January 1992 to March 1993; the engineering was completed April 1994.

Specifications

Sunsat satellite
Launch: 23 February 1999 by Soyuz 11A511U. Out of service 19 January 2001 due to a hardware failure
Orbit: planned 450 × 840 km, 98.7°, variable
Mass: 61 kg at release
Configuration: 450 × 450 mm 600 mm-high box-shaped body with solar panels on four sides, 4 m boom deployed from top face for gravity gradient stabilisation
Power supply: 20 W orbital average
OBC & Data Handling: 80C188EC + 80C386EX redundantly handle flight management tasks and digital communications, and access a 64 Mbyte RAM disk
Attitude control: 2 m boom with 5 kg tip mass provides gravity gradient stabilisation. Magnetorquers damp libration and provide slow yaw spin for thermal control. Redundant attitude determination using Kalman filtered 3-axis magnetometer data, linear CCD Sun/horizon sensors (0.5 mrad) and CCD star camera. T800 transputer used as a processor; back-up by flight computer. When imaging, RW stops satellite spin and re-orients Sunsat
Communications: 10 W transmit allows 1,200/9,600 baud. 5–10 W 2.3 GHz wideband downlink for imager. Adding 1.2 GHz receiver creates a wideband transponder capable of 1 Mbyte/s to 2 m dishes
Payload: Stereo 3-colour Earth imager providing 15 m resolution. Linear 3,500-pixel CCD plus 10 cm aperture optics (by CSIR in Pretoria). Real-time downlink to Stellenbosch. Swath width 50 km. Images up to 50 × 50 km stored in RAM for downloading by amateur operators. For stereo imaging, after 50–500 km strip recorded, camera attitude reversed for second view.

Voice/digital communications package at amateur frequencies.

SA Amateur satellite project. 2 m up/down + 2/0.70 m up/down both used. 'Parrot' mode for novice users uplinks speech on 2 m and re-transmits on same frequency. 2 m also distributes SA Radio League's weekly bulletin. SA Amateur satellite also provided linear 2/10 m Mode A transponder.

Precision receiver and laser retro-reflectors. Intended to reveal fine orbital perturbations for gravity field mapping in conjunction with Ørsted.

Spain

Minisat Series

Background

INTA manages a programme under the Interministerial Commission for Science & Technology (CICYT) to produce a small, low cost, rapid-delivery platform for science, communications and Earth observation missions. When the first bus design was complete, INTA signed a US$10.2 million launch services contract with Orbital Sciences Corp on 26 October 1994 for the L-1011/Pegasus XL to depart from Torrejon Air Base in September 1996. It actually flew on 21 April 1997 from Gando, making it the first orbital mission to depart from western Europe.

CASA is responsible for Minisat's platform and subsystems, Sener for systems/mission analysis, and Inisel for the Earth segment. Buying the unit as subsystem components (a technical assistance

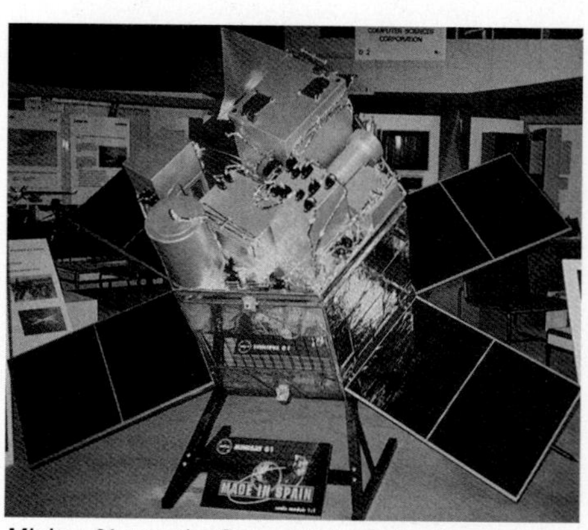

Minisat 01 was the first orbital mission from western Europe (Theo Pirard/Space Information Center)
0517281

Minisat 01 was launched in 1997 to return 1 m-resolution imagery (INTA) 0516737

agreement was signed with Ball Aerospace 9 October 1992) held down development costs and by using one of Spain's existing three sites for the ground station.

Minisat 01, the first in the series, was launched 21 April 1997 by a Pegasus XL from Canary Islands. Although early plans suggested Minisat 01 would enter a 28.5°, 600 km orbit, the one actually used was 151°, 560–581 km. It is possible that early reports of the planned inclination mis-interpreted the true value (180 –28.5° = 151.5°). It carried three science payloads, the EURD Extreme Ultraviolet Radiation Detector diffuse background study (INTA/ University of California at Berkeley), the CPLM to study behaviour of a liquid bridge in µg (also on UPM-Sat 1; Polytechnic University of Madrid) and the LEGRI 10–100 keV Low Energy Gamma Ray Imager (INTA, University of Valencia, Spain's CIEMAT Lab, UK's Rutherford Appleton Lab). Mass 201 kg (platform 100 kg; payload 100 kg); orbital life one to two years first mission; power 160 W EOL (100 W required, 40 W by payload); 32 Mbyte storage; 1 Mbit/s down by 5 W S-band; 3-axis stabilisation (±3° accuracy). The satellite decayed 26 February 2002.

The 550 kg Minisat 01 was to have used the platform to its maximum capability for a 1998 remote sensing mission (1 m + 20 m resolution). 330 kg payload, 480 W, 150 Mbit storage, 3-axis control (0.1° pointing error), N_2/N_2H_4 100 m/s propulsion.

UPDATED

UPM-Sat 1 Satellite

Background

This science and educational micro-satellite was designed and integrated by the CIDE/UPM Centro de Investigación y Desarrollo Espacial de la Universidad Politécnica de Madrid (UPM) for launch as an Ariane ASAP passenger. Another satellite project is proposed, with several universities and research institutes from Europe and Latin America.

UPM-Sat 1 is the first satellite built in Spain for more than 20 years (CIDEUPM) 0517115

An Ariane 4 launched the satellite on 7 July 1995 from Kourou, French Guiana into a 665 × 676 km, 98.07°, 98.2 min orbit. It weighed 47 kg, including separation system, packed into a 530 × 450 × 450 mm box-shaped body with solar panels on four sides. The main structural elements are four aluminium 7075 alloy shelves joined by four L-section beams. Four closing panels, supporting the solar panels, are screwed to the beams. Equipment is divided into three modular layers. For attitude control, the satellite uses active magnetorquing from 3-axis fluxgate magnetometer for attitude data, to provide spin for thermal control and to hold spin axis parallel to geomagnetic dipole axis.

The communications and datalinks provide 9.6 kbit/s downlink on single 400 MHz 10 W UHF dipole antenna. Data goes to a ground station at CIDE (Madrid). To supply the system with power, one GaAs experimental panel and three Si panels provide 20 W orbit average processed from 696 cells. Two sets of 14 Ni/Cd 6 Ah cells support eclipse operations.

In the payload bay, experimenters gather data on fluid behaviour in space with the LBE Liquid Bridge Experiment, at liquid mass held by surface tension forces between two solid supports. The ALE Al interconnector solar panel Experiment (with DASA/ESTEC) allows research in the efficient connection of solar cells with a panel of Si cells.

United Kingdom

United Kingdom Science Satellite Programme

Background

By early 1999, the UK had flown 21 indigenously-built national satellites. This total comprised one launched by Black Arrow, 13 by the USA and seven by Ariane. The total includes the five UoSATs of the University of Surrey. These are in addition to craft produced for international and commercial programmes, such as Giotto and Inmarsat. Prospero remains the only UK satellite launched on a UK vehicle. Like the original USA/USSR programmes, the national space effort depended initially on the conversion of a ballistic missile into a launcher. This was the Hawker Siddeley Blue Streak, finally cancelled as a missile in 1960 though continued until 1973 as first stage for ELDO's Europa. A more modest attempt at a national launch vehicle was announced in September 1964, developing a small three-stage vehicle from the successful Westland Black Knight research rocket. Construction of three Black Arrows was ordered in March 1967. Success in October 1971 with only the second complete vehicle was contrasted by the programme's previous cancellation in July 1971. With Prospero, the UK became the sixth nation to achieve orbital capability – after the USSR, USA, France, Japan and China.

The Prospero technology demonstration satellite was placed in 557 × 1,598 km, 82° orbit. Its sole science payload was Birmingham University's micrometeoroid counter. Prospero made 812 passes, which were monitored, out of the 4,960 orbits in first year. The science portion ended when the tape player failed, after replaying 730 times, on 24 May 1973. RAE Lasham contacted the satellite annually until 1989, when the site's VHF system closed down. The Black Arrow programme was cancelled July 1971 before Prospero's success. NASA Scout launched X-4 Miranda. Its intended launcher, R-4, is displayed at The Science Museum, London.

The UK also provided a 74 kg, 1 m diameter sub-satellite, known as the UKS, to the AMPTE project in a joint effort with the US and West Germany to study how energy carried by the solar wind is transferred to Earth's radiation belts and magnetosphere out to 100,000 km. The three-satellite mission began with a Delta launch in August 1984. The Rutherford Appleton Lab collected data for 4 h on each 44 h orbit for five months until UKS failed unexpectedly 16 January 1985.

The UK science ambitions suffered a serious setback in June 1996, with the uninsured loss of the Cluster solar monitoring missions on Ariane 501. Nonetheless, the country presses ahead by participating in current international missions such as ESA's Ulysses, ISO and SOHO, the ESA-NASA Hubble Space Telescope and the US/German/UK ROSAT and future missions with a UK contribution such as Russia's Spektrum-RG, and ESA's Rosetta cometary probe. The loss of Ariane 501, carrying Cluster, was immediately subjected to an ESA investigation, from which it was clear that no UK components were at fault. Replacement Cluster satellites were launched in 2000.

Specifications

Ariel 1
Launched: 26 April 1962 by Delta from Cape Canaveral and decayed 24 May 1976
Mass: 60 kg
Orbit: 389 × 1,214 km, 54°, 100.8 min
Comment: Transmitted ionosphere and x-ray data until November 1964. First international satellite, evolving from 1959 NASA offer on payloads of mutual scientific interest
Re-entered: 24 May 1976.

Ariel 2
Launched: 27 March 1964 by Scout from Wallops Island
Mass: 60 kg
Orbit: 285 × 1,362 km, 52°, 101.3 min
Comment: Similar mission to Ariel 1. Returned data until November 1964
Re-entered: 18 December 1967

Ariel 3
Launched: 5 May 1967 by Scout from Vandenberg
Mass: 90 kg
Orbit: 498 × 606 km, 80°, 95.6 min
Comment: First all-UK satellite. Observed radio noise from lightning and galactic sources at frequencies too low to penetrate the atmosphere. Spin-stabilised, operated for more than two years
Re-entered: 14 December 1970

Ariel 4
Launched: 11 December 1971 by Scout from Vandenberg
Mass: 100 kg
Orbit: 485 × 599 km, 83°, 95.3 min
Comment: Electron density studies, radio noise and other monitoring. Mission ended April 1973, but the satellite is re-activated occasionally to support sounding rocket experiments
Re-entered: 12 December 1978

Ariel 5
Launched: 15 October 1974 by Scout from San Marco
Mass: 129 kg
Orbit: 504 × 549 km, 3°, 94.9 min
Comment: Six instruments to study x-ray sources. Working with Dutch ANS and NASA Explorer 42 it identified 12 rapid x-ray bursters. Operated for 5 years
Re-entered: 14 March 1980

Ariel 6
Launched: 2 June 1979 by Scout D from Wallops Island
Mass: 154 kg
Orbit: 600 × 654 km, 55°, 92.1 min
Comment: Observations ended in February 1982 on exhaustion of ACS gas after having studied 30 x-ray objects in detail, including several black hole candidates
Re-entered: 23 September 1990.

United States

Advanced Composition Explorer (ACE)

Current Status

ACE has been at the L1 station since 1998 and with the exception of the SEPICA instrument the satellite continues to work well. As of mid 2002 more than

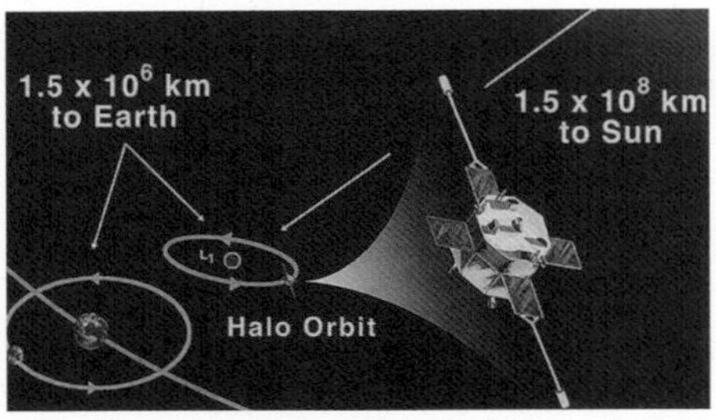

ACE will study charged particles ranging from the solar wind to galactic matter (NASA) 0517090

200 peer reviewed papers had been published by ACE science team members.

Background

ACE was conceived at a June 1983 meeting at the University of Maryland. NASA selected the Advanced Composition Explorer in October 1989 for Phase B, as an intermediate class Explorer, to study the composition of energetic particles. It has a collecting power 10 to 1,000 times greater than past missions. NOAA agreed in 1994 to pay US$450,000 for three instruments to return data continuously, instead of NASA's planned 4 h daily, to act as a Real-Time Solar Wind Monitor. NOAA wanted this capability to provide 1 hour warnings of solar storms to satellite operators and power companies.

Specifications

Contractors: Goddard (project management, ground system, mission operations, orbit determination), Johns Hopkins Applied Physics Lab (spacecraft, including integration and launch support), California Institute of Technology (science payload management), Seakr Engineering (SSRs), Olin Aerospace (propulsion), Ball Aerospace (star tracker), Adcole (Sun sensors), SAFT (battery), Tecstar (solar arrays)
Launch: 25 August 1997 by Delta 7920 from Cape Canaveral
Mission life: 2 years (5 year goal)
Mass: 765 kg (including 189 kg hydrazine)
Orbit: 150,000 × 300,000 km halo orbit around the L1 Sun-Earth Lagrangian point (arriving 3 months after launch)
Spacecraft: octagonal prism 160 cm diameter main body with four 86 × 152 cm deployed solar panels, providing 500 W; 12 Ah Ni/Cd battery. Two panels carry magnetometer booms. 5 rpm spin-stabilised, Earth-pointing spin axis. AOCS by 10 hydrazine 4.5 N thrusters, solid-state star tracker + digital Sun sensors. Solid-state storage capacity is >80 h; data stored for 24 h at 7 kbit/s interleaved with real-time data and returned to Deep Space Network via 66 cm parabolic antenna for up to 4 h daily at 77 kbit/s (but RTSW data returned continuously)
Payload: CRIS Cosmic Ray Isotope Spectrometer. Four identical Si solid-state telescopes with common scintillating optical fibre hodoscope. Few × 100 MeV/nucleon. CIT, GSFC, JPL, Washington University (St Louis).

SIS Solar Isotope Spectrometer. Two identical solid-state telescopes, 10–100 MeV/nucleon. CIT, GSFC, JPL.

ULEIS Ultra Low Energy Isotope Spectrometer. Time of flight *vs.* energy spectrometer. Few × 100 keV/nucleon. University of Maryland (UMD)/APL.

SEPICA Solar Energetic Particle Ionic Charge Analyser. Electrostatic deflection *vs.* energy loss charge state analyser. 1 MeV/nucleon. University of New Hampshire/Max Planck Institute.

SWIMS Solar Wind Ion Mass Spectrometer. Time of flight spectrometer. Few × 100 km/s. UMD/University of Bern.

SWICS Solar Wind Ionic Charge Spectrometer. Electrostatic deflection, time of flight and energy. Few × 100 keV. UMD, University of Bern and Max Planck Institute.

EPAM Electron, Proton and Alpha Particle Monitor. Two telescope assemblies have five apertures. Particle fluxes from 50 keV to several MeV. APL.

SWEPAM Solar Wind Electron, Proton and Alpha Particle Monitor. Separate electron (1–1,450 eV) + ion (260 eV–35 keV) electrostatic analysers. Los Alamos National Lab/Sandia National Lab.

Magnetometer. Two triaxial fluxgate magnetometers. 0.004–65,000 nT. Bartol Research Institute University of Delaware/GSFC
Secondary payload: RTSW Real-Time Solar Wind monitor for solar storm prediction. EPAM, SWEPAM and MAG data broadcast 20 h/day via a 434 bit/s selectable telemetry format to NOAA/USAF-provided ground stations. Data taken during 4 h ACE ground contact sent to NOAA in real time.

CATSAT (STEDI Series)

Current Status

Due to cost escalation NASA withdrew financial support for CATSAT. Low level tests continued while future launch options are evaluated.

Background

The University of New Hampshire's Co-operative Astrophysics and Technology Satellite will investigate the distance and polarisation of gamma bursts. Though high density bursts occur daily, their source remains a mystery. It is not even known if they originate from near our own galaxy or from cosmological distances, leading to an uncertainty in their source energy of 26 orders of magnitude. CATSAT will measure the 500 eV to 10 MeV burst energies to derive the photoelectric absorption along the line of sight. The point of cut-off below a few keV in this absorption depends on the column depth of material traversed. This depth will be compared with known line of sight column densities from radio surveys of galactic neutral hydrogen, thus yielding the burster distance. In its secondary science role, CATSAT will be the only all-sky monitor in the critical 500 eV to 2 keV range, directly complementing MOXE, HETE and RXTE by filling the gaps at 500 eV–2 keV and 1–10 MeV. NASA had scheduled a launch in 2002 but the project was cancelled.

Specifications

Contractors/Institutes: UNH, Weber State University (satellite), University of Leicester (UK)
Launch: planned for December 2001 aboard a Delta 7310 with the NASA ICESat satellite
Planned life: 1 year
Orbit: planned 550 km, 97° Sun-synchronous
Mass: 170 kg
Spacecraft: 53 × 53 × 56 cm. 3-axis control, nadir pointing to ±5°. 100 W nominal power. 1.5 Mbaud S-band science data down to UNH; VHF/UHF up/down command at WSU
Payload: soft X-ray spectrometer, hard X-ray spectrometer, directional γ-ray spectrometer, X-ray albedo polarimeter.

UPDATED

CHANDRA (AXAF)

Current Status

NASA renamed the Advanced X-ray Astrophysics Facility (AXAF) after the renowned astrophysicist and Nobel-Laureate Subrahmanyan Chandrasekhar, giving it the name CHANDRA in late 1998. CHANDRA was launched by STS-93 on 23 July 1999.

Following a series of five engine firings Chandra reached its desired orbit of 9, 676 × 139, 141 km on 7 August.

The official first light image was received from the Advanced CCD Imaging Spectrometer, a picture of Cassiopiea A on 19 August 1999.

On 23 July 2004 scientists celebrated five years of almost continuous use and Chandra continues to provide solid science data with the spacecraft performing well.

Background

The Advanced X-ray Astrophysics Facility (CHANDRA) constitutes the third in NASA's Great Observatories Series, following the Hubble Space

The mirror for the Chandra telescope is subject to environmental tests (NASA) ***NEW***/1047481

The Chandra astronomy observatory satellite during final assembly and checkout (TRW) 0572504

Telescope and the Compton Observatory. AXAF-I (Advanced X-Ray Astrophysics Facility) will resume the work started by Uhuru in 1970 and continued by HEAO 2 until re-entry in 1981. AXAF-I's four pairs of nested mirrors for focusing 0.1–10 keV high-energy x-rays will offer about 10 times the spatial resolution and 100 times the sensitivity of HEAO 2. With 1.2 m outer diameter, the 10 m focal length mirrors will provide 1,100 cm^2 collecting area (instead of the 1,700 cm^2 of the original six pairs).

In March 1985, NASA announced 11 investigators who will provide scientific and technical guidance from initial design to on-orbit operation; nine were US, one Netherlands and one UK. To initiate the programme, Congress gave NASA Marshall US$29 million start-up funding in FY89. In subsequent years NASA allocated US$64 million in FY90, US$113 million in FY91, US$141 million in FY92, US$158 million in FY93. The major contracting work began with the authorisation of US$224 million for AXAF-I in FY94 (reduced from the original request of US$260 million for both AXAF-I/AXAF-S), 234.3 FY95, 237.6 FY96, 178.6 FY97, 92.2 FY98. Congress terminated funding for AXAF-S in November 1993 as part of a far-reaching NASA cost-reduction exercise. The US$60 million shortfall in FY92's request, and Congress's declaration of no FY93 increase, slipped the launch a year to 23 July 1999.

NASA approved a facility design in 1991 for the US$86.7 million contract to design and operate the AXAF Science Center in Cambridge, Massachusetts. Under a US$40.7 million contract to the MIT Center for Space Research, in support of the Principal Investigator at PSU, Martin Marietta was selected in 1986 as subcontractor for the CCD Imaging Spectrometer; the company's Space Systems division was awarded the US$13.1 million contract in mid-1990. The HRC instrument is primarily an in-house effort at SAO, utilising a few small subcontracts. Ball Aerospace Systems Group in Broomfield, Colorado was awarded the US$24 million contract in September 1990 by GSFC for the XRS 490 l helium dewar to cool the instrument to 1.3 K. Fokker Space is providing the solar arrays.

Preliminary plans called for the US$1.4 billion 5,200 kg AXAF-I (imaging) to be launched by the Shuttle with an unspecified upper stage in September 1998 into a 10,000 × 140,000 km, 28.5° orbit. The US$320 million 2,700 kg AXAF-S (spectroscopy) would follow in December 1999 aboard Delta into polar orbit. NASA wanted both to last at least five years in orbit. AXAF-I consists of a four-mirror pair-shell grazing incidence tele scope (instead of the original six parabola/hyperbola pairs), two gratings and two focal plane instruments. Boeing's agreed to provide an IUS upper stage in May 1994 under a US$25 million contract to establish the final orbit. AXAF-S changed the replicated optics telescope with the supercooled x-ray spectrometer in the focal plane.

NASA adopted a phased programme to avoid HST-type cost overruns. Hughes Danbury Optical Systems Inc began grinding the largest parabola/hyperbola pair (P1/H1) of the four paired nested grazing incidence mirrors in December 1989 under TRW's US$158.1 million contract. Congress, as a measure to check NASA's repeated history of cost escalation, required the agency to make a semi-annual progress report through the completion of testing on 1 October 1991. Testing on P1/H1 was completed that September in MSFC's new US$17 million X-Ray Calibration Facility (XRCF).

All eight flight mirrors were completed in early 1995, followed by coating at Optical Coating Laboratories, Inc (H1 was the last, in January 1996) and delivered to Eastman Kodak. Kodak provides the optical bench and assembles the eight mirror elements (HRMA High Resolution Mirror Assembly) and the telescope. Ball Aerospace Electro-Optics/Cryogenics Div will supply a visible wavelength aspect camera for attitude reference. Ball will also provide the SIM Science Instrument Module, which supports the focal plane instruments, provides focus control and translates the instrument in/out of the focal position as required.

Specifications

Launch: 23 July 1999 by shuttle mission STS-93 from Kennedy Space Centre using IUS as boost stage
Contractors: TRW Space and Electronics Group
Mass: 4,790 kg
Orbit: 9,977 × 138,400 km, 64 hr orbit.
Spacecraft: 13.8 m long, 19.5 m span across deployed solar panels. Mission life >5 years. Power provided by 3-panel, silicon solar arrays (2.35 KW) and 3 × 40 Ah metal/hydrogen batteries for eclipse power. Data recording from solid-state recorder with 1.8 giga-bytes (16.8 hr) capability

Design life: 5 years
Payload: AXAF CCD Imaging Spectrometer, sub-arcsec angular resolution (ACIS, Pennsylvania State University/MIT).

High-Resolution Camera, microchannel plate imager with less than 0.5 arcsec angular resolution (HRC, Smithsonian Astrophysical Observatory).

High-Energy Transmission Grating Spectrometer, 0.4–8.0 keV, resolution 800 at 1.0 keV.

ACIS, images spectral dispersion of grating (HETGS, MIT).

Low-Energy Transmission Grating Spectrometer, 0.1–3.0 keV, resolution 750 at low energies, HRC images spectral dispersion of grating (LETGS, Utrecht Lab for Space Research).

Payload weight 954 kg. Focal length 10 m, outer diameter 1.2m.

UPDATED

Compton Gamma Ray Observatory (Great Observatories Series)

Current Status

Orbit re-boost manoeuvre in 1993 extended mission life by 5 years. It recorded over 2,600 gamma explosions.

Following the failure of a control gyroscope in December 1999, mission managers elected to return the spacecraft to atmospheric re-entry, where it could be destroyed over the Pacific Ocean. De-orbit was safely carried out on 4 June 2000.

Background

Before launch, this satellite was simply known as the Gamma Ray Observatory, but shortly after launch, it was re-named the Compton Gamma Ray Observatory (CGRO) in honour of Arthur H Compton, the US joint-Nobel laureate in physics. One of the largest astronomical satellites ever launched, Compton is providing a deeper understanding of high-energy processes within supernovas, pulsars, quasars and radio galaxies, illuminating the character of the Universe shortly after its creation, and indicating the possible existence of anti-matter in the Universe.

As its scientific priority, the Compton seeks to explain the typically 1 second gamma bursts that sweep across the solar system. BATSE had detected 1,500 by the end of 1995. The brightest occurred on 31 January 1993. The first scientists expected they would concentrate in the galactic disc, since they felt the most likely origin was detonations on the surfaces of neutron stars. However, BATSE has found them uniformly over the sky. This indicates they originate either from some exotic small objects near the Solar System or from extremely powerful distant objects well beyond the galaxy. The latter source is more likely, requiring a new type of celestial body generating unprecedented quantities of energy from previously unknown processes. Such an object would have to channel many times the energy of a supernova into a 1 second gamma-only flash.

In another discovery, Compton's EGRET instrument found that quasar 3C-279 emits 10^8 times the galaxy's gamma energy, a step towards explaining the nature of these exotic objects. In early 1993, Germany's COMPTEL made the first detection of a Ti-44 γ line, in the Cas-A supernova remnant. As Ti-44 has a half-life of only 70 years, this line can now be used to look for nearby but obscured supernovas in our galaxy. COMPTEL also found Al-26 emissions from the 10,000 year-old Vela SuperNova (SN). Al-26's 10^6 year half-life means it can be used to look for traces of very old nearby SNs.

NASA's Compton Observatory is probing the Universe's higher energy processes. COMPTEL is the central cylinder, EGRET is to the right and OSSE is housed in the unit at left. The eight BATSE burst detectors are mounted on the platform's corners (NASA)
0516739

COMPTEL's third major 1993 discovery was to identify the Orion nebula as a source of cosmic rays. Such particles filling the galaxy were discovered in 1911 but their origin was a mystery. Following a 17 February 1994 gamma burst, EGRET detected high energy gamma rays for more than 90 minutes. OSSE has detected a cyclotron spectral line from the galactic neutron star A0535 + 26, indicating a surface magnetic field of more than 5×10^{12} gauss.

Specifications

Launch: 5 April 1991 by Shuttle from KSC; deployed 7 April 1991
Contractors: TRW Space & Communications Co
Mass: 15,623 kg at release
Orbit: 445 × 459 km, 28.45° at release; decayed to 345 km by mid-1993; reboosted to 441 × 452 km October to December 1993; decayed to 386 km by March 1996
Spacecraft: 7.6 m long, 3.8 m diameter, solar array span 21.3 m. 3-axis stabilised. Based on 6.30 m long, 3.15 m wide, 61 cm deep platform of aluminium sill/keel beams. Mission life 21/4 years but four tanks carry 1,814 kg hydrazine sufficient for 10 years AOCS via 4 × 445 N orbit adjust + 8 × 22.2 N attitude control thrusters (although not planned, can be refuelled in orbit; total impulse 4.03 MNs). 500 kg hydrazine is reserved for controlled re-entry at mission's end. Control torques and momentum storage provided by four reaction wheel assemblies. Inertial Reference Unit is a self-contained strapdown three-axis dual redundant attitude rate sensing unit, providing reference for high rate (±2°/s) manoeuvring and low rate (±400 arcsec/s) for pointing; associated with fixed head star trackers, fine/coarse Sun sensor assemblies and 3-axis magnetometer. Two solar wings of four 137 × 336 cm panels each (total of 40,488 Si cells) provide 4,500 W BOL/3,980 W EOL to two modular power systems, each supported by three 50 Ah Ni/Cd batteries (one battery has failed; three are adequate). Recorded data were returned at 256/512 kbit/s via 2.2875 GHz S-band 152 cm diameter high gain antenna on 442 cm boom working through TDRSS; both tape recorders have failed (the first in 1991). TDRS 1 was moved to 85° E February 1994 to provide a dedicated 32 kbit/s real-time link through Australia's Tidbinbilla for relay to White Sands via Intelsat, increasing coverage from 65 to 90%. TDRS 3 took over in June 1995. Command rate 1 kbit/s, backed up by 0.125 kbit/s through two omni antennas at 2.1064 GHz
Payload: Oriented Scintillation Spectrometer Experiment (OSSE). Four identical 33 cm diameter NaI(Tl)-CsI(Tl) phoswich detectors for 0.05–10 MeV measurements. Operated in pairs, they can be turned through 180° for background and secondary target monitoring. Tungsten collimator provides 3.8 × 11.4° FOV (resolution). Naval Research Lab; Ball Aerospace.

Imaging Compton Telescope (COMPTEL). 2° resolution over 1.0–30 MeV using two planes of scintillators 1.5 m apart. Compton's most sensitive instrument. Max Planck Institute (Germany); MBB prime.

Energetic Gamma-Ray Experiment Telescope (EGRET). Operating at top end of Compton's energy range, 20–30,000 MeV, using 6,400 cm^2 spark chamber and providing best resolution (0.5° at 2 GeV). NASA Goddard.

Burst and Transient Source Experiment (BATSE). Eight wide field detectors covering the whole celestial sphere (4π radians) each with 50.8 cm diameter, 1.3 cm thick NaI(Tl) scintillation crystal to register 0.20–100 MeV transient events of a few μs and locating positions to within a few degrees.

Cosmic Background Explorer (COBE)

Background

Cosmic Background Explorer (COBE) set out to answer fundamental questions about the universe. Its microwave/infra-red sky maps at 100 wavelengths are helping to determine if the universe is rotating, is distributed homogeneously and is uniformly expanding. In addition, present interstellar and interplanetary dust clouds were mapped. COBE was conceived in the mid-60s after the Big Bang theory of creation received a strong verification in 1964 when Arno Penzias and Robert Wilson serendipitously discovered the remnant radiation resulting from the bang.

COBE: the Cosmic Background Explorer was the first of the new Explorer series (NASA)
0516745

COBE carried three main instruments. The Differential Microwave Radiometer (DMR) mapped the sky at 3.3/5.7/9.6 mm, where the galaxy is more transparent, using standard microwave hardware to compare the emissions from two patches of sky, 60° apart, at a time. If the early universe was thermally and structurally uniform the background would be equal around the sky. If not, the map will highlight the variations. The detectors for the FIRAS Far Infra-Red Absolute Spectrometer were cooled to less than 2K by liquid helium inside a cryostat and protected against Sun/Earth radiation by the conical shield. The 95.7 kg of cryogen limited the far-infra-red mission to a 10-month life, sufficient for 1.6 surveys.

Operating with a sensitivity two orders of magnitude greater than previous balloon/rocket instruments, FIRAS collected radiation with a horn antenna viewing along the spin axis. Like DMR, it provided an angular resolution of 7°. An electrically heated calibrator was inserted into the horn and warmed until its emission matched that of the sky in the field of view for measurements with an accuracy of 0.001°C over 100 μm-1 cm. The Diffuse Infra-Red Background Experiment (DIRBE) searched for the remnant infra-red radiation, previously undetected, from the initial burst of radiation from the primordial galaxies as they formed. Ten filters studied selected bands over 1–300 μm, using cooled bolometers, photoconductors of Ge and Si, and photovoltaic detectors of In and Sb; observations at the shorter wavelengths were still possible after cryogen exhaustion.

FIRAS, measured the background temperature at 2.728 ± 0.004K, varying 0.01 per cent from the blackbody spectrum and indicating that 99.99 per cent of the universe's early radiant energy was released within one year. FIRAS' results mean that the Big Bang theory has been tested against observation to a fine degree of precision. Another instrument, DIRBE mapped the radiation from the interstellar medium, solar system dust and stars; it is hoped that extensive modelling now under way to extract the foreground infra-red will reveal the glow from the universe's first stars and galaxies. DMR's momentous discovery of the predicted background ripples in support of the Big Bang theory was announced April 1992. These 30×10^{-6}°C variations, required for the appearance of galaxies, date back to when the was only 10^6 years old. They also support the theory that most of the universe is composed

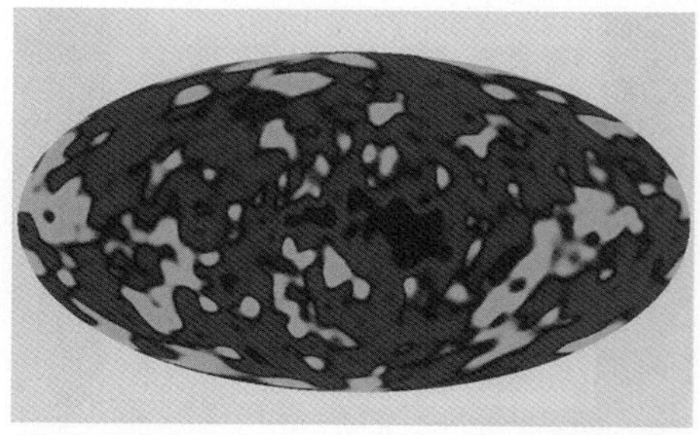

All-sky microwave map generated by COBE's DMR instrument. Our Galaxy's radiation has been processed out, to leave the 0.02 per cent variations in the 2.73 K background. Detailed analyses show that fainter cosmic signals are present from the earliest clumping of material in the Universe
0517102

of dark matter. FIRAS mapped interstellar ionised nitrogen (the first detection at 205 μm), carbon and dust for the first time, atomic emissions that control the cooling of interstellar gas. DIRBE's continuing survey at shorter wavelengths, although six times less sensitive than originally, was primarily aimed at solar system dust/gas mapping. Repeated coverage could show temporal variations (the analysis has yet to be performed).

COBE completed a full-sky survey in mid-June 1990 and was performing its second when the cryogen was exhausted 21 September 1990. FIRAS' observations therefore ended, but the un-cooled DMR continued and DIRBE could still operate at its shorter IR wavelengths with the cryostat's interior warmed to 72 K. Science operations ended 23 December 1993 and EOL engineering tests were completed in January 1994 before COBE was handed over to the Wallops Flight Facility for testing WFF's ground tracking system two to three times monthly, evaluating telemetry reception, command transmission and the angle and Doppler tracking capabilities. TDRSS and JPL similarly access it when WFF work allows. Commanding is limited to basic transponder functions and playback of tape data recorded when COBE was last fully operational. No other housekeeping functions are exercised but, with the high power margin, COBE should continue for some years. It is commanded locally by any of WFF's stations. The original flight engineers are occasionally given data for a quick checkup.

Mission cost was approximately US$230 million. Results from the COBE DMR experiment led directly to the development and launch of the Microwave Anistrophy Probe.

Specifications

Launched: 18 November 1989 by Delta 5920 from VAFB
Lifetime: 1 year nominal
Mass: 2,205 kg BOL, 2,109 kg EOL
Orbit: 887 × 898 km 99.0° Sun-synchronous
Spacecraft: hexagonal bus spin stabilised at 0.8 rpm normal to Sun by RWs (supported by Sun sensors, magnetometers and three scanning infra-red Earth sensors) for scanning the instruments, 95.7 kg/600 litre liquid helium initially carried in cryostat, protection against solar/Earth radiation provided by conical shield. Three 3-segment solar wings totalling 31 m^2 and spanning 8.5 m generate 1,050 W BOL/905 W EOL (615 W EOL eclipse). TT&C provided through NASA's TDRS system but data dumped to Wallops in 10 min each day for relay to the COBE Science Data Center at Goddard.

Discovery Series

Background

NASA's strategic plan, formulated in 1991 for solar system exploration, included the Discovery programme of near-Earth missions. These are rapid turnaround missions each costing less than US$150 million (US$75 million in any one year) at 1992 rates for spacecraft development; total cost must be less than US$245 million. Launch must be within three years of approval. Costs are minimised by emphasising focused science return from mature instrument and spacecraft technology providing simplicity and reliability. Only flight proven launchers (no larger than Delta 2) are employed. Each proposal requires a consortium of a NASA centre (or other federally funded research centre), industry and university partners.

The first two missions are the NEAR (Near-Earth Asteroid Rendezvous) in February 1996 and the Mars Pathfinder in December 1996. US$66.6 million and US$60.8 million was approved by Congress for FY94 to begin NEAR and Mars Pathfinder, respectively. US$129.7 million was provided FY95; US$102.2 million FY96; US$74.8 million requested FY97.

NASA received 28 proposals for round two by October 1994 in response to August 1994's request. NASA announced February 1995 that mission 3 would be the US$59 million Lunar Prospector, to be launched October 1997 to map the Moon's chemical composition and possibly identify polar water ice. The proposal was sufficiently mature to proceed to development after final technical definition. See the separate entry. Three other missions were selected for studies before the final choice in November 1995. Stardust was launched February 1999, which flew through the coma of active comet Wild 2 in January 2004, returning comet and interstellar dust samples on aerogel plates to Earth January 2006. Total cost to NASA will be US$199.6 million. See the separate entry. Unsuccessful were Venus Multiprobe Mission (16 23 kg capsules into Venusian atmosphere to study complex circulation) and Suess-Urey (collect solar wind samples outside of Earth's magnetosphere).

The next round for Discovery missions started in June 1996 and, in April the following year, NASA announced the five proposals for further detailed studies as part of the series. Feasibility studies for the five missions submitted by 15 August 1997, and one or two will be selected the following October for full development in the Discovery programme. The five proposals selected in April 1997 were:

Aladdin, a mission to gather samples of the small Martian satellites Phobos and Deimos, plans to fire four projectiles into the satellites' surface and gather the ejecta during slow fly-bys, and then return the samples to Earth for detailed study;

The Comet Nucleus Tour (CONTOUR), a mission to take images and comparative spectral maps of at least three comet nuclei and analyse the dust flowing from them;

Genesis, a mission to collect a sample of the solar wind and return it to Earth for detailed analysis. It had been one of the finalists in the fourth round of Discovery programme selections under the name Seuss-Urey;

The Mercury Surface, Space Environment, Geochemistry and Ranging mission, or MESSENGER, an orbiter spacecraft carrying seven instruments to image and study the closest planet to the Sun;

The Venus Environmental Satellite (VESAT), an orbiter spacecraft to study the atmospheric chemistry and meteorology of Earth's cloud-covered neighbour using an imager, near-infra-red spectrograph, a temperature mapper and an X-band radar.

The spacecraft must be ready for launch no later than 30 September 2002, and meet the Discovery Programme's development cost (launch plus 30 days) cap of US$183 million in FY97 dollars over 36 months.

The two missions selected in October 1998 were Genesis and Contour. Genesis, the solar wind sample return mission was launched on 8 August 2001 followed by Contour, the mission to fly by three comets on 3 July 2002. In July 1999, NASA announced selection of two more Discovery series missions: Messenger, the Mercury orbiter, launched 3 August 2004; and Deep Impact, a fly by of comet P/Tempel 1 and release of an impact projectile, to be launched in a window that opens 30 December 2004. Three candidate proposals were unsuccessful. Aladdin, a mission to Mars moon Phobos and Deimos; inside Jupiter, a Jupiter orbiter; and Vesper, a Venus orbiter to study the middle atmosphere of that planet. Other Discovery missions selected are; Dawn, the first purely scientific mission propelled by ion engine, will launch in May 2006, reaching the asteroid Vesta in 2010 and Ceres in 2014, examining characteristics of each; Kepler, launched 2001, designed to search for earth-size planets, using a photometer to measure the brightness of 100,000 stars and looking for transits of planetary bodies. A Discovery Mission of Opportunity, ASPERA-3 is one of seven science instruments carried aboard Mars Express, the ESA Mars mission launched in June 2003.

UPDATED

ERBS

Current Status

Lowered in July 2002 for a controlled re-entry.

Background

The 2,240 kg Earth Radiation Budget Satellite (ERBS), launched 5 October 1984 by Shuttle 41G into 610 km, 57° orbit remains operational. Ball Aerospace as part of the three-satellite Earth Radiation Budget Experiment managed by NASA Goddard built ERBS; NASA Langley was responsible for instrument development and data reduction. Earth's radiation budget is probably the weather's principal driver. Energy balances for the entire planet resolve around the energy exchange among the poles, the tropics and outer space. The tropics absorb more energy than they emit, and seasonally the polar regions receive no solar radiation, though they continue to emit. The difference in absorbed and emitted energy causes the atmosphere and oceans to circulate and transfer energy, creating weather systems.

ERBS's instruments were supplemented by similar packages on NOAA 9/10 in near-polar orbits to monitor thermal and solar radiation of the entire Earth at some time each day, and most of the planet

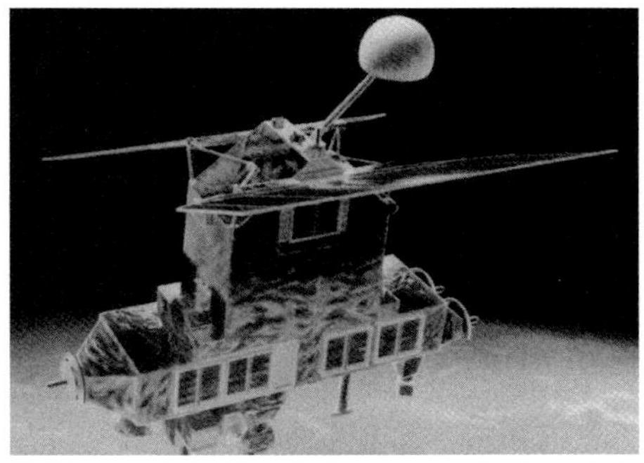

ERBS was released by Challenger's manipulator. Sally Ride had to use the arm to shake the satellite before the 7 m solar array would deploy. Its thrusters later fired for over 10 hours to attain a 610 km orbit 0516740

at all times of day during every month. All three scanning instruments have failed: NOAA 9's on 20 January 1987; NOAA 10's on 22 May 1989 and ERBS' on 28 February 1990. The eight radiometers of each satellite were divided into two separate instruments: a scanning and a non-scanning radiometer (wide/medium FOV). They monitor the incoming solar ultra-violet/visible and subtract the component reflected back into space (principally by clouds/snow) and the infra-red/thermal infra-red emission from the heated atmosphere and surface. The result is Earth's net radiative heating. The observations are indicating variations in the Sun's output, the veiling of Earth by volcanic dust, increases in atmospheric CO_2 from fossil fuel burning.

ERBS demonstrated for the first time that cloud cover is responsible for a net global cooling, depressing surface temperature by an average of 11°C during April 1985. ERBS observations also provided the first direct evidence that a volcanic eruption can lead to significant changes in the global radiation balance. A 3.8 per cent increase in reflected sunlight was detected for several months following Mt Pinatubo's 1991 eruption.

In separate experiments, ERBS carries Langley's SAGE (Stratospheric Aerosol/Gas Experiment) to monitor the vertical distribution of stratospheric aerosols, O_3 and N_2O_4. SAGE and Nimbus 7 showed that Antarctica's 1987 ozone hole, which disappeared during 1988, re-appeared almost as deeply in 1989 and worse in each subsequent year.

In order to deploy ERBS, NASA launched the Shuttle into a 57° inclination for only the second time. It was released by the remote manipulator and employed an integral propulsion system to reach the required orbit. 3-axis control is normally maintained to within 1° by a magnetorquing system, backed up by the hydrazine thrusters, employed primarily for orbit adjust and yaw turns. ERBS must be yawed 180° every four to five weeks to maintain solar array illumination. It requires 150 real-time and 550 stored commands each day for normal operations. The solar arrays (470 W average load) are supported by a 50 Ah Ni/Cd battery (2 of 22 cells failed; the other battery failed). Commands and telemetry are routed through TDRSS by the electronically steerable spherical array antenna, backed up by two omni antennas via TDRSS and ground stations. Data are recorded at 1.6/12.8 kbit/s and dumped at 128 or 32 kbit/s on alternate orbits; normal real-time rates are 1.6/12.8 kbit/s.

Spacecraft systems and the SAGE and ERBE non-scanner instruments continue to operate normally except for a failed battery and some failed gyros. The gyro losses do not affect operations apart from the yaw turns, which are performed by timed thruster firings. As of 5 February 1996, after 4,110 days on orbit, ERBS had completed 61,775 orbits, 114 manoeuvres and more than120,000 SAGE science events.

Explorer Series

Background

Explorer is the oldest US satellite series and will continue to fly science payloads in Earth orbit into the next century. As of the end of 2000, 78 Explorer missions had been launched since Explorer 1 became the first US satellite in February 1958. The Advanced Composition Explorer (ACE) was the last of the Delta-class Explorers, before FUSE appeared as the pathfinder for the smaller and cheaper MIDEX Medium Explorer line. FUSE was launched on 24 June 1999. The first two MIDEX missions were selected in April 1996 for definition studies

SWAS will investigate the chemistry of interstellar clouds (Ball Aerospace) 0517128

leading to 1999 and 2000 Med-Lite launches: MAP Microwave Anistropy Probe and IMAGE Imager for Magnetopause to Auroral Global Exploration. The SMEX Small Explorer programme is providing a return to lower cost, rapid turnaround LEO missions emphasising space physics and astrophysics. Three are flying under the first cycle: Solar, Anomalous & Magnetospheric Particle Explorer (SAMPEX, launched 1992), Fast Auroral Snapshot Explorer (FAST, 1996) and Sub-millimetre Wave Astronomy Satellite (SWAS, 1997). Two, Transition Region & Coronal Explorer (TRACE, 1997) and Wide-Field Infra-red Explorer (WIRE, 1999), were selected in 1994 as the second set.

NASA also plans the UNEX University Explorer series, giving the PI authority and management responsibility for less complex missions than SMEX. It will be managed under a fixed cost ceiling, with NASA retaining oversight authority sufficient only to discharge its responsibility for handling public money. The SNOE and Terriers missions of the USRA Universities Space Research Association are UNEX precursors.

The Explorer programme is headed by NASA's Goddard Space Flight Center for the Office of Space Science.

Extreme Ultra-Violet Explorer (EUVE)

Current Status

In August 2000, the NASS Office of Space Science Senior Review recommended the termination of EUVE on 30 September. The all-survey stay catalogue listed 801 objects and EUVE mode the first EUV detection of extragalactic objects. The satellite operated until 31 January 2001 and was deorbited in January 2002.

Background

EUVE carries four ultra-violet telescopes covering 60–800 Å for the first detailed all-sky observations at these wavelengths. The initial all-sky mapping required six months, 24 July 1992 to 21 January 1993, pinpointing sources with 0.1° accuracy for more detailed subsequent observations over the following year. Some 700 bright EUV sources appeared in EUVE's instruments and more than 40 extragalactic sources have been detected.

The observatory phase began 22 January 1993, looking at one to four targets weekly. Most EUV objects discovered to date have been stars at advanced stages of evolution; white dwarfs and stars with active coronas, although 30 per cent have not been identified with an optical counterpart.

The Challenger accident delayed and forestalled plans for the EUVE and in October 1986 Goddard began procurement of an Explorer Platform, launchable by either Shuttle or Delta, that would permit these payloads to be exchanged in orbit. It was decided, in 1991, that launches would be on expendables to maximise science return. EUVE is the first payload flown on the new Explorer Platform, based on the NASA/Fairchild Multi-Mission Spacecraft (MMS). The EUVE's four instruments are 3 to 188 kg grazing incidence telescopes and a 323 kg deep survey/EUV telescope/spectrometer. The fourth instrument took measurements in the anti-Sun direction along the ecliptic, taking advantage of the low background

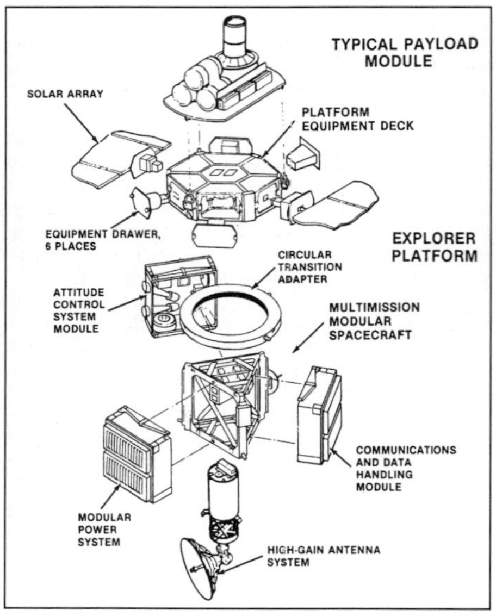

EUVE is undertaking detailed extreme Ultra-Violet observations 0516746

to complete a survey in two bands in 60–400 Å. Its three spectrometers spanning 60–760 Å to study the brightest sources with a resolving power of 350. Each instrument incorporates 40 cm diameter Wolter Schwarzschild Ni-coated aluminium mirrors: two Type I covering 60–400 Å (both for the sky survey) and two Type II for 400–800 Å (one for the deep survey and spectrometer). Detectors consist of Z stacks of micro-channel plates mounted in windowless assemblies that are read out in an array of 1,024 × 1,024 pixels.

EUVE is managed from Goddard's POCC Payload Operations Control Center and the science payload from the Center for EUV Astrophysics at the University of California.

Specifications

Contractors: Goddard (project management; development/integration of payload module), Fairchild Space Corp (prime; communications/data handling MMS module), General Electric (attitude control system MMS module), McDonnell Douglas (power systems MMS module), University of California Space Sciences Lab (instruments, Science Operations Center)
Launch: 7 June 1992 by Delta 2 from Cape Canaveral
Lifetime: 42 months nominal mission; 20 month/US$10 million extended mission funded from January 1996. No degradation of spacecraft or instruments has occurred
Mass: 3,256 kg at launch
Orbit: 514 × 529 km, 28.5°
Spacecraft: based on NASA/Fairchild MMS, with Platform Equipment Deck mated with the Explorer Platform by motorised connectors. MMS carries modular power system (working with two solar wings providing about 300 W to payload from >1 kW BOL total), command/data handling module, the Modular Attitude Control System (MACS) and the signal conditioning unit. 3-axis stabilisation. Data rate of 16 kbit/s shared by the four telescopes, recorded with engineering data at 32 kbit/s on a 10E9 tape recorder for dumping to ground at 512 kbit/s through TDRS system. Platform designed for 10-year life; degrading solar arrays can be replaced in orbit (but not planned).

Far Ultra-Violet Spectroscopic Explorer (FUSE)

Background

NASA initially studied FUSE as a 1 m telescope for very high-resolution spectroscopy over 900–1,200 Å, expanding the preliminary investigations made by OAO 3 and complementing HST and the x-ray missions. FUSE was planned for launch by Delta in 2000 in a total US$254 million programme, but in 1994, Congress reduced its total budget. NASA reduced the budget again in 1998 and renamed it Med-Lite. As with the Hubble Space Telescope, NASA decided to save money by lowering the apogee of the orbit at the expense of the scientific mission. With the FUSE, the Agency lowered the original 71,000 km apogee to LEO, reducing the

FUSE will study ultra-violet spectra below IUE's range 0517284

number of observations but not affecting the basic science.

Instead of NASA Goddard managing the programme and building the spacecraft for the PI's payload, PI Warren Moos of JHU's Center for Astrophysical Sciences has full responsibility for the mission, selecting Orbital Sciences Corp in August 1995 for the US$37 million bus contract. McDonnell Douglas is providing the Delta 7320 at the same cost as planned for Delta-Lite, which it now does not plan to develop.

One FUSE goal is studying trace species in interstellar and intergalactic gas using absorption spectroscopy of distant faint sources. Scientists believe the primordial deuterium/hydrogen ratio is a critical parameter in understanding the first seconds of the Big Bang. In order to conduct its investigations four coaligned mirror telescopes, gather spectra on an individual focal plane assembly carrying four slits: 1.5 × 20 arcsec (highest resolution), 4 × 20 (extended objects), 30 × 30, 0.2 diameter pinhole (bright objects that would overwhelm the detector). Each telescope illuminates a holographic diffraction grating and the output from all four is detected by two 10 × 179 mm microchannel plates.

On 10 December 2001 FUSE went into safe mode when the second of four reaction wheels stopped. In February 2002 partial usability was restored with the use of the magnetic torque technique. Full operations were restored in March 2002. The spacecraft continues to perform well and has made major discoveries in the five years since launch.

Specifications

Launch: 24 June 1999 by Delta 7320 from Cape Canaveral
Lifetime: 3 year nominal mission
Mass: 1,360 kg (bus 500 kg, payload 780 kg, no propellant)
Orbit: 800 km, 25°
Spacecraft: 3-axis 0.9 × 0.9 × 1.3 m box-shaped bus with simple mechanical/thermal links to payload. 0.5 arcsec pointing (1σ) with FES (2° without) using four MWs (16.6 Nms at 5,100 rpm), 3 × 100 Am^2 torque rods, Sun sensors and magnetometers. No propulsion. 1 Mbit/s on 5 W S-band downlink (2 kbit/s command link) from 240 Mbyte memory. Twin GaAs 3.5 m^2 solar wings provide 500 W EOL (payload requires 340 W); 40 Ah Ni/Cd battery.

Fast Auroral Snapshot Explorer (FAST)

Background

FAST is the second of five missions in NASA's Small Explorer (SMEX) series developed by NASA GSFC. The Fast Auroral Snapshot Explorer (FAST) is investigating plasma phenomena in auroral processes discovered in previous satellite and sounding rocket missions. Its orbit at around 100 km takes it through the source region for much of the energy that appears as auroral light. A transportable 8 m antenna is positioned in Alaska to collect real-time data as FAST passes through the northern aurora. These observations are being complemented by data from higher altitude satellites, placing FAST's data in a global context.

FAST had difficulty making it into space because August 1994/5's six-week launch windows were both missed because of June 1994/5's Pegasus failures. FAST was placed in storage waiting its mid-1996 launch in order to make it synchronous with peak northern auroral activity.

FAST is investigating how electrical and magnetic fields accelerate charged particles in the auroral regions. It is shown during spin balancing at Goddard (NASA) 0517113

Specifications

Contractors: GSFC provided the spacecraft for the UCB University of California at Berkeley's payload
Launch: 21 August 1996 by Pegasus XL/Lockheed L-1011 from VAFB
Lifetime: 1 year minimum
Mass: 180 kg (51 kg science)
Orbit: planned 83°, 350–4,165 km
Spacecraft: bus irregular octagonal prism. 50 W orbit average provided by GaAs body panels (5.6 m^2 total) + one 9 Ah Ni/Cd battery (science instruments require 15 W orbit average, 31 W operating, so frequently turned off). 1 Gbit solid-state memory, 0.90/1.5/2.25 Mbit/s rate via NASA standard S-band 5 W transponder. 2 kbit/s commands up, 4 kbit/s housekeeping down. 12 rpm spin stabilisation (normal to orbit) by two magnetorquers, spin Sun sensor, horizon crossing indicator, magnetometer. Attitude post-knowledge 1°. 86 cm diameter × 1.5 m length launch envelope. Four 30 m radial wire booms, 2 axial booms, 2 magnetometer booms
Payload: EESA. Quadrispherical Electrostatic Electron Analyser (UCB, 11.2 kg, on AMPTE, Giotto, Mars Observer, Wind, Cluster).

TEAMS Time-of-Flight Energy Angle Mass Spectrograph (University of New Hampshire + Lockheed Palo Alto Research Lab, 6.4 kg, on Cluster).

EFPE Electric Field Plasma Experiment (UCB, 13.6 kg, on ISEE 1, CRRES, Polar, Cluster).

Magnetometer (UCLA, 3.7 kg, on Pioneer, ISEE 1/2, Polar, Cluster).

Gravity Probe

Current Status

Gravity Probe B has been considerably delayed beyond its planned launch date of 2001. In March 2003 the schedule anticipates launch in the second half of 2003 by Delta 2 from VAFB, California.

Background

Following launch, GP-B will compare the observed precession of gyroscopes in Earth orbit with that predicted by Einstein's theory of general relativity. Stanford University is leading the experiment for NASA; to put angular inertial measuring instruments on a free-flying 650 km polar satellite to look for the expected 6.6 arcsec/year precession perpendicular to Earth's rotational axis and the 0.042 arcsec/year parallel to it.

In order to distinguish between fine grained effects, the 3,300 kg symmetric spacecraft will slowly roll about the line of sight to the guide star. The spin axis will be measured to 0.1 arcsec by superconducting readout of the gyro's magnetic 'London moment' and compared to the line of sight observations from the reference telescope. Maintaining the proper conditions requires near absolute zero temperatures. All four gyros will be cooled to 1.8K by 2,000 litres of superfluid helium, which should be sufficient to keep them cooled for up to 18 months.

Each gyro is electrically suspended and will spin at more than 100 Hz and operate in a 10^{-14}atm vacuum and a less than 10^{-7} gauss magnetic

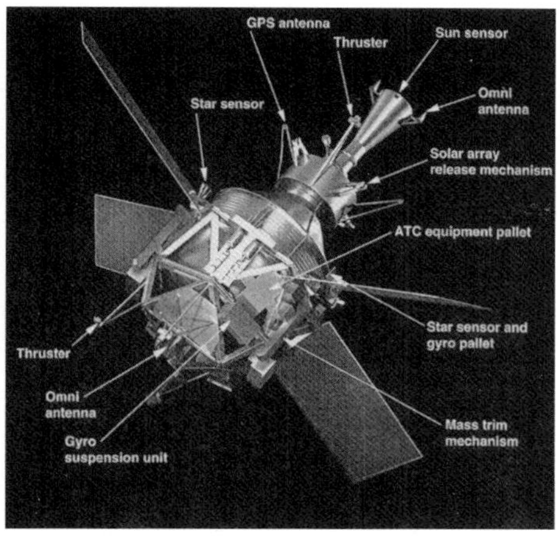

GP-B will test Einstein's theory of general relativity by measuring the precesion of gyroscopes in Earth orbit (NASA Marshall) 0517106

field. To reduce translational contamination, a drag compensation system will reduce the mean gyro acceleration to less than 10^{-10} *g* employing thrusters utilising the Helium boil-off from the dewar. Up to 600 W of power comes from the GaAs wings, spanning 8.8 m and supported by two 35 Ah batteries.

Stanford researchers built a ground-based prototype SIA Science Instrument Assembly with two gyros operating at 1.8K. The second configuration began tests in February 1995, using a flight prototype SIA with four gyros supported by the flight back-up hardware and the engineering development dewar. The third configuration in 1996 tested a protoflight SIA with the flight back-up support hardware and the flight dewar. Ground test of the full flight science payload in 1998 will verify the payload for flight. A Shuttle Test Unit was planned to establish the characteristics of the suspension system in µg, but it was cancelled in early 1994 in a major programme reorganisation.

NASA's cost estimate is US$280 million from FY96. US$10.3 million was expended during FY88 and US$17.9 million in FY89, but GP-B was then deleted from FY90's budget request by the agency. However, congressional review re-instated the programme with US$25 million FY92, 1.9 FY93; 42.4 FY94, 50 FY95, 51.5 FY96 and 59.6 requested FY97. Stanford is supported by two major subcontracts with Lockheed Martin Missiles & Space Co for the spacecraft, dewars and experiment support hardware.

Gravity Probe-A started the series in 1976, when a Scout suborbital package launched it to 10,000 km altitude, on a 116 minute mission over the Atlantic on 18 June 1976, carrying a hydrogen maser clock for telemetric comparison with a duplicate terrestrial version to test the equivalence principle of general relativity.

HETE

Background

MIT's Center for Space Research proposed the High Energy Transient Experiment in 1987 to observe the bright, brief bursts from astronomical objects over UV to gamma energies. NASA suggested in mid-1989 combining HETE's science goals with a proposal from Argentina's space agency for a joint satellite project. That approach was dropped in 1990 and HETE became an MIT project managed from NASA HQ.

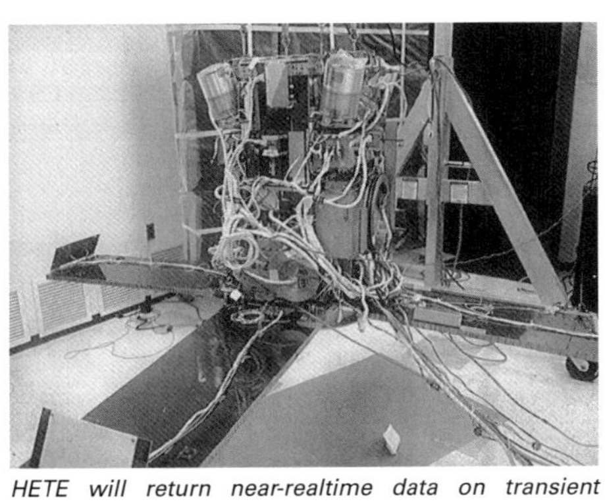

HETE will return near-realtime data on transient astronomical events to observatories worldwide. AeroAstro provided 23 ground stations as part of the turnkey system (Peter Gridley) 0517285

A nuclear test monitor series, called the Vela, first detected gamma bursts in 1969 and they have yet to be precisely mapped. NASA's Compton Observatory is currently providing 10° accuracy of their location, but HETE will locate about 100 to within 10 arcsec. The gamma instruments are derived from the sensors used on the French-Soviet Phobos Mars mission of 1988–89.

HETE went into space along with Argentina's SAC-B on 4 November 1996 on a Pegasus-XL with the L-1011 aircraft taking off from Wallops Island – the first Pegasus launch from there. Although orbit was achieved, neither satellite separated from the Pegasus third stage. Since HETE was contained with the launch vehicle's Dual Payload Attach Fixture there was no opportunity for it to deploy its solar panels and try to obtain some experimental results, and thus it died due to a lack of power. The satellite decayed from orbit and re-entered the atmosphere in April 2002.

Specifications

Contractors/Institutes: NASA HQ (lead), MIT Center for Space Research (science lead, UV instruments), CESR (France, gamma instrument), RIKEN (Institute of Physical & Chemical Research, Japan, X-ray instrument, 2-way ground station), ASI (Italy, primary 2-way ground station), Los Alamos National Lab (X-ray coded mask aperture, software), AeroAstro (bus, payload integration/test, ground stations), Sextant Avionique (GPS receiver)
Launch: 4 November 1996 by Pegasus-XL
Mass: 118 kg (55 kg payload)
Orbit: 38°, 490–555 km (550 km circular planned)
Spacecraft: 66 cm diameter 89 cm high (inclination payload 34 cm high). Momentum biased; ±0.5° attitude control. GPS receiver for position determination. Four 31.5 × 83.2 cm solar wings provide 70 W average. 96 Mbyte mass memory, downlink S-band at 230 kbit/s to three master stations (KPNO Arizona, Sicily and Japan; 7.5 kbit/s uplink) + 20 science receive-only stations worldwide (including Russia). Mission life 1 year, goal 3 years; spacecraft reliability 95 per cent after 18 months
Payload: CESR Omnidirectional Gamma-Ray Spectrometer, CESR, 6 keV-1 MeV.

Wide-Field X-ray Transient Monitor, RIKEN/LANL, 2–25 keV. X-ray emissions of gamma bursts will provide localisation to 10 arcmin.

UV Transient Camera Array, MIT, 4–7 eV (1,800–3,100 Å). Ultra-violet emissions of gamma bursts will provide localisation to 6 arcsec.

Hubble Space Telescope (Great Observatories Series)

Current Status

Following successful repairs and servicing activities carried out during the STS-103 Shuttle mission of 19 to 27 December 1999, Hubble was restored to full operational configuration. Shortly before the launch of Discovery, on 13 November 1999, the HST was placed in safe mode when gyroscope #1 ceased operation leaving only two gyroscopes enabled. Fears that at least one gyroscope would fail had forced mission planners to reschedule an early Shuttle servicing flight. This was carried out by STS-103 but the fourth mission (STS-109) delivered new solar arrays assemblies providing 23 per cent more power than the original arrays. Major changes to the Shuttle programme resulting from the loss of Columbia on 1 February 2003 resulted in a decision announced 21 April 2004 not to fund any further servicing missions although it was expected that this decision could be revoked. A major campaign recruiting support from the scientific community and from the general public for a final servicing mission was partly successful in forcing re-evaluation of this decision. During a Congressional; hearing in April 2004 NASA administrator Sean O'Keefe confirmed that the agency was re-evaluating options in the wake of its decision not to fly the Shuttle to a location from where a safe haven could not be secured. He testified that NASA was examining the possibility of using either NASA's Robonaut developed at the Johnson Space Centre or the University of Maryland's Ranger robot. Robonaut is a human-like android designed by the Robot Systems Technology Branch at JSC and is a collaborative effort with the Defense Advanced Research Agency and was developed as a demonstrator system that can perform similar duties to a space-walking astronaut. Ranger has a series of dextrous manipulators and has already been tested against Hubble servicing

Hubble's first servicing mission (NASA) 0517143

tasks. It is even mooted that one or both of these systems may be able to replace science instruments in a payload performance upgrade. NASA plans to decommission the HST in 2010–2012 but clearly sees an interim servicing and refurbishment capability as crucial to reducing pressure for an early replacement of Hubble by the James Webb Telescope. If an automated repair and instrument replacement capability is demonstrated successfully, the HST may yet receive a longer lease of life.

Background

Renamed the Edwin P Hubble Space Telescope, after the eminent American astronomer in October 1983, this much-postponed project was described by NASA as the most important scientific instrument ever flown. NASA expected the Hubble to resolve objects 50 times fainter and seven times more distant than those visible from Earth-based telescopes. In the most distant views, astronomers would be observing the universe as it was around 14 billion years ago.

NASA began studies of a Large Space Telescope of 3 m aperture in 1971 and 1972. The studies followed closely a realignment of NASA priorities with those of the air force in order to build a space plane according to air force dictates. At the time, the air force planned to launch a new generation of 'Keyhole' close-look spy satellites and, instead of a scientific instrument, the definition studies followed closely the specifications for these satellites.

Astronomers originally expected to launch the Telescope on an expendable booster into an efficient sun synchronous orbit and thereby produce more of a scientific payoff. However, in its final programme review before manufacturing began, NASA shrank the allowable aperture for the Telescope to its current size in May 1975.

ESA agreed to make a 15 per cent contribution in October 1976. In 1976 terms the full ESA cost was projected to be US$595 million, but delays and technical problems caused a rise to more than US$1 billion. NASA promised a return of 15 per cent of observation time for European astronomers. In fact, observing allocation continues to be 20 per cent in open competition.

NASA awarded the US$69.4 million contract for the mirrors and fine guidance sensors to Perkin-Elmer in 1977 following a competition with Kodak Federal System Division of Rochester, New York. Corning Glass, New York cast the 900 kg primary mirror in October 1977. Perkin-Elmer began nine months of fine polishing in August 1980, and aluminised the completed mirror in December 1981. Kodak produced a back-up mirror, although never fully polished it. The Kodak mirror remains in storage.

Hubble deployment by STS-82. Photographed by an aft-mounted wide-angle IMAX camera (NASA) 0006567

The main elements came together during 1985 for the first verification programme to begin, although NASA never asked for an end-to-end optical test. BAe supplied its two solar array wings in 1985 under a 1979 £13 million contract for the 1986 launch. Following the Challenger accident in January 1986 however, the arrays were returned to BAe for a £2 million modification programme. BAe added features to facilitate EVA exchange with new arrays, and the new arrays arrived at Lockheed in April 1989 for installation.

Following the Shuttle accident, NASA delayed Hubble's launch from October 1986 until a possible window in 1988 or 1989. Many astronomers doubted whether the Hubble would have been ready for the 1986 launch anyway. One principal stumbling block was the data transmission of collected images from Hubble to the ground. The Hubble, like its KH-9 counterparts had to use relay satellites or foreign ground stations to receive information. KH-8 satellites used ground stations at remote locations and relays in common orbits with the spy satellites. The TDRS constellation was designed to replace these expensive links. Unfortunately, the air force decided, after the TDRS programme started, that the links needed to be hardened against radio frequency and laser attack from the Soviet Union. The air force argued that the Soviet Union could blind all US intelligence gathering capability with such attacks. In order to make TDRS attack-hardened, the transponders had to be retrofitted at a large cost over-run to the programme and about a two year delay in deployment of the full network. This similarly delayed the relays for the Hubble and without the national security argument behind it, it was unclear how Hubble could use a ground-based data relay network for its image production, let alone do it in a timely fashion. Hence, the Telescope would have remained under-used in orbit while the ground system was brought to full operational capability.

As NASA grappled with this problem, the guide star catalogue for steering the Shuttle had to be finished as well. ST ScI completed the initial Guide Star Catalogue in 1989 after four years of work. When finished, it constituted the largest-ever sky survey. A total of 1,477, 6.4 × 6.4° plates exposed by the US Palomar Schmidt and UK Australian Schmidt telescopes had to be digitised to generate position and magnitude data on 18,819,291 objects for HST's three guidance sensors. Some 15 million objects in the catalogue are stars, most of the remainder are galaxies.

HST arrived at Kennedy Space Center 11 October 1989 aboard a USAF C-5 aircraft for its eventual Shuttle launch. Ground technicians powered it up for the first time on 28 October by remote command from Sunnyvale. Then HST underwent an 11 day GST-8 ground systems test that concluded 8 December 1989, including simulation of early orbit operations and safemode recovery.

After a launch on the Shuttle Discovery, Hubble entered orbit at 20.39 GMT 25 April 1990 619 km over the east Pacific. NASA planned that Orbital Verification (OV) would take three months, followed by five months' Science Verification (SV) in which the instruments would be checked out and calibrated, while allowing some of the Guaranteed Time Observer programmes. Instead the OV took over 19 months.

Once in its lower and less energetic non-sun synchronous orbit, HST passed out of the earth's shadow. Differential heating caused the telescope to ring like a bell for nearly half of the orbit. This ringing prevented precise tracking of the stars that is required for gathering a high resolution image. Lockheed attempted several software 'patches' to correct the problem, but all of them failed to fully alleviate the problem. Later the air force admitted it had known about the problem in advance of the launch, because of similar problems experienced on its KH-9 spy satellites. The air force had refused to alert NASA to the problem, however, for national security reasons. This decision cost the Hubble more than half of its available data gathering time.

The air force nearly cost the United States the use of the HST at another point as well. After release from the Shuttle, Hubble entered its deployment phase in which the solar panels are unfurled. Many problems had developed as the panels unfurled, but the crew, which could have assisted, had moved on to deliver a classified payload to a different orbit.

Cycle 1 of the full observational schedule was to begin towards the end of 1990. The first image, of Carina's NGC 3532 open star cluster, chosen not for

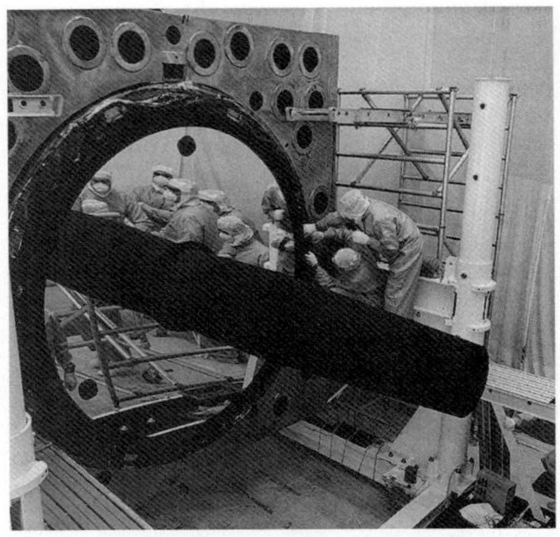

HST's primary 2.4 m {70}, 830 kg mirror with central light baffle installed. The surface is so precise that if it were expanded to cover the entire US its largest bumps would be only a few cm high (Perkin-Elmer) 0517144

scientific reasons, but so that NASA could release a publicity photo, was returned by WF/PC on 20 May 1990. Resolution was 0.7 arcsec, as against the expected 1.5 arcsec, but an improvement by a factor of seven was still expected. This first light observation showed light, which appeared to ground-based telescopes as a single star, was, in fact, composed of two stars. Astronomers had long suspected the fact, but let NASA use the information anyway, for a press release.

After first light was achieved 6 June, it soon became apparent that a large flaw existed in the optics. The first images showed that HST suffers from classical spherical aberration, caused by a 2 μm grinding error around the primary's edge. The mirror's 24 actuators could not compensate and would complicate the 'pure' aberration (the final compromise focus was not decided on until November 1991). Specifications required 70 per cent of the incoming light to fall within a 0.2 arcsec area, and measurements showed 1.4 arcsec. Some 15 per cent covered 0.1 arcsec, with the remainder spread across 1 arcsec (equivalent to ground-based performance). Computer processing for objects seen against a dark background could achieve the target 0.1 arcsec resolution. Low sky background at shorter wavelengths meant that ultra-violet observations were largely unaffected – the visible work suffered most. WF/PC was the most serious loss: it was to have taken 40 per cent of the observing time, but occupied only about 10 per cent.

Hubble was launched with several principal scientific objectives. One key objective was the measurement of the Universe's expansion rate (Hubble constant) by observing the brightness of magnitude 24 and fainter stars (Cepheid variables) in distant galaxies with WF/PC. With the flaws in the optics definitive observations had to wait for WF/PC-II arrived later. ESA's reduced its FOC work by about half after visible observations offered no improvement over ground systems.

The independent six-man HST Optical Systems Board of Investigation established 2 July under JPL Director Dr Lew Allen reported in late November 1990 that the primary mirror had been well figured but to the wrong shape. Perkin-Elmer's Reflective Null Corrector laser interferometer, used to check the figure around 1981, apparently had a lens 1.3 mm out of position. Perkin-Elmer had some indication of the problem, but disregarded the test

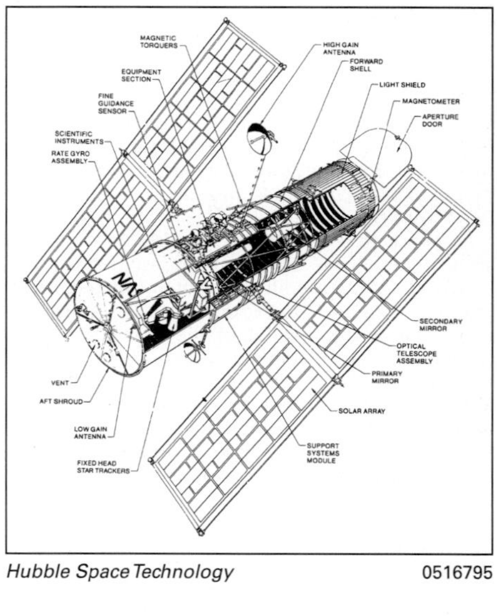

Hubble Space Technology 0516795

Hubble Space Technology 0516796

results. Kodak's bid for the 1977 contract included US$10 million for an end-to-end test, which would have immediately identified the problem.

The Board concluded that the defect could not be solved by a global repair in-orbit but individual instruments could be corrected optically. NASA planned anyway, that WF/PC would be replaced by a second-generation version on SM1. FOC, HRS and FOS are corrected by replacing HSP with the 290 kg 220 cm long 87.6 cm^2. COSTAR Correcting Optics Space Telescope Axial Replacement, carrying a boom to insert 10 mirrors into the light paths (one picks off the light for a second to provide the correction) before they reach the instruments. The coatings call for 56 per cent throughput of 1,216 Å UV. HSP was lost but it was the least-used instrument. Ball Aerospace began work on COSTAR under a February 1991 letter contract, finalised October 1991 at US$30.4 million. ESA considered an on-orbit FOC internal modification and flying a new instrument, although at an estimated US$180 million it proved too costly.

Early observations by the FOC showed that the specified 0.1 arcsec performance was still possible in some circumstances. HST then gathered images of Supernova 87A's ring, Pluto/Charon and of the R Aquarii red giant/white dwarf double star nova. Scientists then made a major step towards determination of the Hubble constant by accurately measuring the distance to the Large Magellanic Cloud. HST/IUE combined to measure the diameter of SN87A's ring at 1.37 light years, translating to an LMC distance of 169,000 light years ±5 per cent (replacing the previous ±30 per cent spread).

As the Telescope settled into its observation schedule, in 1991 it used the HRS to measure the Universe's deuterium content and found the amount supports the Big Bang and ever-expanding cosmos concepts. HRS observed evidence of planetary formation around star Beta Pictoris and HRS discovered huge hydrogen clouds that, theoretically, should have evolved into galaxies long ago. The WF/PC instrument found evidence of a black hole 2.6 billion times the Sun's mass in a region 1–10 times the size of the solar system at the centre of galaxy M87.

In 1992, the HST uncovered strong evidence that discs of planet-forming material surround many stars. WF/PC found 15 in the Orion Nebula; only four were known before HST. Work continued on refining the Hubble constant, but it really required improved optics.

SM1, the first telescope maintenance mission solved a second major problem in December 1993. It corrected the pointing flaws due to the thermal ringing. The ringing produced two modes: a 0.1 Hz end-to-end movement lasting up to 10 minutes after terminator crossing, and a lesser 0.6 Hz transverse action during daylight. The solar array tips flexed up to 1 m under worst conditions. The wobble prevented long exposures. Previous software fixes took up too much of HST antiquated computer memory. SM-1 put in place replacement solar arrays (STSA-2) and an aluminium layer supported by 900 plastic discs in an accordion-like structure covered each modified boom. This reduced thermal gradients by a factor of 20. Then BAe also built a system for countering expansion at the array tips with frictionless springs and immobilised the storage drums by an electrical brake. They were found to reduce jittering by a factor of 10. These arrays should be sufficient until about 2000.

SM1 also replaced two gyro pair packages. The number 6 rate gyro failed 3 December 1990, but one of two back-ups assumed its duties. Then number 4 failed July 1991, possibly from radiation damage of the hybrid circuits generated by high solar activity. Number 5, too, began showing signs of distress, possibly because of dirt in its rotor mechanism. Number 3 failed in November 1992, probably because of a random electronic part failure in the control unit. If they had all failed simultaneously before replacement, HST would have entered a magnetic torque mode to hold safe attitude.

SM1 achieved all objectives in December 1993 during five EVAs totalling more than 35 hours by four crew members.

EVA-1: Replaced Rate Sensing Units numbers 2/3 (each with two gyros) and control units for numbers 1/3, and eight of 12 3 A fuse plugs that protect HST's electrical circuits with 5 A versions.

EVA-2: replaced starboard solar array discarded into orbit. Port wing removed/stowed for return to Earth; two new wings attached.

EVA-3: replaced WF/PC with WF/PC-II. Two new magnetometers installed on top of failed pair at the aperture end.

EVA-4: replaced HSP with COSTAR. Added 386 coprocessor to DF-224 computer.

EVA-5: replaced Solar Array Drive Electronics. Both wings unfurled by 10.35 GMT 9 December; HST switched to array power 15.30 GMT. Installed HRS redundancy kit.

By 1994, NASA had corrected HST's defective optics and observations with the improved optics provided seemingly conclusive evidence for this black hole. WF/PC also began a long-term programme of Mars observations to characterise its climatic processes and surface changes at a resolution of 50 km. 1994 observations showed that Mars' average temperature had dropped 20°C since Viking, but the climate now appears more stable.

In 1994, WF/PC-II observations of 20 Cepheid variables in M100 placed the galaxy at 56 ± 6 million light-years, yielding a Hubble constant equivalent to an age of 8 to 12 billion years for the Universe, instead of the generally accepted 15 billion years. If confirmed, it would require a major reworking of cosmology. However, many more galaxies must be measured because those around the Virgo cluster are mutually perturbed by their concentration. The FOC in 1994 provided the first image of star G1623b, the red dwarf companion in a double star system in Hercules and one of the smallest stars known in our Galaxy. It has only 10 per cent of the Sun's mass and is 60,000 times fainter - it would appear only eight times brighter than the full Moon from 1 AU. Its presence was known from astrometric measurements of its parent but ground-based telescopes were unable to resolve it. Red dwarfs were believed to be the most abundant stars but Hubble's observations show them to be surprisingly rare. They cannot therefore account for much of the Universe's missing mass.

From a study of very young galaxies, HST provided the surprising result that elliptical galaxies formed very early on, but spirals took much longer and most were torn apart. FOC observations of a distant quasar revealed the type and quantity of helium predicted by the Big Bang theory. FOC and WF/PC II observations of 525 km asteroid 4 Vesta in November/December 1994 showed the solar system's oldest terrain at 80 km resolution. The gouged surface revealed details below the crust, lava flows and impact basins. The once-molten interior shows that Vesta should be regarded more as a mini-planet, rather than an inert lump of rock.

As the data gathering entered its full programme, astronomers became disenchanted with the programme as the consequences of working with NASA became apparent. At first, NASA promised to allow scientists unfettered access to the Telescope through the TDRS datalinking network. In fact, NASA had an agreement with the US Air Force to allow the military to have priority linkage through the data relays for nuclear attack planning. NASA also promised the Hubble could be repaired on orbit at least every 30 months. The solar arrays were to be retracted, first every five years, then every three years, and the whole vehicle brought to Earth for refurbishment and relaunch about a year later. By 1987, NASA announced it would avoid Earth return if at all possible and make in-orbit servicing more infrequent.

WF/PC-II observations in 1995 confirmed the existence of the postulated Kuiper Belt of comets 40–500 AU out, beginning just beyond Neptune and the source of comets that orbit the Sun in periods of greater than 200 years. The observations indicate the belt hosts at least 200 million comets that have remained essentially unchanged since the solar system formed. Other 1995 observations strongly suggest that volcanoes are still erupting on Jupiter's moon Io and that Ganymede has a tenuous oxygen atmosphere, created from the bombardment of water ice by Jupiter's charged particles. The first unambiguous brown dwarf was confirmed: Gliese 229B is the faintest object seen orbiting another star. At 20–50 Jupiter masses, it is too large to be classed as a planet but too small and cool (720°C) to shine as a star.

The second HST servicing missions began with a Shuttle launch on 11 February 1997 as the STS-82 flight of Discovery.

When the Shuttle had captured HST and the astronauts were able to look in detail at the telescope's thermal insulation cover, it was found that it had degraded more than expected during the three and a half years since its first servicing mission. As a result, an extra EVA was scheduled so that some running repairs could be done. During the servicing:

EVA-1 (Lee/Smith): removed the High-Resolution Spectrograph and the Faint Object Spectrograph and replaced these with the Space Telescope Imaging Spectrograph and the Near Infra-red Camera and Multi-Object Spectrometer.

EVA-2 (Harbaugh/Tanner): replaced the degraded Fine Guidance Sensor 1 and Engineering/Science Tape Recorder 2 with new models: installed Optical Control Electronics Enhancement Kit. It was during this EVA that the insulation degradation was first noticed.

EVA-3 (Lee/Smith): removed and replaced Data Interface Unit 2 and a reel-to-reel ESTR with a

1	*Scientific Instrument Control and Data Handling*
2	*Faint Object Camera*
3	*Wide Field/Planetary Camera*
4	*Data Interface Unit for Optical Telescope Assembly*
5*	*High Gain Antenna (2)*
6	*Data Management Unit*
7	*DF-224 Computer*
8	*Fine Guidance Electronics (3)*
9*	*Actuator Control Electronics*
10	*Battery (6)*
11	*Fixed Head Star Tracker (3)*
12	*Rate Sensor Unit (3)*
13	*Charge Current Controller (3)*
14	*Data Interface Unit for Support Systems Module*
15	*Electrical Power Thermal Control Electronics*
16	*Power Control Unit*
17	*Power Distribution Unit*
18	*Fuses (14)*
19	*Multiple Access Transponder (2)*
20	*Single Access Transmitter (2)*
21	*Faint Object Spectrograph*
22	*Fine Guidance Sensor (3)*
23	*Reaction Wheel Assembly (4)*
24	*Solar Array (2)*
25	*Low Gain Antenna*
26	*Sun Sensor*

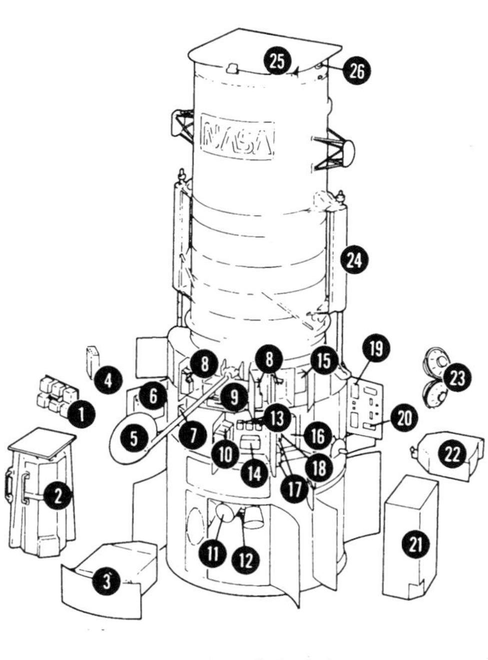

Hubble Space Technology 0516797

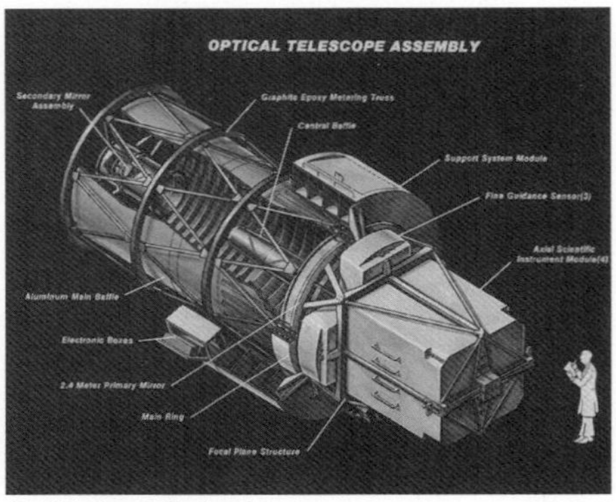

Hubble Space Technology 0517145

new digital Solid-State Recorder that permits simultaneous recording and playback of data.

EVA-4 (Harbaugh/Tanner): replace Solar Array Drive Electronics 2, replace the covers of the Magnetic Sensing System and replaced some of the thermal sets on two areas around the light shield of the telescope.

EVA-5 (Lee/Smith): attached replacement thermal coverings on three equipment compartments at the top of the Support System Module.

While attached to Discovery, the orbiter was used to re-boost the altitude of HST: before being captured it had been in a 590–599 km orbit and when released it was in a slightly higher 599 to 620 km orbit.

Degradation of several key systems and the failure of two crucial gyroscopes out of sync, plus intermittent performance from a fourth, stimulated the HST team to request an interim servicing mission before the planned visit in late 2000. Dubbed mission 3A, the interim visit was assigned to Shuttle Discovery (OV-103) on mission STS-103 launched 19 December 1999. The third STS servicing mission began with the launch of Shuttle Discovery at the start of a mission which lasted almost eight days. After rendezvous and capture, the HST was placed in the payload bay. During the period of servicing several EVA operations were carried out:

EVA-1 (Smith/Grunsfeld): installed six new gyroscopes and six voltage/temperature improvement kits in the telescope. 8 hr 15 min

EVA-2 (Foale/Wicolher): installed a new 486 computer, 20 times faster and with 6 times the memory of the DF-224 computer it replaced; replaced a 250 kg fine guidance sensor, one of three on the telescope. 8 hr 10 min

EVA-3 (Smith/Grunsfeld): replaced failed radio transmitter and installed new digital solid-state recorder capable of 10 times the storage capacity of the unit it replaced; applied new insulation on the equipment bay doors. 8 hr 8 min

Total servicing time on HST was 93 hr 13 min.

Astronauts on servicing mission 3B in March 2002 (Orbiter Colombia) installed the Advanced Camera for Surveys offering 10 × the resolution of on-board cameras. In addition, new high-efficiency solar arrays from Lockheed Martin Missiles and Space generate enough power for simultaneous operation of several science instruments. Following launch, rendezvous and capture, the crew of Space Shuttle Columbia conducted five scheduled EVAs:

EVA-1 (Grunsfeld/Linnehan): installed the V-2 solar array and diode box. 7 hr 01 min.

EVA-2 (Newman/Massimino): activities included the completion of installing the V-2 solar array and the diode box and the installation of a reaction control wheel assembly. 7 hr 16 min.

EVA-3 (Grunsfeld/Linnehan): completed installation of the solar array and the reaction control wheel assembly after a delay due to space suit problems. 6 hr 48 min.

EVA-4 (Newman/Massimino): activites included replacement of the Faint Object Camera with the new Advanced Camera for Surveys, installation of an Electronics Support Module and completion of Power Control Unit cleanup tasks. 7 hr 30 min.

EVA-5 (Grunsfeld/Linnehan): installation of the Near-Infrared Camera and Multi-Object Spectrometer (NICMOS), Cryogenic Cooler and NICMOS Cooling System Radiator. 7 hr 21 min.

SM-4, the fifth servicing mission planned for 2003 or 2004 was cancelled due to the loss of Columbia in February 2003 and plans have been developed to accomplish these objectives using a robotic device.

Observing proposals are still running at five times the level that can be accommodated. Total cost to the end of 15 years of operations was projected in 1989 at US$2,800 million, inclusive of US$1,450 million for development and activities up to launch. In 1990, NASA added US$600 million to the estimate. SM1 was priced at US$674 million, including US$378 million for Shuttle's element. Cost of repairs and correcting for the mirror defects was included at $86 million.

As launch was originally planned for 1983 and full operations were not possible until 1994, advances in the technology of ground-based astronomy have removed some of HST's anticipated unique advantages. It does maintain superiority, however, with its 0.01 arcsec angular resolution (ground telescopes can achieve 0.3 arcsec and then only briefly) and its ability to work with the ultra-violet wavelengths blocked by the atmosphere.

In its first 10 years of operation, the HST has studied 13,670 objects, made 271,000 individual observations, returned 3.5 Tbytes of data and stimulated 2,651 scientific papers.

The fourth HST servicing mission, designated 3B, was made the responsibility of the Shuttle Columbia (OV-102) and its crew on STS-109 launched 1 March 2002. After rendezvous and capture on 3 March several EVA operations were conducted:

EVA-1 (Grunsfeld/Linnehan): installed the first of two new solar array assemblies. 7 hr 1 min

EVA-2 (Newman/Massimino): installed the second (left) of two new solar array assemblies. 7 hr 16 min

EVA-3 (Grunsfeld/Linnehan): installed new Power Control Unit. 6 hr 48 min

EVA-4 (Newman/Massimino): removed Faint Object Camera and installed Advanced Camera for Surveys in its place. Installed Electronic Module supporting a new cooling system to bring NICMOS back to life. 7 hr 30 min

EVA-5 (Grunsfeld/Linnehan): NICMOS cryocooler removed form its location and relocated in aft shroud, cables from the ESM installed on EVA-4 attached. Installed Cooling System Radiator on HST. 6 hr 30 min. Columbia was used to re-boost the HST by about 4 miles.

Specifications

Launched: 24 April 1990 by Shuttle STS-31 from Kennedy Space Center, released 20.39 GMT 25 April. Serviced 4–10 December 1993

Contractors: NASA Marshall (lead NASA centre), NASA Goddard (instrument development management and vehicle TT&C), Lockheed Missiles & Space Co (vehicle integration, Support Systems Module development, ST Operations Control Center), Perkin-Elmer Corp (Optical Telescope Assembly, Fine Guidance Sensors; company now Hughes Danbury Optical Systems), Ball Aerospace (HRS, NICMOS, STIS, HACE instruments), Dornier/Matra (Faint Object Camera), Martin Marietta (Faint Object Spectrograph), British Aerospace (solar arrays, FOC photon detector assembly), Eagle Picher (batteries), Boeing (mirror metering truss), Fairchild Space Co (science instrument command/data handling computer), Honeywell Satellite Systems Operations (RW assemblies), Odetics (three DDS-3100 digital tape recorders, to be replaced by SM2)

Mass: initially 10,843 kg on-station; 10,960 kg after SM1

Orbit: released at 613 × 620 km, 28.5° (altitude to minimise atmospheric drag), 586 × 596 km by time of SM1 capture, raised to 592 × 601 km for SM1 release

Lifetime: 15 years, with 3 to 5 years revisits by Shuttle for maintenance, replacement of instruments and reboost. SM1 revisit December 1993; SM2 planned for STS-82 February 1997, SM3 November 1999, SM4 November 2002, SM5 November 2005

Configuration: HST is 13.1 m long (without aperture door), 4.27 to 4.7 m diameter carrying a 2.40 m diameter primary mirror and optical system held rigidly in an Optical Telescope Assembly (OTA) relaying light to five (four after SM1) aft-mounted focal plane instruments. The Support Systems Module (SSM) encircling the OTA's base and protecting the instruments provides housekeeping functions. 426 kg OTA is required to maintain the 5 m mirror-to-mirror spacing to within 3 μm/0.003 arcsec over 24 h using the 5.08 m long graphite epoxy metering truss. The primary is held in a main support ring inside a large square tooling fixture. Inside the truss is a 3 m diameter aluminium light baffle. Externally, the OTA is covered forward with the SSM light shield and end-aperture door. The twin 33 m^2 solar arrays and primary communications antenna are mounted on the SSM forward shell. The SSM equipment section encircles the metering truss base, accommodating power, control, communications, RWs, rate gyros and other subsystems in its 10 bays, accessible externally for servicing

AOCS: HST avoids thruster exhaust contamination by employing four 3,000 rpm RWs for attitude control (turn rate of 90° in 20 min), with angular momentum offloading by magnetorquers. Periodic reboosting during Shuttle servicing missions combats orbital decay. Six rate gyros (four prime plus two back-up; two pairs replaced by SM1) and three fixed head star trackers in the SSM equipment section provide overall attitude information; three Sun sensors mounted on the aft face prevent pointing within 50° of the Sun (aperture door available for protection). Two magnetometers at the front end sense attitude relative to Earth's magnetic field. Pointing stability requirement is 0.007 arcsec for up to 10 hours, using three interferometric Fine Guidance Sensors on the field of view's periphery. Any two are sufficient for guidance; magnitude +13 was detection requirement, but +15.5 appeared feasible pre-mission. FGS are also operated as a science instrument (see Payload section). The Guidance Star Catalogue, providing position and magnitude data on 18,819,291 objects, was completed in 1989 but continues to be expanded, including proper motion and colour information

Power: two ESA/British Aerospace solar arrays provide 5.0 kW BOL, 4.5 kW after two years and 4.3 kW after five years at 34 V; 5 per cent degradation measured after one year. Each 150 kg 2.83 × 11.8 m wing carries 24,380 Si cells, rolled into a 20 cm diameter drum for launch. Unfurling achieved by Bi-stem actuator units at either end unwinding and interleaving two spring steel elements to form 2.2 cm diameter booms either side of the thermal covers. The arrays can be restowed or jettisoned, and all functions can be achieved manually. The original arrays were replaced by SM1 because incorrect design allowed excessive flexing during day/night cycling. The new arrays reduced jittering by × 10. Each component is qualified for 30,000 thermal cycles of –100/100°C over a five year life. Payload requirement is 2.4 kW. The six 90 Ah Eagle Picher Ni/H_2 batteries, in three modules, can be replaced on each servicing mission

TT&C: 1 Mbit/s data via NASA's TDRS satellite system to White Sands for relay to the ST Operations Control Center at Goddard and to the ST Science Institute at Johns Hopkins. Goddard exercises vehicle control, ST ScI manages the observations. HST is controlled by a triply redundant 50 kg Rockwell Autonetics DF-224 computer (48,000 word total memory; two of six memories have failed, three required for full operation; a co-processor was added by SM1 to increase memSory/speed) and Lockheed's 37.7 kg Data Management Unit. All incoming commands are decoded/relayed by the DMU and it receives the data from the Scientific Instrument Control & Data Handling computer, which relays DMU's commands and formats the science data

Thermal: heaters/coverings maintain mirror at 21°C. Elements such as batteries requiring greater heat dissipation are positioned on the anti-solar side

Payload: Ritchey-Chretien f/24 Cassegrain Optical System comprises an 829 kg 2.40 m diameter primary mirror of ultra-low expansion titanium silicate glass and a 30 cm diameter Zerodur secondary 5 m distant. Effective focal length is 57.6 m. The primary is constructed of 2.5 cm-thick face and

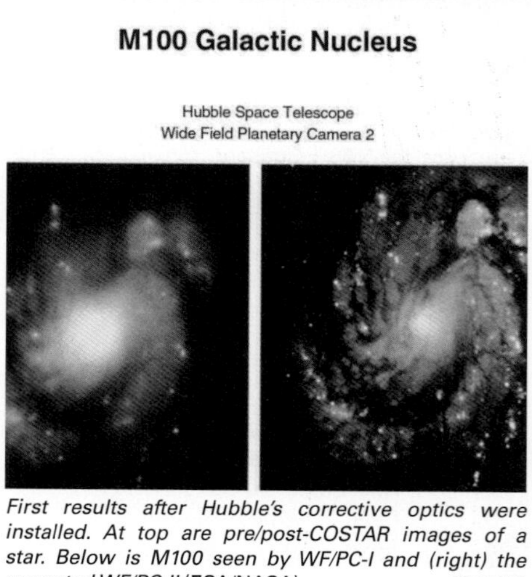

First results after Hubble's corrective optics were installed. At top are pre/post-COSTAR images of a star. Below is M100 seen by WF/PC-I and (right) the corrected WF/PC-II (ESA/NASA) 0516793

rear glass plates separated by a honeycomb layer, and provides l/20 rms at 6,330 Å. Both hyperboloid mirrors are polished to less than 0.05 μm, and coated with a 0.064 μm layer of aluminium and a 0.025 μm layer of protective magnesium fluoride. That outer layer is required for adequate ultra-violet reflectance, although it reduces the aluminium's visible light reflectivity to 85 per cent. Reflection at 1,216 Å is 70 per cent; spectral range is 1,150 Å-1 mm. The secondary directs the light cone through the primary's 60 cm diameter central aperture to a focus 1.5 m behind the face plate for dispersion to the science instruments and FGS (three instruments now receive their light via COSTAR). Optical adjustment by 24 actuators attached to the primary's rear plate and six on the secondary's support structure. The specification required 70 per cent of the incoming light to be focused in a 0.2 arcsec diameter core. The primary mirror spherical aberration discovered in-orbit produces 15% within 0.1 arcsec. COSTAR improved the performance to less than 70 per cent. Four instruments are axially aligned (COSTAR replaced one), the fifth (Wide Field/Planetary Camera) and guidance sensors are in radial bays. All are latched for in-orbit replacement.

Wide Field/Planetary Camera (WF/PC) (designed/built by JPL; US$69 million) for general astronomical far-ultra-violet to near-infra-red applications over 1,200–11,000 Å, including astrometric searches for planetary bodies about other stars (prevented by mirror defect), solar system and galactic/extra-galactic observations. 280 kg. Took light from HST's central FOV for greatest quality. Operated in two modes: f/12.9 2.7 × 2.7 arcmin FOV Wide Field camera (resolution 0.1 arcsec per pixel), and f/30 1.2 × 1.2 arcmin Planetary Camera (resolution 0.043 arcsec per pixel). Light from the f/24 pick-off mirror passed through a wheel of 48 filters, polarisers and gratings, and was divided into four beams by a pyramid mirror for relay to the required camera, selected by pyramid rotation. Each camera employed four CCDs of 800 × 800 pixels (1,515 μm) each. Exposures from 0.110 second to 28 hours. Planetary mode planned for 400 km resolution on Jupiter. WF/PC was most affected by HST's optical defect: was to have taken 40 per cent of observing time but reduced to 10 per cent. Replaced on SM1 by US$101 million 278 kg WF/PC-II: integral optics to correct HST's defect, revised filter set (including far-ultra-violet), updated CCDs reaching into far-ultra-violet, articulation of the pick-off mirror and 3 of the 4 fold mirrors to ensure accurate pointing required by optical correction. 1991 decision reduced detectors from eight to four (one PC plus three WF) on cost grounds.

Faint Object Camera (FOC), provided by ESA (Dornier prime), uses HST's full resolution capabilities for faint objects/structures (magnitude 24–29) using cumulative exposures. 320 kg; 150 W required operating, 75 W standby. Length 2 m, cross section 1 × 1 m. Range 1,150–6,500 Å; most sensitive at blue visible/ultra-violet. Operates in four principal modes: direct imaging at f/48 (22 × 22 arcsec FOV, ×2 magnification), f/96 (11 × 11 arcsec, ×4 magnitude) and f/288 (4 × 4 arcsec, × 12 magnitude), and as a long-slit spectrograph. Four filters provide 12 filters each. Employs two separate imaging photon-counting detectors, each with a three-stage intensifier tube coupled to a TV camera. HST optical defect rendered FOC's visible capability similar to ground-based instruments, but ultra-violet work remained attractive; observing programme was reduced to half. Following SM1, FOC's chains are fed corrected but magnified images: f/96 became f/150 (7 arcsec FOV), f/48 became f/75 (14 arcsec FOV), f/288 became f/4500.

Faint Object Spectrograph (FOS), built by Martin Marietta for University of California. 1,150–8,500 Å; operates in low or intermediate resolution modes plus offers spectropolarimetry capability. Incorporates 1-D Digicon detector.

High-Resolution Spectrometer, built by Ball Aerospace for NASA Goddard. Similar to FOS but optimised for ultra-violet (1,100–3,200 Å, so relatively unaffected by HST optical defect) and high resolution (2,000, 20,000, 120,000) observations. Employs six interchangeable gratings. Began suffering serious power supply fluctuations July 1991; switching to back-up permanently would reduce effectiveness by 30 per cent. SM1 installed relay to allow either detector channel to use either power supply.

High-Speed Photometer (HSP), designed/built by University of Wisconsin. Four photon-counting image dissector tubes, a photomultiplier and >50 focal plane/aperture combinations could detect temporal variations down to 16 μs. 1,150–7,000 Å. Least affected by HST optical defect, but could not begin data return until end-1991 because of HST problems. Sacrificed for COSTAR optics correction module.

Fine Guidance Sensors (FGS) are primarily used for pointing system. Only two are required for 0.007 arcsec pointing goal, but third added for astrometric observations to 0.002 arcsec. Could not begin astrometry until end-1991 because of HST problems.

UPDATED

IMP Explorers

Current Status

Launched on 26 October 1973, IMP 8 (Explorer 50) was retired on 28 October 2001 when the last commands were sent to the spacecraft. Following the failure of the magnetometer in 2000, a review panel advised NASA's office of space science that funds should be diverted to more productive missions. IMP 8 was the last in the series still operating.

Background

Ten Interplanetary Monitoring Platforms monitored Earth's radiation environment during a complete 11-year cycle of solar activity, defining the nature and extent of the magnetosphere. IMPs provided continuous warning of solar flare radiation during the Apollo/Skylab missions, and provided baselines for Pioneer 10/11 and Mariner 10. IMP 8 continues to return data. In December 1992, it provided key data as Galileo passed through Earth's magnetotail. IMP 8's battery is now long since dead (array output 4.40 A) and ranging cannot be performed because the 136.800 MHz transmitter has failed, and insufficient gas precludes manoeuvres. Even so, it remains a principal source of solar wind data and is funded to 2002.

IMP 1, launched 16 November 1963 as Explorer 8 is a 16-sided drum, 135.6 cm diameter, 157 cm high. It has a 22 rpm spin-stabilised control (no AOCS remaining). Data returns from the LAP 2-D electrostatic analyser (4–20 keV electrons and 0.070–20 keV protons), MAP Faraday cup (0.017–7 keV electrons, 0.050–7 keV protons), IOE electrostatic analyser (0.5 eV-50 keV), MAE electrostatic analyser, APP solid-state telescope (fluxes/spectra greater than 220 keV electrons, greater than 300 keV protons, greater than 640 keV alpha particles, and fluxes of more than 1 MeV/n particles), GME cosmic ray telescopes (0.150–15 MeV electrons, 4–80 MeV isotopes), GNF fluxgate sensors, CHE cosmic ray telescopes (0.3–10 MeV electrons; 0.5–1,200 MeV nucleons), CAI multi-element telescope (0.16–2.8 MeV electrons; 0.5–40 MeV isotopes), GWP energetic particles telescopes (30–450 keV electron fluxes and 0.030–8.6 MeV proton fluxes).

Its orbit is 125,000 × 202,000 km, with an on-orbit weight of 62 kg.

Explorer 35, launched 19 July 1967, entered lunar orbit to study the Moon's magnetic field and radiation belt, in addition to Earth's magnetotail.

IMP 8 (Explorer 50), launched 26 October 1973 had an initial orbit of 204,200 × 236,000 km, 28.8°, which has decayed to a now near circular one at 223,000 km, 50°. Its mass at launch was 298 kg and provided 120/42 W to spacecraft and experiments.

Explorer 50 (IMP 8) continued to return fields and particle data from high Earth orbit for 28 years (NASA) 0516742

ISEE Explorer (ICE)

Background

The International Sun-Earth Explorers, part of the International Magnetosphere Study, followed up the IMP series. NASA built ISEE 1 and 3 and ESA provided 2. ISEE 1 and 2 were launched jointly on 22 October 1977. ISEE 1 and 2 decayed 26 September 1987. ISEE 3 followed on 12 August 1978 by Delta from Cape Canaveral (469 kg). ISEE 3 had the distinction of attaining the first halo orbit around the L1 libration point 1.5 million km out on the Earth-Sun line. Eight annual hydrazine thruster firings maintain the orbit and it oscillated 150,000 km about the ecliptic annually so that solar radio emissions did not interfere with tracking and data acquisition.

The ISEE 3 provided 1 hour warning of solar radiation before events reached ISEE 1 and 2. NASA diverted ISEE 3, due to remain as a solar flare sentinel for 10 years, towards Earth in late 1982 so that it could be directed into a series of lunar swing-bys (the last, in December 1983, within 100 km) for the first ever comet encounter. Its speed was increased from 4,679 km/h to 8,278 km/h by sweeping behind the Moon on 22 December 1983, surviving 28 minutes without solar power (its battery failed in 1983).

After the swing-by, ISEE became the ICE, or International Cometary Explorer. On 11 September 1985 it passed 7,860 km behind Giacobini-Zinner's 2.4 km diameter nucleus at a closing speed of 75,300 km/h. Goddard controllers commanding the 422 kg 1.8 m diameter vehicle, built by Fairchild, had feared cometary dust could coat its solar cells and damage its 91 m-span antennas. Throughout the 20 minute coma passage its 5 W transmitter and small antenna, designed to provide data from no further than 1.6 million km, transmitted 70.8 million km to Earth. NASA's Deep Space Network had six large dishes trained on it (Goldstone, Madrid and Canberra), with additional facilities in Japan and Arecibo (Puerto Rico). Though lacking an imaging system, it returned data on plasma densities, flow speeds, temperature and heavy ions in the tail. Initial analysis revealed water ice to be the primary constituent, with cold slow-moving plasma, and water and carbon monoxide ions inside the tail, supporting Whipple's dirty snowball model.

Unexpectedly, ICE began encountering atmospheric ions 1.6 million km out, and the 24,000 km-wide tail was much wider than expected. The first instrument to detect the comet was a European energetic proton instrument. ICE also passed within 40.2 million km of Halley's Comet on 28 March 1986. It was then manoeuvred into an Earth-like orbit with a period of 355 days so that, on 10 August 2014, it will return for capture by the Earth-Moon system, when it could be retrieved for analysis of its cometary dust coating. Ownership has already been formally transferred to the National Air & Space Museum.

Daily data return at 64 bit/s to Goddard continued until December 1995 from the: magnetometer, plasma composition, solar wind plasma, plasma waves, energetic protons, low/medium/high energy cosmic rays and cosmic ray electrons instruments. The telemetry transponders were turned off 19 December for JPL to begin a radio

ICE achieved the first comet interception. It will be recaptured by the Earth-Moon system in 2014 0516743

science mission. The low bit rates means that AOCS cannot be commanded until about 2010. Hydrazine tank 1 is empty (leak), number 2 holds 135 kg; solar array is 6 A. ICE is expected to remain active until at least 2014's return.

IUE Explorer

Background

The International Ultraviolet Explorer followed a modified geosynchronous orbit at 30,221 × 41,349 and 34.9° of inclination, making it the first astronomical satellite in geosynchronous orbit. A joint venture by NASA/ESA/UK, it operated until October 1995 from its 26 January 1978 launch by Delta from Cape Canaveral. At launch it weighed 671 kg (445 kg in orbit) and was placed in a 30,252 × 41,315 km, 34.5°, 1,435.9 min orbit.

IUE provided astronomers with a unique tool and requests for observing time remained three times greater than could be accommodated. Despite the appearance of the Hubble Space Telescope, IUE's 1,150–3,200 Å spectral band covered an entire region not accessible simultaneously to HST's high resolution spectrographs. The fast response time of less than 1 hour provided an unparalleled flexibility in scheduling targets of opportunity. Within one year, IUE had provided 200 investigators from 17 countries with more than 9,000 ultra-violet spectra of objects unobtainable from ground-based instruments. 113,300 spectra had been obtained by the end of 1995 for 2,000 astronomers and 3,300 papers using IUE data had been published in referred journals.

ESA joined NASA and independent researchers from the UK. As part of its programme, to enhance the ESA science contribution to space exploration, European approval for IUE came in 1973, to build the provision of the spacecraft's deployable solar array and a ground station at Villafranca near Madrid. Having passed 18 years of operations, it is responsible for generating more scientific papers than any other astronomical telescope.

Hydrazine propellant is plentiful, with 17.45 kg (75 per cent) remaining by end of 1995; and an annual consumption of only 0.57 kg. On 17 August 1985, a fourth attitude control gyro failed, but software for the remaining two had been prepared in anticipation and only a months' observations were lost. Solar cell degradation had been about 3 to 4 per cent annually (3.31 per cent average 1978–95), with 148 W is required for normal operations.

Still, NASA hinted in January 1994, that it would pull the plug on funding the operations beyond September 1994, and instead dedicate the US$3.5 million annual saving to other science missions. It had to back down from this position, because of European opposition, on 1 October 1995, and since then, science operations have been performed only during the 16 hours low radiation part of the orbit, run from Villafranca. Goddard supplies eight hours of satellite management and housekeeping, while NASA concentrates on the IUE Final Archive.

After the 1994 decision all parties agreed to halt operations on 31 December 1997, but ESA decided in February 1996 that it would shut IUE down by September 1996 at the latest to save ECU5 million annually. The satellite was finally shut down at 18: 42 GMT on 30 September 1996: the day before shutdown the orbit was 35.9°, 29,990–41,620 km.

ESA ran science operations from 1 October 1995 from Villafranca, until the 1996 deactivation. IUE has had a profound impact on the way astrophysical studies are done through the highly flexible operations mode and satellite design, which required continuous contact but allowed for a very efficient science observation schedule. Many programmes of co-ordinated observations with other satellites and ground sites were often performed.

IUE was shut down in 1996 because of funding problems 0516744

NEAR Shoemaker

Current Status

After launch on 17 February 1996, NEAR (Near Earth Asteroid Rendezvous) performed a fly-by of the asteroid Mathilde on 27 June 1997 coming within 1,212 km of its surface. NEAR performed an Earth fly-by on 23 January 1998, passing 540 km above south-western Iran. An attempted engine burn on 20 December 1998 to align the trajectories for the encounter with Eros failed and the spacecraft was placed in 'safe' mode. In a revised mission plan, NEAR performed a fly-by of Eros on 23 December 1998 coming to within 3,830 km and taking 222 pictures. An engine burn on 3 January 1999 ensured that NEAR returned to Eros for the planned 57 sec (10 m/sec) orbit insertion burn on 14 February 2000, about a year later than originally planned. Just over an hour after entering orbit around Eros, NEAR took its first picture of the asteroid from a range of 330 km above the surface. Orbital parameters for NEAR were 327 × 450 km. A series of manoeuvres placed Near in a 204 × 200 km orbit by 3 March until transferred to a 200 × 100 km orbit on 2 April. Also during March, the spacecraft was renamed NEAR Shoemaker after the legendary astrogeologist, Dr Eugene M. Shoemaker, who had died in 1997 after a car accident. Through a series of orbit changes NEAR Shoewmaker entered a circular 50 km orbit of Eros on 30 April 2000 and a circular 35 km orbit by 13 July which was charged to a 35 × 50 km orbit on 24 July and to a 50 km orbit on 31 July. NEAR Shoemaker executed a 16° orbit inclination charge, moving from 90 to 106°, on 8 August and shifted to a 50 × 100 km orbit on 26 August, circularised at 100 km on 5 September. A further series of manoeuvres placed the spacecraft in a 200 km circular orbit on 3 November followed by further manoeuvres which placed it in a circular 35 km orbit on 13 December. During January 2001, several separate manoeuvres placed NEAR Showmaker in several different elliptical orbits around this maximum 35 km trajectory, dipping to 19 km, placing it at times just 2.7 km above the surface (all orbit data are referred to the centre of the asteroid). Reaching the end of its primary mission the spacecraft was brought down to the surface of Eros at 20:01:52 UT, 12 February 2001, at a speed of 1.9 m/s. It continued to return a signal from the surface, but NASA shut down its communications at midnight UT, 28 February 2001. In an attempt to re-activate NEAR, a signal was sent from the 70 m Goldstone antenna at 07:40 UT, 10 December 2002 when an asteroid Eros was 138 million km from earth. No signal was returned and no further re-activation attempts are planned.

NEAR integrated on Delta's PAM kick stage. The science instruments are mounted on the bottom face and the nozzle of the single 445 N thruster protrudes between the solar wings (APL/JHU) 0517273

Background

The Near-Earth Asteroid Rendezvous mission is the first of the Discovery programme. NEAR's first target was the S-type 13 × 14 × 35 km asteroid catalogued as 433 Eros. It is the second largest of the near-Earth asteroids (only three are larger than 10 km). NEAR was expected to orbit the asteroid as a satellite. Scientists assigned NEAR the mission of determining the asteroid's physical and geological properties as well as its chemical and mineralogical composition.

As the first asteroid orbiter, NEAR was designed to evaluate the environment for orbital missions during its lifetime. It was to approach to within 15 km during the one year investigation, and photograph the surface using a camera with about 1 m resolution. NEAR promises to throw light on the nature of planetesimals (the building blocks for the terrestrial planets), the origin of objects impacting Earth, and the relationship between asteroids, comets and meteorites. The data will significantly enhance interpretation of Galileo data from the Gaspra and Ida S-type fly-bys.

En route, NEAR was to fly by the 61 km diameter C-class main belt asteroid Mathilde.

The Applied Physics Laboratory of The Johns Hopkins University was selected in 1992 to build/control NEAR, with the design reflecting simplicity and reliability, and the payload addressing the

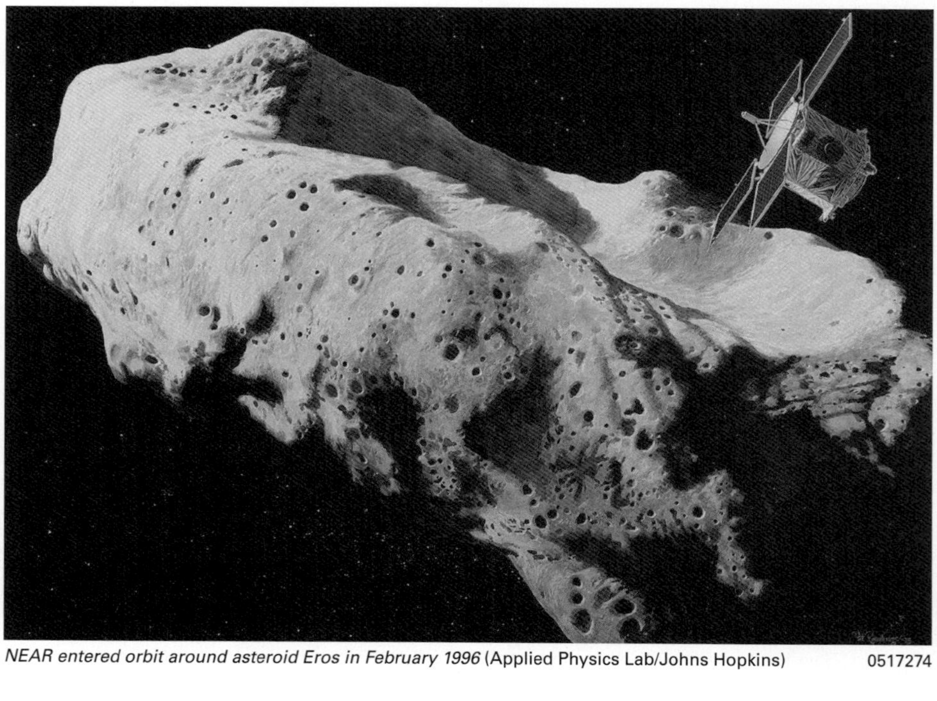

NEAR entered orbit around asteroid Eros in February 1996 (Applied Physics Lab/Johns Hopkins) 0517274

highest priority science issues. It is the first NASA planetary mission to be conducted by a non-NASA centre. APL is conducting mission and science operations from its campus.

Congress capped NASA's cost ceiling at US$150 million (1992 value) for construction, launch and 30 days' operations. When the programme was finished NASA actually paid only US$122 million in real-year dollars. Development began December 1993.

It was launched 17 February 1996 by Delta 7925–8 from Cape Canaveral at 20.43.27 GMT into a two-year Venus Earth Gravity Assist (VEGA) trajectory. By using the Venus fly-by, NEAR only needs a launch energy of 25.9 instead of 42 km^2/s. The 1,200 km 9.9 km/s 253 Mathilde fly-by was scheduled to occur on 27 June 1997. This was to be followed by a 279 m/s deep space manoeuvre 3 July 1997 (in two segments) and a 478 km Earth swing-by 22 January 1998 – cranking the orbital inclination to 10.8°. Then, the mid-space manoeuvres were to take over to nudge NEAR to its final orbital entry around Eros. These manoeuvres, totalling 949 m/s were to start 9 January 1999 to slow approach speed to 5 m/s and enter 500 km Eros orbit 6 February 1999 (2.5 AU from Earth) The Eros orbit is 1.13 × 1.78 AU, at 10.8°. Data gathering was to begin from the initial 200 × 400 km orbit, slowly reduced to 35 to 50 km circular by mid-March 1999. NASA was to shut the mission down on 31 December 1999.

See 'Current Status' above for in-flight changes to this pre-mission plan.

Specifications

Mass: 805 kg at launch, 312 kg propellants, 437 kg dry, 56 kg science payload

Spacecraft: 1.5 m dia hexagonal bus, 3-axis stabilised (1.5 mrad control, 50 µrad knowledge), fixed HGA and solar panels. Four fixed GaAs panels (NEAR's only deployable system, and the first solar cells beyond Mars) provide 1,800 W at 1 AU (81 W required by science instruments). Fixed 1.5 m HGA returns 6–27 kbit/s at X-band; two 512 Mbit SSRs. 445 N Leros NTO/MMH + 12 × 22 N MMH thrusters provide 1,500 m/s total δV. Passive thermal design

Payload: CCD visible imager (95 × 161 µrad resolution, 2.25 × 2.9° FOV, eight-position filter wheel, 10 kg, 15 W, APL, MSX heritage).

X/γ-ray spectrometer (1–10 keV solar monitor, 0.3–10 MeV; Al, Mg, Si, Fe, Ti, Ca, U, Th, K; 26 kg, 10 W, Goddard).

Near-infra-red spectrograph (0.8–2.6 µm, 2–3% spectral resolution, 18 kg, 15 W, APL, DMSP heritage).

Magnetometer (4–65,536 nT, 16-bit resolution, 1 kg, 1 W, Goddard).

Laser altimeter (6 m ht resolution at 50 km, part of attitude/navigation system).

Radio science (2-way Doppler to 0.1 mm/s, part of engineering payload).

UPDATED

New Millennium Series

Current Status

Nasa's New Millennium programme currently supports a total of five missions, one of which has failed: (DeepSpace 2) and one of which (DeepSpace 1) successfully completed a rendezvous with asteroid 9969 Braille on 29 July 1999 and encountered comet Borrelly in September 2001; Earth observing 1, the first of NASA's New Millennium earth observing series, was launched by Delta 7320 from Vandenberg Air Force Base on 17 October 2000; Earth Observing 3 is still in the definition phase and will fly new technology on the NOAA series of GOES; Space Technology 5 will fly three miniature spacecraft high above the earth to evaluate eight innovative technologies in the space environment near the boundary of the magnetosphere following piggy-back launch on a Delta mission in 2003; Space Technology 3 will comprise two separate spacecraft launched on a Delta 7325 in 2005 to test techniques of interferometry and formation flying in space.

Background

Created in 1994, NASA's New Millennium programme covers the identification, development

JPL/NASA Lewis are developing this 30 cm {70} 92 mN Xe thruster under the NSTAR (NASA SEP Technology Application Readiness) project to validate low-power ion propulsion technology. It was carried on DS 1. The thruster is shown before a 2,000 h test at LeRC; a 7,000 h test at JPL began April 1996 (NASA) 0517282

and flight validation of key technologies for incorporation into 21st century science missions without the risks inherent in their first use. NASA projects frequent and affordable science missions with highly focused objectives: numerous micro-spacecraft carrying advanced miniaturised instruments returning a continuous information flow. NM will demonstrate these revolutionary technologies and architectures. These flights also provide opportunities for meaningful science.

Spacecraft development time will be reduced from the eight-year average of the early 1990s to three to four years a decade later. This will help to reduce spacecraft development cost from US$590 million to US$105 million (FY96 rates).

JPL manages the programme, and the advanced technologies for flight validation are identified and developed by integrated product development teams of representatives from NASA, industry, academia and other government centres.

DeepSpace 1

DS 1, the first New Millenium mission, is a comet/asteroid fly-by using solar electric propulsion. The industrial partner for mission definition and implementation is Spectrum Astro, Inc. DS-1 weighs 365 kg at launch, including 125 kg advanced technologies, 23 kg hydrazine, 58 kg Xe. Its design incorporates a fixed 1.5 m X/Ka-band HGA, two LGAs, two star trackers and hydrazine ACS into an aluminium bus, 96 × 112 × 79 cm with a hexagonal spaceframe. Autonomous remote software would plan and execute activities such as autonomous optical navigation. The optical system will image asteroids against star background for position and trajectory information as well as imaging target for fly-by corrections. Primary propulsion is a single JPL/NASA Lewis SEPS 30 cm 20–90 mN, 2.5 kW, SI 3,260 s. Two 152 × 533 cm solar concentrator arrays will generate 2.6 kW.

It carries a payload of the SEPS, solar concentrator array (BMDO), miniature integrated camera/spectrometer, small deep space transponder, autonomous optical navigation, integrated space physics package, beacon mode operations, composite HGA, 3D stack processor. DS1 was launched by Delta 2 Cape Canaveral at 13:08 UT on 24 October 1998. The third stage imparted a velocity of 39.600 km/h sending the spacecraft away from earth but after a successful start-up of the Zenon ion engine on 10 November 1998 a premature shutdown took place. Subsequent operations of the ion engine were successful and over the next several months technology tests were conducted. AK 04:46 UT on 29 July 1999 DS 1 flew to within 15 km of the asteroid 9969 Braille. The primary mission ended on 18 September 1999 but an extended mission has been approved which will take DS-1 to the comet Borrelly for an encounter in September 2001.

DeepSpace 2

DS 2 flew microprobes piggyback on Mars Polar Lander to demonstrate key technologies that will enable network science and *in situ* and subsurface science data acquisition analysis. Within its non-

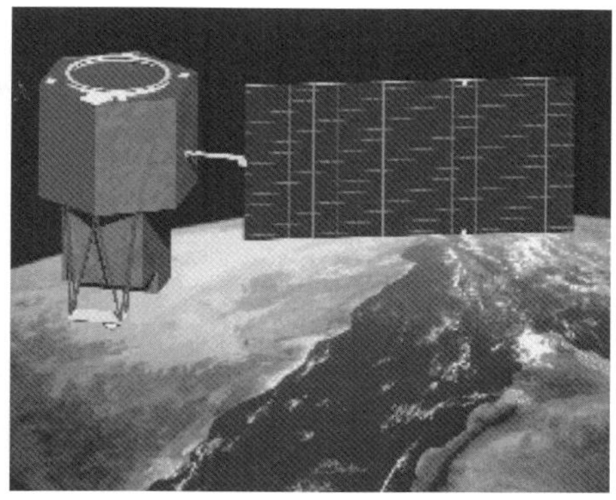

New Millennium's first Earth Observation mission will carry an advanced land imager 0517296

ablative aeroshell, it has a micro-telecommunications subsystem with programmable transceiver, power microelectronics with mixed digital and analogue ASIC, ultra-low temperature lithium battery, microcontroller, flexible interconnects for system cabling, meteorological high-*g* pressure sensor, sample/H_2O experiment soil conductivity and a high-*g* temperature sensor. The DS 2 probes failed to provide information to ground controllers in association with the loss of the carrier, Mars Polar Lander, in December 1999 (see separate entry).

Earth Observing 1

Managed by NASA's Goddard Space flight Center, EO 1 is the responsibility of NASA's Earth Science office and will ensure the continuity of Landsat data. Three land imaging instruments will collect multispectral and hyperspectral data in co-ordination with the Enhanced Therratic Mapper (ETM+) on Landsat 7. EO 1 will provide on-orbit demonstration of spacecraft technologies to enable future spacecraft of this type to be a magnitude smaller and lighter than current versions. EO 1 will be launched by Delta 7320 in October 2000 and inserted into an orbit flying formation with Landsat 7 for comparison of 'paired' images. EO 1 weighs 370 kg without payload and 529 kg at launch to a 98° sun-synchronous, 705 km low earth orbit in a position which follows Landsat 7 by one minute (approx 450 km). Power is provided by three panels of silicon solar cells producing 600 W EOL and 50 Ah super Ni-Cd battery for occultation protection. Data stored at 40 Gbits capacity is transmitted to earth at 105 Mbits/s over x-band with telemetry and housekeeping information sent at 2 Kbyte/s to 1 Kbyte/s selectable. Three axis stabilisation for inertial and nadir positioning has an accuracy of 0.03° in all three axes. For correcting orbit insertion anomalies, orbit maintenance, formation flying and de-orbit Eo-1 has four 1 N thrusters with a 12.3 kg propellant supply of hydrazine. Spacecraft bus is provided by Swales Aerospace and Litton Amecom.

Earth Observing 2

Not a New Millenium mission but a rejected proposal now applied as a Getaway Special for the shuttle programme. It comprises an infra-red laser for measuring the height of atmospheric winds.

Earth Observing 3

NASA carried a Geostationary Imaging Fourier Transform Spectrometer on the GMS satellite planned for launch in 2003. It will test advanced technologies for measuring temperature, water vapour, wind and chemical composition at high resolution. EO3 will expand observations of the atmosphere and its oceans from a few spectral bands to several hundred.

Space Technology 3

Consists of two spacecraft launched by Delta 7325 in late 2005 to test the concepts of the interferometry and formation flying. Candidate propulsion systems for this mission are small 4.5 mN cold gas (GN_2) thrusters or pair of Pulsed Plasma Thrusters at 3 Hz firing rate.

Space Technology 5

Also known as the Narosat Constellation Trailblazer mission, ST5 will comprise three very small satellites, each an octogon 40 cm across and 20 cm high launched together as a piggy-back payload on a Delta mission in 2003. The purpose of ST5 is to test key technologies for future constellations of lightweight narosats weighing about 20 kg. Technologies include:

A miniature communications system to determine the positions of the spacecraft using the Global Positioning System (NASA's Jet Propulsion Laboratory, Pasadena, CA, and Cincinnati Electronics Corp, Mason, OH);

A set of software that automatically operates the spacecraft and determines orbits (Bester Tracking System, Emeryville, CA);

A communications system component that uses one-fourth the voltage and half the power, weighs 12 times less and is nine times smaller than proven technology (Aero Astro, Boston, MA);

A new method of connecting electrical lines that saves weight (Lockheed Martin, Denver, CO);

A new type of microelectronic device that is more reliable and uses 20 times less power than proven technology (Goddard Space Flight Center and the University of New Mexico, Albuquerque);

An electrically tunable coating that can change its properties from absorbing the Sun's heat when the spacecraft is cool to reflecting or emitting heat when needed (Goddard Space flight Center and the Johns Hopkins University Applied Physics Lab, Laurel, MD);

A very tiny microelectromechanical system chip that provides fine attitude adjustments on the spacecraft using 8.5 times less power and weighing less than half as much as proven systems (Marotta Scientific Controls, Montville, NJ);

Development of Lithium Ion Power System for Small Satellites. A rechargeable lithium ion battery that stores two to four times more energy and has a longer life than proven technology (Yardney Technical Products, Pawcatuck, CT).

Space Technology 6

On 31 January 2001 NASA JPC announced that eight teams from industry had been selected to develop new technology concepts such as advanced solar power and optical communications for future missions. Up to five concepts were to have been selected for flights in 2003 and 2004.

Space Technology 7

In March 2002 NASA announced that it had selected Disturbance Reduction System technology scheduled for launch in 2006.

SEDS/SEDSat

Background

SEDS began in 1983 as a Small Business Innovative Research programme. A Delta launched 30 March 1993 carried Seds 1 to a 184 × 746 km, 34° orbit launched to observe the dynamics of a long tether system. It unreeled the 26 kg instrumented end mass on a 0.75 mm diameter polyethylene tether to 20 km in 77 minutes, allowing it to swing like a pendulum for 16 minutes before being detached to burn up. Observations corroborated modelling of the tether dynamics, which introduced a new mathematical description invented by Professor Thomas Kane of Stanford University. Seds 1 decayed 30 March 1993. The 41 kg SEDS 2 flew 10 March 1994 and this time remained attached to stage 2 on its 19.7 km tether. The 25.8 kg 20 × 33 × 41 cm payload included Langley's accelerometer, magnetometer and tensionometer to monitor its motion. Despite its slim diameter, the tether was visible to naked eye ground observers. Placed in a 347 × 352 km, 34.9° orbit. It decayed 8 May 1994.

NASA Marshall planned a third SEDS flight aboard STS-85 in July 1997, releasing the University of Alabama in Huntsville's SEDSat 1. The SEDS will be mounted on top of a Goddard Hitch-hiker carrier in the cargo bay to deploy SEDSat 20 km upwards over 80 minutes from Shuttle's 300 km orbit. The tether will be cut at Shuttle's end first and then at SEDSat's half an orbit later. SEDSat will end up in a 330 × 525 km orbit, while the tether will enter the atmosphere after about one day. SEDSat includes transponders for amateur communications: mode A 145.915–145.975/29.350–29.410 MHz up/down; mode L 1,270/435 MHz up/down at 9.6 or 57.6 kbit/s.

The Shuttle Sed was not flown but Sedsat was launched by Delta 2 on 24 October 1998 to a 574 × 1,078 km, 31.5°, 101 min orbit. Numerous problems plagued the satellite and it was removed from service on 27 October 1998.

SNOE (STEDI Series)

Background

The University of Colorado's Lab for Atmospheric and Space Physics will use the Student Nitric Oxide (NO) Explorer (SNOE) to test possible mechanisms for NO density variations in the lower thermosphere, and to explain why low latitude changes are independent of polar changes. Although only a trace constituent, NO is a major player in many atmospheric processes, mainly ionospheric charge exchange reactions, thermosphere radiative cooling, ozone destruction in the mesosphere and stratosphere.

The compound's density varies with altitude and latitude but nature of the variations and their transport mechanisms are not clearly understood. SNOE will investigate two hypotheses on the energy sources for NO formation. First that solar soft x-rays at low latitude provide the active energy or secondly that secondary auroral electrons in polar regions, where the highest NO levels are found, supply energy. SNOE will simultaneously measure NO density, soft solar x-ray and auroral electron flux.

Performance was excellent and operations continued into 2002.

Specifications

Contractors/Institutes: built inhouse at LASP, collaborating with Ball Aerospace and National Center for Atmospheric Research/High Altitude Observatory
Launch: 26 February 1998 by Pegasus XL/Lockheed L-1011 from Vandenberg AFB
Planned life: one year
Orbit: planned 580 × 550 km, 97.75° Sun-synchronous
Mass: 115.2 kg
Spacecraft: Compact hexagonal structure 91.4 cm high and 99 cm across its widest dimension. Power consumption 35W (average). Data rate: 6 MB/day
Payload: ultra-violet spectrometer, auroral photometer, solar x-ray photometer.

Solar Anomalous and Magnetospheric Particle Explorer (SAMPEX)

Background

SAMPEX was announced 4 April 1989 to study questions in space physics, astrophysics and upper atmosphere science. From 51 submissions the initial missions selected were SAMPEX, FAST and SWAS. SAMPEX is detecting solar/interplanetary charged particles and galactic cosmic rays of energies from 0.4 MeV to hundreds of MeV while over the poles, and observing magnetospheric electrons before they interact with the middle atmosphere. Of the instruments on board, LEICA and HILT were built for GAS (Getaway Special) Shuttle missions but not flown because of delays. Another instrument, MAST/PET, was partially fabricated to fly on the US probe half of Ulysses.

SAMPEX science operations began 10 July 1992. The LEICA initially had power problems, which later resolved themselves. MAST confirmed a belt of anomalous cosmic rays (particles resulting from the Sun's interaction with the interstellar medium) at 600 km within the inner van Allen belt. A US/Russian team in 1991 found the first clear evidence from Cosmos satellite data, but could not determine its location. Because of its orbit, SAMPEX can sample interstellar gas directly from this belt.

SAMPEX was the first Small Explorer 0516741

Specifications

Contractors: GSFC provided the satellite for the University of Maryland's payload
Launch: 3 July 1992 by Scout from Vandenberg AFB
Lifetime: 3 years nominal
Mass: 158 kg (41 kg science)
Orbit: 550 × 675 km, 81.67°, 96 min
Spacecraft: 100 W orbit average provided by four GaAs solar panels (3.4 m^2 total) + one 9 Ah Ni/Cd battery (science instruments require 22 W). 32 Mbyte memory, telemetry by NASA standard 5W S-band transponder: data down at 900 kbit/s for 10 minutes twice daily. Attitude control by MW, three magnetorquers, two-axis fine Sun sensor, four coarse Sun sensors, 3-axis magnetometer. 86 cm diameter × 1.5 m length launch envelope
Payload: LEICA Low-Energy Ion Composition Analyser (8.0 kg, 0.16–25 MeV, University of Maryland).

HILT Heavy Ion Large Telescope (21.0 kg, 3.9–177 MeV, Max Planck Institute Germany).

MAST Mass Spectrometer Telescope (7.5 kg, 7–470 MeV, California Institute of Technology).

PET Proton Electron Telescope (part of MAST, 1–500 MeV).

Spartan

Background

The Shuttle Pointed Autonomous Research Tool for Astronomy was developed by NASA Goddard as a series of short duration low cost free-flyers, meant to extend sounding rocket experiments by using the Shuttle. Spartan 101 used the 100-class carrier, a rectangular 3.2 × 1 × 1.2 m structure, deployed the first time 20 June 1985 by Shuttle 51G with a 136 kg payload and retrieved 45 h later. NRL's x-ray instrument studied the Perseus cluster of galaxies and the central core of the Milky Way galaxy for about 24 hours during 16 orbits at a maximum distance of 192 km from Discovery. It shut down at 40.5 hours, sensing low levels of consumables and battery power. In reality it had used about half its 2,800 m of tape.

Following 101, Goddard developed the 200 carrier for future missions. Three flights were planned for 1986, but the programme was placed on hold after Spartan-Halley, carrying a ultra-violet spectrometer to study Halley's Comet, was lost with Challenger. The Spartan 200 carriers are built from milled aluminium plates bolted together. No welding is used anywhere on the frame. The structure comprises the Instrument Carrier, Upper Structure and Service Module (power, data handling, thermal control, attitude control). The tape recorder were replaced FY96 by a solid-state recorder.

Spartan is deployed by Shuttle's RMS and all subsequent events are pre-programmed with the gathered data stored on the tape recorder. Autonomous free-flying duration is typically 40 hours.

Spartan 201–02 about to be released (NASA) 0517123

Specifications

Spartan 201–02 (general)
Mass: 900 kg Spartan, 500 kg payload (1,000 kg Shuttle support structure); 201–01=1,289 kg deployed
Pointing: 3-axis inertial. Pointing accuracy ±3 arcmin; 0.1 arcsec/s gyro drift; cold gas attitude control
Power: 30 kWh AgZn batteries (28 V DC).

Spartan 201–01
Released: STS-56 April 1993 to a 291 × 298 km, 57°, 90.4 min orbit
Mass: 1,289 kg
Comment: Two solar physics instruments: ultra-violet coronal spectrograph and white light coronograph to probe the physics of solar wind acceleration.

Spartan 201–02
Released: STS-64 September 1994 to a 259 × 269 km, 56.9°, 89.5 min orbit
Comment: Reflight of 201–01 timed to coincide with Ulysses south polar pass.

Spartan 204
Released: STS-63 February 1995 to a 310 × 342 km, 51.6°, 91 min orbit
Comment: Naval Research Lab's FUVIS far-ultra-violet imaging spectrograph observed Shuttle surface glow and thruster firings while attached to RMS, and celestial targets as a free-flyer. Used as large handling mass by EVA astronauts.

Spartan 201–03
Released: STS-69 September 1995
Comment: Reflight of 201–01 timed to coincide with Ulysses north polar pass.

Spartan 206/OAST-Flyer
Released: STS-72 January 1996 to a 302 × 311 km, 28.5° orbit
Comment: Platform for technology experiments. Goddard's Return Flux Experiment characterised return flux contamination to validate computer modelling. Johnson's laser-initiated pyrotechnic characterised space environment effects on reliability. Goddard's Attitude Determination and Control experiment used GPS receivers for navigation and pointing; it was the first satellite to use GPS attitude control. A packet radio system was also demonstrated.

Spartan 207
Released: STS-77 May 1996
Comment: The Inflatable Antenna Experiment deployed JPL's 12.2 m diameter 30 m long aluminised Mylar parabolic antenna structure. It was jettisoned after its surface accuracy was measured, and Spartan was recovered.

Spartan 201–04
Released: STS-87 November 1997 to a 296 km, 28.5° orbit
Comment: Reflight of 201–01.

Spartan 201–05
Released: STS-95 October 1998 to a 550 km, 95.8 min orbit
Comment: Fifth mission to observe solar corona.

STEDI S3eries

Background

In April 2001 NASA decided to prematurely terminate its sponsorship of the STEDI programme in view of delays and cost escalation to the CATSAT project. The USRA Universities Space Research Association is managing the STEDI Student Explorer Demonstration Initiative to launch three small satellites under a US$24 million NASA grant. The first two were selected in February 1995 from 66 proposals for missions costing within US$4.4 million each, excluding launch costs. NASA, in November 1994, awarded OSC the 'Ultralight' launch services contract for two firm Pegasus XL launches plus eight options. The third was selected as a result of intense lobbying by advocates. Cost to USRA is about US$6.7 million per launch, but OSC is offering secondary payload places for US$3.5 million to US$5 million. CATSAT has been pursued at a low level as back-up, but USRA has formally requested full funding for it as the third STEDI. It was planned for launch in December 2001 but attempts to find a replacement funding source to carry the satellite through to launch has been unsuccessful.

Sub-millimeter Wave Astronomy Explorer Satellite (SWAS)

Background

The Sub-millimeter Wave Astronomy Explorer Satellite (SWAS) will observe dense galactic molecular clouds, particularly at water, molecular oxygen, neutral carbon and carbon monoxide wavelengths for information on chemistry, energy balance and structure. Ball Aerospace's Electro-Optics/Cryogenics Division received a US$14.5 million NASA contract in the autumn of 1989 for the primary SWAS instrument. For the first year of operation SWAS concentrated on exploratory observations of more than 70 targets in the galactic plane.

Specifications

Contractors: GSFC provided the spacecraft for the Smithsonian Astrophysical Observatory payload; Ball Aerospace (antenna, star tracker, instrument integration); Millitech Corp (sub-mm heterodyne receiver)
Launch: 9 December 1998 by Pegasus XL/Lockheed L-1011 from VAFB
Lifetime: nominally two years
Mass: 283 kg (science 102 kg)
Orbit: planned 637× 653 km, 70°
Spacecraft: 97 cm diameter. Payload attaches to top of bus as single module. 3-axis stellar pointing to 38 arcsec (<19 arcsec jitter), typically three to five targets per orbit. Nods up to 3° off-target in 15 s increments every 40 s for background reading. Control by three magnetorquers, digital Sun sensor, five coarse Sun sensors, three MWs, magnetometer, three inertial gyros, high accuracy CCD star tracker. 250 W orbit average provided by four deployed fixed + one body panel (3.4 m^2 total GaAs) plus one 21 Ah Ni/Cd battery (science instruments require 50W orbit average). 100 Mbyte bulk memory, 1.8 Mbit/s rate via NASA standard S-band 5 W transponder. 2 kbit/s commands up Telemetry: 2,215 MHz (realtime 18.75 kbps, playlock 1.8 mbps). 5 W transmitter. Command 2,039.65 MHz (2kbps).
Payload: 55 × 71 cm diameter off-axis f/4.8 Cassegrain antenna feeding two sub-mm receivers (cooled to 150K) for University of Cologne's acousto-optical spectrometer (1.4 GHz bandwidth, 1,400 channels, 2.2 GHz centre frequency).

TERRIERS (STEDI Series)

Current Status

Shortly after launch on 18 May 1999 Terriers developed attitude control problems and ground controllers determined that the satellite was unable to orient for solar cell acquisition of the sun-line. Shortly thereafter the satellite ran out of battery power and effectively died. It resided in a 550 km sun-synchronous, 97.7° orbit.

Background

Terriers (Tomographic Experiment using Radiative Recomtinative Ionospheric EUV and Radio Sources) was part of the NASA Student Explorer Demonstration Initiative (STEDI) and carried instruments for creating the first three-dimensional images of the ionosphere using tomographic techniques.

Boston University's Center for Space Physics used the Tomographic Experiment using Radiative Recombination Ionospheric EUV and Radio Sources satellite to create a 2-D, and eventually a 3-D, model of the ionosphere's electron density and photo-emissive components. Terriers employed CAT Computer-Assisted Tomography similar to that developed for medical imaging. Highly directional radiation detectors measure the total radiation in a given spectral band from a particular direction as Terriers spins about a horizontal axis. As Terriers spins and moves along its orbit, the same points in space will be viewed from many different directions, allowing the emission from any one point to be calculated. Visible emission will provide data on chemical sources, such as oxygen ions and nitrogen molecules; EUV emissions from radiative recombinations will be directly related to electron density. Measurements will be co-ordinated with ground-based observations.

Specifications

Contractors/Institutes: Boston University collaborating with Naval Research Lab, MIT Haystack Observatory, University of Illinois, AeroAstro (satellite)
Launch: 18 May 1999 by Pegasus XL/HAPS L-1011 from Vandenberg AFB
Planned life: about 1 year
Orbit: planned 550 km, 97° Sun-synchronous
Mass: 132 kg
Payload: Tomographic EUV spectrographs, two visible photometers, gas ionisation solar spectral monitors, radio beacon.

Transition Region And Coronal Explorer (TRACE)

Background

TRACE is one of the two satellites in the SMEX Mission Set Two series. The SMEX mission matrix compares FAST, SWAS, TRACE and WIRE. TRACE was the third in NASA's Small Explorer series.

The Transition Region And Coronal Explorer studies the connections between fine scale magnetic fields in the solar surface and features in the photosphere, chromosphere, transition region and corona. Scientists want to use it to explore the relation between diffusion of the surface magnetic fields and the changes in heating and structure throughout the solar transition region and corona. The simultaneous ultra-violet films of the 6,000–10,000,000,000 regions at 1 arcsec spatial and 1 second time resolutions will reveal the rate of change of magnetic topology. The launch coincides with the onset of a new solar cycle, so the emerging magnetic flux will be observed in a relatively uncomplicated atmosphere.

TRACE was built at the Lockheed Martin Solar and Astrophysic Laboratory, Palo Alto, California. TRACE's payload consists of a 30 cm diameter ultra-violet/extreme ultra-violet telescope with image motion compensation. Four quadrant normal-incidence coatings on the primary/secondary mirrors form identically-sized and perfectly co-aligned images. A 1,024 × 1,024 CCD collects images over 8.5 × 8.5 arcmin FOV. TRACE's image processor can continuously monitor the data stream and adapt the observing programme and instrument pointing. Guide telescope is integral part of satellite/instrument pointing system.

TRACE and SOHO will co-ordinate their daily observing programme. TRACE is an 'open' mission, in which the PI and co-investigators have waived all priority data rights. The PI will make data available online within 24 hours and science-level data will be released to the PI and general community simultaneously.

Specifications

Contractors: GSFC is providing the spacecraft for the SLISR Stanford Lockheed Institute for Space Research's payload
Launch: 2 April 1998 by Pegasus XL/Lockheed L-1011 from VAFB
Lifetime: 1 year minimum
Mass: 250 kg (59 kg science)
Orbit: 599× 641 km, Sun synchronous 97.8°, 97 min
Spacecraft: bus irregular octagonal prism, 86 cm diameter × 1.5 m length launch envelope. 220 W orbit average provided by four deployed GaAs wings + one 9 Ah Super Ni/Cd battery (science instruments require 83 W). 80C86 instrument processor, 300 Mbyte solid-state memory (66 used by error detection/correction), 2 kbit/s uplink, 0.0234/1.125/2.25 Mbit/s rate via NASA standard S-band 5 W transponder. Four to six daily passes at Wallops/Poker Flat; 320–480 Mbyte science down daily. 3-axis solar 20 arcsec pointing momentum biased around Sun line (stabilised by secondary mirror to <1 arcsec), 3-axis magnetometer, six coarse Sun sensors, bright object sensor, digital Sun sensor, three gyros, guide telescope (fine Sun sensor on instrument), three torque rods, four RWs.

UARS

Background

NASA's Upper Atmosphere Research Satellite (UARS) is investigating the mechanisms controlling the structure and variability of the upper atmosphere. It is the first Mission to Planet Earth satellite and the largest ever flown for atmospheric research. Four instruments (CLAES, ISAMS, MLS, HALOE) provide atmospheric temperature profiles and concentrations of ozone, methane, water vapour and key trace species such as chlorofluorocarbons. Using Doppler shift measurements, two other instruments, HRDI and WINDII, map upper atmosphere wind fields, which shape the global distribution of chemical species. Four other instruments (ACRIM, SUSIM, SOLSTICE, PEM) are monitoring energy inputs from the Sun, which strongly influence chemistry and dynamics.

After gathering data for 4½ years, UARS confirmed that man-made chlorofluorocarbons are responsible. To verify the satellite's findings, mission scientists compare and co-ordinate data, with results from the Solar Backscatter Ultra-Violet (SBUV) spectrometer flying on NOAA meteorological satellites and the Shuttle.

NASA estimates that the total cost through 4½ years of operations and data processing and analysis was US$710 million, with about US$730 million projected through September 1997. The baseline programme authorised 18 months' operations covering two northern winters. Current plans extended those operations through September 1997. Contractors and NASA designers established the 18 month operational window based on the volume constraints in CLAES' coolant supply (exhausted 5 May 1993). All the other instruments had a design life of three years.

The other instruments began failing later in the mission. The ISAMS has not been operational since the mid-1992 failure of the chopper wheel. The MLS 183 GHz radiometer, measuring water vapour, failed April 1993. MLS's vertical scanning activator began to show signs of ageing in 1994. The instrument's operational mode now uses a reduced scan range and duty cycle. MLS focuses its observations on lower stratospheric ClO and ozone, particularly during polar winter. A solar array drive clutch problem halted full science operations 2 June to 20 July 1992. Operations with the degraded primary drive were maintained until October 1993, when the back-up took over. Problems in March 1995 with both drives prompted the May 1995 decision to park the solar array in its safety zone and operate the eight remaining instruments in power-sharing mode. The batteries have given some cause for concern but periodic recharging and revised charge control procedures are allowing full operations.

UARS is based on Fairchild's Multi-mission Modular Spacecraft and employs the refurbished Modular Attitude Control Subsystem package returned from the Solar Max satellite by Shuttle 41C in April 1984. General Electric's Valley Forge centre received the US$221 million prime contract in March 1985 from NASA Goddard. Fairchild's US$41 million contract for the MMS was awarded August 1985.

Trajectory analysis indicates UARS will decay back through the atmosphere between 2016 and 2027 depending on the atmospheric stiffness.

Specifications

Contractors: GE Astro Space (prime), Fairchild (multimission modular spacecraft)
Launched: 12 September 1991 by Shuttle STS-48
Mass: 6,526 kg on-orbit BOL, 2,700 kg science payload; 7,474 kg in Shuttle bay (including support equipment)
Orbit: 571 × 596 km, 57.0°, permitting sampling of all local solar times about every 33 days, 96 min
Spacecraft: 10.36 m long, 4.57 m diameter, 3-axis control, based on Fairchild's MMS. OBC is 64 k NASA Standard Spacecraft Computer (NSSC-1) employing software derived from Landsat
AOCS: 3-axis control provided by inertial reference unit, star trackers, Sun sensors, Earth sensors, RWs, magnetorquers and four monopropellant blowdown hydrazine engine modules, each accommodating 3 × 0.9 + 1 × 22.2 N thrusters; total 204 kg hydrazine. Control to within 36 arcsec, determination to within 20 arcsec
Communications: control is exercised by GSFC at 0.125/1.2 kbit/s on 2.1064 GHz, data return on 5 W 2.2875 GHz 32 kbit/s real-time, 512 kbit/s playback (two NASA Standard 450 Mbit tape recorders); all links via TDRS-based Space Network
Power system: 1.6 kW orbit average provided by single 3.25 × 9.25 m 6-panel Si array and 3 × 50 Ah Ni/Cd batteries
Payload: SUSIM Solar ultra-violet Spectral Irradiance Monitor.

SOLSTICE Solar Stellar Irradiance Comparison Experiment (solar ultra-violet intensity and spectral distribution).

PEM Particle Environment Monitor (magnetospheric energy input to atmosphere from protons/electrons, plus x-ray imaging spectrometer and vector magnetometer).

ACRIM Active Cavity Radiometer Irradiance Monitor (solar output at all wavelengths for long-term measurement of solar constant, continuing work of Solar Max).

CLAES Cryogenic Limb Array Etalon Spectrometer (cryogen depleted).

ISAMS Improved Stratospheric and Mesospheric Sounder (chopper wheel drive system failed).

MLS Microwave Limb Sounder (63/183/205 GHz molecular thermal emission from atmosphere; 183 GHz failed).

HALOE Halogen Occultation Experiment (vertical distribution of key trace species).

HRDI High Resolution Doppler Imager + WINDII Wind Imaging Interferometer (wind speeds in upper atmosphere).

XTE

Background

The X-ray Timing Explorer (also called RXTE) carries three instruments for intensive studies of x-ray source luminosity variations, for the first time ranging from µs to years, to provide information on processes and structures of white dwarfs, black holes, neutron stars and other exotic objects. The large effective area (0.8 m^2) and broadband sensitivity (2–200 keV) are valuable for timing intensity variations and determination of spectra from high energy sources.

The PCA Proportional Counter Array and HEXTE (High Energy X-ray Timing Experiment) form a single telescope, viewing a single source in their common 1° FOV. The large areas and low backgrounds provide high sensitivity to weak sources. The ASM All-Sky Monitor scans 70 per cent of the sky every 1.5 hours to monitor about 75 of the sky's brightest sources. It thus measures long-term variations and identifies changes for PCA/HEXTE to study at short notice. The microprocessor EDS Experiment Data System provides onboard processing/compression of PCA/ASM data from 500,000 bit/s to the telemetry rate.

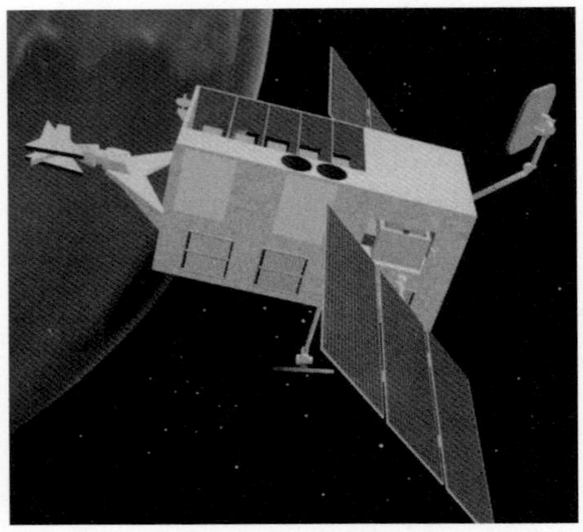

RXTE is aimed at monitoring luminosity variations of x-ray sources (NASA) 0517297

XTE was renamed in February 1996 after x-ray astronomy pioneer Professor Bruno B Rossi. Satellite cost, including ground equipment, was US$195 million; annual operations cost is US$11 million.

Specifications

Contractors: Goddard (project management; development/integration of observatory, development of PCA), MIT (ASM, Experiment Data System), UCSD (HEXTE)
Launch: 30 December 1995 by Delta 7920-10 from Cape Canaveral
Lifetime: 24 months nominal mission (up to 5 years possible)
Mass: 3,045 kg at launch
Orbit: 565 × 583 km, 23°, 96.1 min
Spacecraft: >6°/min manoeuvres via four 75 Nms Ithaco RWs, pointing accuracy <0.1° (knowledge from star trackers + gyros to 60 arcsec). PCA/HEXTE FOV can be pointed to any position if Sun angle >30°. Command/datalinks through TDRS: 32 kbit/s (26 kbit/s science) plus 256 kbit/s for 30 minutes daily. Science Operations Center at Goddard. Two pointable high gain antennas maintain TDRS link while XTE pointed at any target. Twin 3-panel rotating solar wings (total 17.86 m^2) provides 800 W; 2 × 50 Ah batteries
Payload: PCA. Five Xe proportional counters, 6,250 cm^2, 2–60 keV, 1 × 1° FWHM, 1 µs resolution, 0.1 mCrab sensitivity (10 min), 18 kbit/s. Smaller version flown on HEAO 1. NASA Goddard.

HEXTE. Two rocking (every 15 s) clusters of four NaI/CsI detectors, 1,600 cm^2, 20–200 keV, 1 × 1° FWHM (rocking), 10 µs resolution, 1 mCrab sensitivity (10^5 s), 5 kbit/s. University of California at San Diego.

ASM. Three scanning shadow cameras on rotating boom, 90 cm^2, 2–10 keV in 3 channels, 0.2 × 1° FOV, 1.5 h resolution, 30 mCrab sensitivity (1.5 h), 3 kbit/s. MIT.

SEARCH AND RESCUE

International

Nadezhda (COSPAS-SARSAT) Series

Current Status

The Cospas-Sarsat constellation is composed of search and rescue satellites in low earth orbit (LEOSAR) and geostationary orbit (GEOSAR). The nominal system configuration is four satellites, two in LEOSAR and two in GEOSAR. In October 2000 the Council of Cospas-Sarsat announced the gradual phasing out of satellite processing of emergency beacons operating in the 121.5/243 MHz range, encouraging users to adopt 406 MHz beacons.

Background

Canada, France, the US and the former Soviet Union (membership now assumed by Russia) established COSPAS-SARSAT in 1979 to provide an international satellite-based system for search and rescue. The system operates through a constellation of four polar satellites, which detect distress beacons. It locates an activated beacon to within 2 km and the founders of COSPAS-SARSAT credit it with saving more than 5,500 lives by June 1995 in 1,800 search and rescue operations. A Memorandum of Understanding between the signatories governed the actions of the body from 1984 until the International COSPAS-SARSAT Programme Agreement was signed by the four states on 1 July 1988 in Paris.

When a user employs the signal beacon, the distress message arrives via the satellites at a Local User Terminal (LUT) to determine beacon location. The information is passed to a Mission Control Centre to alert the rescue authorities. Globally COSPAS-SARSAT has provided some 550,000 121.5 MHz beacons and made them operational (600,000 are projected by 2000). Because the current system restricted usage to real-time relay by the satellite, a LUT must be within range (about 2,500 km) for position determination to within 10–15 km. The 406.025 MHz units include user identification codes in the message. Even with these systems, though, further localisation results from a measurement of the Doppler shift of the transmitted signal. Mathematically, two-signal localisation results in two possible positions, but the slight change due to the rotational velocity of the earth resolves the ambiguity. If the beacon's frequency stability is sufficient, as with the 406 MHz devices, the satellite can resolve true position on the second pass over the distress beacon. This system, therefore, provides a global service with an average waiting time of 44 minutes, with 94 per cent of signals detected in 90 minutes (results from 1990 exercise).

All satellite downlinks operate at 1.5445 GHz. By end of 1995, 119,000 406 MHz beacons had been produced and distributed in 130 countries. Since the 406 MHz beacons emit a 5 W RF burst of about 500 ms duration every 50 seconds this frequency assures more precise location accuracy (90 per cent within 5 km), while the high peak power increases the probability of detection on a single pass to more than 98 per cent. Typical delay before detection is one to two hours near the equator. Position information from systems such as GPS and Glonass will be encoded in the messages of new 406 MHz beacons after January 1997.

Russia's Nadezhda civil navigation satellites also carry the COSPAS payload to locate search/rescue beacons (AKO Polyot) 0517091

The COSPAS organisation establishes Mission Control Centres in each country operating at least one LUT to disseminate information to the appropriate Rescue Co-ordination Centres. By end of 1995, there were 33 LUTs in 21 countries with others planned. The USSR began deploying the space segment with the launch of Kosmos 1383 in 1982. Designated COSPAS 1 ('Space System for Search of Vessels in Distress'), the 121.5 MHz signal remained operational until March 1988. Kosmos 1447 (1983), Kosmos 1574 (1984), Nadezhda 1 (1989), Nadezhda 2 (1990), Nadezhda 3 (1991) and Nadezhda 4 (1994, in the same orbit as COSPAS 5 as a spare) adopted the roles of COSPAS 2–7, with one more satellite on the ground and in preparation for launch if needed.

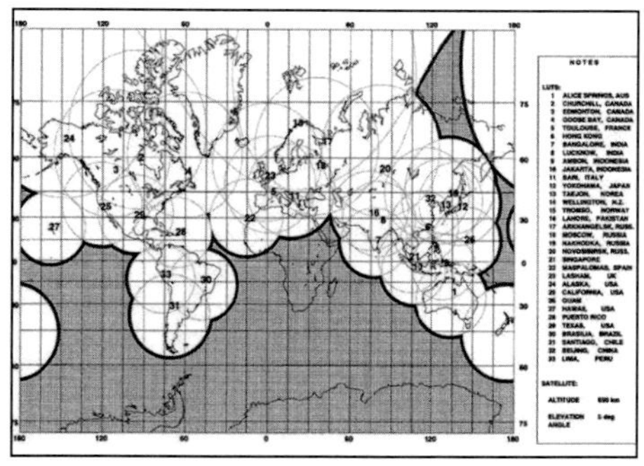

Local User Terminal 121.5 MHz coverage (406 MHz is global) 0517411

NOAA adds a data relay system with each of its meteorological satellites. SARSAT 1 (S&R Satellite-aided Tracking) flew with NOAA 8 in 1983. NOAA 9–11 brought the system to full strength.

Since 1982 more than 11,000 people have been rescued world wide with the assistance of the COSPAS-SARSAT system.

COSPAS-SARSAT satellite launches

	Launch	Orbit	Payload
C1383[1]	29 June 1982	989 × 1,028 km, 83°	COS 1
C1447[2]	24 March 1983	959 × 1,013 km, 83°	COS 2
NOAA 8[3]	28 March 1983	803 × 825 km, 98.7°	SAR 1
C1574[4]	21 June 1984	965 × 1,005 km, 83°	COS 3
NOAA 9[5]	12 December 1984	841 × 862 km, 99°	SAR 2
NOAA 10[6]	17 September 1986	808 × 826 km, 98.7°	SAR 3
NOAA 11	24 September 1988	849 × 865 km, 98.9°	SAR 4
Nadezhda 1[7]	4 July 1989	960 × 1,014 km, 83.0°	COS 4
Nadezhda 2[8]	27 February 1990	956 × 1,021 km, 83.0°	COS 5
Nadezhda 3[9]	12 March 1991	958 × 1,018 km, 82.9°	COS 6
NOAA 13[10]	9 August 1993	850 × 863 km, 98.9°	SAR 5
Nadezhda 4[11]	14 July 1994	954 × 1,005 km, 82.9°	COS 7
NOAA 14[12]	30 December 1994	848 × 863 km, 98.9°	SAR 6
NOAA 15	13 May 1998	808 × 823 km, 98.32°	SAR 7

GOES 7 (May 1986), 8 (April 1994) and 9 (May 1995), Insat 2A (July 1992) and 2B (July 1993), Cosmos 2054 Luch (December 1989), Luch 1 (December 1994) and GMS 5 (March 1995) carried 406 MHz relays into GEO and remain in experimental use.

Key: COS = COSPAS, SAR = SARSAT, C = COSMOS

1) decommissioned March 88 2) decommissioned December 89 3) decommissioned December 1985 4) decommissioned June 90 5) 406 MHz local/global processor mode operational, global mode reactivated October 1991 6) 406 MHz global/local modes not operating since September 88 7) limited availability in Southern hemisphere 8) started Southern hemisphere operations December 1991 9) satellite contact lost 21 August 1993 while S&R testing under way 10) acting as back-up to Nadezhda 3 in same orbit global 11) mode failed soon after launch 12) decommissioned February 1996.

PLANETARY AND SPACE SCIENCE

PLANETARY AND SPACE SCIENCE

International

Cassini

Current Status

Cassini was launched at 08:43 UT on 15 October 1997.

The spacecraft functioned well during test and housekeeping chores including lubricating the reaction wheel assembly mechanisms. Cassini performed a fly-by of Jupiter on 30 December 2000, at a distance of 9.7 million km/h gaining 2 km/s.

Using Jupiter to accelerate the spacecraft on to Saturn, Cassini entered the Jovian magnetosphere on 9 January 2001. It will reach Saturn and enter orbit on 1 July 2004. The Huygens probe will be released about six months later. By early 2003 significant observations had been made of the planet Saturn with several new discoveries made concerning the atmosphere of the planet.

Background

NASA pegged Cassini as the fourth of the 'Core' set of planetary missions sought by its Solar System Exploration Committee in 1983. Cassini was supposed to save funds by employing the sophisticated Mariner Mark II spacecraft, designed to undertake the more demanding deep space missions, by accommodating scan platforms, nuclear power supplies, high data rates and decade-long flight times.

NASA chose Titan as the target for Cassini, partly because the organic processes taking place on Saturn's largest moon provide the only planetary-scale laboratory for studies of a pre-life terrestrial atmosphere. Titan has a dense, predominantly nitrogen atmosphere and the mission will permit intensive scrutiny. Then, suffering from budget constraints due to the Shuttle problems, NASA asked ESA to participate in the programme as well. Following a joint NASA/ESA study, the agencies agreed in June 1985 that NASA would supply a 1,540 kg (dry mass) Saturn orbiter and ESA a 190 kg Titan atmospheric probe.

Originally, NASA set a schedule for a Shuttle/Centaur launch in May 1994 giving the craft the opportunity to study asteroids in fly-bys before Cassini's trajectory returned it to Earth for a swing-by to provide the energy for the 4.5 year journey to Saturn. The mission slipped in 1987 at NASA's insistence and because of the new need to carry Centaur upper stages – then the only possible vehicle for the flight – on expendable launchers.

ESA began to have doubts about the mission in 1988, but at the request of NASA pushed the programme into the phase B programme plan that according to ESA's charter almost ensures a launch date. ESA's Council approved the mission during its June 1990 meeting, planning to begin Phase B in January 1991, although the MoU had still to be signed because of delays in the preceding Soho/Cluster MoU. Aerospatiale received the prime contract (signed with ESA 20 September 1993). Sole bidder General Electric (now Lockheed Martin) was selected by the Department of Energy in mid-1990 to provide the RTGs for both CRAF/Cassini. The mission remained essentially unchanged (calling for Cassini launch 6 April 1996 for 6 December 2002 Saturn arrival) until early 1991 when the CRAF/Cassini launches were swapped and Venus gravity assists added.

The increased mass capacity allowed the addition, for example, of a third RTG, increasing margins and removing the need for batteries. Cassini's 25 November 1995 departure would have produced Saturn arrival 15 May 2004 and allowed an 18 November 1998 fly-by of 40 km diameter F-class asteroid Clarissa in the main belt. CRAF was cancelled in January 1992 and Cassini was slipped to spread out NASA's funding obligations. This two-year delay, which cost ESA contract funds as well, still increased total Cassini funding beyond the original US$1.6 billion Congressional ceiling.

To meet other funding obligations, NASA dropped the scan platforms in 1992. Coupled with a policy of no pre-Saturn science acquisition (an asteroid fly-by would cost 70 m/s delta-V, the equivalent of 7–8 Titan fly-bys), this reduced development cost to US$1.4 billion.

After launch on 15 October 1997, Cassini entered into a Venus/Venus/Earth/Jupiter gravity assist multi-planet swing-by path, thereby reducing launch energy from 80 km^2/s^2 to 22 km^2/s^2. Cassini first encountered Venus on 26 April 1998 when it flew the planet at a distance of 284 km. The 3 December 1998 450 m/s deep space manoeuvre at aphelion set up the second Venus gravity assist for 24 June 1999 at 600 km. Then a 1,171 km Earth fly-by of 18 August 1999 provides the final energy to reach Jupiter. A 115–140 R_J (10 million km) Jupiter swing-by on 30 December 2000 sets up Saturn arrival on 1 July 2004.

Cassini traversed the asteroid belt December 1999 to April 2000, but did not gather data because the science group committee will not turn on their instruments again until two years before Saturn arrival (apart from a 40-day gravitational wave experiment about December 2001). On 23 January 2000 Cassini's cameras took images of the asteroid 2685 Masursky from a distance of 1.6 million km. Too small to image from earth, 2685 Masursky was found to be roughly 15 × 20 km in size.

After getting into orbit about Saturn on 1 July 2004, Cassini also plans a 52,000 km Phoebe fly-by on 11 July 2004, where a 96-minute 621 m/s braking burn establishes an initial 147-day, 1.3 × 176 R_S 17° inclination orbit. This is the closest approach to Saturn during the entire mission and the insertion burn precedes its optimal point (centred on periapsis) by 48 minutes for science observations. A 332 m/s burn 12 September 2004 will raise periapse to 8.2 R_S for Europe's Titan probe to spin up and release itself on 6 November 2004. A 57 m/s deflection burn follows two days later to fly past Titan at 1,500 km during 27 November while the probe descends. The orbiter will fly by Titan on 33 of the planned 63 orbits over the four-year mission, sometimes approaching to within 950 km and eventually raising the orbital inclination to 85° (Voyager photographed the planet from an inclination of less than 30°). Plans allow the Cassini to perform radar mapping during the encounters with the rings if power is available. Fly-bys of Iapetus, Enceladus, Dione and Rhea will also be made, plus more distant encounters with Tethys, Mimas and Hyperion, before the mission formally ends on 1 July 2008.

During the Titan fly-by, Europe's Huygens will take 2.5 h to descend during 27 November 2004. The probe will enter the moon's atmosphere at 6.15 km/s over Titan's day side at about 18° N/200–220° longitude (where 0° is defined as the sub-Saturn point; recent Hubble observations confirmed that Titan is tidally locked to Saturn). ESA plans to use the probe to determine the atmosphere's chemical composition and measure winds, temperature and pressure profiles from 170 km down to the surface. It is possible that

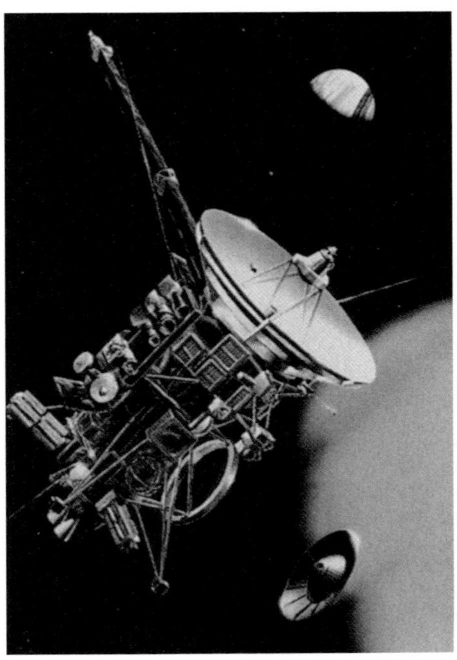

Cassini releases Huygens into Titan's atmosphere 0516789

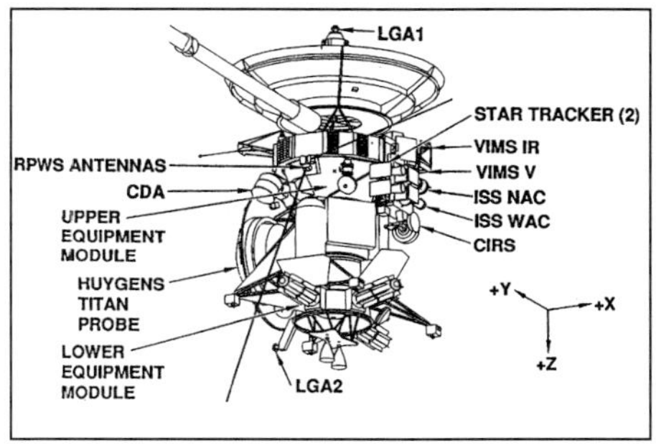

Cassini cruise configuration (NASA/JPL) 0516790

Huygens could continue returning data after touchdown, but planners speculate that, at the very least, accelerometer data will reveal if it has splashed into an alien sea.

As the probe descends, Huygens will jettison a canister when M1.5 has been attained at around 170 km. Then a mortar-deployed 2.6 m diameter pilot parachute will pull off the aft cover to allow deployment of the 8.3 m diameter first main parachute. Following stabilisation, the heat shield will be ejected to free the instruments' inlet ports. After a 15-minute descent, a 3.0 m diameter stabiliser will take over for the final descent. $LiSO_2$ batteries will provide 350 W for 3 h, sufficient for an 8 kbit/s data rate on each of two redundant probe to orbiter S-band links. All data are stored aboard the orbiter for later relay to Earth.

Specifications

Cassini spacecraft

Launch: 15 October 1997 by Titan 4 Centaur from Canaveral

Mass: 2,150 kg dry orbiter, 3,132 kg propellants, 373 kg Huygens (335 kg probe), 165 kg launch adapter. Total injected mass 5,820 kg

Design life: 11 years

Communications: 4 m diameter HGA and GHz X-band 20 W TWTA will return 20 bit/s – 169 kbit/s. Data stored on two 1.8 Gbit SSRs, providing up to 403.2 kbit/s record/playback

Power: Three RTGs provide 815 W BOL (638 W EOL; no batteries)

AOCS: 3-axis attitude control (2 mrad pointing accuracy, 6.98 mrad/s max slewing) maintained via SRU, four gyros, RWs and four clusters of four 1.0–0.6 N hydrazine blowdown thrusters on propulsion module. (primary: two redundant 445 N MMH/NTO bipropellant R-4D thrusters will be used for up to 200 manoeuvres totalling 2,360 m/s)

Orbiter payload:

Cassini Plasma Spectrometer; 12.5 kg, Southwest Research Institute. The Cassini Plasma Spectrometer (CAPS) measures the flux of ions as a function of mass per charge and the flux of ions and electrons as a function of energy per charge and angle of arrival relative to CAPS. Science objectives are: to measure the composition of ionised molecules originating from Saturn's ionosphere and Titan; to investigate the sources and sinks of ionospheric plasma (ion inflow/outflow, particle precipitation); to study the effect of magnetospheric/ionospheric interaction on ionospheric flows; to investigate auroral phenomena and Saturn Kilometric Radiation (SKR) generation; to determine the configuration of Saturn's magnetic field; to investigate the plasma domains and internal boundaries; to investigate

Huygen's final descent sequence was demonstrated with a drop from 36 km over Kiruna, Sweden (Fokker Space/ESA) 0517305

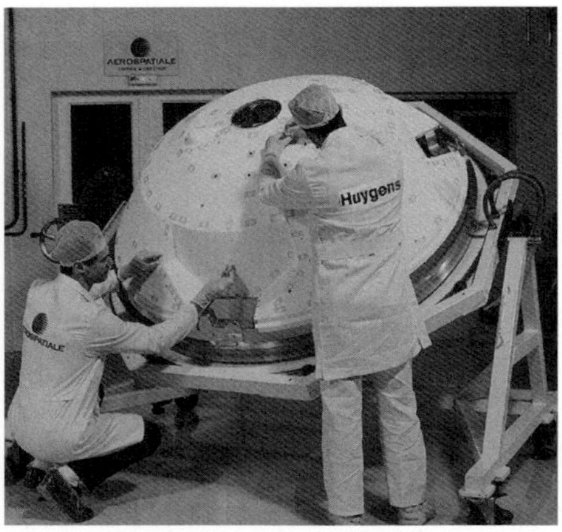

Huygens heat shield and aft cover (ESA/Aerospatiale)
0517146

the interaction of Saturn's magnetosphere with the solar wind and solar-sind driven dynamics within the magnetosphere; to study the microphysics of the bow shock and magnetosheath; to investigate rotationally driven dynamics, plasma input from the satellites and the rings and radial transport and angular momentum of the magnetospheric plasma; to investigate magnetotail dynamics and substorm activity; to study reconnection signatures in the magnetopause and tail; to characterise the plasma input to the magnetosphere from the rings; to characterise the role of ring/magnetosphere interaction in ring particle dynamics and erosion; to study dust-plasma interactions and evaluate the role of the magnetosphere in species transport between Saturn's atmosphere and rings; to investigate auroral phenomena and SKR generation; to study the interaction of the magnetosphere with Titan's upper atmosphere and ionosphere; to evaluate particle precipitation as a source of Titan's ionosphere; to characterise plasma input to magnetosphere from the icy satellites and; to study the effects of satellite interaction on magnetospheric particle dynamics inside and around the satellite flux tube. CAPS incorporates an ion mass spectrometer, ion beam spectrometer, electron spectrometer, data processing unit and a scan motor.

CAPS has a total mass of 12.5 kg, average operating power of 14.5 W and an average data rate of 8 kbps. The CAPS subsystem consists of six major subassemblies: the ion mass spectrometer, the ion beam spectrometer, the electron spectrometer, the data processing unit, a high-voltage power supply and an actuator. The ion mass spectrometer (IMS) provides species-resolved measurements of the flux of positively charged atomic and molecular ions as a function of energy/charge verses aperture entry direction.

The IMS uses a toroidal top-hat electrostatic analyser (for energy/charge data and to create a narrow field of view) combined with a linear electric field time-of-flight mass spectrometer (for mass/charge and other species-resolving data). The IMS consists of an aperture cover/actuator, a toroidal analyser, carbon foils, a time-of-flight spectrometer, microchannel plates, amplifier/discriminators, a time-to-digital converter, a spectrum analyser module and high-voltage power converters contained in high-voltage units 1 and 2. The IMS aperture cover protects the IMS during ground handling and during launch and early flight. The cover is a strip of flexible material that fits over the IMS annular aperture without creating a hermetic seal. To open the cover the

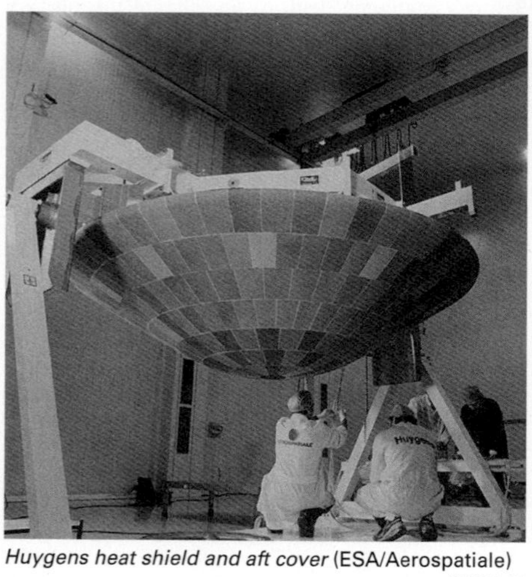

Huygens heat shield and aft cover (ESA/Aerospatiale)
0517147

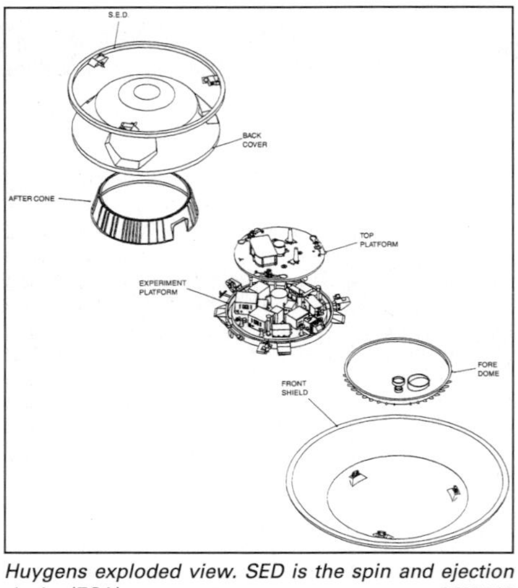

Huygens exploded view. SED is the spin and ejection device (ESA)
0517148

strip is reeled into a container by a spring after the strip is released by a wax thermal actuator. Closing the cover requires ground handling to unreel the cover and re-attach its end to the cover-release mechanism. Thus, in flight, the aperture cover is a one-time use device. The toroidal analyser (toroidal refers to the configuration of the target reflector) consists of a baffled collimator mounted on a toroidal "top hat" electrostatic analyser.

The geometry of the collimator determines the narrow, annular IMS field of view of 12° × 160° which is divided into eight angular pixels of 12° × 20° each. An electrical potential between the inner and outer conductors of the electrostatic analyser allows through this device only ions having energies within a range selected by the analyser potential (i.e., only ions with certain energies will have trajectories, at a given analyser potential, that navigate through the analyser without being stopped by hitting a wall). Energy spectra are taken by stepping the analyser potential through a set of 64 space values. Eight thin carbon foils are arranged in an arc along the exit of the toroidal analyser. The foils at the entrance to the time-of-flight spectrometer chamber formed by the linear electric field (LEF) rings. A –15 kV potential accelerates the positive ions exiting the toroidal analyser into the foils and this permits ions entering the IMS with as little as 1 eV of energy to pass through the foils. Molecular ions travelling through the foils are usually broken into their constituent atoms and/or molecular fragments. Molecular and atomic ions and molecular fragments exit the foils as neutrals or ions and on exiting the neutrals and ions eject secondary electrons into the time-of-flight chamber.

The time-of-flight spectrometer is a cylindrical chamber in the IMS bounded by linear electric field (LEF) rings where particles exiting the carbon foils encounter an electric field with a strength that increases linearly with distance parallel to the LEF axis of symmetry. The linear electric field is generated by a stack of 30 equally spaced aluminium rings along which a network of giga-ohm resistors establishes a quadratic electric potential with a total potential drop across the stack of rings of 30 kV. The LEF focuses positively charged particles with energies up to approximately 15 KeV

Huygens approaches the surface of Titan (ESA)
0517306

independently of the energy and angle with which they exit the carbon foils. Microchannel plate (MCP) multipliers are located at the ends of the stack of LEF rings. Detection by the "start" (ST) MCP of rapidly travelling secondary electrons from the carbon foils is used as a time-of-flight "start" signal and for determining an ion's elevation angle with respect to the IMS aperture. Simple harmonic motion of ions in the LEF together with knowledge of their energy gained from the toroidal analyser setting relates their time-of-flight from carbon foils to the LEF MCP (at the other end of the chamber from the ST MCP) to mass per charge. The two microchannel plates (MCPs) each consist of three circular plates of lead oxide glass with a multitude of microscopic channels running at an incline through the thickness of each plate. Electrons, ions and neutrals striking the outer plate cause secondary electrons to cascade down the semiconductive walls of the channels. With a potential drop of about 1 kV across the thickness of a plate there is a yield of about 300 electrons per incident particle or a total gain of 106 electrons across the three-plate stack. Nine anodes under the ST MCP and one anode under the LEF MCP collect the electrons emitted from the MCPs and pass the signals to two high-speed current amplifiers (one "start" amplifier for the eight 20°-wide anodes and one "stop" amplifier for the LEF anode and the centre ST anode).

Two constant fraction discriminators accept the amplifiers' signals and send digital timing pulses (independent of the amplifier's signal amplitude) to the time-to-digital converter (TDC). Also sent to the TDC is the identification of which of the eight annular anodes was the source of the "start" pulse and whether a "stop" pulse came from the LEF anode or the centre ST anode. The time-to-digital converter (TDC) measures the time interval between the start and stop pulses from the amplifier/discriminators. The TDC outputs an interval length to the spectrum analyser module along with information identifying the "start" and the "stop" anode. The "start" anode identity determines which of the eight 20°-wide IMS elevation resolution elements on ion entered through. A device in the TDC (a pulser) can eject simulated MCP signals just downstream of the MCP anodes via high-voltage isolating capacitors in order to test the signal handling functions of the IMS and related hardware and software in the CAPS data processing unit. The spectrum analyser module (SAM) provides the data gathering, sorting and transfer functions between the TDC and the data processing unit (DPU). Time interval data from the TDC and "binned" or grouped into a pre-selected (by the DPU) set of time channels associated with certain selected ion species, whether atomic or molecular.

The SAM processor performs a deconvolution of the time-of-flight spectra to obtain ion identifications which it passes on to the DPU. The IMS has four programmable pulse width-modulated high-voltage power converters which supply high voltage to the toroidal analyser, the LEF rings and the LEF MCP. The second major CAPS subassembly is the ion beam spectrometer (IBS) which measures the flux of positively charged ions of all species as a function of energy/charge and aperture entry direction and consists of a hemispherical electrostatic analyser, channel electron multipliers, amplifiers/discriminators and high-voltage power converters. The hemispherical electrostatic analyser consists of two conductive hemispheres of slightly different radius mounted concentrically one within the other in such a way that there is a small gap between the two conductors. The electric field in the gap selects the range of energy per charge and angular direction that ions are allowed to have in order to pass through the analyser.

Energy spectra are taken by stepping the analyser potential through a set of spaced values. Three apertures, each defining a fov of 1.5° × 150°, spaced 30° apart, allow particles to enter the analyser. Particles are "focused" into three channel electron multipliers 180° from each of the entrance apertures. Ions entering the channel electron multipliers (CEMs) strike the inner semiconducting surfaces of the devices and spawn secondary electrons which bounce down the curved channel in the CEMs spawning more electrons with each bounce. In this way the entry of an ion into a CEM results in a pulse of approximately 10 × 8 electrons being collected by an anode at the exit of the CEM. The amplifiers/discriminators amplify the electron pulses collected by the CEM anodes and send the signals to the DPU's IBS interface. A device in the IBS (a pulser) can inject simulated CEM signals just downstream of the CEM anodes in order to test the signal handling functions of the IBS and

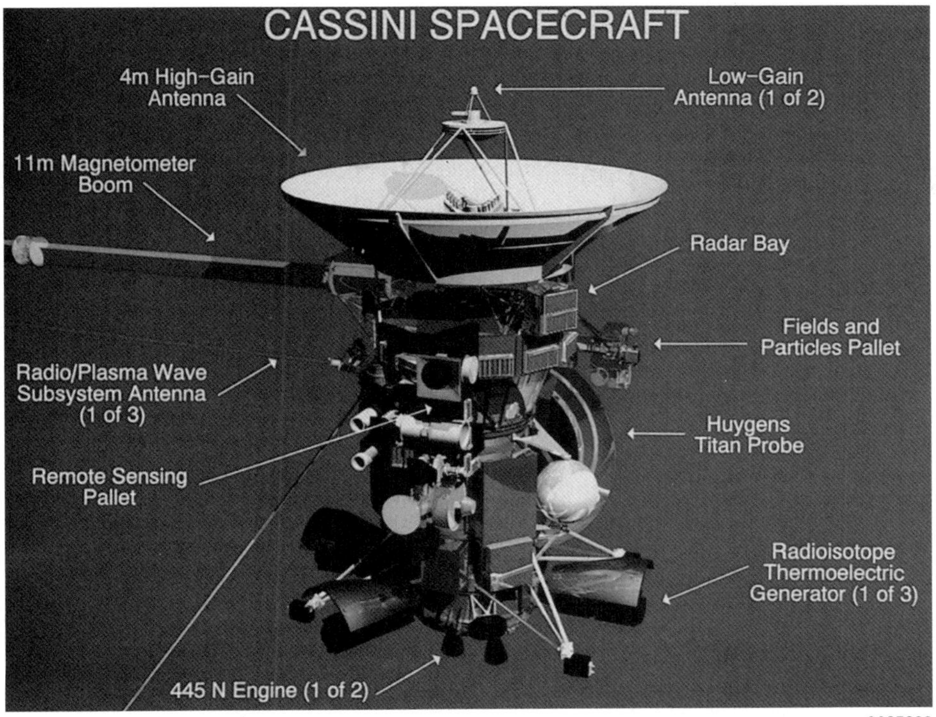

Cassini spacecraft (an image) 0085829

related hardware and software in the DPU. The IBS has two programmable pulse width-modulated high-voltage power converters which supply high voltage to the hemispherical analyser and the CEMS.

The third major CAPS subassembly in the electron spectrometer (ELS) which measures the flux of electrons as a function of energy/charge and aperture entry direction. The ELS consists of a spherical analyser, microchannel plates, amplifiers/discriminators, a sensor management unit and high-voltage power converters. The spherical analyser consists of a baffled collimator mounted on a spherical "top hat" electrostatic analyser. Collimator geometry determines the narrow, annular ELS field of view of 5° × 160° which is divided into eight 20° "pixels". An electric potential between the inner and outer conductors of the electrostatic analyser allows through this device only electrons having energies and angles within a range selected by the analyser potential and top-hat collimation. Energy spectra are taken by stepping the analyser potential through a set of 96 curved values. Two 90° "long" annular (curved) microchannel plates (MCPs) are arranged in a 180° arc along the exit of the spherical analyser, electrons striking the surface of the MCPs are multiplied into signals collected by an arc of eight 20° anodes. The anode signals are passed into eight amplifiers/discriminators via high-voltage isolating capacitors and are then accumulated by the sensor management unit. The anode identity determines which of the eight 20°-wide ELS elevation resolution elements an electron entered through.

The sensor management unit (SMU) controls the voltage level of the ELS high-voltage power converters in accordance with requests from the CAPS data processing unit (DPU) and passes counts of electrons (from each of the eight angular resolution elements) detected by the ELS to the DPU. A device in the SMU (a pulser) can inject simulated MCP signals just downstream of the isolating capacitors in order to test the signal handling functions of the ELS and related hardware and software in the DPU. The ELS has two programmable pulse width-modulated high-voltage power converters that supply high voltage to the spherical analyser and the MCPs. The fourth major CAPS subassembly is the data processing unit (DPU) which manages the acquisition and onboard data processing of all CAPS data and controls sensor and actuator motor functions. The DPU is designed to use two CPUs, in addition to the processor in the spectrum analyser module. The first CPU accumulates IMS time-of-flight spectra and compresses all IMS data. The other CPU controls the IMS, ELS, IBS and the actuator and performs onboard data analysis to determine what measurements will be taken and what data will be placed into housekeeping and science packets.

The DPU consists of the sensor and actuator data control interfaces, a housekeeping analogue-to-digital converter, a safe/arm control, a wax thermal actuator (WTA) driver, CPUs with memory, a bus interface unit, low-voltage power converters, supplemental and replacement heaters, a radiator and the principal structure of CAPS. Through the sensor and actuator data control interfaces the DPU performs several functions: it accumulates IBS ion flux counts; it collects ELS and IMS data products from the SMU, the TDC and the SAM; it directly controls the IMS and IBS pulsers and high-voltage converters. The DPU feeds ion energy/charge data to, and controls, the SAM. In addition the DPU controls the CAPS actuator (ACT) motor stepping and accepts position and status data from the ACT. The output voltage of all nine CAPS high-voltage converters, CAPS low voltages, the actuator position encoder and the temperature at six different locations in CAPS are monitored by a housekeeping analogue-to-digital converter.

In addition, two temperature sensors (one in the DPU, the other in the IMS cover release mechanism) are monitored directly by the spacecraft. These two sensors do not require that CAPS be powered. When not enabled by the DPU (via CAPS software) all high-voltage power converters have zero potential. When high-voltage converters are enabled a safe/arm connector on the DPU can be used during ground handling to limit high voltages to approximately 3 per cent of what the DPU has commanded them to be. The wax thermal actuators (WTAs) in the IMS cover release mechanism and the scan motor launch latch are driven and switched by a WTA driver in the DPU. The DPU uses two nearly identical CPU boards each containing its own RAM, ROM and PACE 1750a microprocessor. CAPS communicates with the command and data subsystem via a bus interface unit (BIU) that is electrically, mechanically and thermally accommodated within the DPU. All CAPS low-voltage power converters are housed in the DPU.

Power at various voltages is supplied for use by the electronics boards in the DPU, including the BIU, and by the amplifiers, D/A converters and other circuitry housed in the IMS, ELS, IBS and ACT. The CAPS supplemental heater (controlled by the CAPS) and replacement heater (controlled by the spacecraft) are mounted to the inside of the DPU's top plate. The CAPS radiator mounts to the back plate of the DPU. Along with the actuator, the DPU box forms the principal structure of CAPS. The IMS, ELS, IBS and the high-voltage power supplies mount to the top plate of the DPU. CAPS mounts to the spacecraft via the actuator which is mounted to the bottom plate of the DPU. The fifth major subassembly of the CAPS instrument is the high-voltage power supply. Two redundant power supplies, HVU1 and HVU2, mount to the top plate of the DPU. The last subassembly of CAPS is the actuator (ACT). The ACT rotates the CAPS instrument at a steady rate over a maximum range of 184° with the acceleration/deceleration over a further 12° (minimum) at either end of this range. The steady rate and acceleration/deceleration range can be adjusted in-flight and the 216° total range of movement will be limited by hard stops to prevent any "wraparound" effects on the CAPS interface cables.

VIMS Visual Infra-red Mapping Spectrometer (320 channels; 40.0 kg, facility instrument). The VIMS comprises a pair of imaging grating spectrometers designed to measure reflected and emitted radiation from atmospheres, rings, and surfaces over wavelengths from 0.35–5.1 micrometers to determine their composition, temperatures and structures. Scientific objectives are: to map the temporal behaviour of winds, eddies and other features on Saturn and Titan; to study the composition and distribution of atmospheric and cloud species on Saturn and Titan; to determine the composition and distribution of the icy satellite surface materials; to determine temperatures, internal structure and rotation of Saturn's deep atmosphere; to study the structure and composition of Saturn's rings; to search for lightning on Saturn and Titan and for active volcanism on Titan; to observe Titan's surface. The sensing instrument has a visible channel (VIMS-V) of 0.35–1.07 μm (96 channels), 32 × 32 mrad fov; and an infrared channel (VIMS-IR) of 0.85–5.1 μm (256 channels), 32 × 32 mrad fov. The instrument weighs 37.14 kg and has a peak operating power of 27.2 W, average operating power of 21.83 W, peak data rate of 182.784 kbps and has dimensions of 78 × 76 × 55 cm. The VIMS subsystem is organised into two assemblies: the optical pallet assembly (OTA) and the main

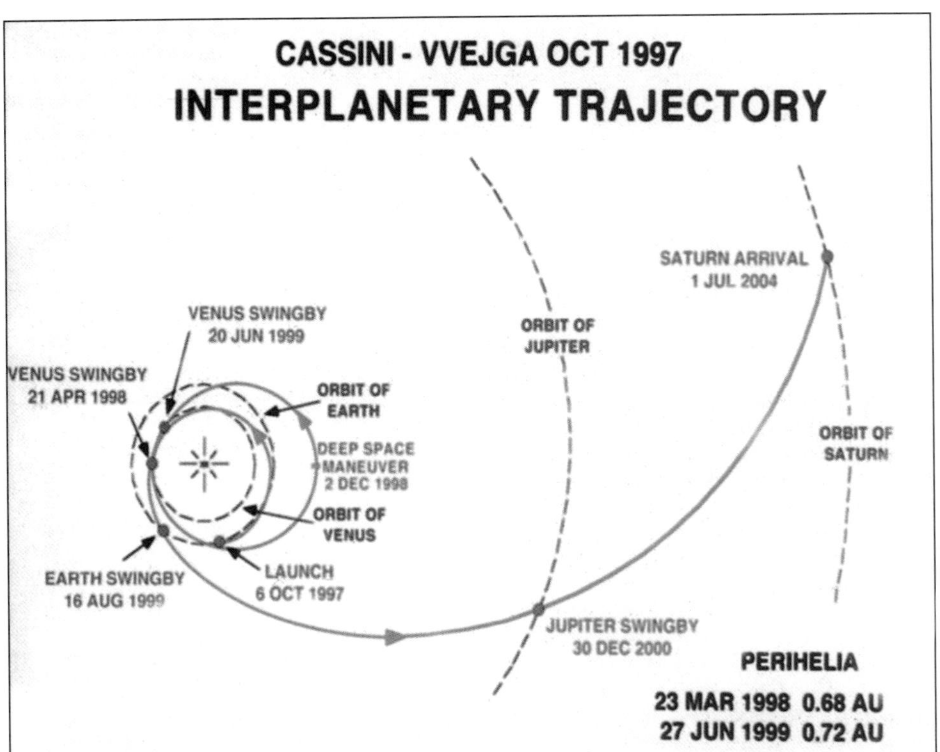

Interplanetary trajectory (an image) 0085785

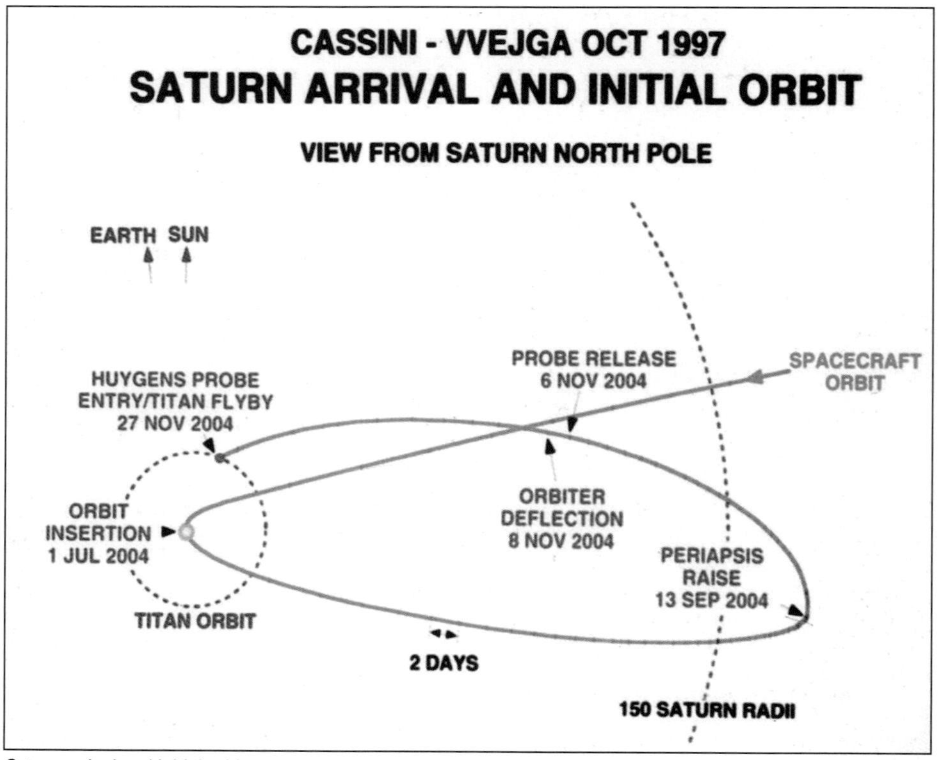

Saturn arrival and initial orbit 0085787

electronics assembly (MEA). The OPA consists of: the infrared channel, the visible channel, the visible channel electronics and the signal processing electronics and has one mechanical interface with the spacecraft and all electrical interfaces are via the main VIMS electronics. The infrared channel (VIMS-IR) is an opto-mechanical subassembly designed to produce multispectral images in the IR range and consists of a Cassegrain telescope, a conventionally ruled spectrometer grating and 256-element linear array focal plane assembly cooled to its required operating temperature by a passive radiator.

The VIMS-IR is configured as a whisk-broom scanning imager where the optical instrument's instantaneous fov is a single pixel. A two-dimensional image is created by scanning along a row of pixels, dropping down a row, scanning that row, etc., using a two-axis scanning mirror. The visible channel (VIMS-V) is an opto-mechanical assembly designed to produce multispectral images in the visible range and consists of a Shafer telescope, a holographic spectrometer grating and a silicon CCD area array focal plane detector cooled to its required temperature by a passive radiator and is configured as a push-broom imager where the optical instruments ifov is an entire line of pixels. This is scanned over the scene with a single-axis scanning mirror to produce a series of contiguous rows which together form a two-dimensional image. The VIMS main electronics is a single assembly that synchronises the visible and IR channel mirror scanning mechanisms and controls the acquisition of data according to the selected operational hardware configuration. The ME serves as the only electronic interface between all elements of the VIMS instrument and the spacecraft including power, data, command and telemetry. During flight operations the ME serves as the interface between ground operations (via the spacecraft) and the VIMS subsystem. During the first 180 after launch all VIMS decontamination heaters were on to prevent contamination from outgassing products and during cruise conditions the VIMS is turned on once for monitoring of instrument health and performance and at other times turned off. During off times decontamination heaters are used periodically to keep the optics and radiator surfaces clean; at Saturn VIMS will be used as determined by spacecraft command sequencing to acquire the desired science data.

ISS Imaging (65.1 kg, 250 mm fl wide angle, 2,000 mm narrow angle, 1,024 × 1,024 element CCDs; facility instrument). The Imaging Science Subsystem is a remote sensing instrument that captures images in visible light and some in infrared and ultraviolet light and has a camera that can take broad, wide angle, pictures and a camera that can record small areas in fine detail and the ISS is expected to return hundreds of thousands of images of Saturn and its rings and moons. Designed for exploring the entire Saturnian system, the ISS performed well in delivering unusual images back to earth during its Jupiter fly-by. The ISS consists of a wide angle camera with angular resolution of 60 microradians per pixel and a narrow angle camera with angular resolution of 6.0 microradians per pixel, the sensors being 1,024 × 1,024 CCD arrays. Science objectives are: to map the three-dimensional structure and motions within the Saturn/Titan atmosphere; to study the composition, distribution and physical properties of clouds and aerosols; to investigate scattering, absorption and solar heating within the Saturn/Titan atmospheres; to search for evidence of lightning, aurorae, airglow and planetary oscillations; to study the gravitational interactions between the rings and Saturn's satellites; to determine the rate and nature of energy and momentum transfer within the rings; to determine ring thickness and sizes, composition and physical nature of ring particles; to map the surfaces of the satellites (including Titan) to study their geological histories; to determine the nature and composition of the icy satellite surface materials; and to determine the rotation of the icy satellites. The ISS sensing instruments include a wide angle camera (WAC), incorporating a 20 cm f/3.5 refractor (380–1100 nm) with 18 filters and a 3.5° × 3.5° fov, and a narrow angle camera (NAC), incorporating a 2 Metre f/10.5 reflector (200–1100 nm) with 24 filters and a 0.35 × 0.35° fov. The ISS weighs 57.83 kg with a peak operating power of 55.9 W, a peak data rate of 365.568 kbps with dimensions of 95 × 40 × 33 cm (NAC) and 55 × 35 × 33 cm (WAC). The orbiting imaging experiments encompass a wide variety of targets including Saturn, the rings, Titan, the icy satellites and star fields, and a wide range of observing distances for various scientific purposes. The science objectives include studying the atmospheres of Saturn and Titan, the rings of Saturn and their interactions with the planet's satellites, determining the surface characteristics of the satellites including Titan. The NAC obtains high resolution images of targets of interest while the WAC provides a different scale of image resolution and more complete spatial coverage but the NAC is also used to obtain optical navigation images for the mission with the WAC acting as a functionally redundant backup unit for this purpose. The cameras are CCD imagers that differ in the design of the optics. The NAC has a focal length of 2000 mm and the WAC, which uses optics inherited from the Voyager mission, has a focal length of 200 mm; the cameras each have a focal plane shutter of the type used on both Voyager and Galileo and they have a two-wheel filter-changing mechanism derived from the Hubble Space Telescope Wide Field/Planetary Camera (WF/PC) design. The CCD detector is cooled to suppress dark current (residual current in the CCD beyond that released by incident light), which is dependent upon temperature; it is also shielded from ionising radiation. The CCD detector

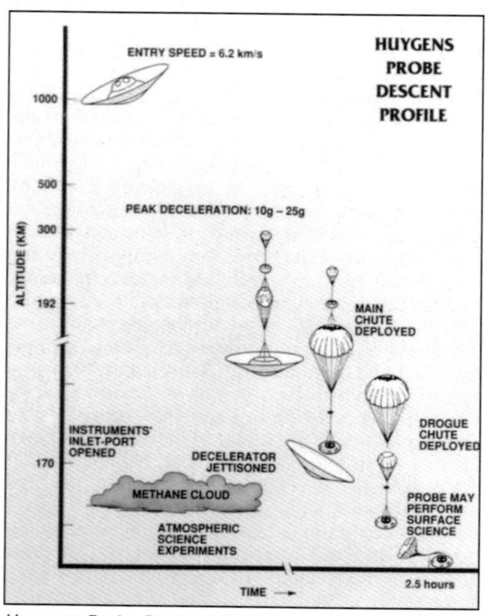

Huygens Probe Descent Profile 0085786

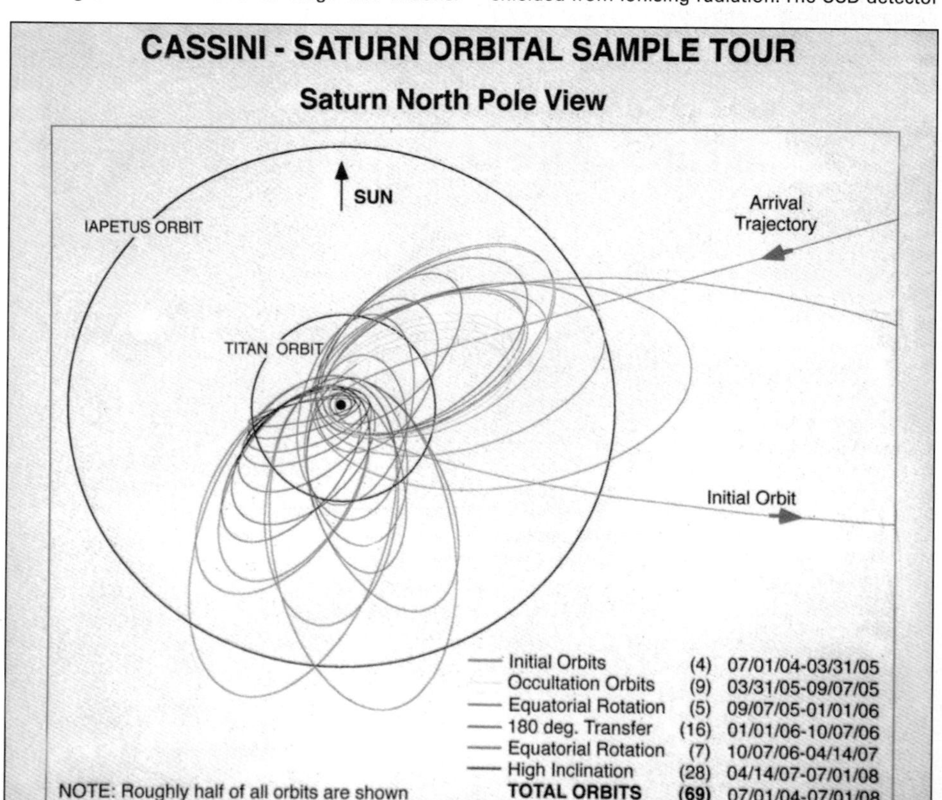

Cassini-Saturn Orbital Sample Tour 0085788

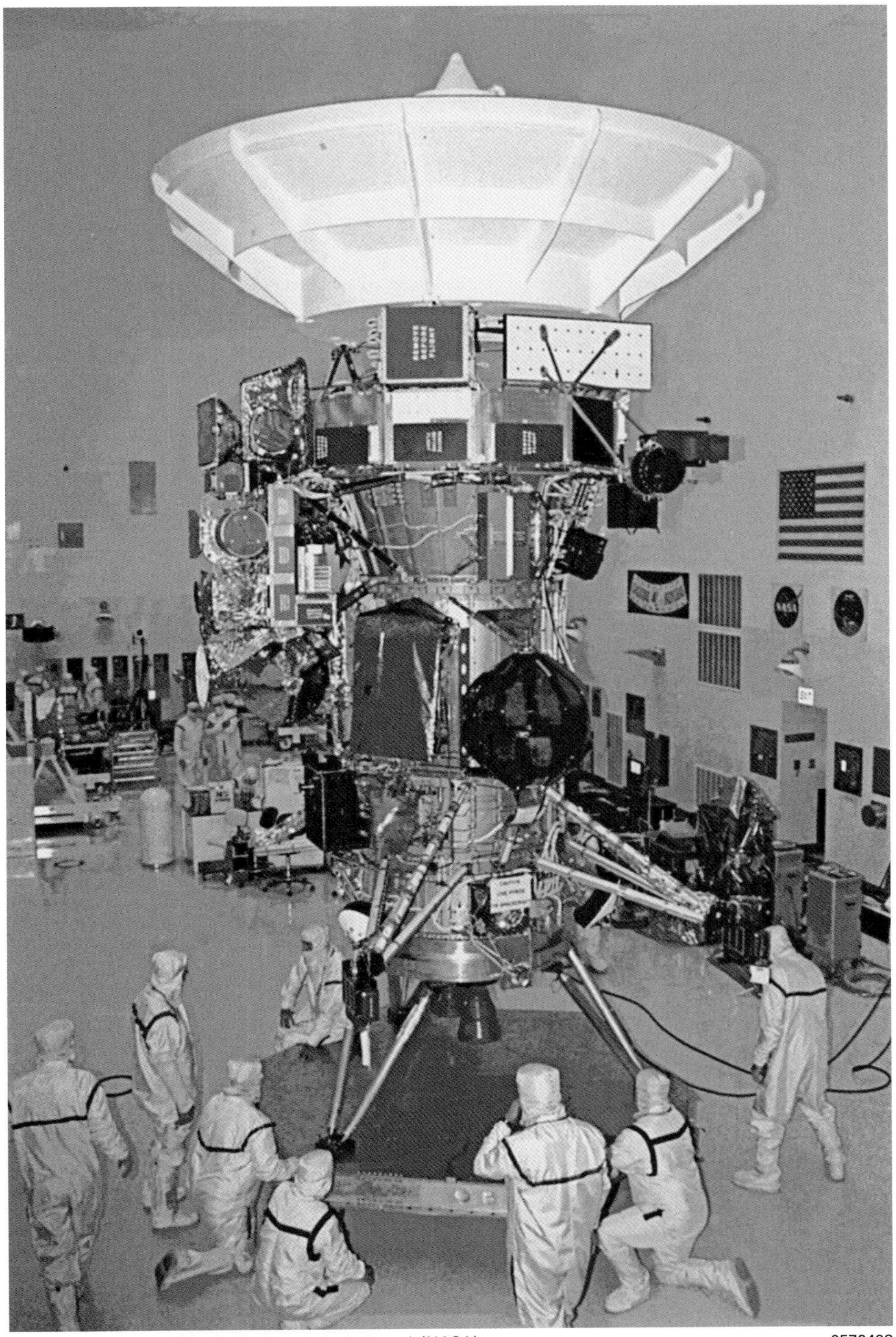

The Cassini spacecraft is checked out prior to launch (NASA) 0572498

design is a square array of 10,242 pixels, each pixel 12 μm on a side and the IC chip uses three-phase, front-side illuminated architecture with a coating of lumogen phosphor to provide ultraviolet response. The detector is passively cooled by a radiator to approximately 10° C below its normal operating temperature (approximately –90° C) and then it is controlled to the operating temperature by a proportional control heater. To minimise radiator size and heater power, the detector/radiator combination is thermally isolated from the rest of the camera head assembly.

The entire NAC is thermally isolated from the remote sensing pallet on which it is mounted in order to minimise the effects of the pallet's thermal variations on NAC image quality; being an inherited design with less stringent imaging requirements, the WAC is not thermally isolated. The electronics for each camera are identical and all ISS command and telemetry functions are handled by the electronics, including receipt of commands from the Command and Data Subsystem, expansion of commands and collection and transmission of imaging data and telemetry to the CDS. The ISS controls the amount of power it draws from the spacecraft during operations and to accomplish this the profile of ISS command timing is structured to reduce the power the ISS requires for certain internal functions (e.g., shutter or filter wheel movement) and when the filer is moving the power from the optical heater (if present) in the active camera is turned off. When the movement is complete the optical heater is turned on, if needed, but simultaneous filter positioning within a single camera, either the WAC or NAC, is not permitted. During cruise the cameras are periodically turned on for maintenance, calibration and monitoring of instrument health and performance but other than these specified times the ISS is off and replacement heaters are on with decontamination/radiation heater number 1 on throughout most of the cruise portion of flight. When Cassini arrives at Saturn the cameras will be turned on most of the time with spacecraft power limitations the controlling parameter determining whether the ISS will be turned off or put into a low-power state. During the Saturn tour the high activity for periods for imaging Saturn and its rings will be clustered around periapsis but for the satellites the high activity periods will be when the spacecraft is closest to them and at these times high resolution images of all targets will be acquired through various camera filters and the data stored in the spacecraft solid-state recorder; during low activity periods, long term atmospheric and ring monitoring will take place and ISS calibrations will be performed.

RADAR (58.0 kg, 13.8 GHz, 120 W raw power, employs HGA; facility instrument). The Radio Detection and Ranging Instrument is a remote active and remote passive sensing instrument that is designed to produce maps of Titan's surface and measures the height of surface objects. The RADAR will facilitate imaging, altimetry and radiometry. Each mode allows for the collection of different types of data from straightforward imaging and 3-D modelling to passive collection of information such as recording energy emanating from the surface; an important aspect is the opportunity it affords to observe and map the synchrotron emissions at a new frequency (13.8 GHz) not possible from earth-based telescopes, this opportunity stemming from the Cassini radiometer's ability to separate atmospheric thermal emission from synchrotron emission. The Cassini RADAR experiment uses the five-beam Ku-band antenna feed assembly associated with the spacecraft high gain antenna to direct radar transmissions toward targets and to capture blackbody radiation and reflected radar signals from targets. Scientific objectives aim: to determine whether oceans exist on Titan and if so to determine their distribution; to investigate the geologic features and topography of the solid surface of Titan; to acquire data on the rings and icy satellites as conditions permit. The RADAR sensing instruments include: Synthetic Aperture Radar (SAR) imager operating at the 13.78 GHz Ku-band with 0.35–1.7 km resolution; Altimeter operating at 13.78 GHz with 24–27 km horizontal and 90–150 m vertical resolution; Radiometer with 13.78 GHz passive Ku-band with 7–310 km resolution. The instrument has a mass of 41.43 kg with peak operating power of 108.4 W and a peak data rate of 364.8 kbps. The RADAR will be used to investigate the surface of Saturn's moon Titan by taking four types of observations: imaging, altimetry, backscatter and radiometry. In the imaging mode the radar will bounce pulses of microwave energy off the surface of Titan from different incidence angles and record the time it takes the pulses to return to the spacecraft and when converted to distance these measurements will allow the construction of visual images of the target surface. In the backscatter mode the RADAR will act as a scatterometer by bouncing pulses off Titan's surface to measure the intensity of the energy returning for a determination of the surface composition. In the radiometry mode the RADAR will act as a passive instrument recording emanations for information on latent heat in the atmosphere, a factor that has an impact on the precision of the other measurements taken by the instrument.

During imaging, altimetry and backscatter operations the RADAR instrument will transmit linear frequency-modulated Ku-band pulsed signals toward the surface using the HGA (High Gain Antenna) and these signals will be captured by the RADAR Radio Frequency Electronics Subsystem (RFES); during radiometry operations the instrument will not transmit radio signals but the HGA will again be used for radiometric observations. To improve surface coverage by radar imaging a switched multiple Ku-band antenna feed array structure is part of the HGA and permits the formation of five antenna beam patterns, each beam of which will have a different pointing angle relative to the antenna reflector's focal axis. The major functional components of the RADAR subsystem are the Radio Frequency Electronic Subsystem (RFES), the Digital Subsystem (DS) and the Energy Storage Subsystem (ESS). The RFSE comprises: the Frequency generator which contains an ultra-stable oscillator that is the system timing source for the RADAR instrument; the digital chirp generator, which generates the low-power, baseband frequency, modulated pulse upon request from the RADAR Digital Subsystem; the chirp up-converter and amplifier (CUCA) which converts the baseband chirp pulse to Ku-band and provides the up-converted pulse to the high-power amplifier; the high-power amplifier which receives a low-power Ku-band chirp pulse from the CUCA and amplifies that pulse to the required power level for transmission; the front-end electronics (FEE), which routes the high power transmission pulses, the returning low energy echoes and radiometric signals (the FEE receives the high power pulse from the HPA and routes the signal to one of five different antenna ports on the RFES via an antenna switch module; the echo returns and radiometric signals are routed from one of the five antenna ports to the RFES microwave receiver – the FEE also steers the selected calibration signal to the microwave receiver during periods of calibration mode operation); the microwave receiver which receives signals at Ku band and down-converts these to baseband so that they can be properly sampled, the source of these signals being the echo returns, the radiometric signals and the calibration signals routed through the FEE (the MR receives the re-routed chirp calibration signal from the CUCA and passes that signal to the FEE for proper routing but the MR is also the source of the noise diode calibration signal that is provided to the FEE for routing, the MR gain and bandwidth information being provided to the MR from the DSS); and the RFES power supply which converts the (approximately 30 vdc input from the Power and Pyrotechnic Subsystem to the required voltages for the RFES.

The Digital Subsystem performs three principal functions: reception and depacketisation of RADAR commands from the Command and Data Subsystem (CDS), configuration control and timing signal generation for RADAR and the packetisation of RADAR housekeeping data and science data for transfer to the CDS. DSS subassemblies are contained within a spacecraft bay and are supported in shear by shear plates and the top and bottom rings of the Cassini spacecraft bus. Electronic harnesses, which face inboard on the spacecraft and be supported by the inboard shear plate, are used to provide interconnections between the RADAR subassemblies and the spacecraft.

The DSS uses two primary modes of heat transfer on its design: the conduction of heat from the electronic components to the subchassis and the outboard shear plate, and the radiation of heat from the outboard shear plate to the space environment; high power heat dissipation components are mounted on a special heatsink bracket, which is bolted directly to the outboard shear plate to optimise heat transfer (thermal components are applied between the components and the heatsink to minimise contact thermal resistance). The DSS consists of: the bus interface unit (BUI), the interface between RADAR and the CDS (on the RADAR side the BIU interfaces to the flight computer unit for command, software and data transfers); the flight computer unit (FCU) which receives commands and software from the CDS and sends data and status to CDS by the BIU, depacketises the commands and provides the RADAR configuration and timing information to the control and timing unit (it also receives housekeeping values in a predetermined order from the low-speed A/D converter and packetises the housekeeping and science data to be passed to the CDS by way of the BIU – in addition the FCU receives spacecraft time broadcasts and RADAR software uploads from CDS by the BIU) – the FCU is built around an engineering flight computer (EFC) with additional banks of ROM and internal interface circuitry; the control and timing unit (CTU) which controls the hardware configuration and the timing control signals within RADAR and the parameters for determining RADAR configuration are passed to the CTU from the FCU, the CTU providing the configuration and timing control signals to the RFES and to other portions of the DSS (the CTU is responsible for updating to millisecond resolution the spacecraft time received from the CDS); the signal conditioning unit (SCU) which consists of a science data buffer and high- and low-speed A/D converters comprising the science data buffer (SDB) which exists to receive and store the high-rate digital science data from the high-speed A/D converter during the proper receive window period (as determined by the CTU) and then to provide this data upon request to the FCU at a slower rate while the high-speed A/D converter digitises the imaging data output from the RFES microwave receiver and provides the data to the SDB for buffering (the low-speed A/D converter performs two tasks: it digitises the analogue housekeeping telemetry values from throughout RADAR at predetermined times and provides these digitised values to the FCU on request and also digitises the radiometer output from the RFES microwave receiver and provides those values to the FCU on request); the DSS power supply which converts the (approximately) 30 vdc input from the Power and Pyrotechnic Subsystem to the voltages required for the DSS; the RADAR energy storage subsystem (ESS) which converts the 30 vdc input from the PPS to a higher voltage, stores energy in a capacitor bank and provides regulated voltage to the high-power amplifier (HPA) of the RFES – the ESS subassemblies are contained within a spacecraft bay and are supported in shear by shear plates and the top and bottom rings of the Cassini spacecraft bus, high strength fasteners being used to tie the electronics assemblies to the spacecraft and supported by the inboard shear plate used to provide interconnections between the RADAR assemblies and the spacecraft (the ESS uses two primary modes of heat transfer in its design identical to that engineered into the Digital Subsystem outlined above).

The ESS consists of boost circuitry, the capacitor bank and a buck regulator: the boostcircuitry increases the 30 vdc input power to approximately 85 vdc for more efficient storage by the capacitor bank, soft-start circuitry limiting the current drawn from the power source and an input voltage filter prevents electromagnetic interference from being conducted back into the source; the capacitor bank stores energy to supply to the buck regulator and the HPA during RADAR pulse bursts, the capacitor bank voltage dropping during each burst which returns to normal before the next burst; the buck regulator regulates the varying capacity bank voltage for the HPS.

Ion Neutral Mass Spectrometer (10.8 kg facility instrument). The Ion and Neutral Mass Spectrometer (INMS) is intended to measure positive ion and neutral species composition and structure in the upper atmosphere of Titan and magnetosphere of Saturn and to measure positive ion and neutral environments of Saturn's icy satellites and rings, study the interaction of Titan's upper atmosphere with the magnetosphere and the solar wind and to measure ion and neutral species composition during ring plane crossings. The sensing instruments include a closed source (neutrals only), an open source (ion and neutrals), a lens system and a quadrupole mass analyser and detector, the INMS having a mass of 9.25 kg with average operating power of 27.7 W and an average data rate of 1.5 kbps. The open ion source produces ions by ionising neutral gases and includes an ion trap/deflector that forms trapped ions into a beam which minimises interaction effects between the gas environment and the open source surface as the source directly samples the gaseous species. The closed source also produces ions by ionising neutral gas and uses ram density enhancement to provide measurements of higher accuracy and sensitivity for the more inert atomic and molecular species than provided by the open ion source, achieved by maintaining a high input flux to an enclosed antechamber and then limiting the gas conductance or output from the antechamber by the use of an orifice. Ions are directed to the mass analyser from the selected ionsource by changing the potentials on a 90° quadrupole deflector that allows both sources of ions to be sequentially switched into a common exit lens systems. The quadrupole mass analyser consists of four precision ground hyperbolic rods mounted in a rigid mechanical assembly. The transmitted mass, the resolution and the ion transmission are controlled by variations in RF and dc electric fields between adjacent rod pairs while opposite rod pairs are kept at the same potential. The ion dual detector system amplifies and detects the input from the mass analyser by the use of two continuous dynode multipliers.

Cosmic Dust Analyser (15.1 kg, Max Planck Institute für Kernphysik, Germany). The Cosmic Dust Analyser is designed to provide direct observations of particulate matter in the Saturnian system, to investigate the physical, chemical and dynamical properties of these particles and to study their interactions with the rings, icy satellites and the magnetosphere of Saturn. Specific science objectives being to: extend studies of interplanetary dust (sizes and orbits) to the orbit of Saturn; define dust and meteoroid distribution (size, orbits, composition) near the rings; map the size distribution of ring material in and near the known rings; analyse the chemical composition of ring particles; study processes (erosional and electromagnetic) responsible for the E ring structure; search for ring particles beyond the known E ring; study the effect of Titan on the Saturn dust complex; study the chemical composition of icy satellites from studies of ejecta particles; determine the role of icy satellites as a source for ring particles; determine the role of icy satellites as a source for ring particles; determine the role that dust plays as a magnetospheric charged particle source/sink.

The instruments comprise a dust analyser (DA) and a high rate detector (HRD) with a total experiment mass of 16.36 kg, a peak operating power of 18.38 W, an average operating power of 11.38 W, a peak data rate of 0.524 kbps and an overall size of 81 cm × 67 cm × 45 cm. The four major functional elements of the CDA are the dust analyser, the main electronics, the articulation mechanism and the high-rate detector assembly. The dust analyser consists of: four charge pick-up grids; a hemispherical target; an ion collector; an electron multiplier; and the sensor electronics. The charge pick-up grids collect the initial impact particles and are mounted at the entrance of the sensor; the hemispherical target is divided into two parts – a ring shaped impact ionisation target and a chemical analyser target in the middle of the ionisation target – and the chemical analyser has an acceleration grid mounted 3 mm in front of it; the ion collector has a grid that is negatively biased in order to collect the positively charged plasma ions produced at the impact ionisation target; the electron multiplier is located in the centre of the hemispherical ion collector target and amplifies the signal produced by ions capable of penetrating the ion collector grid (ions originate from plasma produced by particle impact either on the impact ionisation target or the chemical analyser target) and the output signal from the multiplier differs depending upon the target from which the impacts are being measured; the sensor electronics are contained in an electronics box attached to the DA sensor chassis and among other components this box contains charge-sensitive amplifiers that measure the signals from all the grids in the DA.

The CDA main electronics includes amplifiers and transient recorders, a control and timing unit, a microprocessor unit, a bus interface unit, a power input circuit, a low-voltage converter and a housekeeping system. All CSA and electron multiplier signals are separately amplified by logarithmic amplifiers and then digitised by an analogue-to-digital converter, the data being stored on transient recorders with only the recorder connected to the pick-up grids operated continuously, all others being activated by a signal detected at a target or the acceleration grid. The control and timing unit stores and decodes information received from the microprocessor and produces all timing and synchronisation signals required for instrument operation. The microprocessor samples and collects the buffered measurement data, co-ordinates the subsystem measurement cycle, controls the instrument operating modes, processes the data according to a programme loaded in its memory and outputs data to the spacecraft on reques through the bus interface unit (BIU), which is the interface circuit between the spacecraft and the microprocessor and is powered by the CDA instrument.

The power input circuitry is the interface with the spacecraft power and pyrotechnics subsystem and contains a filter circuit and a regulator to produce a DC voltage to feed the low-voltage converter, which provides different regulated low voltages for the high-voltage converters. The converters are synchronised to the 100 kHz clock provided through the BIU from the command and data subsystem. The CDA housekeeping system is a data system that multiplexes, digitises and stores information on the instrument current, the low voltages, the high voltages and the temperature measurements. The articulation mechanism allows the entire CDA instrument, including the high-rate detectors, the dust analyser, the main electronics and the articulation mechanism electronics to be rotated or repositioned with respect to the spacecraft co-ordinate system. The high-rate electronics HRDs) are two redundant independent sensors, the electronics for these being contained in the HRD electronics box and each sensor having its own electronics independent of the other sensor. The HRD is operated in two modes: "normal" mode and "calibrate". In normal mode the operational HRD continuously collects dust particle data; in the calibrate mode a calibration cycle is initiated which consists of a sequence of pulses sent to the HRD by the in-flight calibrator to verify the stability of the electronics.

Radio Plasma Wave Spectrometer (37.4 kg, University of Iowa). The major functions of the Radio and Plasma Wave Science (RPWS) instrument are to measure the electric and magnetic fields and electron density and temperature in the interplanetary medium and planetary magnetospheres. The scientific objectives of the mission are to: study the configuration of Saturn's magnetic field and its relationship to Saturn kilometric radiation (SKR); to monitor and map the sources of the SKR; to study daily variations in Saturn's ionosphere and search for outflowing plasma in the magnetic cusp region; to study radio signals from lightning in Saturn's atmosphere; to investigate Saturn's electrical discharges; to determine the current systems in Saturn's magnetosphere and study the composition, sources and sinks of magnetospheric plasma; to investigate the dynamics of the magnetosphere with the solar wind, satellites and rings; to study the rings as a source of magnetospheric plasma; to look for plasma waves associated with ring spoke phenomena; to determine the dust and meteoroid distributions throughout the Saturnian system and interplanetary space; to study waves and turbulence generated by the interaction of charged dust grains with the magnetospheric plasma; to investigate the interactions of the icy satellites and the ring systems; to measure electron density and temperature in the vicinity of Titan; to study the ionisation of Titan's upper atmosphere and ionosphere and the interactions of the atmosphere and exosphere with the surrounding plasma; to investigate the production, transport and loss of plasma from Titan's upper atmosphere and ionosphere; to search for radio signals from lightning in Titan's atmosphere, a possible source for atmospheric chemistry; to study the interaction of Titan with the solar wind and magnetospheric plasma; to study Titan's vast hydrogen torus as a source of magnetospheric plasma; to study Titan's induced magnetosphere. The RPWS sensing instruments include an electric field sensor, a magnetic field sensor, a Langmuir probe and high, medium and wide band receivers, the collective suite weighing 37.45 kg with an average operating power of 7 W and a data rate of 0.9 kbps.

The major components of the RPWS subsystem are the electric field sensor, the magnetic search coil sensor assembly, the Langmuir probe sensor assembly and the instrument main electronics. The electric field sensor is made up of three deployable antenna elements, an associated preamplifier and antenna deployment mechanism drive electronics, the antennas composed of interlocking sections made from beryllium copper and each antenna element is deployable separately to 10 metres with its own 400-Hz AC motor. The electric field preamplifier is used to add gain to the output signals from the antennas; the antenna deployment mechanism electronics convert +/–15 volt primary power to 400-Hz AC power for the antenna drive motors. The magnetic search coil sensor assembly is composed of a triaxial sensor assembly and an associated preamplifier. The triaxial sensor consists of three orthogonal metallic alloy cores with two sets of windings each, one to produce flux in the core and another to detect the flux. The magnetic search coil amplifier adds gain to the output signal from the sensor assembly. The Langmuir probe sensor assembly consists of a sensor, a preamplifier and associated control electronics, the probe sensor consisting of a 5 cm diameter sphere located at the end of a rod approximately 1 metre in length. The sensor rod is folded in a stowed state until deployment in flight. The probe sensor preamplifier adds gain to the output from the probe.

The RPWS main electronics includes a digital data processing unit, a high frequency receiver, a wideband receiver, a medium frequency receiver, a low frequency five-channel waveform receiver, the Langmuir probe bias circuitry and a power converter. The data processing unit controls all instrument functions and handles all communications with the orbiter; it contains a large block of RAM to be used as waveform storage for the five-channel waveform receiver. Software in the DPU is used to enhance the scientific return of the instrument by performing various analysis and data compression operations. The high-frequency receiver is a digital waveform processor that operates by digitising a portion of the bandwidth received from the electric field sensor antennas and deriving spectral and waveform vector information from the waveforms using digital signal processing techniques. The wideband receiver obtains very high resolution electric or magnetic field waveforms for selected time intervals that vary from under a minute to as much as an hour or more. The receiver has two selectable passbands: 50 Hz to approximately 10 kHz and 10 kHz to approximately 80 kHz. The wideband receiver uses high-rate telemetry to transfer waveform information in a given bandwidth from a selected sensor directly. The input signal is selectable from one of five inputs: two electric, one magnetic, a frequency converted output from the high frequency receiver and the Langmuir probe. The medium frequency receiver provides spectrum measurements over the frequency range from 25 Hz to 12.6 kHz, this receiver being attached to one of four sensor inputs (two electric and two magnetic) and uses double frequency conversion to convert the input bandwidth down to a low frequency constant frequency band where it is detected by an amplitude detector. The low frequency five-channel waveform receiver provides high resolution spectral measurements of electric and magnetic fields over the frequency range from 0.1 Hz to 2.5 kHz and provides simultaneous waveforms from all five antennas (three magnetic axes and two electric axes). This receiver captures blocks of waveform data simultaneously from the five sensors and maintains a high degree of phase and amplitude accuracy. The data is processed through five parallel amplifier and filter channels. The bias converter converts DC power from the spacecraft power supplies to ac power to operate the instrument, the conversion frequency fixed at 100 kHz by locking onto a signal from the spacecraft bus interface unit.

Ultra Violet Imaging Spectrograph (14.9 kg, University of Colorado). The Ultra Violet Imaging Spectrograph (UVIS) is a set of detectors designed to measure ultraviolet light reflected or emitted from atmosphere, rings and surfaces over wavelengths from 55.8 to 190 nanometres to determine their compositions, distribution, aerosol content and temperatures. Science objectives aiming to: map the vertical/horizontal composition of Titan's and Saturn's upper atmosphere; determine the atmospheric chemistry occurring in Titan's and Saturn's atmospheres; map the distribution and properties of aerosols in Titan's and Saturn's atmospheres; infer the nature and characteristics of circulation in Titan's and Saturn's atmospheres; map the distribution of neutrals and ions within Saturn's magnetosphere; study the radial structure of Saturn's rings by means of stellar occultations; study surface ices and tenuous atmospheres associated with icy satellites. The UVIS contains four separate instruments: Far Ultraviolet Spectrograph (FUV) in the 110–190 nm range [0.75, 1.5, 6] × 64 mrad fov; Extreme Ultraviolet Spectrograph (EUV) in the range 55.8–118 nm [1, 2, 6] × 64 mrad fov; High Speed Photometer (HSP) in the 115–185 nm 6 × 6 mrad fov, 2 msec time res; Hydrogen-Deuterium Absorption Cell (HDAC) at 121.5 nm, 55.8 circular fov. The UVIS has a mass of 14.46 kg with peak operating power of 11.83 W, peak data rate of 32.096 kbps and dimensions of 48 cm × 30 cm × 23 cm. The UVIS will measure the fluctuations of starlight and sunlight as the sun and stars move behind the rings and the atmospheric concentrations of hydrogen and deuterium, data which will be used for studies of the atmospheres, the magnetospheres and the rings of the Saturnian system. The UVIS has two channels, the extreme UV channel and the far UV channel; the UV channels are built into weight relieved aluminium cases and each contains a reflecting telescope, a concave grating spectrometer and an imaging pulse-counting detector.

The UVIS also includes a high-speed photometer channel, a hydrogen-dueterium absorption cell channel and an electronics and control subassembly. The Extreme Ultraviolet Channel (EUV) will be used for imaging spectroscopy and spectroscopic measurements of the structure and composition of the atmospheres of Titan and Saturn. The EUV consists of a telescope with a three-position slit changer, a baffle system and a spectrograph with CODACON microchannel plate detector and associated electronics. The telescope consists of an off-axis parabolic section with a focal length of 100 mm, a 22 mm × 30 mm aperture and a baffle with a fov of 3.67° × 0.34°; a precision mechanism positions one of the three entrance slits at the focal plane of the telescope, each translating to a different spectral resolution. The spectrograph uses an aberration-corrected toroidal grating that focuses the spectrum onto an imaging microchannel plate detector to achieve both high sensitivity and spatial resolution along the channel slit. The microchannel plate detector consists of a low-voltage power supply, a programmable high-voltage power supply, charge sensitive amplifiers and associated logic. The EUV channel also contains a solar occultation mechanism to allow solar flux to enter the telescope when the sun is still 20° off-axis from the primary telescope. The far ultraviolet channel (FUV) is used for imaging spectroscopy and spectroscopic measurement of the structure and composition of the atmospheres of Titan and Saturn and of the rings and is similar to the EUV channel except for the grating ruling density, optical coatings and detector details (the FUV electronics are similar to those for the EUV except for the addition of a high-voltage power supply for the ion pump). The high-speed photometer channel (HSP) performs stellar occultation measurements of the structure and density of material in the rings and resides in its own module measuring undispersed light from its own parabolic mirror with a photomultiplier tube detector. The electronics consists of a pulse-amplifier-discriminator and a fixed-level high-voltage power supply. The hydrogen-deuterium absorption cell channel (HDAC) will be used to measure hydrogen and deuterium in the Saturn system using a hydrogen cell, an oxygen cell, a deuterium cell and a channel electron multiplier detector to record photons not absorbed in the cells. The hydrogen and deuterium cells are resonance absorption cells filled with pure molecular hydrogen and deuterium and are located between an objective lens and a detector. Both cells are made of stainless steel coated with Teflon and are sealed at each end with MgF2windows. The electronics consists of a pulse-amplifier-discriminator, a fixed-level high-voltage power supply and two filament current controllers. The UVIS microprocessor electronics and control subassembly consists of input-output elements, power conditioning, science data and housekeeping data collection electronics and microprocessor control elements.

Magnetospheric Imaging Instrument (24.9 kg, The Johns Hopkins University). The Magnetospheric Imaging Instrument (MIMI) is designed to: measure the composition, charge state and energy distribution of energetic ions and electrons; and to conduct remote imaging of the Saturn's magnetosphere, information which will be used to study the overall configuration and dynamics of the magnetosphere and its interactions with the solar wind, Saturn's atmosphere, Titan, rings and icy satellites. The MIMI is designed to accomplish the following scientific objectives: to determine the global configuration and dynamics of hot plasma in the magnetosphere of Saturn; to monitor and model magnetospheric substorm-like activity and correlate this with Saturn kilometric radiation (SKR) observations; to study magnetosphere/ionosphere coupling through remote sensing of aurora and measurements of energetic ions and electrons; to investigate plasma energisation and circulation processes in the magnetotail of Saturn; to determine through imaging and composition studies the magnetosphere/satellite interactions at Saturn and understood the formation of clouds of neutral hydrogen, nitrogen and water products; to measure electron losses due to interactions with whistler waves; to study the global structure and temporal variability of Titan's atmosphere; monitor the loss rate and composition of particles lost from Titan's atmosphere due to ionisation and pickup; to study Titan's interaction with the magnetosphere of Saturn and the solar wind; to determine the importance of Titan's exosphere as a source for the atomic hydrogen torus in Saturn's outer magnetosphere; to investigate the absorption of energetic ions and electrons by Saturn's rings and icy satellites; to analyse Dione's atmosphere.

The MIMI has a mass of 16 kg, an average operating power of 14 W and an average data rate of 7 kbps. It consists of one set of electronics, the MIMI electronics box, servicing three detector heads that perform the various measurements: the low-energy magnetospheric measurements system (LEMMS), the charge-energy-mass spectrometer (CHEMS), and the ion and neutral camera (INCA). The MIMI electronics box contains the data processing unit (DPU) and the digital processing electronics for all three detector heads. The low-energy magnetospheric measurements system (LEMMS) detector head will measure low- and high-energy proton, ion and electron angular distributions; the LEMMS head is mounted on a scan platform capable of 180° rotations and the platform is mounted so that the rotation axis is orientated perpendicular to the spacecraft X axis so that its extrapolation intersects the spacecraft Z axis. The charge-energy-mass spectrometer (CHEMS) head will measure the charge state and composition of ions in the most energetically important portion of the Saturnian magnetospheric plasma. The ion and neutral camera (INCA) will make two different types of measurements and obtains very high sensitivity the three-dimensional distribution, velocities and rough composition of magnetospheric and interplanetary ions for those regions in which the energetic ion fluxes are very low. The INCA instrument will also obtain remote images of the global distribution of the energetic neutral emission of hot plasmas in the Saturnian magnetosphere, measuring the composition and velocities of those energetic neutrals for each image pixel.

Dual Technique Magnetometer (8.7 kg, Imperial College, UK). The primary objective of the Dual Technique Magnetometer (MAG) is to determine the planetary magnetic fields and the dynamic interactions of the planetary environment. Its science objectives being: to determine the internal magnetic field of Saturn; to develop a three-dimensional model of Saturn's magnetosphere; to determine the magnetic state of Titan and its atmosphere; to derive an empirical model of the Titan electromagnetic environment; to investigate the interactions of Titan with the magnetosphere, magnetosheath and the solar wind; to survey the ring and dust interactions with the electromagnetic environment; to study the interactions of the icy satellites with the magnetosphere of Saturn; to investigate the structure of the magnetotail and the dynamic processes therein. The MAG sensing instruments include a Vector/Scalar Helium Magnetometer and collectively they have a mass of 3 kg, an average operating power of 3.1 W and an average data rate of 3.6 kbps. The Cassini MAG measures magnetic fields during the Titan and Saturn encounters and consists of a vector/scalar helium magnetometer sensor, a fluxgate magnetometer sensor, a data processing unit, three power supplies and operating software and electronics associated with the sensors. The vector/scalar helium magnetometer (V/SHM) sensor is used to make both vector (magnitude and direction) and scalar (magnitude only) measurements of magnetic fields, the electronics being supplied by JPL (the sensor and its electronics are provided by Imperial College, London). The instrument data processing unit (DPU) is provided by the technical university of Braunschweig and interfaces with the spacecraft Command and Data Subsystem through the JPL designed bus interface unit (BIU) which processes all commands, data and processor programme changes. The MAG components are powered by three power supplies plus the 30-volt spacecraft bus; power supply 0 powers the BIU and the DPU, power supplies 1 and 2 are redundant and power the V/SHM electronics (the FGM electronics are powered by the spacecraft bus). The FGM sensor is located midway out on the Cassini magnetometer boom and the V/SHM sensor is located at the end of the boom which, composed of thin, non-metallic, rods is deployed only after the spacecraft has separated from the launch vehicle.

Radio Science Subsystem (13.2 kg). The Radio Science Subsystem (RSS) uses the spacecraft X-band communication link as well as S-band downlink and Ks-band uplink and downlink to study compositions, pressures and temperatures of atmospheres and ionospheres, radial structure and particle size distribution within rings, body and system masses and gravitational waves. The RSS science objectives are: to search for and characterise gravitational waves coming from beyond the solar system; to study the solar corona and general relativity when Cassini passes behind the Sun; to improve estimates of the masses and ephemerides of Saturn and its satellites; to study the radial structure and particle size distribution within Saturn's rings; to determine temperature and composition profiles within Saturn's/Titan's atmospheres; to determine temperatures and electron densities within Saturn's/Titan's ionospheres. RSS transmitting and receiving frequencies: Ka-band (T: 32.02344[NC] or 32.02860[CK] or 32.03377[CX] GHz; R: 34.31636 GHz); X-band (T: 8.4272[NC] or 8.4299[CX] GHz; R: 7.175 GHz); S-band (T: 2.29833[NC] or 2.29907[CX] GHz) The RSS instrument has a mass of 4.38 kg with peak operating power of 80.7 W and consists of a Ka-band TWTA, a translator, an exciter, an S-band transmitter and various microwave components. The Ka-band TWTA consists of a TWT and an electronic power conditioner (EPC). The non-redundant TWT is a signal amplifier which can be commanded into a standby mode for low DC power consumption.

The EPC converts DC power from the Power and Pyrotechnic Subsystem (PPS) to the voltages required to operate the TWT; it can power the TWT in the standby mode or power down the TWT in case of TWT or EPC fault detection (the EPC also supplies engineering telemetry to the RFS and provides direct-access signals. The Ks-band translator (KAT) subassembly receives the 34-GHz uplink carrier from the high-gain antenna and translates it by a factor of 14/15 for retransmission back to the DSN; the phase and phase-shift of the signal are used for the actual science observations and measurements. The KAT contains a power converter that allows it to operate from the 30-volt DC power bus and also supplies engineering telemetry data to the RFS and provides for direct access. The Ka-band exciter (KEX) generates a stable 32-GHz signal and provides an RF power combiner to combine the RF signal generated by the KAT with its own signal and is powered by the 30-volt spacecraft bus which supplies telemetry data to the RFS and provides direct access. The S-band transmitter (SBT) receives a 115-MHz signal from the RFS, multiplies it by 20, amplifies it to 10 watts and supplies the resultant signal at approximately 2290 MHz to the high-gain antenna. This carrier signal is used for radio science experiments and the transmitter contains a power converter to allow operation from the 30-volt power bus and it supplies telemetry data to the RFS and provides direct access. The microwave components consist of two band pass filters (BFPs) and waveguide components, in this subsystem used for all Ka-band microwave component interconnections.

Composite Infra-Red Spectrometer (36.2 kg, NASA Goddard) The Composite Infra-Red Spectrometer (CIRS) evolved from the Voyager Infrared Interferometer Spectrometer (IRIS) exhibiting significant improvements with spectral resolution 10 times higher than the Voyager IRIS, and a larger wavelength with more closely spaced data points greatly increasing the detail seen in the spectrum. The CIRS consists of dual interferometers that measure infrared emission from atmospheres, rings and surfaces with wavelengths from 7 to 1000 micrometers (1400 to 10-cm^{-1}) to determine their composition and temperatures. CIRS science objectives are: to map the global temperature structure within Titan's and Saturn's atmospheres; to map the global composition within Titan's and Saturn's atmospheres; to map global information on hazes and clouds within Titan's and Saturn's atmospheres; to collect information on energetic processes within Titan's and Saturn's atmospheres; to search for new molecular species within Titan's and Saturn's atmospheres; to map the global surface temperatures at Titan's surface; to map the composition and thermal characteristics of Saturn's rings and icy satellites. Sensing instruments include: Far-Infrared Focal Plane [FP1] (16.67 to 1000 μm; 4.3 mrad circular fov); Mid-Infrared Focal Plane [FP3] (9.09 to 16.67 μm; 1 × 10 array of 0.273 mrad squares); Mid-Infrared Focal Plane [FP4] (7.16 to 9.09 μm; 1 × 10 array of 0.273 mrad squares). The CIRS instrument has a mass of 39.24 kg, a peak operating power of 32.89 W, an average operating power of 26.37 W and a peak data rate of 6 kbps. The CIRS with its 50 cm diameter telescope has dimensions of 89 cm × 76 cm × 52 cm. The CIRS instrument consists of two assemblies, optics and electronics. The optical assembly comprises the telescope, a far infrared interferometer (FIR), a mid infrared interferometer (MIR), a reference interferometer, a reference interferometer, a moving scan mechanism, a cooler, thermal control equipment, an optics assembly mount, covers for the telescope and cooler and a calibration shutter mechanism.

The telescope is composed of a 50.8 cm diameter paraboloid primary mirror and a hyperboloid secondary mirror. A sun shade will be present around the primary mirror. This shade will also serve as the primary mirror radiator. A cylindrical tube extends from the central portion of the primary mirror to support the secondary mirror which has its own radiator. The CIRS science data is collected by two of the instrument's three interferometers. The CIRS FIR covers a spectral range of 10–600 cm and is a polarising interferometer that uses substrate-mounted wire grid polarisers to polarise and analyse the radiation. The interferometer operates by first polarising the radiation and then modulating its polarisation. The FIR interferometer has its moving mirror mounted on one end of the moving scan mechanism which it shares with the mid-infrared interferometer. The fixed and moving mirrors are roof mirrors and the FIR focal plane consists of a matched pair of thermopile detectors, each with a concentrator. The CIRS Michelson MIR covers a spectral range of 600–1500 cm and has its moving mirror mounted on the opposite end of the moving scan mechanism from the FIR mirror. The fixed and moving mirrors are cube corners and the MIR uses a germanium lens to focus the interferometer output onto focal planes FP3 and FP4. The reference interferometer is designed to provide timing correlation of the science data sampling to the scan mechanism position. The scan mechanism's motion produces a variable reference interferometer signal used to generate the timing signals necessary to time the recording of science data. The reference interferometer is a Michelson instrument used on-axis at the optic centre of the MIR interferometer and includes laser diode and LED sources, a quartz beamsplitter/compensator, optics and a silicon detector. The reference interferometer uses cube corners for the fixed and moving mirrors. The moving scan mechanism subassembly includes the optical and mechanical components in the optics assembly required for moving the interferometer mirrors to permit controlled sampling in the optical path difference.

This subassembly consists of a common carriage with a moving shaft for the three interferometer mirrors, a motor to drive the shaft, a cantilever-spring motor mount and a velocity transducer. The scan mechanism includes a launch lock that can be locked and unlocked repeatedly without refurbishment by remote command through the onboard computer. A single-stage passive cooler, radiating to space, provides a 70–80 K cold finger with four discrete, commandable set points within that range; nominal set point temperature is 80 K. The cold finger has heaters for decontamination and detector annealing. A system of thermal control equipment including temperature sensors, electrical heaters, proportional heater controllers and radiators used to maintain the thermal control of the optics assembly. The temperature of the instrument is monitored by sensors located at appropriate places in the instrument, including, but not limited to, the telescope mirrors, the interferometers, the detectors, the optics housing, the radiating surface and the Michelson motor. The optics assembly mount thermally decouples the optics assembly from the remote sensing pallet. The telescope and 80 K cooler is protected with covers until the orbiter leaves the solar system, at which point the covers are separately jettisoned by wax thermal actuators. Each cover has two redundant paraffin actuators, either one of which can initiate the action of a pin-puller assembly, which in turn initiates the action of an ejection spring assembly. The calibration shutter mechanism is used to interrupt the MIR beam, causing the MIR detectors to view a controlled 170 K black surface inside the instrument. The shutter is commandable through the onboard computer. The CIRS electronics assembly includes front-end electronics, scan mechanism electronics, reference interferometer electronics, temperature control and monitor electronics, the instrument data system and power converter electronics for conditioning spacecraft power as required by the instrument. The front-end electronics provides analogue and digital processing of the detector signals and this includes filtering, multiplexing, amplification and digitisation to achieve the science data requirements. The scan mechanism electronics provides the operation and control of the linear moving scan mechanism. The control system provides a constant velocity travel during the forward (scan) direction and a fast flyback during the reverse (flyback) direction. The control is a function of the reference interferometer timing signals and other sensors, as needed. The scan mechanism electronics also includes circuitry to actuate the calibration shutter. The reference interferometer electronics provides timing signals for accurate sampling of the scan mechanism velocity.

The CIRS instrument has four zones that are independently controlled by the temperature control and monitor electronics. These zones are the primary mirrors, the secondary mirrors, the interferometers and the 80 K cooler. The temperature is monitored by temperature sensors located on the four zones which provide accurate temperature values at the cooled temperatures. The sensors located on the optics assembly provide course values at the decontamination temperature and the monitored temperatures transmitted to the spacecraft. The instrument data system (IDS) is a microprocessor circuit that provides the instrument data processing function and communication with the spacecraft. The IDS receives and processes commands, data and timing information from the spacecraft and configures and controls the instrument's operational states. It processes the science data from the front-end electronics and the housekeeping data from all of the subassemblies and then it transmits these data to the spacecraft. The power converter electronics conditions the regulated power received from the spacecraft and provides direct power to the instrument data system and the temperature controller and monitor subassemblies. The PCE also provides, on command, power to other subassemblies and release of the scan mechanism launch lock, the 80 K cooler cover and the telescope cover.

Huygens payload: HASI Huygens Atmospheric Structure Instrument: 6.7 kg, temperature and pressure of Titan's atmosphere, winds and turbulence, atmosphere electricity. (University of Rome)

GCMS Gas Chromatograph Neutral Mass Spectrometer: 19.5 kg, atmosphere composition. (NASA Goddard)

ACP Aerosol Collector/Pyrolyser: 6.7 kg, atmosphere aerosol composition. (France)

DISR Descent Imager/Spectral Radiometer: 8.5 kg, image clouds/surface and UV-NIR spectroscopy. (University of Arizona)

SSP Surface Science Package: 4.2 kg, surface state after touchdown. (University of Kent, UK)

DWE Doppler Wind Experiment 2.1 kg, high accuracy zonal wind characteristics. (University of Bonn, Germany)

Huygens probe

Mass: 335 kg probe containing an 87 kg 2.7 m diameter decelerator. Science instrumentation: 48 kg.

Payload: 15.4/15.8 GHz phased array radar to show altitude and surface reflectivity data from 10 km to 150 m.

VERIFIED

Ulysses (ISPM)

Current Status

Ulysses made a Jupiter's gravity assist manoeuvre in early 1992 to swing south of the ecliptic towards the Sun, becoming the first probe to undertake detailed investigations of the heliosphere's third dimension. On 9 June 1993, it became the first spacecraft to exceed 32°, surpassing Voyager 1's record. The trajectory brought Ulysses over the Sun's south pole during 1994, with 132 days spent above 70° solar latitude between 26 June and 6 November. A maximum 80.2° was reached on

Ulysses continues to monitor activity (ESA) 0517150

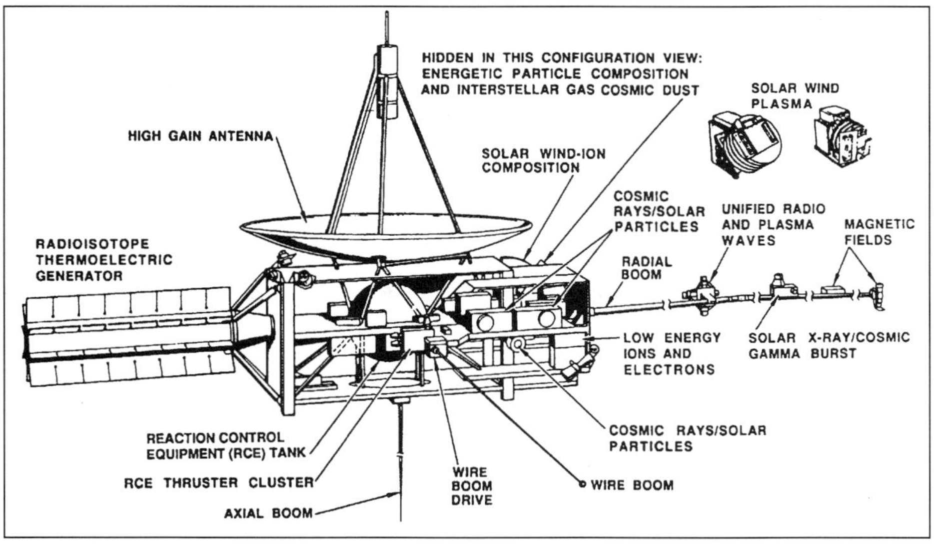

Ulysses principal features (ESA) 0516811

13 September 1994, at a distance of 2.3 AU from the Sun. The equator was crossed on 5 March 1995. Passage over the north pole came during 19 June to 30 September 1995; 80.2° was attained on 31 July. Ulysses arrived during a period of minimum solar activity, instead of the original schedule's maximum.

Ulysses is tracked 8 hr per day throughout the mission by the 34 m dishes of NASA's Deep Space Network. A joint NASA/ESA team at JPL's Mission Control and Computing Center conducts operations.

Ulysses made a second pass near the next solar maximum, returning in 2000–2001 after reaching 5.4 AU aphelion in April 1998. Its second southern polar pass above 70° occupied September 2000 to January 2001, followed by perihelion in May 2001 and the northern pass September to December 2001. Power-saving measures are necessary. This extension, beginning 1 October 1995, is formally known as the Second Solar Orbit and cost ECU14 million through December 2001. The original mission, ended 30 September 1995, cost ECU168 million (1995 rate). Ulysses is in excellent engineering health. At its June 2000 meeting ESA's Science Programme Committee approved the continuation of the mission until September 2004 when Ulysses will have completed two full out-of-elliptic orbits of the Sun. During the first quarter of 2004 Ulysses was again in the vicinity of the Jovian environment performing measurements. The mission is funded through October 2008.

Background

Until 1981, the International Solar Polar Mission was a two-craft project to be launched by a single Shuttle and three-stage IUS in February 1983. ESA had the responsibility to provide one spacecraft and about half of the 16 experiments, while NASA promised to provide the other craft, the rest of the experiments, RTGs and Shuttle launch. The two agencies signed the MoU 29 March 1978.

The original plan called for the two spacecraft to fly for 15 months to Jupiter, one passing above and the other below the planet at distances of 450,000 km, causing both to enter high inclination, but opposite (posigrade and retrograde solar) orbits. Four years after launch, they would have passed over opposite solar poles at 250 million km, through the ecliptic plane and then over the other poles. The five-year mission would have been the first to investigate the Sun in three dimensions, providing an accurate assessment of the solar magnetic field and solar wind and studying solar flares, coronal holes and loss of mass/energy.

Then NASA's decision in 1980 not to proceed with the three-stage IUS raised doubts about the feasibility of the dual launch plan. The project switched to the cryogenic liquid Centaur upper stage, but US budget cuts in February 1981 forced NASA to save US$250 million to US$300 million on ISPM by cancelling its craft. Shuttle delays and other problems had already delayed the launch to 1985. It was finally agreed that NASA would continue to provide launch/tracking services and an RTG. With ESA's stepped up prominence as the majority partner in the project, it was renamed Ulysses (from ISPM), at ESA's request to honour both Homer's mythological hero, and a reference in Dante's *Inferno* of Ulysses' urge to explore 'an uninhabited world behind the Sun'.

Doubts again crept into the programme plan when controversy arose about Shuttle/Centaur threatening the planned May 1986 launch even before the Challenger accident in January 1986. After the accident, the earliest possible launch window became September/October 1989, with subsequent windows available only every 13 months because of Jupiter's motion. Following NASA's June 1986 decision to prohibit the use of the liquid Centaur in the Shuttle, the ESA and NASA agreed to move the craft to an IUS/PAM-S combination. Each had been employed separately from the Shuttle, but never in combination.

There was also concern over launch/injection loads and Ulysses' RTG. A Centaur launch would have restricted peak loads to 5.8 *g*, whereas the substituted solid rocket upper stages could have imposed up to 11 *g*. However, Ulysses was designed to withstand 11.5 *g* because of uncertainty over upper stages in the programme's early days.

Following the closure of the US Department of Energy plant that produced the RTG fuel, 1989 marked the last year a launch could be made, without reductions in the power available for the experiments. These issues added complexity to the competition for Shuttle launch slots with Galileo and Magellan, all requiring the October 1989 window. In the end, the ESA agreement with NASA for an October 1990 launch was signed April 1987, Galileo took the Autumn 1989 slot and Magellan departed the previous May.

Following Challenger's loss, Ulysses was returned from Kennedy Space Center to ESA's ESTEC facility in The Netherlands on 16 June 1986. ESA removed its instruments and returned the majority to the Principal Investigators. The spacecraft itself spent 1987 in storage at prime contractor Dornier in Friedrichshafen, with some units, such as the tape recorders and axial boom, which required special treatment, removed.

By February 1989, a new schedule was issued after Ulysses was brought out of storage in October 1988. It called for it to be reintegrated with its instruments starting in May 1989, tested during July 1989 to February 1990, and delivered to Cape Canaveral in May 1990 for mating with its upper stages in July 1990 in preparation for installation in Shuttle's cargo bay in September 1990. Some 200 microprocessor memory chips were replaced in the decoder, data handling and experiment units beginning April 1988 following the discovery of a potential generic fault.

NASA launched the Ulysses by Shuttle during the 5–23 October 1990 Jupiter window, 13 years after ESA's Science Program Committee approved the mission. The Inertial Upper Stage and PAM-S motor accelerated it to a record 15.4 km/s, exceeding the velocities of NASA's Pioneers and Voyagers. After launch, the probe had to arrive at Jupiter within a 160 km corridor, forcing an IUS guidance system in orbit from Discovery's more accurate unit. Some 45 minutes after release, at 18.53 GMT, IUS stage 1 ignited for a 148-second burn. The 125-second coast was followed by stage 2's 108-second firing. The IUS provided final pointing and separated, leaving four small Atlantic Research Corp solids around PAM's waist to spin the craft up to 70 rpm. They were released and PAM fired for about 88 seconds to add 4.3 km/s for a record 15.4 km/s departure speed, outstripping the four Pioneers and Voyagers and crossing the Moon's orbit in about 8 hr. Two 0.9 kg masses on 12 m wires were freed to reduce the spin to 6.8 rpm and Ulysses separated about 10 seconds after PAM's burn. The deployment of Ulysses' 5.55 m experiment boom on 7 October cut the spin to 4.7 rpm. The RTG was found to be providing 283 W, 2 W less than expected.

The Ulysses deployed its 7.5 m boom on 4 November and nutation peaked after 11 days at 6.5°, before decreasing to about 3°. Such motion must be <0.75° for the narrow-beam X-band link to lock on and when Ulysses' automatic Conscan system was activated on 12 December (designed to cope with up to 1.5° nutation) it fell to 0.5°. The cause was solar heating of the 7.5 m boom as Ulysses rotated. The effect disappeared as the craft moved further from the Sun.

Ulysses' outbound path would have taken it to 17 AU, almost to Uranus's orbit, if Jupiter had been avoided. A 0.29 m/s tweak 8 July 1991 set up the Jupiter encounter. After travelling 993 million km and already collecting important information on the interplanetary medium, Ulysses began the Jupiter encounter by detecting the bowshock crossing at 17.33 GMT 2 February 1992, some 113 R_J out. This was more distant than expected: Pioneer's was 100 R_J and Voyager's 60 R_J. It approached through the late morning region of the magnetosphere at about 30° N, came within 378,400 km of the cloud tops (5.3 R_J) at 12.02 GMT 8 February and then exited unscathed on the previously unexplored evening side, at high southern latitudes, having spent more than a week in the magnetosphere.

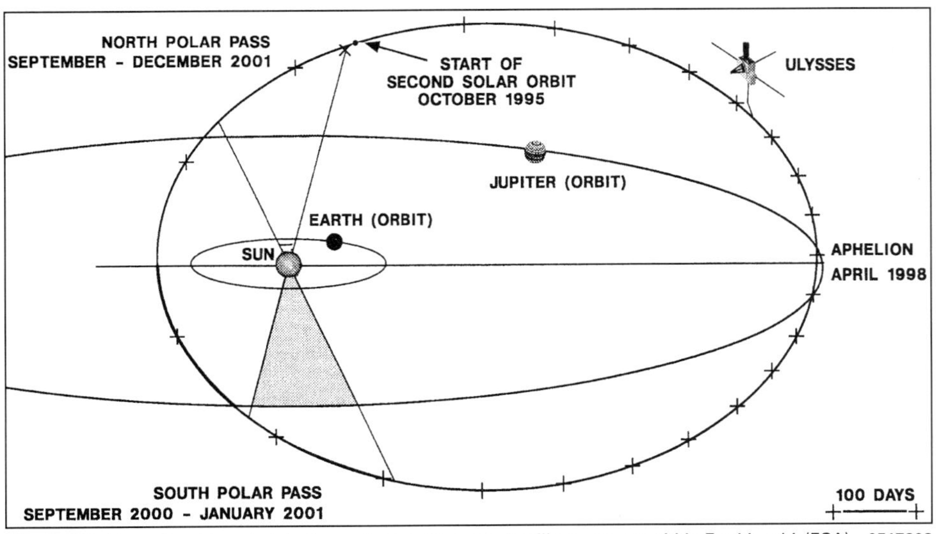

Ulysses' flight profile will again take it over both solar poles. It will never pass within Earth's orbit (ESA) 0517303

Preparing to insert the RTG's fuel elements on the Shuttle pad (NASA) 0517304

Just before closest approach, Ulysses entered the magnetosphere's polar cap. The close 13.5 km/s fly-by was necessary to produce the high ecliptic inclination - IUS/PAM could have reached only 23° without the assist. Ulysses found that Jupiter's intense radiation belts reach only 40° latitude, whereas Earth's extend to 70°. Ulysses then passed directly through the Io plasma torus (only Voyager 1 preceded it), which plays a key role in refuelling the magnetosphere with plasma. The overall density was about as expected, implying injection from Io has not changed dramatically since Voyager, but significant azimuthal asymmetries were found.

At end-February 1992, the Sun, Earth and Ulysses were in direct opposition, the best time to detect gravitational waves by observing a minute shift in the craft's position. Although no scientist could positively identify gravity waves, this radio science experiment, set new upper limits with a factor of 20 improvement in sensitivity on their value. Other important Ulysses' observations included the first direct detection of ionised O, N and He (and neutral He) atoms arriving from interstellar space and the measurement of μm-sized dust grains from interstellar space.

By mid-1993, Ulysses was permanently immersed in the region of space dominated by the Sun's southern pole. This could be seen in the consistently negative polarity measured by the magnetometer since April 1993. Increasing latitude also reduced the intensity of charged particles.

Ulysses monitored radio science events as Jupiter was impacted 16–22 July 1994 by the 21 large objects from the disrupted Comet Shoemaker-Levy 9. No evidence for radio emission from the impacts was found.

Now in high inclination heliocentric orbit, Ulysses has passed over the solar south pole and conducted the first long-term in situ observations of high speed solar wind flowing from the large coronal hole that covers that polar cap at solar minimum (the Sun's current phase). Ulysses measurements of ionised interstellar neutralised gas particles, so-called interstellar pick-up ions, have led to a major advance in our understanding of the processes affecting this component of our heliospheric particle population.

Unexpectedly, the magnetometers detected a wide variety of fluctuations at many spatial scales in the Sun's high latitude field. They are believed to represent relatively unevolved turbulence originating at the southern polar coronal hole. Another important finding was that the magnetic field radial component varies relatively little with latitude. This is contrary to the expectation that the imprint of the underlying dipole-like magnetic field at the Sun's surface, showing an increase in the radial field at the poles, would be detected at Ulysses' position. In simple terms, a south magnetic pole was absent at high latitudes. Undoubtedly related to the presence of the large-scale fluctuations in the polar magnetic field was the detection of an unexpectedly small increase in the measured influx of cosmic ray particles over the poles compared with model predictions.

Moving back towards the equator, Ulysses left the fast solar wind (750 km/s) in which it had been immersed for more than 18 months, crossed briefly into the equatorial region dominated by slow wind (400 km/s) and once again became immersed in the fast wind, this time from the North pole. The boundaries were strikingly symmetric: 22° S/ 21° N. The magnetic field changed from negative (directed inwards) to positive (outwards), reflecting the current configuration of the Sun's dipole-like surface field. The radial component's strength continues to show no change with latitude. With the exception of the north-south solar wind asymmetry and cosmic ray intensity, the north pass was generally similar to the S pass.

Observations as Ulysses heads slowly south are now being correlated with those from Soho. During 1997–98, it will remain at almost constant distance from the Sun during aphelion, allowing temporal effects to be separated from spatial.

Ulysses in final assembly 0085789

Specifications

Ulysses satellite

Mission: fields/particles and cosmic dust investigations of the heliosphere up to high solar latitudes.
Launch: 6 October 1990 aboard Shuttle Discovery from Kennedy Space Center.
Contractors: Dornier (prime), BAe (AOCS, HGA), Fokker (thermal, nutation damper), FIAR (power), Officine Galileo (Sun sensors), Laben (data handling), Thomson-CSF (telecommand), MBB (thrusters).
Mass: 371 kg at launch; 366.6 kg at PAM separation. Science payload 55 kg.
Orbit: heliocentric, including Jupiter swing-by, perihelion 1.34 AU.
Description: spin-stabilised at 4.99 rpm with the 1.65 m diameter HGA pointed continuously towards Earth. Powered by a single DoE RTG providing 284 W BOM and 255 W EOM October 1995 (4.5 kW heat). Double-hinged 5.55 m radial boom carries four sensors for HED, STO and HUS. 7.41 m axial boom as monopole for STO. two wire booms 72.5 m tip-to-tip as dipole for STO.
Communications: X-band (20 W, 8,408.209 MHz, 2° 3 dB beamwidth); 5 W 2 GHz S-band 2,111.607/ 2,293.148 MHz up/down is used for dual frequency radio science investigations and early manoeuvres. During contact periods, real-time data at 1,024 kbit/s are interleaved with playback of stored data. Two redundant tape recorders can each hold 45.8 Mbit, stored at 128/256/512 bit/s.
Propulsion: 33.5 kg high purity hydrazine in single bladder tank with dual manifold feeds two clusters of four 2 N catalytic thrusters (total delta-V 185 m/s)
Payload: BAM solar wind plasma (two electrostatic analysers with channel electron multipliers; US)

GLG solar wind ion composition (electrostatic analyser with time-of-flight/energy measurement; Switzerland/US)

HED magnetic fields (triaxial vector helium and fluxgate magnetometers; UK)

KEP energetic particle composition/neutral gas (four solid state detectors; Germany)

LAN low energy charged particle composition/ anisotropy (two sensor heads with five solid state detectors: US)

SIM cosmic rays and solar particles (five solid state detectors, US)

STO radio/plasma waves (72.5 m radial dipole plus 7.55 m axial monopole antennas with 2-axis search coil; US)

HUS solar X-rays and cosmic ν bursts (two silicon plus two cesium iodide detectors; France/Germany)

GRU cosmic dust (multi-coincidence impact detector with channeltron: Germany).

UPDATED

Japan

Lunar-A (SS-17) Satellite

Current Status

The launch was scheduled for 24 August 1999 but has been rescheduled to 2004 because a penetrator failed a 300 m drop test and must be redesigned. The original mission plan remains unchanged.

Background

ISAS selected a lunar orbiter/penetrator mission in 1990 for its launch vehicle M-5's second mission. It then chose NEC to act as system integrator. As it has laid out the programme to date, ISAS plans for an orbiter to fire three 13.6 kg, 13.8 cm diameter, 71.4 cm-long (1.44 m with motor attached)

The Lunar-A 1997 lunar orbiter/penetrator mission is selected for the second M-5 (ISAS) 0516736

penetrators carrying seismometers and heat flow probes 1–3 m into the surface. The probes will penetrate both the far side and the near side of the moon. Japan has designed the seismic data experiments to clarify long-standing questions on the far side crustal and deep mantle structure and existence of the lunar core. The orbiter will act to relay data to Earth, fleshing out the Apollo programmes near side measurements. In order to test its plans, ISAS has set up an N_2-powered gas gun at ISAS' Noshiro facility to fire penetrator designs into a simulated lunar surface. Lower speed drop tests began in August 1986. A one-stage S-520 sounding rocket, launched on 17 September 1995, also tested penetrator separation and attitude control plans but during the test the ground station lost telemetry after 25 seconds.

Physically, Lunar-A is a 540-kg spin-stabilised unit, 2.2 m in diameter and 2 m high vehicle. The three-stage M-5 plus two solid kick stages will inject it into a Geotail-type highly eccentric Earth orbit with an apogee of about 500,000 km. N_2O_4/N_2H_4 thrusters account for 189 kg of the weight budget and the penetrator modules about 124 kg. Six 20 N and four 1 N/N_2H_4 thrusters augment the main 500 N bipropellant engines.

In order to reach the final trajectory, Lunar-A will fly-by the moon to obtain an apogee of 1,185,000 km and reduce the delta-V to 550 m/s for the main engine to establish a 200 × 200 km, 20° orbit during the second encounter.

For each penetrator release, the orbiter will descend to 45 × 200 km before returning to its higher orbit to use its LIC Lunar Imaging Camera (with an image compression chip provided by CNES). The orbiter will spin up to 120 rpm for release from its normal 30 rpm. A 30 cm diameter spherical solid motor (20 second burn, 5 kN max thrust) will brake the penetrator out of orbit. Its N_2 attitude control system will ensure a head-on impact at 8° and a velocity of 250–300 m/s, a deceleration of 8,000 g, burrowing 1–3 m into the surface. One penetrator is planned for the Apollo 12 and 14 region, one into the far side and one into the far/near side boundary. The orbiter will then rise to a 300 × 300 km orbit for data relay. Data transfer from penetrator to the orbiter is at 256 bps and from the spacecraft to earth at 8 kbits/sec.

Planet-B (Nozomi)

Current Status

Japan launched the Planet-B spacecraft on 3 July 1998 aboard an M-5 rocket built by ISAS and launched from Kagoshima. After lift-off, ISAS renamed Planet-B Nozomi, or 'Hope'. Nozomi performed two lunar fly-bys on 24 September and 18 December (2,809 km) and then a first Earth fly-by (1,003 km) on 20 December 1998 for a gravity assist to trans-Mars trajectory. However, a perigee burn planned at this time provided insufficient velocity and an extra burn a day later used more propellant than planned. Instead of reaching Mars on 11 October 1999 and entering a 300 × 47,500 km orbit, Nozomi conducted a second earth swing-by in January 2002 to an out of plane trajectory providing a third earth fly-by in June 2003. Mars arrival is scheduled for January 2004. Japan launched Nozomi on 3 July 1998 specifically to acknowledge the first anniversary of the Mars Pathfinder mission. Japan has spent approximately ¥11 billion (US$80 million) for the orbiter, which will examine the planet's atmosphere for details of the Aerian atmosphere's interaction with the solar wind and the escape trajectories of oxygen molecules from the thin atmosphere of Mars.

At the time of launch in 1998 Japan was only the third country to launch a planetary mission.

Background

Planet-B, operating for at least a full Mars year, will provide synoptic data on turbulence, motion and seasonal variation of the upper atmosphere, interaction with the solar wind and map Mars' magnetic field. ISAS has overall responsibility for the spacecraft and it contracted with NEC to perform the systems integration and MHI to build the 500 N thruster. The spacecraft has a mass of 536 kg at launch of which 258 kg is the dry mass carrying a 33 kg payload. After launch, Planet-B enters a 7000 × 400,000 km Earth orbit for two months before a lunar swing-by now scheduled for the middle of October. When the spacecraft completes 4.5 orbits, then a 420 m/s burn, it will be on its planned trajectory for Mars. Upon arrival it will then perform a 1,257 m/s Mars injection burn originally planned for 11 October 1999 to create an initial 300 × 47,500 km 170° inclination (ecliptic) orbit further refined and adjusted to 150 × 47,500 km (at times lowered to 130 km), but see current status.

Specifications

Planet B

Launch: 3 July 1998 by M-5 from Kagoshima
Description: 2 m diameter irregular octagonal prism body, 7.5 rpm spin
Mass: 536 kg; 258 kg dry
Telemetry: fixed 1.6 m diameter Earth-pointing X/S-band HGA. LGA provides S-band link during non-Earth pointing, 1.7 m boom and four 25 m sounder antennas deployed in Mars orbit. 2–32 kbit/s X-band + 64-1024 bit/s S-band, depending on Earth distance (2 kbit/s X-band minimum), 16 + 125 bit/s command uplink; 128 Mbit SSR (1 kg).
Power: 2 fixed solar panels provide 200 W min from 20 × 40 mm Si cells (18 per cent efficiency). 2 × 15 Ah Ni-MH batteries. 5 m magnetometer boom
Propulsion: Trans-Mars injection and Mars orbit insertion by 500 N NTO/N_2H_4 thruster (10 rpm spin for stability). RCS by 4 axial, 2 radial, 4 tangential 2.3 N N_2H_4 thrusters.
AOCS: Attitude data from V-slit star scanner, spin Sun aspect sensor (170° FOV), and rate gyros for active nutation control in launch phase.
Payload: MIC visible camera (with data compression chips provided by CNES)
MGF magnetometer (3-axis 0.1 nT accuracy)
ESA energetic electrons (5 eV-22 keV)
ISA energetic ions (10 eV-20 keV/q)
IMI energetic ion mass (0.5 eV-40 keV/q; Swedish Institute of Space Physics)
EIS high-energy particles (40–500 keV)
TPA thermal ion drift (0.1–100 eV; Canadian Space Agency)
PET electron T, UVS UV spectrometer (H, O, CO, CO_2, imaging)
PWS sounder/HF waves (electron density profile and HF waves)
LFA plasma waves (VLF/ELF waves)
NMS neutral gas mass spectrometer (NASA)
MDC dust counter (Munich Technical University, Germany)
XUV EUV spectrometer (helium)
USO (ultra-stable oscillator for radio science, NASA).

Russian Federation

Mars Series

Current Status

Russia abandoned its ambitious planetary exploration programme after the failure of the Mars–96 mission on 16 November 1996. IKI still discusses the Mars Together programme with JPL Russian Space Agency, although funding for any planetary mission remains tight and the budget is being drawn away from the exploration programme to finance the Russian contribution to the International Space Station. Russia continues to discuss participation in US and European planetary exploration initiatives, but has no plans to conduct an independent launch programme. Several research institutes have developed rover vehicles and sample return vehicles and continue to press for a Russian Mars surface earth return sampler.

Background

Russia finally launched Mars-96 on 16 November 1996 at 20.48 GMT from pad 200L at Tyuratam/Baikonur aboard a four-stage Proton-K using a Block D-2 fourth stage. The Block D-2 (11S824F) stage had previously been used for the two 1988 Phobos launches. Initially, the launch went well with the Block D-2 performing its first burn to place itself and the attached Mars probe – re-named Mars 8 after launch - into an initial 51.5°, 139–155 km parking orbit. The flight plan called for the Block D-2 to fire again about an hour after launch to place itself and Mars 8 into a highly eccentric Earth orbit. (Block D2 should have provided a delta-v of 2.850 km/s). After this firing Mars 8 would fire the Fregat-class propulsion module for the final burn to a heliocentric, trans-Mars orbit.

At the time of the Block D-2 re-ignition, an incorrectly aligned assembly caused the engine to shut down after 20 seconds on command from Mars 8. Stranded in a low Earth orbit, the Mars 8 still separated from the Block D-2. Some reports suggest that the Fregat module then burned as planned, placing the Mars 8 spacecraft on an orbit with a perigee of about 87 km and apogee of 1,500 km or more and the low perigee quickly brought the stage back to Earth. Initially the Russians thought that it had re-entered over the southern Pacific Ocean, but later debris tracks indicated it survived long enough to possibly land in Chile or Bolivia, raising public health concerns over the spacecraft's plutonium power source. Sweeps by the Chilean Air Force failed to turn up any sign of radioactive contamination. Meanwhile, the Block D-2 continued in orbit until 18 November. The Russians stated that the stage re-entered at 01.13 GMT with debris falling into the southern Pacific Ocean close to 50.9° S, 191.9° E. USSPACECOM challenged this, stating the decay time was actually 01.32 GMT over 32° S, 264° E.

Mars-96

Mars-96 consisted of a single spacecraft based on the Phobos design. It was meant to follow a Type II trajectory to arrive at the Red Planet in September 1997 after 315 days. At launch, the spacecraft assembly had a mass of 6.2 tonnes. If it succeeded, the orbiter would have remained in its initial 12 h orbit for three to four weeks for regular data contact with surface elements. Then it would have released two small landers before rising to the mapping path that allows surface contact only every three to four days. The mapping was planned to last a full seasonal cycle or roughly two Earth years from 250 × 18,000 km 101°.

The imaging platform known as ARGUS had many development problems at VNII TransMash as Mars 8 neared completion. The problems prompted Russia to propose ARGUS' deletion, but DARA/DLR strongly resisted the move because body-mounting the instruments instead would cost at least 60 per cent of the science return and compromise the project goal of 10 m resolution (from 300 km) stereo colour imagery.

MARS-96's orbiter would have sent the two autonomous stations on a trajectory to the surface three to five days after achieving a final orbit. The small stations would have followed a descent path with a 5.6 km/s entry. Just before landing the re-entry capsule would have ejected a 65 cm diameter heatshield leaving the lander suspended on 130 m line and protected by a 1.35 m inflated shell. Upon touching down, the shell would have split around the equator and cut the parachute and the carrier free. Russia expected the final landing to impact at between 15–20 m/s activating four petals à la Luna 9 through 13. After a final checkout, the lander would have uplinked data from a 32 Mbit memory at 8 kbit/s on 401 MHz. The power supply could provide 2 W for 20 minutes daily initially, then every three days. A 9 kg payload stereo panoramic camera provided by Russia would use most of the telemetry band. NASA and RKA signed a contract in April 1993 for the US to fly a duplicate Mars Oxidiser experiment on two landers. The MOX would measure the soil to determine if it is strongly oxidising. It would have exposed bundles of optical fibers with coatings sensitive to peroxides and superoxides to the air and soil and then measured changes in the light transmission characteristics.

Russia based its designs for the independent surface penetrators on those planned for the Vesta mission. As envisioned, the 75 kg penetrators would directly enter the Martian surface, by spinning up to 95 rpm in orbit and upon release, igniting a four-nozzle solid propellant motor to provide 4.6 km/s entry. The 45 kg 1.5 m long 11 cm diameter titanium body would impact at 76 m/s, reaching 6–8 m depth. The deceleration peak of 500 *gs* breaks a forebody free to penetrate deeper. Fore and aft bodies are connected by cable. Once in place, a small isotope generator provides one year of life. The penetrators and surface stations will create a seismic network.

Specifications

Mars-96 (Mars-8)

AOCS: 3-axis controlled to 1° per axis
Data rate: 64 kbit/s when Earth distance is greater than 200 million km, 128 kbit/s when less. 0.5 Gbit/s daily downlink capacity
Mass (science payload): 250 kg
Mass (dry): 1,750 kg
Orbiter sensors (Plasma science): ASPERA-S energy-mass (Sweden).
FONEMA omni non-scanning energy-mass ion analyzer (UK)
DYMIO omni ionosphere energy mass spectrometer (France)
MARIPROB ionosphere plasma spectrometers (Russia)
MAREMF electron analyser/magnetometer (Russia)
ELISMA wave complex (France)
SLED-2 low energy charged particle spectrometer (Ireland)
Orbiter sensors (Cruise science): PGS precision γ spectrometer (Russia)
LILAS-2 spectrometer (cosmic/solar ν bursts, France)
EVRIS photometer (stellar seismology, France, on PAIS 2-axis platform)
SOYA photometer (part of SPICAM, solar oscillations, France)
RADIUS-MD dosimeter (Bulgaria)
MORION-S science data acquisition system (Russia), 2 Gbit mass memory (ESTEC)
Imaging Sensors: ARGUS 3-axis scan platform with four imaging instruments
WAOSS Wide Angle Opto-electronic Stereo Scanner (DLR Germany). WAOSS: focal length 21.7 mm, f/5.6, 0.4–0.76 μm, each push-broom linear CCD 5,184 elements, swath 80°, viewing nadir + 20° back/forward, 500 kbit/s; resolution 100 m.
HRSC High Resolution Stereo Camera (DLR Germany). HRSC: fl 175.0 mm, f/5.6, nine push-broom linear CCDs, 3,950–9,250 Å, nine channels (4 XS, 10 m resolution PAN, 2 stereo PAN, 2 photometric), swath 11°, 1.6 Mbit/s. HRSC/WAOSS total 47.7 kg, 75 W; data stored in 1 Gbit mass memory shared with Omega.
Omega/VIMS 0.5–5.2 μm visible/IR mapping spectrometer (CNES; Italy contributing to visible channel)
NK navigation camera (Russia). 4.0 kg, 520 × 580 CCD, 100 mm fl, f/3.5, 5 × 8° FOV.
Other planetology instruments: TERMOSCAN mapping radiometer (Russia)
SVAT mapping spectrometer (Russia)
SPICAM multi-channel spectrometer (on PAIS platform, Russia)
UVS-M UV spectrometer (Russia)
LWR longwave radar (Russia), Photon ν spectrometer (on PAIS platform, Russia)
NEUTRON-S neutron spectrometer (Russia), MAK quadropole mass spectrometer (Russia).

SAS small autonomous stations

Mass: 90 kg each (34-kg surface mass).
SAS sensors: DESCAM descent camera (France, 200 images during descent from 10 km at 1 s intervals, 544 × 240-pixel CCD, 10 mm fl)
Optimism vertical axis seismometer (France, 0.350 kg, 9 cm cube, 0.01-4 Hz, 10^{-8} m/s^2 sensitivity), magnetometer (on deployed arm, France)
Metegg meteorology package (FMI Finland, on deployed arm)
Mars Oxident Experiment (MOx, NASA), carrying (a) a neutron-proton spectrometer (on deployed arm, Germany); (b) optical depth analyser (France). 1 year surface life provided by two 150 mWe RTGs and Ni/Cd battery (four RTGs in total generate 34 W heat) (c) Oxidiser/superoxidiser experiment
Penetrator sensors: TV cameras (1.2 kg Kassei + 0.15 kg Mikrotel; RFAS, Japan, Czech Republic)
Meteorology unit (0.4 kg Mekom; RFAS, Finland)
Gamma-ray spectrometer (1.2 kg Pegas, with 3 cm tungsten shield against RTG; RFAS, Czech Republic)
X-ray spectrometer (0.3 kg Angstrem; RFAS, Germany),
Alpha-particle spectrometer (0.3 kg Alpha; RFAS, Germany)
Neutron detector (0.2 kg Neutron-P; RFAS, Germany)
Accelerometer (0.2 kg Grunt; RFAS, UK)
Thermometer (0.3 kg Termozond; RFAS, UK)
Seismometer (0.3 kg Kamerton; RFAS, UK)
Conductometer(0.1 kg Epsilon; RFAS, UK)
Magnetometer (<0.1 nT resolution, 0.2 kg IMAP-6; RFAS, Bulgaria).

Zond

Background

The Zond missions can be split into two classes, the original series which comprised Zonds 1–3 (plus the launch failure identified as Cosmos 27) and the second series was initiated by Cosmos 146 and Cosmos 154, launched in 1967 as test flights of the L-1 circumlunar spacecraft (called L-1P) derived from Soyuz.

All first series Zonds were dedicated at specific solar system research and planetary studies, but the Zond name covered the general target planet because of the poor performance of previous planetary spacecraft. Zond 1 aimed at Venus, Zond 2 launched 30 November 1964 at Mars and Zond 3 - rumoured to have been a delayed Mars probe - flew past the Moon into heliocentric orbit, taking photographs of the lunar far side on its way out of the Earth-Moon system.

As the moon race with the Americans heated up, the Russians chose to concentrate on lunar exploration missions with the second series of Zonds. After launch failures in September and November 1967, Zond 4 was launched on 2 March 1968 for a communications and high-velocity re-entry test from elliptical orbit. The spacecraft possibly reached out to approximately the lunar orbit, but in a different direction. A further launch failure occurred in April 1968 and the Soviets cancelled a July 1968 launch which may have been manned if tests had gone according to plan after a Proton-K Block D propellant tank ruptured before launch.

The Soviets achieved a first with Zond 5, when they recovered it after a circumlunar mission, albeit unmanned, following launch on 14 September 1968, beating NASA's manned Apollo 8 mission by several months. Launched 10 November 1968, Zond 6 tried to repeat the success but crashed on landing. Zond 7 launched 7 August 1969 and appears to have been the only fully successful mission in this series when it returned to earth recovery following lunar fly-by. Zond 8 closed down the flights in October 1970 with a repeat of the Zond 7 mission.

The Soyuz-based Zond 4-8 flew in preparation for a Soviet manned circumlunar programme 0516728

Zond 5 awaiting recovery in the Indian Ocean. It is clearly a Soyuz descent module 0516729

In their desperation to compete with the Americans, the Soviets cannibalised the last two L-1 spacecraft, intended for manned missions around the Moon for the November 1969 and December 1970 L-1E missions. These were primarily Block D propulsion tests with the spacecraft heavily instrumented for launch vehicle monitoring. The first launch failed and the second was named Kosmos 382 indicative of a failure. The Zond/L-1 spacecraft – called L-1S – went aloft on the first two N-1 lunar boosters, launched February and July 1969, but neither reached orbit.

United States

Clementine Satellite

Current Status

Over the course of 71 days in lunar orbit, Clementine mapped the 38 million sq km of the moon in 11 colours and took 620,000 high resolution and about 320,000 mid-IR thermal images. Clementine departed lunar orbit on 5 May 1994 to begin a series of earth and lunar fly-bys before rendezvous with asteroid 1620 Geographos. A computer problem 7 May exhausted all the hydrazine attitude control propellant and created an 80 rpm spin. Engineers determined that spin could be reduced to 30 rpm – still too high for worthwhile asteroid data return. Instead, the main engine on 19 May made the first of three burns to establish a 44,769 × 388,197 km orbit for prolonged radiation exposure studies. BMDO attempted to contact the satellite in February 1995 but it had insufficient power for any significant telemetry. However, since the solar array Sun angle improved daily Clementine responded to commands 10 April 1995. Science data returned by Clementine forms the basis for an early evaluation of potential resource materials on the moon. Determinations from preliminary geochemical analysis will help shape requirements for the next generation of lunar exploration vehicles, the first of which is tentatively scheduled for launch in 2008.

Background

In September 1990, NASA administrator Richard Truly approached Deputy Secretary of Defence Donald Atwood with a view to utilising advanced, lightweight, technologies being developed by the Department of Defence in a joint space mission. NASA proposed using technologies developed at its own field centres with those from various military research projects. The proposed mission envisaged an asteroid fly-by and it was subsequently incorporated into the Ballistic Missile Defence Organisation Sensor Integration Programme. Design of the spacecraft began in March 1992 and an additional two-month lunar mapping phase was introduced to demonstrate spacecraft and sensor performance prior to the demanding asteroid fly-by. The interstage adapter housing the solid rocket motor was also designated as a platform to remain in a highly eccentric earth orbit as a radiatonexperiment and additional sensor target.

The Clementine spacecraft was built and integrated at the US Naval Research Laboratory in Washington, DC, and carried sensors, attitude control systems and software designed and built by the Lawrence Livermore National Laboratory. The US Air Force supplied lightweight composition structures and the Titan 2G launch vehicle. Total cost came to about US$75 million including US$17 million paid to the USAF for the refurbished Titan 2 ICBM as launch vehicle. NASA supplied the communications network including the services of the Deep Space Network and orbit determination and operations support from both the Goddard Space Flight Centre and the JPL.

Clementine was launched at 16.34 UT on 25 January 1994 by Titan 2G. The InterStage Adapter (ISA) remained in the 189 × 126,990 km 66.77° inclination 51.2 hr Earth orbit with its package of radiation detectors while the satellite entered moon orbit to conduct a radar mapping mission. With Clementine 1 spinning at 60 rpm, the Thiokol Star 37 FM solid ignited 06.29 GMT 3 February 1992 during a north bound pass over the Indian Ocean for a 2,959.7 m/s delta V burn. The Star/ISA dropped off and during perigee 1, at 10.01 UT on 5 February, the 490 N motor added 213.5 m/s to stretch the apogee out to the Moon, with a 31.9 m/s

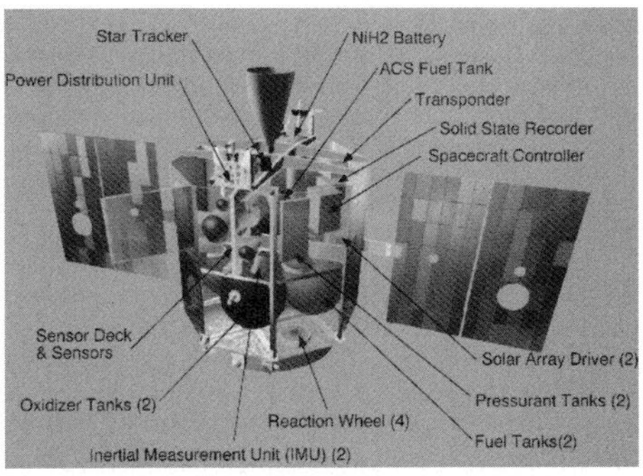

Clementine 1 in February 1993 became the first mapping lunar orbiter since Luna 22 in 1974. It was the first US lunar mission since 1972. Note the instrumented interstage adapter below the main craft (Naval Research Lab) 0516827

trim at 10.00 UT 6 February at the Moon. A few other trim manoeuvres occurred and the spacecraft performed a 460.3 m/s burn at 12.52 UT 19 February, 103.5 m/s at 12.16 21 February and a 5.3 m/s at 12.17 UT 22 February to establish a 382 × 2,971 km, 89.3° inclination lunar orbit. This provided global mapping over two lunar months; while data were taken at 15 hours around perigee and the rest of the orbit was used to relay the data to Earth. Clementine returned a total of 1.8 million images from lunar orbit.

Two major burns on 26 March 1994 to change the argument of perigee carried Clementine up to 1,009 × 3,560 km and back. When the mission concluded another major burn on 5 May, this started the return to high Earth orbit for a series of earth and lunar fly-bys prior to a rendezvous with asteroid 1620 Geographos on 31 August 1994.

The spacecraft consists of an octagonal prism 1.8 m high and 1.14 m across. A 110 lb thruster for delta-V manoeuvers is on one end of the prism and a high-gain fixed dish antenna is on the other end. The spacecraft propulsion system consists of a non-propellant hydrazine system for attitude control and a bipropellant nitrogen tetraoxide and monomenthyl hydrazine system for the manoeuvers in space. The bipropellant system had a total capability of about 1.9 km/s with about 550 m/s required for lunar insertion and 540 m/s for lunar departure. The power system consists of a gimballed, single axis, GaAs/Ge solar array providing a total spacecraft power of 360 W at 30 V DC, with a specific power of 240 w/kg, based on lightweight construction. Two arrays of the rotating solar panels protude from opposite sides, by rolling the spacecraft and rotating the panels, full solar illumination of the panels would be achieved. The solar array was used to charge a 15 A-h, 47 W hr/kg, Nihau common pressure vessel battery. The total mass of the spacecraft in the launch configuration was 1,690 kg, with most of the weight in the solid rocket motor required for translunar insertion. The spacecraft dry mass is about 227 kg, with a roughly equal mass for liquid fuel. This weight was achieved by incorporating many of the previously discussed lightweight technologies.

The main instrumentation on Clementine consists of four cameras, one with a laser-ranging system. The cameras include an UltraViolet-Visual (UV-VIS) camera, a Long-Wave Infra-Red (LWIR) camera, the laser-ranger (LIDAR) High-Resolution (HiRes) camera, and a Near-Infra-Red (NIR) camera. The spacecraft also has two star tracker cameras, used mainly for attitude determination, but they also serve as wide-field cameras for various scientific and operational purposes. The sensor package has a mass of 8 kg and represents the state-of-the-art at the initiation of the Clementine program. The sensors are all located on one side, 90° away from the solar panels. Clementine has 12 small attitude control jets that were used to orient the spacecraft to point the cameras to desired targets. The attitude control system includes the two start tracker cameras and two lightweight inertial measurement units, based on a ring laser gyro and an interferometric fiber optic gyro. During initial deployment at low-earth-orbit, the full size spacecraft was 3-axis stabilised. The spacecraft was 3-axis stabilised in lunar orbit via reaction wheels, with a precision of 0.05° in control and 0.03 in knowledge.

The spacecraft data processing was performed by 3 computing systems. A MIL-STD-1750A computer with a capacity of 1.7 million instructions per second was used for save mode, attitude control system, and housekeeping operations. A Reduced Instruction Set Computer (RISC) 32-bit processor with 18 million instructions per second was used for image processing and autonomous operations. The Clementine mission represents the first long duration flight of a 32 bit RISC processor. Also incorporated is a state-of-the-art image compression system provided by the French Space Agency CNES. A data-handling unit with its own microcontroller sequenced the cameras, operated the image compression system and directed the data flow. During imaging operations, the data were stored in a 3 kg, 2 Gbit dynamic solid state data recorder and later transferred to the ground stations using a 128 kb/s downlink. The spacecraft was commanded from the ground using a 1 kb/s uplink from the NASA Deep Space Network and DOD stations. Demonstration of autonomous navigation including autonomous orbit determination was a major goal of the Clementine mission. Autonomous operations were conducted in lunar orbit.

The Scientific Payload

UV-Visible Camera: The UV-Visible camera has a catadioptic telescope using fused silica lenses focusing onto a metachrome-coated CCD imager. Active wavelength response is limited on the short wavelength end by the transmission of fused silica and the optical blur of the lens.

Wavelength response on the long end is limited by the response of the CCD. Six spectral bands can be selected from a filter wheel which is controlled throught the same Serial-Addressable Synchronous Interface (SASI).

The Thomson Focal Plane Array (FPA) used is a frame-transfer device, accomplishing electronic shuttering by rapidly shifting the active pixel area into the storage area, pausing for the (13-bit programmable) integration time, then rapidly shifting the captured image into a storage buffer from which the image is read out. Post-FPA electronics allow three gain states followed by 5 bits of offset that span 248 counts in the analog regime to augment the basic 8-bit A/D conversion. Gain is A/D digitisation noise limited, so proper exposure is critical. Working against the day side of the Moon as a target, typical integration times were as short as several milliseconds in the lowest gain state (1,000 electrons/bit) near sub-solar illumination points at the brighter spectral bands, increasing to 40 msec near the polar regions in the mid-gain setting for the weaker 415 and 1,000 nm spectral bands.

Wide Field of View Star Trackers: The star tracker cameras have a concentric optics design with a fibre optic field flattener to couple the image surface to a CCD array. The CCD is kept cleared by continuously shifting lines and reading out pixels at the 5 MHz base rate, which is approximately 100 microseconds per line. Integration is accomplished by stoppping this clearing process for the specific time (13 bit programmable), then proceeding with a second line-shifting operation into a readout buffer line and completing readout at 5 MHz. Camera electronics are identical to those of the UV-VIS camera, with the gain circuit resistors selected to be slightly more sensitive.

The primary function of the star tracker camera was to provide stellar images that were processed against an onboard star catalogue to find pointing, thus establishing absolute angular references for navigation. The scientific uses of the cameras were secondary. Only broad band operation was available. Owing to the line-transfer electronic shuttering, imaging was limited to dim targets such as the lunar surface illumination by earth shine. System noise is about 1.0 bits rms at the highest sensitivity setting.

High Resolution Imaging (HiRes) camera: The HiRes camera combines a lightweight beryllium telescope with an image intensifier-coupled frame transfer CCD imager. Image shuttering is accomplished through voltage gating of the image intensifier. maximum integration time in 733 milliseconds in 10.67 microsecond increments. Spectral response is limited in the system by the S-2 photocathode between 0.4 and 0.8 microns. Five spectral bands can be selected from a filter wheel which is controlled through the same SASI interface. A sixth filter position is allocated to an opaque filter for the image intensifier's protection.

The post-FPA electronics circuitry is identical to that used in the UV-VIS camera. Images of the day side of the moon used intensifier gate times with relatively low gain settings on the order of 1 millisecond. Lifetime concerns about the photocathode and micro channel plates in the intensifier unit drove operational settings to low exposures. This resulted in photon shot noise significantly contributing to the overall noise in the HiRes sensor.

Laser Range-finder (LIDAR): The LIDAR unit shares the telescope of the HiRes camera, splitting the 1,064 nm return signal from the NdYag source off to an avalanche photodiode (APD) detector with dichroic filter. The optics are non-imaging, providing an exit pupil through simple relay optics at the APD. The APD electronics includes a temperature compensation feature for the APD bias voltage and programming thresholding of the output signal. The APD curent is amplified and inverted to a voltage by a transimpedance amplifier with a gain of 230X, a low frequency cutoff of 3 MHz, and a high frequency cutoff of 23 MHz. The APD current is amplified, then discriminated for changes (increase) through 14 MHz discriminator. Voltage changes exceeding the programmed threshold are flagged as returns.

Range value is determined by the clock cycles since a laser output sampled start pulse is received. The clock counter has only 14 bits owing to the hardware availability limitations. In order to allow returns up to the 640 km maximum range required in the lunar mission, returns from the discriminator are binned 4 to a clock count, turning the 23 MHz response into a 40 m high bin. Internal memory in the LIDAR unit saves up to 6 'returns' per laser firing, with up to 4 saved between programmable search range minimum/maximum values. Threshold is set for the best compromise between missed detection and false alarms.

Near Infra-Red Camera (NIR): The NIR camera uses a catadioptric lens with a 256 × 256 indium antimonide (InSb) FPA mechanically cooled to cyrogenic temperature. The FPA operated at 70 plus-or-minus 0.5 K at the Moon and showed excellent stability over more than 500 hours of operation in space. The lens design features all ZnSe refractive elements with a relay to provide an external pupil for 100 per cent efficient cold shielding. This lens design was chosen for image quality and focus stability. Wavelength range was constrained by the optics and the InSb response to somewhat less than 1.0 to 5.5 microns. Six wavebands were selected by the NASA advisory committee, all falling well inside this range.

Camera electronic programmable variables allow 4 integration times, 5 bits of gain, and 8 bits of offset. Gain states are spaced approximately evenly from 0.5 to 36 factors of voltage multiplication. Offset is subtracted before gain is applied with 0 V to full well range that can be set in 1/255 full well increments. The NIR camera performance specifications are shown in Table 6.

Longwave Infra-Red Camera (LWIR): The LWIR camera uses a catadioptric lens with a 128 × 128 mercury cadmium telluride (HCT) FPA. The FPA was mechanically cooled to cyrogenic temperatures during operation with a temperature set point goal of 65 K. The optics design incorporates an internal relay with 100 per cent cold shield efficiency. Wavelength ranges was controlled by the cold filter to 8.0 to 9.5 microns.

UPDATED

CONTOUR

Current Status

The CONTOUR comet-chasing science satellite was launched from Cape Canaveral, Florida, at 06:47 UT on 3 July 2002 by Delta 2, delayed two days in to a 25 day launch window that began 1 July to check the spacecraft for suspected dust contamination. A modest 3 m/second over-speed on the Delta first stage was compensated by a 2 second under-burn on the Aerojet powered second stage resulting in a 185 × 196 km, 29.7° orbit, close to nominal. After a 47 minute coast, the second stage fired a second time for 4 second after which the third stage was spun up to 49 rpm and its Star 48B solid rocket motor fired for 84 second placing the combined third stage and spacecraft into a 211 × 108,986 km orbit. About 2 minute after burn termination CONTOUR was spring-ejected from the third stage and coasted during the two elliptical phasing orbits designed to set it up for the injection burn. This would set the spacecraft on course for an earth gravity assist on 15 August 2003 and encounter with Comet Encke on 12 November 2003 followed by three more earth gravity assists and rendezvous with comet Schwassmann-Wachmann 3 (SW3) in June 2006. At 08:49 UT on 15 August the Star 30BP solid rocket motor fired for a 50 second burn designed to increase the velocity of the spacecraft by 1,920 m/second. This would propel CONTOUR from its perigee altitude of 224 km to a sun-orbiting earth-return path for a gravity assist one calendar year later. Ignition of the one and only firing for this boost motor took place with the spacecraft over the Indian Ocean and out of contact with NASA's DSN stations. Contact with the spacecraft was not re-established after this burn and subsequently ground-based telescopes observed three fragments in a path very close to that predicted for a post-burn trajectory but 1,000 km behind the expected position of the spacecraft. This allowed engineers to calculate that the burn probably terminated 2 second early and that the spacecraft may have broken up.

Background

Selected in 1998 as the sixth mission in NASA's Discovery series of lower cost, highly focused space science investigations, CONTOUR is NASA's second mission after Stardust to investigate a comet and is to be followed by the third, Deep Impact, scheduled for launch in 2005. Its name is an acronym of Comet Nucleus Tour, the first mission to study comet diversity by looking at the nucleus of a cometary body on a four-year plan aimed at examining two short-period comets, 2P/Encke and 73P/Schwassmann-Wachmann 3. The mission was also designed to be sufficiently flexible to include a newly discovered and scientifically valuable comet should one appear. CONTOUR was designed to pass within 100 km of the comet nucleus, taking it closer than any previous spacecraft, during which four science instruments would provide images and spectral maps of the nucleus and analyse the surrounding gas and dust. Spacecraft was designed and built by the Applied Physics Laboratory (APL) of the John Hopkins University with much of the design and components from past or operating missions such as NEAR, TIMED, Cassini and Stardust.

To achieve this, CONTOUR would be launched by a Delta 2 (7425) during a 25 day launch window beginning 1 July 2002 with daily opportunities each lasting a mere 6 second. The first mission to use the Indirect Launch Mode (ILM), CONTOUR would be placed in a highly elliptical orbit for several weeks An onboard Star 30 solid rocket motor fired on 15 August 2002 would propel it to a velocity of almost 13 km/second and a sun-orbiting, earth-return trajectory. Exactly one year later the spacecraft would return to earth for a 58,000 km gravity-assist fly-by and instrument calibration taking CONTOUR to its encounter with comet Encke on 12 November 2003 followed by a second earth gravity-assist fly-by at 40,180 km on 14 August 2004. Successive earth fly-by encounters of 218,770 km on 10 February 2005 and 30,000 km on 10 February 2006 would set up the path of the spacecraft for an encounter with Schwassmann-Wachmann 3 on 19 June 2006. The baseline mission was scheduled to end on 30 September 2006. Between comet enounters and during the four cruise periods lasting 120 to 300 days, CONTOUR would be in a state of spin-stabilised hibernation designed to help reduce spacecraft operations and DSN tracking costs. Mission operators would turn off CONTOUR's instruments and most subsystems leaving only the command receivers, thermostatically controlled heaters and critical core components switched on. The command systems automatically monitor spacecraft status and correct potential faults. Ground controllers wake the spacecraft by sending "active spin mode" commands 35 days prior to each earth swing-by, giving time to track the spacecraft, calibrate the instruments and prepare for the swing-by manoeuvre.

Comet encounters are conducted on a standard operational format common to each and typically begin 60 days before the fly-by with spacecraft orbit determination and systems and instrument checks completed 10 days prior to encounter. From this date and up to 12 hours prior to encounter, CONTOUR is programmed to take pictures and make spectral observations of the coma. Optical navigation images used to determine the relative positions of spacecraft and comet are taken daily from up to 5 days before and thereafter twice daily. From 12 hours before, to 12 hours after encounter, CONTOUR is in full encounter mode with all instruments turned on and data recorders operating. From 12 hours to 15 days after encounter CONTOUR plays back its data and a brief thruster firing five days after encounter changes the spacecraft trajectory. CONTOUR was to encounter Encke at a fly-by speed of 28.2 km/second and SW3 at a speed of 14 km/second. The spacecraft's earth-return trajectory makes it possible to redirect the spacecraft toward a new comet and several additional targets were available after the SW3 encounter including comets 6P/d'Arrest and 46P/Wirtanen. Mission flexibility was built in to make possible rare opportunities to catch bright, long period, comets like Hale-Bopp that approached the sun from the Oort cloud in 1997. To find and locate suitable candidates as early as possible, CONTOUR supported a global comet watch in which amateur and professional astronomers search for previously undetected cometary bodies approaching from the fringes of the solar system.

CONTOUR's data was expected to cover much of the ground addressed by ESA's Giotto at Halley in 1986 and NASA's Deep Space 1 at Borelly in 2001, providing insights into the differences between Halley-like comets and shorter period comets such as Encke and SW3. CONTOUR and NASA's Stardust spacecraft carry the same dust analyser offering a direct comparison of mission data.

Specifications

Mass: 970 kg at launch, 387 kg dry with instruments, 503 kg STAR-30 SRM, 80 kg hydrazine propellant. 60.4 kg science instruments.

Spacecraft: 1.8 m tall octagonal main structure, 2.1 m wide, spin stabilised in cruise and three-axis stabilised for comet encounters, incorporating nine body-mounted GaAs solar panels with NiCd battery backup. Layered, 25 cm thick shield of Nextel with Kevlar backing to protect against impacting dust grains mounted as a flat surface to one end of octagonal structure and incorporating CIDA, NGIMS and CFI instruments.

Electronics: APL developed Integrated Electronics Module (IEM) incorporating core avionics onto small circuit cards in one box. IEM incorporates 10 cards for command system, data collection and formatting system, data recorder, guidance and control processor and X-band receiver and transmitter. Backup IEM included in spacecraft.

Power: Body mounted nine-panel GaAs solar array providing 670 W peak power at 1 AU from sun; 9 ampere-hour super CiCd backup battery to store power in event solar panels point too far off sun.

Propulsion: STAR-30 solid rocket motor provides 1,922 m/second velocity change for boost to earth-return heliocentric trajectory. Hydrazine system for 16 thrusters in four clusters of four.

Telecommunications: Transceiver based X-band communications system including 46 cm directional high-gain dish, two low-gain antennas and one pancake-beam antenna. HGA used for communicating during three-axis stabilisation; low-gain during spin-stabilisation. Can receive and send commands and data simultaneously. Non-coherent Doppler system uses transceiver in which uplink and downlink frequencies are independent with onboard oscillator to determine spacecraft velocity relative to earth utilising simpler and more flexible hardware.

Command and data handling: Radiation hardened, high-performance 32-bit Mongoose V processor receives time-tagged commands from the ground at least a week before a scheduled manoeuvre. Commands are usually uplinked at 500 bps but rates of 7.8 bps and 125 bps are also possible. Two solid-state recorders are each capable of storing up to 5 Gigabytes and data and telemetry can be downlinked at 11–85 kbps depending on earth distance and whether low- or high-gain antenna is employed.

Guidance and control: System includes an earth-sun sensor, advanced stellar compass (star tracker) and a gyroscope. When 3-axis stabilised, Mongoose V flight computer processes location and position information from the sensors to carry out specific sun, earth or comet-pointing instructions from mission operators. Ephemeris data on positions of earth, sun and target comets is regularly uploaded into CONTOUR's flight computer. No internal reaction wheels. Hydrazine thrusters employed for pointing, spin-up, spin-down or spacecraft movements or trajectory changes. Processor in primary digital camera also takes signals from flight computer during comet manoeuvres directing spacecraft in roll to maintain camera tracking mirror lock on comet nucleus.

Payload: CONTOUR Remote Imager/Spectrograph (CRISP) combines narrow-angle optical imager with near-IR spectrometer, offers top pixel scale of ~4 m with 10 filters for visible colour of the nucleus. Spectrometer covers 780 to 2,500 nm, spatial pixel scale about three times that of the imager. CRISP points out from the side of the spacecraft and remains protected by the dust shield. Tracking mirror keeps the nucleus in the field of view and guides the spectrometer slit across the surface building an IR compositional map. Seven different imaging sequences are loaded into CRISP computer prior to encounter and selects appropriate sequence for comet's location as comet appears behind dust shield. CRISP directs spacecraft computer to roll onto optimal lock and to track nucleus for a full 30 degrees. Telescope aperture is 100 mm, focal length is 683 mm, fov 1.2° × 1.2°, pixel fov 20 mrad across wavelength range 450 to 770 nm. 10 spectral channels with a TH7888A detector with 1,024 × 1,024 pixel per image. CRISP weighs 36.7 kg and draws on 45 W of electrical power. It is supplied by the Applied Physics Laboratory of the Johns Hopkins University.

CONTOUR Forward Imager (CFI) locates and images comet beginning several days prior to encounter through opening in dust shield providing navigation information for terminal position guidance and scientific images of coma. CFI telescope images from the mirrored side of a cube and when obscured by dust impact on encounter cube is rotated to present the next mirrored surface for the next encounter. Telescope aperture 60 mm, focal length 300 mm, fov 2.5° × 2.5° with a pixel fov of 43 mrad in a wavelength range of 308–920 nm in 10 spectral channels with a Marconi 47–20 detector, 1,024 × 1,024 pixels per image. CFI weighs 9.7 kg and draws 9 W electrical power. It is supplied by the Applied Physics Laboratory of the Johns Hopkins University.

Neutral Gas and Ion Mass Spectrometer (NGIMS) measures abundance and isotope rations of neutral gas and ion species in each comet coma. IGIMS is a quadrupole mass spectrometer employing two separately optimised ion sources. Using both sources NGIMS is built to rapidly switch between measurements of neutral gas and ambient ions in the coma. It measures the chemical and isotopic composition of neutral and ion species over a range of 1–294 atomic mass units. The NGIMS weighs 13.5 kg, draws 36 W of power and is supplied by the NASA Goddard Space Flight Center.

Comet and Interstellar Dust Analyser (CIDA) analyses elemental and chemical composition of dust and ice grains in coma. Consist of an inlet, target, ion extractor, time-of-flight mass spectrometer and ion detector.

UPDATED

Galileo

Current Status

The end of the Galileo mission on 21 September 2003 after almost eight years in the Jovian environment conducting surveys of the planet and its major moons has stimulated support for the Jupiter Icy Moons Orbiter (JIMO) mission proposed by NASA for launch in 2012. Responding to searching questions arising from the Galileo results, JIMO would be powered by a nuclear electric engine and produce electricity through a nuclear fission reactor. It would take five to eight years to reach Jupiter and spend four to six years orbiting Europa, Gannymede and Callisto in turn.

Background

Plans to follow up the Voyager fly-bys started in the mid-1970s with a formal proposal from the NASA scientific community for what was then known as Jupiter Orbiter-Probe (JOP) in 1976 and was approved by Congress for an October 1977 start. By 1981, NASA had invested US$300 million in the project. As first approved, NASA envisioned the JOP spacecraft would be launched in January 1982 from the Shuttle with a three-stage USAF Inertial Upper Stage (IUS) providing the energy to depart earth orbit for a trajectory to Jupiter using gravity assist from a Mars fly-by. By January 1980 NASA was forced to change this plan when it became apparent that the IUS would lack the energy to propel the spacecraft to Jupiter and it was decided to send the Orbiter and the Probe on separate Shuttle-IUS combinations. Launch dates were reset to February 1984 with arrival at Jupiter in July 1986 for a total mission cost of US$650 million. In November 1980, with a further modification to the plan, Hughes Aircraft received a US$40 million contract to develop the probe carrier, to be launched by Shuttle in March 1984 on a trajectory to reach the planet 1,200 days later in July 1987. Some of the delay in the programme was caused as much by the delay to Shuttle flight operations as by the uncertainty over upper stage and launch opportunities, with the first fight of Columbia not taking place until April 1981.

As NASA envisioned the project, the probe would separate 100 days before planet arrival and enter Jupiter's atmosphere on the sunlit side at 160,000 km/h. After slowing in a few seconds to a few hundred km/h, a parachute would open to allow the instruments to measure the vertical profile of the atmosphere. The carrier would relay the probe's data back to Earth, from an orbit similar to that of Ganymede, at about 1,002,400 km from Jupiter. It would use that body's gravity to skew its orbit at each near approach so that in 11 passes in 20 months it would cover all regions of the Planet. Powered by two RTG nuclear sources used on previous outer planet missions, Galileo would incorporate a 5 m diameter furlable antenna to relay 30–50 m resolution imagery of the moons.

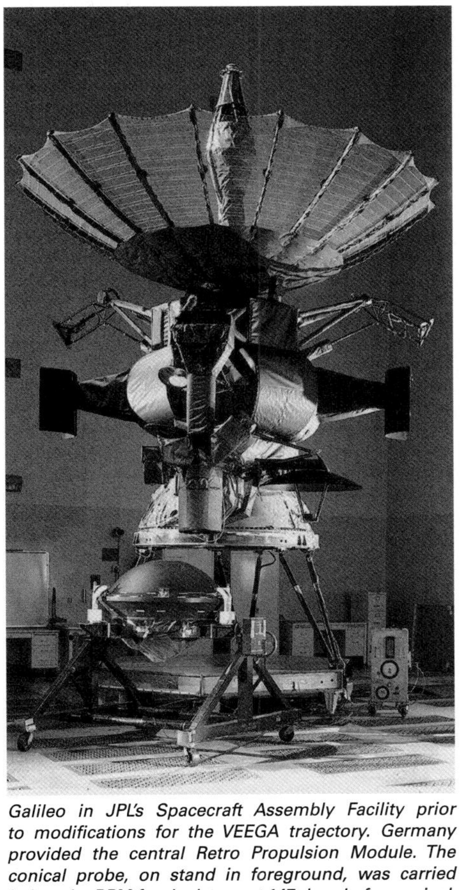

Galileo in JPL's Spacecraft Assembly Facility prior to modifications for the VEEGA trajectory. Germany provided the central Retro Propulsion Module. The conical probe, on stand in foreground, was carried below the RPM for deployment 147 days before arrival. Galileo is topped by the 4.8 m {70} High Gain Antenna, which has not deployed because of three stuck ribs (NASA) 0517112

As NASA faced financial restrictions, the project suffered many delays and changes in the original plan. At the urging of Congress, in January 1981 the three-stage planetary IUS was cancelled in favour of a wide-body cryogenic Centaur stage, known as Centaur G-Prime, and the dual-launch hitherto conceived for the Shuttle-IUS combination reconverted to a single launch scheduled for flight in April 1985 and arrival at Jupiter in mid-1987 for a revised total mission cost of US$695 million. In January 1982 the Office of Management and Budget forced NASA to cancel the Shuttle-launched Centaur declaring that NASA did not have a sufficient number of heavy deep-space missions to justify its development cost, forcing NASA to readopt the IUS utilising a trajectory that envisaged a launch in May 1985, a fly-by of Earth in July 1987 to acquire the necessary energy for a slingshot to Jupiter, arriving at Jupiter at the end of 1989 for a two-year tour on 11 encounters with the planet and its Galilean moons. The mission cost had now risen to US$864 million. In September 1982 Congress vetoed the decision to cancel the wide-body Centaur, reinstating it for a two-year journey beginning May 1986, a launch delay made necessary because of time lost when the stage had been cancelled in the January of that year. Mission cost had now increased to US$874 million.

By 1983, NASA set the total cost already expended at US$630 million, but said that spare parts already acquired could be used for a US$100 million Galileo Saturn orbiter. This would have been launched in 1987–88, using Earth gravity assist for an eight-year transit. That project was cancelled and eventually emerged as Cassini-Huygens.

There were growing concerns even before the Shuttle accident of January 1986 over safety considerations with the Centaur G-Prime in the payload bay. When the Shuttle version of Centaur was finally cancelled in June 1986, some Galileo proponents proposed dumping a Shuttle launch altogether in favour of a Titan 4 Centaur G-prime combination in 1991 as an alternative to competing with Ulysses for a Shuttle launch. NASA refused to entertain the notion even at the risk of delaying the project until it might be ultimately cancelled. Ultimately, a Shuttle launch combined with IUS #19, previously assigned to a DoD mission, in conjunction with a Venus double-Earth swing-by (VEEGA: Venus-Earth-Earth Gravity Assist), emerged as the most promising solution to the launch dilemma for a new mission cost estimate of US$1.3 billion. The launch was now assigned to the October-November 1989 launch window for a six-year VEEGA trajectory with encounter in 1989.

Because a Shuttle/IUS combination is insufficient to fling Galileo into a direct Jupiter trajectory, JPL's mission design section identified a multiple gravity assist scheme leading to arrival in December 1995. JPL identified no additional benefit with a Mars swing-by, but it was found that the six-year VEEGA route would generate the required energy. On that basis, NASA set the launch of the 2,561 kg craft (including 925 kg propellant and 339 kg probe) for the window of 8 October-24 November 1989, with 30-minute daily windows. The lengthy period within Earth's orbit presented the greatest problem and Galileo underwent modifications to combat the almost doubled solar flux and the effect it would have on delicate instrumentation. The modified spacecraft, even with an additional 100 kg of shielding and equipment, emerged slightly lighter. Modifications included the addition of a large thermal disc above the bus and an LGA on the RTG boom for Earth communications during the Venus/Earth fly-bys.

By the end of 1987, delays and modifications had increased the total cost of development, launch and 30 days of space operations from US$673.5 million to US$893 million, with the rest of the mission rising from US$200 million to US$500 million (assuming a 7 per cent inflation rate). Added to the fact that Galileo's design would represent 20-year-old technology when it reached Jupiter, this generated speculation during 1987 that the mission would be cancelled. Galileo competed with Ulysses for a Shuttle launch in a Jupiter window that recurs only at 13-month intervals. ESA and NASA finally agreed in April 1987 that Galileo would be launched in the 45-day October-November 1989 window, with Ulysses following in October 1990. There was to be no repetition of the 1986 plan (foiled by Challenger) of attempting both during the same window.

Many observers doubted that departure would occur at all, because the spacecraft suffered a legal challenge prompted by concerns that the 22 kg of plutonium dioxide carried by Galileo's two RTGs (the first new-model RTGs referred to as GPHS General Purpose Heat Source RTG) might spill in an explosion. The RTGs were not inserted until 3 October. Atlantis began its fifth mission at 16.53.40 UT 18 October 1989.

Galileo was the first to make in situ observations of an outer planet's atmosphere (Hughes) 0517111

Checkout of the Galileo/IUS combination began 2 h into the mission and after 5 h 42 min the crew reported the IUS table was raised to 29°, followed 10 minutes later by the required 58°. Mission Specialist Shannon Lucid commanded spring deployment at 6 h 21 min 23 s. The IUS automatically ignited its first stage 60 minutes later over Borneo, following the 149 second burn with stage 2's 105 seconds. At this stage, the vehicle was being monitored by the USAF Consolidated Space Test Center at Onizuka AS, California via their ARIA Advanced Range Instrumentation Aircraft. IUS then pointed Galileo within 9.5° of the Sun for thermal shading, spun it up to 2.8 rpm and deployed the craft's magnetometer and RTG booms, before separating at 01.05 GMT 19 October.

The probe and its instruments underwent 6 h of tests on 26 October 1989 before being closed down until its next annual test during December 1990's Earth fly-by. The orbiter's instruments were also checked and the Heavy Ion Counter and magnetometer left on to return cruise data. The HIC was conceived in 1984 to answer concerns over Single Event Upsets that were being found on Earth-orbit satellites. The test model of Voyager's Cosmic Ray Science instrument was reworked to monitor the rich heavy ion environment found around Jupiter. Ironically, HIC's older components are more SEU-resistant than Galileo's more modern elements. Four large solar events occurred soon after launch but the detected impacts caused no damage.

The first mid-course correction sequence TCM-1 was executed over 9–11 November, yielding a 17 m/s delta-V. 10 N thrusters on the spin section fired >5,500 times in typically 1 second pulses. TCM-2 on 22 December provided a 0.8 m/s trim for a 16,000 km Venus fly-by. TCM-3 was not required. During its passage over Venus' morning terminator, Galileo employed its NIMS infra-red mapper to observe cloud features on the dark side for the first time and map airglow at 2.7 μm. The instrument provided the first views of small scale features over the mid-latitude and equatorial regions, in addition to performing spectroscopy of the middle/deep layers of the thick atmosphere. Features matching those on radar images suggest the surface may be visible in the 1 μm carbon dioxide window. The plasma wave instrument confirmed the presence of lightning. A closest approach of 16,106 km was made over 41° S at 05.58.48 GMT on 10 February 1990, gaining 8,030 km/h. For a day before and for up to seven days after, the charged particles, dust and magnetic fields, infra-red/ultra-violet instruments were active but most of the data were recorded for playback mid-November shortly before the first Earth fly-by. The link went via one of the LGAs (thermal considerations required the HGA to remain furled until April 1991). However, three TV images and some NIMS mapping were returned. As planned, 81 TV images were recorded but a software timing problem led to the system shuttering an additional 452 times (without recording) in the 5 hr after closest approach. Final imaging took place 17 February.

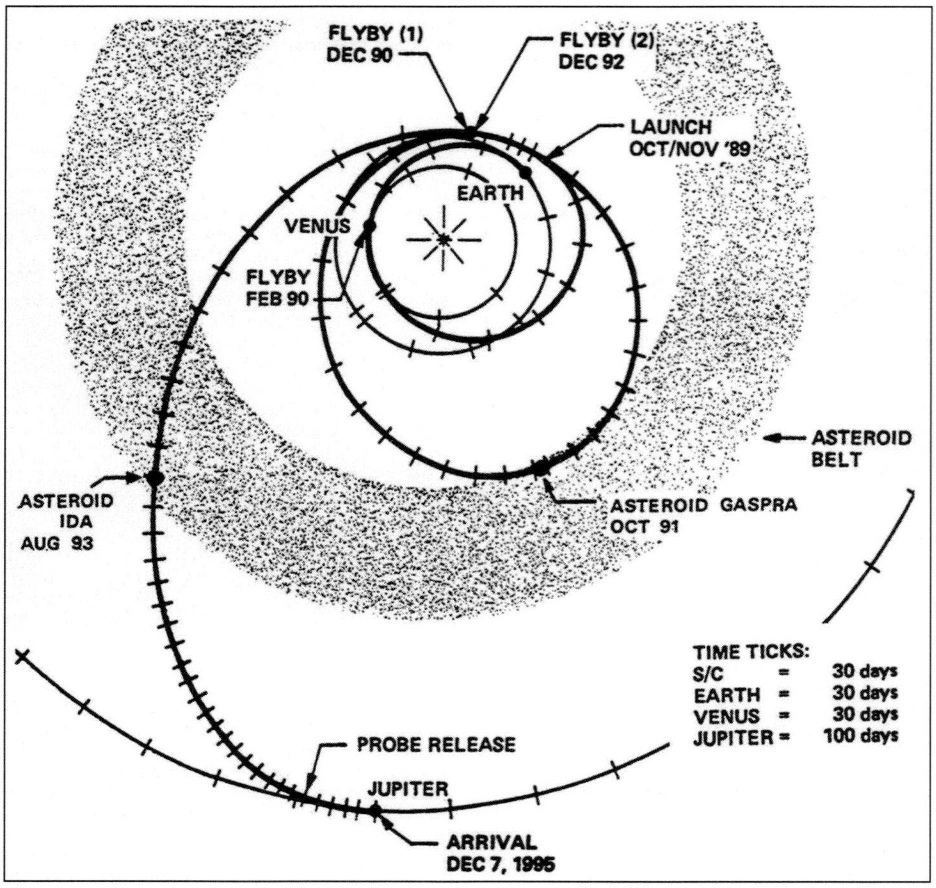

Galileo's unique Venus-Earth-Earth swingby path enhanced the Inertial Upper Stage's Jupiter capability and permitted dual asteroid encounters 0516747

Inserting the probe's descent module into the heatshield (Hughes) 0517109

Following the fly-by, Galileo's heliocentric orbit was 0.699 × 1.28 AU, heading towards first perihelion 25 February and first aphelion 23 August 1990. Galileo performed its largest course adjustment before Jupiter in April/May 1990, to patch its post-Venus path into the required Earth fly-by trajectory. The VEEGA route initially identified in July 1986 required a November 1989 launch but the insertion of this large adjustment expanded the window into October. TCM-4's first set of pulses, over 9–12 April, yielded a 24 m/s (24.8 m/s planned) change, followed by 11.3 m/s on 11/12 May, using the lateral thrusters.

After the Venus fly-by, mission planners approved 1991's encounter with asteroid Gaspra. Propellant margins had eased with IUS' accurate injection, further trajectory calculations and the discovery that Galileo's thrusters were providing 2 per cent greater efficiency than expected all encouraged confidence in the mission.

Observations were made of Comet Austin during May 1990 and the following August the Extreme UV instrument registered hydrogen in the coma and tails of Comet Levy. Only small corrections were required after TCM-4 to adjust the trajectory for the first Earth fly-by: TCM-5 provided 0.9 m/s on 17 July; TCM-6 0.5 m/s on 9 October, TCM-7 1.3 m/s on 13 November and TCM-8 0.05 m/s on 28 November. The tape recorders' stored Venus data dumped their data to earth station on the way by on 19–21 November.

Passage through the Earth-Moon system provided the opportunity to certify the instruments and undertake the first terrestrial observations from a planetary craft. Galileo flew within 960 km (8 km above target) of Earth at 20.34.34 GMT 8 December 1990, detecting clear evidence of life in atmospheric trace constituents. The cameras made one exposure each minute for a day for a 1,500-frame film sequence to be assembled, showing Antarctica from a unique perspective. Lunar observations were aimed at composition of the far side crust, Mare Orientale's structure and imaging little-known regions around the south pole. Galileo confirmed the existence of the suspected (from Apollo 15 data) 1,940 km diameter Aitken Basin in the south, underlying the younger Orientale impact feature. The entire Earth/Moon encounter returned 2,675 images. It also set up October 1991's Gaspra fly-by, by increasing Galileo's heliocentric period from one to two years and increasing ecliptic inclination to 4.5°. The TCMs before Gaspra fine-tuned this gravity assist.

Previously furled because of thermal requirements, the 4.8 m diameter HGA was commanded to open 11 April 1991 but jammed less than half way, with three of the 18 ribs still slotted in their launch positions on the central mast. Analysis and tests with the spare flight model suggested that insufficient lubricant, aggravated by chafing during road transport caused the problem. JPL turned Galileo in late May 1991 to use solar heating induced thermal expansion to unjam the mast. Then JPL engineers realised contraction and not expansion would be the best solution. The mast was held in shadow for 32 hr in July and then for 50 hr in August while Galileo was 2.0 AU from the Sun. These attempts failed. The HGA had been scheduled for the Gaspra fly-by, but the data had to be recorded instead for replay during the December 1992 Earth fly-by. The fifth and most aggressive cold soak was tried in December 1991 2.2 AU out around aphelion but the ribs remained stuck. As Galileo approached 1 AU during 1992 for the Earth fly-by, hot/cold cycling was attempting to 'walk' the ribs out. Rotating the craft cost 4 kg of valuable propellant.

Galileo, meanwhile, achieved the first close encounter with a known asteroid during 29 October 1991. Following Earth's December 1990 gravity assist, five TCMs fine-tuned the approach to 951 Gaspra. Ground-based observations had located Gaspra to within 200 km and Galileo's own navigation images during September and October 1991 refined the ephemeris to 50 km. To avoid any debris cloud, the probe approached no closer than 1,600 km (actual 1,604 km) during the 8.0 km/s fly-by at 22.37 GMT, some 410 million km from Earth. In fact, no dust impacts were registered. Even with the relatively large separation, the gravitational deflection of Galileo's trajectory was sufficient to yield a good estimate of Gaspra's mass and hence, average density. Field reversals detected by the magnetometer indicate a strong magnetic field, requiring Gaspra to be iron or nickel.

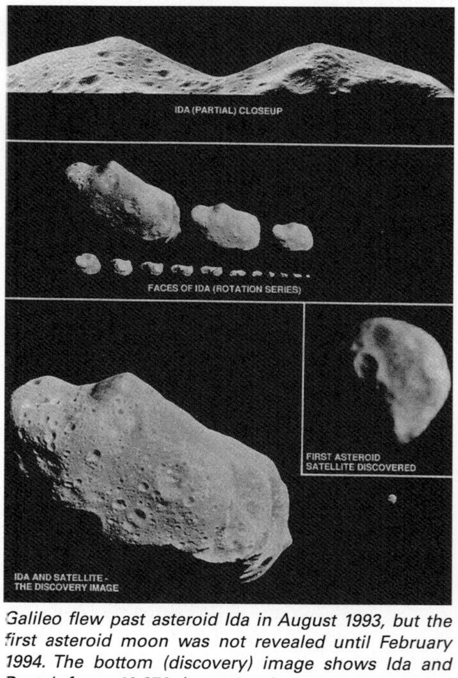

Galileo flew past asteroid Ida in August 1993, but the first asteroid moon was not revealed until February 1994. The bottom (discovery) image shows Ida and Dactyl from 10,870 km, 14 minutes before closest approach. Inserted is the best image of Dactyl, 10 minutes later, from 3,900 km. At the top is the highest resolution achieved at Ida, some 25 m/pixel 46 s after closest approach. The images remained stored until 1994 when Earth was close enough again for the 40 bit/s relay (NASA/JPL) 0517110

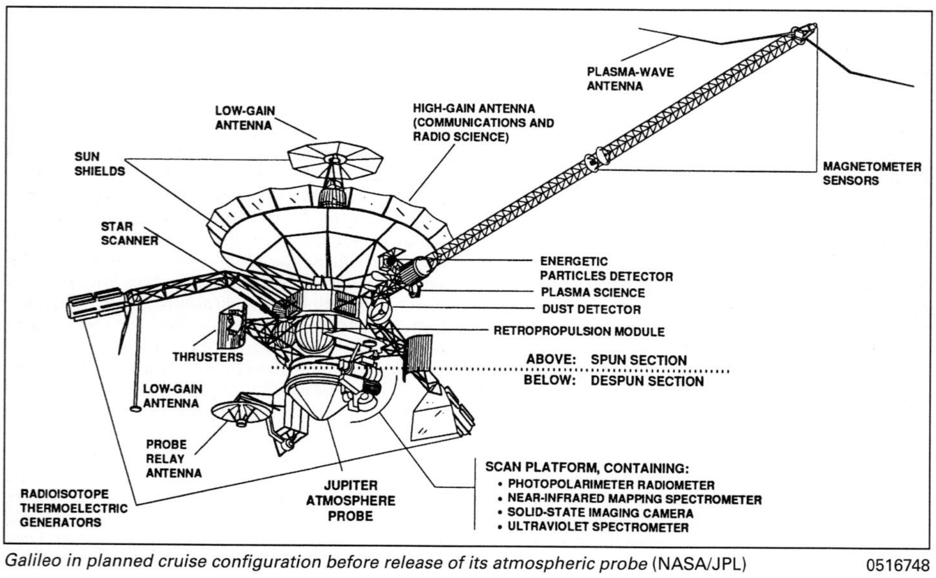

Galileo in planned cruise configuration before release of its atmospheric probe (NASA/JPL) 0516748

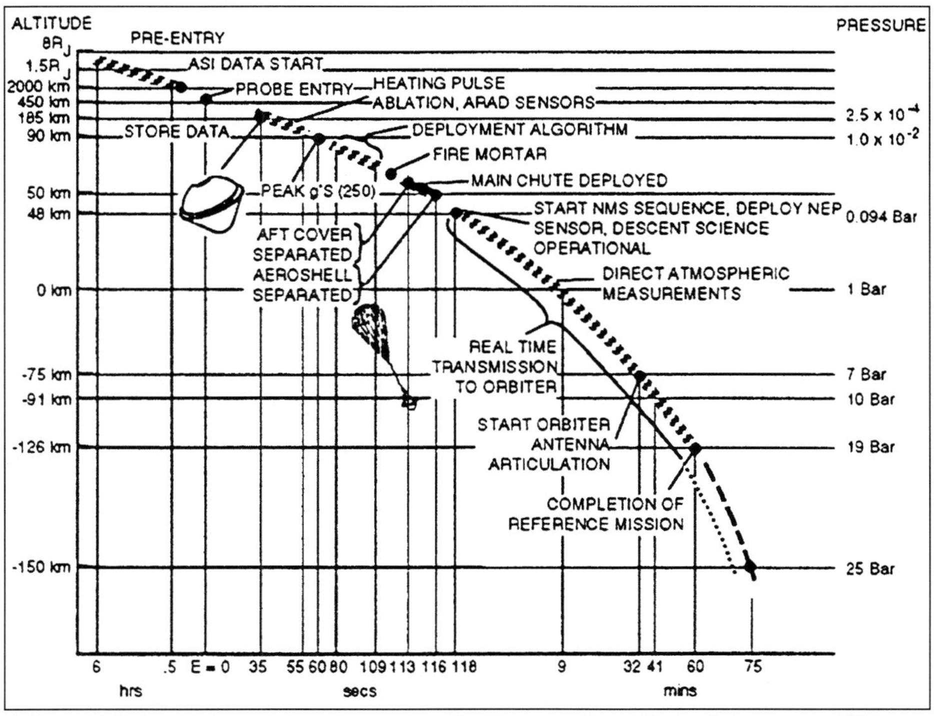

Galileo probe's descent sequence was pre-programmed, with events triggered by accelerometer cues (JPL)
0517108

A total of 150 images were recorded during the encounter, 126 of them during the last hour before closest approach. From 35–29 minutes before closest approach, a mosaic of nine colour frames (that is 27 frames through three filters) was acquired to ensure Gaspra was seen. Similarly, a 49-frame black/white mosaic was recorded 29–16 minutes before closest approach. No frames were taken at closest approach because of the angular speed. The 800 × 800 element Si CCD camera has a broader spectral range and faster response than the vidicon tubes aboard Voyager. The Near-IR Mapping Spectrometer (NIMS) is a powerful diagnostic tool never flown before, capable of classifying the surface mineralogy through the solar reflection spectrum, highlighting silicates, carbonates and nitrates, for example. The photopolarimeter radiometer studied the photometric and thermal surface properties. If there is an associated dust field, the German dust detector would have measured the size distribution, spatial extent and trajectories of its sub-μm particles. With a sensitive area of 1,000 cm^2,= particles are registered after impact with a gold target which generates a small plasma cloud. All of the data were recorded for playback when the HGA was released or during end-1992 Earth's fly-by, whichever came sooner.

In fact, mankind's first close-up view of a known asteroid came on 7 November 1991 when controllers retrieved the second-highest resolution image (comprising four frames for full colour), providing 160 m/pixel from 16,000 km, at 40 bit/s via the LGA. The 17 × 12 × 11 km S-type Gaspra was surprisingly rounded, suggesting a considerable dust layer held in place by the 0.0005 *g* gravity.

Release of Galileo from Shuttle Atlantis in October 1989 0085791

Craters ranged up to around 2 km diameter. Grooves similar to those seen on Phobos were at the edge of resolution. The highest resolution image (50 m/pixel) was retrieved 18 May – 5 June 1992, and all the rest in late November 1992. The International Astronomical Union in 1994 approved names for Gaspra's features based on the world's spas and resorts, as the asteroid itself was named after a resort on the Crimean Peninsula.

November 1991's TCM-13 was not required but four sequences were executed for the second Earth/Moon fly-by: TCM-14 21.0 m/s 4 August, –15 0.7 m/s 9 October, –16 0.9 m/s 14 November and –17 0.03 m/s 28 November. Mass was 2,415 kg after TCM-14. The probe was put through its first Mission Sequence Test (MST) during 20 November, taking it through its actual mission sequence. Performing observations similar to those on the first passage, Galileo flew past the Moon at 110,000 km at 03.58 GMT 8 December 1992 and then Earth at 303.1 km (303.8 km planned) at 15.09.25 GMT over 34° S/6° W, adding 3.7 km/s to enter the direct Jupiter route. The passage permitted high resolution (and the first multi-spectral) observations of the lunar north pole. UVS discovered that Earth's geocorona is far more extensive than previously believed. As last time, time lapse films were made, including the first-ever showing of a full 24 hr rotation of Earth.

After thermal walking failed to release the HGA by July 1992, engineers switched to a 'hammering' technique, pulsing the two electric motors to drive the ballscrew mechanism further. They waited until December 1992 around perihelion when the antenna was warmed to its assembly temperature. The first attempt 29–30 December did move rib No 2 by 8° but the 5–11/14–19 January 1993 sequences failed. After 13,320 pulses, they gave up hope of deploying the HGA on this first Jupiter orbiter. A final attempt by spinning Galileo up to 10.5 rpm 11 March 1993 while hammering, also failed. 312 pulses of a 10 N thruster were required for the spin up from 2.89 rpm and 313 pulses for the spin down. It was next hammered 1,080 times again 29 August 1994, as a last try before arriving at Jupiter, although there was little hope of it working. It was hammered 1,080 times for the last time 22 March 1996.

The LGA officially became the baseline mission in March 1993, although software and ground hardware improvements should yield around 70 per cent of the originally planned science return (80 per cent atmosphere, 100 per cent probe, 70 per cent satellites and 60 per cent magnetosphere). 1,500–2,000 images will be returned instead of the 50,000 expected. The 40 bit/s has been improved to an equivalent of more than 1 kbit/s (160 bit/s are actually returned, but data compression multiplies the effective rate). By using data compression and new error correction algorithms, arraying Earth antennas and increasing their sensitivities (particularly of the Canberra station, the most important, as Jupiter is then in the southern sky).

Galileo made history again, when in conjunction with the Mars Observer and Ulysses, JPL used the craft as along baseline gravity wave detector between 21 March and 11 April 1993. The results of this experiment were negative.

Approval for the asteroid Ida fly-by on 28 August 1993 was required before the Earth-2 TCMs. Although it cost 30 kg propellant and earlier in the mission looked likely to be rejected (principally because the six-day launch delay imposed a 58 kg penalty), Galileo's performance and the science return from Gaspra prompted approval from NASA HQ 25 June 1992. Earth-2 fly-by was so accurate that TCM-18 13 days later was not required. TCM-19 (2.1 m/s) on 19 March 1993 provided the bulk of retargeting to Ida and the 0.6 m/s TCM-20 trim 13 August cancelled the need for TCM-21 26 August. Galileo followed a Gaspra-type sequence and made its closest approach of 2,410 km at 16.51.59.0 GMT 28 August 1993. A potentially disastrous event hit Galileo 256 minutes out: for some reason, it turned off its gyros and returned to cruise mode. Engineers determined the encounter could still be successful in this mode by pointing the scan platform using the actuators' encoders. However, the mode had stowed the platform, so a precisely timed command was sent 198 minutes before arrival to restore proper pointing. 150 image frames were recorded, about half of them to ensure capture. 13 colour and 8 black and white images were stored, to be transmitted over 1,300 h at 40 bit/s via the LGA September 1993 and 16 February – 26 June 1994 when Earth was close enough. The highest resolution five-frame mosaic was recovered over about 30 h by 22 September 1993 (the last day before Earth was too distant), showing the 52 km-long Ida at 25 m resolution. The battered surface shows it to be an old body. Ann Harch of the SSI team discovered a 1.2 × 1.4 × 1.6 km diameter moon, subsequently named Dactyl, while examining image slices on Galileo's recorder 17 February 1994. It was found in the NIMS data 23 February. The image return completed 26 June 1994 showed it to be as battered, and thus as old, as Ida. They are similar bodies but NIMS revealed different compositions, suggesting that Dactyl was captured and did not break off Ida. The pair appeared together in 47 SSI images, allowing Dactyl's orbit and hence Ida's density to be modelled. The orbit could not be calculated with certainty, but gave an Ida density range of 1.9–3.2 g/cm^3, suggesting a light and/or porous structure.

The 38.6 m/s TCM-22 was performed 4–8 October 1993 using 35.2 kg propellant, for the first time slotting Galileo into the Jupiter atmosphere entry corridor. The almost 5,000 pulses on each of the two lateral thrusters provided the most accurate burn yet – only a 0.15 per cent underburn. The 10 N thrusters will not be used for such a large delta-V again. 15 February 1994's 0.1 m/s TCM-22A corrected the error. Five instruments provided astronomers' only direct observations of the disrupted Comet Shoemaker-Levy 9 impacting Jupiter 16–22 July 1994. Data return was completed 29 January 1995.

TCM-23 (12 April) was perfect, so TCM-24 (23 June) was cancelled. The probe was released on 13 July 1995, 147 days and 80 million km from Jupiter. The umbilical was cut 11 July, meaning that it could never again receive commands and would not transmit telemetry until after its parachute opened. Galileo assumed the release attitude, spun up to 10.5 rpm and fired the three explosive nuts on the probe's tripod assembly. Springs ensured a 30 cm/s separation. Galileo avoided following the probe into the planet by making the 62 m/s 308 second TCM-25/ODM Orbiter Deflection Burn 27 July, firing its 400 N engine for the first time. It also set up unique inbound fly-bys of Europa and Io, the reason for selecting arrival 7 December. The 10 N thrusters could have made the ODM, but the larger engine's performance was demonstrated for Jupiter Orbit Insertion. Galileo's approach 6° below the satellites' orbital planes provided the mission's only views of Europa's south pole region, from 32,500 km (six times closer than Voyager) within 9 h of perijove. Then, 4 hr before perijove, came Galileo's only close (900 km) fly-by of Io because of the harsh radiation environment, providing 20–30 m resolution. The encounter reduced Galileo's JOI delta-V by 175 m/s and provided passage through Io's plasma torus. However, the tape recorder jammed 11 October 1995 and all of the inbound remote sensing was cancelled. Instead of rewinding after recording an image, the tape continued slipping on a capstan for 15 h before a ground command shut it down. Fearing the tape was weakened, that spot was hidden under 25 turns on the takeup reel. As the tape was essential for recording the probe's unique descent data, it was deemed safer to cancel further remote sensing until April 1996 after the last probe data had been retrieved. In fact, 3 h of high-resolution

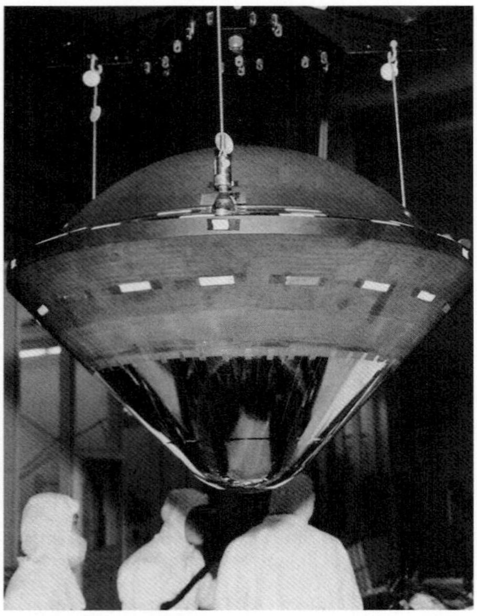

Galileo entry probe fully assembled prior to launch
0085822

fields/particles data was recorded during Io plasma torus passage.

As Galileo hit perijove 130,000 km above the clouds at 21.53 GMT 7 December 1995, the probe entered the atmosphere at 6.5° N/4.4° W at 22.04 GMT and transmitted data for 57 minutes the descent lasted 156 km after its parachute opened 53 seconds/26 km late (which meant data began below the cloud tops. The cause was a faulty *g*-switch). Timers activated the probe 6 h before entry. It was expected to survive down to 20 Earth atmospheres following the 47 km/s, 250 *g* entry (designed to withstand 400 *g*), but it achieved 22 atmospheres and 425K. The full data were stored on Galileo's tape recorder and in partial back-up form in its solid-state Command & Data Subsystem. The back-up data were read out at intervals in December/January and the tape January-April 1996. Preliminary findings showed an intense radiation belt 50,000 km above the clouds, a sun-like helium to hydrogen ratio, few organic compounds, one cloud layer (three expected) and 640 m/s winds below the clouds (powered by internal heat instead of the Sun).

Regardless of the probe's survival at 75 minutes, the orbiter had to break off to prepare for JOI. The 643 m/s, 48.98 minutes 400 N burn starting 00.27 GMT 8 December 1995 and consuming 377 kg propellant established Galileo in a 198-day closed path about Jupiter – the first man-made satellite of the giant planet. Perijove was raised from 4 R_J to 10 R_J at first apojove on 14 March 1996 in a 377 m/s burn to avoid the radiation hazards. The planned 11 tours over 23 months with Europa, Callisto and Ganymede close encounters thus began. The 400 N engine is not needed after this third firing. The 185 kg consumed left 105 ±5 kg, leaving a 5 kg margin to complete the mission with 90 per cent confidence (13 kg is still held aside for the project manager to allocate). Galileo's total delta-V capacity is 1.6 km/s; the tour would require 6 km/s without gravity assist, whereas 146 m/s is allocated.

On 15 April, the play-back of the tape-recorded science data from the Atmospheric Probe was completed. Other experiments on the spacecraft were gathering and transmitting data about the jovian magnetic field during the period leading up to the first Ganymede encounter. On 3 May, an orbital trim manoeuvre was completed to adjust the arrival time and geometry for the encounter. On 3 June, the spacecraft began to transmit data which had been recorded during its fly-by of the satellite Io six months earlier on 7 December. Images showed how much the satellite's geography had changed due to volcanism since the jovian fly-by visits of the Voyager spacecraft.

A final orbital trim manoeuvre took place on 24 June and the first fly-by on Ganymede came at 09.29 GMT on 27 June, when Galileo passed 844 km above the satellite's surface. On 24 August, Galileo unexpectedly put itself into a safe mode due to processing time allocation being exceeded in a central computer on the spacecraft: control was switched to a back-up computer which performed basic engineering functions. Four days later, the main on-board computer was brought back on-line. In the meantime, on 27 August the spacecraft had fired its thrusters once more to fine-tune its trajectory for the second Ganymede fly-by. Encounter came at about 19.00 on 6 September, this time the minimum distance being 262 km. Once more, instruments were not only taking photographs of the satellite's surface, but also probing the magnetic and particle environment and gravity characteristics.

The first Galileo fly-by of the satellite Callisto took place at 13.34 GMT on 4 November, the minimum distance being 1,104 km. On 26 November, a trajectory trim was performed in advance of the first encounter with Europa. The encounter took place at 06.53 on 19 December at a minimum distance of 692 km. There was a lull in Galileo's encounters with the Jovian moons while Jupiter was experiencing solar conjunction, so the revolution 5 encounter was with the planet Jupiter itself, closest approach of 646,000 km coming at 09:36 GMT on 19 January 1997. Then it was back to the moons of Jupiter beginning again on revolution 6. On 20 February 1997 at 17.06 GMT, the second encounter with Europa took place with the minimum distance being within 1 km of the planned 586 km. Further trim manoeuvres took place on 13 and 31 March in advance of the next satellite encounter, the third with Ganymede. This fly-by came at 07.11 on 5 April, the distance being 3,102 km.

Galileo began its investigation of the Jovian moons with two encounters at Ganymede on successive orbital passes, each elliptical swing around Jupiter taking several weeks while the spacecraft drifted several million kilometres out to the apojove of its orbit ensuring protection from the planet's intense radiation belts. Revolutions 7 and 8 also provided two successive encounters with Ganymede in April and May respectively, the last time this Jovian moon would be visited by Galileo. Following this, two close passes of Callisto where Galileo swooped to within 415 kilometres of the surface preceded a third encounter with Europa to close out the 11 fly-by encounters that formed the primary mission for which the spacecraft had been launched. Successful performance in the Jupiter system and equally successful lobbying from the Galileo team gained approval for a two-year extension to be known as the Galileo Europa Mission (GEM).

GEM would double the science time at Jupiter, extending the mission from two to four years and extend the number of encounters from 11 to 25. Originally scheduled to end on 7 December 1997, the mission would continue to the end of 1999. Divided into three tightly focused periods of observation known as 'Ice', 'Water' and 'Fire', GEM would build upon findings from encounters at Europa to investigate key aspects of the moon and its interaction with Jupiter. The 'Ice' phase would require eight successive passes of Europa (December 1997 to February 1999) during which the spacecraft would search for further evidence of a salty ocean beneath an ice crusted surface. It would look for volcanic activity to indicate a geologically active interior and obtain images with a resolution of just six metres during a close pass of just 200 km.

The second, or 'Water' phase would call for four successive swings round Callisto (May, June, August, September 1999) to halve the distance of closest approach to Jupiter (perijove reduction) for investigations into water circulation in cloud tops that surround the giant planet. On each encounter, Galileo would fly through the ring-shaped torus of charged particles that links innermost major moon Io with Jupiter itself, mapping the density of sulphur spewing from the surface.

The most dramatic phase, 'Fire', provides a unique opportunity on the last two GEM passes (October and November 1999) for a close encounter with Io. On the last one, the spacecraft dives down to a distance of just 300 km as Galileo swoops through the active plume of frozen sulphur from Io's Pillan Patera volcano taking pictures with a resolution of six metres while intense radiation peppers the camera's light detector with flashes visible in the images. GEM formally ended on the last day of January 2000. Galileo began 2000 with an intensive series of operations of which a fly-by of Europa at a distance of 351 km was the centrepiece. Galileo then performed observations of Jupiter's smaller moons Amalthea, Thebe and Matis and of the large inner moon Io. These observations began 3 January and lasted until 4 January. Galileo's multi-suite of fields and particles instruments observed the plasma torus, recording data from 357,000–429,000 km from Jupiter-the third deepest set of torus measurements since Galileo arrived at the planet in December 1995. Galileo returned to the Jovian moons to make its third close encounter with Io at 14:32 UT 22 February 2000 at a distance of 198 km. During this penetration of Jupiters inner, and intense, radiation belts, Galileo encountered several computer resets but software, written to handle these, prevented data loss. Galileo successfully completed its GEM extension phase with an encounter with Europa on 3 January 2000 at 351 km. NASA headquarters has agreed in principle to a so called Galileo Millennium Mission which anticipated encounters with Io on 22 February, Ganymede on 20 May and 28 December and for Joint observations of Jupiter in conjunction with the Cassini spacecraft en-route to Saturn but scheduled to fly by Jupiter in December 2000. The 22 February 2000 fly-by of Io was at a closest ever distance of 198 km, taking pictures of the bizarre moon and bringing to about 14,000 the number of images taken by Galileo. Shortly after the encounter with Europa on 3 January, Galileo imaged inner moons Amalthea, Thebe and Metis and then took pictures of Io.

Mission planners were given tentative approval for further satellite encounters: with Io on 22 February and Ganymede on 20 May and 28 December 2000. On 8 September 2000, Galileo reached an apojove of 20.7 million km, the longest and largest orbit since reaching the planet in December 1995, placing it on a trajectory to support co-operative operations with Cassini as it paned Jupiter on 30 December en-route to an encounter with Saturn in July 2000. In June 2000, Galileo left Jupiter's magnetosphere for the first time since 1996.

In March 2001, NASA finalised definitive plans for remaining mission time with Galileo. Anticipating the arrival of Cassini in Saturn orbit in 2004 and balancing the needs of the DSN, Galileo would perform a final series of encounters including scans of Io, Callisto and Amalthea before completing a direct plunge into Jupiter in September 2003.

Activating this plan, four encounters in 2000 visiting Europa, Io and Ganymede (the latter twice) preceded one encounter with Callisto and three encounters with Io in 2001. In 2002, a sixth and last encounter with Io in January preceded a long, elliptical flight taking Galileo 24.9 million km from Jupiter, the most distant it had been since approaching the planet in 1995. Descending to perijove, Galileo performed its first and only encounter with the tiny asteroid-like moon Amalthea in November 2002. On 14 April 2003 Galileo once again broke its distance record by reaching apojove at 26.4 million km from the giant planet. This final looping path for the ageing spacecraft took it on a trajectory which would result in its collapse into Jupiter's atmosphere on the dark side of the planet. Due to the fact that Galileo had not been intended to search for life, or land on any of Jupiter's moons which the spacecraft showed may actually carry living organisms, Galileo had to be destroyed in the planet's turbulent atmosphere to prevent contamination in the event it collided with a natural Jovian satellite. The final hours of Galileo's 14-year sojourn through space, during which it travelled 4,631,778,000 km, optimised the telecommunications and data-handling systems to send science information without buffering, thus ensuring any data it was able to gather left the spacecraft before it was destroyed. The end came at 18:57 UT on 21 September 2003 when Galileo sliced into the Jovian atmosphere slightly south of the planet's equator at a speed of 48.26 km/sec, completing almost eight years of outstanding performance and discoveries that will keep scientists busy for many years.

The accompanying table identifies all the encounters by Galileo accomplished to date and those scheduled for the remainder of the mission.

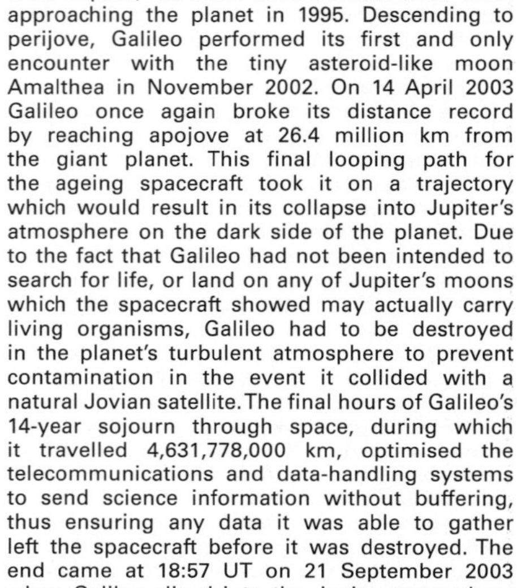

Asteroid Gaspra imaged by Galileo in November 1991
0085823

Primary Mission:

Orbit	Moon (sequence)	Date	Time (UT)	Distance from objective (km)
1	Ganymede (1)	27 June 1996	06:29	835
2	Ganymede (2)	6 September 1996	19:00	262
3	Callisto (1)	4 November 1996	13:34	1,118
4	Europa (1)	19 December 1996	06:53	698
5	Jupiter (1)	19 January 1997	09:36	646,000
6	Europa (2)	20 February 1998	17:06	586
7	Ganymede (3)	5 April 1998	07:10	3,102
8	Ganymede (4)	8 May 1998	15:57	1,596
9	Callisto (2)	25 June 1997	13:48	415
10	Callisto (3)	17 September 1997	00:19	538
11	Europa (3)	6 November 1997	20:32	2,042
Galileo Europa Mission:				
12	Europa (4)	16 December 1997	12:05	200
13	Europa (5)	10 February 1998	17:57	3,562
14	Europa (6)	29 March 1998	13:23	1,649
15	Europa (7)	31 May 1998	21:12	2,521
16	Europa (8)	21 July 1998	05:07	1,837
17	Europa (9)	26 September 1998	03:54	3,582
18	Europa (10)	22 November 1998	11:47	2,281
19	Europa (11)	1 February 1999	02:10	1,495
20	Callisto (4)	5 May 1999	13.56	1,315
21	Callisto (5)	30 June 1999	07:47	1,047
22	Callisto (6)	14 August 1999	08:31	2,299
23	Callisto (7)	16 September 1999	17:26	669
24	Io (1)	11 October 1999	05:01	611
25	Io (2)	26 November 1999	03:40	300
Galileo Millennium Mission:				
26	Europa (12)	3 January 2000	351	
27	Io (3)	22 February 2000	14:32	198
28	Ganymede (5)	20 May 2000	11:10	808
29	Ganymede (6)	28 December 2000	08:25	2,337
30	Callisto (8)	25 May 2001	12:13	123
31	Io (4)	6 August 2001	05:48	200
32	Io (5)	16 October 2001	02:04	181
33	Io (6)	17 January 2002	14:44	101
34	Amalthea (1)	5 November 2002	07:02	160

Orbiter description

The 2,222 kg orbiter is divided into an upper section incorporating an HGA, an 8-sided bus and retro-propulsion module (RPM) spun at 3.15 rpm (increased to 10.5 rpm for probe deployment). This was also used March 1993 in an attempt to free the HGA, and a despun lower section of probe, its data relay antenna and scan platform (0-1°/s slew rate, 0.2° pointing accuracy) with the four remote sensing instruments. The five fields/particles instruments are mounted on or near a single 11 m boom. Science payload is 118 kg.

Two new-model RTGs on 5 m booms provided 570 W at launch, projected to decay to 480 W in 1997 (496 W, 31 W margin, as of September 1995); each carries 10.9 kg of plutonium dioxide. Germany's 1,120 kg RPM provides all attitude control and retro capability from two independent networks of six 10 N thrusters, each mounted on two 2 m arms and central 400 N motor blocked by the probe until release. A usable total of 925 kg of MMH/NTO is divided between four tanks, with two associated helium pressurant spheres. The three major burns (planet avoidance after probe release, orbit insertion and orbit raising) were performed by the large motor with the entire spacecraft spinning at 10 rpm, but four of the 10 N thrusters are positioned as back-ups. With all of the tankage emptied the probe has a total delta V capability of 1,600 m/s.

The 4.8 m diameter TDRS-type HGA, comprising gold-plated molybdenum wire mesh stretched between 18 graphite epoxy ribs, is designed for X-band (8.415 GHz) 115.2 kbit/s real-time and 134 kbit/s playback datalinks at Jupiter. The LGA gives a 40 bit/s in S-band at 2.296 GHz. Uplinking occurs at 34 kbit/s on 2.115 GHz S-band. Probe data were to be relayed at 28.8 kbit/s in real-time, received by the dedicated despun dish, and also recorded on the single Odetics DDS-3100 900 Mbit digital tape recorder, able to hold the equivalent of about 200 images. Instead, the LGA required 3 months to transmit the stored probe data between January and April 1996.

Probe description

The 339 kg probe, built by Hughes Aircraft under NASA Ames' management, was activated by timer 6 hr before entry to follow a pre-programmed descent with sequences triggered by accelerometer cues. Pre-programmed system testing was undertaken annually during coast and before release. The Probe contained two main elements: the external 220 kg Deceleration Module, and the Descent Module uncovered after entry.

Some 90 kg of the 145 kg carbon phenolic heatshield was lost during the 43 kW/cm^2 peak heating entry; a phenolic nylon shield protected the upper surface. A 2.5 m Dacron parachute pulled the Descent Module free. The DM was not a pressure vessel (as were Pioneer Venus' capsules) but the 28 kg of science instrumentation, mounted on a 66 cm diameter aluminium honeycomb shelf, was protected inside individual housings expected to survive up to 20 atmospheres. Data were transmitted on 1.3870/1.3871 GHz L-band at 28.8 kbit/s for orbiter storage. A 21 Ah $LiSO_2$ battery provided power, with 72 W peak allocated for science instruments.

NASA did not sterilise the probe since Jupiter is a Category 2 planet under NASA's 1984 quarantine guidelines.

Specifications

Orbiter payload

Solid-State Imaging: 28 kg, 1,500 mm, f/8.5, 800 × 800 CCD, map Galilean moons at 1 km resolution and monitor atmospheric circulation, 115 kbits/s.

Near-IR Mapping Spectrometer: 18 kg, 0.7–5.2 µm, satellite surface composition, atmosphere T/composition.

UV Spectrometer: 4 kg, 1,150–4,300 Å, atmospheric gases/aerosols. 544–1,280 Å Extreme UV detector added later.

Photopolarimeter-radiometer: 5 kg, visible/near-IR bands, radiometry to greater than 42 µm, atmospheric particles, thermal radiation.

Magnetometer: 7 kg, 32–16,384 ν, magnetic fields.

Energetic Particle Detector: 9 kg, 0.02–55 MeV ions, 0.015–11 MeV electrons, and high energy ions in magnetosphere.

Plasma Detector: 12 kg, 1 eV-50 keV in 64 bands, energy composition/distribution of low energy ions.

Plasma Wave: 6 kg, 6–31 Hz, 50 Hz-200 kHz, 0.1–5.65 MHz, electromagnetic waves and wave/particle interactions. Dust Detector: 4 kg, 10^{-16}–10^{-6} g, 2–50 km/s, measures γ and charge.

Heavy Ion Counter: modified from Voyager Cosmic Ray Science detector, added in 1985 to monitor spacecraft exposure to damaging radiation in addition to observing solar flares and Jupiter's environs. Five single Si crystal wafers 30–2,000 µm thick, 6–200 MeV.

Radio system: determine planet/moon masses and radio/atmospheric structures.

Probe Payload

Atmospheric Structure Instrument: 4 kg, 0–540 K, 0–28 atmosphere, temperature and pressure, density and molecular weight with altitude

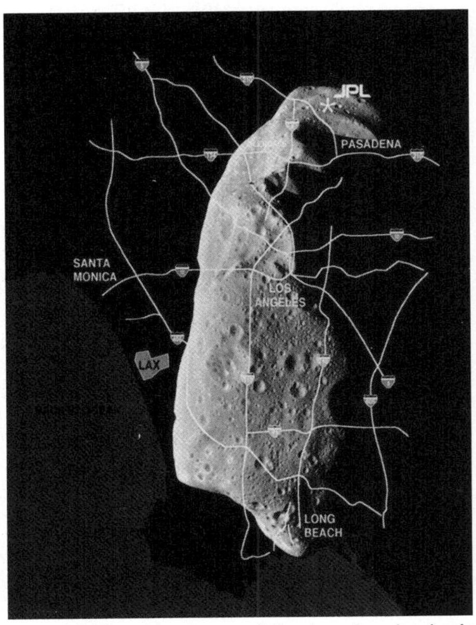

Asteroid Ida with a map of the Los Angeles basin superimposed to scale 0085824

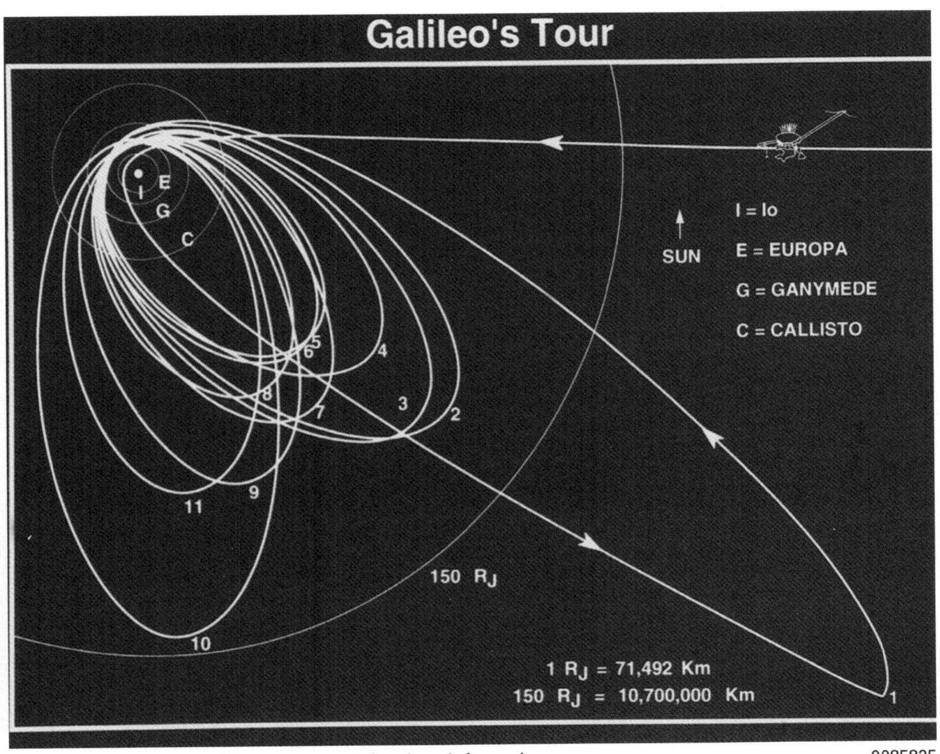

First Io orbits of Jupiter with encounters at the planet's four primary moons 0085825

Neutral Mass Spectrometer: 11 kg, 1-150 amu, atmospheric chemical composition

Helium Abundance Detector: 1 kg, 1% accuracy

Net Flux Radiometer: 3 kg, 0.3 to more than 100 μm, ambient thermal and solar energy.

Nephelometer: 5 kg, 0.2–20 μm cloud particles more than 3/cm^3

Lightning/Energetic Particles: 2 kg, fisheye lens, 1 Hz-100 kHz, lightning, measure energetic particles.

UPDATED

Lunar Prospector

Current Status

Following a manoeuvre performed at 09:16 UT on 31 July 1999 Lunar Prospector impacted the moon as expected at 09:52 UT just over the southern horizon as seen from earth. It struck the surface in a permanently shadowed crater at –87.7° latitude and 42° longitude selected because of the possibility that it contained surface ice which would vapourise when hit by the impacting spacecraft. The de-orbit manoeuvre involved firing two thrusters for 276.5 second burning 3.38 kg of propellant and decreasing spacecraft velocity by 44.3 m/sec. There were no reported observations of an impact debris plume and no recorded observations from earth of water vapour. Lunar Prospector had been in lunar orbit since 11 January 1998 and over a period of 18 months performed all its mission objectives, returning a unique sequence of data from low altitude. Contributions from the science results of Lunar Prospector will feed into the requirements and specifications for the new generation of lunar exploration vehicle NASA is charged with developing over the next several years under the new initiative announced by President George W. Bush in January 2004.

Background

Following NEAR and Mars Pathfinder, Lunar Prospector was the third mission in NASA's Discovery series formulated in 1992 for solar system exploration. Following an announcement of opportunity in August 1994 resulting in 28 proposals for round two of the Discovery selection process by October that year, Lunar Prospector was selected in February 1995. The primary mission objective was to map the moon's elemental abundances, gravity field, magnetic fields and outgassing events for one year from a near circular, 100 km, polar orbit. As one of the first missions in NASA's mantra of "faster, better, cheaper", Lunar Prospector was put out to competitive bids and built by Lockheed Martin in just 22 months. The total budget was constrained to US$63 million including the launch vehicle. Discovery missions are performance-based and reverse the traditional concept of specifying development details by introducing the concept of buying the mission from the principal investigator. NASA Ames Research Center managed the mission.

The objectives set for Lunar Prospector originated within the Apollo missions science data set and were a legacy of repeated but unsuccessful attempts to fund a lunar orbiting spacecraft for chemical and physical analysis of the moon to follow the manned landings of 1969–72. Launched on 25 January 1994, the Clementine Deep Space Program Science Experiment funded by the US Department of Defense was the first scientific attempt to provide further analysis of the moon from close orbit remote sensing and the first science mission to the moon since the early 1970s. Results from the Clementine radar data provided information leading some scientists to conclude that water may be present in the form of ice in permanently shadowed areas. Accordingly, the Discovery Lunar Prospector spacecraft was given a neutron spectrometer in an attempt to resolve the matter. The presence of water on the moon, in whatever form, would have profound impact on design criteria for lunar bases designed for human habitation. Beyond the first year of operation, mission planners hoped to lower the 100 km orbit used by Lunar Prospector for the first phase of data gathering to as low as 10 km for high fidelity data.

The science payload selected for Lunar Prospector addressed key questions raised by earlier missions and extended the characterisation of the lunar environment by analysing the lunar crust and atmosphere for potential resources. This included the distribution of minerals, possible water ice and certain gases, mapping the gravitational and magnetic shape of the moon and understanding more about the size and content of the moon's core. In addition to the neutron spectrometer capable of detecting hydrogen and determining the probability of water ice deposits the spacecraft also carried a gamma ray spectrometer for determining elemental composition. Additionally Lunar Prospector carried an alpha-particle spectrometer for measuring outgassing activity, a magnetometer and electron reflectometer to map magnetic fields and a doppler velocity experiment to map the lunar gravity field. The ability to conduct an extended mission at lower altitude enhanced the opportunity for adding valuable data in all science objectives, particularly the gravity mapping objective.

Lunar Prospector was launched by Lockheed Martin Athena 2 at 02:28:44 UT on 6 January 1998 from pad 46 at Cape Canaveral. After a 13 min flight the spacecraft and Trans Lunar Injection (TLI) Module were placed in a 100 km interim earth orbit. Following a 42 min coast the Star 37FM TLI motor fired for 64 second applying a nominal 3,142 m/s velocity increment boosting the spacecraft toward the moon. The first course correction manoeuvre came at 12:30 UT 7 January when thrusters performed a 50.2 m/s velocity change and this was followed by a second manoeuvre at 08:45 UT the following day producing a 7.4 m/s burn. Just over three days later the spacecraft passed within 71 km of the lunar surface. In a continuous 32.2 min firing of the two aft thrusters that began at 11:45 UT on January 11 Lunar Prospector was placed in an elliptical orbit with an aposelene of 8,500 km and a period of 11.8 hr. At 10:58 UT on 12 January the thrusters were fired again, for 27 min, lowing aposelene to about 1,870 km (periselene 83 km) and reducing the orbital period to 3.5 hr. The third orbit burn began at 11:38 UT on 13 January, lasted 27 min and placed the spacecraft in a 92 × 153 km, 90.1°, orbit with a period of about 2 hours. Lunar Prospector reached its final mapping orbit on 15 January in a sequence of delicate orbital adjustments. Those began at 21:44 UT with a 76.6 second firing to lower aposelene to 100 km, followed at 22:32 UT by a 16.4 second burn to raise periselene to 99 km. At 23:57 UT a series of 139 pulsed thruster firings began to reorientate the attitude and to precess the spin axis 31.5°. From its 99 × 100 km orbit the spacecraft began its mapping science. Over several orbits the anomalous gravitational field of the moon changed orbital altitude considerably. The lunar gravity maps thus obtained were used to plan an extended mission which would involve lowering the spacecraft to a mean altitude of 30 km above the surface increasing instrument resolution as well as conducting a more detailed gravity anomaly map.

On 19 December 1998, with the spacecraft orbit degraded by gravitational influence to 77.5 × 122.5 km, twothrusters were fired at 18:11 UT for46.8 second lowering periapsis to 50.5 km followed by another firing at 19:11 UT for 150.4 second lowering apoapsis. The resulting orbit of 25 × 55 km (average 40 km) with a period of 112 min was an interim one prior to the final extended mission orbit with an average altitude of 30 km. The final altitude reduction took place in a series of manoeuvres conducted on January 29, 1999 (UT). They began with a 40.4 second firing of two thrusters at 06:55 UT resulting in a 43 × 63 km orbit, followed by a 71.6 second thruster firing at 07:43 UT resulting in a 15 × 45 km orbit. Over the next several months the spacecraft continued to operate from this low average altitude of 25–30 km, at times dropping to within 10 km of lunar mountains and crater walls greatly enhancing the resolution of the scientific instruments. Frequent thruster firings were necessary to maintain a safe altitude and to reconfigure the orbit geometry for optimised science observations.

Specifications

Fuelled mass: 295 kg

Description: Lunar Prospector is a small, spin-stabilised, orbiter. It consists of a graphite-epoxy bus 1.3 m diameter by 1.4 m tall supporting three 2.5 m science masts displaced at 120° intervals for deploying the five science instruments. Spacecraft was designed with no computer other than a simple command and data handling unit. Data transmission rate was 3,600 bits/sec. Electrical power is provided by solar cells attached to the drum-shaped exterior surface. The spin stabilised spacecraft carried four thrusters, two at the top and two at the bottom, with 137 kg of hydrazine propellant.

Payload: Consists of Neutron Spectrometer (NS), Gamma Ray Spectrometer (GRS), Alpha Particle Spectrometer (APS), Magnetometer and Electron Reflectometer (MER) and the Doppler Gravity Experiment (DGE).

The NS consists of two canisters each containing helium-3 and an energy counter. Neutrons colliding with helium atoms will give an energy signature which will be detected and counted. One of the canisters is wrapped in cadmium and one in tin. The cadmium screens out thermal neutrons (low energy or slow moving neutrons cosmic-ray generated with lost energy from collisions with hydrogen atoms) while the tin does not. Differences in counts between the two canisters indicate the number of thermal neutrons detected which in turn indicates the quantity of hydrogen on the moon's crust at a given location. Large quantities would be due to the presence of water. The NS is a thin cylinder co-located with the APS at the end of one of the three radial science booms. The instrument has a surface resolution of 150 km. For polar ice studies the NS examined the poles to 80° latitude with a sensitivity of at least 10 ppm of hydrogen. For implanted hydrogen studies the NS examines the entire lunar globe with a sensitivity of 50 ppm.

The GRS was designed to provide global maps of elemental abundances on the lunar surface by recording the spectrum of gamma rays emitted by the radioactive decay of elements (U, Th, and K) in the lunar crust and elements in the crust bombarded by cosmic rays and solar wind particles (Fe, Ti, O, Si, Al, Mg and Ca). The uranium, thorium and potassium in particular is used to map the location of KREEP (potassium, rare-earth element, phosphorous containing material believed to have

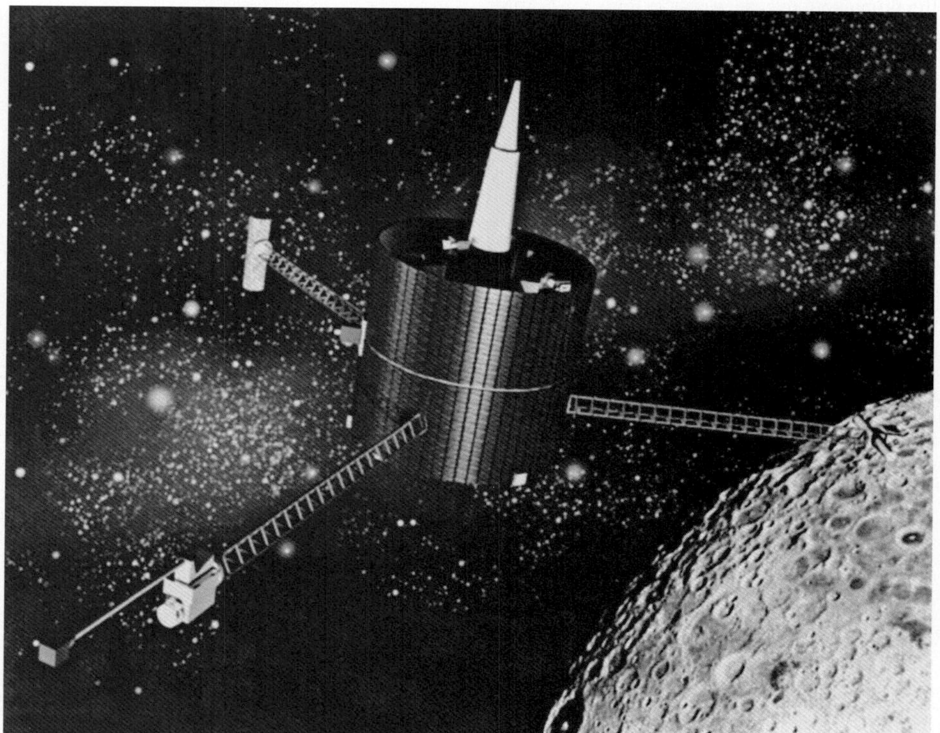

The Lunar Prospector spacecraft in an artists view depicting full deployment (NASA) 0572508

developed late in the formation of crust and upper mantle). The instrument also detects epithermal neutrons which complements the NS in the search for subsurface water. The GRS is a small cylinder mounted on the end of one of the three 2.5 m radial booms and consists of a bismuth germanate crystal surrounded by a shield of borated plastic. Gamma rays striking the bismuth atoms produce a flash of light with an intensity proportional to the energy of the gamma ray which is recorded by detectors. The energy of the gamma ray is associated with the element responsible for its emission. Due to a low signal to noise ratio, multiple passes are required to generate statisically significant results. At nine passes a month it will take about three months to confidently estimate abundances of thorium, potassium and uranium and 12 months for the other elements. The precision varies according to element measured. For U,Th and K the precision is 7 to 15 per cent, for Fe 45 per cent, forTi 20 per cent and for the overall distribution of KREEP 15 to 30 per cent. The borated platsic shield is used in the detection of fast neutrons.The GRS will achieve global coverage from an altitude of approximately 100 km and with a surface resolution of 150 km.

The APS is designed to detect radon outgassing events on the lunar surface and records alpha particle signatures of radioactive decay of radon gas and its daughter product, polonium. These putative outgassing events in which radon, nitrogen and carbon dioxide are vented are hypothesised to be the source of the tenuous lunar atmosphere and may be the result of low-level volcanic/tectonic activity. The APS is a cube approximately 18 cm on a side colocated with the NS on the end of one of the 2.5 m radial booms. It contains 10 silicon detectors sandwiched between gold and titanium discs arranged on five sides of six sides of the cube. Alpha particles produced by the decay of radon and polonium leave tracks of charge on the silicon wafers when they impact. A high voltage is applied to the silicon and the current amplified by being funnelled along the tracks to the aluminium disc and is recorded for identification. The APS makes a global examination of gas release events and polonium distribution with a surface resolution of 150 km and a precision of 10 per cent.

The Magnetometer is designed to map weak remanent lunar magnetic fields, either paleomagnetic or due to meteor impacts, secondarily to provide information on their origin and to allow estimates of the size and composition of the lunar core and to provide information on the lunar induced magnetic dipole. The magnetometer is located at the end of one of the three radial booms on a 0.8 m pole extending from the electron reflectometer. It is 2.6 m from the spacecraft in order to isolate it from generated magnetic fields. The instrument is a triaxial fluxgate magnetometer similar in design to the instrument used on the Mars Global Surveyor and the magnetometer and the electron reflectometer have a combined mass of 5 kg and use 4.5 W of power. The Magnetometer measures the magnetic field amplitude and direction at spacecraft altitude with a spatial resolution of about 100 km when ambient plasma disturbances are minimal.

The Electron Reflectometer and its associated electronics package are located at one end of one of the three radial science booms and provides a support for the magnetometer which is further mounted on a 0.8 m boom. The ER works by measuring the pitch angles of solar wind electrons reflected from the moon by lunar magnetic fields. Stronger local magnetic fields can reflect electrons with larger pitch angles. The ER measures field strengths as small as 0.01 nT with a spatial accuracy of about 3 km at the lunar surface.

The DGE is designed to provide the first lunar gravity data from a low polar orbit in a scientific investigation into the surface and internal mass distribution of the moon. This is accomplished by measuring the Doppler shift in the S-band tracking signal as it reaches earth, converted to spacecraft accelerations processed to provide estimates of the lunar gravity field. Estimates of the surface and internal mass distribution give information on the crust, lithosphere and internal lunar structure. Because line-of-sight tracking is required, only the near-side gravity field can be estimated using the Doppler method. The experiment is a by-product of the spacecraft S-band tracking and so has not specific weight or power requirements. The experiment will give near-side gravity field surface resolution of 200 km and precision of 5 mgal in the form of spherical harmonic coefficients. The downlink telemetry signal is transmitted at 2273 MHz over a ±1 MHz bandwidth as a right-hand circularly polarised signal at a nominal power of 5 W and a peak power of 7 W. Command uplinks are sent at 2093.0542 MHz over a ±1 MHz bandwidth. The transponder is a standard Loral/Conic S-band transponder. An omnidirectional antenna can be used for uplink and downlink or a mediu, gain helix antenna can be used for downlink only. Because the spacecraft is spin-stabilised the spin results in a bias in the Doppler signal due to the spacecraft antenna pattern spinning with respect to the earth station of 0.417 Hz (27.3 mm/s) for the omnidirectional antenna and –0.0172 Hz (–1.12 mm/s) for the medium gain antenna. LOS data is sampled at 5 second to account for the approximately 5 second spin rate of the spacecraft leaving a residual of less than 0.1 mm/s.

UPDATED

Magellan

Current Status

One of the most successful planetary missions in history ended in October 1994 when the NASA/JPL Magellan Venus orbiter fell silent and then burned up in the Venusian atmosphere. Having completed the planet's first detailed radar and gravity mapping, Magellan lowered itself into the atmosphere to generate unique aerodynamic data for future aerobraking missions. The US$800 million mission completed the first 243-day cycle of radar mapping Venus on 15 May 1991, providing the first clear views of 83.7 per cent of the surface at 120–300 m resolution. The formal project goal of 70 per cent coverage passed as a milestone on 2 April 1991. Magellan returned 1,200 Gbit of data and far surpassed the 900 Gbit from *all* of NASA's other planetary missions.

Mapping cycle 2 ended 15 January 1992, increasing coverage to 96 per cent, particularly adding high southern latitudes by changing attitude to look right. Cycle 3, ending 13 September 1992, concentrated on stereo imaging (achieving 21 per cent coverage and increased mapping to 98 per cent. No further radar mapping was performed because of communications problems. Cycle 4, ending 25 May 1993, emphasised

Magellan Spacecraft in final assembly prior to launch 0085839

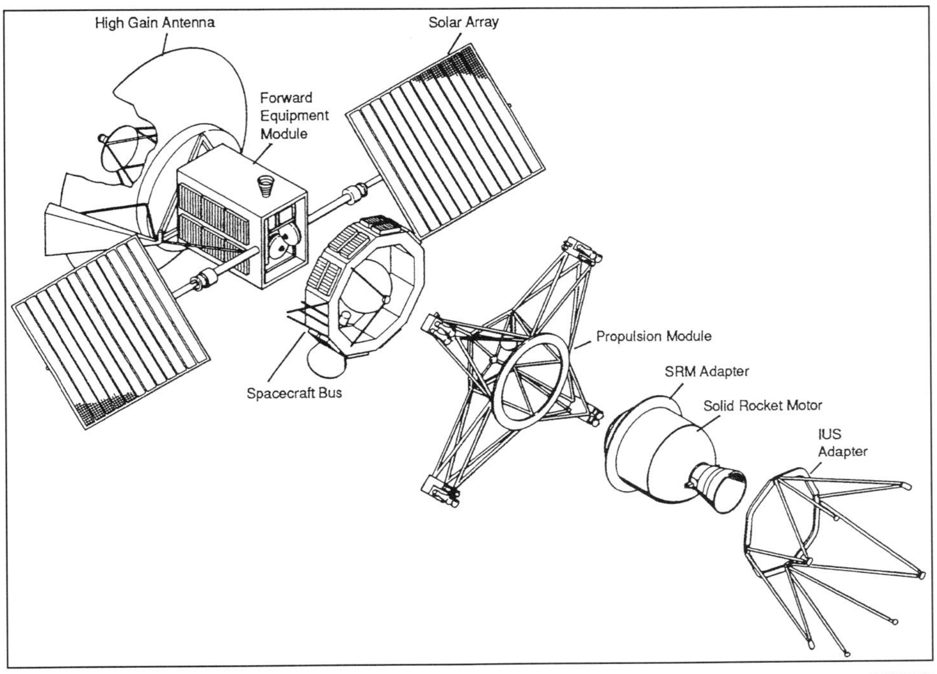

Magellan spacecraft 0085840

Elliptical orbit of Venus permitted radar mapping at low altitude 0085842

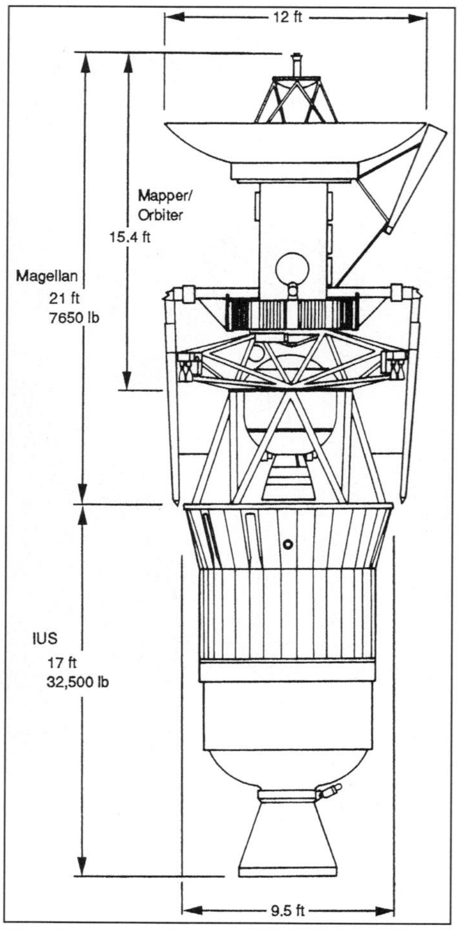

Magellan Spacecraft in launch configuration attached to IU5 0085841

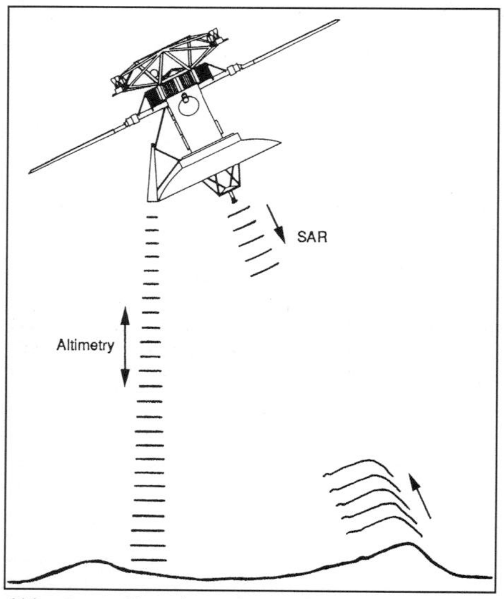

Altimetry and imaging radars on Magellan spacecraft 0085843

gravity data between ±30° latitudes. The first ever extended aerobraking sequence, 25 May to 3 August 1993, then lowered the orbit. Cycles 5/6, covering 3 August 1993 to 10 October 1994, thus collected 200–400 km high resolution gravimetry data over 95 per cent of the planet with 1.5 milligals sensitivity – better than so far acquired for Earth. In its final death throes Magellan was used to evaluate atmospheric torque effects at Venus. By turning the solar cell arrays in opposite directions and observing the way the outer atmosphere interacts with them, information was obtained about turning moments. Atmospheric destruction came on 14 October 1994.

VERIFIED

Mars Climate Orbiter

Current Status

Just one day late, Mars Climate Orbiter was launched by Delta 7425 at 18:45:51 UT on 11 December 1998 from launch pad 17A at the Cape Canaveral Air Force Station, Florida. Before the flight, engineers discovered a potential problem with a charge control unit that regulates the flow of current from the solar cell arrays to the battery. If it had failed in flight, the battery would have overcharged and been rendered useless. Just 63 seconds after lift-off, the four solid rocket boosters burned out and were jettisoned, two at a time, with first stage cut-off at approximately 4 minutes 24 seconds. Second stage burn lasted about 11 minutes 22 seconds and put the spacecraft in a low Earth orbit at about 189 km. Second stage restart came at 34 minutes and lasted 19 seconds before third stage ignition at 40 minutes 30 seconds for 1 minute 28 seconds pushing the MCO out of Earth orbit and on its way to Mars. This was followed by a de-spin operation using extended lines and yo-yo weights with final dispersions nulled by spacecraft thrusters. MCO's first trajectory correction manoeuvre at 21:33 UT on 21 December to remove an intentional bias placed in the launch vehicle's guidance steering to prevent the unsterilised third stage of Delta from following on to Mars and entering the atmosphere. The spacecraft's thrusters were fired for 2 minutes 48 seconds effecting a 19.1 m/sec change in velocity. Two days later, engineers successfully conducted a health check on the MCO's two science instruments. On 4 March 1999, the MCO performed its second course correction manoeuvre with a burn involving its four thrusters for 8.2 seconds (0.86 m/sec) 17 million km from Earth. The next course correction took place on 15 September 1999 with a thruster burn lasting 15 sec which controllers predicted would put the spacecraft 193 km above the north pole of Mars on 23 September 1999 ready for the Mars orbit insertion burn. At approximately 0700 UT on 23 September the orbiter main engine fired but tracking over the preceding several days had shown ground controllers that the MCO was heading for a closest approach to Mars of 60 km which is well within the minimum survivable fly-by altitude of 85 km. Consequently, the MCO spacecraft entered the tenuous outer layers of Mars' atmosphere and was decelerated too highly to remain in orbit and it burned up on its way to a crash at the surface. In a subsequent investigation it was revealed that an error was made by Lockheed Martin engineers and flight controllers in the unit of measurement employed to pass along flight manoeuvre data to the software engineers writing instructions for the spacecraft event sequencer; in effect, one group using English units, the other using SI units. This, together with the loss of Mars Polar Lander less than three months later, has brought a radical re-evaluation of Mars mission strategy. As a result Mars mission planning changes eliminated the lander from the Mars 2001 plan, leaving one orbiter, called Odyssey, successfully launched by Delta 2 on 7 April 2001. Orbital images from Mars Global Surveyor and Mars Odyssey have been used in an attempt to locate remains of MCO on the surface, without success.

Background

NASA developed the Mars Surveyor programme in 1994 as a commitment to sending low-cost spacecraft to Mars at every launch window, which take place roughly every 26 months. The first mission in this series was Mars Global Surveyor launched in 1996 and under the mission opportunity for 1998, Mars Climate Orbiter (MCO) and Mars Polar Lander (MPL) were developed collectively as the next two missions under the Mars 98 umbrella. All of NASA's Mars missions planned for the first decade of the 21st century come under the Surveyor programme and in efforts to cut costs, all involve a single industrial partner with the Jet Propulsion Laboratory. In support of this, JPL set up the Mars Surveyor Operations Project. Along with Mars Polar Lander, MCO was one of two Mars 98 missions and, like Mars Global Surveyor, was designed for aerobraking into the optimised mission orbit following initial capture by retro-propulsion. Synchronised from a staggered launch with MCO being sent off first, both spacecraft were scheduled to begin 65 days of simultaneous operations in orbit and at the surface in December 1999.

Excluding launch vehicle costs, the Mars 98 Surveyor orbiter and landing funding was capped at US$193.1 million for all project management and spacecraft, instrument and mission system development activities from the authority to proceed (November 1995) through lander launch +30 days (February 1999). This sum included a US$24.7 million contingency fund. In addition, launch costs were budgeted at US$91.7 million and missions operations costs at US$42.8 million, excluding the Deep Space 2 costs for Mars Polar Lander. The schedule required a preliminary design review in March 1996, critical design review in January 1997, MCO integration and test in May, instrument delivery in August and delivery to the Cape Canaveral Air Force Station in September 1998 for the launch in December 1998. MCO and MPL shared 70 per cent in hardware and 89 per cent in software modules.

While Mars Global Surveyor is dedicated to topographic mapping and geomorphologic studies, Mars Climate Orbiter was dedicated to better understanding of the atmosphere and climate. Lockheed Martin was selected for Phase B definition studies in March 1995 as the successful respondent to an industry-wide competitive solicitation for the Mars 98 Surveyors. MCO was designed for launch by Delta 2 (7425) launch vehicle, a more powerful version of the rocket used to send Near East Asteroid Rendezvous (NEAR) on its way. The primary launch window was 10–17 December 1998 with a secondary window 18–25 December. Launch during the initial period insured a higher probability that the MCO would complete aerobraking before Mars Polar Lander, the second Mars 98 mission, arrived at the planet. Launch during the secondary period would result in a higher capture orbit at arrival requiring more aerobraking to bring the spacecraft into its final science orbit. Launch after 25 December was possible but would result in further compromise to the capture orbit.

MCO would take an estimated 286 days from launch to encounter, reaching Mars on 23 September 1999. Between two and four mid-course corrections can be made to refine the trajectory during interplanetary cruise, the first of these to be carried out 10 days after launch. This manoeuvre was expected to be the largest and longest and would be used to correct injection errors and to adjust the Mars arrival aimpoint. The remaining manoeuvres would be used to direct the spacecraft to the proper aimpoint and were scheduled at 45 days after launch and 60 days and 10 days before arrival at Mars. During cruise the science instruments were turned on, calibrated and tested, the Pressure Modulator Infra-red Radiometer and Mars Colour Imager was calibrated during a week-long checkout 80 days after launch. Eighteen days before arrival, the imager was to take pictures of Mars that would be transmitted to Earth over the following three days.

Mars orbit insertion was scheduled for 23 September 1999. As the spacecraft nears its closest point to the planet, coming in over the northern hemisphere, it was to have fired its 640 N thrust main motor for 16–17 minutes to brake into an elliptical capture orbit, reducing speed from 5.9 km/sec to 4.7 km/sec. In an initial orbit of 160 × 39,000 km, the spacecraft would have looped around Mars roughly once every 29 hours, but about 22 minutes after completing the burn the spacecraft would be turned to point its high-gain antenna at Earth. Continuous communication with the Deep Space Network's 70 and 34 m antennas was to have been maintained for two days after orbit insertion. Based on the vehicle's state vector, the spacecraft was to have fired its thrusters during periapsis to lower the orbit and reduce the period by 2–4 hours. The aerobraking phase, similar to that planned for Mars Global Observer, would have begun when periapsis had been lowered to 100 km and was to have lasted 200 aeropasses or 57 days. At the end of this phase, on 22 November 1999, MCO was to have been in an orbit of 90 × 405 km and periapsis would have shifted north from 34° latitude to 89° latitude, almost directly over the planet's north

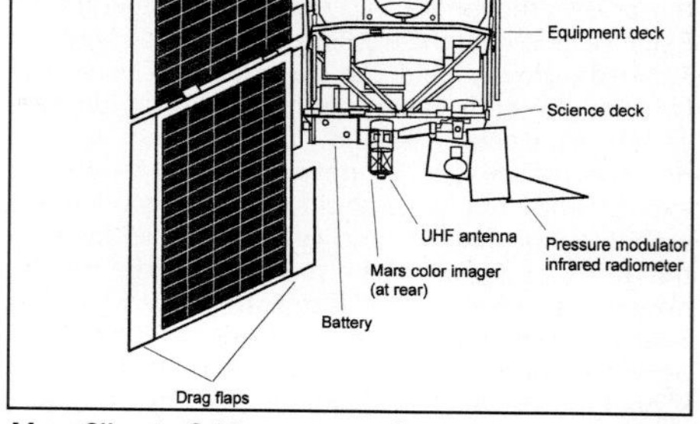

Mars Climate Orbiter spacecraft 0085848

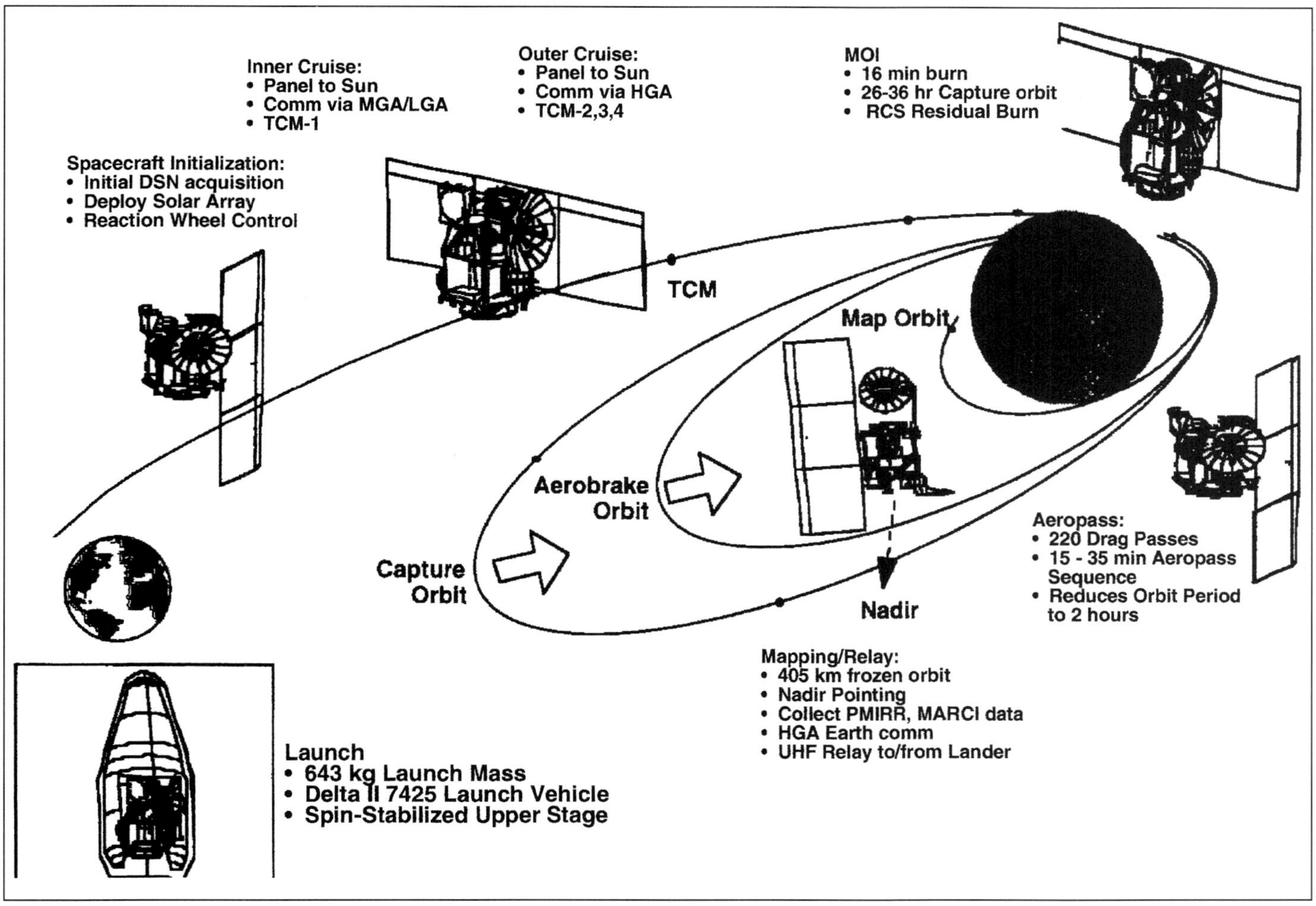

Mars Climate Orbiter depicted in mission phases prior to the commencement of mapping operations 0085847

pole. Small thruster firings would have kept the MCO at the desired level for heating and dynamic pressure, a maximum 0.46 W/cm^2.

The operating orbit was designed to be one in which MCO passed over Mars' equator at 4:40 pm local time over the sunlit side and 4:30 am on the night side. It was scheduled to achieve this when the MCO manoeuvred to a 405 km near circular orbit, a Sun-synchronous path, on 1 December and the high-gain antenna had been deployed. For the first three months, the MCO was to have acted as a radio relay for the Mars Polar Lander, relaying commands from Earth to the lander as well as sending data from the lander to the Earth via the UHF two-way relay link. The orbiter was to overfly the lander typically 5–6 minutes 10 times a day and in these periods communicate with the lander at 128 kbits/s with X-band telecommunications systems. This support was scheduled to take place for 10 hours each day through to the end of February 2000. From 3 March 2000 until 15 January 2002, the MCO was to perform its primary mission, making systematic observations of the atmosphere and the surface of Mars using its two science instruments, the Pressure Modulator Infra-red Radiometer and the Mars Colour Imager. The duration of the science mission was designed to be one Mars year (687 Earth days) to capture a full cycle of seasonal changes in Martian weather. After this the MCO would have been allowed to drift in a low-maintenance orbit awaiting the arrival of the Mars 2001 spacecraft in January 2002 when it was once again to have operated as an orbital relay station.

Specifications

Description: The Mars Climate Orbiter is a three-axis stabilised spacecraft comprising a bus 2.1 m tall, 1.6 m wide and 2 m deep. Framework is made of combined graphite composite facesheets on aluminium honeycomb with a gusset plate construction propulsion module (scaled down from the Mars Global Surveyor design) and a truss construction equipment module.

Mass: 629 kg at launch; 338 kg dry plus 291 kg of fuel.

Computation: RAD6000 32-bit processor with Mars Pathfinder heritage in the Orbiter Command and Data Handling subsystem, provides a central processing capability for all spacecraft systems including payload elements. It can be switched between clock speeds of 5, 10 or 20 MHz and includes 128 Mbytes of RAM of which about 20 per cent is used for running spacecraft programmes. The rest of the space is used for science and for storing data which will be relayed to Earth or the Mars Polar Lander. Unlike many other spacecraft it does not have an on-board tape recorder or solid-state data recorder, but instead uses its RAM for transmission to Earth. The computer also has 18 Mbytes of flash memory which will be used to store data even when the computer is off, of which 8 Mbytes is for storing triplicate copies of high priority data. When the computer accesses this triplicate data, it checks all three copies of each byte to ensure there has been no corruption.

Telecommunications: Primary communication between the spacecraft and the Earth is via the Deep Space Network's X-band (up/down) link and the orbiter's deep space transponder developed for the Cassini spacecraft. MCO has 15 W RF solid-state power amplifiers and a 1.3 m diameter articulated, two-axis, high gain antenna, one transmit only medium gain antenna and one receive-only low gain antenna. A two-way 10 W RF UHF link is provided for on-orbit communication with the lander while it is on Mars.

Power: The two-axis articulated, gallium arsenide, three-panel solar array (7.4 m^2 cell area; 11 m^2 total wing area), 5.5 m tip-to-tip, provides electrical power and serves as the most significant drag brake during aerobraking. Nickel-hydrogen (NiH_2) 16 Ah common pressure vessel batteries provide power during eclipses and during peak power needs. The electrical power control electronics are derived primarily from the Small Spacecraft Technology Initiative (SSTI).

Thermal control: The TC system uses passive methods and louvres to control the temperature of the batteries and the solid-state power amplifiers. Passive coatings as well as multi-layer insulation blankets are used to control temperatures along with thermostatically controlled and computer controlled heater circuits. Where needed, radiators are used to take the excess heat out of the spacecraft components to maintain proper operating temperatures.

Attitude control: The orbiter is three-axis stabilised in all mission phases except following separation from the launch vehicle. Primary attitude determination is via star camera derived from the Clementine spacecraft star camera and an inertial measurement unit with analogue star sensors as back up. Reaction wheels provide primary attitude control during most mission phases, desaturated via RCS thrusters. Because of lifetime concerns the IMUs are switched off during significant portions of cruise and mapping and the vehicle operated in an all-stellar mode except during manoeuvres. RCS thrusters also provide attitude control during trajectory correction manoeuvres, orbit insertion, aerobraking drag passes, orbital trim manoeuvres and safe mode. The MCO has four 22.24 N thrusters used for trajectory correction and for pitch and yaw control and four 0.89 N thrusters for roll control.

Propulsion: Derived from the propulsion subsystem for Mars Global Surveyor the MCO has a LEROS 640 N main engine operating with hydrazine and nitrogen tetroxide bi-propellants for Mars orbit insertion. Following this burn the oxidiser is gone and the attitude thrusters use the remaining fuel (hydrazine) as a monopropellant.

Payload: Consists of a rebuilt version of the Mars Observer Pressure Modulated Infra-red Radiometer (PMIR) and the Mars Colour Imaging (MARCI) system. The PMIR weighs 42 kg and uses 41 W of power in an instrument box 23 × 30 × 74 cm. A smaller cooling unit 58 × 65 × 30 cm is attached to the box. PMIR has nine channels, visible light + IR wavelengths 6–50 microns. These will construct vertical profiles of the atmosphere from 80 km to the surface. Bands of water vapour and CO_2 will be detectable at a vertical resolution of 5 km. The two MARCI wide-angle and medium-angle cameras weigh 2 kg and each measures 6 × 6 × 12 cm and use 4 W power. The wide-angle camera detects light in the visible (5 × 425–750 nm bands) and UV (2 × 250–350 nm) with a resolution of 1 km when large numbers of pictures are being sent or 7.2 km at other times when averaging adjacent pixels. The camera can also be used to image the planet's limb for cloud structure and haze patterns at a resolution of 4 km. The medium-angle camera detects light in eight spectral bands from violet to near-IR (425–930 nm) with a maximum resolution of about 40 m across a 6° Field of View (FoV) covering 40 km.

UPDATED

Mars Exploration Programme

NASA's strategic plan for the robotic exploration of Mars evolved as a result of recommendations from the Solar System Exploration Committee (SSEC) in

1983. It sought to resume the exploration of Mars after completion of the Viking programme which had put two spacecraft in Mars orbit and two on the surface in 1975. The strategy was substantially modified after the loss of Mars Observer in August 1993 and was changed again in response to the dual loss of Mars Climate Orbiter in September 1999 and Mars Polar Lander in December 1999. The first launch in this revised strategy was Mars Odyssey launched in 2001.

In 1983, the SSEC recommended a series of low-cost Planetary Observer missions beginning with Mars Geoscience/Climatology Orbiter (subsequently renamed Mars Observer), but budget restrictions delayed launch from 1990 to 1992. Based on tried and tested hardware, the spacecraft bus was built up from the RCA Satcom K design using electronics derived from the company's Tiros/DMSP meteorological satellite programme. Originally designed to be carried into low earth orbit by Shuttle and boosted on its way to Mars by an Orbital Science Corporation TOS upper stage it (and TOS) was assigned a Titan 3 launcher following the Challenger disaster of January 1986 and sent on its way in September 1992.

Prior to the loss of the 2,565 kg Mars Observer in August 1993, when it was believed an explosion occurred in orbit-insertion propellant lines, NASA had proposed a follow-on programme of small lightweight landers under what it called the Mars Environmental Survey (MESUR) programme. MESUR would place 12–16 100 kg landers on the surface with proof-of-concept performed by a precursor Pathfinder mission funded under NASA's Discoverer programme. With the loss of Mars Observer, NASA cancelled the MESUR programme and re-examined exploration strategy along more fiscally stringent lines with Pathfinder and its Sojourner rover retained as a spring-board to a new generation of low-cost orbiters and landers. As part of the revised Mars exploration strategy NASA began a series of Surveyor orbiters and landers, placing the newly defined Mars Global Surveyor (MGS) alongside Mars Pathfinder for the 1996 launch opportunity. MGS would refly some instruments previously carried on the failed Mars Observer spacecraft and utilise the same launch window as Russia's Mars 96 mission. This would involve a single 6,000 kg orbiter, two small 50 kg lander stations and two 65 kg penetrators. The landers would be released during the final orbiter approach phase and the penetrators from Mars orbit.

For future extended mission operations NASA decided on a strategy shaped by a series of scientific goals based on studying Mars in three primary areas: evidence of past or present life; climate (weather, processes and history); and resources (environment and utilisation). The Mars Science Working Group laid out a strawman strategy for fitting science goals into a set of missions which evolved throughout 1995 and would begin with Mars Pathfinder and MGS launched in late 1996 in the same window as Russia's Mars 96 mission. NASA decided to launch in 1998 Surveyor orbiter and lander spacecraft, the former reflying another instrument from Mars Orbiter and the latter searching for life near the south pole of Mars. The final element lost with Mars Observer would fly on a Surveyor orbiter launched in 2001, along with a lander which would analyse rocks and surface geology. Original plans were to launch the lander on a Delta and the orbiter on a Russian Molniya.

For 2003, NASA wanted to join forces with ESA and add three surface landers to a European Mars orbiter launched on a single Ariane 5 in a joint endeavour called InterMarsNet. The 415 kg landers would explore the interior of the planet using seismology to detect 'Marsquakes' and study geochemistry at three sites. In addition, NASA contemplated sending a network of small and complementary weather stations around the planet. The earliest opportunity to retrieve samples from the surface of Mars and return them to Earth would come, said NASA, with a possible launch in 2005 and, because it would contravene the stringent low-cost standards of the Surveyor missions, involvement with either Russia or Europe. Beyond 2005 (a decade after the 1995 planning date) NASA wanted to fill the gap between this published sequence of missions and possible human expeditions which it believed it could mount from 2018, with precursor robotic paving the way for the first scientific research base.

Of the three Mars 96 missions, only the two NASA spacecraft (Mars Global Surveyor and Mars Pathfinder/Rover) were successful, with Russia's Mars 96 being destroyed in a launch failure. MGS reached Mars orbit in September 1997 but experienced aerobraking difficulties due to a technical problem with the –Y solar cell array. A modified orbit construction sequence delayed the start of the mapping mission until 9 March 1999. Meanwhile, Pathfinder/Sojourner rover landed on the surface in July 1997 and carried out a spectacular survey of the landing site. During 1998, and in direct response to the successful return to Mars, NASA reconsolidated its revised Mars exploration strategy, reaffirming its intention to send Mars Climate Orbiter and Mars Polar Lander at the 1998 window. It also reaffirmed its decision to launch identical sample collecting rovers in the 2001 and 2003 windows complemented by respective orbiters. Extant plans for the 2001 and 2003 missions included surface sample collection by rovers that could cover sufficient ground to ensure a diverse suite of rock and soil samples delivered to the prospective landing site for the 2005 surface sample return mission. The two rovers, identical to save costs, were also to have carried an Athena payload designed to locate, select, collect and store multiple fresh samples. The 2005 sample collection mission envisaged Mars-orbit-rendezvous from the surface instead of direct ascent and anticipated delivery of those samples to earth in 2008.

Following the loss of Mars Climate Orbiter in September 1999 and Mars Polar Orbiter in December 1999, NASA conducted several major internal reviews of its 'faster-better-cheaper' strategy which underpinned plans developed after the SSEC review in 1983. Criticism from scientific institutions and determined investigations by Congressional subcommittees responsible for science and budgetary oversight claimed NASA had cut corners in an attempt to do too much with too few resources. As a result, a refined strategy slowed the pace of the robotic exploration of Mars and plans to move from the robotic to the human exploration of Mars were deferred indefinitely. The existing plan for Mars 2001 was cancelled and the lander was eliminated, but the orbiter Odyssey was allowed to continue. The following mission structure for 2003 and subsequent launch opportunities was tentatively agreed and modified by the new Exploration Initiative announced January 2004.

2003
Mars Exploration Rovers: Launched separately between May and July 2003 twin rovers, able to travel almost as far in one Martian day as Sojourner did over its entire lifetime, landed at separate sites and set out to determine the history of climate and water on the planet where conditions may once have been very favourable for life. By means of sophisticated sets of instruments and access tools the twin rovers evaluated the composition, texture and morphology of rocks and soils at a broad variety of scales, extending from those accessible to the human eye to microscopic levels. The rover science team will select targets of interest such as rocks and soils on the basis of images and infra-red spectra sent back to earth. Two different Martian sites will be chosen on the basis of an intensive examination of information collected by the Mars Global Surveyor and Mars Odyssey orbiters as well as previous missions.

2005
Mars Reconnaissance Orbiter: Launched in August 2005, this scientific orbiter will attempt to bridge the gap between surface observations and measurements taken from orbit. It will reach Mars in 2006 and focus on analyzing the Martian surface at new scales in an effort to follow the tantalizing hints of water from the Mars Global Surveyor images. For example, the Mars Reconnaissance Orbiter will measure thousands of Martian landscapes at 20 to 30 cm (8 to 12 in) resolution, which is adequate to observe rocks the size of beach balls. In addition, maps of minerals diagnostic of the role of liquid water in their formation will be produced at unprecedented scales for thousands of potential future landing sites. A specialised, high-resolution sounding radar will probe the upper hundreds of metres of the Martian sub-surface in search of clues of frozen pockets of water or other unique layers. Finally, the Mars Reconnaissance Orbiter will finish the job of characterising the transport processes in the present-day Martian atmosphere, including the planet's annual climate cycles, using a unique infra-red sounding instrument, originally carried to Mars on the ill-fated Mars Observer, and then again on Mars Climate Orbiter.

2007
Scout Mission (Phoenix): NASA proposed to create a new line of small 'scout' missions that would be competitively selected from proposals submitted by the scientific and aerospace community. Exciting new vistas could be opened by means of this innovative approach, either through observations made from airborne vehicles, networks of small surface landers, or from highly focused orbital laboratories. NASA aims to compete these scout missions as often as possible, and potentially every four years, depending on resource availability. Selected in August 2003, the first Scout Mission (Phoenix) will send a copy of the failed Mars Polar Lander to an ice-rich region of northern Mars. It will scoop up soil to analyse at the landing site and radio evidence about the history of Martian water and the possibility of current or past life. Phoenix will land in May 2008. A stereo colour camera and a weather station will study the surrounding environment while other instruments check excavated soil samples for water, organic chemicals and conditions that could indicate whether the site was ever hospitable to life. NASA plans to select a second Mars Scout mission from a future round of proposals to fly in 2011.

2009
Mars Science Laboratory: NASA has proposed to develop and launch a next-generation 'mobile surface laboratory' with potentially long-range roving capabilities (greater than 10 km (about 6 miles)) and more than a year of surface operational lifetime as a pivotal step toward a future Mars sample return mission. By providing a major leap forward in surface measurement capabilities and surface access, this mission will also demonstrate the technology needed for accurate landing and surface hazard avoidance in order to allow access to potentially compelling, but difficult to reach, landing sites. Its suite of scientific instruments could include new devices that will sample and probe the Martian subsurface of organic materials. Launch October 2009 with arrival at Mars in 2010.

2009
Telesat: A multiband (X, Ka, UHF) telecommunications relay satellite in an optimal orbit to maximise coverage of orbital, sub-orbital and surface assets. Launched in 2009 it will support the 2009 MSL and provide a significant increase in the quantity of data relayed from the surface of Mars to the DSN on earth for this and successor missions.

2011
NASA plans to launch the first in a series of human precursor missions to Mars, robotic testbeds which will demonstrate technologies such as improved aerodynamic entry, Mars orbit rendezvous and docking, high precision landing and resource extraction and utilisation. These missions will also obtain critical data for future human missions on chemical hazards, resource locations and research sites and may prepare resources and sites in anticipation of human landings.

Mars Sample Return: NASA is studying additional scientific orbiters, rovers and landers, as well as approaches for returning the most promising samples of Martian materials (rocks, soils, ices and atmospheric gases/dust) back to Earth. While current schedules call for the first of several sample return missions to be launched in 2014 with a second mission in 2016, options that could move the date sooner to 2011 are presently under detailed examination. Technology development is underway for advanced capabilities including a new generation of miniaturized surface instruments such as mass spectrometers and electron microscopes, as well as deep drilling to 20 m or more.

UPDATED

Mars Global Surveyor

Current Status

Mars Global Surveyor continues to operate well. It has far exceeded the operating record of the three previous Mars orbiters, all NASA spacecraft, which were able to perform useful science. Mariner 9 entered Mars orbit in November 1971 and operated for just over eleven months; Viking 1 Orbiter entered Mars orbit in June 1976 and operated for almost four years and two months; and Viking 2 Orbiter reached Mars orbit in August 1976, operating for almost eleven months. At the time of writing MGS has operated in Mars orbit for five years eight months and supported the landings of rovers Spirit and Opportunity which reached their landing sites in January 2004. A mission aimed primarily at providing global visible and topographic and thermal emission spectroscopy of Mars has been transformed into a flight support operation backing up the science assignments of NASA's Mars Odyssey and ESA's Mars Express and when called upon serving as a relay station. Current plans are to continue to operate MGS until the end of September 2006 when, along with Mars Odyssey, it will be replaced by Mars Reconnaissance Orbiter. By September 2003, MGS had returned more than 137,000 pictures in the six years since reaching Mars orbit on 12 September 1997 and continues to operate, having captured about 3 per cent of the planet's surface representing nearly 5 million km^2 of the surface. After an unsuccessful attempt to locate the Mars Polar Lander, contact with which was lost on 3 December 1999, Mass Global Surveyor continues to return good data and high-quality images. In the absence of data from Mars Climate Orbiter and Mars Polar Lander following the loss of both in 1999, MGS has continued to provide high-resolution images stimulating renewed conjecture about the origin and evolution of surface features. The mission has been formally extended and has been supplemented by Mars Odyssey, which arrived at Mars in October 2001. The first year of Extended Mission operations began 1 February 2001 and a second EM began in February 2002.

Background

NASA began a new initiative, the Mars Surveyor programme in the wake of the Mars Observer loss in 1993. In effect the administration intended to demonstrate a new commitment towards cheaper, faster and more frequent missions. Rather than flying a duplicate mission of the lost Mars Observer, NASA shook up its approach by including US$78.4 million in FY95's request (receiving $59.4 million) to begin the Mars Surveyor series of small orbiters/landers. Congress capped the programme to US$100 million annually, and contrary to NASA's track record the budget cap seems to be holding.

The 1 tonne first orbiter, Mars Global Surveyor, carries five of Mars Observer's seven experiments. The Mars Pathfinder orbiter carries one of MO's two remaining instruments and acts as a relay for the separate lander. 2001's mission will carry MO's last instrument. NASA hopes to return 2 kg of samples in 2005, while at the same time remaining within the US$200 million ceiling, and in situ propellant production may be required to meet the mass target.

Lockheed Martin Astronautics built MGS, employing Mars Observer's spare thrusters and HGA but with a graphite composite bus instead of aluminium, because of mass constraints. Major elements draw on MO spares: most major electronic assemblies are MO spares, retrofitted to eliminate MO-identified discrepancies. The dual-mode propulsion system draws on Cassini designs.

NASA launched MGS at 17.00.50 GMT on 7 November 1996 and initially the Delta-2 launch vehicle placed the satellite and PAM-D third stage into a near-circular orbit at 28.5°, 185 km. At 17.41, the second stage of the Delta-2 re-ignited to place the assembly into an orbit reaching out to 4,700 km and then separated from the combined PAM-D/MGS. PAM-D ignition at 17.44.30 put the spacecraft into a heliocentric, trans-Mars orbit.

At about 17.52, MGS spread out its solar array, but telemetry indicated that one of the solar panels (–Y) did not fully open, tilting instead about 20° from the expected angle relative to the spacecraft's main bus. Both solar panels were receiving sunlight, while still fully charging the batteries. The orientation of the (–Y) solar panel is expected to have little effect during the cruise phase to Mars.

On 10 November, the spacecraft used its star sensor to establish its orientation in space and later that day it was put into an 'array normal spin' orientation which was used for the rest of the trans-Mars coast. On 21 November, MGS fired its main engine and performed the first of its four planned flight path correction manoeuvres. The 44 s burn achieved a change in spacecraft velocity of about 27 m/s. By performing the burn, the third stage of the Delta-2 launch vehicle is steered away from a Mars collision.

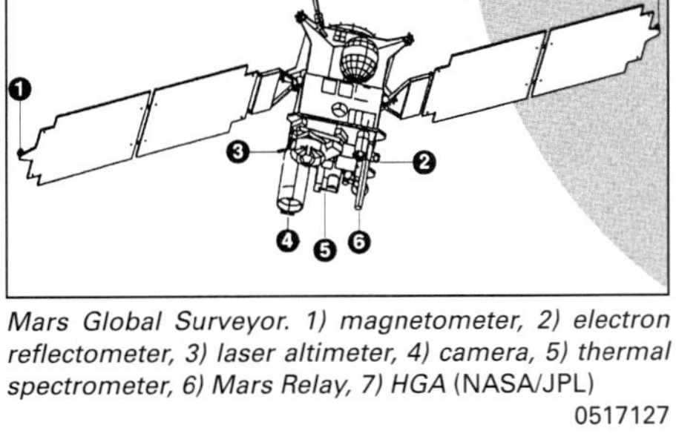

Mars Global Surveyor. 1) magnetometer, 2) electron reflectometer, 3) laser altimeter, 4) camera, 5) thermal spectrometer, 6) Mars Relay, 7) HGA (NASA/JPL) 0517127

On 22 January 1997, the spacecraft's flight computer activated a 53 W heater in the Mars Orbiter Camera to begin 'baking' the instrument and removing residual moisture. Without this 14-day bakeout period, the moisture in the camera's tube-like structure would leak into space at a slow rate and cause a gradual shift in the camera's focus. The bakeout will remove all of the moisture at once and stabilise the focus of the camera. Flight controllers also performed a series of very slight manoeuvres on 22, 23 and 24 January to attempt to manipulate the MGS solar array and characterise the exact condition of the debris that is preventing the panel from fully deploying.

A 26 second burn on 20 March 1997, refined MGS's flight path to Mars and achieved a change in spacecraft velocity of about 3.87 m/s. The burn was performed in two stages, in which flight controllers first commanded the spacecraft to fire its small thrusters for 20 seconds, then to fire its main engine for another 6 seconds. During April the spacecraft was in a quiet phase and conducted a search campaign to detect gravity waves.

MGS arrived at the Red Planet on 11 September 1997, and made a 973.03 m/s delta-V burn of 22 minutes 39 seconds to put it into an initial 250 × 54,000 km, 48 hour orbit. Aerobraking over a five-month period was to have changed the initial, elliptical path to a near circular 378 km orbit with an inclination of 92.9°. This would have provided an orbital period of 117.65 minutes and a 2 pm equatorial sun-synchronous repeat cycle of 7 days (88 revolutions) so that full mapping could begin in March 1998. On 1 October 1997, when the spacecraft descended through periapsis of 110 km, engineers observed movement in the –Y solar cell array that had given trouble shortly after launch. This generated concerns about possible damage caused by successive aerobraking passes through the tenuous outer layers of the Mars atmosphere and the periapsis altitude was temporarily raised back up to 121 km while an engineering analysis was performed. On 11 October, it was decided to raise periapsis to 171 km and on 27 October a management meeting was held to discuss options; at this date apoapsis was about 45,135 km and the orbital period 35.4 h. Periapsis reduction recommenced 7 November in a burn that put the low point of the orbit down at 134.8 km but aerobraking would be performed at a much slower rate than originally planned, enabling engineers to carefully monitor any stress on the –Y panel. Engineers worked out that during initial deployment, a damper arm had fractured resulting in structural damage to one end of the solar cell frame.

The revised aerobraking plan reduced the periapsis pressure exerted on the solar panels to 0.2 N/m^2, one-third the level assumed in the original plan. Global Surveyor would reach a circular mapping orbit in March 1999 after a total of more than 900 orbits. To achieve that, aerobraking continued until 23 March 1998 when it was suspended for four months during the period covered by solar conjunction. Four days later, the science instruments were turned on for a period of observation and imaging. On 23 September, Global Surveyor performed the first of three propulsive manoeuvres, completed two days later, totalling 11.62 m/sec to step back into the atmosphere, lower periapsis to 121 km and settling it into a pressure regime of 0.14 N/m^2. On 26 and 27 September, further manoeuvres of 0.18 m/sec lowered periapsis to 116 km but engineers recorded peak aerobraking pressures of 0.25 N/m^2. As the spacecraft continued to reduce its orbital period it experienced peak pressures of 0.28 N/m^2. At 12:40 UT on 12 December it passed within 300 km of the moon Phobos.

On 29 January 1999, with apoapsis altitude down to 1,000 km and periapsis of 103 km, the spacecraft conducted the first of four tiny 'walk-out' manoeuvres, each less than 1 m/sec, allowing it to step out of the atmosphere and stabilise in the desired orbit. Having slowed the spacecraft by more than 1,200 m/sec, aerobraking was completed at 20:11 UT on 4 February when a 61.9 m/sec burn of the main propulsion engine raised periapsis and put the spacecraft in a 1.97 hour orbit with a local solar mean time of 2:04 pm. At 22:00 UT on 19 February, a further burn marginally lowered the newly raised periapsis from 405 to 367 km establishing a fully sun-synchronous orbit.

On 9 March the mapping mission began and on 28 March the high-gain antenna was successfully deployed for optimised data return to Earth. On April 15, the spacecraft put itself into an automatic contingency mode when the azimuth hinge on the high-gain antenna failed to point the antenna toward Earth. The entire spacecraft was turned to align the antenna with Earth and normal mapping operations resumed 5 May. Having completed an 88-orbit repeat pattern, a propulsive manoeuvre of 3.5 m/sec on 7 May started the spacecraft 'walking' round the planet by 58 km on each successive revolution for the next mapping phase. Primary objective of initial mapping operations during June and July was the Mass Polar Lander sector near the South Pole. On 4 November the MGS control team reported success with a test of the DeepSpace 2 microprobes relay system, which was to relay telmetry from the probes when they reached the surface of Mars on 3 December 1999. When contact with Mars Polar Lander was lost, the MGS spacecraft supported attempts to communicate with the spacecraft via UHF. On 22 December, MGS completed 16 of 19 planned imaging targets in the hope of finding the spacecraft on the surface but this was unsuccessful. MGS continues to return excellent data and high-quality imagery and performed science mapping operations during its second extended mission phase after which it served as a communications relay during the entry, descent and landing phase of the 2003 Mars landers when they reached the planet in January 2004.

Specifications

Mars Orbiter

Mass: 1,062.1 kg at launch (388.3 kg propellant, 75.8 kg science, 598 kg bus); body approx 1.5 m square by 3 m high.

Attitude Control: 3-axis pointing to 10 mrad (3 mrad knowledge, stability 1 mrad for 0.5 s and 3 mrad for 12 s) using 4 RWs, Sun sensors, IRU, star sensor, Mars Horizon sensor. 596 N bipropellant thruster for large delta-Vs; 12 × 4.45 N for momentum unloading, small delta-Vs and TVC.

Communications: 2 × 1.5 Gbit SSRs. 25 W X-band transmitter provides max 85.3 kbit/s via 1.5 m parabolic articulated boom-mounted HGA. Command uplink 7.8–500 bit/s.

Power: Twin 2-panel solar wings (12 m top-to-top extended) provide 940/660 W aphelion/perihelion (GaAs inner panel, Si outer); 2 × 20 Ah Ni/H_2 batteries

Payload: (all surface instruments are nadir body-mounted.)

MOC Mars Orbital Camera. CCD line-scan camera for wide angle (f/6.5, 11.3 mm focal length, 140° FOV, 3,456 CCD array length, resolution 7.5 km/240 m) and narrow angle (f/10, 350 cm fl, 2,048 CCD array length, resolution 1.4 m) imaging to monitor surface and atmospheric changes. 20.5 kg, 6 W. Imaging systems include 96 Mbit solid-state memory. (California Institute of Technology)

MOLA Mars Orbital Laser Altimeter. 10 Hz pulses, surface topography within 30 m (1–10 m relative) and surface reflectance at 1.06 μm. 160 m diameter footprint. NASA Goddard

TES Thermal Emission Spectrometer. Mapping of rock mineral composition, frost coverage and cloud composition at up to 3 km resolution across 9 km FOV using 6.25–50 μm Michelson interferometer

MAG/ER Magnetometer/Electron Reflectometer. Two NASA Goddard triaxial fluxgate magnetometers and French CNES ER to map magnetic field and monitor solar wind-Mars plasma interactions; resolution ±0.004 to ±16 nT

RS Radio Science. Uses radio system's X-band transponder + ultrastable oscillator to measure vertical profile of atmospheric refractive index

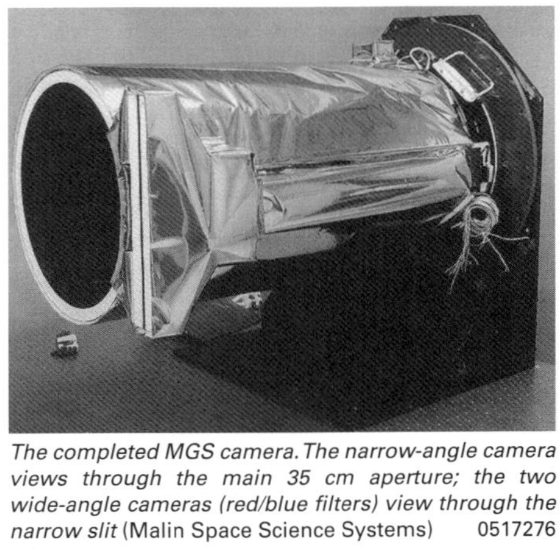

The completed MGS camera. The narrow-angle camera views through the main 35 cm aperture; the two wide-angle cameras (red/blue filters) view through the narrow slit (Malin Space Science Systems) 0517276

(indicating pressure, temperature, density) by monitoring Earth-received signal after atmospheric passage; map gravity field by spacecraft Doppler tracking. JPL

MR Mars Relay. Relays data from NASA Mars Surveyor landers. Data stored in MOC memory. Downlink beacon activates data uplink.

UPDATED

Mars Odyssey

Current Status

Mars Odyssey was launched by Delta 2 on 7 April 2001 and arrived at Mars on 24 October 2001 at 02:30 UT. The primary science mission lasts from February 2002 through August 2004. The spacecraft supports the other missions and serves as a relay for the two Mars Explorer Rover spacecraft launched in mid-2003 and successfully landed on the surface in December 2003 and January 2004. Odyssey will also assist with site selection for future landers. Mars Odyssey is scheduled to remain operational until the end of September 2006 at which time, along with Mars Global Surveyor, it will be succeeded by the Mars Reconnaissance Orbiter.

Background

Mars Odyssey is an orbiter carrying three packages of science experiments designed to make global observations of Mars to improve our understanding of the planet's climate and geologic history including the search for liquid water and evidence of past life. The mission will extend across a full Martian year, or 29 earth months.

The orbiter launch period extends for 21 days, opening on 7 April 2001 and closing on 27 April. The first 12 days make up what is considered the primary launch period; a secondary launch period runs from 19 April to 27 April. If Odyssey is launched during the secondary period, science data return at Mars may need to be reduced slightly because of higher arrival speeds and longer aerobraking periods. Arrival dates at Mars vary with launch dates, and range from 17–28 October 2001.

Odyssey will lift off from Space Launch Complex 17 at Cape Canaveral Air Station, Florida. Sixty-six seconds after launch, the first three solid rocket boosters will be discarded followed by the next three boosters one second later and the final three boosters one second after that. The final three boosters are jettisoned two minutes, 11 seconds after launch. About four minutes, 24 seconds after lift-off, the first stage will stop firing and be discarded eight seconds later. About five seconds later, the second stage engine ignites. The fairing or nose cone will be discarded four minutes, 41 seconds after launch. The first burn of the second stage engine occurs at 10 minutes, three seconds after launch.

At this point the vehicle is in low Earth orbit at an altitude of 189 km (117 miles). Depending on the actual launch day and time the vehicle will then coast for several minutes. Once it is in the correct point in its orbit, the second stage will be restarted at 24 minutes, 32 seconds after launch.

Small rockets will then be fired to spin up the third stage on a turntable attached to the second stage. The third stage will separate and ignite its motor, sending the spacecraft out of Earth orbit. A nutation control system (a thruster on an arm mounted on the side of the third stage) will be used to maintain stability during the third stage burn. After that, the spinning upper stage and the attached Mars Odyssey spacecraft must be despun so that the spacecraft can be separated and acquire its proper cruise orientation. This is accomplished by a set of weights that are reeled out from the side of the spinning vehicle on flexible lines, much as spinning ice skaters slow themselves by extending their arms. Odyssey will separate from the Delta third stage about 33 minutes after launch. Any remaining spin will be removed using the orbiter's onboard thruster. About 26 minutes after launch the solar array is unfolded and about eight minutes later it is locked in place. Then the spacecraft turns to its initial communication attitude and the transmitter is turned on. About one hour after launch the 34 m (112 ft) diameter antenna at the Deep Space Network complex near Canberra, Australia will acquire Odyssey's signal.

The interplanetary cruise phase is the period of travel from the Earth to Mars and lasts about 200 days. It begins with the first contact by the DSN after launch and extends until seven days before Mars arrival. Primary activities during the cruise include check out of the spacecraft in its cruise configuration, check out and monitoring of the spacecraft and the science instruments and navigation activities necessary to determine and correct Odyssey's flight path to Mars.

There are science activities planned for the cruise phase including payload health and status checks, instrument calibrations, as well as data taking by some of the science instruments according to spacecraft limitation.

Odyssey's flight path to Mars is called a Type 1 trajectory that takes it less than 180° around the Sun. During the first two months of cruise, only the Deep Space Network station in Canberra will be capable of viewing the spacecraft. Late in May California's Goldstone station will come into view, and by early June the Madrid station will also be able to track the spacecraft. The project has also added the use of a tracking station in Santiago, Chile, to fill in tracking coverage during the first seven days following launch.

The orbiter will transmit to Earth using its medium-gain antenna and it will received commands on its low-gain antenna during the early portion on its flight. At some point during the first 30 days after launch, the orbiter will be commanded to receive and transmit through its high-gain antenna. Cruise command sequences are generated and uplinked approximately once every four weeks during one of the regularly scheduled Deep Space Network passes.

The spacecraft will determine its orientation in space chiefly via a star camera and a device called an inertial measurement unit. The spacecraft will fly with its medium or high-gain antenna pointed towards the Earth at all times while keeping the solar panels pointed toward the Sun. The spacecraft is stabilised in three axes and will not spin to maintain its orientation, or attitude.

The spacecraft's orientation will be controlled by reaction wheels, devices with spinning wheels similar to gyroscopes. These devices will be occasionally 'desaturated', meaning that their momentum will be unloaded by firing the spacecraft's thrusters.

During interplanetary cruise, Odyssey is scheduled to fire its thrusters a total of five times to adjust its flight path. The first of these trajectory correction manoeuvres is schduled for eight days after launch, and will correct launch injection errors and adjust the Mars arrival aim point. It will be followed by a second manoeuvre 90 days after launch.

The remaining three trajectory correction manoeuvres will be used to direct the spacecraft to the proper aim point at Mars. These manoeuvres are scheduled at 90 days after launch, 12 days before arrival and seven hours (24 October) before arrival. The spacecraft will communicate with Deep Space Network antennas continuously for 24 hours around all of the trajectory correction manoeuvres. Manoeuvres will be conducted in what engineers are calling a 'constrained turn-and-burn' mode in which the spacecraft will turn to the desired burn attitude and fire the thrusters, while remaining in contact with Earth.

Navigation tracking during cruise involves the collection of two-way Doppler and ranging data. In order to provide additional information for navigation, the project has added a program of delta differential one-way range measurements, called delta DOR, which will be taken periodically during cruise and Mars approach. Delta DOR measurements are interferometric measurements between two radio sources. In this case, one of the radio sources is the DOR tones or telemetry signal coming from Odyssey. The second source will be either a known, stable natural radio source like a quasar or the telemetry signal from the Mars Global Surveyor spacecraft. Each source is recorded simultaneously at two radio antennas. The triangulation achieved through this method provides navigators with much more refined knowledge of the spacecraft's position. With this information, spacecraft operators can more precisely adjust Odyssey's flight path. Delta DOR measurements will be collected and processed for system testing during early and mid-cruise and weekly during the Mars approach phase to provide additional data to the navigation team. For the first 14 days after launch, the Deep Space Network will continuously track the spacecraft. During the quiet phase of cruise when spacecraft activity is at a minimum, only three 8-hour passes per day are scheduled. Continuous tracking will resume for the final 50 days before Mars arrival.

Science instruments will be powered on, tested and calibrated during cruise. The thermal emission imaging system will take a picture of the Earth-Moon system about 12 days after launch if the spacecraft is operating normally. Star calibration imaging is also planned 45 days after launch, while a Mars approach image is planned about 12 days before arrival if the Earth-Moon calibration image is not taken.

Two calibration periods are planned for the gamma ray spectrometer during cruise. Each of the spectrometer's three sensors may be operated during the calibration periods depending upon spacecraft power capabilities. The Mars radiation environment experiment is designed to collect radiation data constantly during cruise to help determine what the radiation environment is like on the way to Mars.

A test of the orbiter's UHF radio system is planned between 60 and 80 days after launch. The 45 m (150 ft) antenna at California's Stanford University will be used to test the UHF system ability to received and transmit. The UHF system will be used during Odyssey's relay phase to support future landers, it is not used as part of the orbiter's science mission.

Odyssey will arrive at Mars on 24 October, 2001. As it nears its closest point to the planet over the northern hemisphere, the spacecraft will fire its 640 N main engine for approximately 22 minutes to allow itself to be captured into an elliptical, or egg-shaped, orbit. If the launch occurs early in the period, Odyssey will loop around the planet every 17 hours. About three orbits after insertion, the spacecraft will fire its thrusters in what is called a period reduction manoeuvre so that it orbits the planet approximately once every 11 hours.

Aerobraking is the transition from the initial elliptical orbit to the science orbit where Odyssey will circle Mars at a uniform altitude. It is a technique that slows the spacecraft down by using frictional drag as it flies through the upper part of the planet's atmosphere.

During each of its long, elliptical loops around Mars, the orbiter will pass through the upper layers of the atmosphere each time it makes its closest approach to the planet. Friction from the atmosphere on the spacecraft and its wing-like solar array will cause the spacecraft to lose some of its momentum during each close approach, known as 'a drag pass'. As the spacecraft slows during each approach, the orbit will gradually lower and circularise.

Aerobraking will occur in three primary phases that engineers call walk-in, the main phase and walk-out. The walk-in phase occurs during the first four to eight orbits following Mars arrival. The main aerobraking phase begins once the point of the spacecraft's closest approach to the planet, know as the orbit's periapsis, has been lowered to within about 100 km (60 miles) above the Martian surface. As the spacecraft's orbit is reduced and circularised during approximately 273 drag passes in 76 days, the periapsis will move northward, almost directly over Mars' north pole. Small thruster firings when the spacecraft is at its most distant point from the planet will keep the drag pass altitude at the desired level to limit heating and dynamic pressure on the orbiter. The walk-out phase occurs during the last few days of aerobraking when the period of the spacecraft's orbit is the shortest.

The aerobraking drag pass events will be executed by stored onboard command sequences. The drag pass sequence begins with the heaters for the thrusters being warmed up for about 20 minutes. The transmitter is turned off to conserve power during the drag pass. The spacecraft then turns to the aerobraking attitude under reaction wheel control.

Following aerobraking walk-out, the orbiter will be in an elliptical orbit with a periapsis near an altitude of 120 km (75 miles) and an apoapsis – the farthest point from Mars – near a desired 400 km (249 mile) altitude. Periapsis will be near the equator. A manoeuvre to raise the periapsis will be performed to achieved the final 400 km (249 mile) circular science orbit.

The transition from aerobraking to the beginning of the science orbit will take about one week. The high-gain antenna will be deployed during this time and the spacecraft and science instruments will be checked out.

NASA's Langley Research Center in Hampton, Virginia will provide aerobraking support to JPL's navigation team during mission operations. Langley's role includes performing independent verification and validation, developing simulation tools and assisting the navigation team with trade studies and performance analysis.

The science mission begins about 45 days after the spacecraft is captured into orbit about Mars. The primary science phase will last 917 Earth days. The science orbit inclination is 93.1°, which results in a nearly Sun-synchronous orbit. The orbit period will be just under two hours. Successive ground tracks are separated in longitude by approximately

29.5° and the entire ground track nearly repeats every two sols, or Martian days.

During the science phase, the thermal emission imaging system will take multispectral thermal-infra-red images to make a global map of the minerals on the Martian surface, and will also acquire visible images with a resolution of about 18 m (59 ft). The gamma ray spectrometer will take global measurements during all Martian seasons. The Martian radiation environment experiment will be operated throughout the science phase to collect data on the planet's radiation environment. Opportunities for science collection will be assigned on a time-phased basis depending on when conditions are most favorable for specific instruments.

The relay phase began at the end of the first Martian year in orbit (about two Earth years). During this phase the orbiter provided communication support for US rovers *Spirit* and *Opportunity*.

Specifications

Description: The Mars Odyssey orbiter is a three-axis stabilised spacecraft consisting of a bus 2.2 × 1.7 × 2.6 m in size composed of a frame fabricated from aluminium with some titanium. Most spacecraft systems are fully redundant, the exception being the memory card that collects imaging data from the thermal emission imaging system. Spacecraft structure is divided into two modules: propulsion module containing tanks, thrusters and associated plumbing and an equipment module composed of an equipment deck, which supports engineering components and the radiation experiment, and a science deck connected by struts. The top side of the science deck supports the thermal emission imaging system, gamma ray spectrometer, the high-energy neutron detector, the neutron spectrometer and the star camera, while the underside supports engineering components and the gamma ray spectrometer's electronics box. The structures subsystem weighs 81.7 kg.

Mass: 725 kg at launch; 331.8 kg dry plus 348.7 kg of propellant and 44.5 kg of science instruments.

Command and Data Handling: RAD6000 32-bit processor with Mars Pathfinder heritage in the Orbiter Command and Data Handling subsystem, provides a central processing capability for all spacecraft systems including payload elements. It can be switched between clock speeds 5, 10 or 20 MHz and includes 128 Mbytes of RAM of which approximately 20 per cent is used for running spacecraft programmes. The rest of the space is used for science and for storing data, which will be relayed to Earth. Communication with Odyssey's sensors that measure the spacecraft's orientation in space, or attitude, and its science instruments is done via another interface card. A master input/output card collects signals from around the spacecraft and also sends commands to the electrical power subsystem. The interface to Odyssey's telecommunications subsystem exists through another card called the uplink/downlink card. Two other boards are located in the data handling subsystem. The module interface card controls when the spacecraft switches to back-up hardware and serves as the spacecraft's time clock. A converter card takes electricity produced by the power subsystem and converts it into the proper voltages for the rest of the command and data handling subsystem components. The last interface card is a single, non-redundant, 1Gbyte mass memory card used to store imaging data. The command and data handling subsystem weighs 11.1 kg.

Telecommunications: Comprises a radio system operating in the X-band microwave frequency range for direct communications with Earth and a system that operates in the UHF range for communication with future landers. System weighs 23.9 kg.

Power: The two-axis, articulated, gallium arsenide, three-panel, solar array (7.4 m^2 cell area) provides electrical power and serves as the most significant drag brake during aerobraking. Nickel-hydrogen (NiH2) 16 Ah common pressure vessel batteries provide power during eclipses and during peak power needs.

Thermal Control: The TC system uses passive methods and louvres to control the temperature of the batteries and the solid-state power amplifiers. Passive coatings as well as multilayer insulation blankets are used to control temperatures along with thermostatically controlled and computer controlled heater circuits. Where needed, radiators are used to remove excess from spacecraft components to maintain proper operating temperatures. The TC system weighs 20.1 kg.

Attitude control: The orbiter is three-axis stabilised in all mission phases except following separation from the launch vehicle. Primary attitude determination is via star camera derived from the Clementine spacecraft star camera and an inertial measurement unit with analogue star sensors as back-up. Reaction wheels provide primary attitude control during most mission phases, desaturated via RCS thrusters. Because of lifetime concerns the IMUs are switched off during significant portions of cruise and mapping and the vehicle operated in all-stellar mode except during manoeuvres, orbit insertion, aerobraking drag passes, orbital trim manoeuvres and safe mode. Mars Odyssey has four 0.98 N thrusters for roll control and four 22.56 N thrusters for pitch and yaw control. Using three redundant pairs of sensors, the guidance, navigation and control subsystem determines spacecraft orientation. A sun sensor is used to detect the position of the sun as a backup to the star camera, used primarily to locate star fields. Between star camera updates the inertial measurement unit collects information orientation is maintained or changed using four reaction-control wheels, three used for primary control and one as a backup. The guidance, navigation and control subsystem weighs 23.4 kg.

Propulsion: Derived from the propulsion system for Mars Global Surveyor, the Mars Odyssey spacecraft has a single LEROS 640 N main engine operating on hydrazine and nitrogen tetroxide bipropellants for Mars orbit insertion. In addition to miscellaneous tubing, pyrotechnic valves and filters, the propulsion subsystem also includes a single gaseous helium tank to pressurise the fuel and oxidiser tanks. The propulsion subsystem weighs 49.7 kg.

Payload: One of the chief scientific goals that Mars Odyssey will focus on is mapping the chemicals and minerals that make up the Martian surface. As on Earth, the geology and elements that form the Martian planet chronicle its history, and while neither elements, the building blocks of minerals, nor minerals, the building blocks of rocks, can convey the entire story of a planet's evolution, both contribute significant pieces to the puzzle. These factors have profound implications for understanding the evolution of Mars' climate and the role of water on the planet, the potential origin and evidence of life, and the possibilities that may exist for future human exploration.

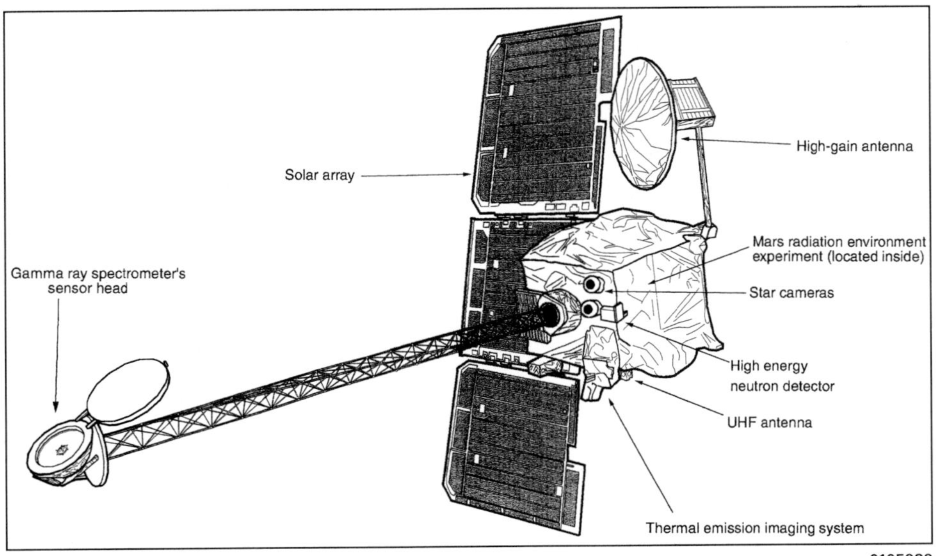

Mars Odyssey spacecraft 0105939

Other major goals of the Odyssey mission are to determine the abundance of hydrogen, most likely in the form of water ice, in the shallow subsurface; globally map the elements that make up the surface; acquire high-resolution thermal infra-red images of surface minerals; provide information about the structure of the Martian surface; and record the radiation environment in low Mars orbit as it relates to radiation-related risk to human exploration.

During the 917-day science mission, Odyssey will also serve as a communication relay for US or international scientific orbiters and landers in 2003 and 2004. After this period, the orbiter will be available as a communication relay for an additional 457 days, making for a total mission duration of 1,374 days, or two Martian years. Science operations may still continue during the communication relay-only phase depending on remaining orbiter resources.

The orbiter carries three science instruments; a thermal infra-red imaging system, a gamma ray spectrometer and a radiation environment experiment. These are all calibrated during the spacecraft's cruise phrase on its way to Mars. Opportunities for data collection are assigned on a time-phase basis depending on when conditions are most favourable for specific instruments.

The thermal imaging system is responsible for determining Mars' surface mineralogy. Unlike our eyes, which can only detect visible light waves, a small portion of the electromagnetic spectrum, the instrument can see in both visible and infra-red, thus collecting imaging data that has been previously invisible to the scientists. In the infra-red spectrum, the instrument uses 10 spectral bands to help detect minerals within the Martian terrain. These spectral bands, similar to ranges of colours, serve as signatures, or spectral fingerprints, of particular types of geological materials. Minerals, such as carbonates, silicates, hydroxides, sulfates, hydrothermal silica, oxides and phosphates, all show up as different colours in the infra-red spectrum. This multispectral method allows researchers to detect in particular the presence of minerals that form in water and understood those minerals in their proper geological context. Remote-sensing studies of natural surfaces, together with laboratory measurements, have demonstrated that 10 spectral bands are sufficient to detect minerals at abundances of five to 10 per cent. In addition, the use of 10 infra-red spectral bands can determine the absolute mineral abundance in a specific location within 15 per cent. The instrument's multispectral approach will also provide data on localized deposits associated with hydrothermal and subsurface water and enable 100 m (328 ft) resolution mapping of the entire planet. In essence, this allows a broad geological survey of the planet for the purpose of identifying minerals, with 100 m (328 ft) of Martian terrain captured in each pixel, or single point, of every image. It will also allow the instrument to search for thermal spots during the night that could result in discovering hot springs on Mars. Using visible imaging in five spectral bands, the experiment will also take 18 m (59 ft) resolution mineralogical and structural measurements specifically to determine the geological record of past liquid environments. More than 15,000 images each 20 by 20 km (12 by 12 miles) will be acquired for Martian surface studies. These more detailed data will be used in conjunction with mineral maps to identify potential future Martian landing sites. These images will provide an important bridge between the data acquired by the Viking missions and the high-resolution images captured by Mars Global Surveyor. The instrument weighs 11.2 kg; is 54.5 cm long, 34.9 cm tall and 28.6 cm wide; and runs on 19 W of electrical power. The principal investigator for the instrument is Dr Philip Christensen of Arizona State University in Tempe.

The gamma ray spectrometer plays a lead role in determining the elemental makeup of the Martian surface. Using a gamma ray spectrometer and two neutron detectors, the experiment detects and studies gamma rays and neutrons emitted from the planet's surface. When exposed to cosmic rays, all chemical elements emit gamma rays with distinct signatures. This spectrometer looks at these signatures, or engines, coming from the elements present in the Martian soil. By measuring gamma rays coming from the Martian surface, it is possible to calculate how abundant various elements are and how they are distributed around the planet's surface.

By measuring neutrons, it is possible to calculate Mars' hydrogen abundance, this inferring the presence of water. The neutron detectors are sensitive to concentrations of hydrogen in the upper metre of the surface.

Gamma rays, emitted from the nuclei of atoms, show up a sharp emission lines on the instrument's spectrum. While the energy represented in these emissions determines which elements are present, the intensity of the spectrum reveals the elements' concentrations. The spectrometer will send a reading to Earth every 20 seconds. This data will be collected over time and used to build up a full-planet map of elemental abundances and their distributions.

The spectrometer's data, collected at 300 km resolution, will enable researchers to address many questions and problems regarding Martian geoscience and life science, including crust and mantle composite, weathering processes and volcanism. The spectrometer is expected to add significantly to the growing understanding of the origin and evolution of Mars and of the processes shaping it today and in the past.

The gamma ray spectrometer consists of two main componets: the sensor head and the central electronics assembly. The sensor head is separated from the rest of the Odyssey spacecraft by a 6 m boom, which will be extended after Odyssey has entered the mapping orbit at Mars. This is done to minimize interference from any gamma rays coming from the spacecraft itself. The initial spectrometer activity, lasting between 15 and 40 days, will perform an instrument calibration before the boom is deployed. After 100 days in orbit, the boom will deploy and remain in this position for the duration of the mission. The two neutron detectors, the neutron spectrometer and the high-energy neutron detector are mounted on the main spacecraft structure and will operate continuously throughout the mission. The instrument weighs 30.3 kg and uses 32 W of power. Along with its cooler, the gamma ray spectrometer measures 46.8 cm long, 53.4 cm tall and 60.4 cm wide. The neutron spectrometer is 17.3 cm long, 14.4 cm tall and 31.4 cm wide. The high-energy neutron detector measures 30.3 cm long, 24.8 cm tall and 24.2 cm wide. The instrument's central electronics box is 28.1 cm long, 24.3 cm tall and 23.4 cm wide. The principal investigator for the gamma ray spectrometer is Dr William Boynton of the University of Arizona.

The Martian radiation environment experiment characterises aspects of the radiation environment both on the way to Mars and the Martian orbit. Since space radiation presents an extreme hazard to crews of interplanetary missions, the experiment will attempt to predict anticipated radiation doses that would be experienced by future astonauts and help determine possible effects of Martian radiation on human beings. Space radiation comes from two sources – energetic particles from the Sun and galactic cosmic rays from beyond our solar system. Both kinds of radiation can trigger cancer and cause damage to the central nervous system. A spectrometer inside the instrument will measure the energy from these radiation sources. As the spacecraft orbits the red planets, the spectrometer sweeps through the sky and measures the radiation field. The instrument, with a 68° field of view, is designed to continuously collect data during Odyssey's cruise from Earth to Mars. It can store large amounts of data for downlink whenever possible, and will operate throughtout the entire science mission. The instrument weighs 3.3 km and uses 7 W of power. It measures 29.4 cm long, 23.2 cm tall and 10.8 cm wide. The principal investigator for the radiation environment experiment is Dr Gautum Badhwar of NASA's Johnson Space Center.

UPDATED

Mars Pathfinder

Current Status

Scientists at the JPL in Pasadena, California lost contact with the Mars Pathfinder in October 1997. During the four-month life of the mission, Pathfinder transmitted 2.3 billion bits of data from the six-wheeled Sojourner rover and the Pathfinder lander. First impressions from the scientific team sifting through the data indicated that all the rocks studied by Pathfinder's alpha proton X-ray spectrometer, contain a high-silicon content and resemble volcanic rock on Earth known as andesite. Mars Global Surveyor and Mars Odyssey have photographed the lander region for added verification of horizon features on images taken from the surface.

Background

NASA approved the Mars Pathfinder in 1993 as one of the first two Discovery missions, which feature low cost planetary missions. To ensure a cost effective mission, NASA set a US$150 million FY92 development cost cap with fast development schedules of less than three years and focused science objectives.

NASA wants Pathfinder to demonstrate a low cost cruise, entry, descent and landing system for exploration of the Martian surface, with the additional objectives of the deployment and operation of three science instruments and operation of the small rover (Sojourner Truth, named after a pre-Civil War slavery abolitionist).

The spacecraft and instruments cost US$171 million (excluding the separately funded US$25 million MFEX). It cost NASA US$61 million to launch the craft and US$14 million for one year of tracking and mission operations.

Mars Pathfinder was originally proposed as a technology demonstration for the MESUR Mars Environmental Survey Network mission, but after the loss of the Mars Observer, NASA shelved MESUR and kept Mars Pathfinder as the flagship mission for the Mars Surveyor and Discovery programmes.

Sojourner is a NASA Office of Advanced Concepts & Technology experiment for mobile vehicle technologies, whose primary mission is to determine microrover performance in the poorly understood Martian terrain. The total rover budget was US$25 million including development and operation.

To save time and operational costs, the spacecraft performed no science experiments on the entry, which was a direct hyperbolic approach. The rover landed about 850 km southeast of Viking 1, in the Ares/Tiu Valles. This site elicited special interest among project planners because of its outflow of a catastrophic flood system.

JPL built the spacecraft and handled the mission operations. NASA launched the Pathfinder on a McDonnell Douglas Delta. Lockheed Martin contributed the aeroshell and the IMP optics, Loral Federal Systems the flight computer, Motorola the transponder, ILC Dover the airbags, Pioneer Aerospace the parachute, and Thiokol the Rocket Assisted Deceleration (RAD) subsystem, airbag gas generators.

The Mars Pathfinder was launched at 06:58:06 on 4 December 1996. Unlike the Mars Global Surveyor, the solar panels on MPF deployed successfully and performed some 10 per cent better than predicted. On 11 December, the spacecraft spin rate was reduced from 12.3 rpm down to just 2 rpm and the star scanner was activated to establish full three-axis knowledge of orientation in space. On 17 December, the Sojourner rover, encased within the MPF spacecraft, awakened on a ground signal and performed an internal health check.

MPF underwent its first trajectory correction manoeuvre (TCM) on 10 January 1997 with two of the spacecraft's eight 0.4 kg thrusters firing continuously for 90 minutes, changing the spacecraft's velocity by 31 m/s. With the spacecraft 19 million km from Earth, the second TCM was performed on 3 February, a two-part manoeuvre designed to null out dispersions from TCM-1 and shift the trajectory closer to a planetary encounter. For quarantine reasons, Mars Pathfinder would not be placed on a direct Mars intercept path until TCM-3. The first component of TCM-2 involved the two small forward-firing thrusters operating for five minutes, changing velocity by 1.5 m/sec. The second burn required all four thrusters on one side to fire for five seconds and change velocity by 0.5 m/sec. Conducted on 6 May, TCM-3 was a three-component manoeuvre consisting of a 0.4 m/sec lateral burn away from Mars, a 0.1 m/sec axial burn to correct arrival time and a 0.5 m/sec burn back toward Mars. About 17 million km from Mars, activity during the last week in May focused on charging up the flight battery. About 22 Ah had been taken out of the 56 Ah capacity through degradation and by re-charging the level was returned above the 40 Ah level felt necessary for the remainder of the mission.

TCM-4, the last scheduled course adjustment, was accomplished on 25 June with the spacecraft 180 million km from Earth, about 3.5 million km from Mars and just nine days to go before direct entry. The first component involved firing four thrusters on one side of the vehicle for 1.6 seconds while the second burn involved firing two thrusters close to the heat shield for 2.2 seconds. The combined velocity change was a mere 0.018 m/sec and aligned the trajectory for a direct entry on target to a landing in the Ares Vallis flood plain on 4 July 1997. On 30 June, controllers turned the spacecraft into its proper orientation for entry and switched to the 'entry, descent and landing' (EDL) computer sequence. An optional TCM-5 at either 12 or six hours prior to entry was not necessary, tracking indicating the spacecraft was aligned with the centre of the pre-planned entry path and at an angle of 14.06° instead of 13.9°.

Mars Pathfinder will demonstrate a low-cost approach to delivering payloads to the surface of Mars. The 9 kg Sojourner rover will probe the elemental composition of rocks and soil. The flood plain should provide a grab bag of rocks washed down from the highlands (NASA/JPL)
0517292

The EDL phase began about 35 minutes prior to landing (L–35) with cruise stage separation followed by entry at L–5 at an altitude of 130 km, a velocity of 7,470 m/sec and peak deceleration of 25 *g*. At L–134 sec and 9.4 km, with velocity reduced to 370 m/sec, the parachute deployment was triggered by an accelerometer sensing 16 *g* and diminishing. Heatshield release occurred 20 seconds later followed by the lander separating, so that it was suspended beneath the backshell by a 20 m kevlar bridle, 20 seconds after that (L–94 sec). Ground radar acquisition came at L–28.7 sec at a height of 1.6 km and a descent rate of 68 m/sec followed at L–10.1 sec (355 m altitude) by inflation of the airbag. Just four seconds later (L–6.1 sec) and at an altitude of 98 m with the spacecraft falling at 61.2 m/sec, the descent rate was dramatically arrested by ignition of the rockets in the backshell which fired for 2.5 seconds. At L–3.8 seconds and at a height of 21.5 m, the bridle holding the spacecraft to the backshell was cut. Encapsulated by the multi-lobe airbag, in free-fall the lander struck the surface at 16:56:55 UT event time, 17:07:25 UT Earth receive time, on 4 July 1997, at a true solar time of 2:58 am. Impact velocity of 10.5 m/sec and 18 *g* was sufficient to cause Pathfinder to bounce about 15 m off the surface with a second bounce of 9 *g* and about 7 m. This was followed by a third impact of 11 *g*, probably caused by the airbag/lander combination spinning up as it bounced. The spacecraft bounced at least 15 times before rolling up a slight incline and some way back down again before coming to a stop. Landing co-ordinates are 19.30° N by 33.52° W, 19 km south-west of the planned spot with the spacecraft at a slight tilt of 2.5°.

Named the Sagan Memorial Station after the late and eminent planetologist and astronomer Carl Sagan, the Mars Pathfinder site was explored by

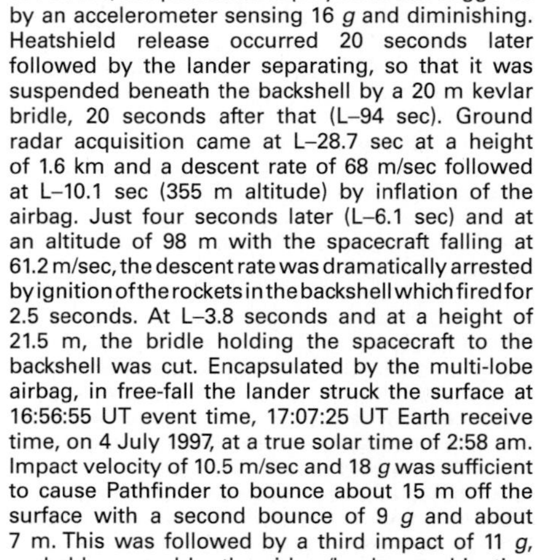

Mars Pathfinder exploded view (JPL) 0517116

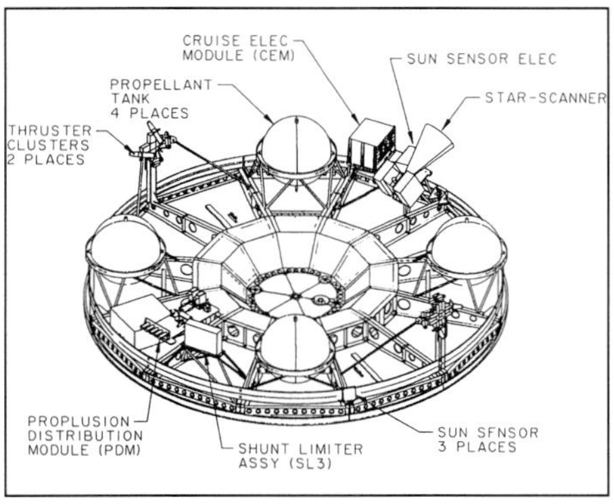

Mars Pathfinder cruise stage. Separated 30 min before entry, it will impact the surface (JPL) 0517117

the Sojourner rover deployed to the surface down extended metal tapes serving as runners for the rover wheels. Before that the airbag was partly reinflated and recollapsed in an effort to rearrange the folds from obstructing the rover's traverse down to the surface. An extensive panorama of images secured the preliminary survey of local topography and pictures taken of Sojourner on the lander helped verify the condition of the 10.5 kg rover. Minor software problems between Pathfinder and Sojourner were cleared up and the rover moved halfway down the ramps, one of which was inclined 20°, before taking pictures. Resuming its movement down to the surface, which took four minutes, the first TV pictures showing all six wheels on Mars were received at 05:59 UT on 6 July. It was the first vehicle to roam the surface of another planet. The APXS designed to analyse surface materials for their chemical content was activated on sol 2 (Mars day 2 of the mission). Data assessment at the end of the mission indicated that local rocks resemble volcanic types with high silicon content, a rock known as andesite on Earth, covered with a fine layer of dust.

Mars Pathfinder was as much a proof-of-concept mission as a scientific exploration of Mars and proved that low-cost technology and innovation can dramatically reduce the price of planetary missions. It demonstrated that the small rover is an ideal mobile vehicle for placing instruments up against rocks and soil samples and for conveying a camera around the surface for topographic, geologic and morphologic studies of surface features, in both mono and stereo images. The last signals from the lander were received on 27 September 1997 after which 2.6 Gbits of data had been transmitted including more than 16,000 images from the lander camera and 550 from the rover. Approximately 8.5 million temperature, pressure and wind measurements had been made. The rover travelled a total 100 m in 230 commanded manoeuvres, exploring about 250 m^2 of the local area. The lander had been in contact for 83 days, nearly three times its design life of 30 days and the Sojourner rover operated for 12 times its design life of seven days.

Specifications

Mass: 870 kg at launch; 785 kg dry, 85 kg hydrazine. 556 kg at entry, 325 kg lander mass includes 25 kg payload

Cruise stage

Description: 2.65m diameter, 2 rpm spin stabilised during cruise, attitude control knowledge to less than 0.5° using Magellan-heritage star scanner and Mars Observer-heritage Sun sensors.
Telemetry: Minimum command rate (from DSN 34 m High Efficiency net) 125 bit/s; min telemetry rate 40 bit/s to DSN 34 m net.
Propulsion: Four hydrazine spheres with total 100 kg capacity feed propulsion system for TCM manoeuvres and attitude control by eight 4.45 N thrusters. TCMs in Earth-line vector mode or turn/burn.
Power: 4.4 m^2 GaAs solar array covering the top face provides power for cruise ops.

Entry vehicle

Description: 2.65 m diameter aeroshell, parachute (based on Viking design), radar altimeter (operates from 1,500 m), RAD solid retros and 20 m^3 double-layer airbags. Aeroshell, based on Viking-heritage, employs SLA-561 ablative material. 11.5 m diameter parachute deployed through backshell; three modified Star 5 solid retros fired to reduce landing speed. Airbag mounted on each face of tetrahedral lander limits lander deceleration to 50 g. Communication is performed via lander RF equipment and backshell LGA. Telemetry rate to DSN 70 m antenna 40 bit/s.

Lander

Description: 1.1 m isosceles tetrahedron with deployable sides for self-righting.
Power: 2.9 m^2 GaAs array on petal provides power during day. 40 Ah AgZn secondary battery allows night operations.
Computation: A single board Loral R6000 computer (20 Mips max). Data storage by 128 Mb DRAM plus 4 Mb EEPROM. Flight software encoded in 'C' using commercial operating system (VxWorks). Computer and associated command/data handling hardware use standard VME architecture.
Communication: Cassini-heritage deep space transponder plus 13 W X-band SSPA for telecoms (including 5 W X-band back-up). 125 bit/s command rate via HGA from DSN 34 m HEF net. 5.0 kbit/s telemetry to DSN 70 m net. HGA is a mechanically steered slotted plate in thermal enclosure. Command rate through lander LGA (whip antenna on bottom, used during final descent) is 7.8 bit/s, telemetry 40 bit/s.
Lander payload: IMP Imager for Mars Pathfinder. Stereo colour imager mounted on 80 cm deployable mast. Stepper motors provide camera head steering, ±178° azimuth, −72° to +83° elevation. Loral CCD mounted at foci of two optical paths divided into two 256 × 256 square frames; stereo optical paths separated by 120 mm. f/10, stepped down to f/18 with 23 mm effective fl, 14.4° FOV, 0.5–32,000 ms exposures. Pixel IFOV 1 mrad. Fused silica window at each path entrance prevents dust intrusion. Each path's 0.5–1.05 μm filter wheel has 12 positions: eight for geology, two for water vapour, blue for atmospheric dust, and a broadband for stereo. IMP also carries magnets for magnetic properties experiment (Denmark; infra-red imaging may identify magnetic minerals) and wind socks for 5–50 m/s wind direction observations. 5.20 kg, 2.6 W, US$6 million cost cap.

APXS Alpha/proton/X-ray Spectrometer. Alpha-particle source/detector for backscattered alpha, protons and X-rays to determine sample elemental composition. APXS electronics mounted in rover WEB; sensor head mounted on rover deployment mechanism. 0.74 kg, 0.8 W. (Max Planck Institute/University of Chicago)

ASI/MET Atmospheric Structure/Meteorology Package. JPL facility instrument assesses atmospheric structure and surface meteorology. Surface meteorology hardware includes 1 m weather boom with a set of temperature and pressure sensors and wind speed/direction sensors. Acceleration data acquired throughout entry using range switchable accelerometers. temperature and pressure measurements after heatshield separation, using surface temperature and pressure sensors. 2.04 kg, 3.2 W.

Sojourner (MFEX Microrover Flight Experiment). 9 kg free-ranging rover for autonomous ops in lander vicinity; nominal life 7 sols but >30 sols possible. 63 cm long, 48 cm wide, 28 cm deployed height. Six 13 cm diameter wheels, independently powered (front/rear independently steerable), on rocker bogie suspension; top speed 0.4 m/min. Power provided by 0.22 m^2 GaAs top face array plus 150 Wh $LiSOCl_2$ primary battery. Peak solar panel power production 15 W; mobility requires 10 W. Sensitive electronics housed in insulated WEB warm electronics box; no night heating. Control by integrated set of computing/power distribution electronics. 80C85 computer rate at 100 Kips; memory includes 176 kbyte PROM plus 576 kbyte RAM. Motion control partially autonomous using feedback from onboard sensors: three CCD monochrome cameras (stereo at front; rear for 1 mm resolution of APXS sites), laser stripers, accelerometers, potentiometers. Distance travelled and heading from averaged odometry and gyro. Deployed from lander over ramps, deployable metal reels providing path over airbags. Communicates to lander via UHF modems at up to 9,600 bit/s.

Mars Pathfinder airbag testing (NASA) 0517275

UPDATED

Mars Polar Lander

Current Status

Mars Polar Lander was selected as the vehicle design to send to Mars as part of its new Scout missions programme which aims to send relatively low cost spacecraft to selected targets of interest. The original MPL concept will be re-scoped into Phoenix, the name of the Mars lander set to reach Mars in 2007. The lack of success with the original Mars Polar Lander does not reflect negatively on the basic design and on its application to the 2007 opportunity. Several efforts have been made to locate the remains of the original MPL spacecraft on the surface of Mars using Mars Global Surveyor and Mars Odyssey but without success.

Mars Polar Lander was launched by Delta 2 (7425) at 20:21:10 UT on 3 January 1999 from Launch Complex 17B at the Cape Canaveral Air Force Station, Florida. The four solid rocket boosters were jettisoned at 60 seconds followed by first stage cut-off 3 minutes 24 seconds later and separation 8 seconds after that. The second stage lit up 5.5 seconds after first stage separation and burned for 6 minutes 44 seconds placing the vehicle in a low Earth orbit of 119 km. After a 23-minute coast over the Indian Ocean, the second stage fired briefly a second time followed by the third stage which fired for 88 seconds to inject the MPL to a Type 2 Mars transfer trajectory. Shortly thereafter the spacecraft separated from the third stage and deployed the cruise solar panels. The spacecraft signal was acquired by the DSN station at Canberra, Australia. The first trajectory correction manoeuvre was performed on 21 January when the four manoeuvring engines fired for almost 10 seconds to change velocity by 0.89 m/s. The second correction manoeuvre was performed on 15 March, a burn using the same engines and with the same parameters. The next scheduled course correction manoeuvre was set for 1 September, three months before the spacecraft was expected to arrive at the planet to align the flight path for a specific landing zone near the south pole of Mars. On 25 August 1999 NASA revealed the final landing site selected by the evaluation committee, a swath of terrain 4,000 km^2 located at 76° south latitude by 195° west longitude, a region where slopes were no steeper than 10°. The landing was targeted to the centre of the site, a rectangular area 200 km long and 20 km wide. The site had been selected following extensive mapping and altimeter information from Mars Global Surveyor (which see). The landing would take place toward the end of spring in the southern hemisphere when the spacecraft would have 90 days of continuous sunlight until the seasons changed and the lander's mission ended around the beginning of March 2000. On 1 September 1999, at 17:07 UT, MPL fired its thrusters for 30 seconds increasing speed by 2.3 m/s, refining the flight path. After the loss of Mars Climate Orbiter on 23 September extensive analysis was conducted to ensure that the MPL did not have the same kind of unit conversion error that doomed its sibling spacecraft. However, during the evaluation it was discovered that a potential problem with the PL's descent engine could cause catastrophic failure if left uncorrected and a workaround procedure was devised. A third trajectory correction burn was performed at 17:28 UT 30 October 1999 for 12 seconds changing the speed of the spacecraft by about 1 m/s. On 1 November the spacecraft's landing radar was turned on for successful test, the first time since launch and, at 18:00 UT on 30 November, the fourth course correction burn was performed, lasting 12.6 sec and adjusting speed by 0,06 m/s. With this the MPL was aligned to enter a corridor 10 km wide and 40 km long. Flight controllers opted to perform a fifth course correction carried out at 13.39 UT 3 December for 8 sec, to increase the flight path angle from 13° to 13.25°. The MPL was calculated to begin its 5 min 30 sec descent to the surface at about 20:10 UT on 3 December and to arrive at its assigned

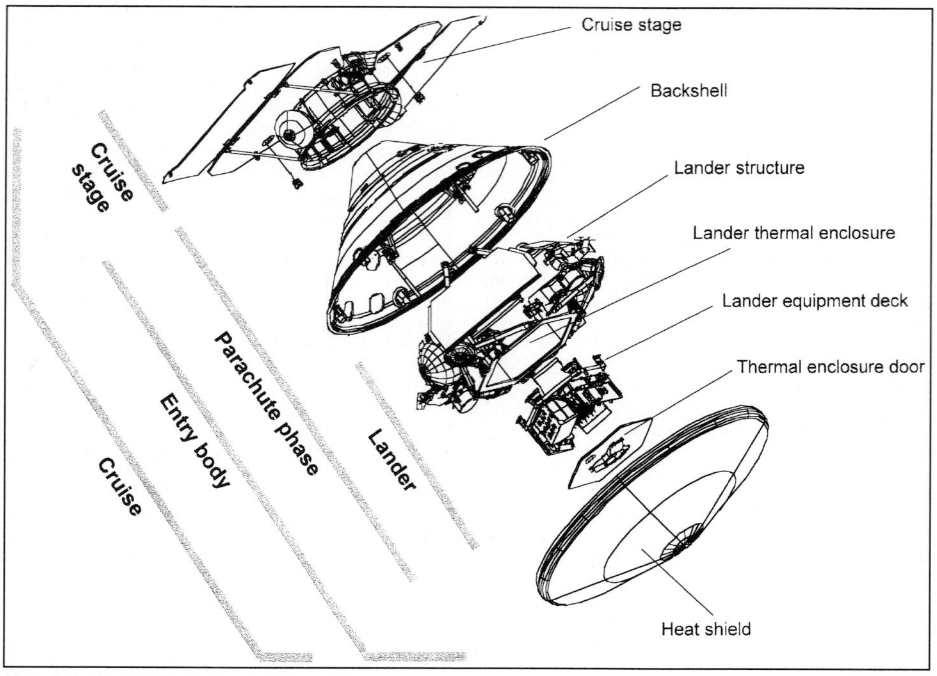

Mars Polar Lander flight encapsulation structures 0085846

landing site at 20:15 UT; not before 20:39 UT would the signal be heard on earth due to the 19 min time lag at the speed of light. No further word was heard from the spacecraft or the DS2 probes. An extensive search was made using alternative command configurations and the Mars Global Surveyor was used to search for the spacecraft on the surface, although an object the size of the lander was right on the limit of the MGS imaging system. In the extensive investigation that followed questions were raised about the future of Mars exploration and plans for mission opportunities in 2001 and 2003 were scaled down. NASA would launch only one orbiter at the 2001 opportunity and two landers in 2003 with at least one orbiter or one lander at 26 month launch windows thereafter.

Background

Mars Polar Lander (MPL) was one of the Surveyor missions instituted by NASA in 1994 as a series of small spacecraft for the sustained exploration of Mars. MPL is one of two spacecraft developed for the Mars 98 mission opportunity. The other is Mars Climate Orbiter (MCO). See Mars Climate Orbiter (this section) for general Mars 98 background and costs. MPL fills the scientific need for further information about the planet's past and present water resources and will seek answers by probing layered terrain near Mars' south pole. It will deploy two penetrators before impact and a 2 m robotic arm on the main spacecraft will dig into the top of the terrain. This is the first time a lander has examined the surface of Mars at such a high latitude and it is the second lander mission in the new strategy of sending spacecraft to this planet at every launch window approximately 26 months apart, the first being Mars Pathfinder. Developed and built by Lockheed Martin, the MPL will also carry two Deep Space 2 microprobes, released for ballistic descent to the surface shortly before the lander performs touch down. Scientific objectives specifically include a search for ancient and modern ground and surface water environments.

The mission plan required MPL to be launched less than one month after Mars Climate Orbiter by Delta 2 (7425) identical to the one used to send MCO on its way and similar to the one used to launch NEAR. The trajectory selected for MPL was a Type 2 path where the spacecraft would travel more than 180° around the Sun, launching marginally inward towards the Sun before spiralling out towards Mars. This would have enabled MPL to reach the appropriate landing zone selected close to the south pole. Because it flew a Type 1 trajectory (travelling less than 180° round the Sun before encounter) Mars Climate Orbiter would reach Mars more than two months ahead of Polar Lander. In the intervening period it would use aerobraking to achieve its operational (near circular) orbit by 1 December 1999, two days ahead of the date Polar Lander was scheduled to reach the surface. This configured the relative position of MCO to operate as a relay to the Lander for approximately three months. The primary launch window of 3–10 January 1999 provided the best opportunity for MPL to reach the surface of Mars on 3 December 1999 in the target area 73–77° S latitude. The secondary launch window of 11–16 January would incur a northward drift in the landing site to about 71° S and extend the landing date to no later than 17 December. Launch times within both periods were almost instantaneous to the second allowing no room for unexpected holds in the countdown.

The lander portion of the MPL was encapsulated in a protective aeroshell to the backshell of which was attached the cruise stage providing power, propulsion and communications and housekeeping duties until shortly before entry into the Martian atmosphere. Less than one hour after launch with the spacecraft on its Type 2 trajectory, the cruise stage solar panels were deployed and the vehicle orientated towards the Sun at which point the 34 m DSN antenna at Canberra, Australia, acquired it. Interplanetary cruise lasted 11 months during which communication was maintained via X-band link and the medium gain antenna on the cruise stage. Up to six course correction manoeuvres were built in to the cruise plan, the first conducted 15 days after launch to adjust the Mars arrival aimpoint. The remainder were scheduled at 45 days after launch and at 60 days and 10 days before Mars arrival with two manoeuvres possible between 3 days and 7 hours before arrival. Tracking increased 45 days before encounter and the precise aimpoint calculated while science instruments that gathered data during the cruise phase transmitted to the 70 m DSN antennas on Earth. Selection of the precise aimpoint within the preferred landing zone made between June and August 1999 using images of the region transmitted to Earth by Mars Global Surveyor. Mission managers designated that spot within four months of arrival. Unlike the Viking landers that touched down in 1976 the MPL was to have conducted a ballistic entry rather than first going into Mars orbit but like the Viking landers and unlike the Mars Pathfinder of 1997 the MPL was designed to perform a controlled descent using retro-motors almost to touchdown.

Entry events commenced about 14 hours before encounter when the final four-hour tracking phase began, the final opportunity for navigation data which could signal the need for a trajectory correction manoeuvre. A final 30-minute tracking session started 7 hours 25 minutes before entry, at which point the final trajectory correction command was issued. At E-5 hours, a one-hour tracking session provided spacecraft health data and status checks following any last course correction and at E-4 hours 40 minutes a series of valves opened to vent the descent motors. Final contact with the spacecraft began at E-25 minutes and was expected to last 15 minutes. At E-20 minutes the lander's descent motor heaters were to have turned on five minutes before the software controlling the Mars descent imager was initialised. At E-10 minutes the MPL was to have switched to inertial navigation computations obtaining data from gyroscopes and accelerometers and at E-6 minutes the spacecraft was to have fired its thrusters for 80 seconds to put the vehicle in the correct orientation. At E-5 minutes (10 minutes to touchdown) with the MPL travelling at 6,200 m/s and about 2,300 km from the surface of Mars, the cruise stage and solar panels were to have jettisoned followed 18 seconds later by the two Deep Space 2 microprobes. Five minutes later, at 6,900 m/s and an altitude of 125 km, the aeroshell encountered the atmosphere with just 4 minutes 33 seconds to touchdown. At maximum deceleration, the aeroshell experienced 12 *g* and a temperature of 1,650° C

About two minutes before landing when the MPL was descending at 493 m/s and just 8.8 km from the surface, the parachute was to have been released by mortar followed 10 seconds later by release of the heat shield from beneath the lander, which is suspended under the backshell. At this point the descent imager would have been switched on and the first picture taken 0.3 seconds before shield separation, the first of 10 pictures taken during descent. The first pictures would have shown a 9 × 9 km area at 7.5 m resolution while the final picture would have shown a 9 × 9 m area with a resolution of 9 mm per pixel. Between about 70 and 100 seconds before touchdown, the landing gear deployed followed 1.5 seconds later by activation of the doppler velocity landing radar. This determined the lander's altitude approximately 44 seconds later about 2.5 km above the surface descending at 80 m/s with 30 seconds to touchdown. At 1.4 km the manoeuvre thrusters were to have been turned off and lander was to have separated

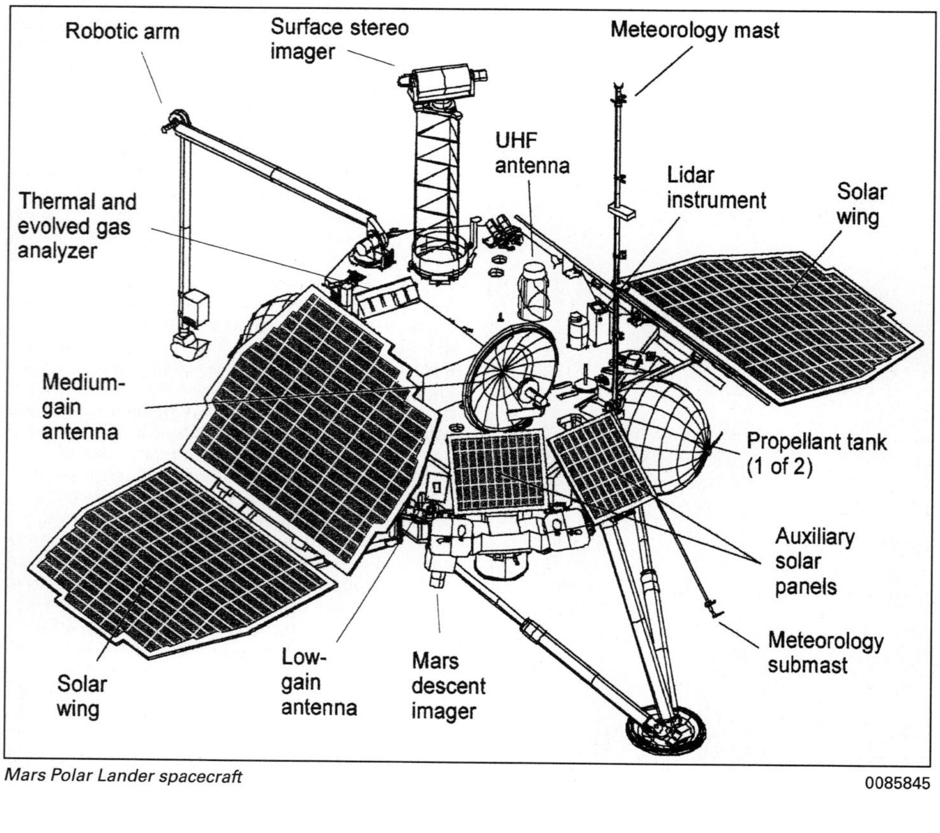

Mars Polar Lander spacecraft 0085845

from the backshell and parachute assembly. Just 0.5 seconds later the descent engines were to have been turned on for deceleration, attitude and roll control to turn the descent path vertical and put the lander in the proper orientation so that it would sit on the surface with the solar panels facing the Sun as it moves across the sky. The radar was to have turned off at a height of 40 m and the lander on inertial navigation. At 12 m or a descent rate of 2.4 m/s, whichever came first, the lander was to have lowered itself at a constant velocity (<1 m/s horizontal velocity) and the descent engines were to have been turned off when pad sensors touched the surface.

The lander was expected to contact the surface at 5:00 am local time, 21:03 UT, on 3 December 1999. About two minutes later, the solar panels would have deployed and three minutes later the medium-gain antenna would have aligned with the Earth. Two-way communication with Earth would have begun about 20 minutes after landing and much of that Mars morning would have been spent diagnosing spacecraft health and condition, sending meteorological data, some descent imager pictures and some from the surface stereo imager via the high-data rate UHF transmitter. When the lander was ready to begin science operations, the robotic arm would dig a shallow trench a few centimetres deep, probably taking the sunlight hours of two Mars days (sols) to complete the initial task. A sample would have been scooped up and dropped into the Thermal and Evolved Gas Analyser where a low temperature cooking sequence would have heated it to 27° C followed by a high-temperature cycle to 1,027° C a day later. Each day the MPL would have had 3 to 4 opportunities to send engineering and science data to Earth via the Mars Climate Orbiter, when the latter would be at least 20° above the horizon.

Meanwhile, the two Deep Space 2 probes (named Amunsden and Scott after the polar explorer) released from the MPL about 5 minutes before entry (encapsulated in an aeroshell to protect them from the 2,000° C of re-entry) would have impacted the surface at 200 m/s about 50 to 85 seconds before the lander touched down 100 km distant. Each penetrator comprised a main surface body and a probe connected by a flexible line protected by the aeroshell during descent. When it strikes the surface the aeroshell shatters and the shock of sudden deceleration (30,000 *g* on the probe and 60,000 *g* on the main surface body) drives the probe into the surface still connected to the main surface body, which remains above ground. The two probes, up to 100 km apart, would have each obtained a small sample of subsurface soil using a small electric drill extending sideways from the probe pulling less than 100 mg into a small cup. The sample would have been sealed and heated turning any water ice into vapour through which a tiny laser beam is shone to record the quantity. The probes were designed to measure the rate at which heat dissipates into the surrounding soil. The primary mission of the probes was expected to last no more than 36 hours although they would have continued to obtain data for as long as their batteries lasted. The prime mission of the lander was due to finish on 1 March 2000, releasing the Mars Climate Orbiter from its role as relay and enabling it to begin systematic observations of the atmosphere and the surface.

Specifications

Description: The Mars Polar Lander stands 1.06 m tall from the ground to the tip of the science deck and measures 3.6 m wide. The spacecraft is constructed of a composite material with honeycomb aluminium core and graphite epoxy facesheets bonded to each side of the bus. A thin aluminium sheet is bonded to the composite to provide a Faraday cage around the thermal enclosure. The landing legs are made of aluminium and equipped with compression springs to deploy them from the stowed position. Tapered, crushable, aluminium honeycomb inserts in each leg provide shock absorption at landing. The central enclosure houses the onboard computer, power distribution, the 16 Ah, 12-cell, nickel-hydrogen battery, a battery charger and radio equipment. A separate thermally-isolated component deck outside of the central electronics enclosure contains gyroscopes, electronics to fire pyrotechnic devices for instrument deploy and radar equipment for entry, descent and landing (EDL). Embedded within the component deck are the evaporators, which transfer heat from the aluminium facesheets to the radiators outside the thermal enclosure.

The lander flight system consists of a separable cruise stage with a MDAC V-band launch vehicle separation interface and propulsive lander/entry assembly. The cruise stage is jettisoned before EDL providing a clean aerodynamic shape for entry and a reduced ballistic coefficient of 58–62 kg/m^2 across the 2.4 m ablator heat shield. Cruise stage operational components include star cameras and sun sensors for attitude determination, two solar array wings (2.6 m^2 area) for power generation, X-band medium gain transmit/receive horn antenna and one low gain receive patch antenna and a redundant pair of SSPAs for cruise telecom. Three-axis attitude control is achieved using redundant IMUs and four cruise reaction engine modules (REMs) on the lander, each REM containing one aft-facing 22.2 N TCM thruster and one 4.45 N RCS thruster (canted 20° out, 15° aft). The EDL stage consists of a 2.4 m diameter ablator aeroshell utilising tooling originally developed for Mars Pathfinder and sharing the same nose radius and cone angle. The parachute is also from the Pathfinder design to lower cost and is deployed based on an on-board navigation velocity estimate eliminating the long range radar used on the Viking landers in 1976.

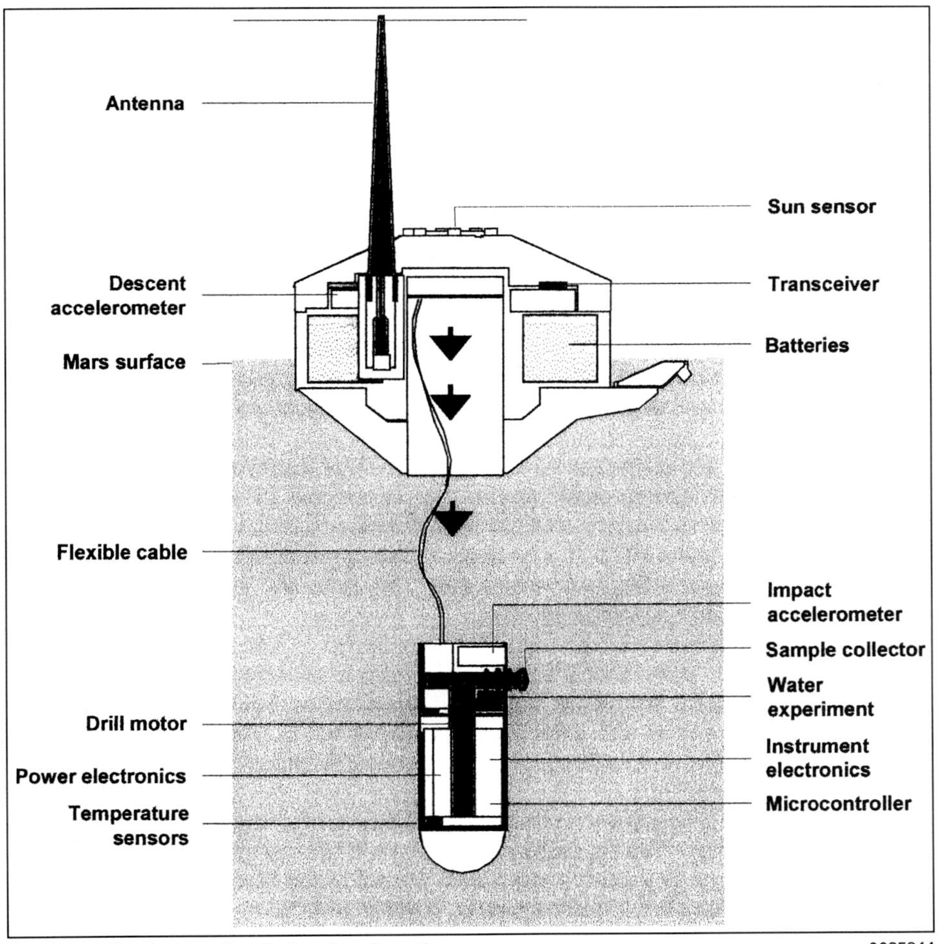

DeepSpace 2 probe in surface deployed configurations 0085844

Mass: 576 kg at launch; lander 290 kg; cruise stage 82 kg; aeroshell/heatshield 140 kg; propellant 64 kg.

Computation: The lander employs the same SSPA and RAD6000 32-bit processor as the orbiter (see Mars Climate Orbiter).

Telecommunications: The primary telecom link for science data relay and spacecraft commands is the UHF relay through the Mars Climate Orbiter. A UHF link to Mars Global Surveyor is also available for data relay only. The maximum duration of an X-band transmission is limited to one hour by the capability of the Loop Heat Pipe to transfer heat energy from the X-band SSPA out of the thermal enclosure. The number of UHF and X-band cycles is limited by the amount of daytime power available. The lander can communicate direct with Earth via X-band at 5.7 kbits/s through the DSN's 70 m antennas or at 1.4 kbits/s via the 34 m antennas. Data relayed via Mars Climate Orbiter or Mars Global Surveyor can be transmitted at 128 kbits/s.

Power and thermal: For the landing footprint of 73–77° S latitude, the Sun does not go below the normal horizon for the season of the prime mission. However, a 10° terrain mask is assumed for power analysis and results in a defined day and night interval when the Sun goes below that horizon mask. The solar array consists of six panels (four fixed and two deployed) to provide daytime power for payload operations and for charging the 16 Ah batteries which provide night-time heater power for the thermal enclosure. The two deployed auxiliary solar panels were necessary because the fixed panels constrained in size by the launch vehicle payload shroud were of insufficient area to provide the required amount of electrical power. The C&DH and Power Distribution electronics have a low-power sleep mode to reduce night-time energy consumption. The lander lifetime is limited by the size of the batteries, as are daytime operations, and by the amount of power available to recharge them. At time of landing the lander can perform 8–9 hours of daytime operations but this decreases as the sun gets lower in the sky. The power system is designed for full operation in a dusty atmosphere of opacity (Tau) = 0.5. The rechargeable NiH_2 battery will keep temperatures in the central electronics enclosure above –30° C during –90° C night-time temperatures at the south pole.

Propulsion: The lander is equipped with eight thrusters in four clusters. Each cluster consists of one 22 N thruster and one 4.4 N thruster. The lander's final descent is controlled by 12 pulse-modulated, 266 N thrust, descent engines (retro-rockets) arranged in three groups of four on the underside of the lander. Two spherical diaphragm propellant tanks underneath the lander's solar arrays carry 64 kg of hydrazine for all 20 thrusters and retro-rockets. This is a pressure-regulated system with serially-redundant pressure regulators utilising helium gas as pressurant. For final descent doppler radar provides accurate altitude and 3D velocity estimates. As the spacecraft descends to within 12 m of the surface, the spacecraft control system begins the 2.4 m/s terminal velocity descent phase. The AACS subsystem controls the orientation of the vehicle on landing, placing the X-axis within 5° of the desired azimuth (45° W of north) to maximise solar array efficiently and minimise direct-to-Earth antenna blockage.

Payload: The science payload for the Mars Polar Lander consists of six instruments: the four instruments comprising the Mars Volatile And Climate Surveyor (MVACS) suite, the Mars Descent Imager (MDI) and the LIght Detection And Ranging (LIDAR) instrument. The MVACS consists of the Stereo Surface Imager (SSI), the robotic arm, the meteorology package and the Thermal and Evolved Gas Analyser (TEGA). The SSI is mounted at the top of a 1.5 m mast that will deploy upward from the lander deck and is identical to the imager on the Mars Pathfinder lander. It has a spectral range from violet to IR (400–1,100 nanometer) and two sets of lenses mounted slightly wider apart than a normal pair of human eyes so as to capture stereo images. Two sets of optics share a single CCD detector and two filter wheels allow the camera to take images in selectable spectral ranges. The

2 m robotic arm is attached to the lander deck and has an articulated member on the other end with a digging scoop, camera and temperature probe. The scoop will dig trenches and deliver soil samples to the TEGA. The temperature probe will measure the ambient temperature and thermal conductivity of the soil. The meteorology package includes a 1.2 m mast with wind speed and direction sensor, temperature sensors and tunable diode lasers that detect water vapour and isotopes of water and CO_2. A 0.9 m submast with a wind speed sensor and two temperature sensors points downward from the lander deck and is designed to study atmospheric effects in the zone 10–15 cm above the surface to determine the threshold wind speed required for dust storms to start. The TEGA heats soil samples and analyses them to determine concentrations of volatiles such as water and CO_2, whether present as ice or in volatile-bearing minerals. The arm deposits 0.1 g soil samples at a time in a receptacle which is mated with a cover to form a small oven. Heater wires heat the sample across a range of temperatures up to 1,027° C and a tunable laser emits beams which pass through the gases to a detector which measures the amount of light to determine the presence of vapour. The instrument can perform eight soil analyses. The MDI consists of a single camera head with an electronically shuttered CCD to capture black and white images in 1,000 × 1,000 pixel format. The LIDAR uses a gallium aluminium arsenide laser diode to emit 2,500 pulses of near-IR energy per second upward from the lander deck. A detector shows how long it takes the pulses to return following reflection from ice and dust haze in the lower part of the atmnosphere (2–3 km maximum altitude). provided by Dr V S Linkin of IKI, this is the first Russian experiment flown on a US planetary spacecraft.

UPDATED

Pioneer Series

Current Status

Pioneer Venus Orbiter burned up in October 1992 and project funding ended September 1994, but the Venus Data Analysis Program continued until December 1995. Pioneers 6 to 8 continue their occasional operations.

Tracking of Pioneers 6, 7, and 8 takes place once or twice a year for an hour, mainly out of curiosity about system functionality. The solar arrays continue to deteriorate and the transmitters can turn on only around perihelion where there is enough solar flux to provide sufficient power. If power is adequate, NASA activates some instruments, although the scientific value is insignificant. Three of Pioneer 6's six instruments, two Plasma Analysers and the Cosmic Ray Detector, can still return real-time data. It was tracked 29 July and 15 December 1995 (after 30 years), returning 1 h science data each time. Successful contact was made on its 35th launch anniversary (8 December 2000) for about two hours, making it the largest listed NASA spacecraft extant.

As of February 2003 Pioneer 10 is travelling at about a constant 12.224 km/s relative to the Sun and round-trip communications time with Earth takes 22 h. It is heading towards Aldebaran, but it will take two million years to reach that point and by then the star will have moved on. Pioneer 10 should encounter another system about once every million years. As of May 2000 Pioneer 10 RTGs were producing 65 W of power (42 per cent of the 155 W available at launch in 1972). On 11–12 February 2000 ground controllers fired Pioneer 10's attitude thrusters for the biannual pointing manoeuvre toward earth. Platform temperature is –40°C and the strength of the received signal on earth is –178 dBm. Scientific data from Pioneer 10 was retrieved on 5 and 6 August 2000, reporting no change in the cosmic ray intensity and therefore confirming that it has yet to pass through the heliospheric boundary.

The flight of Pioneer 10. Its firsts include: Asteroid Belt crossing January 1973, Jupiter fly-by December 1974, crossing of Saturn's orbit February 1976, crossing of Uranus' orbit August 1979, crossing of Neptune's orbit and leaving the solar system June 1983 (NASA)
0516751

NASA Ames officially ended routine contact attempts with Pioneer 11 on 30 September 1995. A few minutes of healthy engineering data was received 24 November 1995. This was the last communication from Pioneer 11 and shortly afterwards the Earth moved out of view of the spacecraft's antenna. Pioneer 10 will be dominated by the gravitational field of the sun for the next 126,000 years until it reaches a distance of 46 billion km – about seven times the distance from the earth to the outermost planet – when it will enter the gravitational field of the galaxy. On 7 March 2001, Pioneer 10 was 77.3 AU from the sun, 11.56 billion km from earth, travelling at 12.24 km/s relative to the sun with a round-trip light time to earth of 21 hour 26 minutes. On 20 February 2003 it was 12.21 billion km from earth, 82.19 AU from the Sun with a round trip light time of 22 h 38 m.

There was a contact with Pioneer 10 on 5 December 2002 via the DSS-63 Madrid station, but no lock on the receiver and therefore no telemetry. A signal was again received on 23 January 2003.

Background

The relatively simple spin-stabilised Pioneer craft, managed by NASA Ames, paved the way for more sophisticated subsequent explorations of the Sun, Jupiter, Saturn and Venus. Pioneer 1, launched in October 1958, was nominally the first NASA deep space probe, having been transferred by the USAF. Pioneers 1–4 failed to attain their lunar target but probed Earth's radiation belts. Three more Pioneer launches by Atlas Able in November 1959, September 1960 and December 1960 and what should have been Pioneer 10, launched by Delta in August 1969, were failures.

NASA Ames' deep space and interplanetary Pioneers began with number 5, which, with its successors, monitored solar particle radiation and provided up to 15 days' warning of solar flares. Originally designed for six months' operation, Pioneers 6 through 9 provided coverage from widely separated points during an entire solar cycle of 11 years and were all still predicting solar storms at the end of 1981. Pioneer 9 was the first to fail, in 1983. Pioneer 6 marked a quarter century of operations in December 1990, and could still be re-activated again at any time after 30 years in orbit.

Pioneers were the first spacecraft to detect the Earth's magnetotail, measuring some 5.6 million km on the anti-solar side.

Beginning in 1972 and continuing through 1974, Pioneer first began to fly to Jupiter. These flights introduced a new generation of planetary probes designed as precursors to more sophisticated vehicles such as Voyager. Pioneer 10 placed on a sidereal trajectory became the first man-made object to exit the solar system, the first to fly beyond Mars, the first to enter the asteroid belt and the first to reach Jupiter. It proved that the asteroid belt contains fewer dispersed particles than at one time believed and showed that traversing vehicles faced little hazard from collisions.

When Pioneer 10 retired after 25 years of deep space exploration, only a few weeks' of data have been lost due to spacecraft problems. It became the first spacecraft to reach 50 AU from the Sun, on 22 September 1990.

To date, Pioneer 10 flies on a predictable trajectory indicating that no gravitational effects from a 10th planet occur in its present vicinity. However, there is an unaccounted constant acceleration of 8×10^{-10} m/s^2 towards the Sun acting on both craft. It is not caused by a 10th planet, a dark star companion to the Sun or an unseen disc of outer solar system material; alternative causes are under investigation. In view of its potential importance, The Aerospace Corp is using independent software to check the discovery.

Pioneer 10 continues to move outward at about 2.7 AU annually in a direction generally away from the Galactic centre and nearly opposite the direction of the basic solar system motion with respect to the nearby stars. Pioneer 11 continues to depart at about 2.48 AU per year, towards the Galactic centre in the general direction of Sagittarius.

The two Pioneers have shown the heliospheric boundary with the interstellar medium to be significantly more distant than expected. Measurements of the solar wind indicate the boundary of the heliosphere may lie some 90–120 AU

Pioneer 6 is the first spacecraft to operate for more than a quarter century. This is the similar Pioneer 8 undergoing final balance checkout at the Cape
0516750

from the Sun. Inference on the heliosphere's extent is also provided by monitoring the solar modulation of cosmic ray intensities and from UV observations of changing atomic interactions with sunlight that result from incursion of extra-solar particles into the heliosphere.

In October 1990, Pioneer 11's primary receiver developed a problem that required very high power commands and prevented measurement of spin axis orientation. The radio science search for extra sources of gravitation ended with that failure.

Pioneer Progress: distance from Sun in Astronomical Units on 1 January each year

Year	Pioneer 10	Pioneer 11
1987	40.0	22.4
1988	42.7	24.9
1989	45.4	27.3
1990	48.1	29.8
1991	50.7	32.2
1992	53.4	34.8
1993	56.0	37.3
1994	58.6	39.8
1995	61.2	42.2
1996	63.8	44.7
1997	66.5	47.2
1998	69.1	49.6
1999	71.7	52.0
2000	74.3	54.5
2001	76.9	57.1
2002	79.5	59.5

Pioneers 10 and 11 ventured into the outer solar system by design as the first man made objects to ever leave the Solar System. NASA built them to last for seven years of data transmission. In their 258 kg of mass, these Pioneers carried 30 kg of science instruments and 27 kg of propellant (18 kg remained in January 1996, the only drain on this supply is an insignificant periodic precession manoeuvres).

Six 4N thrusters each provide control and mid-course corrections and using the original propellant supply can produce a total delta-V of

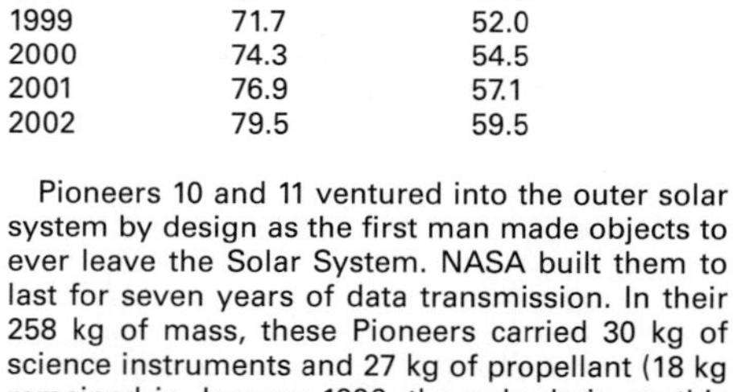

Pioneer plaque. The key is the hydrogen atom (top left), the most common in the Universe. 14 lines from Sun indicate distance to pulsars. Human figures, with the man's hand raised in goodwill gesture, are compared with spacecraft to give scale. Bottom, plan of solar system, showing Pioneer originating from Earth
0516753

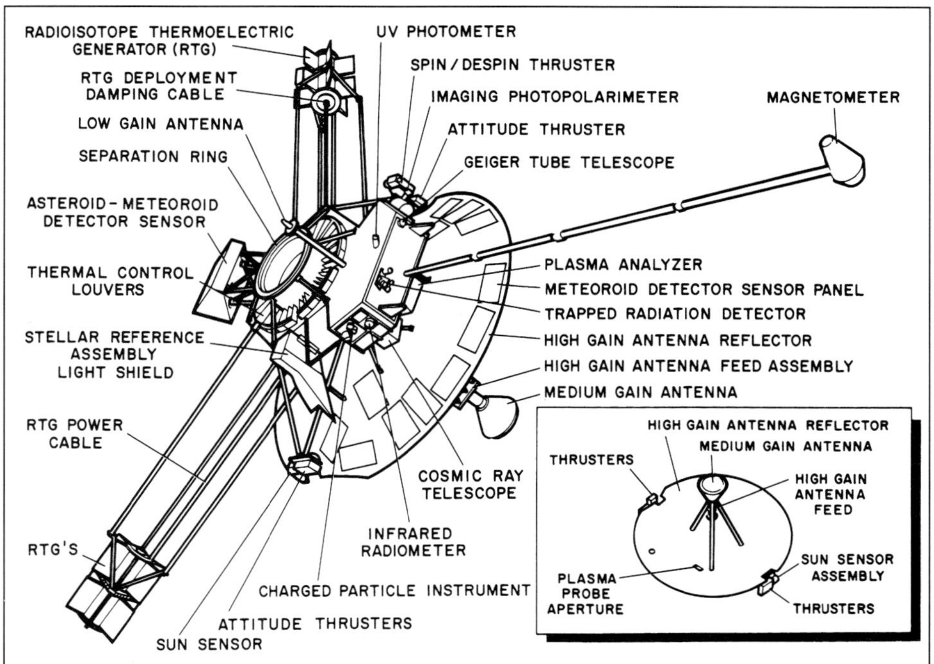

Pioneer 10/11 principal features (NASA) 0516752

670 km/h. The Pioneers retain their attitude with respect to the fixed stars by two Sun and one star sensors mounted on a spin-stabilised platform having a full-circle scan rate of about four times a minute. (The spin rate is now decreasing slowly; the cause is unknown but has been theorised as a build-up of electrostatic charge interacting with the interplanetary magnetic field). Four nuclear units, held on 2.7 m booms to avoid interference with the science experiments provide power.

The heart of the communications system is the fixed 2.7 m diameter dish antenna and redundant 8 W S-band transmitter (2,110/2,292 MHz up/down through NASA's Deep Space Network).

Pioneer's instruments have enjoyed a remarkable and unpredicted longevity and continue returning data on the solar atmosphere from the outer reaches of the solar system. Rather than instrument death, it now appears operations will fail due to the decreasing power, which diminishes by 81 mA annually. About 140 mA was available for science as of 17 January 1996 (the actual level is unknown as the telemetry provides 50 mA steps). Only those instruments with the lowest power consumption can now be operated and even then not all simultaneously.

The Plasma Analyser (179 mA required) could be operated until 9 September 1995. The Cathode Ray Telescope (100 mA) was dropped in 1996 and the Charged Particle Instrument was turned off on 30 December 1996, leaving just the Geiger Tube Telescope and the Ultraviolet Photometer operating as 1997 opened. On 31 March 1997 – more than 25 years after launch and nearly 10 billion km from Earth, Pioneer 10 retired and fell silent.

At 20.19 GMT on 22 September 1990, Pioneer 10 became the first spacecraft to penetrate to 50 AU, although it had still not reached the heliosphere's boundary (nor had it by the beginning of 1996).

Each craft carries a 15 × 23 cm plaque indicating its origin should it be recovered by an intelligent species.

Specifications

Pioneer 6
Launched: 16 December 1965 by Delta from Cape Canaveral
Orbit: 0.814 × 0.985 AU solar orbit.
Characteristics: Mass: 63 kg cylinder, 1 m diameter, 0.9 m high.
Mission Description: Pioneer 6 returned the first detailed data on the tenuous solar atmosphere and recorded passage of Comet Kohoutek's tail. By December 1990, having circled the Sun 29 times and covered 24,800 million km, it had been operational for 25 years – a record for an interplanetary spacecraft – compared with planned six months. With Pioneers 7–9 it formed a network of solar weather stations spaced along Earth's orbit. On 26 November 1988, it made its closest approach to Earth since launch, the 18.66 million km encounter increasing its heliocentric orbital period from 311 to 317 days.

Pioneer 7
Launched: 17 August 1966 by Delta from Cape Canaveral
Orbit: 1.010 × 1.125 AU solar orbit
Characteristics: 63 kg.
Mission Description: In 1977, Pioneer 7 registered Earth's magnetic tail 19.3 million km out, three times deeper into space than before. Pioneer 7 was tracked 17 April 1994 but insufficient power prevented the transmitter operation. On 31 March 1995, 2 hours of contact was made and the Plasma Analyser turned on.

Pioneer 8
Launched: 13 December 1967 by Delta from Cape Canaveral
Orbit: 1.0 × 1.1 AU solar orbit
Mission Description: Pioneer 8 defined Earth's magnetotail. NASA tracked it on 23 July 1995 but could not turn on the transmitter, either because of spacecraft failure or too great a Sun distance for the solar cells.

Pioneer 9
Launched: 8 November 1968 by Delta from Cape Canaveral.
Orbit: 0.75 × 1.0 AU solar orbit.
Characteristics: 66.6 kg
*Mission Description:*This spacecraft was the fourth of what was to be a series of five solar monitors (Pioneer E, intended to be Pioneer 10, failed to orbit 27 August 1969). NASA contacted Pioneer 9 last on 18 May 1983, but it was not officially declared inactive until March 1987 after attempts to detect it using SETI equipment with a very low signal detection threshold failed on 3 March.

Pioneer 10
Launched: 3 March 1972 by Atlas Centaur from Cape Canaveral.
Mission Description: After launch Pioneer 10 had an initial 51,800 km/h velocity, faster than any previous man-made object. Its trajectory took it within 130,300 km of Jupiter's cloud tops on 4 December 1973. On 13 June 1983, it crossed Neptune's orbit; further from the Sun than any known planet and effectively it left the solar system at that point.

Pioneer 11
Launched: 5 April 1973 by Atlas Centaur from Cape Canaveral.
Mission Description: NASA added a second magnetometer to measure Jupiter's intense fields. With the precedent of Pioneer 10, NASA knew how to retarget Pioneer 11 to approach three times closer to Jupiter and then proceed to the first encounter with Saturn. At 17.22 GMT 3 December 1974, travelling faster than any previous man-made object, 171,000 km/h, Pioneer 11 passed 42,940 km below the Saturian South pole. Its closest approach to Saturn, occurred on 1 September 1979 after a six-year and 3,200 million km journey. At that time it was only 20,800 km from the cloud tops and travelling at 114,500 km/h. Pioneer 11 crossed the orbit of Neptune, then the most distant planet, during 23 February 1990, becoming the fourth spacecraft to exit the solar system. By 1995, Pioneer 11 could support only two science instruments, the Geiger Tube Telescope and UV Instrument. JPL turned off the Charged Particle Instrument by the end of 1994.

STARDUST

Current Status

The first launch window was at the precise time of 21:06:42 UT on 6 February 1999. A problem with the Delta launch vehicle's C-band beacon caused a one-day postponement to the next launch window at 21:04:15 UT 7 February 1999 when Stardust was successfully launched by Delta 7426 from SLC-7 at Cape Canaveral. Spacecraft signal was successfully acquired by DSN Canberra, Australia, at 21:55 UT. All injection parameters were within a 3-sigma dispersion. At 17:16 UT the next day, Stardust passed the moon at the distance of 50,000 km. Initial tracking indicated that the injection parameters were so close to the optimum that no course correction manoeuvre would be needed, as optional in the flight plan on 22 February 1999. Early in March 1999 the spacecraft switched from low to medium gain as distance to earth exceeded 12 million km. Spacecraft systems checked out well but some problems with the solid-state power amplifier were observed in early April indicating problems with trapped charges, but this and other minor system anomalies were corrected.

The first trajectory correction manoeuvre was scheduled for 28 December 1999, an 11 m/sec burn conducted for 5 minutes 11 seconds at 15:30 UT, and this was carried out as planned. The next scheduled course change, the deep space manoeuvre was scheduled to take place in three parts. The first burn would come on 18 January 2000 at 19:00 UT (58 m/sec), the second on 20 January at 14:00 UT (52 m/sec) and the third on 22 January at 18:00 UT (48 m/sec). Each burn lasted about 30 minutes in duration, total of 5,884.4 seconds, and used a total 29.341 kg of propellant. The aerogel collector was opened up toward the end of February 2000 to begin its interstellar collection activity until the collector was retracted again on 1 May 2000. The next trajectory correction manoeuvre was carried out on 24 May 2000, a 72 second burn imparting a velocity change of 2 m/s, fire turning the trajectory for a rendezvous with Wild-2 in 2004. Before that, Stardust performed a fly-by of earth on 15 January 2001, at a distance of 6,008 km followed 17 hours later by the closest pan of the moon at 108,000 km, taking 23 images in the process. On 18 April 2002 Stardust reached its aphelion, a distance of 2.72 AU (407 million km) from the Sun and near the middle of the asteroid belt. Between August and December 2002 the second period of interstellar dust collection was performed. On 3 November 2002 Stardust performed a 3,300 km fly-by of the asteroid Annefrank, a full scale rehearsal for its encounter with Wild 2 on 2 January 2004. The spacecraft survived the encounter and obtained valuable and extensive data about the immediate environment of the comet nucleus. Stardust is expected to return to earth on 16 January 2006.

Background

Stardust is one of NASA's Discovery Programme missions which sponsor low-cost solar system exploration projects with highly focused scientific goals. Created in 1992, the Discovery Programme competitively selects proposals submitted by teams led by principal investigators and supported by organisations which provide project management and build and fly the spacecraft (see

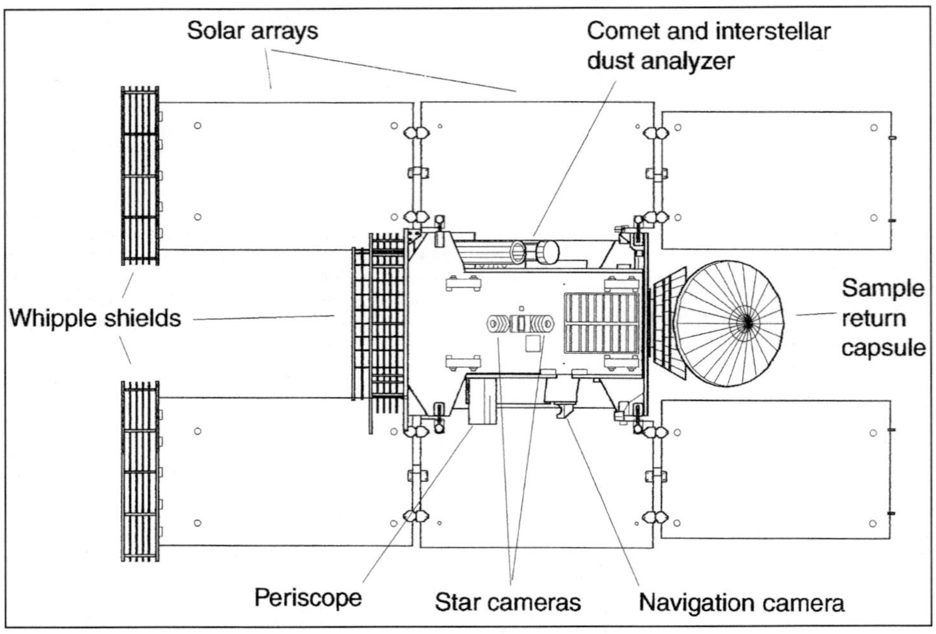

Stardust spacecraft layout 0089026

Discovery Series). Stardust was selected in 1995 as a mission designed to obtain samples from the cloud of dust that surrounds the nucleus of comet Wild-2 and bring it back to earth for analysis. Comets are thought to hold many of the original ingredients of the mixture of materials that formed the solar system almost 5 billion years ago, rich in organic material which provided the earth with many of the molecules that gave rise to life. Due to their unique cocktail of organic and inorganic molecules, samples obtained directly from a comet will contain important chemical and physical information about the earliest history of the solar system.

The mission envisaged a launch in February 1999 from Cape Canaveral to a solar orbit which would loop it three times round the sun in seven years. It would collect interstellar dust on two different orbits, flying through the coma, the dust and debris cloud that surrounds the nucleus, before approaching each and ejecting a re-entry capsule that would descend to the desert in Utah, USA, with its cometry and interstellar dust samples. The significance of this mission is enhanced when considering that this is the first time since Apollo that a spacecraft will have returned to earth carrying primordial materials collected in space. Samples returned from earth, orbiting satellites or space stations are not primordial and are gathered from locations within the magnetosphere.

Launch was assigned to a Delta 7426 at Cape Canaveral Air Force Station with a single launch window of just one second per day between 6 February 1999 and 25 February 1999. One of the new Med-Lite series of Delta launch vehicles, the first stage of the Delta 2 is augmented by four strap-on solids which are jettisoned just 66 seconds after launch, 3 seconds after burnout at an altitude of 21.9 km and a velocity of 3,862 km/hr. Main engine cut-off is scheduled for 4 minutes 24 seconds at an altitude of 114.5 km and a velocity of 20,096 km/hr followed by ignition of the liquid-propellant second stage 13.5 seconds later. The fairing protecting the spacecraft is jettisoned at 4 minutes 44 seconds at a height of 126.9 km and a velocity of 20,171 km/hr followed by second stage cut-off at 9 minutes 57 seconds at a height of 189 km and a velocity of 28,055 km/hr. The combined upper stages and Stardust spacecraft are now in a temporary earth orbit of 185 × 189 km and coast for 11 minutes 52.5 seconds. Then, the second stage ignites for a second time and burns for 2 minutes 37 seconds propelling the stack to a height of 178.3 km and a velocity of 32,593 km/hr. The second stage separates 53 seconds later and 37 seconds after that the solid-propellant third stage ignites and burns for 1 minute 5 seconds. Some 1 minute 10 seconds later, 27 minutes 19 seconds after launch, Stardust separates and is on its way to comet Wild-2.

The cruise phase begins with the spacecraft firing thrusters to stop the spinning rotation imparted by the third stage and about 4 minutes after separation the 34 m diameter antenna at the DSN complex in Canberra, Australia, acquires the spacecraft. In March 2000 when Stardust is between the orbits of Mars and Jupiter, the greatest distance it will be from the sun, it will fire its thrusters to put itself on an earth fly-by trajectory on 15 January 2001 at an altitude of 5,964 km. Before that, between March and May 2000, Stardust would collect interstellar particles on one dedicated side of the sample collector to separate the samples from those obtained later during the fly-through of Wild-2. Stardust will use aerogel, a silicon-based solid with a porous, sponge-like structure in which 99 per cent of the volume is empty space, to collect interstellar and cometary particles. Originally invented in 1933, aerogel is made from fine silica mixed with a solvent set in moulds of the desired shape and thickness and then pressure-cooked at high temperature. It is the lightest and lowest mass solid known and ideal for capturing tiny particles in space. The earth fly-by would set up the trajectory for a further 2.5 loops of the sun but as Stardust travelled back toward the sun it would again collect interstellar particles between July and December 2002.

On about 25 July 2003, Stardust is scheduled to fire its thrusters again to fine tune the flight path through the comet's coma, based on information supplied by the spacecraft's navigation camera. Encounter with Wild-2 will occur on 1 January 2004 at a radius of 1.86 AU (Astronomical Unit = mean distance between the earth and the sun) with Stardust travelling at a velocity of 6.1 km/sec relative to the comet. The velocity has been selected to optimise particle capture by the aerogel collector and passage through the most intense region will last about eight minutes. Wild-2 will be far from its peak period of activity and relatively safe for a close fly-by about 150 km from the comet's nucleus when Stardust's navigation camera will take images of the core body. The mission's primary period of data gathering begins about five hours before encounter at a distance of about 100,000 km. Just before closest approach the spacecraft will deploy the second side of the aerogel collector facing the direction of incoming particles. When Stardust has completed its passage through the coma, the aerogel collector will be stowed for the third and final time, sealing the interstellar and cometary particle collection inside the re-entry capsule.

Stardust is scheduled to fire its thrusters three times to fine tune the flight path to earth. The first of these will take place 13 days prior to entry, the second at three days and the final manoeuvre three hours before entry. Shortly after this final burn, at 110,728 km from earth, Stardust will release the sample return capsule. A spring mechanism will impart a spin to the capsule as it is pushed away from the spacecraft to stabilise it, after which the main spacecraft will perform a manoeuvre to avoid the earth's atmosphere enabling it to remain in solar orbit. The capsule will enter earth's atmosphere at a velocity of about 12.8 km/sec, stabilised nose down. It will slow to a speed of about Mach 1.4 at 30 km altitude and deploy a small drogue parachute. After descending to about 3 km a line holding the drogue parachute will be cut, allowing the drogue to pull out a larger parachute that will carry the capsule to a soft landing at 10:00 UT on 15 January 2006. At touchdown, the capsule will be travelling at about 4.5 m/sec just 10 minutes after entry interface. The landing site is a 30 × 84 km footprint at the Utah Test and Training Range near Salt Lake City, Nevada, an ample area to accommodate aerodynamic uncertainties and winds that might affect terminal descent. To land the capsule within the footprint, the entry angle of the flight path must be within 0.08°. For locating the capsule a UHF radio beacon on the capsule will transmit a signal as it descends to earth and can be tracked by radar. A helicopter will fly the recovery crew to the capsule. The canister will eventually be transported to the curatorial facility at NASA's Johnson Space Center.

Specifications

Stardust satellite

Mass: 385 kg at launch including propellant and 45.7 km sample return capsule including parachute.

Description, main bus: Rectangular deep-space bus called SpaceProbe developed by Lockheed Martin Astronautics, Denver, Colorado, USA. Main bus 1.7 m high, 0.66 m wide and 0.66 m deep. Panels fabricated from aluminium honeycomb core with layers of graphite fibres and polycyanate face sheets. Cometry particle impact protection provided by Whipple shields, comprising two bumpers at the front of the spacecraft which protect solar panels and a third to protect main spacecraft bus. Each shield is built around composite panels designed to disperse particles as they impact augmented by Nextel ceramic cloth blankets that further dissipate and spread particle debris.

Description, capsule: Sample return capsule is a cone with a diameter of 81 cm. It has five major components: heat shield, back shell, sample canister, parachute systems and avionics. A hinged clamshell mechanism opens and closes the capsule. Dust collector fits inside extending on hinges to collect samples and retracting to fold back down inside. Capsule is encased in ablative graphite-epoxy composite covered with a Thermal Protection System (TPS) made of phenolic-impregnated carbon ablator developed by NASA Ames Research Center for use on high-speed re-entry vehicles.

Backshell structure is made of graphite-spoxy composite covered with a TPS made of cork-based SLA 561V developed by Lockheed Martin for the Viking missions to Mars in the 1970s.

Sample canister is an aluminium enclosure that holds the aerogel and the mechanism used to deploy and stow the aerogel collector during the mission. Canister is mounted on an equipment deck suspended between the backshell and the heat shield. A parachute system incorporates drogue and main assemblies in a single canister. A gravity switch sensor and timer will trigger a pyrotechnic gas cartridge that will pressurise a mortar tube and expel the drogue chute. Drogue deploys at 30 km and a velocity of about Mach 1.4.

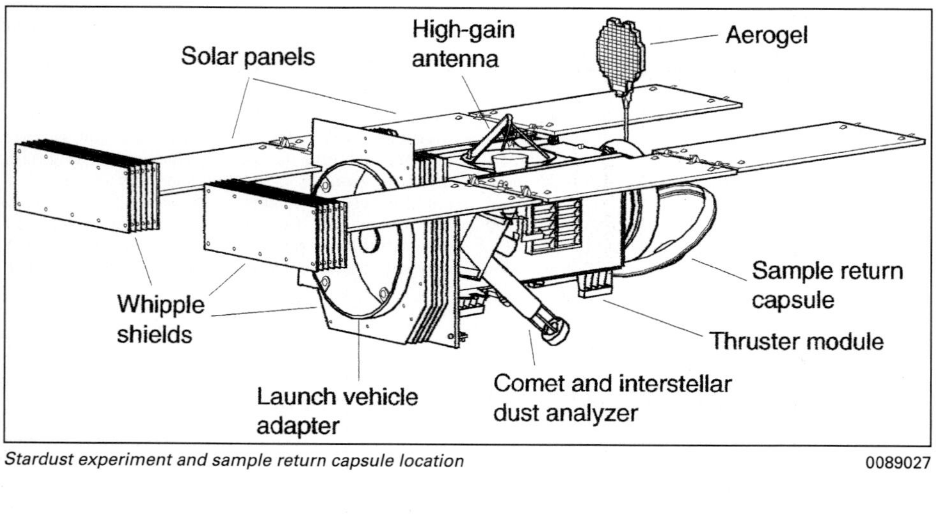

Stardust experiment and sample return capsule location 0089027

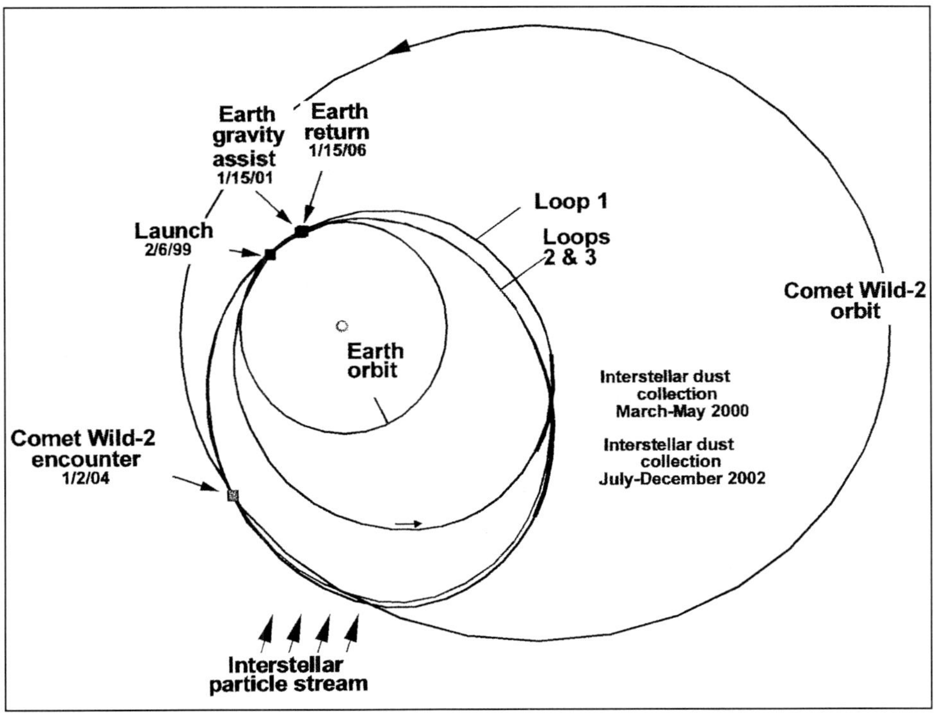

Stardust solar tour 1999–2006 0089028

Based on information from the timer and backup pressure transducers, a small pyrotechnic device will cut the drogue chute from the capsule at an altitude of approximately 3 km and as the drogue moves away it will extract the 8.2 m diameter main chute. Cutters cut main chute cables to prevent winds dragging capsule across the ground.

Capsule carries a UHF radio-locator beacon used in conjunction with locator equipment on recovery helicopters. Beacon is turned on at main chute deploy and is powered by redundant sets of lithium sulphur dioxide batteries with 40-hour operating life.

Power: Two solar array panels, total 6.6 m^2, using high-efficiency silicon solar cells. One 16 Ah N-H_2 battery for peak power operations.

Computation: Single 128 Mb RAD6000, radiation hardened version of the PowerPC chip, switchable clock speeds of 5, 10 or 20 Hz. RAM storage in lieu of TR includes 75 Mb for nav-camera images, 13 Mb for dust analyser, 2 Mb for dust flux monitor.

Communications: Xponder developed for Cassini spacecraft with 15 W RF SSPA. One medium and three low-gain antennas with data rates from 40 to 4 K bits/sec. One 0.6 m diameter HGA for encounter transmissions.

Thermal control: Stardust TCS uses louvres to control temperature of the IMU and the telecommunications systems's SSPAs with thermal coatings and multilayer insulation blankets and heaters to control bus elements.

Propulsion and attitude control systems: Two sets of eight thrusters operating on a total 85 kg of hydrazine as a monopropellant. Eight 4.4 N large thrusters for trajectory correction manoeuvres or turning the spacecraft. Four 0.9 N small thrusters for attitude control. Thrusters in four clusters of two large and two small each on the opposite side of the spacecraft to the deployed aerogel device. Stardust is three-axis stabilised and orientation is determined using a star camera or one of two IMUs each of which has three ring-laser gyroscopes and three accelerometers. IMUs only needed during trajectory correction manoeuvres and during cometary fly-by. Otherwise, spacecraft operated using stellar guidance for spacecraft positioning. Two sun sensors serve as back-up units by the rest of the attitude control system's elements.

Navigation Camera: Capable of acquiring images with 6 m resolution per pixel at 100 km distance. Main camera is a spare wide-angle camera from the two Voyager spacecraft and uses a single eight-position filter wheel, thermal housing and spare optics and mechanisms. New thermal radiator added. Combined with camera is a modernised sensor head using existing Galileo design updated with a 1,024 × 1,024 pixel array CCD from Cassini spacecraft modified to use miniature electronics. Eigth position filter wheel fitted with three gas filters, three dust filters, a polarising filter and a clear filter. Gas filters will permit study of the comet's dust jets as possible sources of gas. Three dust filters and polarising filter will allow study of colour and scattering properties of the dust. During distant imaging, the camera will image through a periscope to protect primary optics as spacecraft enters coma. In periscope light is reflected off mirrors made of highly polished metals designed to minimise image degradation while with standing particle impacts. During close approach, nucleus is tracked and several images taken with rotating mirror and not the periscope.

Payload: Consists of Aerogel Dust Collector (ADL), Comet and Interstellar Dust Analyser (CIDA) and Dust Flux Monitor (DFM).

The ADL collects particles ranging in size from 1–100 microns and comprises a dust collection module with two active particle acquisition sides, the cometary dust collection side Side-A and the interstellar dust collection side Side-B, each with at least 1,000 cm^2 of collection area. Dust collectors each have 130 rectangular blocks of aerogel 2 × 4 cm in area plus two slightly smaller rhomboidal blocks coated with aerogel varying in thickness from 3 cm on Side-A to 1 cm on Side-B. Aerogel density is graded, being less dense at the point of particle entry and progressively denser in the material. Each block of aerogel is retained by a thin aluminium frame.

The CIDA is derived from an instrument that flew on ESA's Giotto mission and the Russian Vega spacecraft which conducted a fly-by of Comet Halley in 1986. The purpose of the analyser is to intercept and perform instantaneous compositional analysis of dust as encountered by Stardust. The CIDA comprises a mass spectrometer to enable Principle Investigators to determine chemical composition. The instrument consists of a particle inlet, a target, an ion extractor, the mass spectrometer and an ion detector.

The DFM measures the size and frequency of dust particles in the comet's coma and the instrument comprises two film sensors and two vibration sensors. The film material responds to particle impacts by generating a small electrical signal when penetrated by dust particles. Their mass is determined by measuring the size of the electrical signals. The number of particles is determined by measuring the size of the electrical signals. By using two film sensors with different diameters and thicknesses the instrument will provide data of particle sizes and size populations.

Voyager

Current Status

During 2003 several investigators reported conclusions from Voyager science instruments to the effect that the spacecraft had passed through the bow shock wave of the heliosphere and that it was in interstellar space. This has been contested by other scientists who expected to see much stronger indications than those recorded from instruments on board.

Background

Dr Gary Flandro of the University of Utah, first noticed that a rare alignment of the planets late in the 1970s allowed a single spacecraft to visit all of the major outer planets of the solar system. Working with NASA's JPL, he and JPL engineers conceived of the 'Grand Tour' in the 1960s. In the mission profile, gravitational forces of the outer planets served to swing one spacecraft in turn from Jupiter to Saturn, Uranus and Neptune and another from Jupiter to Saturn and Pluto. Financial cuts at NASA required to pay for the burgeoning budget of the Shuttle pared the project to a mere extension of the Mariner programme, which had explored the inner planets. Work began in January 1972 but the project underwent many changes. For example, NASA lacked the funds for Jupiter orbiters and postponed a visit until Galileo. The craft were designed, built and operated by JPL.

As the flight date neared, JPL prepared the craft for launch. But following the four-day road journey from JPL to Cape Canaveral, engineers discovered faults in the attitude and articulation control and flight-data subsystems of Voyager 2, due to depart first because its trajectory included the Uranus option. The two Voyagers thus had to be interchanged, with consequent swapping of equipment.

The overall project cost about US$320 million not including launchers and flight support activities and NASA sources indicated the entire programme cost after 1986's Uranus encounter was $600 million. Voyager 1 left on a Titan-Centaur IIIE/Centaur on

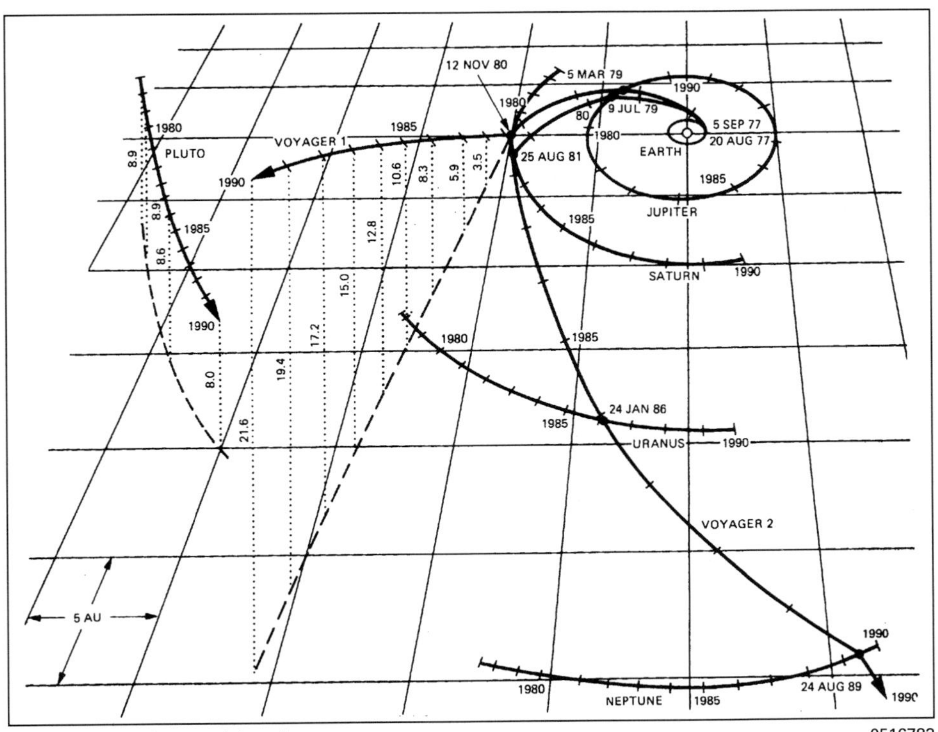

Voyager trajectories out of the solar system 0516783

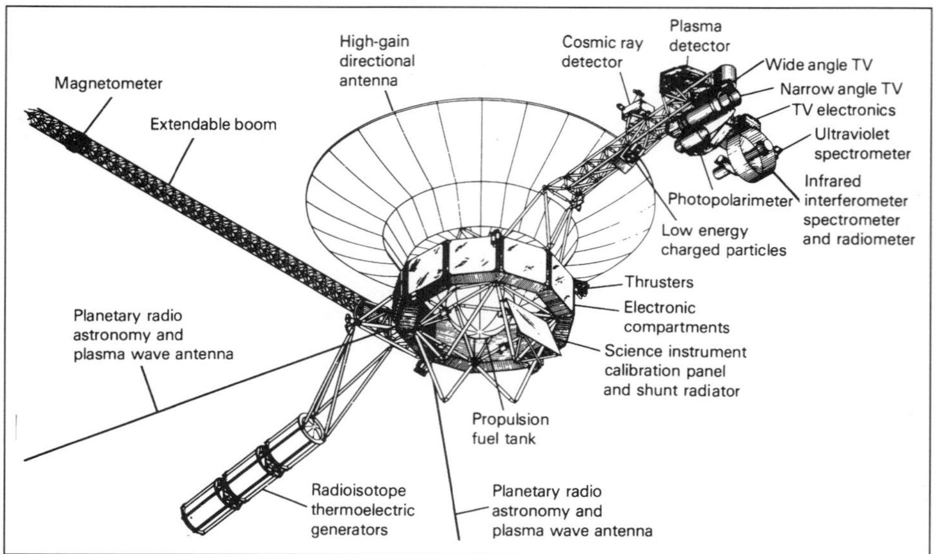

Voyager principal features 0516784

Close-up view of Triton (image area about 450 km across) 0085828

High Resolution image of Ariel, moon of Uranus during Voyager 2 fly-by in 1986 0085826

5 September 1977 on a Titan IIIE/Centaur and coming within 278,000 km of Jupiter on 5 March 1979 and 124,000 km of Saturn 12 November 1980. Voyager 2 left on 20 August 1977, coming within 650,000 km of Jupiter 9 July 1979, 101,300 km of Saturn 25 August 1981, 71,000 km of Uranus 24 January 1986 and 5,016 km of Neptune 25 August 1989.

Voyager 2's fly-past of Neptune formally ended 2 October 1989, the sixth and last planetary encounter of the two-craft programme. When this happened, NASA formally renamed the project Voyager Interstellar Mission and estimated the programme had cost US$863 million. Under its new designation, VIM cost US$70 million through 1995, exclusive of tracking. Voyager's two Jupiter, two Saturn and the first Uranus and Neptune fly-bys sustained NASA's planetary programme during an 11-year dearth of launches. Voyager 1 captured the last of some 67,000 images in February 1990 to create a mosaic portrait of the solar system.

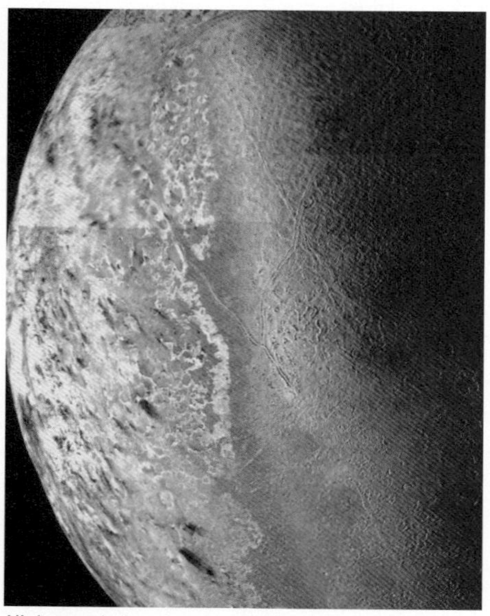

High resolution image of Neptune's moon Triton viewed by Voyager 2 in 1989 0085790

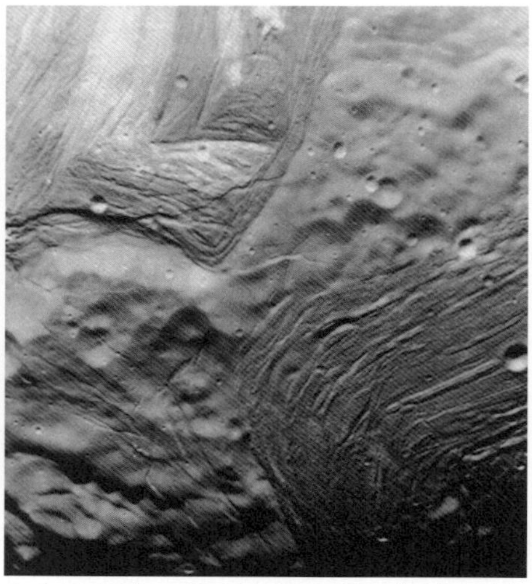

Surface detail on Uranus moon Miranda viewed by Voyager 2 0085827

The Voyagers and two Pioneers are exiting the Sun's system in different directions, heading towards the hypothetical heliopause, where the solar wind is turned back by the interstellar wind. Voyager 1 has been heading 35° N of the ecliptic since its November 1980 Saturn encounter. Voyager 2 on a 48° S trajectory since Neptune. Instruments no longer collecting data and non-essential heaters on Voyager 2's scan platform were turned off in November 1998 as a power conservation move. Five instruments continue to provide data. Voyager 2 is now more than 9 billion km (8.75 light hours) from Earth travelling at 15.8 km/s. The Voyager 1 spacecraft scan platform was turned off in mid-2000 to conserve power by saving 43.9 W. The most distant object made by humans, on 9 February 2001, Voyager 1 was 12,036 billion km (11 hours, 9 minutes, 6 seconds light time) from Earth travelling at 17.256 km/s. Since launch, in September 1977, Voyager 1 has travelled 13.8 billion km and, since launch in August 1977, Voyager 2 was 9,566 million km (8 hours, 52 minutes, 4 seconds light time) from earth travelling at 15,763 km/s has travelled 12.97 billion km. Both spacecraft confirmed the earlier Pioneer finding that the heliopause exists further than this earlier estimate of its position. Based on the rate of change of solar wind material, the Voyager science data sets the current estimate for heliopause at 80–90 AU. The satellites scan the path ahead for low energy cosmic rays penetrating the outer reaches of the solar system from nearby supernovae remnants. The plasma wave receivers have been recording intense 2–3 kHz emissions since August 1992, believed to originate at the heliopause where the solar wind interacts with the cold interstellar plasma.

JPL's Deep Space Network continues to receive daily ultra-violet and fields/particles data. The Laboratory's scientists turned off the imaging cameras, photopolarimeter and IR spectrometer on both craft after Voyager 2's Neptune fly-by as they are of no further scientific use. The remaining seven instruments (UV spectrometer, cosmic ray telescope, low energy charged particle experiment, magnetometer, plasma and plasma wave subsystems and the planetary radio astronomy instrument) continue to return data at a combined rate of 160 bit/s. Their data return and communications could be maintained to around 2017, when power levels reduce to the minimum 230 W required for critical life functions. Voyager 1 and 2 will then be 138/113 AU out. Electrical power outputs by April 2000 were 318/220 W (480 W at launch). With sufficient power, contact would be lost with their 20 W transmitters around 2033 because of the distances (29,000 million km for Voyager 1). Attitude control propellant is not a limiting consumable: the 32.05/34.07 kg of hydrazine remaining at April 2000 (104 kg at launch) is sufficient for operations well beyond 2020. On 3 January 2003, Voyager 1 was 13.015 billion km from the Sun travelling at 17.216 km/sec relative to the Sun with a round trip flight time of 24 hours 18 minutes 44 seconds. Voyager 2 was 10,335 billion km from the Sun moving out at 15, 696 km/sec relative to the Sun with a round trip flight time of 19 hours 23 minutes 34 seconds.

Specifications

Launch date: Voyager 1 5 September 1977 on a Titan IIIE/Centaur from Cape Canaveral; Voger 2 20 August 1977 of a Titan IIIE/Centaur from Cape Canaveral.

Mass: 2,016 kg launch mass, Voyager 792 kg.

Communications: 3.7 m antenna dish, normally Earth-pointing (largest flown on a planetary mission). Transmitter power is 20 W.

Power: 3 RTGs, each 39 kg, encased in beryllium on a deployed boom. (To avoid their radiation, the most sensitive science instruments are mounted on an opposite 2.3 m boom).

Computation: 536 Mbit tape recorder memory (allowed the storage of the equivalent of 100 full-resolution images) Six computers (three back-up), total 32 kbit capacity, reprogrammable to meet the needs of each encounter.

Propulsion: Spherical propellant tank at the centre of the basic 10-sided aluminium bus supplies hydrazine (104 kg loaded) for the 12 attitude control and four trajectory correction thrusters.

Payloads: TV: 200 mm f/3 and 1,500 mm f/8.5 cameras.

IR spectrometer: 51 cm diameter gold-plated telescope covering 2,000 wavelengths over 4–50 μm ultra-violet spectrometer: 128 wavelengths over 500–1,700 Å.

Photopolarimeter: 15 cm Cassegrain telescope.

Cosmic ray telescope: covering 0.5–500 MeV.

Low energy charged particle experiment: 0.01–11 MeV electrons and 0.015–150 MeV protons/ions in the planets' magnetospheres.

Planetary radio astronomy: radio emissions.

Plasma particles: 10–6,000 eV charged particles in the solar wind and the planet/moon magnetospheres.

Plasma waves: 0.01–56 kHz oscillations in surrounding plasma to determine temperature and density.

Magnetometers: magnetic fields around planets/moons and in interplanetary space; $20^{-2} \times 10^{-8}$ gauss.

Comment: Each carries a 30 cm copper 'Sounds of Earth' record containing greetings in 60 languages, sounds of birds, whales and other animals, samples of music and other information in case they are ever retrieved by intelligent beings. The record's aluminium jackets carry playing instructions using the cartridge and needle provided.

UPDATED

HUMAN SPACE FLIGHT

HUMAN SPACE FLIGHT

China

Project 921

Background

China became the third nation to launch its own manned space mission in October 2003. The overall programme designator is Project 921, while the manned spacecraft is called *Shen Zhou* (SZ).

Project 921 came to fruition almost three decades after the cancellation of China's first attempts at a manned space programme. In 1966 a programme called *Shug Yuang* ('dawn') was initiated. There has been Western speculation that the *Shug Yuang* spacecraft would have been a manned variant of the *Fanhui Shi Weixing* /*Jian Bing* (FSW/JB) photo reconnaissance satellite, but this is not confirmed from reliable Chinese sources. If correct, then the spacecraft would have been similar to the United States Mercury 'man in a can' concept, allowing a single person to fly in orbit for a maximum of 1 to 2 days. The original FSW satellite had a mass of 1.8 tonnes and was launched aboard the CZ-2A launcher.

In 1971, 19 trainee 'astronauts' were selected for the initial *Shug Yuang* missions, which were then expected to begin in 1973. Before test flights could take place, politics and budgets led to the cancellation of the programme in 1972.

China's second attempt at a manned space programme is an offshoot of **Project 863** (for example the third major government project in 1986) which was intended to improve China's science and technology capabilities and standards. Project 921 is a direct derivative of Project 863, and it is acknowledged by China to be the largest and most expensive project ever undertaken by the country. While designators like Project 921-1, 921-2 and so on, litter Internet websites, none of these reflect the true Project 921 organisation.

Seven Elements of the *Shen Zhou* Programme

When authorised in 1992, Project 921 comprised seven elements which were designated Project 921-1 to 921-7, although in the public domain it is not known which program element refers to which designator. The seven elements are:

- launch site system
- astronaut system
- measurement and control system
- manned delivery rocket system
- manned spacecraft system
- landing system
- spacecraft application system

By the end of the second test flight of *Shen Zhou* China had demonstrated all of these elements.

Jiuquan Launch Facility

The launch of SZ 1 introduced the new launch pad constructed at Jiuquan to operational launches. The pad was associated with a nearby assembly building, akin to the Vehicle Assembly Building at the Kennedy Space Center, constructed for the stacking of Saturn-5/Apollo vehicles and now used for stacking space shuttles prior to roll-out to the launch pad. The Chinese assembly building is large enough to allow the simultaneous assembly/integration of two, possibly three, launch vehicles and spacecraft.

The new assembly building and launch pad brought a new philosophy to China's preparations for launches. Previously, all launch vehicles had been transported to the launch pads stage-by-stage and then stacked on the launch pad. For the manned programme the launch vehicles would be stacked and integrated in the assembly building, and then rolled out to the launch pad a few days before the planned launch.

This change makes sense when it is considered that at present there is only one launch pad available for the manned programme. If the Chinese want to launch rendezvous/docking missions – like the Soviet-era Soyuz 4 and Soyuz 5 missions, launched on consecutive days but using two pads – then they need to have a minimum on-the-pad time for their launch vehicles. Also, this approach allows for a simplified launch pad design and construction – so that if the pad is damaged or destroyed during a launch mishap it should not be a major job to repair or rebuild it.

In 2002 it was reported that the Chinese were working on a second launch pad close to the one used for the CZ-2F launches. The new pad was smaller than the one used for *Shen Zhou* and it had no rail links to the assembly building. Completed by early 2003, the new pad saw its first launch in November 2003 when the first Uprated CZ-2D carried the first of a new series of recoverable *Jian Bing* reconnaissance satellites into orbit.

The Hangtianyuan Team

Independently, Chen Lan and Chiew Lee Yih "invented" the term *taikonaut* for a Chinese space traveller, and this term appears to have come into popular use, even though it is unofficial. Official Chinese sources have used the words *yuhangyuan*, *taikongren* and *hangtianyuan* – the latter apparently being the most "official" of the various terms.

The first Chinese *hangtianyuans* to be named were Li Tsinlung and Wu Tse. The Russians announced in late 1996 that these two men would be undertaking a year's cosmonaut training at the Zvezdny Gorodok centre, and they were met and photographed by the British Neil Da Costa and Rex Hall when they visited the training centre in April 1997. Following the successful SZ 4 mission the name "Chen Long" circulated as a possible candidate for China's first piloted space mission, although in reality there has never been a Chinese trainee with this name.

The Russian announcement referenced above noted that the two Chinese trainees had been selected by the Chinese as "national astronauts" in March 1995 and had been given the designation "instructor astronauts" by the Chinese. They were to complete a one-year long general space training course, which would include three months learning Russian, physical training and familiarisation with the technical details of piloted space missions. The Russians said that the two trainees would return to China "where the launch of the first national manned spacecraft carrying a crew of two is scheduled for 1999". Of course, this target was not met, and it is possible that the Russians misunderstood Chinese plans which called for the maiden unmanned launch of a Shen Zhou during the 1998–1999 period.

The training at Zvezdny Gorodok was conducted on a purely commercial basis, with the Chinese paying "several hundred thousand dollars" for the two men to complete the training course. Privately, the Russians said that they believed that some of the "trainers" who accompanied Li Tsinlung and Wu Tse were also astronaut-candidates, although the Chinese said nothing to confirm this.

The Chinese had announced that they had 14 *hangtianyuans* in training but did not immediately release the names of the candidates. It was not until April 2003 that the authoritative *Novosti Kosmonavtiki* web site listed 14 names of *hangtianyuan* s which had been quoted in the German publication *Fliegerrevue*, issue 5, 2003:

- Chen Quan
- Deng Qingming
- Fei Junlong
- Jing Haipen
- Li Qinglong
- Liu Buoming
- Liu Wang
- Nie Haisheng
- Pan Zhanchun
- Wu Jie
- Yang Liwei
- Zhai Zhigang
- Zhao Chuandong
- Zhang Xiaoguan

The names Li Qinglong and Wu Jie are clearly different transliterations of Li Tsinlung and Wu Tse whom Rex Hall and Neil Da Costa met in April 1997.

The Chinese have described the basic requirements for the members of the *hangtianyuan* team: these are:

- age approximately 30 years old
- height approximately 1.7 metres
- weight approximately 50 kg (65 kg is quoted in some sources)

All of the trainees were all said to hold a master's degree in "mathematics, physics, engineering and biology, etc". They have a strict training regime of ten hours per day in the "aerospace city" centre, located in the suburbs of Beijing. The training includes training on "specific" training relating to the aerospace environment and spaceflight simulation – general knowledge of "astrology" (sic – astronomy ?), geography, medicine, flight dynamics, and rocket and spacecraft technology.

We now know that Yang Liwei became China's first *hangtianyuan* on October 15, 2003 and that his back-ups for the flight were Zhai Zhigang and Nie Haisheng – and all three names had appeared in the *Fliegerrevue* listing earlier in 2003.

In March 2004 the Chinese press started to talk about the future selection of women *hangtianyuans*: like their male counterparts, the women would be drawn from pilots within the PLA Air Force. Additionally, a second team of male *hangtianyuans* is said to be in the process of being recruited.

Yuan Wang Tracking Ships

There are four *Yuan Wang* ships operated by the Chinese for the tracking of satellite and missile launches, tracking newly-launched payloads into orbit and assisting in the tracking and recovery of missile re-entry vehicles. The first two *Yuan Wang* ships were operating in 1980 (possibly accepted for operations in 1979): their first operations appear to have been in connection with the DF-5 missile test on May 18, 1980 which fired a re-entry vehicle from Jiuquan into the Pacific Ocean, near the Solomon Islands. The *Yuan Wang* 3 was delivered in March 1995 and *Yuan Wang* 4 in July 1999.

The original three ships underwent a major systems upgrading which was announced in January 1999. The work had been carried out in Shanghai and had taken 16 months to complete. The ships were now carrying more advanced equipment on board than they had previously used, and this ensured that they were capable of global tracking and control. The data transfer speed had been increased by a factor of 400.

At the times of the *Shen Zhou* flights the ships are deployed as follows:

Yuan Wang 1: North Pacific (off Chinese coast)
Yuan Wang 2: South Pacific (east of Australia)
Yuan Wang 3: South Atlantic (west of Angola)
Yuan Wang 4: Indian Ocean (west of Australia)

CZ-2F Launch Vehicle

The CZ-2F launch vehicle used for the *Shen Zhou* missions is an uprated version of the CZ-2E vehicle, and it is described in the launch vehicle section of the *Directory*. The 2F has improved propulsion systems, improved control and computer systems and carries a payload shroud with an escape tower at the top for use in emergencies.

The manned spacecraft system is *Shen Zhou* itself, which described in more detail in the following text. The landing system, of course, is incorporated into the *Shen Zhou* design, but the Chinese probably include the ground recovery vehicles and associated systems as part of the landing system.

The spacecraft applications system relates to the equipment carried aboard the *Shen Zhou* orbital module which functions in orbit as an independent satellite once the descent module has returned to Earth.

Shen Zhou Design

The name *Shen Zhou* is said to have been chosen personally by President Jiang Zemin, and it is capable of different translations into English: *shen* can translate as "God", "heavenly" or "divine", while *zhou* can translate as "ship" or "vessel".

The prime contractor for the *Shen Zhou* spacecraft is the China Research Institute for Space Technology, and the spacecraft's chief designer is Qi Faren.

Whenever the Chinese talked about possible manned space missions during the 1990s, the goal was always said to be the establishment of an orbiting laboratory. As a result, the Chinese have based the design of *Shen Zhou* on the World's most-used space station ferry, the Soviet Soyuz which first flew a space station mission in April 1971: subsequent variants have been Soyuz-T, Soyuz-TM and now Soyuz-TMA which is used for supporting the International Space Station. The Chinese took the basic Soviet three-module concept and modified it for their programme.

This has often been either misunderstood or completely misrepresented in the West, with even the official US Cox Report claiming that when China started its manned space programme it would be using no more than surplus Soyuz spacecraft, purchased from Russia. Overall, the design of *Shen Zhou* is very much like Soyuz, but physically it is slightly larger and about half a tonne more massive. In the detail there are major design differences, the most obvious being that the forward orbital module on *Shen Zhou* carries its own set of solar panels and its own orbit manoeuvring system: on the SZ 2, 3 and 4 missions the orbital module has operated as an independent satellite for six months or more after the descent module returned to Earth. The Soyuz orbital module was never designed for this.

On the other hand, the Chinese have clearly benefited from the Russian experiences. They have purchased single items of Russian technology to see how the Russians solved specific problems and then incorporated those solutions into the *Shen Zhou* design. Examples of the technology purchases are that the Chinese obtained a Soyuz life support system, the shell of a Soyuz descent module (with most of the interior instrumentation and equipment stripped out), an androgynous docking unit and a Soyuz pressure suit (of the type worn at launch and landing – not an EVA suit). Details of the *Shen Zhou* spacecraft and the missions flown are included in this entry.

Physically, the Chinese spacecraft is slightly larger than the Soyuz, and a comparison is given in the Table below.

Russia's Soyuz-TM is the variant designed to take crews to and from the Mir Complex: the improved Soyuz-TMA is now used as the "lifeboat" spacecraft for the International Space Station.

The return of the Shenzhou spacecraft follows the philosophy of the Russian Soyuz closely, and was described in general terms for the flight of *Shen Zhou* 1.

The orbital module separates in orbit and approximately ten minutes later the combined descent and instrument modules are oriented as they as passing over the *Yuan Wang* 3 ship off the coast of Angola and the main engine on the instrument module fires for a last time to bring the assembly out of orbit. Atmospheric re-entry is considered to begin at an altitude of 80 km. Once the craft had descended to an altitude of about 40 km a communications blackout started. Once the blackout was ended the descent was followed by radar with helicopters converging on the predicted landing area. At a distance of 30 km from the; landing site the spacecraft's electrical power is turned on, followed by the separation of the drogue parachute and the deployment of the main parachute with an area of 1,200 square metres was completed. As the landing draws close the descent module's base falls away, exposing the four solid-propellant motors which ignite at an altitude of 1.5 metres to cushion the landing.

Future Manned Programmes

Even before the first manned *Shen Zhou* is launched, China has made it clear that the immediate goal is the establishment of a small space laboratory, to be followed by something similar to the Russian Mir modular space station by around 2010. And a manned lunar programme is at the very preliminary stage.

Wang Yongzhi, described as one of the people behind China's manned space programme, has described a three-step programme for manned operations:

- launch unmanned and manned spacecraft: conduct suitable Earth observations and scientific experiments
- EVA (space walk), rendezvous and docking activities: launch a space laboratory
- Build large, permanent manned space station.

Since the *Shen Zhou* spacecraft has been designed from the outset for docking operations, an initial docking mission could come relatively early in the flight programme. *Shen Zhou* 6 is planned for late 2005 and is said to be scheduled to carry two men on a flight lasting for about a week. The Chinese have specifically said that no docking is planned for this mission, but it is possible that the *Shen Zhou* 6 orbital module could be used as a docking target for *Shen Zhou* 7 if that can be launched in early 2006, within 3–4 months of the previous mission.

An initial space station module can be launched using the CZ-2F vehicle, and this has been depicted in Chinese animations released following the *Shen Zhou* 5 mission. Without the need for the heavy payload shroud and escape tower, the CZ-2F should be capable of orbiting 9.5–10 tonnes. This squares with the animations of a *Shen Zhou* docking with a two-cylinder space laboratory which is slightly larger than the piloted spacecraft. Such a mission could come around 2007–2008, linked for propaganda value with the Beijing Olympic Games.

Slightly further ahead, the Chinese have hinted at having their equivalent of the Mir orbital station in orbit by 2010, and this could be based upon 20 tonnes modules launched by the CZ-5/504 vehicle (*qv*).

It is possible that the orbital module left by one *Shen Zhou* mission could be used as an expansion module for the crew on the next *Shen Zhou*, and a 'train' of orbital modules could be assembled over a period of time. One problem with this scenario is that there is a docking unit at only one end of the module and it would therefore be impossible to build up a 'train' of such modules as an interim space station.

Once the initial piloted test flights of *Shen Zhou* are complete, a logical goal will be the docking of two spacecraft in orbit, much as the Soviets did with Soyuz 4 and Soyuz 5 in January 1969. Unlike the Soyuz of that era, *Shen Zhou* has an internal transfer system built into it, so the *yuhang yuans* would be able to move from one craft to another with ease. If one of the *Shen Zhou* spacecraft for such a mission were to be launched unmanned the two or three *yuhang yuans* on the other spacecraft could fly a longer mission with the same consumables that could be undertaken if two crews were launched.

A larger space laboratory could be launched if the development of the CZ-2E(A) launch vehicle proceeds: that vehicle has been described as having a payload capability of around 14 tonnes, enough for a laboratory significantly larger than either the *Shen Zhou* orbital module or a complete unmanned *Shen Zhou* spacecraft. With an advertised payload capability of 23 tonnes, the planned CZ-5 launcher can orbit space station modules similar to those of the Soviet Salyut programme and which later comprised the Mir Complex. The goal for China's 'Mir' seems to be around 2010.

The Chinese have strongly indicated that they are interested in eventually flying men to the Moon, but such a mission is unrealistic for the next decade or so. None of the announced launch vehicles in development has the payload capacity for a lunar landing mission.

One scenario presents itself for *Shen Zhou* loops around the Moon, akin to the Soviet Zond/L-1 missions launched during 1967–1970. The CZ-5/504 variant with the planned LOX/hydrogen third stage could easily launch a complete *Shen Zhou* spacecraft on a circumlunar mission, although without a larger launch vehicle for follow-on lunar orbit and lunar landing missions such a project would be a dead-end.

In 1992 a Chinese paper presented at the International Astronautical Federation Congress described a possible family of modular launch vehicles which could be developed. With hindsight, this seems to have been one of probably a series of such studies which were underway at the time and which have led to the current development of the modular CZ-5 launch vehicle.

The largest of the 1992 proposals was a vehicle which could place 70 tonnes into low Earth orbit. While not enough in itself for a manned lunar landing mission, such a vehicle could be used to support such a mission. One possibility is that the manned lunar assembly could be launched on two of these vehicles with Earth orbit rendezvous before trans-lunar injection. Another possibility is that a single 70 tonnes rocket stage could be launched, followed by the manned uprated *Shen Zhou* and the lunar lander on separate launches of CZ-2F class launchers.

It is far too early for even the Chinese to have decided upon the exact mission profile and thus launch vehicle requirements for manned lunar landing programme, and any analysis like that in this entry can only be guesswork at present. A realistic pencilled-in date for China's first crew on the Moon would be around 2020: possibly July 2019, to celebrate 50 years since Apollo 11. And unlike Apollo, the Chinese have indicated that their plan will escalate to include an eventual permanently manned base on the Moon.

Spacecraft	Soyuz-TM	Shen Zhou
Overall spacecraft		
Launch mass, kg	7,250	7,800
Length, metres	7.48	8.80
Base diameter, metres	2.72	2.80
Instrument/propulsion module		
Mass, kg	2,950	3,000?
Propellant mass, kg	800	
Length, metres	2.60	2.94
Diameter (cylindrical), metres	2.17	2.50
Diameter (base), metres	2.72	2.80
Area of solar panels, m^2	10	24
Descent module		
Mass, kg	3,000	3,200?
Length, metres	1.90	2.059
Diameter, metres	2.17	2.50
Orbital module		
Mass, kg	1,300	2,000?
Propellant mass, kg	0	40
Length, metres	2.98	2.80
Diameter, metres	2.26	2.25
Area of solar panels, m^2	0	12

Specifications

Shen Zhou Spacecraft

Prime contractor: China Research Institute for Space Technology
Launch vehicle/site: CZ-2F from Jiuquan
Dimensions:
Overall: 2.5 × 7.9 m
Instrument/propulsion module: 2.5 × 2.8 m
Descent module: 2.5 × 2.059 m
Orbital modules: 2.25 × 2.8 m
Power: instrument/propulsion module: two vanes of solar panels with a total surface area of 24 sq m
Orbital module: two vanes of solar panels with a total surface area of 12 sq m

Shen Zhou 1

Launch date/time: 20 November 1999 at 22.30 GMT
Descent module recovery: 21 November 1999 at 19.41 GMT
Orbital module decay: 1 December 1999

Western observers were ready for the expected maiden launch of what would become China's manned spacecraft in mid-November 1999, since the four *Yuang Yuan* tracking ships were deployed around the World for the first time. China announced nothing until the descent module had returned to Earth, although Western observers had been following the mission using orbital data issued through the NASA/Goddard Space Flight Center. The first orbital data for what would be announced as *Shen Zhou* 1 showed the spacecraft in a 42.6°, 197–325 km orbit. No orbital manoeuvres took place until retrofire brought the spacecraft out of orbit after less than a day in flight.

SZ 1 was a very basic test flight with the objectives being the successful launch and recovery of the spacecraft. No life support system was carried and the orbital module appears to have carried only 'dummy' solar panels which remained folded throughout the flight. SZ 1 is the only flight where the orbital module apparently had no orbital activities after the descent module's return to Earth, and it simply decayed from orbit.

Shen Zhou 2

Launch date/time: 9 January 2001 at 17.00 GMT
Descent module recovery: 16 January 2001 at 11.22 GMT
Orbital module decay: 24 August 2001

The launch of SZ 2 was originally planned for October 2000, but nothing was heard until December when a 5 January 2001 launch date was announced: the day before the planned launch a delay was announced with launch then planned for 10 January (Beijing time). When the launch took place it was the first launch of the New Millennium (which, of course, started on 1 January 2001): the final launch of the Old Millennium had been China's *Bei Dou* 2 navigation satellite.

SZ 2 was launched into a 42.6°, 197–336 km orbit and at about 13.23 GMT on 10 January (as the spacecraft was making the same geographical orbital pass as had SZ 1 at the time of retrofire) the spacecraft's propulsion system fired to circularise the orbit at 329–334 km. This orbit was an almost perfect 31-circuit repeater, the kind of orbit which will be useful when rendezvous/docking missions begin. A further manoeuvre on 12 January raised the orbit to 331–340 km and a final one on 15 January resulted in a 330–346 km orbit. The orbital module separated from the combined instrument/propulsion + descent modules on 16 January at 10.23 GMT and retrofire took place ten minutes later. The descent module landed in Inner Mongolia.

Ever since the SZ 2 descent module's return to Earth there have been persistent reports that something had gone wrong with the landing. Unlike the SZ 1 recovery, there were no photographs or television coverage of the SZ 2 descent module at the landing site or after its return to Beijing. Although the Chinese have denied that there was a serious problem, reliable sources suggest that one of the parachute attachments failed during the descent, and as a result the descent module made a hard landing.

Various animals, plants and seeds were carried on board SZ 2, and the Chinese said that these were recovered intact. According to a *Reuters* report after the landing, the descent module had carried a monkey, a dog and a rabbit, along with an unspecified number of snails.

Even though the descent module was back on Earth, the SZ 2 mission was not over. On this flight a fully functional orbital module had been carried. On 17 January the module used its thrusters to raise its orbit from 330–345 km to 388–404 km. Two further manoeuvres took place: on 20 February to raise the orbit from 375–391 km to 389–403 km and on 15 March to raise the orbit from 382–390 km to 394–405 km. On 11 July the Chinese announced that the module was still working after 170 days of operations. The module finally decayed from orbit on 24 August.

SZ 2 was still not a man-rated spacecraft but the flight had tested far more of its systems that had the SZ 1 mission. The spacecraft had remained in orbit for a week and had performed all of the manoeuvres necessary to fly a docking mission. A total of 64 scientific experiments were carried aboard the spacecraft: 15 were inside the descent module, 12 inside the orbital module and 37 in what was called the 'attached segment' – possibly the assembly on the orbital module in place of the docking system.

While details of the orbital module's operations have not been disclosed in detail, it is believed that an Earth imaging system was carried, although whether this was intended for civil remote sensing or military reconnaissance observations is not known.

Shen Zhou 3

Launch date/time: 25 March 2002 at 14.15 GMT
Descent module recovery: 1 April 2002 at 08.51 GMT
Orbital module decay: 12 November 2002
Mass: 7,800 kg (orbital module 2,000 kg)

The original plan was for China to launch SZ 3 in August – September 2001, and indeed the launch vehicle was seen by Western imaging satellites as it travelled to and from the launch pad during this time period. The Chinese admitted that the launch vehicle had to be returned to Beijing for modifications. In 2002 it was said that some modifications to the spacecraft were not found to be improvements during 2001, and the launch was delayed for some real improvements to be made to the spacecraft.

SZ 3 was not planned as a simple re-run of SZ 2. For the first time, SZ 3 would carry a fully functioning life-support system, and 'dummy' *yuhang yuan* would be carried to test these systems. When previewing the flight it was explained why a simulated man was being carried rather than a live animal: if a monkey were to be launched then there was the danger of it breaking loose and it would be "sure to make some troubles, skipping and fumbling about when loosened" (*sic*). One might wonder whether this statement might be based upon some experiences aboard SZ 2 ?.

The SZ 3 launch was originally rumoured for mid-January 2002, but it did not come until 29 March: the initial orbit was not catalogued by USSPACECOM, but they tracked the CZ-2F final stage in a 42.4°, 198 to 326 km orbit. The slight shift in orbital inclination appears to be linked with the new inclination allowing better rendezvous and repeating orbits to be maintained.

SZ 1 had performed its retrofire 20 hour 19 min after launch: SZ 2 had performed its orbit-raising manoeuvre 20 hour 23 min after launch. However, SZ 3 performed its orbit-raising manoeuvre far earlier in the flight, only 7 hours after launch. Calculations showed that this would give a perfect rendezvous orbit for a hypothetical second spacecraft launched about two days later. After the first manoeuvre the SZ 3 orbit was 332 to 338 km. As with SZ 2, two further manoeuvres followed: on 29 March to raise the orbit from 331–336 km to 331–338 km and on 31 march to raise the orbit from 330–337 km to 330–340 km. After flying for nearly seven days – like SZ 2 – the SZ 3 descent module returned to Earth on 1 April. This time the Chinese media immediately released photographs and television pictures of the scene at the landing site.

On SZ 2 the orbital module had waited for just over a day after the descent module recovery, but on SZ 3 the first orbital module manoeuvre came only a few hours after the descent module's return: on the afternoon of 1 April the orbit was raised from 330–340 km to 353–358 km. There were four further manoeuvres of the orbital module with the following post-manoeuvre orbits: 25 April – 382 to 388km, 13–14 June – 375 to 385 km, 16 July – 373 to 378 km and 17 July 377 to 382 km. After this, the orbital module was allowed to decay naturally from orbit.

It will be noted that the SZ 3 orbital module orbits are lower than those seen on SZ 2: it is possible that this was because either a different imaging system was being carried or the same imaging system was being carried but the lower altitude was allowing higher resolution images.

Shen Zhou 4

Launch date/time: 29 December 2002 at 16.40 GMT
Descent module recovery: 5 January 2003 at 11.16 GMT
Orbital module decay: 9 September 2003
Mass: 7,794 kg (orbital module 2,000 kg)

In an interview before the launch of SZ 4 but published once the spacecraft was in orbit, Yuan Jiajun, a general director of the manned space programme, previewed the forthcoming flight. For the first time, the spacecraft would be identical with the version which would carry crews into orbit. It carried both automatic and manual controls, as well as all of the emergency systems and equipment which a crew might need. The descent module's instrument panel now used 'sound prompts' for data to be input or to alert crewmembers of tasks. The spacecraft was carrying a fully functioning life support system, sleeping bag, food, medicine, a fire extinguisher and 'hygiene articles and other necessities'.

Prior to the spacecraft's launch, *yuhang yuan*s completed training exercises using the spacecraft on the launch pad, simulating emergency escape from the spacecraft.

Should inclement weather prevent a landing at the primary site in Inner Mongolia then the descent module could land at Jiuquan.

SZ 4 was launched on 29 December, 2002, and like SZ 3 the orbit-raising manoeuvre took place about seven hours after launch: the second stage rocket body was tracked in a 42.4°, 198 to 332 km orbit, while SZ 4 itself was in a 42.4°, 331 to 337 km orbit. On this flight, smaller orbital manoeuvres took place, probably because the solar activity was not affecting the Earth's atmosphere and thus the spacecraft's decay rate as much as on the SZ 2 and SZ 3 missions. Small manoeuvres were noted on 2, 4 and 5 January, and the descent module was recovered after another flight lasting for seven days.

Again like SZ 3, the SZ 4 orbital module performed its first orbital manoeuvre only a few hours after the descent module's return to Earth: on 5 January the orbit was raised from 331–339 km to 354–366 km. Subsequent manoeuvres were as follows: 9 February to 359–366 km, 2 March to 359–373 km, 17 April to 345–366 km, 22 April to 346–382 km and 27 April to 345–381 km. The orbital module was then allowed to decay from orbit.

Shen Zhou 5

Launch date/time: 15 October 2003 at 01.00 GMT
Descent module recovery: 15 October 2003 at 22.23 GMT
Orbital module decay: Still in orbit in March 2004
Mass: 7,790 kg (orbital module 2,000 kg)

Leading up to the launch

The official goal for China's first manned space launch was the end of 2005, when the current Five Year Plan comes to an end, but it was clear as 2002 drew to a close that – barring major failures – the first manned flight would come during 2003. When *Shen Zhou* 4 was launched it was clearly stated that this would be the final test flight, and that *Shen Zhou* 5 would carry China's first crew around October 2003.

The number of people to be carried on the first flight was initially unclear. It was known that Shenzhou was capable of carrying three people in the descent module, but it seemed unlikely that the full compliment would fly on the maiden manned mission. On the other hand, it was thought that China might want to launch more than one person on *Shen Zhou* 5, simply because both the United States and Soviet Union launched a single crewman on their debut missions. Therefore, two crewmen was considered to be a distinct possibility.

The Chinese themselves were ambiguous concerning the number of people for the first flight, and even in summer 2003 the official People's Daily was saying that it had still to be decided whether to fly one or two people on the maiden flight.

The duration of the *Shen Zhou* 5 mission was also uncertain to western observers. Since the second, third and fourth *Shen Zhou* test flights had all lasted for seven days, and the spacecraft on the fourth mission was said to be identical in all respects to the future manned spacecraft, there was the possibility that *Shen Zhou* 5 might fly for about seven days.

If China had flown two men on a flight lasting for nearly seven days on the first manned mission, then that flight would have virtually accomplished the equivalent of the United States August 1965 Gemini 5 flight which saw two men in a single spacecraft for eight days.

The Chinese decided to err on the side of caution, and in September 2003 it was confirmed that the first flight would carry a single man and last for approximately a day. The launch period was expected to be during the ten days or so starting 10 October, and it was announced that StateTelevision would be carrying the launch "live" on television.

In early October it was announced that the target launch date would be 15 October and the launch time would be approximately 9.00am Beijing Time. The four unmanned tests had all been launched during local darkness and recovered during local darkness, the rationale behind this being that it would ease visual tracking of both the launch and re-entry phases. For *Shen Zhou* 5 this rationale was reversed – the launch and landing times were chosen so that both events would come during daylight, although landing would be made very close to local dawn: in the case of an emergency it would be easier to find and recover the descent module during daylight.

As the announced launch date approached, somewhere within the Chinese bureaucracy, nerves were lost. This can be understood when the various parties involved with the *Shen Zhou* programme are considered. Being a major government programme, *Shen Zhou* is run by the People's Liberation Army: there are also the Chinese government, the scientists and technicians which construct the spacecraft and launch vehicle and plan the missions and the Chinese media. One can imagine that the PLA was horrified with the thought of live launch coverage when something could go seriously wrong. A few days before the launch it was announced that the launch would be shown on Chinese television shortly after it took place, but not live.

There were also Chinese press reports that the flight would only last for about 90 minutes: these could be quickly discounted because it would have meant a landing far to the west of the normal landing site. These reports – quickly corrected by official Chinese media – were possibly a misunderstanding of the spacecraft entering an orbit which would take 90 minutes to circle the Earth.

Also the Chinese started to hedge their bets as far as the launch date and time were concerned: rather than being on 15 October the launch became sometime during 15–17 October, and even the 9.00am launch time seemed to disappear from the media. However, the Chinese media continued to build up as if the launch would be taking place as previously announced.

The spacecraft and launch vehicle had been delivered to the VAB at Jiuquan from Beijing on 8 October, they were integrated and checked, and the roll-out to the main launch pad took place on 11 October.

The complete *hangtianyuan* group arrived at Jiuquan during the days leading up to launch and all of the members got a chance to practice inside the spacecraft while it stood on the launch pad. Although there were fourteen men in the group, it was known that five unnamed men had been selected for the final *Shen Zhou* 5 training. On the day of launch this had already been reduced to three men: Nie Haisheng, Yang Liwei and Zhai Zhigang. Finally, on the morning of launch Lt-Col Yang Liwei was introduced to the Chinese media as China's first *hangtianyuan* with Nie and Zhai named and identified as his two back-ups.

Mission Report

Yang Liwei was driven to the launch pad and entered the *Shen Zhou* 5 spacecraft via the orbital module hatch (then dropping down into the descent module) approximately two and a half hours before launch. Even though there was no official word when the launch would take place,

Beijing		**GMT**		**Mission Elapsed Time**	**Event**
Date	Time hh:mm	Date	Time hh:mm		
Oct 15	05:30	Oct 14	22:30		Yang Liwei leaves for the launch pad
	09:00	Oct 15	01.00	00h 00m 00s	Launch
				2m 00s	Launch escape tower separates
				2m 16.3s	Strap-on boosters shut down
				2m 16.6s	Strap-on boosters separate
				2m 38.9s	First stage core shut down
				2m 39.4s	First stage core separates
				3m 20.4s	Payload shroud/escape system separates
				7m 41.6s	Second stage main engine shut down
				9m 43.6s	Second stage vernier engines shut down: orbital injection
	09.34		01.34	34m	Yang reports that he "feels good"
	09.42		01.42	42m	Li Jinai reports successful launch of Shen Zhou 5
	10.31		02.31	1h 31m	Yang Liwei takes off his gloves and frees his legs
	11.08		03.08	2h 8m	Yang Liwei has his first meal in orbit
	11.12		03.12	2h 12m	Yang Liwei has begun his second period of sleep
	14.15		07.15	6h 15m	Yang Liwei ends his first period of sleep
	15.57		07.57	6h 57m	Shen Zhou 5 performs orbit circularisation burn
	17.05		09.05	8h 5m	Yang Liwei transmits pictures of Earth from orbit
	17.30		09.30	8h 30m	Cao Gangchuan – vice-chairman of CPC military commission and defence minister – speaks with Yang Liwen
	20.00		12.00	11h 0m	Yang Liwei speaks with his family from orbit
Oct 16	00.18		16.18	15h 18m	Yang Liwei has begun his second period of sleep
	05.04		21.04	20h 4m	Yuan Wang 3 ship transmits retrofire commands to Shen Zhou 5
	05.36		21.36	20h 36m	Shen Zhou 5 orbital module separates from descent/instrument modules
	05.38		21.38	20h 38m	Retrofire takes place
	05.59		21.59	20h 59m	Descent and instrument modules separate
	06.07		22.07	21h 7m	Helicopters receive signals from descending spacecraft
	06.08		22.08	21h 8m	The main recovery switch was turned on
	06.11		22.11	21h 11m	First parachute deployed
	06.14		22.14	21h 14m	Heat shield at base of descent module jettisoned
	06.16		22.16	21h 16m	Main parachute deployed
	06.23		22.23	21h 23m	Four retrorockets fire and the descent module lands
	06.24		22.24	21h 24m	Helicopter sees main parachute on the ground
	06.25		22.25	21h 25m	Main parachute discarded from descent module
	06.27		22.27	21h 27m	Recovery team 3 report that the descent module is 7.05 km away
	06.31		22.31	21h 31m	Premier Wen Jiabao speaks with Yang Liwei, congratulating him on completion of his flight
	06.33		22.33	21h 33m	Recovery helicopter finds descent module in grass
	06.38	Oct 15	22.38	21h 38m	Rescue workers find Yang Liwei is alive and well and he climbs outside his descent craft

Based upon data published by the People's Daily Online on October 16 and Xinhua (date of original posting not known).

everything was going to the previously-announced schedule for a launch at 9.00am Beijing Time (01.00 03.497s GMT). The Chinese press witnessed the launch, as did local people and sightseers, even though they were kept well outside the launch base. Launch was rapidly reported to the western media about four minutes after the event.

As the CZ-2F launch vehicle climbed away from the pad, debris was seen to be falling off the vehicle: while this initially caused some consternation for western observers, in fact the debris from the interstage area between the first and second core stages was perfectly normal – but it had not been so easily visible on the previous night-time launches of the vehicle.

The Chinese media quickly showed pictures of the launch, showing the vehicle climbing away with the four main engines on the first stage core and the four strap-on booster engines (one per strap-on) clearly firing. It took nearly ten minutes for *Shen Zhou* 5 to reach orbit. China had indicated that the initial orbit would be approximately 200–350 km: the first data from United States Space Surveillance Network (USSSN) showed that the second stage of the CZ-2F was in a 197–328 km orbit inclined at 42.4 degrees to the equator approximately ten hours after launch.

(As a first aside, starting with *Shen Zhou* 3 when the spacecraft introduced the orbit-raising burn approximately seven hours after launch, USSSN has tended not to have officially catalogued anything from the launch until after the manoeuvre has taken place. A second aside is that China and USSSN are using different "Earth models" when preparing orbital data: while the Chinese model is not known, the USSSN data are based upon a spherical Earth with a radius of 6,378 km. This is why "western" orbital data differ from those announced by China.)

Yang reported that he was feeling fine some 34 minutes after launch and as the second orbit started he received his orders to remove his gloves and release his leg restraints. Throughout the flight Yang would remain in the descent module, even though the orbital module allowed extra living area. For this first flight China wanted to have everything as simple as possible and since the descent module was designed to accommodate three people at launch, there was deemed to be sufficient room for Yang to stay there throughout his flight.

While Yang was safely in orbit, on the ground the search was underway for the "black box" magnetic recorder which held all of the launch vehicle's flight data which could not be obtained from the ground monitoring stations. The box was expected to be found near Otog Banner in Inner Mongolia.

The successful orbiting of the spacecraft also meant that three emergency rescue ships – Beihai 102, De Kun and De Yi – could return home. They had arrived under the launch vehicle's trajectory over the ocean (the actual locations were note stated) about 8am Beijing Time on October 14, and they were capable rescuing the hangtianyuan and his decent module if there had been a launch aboard and he had to ditch in the ocean.

A major event for the mission was the circularisation of the *Shen Zhou* 5 orbit which was completed 6h 57m after launch: China said that post-manoeuvre the spacecraft was in a 343 km circular orbit, while finally USSSN started to issue orbital data for the spacecraft – showing it to be in a 331–338 km orbit.

The rationale behind this manoeuvre is the planned mission of the *Shen Zhou* 5 orbital module. This carries 40 kg of propellant, allowing manoeuvres totalling 60 m/s after the manned phase of the flight is over. On the *Shen Zhou* 2–4 flights the module was left in the 335 km circular orbit by the returning spacecraft. If *Shen Zhou* 5 had not manoeuvred to the same circular orbit then the orbital module would have had to use 60–70 per cent of its propellant to reach the circular orbit, and this would mean that there was insufficient propellant left for the orbital module to remain in orbit for the planned 6–8 months.

While in orbit Yang had the traditional conversations with Chinese political leaders as well as his family, he had two periods of rest and he ate traditional Chinese food, especially prepared for his flight. He also found time to take excellent-quality video images of the Earth from his spacecraft, showing one of the solar panels at the right of the image: it was not clear whether the video was showing one of the orbital module solar panels ahead of the descent module or one of the instrument module panels behind the module.

As happened on the four unmanned *Shen Zhou* flights, the *Yuan Wang* 3 space tracking ship was located off the west coast of Africa, close to the equator: it would have the honour – as always – of transmitting the command to bring the spacecraft back to Earth. The commands were transmitted to the ship on 16 October at 5.04am Beijing Time. As *Shen Zhou* 5 rose in the sky as seen from the ship, the orbital module was first separated and then two minutes later the command for retrofire was sent to the spacecraft and executed. As the orbital module continued in solo flight, the manned part of *Shen Zhou* 5 started its return to Earth.

The landing profile was familiar to those who follow the Russian space programme. After atmospheric entry and separation from the instrument module, a drogue parachute opens from the descent module, is discarded and then a main parachute opens and continues to slow the descent module. The base plate of the heat shield is discarded, exposing four solid-propellant retro-rockets. At an altitude of 1.5 metres these fire to cushion the final landing on terra firma. Landing came in the Dorbod Xi area of Inner Mongolia which is located at 41.3° N, 111.4° E: the descent module had landed a mere 4.8 km from the planned touchdown point – remarkably accurate for the first piloted recovery.

Chinese television was showing programmes devoted to the flight, and it was interesting to note that the television transmissions from the recovery crew as shown by CCTV4 had a delay of about 6–7 minutes. This continued until the camera team had reached the spacecraft and the transmissions suddenly switched to "live" once the hatch was opened and Yang was seen sitting in the descent module.

The landing must have been more than a little rough on Yang. After the retrorockets fired the parachute must have toppled the descent module onto one side. The result was that Yang was initially hanging upside-down inside his descent module until he was able to carefully release himself and prepare for the arrival of the rescue team.

Orbital Module Operations

The flight of *Shen Zhou* 5 appears to have been an unqualified success. The orbital module remains in orbit, and that is equipped with a photoreconnaissance camera carrying a reported ground resolution of 1.6 metres and an electronic intelligence-gathering package. There is also expected to be a retinue of scientific instruments on board the module as well. The module performed its first orbit-raising manoeuvre close to midnight (GMT) of October 21–22, manoeuvring from a 330–332 km orbit to 339–346 km. Subsequent manoeuvres resulted in the following post-manoeuvre orbits: 3 November – 343–354 km, 12 November – 357–367 km, 24 December – 343–358 km and 31 December – 341–387 km.

***Shen Zhou* 6**

Launch date/time: late 2005

It was a surprise when the Chinese stated that the second piloted mission would not come until approximately two years after the *Shen Zhou* 5 flight. One may speculate that this is because the current Chinese Five-year Plan only called for two piloted flights, and the test programme was so successful that the initial piloted flight came earlier than expected.

The Chinese have discussed various scenarios for the *Shen Zhou* 6 flight, discussing a flight of up to three *hangtianyuans* lasting for more than one day. By the end of 2003 these reports had settled to there being two people of the second piloted mission, with the flight lasting about a week. It has been said specifically that the *Shen Zhou* 6 flight does not call for an orbital docking, but it is likely that this refers to the *piloted* phase of the mission. The separated *Shen Zhou* 6 orbital module could be used as a docking target for the *Shen Zhou* 7 mission of China can launch its third piloted mission within four months of the second. There has also been speculation that *Shen Zhou* 6 might include China's first spacewalk, but once more this might be something planned for *Shen Zhou* 7 in 2006.

International

Spacelab

Current Status

The last Spacelab flight was launched on 17 April 1998 aboard Columbia on STS-90. In 22 flights, the pressure module has accumulated 244 days in space over more than 14 years of operations. The 10 pallet-only flights between November 1981 and February 1996 accumulated 89 days of mission time. Experience with developing and managing Spacelab allowed Europe access to technology and operations experience essential to admitting ESA to the International Space Station programme. The research base and experimental programmes conducted aboard Spacelab provided technical expertise for Columbus and ISS participation.

Background

Spacelab was Europe's major contribution to the US Space Shuttle programme, and ESA's most important long-term project for the nine-year span between 1974 and the first flight in November 1983, about four years behind schedule because of delays in the Shuttle programme. NASA and ESA jointly ran the programme until the third Spacelab flight (STS-51F/Spacelab 2) in 1985. Later that year, ESA delivered to NASA the second Spacelab, including IPS, for which NASA paid about 200 MAU, formally ending the joint programme. Europe's US$1 billion development of Spacelab at no cost to NASA did not provide ESA with continuing right of access to or use of the laboratory once the joint programme ended. By the end of 1985, four Spacelab missions had been completed, the fourth being the German-funded/controlled D1 mission.

Discussions that eventually led to Spacelab, began in 1969, when NASA invited Europe to participate in the post-Apollo programme. Agreements on Spacelab between NASA/ESA and between the US and nine European governments followed in September 1973. ESA, basing its plans upon American claims of space launch affordability believed US would put the Spacelab into orbit at least 50 times during its lifetime. To date, the flight rate is less than half that.

NASA also promised the Europeans that Spacelab would be a first step towards internationalising the US manned programme, however NASA would control all of its activities. NASA would permit European flight crew opportunities in conjunction with flight projects sponsored by ESA or by governments participating in the Spacelab programme. West Germany contributed the most in percentage terms. The breakdown of contributions is: West Germany 54.94 per cent, Italy 15.57 per cent, France 10.29 per cent, UK 6.51 per cent, Belgium 4.32 per cent, Spain 2.88 per cent, Netherlands 2.16 per cent, Denmark 1.54 per cent, Switzerland 1.00 per cent and Austria 0.79 per cent.

An initial six-year contract worth US$226 million and covering Spacelab design/development was awarded in June 1974 to VFW-Fokker/ERNO of Bremen, whose major subcontractors were Bell Telephone in Belgium, Aeritalia (pressure shell) in Italy and BAe Dynamics (Pallets). In all, the industrial consortium finally embraced 50 companies in 10 European countries. TRW provided technical assistance to ERNO.

During 1979, the NASA/ESA partnership nearly dissolved. Spacelab integration tests brought to light numerous mismatches of hardware and incompatibilities between hardware and software. New cost projections to fix the problems showed the project would break the 120 per cent ceiling and more funds had to be sought from the participating countries. By then, the original estimated cost of US$515.7 million had grown to US$850 million.

With additional funding, however, in November 1980, after more than six years of development, the EM was delivered to NASA at the Cape, where it was used for interface verification and training. Then, US Vice-President Bush formally took delivery of the first flight version of Spacelab on 5 February 1982, together with an EM, ground support equipment and some computer software. ESA supplied a Long Module, five Pallets, one igloo, an Instrument Pointing System and two sets each of EGSE/MGSE.

In 1980, NASA ordered a second Spacelab worth US$144 million and subsequently a second IPS worth US$18.4 million for delivery in 1983. NASA would have preferred a better deal under which it would have provided flight facilities in return for the hardware. NASA dashed ESA hopes of continuing the project by dropping an option for a third Spacelab.

Beginning with 1990's Astro mission, NASA took over control, with ESA participation, of all science activities from Marshall's Spacelab Mission Operations Control Center, replacing Johnson's Payload Operations Control Center. McDonnell Douglas, the American aerospace company,

Spacelab 1 in Columbia's payload bay. The Z-shaped docking tunnel at the far end (top right) matches the levels of the Orbiter and Spacelab hatches and allows longitudinal movements during launch and re-entry to reduce stress (NASA) 0516807

provides Spacelab processing and support. NASA extended the contract in 1992 by four years at US$163 million, plus a four-year US$120 million option. Germany dropped out of Spacelab citing the soaring costs related to American financial management of the overall programme. In 1993, ESA dropped plans for its own Spacelab-E flight on cost grounds.

Spacelab flies in the cargo bay of the Shuttle and relies on the Orbiter for life support during the 9–16-day missions. Columbia accommodates most Spacelab flights and, from 1992, its Extended Duration Orbiter capability has permitted 13-day missions. In 1995, NASA adopted the Mir Shuttle visits, greatly reducing the need for Spacelab sorties.

Spacelab comprises three principal units: a 75 m^3 Pressure Module, Pallets for mounting external payloads and an 880 kg tunnel linking the PM to the Orbiter airlock. In preparing the Spacelab for flight, the ESA sets a landing mass and centre of gravity restriction that require aftward mounting in the cargo bay of the Shuttle, but a five-Pallet train can be accommodated along the full length. The total load limit on the Spacelab upper mass limit is 14.5 tonnes. British Aerospace constructs each 725 kg 3 m long 4 m wide Pallet with conventional aircraft-type designs and aluminium skin over an aluminium frame. The final pallet configuration is capable of carrying up to 3 tonnes of payload. Data, power and cooling links are routed through the PM. The first Pallet, an engineering model, flew on STS-2. If the PM is absent, a 640 kg 2.2 m^3 2.4 m high 1.1 m diameter aluminium igloo cylinder at the front provides a pressurised environment for the avionics.

The Pressure Module consists of a single shell of 0.16–0.35 cm thick aluminium capped at both ends by an Al cone. Each 4.06 m diameter 2.7 m long segment consists of eight butt-welded panels ending in 24 cm wide 4.10 cm outer diameter aluminium load-bearing rings. A two-segment Long Module is created by bolting two units together, with O-rings providing the seal; a single segment Short Module configuration is also available (but has not yet flown). Two longerons fastened to the aluminium rings attach Spacelab to the Orbiter's longerons via a single 82 mm diameter Inconel pin on each side and a keel point underneath.

The 78.2 cm deep 4.11 m diameter cones are bolted to the cylinder only when the internal equipment is in place and thereafter provide access through central 1.3 m circular openings. ESA removes the rear hole for flight and uses the forward opening for the tunnel. The LM's overall length is about 7 m. Each cone includes three 40 cm holes: two at the base for Orbiter supplies and Pallet connections and one near the top. On the rear cone, the latter is always a 30 cm diameter 2.6 cm thick window for viewing the Pallets, whereas the forward version carries pressure relief valves (the structure is stressed for a maximum pressure differential of 1.1 atmosphere).

Dual-triangular aluminium beams run longitudinally inside each PM segment and are attached to the end rings as support for the floor and experiment racks. Prior to flight, experimenters place their packages and integrate them into the Spacelab offline and then the racks are slid with the floor into place on the beams. The aluminium racks are designed to accept standard lab trays 48.3 cm wide: the 42 kg Single Rack can hold 0.9 m^3/290 kg of equipment and the Double Rack 1.75 m^3/580 kg. Each Spacelab segment can hold two DR and one SR on both sides, creating a central aisle similar in size to Mir. Typically, a rack is devoted to a specific discipline or reserved as a storage unit. The racks are attached to two overhead Al beams that also support up to eight storage containers. The Al honeycomb floor panels can be lifted to expose

The Spacelab long module continues to be used for science missions (NASA) 0517298

Spacelab Mission History

Date	Configuration	Mission
November 1983	LM+1P	Spacelab 1[1]
April 1985	LM+MPESS	Spacelab 3[2]
July 1985	IG+3P	Spacelab 2[3]
October 1985	LM+MPESS	Spacelab D1[4]
December 1990	IG+2P	Astro-1 + BBXRT[5]
June 1991	LM	SLS-1[6]
January 1992	LM	IML-1[7]
March 1992	IG+2P	Atlas-1[8]
June 1992	LM	USML-1 (1st EDO)[9]
September 1992	LM	Spacelab-J (Japan)[10]
April 1993	LM	Spacelab-D2 (Germany)[11]
April 1993	IG+P	Atlas-2[12]
October 1993	LM	SLS-2 (EDO)[13]
July 1994	LM	IML-02 (EDO)[14]
November 1994	IG+P	Atlas-3[12]
March 1995	IG+P	Astro-2 (EDO)[15]
June 1995	LM	SL-M Mir visit
October 1995	LM	USML-2 (EDO)[16]
June 1996	LM	LMS (EDO)
April 1997	LM	MSL-1 (EDO)[17]
July 1997	LM	MSL-1 (EDO)[17]
April 1998	LM	Neurolab (EDO)

Key:
EDO: Extended Duration Orbiter; IG: Igloo; LM: Long Module (pressurised);LMS: Life & Microgravity Spacelab; MPESS: Mission Peculiar Experiment Support Structure; P: Pallet; Atlas: Atmospheric Lab for Applications & Science; SLS: Spacelab Life Sciences; IML: International Microgravity Lab; MSL: Microgravity Sciences Lab; USML: US Microgravity Lab.
The first eight PM missions (through D2) involved 387 experiments covering 323 PIs from 148 institutes in 26 countries.
1 38 NASA/ESA experiments (16 Pallet/22 PM), to demonstrate basic Spacelab capabilities.
2 10 μg experiments in LM, including Drop Dynamics Module and animal holding units with 2 monkeys & 24 rats, 2 on MPESS (Atmos for monitoring upper atmosphere trace constituents and Indian Ions for studying cosmic rays).
3 carried IPS with four solar telescopes, plus x-ray, cosmic ray & IR telescopes and retrievable Plasma Diagnostics Package first flown on SM3.
4 first 8-man crew, 76 μg experiments: life sciences, Biorack, Fluid Physics Module, Sled, materials processing, plus Materials Experiment Assembly on MPESS. Sponsored by FRG BMFT & managed by DFVLR.
5 7,836 kg Astro telescopes on ESA/Dornier IPS (HUT Hopkins UVTelescope, The Johns Hopkins University; UIT UV Imaging Telescope, GSFC; WUPPE Wisconsin UV Photopolarimeter Experiment, University of Wisconsin). 3,294 kg BBXRT BroadBand X-RayTelescope, GSFC.
6 most intensive life sciences research mission flown, first three-woman crew, 29 rats and 2,478 jellyfish, plus 12-can GAS Bridge.
7 55 materials and life sciences experiments, plus 10-can GAS Bridge.
8 11 instruments on 2 pallets for atmosphere, solar, space plasma and astronomy.
9 31 experiments in four broad areas: materials, fluid physics, combustion and biotechnology.
10 43 materials and life sciences experiments (34 from Japan).
11 90 experiments including Anthrorack, Advanced Fluid Physics Module, Medea, Holop, Biolabor, MOMS-02 Earth observations, Gauss galactic plane UV film imaging. Funded by BMFT, managed and payload controlled by DLR.
12 six instruments on one pallet plus seventh in two side cans, for atmosphere and solar studies.
13 14 experiments continuing SLS-1 life sciences studies.
14 >80 experiments, including Large Isothermal Furnace, Critical Point Facility, Advanced Protein Crystallisation Facility, Free-Flow Electrophoresis Unit, Biorack, Aquatic Animal Experiment Unit, Bubble Drop & Particle Unit.
15 as Astro-1 (note 5), except no BBXRT.
16 followed on and repeated USML 1 (note 9).
17 Mission terminated early and to be reflown July 1997.

additional space in the rear segment and equipment can be bolted down on the 60 cm wide central aisle. This was employed, for example, during Spacelab D1 for the Sled chair/track experiment.

The two forward DRs house connections for standard functions, even on a Short Module flight. The port side workbench rack carries a work surface, tools, high-quality lighting and writing equipment. The starboard unit controls Spacelab's systems. Three identical IBM AP-101SL (replacing the Matra 125/MS 64 kbit) computers handle either life support or experiments, retaining one as a back-up. The Orbiter's flight deck hosts a duplicate terminal and a third can be added to the second Spacelab segment. The same rack holds the High Data Rate Recorder for storing data at up to 32 Mbit/s. Data can be transmitted in real time through NASA's TDRS satellite system at up to 50 Mbit/s via the steerable Ku-band dish on the Orbiter's forward cargo bay starboard side. The recorder is employed when TDRS is unavailable.

The roof of each segment incorporates a 1.3 m diameter opening for windows or, for the rear opening only, a science airlock. The window unit contains a general viewport and a high-quality optical 4.1 cm single thickness 40.9 × 55.2 cm remnant Skylab pane. ESA's large format Metric Camera employed this port during Spacelab 1. An external aluminium cover provides protection against radiation, contamination and micrometeoroids. A simple viewport can replace this unit as required. D1 used the simple window.

Dornier's Instrument Pointing System (IPS) provides three-axis control for a payload of up to 3 tonnes, although with modifications the system can handle 7 tonnes. It carries all inertial sensors, data and power electronics and the dedicated software for control via the Spacelab computers in a unit that weighs 1,184 kg. The nominal payload is 2 tonnes, on a 0.5 to 3 m diameter. Pointing accuracy (arcsec) for a 2 tonnes payload is 0.4 lateral/11.2 roll under star tracker (3) control; 0.5 lateral 41.0 roll in the Sun mode. Pointing stability is sub-arc/sec. The available control torque is 30 Nm/axis. Slewing rate is 1°/s maximum in inertial mode, power required during pointing is 500 W. Payload power provided is 1.25 kW. Data rate 16 Mbit/s on six channels. The control system has flown on Spacelab 2 and Astro 1/2.

Spacelab's life support system is carried under the forward segment's floor. The twin holes each side in the main floor beams blow cooling air up behind the equipment racks. The racks and second segment await mating in the background 0516809

The 1 m long 1.05 m diameter airlock can accommodate 98 cm diameter 99 cm long packages bolted to a side-mounting table, which is then hand-cranked 96 cm out into open space. When astronauts require easy access they may remove the inner hatch. D1 deleted the airlock, for instance. It carries an 18 cm diameter off-centre porthole for observing experiment progress. Spacelab 1/3 used it to place an ultra-violet Very Wide Field Camera outside the main structure for celestial imaging. Spacelab carries sufficient N_2 for up to seven airlock repressurisations on a normal mission. Should the outer hatch jam open, preventing the Orbiter's door from closing, it can be ejected or removed during EVA.

Astronauts perform EVAs as needed through a top hatch on the transfer tunnel adjacent to the Orbiter's airlock. The 1.3 m diameter aluminium tunnel, built by McDonnell Douglas, provides a 1.02 m internal diameter for passage of a suited astronaut and accommodates the 1.07 m height difference between the Shuttle rear hatch and the Spacelab central opening. Only the 5.75 m long six-segment tunnel has been employed so far but a 2.66 m five-segment version is available for the Short Module (which can be carried further forward). Flexible sections at both ends allow some movement in the bay.

The Orbiter and Spacelab work in concert to perform life support activities and to maintain a 1.013 atmosphere 18 to 27°C 30 to 70 per cent humidity environment. A ventilation unit incorporating an old Skylab fan at the bottom of the tunnel bend draws Orbiter cabin air into Spacelab through a floor conduit and returns it via the main volume. A charcoal scrubber converts CO to CO_2 and removes other trace materials. When the end hatch is closed (the Spacelab end does not carry a door) during ascent and landing,

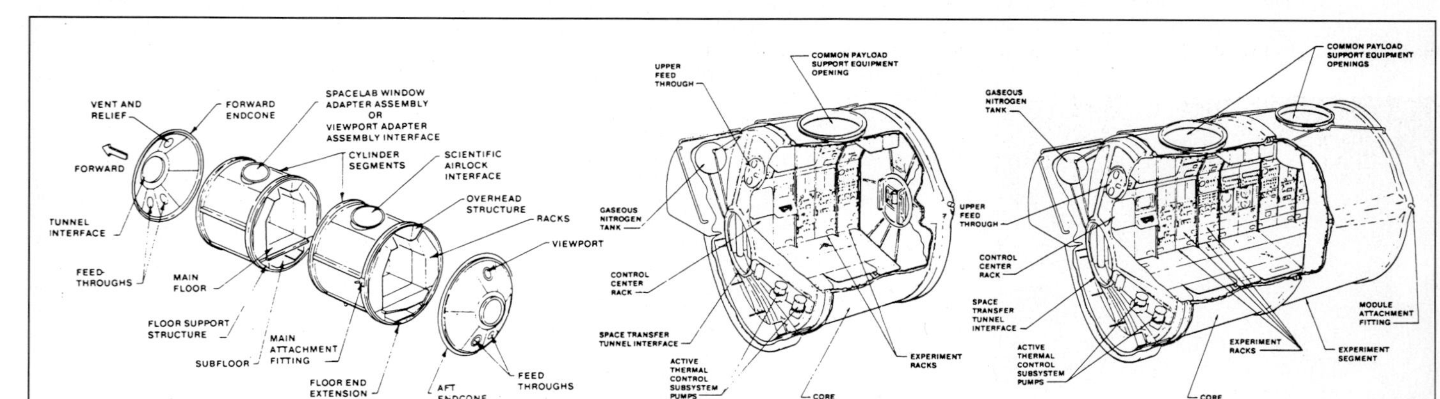

The Pressurised Module can be flown in either long or short form. It has always flown as a Long Module so far 0516808

the system continues to ventilate the tunnel/lab. The air is cooled in heat exchangers before being blown through roof-mounted diffusers. Air is also forced up inside the racks for cooling and drawn off at several points inside the curved wall.

The payload is limited to about 8 kW (28 V DC) from the fuel cells because of the Orbiter's heat rejection capacity. To maintain thermal control some experiments and avionics are mounted on cold plates using the Orbiter's radiator system. The whole module carries an external jacket of 39 interleaved layers of Dacron and gold-covered Kapton topped by Teflon-coated beta cloth, the whole held in place by studs.

Before a launch, researchers normally insert their experiments while Spacelab is horizontal and the whole configuration is slotted into the Cargo Integration Test Equipment (CITE) at KSC for Orbiter simulations. Then the NASA ground crew installs Spacelab in the Orbiter Processing Facility before the spaceplane is stacked and transferred to the pad.

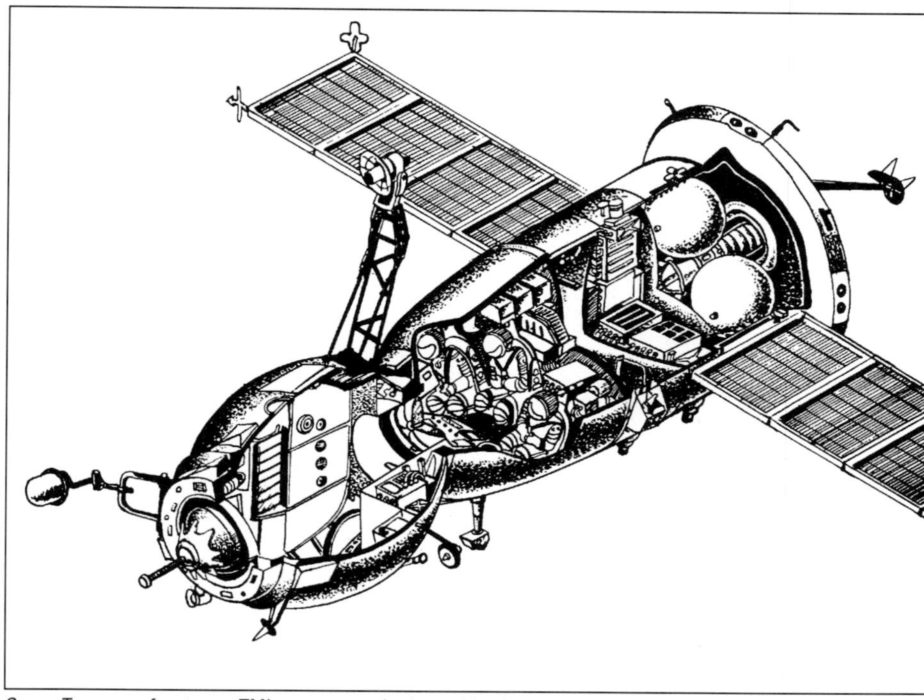

Soyuz-T spacecraft cutaway. TM's appearance is almost identical 0516698

Russian Federation

Soyuz

Current Status

An improved Soyuz TMA was selected 1 June 1996 as the initial Assured Crew Return Vehicle (ACRV) for the international Space Station. NASA signed an agreement with RKK Energia in December 1992, when the Phase A study ended, affirming that Soyuz is technically suitable. Plans originally called for two docked three-man Soyuz-TMs to be used as the emergency escape vehicle in much the same role as the spacecraft was used aboard the Mir for American astronauts. Each Soyuz costs US$30 million, plus enhancements such as increasing service life from six months to three years. The vehicle must be able to land within 20 km of a designated landing spot. Crews will not wear pressure suits. The Soyuz TMA is modified to accommodate the larger torso size range of US astronauts. For example, in 1995, Scott Parazynski stood down from May 1997's Mir visit because of his torso length. Although he was to fly up and down on Shuttle, he had to be able to return in Soyuz in an emergency. It also has improved parachutes and two new Amber display units. Because the ISS crew size is limited to a maximum of 3 people for the foreseeable future, only one Soyuz TMA is needed on the station at any one time. Also, since the loss of Columbia on 1 February 2003, and the grounding of NASA orbiters Discovery, Atlantis and Endeavour, Soyuz TM series spacecraft are the only means of maintaining an ISS caretaker crew.

Background

Soyuz is the most-utilised manned spacecraft and is now into its fourth decade with a flight total comparable to the aggregate for Mercury, Gemini, Apollo and Shuttle. There have also been dozens of unmanned orbital tests and it probably has a version that houses a photo-reconnaissance satellite. Despite its age, Soyuz operated as a Mir ferry. RKK Energia in Kaliningrad is responsible for design and production. Russian planners have long floated plans for an upgraded version of the Soyuz but, in 1991, Energia halted design work on the 13 tonne Zarya super-Soyuz, carrying five to six crew to be launched from a Zenit booster, because of funding difficulties. The overall form of the descent module was similar to that of Soyuz, but with a 3.7 m base diameter. The manned version would have included 1.5 tonnes of cargo and the unmanned 3.8 tonnes.

TM16 displays the forward-looking cupola (arrowed) and androgynous docking system 0517094

Originally, Chief Designer Sergei Korolev intended to build several versions of Soyuz. His ideas called for a full three-module craft for Earth-orbit operations, a stripped-down two-module Zond for demonstrating lunar mission techniques, and a lunar-landing orbital version. As flown, Soyuz has had three distinct mission categories. Soyuz 1–9 (1967–69) carried three-man LEO solo missions of durations up to 2½ weeks. Then when it became evident to the Soviets that they had lost the moon race, design work began on the Salyut space station in late 1969 and Soyuz 10 and 11 functioned as crew ferries with integral transfer tunnels. After the Soyuz 11 failed, killing its crew during descent, the Soviets initiated a thorough redesign and Soyuz became a two-man station ferry capable of only 2½ days' independent flight. At this time, the Soviet space program evolved towards a more automated program than its American counterpart and as a result, the unmanned Progress cargo version of the Soyuz capsule came about.

The post-Soyuz 11 version of the craft was 24 times in its manned configuration when Soyuz 40 retired the model in May 1981. After Soyuz 40, the Soviet Union introduced the Soyuz-T with its updated systems and three-man crews. This capsule took over for Salyut 7 operations until a further reworking produced the 'TM' Mir ferry. The 'TM' designates all Russian craft as modified transports. Soyuz has thus been transformed from a spacecraft with lunar missions to a dedicated space station ferry, through several generations of space exploration objectives.

The Soyuz Orbital Module (OM) provides space for cargo, food, waste management services, orbital experimentation on solo missions and EVA operations through the side hatch which is employed primarily for crew boarding on the pad. The forward hatch houses the docking apparatus of a central cone and eight metal blades inside a 1.3 m diameter docking collar that has eight mechanical latches (Soyuz 16/19/TM16 carried androgynous docking units). Two crew use sleeping bags in the OM, while the third sleeps in the Descent Module (DM) for safety (a rule not always adhered to). The toilet swings out from behind a panel, using pressurised airflow as a gravity substitute.

When delivering payloads to space, the Soyuz uses built- in connections for electrical, communications and hydraulics links. The original Soyuz design included four OM waist portholes, reduced to two on Soyuz-T and one on TM. Since 1991 and possibly before, TM has carried at least one front-viewing cupola in the OM, with controls for flyaround operations using direct vision. The OM is jettisoned by 12 pyrotechnic charges after retrofire. Until the TM5 failure of September 1988, cosmonauts ejected the OM before retrofire. This sequence saved about 10 per cent of the propellant. Failure investigations proved that crew safety required the post-retrofire jettisoning operation.

The Soyuz Descent Module (DM) accommodates a crew of two or three, with the Cdr occupying the central couch, the flight engineer the port side and the cosmonaut-researcher the starboard side. Alternatively, a 100 kg cargo pod can replace one crew member. These Kazbek-Y couches are supported on shock-absorbing struts and can be lowered to provide increased space. Ten minutes before landing, compressed N_2 raises the couch 10 cm to help absorb the shock. Each seat liner is customised for the occupant and when cosmonauts return in different craft from their delivery craft, then the liners must be changed.

The commander provides manual thruster control via translational and rotational handles to his left and right, respectively, performing rendezvous and docking operations using the range and range-rate radar data and the 15° Field of View (FoV) VSK-4 periscope below the centre of the main control panel. From the introduction of Soyuz-T, the onboard computer has been capable of performing a fully automatic docking. The DM's separate 67 N H_2O_2 thruster network is activated only for re-entry: two nozzles near the apex some 60° apart provide roll control, with pitch control from a pair located between, and two at the base for yaw.

The service module completes the vehicle, supporting the DM on its forward face. The Service Module (SM) carries the propulsion unit. In the old Soyuz, the propulsion system employed a KTDU-35 assembly at the rear of the SM. Pumps fed propellant and oxidisers to a single main engine, which fired for 2½ minutes to initiate re-entry. A single start twin-nozzle engine served as a back-up unit. AOCS thrusters distributed around the SM's forward and aft faces provided guidance and control of the vehicle. Soyuz-T introduced the ODU unified propulsion system in which all engines were supplied from the same tank. Soyuz T's smaller AOCS provided a back-up re-entry capability. The Russians have revealed less about the Soyuz-TM system. Jane's believes the primary engine provides 3.1 kN of thrust. Mir incorporates similar engines and the timeframe of the development programme, the contractors and the history of Russian design practices suggest the two engines would be used for both projects.

As a safety measure, the propulsion system tanks now use metallic membranes to avoid pressurising N_2 which would have leaked through the previously elastic barrier. Some 150 kg of the propellants are reserved for rendezvous and docking and the retrofire operation consumes an additional 200 kg (400 kg allocated) for Earth return.

In the event of an emergency, an escape tower can fire the capsule away from the immediate area. From the top of the launcher on the pad, the capsule would reach an altitude of 1 km and land some 2½ km away, using the 27 m diameter reserve parachute. On a normal mission, descent is initiated by a 3 to 4 minute ~155 m/s retroburn by the main engine over the south Atlantic, followed by OM and then SM ejection. Re-entry occurs over Africa at an angle of attack of about 30°; a lift to drag ratio between 0.25–0.30. A 0° bank (heads-up) generates maximum lift and a deceleration of 3 *g*, 60° produces 4 *g* and the no-lift and a 90° bank leads to 8 *g*. The Soyuz TM can land cross-range of ±65 km, with a landing accuracy of 20 × 30 km ellipse. During re-entry, the external temperature reaches 3,000°C,

Soyuz-TM: the version of the Soyuz manned craft initially used for Mir duties. This full-scale representation at 1987's Paris Salon was the first public appearance. Note the SM's two Earth horizon sensors, the DM's periscope and umbilical panel and the OM's hatch for on-pad access (Andrew Wilson) 0517095

but internal temperature maintains a 25–30°C constraint.

When landing, a pressure switch, activated at 9 to 11 km altitude and 850 km/h, unfurls two sequential stabilising drogues releasing the 4.25 m braking parachute from the port compartment. The primary canopy is deployed reefed at 8 km some 40 to 45 minutes after retrofire. It is freed to its full 35.5 m diameter at a descent rate of around 35 m/s to reduce sink rate to 8 m/s. Soyuz-TM retains the same size parachutes but their lighter material accounts for 110 kg of the craft's increased lift capacity. The 90 kg reserve system yields 10 m/s descent rate with its 25 m parachute, activated at 6 km. The heatshield is dropped at about 3 km some 5 minutes before landing to clear the base retromotors for a soft landing.

In the final burn, four solids initiate upon command from a radar altimeter about 2 m above the ground to cushion the impact. Soyuz-TM carries an improved system that has reduced contact speed from 3 to 4 m/s to 1 to 2 m/s. Following the Soviet design, the Soyuz generally touches down on land but the craft is equipped for water landings. Coping with splashdown is a routine element of cosmonaut training.

Each capsule carries standard Granat-6 survival kits of Forel nylon flotation suit, Neva self-inflating life preserver, TZK cold weather suit, dried food, medical supplies, flares, radio, machete and canteen. The equipment is designed to sustain the crew for up to 3 days in the capsule in severe conditions and up to 12 hours in 2°C water with −10°C ambient air. A recovery beacon automatically activates on touchdown from behind a small ejectable cover to the left of the parachute compartment. In the event of a water landing, a balloon inflates from the same location. A radioactive source also acts as a beacon, something the recovery crew has to avoid after landing.

Specifications

Cargo/Manned Unit

First manned launch: 23 April 1967 Soyuz; 5 June 1980 Soyuz-T; 5 February 1987 Soyuz-TM; 30 October 2002 Soyuz TMA
Number manned launches: 40 Soyuz, 15 Soyuz-T, 34 Soyuz-TM (to end-2002); 1 Soyuz TMA (to end 2002)
Principal applications: space station ferry, LEO solo operations
Cost: Rb95 million (plus Rb56 million launcher) quoted in November 1992. Rb5 million quoted during Bulgarian TM5 mission of 1988, plus Rb4 million for Soyuz-uluz launcher.
Availability: typically 2–4 launched annually.
Performance: TM can accommodate a crew of three and 50 kg cargo for Mir-type operations, returning a full crew and 50 kg to Earth (no crew allows 450 kg up and 300 kg down).
Flight longevity: Independent 3.2 days and attached to Mir 180 days. Capabilities are constrained partially by launch escape system.
Cargo size: 45 × 60 × 100 cm up and 30 × 40 × 50 cm down.
Length: 6.98 m with docking probe withdrawn (on old Soyuz retraction was 43 cm)
Mass: 7,070 kg at launch (Soyuz-T 6,850 kg). Typically 6,790 kg at docking. Soyuz 19 mass breakdown was 1,224 kg OM, 2,802 kg DM, 2,654 kg SM.
Habitable volume: ~10 m^3 (T/TM), previously ~9 m^3
Landing Mass: about 3,000 kg

Soyuz Orbital Module

Mass: 1.2 t
Size: 2.2 m diameter spheroidal
Volume: 6 m^3
Hatches: 65 cm diameter side hatch, 80 cm forward hatch

Soyuz Descent Module

Size: 2.2 m base diameter, 2.2 m high
Mass: 2.85 t re-entry capsule
Usable volume: ~4 m^3
Standing height limit: 1.86 m, 94 cm when seated. Some cosmonauts have exceeded the height limit while still within the seat range.
Hatches: A single 70 cm diameter overhead hatch, opened from either side and two 70 cm diameter side hatches for the primary and reserve parachute systems. The hatch lips extending from the 7° slope wall add to the DM's lift during re-entry and help to flip the hatches away in the airstream. The DM provides occupants with two 20 cm diameter portholes, one either side of the control panels; the discoloured outer pane is ejected after entry to restore vision. It is possible the aluminium airframe has been reworked and lightened.

Soyuz Service Module

Size: 2.3 m long, 2.2 m diameter (flaring to an aft 2.72 m)
Power systems: Twin Si solar wings (spanning 10.6 m).
Thermal Control: ~8 m^2 of thermal radiators.

KTDU35

Fuel/Oxidiser: ~500 kg of UDMH/nitric acid in four 80 cm diameter spheres.
Engine: A single 4.09 kN main engine with back-up of a single start twin-nozzle engine generating 4.03 kN,
AOCS thrusters: 14 with 4 back-up providing 98 N and 8 with 4 back-up providing 9.8 N, supplied by H_2O_2 from spheres in the SM structure below the DM.

Soyuz-T

Fuel/Oxidiser: ~700 kg of UDMH/NTO (150 kg reserved for retrofire)
AOCS thrusters: 14 providing 137 N and 12 providing 24.5 N. Capable of acting as retrofire back-up.

Soyuz-TM

Thrust: 3.1 kN
Fuel Oxidiser: 800 kg supply of UDMH/NTO. 150 kg reserve for rendezvous/docking.

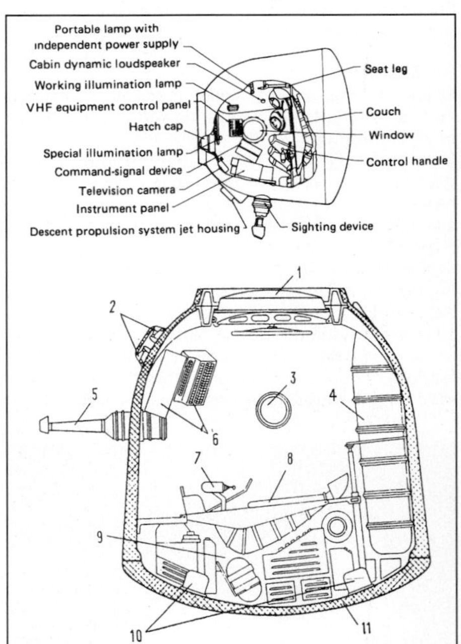

Soyuz descent module. Key: 1 crew hatch, 2 pitch thrusters, 3 porthole (one of two), 4 'chute container (one of two), 5 periscope, 6 control panels, 7 hand controller, 8 couch, 9 life-support equipment, 10 soft-landing thrusters, 11 detachable heatshield
0516699

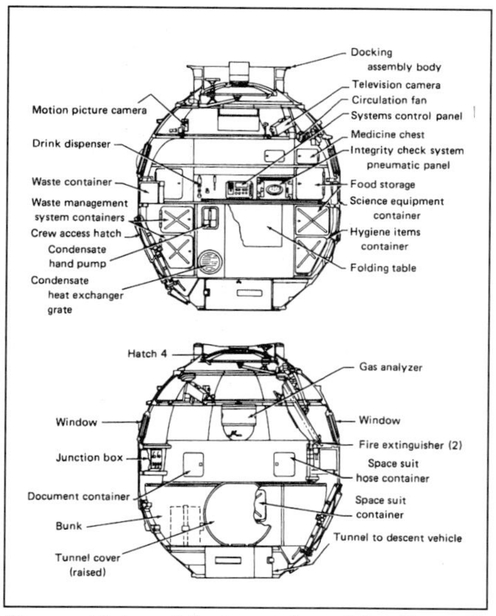

Soyuz orbital module, as configured for Soyuz 19. Little has been seen of the interior during Soyuz-T/TM missions, and no comparable illustrations have been released (NASA) 0516700

Power
Soyuz-T
Type: Twin 4-segment solar wings 1.4 × 4.4 m with batteries in the service module.
Output: ~1.3 kW, with a significant proportion lost in cabling/processing.

Soyuz-TM
Type: Solar cells with a span of 10.6 m. Soyuz locks on to the Sun with a sensor and slowly cartwheels around the Sun axis. DM includes a small battery supply for power after SM separation.

Thermal protection
Soyuz T
Type: Hand-cut asbestos sheets impregnated with ED-5 or ED-20 epoxy resin laid on to the cleaned aluminium structure. The base was made separately and attached mechanically before the whole module was baked at 200°C and then machined to the required dimensions on a large lathe. From 1969, a 1–2 cm thick titanium honeycomb layer injected with an asbestos fibre/binder mixture provided base protection.

Soyuz-TM
Type: 8 m^2 of SM radiators with the rest of the surface area covered by green thermal blanketing.

Descent Module
Type: Eight blankets held by apex and base rings released when the other modules separate.
Comment: Three of these blankets broke free of the base on 1990's TM9 at shroud separation, requiring EVA repairs.

Life support systems
Soyuz-TM
Pressure: Hermetically sealed on the pad under sea level conditions at 184–263 mbar partial O_2 pressure.
Regeneration: Potassium superoxide and LiOH cylinders.
Suits: Sokol-KV2 (Falcon) pressure suits. Wearer has to be minimum of 1.64 m tall.
Contractors: NPO Zvezda is responsible for all Russian life support systems and suits.
Comment: The bulk of the system is housed in the OM but the DM carries a smaller, independent unit under the couches. O_2 bottles provide emergency protection. Waste management and eating is handled entirely in OM; DM offers air sufficient for 48 hours and only food/water packs for landing emergencies.

Avionics/control
Soyuz
Type: Ground controlled by command and/or pre-programmed analogue sequencers activated from switch panels either side of the main display.

Soyuz-T
Type: 16 kbyte capacity Argon computer, capable of completing a fully automatic rendezvous and docking. (Cdrs previously assumed manual control 180 m from the target.) Its Russian acronym was BTsVK, 'onboard numerical calculation complex'.
Attitude: Provided by infra-red Earth horizon plus Sun sensors in conjunction with an improved strapped-down inertial platform. Range/range-rate information comes from radar deployed on a 2.5 m pylon.

Soyuz-TM
Docking type: S-band Kurs ('course') docking system. Allows approach to Mir from any angle without the entire station having to rotate, as under the previous Igla ('needle') system.
Sensors: 15° FoV VSK-4 periscope for final approach. Two external 625-line 25 pictures/s TV cameras. The OM's waist carries VHF TV, radio and telemetry/command aerials, T-shaped telemetry/command antennas encircle its base.
Communication channels: Direct voice communications on 121.75 MHz. Data relay performed by the Rassvet system working via Mir through GEO satellites providing simultaneous dual-voice channels.

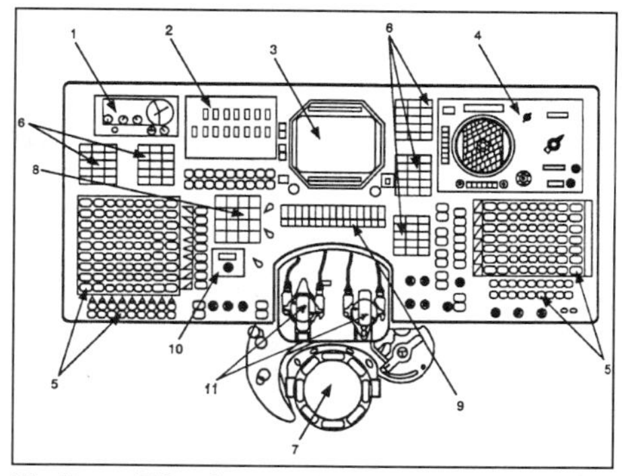

The Soyuz-TM main control panel in Star City's primary TM trainer. The commander's abort column is at bottom left 1 clock, 2 computer keyboard, 3 computer display, 4 Globus Earth position display, 5 control panels and status lights (81 on left, 62 right), 6 systems status lights, 7 periscope screen, 8 ignition panel (protected), 9 keyboard for information selection, 10 propellant supply indicator, 11 oxygen supply valves 0517096

Landing/escape system
Soyuz/SL-4 launcher
Contractor: NPO Iskra's
Type: Tractor escape motor system providing protection for the first 115 s (the fairing's own separation rockets can then separate with Soyuz attached up to 165 s).

Soyuz TM
Type: 1,971 kg solid propellant tower, 6.680 m long, 1.415 m diameter.
Operation: Pyrotechnics separate below the DM and halfway down the shroud for the main 441 to 715.4 kN solid (1,607 kN total impulse) to pull the DM/OM combination free.
Peak acceleration: about 14 *g*. The motor attains 70 per cent thrust in less than 0.07 seconds. The main escape motor has two components, a larger motor with four nozzles at its apex, plus an upper smaller motor with four smaller nozzles at its base. Both fire for aborts up to 20 seconds, but from 20 seconds only the larger one operates. The firing sequence lasts for about 5 seconds and then four shroud panels pivot outwards to slow the ascent. Throughout the escape, a small apex unit with four nozzles curves the vehicle onto its recovery path.
Engines/Motors: 150 kg, 98 to 171.5 kN (98 kN total impulse) 12 nozzle tower separation motor.

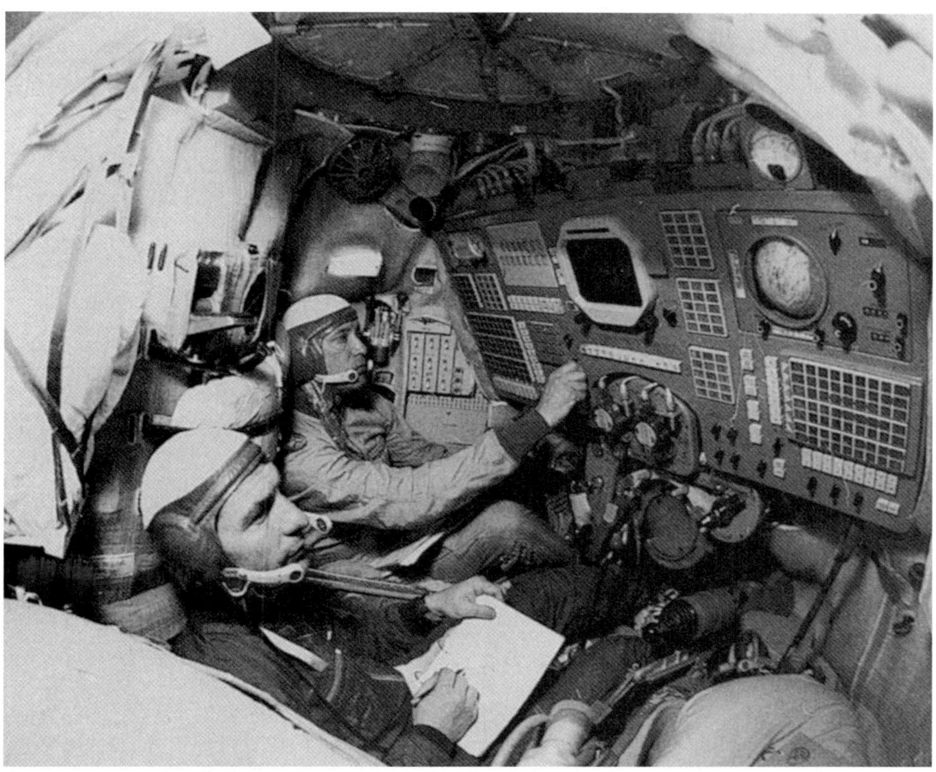

Cosmonauts Malyshev (left) and Aksynov train in Star Town's simulator. The attitude controller can be seen by Malyshev's right knee, the periscope below panel centre, with oxygen controls above, and the Globus Earth position mechanical display at top right. Globus' main function is to show the landing site at any instant if retrofire begins immediately; pushing a button turns the globe from the current orbital position to the landing co-ordinates. Rods are used for pressing buttons difficult to reach in the cramped conditions. The commander holds a flimsy column bearing an abort button during launch. Note the hatch at top and the parachute container bulging between the cosmonauts' positions 0516701

Recent missions
Soyuz-TM 21
Launched 06.11 GMT 14 March 1995 with the eighteenth main crew and NASA astronaut Dr Norman Thagard with a primary interest in life sciences. Commander Vladimir N Dezhurov and engineer Gennadi Strekalov (the first Russian to six launches). This was the 100th space launch of a US astronaut; all planned to return aboard NASA Shuttle Atlantis. The launch created a record (13) for the number of people in space at one time. Docked automatically with Kvant 1 07.46 GMT 16 March at a final rate of 27 cm/s. Thagard began his experiments 17 March with body mass measurements but his full programme had to wait for Spektr's delivery of his equipment. Progress-M 27 docked 21.01 GMT 11 April, bringing Germany's 20 kg 21.5 cm diameter GFZ 1 laser-reflecting geodetic satellite for release 19 April from Mir's core garbage airlock (see German entry for other details). 48 Japanese quail eggs, fertilised just before launch, also arrived. Thagard 'fixed' them at various points in their development, the fourth and last time on 29 April at full term, for return on Atlantis. Dezhurov/Strekalov began 18 April removing Kvant 2's unsuccessful shower unit (unused as a shower for two years) to make space for more gyrodynes, breaking it up to fit into Progress-M 27 for disposal. M27 undocked 23.40 GMT 22 May (leaving the axial port clear for Spektr) and de-orbited 03.17 23 May.

Dezhurov/Strekalov made three EVAs to move Kristall's solar wings before Shuttle's arrival. EVA-1: 12 May, 6 hours 15 minutes, Kvant 2 hatch opened 04.20 GMT; planned to transfer one wing to Kvant 1, but retraction problems. EVA-2: 17 May, 6 hours 52 minutes, exited Kvant 2 02.38. Moved wing to Kvant 1 using Strela, but unable to complete connections in time. EVA-3: 22 May, 5 hours 15 minutes, Kvant 2 hatch open 00.10. Completed wing-1 installation and extension. Wing-2 was partially retracted to give clearance while still helping the power shortage.

Mir's 19.6 tonne Spektr module was launched 03.33 GMT 20 May on Proton, although Mir was

TM3's descent module. A rare view of the capsule's underside. Note the scorch marks from the soft-landing rockets 0516702

not yet ready to receive it. Dezhurov commanded the node's Lyappa arm to start swinging Kristall from its –Y/Earth-pointing position at 23.28 26 May to the –X/axial face, taking about 30 minutes. A 28 May internal EVA in the depressurised node moved the docking cone to the -Z/starboard port; Kristall moved there 30 May. Spektr docked 00.58 GMT 1 June – the first time a station module had arrived at the first attempt. It included 755 kg of NASA equipment for Thagard and his NASA successors. Spektr added 62 m^3 of habitable volume. Its two forward solar wings remained stowed in transit, but one deployed only partially 5 June; a second attempt was unsuccessful 8 June. It was believed that a launch restraint was blocking it, so a July EVA was planned to cut it free using one of two cutters (one Russian and one US) brought up by Atlantis. Another array was to be checked because it had problems tracking the Sun.

Another internal node EVA, 1 June, starting 22.05 GMT moved the cone to –Y for Spektr to swing there 19.52–21.50 GMT 2 June. With Spektr in its final location, Kristall returned to the axial face 10 June to await Atlantis, which docked at 13.00 GMT 29 June 395 km over Lake Baikal, creating a ~223 tonne complex. Dezhurov and Shuttle commander Gibson shook hands some 2 hours later and the record 10-man team collected in Mir's core for photographs. Mir was formally handed over to the fresh cosmonauts. Most of the joint mission was devoted to investigating the long-term physiological effects on the Mir 18 crew in Spacelab. Some 400 saliva, urine and blood samples were returned from Mir/Spacelab. Spacelab's treadmill and Lower Body Negative Pressure unit helped to prepare for Earth return. 200 kg of food, clothing and science equipment and <500 kg of water were delivered, and Mir's pressure was raised from 0.92 atmosphere to 1.02 atmosphere. Shuttle provided most of the attitude control, holding the complex for optimuum power generation during working hours. It would then return to the local vertical, with its belly pointing at Earth, as had been planned before Mir's power shortage occurred.

Kristall's hatch was closed at 19.32 GMT 3 July and Shuttle's at 19.48. Solovyov/Budarin undocked from Kvant 1 in Soyuz-TM 21 at 10.55 4 July to photograph Shuttle's 11.10 undocking and flyaround. However, Mir began drifting when its main computer failed so TM21 redocked 5 minutes early at 11.39. Mir remained drifting until they could replace the hardware. Gibson fired his separation thrusters at 12.35 4 July to end the joint phase. Thagard reported that his greatest challenge had been cultural isolation – sometimes going days without talking to a native English-speaker. He broke the US endurance record of 84 days 1 hour at 04.28 6 June, eventually achieving 115 days 8 hours.

Soyuz-TM 22

Launched 09.00.23 GMT 3 September 1995 as the 20th main crew; commander Yuri R Gidzenko and engineer Sergei Avdeyev, with ESA researcher Thomas Reiter (back-up Christer Fuglesang) for the EuroMir-95 mission. For the first time, a visitor trained (3,500 hr) as a flight engineer, working with Mir systems such as thermal control for 2 h daily and qualified for EVA. Fuglesang remained at TsPK until mid-1996 to qualify for landing a Soyuz for the Space Station era; he could also fly EuroMir-97/98. TM22 docked automatically at Mir's forward node 10.29.54 GMT 5 September, and the hatches were opened at 12.01. Most of ESA's equipment was delivered by Progress-M 28, but number 29 arrived at 20.33 GMT 10 October at Kvant 1 with another 85 kg (including Reiter's EVA cassette). 10 kg went up with TM22. M29's hatch was opened 21.57 GMT.

It became clear in late September that TM23's Soyuz booster would be delayed because of funding problems. It was confirmed with TM22's crew 17 October that they would not return until 29 February 1996, instead of 16 January. The extension had been an option before launch and allowed Reiter a second EVA and increased science return. It meant that Progress-M 30 in December 1995 needed to carry 58 kg of ESA science consumables for the extension (ESA paid US$900,000 for the cargo plus US$700,000 for EVA 2). Before the extension, 41 experiments were planned, mostly in Spektr and Mir's core: 18 life sciences, 10 technology, 8 materials science and 5 astrophysics, totalling 450 h, or 4.5 h daily. Germany's TITUS (Tubular furnace with Integrated Thermal analysis Under Space conditions), developed for EuroMir from CSK-1 experience, was used for solidification experiments (see EuroMir-94 mission for CSK-1 failure). Germany's Oberpfaffenhofen facility provided EuroMir operations control.

ESA's first EVA began 11.50 GMT 20 October when Avdeyev/Reiter opened Kvant 2's hatch with the main goal of installing a dust collection experiment on Spektr's exterior. This European Space Exposure Facility (ESEF) was launched attached to Spektr and Reiter added four boxes (delivered by Progress-M 29) opened by remote control: two to collect dust particles (UK's University of Kent), one for environment monitoring and the fourth for electronics. Reiter and the cassettes rode to the site on the Strela arm, operated by Avdeyev, who then climbed along it. ESEF's four covers were removed and retained as they were experiments themselves, having collected radiation data since launch (they were brought back by STS-74). The pair then exchanged cassettes on the nearby Russian-Swiss Komza experiment, finishing at 15.23 GMT, and returned to Kvant 2, closing the hatch at 17.06 GMT. The dust collectors were soon opened for Leonid micrometeoroids. One remained open for most of the time until it was retrieved 8 February 1996 (see below), closing briefly to avoid contamination from approaching and departing craft; the other was opened only during passage through cometary dust streams.

Kvant 1's Vozdukh CO_2-scrubbing molecular sieve was discovered to be leaking glycol coolant 1 November. Kvant 2 carries another, but Mir otherwise then had to rely on the 25 to 30 day supply of chemical canisters aboard the core. Kvant 1's leak was plugged and tested satisfactorily 3 November. Nevertheless, STS-74 still delivered 20 LiOH canisters and an adaptor to plug them into the foreign system. By coincidence, ESA's Biokin scrubber was tested for a week starting 9 November, operating it in an enclosed bag to see if microbes could clean out a preset level of noxious gases. The membrane bioreactor was then frozen and returned on STS-74 for analysis.

STS-74 made Mir's first radial docking at 06.27.39 GMT 15 November 1995, attaching Energia's 2.2 m diameter 4.7 m long 4,087 kg docking module to Kristall so that future dockings will adequately clear Mir's solar wings. Without it, Kristall would have to be shifted to the forward face each time. The hatches were opened at about 09.00 when the pressures had equalised. Atlantis delivered 275 items totalling 968 kg and took away 195 items (370 kg). Delivered were 400 kg of water in containers,

Few details are available of Soyuz-TM's propulsion system. This mockup at 1987's Paris Salon suggests a single main engine, with protective cover and four peripheral thrusters. The large backup retro of previous designs appears to be absent 0516703

The 8 kg Sokol-KV pressure suit flew for the first time on Soyuz-T 2. It was improved to the KV2 for Soyuz-TM 0516704

65 kg of air, experimental equipment, fresh food (including vegetables, steak and ice cream), gifts of flowers and sweets and a collapsible classical guitar. It departed with 20 kg of ESA cargo from Reiter's activities: frozen medical samples, furnace material samples and science data on videotape and computer disk. Left behind on the DM were two stowed solar wings, which were planned for deployment by EVAs during 1996 on Kvant 1. Atlantis undocked 08.16 GMT 18 November to circle Mir at 130 m for 45 minutes while the crew took detailed engineering photographs.

An internal 'EVA' by Gidzenko/Avdeyev 19.23–19.52 GMT 8 December in Mir's node moved a docking cone to prepare for Priroda's April 1996 arrival. Progress-M29 departed 09.15 19 December, leaving Kvant 1's port free for M30's 16.30 20 December docking. A 182 minute EVA starting 14.03 GMT 8 February 1996 from Kvant 2 with Gidzenko allowed Reiter to retrieve the two ESEF dust collectors and install a third. They began by removing the old manoeuvring backpack, unused since 1990 and cluttering up the airlock, to an outside storage position on Kvant 2. The crew rode back aboard TM22, which undocked 07.20 GMT 29 February, began its 261.6 second 115.2 m/s retro burn 09.47 and landed 105 km northeast of Arkalyk at 10.42.08, only 2 km off target.

Soyuz-TM 23

Launched 12.34.05 GMT 21 February 1996 with the 21st main crew, Commander Yuri Onufrienko and flight engineer Yuri Usachev, docking with the Kvant 1 port 14.20.36 23 February: Progress-M 30 had vacated this port at 07.26.47 22 February. Soyuz-TM 23 had been planned for 25 December 1995, but funding problems and the unavailability of a Soyuz-U2 launch vehicle required an extension of the TM 22 crew's occupation of Mir. The 351 minute EVA, starting 01.04 15 March, installed a new core crane for moving the new Docking Module solar wing packages to Kvant 1.

Third docking of Atlantis to Mir (STS-76 mission) brought NASA astronaut Shannon Lucid to join the main crew until the STS-79 visit in August. Atlantis docked at 02.34 24 March, attaching to the Docking Module delivered during the previous visit of Atlantis. An EVA lasting for 362 minutes by Clifford and Goodwin exiting from the Shuttle's tunnel, began 6.34 27 March (the first US EVA outside Mir) attached the MEEP to Mir to gather space debris and materials data (to be retrieved by an EVA during the STS-86 mission) and also tested common US/Russian tether hooks and foot restraints. Some 640 kg of water was transferred from the Shuttle to Mir, along with 880 kg of Russian supplies (including gyrodynes and batteries), 740 kg US science equipment and 93 kg of other items. More than 60 per cent of the Spacehab's cargo of 2,010 kg comprised supplies for Mir. Of these supplies, 450 kg of material was returned to Earth aboard Atlantis, including 110 kg science payload. Mir's radio and electric field environment was mapped. Atlantis undocked 001.08 29 March.

The final module planned for Mir finally arrived at the station following launch on 23 April. The 19.7-tonne Priroda remote sensing module carrying a tonne of US equipment docked with the forward port of Mir 12.43 26 April and the following day it was relocated to the portside position. The battery-powered module was the first large module not to carry solar panels (the smaller Kvant 1 had not carried its own panels either), although early in its design a single large solar panel had been planned for the module.

Progress-M 31 was launched 07.04 5 May and docked at the front port 08.54 two days later. Cargo freighter carried 1,140 kg of propellant for Mir plus 1,700 kg of other supplies.

On 20 May, Onufrienko and Usachev began a series of EVAs: beginning at 22.50, the cosmonauts exited via the Kvant 2 airlock. They used the telescopic lifting device installed during their earlier EVA to transport a folded solar battery (manufactured in the USA) to outside Kvant 1: the crew also filmed an advertisement for the Pepsi-Cola soft drink company while outside Mir, using a larger-than-life mockup of a newly designed can. The EVA lasted for 320 minutes. Three days later at 20.47 the two men were outside Mir again for 347 minutes, during which they completed the installation of the solar batteries on Kvant 1. At 18.20 on 30 May, the fourth EVA began, during which Onufrienko and Usachev mounted the MOMS-2P multispectral scanner (intended for geophysical studies) outside the Priroda module. They then installed a new handrail outside Priroda, before returning inside Mir after spending 260 minutes outside. The mission's fifth EVA began on 6 June at 16.56 and the cosmonauts used the telescopic crane to move to the Spektr module. They replaced equipment for a Swiss-Russian experiment to measure interstellar gas and then returned to Kvant 2. They installed two US experiments for the detection of micrometeorites and a Russian experiment to measure the effects of the space environment on various materials. The EVA lasted for 214 minutes. The sixth and final EVA of the mission began at 12.45 on 13 June when the prime objective would be to replace a truss on Kvant 1 with a new 5.9 m model. The old truss (named Rapana) would later be attached to the new one, allowing additional experiments to be deployed outside the station. Onufrienko and Usachev also filmed the second part of the soft drink commercial.

On 21 June, it was announced that Onufrienko and Usachev would remain on board Mir until around 30 August, 40 days longer than planned, since the Soyuz-TM 24 mission was being delayed. On 2 and 4 July, two orbital manoeuvres reduced the orbital altitude of Mir by 11 km, to a level at which Soyuz-TM 24 could dock with the station after being launched using a Soyuz-U vehicle. Onufrienko and Usachev continued working with Lucid on Mir: in June they assembled a new Crater-5M furnace in the Kristall module, this being planned for use in producing improved-quality semiconducting materials.

The launch of the Progress-M 32 cargo freighter was cancelled three times in July before it was launched, first on 22 July, then 25 July and finally on 27 July. Launch took place at 20.00 on 31 July and the freighter docked at the front port of Mir at 22.04 2 August. In the meantime, Progress-M 31 had undocked 16.45 on 1 August and had been de-orbited later that day. The 2.4 tonnes of supplies carried to Mir included 400 kg of propellant, 300 kg of food and 340 kg of equipment for the forthcoming French Cassiopeia mission. At 09.34 on 18 August, Progress-M undocked from Mir to free the port for the new Soyuz-TM, but it remained in orbit to redock after the return of Soyuz-TM 23.

TM8's Viktorenko (left) and Serebrov with the MKF-6MA camera in the Kvant 2 trainer at Zvezdny Gorodok, seen through the 1 m EVA hatch 0516706

After the arrival of Soyuz-TM 24 on 19 August, Onufrienko and Usachev worked with the new crew on the French Cassiopeia experiments. Soyuz-TM 23 undocked from Mir at 04.20 2 September with Onufrienko, Usachev and Andre-Deshays (who had been launched aboard Soyuz-TM 24) and landed south west of Akmola at 07.41 the same day.

Soyuz-TM 24 (Cassiopeia)

Russian Space Agency informed CNES that the planned launch of Soyuz-TM 24 with French spationaut Claudie Andre-Deshays (for the Cassiopeia mission), originally planned for June 1996 and then delayed until 6 to 22 July would be further delayed until 14 to 30 August. A further change to the mission was announced on 7 August when the prime Russian crew of Gennady Manakov and Pavel Vinogradov would be grounded because doctors had found minor heart troubles with Manakov. Rather than split the Russian team, RSA decided to ground the prime crew and fly with the back-ups, Valery Korzun and Aleksandr Kaleri. The prime spationaut crew assignment would not be changed. At the same time, it was announced that the Soyuz-TM 24 mission would begin on 17 August.

At 13.18 GMT on 17 August 1996, Soyuz-TM 24 was finally launched using a Soyuz-U vehicle to begin the 22nd residency aboard Mir. The spacecraft was manually docked with Mir at the front port two days later at 14.50.

On board Mir the French Cassiopeia research programme included five main experiments. Cognilab was a laboratory which is dedicated to neuroscience and robotics and was used to study the mechanisms used by the central nervous system and to evaluate the adaptive capabilities of the brain while in a microgravity environment. Fertile was a biology experiment, devoted to the study of

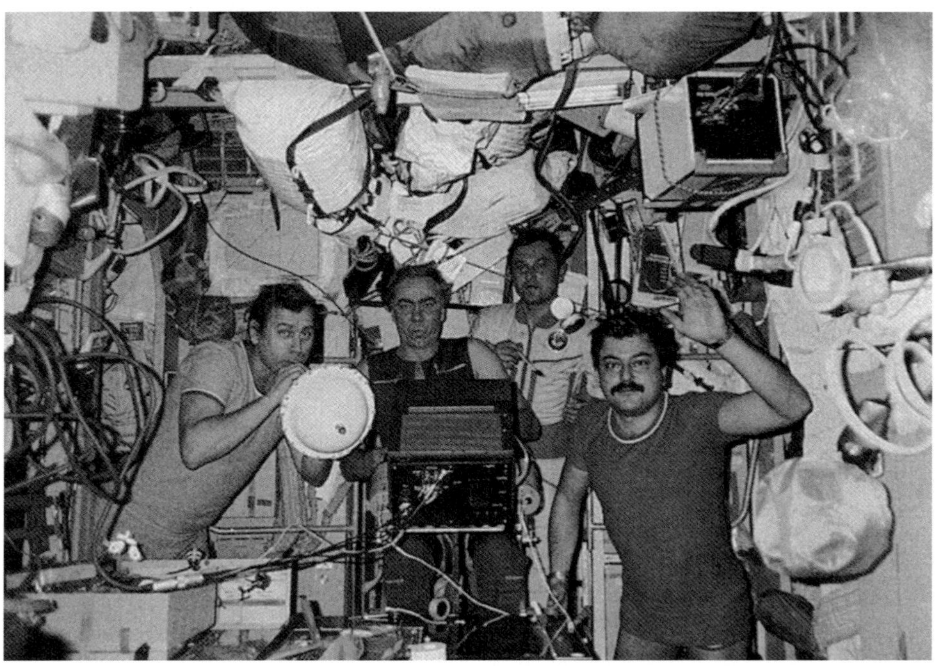

Alexander Alexandrov (left) aboard Mir June 1988 with, from left, Levchenko (on veloergometer), Titov and Manarov. Compare the cluttered appearance with the photograph of the same main living area in the Mir space station section. Titov and Manarov became the first to spend a year in space. Levchenko was a candidate for the first manned shuttle test but died of a brain tumour soon after 0516705

TM8 commander Viktorenko tests the new manoeuvring backpack 5 February 1990. Note Kvant 2's 1 m hatch, the first outward-opening hatch on a Soviet craft. It was stored on Kvant 2's exterior 8 February 1996 to move it out of the way 0516707

the development of embryos of amphibians under microgravity conditions. Castor was a technical experiment, intended to measure the vibrations of the Mir orbital station complex, analyse its modes of deformation and study the dynamic behaviour in the microgravity environment of a trellis equipped with shock absorbing systems. Physiolab was designed to study the cardiovascular system with two planned goals. First, the study of the fundamental mechanisms that regulate blood pressure and second, to be used as a tool for medical monitoring. Finally, ALICE-2 (Analyse des LIquides Critiques dans l'Espace) covered scientific research into the behaviour of fluids in microgravity. The French had paid FFr82 million for the mission.

With the completion of her part of the mission, Andre-Deshays returned to Earth with Onufrienko and Usachev aboard Soyuz-TM 23 on 2 September (see Soyuz-TM 23 above).

On 3 September at 09.35, Progress-M 32 redocked with Mir, this time at the rear port, which had been vacated by Soyuz-TM 23.

The fourth visit of Atlantis to Mir for the STS-79 mission began on 19 September, with the Orbiter docking at 03.13. Atlantis carried more than 2.1 tonnes of supplies to Mir. Shannon Lucid ended her residency aboard Mir and her place was taken by John Blaha. Atlantis undocked from Mir 23:33 23 September and returned to the Kennedy Space Center three days later at 12:13.

Following the departure of Atlantis, the Mir cosmonauts and astronauts settled into the routine of space station operations. They undertook experiments using the astrophysical equipment aboard Kvant 1, the Mariya magnetic spectrometer was used for measurements of high-energy charged particles and Earth photography using the Priroda-5 equipment was undertaken.

The launch of Progress-M 33 had been scheduled for 15 October, but because there was no Soyuz-U launch vehicle available for that date it was delayed until November. Launch finally came at 23.21 on 19 November and docking at the rear Mir port took place at 01.01 22 November. In the meantime, Progress-M 32 had undocked from Mir 19.44 21 November and was de-orbited later that day. Progress-M 33 brought about 2.4 tonnes of supplies to the station.

On 26 November, preparations began for the first EVA during this residency on board Mir, as Korzun and Kaleri checked the EVA suits. On 2 December at 15.54, the Kvant 2 airlock hatch opened and the two cosmonauts began their EVA to install a new solar battery: the EVA lasted for nearly 6 h. A second EVA took place on 9 December, lasting for more than 6 hr (an hour longer than planned). During the EVA, the cosmonauts completed the connections of the new solar battery and installed a new Kurs antenna on the docking Module, intended to ease future dockings of the Atlantis shuttle orbiter.

The New Year of 1997 saw the Mir crew looking forward to the arrival of Atlantis for its fifth visit to the station. Launched 12 January on the STS-81 mission, Atlantis docked with the station at 03.55 15 January. While the Russian and American crews were on board Mir they undertook joint biomedical experiments, assessing adaptation to the space environment and radiation studies. A further crew rotation took place, as Blaha prepared to return to Earth aboard Atlantis and Jerry Linenger prepared for four months in orbit. Atlantis undocked from Mir 02.16 20 January and landed at the Kennedy Space Center two days later at 14.22.

In preparation for the forthcoming Soyuz-TM 25 launch, Progress-M 33 undocked from Mir 12.14 6 February and remained in orbit for a planned redocking after the return of Soyuz-TM 24. At 16.28 on 7 February, Soyuz-TM 24 containing Korzun,

Tokyo Broadcasting Service paid ¥12 million for its journalist, Toyohiro Akiyama, to travel to Mir aboard Soyuz-TM 11 0516708

Kaleri and Linenger undocked from the front port of Mir and redocked at the rear port 23 minutes later.

With the arrival of Soyuz-TM 25 on 12 February, the Soyuz-TM 24 residency drew to a close. Joint work with the new Russian crew and the German researcher Ewald was undertaken before the return to Earth. Soyuz-TM 24 with Korzun, Kaleri and Ewald undocked from Mir 03.25 2 March and landed near Arkelyk at 06.44.

Soyuz-TM 25 (Mir-97)

The launch of Soyuz-TM 25 with a Russian-German crew had originally been planned for 15 December 1996, but delays in the delivery of a Soyuz-U launch vehicle for the flight meant that on 7 November that year, it was announced that the flight would be delayed until February 1997. Soyuz-TM 25 was actually launched 14.09 GMT 10 February 1997 with two Russian cosmonauts, Vasily Tsibilev (Cdr), Aleksandr Lazutkin (flight engineer) and German cosmonaut Reinhold Ewald.

Docking at the front port of Mir came two days later at 15.51. Once more the docking had to be conducted manually after the automatic Kurs system failed. Russia noted when reporting the docking that there would be only one further flight using Kurs, presumably Soyuz-TM 26.

During the handover period, there was a fire on board Mir. While there are more than three people on Mir, the oxygen supply is maintained by the burning of oxygen generators using lithium cartridges: when combusting, the cartridges release oxygen into the station's atmosphere, a process that is accompanied by a great deal of heat. During the evening of 23 February, the fourth and final cartridge of the day was burning. Without warning, the cartridge exploded and some of the panels inside Kvant 1 caught fire. The crew used three fire extinguishers to fight the fire, working in gas masks for 15 minutes. When the fire had been put out, the crew dismantled the air line and retreated to their respective Soyuz-TM spacecraft, while ground controllers considered the emergency evacuation of the station. It was decided that the mission could continue, although the cosmonauts continued to wear breathing masks until 25 February as they conducted repairs. The Russians noted that this was the second time that a lithium cartridge had caught fire: Valery Poliakov had to extinguish a similar fire in October 1994.

Ewald returned to Earth on 2 March aboard Soyuz-TM 24, leaving Tsibilev and Lazutkin in orbit with Linenger. On 4 March, the attempt to redock Progress-M 33 at the rear port of the station failed and the spacecraft had too little propellant available for further docking attempts: the spacecraft was de-orbited on 12 March. On 19 March, the primary angular rate sensor in the Spektr module on Mir failed and 3 minutes later, the motion control system computer switched to a backup. In the interim, the gyrodynes started manoeuvring Mir in all three axes at rates beyond the capacity of the gyrodynes to compensate. The oven switched off the attitude control system relaxing the station into 'free drift' and used thrusters to stabilise it. With Mir in gravity gradient attitude the crew installed a spare Omega attitude sensor in the Kristall module and its sensor served as a backup.

On 22 March, the United States announced that they had achieved a year-long presence in space, thanks to astronauts working on board Mir: however, since Mir had been permanently occupied by Russians since September 1989, the American record paled into insignificance.

As April approached it became clear that Mir was showing its age: launched more than eleven years earlier and expected to remain operational only until around 1992, the core module's life support system was causing serious problems. In November 1996, there had been problems reported with the recycling of water, as 'industrial' water from the waste management system had begun to enter the station's cooling system. The two Elektron oxygen supply systems broke down in March 1997, meaning that the crew had to rely on burning the lithium cartridges to maintain an oxygen supply.

The launch of Progress-M 34 came at 16.04 on 6 April and it docked at the rear port of the station at 17.30 two days later. Among the supplies being taken to the station were spare parts for the repair of the two Elektrons, a fresh supply of lithium cartridges, which could supply oxygen in the meantime, and 50 kg of fresh oxygen. The day following the docking, the Elektron spare parts had been unloaded from Progress-M and the crew began their repairs of the Elektrons. By the end of the month the Elektrons were once more operating.

One of the flight tasks conducted by NASA astronaut Linenger was participation in a year-long sleep study involving three experiments put together between Pittsburgh University, Harvard University and the University of Toronto together with Moscow's Institute of Bio-Medical Problems. The North American work was sponsored by the Canadian Space Agency and co-ordinated between NASA and the Russian Space Agency. The results of the research were said to be important for the clinical understanding of several disorders such as stress and depression which it appeared to the experimenters could be linked to altered immune functions brought about by sleep disorder. The difficulty astronauts face in matching the circadian rhythm cycle of living things on Earth seriously affects their immune systems. In attempts to understand this, Linenger was to be assessed at 81 and 141 days and again when he returned to Earth and the same process would be applied to Russian cosmonauts Lazutkin and Tsibilev as well as their backups, Musabayev and Budarin.

Linenger was one of the first in space to have use of the new Orlan M pressure suit, brought by Progress M 34, which weighs 80 kg and is designed to afford easier access through the rear. Linenger used this suit to conduct an EVA beginning 05:10 UT, 29 April, and ending 4 hours 58 minutes later. During this, Linenger collected the Partial Impact Experiment and the Mir Sample Experiment from their respective locations on the outside of the Kvant-2 module where they had been placed in 1996 and installed the Optical Properties Monitor Experiment on the Docking Module at the end of the Kristall module. Linenger was moved between locations by the Strela telescopic boom and worked with Tsibilev to place the Benton radiation dosimeter on Kvant-2. In the first two weeks of May, the crew tidied Mir and stowed equipment ready for the arrival of Shuttle Atlantis (STS-84) and its seven-person crew including Michael Foale, replacement for Linenger. While the Elektron oxygen generators were working satisfactorily, the 50 kg of oxygen brought up by Progress-M 34 was to be used first. There was still a problem with fumes getting into the station atmosphere.

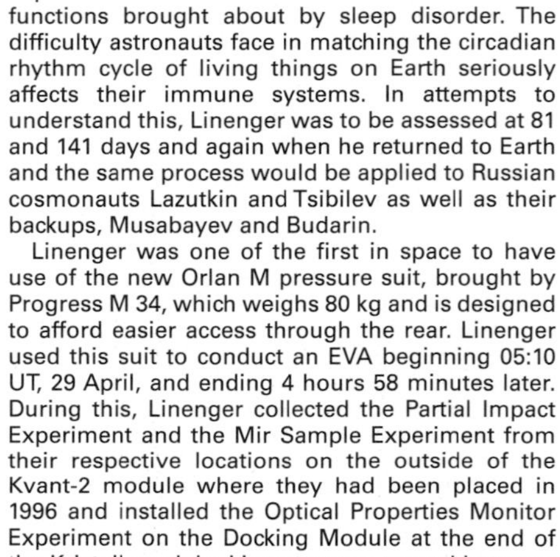

The UK's Helen Sharman aboard Mir with (from left) Afanasyev, Artsebarsky, Manarov and Krikalev 0516709

Soyuz-TM 14 training: test pilot Klaus-Dietrich Flade (left) and physicist Dr Reinhold Ewald (right), with Kazakh candidate Talgat Musabayev (flew 1994 aboard TM19) (DLR) 0516710

Launched on 15 May 1997, Atlantis docked with Mir for the sixth time at 14:33 UT two days later, bringing more than 2,800 kg of food, water and experiments in addition to a new Elektron unit installed as a backup to the main device on Kvant-2. During joint operations involving the 10 Mir-Shuttle crewmembers, the KOB-2 cooling loop was reactivated although a leak in a supplementary cooling loop, the VGK, awaited repair. With Linenger in Atlantis and Foale left aboard Mir, the Shuttle undocked at 01:04 UT 22 May and tested a European laser docking sensor from a distance of about 900 m, returning to Earth two days later. Foale integrated quickly with Lazutkin and Tsibilev aboard Mir and helped track down various coolant leaks and check out experiments before packing redundant equipment in Progress-M 34 prior to its departure from the rear docking port.

M 34 undocked from the Kvant port at 10:22 UT 24 June under a plan to have Progress perform a one-day autonomous flight for a manual re-docking test. Next day, out of radio contact with the ground, Tsibilev attempted to re-dock the Progress using the dual-band controllers of the TORU system and a TV link when the cosmonaut noticed that the vehicle was deviating from its intended course. It missed the docking port, travelled down the side of Kvant and the main Mir station block and struck one of the four solar arrays on the Spektr module causing it to tilt and collide with one of the radiator panels sending a pronounced shock throughout the complex. Physical damage breached the hull causing a pressure leak audibly detectable to the crew, who quickly severed electrical cables across the open hatch to Spektr and sealed the module to secure the leak. Damage to the solar panel cut Mir electrical power by 50 per cent but this was partially restored later that day through manual tracking of the arrays. However, the Elektron in Kvant-2 was shut down and the Vozdukh CO_2 removal system turned off. For the time being, CO_2 levels would be kept to acceptable values by lithium hydroxide canisters brought up by STS-84. Meanwhile, discussions began about entering Spektr wearing space suits to connect electrical cables from the solar arrays to batteries elsewhere on Mir and restore power.

Early on 27 June, the Kvant-2 batteries ran down and thrusters aboard Soyuz TM-25 were used to adjust station attitude and by building battery power back up it was hoped that the gyrodones could be brought back on the following day. Next day with power being slowly restored, the Vozdukh CO_2 scrubber was turned back on and the crew activated the new Elektron unit brought aboard by the STS-84 crew. By early July, plans were made to bring special repair equipment to Mir in Progress-M 35 and for the two Russian cosmonauts aboard Mir to perform the suited inspection of Spektr with access to all other modules sealed and Foale in Soyuz TM-25. Progress-M 34 was de-orbited on 2 July and its successor was launched at 04:11:54 UT on 5 July, carrying 2.4 tonnes of supplies and equipment. M 35 docked at the Kvant port at 05:59:24 UT 7 July, a day in which the Mir crew succeeded in switching on 10 of the 11 operational gyrodones (one had not been working for some time). Most of 8–10 July was spent unloading the Progress cargo-tanker but conditions in the Kristall and Priroda modules was getting worse. The crew used towels to mop up condensation caused by high humidity and Foale abandoned his traditional sleep location in the docking node and moved into Kvant-2.

On 13 July, Tsibilev displayed a cardiac arrhythmia, not unknown among Mir crewmembers (including NASA astronauts) but one which under the stress

Tognini, Kaleri and Viktorenko (from left to right) aboard Mir during Soyuz-TM 15 (CNES) 0516711

of the circumstances caused concern. On 16 July, during lengthy cable disconnection procedures, a line to the Omega attitude control system computer was inadvertently removed causing the station to slowly lose orientation on the Sun and most of the power from the solar arrays. When the crew woke next day to find the situation they manually rotated the Kvant-2 module arrays and gradually re-charged the batteries. Once again the crew had to go through lengthy power-down procedures similar to those following the collision three weeks earlier and the internal inspection of the crippled Spektr module was postponed. On 19 July, the crew were informed that the suited activity would be performed by Soyuz TM-26 cosmonauts Solvyev and Vinogradov, due to replace them in early August.

Soyuz TM-26

Launched at 15:36 UT on 5 August 1997, Solovyov and Vinogradov carried 96.5 kg of extra equipment for Mir, including 40 kg of oxygen along with tools and EVA equipment, instead of French astronaut Leopold Eyharts who would fly to the space station on Soyuz TM-27. Next day, at 11:46 UT, Progress-M 35 was undocked from the aft port and removed to a safe station-keeping distance to make room for the unmanned vehicle. At 17:02 UT on 7 August, Solovyov and Vinogradov steered Soyuz TM-26 to a manual docking at the aft Kvant port. Cosmonauts Tsibilev and Lazutkin returned to Earth in Soyuz TM-25, touching down 168 km from Dzezkagan, Kazakhstan, at 12:17 UT on 14 August, leaving Foale to support Solovyov and Vinogradov in repairs to the Mir complex. Next day, the Mir crew conducted a fly-around inspection of the station exterior from Soyuz TM-26, undocking at 13:30 UT and re-docking at 14:14 UT. Progress-M 35 was re-docked to the aft port at 12:53 UT where it would remain as a refuse bin until shortly before the arrival of Progress-M 36 in October.

The much delayed suited internal inspection of Spektr began with the opening of the hatch at 11:10 UT on 22 August with Foale in the Soyuz. Retrieving a vacuum cleaner, Foale's personal ditty bag and effects and after having video documented the state of the interior and installed the power line patch, Solovyov and Vinogradov returned to the node and at 14:30 UT, closed the hatch on the unpressurised module. With more power flowing, systems were turned on, sequentially over the next several days, in Kristall, Kvant-2 and Prioroda modules respectively. For the next two weeks the station complex was slowly restored to its former power levels and plans were made for a Spektr inspection EVA involving Foale in attempts to locate the rupture which had caused depressurisation after the impact.

The inspection EVA commenced at 01:07 UT on 6 September when Foale opened Kvant-2's airlock hatch. In the early part of the excursion, Foale assisted Solovyov as he was manoeuvred by the Strela crane into position over Spektr. Cutting tools were used to remove thermal blankets to inspect the hull before moving to the damaged solar arrays where they installed handrails to aid later cosmonauts on full-scale repair tasks. Moving them to undamaged Spektr arrays they eased the panels into a new orientation to increase solar exposure and improve the electrical supply. Foale removed the Benton dosimeter left by Linenger before the EVA ended after exactly 6 hours, Solovyov's eleventh space walk.

On 8 September, the crew were busy correcting a major failure with the Mir main command computer, vital for station orientation and maintain the correct attitude alignment for optimum solar energy, but general maintenance was now taking increasing slices of work time. A new fluid unit for the back-up Elektron brought up by STS-84 was installed in Kvant-2, the old one being returned aboard Atlantis after its late-September visit and water purification and elements on the condensate recovery system were replaced. On

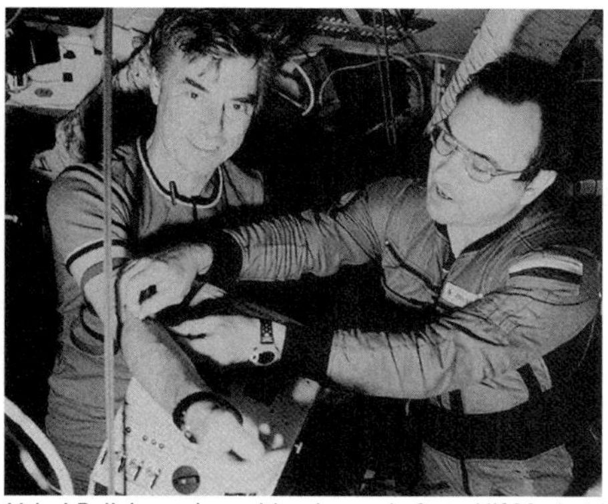

Valeri Poliakov takes a blood sample from Ulf Merbold (ESA) 0517092

14 September, the main computer failed again and power requirements were cut. In an operation similar to that when the computer failed in 1995, the cosmonauts stripped two failed machines to make one good working computer, taking the central data processing unit from the failed unit and the central processing unit from the back-up. Two days later, the solar power was back to normal and the gyrodones were switched back on, one by one. However, the planned launch of Progress-M 36 on 20 September was postponed to allow time for a new computer to be produced and delivered by the cargo-tanker in October. The third computer failure on 22 September and again the gyrodynes stopped and systems were cut until the crew had repaired the faulty item and begun to restore the station once more.

Shuttle Atlantis was launched to Mir at 02:34:19 UT on 26 September at the start of STS-86 during which Russian cosmonaut Titov and NASA astronaut David Wolf would be left at Mir and Foale would be returned to Earth. Docking took place at 19:28 UT on 28 September and over the next two days, the crew, including veteran visitor Jean-Loup Chretien, offloaded 3,160 kg of supplies and equipment to the Mir complex. A late addition to the cargo manifest was a new interim computer for Mir mounted to the aft bulkhead on Spacehab via the Module Vertical Access kit. The computer was installed by Solovyov and Vinogradov while Parazynski and Titov performed a 5 hour 1 minute EVA from Atlantis to retrieve four Mir Environmental Effects Packages and test the Simplified Aid for EVA Rescue (SAFER). They also installed a 121 kg Solar Array Cap to the Docking Module which cosmonauts could use on a future EVA to seal the leak on Spektre. With astronaut Foale on board, Atlantis undocked from Mir at 17:28 UT on 3 October and returned to a landing at 21:55 UT.

At 15:09 UT on 5 October, the Progress-M 36 supply ship was launched toward Mir and, after Progress-M 35 was undocked at 12:04 UT two days later (following some unexpected technical difficulties), a docking was achieved at 17:08 UT on 8 October. Progress had 2,500 kg of equipment including a back-up Motion Control Computer, 1,000 kg of propellant and 100 litres of water which the crew, including Wolf, offloaded to Mir. On 20 October, Solovyov and Vinogradov performed a second suited inspection of Spektr's interior and in an operation lasting 6 hours 38 minutes routed cables to partially restore orientation control for the remaining healthy solar arrays. Two more maintenance EVAs to install a new solar array brought up aboard STS-84 were conducted over the next few weeks.

On 3 November, Solovyov and Vinogradov went outside to remove the old array from Kvant-2 with the aid of the Strela device. While outside they released a small model of Sputnik 1. On trying to return through the hatch, they found it buckled, requiring them to enter through the second Kvant-2 module and use its inner hatch as the pressure seal. In all the EVA lasted in excess of 6 hours. The second EVA 3 days later successfully installed the replacement arrays to Kvant-2 but they had to be manually deployed. After installing 10 clamps around the failed airlock hatch, the 6 hour 17 minute EVA ended.

On 14 November, the Mir complex sustained another major setback when ground controllers failed to command rotation of the Kvant-2 solar arrays to prevent a power surge. This caused the main computer to fail again, once more requiring the crew to switch off all non-essential systems including the Vozhdukh and Elektron systems. To supplement diminished power, the crew transferred charged batteries from the Kristall module to the main core module to restore power to five gyrodynes which spun down after the failure. Following yet another failure on 21 November, the crew began to replace the interim computer brought up in Atlantis with the completely new computer delivered by Progress-M 36 and the recently removed Kristall batteries ensured power levels prevented the gyrodynes spinning down as they had on earlier occasions. Despite frequent problems with systems the crew were able to conduct some science tasks and plans were laid for two space walks in January 1998.

On 17 December at 06:02 UT, Progress-M 36 was undocked at the start of an exercise to deploy a small German Inspector sub-satellite designed as a test precursor to a fly-around TV inspection satellite for use with the International Space Station. Inspector was released from the cargo section of Progress at 07:37 UT but a software problem prevented it responding to commands from a laptop computer on Mir although it did take a few TV shots of the complex. Next day, the crew reported a leak of freon gas from the air conditioning unit and concerns were expressed over the quantity of water from condensation gathering because of temperature differences between the modules and the base block. Progress-M 36 was commanded to destructive re-entry over the Pacific Ocean at 13:20 UT 19 December and at 08:45 UT next morning Progress-M 37 was launched from Baikonur. It docked with the aft Kvant-1 port at 10:22 UT on 22 December.

The first day of 1998 was marked by another main computer failure triggering the now routine power-down procedure, reversed when a new central exchange unit was hooked up. Solovyov and Vinogradov conducted an EVA lasting little

From left: Malenchenko, Musabayev, Viktorenko and Kondakova, during EuroMir 94. At bottom right is the Chibis suit (ESA) 0517093

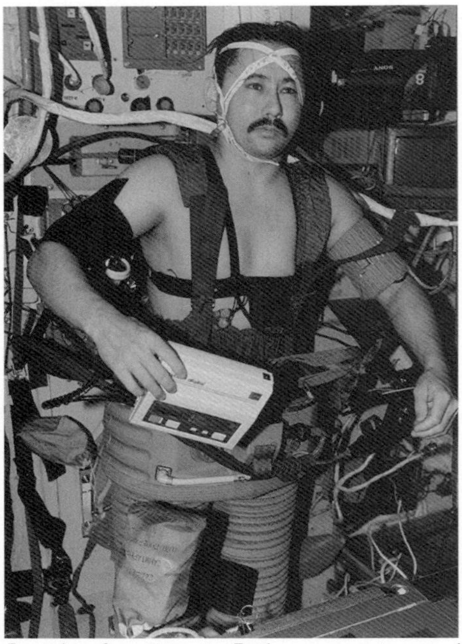

Preparing for return to Earth, cosmonauts spend increasingly long periods in the Chibis lower body negative pressure suit. Shown is TM19's Talgat Musabayev 0517140

more than 4 hours during which they retrieved the OPM experiment and examined the outer airlock hatch before successfully locking all latches. By 12 January it was found to be slowly leaking again. On 14 January, Wolf and Solovyov conducted a 3 hours 52 minutes EVA to use the NASA Space Portable Spectroreflectometer, a device for measuring the effects of the space environment on several materials in different places.

The seventh and penultimate Shuttle-Mir docking began with the launch of STS-89 on 23 January carrying Andrew Thomas, the last NASA astronaut to serve aboard Mir. Endeavour docked with Mir at 20:14 UT next day. More than 3.6 tonnes of equipment and supplies were transferred to Mir over 5 days of joint activities, returning with Wolf after separating from the Docking Module at 16:57 UT 29 January leaving Thomas to await the arrival of a new crew Baikonur. To vacate the aft docking port, Progress-M 37 was undocked at 12:53 UT on 30 January and pulled back to a safe station-keeping distance for re-docking after the next manned visit.

Soyuz TM-27

The 34th manned spacecraft to dock with the Mir space station, Soyuz TM-27 was launched at 16:33:20 UT on 29 January 1998 and mated to the Kvant-1 port at 17:54 UT two days later. The mission was a repeat of the Cassiopeia flight of 1996 with Pegasus experiments occupying the full attentions of CNES astronaut Leopold Eyharts using cosmonauts Talgat Musabayev and Nikolai Budarin as test subjects when time permitted. A dubious 'first' of a questionable nature, Solovyov and Vinogradov appeared live on TV selling space pens through a US cable shopping channel before returning home, with Eyharts, in the Soyuz TM-26 spacecraft.

After undocking from the front port of the Mir complex at 05:52:50 UT on 19 February, the three men landed at 09:10 UT, 30 km south-east of Arkalykh in the middle of a blizzard. Next day Musabayev, Mudarin and Thomas boarded TM-27 and, at 08:47 UT, undocked from the aft port and backed away 60 m while Mir rotated 180° facilitating a manual docking to the forward port at 09:32 UT. Using the Kurs automated system Progress-M 37 re-docked to the aft port at 09:42:27 on 23 February.

In an EVA attempt on 3 March, Musabayev and Budarin were unable to release all 10 manual latches on the troublesome Kvant-2 airlock module hatch because the tool they needed could not be found and the attempt was aborted. Science was interspersed with maintenance duties although on 13 March, a problem with the air conditioning unit cleared up when new parts brought by STS-89 were fitted. Science tasks included operations with the MOMS-2P multispectral scanner on the Priroda module. At 22:45:55 on 14 March, the Progress-M 38 cargo tanker was launched from Baikonur. At 19:16 UT next day, the Progress-M 37 ship was undocked from the aft port allowing its successor to dock at 00:31:17 on 17 March. Musabayev took over manual control with the TORU system after the automatic Kurs unit failed.

On 27 March, the two cosmonauts spent several hours in the Kvant-2 airlock replacing broken latches on the troublesome hatch and freeing up others in preparation for several major EVAs to repair the Spektr solar arrays. They would also search for possible leaks on the module itself and change a VDU thruster module on the outside of the station. The first EVA began at 13:35 on 1 April and lasted 6 hours 40 minutes during which Musabayev and Budarin worked at the base of the array using the Strela access device. A second EVA took place on 6 April but it ended after 4 hours 23 minutes when Mir appeared to ground controllers to be drifting off alignment with the Sun, and jets on the Priroda module had to be used to restore attitude. The third EVA started at 09:55 UT on 11 April when Musabayev and Budarin exited the Kvant-2 module, moved up the Sofora girder to remove the VDU block and pushed it away into space from where its independent orbit would decay within a year. They returned to Kvant-2 after 6 hours 25 minutes but again experienced problems with the hatch latches.

The fourth EVA in this work series started at 07:40 UT on 17 April and in 6 hours 33 minutes set up the Sofora girder for attachment of the new VDU which was raised by ground command from its special mounting on the modified Progress cargo ship. A science support girder, Rapana, which the crew had originally planned to jettison was stowed for possible later use and the VDU was locked in place at an angle to facilitate further work on a later EVA. The fifth EVA got underway at 05:34 UT on 22 April and the new VDU was attached to the Sofora plate with several cables. It took several hours to complete the assembly, 14 m long, with the VDU block on top of the Sofora in its working position. Inside Mir, Thomas activated heaters to prepare the VDU for operations. The 6 hour 21 minute EVA brought the total amount of work time outside the complex during April 1998 to 30 hours 22 minutes.

About this time, space officials began to plan the demise of Mir and with it the historic sequence of Soyuz missions to the orbiting complex that had kept it permanently manned since 1989. NASA wanted the Russians to end the Mir programme to ease the pressure on limited funds for their contribution to the International Space Station (ISS). Efforts to find commercial sponsors began in the hope of maintaining a human presence aboard the station until the ISS could be permanently manned early in the next decade. The final Shuttle visit to Mir – STS-91 – would complete what NASA called Phase 1 of the ISS, but the Russians wanted NASA to help them maintain Mir by using the Shuttle for two additional visits. NASA refused to consider this and the Russians postponed plans to lower Mir in May 1998, essential for a controlled de-orbit to an unpopulated part of the Pacific Ocean in 1999.

Progress-M 39 was launched at 22:13 UT on 14 May carrying 1.5 tonnes of water, food, scientific equipment, instruments and personal items for the crew. Progress-M 38 was undocked from the aft axial port at 18:41 UT next day clearing the way for its successor to dock at 23:51 UT on 16 May. Routine servicing and technical checks followed the transfer of cargo and equipment from Progress but at the end of the month the main core computer once again failed, Mir lost Sun-lock and the manoeuvring engines on Soyuz TM-27 were used to control the complex. This occurred three days before the launch of STS-91, the last Shuttle visit to Mir, coming to deliver supplies and to collect NASA astronaut Andrew Thomas. On board with five NASA astronauts was the veteran cosmonaut and head of the Russian management team on the Shuttle-Mir programme, Valeri Ryumin, who would conduct a full inspection of Mir and report on its potential future use.

STS-91 was launched at 22:06:18 on 2 June when mission managers were assured that the computer problem aboard Mir had been solved and attitude control, an essential prerequisite for docking, was assured. After a nominal rendezvous, Discovery docked with the DM at 16:58 UT two days later carrying 500 kg of water and 2,100 kg of cargo experiments and supplies. The visit was a busy one with removal from Mir and stowage in Discovery of equipment supporting the eight Shuttle visits, which had begun with STS-71 in September 1995. Ryumin performed his inspection and videoed the interior, claiming that the station was undermanned and that it needed at least six cosmonauts on permanent duty to cope with all the work required. Leaving Musabayev and Budarin aboard Mir, Thomas said his farewells and joined Ryumin and the other five NASA astronauts aboard Discovery for the ride home, undocking at 16:01 UT on 8 June, landing at KSC on 12 June.

Three days later, mission control formally announced that the orbit of Mir had been lowered by 6 km to start the process whereby a specially configured Progress vehicle could, at some date in 1999, perform a retrograde burn to destruction in the atmosphere. While the crew set about the busy scientific and engineering tasks aboard Mir, the fate of the complex was discussed at length and resistance to its demise stiffened as potential backers supported the possibility of a commercial purpose to its continued operation in orbit. This even extended to the Menatep bank which had been sending souvenirs up to Mir for some potential commercial gain in the future. Severe funding problems were causing difficulties with mission life planning. After a representative from the Russian Finance Ministry stepped in, the desire to keep Mir aloft until the end of 1999 was summarily cut by six months.

This, plus difficulties with paying wages and utilities bills at the Baikonur cosmodrome in Kazakhstan, delayed the launch of the next Progress and, in turn, the next crew to Mir. Finally, it was agreed that the next Soyuz crew would be launched in August 1998 with the last crew going aboard in February 1999 before their return, and de-orbit of the station, in June 1999. The launch of Progress-M 40 would be delayed by the inability of the space agency to buy a launch vehicle until later in the year.

Soyuz TM-28

In preparation for the next Mir visit, Progress-M 39 was undocked from the aft port at 09:28 UT on 12 August and displaced several kilometres to a safe station-keeping position. On 13 August 1998 at 09:43 UT, the Soyuz TM-28 carrying Padalka, Avdeyev and Baturin was launched from Baikonur

Solovyov (left) and Dezhurov work inside Mir's core (NASA) 0517259

Solovyov and Budarin close Kristall's hatch before STS-71's departure. Note the docking system petals at top/bottom, and the docking target (NASA) 0517260

and only a relatively minor problem with the Kurs docking system marred a smooth link-up, which took place under Padalka's control at 10:56:52 two days later. Once a presidential aide, Yuri Baturin would conduct physics, technology and medical experiments during his relatively brief stay aboard Mir. During the 10 days of changeover, the three cosmonauts transferred several new items including a new furnace for materials processing activities. With five crewmembers, it was possible to perform critical maintenance tasks, especially on the thermal control system and the Vozdukh CO_2 removal equipment. Musabayev, Budarin and Baturin undocked the Soyuz TM-27 from the forward port at 02:05 UT on 25 August and landed 40 km from Dzhezkazgan 3 hours 17 minutes later.

At 16:44 UT 27 August, Padalka and Avdeyev withdrew from Mir in Soyuz TM-28 while the station complex rotated 180° to allow the spacecraft to dock at the forward port 23 minutes later. Finally, two days later, Progress-M 39 was returned to a docking with the aft port at 05:35 UT where it would remain until its successor came up with fresh cargo and equipment. At 20:00 UT 15 September, Padalka and Avdeyev performed a suited inspection inside the depressurised Spektr module and within minutes had successfully reconfigured cables to allow orientation of three operable solar panels on the crippled module; the fourth had been damaged beyond repair.

Plans to send Progress-M 40 to Mir were seriously delayed by funding difficulties and activities aboard Mir were severely curtailed by the absence of the supply ship. Most of October 1998 was spent trading maintenance time with repeating experiments already carried out. Signs of friction with senior space agency managers flared when Padalka and Avdeyev sympathised with striking science and engineering workers in Moscow but declared they would continue working as usual, much of which was devoted to repairs and continued tinkering with station systems.

Progress-M 40 was finally launched at 04:17:57 UT on 25 October carrying the second Znamya 'space-mirror' device similar to the one carried in 1993. The plan was to unfurl the 25 m reflector after Progress had undocked following the transfer of cargo and fresh supplies to Mir. In preparation for its arrival, on the same day at 23:03 UT, Progress-M 39 was undocked from the aft port and the cosmonauts rehearsed the manoeuvres they would apply to M 40 during the Znamya experiment. M 40 docked with the aft port at 05:43:42 on 27 October and two days later, M 39 was de-orbited over the Pacific Ocean.

To retrieve particles from the Leonid meteor shower the two cosmonauts performed an EVA starting at 19:24 UT 10 November, during which they attached sets of equipment to the exterior of the complex including a US-built experimental solar array, a commercial experiment paid for by the development company. One of their first tasks was to release a small miniature of Sputnik 1 built by French students and Energia similar to the one released one year and one week earlier. The EVA ended after 5 hours 54 minutes. With mixed emotions on 20 November, the crew acknowledged the launch of Zarya, Russia's contribution to the International Space Station and the first element sent into space. The ISS would remove all need to maintain Mir in orbit but Energia claimed potential investors were again discussing commercial uses for the station.

On 24 December, station perigee was raised by 14 km to preserve it for the last scheduled visit to Mir, placing it in a 373 × 361 km orbit. Early in January 1999 it was announced that the Soyuz TM-29 flight would be launched on 20 February, the 13th anniversary of the launch of the Mir core block. This last scheduled crew would include the first Slovak, Ivan Bella and the French cosmonaut J-P Hagniere flying as the first non-Russian flight engineer on Soyuz. If investors could be found, the mission would last until 21 August anticipating further visits but if talks failed, the crew was to return on 1 June. Three weeks later, Prime Minister Primakov declared that if investors could be found, Mir would continue in orbit for a further three years.

On 27 January, Progress-M 40 was used again to raise the orbit of Mir and on 2 February at 09:59:32 UT, it undocked from the aft port and backed away 400 m from the station in preparation for the Znamya tests which were to reflect sunlight to specified locations on Earth. At 11:33 UT, commands were sent to deploy the mirror assembly, but as it began to unfurl, a portion of the reflective fabric snagged a Kurs docking antenna which controllers had unwisely commanded to deploy. The material failed to break free despite thruster firings to jerk it away from the antenna and Progress-M 40 was commanded to de-orbit at 10:16 UT next day. At 11:23 UT on 8 February, Soyuz TM-28 carrying Padalka and Avdeyev backed away from Mir as it rotated for a re-docking at the aft port, completed 16 minutes later.

Just three days later, it was announced that commercial investors had gone away and that the Soyuz TM-29 crew would be the last aboard Mir. Now the plan was to use the propulsion system on Progress-M 41 to de-orbit the station on 28 August just five days after its crew returned home, little more than one week short of a record 10 years of continuous human occupation.

Soyuz TM-29

Russian cosmonaut Viktor Afanasyev, Slovakian cosmonaut Ivan Bella and French comsonaut Jean-Pierre Haigneré were launched to the Mir complex at 04:18:01 UT on 20 February 1999. Haigneré was to be part of the long term-crew while Bella would return after a few days. Soyuz TM-29 docked to the forward Mir port at 05:36:16 UT on 22 February and moved into the main complex to be greeted by resident crew Padalka and Avdeyev. Bella carried out medical experiments under the name Stefanik investigating metabolic and cardiovascular reactions to space flight, the weightless development of quail eggs and radiation measurements at various phases. After undocking from the Mir on 28 February the Soyuz TM-28 spacecraft returned to earth with Padalka and Bella, landing 58 km north of Arkalyk in Kazakhstan at 02:14, about 3 hours 22 minutes after undocking. Several EVA and two visits from Progress cargo-tankers were planned during the remaining period of Mir habitation.

On 3 March 1999, Avdeyev and Afanasyev turned the Buket X-ray spectrometer and Maria magnetic spectrometer on the earth's oceans in search of internal wave forms and aligned these observations with tests using equipment brought to Mir as part of the French Perseus earth observation programme from the Priroda module. The Roentgen X-ray equipment in the Kvant module was also used for the first time, beginning a series of observations that would continue for the remainder of the flight. Over the next several days Hagnieré planted seeds in the Svet greenhouse, some of which were from plants grown aboard Mir a few months earlier. Some science experiments were curtailed, reducing the number of EVAs planned for the mission, when budget cuts restricted operational support from the ground. Also, a problem with data relay via the Luch-1 geostationary communications satellite reduced the downlinking of scientific information. There was time, however, for Afanasyev to call up a Welsh school in Penarth to start an internet link with 2,000 schools around the world, an activity that also involved several NASA personnel.

Progress-M 41 was launched from Baikonur at 11:28:43 UT on 2 April 1999 with 2,438 kg of oxygen, propellant, science equipment (including 18 live Salamanders for a French experiment) and spare parts for Mir complex. It docked at Mir's rear docking port at 12:46:19 on 4 April. Further restrictions on communications with the ground came on 12 April when the Mir crew was informed that the Luch-1 satellite had failed and that no contact would be possible unless the complex was over Russian land stations. On 16 April 1999 Afanasyev and Haigneré performed an EVA which started at 04:37 UT and lasted 6 hours 19 minutes. Problems with Afanasyev's Orlan space suit delayed the start of activities but when the crew tried to demonstrate a special sealant which could be used to repair punctures in the hull by applying it to a simulated leak on Kvant 2, the valve on the sealant stuck. Due to delays the planned deployment of a Sprut-6 radiation detector was cancelled but the two cosmonauts did retrieve a gaseous pollution detector and a micrometeoroid detector leaving a French exobiology experiment for retrieval on a later EVA. Their final act was to hand-deploy a small amateur radio satellite, a replica of Sputnik 1.

Beginning in February there were several orbit-raising manoeuvres to preserve the life of Mir. In the first of these the orbit was raised from 347 × 365 km to 348 × 368 km and, in early April, from 335 × 351 km to 337 × 357 km, in late April from 336 × 356 km to 337 × 363 km. Early in May the Mir orbit was raised again in two manoeuvres, the first from 336 × 361 km to 343 × 361 km and the second from 342 × 361 km to 352 × 362 km. In late June two more orbit-raising manoeuvres took place, from 345 × 353 km to 347 × 355 km and from 347 × 354 km to 350 × 355 km. The net effect was to reduce the eccentricity of the orbit and extend the projected lifetime of the complex. This has a direct impact on perceptions about renewed emphasis placed by the Russians on preserving their options for continued operations with Mir. In the United States concern over delays to Russian-built hardware for the International Space Station (see separate entry) raised Congressional anger over limited Russian funds seemingly diverted to maintaining their national icon at the cost of delivering promised elements for the ISS. On 1 June the Russians announced that Afanasyev, Avdeyev and Haigneré would return to earth on 28 August and leave Mir unmanned for about six months when a fresh crew might be launched in February 2000 to prepare the complex for the de-orbit sequence. This would be carried out in conjunction with a special Progress M vehicle equipped with eight propellant tanks for de-orbiting the 125 tonne complex. The prospect of sending another crew rather than the Progress M alone was considered more likely.

On 5 July 1999 the failure of a Proton launch vehicle carrying the first Breeze-M upper stage resulted in wreckage falling to earth near populated area of Kazakhstan. Local authorities banned all flights from Baikonur – operated by the Russians and leased from Kazakhstan – until safeguards were put in place. This prevented the scheduled launch of Progress-M 42 on 14 July, which should have carried supplies and a new control unit to enable Mir to be operated unmanned between August 1999 and February 2000. Although there was sufficient consumables to last the crew through to the middle of August, concern about the survivability of the orbiting complex, and the immediate inability of Russian space officials to set a launch, had its effect on the crew. A psychological difference between the French and Russian cosmonauts on board had already resulted in friction and a daily air leak of 1.5 mm pressure did little to alleviate the tension. On 11 July, however, the crew finally located the leak by isolating each element in turn and checking

Russia's Mir Space Station viewed from the Shuttle Atlantis on the STS-86 mission in 1997 0085776

for pressure drop, finding it to be in the Kvant 2 module in an equipment bay near the EVA hatch.

On 12 July, despite a refusal by the Kzakh authorities to allow launches to proceed, the Progress-M 42 and carrier rocket were rolled to the pad. With only four days to go in the launch window, on 16 July the authorities finally gave permission for the flight to go ahead. At 16:37:33 UT the Progress cargo-tanker lifted off into orbit, arriving to dock with the rear port on Mir at 17:53:23 UT on 18 July. Progress-M 41 had undocked at 17:47 UT on 17 July. The second EVA of the mission began at 11.06 UT on 23 July at the beginning of a sequence of activities scheduled to keep Afanasyev and Haigneré outside for 5 hours 39 minutes. The Strela crane was used to transport the cosmonauts to the Sofora girder assembly to which a folded reflector was attached and electrical cables connected. Built by the Georgian Polytechnic Intellect, the reflector should have unfolded to an elliptical shape with a height of 5.2–6.4 m. Despite strenuous pulling and tugging for 40 minutes the crew could not get the reflector to unfurl and the attempt was abandoned on the orders of ground control. External experiments were retrieved and a hurried return inside was mandated by a sudden rise in ventilator temperature in Afanasyev's suit. The EVA had lasted 6 hours and 7 minutes.

On 25 July 1999 Haigneré spoke by radio with fellow Frenchman Michel Tognini, 12,000 km away aboard the Shuttle Columbia on behalf of the European Space Agency for the release of the Chandra X-ray telescope (see Shuttle Mission STS-93). The third and final EVA of the mission began at 09:37 UT on 28 July and once again the cosmonauts made their way to the Sophia girder by way of the Strela crane. This time they successfully unfurled the reflector, at 11:15 UT, releasing it into space at 12:44 UT to become Georgia's first artificial earth satellite. Further exchanges of experiments attached to the exterior and left in their place occupied the rest of the EVA until the cosmonauts returned inside Mir at an elapsed time of 5 hours 22 min, eight minutes short of the schedule. A minor problem with air pressure in the Kvant 2 hatch seal required the crew to close the pressure equalisation valve and prevent a continuous leak through that element. Over the next several days batteries were changed and science work was carried out but an incorrect command sent to the main computer resulted in a temporary loss of attitude alignment.

The final two weeks of their stay abroad Mir saw Afanasyev, Avdeyev and Haigneré stow equipment, configure the systems for six months of unmanned flight and generally prepare the complex for an inert period. But if the significance of this event, leaving the station unmanned for the first time in almost 10 years, was felt by the Mir leadership there were plenty who were reluctant to mourn the loss of this ageing facility. When asked by cosmonaut Aleksandrov to gather in the control centre for a final farewell to Mir, few of the former crewmembers bothered to turn up. Advising then that by now there was little probability that a replacement crew would be launched in February 2000, controllers kept options open by raising the orbit to 353×359 km. The crew sequentially closed off each module as they progressively worked their way back toward the Soyuz-TM 29 spacecraft and, at 18.15 UT on 27 August, the last hatch was finally closed. Undocking took place at 21:17 UT and two revolutions of the earth later the retro-engine was fired. At 00:34:54 on 28 August 1999 the crew landed 70 km north-east of Arkalyk in Kazakhstan, setting fire to grass in the vicinity. On touchdown the descent module rolled over, the three crewmen suspended from their straps for 15 minutes until it was put to the base-down position. The mission had lasted 188 days 20 hours 16 minutes but Avdeyev had logged 379 days 14 hours 52 minutes since his launch aboard Soyuz-TM 28 on 13 August 1998.

During 1999, intensive efforts were made by politicians in Russia and commercial entrepreneurs in the United States and Europe to resurrect Mir as a viable facility and there were several false starts towards achieving that. Just before the launch of Soyuz-TM 29 a deal involving an unnamed sponsor was said to have fallen through and, in April 1999, a Canadian businessman was rumoured to have put up the necessary funds for a flight to Mir but this came to nothing. Meanwhile, crews for two or three additional flights to Mir were training in case a way was found to keep flying, the first of these being Sergei Zalyotin and Aleksandr Kaleri. In yet another false hope, British businessman Peter Llewelyn was reported to be willing to pay US$100 million for seven days on Mir but Llewelyn claimed he had been offered the opportunity as a publicity stunt. When questions were raised about the validity of his wealth and about former business dealings he left Mosocw and the flight was cancelled.

During the latter months of 1999 a company called MirCorp was formed to handle fund-raising among sponsors in the United States and Europe and substantial funds were raised through offices in the Netherlands. On 17 February 2000 MirCorp made its official debut at a press conference in London and unveiled plans to return Mir to operational condition claiming to have substantial funds which it would use to resume Soyuz operations. MirCorp President Jeffrey Manber, a long standing space advocate from the United States, signed the lease on Mir with DrYuri Semenov, General Designer and President of RSC Energia and MirCorp's Chairman. MirCorp is 60 per cent owned by Energia and 40 per cent by two capital investors. Andrew Eddy began duties as MirCorp's Senior Vice-President for Business Development on 1 March 2000. Dr Semenov claims that Mir has 60 per cent of its operational life remaining and Energia concluded a deal with the Russian government allowing the company to acquire rights over the station for the rest of its orbital activity. MirCorp already has the US$20 million needed for the first round of financing and has said it would go to the capital markets for the US$150 million needed for improvements to Mir and full operations.

As part of the preparations for commercial activity with Mir, initial funding was made available to start preparing the station for a resumption of flight operations with the launch of the first in a new generation of Progress M-1 enhanced cargo-tankers. Called Progress M-1 it was launched at 06:47 UT on 1 February 2000 (see separate entry under Progress) to deliver air, propellant and supplies to the inert Mir. Developed specifically for the International Space Station, the enhanced M-1 version was assigned to Mir. Progress-M 42 was undocked from Mir's Kvant module at 03:11:52 UT on 2 February and deorbited over the Pacific Ocean the same day at 06:10:40 UT with a retro-burn lasting eight minutes. Progress M-1 1 docked with the unoccupied Mir at 08:02:20 UT on 3 February and two days later made the first in a series of propulsive manoeuvres to raise the orbit of the now privately owned space station.

Soyuz-TM 30

To prepare Mir for human habitation Serigei Zalyotin and Aleksandr Kaleri were launched aboard Soyuz-TM 30 at 05:01:29 UT on 4 April 2000, the first commercial manned space flight. Their spacecraft was the first in an evolved 200-series design with improvements and upgrades originally designed in for use with the International Space Station but now brought forward for flights to Mir as well. TM 30 docked with Mir at its forward axial port at 06:31 UT on 6 April. Since August 1999, Mir had been controlled autonomously through the analogue computer brought up by Progress-M 42. The gyrodynes had been turned off at 13:30 UT on 7 September 1999 followed by the main digital control computer at 21:15 UT the same day. Progress M-12 was launched from Baikonur at 20:08 UT on 25 April and Progress M-1 1 undocked from Mir at 16:33 UT the following day, de-orbiting over the Pacific Ocean at 19:27 UT on 26 April. Progress M-12 docked with the rear Kvant port at 21:28 UT on 27 April. On 12 May 2000, cosmonauts Zalyotin and Kaleri conducted an EVA via the airlock on the Kvant-2 module. It began at 10:44 UT and included tests with the Germatizator experiment, a device to dispense a special adhesive to seal cracks and small orifices on the exterior. A failed solar array on the Kvant-1 module was inspected and shown to have a rotor-steering cable burned through due to short circuit. Zalyotin and Kaleri took apart an experimental lightweight solar battery from the main docking compartment. Finally, a panoramic photographic survey was conducted of the entire complex to determine the stations ageing status. The EVA lasted 4 hours, 52 minutes.

During early June the crew prepared to return to earth and hatches were closed at 18:17 UT, 15 June 2000, with undocking at 21:24:23 UTC. Soyuz-TM 30 landed 45 km south-east of Arkalyk at 00:44 UT on 16 June. Progress M 43 was launched by Soyuz from Baikonur at 21:27 UT on 16 October 2000, carrying food and fuel for Mir, taking four days rather than two to rendezvous with the space station thus conserving 150 kg of propellant for raising Mir's orbit.

Soyuz-TM 31

Originally scheduled as the first of two missions to Mir for early 2001, Soyuz-TM31 was launched at 07:53 UT on 31 October 2000, carrying the Expedition 1 crew to the International Space Stations comprising Commander Yuri Gidzenko, flight engineer 1 Sergey Krikalov and flight engineer 2 Bill Shepherd of NASA. Earlier, Progress M-13 had been launched from Baikonur at 18:27 UT on 6 August 2000 and docked at the near Zvezda port at 20:13 on 8 August. It undocked at 04:05 UT, 1 November 2000, and was de-orbited over the Pacific Ocean at 07:05 UT, thus vacating the Zvezda docking port to which Soyuz TM-31 was mated at 09:21 UT, 2 November. Once on board the ISS, NASA astronaut Shepherd became the mission commander. Soyuz-TM 31 returned to earth with Musabayev, Baturin and Tito (launched in TM 32 on

28 April 2001) on 6 May 2001, the spacecraft having been in space for 186.9 days.

Soyuz-TM 32
Designated ISS flight 2S by NASA and EP-1 by Energia, the mission was designed to deliver three crewmembers and provide a new station lifeboat. The spacrcraft was launched at 07.37 UT on 28 April 2001 carrying Commander Talgat Musabayev, Flight Engineer Yuri Baturin and space tourist Dennis Tito. The spacecraft docked at the -Z port on the Zarya module at 07:58 UT two days after launch. The three crewmembers returned to earth in Soyuz-TM 31 in orbit since 31 October 2000 after removing their couches from TM 32 and placing them in the homebound spacecraft. Musabayev, Baturin and Tito returned to earth after a flight lasting 7 days, 42 hours 4 minutes. Soyuz-TM 32 returned to earth 31 October 2001 after 185.9 days in space carrying returning Expedition 2 crewmembers Afanaseyev, Haignere and Kozeyev to a safe landing.

Soyuz-TM 33
Launched as an ISS lifeboat carrying two Russian cosmonauts and one French cosmonaut TM 33 lifted off atop a Soyuz 11A511U at 08:59 UT on 21 October 2001 from the Baikonur cosmodrome, Kazakhstan. Commander Viktor Afanaseyev, Flight Engineer Konstantin Kozeyev and Claudie Haignere docked with the ISS at 10:00 UT on 23 October and supported eight days of activity with the ISS crewmembers before returning to earth in TM 32 on 31 October 2001. Three crewmembers from TM 34, Gidzenko, Vittori and South African space tourist Mark Shuttleworth returned in TM 33 which has been in space 195.8 days, landing on 5 May 2002.

Soyuz-TM 34
Launched at 06:26 UT on 25 April 2002 by Soyuz 11A51U from Baikonur, TM 34 carried Russian cosmonauts Gidzenko and Vittori and South African space tourist Mark Shuttleworth. The flight had been delayed from 10 April. It docked with the ISS Zarya module at 07:55 UT on 27 April. This short-duration ISS visiting crew transferred their seats into the Soyuz-TM 33 spacecraft which had been docked to the station since 23 October 2001 and returned to earth in TM 33 on 5 May 2002. TM 34 brought cosmonauts Zalyotin, De Winne and Lonchakov (previously launched on Soyuz TMA-1) back to earth on 10 November 2002, the spacecraft mission having lasted 198 days 17 hours 37 minutes 45 seconds.

Soyuz TMA-1
Essentially a modified Soyuz-TM configured to satisfy requirements for an ISS lifeboat and to agreed specifications approved by the ISS partners. Development followed a 1 June 1996 decision to adopt the modified Soyuz as an interim ISS lifeboat pending availability of NASA's X-38 Crew Return Vehicle. In this plan, two Soyuz TMA spacecraft were to be docked to the ISS at all times providing the full ISS crew complement with immediate abort and return capability. When the Bush Administration came to power in January 2001 the X-38 CRV was cancelled along with the NASA Habitation Module, which would have increased ISS crew size to the 5 or 6 originally specified in the Space Station plan. With reduced manning levels, the ISS would require only one Soyuz TMA lifeboat. Incorporates provision for taller crewmembers, has two new amber display units and improved parachute recovery system. Manufacturers designation is 7K-STMA, TMA-1 being designated factory number 11F732. Because of an investigation into the failure of a Soyuz launch vehicle on 15 October 2002, TMA-1 was delayed eight days before finally lifting off on a Soyuz FG with cosmonauts Sergei Zalyotin and Yuri Lonchakov and ESA Belgian astronaut DeWinne on 30 October 2002. It docked with the ISS two days later and remained at the station to serve as an emergency lifeboat and scheduled return vehicle until it brought back the ISS long-duration crew in May 2003. After transferring their crew couches to the return craft Zalyotin, Lonchakov and DeWinne returned to earth on 10 November 2002 aboard the Soyuz-TM 34 spacecraft which had been in orbit since 25 April 2002 (198.7 days). Soyuz TMA-1 returned to earth with Bowersox, Budarin and Pettit landing at 02:04:25 UT on 4 May 2003. TMA-1 mission duration totalled 185 days 22 hours 53 minutes 14 seconds.

Soyuz TMA-2
Carrying the Russian commander Yuri I. Malenchenko and NASA flight engineer Edward T. Lu, the second modified Soyuz TMA spacecraft was launched from Baikonur at 03:53:52 UT on 26 April 2003 with a mass of 7,136 kg. Its mission objective (ISS-6S) was to deliver the replacement crew that would tend ISS systems as part of the caretaker operation implemented on the grounding of the NASA Shuttle fleet necessitating a halt to ISS assembly until Shuttle vehicles are cleared for a return to flight. The hatches between TMA-2 and the ISS were opened at 07:27 UT on 28 April allowing ingress to the station. The TMA-2 spacecraft was left at the ISS while Budarin, Bowersox and Pettit returned to earth in Soyuz TMA-1 on 3 May 2003. Soyuz TMA-3 was launched 18 October 2003 and docked to the ISS two days later carrying Kaleri, Foale and Duque. After seven days of joint activity and handover, Malenchenko and Lu took Duque into Soyuz TMA-2, docked to the ISS since 28 April, and closed the hatches at 20:14 UT, 27 October, leaving Kaleri and Foale to keep the ISS in working order and to monitor a limited number of science experiments. TMA-2 undocked from the Pirs module at 23:17:09 UT. At 01:46:51 UT, 28 October the retrofire burn sent the spacecraft toward the atmosphere and at 02:40:20 UT the re-entry module landed 42 km south of Arkalyk. TMA-2 mission duration totalled 184 days 22 hours 46 minutes 28 seconds.

Soyuz TMA-3
The Soyuz TMA-3 spacecraft was launched from Baikonur cosmodrome at 05:38:03 on 18 October 2003 carrying Russian cosmonaut Aleksandr Y. Kaleri, NASA astronaut Michael C. Foale as the Expedition 8 contingency caretaker crew pending the return to flight of NASA's grounded Shuttle fleet. Also on board was Spanish (ESA) astronaut Pedro Francisco Duque previously flown into space on Shuttle mission STS-95 in 1998. Flying mission ISS-7S, TMA-3 was inserted to a 193 km × 227 km orbit and two days later docking took place with the Pirs module at 07:15:58 UT on 20 October. The hatches between the spacecraft and the ISS were opened little more than three hours later allowing the TMA-3 crew to join Malenchenko and Lu.

UPDATED

United States

Apollo

Current Status
Potential contractors and NASA engineers are reviewing the technology developed for the Apollo spacecraft to determine the feasibility of resurrecting some of the design principles behind the ballistic recoverable spacecraft for application with the Crew Exploration Vehicle, the successor to NASA's Shuttle in the new exploration initiative announced January 2004. Use of Apollo recovered Lunar samples continues in laboratories around the world with increasing emphasis on using this unique material to unravel the early evolution of the earth-moon bi-planetary system.

Use of the laser reflectors deposited by Apollos 11/14/15 in 1969–72 continues for measurement of the Earth-Moon distance to within 2.5 cm. Laser beams are routinely fired through optical telescopes sited at McDonald Observatory in Texas and near Grasse in southern France at the 100 fused silica half cubes of each reflector. The research has identified that the Moon is receding at 3.8 cm/year, has measured the combined Earth-Moon mass within one part in 100 million, determined the character of small scale variations in lunar rotation and identified variations of 0.001 s/year in the length of a terrestrial day. Lunar ranging has also yielded a large improvement in knowledge of the Moon's orbit and has more recently been utilised for plate tectonics migration rate analysis as an analytical tool in earth sciences.

Background
NASA protects and scientifically investigates Apollo's lunar samples in NASA Johnson's Lunar Sample Building (number 31N), built in 1979 specifically to provide permanent storage for the moon rocks. In all, Apollo astronauts returned with a total of 381.7 kg, comprising 2,196 individual specimens from the Moon. NASA scientists and other researchers have processed the samples into more than 97,000 individually catalogued samples.

Four of the 54 4 cm diameter 40 cm long core tube sections collected at 27 localities remain to be opened and analysed. As of 27 October 1995, specimens were widely distributed. JSC held 298 kg, Brooks AFB remote vault 52 kg, 18.8 kg was out with researchers, on display, being used for educational purposes or given as gifts, while 13.2 kg had been consumed during research. Only 33 g ended up classified as 'lost'.

Apollo's lunar samples are stored and processed in NASA Johnson's Lunar Sample Building. They are handled in cabinets filled with high-purity nitrogen (NASA) 0517114

Two vaults store the samples: the 'pristine' vault houses those that have never been outside the facility (270 kg as of October 1995) and a second accommodates those that have been returned (27.6 kg by October 1995) by researchers. More than 60 laboratories worldwide actively pursue sample studies and some 1,100 samples are sent out to researchers annually. They are prepared in stainless steel environmental cabinets purged by high-purity nitrogen that is continuously monitored to keep oxygen and moisture contents within 6 ppm. Samples that are not consumed in analysis are returned and can be recycled to other users, but are stored separately from the pristine samples. The facility also includes rooms to support sample examinations and experiments by visiting scientists. The building features steel-lined storage vaults with thick reinforced concrete walls and stands above projected storm-surge sea level heights. Bank-type vault doors remain closed except for sample transfer from storage to work areas. All pipes and openings into the vault close automatically in response to fires or attempted intrusions. Other parts of the building, including separate laboratories where samples are processed, are also heavily alarmed. All materials used in constructing and equipping the building, including floor coverings, walls, plumbing, light fixtures and paint, were screened to exclude chemical elements posing sample contamination threats (particularly lead and gold).

UPDATED

Space Shuttle

Current Status
On 1 February 2003 NASA's first Shuttle orbiter specifically built from the outset for operational space missions was destroyed on re-entry when its thermal protection systems failed to defend its aluminium structure from the searing heat of kinetic friction during atmospheric descent. Carrying a crew of seven returning from a 16 day solo science mission that began with launch on 16 January 2003, Columbia was conducting its 28th orbital flight in a series that began with the first launch of a Shuttle on 12 April 1981. For two years Columbia (OV-102) was the only Shuttle orbiter flying and had conducted five flights before Challenger (OV-099) joined it in April 1983. Challenger had been converted to operational flight status from its original role as a structural test article originally built as a ground vehicle only. Columbia completed the Shuttle flight test programme with its fourth flight in June 1982, conducted the first in-orbit satellite launch in November 1982 and carried the first European built Spacelab science laboratory into space during November 1983. After a period of modifications conducted after Challenger began flights operations in 1983, Columbia had a large number of items removed or changed, including the ejection seats, fully adapting the orbiter from a test and development role to a fully fledged

Typical Space Shuttle launch. Note the 'white room' (centre bottom), which provides crew access and the only pad escape route (NASA) 0517287

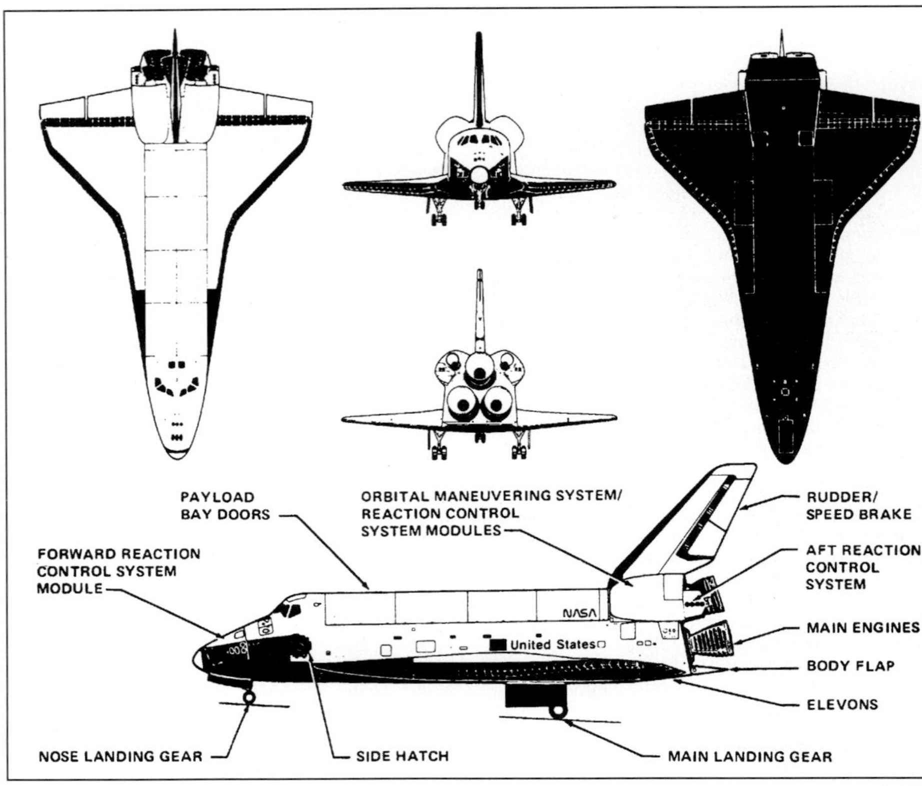

Diagram of Space Shuttle 0516785

operational spaceship. In 1984 Discovery joined the fleet followed by Atlantis in 1985, completing the approved inventory of four flight orbiters with funds for a fifth flight vehicle not approved by Congress.

On returning to flight in early January 1986 Columbia performed the last Shuttle mission before Challenger was destroyed during launch at the end of that month. Columbia returned to space in August 1989 when it launched a DoD mission and in 1990 recovered the Long Duration Exposure Facility from orbit. Columbia continued to support science missions and Spacelab flights, launching the Chandra X-ray observatory in July 1999 before stand-down for extensive modifications and refurbishment. As the first flight orbiter, Columbia carried a lot of electrical wiring and data lines unnecessary for operational flights. Although much had been removed during its earlier refurbishment, much wiring built in for test and operational monitoring equipment used only on the first few flights remained in the vehicle. Columbia returned to flight in March 2002 supporting the fourth Hubble Space Telescope servicing mission (Servicing Mission 3B) before preparing for its science mission, STS-107, during which it was destroyed. Before the catastrophic STS-107 flight, Columbia had completed the equivalent of 284 days 19 hour 19 minutes 1 second in flight, more than 40 days more than Discovery, 65 days more than Atlantis and 78 days more than Endeavour (Challenger had totalled more than 62 days when it was destroyed).

Investigation into the cause of the disaster is the responsibility of an independent Columbia Accident Investigation Board (CAIB) chaired by Admiral (retired) Harold Gehman. NASA has conducted a parallel, internal, investigation. The CAIB issued its preliminary findings in August 2003 but the long-term implications are potentially quite important for the future of the Shuttle programme and for the International Space Station which is ultimately dependent on the Shuttle for continued assembly and operation. The future of the Shuttle has long been debated by NASA and Congress and the development money for a successor has had to wait for a reduction in ISS funding peaks, a reciprocal role for that played by the ISS waiting for the decline from peak Shuttle development funding. Prospects for an early replacement evaporated when ISS funding crises and the failure of concepts such as the X-33 necessitated upgrades and improvements to the Shuttle fleet. Various future launch vehicle studies embracing expendable and reusable concepts have come and gone and NASA focused those requirements into the Integrated Launch System, which was to develop new technologies and synthesise a wide range of operational roles for the post-2012 period. Congress has already warned NASA that it expects to take a leading role in determining not only the future of the Shuttle but of the entire concept of human space flight and what it should be shaped towards in the years ahead. The Shuttle is not expected to return to flight before the second quarter of 2005 although predictions are uncertain as changes mandated by the CAIB have yet to be implemented. In a pre-planned announcement made from NASA Headquarters President Bush charged NASA to retire the Shuttle by 2010 and redirect efforts toward a new human exploration initiative resulting in a return to the moon by 2015.

Background

Shuttle manifest changes

A significant shift in the Shuttle flight manifest was forced upon NASA by a presidential decision following the loss of orbiter Challenger in January 1986, not to fly commercial satellites on future missions.

Columbia approaches touchdown at the conclusion of STS-73 0517288

The recovery programme cost the United States an estimated US$2.4 billion, with an additional US$1.8 billion spent on the replacement vehicle Endeavour. Rockwell delivered the Orbiter in 1991 for its 1992 debut. NASA currently aims to maintain an annual flight rate of seven Shuttle launches, spending no more than 88 days processing each Orbiter after return from orbit. In this 88 day objective, NASA budgets 23 days on the pad.

In order to make operations run more smoothly, NASA removed all commercial satellites from Shuttle manifests, leaving only a backlog of science and defence missions and beginning in November 1998, Space Station assembly and utilisation launches. STS-53 in December 1992 entered history as the last dedicated military mission. The Department of Defense has dramatically reduced its Shuttle requirements and transferred most of its launches to new Atlas, Titan and Delta launchers. One of the last DoD entries on the Shuttle mission docket was the September 1999 Shuttle Radar Topography Mission, added in July 1996 after struggling for years to find a slot.

While the Shuttle was able to maintain most of its target launch dates (apart from minor delays generally due to the weather), NASA has announced repeated delays in the first missions dedicated to the International Space Station. NASA tried to blame these delays on Russian contributions to the Station, but later NASA's own contractors admitted they were behind schedule, as well, and over budget. Initially, NASA delayed Shuttle missions starting with STS-88 by about eight months. After the May 1998 release of a study that showed that NASA had provided unrealistic estimates of the Station's cost to the American public, all Shuttle missions related to the Space Station's construction were rescheduled.

When conceived in the mid-1960s, the Space Transportation System was to be fully recoverable, but the concept was abandoned because of its high development cost in favour of the present system. This involves the Solid Rocket Boosters parachuting into the sea for recovery and the External Tank being irrecoverably jettisoned. This compromise, it was intended, would still enable the system to reduce overall programme costs by 5/6 compared with the then available expendable launch vehicles, which were to be phased out. This was not achieved because manpower requirements and turnaround times were not reduced to the expected levels.

Shuttle components

Shuttle Orbiters

The Orbiters are about the size of a DC-9 airliner. Long-term initial planning was based on production of five, each capable of at least 100 flights. Before being named, what was intended to be the structural test article Orbiter Vehicle was designated STA-099 and the five flight-standard vehicles OV-101 to OV-105. Shortly before the Approach & Landing Tests (ALT) began in 1977, NASA intended to name OV-101 Constitution, but

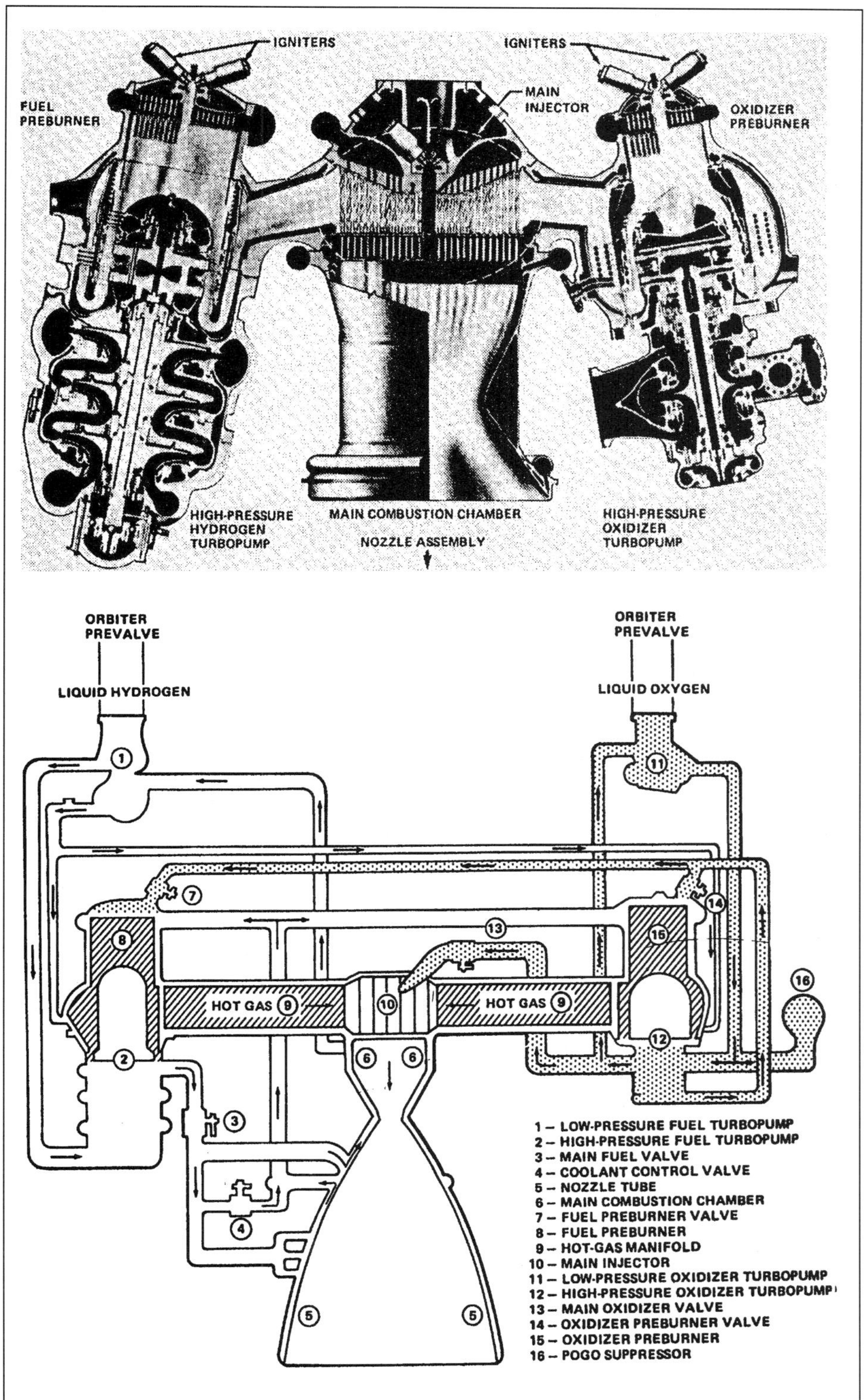

SSME: the main engines operate at a LOX/LH$_2$ mixture ratio of 6:1, generating a sea level thrust of 1,670 kN. Thrust can be throttled over 65 to 109 per cent, allowing thrust reduction to limit maximum dynamic pressure and ensure that astronauts and equipment endure no more than 3 g. Following Challenger, the upper limit was established at 104 per cent to reduce engine wear, although 109 per cent would be used in the event of a trans-Atlantic abort. Improvements will allow Space Station assembly missions currently beyond Shuttle's capability to fly at 106 per cent. Total engine testing exceeds 650,000 s
0516754

a letter campaign to President Ford produced a change to Enterprise after the starship in the 'Star Trek' TV series. However, it was decided shortly afterwards to upgrade STA-099 as the second flight vehicle, replacing it with Enterprise as the ground test vehicle after the ALT programme. The first four operational Orbiters were then named after pioneer sailing ships. Columbia (OV-102) is named after one of the first navy frigates to circumnavigate the world (it was also the name of the Apollo 11 Command Module). Challenger (OV-099) is named after the ship that explored the Atlantic/Pacific 1872–76 (also the name of the Apollo 17 Lunar Module). Discovery (OV-103) was named after two ships, the vessel that discovered Hudson's Bay in 1610–11 and the one in which Captain Cook discovered the Hawaiian Islands, and Atlantis (OV-104), after the ketch that performed oceanographic research 1930–66. Challenger's OV-105 replacement was named in May 1989 as Endeavour after Cook's first command; it was also used by Apollo 15's Command Module.

About 210 significant hardware changes were made to each Orbiter between missions 51L & STS-26, in addition to 100 software and 35 SSME changes. The system now contains some 800 Criticality 1 items – that is, equipment without back-up and creating significant mission impact should they fail. Improvements included RCS rewiring so that thrusters cannot fail into the 'on' position, replacement of the limited-life APUs (the retrofitted models offer 75 h lives), and rerouteing of fuel cell waste water outlets so that freezing cannot disable the whole power production system. The main axles of the landing gear were thickened and the capacity for nosewheel steering, introduced on Columbia for mission 61C, was added to the other two vehicles. Higher temperature thermal protection was added in the wing elevon cove areas, along with RCC protection between the nose cap and landing gear door. Columbia's low-T thermal protection along its mid-fuselage, payload doors and tail was replaced by Advanced Flexible Reusable Surface Insulation.

OV-103/104 now have strengthened wings after analyses showed that the weight reduction programme during construction left them vulnerable to flight loads higher than predicted. Closure of the flapper valves on the 43 cm External Tank propellant lines during SSME burn was a major concern before Challenger's loss because of the fatal consequences, but latching mechanisms were added to ensure they cannot close while liquid is passing through to the engines.

Enterprise (OV-101)

Used for the Approach & Landing Tests (ALT) of 1977–78, Enterprise subsequently served as a ground test vehicle, ending at Vandenberg in 1985. It was used there for preflight test matings on the SLC-6 complex, later closed. It now belongs to the Smithsonian's National Air & Space Museum at Washington DC's Dulles Airport. In June 1987, NASA borrowed 101 for arrester net tests at Dulles. The need for such nets at emergency landing sites was under consideration before the Challenger accident and particularly at Dakar (Senegal) and the Pacific Ocean runways on Easter Island and Hoa. The net is based on those already in use at military airfields.

Columbia (OV-102)

After 27 successful flights into space, Columbia was destroyed during re-entry on 1 February 2003.

As the heaviest flying Orbiter, Columbia's utilisation potential was enhanced by the increase in permitted landing weights. It was modified for longer duration missions of nine to twelve days, instead of the normal five to seven days, and flew most of the Spacelab missions. It was taken out of service after STS-40, in June 1991 for about a year, for conversion in Rockwell's Orbiter final assembly facility as the first Extended Duration Orbiter to accommodate missions of up to 16 days (plus two days' contingency); dry mass is 82,267 kg. Additionally, improved APUs, GPCs, carbon brakes and a drag parachute were fitted and a major structural inspection recertified the airframe for a further three years. Obsolete development instrumentation was removed and provision made for later installation of the new Collins' TACAN landing navigation system. Its third inspection/recertification took place October 1994–April 1995.

As the first Orbiter in space, Columbia had made six flights by STS-9 in November 1983, had flown 18 astronauts and launched two commercial satellites. Columbia returned to service for STS-61C in January 1986 after 18 months' major modifications to bring it to an operational configuration. It should have returned to Palmdale for removal of its ejection seats and overhead escape windows after STS-5 in late 1982, but DoD insisted that two Orbiters should be available in case of loss, so the work was delayed until after STS-9/Spacelab 1. When it was returned to Palmdale, the planned 24 modifications became >240. Because of the greater dry mass, monitoring instrumentation and extended mission facilities such as additional cryogen storage, Columbia's maximum payload capacity for a 204 km, 28.5° orbit was 3,800 kg less than that of OV-103/104. Work included structural strengthening, to enable it to withstand higher aerodynamic loading, wiring changes and avionics updating and a reinforced carbon-carbon nose cap, incorporating pressure sensor inlets as part of the programme to collect data on dynamic pressure during re-entries. A pod was also installed on the vertical stabiliser to obtain detailed infra-red (IR) imagery of upper surface heating of the port wing and fuselage during re-entry. A plan to take Columbia to Vandenberg for Shuttle handling tests before the facilities were shut down, was abandoned in October 1986. NASA estimated that each of Columbia's first six flights cost US$300 million.

As of end 2002, OV-102 had flown 27 missions totalling 284 days, 19 hours, 19 minutes and 01 seconds.

Challenger (OV-099)

The second Orbiter to fly, in April 1983. So much was learned during construction that it was upgraded from the Structural Test Article and replaced by Enterprise in that role. By November 1985, it had made nine flights, flown 45 astronauts and launched five commercial satellites. Its milestones included the first 7/8-member crews, the first EVA (including the first untethered), and the first capture, repair and redeployment of a disabled satellite. Following STS-41G in 1984, 13,984 thermal tiles had to be replaced in an emergency operation involving 200 Rockwell tile specialists being sent to KSC, after 2,500 new tiles had been manufactured at Palmdale. Challenger was lost in the launch explosion of 28 January 1986;

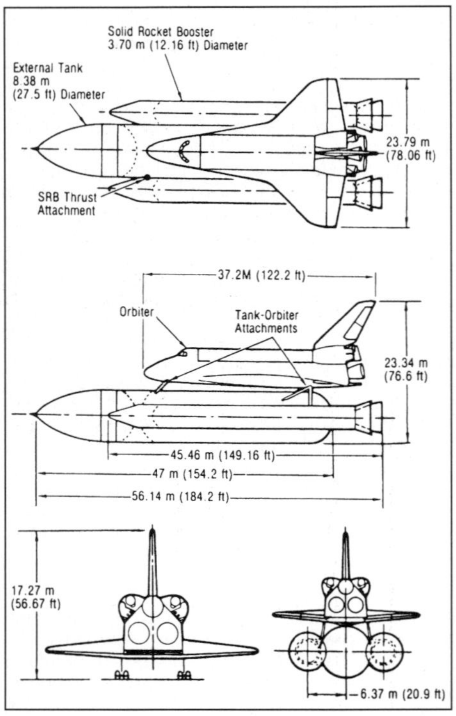

Space Shuttle 0516786

the debris is sealed in two Minuteman silos at Cape Canaveral. Average cost of each of its eight flights was estimated at US$290 million. OV-099 performed 10 flights totalling 62 days, 7 hours, 56 minutes and 22 seconds.

Discovery (OV-103)
The third Orbiter to fly, OV-103 undertook STS-26, the first mission of the new post-Challenger series. It debuted in August 1984 and by November 1985, it had made six flights, flown 35 astronauts and launched 13 commercial satellites. Its firsts included the first capture/return to Earth of two disabled satellites, the first orbital test of a laser tracking system, the first trial of a 32 m solar array and the first classified military flight. Before the Challenger accident it was to have been based permanently at Vandenberg from its seventh flight for polar operations, but its delivery from Kennedy Space Center had been delayed by problems with the new lightweight SRBs, followed by delays with completion of the Vandenberg facilities. Between the accident and the cancellation of Shuttle/Centaur, Discovery was modified, at a cost of US$5 million, to enable it to carry the Centaur payloads in place of Challenger. Average cost of each of Discovery's first six flights was estimated at US$170 million. Capacity of the 78,700 kg inert-mass Orbiter into a 204 km, 28.5° orbit is 24,950 kg. Discovery was the first to use carbon-carbon brakes, on STS-31 during 1990. It was removed from service after STS-42 in January 1992 for extensive modification (the 78 major changes included addition of a drag' chute and nosewheel steering) and inspection at Kennedy Space Center, returning for STS-53 in December 1992. It returned to Palmdale following the July 1995 STS-70 mission for its regular three year major service, this lasting from September 1995 to June 1996. During the servicing a complete structural inspection was made of the orbiter, which found no major problems.

More than 750 kg of performance-enhancement weight savings were achieved to reach the requirements for International Space Station missions, the first of which was then planned to be STS-94 in November 1998 (this mission has now been cancelled). AFT ballast was added, as well as a fifth cryogenic tank which set provisions for additional on-orbit capabilities, additional nitrogen supply, installation of lightweight seats and a reinforcement of the crew module's floor under seats 1, 2 and 5 to meet the requirements for a 20 *g* crash overload. Modifications in support of the International Space Station and Shuttle/Mir missions included air-cooling provisions for mid-deck locker payloads, installation of a UHF space station communications system, orbiter provisions for ISS assembly power conversion units, removal of the internal airlock and the installation of a new airlock which provides a Mir docking capability. Design improvements included a redesign of the main propulsion system gaseous hydrogen flow control valves and a wing doubler modification. Four next-generation toughened single-piece fibrous insulation tiles were installed on Discovery's side hatch. Discovery returned to flight for the STS-82 second Hubble Space Telescope servicing mission in February 1997.

The 1995–1996 Discovery servicing marked the first use of a new vehicle automated power-on checkout system at the Palmdale facility. This demonstrated programme savings which reduced the number of data storage tapes from 1,060 to 10, cutting the number of checkout support staff from 21 to 3 and cutting the start-up time from four hours to five minutes.

As of end 2002, OV-103 had flown 30 missions totalling 241 days, 23 hours, 19 minutes and 54 seconds.

Atlantis (OV-104)
OV-104 performed STS-27, the second mission of the post-Challenger series, reportedly deploying the new Lacrosse imaging radar reconnaissance satellite. Its first mission, the classified STS-51J for DoD in September 1985, was claimed to be the most trouble-free of all Shuttle flights until then. It had been used for only one more, STS-61B in November 1985, before the 2 year shutdown. The scheduled December 1984 delivery to Kennedy Space Center was delayed until 13 April 1985 for modifications to enable it to carry the Centaur upper stage in the payload bay. These modifications included extra plumbing to load/vent cryogenic propellants and controls in the aft flight station for loading, sensing and monitoring the cryogenics. When delivered, Atlantis' inert weight was 76,658 kg – slightly above Discovery's, which had not undergone wing spar modifications. Atlantis was protected with 21,801 high-temperature reusable insulation tiles and 1,977 thermal protection tiles. Final assembly began at Palmdale in 1983, with experience of earlier Orbiters enabling it to be completed with a 49.5 per cent reduction in man-hours. In October 1986, it was rolled out to Pad 39B and fully stacked with ET/SRBs for tests of a new weather protection system, emergency evacuation and slide wire tests. Following the post-Challenger modifications, its dry mass was 78,255 kg, with a 24,950 kg capacity into a 204 km, 28.5° orbit; dry mass is now 78,372 kg. Atlantis was the first to fly the upgraded GPCs, on STS-37 in 1991. It was returned to Palmdale following return from STS-46 in August 1992 for conversion as the second EDO (although it has no scheduled EDO missions) and modifications to allow it to carry an RKK Energia docking unit for the flight to Mir in June 1995. It returned to KSC 29 May 1994.

As of end 2002, OV-104 had flown 26 missions, totalling 219 days, 21 hours, 27 minutes and 17 seconds.

Endeavour (OV-105)
Work began 1 August 1987, at Rockwell's Palmdale and Downey facilities, on construction of the Challenger replacement. Final assembly was performed in Palmdale Building 150. NASA awarded a US$1.3 billion contract to Rockwell (total cost about US$1.8 billion), which at peak production employed 2,000. OV-105 was rolled out, on schedule, 25 April 1991 and formally handed over to NASA. It arrived at Kennedy Space Center 6 May minus its main engines, forward RCS module and OMS pods. The 20 seconds on-pad Flight Readiness Firing (FRF) standard for all new Orbiters was made 6 April 1992. While the FRF was considered successful, all three of the new engines were replaced because of minor irregularities. Rockwell completed mating of the major structural elements, apart from the OMS pods, in January 1991. Onboard systems were powered up for the first time 6 July 1990. Endeavour made its debut as an Extended Duration Orbiter in March 1995 for STS-67. If NASA ever decides on a 28-day Orbiter, then OV-104/105 will provide the capacity.

While duplicating the original OV-103/104 configurations in most respects, post-Challenger modifications included the new GPCs with higher-density memories. They also included quick disconnect valves from ET to Orbiter; carbon brakes

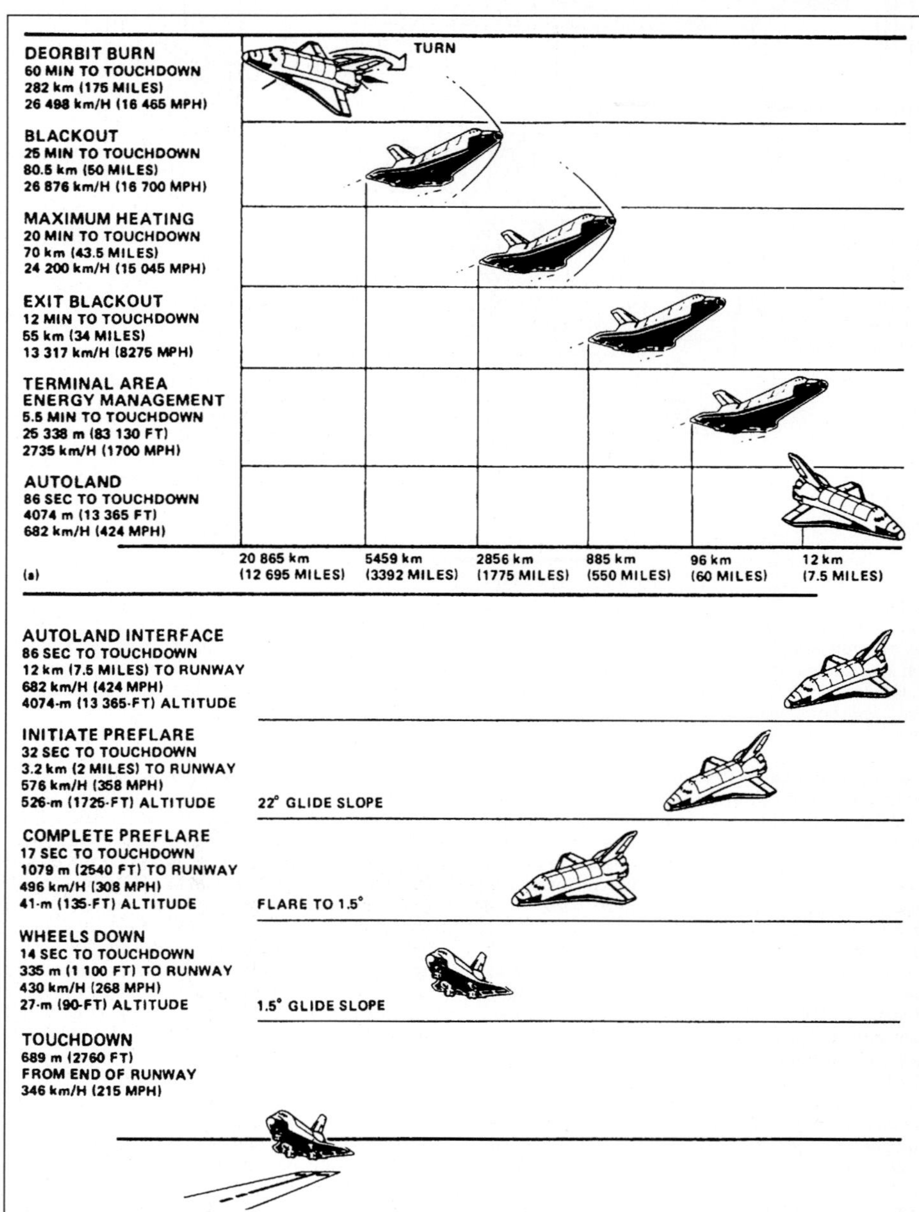

Space Shuttle re-entry and landing sequence 0516755

View of the original Flight Deck layout, showing the 1,800 switches, push buttons, circuit breakers and other controls. Compare with the 'glass cockpit' Multifunction Electronic Display Subsystem (right), due to have been retrofitted 1997–98, which makes use of eleven 17 × 17cm flat LCD colour screens. See Rockwell's Shuttle Orbiter entry in the Launchers section for further details (Honeywell) 0516756

to resolve the overheating problems experienced on most landings; modified RCS valves to prevent them sticking open and operating continuously or dumping fuel overboard; and crew escape hatch. A manually deployed 12.2 m diameter nylon/Kevlar drag parachute on a 26 m line is carried at the tail base to shorten rollout by 760 m and cut nosewheel contact speed by 37 km/h to 260 km/h, reducing tyre/brake wear. It also improves handling characteristics, particularly in crosswinds and on wet runways. The system has been retrofitted to the other Orbiters. Eight 'chute trials were conducted July to October 1990 at the Dryden Flight Research Center by NASA's B52, employed the previous April to launch OSC's orbital Pegasus vehicle. OV-105 tested the 'chute on its six initial missions under a US$33 million Rockwell contract.

OV-105 carries the same thermal tile patterns as 103/104 but employs a new type of gap filler: Ceramic Ames Gap Filler (CAGF), developed by Rockwell. The older rubber-impregnated quartz fabric leaves a whitish residue on the tiles as it burns away. CAGF avoids that build-up and can undertake at least 30 missions. It will gradually appear on the other Orbiters as attrition occurs.

Five assemblies (mid-fuselage, body flap, wing gloves, vertical stabiliser and main landing gear doors) were already available, having been built as operational spares under a 1983 NASA contract. Structural assemblies remaining to be built included wings, forward/aft fuselage sections, crew module, payload bay doors and the forward RCS module. Endeavour's waste management system is based on the Space Station design. Previous Orbiters carry nine computer-controlled 23 × 56 cm vent doors along each fuselage side to manage pressure variations during launch/landing, but assemblies 4/7 were eliminated on OV-105 because mission experience showed the original design to be conservative. Following the STS-77 mission in May 1996, Endeavour was returned to Palmdale for its first servicing since being introduced to flights in 1992. OV-105 returned to flight with the STS-89 mission to the Mir space station in January 1998 and flew the first US International Space Station mission (STS-88) in December 1998.

As of end February 2002, OV-105 had flown 19 missions, totalling 206 days, 14 hours, 12 minutes and 17 seconds.

OV-106

The Augustine committee report of December 1990 was against a sixth Orbiter, preferring the funding for a heavy-lift unmanned vehicle. President Bush signed a domestic launch policy in Summer 1991 that rejected any new Orbiters. As a follow-on to Endeavour's August 1987 contract, NASA and Rockwell began discussions on replacing the spares used for 105. Although for use primarily as spares to replace possible damage on existing Orbiters, they could also be used for construction of a new Orbiter. Rockwell was awarded a 4.5-year, US$375 million contract in late 1989. The contract included US$35 million for FY90, covering fabrication of major structural spares: aft fuselage, crew compartment, lower/upper forward fuselage, mid-fuselage, wings (including elevons), vertical stabiliser (including rudder and speed brake), body flap, forward RCS system and one set of OMS pods. The spares were to be of the same configuration as Endeavour. However, NASA decided, in January 1994, to cancel the work on cost grounds. In Summer 1989, NASA was planning to include, in its FY91 budget request, funding for two further Orbiters, partly to support demands on the fleet from Space Station assembly and operations (the funding was not requested). It was then considered for the FY92 request, aiming for 1997 delivery, but again was not included.

External Tank (ET)

The largest Shuttle component (46.88 m high, 8.40 m diameter), the ET dominates the cluster and carries typically 730 tonnes of propellant for the Orbiter's three main engines. It remains the only part of the stack not recovered. As its 29,930 kg dry mass equates with that of the Shuttle's original theoretical maximum payload, there are continuing proposals for it to be taken into orbit. It is jettisoned shortly before orbital speed to ensure a controlled re-entry. Releasing it at the most economic moment would leave it tumbling in low orbit, to re-enter within days.

The ET provides the cluster's backbone at lift-off, absorbing the SRB/SSME thrust. It incorporates three units, the LOX tank forward, containing 625,850 kg at −183°C, the intertank, with instrumentation, and aft, the LH_2 tank, containing 104,308 kg at −253°C. Total loaded mass, which varies for each mission, is about 760 tonnes. External insulation is provided by a 2.5 cm multilayered polyurethane foam coating to withstand the extreme thermal differential. At launch, the propellants are pressure-fed at a combined rate of 3,900 litres/s (1,407 kg/s) through 43.2 cm diameter feedlines to the three SSMEs. The tanks are self-pressurised: 3.0 atmosphere LH_2; 1.43 atmosphere LOX. The ET is jettisoned after 8.5 minutes, at about 114 km altitude. Venting residual LOX was used to begin a slow tumble, preventing skipping off the atmosphere and promoting break-up over a remote ocean area. The valve has been omitted since STS-65 as the tank tumbled anyway when venting was deliberately not used on STS-31. A 'beanie cap' on the launch tower was added, to siphon off vented oxygen from the apex during countdown, to prevent ice formation and possible launch damage to the thermal tiles. It was first fully implemented for STS-2. The ET carries typically 770 kg of residual propellants at separation, intended as a safeguard against premature depletion before main engine cut-off. Early exhaustion of LH_2 would be particularly serious as running the SSMEs on an oxygen-rich mixture would cause erosion.

NASA decided, in 1994, to proceed with development of an SLWT Super LightWeight Tank version, awarding Lockheed Martin Manned Space Systems US$172.5 million for the work. Replacing 2219 aluminium with the Weldalite 2195 aluminium–lithium alloy, changing to an orthogrid design and refining the foam insulation spraying technique will save 3,600 kg. The alloy is 30 per cent stronger and 5 per cent less dense than the current material and comprises 1 per cent Li, 4 per cent Cu, 0.4 per cent Ag, 0.4 per cent Mg and 94.2 per cent Al. The saving trades directly to payload capacity, necessary for Space Station missions. A 12.2 m long full-diameter version, essentially a segment of the hydrogen tank with a LOX tank dome at one end, was delivered to NASA Marshall 1 February 1996 for 6 months' dynamic and pressure testing to verify the structural design. A one-piece graphite phenolic nose cap that does not require the spray-on foam is already under development for first delivery and flight in 1997. Studies showed it was possible for the insulation to break off and impact an Orbiter window, with fatal consequences. The new design also eliminates 961 fasteners, requiring only about 50. It was used for the first time on the STS-91 flight launched in June 1998.

Computer-generated image of the 'glass cockpit' Multifunction Electronic Display Subsystem incorporating nine 17 × 17 cm flat LCD colour screens, due to have been retrofitted 1997–98 (the aft panels providing another two). They can also show video. Other functions can be added because of the increased computing Power: one candidate is landing simulation to refresh pilot training during longer missions. See Rockwell's Shuttle Orbiter entry in the Launchers section for further details (Honeywell) 0516757

Endeavour's cockpit, minus seats (Rockwell) 0516758

The major post-Challenger modification involved the installation of a mechanical latch on the Orbiter side of the 43 cm feed lines to prevent the fatal closing of either flapper valve before separation. These 43 cm disconnect mechanisms were replaced by simplified 35.6 cm versions in 1993, with flapper valves folded against the walls (instead of opening sideways). This cleaner design, using 51 per cent fewer parts, permits the same propellant flow; the rest of the 43 cm plumbing remains. NASA also plans to redesign the systems tunnel. The projected cost of an ET rose from US$1.8 million in 1971 to US$10.1 million 10 years later. A September 1983 contract for 26 ETs plus material for another 21 (following the existing contract for 15 ETs) was worth US$505 million. ETs are manufactured by Lockheed Martin Manned Space Systems at NASA's Michoud Assembly Facility in New Orleans for NASA Marshall and shipped by ocean-going barge to Kennedy Space Center. After the 1986 accident, production of the 59 contracted tanks was halted with seven available, with a 1985 contract for a further 60 held in abeyance. The US$1,797 million contract was finally awarded in August 1989; the first was completed in late 1991 and the 60th is planned for 2000.

Solid Rocket Boosters (SRB)

The SRBs produce their thrust rapidly upon ignition and burnout in about 130 seconds, having provided the main thrust to lift the Shuttle cluster up to about 45 km. Thiokol is the motor prime contractor, selected in 1974. Other contractors include McDonnell Douglas for the structures and USBI for SRB checkout, assembly, launch and refurbishment (USBI's contract was most recently extended by two years through September 1999. Each booster generates a sea level thrust of 14,678 kN at launch and together they provide 71.4 per cent of the thrust at lift-off and during first stage ascent. After separation, they coast upwards to about 65.9 km. They then fall in a ballistic trajectory for four minutes; at 4,700 m, a barometric switch causes the 3.5 m drogue first and then three 41.4 m parachutes to deploy; the SRBs splashdown at about 95 km/h for recovery in the Atlantic about 227 km downrange. Each SRB is 45.46 m long, 3.7 m diameter and (from STS-6) weighs about 589,670 kg, of which about 503,950 kg is propellant – a mixture of ammonium perchlorate (oxidiser, 69.6 per cent by mass), aluminium (fuel, 16 per cent), iron oxide (a catalyst, 0.4 per cent), a polymer binder and an epoxy curing agent (1.96 per cent). It is configured to provide high thrust at ignition, reducing by about one-third after 50 seconds to avoid overstressing the vehicle during maximum dynamic pressure. The SRBs are interchangeable matched pairs, each comprising four motor segments. SRB ignition can occur only when the ground crew manually removes a lock pin from the Safe & Arm device. The computer launch sequence starts the three SSMEs in sequence (each within 1/4 second), at −6.6 second; only when they have all attained 90 per cent, rising to 104 per cent, will the four onboard computers command SRB ignition. Two Orbiter-type APUs provide hydraulic power from T-28 seconds until separation for TVC. Each APU is fuelled by 10 kg of hydrazine.

On 1 February 1987, a year after the Challenger accident, NASA reported that the PDR of the Redesigned Solid Rocket Motor (called the Reusable SRM since 1995) was complete; CDR followed in October 1987. The field joints now incorporate a tang capture feature to reduce joint deflection, a third O-ring, changes to the design of the joint insulation, sealing without putty and external heaters. The heaters maintain seal temperature at greater than or equal to 24°C in all weather conditions and a weather band prevents

The escape system retrofitted on each Orbiter consists of a telescopic pole extended from the side hatch. The astronauts use a lanyard to slide along it to clear the Orbiter's wings and tail. This picture was taken from a C-141 aircraft as a Navy parachutist demonstrated the technique 0516759

water penetration. The nozzle/casing joint, which showed evidence of erosion on some flights, was similarly redesigned. Five full-scale RSRM firings were made, culminating in the PVM-1 (Product Verification Motor) test at Thiokol's Brigham City facility on 18 August 1988. At the time of Challenger's loss, 11 motors remained to be used. Thiokol fired these over November 1988 to late 1991 to empty them for conversion to the current design; the firings also generated further engineering data. NASA annnounced July 1991 a US$2.6 billion Buy 3 contract extension to Thiokol for 67 further flight pairs to 1999/STS-105, plus eight ground test motors.

Proposals by contractors for a second-generation SRM design led to the award, in August 1987 by NASA Marshall of study contracts to five companies for an Advanced SRM, providing the Shuttle with payload increases of up to 5,440 kg. This almost makes the Shuttle deliver as much payload as NASA originally promised, being close to the original 29.5 tonnes (65,000 lb). Capacity was to be 27.67 tonnes into a 280 km orbit and raised Space Station 28.5° orbit capability from 20.9 tonnes to 26.3 tonnes, essential for station assembly. Thrust profile was designed to greatly reduce Orbiter main engine throttling during maximum dynamic pressure, removing some 175 SSME and APU critical failure modes associated with the potential inability to throttle up. Lockheed Missile Systems Division, teamed with Aerojet Space Booster Co, was selected April 1989 to develop ASRM. An associated 3½-year US$550 million contract required construction of a NASA production facility at Yellow Creek in Iuka, Mississippi capable of handling 32 motors annually. Programme cost was estimated at US$3.9 billion, but Congress cancelled it in October 1993 after about US$2 billion had been expended. Cost of cancellation is estimated at US$240 million; and Lockheed/Aerojet handed over the Yellow Creek facility to NASA in December 1994. Thiokol then planned to transfer its RSRM nozzle manufacturing there, but was directed 2 May 1995, by NASA, to discontinue work.

Studies have also been conducted into liquid boosters for future Shuttle and other launch vehicle use. They would increase Shuttle payload capability by 5.5 to 9 tonnes and add an abort capability during early ascent, but would require US$5 billion and eight years to develop, plus US$500 million for pad modifications. NASA's initial studies ended in Spring 1989, with General Dynamics and Martin Marietta completing their studies in December 1989.

The intention was to order five SRB pairs, at about US$50 million per pair, for each Orbiter and reuse them 20 times. However, one pair was lost on STS-4 as a result of excessive water impact speeds, leading to enlargement of the main parachutes from the original 35 m. A second pair was lost with Challenger on STS-51L. After recovery, each is broken down into 14 major components for refurbishing, with the sections being reused separately. Nine is the record held to date for reuse. Total mass is being steadily reduced – for example, reducing wall thickness in some areas produced a 1.8 tonne saving. A proposal was also made to use the SRBs as sounding rockets on each mission, since each could carry 90 kg of instruments without loss of performance. A US$46 million programme was planned (but cancelled when Orbiter weight reduction proved cheaper) to reduce the weight of each SRB by 2,720 kg, adding 540 kg capacity into the planned 51.6° Space Station orbit. A further 1,025 kg capacity would be created if the 'chutes and various recovery structural items were omitted, saving 5,125 kg per booster.

Space Shuttle Main Engines (SSME)

The SSMEs have been undergoing upgrades since 1983 under a 10-year US$1 billion improvement programme. Improved turbopumps were introduced for STS-26. In the 18 months before that mission, SSME firing trials accumulated the equivalent of more than 40 missions. On 15 November 1991, total testing and flight time passed 500,000 seconds, the equivalent of 320 missions; it exceeded 600,000 seconds in December 1993. New Pratt & Whitney high-pressure turbopumps, allowing 7.5 hr of firing (55 mission equivalents) between major overhauls are replacing Rocketdyne's versions. The older pumps require refurbishment after 33 minutes use (three flights). Problems increased development cost to US$504 million, so NASA halted fuel pump work from December 1991 in order to focus on the more critical oxygen unit for the Block 1 upgrade.

It restarted work May 1994 for 1997 debut as part of the Block 2 upgrade once the LOX pump was on schedule for its July 1995 flight debut. The Block 1 upgrade includes a single coil heat exchanger (eliminating seven weld joints) and a two-duct hot gas manifold, which allows the fuel pump to run at lower temperatures and pressures. A single Block 1 upgrade engine (number 2036) flew successfully on STS-70 in July 1995. All three engines were upgraded for STS-77 in May 1996. Testing of the Block 2 upgrade began at NASA Stennis on number 0521, once Block 1 certification was completed 26 May 1995. Block 2 includes P&W's fuel pump and a 10 per cent diameter larger throat combustion chamber that will reduce wear, reducing pump discharge pressure by 4.5 to 6 per cent and pump operating temperature by 7 to 12 per cent. All three engines on STS-86 in September 1997 introduced the Block 2. STS-85 will be the last mission to fly with an old engine; Block 1 or 2 will be used on all subsequent missions.

Computer and automated control systems

Each Orbiter's five IBM AP-101 General Purpose Computers of 1972 vintage are being replaced by IBM AP-101S models designed in 1984 and offering 2.5 times the capacity and three times the speed for half the size and mass. At peak, they will be used at more than 50 per cent of their capacities, contrasted with 80 per cent for the older models. The new GPCs first flew aboard Atlantis on 1991's STS-37. New Block 2 Honeywell SSME controllers were initially installed on OV-105 and then retrofitted to the others, providing four times the capacity with the first space-qualified 68000 series microprocessors. They also eliminate a separate 13.6 kg vibration monitoring box and add 20 engine parameter monitoring channels to the previous 120. The last of 30 ordered was delivered April 1993. The KT-70 Inertial Measurement Units are being augmented and eventually replaced by the High Accuracy Inertial Navigation System, of similar configuration but with increased accuracy and reliability (the first one flew on Atlantis during STS-44 in 1991). A new TACAN system is also due. Endeavour was delivered with the new items installed. Replacing the copper wire payload links in the cargo bay with a fibre optic system would save 545 kg.

Remote Manipulator System (RMS)

Spar Aerospace developed the RMS as Canada's principal Shuttle contribution. 38 cm diameter, 15.3 m long, mass of 408 kg, capable of handling 29.5 tonne payload. Shoulder, elbow and wrist joints are driven by DC electric motors and can be controlled in five different modes, ranging from manual hand controllers for the crew, using direct observation and TV cameras on the elbow/wrist, to full computer control. Stowed on the bay's port side and latched into three cradle pedestals, it can be restowed by EVA if it jams in use or, in the last resort, jettisoned by pyrotechnics. Each RMS is designed for a 10 year, 100 mission life. The first, the number 201 DDT&E arm and associated ground equipment, was delivered April 1981 at a cost to Canada of C$100 million. The operational 301/302/303 models were delivered May 1982, 14 December 1983 and 29 March 1985, respectively. 302 were lost on Challenger, but were not replaced until the refurbished 202 qualification model debuted on STS-66 in November 1994. RMS' highly successful use as a mobile crew work platform, in addition to its primary role as a manipulator, established it as an indispensable tool. A servicing centre based on RMS derivatives is Canada's primary Space Station contribution. Spar was contracted in 1991 to upgrade the arms to handle up to 265 tonne for berthing the Orbiter to Space Station. This includes a C$30 million redesign of arm control electronics, followed by a refurbishment and overhaul of various critical components completed in 1996.

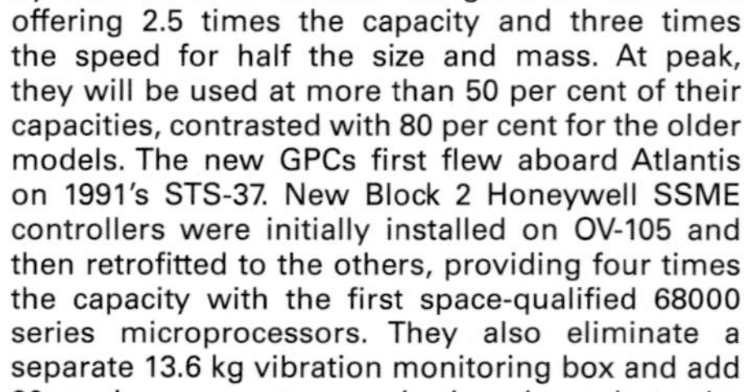

Four Gulfstream IIs, powered by Rolls-Royce engines, are fitted out as Shuttle flight decks for pilot training. They simulate the powerless handling characteristics and steep descent during approach and landing (Grumman) 0516760

Crew module for replacement Orbiter Endeavour 0516761

Modifications to the RMS will be certified to handle up to 265 tonnes and a remotely operated power/data connector will be carried on one side of the cargo bay for early Station flights until the pressurised docking system can be used. The existing internal airlock will be refurbished and mounted outside, attached to the cabin by a short flexible tunnel. On top will be an ASTP-type androgynous docking unit, with integral power/datalinks, and crew hatch. Three docking units will be built and carried as required.

Environmental Mobility Unit (Spacesuit)

EVAs have played a major role in Shuttle activities. Through February 1996, 35 different astronauts had conducted 60 EVAs totalling 378.2 hours, had repaired and redeployed four satellites and brought back to Earth two more (see EVA table at front of Directory). Crew members must be able to leave the Orbiter for routine work and emergencies. Three of the 14, 117 kg, suits are normally carried on each EVA mission (four were carried for the first time on STS-49 May 1992 and again STS-61 December 1993) and two for emergencies on non-EVA missions. They are stowed in the airlock, where they are donned/doffed, their size adjusted to fit the crewmembers trained for EVA on the particular mission. The crew uses 30.5 m tethers when working in the payload bay, unless they are working in the MMU or on the RMS foot restraint.

Cheaper and more flexible than Apollo's lunar suits, donning the suit begins with a zip-on one-piece Spandex liquid cooling/ventilation garment. Weighing 3 kg and with a projected life of 15 years, it includes urine collection facilities for up to 950 ml and a drink bag containing 0.6 litres of water. A Snoopy cap with headphones and microphone fits over the head and chin, and also provides caution and warning tones. Over this fits the two-piece spacesuit itself, comprising: hard upper torso, made in five sizes with hard waist ring. The lower torso comes in various sizes with hard waist ring; the helmet (one size only), and the visor assembly, which protects against micrometeoroids and the Sun's UV/IR radiation fits over the helmet. A more dextrous glove was tested during STS-37 in 1991. Expected life is 15 years for the hard sections and six years for the softer portions. There are no zippers on the suit; the components connect with hard snap-rings. The suit is composed of several bonded layers, beginning with a polyurethane-on-nylon pressure bladder, followed by Kevlar layers with folded/tucked joints (for mobility) and ending with a Kevlar, Teflon and Dacron anti-abrasion layer. The hard upper torso has a glass fibre shell. The materials are designed to prevent fungal or

Endeavour configuration by late 1989 0516762

bacterial growth: unlike the individual Apollo suits, they are used by different astronauts and must be cleaned and dried after each use. Each suit can be used up to 25 times during a mission.

The Portable Life-Support System (PLSS), attached to the back of the upper torso, provides sufficient oxygen and water for 7 h inside the suit, including 6 h for EVA and a 30-minute reserve. There is also a secondary pack to supply O_2 and maintain suit pressure for 30 minutes if the primary system fails. Astronauts can plug in to the Orbiter airlock for fresh supplies of O_2 and water. A liquid crystal display/control module on the front of the upper torso indicates remaining EVA capacity, and warns of excessive consumption or problems with the water pressure and T in the cooling garment. A suit vents 0.7 kg of water/O_2 hourly, a potential source of contamination. Hamilton Standard was awarded a US$97.3 million contract in 1991 for two further PLSS (adding to the 15 built) and continued engineering support through September 1997, bringing contract total value to US$212 million. Astronaut training and suit maintenance costs NASA some US$100 million annually.

It was planned to provide suitless crew members with a Personal Rescue Enclosure, an 86 cm diameter inflatable sphere made of suit material. Sitting inside with a portable oxygen system, the occupant would have had a communications system and viewing port, and sufficient pressurising gas for a 1 hr rescue. If an Orbiter became disabled, suited crew members could carry the others across to a rescue Orbiter; alternative methods included use of the RMS and passing them along a line as in a ship-to-ship rescue. The PREs were not produced.

The initial disadvantage of Shuttle's suit was that, as it is pressurised to only 0.29 atmosphere (4.3 psi) of pure O_2, the wearer had to prebreathe pure O_2 for 3.5 hr prior to EVA to prevent N_2 bubbles forming in the joints. Pre-breathing now begins with an hour while cabin pressure is reduced from 14.7 psi to 10.2 psi. After 12 hr, the EMU is donned and, during checkout, another 40 minutes is spent breathing pure O_2 before depressurizing the airlock. To overcome the prebreathing necessity, which might delay an EVA to meet an emergency, NASA/Hamilton Standard began work on a zero prebreathe high-pressure hard suit. NASA decided later because of the cost (estimated in 1995 at US$700 million) to use Shuttle-type suits on Space Station. The Station will use 1 atmosphere, requiring the full 3.5 h prebreathe. Hamilton Standard has proposed a US$10 million minor upgrade that would operate the suit at 0.4 atmosphere, the same as Russia's Orlan suit, which requires only 30 minutes prebreathing. Thermal modifications are under way as Station EVA work will be away from Shuttle's warm cargo bay. STS-69 in September 1995 tested fingertip heaters and the body cooling loop could be shut off.

Manned Manoeuvring Unit (MMU)

The 153 kg Manned Manoeuvring Unit, which snapped on to the back of the PLSS, enabled an astronaut to operate free of tethers. Two hand controllers, operating 24 × 7.6 N N_2 thrusters, provided 6 DoF and a total 6,190 Ns impulse. Two were carried in the forward cargo bay when required. Martin Marietta built only three at US$10 million each. On missions 10/11/14 they were used nine times, with an accumulated flight time of 10 hr 22 minutes. NASA began a US$600,000 study in mid-1990 to determine the cost of restoring them

A completed ET leaves NASA's Michoud plant 0517266

to flight readiness, possibly for a flight in late 1991 carrying an IMAX camera and other detectors to observe Orbiter thruster firings and water dumps. A total of about US$5 million was projected. However, NASA management ordered them returned to storage in 1991 as they were developed for demonstration purposes and are too limited for effective deployment.

The 37.6 kg SAFER Simplified Aid for EVA Rescue was tested September 1994 during STS-64 to show its value in an emergency, such as when an astronaut comes adrift from Space Station. At the moment, astronauts have to work in pairs, carefully tethered, and with the Orbiter ready to provide rescue. A drifting astronaut would activate SAFER, to stop any tumbling, using its automatic attitude hold mode and then jet back to safety. SAFER is attached below the PLSS, providing a 3 m/s total impulse from a 1.4 kg pressurised N_2 supply via 24 3.5 N thrusters. It was first used operationally during March 1996's STS-76 EVA at Mir, as it would have taken Atlantis 20 minutes to undock for a rescue.

Shuttle support equipment, maintenance and reliability

Shuttle carrier aircraft

Boeing 747-123 N-9668, purchased by NASA from American Airlines in June 1974 after logging 8,999 hr, was modified (and registered as N-905NA) to ferry Orbiters between Shuttle facilities and to carry Orbiter *Enterprise* for the original ALT programme. For ferry flights, a tail cone fairing is installed on the Orbiter to reduce aerodynamic drag and buffet and aerosurface control locks are added to the Orbiter's elevons. The Orbiter is carried unmanned with its systems inert. A bailout system is carried for the 747 crew. With use of the Kennedy runway limited, the aged lone 747 was a vital element of the Shuttle system. Had it been damaged or lost, the Shuttle fleet would have been grounded. In September 1987, NASA signed a letter of agreement with Boeing for a second aircraft. An ex-Japan Air Lines 747-SR was delivered to Boeing's Wichita plant 17 April 1989 under a US$55 million contract to undergo conversion; NASA accepted it in November 1990. Designed to carry 109 tonnes (22 tonnes more than the older aircraft), N-910NA has a 1,700 km range and is designed for 265 ferry flights over 15 years. The aircraft may be used for carrying the X34 hypersonic test vehicle.

The USAF's 433rd Military Airlift Wing at Kelly AFB, Texas, operates two Lockheed C5As modified by the company under an October 1985 US$133 million contract for carrying a loaded Shuttle cargo container and its towing tractor. These C5A SCM (Shuttle Cargo Modification) aircraft, delivered November 1988/Oct 1989, can handle a full cargo bay load such as the Hubble Space Telescope.

Shuttle operations

Processing/turnaround

NASA plans an annual flight rate of seven for the full four-Orbiter fleet through the Space Station era. The processing goal for each mission is 88 days, covering 60 days in the Orbiter Processing Facility, 5 days in the VAB and 23 days on the pad (Endeavour set the current 52-day OPF record preparing for STS-57 in 1993, beating the previous best, of 1993's STS-45, by 3 days). Platforms and equipment from the mothballed Vandenberg site are installed in KSC's Orbiter Maintenance & Refurbishment Facility, creating a third OPF position from late 1991 and allowing simultaneous processing of three Orbiters.

In September 1983, it was announced that Lockheed Space Operations Co (now Lockheed Martin Space Operations), a consortium including Grumman, Thiokol and Pan American World Services, had won the NASA/USAF Shuttle processing contract, effective for three years from 1 October 1983, with four three-year options. NASA exercised the US$1.60 billion second option in September 1989, extending the contract to 30 September 1992. The third was exercised in 1992, worth US$1.830 billion through September 1995, bringing the total for 1983–95 to US$6.3 billion. A one year US$638 million extension began 1 October 1995, in preparation for the new approach of awarding Shuttle operations to a single contractor. LMSOC assumed all the work from October 1995. Before then, Lockheed Martin's share as consortium leader was 75 per cent. Grumman, responsible for launch processing, had a 13 per cent stake. Thiokol's task of processing the ET and SRB (including recovery after launch) was worth 11 per cent, and Johnson Controls World Services (who bought out Pan American), responsible for operations analysis, process planning/control and logistics engineering, received 1 per cent.

NASA's policy was to hand over responsibility for the whole STS processing operation at the Cape and Vandenberg, thus freeing itself for development work on Space Station. It was also hoped that the single processing contract would improve safety and mission effectiveness while reducing costs and turnaround time. Turnaround time was down to 69 days between STS-6/7. Processing had taken 649 days for STS-1, compared with 198, 114, 80 and 120 days for the next four missions. With the near achievement of 10 flights in 1985, turnaround time was down to an average of 36 days.

NASA's STS Operations Contract (STSOC), a four-year award worth an estimated initial US$685 million, was awarded to Rockwell's Shuttle Operations Co in September 1985, effective 1 January 1986. A US$2.3 billion extension through December 2000 was granted Spring 1991, bringing total contract value to US$4.8 billion. RSOC became responsible for managing 22 Shuttle operations functions previously performed by 17 contractors, involving mission planning, flight design, mission data production, flight crew/ground controller training, and ground facility engineering. The work complements Rockwell's Florida logistics contract, extended in October 1991 to 1994 at a cost of US$453 million. The 25,000 m^2 NASA Shuttle Logistics Depot (NSLD) became fully operational in 1993, certified to handle >3,200 of Shuttle's 4,000 Line Replaceable Units.

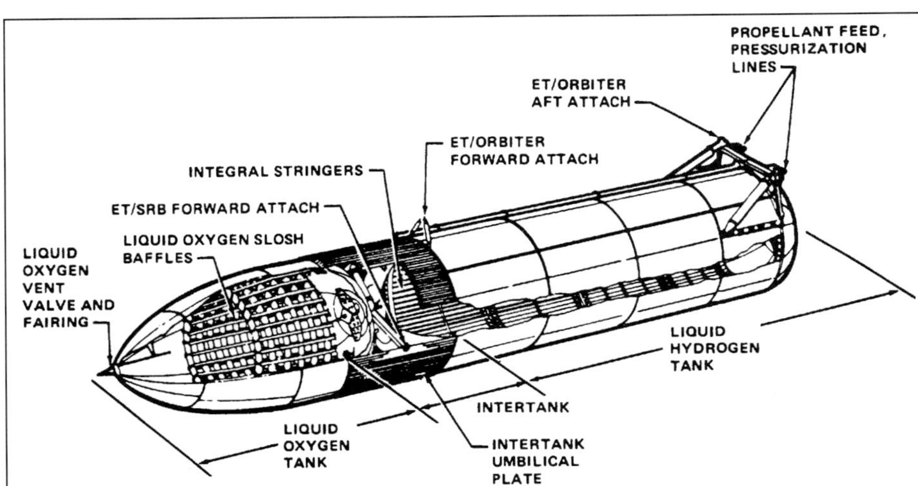

External Tank principal features (Rockwell) 0516763

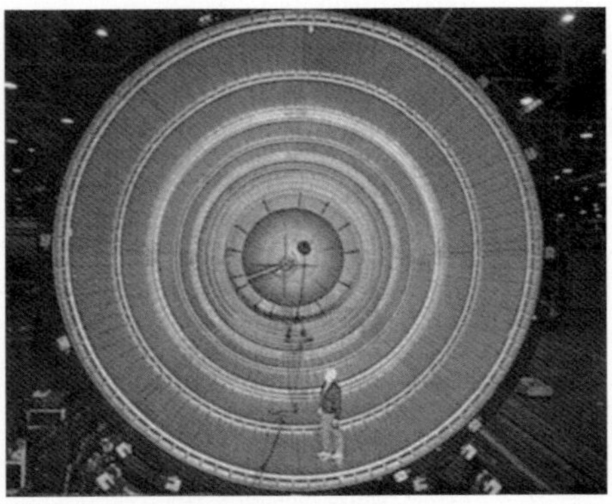

ET hydrogen tank (Lockheed Martin) 0516764

Orbiter pilot training

Pilot training includes at least 30 flights on Shuttle Training Aircraft (STA), with about 900 practice landing approaches. Some 10 training sorties are made weekly at the White Sands Space Harbour. The aircraft are housed in the NASA hangar at El Paso International Airport, Texas. STAs are Gulfstream II executive jets, with Rolls-Royce engines and extensive structural, wiring and cockpit changes and with installation of direct lift control systems and thrust reversers to simulate Orbiter flight characteristics. In early 1985, NASA ordered a third STA from Grumman, the modifications costing US$14 million. A fourth was being considered before the Challenger accident to support a launch rate of about 10 per year. It became operational in April 1991. Honeywell's glass cockpit displays will be fitted from 1997 as they are installed on the Orbiters. The two pilots during STS-58 in 1993 introduced a new training aid designed to refresh landing skills on long missions: the PILOT Pilot In-flight Landing Operations Training unit is a portable computer with an Orbiter-type hand controller. It is mounted on the console in front of the pilot's flight deck seat and the controller attached to the actual Orbiter controller.

Shuttle maintenance and reliability

All of the Orbiters return to Palmdale every three years for a major inspection, recertification and maintenance. Realistic analyses of the chance of a Shuttle loss resulting from a 1988 NASA analysis, state the chance of loss at a 1 in 78 median probability of loss during ascent. In 1989, NASA recalculated the values at 1 in 94. Then in 1995, NASA again upgraded safety projections for any Shuttle mission as 1 in 145 (estimates ranged from 1 in 76 to 1 in 230), made up in part of a factor of 1 in 248 for ascent only. The main engines are the major contributors with a mean probability of failure of 1 in 410. With the upgraded engines, the chance will level off at less than 1 in 200 for the whole mission and less than 1 in 300 for ascent. The first estimates that NASA provided to the American public in 1975 indicated that the loss rate would be less than 1 in 1,000,000.

Shuttle orbiter modifications

Through the first several years of launch operations, Shuttle in-orbit missions usually lasted less than 10 days. Columbia and Atlantis have undergone major modifications to provide additional fuel cell cryogenic supplies and improved life support systems for 16-day missions and, as a result, are now referred to as Extended Duration Orbiters (EDO). The Columbia retrofit was completed in 1992. The 13-day US Microgravity Lab STS-50, in June 1992, used the EDO for the first time. *Atlantis* underwent the same conversion following return from STS-46 in August 1992 (although it has no scheduled EDO missions yet) and was also equipped for docking with Mir in June 1995. Endeavour was built with the required plumbing and electrical connections and flew its first EDO mission as STS-67 in March 1995.

In order to make the modifications, Rockwell added a cryogenic pallet in the aft cargo bay to extend fuel cell capacity and enlarged the storage space in the cabin for additional food and clothing. In addition, they installed an improved waste collection system (first flown operationally STS-65 July 1994) and a reusable CO_2 scrubber to replace the previous lithium hydroxide canisters. Atlantis and Endeavour are capable of carrying EDO kits allowing up to 28-day missions, although none is yet planned.

During the preliminary planning for the Shuttle, NASA announced that the spaceplane would have a mission duration of up to 30 days. When it actually flew, NASA could not promise an on-orbit duration of more than eight days and returned to Congress to request additional funds for modification. In November 1987, NASA provided Congress with a report on methods of extending the in-orbit capability of at least one Orbiter to 16 days. This extended the time from the pre-Challenger limitation by fuel-cell life of 9–10 days. In May 1983, Rockwell proposed a two-stage plan to extend stay-time to 15 to 18 days by 1985 and to 45 days by 1988–89, but NASA did not proceed with it. The first stage, comparable to the current improvements, was projected to cost US$200 million. The second stage, costing US$200–250 million, would have required an orbiting solar array and battery system with which the Shuttle could dock. This required relocation of the airlock and mid-deck modifications and gained some DoD support as a cheaper alternative to space station development. NASA agreed to Congressional suggestions in early 1989 of providing OV-105 with a 28 day mission capability at a FY90 cost of US$52 million.

Rockwell agreed in March 1990 to contribute US$15 million to the US$53 million cryogenic pallet element under a semi-commercial agreement. If Rockwell signs up a non-government user, the company will receive US$1.85 million for each day beyond an 8-day mission, thereby recouping the investment in a single 16-day flight. Similar missions up to end-1997 would provide profit. NASA management indicated in January 1990 that a 28 day modification is unlikely because of the associated risks.

For EDO, the 2,926 kg cryo pallet plumbs four tank sets to the standard five, each capable of supporting two days' activities. The pallet could carry four further sets on its aft side but Columbia is too heavy to be converted to a 28 day Orbiter. Hamilton Standard's 147 kg Regenerative CO_2 Removal System, forward of the mid-deck lockers, operates two beds in a 15-minute cycle, one absorbing the gas while the other is exposed to vacuum. A similar system was flown on the Skylab space station. Some N_2 is also recovered for addition to the two extra tanks carried under the payload bay against the aft bulkhead. STS-50, the first EDO flight, still carried 50 LiOH canisters as back-up (astronauts had to repair the RCRS in flight). Although RCRS is now fully qualified, EDO missions continue to carry sufficient canisters for four days.

The EDO 22 kg trash compactor was demonstrated aboard 1990's STS-35, carried in place of a mid-deck locker. Rubbish is loaded in a 0.014 m^3 plastic bag, equivalent to one person's daily waste and compacted by pistons at 41 N/cm^2 into quarter the volume. The bag is then stored under the mid-deck floor. Some items, such as the rehydratable food package, have been redesigned to allow crushing. STS-35 also demonstrated the new Waste Containment System. The old toilet has to be removed/cleaned after each flight, but Hamilton Standard's newer design collects the waste in bags, which are then compacted into storage tubes with odour/bacterial filters. Using removable bags means the toilet has unlimited capacity. Development versions were carried by Endeavour on STS-54/57 in 1993, allowing the then record-breaking STS-65 July 1994 to be the first operational mission.

NASA has studied the Longer Duration Orbiter (LDO) modification for longer missions. The preliminary results indicated that 30 days are possible and 60 days appear feasible if the Shuttle remains attached to the Space Station for mutual support. Power (fuel cells) is the limiting factor beyond 30 days. A solar array module has been proposed, which would also make it independent of Space Station. LDO was deleted from FY94's budget to save US$43 million, particularly as long missions are now available aboard Mir.

Beginning in 1994, NASA and its contractors modified Atlantis to fly seven Mir docking missions between 1995 and 1997 and Discovery for two missions. Engineers believe the Orbiter structures are good for more than 100 missions, allowing flights until around 2030. The thrust structures are the principal limiting factor on a Shuttle lifetime. Discovery, Atlantis and Endeavour are modified for Space Station operations. Orbiter Atlantis stood down for 10 months while engineers and technicians replaced electromechanical cockpit instruments with 11 colour flat-panel screens, improving pilot efficiency and reducing costs of eliminating the need to maintain outdated equipment. Atlantis arrived at Boeing's orbiter processing facility at Palmdale, California, on 14 November 1997 and returned with more than 130 major modifications on 27 September 1998.

Space Shuttle Orbiter specifications

Shuttle payload capacity

Structural analyses and reviews of landing stresses following Challenger's loss showed that landing weight could be increased by 8,619 kg to 104,328 kg. Abort landings, which constrain launch mass, are limited to 108,860 kg. Discovery, Atlantis and Endeavour can deliver 24,990 kg into 204 km orbits; they are referred to as 55,000 Orbiters because of their 55,100 lb capacities into NASA's standard 120 n mile (220 km) orbit. Originally NASA promised the Shuttle would deliver 65,000 to the standard orbit. Columbia can handle 3,152–3,840 kg less because of its greater dry mass. As a result, it will not be used for Station assembly. Advanced solid boosters, available from 1999 at the end of a development cycle cancelled in October 1993, would have added 5,440 kg capacity. Some Station assembly flights might require SSME levels of 106 per cent, allowed by pump improvements and other changes.

Shuttle costs

NASA's own figures for the first 20 flights made by Columbia, Challenger and Discovery showed an average cost of US$257 million. The improved Discovery, new Atlantis and modified Columbia were expected to reduce the average, but not to the US$74 million per 'dedicated flight' (use of the whole payload capacity) which NASA intended to charge after 1988. NASA quoted a cost per mission of US$286 million in 1990, but charges only US$130 million at 1988 rates. True amortised rates suggest that the real cost of Shuttle missions is probably in excess of about US$350 million (1986 rates). The true cost is debatable, depending on what costs are included. NASA in 1992 estimated the *marginal* cost (that is, the saving by reducing the flight rate from eight to seven) at only about US$40 million. But NASA's Shuttle operations budget produces a figure of US$450 million for FY95. Adding other programme costs such as TDRS, ASRM and civil service salaries puts the figure at more than US$500 million in 1992 terms.

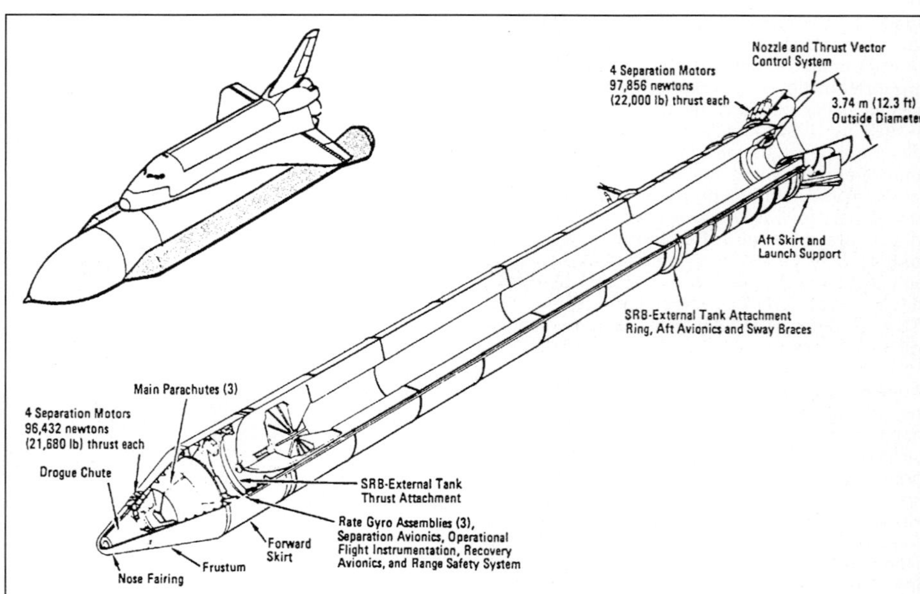

Space Shuttle Solid Rocket Booster (NASA) 0516765

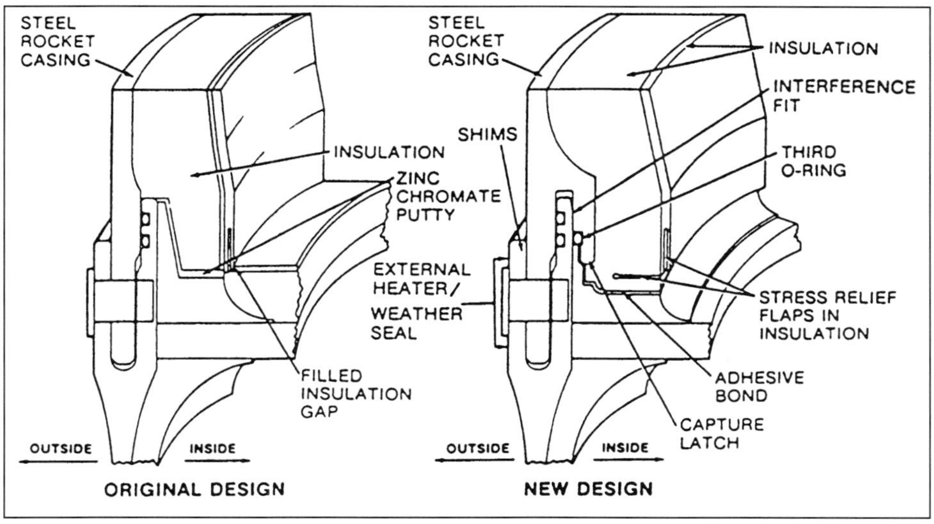

The post-Challenger SRB field joint includes a tang capture feature, with a third O-ring, to reduce joint deformation and a combined weather seal/heating band to exclude water and maintain joint temperature at 24°C 0516766

NASA estimates the Shuttle's original development at US$5,150 million (FY71 rate). The actual cost was US$10.1 billion (real year cost), or US$6,744 million (FY71 rate). The 1971 estimate of US$250 million for each Orbiter rose to about US$2 billion in 1983 dollars, though this included, for the first time, engines and equipment, Remote Manipulator System, galley and the closed-circuit TV. The major Shuttle contracts, with their values in 1982 US$ million were:

Rockwell (Shuttle Orbiter)	3,560
Grumman (wings)	45
McDonnell Douglas (OMS/RCS pods)	85
McDonnell Douglas (Support)	52
Rocketdyne (Main Engines)	1,546
Thiokol (Solid Rocket Boosters)	206
USBI (SRB assembly/retrieval)	89
Martin Marietta (External Tank)	529

Following the Challenger accident, NASA submitted, in February 1987, a series of proposed actions in response to the Rogers Commission recommendations, with a total estimated cost of US$1,379 million. The estimated total expenditures to meet this goal were actually US$2.4 billion at the time of STS-26.

Other costs may be estimated from subsequent actual mission costs. As an example, President Reagan's 15 point policy, issued in March 1988, was designed to create opportunities for US commercial space interests. The proposals emphasised the Shuttle-launched Industrial Space Facility as a man-tended microgravity platform and the use of Spacehab inside the payload bay. It was projected that the former price of US$74 million for the use of the whole cargo bay could go as high as US$255 million by the early 1990s. As the 1990s began, NASA cited a new price of around US$180 million, which represented full cost recovery if there were 14 flights annually. NASA reached a *quid pro quo* agreement with the DoD in October 1987 citing US$115 million (FY89 rate) for a full cargo bay but, in 1996, NASA quoted US$200 million to US$250 million for DoD.

President Reagan's August 1986 edict, prohibiting NASA from using the Shuttle for commercial satellite launches, made his earlier Shuttle pricing policy. The intention was to charge a minimum of US$74 million (1982 US$) for a full payload bay and to offer three such missions by auction two years in advance for commercial and foreign use. At less than half the cost of a Shuttle mission, NASA was hoping that demand would increase the price.

NASA is consolidating more than 80 Shuttle processing and operations contracts, saving US$770 million annually, into a single prime contractor. The final contract, lasting for six years and worth US$7 billion, was awarded to United Space Alliance. This consolidated ground processing and in-flight operations within a single company. United Space Alliance will suffer penalties if it fails to meet cost-cutting targets, but it will be awarded 35 per cent of additional costs saved after an initial target of US$400 million has been reached. NASA retains the responsibility for safety, scheduling and budgeting. 1995's operating budget was already 25 per cent lower than 1992's. US$3.149 billion was allocated to Shuttle operations and hardware for FY96; US$3.151 billion was requested for FY97. For other years the requests were FY95 US$3.155 billion, FY94 US$3.549 billion, FY93 US$4.061 billion, FY92 US$4.271 billion, FY91 US$4.066 billion, FY90 US$3.682 billion.

Aborts

Four basic abort modes are available to Shuttle controllers, the precise scheduling of each varying with the mission ascent profile. Return To Launch Site (RTLS) would be used in the event of a main engine failure in the first 260 seconds. The remaining engines and RCS thrusters would achieve a pitch-around, enabling the Orbiter to jettison the ET 45 km from the coastline and glide back to the KSC runway. RTLS is considered to be a very risky manoeuvre.

The Transatlantic Landing (TAL) abort is required when an SSME fails after the last opportunity for RTLS and before an Abort to Orbit is possible. There are four TAL sites where an Orbiter would land about 25 minutes after launch: Moron Air Base (Spain), Yundum International Airport (Banjul, Gambia), Zaragoza Air Base (Spain, prime site for high-inclination flights) and Ben Guerir Airfield (Morocco). All sites provide Shuttle visual and navigational aids (principally TACAN and a Microwave Landing System); two or three are activated for each ascent and one (usually Ben Guerir) remains active throughout in case of emergency de-orbit. There are also two ACLS (Augmented Contingency Landing Site), at Hickham AFB (Honolulu) and Andersen AFB (Guam). They are similar to TAL sites in facilities and equipment but used instead for contingency de-orbits and not ascent aborts.

Abort Once Around (AOA) would be used from about 2 minutes after SRB separation, again in the event of a main engine failure. The procedure relies on two OMS firings after ET jettison to place the Orbiter in a sub-orbital coast and free return orbit for re-entry and glide back to the runway at KSC, Edwards or White Sands. Abort To Orbit (ATO) is available in the event of a main engine failure after passing the AOA point. Again the procedure relies upon two OMS firings, one for orbit insertion and the other to circularise the path. An alternative mission might then be possible, depending on mission rules. An ATO occurred on STS-51F in July 1985, but most of that mission was completed. Each mission has its emergency landing sites, which vary with the launch profile, but they carry no Shuttle equipment or aids. In the case of STS-51F, a landing could have been made at Zaragoza, although it would have taken at least 5 weeks to recover Challenger. The AOA site was White Sands Space Harbour, New Mexico.

One abort site safety device is a 13.7 m high arresting net attached to an energy-absorbing tape drum, capable of stopping an Orbiter travelling at 185 km/h. Tests of this Shuttle Orbiter Arresting System (SOAS) were made with Enterprise in June 1987 at Washington DC Dulles airport. Nets were installed at the four TAL sites and at Hickham AFB. Similar abort procedures were devised for Vandenberg launches and arrangements were made for emergency landings on several Pacific Islands, including Chile's Easter Island.

Crew escape system

Following Challenger's loss, each Orbiter was retrofitted with a telescoping pole escape system to permit evacuation while the craft is in a controlled glide descent. As such, it provides very limited cover and would primarily be called upon in the event of a water ditching, which would probably be non-survivable for a crew remaining with the craft. The system comprises a 109 kg, 7.5 cm diameter aluminium/steel pole that would be extended 2.67 m through the jettisoned hatch to guide evacuating crew 45° down and 15° back to avoid the port wing. The abort decision must be made below 18,300 m, with the Cdr and GPCs achieving a controlled glide descent before the designated jumpmaster, occupying the mid-deck seat closest to the door, manually depressurises the cabin and blows the hatch pyrotechnically at 6,710 m. Bale-out begins at 6,100 m, with each of up to eight crew members attaching the suit lanyard to a hook to slide down

Stacked SRBs await ET mating in the Vehicle Assembly Building at Kennedy Space Center 0516767

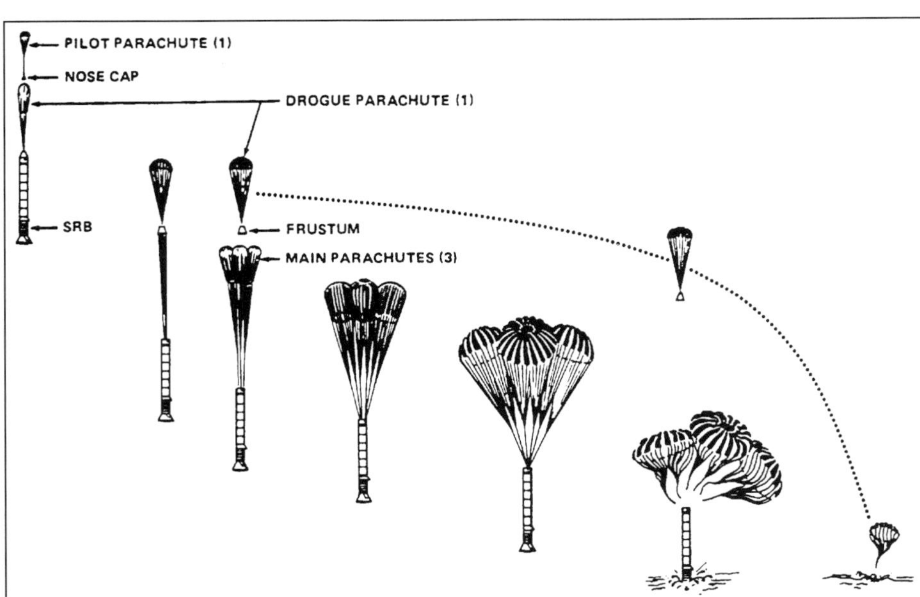

SRB recovery: after separation, the motors follow a 4-minute ballistic path before the recovery sequence begins 0516768

The small SAFER (seen below the main backpack, with chest controls) carries rescue thrusters should the wearer drift free of Mir or Alpha (NASA) 0517101

the pole. The main parachute opens automatically within 6 seconds. All crew members now wear 32 kg high-altitude pressure suits carrying emergency oxygen, liferaft and other survival equipment.

The transcript of onboard crew conversation, plus the fact that at least two of the air-breathing packs were activated before Challenger fell into the Atlantic, convinced the astronauts that even though it was a 'worst case' accident, at least some crew members could have survived if there had been an escape system.

Recent missions

STS-66 ATLANTIS (66)

Orbiter flight: 13
Mobile Launch Platform: 3
Launch pad: KSC 39B
Launch date/time: 3 November 1994 at 16.59.43 GMT
Landing date/time: 14 November 1994 at 15.33.45 GMT
Mission duration: 10 days 22 hours 34 minutes 2 seconds
Landing site: EAFB
Main engines: 1: 2030 2: 2034 3: 2017
Launch Mass: =~2,045.0 t
Landing Mass: 95,184 kg
Crew: Donald R McMonagle (Commander), Curtis L Brown Jr (Pilot), Ellen Ochoa (Payload Commander/Mission Specialist-1), Joseph R Tanner (Mission Specialist 2/EV2), Jean-François Clervoy (Mission Specialist 3/ESA), Scott E Parazynski (Mission Specialist 4/EV1).
Payload (Primary):
Atlas 3 Spacelab (3,759 kg)
Crista-SPAS (3,263 kg)
Payload (Secondary): cargo bay,
ESCAPE II Experiment of the Sun for Complementing the Atlas Payload and for Education
Payload (Secondary): mid-deck,
PCG Protein Crystal Growth
HPP 2 Heat Pipe Performance
PARE R1 Physiological & Anatomical Experiment – Rodents (see STS-56)
SAMS Shuttle Acceleration Measurement System (supporting PCG)
STL Space Tissue Loss (flew STS-45/53/56/59)

Mission summary

The third Atmospheric Lab for Applications and Science (Atlas) Spacelab carried six instruments on a Spacelab pallet, plus a seventh in two sidewall cans, to study the atmosphere's chemical composition over 15–600 km, measure sunlight's total energy and variation and how these factors affect ozone levels. Atlas was intended to fly every one to two years throughout the Sun's 11-year activity cycle, complementing UARS observations and providing a baseline for discovering changes by flying highly calibrated instruments. NASA has dropped the original schedule and Atlas 3 was the last manifested. Six of the instruments (ATMOS, MAS, SSBUV, ACRIM, SOLCON, SOLSPEC) flew on STS-45/56 Atlas 1/2. SUSIM, added to measure the Sun's UV spectrum, also flew on Atlas 2.

Although MAS failed on day 2, Atlas 3 allowed investigators to study the northern hemisphere atmosphere as it moved from its relatively quiet Summer state to its more active Winter conditions. It also observed the Antarctic ozone hole near its maximum for comparison with the March and April data of Atlas 1/2.

Germany's retrievable Crista-SPAS complemented Atlas with the first study of global atmospheric variability by limb sounding. It returned 51,000 vertical scans, measuring 15 trace gases every one second. A GPS receiver provided the required position information. A secondary SPAS payload was the US Naval Research Lab's MAHRSI Middle Atmosphere High-Resolution Spectrograph Investigation for nitric oxide and hydroxyl measurements. OH is a key compound in ozone's destruction. MAHRSI returned four global maps of its distribution and concentration. ESA's Clervoy released SPAS from the robot arm at 12.50 GMT 4 November for it to be controlled by a German team at KSC.

SPAS' recovery demonstrated a new rendezvous technique for Shuttle missions to Mir. The Orbiter approached from below, firing thrusters away from its target. Ochoa retrieved the free-flyer at 13.05 GMT 12 November. Crista-SPAS was planned for reflight aboard STS-85 in July 1997.

ESCAPE obtained digital images of the Sun at 1,216 Å to help correlate how the Sun's extreme UV affects the temperature and chemical composition of Earth's upper atmosphere. PARE carried 10 pregnant rats in two Animal Enclosure Modules for NASA/National Institute of Health studies into µg effects on living organisms.

Clervoy also tested a special reclining seat for returning Mir's long-duration crews. The landing was 1994's fourth diverted to Edwards because of bad weather at KSC. Each diversion costs about US$1 million to return the Orbiter to Florida.

STS-63 DISCOVERY (67)

Orbiter flight: 20
Mobile Launch Platform: 2
Launch pad: KSC 39B
Launch date/time: 3 February 1995 at 05.22.04 GMT
Landing date/time: 11 February 1995 at 11.50.19 GMT
Mission duration: 8 days 6 hours 28 minutes 15 seconds
Landing site: KSC
Main engines: 1: 2035 2: 2109 3: 2029
Launch Mass: =~2,046.4 t
Landing Mass: 95,835 kg
Crew: James B Wetherbee (Commander), Eileen M Collins (Pilot), Bernard A Harris Jr (Payload Commander/Mission Specialist 1/EV2), C Michael Foale (Mission Specialist 2/EV1), Janice E Voss (Mission Specialist 3), Vladimir G Titov (Mission Specialist 4, Russia).
Payload (Primary):
Spacehab 3
Spartan 204
Payload (Secondary): cargo bay,
CSE Cryo Systems Experiment
GLO 2 (flew STS-53)
ODERACS 2 Orbital Debris Radar Calibration Spheres (flew STS-60)
IMAX
Payload (Secondary): mid-deck,
SSCE Solid Surface Combustion Experiment (flew STS-41/40/43/50/47/54/64)
AMOS AF Maui Optical Site

Mission summary

This was Shuttle's first launch into 51.6° and with the five minute window that will be standard for Alpha missions. NASA had to fly the high-inclination orbit with its dogleg ascent to complete the Mir rendezvous and flyaround as a final dress rehearsal for the series of dockings beginning with STS-71. Discovery approached to within 10 m of Mir for 10 minutes without disturbing the delicate solar wings. Russian controllers evinced some concern during the three-day rendezvous because an aft Mir RCS thruster began leaking oxidiser soon after launch. The problem was solved by turning off the entire group of three leaving no redundancy in

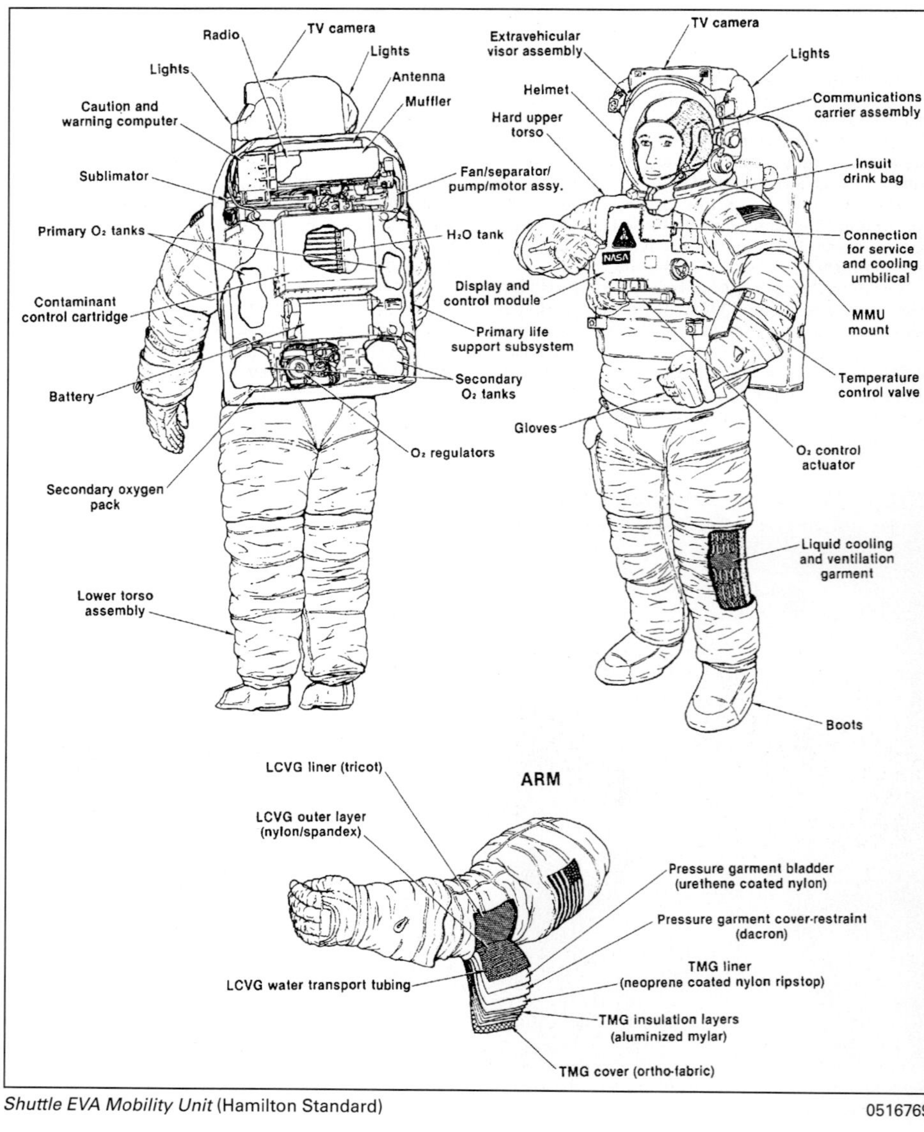

Shuttle EVA Mobility Unit (Hamilton Standard) 0516769

Harris (left) and Foale prepare to leave the airlock for STS-63's EVA. The new electronic checklist can be seen on Foale's right arm; it failed, probably because of the cold conditions, but has since been successful. This was Shuttle's first EVA to exit from a hatch in the tunnel leading to Spacehab (NASA) 0517124

the event of an emergency. A nose thruster began leaking 4 February but the problem was cured by warming it in the Sun (a standard technique that failed on the aft unit).

Shuttle astronauts Wetherbee and Collins took over manual control from 600 m out using the aft flight deck, slowing the closing rate down at 100 m to 30 cm/s. A camera in Spacehab's top window provided a direct view. Foale operated two laser ranging systems, namely the cargo bay Trajectory Control System and a hand-held unit. Discovery halted within 10 m at about 19.20 GMT 6 February. 10 minutes later it began backing off to 150 m and spent 45 minutes photographing Mir in detail. Discovery finally departed 21.13 GMT.

Spartan 204 carried the Naval Research Lab's FUVIS Far-UV Imaging Spectrograph. Titov used it on day 2 on the RMS to observe oxygen glow on Discovery's tail and observed thruster firings. He released it at 12.26 GMT 7 February for autonomous astronomical observations. As Voss recovered it at 11.53 GMT 9 February, Harris and Foale waited in the airlock to begin their EVA. They began their 278 minute EVA on the end of the RMS away from the warm cargo bay to test suit modifications. One modification disconnected the arm tubes in their liquid-cooled undergarment. While they were holding Spartan during a night pass to investigate the handling of large masses, they reported they were both very cold and the EVA was ended. Sensors showed Foale's glove fingers had fallen to 4°C. New electronic cuff checklists failed, apparently because of the cold.

Spacehab 3 carried 20 primarily μg experiments (some were actually housed in the mid-deck). McDonnell Douglas' robotic box-like Charlotte moved along cables extending from each of its eight corners to Spacehab attach points. Reels allowed it to move anywhere inside Spacehab, demonstrating remote control and observation of experiments. Its arm, equipped with a simple gripper, can flip switches and exchange samples and data cartridges.

Goddard's Hitchhiker bridge carried four experiments plus the IMAX camera. Among the experiments the CSE evaluated an integrated Stirling 65K cryocooler and oxygen diode heat pipe, the GLO 2 made glow observations of Orbiter surfaces, airglow and contaminating events at 1,150–11,590 Å. The Hitchhiker also contained two SAAMD stand-alone acceleration measurement devices and ODERACS 2.

On day two, ODERACS ejected 5, 10 and 15 cm diameter spheres and three dipoles (two 13.3 cm long and one 4.42 cm) to help calibrate radar and optical tracking systems for debris monitoring, particularly the Haystack Radar.

At the end of the mission, the Shuttle performed the first KSC landing since the runway was resurfaced in September 1994 to decrease tyre wear and increase crosswind tolerances.

STS-67 ENDEAVOUR (68)
Orbiter flight: 8 (5th Extended Duration Orbiter)
Mobile Launch Platform: 1
Launch pad: KSC 39A
Launch date/time: 2 March 1995 at 06.38.13 GMT
Landing date/time: 18 March 1995 at 21.47.47 GMT
Mission duration: 16 days 15 hours 9 minutes 34 seconds
Landing site: EAFB
Main engines: 1: 2012 2: 2033 3: 2031
Launch Mass: =~2,050.6 t
Landing Mass: 98,879 kg
Crew: Stephen S Oswald (Commander), William G Gregory (Pilot), John Grunsfeld (Mission Specialist 1/EV2), Wendy B Lawrence (Mission Specialist 2), Tamara E Jernigan (Payload Commander/Mission Specialist 3/EV1), Ronald A Parise (PS1), Samuel T Durrance (PS2).
Payload (Primary):
Astro 2
Payload (Secondary): cargo bay,
Endeavour UV telescope (flew STS-42)
Payload (Secondary): mid-deck,
MACE Mid-deck Active Control Experiment
PCG Protein Crystal Growth (flew STS-66)
CMIX 3 Commercial Materials
ITA Experiment (flew STS-52/56)
SAREX-II Shuttle Amateur Radio Experiment

Mission summary
Astro's set of three broadband ultra-violet telescopes, designed to observe astrophysical targets such as quasars and active galactic nuclei, first flew aboard SM38/STS-35 in December 1990. Astro 2 was cancelled in January 1990 to save US$10 million to US$20 million and the telescopes were stored at KSC. However, NASA announced in May 1991 that an Astro 2 mission would be funded, although Goddard's complementary Broad Band X-ray Telescope (BBXRT) was not re-flown because Japan's 1993 Asuka satellite carries a similar telescope.

Carried by ESA's Instrument Pointing System (IPS) on two Spacelab pallets, Astro's payload carried three instruments. The most important is the Hopkins UV Telescope (HUT),The Johns Hopkins University, 90 cm f/2 silicon carbide-coated paraboloid mirror optimised for 900–1,200 Å but extending to 425–1,850 Å. It has a prime focus Rowland Circle spectrograph, 3.0 Å spectral resolution and weighs 789 kg. Next is the UIT UV Imaging Telescope and film cameras, a 38 cm f/9 Ritchey-Chretien optics, 1,200–3,200 Å, 40 arcmin FOV, 2 arcsec resolution, device built by NASA Goddard. It can see down to a magnitude limit of +25 with two image intensifiers and 70 mm film. The cameras can take 1,000 frames each, with exposures up to 30 minutes. It weighs 474 kg. A University of Wisconsin research group built the Wisconsin UV Photo-Polarimeter Experiment (WUPPE), with 50 cm f/10 Cassegrain optics, 1,400–3,200 Å, and detector spectropolarimeter with dual electronic diode array detectors. The device studies stars (particularly white dwarfs), interstellar medium, asteroids, comets, nebulas and galaxies with 3.3 × 4.4 arcmin resolution and a magnitude limit of +16, and spectral resolution of 6 Å. It weighs 446 kg.

The first Remote Manipulator System, being assembled in Toronto (Spar Aerospace) 0516770

STS-48 astronauts Brown (left) and Buchli assemble a Space Station truss model to study its vibration characteristics. An improved version flew on STS-62 in March 1994 (NASA) 0516771

With the UIT/WUPPE Spacelab's IPS stabilisation improves to less than 1 arcsec by a gyro-based image motion compensation system adjusting the secondary mirrors. IPS activation began 200 minutes after launch.

MACE was a 1.5 m long flexible beam with scanning and pointing payloads at each end and a central control system with a rate gyro and reaction wheels to test active damping techniques.

Endeavour's first EDO mission (and the 99th US manned launch) set a Shuttle duration record.

STS-71 ATLANTIS (69)
Orbiter flight: 14
Mobile Launch Platform: 3
Launch pad: KSC 39A
Launch date/time: 27 June 1995 at 19.32.19 GMT
Landing date/time: 7 July 1995 at 14.54.34 GMT
Mission duration: 9 days 19 hours 22 minutes 15 seconds
Landing site: KSC
Main engines: 1: 2028 2: 2034 3: 2032
Launch Mass: =~2,046. 4 t
Landing Mass: 97,447 kg
Crew: Robert L Gibson (Commander), Charles J Precourt (Pilot), Ellen S Baker (Mission Specialist 1/EV2), Gregory J Harbaugh (Mission Specialist 2/EV1), Bonnie J Dunbar (Mission Specialist 3), AnatoliY Solovyov (Russia, Mir 19; up only), Nikolai M Budarin (Russia, Mir 19; up only), Vladimir N Dezhurov (Russia, Mir 18; down only), Gennady M Strekalov (Russia, Mir 18; down only), Norman E Thagard (Mir 18; down only).
Payload (Primary):
Spacelab
Payload (Secondary): cargo bay,
ODS Orbiter Docking System
Payload (Secondary): mid-deck,
SAREX-II Shuttle Amateur Radio Experiment
IMAX In-Cabin Camera

Mission summary
Atlantis departed at the beginning of a 10-minute window on the 100th US manned flight and the first of seven planned dockings with Mir in preparation for the International Space Station. After launch, the pilot Precourt halted 350 m directly underneath Mir by 10.50 GMT 29 June and then used a mean distance navigation strategy for final approach. This allowed thruster firings *away* from the passive Mir to avoid damage to the delicate solar wings, which remained in a feathered configuration as a further precaution.

Precourt station-kept at 82 m until Mir control gave permission to proceed. He used a camera inside the ODS to assist with alignment by aiming it through the hatch window at Kristall's hatch target. He again halted at 10 m until the final approach at 12.55 GMT. The ODS docking module completed capture at 13.00.15 GMT at 395 km altitude over Lake Baikal. The ODS capture ring carried three guide petals, each with two latches for soft docking. The ring was held 33 cm out from the ODS base ring by six 6-DoF shock absorbers. At contact, the closing rate was 3.26 cm/s (5.1 cm/s allowed), displacement error 1.8 cm, angular error 0.4° (4° allowed). At that point, Precourt triggered a 2.46 second computer sequence firing Shuttle's aft thrusters 14 times and the forward thrusters six times to push the two vehicles together. Harbaugh retracted the ring for the 12 base latches to hook on to Kristall, creating a 223 tonne space complex.

Precourt & Dezhurov shook hands some 2 h later and the record 10-man team collected in Mir's core for photographs. Mir was formally handed over to the fresh cosmonauts. The Shuttle delivered 200 kg

STS-49 astronauts Hieb (left), Akers (centre) and Thuot (right) capture Intelsat 603 during the first-ever three-man EVA (NASA) 0516772

of food, clothing and science equipment and more than 500 kg of water were delivered and Mir's pressure was raised from 0.92 atmosphere to 1.02 atmosphere. Shuttle provided most of the attitude control, holding the complex for optimum power generation during working hours.

Most of the rest of the joint mission investigated the long-term physiological effects on the Mir 18 crew, aboard Mir since March 1995. Some 400 saliva, urine and blood samples were returned from Mir/Spacelab. Spacelab's treadmill and Lower Body Negative Pressure unit helped to prepare the cosmonauts for return to Earth.

Kristall's hatch was closed at 19.32 GMT 3 July and Shuttle's at 19.48 GMT. Solovyov and Budarin undocked from Kvant 1 in Soyuz-TM 21 at 10.55 GMT 4 July purely to photograph Shuttle's 11.10 GMT undocking and flyaround. However, Mir began drifting when its main computer failed, so TM21 redocked 5 minutes early at 11.39 GMT. Mir remained drifting until they could replace the hardware. Precourt fired his separation thrusters at 12.35 GMT 4 July to end the joint phase. The three Mir 18 returnees occupied reclining mid-deck seats designed to help their adaptation to 1 *g*. US record-holder Thagard insisted on standing up 10 minutes after landing, while his Russian colleagues were stretchered out to three weeks of biomedical tests. Thagard reported that his physical condition was excellent, but his greatest challenge had been cultural isolation – sometimes going days without talking to a native English speaker.

STS-70 DISCOVERY (70)
Orbiter flight: 21
Mobile Launch Platform: 2
Launch pad: KSC 39B
Launch date/time: 13 July 1995 at 13.41.55 GMT
Landing date/time: 22 July 1995 at 12.02.00 GMT
Mission duration: 8 days 22 hours 20 minutes 5 seconds
Landing site: Kennedy Space Center
Main engines: 1: 2036/B1 2: 2019 3: 2017
Launch Mass: =~2,051.1 t
Landing Mass: 88,411 kg
Crew: Terrence T Henricks (Commander), Kevin R Kregel (Pilot), Donald A Thomas (Mission Specialist 1/EV2), Nancy J Currie (Mission Specialist 2), Mary E Weber (Mission Specialist 3/EV1).
Payload (Primary):

TDRS 7

Payload (Secondary): mid-deck,

BRIC Biological Research in Canisters (flew STS-64/68)

BDS Bioreactor Development System

CPCG Commercial Protein Crystal Growth block III

PARE R2 Physiological & Anatomical Experiment-Rodents

STL Space Tissue Loss (flew STS-45/53/56/59/66)

HERCULES Hand-held Earth-oriented Real-time Co-operative User-friendly Location-targeting & Environmental System (flew STS-53/56)

MIS Microcapsules in Space (flew STS-53)

MAST Military Application of Ship Tracks (flew STS-65/64/68)

RME-III Radiation Monitoring Experiment (flew STS-48/44/42/45/53/56/51/64)

SAREX-II Shuttle Amateur Radio Experiment

WINDEX Window Experiment

VFT Visual Function Tester 4 (flew STS-27/28/36/44/45/53/59)

Mission summary

NASA wanted to put STS-70 in space on 8 June but cancelled it on 2 June after a woodpecker attacked the ET's insulation. It left more than 100 holes of up to 4 cm, which NASA thought, would delay the tight schedule of STS-71. So STS-70 was rolled back to the VAB on 8 June and STS-71 became the next mission. Only six days separated the launch of 70 and 71, a NASA record.

NASA described the debut of the Block 1 improved SSME (#2036) as flawless. After launch, STS-70 released TDRS-7 for orbital insertion at 19.55 GMT 13 July. Houston's new-generation Flight Control Room handled, albeit partially (the orbital phase after IUS deployment). NRL's HERCULES camera, which first flew on STS-53 to photograph Earth features and simultaneously record their latitude and longitude to within 3.7 km added the capability of transmitting images and location data to the ground for analysis during flight with STS-56. The STS-70 version added a 10 m resolution multispectral night vision army package, although there were IMU alignment problems. HERCULES also transmitted an image of a debris crater discovered in one window. Following STS-70, OV-103 stood down from service before its nine-month maintenance period at Palmdale from October 1995.

STS-69 ENDEAVOUR (71)
Orbiter flight: 9
Mobile Launch Platform: 1
Launch pad: KSC 39A
Launch date/time: 7 September 1995 at 15.09.00 GMT
Landing date/time: 18 September 1995 at 11.37.56 GMT
Mission duration: 10 days 20 hours 28 minutes 56 seconds
Landing site: KSC
Main engines: 1: 2035 2: 2109 3: 2029
Launch Mass: =~2,050.3 t
Landing Mass: 99,740 kg
Crew: David M Walker (Commander), Kenneth D Cockrell (Pilot), James S Voss (Payload Commander/Mission Specialist 1/EV1), James H Newman (Mission Specialist 2), Michael L Gernhardt (Mission Specialist 3/EV2).
Payload (Primary):

Wake Shield Facility. 2×1,950 kg Spartan. 1,288 kg

Payload (Secondary): cargo bay,

GAS Bridge Assembly including CAPL Capillary Pumped Loop 2 (flew STS-60, ESA)

IEH International Extreme UV Hitchhiker

ITEPC Inter-Mars Tissue-Equivalent Proportional Counter

Payload (Secondary): mid-deck,

APE-B Auroral Photography Experiment (flew STS-43/51/60/62)

BRIC Biological Research in Canisters (flew STS-64/68/70)

CGBA Commercial Generic Bioprocessing Apparatus (flew STS-50/54/60/62/63)

CMIX 4 Commercial Materials

TSS in Alenia Spazio's anechoic chamber during final tests in Turin before delivery to Kennedy Space Center 0516773

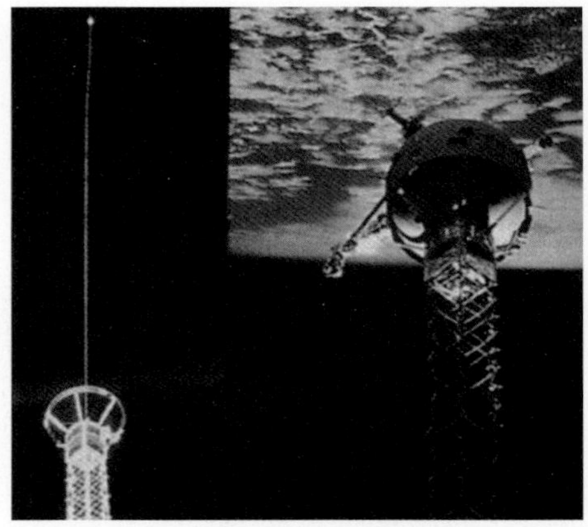

Beginning deployment of Italy's tethered satellite during STS-46. The 2.54 mm 70 tether repeatedly jammed in the reel mechanism (the box unit at the boom's apex). At left, the sphere approaches its maximum altitude (NASA and Alenia Spazio) 0516774

ITA Experiment (flew STS-52/56/67)

EPICS Electrolysis Performance Improvement Concept Study

STL Space Tissue Loss (flew STS-45/53/56/59/66/70)

Mission summary

Engineers delayed STS-69 from its 5 August launch when they discovered hot gases had damaged O-rings on STS-70 and 71's SRB nozzle joints. Thiokol reinsulated STS-69's.

Spartan 201–03 was a reflight of STS-56's 201–01 and 201–02 on STS-64. Gernhardt released it from the RMS 15.42 GMT 8 September, leaving Endeavour to move 105 km behind. The RMS retrieved the satellite at 15.02 GMT 10 September. This time the observatory viewed the Sun while ESA's Ulysses made its northern polar pass.

WSF, developed by the Space Vacuum Epitaxy Center under a NASA grant, investigated the commercial possibilities of producing new electronic, magnetic and superconducting thin-film materials and devices in the 10^{-14} torr ultravacuum of space. In orbit, however, technical problems prevented the first WSF from being released from the Shuttle during STS-60. Still it grew five 7 cm diameter GaAs wafers under 10^{-10} torr (about the best achievable on Earth). Endeavour's astronauts released WSF 2 at 11.25 GMT 11 September. The satellite used its own 0.6 N N_2 thruster to move 25 km behind Endeavour (making it the first payload to perform its own separation manoeuvre). It shut down during the fifth of seven planned runs when its batteries depleted. Despite cooling, attitude control and communications problems, SVEC declared that WSF had achieved its primary objectives. Newman grappled it 13.59 GMT 14 September after the approaching Orbiter deliberately hit it with thruster exhaust to test plume effects on rendezvous targets. During day 9, Newman hoisted it to various positions to map Endeavour's plasma wake for 5 h.

Voss and Gernhardt performed a 406 minute generic EVA during 16 September to test new glove heaters as they worked on typical station tasks. Newman lifted them away from the cargo bay's warmth and they could also shut off their body cooling loops to examine extreme conditions.

STS-73 COLUMBIA (72)
Orbiter flight: 18 (6th Extended Duration Orbiter)
Mobile Launch Platform: 3
Launch pad: KSC 39B
Launch date/time: 20 October 1995 at 13.53.00 GMT
Landing date/time: 5 November 1995 at 11.45.21 GMT
Mission duration: 15 days 21 hours 52 minutes 21 seconds
Landing site: KSC
Main engines: 1: 2037/B1 2: 2031 3: 2038/B1
Launch Mass: =~2,051.5 t
Landing Mass: 104.400 kg (heaviest landing to date)
Crew: Kenneth D Bowersox (Commander), Kent V Rominger (Pilot), Catherine G Coleman (Mission Specialist 1/EV2), Michael E Lopez-Alegria (Mission Specialist2/EV1), Kathryn C Thornton (Payload Commander/Mission Specialist 3), Fred W Leslie (PS1), Albert Sacco Jr (PS2).
Payload (Primary):

US Microgravity Lab-2 Spacelab 10,308 kg of experiments composed of Crystal Growth Furnace-4 experiment, Drop Physics Module, Surface Tension Driven Convection Experiment, Geophysical Fluid Flow Cell, ESA Advanced Protein Crystallisation Facility, ESA Glovebox-7 experiment, SAMS 07

Training with the HERCULES camera, which includes a ring laser gyro for accurate pointing information 0516775

Space Acceleration Measurement System, 3-Dμg Accelerometer, STABLE Suppression of Transient Accelerations by Levitation Evaluation.
Payload (Primary): mid-deck,
Astroculture
CGBA Commercial Generic Bioprocessing Apparatus
CPCG Commercial Protein Crystal Growth
ZCG Zeolite Crystal Growth
SPCG Single Locker Protein Crystal Growth
Payload (Secondary):
OARE Orbital Acceleration Research Experiment

Mission summary
USML 2 followed up and repeated experiments flown aboard STS-50/USML 1, conducting 16 experiments in materials, fluid physics, combustion and biotechnology. The crew divided into two shifts to pursue the experiments 24 hours a day, laying the foundations for μg research aboard the International Space Station. It was Columbia's first flight since its most recent inspection at Palmdale after STS-64 and the first to carry two Block 1 engines. The Orbiter should have been the first to fly the full set of new engines, but #2039 was replaced by the old #2031 when technicians feared that a seal in the upgraded LOX pump was improperly seated. It took six attempts to get the Columbia launched.

STS-74 ATLANTIS (73)
Orbiter flight: 15
Mobile Launch Platform: 2
Launch pad: KSC 39A
Launch date/time: 12 November 1995 at 12.30.43 GMT
Landing date/time: 20 November 1995 at 17.01.27 GMT
Mission duration: 8 days 4 hours 30 minutes 44 seconds
Landing site: KSC
Main engines: 1: 2012 2: 2026 3: 2032
Launch Mass: =~2,046.5 t
Landing Mass: 92,700 kg
Crew: Kenneth D Cameron (Commander), James D Halsell Jr. (Pilot), Chris A Hadfield (Mission Specialist 1, Canada), Jerry L Ross (Mission Specialist 2/EV1), William S McArthur Jr (Mission Specialist 3/EV2)
Payload (Secondary): cargo bay,
IMAX camera
GLO 4
PASDE Photogrammetric Appendage Structural Dynamics Experiment (measured motion of Mir solar arrays)
ITEPC Inter-Mars Tissue Equivalent Proportional Counter (flew STS-69)
Payload (Secondary): mid-deck,
SAREX-II Shuttle Amateur Radio Experiment

Mission summary
RSA and NASA planned to use this second docking mission to Mir to attach RKK Energia's 4,087 kg docking adaptor and to perform two days of life sciences experiments. After installation, the 2.2 m diameter 4.7 m long androgynous docking module attaches to Kristall at Mir's starboard position so that future dockings will adequately clear Mir's solar wings. Without it, Kristall has to be shifted to the forward face during each Shuttle visit. In order to perform the final docking, four DM retroreflectors guide the Shuttle's Laser Trajectory Control Sensor to generate 6 Degrees of Freedom (DoF) information about position between the two vehicles.

During 14 November, Hadfield unberthed the DM from the aft cargo bay and held it 12 cm above 104's Orbiter Docking System, reused from STS-71. With RMS in 'limp mode', Cameron thrusted Atlantis into contact for retraction at 07.17 GMT. Atlantis docked at 06.27.39 GMT 15 November and astronauts opened the seven hatches at 09.00 GMT when the pressures had equalised.

STS-74 delivered 275 items totalling 968 kg and took away 195 items (370 kg). 400 kg of water in containers, 65 kg of air, experimental equipment, fresh food (including vegetables, steak and ice cream), gifts of flowers and sweets, and a collapsible classical guitar were among the goods brought aboard. The Shuttle departed with 20 kg of ESA cargo from Reiter's activities, including frozen medical samples, furnace material samples and science data on videotape and computer disk.

Atlantis undocked 08.16 GMT 18 November to circle Mir at 130 m for 45 minutes while the crew took detailed engineering photographs. STS-74 carried the 200th individual to fly in the Shuttle.

STS-72 ENDEAVOUR (74)
Orbiter flight: 10
Mobile Launch Platform: 1
Launch pad: KSC 39B
Launch date/time: 11 January 1996 at 09.41.00 GMT
Landing date/time: 20 January 1996 at 07.41.41 GMT
Mission duration: 8 days 22 hours 0 minutes 41 seconds
Landing site: KSC
Main engines: 1: 2028 2: 2039/B1 3: 2036/B1
Launch Mass: =~2,048.0 t
Landing Mass: 98,553 kg
Crew: Brian K Duffy (Commander), Brett W Jett Jr (Pilot), Leroy Chiao (Mission Specialist 1/EV1), Winston E Scott (Mission Specialist 2/EV3), Koichi Wakata (Mission Specialist 3, Japan), Daniel T Barry (Mission Specialist 4/EV2).
Payload (Primary):
Spartan 206 OAST-Flyer
Payload (Secondary): cargo bay,
SSBUV/A Shuttle Solar Backscatter UV
SLA Shuttle Laser Altimeter
5 GAS
Payload (Secondary): mid-deck,
PARE R3 Physiological & Anatomical Rodent Experiment (three female rats, each with 10 newborn)
PCG/STES 4 Protein Crystal Growth Single Locker Thermal Enclosure System
CPCG Commercial Protein Crystal Growth
STL Space Tissue Loss (flew STS-45/53/56/59/66/70/69)

Mission summary
NASDA's astronaut Koichi Wakata captured Japan's SFU (Space Flyer Unit) with the RMS at 10.57 GMT 13 January after one orbit caused by the free-flyer's solar wings refusal to latch down after retraction. They had to be pyrotechnically jettisoned. Wakata released Goddard's OAST-Flyer from the RMS 11.32 GMT 14 January for two days of separate operations, regrappling it 09.47 GMT 16 January. The satellite conducted four experiments, namely: surface contamination investigations; the first attitude control of a satellite using GPS data; a laser-triggered pyrotechnic device firing and amateur packet radio.

Astronauts Chiao and Barry performed a 369-minute EVA 1 15 January, to test generic Space Station operations such as moving equipment, the strength of a flexible foot restraint and connecting simulated fluid lines. Chiao and Scott followed with a 414-minute EVA on 17 January performing similar duties and including Scott remaining still for 30 minutes in the shadowed cargo bay to test suit modifications. The test proved the suit could cope with –60°C conditions during Station work.

SLA began a series of four flights sampling land topography and vegetation data from a spaceborne laser altimeter.

STS-75 COLUMBIA (75)
Orbiter flight: 19 (7th Extended Duration Orbiter)
Mobile Launch Platform: 3
Launch pad: KSC 39B
Launch date/time: 22 February 1996 at 20.18.00 GMT
Landing date/time: 9 March 1996 at 13.58.21 GMT
Mission duration: 15 days 17 hours 40 minutes 21 seconds
Landing site: KSC
Main engines: 1: 2029 2: 2034 3: 2017
Launch Mass: =~2,053.2 t
Landing Mass: 103,160 kg
Crew: Andrew M Allen (Commander), Scott J Horowitz (Pilot), Jeffrey A Hoffman (Mission Specialist 1), Maurizio Cheli (Mission Specialist 2/ESA), Claudie Nicollier (Mission Specialist 3/ESA/EV2), Franklin R Chang-Diaz (Payload Commander/Mission Specialist 4/EV1), Umberto Guidoni (PS1, Italy)
Payload (Primary):
TSS 1R Tethered Satellite System (system mass 5,506 kg)
USMP 3 US Microgravity Payload
Payload (Secondary):
OARE Orbital Acceleration Research Experiment
Payload (Secondary): mid-deck,
Glovebox (combustion studies)
CPCG 09 Commercial Protein Crystal Growth

Mission summary
The joint Italian and US TSS 1R reflew STS-46's tethered satellite system. In the first flight the tether jammed at only 256 m instead of the planned 20 km. When fully deployed, the Italian designers expected that 2.3 kW should have flowed along the 2.5 mm tether's copper core down to the Shuttle. In this test, the researchers expected to deploy the 1.6 m diameter sphere from a 12 m boom to 20.7 km. They planned for the deployment to take 5.5 h with no stops. Then, after 22 h on station, the satellite

STS-57's Wisoff on the RMS watches David Low working on Eureca's antenna. Spacehab is in the foreground (NASA) 0516776

Endeavour makes the 16th Shuttle landing in Florida to conclude STS-57 (NASA) 0516777

NASA's ACTS communications satellite begins the journey out of Shuttle's cargo bay (NASA) 0516778

retrieval would require 17 h, including a stop at 3.2 km for experiment operation.

Unreeling began 20.45 GMT 25 February. Unfortunately, the tether snapped near the top of the boom without warning at 01.30 GMT 26 February while TSS was 19.7 km out, leaving the sphere in a 316 × 414 km, 28.47° orbit. Since the original orbit began at 294 × 304 km, the Italians at least demonstrated orbit raising without propellant. At the time of failure, the cable experienced only 70 N of tension against a testing breaking strength of 1.78 kN. NASA Marshall controllers continued taking TSS data for 4 days until the batteries were exhausted early 1 March. Ground observers clearly witnessed the stable tether before the 00.55 GMT re-entry 19 March over the Atlantic.

The two missions cost US$404 million, but no more are planned. A failure review board concluded in June 1996 that the cable insulating layer had been penetrated, possibly by debris squeezed in while it was still wound on the reel. Electric arcing began while it was in the lower tether control mechanism, sputtering intermittently for 9 seconds as the moving tether passed through the deployer mechanisms and into the boom area. The remaining Kevlar strands could not withstand the small deployment forces.

STS-76 ATLANTIS (76)
Orbiter flight: 16
Mobile Launch Platform: 2
Launch pad: KSC 39B
Launch date/time: 22 March 1996 at 08.13.04 GMT
Landing date/time: 31 March 1996 at 13.28.57 GMT
Mission duration: 9 days 5 hours 15 minutes 53 seconds
Landing site: EAFB
Main engines: 1: 2035 2: 2109/B1 3: 2019
Launch Mass: =~2,045.5 t
Landing Mass: 95,400 kg
Crew: Kevin P Chilton (Commander), Richard A Searfoss (Pilot), Ronald M Sega (Mission Specialist 1), Michael R Clifford (Mission Specialist 2/EV2), Linda M Goodwin (Mission Specialist 3/EV1), Shannon W Lucid (Mission Specialist 4, up only).
Payload (Secondary): cargo bay,
MEEP Mir Environmental Effects Programme
TRIS Trapped Ions in Space
Payload (Secondary): mid-deck,
Kidsat
SAREX-II Shuttle Amateur Radio Experiment

Mission summary
NASA intended to use this flight to deliver Shannon Lucid to the Mir in its third docking with the space station, to perform an EVA and deliver supplies. Atlantis docked at 02.34 GMT 24 March with Kristall's adaptor module (attached by STS-74). Clifford and Goodwin's 363-minute EVA, beginning 06.34 GMT 27 March from Shuttle's tunnel hatch, attached the MEEP to Mir to gather space debris and materials data. It also tested common US/Russian tether hooks and foot restraints. The Shuttle transferred some 640 kg of water along with 880 kg of Russian supplies (such as one gyrodyne and three batteries returned by STS-71), 740 kg US science equipment and 93 kg other items. About 60 per cent of Spacehab's 2,010 kg cargo were Russian supplies. 450 kg returned, with the Shuttle including 110 kg science payload. One experiment mapped Mir's radio and electric field environment. Then the Shuttle undocked at 01.08 GMT 29 March. Because of the orbit inclination, the more stringent landing conditions resulted in a switch to EAFB for the return.

Musgrave and Hoffman ride the RMS near the end of 35 hr of EVA servicing of the Hubble telescope in Endeavour's cargo bay (NASA) 0516779

STS-77 ENDEAVOUR (77)
Orbiter flight: 11
Mobile Launch Platform: 1
Launch pad: KSC 39B
Launch date/time: 19 May 1996 at 10.30.00 GMT
Landing date/time: 29 May 1996 at 11.09.18 GMT
Mission duration: 10 days 0 hours 39 minutes 18 seconds
Landing site: KSC
Main engines: 1: 2037/B1 2: 2040/B1 3: 2038/B1
Launch Mass: =~2,049.8 t
Landing Mass: 100,420 kg
Crew: John H Casper (Commander), Curtis L Brown Jr (Pilot), Andrew S W Thomas (Mission Specialist 1), Daniel W Bursch (Mission Specialist 2/EV2), Mario Runco Jr (Mission Specialist 3/EV1), Marc Garneau (Mission Specialist 4, Canada)
Payload (Primary):
Spacehab 4 Spacehab carried commercial space product development payloads made up of the Advanced Separation Process for Organic Materials, CGBA Commercial Generic Bioprocessing Apparatus, Plant GBA, Fluids GBA 2, Hand-Held Diffusion Test Cell, Immune 03, CPCG Commercial Protein Crystal Growth, GPPM Gas-Permeable Polymeric Membrane, Commercial Float Zone Furnace, Space Experiment Facility, Space Tissue Loss/NIH-Cells 07.
Spartan 207/IAE The Spartan 207 Inflatable Antenna Experiment deployed JPL's 12.2 m diameter 60 kg aluminised mylar parabolic antenna structure on three 28 m inflatable struts. It was jettisoned after its surface accuracy was measured and Spartan was recovered.
TEAMS 01 Technology Experiments Advancing Missions in Space. TEAMS' PAMS released a small passive satellite from its Hitchhiker canister, illuminating it with a laser to observe its attitude as it was stabilised aerodynamically and magnetically.
GANES GPS Attitude & Navigation Experiment
Liquid Metal Test Experiment
PAMS Passive Aerodynamically Stabilised Magnetically Damped Satellite
VTRE Vented Tank Resupply Experiment
Payload (Secondary): cargo bay,
BETSCE Brilliant Eyes Ten-Kelvin Sorption Cryocooler Experiment,
GAS Bridge (12 cans)
RFTPCE Reduced-Fill Tank Pressure Control Experiment
Payload (Secondary): mid-deck:
ARF Aquatic Research Facility. The Canadian Space Agency/NASA ARF debuted as a modular facility that can support different biological experiments in two standard mid-deck lockers. This flight carried mussels, starfish and sea urchins.
BRIC 07 Biological Research in Canisters (flew STS-64/68/70/69)

STS-78 COLUMBIA (78)
Orbiter flight: 20 (8th Extended Duration Orbiter)
Mobile Launch Platform: 3
Launch pad: KSC 39B
Launch date/time: 20 June 1996 at 14.49.00 GMT
Landing date/time: 7 July 1996 at 12.37.45 GMT
Mission duration: 16 days 21 hours 47 minutes 45 seconds
Landing site: KSC
Main engines: 1: 2041/B1 2: 2039/B1 3: 2036/B1
Launch Mass: =~2,049.3 t
Landing Mass: 103,424 kg
Crew: Terrence T Hendricks (Commander), Kevin R Kregel (Pilot), Richard M Linnenhan (Mission Specialist 1/EV2), Susan J Helms (Payload Commander/Mission Specialist 2/EV1), Charles E Brady Jr (Mission Specialist 3), Jean-Jacques Favier (PS1, France), Robert Brent Thirsk (PS2, Canada)
Payload (Primary):
LMS 1 Life and Microgravity Sciences.
Payload (Secondary): payload bay,
Orbital Acceleration Research Experiment
Payload (Secondary): mid-deck,
SAREX-2 (Shuttle Amateur Radio Experiment). Flown in configuration C which operates in either voice or data mode in communications with amateur stations within line of sight of the orbiter).
BRIC-08 (Biological Research in Canisters, block 2 configuration)

Mission summary
After a successful launch, the crew of the 78th Shuttle mission successfully activated the Spacelab, which housed more than 40 life science and microgravity experiments in the Shuttle's cargo bay. The experiments measured the physiological changes that take place in the human body during space flight and also on utilising the unique microgravity environment to study materials processing techniques. On the first day in orbit, the flight controllers dumped the back-up flight software that had experienced intermittent transients during Columbia's launch. The day after launch, Henricks worked with the Bubble Drop Facility while Kregel laboured over a computer workstation to assess human behaviour changes during long-duration

Thomas (left) and Cabana at the conclusion of STS-65 (NASA) 0517098

Mark Lee tests the SAFER rescue propulsion pack during STS-64. SAFER sits below his suit's backpack, with the controls at front. Behind him resting along the bay's sill is SPIFEX (NASA) 0517097

space flight. The other crewmembers worked with neurological and cardiovascular experiments. Human physiology tests on the flight included the Direct Measurement of the Initial Bone Response to Space Flight investigation, which scientists used to explore ways to counter this detrimental effect of space flight.

On 22 June, flight controllers assisted the astronauts in a procedure to flush an ice blockage out of the Flash Evaporator System (FES), used to dissipate heat from the shuttle. The procedure, which ran warm freon through the FES, succeeded in making the system once again fully operational. To supplement the FES, flight controllers also deployed one of two sets of radiators mounted on the Shuttle's payload bay doors to further radiate heat from the orbiter. Henricks and Kregel tested a device designed to permit voice control of the Shuttle's closed circuit television system. The voice command system should allow astronauts to command the CCTV system verbally while performing other tasks. Kregel reported that the system responded to at least 90 per cent of his verbal commands. On 24 June, the STS-78 crew members contacted friends and family members on the ground with the SAREX.

Three days later, Brady held an unlit Olympic Torch aloft while exercising on the bicycle ergometer in the Spacelab. The other astronauts on board joined him in that informal salute to the 1996 Olympics, which opened in Atlanta on 19 July.

On 28 June, the crew performed a second successful flush of a small buildup of ice from orbiter's FES after the ice formed during a dump of the supply water tanks. The procedure took about 20 minutes to complete. Canadian payload specialist Bob Thirsk observed Canada Day 1 July as the other crewmembers spent their twelfth day in orbit. The Americans then had the opportunity to observe their own 4 July holiday three days later.

After a one-day extension from the ground, on 6 July the astronauts conducted routine firings of the orbiter's reaction control system jets and checked out its flight control systems and aero surfaces in anticipation of the forthcoming landing at the Kennedy Space Center. Henricks and Kregel pulsed Columbia's vernier reaction control system jets to gently raise Columbia's altitude. These firings tested the manoeuvres needed by Discovery's next mission (STS-82) to service the Hubble Space Telescope. The vernier jet firings must be gentle enough to raise the orbit without disturbing any payloads on board, or in the case of the Hubble Space Telescope, without placing any force on the telescope's fragile solar arrays.

The return on 7 July ended the longest flight to that date in the Shuttle programme.

STS-79 ATLANTIS (79)
Orbiter flight: 17
Mobile Launch Platform: 1
Launch pad: KSC 39A
Launch date/time: 16 September 1996 at 08.54.49 GMT
Landing date/time: 26 September 1996 at 12.13.15 GMT
Mission duration: 10 days 3 hours 18 minutes 26 seconds
Landing site: KSC
Main engines: 1: 2012 2: 2031 3: 2033
Launch Mass: =~2,045.9 t
Landing Mass: 97,603 kg
Crew: William F Readdy (Commander), Terrence W Wilcutt (Pilot), Jay Apt (Mission Specialist 1/EV1), Thomas D Akers (Mission Specialist 2), Carl E Walz (Mission Specialist 3/EV1), John E Blaha (Mission Specialist 4, up only), Shannon W Lucid (Mission Specialist 4, down only)
Payload (Primary):
SPACEHAB 05 Double module, orbiter docking system.
Payload (Secondary): mid-deck,
SMM-04
SAREX-2 (configuration M)
IMAX in-cabin camera
MSX Mid-course Space Experiment

Mission summary
On 1 July 1996, NASA transferred Atlantis from the VAB to pad 39A. As a precautionary measure, NASA, on 10 July, rolled the Shuttle stack back to the VAB to avoid any damage from the approach of Hurricane Bertha. Once retrieved, NASA decided to delay the launch until September, as a result of the discovery that hot gases had seeped into the field joints of the STS-78 SRBs. The ground crew destacked Atlantis from its SRBs and re-stacked it with new solids. Then NASA rolled the stack back to the pad on 20 August. Hurricane Fran forced NASA to return the stack to the VAB on 4 September as the launch awaited a one day delay. Launch finally occurred on 16 September.

After reaching its high-inclination orbit, Atlantis docked with Mir at the Docking Module attached to the Kristall on 19 September at 03.13 GMT. After two hours of pressure and leak checks, astronauts opened the hatches between the two spacecraft at 05.40 GMT. The two crews greeted one another to begin five days of joint operations. Soon after the crewmembers completed their welcoming ceremony, they hauled bags of water and other supplies from the Shuttle's SPACEHAB module into the Mir, eventually transferring more than two tonnes of equipment and logistical supplies to Mir.

During the first night that Atlantis was docked with Mir, scientists on the ground continued work with the Active Rack Isolation System in the SPACEHAB module, completing several successful tests. The system experimented with dampening crew motion and engine firings to protect the microgravity environment on the International Space Station.

During the first full day of joint experiments, the Biotechnology System, an investigation to study the long-term development of cartilage cells in microgravity, moved from the Shuttle to Mir. During his stay on Mir, John Blaha took weekly samples of the culture, to determine the feasibility of engineering cartilage cells for possible use in transplantation.

On 20 September, Carl Walz worked with the Mechanics of Granular Materials experiment in the double SPACEHAB module while Mission Specialist Shannon Lucid had a workout on the Mir treadmill, continuing physical conditioning exercises designed to prepare her body for the return to Earth's gravity after six months in orbit. Jay Apt monitored the ARIS experiment as Readdy and Korzun fired manoeuvring jets on their spacecraft to test the ability of the International Space Station prototype experiment rack to dampen the resulting vibrations.

Lucid's stay on Mir ended on 23 September at 23.33 GMT when Atlantis undocked from the orbital station to begin the return home, leaving John Blaha behind. After undocking, Atlantis flew around Mir 1½ times at a distance of over 100 m, before the orbiter performed its final separation manoeuvre. When Atlantis returned to KSC on 26 September, Lucid claimed the record for the longest flight by a non-Russian (non-Soviet), having been in orbit for 188 days 4 h.

STS-80 COLUMBIA (80)
Orbiter flight: 21 (9th Extended Duration Orbiter)
Mobile Launch Platform: 3
Launch pad: KSC 39B
Launch date/time: 19 November 1996 at 19.55.47 GMT
Landing date/time: 7 December 1996 at 11.49.05 GMT
Mission duration: 17 days 15 hours 53 minutes 18 seconds
Landing site: KSC
Main engines: 1: 2032 2: 2026 3: 2029
Launch Mass: =~2,052.7 t
Landing Mass: 103,365 kg
Crew: Kenneth D Cockrell (Commander), Kent V Rominger (Pilot), Tamara E Jernigan (Mission Specialist 1/EV1), Thomas D Jones (Mission Specialist 2/EV2), F Story Musgrave (Mission Specialist 3)
Payload (Primary):
WSF 3 (Wake Shield Facility)
ORFEUS-SPAS 2 (Orbiting and Retrievable Far and Extreme Ultraviolet Spectrograph – Shuttle Pallet Satellite). 2 free-flying spacecraft
Payload (Secondary): payload bay,
ITEPC Inter-Mars Tissue Equivalent Proportional Counter
SEM 01 Space Experiment Module/Getaway Special
Payload (Secondary): mid-deck,
PARE Physiological and Anatomical Rodent Experiment
MDA Materials Dispersions Apparatus
VIEW-CPL Visualisation in an Experimental Water Capillary Pumped Loop
CCM Cell Culture Module
BRIC 09 Biological Research in Canisters

Mission summary
NASA employed the Mobile Transporter to move The STS-80 stack from the VAB to launch pad 39B on 16 October 1996, ready for launch the following month. The space agency had to delay the launch on 4 November until no earlier than 15 November to allow additional time to study unexpected SRB nozzle erosion which was observed following the STS-79 launch. Then on 13 November, NASA announced that the launch needed an additional five-day delay until 19 November because of inclement weather. The final launch had a two-minute hold to monitor the hydrogen concentration in the orbiter's aft compartment.

Columbia released its first satellite, the ORFEUS-SPAS 2, from the orbiter's RMS at 04.11 GMT 20 November. The release occurred just over an hour late as a result of a longer-than-planned predeployment checkout. During its mission, the science satellite gathered data on the origin and makeup of stars.

Two days later, Columbia released WSF 3 by the Shuttle's robot arm at 00.38 GMT 22 November. Within 48 h ground controllers feared that ORFEUS-SPAS might be closing in on WSF at a higher rate than planned and they decided to retrieve the latter satellite about a day earlier than the original plan. The remote manipulator latched on to the satellite

ESA's STS-66 mission specialist Clervoy tested a new reclining seat for returning long-duration occupants from Mir. Behind him is the airlock hatch; to his right are sleep cubicles (NASA) 0517133

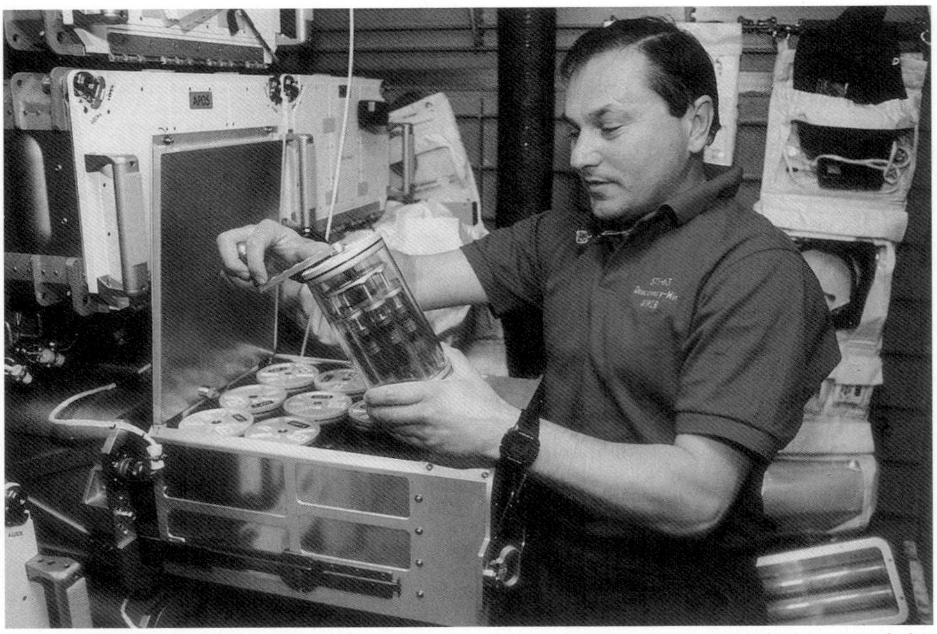

Russian Vladimir Titov handles samples for Spacehab 3's Commercial Generic Bioprocessing Apparatus during STS-63 (NASA) 0517122

at 01.01 GMT on 26 November and reberthed the satellite in the orbiter's payload bay. WSF scientists reported they completed seven thin film growths of semiconductor materials, the maximum capability for the satellite. At 08.23 GMT the Shuttle retrieved ORFEUS-SPAS as well.

NASA cancelled a planned EVA activity for STS-80 on 29 November. Depressurisation had already begun and Jernigan and Jones had to leave Columbia's airlock because, despite repeated attempts, they were unable to open the airlock due to a stuck handle. The EVAs had been planned as tests of equipment and procedures for International Space Station operations.

Bad weather cancelled the first planned landings on 5 and 6 December, and Columbia finally returned to KSC on 7 December, setting a new Shuttle duration record.

STS-81 ATLANTIS (81)
Orbiter flight: 18
Mobile Launch Platform: 2
Launch pad: KSC 39B
Launch date/time: 12 January 1997 at 09.27.23 GMT
Landing date/time: 22 January 1997 at 14.22.55 GMT
Mission duration: 10 days 4 hours 55 minutes 32 seconds
Landing site: KSC
Main engines: 1: 2041/B1 2: 2034 3: 2042/B1
Launch Mass: =~2,046.1 t
Landing Mass: 97,275 kg
Crew: Michael A Baker (Commander), Brett W Jett Jr (Pilot), Peter J K Wilsoff (Mission Specialist 1), John M Grunsfeld (Mission Specialist 2), Marsha Ivins (Mission Specialist 3), Jerry M Linenger (Mission Specialist 4, up only), John E Blaha (Mission Specialist 4, down only)
Payloads (Primary): mid-deck,
SMM-05
SPACEHAB 06 Double module, orbiter docking system.
Payloads (Secondary): mid-deck,
CREAM Cosmic Radiation Effects and Activation Monitor
KidSat

Mission summary
On 10 December 1996, the ground crew moved the STS-81 stack from the VAB to pad 39B. After launch, one month later, the orbiter docked with the Docking Module attached to the Kristall module of Mir to begin joint operations as part of the fifth Shuttle-Mir Mission. The docking occurred at 03.55 GMT on 15 January. When the Americans entered Mir the handover between John Blaha and Jerry Linenger as a resident Mir crewmember took place. The Shuttle carried three tonnes of supplies to Mir, including equipment to be used by Linenger for experiments.

In the morning on 17 January, one of three Inertial Measurement Units (IMU) which provide navigation information to the Shuttle's general purpose computers showed significant drifting and Mission Control put it in standby mode. The IMU returned to service for use during Atlantis' landing.

Undocking from Mir came at 02.16 GMT on 20 January and Blaha returned to Earth with the remaining STS-81 crew two days later. Linenger stayed in orbit, awaiting collection by the STS-84 (SMM-06) mission due in May 1997.

STS-82 DISCOVERY (82)
Orbiter flight: 22
Mobile Launch Platform: 1
Launch pad: KSC 39A
Launch date/time: 11 February 1997 at 08.55.17 GMT
Landing date/time: 21 February 1997 at 08.32.37 GMT
Mission duration: 9 days 23 hours 37 minutes 20 seconds
Landing site: KSC
Main engines: 1: 2037/B1 2: 2040/B1 3: 2038/B1
Launch Mass: =~2,047.8 t
Landing Mass: 97,076 kg
Crew: Kenneth D Bowersox (Commander), Scott J Horowitz (Pilot), Joseph R Tanner (Mission Specialist 1/EV4), Steven A Hawley (Mission Specialist 2), Gregory J Harbaugh (Mission Specialist 3/EV3), Mark C Lee (Mission Specialist 4/EV1), Steven L Smith (Mission Specialist 5/EV2)

Mission summary
On 17 January 1997, NASA transported the STS-82 stack with the Shuttle orbiter *Discovery* from the VAB to pad 39A for launch the following month. NASA dedicated this flight to the second servicing mission for the Hubble Space Telescope, the installation of Near-Infra-red Camera and Multi-Object Spectrometer (NICMOS), the Solid-State Recorder (SSR) 1 and the Space Telescope Imaging Spectrograph (STIS). The crew also replaced the Fine Guidance Sensor (FGS) 1, Reaction Wheel Assembly (RWA) 1. As a secondary objective, the crew planned to install the Optical Control Electronics Enhancement Kit (OCE-EK), Magnetic Sensing System (MSS) covers and to replace the Data Interface Unit (DIU) 2, the Engineering/Science Tape Recorder (ESTR) 2 and the Solar Array Drive Electronics (SADE) 2.

The launch occurred on time. Two days after launch, at 08.34 GMT on 13 February Steven Hawley – who had first deployed HST during the STS-31 mission – used the orbiter's robotic arm to capture the telescope. Within half an hour it was berthed in Discovery's payload bay. The first EVA to service HST began at 04.34 GMT on 14 February when astronauts Lee and Smith left the Shuttle proper for 402 minutes. Once outside, the astronauts opened a hatch on the aft side of HST and removed the Goddard High-Resolution Spectrograph and the Faint Object Spectrograph and replaced them with the STIS and the NICMOS.

On 15 February, the second EVA took place. Astronauts Harbaugh and Tanner began their EVA at 01.25 GMT and it lasted for 442 minutes. They replaced a degraded FGS and a failed ESTR with new models. They also installed the OCE-EK, which will increase the capability of the new FGS. During the EVA activities the astronauts and flight controllers noticed that the thermal insulation showed signs of cracking and general wear and suggested an additional EVA to allow extra unplanned repair work to be completed. As the EVA drew to a close, Discovery's RCS fired to slightly raise the orbit of the assembly, in advance of a larger re-boost later in the mission. On 16 February, NASA decided that a fifth EVA would be added to the four planned for the STS-82 mission.

EVA number three started at 01.53 GMT 16 February. During this effort, astronauts Lee and Smith removed and replaced a DIU and a reel-to-reel style ESTR with a new digital SSR that will allow the simultaneous recording and playback of data. Then they replaced an RWA, and ended the EVA after 431 minutes.

The fourth EVA began at 03.45 GMT on 17 February with Harbaugh and Tanner outside Discovery. They first replaced a SADE used to control the positioning of HST's solar arrays. Next, the astronauts moved to the top of the telescope and replaced the covers of the MSS. They then placed thermal protection over two areas of degraded insulation around the light shield part of the telescope. Harbaugh and Tanner returned inside Discovery after an EVA lasting for 394 minutes.

The fifth (unplanned) and final EVA began at 03.15 GMT on 18 February, when astronauts Lee and Smith left Discovery for the third time. They attached several thermal insulation patches to three equipment compartments at the top of the Support Systems Module of HST which contain key data processing, electronics and scientific instrument telemetry packages. Following the completion of that work, Lee and Smith briefly returned to the airlock while flight controllers evaluated a possible problem with one of four RWA units used to manoeuvre the telescope for its scientific observations. After determining that further analysis of the Reaction Wheel Assembly would be required, ground control directed the astronauts to end their spacewalk and re-enter the airlock for the final time.

With the HST servicing complete, Discovery raised the orbit of the assembly in anticipation of the re-deployment of the telescope the following morning. Additional orbital 'tweaking' resulted from an orbital prediction that HST might collide with a piece of space debris (1994-029HU). Before being captured, HST orbited at 590–599 km and when it had been released, its new orbit was 599–620 km. The Shuttle released the HST on 19 February at 06.41 GMT. Within minutes, thrusters on Discovery had fired to begin the retreat from the telescope. Two days later, Discovery returned to Earth.

With the mission complete, testing of the new HST instruments showed that one of the three cameras on the NICMOS is not focusing properly. Controllers hope that this problem might cure itself with time.

STS-83 COLUMBIA (83)
Orbiter flight: 22 (10th Extended Duration Orbiter)
Mobile Launch Platform: 3
Launch pad: KSC 39A
Launch date/time: 4 April 1997 at 19.20.32 GMT
Landing date/time: 8 April 1997 at 18.33.23 GMT
Mission duration: 3 days 23 hours 12 minutes 51 seconds
Landing site: KSC
Main engines: 1: 2012/B2 2: 2109/B2 3: 2019/B2
Launch Mass: =~2,051.7 t
Landing Mass: 96,643 kg
Crew: James D Halsell Jr (Commander), Susan L Still (Pilot), Janice S Voss (Payload Commander/ Mission Specialist 1), Michael L Gernhardt (Mission Specialist 2/EV1), Donald A Thomas (Mission Specialist 3/EV2), Roger K Crouch (PS1), Gregory T Linteris (PS2)
Payload (Primary): payload bay,
MSL 1 Microgravity Science Laboratory

STS-71 cargo bay configuration for Mir docking (NASA) 0517261

Gibson (foreground) and Dezhurov shake hands across the Mir-Shuttle docking plane. At bottom is one of the androgynous docking unit petals. The mission marked the first US-Russian docking since Apollo-Soyuz in 1975 (NASA) 0517257

Payload (Secondary): payload bay,
CRYOFD Cryogenic Flexible Diode Experiment
OARE Orbital Acceleration Research Experiment
Payloads (Secondary): mid-deck,
SAREX 2 Shuttle Amateur Radio Experiment

Mission summary

The ground crew moved the stack for STS-83 from the VAB to pad 39A on 11 March 1997 in advance of the launch planned for the following month. Launch came without any delays on 4 April. Two hours after reaching orbit, astronauts Voss and Crouch entered SPACELAB for the first time.

Shortly after the mission began, Mission Control discovered that one of the three fuel cells suffered some degradation. The loss of a fuel cell usually prompts a flight termination, but initially controllers decided to simply monitor the problem. Finally, on 6 April, Shuttle programme managers decided that the fuel cell problem meant that the mission would have to end. Astronauts shut down the faulty fuel cell that afternoon but, with careful management of the available electricity, the astronauts continued to conduct some experiments as part of the MSL 1 programme. The science activities ended on 7 April as preparations for the early return continued. The following day Columbia landed at KSC.

Once on the ground, NASA considered the delays in the Shuttle flights related to the International Space Station programme and decided to immediately refurbish Columbia for relaunch around 1 July for a mission designated STS-94 (originally the reflight was tentatively designated STS-83R).

STS-84 ATLANTIS (84)

Orbiter flight: 19
Launch pad: KSC 39A
Launch date/time: 15 May 1997 4:07:48.62 EST
Landing date/time: 24 May 1997 09:27:44 EDT
Mission duration: 9 days, 5 hours, 19 minutes 56 seconds
Landing site: KSC, Runway 33
Main engines: SSME-1: SN-2032, SSME-2: SN-2031, SSME-3: SN-2029
Launch Mass: = ~2,052.7 t
Landing Mass: 103,365 kg
Orbit: Altitude, 184 statute miles. Inclination: 51.6°
Orbits: 145
Crew: Charles J Precourt (Commander), Eileen M Collins (Pilot), C Michael Foale (Mission Specialist-1), Carlos I Noriega (Mission Specialist-2), Edward T. Lu, (Mission Specialist-3), Jean-François Clervoy (Mission Specialist-4), Elena V. Kondakova (Mission Specialist-5)
Payload (Primary):
SpaceHab-DM
LME
SAMS
CGEL

Mission summary

The STS-84 mission was the 6th Shuttle/Mir docking mission and is part of the NASA/Mir programme, which consists of nine Shuttle-Mir dockings and seven long duration flights of US astronauts aboard the Russian space station. STS-84 transferred 7,314 lb of water and other material to and from the Mir. During the docked phase, 1,025 lb of water, 844.9 lb of US science equipment, 2,576.4 lbs of Russian logistics along with 392.7 lb of miscellaneous material passed between the vehicles. 897.4 lb of US science material, 1,171.2 lb of Russian logistics, 30 lb of ESA material and 376.4 lb of miscellaneous material returned to Earth aboard the Shuttle. C Michael Foale stayed aboard Mir, replacing Jerry M Linenger who will have arrived on Mir from STS-81.

The Shuttle missed the first landing opportunity due to low cloud cover over the KSC. At 08:19 EDT, the STS-84 crew received approval for the de-orbit burn of 3 minutes 7 seconds on the second KSC landing opportunity for 24 May 1997. The burn occurred at 08:24 EDT. This allowed Atlantis to enter the atmosphere over the Pacific Ocean and travel in a northwest to southeast path toward the Kennedy Space Center.

STS-94 Columbia (85)

Orbiter flight: 23
Launch pad: KSC LC-39A
Launch date/time: 1 July 1997/14:02:02 EDT
Launch window: 2 hours 30 minutes
Main engines: SSME-1, 2037; SSME-2, 2034; SSME-3, 2033
Lift-off weight: 2,051,809 kg
Orbitor weight at lift-off: 118,057 kg
Orbitor altitude/inclination: 341 km/28.45°
Mission duration: 15 days 16 hours 45 minutes 29 seconds
Landing site: KSC (38) Runway 33
Landing date/time: 17 July 1997/06:47:29 EDT
Rollout distance: TBD
Rollout time: 55 seconds
Crew: James D Halsell (Commander), Susan L Still (Pilot), Janice E Voss (Payload Commander/Mission Specialist), Donald A Thomas (Mission Specialist), Michael L Gernhardt (Mission Specialist), Roger K Crouch (Payload Specialist), Greg T Linteris (Payload Specialist)
Payload (Primary):
First Microgravity Science Laboratory (MSL-01)
Spacelab module with long crew transfer tunnel
EDO Cryogenic Pallet
Cryogenic Flexible Diode Heat Pipe Experiment (CRYOFD)
Orbiter Acceleration Research Experiment (OARE)
Payload (Secondary):
Protein Crystal Growth (PCG)
Single-Locker Thermal Enclosure System (STES)
Shuttle Amateur Radio Experiment (SAREX II)
Payload weight: 11,332 kg (up); 11,269 kg (down)

Mission summary

Flow: A
OPF: 9 April 1997
VAB: 4 June 1997
Pad: 11 June 1997

STS-94 marked the first time the same vehicle had been reflown with the same crew and the same payload and followed STS-83 flown 4–8 April 1997 which executed a premature return when one of three fuel cells (#2) failed on the first day of that mission. Less than three months separated the two flights, with a nine day Shuttle for Atlantis between the two. On orbit the crew maintained 24 hour/two-shift operations and using the Spacelab as a test bed for facilites, equipment and procedures that will be used on the International Space Station. The 33 investigations yielded new knowledge on the principal scientific fields of combustion, biotechnology and materials, processing. Combustion experiments resulted in the discovery of a new mechanism of flame extinction caused by radiation of soot and in the ignition of the weakest flame (about 1 watt) ever burned under laboratory conditions as well as the longest burning flame in weightless space (500 seconds). Although only 144 combustion experiments were scheduled more than 200 were actually completed.

The MSL featured 19 science investigations in four major facilities: the Large Isothermal Furnace, the Expedite the Processing of Experiments to the Space Station (EXPRESS) pack, the Electromagnetics Containerless Processing Facility (TEMPUS) and the Coarsening in Solid-Liquid Mixtures (CSLM) Facility. Additional technology experiments were also performed in the Middeck Glovebox (MBGX) developed at NASA's Marshall Space Flight Center.

The Large Isothermal Furnace was developed by the Japanese Space Agency (NASDA) for the STS-47 Spacelab-J mission and was also flown on STS-65 IML-2 mission. It houses the Measurement of Diffusion Coefficient by Shear Cell Method Experiment, the Diffusion of Liquid Metals and Alloys Experiments, the Diffusion in Liquid Lead-Tin-Telluride Experiment, the Impurity Diffusion in Ionic Melts Experiment, the Liquid Phase Sintering Experiment (LIF) and the Diffusion Processes in Molten Semiconductors Experiment (DPIMS). The Combustion Module-1 facility from NASA Lewis (Now Glenn) Research Center houses experiments on Laminar Soot Processes Experiment and the Structure of Flame Balls at Low Lewis-Number Experiment (SOFTBALL). The Droplet Combustion Experiment (DCE) is designed to investigate the fundamental combustion aspects of single, isolated droplets under different pressures and ambient oxygen concentrations for a range of droplet sizes varying between 2 and 5 mm. The DCE apparatus is integrated into a single MSL Spacelab rack in the cargo bay.

The STS-71/Mir crew, creating a record for the largest crew of a single complex, mark the accomplishment inside Spacelab (NASA) 0517258

STS-69 prepares to release the Wake Shield Facility (NASA) 0517269

The EXPRESS rack replaced a Spacelab Double rack and special hardware provided the same structural and resource connections the rack will have on the International Space Station. It houses the Physics of Hard Spheres (PhaSE) experiment and the Astro/PGBA experiment. The Electromagnetic Containerless Processing Facility was used for the Experiments on Nucleation in Different Flow Regimes,Thermophysical Properties of Advanced Materials in the Undercooled Liquid State Experiment, Measurements of the Surface Tension of Liquid and Undercooled Metallic Alloys by Oscillating Drop Technique Experiment, Alloy Undercooling Experiments, the Study of the Morphological Stability of Crowing Dendrites by Comparative Dendrite Velocity Measurements on Pure Ni and Dilute Ni-C Alloys in the Earth and Space Laboratory Experiment, the Undercooled Melts of Alloys with Polytetrahedral Short-Range Order Experiment, the Thermal Expansion of Glass Forming Metallic Alloys in the Undercooled State Experiment, the AC Calorimetry and Thermophysical Properties of Bulk Glass-Forming Metallic Liquid experiment and the Measurement of Surface Tension and Viscosity of Undercooled Liquid Metals experiment.

Other experiments included the Space Acceleration Measurement System (SAMS), the Microgravity Measurement Assembly (MMA), the Quasi-Steady Acceleration Measurement System and the Orbital Acceleration Research Experiment (OARE). The Middeck Glovebox facility supported the Bubble and Drop Nonlinear Dynamics (BDND) Experiment, the Study of the Fundamental Operation of a Capillary-driven Heat Transfer Device in Microgravity Experiment, the Internal Flows in a Free Drop (IFFD) experiment and the Fiber Supported Droplet Combustion Experiment (FSDC-2).

The 25 primary experiments, four glovebox investigations and four accelerometer studies on MSL-1 were contributed by scientists from NASA, ESA, the German Space Agency and NASDA of Japan. More than 35,000 commands were sent to MSL-1 from the Spacelab Missions Operations Control Center at the NASA Marshall Space Flight Center, constituting an all-time record.

STS-85 DISCOVERY (86)
Orbiter: Discovery (OV-103), 23rd flight
Launch date: 7 August 1997
Launch pad: KSC 39-A
Launch time: 10:41 EDT
Launch window: 1 hour, 39 minutes
Orbit Altitude/Inclination: 160 n miles, 57°
Mission duration: 10 days, 20 hours, 24 minutes
Landing date: 18 August 1997
Landing time: 07:05 EDT
Primary Landing site: Kennedy Space Center, Florida
Launch Mass: =~2,028.4 tonnes
Landing Mass: 97,919 kg
Crew: Curt Brown, (Commander), Kent Rominger (Pilot), January Davis, (Payload Commander/ Mission Specialist-1), Robert Curbeam (Mission Specialist-2), Steve Robinson (Mission Specialist-3), Bjarni Tryggvason (Payload Specialist-1)
Payload (Primary): cargo bay,
CRISTA-SPAS-02. Cryogenic Infra-red Spectrometers and Telescopes for the Atmosphere-Shuttle Pallet Satellite-2
TAS-01 Technology Applications and Science-01
IEH-02 International Extreme Ultraviolet Hitchhiker-02
MFD

STS-79 was the first to carry the Spacehab Double Module, providing 2.7 t cargo capacity (Spacehab) 0517271

Payload (Secondary): crew cabin,
BDS-03 Bioreactor Demonstration System-3
BRIC
PCG-STES-05
ACIS
MSX
SIMPLEX
SWUIS Southwest Ultraviolet Imaging System
SSCE

Mission summary
STS-85 had the prime objective of launching the CRISTA-SPAS-2 which is making its second flight on the Space Shuttle and is the fourth mission in a co-operative venture between the German Space Agency (DARA) and NASA. The RMS will deploy the satellite and then the CRISTA-SPAS system will free-fly for over 200 hours and will use three telescopes and four spectrometers to measure infra-red radiation emitted by the Earth's middle atmosphere. Data gathered will help investigators from 15 countries to understand how small-scale tracer 'filaments' in the stratosphere contribute to transport of ozone and chemical compounds that affect the distribution of ozone.

NASDA, the Japanese Space Agency, plans to test the RMS with the MFD series of experiments. MFD consists of three separate experiments located on a support truss in the payload bay. It is designed to demonstrate applications of a mechanical arm for possible use on the Japanese Experiment Module of the future International Space Station.

The TAS holds seven separate experiments that provide data on the Earth's topography and atmosphere study the sun's energy and test new thermal control devices. It is joined by the four experiments comprising the IEH payload. NASA intends to study ultraviolet radiation from the stars, the Sun and other sources in the Solar System with the IEH.

Payload and experiments flying in the crew cabin area include the (SWUIS), a 7 in imaging telescope that will be pointed at the Hale-Bopp comet. The BDS-3 payload performs cell biology experiments under controlled conditions on small samples of material.

STS-85 carries several payloads to continue NASA's study of natural and human-induced changes in Earth's environment as part of the Mission to Planet Earth programme. Mission to Planet Earth programme investigations include three experiments - SOlar CONstant Experiment (SOLCON), Infra-red Spectral Imaging Radiometer (ISIR) and Shuttle Laser Altimeter (SLA), that are a part of the Technology Applications and Science-1 (TAS-1) payload.

The Middle Atmosphere High-Resolution Spectrograph Investigation (MAHRSI) is another NASA payload to obtain data on the natural environment. During STS-85 MAHRSI will obtain new vertical profile data on the distribution of hydroxyl (OH) in the mesosphere and upper stratosphere under different conditions from its previous flight on STS-66. At the end of the flight, NASA plans to develop global maps of the vertical distribution of OH and measurements of nitric oxide in the middle atmosphere.

STS-86 ATLANTIS (87)
Orbiter: Atlantis, 20th flight
Launch date: 25 September 1997
Launch pad: KSC 39-A
Launch time: 22:34 EDT
Launch window: 10 minutes
Orbit altitude/Inclination: 184 statute miles, 51.6°
Orbits: 169
Mission duration: 10 days 19 hours 22 minutes 12 seconds.
Landing date: 6 October 1997
Engines: SSME-1: SN-2012, SSME-2: SN-2040, SSME-3: SN-2019
Landing time: 17:55 EDT
Primary Landing site: Kennedy Space Center, Florida, Runway 15
Crew: James D Wetherbee (Commander), Michael J Bloomfield (Pilot), Vladimar G Titov (Mission Specialist), Scott E Parazynski (Mission Specialist), Jean-Loup J M Chretien (Mission Specialist), Wendy B Lawrence (Mission Specialist), David A Wolf (Mission Specialist)
Payload (Primary): cargo bay,
SpaceHab-DM
MEEP-R
EDFT-06
SEEDS-II
CCM-07
MSX-09
CREAM-09
KidSat-03
RME-III-21
SIMPLEX-02
Payload (Secondary): crew cabin,
GAS(G-036)

Mission summary
After receiving final clearance to launch from NASA Administrator Daniel Goldin earlier in the day, Atlantis' astronauts lifted off from the Kennedy Space Center at 21:34 Central time to begin the seventh docking mission to the Mir Space Station.

Shortly after reaching orbit, the astronauts began to activate Shuttle and Spacehab systems and

The new Multifunction Electronic Display Subsystem (MEDS) is shown in the cockpit of Orbiter Atlantis 0054448

The Unity connecting module is moved into the payload bay of Orbiter Endeavour prior to STS-88
0054449

planned a minor firing of the orbital manoeuvring system engines to refine Atlantis' path to the Mir, which was flying over southern Europe at launch time at an altitude of about 215 n miles.

On 27 September 1997, 14:58 EDT, for the seventh time in two years, the Shuttle Atlantis gently docked with the Russian Mir Space Station. Operating from Atlantis' aft flight deck, Cdr Jim Wetherbee guided Atlantis to a smooth docking with the Mir at 14:58 Central time as the two vehicles flew north of the Caspian Sea east of the Russian-Kazakh border at an altitude of about 215 n miles.

Astronauts Scott E Parazynski and Vladimir G Titov performed an EVA outside the Atlantis-Mir complex to retrieve four Mir Environmental Effects Packages which had been mounted on the docking module last March by astronauts Linda Goodwin and Rich Clifford. Parazynski and Titov tethered the Spektr Solar Array Cap to the docking module. The Solar Array Cap may be used in a future Mir spacewalk to seal any hole found in the hull of the damaged Spektr module. The EVA began on 1 October at 12:29 Central Time.

Atlantis' landing with astronaut Mike Foale was postponed for 24 hours, after clouds at the Kennedy Space Center in Florida prevented a landing on the two opportunities available at the Florida spaceport. The Space Shuttle Atlantis finally landed on Kennedy Space Center, Florida at 16:55 CDT 6 October.

STS-87 COLUMBIA (88)
Orbiter flight: 24 (Extended Duration Orbiter)
Launch pad: KSC LC-39B
Launch date/time: 19 November 1997/14:46:00 EST
Launch window: 2 hours 30 minutes
Main engines: SSME-1, 2031; SSME-2, 2039; SSME-3, 2037
Lift-off weight: 2,051,833 kg
Orbiter weight at lift-off: 118,298 kg
Orbit altitude/inclination: 278 km/28.45°
Mission duration: 15 days, 16 hours, 34 minutes 04 seconds
Landing site: KSC (41) Runway 33
Landing date/time: 5 December 1997/07:20:04 EST
Rollout distance: 2,440 m
Rollout time: 57 seconds
Crew: Kevin R Kregel (Commander), Steven W Lindsey (Pilot), Kalpona Chawla (Mission Specialist), Winston E Scott (Mission Specialist), Takao Doi, NASDA (Mission Specialist), Leonid Kadenyuk, National Space Agency of Ukraine (Payload Specialist)
Payload (Primary):
United States Microgravity Payload-4 (USMP-4)
Spartan 201–04
Payload (Secondary):
Shuttle Ionospheric Modification With Pulsed Local Exhaust (SIMPLEX)
EVA Demonstration Flight Test 05 (EDFT)
Shuttle Ozone Limb Sounding Experiment (SOLSE)
Loop Heat Pipe (LHP)
Sodium Sulphur Battery Experiment (NaSBE)
Turbulent Gas Jet Diffusion Flames (TGDF)
GAS 036
Collaborative Ukraine Experiment (CUE)
Autonomous EVA Robotic (AER) Camera/Sprint (RME 1323)
Midcourse Space Experiment
Payload weight: 10,038 kg

Mission summary
OPF: 17 July 1997
VAB: 24 October 1997
Pad: 29 October 1997

Following a near flawless countdown, STS-87 was launched on time and, for the first time, on to a new attitude profile for ascent. Approximately 6 minutes after launch, the orbiter/ET combination was rolled from 'heads-down' to 'heads-up' allowing the antennas to communicate through the TDRSS network of geosynchronous satellites about 2 minutes 30 seconds earlier than would otherwise be possible. This profile will be flown on all subsequent low-inclination trajectories to eliminate the need for the Bermuda tracking station.

The microgravity science and technology research aboard Columbia performed well throughout the mission. Experiments conducted aboard USMP-4 included: Advanced Automated Directional Solidification Furnace; Confined Helium Experiment; Isothermal Dendritic Growth Experiment; Material for the Study of Interesting Phenomena of Solidification on Earth and in Orbit (MEPHISTO); Microgravity Glovebox Facility with several unique experiments; Enclosed Laminar Flames; Wetting Characteristics of Immiscibles; Particle Engulfment; Pushing by a Solid/Liquid Interface; Space Acceleration Measurement System; Orbital Acceleration Research Experiment.

The plan to deploy the Spartan was not a success. Following deployment at 16:04 EST on 21 November Spartan executed a pirouette manoeuvre, indicating it might have had an attitude control problem. On attempting to re-capture the grapple fixture, Chawla failed to obtain a proper attachment and on trying to re-grapple the platform inadvertently imparted a 2°/s spin to Spartan. An attempt to synchronise the rotation of the orbiter with that of the platform was called off by the Flight Director. During the first EVA on 24 November, Scott and Doi managed to grab the platform and stow it aboard the payload bay. The EVA ended after 7 hours 43 minutes. The second EVA lasted 4 hours 59 minutes during which Scott and Doi completed unfinished tasks from the first EVA.

Columbia landed at KSC's runway 33 on orbit 252 at 07:20:04 EST.

STS-89 ENDEAVOUR (89)
Orbiter flight: 12
Launch pad: KSC LC-39A
Launch date/time: 22 January 1998/21:48:15 EST
Launch window: ~10 minutes
Main engines: SSME-1, 2043; SSME-2, 2044; SSME-3, 2045 (all Block II engines)
Lift-off weight: 2,046,919 kg
Orbiter weight at lift-off: 114,167 kg
Orbit altitude/inclination: 394 km/51.6°

Russian cosmonaut Sergei Krikalev and NASA astronaut James Newman (right) sit inside Orbiter Endeavour during a terminal countdown demonstration test
0054450

Endeavour lifts off into a night sky for the start of mission STS-88, 4 December 1998
0054451

Mission duration: 8 days 19 hours 46 minutes 54 seconds
Landing site: KSC (42) Runway 15
Landing date/time: 31 January 1998/17:35:09 EST
Rollout distance: ?
Rollout time: ?
Crew: Terrence W Wilcutt (Commander), Joe Frank Edwards, Jr (Pilot), James F Reilly II (Mission Specialist), Michael P Anderson (Mission Specialist), Bonnie J Dunbar (Mission Specialist (Payload Commander)), Salizhan Shakirovich Sharipov (Mission Specialist), Andrew S W Thomas (Mission Specialist (up)), David Wolf (Mission Specialist (down))
Payload (Primary):
Spacehab 08-DM, transfer tunnel, transfer tunnel extension, orbiter docking system
Payload (Secondary):
GAS 093, 141, 145 and 432
CREAM
SIMPLEX
EarthKAM (Kidsat)
MPNE
HP
CEBAS
Payload weight: 9,952 kg (up); 8,858 kg (down)

Mission summary
OPF-1: 28 March 1997
VAB: 8 April 1997 (temporary stowage)
OPF-3: 21 April 1997
VAB: 23 May 1997 (temporary stowage)
OPF-1: 4 June 1997 (begin preflight processing)
VAB: 12 December 1997
Pad: 19 December 1997
Mission managers announced 22 May 1997 orbiter Endeavour would fly STS-89 instead of Discovery and that it would become the first orbiter other than Atlantis to dock with Mir. Launch was originally scheduled for 15 January but this was changed first to 20 January and then to 22 January at the request of the Russians who needed extra time in orbit to prepare for the visit. Following the retirement of veteran launch director Jim Harrington, this was the first launch overseen by Dave King, one of two rotational launch directors.

Following nominal ascent to orbit, the crew activated secondary experiments. Nominal rendezvous following standard ascent to orbit with docking at Mir at 15:14 EST 24 January. The hatches were opened at 17:25 EST at the start of five days of joint activities. Endeavour carried 3,629 kg of scientific equipment, logistical hardware and water for the Russian space station. Astronaut Thomas was switching positions with astronaut Wolf and would remain aboard Mir as the last NASA astronaut to conduct a long duration stay aboard the Russian space station. Wolf had spent 119 days aboard Mir and a total on-orbit time of 128 days.

The Shuttle-Mir science programme focused on 27 areas of study: advanced technology, in which commercial development of research tasks was mooted as a test bed for the ISS; Earth sciences, including research in ocean biochemistry, land surface, hydrology and meteorology; human life sciences, focusing on crewmembers' adaptation to zero-g; and ISS risk mitigation, involving crew health and safety.

Hatches between Mir and the Shuttle were closed 17:34 EST on 28 January and the two spacecraft undocked at 11:57 EST the following day. Orbiter landed on orbit 139 at 17:35:08, 31 January, on

Shuttle Discovery lifts off on 27 May 1999 to deliver supplies and cargo to the ISS 0054452

runway 15 at the Kennedy Space Center, the 13th time in succession and the 20th time in the previous 21 landings, having travelled 5.8 million km.

STS-90 COLUMBIA (90)
Orbiter flight: 25
Launch pad: KSC LC-39B
Launch date/time: 17 April 1998/14:19:00 EDT
Launch window: ~2 hours 30 minutes
Main engines: SSME-1, 2041; SSME-2, 2032; SSME-3, 2012
Lift-off weight: 2,051,982 kg
Orbiter weight at lift-off: 119,005 kg
Orbit altitude/inclination: 278 km/39°
Mission duration: 15 days 21 hours 49 minutes 59 seconds
Landing site: KSC (43) Runway 33
Landing date/time: 3 May 1998/12:08:59 EDT
Rollout distance: 3,047 m
Rollout time: 58 seconds
Crew: Richard A Searfoss (Commander), Scott D Altman (Pilot), Kathryn 'Kay' Hire (Missions Specialist), Dafydd (Dave) Rhys Williams, Canadian Space Agency (Mission Specialist), Dr Jay C Buckey (Payload Specialist), Dr James A Pawelczyk (Payload Specialist)
Payload (Primary):
Spacelab LM Neurolab
EDO cryogenic pallet (5 full and 4 extended duration cryo tank sets)
Payload (Secondary):
GAS 197, 744 and 722
Shuttle Vibration Forces
Bioreactor Demonstration System 04
Payload weight: 11,862 kg

Mission summary
OPF: 5 December 1997
VAB: 16 March 1998
Pad: 23 March 1998
Launch originally planned for 16 April 1998 but delayed for 24 hours because of difficulties with Network signal processor No 2, one of two processors which format data and communications between the ground and the Shuttle and which was subsequently replaced. Lift-off occurred on time a day later and the direct insertion ascent was nominal with only one OMS thrusting manoeuvre to reach the operational orbit. Neurolab mission was the last flight of the Spacelab pressurised modules to fly.

Mission was to combine biomedical techniques with new and sophisticated measurement and recording devices to advance neuroscience as NASA's contribution to the Decade of the Brain designated by President Bush and Congress. The experiments were to examine the effects of space flight and microgravity on the nervous system. Neurolab's 26 experiments targeted basic research in attempts to understand the role played by gravity in the development of the nervous systems in humans and animals. Indications are strong that critical early development periods shape the way the brain interprets stimuli in later life. These tests were conducted with the astronauts and with rats, mice, crickets, snails and two kinds of fish.

About one week into the mission, the crew had to jury-rig a bypass system for a suspect regulator valve in the CO_2 removal system, preventing the need for an earlier return to Earth than planned. Consumables resources permitted consideration of a one-day extension but this was not adopted because of a forecast for deteriorating weather at KSC. Columbia landed on orbit 256 after logging almost 10.26 million km making the 14th consecutive touchdown at KSC.

STS-91 DISCOVERY (91)
Orbiter flight: 24
Launch pad: KSC LC-39A
Launch date/time: 2 June 1998/18:06:24 EDT
Launch window: ~10 minutes
Main engines: SSME-1, 2047; SSME-2, 2040; SSME-3, 2042
Lift-off weight: 2,047,782 kg
Orbiter weight at lift-off: 117,861 kg
Orbit altitude/inclination: 320 km/51.6°
Mission duration: 9 days 19 hours 53 minutes 57 seconds
Landing site: KSC (44) Runway 15
Landing date/time: 12 June 1998/14:00:18 EDT
Rollout distance: 3,576 m
Rollout time: 1 minute 4 seconds
Crew: Charles (Charlie) J Precourt (Commander), Dominic (Dom) L Gorie (Pilot), Franklin Chang-Diaz (Mission Specialist), Wendy Lawrence (Mission Specialist), Janet Kavandi (Missions Specialist), Valeriy Ryumin, Russian Space Agency (Mission Specialist), Andrew (Andy) S W Thomas (down (Mission Specialist))
Payload (Primary):
Spacehab 09-SM, transfer tunnel, transfer tunnel extension, orbiter docking system
Payload (Secondary):
GAS 090, 743, 765, 648, 2 × SEM, 2 × Phase 1 Program Support Packages
Alpha Magnetic Spectrometer
Phase 1 Requirements
Commercial Protein Crystal Growth
Solid Surface Combustion Experiment
Shuttle Ionospheric Modification with Pulsed Local Exhaust
SIMPLEX
Payload weight: 11,782 kg (up); 11,843 kg (down)

Mission summary
OPF-3: 19 August 1997 (temporary storage)
OPF-2: 1 October 1997 (temporary storage)
OPF-2: 30 October 1997 (begin preflight processing)
VAB: 21 April 1998
Pad: 2 May 1998
With few exceptions, the countdown proceeded smoothly and final launch time was determined at the T–9 minutes built-in hold to synchronise the orbit insertion point with the Mir orbital state vector. This was the first time Discovery had been used on a Shuttle-Mir mission and the docking took place on 4 June at 12:58 EDT at an altitude of 334 km. At hatch opening (14:34 EDT), NASA astronaut Andy Thomas officially became a member of the STS-91 crew, ending 130 days of living and working aboard Mir and bringing to 907 days the total amount of time spent on Mir by seven US astronauts as long-duration crewmembers.

During the next four days, the crews transferred more than 499 kg of water and 2,132 kg of supplies between the two spacecraft. Long-term US experiments were transferred to the mid-deck locker area and into the Spacehab single module in the orbiter payload bay including the Space Acceleration Measurement System and the tissue engineering co-culture investigations, as well as two crystal growth experiments. The crews also conducted Risk Mitigation Experiments and Human Life Sciences investigations. Carried for the first time, the Alpha Magnetic Spectrometer was designed to look for 'dark matter' in the universe and was switched on during Flight day 1 but data that should have been sent to the ground via the Ku-band antenna was recorded onboard because of a problem with that antenna. On

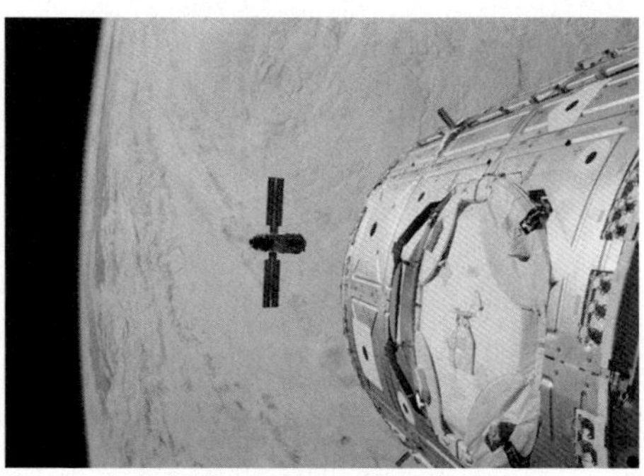

Russian-built FGB approaches Shuttle Endeavour and Unity module (NASA) 0062166

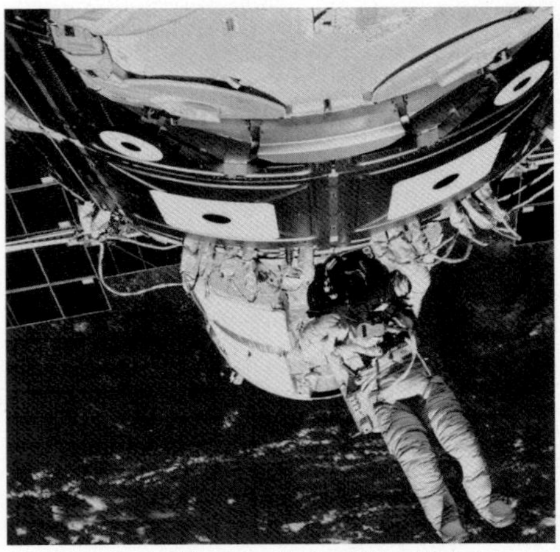

NASA astronaut Newman during EVA at Unity module (NASA) 0062167

3 June, the crew were able to set up a bypass system allowing AMS data to be downlinked through the S-band/FM system. The Ku-band failure prevented the transmission of TV from Discovery and a fault between ground stations and mission control in Moscow prevented TV from being broadcast out of Mir.

Hatches between Mir and Discovery closed at 09:07 on 8 June 1998 and the two undocked at 12:01 EDT ending Phase 1 of the ISS programme. Discovery landed on orbit 155 having logged 6.1 million km making the 15th consecutive landing at that facility and the 22nd in the last 23 missions.

STS-95 DISCOVERY (92)
Orbiter flight: 25
Launch pad: KSC LC-39A
Launch date/time: 29 October 1998/14:20:19 EST
Launch window: 2 hours 30 minutes
Main engines: SSME-1, 2048; SSME-2, 2043; SSME-4, 2045
Lift-off weight: 2,051,142 kg
Orbiter weight at lift-off: 103,061 kg
Orbit altitude/inclination: 471 km/28.45°
Mission duration: 8 days 21 hours 43 minutes 57 seconds
Landing site: KSC (45) Runway 33
Landing date/time: 7 November 1998/12:04:16 EST
Rollout distance: 2,898 m
Rollout time: 59 seconds
Crew: Curtis L Brow, Jr (Commander), Steven W Lindsey (Pilot), Scott E Parazynski (Missions Specialist), Stephen K Robinson (Missions Specialist), Pedro Duque (NSDA, Japan (Mission Specialist)), Chiaki Mukai (Payload Specialist), Senator John H Glenn (Payload Specialist)
Payload (Primary):
Spacehab
Spartan 201–5
Hubble Space Telescope Orbital Systems Test Platform
International Extreme Ultraviolet Hitchhiker
Payload (Secondary):
Cryogenic Thermal Storage unit
Space Experiment Module
GAS
Biological Research in Canisters
Electronic Nose
Payload weight: 12,937 kg (up); 12,867 kg (down)

Mission summary
OPF: 15 June 1998
VAB: 14 September 1998
Pad: 21 September 1998
The launch was scheduled for 14:00 EST but an unscheduled hold of 8 minutes 30 seconds at the T–9 minute built-in hold period was made in order to investigate a master alarm during cabin leak checks after hatch closure. The Range Safety Officer requested a further hold at T–5 minutes when an unidentified aircraft entered the restricted airspace around KSC. After main engine ignition but before solid rocket booster ignition, the drag chute compartment door came off without any effect on the lift-off or ascent to orbit. This was the first time veteran Mercury MA-6 astronaut John Glenn had been into space since 20 February 1962 when he became the first American to orbit the Earth 36 years 8 months and 9 days earlier.

The nine days in space were filled with concentrated scientific activity in a variety of disciplines and involved the independent flight of the free-flying Spartan on its fifth flight into space. During STS-95 it would perform a repeat of the failed Spartan 201–4 mission on STS-87 when it had to be retrieved by a contingency EVA (see STS-87 Columbia). Some of the on-board science supported studies into ageing

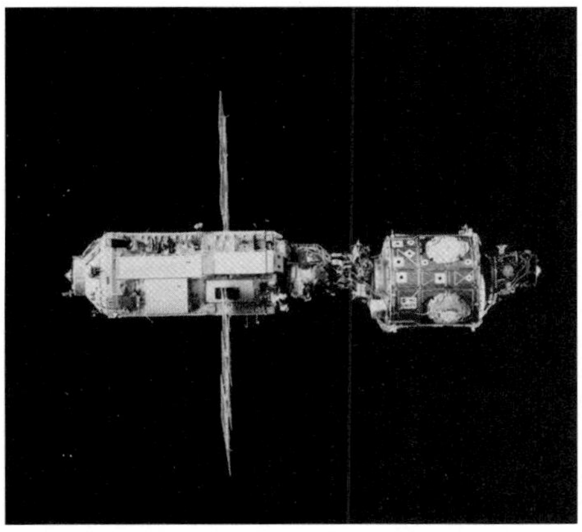

Zarya and Unity modules viewed from Endeavour (STS-88) (NASA) 0062168

and the 77-year old Glenn was a test subject required to deliver 10 blood and 16 urine samples over the period of the mission. On 30 October, the crew released into orbit the 70 kg Petite Amateur Naval SATellite (PANSAT), a small telecommunications satellite developed by the Naval Postgraduate School in Monterey, California. Next day Glenn began giving the cycle of blood and urine samples.

Spartan was released from the remote manipulator arm at 13:59 EST on 1 November and for about 9 h, the Shuttle remained 9.6–16 km away before moving to a distance of around 48 km. Spartan 201–5 carried instruments to study the Sun and the solar wind and after a series of rendezvous manoeuvres, it was back on the remote manipulator at 15:45 CST on 3 November and berthed back in the payload bay shortly thereafter. A further use of Spartan on 4 November extended it above the Shuttle for additional measurements. In all, STS-95 completed some 80 experiments.

Preparations for the return to Earth included closing the payload bay doors at 08:17 EST on 7 November followed by the 4 minute 40 seconds firing of the deorbit engines on orbit 135 starting at 10:53 EST. Entry began at 11:30 EST followed by touchdown at 12:04 EST, the 16th consecutive landing at KSC. The drag chute was not deployed.

STS-88 ENDEAVOUR (93)

Orbiter flight: 13
Launch pad: KSC LC-39A
Launch date/time: 4 December 1998/03:35:34 EST
Launch window: 5–10 minutes
Main engines: SSME-1, 2043; SSME-2, 2044; SSME-3, 2045
Lift-off weight: 2,049,542 kg
Orbiter weight at lift-off: 119,717 kg
Orbit altitude/inclination: 320 km/51.6°
Mission duration: 11 days 19 hours 17 minutes 55 seconds
Landing site: KSC (46) Runway 15
Landing date/time: 15 December 1998/22:53:29 EST
Rollout distance: 2,543 m
Rollout time: 44 seconds
Crew: Robert D Cabana (Commander), Frederick (Rick) Sturckow (Pilot), Nancy Currie (Mission Specialist), Jerry Ross (Mission Specialist), Jim Newman (Mission Specialist), Sergei Krikalev, Russian Space Agency (Mission Specialist)
Payload (Primary):
Unity Module (Node 1/PMA1/2)
Imax Cargo Bay Camera
Argentinian Scientific Applications Satellite-C
Payload (Secondary):
MightySat 1 Hitchhiker
Space Experiment Module-07
GAS 093
Payload weight: 11,612 kg (Unity)

Mission summary

OPF: 1 February 1998
VAB: 13 October 1998
Pad: 21 October 1998
STS-88 was the long-awaited first US assembly flight in support of the ISS, the culmination of almost 15 years of design, redesign and preparation in an effort pledged by President Ronald Reagan in his state of the union address on 25 January 1984. The ISS was re-scoped by President Bill Clinton during the mid-1990s to include the Russians as well as the Europeans, the Japanese and the Canadians. STS-88 followed the launch of the first element, the Russian-built Zarya (or SUV, Space Utility Vehicle), on 20 November 1998 in ISS Flight 1A/R. The STS-88 mission was known as Flight 2A.

The first launch attempt aimed for a lift-off at 03: 58 EST on 3 December 1998 but at T–4 minutes 24 seconds, a master alarm halted the clock 24 seconds later. The alarm had been triggered by pressure in the orbiter's hydraulic system #1 temporarily registering a low of 19,306 kPa. The countdown was held at T–31 seconds to assess the situation, but engineers ran out of time before constraints governed by LOX drain-back redlines closed the launch window. Next day the countdown proceeded without major incident and Endeavour was launched at 03:35 EST. Mission objectives focused on attaching the Unity (Node 1) module and PMA attachments to Zarya, the first (Russian-built) element in the ISS launched 20 November 1998.

Following a trouble-free ascent and during the early stages of rendezvous, Endeavour was reconfigured. Using the remote manipulator, the crew manoeuvred the Unity module from its cradle at the rear of the payload bay to a vertical position and placed it on top of the external docking module and airlock assembly in the forward part of the payload bay. Mating occurred at 18:45 EST on 5 December. Next day Endeavour rendezvoused with Zarya and, using TV cameras and an Orbiter Space Vision System tested on earlier flights, drew Zarya down on the PMA#1 docking ring at 21:07. Initial attempts at a docking were unsuccessful and only after Currie ungrappled the module and tried again was the Shuttle assembly docked to the Russian element, at 21:48 UT on 6 December.

The first of two EVAs was conducted by Ross and Newman starting at 17:10 EST on 7 December and lasted 7 hours 21 minutes during which the two astronauts hooked up external power and data cables, installed handrails to be used later during assembly flights and checked out the exterior. Power was applied to Unity at 22:49 EST to verify the integrity of the connectors before the crew completed their EVA. Newman was moved by the manipulator arm to inspect Zarya's two TORU rendezvous antennas which did not fully deploy after launch. Shortly before the EVA ended, Ross broke the record for the most cumulative EVA time (29 hours 41 minutes) previously held by astronaut Akers during five EVAs on STS-49 and STS-61. He ended his EVA with 30 hours 8 minutes accumulated time.

On 8 December, the orbiter propulsion system was used in a sequence of pulsed firings lasting 22 minutes to raise the orbit of the docked configuration giving it greater orbital life before the next visit in 1999. The second EVA began at 15:33 EST 9 December with Ross and Newman setting out to install two systems health monitoring boxes on the exterior of Unity which double as a video-conferencing system. Then they removed restraint pins on the four Unity hatchways which will be used by future crews during the assembly phase. Newman was raised to one of the Zarya TORU antennas which he freed with a grappling hook, ending the EVA at 7 hours 2 minutes.

Cabana and Krikalev were first to begin the occupation of the new complex and moved into the pressurised Unity module at 14:54 EST on 10 December, the rest of the crew following. Turning on lights, unstowing gear and checking equipment, Cabana and Krikalev opened the hatch to Zarya at 16:12 EST and were followed inside by the remainder of the crew. Several work tasks were completed before the ensuing rest period. Following further work sessions, Cabana and Krikalev closed the hatch on Zarya at 17:41 EST on 11 December and shut the hatch on Unity at 19:26 EST completing the first 28 hours 32 minutes visit to the docked assembly.

The third and final EVA began at 15:33 EST on 12 December during which Ross and Newman disconnected umbilicals, stowed an exterior tool work bag for later astronauts and took an extensive series of photographs to document the exterior configuration. Ross then freed the second jammed TORU antenna and completed a total 21 hours 22 minutes EVA on this mission. Ross has now completed seven EVAs totalling 44 hours 9 minutes. Endeavour was undocked from the PMA at 15:25 EST on 13 December for a complete fly-around and photograph exercise after which, at 16:49 EST, the final separation burn was made. At 22:31 EST the crew deployed the Argentinian satellite.

Endeavour executed the de-orbit burn at 21:48 EST, 15 December and landed at 22:53 EST making the 10th night landing, the 5th night landing at KSC and the 18th consecutive landing in Florida.

STS-96 DISCOVERY (94)

Orbiter flight: 26
Launch pad: KSC LC-39B
Launch date/time: 27 May 1999/06:49:42 EDT
Launch window: 5–10 minutes
Main engines: SME-1, 2047; SSME-2, 2051; SSME-3, 2049
Lift-off weight: 2,047,756 kg
Orbiter weight at lift-off: 118,859 kg
Orbiter altitude/inclination: 320 km/51.6°
Mission duration: 9 days 19 hours 13 minutes 1 second
Landing site: KSC (47) Runway 15
Landing date/time: 6 June 1999/02:02:43 EDT
Rollout distance: ?
Rollout time: 56 seconds
Crew: Kent V. Rominger (Commander), Rick D Husband (Pilot), Tamara E Jernigan (Mission Specialist), Ellen Ochoa (Mission Specialist), Daniel T Barry (Mission Specialist), Julie Payette, Canadian Space Agency (Mission Specialist), Valery Tokarev, Russian Space Agency (Mission Specialist)
Payload:
Spacehab DM
Integrated Cargo Carrier
STARSHINE
Shuttle Vibration Forces (experiment)
Integrated Vehicle Health Monitoring
Payload weight: 9,047.4 kg

Mission summary

VAB: 15 April 1999
Pad: 23 April 1999
VAB: 16 May 1999
Pad: 21 May 1999
STS-96 was the second manned mission in the assembly sequence for the ISS and was known as Flight 2.A1. Following a countdown marred only by the potential hold waiting for a sailboat to clear restricted waterways, STS-96 was launched on time at 06:49 EDT and cleared to orbit at an initial altitude of 315 km. At orbit insertion, the ISS was in a near circular orbit of 389 km and over the next two days, the two vehicles converged through a sequence of rendezvous manoeuvres actioned by Discovery. Contact between the orbiter and the docking unit on PMA-1 attached to the Unity module launch by STS-88 came at 00:24 EDT on 29 May with docking 15 minutes later.

The object of the mission was to transfer 1,618 kg of supplies and equipment in 115 separate

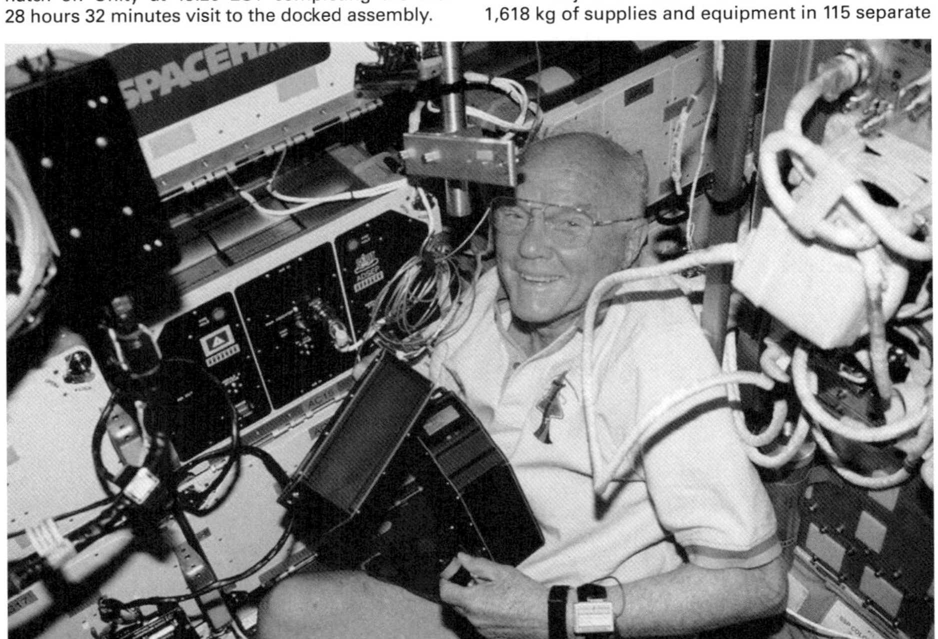

John Glenn returns to space aboard Shuttle STS-95 (NASA) 0062169

STS-95 crew commanded by Curtis L Brown Jr (right, seated) and including John Glenn (right, standing) (NASA) 0062170

itemised units from Discovery to the interior of the ISS modules. The crew would also move 18 items weighing 89 kg from the ISS to Discovery and remove 300 kg of equipment from the integrated Cargo Carrier and attach it to the outside of the ISS. The first EVA began at 06:51 EDT on 30 May. Jernigan and Barry retrieved operator posts and grapple fixture plates for the Strela crane which will be assembled at the station on later flights and placed on the PMA-2 adapter. The EVA ended after 7 hours 55 minutes, the 45th Shuttle-based EVA but the fourth ISS EVA raising that total to 29 hours 17 minutes.

At 21:14 EDT on 30 May, Jernigan and Tokarev opened the hatch into Unity and at 22:07 EDT moved across into Zarya followed by the rest of the crew for more than three days of busy work. They replaced 12 of 18 MIRTS battery recharge controllers, three to each of the six batteries which had shown signs of under-charge during the months since Zarya's launch six months earlier. On 31 May, the crew installed sound mufflers inside Zarya and completed the transfer of 318 litres of water for the use of the first resident crew. The hatch to Zarya was finally closed at 02:40 EDT on 3 June followed by the hatch to Unity at 03:12 EDT and the hatch to PMA-2 at 04:44 EDT completing a 79 hour 30 minute period of occupation.

A series of 17 pulsed firings over 37 minutes raised the orbit of the docked configuration to 396×388 km. By the time the Service Module (Zvezda) is launched late in 1999, this orbit will have degraded to about 357 km. Undocking came at 18:39 EDT on 3 June after 5 days 18 hours 13 minutes in the docked configuration. A photographic fly-around was performed before the separation burn at 20:53 EDT that day. At 03:31 EDT on 5 June, the STARSHINE satellite was ejected to independent orbit. Designed and built by the Naval Research Laboratory, the 160 kg satellite carries 878 reflective mirrors polished by school children in 18 countries enabling them to track it for calculations on the atmospheric density. Discovery landed at 02:03 EDT on 6 June performing the11th Shuttle night landing, the 6th at KSC.

STS-93 Columbia (95)
Orbiter flight: 26
Launch pad: KSC LC-39B
Launch date/time: 23 July 1999/00:31 EDT
Launch window: 1 hour 56 minutes
Main engines: SSME-1, 2012; SSME-2, 2031; SSME-3, 2019
Lift-off weight: 2,052,527 kg
Orbiter weight at lift-off: 122,647 kg
Orbiter weight at landing: 91,780 kg
Orbiter altitude/inclination: 283 km/28.45°
Mission duration: 4 days 22 hours 50 minutes 18 seconds
Landing site: KSC (48) Runway 33
Landing date/time: 27 July 1999/11:20:35 EDT
Rollout time: 43 seconds
Crew: Eileen M Collins (Commander), Jeffrey S Ashby (Pilot), Steven A Hawley (Mission Specialist), Catherine G Coleman (Mission Specialist), Michel Tognini (Mission Specialist)
Payload (Primary):
Chandra X-ray Onservatory
Inertial Upper Stage
Payload (Secondary):
Plant Growth Investigations in Microgravity
Payload weight: 22,584 kg (up); 3,043 kg (down)

Mission summary
Flow: A
VAB: 1 February 1999
Pad: 7 June 1999
STS-93 launched the last of the great space observatories which evolved during the 1980s and follows the Hubble Space Telescope launched by STS-31 in April 1990 (delayed from 1986 by the Challenger accident) and the Compton Gamma Ray Observatory launched by STS-37 in April 1991. STS-93 also carried the Midcourse Space Experiment (MSX), Shuttle Ionospheric Modification with Pulsed Local Exhaust (SIMPLEX), Southwest Ultraviolet Imaging System (SWUIS), Gelation of Sols: Applied Microgravity Research (GOSAMR), Space Tissue Loss-B (STL-B), Light Weight Flexible Solar Array Hinge LFSAH), Cell Culture Module (CCM) and the Shuttle Amateur Radio Experiment (SAREX-2), EarthKam, Plant Growth Investigations in Microgravity PGIM), Commercial Generic Bioprocessing Apparatus (CGBA), Micro-Electrical Mechanical System (MEMS) and the Biological Research in Canisters (BRIC) experiment. The MSX required Shuttle orbiter thruster firings to be used as a sensor calibration and evaluation target for the space-based ultraviolet, infra-red and visible sensors on the MSX satellite, which was to be placed in a 1,037 km, 99° orbit.

The launch of STS-93 was delayed three days beyond its scheduled launch date of 20 July 1999, first by a spike detected in one of the sensors designed to measure concentrations of hydrogen in the orbiter's aft engine compartment. A new launch date was set for 22 July at 00:28 EDT but storm cells moved into the area and lightning was reported within 15 km of the launch pad forcing cancellation of the attempt. The flight began at the third attempt just 31 minutes into 23 July 5 seconds after liftoff the crew noticed a voltage drop one electrical bus and one of two redundant main engine controllers on two of the three engines shut down. The first female Shuttle commander Eileen Collins and pilot Jeffrey Ashby handled the situation impeccably and Columbia reached orbit on schedule.

Precisely on time, the Chandra observatory was spring-ejected from its cradle at 06:47 CDT as Columbia flew across the Indonesian islands in a 283 × 301 km orbit. The first stage of the IUS boost stage was ignited at 07:47 CST. After the IUS second stage burn Chandra's solar arrays were deployed at 08:22 CDT and the observatory separated from its booster stage 27 minutues later. After a sleep period the crew were wakened at 18:31 CDT 23 July to begin their second day in space during which secondary payloads and experiments were activated. Meanwhile, the Chandra X-ray Observatory was in a 1,245 × 71,761 km orbit imparted by the two-stage IUS. Five orbital manoeuvres over two weeks using on-board thrusters would put Chandra in its planned orbit of 9,977 × 138,400 km orbit.

On Columbia, the crew tested a procedure which would be used on the STS-99 flight carrying a 61 m long mast deployed from the cargo bay on the Shuttle Radar Topography Mission. Known as the 'flycast' manoeuvre, it involved the Shuttle autopilot and multipilot thruster firings to maintain rigidity in the mast. At 20:16 CDT on 24 July the first of four apogee burns was performed by the Integral Propulsion System on Chandra putting the observatory in a 1,200 × 72,023 km path. When Columbia landed at 11:20:35 UT on 27 July it was the 19th consecutive landing at the Kennedy Space Center and the 12th night landing in the Shuttle programme. The landing was preceded by a spectacular light show as it streaked across toward Florida at a height of 61 km and a speed of Mach 15.

STS-103 Discovery (96)
Orbiter flight: 27
Launch pad: KSC LC-39B
Launch date/time: 19 December 1999/19:50:00 EST
Launch window: 42 minutes
Main engines: SSME-1, 2053; SSME-2, 2043: SSME-3, 2049
Lift-off weight: 2,044,112 kg
Orbiter weight at lift-off: 112,565 kg
Oriter altitude/inclination: 587 km/28.45°
Mission duration: 7 days 23 hours 11 minutes 34 seconds
Landing site: KSC (49) Runway 33
Landing date/time: 27 December 1999/19:01:34 EST
Rollout time: 47 seconds
Crew: Curtis L Brown (Commander), Scott J Kelly (Pilot), Jean-Francois Clervoy (ESA Mission Specialist), Steven L Smith (Mission Specialist), C Michael Foale (Mission Specialist), John M Grunsfeld (Mission Specialist), Claude Nicollier (ESA Mission Specialist)
Payload (Primary):
Hubble Space Telescope Servicing Mission 03-A
Payload weight: 5,9991 kg (up); 5,954 kg (down)

Mission summary
OPF: 6 June 1999
VAB: 4 November 1999
Pad: 13 November 1999
Originally scheduled for launch at 13:10 EST on 9 December at the start of launch window lasting 42 minutes, STS-103 was delayed by a dented main propulsion system hydrogen line discovered during closeout inspections on 8 December. The recirculation line was replaced and the mission re-scheduled for 16 December but this was put back a day to check quality control paperwork when a routine manufacturing inspection on the External Tank gaseous pressurisation line revealed welds made with an improper welding rod. Then the launch was delayed until 18 December and finally to 19 December to ensure better weather. STS-103 was the first component of the third Hubble servicing mission (SM-3A), advanced because of problems with the gyroscopes essential for attitude control with the observatory. It was a timely decision to split the third visit in two and bring it forward. On 13 November the HST was effectively disabled with the loss of the fourth gyroscope.

Following a successful launch on 19 December 1999, Discovery performed a rendezvous with the HST, attaching the remote manipulator on the orbiter to the observatory's grapple fixture at 18:34 CST, 23 hours 44 minutes into the mission. Clervoy used the arm to rotate the HST and bring its base down into the cargo bay, firmly attaching it to the Flight Support System restraints less than one hour later. A specially choreographed survey of the HST using the camera on the robotic arm was conducted both before and during the crew sleep period which began 00:50 CST 22 December. Before the sequence of delays that prevented launch on 9 December, NASA had scheduled the STS-103 mission to last 9 days 9 hours 31 minutes. With the launch 10 days late the mission was shortened to just under 8 days, cutting the number of EVAs from four to three, while achieving the same objectives.

Under the original plan, on EVA-1, Smith and Grunsfeld would install three new rate sensor units (RSUs) each containing two gyroscopes, install voltage/temperature improvement kits, open coolant valves and remove caps on the Near Infrared Camera and Multi-Object Spectrometer which had been installed on the second servicing visit. Opening the coolant valves would allow residual coolant frozen in the line to dissipate to space prior to the next servicing visit (SM-3B) in 2001. On EVA-2 Foale and Nicollier would replace one of the Fine Guidance Sensors. On EVA-3 Smith and Grunsfeld were to have replaced the S-band

The new cockpit upgrade was flown for the first time on STS-101 0085851

transmitter, installed another Solid state Recorder and attached new insulation on equipment bays midway up the side of the telescope. On EVA-4, Foale and Nicollier would finish up uncompleted tasks from previous EVAs, lay up more insulation around the top of the telescope and on handrails with peeling paint. Under the revised EVA plan, some EVA-4 tasks would be attempted on EVA-1 although the priority would be replacing all six gyroscopes by changing out the three RSUs. To recognise astronauts outside during EVA, Smith would wear solid red stripes, Grunsfeld would have no special markings, Foale would wear broken red stripes and Nicollier would wear diagonal broken red stripes.

On 22 December Smith and Grunfeld performed EVA-1 which lasted 8 hours 15 mintues, extended in duration due to problems fitting one of the old RSUs into its return box and in removing caps from the Near Infrared Camera and Multi-Object Spectrometer. On EVA-2, conducted on 23 December, Foale and Nicollier spent 8 hours 10 minutes installing a new advanced 486 computer, 20 times faster and with six times the memory of the DF-224 it replaced. They also replaced one of three 250 kg Fine Guidance Sensors on board. On the 8 hour 8 minute EVA-3 conducted 24 December Smith and Grunsfeld replaced a failed radio transmitter and installed a new solid-state recorder. The transmitter replaced one that failed in 1998, leaving a second transmitter to carry the load. The transmitters were not designed to be replaced in orbit and special tools had to be developed to allow this work to be carried out on EVA. The digital recorder replaced a conventional reel-to-reel, version and has more than 10 times the storage capacity of the unit it replaced. The astronauts also replaced insulation on two equipment bay doors. The EVA was extended somewhat by a problem hooking Grunsfeld's suit to orbiter power after he returned to Discovery's airlock. The last EVA brought the total amount of time spent servicing Hubble to 93 hours 13 minutes. Total Shuttle programme EVA now totals 317 hours 3 minutes. With 35 hours 33 minutes, Steve Smith now has the second longest combined EVA time, behind only Jerry Ross with 44 hours 11 minutes.

On board Discovery, the astronauts became the first Shuttle crew to remain in space over Christmas Day, which began with an uplinked wakeup song from Bing Crosby. At 17:03 CST, while flying over South Pacific's Coral Sea north-east of Australia at an altitude of 595 km, Clervoy used the robot arm to release the HST to free orbit, its life restored and enhanced by 24 hours 33 minutes of repairs conducted during this mission. At 17:30 CST controllers at the Space Telescope Operations Control Center in Maryland reported the telescope in normal operating mode. At 17:39 CST Brown executed a steering jet burn lowering Discovery's orbit slightly so that the Shuttle would begin to move ahead of the HST at about 9.6 km/h. The de-orbit burn began at 16:48 CST (17:48 EST) 27 December following a one orbit postponement due to crosswinds at the Kennedy Space Center landing strip and Discovery touched down on runway 33 at 19:01:34 EST.

STS-99 Endeavour (97)
Orbiter flight: 14
Launch pad: KSC LC-39A (68)
Launch date/time: 11 February 2000/12:43:40 EST; 17:43:40 UT
Main engines: SSME-1, 2052; SSME-2, 2044; SSME-3, 2047.
Lift-off weight: 2,050,460 kg
Orbiter weight at lift-off: 116,375 kg
Orbiter altitude/inclination: 233 km/57°
Mission duration: 11 days 05 hours 38 minutes 43 seconds
Landing site: KSC (50) Runway 33
Landing date/time: 22 February 2000/18:22:23 EST; 23:22:23 UT
Rollout time: 62 seconds (3,030.6 metres)
Crew: Kevin R. Kregel (Commander), Dominic L. Pudwill Gorie (Pilot), Janet L Kavandi (Mission Specialist), Janice E. Voss (Mission Specialist), Mamoru Mohri (Mission Specialist), Gerhard P. J. Thiele (Mission Specialist)
Payload (Primary):
Shuttle Radar Topography Mission
EarthKAM
DTO 690 Urine Collection Device
DTO 686 Heat Exchange Unit Evaluation
Payload weight: 13,152 kg

Mission summary
OPF: 15 December 1998
VAB: 11 July 1999; 2 December 1999
Pad: 13 December 1999
Accomodating work to repair wiring defects found in orbiters Discovery and Endeavour, NASA managers flipped the original schedule for STS-103 and STS-99, preferring to launch the latter after rather than before the former Hubble servicing mission as originally scheduled. Following alarms raised by extensive wiring checks on orbiter Columbia after the shutdown of two main engine computers during launch on 23 July 1999, Endeavour was found to have many more wiring defects than Discovery. STS-99 had been scheduled for September 1999. Because the HST was experiencing major control moment gyroscope failure NASA was concerned not to delay the repair mission any longer than possible and sent Discovery up first. On 7 October 1999 NASA managers decided to slip the STS-99 mission to 13 January 2000, eliminating original hopes of getting the SRTM flight up in September, the HST repair mission in October and a ISS logistics mission in December 1999. By the end of December STS-99 had slipped to the end of January and on 13 January NASA and the USAF agreed on a flight attempt scheduled for 31 January with a back-up opportunity the following day. If delayed the launch opportunity would slip to mid-February as a result of scheduling problems for tracking and communications requirements at the Cape. During late January NASA ordered inspections of Endeavour's SSMEs after inspections on Discovery revealed de-bonded blade tip seal material on a high-pressure hydrogen turbopump. Low clouds and rain coupled to ongoing technical problems halted launch day preparations for flight at T-29 minutes and a suspect avionics box prevented launch the following day. An attempt to launch on 9 February failed when a suspect avionics box scrubbed the countdown and the mission was re-scheduled for 12:30 EST (17:30 UT) on 11 February. During the launch attempt on 11 February almost 14 minutes of holds accumulated through a series of minor technical problems. During ascent sections of foam insulation came away from the External Tank intertank section, observed in images from Endeavour following separation.

The mission objective for STS-99 was supported by the Shuttle Radar Topography Mission payload, a radar remote-sensing technology capable of producing topographic maps of earth 30 times as precise as the best global maps available heretofore. The imaging systems constructs its maps through two radar antennas mounted in the orbiter payload bay and two antennas extended on a 60 m long mast deployed on orbit. The system was designed to construct high-resolution images of earth's surface between 60° north latitude and 56° south latitude, mapping areas populated by 95 per cent of the world's people to an accuracy of better than 30 m. The genesis of the SRTM lay in the STS-59 and STS-68 missions flown in 1994 and carrying the Spaceborne Imaging Radar C/X-band Radar in 1994. Modifications to the radar provides new capabilities compared with predecessors. The cornerstone of the innovation was the C-band and an X-band antenna at the extremity

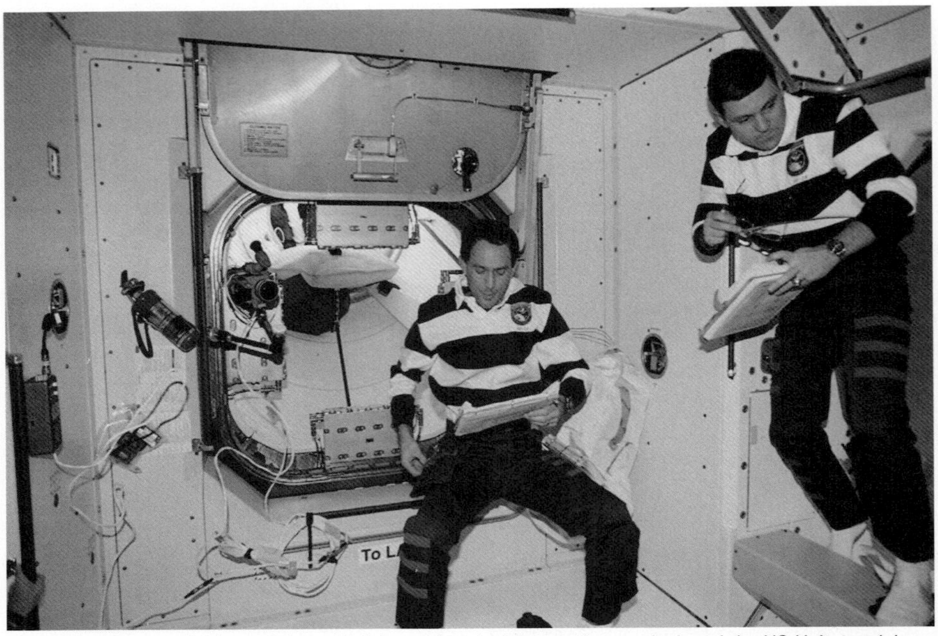

Astronauts Newman (left) and Cabana (right) examine checklists as they work aboard the US Unity module on STS-88 0085782

of the deployed mast, the longest rigid structure ever flown in space. This was the first time a dual-antenna imaging radar had been flown and utilised interferometry to map terrain elevation and produce digital elevation models. The mission was a partnership between NASA, the National Imagery and Mapping Agency, the German Deutsches Zentrum fur Luft-und Raumfarht, the Italian Space Agency which provided the experimental X-SAR radar system. The SRTM was the latest in a technology development application that began with the first civilian spaceborne imaging radar, Seasat, launched in 1978 and carrying a single-frequency 24 cm L-band single polarisation, fixed look angle radar designed for ocean studies. The first of the Shuttle imaging radars, SIR-A, flew on STS-2 in November 1981, also an L-band with fixed look angle. The second, SIR-B, flew on STS-41G in October 1984 and used an L-band, single-polarisation, radar with adjustable look angle.

The third Shuttle Imaging Radar, SIR-C, flew on STS-59 and STS-68 in April and September 1994 respectively. It was capable of imaging at three microwave wavelengths: L-band with quadruple polarisation; 6 cm C-band with quadruple polarisation; and 3 cm X-band with single polarisation. SIR-C/X-SAR also had a variable look angle and could image at incidence angles between 20° and 65°. Typical image sizes were 50 km × 100 km with resolutions of between 10 m and 25 m. SIR-C tested critical technologies leading to the development of the SRTM: active, phased array antennas and ScanSAR. The active SIR-C antenna provided the capability to steer to any angle through electronic manipulation of the radar beam and ScanSAR for STS-99 was derived from that capability. The radar beam is scanned back and forth painting a much wider swath than possible before. The earlier swath of 50 km was increased to 225 km, sufficient for total earth coverage between specified latitudes in one 11 day mission. The power required to operate the SRTM and associated equipment averaged 900 kW h. Data for the C-band radar was to be recorded at 180 mbps and for the X-band radar at 90 mbps. Total data was expected to be almost 10 terabytes recorded on 300 high-density tapes. The 609 m antenna mast is deployed and retracted by a motor-driven nut within the mast canister which pulls the mast from its stowed configuration, taking about 20 minutes to reach full deployment. The mast supports a 360 kg antenna structure at its tip and carried 200 kg of stranded copper, coaxial, fibre optic cables and thruster gas lines along its length. A secondary payload, EarthKAM is a NASA sponsored student programme involving a modified electronic Kodak DCS-460 still camera bracket-mounted to the overhead starboard window on the orbiter aft flight deck. Operated from a special facility at the University of California at San Diego, it requires no crew participation other than set-up and stowage. DTO 686 was designed to test the effectiveness of a portable heat exchange unit that uses the vacuum of space to chill nearby items by controlling the evaporation of water. DTO 690 involved five crewmembers evaluating the fit of different sizes of urine collection device, evaluating adapter (anatomical interface) and valve design, evaluating hygiene aspects involving seepage and leakage and testing techniques for returning urine to the waste collection system.

Continuous, round-the-clock, SRTM operation began within hours of Endeavour reaching orbit. Only two minor anomalies plagued early parts of the mission. A damper system designed to minimise forces acting on the mast could be activated because of temperature issues and a thruster on the mast designed to counteract the miniscule force of gravity that over time would torque the orbiter around was consuming excess nitrogen propellant. This caused the Shuttle orbiter to experience a doubling of propellant usage from 0.07 per cent/h to 0.15 per cent/h. Neither became a serious problem and the anomalies were resolved. Scientists were able to gain approval for an extra nine hours of radar data acquisition before mast retraction, finally logging 222 hours 23 minutes of almost continuous radar mapping when it ended at a mission elapsed time of 9 days, 18 hours 10 minutes during which scientists reported complete success. Mast retraction took place at 10:50 EST (15:50 UT) on 21 February after three previously unsuccessful attempts to achieve full retraction where the mast stopped just 3 cm from latch-up. Landing was delayed one orbit to allow relatively high crosswinds to dissipate, touchdown taking place at 18:22:23 EST on 22 February, making the 50th landing at KSC. Due to high electrical power usage from the SRTM the orbiter had little in reserve and almost no contingency for repeated wave-offs. Scientists were delighted with the extent of the radar data acquired during the mission, accomplishing 94.6 per cent of the pre-flight goal in collecting dual elevation measurements over 112.665 million sq km. The astronauts actually mapped 99.96 per cent of their objectives covering 118.881 million sq km. Technicians discovered that the radar system performed better than expected, making the one-pass data almost as good as the dual scan areas. Radar images filled 332 high-density tapes.

STS-101 Atlantis (98)

Orbiter flight: 21
Launch pad: KSC LC-39A
Launch date/time: 19 May 2000/06:11:10 EDT/10:11:10 UT
Launch window: 5 minutes 20 seconds
Main engines: SSME-1, 2052; SSME-2, 2044; SSME-3, 2047
Lift-off weight: 2,050,111 kg
Orbiter weight at launch: 119,099 kg
Orbiter weight at landing: 101,835 kg
Orbiter altitude/inclination: 332 × 341 km/51.6°
Mission duration: 9 days 20 hours 9 minutes 8 seconds
Landing site: KSC (51) runway 33
Landing date/time: 29 May 2000/02:20:18 EDT/06:20:18 UT
Rollout time: 62 seconds
Crew: James D. Halsell (Commander), Scott J. Horowitz (Pilot), Mary Ellen Weber (Mission Specialist 1), Jeffrey N. Williams (Mission Specialist 2), James S. Voss (Mission Specialist 3), Susan J. Helms (Mission Specialist 4), Yuri V. Usachev (Mission Specialist 5)
Payload (Primary):
ICC; MARS
Payload weight: ICC 1,678 kg; MARS 122.5 kg

Mission summary
OPF: 27 September 1998; 17 February 1999; 24 September 1999
VAB: 10 December 1998 (storage); 25 August 1999 (transfer aisle)
Pad: 25 March 2000
Returning to flight after a 10 month period at Boeing's Palmdale, California, facility, two years eight months after its last mission (STS-86 in September 1997), orbiter Atlantis carried more than 100 significantly improved and upgraded systems and features. These included relocating the airlock from the mid-deck area to the payload bay prior to ISS assembly flights, updated communication system, several weight reduction measures, enhancements to adding protection to the cooling system and a strengthened crew cabin floor. The flight deck incorporated 11 new full-colour, flat-panel, display units replacing 32 gauges and electromechanical displays and four CRT displays. This new Multifunction Electronic Display Subsystem (MEDS) is 34 kg lighter and uses less power than the subsystems it replaces and the colour displays facilitate pilot recognition of key features. All four orbiters would have the MEDS by 2002.

Inspection of Columbia in its extended refurbishment period revealed electrical wiring problems that instigated similar inspections of Discovery and Endeavour. During October 1999 inspections on Atlantis began and on 29 October NASA managers decided to launch STS-101 on or after 16 March 2000. Unplanned replacement of the ammonia boiler was found necessary after corrosion was found during the inspection and repairs to a damaged eleven were carried out. Due to a failure to a Proton launch vehicle of the type that would be used by the Russians for lifting the ISS Service Module, Proton launchers were grounded pending an investigation. The STS-101 mission was designated 2A.2 and was to have been launched after the Service Module had been sent to the ISS to install new equipment on Zarya and provide logistical support to the SM. The Russian Zarya module was certified operational through end March 2000 but it had been showing signs of wear and beyond that date it would need additional support. As a result, because of the delay to the SM launch, NASA decided to break up the mission tasks into two flights. STS-101 would fly on or after 16 March 2000 and, designated 2A.2a, would service Zarya, leaving STS-106 to fly mission 2A.2b to the Service Module and deliver its supplies. Atlantis carried the Integrated Cargo Carrier (ICC) for the first time, an open structure 2.44 m long, 4.57 m wide weighing 1,383 kg. It could accommodate six 181 kg capacity boxes on a pallet surface. For this mission it carried the 95 kg US Spacewalkers Crane and the 89.8 kg Russian crane Strela. STS-101 would install a new battery and other electrical equipment in Zarya. In January 2000 four of its six batteries were working normally. An EVA would be conducted to move the cargo boom from Zarya to a stowage point on the Unity module. In late January 2000 engineers replaced a defunct thruster on Atlantis and performed ammonia system and fuel cell tests while continuing with extensive wiring inspection. At the end of January managers set a launch date of 13 April. Main propulsion leak checks were conducted early February and work to install the transfer tunnel adapter in the payload bay began.

Because the original 2A.2 mission was being split into two, some crew changes were made and astronaut Edward Lu and cosmonauts

Shuttle Endeavour's remote manipulator arm re-positions the Unity module prior to docking with the Russian Zarja module during STS-88 0085781

EVA astronauts Newman and Ross work on the Zarja and Unity module during STS-88 0085780

Yuri Malenchanko and Boris Morukov were removed from STS-101, assigned to STS-106 and replaced by Voss, Helms and Usachev. Its original launch date was shifted to April 17 for scheduling reasons associated with other launches. On 15 March managers ordered replacement of SSME-1 (2045) because of concerns about the tip seals in the high-pressure hydrogen turbo-pump. Atlantis was moved to the launch pad on 25 March. Its 17 April launch date had shifted to 06:37:18 EDT on 18 April and then a week to 16:14:50 EDT on 24 April allowing astronaut Halsell time to recover from an ankle sprain. Flight scheduling at the Cape would give NASA just three days to launch Atlantis or defer until mid-May. Just eight minutes before the scheduled launch at 16:15 EDT SLF cross wind violations prompted a scrub, as they did on 25 April one hour 34 minutes prior to the planned lift-off time of 15:52 EDT. A third launch attempt scheduled for 15:29 EDT on 26 April was cancelled about one hour prior to launch due to poor weather at the TAL sites. A fourth launch attempt was scheduled for 18 May, a date driven by tracking and range safety schedules involving a flurry of other launches at the Cape. On orbit, the ISS was losing altitude at the rate of about 3 km per week. On 18 May the ISS (comprising Zarya and Unity) would be in a 335 × 326 km orbit; on 30 May the ISS would be in a 328 × 320 km orbit. Managers wanted to conserve Zarya on-board propellant as a contingency against possible further delays and aimed to avoid an orbit altitude boost. To conserve orbiter propellant for the much needed altitude boost of almost 40 km the Shuttle launch window was cut from 10 minutes to 5 minutes, reducing the plane-change propellant requirement. Because of high winds predicted for 18 May the launch of an Atlas 3A was delayed, shifting the Shuttle attempt set for 06:37:58 EDT to 19 May.

Atlantis rocketed toward orbit following a perfect launch at 06:11:10 EDT/10:11:10 UT on 19 May, a OMS burn 44 minutes later placing the Shuttle in a 324.7 × 157 km orbit. The NC-1 rendezvous burn took place at 09:54 EDT/13:54 UT putting Atlantis in a 325 × 200 km orbit. The only anomaly was a suspect propellant shut-off valve on the left side OMS. Managers decided to use only the right side OMS for further rendezvous burns, the next of which occurred at 21:51 EDT 19 May/01:51 UT 20 May after a crew rest period. After several more orbital manoeuvres reconfigured due to the single-OMS operation, Atlantis docked to Zarya at 00:31 EDT/04:31 UT 21 May following which a sleep period preceded preparations for the mission EVA from the Shuttle to install urgent equipment prior to crew transfer to Zarya. The EVA began at 21:48 EDT 21 May/01:48 UT 22 May and in 6 hour 44 minutes astronauts Voss and Williams completed all their scheduled tasks. These included moving the Russian Strela cargo crane to the Unity Module, installing eight handrails on Unity, moving a cable bundle obscuring an optical target and attaching a TV camera cable for future Shuttle dockings. This was the fifth EVA for ISS support bringing to 304 hours 37 minutes the time spent on Shuttle EVAs during 49 excursions. The Atlantis crew entered Zarya at 20:03 EDT 22 May/00:03 UT 23 May, twelve months after the last crew to visit the station. Extensive housekeeping duties were performed and the crew reported improved atmospheric conditions; the previous crew in 1999 reported bad odours and nauseating smells from materials outgassing.

Between 20:02 EDT 23 May/00:02 UT 24 May and 21:01 EDT 23 May/01:01 UT 24 May astronauts Halsell and Horowitz performed 27 propulsion burns to raise the orbit from 326.7 × 334.7 km to 353.2 × 341.6 km. Despite some concerns about the state of a battery, four new batteries were installed in Zarya (nos 1, 2, 3 and 5) leaving two existing batteries in place. Late on 24 May a further 27 propulsion burns over a 56 minute period raised the ISS to a 363.5 × 360 km orbit. At 19:36 EDT/23:36 UT 25 May the first of 26 propulsion burns in a 54 minute period increased the ISS orbit to 382.5 × 369.1 km, a total average altitude increase of 44.4 km since arriving at the station. After further work configuring Zarya and Unity for the next component of the 2A.2 mission, to be flown by STS-106, Atlantis undocked at 19:03 EDT/23:03 UT 26 May, moved away 140 m before beginning separation manoeuvres. Russian flight controllers began activation of the ISS automatic attitude control system aboard Zarya. The 3 minute 5 second Shuttle de-orbit burn began at 01:12 EDT/04:12 UT 29 May slowing Atlantis by 340 km/h setting up the descent path for a landing approach to KSC SLF runway 15. Touchdown took place at 02:20:18 EDT/06:20:18 UT bringing the 9 day 20 hour mission to a successful conclusion. During the mission the Atlantis crew replaced four of Zarya's six batteries, installed new battery control equipment, replaced a faulty radio antenna on Unity, installed a replacement telemetry unit in Zarya and transferred 1.4 tonnes of supplies and equipment for future missions including the long duration crew planned for late 2000.

STS-106 Atlantis (99)
Orbiter flight: 22
Launch pad: KSC LC-39B
Launch date/time: 8 September 2000/08:45:47 EDT
Launch window: 3 minutes 57 seconds
Main engines: SSME-1, 2052; SSME-2, 2044; SSME-3, 2047
Lift-off weight: 2,050,111 kg
Orbiter weight at lift-off: 115,259 kg
Orbiter weight at landing: 100,368 kg
Orbiter altitude/inclination: 320 km/51.6°
Mission duration: 11 days 19 hours 11 minutes 1 second
Landing site: KSC SLF runway 15
Landing date/time: 19 September 2000/03:56:48 EDT/07:56:48 UT
Rollout time: 64 seconds
Crew: Terrence Wilcutt (Commander, Scott D. Altman (Pilot), Edward T. Lu (Mission Specialist 1), Richard A. Mastracchio (Mission Specialist 2), Daniel C. Burbank (Mission Specialist 3), Yuri Malenchenko (mission Specialist 4), Boris V. Morukov (mission Specialist 5)
Payload (Primary):
Integrated Cargo Carrier plus cargo; Logistics Double Module
Payload weight: ICC plus cargo 2,054 kg; LDM 8,165 kg

Mission summary
OPF: 29 May 2000
VAB: 7 August 2000
Pad: 13 August 2000
Mission 2A.2 originally to a single flight was split into two with 2A.2a flown by STS-101 launched 19 May 2000 and 2A.2b flown by STS-106 launched 8 September 2000, both involving orbiter Atlantis. The plan was formally agreed 18 February 2000 at which time STS-101 was slated for 13 April and STS-106 for 19 August. Delays in getting STS-101 spaceborne re-scheduled the STS-106 mission for no earlier than 8 September. Following its successful landing 29 May 2000 from the STS-101 mission, Atlantis was delivered straight to the OPF in preparation for its turnaround to STS-106. During the first week in June 2000 technicians removed the three SSMEs and commenced intensive examination and scrutiny, re-configuring the orbiter for its return to space. Clearance for launch of Atlantis awaited the successful flight of the Zvezda Service Module, its flight delayed from the original date (and the reason for splitting Mission 2A.1) on a Proton booster. Launch took place at 04:56 UT 12 July when Zvezda was lifted to a 365 × 372 km, 51.6°, orbit for two weeks of checkout prior to docking. Link-up to the Zarya control module took place at 00:45 UT 26 July thus opening the way for STS-106, the precursor flight to permanent habitation by a three-person crew on Expedition 1. On 6 August at 18:27 UT a Soyuz-U launched from Baikonur carried the Progress M1–3 cargo tanker into space carrying 588 kg of stores and supplies from where it docked to aft port on Zvezda at 20:13 UT on 8 August. With the addition of these two Russian elements, the ISS was almost 44 m in length and weighed almost 61 tonnes with interior pressurised volume of 246.4 cm^3, considerably bigger than when the STS-101 crew visited in May 2000.

Astronaut Newman waves to the camera while working on Space Station modules during STS-88 0085779

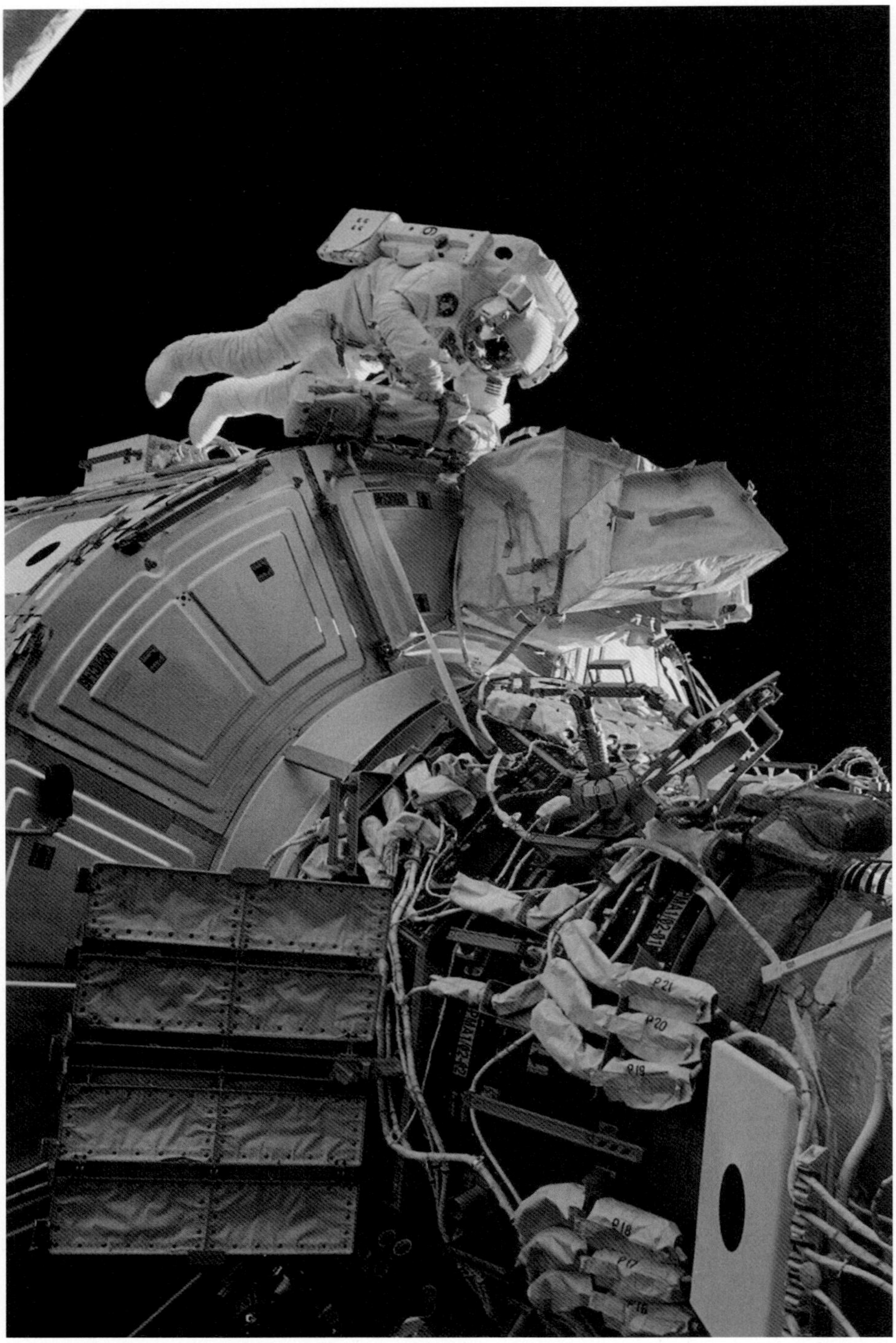

Astronaut Barry works on the Space Station during STS-96 0085778

The payload for STS-106 reflected the ISS support role this mission was designed to fulfil. The ICC was attached to the aft side of the airlock module and the Spacehab Logistics Double Module was attached to the aft end of the ICC, making up the rear 40 per cent of the orbiter payload bay. Access from the LDM was through a continuous tunnel between elements of the ICC and into the airlock module. The LDM carried approximately 3,675 kg of hardware, equipment and logistical supplies to the ISS. Essentially an externally mounted, unpressurised, aluminium flat-bed pallet, the ICC on STS-106 carried an additional 1,300 lb of cargo. The ICC also carried the Spacehab-Oceaneering Space System (SHOSS) box, an unpressurised container attached to the top of the ICC with capacity for up to 181 kg of tools and other flight equipment. Atlantis would carry food, fresh water, additional exercise equipment and two replacement batteries for Zarya (four had been replaced by STS-101). Moved to LC-39B on 13 August 2000, Atlantis was threatened by Hurricane Debby the threat alleviated 24 August when it veered to a more westerly course. At 17:56 EDT 5 September the lightning mast experienced a strike but the vehicle was unaffected although the thunderstorms did delay some work at the pad.

Despite concerns over anomalous weather in the KSC area Atlantis was launched on time at 08:45:47 UT on 8 September 2000. Closing the 9,600 km lead to achieve a nominal rendezvous, Atlantis docked with the PMA-2 adapter at 01:51:37 EDT/05:51:37 UT on 10 September. The EVA began when astronauts Lu and Malenchenko switched their suits to internal power at 00:47 EDT/04:47 UT on 11 September. At an EVA elapsed time of 45 minutes the two EVA astronauts rode the Shuttle RMS about 23 m up the side of the station to a position adjacent the Zarya module. From there the two worked their way along hand rails a further 27 m to install a magnetometer boom. Lu freed a jammed docking target on the Zvezda module which had failed to deploy properly when originally needed during rendezvous and docking. Engineers were worried the spring-loaded boom may deploy without warning during a future EVA, endangering the spacewalkers. Then, getting appreciably ahead of schedule, the two astronauts worked their way back down the stack to start connecting nine power, telemtry and data cables between Russian modules Zvezda and Zarya. Little more than 4 hours into the EVA the wiring work was complete and they were working their way down the hull to stow tools, empty cable spools and associated equipment on the ICC before ingress through the payload bay airlock module. The EVA had lasted 6 hours 14 minutes. Next day, 11 September, the crew began preparations for opening up the station, now comprising Unity, Zarya, Zvezda modules to the extreme end of which was attached a Progress cargo tanker.

Wilcutt opened the hatch into the Unity module at 10:47 EDT 11 September/02:47 UT 12 September and the crew floated in. About one hour 15 minutes later they opened up Zarya and less than two hours after that they entered Zvezda wearing protective goggles and face masks to prevent adverse effects of materials outgassing since the module docked with Zarya almost seven weeks earlier. Zvezda's outer hatch was opened into the Progress cargo tanker at 02:08 EDT/06:08 UT on 12 September. After a rest period the crew set about installation and logistical stowage, equipment configurations and subsystems replacements. Four of Zarya's six batteries had been replaced during STS-101 and batteries 4 and 6 were to be replaced by STS-106. To save weight, Zvezda was sent into orbit with only five of its eight batteries installed and the other three had been carried up by Atlantis along with battery control equipment. After Zarya's battery No 6 had been replaced by Burbank and Morukov and the charge-discharge controller installed a bracket was discovered to foul the voltage controller known to the Russians as the BUPT. Nut plates blocked access to screws that held the bracket in place and the two astronauts had to resort to a hammer and chisel to strip the fixture. The following day the crew set about installing the ISS permanent zero-g toilet facility which would not be used during this visit but plumbed in and activated during the next visit. The toilet would deliver human waste products to containers periodically taken to a Progress vehicle for atmospheric burn-up. The crew also replaced the last of six Zarya batteries, transfered supplies and equipment to the ISS from Atlantis and Progress, transfered almost 180 kg of water and installed a battery charger/discharger which would be utilised by future Soyuz and Progress vehicles. During this activity Russian controllers reported a problem with No 5 of eight batteries on Zvezda but stowage and storage proceeded as planned, including transfer of an IMAX camera for future crews to shoot a space station film for public presentation at IMAX theatres. The crew also performed the second of four planned re-boost manoeuvres.

Over the next several days battery troubleshooting shared time with tests of Zvezda communications and data systems, evaluating voice downlink with Russian controllers and testing power converters which will allow US built solar arrays to feed electrical energy to Russian modules and equipment. A Russian Elektron oxygen generator was assembled for activation by the Expedition 1 crew which, when operational, would use electrolysis to break down water into its constituent chemical components oxygen and hydrogen. Oxygen would be used for breathing but hydrogen would be dumped overboard. By 15 September the crew were ready to install the treadmill in Zvezda, a device carrying an active isolation system so that vigorous users would not disrupt microgravity experiments. When the installation ran into alignment problems controllers suggested the crew disassemble then reassemble the equipment, an activity that solved the unidentified problem. A fourth and final set of orbit re-boost manoeuvres increased altitude by an average 22.5 km leaving the station in a 386 × 377 km orbit. After a total occupation time of 5 days 9 hours 21 minutes the STS-106 crew closed the last hatch and prepared for undocking, which came at 11:46 EDT 17 September/03:46 UT 18 September. The 375 km/h de-orbit burn was effected through a 3 minute 16 seconds firing that began at 02:49:17 EDT/06:49:17 UT 20 September resulting in a precise night-time touchdown at KSC at 03:56:48 EDT.

STS-92 Discovery (100)
Orbiter flight: 28
Launch pad: KSC LC-39A
Launch date/time: 11 October 2000/19:17:00 EDT
Launch window: 4 minutes 12 seconds
Main engines: SSME-1, 2045; SSME-2, 2053; SSME-3, 2048
Lift-off weight: 2,050,542 kg
Orbiter weight at lift-off: 115,127 kg
Orbiter weight at landing: 92,740 kg
Orbiter altitude/inclination: 320 km/51.6°
Mission duration: 12 days 21 hours 42 minutes 41 seconds
Landing site: EAFB runway 22
Landing date/time: 24 October 2000/14:59:41 EDT
Crew: Brian Duffy (Commander), Pamela A. Melroy (Pilot), Leroy Chiao (Mission Specialist 1), William S. McArthur (Mission Specialist 2), Peter J. K. Wisoff (Mission Specialist 3), Michael E. Lopez-Alegria (Mission Specialist 4), Koichi Wakata (Mission Specialist 5)
Payload (Primary):
Z1 Integrated Truss Segment; PMA-3
Payload weight: 12,704 kg
Payload weight down: 2,859 kg

Mission summary
OPF: 27 December 1999
VAB: 21 August 2000
Pad: 11 September 2000

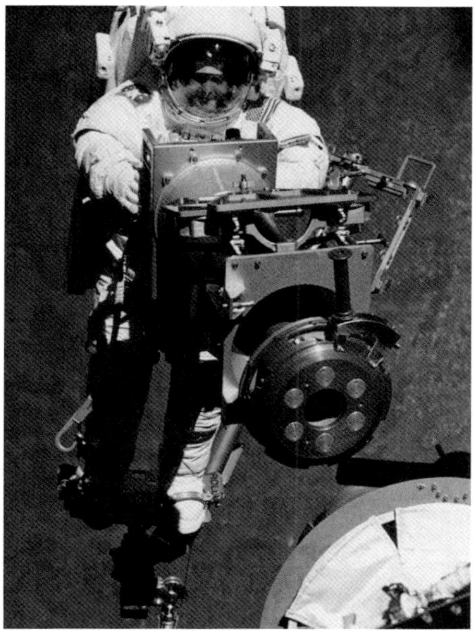

Astronaut Jerrigan moves part of the Russian-built crane Strela on STS-96 0085777

Designated ISS assembly flight 3A, the STS-92 mission was structured to deliver the Z1 truss assembly, Pressurised Mating Adapter 3 and perform five DTOs and cargo transfers from the orbiter to the ISS. The launch time was set for 21:33:27 EDT 5 October/01:33:27 UT 6 October 2000 and preparations continued apace on launch day. Fuelling was held up while engineers consulted about an event just observed in launch film from the STS-106 launch which showed one of two ET-orbiter aft attachment pins failed to fully retract as it should have done. There was concern that a hung pin could cause the ET to rotate and collide with the orbiter. The 35 cm long, 1 cm diameter, bolt extends from the ET into the underside of the orbiter and is secured with explosive bolts torqued with 1,223.2 kN force. By mid-afternoon local time management had decided to scrub the countdown and target the flight for 21:16:10 EDT 6 October giving time to assess records and reach a decision, also affording time to clear another minor technical problem associated with a POGO suppression valve in the main propulsion system. A decision to completely replace the valve re-set the launch to 20:05:17 EDT 9 October. Work to replace the valve was completed 6 October and the bolt issue was resolved two days later but high winds behind a cold front moving through the Cape area forced a further delay in the scheduled launch to 19:39:36 EDT 10 October. Shortly after fuelling on that day and with just three hours to launch the ice inspection team discovered a pip pin found lodged in a strut connecting the orbiter to the ET in an inaccessible region. The 227 g pin had apparently fallen from a temporary platform which had been dismantled some time before final flight preparations. Fear that it could fall into the ET or the orbiter causing damage during launch forced mission controllers to scrub the attempt for a fourth time, re-scheduling the flight to 19:17:00 EDT 11 October. Scheduling requirements were in jeopardy. Discovery had to be launched by 15 October because the Expedition 1 crew was set to launch to the ISS from Baikonur on 30 October and two days were needed after the flight of Discovery to get mission-critical STS-92 personnel over to Russia for the launch of Soyuz TM-31. This time the countdown proceeded on schedule and Discovery was launched as planned at the fifth try.

As Discovery closed on the orbiting ISS engineers tracked a problem which left the orbiter Ku-band antenna system inoperable and unable to send radio and television signals through that route. Instead of live colour TV of the approach and docking, NASA deferred to a grainy black and white picture from a Russian camera attached to the ISS. Communications would have to go through the slower S-band system causing some communications log-jam and reduced data flow rates. Without a rendezvous radar, Shuttle pilots flew the orbiter to a near perfect docking using visual techniques using very little additional propellant compared to a radar-based approach. Docking took place at 13:45:15 EDT (17:43:15 UT on 13 October with Discovery attached to the PMA 2 adapter on the forward common berthing mechanism of the Unity module. At 08:57 EDT on 14 October, just as astronaut Wakata was unlatching the RMS, a 14 amp electrical spike on bus #3 knocked out critical payload support systems. This included an upward pointing payload bay keel camera, part of a Canadian computerised TV system essential for visually manipulating the massive Z1 truss from the rear of the orbiter payload bay to the top of the Unity module. While Mission Control deliberated over the anomaly the astronauts deferred the planned Z1 truss deployment and cut to other activities. On recommendations from engineers on the ground, the crew successfully hooked up a spare computer and powered it from an alternate source. As additional, visual, backup astronaut Lopez-Alegria would position himself by a viewport in the Unity module and astronaut Melroy would work from another laptop computer to command the zenith port common berthing mechanism. The Z1 truss was a vital element for future growth of the ISS. Rigidly attached to the top of the Unity module it would be a strong-back mounting structure to which truss assemblies would be installed to port and starboard like wings, supporting the large solar arrays. Initially the Z1 would also support a huge set of solar arrays to be delivered by STS-97 in December 2000. Throughout the life of the ISS, the Z1 truss structure would support the station's primary communication system with a large dish antenna and associated electronic support boxes. The crew began unlatching the Z1 truss from its carrier pallet at 11:30 EDT on 14 October and found it much easier to manipulate the massive structure from its berthing position to the zenith port on Unity than anticipated.

The first of four EVAs scheduled for STS-92 began at 10:27 EDT 15 October and in 6 hours 28 minutes Chiao and McArthur completed all assigned tasks. Astronaut Wakata operated the RMS, moving the two EVA crewmembers around the vicinity of the Z1 truss attaching wires, cables and fixtures. The early part of the EVA was busy deploying tools, restraints and tethers and then connecting the six umbilical cables between Unity and the Z1 structure. They then removed the S-band Antenna Subassembly and placed it in a temporary position exposing the place where two power converters would be attached on the third EVA. Another set of cables was attached before the Space to Ground Antenna was installed and its dish deployed. Chiao worked late in the EVA on the large ISS Ku-band antenna, erecting it to its deployed position. In Mission Control on 16 October as Wisoff and Lopez-Alegria were preparing to conduct the mission's second EVA the Power, Heating, Articulation, Lighting and Control Officer (Phalcon) successfully completed work on the two Plasma Contactor Units mounted on the Z1 truss. Designed to discharge electrical current on the station and to its immediate environment, the PCUs were launched with their valves open to allow a continuous discharge of xenon gas to prevent contamination of the units. Phalcon powered up and purged the units, closed the valves and deactivated the PCUs. They would be powered on again during STS-97 after the solar arrays had been deployed and started generating current.

The second EVA began at 10:15 EDT on 16 October, about 15 minutes ahead of schedule, and lasted 7 hours 7 minutes. Wisoff and Lopez-Alegria supported transfer of the PMA 3 from the cargo bay to its position on Unity. They began by unlatching the device before Wakata moved it with the RMS as Wisoff and Lopez-Alegria released latches on the top of the Z1 truss in preparation for the large solar array to be carried aboard STS-97. They then moved back to Unity and to visually observe the RMS, controlled by Wakata, latching to PMA 3. Inside Discovery, Melroy again used her laptop to command the first stage of securing capture latches and locking bolts. The final commands would be performed later by flight controllers after MCC-H confirmed temperatures in the docking port and Unity's common berthing mechanism had stabilised. In these two days of activity the ISS had grown in mass by about 9.5 tonnes. As the EVA ended Astronaut Duffy performed the first of three re-boost manoeuvres in 18 pulses of 1.4 seconds each over a 30 minute period. Later that night flight controllers completed the clamp-down operation for securing PMA 3 to the Unity module, latching all 16 bolts by remote command from Houston. PMA 3 would be the docking port for Endeavour on the STS-97 mission two months hence. The third mission EVA began at 10:30 EDT, 17 October, when Chiao and McArthur started work to configure the Z1 truss for upcoming missions. Lopez-Alegria took control of moving the duo around on the RMS for the first half of the space walk and then Wakata took over for the second half. Two 58.5 kg DDCU (DC to DC Converter Unit) hardware sets were installed on top of the Z1, power cable connectors were hooked up to both Z1 and PMA 3 elements, power cables were re-routed from PMA 2 to PMS 3 for the arrival of Endeavour in December and a second tool box was fixed to the Z1 truss. This would support future EVAs and included tools and hardware Jernigan and Barry had left on the Unity module during their 30 May 1999 EVA during STS-96. After the 6 hour 48 minute EVA the Shuttle pilots performed another re-boost manoeuvre, firing thrusters in a series of 1.4 second pulses over a 30 minute period.

The fourth and last EVA began at 11:00 EDT 18 October when Wisoff and Lopez-Alegria began a series of clean-up tasks configuring equipment and hardware for the next stage in ISS assembly and build-up activities. They removed a grapple fixture from the Z1 truss, opened and closed a latch assembly that would hold the P6 solar array truss scheduled for STS-97and deployed a tray that would be used to power the US laboratory Destiny. They tested the manual berthing mechanism latches that would support that module. Wisoff opened and closed the latches on the capture assembly for the P6 array using a pistol grip tool with which he made more than 125 turns to open the latches, then closed and reopened them. He left the capture latch, called the claw, ready to receive the P6 arrays. The EVA astronauts evaluated the nitrogen powered SAFER (Simplified Aid for EVA Rescue) by each jetting a 15 m distance demonstrating the ability of an astronaut to regain the safety of the spacecraft in the event of a severed or disconnected tether. Time constraints prevented a low priority evaluation of one astronaut, simulating a disabled crewmember, being hauled back by the other to the airlock along hand rails. Nevertheless, the EVA had been a stunning success and justified the months of pre-flight preparation. After a day of packing up and stowing, leaving the ISS in an appropriate configuration for the next crew to visit Discovery was made ready for departure.

On 20 October at 11:08 EDT, after 6 days 21 hours 23 minutes docked to the ISS, Discovery unlatched from the ISS and drifted free. The landing was

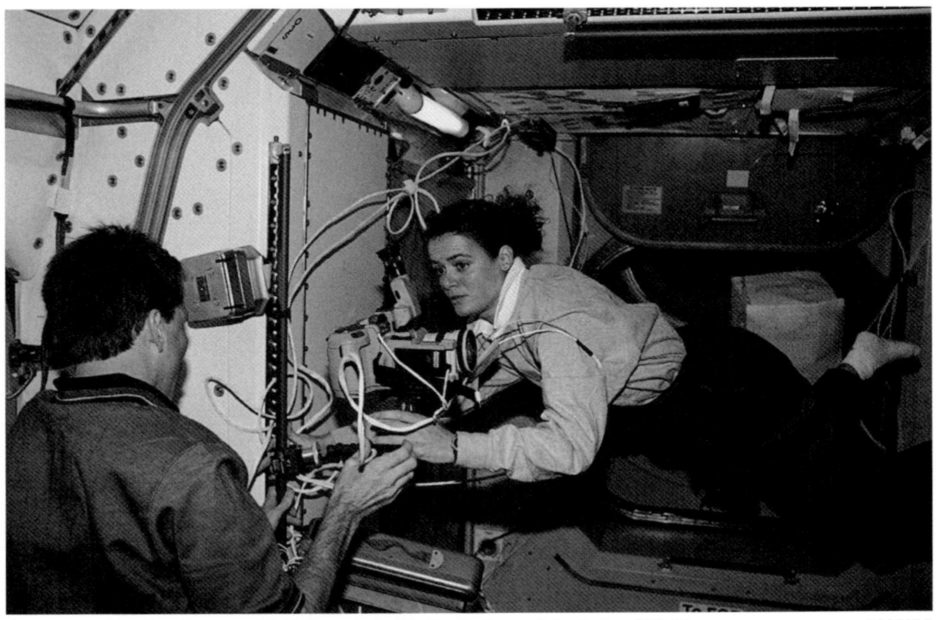

Astronauts Kent Rominger and Julie Payette inside the Unity module during STS-96 0085852

From the left, astronauts Smith, Gunsfeld, Foale and Nicollier prior to the STS-103 mission 0085853

scheduled to take place on 22 October with the de-orbit burn at 13:06:57 EDT for a landing at KSC at 14:14:20 EDT. High crosswinds at the landing facility forced a cancellation and one-day wave-off with a second attempt to be made 23 October with the de-orbit burn at 13:43:00 EDT and landing at 14:51:00 EDT. Several alternate landing options for EAFB in California came and went as rain showers affected the Edwards AFB area, causing another delay until 24 October. NASA had wanted Discovery back at KSC to ensure it would be ready for its next flight in 2001, the STS-102 mission assigned to return with the Expedition 1 crew launched aboard a Soyuz on 31 October 2000. Poor weather at KSC, however, forced controllers to order a EAFB landing and at 15:52 EDT the 3 minute 6 second de-orbit burn slowed the orbiter by 532 km/h and Discovery sliced through the atmosphere to a landing at 16:59:41 EDT, 24 October. Duffy had taken over manual control at 15,000 m. This was the first Shuttle touchdown in California since the STS-76 landing on 31 March 1996 breaking a straight 23 touchdowns at KSC. For a mission that took five launch attempts to gain orbit a three-day wave-off for landing seemed appropriate.

STS-97 Endeavour (101)
Orbiter flight: 15
Launch pad: KSC LC-39B
Launch date/time: 30 November 2000/22:06:01 EST; 1 December 2000/03:06:01 UT
Main engines: SSME-1, 2043; SSME-2, 2054; SSME-3, 2049
Lift-off weight: 2,052,272
Orbiter weight at lift-off: 120,741 kg
Orbiter weight at landing: 89,766 kg
Orbiter altitude/inclination: 320 km/51.6°
Mission duration: 10 days 19 hours 57 minutes 24 seconds
Landing site: KSC SLF runway 15
Landing date/time: 11 December 2000/18:03:25 EST
Rollout time: 55 seconds
Crew: Brent W. Jett (Commander), Michael J. Bloomfield (Pilot), Joseph R. Tanner (Mission Specialist 1), Marc Garneau (Mission Specialist 2), Carlos I. Noriega (Mission Specialist 3).
Payload (Primary):
PVAA; IEA, The Long Spacer
Payload weight: 17,000 kg up; 871 kg down

Mission summary
OPF: 24 February 2000
VAB: ?
Pad: 31 October 2000
The STS-97 flight supported ISS mission 4A designed to lift the P6 photovoltaic module and temporarily install the P6 Integrated Truss Structure on the Z1 truss until re-located to its permanent location on the P5 truss during a later assembly. The P6 ITS was the primary payload and contained the Photovoltaic Array Assembly, the Integrated Equipment Assembly and the Long Spacer. STS-97 would deliver the first of eight sets of solar arrays. The P6 had two identical PVAAs each of which comprises a Solar Array Assembly (SAA) containing two Solar Array Wings (SAW folded onto a mast situated in a Mast Canister from where it would be deployed. The Beta Gimbal Assembly (BGA) consisted of the following: a Mast Canister to house the folded mast; the Bearing, Motor and Roll Ring Module (BMRRM) used to rotate the SAW and transfer power; the Electronic Control Unit (ECU) used to control the BGA motor and mast rotation; and the Sequential Shunt Unit (SSU) designed to coarsely regulate the SAW output voltage. Power generated by the PVAA would be routed via the SSU to the IEA. Two solar array wings were to be deployed from the P6 module, each deployed in the opposite direction to the other. Each SAW was made up of two solar panels mounted to a common mast. Prior to deployment each panel was folded into a Solar Array Blanket Box (SABB) measuring 51 cm high and 457 cm in length. When fully deployed the SAW extends 35 m and spans 11.6 m and since both oppose in different directions the total wing span is 73.1 m. Each solar array wing, the largest ever deployed in space, weighs more than 1,090 kg and carries 32,800 solar cells, each measuring 8 cm^2 with 4,100 diodes.

When deployed, the P6 tower would have port and starboard array wings, each wing comprising two flexible blankets of solar cells separated by the central open frame mast. Each blanket comprises 82 panels each measuring 38 cm × 4.3 m and each panel carries 200 solar cells each of which is 20.6 cm^2. Solar flux in earth orbit is about 1.4 kW/m^2 and cells convert sunlight to electricity with a 12–14 per cent efficiency. Two blanket panels on each solar array make up one circuit that gives each array 41 circuits. As each circuit has 400 solar cells in series it produces about 0.5 V at 2.7 A. At beginning of life each cell produces 1 W of power giving a theoretical maximum 32.8 kW, degrading to 24.6 kW after the design life of 15 years. The two arrays produce a combined power generation capability of almost 64 kW h. The two solar arrays will produce about 30 kW more than needed to re-charge the batteries in the electronics assembly and operate the array's other internal systems. By comparison, the two solar arrays on Zvezda generate about 10 kW while the two arrays on Zarya provide an additional 7 kW. Power generated by the P6 is sent direct to the current switching unit in the IEA while the switching unit would direct power to the station and to 12 batteries during orbital daylight. Direct current converters step the voltage down from 160 V DC generated by the arrays to 120 V for station current. In darkness power would be provided by twelve 170 kg nickel hydrogen batteries in the IEA. The batteries will be replaced every five years but the solar arrays are designed for 15 years of in-orbit use. The 7,711 kg Integrated Equipment Assembly is 4.9 × 4.9 × 4.9 m and is designed to condition and store the electrical power collected by the photovoltaic arrays. The P6 Long Spacer measures 8.5 m long, 4.9 m wide and 4.9 m deep. It is designed to physically separate the P6 solar arrays from the P4 solar arrays when the P6 is re-located on a later mission and it provides temporary cooling for the Destiny laboratory after it was delivered by STS-98.

Preparations for launch were completed without major problems, the only shift being a modest delay in fuelling to remove a loose bracket from the side of the 'white room' on the launch structure. With near perfect weather Endeavour was launched at 22:06:01 EDT into a night sky over the coast of Florida making the 25th Shuttle night launch. One of two explosive cartridges holding the left SRB to the ET failed to fire although a detonator inside the strut did fire and the SRB separated without incident. Analysis of the problem would threaten the assembly of Shuttle Discovery in the VAB. It would later be discovered that a frayed wire inhibited activation. Within minutes of reaching orbit the STS-97 crew unstowed equipment and configured the orbiter for the next several days of busy activity. Almost all of the mission would be conducted from Endeavour, its crew only spending one day at the end of the flight with the Expedition 1 crew aboard the ISS, hatches remaining firm in place until then due to pressure differentials between orbiter and station to facilitate EVA operations. Since the launch of the previous Shuttle mission two key flights had changed activity at the ISS. On 31 October the Expedition 1 crew of NASA astronaut Bill Shepherd and Russian cosmonauts Yuri Gidzenko and Sergei Krikalev were launched on Soyuz TM-31 for a rendezvous and docking with the ISS two days later. On 16 November the Progress M1-3 cargo tanker was launched from Baikonour and docked under manual control with the Zarya nadir port at 03:48 UT on 18 November. At 16:23 UT on 1 December Progress undocked from Zarya and moved away at a rate of 10 km per orbit with a view to its reuse in 2001 as a trash disposal craft. During the initial approach on 18 November the KURS automated docking system failed and Gidzenko had to take over remote control with the TORU back-up system for the final 200 m. The Russians wanted to test a software patch devised to get round the technical fault which prevented automated link-up by re-docking it to the station. NASA was concerned at flight safety issues and no firm decision had been made as to its interim fate.

Terminal rendezvous phases were different from those used on previous ISS missions and similar to those used for Mir rendezvous operations. Endeavour would approach the ISS along the plus R-bar (below the radius vector) with the station flying in the XVV mode (long axis along the velocity vector). Terminal sequences commenced with Endeavour performing a propulsion burn at approximately 15 km separation below and behind. After catch-up Endeavour turned through 180° with the tail pointing in the direction of motion. Because of the relative positional attitude of the orbiter to the station, unlike prior ISS missions the centre of gravity offsets were considerable necessitating extremely accurate mating alignments to prevent tipping the station off axis and fouling the docking mechanism. Jett had a margin of just 7 cm when aligning the Shuttle docking device with the PMA 3 ring. The crew successfully accomplished docking with PMA 3 at 14:59:40 EST, 12 November, secured minutes later when commands were sent to retract Endeavour's docking ring. Initial exchanges between the Endeavour and ISS crewmembers acknowledged that phase of the operation complete through a communications link that would be the only contact the two crew would have with each other for several days. Following docking the crew re-orientated the alignment putting the docked configuration into a X-POP attitude (x-axis perpendicular to the orbital plane), this to ensure proper heating on the Z1 truss. The truss carries four massive gyroscopes which on a later mission would be activated for control-moment attitude stabilisation. Tests indicated that gyroscope speed sensors could be damaged by lengthy exposure to cold temperatures. After the P6 solar arrays had been installed there would be sufficient electrical power to operate internal heaters. Before entering a rest period Canadian Marc Garneau used the Shuttle RMS to delicately lift the massive P6 truss from the cargo bay at 17:30 EST and slide it out and under from a station docking port much like sliding a foot from a shoe. By 18:30 EST the P6 was parked outside the bay at a pitch of about 30°, still attached to the RMS overnight to preserve temperature limits and prevent thermal damage to its delicate systems. Final launch day activities included opening up the PMA 3 docking port and placing a new laptop computer and other essential equipment inside before buttoning it up again.

The first EVA began at 13:35 EST on 3 December and lasted 7 hours 33 minutes during which Tanner and Noriega manipulated the 17 tonne P6 module from its parked position above the cargo bay to the hook on top of the Z1 truss. Noriega engaged the latching mechanism to pull the array down onto the truss at which point both crewmembers tightened four large bolts locking the two together at about 14:32 EST. He plugged power lines from Endeavour to the P6 assembly and both men moved up the P6 tower to prepare the solar wings for deployment by extending attachment joints into the proper position. There was some minor problem deploying bars on the attachment joints before both astronauts extended the solar blanket boxes to the deploy position but there was a further problem with commands to open latches on one of the two boxes. After several further attempts the commands got through but one of 28 latch pins on the port array failed to move. Evaluating that problem, controllers moved the two astronauts to deployment of the starboard wing that began to open up at 20:23 EST but the blankets of solar cells were slack and did not have the tension expected. The next step was to have closed the blanket boxes which would have tensioned the array but the peculiarity of the deployment as seen on video replay concerned controllers and that was postponed pending further analysis. As the EVA was ending Jett successfully commanded open the errant latching pin on the port array, freeing that array for deployment on the second EVA. The crew also commanded to the deployed position the 20 m long radiator designed to removed excess heat from internal electronics on the P6 array. Before day's end the crew successfully commanded the port array to open in a perpetual start-stop-start sequence that allowed the tension cables not to be defeated by the speed of deployment. In an operation that lasted one hour 54 minutes the two port array wings were successfully out and fully tensioned by 21:46 EST.

The second EVA began at 12:21 EST on 5 December and first involved Tanner and Noriega configuring power cables between the Z1 truss and the P6 array. Heat shrouds were removed and Noriega removed the S-band antenna array from its temporary location on the Z1 truss and installed it on top of the P6 tower almost 30 m above the Shuttle cargo bay. Meanwhile, using cameras on his helmet, Tanner completed a full TV inspection of the starboard solar array wing that had slack tension after deployment the previous day. Tanner found that tension cables had jumped off their guide pulleys during deployment. In California, astronaut Jerry Ross worked with Lockheed Martin solar array engineers to trouble-shoot the problem and work a solution. Combined with the TV images engineers worked a procedure to fix the slack array during the third EVA. The second EVA ended at an elapsed time of 6 hours 37 minutes. On 7 December the third EVA began when Tanner and Noriega switched to internal suit power at 11:13 EST. After breaking out tools and equipment they set about the starboard solar array wing tensioning procedure worked up over the preceding day by flight controllers and contractor specialists. Then they set about a complicated but functional procedure to re-thread cable reels on the slack array in an operation EVA planners had scheduled 3 hours to complete. The job was done in one hour 20 minutes, the arrays held taut at their design tension. The two astronauts then set up a TV cable for live transmissions during the docking of the US Destiny laboratory during STS-98 in February 2001. Installation of the floating potential probe facilitated measurement of electrical charge which could build up on the arrays and endanger future EVA operations. Two plasma contactors released a constant stream of xenon atoms to close the circuit and prevent static charge but because they carry no means of registering the electrical potential the contactors would monitor the rates. The third EVA lasted 5 hours 10 minutes after which preparations were made to raise the Shuttle interior pressure allowing hatches into the ISS to be opened after equalisation.

At 09:36 EST on 8 December the Endeavour crew opened the hatch to Unity and joined the Expedition 1 crew in the ISS for the first time since docking almost six days earlier. The station crew were taken up to the Endeavour flight deck for their first look at the enormous array extended successfully by the two EVA astronauts. The Endeavour crew was then given a tour of the Zarya and Zvezda modules before a safety procedures briefing preceded transfer of equipment and supplies into the ISS. The brief visit lasting barely 25 hours 15 minutes ended with PMA 3 hatch closure at about 10:51 EST on 9 December. Undocking took place at 14:13 EST ending a docked duration of 6 days 23 hours 13 minutes. A fly-around ISS inspection was then performed but a planned IMAX shoot could not take place because the camera had failed earlier in the flight so a series of more conventional still, video and movie cameras were used instead. The 2 minute 45 second de-orbit burn took place at 16:58 EST on 11 December and touchdown on runway 15 at KSC came at 18:03:25 EST.

VERIFIED

RESUPPLY VEHICLES

Russian Federation

Progress Series

Current Status

The first Progress cargo tanker launched after the loss of Shuttle Columbia on 1 February 2003, M-47 lifted off on 2 February on a planned logistics flight with docking at 1449 UT on 4 February. Progress M1-10 was launched on 8 June 2003 carrying additional water to maintain the emergency caretaker crew aboard the International Space Station. It successfully docked with the Pirs port at 1115 UT on 11 June.

Background

The first Progress (11F615A15, the A15 mod version of the 11F615 Soyuz class vehicle) had been launched in 1978 followed by the 11F615A55 Progress M vehicle in 1989 with improved systems. Vehicle designation for the basic Progress is 7K-T9 and for the Progress M 7K-T9M. The latest version is officially designated 7K-TGM1 and was developed for use with the International Space Station. The constant demand by Salyut and its Mir successor for consumables sparked the Soviet Union to build an automatic resupply vessel. RKK Energia designed and constructed the Progress vessel based on its manned Soyuz vehicle. In the new configuration, the Soviets replaced the manned descent module with a compartment for transporting up to 940 kg of propellants, other liquids and compressed gases. The new design features a fully autonomous docking port with pipes to carry the consumables into the station.

The Soyuz orbital module accommodates up to 1.8 tonnes of removable cargo. It has been used to carry experimental equipment, food, film and air regeneration cylinders. For the return trip, the crew fills this module with unwanted materials. A Progress may also raise the host station's orbit to combat atmospheric drag. Ground controllers command the ferry to undock and retrofire and then target it to re-enter that atmosphere over Russia for destructive re-entry along a Pacific corridor. Cargo accounts for about 35 per cent of a Progress launch mass, the high proportion made possible by the absence of escape systems and re-entry thermal protection.

All 43 vehicles in the series successfully docked with their intended targets, delivering a total of 99 tonnes. Progress delivered a total of 24.97 tonnes to Salyut 7 by 12 named flights. Kosmos 1669 added 2,254 kg to this total. Mir's first 17 deliveries totalled more than 40 tonnes, or double the station's original mass. The first design lacked solar

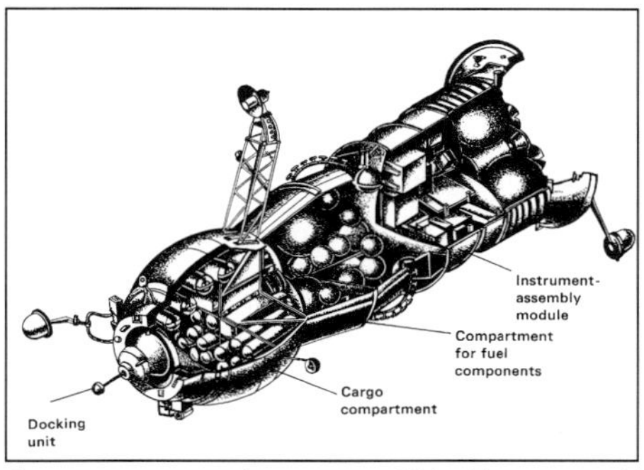

Progress space station cargo vehicle. The current M version incorporates solar panels 0516694

panels, which limited it to an independent flight time of at least three days, although the capsule itself remained docked to Mir for two or three months. The vehicle stretches to a slightly greater length than Soyuz because RKK Energia added an additional pressurised instrument section to the rear unit.

Progress-M flew for the first time in August 1989. It incorporates Soyuz-TM's Kurs rendezvous/docking system, computer, propulsion unit and solar arrays. Kurs allows forward port docking, whereas the previous Progress could only approach a space station through the aft docking unit. The Russians at first planned to build an improved version for delivering supply modules to the ISS, but Russia now expects to improve the current capacity by 200 kg and use a Proton-launched modified FGB for large deliveries. A Zenit 2 booster would have launched the 13.3 tonne Progress-M2 and might have delivered as much as 2.3 tonnes of propellant. Four of these loads would satisfy the Station's projected annual requirement.

The last original Progress spacecraft flew between 5 and 27 May 1990. Progress-M 7 was the first to suffer docking failure, in March 1991, because of a broken antenna on Mir itself. Consequently, it became the first to take propellant from a station for orbit raising. Progress-M 4 carried an open cargo carrier in place of the refuelling mid-section for the first time.

Russian space technology has made great strides in docking compared to the West. Manakov Poleshchuk demonstrated this when he undocked Progress M16 on 26 March 1993, took it out to 70 m and brought it back in for docking. In September 1994, TM19's Malenchenko brought in Progress-M 24 (with its malfunctioning Kurs) manually from 150 m using TV signals from the unmanned craft. Remote-control docking from the ground may be added, with a cosmonaut in the Flight Control Center controlling Progress using Soyuz-type controls and a TV monitor.

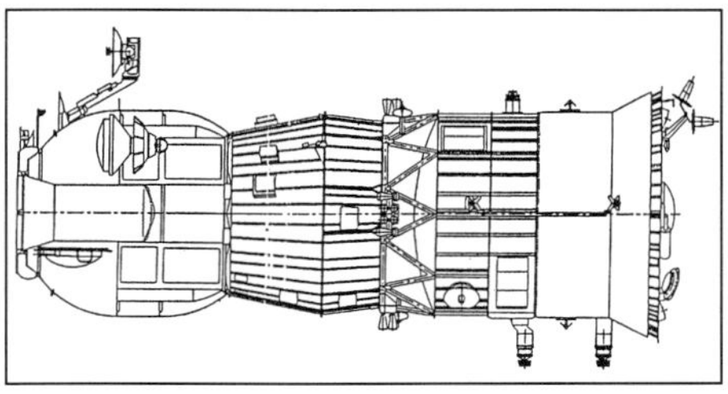

Progress-M side view, configured for undocking with the Raduga return capsule positioned for ejection (RKK Energia) 0516695

Some Progress's now deliver a coni-cylindrical re-entry capsule (VBK: ballistic return capsule) among their cargo. Loaded with 80–150 kg of film and material samples, the capsule mounted in Progress's hatch by the crew, releases itself after retrofire at about 120 km for recovery by parachute (initially in Kazakhstan's normal manned landing area but beginning with M18 within Russia to avoid political problems). M5 carried the first of these configurations and landed with at least some crystals among its 113 kg payload. Unfortunately, the M7's Raduga and 94 kg payload was lost.

After separation, the capsule retains Mir's 1 atmosphere pressure and the outside nose temperature reaches 2,200°C during entry. The braking chute deploys barometrically at 15.5±2.5 km (Mach 1.5±0.5), the main chute spreads to a diameter of 5.6 m. Antennas on 'chute lines and clouds of dipole reflectors released during the descent pinpoint the location. The reflectors are the same ones used as penetration aids for Russia's nuclear re-entry vehicles. A balloon provides buoyancy for water landings.

The first Progress M1 enhanced cargo ferry was launched to Mir from Baikonur on 1 February 2000. It was deorbited 26 April 2000 and was followed by M1-2 launched 25 April 2000, deorbited on 15 October 2000. Progress M1-3 became the first Progress vehicle to service the International Space Station and was launched 6 August 2000. It was followed to the ISS by M1-4 on 16 November 2000 and by M44 on 26 February 2001. Progress docked with the -Y port on Zvezda, that port having been vacated by Soyuz TM-31 when it moved to the Zarya module. Docking took place at 0947 UT on 28 February. It undocked at 0848 UT on 16 April 2001 and was deorbited at 1323 UT the same day. Having re-supplied and replenished three Russian space stations and begun the process of sustaining the International Space Station, more than 23 years after it was first used, a Progress (M1-5) was used to de-orbit the Mir station on 23 March 2001. Having dispatched Mir, Progress was turned fully and exclusively to support for the International Space Station. Progress M1-6 launched 20 May 2001 was

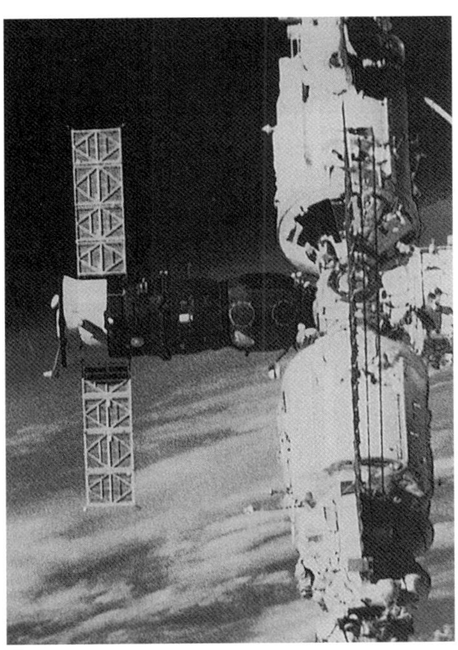

Progress-M 18 attached to Mir's forward port 0517100

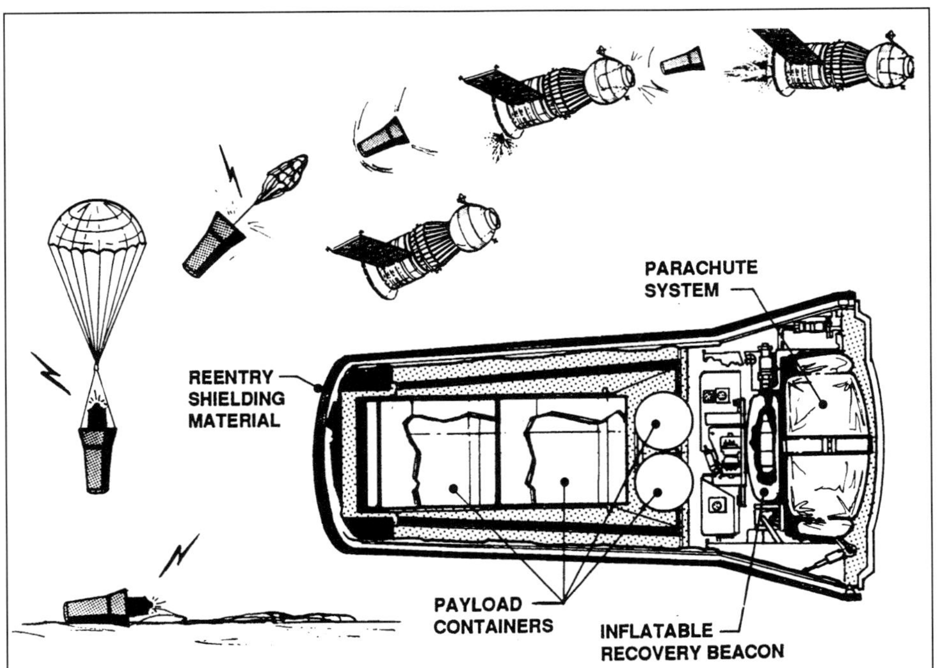

The Raduga return capsule is mounted in Progress' docking hatch and de-orbited by its host for a ballistic re-entry (Teledyne Brown Engineering) 0516696

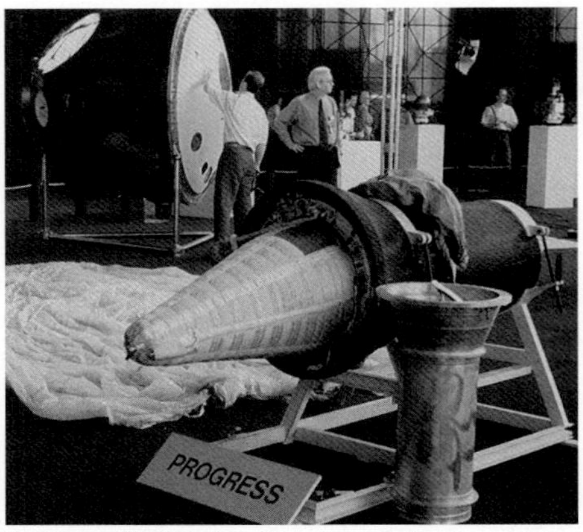

Raduga displayed with its payload container (Theo Pirard/Space Information Center) 0517286

launched by the first Soyuz FG as ISS supply flight 4P carrying 2.5 tonnes of food, fuel, water, life-support equipment and spares including redundant computer equipment for the Destiny Module. Progress docked with the -Y port on the aft end of Zvezda at 0024UT 23 May and undocked at 0601 UT on 22 August 2001, deorbiting at 0900 UT that day. Progress M-45 had been launched the previous day and docked to the ISS at 0951 UT on 23 August, utilising the -Y Zvezda port vacated by M-44 the previous day. It undocked and was deorbited on 22 November 2001. An adapted version of Progress, the SO1 vehicle carried the Pirs docking and airlock module instead of standard cargo and fuel sections, launched on 14 September 2001 and docked to the Zvezda nadir port three days later. Instrument Module separated and deorbited 26 September 2001. Launched 26 November 2001, Progress M1-7 was another ISS servicing flight which docked with Zvezda on 3 December during an EVA conducted by cosmonauts from the ISS to clear a rubber seal left by the departing Progress M-45. The vehicle udocked from Zvezda on 19 March before a Kolibri-2000 microsatellite was released from Progress M1-7. Progress deorbit took place a few hours later in the early morning of 20 March. Progress M1-8 was launched on 21 March 2002 after a five week delay as ISS servicing mission 7P. It docked with the Zvezda module on 24 March and remained attached until early on 25 June, the spacecraft returning to destructive re-entry later that day. Progress M-46 launched 26 June 2002 docked to the ISS as supply mission 8P three days later after conducting tests with the Kurs rendezvous system on 28 June. Progress M1-9 was launched 25 September 2002 as ISS servicing mission 9P and docked to Zvezda on 29 September.

Specifications

First launch: 20 January 1978; 19 July 1985 for original Mir design; 23 August 1989 for Progress-M (27 September 1990 for M with return capsule); 1 February 2000 for Progress-M1 variant
Launch sites: Tyuratam
Launch vehicle: Soyuz SL-4 (without escape system)
Principal applications: unmanned LEO space station ferry
Cost: reported at about Rb10 million/mission
Performance: up to 2,750 kg total. 1,500–1,800 kg in crew-accessible 7.0 m^3 Orbital Module and 940 kg in mid-section (870 kg UDMH/NTO in two spheres each; 50 kg O_2). Progress-M can also transfer its own excess propellant. The preceding design carried up to 2,480 kg (up to 1,400 kg in OM + up to 1,200 kg refuelling section). Raduga, when carried, can return up to 150 kg. Independent flight time 30 days (3 days without solar arrays), attached to Mir max 180 days
Total length: 7.23 m; 2.2 m typical diameter
Launch Mass: 7.3 t; 7,240 kg in original Mir design, 7,020 kg w/2,300 kg cargo for Soyuz.
Propulsion: derived from ODU unified propulsion system of Soyuz-T/TM. Hypergolic UDMH/NTO is employed with the 3.1 kN main engine and a network of 14 × 98 N (four back-up) plus 12 × 9.8 N (four back-up) thrusters. It is not known if a back-up to the main engine is carried (the smaller thrusters provide de-orbit capability). The system is used for Mir orbit raising to combat drag effects before Progress departs. Progress-M can transfer its excess propellant to Mir (M7, because of its docking problems, became the first to take propellant from Mir for orbit raising)
Power: TM-type solar arrays generating 1.3 kW from 10 m^2; batteries carried in pressurised instrument compartment. Augments Mir supply when docked
Avionics/control: Progress M docks with Mir/Kvant's forward or aft port, employing the Kurs ('course') approach system. Two external TV cameras are carried and range/range-rate data are transmitted to both mission control and to the station crew, who can assist in the docking operation if necessary. Attitude reference provided by Earth horizon sensors + gyro package. Soyuz carries its avionics in its manned descent module, for Progress they are repackaged into the instrument unit. Navigational computations are performed by mission control and uplinked for execution.

Launch and re-entry dates up to the time of going to press were as follows:

Salyut 6

1	20 January to 7 February 1978
2	7 July to August 1978
3	8 August to 24 August 1978
4	4 October to 26 October 1978
5	12 March to 5 April 1979
6	13 May to 9 June 1979
7	28 June to 18 July 1979
8	27 March to 26 April 1980
9	27 April to 22 May 1980
10	29 June to 19 July 1980
11	29 September to 11 December 1980
12	24 January to 19 March 1981

Mir

25	19 March to 21 April 1986
26	23 April to 23 June1986
27	16 January to 25 February 1987
M6	14 January to 16 March 1991
28	3 March to 28 March 1987
29	21 April to 11 May 1987
30	19 May to 19 July 1987
31	3 August to 22 September 1987
32	23 September to 19 November 1987
33	20 November to 19 December 1987
34	20 January to 2 March 1988
35	25 March to 5 May 1988
36	13 May to 5 June 1988
37	18 July to 12 August 1988
38	9 September to 23 November 1988
39	25 December 1988 to 7 February 1989
40	10 February to 5 March 1989
41	16 March to 25 April 1989
M1	23 August to 1 December 1989
M2	20 December 1989 to 9 February 1990
M3	28 February to 28 April 1990
42	5 May to 27 May 1990
M4	15 August to 20 September 1990
M5*	27 September to 28 November 1990
M6	14 January to 16 March 1991
M5*	27 September to 28 November 1990
M8	30 May to 15 August 1991
M9*	20 Aug-30 September 1991
M10*	17 October to 20 January 1992
M11	25 January to 13 March 1992
M12	19 April to 28 June 1992
M13	30 June to 24 July 1992
M14*	15 August to 21 October 1992
M15	27 October to 7 February 1993
M16	21 February to 27 March 1993
M17	31 March to 1993 to 3 March 1994
M18*	22 May to 3 July 1993
M19*	13 August to 13 October 1993
M20*	11 October to 21 November 1993
M21	30 January to 23 March 1994
M22	22 March to 23 May 1994
M23*	22 May to 2 July 1994
M24	25 August to 4 October 1994
M25*	11 November to 16 February 1995
M26	15 February to 15 March 1995
M27	9 April to 23 May 1995
M28	20 July to 4 September 1995
M29	8 October to 19 December 1995
M30	18 December to 22 February 1996
M31	5 May to 1 August 1996
M32	31 July to 21 November 1996
M33	19 November 1996 to 12 March 1997
M34	6 April to 2 July 1997
M35	5 July 1997 to 7 October 1997
M36	5 October 1997 to 19 December 1997
M37	20 December 1997 to 15 March 1998
M38	14 March 1998 to 15 May 1998
M39	14 May 1998 to 29 October 1998
M40	25 October 1998 to 3 February 1999
M41	2 April 1999 to 17 July 1999
M42	16 July 1999 to 2 February 2000
M1-1	1 February 2000 to 26 April 2000
M1-2	25 April 2000 to 15 October 2000
M43	16 October 2000 to 29 January 2001
M1-5	24 January 2001 to 23 March 2001

Salyut 7

13	23 May to 6 June 1982
14	10 July to 11 August 1982
15	18 September to 16 October 1982
16	31 October to 13 December 1982
17	17 August to 18 September 1983
18	20 October to 16 November 1983
19	21 February to 2 April 84
20	15 April to 7 May 1984
21	8 May to 26 May 84
22	28 May to 15 July 1984
23	14 August to 26 August 1984
24	21 June to 15 July 1985
C1669	19 July to 30 August 1985

1SS

M1-3	6 August 2000 to 1 November 2000
M1-4	16 November 2000 to 8 February 2001
M44	26 February 2001 to 16 April 2001
M1-6	21 May 2001 to 22 August 2001
M45	21 August 2001 to 22 November 2001
M-S01	14 September 2001 to 26 September 2001
M1-7	26 November 2001 to 20 March 2002
M1-8	21 March 2002 to 25 June 2002
M46	26 June 2002 to 24 September 2002
M1-9	25 September 2002 to1 February 2003
M47	2 February 2003 to 28 August 2003
M1-10	8 June 2003 to 3 October 2003
M48	29 August 2003

M17 undocked 11 August 1993 but was held in orbit for longevity tests.
* return capsule carried. Progress Mir cargo (kg): 25 2,482, 26 2,405, 27 2,406, 28 2,084, 29 2,227, 30 1,856, 31 2,441, 32 2,341, 33 2,082, 34 2,324, 35 2,283, 36 2,337, 37 2,305, 38 2,282, 39 2,242, 40 1,993, 41 2,238, M1 2,682, M2 ?, M3 2,643, 42 2,409, M4 2,689, M5 2,594, M6 2,546, M7 2,542, M8 2,693, M9 2,730, M10 2,623.5, M11 2,576, M12 2,748.5, M13 2,752, M14 2,532, M15 2,558, M16 2,598

SPACE STATIONS

Russian Federation

Mir

Current Status

The Mir 1 space station was launched on 20 February 1986 and remained in space for more than 15 years until intentionally de-orbited to a relatively safe region of the Pacific Ocean on 23 March 2001. It outlived the Communist regime that paid for its development and found new life in a post-Cold War partnership with the United States. The legacy of Mir is enshrined within the core of Russian participation in the International Space Station. The Mir programme was authorised in February 1986 when approval was given for two Mir space stations each of which had a design life of 5 years. When Mir was originally built it had a back-up, the DOS-8 core module, and when the International Space Station programme embraced Russian participation as a full partner DOS-8 became the first Russian segment of the station known as the Service Module. Under the original (revised Clinton administration) ISS programme the Service Module was to have been launched in September 1997 following the Zarya tug in June of that year. Financial troubles delayed launch for nearly three years until it finally went to dock with the Zarya module in July 2000. It had already been around for almost 14 years.

The DOS-8 serial number 128 spaceframe was completed in February 1985 and all internal systems had been installed by October the following year. After President Reagan's announcement of the Strategic Defense Initiative (Star Wars) in 1983 Soviet Premier Andropov authorised a major expansion of Soviet space station operations. The draft programme for Mir-2 was completed in December 1987. It was to be an assembly of massive, 90 tonne, modules based on the DOS-8 which would be launched in 1993. The original Mir outlasted expectations and as the USSR disintegrated down-sizing kept Mir-2 as an optional possibility for a sustained human space flight programme, albeit drastically limited in scope. The massive add-on modules from Energia were eliminated and in 1991 the Mir-1.5 concept emerged which was to be equipped with power modules and two operational biotechnology modules launched by Buran, the Soviet Shuttle. The following year Buran was abandoned and revised, scaled down, plans were approved by the Council of Chief Designers in November 1992. The DOS-8 module would now have a cross-beam science and energy production platform, the additional modules lifted to orbit by Soyuz or Zenit launchers. At the end of 1992 a suspicion from European partners that the US Freedom space station might be abandoned under the new Clinton administration led to talks between ESA and Russian Space Agency over joining with Mir-2 to create a Euro-Russian space station. Reacting quickly, Energia briefed US officials in mid-1993 about Mir-2 plans and by the end of the year Russia had agreed to become a full partner merging Freedom and Mir-2 into the International Space Station. DOS-8 was finally launched to orbit on 12 July 2000 and docked to the Zarya module at the ISS on 26 July.

Background

Mir ('Peace'), a close cousin of the Salyut 7, replaced that programme in 1986. It was configured as the core habitation module of a 135 tonne multivehicle complex. The Soviets launched Mir with only 20 days' provisions on board and Progress 25 had to resupply the consumable by Progress 25. The Soyuz-T15 cosmonauts occupied it first during 1986, Vladimir Solovyev having the honour of entering the spacecraft as the first person aboard. Despite this inauguration, science module delays deferred the beginning of routine occupation until February 1987. Resupply vessels routinely visit Mir every two to three months when it is occupied.

TM2's Yuri Romanenko established a space endurance record of 326 days on his return to Earth 29 December 1987, surpassed by the 365.94 days of TM4's Titov/Manarov during 1988 and then the 437.7 days of TM18's Poliakov in 1995. Manarov's TM11 mission 1990-91 made him the first person to exceed 500 days total in space.

Mir hosted three international visiting crews during 1988 (Bulgarian, Afghani and French), but continuing module delays and the ground test failure of Soyuz-TM 8's service module resulted in it being left unmanned from April to August 1989. Kvant 1's arrival in April 1987 allowed the beginning of detailed astrophysical observations, but without other modules Mir was limited to basic medical, Earth observation and μg research.

Official Soviet sources stated that the Mir programme up to the return of Soyuz-TM7 27 in April 1989 cost Rb1,471 million, including the development, launch and operating costs of the core, Soyuz, Kvant 1 and Progress, and the development costs of the first two awaited modules. Kvant 1/2, Kristall and all Progress through M16 delivered 87 tonnes of cargo.

Recycling reduced the cargo needs of the Progress vehicles. Water and urine recovery has proved a valuable method of reducing cargo deliveries. Through mid 1993, some 7,500 litres of water had been recovered from the atmosphere. Since a Progress can deliver a maximum of 420 litres of water, this saved 18 Progress flights alone. Shuttle crews now carry across their waste water, which would otherwise be dumped, in plastic bags holding up to 43 kg. Kvant 2 had recycled 2,500 litres of urine by electrolysis, providing enough oxygen for 750 days and saving 4 tonnes of Progress cargo capacity. The molecular sieves of Kvant 1/2 for scrubbing out CO_2 had saved the delivery of 450 chemical canisters (3,600 kg cargo). By comparison a Progress can handle 40 canisters.

NASA Administrator Dan Goldin and RKA General Director Yuri Koptev signed a $400 million protocol on 16 December 1993 that includes $305 million for up to 10 American astronaut visits to Mir. The Spektr and Priroda modules carry some US equipment for a broad programme of materials science, biotechnology, life sciences, Earth observations and space technology. The agreement covers upgrading and maintaining Mir to ensure it remains operational through late 1999.

NASA/RKA, in 1993, began joint development of a solar power dynamic power system, for testing aboard Shuttle-Mir in 1997/98 and environmental control and life support systems. NASA has since dropped the plans for the ISS. When NASA signed the agreement, Mir managers told NASA they could operate their vehicle for at least another five years, and possibly 10 years, after space station occupation begins. NASA asked them to refrain from planning to avoid competition with its space station.

First aboard Mir was Dr Norman Thagard, delivered for a three and a half-month stay by Soyuz-TM 21 in March 1995 (Sergei Krikalev flew on STS-60 in February 1994 and Vladimir Titov was aboard STS-63 in February 1995 when it flew around Mir to demonstrate rendezvous and docking procedures). STS-71 docked June 1995, with Spacelab carried aboard Atlantis, delivering the new Russian crew and taking home TM21's three-man team. STS-74 in November 1995 delivered a docking module for Kristall, to hold Shuttle away from Mir's solar arrays and two new solar wings. It also brought 967 kg of cargo and returned with 370 kg of mainly science hardware and results.

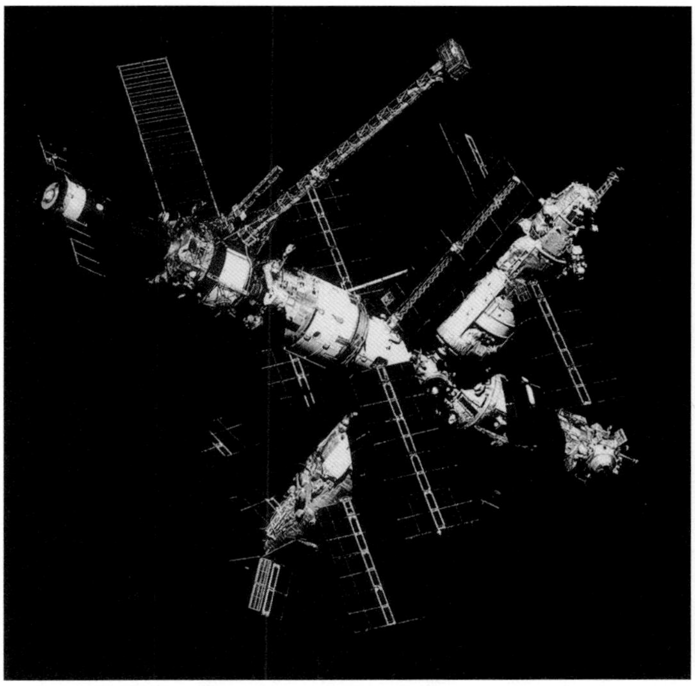

Mir configuration as STS-74 (left) and STS-71 (right) approached for docking. The main difference is that STS-71 docked with Kristall at Mir's axial node; STS-74 brought a docking module that allows Kristall to remain permanently at a radial face. Kvant 2 is at top and Spektr at bottom. Protruding from Kvant 1 is the 14 m Sofora tower carrying the roll control thruster package. Kvant 1 also carried the single solar wing transferred from Kristall (NASA) 0517278

Mir configuration as STS-74 (left) and STS-71 (right) approached for docking. The main difference is that STS-71 docked with Kristall at Mir's axial node; STS-74 brought a docking module that allows Kristall to remain permanently at a radial face. Kvant 2 is at top and Spektr at bottom. Protruding from Kvant 1 is the 14 m Sofora tower carrying the roll control thruster package. Kvant 1 also carries the single solar wing transferred from Kristall (NASA) 0517277

Mir's main living area, looking towards the rear tunnel leading to Kvant 1. At bottom is the main table, at right are the refrigerator (white circle), individual cabin and food store 0516716

Four radial science modules completed attachment in 1996, after the core had already exceeded its planned life. The first, Kvant 2, houses gyrodynes to help control the initially asymmetric complex, water electrolysis equipment, crew shower (removed in April 1995) and manned manoeuvring unit (disassembled July 1991) for use in conjunction with a large EVA hatch of at least 1 m diameter. This general purpose unit was followed in 1990 by the technology Kristall module, providing materials processing and biotechnology facilities. The Spektr remote sensing module appeared May 1995 and the Priroda Earth observation unit April 1996.

Soviet space planners saw Mir as a transition element to an even larger space station, which they would build in the 1993 to 1995 time frame. Much as the United States had, they wanted to use their Shuttle, the Buran to build this new station. After the first flight of the Buran, the Soviets realised that building a space station by Shuttle launch was a highly inefficient process and dropped any subsequent plans. By that time they had a capability that the United States has still not been able to match in the Progress and Merkur space tugs.

Such a grandiose and costly Mir 2 was finally scrapped in early 1991. The new RSA evolving out of the space activities of the former Soviet Union wanted instead a more modest new core installed in 1997 by Proton to dock with Mir, for Buran's manipulator to transfer the newer Spektr and Priroda modules before the old units were de-orbited. In 1994, plans indicated a new Mir one-size core might be launched late 1996 for immediate manning, awaiting a second core and resource, docking and user modules. A truss beam would accommodate solar arrays and possibly solar dynamic collectors. RSA even planned to move the launch site away from Tyuratam and use instead a Proton KM vehicle from Plesetsk into 65° orbit.

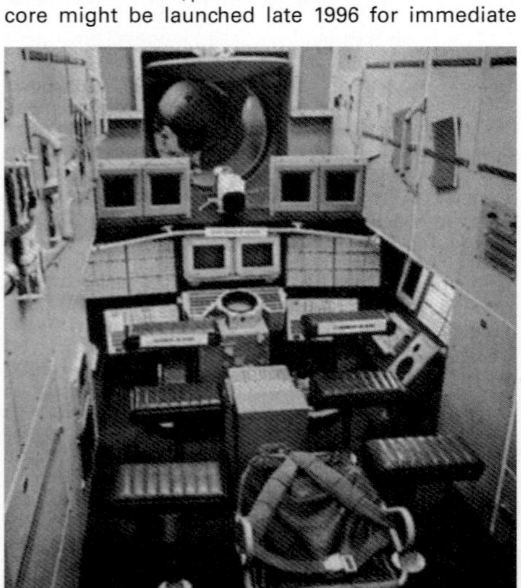

Inside Mir: looking forward from the living section across the main control panels towards the docking node. TV and computer screens face the crew. The pictures were taken from the same perspective, but November 1994's (right) shows how cluttered Mir has become. Work is affected because of the lack of a proper inventory; crews sometimes have to radio previous occupants to find stored items (Graham Turnill/ESA) 0516717

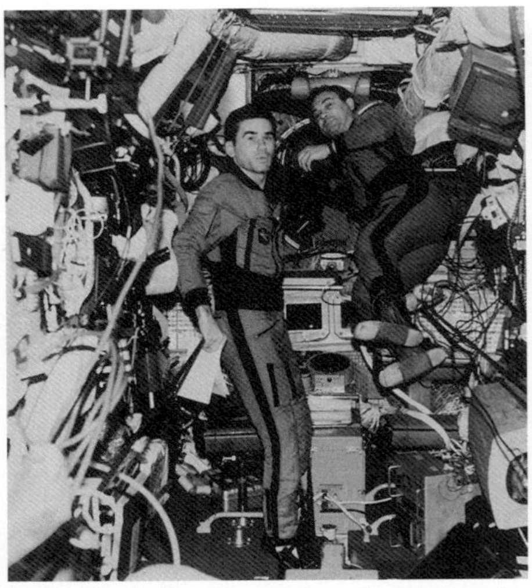

Inside Mir: looking forward from the living section across the main control panels towards the docking node. TV and computer screens face the crew. The pictures were taken from the same perspective, but November 1994's shows how cluttered Mir has become. Work is affected because of the lack of a proper inventory; crews sometimes have to radio previous occupants to find stored items (Graham Turnill/ESA) 0517125

Paradoxically, although RKK Energia leads the Mir programme, the core, heavy modules and their Proton launchers come from the old Chelomei bureaus. Station designer Konstantin Feoktistov commented in 1993 that the original Mir design called for 7 tonne (Progress-sized) add-on modules but 'higher authority' imposed 20 tonne units.

Mir began hosting its 21st long-stay crew in February 1996, with crew exchanges at four to six monthly intervals allowing visits by foreign passengers. CNES/Energia signed a memorandum 29 July 1992 for four missions, in July 1993, August 1996 (Cassiopeia), 1998 and 2000 to exploit Mir's facilities. They paid FFr126 million for the 1993/96 missions, plus equipment. The 2000 Mir visit cannot now occur, but 2000's visit may, in fact, be made to Russia's portion of Alpha. RKA/DARA signed an agreement December 1995 for a second German visit, aboard TM25 in February 1997. ESA's two visits were the 30 day EuroMir-94 from October 1994 and the six-month EuroMir-95 from September 1995, paying ECU45 million for the pair. RKA acknowledged in February 1996 that South Africa and South Korea were interested in visits.

By 1997, NASA began to raise serious questions about the fitness of Mir to continue in orbit. With the fire in February 1997 and the failure of the oxygen-generating equipment the following month, the cosmonauts spent more and more time simply doing repair work to keep the station operating. Then the catastrophic collision late in the Summer caused an almost daily media assault on the station.

On 22 April, the first authoritative Russian comments appeared which raised the possibility of abandoning Mir. Viktor Blagov at the Mission Control in Korolyov said that at the moment there was no reason to evacuate Mir, but added: "The team is continuing its repairs. It's a difficult task and if it cannot be solved Mir might be abandoned by the cosmonauts". Blagov, the deputy head of the flight programme, said the cosmonauts had repaired two leaks in the cooling system with a special glue but had yet to find a third through which antifreeze was escaping into Mir's air supply. "It's on the limit of what is acceptable," He told a news conference, adding that it was not clear how dangerous the concentration of antifreeze vapour was. He said an urgent evacuation of the cosmonauts could be ordered from ground control or by the space station commander. "That could happen if there's a big fire on the station or if there is decompression," Blagov said, adding that the men could reach Earth in an hour using the Soyuz-TMM spacecraft docked with Mir.

Valeri Poliakov observes the 6 February 1995 rendezvous by STS-63 Atlantis from Mir's core (NASA) 0517119

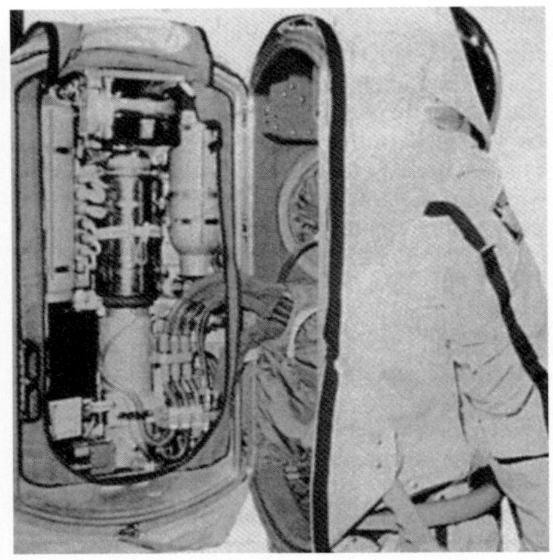

Entry into the Mir EVA suit is via the open backpack. Note the two 1 kg oxygen bottles 0516718

It was argued that if Mir had to be left unmanned for any length of time then it is questionable whether it could ever be used again. In 1985, the Soyuz-T 13 crew brought the Salyut 7 out of mothballs establishing a precedent for such activities. That station was nowhere near as old as is the core module of Mir. For the last several years of its life, cosmonauts on Mir spent at least half of their working days doing repairs to keep the station operational.

If Mir had to be abandoned, it was said, then the chances are that it will re-enter the atmosphere without any control over where debris might land. With the radial modules breaking off and coming down independently along the same ground track the chances of serious damage at ground level cannot be ignored. Accordingly, plans emerged during 1998 for a Progress cargo-tanker to de-orbit the station several days after its last crew planned to leave in August or September 1999. In this way it could be brought down in a controlled manner far from populated regions of Earth. However it appeared likely that Mir would continue to host occupants for at least the next year. If commercial justification could be found and private money forthcoming, it was projected that the orbiting facility probably had another 15 years of active life remaining.

Mir EVA/manoeuvring unit. Note the paired thrusters on the corner units 0516719

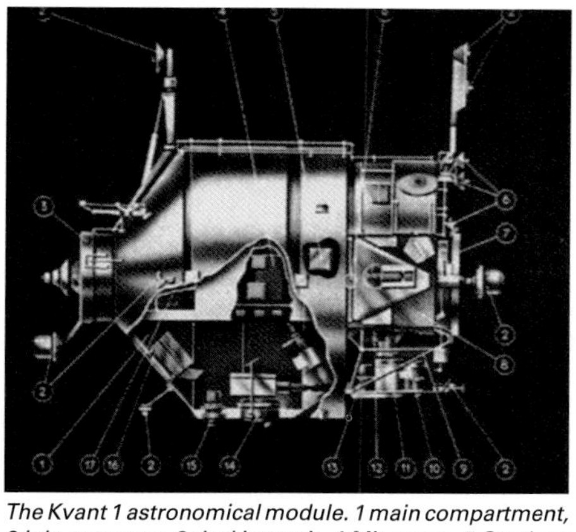

The Kvant 1 astronomical module. 1 main compartment, 2 Igla antennas, 3 docking unit, 4 Mir cargo, 5 Svetlana electrophoresis unit, 6 Kurs antennas, 7 drogue docking unit, 8 science compartment, 9 HEXE x-ray telescope, 10 Sirene 2 spectrometer, 11 Glazar ultra-violet telescope, 12 TTM x-ray telescope, 13 Pulsar X-1 x-ray telescope, 14 optical sight, 15 star tracker, 16 main control panel, 17 life support units 0516720

Mir was reactivated as part of a programme co-ordinated by the Energia corporation and MirCorp to return the orbiting complex to full operational status. Soyuz TM-29 returned the last manned crew in august 1999 and the station was unsupplied until the launch of Soyuz TM-30 in April 2000 (see relevant entries under Soyuz).

A continuous human presence aboard Mir lasted 3,640 days 22 hours 52 minutes between 5 September 1989 and 27 August 1999. Mir had been in space for more than 13 years and fully occupied for almost a decade.

In early 1999 the first commercial manned space flight began with the launch of Soyuz TM-30, paid for by Energia and by MirCorp. Energia has taken over responsibility for Mir from the Russian government in 1999 leaving this orbiting complex as the world's first privatised space station.

Further attempts to maintain Mir for additional manned flights failed and the RSA planned for the deliberate de-orbiting of this 130 tonne complex using a specially adapted Progress vehicle to provide propulsion for that event, which took place on 23 March 2001. Mir had been the 10th space station launched by Russia but far exceeded its predecessors, being visited by 111 spacecraft and occupied for 4,591 days during which 79 EVA were performed.

Specifications

Mir

Launched: 19 February 1986 by Proton from Tyuratam

Mass: 20.9 t at launch; about 51 t for Soyuz/Mir/Kvant 1/Progress; 72.8 t with Kvant 2 added; 90.0 t with Kristall added; up to 135 t fully assembled with four radial modules

Orbit: typically 300–400 km, 51.6°

Crew size: two-six (two long stay typical)

Length: 13.13 m core, about 32.9 m for Soyuz/Mir/Kvant 1/Progress

Diameter: 4.15 m max cylindrical diameter

Pressurised volume: 90 m^3 for core

Configuration: derived from Salyut 6/7 but the interior reworked as a habitation core vehicle and a five-port docking unit replaced the previous single-port forward unit. The 2.2 m diameter node can accommodate a Soyuz-TM/Progress ferry or large module (axial face) and four science modules at the radial positions. Each module docks first at the axial port drogue, inserts a Lyappa manipulator arm into one of two node sockets and swings around mechanically to its allocated site. A conventional docking approach from the side would impose unacceptable loads on the structure (STS-74 made first radial docking 15 November 1995, its 2 cm/s contact within rates).

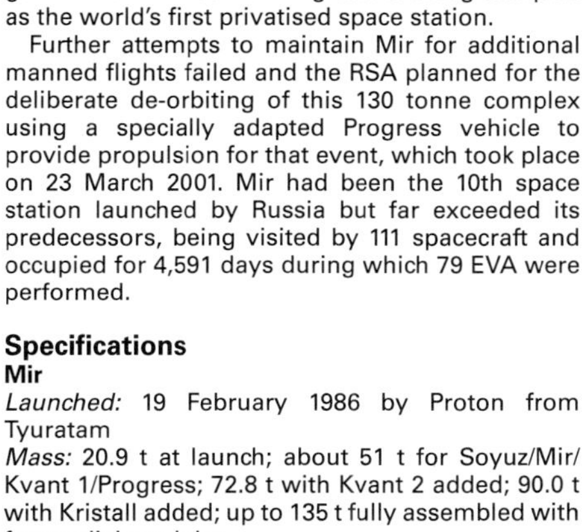

Kvant 2, featuring its ASPG scan platform at top left (NASA) 0517280

All ports incorporate the standard range of electrical, mechanical and hydraulic connections of all the 80 cm diameter hatches. Limited space meant that only two docking drogues were included, the others using blank doors. Swapping doors with drogues to accommodate module movements has required some 'internal' EVAs (11 January 1990, 28 May 1995, 2 June 1995, 8 December 1995). Length of the sphere and its frustum collar is 2.84 m. A hatch leads into the first, 2.9 m diameter, habitation cylinder, which houses the primary station control facilities. The wall units either side of the main control panel carry the air regeneration, water recovery and thermal control systems; this section's exterior provides a mounting for the thermal radiators and the three solar arrays. The habitation section flares to a 4.15 m diameter towards the rear.

The instrumentation carried by every Salyut, which viewed Earth through a large underbody aperture, was removed and floor lockers added. There is also a small airlock in the floor under the wardroom table for experiments. This large table protrudes from the starboard wall with two seats either side to create a dining/recreation area. Heating and drinking facilities are provided, including aft food lockers and a 50 litre portside refrigerator (there is no freezer – ESA had to bring its own). Two personal cubicles are provided behind the table on either side, each with a sleeping bag, porthole and communications panel. On the floor between them is the treadmill aligned with the longitudinal axis. Starboard is a toilet cubicle that includes a spherical vessel for washing hands in running water. Aft of the toilet, a hatch completes the 7.67 m long main cylindrical sections. A 2.26 m long, 4.15 m diameter propulsion unit that provides access to aft-docked vehicles through a 1.67 m long, 2 m diameter tunnel caps Mir.

Power: 9 kW was generated by two Sun-following 38 m^2 4-segment 2 × 2 cm-cell Si solar wings with a rigid central spine; the two outer segments are slightly larger. Span is 29.73 m. The top Salyut attachment point was retained and a 340 kg third array, delivered by Kvant 1 in April 1987, was erected to bring the total area to 98 m^2 and supply to 10.1 kW. Voltage is 28.5 ± 0.5 V DC. 12 74 kg Ni/Cd batteries provide storage. Appears that Kvant 2 added to the power supply but some recabling was performed during 1991 to allow pooling of all modules' power. With Soyuz/Progress attached, total array area 250 m^2. This would provide 28 kW BOL, but shading/ageing reduced it to 10 kW by mid-1992. Power shortage is a persistent Mir problem; ageing reduces the supply by 5% annually. STS-74 in November 1995 delivered two new solar arrays for attaching to Kvant 1 in two 1996 EVAs to add 12 kW. Both built by Energia, but one carries 84 Lockheed Martin panels of 80 Si cells each, totalling 42 m^2 and 6.72 kW.

Propulsion: twin aft-mounted 2.94 kN main engines gimballed ±5° and 32 (in two independent networks of 16 each) 137 N attitude control thrusters. Pressure-fed with hypergolic UDMH/NTO from four tanks incorporating metallic membranes and offering 800 kg capacity. System derived from Soyuz-T/TM. Refuelled by Progress ferries via lines around Kvant; the primary engines cannot be used while Kvant 1 remains attached. Progress' own engines are employed for orbit raising/changing before departure (M7, because of its docking problems, became the first to take propellant *from* Mir). AOCS requirements were reduced by adoption of control moment gyros (see avionics/control). 700 kg thruster package added to 14 m Kvant 1 Sofora mast September 1992 to reduce roll control consumption eight-fold.

Life support: pressure maintained at 1,053–1,276 mb, T 18–28°C, humidity 20–70%, O_2/N_2 mixture with O_2 max content 40%, CO_2 3%. Pressurised air delivered by Progress + Shuttle (20 litres stored for atmospheric makeup), CO_2 scrubbed chemically by LiOH (now used as back-up to molecular sieves on Kvant 1/2; 15 cartridges retained for 3.3 man-days capacity). 40 cartridges of O_2-generating $NaClO_3$ sodium chlorate retained as back-up to Kvant 1/2's electrolytic system and used when >3 people aboard. A 1,000–1,100 K charge ignites the reaction. Other modules provide additional life support capabilities. Salyut's systems were extensively reworked to accommodate crews of up to six. Long-stay cosmonauts remain aboard for typically 6–12 months (more recently six months), punctuated by 8–30 day visiting missions and working 8.5 h days with weekends off. It was discovered that a three-man crew accomplishes little more work than a two-man. Up to 90 min are spent exercising daily. General equivalent protection inside Mir against radiation is typically 1 g/cm^2. Background noise level is 75 dB; Manarov reportedly required treatment for deafness caused by Mir's high levels. Mir's life support system is the responsibility of NPO Zvezda.

Avionics/control: ODCC On-board Digital Computer Complex in conjunction with PINS Platform-less Inertial Navigation System, offering 11/2° automatic/15 arcmin manual pointing accuracy. Eight computers have been referred to aboard Mir's core but the two primary units are: a Soyuz-T Argon 16B derivative capable of performing 200,000 short ops/s and the more powerful Salyut 5B computer (500,000 ops/s) that is employed now as the Mir complex develops. Limited memory means that activities software has to be loaded daily. As modules are added, new control software has to be integrated to accommodate the configuration change; the L-shaped assembly formed after Kvant 2's arrival proved more difficult to control than expected. Salyut 5B was brought into operation during 1990 but difficulties with it delayed the launch of Kristall. There are two CRTs in Mir's forward command position; Kvant 1 includes a third, permitting station control during observing sessions. Attitude/navigation information is provided by: IR Earth horizon scanning sensors (local vertical), solar sensors, star sensors, Sun-presence sensors (4π field of view, signalling Earth shadow entry/exit), automatic/manual star sextant, magnetometer, gyros and linear accelerometers. The Kvant 1/2 modules each carry six gyrodyne control moment gyros, providing 15 arcmin pointing accuracy during astronomical observations. Each 10,000 rpm, 1,000 Nms/200 Nm unit requires 90 W and masses 165 kg with its electronics packages. Mir carries the S-band Kurs ('course') approach/docking system forward/aft, and the older Salyut Igla ('needle') only at the rear. Kurs holds Mir stationary, thus conserving propellant, while the approaching vehicle conducts all the manoeuvres. Soyuz-TM, Progress-M and the new science modules employ Kurs, while the older Progress used Igla. Both generate range, closing rate, line-of-sight angular velocity, perpendicular deviation from the required approach and, from 200 m out, roll deviation. Unmanned vessels are docked under ground control; crews originally occupied their descent craft during such operations on safety grounds. Malenchenko docked progress-M 24 from Mir using TV relayed from the ferry. Mir carries an aft steerable 11–14 GHz antenna for continuous links through GEO telecom satellites at 95/200/344° E (Cosmos 2054 at 344° E, Luch 1 at 95° E and Luch-1 1 at 77° E were in position at the beginning of 1996). C1700 at 95° E permitted 40 min contact/rev during Mir's early life, in contrast to Salyut's 15–20 min, although the satellite subsequently drifted off station.

Mir components

EVA suit/manoeuvring unit

Description: The Zvezda Factory under the direction of Chief Designer Gai Severin is responsible for EVA equipment and Mir's life support system. October 1988's EVA of Titov/Manarov introduced a modified Orlan-DM ('Bald Eagle') suit that did not require power or communications umbilicals; a simple tether was provided for safety. The suits were delivered by Progress 37 in July 1988. Improved Orlan-DMA versions were delivered by Kvant 2 with the manoeuvring unit. Standard duration is limited to 6.5 h by the capacity of 2.2 kg LiOH 120 mm diameter cylinders for scrubbing out CO_2, but can cope with further 90 min. However, the suit can be linked to Mir for up to a week in emergencies. Adjusting via cables/pulleys accommodates different arm/leg lengths. There are two glove sizes. The cosmonaut wearing a liquid cooling garment climbs into the suit from the rear, through the open backpack and inserts his head into the integral helmet. The rear door is closed by pulling a chest-mounted wire and sealed by a lever on the right waist. The aluminium hard upper torso carries a display/control panel at upper right visible from the helmet; bottom left are temperature regulation and emergency oxygen controls with reversed lettering read by a left-wrist mirror. The rear hatch prevents there being a waist bearing (unlike Shuttle's), making shoulder twisting difficult. No food, drink or urine collection is provided. Each suit can undertake 10 EVAs over four years.

Suit Mass: 105 kg loaded

Oxygen system: 1.0/1.0 kg O_2 stored primary/back-up. Primary 0.206–0.255 kg/h maintains 0.4 atmosphere pressure (30 min pre-breathing

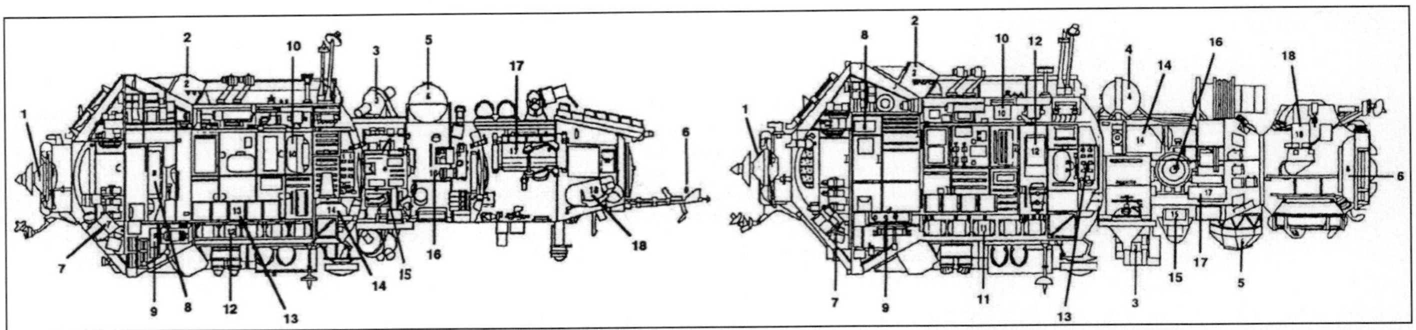

Kvant 2 (left) and Kristall cutaways. Kvant 2 key: 1 docking unit, 2 thruster packages, 3 gyrodynes, 4 star camera, 5 Rodnik water tanks, 6 EVA hatch, 7 control station, 8 shower/hygiene unit, 9 water regeneration unit, 10 Progress water units, 11 attitude control system, 12 power system, 13 food/supply containers, 14 payload container, 15 MKF-6M camera, 16 VIKOD equipment, 17 EVA manoeuvring backpack (stored outside since February 1996), 18 EVA suit. Kristall key: 1 docking unit, 2 thruster packages, 3 Glazar 2 ultra-violet telescope, 4 Rodnik water system, 5 Marina γ-ray telescope, 6 androgynous docking unit, 7 control station, 8 Krater 5 control panel (unit is on opposite wall), 9 running track, 10 attitude control system, 11 power system, 12 SVET plan cultivation unit, 13 Rodnik system control panel, 14 freezer, 15 Ksenia device, 16 reusable solar wing drive, 17 Kristallisator CSK-1 furnace, 18 Priroda 5 Earth camera assembly. The Krater 5, Optizon 1 and Zona 2/3 units are on the starboard wall (The Cosmos Center)
0517107

required); back-up 2 kg/h activated manually or automatically if pressure drops to 0.22 atmosphere, routeing some flow directly to the helmet. If a glove leak is responsible, cuffs inflate to provide 15 mins' grace (unlike the rest of the suit, the gloves do not have a back-up bladder). Mir can also supply 2–3 kg/h through umbilical.
Cooling garment: 3.6 kg water circulated at 60–180 kg/h at 16–22°C, removal capacity 300/600 W average/max. Also available from Mir, in addition to 0.15–0.25 m^3/min cooling gas.
Power: 42 W required. Can be provided by Mir or the manoeuvring pack, along the same cable that relays suit parameters to the pack's telemetry unit.
Communications: 100.125 MHz voice transmit; 23 1/247 MHz suit parameters.
Comment: The new Orlan-M was delivered to Mir in 1996 and will be used in the Space Station era. 110 kg loaded, warranty for 12 sorties in 4 years. 54 W consumed power. It provides easier donning/doffing, improved visibility, increased reliability and partial interoperability with Shuttle EVA suits. The entry hatch is raised 6 cm to eliminate the awkward head insertion into the helmet. The torso provides increased volume. Shortening the hard upper torso and making the longer soft lower torso adjustable (the DMA allows only leg/arm adjustment) accommodates the larger range of US/Western European wearers. New bearings between the shoulder and elbow and above the ankle provide greater flexibility. A small top visor allows overhead viewing. A 'hot' back-up water pump in series increases cooling reliability. The 0.22 atmosphere back-up pressure mode has been eliminated – it has never been used intentionally (but has been activated accidentally) and would be of limited value. The 1 kg emergency oxygen supply cuts in at 0.27 atmosphere (instead of 0.22 atmosphere) and the redundant bladder comes into effect at 0.32 atmosphere instead of 0.24 atmosphere. The LiOH cartridge provides 9 hr operation, allowing EVAs from Shuttle's airlock (the current 6.5 hr is acceptable because Mir provides scrubbing while airlocking).

The Icarus (Ikar) equivalent to NASA's Manned Manoeuvring Unit, known generically as the YMK, was introduced during Serebrov's 1 February 1990 EVA. The original version, incorporating 14 air thrusters and some *solid* motors, was completed in 1968 but cancelled because of lack of missions. The new unit (Zvezda built two) was housed in Kvant 2 and could be used only through that module's 1 m diameter hatch. It was apparently developed for Buran tile inspection/repair and after 1990's two demonstrations was disassembled into six pieces and stored inside Kvant 2, with no plans for further use; it was attached to Kvant 2's exterior in the EVA of 8 February 1996 to move it out of the way. A rigid waist belt, that also provides attachment to the parent spacecraft and payload mounting, holds in the occupant. Two adjustable arms provide consumables status reports and rotational (right) and translational (left) thruster controllers. Four T-shaped corner units each incorporate eight thrusters in redundant pairs powered by compressed air from two titanium tanks; total 30 m/s V (NASA's MMU 20.1 m/s). Fresh tanks delivered by Progress ferries. Two independent gyro systems provide rotational reference and an automatic stabilisation to ±2°. Range for Mir operations limited to 60 m by the safety tether, carried because, unlike the Shuttle, there is no immediate rescue capability; 100 m is the limit without it. Serebrov's initial demonstration involved only translational movements 5 m out from Mir; Viktorenko's four days later included 3-axis variations. Both were tethered: an electric winch on the chest took up slack and could provide propulsion back to the station. Potential improvements include a navigation system, return of an incapacitated occupant by remote control from the station and an uprated propulsion system.
Mass: 218 kg loaded, about 400 kg with suited occupant.
Power: single 18.0 Ah primary + 8.5 Ah back-up 27 V DC AgZn batteries. Power can be provided to suit.
Communications: about 100 unit + suit parameters transmitted automatically. Audio warnings provided in the cosmonaut's helmet.
Total impulse: 12 kNs (δV 30 m/s)
Propulsion system: two independent systems of 16 × 5.0 N thrusters each, mounted on four corner units in redundant pairs. Each system has two titanium tanks holding 28 litres of air compressed at 345 atmospheres, reduced by valves to 12.3 atmospheres. One system utilised until tank pressure drops to 108.5 atmospheres; system number 2 then used to exhaustion, leaving number 1's as an emergency reserve. Crossover lines allow each tank to serve the other system. In economy mode, for proximity operations, the translational thrusters fire for 1 s and rotational provides 3°/s. In boosted mode, values are 4 s and 8°/s, respectively.

Kvant 1
Description: Mir's first additional module was Kvant ('quantum'), comprising a laboratory bay and transfer compartments totalling 40 m^3 and an unpressurised aft payload bay accommodating a large array of high energy telescopes. The principal bay houses the control systems, life support systems, including T and air composition regulators, plus radio, TV and command/display units for Mir's computer system. Six gyrodynes, electrically powered by Mir's solar panels, help to orient the station economically. This bay carries two viewports: one, 43 cm diameter, is intended for a mounted optical device and the other, 22.8 cm diameter, for a visual star tracker. The aft transfer compartment has two 8 cm diameter portholes for visual observations; it also includes an airlock for servicing the Glazar UV telescope. The suite of astronomical instruments can be operated from the ground in untended mode; Glazar's exposed films are delivered to Earth by returning crews. Glycol coolant was discovered to be leaking 1 November 1995 from the Vozdukh CO_2 scrubbing system. It was plugged and tested satisfactorily 3 November.
Launched: 31 March 1987, Mir docking 9 April 1987. Jettisoned service module re-entered 25 August 1988.
Principal application: manned/automatic astronomical observatory, expansion of Mir experimental facilities.
Length: 5.8 m (total length with service module attached about 12 m).

The Spektr module (bottom) emphasises atmospheric remote sensing instruments. Note the jammed solar array. The Kristall docking port faces the camera at top for STS-71's docking (NASA) 0517279

Kvant 2 processing at Tyuratam. The EVA hatch is at right 0516721

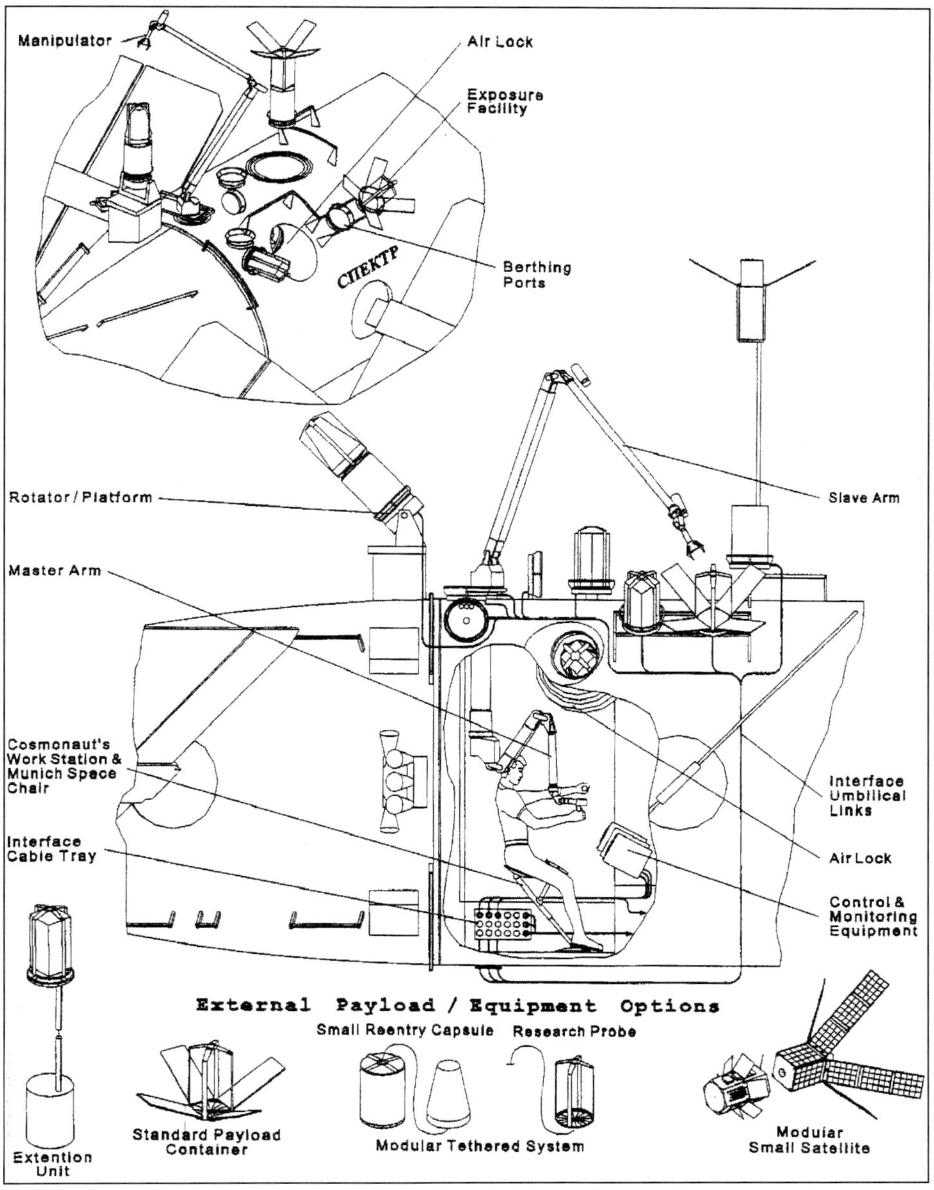

Spektr's manipulator can extract small payloads from an airlock for attachment to the exterior or release into orbit (RKK Energia) 0517267

Diameter: maximum 4.15 m (as Mir core).
Mass: 20.6 t at launch with service module, 11,050 kg docked to Mir (including 1.5 t science payload and 2.6 t Mir equipment such as solar panel).
Pressurised volume: about 40 m^3
Propulsion: none initially, rendezvous/docking operations were performed by the jettisoned service module. Kvant's exterior incorporates plumbing to permit refuelling of Mir's tanks by Progress ferries. 700 kg self-contained thruster package was attached to 14 m Sofora mast by 11 September 1992 EVA to reduce the complex's roll control consumption by up to eight times. Sofora is inclined 11° forward to be closer to centre of gravity.
Power: derived from Mir core. One Kristall wing was transferred by EVAs in May 1995. Two 6 kW wings delivered by STS-74 November 1995 are planned to replace it in 1996.

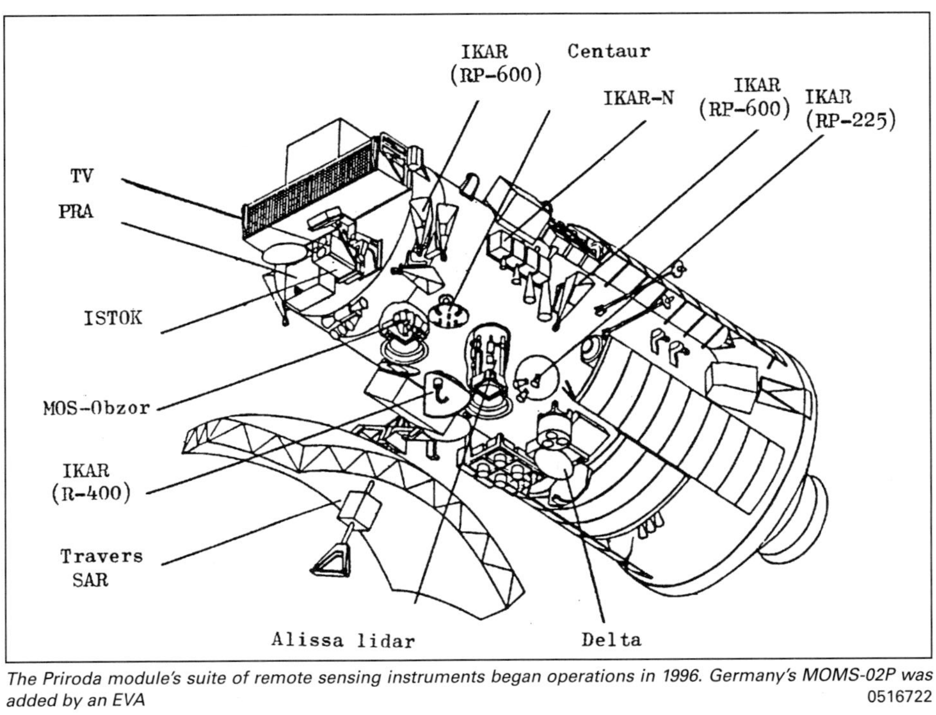

The Priroda module's suite of remote sensing instruments began operations in 1996. Germany's MOMS-02P was added by an EVA 0516722

NASA astronaut Foale operates video camera in Mir's base block (NASA) 0062159

Life support systems: Kvant was hermetically sealed on the pad under sea level conditions. Mir maintains a pressure of 1,053–1,276 mb, T 18–28°C, with max O_2 content 40%, CO_2 up to 3%. Kvant carries experimental unit for O_2 production by electrolytic decomposition of water. 150 kg Vozdukh molecular sieve scrubs CO_2 (280 W daily average requirement).
Avionics/control: derived from Mir (from service module before docking), but Kvant was first to incorporate six gyrodyne control moment gyros, providing 10 arcsec orientation during astronomical observations. Each 10,000 rpm, 1,000 Nms/ 200 Nm unit requires 90 W and with its electronics packages masses 165 kg. Five were reported to be still operational after five years. Kvant also carries two IR Earth horizon sensors, two star sensors, star tracker (two more were added during the EVA of 8 January 1990), Sun sensors and optical sight to contribute to the station's control system; attitude control can be commanded manually from the module during observing sessions. See under propulsion heading above for added roll control thrusters.
Science payload: the unpressurised instrument bay girdles the aft transfer compartment and accommodates the 800 kg 'Roentgen' battery of X-ray and UV telescopes developed in co-operation with the UK, Netherlands, West Germany and ESA. Kvant also houses the Svetlana automated electrophoresis plant in its pressurised volume for biotechnology investigations. Roentgen's detector array comprises: TTM Coded Mask Imaging Spectrometer (Netherlands/UK, 2–30 keV, 7° FOV, 2 arcmin resolution), Sirene 2 Gas Scintillation Proportional Spectrometer (ESA, 2–100 keV, 3° FOV), HEXE High Energy X-ray Experiment (West Germany, 15–250 keV, 1.6° FOV), Pulsar X-1 (USSR, 0–1,300 keV, 3° FOV) and Glazar UV Spectrometer (1,150–1,350 Å, 1–2 Å resolution, 1.3° FOV). Titov/ Manarov undertook two EVAs June/October 1988 to replace TTM's failed 60 kg detector package.

Following launch, Kvant underwent a five-day propellant-efficient rendezvous with Mir and on 5 April 1987, locked on for final approach from 17 km. Mir occupants Romanenko/Laveikin were sealed inside Soyuz-TM2 for safety as Kvant closed in from 500 m but the approach was aborted 200 m out because of control difficulties. The module docked 01.36 GMT 9 April but telemetry indicated that hard dock had not been achieved and the cosmonauts had to perform Mir's first EVA, removing fabric left in Kvant's docking probe area during ground processing. The service module was cast off 13 April to expose the Roentgen observatory and the aft docking unit for visiting Progress/Soyuz. The TM8 cosmonauts attached two 80 kg star trackers during 8 January 1990's EVA to improve the pointing accuracy of Mir's attitude control system. The Sofora roll control thruster package was added September 1992.

Kvant 2

Description: Mir's first radial module, also originally referred to as Module D, incorporates a gyrodyne attitude control system, two water regeneration units, Elektron/Vika oxygen production electrolysis

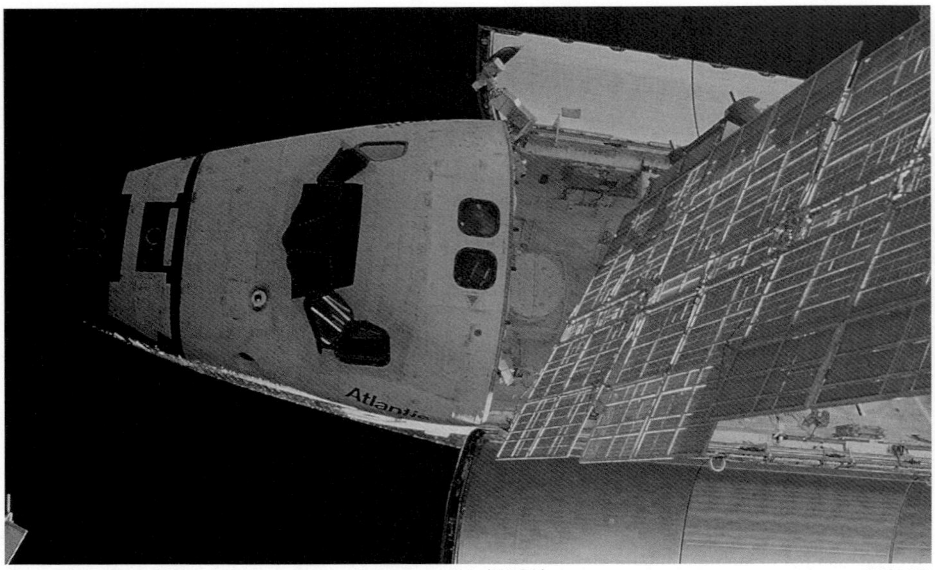

Forward section of Atlantis from a port on Kvant-2 module (NASA) 0062160

Backside of damaged Spektr solar panel (NASA) 0062163

equipment (using recovered urine), shower, MKF-6MA Earth resources camera, 1 m diameter outward-opening airlock (plus suits; Icarus EVA manoeuvring unit was housed here until July 1991), in addition to water (300 litres in external tanks of the Rodnik system), food and propellant supplies. The shower was first used by TM9's crew in February 1990 but was not successful. Air flow substituted for gravity, but the cubicle shape needs to be improved. It was dismantled in April 1995 and disposed of in Progress-M 27; a new design was expected to be delivered. Two portholes in Kvant's central section are used for camera, science instrument and visual observations. Kvant 2's operational life was quoted as three years.

Launch: 26 November 1989 by Proton SL-13 from Tyuratam; docked with Mir's forward axial port 6 December, transferring to radial port 8 December using the Lyappa arm.

Principal application: Mir extension unit, EVA activities.

Length: 13.73 m

Maximum diameter: about 4.35 m

Mass: 18,500 kg

Pressurised volume: 61.3 m^3 in three units: airlock; central science/equipment; base service and cargo.

Propulsion: 2 × 3.9 kN KB Khimautomatiki main engines plus clusters of 400 N KB Khimmach thrusters. Carried sufficient NTO/UDMH propellant for one year of free-flying operations.

Power: 6.9 kW provided by two 26.6 m^2 126 kg solar arrays (40 × 48 mm Si cells). Ni/Cd batteries providing 360 Ah total capacity.

Avionics/control: Kurs approach/docking system. Six external gyrodyne control moment gyros (see Kvant 1) assist complex orientation and pointing (four malfunctioned within three years; two replaced internally in July 1992; more added March 1993). Includes DASA Jena-Optronik Astro CCD star attitude sensor.

Payload: Icarus EVA unit (dismantled 1991; stored outside 8 February 1996), East German MKF-6MA 6-band Earth resources film camera, KAP-350 topographic camera, the 110 kg Czech ASPG-M platform (derived from Vega's scan platform; capable of 3°/s) carrying 115 kg of equipment (ITS-7D IR spectrometer, ARIZ X-ray spectrometer, MKS-M2 optical spectrometer + TV cameras; the Gamma 2 TV/spectrometer package appears to have been added in EVA of 26 January 1990), Sprut 5 charged particle spectrometer installed during EVA of 23 January 1991, Phaza AFM-2 telespectrometer, Spektr-256 spectrometer, 250 kg Volna 2 propellant tank demonstration, Inkubator 2 for hatching birds' eggs, cosmic dust detectors.

Launch was postponed first in early 1989 because of construction delays and again in mid-September when faults were discovered in electronic components of the Kurs rendezvous/docking system. Telemetry indicated after attaining orbit that the starboard solar array's inboard panel had not deployed, leaving the three outer sections free to move during manoeuvres. Full extension/locking was achieved 30 November by rotating the drive mechanism during a vehicle roll. During the final approach to Mir on 2 December for a planned 14.25 GMT docking, Kvant stopped 19 km out, apparently because Kurs detected an excessive closing rate. Docking at Mir's forward axial port came at 12.22 GMT 6 December, bringing the complex's mass to 65.8 t and length to 40 m. On 8 December, Kvant slotted an arm into a socket on Mir's node and took 60 min to swing around to a radial port, the first such procedure in space. Viktorenko/Serebrov then began activating the new section. The first EVA conducted from Kvant 2, beginning with hatch opening at 12.09 GMT on 26 January 1990, prepared for the 1 February demonstration of the manoeuvring pack, followed by a second excursion four days later. Kvant had delivered the suits. Solovyov/Balandin damaged the hatch on 17 July 1990 when they opened it before the compartment was depressurised, preventing its closure at the end of the EVA. They were forced to use the central chamber as an emergency airlock as their suits' capacities were exceeded. The manoeuvring backpack's volume requirements mean that Kvant 2 is the first Russian manned vehicle to carry an outward-opening hatch. The door is equipped with a stop to hold a 1 mm gap for internal pressure to bleed off before full opening, but it was apparently removed prematurely. The hatch was repaired on a 26 July EVA from inside Kvant 2 but the damage was sufficiently serious to warrant further work. Manakov/Strekalov failed to repair the hinge in their EVA of 29 October 1990, leaving Afanasyev/Manarov to replace the hinge unit 7 January 1991. The unused backpack was stored outside on 8 February 1996 EVA.

Radiator damage to Mir's Spektr module (NASA) 0062161

Kristall

Description: Launch of the technology Module T was delayed from 30 March 1990 because of problems with Mir's Salyut 5B control system computer. Equipment includes materials processing furnaces and biotechnology units, initially claimed capable of generating up to 100 kg of material annually. The 2.2 m diameter node carries two APAS-89 Androgynous Peripheral Assembling System docking ports (89 = 1989; 1975's Apollo-Soyuz carried APAS-75) to accommodate Shuttle-type visits (first used by TM16 in January 1993); a third face houses the Priroda 5 Earth resources cameras. Buran was to have used the axial port; the other was reserved for a 1 t X-ray module (housing an uprated version of Kvant 1's Pulsar phoswich telescope) carried up by the spaceplane. Because of Shuttle requirements and as the Spektr Earth observation module is designed to operate opposite Kvant 2, Kristall faced four moves in 1995. On 26 May, it swung from the bottom face to the axial and then to starboard on 30 May, allowing Spektr to move directly to the bottom after docking. On 10 June, it swung around to the axial face, where STS-71 could dock safely away from Mir's solar wings. After STS-71's departure, it moved to its final starboard position 17 July, where STS-74 in November 1995 left it with a 4,087 kg 2.2 m diameter 4.7 m-long docking module, lengthening it for future Shuttles to avoid Mir contact.

Spektr solar array damage (NASA) 0062162

Launch: 31 May 1990 by SL-13 Proton from Tyuratam; docked with Mir's forward axial port 10 June, transferring to radial port 11 June using the Lyappa arm.

Principal application: materials processing units.

Length: 13.73 m

Maximum diameter: about 4.35 m

Mass: 19.64 t in orbit (including 7 t cargo)

Pressurised volume: 60.8 m^3 in two volumes separated by 80 cm diameter hatch

Propulsion: 2 × 3.9 kN KB Khimautomatiki main engines plus clusters of 400 N KB Khimmach thrusters. Secondary system activated for final approach following problems with primary.

Power: 5.5–8.4 kW was provided by two concertina arrays spanning about 36 m, totalling 72 m^2 of 50 × 50 cm Si cells. Each 36-panel wing and drive unit <500 kg and the first designed to be redeployable – the first wing was removed in May 1995 and attached to less shaded positions on Kvant 1. Each wing can be deployed to 5/10/15 m, attaining full extension only after Mir docking. Ni/Cd batteries totalling 360 Ah. Daily average power consumption 0.5–1 kW; peak 3–7 kW.

Avionics/control: Kurs approach/docking system.

Payload: Krater 5, 5 Optizon 1 and 5 CSK-1/Kristallizator units for semi-conductor production (Krater initially used with zinc oxide, Optizon for growing silicon crystals via zonal heating); Zona 2/3 furnaces for other materials processing. The 500 kg of equipment was expected to generate up to 100 kg of materials annually, but output reportedly affected by equipment reliability and power constraint. Ainur electrophoresis unit (also on Priroda module). Glazar 2 UV telescope to extend Kvant 1's Glazar observations. Priroda 5 package of two KFA-1000 film cameras for Earth photography from third face of docking node. Bulgarian Svet plant cultivation unit (inclination cultivating lettuce and radishes). Mariya magnetic spectrometer, Marina #947; telescope, Buket γ spectrometer and Granat astrophysical spectrometer.

Mir's third module finally lifted off aboard an SL-13 from Tyuratam at 10.33 GMT 31 May 1990 into an economical seven-day transfer orbit, planning to dock 11.36 GMT 6 June. However, only 2 h before docking, Kristall's onboard computer detected a thruster problem during the last burn for final approach and abandoned the attempt. The back-up thruster system was activated, initiating final approach with a burn at 08.04 GMT 6 June, followed by docking at Mir's forward axial port 10.47 GMT. The complex's mass was now reported to be 83 t. Restoration of the complex's symmetrical configuration reduced

Spektr station module showing collision damage to solar panels (NASA) 0062164

the attitude control requirements. Transfer using the Lyappa arm system to the face opposite Kvant 2's took an hour the next day before the crew entered and began activating the module.

Spektr

Description: The module's remote sensing instruments are particularly aimed at atmospheric studies. A two-year minimum life is planned. Spektr and Priroda also carry US equipment for a broad programme of materials science, biotechnology, life sciences, Earth observations and space technology. Spektr is equipped with a small external manipulator which can take 'Pelican' packages of up to 33 × 50 cm from a small airlock for positioning at 11 attach points with control, power and telemetry links. The arm can also deploy small satellites.
Launch: 20 May 1995 by SL-13 Proton from
Principal application: remote sensing module, emphasising atmospheric studies.
Length: 14.44 m
Maximum diameter: 4.10 m
Mass: 19,640 kg at launch (11.5 t dry), including 1,260 kg cargo/consumables, 3,240 kg Mir equipment, 2,150 kg science equipment, 700 kg US equipment.
Pressurised volume: 61.9 m^3
Propulsion: 2 × 3.9 kN KB Khimautomatiki NTO/UDMH main engines plus clusters of 400 N KB Khimmach thrusters.
Power: 6.9 kW provided by two 26.6 m^2 Kvant 2-type Si solar wings spanning 23.30 m plus two 38 m^2 angled Si wings adding an estimated 9.3 kW. Ni/Cd batteries providing 360 Ah total capacity.
Avionics/control: Kurs approach/docking system
Payload: Balkan 1 lidar to measure upper cloud altitude, 5,320 Å, 0.15 mJ pulse, 5.5 s intervals, 4.5 m vertical resolution.

Phaza spectrometer for surface studies: 0.340–285 μm, 200–250 km resolution.

Astra 2 atmospheric trace constituents, Mir environment; Taurus + Grif monitors Mir's induced X/γ-ray background.

KOMZA interstellar gas detector.

286 K binocular radiometer.

VRIZ UV spectroradiometer.

EFO 2 photometer.

MIRAS Mir IR Atmosphere Spectrometer monitoring neutral atmospheric composition over at least 1 year using 10 of 32 daily solar occultations for infra-red absorption spectrometry. The Belgian Institute for Space Aeronomy's 205 kg grille spectrometer was launched stowed in two parts inside Spektr, assembled and then installed July 1995 by EVA. 30 cm diameter Cassegrain telescope, 6 m fl. Observing 15 key chemical species using InSb (2.5–5 μm) + HgCdTe (2.5–10.5 μm) detectors, cooled to 77K by Stirling-cycle cooler. Science data stored in 16 Mbyte memory, downlinked at 256 kbit/s in 30 minute daily period to FCC, then transferred to participants via Internet. 230 W power.

Priroda

Description: Mir's fourth module emphasises remote sensing of land, oceans and atmosphere. Bulgaria, Germany, Poland, Romania and Czech Republic have made instrument contributions an Interkosmos project. Spektr/Priroda also carry US equipment for a broad programme of materials science, biotechnology, life sciences, Earth observations and space technology. A two-year minimum life is planned.
Launch: 23 April 1996 by SL-13 Proton from Tyuratam; docked with Mir's forward axial port 12.43 GMT 26 April, transferred to portside position 27 April using the Lyappa arm.
Length: about 13 m
Maximum diameter: about 4.35 m
Mass: 19.7 t in orbit (including 2 t remote sensing payload).
Pressurised volume: 66.2 m^3
Propulsion: 2 × 3.9 kN KB Khimautomatiki NTO/UDMH main engines plus clusters of 400 N KB Khimmach thrusters.
Power: no solar arrays, battery power only.
Avionics/control: Kurs approach/docking system.
Payload: about 2 t of remote sensing instrumentation.

Ikar-N nadir microwave radiometers: R-30 (0.3 μm), R-80 (0.8), R-135 (1.35), R-225P (2.25, polarimeter), R-600 (6.0). Each 60 km/0.15K resolution.

Ikar-D scanning microwave radiometer system viewing 40° off-track in 400 km swath. R-400 (4.0), resolution 30 km/0.15K. Delta system: R-30 (0.3; 5 km resolution), R-80 (0.8; 8 km resolution), R-135 (1.35; 15 km resolution). 8 kbit/s. Bulgaria provided R-400.

Ikar-P panoramic microwave radiometers viewing 40° off-track. Two sets of three RP-225 (2.25) and RP-600 (6.0), all 750 km swath, resolution 75 km/0.15K. 1 kbit/s.

Istok 1 64-channel IR spectroradiometer, 4.0–16.0 μm, 7 km swath, resolution 0.7 × 2.8 km, 0–90° viewing angle. 8 kbit/s. Also used as 0.4–0.75 μm TV system, resolution 6.5 km. Contributions from Czech Republic, Poland and Romania.

Ozon-M spectrometer for ozone/aerosol profiles, 0.257–1.155 μm, 160 channels, 1 km altitude resolution, 50 kbit/s.

Travers SAR, 1.28/3.28 GHz, 50 km swath, 38° look angle, 50–150 m resolution, 16 Mbit/s each channel.

Alissa cloud height, structure and optical properties lidar (second priority: strato/tropo aerosols), four 5,320 Å Nd-Yag Russian lasers, 40 mJ (4 × 10 mJ) pulses at 50 Hz, width 5–15 ns, 150 m vertical resolution, 1 km horizontal resolution, 40 cm diameter 160 cm fl Cassegrain telescope receiver (110 kg in weight). Laser life (25 h or 4.5 million pulses) and 3 kW total consumption limit use. Operates in conjunction with other Priroda instruments and Meteosat. In co-operation with France. Proposed by CNES in 1985 to fly on Salyut (hence Alissa: l'Atmosphere par Lidar Sur Saliout).

MOS-Obzor 17-channel spectrometer, 0.750–1.01 μm, 80 km swath, 700 m resolution, nadir viewing. Contributed by Institute of Space Research of former East Germany. 256 kbit/s. MOS-A: four high resolution (14 Å) channels 7,570–7,664 Å measuring aerosol profile (data also required to correct MOS-B). MOS-B: 13 medium resolution channels 4,080–10,100 Å sampling ocean reflectance signature to identify, for example, chlorophyll and sediments.

NASA astronaut David Wolff sorts microbial samples aboard Mir (NASA) 0062165

Mir space station viewed from Shuttle discovery (NASA) 0062156

MSU-E2 3-channel high resolution (10 m nadir) optical scanner, 0.5–0.6/0.6–0.7/0.8–0.9 μm, 2 × 24.5 km swath, nadir viewing, 11.5 Mbit/s.

DOPI 2.4–20 μm interferometer studying gases/aerosols. With DARA.

Greben ocean altimeter (10 cm height resolution), 13.76 GHz, 2.5 km swath, nadir viewing, 6.5 kbit/s.

Centaur 400 MHz receiver to collect geophysical buoy data.

MOMS-02P 0.440–0.810 μm electro-optical Earth imager with four channels (three stereo). Operates in three modes: multi-spectral, stereo or high-resolution: 37 or 78 km swath, 6 or 18 m res. Provided by Germany, built by DASA Jena-Optronik GmbH; Kayser-Threde responsible for orbit/attitude determination package. MOMS-02 flew on Spacelab D2 Shuttle mission in 1993, returning 5 h data (>6,000 images) at 4.4 m res. The unit was refurbished and added to Priroda (P = Priroda) by EVA 30 May 1996 in a joint venture with RKK Energia. MOMS is the first single unit capable of generating both high-resolution panchromatic and thematic mapping XS images. Imagery transmitted via an intermediate Mir tape recorder.

Mir Chronology

1992

- **20 January:** Progress-M 10 undocks 07.14 GMT.
- **27 January:** Progress-M 11 docks at Mir's forward axial port at 09.31 GMT.
- **20 February:** Volkov/Krikalev perform 252 min EVA, curtailed after Volkov suit problem.
- **13 March:** Progress-M 11 undocks 08.44 GMT.
- **14 March:** Volkov/Krikalev undock in TM13 at 11.43 GMT from Kvant 1 port, docking at forward port after 27 min.
- **19 March:** Soyuz-TM 14 carrying Viktorenko/Kaleri/Flade docks at Kvant 1 at 12.33 GMT.
- **25 March:** Soyuz-TM 13 carrying Volkov/Krikalev/Flade undocks from forward port.
- **21 April:** Progress-M 12 docks at Mir's forward axial port at 23.22 GMT.
- **28 June:** Progress-M 12 undocks.
- **4 July:** Progress-M 13 docks at Mir's forward axial port at 16.55 GMT.
- **8 July:** Viktorenko/Kaleri perform 123 min EVA to install two new gyrodynes on Kvant 2.
- **24 July:** Progress-M 13 undocks 08.49 GMT.
- **29 July:** Soyuz-TM 15 carrying Solovyov/Avdeyev/Tognini docks at Mir's forward axial port at 07.51 GMT.
- **9 August:** Soyuz-TM 14 carrying Viktorenko/Kaleri/Tognini undocks from Kvant 1 aft port.
- **18 August:** Progress-M 14 docks at Mir/Kvant 1's aft port at 00.21 GMT carrying 5th return capsule and Mir Sofora roll control thruster package; it undocks 16.46 GMT 21 October.
- **3 September:** Solovyov/Avdeyev perform 236 min EVA via Kvant 2 to prepare for Sofora thruster package installation.
- **7 September:** Solovyov/Avdeyev perform 308 min EVA via Kvant 2 to attach control cabling to Sofora in preparation for thruster package installation.
- **11 September:** Solovyov/Avdeyev perform 344 min EVA via Kvant 2 to complete Sofora's thruster package installation.
- **15 September:** Solovyov/Avdeyev perform 213 min EVA via Kvant 2 to attach Kurs docking antenna to Kristall.
- **29 October:** Progress-M 15 docks at Mir/Kvant 1's aft port at 19.06 GMT. It undocks 00.45 GMT 4 February 1993 and deploys 20 m test solar reflector.
- **20 November:** Mak 2 satellite released from Mir at 09.00 GMT.

1993

- **26 January:** Soyuz-TM 16 carrying Manakov/Poleshchuk docks 07.31 GMT at Kristall's APAS-89 androgynous axial port.
- **1 February:** Soyuz-TM 15 carrying Solovyov/Avdeyev undocks at 03.00 GMT from Mir forward axial port.
- **23 February:** Progress-M 16 docks at Mir/Kvant 1's aft port at 20.18 GMT. Manakov/Poleshchuk undock it 06.50 GMT 26 March to demonstrate remote control, taking it 70 m and then redocking it. Undocked 04.21 GMT 27 March.
- **2 April:** Progress-M 17 docks at Mir/Kvant 1's aft port at 05.16 GMT. Undocked 15.36 GMT 11 August.
- **19 April:** Manakov/Poleshchuk perform 325 min EVA to prepare for removal of solar arrays from Kristall to Kvant 1. Handle lost from transport boom.
- **24 May:** Progress-M 18 docks at Mir's forward axial port at 08.25 GMT, carrying 6th return capsule. Undocked 3 July. First occasion with two docked Progress'.
- **18 June:** Manakov/Poleshchuk perform 273 min EVA to prepare for removal of solar arrays from Kristall to Kvant 1. New handle installed on transport boom.
- **3 July:** Soyuz-TM 17 carrying Tsebliyev, Serebrov and French researcher Haignere docks 16.24 GMT at Mir's forward axial port.
- **22 July:** Soyuz-TM 16 carrying Manakov/Poleshchuk/Haignere undocks from Kristall axial port.
- **13 August:** Progress-M 19 docks at Kvant 1 at 00.00 GMT, carrying 7th return capsule. Undocked 17.59 GMT 12 October.
- **16 September:** Tsebliyev/Serebrov perform 258 min EVA 1 to install Rapana truss package on Kvant 1.
- **19 September:** Tsebliyev/Serebrov perform 193 min EVA 2 to deploy Rapana truss to 5 m on Kvant 1.
- **28 September:** Tsebliyev/Serebrov perform 112 min EVA 3 to video Mir's exterior to assess condition.
- **22 October:** Tsebliyev/Serebrov perform 38 min EVA 4 to video Mir's exterior to assess condition.
- **29 October:** Tsebliyev/Serebrov perform 252 min EVA 5 to video Mir's exterior to assess condition.
- **14 October:** Progress-M 20 docks at Kvant 1 at 23.25 GMT, carrying 8th return capsule. Undocked 02.36 GMT 21 November.

1994

- **10 January:** Soyuz-TM 18 carrying Afanasyev/Usachev/Poliakov docks at Kvant 1 at 11.50 GMT. Undock 03.12 GMT 24 January and redock Mir forward axial port 05.01 GMT 24 January.
- **14 January:** Soyuz-TM 17 carrying Tsebliyev/Serebrov undocks at 04.37 GMT from Mir forward axial port.
- **30 January:** Progress-M 21 docks at Kvant 1 at 03.56 GMT. Undocked 01.20 GMT 23 March.
- **24 March:** Progress-M 22 docks at Kvant 1 at 06.40 GMT. Undocked 00.57 GMT 23 May.
- **24 May:** Progress-M 23 docks at Kvant 1 at 06.00 GMT, carrying 9th return capsule. Undocked 08.47 GMT 2 July.
- **3 July:** Soyuz-TM 19 carrying Malenchenko/Musabayev docks at Kvant 1 at 13.55 GMT.
- **9 July:** Soyuz-TM 18 carrying Afanasyev/Usachev undocks from Mir forward axial port.
- **2 September:** Progress-M 24 docks at Mir forward axial port. Attempts 27/30 August failed because of Progress Kurs fault. Malenchenko controlled M24 manually from inside Mir for number 3 attempt. Undocked 18.56 GMT 4 October.
- **9 September:** Malenchenko/Musabeyev perform 304 min EVA to examine damage from Progress-M 24 and Soyuz-TM 17 bumps.
- **13 September:** Malenchenko/Musabeyev perform 361 min EVA to continue inspection, including Sofora mast.
- **6 October:** Soyuz-TM 20 carrying Viktorenko/Kondakova/Merbold docks at Mir forward axial port 00.28 GMT.
- **2 November:** Soyuz-TM 19 carrying Malenchenko/Musabayev/Merbold undocks from Kvant 1 10.30 GMT, redocking 11.05 GMT to test automatic Kurs system.
- **4 November:** Soyuz-TM 19 carrying Malenchenko/Musabayev/Merbold undocks from Kvant 1 at 07.29 GMT.
- **13 November:** Progress-M 25 docks at Kvant 1 at 09.04 GMT, carrying 10th return capsule. Undocked 13.05 GMT 16 February 1995.

1995

- **11 January:** Soyuz-TM 20 carrying Viktorenko/Kondakova/Poliakov undocks from Mir's forward axial port 08.57 GMT, redocking 09.23 GMT to test automatic Kurs system.
- **17 February:** Progress-M 26 docks at Kvant 1 at 18.22 GMT. Cargo included 100 kg equipment for Thagard visit. Undocked 02.27 GMT 15 March.
- **16 March:** Soyuz-TM 21 carrying Dezhurov/Strekalov/Thagard docks at Kvant 1 at 07.46 GMT. Undocks 03.31 GMT 11 September 1995, carrying Solovyov/Budarin from STS-71.
- **22 March:** Soyuz-TM 20 carrying Viktorenko/Kondakova/Poliakov undocks from Mir forward axial port 00.40 GMT.
- **11 April:** Progress-M 27 docks at Mir forward axial port at 21.01 GMT, carrying Germany's GFZ 1 laser geodetic satellite. Undocked 23.40 GMT 22 May.
- **19 April:** Germany's GFZ 1 laser geodetic satellite is ejected from Mir at 19.12 GMT.
- **12 May:** 375 min EVA-1 by Dezhurov/Strekalov retracts first Kristall solar wing for move to Kvant 1.
- **17 May:** 412 min EVA-2 by Dezhurov/Strekalov transfer first Kristall solar wing to Kvant 1 portside.
- **22 May:** 315 min EVA-3 by Dezhurov/Strekalov completes connection of first Kristall solar wing to Kvant 1.
- **28 May:** 'internal' EVA by Dezhurov/Strekalov moves docking cone in Mir node for Kristall to move to starboard face 30 May.
- **1 June:** Spektr module docks at Mir's forward axial port 00.58 GMT.
- **1 June:** 'internal' EVA by Dezhurov/Strekalov moves docking cone in Mir node for Spektr to move to lower face 2 June.
- **10 June:** Kristall swings from Mir's starboard face to the forward axial port, to allow Shuttle docking.
- **29 June:** Shuttle STS-71 docks with Kristall androgynous port at 13.00 GMT, delivering 19th main crew Solovyov/Budarin.
- **4 July:** Soyuz-TM 21 with Solovyov/Budarin undocks from Kvant 1 at 10.55 GMT to photograph STS-71 departure; redocks 11.39 GMT.
- **4 July:** Shuttle STS-71 undocks from Kristall at 11.10 GMT, returning 18th main crew Dezhurov/Strekalov/Thagard.
- **14 July:** 334-min EVA-1 by Solovyov/Budarin partially deploys Spektr's jammed solar wing.
- **17 July:** Kristall is moved from Mir's forward axial port to the starboard port.
- **19 July:** planned 300-min EVA-2 by Solovyov/Budarin to install MIRAS spectrometer is shortened to 188 min because of Solovyov suit cooling system problem.
- **21 July:** 335-min EVA-3 by Solovyov/Budarin installs MIRAS spectrometer.
- **22 July:** Progress-M 28 docks at Mir forward axial port at 04.40 GMT, carrying 2,400 kg cargo. Undocked 05.10 GMT 4 September.
- **5 September:** Soyuz-TM 22 carrying Gidzenko/Avdeyev/Reiter docks at Mir's forward axial port at 10.30 GMT.
- **11 September:** Soyuz-TM 21 carrying Solovyov/Budarin undocks from Mir's Kvant 1 port 03.31 GMT.
- **10 October:** Progress-M 29 docks at Kvant 1 at 20.33 GMT. Undocked 09.15 GMT 19 December.
- **20 October:** 316-min EVA by Avdeyev/Reiter installs ESA experiments on Spektr.
- **15 November:** Shuttle STS-74 docks with Kristall androgynous port at 06.28 GMT, using 4.1 t docking module, left attached when Shuttle undocks 08.16 GMT 18 November. Mir's first radial docking - Kristall is at starboard port.
- **8 December:** 19.23–19.52 GMT 'internal' EVA by Gidzenko/Avdeyev moves docking cone in Mir node for Priroda's docking.
- **20 December:** Progress-M 30 docks at Kvant 1 at 16.30 GMT. Undocked 07.26.47 GMT 22 February.

1996

- **8 February:** 182-min EVA by Gidzenko/Reiter recovers ESA experiments from Spektr.
- **19 February:** Mir marks 10 years in space, completing 56,150 revs.
- **23 February:** Soyuz-TM 23 carrying Onufrienko/Usachev docks at Kvant 1's port at 14.20.36 GMT.
- **29 February:** Soyuz-TM 22 carrying Gidzenko/Avdeyev/Reiter undocks from Mir's forward axial port at 07.20 GMT.
- **15 March:** 351-min EVA 1 by Onufrienko/Usachev installs core crane for moving STS-74 docking module solar wing packages to Kvant 1.
- **24 March:** Shuttle STS-76 docks with Kristall adapter at 02.34 GMT, undocks 01.08 GMT 29 March. Lucid remains with main crew.
- **27 March:** 362-min EVA by Clifford/Godwin from Shuttle hatch attaches Mir experiments and tests common EVA equipment.
- **26 April:** Priroda module docks at Mir's forward axial port 12.43 GMT.
- **27 April:** Priroda module swings around to portside position. Mir assembly is now complete.
- **7 May:** Progress-M 31 docks at Mir's forward port.
- **21 May:** Onufrienko and Usachev conduct EVA lasting for 320 min to install a solar panel on Kvant 1. During EVA, the crew photographed a larger-than-life mockup of a soft drinks can to be used by the United States drink manufacturer as part of an advertising campaign.
- **24 May:** Onufrienko and Usachev perform EVA lasting approximately six hours to complete installation of a solar panel to Kvant 1.
- **30 May:** Onufrienko and Usachev conduct fourth EVA lasting for 260 min to mount the optical unit of MOMS-2P multispectral scanner outside the Priroda module. They also installed an additional handrail on Priroda.
- **6 June:** Fifth EVA by Onufrienko and Usachev, lasting for 214 min to remove cassettes from the Swiss-Russian interstellar gas detector on Spektr and installed two US micrometeorite detectors and a Russian experiment for testing the effects of the space environment on various materials.

- **13 June:** EVA number 6 conducted by Onufrienko and Usachev lasting for approximately 330 min: they filmed the second part of the soft drinks commercial and replaced a truss on Kvant 1 with a newer model.
- **21 June:** Russian Space Agency announces that the stay of Onufrienko and Usachev aboard Mir has been extended by 40 days.
- **2/4 July:** Orbit of Mir is slightly lowered to permit the less-powerful Soyuz-U launch vehicle to launch Soyuz-TM 24, due mid-August.
- **1 August:** Progress-M 31 undocks from Mir, de-orbited the same day.
- **3 August:** Progress-M 32 docks with Mir at the front port.
- **7 August:** Announced that back-up Russian crew will fly the next Soyuz-TM mission: prime crew Cdr Manakov is grounded because of heart trouble.
- **18 August:** Progress-M 32 undocks from Mir: remains in orbit.
- **19 August:** Soyuz-TM 24 crew of Korzun, Kaleri and French spationaut Andre-Deshays arrive on Mir: French Cassiopeia mission.
- **2 September:** Soyuz-TM 23 with Onufrienko, Usachev and Andre-Deshays returns to Earth.
- **3 September:** Progress-M 32 re-docks with Mir, using the rear port vacated by Soyuz-TM 23.
- **19 September:** Atlantis docks with module attached to Kristall for the STS-79 mission: undocks 23 Sep, leaving Blaha on Mir and returning Lucid to Earth.
- **5 November:** Russia announces that the Kurs automatic rendezvous system will be dropped due to its unreliability.
- **21 November:** Progress-M 32 undocks from Mir, de-orbited later in the day.
- **24 November:** Progress-M 33 docks with Mir at the rear port.
- **2 December:** Korzun and Kaleri perform EVA lasting for 357 min, installing a new solar panel.
- **9 December:** Second EVA by Korzun and Kaleri lasting more than 6 h installing an antenna on a docking unit to ease docking by Atlantis and also connecting the new solar battery.

1997

- **15 January:** Atlantis docks with Kristall's docking module for STS-81 mission. Undocked 20 January, returning Blaha to Earth and leaving Linenger aboard Mir.
- **6 February:** Progress-M 33 undocks from Mir: re-docking planned following the return of Soyuz-TM 24 to Earth.
- **7 February:** Soyuz-TM 24 undocks from front port of Mir and re-docks at the rear.
- **12 February:** Soyuz-TM 25 docks at front of Mir, carryingTsibilev, Lazutkin and German researcher Ewald flying MIR-97 mission. Docked manually at second attempt after automatic system failed.
- **23 February:** Fire breaks out on Mir: lithium cartridge exploded, but fire soon extinguished. Crew has to wear breathing apparatus until gases vented.
- **2 March:** Soyuz-TM 24 with Korzun, Kaleri and Ewald undocks from Mir and returns to Earth.
- **4 March:** Progress-M 33 fails in re-docking attempt: no further attempt planned and spacecraft is de-orbited 12 March.
- **8 April:** Progress-M 34 docks at rear of Mir carrying supplies and equipment to repair Elektron oxygen-producing systems.
- **29 April:** Linenger performs 4 h 58 min EVA to retrieve/install equipment on exterior of station.
- **17 May:** Atlantis docks with Mir bringing new Elektron unit, collecting Linenger and leaving Foale. Undocked 22 May.
- **24 June:** Progress-M 34 undocks from Mir to perform autonomous operation. Next day during attempted re-docking collides with Spektr module damaging a solar array wing and causing module depressurisation.
- **27 June:** Kvant-2 batteries run down, thrusters on Soyuz TM-25 used for orbit adjust.
- **7 July:** Progress-M 35 docks at Kvant-2 port bringing supplies and equipment.
- **13 July:** Cosmonaut Tsibilev displays cardiac arrhythmia and mission managers decide to control workload to prevent increase in stress.
- **19 July:** Crew given approval to perform a suited entry to depressurised Spektr for situation assessment.
- **6 August:** Progress-M 35 undocked from aft port and removed to a safe distance.
- **7 August:** Soyuz TM-26 docks to aft port brining Solovyov and Vinogradov to replace Tsibilev and Lazutkin.
- **14 August:** Tsibilev and Lazutkin return to Earth in Soyuz TM-25 leaving Foale with Solovyov and Vinogradov.
- **15 August:** Mir crew perform fly-around inspection of complex exterior in Soyuz TM-26 and re-dock to forward port. Progress-M 35 re-docks to aft port.
- **22 August:** Leaving Foale in Soyuz, Solovyov and Vinogradov perform six-hour suited inspection of Spektr.
- **8 September:** Crew replace old back-up Elektron following major power failure.
- **14 September:** Main computer fails again and crew power down while repairs are made. Launch of Progress-M 36 delayed to await new computer it will deliver to Mir.
- **28 September:** Shuttle Atlantis docks with Mir bringing 3.16 tonnes of supplies and effecting an astronaut change.
- **3 October:** Atlantis undocks taking Foale, leaving Wolf.
- **7 October:** Progress-M 35 undocked from Kvant-1 aft port.
- **8 October:** Progress-M 36 docks to Kvant-1 aft port bringing 2.5 tonnes of supplies and equipment.
- **20 October:** Solovyov and Vinogradov perform second suited inspection of Spektr module in 6 h 38 min operation to jury-rig electrical cables for solar arrays.
- **3 November:** Solovyov and Vinogradov perform 6 h EVA to remove old solar arrays from Kvant-2 module and to release a small model of Sputnik as an independent satellite. Hatch locking problem on attempting ingress diverts them to Kvant-2 hatch for entry.
- **6 November:** Solovyov and Vinogradov perform 6 h 17 min EVA to install replacement arrays on Kvant-2. Failed to mend problem hatch.
- **14 November:** Kvant-2 solar array orientation system breaks down necessitating another power-down.
- **21 November:** Following another computer failure crew replace computer with interim unit brought up by Atlantis.
- **17 December:** Progress-M 36 undocked and removed for release of German subsatellite designed to act as precursor TV inspection satellite for International Space Station but suffers software failure.
- **19 December:** Progress-M 36 de-orbited over Pacific Ocean.
- **22 December:** Progress-M 37 docks to aft Kvant-1 port bringing new computer and supplies and equipment for ailing Mir.

1998

- **14 January:** Solovyov and Wolf perform 3 h 52 min EVA to conduct scientific investigations including a NASA environment monitoring device.
- **24 January:** Shuttle Endeavour docked with Mir bringing 3.6 tonnes of supplies. Five days on joint activities.
- **29 January:** Shuttle Endeavour undocks from Mir bringing astronaut Foale back to Earth and leaving Thomas for the final long stay by NASA astronauts aboard the station complex.
- **30 January:** Progress-M 37 undocked from the aft Kvant-1 port and was backed away to a safe station-keeping distance.
- **31 January:** Soyuz TM-27 docks to the aft Mir port and cosmonauts Musabayev, Budarin and French cosmonaut Eyharts to join Solovyov, Vinogradov and Thomas for nearly three weeks of joint activity.
- **19 February:** Solovyov, Vinogradov and Eyharts return to Earth in Soyuz TM-26 leaving Musabayev, Budarin and Thomas aboard Mir.
- **23 February:** Using the Kurs automated system, Progress-M 37 is re-docked to the aft port at the Lvant-1 module.
- **3 March:** Musabayev and Budarin perform an EVA in a futile attempt to repair the sticking hatch on Kvant-2 airlock module.
- **15 March:** Progress-M 37 undocked from Mir aft port and removed to a safe station-keeping distance.
- **17 March:** Progress-M 38 docks to aft Kvant-1 port after Musabayev took over manual control when the Kurs automatic system failed.
- **27 March:** Musabayev and Budarin spend several hours attempting to repair the Kvant-2 airlock module hatch, replacing some latches and freeing up others.
- **1 April:** Musabayev and Budarin perform a 6 h 40 min EVA to repair Spektr module solar arrays and worked at the base of the array using the Strela crane for access.
- **6 April:** Musabayev and Budarin cut short an EVA after 4 min 23 s when ground controllers notice Mir drifting off solar alignment with consequent loss of electrical power.
- **11 April:** Musabayev and Budarin complete third EVA lasting 6 h 25 min working to restore power to Spektr solar arrays and change a VDU thruster module.
- **17 April:** Musabayev and Budarin perform fourth EVA lasting 6 h 33 min and set up Sofora girder for VDU thruster module replacement brought up by Progress-M 38.
- **22 April:** Fifth EVA by Musabayev and Budarin lasts 6 h 21 min and brings month's total to 30 h 22 min by installing VDU block and preparing several experiments for future use.
- **16 May:** Progress-M 38 undocked from Kvant-1 port and backed away to a safe distance and Progress-M 39 docks at the same port 5 h later.
- **4 June:** Shuttle Discovery makes the last docking in the Shuttle-Mir programme bringing 2.6 tonnes of water and equipment and for four days of joint activity as well as a full survey of the station conducted by cosmonaut Ryumin brought up by Discovery.
- **8 June:** Discovery departs the Mir space station carrying astronaut Thomas back from his four-month stay aboard the station complex.
- **15 June:** Moscow announces that the orbit of Mir has been lowered by 6 km so that a specially prepared Progress vehicle launched some time in 1999 can de-orbit the station over the Pacific Ocean.
- **12 August:** Progress-M 39 undocked from Kvant-1 aft port and removed several kilometres to a safe distance.
- **15 August:** Cosmonauts Padalka, Avdeyev and Baturin aboard Soyuz TM-28 dock to the aft port on the Kvant-1 module for nine days of joint activity with Musbayev and Budarin.
- **25 August:** Soyuz TM-27 departs Mir and returns to Earth with Musabayev, Budarin and Baturin leaving Padalka and Avdeyev aboard the space station.
- **27 August:** Padalka and Avdeyev board Soyuz TM-28 and transfer their spacecraft from aft to forward docking ports.
- **29 August:** Progress-M 39 was returned to a docking with the aft Kvant-1 port.
- **15 September:** Padalka and Avdeyev conduct a suited inspection of the Spektr module and reconfigure cables to restore attitude orientation control to the operable solar arrays.
- **25 October:** Progress-M 39 was undocked from the aft port and removed, clearing the way for this replacement and was de-orbited over the Pacific Ocean on 29 October.
- **27 October:** Progress-M 40 docked to the aft Mir port bringing supplies and equipment as well as a repeat Znamaya space mirror experiment designed to reflect sunlight to areas on Earth.
- **10 November:** Padalka and Avdeyev conduct a 5 hr 54 min EVA to attach experiments to the exterior, to catch microscopic particles associated with the Leonid meteor shower and to release a small replica of Sputnik.
- **24 December:** Mir perigee raised by 14 km to preserve its orbit for the last human crew scheduled for launch in February 1999.

1999

- **27 January:** Progress-M 40 engines fired for 1 min 40 s to raise the orbit of the Mir complex to preserve the orbit under the effect of a denser upper atmosphere due to solar activity.
- **2 February:** Progress-M 40 undocks and backs away 400 m from Mir to conduct the Znamya mirror reflector experiment but the thin foil array fouls on a rendezvous antenna.
- **8 February:** Padalka and Avdeyev transfer their Soyuz TM-28 spacecraft from the forward to the aft docking ports clearing the forward port for Soyuz TM-29.
- **22 February:** Carrying Afanasyev, French cosmonaut J-P Hagniere (the first non-Russian

Flight Engineer) and Slovakian researcher Bella, Soyuz TM-29 docks with Mir's forward port as the last Mir crew to visit the space complex.
- **28 February:** Padalka and Bella undock from Mir's aft port in the Soyuz TM-28 spacecraft and return to Earth leaving cosmonauts Afanasayev, Avdeyev and Hagniere as the last crew aboard the station.
- **4 April:** Pro gress-M 41 docks at the aft Kvant-1 port bringing supplies and equipment for the three-man crew aboard Mir which returned to Earth in late August 1999 before the Mir de-orbit burn took place over the Pacific Ocean.
- **16 April:** Atanasyev and Haigneré perform a 6 h 19 min beginning at 04:37 UT. Problems with Atanasyev's suit delay a start to activities. Cosmonauts deploy a small amateur radio satellite during EVA.
- **17 July:** Progress-M 41 undocked from Mir now part to make way for next cargo ferry.
- **18 July:** Progress-M 42 docks with Mir bringing new fuel and supplies to the orbiting complex.
- **23 July:** Atanasyev and Haigneré perform a 6 h 7 min EVA beginning at 11:06 UT and deploy an antenna but efforts to free its folded sections failed.
- **25 July:** Haigneré held a radio conversation with fellow Frenchman Michel Tognini, 12,000 km distant aboard Shuttle Columbia.
- **28 July:** Third and final EVA of this mission began at 09:37UT and lasted 5 hr 22 min, during which Atanasyev and Haigneré successfully deployed the Georgian reflector antenna set out on 23 July.
- **27 August:** Afanasyev, Avdeyev and Haigneré returned to earth aboard Soying TN-29, leaving the station vacant for the first time in almost 10 years.

2000
- **2 February:** Progress M-42 undocked from Mir and de-orbited.
- **3 February:** Progress M11 docks with Mir prior to reactivation.
- **6 April:** New and improved Soying TM-30 docks with Mir forward part bringing cosmonauts Zalyotin and Kalen to Mir for a 45–60 day stay.
- **26 April:** Progress M11 undocks from Mir and is de-orbited later the same day.
- **27 April:** Progress M12 docks with Mir.
- **12 May:** Zalyotin and Kaleri perform a space walk lasting 4 h 52 min.
- **15 June:** Soyuz TM-30 departs Mir with Zalyotin and Kaleri and return to earth again leaving the station unmanned.
- **15 October:** Progress M1-2 undocks from Mir aft docking port at 18:09 UT, retrofire at 22:41 UT followed by destructive re-entry over Pacific east of New Zealand at 23:29 UT.
- **20 October:** Progress M-43 docks with Mir at aft port.

2001
- **25 January:** Progress M-43 undocks from Kvant port to make way for next Progress tanker.
- **27 January:** Progress M1-5 docks with the Kvant port.
- **23 March:** At 00:33 UT Progress M1-5 conducted a propulsion burn, lowering the orbit of Mir from 212 × 218 km to 190 × 219 km. A second burn conducted at 02:01 UT lowering the orbit to 150 × 215 km. The final burn at 05:07 UT lowered perigee to <80 km and at 05:50 fiery re-entry was observed over Fiji. Impact area was centred at 160° W × 40° S.

Three generations of Soviet stations

	Sal 1	Sal 6	Mir
Year of launch	1971	1977	1986
Original mass (t)	19	19	21
Body length (m)	15	15	13.1
Max diameter (m)	4.15	4.15	4.15
Primary crew size	3	2–3	2–6
Design life (month)	3–4	24–36	120?
Docking ports	1	2	6
Main engines	1	2	2
Main thrust (kN)	4.09	2 × 2.94	2 × 2.94
Refuellable?	no	yes	yes
Solar span (m)	11	17	29.73
Panel area (m^2)	<28	51	98
Power (kW)	<2	4	10.1

UPDATED

Salyut/Almaz

Background

The manned programme emphasised long-term occupancy from the beginning when the first Salyut appeared in April 1971. That station was smaller and less capable than NASA's solo Skylab but it had the significant advantage of quantity production with improvements to successive models leading to the current Mir. The first six orbited Salyuts (one was cloaked under a Cosmos designation) were only moderately successful and none was occupied for more than a total of 100 days.

Salyut 7: the 5-year-old station, with T14 at its aft docking port, was photographed by the T13 crew. Note the solar panels added during EVAs 0516723

The parallel military programmes, Almaz, unfolded in 51.6° orbit. The boundaries between civilian and military were sometimes indistinct. The programmes eventually merged and matured with Salyut 6. This single station hosted 16 crews of 27 individuals for a total occupancy of almost 2 years. It demonstrated that a Salyut-type module could support a manned Mars mission. Salyut 7 expanded the record with 812 days' occupation by 19 individuals between May 1982 and June 1986. Its capabilities were also enhanced by the docked Cosmos 1443/1686 modules.

Chelomei's bureau worked on both the military manned station, Almaz, and the civilian station, Salyut. At the time, Korolev's bureau was under the direction of Defence Minister Ustinov in late 1969. Chelomei designed and built the first DOS or 'Permanent Orbital Station' within 18 months just as the United States moved to shut down its own efforts on the MOL or Manned Orbiting Laboratory. Chelomei's failure in the manned moon programme, prompted the Soviet Defence Ministry to move the Almaz to Korolev's bureau. Mishin took over two Almaz vessels. The Salyut programmes remained reasonably autonomous at Chelomei's bureau, although they both used Chelomei's three-stage Proton. NPO Machinostroenye still apparently has four manned Almaz craft in storage, along with five small return capsules (400 kg including 120 kg 0.9 m^3 payload, 85 cm diameter, spin-stabilised).

A slightly different history, however, claims that Almaz was almost completed by Chelomei - with 10 being built at Khrunichev Design Bureau. As the vessels moved down the assembly line the control system lagged behind in development, and the Politburo passed the programme on to a reluctant Mishin, who had no interest in Earth orbital stations. His bureau added the Soyuz elements (particularly the control system, solar panels and propulsion unit) to create Salyut. The programme was known as Zarya but changed shortly before launch as a salute to the 10th anniversary of Gagarin's flight.

Mishin put a kinetic energy interceptor device on his Almaz for defence against US anti-satellite weapons and inspection, although apparently it was never implemented.

Salyut 3 (Almaz 2) and 5 (Almaz 3), with their all-military crews and lower orbits probably originated as military stations. The Soviets may have changed the designation as an answer to the United States' Skylab civilian programme. In contrast, Salyut 4 (DOS 4) and 6 (DOS 5) were unquestionably civil.

Salyut 3 included a 10 m focal length camera, able to resolve Earth surface objects down to 30 cm. Before Popovich/Artyukhin left after their 15-day stay, they prepared Salyut 3 for automatic control and for a possible further manned visit, but Soyuz 15 failed to dock. The return capsule ejected on orbit 1,451 from 253 × 275 km and recovered. The flight ended 24 January 1975 following a re-entry manoeuvre. Salyut had operated for seven months, more than twice as long as originally planned.

The Soyuz 17 crew docked with Salyut 4, 17 days after the launch of the space station and remained aboard until 9 February 1975. It was to have been boarded two months later by the Soyuz 18A crew, whose launch was aborted. It therefore continued under automatic control until the back-up 18B crew arrived for a record 63-day stay from 23 May to 26 July. Much of the subsequent success came from the personal repairs which the Soyuz 17/18B crews accomplished on the solar telescope and cosmic ray equipment, as well as to more fundamental items such as air conditioning. There was also a series of failures of both Soyuz and Salyut equipment between the aborted Soyuz 18A launch and the Soyuz 25 docking attempt, bringing the total success rate to five out of 13 space station missions. Four months after crew 2's departure, the unmanned Soyuz 20 was automatically docked, remaining for 91 days before returning to Earth. It not only demonstrated a resupply capability later employed by Progress, but also the possibility of using an unmanned Soyuz to collect sick crewmembers. On 3 February 1977, only 1.5 days before natural decay would have resulted in re-entry, Salyut 4 entered a re-entry trajectory over the Pacific. It had completed more than 12,000 orbits and two crews had occupied it for a total of 93 days.

The two-man Soyuz 21 crew boarded Salyut 5 14 days after launch, but an acrid odour from the life support system drove them home. Their successors bled off the atmosphere and replaced it from tanks of compressed air. Indications were that, like Salyut 3, the Soviets wanted the vehicle for mainly military purposes. For photo-reconnaissance, the Almaz carried the 6 m focal length Agat camera system, producing 50 × 50 cm negatives. Salyut 5's chief cosmonaut, Gorbatko claimed in 1994 that the manned crew once managed to snap the photographs, develop them on-board and dispatch them to Earth in a capsule all within 30 minutes.

Following the failure of the Soyuz 23 docking attempt, Salyut 5's orbit steadily decayed and re-entry was predicted for March 1977. However, corrections on 14 and 18 January raised it to 256 × 275 km. Then, on 7 February, came the successful boarding and 17 day stay by the Soyuz 24 crew. One day after their departure, the automatic return capsule separated and brought back with it film cassettes and other material loaded via an internal airlock. This capsule was sold for US$48,875 at Sotheby's December 1993 New York Russian space auction. It was the first public sight of the design. On 8 August 1977, after 6,630 orbits, Salyut 5's engines placed it in a descent trajectory over the Pacific.

The steady development of the Progress cargo ferry, also gave the Soviets a unique space station capability still not possessed by the US as it builds the International Space Station. The Salyuts had one major advantage over NASA's Skylab, which descended almost uncontrolled in July 1979. They were equipped with orbital manoeuvring engines that permitted re-entry to be commanded safely over the Pacific, with surviving debris falling harmlessly into the sea (Salyut 7 did not). The manoeuvring motors resulted from a need to initially put the stations in a low orbit to compensate for the limited lifting power of Soviet rockets and, until the T series, for the restricted manoeuvring capability of the Soyuz spacecraft.

Apart from the 'random' Soyuz 25/33 failures, the Soviets realised consistent success with Salyut 6. What appeared to be a routine launch on 29 September 1977 was actually a significant step towards a permanently manned space station. Over 44 months, 30 craft successfully docked. 16 carried a total of 33 crew members and the Progress came into its own with 12 docking. Five long-term missions of 96, 140, 175, 185, 75 days, plus 11 short visits (by crews including nine cosmonauts from Interkosmos countries), meant that Salyut 6 was occupied for a total of 676 days, far surpassing the 171 days' occupation of the US Skylab in 1973–4. Salyut 6 became the home of more than 200 experiments in the material science furnaces and 48 million km^2 of Earth's surface was photographed. The Progress delivered more than 20 tonnes of supplies, enabling crews to replace more than a quarter of the station's equipment and to add to the range of its scientific apparatus.

The station's career began inauspiciously with the failure of Soyuz 25's docking. The conjectured existence of a second port was confirmed when Soyuz 26 docked without difficulty on 19 December at the aft end. Following the Soyuz 33 failure in April 1979, there was a one year gap in manned launchings, but the cosmonauts were so successful in replacing and repairing components that a new lease of life began in April 1980. This phase

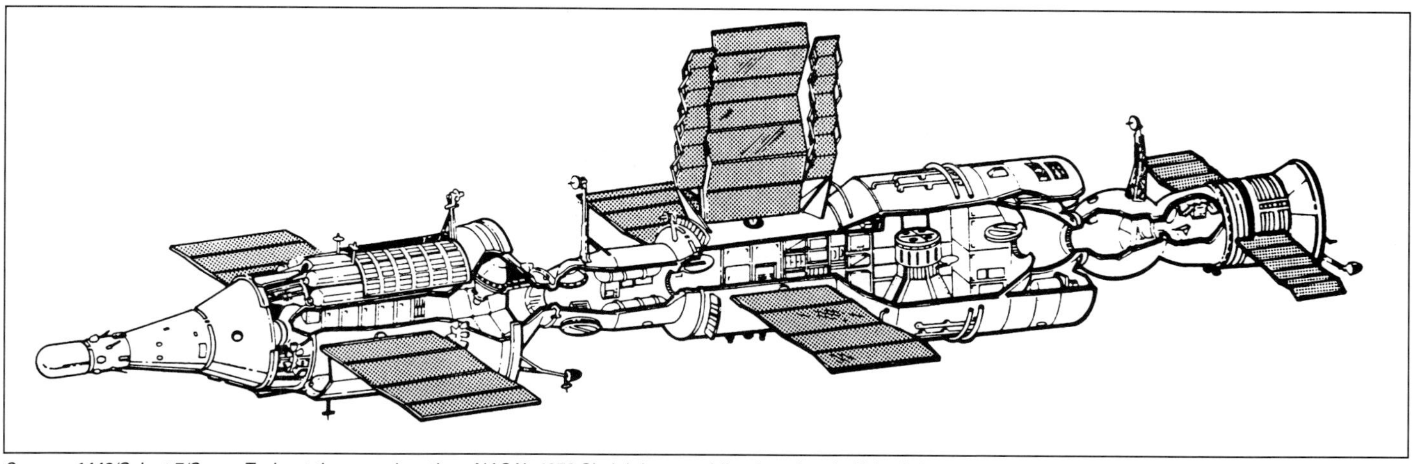

Cosmos 1443/Salyut 7/Soyuz-T: about the same length as NASA's 1973 Skylab but providing less than half the living volume. The TKS/Heavy Cosmos craft (left) was originally intended as a three-man ferry and habitation module for the military Salyuts. The Merkur descent module was derived from Chelomei's small UR500/LK-1 lunar fly-by and later UR700/LK-700 lunar lander designs. C1443's descent module was sold for US$552,500 to an anonymous US collector at Sotheby's in December 1993 space auction in New York (Teledyne Brown Engineering) 0516724

continued through 1981, when four missions were flown, beginning with Progress 12 in January with supplies for the last long-stay crew in Soyuz-T 4. Then came the final docking, on 19 June 1981. Salyut 6 was in a 335 × 377 km, 51.6° orbit and the docking craft was the 15.1 tonne Kosmos 1267, the first use of the TKS/Heavy Kosmos. After manoeuvres on 28 July 1982, Kosmos' 1267 engines were fired the following day to nudge the two vehicles – still joined – into a controlled re-entry over the Pacific.

In its orbit of about 220 × 270 km, periodically raised to about 350 km, Salyut 6 could lose 142 m of altitude every 16 orbits (24 h). This decay depended heavily on solar activity and station attitude. For a long period, Salyut 6's main engines were unusable, but the Soviets compensated for this by showing they could raise the orbit of the whole complex by using the Progress's engines. Soon, spare propellant on Soyuz spacecraft refuelled the stations.

The Soviets expected to follow-on to the Progress with a better vehicle called the Merkur. It became clear in 1991 that the Soviets built the Merkur, otherwise known as the Heavy Kosmos craft (called TKS or 'Logistics transport spacecraft') as a three-man ferry and habitation module for Almaz. However, the Almaz as flown employed only Soyuz and a small payload return capsule ejected from the docking collar.

The Merkur should have made its debut on another military Almaz in 1979–80, to be occupied by a three-man crew flying the TKS/Heavy Kosmos in flight trials. In support of this, Heavy Cosmos Merkur ('Mercury') descent modules were flown two at a time on Proton re-entry tests cloaked under the identifiers Kosmos 881 and Kosmos 882 on 15 December 1976, Kosmos 997/998 on 30 March 1978, and Kosmos 1100/1101 on 22 May 1979. There was also a launch failure 4 August 1977, reusing the capsules from the first two. The tower rescued one but the other was lost. Western reports indicated these were supposed to be small spaceplane trials. Merkur provided three hatches, one on the side for pad ingress, another on top for exit after landing and a base hatch that required removal of the central seat for access to the spacecraft. Kosmos 1443, although it flew unmanned, was equipped as a manned vehicle in one trial.

Salyut 7 was launched amid Western expectations that it would carry three docking ports and form the basis for a 6- to 12-man station. Experience from Salyut 6's career allowed the Soviets to declare a four to five year design life, although major systems failures almost produced an early demise. Ten crews totalling 19 individuals (four paying two visits and Dzhanibekov three) spent a total of 812 days aboard during five periods of occupancy. Repairs, modifications and tests resulted in 13 EVAs and the T13/14 cosmonauts became the first to effect an in-orbit crew exchange.

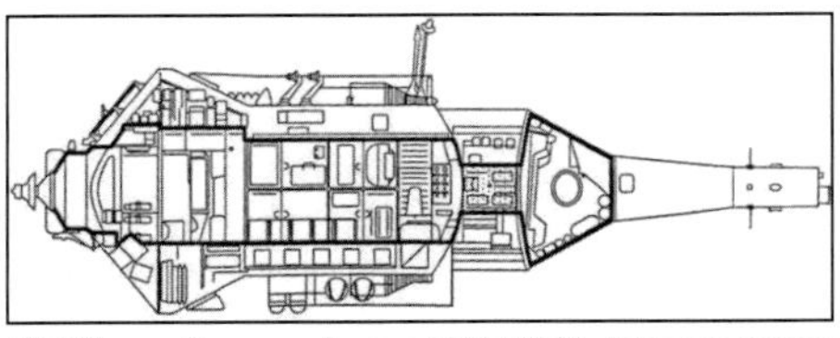

TKS/Heavy Cosmos, from a 1994 NASA Johnson report. The broad line outlines the pressurised volume; details are conjectural. Although a manned launch never occurred, dockings were planned to be controlled by a cosmonaut looking through a port directly over the docking unit (NASA) 0517138

13 Progress's and two Heavy Kosmos (1443/1686) delivered 37.65 tonnes of cargo.

Upgrades included a docking unit strengthened by an additional collar to handle the torques imposed by 20 tonne Heavy Cosmos modules. The portholes included movable external covers to combat the micrometeoroid and UV degradation suffered during Salyut 6. The MKF-6M and Kate 140 Earth resources cameras were again mounted on the floor and allowed each of the first two long-stay crews to return more than 20,000 images. French visitor Chrétien left his Piramig and PCN cameras in July 1982 for sensitive Earth and astronomical imaging at visible/IR wavelengths. The astronomical instrumentation viewing through the large body aperture included the RT-4M X-ray telescope (first flown on Salyut 4), the SKR-02M X-ray spectrometer and the Yelena #947 γ-ray detector (carried by Salyut 6). Materials processing again formed a large portion of Salyut's work, with the Kristall, Magma-F and 136 kg Korund furnaces. Korund produced semiconductor crystals of cadmium selenide and indium arsenide. Evaporator-M used the ShK airlock to deposit metal coatings on surfaces in vacuo and samples were also exposed to study plasma charging effects. The airlock was also employed to launch the 28 kg Iskra 2/3 relay satellites in May/November 1982 on behalf of students at a Moscow engineering institute. A Soviet craft carried electrophoresis equipment for the first time to purify biological materials in a buffer solution subjected to an electric field. Salyut reportedly played a significant role in developing a new flu vaccine. The first run generated material 10–15 times purer than possible under normal gravity.

Some time during late August/September 1983, power was lost from one main solar panel and the life support system struggled to cope under the reduced supply. Heaters maintained 18°C but humidity rose to 100 per cent and condensation dripped off the walls. Fortunately, C1443 had already delivered at least two sets of smaller 1/2 kW panels to satisfy increased loads of the main experiments. Lyakhov/Alexandrov completed two EVAs 1 to 3 November 1983 and attached a 1½ × 5 m GaAs panel either side of the central length. Kizim/Solovyev attached a second pair on 18 May 1984 and brought the supply up to 4 kW, indicating that one main panel was still not contributing. Dzhanibekov/Savinykh added the third pair following delivery by Progress.

Salyut 7's career was further threatened in early September 1983 when oxidiser leaked from the propulsion system during Progress 17's refuelling operations. Lyakhov/Alexandrov retreated to Soyuz in preparation for an emergency departure but were allowed back when it became apparent there was no large breach. The plumbing was shut down, cutting off half of the 32 attitude thrusters and the back-up route to the main engines; the curtailment of attitude manoeuvres seriously affected Earth and astronomical observations. T9's crews could do little but Progress 20 delivered tools and other equipment, and T10's Cdr/engineer arrived trained for a full repair. Kizim/Solovyov undertook four EVAs in April/May 1984 to isolate the leaking pipe and bypass it with new valves and plumbing. They did not complete the work until 8 August after Dzhanibekov arrived with new tools and a videotape illustrating how to pinch off the offending pipe. The duo eventually accounted for eight of Salyut 7's 13 EVAs.

Salyut came closest to orbital death following T10's departure in October 1984. A fault in the radio command system broke contact with ground control and the onboard systems were unable to maintain Sun-lock for the solar panels. The eight Ni/Cd batteries discharged and the station could no longer maintain its vital functions. Experienced cosmonauts Dzhanibekov/Savinykh were dispatched

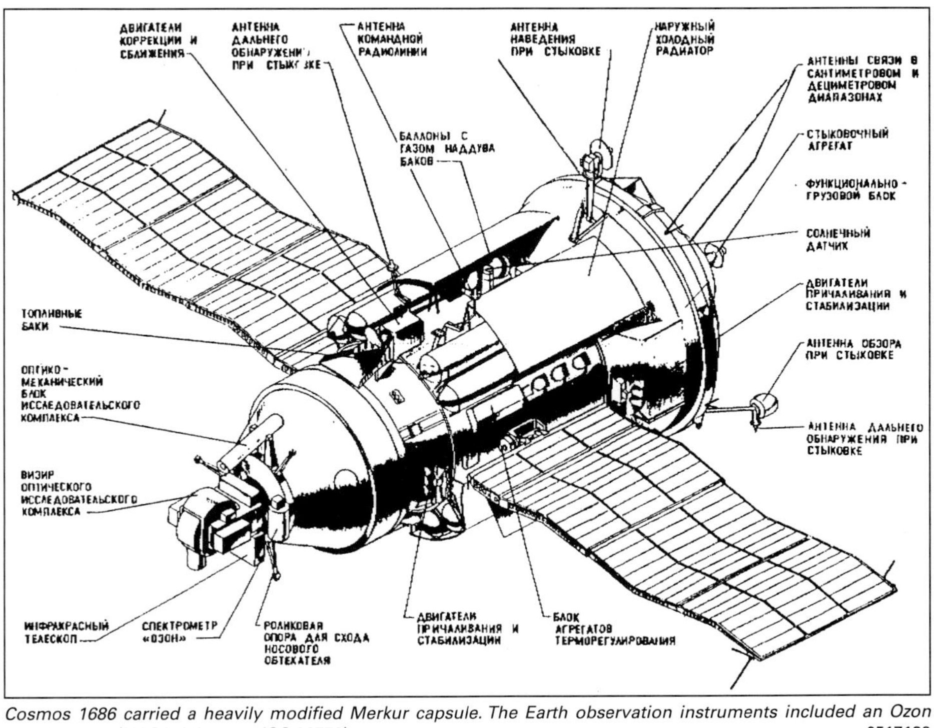

Cosmos 1686 carried a heavily modified Merkur capsule. The Earth observation instruments included an Ozon spectrometer (NASA Johnson, JSC 26770) 0517139

aboard T13 to assess the situation and docked with the inert vehicle using range-speed data from an optical instrument. Salyut's interior was covered in frost and icicles hung from burst pipes, but they discovered that six batteries were serviceable and power could be restored from one panel. Equipment damaged by ice, including eight ventilation fans, was replaced and P24 delivered three new batteries 23 June 1985 so that, within two weeks of boarding, the crew had revived Salyut.

The station saw the beginning of a long occupation by the T14 crew (Vasyutin's illness precipitated an early return) and a brief visit by Kizim/Solovyev from Mir. Salyut was clearly ageing but it and the docked C1686 were still fired into a 474 km circular storage orbit providing at least eight year's life in the hope of a future inspection. During 1988, the Soviets noted that their Shuttle might retrieve it for detailed inspection. However, increased solar activity triggered rapid decay to 445 km by 1 May 1989. Altitude was 413 km by 16 February 1990, 371 × 390 km, 51.61° by 17 April 1990 and dropping by about 1 km per week, leading to a prediction of re-entry by April 1991. It was down to 275 × 305 km by 29 December 1990 and to 176 × 190 km at 12.30 GMT 5 February 1991.

A de-orbiting mission by Progress or Soyuz was mooted during 1989, as a Shuttle retrieval would be both very demanding and too early in the spaceplane's delayed schedule, but it was rejected apparently on cost grounds. The control system had been inactive for two years and only 70 kg propellant remained (500 kg was required for a controlled de-orbit burn). An attitude manoeuvre was attempted the day before entry to direct it by drag modulation into the Atlantic, but it was apparently incomplete. The Soviets predicted that up to 2 tonnes could survive, including 250 lumps of several kg; compensation would be paid for damage or injury. The two largest recovered chunks were reportedly 4 and 8 kg. Several Soviet news releases claimed that C1686 carried a 2.5 tonne 3 m diameter spherical entry capsule; these were the first references to a descent module. Some Soviet reports claimed it had been planned to be separated in January 1986 but Vasyutin's illness during T14 prevented it. Others said Earth return was never planned and the solid retros were not even loaded with propellant. It now appears it was a heavily modified Merkur, carrying Earth observation instruments at its apex instead of the propulsion section. The 40 tonne combination re-entered at about 04.00 GMT 7 February 1991 over 34.9° S/63.8° W, scattering debris over Argentina and Chile as it headed north-east.

Salyut 7 had three habitable compartments totalling 100 m^3, compared with Skylab's 357 m^3. For Salyut 6 and later Mishokin redesigned the original engine system and replaced the single nozzle with twin nozzles on each side of the second docking port added at the rear. He also added more than 20 portholes for experiments, visual observations and photography.

Specifications

Salyut (General)

Forward transfer compartment: The front end, 3 m long, 2 m diameter, was basically an access tunnel with an inner bulkhead enabling it to be sealed off for EVAs from the side hatch; two suits stored here were for the shared use of all visiting crews. There were seven portholes, some carrying astro-orientation sensors.

Working/operations compartment: This main living/working area comprised two cylindrical shells connected by a conical section. The smaller first section was 2.9 m diameter, 3.5 m long, the conical section 1.2 m long and the second section 4.15 m diameter, 2.7 m long. Equipment was arranged in standard racks around the interior's sides. Salyut's central control position in the lower part of the first section had two seats. Further back, a table offered hot food and hot/cold water. On one side was the system that regenerated water from the atmosphere and on the other, the onboard computer (which according to US sources had no back-up). To the outside of the first section were three solar panels, two horizontal and one vertical, spanning 17 m. They rotated 340° to follow the Sun and produce 4 kW of power.

On Salyut 1–3 there were two smaller pairs of horizontal panels, with the Soyuz providing supplementary power from a similar pair. From Salyut 4 onwards the Soyuz began to fly without solar panels as they were able to recharge internal chemical batteries from Salyut's power supplies. This limited Soyuz flying time to 48 h in the event of failure to dock, which happened so often that the solar panels were later restored.

In the rear of the larger-diameter section, on opposite sides, were the crew's bunks and food storage. Aft were a toilet and two small airlocks for ejecting waste. The shower was at the forward end of this section. In the lower centre of the larger-diameter section, the MKF-6M Zeiss camera occupied one of two portholes. Much cosmonaut time was spent nearby on the treadmill, bicycle ergometer and lower-body vacuum suit.

Rear transfer compartment: On the early Salyuts this was an unpressurised service module 2.17 m long, 2.2 m diameter. On Salyut 6, it was widened to the same 4.15 m diameter as the working compartment to take the second docking port. Here the short-term Interkosmos visitors were received when they docked, as were the unmanned Progress ferries, which plugged into the propellant resupply lines. The main engine and thrusters used the same propellants, and refuelling was almost automatic. However, the 1 kW refuelling compressor used so much solar-generated power that refuelling had to be spread over six shifts so that sufficient power remained for other vital systems.

More than 2 tonnes of equipment was used for materials processing, biology, astronomy and Earth resources surveys. The 650 kg BST-1M sub-mm telescope and the MKF-6M multispectral camera dominated Salyut 6's work compartment. The two furnaces (Splav 'Alloy' for smelting metals and Kristall for crystal growth) were delivered by Progress 1/2.

Salyut 1/DOS-1
Launched: 19 April 1971
Orbit: initial 200 × 222 km
De-orbited: 11 October 1971 after about 2,800 orbits
Comment: Following the death of the Soyuz 11 crew on 29 July, the Salyut continuously maintained the temperature and pressure at life-supporting levels. It seemed likely that at least one more Soyuz docking was intended, until the inquiry into the Soyuz failure showed that major modifications would be needed.

Salyut A/ DOS-2
Launched: 29 July 1972
Comment: Proton failure

Salyut 2/Almaz 1
Launched: 3 April 1973
Re-entered: 28 May 1973
Orbit: 215 × 260 km, 51°
Comment: After 130 orbits, the station was said to be stable and the orbit had been raised to 261 × 296 km (now known that main work compartment lost pressure after burn). Shortly after, Western observers reported that it had become unstable. On 14 April, it apparently broke up into 25 fragments; decayed 28 May.

Kosmos 557/DOS-3
Launched: 11 May 1973
Re-entered: 22 May 1973
Orbit: initial 214 × 243 km
Comment: Initial orbit indicated a propulsion or command system failure. Decayed naturally after 11 days. Telemetry suggested civil station. Carried three solar panels, like Salyut 4.

Salyut 3/Almaz 2
Launched: 25 June 1974
Re-entered: 25 January 1975
Orbit: 270 × 219 km
Comment: Modifications included three solar arrays, automatically rotating through 180° for tracking the Sun, instead of the four fixed small arrays on Salyut 1.

Salyut 4/DOS-4
Launched: 26 December 1974 by SL-13 from Tyuratam
De-orbited: 2 February 1977
Mass: 18,900 kg
Orbit: initial 219 × 270 km, 51.6°, raised 11 days later to 343 × 355 km

Salyut 5 /Almaz 3
Launched: 22 June 1976 by SL-13 from Tyuratam
Mass: 19,000 kg
Orbit: 219 × 260 km, 51.6°

Salyut 6/DOS 5
Launch: 29 September 1977
Re-entered: 29 July 1982
Orbit: an initial 219 × 275 km, 51.6°
Mass: 19,825 kg

Salyut 7/Dos 6
Launched: 19 April 1982
Re-entered: 7 February 1991
Mass: 19.92 tonnes
Orbit: 212 × 260 km, 51.6°

United States

International Space Station (ISS)

Current Status

Following years of delay and postponements, assembly of the International Space Station began on 20 November 1998 with the launch of the FGB (Zarya) by a Proton launch vehicle from Baikonur. Known as Flight 1A/R, it put the first element of the ISS into space. It came almost 15 years after President Reagan announced the goal of operating a permanently manned space station, 12 years after the launch of Russia's Mir station and nine years after the Russians began permanent habitation of their laboratory in orbit. The second ISS flight, 2A, was launched on 3 December 1998 when the Shuttle Endeavour (STS-88) carried the Unity module (Node 1) to a docking with Zarya. The third mission, flight 2A.1, was launched on 27 May 1999 when Shuttle Discovery (STS-96) took several tonnes of supplies to the dual complex in preparation for later visits.

The third element, Russia's Service Module, is modelled after the Mir core module and the basic structure comprised DOS-8, which would have been the core module of a Mir 2 which was never built. Known as Zvezda, the ISS Service Module was launched by Proton from Baikonur on 12 July 2000. It docked to the aft end of the FGB, which will serve as a connecting and propellant module between Unity and the Service Module. Unity will form the connecting link to US, Japanese and European elements while the Service Module will be the strongback for Russian elements. The fifth ISS flight, 2A.2 (STS-101), was launched May 19 2000 when Shuttle Atlantis carried additional equipment and a crew to prepare the evolving ISS for later assembly flights and the first occupation crew known as Expedition 1. On 6 August, the Progress M1-3 cargo tanker carried supplies to the ISS, docking to the Zvezda module two days later. Launched 8 September, STS-106 carried the Space Experiment Module and a Spacelab double logistics module on the ISS mission A.2b. Astronauts Lu and Malerchenko conducted a 6 hours 14 minutes EVA on 11 September, during which they lay cable and installed a boom for an exterior navigation unit, moving 33.5 m (110 ft) from the shuttle. The shuttle undocked from the PMA-2 adaptor on 18 September, conducted a photo inspection of the ISS and returned to earth two days later.

On 11 October 2000, the STS-92/3A ISS flight began with the launch of OV-103, carried 14,800 kg of cargo including Z1 truss, PMA-3 located on a Spacelab pallet and the Orbiter Docking System. Four EVA excursions were conducted before STS-92 returned to earth on 24 October. On 31 October, the Soyuz TM-31 spacecraft launched from Baikonur, carried the Expedition 1 crew to the ISS, comprising Russian cosmonauts Yuri Gidzenko and Sergei Krikalev and NASA astronaut William Shepherd.

Space Station after completion in 2002. Russia's elements are clustered to the right; the only point of contact with the rest of Alpha is the node's docking adaptor (John Frassanito & Associates/NASA) 0517129

Space Station configuration in November 1998 after the US Lab module and Canada's RMS are attached (at right). Although occupation was possible from May 1998, the main solar wing was not due to be attached to the US node until September 1998. This wing will be moved later to become one of the four truss-mounted arrays of the completed station (NASA) 0517130

Alpha's FGB, built by Khrunichev under Boeing contract, will provide initial control and power (Theo Pirard/Space Information Centre) 0517294

The US Lab module pressure vessel after completion of welding (Boeing) 0517295

TM-31 docked to the Zvezda port on 2 November, a day after Progress M1-3 vacated that position and de-orbited over the Pacific Ocean, replaced with Progress M1-4 launched 16 November and docked at Zvezda two days later. The STS-97/4A ISS flight began with the launch of Endeavour on 1 December 2000, carrying the P6 photovoltaic assembly and associated solar arrays. A few hours later Progress M1-4 undocked from Zvezda and Endeavour docked to PMA-3 next day. Three EVA excursions were conducted before Endeavour returned to earth on 11 December. On 26 December, Progress M1-4 re-docked to the ISS, un-docked a second time on 8 February 2001 and was de-orbited.

The STS-98/5A assembly flight was launched on 7 February 2001. Atlantis docked to the PMA-3 port on Unity two days later. On 10 February, the Shuttle RMS was used to unberth PMA-2 from Unity and attach it to the Z1 truss at the start of the first mission EVA. The Destiny US laboratory module was removed from the cargo bay and docked to Unity during this EVA, followed by the re-location of PMA-2 to the Destiny end port. Two more EVA's were conducted before Atlantis returned to earth on 20 February. On 24 February, the ISS Expedition 1 crew entered Soyuz TM-31, undocked from Zvezda and re-docked to the -Z port on Zarya. Two days later Progress M44 was launched and docked to the vacated -Y port on Zvezda on 28 February.

Assembly flight STS-102/5A.1 was launched 8 March with the Leonardo logistics module and docked to the USS PMA-2 port two days later. Known as the Expedition 2 crew, cosmonauts Yuri Usachev and NASA astronauts James Voss and Susan Helms, replaced the Expedition 1 crew. The Leonardo module was berthed at Unity's nadir port on 12 March and returned to the shuttle before Discovery landed back at KSC on 21 March. Two EVAs had been conducted.

Background

The evolution, final configuration and Russian involvement in the US International Space Station all came to a conclusion in 1994. NASA had named the assembly Alpha until then, but now simply calls it the International Space Station (ISS). 'Alpha' normally meaning first follows by a decade the permanent manned presence in space that the Soviet Union pioneered.

President Reagan directed NASA to begin development of a permanent Space Station, crewed by six to eight astronauts, in his January 1984 State of the Union Address. Echoing Kennedy's May 1961 Moon landing pronouncement, he proposed that it should be in orbit 'within a decade' and should constitute a bold and imaginative programme to maintain US space leadership into the 21st century. It began as an eight-year, US$8 billion project requiring possibly seven Shuttle flights for initial assembly. NASA administrator James Beggs was requested to discuss participation with European nations, Canada and Japan.

Almost immediately, NASA's various centres suggested methods of building the Station. JSC's Space Station Office advocated an evolutionary facility with 'modular add-ons'. Tasks would include servicing of satellites in low orbits and refuelling of Orbital Transfer Vehicles for GEO satellites. JSC also advocated design of the station as a stepping stone towards a lunar base for mining, military surveillance and communications, and perhaps also for hazardous industrial activities. NASA emissaries seeking international support in Europe in 1983 noted that both polar and LEO stations would be required by 2000, by which time returns

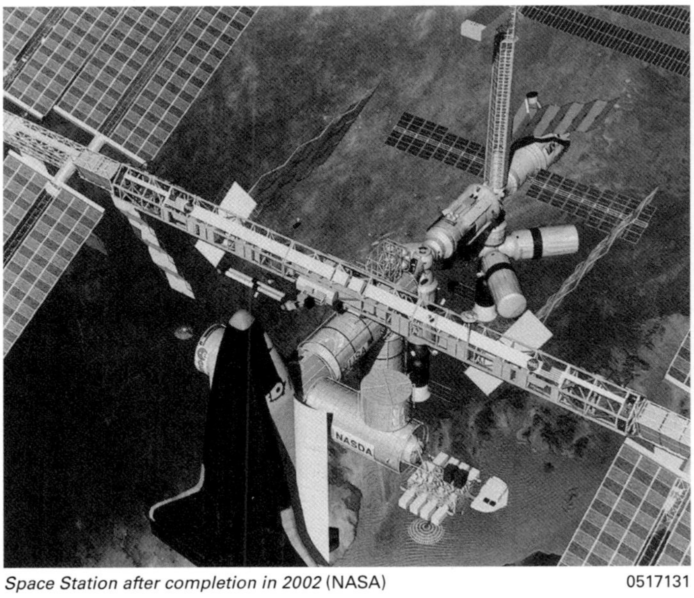

Space Station after completion in 2002 (NASA) 0517131

Space Station after completion in 2002 (NASA) 0517132

The Power Tower concept was abandoned in late 1985 0516780

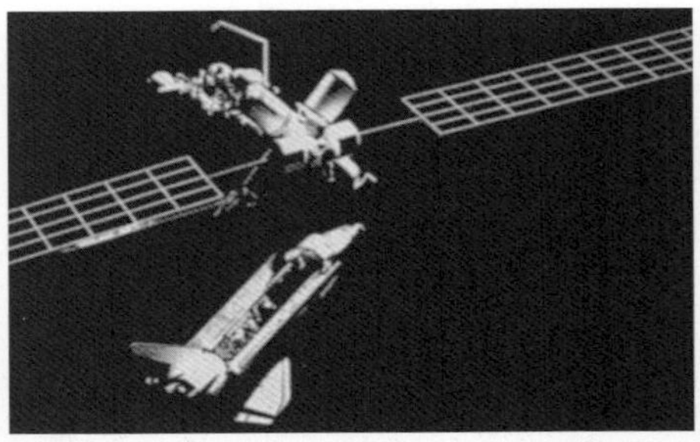

Space Station configuration as presented to President Reagan in December 1983; he used this model at the London Economic summit in June 1984 0516782

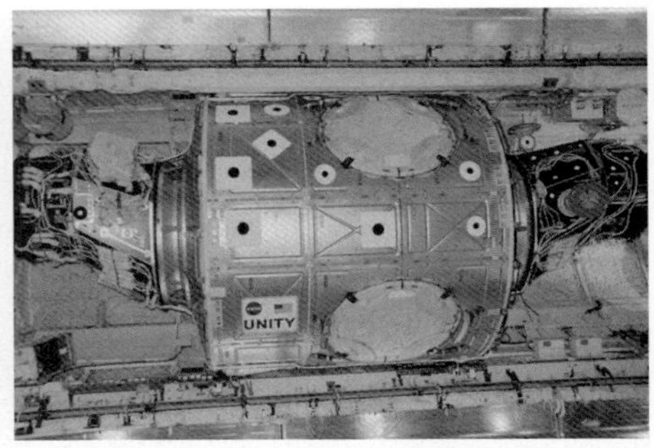

Unity connecting module fitted to payload bay of Shuttle Orbiter Endeavour (NASA) 0054453

from the stations would be worth US$2.4 billion annually.

In its first incarnation, the Station would be placed in a 500 km, 28.5° orbit, with five modules, each requiring three to four hours work per day by crew members. A two-year trial by two crewmen was envisaged in the 1990s, to validate closed-cycle life support systems, ultimately enabling ground supply requirements to be much reduced. The plan specified that another 10 modules would be needed by 1996, with station mass rising from 36 to 94 tonnes by 2000. Included in the US$8 billion price tag were two free-flying platforms. One would accompany the Station, undertaking astrophysics observations, materials processing and other research activities which, if based in the main Station, could be disturbed by crew movements. The second, Shuttle-serviced in a 98° Sun-synchronous polar orbit, would perform Earth observation.

Grumman, under a contractor study, advocated a Power Tower concept, which was eventually accepted by NASA because of the advantages provided by assembling the Station around a 120 m-long central tower. By keeping one end nadir pointing, Earth viewing occurred simultaneously with celestial-viewing. Grumman placed the power generation system between the two sets of instruments, so that it did not obscure the view of either. A massive cross-arm supported either eight solar panels on gimballed joints so that they faced the Sun for about 60 per cent of each orbit; or alternatively, in the solar dynamic concept, four dish reflectors. Further down, above the modules, a pair of large thermal radiators constantly orientated the system to avoid solar illumination as they discharged waste heat. Tether experiments could be simply accommodated, because both Tower and tether orientation stay approximately constant in relation to Earth.

Furthermost from the power system, the Tower offered a large clear area to accommodate the approach and departure of Shuttle Orbiters or OMVs and their payloads. OMVs flew in and out of the lower end for removal and servicing by the RMS. The tower's 2.4 m^2 truss structure provides a convenient track for RMS, as well as supplying a versatile storage area. Finally, the Tower configuration was attractive because it was not sensitive to mass changes from additional modules. By using gravity gradient stabilisation, it inherently accounted for mass changes and therefore did not place tight constraints on the locations of payloads or upon the direction of growth. Attitude control gyros could easily compensate for any stability loss.

Then NASA abandoned the Power Tower approach in October 1985 in favour of the Dual Keel concept. Largely based on Lockheed and McDonnell Douglas designs, it provided the users with a better μg environment (10^{-5} *g* for all modules) by housing the modules centrally. The plan also increased usable area on the structure for attaching external payloads, allowed better pointing accuracy due to the stiffer structure, and reduced traffic through the lab modules. This radically changed plan, announced on 14 May 1986 (and modified again, four months later), was the result of criticisms of the Power Tower expressed in both an internal report and by a group of external engineers and retired astronauts. According to them, the Power Tower design paid insufficient attention to the needs of science and commercial users; in addition, the arrangement of the crew modules, mounted end-to-end in a racetrack pattern, offered poor safety. Individual modules could not be sealed off in emergency – for instance, if punctured by debris. The new design had the modules placed in parallel, and connected with external airlock hubs, enabling any one module to be sealed off without affecting the others. It also provided more room inside the modules.

The design was based on two vertical keels 110 m long, connected by upper and lower horizontal booms 44.5 m long. The central transverse boom upon which the modules would be mounted was 153.3 m long and the truss would consist of graphite/epoxy tubes 5.4 cm in outer diameter, forming a 5 m^2 open structure, designed to provide maximum stiffness for stability during manoeuvres. Astronauts using assisting fixtures during EVAs would erect it. The upper/lower booms were designed for mounting astronomical instruments at the top, and Earth sensors at

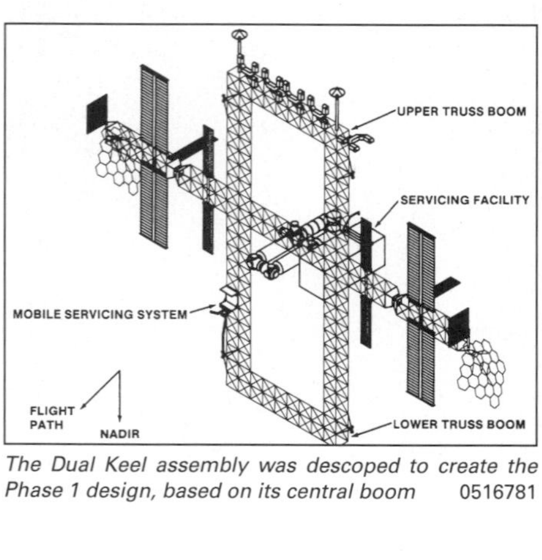

The Dual Keel assembly was descoped to create the Phase 1 design, based on its central boom 0516781

STS-79 was the first to carry the Spacehab Double Module, providing 2.7 tonne cargo capacity (Spacehab) 0007086

US Laboratory module 'Destiny' in space station processing facility at NASA's Kennedy Space Center (NASA) 0054454

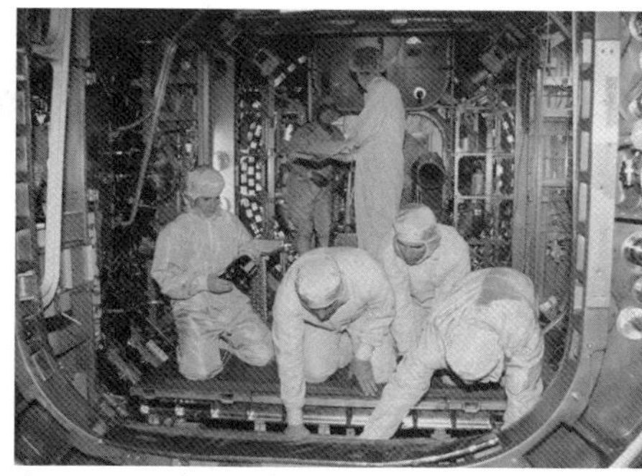

Astronauts take part in a Node 1 systems verification inspection (NASA) 0054457

Unity module pressurised mating adapter (NASA) 0054458

the bottom, with the central transverse boom supporting the pressurised modules. The outer ends of the transverse boom would support the power generation and heat rejection systems, and Canada's mobile RMS would traverse the boom to reach required points of operation.

The new station needed two power generation systems. One was the well-proven Si photovoltaic system, providing 37.5 kW of the total 87.5 kW required. The other 50 kW would come from the newer, and still experimental, solar dynamic system. Large mirrors would concentrate the Sun's energy at a receiver adjacent to the generator, with energy being stored thermally in a molten salt. The solar dynamic collectors must be kept pointing towards the Sun within about 0.1°, whereas solar panels do not have to track the Sun so precisely. In September 1986, the surrounding truss was stripped away to cut costs, leaving the central beam and the configuration in basically the current form.

Before the project was even three years old, NASA began to experience its first real cost overruns. That year NASA spokesmen told Congress the Station cost to completion would be US$21.5 billion, with US$14.5 billion paid by the US (excluding launch costs). By 1989, Station development and initial operations costs by one estimation would be US$24.7 billion at 1989 rates; with launch costs added, the total was around US$30 billion. NASA projected total US cost through FY99 in 1990 at US$38.3 billion. NASA dropped this figure to US$30 billion for its restructured programme unveiled in March 1991. The General Accounting Office estimated in April 1991 US$40 billion, followed by US$78 billion up to 2027.

A major programme review held in the Summer of 1989, under threats from Congress of budget cuts, stated the first element would be in place by March 1995. The Man Tended Configuration (MTC) milestone slipped by five months to April 1996, Permanent Manned Configuration (PMC) by seven months to July 1997 and assembly completion after 29 Shuttle launches by one and a half years to August 1999. NASA issued a new baseline assembly sequence in November 1989, requiring 387 elements to be delivered over four and a half years.

System modifications made as a result of 1989's review included changing from a hydrogen/oxygen propulsion system, involving 4 kW electrolysis of Shuttle waste water, to a modular hydrazine approach. The hybrid AC/DC system was switched to an all-DC network. Development of hard high pressure EVA suits was postponed indefinitely in favour of current Shuttle models. It was decided that various subsystems, such as the closed loop environmental control system, data management, communications and tracking would be incrementally brought to full capability as assembly progressed. It was expected that only 37.5 kW would be available by the time PMC was attained in July 1997. The active thermal control system of fluid/gas heat pipes was replaced by a passive system of body-mounted radiators.

The mass ceiling set for the November 1989 assembly sequence called for a 232 tonne station (including 87 tonne modules, 58 tonne framework, 39.5 tonne power system), with the basic power requirement of 45 kW leaving 30 kW for experimenters. The mass projection peaked at 298 tonnes in June 1990 (and power at 61 kW), before changing the storage tanks' stainless steel to Inconel alloy yielded 283 tonnes the following month. Other measures reduced the power level to 56 kW.

In Autumn 1990, Congress ordered a restructuring of the programme that would reduce expenditure by US$5.7 billion over FY92-96 and peak annual spending from US$4 billion to US$2.6 billion. NASA's 90-day review began in November 1990, encompassing reductions already under consideration. The main US modules were shortened by 40 per cent, and the hexagonal bolted aluminium alloy truss by 50 per cent. In this design, the Shuttle would carry the prices into orbit fully assembled. Assembly elements were reduced from 122 to 17 (an 18th was later added for a node carrying the 2.5 m life sciences centrifuge).

The goal was to reduce in-orbit activity to at most one EVA weekly and preferably monthly. At PMC in late 1999, two years later than planned previously, it would accommodate a permanent crew of four (instead of eight). Mass would be 281 t. Data system capacity was reduced from 300 to 50 Mbit/s and now employing a Shuttle-type Ku-band antenna. The simplified assembly sequence resulted in cancellation of the FTS Flight Telerobotic Servicer, a multi-armed robot for station assembly/maintenance.

An operator viewing via four cameras would control the two manipulators, supported by a leg clamped on the nearest structure. It was to

Laboratory module 'Destiny' undergoes checkout at KSC's space station processing facility (NASA) 0054455

Interior of US Laboratory module 'Destiny' (NASA) 0054456

Interior of Russian built FGB (NASA) 0062158

Zarya and Unity modules viewed from STS-88 0062157

The first two elements of the International Space Station, Russia's Zarya with solar panels and the US-built Unity module 0085849

be available from the first mission, mounted initially on Shuttle's RMS and then the station's Canadian Mobile Servicing System. The project was transferred to NASA's Office of Aeronautics, Exploration & Technology and a late 1993 Shuttle test planned, but it was cancelled in the FY92 budget.

Some modules, such as the US lab/hab, Japan's JEM and ESA's Columbus, required the use of the more powerful Advanced Solid Rocket Motor (ASRM) on Shuttle. It was cancelled from NASA's FY93 budget request in January 1992 on cost grounds. Some US$1.2 billion was allocated through September 1992, with a further US$300–400 million required in termination costs. If it had remained cancelled, then Station's 26 assembly and utilisation flights before PMC would have required an additional two assembly and one utilisation flight using the current Shuttle boosters. ASRM was re-instated September 1992 by Congress, but they finally cancelled it in October 1993.

NASA and Italy's ASI signed an MoU 6 December 1991, for Alenia to supply three Mini-Pressurised Logistics Modules. The MoU included a Mini Laboratory, initially dedicated to life sciences with a 2.5 m tilting centrifuge and three racks. In return for these modules, an Italian astronaut would serve a 90-day tour every six years and one user rack would be allocated 90 days annually (in addition to Italy's ESA entitlement).

FY93's budget allocation US$150 million shortfall in September 1992 prompted further schedule slips: March 1996 first launch (four-month delay), June 1997 MTC (six months) and June 2000 PMC (nine months). The programme and its management were overhauled and by the end of 1993, Russia had agreed to become a fully-fledged international partner. The arrival of a new administration and concern over costs resulted in President Clinton in February 1993 ordering a restructuring to cut cost to completion to US$9 billion (US$9 billion has already been expended), capping annual spending at US$2.1 billion. After years of struggling to survive annual budget procedures, the possibility of cancellation could still not be ruled out. The restructuring aimed to reduce life from 30 years to 10 years (extendable to 15 years) and permanent occupancy was no longer guaranteed. Much simpler schemes, possibly employing Shuttle- and Spacelab-based stations or co-operating on Russia's Mir, were also examined. The 45-member Redesign Team began work 10 March 1993 and submitted its final report 7 June 1993.

The team proposed three options: *Option A* using existing flight proven hardware and cost-effective *Freedom* systems, *Option B* making maximum use of existing *Freedom* designs, *Option C* using a Shuttle-derived vehicle placing a relatively simple station in orbit on a single launch. The White House announced on 17 June 1993, the President's selection of Option A. The plan based on Option A was submitted 7 September 1993 as the 'Alpha Station'. Cost and schedule projections were added 20 September (the US$9 billion goal could not be achieved, but the administration stuck with the US$17.4 billion projection NASA offered for work from 1994 to 2002 (excluding launch costs).

The heads of all the involved space agencies met in Montreal 7 November 1993 to review the outcome of Station's redesign and discuss Russia's participation. The international partners on 7 December formally invited Russia to join the project. Russia's funding problems prompted suggestions in 1995 that Mir should, particularly with the new Spektr and Priroda modules, form the basis for ISS. NASA declined this late suggestion, insisting that any changes must not alter ISS's cost or schedule. Instead, RKA/Energia now plan to keep Mir operational at least through 1999 as

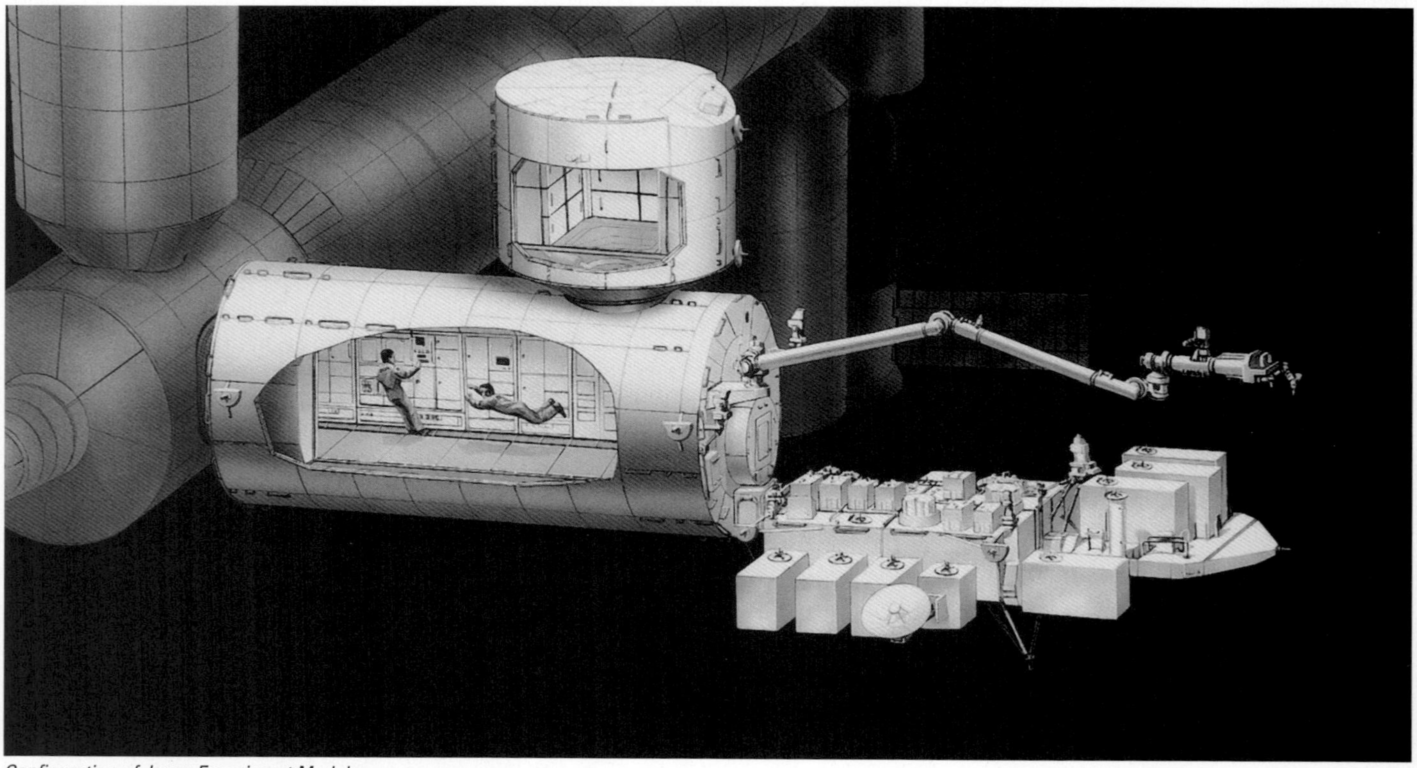

Configuration of Japan Experiment Module 0085784

A view of the International Space Station modules Zarya and Unity during the STS-96 mission 0085850

a much-needed source of revenue, and NASA has added two 1998 Shuttle missions for Mir supplies. Shuttle will also deliver most of Russia's SPP power platform in 1999, saving three Zenit flights.

From 1985 to 1993, NASA spent US$10.1 billion on the station. The US$17.4 billion comprises: US$9.1 billion development, US$4.6 billion operations, US$1.0 billion utilisation plus US$2.7 billion for payloads and Mir support. Operations costs 2003–2012 are projected at US$13.0 billion. The United States plans to pay Russia more than US$1 billion for its contributions, including the US$400 million already agreed on for FY94–97, largely encompassing the Shuttle-Mir missions. Upon completion, NASA will own the FGB 'functional cargo block' under a US$190 million agreement signed by Boeing Defense & Space Group with Khrunichev 15 August 1995 for in-orbit delivery.

Boeing Defense & Space Group received the prime contract for the revised station on 17 August 1993, with the award being US$5.63 billion. Boeing is responsible for delivering the full-up vehicle and for co-ordinating and integrating the US portion with international elements, as well as the design, development, physical and analytical integration, test, delivery and launch of the vehicle. Under the arrangement, Boeing also handles the first one year of sustaining engineering following launch of each package, including spares.

As prime contractor, Boeing oversees three Product Group subcontractors: first, the McDonnell Douglas, truss, external thermal control, command/data handling, communications and tracking, node/cupola integration, GN&C, pressurised mating adaptor components; second, the Rocketdyne, solar arrays, Ni/H_2 batteries, power management/distribution systems; and lastly, Boeing's own, US lab/hab modules, node/cupola primary structures, lab integration, life support systems, international thermal control, internal audio/video, secondary power subsystem.

Assembly will require about 36 Shuttle and nine Russian launches. From the beginning, NASA has had a troublesome time managing the disparate elements of the programme. The FGB was supposed to be on-orbit in November 1997 with the initial Shuttle assembly flight following the next month. However, the Service Module, scheduled to follow FGB by about five to six months, has not been finished. NASA has paid Russia for the manufacturing of the Service Module, but those funds had not been reaching the Khrunichev factory home of the Service Module construction. Because of the lack of funds the work on the Service Module fell behind schedule and it became clear that it would be impossible to meet the scheduled launch in April 1998. As a result it was decided to delay all of the launches for ISS.

Finances and launch delays were not the only problems that ISS suffered during 1996. The first resident crew to ISS was named as A Solovyov and S Krikalev from Russia and W Shepherd from the United States – the latter being backed-up by D Thomas. Both Russian cosmonauts had gained a great deal of experience on multiple visits to the Mir space station, while only Thomas has any space station flight experience. While Solovyov was not questioned as the Soyuz-TM commander, the United States insisted that Shepherd should become the commander once the crew was on board the fledgling ISS. The Russians became reluctant to cede the commander's duties to anybody without space station experience, a matter eventually taken up by the Russian Duma.

Legal precedents were sought out to back up the Russian claim that Solovyov and Krikalev should not be under the command of a foreigner. Of course, the United States is the major financial partner in the ISS programme, but at the time of the flight there would be more Russian units comprising ISS than American ones.

Under pressure from NASA, the RSA removed Solovyov and replaced him with the less experienced Y Gidzenko, while Gidzenko's place on the Soyuz-TM 26 mission to Mir was taken by Solovyov.

In December 1996, ITAR-TASS announced that there were nine Russians in the training group for flights to ISS. From the Russian Air Force, there were Y Gidzenko, Y Malenchenko, Y Onufrienko, V Dezhurov, V Korzun and S Sharipov: the Energiya bureau had supplied N Kuzhelnaya, S Krikalev and M Tyurin. At the time of the report, eight of the nine people were in training at Zvezdny Gorodok for ISS missions: Korzun would join the group at a later date since at that time he was in orbit aboard Mir. Despite having all of these people in training, the Russians had not scheduled any visits beyond Soyuz-TM 29.

A NASA flight director in Houston will control ISS, even when only Soyuz has access in the early phase. The physical control in that Russian-only phase will be from Kaliningrad, but under direction from Houston. NASA will assume full control when the US Lab is attached. English is the working language.

Three Shuttle Orbiters were modified for station operations. The RMS was recertified for loads up to 265 tonnes and a remotely-operated power/data connector was carried on one side of the cargo bay for early flights until the pressurised docking system could be used. The existing internal airlock was to have been refurbished and mounted outside, attached to the cabin by a short flexible tunnel. On top would be an ASTP-type androgynous docking unit, with integral power/datalinks, and crew hatch. Three docking units have been built and carried when required.

Specifications

International Space Station

Mass: 420 t at completion

Crew size: six at completion, including four science (three continuous crew possible from mid-1998). Crews serve six-month tours; typically two will be Russian.

Total pressurised volume: 908 m^3 at completion

Orbit: 426 km circular at 51.6° attained late in assembly sequence; assembled in 355 km, allowing each Shuttle to carry 1.8 t more cargo. The inclination overflies 85% of Earth's landmass and 95% of population.

Microgravity: 180 days quiet μg annually

Configuration: pressurised modules in T-arrangement with transverse 3.7 × 4.9 m hexagonal truss carrying four end-mounted solar arrays for total 110 m length. US will provide two 8.36 × 4.42 m modules outfitted as separate habitation and laboratory facilities, and two connecting nodes (one for storage, other for power control); each 4.42 m long 4.82 m diameter node has six 2 m diameter attachment ports. Node 1 has an attached EVA airlock and a 1.4 m high 2.2 m diameter cupola with seven windows. The 15.5 t 8.364 m long 4.420/4.216 m outer/inner diameter lab (USL) will carry 13 1.054 m wide 2.019 m height 1.50 m^3 usable volume (1.97 m^3 total) standard payload racks. USL hatch 1.3 m^2, 8 latches, central 20 cm diameter window 60 mm thick. 16.1 t hab module includes galley (oven, fridge/freezer, trash compactor, hand washer, water supply) and wardroom (table, 2 × 51 cm windows) at one end and bathroom plus shower in mid section. US modules are principally 2219 aluminium alloy with skin panels featuring external waffle grid.

International Space Station Assembly Sequence

Date	Flight	Launch vehicle	Element(s)
November 20, 1998	1A/R	Russian Proton	Zarya Control Module
			(Functional Cargo Block - FGB)
December 4, 1998	2A	US Orbiter STS-88	Unity Node (1 Stowage Rack)
			2 Pressurized Mating Adaptors attached to Unity
May 27, 1999	2A.1	US Orbiter STS-96	Spacehab – Logistics Flight
May 19, 2000	2A. 2a	US Orbiter STS-101	Spacehab – Maintenance Flight
Jul 12, 2000	1R	Russian Proton	Zvezda Service Module
Sept 8, 2000	2A. 2b	US Orbiter STS-106	Spacehab – Logistics Flight
Oct 11, 2000	3A	US Orbiter STS-92	Integrated Truss Structure (ITS) Z1
			Pressurized Mating Adapter-3
			Ku-band Communications System
			Control Moment Gyros (CMGs)
Oct 31, 2000	2R	Russian Soyuz	Soyuz
			Expedition 1 Crew
Dec 1, 2000	4A	US Orbiter STS-97	Integrated Truss Structure P6 Photovoltaic Module
			Radiators
Feb 7, 2001	5A	US Orbiter STS-98	Destiny Laboratory Module
Feb 26, 2001	4R	Russian Soyuz	Docking Compartment 1 (DC-1) Strela Boom
Mar 8, 2001	5A.1	US Orbiter STS-102	Logistics and Resupply; Lab Outfitting
			Leonardo Multi-Purpose Logistics Module (MPLM) carries equipment racks
Apr 19, 2001	6A	US Orbiter STS-100	(Lab outfitting)
			Ultra High Frequency (UHF) antenna
			Space Station Remote Manipulator System (SSRMS)
Jun 14, 2001	7A	US Orbiter STS-104	Joint Airlock
			High Pressure Gas Assembly
Jul 12, 2001	7A.1	US Orbiter STS-105	Donatello Multi-Purpose Logistics Module (MPLM)
Nov 1, 2001	UF-1	US Orbiter STS-109	Multi-Purpose Logistics Module (MPLM)
			Photovoltaic Module batteries Spares Pallet (spares warehouse)
Jan 17, 2002	8A	US Orbiter	Central Truss Segment (ITS SO) Mobile Transporter (MT)
Mar 14, 2002	UF-2	US Orbiter	Multi-Purpose Logistics Module (MPLM) with payload racks Mobile
			Base System (MBS)
2002	9A	US Orbiter	First right-side truss segment (ITS S1) with radiators
			Crew & Equipment Translation Aid (CETA) Cart A
2002	11A	US Orbiter	First left-side truss segment (ITS P1)
			Crew & Equipment Translation Aid (CETA) Cart B
2002	9A.1	US Orbiter	Russian-provided Science Power Platform (SPP) with four solar arrays
2002	12A	US Orbiter	Second left-side truss segment (ITS P3/P4)
			Solar array and batteries
2003	12A.1	US Orbiter	Third left-side truss segment (ITS P5)
			Multi-Purpose Logistics Module (MPLM)
2003	13A	US Orbiter	Second right-side truss segment (ITS S3/S4)
			Solar array set and batteries (Photovoltaic Module)
TBD	3R	Russian Proton	Universal Docking Module (UDM)
TBD	5R	Russian Soyuz	Docking Compartment 2 (DC2)
Feb 2003	10A	US Orbiter	US Node 2
May 2003	10A.1	US Orbiter	Propulsion Module
Jun 2003	1J/A	US Orbiter	Japanese Experiment Module Experiment
			Logistics Module (JEM ELM PS)
			Science Power Platform (SSP) solar arrays with truss
Sep 2003	1J	US Orbiter	Kibo Japanese Experiment Module (JEM)
			Japanese Remote Manipulator System (JEM RMS)
Oct 2003	UF-3	US Orbiter	Multi-Purpose Logistics Module (MPLM)
			Express Pallet
Jan 2004	UF-4	US Orbiter	Express Pallet
			Spacelab Pallet carrying 'Canada Hand' (Special Purpose Dexterous Manipulator)
Feb 2004	2J/A	US Orbiter	Japanese Experiment Module Exposed Facility (JEM EF)
			Solar Array Batteries
TBD	9R	Russian Proton	Docking and Stowage Module (DSM)
May 2004	14A	US Orbiter	Cupola
			Science Power Platform (SPP) Solar Arrays
			Zvezda Micrometeroid and Orbital Debris (MMOD) shields
Jun 2004	UF-5	US Orbiter	Multi-Purpose Logistics Module (MPLM)
			Express Pallet
Sep 2004	20A	US Orbiter	US Node 3
Oct 2004	1E	US Orbiter	European Laboratory - Columbus Attached
			Pressurized Module (APM)
TBD	8R	Russian Soyuz	Research Module 1
Jan 2005	17A	US Orbiter	Multi-Purpose Logistics Module (MPLM)
			Destiny racks
Feb 2005	18A	US Orbiter	
Mar 2005	19A	US Orbiter	Multi-Purpose Logistics Module (MPLM)
May 2005	15A	US Orbiter	Solar Arrays and Batteries (Photovoltaic Module S6)
TBD	10R	Russian Soyuz	Research Module 2
Jun 2005	UF-7	US Orbiter	Centrifuge Accommodation Module (CAM)
Jul 2005	UF-6	US Orbiter	Multi-Purpose Logistics Module (MPLM)
			Batteries
Sep 2005	16A	US Orbiter	Habitation Module

Notes: Additional Progress, Soyuz, H-II Transfer Vehicle and Automated Transfer Vehicle flights for crew transport, logistics and resupply are not listed.

Canada's Remote Manipulator System will move along truss on mobile transporter carrying 17.6 m long seven-jointed manipulator with 265 t load capacity, tactile feedback, a machine vision system and, if required, Special Purpose Dexterous Manipulator (SPDM) with two 2 m arms. RMS applications include station assembly/servicing, Orbiter docking/undocking, payload manipulation and EVA assistance. Russia will provide 19.3 t FGB, 21 t Mir Service Module; two modified Soyuz-TM as interim Assured Crew Return Vehicles; Science Power Platform to provide 20 kW (US solar cells), attitude control (gyrodynes) and thermal control for early Station; science modules; Progress M and further FGBs to deliver propellant. FGB will provide initial AOCS, power and control; after SM docks, its primary function is as propellant depot/distribution. It also forms a link in Alpha's overall power and data distribution system. Japan's JEM is a 9.7 m long 4.2 m internal diameter pressurised module with attached remote manipulator and exposed platform; a 4 m long pressurised Experiment Logistics Module will be delivered once annually.

ESA's descoped COF will carry 2 t payload in four double racks for materials sciences, fluid physics and life science research. As its contribution to operations costs, ESA will provide servicing and reboost using its ATV Automated Transfer Vehicle launched on Ariane 5; it receives 5.3 per cent of Alpha's resources in return. Italy is providing three 16-rack (including up to five powered refrigerators/freezers) Shuttle-launched Mini Pressurised Logistics Modules: 9.1 t payload, 4.7 t empty, 6.17 m long, 4.5 m diameter. ESA is providing COF's ECLSS. Italy might also provide a Mini Laboratory initially dedicated to life sciences with 2.5 m tilting centrifuge plus 3 racks.

Power: 110 kW at completion, including 46 kW for science. Single US array will provide 23 kW during initial assembly, in addition to the small Russian arrays. At completion, NASA's four photovoltaic Si solar arrays of 32,800 cells

Japan Experiment Module engineering model 0085783

each will provide 92 kW. Each thermal cover 9.8 × 29.3 m. Supported by six Ni/H_2 batteries on each array module. Russian SPP Science Power Platform will add 20 kW. All DC system: 160 V primary, 120 V secondary.
AOCS: initially by FGB then by SM, plus ATV reboost. FGB stores 6.14 t UDMH/NTO in eight external cylinders, refuelled via SM. Attitude determination by US, principally by GPS receivers.
Life support: SM supports three crew, US Hab four crew.
Communications: data 43 Mbit/s Ku-band downlink; 70 kbit/s S-band uplink (300 Mbit/s down projected for future).
Thermal: US will provide six 1,045 kg 22.7 m-long liquid ammonia radiators.
Command/control: from NASA Johnson's 9,500 m^2 five-storey SSCC Space Station Control Center. Working language is English.

UPDATED

MILITARY SATELLITES, SATELLITE SERIES AND SATELLITE CONSTELLATIONS

ASAT
Ballistic Missile Defence
Communications
Data Relay
Early Warning
ELINT
FOBS
Geodetic
Meteorological
Minor Military
Navigation
Nuclear Surveillance
Ocean Surveillance
Photo Reconnaissance
Radar Calibration
Radar Imaging
Science and Engineering Test
Unknown

ASAT

Russian Federation

Polyot Series

Background

In 1992, some Russian scientists admitted the Polyot manoeuvring vehicle tests conducted in November 1963 and April 1964 tested ASAT propulsion systems. They were planned for launch on Chelomei's UR200 but booster delays forced the Soviets to switch the tests to an SL-3 Vostok minus its upper stage.

At the time, the Soviets suggested that manoeuvres supported their manned space programme objective of rendezvous and docking. With cancellation of UR200 in late 1964, the ASATs were switched to the Tsyklon derivative of the R36/SS-9 Scarp missile. TsNPO Kometa had overall responsibility for ASAT development and Chelomei (now NPO Machinostroenye) provided the bus. The 1,950 kg P1, placed in initial 339 × 592 km, manoeuvred several times, ending in 343×1,437 km. It decayed on 16 October 1982. P2 performed several similar manoeuvres and also changed its inclination from 58 to 60°, decaying finally on 6 June 1966. Interest in an advanced autonomous rendezvous, manoeuvre and docking vehicle capable of moving about several separate payloads has been designed on the basis of Polyot tests.

Soviet ASAT Series

Current Status

Inactive. However, the extreme reliance placed on military space-based assets by the US and other NATO countries has revitalised interest in a simple air-launched ASAT. Level of resource allocation is unknown but several Russian defence companies and government organisations have proposed air-launched systems for ASAT roles similar to the US air-launched ASAT programme and with a similar performance.

Background

In 1982, the Soviet Union built and deployed a satellite system which destroyed other satellite systems in space by drawing near and detonating a conventional warhead. Contemporary administration officials and particularly the Department of Defence, often cited this system to claim that the RFAS possesses the world's only operational ASAT system. During the Reagan administration various officials cited this fact while seeking public funding for a new American ASAT system. But the claim still has merit. The RFAS Tsyklon SL-11 booster, basis for the RFAS ASAT system, still serves as an active part of the RFAS launch vehicle inventory. Sources within the RFAS military also point out that the SH-11 Gorgon (Russian name: Topol) ABM, deployed as part of Moscow's ABM defences, has enough energy to launch a LEO ASAT weapon.

The Soviet Union introduced its first ASAT without specifically identifying the purpose in October 1968. The country conducted a series of tests by launching target satellites first from Tyuratam and then from Plesetsk, followed by the interceptor from Tyuratam. The Soviet Union conducted 20 tests in 1982 using radar/optical sensors, to track and close on the target. Sources later revealed that only the radar devices succeeded. The pattern of orbits indicated that the interceptor satellites were expected to kill the targets within one or two orbits. Only two of the four single orbit tests succeeded. Overall, the interceptor hit the target nine of the twelve times. No interceptor moved specifically toward the target if it missed, and observers differ on whether the ASAT has this capability.

ASAT interceptions tended to take place over East Europe or the Soviet Union in the early hours, local time, apparently to allow optical tracking of the ASAT and the target during the closing manoeuvres. Western observers studied the test between 20 and 30 October 1970 by observing Kosmos 373–375. Kosmos 373 acted as the target.

Approximately 3.9 m long and 2.1 m in diameter, it probably had sensors to determine miss distance. Kosmos 374 trailed it three days later, passed very close on its second orbit and then exploded into more than 100 tracked pieces. Kosmos 375 followed on 30 October and also passed very close to Kosmos 373 on the second orbit, 230 minutes after launch. It exploded into almost 50 pieces at its closest approach.

The US DoD estimates the Soviets had an operational system for intercepting and destroying enemy satellites up to altitudes of about 5,000 km since 1972. In 1985 the Soviets publicly announced their system in part as a response to the SDI. As part of the statement, the Soviet Union also announced it had a direct ascent ASAT which would provide a launch on warning capability attack posture towards the SDI satellites the USA expected to deploy.

The last Soviet test of the series of 20, in June 1982, preceded the 'unilateral moratorium' on tests, declared by President Andropov in August 1983. The Soviet Union called for a negotiated ban on space weapons in June 1984. The USA responded with a congressional ban on USAF tests of the ASAT system in effect until October 1988. General Herres, commander in chief of the US Space Command, objected to the congressional ban and warned that almost all near-Earth payloads were within Soviet ASAT range. A publication of the DoD, Soviet Military Power 1985, widely dismissed as a propaganda publication of the Pentagon, noted the Soviets could reach satellite targets orbiting at more than 5,000 km.

The highest observed Soviet era interception occurred at 1,575 km. US estimates suggested the ASAT could reach about 20 US surveillance, weather and navigation satellites known to orbit at or below 1,000 km. However, because of the launch azimuths available, satellites below 45° latitude are probably not vulnerable. NASA pointed out that with the usual 28.5° inclination orbit of the Shuttle, it is not at risk. (Military Shuttle missions have flown up to 62° inclination). Early warning, communication and spy satellites in geosynchronous orbits remain out of RFAS/Soviet ASAT reach; a larger booster, such as Proton, could be used, but would be easily detectable and become a vulnerable target in itself.

The Soviet Union conducted its seventh interception test by the end of 1971 (of which five are believed to have been successful). It concluded this initial testing period with Kosmos 404 launched 4 April 1971, which successfully engaged Kosmos 400 at 1,005 km altitude. Kosmos 462, launched 3 December 1971, did equally well against target satellite Kosmos 459 at only 230 km altitude. Then a four year gap followed these successes until Kosmos 803 went into orbit against Kosmos 462 in February 1976. Some observers have stated the Soviets intentionally conducted this test to coincide with the Salt talks. In April 1980, the test resumed when Kosmos 1171 was launched. The interceptor missed its target by 60 km.

Kosmos 1174 apparently used optical-thermal detection of targets in place of the older radar systems in two tests, one in 1981 and one in 1982. After the collapse of the Soviet Union, researchers in to the Soviet era archives discovered in 1992 that the Polyot missions of November 1963/April 1964 carried ASAT propulsion systems. They were scheduled to fly aboard Chelomei's UR200 but booster delays forced a switch to an SL-3 Vostok. When the Soviet Union cancelled UR200 in late 1964, the ASATs were switched to the SL-11 Tsyklon derivative (first launch 1966) of the R36/SS-9 Scarp missile.

Besides direct hit or kinetic kill ASATs, both the Soviet Union and the United States expended considerable sums on ground-based lasers as a means to kill satellites. For years, Western observers have held that the Sary Shagan and Dushanbe lasers may be capable of damaging passing satellites' sensitive components. Although weather and atmospheric beam dispersion may limit such ground-based lasers, they have the advantage of being able to refire and thus disable several satellites. The large-scale ballistic missile defence research into laser, particle beam, RF and kinetic energy technologies could also provide significant advances in space-based ASATs.

In 1986, some reports suggested the Soviets used the Dushanbe laser, north of the Afghanistan border to disable some US satellites. Secretary of Defense Caspar Weinberger refused to confirm the story; the Soviets made no specific refutation. Interestingly, cosmonauts Kizim and Solovyov mentioned Dushanbe earlier in the year in connection with a 'geophysical survey' they conducted from the Mir space station. The high elevation around Dushanbe permits its laser beam to be particularly effective as an ASAT due to low atmospheric attenuation of the beam strength. The longitudinal spacing of what could well be ASAT facilities at Tyuratam, Dushanbe, Sary Shagan and Semipalatinsk make it possible to attack low-altitude satellites on at least two passes daily.

The Swedish Space Media Network, using the French Spot satellite in October 1987, produced high-resolution photographs that pinpointed

Russia's ASAT vehicle was publicly revealed in 1993. The main radar antenna is at right 0516850

The Dushanbe facility was described as a space tracking centre but its elevation is ideal for siting directed energy weapons (Soviet Military Power) 0516814

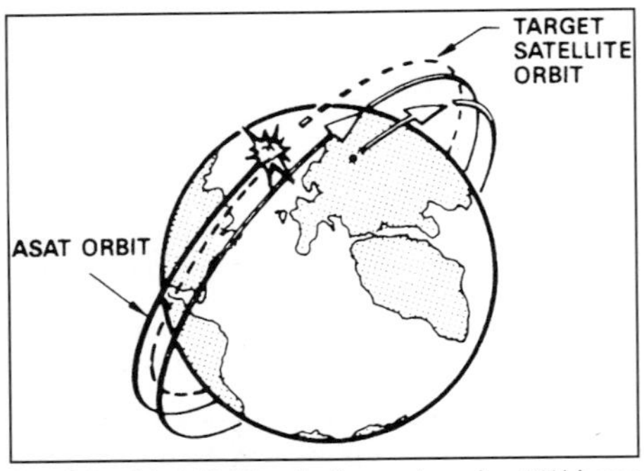

Russia's orbital ASAT typically requires about 3½ hours (2 revolutions) to intercept its prey
(Teledyne Brown Engineering) 0516815

new construction work at Nurek, near Dushanbe. Purportedly, the Soviets were building four domed laser installations, with direct access to the 2,700 MW Brezhnev hydroelectric plant. The Soviets had once described a separate facility nearby as an acquisition/tracking system that could operate in conjunction with the lasers. In response to accusations of SDI research at Dushanbe, Soviet officials said that facilities near both Sary Shagan and Dushanbe were used for satellite tracking, but not SDI-related activities. On 2 January 1988, *Pravda* issued a specific denial that the Sanglok observatory, near Nurek, was part of the SDI-type programme.

A US Congressional delegation received a tour from the Soviets in 1989 of Sary Shagan. According to this delegation, a single 1 m diameter beam director supported two different systems: a very low-power ruby laser and a more powerful CO_2 laser. The 0.7 µm ruby laser housed in the main building adjacent to the beam director consists of 19 individual 5 W lasers whose outputs were merged into a single beam. In a nearby building, a 20 kW CO_2 laser operating at 10.6 µm is linked via a tunnel to the director for a final output of 1 to 2 kW. The Soviet guides claimed the devices had been employed only for tracking, with the most recent exercise in August 1988.

United States

Army ASAT

Current Status

In 1997, the US Army spent US$50 million to develop an anti-satellite weapon, up from US$30 million last year. The principal target of the new army ASAT is foreign photo-reconnaissance satellites. In May 1998, the Pentagon withdrew a funding request to continue the programme.

The army programme bases its kill strategy on a direct ascent kinetic vehicle, launched from Vandenberg AFB, which hits the enemy satellite in a battle space within the immediate area, rather than as a co-orbital or partial orbital system. In an attempt to reduce space debris, the army ASAT releases a 6 m^2 Mylar bag to trap debris created by the impact. The plan has earned the programme the nickname "Green Wars".

Studies have been conducted on directed energy weapons for ASAT purposes but no significant levels of support have been forthcoming. Renewed efforts are redirecting efforts toward a joint endeavour for combined ASAT and ABM systems.

United States ASAT Series

Current Status

Significant progress has been made since 2000 on the integration of ASAT and BMD technologies and although no early linkage in operational plans is sought the development of common systems is a step in that direction.

Background

Just shortly after the Space Age dawned, the US services began plans to shoot satellites out of space. In 1960, one military proposal gained funding and the first test of a system framework came out of Project Bold Orion where a US air-launched missile flew from the atmospheres up to the orbital altitude of NASA's Explorer 6, purposely missing by the satellite by just a few miles. Then the US Air Force began a project named SAINT, variously reported to be an acronym for SAtellite INspector, SAtellite INTerceptor, or SAtellite Inspection & NegaTion. Congress cancelled the programme in 1962 without any flight tests, but the US Air Force followed up quickly with a proposal to knock satellites out of the sky with nuclear-armed Thor missiles fired directly into orbit. The air force conducted tests and began plans to build a base on Johnston atoll and President Johnson declared the system operational in 1964. The Nixon administration dismantled it in 1975, although by then Soviet ASAT tests had been under way for seven years.

Development of the USAF ASM-135 system began under President Ford following concern that the Soviet system was by then operational. But ASM-135 first had to win a bureaucratic and then a technical battle to stay alive. The USAF did get a missile into the air for the first time when it tested the two-stage hit-to-kill ASM-135 from a modified McDonnell Douglas F-15 fighter on 21 January 1984 over the Western Test Range. It was aimed at a point in space, not at a target. The USAF called the second test, in November 1984, 'a partial success'.

On 13 September 1985, the USAF conducted its third test. An F-15 aircraft, piloted by Major Wilbert Pearson, director of the ASAT test force at Edwards AFB, launched an ASM-135 from a position 320 km off the Californian coast, at the DARPA P78–1 Solwind gamma ray spectrometer satellite. The attack followed a 60 to 65° climb out at just below M1.0, and was aimed at 55 m to the right and 25 m above the planned altitude of 11,582 m within a 12.8 km geographic window. The target, launched 24 February 1979, was then 8,600 km distant at 525 km altitude. The ASAT hit it as planned and the impact broke it into at least 285 catalogued pieces (more than 10 pieces are still in orbit). Solwind, it was subsequently revealed amid controversy, had been operational at the time it was hit and was still returning scientific data, returning images of the Sun and corona; it had also recorded comets diving into the Sun.

In the mid-1980s, as the Soviet Rorsats and Eorsats began to orbit, the US Space Command justified funding to Congress on the basis that these satellites could neutralise the US Navy in a time of hostilities. In July 1985, the USAF postponed the launch of two ITVs – the 1.98 m diameter balloons target vehicle which the ASAT would hit – because of reported problems with the ITV's miss-distance signalling system. While the maker, Avco Systems of Wilmington, Massachusetts, was studying the problem, the Summit talks got under way in Geneva. At the talks, the Soviet Union made a dramatic proposal to halt all ASAT testing. However, US$65 million had by then been provided by Congress to start ASAT production, and 58 missiles had been ordered for delivery by 1989.

Space Command continued to argue vigorously for the ASAT but the Congressional Budget Office in February 1987 proposed cancellation. The office contended the move would save US$2.6 billion over five years and would not affect US security because direct ascent technology was rapidly becoming obsolete in light of emerging SDI technologies.

Some details of the ASAT itself became available in September 1981 when a full-size mockup was fitted to an F-15. The interceptor, based on Boeing's SRAM missile, was 5.4 m long, 50.8 cm in diameter and weighed 1,179 kg. A simple infra-red seeker spinning at 1,200 rpm and guided by a laser gyro roll sensor indicated the intercept trajectory. The USAF wanted a direct ascent attack to allow less time for the Soviets to move their satellite out of the way.

Space Command claims that, in addition to ocean reconnaissance satellites, the F-15 ASAT could attack Russian reconnaissance, weather and navigation satellites, in addition to manned space stations and associated spacecraft. In order to reach the Glonass satellites it would have to fit on to a Minuteman missile. The development and procurement of the F-15 low-altitude system is reported to have cost US$3.850 billion.

The air force laid plans to base F-15 ASAT squadrons at Langley AFB (Virginia) and McChord AFB (Washington) because of the orbital inclinations of Soviet target spacecraft and the need for booster debris from tests to fall into the Atlantic or Pacific Oceans. They conducted ASAT command and control functions from a facility developed since 1979 in Space Command's Cheyenne Mountain complex. Full cancellation of the USAF aircraft-based ASAT activities came in March 1988.

The Congressional ban imposed in year-long blocks since September 1985 against ASAT testing on space targets, automatically ended on 30 September 1987. At this time the USAF hoped to launch three intercept demonstrations during 1988 and it launched the ITVs – (Instrumented Target Vehicle) by Scout in December 1985. They re-entered August 1987 and May 1989 respectively. NASA's October 1987 manifest showed that further Scout/ITVs were planned for May in each year between 1988 and 1991. The Congressional ban was extended, however, and all the launches were deleted by the time March 1988's ELV listing was released. The USAF's initial projection of a fully operational system by 1989, and then by the early 1990s, was untenable and the project's cancellation was announced in March 1988. Five F-15 ASM-135 interceptors are now stored in Dallas.

The USAF argued in 1988 that Congress should fund the resumption of tests because the target vehicles would soon be lost. The service argued the instrumented target vehicle balloons cost US$20 million and the opportunity should not be lost to continue the multi-billion dollar programme. DoD followed up the US$183 million FY86 funding and requested US$322 million FY87 funding, with requests for US$448.7 million in FY88 and US$784.2 million in FY89 for its 'restructured' ASAT programme.

Incoming US President George W Bush II is believed to be discussing the possibility of acquiring ASAT technology from the National Missile Defence programme.

BALLISTIC MISSILE DEFENCE

United States

Delta 180 Satellite

Background
Delta 180, conceived by the SDIO, planned to demonstrate that small kinetic energy weapons could destroy ballistic missiles during the launch phase, before the release of multiple warheads. The complex exercise, costing US$150 million, lasted only 205 minutes, and involved six aircraft, 38 radars, 31 satellite communications links and the co-ordination of the White Sands and Kwajalein missile ranges with the Cape and Vandenberg test ranges. The orbiting Delta 3920 second stage was equipped with four forward-looking sensors to assess the target's rocket plumes and four others pointing aft to assess its own. The test began with the launch of the Delta 180 Sensor Module on 5 September 1986 combining the SDI test and an ASAT test. From its initial 222 × 220 km, 28.5°, stage 2 released a satellite mounted on a new McDonnell Douglas Payload Assist System (PAS) powered by an Aerojet 201 engine. The SDI satellite carried not only advanced infra-red and other sensors but also, mounted on a mast, the first laser radar flown in space. After stage 2 had manoeuvred to a separation of about 200 km, they crisscrossed and obtained data simulating observations of ICBMs against various Earth horizon and terrain backgrounds. At 92 minutes a Minuteman stage 2, fired from White Sands, was detected by the infra-red system (modified from the front end of a Maverick anti-tank missile). Finally, the satellite used its radar (modified from a Phoenix air-to-air missile) to impact the burning stage 2 at 10,450 km/h over Kwajalein. TV images relayed by aircraft showed two plumes merging into a flash as a conventional warhead on one exploded to ensure complete destruction of the classified sensor packages; impact was within 75 cm of dead centre. SDI director General Abrahamson claimed it as the most complex command and control mission undertaken by the USA. The Delta 180 SDI test vehicle decayed 28 September 1986 and the target decayed 2 November 1986.

Delta 181 Satellite

Background
In this second satellite of the SDIO series, the Office demonstrated that space-based trackers could identify hostile objects and provide the information to homing seekers. At a cost of US$250 million, the SDIO reported that the Delta 181 mission met all of its scientific objectives. Launched 8 February 1988 by Delta to a 223 × 333 km, 28.6°, 90.1 min orbit, the 1.8 tonne payload module ejected 14 objects for seven sensors (infra-red, visible, ultra-violet and a single LIDAR) to track against the cold background of deep space, then against the more cluttered infra-red background of Earth's surface and finally against the horizon. During the test, four of the target objects also ignited solid-propellant motors to simulate missile signatures. The test concluded with the sensor observation of a Strypi sounding rocket launch.

SDIO states the mission succeeded in tracking cold objects against a variety of backgrounds and demonstrated that the laser radar's ability could discriminate between warheads and decoys by determining the shapes and masses of potential targets for the first time.

Delta Star Satellite

Background
A Delta 3920 launched SDI's third major orbital mission from Canaveral 24 March 1989 into 482 × 504 km, 47.7°, 94.5 min orbit. The 2,720 kg craft had a 698 kg package of eight visible, infra-red, ultra-violet

Delta Star principal features (McDonnell Douglas)
0516825

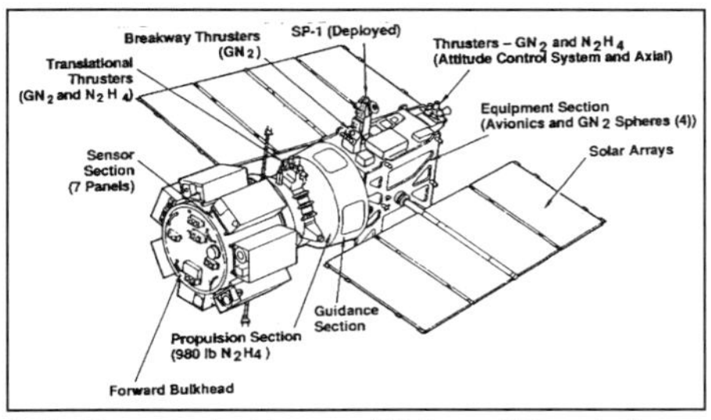

Delta Star principal features (McDonnell Douglas)
0516826

and laser radar sensors that were intended to collect signature data on hot rocket motors and other targets against a variety of space and terrestrial backgrounds. In the second primary set of experiments, the payload returned real-time information on materials degradation in space. The US$140 million vehicle failed to observe Delta's stage 2 separation because of a slow opening sensor door.

Mission planners wanted to observe the Delta, Titan and Trident launches from the Cape in order to establish the feasibility of identifying boosters from their plumes. The sensors would also have trained on at least six Black Brant sounding rockets from a variety of US sites. McDonnell Douglas provided the bus accommodating the Johns Hopkins Applied Physics Lab sensors. Apparently, the manoeuvres to observe these launches did not occur as planned.

The USAF controlled the mission from its Consolidated Space Test Centre in Sunnyvale. SDIO announced in March 1990 that Delta Star shut down on command in December 1989 after propellant exhaustion and 4,253 orbits. Orbital decay was predicted after six years.

The mission did succeed in returning 110 observations, including the return of 1,000 minutes of video spectral data at 30 frames/s. The vehicle re-entered the atmosphere on 23 June 1992.

Losat-X Satellite

Background
SDIO funded Ball Aerospace to create the 'Low-Altitude Satellite Experiment' to demonstrate small satellite and complementary sensor technologies. Losat carried an integrated avionics suite built around two 80C86 processors, a 0.25 Gbit mass memory, Ball reaction wheels and new wide field of view camera. Losat's mission included collection of multispectral data on rocket launches and the space environment. It was originally planned to fly on the RME/LACE Delta but was not ready in time. Carried on a Delta 7925 from CCAFS, the Losat-X satellite was launched on 7 July 1991 to a 405 × 417 km, 40° orbit. In order to reach orbit the Delta 7925 ejected the satellite from stage 2 adaptor after release of primary GPS payload. Losat-X weighed 75.3 kg.

MSTI Series

Background
BMDO started the Miniature Sensor Technology Integration programme to provide realistic tests of miniature sensors on low-cost and rapidly-built satellites in LEO. MSTI-1 reached orbit a within a year of the programme start. The Office wanted to launch two to three satellites annually over five years following what they called a 'build a little, test a little' approach. BMDO transferred MSTI to the AF Space and Missile Systems Center 1 October 1994. MSTI 1 cost US$15 million plus an additional US$3 million for initial costs such as ground equipment. Each follow on flight cost about US$10 million each.

Launched 21 November 1992 to a 289 × 366 km, 96.7° orbit by Scout G, MSTI 1 started the test series to track and guide a homing vehicle against a detected track. Rockwell's 4.4 to 4.8 micrometer mid-infra-red 256 × 256 element HgCdTe staring array detected the target and provided 1.4 arc/min resolution with the 100 mm fl telescope. Observations included a static solid motor firing. MSTI spun up after two leaking thrusters lost all the nitrogen propellant on day one, some imaging still occurred. Imaging continued until April 1993. AF Philips Lab directed the project; JPL was responsible for structure, power, thermal and attitude control systems.

Launched 9 May 1994 by another Scout G to a 415 × 433 km, 93.1° orbit, MSTI 2 examined Loral Electro Optical Systems' 2.7 to 3.0 micrometer platinum silicide camera and Amber Engineering 3 to 6 micrometer InSb instrument with six filters. The new technology introduced on MSTI 2 included a Ni/H_2 CPV battery, lightweight GaAs solar array, high-performance Sun/horizon sensors and a lightweight low power erasable disk magnetic memory recorder. More capable than MSTI 1, the two-camera system with a gimbaled tracking mirror could follow both fixed targets and ballistic missiles. MSTI 2 acquired and tracked Minuteman 3 from Vandenberg, observed two Sergeant launches from Wallops and made multiple tracking observations of ground test objects. In addition to its primary TMD space-based sensor demonstration role, it achieved connectivity with US Navy shipboard assets in a theatre space-based queuing demonstration.

Launched 17 May 1996 by Pegasus to a 420 × 432 km, 97.1° orbit, MSTI 3 carried a SWIR/MWIR 256 × 256 pixel InSb array to observe ground and atmospheric features to determine their variations with aspect, time of day and season. BMDO intended to use it to test future early warning infra-red instruments. The visible wedge 499 × 768 pixel CCD spectrometer provides comparison data for a variety of target signatures. Payload provided by SAIC Science Applications International Corporation under contract to AF Phillips Laboratory.

MSX Satellite

Current Status
The Midcourse Space Experiment Program successfully completed demonsrations of discrimination, tracking and surveillance from space, as well as data collection to characterise stressing backgrounds against which SBIRS must operate. During the lifetime of the primary infra-red sensor (April 1996 to February 1997), MSX collected approximately two terabytes of data. MSX science teams are in the process of analysing this data in support of the SBIRS and GBI programmes.

Targets tracked include a dedicated STARS rocket and two sounding rockets with a variety of deployed objects; co-operative target launches with the UK (Red Tigress II), Australia (DUNDEE) and Russia (AGRE) and co-operative tracking demonstrations of US, UK and Australian aircraft Aurora, airglow, clouds and other phenomena were observed; stellar occulation, a measurement technique for determining atmospheric composition by observing stars as they go below the horizon, has been successfully demonstrated. Space-based space surveillance was successfully demonstrated.

MSX is demonstrating SMTS-type sensors for discriminating/tracking mid-course targets (APL/JHU) 0516830

BMDO data collection with MSX has been completed, although the spacecraft and remaining sensors have three years of life remaining. BMDO is maintaining the capability to collect data and pursuing co-operative data collection efforts with other agencies.

Based on the successful space surveillance demonstration, BMDO has entered into a joint Advanced Concept Technology Development (ACTD) programme with Air Force Space Command. AFSPC will use the MSX Space-based Visible (SBV) sensor as a pathfinder for operational space-based space surveillance. The ACTD is a three-year, jointly funded effort.

Background

The US$866 million Midcourse Sensor Experiment mission is demonstrating SMTS-type sensors for discriminating and tracking mid-course missile warheads. Johns Hopkins' Applied Physics Lab directed the construction of the satellite. The main instrument was projected to last for 20 months, only limited by the cryogenic coolant supply. All other sensors will operate for five years. MSX is also gathering global environmental data, such as ozone levels, in co-operation with NASA. The APL was originally scheduled for launch on 18 November 1994 but a Dewar vacuum leak allowed the frozen hydrogen to boil off. The cause (incompatible aluminium types attaching equipment to the Dewar developing microcracks) took months to determine.

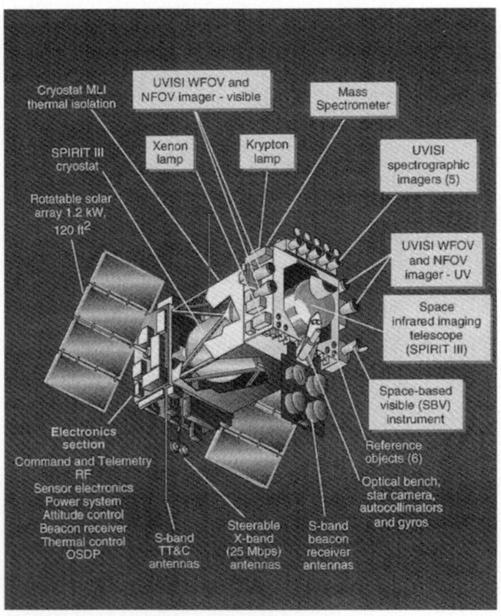

MSX principal features (APL/JHU) 0517151

The completed MSX at Johns Hopkins' Applied Physics Laboratory before it was shipped to Vandenberg 0517152

To help improve orbital debris models, MSX will track along the orbits of Kosmos 2227 in LEO, Kosmos 1278 in GTO and a Titan transtage in GEO looking for debris from their fragmentation. Aurora, airglow, clouds and other phenomena are being observed; key trace constituents will be measured to help establish a database for climate change research. MSX will gather data for two years before NASA's first Earth Observing System (EOS) will take its place. The satellite also performs many astronomy experiments including the UVISI instrument, which looks for photons emitted by decaying neutrinos in a search for the Universe's 'missing mass'. MSX released MIT's 2.0 cm reference spheres for instrument calibration.

The design life of the MSX satellite if 4 years with 18 to 20 months of coolant for IR sensors. Initial project cost US$800 million, scaled down to US$350 million.

Specifications

Launch: 24 April 1996 by Delta 7920 from Vandenberg AFB
Orbit: 897 × 906 km, 99.4° Sun-synchronous
Mass: 2,680 kg at launch
Configuration: 510 cm high bus with three 150×150 cm cross section modules consisting of electronics, Dewar support module and instrumentation (11 optical sensors precisely aligned). Attitude control by four RWs plus three magnetorquers to avoid thrusters contaminating the sensors' environment: 0.1° real-time pointing accuracy (post-processing knowledge 9 μrad). Attitude sensors include two three-axis RLGs, star camera, two horizon sensors, five digital Sun sensors plus 3-axis magnetometer. Line of sight jitter held to ±9 μrad over instrument integration periods of 1 s. Background data collected for 20 min/rev, plus 35 min for target tracking. Data recorded at 5/25 Mbit/s on two 54 Gbit Odetics tape recorders and later dumped at 25 Mbit/s X-band to the primary ground site at Johns Hopkins. 1,200 W provided from twin 11.16 m^2 solar wings, 28 VDC; supported by 50 Ah Ni/H_2 battery. The 200 cm long mid-section graphite epoxy truss supports the Dewar, thermally isolating the 200K outer surface from the much warmer bus
Sensors: SPIRIT III (Space Infra-Red Imaging Telescope) scanning radiometer and interferometer/spectrometer: five radiometer bands collect data over 4.2–26 μm and six interferometer channels over 2.5–28 μm. 33.7 cm aperture telescope, 1 × 3° FOV, 90 μrad resolution. 967 kg, 334 W. Nominal life 18–20 months, limited by the solid hydrogen (8.5°K) supply. Utah State University. Space Dynamics Laboratories.

UVISI (Ultra-violet Visible Imagers & Spectrographic Imagers). Five 1,100-9,000 Å 1.0 × 1.0° spectrographic imagers plus four 1,100-3,000/3,000-9,000 Å ultra-violet/visible narrow (1.6 × 1.3°) + wide (10.5 × 13.1°) FOV imagers. 207 kg, 105 W. APL

SBV (Space-Based Visible) sensor. 15.2 cm 4,000–10,000 Å visible telescope. 1.4 × 6.6° FOV, 60 μrad resolution. 78 kg, 68 W. MIT Lincoln Labs
Experiments: CE Contamination Experiment: mass spectrometer, quartz crystal microbalance, and Xe/Kr flashlamps to measure contamination around MSX and on optical surfaces. Separate balances measure mass deposition on MSX's external surface and on Spirit's main mirror.

Skipper Satellite

Background

This 230 kg satellite flew 28 December 1995 as a secondary payload on a Russian Molniya 8K78M from Tyuratam to investigate aerothermochemistry and aerobraking by dipping into the atmosphere. The satellite was placed in a 804 × 813 km, 98.6° orbit. Unfortunately, the solar arrays had a wiring problem and the battery was drained within a day. The SDL/USU Space Dynamics Labs of Utah State University meant Skipper to conclude the three bow shock studies flights conducted. Bow Shock 1 and 2 sounding rocket missions investigated ultra-violet emissions from bodies at 3.5 to 5.1 km/s and 40 to 110 km. These new data prompted Skipper to look at the 7 km/s region. The elliptical orbit had a perigee of 130 to 150 km for aerodynamic studies.

Skipper's payload included two scanning spectrometers and 18 photometers. The spectrometers when powered, scan every 0.25 second over 0.200 to 0.420 micrometers in 10 angstrom steps and the photometers view ultra-violet and VUV, with emphasis on atomic oxygen, Lyman-alpha, NO, OH and N_2.

COMMUNICATIONS

China

Feng Huo (Zhongxing 2X Series)

On January 25 2000, a new communications satellite usually designated *Zhongxing*-22 was launched and located over 98° E. Prelaunch reports had stated that the new satellite would be called *Feng Huo*.

The name *Feng Huo* means 'fire and smoke', and is taken from the ancient Chinese Army's practice of communicating an impending invasion by lighting a series of beacons along the Great Wall. This allowed warning to be transmitted in only a few days along the length of the Wall, while a human with a verbal or written warning would take far longer to cover the distance.

Based upon this derivation of the name and its military implications, it seems likely that *Feng Huo* 1 is being used primarily for military communications. It is believed that the satellite is being used for mobile communications work. Chinese literature has confirmed that the satellite, launched aboard a CZ-3A, uses a standard DFH-3 satellite platform.

The use of the *Zhongxing*-22 (ChinaSat-22) location for the first *Feng Huo* satellite suggests that a constellation of five such satellites might be planned: the full *Zhongxing*-2X locations are: 21–103° E, 22–98° E, 23–75.5° E, 24–127.5° E and 25–117.5° E.

NATO

NATO Series

Background

Before dedicated international geosynchronous satellites appeared, NATO paid the US government about US$60,000 annually to utilise IDSCS. However, beginning in the late 1960s, NATO began to assemble its own satellite communication network, generally referred to as the second-generation system. Philco-Ford built the NATO 2A and 2B, 243 kg GEO communications satellites, which operate about 5,950 km apart at 18° west and 26° west. These two initial satellites covered the northern hemisphere from Ankara (Turkey) to Virginia in the US, providing 57 voice and 100 telegraph point-to-point circuits through a dozen 12.8 m diameter fixed Earth stations. NATO retired 2A in May 1972 and 2B June 1976, after a total operational expenditure of US$15 million. NATO Communications and Information Systems Agency (NACISA) in Brussels manages NATO communication traffic and has had this responsibility since the inception of the second-generation system.

In February 1973, Ford Aerospace received a US$27.7 million contract to build three NATO 3 satellites and continue the programme into its third generation. Ford's new satellite weighed 720 kg and had a seven-year life. Each could handle hundreds of voice, telegraph, facsimile and wideband datalinks simultaneously on its three X-band channels. NATO manoeuvred NATO 3A into position, following its launch on 28 April 1976 by Delta from Canaveral, and stationed it between Africa and South America. The satellite retired December 1982. NATO 3B followed shortly after the NATO 3A launch on 28 January 1977 where it took up position off the US West Coast above the east Pacific. This satellite also filled in for the traffic that began overwhelming the US DSCS system after 1979. It retired from service in October 1986 and was boosted above GEO from 60° west in July 1993. NATO 3C joined 3A on 19 November 1978 in the slot midway between Africa and South America as an in-orbit spare.

Because NATO 3B continued to perform flawlessly beyond its expected lifetime, NATO 3C remained in orbital storage for seven and a half years, the longest time on record. In October 1986, 3C emerged from storage to begin communication on command from USAF's Sunnyvale facility, relayed through a New Hampshire link. It remained in service until retirement in October 1991, ending its life at 21° west. A small boost command lifted it above GEO in mid-1992. NATO 3D, launched 13 November 1984 by Delta from Canaveral completed the series. 3D cost US$80 million, including US$40 million for the satellite. It was placed initially at 21° west inclined before moving to its operational position at 18.5° west.

At the end of the NATO 3D lifetime, NATO began considering a more advanced, follow-on satellite, but deferred the plans because it normally needs only one satellite, and was owed two years' service from the US DSCS system in return for its use of NATO satellites. NATO 3D moved from 30° west to 21° west in mid-1993, where it remains as a back-up. At the beginning of 1996, the satellite had drifted into an inclination of 3.70°. NATO 3 operational control transferred on 11 July 1991 from the USAF CSTC Consolidated Space Test Center in Sunnyvale to the 3rd Satellite Control Squadron at the CSOC Consolidated Space Operations Center in Colorado, now referred to as the 3rd Space Operations Squadron of Space Command's 50th Space Wing. The control transferred again in 1994 to 5th SOPS.

NATO ordered two NATO 4 satellites, almost identical to the UK's Skynet 4 in February 1987 at a total cost of about £150 million. The total programme will ultimately cost NATO US$364 million at 1987 rates. 4A entered orbit from a Delta 7925 from Canaveral on 7 January 1991 for positioning at 18° west, where it remains as the principal satellite. Like Skynet, it has three 40 W SHF transponders to provide four channels of 60–135 MHz traffic and two UHF transponders which each carry a single 25 kHz channel on a full-Earth helix antenna. The payload includes signal processing and anti-jamming features that are normally associated with electronic intelligence

The NATO IV satellite employs the same basic configuration as the Skynet 4 and the NATO IV satellites for military communications (Matra-Marconi Space/BAe) 0019684

satellites. 4B departed 8 December 1993, also aboard a Delta 7925 and NATO located it at 5° east, where it served as a hot standby for 4A until it was moved to 20° west. The United Kingdom manages the NATO 4 programme on behalf of NACISA.

British Aerospace and Marconi Space Systems act as prime and payload contractors respectively, while the operational control emanates from No 10001 Signals Unit at RAF Oakhanger in the UK. Since the Skynet 4 programme also originates at that facility, speculation is rife that the Zircon capability also exists on the NATO 4 satellites.

Russian Federation

Molniya (Lightning) Constellation

Background

Molniya-1

The Soviet Union built a continuous communication constellation in a sub-geosynchronous orbit by placing eight satellites in orbital planes separated by 45°. Though the Soviet Union never commented publicly on the satellites and the new openness in Russia has added little new information, most observers believe Molniya-1 served primarily for government and military communications. Gorbachev's government made a substantial financial investment and replaced half of the satellites, which have operating lifetimes of around two to three years, in 1988. In 1989, two more launches continued the replacement programme, three in 1990, three in 1991, two in 1992, three in 1993, one in 1994, none in 1995 and one in 1996. The Molniya-M launch vehicle places the Molniya-1 satellites in orbit mainly from the launch field from Plesetsk but very occasionally from Tyuratam. Tyuratam's last launch, Molniya-1, occurred in February 1989. The sole Molniya 2 series launched 25 November 1971 with the first Molniya 2 series on 21 November 1974. The latest Molniya 3 was launched 25 October 2001.

Raduga (Rainbow) Series

Background

Raduga satellites provide domestic telecommunications links and include an X-band military and government transponder for the Gals service dating to the 1970s. The L-band Volna maritime/aeronautical module and Ku-band Luch transponders also occupy spots on the Raduga satellites. Confusion arose in Western circles when Soviets introduced the Gals and Luch satellites, since they seemed at first to be identical to the supplementary payloads carried on Statsionar system satellites.

Raduga satellites have used the Statsionar geosynchronous slots as one of the three types of satellites to operate in these positions. (The other satellite types are Ekran and Gorizont. The Russians have refused to issue any illustrations that can be definitely identified as Raduga satellites, although some pictures have been released claiming to be Raduga. Careful examination of the photographs indicates that they have been Ekran or Gorizont payloads, which have been misidentified. Thirty-nine Raduga satellites have orbited with the first orbit starting in December 1975. 38 conventional Radugas have reached their intended GEO orbits, one (Raduga 33) reached geosynchronous transfer orbit but an explosion of the Block DM-2 fourth stage destroyed the satellite at geosynchronous orbit injection. The Soviets lost only one Raduga to a launch failure on 24 December 1982. Russian statements released with Raduga's 1–3 launch in February 1994 indicated that this was an improved type of satellite intended to replace the existing Raduga satellites with Raduga-1 class payloads. Press releases accompanying the three subsequent launches identified these satellites as the Raduga variant, not Raduga-1.

In February 1992, Italy joined Russia in the joint Astelit venture between Italy's Telespazio and Russia's NPO Astra, which has held responsibility previously for most of the Soviet ground segment, including the Moskva and Orbita stations. Astelit formed to provide Russia's first reliable telephone links into the international telephone network. Ground stations in Moscow and St. Petersburg uplink calls via a single transponder on a Raduga to Fucino in Italy and from there on to the network.

Specifications

Launches: typically one to two annually using a four-stage Proton-K from Tyuratam (Baikonur)
Design life: two to three years, although operational lifetimes well in excess of this have been demonstrated
Contractors: NPO PM (spacecraft), NPO Radio (payload)
Transponders: 6 × 15 W TWTA, 5.75–6.25/3.42–3.92 GHz up/down C-band global, zonal and spot beams, 34 Mhz bandwidth, edge EIRP 26 dBW global, 35 dBW zonal, 45 dBW spot circular polarisation. Single 7.9–8.4/7.25–7.75 GHz up/down X-band beam.
Configuration: unknown, but probably similar to Ekran and Gorizont
Mass: 1,965 kg
AOCS: three-axis, east-west station-keeping to within 0.5° by liquid thrusters

Strela (Arrow) Series

Current Status

Strela-3 continues to be the prime Russian series for military store – dump satellites.

Background

Strela 1 series

Soviet direct communications between ships, aircraft and bases used to be passed over a constellation of eight 61 kg 100 cm diameter Strela 1 ('Arrow') satellites put in orbit by a single SL-8 Kosmos launched from Plesetsk into 1,500 km circular orbits at 74° with 115-minute periods. Strela 1s offered medium-range links between forces in the field, and possibly at sea, over VHF/UHF bands. The system was developed in the 1960s with 26 flight tests of the Strela 1 from 18 August 1964 to 18 September 1965.

At any one time, the Soviets had to keep about 24 satellites operating in order to ensure global coverage. Each had an operational life of about two years, but as their orbital life is about 10,000 years, with a 20,000-year life for the rocket stages, the orbits had to accommodate more than 350 satellites and stages. These orbits are now largely avoided because of the debris left behind from the Strela-1.

The Strela 1M was used with Strela 2M for Soviet military and intelligence communications. The first was launched on 25 April 1970 and 360 were launched before the system was phased out with the launch of Mozhaets on 3 June 1992.

Strela 2 series

The Soviets conducted longer-range military communications on a constellation of three 875 kg Strela 2 satellites, built by AKO Polyot for NPO PM. The practice remains in place with the RFAS. An SL-8 Kosmos places them in 786–810 km orbits at 74° inclination and separated in spacing on the orbital plane by 120°. They apparently receive information from low-power spy and other clandestine transmitters around the world, and store the information until a suitable pass enables it to be dumped to RFAS receiving stations. Their ground tracks repeat approximately every 85 revolutions. Two or more satellites have been operated in recent years in each plane of the system, which began with Cosmos 103 on 28 December 1965, the first of five launched by 27 August 1968. The first operational Strela 2M flew on 27 June 1970, the last of 52 on 20 December 1994.

Strela 3 series

Analyses of a more recent system, which carries six 225 kg Strela 3 satellites on each launch to orbits of 1,400 km at 82.6° inclination, suggest it is a second-generation replacement for the octets. AKO Polyot builds the satellites for NPO PM. Store-dump second generation series developed from 1973, six launched on each Cosmos 3M instead of eight, an operational constellation consisting of 12 in two orbital planes, 90° apart. Replaced Stela 1M in 1992 and 2M from 1994 in strategic communications role. Commercial version marketed as Gonets. First launch on 15 January 1985, last of 126 launched 8 July 2002.

United Kingdom

Skynet Series

Current Status

The two most recent satellites in this series were Skynet 4D launched by Delta 7925 on 10 January 1998 and Skynet 4E launched by Ariane 44L on 26 February 1999. Early in 2002 the UK MoD decided in favour of the Paradigm team for development and production of the Skynet 5 system in the largest UK government Priate Finance Initiative of its kind. Paradigm Secure Communications supports the prime system manufacturer Astrium and the service company, Paradigm Services. Skynet 5 is expected to debut in 2006.

Background

The UK began planning to build a network of communication satellites for its far-flung outposts even before the country had begun to withdraw from Asia and the Middle East. Britain needed secure voice, telegraph and fax links for strategic and tactical communications between UK military headquarters and ships and bases around the globe. However, the first attempts to obtain this capability proved unsatisfactory and the UK began to lose interest.

Skynet 1 was launched on a Delta launcher in November 1969 from Cape Canaveral almost as soon as geosynchronous orbits became accessible. But Skynet 1's TWTA failed within a year. Skynet 1B followed shortly, but it failed to reach the correct orbit and remained stranded in a highly elliptical orbit. The first three satellites in this series offered two UHF global channels for submarine and other mobile users, four SHF channels ranging from global to 3° spot beam coverage, and an experimental EHF uplink channel for propagation studies.

For its second-generation satellites the UK sought to build an indigenous capability. Ironically, the UK, with the most sophisticated electronics industry in the world, contracted with Philco-Ford to build the first Skynet. With help from Philco, the UK's Marconi Space & Defence Systems built the first communications satellite produced outside the US/USSR. A launch failure set the Skynet 2A tumbling through space until it plunged back into the atmosphere six days after being launched. Skynet 2B succeeded and broke the string of failures for the UK. It linked the UK directly with Western Australia for the first time in history.

The original Skynet system included 17 ground stations, with a 12.8 m master terminal at RAF Oakhanger, air-transportable terminals for use on land and shipborne terminals. By the time the system became operational, successive defence cuts and British withdrawals from Asia and the Near East made it seem like an unnecessary luxury. The British quietly cancelled the planned Skynet 3 series in 1975 and replaced it by leasing capacity from US/NATO satellites.

In the decade before the Falklands conflict of 1982, the UK's military leased channels from the US and the plan worked well for all UK defence needs. Nonetheless, the Falklands conflict itself ironically proved the difficulty of routeing military communications through the good offices of a foreign provider, in this case the US DSCS circuits. The UK revived attempts shortly thereafter to establish an independent and secure communications network for the Royal Navy and other UK forces. The idea passed in Parliament and the budget included funds for the Skynet 4 programme.

After several delays, the UK finally scheduled Skynet 4A and 4B for launches in June and

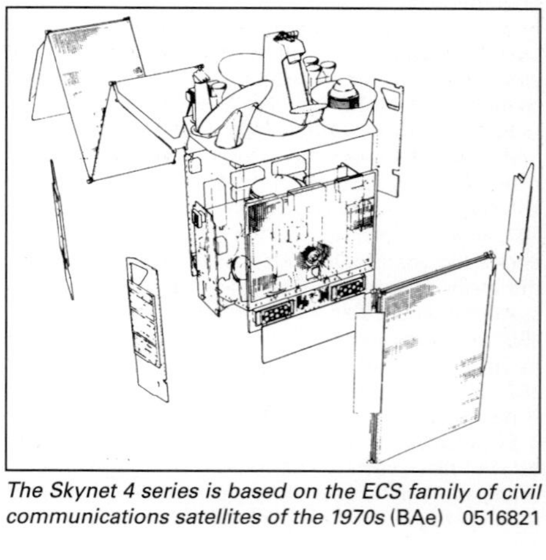

The Skynet 4 series is based on the ECS family of civil communications satellites of the 1970s (BAe) 0516821

Skynet 4: the series of UK milcom satellites will continue beyond 2000 (MMS) 0517317

Skynet's 14 m Earth stations at RAF Oakhanger. The site also controls NATO 4. Siemens Plessey provided the turnkey station under a £40 million contract; a £60 million contract covers a second large anchor station, a Satellite Network Control Centre and two large transportable terminals
(Siemens Plessey Defence Systems) 0516822

Skynet 2B was the first communications satellite built outside of the US/USSR. Launched in 1974, it can still be used (MMS) 0516823

December 1986, both by Shuttle despite strong French pressure to use Ariane. The UK finally agreed to the Shuttle launches because of NASA's standing offer to let a British astronaut go along and the hope that the US would help the UK make the Skynet 4 the standard NATO system. By early 1988, the Skynet 4B/C had berths on the Ariane 4 and in late 1988 and early 1990, and 4A had a Shuttle booking.

The continuing problems with the Shuttle moved Skynet 4A to a Titan 3 launch in August 1989 at a reported cost of US$55 million. With the launch vehicles chosen, the UK government ordered Skynet 4C from BAe in May 1985.

The programme, popular in Parliament, seemed to have assured backing but, in January 1987, a journalist reported that the Skynet actually cloaked an electronics intelligence programme known as Zircon. In a secret intelligence sharing deal with the United States, the UK would provide all the message traffic that its Skynets snatched from their Asian orbital slots in return for US information. The United States benefited from this by averting suspicion from electronic intelligence satellites that it needed to have in the region through the mechanism of the false flag. The satellite eavesdropped on other electronic communication at geosynchronous. In the published article, the author cast suspicion on the location registered for the satellite, which was far from any conceivable orbit for communications. The Ministry of Defence eventually admitted its role in Project Zircon and confessed to spending £70 million on the programme.

In operation it would resemble the US Rhyolites and the later Magnum signal intelligence satellites. Reports indicate that the UK would share intelligence with the US via the UK government's monitoring centre GCHQ Cheltenham. Zircon was abandoned following disclosure amid concerns over its ageing technology. By the time Zircon was ready to go though, the ageing satellites, delayed after years of waiting for a Shuttle ride, were simply obsolete.

Skynet suffered from another American problem. Originally, the British military projected Zircon would cost about £400–500 million to build, launch and operate two to three satellites to provide Elint reportedly from GEO at 53° east. The UK Parliamentary Space Committee received a report in June 1990 that the entire Skynet series had cost almost £1 billion, even before Skynet 4D-4F came along.

The Skynet 4 generation of telecom satellites assures continuity of UK tactical and strategic defence communications to at least 2006. The fourth-generation Skynet adds a considerable capability to the communication system with steerable antennas for SHF spot beams, increased power, improved anti-jamming and a fully tunable UHF system for increased flexibility.

Specifications

Skynet 1

Launched: 22 November 1969 from Cape Canaveral by Delta
Launch mass: 243 kg, 129 kg on-station. Payload of two 3 W X-band TWTAs. Cylindrical bus 137 × 81 cm high (157 cm with solid AKM), 90 rpm spin-stabilised with a despun antenna. 7,236 solar cells supported by two 16 Ah Ni/Cd batteries. Prime contractor Philco-Ford; same design as NATO 2A/B. Positioned in GEO over the Indian Ocean but its TWTAs failed within a year.

Skynet 1B

Launched: 19 August 1970 by Delta from Cape Canaveral. Abandoned in 270 × 36,058 km GTO after AKM failure. Used for meteorology and communications tests. Other details as Skynet 1

Skynet 2A

Launched: 19 January 1974 by Delta from Cape Canaveral. Launch mass 435 kg; 235 kg on-station. Built by the UK's Marconi Space & Defence Systems (with Philco-Ford providing technical direction) as the first communications satellite produced outside US/USSR. Payload of two 16 W X-band TWTAs provided redundant 2 + 20 MHz channels. Cylindrical bus 191 × 208 cm high (with solid AKM), 90 rpm spin-stabilised with despun antenna. Launch failure left it tumbling in 120 × 1,857 km orbit; re-entered after six days

Skynet 2B

Launched: 23 November 1974 by Delta from Cape Canaveral, placed in GEO over Indian Ocean. With a design life of at least three years, it provided links from the UK to W Australia. Other details as 2A. It began drifting in late 1992, indicating its propellant was exhausted. Tests in 1994 showed that its two 16 W TWTAs still met specifications. It remains usable, inclined at <13° and slowly drifting. The original Skynet system included 17 ground stations, with 12.8 m master terminal at RAF Oakhanger, air-transportable terminals for use on land and 1 m shipborne terminals. By the time the system was operational, successive defence cuts and British withdrawals from Asia and the Near East had rendered it largely unnecessary

Skynet 4A

Launched: 1 January 1990 by Titan 3 from Cape Canaveral
Location: 6° E (1990); 29° E (1991); 65° E (1991); 34° W (1992 to 1999)
Other specifications as for Skynet 4B, except McDonnell Douglas PAM-D2 provided perigee kick

Skynet 4B

Launched: 11 December 1988 by Ariane V27 from Kourou, French Guiana, declared operational 22 February 1989
Location: 1° W (1989 to 1990); 53° E (1990 to 1998)
Design life: 7 years
Contractors: BAe (prime), Marconi Space Systems (payload), and CAL Ltd. (UHF antenna)

Transponders: 3 × 7.250–8.400 GHz X-Band 40 W TWTAs providing four SHF channels made up of channel 1, 31 dBW EIRP 17.0 dB gain global 135 MHz bandwidth; channel 2, 34 dBW EIRP 24.7 dB gain narrow (Europe) 85 MHz beam; channel 3, 35 dBW EIRP 21.7 dB gain wide (hemispherical) 60 MHz beam; channel 4, 39 dBW EIRP 29.5 dB gain 3° spot (central Europe) 60 MHz beam

2 × UHF 305–315/250–260 MHz up/down global 40 W transponders, 25 kHz bandwidth, 26 dBW EIRP, working through single helical antenna. With EMP hardened anti-jamming and ECM interference features
Principal applications: UK Ministry of Defence. Submarines, aircraft, maritime patrol, tankers, AWACs, transportable Earth terminals, 1.7 m mobile terminals, manpacks
Configuration: based on MMS ECS/OTS bus, 2.1 m high, 1.9 m deep, 1.4 m wide; 16.0 m span with solar wings deployed. The Communications Module is a U-shaped structure sitting on the Service Module; major structural elements are predominantly aluminium alloy honeycomb with aluminium or CFRP skins. Payload components are mounted on both sides of the upper floor and on the sidewalls. The central 2.2 m diameter thrust cone is CFRP honeycomb with aluminium support struts
Mass: 1,433 kg at launch, 790 kg after AKM firing, 770 kg on-station BOL, 655 kg dry
AOCS: 3-axis control with ±0.1° E-W station-keeping incorporating IR Earth sensors (Officine Galileo, accuracy ±0.04°) and Sun acquisition sensors (TNO/TDP, accuracy ±2° coarse) maintained by two 25 Nms and one 16 Nms Teldix MWs and 20 catalytic hydrazine thrusters. (0.5 N roll/yaw, 2 N east-west station-keeping, 5 N pitch, 20 N active nutation damping). The thrusters draw on 70 kg propellant in two N_2-pressurised tanks. Pointing accuracy 0.07° roll/pitch, 0.35° yaw
TT&C: through RAF Oakhanger in Hampshire, UK, which also acts as the primary communications

node using 14 m dish. A dedicated Skynet 4 in-orbit checkout facility is also maintained at Oakhanger. The TT&C subsystem, working through the SHF receivers with S-band utilised during launch and as back-up, can handle 512 8-bit words about every 16 s; a total of 750 telecommands is available
Power system: twin 228 × 730 cm 3-segment (each 128 × 205 cm) Si solar wings and auxiliary body panels, providing 1,200 W at regulated 42 V in sunlight; eclipse power provided by two banks of 14-cell 35 Ah Ni/Cd batteries with 30–37 V unregulated output. Eclipse power requirement 940 W, maximum requirement 1,083 W (equinox). Degradation of AEG back surface reflector cells 9% over 7 years
Thermal control: heaters plus passive radiators, including Secondary Surface Mirrors and 10–15 layers
Apogee kick motor: Thiokol Star 30E solid propellant, burn time 49 s, specific impulse 290 s

Skynet 4C
Launch: 30 August 1990 by Ariane V38 44LP from Kourou, French Guiana. AKM fired 2 September, 69 h after launch; solar wings deployed 3 September; UHF antenna deployed 4 September
Location: 1° W geostationary (moved from storage position at 53° E in 1992, where it was ideally placed to provide links during the 1990–91 Gulf conflict); inclination 1.1°

Skynet 4D
Launch: 10 January 1998 by Delta 7925 from Cape Canaveral
Location: 53° E (1998 to 1999)
Life: 7 years
Platform: as 4A-C but reworked TT&C (including increased autonomy); improved power system; improved reaction control system
Transponders: 3 × 7.250–7.725 GHz X-Band 50 W TWTAs providing four SHF channels: channel 1, more than 31 dBW EIRP 17.0 dB gain global 125 MHz bandwidth; channel 2, >34 dBW EIRP 26 dB gain narrow (Europe) 75 MHz beam; channel 3, >36 dBW EIRP 21.5 dB gain wide (hemispherical) 75 MHz beam; channel 4, >40 dBW EIRP 30 dB gain 3° steerable spot (central Europe) 60 MHz beam; two UHF 295–318/254–258 MHz up/down global 50 W transponders, 25 kHz bandwidth, >25 dBW EIRP, working through single helical antenna (up/down frequency selection independently tunable). With EMP-hardened anti-jamming and ECM interference features
Mass: 1,510 kg at launch, 868 kg after AKM firing, 851 kg on-station BOL, 256 kg payload
Power system: as Skynet 4A-C, except 2 × 29 Ah Ni/Cd batteries
Other specifications as Skynet 4A-C

Skynet 4E
Launch: 26 February 1999 by Ariane 44L from Kourou, French Guiana
Location: 53° E
Mass: 1,510 kg at launch, 873 kg after AKM firing, 851 kg on-station BOL, 256 kg payload
Other specifications as Skynet 4A-C

Skynet 4F
Launch: 7 February 2001 by Ariane 44L from Kourou, French Guiana
Location: 16° E
Mass: 1,510 kg at launch, 851 kg on-station BOL, 256 kg payload
Other specifications as Skynet 4A-C.

United States

DSCS II Series

Current Status
Launched 4 September 1989, DSCS-II F15 is still operational despite it having a design life of 7.5 years.

Background
The IDSCS, often referred to as phase I of military defence communication of the defence Satellite Communications System, emphasised the Cold War needs of nuclear hardening and anti-jamming protection. In phase II, building a network made up of 16 satellites, the military emphasised an increased capacity and system flexibility that reflects the importance of the initial system. The satellite system can handle 1,300 voice channels or 100 Mbit/s data. Each satellite carries two steerable narrowbeam antennas to work through small ground terminals. Self-contained and mobile, these small antenna enable US Army units to operate full communications through DSCS within 20 minutes of arriving at a new site. The satellite can redirect its beams to different locations within minutes, and the army can move the satellite within a few days to a new GEO position to meet defence contingencies in any part of the world. Antenna nodding reduces north-south station-keeping, a measure that reduces the propellant requirements and extends useful lifetimes.

DSCS 1 and 2, launched 3 November 1971, ceased operations in 1973 and 1980 respectively, but DSCS 3 and 4, launched 14 December 1973, continued to operate over the west Pacific while 1 and 2 shut down. DSCS 3 was retired in 1982 Number 4 was raised above GEO from 63° west in March 1993 and de-activated 17 December 1993. DSCS 5 and 6, launched 20 May 1975, would have completed world coverage except for polar areas, but they fell into the Pacific after six days, due to a Titan guidance failure. A second set of six satellites offering five-year design lives, began with DSCS 7 and 8, launched 12 May 1977. DSCS 7 was shut down on 20 December 1981 followed by DSCS 8 in February 1990. The DSCS system still did not reach completion as projected when DSCS 9 and 10 were lost 25 March 1978 following Titan stage 2 failure. During this period NATO provided a 'loaner', repayable in similar services when needed, of the use of its own system. DSCS 11 and 12 were launched on 14 December 1978 and both were shut down on 13 December 1993. DSCS 13 and 14 were launched on 21 November 1979, the former shut down on 19 July 1994, the latter on 19 June, 1995.

DSCS-2 16 was launched with the first DSCS-3 satellite on October 30 1982 and number 15 (the last one) with the second DSCS-3 September 4 1989. In March–April 1997, the orbital period of DSCS-2 15 was raised from 1,436 minutes to 1,451 minutes and the satellite began to drift off-station: it is unclear whether the satellite is being moved to a new location or – more likely – being retired. Orbital data for DSCS-2 16 has been classified and therefore its location history is not known. It was retired on 23 May 1997. All except DSCS 16, carried by an IUS, were launched on Transtage upper stages.

A full complement of four operational satellites occurred when the successful launch of DSCS 11 and 12 on 14 December 1978 placed them over the east and west Pacific. All DSCS II series satellites remained operational 10 years later but are no longer in service.

Specifications
Contractors: TRW Space & Defence Group
Transponders: numbers 1–12. 2 × 20 W TWTA 7/8 GHz up/down X-band 410 MHz bandwidth circular polarisation transponders employable as 28 dBW global (18° beamwidth), 32 dBW regional (6.5° beamwidth, 4,100 km diameter coverage) and 43 dBW spot (2.5° beamwidth, 1,600 km diameter coverage). numbers 13–16 increased TWTA to 40 W
Principal applications: 1,300 duplex voice channels or up to 100 Mbit/s data
Configuration: 2.7 m diameter, 3.95 m deployed height, spin-stabilised cylindrical bus with de-spun platform incorporating 2 global coverage horns and two narrow coverage steerable (±10°) dishes
Mass: 522 kg numbers 1–5, 567 kg numbers 7–12, 612 kg numbers 13–16 fuelled
AOCS: spin-stabilised using infra-red Earth horizon sensors and hydrazine thrusters; small solid thrusters at base of drum provided initial spin-up. Inserted into GEO by launcher final stage. NSSK not provided.
TT&C: via base S-band biconical horn antenna with 32° toroidal beamwidth
Power system: eight solar array panels on cylindrical bus provide 535 W BOL and 358 W after five years; 235 W required by communications payload. 3 × 12 Ah Ni/Cd batteries provide full eclipse protection for minimum of five years.

DSCS III Series

Current Status
The DoD launched an improved 1,040 kg DSCS 3 B13 from Cape Canaveral on 25 October 1997. The DoD code-named the satellite USA 135. Currently, F-4, F-5, F-6, F-7, F-8, F-9, F-10 and F-11 (launched 21 January 2000) are operational. The 13th DSCS 3 series (A-3) was launched by Delta 4M on 11 March 2003, the first commercial Delta 4 and the second launched. DSCS III-14 was launched by Delta on 29 August 1993. It was the last satellite in the series.

Background
DSCS Phase III satellites have 10-year lives and are three-axis stabilised. The DoD ordered 14 in total when the programme began in 1975. Their multiple-beam antennas employ advanced encryption to protect their data and 1,300 duplex voice links for both tactical and strategic users. In addition to nuclear hardening, sensors can detect a jamming attempt, report it and wait for ground control to plot the location of the jammer and dispatch instructions on how to use its steering capacity to null the jammer. The nuclear hardening cannot protect against direct attack, but enables it to withstand low-level flash effects, X-rays, gamma rays and electromagnetic pulses (EMPs). The system provided more than 80 per cent of all communications during Operation Desert Storm/Shield in 1990 and 1991.

A complete DSCS III ground terminal can be transported by a C-5 Galaxy. The first operational satellite in this series weighed 1,042 kg. and orbited with II-15 from the same Titan 34D/IUS on 30 October 1982 from Canaveral. The entire Phase II/III system had a network of six satellites, four active and two spares, which operated at 7/8 GHz X-band. They provided secure strategic and tactical voice and data transmission, military command and control and ground mobile communications. In the event of a nuclear war, the National Command Authority, the White House Communications Agency and the Diplomatic Telecommunications Service all had priority usage for the DSCS.

The constellation had seven satellites by end-1985, (the latter ones launched cost US$150 million each). Four DSCS IIIs were required on-station in GEO to provide full coverage; and the first two (B-4/B-5) left aboard the single IUS from STS-51J on 4 October 1985. The Shuttle Atlantis had to fly to an extremely high orbit of 515×475 km, to help the IUS deliver its heavy cargo into GEO. The next DSCS III number 4 (A-2), departed 4 September 1989. The third IIIB created, for the first time, a five-satellite DSCS III network.

The operational system requires four active satellites plus two in-orbit spares, and the army acknowledged that five primary and four back-up satellites were in place by September 1995. The ban on releasing orbital elements for the post-1982 satellites prevents a comprehensive identification of these satellites. The available elements indicated the army still actively controlled one of the last three IIs (II-16 is post-1982) and the first IIIs were still controlled at the beginning of 1996. A tentative identification locates these satellites at 64° east (II-15), 130° west (III-1). Space observers identified II-16 as still operational in September 1995. II-13 retired in mid-1993 and boosted itself above GEO. II-4, II-11 and II-12 left active service in December 1993 and the army took II-14 off-line in mid-1995. One source indicated in 1992 that II-13 and II-14 were severely degraded and barely usable, as was III-1.

The primary DSCS III positions are 12° west Atlantic, 136° west (east Pacific), 175° east (west Pacific), 60° east (Indian Ocean). The army has also registered slots with the ITU for 130° west, 52.5° west, 42.5° west, 57° east, 180° east.

The new US Army Space Command, created in 1988, collocated at Colorado Springs with Space Command/NORAD, accepted full authority for the system in the summer of 1988. Operational control was transferred 11 July 1991 from the Consolidated Space Test Center at Sunnyvale to the 3rd Satellite Control Squadron at the Consolidated Space Operations Center in Colorado (now referred to as the 3rd Space Operations Squadron of Space Command's 50th Space Wing).

The air force planned to award Lockheed Martin a US$100 million to US$125 million contract in March 1996 to upgrade at least four of the five satellites in storage in order to prolong full system operability until 2008. The IIIs, despite a design goal of 10 years, only average a seven year lifetime. Improved thrusters will stretch the station-keeping propellant from 9 to 11½ years, and replacing the Si cells on the two inboard panels with GaAs will combat degradation. The 40 W TWTAs of channels 1/2 will be replaced with either SSPAs or improved TWTAs. The company has already replaced amplifiers with upgraded versions capable of handling 50 per cent more traffic under a US$9.6 million contract in early 1990.

1995's DCSC IIIB B-7 carries STP's CHARGECON-GEO (Satellite Charge Control Experiment at GEO) to test the ability to detect and discharge an electrostatic build-up while still small.

Specifications
Contractors: Lockheed Martin Astro Space
Transponders: 6 × SHF 7.900–8.400/7.250–7.750 GHz up/down channels: 1 40 W 50 MHz; 2 40 W 75 MHz; 3/4 10 W 85 MHz; 5 10 W 60 MHz; 6 10 W 50 MHz. Two (plus two back-up) 40 W TWTAs + four (plus two back-up in 1-for-2) 10 W SSPAs.

Working through: One high gain (30.2 dB min) steerable 3.0° spot dish for channel 4 (EIRP 37.5 dBW) + channels 1/2 (EIRP 44 dBW).

Two 19-beam waveguide lens Multi-Beam Antennas (MBAs) for ch1-4 with beam-forming networks to produce selected antenna patterns accommodating the ground receiver network. EIRP: Earth coverage channels 1/2 29.0 dBW, channels 3/4 23.0 dBW; 1° narrow coverage channel 1/2 40.0 dBW channels 3/4 34.0 dBW. Receive through 61-beam waveguide lens MBA, with associated beam-forming network to provide selective coverage and jamming protection.

Two global horns receive and two transmit for channels 3–6 (EIRP 25.0 dBW)

UHF bow tie 300–400 MHz receive and cross dipole 225–260 MHz SSPA transmit for Single Channel Transponder to support AFSATCOM. AFSATCOM II signals handled at either UHF or SHF (channels 1)

Configuration: 206 cm long body (279 cm across antennas), 193 cm wide, and 196 cm deep to antenna tips. Transponder amplifiers and associated components are mounted on the north panel. Modular structure, 3-axis stabilised

Mass: about 1,170 kg at launch

AOCS: 3-axis, pointing accuracy 0.09° pitch/roll, 1.0° yaw, 16 × 4.45 N thrusters, 276 kg capacity hydrazine system utilising four spheres. Inserted into near-GEO by IUS stage 2 (III on Shuttle) or IABS (IIIB)

TT&C: via S-band (1,807.764 and 1,823.779 up/ 2,257.5 and 2,277.5 MHz down) and SHF (channels 1/5 up/7,600.0 and 7,604.70588 MHz down). The SHF link's primary function is to control channel and antenna configuration in real time through the defence Information Systems Agency's SCCE Satellite Configuration Control Elements

Power system: 2-panel Si dual solar wings spanning 11.62 m, totalling 11.72 m^2, providing 1,240 W BOL, 980 W after 10 years, supported by 3 × 35 Ah Ni/Cd batteries. 190 W provided by stowed array during spin transfer phase.

DSCS Phase III: these models are the primary user of the Atlas 2 launcher. At bottom is the large 61-beam receive array; the 19-beam transmit antennas are at top, below the single steerable spot. The two UHF antennas are central. At left are the global horns (Lockheed Martin Astro Space) 0516831

Lockheed Martin produces the third-generation Defense Satellite Communications System (DSCS III). It features high-data rate and around the globe, secure communications. Each satellite has six independent SHF transponders with a 500 MHz bandwidth (Lockheed Martin) 0022567

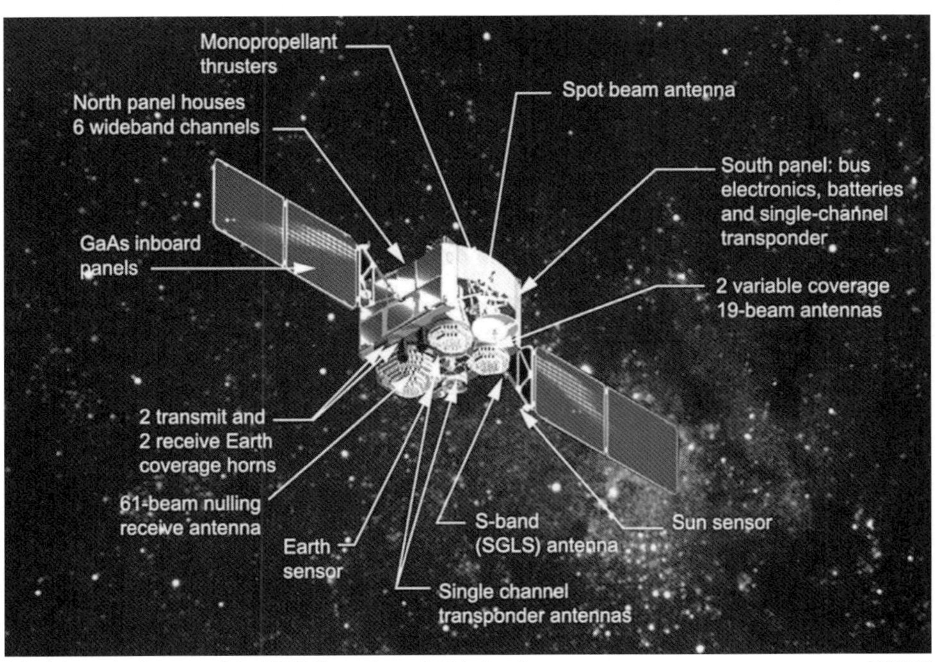

All of the major features of the DSCS III are shown in this drawing 0022568

FltSatCom/AFSATCOM Series

Current Status

All FltSatCom satellites have now been retired but four of these were required to remain in service for twice the duration of their design lives. FSC-7 and FSC-8 carried experimental EHF 20 W transponders to evaluate new ground terminals.

Background

When it was the backbone of navy communications, FltSatCom handled 90 per cent of the USN's communications and provided two-way UHF links for the Strategic Command and the US Army/ Marine components of the Rapid Deployment Force. The navy placed its satellites at 110° west, 23° west, 72.5° east and 172° east, although the service has also recently used 100° west, 23° west, 15° west and 171° east. Declared operational 1 January 1981. The four operational satellites dropped out of service between 1997 and 2000 as the last of the onboard propellant is exhausted, but they will still exceed their planned seven-year lifetimes by a good measure. The NAVSOC Naval Satellite Operations Center at Point Mugu in California operates the system after taking over from the USAF June 1996.

During its operational lifetime, FltSatCom provided high-priority UHF communications between the US President and commanders in the field. The system began full operations in January 1981, providing global links for command and control of US nuclear forces. It handled between 900 ships, submarines and aircraft of the navy and related fleet ground stations, while also supplementing communication between 1,000 USAF aircraft and air-to-ground terminals for Strategic Command and other air force operations. The coverage extended into the high latitudes and only the polar regions remained without service.

Comprehensive coverage up to 70° latitude required four satellites and generally five were in orbit because the navy scheduled one in-orbit spare. By 1988, five of the seven placed in orbit up to that time still remained operational and Fl had passed its tenth anniversary in space. F8 completed the series on 25 September 1989. FltSatCom experienced a spate of launch failures between 1987 and 1988 when on 27 March 1987 an Atlas Centaur carrying FltSatCom F6 suffered a lightning strike and had to be destroyed by the range safety officer 51 seconds after lift-off from Canaveral.

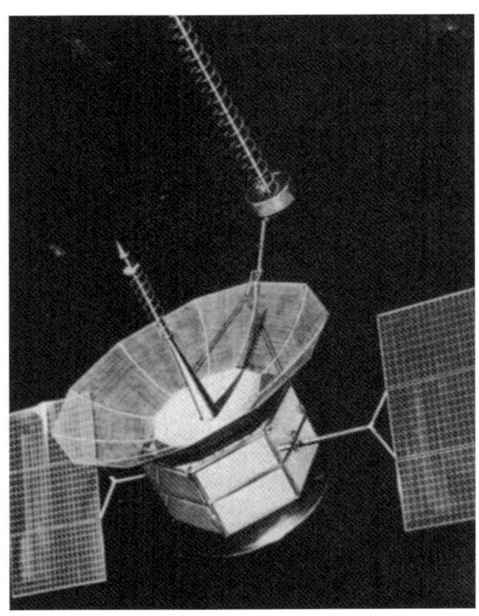

FltSatComs provide global communications for command and control of US nuclear forces. They will remain operational into the late 1990s (TRW) 0516832

FSC has 1,005 kg on-station mass and 13 to 19 m span with solar wings extended. It carries enough propellant to remain operational for 10 years. Two redundant networks of eight (2 roll, 2 pitch, 4 orthogonal) thrusters maintain the satellite position and attitude. FltSatCom's solar panels generate more than 1,400 W. The antennae consist of a deployable mesh dish 4.9 m in diameter with a 3.4 m long helical receive antenna. Each provides nine 25 kHz bandwidth channels and 12 × 5 kHz channels for small mobile users, plus a 25 kHz fleet broadcast channel and a 500 kHz National Command Authority channel.

Two onboard FltSatCom processors, one each for the air force and the navy, protect against uplink UHF/SHF signal jamming. The older satellites provide 75 bit/s UHF down with limited anti-jamming capability. AFSATCOM-II provides 150 bit/s at UHF/SHF. The system would handle nuclear weapon release authorisation codes, for example. Air force Space Command assumed AFSATCOM responsibility 1 June 1992 from Strategic Command at Offutt AFB, Nebraska. During 1990, the Single Channel Transponder on DSCS III entered AFSATCOM service.

Specifications

FSC-1

Launched: 9 February 1978, took up a position at 100° west over Midway Island for coverage across the US to the Azores and Atlantic. It was moved in October 1992 over a 25-day period to provide more capacity over the Atlantic at 15° west near FSC-8, which is inclined at 12.5°. The navy manoeuvred FltSatCom-1 off-station in December 1996 over approximately 344–345° E and the satellite had its location restabilised over 71–72° E at the end of January 1997.

FSC-2

Launched: 4 May 1979, filled the slot at 71.5° east to cover the Indian Ocean from Africa to the Philippines. By 1992, it was clearly out of service, and drifting rapidly.

FSC-3

Launched: 17 January 1980 into 23° west, covered the Atlantic and Mediterranean; it has been drifting slowly since 1991.

FSC-4

Launched: 30 October 1980 to 172° east over the Pacific, provided coverage from East Asia to the US west coast. It remained there in an inclination of 10.6°.

FSC-5

Cargo/Manned Unit

Launched: 6 August 1981, had been scheduled for 93° west near FSC-1, the busiest region, to act as an instantly available spare, but was damaged by implosion of the Centaur's payload fairing during launch.

FSC-6

Cargo/Manned Unit

Launched: 27 March 1987 and disappeared when lightning struck its Atlas Centaur. The satellite was reported to be worth US$83 million, and the launcher US$78 million.

FSC-7

Cargo/Manned Unit

Launched: 4 December 1986 by Atlas Centaur from Canaveral after eight delays from the original 22 May date. The more advanced FSC-7, equipped with the jam-resistant EHF, actually preceded FSC-6 when the latter needed to have suspect components replaced. It was positioned at 105° west and later moved to 100° west January 1996, in an inclined orbit of 2.55'.

FSC-8

Cargo/Manned Unit

Launch scheduled: 24 July 1987, got bumped on 13 July when its Centaur upper stage, the last of its type, was irreparably punctured by a pad work platform. The mission departed on 25 September 1989 instead. By January 1996, it held the 23° west slot, inclined at 2.0°.

IDCSP Series

Background

Lack of launcher power, together with political and economic arguments, delayed the first deployment of a robust communication satellite architecture for the US military until 16 June 1966, when a Titan 3C deployed eight satellites, including the first seven of the Initial Defense Satellite Communications System (IDSCS). Built by Ford Aerospace, these 45 kg, 26-sided 86 cm diameter polygons served as signal repeaters. When the Titan released its payload, the eight individual satellites dispensed themselves over six hours at slightly different orbital velocities to provide global coverage at points just below GEO altitude, at 33,915 km. Because they were not perfectly geosynchronous, the satellites drifted about 30 minutes relative to Earth daily, but stayed in view of an equatorial station for four and a half days so that even when one malfunctioned another drifted over the horizon to take up the traffic. The satellites maintained attitude control by spin-stabilisation. Over the years the network of IDSCS satellites slowly grew to total 26 satellites on 13 June 1968. IDSCS transmitted reconnaissance photos and other data during the Vietnam war. Succeeded by NATO and DSCS, truly geosynchronous.

The system provided five high-quality voice, 1,550 Teletype or 11 tactical voice channels using 20 MHz bandwidth in the 7.266–7.286 GHz X-band. They linked users 16,000 km apart and, from 1967, provided the military with direct communications between South Vietnam and Hawaii, and Hawaii to Washington. The military intelligence services in Vietnam transmitted high quality reconnaissance photographs. Designed to last 18 months, five were still operational after eight years and two after 10 years. New studies of cost-effective communication satellites indicate IDSCS-type satellites may appear again since it has been estimated that 200 could be launched at the same cost as a single Milstar.

Leasats supplement the FltSatCom network 0516833

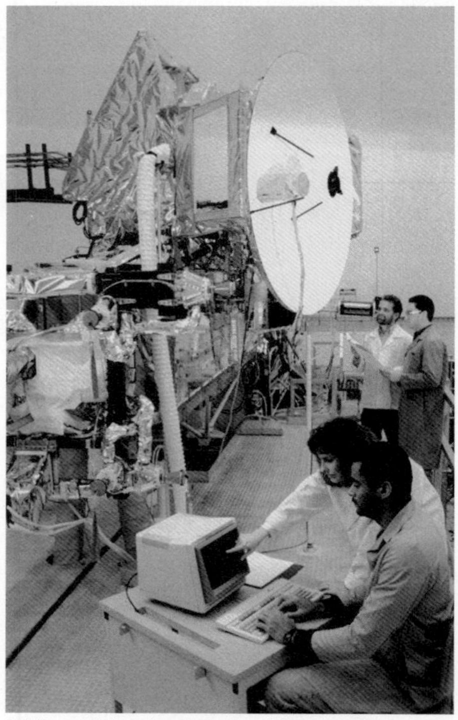

TRW technicians integrate the LDR payload on to DFS 2 in Lockheed Martin's facilities in Sunnyvale. Prominent is the 60 GHz inter-satellite antenna at top (TRW) 0517308

Leasat/Syncom 4 Series

Current Status

Only Leasat 5, launched 9 January 1990, is currently operational, 4 years beyond its design life.

Background

Also known as Syncom IV, Leasats took full advantage of the Shuttle's wide payload bay, the first satellites to require the full capacity of the Shuttle. A Leasat launch uses the unique 'Frisbee' or rollout method. The satellite has a built-in upper stage that remains in place on station. Each satellite sits 6.1 m high with deployed antennas and has a 4.2 m diameter. Two large helical UHF antennas provide receive and transmit signals at 240 to 400 MHz. The principal Fleet Broadcast function includes an SHF uplink and SHF/UHF downlinks.

Owned and operated by Hughes Communications Inc, the system has five HS-381 satellites (which included one ground spare) which DoD leases for worldwide UHF links between ships, aircraft and fixed facilities. DoD and the navy specified that US$83.75 million would be paid annually for each satellite once it was operational in its GEO position – but not before. This clause served the US Navy well when Leasat 3 became stranded in the wrong orbit for eight months and Leasat 4 failed completely after attaining GEO.

Leasat's original design kept it unpowered in the payload bay. Unlike other Shuttle-launched satellites, astronauts could not perform a final electrical checkout before release. As they left the bay, an arming lever automatically initiated a timing sequence which first raised the 2.4 m omni antenna after 80 seconds, then at 6.5 minutes commanded 22 N thrusters to increase the spin rate to 2 to 33 rpm. At 45 minutes the sequencer fired the 131 kN Minuteman 3 stage 3 solid motor for 64 seconds to raise apogee to 15,450 km. Three additional firings by two 448 N liquid thrusters then took them into GEO. After the failure of Leasat 3 however, NASA required a redesign of this automatic arming feature.

Leasats occupied four operational slots at 177° west, 105° W, 15° W and 75° E. The ground segment operates from Hughes Operations Control Center in Los Angeles working through 10 m antennas in Italy (NATO base at Lago Patira, near Naples), California (Stockton), Hawaii (Wahiawa), Guam (Finegayan) and Norfolk, Virginia. Guam and Norfolk also have transportable stations originally used during Leasat deployment but now held as back-ups. Although controlled from Hughes in Los Angeles, each station can assume control of the whole fleet if required. Operations are co-ordinated via dedicated leased terrestrial lines to the US Naval Space Command Operations Center in Dahlgren, Virginia.

Leasat Launch History

Satellite	Launch Date	Vehicle	Position	Comments
Leasat 1	8 November 1984	STS-51A	105° W /15° W	In October 1985, the only fully operational Leasat. Moved to 33,093 × 39,532 km early 1993.
Leasat 2	30 August 1984	STS-41D	177° W	Operational until September 1985, when its wideband channel began to malfunction
Leasat 3	12 April 1985	STS-51D	107° W	Automatic timing sequence failed to activate after deployment. Hughes claimed US$85 million insurance and paid NASA US$8.5 million for a repair operation. NASA left the satellite in low orbit until 27 October to warm up the solid motor.
Leasat 4	29 August 1985	STS-51L	177° W	The primary communications system failed eight days after deployment. Hughes claimed the US$85 million insurance.
Leasat 5	9 January 1990	STS-32	72° E	

LES Series

Current Status
Powered by RTGS, both satellites LES 8 and LES 9 remain usable. LES 9 provided links with a polar exploration team in 1995 while LES 9's AFSATCOM transponder carried Gulf War communications. LES 8 went out of operation in 1998 but LES 9 remains operational.

Background
Tactical communication satellites require much higher power levels than conventional satellites because they must work through small ground terminals. Lincoln Laboratories acted as the prime contractor for the US military to study methods to bring down the costs of these communication satellites. Launched on 11 February 1965, LES 1 began the series with a Titan 3A. LES 2 on 6 May 1965, also on a Titan 3A, was propelled to a 2,771 × 14,810 km orbit while LES 4 on 21 December 1965 and LES 3 on the same launcher, were lifted by Titan 3C.

LES 5 (Lincoln Experiment Satellite) orbited 1 July 1967 with a six-satellite payload, which included IDSCS 16/17/18. It employed a 225–400 MHz military UHF band. Two days after manoeuvring into a 33,360 km near-synchronous orbit, the satellite conducted the first communications between US aircraft, a US Navy submarine and surface vessel and army ground units.

LES 6 continued these investigations after September 1968. With the success of the two LES satellites the US DoD launched the follow-on Tacsat 1 (LES 7) on 9 February 1969. A 7.6 m tall, 2.3 m diameter satellite weighing 726 kg, it maintained its attitude with spin stabilisation on a cylinder bearing a despun platform. It targeted 30 cm diameter UHF/X-band land-based antennas using about 40 coded UHF voice channels for mobile antennas; and in X-band either the same or up to 700 teletype channels which it transmitted to 1 m diameter dishes.

LES 8 and 9 launched together on 15 March 1976, took up positions in 25° GEO-altitude orbits to conduct inter-satellite K-band links at ranges up to thousands of km, plus UHF downlinks. LES 8/9 appear to be stationed over 103° and 105° W respectively now inclined at 15.5°.

MACSAT Series

Background
MACSAT, the Multiple Access Communications Satellites, M1 and M2, built by Defence Systems Inc, demonstrated tactical UHF data store/forward techniques. NAVSOC operated the satellites from its operations centre at Naval Satellite Operations Center Pt Mugu, California until the satellites failed in 1994.

A single Scout carried the two 68 kg satellites into space on 9 May 1990 to a 613 × 769 km, 89.90°. 98.5 min, and a 614 × 767 km, 89.9°, 98.5 min, orbit respectively. DARPA provided the initial seed money, to build the satellites, through its 'lightsat programme'. Each satellite has a diameter of 61 cm, and sits 35.6 cm high. MACSAT incorporated two 80C86 processors, each with 1.2 Mbyte RAM providing 8.64 kbyte mailboxes in eight, 192 kbyte increments. A 64 kbyte (32 page) document could be relayed at 2.4 kbit/s from a given footprint during one pass. Two digitally tuneable 10 W transmitters, a 65 W high-power auxiliary receiver. 54 blocks of solar cells generated 14 W orbital average power to supply the one 10 W and one 60 W transmitters. The design called for the MACSAT to maintain stabilisation by deploying two gravity gradient booms. Radar images indicated on boom remained deployed.

Originally designed for demonstration purposes only, the DoD pressed M2 into service by US Persian Gulf forces from 20 August 1990 until 5 April 1991. MACSAT also provided links between US Navy personnel in Antarctica, New Zealand and California. RADM John Weaver, Commander of USN Space & Naval Warfare Systems Command noted, during 1990, that USN plans included several (described in 1991 as a network of six) polar UHF 'Arcticsat' relays based on MACSAT. Apart from improving high-latitude communications, they would be available for rapid launch and replacement in times of need. Arcticsat failed to acquire the US$37 million requested for FY92 and was cancelled from FY93.

The NAVSOC Naval Satellite Operations Center at Point Mugu in California operated them until their failures in 1994.

Milstar Series

Current Status
The first Milstar 2 was launched by Titan-4B/Centaur on 30 April 1999 from Cape Canaveral. Designated DFS 3 (or USA 143 in the official announcement of launch) Milstar 2 has an added medium data rate data transmission system. A failure of the Centaur upperstage to fire properly left the satellite stranded in a useless orbit of 1,097 × 5,149 km, 20.2° inclination orbit instead of the geostationary orbit planned. Fault was traced to an incorrect mathematical constant that steered the Centaur off course shortly after the first of three planned burns.

The second Milstar 2 was launched by Titan 4B on 27 February 2001, the first with higher data rates and more secure anti-jamming capability. The third Milstar 2 was launched 16 January 2002.

Background
The troubled Milstar programme, providing secure tactical links via hardened satellites, is now operational. The first of six Milstar 1 satellites left for orbit on 7 February 1994 aboard the first Titan 4 Centaur. At first DFS 1 (Development Flight Satellite) took up a position over the 90° west test slot until the air force moved it about September 1995 to 120° west. DFS2 was launched 6 November 1995 for testing at 90° west, before the air force put it in the planned operational slot at 4° east, inclined at 4°.

The air force initiated the Milstar programme in 1983 as an all-service, jam-resistant and survivable system to become operational in the second half of that decade. The 4.67-tonne satellites offer tactical communications via manpacks that access the 44.5 GHz uplink and 20.7 GHz downlink frequencies. The EHF band ensures communications continue during nuclear conflicts and for the first time provides interservice links. Frequency-hopping techniques eliminate jamming to some degree and EHF inter-satellite transmissions permit users to communicate around the globe without routeing via intermediate ground stations.

Each satellite has a 0.5 Mbit/s total capacity, which is limited because of the nuclear countermeasures, spread over 192 channels at 752,400 bit/s, with onboard processing and five agile antennas covering 185 service areas. The first two satellites, Milstar 1A, carry only this LDR Low Data Rate payload. Tactical forces will carry 1,050 terminals, strategic forces have 279 and the USAF EC-135 aircraft has 138 installed terminals. The services had built some 200 army, USN and air force terminals by early 1996.

DoD notified the International Telecommunications Union it would use 148° west, 120° west, 90° west, 68° west, 16° west and 9° west. In the eastern hemisphere it would use 14° east, 19° east, 30° east, 55° east, 90° east, 133° east, 150° east, 152° east, 177.5° east.

Congress ordered a restructuring of DoD's Military Strategic & Tactical Relay programme in 1990 to reduce cost and refocus the satellite communication needs from strategic applications during and after nuclear wars to tactical uses. The 4th Space Operations Squadron (4th SOPS) of Space Command's 50th Space Wing, operates Milstar after commissioning by 3rd SOPS. The Congress announced that the total cost between 1991 and 2011 for development, launch, 20 years of operations and user terminals is projected at $17 billion. Each Milstar 2 unit costs $800 million. The helical antennas provide US Navy UHF links (larger for receive, smaller transmit) and the two dishes are the source of the EHF/SHF spot beams.

At one time, provisions on Milstar allowed it to relay encrypted KH11 imagery, but that apparently has been dropped. Milstar's space testing phase began with the launch of FltSatCom 7 in December 1986 which carried its highly jam-resistant 44 GHz EHF transmitter. The US headquarters in Riyadh, Saudi Arabia used a prototype Milstar terminal during 1991's Desert Storm to provide a demonstration link with the Pentagon.

Milstar is the first system to allow all three armed services to communicate with each other on

Milstar 1 LDR payload final test/integration at TRW during 1989. The helical antennas provide US Navy UHF links (larger for receive, smaller transmit) and the two dishes provide EHF/SHF spot beams. The smaller arrays provide agile EHF links; missing at centre is the SHF agile unit. Milstar's other payload wing, originally planned for the Science Prime package to relay KH11 imagery, will carry the MDR package from number 3
0516834

The USAF established the Milstar programme in 1983 to provide military communication before and after a massive nuclear exchange. Each satellite carries 13 assorted antennas. The US Air Force paid a reported US$1.5 billion for the first Milstar. A total constellation requires at least four satellites (TRW) 0022565

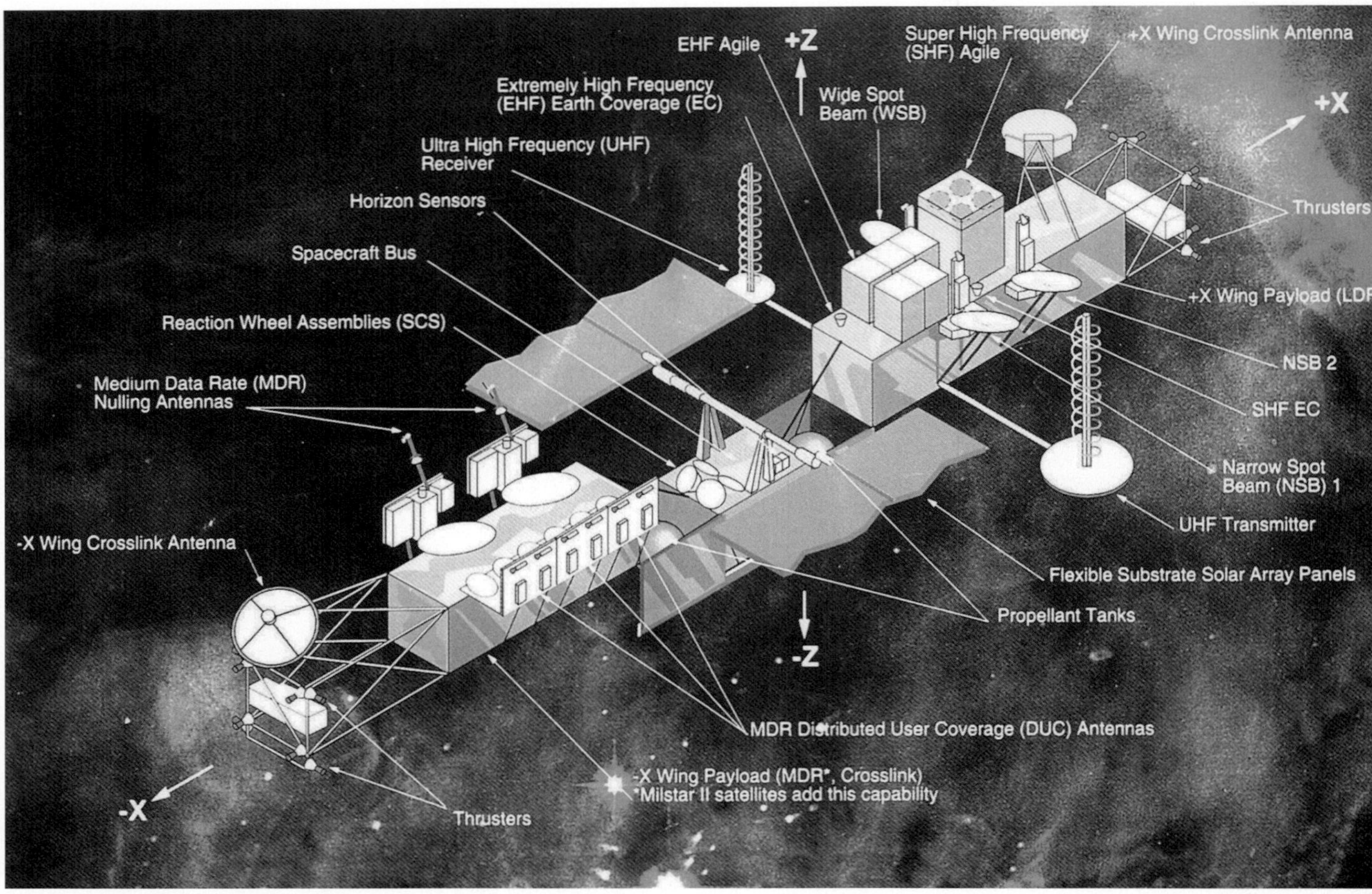

A drawing showing the major features of the Milstar satellite (Lockheed Martin) 0022566

the same network. The extensive onboard signal processing and resource management allows it to automatically set up, maintain and reconfigure voice/data networks in real time. It is the first payload to operate at EHF using unique frequency hopping to avoid interception and interference. Crosslinking between satellites eliminates the need for costly, politically sensitive and vulnerable ground stations.

In his testimony in 1988, Air Force Secretary Edward Aldridge admitted that technical problems had raised the cost for each satellite and its Titan Centaur launch from Cape Canaveral to US$1 billion, but in 1992, other air force sources quoted a price of US$515 million per satellite. As it debated a funding bill in 1990, the Senate Armed Services committee estimated that the entire system would cost US$35–40 billion by the time it was complete. The Congress gave the air force US$345 million in FY90, with US$956 million in FY91, US$1.6 billion FY92, US$1.14 billion FY93, US$918 million FY94 and US$647 million FY95. Milstar's redirection reduced the number of satellites from 10 to four and cut the numbers of ground stations and user terminals.

Specifications

EHF/SHF Services: EHF 44.5 GHz (2 GHz bandwidth) up/SHF 20.7 GHz (1 GHz bandwidth) down
Data rates: LDR 75–2,400 bit/s; MDR 4.8–1,544 kbit/s
Channels: LDR 192 (100 at 2,400 bit/s); MDR 32
Antenna coverage: LDR 1 up/down (Earth), 5 up/1 down agile, 2 up/down narrow spot, 1 up/down wide spot; MDR 2 up nulling spots (allowing operation in presence of in-beam jammers), 2 coincidental down spots, 6 up/down spots (distributed user coverage)
UHF Services: 4 transmit/rx 75 bit/s AFSATCOM IIR channels (Earth coverage) and 1 transmit-only 1.2 kbit/s Fleet Broadcast channel (Earth coverage)
Crosslinks: (60 GHz) 2 per satellite (1 each direction), compatible with LDR/MDR

Polar Bear Satellite

Background

The Polar Beacon Experiment and Auroral Research satellite was a joint USAF/defence Nuclear Agency mission. The satellite was built up from a Transit navigational satellite retrieved from the Smithsonian National Air and Space Museum. The air force launched the 122 kg craft by Scout G on 13 November 1986 from Vandenberg into a 962 × 1,019 km at 89.5° inclination with a period of 104.9 min. It was also known as STP P87-1, and the air force intended to use it to study communications interference caused by solar flares and increased auroral activity. DAN contributed to the effort in order to estimate the effects of nuclear bursts on communication signals. Gravity gradient stabilisation by deployable boom with tip masses. Design life 3 years. The satellite had a length of 1.3 m, a diameter of 0.4 m and a mass of 122 kg. The USAF recycled an existing Transit from the navy programme and saved US$2 million on the cost of the US$13 million satellite. The satellite had originally been displayed from 1976 to 1984 at the National Air and Space Museum, Washington DC. Johns Hopkins University discovered that only one solar cell out of thousands was marginal and required replacement. The Auroral Imaging Remote Sensor, sponsored by the Air Force Geophysics Laboratory, imaged the aurora borealis; the Beacon Experiment, sponsored by the DNA, monitored ionosphere propagation over the poles. The satellite lost most of its battery capacity in early October 1992, leaving insufficient power for the remaining SIDEX Signal Identification Experiment. The air force deactivated Polar Bear 2 November 1992. Data provided by the satellite was complementary to the mission's predecessor, HILAT.

UHF Follow-On Series

Current Status

Boeing Satellite Systems has been awarded a contract to build UFO F-11, for launch in 2003.

Spacecraft	Blk	Weight (lbs)*	Power (w)**	L (ft)	W (ft)	Payload
F1-F3	I	2,600	2,500	60	23	UHF/SHF
F4-F7	II	3,000	2,800	60	23	UHF/SHF/EHF
F8-F10	III	3,400	3,800	86	22	UHF/EHF/GBS

*Begining of life **End of life

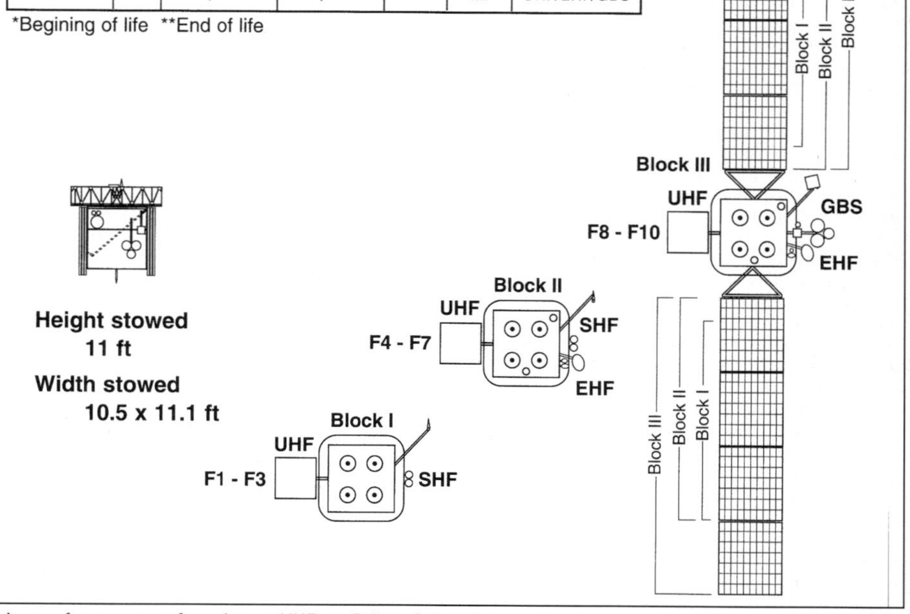

Major features of the UHF Follow-On satellite are shown in this diagram (Hughes Space and Communications Division) 0022570

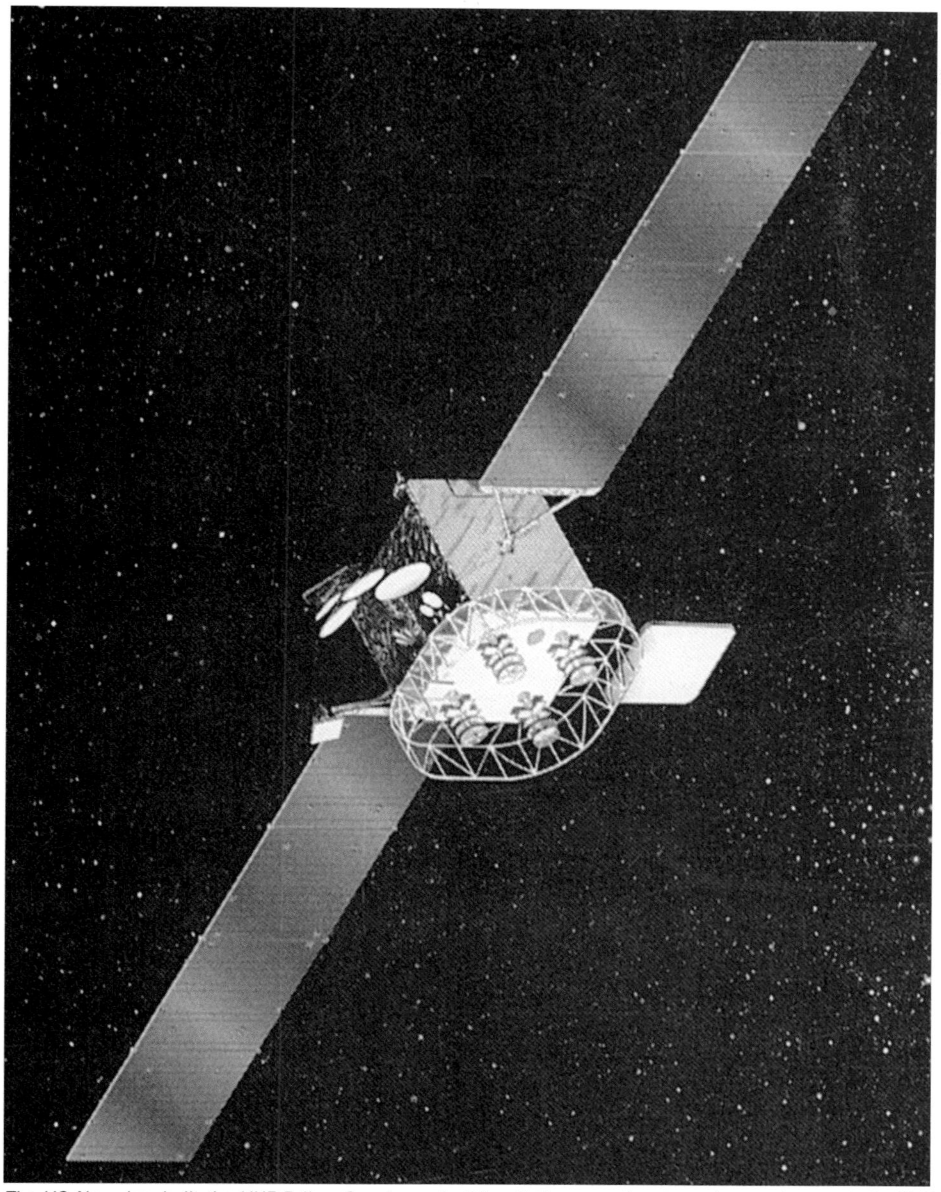

The US Navy has built the UHF Follow-On, shown in this artist's conception, to replace and upgrade its UHF (Ultra High Frequency) satellite communications network. Navy communications needs have prompted it to order 10 satellites in the series through January 1994 (Hughes Space and Communications Division) 0022569

FO-10: Launched 23 November 1999 by Atlas-2A as UFO-10 to geosynchronous orbit (incln: 6°) stationed at 170° west.

FO-9: Launched 20 October 1998 by Atlas-2A as UFO 9 to geosynchronous location at 186° E.

FO-8: The navy had an 80-minute launch window on 16 March 1998 to put the UFO FO-8 satellite in orbit. Hughes contracted with Lockheed Martin for launch on board an Atlas II rocket. The satellite is a commercial-based HS 601 spacecraft and is the first in the fleet to carry Global Broadcast Service (GBS). GBS uses high-power satellite transponders to provide high-speed wideband, simplex broadcast signals. The GBS payload utilises four 130 W, 24 Mbit/s military Ka-band transponders operating in the 30/20 GHz frequency. When the navy's third GBS satellite, UHF FO, is launched in 1999, the DoD will have near-global GBS coverage. UBF FS, F9, and F10 will carry the GBS payload, in addition to the existing IW and EHF payloads. UFO FO-8 will be stationed at 172° east longitude over the Pacific Ocean. The UBF Follow-On satellites have replaced the Fleet Satellite Communications (FltSatCom) and the Hughes-built Leasat spacecraft. In March 1996, the navy ordered the special GBS payloads for FS, F9, and F10, bringing the total contract value to US$1.85 billion.

The US Navy has contracted with Hughes Space and Communications Company to add GBS capabilities to the UHF Follow-On (UHF F/0) satellites 8, 9, and 10, the last three in a series it is building under a total US$1.9 billion contract. Hughes derives the space segment GBS package from its commercial system. The GBS payload will replace the current SHF X-band payload with four 130 W, 24 Mbit/s military Ka-band (30/20 GHz) transponders. This modification results in a 96 Mbit/s capability per satellite.

Data uplinks to the satellite pass through a fixed receive antenna from a broadcast management centre and a steerable receive antenna from theatre injection points. Each of the four transponders can be accessed through either of the receive paths, configured by ground command. Data is transmitted on three steerable spot beam antennas per spacecraft into 22 in receive antennas. Each of two spot beams covers an area of 500 n miles in diameter at the sub-satellite point. The third downlink spot beam covers an area of 2,000 n miles in diameter at the sub-satellite point.

To integrate the GBS payload requires several spacecraft modifications. The payload's high-power amplifiers are integrated on to a larger south radiator panel with heat pipes added to accommodate the increased thermal dissipation. The GBS fixed receive antenna together with the forward SGLS omni antenna will be mounted on the structure supporting the current SHF antennas. The GBS steerable receive antenna will be mounted on a deployable boom. A new pallet structure will be added to integrate the three GBS transmit spot beam antennas. The power subsystem requires modification to provide additional capacity to support the high-power GBS payload.

More efficient Gallium Arsenide (Ga As) solar cells will replace silicon-type cells on four solar wing panels. The battery grows in capacity to support eclipse operations. All the spacecraft bus changes can be made with commercially proven space hardware.

Background

UFO offers double the communications capacity over FltSatCom for about the same amount of weight in orbit. An EHF package provides 11 channels on some satellites and is distributed between a 5° spot beam and global coverage.

The first of FltSatCom's UFO follow-ons, with 14-year lifetimes, departed March 1993, but its Atlas 1 launcher left it in an unusable low orbit. UFO 2 departed successfully in September 1993 and the USN formally accepted command authority on 2 December. The satellite occupies a position over the Indian Ocean, collocated with Leasat 5. Two more UFOs departed successfully by March 1995 and a further six were ordered.

Block 1 (F1-F3) weigh 1,180 kg and produce 2.5 kW of electrical power, block 2 (F4-F7) weigh 1,270 kg and produce 2.8 kW, block 3 (F8-F10) weigh 1,720 kg and produce 3.8 kW. The sole block 4 (F11) is a block 2 configuration.

UFO Satellites numbers 1 through 3 cost US$138 million and each of the successors will cost US$198 million. Hughes builds the initial 1 tonne satellites, using its HS-601 three-axis design. They have 21 narrowband (5 kHz), 17 relay (25 kHz) and one fleet broadcast (25 kHz) channel working through 11 SSPAs using TDMA. UFO features enhanced anti-jamming and EMP hardening. It can function for up to 30 days without ground contact of any kind. The 3rd Space Operations Squadron of the 50th Space Wing has command of the satellites for the Naval Space Command.

The US government made the contract require a first launch before 29 July 1992 and the military would levy fines against Hughes up to US$80,000 daily for late launch, up to a maximum of US$16 million. The USN had a problem with a computer it planned to supply to the project and the first departed March 1993 but was left in an unusable low orbit by its Atlas 1. Hughes still had to pay the navy a US$137 million penalty.

The initial 1 tonne satellite, based on Hughes' HS-601 three-axis design, previously selected only for Australia's Aussat 2, replaces ageing Leasat/FltSatCom families with 21 narrow band (5 kHz), 17 relay (25 kHz) and one fleet broadcast (25 kHz) channel working through 11 SSPAs using TDMA.

UFO Satellite Launch History

Satellite	Launch Date	Position	Comments
UFO 1	March 1993		In an unusable low orbit because of an Atlas 1 failure. Hughes was required to pay the navy a US$137 million penalty.
UFO 2	3 September 1993	174° west/ 71.5° east	Navy and Hughes originally tested it at 174° west. Hughes formally handed it over 2 December 1993.
UFO 3	24 June 1994	15° west	
UFO 4	29 January 1995	161° west / 177° west	
UFO 5	31 May 1995	17.1° east / 71.5° east	
UFO 6	22 October 1995	106° west	Inclination of 4 to 5°.
UFO 7	25 July 1996	337° east	
UFO 8	16 March 1998	188° east	
UFO 9	20 October 1998	186° east	
UFO 10	23 November 1999	170° west	

DATA RELAY

Russian Federation

Potok Series

Current Status
As of 1999, a single satellite occupies each of the formerly occupied Potok slots at geosynchronous, Kosmos 2319 over 80° E and Kosmos 2291 over 346.5° E.

Background
Izvestia noted in a 5 February 1992 story that the Russian armed forces employ about 10 satellites in geosynchronous orbits, but these satellites utilise only 30 per cent of their available 1,500 telephone and 1,000 telegraph channels. The remainder lie idle, except when the Russian military conducts exercises. Then the network usage swells to 60 per cent for 60 days annually. Izvestia suggested the military turn this capacity over to more productive, non-military applications.

Within this network the Russian military has registered three geosynchronous orbit locations for the Potok military data relay system. Potok uses NPO PM's Geizer satellites launched within the Kosmos programme. Potok 1 resides at 346.5° E, Potok 2 at 80° E, and Potok 3 at 192° E, all potential useful orbits for civilian communication systems or data relay networks. The Russians have never occupied the third location Each Potok employs C-band transponders at frequencies up and down of 4.40–4.68 and 3.95–4.00 GHz. NPO ELAS adopted Potok's phased-array antenna system for the commercial Kupon satellite, but the Russians have thus far refused to publish an illustration of the satellite.

Intelligence sources report that the Geizer satellites permit communications with the fifth-generation photo-reconnaissance satellites using a ground-space-space link to provide near real-time imagery. Launch dates correspond to this theory since the first Geizer satellite went into space in May 1982 and the first fifth-generation launch came in December of that year. Amateur satellite observers have been puzzled for years by the lack of transmissions from fifth-generation photo-reconnaissance payloads and an uplinked transmission to geosynchronous would explain the absences of direct links to a ground station. In this vein, speculation also holds that the Geizer satellites may perform the same function for continuous communications with the EORSATs.

The first Potok (Cosmos 1366) was launched by Proton 8K82K on 18 May 1982 and stationed at 80° E and the second (Cosmos 1540) was launched by Proton on 2 March 1984 also to 80° E. Three more followed in 1986, 1987 and 1988 and a sixth was launched in 1990 with a seventh in 1991. Two more followed in 1994 and 1995 with the most recent in July 2000.

The relay network has no fixed boundary in space. For example, from March 1984 to November 1987, both Kosmos 1366 and Kosmos 1540 operated at 80° east. When Kosmos 1366 died, Kosmos 1888 replaced it and operated with Kosmos 1540 from October 1987 to December 1988 when Kosmos 1540 died. In July 1990, Kosmos 2085 replaced Kosmos 1888 and, from April 1992 to mid-1993, Kosmos 1961 operated with Kosmos 2085 until the former satellite died. During September 1994 and continuing until February 1995, Kosmos 2085 operated in tandem with Kosmos 2291, until Kosmos 2085 died. Finally in September 1995, Kosmos 2319 replaced Kosmos 2291.

A similar pattern occurred at the other Potok slot, 346.5° E. From August 1988 to August 1989, Kosmos 1738 operated with Kosmos 1961 until the older satellite died. From September 1990 to March 1992, Kosmos 1888 and Kosmos 1961 operated together and during December 1991 to March 1992, Kosmos 2172 joined the two older satellites, so that for a few months there were three satellites operating at this position. Kosmos 1888 and Kosmos 2172 continued to operate together, after the departure of Kosmos 1961 (which manoeuvred to 80° E), until December 1994 when Kosmos 1888 died. And fromNovember 1995 to March 1996, Kosmos 2172 and Kosmos 2291 operated together until Kosmos 2172 died.

United States

Glomr Series

Background
Defence System Incorporated built the 68 kg Global Low Orbiting Message Relay satellite, a 62-sided polyhedron, designed for the Defence Advanced Research Projects Agency. The satellite launched itself on a spring ejection from a Getaway Special canister in the payload bay of STS-61A on 30 October 1985. It was the second launch attempt for the satellite after a failure to deploy it on STS-51B in April 1985 because of a battery problem. Glomr had a one-year life in a 317 × 332 km, 57° orbit and DARPA wanted to use it to demonstrate techniques for locating, commanding and relaying data from oceanographic and other sensors such as submarine detectors dispersed on Arctic pack ice. A follow-up satellite was launched aboard the first Pegasus vehicle on 5 April 1990 to a 477 × 642 km, 94.1° orbit.

SDS Series

Current Status
Only SDS 8 remains operational. Launched on 2 July 1996 it had a design life of 7–10 years.

Background
The Satellite Data System craft operate in highly elliptical orbits similar to those of the Russian Molniya, relaying imagery from KH 11 and other reconnaissance satellites, to ground stations in the United States. They also provide AFSATCOM communications links with US nuclear and other forces in the polar regions that have spotty coverage from GEO satellites. The indistinguishable Jumpseat versions act as communications intelligence satellites. DoD refuses to release any details about the programme or its associations with the KH-11 reconnaissance satellites.

Trade reports in 1994 claimed that the 'Heritage' sensor being proposed for DSP's follow-on has long flown aboard the Jumpseat and SDS to augment the DSP system. Each satellite weighs about 700 kg. Hughes builds them at a cost to the DoD of about $100 million each. Until the Shuttle became available the satellites had to be launched by Titan 3B Agena D from Vandenberg into 63.4° inclination orbits. In association with each new reconnaissance satellite launches of SDS 1 series are SDS-1 2 June 1976, SDS-2 6 August 1976, SDS-3 5 August 1978, SDS-4 13 December 1980. SDS 2 series began with the launch on 28 August 1984 followed by launches on 6 February 1985 and 12 February 1987. Successor series followed with launches on 8 August 1989, 2 December 1992, 12 May 1996, 3 July 1996, 6 December 2000 and 11 October 2001. A review of the Shuttle manifest and other orbital details indicates that Shuttle STS-28 in August 1989 and STS-53 in December 1992 both released new versions of the Jumpseat. The first two Advanced KH 11s (KH 12 in some sources), appearing February 1990 and November 1992, and the launches of the advanced Jumpseat coincide with these departures.

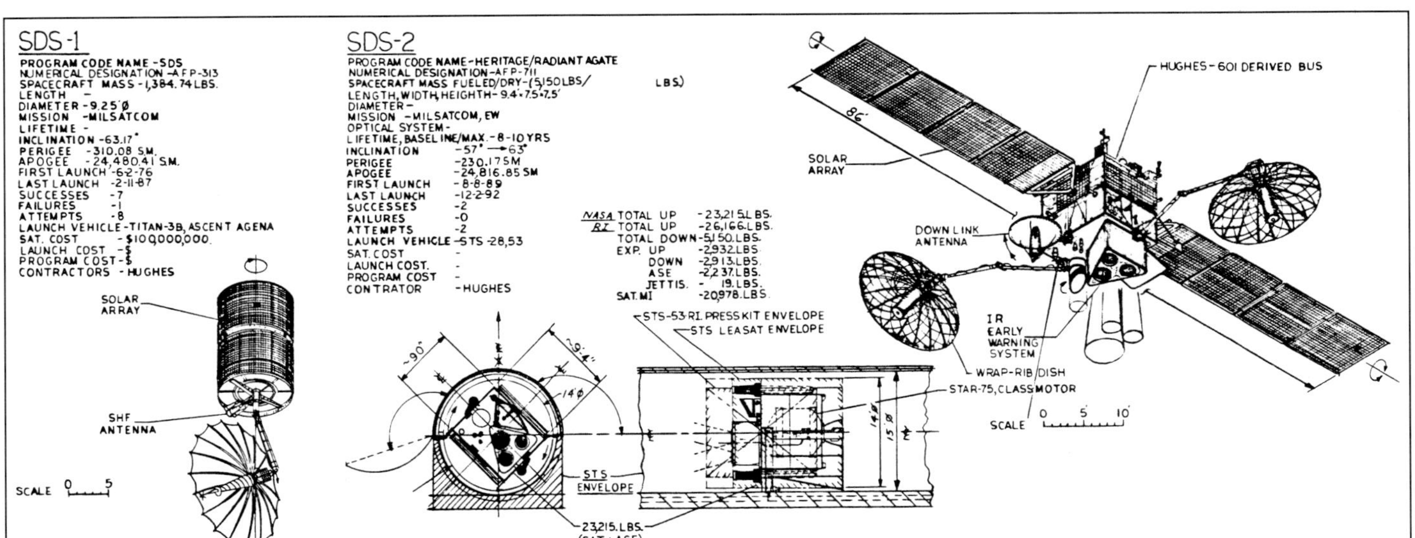

SDS conceptual studies using information from open sources. The current version, suggested as being launched by STS-28/53, is at right. Trade reports in 1994 claimed that the 'Heritage' sensor being proposed for DSP's follow-on has long flown on Jumpseat/SDS to augment the DSP system. Dimensions are in feet/inches (C P Vick) 0517318

EARLY WARNING

Russian Federation

Oko (Eye) Constellation

Background

The system of five Oko satellites roughly corresponds to the US DSP system. The Soviets intended for the system to produce an early warning of a ballistic missile attack. Work on the system began in 1967 with TsNII Kometa as prime contractor on a 15 to 30 minute warning system. It took 15 difficult years for the Soviets to place the system and get it fully operational in 1987. US military observers claim that 'it is capable of providing about 30 minutes warning of a US attack and of determining the general area from which it originated'. Okos probably also monitor nuclear tests, as do the US systems.

The Soviets began a four-year test programme for the first-generation system in 1972 with launch of Cosmos 520 on 19 September 1972, and operationally deployed the system starting in 1976. An SL-6 Molniya, launched from Plesetsk, places each Oko in a highly elliptical (about 600 × 40,000 km) semi-synchronous orbit at 40° intervals. A full system requires nine satellites operating 160 minutes apart. With apogees over Western Europe and the Pacific and overlapping coverage, each views ICBM sites in the US for 5 to 6 h/revolution (12 hours daily) and transmits data to RFAS bases at the same time. Oko also monitor routine launches from the US and other countries.

When Kosmos 1849 reached orbit on 4 June 1987, the Soviets filled all their requested operational slots for the first time. The RFAS keeps one or two older satellites on a more westerly ground track, possibly for better viewing of the western US. Despite its age and the lessening of tension with the United States, Russia shows no sign of abandoning the system and, during the last days of the Soviet Union, Oko established a new orbital plane, in late 1990, by Kosmos 2105 (20 November 1990). Western analysts still have not determined the reason for this new plane. The Soviets also created full-time global coverage by launching four unmodified Okos in geosynchronous orbit at 335° east to provide Pacific Ocean coverage: Kosmos 775 (1975), Kosmos 1546 (1984), Kosmos 1629 (1985) and Kosmos 1894 (1987). The last one finally shut down in January 1992 when Prognoz took over.

In the late 1990s, a decline in launches has caused the constellation to fall below its full population. In 1990, there were six launches to replenish the system, then none in 1991, four in 1992, three in 1993, one in 1994, one in 1995 and none in 1996. The Russians launched a new Oko satellite in April 1997 with the new Kosmos 2340. Kosmos 2222 apparently died in the second half of 1996 and a standard manoeuvre reduced the orbital period by 0.4 minute. Through April 1997, no further manoeuvres appeared, indicating that the satellite is no longer operational.

Interfax News Agency initially stated that the planned launch of a Kosmos satellite aboard a Molniya-M on 18 December 1996 had been delayed without giving a specific reason. After the Russians rescheduled the launch attempt for 10 January 1997, it was again postponed 'for technical reasons'. After a launch on 9 April 1997, the Russians named the satellite Kosmos 2340 once in orbit. It was followed within a month by Cosmos 2342, with further launches of one each in 1998 and 1999. The most recent Oko launch took place on 24 December 2002, the second that year after a gap in 2000 and 2001.

Russian drawings have depicted the Oko satellites, showing a cylindrical drum about 2 m diameter and 2 m long, plus two sets of solar panels and a telescopic housing for the instruments. They appear to weigh about 1,250 kg. Oko is the responsibility of TsNPO Kometa, with NPO Lavochkin responsible for the satellite bus. The latest Oko satellite was launched on 27 December 1999 (Oko 38).

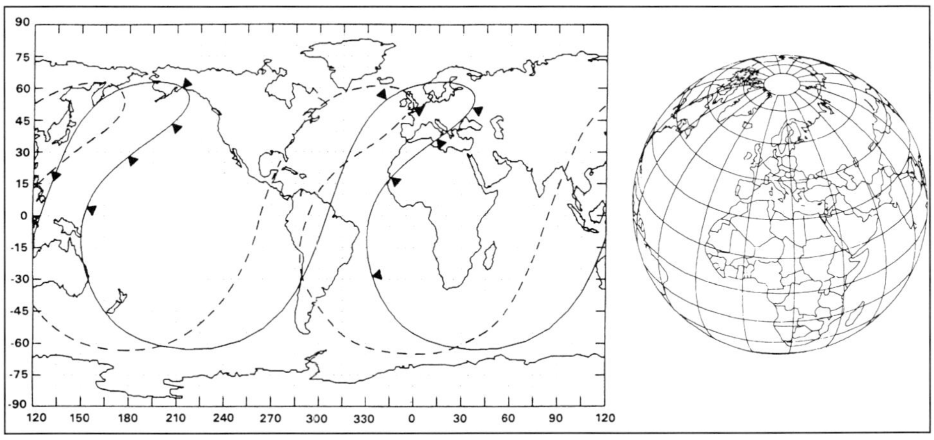

The Oko early warning satellites typically operate with apogees that provide surveillance of the US while remaining in contact with the RFAS/USSR. The globe illustrates the view from one apogee. The solid groundtrack depicts the daily path of the primary satellite, while one or two satellites are usually maintained in the more westerly dotted track (Teledyne Brown Engineering)
0516818

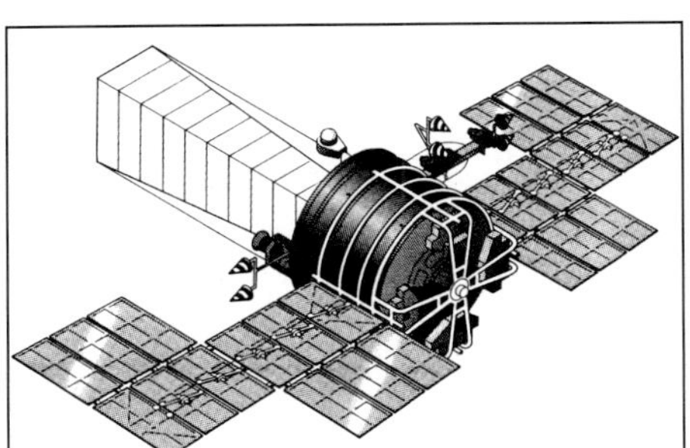

The original Oko early warning satellite
(Europe and Asia in Space, 1991–92; USAF Phillips Lab)
0517154

Prognoz (Forecast) Series

Current Status

Cosmos 2282 was launched in July 1994 followed by Cosmos 2345 in August 1997, Cosmos 2350 in April 1998 and Cosmos 2379 in August 2001.

Background

As a follow-on plan to Oko, the Soviets requested seven GEO positions in 1981 and named the satellites Prognoz. The Soviets claimed the Prognoz would study the Earth's oceans and natural resources beginning in 1982. The Soviets requested slots at: 12° east, 35° east, 80° east, 130° east. 166° east, 159° west, and 24° west. The first Prognoz satellite took up its position 26 April 1988 named Kosmos 1940. An SL-12 Proton carried it into orbit, but the flight plan left it in GTO for several weeks prior to transfer.

For 14 years before this, the Soviets inserted all their GEO satellites close to 90° east and then shifted them into final operating positions. Kosmos 1940 instead remained in its low parking orbit for an additional 6 h and then transferred directly into 15° west before moving to 24° west. The Soviets moved it to 12° east during July/August 1988 to support Buran's maiden flight. It failed on 14 September 1988. In 1993, the Russians finally identified Prognoz as an early warning system.

Kosmos 2133 (14 February 1991) established the series as the second vehicle launched under the Prognoz programme. It occupied two slots first at 80° east and Western observers claimed it handled Potok communications at the time. Then it manoeuvred first to 35°, 12°, 336° and 80° east – each of the Prognoz locations. In August 1995, it drifted off its assigned longitude and appears to be no longer operational.

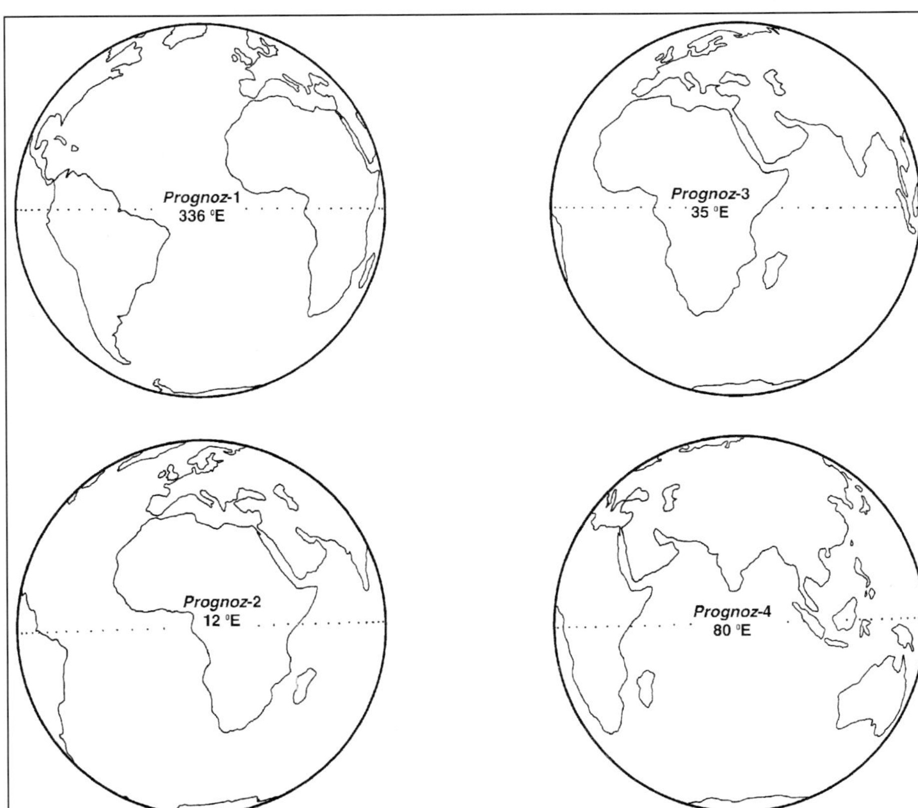

Perspectives from four of the Prognoz locations. The positions allow the satellites to monitor launches from the US east coast through to the east of China (Nicholas Johnson)
0516817

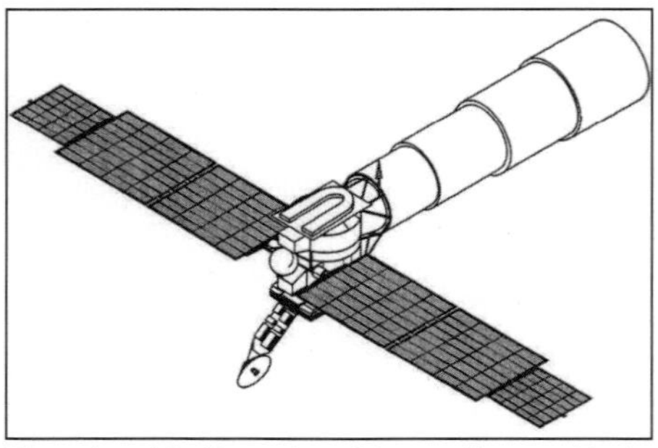

The only public illustration of the satellites used in the Prognoz system appeared in March 1993. This version is drawn by Ralph Gibbons. Compare it with the Spektrum-UV artwork in the Spektrum section 0516816

Prognoz 2 was joined at 336° east by Kosmos 2209, launched 10 September 1992 then Kosmos 2224 17 December 1992 at 12° east, and later moved to 336° east. It returned to 12° east in April 1994. Kosmos 2282 appeared 6 July 1994 and settled at 335° east; but began drifting in October 1995.

The satellite is made up of a 2 m diameter main instrument section, two large solar wings and a large telescope tube housing a beryllium mirror. The PbS linear CCD can apparently detect aircraft on afterburners. The same bus will be used for the Spektrum science mission in 1998; and the sole published illustration closely reflects the Spektrum-UV artwork the Russians have released. Proton GEO performance indicates a mass of about 2.5 tonnes. Oko and Prognoz are both the responsibility of TsNPO Kernels; NPO Lavochkin provides the bus.

The Russians hinted at the close association of the Prognoz and Oko programmes because, though they registered the Prognoz system in 1981, no Prognoz launches took place until 1988. The three geosynchronous Oko satellites launched during 1984–1987 all took up orbital slots registered at Prognoz longitudes.

The first Prognoz satellite apparently failed because, after one orbital relocation, during July of that year, the satellite was boosted off-station in September and never had its geosynchronous longitude restabilised. The satellite probably suffered some failure during the attempt to relocate it around the geosynchronous orbit band. Kosmos 2155 appears to have been another failed mission in terms of operating time since the Russians manoeuvred the satellite off-station nine months after launch.

The launches in 1992 turned out to be two of the most successful Prognoz satellites. Kosmos 2209 operated for just over four years, while Kosmos 2224 is still operational. Kosmos 2282 in July 1994, only operated for 15 months, drifting off-station in October 1995.

United States

Defense Support Program (DSP) Series

Current Status

A total of 22 satellites has been launched and TRW is under contract to build a total of 23. The most recent was DSP-1 launched by Titan 4B on 14 February 2004. DSP-18 was launched on 24 February 1997, followed by DSP-19 on 9 April 1999 and DSP-20 on 8 May 2000. As of March 2001, only DSP 17, DSP 18, launched 22 December 1994 and 22 February 1997 respectively, and DSP-20 are operational.

DSP Block 3 were launched 14 June 1989, followed by others on 13 November 1990, 25 November 1991, 22 December 1994, 9 April 1999 and 8 May 2000. The most recent was DSP-21 launched 6 August 2001.

Background

The Defense Support Program (DSP) is the air force catch-all phrase invented in 1966 to hide the placement of geosynchronous satellite monitors of Soviet ballistic missile launches. The air force did not acknowledge the existence of the programme until the launch of DSP 16 aboard the Shuttle. It still refuses to reveal which DSP satellites remain operational. Other sources indicate that DSP 17 is operational at 37° west Atlantic, DSP 16 at 69° east, DSP 15 at 10° east, and DSP 14 at 152° west. DSP 13 is probably still operational in a limited capacity at the 110° east back-up position, to cover Russian ICBM/SLBM launch areas. DSP 16 probably replaced number 14. Another five satellites sit in storage at the manufacturer and the air force cancelled the manufacture of DSP 24 and 25 to help pay for the next generation. This next generation known as the SBIRS Space-Based Infra-Red System will eventually cost $22 billion, but its overlappping coverage will provide stereo views in order to show three dimensional track files for ballistic missiles thereby improving warnings of theatre launches. The more agile system will work in conjunction with the JTAGS vehicle to provide launch site and target data directly to tactical commanders and ballistic missile defence batteries.

The Space Command's 50th Space Wing at Falcon AFB controls the DSP for its user, the 21st Space Wing at Peterson AFB, working through stations in Guam, Pine Gap and Nurrangar. Loral Federal Systems received a five-year, US$150 million contract July 1995 to provide software and engineering support for DSP ground stations. Observational data are received and processed at Buckley Air National Guard Base in Aurora, Colorado, Nurrangar and Kapuan (Germany). The United States has plans to close down the two foreign sites and consolidate operations, from 1999, at a new central processing station at Buckley. Pine Gap, 19 km southwest of Alice Springs in Australia, controls all military satellites when they are in the southern hemisphere. The 21st Space Wing's 5th Space Warning Squadron (jointly with the Australian 1st Joint Communications Squadron) acts as the control authority. When the United States publicly renewed the leases on the facilities in November 1988, the public knew the functions of the bases for the first time.

Nurrangar acts as the principal data and imaging centre and it became operational in early 1971 to take over the early warning satellite system from Pine Gap. The United States had to upgrade the facility in order to use the current Block 14 satellites, and it can now receive, process and re-encrypt data at 100 Mbit/s.

Some concern occurred after the loss of the Shuttle Challenger that the new generation of DSP would be stranded until a new launch vehicle could be found. As the older second-generation satellites began to go offline there seemed to be no interim solution to keep an early warning system in place for the United States. Instead, the USAF launched the last of the second-generation satellites in November 1987 and the Shuttle became available again to handle the new Block 14 DSPs. The first Block 14 orbited aboard the new Titan 4 launch vehicle on 14 June 1989. The third Block 14, DSP 16, appeared aboard Shuttle mission STS-44 in November 1991. This was the first advance notice of a DSP departure, given by NASA in August 1990 on USAF clearance. These enhanced spacecraft incorporate more sensitive IR and Vela-type nuclear detectors for improved discrimination. They also employ two infra-red wavelengths to combat laser jamming.

Acquisition of DSPs continues apace. TRW received a US$743 million contract by USAF's Space and Missile Systems Center in mid-1987 to build a second batch of third-generation satellites, numbers 18–22, to follow on from 14–17. Then the USAF ordered numbers 23 to 25 in June 1993 under an eight-year US$724 million contract to TRW, plus US$501 million to Aerojet for the payload.

The DSPs monitored 166 exchanges of Soviet SS-1 Scud medium-range missiles between Iraq and Iran during the missile exchange known as the 'War of the Cities'. During 1991's allied Gulf War against Iraq, the last two second-generation DSPs – the first to carry the 6,000 element detectors – moved their orbits from their ordinary positions to give stereo coverage. The 69° east slot provided an ideal position to view the launch field of the H2 airfield in Iraq. Reports from field commanders indicate the satellites provided launch information within two minutes of the launch and therefore could give the Patriot batteries information for targeting well before the end of the Scud's seven-minute flight. Target field identification was said to be less than 6 km of uncertainty.

Prompted by the war, two programmes have developed ground stations for tactical commanders to receive DSP data directly. A prototype of the army/Aerojet Tactical Surveillance Demonstration station began operating in Germany in 1993 and was transferred to South Korea in 1994 as a precaution against North Korean missiles. The station merges data from two or more satellites. The army plans to acquire six operational JTAGS Joint Tactical Ground Stations. The USAF built its own version despite the 'joint' claim and called it Talon Shield. It became operational in October 1994. The 21st Space Wing's 11th Space Warning Squadron at Falcon AFB provides theatre commanders with continuous tactical surveillance and alert information.

DSP 16 was released by Shuttle STS-44. Clustered below the main barrel are optical sensors looking for the characteristic double flashes of terrestrial nuclear explosions (NASA) 0516835

Although DSP provides launch location, some partial information about missile type and approximate azimuth, contrary to popular reports it cannot pinpoint accurate range information. After a launch signal is passed to the ground station a BMEWS or ground-based radars must calculate the range with supplementary information. DSPs have been used formally since 1993 to log asteroid atmospheric impacts. There are 10–30 annually, equivalent to 1 kT explosions.

The second-generation DSP busses extend 2.78 m to house the optics on a 2.91 m high cylindrical bus. The bus provides power from its solar cells on its outer surface supplemented by four small panels, generating a total of 680 W. The air force hardened the satellites against laser jamming and nuclear blasts.

DSP 3s have a 3.63 m long Schmidt telescope with 92 cm aperture, to collect IR radiation and permit target discrimination by ground controllers within 1.20 seconds. The satellite spins at 5.7 rpm to maintain orientation. The telescope's optical axis points at 7.5° off the spin axis to generate a conical scanning pattern as the vehicle rotates. From GEO, each of the Aerojet IR array's 2,000 lead sulphide cells views a region on Earth about 3 km in diameter. The USAF introduced a new-generation sensor and retrofitted the last two in the primary series – DSP 12 and 13 – with this new

The second-generation DSP was first launched and entered orbit in 1975; launches continued until 1987 (TRW) 0516836

First of the new DSP early warning satellites was launched by Titan 4 in June 1989 (TRW) 0516837

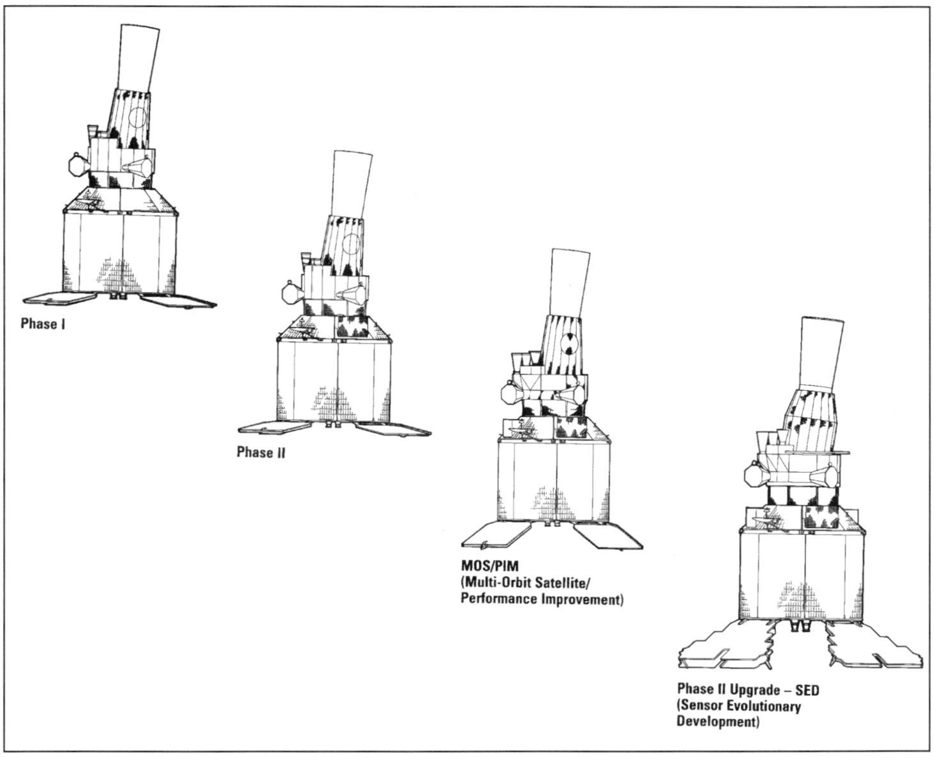

The Thermatic Mapper aboard Landsat 4 produces processed images in seven spectral bands with a resolution of 20 m. False colour images reveal topographical details of the mountain range traversing White Sands Missile Range, New Mexico 0022562

sensor. Each new sensor has about 6,000 elements commensurately increasing the resolution. DSP 12 carries HgCdTe detectors as well. DSP 12 and 13 provided the primary coverage of the Gulf to detect and warn Patriot batteries of Scud launches.

The earliest DSP satellites weighed 2,000 lbs, had 400 W of power, 2,000 infra-red detectors and a design life of one and a quarter years. In the 1970s, the satellites were upgraded. The satellite weight grew to 3,690 lbs, the power to 680 W, and the number of detectors increased to 6,000 and the design life to three years.

Late in the decade, TRW upgraded the sensor to improve its survivability against blinding attacks from the Soviet Union. The company added a second infra-red colour to protect against attack from lasers and particle beams. The second colour or medium-wavelength infra-red resulted, for a focal plan array made up of mercury cadmium telluride, the first application of the material in a space system.

Evolutionary growth has continued to improve satellite capability, survivability and life expectancy without major redesign. TRW is currently building DSP spacecraft that accommodate 6,000 detectors. Today's DSP satellite weighs 5,200 lbs and uses 1,274 W of power. At its peak, in the late 1980s, DSP employed more than 800 people at TRW.

Early in the programme, engineers used a so-called 'zero momentum' approach that allows for spacecraft attitude control with a minimum of fuel expenditure. The DSP satellite is spun about its Earth-pointing axis to provide a scanning motion for the infrared sensor. To reduce the satellite spin momentum to a nominal value of zero, engineers introduced a reaction wheel that achieved an equal and opposite momentum. To maintain the satellite's pointing accuracy throughout its lifetime, the DSP has an Inertial Properties Adjustment Device (IPAD) which balances the changes in moments of inertia as fuel and other consumables are depleted.

DSP Block 1

In 1966, TRW and Aerojet began a joint programme to build the first DSP satellites. Initially designated Program 266, then Program 949 and finally, in June 1969, Program 647 to hide its true purpose. The design used a 400 W panels spanning about 7 m with 3-year design life. Titan 3s placed the satellites in orbit. Each focal plane array carried 2,000-element infra-red detectors. The original specification for the satellites called for them to include nuclear test detection among their missions, thus making the Velas obsolete. The first DSP (DSP 1) went into orbit 6 November 1970 from Cape Canaveral. It weighed 820 kg and entered a 26,070 × 36,050 km 7.8° inclination orbit. By March 1974, it had exhausted all of its propellant. DSP 2 and 3 flew 5 May 1971 and 1 March 1972 respectively into synchronous orbits at 0.1 and 0.2° inclinations. The USAF placed DSP number 2 over the Panama Canal to watch for submarine missiles. Number 3 sat over the Indian Ocean to warn of an attack by Soviet land-based missiles in addition to providing information on Soviet and Chinese missile tests. DSP 4 followed on 12 June 1973. The four satellites detected 1,014 missile launches by 30 June 1973.

DSP Block 2

The second-generation satellites completed their launch programme in 1987. DSP 5 entered orbit on 14 December 1975, but its hydrazine system ruptured four days after launch. DSP 6 followed on 26 June 1976 and replaced number 3 over the Atlantic. The USAF moved it March 1981 to 75° east on standby; and removed it from GEO March 1985. Then DSP 7 flew on 6 February 1977, DSP 8 on 10 June 1979 (retired by the USAF August 1986), DSP 9 on 16 March 1981 (removed from GEO April 1988), DSP 10 on 6 March 1982. The air force added DSP 11 on 15 April 1984 (retired 1992 from 110° east), DSP 12 on 22 December 1984 (operational since May 1985 at 152° west, probably retired 1995) and, concluding the series, DSP 13 on 29 November 1987 (operational at 37° west).

The USAF planned to give the second- and third-generation satellites the capability to fly in Molniya-type orbits if required. When launched, they expected the satellite to last three years, but hoped for five years. The loss of the Challenger forced the air force to keep them in service much longer.

India's latest IRS satellite operates with a special resolution of better than 20 m and has four spectral bands. It resembles the Landsat 4 in performance characteristics (Radarsat) 0022563

DSP Block 3

By the late 1980s, the air force exhibited some anxiety about getting the DSP third-generation satellites into space. It had scheduled with NASA the first launch of the new series in December 1986, but the Challenger disaster ended the plan. Then discussions of a modified Titan launch changed when a Titan, flying from Vandenberg, failed in April of that year. The first DSP had to wait until the beginning of 1989, to be launched aboard the maiden Titan 4 in March 1989. The first Shuttle launch didn't follow until a year later. The air force will eventually launch 10 new DSPs, in intervals of roughly 18 months. The Titan 4 launch on 14 June 1989 from Canaveral's pad 41 to 152° W, replaced DSP 10.

Each third-generation DSP has a seven- to nine-year lifetime and senses targets at two IR wavelengths (with HgCdTe and PbS detectors) to avoid laser jamming and improve discrimination. The air force claims this new two-colour system has the sensitivity to detect aircraft operating on afterburners. The satellites also carry nuclear explosion detectors for the Department of Energy. Since each one between numbers 14 and 23 carries a Magnetic Plasma Analyser and a Synchronous Orbit Particle Analyser they can monitor spacecraft charging and radiation environments. They thus coincidentally provide unique 3-D observations of Earth's near-equatorial inner magnetosphere. The air force makes data available to qualified researchers through the IASTP Inter-Agency Solar Terrestrial Physics programme. Los Alamos

National Laboratory makes the instrument on funding from the Department of Energy's Office of Intelligence and National Security. The first result to come from this scientific investigation came from Los Alamos when it announced in 1994 that electron levels observed back to 1979 exhibit a solar-like 11-year cycle but, surprisingly, peak one year before the Sun's minimum activity.

The third-generation DSP weighs 2,360 kg and is 10.0 m long by 6.74 m across the four deployed solar panels. The solar arrays produce 1,274 W of power, almost double the preceding generations. Impact detectors monitor physical interference and provide the option for manoeuvring away from the source in case a kinetic kill vehicle attacks the satellite.

The first spacecraft (number 14) carried a piggyback SDI visible/ultra-violet sensor to characterise targets against bright backgrounds such as Earth's limb. Number 15 launched by Titan 4/IUS 13 November 1990 from Canaveral to 100° east, replaced #11. DSP 16 left for geosynchronous aboard Shuttle mission STS-44, and probably replaced number 12 at 69° E. NASA announced this launch the first ever for the classified programme on USAF clearance. The USAF named this DSP Liberty and the NASA launch cost was quoted at US$300 million in contrast to the standard US$180 million. DSP 17 appeared 22 December 1994 on Titan 4, probably replacing DSP 13 at 37° W. Many observers say DSP 17 is the satellite that witnessed the 10 December 1995 re-entry, some 300 km north of the Falkland Islands, of Russia's Kosmos 398 lunar module test vehicle, launched 26 February 1971. DSPF 18 launched 23 February 1997.

At the beginning of 1994, both DSP 14–16 remained operational but only #16 offered full capabilities. DSP 14 and 15 had both suffered failures in the thermal control of their focal plane arrays because of blown fuses, reducing their sensitivity to smaller missiles.

DSP Block 4

Phase II Upgrade programme, 1984–1987 with two satellites launched.

DSP Block 5

DSP-1 series beginning 1989 to the present with a total 8 satellites launched to date.

UPDATED

The USAF was planning to spend a total of US$2.6 billion between 1998 and 2001, dependent upon Congressional approval of the deployment of SBIRS (Boeing) 0022578

Midas Series

Background

As both sides in the Cold War built ballistic missiles, the United States felt an acute need to augment its DEWS radar tracking stations with a space-based missile detection system. The USAF called the first design for this system the MIDAS. Each MIDAS satellite used IR sensors capable of detecting the hot exhaust plume of a ballistic missile against the background of the Earth.

Ground-based BMEWS had to wait until missiles rose above the horizon and could only provide the US with a 15-minute attack warning. 12–15 Midas satellites spaced equally around a polar orbit detected an attack soon enough to provide the US with 30 minutes warning. The USAF launched 12 satellites in three series.

Midas 4, launched 21 October 1961 aroused an international furore when details of Project Westford became known. Westford ejected a 36 kg canister containing 350 million hair-like copper dipoles each 21 mm long. After separation, the spinning canister should have dispensed the dipoles in a belt 3,220 km high, 8 km wide and 40 km deep to act as passive reflectors to relay military communication messages. Radio astronomers protested the possibility of interference with their work. Midas 4 ejected its canister but the dipoles failed to disperse.

Midas suffered from a large number of false alarms, interpreting Sun reflections off cloud tops as missile launches. The USAF dropped the name Midas and the satellites became simply Program 461. However, by 1963 President Johnson, apparently referring to the 7 and 9 launches claimed that Midas detected both liquid and solid ICBM launches.

Midas Launch History

Satellite	Launch date	Outcome
Midas 1	25 February 1960	Failed to reach orbit
Midas 2	24 May 1960	484 × 511 km orbit. Telemetry link failure
Midas 3	12 July 1961	3,428 km × 91° inclination
Midas 4	21 October 1961	Ejected its canister but the dipoles failed to disperse
Midas 6	17 December 1962	Failed
Midas 8	9 May 1963	Successfully released a Westford canister
Midas 9	13 June 1963	Failed
Midas 10	9 June 1966	Failed
Midas 11	19 August 1966	
Midas 12	5 October 1966	

SBIRT Series

Current Status

The SBIR programme will cost US$17 billion for the high satellites, plus US$11 billion for SMTS.

Background

In 1992, the USAF Space & Missiles Systems Center awarded US$240 million 24-month contracts for the next generation of early warning satellites. The USAF projected its expenditures at US$1.6 billion through the end of 1994 and asked Congress to provide US$84 million FY92, US$251 million FY93, and US$215 million requested FY94. However, the Congress cancelled the FEWS (Follow-On Early Warning System), programme in December 1993. Instead, the air force proposed the ALARM (Alert Locate And Report Missiles) approach for deployment beginning in 2004. To accompany the proposal, the air force requested US$150 million for FY95 and projected US$1 billion for 1995–99. The air force planned to reduce the cost of FEWS by reduced onboard processing and tactical links in the first years of the programme and then gradually build to FEWS capabilities in the out years.

Many observers expected the air force to cancel DSPs 24/25 to help pay for ALARM, and DSP 23 production slowed. As the Ballistic Missile Defense Office sought to incorporate its own planning into the programme ALARM disappeared into the broader-based SBIR Space-Based Infra-Red programme in December 1994. SBIR requested US$283 million FY96, US$319 million FY97, with US$2.6 billion projected between FY98-01.

As proposed by BMDO, SBIR will initially cover four GEO satellites (the first 2002) and two highly inclined elliptical satellites. Trade reports in 1994 claimed SBIRs would use the 'Heritage' sensor now included or potentially included on the Jumpseat/SDS surveillance satellites. BMDO will add LEO satellites beginning in 2006, based on the Brilliant Eyes (now called the Space and Missile Tracking System or SMTS). The air force picked up the programme in 1994 from BMDO and selected two teams headed by Lockheed Martin and Hughes/TRW in August 1995 for 15-month US$80 million pre-engineering and manufacturing development contracts.

If SMTS is cancelled, then SBIR will exclude tracking after burnout, manoeuvring, decoy discrimination, accurate impact prediction and cueing ground/space defences. TRW/Aerojet proposed adding a small field of view sensor (developed for a cancelled classified mission) in place of Trailblazer on a current DSP to provide coverage of theatre-sized areas, significantly improving tactical value.

ELINT

Introduction

Current Status

Elints record radio and radar transmissions from areas of military activity. When replayed to ground stations the radar signatures (characteristics such as pulse repetition frequency, pulsewidth, transmitter frequency and modulation) enable the likely function and method of operation of a particular centre to be identified. The number and type of electronic systems at a particular site, and subsequent changes in the signals, provide a valuable indication of its purpose and capability. The ability to intercept and decode satellite and ground communications, and to interfere with them by rival satellite activities, is likely to be a decisive factor in any large scale hostilities. In times of hostilities, even if the contents of the messages are not decoded in time, their interception can indicate the locations and identities of the communicators. Together with the movements of the mobile transmitters, an Electronic Order of Battle (EOB) can be constructed. It is continued monitoring of large ground radars such as ballistic missile early warning systems, with their specific ranges, viewing angles and other operational restraints, that enable details of their modernisation and upgrading to be logged and analysed. The nature of elint gathering has changed dramatically as technology has advanced and shifted preferred orbit type as operational requirements have changed with improved capabilities.

China

Chang Kong Satellites

Background

The main series of satellites to be flown atop the FB-1 booster has been virtually written out of Chinese space history. No Chinese description of the programme is known to exist and Chinese launch listings now exclude these FB-1 launches, possibly on the grounds that the FB-1 has never been offered for commercial satellite launches.

When these satellites appeared in Chinese listings during the 1980s they were identified by the generic *Ji Shu Shiyan Weixing* (JSSW – "technical experiment satellite") name but it has been reported in recent years that the programme name was actually *Chang Kong* (variously translated as "long sky" or "wide and high sky").

At the time of launch no details were given of the programmes being undertaken, other than they were connected with the "preparedness against war" – a common claim made for the early Chinese satellites and not exclusive to this series. To all intents and purposes, these satellites were simply launched and then allowed to decay from orbit: no telemetry was picked up by open source observers in the West, suggesting that data transmissions were limited to passes over the Chinese mainland when in line-of-sight of the control centre.

These satellites are assumed in the West to have been primarily military in purpose almost by default, possibly undertaking ELINT work. It is possible that this was a Shanghai Bureau imaging satellite, intended to rival the Beijing *Jian Bing*-1 (FSW) satellites (*qv*) being developed separately (although the high apogee seen on the August 1976 payload would not square with such an application).

No Chinese publication has shown any representation of the *Chang Kong* satellites, but American visitors to China in 1979 are believed to have seen a flight spare from the programme, and the description which follows is based upon this.

Specifications

***Chang Kong* (1973)**
Launch: 18 September 1973 using an FB-1 from Jiuquan
Design life: 1 month ?
Contractor: Shanghai Bureau of Astronautics
Orbit: Failed to reach orbit
Configuration: Cylinder with hemispherical top, 1.7 metres diameter and 2.5 metres high
Mass: 1,138 kg
Power system: Probably on-board batteries, although the satellite carried some experimental solar cells on the body.

***Chang Kong* (1974)**
Launch: 12 July 1974 using an FB-1 from Jiuquan (some Chinese sources give the launch date as 14 July)
Orbit: Failed to reach orbit
Mass: 1,108 kg
Other specifications: as *Chang Kong* (1973)

***Chang Kong* 1**
Launch: 26 July 1975 using an FB-1 from Jiuquan
Orbit: 69.03°, 183–460 km
Orbital decay: 14 September 1975
Mass: 1,107 kg
Other specifications as *Chang Kong* (1973)

***Chang Kong* 2**
Launch: 16 December 1975 using an FB-1 from Jiuquan
Orbit: 69.00°, 186–387 km
Orbital decay: 27 January 1976
Mass: 1,109 kg
Other specifications as *Chang Kong* (1973)

***Chang Kong* 3**
Launch: 30 August 1976 using an FB-1 from Jiuquan
Orbit: 69.15°, 198–2,145 km
Orbital decay: 25 November 1978
Mass: 1,108 kg
Other specifications as *Chang Kong* (1973)

***Chang Kong* (1976)**
Launch: 10 November 1976 using an FB-1 from Jiuquan
Orbit: Failed to reach orbit
Mass: 1,208 kg
Other specifications as *Chang Kong* (1973)

Russian Federation

Tselina Constellations

Current Status

Launches of the Tselina-2 satellites were planned at a rate of one to two per year. The latest was launched 3 February 2000.

Background

Tselina/Tselina-O

The Soviet Union apparently named its electronic intelligence programme Tselina and initiated it with some Elint payloads on the small Kosmos-1/2 class launch vehicle. Starting in the late 1960s, the Kosmos-3M booster launched the Tselina-O family of Elint satellites which appear to be the payloads primarily launched into 74°, 540 km orbits. The first Tselina was launched by Kosmos 3M on 26 June 1967 with the last of 60 flown on 31 March 1982. These satellites, like all Elints, pinpoint to within about 10 km, the sources of radar and radio emissions and make it possible to identify command and control centres, forward battle elements, air defence units and, even more importantly, reveal movements and changing tactics.

Tselina 2

The Soviets built a variety of electronic intelligence-gathering satellites over the years, but the Russians now use only two varieties: Tselina D and Tselina 2, both from NPO Yuzhnoye. The new Tselina 2, 3.2 tonne Elint type appeared as Kosmos 1603, launched 28 September 1984, and Kosmos 1656, launched 30 May 1985, by Protons from Tyuratam. Both performed very extensive manoeuvres. Starting from an initial 198 by 186 km orbit in an inclination of 51.6°; the satellite and its upper stage remain in this parking orbit for only about 30 minutes until it uses the booster to raise the apogee to 860 km. Some 50 minutes later, another manoeuvre raises the perigee to 850 km and changes the inclination to 66.6°.

On 22 October 1985, the Soviets launched the third satellite in the series, Kosmos 1697. It left no debris in the 51.6° inclination and 66.6° inclination orbits it used as it adopted its final position. A re-examination of the programme led Western observers to conclude that, for the first time, the new SL-16 Zenit launcher carried this satellite into orbit. A second similar satellite followed on 28 December 1985, but this time the Zenit upper stage failed to re-ignite and the Tselina re-entered in February 1986.

Flights resumed in 1987 with Kosmos 1833 (18 March) and Kosmos 1844 (13 May), and two more in 1988: Kosmos 1943 (15 May) and Kosmos 1980 (23 November). Kosmos 2082 appeared next on 22 May 1990. The Soviet attempt to complete a constellation of four satellites in orbital planes spaced 45° apart failed 4 October 1990 when the Zenit exploded within seconds of launch destroying both the payload and pad. The Russians experienced another major setback 30 August 1991 when the next Zenit exploded after launch from a new pad. The launch authority issued a statement saying: it was a 'military-technical satellite to verify the fulfilment of disarmament treaty commitments.' A third Zenit failure followed on 5 February 1992, but the Zenit of 17 November 1992 carried Kosmos 2219 into an orbit almost coplanar with 1988's Kosmos 1943. The second Zenit in 6 weeks delivered Kosmos 2227 on 25 December 1992, into an orbital plane 90° separated from Kosmos 2219's. Kosmos 2237 appeared 26 March 1993, apparently replacing Kosmos 1980. Kosmos 2263 appeared 16 September 1993. These most recent additions form a 3-satellite system with planes 120° degrees apart. Kosmos 2278 appeared 23 April 1994 with its plane 40° west of Kosmos 2263's and Kosmos 2297 appeared 24 November 1994 with its plane 40° west of Kosmos 2278's. Kosmos 2322 appeared 31 October 1995. A study of the orbits indicated in 1995 that Tselina 2's mass is only 3.2 tonnes, rather than the 9 tonnes previously believed, from Zenit's capacity in that orbit.

The tests using the Proton-K launch vehicle were possibly a result of delays in flying the new Zenit-2 and when that vehicle started to fly problems arose. The first Zenit-2 launch in April 1985 carried a mockup Tselina-2 intended for a 64.4°, 850 km orbit, but for unknown reasons the launch failed. In June, the second test flight failed when a second stage vernier engine exploded 511 seconds after launch. Three pieces of debris reached short-lived orbits with inclinations of 64.4°. The launch of Kosmos 1714 marked the third Zenit-2 failure in four launches, (Kosmos 1697 being a success), when the payload shroud failed to separate from the second stage and a second stage vernier engine exploded. The combined second stage shroud and encapsulated satellite reached an intermediate orbit with an apogee of 850 km without circularisation.

Unlike the earlier Tselina constellations, Tselina-2 series show no regular orbital plane spacing indicative of a constellation.

Tselina D

Another Soviet era operational electronic intelligence system, consists of a constellation of six 2.5 tonne satellites, spaced at 60° intervals in roughly 630 km, 81° inclination orbits. The Soviets used these satellites to identify and pinpoint military radio/radar stations. An SL-3 carried the first of these Elints into space on Kosmos 389, launched 18 December 1970 from Plesetsk (642 × 687 km). The constellation continued to be upgraded at a launch rate of approximately two each year until 1976, when the launch rate suddenly increased to four per year. Each had an average operational life of 18 months. Western observers who studied the orbits, believe the orbits suggest that the satellite is based on the Meteor bus configuration; and when the series began, many initially believed them to be Meteor failures.

In 1978, the SL-14 Tsyklon 3 began launching Kosmos satellites into 635 × 665 km, 82.5° inclination orbits. The Soviets named the first in the series the Kosmos 1025 (although this might be Okean-related). Various statements by Soviet era officials called a few of these satellites (Kosmos 1076, Kosmos 1151, Kosmos 1500, Kosmos 1602) oceanographic satellites. But their true mission as an Elint became apparent in 1984, when a complete constellation of six satellites spaced in planes 60° apart, emerged just as the Soviets announced the termination of the SL-3 programme. The coincidence of these events indicated that the SL-3 assembly line remained open just to launch the last Elint Kosmos 1441 (launched 16 February 1983), the 35th in the series. Built by NPO Yuzhnoye, the Soviets call the Tsyklon ELINTS the Tselina D.

The last of 1986's six Tsyklon Elints (Kosmos 1805, 10 December, 635 × 662 km, 97.7° inclination) entered a plane exactly between two established planes. These clues indicated that the constellation experienced an enlargement or reconfiguration, although a supplementary satellite did not arrive until Kosmos 2242 in 1993 (between Kosmos 2221 and Kosmos 2228). 1987's five Elints routinely replaced existing satellites and this pattern continued through the four missions of 1988. Kosmos 1805 transmitted Elint-type signals throughout 1988. No new Tselina's appeared during 1989. Then the single Kosmos 2058 orbited on 30 January 1990, replacing Kosmos 1812. Another long break occurred until Kosmos 2151 appeared on 13 June 1991, replacing Kosmos 1908. Kosmos 2221 and Kosmos 2228 appeared in 1992, replacing Kosmos 1842 and Kosmos 2058, respectively. Then 1993's single Tselina D Kosmos 2242, slipped into a new plane midway between them. A launch failure on 25 May 1994 may have been a Tselina-D and that is the last recorded launch of the series.

United States

Canyon (Program 827) Series

Background

The first large US signals intelligence satellite operating from near-geosynchronous orbit. The air force built Canyon to monitor radio communications. They began to go into orbit, with the Atlas Agena D launch of 6 August 1968, in a 10° inclination orbit that traced a figure-eight over each hemisphere improving northern coverage. Each satellite weighed about 700 kg. The other six launches were 12 April 1969, 31 August 1970, 4 December 1971 (launch failure), 20 December 1972, 18 June 1975 and 23 May 1977.

Chalet/Vortex (Program 366) Series

Background

Chalet succeeded Canyon and weighed 820 to 1,000 kg. It first flew on 10 June 1978 following the US loss of ground stations in Iran. Five more were launched by Titan 3C as follows: 1 October 1979, 31 October 1981, 31 January 1984, 2 September 1988 and 10 May 1989. The 1988 launch apparently suffered a Transtage failure. Sources hinted that the 15th and last Titan 34D launch, of 4 September 1989, may have been a Chalet mission but it was subsequently confirmed as a DSCS/IUS flight. USA 67 on STS-38, in November 1990, has also been suggested as a Vortex, although trade reports at the time claimed it was a Magnum. Information released following 1988's failure indicated that Chalet had been renamed Vortex and that the two first craft remained operational at 45° east and 115° east. Chalet operated in an elliptical orbit of about 30,000 × 40,000 km with an inclination of about 29°.

Heavy Ferret Series

Background

The United States achieved its first significant Elint capability starting 18 June 1962 when a Thor Agena B launch carried a Heavy Ferret weighing 1,500 kg from Vandenberg into 370 × 411 km 82° inclination orbit. Eventually the United States launched 18 satellites, at a rate of two to three annually. The Elint programme apparently began with test packages aboard at least the early Samos satellites. As they were developed, the Thrust Augmented Thors and Agena Ds then later the Long Tank Thrust-Augmented Thors and Agena Ds, carried these satellites into orbit. Known variously as Program 102, Program 770 and the designation 698BK. The NRO launched its last Heavy Ferret apparently on 16 July 1971 into 488 × 508 km 94.5° orbit. The programme eventually became obsolete as it was succeeded by Canyon.

Jumpseat Series

Background

Jumpseat record radio and radar transmissions from areas of military activity. When replayed to ground stations the radar signatures (characteristics such as pulse repetition frequency, pulsewidth, transmitter frequency and modulation) enable the likely function and method of operation of a particular centre to be identified. The number and type of electronic systems at a particular site, and subsequent changes in the signals, provide a valuable indication of its purpose and capability. The ability to intercept and decode satellite and ground communications, and to interfere with them by rival satellite activities, will most likely help determine the course of future wars. In times of hostilities, even if the contents of the messages are not decoded in time, their interception can indicate the locations and identities of the communicators. Together with the movements of the mobile transmitters, an Electronic Order of Battle (EOB) can be constructed. It is continued monitoring of large ground radars such as ballistic missile early warning systems, with their specific ranges, viewing angles and other operational restraints, that enable details of radar modernisation and upgrading to be logged and analysed. Jumpseat are 2 × 2 × 4 m in dimension with two large solar arrays and a large primary parabolic antenna.

Jumpseat have a 12-hour, 63.4° inclination Molniya-type orbit to provide polar coverage. They use almost the same orbits as SDS and use the same Titan 3B launcher. Jumpseat probably began to be manufactured in the late 1960s by either TRW or Hughes under the air force, reportedly to monitor Moscow ABM site radar emissions. Other targets include space tracking radars, SAM sites, airbases and SLBM submarine bases. The air force ordered six satellites. They were launched as number 1 on 21 March 1971, number 2, 21 August 1973, number 3, 21 August 1973 (failure), number 4, 10 March 1975, number 5, 25 February 1978, number 6, 24 April 1981, number 7, 31 July 1983. Advanced Jumpseat have followed. The Titan failure of 16 February 1992 probably carried a Jumpseat. The first Titan 4 Centaur launch into an apparently Molniya-type orbit, 3 May 1994, has been suggested as carrying a Jumpseat successor. The Titan 4 Centaur of 10 July 1995 was apparently similar. Shuttle STS-28 and STS 53 flew similar profiles with IUS, but their payloads have generally been attributed to advanced SDS.

Magnum/Orion/ Intruder Series

Background

The CIA initiated development of these advanced Ferrets to replace Rhyolite/Aquacade following the complete loss of Iranian ground stations by early 1979, but they suffered deployment delays because of launcher problems. Designed to take advantage of Shuttle launches, and pick up signals across the whole radio frequency range, the CIA had to pull together plans to use the Titan/IUS launch as the Shuttle fell behind schedule. Although NASA finally scheduled the first Magnum for Shuttle/IUS launch STS 10, the mission did not take place until STS-51C on 25 January 1985. The satellite started at a 28.5° GTO orbit. It attained GEO despite the IUS' small liquid thrusters being required to make up for I's delta velocity shortfall of some 1.6 m/s. The Magnum, many believe used STS-33 to travel to LEO during 23 November 1989. The same speculation surrounds STS-38's payload a year later. Magnums weigh 2.7 tonnes. Each spacecraft cost US$300 million. The Titan 4 Centaur of 14 May 1995 headed due east from Canaveral, indicating a 28.5° orbit, preparatory to departure for GEO. This may have been the first Intruder, which combines the functions of Magnum and Vortex. The new satellite weighs about 4.5 tonnes and has a 100 m diameter antenna. Extensive building work at Menwith Hill supports the debut of a new vehicle. The 24 April 1996 launch of USA 118 probably included the same type of satellite. A new advanced Orion was launched on 9 May 1998 on a Titan 4B.

The satellites are controlled from Pine Gap, which was expanded 1983–85 for Magnum operations.

The Magnum/Orion type tend to specialise in datalinks, such as missile telemetry transmitted during tests, while Vortex's voice communications speciality may have been added to Magnum's successor. Chalet was originally tasked with communications intelligence but from the second vehicle it was modified to encompass signals intelligence. Jumpseat is targeted at radar installations.

Rhyolite/Aquacade Series

Background

Electronic intelligence satellites, also known as Ferrets and designated Program 720 or Program 472 and code named Aquacade, matured with Rhyolite in the 1970s. Earlier Elints were usually launched into circular 500 km orbits. The first Rhyolite was launched 19 June 1970 followed by Rhyolite 2 on 6 March 1973, by Atlas Agena D into geosynchronous orbit at 0 to 2° inclination. It only weighed about 275 kg, but it became the first operational GEO Ferret. Rhyolites pick up telemetry from Soviet missile tests, monitor weapons development and check compliance with arms control treaties. Built by TRW for the CIA, two more were launched: 11 December 1977 and 7 April 1978. Satellites weigh about 700 kg. The CIA originally funded and then dropped project Argus, a more sophisticated follow-on, in 1976.

The GEO satellites return their data to Pine Gap (Australia), Bad Aibling (Germany), Menwith Hill (UK), and Ft Meade, Maryland for DSCS relay to Buckley Air National Guard Base, Aurora, Colorado. Pine Gap, near Alice Springs, is a 2.25 km^2 base under CIA/National Range Operations Control. Operational from 1970 in time for Rhyolite, two new radomes were added during the expansion of 1983–85 for Magnum operations, just as others were added in the early 1970s for Rhyolite. There were 11 radomes by 1991, operated by almost 700 personnel. Data tapes are sometimes airlifted out of Alice Springs, Australia to avoid interception.

Subsatellite Ferret Series

Background

A series of small subsatellites, launched piggyback with photo-reconnaissance satellites, identifies possible targets for detailed examination by the heavy Ferrets. The first of these, weighing 50 kg, left Vandenberg on 29 August 1963 from Vandenberg aboard a Thor Agena D. It adopted a 310 × 431 km 82° inclination orbit and had a 30 day life. By 8 June 1975, 35 had been identified; Big Bird took over as their launcher from 20 January 1972. There were interesting variations on 12 December 1968, 5 February 1969, 10 October 1972, 10 November 1973, 8 June 1975, 18 June 1980 and 20 June 1983, when the Ferrets went as high as 1,400 km. They monitored the Sary Shagan research and development centre at the time of Soviet ABM tests. Information from these subsatellite ferrets has been crucial to determining Russia's level of research into directed energy weapons and its ongoing development programmes.

Titan 2 Ferret Series

Background

Russia's Tass news agency reported on 22 April 1992 that the USA planned to launch a third new type of Elint satellite on a Titan 2. The first was USA 32 on 5 September 1988 in 185 × 292 km 85°

inclination. Titan 2 had been expected to continue the Whitecloud series but these satellites clearly did not fit the Whitecloud mould. The second in the series, USA 45 on 5 September 1989, suffered technical failures according to Tass. USAF claimed a successful launch, indicating the fault lay with the payload. USA 81 ascended 25 April 1992 into 784 × 805 km, 85.1° inclination orbit and appeared to continue the series.

Trumpet Series

Current Status

The NRO launched a Trumpet class electronic intelligence satellite on 8 November 1997, from Cape Canaveral aboard a Titan 4A to a 1,000 × 39,059 km, 63.3° orbit. The Intelligence Agency code-named the satellites USA 136. The launch was the third in the series for Trumpet satellites, following USA 103, launched on 3 May 1994 to a 518 × 537 km, 55.1° orbit, and USA 112, launched 13 July 1995. Another Trumpet was launched 8 November 1997 to a 1,000 × 39,000 km, 64° orbit. The three satellites are all operational providing continuous coverage of the Russian signal spectrum. The satellites are equipped with a large deployable mesh antenna designed to monitor Soviet communications and missile tests. They were all deployed to Molniya-type orbits.

Background

The Trumpet is a massive Hughes electronic intelligence satellite with the more specific objective of signal intelligence (signals intelligence). To eavesdrop on foreign military signals it unfolds an antenna about the size of a football field. The spacecraft cost roughly a billion dollars, and cost an additional US$200 million to US$300 million to launch.

It carries a special phased-array broadband eavesdropping antenna that measures roughly 300 ft across when it is unfolded. In order to keep the satellite placed over the proper point for the longest period of time, the spacecraft flies a Molniya orbit with an apogee of 23,000 miles, and a perigee of a few hundred miles, at an orbital inclination of 63°. The spacecraft uses its broadband capability to monitor more than 2,000 ground sources simultaneously, including communications between Russian generals and the country's nuclear submarine fleet.

The antenna demonstrates a new technology of construction that differs from the Magnum, Mentor and Jumpseat, 'wraprib' umbrella type under girding structure. The Trumpet has a mesh antenna and small pulleys and tiny motors unfurl it when it reaches its assigned orbit. Like the TDRS and Jumpseat systems it carries an Extremely High-Frequency (EHF) relay system.

In order to achieve final orbit the TITAN/CENTAUR must fire four separate times, two to achieve the inclination change and two to change the eccentricity of the orbit.

FOBS

Russian Federation

Fractional Orbital Bombardment System (FOBS) Series

Background
The Soviets provided their answer to the United States deployment of a missile tracking system on the northern perimeter of the western hemisphere when they launched The Fractional Orbit Bombardment System (FOBS). They demonstrated its potential in up to 29 tests conducted between 16 December 965 and 9 August 1971. The purpose of FOBS is to fly a nuclear warhead into the United States by using a southern trajectory, essentially coming into the western hemisphere over the south pole instead of the north pole.

FOBS are fired into an orbit of 160 km and then braked by retro for re-entry, their nuclear warheads (though none was ever carried during the tests) falling on the target before completion of the first revolution (hence the name fractional orbit). In these series of tests, the Soviets purposely launched the satellites on a track to miss the United States, with the test warhead being called down on to Soviet territory shortly before the end of the first orbit. From the Soviet point of view, FOBS forced the West to invest in more complicated defences such as Over-The-Horizon radar (OTH). But they paid a price in reduced accuracy and payload capacity.

All FOBS tests used a SL-11 Tsyklon (SS-9 Scarp) from Tyuratam, with a retro stage fitted to the warhead. The 1966 tests 'unannounced' by the Soviet Union and allocated a 'Kosmos-U' designation by RAE Farnborough, initiated the series and the response in the West prompted the Soviets to continue with nine in 1967: Kosmos 139, 160, 169, 170, 171, 178, 179, 183, 187. Then the Soviets continued with two in each of the following three years, and the last, Cosmos 433, came on 9 August 1971. Opinion in intelligence services is divided. Some feel the lack of tests indicate the system is operational and others feel it means the system was abandoned. Under Article VII of the unratified Salt 2 Treaty and the accompanying MoU, all FOBS must be dismantled. Specifically, the acknowledged 18 FOBS launchers at Tyuratam were to be either destroyed or converted.

GEODETIC

Russian Federation

Etalon Series

Background
Etalon 1, launched 10 January 1989 as Kosmos 1989 and Etalon 2 launched 31 May 1989 as Kosmos 2024 are 1,415 kg, 1,294 mm diameter geodetic satellites covered with 306 antenna arrays each with 14 corner cubes for laser reflection. Etalon reflects laser light in a passive role. The first was placed in 19,102 × 19,149 km at 64.9° along with two Glonass satellites. Etalon carries 306 laser retro-reflector assemblies for international geodetic studies; six germanium reflectors may be used for future IR interferometric measurements. Etalons supplement Glonass information, aiming to improve the navigation system's accuracy. Three laser stations at Ternopol, Yevpatoria and Maydanak provide 25 cm ranging accuracy, and were later joined by 10 foreign sites.

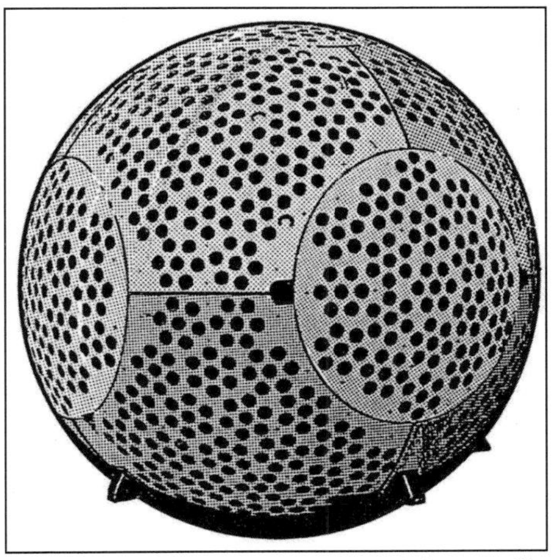

The Etalon laser-reflecting geodetic satellite complements the Lageos, Starlette and EGP spheres (Teledyne Brown Engineering) 0516693

Geodetic Type 1 Series

Background
Three satellites have been launched using the Kosmos-3M into the normal Parus/Tsikada navigation satellite orbital regime but their orbital planes and transmissions indicated that they were not part of the navigation satellite system. Speculation centres upon navigation calibrations as their ultimate mission. Kosmos 842 launched 21 July 1976 to a 965 × 1,006 km, 83° orbit and Kosmos 911 launched 25 May 1977 to a 965 × 998 km, 82.9° orbit had orbital planes 180° apart.

Kosmos 2285 remains an isolated mission. It was launched 2 August 1994 to a 974 × 1,013 km, 74° orbit. Although it shared the orbital altitude of the navigation satellites, the orbital inclination of 74° had not been used in that programme since 1972. This mission has not been explained and geodesy is perhaps the most likely explanation.

There is also a small group of satellites which are geodetic but their orbital parameters differ from the other acknowledged geodetic missions. These are part of the Sfera series or a different type of satellite as yet unidentified.

GEO-IK Musson (Monsoon) Series

Current Status
None in this series have been launched since 1994.

Background
The Soviets publicly discussed this series for the first time in 1989. They stated it used a design based on the Parus/Tsikada military/civil navigation satellites. NPO PM builds the satellite bus, and the NIIKP Institute of Space Device Engineering provides the payload. Soviet era reference call them the 'Musson' (Monsoon). Solar cells power the instrumentation through 10 deployed petals, working with Ni/H_2 batteries. GEO-1K is the final KAUR-1 bus with panels wrapped on a cylindrical body. The satellites takes measurements five days a week using a 150 to 400 MHz Doppler transmitter which operates for up to 12 hours daily, providing 1–3 cm/s accuracy. Musson uses 9.4 GHz radar altimeter to measure height over sea with 3 to 5 m accuracy, and a 5.7/3.4 GHz C-band up/down transponder measures satellite range within 3 to 5 m. Laser retro-reflectors calibrate the range measurement to within 1.5 m. A flashlight system emits a sequence of nine 800 to 1,200 Joule flashes up to 55 times daily, limiting positioning errors to 15 m for a satellite at 1,500 km.

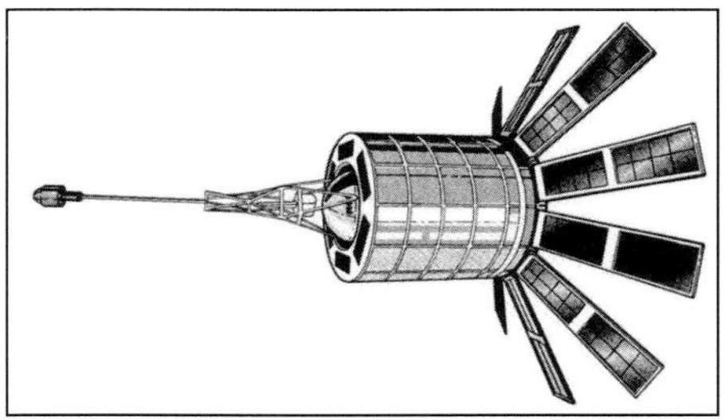

The GEO-IK class of geodetic satellites, derived from NPO PM's series of low altitude navigation satellites, was acknowledged in 1989. 1994's was actually named GEO-IK 1 (Teledyne Brown Engineering) 0516692

GEO-1K 1 became the first satellite in the series to carry a name and not a Kosmos serial number. It included an Elekon payload to demonstrate a location and data relay service for Russian cargo transport services. A joint venture between NPO PM and Germany's Elbe Space and Technology (Dresden) GmbH, the operational Elekon system will comprise payloads on sevenTsikada navigation satellites.

Launch date	Cosmos	Incl deg	Period min	Perigee km	Apogee km
23 January 1981	Failed to reach orbit				
23 September 1981	1312	82.6	116.0	1,493	1,505
24 September 1982	1410	82.6	116.0	1,495	1,503
24 November 1983	1510	73.6	116.1	1,481	1,526
8 August 1984	1589	82.6	116.0	1,496	1,504
14 June 1985	1660	73.6	116.1	1,482	1,526
11 February 1986	1732	73.6	116.1	1,480	1,526
2 December 1986	1803	82.6	116.0	1,498	1,503
20 February 1987	1823	73.6	116.0	1,479	1,526
30 May 1988	1950	73.6	116.1	1,485	1,522
28 August 1989	2037	73.6	116.1	1,484	1,525
30 July 1990	2088	73.6	116.1	1,483	1,525
22 December 1992	2226	73.6	116.0	1,482	1,526
29 November 1994	GEO-IK 1	73.6	116.1	1,480	1,527

Sfera (Sphere) Series

Background
The Soviet Union built the preponderance of geodetic satellites and launched them as part of their Kosmos series. They called the first generation the Sfera geodetic satellites and put them into three orbital inclination slots and two orbital altitude slots. Sfera has a length of 3 m, a maximum diameter of 2 m with a total mass of 800 kg. The Soviets launched two each year during the period 1968–78. The satellites provided improved accuracy for long-range weapons and each was built on the basic KAUR-1 bus and launched by Cosmos 3M. Col Ye S. Shchapov was in charge of Sfera design and production. The satellites comprised a cylindrical body with solar cells and radiators of the thermostatic temperature regulating system attached to the exterior. Orientation was by single-axis gravity gradient boom. NPO Prikladnoi Mekhaniki built them.

Launch date	Cosmos	Incl deg	Period min	Perigee km	Apogee km
20 February 1968	203	74.1	109.2	1,178	1,208
4 June 1968	Failed to reach orbit				
30 November 1968	256	74.1	109.5	1,175	1,227
17 March 1969	272	74.0	109.4	1,181	1,211
24 November 1969	312	74.0	108.6	1,144	1,179
28 April 1971	409	74.0	109.4	1,177	1,216
20 November 1971	457	74.0	109.5	1,185	1,221
25 March 1972	480	83.0	109.2	1,175	1,203
21 December 1972	539	74.0	113.0	1,343	1,383
8 September 1973	585	74.0	113.6	1,368	1,416
29 April 1974	650	74.0	113.5	1,369	1,402
29 August 1974	675	74.0	113.7	1,365	1,426
12 February 1975	708	69.2	113.6	1,369	1,413
24 September 1975	770	82.9	109.2	1,169	1,210
24 November 1977	963	82.9	109.4	1,182	1,210
26 December 1978	1,067	83.0	109.1	1,158	1,208

METEOROLOGICAL

United States

Defense Meteorology Satellite Program (DMSP) Series

Background

The US military launched 39 military meteorological satellites beginning in 1965 through 1999. The military uses these satellites to forecast, identify and track developing patterns of severe weather, hurricanes and typhoons and to prevent imaging reconnaissance satellites and aerial surveys from wasting film and flying time on targets that are covered by clouds. In recent years, the USAF has also begun to build three dimensional cloud profiles to optimise aircraft route planning, and during the Cold War and since, these same satellites have watched the atmosphere for telltale signs of nuclear blast effects. The military provides its information about the atmosphere using two satellites in near-polar Sun-synchronous orbits to both ground and shipborne stations to give warning of thunderstorms, typhoons and hurricanes.

After the first reconnaissance satellites returned many useless photos because of cloud cover at the target, the air force proposed the DMSP system to sample the target ahead of photo-reconnaissance satellites and only use the film if the conditions at the target warranted it. The mission requires two satellites in Sun-synchronous orbits. The first crosses the equator at 6 am local time and the second at 10.30 am. When the satellites go off line a replacement can wait no longer than 90 days.

When the Thor Burner booster became available in 1965 the air force pressed ahead with its plans for the Block 4A DMSP. The first of the satellites appeared in 1965 on Thor Burner. The air force launched 10 satellites on 19 January 1965, 18 March 1965, 20 May 1965, 10 September 1965, 8 January 1966, 31 March 1966, 16 September 1966, 8 February 1967, 23 August 1967, 11 October 1967, 22 May 1968, 22 October 1968 and 22 July 1969. The Block 5A series were launched on 11 February 1970, 2 September 1970, and 16 February 1971. The next generation, the Block 5B, switched to a more powerful Thor Burner combination: 14 October 1971, 24 March 1972, 9 November 1972, 17 August 1973 and 16 March 1974. Block 5C satellites were launched on 9 August 1974, 24 May 1975 and19 February 1976.

The five Block 5D-1, Advanced Meteorological Satellites, placed in 850 km, 98° inclinations with 1½-year operational lifetimes had a spotty record: one failed after launch and another never achieved full operational status. The air force launched AMS-1, 11 September 1976; AMS-2, 5 June 1977; AMS-3, 1 May 1978; AMS-4, 6 June 1979; AMS-5, 14 July 1980 but this last one failed to reach orbit. When Block 5D-2 appeared it offered additional redundant systems and three-year operational life, adding 155 kg to the 613 kg mass. This last mass addition required the air force to switch to the more powerful Atlas. The first 5D-2 (2–01 SI) orbited on 21 December 1982 by Atlas E from Vandenberg into a 724 km Sun-synchronous orbit; the second a 751 kg. 2–02 S2 on 18 November 1983 by Atlas E from Vandenberg, and the next also a 751 kg satellite launched 20 June 1987 went into a 816 × 833 km, 98.7° orbit. Its primary sensor failed after 47 months.

DMSP has grown in complexity and sophistication in several stages. The air force called the next version of the satellites Block 5D-2. The air force purchased nine (S6–14) of these satellites. Both of the original DMSP 5D-2s are now well beyond their three-year design lives, but an upgraded 823 kg DMSP 5D-2 (S8) replaced the original satellites after being launched by Atlas E from Vandenberg SLC-3W in June 1987. S8 and S9 still operated well when S10 joined them on 1 December 1990, guaranteeing DMSP services during Operation Desert Storm in the Gulf. S9's appeared about to fail during this period and the air force hoped to use S10 to maintain services, but S10's eccentric orbit reduced its effectiveness. S8 failed at the end of 1991 and the replacement S11 arrived promptly after launch 28 November 1991. S12 appeared on 12 August 1994 and S13 in March 1995, on the last Atlas E in the inventory. Both remained fully operational at the end of 1995, supported by three on standby with degraded capabilities. Finally, S14 was launched 4 April 1997 by Titan 2.

In May 1994, President Clinton approved a long debated plan to combine DMSP and the civil NOAA system, halving the number of operational satellites to two. The first joint satellite is expected in 2005, controlled by NOAA from its Suitland, Maryland centre. Suitland will take over DMSP control from 1998. DoD will close its own stations, but retain one as backup.

The USA instituted an Integrated Program Office on 1 October 1994, headed by James T Mannen to set up the joint operations. The plan projects a US$300 million saving through 1999, and then at least US$1 billion over the next 15 years. Until March 1996, it was planned that two US$100 million Phase 1 contracts for system demonstration, risk reduction and prototyping would begin August 1996,

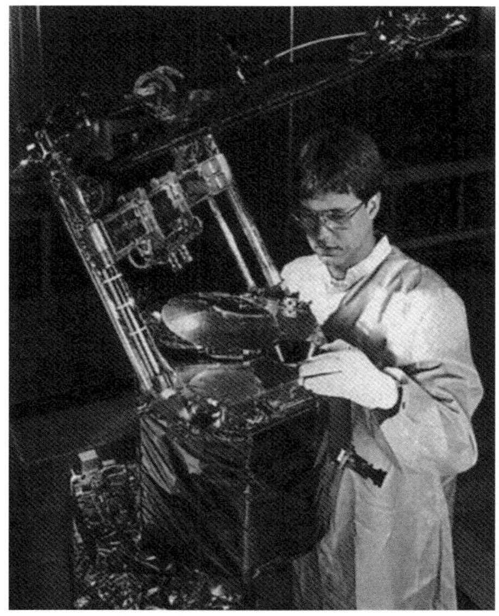

Hughes' DMSP microwave imaging sensor. NASA's Tropical Rainfall Mapping Mission has an enhanced version 0516839

but the programme is being restructured as a result of budget constraints and performance of current generations. Five or six satellites may be ordered, launched on Delta 2-class vehicles and offering 5½-year lives. Phase 2, starting in 1999, was to cover two satellites, and Phase 3 three or four. The USA will also fly instruments on Europe's Metop satellites, due 2001 and 2006.

Lockheed Martin Astro Space received the contract for the new generation of satellites valued at US$228.2 million in July 1989 for five satellites (S15–19). They are expected to feature increased onboard processing and anti-jamming protection. Similar in appearance to their predecessors, some US$40 million of the US$60 million unit cost is attributed to the instrument package, which Aerojet will provide for US$62.1 million for three SSMIS Special Sensor Microwave Imager Sounders. The sounders combine the functions of the current Barnes Engineering humidity profiler, Aerojet's own temperature sounder and Hughes' microwave imager. The air force US$7.9 million option was exercised in January 1990 for a fourth sensor, the US$8.8 million fifth followed in August 1991. Each 57 kg, 70° west sensor will provide 19 microwave channels over 19 to 183 GHz for observations from 30 km to ground level, including rainfall to an accuracy of 5 mm/h and ocean wind speeds to within 2 m/s. Temperature profiles up to 73 km will also be generated.

S15 will carry STP's POGS II Polar Orbiting Geomagnetic Survey magnetometer on a 5 m boom to update geomagnetic maps. S16–19 will also carry two ionosphere instruments for observing UV emissions for improved prediction of disruptions to communications and Over-the-Horizon radar systems. The Space Forecast Center at Falcon AFB will be responsible for processing

DMSP 5D-2, the current model. Among other applications, DMSPs prevent reconnaissance satellites wasting resources when targets are cloud-covered (Lockheed Martin Astro Space) 0516838

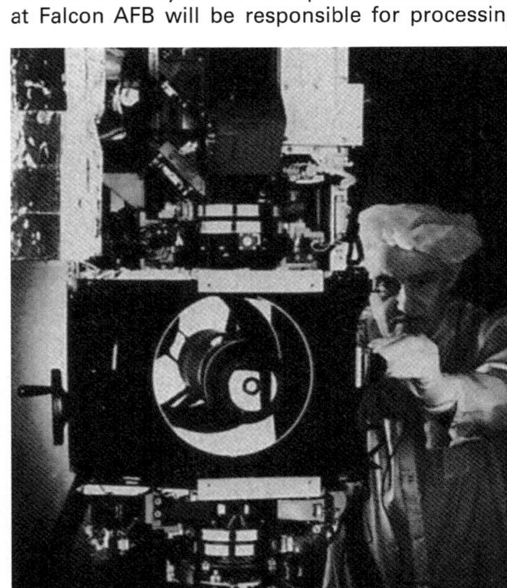

DMSP's primary imager is the Operational Linescan System (Westinghouse Electric Corporation) 0516840

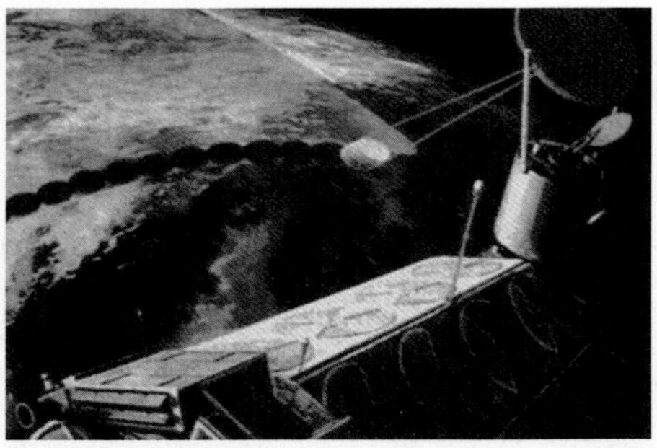

DMSP Block 5D3 will incorporate the first Special Sensor Microwave Imager Sounder instrument
0516841

the data into warnings. Johns Hopkins University's Applied Physics Lab is providing the USAF instrument for nadir viewing, while the USN's own Naval Research Lab is building the complementary horizon-viewing unit. The NRL's RAIDS Remote Atmospheric & Ionospheric Detection System test version was planned for NOAA 13 in 1994 but it had to be removed and may now never fly.

S15 finally launched on 12 December 1999 by Titan 2 to a 837 × 851 km, 98.9° orbit.

The air force is expected to issue a Block 6 contract in 1998 for first spacecraft delivery in 2004–05. Martin Marietta Astro Space and Lockheed Missiles & Space were awarded US$8.2 million and US$8.3 million contracts, respectively, July 1991 for sensor risk reduction experiments. A multispectral instrument was under consideration in addition to meteorological sensors. A radar altimeter would have continued the ocean topography work of the GFO Geosat Follow-On. Block 6 is now submerged within the joint DMSP/NOAA programme.

The air force, through the USAF Space Command, manages the defence Meteorological Satellite Program. The AF Space Command's 6th Space Operations Squadron (part of 50th Space Wing, Falcon AFB, performs TT&C at Offutt AFB's Satellite Operations Center. Each satellite uses a visible/IR OLS Operational Linescan System. An 85.5 GHz Microwave Imaging Sensor (an enhanced version is on NASA's TRMM) measures ocean surface wind speed, ice coverage/age, areas and intensity of precipitation, cloud water content and land surface moisture; the microwave temperature sounder returns cloud temperature profile data. The precipitating electron spectrometer monitors electron and proton densities at different energy levels, and the electrostatic ionosonde locates auroral boundaries to aid radar and long-range radio operations. There is also a Gamma/X-ray Spectrometer and Magnetometer on board.

Data are provided in real-time to tactical ground stations and navy ships worldwide. They can be encrypted at times of hostility. They are also stored onboard for transfer via US readout stations and then commercial telecommunication satellites to the US Air Force Global Weather Center at Offutt AFB, Nebraska and the US Navy's Fleet Numerical Meteorology and Oceanography Center in Monterey, California. The command readout stations are sited at Fairchild AFB (Washington), Keene Pt (Hawaii) and Loring AFB (Maine; Loring's TT&TC facility closed FY89); Thule AFB in Greenland provides TTETC support. DMSP information is also made available for civil use through NOAA, and much of it is kept on file at NOAA's depository at Colorado University. Forces equipped with Mark 4 truck-mounted terminals receive the imagery.

Ground forces used six mobile receiver units during the Gulf War but their bulk (1.3 tonnes) required 75 per cent of a C-141 cargo transport plane and so slowed their introduction into the theatre. In an ironic twist, tactical users in the forward lines of the war carried commercial receivers for civil satellites.

Specifications

DMSP 2–03 (S8)
Carries a microwave imaging sensor intended to identify tropical storms earlier than previous systems. The sensor can penetrate clouds to measure storm intensity, instead of measuring only the cloud tops; the system can also determine wind speeds over the ocean, measure ground and snowpack moisture content, determine ice coverage and indicate whether rain is falling over land.

DMSP 2–04 (S9)
Launched: 3 February 1988 by Atlas E from Vandenberg to replace 2–02. 823 kg; 820 × 829 km, 98.77°. Was to be replaced by 2–05 but operations continued. See 2–03 for comments.

DMSP 2–05 (S10)
Launched: 1 December 1990 by Atlas E from Vandenberg SLC-3W at 15.57 GMT into 732 × 849 km, 98.9°. Low-perigee (almost 50 trackable fragments indicate solid motor explosion) reduced effectiveness. Intended to replace 2–04, guaranteeing services during Operation Desert Storm in the Gulf.

DMSP 2–06 (S11)
Launched: 28 November 1991 by Atlas E from Vandenberg SLC-3W into 840 × 857 km, 98.9°. Replaced 2–03.

DMSP 2–07 (S12)
Launched: 29 August 1994 by Atlas E from Vandenberg SLC-3W at 17.38 GMT into 843 × 859 km, 98.9°. Replaced 2–04/2–05.

DMSP 2–08 (S13)
Launched: 24 March 1995 by Atlas E from Vandenberg SLC-3W at 14.05 GMT into 847 × 854 km, 98.83°. Replaced 2–06.

DMSP 2–09 (S14)
Launched: 4 April 1997 by Titan 23G from Vandenberg SLC-3W into 844 × 855 km, 98.94° orbit.

DMSP 2–10 (S15)
Launched: 12 December 1999 by Titan 23G from Vandenberg into a 837 × 851 km, 98.9° orbit.
Contractors: Lockheed Martin Astro Space (prime), Westinghouse Electric Corp (Operational Linescan imaging system), Hughes Aircraft (microwave imager), Aerojet (microwave temperature sounder), Barnes Engineering (infra-red temperature/moisture sounder)
Resolution: typically 0.5 km
Design life: 48 months
Mass: 823 kg at launch, including 180 kg sensor payload
Orbit: typically 820–860 km Sun-synchronous at 98.8°, providing repeated passes at either 6 am or 10.30 am local time. Scan width 2,960 km, creating continuous coverage at equator and global coverage in 12 h
Spacecraft: based on Astro Space's NOAA Tiros bus, 1.22 m wide bus, 6.4 m long with single solar wing (1 kW from 8 panels of 12,500 Si cells) deployed. 127 kg structure provides 5.4 m^2 of payload mounting area. Zero momentum 3-axis control by MWs plus magnetorquers; 4 hydrazine (15.9 kg propellant) plus 8 N_2 (2.3 kg propellant) thrusters. Earth pointing accuracy of 0.01° provided by strapdown star sensor and gyros; 0.12° accuracy backup system of static Earth sensor, Sun sensor, gyros. Final orbit insertion provided by Thiokol Star 37S solid. TT&C assigned to AF Space Command's 6th Space Operations Squadron (part of 50th Space Wing, Falcon AFB, which can also provide control) at Offutt AFB's Satellite Operations Center. Data downlink on 2,207.5, 2,252.5, 2,267.5 MHz; telemetry downlink 2,237.5 MHz; command uplink 1,792 MHz at 2 kbit/s (minimum of 300 stored commands); data stored on Odetics DDS-5000 tape recorder. Passive thermal control provided by finishes and radiators; active control by electrically-controlled louvres and heaters (thermal subsystem mass 24 kg)
Sensors: primary meteorological imaging is provided by the visible/infra-red OLS Operational Linescan System. The system includes an 85.5 GHz Microwave Imaging Sensor (SSM/I, an enhanced version rides on NASA's TRMM) measures ocean surface wind speed, ice coverage/age, areas/intensity of precipitation, cloud water content and land surface moisture, the microwave temperature sounder (SSM/T-1) returns cloud temperature profile data, the SSM/T-2 microwave water vapour sounder will be added later for atmospheric humidity monitoring to improve global forecasting. The precipitating electron spectrometer (SSJ/4) monitors electron plus proton densities at different energy levels, and the electrostatic ionosonde locates auroral boundaries to aid radar and long-range radio operations. Gamma/X-ray Spectrometer (SSB/X2) and SSM Magnetometer also carried.

DMSP 2–10 (S15)
Launched: 12 December 1999 by Titan 23G from Vanderberg SLC4W into 837 × 851 km, 98.9° orbit.
Contractors: Lockheed Martin Astro Space (prime), Westinghouse Electric Corp (Operational Linescan imaging system), Aerojet (Special Sensor Microwave Imager Sounder)
Resolution: typically 0.5 km
Design life: 60 months
Mass: 1,154 kg. on-station
Orbit: as 5D2
Spacecraft: based on 5D2 bus, 1.22 m wide bus, 6.4 m long with single solar wing (10 panels of Si cells) deployed. 187 kg structure provides 6.5 m^2 of payload mounting area. Zero momentum 3-axis control by MWs + magnetorquers; 4 hydrazine (21.8 kg propellant) plus 8 N_2 (3.6 kg propellant) thrusters. Earth pointing accuracy of 0.01° provided by strapdown star sensor plus gyros; 0.12° accuracy backup system of static Earth sensor, Sun sensor, gyros. Final orbit insertion provided by Thiokol Star 37FP solid (19 per cent offload). TT&C and frequency details as 5D2. Passive thermal control provided by finishes and radiators; active control by electrically-controlled louvres and heaters (thermal subsystem mass 29.5 kg)

MINOR MILITARY

Russian Federation

Introduction

Current Status

Russian literature has now identified three classes of satellites within the 'minor military' series: Yug (south), Vektor (vector) and Romb (rhombus). All three appear to have been designed by a laboratory at the Mozhaisky Military Engineering Space Academy (VIKA) in St. Petersburg. After subtracting the Yug, Vector and Romb satellites from the list of satellites generally classified as 'minor military' or 'radar calibration' in the west, there is still a large number of satellites not identified with these groups. None of these satellites produced the additional sub-satellites which are associated with Romb missions although in many cases the satellites were flown to Romb-type orbital altitudes.

Background

Type 1

The late Dr Charles Sheldon coined the term 'minor military' to describe families of Soviet satellites for which no other information had leaked to the west, but which appeared to have a military purpose. Dr Sheldon filed his reports with the US Library of Congress, and so the coinage appears in most official US documents describing the Soviet era space programme. Most minor military satellites fly at low altitude and their purpose has been suggested (but not confirmed) to include radar calibration, atmospheric drag measurement and spacecraft technology experiments. In 1993 one Soviet era official acknowledged that some dedicated missions each carry a 2 m sphere for passive atmospheric monitoring. Others carry 1 m spheres as sub-satellites on other missions.

The categorisation of Dr Sheldon covers four orbital inclinations with two subclasses. The Soviets placed the first subclass in eccentric orbits about twice a year. Between 1980 and 1984 they launched six by the Kosmos booster into 300 to 400 × 1,600 to 2,000 km orbits. Then a hiatus occurred until the next such satellites appeared on 28 August 1990 as Kosmos 2098 entered a 396 × 1,977 km, 83.0° inclination orbit. Kosmos 2265 (26 October 1993) released a 2 m sphere into 291 × 1,574 km, 82.9° orbit for passive atmospheric measurements, and the follow on satellites Kosmos 2292 (27 September 1994) and Kosmos 2332 (24 April 1996) performed similar functions. The Soviets tinkered with the basic design before launching a new variant in 1989, on Kosmos 2002 (14 February 1989) placed in 186 × 2,307 km, 65 to 85° orbit. Kosmos 2002 for the first time, released 10 objects shortly after launch and then itself decayed naturally during 15 October 1989. Kosmos 2059, 6 February 1990 performed a similar operation.

Type 2

The second sub-class has less eccentric, more circular orbits having 92–95 minute periods. These satellites decay naturally in only a few years but many operate simultaneously. Some minor military satellites – C 1501 (30 September 1983, 468 by 614 km. 82.94 degree inclination), Kosmos 1601 (27 September 1984, 475 × 516 km, 65 to 84° inclination), and Kosmos 1662 seem to be cylinders about 4 m long and 2 m in diameter. They periodically jettison small fragments about 0.1 m^2 in radar cross-section. Originally, they dropped 22 to 24 objects, but after Kosmos 1601 the number rose to 28. Kosmos 2075, launched 25 April 1990 appears to be the last of this series.

The released objects decay much more rapidly than the parent body and may simulate MIRVs (Multiple Independent Re-entry Vehicles) to give the Russians radar detection system practice for engaging possible US nuclear warheads. Kosmos 2306 (2 March 1995, 469 × 517 km, 65.9° inclination) immediately released four sub-satellites. Three minor military launches occurred in 1991, all using SL-8 Kosmos. Kosmos 2137 entered 449 × 495 km, 65.8° inclination orbit on 19 March and Kosmos 2164 entered 286 × 708 km, 74° inclination orbit. The launch of 25 June should have placed its payload into 485 × 515 km, 74° inclination orbit but stage 2 failed to re-ignite and prevented orbital injection.

Type 3

A new larger type of minor military satellite, possibly an Elint, debuted in late 1988. The launch on 23 December had the name Kosmos 1985. It had a 622 × 633 km, 73.6° inclination and was the first minor military satellite ever launched by SL-14 Tsyklon. Four small sub-satellites soon appeared from the main bus and two more emerged in late January 1989. By 1990, a total of 36 objects had appeared on radar tracks. The satellite Kosmos 2053 (27 December 1989) appears to be the second of this type ever launched. The third, Kosmos 2106, entered a higher inclination (82.5°) 28 November 1990.

Launch Date	Cosmos	Incl deg	Period min	Perigee km	Apogee km
24 July 1975	752	65.9	94.6	481	515
17 December 1976	885	65.8	94.4	467	512
2 February 1977	891	65.8	94.5	473	516
22 July 1977	933	65.8	92.5	384	408
12 May 1978	1006	65.8	92.4	382	408
8 February 1979	1075	65.8	94.5	473	516
4 December 1979	1146	65.8	94.0	444	494
27 March 1980	1169	65.8	94.5	477	515
23 September 1981	1310	65.8	94.5	477	518
21 October 1982	1418	50.7	92.3	371	413
29 December 1982	1427	65.8	94.0	445	499
25 January 1983	Launch failure				
6 April 1983	1450	65.9	94.4	472	512
5 October 1983	1502	65.8	92.3	369	412
26 January1984	1534	65.8	94.5	469	517
20 December 1984	1615	65.8	94.0	441	501
27 February 1985	1631	65.8	94.5	472	512
27 October 1986	1788	65.8	94.5	468	517
15 December 1987	1902	65.8	92.4	368	411
14 July 1988	1958	65.8	92.4	371	413
14 June 1989	2027	65.8	94.6	482	513
19 March 1991	2137	65.8	94.0	449	495

NAVIGATION

China

Navigation Satellite Programme

Background
In the late 1980s an undated and untitled brochure issued by China Great Wall Industries Corporation included a photograph of an early *Shi Jian* class satellite which was simply captioned 'Navigation satellite'. The satellite appeared to use the same satellite bus as *Shi Jian* 2, but with the solar cells mounted on the body of the satellite rather than on folding solar panels. At the time the Chinese had no acknowledged navigation satellite programme, and when their representatives were asked at the 1989 Paris Air Show the response can best be described as a 'collective shrug of their shoulders'. It is not known whether the photograph was simply incorrectly captioned or whether it represented a Chinese navigation satellite programme which never reached flight status.

A paper presented at the International Astronautical Federation Congress in 1994 stated that two communications satellites in the DFH-2 and DFH-2A series had been used for some navigation satellite related tests in 1989. The specific satellites were not identified, but there are three candidates which were operating at that time:

STTW 1	(DFH-2)	over 103° E
STTW 2	(DFH-2A)	over 87.5° E
STTW 3	(DFH-2A)	over 110.5° E

Since the paper specified that a DFH-2 satellite was used then one of the satellites must have been STTW 1: if STTW 2 were to be the second satellite then a longer baseline would result than if STTW 3 had been used.

The 1994 paper described a planned navigation satellite system called Twin Star, which would use satellites deployed in geosynchronous orbit. The ground user would transit a signal to the two satellites and this is re-transmitted to a ground control centre. Using the phase difference in the signals as received by the satellites the ground control system will automatically calculate the location of the user, transmit this information to the satellites and this will be re-transmitted to the user.

The 1994 paper discussed two satellites in geosynchronous orbit with their longitudes 40° or more apart: actual longitudes were not quoted, but a sketch map seemed to suggest locations over 85° E and 132.5° E approximately. Specific accuracy figures for the system were not quoted, but it was stated that it would be dependant upon the distance between the two satellites (the greater the distance the greater the accuracy), the accuracy of range measurements and the altitude of the user.

Bei Dou
With no advance warning, the first *Bei Dou* (BD) satellite was launched on October 30, 2000: it was not clear whether this was BD 1 or the first in the BD-1 series. The name *Bei Dou*, meaning 'north dipper', comes from the seven stars which are called The Plough in the UK and The Big Dipper in the US, these stars being part of the larger constellation Ursa Major. Two stars in The Plough point almost directly to Polaris, the North Pole Star, and therefore are a famous navigation aid known as The Pointers.

The location of BD 1 over 140° E was registered by the Chinese as ChinaSat-32, and therefore it was expected complete set of three BD satellites will use the ChinaSat –31, –32 and –33 locations. This was verified with the launch of a third BD satellite in 2003.

The Chinese have also registered four geosynchronous orbit locations named *Compass* with the International Telecommunications Union, and this name appears to be indicative of a navigation role. The four *Compass* locations are simply identified by their longitudes and these are: 58.75° E (registered via the ITU as *Compass*-58.75), 80° E, 110.5° E and 140° E. It will be noted that the second, third and fourth locations are the ChinaSat-3X series and used by the BD satellites. It remains to be seen whether a fourth BD satellite will go to the unoccupied *Compass*-58.75 location or replace/work alongside one of the existing satellites.

Specifications

Bei Dou 1
Launch: 30 October 2000 using a CZ-3A from Xi Chang
Location: 140° E (not announced by China: this is derived from US tracking data): Chinasat-32 location
Design life: >8 years?
Contractor: China Academy of Space Technology
Configuration: Uses the DFH-3 satellite bus: 3-axis stabilised, 2.2 × 1.72 × 2 metres box plus two vanes of solar panels: on communications satellites. The solar panels span 18.1 metres.
Mass: ~2,300 kg at launch, ~1,150 kg on-station at the beginning of operations
Power system: solar panels provide ~2 kW BOL, 1.7 kW at the EOL

Bei Dou 2
Launch: 20 December 2000 using a CZ-3A from Xi Chang
Location: 80° E (not announced by China: this is derived from US tracking data): Chinasat-31 location
Other specifications as BD 1

Bei Dou 3
Launch: 24 May 2003 using a CZ-3A from Xi Chang
Location: 110.5° E (not announced by China: this is derived from US tracking data): ChinaSat-33 location. Other specifications as BD 1.

Bei Dou 4
Launch: Scheduled for 2004 from Xi Chang using CZ-3A
Location: Not known whether BD 4 will replace/supplement an existing satellite or use the *Compass*-58.75 location.
Other specifications as BD 1.

Russian Federation

Introduction

Current Status
Russia maintains three navigational satellite programmes, one low altitude military and one low altitude civil and the Russian equivalent of GPS known as Glonass. The first two employ satellites in 1,000 km, 83° inclination orbit and give positions to within 80 to 100 m, or roughly Transit-type positioning, for navy/merchant marine vessels equipped with Shkhuna terminals. Both programmes are inter-related in that they can use common transceivers and ground control equipment.

Both the civil and military systems use 810 kg, 2 m in diameter gravity gradient satellites which fly from Plesetsk on SL-8 Kosmos vehicles. A receiver establishes its position from the Doppler shift of various satellites in the constellation employing the same method as Transit and uses similar VHF frequencies (150 to 400 MHz) as Transit. More than 1,400 ships around the world have terminals to receive both systems' signals. Extra accuracy comes from the military's six satellite constellation as opposed to the four satellite civil equivalent.

Glonass Series

Current Status
In July 2000 Russia offered China the chance to participate in the Glonass programme. At this date there were only 14 satellites available and only nine fully operable. The system would require Rb1.5 billion a year to operate and replenish the system. Further blows to Russia's attempt to make Glonass the cornerstone of a new European competitor to the US Navstar system came with the European decision to develop the independent system Galileo. China has snubbed Russia's approach to join the Glonass programme by agreeing to provide financial support to the Galileo system.

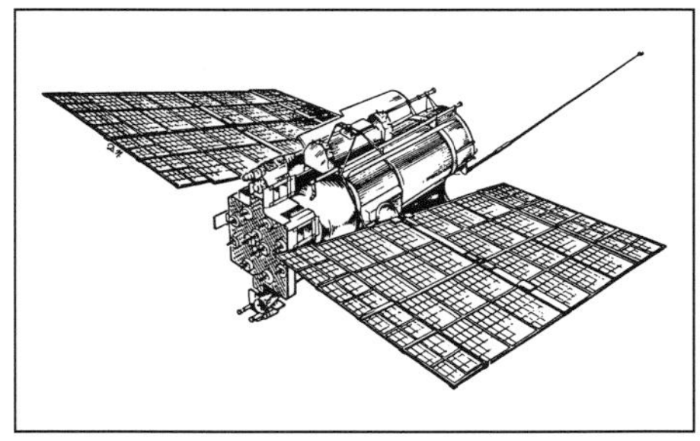

Glonass navigation satellites are launched by Proton, typically in triplets. Height is 7.840 m with magnetometer boom deployed; 7.230 m span across solar wings (Nicholas L Johnson/Kaman Sciences)
0517156

Background
The GPS equivalent Global Navigation Satellite System, Globalnaya Navigatsionnaya Sputnikovaya Sistema (Glonass) system has been under deployment since 1982 and became fully operational in December 1995. Glonass follows in a long tradition of Soviet navigation satellites going back to 1967's Kosmos 192 'Tsyklon'. The Soviet Union refused to publicly acknowledge it had a navigation satellite system until the launch of Kosmos 1000 in 1978. Then the Soviet Union informed the International Telecommunications Union it would inaugurate the Glonass service in 1982. On 12 October 1982, an SL-12 Proton launched the first three satellites on a single booster from Tyuratam and identified them as Kosmos 1413–1415. In order to achieve the constellation spacing the satellites first entered a low altitude parking orbit at 51–60 kms. Then the upper stage performed two manoeuvres to place the three payloads into their required orbits. The Soviets repeated this pattern on their next five launches during the next three years. Curiously, each trio had only one or two Glonass pre-operational satellites and the remaining seven were dummies, although still called Kosmos and given Kosmos numbers. The 10 operating satellites weighed 1,260 kg and had one-year design lives. But they achieved an average of 14 months. During this pre-operational testing phase the Soviets only had six satellites simultaneously operating at any one time. The last launch of the pre-operational phase included the first Block 2a, transmitter with improved clocks. Six Block 2a satellites eventually appeared and each had an average life of 17 months.

With the pre-operational phase complete, the Soviets entered into a new phase and launched the satellites on a more direct profile, to distinguish it from the earlier phase. Each Proton carried the new satellites directly into an initial inclination of almost 65°. This track surprised long time space observers because it was the first time the Soviets had used it in almost 20 years. Again, two manoeuvres by the upper stage spaced the satellites properly, but no orbital plane change occurred during this manoeuvre. All subsequent missions have used this more direct route.

With September 1986's launch, the Soviets put up the first payload in which all three satellites were live. But a launch failure marred the debut of the Block 2b, which the Soviets claimed had a 24-month design life. The Soviets launched 12 Block 2Bs between 1987 and 1988, but only half reached orbit. These satellites achieved a 22-month average life. The 1,415 kg Block 2v first appeared in September 1988. The new satellite has a reported three-year lifespan and is the basis of the current Glonass system.

Two 1990 missions raised the number of operational craft to 11 and 1992's first launch maintained the active number at 11. There were 13 active satellites by September 1992. Lifetimes began to improve significantly following the launch on April 1994, which had a radiation, hardened satellite. This launch also filled every slot in planes one and three at some time. Then August 1994's launch began the process of filling plane two.

Glonass planners wanted to have a Phase 1 system of 10 to 12 operational satellites ready by the end of 1990, but didn't realise that goal until 11 March 1992. From this original deployment the Russians planned to grow the system to a full operational capability of 21 craft, plus three in-orbit spares by the end of 1995. Still, by end of 1993 the

Glonass satellites	Launch		Plane	Slot	Block
1	12 October 1982	Cosmos 1414	1	1	1
2	10 August 1983	Cosmos 1490	1	3	1
3		Cosmos 1491	1	2	1
4	29 December 1983	Cosmos 1519	3	18	1
5		Cosmos 1520	3	17	1
6	19 May 1984	Cosmos 1554	3	19	1
7		Cosmos 1555	3	18	1
8	4 September 1984	Cosmos 1593	1	2	1
9		Cosmos 1595	1	3	1
10	17 May 1985	Cosmos 1650	1	1	1
11		Cosmos 1651	1	2	2a
12	24 December 1985	Cosmos 1710	3	18	2a
13		Cosmos 1711	3	17	2a
14	16 September 1986	Cosmos 1778	1	2	2a
15		Cosmos 1779	1	3	2a
16		Cosmos 1780	1	8	2a
17*	24 April 1987	Cosmos 1838	3	2b	
18*		Cosmos 1839	3	2b	
19*		Cosmos 1840	3	2b	
20	16 September 1987	Cosmos 1883	3	18	2b
21		Cosmos 1884	3	17	2b
22		Cosmos 1885	3	24	2b
23*	17 February 1988	Cosmos 1917	1	2b	
24*		Cosmos 1918	1	2b	
25*		Cosmos 1919	1	2b	
26	21 May 1988	Cosmos 1946	1	8	2b
27		Cosmos 1947	1	7	2b
28		Cosmos 1948	1	1	2b
29	16 September 1988	Cosmos 1970	3	24	2v
30		Cosmos 1971	3	18	2v
31		Cosmos 1972	3	19	2v
32	10 January 1989	Cosmos 1987	1	2	2v
33		Cosmos 1988	1	3	2v
34	31 May 1989	Cosmos 2022	3	19	2v
35		Cosmos 2033	3	24	2v
36	19 May 1990	Cosmos 2079	3	17	2v
37		Cosmos 2080	3	19	2v
38		Cosmos 2081	3	20	2v
39	8 December 1990	Cosmos 2109	1	4	2v
40		Cosmos 2110	1	7	2v
41		Cosmos 2111	1	5	2v
42	4 April 1991	Cosmos 2139	3	21	2v
43		Cosmos 2140	3	22	2v
44		Cosmos 2141	3	24	2v
45	29 January 1992	Cosmos 2177	1	3	2v
46		Cosmos 2178	1	8	2v
47		Cosmos 2179	1	1	2v
48	30 July 1992	Cosmos 2204	3	24	2v
49		Cosmos 2205	3	18	2v
50		Cosmos 2206	3	21	2v
51	17 February 1993	Cosmos 2234	1	3	2v
52		Cosmos 2235	1	2	2v
53		Cosmos 2236	1	6	2v
54	11 April 1994	Cosmos 2275	3	17	2v
55		Cosmos 2276	3	23	2v
56		Cosmos 2277	3	18	2v
57	11 August 1994	Cosmos 2287	2	12	2v
58		Cosmos 2288	2	16	2v
59		Cosmos 2289	2	14	2v
60	20 November 1994	Cosmos 2294	1	3	2v
61		Cosmos 2295	1	6	2v
62		Cosmos 2296	1	4	2v
63	7 March 1995	Cosmos 2307	3	20	2v
64		Cosmos 2308	3	22	2v
65		Cosmos 2309	3	19	2v
66	24 July 1995	Cosmos 2316	2	15	2v
67		Cosmos 2317	2	10	2v
68		Cosmos 2318	2	11	2v
69	14 December 1995	Cosmos 2323	2	9	2v
70		Cosmos 2324	2	–	2v
71		Cosmos 2325	2	13	2v
72	30 December 1998	Cosmos 2362	1	–	2v
73		Cosmos 2363	1	–	2v
74		Cosmos 2364	1	–	2v
75	13 October 2000	Cosmos 2376	1	–	2v
76		Cosmos 2374	1	–	2v
77		Cosmos 2375	1	–	2v
78	1 December 2001	Cosmos 2380	2	–	2v
79		Cosmos 2382	2	–	2v
80		Cosmos 2381	2	–	2v
81	25 December 2002	Cosmos 2394	3	–	2v
82		Cosmos 2395	3	–	2v
83		Cosmos 2396	3	–	2v

* Launch failure. The slot listing is the initial position for each; some later transferred to other slots.
Note: 1: spare. **Active satellites** January 1996: C2111, C2178, C2179, C2204, C2205, C2235, C2236, C2275–77, C2287–89, C2294–96, C2307–09, C2316–18, C2323–25 (C2324 spare). Active satellites as of 15 December 1993 were: C2079, C2080, C2110, C2111, C2140, C2178, C2179, C2204, C2205, C2206, C2234, C2235, C2236; C1987 was held in reserve.

Russians had 13 active satellites (the oldest from 1990) plus one reserve. The system continued to grow in 1994, building to 16 satellites before the end of that year. When the Russians announced a full deployment of 24 satellites after the launch of three in December 1995 they had thus made up for schedule slippage and achieved a considerable success. At least part of the urgency centred on keeping pace with the west in having a commercial navigational system in place to compete with GPS.

Russia's own Institute of Radionavigation & Time estimated in 1993 that a scant 200 to 500 civil Glonass receivers had made it into widespread use compared with the hundreds of thousands of contemporary GPS receivers. As the Soviets originally conceived the system, maritime and aeronautical users would be the principal consumers of signals allowing positional accuracy within 100 m surface location and 150 m altitude and 15 cm/s velocity readings. Most civil users now report the actual figures are about 20 m of accuracy, better than original expectations primarily because phantom signal ghosting from multiple paths has not proven a serious problem. Unlike GPS, Glonass cannot degrade with selective availability. The P-code yields accuracy about twice as good as the civil code. Although the P-code encryption is being rewritten following the demonstrated ease with which Professor Peter Daly of Leeds University, England was able to crack it. Glonass now offers better coverage than GPS at higher latitudes, because of the constellation configuration but GPS is better nearer the equator.

The system has an operating philosophy that closely mirrors GPS. One difference arises from the geoid model upon which the two systems base their calculations. GPS uses the WGS-84 geodetic systems, Glonass uses SGS-85 (Soviet Geocentric System), adding an error of a few metres whenever a receiver averages between signals from the two systems. Satellites transmit in narrow bands centred around 1.250 and 1.6035 GHz. The higher frequency band creates complaints from radio astronomers that cite its interference with the hydroxyl line at 1.612 GHz. In response the Russians have begun changing the frequencies of some satellites beginning in September 1993. Glonass itself will soon suffer interference in its own upper L1 band from LEO satellite communication systems, such as Iridium, because these systems have been allocated 1.610–1.6265 GHz frequencies. Both problems force Glonass to use only 12 channels, instead of the planned 25, because satellite transmissions are doubled at the poles. By January 1996, nine pairs were doubling up on channels.

In terms of constellation the Glonass does differ from the US system, which uses circular semi-synchronous orbits near 20,000 km (718-minute period). Glonass satellites fly at a slightly lower altitude of 19,100 km (676 minute period) at inclinations of 64.8° which gives a precise retrace of the ground track every 17 revolutions (eight days). The VKS Military Space Forces launch and operate Glonass and monitors it through five principal monitoring and command stations: Moscow, St. Petersburg, Eniseyisk, Komsomolsk-on-Amur and Balkhash. The VKS performs control functions from a single location at Golitsyno-2 in the Moscow region, the core of VKS's command system.

NPO PM has responsibility to produce the satellites which are manufactured at AKO Polyot, using the NIIKP Institute Space Device Engineering's payloads. NIIKP has responsibility for manufacturing both the civil and military receivers. The Institute of Radionavigation and Time maintains the ground time/frequency standards. Individual satellites are called Uragan (hurricane).

The 1,480 kg Glonass M improves stability of the three cesium clocks from 5×10^{-13} daily to 1×10^{-13} and extending the series lifetimes to five to seven years per satellite. The new civilian band includes an ionospheric correction transmission, which allows for compensation based on the comparison of effects on signals of two different frequencies. Passage through the ionosphere generally creates the largest error in a receiver's position. Pending military review the entire system may be uprated again providing inter-satellite links and 60-day satellite autonomy.

Parus Sail Series (Generation 1 and 2)

Current Status

The latest Parus satellite was launched as Cosmos 2389 by Kosmos 11K65M Plesetsk on 28 May 2002.

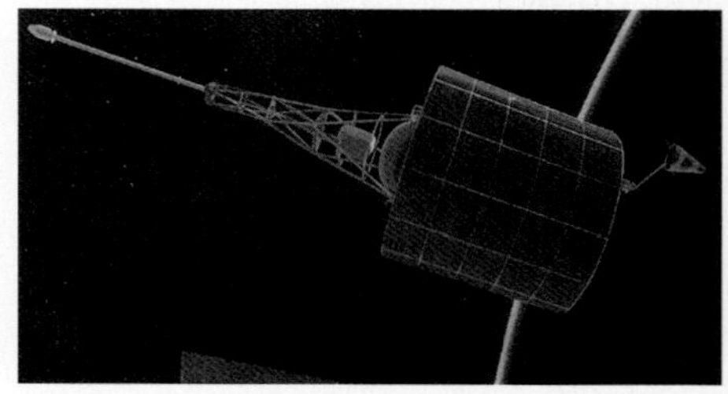

The Parus military navsats are probably almost identical to this Nadezhda civil version. See the Cospas-Sarsat entry for an illustration showing Nadezhda under construction 0517157

Launches of Parus Navigation satellites 1990–2002

Launch Date	Satellite	Node	Plane °E
20 March 1990	Cosmos 2061	90	3
20 April 1990	Cosmos 2074	60	2
14 September 1990	Cosmos 2100	120	4
16 April 1991	Cosmos 2142	150	5
22 August 1991	Cosmos 2154	90	3
27 November 1991	Cosmos 2173	120	4
17 February 1992	Cosmos 2180	180	6
15 April 1992	Cosmos 2184	60	2
1 July 1992	Cosmos 2195	30	1
29 October 1992	Cosmos 2218	90	3
9 February 1993	Cosmos 2233	150	5
1 April 1993	Cosmos 2239	120	4
2 November 1993	Cosmos 2266	30	1
26 April 1994	Cosmos 2279	180	6
22 March 1995	Cosmos 2310	60	2
6 October 1995	Cosmos 2321	[30]	[1]
16 January 1996	Cosmos 2327	30	1
5 September 1996	Cosmos 2334	30	1
20 December 1996	Cosmos 2336	120	4
17 April 1997	Cosmos 2341	60	2
23 September 1997	Cosmos 2346	30	1
24 December 1998	Cosmos 2361	–	–
26 August 1999	Cosmos 2366	90	3
8 June 2001	Cosmos 2378	–	4
28 May 2002	Cosmos 2389	–	1

The nodes are referred to an arbitrary zero point. Cosmos 2321 was left in its transfer orbit due to a malfunction with the second stage of the Cosmos-3M launch vehicle.

Background

The Soviet Union built a low altitude military system similar to the US Transit system to locate and guide its ballistic missile submarines. The first trace of this system appeared on an SL-8 Kosmos launch in 1967. Following a two-phase test programme at 74° inclination between 1967 and 1972, the Soviets declared the system operational following the launch of Kosmos 514, which completed a three-satellite constellation at 83° inclination. Eventually the Soviets phased this system out in 1978 (the last launch, Kosmos 1027, occurred on 27 July 1978). In its place they established the current constellation of six Parus 'sail' satellites in orbital planes spaced 30° apart and inaugurated by Kosmos 700 (26 December 1974). Kosmos 2061 appeared on 20 March 1990, apparently replacing Kosmos 1959; and reconfiguring the constellation as Kosmos 2026, 1904, 2061, 2016, 2034, 2004 in order of the orbital planes. Kosmos 2074 appeared 20 April 1990, replacing Kosmos 1904, and Kosmos 2100 14 September 1990, replaced Kosmos 2016. In 1991, Kosmos 2135 (26 February) replaced Kosmos 2026, Kosmos 2142 (16 April) took up the duties for Kosmos 2034, Kosmos 2154 (22 August) for Kosmos 2061 and Kosmos 2173 (26 November) for 2100. In 1992, Kosmos 2180 (18 February) replaced Kosmos 2004, Kosmos 2184 (15 April) substituted for Kosmos 1904, Kosmos 2195 (1 July) for Kosmos 2026 and Kosmos 2218 (29 October) for Kosmos 2154. By the beginning of 1993, the operational system consisted of Kosmos 2142, Kosmos 2173 and 1992's four satellites. In 1993, Kosmos 2233 (9 February) replaced Kosmos 2141, Kosmos 2239 (1 April) replaced Kosmos 2173 and Kosmos 2266 (2 November) replaced Kosmos 2195. In 1994, Kosmos 2279 (26 April) replaced Kosmos 2180. There were no more launches through 1994, leaving the operational system at the beginning of 1995 as Kosmos 2184, Kosmos 2218, Kosmos 2233, Kosmos 2239, Kosmos 2226. Kosmos 2285 (2 August 1994) remains to be identified although it resembles an original generation Parus, the 74° inclination has been vacant since 1972. In 1995, Kosmos 2310 (22 March) replaced Kosmos 2184, but Kosmos 2321 (6 October) was stranded in 258 × 793 km and 82.94° inclination, because of a stage 2 sticking fuel valve. The orbital plane suggests that it was intended to replace 1993's Kosmos 2266. It was claimed to be usable, but Kosmos 2327 (16 January 1996) then replaced Kosmos 2266.

Tsikada 'Cicada' Series

Background

The system Soviet sources refer to as both Tsikada 'Cicada' and Nadezhda ('Hope', when applied to the Cospas-related satellites) provides location information and warning and distress signal identification. As the Russian republic inherited some of the existing Soviet launch infrastructure, the country has sought to maintain the Tsikada network. It still launches replacement satellites into each civil plane approximately every 26 months, and the payload sometimes includes a Cospas/Sarsat transponder. The original Tsikada system began with the launch of Cosmos 883 on 15 December 1976, a total of 15 including the last on 23 June 1987. After the original constellation no new satellites appeared in 1988 but a replacement took up an orbit at 960 × 1,014 km, 82.96° on 4 July 1989 with the Cospas 4 transponder. Nadezhda 2 (Cospas 5) appeared 27 February 1990 in 956 × 1,021 km, 83.0° inclination orbit. Nadezhda 3 (Cospas 6) followed on 12 March 1991 at 958 × 1,018 km, 82.9° inclination, then Kosmos 2181 on 9 March 1992 at 973 × 1,014 km, 83.0° inclination replaced Naddezhda 1. Kosmos 2230 on 12 January 1993 at 973 × 1,007 km, 82.9° inclination replaced Kosmos 2181. Nadezhda 4 (Cospas 7) on 14 July 1994, replaced Nadezhda 2. The first named Tsikada appeared in 1995 when the Russians launched Tsikada 1, on 24 January 1995, probably replacing Kosmos 2123. Kosmos 2315 appeared 5 July 1995, replacing Kosmos 2230, and including the Kurs marine location/control system.

Guided by hints from Russia and press reports, space analysts expect that the Russians will eventually launch seven Tsikada satellites, the first two in 1996, carry the Elekon payload to provide location and data relay for global store/forward services.

Launch date	Satellite	Node deg E	Plane
27 February 1990	Nadezhda 2	0	14
5 February 1991	Cosmos 2123	315	13
26 February 1991	Cosmos 2135	0	14
12 March 1991	Nadezhda 3	270	12
9 March 1992	Cosmos 2181	235	11
12 January 1993	Cosmos 2230	235	11
14 July 1994	Nadezhda 4	0	14
24 January 1995	Tsikada 1	315	13
5 July 1995	Cosmos 2315	235	11
6 October 1995	Cosmos 2321	–	–
16 January 1996	Cosmos 2327	–	–
5 September 1996	Cosmos 2334	–	–
20 December 1996	Cosmos 2336	–	–
17 April 1997	Cosmos 2341	–	–
10 December 1998	Nadezhda 5	0	14

The nodes are referred to an arbitrary zero point.

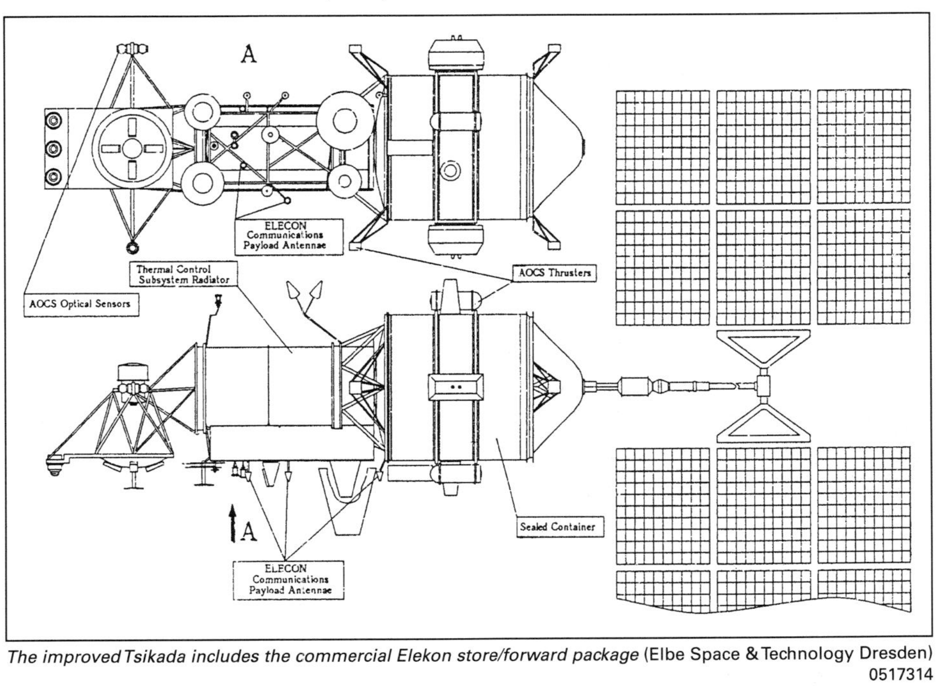

The improved Tsikada includes the commercial Elekon store/forward package (Elbe Space & Technology Dresden)
0517314

Tsiklon (Cyclone) Series

Background

Some observers believe the original Soviet navigation programme began with this series of satellites operated under the name Tsiklon (Cyclone). Satellite uses the basic KAUR-1 bus with a 2,035 m diameter body, radar cells and radiators of thermostatic temperature regulating system on the exterior. Orientation by passive gravity gradient boom. Mass 800 kg. Flights started with the launch of Kosmos 158 (15 May 1967) into a 74°, 740 × 820 km orbit: the series continued with Kosmos 192 (1967), 220 (1968), 292, 304 (both 1969), 332, 358 and 371 (1970), all using a slightly lower orbital altitude. Starting with Kosmos 385, flights used a higher orbital regime, but the same inclination – 74°, 980 × 1,020 km: later flights were Kosmos 422, 465 (both 1971), 475 and 489 (both 1972). Flights then switched to 83°, using the same altitude regime: Kosmos 514 (1972), 574 (preceded by a launch failure on 25 May), 586, 627 (all 1973), 628, 663, 689 (all 1974), 729 (1975), 800, 823, 846 (all 1976), 890, 962 (both 1977), 994 and 1027 (both 1978). By the time that the last Tsyklon series navigation satellite had been launched the Parus and Tsikada launches were underway.

United States

GPS Constellation

GPS: the first Navstar Block 2 satellite (Navstar 14) undergoes solar panel deployment testing at Rockwell's Seal Beach facility. It was delivered to the US Air Force 27 April 1987 as a Pathfinder in the Navstar Processing Facility at Cape Canaveral 0517316

Current Status

As presently planned, GPS III will be the fifth generation system and continue to deliver the new civil signals and improved military codes begun with the GPS IIIM and IIIF satellites. The first launch is planned for no earlier than 2010.

Background

The Navstar Global Positioning System (GPS) is a space-based radio positioning, navigation and time transfer system operating at L-band frequencies, providing 16 m, 0.1 m/s and 0.1 µs accuracy to military users. The operational space segment consists of 24 Navstar (Navigation System using Timing & Ranging) satellites, including three hot in-orbit spares, in six orbital planes (A-F). Each carries four atomic clocks, two caesium with a stability of 1 second in 300,000 years and two rubidium with about half that stability (Block 1 satellites carry only rubidium clocks). The satellites fly in circular 20,182 km, 55° inclination, 11.967 hour (half a sidereal day) orbits, spaced so that a minimum of four appear to a receiver worldwide. They transmit information on L1 (1,575.42 MHz) and L2 (1,227.60 MHz), phase modulated by the data they carry and by spread spectrum pseudorandom number codes.

The GPS programme began in 1973 as a navy initiative to replace the 200 m accuracy of the Transit system. Successful as the Transit had been it did not allow the navy to build accurate SLBMs for its submarine fleet. The navy awarded Rockwell the prime contract in 1974 to build the first 12 Navstars, but then delayed the programme by deciding to harden the satellites and to re-designate the 11 already launched by Atlas as prototypes. So number 12 became a ground qualification unit to prepare for the Block 2 production series. As part of the Cold War the navy then added funding in 1979 to develop an onboard sensor able to report whether the satellite had been illuminated by laser energy or interfered with by another spacecraft. The design review completed in mid-1982 resulted in the Block 2 being one-third larger, with lifetime increased from 5 years to 7 years 6 months, larger solar panels increasing power supply from 400 to 700 W, and the adoption of improved atomic clocks.

An L1 carrier provides the C/A coarse/acquisition code at 1.023 MHz and both LI/L2 are modulated by the P precise code at 10.23 MHz. These codes contain a message at 50 bit/s giving details about the satellite ephemeris, atmospheric propagation correction and clock bias information. By measuring the time interval between transmission and reception, the receiver calculates the distance to a satellite. Using the distance to at least three satellites in an algorithm provides the position fix. A minimum of three satellites signals provides latitude/longitude; the fourth adds altitude.

With the C/A code Standard Positioning Service (SPS) yields about 100 m civil accuracy and the P-code the PPS Precise Positioning Service, intended to provide 16 m accuracy on 50 per cent of occasions. Commercial C/A SPS receivers originally provided 20 to 40 m accuracy, but the C/A code is now deliberately degraded by SA Selective Availability to 100 m by broadcasting slightly erroneous clock and ephemeris information. Differential GPS (DGPS) overcomes SA and provides increased overall accuracy. DGPS places a transmitter at a known location and its position information is used to calculate corrections communicated to other receivers in the area. With DGPS, civilian users get real-time accuracy of 10 to 15 m, with 3 m possible for post-processing calculations in static positions.

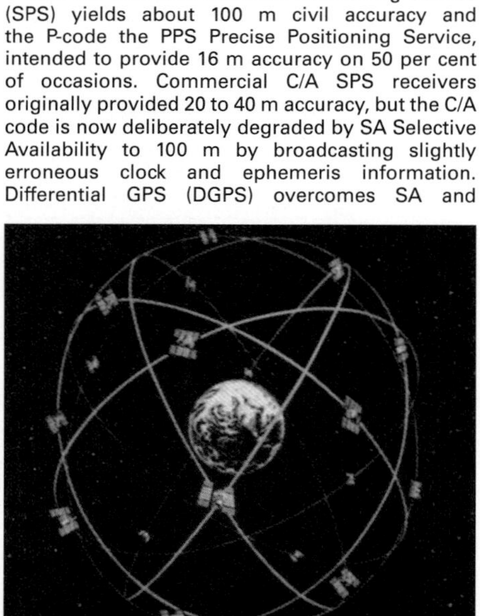

The operational GPS network comprises 24 satellites in six orbital planes. The system can fix the positions of military users to within 16 m and offers a downgraded civil facility (Rockwell) 0516842

GPS has an additional A-S (anti-spoofing) signal to prevent friendly forces being deceived by an enemy deliberately transmitting inaccurate signals. A-S encrypts the P-code to become the Y-code, which is accessed by an auxiliary output chip in the receiver. The military classified early P-code receivers once they had been loaded with the cryptographic material. The new generation incorporate a PPS-SM Precision Positioning Service-Security Module which isolates the security aspects into a sealed tamperproof module so that the receiver as a whole is unclassified when keyed. Current military GPS receivers require a hand-held, hand-inserted cryptographic code. With thousands of users, handling that cryptographic key becomes very cumbersome and expensive. So the United States has adopted a national policy of an electronic key management system – which transmits the code to a user without involving paper.

Four Monitor Stations on Hawaii, Kwajalein (Pacific), Diego Garcia (Indian Ocean) and Ascension Island (Atlantic) continually track the health of the system and route data through to the Master Control Station of the 2nd Space Operations Squadron of the 50th Space Wing, Falcon AFS, Colorado Springs.

The DoD has ordered about 27,000 receiver sets, ranging from man-packs to F-16 navigation systems. The military had to rush 10,000 hand-held units to the Gulf in 1990 and 1991. Currently, the military has outstanding orders for 13,999 receivers and accessories, with options that could bring the total purchase to 93,999 units through 1998. The 50,000th unit was delivered October 1995.

Each Block 2 incorporates the Integrated Operational Nuclear Detection System (IONDS) devices to detect nuclear explosions for assessment of nuclear attacks. Detonations can be pinpointed with 100 m accuracy, with every point on the globe within view of at least four satellites. The Dept of Energy's Sandia National Labs provide the Earth-facing 'bhangmeter' optical sensor that detects the characteristic double flash of an Earthbound nuclear explosion, while the Los Alamos National Lab's X-ray/neutron detectors are spread around each satellite to monitor for space events. Rockwell received the US$1.2 billion production contract in 1983 for 28 satellites of these Block 2s. The navy planned to deliver them to orbit aboard the Shuttle (10 planned for 1987 alone) in conjunction with McDonnell Douglas' PAM-D2 solid upper stage. These plans had to be dropped after the Challenger accident and the satellites had to be retrofitted for the Delta 2 expendable launcher.

In practice the GPS Navstar system has been operating at full strength (24 Block 2/2A satellites, including three hot spares) since March 1994. One ground spare remains in the military's inventory to be launched as an emergency replacement. Another can be built with two months' notice.

The Department of Transportation formally became part of the GPS management system in 1994 to accommodate civil interests. The DoD is thus officially operating the system in part for the civil community and must ensure that further testing and development has minimum impact on those users. With DoT on board pilots have now been given authority to use GPS as primary guidance for non-precision approaches without monitoring the ground navigation aid, providing they have autonomous integrity monitoring. Improvements in military cryptography may lead to at least partial civil access to the military PPS signal, improving accuracy in the near future.

President Clinton, on 29 March 1996, declared his intention to halting civil signal degradation within 10 years. March 1994's launch brought the system up to full Block 2/2A strength for the first time when it became operational on 28 March, but he left the USAF Space Systems Division to the USAF Space Command in Colorado with operational control of the system. The DoD is thus officially operating the system in part for the civil community and it must seek Presidential approval and then notify the FAA and Coast Guard (USCG) at least 48 hours before any planned reduction in satellite numbers. GPS became integrated 17 February 1994 into the US National Airspace System and the FAA plans the US$500 million Air Traffic Management system to be operational in late 1997, using GPS and 24 ground stations for aircraft.

In FY97, Congress gave the USAF a GPS budget of US$278 million (which supports procurement of first three Block 2F satellites). Other budget requests have been FY96 US$201 million, FY95 US$226.5 million, FY94 US$94.2 million, FY92 US$66.41 million.

Specifications

Navstar Block 1

Navstar 1–6 used the Atlas F from Vandenberg to achieve a 20,000 km circular orbit (all planes A through C) at 63° inclination. Navstar Block 1 were launched in February, May, October, December 1978, February and April 1980, December 1981, July 1983, June, September 1984, October 1985. Number 3 suffered a clock problem. Number 6 took over the detection of nuclear explosions from the Vela Hotel satellites. Number 3 had to be taken out of service at the height of the Gulf Conflict on 11 December 1990 because a reaction wheel problem affected attitude control. The satellite returned to partial service 16 January 1991. Number 7 burned up in the Atlas launch failure. Navstar 8 was deactivated in May 1993, because of battery and attitude problems. Number 31 replaced it. Number 9 had a similar battery problem and left service in February 1994. Number 10 retired in 1996 as the last in-service Block 1 satellite. Number 11 had already retired from service February 1994 because of clock problems and removed from the orbit 14 April.

Signal strength: +23.8 dBW P signal, +26.8 dBW C/A signal (+28.8 dBW in high power mode)

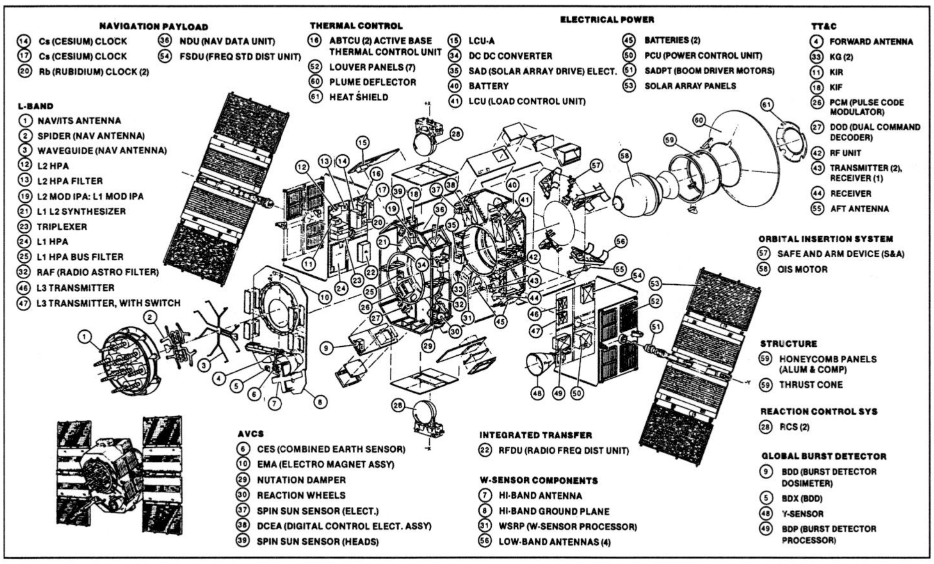

Navstar Block 2A satellite configuration; Block 2 was very similar (Rockwell) 0516843

Atomic clocks: three rubidium, one caesium
Mass: 760 kg at launcher separation, 455 kg initial on-orbit
Span: 5.34 m across deployed solar panels
Design life: 5 years (ACS propellant sufficient for 7 years)
Structure: aluminium honeycomb panels
AOCS: spin-stabilised. 18 hydrazine thrusters in two modules: 2 × 22 N + 16 × 4.5 N. Orbital injection by 28.8 kN maximum Thiokol TE-M-616 solid (25 kg burnt-out motor remains attached)
Power: 410 W EOL from 5.02 m^2 array of two Si cell panels; 3 × 15 Ah Ni/Cd batteries

Navstar Block 2
Navstars 13–21 weigh about 840 kg in orbit and each unit cost US$38.8 million in FY87 dollars. Block 2 and 2A launched 1989 (5), 1990 (5), 1991 (1), 1992 (6), 1993 (6), 1994 (1), 1996 (3), 1997 (1). All operated throughout 1996, but the aging GPS 16 and GPS 13 had already been replaced by GPS 33 and GPS 40 respectively.

Navstar Block 2A
Navstars 22–40 have enhanced survivability in the event of a nuclear attack and weigh about 1,881 kg at launch. Navstar 40, the last of the contracted 28 satellites, was delivered in 1992 and launched by Delta 7925. The first (Navstar 22 or GPS 23) appeared in November 1990, increasing the operational system to 16 satellites. However, a problem developed in the circuit board of one solar array's control system 12 December just two days after launch. This glitch turned out to be a design fault and delayed the launch of Navstar 23 (GPS 24) from February to July 1991 while it was corrected. Then number 24 (GPS 25) experienced a further delay from August 1991 to February 1992 after a reaction wheel failed in a ground test because its extended storage had affected lubrication. Navstar 40 was launched 12 September 1996.

Navstar Block 2/2A
Signal spread: –160 dBW C/A & –163.0 dBW P on L1; –166.0 dBW C/A or P on L2, –165.2 dBW C/A on L3
Atomic clocks: two rubidium and two caesium clocks
Mass: 843/930 kg on-orbit BOL (1,667/1,881 kg launch)
Span: 5.34 m across deployed solar panels
Design life: 7.5 years
Structure: aluminium honeycomb panels, 1.95 m^2 radiating array
AOCS: spin-stabilised, 22 hydrazine thrusters in two modules with 60 kg supply sufficient for >10 years of normal operations. Orbital injection by 47.1 kN (2,097 kN total impulse) Thiokol Star 37XFP
Power: 710 W EOL from 7.26 m^2 array of two Si cell panels; 3 × 35 Ah Ni/Cd batteries.

Navstar Block 2R
Navstar 41–60 became the first round of replenishment satellites to maintain the fully occupied constellation. Each has been improved to provide six months of autonomous satellite operations without ground control corrections, increased service life and improved survivability. The atomic clocks, normally updated daily, offer five times the stability of their predecessors. The design phase contract cost US$119 million: the first 20 satellites will then cost US$575 million. First delivery was planned for October 1995 for launch March 1996, although payload development problems delayed first launch to 23 January 1997. First Block 2R satellite was launched aboard a Delta-2 (7925) 17 January 1997, but following a fault with one of the GEM strap-on boosters the launch vehicle self-destructed 21 s after launch, resulting in the loss of the satellite. The satellite had been intended to fill the vacant F5 slot. The second satellite (No 44) was launched on 23 July 1997 followed by the third on 7 October 1999, the fourth on 11 May 2000, fifth on 16 July 2000, sixth on 10 November 2000, seventh on 30 January 2001, eighth on 29 January 2003, ninth on 31 March 2003 and tenth on 21 December 2003. GAO estimated during 1990 that each will cost US$200,000 annually to store. Launch will be on USAF's MLV-3 medium launch vehicle: Delta 7925 was selected April 1993. Cost of design and development was acknowledged in March 1993 to have risen from the planned US$99. 1 million to US$119 million, primarily because of the requirement that the satellites should remain in continuous service throughout a nuclear war.
Mass: 2,032 kg at launch; 1,075 kg BOL on-station
Deployed dimensions: 152 × 193 × 191 cm body, 19.3 m span across solar arrays
Design life: 10 years minimum
Structure: equipment is mounted on six aluminium honeycomb structural panels mounted on central aluminium core. Communications equipment and some housekeeping units are mounted on the N/S faces; antennas and Earth sensors are on the Earth-facing panel
Power: twin 2-panel solar wings (totalling 13.4 m^2) using N-on-P Si cells providing 1,136 W EOL (970 W requirement); Ni/H^2 batteries provide 100 per cent eclipse protection
Pointing: not available
TT&C: at 2,227.5 MHz S-band
AOCS: 3-axis by momentum wheel, magnetorquing and 16 blowdown monopropellant hydrazine thrusters. The thrusters are divided into two redundant sets of eight: six 0.89 N catalytic Reaction Engine Assemblies (REAs) accompanied by two 22.2 N Electrothermal Hydrazine Thrusters (EHT) for increased performance NSSK, drawing on four tanks (94.8 kg total hydrazine). Insertion by Thiokol Star 37FM solid
Thermal systems: passive and heaters.

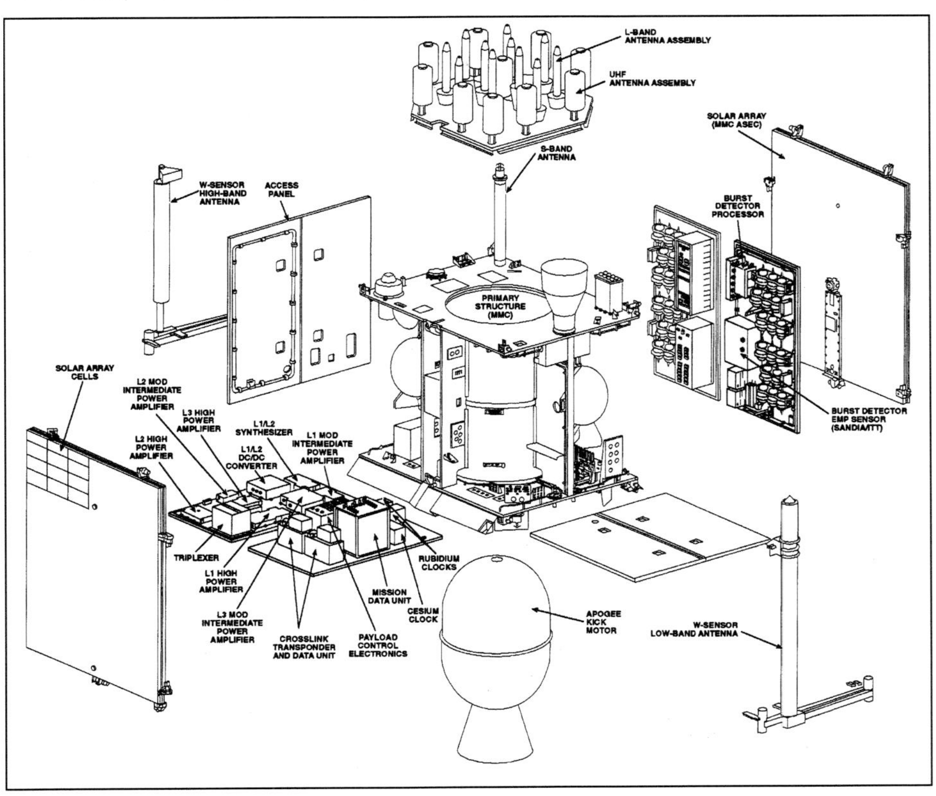

GPS Block 2R principal features (ITT) 0517310

Navstar Block 2F
Block 2F was planned as the fourth generation system. Navstar 61–93 got underway with the initial US$382 million award for the next generation of satellites made on 22 April 1996 to Rockwell, the single team which initially aimed for first delivery April 2001 and launch third quarter of 2002 on the USAF Evolved ELV. Total programme cost over its life cycle is US$3.5 billion. The first award for six satellites was followed by a contract for 15 and in 2002 for 12, totalling US$1.3 billion. The original intention was for 51 satellites, but that was reduced to 33 in June 1995. DoT is funding a civil addition to provide the L5 signal for improving civil accuracy. The satellite network will be all-2F by 2010. Individual minimum life will be 10 years; design life is 15 years. Rockwell teamed with Computer Sciences Corp to bid; Lockheed Martin Astro Space again teamed with ITT Aerospace/Communications and Loral Federal Systems; Hughes' team covered Hughes Information Technology Co, Space Applications Corp and Stanford Telecom. Hughes had not bid on GPS contracts before. The Nuclear Detection System will include new technology demonstrated by the FORTE satellite launched in 1997. It will look for RF pulses from low technology explosions, such as might result from a nuclear test by a proliferant nation. Delays to the programme result in an initial launch no earlier than 2005. Only 12 will now be built.

Nova Series

Background
After the commercial success of the original Transit (US Navy Oscar) series, the navy introduced Nova, an improved version of Oscar, designed by Johns Hopkins University and built by RCA. Each Nova includes a more powerful transmitter, improved reference clock, greater computer capacity, an onboard ability to compensate for orbital disturbances and a 7.6 m long gravity gradient boom for stabilisation. Nova 1 launched 15 May 1981 by Scout from Vandenberg; weighed 136 kg and went into a 1,170 × l,187 km, 109° inclination orbit. It failed March 1991. Nova 3 launched

12 October 1984 into 1,150 × 1,198 km, 90.1° orbit failed November 1993 and Nova 2 launched 16 June 1988 into 1,150 × 1,199 km, 90° inclination orbit is still active. Nova 2 and Nova 3 carried pulsed plasma thrusters.

Timation Series

Background

Time Navigation, later called Timation (Time Navigation), began in 1964 as a Naval Research Lab project to demonstrate the principle of navigation by ranging using synchronised space and ground clocks as opposed to Doppler navigation used by Transit. The 700 kg Timation 1 appeared 31 May 1967 as a piggyback satellite released byThor Agena D into 906 × 915 km 70° inclination orbit. Timation 2 followed on 30 September 1969 to a 898 × 925 km orbit transmitting at 150 to 400 MHz. The USAF had its own version of time ranging Project 621B, calling for 12 satellites in eccentric 40,800 km orbits. The DoD combined the two programmes in 1973 creating GPS Navstar. NRL's 362 kg Timation 3, built before the merger, appeared in orbit after the merger on 14 July 1974 launched by Atlas F to a 13,442 × 13,777 km, 125° orbit and received the name NTS 1 (Navigation Technology Satellite). The 430 kg NTS 2 launched 23 June 1977 into 20,181 × 20,181 km, 63.3° inclination orbit actually served as the first Navstar verifying the launcher, onboard systems, ground control software and tracking techniques. It carried the first caesium clock in space, now flying on Block 2/2A GPS.

Transit/NNSS (US Navy Oscar) Series

Current Status

Three Nova satellites are currently stored in orbit.

Background

The Applied Physics Laboratory of Johns Hopkins University originally proposed and designed Transit to update the inertial navigation systems it had helped build on Polaris submarines. Eisenhower approved initial project funding in December 1958, but the first Polaris navigation fix did not occur until January 1964. After this success the US Navy opened up the sitemaps to civil users in 1967. Transit can be used in any situation where dead reckoning can be used for about an hour between updates. The navy did not shut the system down until 1996. The satellites comprised an octagonal prism, 0.25 m high, 0.37 m across and four blades of solar cells, each 1.67 m long and 0.25 m wide. The 65 kg carried two transmitters for telemetering doppler signals and memory readout.

Even though more than 400 military users relied on the system, the commercial system found the greatest audience by far, with an estimated 80,000 commercial sets available at one time.

Transit operates by measuring the Doppler shift in the carrier signals from the satellite as it passes overhead. Each Doppler sample requires 10 to 15 minutes for proper accuracy. The satellites transmit timing marks and orbital information on two stable coherent frequencies (149.99/399.97 MHz) that permit corrections for ionospheric refraction.

The navy created the Naval Satellite Operations Center (NAVSOC) in 1962 to operate Transit from Point Mugu, California. It has TT&C stations at Prospect Harbor (Maine), Rosemount (Minnesota), Laguna Peak (California) and Wahiawa (Hawaii). Wahiawa closed in September 1993 as a step towards phasing out Transit.

Using data from these stations the Pt Mugu computing centre refines the orbital model of each satellite every 24 hours and uplinks new ephemeredes typically every 12 hours. Sample tests indicate Transit locates a receiver to within an RMS accuracy of 15 to 25 m utilising a dual-frequency receiver. Single frequency reception yields 80 to 100 m. Because the speed of the vessel affects the signal, an error of 370 m is introduced for each 0.5 m/s of unknown vessel speed. As the current constellation has reached the end of its useful life, intervals between fixes now vary over 35 to 100 minutes. For stationary users, such as surveying and oil platform location, integration of measurements from 25 passes yields 5 m RMS accuracy. If a second receiver can be located several hundred km distant, then simultaneous tracking over 10 passes generates 0.5 m RMS accuracy.

After an initial launch failure of Transit 1A on 17 September 1959, Transit 1B, launched 13 April 1960 into a 373 × 745 km, 51° inclination orbit inaugurated a test for the service by transmitting for three months. Over the next six years, 22 more Transit satellites joined the constellations. Most operated for about one year. However, many far exceeded their projected three-year lives. The most long-lived satellites included number 12 151 months, number 13 260 months and number 14 199 months.

In September 1983, the navy announced that of 15 satellites built by RCA (now Lockheed Martin Astro Space) in the 1970s, 12 remained in store because the three launched (2 March 1968, 27 August 1970, 30 October 1973) had far exceeded their design lives. RCA received a US$9.9 million contract to modify eight for dual launches by Scout from 1985 under the Stacked-Oscar-on-Scout (SOOS) programme. The navy intended to store these satellites in orbit rather than on the ground. The first pair, Oscar 24 and Oscar 30 went into orbit 3 August 1985 on a Scout.

At 55 kg, the Scout could put them into 1,005 × 1,263 km, 89.8° inclination orbits. The 59 kg Oscar 27/29 pair joined the predecessors on 16 September 1987 from Vandenberg in a 1,017 × 1,183 km, 90.3° inclination orbit. They brought the total of Transits in orbit at that time to nine. They were joined by the SOOS-3 pair (Oscar 23/32) on 26 April 1988 and then SOOS-4 (Oscar 25/31) on 25 August 1988.

The Defense Nuclear Agency purposely destroyed one of the four remaining satellites – Oscar 22 – on the ground by a 1.50 g aluminium pellet travelling at 6 km/s to simulate a space debris strike. The navy gave Oscar 26 to The Naval Post-graduate School in Monterey, California. The other two (Oscar 21/28) remain in storage at APL.

The US Navy shut Transit down in December 1996 now that GPS is in place. Three Oscars and one Nova were active at end-1995 in the system, plus three Oscars were stored on-orbit.

NUCLEAR SURVEILLANCE

United States

Alexis (STP P891-B)

Current Status
Launched 25 April 1993, the US$17 million Department of Energy-funded satellite has lasted well beyond its nominal one-year mission to demonstrate its telescope and radio-receiver technology for nonproliferation applications and past a three-year lifetime engineering estimates gave it. After five years it still has most of its capabilities intact, even though the satellite is starting to show signs of wear and tear. The Blackbeard experiment has been superseded by Los Alamos' FORTE satellite.

A broken solar panel at launch imparted a wobble to the spin-stabilised spacecraft. Los Alamos scientists modelled the wobble and mapped the actual positions of the photons captured by the six telescopes to their originating location in the sky. The final computer software for decoding the telescope data was recently completed and used to process all 49 months' worth of archived data, stored on more than 150 CDs.

Background
LANL's Alexis (Array of Low-Energy X-ray Imaging Sensors) satellite performs an all-sky survey of the Universe's soft X-ray background glow at 62/72/93 electron Volts. The X-rays, possibly generated by interstellar gas, are detectable by Alexis's three telescope pairs, each with 30° fields of view. Los Alamos National Laboratory designed the project to demonstrate that low cost complex sensors can be built quickly for verification of future arms control agreements. Although Alexis's telescopes perform an astrophysical survey, they really monitor space for nuclear flashes as revealed by the fact that the Dept of Energy's Office of Intelligence and National Security funded the US$17 million programme.

In order to put the satellite in space the USAF provided launch services through its Space Test Programme and called the satellite P89-IB. AeroAstro provided the platform and the Apple Macintosh-based ground station. The launcher's avionics deck carries the VHF Spectrum Utilisation Measurement Experiment (VSUME) to research VHF usage and densities in specified military frequencies for the Naval Space Systems Activity. Alexis also carries the Blackbeard radio propagation technology development experiment for ionospheric research, Blackbeard can take 150 million measurements/second and detect radio noise from everyday sources such as car ignitions and toy walkie-talkies. This survey could lead to detection by future satellites of RF pulses from nuclear detonations.

It took the ground station several months to gain full control of Alexis because one solar panel broke loose during launch. Blackbeard began operating 11 July and the first X-ray telescope 26 July. Blackbeard has discovered bright and rapid near-surface RF flashes: hundreds of transionospheric pulse pairs. Although associated with thunderstorms, they are not regular lightning, but they are similar to nuclear pulses and so must be characterised. They were explained in 1995 as resulting from a runaway process creating an upward avalanche of electrons in the strong electric fields above thunderstorms. The initial electrons are provided by cosmic gamma rays colliding with

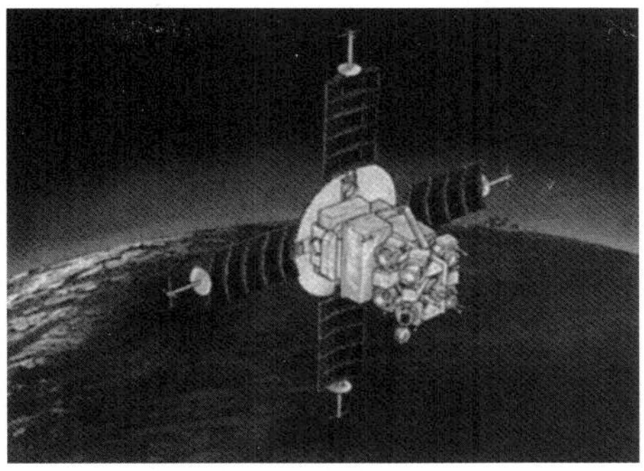

Alexis carries an X-ray telescope array and radio propagation experiment (AeroAstro) 0516844

air molecules. The second pulse arises from the weakening electric field still being strong enough in the very thin air at high altitude to generate another avalanche. The theory also explains the blue and red flashes discovered shooting up from thunderstorm clouds: the electrons cause air molecules to fluoresce, blue in thicker air and red in thinner.

Specifications
Launch: 25 April 1993 by Pegasus
Mass: 109 kg (bus 45 kg, payload 64 kg)
Orbit: 747 by 836 km, 69.93°
Spacecraft: 61 cm diameter 104 cm high. Spun around Sun-pointing axis (to within 2°) at 3 rpm; 0.25° attitude determination. Four solar panels provide 50 W average to payload and four NiCd batteries. Redundant 80C86 processors. S-band communications. Bus points the payload in anti-solar direction. 96 Mbyte mass memory, downlink S-band at 750 kbit/s to 1.8 m Los Alamos station daily, uplink 9.6 kbit/s. Mission life 1 year, goal 3 years; spacecraft reliability 95% after 18 months.

FORTE (STP P94-1)

Current Status
Los Alamos National Laboratory's FORTE (Fast On-orbit Recording of Transient Events) satellite has detected many thousands more radio bursts from lightning strikes and other phenomena than previously reported. Los Alamos scientists are using another major FORTE instrument, an event classifier, to distinguish the inherent structures of radio frequency signals in the 30 to 300 MHz range, which includes commercial television, FM radio, aircraft navigation and communication bands. Los Alamos claims a library of these sources will assist it in sorting out suspected nuclear detonations from ordinary events. Essential for monitoring nuclear test ban treaties. Instruments developed through the Forte programme will be incorporated into the GPS III satellites.

Background
Sandia National Laboratories launched the Fast On-orbit Recording of Transient Events satellite on a Pegasus XL on 29 August 1997 into 799 × 833 km 70° to test the detection of nuclear detonations by RF pulses. The mission is sponsored by the Department of Energy and jointly executed by the Los Alamos and Sandia National Laboratories. The purpose of the mission is to detect and distinguish between electromagnetic pulses caused by events such as low-technology nuclear detonations and lightning-strikes. This

Forte will study RF pulses from nuclear detonations (Los Alamos National Laboratory) 0517153

characterisation will be aided by optical sensors in order to correlate the optical signals with the RF signals. The satellite will also provide a proof of concept for an advanced composite bus made of graphite epoxy and cut into shape with a water jet. The advanced structure allows the satellite to carry an additional 23 kg of payload. The satellite is three axis stabilised, and most of the exterior panels are covered with solar cells to provide 160 W of power.

The satellite has an expected lifetime of at least one year (three-year goal). SNL supplied an optical monitor to eliminate lightning flashes. The satellite serves as a prototype for the GPS 2F-satellite generation, which the air force intends to use to look for the pulses from low technology explosions. The 10.7 m antenna, built by Astro Aerospace and comprising two log-periodic orthogonal arrays, is held pointing Earthwards by magnetorquers. Sandia is contributing the data to a database of lightning's global distribution. The optical monitor uses a 10 km^2 resolution wide field of view imager to locate lightning and a 50 kHz sampling rate to record separate flashes and trigger RF recording. Forte will test if it can sort out lightning automatically.

Ionds Sensors

Background
The United States monitors the world for nuclear detonations represented by X- and gamma ray wavelengths with the Integrated Operational Nuclear Detection System (IONDS) packages aboard the DSP early warning and GPS navigation satellites. Detonations can be pinpointed with 100 m accuracy now that the Navstar system is completed, with every point on the globe within view of at least four satellites. The Dept of Energy's Sandia National Labs provide the Earth-facing 'bhangmeter' optical sensor that detects the characteristic double flash of an Earthbound nuclear explosion, while the Los Alamos National Lab's X/neutron detectors are spread around each satellite to monitor for space events. Rockwell equipment also looks for electromagnetic pulses. USAF in 1993 decided to delete detectors from the FEWS follow-on to DSP in order to save mass and money (FEWS was cancelled anyway later that year), relying instead on GPS after DSP has expired. LANL/SNL are flying a series off three satellites to demonstrate techniques for monitoring of nuclear detonations. The technical requirements have been considerably expanded and are being applied to a variety of platforms and indigenous to new satellites.

OCEAN SURVEILLANCE

Russian Federation

Eorsat Series

Eorsats appeared for the first time at the end of 1974 with the launch of Cosmos 699 on 24 December. They differ from the actively transmitting Rorsats and instead have the passive task of identifying the type and role of ships of the Western navies by noting transmission frequencies of their radio and radar. Consequently Eorsats do not have the same huge power requirements that the Rorsats have. As they gather information they manoeuvre extensively in low orbits and consistently maintain an exact orbital period of 93.3 minutes in circular orbits at 65°.

During the Falklands War, Eorsats monitored the progress of the British Task Force's journey to the Southern Hemisphere. A British commentator asked one commander how he avoided being tracked by Soviet satellites. He replied that the British Navy had countermeasures against the satellite monitoring. In 1982, the Soviets had three Eorsats in orbit, Kosmos 1286 (4 August 1981, 432 × 434 km, 93° inclination); Kosmos 1306 (14 September 1981, 409 × 462 km, 93° inclination) and Kosmos 1337 (11 February 1982, 429 × 447 km, 93° inclination). The Soviets did not have a paired Rorsat in orbit at the time of the movement of the British fleet, but did launch one when the fleet arrived in the Falklands.

Following their testing phase, the Soviets since 1979 usually used Eorsats in pairs, like the Rorsats. The Eorsats enjoy longer lifetimes than the Rorsats however, with some having operational lives of more than a year. Curiously, Eorsats usually exploded just before re-entry. The Soviets refused to announce whether the explosions were intentional or not, but characteristically ceased after 1987, when the space debris issue became a widespread concern. In 1985, the Soviets had three Eorsats active simultaneously for the first time, Kosmos 1567, Kosmos 1588 and Kosmos 1646 between 18 April and 14 July, and then again with Kosmos 1567, Kosmos 1646 and Kosmos 1682 beginning 19 September 1985. But the simultaneous operation ended 19 November 1985 when Kosmos 1567 left its tightly controlled orbit after a record 538 days on station.

The Eorsat programme began to evolve in 1986 when the first Eorsat of the year, Kosmos 1735 (27 February), entered a slightly lower orbit with a three day repeating pattern instead of the standard four day repetition. When the second satellite of the series that year, Kosmos 1737 (25 March) it followed the new orbital pattern, but at the higher inclination of 73.4°. All previous Rorsats and Eorsats had flown only at 65° inclination. But more importantly, Kosmos 1735 and Kosmos 1737 had exactly the same ascending node indicating a connection between the information gathered by the satellites. The two flew together until the Soviet ground control commanded Kosmos 1737 to re-enter. The ocean surveillance programme had never before witnessed an intentional re-entry.

By 1988, all Eorsats manoeuvred to achieve the three day repeating pattern. Those that had a 65° inclination however remained at that inclination with no repeat of the Kosmos 1737 profile. The first launch of 1988, Kosmos 1949 (28 May), established the first complete constellation of evenly-spaced Eorsats. The Soviets maintained that the constellation Kosmos 1834 terminated its mission on 29 August. A new constellation formed when Kosmos 1979 joined the other two in November,

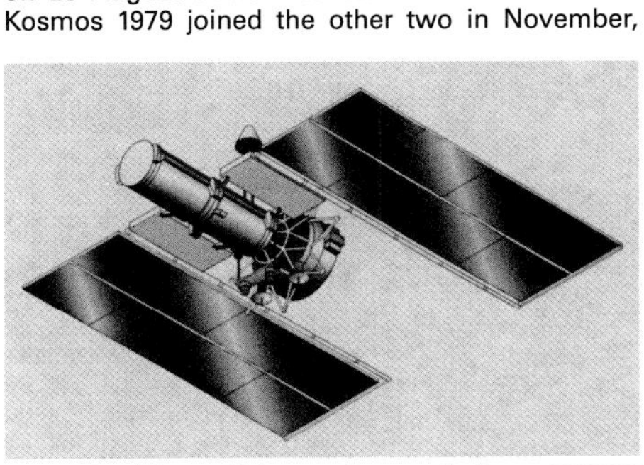

Eorsat bus (Kaman Sciences Corporation) 0516851

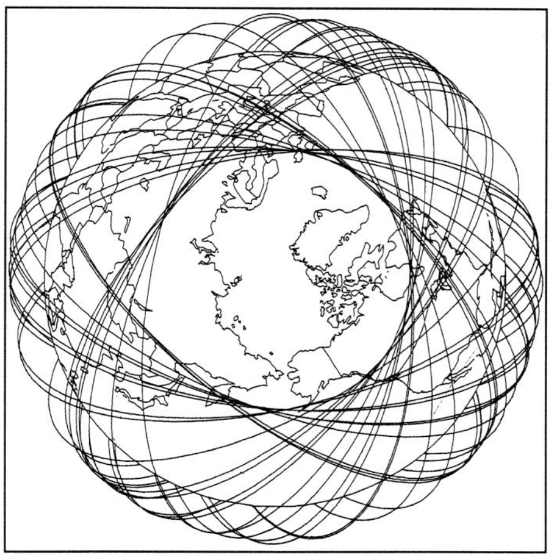

A considerable number of spent nuclear reactors and fuel cores are now in long-lived orbits about the Earth (Teledyne Brown Engineering) 0516819

but by 8 December Kosmos 1890 completed its mission, leaving only the two 1988 satellites still operating. In a surprise move, Kosmos 1834 and Kosmos 1890 were both commanded to lower their orbits to accelerate natural decay. The two satellites re-entered without incident, signalling a departure from earlier fragmentation practices.

By 1989, the Soviet Union decided to expand the highly successful Eorsat programme and add new operational characteristics. Three Eorsats appeared as the year progressed. Kosmos 2033 went up on the 24 July, Kosmos 2046 on the 27 September, and Kosmos 2051 on 24 November enlarged the constellation to four active craft and for the first time inaugurated a multiple plane system to permit faster revisits to ocean areas of interest. Kosmos 2033 completed the plane started by Kosmos 1949 and Kosmos 1979. When Kosmos 2046 orbited it occupied a position exactly 172° behind the earlier plane. The orbit is significant in that its plane allows it to repeat exactly Kosmos 2033's ground track just 11 hours later. Kosmos 2051 joined Kosmos 2046 in the new plane but immediately experienced a propulsion problem. Several weeks of small manoeuvres saved the mission, although Kosmos 1979 had come to the end of its own useful life in the meantime. The replacement philosophy already in place repeated itself in March 1990 when Kosmos 2060 (14 March 1990) completed the plane started by Kosmos 2046 and Kosmos 2051, while Kosmos 1949 finally failed after a record mission of 22 months.

The Eorsats programme continued in 1990 with three new launches: Kosmos 2096 (23 August), Kosmos 2103 (14 November) and Kosmos 2107 (4 December), creating a record network of six satellites. Kosmos 2103 failed during the first few days of 1991, but Kosmos 2122 (18 January 1991) quickly replaced it. The CIS allowed the Soviet network to gradually deplete itself during 1991 as satellites ended their careers and no replacements appeared: Kosmos 2046 re-entered in April, and Kosmos 2060 retired in August and re-entered during September. At the end of 1991, only Kosmos 2096, Kosmos 2107 and Kosmos 2122 still operated. Kosmos 2107 manoeuvred off-station 13 March 1992 and decayed 6 April. Kosmos 2096 ended operations 2 August 1992 and re-entered 30 August, leaving only Kosmos 2122 at the beginning of 1993. Even this satellite re-entered 28 March 1993 after operating for 776 days.

Then Kosmos 2238 appeared 30 March 1993, and was joined by Kosmos 2244 on the 28 April 1993, in a co-planar orbit but trailing 120° behind. Kosmos 2258 appeared 7 July, coplanar with the previous two but 120° ahead of Kosmos 2238 and 120° behind Kosmos 2244. Kosmos 2264 started a new triplet 17 September 1993 with its plane 145° away and Kosmos 2293 (2 November 1994) opened up a third plane.

Kosmos 2244 moved off-station about 14 February 1995 and decayed 18 March. Kosmos 2258 moved off-station about 2 March 1995 (decaying 8 June), leaving only Kosmos 2264 launched 17 September 1993 and Kosmos 2293 launched 2 November 1994. Kosmos 2264 then deactivated 4 April and decayed 7 August 1995. Kosmos 2313 appeared 8 June 1995, joined on 20 December 1995 by Kosmos 2326. So at the beginning of 1996 the CIS had Kosmos 2293 Kosmos 2313 and Kosmos 2326 operating. Kosmos 2293 decayed May 1996. The role of Eorsats expanded when the Ioffee Physics Institute of St. Petersburg included a 131 kg Konus, a gamma ray burst detector from Kosmos 2326. The Granat satellites also carry the Konus launched 11 December 1996, Kosmos 2335 was joined by Kosmos 2347 launched 15 December 1997. Kosmos 2383 launched 21 December 2001 is a signals intelligence satellite with the launch vehicle placing it in a 145 × 405 km orbit at 65° inclination later circularised at 405 km by a 70–80 m/sec velocity increment at first apogee.

Okean Series

Current Status

Okean-0 No 1 was launched on 17 July 1999 by a Zenit 2 from Baikonur to a 661 × 663 km, 98.06° orbit. First in a new generation of Okean-0 oceanographic satellites, Okean-0 No 1 has a mass of about 6,300 kg and a payload which includes side-looking radar, multispectral scanner, radiometers and spectro radiometer.

Background

The precursor technology development programme began with the launch of Cosmos 1076, an Okean-E type with a mass of 1,950 kg, launched 12 February 1979, followed by successors on 23 January 1980, 28 September 1983 and 28 September 1984. During the summer of 1985, US Intelligence agencies reported that the Soviets had recently deployed a radar carrying satellite system. The satellite according to Soviet sources mapped ice formations in polar regions, and thereby enhanced the ability of the Soviet Navy to operate in icebound areas. The system aided navigation of northern sea routes, in order to assist in moving naval ships from construction yards in the western Soviet Union to new ports in the Pacific.

Radar mapping of ice floes certainly appeared to explain those satellites belonging to the Kosmos 1500 type, with its side-looking radar, and later emerging in a civilian guise as the Okean satellites. The system uses a wavelength of 3.2 cm, a swath width of 460 km, and a resolution of 1.5 km. Since 1985, the satellites transmit their data to Moscow, which uplinks it to ships via the Ekran geosynchronous satellites.

Kosmos 1602 followed by Kosmos 1500 on 28 September 1984, in the traditional 636 × 669, 82.5° inclination orbit. The third oceanographic satellite, Kosmos 1766, appeared on 28 July 1986 from Plesetsk into 635 × 666 km, 82.5° inclination orbit. The orbital plane for 1766 was 60° away from Kosmos 1602 and gave the impression that the Soviets were in the process of establishing a fixed constellation.

The next oceanographic satellite, Kosmos 1869, launched 16 July 1987, assumed a position 60° from its predecessors. The satellite failed to deploy its antenna correctly rendering the SLR unusable. In 1988, the full programme became operational with the 5 July launch of Okean 1.

Rorsat Series

Current Status

Functions previously carried out by Rorsat have been dispersed to other satellites such as Eorsat. No launches of Rorsat since 1988.

Background

During the Cold War the Soviet Union sought to monitor US aircraft carriers in the proximity of the USSR and to maintain the capability 'to oppose them in the event of war'. To carry out this plan they invested in a suite of ocean surveillance satellites. Hardware tests were initiated with the launch of Cosmos 102 on 28 December 1965 with a second flight on 20 July 1966 and the first nuclear powered Rorsat launched as Cosmos 198 on 27 December 1967. The US Navy viewed the development of this capability as a severe threat to its interests and has justified the existence of the US ASAT programme based on the need to eliminate the Soviet monitoring network. The satellites as configured by the Soviets and carried on by the Russians apparently relay the position, heading and identification of Western naval forces to RFAS forces in real time.

Each nuclear-powered Rorsats (radar ocean reconnaissance satellites) or Eorsats (Elint ocean reconnaissance satellites) works as part of a pair and the corresponding interferometric radar return allows the RFAS to build up a comprehensive view of surface activity. The satellites maintain the required separation accuracy and height by very low thrust engines. If a vessel attempts to use electronic countermeasures against the low altitude (255 km) Rorsats the radio activity probably exposes the existence of the ship to the accompanying Eorsat, orbiting above at about 435 km. Significant as this threat is to US Naval operations, the Russians are apparently phasing the programme out and the country has launched no identifiable Rorsat launches since 1988.

Viadimir Vasilkovsky, Deputy Chief of Department for nuclear reactors of the Russian Atomic Energy Ministry, said in January 1994 that the Russians have no plans to launch nuclear power sources in the next few years.

The Soviet Union launched two Eorsats and two Rorsats in 1987. Kosmos 1834 (April 1987) joined two similar Eorsats, Kosmos 1735 (27 February 1986) and Kosmos 1769 (4 August 1986), at the same time that five Soviet Victor 111-class nuclear attack submarines conducted simulated attacks against New London and Norfolk. Another Eorsat, Kosmos 1890 (10 October 1987) soon followed Kosmos 1735 after Kosmos 1769 ceased functioning. Kosmos 1769 and another Eorsat, Kosmos 1646 (18 April 1985), broke up in September/November, respectively, continuing a pattern set with the very first Eorsat, Kosmos 699 (1974).

Of the two Rorsats launched in 1987, the first lasted for only a brief period. The second had an unusual orbit. Kosmos 1860 (18 June 1987) flew shortly after the Soviet Union publicly complained of US violations of Soviet territorial waters and roughly coincident with the build-up of US naval forces in the Persian Gulf after the Iraqi attack on the USS Stark. During the 1980s, most Rorsats lasted in orbit about 80 days (Kosmos 1365 lasted 136 days, the maximum of any in the series), but Kosmos 1860 ended its mission after only 40 days when its nuclear reactor separated and manoeuvred into a higher altitude storage orbit. Kosmos 1900 (12 December 1987) deviated from the 20-year-old orbit unique to the radar satellites when it entered an orbit about 8 km higher than its predecessors.

In 1988, Kosmos 1900 experienced a control malfunction and almost re-entered with its nuclear power supply. Soviet ground controllers attempted to change its orbit to no avail and then on 30 September 1988, only five days before re-entry a back-up system propelled the reactor to a safe storage orbit. The Soviet's launched their last mission Kosmos 1932 on 14 March 1988. Lincoln Lab's Haystack radar reports the Rorsats appear to be leaking coolant, according to orbital debris research.

While the Rorsat programme has been inactive since Kosmos 1932, the Eorsat operations continued to expand. The number of operational craft in working constellations has grown from an average two to three operational craft in 1987. The constellation grew steadily until a record six satellites operated simultaneously in late 1990. Perhaps to compensate for the absence of Rorsats, the CIS launched five Eorsats between March 1990 and January 1991. Three remained operational at the end of 1991 and only Kosmos 2122 still operated at the end of 1992. That re-entered 28 March 1993 after operating for 776 days, leaving no operational satellite for the first time in a decade.

The powerful pulsed X-band radar signals, used to locate naval ships in all weathers with resolution of tens of metres, easily indicate the purpose of the Rorsat satellites. These satellites can identify formations of ships and their headings/speeds. Then they downlink targeting date within one revolution to ships or other weapons platforms carrying long range anti-ship missiles. Ships of the western powers therefore have less than 90 minutes to take evasive action or otherwise prepare for attack.

The paired satellites flew out of Tyuratam on SL-11 into 270 km circular orbits. Each satellite is about 10 m long, 1.3 m in diameter. A 2 kW Bouk multi-cell thermionic reactor built by Kraznaya Zvezda containing 31.1 kg of 90 per cent enriched U235 divided into 37 fuel elements in beryllium reflectors powers each of the X-band phased array radars. The USAF uses its visible/infra-red telescopes, such as those at the Maui Optical Station on Hawaii, to determine if the satellite's reactor remains active by its glow. The Soviets and then the RFAS normally separates the reactor and boosts it to a 900 km orbit with a 500- to 600-year lifetime at the end of its 60- to 70-day mission.

The Soviet Union placed 33 Rorsats in orbit (plus two launch failures), although some early flights might have been reactor tests only with no radar. Kosmos 198, launched 27 December 1967 (249 × 270 km, 66° inclination), began the series. In the 1970s, a Rorsat created an international furor when it failed and its nuclear reactor began an unpredictable re-entry sequence. The satellite Kosmos 964 launched 18 September 1977 into 251 × 265 km, 64.98° inclination orbit malfunctioned and started to decay on 1 November. The Soviet controllers were unable to boost its nuclear power pack to a safe high orbit, and the world awaited with grave speculation wondering which country the radioactive debris would impact. 92 days after launch, on 24 January 1978, Canada became the site and main debris fell west and south of Yellowknife along an 800 km strip of the Northwest Territories. In one of the first international applications of the US Nuclear Emergency Search Team the American nuclear specialist set up operations known as Operation Morning Light to recover the debris.

Still, the US withheld protesting too vigorously because its own nuclear powered US Navy satellite launched from Vandenberg and powered by a Snap-9 nuclear generator with plutonium 238 had failed to reach orbit, broken up and released radioactive material over the Indian Ocean. In the end the Canadian government estimated that the Kosmos 954 search had cost C$6 million, the Soviet Union contributing about C$3 million.

In April 1979, the Soviet Union used a pair of non-nuclear satellites for ocean survey, but then quickly resumed the nuclear powered ones a few months later. The US State Department issued a 'statement of regret' to protest the resumption of Rorsat nuclear power when Kosmos 1176 orbited on 29 April 1980. Despite this, the programme continued in 1981 with Kosmos 1249, launched 5 March into 252 × 265 km. At the end of its life the Soviet Union boosted its nuclear power pack to 898 × 985 km. Kosmos 1266, launched 2 April 1981, operated for only eight days before its manoeuvring rocket and platform detached and manoeuvred to an 891 × 965 km with 600-year life. Apparently the satellite malfunctioned causing the Soviets to abandon it early and a similar incident prematurely ended the life of Kosmos 1299, launched 24 August 1981 into 248 × 269 km and boosted to 910 × 984 km after only 12 days.

For the first time since the Kosmos 954 drama, a pair of Rorsats, Kosmos 1365 and Kosmos 1372, operated as planned for 136 and 71 days respectively before being boosted into higher orbits. They were thus able to work together during June, July and early August, surveying the South Atlantic and, presumably, giving the Soviets information about the Falklands war.

Two more satellites followed, identified as Kosmos 1402 and Kosmos 1412, launched 30 August and 2 October. They operated for 118 and 39 days respectively. They too monitored events in the South Atlantic together during October and early November. Then, exactly five years after Kosmos 964, another Rorsat Kosmos 1402 created the same heart-stopping drama when the Kettering Group reported that between revolutions 1,925 and 1,926 the separation and boosting events failed. The nuclear reactor would remain stuck in a lower orbit that would eventually decay.

The Soviets expressed confidence that their new design would ensure that the reactor would burn up upon re-entry. The safety measure succeeded: the less radioactive reactor vessel and instrument section re-entered over the Indian Ocean in January 1983, and the core disintegrated over the South Atlantic on 7 February. No increase in radioactivity was detected in either area.

Other Soviet end of life operations proceeded successfully and the booster put Kosmos 1412 in the proper end of life orbit at 901 km. The Soviets may have again modified the units after the Kosmos 1402 incident, because no new Rorsats entered service in 1983, but two, Kosmos 1579 and Kosmos 1607, orbited in the second half of 1984. Both operated for about three months and then their nuclear power supplies separated and entered higher orbits.

In 1985, the Soviets launched two more Rorsats Kosmos 1670 (1 August) and Kosmos 1677 (23 August). The timing of these launches as two other Eorsats already operated in orbit, appeared to coincide with the large NATO naval exercise Ocean Safari 85. Both reactors separated during 22–23 October 1985 after missions of 82 and 61 days respectively. They then manoeuvred into the standard storage orbits. The 1986 Rorsats, Kosmos 1736 (21 March) and Kosmos 1771 (20 August), completed separate missions of 92 and 56 days. At the end of 1987, only Kosmos 1900 still operated. During the second week of April, Kosmos 1900 failed to respond to radio commands and began to decay slowly from its 265 km mean attitude. On 13 May, Tass acknowledged the satellite was in trouble and predicted a re-entry in August or September. Meanwhile, the spacecraft's nuclear reactor and attitude control system continued to operate which prevented any of the backup safety systems from being activated and sending the reactor to a higher orbit. On 30 September the satellite apparently lost all stabilisation when it used the last of its propellant supply. The depletion of propellant triggered a stored computer command to separate the reactor and propel it to a higher storage orbit.

In 1988, the Soviets followed up on the programme with only one Rorsat launch, Kosmos 1932, which operated for 66 days. Because of the malfunction of Kosmos 1900 and a growing worldwide concern about nuclear reactors in near Earth orbit, the Rorsat programme apparently ended with this launch. Nonetheless, Kraznaya Zvezda, which developed the 5 to 6 kW Topaz multi-cell thermionic reactor, has discussed sales to the United States and provided prototype reactors to the US Department of Energy.

United States

Geosat FO Series

Current Status

A Taurus rocket successfully deployed the US Navy GEOSAT Follow-On (GFO) satellite on 9 February 1998 at 5.20 am Pacific time. The launch from Vandenberg placed the satellite released from the Taurus's fourth stage into a 775 km circular, 108° inclination orbit. The GFO has a final configuration of 813 lb. Ball Aerospace and Technologies Corporation built it for the US Navy's Space and Naval Warfare Systems Command. The satellite uses a radar altimeter and a water vapour radiometer to measure differences in sea surface height associated with ocean currents and eddies. The 13.5 cm radar altimeter provides 3.5 cm height precision. The 22/37 GHz dual frequency water vapour radiometer with path corrective accuracy of 1.9 rms.

Background

The US Navy awarded a US$46 million contract to Ball Corp by their Space & Naval Warfare Systems Command in August 1992 to build the Geosat Follow-On. This satellite is a radar altimeter, which provides accurate measurement of ocean topography for the Naval Meteorology & Oceanographic Command in Bay St. Louis, Missouri. By measuring sea height the satellite provides the navy with information on sea floor topography and water density, surface roughness and by association, an index of wind speed. GFO when properly interpreted forms a vital link to the scientific understanding and accuracy of global ocean circulation models under development for the navy's new class of supercomputers. It also identifies warm and cool water masses, which is important knowledge for anti-submarine warfare. For the first time, the navy may also use the satellite for real-time direct encrypted readout to 65 ship/shore terminals. GFO has an important role

Geosat Follow-On will provide accurate sea height measurements. It will directly support naval operations such as ship routeing, anti-submarine warfare and amphibious operations (Ball Aerospace) 0517319

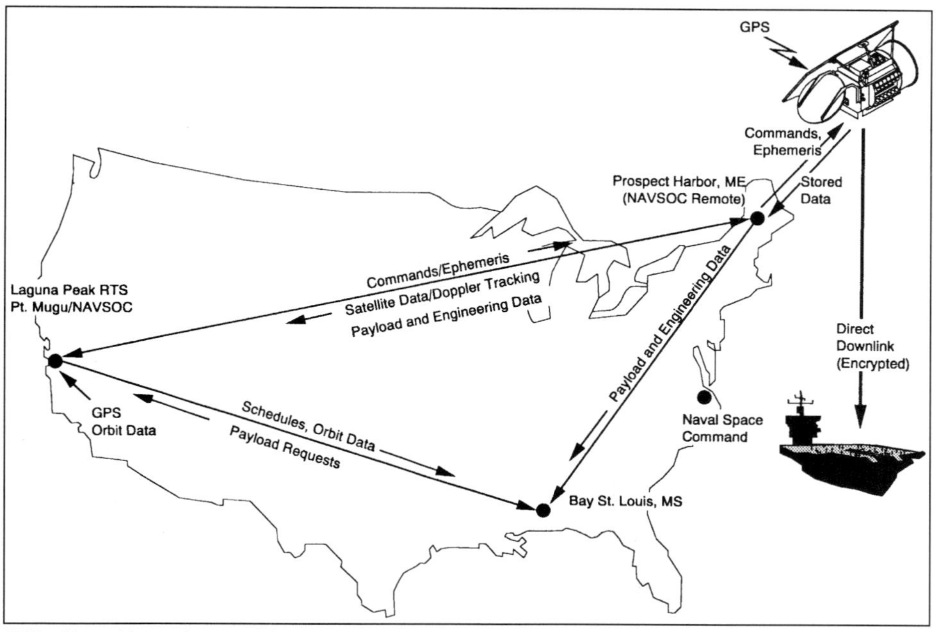

GFO will provide continuous global oceanographic data to ships and shore facilities (Ball Aerospace) 0517159

for NOAA as well as filling in gaps in the data for climate and long range weather forecasting and global change predictions.

Specifications

Launch: 9 February 1998 by Taurus from VAFB
Orbit: 785 × 788 km 108° polar, 17 day repeat
Design life: 8-year minimum
Mass: 347 kg at launch
Payload:
Pulse radar altimeter, measure sea surface height with <3 cm accuracy
Passive radiometer to correct altimetry data for tropospheric water vapour effects
AOCS: 3-axis control. GPS will provide precise orbit determination (altitude precision <5 cm). Eight hydrazine thrusters drawing on two tanks
TT&C: by NAVSOC Naval Satellite Operations Center at Pt Mugu, California.

Geosat Series

Background

The Navy launched Geosat on 13 March 1985 by Atlas F from Vandenberg into 775 × 779 km, 108.05° inclination orbit. Currently the Johns Hopkins Applied Physics Laboratory, developers of the satellite, operate it for the US Navy. Geosat houses a radar altimeter to measure sea surface height and map wave action. The navy has been largely without such data of 3 to 4 m accuracy since Seasat's failure in 1978. Radar altimetry and the instruments to conduct it are also carried by the navy NROSS programme. The satellite has a mass of 635 kg and a life of 3 years.

When it was launched, the navy hoped to get three years out of Geosat, but RADM Richard Pittenger, Oceanographer of the Navy, indicated in April 1989 that it remained operational and would probably remain so for several years more. However, its onboard recording facility failed in October 1989 and the altimeter failed 5 January 1990. In June 1992, the navy declassified all Geosat data covering south of 30° south and followed in July 1995 by declassifying the remainder. NOAA is responsible for public distribution.

NOSS (White Cloud)

Background

The advent of NOSS-1, launched 30 April 1976 by Atlas from Vandenberg into 1,092 × 1,128 km, 63.5° inclination orbit, marked a new capability in US monitoring programmes because of the much larger main satellite. This programme became known as Navy Ocean Surveillance Satellite programme or NOSS. In orbit NOSS-1 and its cylindrical successors released three box-shaped subsatellites into similar orbits. 1993's Russian article claimed the satellites flew in a triangular formation around the parent, separated by 50 to 240 km. Actually, the new system flies with 30 to 110 km separations. NOSS locates targets by passive radar interferometry (hence the subsatellites), and is able to track surface ships in all weathers, with RF facilities for listening to their radar and communications. It may incorporate infra-red sensors to track submerged nuclear submarines, detecting their warm water wakes. Some have suggested the infra-red detectors also find low flying missiles.

NOSS-2 was launched 8 December 1977, followed by NOSS-3 on 3 March 1980, suggesting a three- to five-year operational life. A launch failure of 9 December 1980 was followed by NOSS-4, on 9 February 1983 by Atlas from Vandenberg into 1,063 × 1,186 km, 63.4° inclination orbit. It released four box-shaped subsatellites, designated SS (for surveillance satellite) A-D into similar orbits. NOSS-5, launched 10 June 1983, released three subsets designated GB1-3; and NOSS-6, launched 5 February 1984, released three subsets designated JD1-3 into similar orbits. NOSS-7 followed 9 February 1986 and NOSS-8 concluded the series of five Atlas H launches on 15 May 1987, heading into 1,050 × 1,170 km 63.40° orbits. The Titan 2 launch of 5 September 1988 was initially believed to have carried NOSS-9 but the lack of subsets and the subsequently announced orbit of 185 × 292 km, 85.0° seemed incompatible with the programme. No orbit is known for the second Titan 2 launch, of 5 September 1989, but again no subsatellites appeared. These are now believed to have been Elints.

Since 1990 the NOSS (White cloud) system has been replaced by the more capable Space Based Wide Area Surveillance System. With a total mass of almost 8,000 kg, the SB-WASS series is launched by Titan 4.

PARCAE/Whitecloud Series

Background

As Soviet Naval power grew during the 1960s, the United States Navy responded by developing a Naval surface ship detection programme to track and locate the Soviet fleet. Satellites offered the only hope because US aircraft had become woefully inadequate at tracking and monitoring the newly expanded fleet. The resulting programme called Whitecloud in unclassified documents is also known by the classified name Parcae, the three daughters of Zeus and the goddess Themida. Experiments on the space reconnaissance of radio signals were conducted in the early 1960s using small Elint satellites. The name Whitecloud surfaced in October 1984, when it was reported that postal covers carrying the name and some details were on sale at NASA Johnson Space Center's gift shop.

Whitecloud/NOSS: a rare sight of the US Navy's Ocean Surveillance Satellite. Three subsatellites cluster around the parent vehicle before release (Aviation Week) 0516845

Whitecloud resembles the Russian Rorsat programme in that it passively measures signals emitted from naval ships. The Naval Research Lab developed and built the first two prototype satellites, and the navy launched the first one on 14 December 1971 by Thor Agena developing three 123 kg SSU satellites. Then contractor Martin Marietta took over manufacture along with E-Systems of Dallas, which built the payload itself. These satellites provided the basis for all US surveillance of Russian shipping until about 1976 when other programmes took their place.

At one time the navy planned for a programme equivalent to the Russian Eorsat called Clipper Bow. Congress killed the programme in 1980, because among other reasons the navy could not offer a sufficient justification for duplicating the capability already available on the LaCrosse. Some reports still indicate that Clipper Bow is alive and the navy will deploy it sometime in the 1990s.

When the US Navy launched the second Titan 4 vehicle off the assembly line into an unusual orbit, space observers speculated that the navy had finally completed its Clipper Bow programme. The launch on 8 June 1990, originally put the satellite into a 448 km, 61° inclination orbit. The satellite, a single USA-59 surveillance satellite went into this orbit, but the principal satellite then manoeuvred into a 1,116 km 63.4° inclination orbit and released three payloads (USA 60, USA 61 and USA 62). Then Vandenberg's 8 November 1991 Titan 4 launch of USA 72 followed a similar pattern with objects (USA 74, USA 76 and 77) released into 1,053 × 1,165 km 63.4° inclination orbits. The Titan 4 failure of 2 August 1993 probably carried the same platforms. Titan's 5 December 1995 launch of USA 116 may have been the first replacement for the original payload, but the navy did not give this satellite a USA designation which is generally the surest indication of the nature of the payload. The May 1996 launch carried the pattern forward with the first object being identified as USA 119 and three subsets as USA 120, USA 121 and USA 122.

An article published in July 1993's Foreign Military Review by Russia's Ministry of Defence, provided the first generally available information that these satellites are the advanced version of the US Whitecloud programme satellites. According to the article, the main satellite weighs 7 tonnes, compared to the initial Whitecloud which weighed only 600 kg. A fully operational Whitecloud uses four clusters of satellites, one transmitter and three interferometric receivers. The detection of ocean going vehicles requires 1.5 m of resolution or better a level of radar imaging that has only been available on the Vortex satellites to date.

In May 1987, the navy launched its tenth Whitecloud and the last (fifth) by Atlas H.

PHOTO RECONNAISSANCE

China

Jian Bing Satellites

Background
China's original photo reconnaissance satellite programme was the *Fanhui Shi Weixing* (FSW) series: within Chinese Government circles the programme was called *Jian Bing* which means 'pathfinder'. The designator *Jian Bing* was only applied to the specific FSW satellites were launched on photo reconnaissance missions: other FSW satellites which flew on nominally civil remote sensing and microgravity missions were *not* given *Jian Bing* designators, although such designators appear to have been applied throughout the FSW programme by Western intelligence circles.

From the history of the FSW programme, it would be reasonable if the FSW-0 series of satellite comprised the *Jian Bing*-1 programme and that the specific photo reconnaissance missions in the FSW-1 programme were given the *Jian Bing* -2 designator. It is possible that the latter group might be further divided into *Jian Bing*-2A and -2B series, although such a division might be a figment of western intelligence designator systems. Open literature does not allow a definitive decision concerning which FSW satellites had which *Jian Bing* designators.

The recoverable FSW-0 and -1 satellites were designated *Jian Bing*-1, while the FSW-2 series was an improved *Jian Bing*-1. The designator *Jian Bing*-2 refers to an as-yet unflown satellite design (and not the FSW-2 series as some western sources have speculated).

The China-Brazilian Earth Resources Satellite (CBERS) programme is discussed elsewhere: the Chinese name for the programme is *Zi Yuan* (ZY) which means 'resource', with the CBERS satellites specifically comprising the ZY-1 series.

With no prior announcement, China launched the first *Zi Yuan*-2 satellite on 1 September 2000. While the first ZY-1/CBERS satellite would operate in a 774 to 775 km orbit, ZY-2 1 manoeuvred to a much lower 484 to 500 km orbit. The Sun-synchronous orbit was co-planar with that of the first ZY-1/CBERS and the orbital periods were in the ratio (ZY-2:ZY-1) 17:16.

Unlike the CBERS satellite system, China has released no information concerning the design of the ZY-2 satellite or the imaging system which it is carrying. Officially, the satellite is a Chinese domestic equivalent of the ZY-1/CBERS satellite, but western observers believe that it is a photo reconnaissance satellite in the *Jian Bing*-3 series, returning images through a digital link.

An improved version of the FSW recoverable satellite series made its debut in November 2003: western reports indicated that the satellite was the first in the FSW-3 series and that it bore the designator *Jan Bing*-4, but the Chinese have not confirmed this speculation.

Specifications
***Zi Yuan*-2 1 (*Jian Bing*-3)**
Launch: 1 September 2000 using a CZ-4B from Taiyuan
Design life: 2–4 years?
Contractor: China Academy of Space Technology
Orbit: initially 97.42°, 475 to 493 km, initial operational orbit 485 to 499 km, standard operational orbit 490 to 495 km
Configuration: no illustrations of the ZY-2 satellite have been released, but it is believed to use the same satellite bus as the ZY-1/CBERS satellite. This implies a 1.8 × 2.0 × 2.2 m body plus a single vane of solar panels 2.6 × 6.3 m: 3-axis control by RWs and hydrazine thrusters. Orbit-raising manoeuvres take place at ~2 week intervals (with a large amount of variation in this interval)
Mass: ~1,500 kg at launch
Power system: solar panels provide ~1.1 kW BOL
***Zi Yuan*-2 2 (*Jian Bing*-3)**
Launch: 27 October 2002 using a CZ-4B from Taiyuan
Orbit: initially 97.40°, 471 to 483 km, initial operational orbit 478 to 506 km.
Other specifications as ZY-2 1
***Zi Yuan* -2 3 (*Jian Bing*-3)**
The launch of a third *Zi Yuan*-2 electro-optical reconnaissance platform is expected during 2004.
***Fanhui Shi Weixing*-3 1 (*Jian Bing*-4 ?)**
Launch: 3 November 2003 using an uprated CZ-2D from Jiuquan
Contractor: China Academy of Space Technology
Orbit: 62.99°, 194–324 km
Configuration: Three module spacecraft, comprising conical instrument/propulsion module plus retrorocket module plus dome-shaped recoverable module, maximum (base) diameter 3.35 metres, length ~5.5 metres. Recoverable module is approximately 1.75 metres base diameter and 1.5 metres high.
Mass: ~3,500 kg at launch ?

The Chinese referred to the satellite as being "scientific": although scientific experiments could have been carried the primary mission is believed to be military photo-reconnaissance, possibly including some "civil" remote sensing work as well. Satellite manoeuvred six times in orbit before the descent module was recovered on 21 November in Sichuan Province. The instrument/propulsion module continued in orbit until 18 December.

The launch was the first from the new launch pad close to the one used for the piloted *Shen Zhou* launches and it was the first to use a stretched version of the CZ-2D vehicle.
***Fanhui Shi Weixing*-3 2 and 3 (*Jian Bing*-4?)**
Chinese sources have hinted that there could be two launches in the FSW-3 (western designator) programme during 2004. The first would be the long-promised "seeds satellite" which will carry a package of seeds and other biological experiments to study the effect of space radiation and other aspects of the space environment on the specimens. The second would be a mission like the first FSW-3, dedicated to photo-reconnaissance and land survey work.

France

Helios 1A Satellite

Current Status
In March 1998, Germany announced it was withdrawing from the Helios 2 partnership and France began to re-evaluate its continuation of the programme. Helios 1A continues to operate 9 years after launch by Ariane 40. Helios 1B was launched by Ariane 40 on 3 December 1999 and continues to operate.

Background
France joined the rank of nations capable of performing autonomous military reconnaissance from space when it launched the Helios 1 surveillance satellite in July 1995 aboard an Ariane. On a platform based on the civil Spot Earth resources observation satellite, Helios provides the fourth independent military surveillance capability after those of the USA, RFAS and China. During the 1987 parliamentary season, France announced it had paid FFr7.6 billion for the programme, and had saved FFr300 million by

Helios 1A platform testing at Intespace. See France's National entry for an artist's impression of the satellite in orbit (Ministère de la Défense) 0517311

joint development with Spot 4. France claims the satellite allows it autonomy from the US decision-making process in the western alliance, by giving it an independent source of corroboration for US Intelligence claims. France justified the decision at the time by saying it needed to upgrade its data gathering ability to prepare for the expected improvements in Soviet strike capability in response to the US Strategic defence Initiative. Germany for a while agreed to participate in the programme but then later backed out, citing a desire to build a radar imaging system rather than a visible spectrum satellite. After this, CNES proposed that the Helios and Spot 4 production programmes be merged to reduce costs. The world recognised France's sophistication in this new field when it first published Spot photographs of 10 m resolution, showing Soviet naval, aircraft and nuclear weapon storage facilities at Murmansk and Severemorsk in 1987. The USA and the CIA immediately challenged France's right to publish such data at the time. The White House Senior Interagency Group for Intelligence had to file a report on the national security aspects of civil remote-sensing programmes.

The Italians and Spanish share part of the financial burden and the image products with France. Helios 1 is controlled from the CMP (Centre de Maintien Poste) at Toulouse, and has a back-up control facility at Francazal Air Base. The new Fresnel building at CNES Toulouse handles both Helios 1 and Spot 4 control and schedule uplinking. The images are downlinked to three CRI (Centre de Reception d'Images): Colmar in northeast France, Lecce in southeast Italy and Maspalomas in Spain. Each of the three capitals has an individual processing centre. Controllers can select onboard encryption so that imagery is restricted to one or more of the three principal nations or is more widespread. Initial reports declared Helios 1A imagery to be excellent. Helios I initially flew into a 673 × 676 km orbit, but the French adjusted it 14 July 1995 to 680 × 682 km, 98° inclination Sun-synchronous orbit that has a spot revisit time of 48 hours for one satellite.

Aerospatiale is working on the IHR Instrument Haute Resolution, a second-generation imaging system, increasing resolution to 50 cm and adding an infra-red capability.

Specifications
Helios
Launch: 1A 7 July 1995, by Ariane 4 from Kourou, French Guiana 1B 3 December 1999 by Ariane 40
Lifetime: 5 years
Orbit: 1A: 679 × 682 km, 98.1°; 1B: 679 × 681, 98.1°
Resolution: 1 m
Lifetime: 4 years
Mass: 2,537 kg.
Comment: Italy is providing 14.1 per cent funding and Spain 7 per cent. Based on the extensive experience derived from the civil Spot, Helios provides the 4th independent military surveillance capability after those of the USA, RFAS/USSR and China. Programme cost about FFr10 billion. Helios 1B is planned for launch by Ariane to back-up and complement 1A operations; orbits will be phased to provide 24-hour revisit periods.
Contributors: Délégation Générale pour l'Armement/Direction des Missiles et de l'Espace (DGA/DME, executive programme management under AGEX Agence Exécutive), CNES (system/spacecraft architect), Aerospatiale (system industrial architect, EPV imaging system, GSR-3 solar array), Matra Marconi Space France (satellite, processing centre), SEP (processing centre in collaboration with Matra), Alenia/Inisel (ground stations), Alcatel Espace (payload electronics, command/communications systems), Sodern (DTA 04 cameras).
Orbit: 1A released into 673 × 676 km, adjusted by 14 July 1995 to 680 × 682 km, 98.07° Sun-synchronous; revisit time 48 hours for one satellite. 1B released into a 679 × 681 kn orbit, 98.1° Sun-synchronous.
Comment: Spacecraft: based on Matra's Spot 4 bus 3-axis control: Spot 4's 0.05° RMS pointing accuracy improved to 0.005° RMS by addition of star tracker. 5-panel GSR-3 solar wing.
Sensors: EPV Ensemble de Prises de Vues imaging system using DTA 04 multispectral CCD cameras, and electronic intelligence equipment.

UPDATED

Russian Federation

Introduction

Current Status

The Russians launched a Yantar 4K class (Kobalt) close look photo-reconnaissance satellite from Plesetsk on 15 December 1997. The spacecraft carries two or possibly four externally mounted data return capsules which are ejected for earth return while the main payload remains in orbit.

Kosmos 2349a Yantar 1KFT/Kometa topographical mapping satellite orbited on 17 February 1998, aboard a Soyuz-U launched from Baikonur. The satellite had an expected lifetime of 45 days at launch.

Background

In the more than 40 years since the Soviet Union orbited a Zenit 2 photo-reconnaissance satellite aboard Kosmos 4 on 26 April 1962, the country has launched more satellites for this purpose than any other. With almost 800 launches having reached orbit, no other country even comes close to this dedication to photo surveillance.

But since the break-up of the Soviet Union and the introduction of new technology, the new Russian Republic has greatly decreased the number of its photo-reconnaissance launches. In part, simple economics influenced this trend, but another factor is that the third-generation satellites which remained in orbit only about two to three weeks are being replaced by the fourth- and fifth-generation satellites, with lifetimes of around two months and around a year respectively.

In 1987, photo-reconnaissance satellites reached a peak in orbital days of operation by completing more than 1,400 days in orbit. In the 1990s this has dropped to: 1990 – 962 days, 1991 – 896 days, 1992 – 808 days, 1993 – 840 days, 1994 – 940 days, 1995 – 745 days and 1996 – 361 days. Part of the lost operational time resulted from launch failures of fourth-generation satellites in May and June 1996. Following the de-orbiting of Kosmos 2320 on 28 September 1996 the Russians have not had a photo-reconnaissance satellite in orbit through April 1997.

The Russians launched only two of their new fourth-generation Yantar-2K satellites in each of 1994 and 1995, and so, if the two launch failures of 1996 each carried a Yantar-2K, the Russians may simply have depleted their inventory of satellites for the year. Normally the Russians also launch only one (sometimes two) Kometa satellites each year, but failed to put one in orbit during 1995.

What intrigued Western observers is the fact that the Russians did not replace the fifth generation Zenit satellite after the de-orbit of Kosmos 2320. The Russians did phase out the Soyuz-U2, so it could be simply that no method to put the satellite in orbit currently exists.

Zenit 2 (Generations 1–3) Series

Background

The Soviets' first of the type of photo-reconnaissance satellite with origins prior to Sputnik and only their 18th launch, Kosmos 4 on 26 April 1962, carried the 4,600 kg Zenit 2. After the fall of the Soviet Union, various commentators revealed the name and some operational details of the programme in 1990. More recent accounts in the 1995 book Corona fill in some of the details. Sergei Korolev developed the Zenit 2 at his design bureau OKB-1 and fitted the film capsule into the manned Vostok platform which could be retrieved from space automatically. Zenit originated in 1956 in a Soviet military requirement for a photo-reconnaissance satellite. In November 1958 Sergei Korolev, largely responsible for Sputnik 1, received approval of a combined spy satellite and manned spacecraft concept where the same design could serve both requirements. Zenit as a spy satellite and Vostok as a manned spacecraft. Cosmos launches of 1K prototypes would lead to the 2K and 4K unmanned reconnaissance satellite and the 3K the manned Vostok. Development began in 1959. The affiliate Central specialised Design Bureau took over the programme in 1964 and continued to develop new and better film canisters. As the programme became more sophisticated, the film canisters became able to carry more film and the vehicle itself could manoeuvre more in space. The satellites referred to in the West as Generations 3 through 5 grew in mass to about 6 to 7 tonnes. The Soviets launched roughly 30 to 35 a year through the 1980s, but cut back sharply to 21 in 1990 as the more advanced longer-lived models took over. 1991 and 1992 saw the Russians only launch 11 Zenits in each year. In 1993 and 1994, they cut back further to seven each, and in 1995 the number of launches fell to only three.

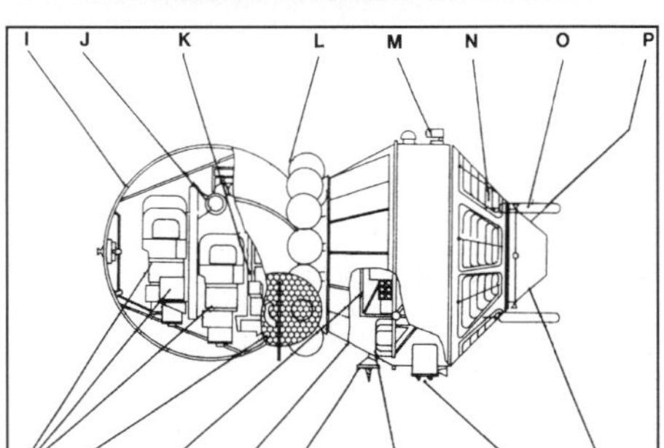

This first drawing of the Zenit 2 photo-reconnaissance satellite was released in 1993. It is very similar to the manned Vostok except principally for the cylindrical section in the service module. This carried IR/Sun sensors for the improved orientation required for imaging, and also possibly additional batteries. A cameras, B sigint antenna, C electronics, D service module, E radio antenna, F/K control system, G IR vertical sensor, H retrorocket, I parachute hatch, J self-destruct system, L N_2 spheres for cold gas attitude thrusters, M solar sensor, N thermal radiators, O/P radio antennas 0517312/0516820

The Russians continue to rely on film returned from space at a time when the United States has long abandoned the practice preferring instead digital imaging. Film canisters offer higher resolution than the digital equivalent and some sources say the Russian photographic resolution is 20 cm. Electro-optical digital imaging provides sub metre resolution. The Russians have moved into digital processing however, to receive more timely the information than film. The increasingly-important Generation 5 Zenit returns digital data and the electronic emission form the Generation 6 and Generation 7 satellites indicate this is their method as well.

Initially, the Soviets snapped both high- and low-resolution pictures from orbits of about 200 × 250 km. They increased the image quality, beginning in 1974, by lowering the perigee of the satellites to about 170 km. The Generation 4 Zenits can lower their orbit to 150 km or lower over areas of military or geographical significance. They also co-ordinate these changes with information from weather satellites to check whether the target cloud cover would waste film.

The Soviets claim their satellite, Kosmos 932, launched 20 July 1977, 180 × 342 km, 65° inclination, warned the world that South Africa verged on detonating a nuclear device in the atmosphere. As a result of this the Western Alliance joined with the Eastern Bloc to urge the South Africans to desist.

Photo-reconnaissance satellites found other uses beyond their primary missions. The Soviets manoeuvred some Generation 3 low-resolution photo mappers into the unusual orbit of 350 × 420 km early in their lifetimes. From this altitude the satellites composed photo-mosaics of the earth in 11 days, covering the globe in strips of 2° of longitude. The Soviets launched their early Zenit satellites from Plesetsk at an inclination of 73° or of 70° from Tyuratam. By the fifth generation all launched activity had switched exclusively to Tyuratam.

Several marketing firms have begun to sell old photographs with 5 m resolution to foreign customers. Even the US Air Force has become a potential customer. It has its own interest in buying Generation 5 satellite imagery. Priroda plans to distribute up to the minute digital 3 m resolution panchromatic imagery via their own ground stations from a military satellite, presumably a Generation 5 Zenit.

Operationally, some Russian and former Soviet spy satellites use a beam-splitter mirror to locate themselves extremely precisely by photographing a star background. This triangulation technique is often used to find exact locations of US ICBM silos and may even be used operationally to increase the accuracy of SLBMs. The Soviets adopted the practice of destroying their spy satellites at the end of their useful life in order to prevent them from falling into western hands if they did not completely disintegrate upon re-entry. The Russians apparently ended this practice after the detonation of Kosmos 2262, 102 days after its launch 7 September 1993.

The Soviets have co-ordinated launches and orbits with events unfolding internationally. When Iraq invaded Iran in 1980, the Soviets quickly launched Kosmos 1210, into a 187 × 320 km 82° inclination orbit. The satellite passed over the battlegrounds at noon each day and was recovered on the 14th day, 3 October Kosmos 1211 left for orbit before Kosmos 1210 returned and this new satellite followed the exact orbit of its predecessor.

The venerable Zenit-class photo-reconnaissance satellites ended their tenure as the workhorse of photo reconnaissance with Kosmos, launched in June 1994 – the last of the series.

Generation 4 (Yantar/Kometa) Series

Background

At the same time as the United States moved toward extreme close up photography, the Soviet Union had similar projects in the works. One was the Yantar (Amber) 'Close Look' satellites also known as Generation 4. Code name Feniks, the Yantar 2K was the second series to the Zenit generations (identified outside the USSR as 'Fourth Generation' systems). Yantar had orbit manoeuvre capability, the small re-entry capsules in addition to the main re-entry module and a design life of 30 days opposed to the 12 days of Zenit. The small SpK capsules would make deliveries typically on the 10th and 18th days of flight. The Series got underway with the launch of Kosmos 697, on 13 December 1974. At 6,700 kg, the satellite flew originally into orbits resembling the premier mission conditions of orbits of 174 × 326 km at 67° inclination. Some Soyuz hardware has found a place on the Yantar, but it carries at least two film capsules. At the end of its lifetime, the Soviets de-orbited the film canister module and often reused the hardware. In those cases where platform did not de-orbit, the Soviets commanded its destruction from the ground.

The Russians planned to operate one Yantar at all times but later abandoned the plans after 1994, when only two launches took place, an insufficient number to keep up continuous surveillance. But the lifetimes of Yantars increased from 58–59 days to 70 days and then to 89 days with the flight of Kosmos 2331. When that satellite returned on 11 June 1996, the Russians tried to put a replacement in orbit within nine days, but a failure of the payload shroud caused the launch to fail 50 seconds after take-off when the Soyuz-U vehicle's strap-on and core engines shut down.

The Soviets experienced some launch failures and acknowledge that vehicles carrying Yantars failed on 23 May 1974 from Plesetsk and 28 March 1981 from Tyuratam. Other failures carrying photo-reconnaissance satellites occurred at Tyuratam on 12 June 1982, 26 March 1986, 9 July 1988 and 11 November 1988. Speculation holds that these were probably third or fourth generation satellites.

At the time of publication, the Russians have not attempted another in this particular series following this launch failure. Some Russian sources state that the country is out of Yantars in its inventories with no prospects for more because of budget cutbacks.

One subgroup of the Generation 4 Yantar performs topographic and mapping functions and is known in its civilian mode as Kometa. The first of these satellites appeared with Kosmos 1246 in February 1981. Western observers easily identified it as a new mission class because of its unusual orbit. Most Yantar-2K close look satellites orbit around 170 to km. Kosmos 1246 instead manoeuvred to a more circular 210 to 265 km orbit. Other similar satellite orbits soon followed, typically with operational perigees of around 210 km and apogees within the range 270 to 300 km. The character of the electronic emissions from these satellites identified them as Yantar – like and also resembling the transmissions from known to be mapping and surveying topography. Much later the Russians confirmed this deduction, identifying the name of the satellite series as Kometa (Comet) and the series Yantar 1KFT. Some Russian reports indicate the original name was Yantar-1KFT, but this is slightly out of character since the Yantar-2 designator appeared six years prior to the flight of Kosmos 1246.

The Kometa launches occur less frequently than the Yantar-2K missions. Kometas go into space only one or two times a year and all have originated from the Tyuratam launch site. Western journalists witnessed the launch of Kosmos 1944 in May 1988 and their Soviet hosts described the mission as cartographic in nature.

Kosmos 2243, probably a Kometa mission exploded at orbital injection because the satellite's on-board self-destruct system inadvertently activated, causing the satellite to explode. 172 pieces of debris entered orbit, but the main object decayed after nine days. In its first commercial application, the Russians attempted to launch a Kometa in May 1996 to begin the SPIN-2 (Space Information – 2 m) imagery programme for a United States consortium under an agreement with Sovinformsputnik. The payload failed to reach orbit. The payload shroud failed 49 seconds after launch due to inferior bolts and materials having been used in its construction and debris from the launch crashed at Tyuratam 310 seconds after launch.

Sovinformsputnik (whose dominant share-holder is the Kosmosentral specialised Design Bureau) holds a licence to sell imagery from the Kometa's KVR-1000, TK-350 and DD-5 camera systems of military satellites. Orders can be directed through Space Liaison and Imaging Kosmosorp and the WorldMap consortium. The company has released details of the camera systems aboard the satellite. The KVR-1000 is a panoramic low-angle camera flying aboard Kometa since 1983, usually in conjunction with TK-350. The focal length is 1,000 mm, and has a ground resolution of 75 cm. Commercially the photos only have a resolution of 2 m. The frame size is 18 × 72 cm covering a swath of 37 × 165 km; cut into three portions and a scale 1:220,000. The image has been enlarged to 1:10,000 without loss of detail. The TK-350's topographic cameras have a focal length of 350 mm, with 10 m resolution, a frame size of 30 × 45 cm for a scale of 1:660,000 and 60 to 80 per cent stereo overlap. The digital DD-5 imaging system makes high-resolution data, which the Russians degrade, to 2 m resolution in the digitising process. DD-5 does not refer to a specific camera system but is a catch-all designation for digitised data from military satellites.

Generation 5 Series

Background

The fifth generation photo-reconnaissance satellites remain unidentified as such and the title derives from Western observation of their behaviour based on a model of a digital imaging remote-sensing satellite identified at the 1995 Paris Air Show as Resurs-Spektr-V. This was a modified Yantar 4KS1 and could claim a 3 to 5 m resolution. Typically Generation 5 satellites fly at 230 km perigee and 280 to 320 km apogee. Kosmos 2320, which de-orbited on 28 September 1996 after 365 days in orbit, did not have a replacement. Unlike the budgetary problems alluded to for Generation 4 satellites, the likely reason the Generation 5s have stopped flying is the lack of a launch vehicle. When a factory involved with the production of SYNTIN fuel closed, Russia had to abandon the Soyuz U2 vehicle, which relied on the fuel.

Generation 6 Series

Background

On 18 July 1989, Kosmos 2031 entered a rarely used orbital inclination of 50.6° from Tyuratam and during its lifetime it manoeuvred in a manner different from the fourth and fifth generation satellites. As planned, perigee would be maintained at 210 to 220 km while allowing apogee to be adjusted to a maximum 350 km. Ground controllers destroyed the satellite after 44 days in orbit. In October of the following year, Kosmos 2101 entered a routine 64.8° inclination slot, with apogee and perigee similar to Kosmos 2031. The Soviets destroyed this satellite too after 60 days. Whether coincidentally or by intention, these lifetimes matched the standard ones being seen for Kometa and Yantar-2K missions respectively. Three further missions took place at approximately yearly intervals between 1991 and 1993, with Kosmos 2262 remaining in orbit for 101 days before being destroyed.

Generation 7 Series

Background

To all indications, the Russians introduced a new type of satellite on 26 August 1994 when a Zenit-2 flew from Tyuratam carrying Kosmos 2290. It had a mass of 10.6 tonnes and entered a 212 × 293 km, 64.81° inclination orbit. Extensive manoeuvres produced a 182 × 577 km orbit by the end of March 1995. The Russians commanded it to re-enter on 4 April 1995. No further missions had appeared through May 2000.

United States

Introduction

Current Status

The National Reconnaissance Office set up the new National Imaging and Mapping Agency to facilitate the distribution of images from the Discoverer, Argon Corona, and Lanyard satellites. Gambit imagery remains classified. The NRO began a study in 1998 of a Future Imagery Architecture as a roadmap to the design and development of future satellite systems and associated ground systems and facilities. The Department of Commerce still places tight restrictions on the export of remote-sensing technology. DoC estimates that the current US$400 million annual market for remote-sensing data will grow to US$2 billion by 2000. The National Archives holds more than 800,000 images from some reconnaissance satellites flown between 1960 and 1972. In February 1995 President Clinton authorised the public release of more than 860,000 earth images acquired by Corona, Argon and Lanyard satellites. In November 1997, the National Reconnaissance Office (NRO) released about 38,000 newly declassified records associated with early photosatellite programmes, plus historical videotape. The records involve the development, construction, launch and operation of the CORONA, ARGON and LANYARD photo-reconnaissance satellite systems. This release added no satellite imagery to that already available.

Background

President Eisenhower approved the first programme to build photo-reconnaissance satellites in the United States in 1958 and heartily endorsed the idea behind the programme with a policy initiative he referred to as the 'Open Skies'. The CIA brought Eisenhower's dream into action in developing the Discoverer satellites series by drawing on commercial industry in America. From the earliest days, the programme considered two technologies: 1) snapping a picture of the area of interest and then returning a film canister or 2) developing the picture on the satellite and scanning it for transmission to the earth. Both had strengths and liabilities. The returned film capsules offered, by far, the highest resolution and best quality, but it often took as much as two weeks to get the actual photograph back from space and into the hands of the President. The second technique offered immediate transmission so the President could follow unfolding events, but often the quality of the imagery gave little information.

To carry out Eisenhower's vision, the Secretary of Defense set up the National Reconnaissance Office on 25 August 1960. The NRO introduced the 'Keyhole' designation in 1962 and thereafter all American reconnaissance satellites carried the moniker 'KH'. The first officially named was KH-4, fourth in the series. The name came about not because of the satellite but because of the camera it carried, the fourth different type used in Corona/Discoverer by 1962. Although DoD did not officially acknowledge the NRO even existed until 18 September 1992, most Pentagon insiders found it by looking up the name of the long time director who had a cover title in the DoD phone book. NRO co-ordinated all overhead reconnaissance operations, including those from satellites and aeroplanes. Today it runs on a US$5 billion annual budget, reports to the Secretary of Defense but allows the CIA to target and control its assets.

Until 1992, the inter-agency Comirex (Committee on Imagery Requirements & Exploitation) met daily to review requests and direct targeting. But the DoD Central Imagery Office, created in June 1992 received authority to take over the obligations in the aftermath of the Gulf War, which revealed a clear need for more efficient dissemination of wartime imagery.

Imagery is relayed to Fort Belvoir in Virginia, from where it is passed to the National Photographic Interpretation Center in Washington DC established 1961 and run by the CIA for all of the intelligence community. Another major function of overhead imagery, defence mapping, became the responsibility of the new NIMA National Imaging & Mapping Agency (NIMA) created 1 October 1996 to co-ordinate imagery collection, processing, exploitation, analysis and distribution. NIMA consolidates all defence and intelligence imagery and mapping functions, absorbing the defence Mapping Agency, CIA, NPIC, imagery analytical support elements from the Defense Intelligence Agency, and dissemination and exploitation assets from the National Reconnaissance Program. NRO will continue to build the satellites.

KH-1-KH-3 (Corona/ Discoverer) Series

Background

After Eisenhower released the authority to build a spy satellite to the CIA, the Agency built an entire front programme, putatively to conduct important civilian scientific work. Ironically, Congress never explained to the American public why it had a separate scientific satellite programme in the Pentagon just as it was in the throes of setting up the new supposedly all-civilian agency NASA, which was to conduct all American space science. Corona began from the CIA offices as a high-risk programme funded from an almost unlimited intelligence budget to prove that a satellite could be built and controlled adequately enough for photography in the first place.

Even with the first disappointing results, Corona clearly had the potential to revolutionise the collection of intelligence. The United States had suffered a major embarrassment when the Soviet Union shot down CIA U-2 pilot Gary Powers on 1 May 1960. But flights like Powers proved that the Soviet missile build-up was largely posturing on the part of the Soviets who had maybe four satellites all together instead of the hundreds being hinted at in the Kennedy presidential campaign. After the loss of the U-2, for reasons of pure practicality, the United States had to continue with the Discoverer programme since without the spy plane the USA had no means to keep abreast of Soviet developments.

The CIA rushed the Corona 1 (KH-1) satellite through production and testing and launched the first in the series on 28 February 1959, only 13 months after the launch of Explorer 1, America's first satellite. For 18 months failure dogged the programme.The air force launched the first satellite into a polar orbit, but it tumbled and re-entered only five days later. Discoverer 2 carried the first RV but it re-entered at the wrong time and landed in Spitsbergen touching off a behind-the-scenes diplomatic flap as the CIA sought to recover the satellite before the Soviets found it. They did not succeed and the Soviets retrieved it ahead of the American search team.

Missions then suffered a continued series of failures, the most obvious occurring because the cold conditions in orbit led to the loss of capsule batteries and solid rocket motors. The acetate film also degraded in the vacuum of space and jammed the camera. When this problem appeared the CIA turned to Kodak to develop polyester film base. Discoverer 11 carried this new film for the first time, but that mission failed when the solid spin rockets exploded. Nonetheless, Kodak inaugurated a new civilian line of film based on the new substrate. To prevent failure of the solid rocket motors, Lockheed designers introduced new cold gas thrusters for attitude control. To test the concept, the CIA launched the next two Discoverers numbers 12 and 13 with no film canister on board, just a dummy mass. Because of this these two satellites have never carried the venerable KH designator. To improve the recovery procedure, which until then had a spotty record, the CIA used Discoverer 13 on 10 August 1960 to test new techniques. They retrieved it intact from the ocean. Even with progress clearly being made, Eisenhower sometimes lost hope for the satellite programme and considered cancelling before the end of his term. The first KH-2 was launched on 26 October 1960 but the mission failed, the first success in this series launched 7 December 1960.

Then with D14, all problems suddenly evaporated. Occurring immediately after the loss of the U-2, Discoverer 14 left Vandenberg on 18 August 1960, flew a perfect mission over the Soviet Union and returned on its 18th orbit. This one satellite provided more photographic coverage of the USSR than all the previous U-2 missions combined. It returned from space with 1,082 m of film covering 4.27 million km^2 of the Soviet land mass. It succeeded in imaging all Soviet MRBM, IRBM and ICBM complexes, revealing there to be no missile gap.

Discoverer 14 changed its recovery mode to an air capture when a C 119 transport snatched its parachute from mid-air over the Pacific Ocean. Discoverer 14 aimed its cameras at a military airfield near Mys Schmidta on the Chukchi Sea in Northeast Russia, which showed up clearly in the 1.5 m resolution photographs. The CIA declared Corona operational. The first KH-3 was launched on 12 September 1961 and was a success.

From 1961 onwards, Corona provided regular imagery of the USSR, and eventually led the Pentagon to make the public admission less than one month into Kennedy's administration that one of the issues that elected him – the missile gap – did not exist, nor had it ever. Corona imagery had brought the estimates of Soviet ICBMs down from hundreds to tens. Eventually Corona provided the United States with images of the Plesetsk test range, the Severodvinsk submarine construction facility and ABM deployments such as Galosh around Moscow, and their radars. Corona became almost the sole source of the defence Mapping Agency's data. Then casting the veil of secrecy back over the programme starting on 22 November 1961 the DoD classified all military missions.

In practice, a specially modified air force Thor IRBM put the satellite into orbit and then the Agena's 66.7 kN second stage engine fired to put the whole stage in a circular orbit. Once in orbit, the booster used its attitude control thrusters to turn the camera in the nose cone around to view the earth. This attitude manoeuvre often caused the re-entry capsule to enter at the wrong angle far overshooting the projected return zone. Sometimes the capsule even boosted itself into a higher orbit. One such incident resulted in a false alarm in February 1960, when the US Navy's early warning detection system found an unknown satellite, believed to be Soviet, in near-polar orbit. It was eventually established to be an errant Discoverer 5 re-entry vehicle. Depending on the quantity of film onboard, explosive bolts separated the capsule for re-entry on either orbit number 17 or orbit number 33. At 15,240 m a parachute pulled the recovery package clear of the heatshield, which fell into the sea, usually the Pacific near Hawaii. Beginning with Discoverer 30 in September 1961, the air force began using C-130s for mid-air recovery.

The 10 KH-1 missions used the C camera with an f/5 Tessar 61 cm focal length lens. D16 unsuccessfully introduced the C camera for the KH-2 missions, although it was little different from C/KH-1. But Discoverer 16 also flew aboard the more capable Agena B upper stage, allowing it to almost double the mission time by using an 18 kg film payload. The last KH-3 mission was launched on 13 January 1962.

Discoverer 29 introduced the C/KH-3 camera with the f/3-5 Petzval 61 cm focal length lens. It should have doubled resolution to 4 to 6 m, but a design fault blurred all of Discoverer 29's exposures. Discoverer 38 in February 1962 introduced the radically new Mural camera system made up of two KH-3 cameras angled 15° fore/aft from the vertical for stereoscopic photography. These cameras flew successfully on 20 of the 26 KH-4 missions. Mural also added a wide angle index camera and a stellar location and triangulation system to help identify target areas more precisely. The next major upgrade came when the CIA put the KH-4A camera into service in August 1963. This camera designated the J1 retained Mural but the 73 kg film load split out between two RVs, so part of the secret data could be retrieved more quickly.

In public bland announcements from various front organisations said that the Discoverer made important discoveries every day about 'radiometric' conditions in space.

The pioneering Discoverer 14/Corona imaged its first target 18 August 1960, showing a military airfield near Mys Schmidta on the Chukchi Sea in NE Russia. It was a remarkable achievement; before then, the best had been simple cloud cover images scanned by Tiros. Resolution is about 7 m. The airfield became a regular Corona target (CIA) 0517158

KH-4A/B Corona Series

Background

The first KH-4 was launched on 27 February 1962 and the last of 26 on 21 December 1963. As the spy satellite programme moved beyond its frontier days and matured into a significant technical and political asset, the CIA introduced the more sophisticated KH-4A/B cameras on their Agena carrier. At the very end of the Discoverer programme – on Discoverer 38 – the stereo KH-4 flew for the first time in February 1962 providing the most accurate mapping data ever returned from space.

The upgraded KH-4A took over as the programme lost the Discoverer designation in August 1963, with dual film capsules intended to double the mission duration. The first KH-4A launch took place on 24 August 1963 with the last of 52 on 22 September 1969. The programme experienced the same start-up glitches that beset the Discoverer in its earlier days. When the CIA first tried to retrieve the dual canister, they discovered that only one capsule could be recovered from the first two missions. The first true double recovery came on the third mission, launched 15 February 1964. Eventually, 50 of 52 launches returned RVs and most succeeded in delivering the double RVs. The KH-4A provided resolution of about 3 m from the satellite's altitude of around 175 km. Each satellite lifetime lasted about 1–4 weeks, complementing the new close look KH-7.

Once again the introduction of a new booster extended the amount of film the KHs carried when the Thorad Agena D launched a KH-4A on 9 August 1966. The KH4B camera, first launched in 1967, substantially improved resolution to 2 m and had selectable exposure and filters. It used the DISIC Dual Improved Stellar Index Camera for target location identification and attained orbit 16 out of 17 times. A more flexible control system also allowed the spacecraft to power down for up to 21 days. The first KH-4B was launched on 15 September 1967, the last of 17 on 25 May 1972.

102 of the 144 missions succeeded when the Corona programme formally ended by putting a last KH-4B camera in space in May 1972.

The KH-4B payload carried two return capsules (the central one is omitted here), fed film streams from the two rotating panoramic cameras. The film cassette is at far right (CIA) 0517313

KH-5 Argon Series

Background
As the Soviet Union deployed more missiles in hardened underground silos during the early 1960s, the United States' accuracy for its own missiles had to improve in order to destroy the silos. After a successful programme to advance missile guidance systems and the re-entry vehicle itself, the United States arrived at a point where its missiles had become more precise than its knowledge of the target locations. To support the new accuracy the air force needed more exact mapping co-ordinates. The defence Mapping Agency provided all targeting maps to the air force, and so the US Army devised its own overhead mapping project to be flown within Corona. It called the programme Argon and the new camera the KH-5. In all it had seven successful flights out of 12 attempts.

When the army finished its programme it could locate targets to within 300 m using Itek's 7.5 cm focal length camera which covered 550 × 550 km on each 11.4 × 11.4 cm frame at 140 m resolution. The first KH-5 orbited as Discoverer 20 on 17 February 1961, but the camera failed. The first success came May 1962 and the last of the 12 Argons flew on 21 August 1964, when the entire Soviet Union had been mapped.

KH-6 Lanyard Series

Background
Perhaps the least used of the cameras, Itek's 168 cm focal length panoramic camera inaugurated the KH-6 programme. Lanyard was the first determined effort to gain high-quality, high-resolution images from orbit and the satellite carried a single panoramic camera. Film was to be returned in a re-entry capsule. The CIA originally intended to use this device on the Samos. It was designed to acquire 60 cm resolution images of targets such as the Tallin (Estonia) ABM site. But the first, on 18 March 1963, failed to achieve orbit; then the Agena upper stage failed on the second, launched 18 May 1963. The third attained orbit 31 July 1963 but this time the camera failed after 32 hours. Then when the first successful photographs appeared, investigators realised the lens had a serious focus problem, later traced to thermal factors. The programme ended after these three attempts because of KH-7's successful introduction.

KH-7 Gambit Series

Background
The USAF built its own photo-reconnaissance system to provide detailed assessments of certain installations in order to determine the effectiveness of its targeting programme. It named the system the Gambit. Gambit returned 0.5 m resolution film by flying at a perigee as low as 137 km. Often flown in conjunction with a KH-4A, one system provided detailed images while the other provided a more panoramic view. Between 12 July 1963 and 4 June 1967 when the programme ended, 36 out of 38 KH-7s launches achieved orbit. Its principal contribution to intelligence lore was to photograph the new SS-7 and SS-8 ICBMs silos in the Soviet Union. Like KH-4A, film came back in two capsules and the satellite had an approximate lifetime of six days. Of all the early generation military reconnaissance satellite systems, Gambit is one of the most tightly guarded because its technical legacy has been to advance the capabilities of successor generations. No images have been released.

KH-8 Gambit Follow-on Series

Background
As the KH-7 provided the test bed for new cameras, the KH-8 brought them into being. A single overwhelming advantage of the KH-8 was its long orbital operations life, cutting costs of repeated and frequent launches and maintaining consistency of operation. The first KH-8 flew on a Titan 3B Agena D on 29 July 1966. It had a longer life than its KH-7 predecessor and carried first the KH-4B and then KH-9 camera which had a resolution down to 15 cm. Agena's restartable engine enabled perigees as low as 135 km to be maintained for more than 50 days, so fewer of the expensive satellites had to be flown. In the early years from 1960 on, the CIA flew about nine spy flights a year. The KH-8 allowed them to reduce this number to two or three. Even though the Big Bird entered service in the mid-1970s, the CIA continued to fly the KH8 apparently to fill gaps in Big Bird launches. Finally in April 1984 the CIA depleted the stockpile of KH-8 cameras and the programme ended after 53 launches. The production stopped in 1980 after Congress became alarmed about cost overruns, reported at about US$400 million, in the Keyhole programmes.

KH-9 Hexagon (Big Bird) Series

Status
The first KH-9 was launched by Titan 3D on 15 June 1971 from the VAFB to a 180 × 300 km, 96.4° orbit. The second KH-9 was lofted on 20 January 1972 followed by the third on 7 July 1972. The last of 20 was launched on 18 April 1986 but lost in launch failure.

Background
By the mid-1960s, the air force desired its own completely independent method to observe Soviet ground installation. It had proceeded with the Manned Orbiting Laboratory programme, which had Earth observation as one of its primary objectives. To support the plan, the air force built the KH-10 camera. But when Congress cancelled the MOL in 1969, the air force still made its new camera available for imaging principally as the Big Bird. Instead of broad swaths of countryside the Big Bird is a 'close look' operation, intended to make detailed photos. Some programme managers refer to it by its service acronym as LASP (Low Altitude Surveillance Platform).

The orbiting spacecraft weighs more than 11 tonnes. Lockheed built it for the air force in a cylindrical canister 15.2 m long, 3.05 m in diameter and powered by two large solar arrays along with a 6 m diameter antenna. The Big Bird expanded the number of film canisters to four and it could eject them as required one at a time in times of crisis. Big Bird truly serviced two audiences, the camera combined KH-8's 45 cm resolution with KH-4B's 64 km swath width and achieved a 60 cm resolution across a 130 km swath. In addition to their primary mission, many Big Birds carried 60 kg Ferret subsatellites (some two), signals intelligence antennas and covert communications relays.

Lockheed enjoyed a boom in its launcher business and developed the Titan 3D specifically to carry the Big Bird into space. Most observers think that the air force tried to clandestinely launch the first Titan 3D without registering it. Each satellite entered a Sun-synchronous orbit to pass regularly over the targets at the same time of day. In this manner each target in subsequent frames should have had exactly the same size shadows, so if new construction occurred, an analyst could follow the progress by watching the shadow lengthen. Lockheed overcame the drag that such a vehicle incurred, normally enough to make it re-enter within 7 to 10 days, by periodic Agena firings. At the end of the satellite's lifetime, the air force de-orbited it directly into the ocean rather than allowing it to decay naturally. Big Bird production ended in 1981, with a number held in store for annual launches until later systems made them unnecessary.

After an initial start-up period the air force managed to get a Big Bird into orbit about once every 90 days. Only one six-month gap occurred between number 8, launched 10 April 1974 and number 9 launched 29 October 1974. The air force may have been using the time to modify the camera in order to counter Soviet efforts to camouflage missile silos and control centre construction. Number 9 provided surveillance at a time of increased tension in the Middle East and witnessed 16 Soviet ships unloading crated

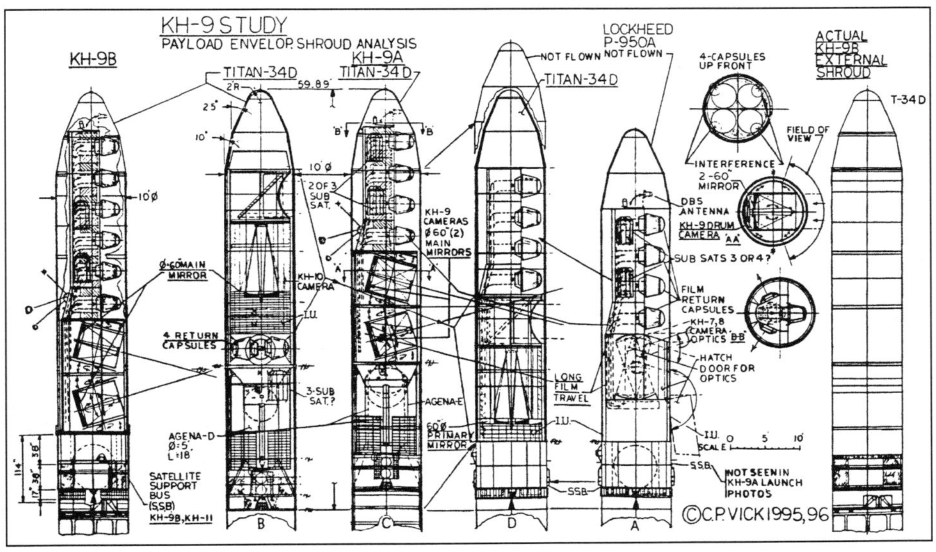

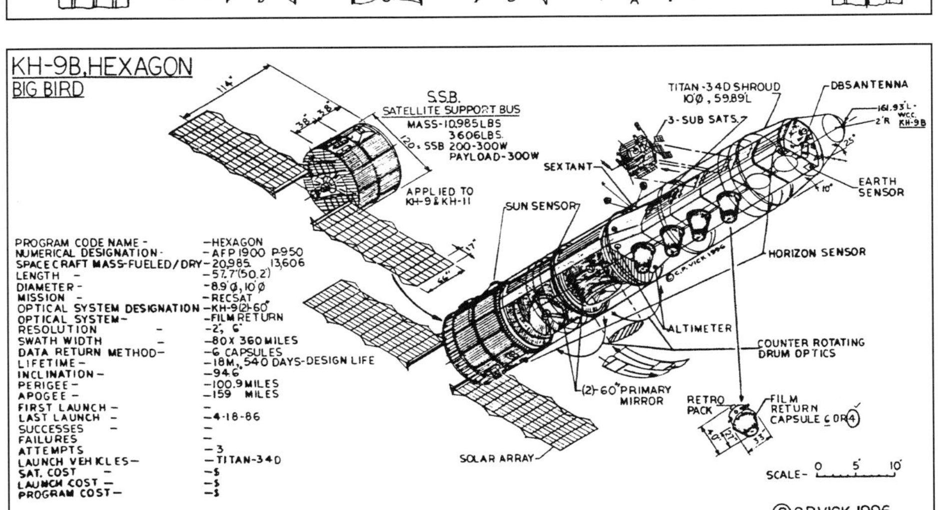

KH-9 conceptual studies. The last three KH-9s were retrofitted with the SSB Satellite Support Bus and two more return capsules. Dimensions are in feet (C P Vick)
0003433/0003434

materials, believed to include SA-6 spare parts and components for Scud surface-to-surface missiles.

Big Bird 16 returned film over a period of 261 days, watching the Iran-Iraq war. Big Bird 17 launched 11 May 1982 watched the Falklands War. A Big Bird probably identified the new Soviet missile radar at Abalakova, central Siberia, which became important to the Reagan administration's claims that the Soviets violated the ABM treaty.

KH-11 Kennan/Crystal Series

Current Status

No KH-11 Kennan/Crystal is operational today and the lack of a shuttle launch facility for polar orbit flights undoubtedly had a part to play in the retirement of this type. Vandenbeng AFB, Calif, was to have been the base from which shuttle missions would have been available to extend the life of the satellite. It was the concept of in-orbit-servicing for military satellites that inspired the mission design for the Hubble Space Telescope.

Background

KH-11 introduced electro-optical imaging to the photo-reconnaissance inventory of the United States. Instead of returning a film canister several days to weeks after a launch an electro-optical imaging system takes the equivalent of a high resolution television picture of the earth which is then transmitted to a ground station. KH-11, despite a lower resolution than its film bearing predecessors, became more practical for routine use, and the 1970s semi-annual launch rate for espionage satellites with film aboard dropped to a few a decade.

KH-11s also occupied a higher orbit than the Big Bird because their lengthy mission duration did not allow for the frequent orbit adjustments that atmosphere skimming satellites require. A usual KH-11 orbit had a perigee of about 300 km and an apogee of 500 km with some as high as 1,000 km. Each KH-11 weighs more than 13 tonnes, is about 19.5 m long and 3 m in diameter. TRW builds the satellites for the USAF who uses the Titan 3D/34D to place them in 97° inclination orbits. Even at the higher altitude the KH-11 must raise its altitude about every three months during a three-year lifetime. Its real-time sensing systems (8–12 images per minute) and high resolution CCD cameras enable it to distinguish military from civilian personnel on the ground. The infra-red and multi-spectral sensing devices of Advanced KH-11 can locate mobile missiles, trains and launchers by day or night, and distinguish camouflage and artificial vegetation from real plants and trees. Each satellite has a side-looking

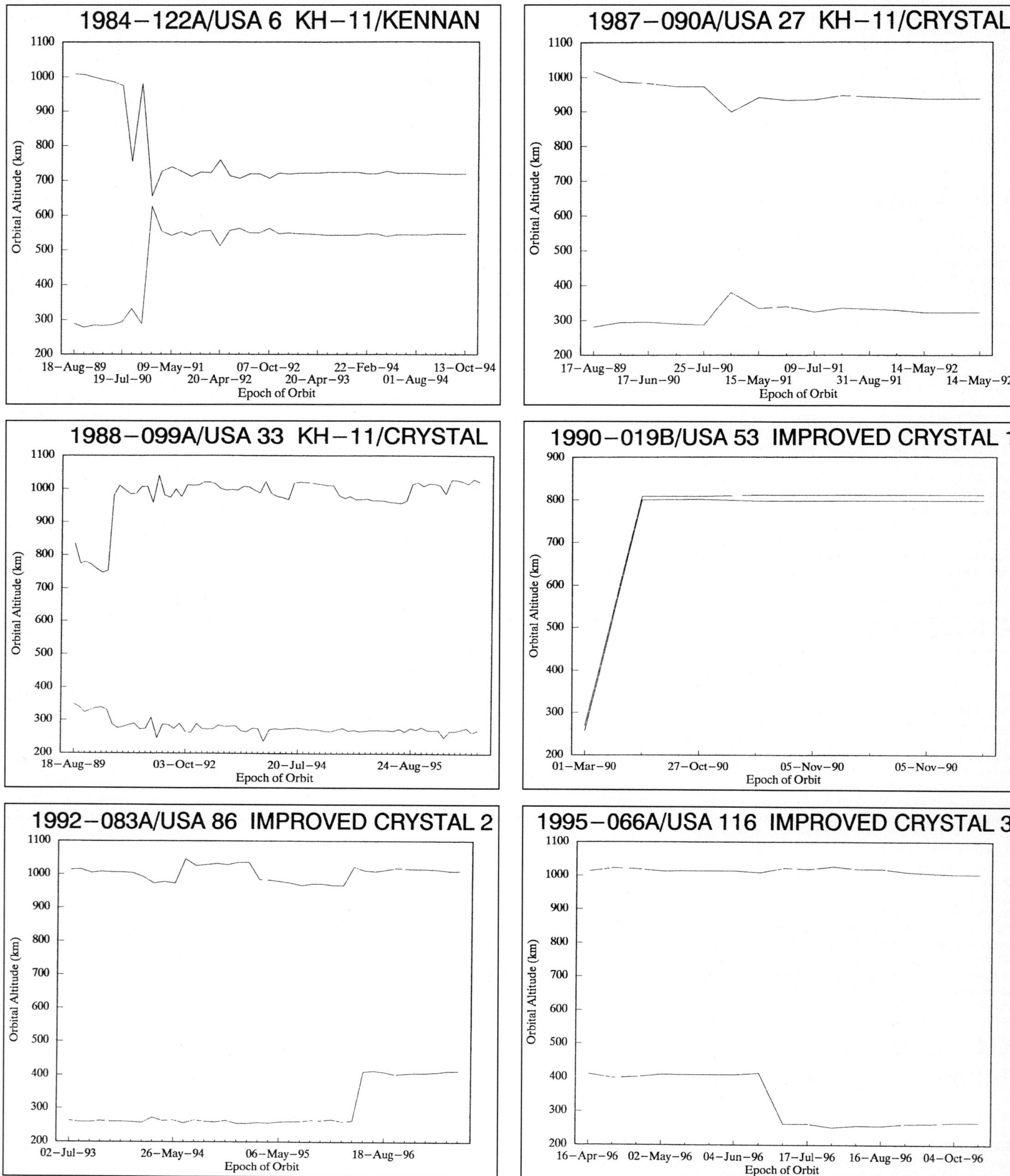

Graphs showing the orbital altitude regimes for photo-reconnaissance missions operating during the 1980s–1990s. USA 6 was a KH-11/Kennan satellite, USA 27 and USA 33 were KH-11/Crystal satellites and USA 53, USA 86, USA 116 and USA 129 were Improved Crystal satellites. These orbits are based upon amateur observations of the satellites, since the actual operating orbits are classified

0003400/0003401/0003439/0003440/0003441/0003442

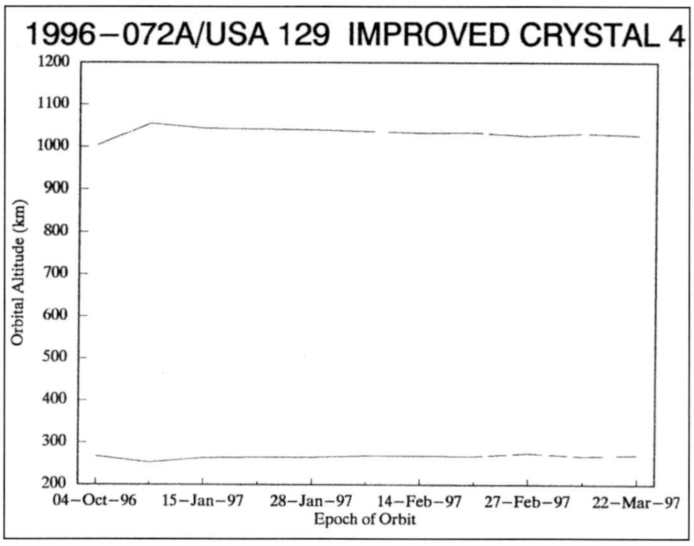

Graphs showing the orbital altitude regimes for photo-reconnaissance missions operating during the 1980s–1990s. USA 6 was a KH-11/ Kennan satellite, USA 27 and USA 33 were KH-11/Crystal satellites and USA 53, USA 86, USA 116 and USA 129 was an Improved Crystal satellite. These orbits are based upon amateur observations of the satellites, since the actual operating orbits are classified
0003443

capability, doubling the number of daily passes over a site from two to four.

An SDS helper satellite relays the data to Fort Belvoir in Virginia as the KH-11 takes each picture. As configured by the USAF, the KH-11 operational system consists of a constellation of two satellites in planes 48.7° apart so that one covers a target in mid-morning while the second follows mid-afternoon. Each satellite repeats its track every four days.

There had been no public reference to the KH-11 designation on the part of the CIA until a CIA agent sold an operating manual to a KGB agent in Greece in 1978. The KH-11 scored its largest intelligence coup by monitoring a shed in early 1980 at the Severodvinsk naval yard on the White Sea. A submarine larger than the US Trident boats finally emerged from this shed a few months later. In September 1980 the Pentagon purposefully leaked information that a KH-11 had photographed side by side emplacements of SS-20 and SS-16 missiles. The Pentagon claimed the Soviets planned to use their own reconnaissance satellites as surrogate KH-11s to determine how much camouflage each missile required to be mistaken for the other. In short order the Pentagon cited this as evidence that Soviet deception would not allow the US to verify adherence to the Salt 2 agreement on ballistic missile limitation.

KH-11-1, launched 19 December 1976, and KH-11-2, launched 14 June 1978, worked in pairs during part of their lifetimes. Both made their observational passes in the morning. After KH 11-3 joined them on 7 February 1980, the pairs alternated observations in mornings and afternoons to watch the early stages of the Iran-Iraq war in conjunction with Big Bird.

KH-11-4, launched 3 September 1981, set an operational record at 1,177 days, KH-11-5, launched 7 November 1982, and later joined by KH-11-6 on 4 December 1984, worked as a pair until 13 August 1985, when KH-11-5 re-entered. Two weeks later, on 28 August 1985, its KH-11-7 replacement disappeared when the Titan 34D launcher exploded shortly after lift-off, seriously damaging the Vandenberg pad. Because of the loss of the Shuttle Challenger and the Titan explosion, the United States had lost a significant portion of its photo-reconnaissance ability and heroic measures prolonged KH 11-6's life by using it to only relay images of the highest priority intelligence targets.

The satellite executed few manoeuvres in order to conserve fuel, but amateur observations revealed a manoeuvre in mid-1989 into a still unexplained 350 × 900 km orbit. The CIA allowed the orbit to decay back to 336 × 732 km when its perigee was raised to 559 km in November 1990, possibly to support Persian Gulf War operations. No further manoeuvres were seen before it apparently de-orbited in late 1994 after an impressive 10-year life. KH-11-8 was joined by KH-11-6 on 26 October 1987 in an afternoon orbit. The Titan 34D launch of 6 November 1988 was the last launch of a KH-11, arriving at the morning slot in unison with KH-11-8. It was initially placed in a 156 × 1,012 km, 97.9° orbit by Titan 34D.

Even with the lengthened lifetimes, many in Congress talked of killing the KH programme. Total cost overruns in the KH-11 had come to more than US$1 billion and the Soviets had thwarted it by introducing an elaborate camouflage system. To overcome the objection of cost the air force at one time planned to refuel each satellite in orbit to stretch its three-year lifetime. Some observers have suggested that military astronauts made a practice refuelling attempt using the Space Shuttle. Others have indicated this was just a plan that was never tried.

The first in a series of Improved Crystal was launched by Titan 4 on 28 November 1992, followed by USA-116 on 5 December 1995 and USA-129 on 20 December 1996.

KH-12 Ikon

Current Status

First in the series was launched by Titan 4 on 28 November 1992. Advanced KH-11 presently in operation was KH-11-3 launched by Titan 4 on 20 December 1996.

Background

In 1989, the CIA had clearly dropped the KH-11, but while some observers called the new series the KH-12 others claimed it is really an upgraded KH-11. Electro-reconnaissance with some signals intelligence payloads and wider spectral band sensitivity. Originally designed for launch from the Shuttle the USAF switched to Titan 4 and it was lightened. Data is transmitted via the SDS relay satellites. In August 1989, the Shuttle placed a payload called USA-40, into a 303 × 315 km, 57° orbit, which later raised itself to 460 km, although observations by amateur astronomers indicated initial tumbling problems. This new satellite obviously differed from the KH-11 since it had a mass of only 9,350 kg, which included a large propellant load for manoeuvring.

The high orbit requires improved optics to maintain the resolution of earlier KH-11s, and some circumstantial evidence exists for such a change. As part of its promotion for using the 'Bus I' service module for the International Space Station, Lockheed claims the bus is operational on classified missions. Advanced KH-11 is a prime candidate. Its 4.04 m diameter and 2.67 m length accommodates a larger mirror than KH-11's. (Compare this to the 2.40 m diameter mirror on the Hubble Space Telescope and NASA's claim that HST has the largest mirror in space is suspect.)

Some observers suggest USA-40 is an improved SDS satellite intended to relay data from a Molniya type orbit. If true, Shuttle STS-36 actually deployed the first upgraded KH-11 when it released the 16.9 tonne AFP-731 into a 204 km 62° orbit on 28 February 1990. The fact that this new satellite is significantly larger than KH-11, possibly up to 18 tonnes fully fuelled, lends credence to the belief USA-40 is actually an SDS. In order to launch this satellite from Vandenberg the Shuttle would have had to use 109 per cent main engine thrust - a condition no longer permitted. During 1986, the lack of appropriate photo-reconnaissance satellites prompted some discussion of an emergency Shuttle flight to orbit an advanced KH-11 (KH-12). Since the flight would have been unmanned, the military astronaut corps fought it, because it would have indicated they were not needed to pilot the spaceplane. The first advanced KH-11 should have flown on Shuttle 62C – the first scheduled flight from VAFB in September 1986 – but when the air force mothballed the west coast's US$3 billion Shuttle facilities it had to wait until a 1989 Shuttle departure from the Kennedy Space Center.

Each advanced KH-11 carries both digital imaging sensors and signals intelligence receivers covering a wide range of transmissions over the Soviet Union, from telephone to TV. Each also supposedly carries a laser altimeter and the ICMS Improved Crystal Metric System for the defence Mapping Agency, adding fiducial marks to imagery.

Publication of commercial imagery of Chernobyl and Soviet military bases brought the subject of space-based observations into the open so heavily by April 1986 that Keyhole, and even KH-12, were for the first time discussed in the US House of Representatives. Representative George Brown called for relaxation of the existing limit of 10 m resolution on civilian remote sensing satellites (which President Reagan subsequently agreed to). Earth resources satellites being developed by Europe, Japan and China do not adhere to such limits.

Samos Series

Background

As the Corona/Discoverer programme developed, a competing philosophy struggled to find a voice in the United States bureaucracy. The USAF had long proposed and funded background studies in direct transmission of data from space in an existing RAND programme named WS-117L. But James Killian and Edwin Land, two of Eisenhower's advisors on photographic intelligence, still recommended recovering a film canister from space. The air force refused to let the direct readout transmission idea die and continued funding WS-117L on a parallel track with Corona. Eisenhower's National Security team actually encouraged this programme as a public shield for the Corona programme. From its press releases the public believed the air force programme constituted the sum of the United States investigations into satellite reconnaissance. Eventually the programme became known as Sentry. Many historians feel the name SAMOS stood for Space and Missile Observation Satellite, but in fact the name was chosen to honour the Greek island where Aristarchus lived.

The Polaroid corporation provided the onboard film processing procedure, which NASA adapted

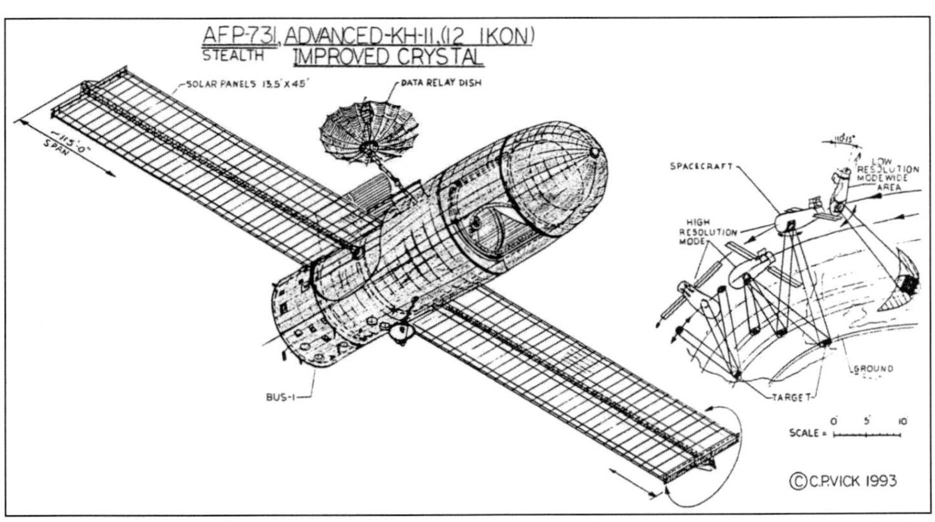

Advanced KH-11/Improved Crystal conceptual study. Dimensions are in feet (C P Vick) 0003438

for the Lunar Orbiters of 1966–68. Samos itself though was apparently a dismal failure and had to be cancelled in 1962, although there have been references to missions beyond the five admitted to at the end of 1961.

Samos could accommodate four different payload configurations: E-1 frame readout camera with 1.83 m focal length, 30 m ground resolution, 161 × 161 km frame coverage; E-2 frame readout camera with 0.91 m focal length, 6 m ground resolution, 27 × 27 km frame coverage; E-5 recoverable panoramic camera with 1.67 m focal, 1.5 m ground resolution, 98 km swath; E-6 recoverable panoramic panorama with 0.7 m focal length, 2.4 m resolution, 280 km swath.

The publicly acknowledged missions are Samos 1 launched 11 October 1960 by Atlas Agena A from Vandenberg, which failed to achieve orbit. Samos 2 launched 31 January 1961 made it into a 476 × 554 km 97° orbit, but Samos 3 exploded on the pad 9 September 1961. Samos 4 also failed to attain orbit, 22 November 1961. Samos 5 succeeded 22 December 1961. Some observers hold that 7 March 1962 may have been a Samos. One author who has studied the programme claims that the entire series returned only a single set of images, with resolution about 30 m, and that analysts were unable to decide if it showed China or the USSR. A second Samos, E-6, series apparently flew five times in 1962 (26 April, 17 June, 18 July, 5 August, 11 November), returning film aboard capsules and using Agena to initiate re-entry.

RADAR CALIBRATION

Russian Federation

Romb Series

Current Status
No launches have taken place since 1995.

Background
Manufactures by Reshetnev, Romb were subsatellites released from Taifun-1 spacecraft for the calibration of PVO air and space defence radars.

Taifun Series

Background
Taifun (code name Vektor) are produced in two types, Taifun 1 operates independently while Taifun 2 carry up to 25 Romb subsatellites. Satellites fly to allow ground-based radars perform experiments for the determination of a satellite's characteristics. Unlike Yug, Taifun is an active satellite and carries solar cells over its 2 m diameter spherical surface, as well as four antennas. Two classes of Taifun satellites have appeared. The first group started with Kosmos 660 and flew into approximately 400–2,000 km orbits while the second group started with Kosmos 687 and flew into approximately 300–700 km orbits. The Russians have not revealed whether satellites of slightly different designs fly the two different orbital regimes or whether all of the Vektor satellites were identical.

Yug Series

Background
Yug is the code name for Taifun 2, equipped to carry up to 25 Romb calibration subsatellites. The system was integrated into military service in March 1981. Satellites are spherical in shape, 2 m in diameter. Mass 950 to 990 kg.

Romb Series

Launch date	Cosmos	Incl (degrees)	Period	Perigee	Apogee
28 April 1976	816	65.8	94.6	481	515
27 April 1977	906*	50.6	94.3	463	515
30 May 1977	913	74.0	94.5	472	520
19 July 1977	930*	74.0	94.6	481	514
8 December 1977	965	74.0	94.4	465	516
22 December 1978	1065	50.7	93.4	344	548
6 July 1979	1112**	50.7	93.4	344	542
6 June 1980	1186	74.0	94.5	473	519
31 July 1980	1204	50.7	93.3	345	538
28 September 1981	1311	83.0	94.5	463	519
29 January 1982	1335	74.1	94.6	482	518
21 April 1982	1351	50.7	93.5	348	548
29 July 1982	1397	50.7	93.4	345	540
19 April 1983	1453	74.0	94.4	463	517
26 May 1983	1465	50.7	93.4	349	543
31 August 1983	1494	50.7	93.4	345	550
3 September 1983	1501	82.9	94.5	468	514
27 September 1984	1601	65.8	94.5	475	516
19 June 1985	1662	65.8	94.5	476	513
2 October 1985	1688***	50.7	93.5	347	548
3 September 1986		74.0	94.6	474	516
22 January 1987	1815***	50.7	93.5	345	550
28 July 1988	1960	65.8	94.5	473	513
23 December 1988	1985	73.6	95.2	522	533
15 February 1989	2002	65.8	110.3	186	2,299
27 February 1989	2053	73.5	95.1	518	526
6 February 1990	2059	65.8	110.2	189	2,280
24 April 1990	2075	74.0	94.6	484	514
28 November 1990	2106	82.5	95.2	518	537
25 June 1991		74.0	Launch failure		
2 March 1995	2306	65.8	94.5	469	517

The inclusion of some satellites in this listing is based upon their orbital parameters rather than the existence of sub-satellites deployed in orbit. The launch failure in June 1991 might not have been a Romb mission, but this seems to be the most likely satellite group for the failure.
* Cosmos 906 and Cosmos 930 remained attached to their respective Cosmos-3M launch vehicle second stages, so could have been Romb failures (no additional sub-satellites were tracked).
** The Cosmos-3M second stage from the Cosmos 1112 mission exploded in orbit and a determination of any Romb mission-related sub-satellites cannot be confirmed.
*** These two satellites are the only ones to have used this particular orbital regime during 1980–1990 which are not identified as Romb missions: it is possible that they were mission failures (or a totally different class of satellite).

Taifun Series

Launch date	Cosmos	Incl	Period	Perigee	Apogee
18 June 1974	660	83.0	109.1	397	1,972
12 March 1976	807	83.0	109.1	398	1,973
16 January 1981	1238	83.0	109.0	406	1,958
9 April 1981	1263	83.0	109.1	397	1,970
11 November 1983	1508	82.9	109.1	400	1,966
28 August 1990	2098	83.0	109.2	396	1,977
27 September 1994	2292	83.0	108.9	400	1,954
Taifun Series: Class 2 Missions (Cosmos 687 Type)					
11 October 1974	687	74.0	94.5	286	698
28 May 1976	822	74.0	94.5	280	711
14 July 1987	1868	74.0	94.6	284	709
10 October 1991	2164	74.0	94.5	285	708

RADAR IMAGING

United States

Lacrosse/Onyx Series

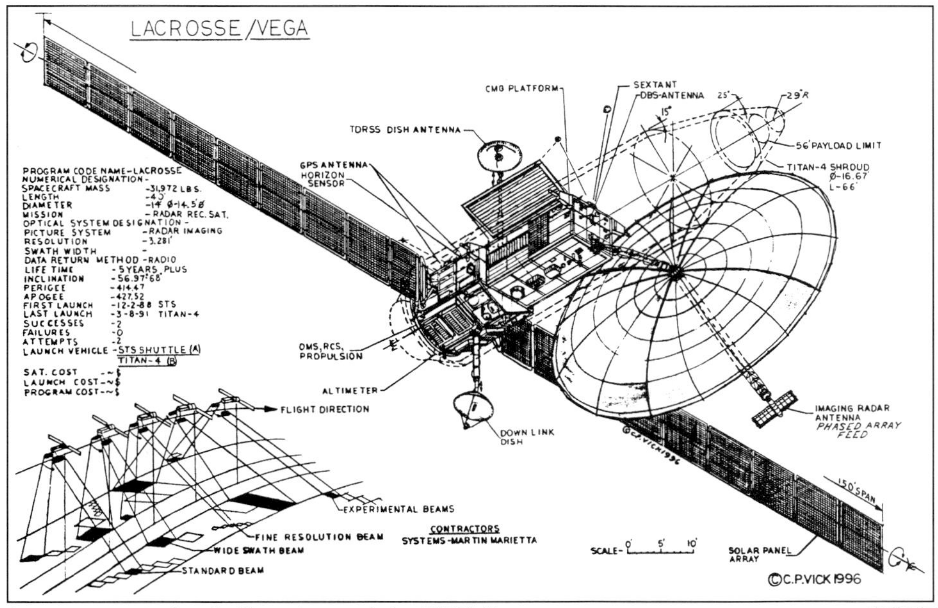

Lacrosse conceptual study. Dimensions are in feet (C P Vick) 0003435

Current Status

The Central Intelligence Agency launched its third Lacrosse/Vega high-resolution imaging satellite on 24 October 1997. The Agency code-named the satellite USA 133. It arrived in a higher than expected 674 km orbit at 57° inclination. The new Lacrosse has a higher resolution than the 1 m capability of its two predecessors. This Lacrosse had to go into the 57° orbit because the CIA planned to operate it in conjunction with an earlier version to obtain stereo coverage. Because the CIA chose to launch the earlier satellite on the Shuttle in 1988, the mission could not achieve a higher inclination.

The Lacrosse Onyx 4 satellite was launched by Titan 403B on 17 August 2000.

Background

The first high resolution imaging radar satellite and the first US military active radar craft of any type left the Shuttle cargo bay as USA-34 from Atlantis on 2 December 1988 during the STS-27 mission the first Shuttle flight after the loss of Challenger on 28 January 1986. The air force used it to provide all weather day/night coverage of Soviet-bloc armour movements. Lacrosse evolved as a side looking radar all-weather reconnaissance satellite. To keep track of armour requires that the satellite images at 1 m resolution or better. By comparison NASA's Shuttle Imaging Radar (SIR) has a 10–20 m resolution. Some sources feel that the Lacrosse also carries optical sensors. If true the sensors probably only identify targets optically for the Radar to scan. The US military probably proposed that Lacrosse act to detect Soviet mobile missiles, but the end of the Cold war required the DoD to dream up another mission to keep the programme afloat. One candidate often cited by sources is to verify adherence to mobile ICBM treaties. Lacrosse returns data via TDRS to the White Sands, New Mexico ground station.

Lockheed Martin Astronautics in Denver acted as the prime contractor for the Vega/Lacrosse satellite, which reportedly cost in excess of US$500 million. The CIA originally named the programme Indigo. In its present version it somewhat resembles the nuclear-powered RORSAT, utilised by the Soviet Union for monitoring the world's navies.

Atlantis's remote manipulator arm released the Lacrosse during orbit number 5 some 7 hours after launch but the solar panels initially failed to deploy to their 45.7 m span. The satellite subsequently raised its orbit by some 240 km to 668 × 703 km at 57° inclination. The Lacrosse cost the US government more than US$500 million. Some sources say it was originally called Indigo.

Vandenberg's first Titan 4 launch on 8 March 1991 carried Lacrosse. Amateur observations of USA-69 during March 1991 identified a 683 km 68.0° orbit, indicative of a radar satellite. Titan 4's 8 November 1991 was an advanced Whitecloud. USA-116 of 5 December 1996 may have been a Lacrosse. It has been claimed that a Titan 3B Agena mission of 21 January 1982 was launched into 553 × 646 km, 97.3° orbit in order to demonstrate the Lacrosse technology. Orbit manoeuvres on the part of several nearby satellites indicate they acted as calibration targets for the radar satellite. Lacrosse 3 was launched by Titan 4A on October 24, 1997 followed by Lacrosse 4 on 22 May 1999.

This orbit was maintained through to the end of 1996. In March 1997 amateur satellite observers found that USA-34 had disappeared from orbit, presumably having been de-orbited that month.

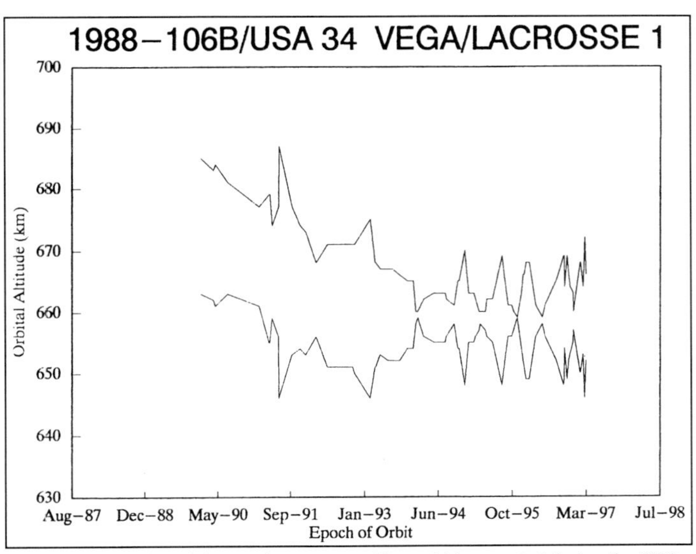

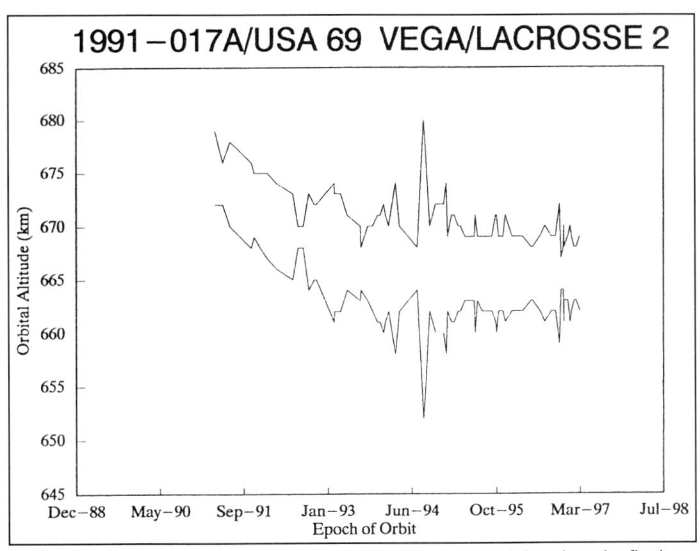

The orbital altitudes of the two Lacrosse satellites which operated during the 1990s. It will be noted that each satellite started life in an orbit slightly higher than the final one and was allowed to slowly drift down until the altitude was stabilised at around 660 to 670 km. USA 34/Lacrosse 1 disappeared from orbit in March 1997. These orbits are based upon amateur observations of the satellites, since the actual operating orbits are classified 0003436/0003437

SCIENCE AND ENGINEERING TEST

France

Cerise

Background
The 50 kg Cerise (Caractérisation de l'Environnement Radioélectrique par un Instrument Spatial Embarqué) was carried into orbit in order to characterise the Earth's radio environment in research that could lead to a 'Zenon' Elint satellite during the next decade. It came with Ariane 40 on 7 July 1995. Alcatel Espace built the satellites for DGA, using the University of Surrey's UoSAT bus. The French Ministry of Defence expects to use the satellite for about 2.5 years. In 1993, the MoD requested industry proposals for 1998's Clementine Cerise follow-on to investigate environmental interaction at different radio frequencies.

Cerise collided with a piece of debris from Ariane third stage, which orbited Spot 1 in 1986, making it the first proven casualty of space debris. The collision with the debris 1986-019RF occurred at 09.48 GMT on 21 July 1996 breaking off the upper portion of the Cerise gravity gradient boom. French ground controllers regained control over the satellite following the collision.

United States

APEX (P90-6 Advanced Photovoltaic and Electronics Experiment)

Background
A Pegasus launched the 261 kg APEX (Advanced Photovoltaic & Electronics Experiment) 3 August 1994 to determine the long-term effects of harsh radiation and plasma environments. The satellite achieved a 361 × 2,528 km, 70° inclination orbit that passed repeatedly through Earth's radiation belts at apogee and through the upper atmosphere at perigee.

The USAF awarded Orbital Sciences Corporation a US$9.6 million contract on 18 March 1991 to build the APEX P90-6 mission in the STP series of experiments. Orbital Sciences built the test bed that housed the many experiments on board. It is based on the PegaStar bus which integrates the booster upper stage avionics and spacecraft bus functions. They included the PASP Plus (Photovoltaic Array Space Power Plus Diagnostics) which tested 12 solar cell samples with Fresnel lens and Cassegrain concentrators, the CRUX (Cosmic Ray Upset Experiment) which studied microelectronics single event upsets, and the radiation degradation of ferro-electric devices with the FERRO (Thin Film Ferro-electric Experiment).

APEX is the first application of OSC's PegaStar bus (OSC) 0517162

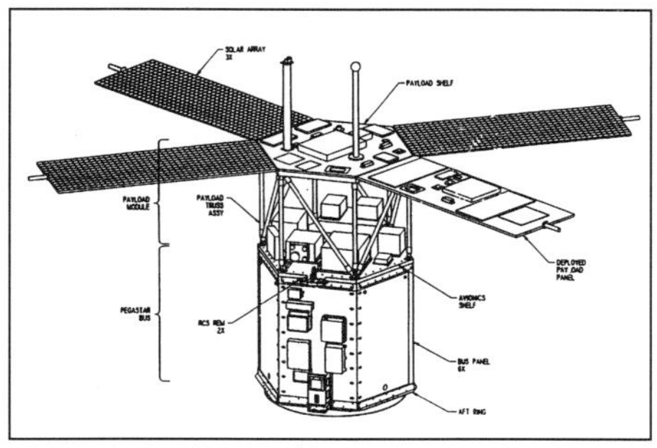

APEX principal features (Orbital Sciences Corporation) 0516824

PASP Plus ceased operating 11 August 1995 after collecting 10 times more data that all such previous space experiments combined. CRUX/FERRO remains operational. The total mission cost to the air force is about US$30 million, including about US$12 million for the spacecraft and US$7.8 million for the Pegasus.

ARGOS (P91-1, Advanced Research and Global Observation Satellite)

Current Status
Argos (P91-1) was launched on 23 February 1999 by Delta 7925 from Vandenberg AFB pad SLC2W to a 831 × 847 km orbit at 98.7° inclination. Co-orbited by the same launch vehicle were the magnetophonic mapping satellite Orsted and the South African student satellite Sunsat.

Background
The ARGOS mission is to fly and operate joint service payloads which include two technology demonstrators and seven experiment payloads for global and celestial observation. It is categorised as USAF Space Test Programme P91-1 built by Boeing at its Seal Beach facility. ARGOS weighs approximately 2,700 kg and has a goal of three years of on-orbit operations. ARGOS incorporates an accurate attitude determination and control system with an embedded Global Positioning System (GPS) receiver for position determination and time reference and a solid-state recorder, providing 2.6 Gbits of onboard data storage. ARGOS also includes an automated mission planning system, developed by the University of Colorado's Laboratory of Atmospheric and Space Physics, to optimise ground contacts as well as onboard power and data storage requirements. The automated mission planning tool adds tremendous capability to manage the ARGOS downlink of 9.6 Gbits each day. On board experiments are:

The High Temperature Superconductivity Space Experiment, or HTSSE 2, developed by the Naval Research Laboratory (NRL), Naval Center for Space Technology. HTSSE 2 will space qualify super-conducting digital and RF subsystems to demonstrate an operational space capability. Expected performance of super-conducting components include factors of 100–1,000 in power reduction, more than 10 times higher speed and similar weight reduction than today's silicon or gallium arsenide (GaAs) based electronics.

The Extreme Ultra-Violet Imaging Photometer (EUVIP) payload is sponsored by the US Army Space and Strategic Defense Command and built by the University of California at Berkeley Extreme Ultra-Violet Astrophysics Laboratory. EUVIP will establish the behaviour of the upper atmosphere and plasmasphere as needed for Army RF secure communication systems design, prediction of magnetic storms, and characterisation of the aurora.

The Naval Research Laboratory, Space Science Division, sponsors the Unconventional Stellar Aspect (USA) payload. ARGOS will be one of the first research satellites to fly an embedded Global Positioning System (GPS) receiver, while characterising astronomical x-ray sources for potential use as autonomous position, attitude and time-keeping references for military space systems. The augmentation of a proven x-ray detector design with advanced computers will help to demonstrate the feasibility of autonomous satellite navigation using x-ray pulsars in place of GPS timing and navigation signals.

Electric Propulsion Space Experiment or ESEX, developed at TRW's Space and Electronics Group facilities will demonstrate reliable high powered arcjet thruster operations in space. ESEX will also execute orbit transfers and verify compatibility with spacecraft subsystem controls. Electric propulsion is projected to double the payload-to-orbit capability of current space propulsion systems. The managing organisation for this payload is the Air Force Phillips Laboratory, Electric Propulsion Laboratory.

The Space Dust Experiment of SPADUS will provide a 3-D survey map of the present dust distribution in low Earth polar orbit and allow prediction of orbital debris 'showers' which could affect spacecraft such as the Space Shuttle. SPADUS will also obtain early flight experience for sensors and electronics, which are planned for International Space Station and the Cassini mission to Saturn in 1998. This payload is sponsored by the Office of Naval Research, Electronics Division and built by the University of Chicago Enrico Fermi Institute, Laboratory for Astrophysics and Space Research.

The Critical Ionisation Velocity experiment of CIV will release xenon and carbon dioxide to study ionisation caused by plasma and collision processes in the upper atmosphere. The results will be used to help identify plumes and atmospheric wakes of launch and orbital vehicles. CIV is developed and sponsored by the Air Force Phillips Laboratory, Satellite Assessment Division.

The High Resolution Airglow Aura Spectroscope (HIRAAS) payload, developed and sponsored by the Naval Research Laboratory, Space Science Division, consists of three high spectral resolution ultra-violet spectrographs designed to measure the naturally occurring thermospheric and ionospheric airglow. Measurements of the ionosphere can improve a number of DoD systems that depend on radio and microwave propagation through the upper atmosphere and ionosphere. Neutral density measurements also support NORAD operational requirements for satellite drag forecasting and the ability to predict orbital life and re-entry impact locations.

Global Imaging Monitor of the Ionosphere or GIMI will demonstrate operational charged coupled device (CCD) sensor technology for environmental monitoring of upper atmospheric perturbation due to meteors, rocket exhausts, and aurora. Use of wide-field sensors in three separate wavelengths

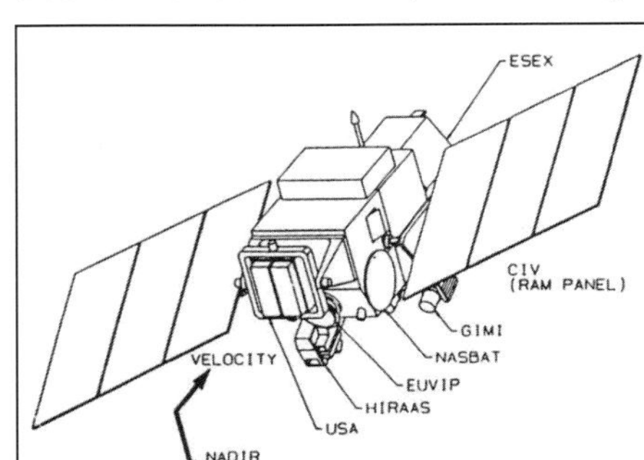

Argos: the artist's impression highlights the operating ammonia arcjet (TRW) 0516847/0516848

will enable GIMI to continuously image 560 m^2 of the Earth's limb. GIMI is developed and sponsored by the Naval Research Laboratory, Space Science Division.

The Coherent Radio Topography Experiment Payload (CERTO) consists of a three-frequency radio beacon and radiating antenna mounted on the space vehicle. Receivers on the ground use the differential phase techniques to derive the integrated electron density. Characterising CERTO data plays a critical role in assessing impacts on navigational accuracy, communication systems, and remote sensing by radars. CERTO is managed by the Naval Research Laboratory Plasma Physics Division.

Co-manifested on the Delta-2 launch vehicle are two secondary payloads from Denmark and South Africa. Assembled is a coalition of international test teams to support space test programmes from the United States, Denmark and South Africa. The Danish satellite ORSTED will map the Earth's magnetic field and the South African satellite SUNSAT will map the Earth's gravitational fields. The integration and development for these micro-satellites are managed by the NASA Orbital Launch Services Project.

JAWSAT (P98-1, Joint Academy, Weber State University Satellite)

Background

STP continued to prepare for the JAWSAT (P98-1), Joint USAF Academy/Weber State University Satellite. JAWSAT, along with the Air Force Academy's FalconSat, was due to be launched on the first Orbital/Suborbital Program (OSP) launch vehicle in September 1999 but the flight was delayed to 27 January 2000 by Minotaur from Vandenberg AFB, pod CLF (Commercial Launch Facility) on a leased site. The US Air Force refers to Minotaur on the Orbital/Sub-orbital Program Space Launch Vehicle (OSPSLV), it comprises the Minuterran and Taurus stages in a configuration put together by Orbital Services. JAWSAT was placed in a 746 × 810 km, 100.2°, polar orbit. JAWSAT carries three experiments: a Thermospheric Temperature and Nitric Oxide Spectrograph, an Ionospheric occultation Experiment, and the Coherent Electromagnetic Radio Tomography Probe (CERTO) series experiment. Success with this flight confirmed the JAWSAT concept for future science platform applications, a secondary objective of the P98 effort.

MightySat Satellite

Current Status

MightySat was launched from shuttle Endeavour on STS-88 which lifted off from the Kennedy Space Center on 4 December 1998 placing MightySat 1 in a 381 × 395 km, 51.6° orbit followed by MightySat 2 on 19 July 2000 to a 547 × 581 km, 97.8° orbit.

MightySat 1 composite mockup, with four 45° antennas for the UHF half duplex TT&C system (Phillips Laboratory) 0517309

MightySat 2 carries a hyperspectral imager (Spectrum Astro Inc) 0517315

Background

The Space Experiments Directorate of AF Phillips Lab developed a series of MightySat micro-satellites carrying experimental payloads. Phillips' Program Office was created in January 1994 to provide a standard small platform for testing new technologies without placing larger missions at risk. Phase 1 covers the refurbishment by CTA under a US$2 million contract of an existing Space & Missiles Systems Center bus from an earlier programme. Launch took place from Shuttle STS-87 in July 1998. Spectrum Astro was awarded the contract 22 November 1995 for the five-year Phase 2, for at least three micro-satellites, with options for two more, worth up to US$23.5 million. First launch was planned for March 1998 and then at one-year intervals. Payload includes pulsed plasma thrusters, Kestrel Corp's Fourier transform hyperspectral imager (256–512 channels, 1.7 nanometer resolution, 0.350–1.050 μm), and NRL/Space Instruments' solar output Total and Ultra-Violet Irradiance Radiometer.

The air force used Phase 1 to prove the concept effort, retrofitting the bus with an advanced composite structure and high-efficiency, multijunction solar cells. It will carry an experimental release mechanism using shape memory devices and a reliability experiment space qualifying a new set of advanced electronics and electronic packaging techniques.

MTI (P97-3, Multispectral Thermal Imager)

Background

The Multispectral Thermal Imager is a Department of Energy (Sandia National Laboratory, Los Alamos National Laboratory, and Savannah River Technology Center) and STP sponsored mission. MTI was placed into a 574 × 609 km circular, 97.5° inclination Sun-synchronous orbit on 12 March 2000 by a Taurus launch vehicle from VAFB. The 1,610 kg spacecraft is 3-axis stabilised, provides an average of 540 W of power, and is designed for a 14-month mission life with a three-year goal. The objective of the MTI mission is to demonstrate and evaluate multispectral and thermal imaging technology of nuclear proliferation monitoring and for other military and civilian applications. The MTI sensor will provide multispectral data with high ground resolution and radiometric accuracy. Ball Aerospace is building the MTI spacecraft bus and Hughes Santa Barbara Research Center, Hughes Danbury Optical Systems, TRW, and Sandia National Laboratory are building the sensor.

PICOSat (P97-1) 7/8

Background

Surrey Satellite Technology Ltd (SSTL) is building PICOSat. It will provide a test of an off-the-shelf micro-satellite to determine if cost effective spaceflight of DoD experiments can be achieved in this manner. The four experiments to be flown on the satellite are: the Polymer Battery Experiment, the Ionospheric Occultation Experiment (IOX) to demonstrate the use of occultation of GPS signals to characterise the ionosphere, Coherent Electromagnetic Radio Tomography Ionospheric (CERTO), and Optical Precision Platform Experiment (OPPEX). P13EX is an all plastic battery that has virtually limitless charge/discharge cycles and other space-favourable characteristics. IOX and CERTO are experiments measuring electron content and densities in the ionosphere. OPPEX is an experimental platform that demonstrates active and passive vibration isolation for future precision instrument applications. The carrier satellite MightySat 2 was launched 19 July 2000 to a 547 × 581 km, 97.8°. orbit. PICOSat 7/8 were launched from MightySat 2 on 7 September 2001 into a 511 × 539 km, orbit.

RADCAL (P92-1 Radar Calibration Satellite)

Background

A Scout booster launched the 95 kg RADCAL 25 June 1993 from Vandenberg into 759 × 888 km, 89.5° orbit. It provides radar calibration via two-C band transponders for more than 13 agencies. The Doppler transmitter permits use of ground processed ephemeris (by Naval Surface Warfare Center) as a standard reference for the radars. Two onboard GPS receivers using 4 patch antennas demonstrate a new technique for generating ephemeredes using ground processing. The USAF Phillips Laboratory Small Satellite Power System Regulator (SSPSR) experiment is also testing improved methods of battery charging. The RADCAL experiments are housed on a modified X-Sat bus. Defense Systems Inc (then CTA Space Systems, now Orbital Sciences Corporation) was prime contractor for the 87 kg spacecraft at a total cost of less than US$4 million. Both satellite bus and payload integration were used as a potential template for future missions of this type.

Future Payloads

POGS 2 (S92-1, Polar Orbiting Geomagnetic Survey 2)

The experiment uses a Special Sensor Magnetometer to collect data, which would be used to update the geomagnetic maps of the Earth. POGS 2 was due to be carried onboard defence Meteorological Satellite Program spacecraft S-15, which had a predicted launch need date of 1998. The magnetometer was to be placed on an extendible 5 m boom in order to reduce the effects of electromagnetic noise from the spacecraft, planned to fly on DMSP's S15-S20. First was planned for the second quarter of 1997.

CERTO PLUS (S97-2, Coherent Electromagnetic Radio Tomography/Profiling the Limb with Ultra-violet Sensors)

CERTO PLUS will provide measurements of the integrated electron density of the ionosphere in the satellite orbit plane. It will also provide a database for global models of the ionosphere and will test tomographic algorithms for reconstruction of ionospheric densities and irregularities. CERTO PLUS is scheduled to fly on STRV ID, a spin stabilised, 100 kg class spacecraft to be launched into a geosynchronous transfer orbit by an Ariane 5 launch vehicle.

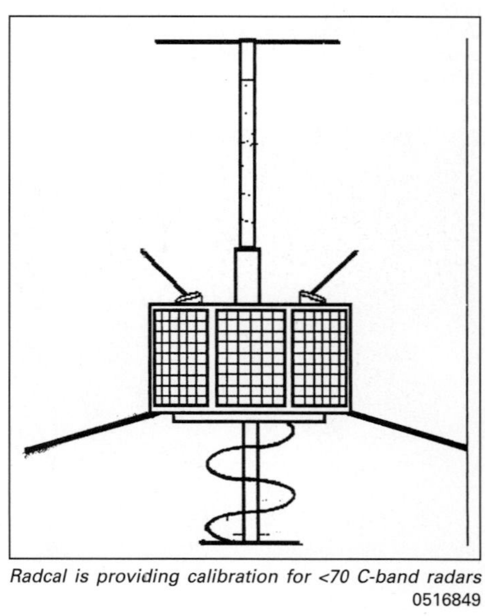

Radcal is providing calibration for <70 C-band radars 0516849

REX II (P94-2) Satellite

Background

STP's follow-on mission to 1991's REX (PS9-1) for research into the physics of electron density irregularities that cause disruptive scintillation effects on radio signals. The payloads include a Rome Laboratory Communications experiment and an OPS Attitude Determination and Control System (ADACS) experiment to perform on-orbit attitude determination and direct feedback to the control system using the Global Positioning System. REX I was launched by Scout G on 29 June 1991 to a 766 × 869 km, 89.6° orbit.

The air force successfully launched the spacecraft by a hybrid Pegasus on 9 March 1996 from Vandenberg into a 799 × 835 km, 90°, orbit. Built by CTA Space Systems, the 113 kg, gravity gradient stabilised with three-axis magnetotorque coils and a momentum wheel for pitch momentum bias, with four deployable solar arrays and gravity gradient boom. A failure occurred in the REX payload in July 1996, which prevents normal operations, but efforts continue to identify and compensate for the cause of the failure. The spacecraft and ADACS payload continue to operate nominally. The air force paid less than US$14 million for REX, including US$9 million for Pegasus.

STEP Series

Current Status

STEP M5 was launched on 7 June 2000, by Pegasus XL to a 405 × 1,689 km, 69°, orbit.

Background

TRW signed a US$5.48 million contract 17 April 1990 for STP's Space Test Experiments Platform (STEP). The STEP designs and flies satellites below 450 kg, with lifetimes of one to three years, which communicate with small ground stations. The bus upon which the air force places its experiments provides up to 300 W power and is 3-axis, or gravity gradient stabilised. The TRW contract allows for options on a further 11 satellites. OSC's Pegasus became the prime launch vehicle after selection in April 1991 for one firm (APEX) and 39 optional launches under the AFSLV Air Force Small Launch Vehicle Program.

STEP MO (P90-5)

STEP MO carried a payload of ten TAOS Technology for Autonomous Operational Survivability experiments, including seven computers. It weighed 503 kg at launch on 13 March 1994 into 539 to 560 km, 105° on Taurus with the classified DARPAsat. The air force declared it fully operational on 31 March 1994. The satellite package cost US$56 million, including US$16 million for launch. It demonstrated Phillips' Lab project for satellite autonomy from ground controllers using a 6-channel GPS receiver, horizon scanner and high speed computer for 400 m and 0.03° accuracy position determination. It had a design life of 18 months and two articulated solar panels, generating 195 W (100 W available to payload).

TAOS is the first STEP mission (TRW) 0516846

STEP's first four missions: M0 TAOS (bottom), M1 (centre), M2 SIDEX (top left) and M3 (top right) (TRW) 0517161

MO suffered a serious blow on 20 July 1994 when the IMU failed and full capability was not restored with new software circumventing the problem until 28 February 1995. Responsibility for operating the MO spacecraft has now been transferred to Air Force Space Command.

STEP M1 (P90-1)

The air force originally planned for a 500 km 90° orbit on a Pegasus-XL to investigate HF radio propagation in the ionosphere below 250 km, atmospheric drag over 195–400 km, and plasma instabilities and electric fields. The satellite weighed 348 kg, had 3-axis control and 105 W of power from deployed solar panels. It carried payloads of Ducted HF Propagation; Satellite Electrostatic Triaxial Accelerometer (SETA); Absolute Density Mass Spectrometer (ADMS); Composition and Density Sensor (CADS); Coordinated Heating and Modification Process in the Ionosphere (CHAMPION), Plasma Environment Analyser (PEA). The satellite was lost in the 27 June 1994 Pegasus launch failure.

STEP M2 (P91-2)

The air force launched the 180 kg SIDEX (Signal Identification Experiment) on 19 May 1994 for the USAF Rome Laboratory Air Development Center into 603–821 km, 81.96° orbit (830 km circular planned) aboard Pegasus. The low orbit degraded data collection. The STEP M2 spacecraft failed December 1995 and investigators never determined the cause. Full operations of the spacecraft did not begin until October 1994 because of several anomalies, and then suffered power problems because of overheating. The satellite payload consisted of a narrow beam parabolic gimbaled antenna; wide beam fixed helix antenna, and a GPS antenna. The experiments were designed to evaluate communications in dense signal environments. Controllers flipped the M2 180° in yaw 22 December 1994 to position the battery on the anti-Sun side to resume normal operations after the over-heating.

STEP M3 (P92-2)

The fourth STEP was lost 22 June 1995 due to failure of the Pegasus-XL launch vehicle. It carried five experiments: ACTEX advanced controls technology experiment (BMDO), EDMM erasable disk mass memory, SAMMES space active modular materials experiment (BMDO), SAWAFE strategic attack warning and assessment flight experiment ('smart skin', BMDO), SQUOD space qualified optical disk. The mission cost the air force about US$40 million. M3 was completed in August 1994 but then had to be stored because of Pegasus-XL's 1994 failure.

STEP M4 (P95-1)

The fifth and final STEP Mission is a 402 kg spacecraft launched on 22 October 1997 aboard a Pegasus Xl, into a 430 × 511 km parking orbit at 45°. Ground radar confirms that the solar arrays are not deployed, leaving the satellite in a deep under-voltage condition. STEP M4 hosts three experiments. They are an Orbiting Ozone and Aerosol Measurement (OOAM) experiment, which measures global ozone depletion and water vapour in the middle atmosphere to help gauge infra-red sensor performance, the Electromagnetic Propagation Experiment (EMPE), which characterises ionospheric effects and the Digital Ion Driftmeter (DIDM), which measures the flow of ionospheric ions. Mission requirements demand that the attitude control system operate in both spin and 3-axis modes. In July 1996, CTA Space Systems delivered the spacecraft to TRW S&EG in Chantilly, Virginia for final integration and test. The mission cost the air force an estimated US$80 million.

STP (Space Test Programme) Payloads

Current Status

Since inception in 1966, STP has flown more than 420 experiments on over 130 missions supporting scientific research and operational satellite programs.

Background

The Air Force Space and Missile Systems Center, created in May 1965, runs a programme of science and engineering test payloads known as the Space Test Program. The Department of Defense relies on the STP (formerly called STEP) programme to test new concepts and gather information about the space environment. STP put its first payload in space in 1967 and since that time had either launched or participated in the launches of 415 different experiments. DoD provided US$38.3 million for STP operations and experiments this year. In past years the budget has been FY97 US$43.1 million, FY96 US$43.8 million, FY95 US$53.l million, FY94 US$73.1 million, FY93 US$77.9 million, FY92 US$69 million, FY91 US$62.4 million. STP participates in space experiments in three ways: free-flyers, Shuttle pallets and piggyback flights on other experimental packages. For experiments with unique orbital requirements that can best be met by free-flying spacecraft, STP contracts for spacecraft development, experiment integration, and launch service. A recent example is TRW's Space Test Experiments Platform. These experiments usually have a 'P' designation and carry a numerical identifier closely related to the fiscal year of their completion. STP also flies experiments as piggyback payloads on spacecraft of various agencies of the US (NASA and DoD) and other countries (CNES-France, DRA-UK, Russia), usually carrying an 'S' designation. STP also makes use of the Space Shuttle mid-deck lockers, the Hitchhiker, the Getaway Specials, Spartan and other low cost available space with the NASA programme.

Current and recent payloads

STS-83, Columbia (OV-102)

STS-83 launched on 4 April 1997 carried the Cryogenic Flexible Diode (CRYOFD), a DoD Cargo Bay payload. CRYOFD is designed to determine the behaviour of Cryogenic 2-phase thermal control components in microgravity, demonstrate oxygen and methane heat pipe startups from a super-critical condition, demonstrate operations, verify analytical performance models, and establish the correlation between one *g* and microgravity thermal performance. A secondary objective is to validate the performance of an American Loop Heat Pipe with Ammonia (ALPHA).

STS-84, Atlantis (OV-104)

STS-84 launched on 15 May 1997 as the sixth Shuttle/Mir docking mission. Atlantis carried two DoD payloads, which were the first DoD sponsored experiments to operate inside the Mir Space Station. The Radiation Monitoring Equipment – RME III – attempted the first ever direct measurement of the East-West effect, a theory that predicts higher radiation levels (by a factor of 2 to 4) on the West facing side of the Mir. Data was collected throughout the docked phase of the mission and also from inside the Spacehab on the way back to Earth. The Cosmic Radiation Effects and Activation Monitor is composed of passive and active radiation detectors that work together to better understand the effects of cosmic radiation on both the crew and electronics. CREAM recorded over 151 hours of data.

STS-94, Columbia (OV-102)

STS-94, launched on 1 July 1997, again carried a Cryogenic Flexible Diode (CRYOFD). It performed successfully.

Argos was STP's next major satellite, planned for 1997
(Naval Research Laboratory) 0517160

STS-85, Discovery (OV-103)
STS-85 was launched from KSC on 7 August 1997 carrying the COOLLAR Flight Experiment activated nine hours into the mission aboard the TAS-01 bridge. CFE completed all of its planned activities and was able to run for an additional 18 hours to gather secondary data.

STS-86, Atlantis (OV-104)
STS-86 launched on 25 September 1997 carried three DoD STP payloads, the Cell Culture Module (CCM), Cosmic Radiation Effects and Activation Monitor, and Shuttle Ionospheric Modification with Pulsed Localised Exhaust (SIMPLEX). CCM performed very well throughout the mission. CREAM was deployed and activated inside the Mir Space Station on 28 September. The experiment ran flawlessly until its deactivation and partial stowage in the Shuttle on 2 October. The remainder of CREAM's components (all passive) remained deployed in the Mir module, Kvant-II module, and Kristall module until STS-89 (January 98). NASA supported SIMPLEX with a pass over the Jicamarca, Peru ground station at approximately 15:30 CDT on 4 October 1997. The pass consisted of a 10-second OMS burn producing an exhaust plume northward. The orbiter aft cameras focused on the starboard OMS engine throughout the burn which produced excellent video.

STS-87, Columbia (OV-102)
STS-87 lifted off from KSC on 19 November 1997 with the Midcourse Space Experiment (MSX) and Shuttle Ionospheric Modification with Pulsed Localised Exhaust (SIMPLEX). A primary objective of this mission was the deployment and retrieval of a Spartan free-flyer. Due to problems with the Spartan deploy and retrieve operations, propellant was at a premium for this mission causing SIMPLEX to 'stand down' halfway through the mission. MSX was able to observe a 10-second, 2-engine, RAM burn (from the Orbiter's rear engines). The ram burn was visible using the LTV narrowband sensor, and the spectrographic imagers. The result was excellent data, making the experiment a success.

STS-89, Endeavour (OV-105)
STS-89 launched on 22 January 98 with the Cosmic Radiation Effects and Activation Monitor (CREAM) active monitors on board the Spacelab Payload Processing Facility (SPPF). Some modifications took place in order to extend the data collection capacity during the Mir phase of the flight. CREAM was successfully deployed and activated inside the Mir Space Station on 25 January. CREAM's active and passive components were deployed in the Mir, Kvant-II module, and Kristall module. All of the CREAM components from STS-86 aboard the Mir were returned to Endeavour at the same time that the STS-89 components were deployed.

CHARGECON-GEO (SP90-3)
The Satellite Charge Control Experiment at GEO flies on the DSCS 3B B-7 communications satellite. The satellite tests the build-up of an electrostatic charge on the spacecraft and methods to then dissipate the charge by releasing an ionised xenon plasma. The experiment started gathering data in August 1995 and through July 1996 the experiment had completed 20 automatic charge dissipation operations. The experiment continues to run.

SWIM (S91-4)
The Solar Wind Interplanetary Measurements instrument on NASA's Wind satellite, launched 1 November 1994.

POAM 2 (S88-1)
NRL's 25 kg POAM Polar Ozone and Aerosol Measurement package orbits attached to France's SPOT 3 launched 26 September 1993. Thermo Trex of La Jolla, California built the instrumentation. Nine UV-IR channels observe sunsets/rises over the poles to provide vertical profiles of polar ozone, aerosols, water vapour and nitrogen tetroxide, and stratosphere and upper troposphere density and temperature. POAM 2 characterises and clarifies the intense turbulence in the middle atmosphere that adaptive optics in systems such as laser communications would have to accommodate. Two of POAM's channels are at important laser wavelengths: 0.353 and 1.059 µm. The other frequencies are 447.4, 452.5, 550.0, 600.0, 763.4, 780.0 and 945.0 µm. When NASA's TOMS instrument aboard Russia's Meteor-3 failed in December 1994, POAM became the only operating US ozone sensor, returning data in unprecedented detail on the ozone hole. The French announced the end of the SPOT 3 mission in December 1996 after an unexplained failure rendered the spacecraft inoperative.

POAM 3 (S95-A)
The Naval Research Laboratory Polar Ozone and Aerosol Measurement 3 is a congressionally mandated follow-on to the POAM 2 project in support of an integrated NRL effort to globally monitor environmental parameters. NRL contracted with the Centre National d'Etudes Spatial (CNES) to integrate, test and launch POAM 3 on the SPOT 4 spacecraft. An Ariane 5 launch vehicle boosted POAM 3/SPOT 4 to an 833 km Sun-synchronous orbit from Kourou, French Guiana.

BINRAD (S93-1, Beryllium-7 Induced Radiation Experiment)
STP conducted a third reflight of the experiment from 14 March 1996 to 15 June 1996. The experiment first flew with Kosmos 2331, launched on a Soyuz-U launch vehicle. The analysis of the foil showed a Be-7 level only 1 per cent of the expected level and analysts are still struggling to explain these results. The experiment followed-up on NASA's LDEF to help explain how LDEF accumulated Be-7 at a rate several thousand times greater than modelled for that altitude. Some researchers have theorised a heretofore unknown transport mechanism from a lower altitude exists. Two previous reflight attempts in 1992 and 1993 on Resurs-F spacecraft failed when the aluminium foil remained unexposed because of improper release from the experiment containment system.

MPTB (S96-1)
The Naval Research Laboratory Microelectronics & Photonics Test Bed is a piggyback experiment that was launched into a high radiation orbit in Autumn 1997. MPTB is presently returning data vital to predicting the performance of advanced microelectronic and photonic devices. Once proven, these off-the-shelf devices will be used in military SPO procurements. The experiment contains 24 individual experiment boards and a charged particle telescope. The decline of the Cold War means that a full complement of radiation hardened parts is no longer available for space systems, leaving commercial unhardened devices to be employed.

TSX-5 (P95-2) Tri-Service Experiments Mission-Mission 5

Background
The air force launched TSX-5 on a Pegasus XL launch vehicle on 7 June 2000. It was placed in a 405 × 1,689 km, 106.1 min, orbit at a 69° inclination controlled by the Research and Development, Test and Evaluation Support Complex (RSC) at Kirkland AFB NM. TSX-5 carries the Space Technology Research Vehicle II (SRRV 2) experiment package and the USAF Research Laboratory's Compact Environment Anomaly Sensor (CEASE).

TSX-6 (P97-x) Tri-Service Experiments Mission-Mission 6

Background
Tri-Service Experiment Mission 6 is expected to carry the Optical Reflection Experiment Spacecraft Evaluation Model (OSEM) and other, yet to be determined, experiments into an 835 km orbit at an inclination of 82–98°. The launch date has yet to be determined.

UNKNOWN

Russian Federation

Kosmos Series

Background

More than 90 small Kosmos launches have taken place since 1964 whose purpose has not been identified. They share the common feature of Kosmos 3 launch vehicle based on the R-12 designed by Yangel in the 1950s. The last in the series seems to be Kosmos 919 in 1977. Western observers have classed them as military because they emit no characteristic communications signals and no scientific results have been published at international scientific conferences that reference these launches.

One series, with 62 total launches, used Kapustin Yar as a spaceport and had an orbital parameter of 280 × 500 km, 49° inclination. These satellites lasted until 1967–68. When Plesetsk opened, the satellites came from that spaceport and flew into 71° inclination orbits. A second series, adopting an orbit of 200 × 1,100–1,500 km and 49° inclination, also flew from Kapustin Yar beginning in 1967, and then switched to Plesetsk in 1968. A third series also left for orbit from Kapustin Yar into 260 × 2,100 km at 49° inclinations. The fourth series used a 280 × 840 km, 71° inclination orbit from Plesetsk. The constant replacement of one satellite by another suggested that the Soviets were making continuous observations. Another 40 Kosmos satellites used similar orbits, but these could be identified as scientific payloads from publication of their findings.

The Soviets conveniently announced launches under the Kosmos title, accompanied by serial numbers, orbital parameters and a vague description of purpose from the time the series began on 16 March 1962. Kosmos designations have also been used to cloak inevitable failures during the development of a new series. Interplanetary probes, to cite one example, received a name in the Venera, Mars or other series only after successful injection. Otherwise they were allocated the next Kosmos number and a routine announcement released. A typical launch failure disguised by the Kosmos label was Kosmos 1612 in November 1984. Orbital analysis suggests that this satellite, stranded in an unusable transfer orbit when its upper stage failed to re-ignite, was the first of the new-generation Meteor 3 weather satellites. Similarly, a geosynchronous launch failure in January 1987 received the designation of Kosmos 1817 before it re-entered the following day.

CONTRACTORS

CONTRACTORS

Argentina

Government – Ministry/Agency

Instituto de Astronomia y Fisica del Espacio (IAFE)

Ciudad Universitaria, 1428 Buenos Aires, Argentina
Tel: (+54 1) 783 26 42
(+54 1) 781 67 55
(+54 11) 47 83 26 42
Fax: (+54 1) 786 81 14
e-mail: difusion@iafe.uba.ar
Web: http://www.iafe.uba
Director: Dr Marta Rovira
Contact: Ms Gloria Nazar

UPDATED

Industry – Manufacturing

Nahuelsat SA

Lavelle 472, No. 1 Piso, 1047 Vapital Federal, C1047AAJ Buenos Aires, Argentina
Tel: (+54 11) 58 11 26 00
Fax: (+54 11) 58 11 26 88
e-mail: info@nahuelsat.com.ar
Web: http://www.nahuelsat.com.ar
General Manager: Daniel Salzer

UPDATED

Australia

Government – Ministry/Agency

Department of Industry, Tourism and Resources

GPO Box 9839, Canberra, Australian Capital Territory, 2601, Australia
Tel: (+61 2) 62 13 60 00
Fax: (+61 2) 62 13 70 00
Web: http://www.isr.gov.au
http://www.minister.industry.gov.au
http://www.ausindustry.gov.au
Minister for Industry, Tourism and Resources: Ian E McFarlane
Minister for Sport and Tourism: Ms Jackie Kelly

UPDATED

Centres of Learning & Research

Institute for Telecommunications Research (ITR)

University of South Australia, The Levels, South Australia, 5095, Australia
Tel: (+61 8) 83 02 33 36
Fax: (+61 8) 83 02 38 73
Web: http://http://www.itr.unisa.edu.au
Director: Andrew Parfitt
Tel: (+61 8) 83 02 33 10
Business Manager: J Kasparian
Tel: (+61 8) 83 02 38 91
Operations Manager: Richard Akroyd
Mobile: (+61 0407) 97 51 99
Director: Prof William Cowley

- Satellite communications systems and telecommunication services.

UPDATED

Association

Astronautical Society of Western Australia Inc.

PO Box 278, Perth, South, Western Australia, 6151, Australia
Tel: (+61 8) 94 57 36 20
(+61 8) 94 57 30 16
e-mail: astrosocwa@hotmail.com
Editor and Public Relations: Jos Heyman
Publications: News Bulletin

UPDATED

Australian Space Industry Chamber of Commerce (ASICC)

GPO Box 7048, Sydney, New South Wales, 2001, Australia
Tel: (+61 2) 99 88 02 52
Fax: (+61 2) 99 88 02 62
Chairman: Kirby Ikin
Treasurer: Tim McEgan
Tel: (+61 413) 54 71 50

- Industry association promoting growth of Australian space industry.

UPDATED

National Space Society of Australia (NSSA)

GPO Box 7048, Sydney, New South Wales, 2001, Australia
Tel: (+61 2) 99 88 02 52
Fax: (+61 2) 99 88 02 62
e-mail: nssa@nssa.com.au
Web: http://www.nssa.com.au
Chairman: Kirby Ikin
President: Philip Young
Tel: (+61 412) 01 84 10
Director/Financial Controller: Tim McEgan
Tel: (+61 2) 95 67 14 61
Public Relations Director: Tony James
Chief Operating Officer: Jeffery Candiloro

- Promotion of space development, education and industry liaison.

UPDATED

Industry – Manufacturing

Codan Limited

(Head Office)
PO Box 96, Campbelltown, South Australia, 5074, Australia
Tel: (+61 8) 83 05 03 11
Fax: (+61 8) 83 05 04 11
e-mail: info@codan.com.au
Web: http://www.codan.com.au
Managing Director: Michael K Heard
Marketing Manager: Kent Parkin
Marketing Publisher: Adam J Diggens
Tel: (+61 8) 83 05 05 06

UPDATED

Electro Optic Systems Pty Ltd (EOS)

Locked Bag 2, Post Office, Queanbeyan, New South Wales, 2620, Australia
Tel: (+61 2) 62 98 80 00
Fax: (+61 2) 62 99 76 87
e-mail: eos@eos-aus.com
Web: http://www.eos-aus.com
Managing Director: Ron Thompson
Chairman and Executive Director: B Greene
Business Development Manager: Bob Quodling
Tel: (+61 2) 61 61 21 57
Mobile: (+61) 24 14 36 56 57
Fax: (+61 2) 61 61 21 57

- Aerospace lasers and electronics. Turnkey systems or sub-systems in multi-disciplinary project areas. Spaceborne or airborne systems. Military rangefinders.

UPDATED

Industry – Service

Asia Pacific Aerospace Consultants Pty Ltd (APAC)

PO Box 98, Hawker, Australian Capital Territory, 2614, Australia
Tel: (+61 2) 99 88 02 52
Fax: (+61 2) 99 88 02 62
e-mail: info@apac.com.au
Web: http://www.apac.com.au
Managing Director: Kirby Ikin
Managing Director: Dr Bruce S Middleton

- APAC is a commercial space consultancy specialising in the Asia Pacific region. Its expertise covers satellite-based communications, launch services, remote sensing and related areas. It provides support for commercial space companies interested in growing their business in the Asia Pacific, and to companies located in that region that are interested in growing global space business. APAC generally works at the strategic business planning level. Its services include identifying market trends, undertaking market research, locating and researching prime marketing targets, identifying critical success factors, and advising on objectives and strategies. Assessing the risks of space-related investments, contributing to feasibility studies, advising on acquisitions, and identifying and making contact with potential investors and customers are all services APAC offers.

UPDATED

Auspace Ltd

50 Hoskins Street, PO Box 17, Mitchell, Australian Capital Territory, 2911, Australia
Tel: (+61 2) 62 42 26 11
Fax: (+61 2) 62 41 66 64
e-mail: admin@auspace.com.au
Web: http://www.auspace.com.au
Managing Director: Roger Franzen
Tel: (+61 2) 62 42 26 14
Marketing Director: Ms Nicola Porecki
Finance and Commercial Manager: Mark Childs
Public Relations Officer: Ms Sue Callow

- Systems studies and project management for satellite communications, remote sensing and science applications in the following areas: thermal analysis and design; mechanical/structural design; electro-optics (Vis, IR, UV); assembly, integration and test in Class 100 clean conditions; instrumentation design and development; electronic, synthetic aperture radar; advanced ground support equipment; satellite earth station design and integration; real-time software and light satellites.

UPDATED

Co-operative Research Centre for Satellite Systems (CRCSS)

GPO Box 1483, Canberra, Australian Capital Territory, 2601, Australia
Tel: (+61 2) 62 16 72 70
Fax: (+61 2) 62 16 72 72
e-mail: satsys@crcss.csiro.au
Web: http://www.crcss.csiro.au
Executive Director: Dr Brian Embleton
Tel: (+61 2) 62 16 72 80
Deputy Executive Director: Prof. Michael Miller
Tel: (+61 8) 83 02 33 10
Centre Manager: Jeff Kingwell
Tel: (+61 2) 62 16 72 78
Office Manager: Ms Karen Doull
Tel: (+61 2) 62 16 72 71

- Development and launch of a small scientific satellite FedSat. Research and development in areas of space science, communications, remote sensing and satellite systems.

Core partners
Auspace Limited, CSIRO, Queensland University of Technology, University of Newcastle, University of South Australia, University of Technology, Sydney, VIPAC Engineers and Scientists Ltd.

UPDATED

Geoimage Pty Ltd

(Head Office)
PO Box 789, Indooroopilly, Queensland, 4068, Australia

Tel: (+61 7) 38 71 00 88
Fax: (+61 7) 38 71 00 42
e-mail: geoimage@geoimage.com.au
Web: http://www.geoimage.com.au
Managing Director: Bob N Walker
Director: Ms Sylvia Michael
Satellite Image Processing Manager: Michael Peters
Marketing Manager: Bernie Fitzpatrick
Systems Manager: Don Lindsay
Technical Officer: Mark Winterbotham
Administration Officer: Ms Lee Anderson
■ Advice, purchase and supply of satellite image data, image processing and remote sensing, production of hard-copy imagery, geophysical data processing services, sales of airborne geophysical data, production and sale of multiclient satellite image products, multiclient geophysical image products, sale, support and training of ER Mapper, geographic information systems services and data archiving services.

UPDATED

Geoimage Pty Ltd
(Darwin Office)
PO Box 36289, Winnellie, Northern Territory, 0820, Australia
Tel: (+61 8) 89 47 17 55
Fax: (+61 8) 89 47 17 88
e-mail: darwin@geoimage.com.au
Web: http://www.geoimage.com.au
Office Manager: Ms Hazel Morgan
General Manager: Paul Ryan
Administration Officer: Ms Michelle Mcdonnell

UPDATED

Helitech Pty Ltd
(ex-BH Pacific Pty Ltd)
(ex-Helitech Industries Pty Ltd)
PO Box 1374, Eagle Farm, Queensland, 4009, Australia
Tel: (+61 7) 36 32 70 00
Fax: (+61 7) 32 60 16 16
e-mail: helibne@helitech.com.au
Web: http://www.helitech.com.au
Group Managing Director: Joseph V Moharich
Chief Operating Officer: Les Healey
Financial Director: Glen Battershill
Sales and Marketing Director: Peter Crook
Service Director: Alan Crawford
Purchasing Manager: Pam Harris
Customer Support Representative: N Hetherington
■ Complete range of Bell Helicopters and spare parts. Helicopter overhaul and maintenance services.
Representing: Bell Helicopter Textron Inc., US

UPDATED

Jane's Information Group (JIG)
(Australia)
PO Box 3502, Rozelle, Delivery Centre, New South Wales, 2039, Australia
Tel: (+61 2) 85 87 79 00
Fax: (+61 2) 85 87 79 01
e-mail: info@janes.thomson.com.au
oceania@janes.com
Web: http://www.janes.com
Business Manager: Dr Pauline J Roberts
Representing: Jane's Information Group, UK

UPDATED

KEL Aerospace Pty Ltd
231 High Street, Ashburton, Victoria, 3147, Australia
Tel: (+61 3) 98 89 00 22
Fax: (+61 3) 98 89 00 06
e-mail: kelaero@kel.com.au
Web: http://www.kel.com.au
Managing Director: S Gruca
Export Manager: Ms Sue Smith
■ Supply of HF radar receivers. HF amplifiers and mixers, over-the-horizon radar receivers, ionospheric monitoring systems, wideband COMINT receivers, high voltage electrical transient monitoring systems, digital ionosondes, automatic ionogram interpretation and scaling.

UPDATED

Industry – Distributor

Spot Imaging Services Pty Ltd
(ex-SPOT Imaging Systems)
Deakin Chambers, PO Box 9141, Epping, New South Wales, 2600, Australia
Tel: (+61 2) 62 32 51 71
Fax: (+61 2) 62 32 59 92
Web: http://www.spotimage.com.au
Managing Director: Rob Lees
■ Distribution of RS data and products from Spot satellites.

UPDATED

Industry – Sales Agent/Office

J C Aviation International Pty Ltd (JCAI)
5b Jubilee Avenue, Warriewood, New South Wales, 2102, Australia
Tel: (+61 2) 99 79 17 77
Fax: (+61 2) 99 79 17 88
e-mail: jc@jc-aviation.com
Web: http://www.jc-aviation.com
Managing Director/Aviation Director: Jack Cairns
■ Supplier of Ni-Cad batteries, borescopes, diesel and electric GPU's, frequency changers, rechargeable hi-intensity flashlights, ground support equipment, aircraft spares and tooling, air-conditioning units and aircraft tugs.

UPDATED

Austria

Government – Ministry/Agency

Austrian Space Agency (ASA)
(Österreichische Gesellschaft für Weltraumfragen Ges.mbH)
Canovagasse 7, A-1010 Wien, Austria
Tel: (+43 1) 40 38 17 70
Fax: (+43 1) 405 82 28
Web: http://www.asaspace.at
Managing Director: Dr Klaus Pseiner
Tel: (+43 1) 403 81 77 12
Telecommunication and Industrial Policy: Harald Posch
Tel: (+43 1) 403 81 77 14
Manned Spaceflight and Migravity, Technology Transfer: Dr Werner Balogh
Tel: (+43 1) 403 81 77 20
Galileo Contact Point Austria, Satellite Navigation: Dr Stephan Mayer
Earth Observations, Space Transportation: Dr Erwin Mondre
Tel: (+43 1) 403 81 77 16
Treasury and Accounting: Ms Ernestine Stöger
Tel: (+43 1) 403 81 77 13
Aeronautics Programme: Mag Andreas Geisler
Tel: (+43 1) 403 81 77 31
Nanotechnology Unit: Dr Emmanuel Glenck
Tel: (+43 1) 403 81 77 27
European Space Policy Institute Point of Contact: Michel Jakob
Tel: (+43 1) 403 81 77 26
Austrian Space Applications Programme: Wolfgang Würz
Tel: (+43 1) 403 81 77 25
Administration, Public Relations and Education: Ms Michaela Gitsch
■ Information and promotion of contracts to Austrian research institutions and industry. Advises the Federal Government in matters concerning space research and technology. Contracts with foreign institutions. Promotion of training of experts. Co-ordination of space activities.

UPDATED

Transport Administration – Space

Office for Outer Space Affairs (UNO)
Vienna International Centre – Room EO954, PO Box 500, A-1400 Wien, Austria
Tel: (+43 1) 260 60 49 50
Fax: (+43 1) 260 60 58 30
Contact: Ms Natercia Rodrigues
Publications: Annual Reports, Intergovernmental & Expert Group Meetings

UPDATED

Industry – Manufacturing

Andritz AG
18 Stattegger Strasse, A-8045 Graz, Austria
Tel: (+43 316) 690 20
Fax: (+43 316) 690 24 15
e-mail: welcome@andritz.com
Web: http://www.andritz.com
Head of Corporate Communications and Investor Relations: Michael Buchbauer
Tel: (+43 310) 69 02 29 79
■ Components for Ariane-5 including booster attachments.

UPDATED

Austrian Aerospace GmbH
(ex-Schrack Aerospace/ORS)
Stachegasse 16, A-1120 Wien, Austria
Tel: (+43 1) 801 99 30 00
Fax: (+43 1) 801 99 69 50
Web: http://www.space.at
Assistant, Marketing and Sales: Ms Gabriele Riedl
■ Electrical and mechanical equipment and thermal hardware for communications, earth observation, navigation and scientific satellite missions, electronic satellite check out equipment. Thermal protection, multi-layer insulation, cryogenic insulation.

UPDATED

Böhler Schmiedetechnik GmbH & Co. KG
Mariazeller Strasse 25, PO Box 96, A-8605 Kapfenberg, Austria
Tel: (+43 3862) 20 61 70
Fax: (+43 3862) 20 75 70
e-mail: office@bstg.buag.co.at
Web: http://www.bohler-forging.com
http://www.bohler-schmiedetechnik.com
Chairman: Dr Knut Consemüller
Managing Director: Ing Heinz-Fritz Romen-Kierner
Managing Director: Dipl Ing Peter Bojdunyk
■ Precision forged and machined parts for engines, compressors and aerospace applications.

UPDATED

Fischer Advanced Composite Components GesmbH (FACC)
Fischerstrasse 9, PO Box 192, A-4910 Ried, Austria
Tel: (+43 7752) 61 60
Fax: (+43 7752) 61 63 50
e-mail: info@facc.co.at
■ Composite components. Aircraft interiors, furnishings, aircraft access doors, fairings, structural aircraft assemblies, galley equipment, landing gear doors.

UPDATED

Plansee AG
(Plansee Aktiengesellschaft)
Postfach 74, A-6600 Reutte, Austria
Tel: (+43 5672) 60 00
Fax: (+43 5672) 60 05 00
Web: http://www.plansee.com
Marketing Manager: Alfred Troy
Communications Manager: Ms Virginie Amanieu
Marketing: Dr A Eichwalder
Tel: (+43 5672) 600 24 30
■ Manufacture of powder metallurgical products and components including refractory metal components for launchers and satellite engines; thermal protection systems for reuseable space transport systems; heavy metal materials for aerojet balance weights and radiation protection shieldings; hard metal cutting tools for the machining of structural components. Industry development and engineering services.
Accreditations: ISO 9001, BSI, GAZ

UPDATED

Azerbaijan

Transport Administration – Space

Azerbaijan National Aerospace Agency (ANASA)
159 Azadlyq Prospect, 1106 Baku, Azerbaijan
Tel: (+994 12) 62 93 87
Fax: (+994 12) 62 17 38
Director General: Aflatun Hasanov

■ State policy and co-ordination of national aerospace and space programmes focusing on two areas: aerospace environmental monitoring; thematic mapping, developing GIS, and space applications. Remote sensing of the earth's surface using airborne and satellite systems

Managing the following:
Ecology Institute
Experimental Industry Work
Institute of Aerospace Informatics
Institute for Space Research of Natural Resources
Special Design Unit

UPDATED

Belarus

Industry – Manufacturing

Peleng JSC
23 Makayonok Street, 220023 Minsk, Belarus
Tel: (+375 17) 263 82 04
Fax: (+375 17) 263 65 42
e-mail: market@belomo.minsk.by
■ Space photographic systems, photogrammetric instruments and techniques, stabilised night vision devices and binoculars; anti-tank weapon systems.

UPDATED

Belgium

Government – Ministry/Agency

Royal Meteorological Institute of Belgium (RMIB)
3 Avenue Circulaire B, B-1180 Bruxelles, Belgium
Tel: (+32 2) 373 06 01
(+32 2) 373 05 20
Fax: (+32 2) 374 67 88
(+32 2) 373 05 28
e-mail: uidoma.be
rmiinfo@oma.be
Web: http://www.meteo.be/irm-kmi
Head of Section (Remote Sensing from Space): Dr Alexandre Joukoff
Scientist: Dr Steven Dewitte
Tel: (+32 237) 306 24
Information Officer: Luc Debontridder
Tel: (+32 237) 305 21
■ Research and development study services.

UPDATED

Government–Intergovernmental

European Space Agency (ESA)
(Brussels Office)
Avenue de Cortenbergh 52, 4th Floor, B-1000 Bruxelles, Belgium
Tel: (+32 2) 743 30 70
Fax: (+32 2) 743 30 71
Head of Office: Michel Praet

UPDATED

Western European Union (WEU)
(Union de l'Europe Occidentale)
15 rue de l'Association, B-1000 Bruxelles, Belgium
Tel: (+32 2) 500 44 12
(+32 2) 500 44 15
Fax: (+32 2) 500 44 70
e-mail: ueo.secretariatgeneral@skynet.be
Web: http://www.weu.int
Secretary General: Dr Javier Solana Madariaga
Head of Secretariat: A Jacomet
Acting Head of Council Secretariat: Thomas Evenson
■ European intergovernmental defence and security organisation created by the modified Brussels Treaty in 1954. Member states: Belgium, France, Germany, Greece, Italy, Luxembourg, Netherlands, Portugal, Spain and the United Kingdom. Associate members (also members of NATO): Czech Republic, Hungary, Iceland, Norway, Poland, Turkey. Observers (also members of the EU): Austria, Denmark, Finland, Ireland, Sweden. Associate partners (all are signatories of agreement with EU): Bulgaria, Estonia, Latvia, Lithuania, Romania, Slovakia, Slovenia.

UPDATED

Centres of Learning & Research

Centre Commun de Recherche des Communautes Européennes (JRC)
(Directorate For Science Strategy)
SDME 10/53, B-1049 Bruxelles, Belgium
Tel: (+32 2) 295 80 43
Fax: (+32 2) 295 01 46
Telex: 21877 comeu b
Web: http://www.jrc.cec.eu.int
Director General: B McSweeney
Head of Co-operation Strategy and Technology Transfer: Robin Miège
■ Joint research centre acting as scientific and technical body for the CEE in the fields of remote sensing applications, environment, safety technology, prospective technological studies, advanced materials, reference materials and measurements, transuranic elements, systems engineering and informatics, health and consumer protection.

UPDATED

Laboratoire d'Hyperfréquences
Université Catholique de Louvain, 3 Place du Levant, B-1348 Louvain la Neuve, Belgium
Tel: (+32 10) 47 80 95
Fax: (+32 10) 47 87 05
Telex: 59315 telhy
e-mail: secretariat@emic.ucl.ac.be
Web: http://www.emic.ucl.ac.be
Deputy Director: J P Raskin
■ Development of hardware and software products for industrial partners in the field of satellite communications.

UPDATED

Association

Belgian Inter-Trade Association for Space Activities (BELGOSPACE)
80 Building A. Reyers, B-1030 Bruxelles, Belgium
Tel: (+32 2) 706 81 48
Fax: (+32 2) 706 79 52
e-mail: belgospace@agoria.be
Web: http://www.agoria.be/belgospace
Chairman: P Verhaert
Managing Director: J C Lacroix
■ Association of Belgian companies working in space technology.

Association Members
Alcatel Bell Space, Alcatel-ETCA, Cegelec, E.N.E, Gillam Fei, Newtec Cy, Nexan HARNESSES, SABCA, SONACA, Space Applications Services, Spacebel, Techspace Aero, Verhaert Design & Development

UPDATED

Industry – Manufacturing

Advanced Products NV (APNV)
Industrieterrein Krekelenberg, Rupelweg 9, B-2850 Boom, Belgium
Tel: (+32 3) 880 81 50
Fax: (+32 3) 888 48 62
e-mail: sales@advpro.be
Web: http://www.advpro.com
Sales Director: P Amos
Technical Director: E Vanswijgenhoven
Administration and Finance Manager: P Van Riet
■ Manufacture of seals for extreme environmental conditions. Markets: aircraft and aerospace, automotive, chemical process and equipment manufacturing industries.

UPDATED

Britte SA
27 rue de Cheratte, B-4683 Vivegnis, Belgium
Tel: (+32 4) 256 90 69
Fax: (+32 4) 264 08 63
e-mail: info@britte.be
Web: http://www.britte.be
Managing Director: Vincent Pissart
Tel: (+32 4) 256 90 65
Financial Director: Mrs Josiane Willems
Commercial Director: Jean-Marie Agnessen
Tel: (+32 4) 256 90 72
Plant Director: Michel Pondant
Purchasing Manager: Ms Ghislaine Andrien
Tel: (+32 4) 256 90 92
■ Machining of high precision mechanical parts for defence, space and aerospace applications.

UPDATED

German Aerospace Center
(Brussels Office)
(Deutsche Forschungsanstalt für Luft- und Raumfahrt)
Rue du Trône 98, B-1050 Bruxelles, Belgium
Tel: (+32 2) 500 08 41
(+32 2) 500 57 81
Fax: (+32 2) 500 08 40
(+32 2) 500 57 83
e-mail: internet@dlv.de
Web: http://www.dlv.de
Contact: Dr Uwe Möller

UPDATED

Nexans Harnesses NV
(ex-Alcatel Fabrisys NV)
Heideveld 1, B-1654 Huizingen, Belgium
Tel: (+32 2) 363 17 11
Fax: (+32 2) 363 17 87
e-mail: mkt.sales@nexans.com
Web: http://www.nexans.com
Financial Director: Vincent Caenepeel
Sales and Marketing Manager: Eric Leurquin
Purchasing Manager: Francoic Didden
Development: Eric Cochez
■ Wiring harnesses, electrical system and transport vehicles (armoured and non-armoured).

UPDATED

OIP Sensor Systems
(ex-Delft Sensor Systems)
(ex-OIP Optronic Instruments & Products)
Westerring 21, B-9700 Oudenaarde, Belgium
Tel: (+32 55) 33 38 11
Fax: (+32 55) 33 38 02
e-mail: sales@oip.be
Web: http://www.oip.be
Managing Director: Freddy Versluys
Finance and Administration Director: M Allegaert
Research and Development: W Camphyn
Purchasing Manager: K Segers
Quality Assurance Manager: G Brandt
Operations Manager: Robert Van Acker
Manager Space Division: L De Vos
■ Portable display systems, thermal observation equipment (handheld and vehicle mounted), night vision systems, fire control systems for

armoured vehicles, space projects, advanced and holographic optics, customised research and development projects, equipment qualification.

UPDATED

Pedeo Techniek NV
Martijn van Torhoutstraat 15, B-9700 Oudenaarde, Belgium
Tel: (+32 55) 31 35 61
Fax: (+32 55) 31 26 85
Web: http://www.pedeotechniek.be
■ Design and manufacture of molds and dies, precise equipment and tool parts for industries including automation, medical, electronics, aviation, aerospace and defence.

UPDATED

Radio Holland Belgium NV
(ex-Sait Marine NV)
Noordersingel 17, B-2140 Borgerhout, Antwerpen, Belgium
■ Marine communication and navigation equipment.

UPDATED

Rhea System SA
New Tech Center, Avenue Einstein 2a, B-1348 Louvain la Neuve, Belgium
Tel: (+32 10) 48 72 50
Fax: (+32 10) 45 25 07
e-mail: sales@rheagroup.com
Web: http://www.rheagroup.com
Managing Director: David Kenyon
Technical Writer: Richard Swifte

UPDATED

Siemens Atea NV
(ex-ATEA NV)
Gent, Industriepark Klein Gent, Atealaan 34, B-2200 Herentals, Belgium
Tel: (+32 14) 25 21 11
Web: http://www.ic.siemens.be/eng/default.shtm
■ Design, production, marketing, installation and maintenance of voice-data communications, PABXs (ISDN) networks, mobile PABXs, telephone sets, virtual private networks (VPN) systems integrator, AWHQ communications, ASM communications, CCIS and C3I communications, optical fibre LANs for headquarters, peace-keeping systems and shipboard communications.

UPDATED

Société Anonyme Belge de Constructions Aéronautiques (SABCA)
1470 Chaussée de Haecht, B-1130 Bruxelles, Belgium
Tel: (+32 2) 729 55 11
Fax: (+32 2) 705 15 70
e-mail: sabca.secr@sabca.be
Web: http://www.sabca.com
■ Design, development and manufacture of aircraft structures and components for F-16, Mirage 2000, Mirage 5, Mirage F-1, Alpha Jet, Fokker F-27/F-50, Airbus A330-A340, Dornier 728, Dassault Business Jets, Breguet Atlantic 2 and helicopters. Design, development, manufacture, repair and overhaul of hydraulic servo-actuators for F-16 and Ariane launcher. Design and development of digital and direct drive servo-actuators. Design, development and manufacture of electronic fire control systems for Leopard tanks and training simulators for main battle tanks. Development of low altitude air defence fire control systems. Development of advanced systems of automatic data processing for field artillery. Participation in space programmes for Ariane (parts, structures and servo-actuators), Spacelab, Eureca, ERA. Design, development and manufacture of composite structures such as glass, aramid and carbon fibre. Co-operation with Belgian universities. Final assembly, systems integration, ground and flight tests of aircraft and helicopters (F-16, Mirage, Alouette, Puma, Agusta 109, Sea King, Mirage F-1, Northrop F-5). Overhaul and repair of electrical, mechanical and hydraulic systems. Integration, testing of weapon systems and ECM devices on F-16, A-10, Mirage and Northrop F-5 aircraft. Repair, overhaul and integration of airborne electronic equipment and missile electronic devices. Study of improvements of present aircraft systems.

UPDATED

Sonaca SA
(Société Nationale de Construction Aérospatiale SA)
Aérodrome de Charleroi, Route National 5, B-6041 Gosselies, Belgium
Tel: (+32 71) 25 51 11
Fax: (+32 71) 34 40 35
Web: http://www.sonaca.com
Chairman: Pierre Sonveaux
Managing Director: Christian Jacqmin
Strategic Development and Business Director: Michel Milecan
Chief Financial Officer: Yves Delvigne
Technical Director: Claude Nyssen
Production Director: Thierry Dufour
Personnel Director: Pierre Grenier
Marketing and Programs Director: Marcel Devresse
Purchasing and Logistics Director: Michel Bilocq
Commercial Department: Ms Nathalie Pourignaux
Tel: (+32 71) 25 53 77
Fax: (+32 71) 25 50 82
Research and Development, Space Contracts Manager: Pierre Taquet
■ Design, development, manufacturing, assembly and testing of aerospace structures and their associated systems.
Accreditations: AQAP 110, ISO 9001, JAR/FAR 145, BCAR

UPDATED

Techspace Aero SA
121 Route de Liers, B-4041 Herstal Milmort, Belgium
Tel: (+32 4) 278 81 11
Fax: (+32 4) 278 52 07
Telex: 42075 techsp
e-mail: info@techspace-aero.be
Web: http://www.techspace-aero.be
Chief Executive Officer and President: Jean-Lin Fournereaux
Tel: (+32 4) 278 84 33
Chief Finance Officer: Claude Lemaire
Tel: (+32 4) 278 86 77
General Manager, Commerce and Programs: Jean Christophe Dalla Toffola
Tel: (+32 4) 278 87 02
General Manager, Engineering: Pierre Guillaume
Tel: (+32 4) 278 82 20
General Manager, Human Resources: Georges Massart
Tel: (+32 4) 278 83 05
General Manager, Manufacturing: Vincent Verleye
Tel: (+32 4) 278 81 81
General Manager, Economics and Cost Management: Jean Claude Morin
Tel: (+32 4) 278 87 10
General Manager, Quality: Gilles Debyser
Tel: (+32 4) 278 88 16
General Manager, Information System: Christian Thonon
General Manager, Services: Thierry Chantraine
General Manager, Purchasing: Philippe Menu
Tel: (+32 4) 278 84 37
General Manager, Operations: Jean-Pierre Sainty
General Manager, Test Cell Engineering Business Unit: Pascal Brochard
General Manager, Facility Management: Xavier Naveau
Tel: (+32 4) 278 84 76
Communication Manager: Joëlle Wathelet
Tel: (+32 4) 278 87 28
Fax: (+32 4) 278 80 25
■ Design, development and production of components and equipment for aircraft engines, lubrication systems, boosters (low pressure compressors), critical engine components including casings, bearing supports, disks, drums and spool assemblies; turbine blades and vanes; compressors and turbine disks. Design and production of components for space propulsion systems, control cryogenic valves, regulators and leak control systems. Assembly and testing of complete turbo engines. Repair of assemblies for commercial engines and full overhaul of F100 military engines. Design and installation of turbo engines and equipment.

UPDATED

Thales Communications Belgium SA (TCB)
(ex-Alcatel Bell Space and Defence)
(ex-Thomson-CSF Electronics Belgium)
(ex-Thomson-CSF Sysbel SA)
(ex-Thomson-CSF Systems Belgium)
Rue des Frères Taymans 28, B-1480 Tubize, Belgium
Tel: (+32 2) 391 22 11
Fax: (+32 2) 391 23 00
Managing Director: Jean-Louis Blanchart
Technical Director: Ms Etienne Pourbaix
Tel: (+32 3) 820 50 60
Marketing and Business Development: Pierre de Ponthiere
Tel: (+32 2) 391 24 69
Finance: Alain Malingreau
Tel: (+32 2) 391 22 18
Operations: Carlo Liccardo
Tel: (+32 2) 391 24 45
Purchasing: Patrice Partoune
Tel: (+32 2) 391 22 49
External Communications Manager, Marketing and Business Development: Gianni Butera
Tel: (+32 2) 391 22 23
Fax: (+32 2) 391 24 56
■ Development and manufacture of defence and aerospace C3 systems and navigation aids.

UPDATED

Verhaert
Hogenakkerhoekstraat 21, B-9150 Kruibeke, Belgium
Tel: (+32 3) 250 14 14
Fax: (+32 3) 253 14 64
e-mail: info@verhaert.com
Web: http://www.verhaert.com
General Director: Paul Verhaert
Manager of Satellites and Platforms: Jo Bermyn
Manager of Space Instruments: Frank Preud'Homme
Communications: Ms Ina van Hoye
Marketing: Dany Robberecht
■ Instruments and laboratories for microgravity research services in life, material and fluid sciences and biotechnology. Development, construction, launch and operation of small autonomous satellite systems from 10 to 400 kg, satellite subsystems and attached platforms, planetary research. Development and construction of satellite-related user terminals. Instruments and professional equipment for military, life science, environmental and automotive applications.

UPDATED

Industry – Service

Spacebel SA
Rue des Chasseurs Ardennais, Liege Science Park, B-4031 Angleur, Belgium
Tel: (+32 4) 361 81 11
Fax: (+32 4) 361 81 20
e-mail: info@spacebel.be
Web: http://www.spacebel.com
Managing Director: Bernard Stevenot
Finance and Administration Manager: Ms Nathalie Winandy
Sales and Marketing Manager: Luc Halbach
General Manager: Thierry du Pre Werson
Healthcare Business Unit Manager: André Bonjean
Public Relations and Communication Manager: Ms Ingrid Leonard
Tel: (+32 4) 361 81 13
■ Study, design and realisation of software systems for scientific, technical and industrial applications in the field of aerospace, healthcare and geomanagement.

UPDATED

Bermuda

Intelsat, Ltd
Dundonald House, 14 Dundonald Street West, Suite 201, Hamilton, HM 09, Bermuda
Tel: (+1 441 294) 16 50
Fax: (+1 441 292) 83 00
Web: http://www.intelsat.com
■ Provide telephony corporate network, video and internet solutions around the globe.

UPDATED

Brazil

Centres of Learning & Research

Instituto de Aeronáutica e Espaco (IAE)
Prac Mal.Eduardo Gomes, 12228-904 São José dos Campos, São Paulo, Brazil

Tel: (+55 12) 347 65 55
Fax: (+55 12) 341 25 22
Web: http://www.iae.cta.br

UPDATED

Industry – Manufacturing

Avibras Indústria Aeroespacial SA

14, Rodovia dos Tamoios km 14, Jacre, PO Box 278, 12300-000 São Paulo, Brazil
Tel: (+55 12) 39 55 60 00
Fax: (+55 12) 39 51 62 77
e-mail: govsales@avibras.br
Web: http://www.avibras.com.br
Secretary: Ms Maria Clara Paes Tavares

■ Rocket research and sounding systems and equipment. Meteorological sounding rockets, command post, radio vehicles, gendarmerie and utility vehicles. Ground-based weapon fire control systems. Data processing systems. Missiles and missile systems.
Products include AV-BAFG, AV-BI, AV-BP 250, BFA-230/2, BADI, BLG-120, BLG-252 bombs; gun pods; SBAT-37, SBAT-70, SBAT-127, Skyfire AV-SF-70, LM-70/19 rockets.

UPDATED

Bernardini SA Indústria e Comércio

(Bernardini SA)
Rua Hipólito Soares 79, 04201 São Paulo, Brazil
Tel: (+55 11) 274 80 33
(+55 11) 272 33 23
(+55 11) 67 15 78 15
Fax: (+55 11) 274 81 86
(+55 11) 274 85 67
President: Richard B Cury
Vice President: Flavio M Bernardini

■ Armoured vehicles and systems. Armoured vehicle bridge-laying unit. Military jeeps. Upgrading of tanks and military trucks. Parts, casings, nozzles, nutation dampers and magnetic coils for satellite attitude control systems.

UPDATED

Industry – Service

Aeromot Group

(ex-Aeroeletrônica Indústria de Componentes Aviônicos SA)
(ex-Aeromot Mechânico-Metalúrgica Ltda)
Aeroporto Internacional Salgado Filho, PO Box 8031, 90201-970 Porto Alegre, Rio Grande do Sul, Brazil
Tel: (+55 51) 33 71 16 44
(+55 51) 33 37 29 77
Fax: (+55 51) 33 71 16 55
Web: http://www.ximango.com
http://www.aeromot.com.br
President: Cláudio B Viana
Tel: (+55 51) 371 27 89
Managing Director: João Cláudio Jotz
Financial Director: Nestor Rancich
Technical and Operations Director: Luís Castilho
Purchasing Manager: Ms Paula Cruz
Marketing Assistant: Fabiane Dorneles

■ Aircraft sales and maintenance, overhaul and major repairs.

UPDATED

Bulgaria

Centres of Learning & Research

Space Research Institute (SRI-BAS)

(ex-Central Laboratory for Space Research)
6 Moskovska Street, PO Box 799, 1000 Sofia, Bulgaria
Tel: (+359 2) 988 35 03
(+359 2) 986 16 83
(+359 2) 986 28 78
Fax: (+359 2) 981 33 47
e-mail: office@space.bas.bg
Web: http://www.space.bas.bg
Director and Head of Aerospace and Department of Aerospace Control Systems: Associate Prof Dr Peter Getsov
Tel: (+359 2) 71 83 51
Deputy Director: Associate Prof Dr Plamen Angelov
Deputy Director and Department of High Energy Astrophysics: Associate Prof Dr Latchezar Filipov
Tel: (+359 2) 713 34 22
Scientific Secretary and Department of Experimental Methods of Space Physics: Associate Prof Dr Stefan K Chapkunov
Tel: (+359 2) 713 34 74
Scientific Council Chairman and Department of Earth Remote Sensing: Prof DSc Hernany Spiridonov
Tel: (+359 2) 713 33 45
Department of Space Physics: Associate Prof Dr Georgi Stanev
Tel: (+359 2) 70 05 70
(+359 2) 979 34 28
Department of Remote Sensing and GIS: Associate Prof Dr Anton Stoimenov
Tel: (+359 2) 713 33 13
(+359 2) 70 92 20
Department of Space Biotechnology: Associate Prof Dr Tania Ivanova
Tel: (+359 2) 713 34 67
(+359 2) 70 04 72
Department of Vacuum Experimental Studies: Associate Prof Dr Yulika Simeonova
Tel: (+359 2) 979 32 05
(+359 2) 979 32 06
Department of Aerospace Control Systems: Prof Dr Petar Getsov
Tel: (+359 2) 71 83 51
Department of Onboard Optical Systems: Associate Prof Dr Dimitar Jordanov
Tel: (+359 2) 77 03 05
(+359 2) 714 48 85
Department of Aerospace Technics and Technology: Associate Prof Dr Petar Genov
Tel: (+359 2) 979 34 14
(+359 2) 979 34 86
Department of Space Material Science: Associate Prof Dr Stavri Stavrev
Tel: (+359 2) 975 34 43
Avionics Applied Research Laboratory: Associate Prof Dr V Patov
Tel: (+359 2) 77 03 05
Applied Research Division, Shoumen: Associate Prof Dr Zh Zhekov
Tel: (+359 54) 340 18
Applied Research Division, Plovdiv: Dr S Avramov

UPDATED

Canada

Centre for Research in Earth and Space Technology (CRESTECH)

(Corporate Office)
4850 Keele Street, Toronto, Ontario, M3J 3K1, Canada
Tel: (+1 416) 665 33 11
Fax: (+1 416) 665 20 32
Web: http://www.crestech.ca
Theme Director, Controlled Environment Systems: Andrew M Gordon
Tel: (+1 519) 824 41 20
Fax: (+1 519) 837 04 42

UPDATED

National Research Council of Canada (NRC)

(Herzberg Institute of Astrophysics)
5071 W Saanich Road, Victoria, BC, V9E 2E7, Canada
Tel: (+1 250) 363 00 07
Fax: (+1 250) 363 00 45
Web: http://www.hia.nrc.ca

UPDATED

Ryerson University

(Mechanical, Aerospace and Industrial Engineering)
350 Victoria Street, Toronto, Ontario, M5B 2K3, Canada
Tel: (+1 416) 979 53 03
Fax: (+1 416) 979 52 65
Web: http://www.ryerson.ca
Professor and Chair: Sanjeev D Bhole

UPDATED

University of Toronto

(Institute for Aerospace Studies)
4925 Dufferin Street, Toronto, Ontario, M3H 5T6, Canada
Tel: (+1 416) 667 77 00
Fax: (+1 416) 667 77 99
e-mail: info@utias.utoronto.ca
Web: http://www.utias.utoronto.ca
Director: A A Haasz
Tel: (+1 416) 667 77 17
Library Technician: Ms Nora Burnett
Tel: (+1 416) 667 77 12

■ Research and education in selected areas of aerospace science and engineering including: aircraft flight dynamics, control and simulation, systems integration and virtual aircraft, aerospace propulsion and gas turbine combustion, aerodynamics, structures and multidisciplinary optimisation, space systems engineering, including spacecraft dynamics and control, intelligent and autonomous space robotics and microsatellite technology. Spin-off technologies, such as plasmas and fusion energy, are also pursued.

UPDATED

Industry – Manufacturing

Andrew Canada Inc.

(ex-Andrew Antenna Co. Ltd)
606 Beech Street, PO Box 177, Whitby, Ontario, L1N 5S2, Canada
Tel: (+1 905) 668 33 48
Fax: (+1 905) 668 85 90
Web: http://www.andrew.com

■ Microwave terrestrial antennas, HELIAX coaxial cable and elliptical waveguide, earth station antennas, radar and navigation aid systems, HF antennas, direction finding tactical antennas, weather radar. Radiax cables, towers, weather sensors and broadcast antennas.

UPDATED

Aut Inc.

300, rue Cormier, Sorel-Tracy, Québec, J3R 1V2, Canada
Tel: (+1 450) 743 31 31
e-mail: aut@qc.aira.com
Web: http://www.aut.qc.ca
Chargé de Projet: Ing. Richard Payette

UPDATED

Bristol Aerospace Ltd (BAL)

PO Box 874, Winnipeg, Manitoba, R3C 2S4, Canada
Tel: (+1 204) 775 83 31
Fax: (+1 204) 775 74 94
Web: http://www.bristol.ca

■ CRV-7, 70 mm (2.75 inch) air-to-ground rocket weapon system, Black Brant family of sub-orbital research rocket vehicles. Wire Strike Protection Systems (WSPS) for helicopters, large airframe structures, composites and gas turbine engine 'hot section' components. Repair, overhaul and modification of high performance fixed-wing fighter aircraft.

UPDATED

COM DEV International Ltd

(Corporate Headquarters and R & D Centre)
155 Sheldon Drive, Cambridge, Ontario, N1R 7H6, Canada
Tel: (+1 519) 622 23 00
Fax: (+1 519) 622 16 91
e-mail: investor.relations@comdev.ca
media.relations@comdev.ca
webmaster@comdev.ca
Web: http://www.comdev.ca
Chief Executive Officer: John Keating
Chief Financial Officer: Gary Calhoun

UPDATED

Eidetic Digital Imaging Ltd

1210 Marin Park Drive, Brentwood Bay, British Columbia, V8M 1G7, Canada
Tel: (+1 250) 652 93 26
Fax: (+1 250) 652 52 69
e-mail: eidetic@eidetic.bc.ca
Web: http://www.eidetic.bc.ca/~eidetic/homepage.htm
President/Physicist: Dr Fred G Peet

■ Development of digital image processing systems for the analysis of remotely sensed images of the Earth's surface.

UPDATED

EMS Technologies Inc.
(Space & Technology Group, Canada)
(ex-Satellite and Communications Systems Group)
(ex-Sovcan Star Satellite Communications Inc.)
(ex-Spar Space Systems)
21025 Trans Canada Highway, St Anne-de-Bellevue, Québec, H9X 3R2, Canada
Tel: (+1 514) 457 21 50
Fax: (+1 514) 457 27 24
Telex: 05822792
e-mail: marketing@ems-t.ca
Web: http://www.ems-t.com
President: Gerry Bush
Senior Vice President and General Manager: Don Osborne
Marketing Director: Gareth Lewis
Public Affairs Director: Graeme Maag
Tel: (+1 514) 425 30 79

- Design and manufacture of products and systems for space-based communications, remote sensing and manned-space missions. Technological achievements in satellite equipment include: the Alouette payload in 1961, various payload products and systems for the ANIK series of satellites, the MSAT mobile communications satellite, the Radarsat 1 satellite and the radar antenna and antenna control electronics for the Radarsat 2 programme. Also provides products and systems from the following product lines: antennas, microwave and power products, digital command and control products, optical products and broadband products.

UPDATED

EMS Technologies Inc.
(Satcom Division)
(ex-CAL Corporation)
1725 Woodward Drive, Ottawa, Ontario, K2C 0P9, Canada
Tel: (+1 613) 727 17 71
Freephone: (+1 800) 600 97 59
Fax: (+1 613) 727 12 00
e-mail: info@emssatcom.com
Web: http://www.emssatcom.com
President and Chief Executive Officer: Al Hansen
Senior Vice President and General Manager, SATCOM Division: Neil MacKay
Tel: (+1 613) 727 62 77
Vice President, Engineering and Business Development: Gary Hebb
Vice President, Emergency Management Product Group: Stephen R Edgett
Director, Sales and Marketing – Land Mobile Products: James McMillan
Director, Sales and Marketing – Aeronautical Products: Ray Larkin
Public Relations, Manager: Ms Halina Sejdak-Rydel
Tel: (+1 613) 727 62 77
Public Relations and Communications Assistant: Kelly Lannon
Tel: (+1 613) 727 62 77

- Specialise in the design and development of satellite-based terminals and antennas for the aeronautical, land-mobile, maritime, and search and rescue markets. Development of steerable antenna systems to provide live television to commercial aircraft by a wireless link to a direct broadcast satellite.

UPDATED

Evans
(Head Office)
1616 27th Avenue NE, Calgary, Alberta, T2E 8W4, Canada
Tel: (+1 403) 291 44 44
Fax: (+1 403) 250 65 49
e-mail: info@evansonline.com
Web: http://www.evansonline.com
Chief Executive Officer: Greg Smith
Chief Financial Officer: Tom Smillie
Vice President Sales: Adrian Gale
Marketing and Communications: Scott Taylor

- Design and manufacture of technical furniture, including enclosed consoles, open desking and rear projection display walls specifically for control centres.

UPDATED

Fell Fab Limited
2343 Barton Street East, Hamilton, Ontario, L8E 5V8, Canada
Tel: (+1 905) 560 92 30
Fax: (+1 905) 560 98 46
e-mail: cansales@fellfab.com
Web: http://www.fellfab.com
Chairman: D R Fell
President: Glen Fell
Vice President, Sales and Marketing: Brian Ashton
Marketing Director: Marie Darling
Purchasing Manager: Raza Ali Nawabzada

- Aircraft seats, seat covers, interiors, ground support, liners for bulk liquid and dry goods transportation. Thermal insulation, custom packaging and covers, tents and tarps.

UPDATED

Fleet Industries Ltd
1011 Gilmore Road, PO Box 400, Fort Erie, Ontario, L2A 5N3, Canada
Tel: (+1 416) 871 21 00
Fax: (+1 416) 871 27 22
President and General Manager: James S Butyniec
Managing Director: Dan Zanatta
Manager of New Business Development: Brian M Oakley

- Manufacture of complex aerospace structural assemblies, specialising in flight control components with advanced metal to metal bonding techniques.

UPDATED

Irvin Aerospace Canada Ltd
(ex-Irvin Industries Canada Ltd)
PO Box 1510, Belleville, Ontario, K8N 5J1, Canada
Tel: (+1 613) 967 80 69
Fax: (+1 613) 967 04 69
e-mail: marketing@irvincanada.com
Web: http://www.airbornesystems-na.com/main.html
Managing Director: Peter Johnson
Financial Director: Ms Lisa Donovan
Marketing and Sales Director: Mrs Elizabeth Johnson
Purchasing Manager: Ms Christina Martin
Marketing Assistant: Ms Sapphire Crow

- Parachutes for personnel, cargo, weapon and aircraft retardation. Recovery systems for missiles and space vehicles, cargo delivery systems. Safety harnesses and air drop sequence control devices. Inflatable survival products and systems. Anti-g suits, paratrooper parachutes, missile recovery parachutes, flare chutes, jump tower chutes, safety harnesses.

Accreditations: ISO 9001:2000

UPDATED

Magellan Aerospace Corporation
3160 Derry Road East, Mississauga, Ontario, L4T 1A9, Canada
Tel: (+1 905) 677 18 89
(+1 905) 673 32 50
Fax: (+1 905) 677 56 58
(+1 905) 673 53 00
Web: http://www.magellanaerospace.com

UPDATED

Magellan Aerospace Corporation
(Bristol Division)
660 Berry Street, PO Box 874, Winnipeg, Manitoba, R3C 2S4, Canada
Tel: (+1 204) 775 83 31
Fax: (+1 204) 775 74 94

- Manufacture of rocket weapon systems, warheads, rocket launchers and motors.

UPDATED

Optech Incorporated
(ex-Optech Systems Inc)
100 Wildcat Road, North York, Toronto, Ontario, M3J 2Z9, Canada
Tel: (+1 416) 661 59 04
Fax: (+1 416) 661 41 68
Web: http://www.optech.on.ca
President: Dr Allan I Carswell
Vice President: S Szgoric
Program Development Director: Gregg Van Volkenburgh
Marketing Manager: Bill F Matvichuk
Promotions Co-ordinator: Ms Joanne Tassone
Marketing Writer: Joshua Kern
Market Analyst: Sumona Datta

- Manufacture of laser radar systems including military laser systems, lidar bathymetry systems, laser altimeters, high-accuracy laser rangefinders, laser terrain profilers and laser water depth sounders. Remote-sensing systems, lidar pollution-detection systems, remote-sensing optical scanners and wideband logarithmic amplifiers. Airborne ASW and MCM systems.

UPDATED

Routes Astroengineering
303 Legget Drive, Ottawa, Ontario, K2K 2B1, Canada
Tel: (+1 613) 592 07 48
Fax: (+1 613) 592 65 53
e-mail: staff@routes.com
Web: http://www.routes.com
Managing Director: W F Payne
Marketing Director: B Gordon
Technical Director: K Smith
Operations Director: L Piche
Mechanical Engineer: Paul Marchand
Fax: (+1 613) 592 65 53

- Design and development of scientific instruments for space applications. Manufacture of solar panels.

UPDATED

Satlantic Inc.
Richmond Terminal, Pier 9, 3481 North Marginal Road, Halifax, Nova Scotia, B3K 5X8, Canada
Tel: (+1 902) 492 47 80
Fax: (+1 902) 492 47 81
e-mail: info@satlantic.com
Web: http://www.satlantic.com
President and Chief Executive Officer: William Ricketts
Vice President, Marketing Director: Mike Hodgett
Operations Director: Rob McMahon
Sales and Marketing Director: Norm Countway
Sales Director: Ms Marcia Walsh

UPDATED

Seimac Ltd
(ex-Seimac Research Ltd)
271 Brownlow Avenue, Dartmouth, Nova Scotia, B3B 1W6, Canada
Tel: (+1 902) 468 30 07
Fax: (+1 902) 468 30 09
e-mail: info@seimac.com
Web: http://www.seimac.com
Marketing and Sales Manager: Paul Adlakha
Program Manager: Paul Steward
Sales and Marketing: Ms Trisha Hensley
Customer Service Representative: Darren Penney
Tel: (+1 902) 468 72 75

- Provide radio and data acquisition products in the research, military, search and rescue, and oil and gas industries. Design, manufacture and market a variety of products including Emergency Position Indicating Radio Beacons (EPIRBs), GOES satellite transmitters, spread spectrum radio modems, VHF radios and optical beacons and a range of miniaturised ARGOS satellite radio transmitters that are used in tracking objects and animals in marine and terrestrial applications.

UPDATED

Techspace Aero Canada Limiteé
Rue Prince 33, Montréal, Québec, H3C 2M7, Canada
Tel: (+1 514) 397 02 71
Fax: (+1 514) 397 85 85
General Manager: Jean-Pierre Huchin
Tel: (+1 151) 48 66 88 61

- Engineering and development for aero and space propulsion systems.

UPDATED

The Aerospace Consortium Inc.
730 Gana Court, Mississauga, Ontario, L5S 1P1, Canada
Tel: (+1 905) 564 66 01
Fax: (+1 909) 670 16 95
President: William H Reil
Vice President, Business Development: F Davenport
Secretary: Paul MacPherson

- Aerospace and defence components, subsystems and systems: machined components and assemblies, military communications and electronics equipment, whip antennas, composite parts, repair and overhaul, programme management, meteorological equipment.

UPDATED

Ultra Electronics TCS
(ex-BAE Systems Canada Inc.)
(ex-Canadian Marconi Co.)
(ex-CMC Electronics Inc.)
600 Dr Frederik Philips Boulevard, Ville St-Laurent, Québec, H4M 2S9, Canada
Tel: (+1 514) 855 63 97
(+1 514) 748 31 13

Fax: (+1 514) 855 63 54
(+1 514) 748 31 84
Web: http://www.cmcelectronics.ca
Public Relations and Marketing Support Services Manager: Ms Janka Dvornik
Contact: Ms Cathy Babos
Tel: (+1 630) 466 21 12
Fax: (+1 630) 466 43 58
■ Commercial and military aviation electronics: navigation and landing systems (Doppler, GPS, MLS); intelligent monitoring and display systems, adaptive and phased-array antenna systems, airborne satellite communications antenna systems, systems engineering, aeronautical television antennas, flight management systems, cockpit management systems and enhanced vision systems, HUDs, multi-function displays, mission-computers, cockpit systems integration, navigation sensors (GPS and Doppler). Customer support: technical publications, training, field engineering, automated test equipment, logistics systems engineering, spares and repair and support planning services to CMCs worldwide commercial and military aviation customers.
Custom electronics: microelectronics, power conversion and magnetics, specialised electronic components, display components and machined parts, contract manufacturing and circuit packaging.

UPDATED

CMC Electronics Inc.
(Ottawa Facility)
415 Leggett Drive, PO Box 13330, Ottawa, Ontario, K2K 2B2, Canada
Tel: (+1 613) 592 65 00
Fax: (+1 613) 592 74 27
Vice President, Government and Public Affairs: Jean-Michael Comtois
Vice President, Commercial Aviation: Bruce Bailey
Government and Public Relations Co-ordinator: P M Rogers

UPDATED

Industry – Service

Array Systems Computing Inc.
1120 Finch Avenue West, 7th Floor, Toronto, Ontario, M3J 3H7, Canada
Tel: (+1 416) 736 09 00
Fax: (+1 416) 736 47 15
Web: http://www.array.ca
Manager, Business Development: Laszlo Mayer
Business Development and Communications Specialist: Joshua Caplan
■ Specialise in sophisticated software and systems integration in the areas of sonar, radar, Synthetic Aperture Radar (SAR) and remote sensing.

UPDATED

Geomat International Inc.
5160, Decarie Boulevard, Suite 300, H3X 2H9, Montréal, Québec, Canada
Tel: (+1 514) 369 50 00
Fax: (+1 514) 369 50 59
e-mail: geomat@geomat-intl.com
Web: http://www.geomat-intl.com
Geographic Information Systems and Cartography and Imagery: Bill Maag
Remote Sensing: Ms Diane Thibault
Sales: Santiago Forteza

UPDATED

Jane's Information Group (JIG)
(Canada)
1400, 100, Queen Street, Ottawa, Ontario, K1P 1J9, Canada
Tel: (+1 613) 566 36 42
Fax: (+1 613) 566 36 40
Web: http://www.janes.com
Canadian Sales: Geoff Mizen
Representing: Jane's Information Group, UK

UPDATED

Lansdowne Technologies Inc.
1001-275 Slater Street, Ottawa, Ontario, K1P 5H9, Canada
Tel: (+1 613) 236 33 33
Fax: (+1 613) 236 44 40
e-mail: lti@lansdowne.com
sales@lansdowne.com
Web: http://www.lansdowne.com
Managing Partner: Michel Anglehart
Managing Partner: Lorne Schmidt
Director, Business Development: Alex McPhail
■ Project management, planning and control, proposal and logistic management for defence and space industries. Non-tactical software development services.

UPDATED

MacDonald, Dettwiler and Associates Ltd (MDA)
13800 Commerce Parkway, Richmond, British Columbia, V6V 2J3, Canada
Tel: (+1 604) 278 34 11
Fax: (+1 604) 273 98 30
(+1 604) 278 25 33
e-mail: info@mda.ca
invest@mda.ca
Web: http://www.mda.ca
External Relations Manager: Ms Liza Aboud
■ Manufacture information systems for use with monitoring elements of the planet's surface, managing transport systems and mobile assets: provision of information to relevant staff in the field.

UPDATED

MD Robotics
(ex-MacDonald Dettwiler Space and Advanced Robotics Ltd)
(ex-Spar Aerospace Limited)
9445 Airport Road, Brampton, Ontario, L6S 4J3, Canada
Tel: (+1 905) 790 28 00
Fax: (+1 905) 790 44 00
e-mail: info@mdrobotics.ca
Web: http://www.mdrobotics.ca
Vice President and General Manager: Magued Iskander
Vice President, Business Development and Research and Development: Peter Stibrany
Vice President, Finance and Administration: Hiten Makim
Vice President, Operations: Bruce Mack
Public Relations Director: Ms Cheryl Prince
Contracts and Material Director: Allan Luborsky
Manager Public Affairs: Ms Lynne Vanin
■ Development and manufacture of robotics for use in space. Products include the Canadarm robot arm used in the US space shuttles and a robotic servicing system to be used in assembly, maintenance and servicing of the International Space Station.

UPDATED

Radarsat International (RSI)
(Headquarters)
13800 Commerce Parkway, MacDonald Dettwiler Building, Richmond, British Columbia, V6V 2J3, Canada
Tel: (+1 604) 231 50 00
(+1 604) 244 04 00
Fax: (+1 604) 231 49 00
(+1 604) 244 04 04
e-mail: info@rsi.ca
Web: http://www.rsi.ca
President and Director, Business Operations: Dr John Hornsby
Chief Technical Officer, Corporate Planning: Dr Bill Jefferies
Director, Finance: Ms Pam Egan
Director, International Network Stations: Dr Pierre Engel
Director, Worldwide Sales: Adrian Bohane
Marketing Manager: Ms Kate Pilgrim
Director, Public Relations: Ms Kate Stephens
Communications Specialist: Ms Stacey McGregor
Tel: (+1 604) 231 49 40
Fax: (+1 604) 231 49 99
■ Satellite remote sensing data distribution and processing.

UPDATED

Radarsat International (RSI)
(Ottawa Office)
2060 Walkley Road, Ottawa, Ontario, K1G 3P5, Canada
Tel: (+1 613) 238 54 24
Fax: (+1 613) 238 54 25

UPDATED

SED
18 Innovation Boulevard, PO Box 1464, Saskatoon, Saskatchewan, S7K 3P7, Canada
Tel: (+1 306) 931 34 25
Fax: (+1 306) 933 14 86
Web: http://www.sedsystems.ca
President: Ray Basler
Vice President, Operations: Brent McConnell
Director, Communications: Don Epp
Programme Manager: Terry Rohrke
Head of Engineering and Technical: Dennis Akins
Communications Administrator: Ms Karen Davidson
Purchasing Manager: Ms Sharon Carruthers
■ Space and communications systems design, engineering and integration specialising in satellite ground terminals, defence communication systems, satellite test and control and mobile satellite communication systems. Major products include: INMARSAT coast-earth stations, satellite in-orbit test systems, environmental monitoring, remote control systems, HF adaptive beam forming receiver systems.

UPDATED

Smiths Detection
(ex-Barringer Instruments Ltd)
1730 Aimco Boulevard, Mississauga, Ontario, L4W IVI, Canada
Tel: (+1 905) 238 88 37
Fax: (+1 905) 238 30 18
Web: http://www.smithsdetection.com

UPDATED

Telesat Canada
1601 Telesat Court, Gloucester, Ontario, K1B 5P4, Canada
Tel: (+1 613) 748 87 00
(+1 613) 748 01 23
Fax: (+1 613) 748 87 12
e-mail: info@telesat.ca
Web: http://www.telesat.ca
Chairman: Richard Currie
President and Chief Executive Officer: Larry J Boisvert
Vice President, Space Systems: Roger J Tinley
Vice President, Business Development: Dave Lahey
Vice President, Corporate Development: Paul D Bush
Vice President, Network Services: Pat Enright
Vice President, Finance: Ted H Ignacy
Vice President, Human Resources and Administration: Ms Marilynn A Wright
Director, International Marketing and Sales: Rory McCormick
Director, Broadcast Sales and Marketing: Len Lawson
Secretary and General Consul: Ms Jennifer E Perkins
Public Information Officer: Stuart MacMillan
■ Provide specialised engineering consulting service on a worldwide basis to governments and private companies interested in establishing, operating or upgrading satellite systems and associated ground infrastructures. Domestic satellite owner for Canada and operator of ANIK and NIMIQ satellites. Design, implementation, maintenance operation and marketing of satellite telecommunications services for North and South America.

UPDATED

Chile

Centres of Learning & Research

Centro de Estudios Espaciales (CEE)
Universidad de Chile, A. Part 1171, Casilla 411-3, Santiago de Chile, Chile
Tel: (+56 2) 555 34 00
(+56 2) 555 33 71
Fax: (+56 2) 556 83 82
e-mail: infocee@cec.uchile.cl
Web: http://www.cee.uchile.cl
Executive Director: Martin Arluciaga
Tel: (+56 2) 672 18 16
Director of Research: Armando Scheneidewind
Financial Manager: Pedro Ramirez
National Services Manager: Gabriel Soto
Public Relations Officer: Ms Alicia Reyes
Contact: Ms Norma Gornall
■ Satellite tracking station and ground support services. Offers remote sensing, a geographic information system, geomatics, and telecommunications.

UPDATED

Industry – Service

Sisdef Ltda

Parque Industrial Aconcagua, Camino Con Cón-Quintero Km2, PO Box Casilla 5C-Correo Con Cón, Quintero, Chile
Tel: (+56 32) 81 07 77
Fax: (+56 32) 81 11 90
Web: http://www.sisdef.cl
General Manager and Chief Executive Officer: Claudio Niada
Administrative and Financial Manager: Gonzalo Carrasco
Technical Manager: Jorge Pimentel
Marketing and Sales Manager: Eduardo Morales
Research and Development Manager: Fernando Gray
Marketing and Sales Assistant: Ms Rosangela Matoso

■ Systems engineering in electronics and computing, command and control systems, fire control systems, engineering, simulators, communications, aerospace, tactical positioning for land, surface and underwater applications.

UPDATED

China

Government – Ministry/Agency

China National Space Administration (CNSA)

8A, Fucheng Road, 100037 Haidian, Beijing, China
Tel: (+86 10) 88 58 13 77
(+86 10) 88 58 13 79
(+86 10) 88 58 13 80
(+86 10) 88 58 13 86
Fax: (+86 10) 88 58 15 15
e-mail: cnsa@cnsa.gov.cn
Web: http://www.cnsa.gov.cn

UPDATED

Centres of Learning & Research

China Academy of Launch Vehicle Technology (CALT)

No.1 Nan Da Hong Men Road, Fengtai District, 100076 Beijing, China
Tel: (+86 10) 68 38 10 35
(+86 10) 68 75 75 98
(+86 10) 68 38 13 83
e-mail: caltinfo@calt.com.cn
Web: http://www.calt.com.cn

■ Launch vehicle research base. Products include liquid fuel strategic rockets and long march launch vehicles, space craft satellite communications systems, precision measuring and testing instrumentation; applied computer technology and automatic control technology.

UPDATED

Nanjing University of Aeronautics & Astronautics (NUAA)

29 Yudao Street, 210016 Nanjing, Jiangsu, China
Tel: (+86 25) 84 89 27 76
(+86 25) 84 89 24 40
Fax: (+86 25) 84 89 27 76
(+86 25) 84 49 80 69
e-mail: icedao@nuaa.edu.cn
Web: http://www.nuaa.edu.cn
President: Prof Hu Haiyan
Tel: (+86 25) 489 32 78

■ Development of super-light pilotless fixed- and rotary-wing aircraft.

UPDATED

Association

Chinese Society of Aeronautics & Astronautics (CSAA)

5 Liangguochang Road, Dongchen District, 10010 Beijing, China
Tel: (+86 10) 64 02 14 16
Fax: (+86 10) 64 02 14 13
e-mail: mail@csaa.org.cn
Web: http://www.csaa.org.cn

■ Consistsof24technicalcommittees(aerodynamics, flight dynamics and testing, structural design and analysis, propulsion systems, materials technology, manufacturing technology, aviation medicine and human engineering, maintenance engineering, management science, composites, instrumentation and measurements, weaponry and fire control, helicopters, light flying vehicles, avionics, automatics, information, reliability) and organises academic conferences, meetings, forums, technical exhibitions and information exchanges. Over 40,000 members.

Publications: Acta Aeronautica et Astronautica Sinica (with English abstracts)
Aerospace Knowledge Magazine (in Chinese)
Model Airplane (in Chinese with English text)
Regional societies
Beijing Aerospace Society and the Aeronautics and Astronautics Societies of Guangdong, Guangxi, Ghuizou, Heilongjiang, Hubei, Hunan, Jiangsu, Jiangxi, Jilin, Liaoning, Shaanxi, Shanghai, Sichuan, Zhejian, Fujian, Henan

UPDATED

Industry – Manufacturing

China National Electronics Import & Export Corp. (CEIEC)

Electronics Building A23, Fuxing Road, PO Box 140, 100036 Beijing, China
Tel: (+86 1) 821 95 32
Fax: (+86 1) 821 23 52
Web: http://www.ceiec.com

■ Products include air, naval and army radios and radars, air defence systems, navigation systems, optical systems, cryptographic equipment, mine detection equipment, fibre and laser optics, command, control and communications systems, electronic warfare systems, simulators, components and spare parts. Services include the modification, overhaul and upgrading of radar systems, military communication systems, fire control systems and air defence networks, operates a spare parts supply service and export of military electronics systems.

UPDATED

China National Precision Machinery Import & Export Corp. (CNPMIEC)

(ex-China Precision Machinery Import & Export Corp.)
30 Haidian Nanlu, Haidian District, 100080 Beijing, China
Tel: (+86 10) 68 74 88 77
(+86 10) 68 74 88 91
Fax: (+86 10) 68 74 88 44
Web: http://www.cpmiec.com.cn

■ Import and export and production of tactical defensive missiles and related equipment, including anti-ship missiles, air defence missile systems from hand held to vehicle mounted and multilaunch rocket systems. Other products include space equipment, precision machinery, electronic products, space hardware and optical instruments, missile components such as gimbals and gyroscopes. Research and development facilities have wind tunnel, plasma etching technology, IC photolithograph capabilities as well as microwave and infra-red homing guidance simulation laboratories. Deals with joint ventures, compensation trade, contracts for the design and construction of industrial and civil projects.

UPDATED

Ningbo Donglian Mechanical Seal Co. Ltd

Jiangshan Science and Technology Zone, Ningbo, Zhejiang, China
Tel: (+86 574) 88 45 23 48
(+86 574) 88 45 23 54
(+86 574) 88 45 83 54
Fax: (+86 574) 88 45 74 35
e-mail: marketing2@dyseals.com
Web: http://www.dyseals.com

■ Production of ceramic, tungsten carbide and carbon graphite mechanical seals.

Accreditations: ISO 9002, ISO 9001-2000

UPDATED

Industry – Service

China Great Wall Industry Corp. (CGWIC)

30 South Haidian Road, 100080 Beijing, China
Tel: (+86 10) 68 74 88 88
(+86 10) 68 74 88 00
Fax: (+86 10) 68 74 88 65
(+86 10) 68 74 88 76
e-mail: cgwic@cgwic.com
Web: http://www.cgwic.com

■ Trading company acting as the main foreign trade and marketing channel for China Aerospace Corporation (CASC) and is the sole commercial organisation authorised by the government to provide commercial satellite launch services and space technology co-operation to the world market. Development of international activities and aerospace technology collaboration specialising in space technology products. Import and export of equipment, precision machinery, instruments and meters. Technology consulting, project contracting, employment services, technology transfer, joint production, equity, co-operative joint ventures, investment co-ordination, export capital construction, storage and transportation, international exhibition, tourism, satellite communications, property management and domestic trade.

Locations
Jiuquan Satellite Launch Center
Taiyuan Satellite Launch Center
Xian Satellite Control Center
Xichang Satellite Launch Center

UPDATED

Gilat Satellite Networks Ltd

(Beijing Representative Office)
1725, Tower 2, Bright China ChangAn Building, No.7, Jianguomeng Nei Avenue, 100005 Beijing, China
Tel: (+86 10) 65 10 28 38
Fax: (+86 10) 65 10 28 39
Web: http://www.gilat.com.cn
Senior Vice President and General Manager: Osvaldo Bergstein

UPDATED

Industry – Sales Agent/Office

Radyne Comstream Inc.

(China)
(ex-Comstream Corporation)
Room 405, Building B, No. 8 Guanghaua Road, Chaoyang District, 100026 Beijing, China
Tel: (+86 10) 65 83 19 75
(+86 10) 65 83 19 76
(+86 10) 65 83 19 77
(+86 10) 65 83 19 78
Fax: (+86 10) 65 83 19 74
e-mail: chinasales@radn.com
Web: http://www.radn.com
Representing: Radyne Comstream Inc., US

UPDATED

Côte d'Ivoire

Government – Intergovernmental

Regional African Satellite Communications Organization (RASCOM)

(Organisation Regionale Africaine de Communications par Satellite)
2 avenue Thomasset, Abidjan, Côte d'Ivoire
Tel: (+225 20) 22 36 74
(+225 20) 22 36 83
Fax: (+225 20) 22 36 76
(+225 20) 22 36 79
e-mail: info@rascom.org
Web: http://www.rascom.org
Vice President, International Relations: Leke Betechuoh Casimir
Vice President of Marketing: Christian Sagoe Kow
Vice President of Finance: Mrs Membere Kassa

Vice President RASPAC: Sheriff Adams
Commander/Director General/Chief Executive Officer: Goundé Desiré Adadja
Vice President and Director of Engineering Technology: Eliman H Cham
Vice President of Human Resources and Logistics: Machioudi Nassirou

UPDATED

Cyprus

Industry – Service

Imagesat International NV
(Corporate Headquarters)
45 Ayias Paraskevis, Office IA, Yermasoyia, 4044 Limassol, Cyprus
Tel: (+357 25) 82 11 14
Fax: (+357 25) 82 11 15
e-mail: information@imagesatintl.com
Web: http://www.imagesatintl.com
Customer Service Manager: Ms Corieda Kotze
Mobile: (+357 99) 66 12 57

UPDATED

Czech Republic

Industry – Manufacturing

Artisys S.R.O.
Stursova 71, CZ-616 00 Brno, Czech Republic
Tel: (+420 541) 22 48 36
Fax: (+420 541) 22 48 70
e-mail: info@artisys.aero
Web: http://www.artisys.aero
Managing Director: Ms Dana Brhelova
Operations and Technical Director: Jiri Sasek
Financial Director: Petr Bahula
Purchasing Manager: Pavel Jerabek
Project Manager: Ladislav Kracik
■ Software house and system integrator focused on real-time systems and related technologies in the aerospace and space fields. Production of real time information and control systems based on mission critical key technologies. Products and services include: ATC systems and simulators, airport systems, time servers, data engineering, computer technologies consulting, MMI development, Lynxos and Kinesix/Sammi authorised distributor, Linux consulting.
Representing: Kinesix Corporation, US
Lynx Real-Time Systems Inc., US

UPDATED

Denmark

Alcatel Space Denmark A/S
(ex-Alcatel Kirk A/S)
Lautrupvang 2, DK-2750 Ballerup, Denmark
Tel: (+45) 44 86 75 00
Fax: (+45) 44 86 75 01
e-mail: info@alcatel.dk
Web: http://www.alcatel.dk
■ Custom designed power supplies for launchers and satellites. DC/DC converters, servo amplifiers, control electronics, AC power sub-systems and power electronics.

UPDATED

Sailor
(ex-SP Radio A/S)
Porsvej 2, DK-9200 Aalborg, SV, Denmark
Tel: (+45) 96 34 61 00
Fax: (+45) 96 34 61 01
e-mail: sailor@sailor.dk
Web: http://www.sailor.dk
Managing Director: Svend Tottrup
Sales Director: Jørgen Fløe Mikkelsen
Marketing Manager: Henrik Fyhn
Sales Manager, Europe: Kjeld Skovgaard
Sales Manager, OEM: Henning Munch
Sales Manager, Overseas West: Allan Lydersen
Sales Manager, Overseas East: Peter Andersen
■ Maritime radio communication equipment, including VHF, SSB radio-telephones, radio-telex and complete GMDSS equipment. Portable VHF/UHF programme. Satellite communication equipment B, C, Mini-M.

UPDATED

Terma A/S
(Headquarters)
(ex-Terma Elektronik AS)
Hovmarken 4, DK-8520 Lystrup, Denmark
Tel: (+45) 87 43 60 00
Fax: (+45) 87 43 60 01
e-mail: terma.hq@terma.com
Web: http://www.terma.com
Chairman: Svend-Aage Nielsen
President and Chief Executive Officer: Jens Maaløe
Tel: (+45 8743) 62 01
Senior Vice President and Chief Financial Officer: Holger Christensen
Senior Vice President of Radar Systems: Per Borggaard
Senior Vice President of Naval and Ground Systems: Eric Kressel
Senior Vice President of Airborne Systems: Børge Witthøft
Senior Vice President of Space Systems: Claus J Nielsen
Head of Electronic Warfare Systems: Maj Gen Ole Fogh
Tel: (+45 8743) 63 94
Manager, Electronic Warfare Marketing: Henning Hansen
Tel: (+45 8743) 63 35
Sales Manager, SMR: Steen Trier
Tel: (+45 8743) 62 06
Public Affairs Manager: Kasper Rasmussen
Tel: (+45 8743) 60 91
Contract Manager EW: Frank Solhoej
Tel: (+45 8743) 63 16
Area Sales Manager: John Rygaard
Tel: (+45) 86 33 17 43
Mobile: (+45) 40 18 65 12
Public Affairs Assistant: Ms Christina Krikau Petersen
Tel: (+45 8743) 60 39
Mobile: (+45) 40 46 60 39
■ Radar and display systems for communication, navigation, surveillance and velocity measurements. Satellite equipment, antennas and telecommunication systems. Countermeasure dispenser systems.

UPDATED

Terma A/S
(Aerospace and Defence)
(ex-Terma Elektronik AS)
Mårkaervej 2, DK-2630 Tastrup, Denmark
Tel: (+45) 43 52 15 13
Fax: (+45) 43 52 57 58
e-mail: terma.dk@terma.com
Web: http://www.terma.com
Senior Vice President, Head of Division: Eric Kressel
Director, Business Development: Michael Norup
Marketing Manager: Tom Frederiksen
Manager, Business Development: Thomas Blom
Manager, Business Development: Frank B Jorgensen
Sales and Marketing Co-ordinator: Pernille B Sander
Fax: (+45) 43 52 23 80
■ Development of naval C3 systems that integrate, operate and control the entire weapon and sensor inventory on naval ships.

UPDATED

Terma A/S
(Aerospace and Defense)
(ex-Per Udsen Co. Aircraft Industry A/S)
(ex-Terma Industries Grenaa AS)
Fabrikvej 1, DK-8500 Grenaa, Denmark
Tel: (+45 8779) 52 00
Fax: (+45 8779) 52 01
e-mail: ihr@terma.com
Web: http://www.terma.com
Vice President, Business Development: Peter Worsøe
Tel: (+45 8779) 52 03
Operations Director: Jens Nygaard Jensen
Tel: (+45 8779) 52 48
Financial Controller: Birthe H Rask
Tel: (+45 8779) 52 70
Head of Department, Aerostructures: Karl Bendtsen
Tel: (+45 8779) 53 21
Purchasing Manager: Leif Kølleskov
Tel: (+45 8779) 52 07
Engineering Manager: Claus Bratsbjerg
Tel: (+45 8779) 53 34
Executive Secretary: Ms Lisbeth Hesse Rygaard
Tel: (+45 8779) 52 04
■ Design, manufacture and assembly of aircraft parts (airframe). Graphite composite parts.

UPDATED

Thrane & Thrane A/S (T&T)
Lundtoftegaardsvej 93 D, DK-2800 Lyngby, Denmark
Tel: (+45) 39 55 88 00
Fax: (+45) 39 55 88 88
e-mail: info@tt.dk
Web: http://www.tt.dk
Chief Executive Officer: Lars Thrane
Tel: (+45 39) 55 88 10
Chief Financial Officer: Svend-Aage Lundgaard Jensen
Tel: (+45 39) 55 89 33
Chief Operating Officer: Henrik Lunde
Tel: (+45 39) 55 88 79
Marketing Director: Martin Kauffmann
Tel: (+45 39) 55 89 45
Sales and Marketing Manager: Christian L Hammer
Marketing Project Co-ordinator: Nikolaj Hvegholm
Tel: (+45 39) 55 83 26
Logistic Manager: Ms Amanda Jane Federiksen
Commercial Sales, Land Mobile INM-C Systems: Lars Ackermann
■ Manufacture of radio and satellite communication equipment.

UPDATED

Industry – Service

Danish Space Research Institute (DSRI)
(Dansk Rumforskningsinstitut)
30 Juliane Maries Vej, DK-2100 København, Denmark
Tel: (+45) 35 32 58 30
Fax: (+45) 35 36 24 75
e-mail: office@cisri.dk
Web: http://www.dsri.dk
Director: Eigil Friis-Christensen
Tel: (+45) 35 32 57 07
Manager: Mette Moland
Tel: (+45) 35 32 57 30
Head of Department: Carsten Kyhnauv
Tel: (+45) 35 32 57 14
■ Government agency concerned with space research, x-ray astronomy, plasma physics and instrument development.

UPDATED

FLS Aerospace (DK) A/S
Copenhagen Airport South, Hanger 243, DK-2791 Dragør, Denmark
Tel: (+45) 32 45 00 45
Fax: (+45) 32 45 00 12
Web: http://www.flsaerospace.com
■ Undertakes light maintenance of civil aircraft.

UPDATED

Satcom1
(Headquarters)
Sprogovej 6, 2.t.h., DK-2000 Frederiksberg, Denmark
Tel: (+33 145) 75 10 54
Fax: (+33 145) 78 83 67
e-mail: activation@satcom1.com
support@satcom1.com
finance@satcom1.com
info@satcom1.com
webmaster@satcom1.com
sales@satcom1.com
Web: http://www.satcom1.com
Chief Engineering Officer: Rino Ranheim
Chief Financial Officer: Jean Francois Gault
Sales and Marketing Manager: Ms Karina B Larsen
■ Communication provider and integrator for aero, maritime and land mobile applications.

UPDATED

Terma A/S
(Birkerød Office)
(ex-Computer Resources International A/S)
(ex-Terma Elektronik AS)
Bregnerødvej 144, DK-3460 Birkerød, Denmark
Tel: (+45) 45 94 96 00
Fax: (+45) 45 94 96 99
e-mail: terma.spd@terma.com
Web: http://www.terma.com
Sales Manager: Karsten Meier
Division Manager: Claus Jorgen Nielsen
Space Business Development Manager: Jens Guldberg
Tel: (+45 45) 94 96 10

■ Development and deployment of mission-critical space and defence systems, maintenance and engineering systems for airlines and aerospace, systems engineering environments, document and workflow management systems, technical documentation and data delivery systems to various industrial domains.

UPDATED

Ticra
Laederstraede 34, DK-1201 København, Denmark
Tel: (+45) 33 12 45 72
Fax: (+45) 33 12 08 80
e-mail: ticra@ticra.com
Web: http://www.ticra.com
Chairman: Rolf Jorgensen
General Manager: Dr Hans-Henrik Viskum
Executive Secretary: Mads Bue Johnsen

■ Commercial foundation. Engaged in the development of antenna synthesis, analysis and measurement methods on research contracts for international organisations and private companies. Organisation has a library of advanced antenna software which is commercially available. Manufacture of dual polarised probes for near-field measurements.

UPDATED

Industry – Sales Agent/Office

Scandinavian Avionics A/S
Stratusvej 9, Billund Airport, PO Box 59, DK-7190 Billund, Denmark
Tel: (+45) 79 50 80 00
(+45) 40 43 84 22
(+45) 44 64 88 28
Fax: (+45) 79 50 80 99
e-mail: sa@scanav.com
Web: http://www.scanav.com
President and Sales Director: H B Truelsen
Tel: (+45) 79 50 80 01
Financial Director: J K Nielsen
Tel: (+45) 79 50 80 12
Marketing Manager: Ivan Jensen
Tel: (+45) 79 50 80 03
Production Manager: P Albertsen
Tel: (+45) 79 50 80 60
Purchasing Manager: M Pedersen
Tel: (+45) 79 50 80 15
Quality Manager: K Poulsen
Tel: (+45) 79 50 80 10
Instrument Service Manager: E Ubbesen
Tel: (+45) 79 50 80 40
Military Representative: J Thalund
Tel: (+45) 79 50 80 02

■ Sales, installation, repair and overhaul of HF, VHF, UHF, communications and navigation, DME, transponder, radio/radar altimeters autopilots, flight directors, voice/flight data recorder, inertial navigation, GPS, VLF/Omega, instrument, gyro, ground proximity and aircraft electrical systems for civil and military purposes.

UPDATED

Skanti A/S
(ex-Skandinavisk Teleindustri A/S)
Porsvejz, DK-9200 Aalborg, SV, Denmark
Tel: (+45) 44 74 84 00
Fax: (+45) 44 74 84 01
e-mail: skanti@skanti.dk
Web: http://www.skanti.dk

■ Sales of satellite HF and VHF marine radios for ships and ground use.

UPDATED

Ecuador

Association

Association of Andean Community Telecommunications Enterprises (ASETA)
La Pradera 510 y San Salvador, Quito, Ecuador
Tel: (+593 2) 256 38 12
(+593 2) 250 98 21
(+593 2) 250 98 22
Fax: (+593 2) 256 24 99
e-mail: info@aseta.org
Web: http://www.aseta.org
Secretary General: Ing Marcelo López Arjona
International Affairs Director: Ing Eduardo Pichilingue
Studies and Projects Director: Ing Jairo Gomez Malaver
Advisor: Ing Marco Solano

■ Association of telecommunications service operating companies within the Andean community, supporting development of telecommunications, to contribute to their countries' process of integration.

UPDATED

Egypt

Industry – Manufacturing

Arab Organisation for Industrialisation (AOI)
2D Midan al-Abbasia, PO Box 770, Cairo, Egypt
Tel: (+20 2) 482 33 77
(+20 2) 593 24 83
(+20 2) 593 28 22
(+20 2) 593 23 95
Fax: (+20 2) 482 60 10
(+20 2) 401 25 83
Telex: 92014 aoi
92090
Cable: SONARABCAIRO
e-mail: aoihq@idsc.gov.eg
Chairman: Eng Ibrahim Youness
Managing Director: Eng M Samir Abd El-Wahab
Technical Director: Dr Nabil Salama
Operations and Marketing Director: Mahmoud El Refai
Public Relations Director: Mohamed Abdel Azim
Purchasing Manager: El Prince Adel Abd El Rahman

■ Military activities: manufacture of aircraft and aircraft engines, missiles, electronic devices, armoured and military vehicles.

Civil activities: spare parts for miscellaneous capital equipment, bottling lines, sanitary drainage treatment plants, agricultural and medical equipment.

AOI constituent companies
AOI Aircraft Factory
AOI Electronics Factory
AOI Engine Factory
Arab American Vehicles Co. (AAVCO)
Arab British Dynamics Co. (ABDCO)
Arab British Engine Co. (ABECO)
Helwan Factory for Developed Industry
Kader Factory for Developed Industry
Sakr Factory for Developed Industry

UPDATED

Finland

Government – Ministry/Agency

Finnish Space Committee
c/o TEKES, PO Box 69, FIN00101 Helsinki, Finland
Tel: (+358 10) 521 58 56
Fax: (+358 10) 521 59 01
Web: http://www.tekes.fi/space
Ministry of Trade and Industry: Timo Kekkonen
Secretary: Pauli Stigell

■ An interdepartmental advisory body responsible for planning the national space strategy.

UPDATED

Tekes
Ministry of Trade and Industry, PO Box 69, FIN00100 Helsinki, Finland
Tel: (+358 10) 521 57 37
Fax: (+358 9) 521 59 01
Web: http://www.tekes.fi/space
Director General: Dr Veli-Pekka Saarnivaara
Director, Space Unit and European Commission Relations and Satellite Navigation Manager: Esa Panula-Ontto
ANTARES Space Science Programme Manager: Pauli Stigell
Tel: (+358 10) 521 58 56
AVALI Commercial Space Programme Manager and Satellite Telecommunications Manager: Heikki Hannula
Tel: (+358 10) 521 58 57
Earth Observation Manager: Dr Antti Herlevi
European Commission Relations and Satellite Navigation Manager: Esa Panula-Ontto
Technology Transfer Manager: Hakan Sandell
Tel: (+358 10) 521 58 54
US Office (San Jose, California): Kimmo Ahola

■ Governmental agency responsible for planning, promotion, financing and co-ordination of Finnish space activities.

UPDATED

Centres of Learning & Research

Finnish Geodetic Institute
(Geodeettinen Laitos)
Ministry of Agriculture and Forestry, Geodeettinen 2, PO Box 15, FIN02431 Masala, Finland
Tel: (+358 9) 295 55 50
Fax: (+358 9) 29 55 52 00
Web: http://www.fgi.fi
Director General: Prof. Risto Kuittinen
Remote Sensing Director: Prof. Juha Hyyppä

■ Research in geodesy, photogrammetry, remote sensing, cartography and space geodesy. Activities include satellite tracking, GPS network operation, development of digital satellite image interpretation methods and GIS.

UPDATED

Finnish Meteorological Institute (FMI)
(Geophysical Research Division)
15A Vuorikatu, PO Box 503, FIN00101 Helsinki, Finland
Tel: (+358 9) 192 91
Fax: (+358 9) 19 29 46 03
Web: http://www.fmi.fi
Financial Director: Mrs Taimi Siltala
Director: Prof Risto Pellinen
Public Relations Director: Mrs Katrina Palmroth
Head of Planetry Research Group: Walter Schmidt
Tel: (+358 9) 19 29 46 58
Research: Prof Tuija Pulkkinen

■ Scientific and technical research in space-physics, geomagnetic, climatological and air quality. Relevant computer software and instrumentation.

UPDATED

Association

Finnish Astronautical Society
(Sallskapet for Astronautisk Forskning)
(Suomen Avaruustutkimusseura)
PO Box 507, FIN00101 Helsinki, 10, Finland
Tel: (+358 10) 521 58 56
Fax: (+358 10) 521 59 10
Web: http://netlander.fmi.fi/nsats
Chairman: Paul Stigell
Vice Chairman: Bernt Hoffrén
Tel: (+358 19) 34 04 51
Publications: Avaruusluotain (a quarterly journal, in Finnish) relating to interests in space.

UPDATED

Industry – Manufacturing

Metorex Security Products Oy
Nihtisillankuja 5, PO Box 174, FIN02631 Espoo, Finland
Tel: (+358 9) 32 94 15 00
Fax: (+358 9) 32 94 13 02
e-mail: security@metor.net
Web: http://www.metor.net
President: Petri Ikonen
Sales Manager: Jarmo Saari
Tel: (+358 9) 32 94 12 26
Marketing Manager: Ms Nina Hälli
Tel: (+358 9) 32 94 12 05
■ Metal detector gates: Metor 200 multi-zone gate, Metor 200 HD high-discrimination multizone gate, Metor 200 HS high sensitivity multizone gate, Metor 200 WP weather-proof multi-zone gate, Metornet PC-based network for remote monitoring and adjusting multiple Metor gates, Metor 28 handheld metal detector, Metor 160 high discrimination walk-through metal detector.

UPDATED

Space Systems Finland Ltd (SSF)
Kappelitie 6, FIN02200 Espoo, Finland
Tel: (+358 9) 61 32 86 00
Fax: (+358 9) 61 32 86 99
Web: http://www.ssf.fi
Managing Director: Dr Seppo Korpela
Tel: (+358 9) 61 32 86 11
Financial Director: Jari Parviainen
Tel: (+358 9) 61 32 86 14
Space Applications Business Unit: Jyri Heilimo
Tel: (+358 9) 61 32 86 41
New Ventures Business Unit: Fredrick von Schoultz
Tel: (+358 9) 61 32 86 22
■ On-board flight software, system design, software and integration for embedded systems, small satellite technology, software simulations, EO instrument and optical system analysis and consultancy.

UPDATED

Vaisala Oyj
(Head Office)
PO Box 26, FIN00421 Helsinki, Finland
Tel: (+358 9) 894 91
Fax: (+358 9) 89 49 22 27
Web: http://www.vaisala.com
■ Development and manufacture of electronic measurement systems and equipment for meteorology and the environmental sciences, traffic safety and industry. Products include: radiosondes and their ground equipment, dropsonde systems, surface weather observation instruments and systems, along with transmitters and instruments for the measurement of relative humidity, dewpoint, barometric pressure and carbon dioxide.

UPDATED

Ylinen Electronics Ltd
Teollisuustie 9 A, FIN02700 Kauniainen, Finland
Tel: (+358 9) 57 12 17 32
Fax: (+358 9) 505 55 47
Web: http://www.ylinen.fi
Managing Director: Pekka Jukkala
Sales Director: Nicholas Hughes
■ Provision of engineering services including requirements definition, prototype design and verification, design, development and manufacturing. Provision of microwave products from 1GHz to 120 GHz and from single waveguide components to complete systems. Current activities include provision of 70 GHz radiometric receivers for the Planck/Herschel spacecraft for low noise temperature (-30K) and wide bandwidth (14GHz). Provision of L-band sympathetic aperture radiometers and calibration subsystem fo the SMOS earth observation spacecraft. Provision of transmit/receive assemblies for the TerraSAR – X synthetic aperture radar spacecraft.

UPDATED

Industry – Service

Patria Aviation Oy
(ex-Patria Finavitec Oy)
(ex-Valmet Aviation Industries Inc.)
Lentokonetehtaantie 3, FIN35600 Halli, Finland
Tel: (+358 20) 46 91
Fax: (+358 20) 469 33 85
(+358 20) 332 14
(+358 20) 469 36 00
Web: http://www.patria.fi
■ Overhaul, modification and repair of airframes and engines, final assembly. Development and support of aircraft systems. Overhaul of fast diesel engines. Structures for aircraft and spacecraft. Military systems integration.

UPDATED

Patria New Technologies Oy
(Systems Division)
(ex-Patria Finavitec Oy)
Naulakatu 3, FIN33100 Tampere, Finland
Tel: (+358 20) 46 91
Fax: (+358 20) 469 26 90
e-mail: new.technologies@patria.fi
Web: http://www.patria.fi
■ Design and manufacture of surveillance equipment, space electronics and test systems.

UPDATED

France

Government – Ministry/Agency

Centre de Météorologie Spatiale (CMS)
BP 147, F-22302 Lannion, Cedex, France
Tel: (+33 2) 96 05 67 00
Fax: (+33 2) 96 05 67 37
Web: http://www.meteo.fr
Chef du CMS Lannion: J Lafeuille
■ This centre belongs to the French Weather Service. Operational reception and processing of meteorological satellite data. Archiving and distribution. Research activities.

UPDATED

Indian Space Research Organisation (ISRO)
(Technical Liaison Unit)
Embassy of India, 8 rue Halery, F-75009 Paris, France
Tel: (+33 1) 42 66 33 62
Fax: (+33 1) 42 66 32 72
(+33 1) 40 50 09 96
e-mail: itlu@wanadoo.fr
Web: http://www.isyo.org
Minister and Scientific Secretary: V Sundararamaiah
Chairman: Dr K Kasturirangan
Director, Foreign Relations: Jacob Ninan
Liaison Officer: G Narayanan
Public Information Officer: S Krishnamurthy

UPDATED

Japan Aerospace Exploration Agency (JAXA)
(Paris Office)
(ex-National Space Development Agency of Japan)
3 Avenue Hoche, F-75008 Paris, France
Tel: (+33 1) 46 22 49 83
Fax: (+33 1) 46 22 49 32
Director: Jun Gomi
Deputy Director: Hiroyuki Iwamoto

UPDATED

Government – Intergovernmental

European Space Agency (ESA)
(Headquarters)
(Agence Spatiale Européenne)
(Europaische Weltraumorganisation)
8-10 rue Mario Nikis, F-75738 Paris, Cedex 15, France
Tel: (+33 1) 53 69 76 54
(+33 1) 53 69 71 55
Fax: (+33 1) 53 69 75 60
(+33 1) 53 69 75 61
(+33 1) 53 69 75 62
(+33 1) 53 69 76 90
Web: http://www.esa.int
http://www.esoc.esa.de
http://www.esrin.esa.it
http://www.estec.esa.nl
■ Space research and application. Provides and promotes, for exclusively peaceful purposes, co-operation among European states in space research and technology. Also promotes their space applications for scientific purposes and for operational space applications systems.
Member states:
Austria
Belgium
Denmark
France
Germany
Ireland
Italy
Netherlands
Norway
Portugal
Spain
Sweden
Switzerland
UK

UPDATED

EUTELSAT SA
(European Telecommunications Satellite Organisation)
(Organisation Europeenne de Telecommunications par Satellite)
70 rue Balard, F-75502 Paris, Cedex 15, France
Tel: (+33 1) 53 98 47 47
Web: http://www.eutelsat.com
■ Intergovernmental organisation enabling TV and radio broadcasters, internet service providers, telecommunications companies and corporations to transmit and receive content in Europe, Africa, Asia and the Americas.

UPDATED

NASA European Representative
c/o American Embassy, 2 Avenue Gabriel, F-75382 Paris, Cedex 8, France
Tel: (+33 1) 43 12 21 00
Fax: (+33 1) 42 65 87 68
NASA European Representative: Dr J Donald Miller

UPDATED

Centres of Learning & Research

École Nationale Supérieure de l'Aéronautique et de l'Espace
(Supaero)
10 Avenue Édouard Belin, PO Box 4032, F-31055 Toulouse, Cedex 4, France
Tel: (+33 5) 62 17 80 80
Fax: (+33 5) 62 17 83 30
Web: http://www.supaero.fr
Director: Pierre Bascary
Scientific Director: Olivier Prats
Public Relations Manager: Olivier Renne
■ Higher education establishment open to students of all nationalities in the fields of aeronautics and space.

UPDATED

International Space University (ISU)
Parc d'Innovation, 1 Rue Jean–Dominique Cassini, F-67400 Illkirch, France
Tel: (+33 3) 88 65 54 30
Fax: (+33 3) 88 65 54 47
e-mail: info@isu.isunet.edu
Web: http://www.isunet.edu
President: Dr Karl Doetsch
Director, Financial: Jean-Marc Seiler
Director, Marketing: John Egan
Head, External Relations: Ms Caroline Vignard
Tel: (+33 3) 88 65 54 45

UPDATED

Medes-Imps
(Institut de Médecine et de Physiologie Spatiales)
1 avenue Jean Poulhès – BP 4404, F-31405 Toulouse, Cedex 4, France
Tel: (+33 5) 62 17 49 50
Fax: (+33 5) 62 17 49 51
e-mail: contact@medes.fr
Web: http://www.medes.fr
■ Research into medicine and physiology in space. Specialise in life sciences and biomedical research.

UPDATED

Thales Technology & Research France
(ex-Thales Laboratoire Central de Recherches)
(ex-Thomson-CSF Laboratoire Central de Recherches)
Domaine de Corbeville, F-91404 Orsay, Cedex, France
Tel: (+33 1) 60 33 00 00
Fax: (+33 1) 60 33 08 78
Managing Director: Laurie Reinhart
■ Advanced research into new technologies, material science, optics, liquid crystal displays and rapid semiconductor components.

UPDATED

Association

Committee on Space Research (COSPAR)
(Comité pour La Recherche Spatiale)
51 Boulevard de Montmorency, F-75016 Paris, France
Tel: (+33 1) 45 25 06 79
Fax: (+33 1) 40 50 98 27
e-mail: cospar@cosparhq.org
Web: http://www.cosparhq.org
President: R-M Bonnet
Executive Director: Dr I Révah
Associate Director: A Janofsky
■ Promote internationally scientific research through the organisation of scientific assemblies, publications and other activities. Members include 42 National Scientific Institutions engaged in space research, and 12 International Scientific Unions adhering to the International Council for Science (ICSU). Acts as a forum with contributions from most countries engaged in space research, for the presentation of the lastest scientific results, for the exchange of knowledge and also for the discussion of space research problems. Acts as a scientific committee advising, as required, the UN and other intergovernmental organisations on space research matters or on the assessment of scientific issues in which space can play a role. Acts as a panel for the preparation of scientific and technical standards related to space research. Acts as an entity promoting, on an international level, research in space, much of which has grown into large international collaborative programs in the mainstream of scientific research. Strives to promote the use of space science for the benefit of mankind and for its adoption by developing countries and new space-faring nations. Objectives include: promoting on an international level, scientific research in space, with emphasis on the exchange of results, information and opinions, and to provide a forum, open to all scientists, for the discussion of problems that may affect scientific space research.
Publications: Advances in Space Research, 24 times a year
COSPAR Information Bulletin, 3 times a year
Directory of Organization & Associates, biennial
COSPAR Colloquia Series, irregular

UPDATED

European Association of Remote Sensing Laboratories (EARSEL)
(Association Européenne de Laboratoires de Télédétection)
2 Avenue Rapp, F-75340 Paris, Cedex 7, France
Tel: (+33 1) 45 56 73 60
Fax: (+33 1) 45 56 73 61
e-mail: earsel@meteo.fr
Web: http://www.earsel.org
Chairman, Switzerland: Prof Dr Eberhard Parlow
Vice Chairman, Belgium: Prof Dr Rudi Goosens
Tel: (+32 9) 264 47 09
Secretary General, Germany: Dr R Reuter
Treasurer, Spain: Prof Dr José-Luis Casanova
Executive Secretary: Ms M Godefroy
■ Co-ordination of European research in earth observations, from aircraft and satellites through special interest groups. Remote sensing from satellites and aircraft. Aerial photography.
Publications: EARSeL Newsletter (quarterly on the web)
Proceedings GA (annual)
Proceedings Symposium (annual)
Workshop Reports

UPDATED

French Aeronautics and Space Industry Group (GIFAS)
(Groupement des Industries Françaises Aéronautiques et Spatiales)
4 rue Galilée, F-75782 Paris, Cedex 16, France
Tel: (+33 1) 44 43 17 00
Fax: (+33 1) 40 70 91 41
e-mail: com@gifas.asso.fr
Web: http://www.gifas.asso.fr
President, EADS: Philippe Camus
First Vice President, Dassault Aviation: Charles Edelstenne
Vice President, Messier-Bugatti: Yves Le Clere
Vice President, Thales: Bernard Retat
Vice President, SNECMA: Dominique Paris
General Commissioner for Paris Air Show: Louis Le Portz
General Delegate: Guy Rupied
General Secretary for Equipment Group: Olivier Gorge
Director of Operations: Ms Anne-Marie Roy
Export Director: Renaud Favier
Director of Airworthiness: Stéphane Flori
Director of Space Division: Mrs Marie José Le Breton
Director of Communications: Patrick Guerin
Director of Training and Social Welfare: Claude Bresson
Director of Strategy and European Affairs: Mrs Anne-Sophie de la Bigne
Director of Economics and Statistics: Jacques Moureu
Director of Administration, Legal and Business Affairs: Laurence Thuret-Demoulin
Treasurer, Micro Mecanique Pyreneenne: Jean-Marie Virepinte
Export Matters, Equipment Group: Alain Audier
Technical Matters, Equipment Group: Jacques Michel
Communications Department, Equipment Group: Mrs Mireille Soubrier
Accountancy: Mrs Elisabeth Lorin
■ Co-ordinates the industrial and economic activities of its 200 members engaged in the development, manufacture, maintenance and sales of aircraft, helicopters, engines, missiles, space vehicles and their associated components, equipment, accessories, products and operating facilities. Develops and defends interests of the profession in dealings with the Government. Co-ordinates the industrial activity and economic action of its members and organises lectures and exhibitions to promote the image of the French aerospace industry. Organisers of the Paris-Le-Bourget International Air Show.
Publications: GIFAS Annual Report
GIFAS Directory
GIFAS Info News Letter

UPDATED

Industry – Manufacturing

Arianespace SA
(Head Office)
Boulevard de l'Europe, PO Box 177, F-91006 Evry, Courcouronnes Cedex, France
Tel: (+33 1) 60 87 60 00
Fax: (+33 1) 60 87 62 47
Web: http://www.arianespace.com
■ Manufacture, sales and launch operations of satellites.
Subsidiaries and offices:
Arianespace Inc., US
Arianespace, Japan
Arianespace, Singapore

UPDATED

Celerg
(Tactical Propulsion)
Immeuble Jura, Centre d'Affaires La Boursidiers, F-92357 Le Plessis-Robinson, Cedex, France
Tel: (+33 1) 41 07 82 21
Fax: (+33 1) 46 30 22 37
Web: http://www.noxelgroup.com
Chief Executive Officer: David Quancard
Operations Director: Jean Marie Laurent
Sales and Marketing Director: Francis Rodriguez
General Secretary: Ms Sophie-Anne Magdeleine
■ Design and production of solid propellant rocket motors, air-breathing and advanced propulsion concepts.

UPDATED

Dassault Aviation
(Head Office)
78 Quai Marcel Dassault, F-92552 Saint Cloud, Cédex 300, France
Tel: (+33 1) 47 11 86 90
Fax: (+33 1) 47 11 87 40
Web: http://www.dassault-aviation.fr
Honorary Chairman: Serge Dassault
Chairman and Chief Executive Officer: Charles Edelstenne
Vice Chairman: Bruno Revellin-Falcoz
Senior Vice President, International: Bruno Gotté
Senior Vice President, Operations: Christian Decaix
Vice President, Institutional Relations of Corporate Communications: Gérard David
Tel: (+33 1) 47 95 86 90
Information Officer: Luc Berger
■ Scientific research and studies. Manufacture of civil and military aircraft and flight control systems. Maintenance and repair of aeronautical material and space systems.
Corporate structure:
Defence Division:
Dassault Aero Service, France (100%)
European Aerosystems Ltd, UK (50%)
Secbat, France (36%)
Eurotradia International, France (16%)
Sofema, France (6.7%)
Sofresa, France (6%)
Falcon Division:
Dassault Falcon Jet, US (100%)
Dassault Falcon Jet Wilmington (100%)
Aero Precision Inc., US (50%)
Dassault International Inc., US (100%)
Dassault Procurement Services, US (100%)
Dassault Falcon Service, France (100%)
Falcon Training Center, France (50%)
Other activities:
Dassault International, France (100%)
Dassault Assurances Courtage, France (100%)
Dassault Réassurance, France (100%)
Sogitec Industries, France (99%)
Corse Composites Aeronautiques, France (33%)
Embraer, Brazil (5.7%)

UPDATED

Dassault Aviation
(Saint Cloud Factory)
78 Quai Marcel Dassault, F-92552 Saint Cloud, Cedex 300, France
Tel: (+33 1) 47 11 40 00
Fax: (+33 1) 47 11 49 01
Director: Yves Petit

UPDATED

EADS Astrium
(Headquarters)
37 Avenue Louis Bréguet, BP 1, F-78146 Vélizy-Villacoublay, Cedex, France
Tel: (+33 1) 34 88 30 00
Fax: (+33 1) 34 88 43 43
Web: http://www.astrium.eads.net
■ Prime contractor of both space and ground segments and a major supplier of products and technologies; conducts business in science and earth observation, telecommunications and launchers and orbital infrastructure.
Programs and products: Communications Satellites Eutelsat 2, Inmarsat 2, Telecom 2, Hispasat, Orion, Hot Bird, Skynet 4, Nilesat, ST-1, Worldstar, Astra 2B, Intelsat K-TV, EAST, WEST. Communication ground stations and satellite control centres. Earth observation satellites: SPOT 1, 2, 3, 4, 5, ERS 1, 2, HELIOS 1 and 2, METEOSAT, MSG, ENVISAT, METOP. Scientific programs: Foc, Giotto, Hipparcos, Soho, Cluster, Rosetta, XMM, Integral. Launchers and orbital infrastructure: vehicle equipment bay for Ariane, Spacelab, Columbus Orbital Facility (COF), Automatic Transfer Vehicle (ATV).

UPDATED

EADS Astrium
(Toulouse Facility)
31 Avenue des Cosmonautes, F-31402 Toulouse, France
Tel: (+33 5) 62 19 62 19
Fax: (+33 5) 62 54 57 10
■ Development and production of scientific and earth observation satellites, communication satellites, Ariane vehicle equipment bays, optical space instruments, avionics systems flight and ground software.

UPDATED

EADS Defence & Communication Systems
(Headquarters)
6 rue Dewoitine, F-78143 Vélizy-Villacoublay, Cedex, France
Tel: (+33 1) 34 63 70 00
Fax: (+33 1) 34 63 70 70
Web: http://www.eads.com

UPDATED

EADS France
(Headquarters)
(ex-Aerospatiale)
(ex-Aerospatiale-Matra)
37 Boulevard de Montmorency, F-75781 Paris, Cedex 16, France
Tel: (+33 1) 42 24 24 24
Fax: (+33 1) 42 24 21 32
Web: http://www.eads.net

UPDATED

EADS France
(Missiles Division, Châtillon Plant)
2 rue Beranger, PO Box 84, F-92323 Châtillon, Cedex, France
Tel: (+33 1) 47 46 21 21
Fax: (+33 1) 47 46 27 77
Director, Information and Communication Directorate: Patrick Mercillon
Press Relations: Mrs Sophie Thuillier
Tel: (+33 1) 47 46 30 04
Fax: (+33 1) 47 46 33 19
■ Design and production of: strategic weapon systems (ASMP); anti-tank weapons systems (Milan, Hot, Eryx, Tigat LP, Trigat MP); air defence systems (Roland, Aster, future program in co-operation); anti-ship weapon systems (Exocet, AS15TT and MM15TT, and MM40 block 2); air-to-surface missiles (AS30 Laser); drones (C22 and Piver CL289); batteries, rocket motors for all tactical weapons and cruise missiles.

UPDATED

EADS France SAS
(Space Division, Paris)
37 Boulevard de Montmorency, F-75781 Paris, Cedex, France
Tel: (+33 1) 42 24 25 44
Fax: (+33 1) 42 24 25 79
Vice President, Communication and External Relations: Patrice de Lanversin

UPDATED

EADS Space Transportation
(ex-Aerospatiale-Matra Lanceurs)
(ex-EADS Launch Vehicles)
66 route de Verneuil, BP 3002, F-78133 Les Mureaux, Cedex, France
Tel: (+33 1) 39 06 36 01
Fax: (+33 1) 39 06 12 54
e-mail: dir-com@space.eads.net
Web: http://www.space.eads.net
President and Chief Executive Officer: Philippe Couillard
Executive Vice President, Technical Affairs and Programmes: Jacques Jacob de Cordemoy
Vice President, Finance and Control: Jean-Michel Pivin
Vice President, Quality, Process, Improvement and Information: J Jacob de Cordemoy
Vice President, Defence Programmes and Product Business Development: Daniel Buclet
Vice President, Systems, Design and Tests: Serge Petit
Vice President, Research: Gérard Laruelle
Vice President, Industrial Operations: Michel Freuchet
Vice President, Communication and External Relations: Patrice de Lanversin
Vice President, Sales, Marketing and Purchasing: Joël Le Breton
Vice President, Space Programmes: Bernard Humbert
Vice President, Human Resources: Michel Sesques
Manager, Media Relations: Ms Sophie Belaud
Tel: (+33 5) 56 57 37 16
(+33 1) 39 06 14 50
Fax: (+33 5) 56 57 30 19
Military Advisor: P Puy Moutbruu
Executive Assistant to the President: Jurgen Ackermann
Communications and External Relations Directorate: Ms Angelika Teissiere
Contact: Ms Sophie de La Noue
Press Manager: Christian Erles
■ Design, development and manufacture of civil and military space launch systems, space transport systems, re-entry vehicles and spaceborne platforms. Major programmes include: Ariane 4 and Ariane 5. Automated transfer vehicles (ATV) for International Space Station and Soyuz. Prime contractor for strategic missiles M45 and M51
Industrial sites at:
Aquitane, near Bordeaux, France
Les Mureaux, near Paris, France
Guiana Space Centre, Kourou, French Guiana
Branch units in France at:
Brest
Cherbourg
Biscarosse
Toulon

UPDATED

Enertec SA
(ex-Schlumberger Industries DAR Division)
185 Avenue du Général de Gaulle, BP 316, F-92143 Clamart, Cedex, France
Tel: (+33 1) 41 28 87 87
Fax: (+33 1) 41 28 87 00
e-mail: sales@enertecgroup.com
Web: http://www.enertecgroup.com
Chairman: Jean-Paul Vernhes
Chief Operating Officer: Jean Loup Rosseau
Executive Vice President, Marketing and Sales: Bernard Espannet
Business Development Director: Jean François Sulzer
Sales and Export Director: Paul Hartland
Sales Support Manager: Damien Folléa
Tel: (+33 1) 41 28 87 71
Space Market Segments Manager: Jean-Pierre Dauvergne
Tel: (+33 1) 41 28 87 55
Fax: (+33 1) 41 28 87 00
■ Design and manufacture of high-end data and video acquisition, storage and processing systems for aerospace, defence, security and related markets.

UPDATED

ESDT
Centre d'Affaires La Découverte, Bât, Fastnet, PO Box 652, F-56106 Lorient, Cedex, France
Tel: (+33 2) 97 87 20 20
Fax: (+33 2) 97 87 20 21
e-mail: esdt@esdt.com
Web: http://www.esdt.com
■ Manufacture of products for individual protection, night vision, EOD/IEDD, detection, arming and ammunitions, long range observation, telecommunication and roping and extraction.

UPDATED

Framatome Connectors France
145 rue Yves le Coz, F-78035 Versailles, France
■ Manufacture of aircraft components.

UPDATED

Geosys
20 Impasse René Couzinet, BP 5815, F-31505 Toulouse, Cedex 5, France
Tel: (+33 5) 62 47 80 80
Fax: (+33 5) 62 47 80 70
e-mail: info@geosys.fr

UPDATED

Goodrich Actuation Systems
(France Headquarters)
(ex-TRW Aeronautical Systems, Lucas Aerospace)
106, rue Fourny, F-78530, Buc, France
Tel: (+33 1) 39 20 52 00
Fax: (+33 1) 39 20 52 99
e-mail: marketing.buc@goodrich.com
Web: http://www.goodrich.com
Vice President, Military and Civil: Sylvain Prioult
Tel: (+33 1) 39 20 52 01
Business Development Director, Defence Europe: Robert Turpin
Tel: (+33 1) 39 20 52 11
■ Providing a full range of actuation systems for airborne and land based platforms. Products include primary and secondary flight controls; helicopter main and tail rotor, engine and nacelle, utility, precision weapon and land vehicle actuators.
Provision of complete systems including hydraulic, fly-by-wire and power-by-wire controls using technologies such as hydraulic, rotary geared, electro-hydrostatic and electro-mechanical actuators.

UPDATED

Goodrich Actuation Systems
(Cergy-Pontoise)
(ex-Lucas Aerospace)
(ex-Lucas Aerospace FCS Paris)
13 Avenue de l'Eguillette, Saint Ouen 1'Aumone, BP 7186, F-95056 Cergy-Pontoise, Cedex, France
Tel: (+33 1) 34 32 63 00
Fax: (+33 1) 34 32 63 10
Web: http://www.goodrich.com

UPDATED

Jehier
(Jehier Ingénierie Informatique)
Route de Saint Lézin, BP 29, F-49120 Chemillé, France
Tel: (+33 2) 41 64 54 00
Fax: (+33 2) 41 64 54 01
e-mail: info@jehier.fr
Web: http://www.jehier.fr
President: Félix Jehier
Marketing Director: Jacques Maillard
Communication Officer: T Renault
■ Thermal and acoustic insulation for performance applications in aerospace, marine, military vehicles and industry. Manufacture of materials and equipment with metallic or non-metallic cover.
Representing: Inspec, US

UPDATED

Le Cinq Chemins
(Bordeaux – Le Haillan Plant)
Les Cinq Chemins, F-33187 Le Haillan, Cedex, France
Tel: (+33 5) 56 55 30 00
Fax: (+33 5) 56 55 30 01
Telex: 560678
Managing Director: Michel Laroche
Administration Director: Jean-François Gargou
Tel: (+33 5) 56 55 31 76
Operations Director: Jean-Philippe Olier

UPDATED

MAPAERO – Aerospace Coatings
10 Avenue de la Rijole, F-09100 Pamiers, France
Tel: (+33 5) 34 01 34 01
Fax: (+33 5) 61 60 23 30
e-mail: sales.mapaero@mapaero.com
Web: http://www.mapaero.com
Managing Director: Jean-François Brachotte
Deputy Director: Joël Loiseau
Sales and Marketing Director: Patrick Clairefond
Export Sales Support: Ms Maryline Barbe
Personal Assistant and Purchasing: Mrs Eleanor Poujol
■ Development, manufacture and sale of paints for aircraft, space launchers and satellites. Products include aircraft exterior paint; water based primers for aluminium structures; fire retardant water based paints for cabin interiors and special paints for high temperature, anti-skid, anti-static and anti-erosion applications.
Accreditations: ISO 9001

UPDATED

Mecachrome SA
Avenue Eugéne Casella, PO Box 9, F-18700 Aubigney-Sur-Nère, France
Tel: (+33 2) 48 81 22 00
Fax: (+33 2) 48 58 20 84
e-mail: mecachrome@groupe-mecachrome.com
Managing Director: Gérard Casella
Financial Director: M le Pallec
Purchasing Manager: Michel Piolet
Sales Manager: Alain Etienne
■ Mechanical machining and assembly of structural parts and main assemblies; engine parts and components, landing gear.

UPDATED

Novintec
ZA De la Pillardiere, F-45600 Sully-sur-Loire, France
Tel: (+33 2) 38 29 57 10
Fax: (+33 2) 38 29 57 11
e-mail: contact@novintec.com
Web: http://www.novintec.com/
■ Design and manufacture of filtration products and sub-assembly components for aerospace, defence, space, nuclear industries. Manifold filters and accessories, differential pressure indicators for filters and accumulators, visual and electrical reservoir gauges, blocking valves and accumulators, magnetic plugs, check valves, relief valves, drain valves, sampling valves and fuel injectors.

UPDATED

PyroAlliance
(ex-Pyromeca)
(ex-Pyrospace)
139 Route de Verneuil, BP 2052, F-78132 Les Mureaux, Cedex, France

Tel: (+33 1) 34 92 44 44
(+33 4) 94 22 86 86
Fax: (+33 1) 34 74 30 93
(+33 4) 94 22 86 99
e-mail: pyroalliance@snpe.com
Web: http://www.pyroalliance.com
Chairman and Chief Executive Officer: Bernard Zeller
Chief Operating Officer: Jean-Yves Canihac
Financial Director: Philippe Van Bockstaele
Technical Director: Pierre Guy Amand
Sales Engineer: Christian Fritsch
Tel: (+33 1) 34 92 44 22
Sales Engineer, Space: Bernard Lallemant
Tel: (+33 1) 34 92 44 50
Purchasing Manager: Rémi Foudrinier
■ Design, development, manufacture and sale of pyrotechnic devices and components for space, aeronautical, civil, military and commercial programs. Manufacture of safety devices such as pyrotechnically opened and closed valves for use in oil, gas and chemical installations. Manufacture of crew escape pyrotechnical systems and parts. Manufacture of pyrotechnics for satellites and separation of payloads from the launchers. Pyrotechnical chains for separation and destruction missions on launchers.

UPDATED

Pyroalliance
(Toulon Plant)
Quartier Lagoubran, Chemin Charles Battezzatti, BP 2148, F-83063 Toulon, Cedex, France
Tel: (+33 4) 94 22 86 86
Fax: (+33 4) 94 22 86 99
■ System studies and ordnance pyrotechnics for spacecraft, launch pad equipment and civil programmes. Manufacture of pyrotechnical components.

UPDATED

Radiall

101 rue Philibert Hoffmann, F-93116 Rosny sous Bois, France
Tel: (+33 1) 49 35 35 35
Fax: (+33 1) 49 35 35 14
Web: http://www.radiall.com
■ Coaxial connectors, rack and panel connectors, optical connectors and passive microwave devices.

UPDATED

Saft

(International Headquarters)
12 rue Sadi Carnot, F-93170 Bagnolet, France
Tel: (+33 1) 49 93 19 18
Fax: (+33 1) 49 93 19 50
Web: http://www.saftbatteries.com
Chairman and Managing Director, Specialty Battery Group (SBG): John Searle
Managing Director, Industrial Battery Group (IBG): Bertrand Olivesi
Administration and Finance Director: Alain Duwaerts
Human Resources Director: François-Michel Sissung
Purchasing Director: Jean-Claude Bourget
Communications Director: Ms Jill Ledger
Tel: (+33 1) 49 93 17 77
Research and Development Director: Khushrow Press
Primary Lithium Division Director (SBG): Tom Alcide
Space and Defence Division Director (SBG): Tristan Grivel
Rechargeable Batteries Systems Director (IBG): Franck Cecchi
Marketing and Sales Director (IBG): Fred-Eric Hapiak
Aviation Marketing and Sales Manager (IBG): Bernard Weber
Marketing Manager, Stationary and Railway (IBG): Michael Lippert
■ Design and manufacture of high-technology specialist battery solutions for industries including aeronautics, rail, aerospace, defence, automotive, energy, emergency and security installations and mobility and communications.
Subsidiary companies:
Saft AB, Sweden
Saft AS, Norway
Saft America Inc., US
Saft Australia Pty Ltd, Australia
Saft Batterien GmbH, German
Saft Ferak AS, Czech Republic
Saft Ltd, UK
Saft Nife M.E. Ltd, Middle East
Saft Singapore Pte Ltd, Singapore

UPDATED

Sagem SA

(Head Office)
Le Ponant de Paris, 27 rue Leblanc, F-75512 Paris, Cedex 15, France
Tel: (+33 1) 40 70 63 63
Fax: (+33 1) 40 70 66 40
(+33 1) 40 70 68 68
(+33 1) 40 70 66 00
Web: http://www.sagem.com
Chairman of the Supervisory Board: Mario Colaiacovo
Chairman of the Executive Board: Gregoire Olivier
Senior Vice President and Chief Financial Officer: Hervé Philippe
Vice President and Managing Director, Defence and Security Branch: Jacques Paccard
Director, Personnel and Social Affairs: Mrs Dominique Castera
Director, Marketing and Communication, Defence and Security: Jean-Charles Pignot
Tel: (+33 1) 40 70 63 54
Marketing and Communications, Defence and Security: Françoise Hugonet
Tel: (+33 1) 40 70 67 33
Purchasing Manager: Dominique Dupuy
■ SAGEM is composed of two main activities:
The communications activity composed of the Mobile Phone Division and the Internet and Networks Division.
The defence and security activity is composed of
-Navigation and Aeronautics Systems Division
-Optronics and Aeronautical Systems Division
-Security Division

UPDATED

Sagem SA
(Defence and Security)
Le Ponant de Paris, 27 Rue Leblanc, F-75512 Paris, Cedex 15, France
Tel: (+33 1) 40 70 63 63
(+33 1) 40 70 62 68
(+33 1) 53 23 79 00
(+33 1) 40 70 63 54
Fax: (+33 1) 40 70 66 00
(+33 1) 40 70 66 40
(+33 1) 40 70 66 78
(+33 1) 40 70 68 68
(+33 1) 40 70 64 54
Telex: 205255
e-mail: francoise.hugonet@sagem.com
Web: http://www.sagem.com
Senior Vice President, Chief Financial Officer: Hervé Philippe
Managing Director, Defence and Security: Jacques Paccard
Chargé de Missions, Marketing (Defence and Security Branch): Jean-Charles Pignot
Marketing and Promotion: Ms Francoise Hugonet
Tel: (+33 1) 40 70 67 33
■ Guidance and navigation, military avionics, systems and modernisation. Electro-optics.
Defence and Security Sectors
Navigation and Aeronautic Systems
-Avionics
-Military Avionics
-Navigation and Sensors
Optronics and AirLand Systems
-Optical Technology and Simulation
-Sights
-AirLand Warfare
Security Systems
-Government Solutions
-Secure terminals and Transaction

UPDATED

SME GROUP SNPE

(Head Office)
(ex-SNPE Explosives and Propellants)
12 Quai Henri IV, F-75004 Paris, France
Tel: (+33 1) 48 04 66 66
(+33 1) 49 96 74 00
Fax: (+33 1) 48 04 68 87
(+33 1) 49 96 74 01
e-mail: dircom@snpe.com
Web: http://www.explosives-propellants.com
http://www.sme-propulsion.com
http://www.energetic-materials.com
http://www.materialspyrotecs.com
President and Chief Executive Officer: Jacques Loppion
Tel: (+33 1) 48 04 65 84
Executive Vice President, Energetic Materials: Jacques Cardin
Tel: (+33 1) 64 99 10 04
Explosives and Propellants: Jean-Claude Bossy
Tel: (+33 1) 49 96 74 30
Advertising Manager: Régine Rougé-Bouyssou
Tel: (+33 1) 48 04 68 45
Contact: Alain Freche
Tel: (+33 1) 49 96 74 42
Contact: Ms Valerie Simon
Tel: (+33 1) 49 96 74 60
■ Research and development in the field of energetic materials (new molecules, manufacturing processes, solid propellant motor and ramjet operation). Solid propellants for rockets, missiles and launch vehicles propulsion, as well as gas generators (air bags and others). Liquid propellants and energetic materials for spacecraft (MMH, ammonium perchlorate). Pyromechanisms for industry, defence and space sectors. Countermeasure equipment for aircraft and helicopters. Oxygen generation systems. Application of energetic materials for safety devices. Expertise in modelling risk, in the methods used for the complete reliability, availability, maintainability and safety (RAMS), analysis of processes and industrial facilities.
Pyroalliance
Roxel
Structil

UPDATED

SME Propulsion

(Saint Médard Plant)
F-33160, Saint-Médard-en-Jalles, France
Tel: (+33 5) 56 70 50 50
Fax: (+33 5) 56 70 50 51
Web: http://www.sme-propulsion.com
Manager: Jean-Claude Labourroire
■ Solid propellants; military powders and lightweight materials.

UPDATED

Snecma Motors

2 Boulevard du Général Martial-Valin, F-75724 Paris, Cedex 15, France
Tel: (+33 1) 40 60 80 80
Fax: (+33 1) 40 60 81 02
Web: http://www.snecma-moteurs.com
Vice President, Human Resources: Jean-Luc Bérard
Vice President and General Manager, Space Engines: Joel Barre
Vice President and General Manager, Military Engines: Jean-Luc Engerand
Vice President, Engineering: Pierre Thouraud
Vice President, Risk Manager: Jacques Nouailles
Chief Transformation Officer: Jean Piquet
Communications Manager: Sylvie Beaumonte

UPDATED

Snecma Moteurs
(Evry-Corbeil Plant)
Route Nationale 7, PO Box 81, F-91003 Evry, Cedex, France
Tel: (+33 1) 69 87 92 60
Fax: (+33 1) 69 87 89 28
■ Production plant, quality control, purchasing, sales services, military engine support, technical training centre.

UPDATED

Snecma Moteurs
(Gennevilliers Plant)
291 Avenue d'Argenteuil, PO Box 48, F-92234 Gennevilliers, Cedex, France
Tel: (+33 1) 47 60 72 06
Fax: (+33 1) 47 60 73 00
■ Forge, foundry, and machining of mechanical parts.

UPDATED

Snecma Moteurs
(Le Creusot Plant)
Avenue de l'Europe, PO Box 97, F-71203 Le Creusot, Cedex, France
Tel: (+33 3) 85 77 77 12
Fax: (+33 3) 85 77 77 77
Managing Director: Xavier Genthial
■ Automated production of turbine discs.

UPDATED

Snecma Moteurs
(Rockets and Flight Test Department)
Chemin des Bellons, PO Box 644, F-13804 Istres, Cedex, France
Tel: (+33 4) 42 35 90 00
Fax: (+33 4) 42 35 90 68
Managing Director: Eric Nicolet
Tel: (+33 4) 42 47 19 09
■ Engine testing.

UPDATED

Snecma Moteurs
(Space Engine Division – Paris)
2 Boulevard du General Maulial-Valin, F-75724 Paris, Cedex-15, France
Tel: (+33 1) 40 60 80 80
Fax: (+33 2) 40 60 81 02
Managing Director: Joel Barre
■ Development and manufacture of large liquid propulsion for launchers. Undertakes testing of propulsion systems. Also undertakes the prime contractorship in the development and manufacture of large solid propellant rocket motors for ballistic missiles.

UPDATED

Snecma Moteurs
(Space Engine Division)
(ex-Société Européenne de Propulsion)
Large Liquid Propulsion, BP 802, F-27208 Vernon, Cedex, France
Tel: (+33 2) 32 21 72 00
Fax: (+33 2) 32 21 27 01
Web: http://www.snecma.com
http://www.snecma-moteurs.com
Vice President and General Manager: Guy Corteel
Public Relations Attaché: Michel Muszynski

UPDATED

Snecma Moteurs
(Villaroche Nord Plant)
Direction Propulsion et Équipements de Satellites, F-77550 Moissy Cramayel, France
Tel: (+33 1) 64 71 46 97
Fax: (+33 1) 64 09 33 38
General Manager: Bernard Cantaloube
Tel: (+33 1) 64 71 46 01
■ Manufacture of satellite thrusters, propulsion systems and equipment. Solar array drive mechanisms, microgravity furnaces, surface tension propellant tanks, cryogenic engine testing.

UPDATED

Snecma Moteurs
(Villaroche Plant)
Rond Point René Ravaud-Reau, F-77550 Moissy Cramayel, Cedex, France
Tel: (+33 1) 60 59 71 23
Fax: (+33 1) 60 59 71 36
■ Design, development, assembly and test for aircraft engines. Spare parts warehouse. Civil engine advertising and sales.

UPDATED

Snecma SA
(Société Nationale d'Étude et de Construction de Moteurs d'Aviation)
2 Boulevard du Général Martial-Valin, F-75724 Paris, Cedex 15, France
Tel: (+33 1) 40 60 80 18
(+33 1) 40 60 80 80
Fax: (+33 1) 40 60 80 09
(+33 1) 40 60 81 02
Web: http://www.snecma.com
Chairman and Chief Executive Officer: Jean Paul Bechat
Senior Executive Vice President, Propulsion: Dominique Paris
Senior Executive Vice President, Strategy and Development and Equipment: Yves Imbert
Senior Vice President, International Development: Philippe Humbert
Senior Vice President, Management: Philippe Bry
Senior Vice President, Progress Initiative Group: Christian Mari
Senior Vice President, Industrial Relations: Alain Coté
Executive Vice President, Group Industrial Strategy: Dominique Hédon
Executive Vice President, Finance: Alain Marcheteau
General Counsel and Secretary: Alain Bosser
Vice President, Public Relations and Communications: Mrs Fransoise Descheemaker
Tel: (+33 1) 40 60 81 09
Press Officer: Vincent Chappard
■ Manufacture turbojet engines and components for commercial and military aircraft, rocket engines for satellite launchers and missiles.

UPDATED

Société Anonyme de Télécommunication (SAT)
41 rue Cantagrel, F-75631 Paris, Cedex 13, France
Tel: (+33 1) 45 82 31 11

UPDATED

Société Anonyme d'Études et Réalisations Nucleaires (SODERN)
(EADS Sodern)
20 Avenue Descartes, F-94451 Limeil Brévannes, Cedex, France
Tel: (+33 1) 45 95 70 00
Fax: (+33 1) 45 69 14 02
Telex: 265543 sodern
e-mail: com@sodern.fr
Web: http://www.sodern.fr
Chairman and Managing Director: Franck Poirrier
Director of Space and Optics Activities: Philippe Lugherini
Tel: (+33 1) 45 95 71 47
Director of Nuclear Activities: Bernard Aubert
Tel: (+33 1) 45 95 70 02
Director of Optics Activities: Claude Babolat
Tel: (+33 1) 45 95 70 80
Public Relations and Communication Manager: Jean-Luc Guillaume
Tel: (+33 1) 45 95 71 13
■ Space: satellites and spacecraft attitude measurement, star sensing, earth observation cameras. Nucleonics: neutron emission and detection, detection of explosives, chemical materials and narcotics by neutron activation. Bulk material analysis. Optics: optical systems for space instrumentation (IR,UV), flight simulation, nuclear industry.

UPDATED

Thales
(Corporate Headquarters)
(ex-Thomson-CSF)
45 rue de Villiers, F-92526 Neuilly sur Seine, Cedex, France
Tel: (+33 1) 57 77 80 00
(+33 1) 57 77 83 00
e-mail: press@thalesgroup.com
info@thalesgroup.com
dtri@thalesgroup.com
Web: http://www.thalesgroup.com
Chairman and Chief Executive Officer: Denis Ranque
Vice Chairman: Bernard Rétat
Executive Vice President, International Operations and Chief Executive Officer Thales UK: Alex Dorrian
Executive Vice President, Sales and Marketing and Chief Executive Officer Thales International: Jean-Paul Perrier
Senior Vice President, Human Resources and Social Affairs: Yves Barou
Senior Vice President, Finance: Ross McInnes
Senior Vice President, Strategy, Research and Technology: Jean-Loup Picard
Senior Vice President, Operations: Reynald Seznec
Senior Vice President, Aerospace: François Quentin
Senior Vice President, Air Systems and Corporate Secretary (acting): Alexandre de Juniac
Senior Vice President, Naval: Jean-Georges Malcor
Senior Vice President, Services: Jean-Paul Lepeytre
Senior Vice President, Land and Joint Systems: Bruno Rambaud
Senior Vice President, Security: Tim Robinson
Senior Vice President, Deputy Managing Director, International Operations: José Massol
Senior Vice President, Deputy Managing Director, Thales International: François Gayet
Vice President, Communications: Ms Sylvie Dumaine
Press Officer, Corporate Communications: Matt Pothecary
Press Assistant: Ms Tracy-Louise Brooks
Tel: (+33 1) 57 77 86 26
Fax: (+33 1) 57 77 87 44
■ Design, development and implementation of high technology electronic solutions for global communication and information requirements. Major business areas include airborne systems (surveillance and combat systems, missile electronics); avionic systems (military and commercial aircraft, aircraft components, power generation); air security and missile systems (missile systems, air command and control systems, air traffic management, armament and propulsion); naval systems (surface and underwater systems); communications systems, military communications, professional applications); optronics; information systems and services (simulation, training and synthetic training and environment, TV and radio broadcasting, information processing systems, computer services engineering and security); and tubes and components (electron tubes, microelectronics).
Undertakes design, development, manufacture and the provision of in-service support.

International companies:
Africa:
African Defence Systems (Pty) Ltd (ADS) (South Africa)
Asia:
Samsung Thales (joint venture) (South Korea)
Australia:
ADI Limited (JV)
Europe:
Thales Electronic Systems (Greece)
Middle East:
NTS-National Telesystems and Services (UAE)
STESA-Saudi Technical Engineering Systems Associated RL (joint venture)
North America:
Thales Systems Canada, Inc.
Thales Business Units:
Thales Aerospace
-Thales Airborne Systems
-Thales Sensors, UK
-Thales Avionics Systems
-Computers
Air Systems
-Air Defence Systems
-ThalesRaytheonSystems France
-Samsung Thales Co. Ltd, Korea, South
-Thales Sensors, UK
-Air Traffic Management
Land & Joint Systems
-Land & Joint Systems
-Broadcast Solutions
Naval
-Thales Naval
-Warship Prime
-Systems
-Equipment
-Support & Services
-Thales Underwater Systems
Security
-Banking Solutions
-Enterprise Security
-Components
-Homeland Security
-Navigation
-Transport & Energy
Services
-Consulting
-IT Outsourcing
-Systems Integration
-Training & Simulation

UPDATED

Thales Airborne Systems
(Headquarters)
(Thales Systemes Aeroportes)
(ex-Thomson-CSF Detexis)
Centre Charles Nungesser, F-78851 Elancourt, Cedex, France
Tel: (+33 1) 34 81 60 00
Fax: (+33 1) 30 66 79 66
e-mail: customer.service-tasf@fr.thalesgroup.com
info@fr.thalesgroup.com
Web: http://www.thales-airborne.com
Chairman and Chief Executive Officer: José Massol
Chief Executive Officer: Emmanuel Grave
Director, Communications: Michel Roche
Operations Manager: Jean-Christophe Alessandrini
Press Officer: Ms Monique Navarra
Contact: Ms Anne-Marie Nevo
■ Electronic warfare, radars, missile electronics, information technology systems.

UPDATED

Thales Avionics SA (TAV)
(Head Office)
(ex-Thomson-CSF Sextant)
18 Avenue du Maréchal Juin, BP 49, F-92362 Meudon-la-Forêt, Cedex, France
Tel: (+33 1) 39 45 50 00
(+33 1) 39 45 56 27
(+33 1) 39 45 50 60
Fax: (+33 1) 39 45 50 30
(+33 1) 39 45 58 76
Web: http://www.thales-avionics.com
Chairman and Chief Executive Officer: Francois Quentin
Vice President, Communications: Ms Brigitte Porée
Vice President, Engineering Technology: Hervé Girault
Vice President, Finance and Control: Jean-Luc Chevalier
Vice President, Strategy: Ms Dominique Pons
Vice President, Central Procurement: Daniel Delacour
Vice President, Human Resources: René Maisonneuve

Vice President, Legal Affairs: Henry de Moussac
Vice President, Manufacturing Operations: Norbert Dubost
Operation Quality Development: Christian Bernard
Press Attaché: Ms Dominique Huard
Press Assistant: Ms Patricia Soulié
■ Manufacture of avionics and associated components. Communication surveillance systems and inflight entertainment systems.

UPDATED

Thales Avionics Electrical Motors SA (TAEM)
5 rue du Clos d'en Haut, PO Box 115, F-78702 Conflans-Ste.Honorine, France
Tel: (+33 1) 39 19 60 60
Fax: (+33 1) 39 19 96 31
e-mail: yann.ansquer@thales-avionics-es.com
Chairman: Steve Grinham
General Manager: Patrick Philibert
Technical Manager: Didier Renté
Tel: (+33 1) 39 19 11 48
Commercial Manager: Didier Froger
Purchasing Manager: Michel Lizot
Tel: (+33 1) 39 19 16 64
■ Special motors for aerospace applications, space and defence (AC/DC BLDC and PM-VR). Stepper motors, blowers and fans.

UPDATED

Thales Land & Joint Systems
(Thales Land and Joint Systems)
(ex-Thales Communications)
(ex-Thales, Communications Systems, Business Group)
160 Boulevard de Valmy, BP 82, F-92704 Colombes, Cedex, France
Tel: (+33 1) 41 30 30 00
Fax: (+33 1) 41 30 41 86
(+33 1) 41 30 33 57
e-mail: info@tccthomson.fr
Web: http://www.thales-communications.com
Commercial Director: Ms Dominique Simonneau
Communications Director: Ms Anne Lechevranton
Press Assistant: Ms Sylvie Starzinskas
Press Relations Officer: Aline Borne
Tel: (+33 1) 41 30 21 40
■ Prime contractor specialising in radiocommunications equipment and networks, radio navigation, identification, electronic warfare, command systems and security and microwave communications and optronics equipment.
Companies in the Land & Joint Systems Division
-Land & Joint Systems
-Broadcast Solutions

UPDATED

Thales Communications SA
(Gennevilliers Industrial Centre)
66 rue du Fossé-Blanc, BP 156, F-92231 Gennevilliers, Cedex, France
Tel: (+33 1) 46 13 20 00
Fax: (+33 1) 46 13 21 63
■ Mobile radiocommunications, radio surveillance and COMINT systems, information technology security, space communications systems.

UPDATED

Thales Angénieux SA
(ex-Angénieux SA)
F-42570 Saint Héand, France
Tel: (+33 4) 77 90 78 00
Fax: (+33 4) 77 30 78 03
e-mail: angenieux@fr.thalesgroup.com
Web: http://www.angenieux.com
President: Jean François Pernotte
Financial Director: Philippe Cotte
Chief Executive Officer and General Manager: Denis Levaillant
Director Business Unit, Defence: A Piodi
Technical Director: Joel Rollin
Purchasing Manager: Ms Christine Beal
Contact: Gilles Delic
■ Design and manufacture of satellite optical elements.

UPDATED

Thales Electron Devices SA
(ex-Thomson Tubes Electroniques)
2 bis rue Latecoere, F-78941 Vélizy, France
Tel: (+33 1) 30 70 35 00
Fax: (+33 1) 30 70 35 35
Telex: 634280 thomtub
e-mail: info@thales-electrondevices.com
Web: http://www.thalesgroup.com/electrondevices
Chief Executive Officer: Jacques Belin
Vice President, Marketing and Sales: Paul Maisonnier
Space Technical Director: Georges Fleury
Space and Microwave Marketing Director: Regis Cabanel
Microwave Communication Manager: Gilles Khong
Tel: (+33 1) 30 70 36 40
Fax: (+33 1) 30 70 36 50
■ Systems, equipment and services for aerospace.

UPDATED

Weir Valves and Controls France
(Head Office)
(ex-Groupe Sebim)
Zone Industrielle La Palunette, F-13220 Chateauneuf-les-Martigues, France
Tel: (+33 4) 42 07 00 95
Fax: (+33 4) 42 07 11 77
e-mail: coseb@sebim.fr
Web: http://www.weir-valve.com
Managing Director: Christian Goetz
Technical Director: Gerald Schaumburg
Operations, Sarasin/RSBD: Joel Annodeau
■ Safety valves and rupture discs for Ariane launching pads.

UPDATED

Wier Valves and Controls France
(ex-Sarasin RSBD)
Zone Industrielle du Bois Rigault, F-62880 Vendin-le-Vieil, France
Tel: (+33 3) 21 79 54 50
Fax: (+33 3) 21 28 62 00
e-mail: sarasin@weirvalvefr.com
Web: http://www.weir-valve.com
Managing Director: Christian Goetz
Financial Director: David Letrun
Tel: (+33 3) 21 79 54 55
Technical Director: Philippe Duribreu
Tel: (+33 3) 21 79 54 66
Sales Director: Alain Amand
Tel: (+33 321) 79 54 69
Purchasing Manager: Jean-Pierre Cornubert
Tel: (+33 3) 21 79 54 64
Assistant to the Managing Director and Sales Administrator: Ms Michele Curtis
Tel: (+33 3) 21 79 54 53
■ Safety valves and rupture discs for Ariane launching pads.

UPDATED

Zodiac International
2 rue Maurice Mallet, F-92130, Issy-les-Moulineaux, France
Tel: (+33 1) 41 23 23 23
Fax: (+33 1) 46 48 83 87
Web: http://www.zodiacsolas.com
■ Manufacture of inflatable boats, hurricane semi-rigid craft and SOLAS liferafts. Design and manufacture of inflatable products. Superinsulating materials for satellite thermal control systems. Aircraft interiors.

UPDATED

Industry – Service

Centre National d'Études Spatiales (CNES)
(Headquarters)
2 Place Maurice Quentin, F-75039 Paris, Cedex 1, France
Tel: (+33 1) 44 76 75 00
Fax: (+33 1) 44 76 76 76
Web: http://www.cnes.fr
■ French Space Agency. Analysis of long term issues and future course of space activities, and submission of proposals to the French Government. Major development programmes undertaken nationally as part of European Space Agency activities (Ariane).
Publications: CNES Magazine (in French and in English, 4 per year)
CNES subsidiary and participations:
Arianespace
Cerfacs
CLS
DERSI
GDTA
IFRTP
Intespace
MEDES-IMPS
Medias France
Novespace
OST
Prospace
Renater
Satel Conseil
Scot
Semeccel
Simko
Skybridge
Spot Image
Telespace Participation

UPDATED

Cépaduès Editions
111 rue Nicolas-Vauquelin, F-31100 Toulouse, France
Tel: (+33 5) 61 40 57 36
Fax: (+33 5) 61 41 79 89
e-mail: cepadues@cepadues.com
Web: http://www.cepadues.com
President: Jean-Claude Joly
Publishing Director and Sales Manager: Jean-Pierre Marson
Contact: F Boudaille
■ Publisher specialising in aviation and space (SFACT, CNES, SUP'AERO, INP).

UPDATED

Fleximage
(Headquarters)
113 avenue Aristide Briand, F-94117 Cedex, France
Tel: (+33 1) 49 08 76 00
Fax: (+33 1) 49 08 76 02
e-mail: info@fleximage.fr
Web: http://www.fleximage.fr
Marketing and Communication: Ms Camille Deniau

UPDATED

IN-SNEC
5, Avenue des Andes – ZA de Courtaboeuf, BP 101, F-91943 Les Ulis, Cedex A, France
Tel: (+33 2) 31 29 49 01
(+33 2) 31 29 49 49
Fax: (+33 2) 31 80 65 49
Web: http://www.in-snec.com
Vice President, Sales and Marketing: Jean Casati-Ollier
Sales Manager: Michel Calistri
Marketing Manager: François Varieras
Tel: (+33 2) 31 29 49 82
Product Line Manager: Christophe Foret
Product Line Manager: Frederic Saffre
■ Design, engineering and manufacture of complete ground station (telemetry and tracking system) CE-GPS. Mobile satellite communications terminals.

UPDATED

IN-SNEC
(Les Ulis Office)
5 Avenue des Andes, BP 101, F-91943 Les Ulis, Cedex A, France
Tel: (+33 1) 69 82 78 00
Fax: (+33 1) 69 07 39 50
Marketing and Sales Manager: Jean-Claude Dardelet
■ Vehicle borne TM kits or launchers, missile, aircraft. PCM telemetry pre-processing for the ground counterpart. TT and C base band station.

UPDATED

Intespace
(Ingenerie Test and Environmental Spatial)
18 avenue Edouard Belin, BP 4356, F-31029 Toulouse, Cedex 4, France
Tel: (+33 5) 61 28 11 11
Fax: (+33 5) 61 28 11 12
e-mail: marketing@intespace.fr
Web: http://www.dynaworks.com
Chairman and Chief Executive Officer and Acting Sales and Marketing Director: Jean-Louis Marcé
Tel: (+33 5) 61 28 11 50
Testing Department Manager: Jean-Claude Pasquet
Tel: (+33 5) 61 28 12 20
Dynaworks Product Group Manager: Joseph Merlet
Tel: (+33 5) 61 28 12 57
Assurance Product Manager: Charles Blanchard
Tel: (+33 5) 61 28 11 90
■ Environmental test including mechanical, thermal and electrical, for all sectors of industry. On-site assistance for European customers for operating and maintaining test facilities. Engineering and consulting services in environmental technology. Test centre engineering, principally for the space industry.

UPDATED

ISTAR
2600 Route des Crétes, BP 282, F-06905 Sophia-Antipolis, Cedex, France
Tel: (+33 4) 97 23 23 23
Fax: (+33 4) 93 95 83 29
Web: http://www.istar.com
President: Marc Fontaine
Tel: (+33 497) 23 23 69
Sales Director: Jean-Christian Bourion
Marketing Director: Ms Nathalie Pisot
Financial Director: Bruno Fricheteau
Technical Director: Benoît Fleury
■ Cartographic databases. RS satellite image processing, specialising in digital terrain models.

UPDATED

Météo-France
(ex-Meteorologie Nationale)
1 quai Branly, F-75340 Paris, Cedex 7, France
Tel: (+33 1) 45 56 71 71
Fax: (+33 1) 45 56 70 05
(+33 1) 45 56 71 11
Telex: 202876
Web: http://www.meteo.fr
Président and Diréctéur General: Jean-Pierre Beysson
Tel: (+33 1) 45 56 70 00
Director of Communications: Jean-Pierre Macveigh
Tel: (+33 1) 45 56 71 00
Communications Department: Ms Marianne Lyon-Caen
Tel: (+33 1) 45 56 74 05
International Relations Officer: François Duvernet
Tel: (+33 1) 45 56 70 50
International Relations Officer: Denis Lambergeon
Tel: (+33 1) 45 56 74 94
Contact, External Communications: P Cheminot
■ Meteorological services.

UPDATED

Prospace
34 rue des Bourdonnais, F-75001 Paris, France
Tel: (+33 1) 44 88 99 34
Fax: (+33 1) 44 88 99 39
e-mail: prospace@prospace-fr.com
Web: http://www.prospace-fr.com
President: Stephane Janichewski
Executive Manager: Pierre Bescond
Communication – Exhibitions: Ms Dominique Pécontal
Marketing Conferences: Ms Monique Brossaud
Administration – Finance: Ms Patricia Bertron-Chopin
Executive Assistant: Ms Martine Pringault

UPDATED

Satel Conseil (SC)
34, rue des Bourdonnais, F-75001 Paris, France
Tel: (+33 1) 44 88 99 30
Fax: (+33 1) 44 88 99 39
Web: http://www.satelconseil.com
President: Pierre Bescond
Vice President, Administration: Ms Claude Chatellier
Marketing Director: Ms Monique Brossaud

UPDATED

Sevig Press
Rue Bellart 6, F-75015 Paris, France
Tel: (+33 1) 42 73 28 37
Fax: (+33 1) 42 73 20 95
e-mail: sevig.press@noos.fr
Web: http://www.sevig-press.com
Editor: Dick Shirvanian
Assistant Editor: Anayis Khedjian
Publications: Russian Space Directory – published every two years in co-operation with the Russian Space Agency (RSA) and the European Space Agency (ESA).
European Space Directory – Published annually in co-operation with Eurospace and the technical assistance of the European Space Agency (ESA).

UPDATED

Spot Image
5 rue des Satellites, PO Box 4359, F-31030 Toulouse, Cedex 4, France
Tel: (+33 5) 62 19 40 55
(+33 5) 62 19 40 40
Fax: (+33 5) 62 19 40 11
(+33 5) 62 19 40 55
Web: http://www.spotimage.fr
Chairman and Chief Executive Officer: Jean-Marc Nasr
Vice President, Administration and Finance: Prosper Isierte
Vice President, Sales and Marketing: Manuel Lagarde
Vice President, Exploitation and Production: Philippe Delclaux
Vice President, Applications and Development: Antoine de Chassy
Vice President, Systems and Technical Services: Laurent Maggiori
Chief Operating Officer: Jean-Michel Darroy
General Manager: Philippe Munier
Marketing Manager: Laurence Bolopion
Tel: (+33 5) 62 19 42 33
Corporate Communications: Ms Anne-Marie Bernard
Tel: (+33 5) 62 19 40 10
Fax: (+33 5) 62 19 40 54
Customer Department: Ms Jean Bobo
■ Commercial operator of the Spot earth observation satellites. Also offers distribution of worldwide optical and radar satellite data and services.
Representing
Beijing Spot Image, China
Spot Asia, Singapore
Spot Image Corp., USA
Spot Imaging Services, Australia
Tokyo Spot Image, Japan

UPDATED

Starsem
5-7 rue Francois Truffaut, F-91042 Evry, Cedex, France
Tel: (+33 1) 69 87 01 10
Fax: (+33 1) 60 78 31 99
e-mail: communication@starsem.com
Web: http://www.starsem.com
Contact: Andrei Egochkine
■ Marketing, sales and management of commercial launch services using Soyuz launchers; supervision of Soyuz commercial launcher production and development of new European and Russian joint space programs.

UPDATED

Sun Microsystems Inc.
(Defence Team)
13 Avenue Morane Saulnier, BP 53, F-78142 Vélizy, Cedex, France
Tel: (+33 1) 34 03 00 00
Fax: (+33 1) 34 03 00 01
OEM Sales Manager and Defence Team Leader: Jean-Marc Aubry de Maraumont
Tel: (+33 1) 34 03 05 50
Fax: (+33 1) 34 03 05 79
■ Work closely with military organisations and equipment providers to implement the latest technology across all areas of tri-sector military and defence applications.

UPDATED

The Toulouse Space Centre
18 Avenue Edouard-Belin, F-31 401 Toulouse, Cedex 4, France
Tel: (+33 5) 61 27 31 31
Fax: (+33 5) 61 27 31 79
Web: http://www.cnes.fr

UPDATED

Industry – Distributor

Acal SA
(Fontenay sous Bois Office)
(ex-Technitron)
1 avenue Louison Bobet, BP 64, F-94122 Fontenay sous Bois, Cedex, France
Tel: (+33 1) 45 14 73 00
Fax: (+33 1) 48 77 62 30
e-mail: is@acal.fr
Web: http://www.acal.fr
Managing Director: Olivier Mabille
Finance Director: Gérard Fosse
Instrumentation Manager: Gérard Renault
■ Distribution of avionics components and instrumentation dedicated to aerospace, space and transport industries.
Representing:
Ballard, USA
Behlman, USA
KMT GmbH, Germany
Metrum-Datatape, USA
Polytronics, USA
Sycos, USA
Trompeter Electronics, USA

UPDATED

Industry – Sales Agent/Office

Emerson & Cuming Microwave Products Sarl
9 Rue du Colonel Moraine, F-92360 Meudon, France
Tel: (+33 1) 46 01 58 20
Fax: (+33 1) 46 01 58 25
e-mail: contact@emerson-cuming-microwave.fr
Web: http://www.eccosorb.com
Managing Director: A J Geurts
Sales and Marketing Director: J P Heitzmann
Technical Director: B Timmermans
Financial Director: K Geypen
Human Resources and Communications Manager: K De Witte
■ Microwave absorbers, dielectric and shielding materials.

UPDATED

French Guiana

Industry – Service

Arianespace SA
(Kourou Facility)
BP 809, F-97388 Kourou, French Guiana
Tel: (+594) 33 67 07
Fax: (+594) 33 69 13

UPDATED

Centre National d'Études Spatiales (CNES)
(Centre Spatial Guyanais)
PO Box 726, F-97387 Kourou, Cedex, French Guiana
Tel: (+594) 33 51 11
Fax: (+594) 33 47 19
Web: http://www.csg-spatial.tm.fr/english
Head of Press and Information: Ms Marie-Pierre Joseph-Alberton
■ A launch range and test facility of CNES (the French Space Agency) made available to the European Space Agency (ESA) and its member states for the purpose of their programmes under the terms of an agreement with the French Government. Incorporates Ariane launching pads for putting satellites into orbit and all relevant electronic and optical telemetry monitoring equipment.

UPDATED

Germany

Government – Ministry/Agency

Japan Aerospace Exploration Industry (JAXA)
(Bonn Office)
(ex-National Space Development Agency of Japan)
H 1204 Bonn Centre, 2–10 Bundeskanzlerplatz, D-53113 Bonn, Germany
Tel: (+49 228) 91 43 50
Fax: (+49 228) 911 21 50

UPDATED

Government – Intergovernmental

European Organisation for the Exploitation of Meteorological Satellites (EUMETSAT)
Am Kavalleriesand 31, D-64295 Darmstadt, Germany
Tel: (+49 6151) 80 77
Fax: (+49 6151) 80 75 55
Web: http://www.eumetsat.de
Head of Conferences and Publication: Ms Claudia Ritsert-Clark

Communications Manager: Ms Livia Briese
Administrative Assistant: Ms Anita Brumori
Administrative Assistant: Ms Tanja Mollitor
■ European intergovernmental meteorological satellite organisation, representing the national interests of 18 member states and 6 co-operating states. Primary objective is to establish, maintain and exploit European systems of operational meteorological satellites.

UPDATED

European Space Agency (ESA)
(European Space Operations Center)
Robert-Bosch-Strasse 5, D-6100 Darmstadt, Germany
Tel: (+49 6151) 88 61
Fax: (+49 6151) 88 66 11
Web: http://www.esoc.esa.int
Head of EO Applications Department: Stephen Briggs
Head of ESRIN Public and Institutional Relations Office: Ms Simonetta Cheli
■ Satellite operation, reception and processing of transmitted satellite data. Direct links to ground stations across the world.

UPDATED

Defence Force – Air Force

German Aerospace Center
(Deutsches Zentrum für Luft- und Raumfahrt)
DLR – Lampoldshausen, D-74239 Hardthausen, Germany
Tel: (+49 6298) 280
Fax: (+49 6298) 281 12
Web: http://www.dlr.de
University Professor: Prof. Dr.-Ing. Wolfgang Koschel
Tel: (+49 6298) 282 03
Public Relations: Ms Heidelinde Fehrer
Tel: (+49 6298) 282 01
■ European research and testing centre for rocket propulsion systems.

UPDATED

Centres of Learning & Research

German Aerospace Center
(Executive Headquarters)
(Deutsches Zentrum für Luft- und Raumfahrt eV)
Linder Höhe, D-51147 Köln, Germany
Tel: (+49 2203) 60 10
Fax: (+49 2203) 673 10
Telex: 8810-0 dlr
Web: http://www.dlr.de
Chairman: Prof.-Dr Walter Kröll
Vice-Chairman: Dr Bernd Hofer

UPDATED

German Aerospace Center
(Press Office and Public Relations)
(Deutsches Zentrum für Luft- und Raumfahrt)
D-51170 Köln, Germany
Tel: (+49 2203) 60 10
Fax: (+49 2203) 673 10
e-mail: pressestelle@dlr.de
Web: http://www.dlr.de

UPDATED

German Aerospace Center
(Köln-Porz Office)
(Deutsches Zentrum für Luft und Raumfahrt eV)
Linder Höhe, D-51147 Köln-Porz, Germany
Tel: (+49 2203) 60 10
Fax: (+49 2203) 673 10
e-mail: pressestelle@dlr.de
Web: http://www.dlr.de
Chairman: Prof. Sigmar Wittig
Vice Chairman: Prof. Bernd Höfer
Head of Press and Public Relations: Bernhard Fuhrmann
Tel: (+49 2203) 601 21 17
Editor, Transport: Cordula Tegen
Tel: (+49 2203) 601 38 76
Editor, Space Flight: Peter Zarth
Tel: (+49 2203) 601 32 85
Editor, Space Flight: Eduard Muller
Tel: (+49 2203) 601 28 05
Editor, AV-Media, DLR-Nachrichten: Sabine Hoffmann
Tel: (+49 2203) 601 21 33
Editor, Space Missions: Vanadis Weber
Tel: (+49 2203) 601 30 68
Editor, WWW: Marco Trovatello
Tel: (+49 2203) 601 27 26
Editor, Aeronautics, Energy Technology, Echtzeit: Hans-Leo Richter
Tel: (+49 2203) 601 24 25
Assistant Editor, Space Flight: Ms Astrid Foadi
Tel: (+49 2203) 601 21 64
Event Co-ordination: Rudiger Clausen
Tel: (+49 2203) 601 24 74
Events, WWW: Rolf Jansen
Tel: (+49 2203) 601 29 04
Contact: Marlies Stuttgen
Tel: (+49 2203) 601 22 86

UPDATED

Rheinisch-Westfalische Technische Hochschule Aachen (RWTH)
Templergraben 55, D-52062 Aachen, Germany
Tel: (+49 241) 809 43 22
Fax: (+49 241) 809 23 24
e-mail: pressestelle@zhv.rwth-aachen.de
Web: http://www.rwth-aachen.de
Faculty of Mechanical Engineering: Prof. Dr.-Ing. Aloys Krieg
Faculty of Electrical Engineering: Prof. Dr.-Ing. Rolf H Jansen
Public Relations Office: Toni Wimmer
Tel: (+49 241) 80943 23
Editor: Christof Zuerath

UPDATED

Sternwarte Bochum – Planetarium
Castroper Strasse 67, D-44777 Bochum, Germany
Tel: (+49 234) 51 60 60
Fax: (+49 234) 516 06 51
e-mail: planetarium@bochum.de
Web: http://www.bochum.de
Director: Prof.-Dr Johannes Viktor Feitzinger
■ Public observatory and planetarium to promote space sciences and astronomy.

UPDATED

Universität zu Köln
(Institute of Air and Space Law)
Albertus-Magnus-Platz, D-50923 Köln, Germany
Tel: (+49 221) 470 23 37
Fax: (+49 221) 470 49 68
e-mail: sekretariat-hobe@uni-koeln.de
Web: http://www.uni-koeln.de/jur-fak/instluft
University Professor: Prof Stephan Hobe

UPDATED

Association

Bundesverband der Deutschen Luft- und Raumfahrt-Industrie eV (BDLI)
(German Aerospace Industries Association)
Friedrichstrasse 152, D-10117 Berlin, Germany
Tel: (+49 30) 206 14 00
Fax: (+49 30) 20 61 40 90
e-mail: info@bdli.de
Web: http://www.bdli.de
■ Association representing 120 companies organised into four manufacturer groups: aerospace systems, engines, equipment, and materials technology and components. Organisation of Berlin-Brandenburg International Aerospace Exhibition.
Accreditations: A member of the European Association of Aerospace Industries (AECMA) and the Federation of German Industry (BDI)
Ordinary members:
Aero Tech Peissenberg GmbH
Aerodata-Flugmesstechnik GmbH
Airsys Navigation Systems GmbH
AOA Apparatebau Gauting GmbH
ASG Luftfahrtechnik und Sensorik GmbH
Astrium GmbH
Autoflug GmbH & Co.
Base Ten Systems Electronics GmbH
Bayern-Chemie GmbH
Becker Flugfunkwerk GmbH
Behr Industrietechnik GmbH & Co.
Bernd Stephan Stahlhandel GmbH
Bodenseewerk Geraetetechnik GmbH
Böhler Uddeholm Deutschland GmbH
Bosch SatCom GmbH
CAE Electronics GmbH
Cargolifter AG
Carl Zeiss
Corus Aluminium Walzproduct GmbH
Deutsche Titan GmbH
Diehl Avionik Systeme GmbH
Diehl Luftfahrt Elektronik GmbH
Diehl Munitionssysteme GmbH & Co. KG
Doncasters Precision Castings – Bochum
Dornier GmbH
Dornier Flugzeugwerft GmbH
Draeger Aerospace GmbH
EADS Airbus GmbH
EADS Deutschland GmbH
EATON Fluid Power GmbH
E.I.S. Electronics GmbH
Elan GmbH
Elektro-Metall Export GmbH
Elettronica GmbH
Entrak GmbH & Co. KG
ESG Elektroniksystem- und Logistic GmbH
ESW – Extel Systems Wedel Gesellschaft für Ausrüstung
Eurocopter Deutschland GmbH
Europa Fasteners GmbH
EVAC GmbH Kanalisationssysteme
Fairchild Dornier GmbH
Gebr. Hoever GmbH & Co.
GKN Aerospace GmbH
Goodrich Aerospace Europe GmbH
Goodrich Rosemount Aerospace GmbH
Grässlin Präzisionstechnik GmbH
Hella Aerospace GmbH
Holmburg GmbH & Co. KG
Honeywell Aerospace GmbH
Honeywell Regelsysteme GmbH
Honsel AG
IABG Industrieanlagen-Betriebsgesellschaft mbH
Jena-Optronik GmbH
Jenoptik Laser, Optik, Systeme GmbH
Jeppesen GmbH
Kayser-Threde GmbH
KID-Systeme GmbH
Klaus Stegmann GmbH & Co. KG
Krupp VDM GmbH
Labinal Aero & Defence Systems GmbH
Leistritz Turbomaschinen Technik GmbH
Liebherr-Aerospace Lindenberg GmbH
LITEF GmbH
Lockheed Martin GmbH
MAN Technologie AG
Mannesmann Rexroth AG
MST Aerospace GmbH
MTU Aero Engines
Nord-Micro AG & Co. OHG
OHB – Orbital- und Hydrotechnologie Bremen System GmbH
Otto Fuchs Metallwerke
Pace Aerospace Engineering and Information Technology GmbH
Panavia Aircraft GmbH
Parker Hannifin GmbH
Pfalz-Flugzeugwerke GmbH
Pierburg Luftfahrtgerate Union GmbH
Praezisionsmechanik Gilchinp GmbH
Recaro Aircraft Seating GmbH & Co. KG
Rhode & Schwarz GmbH & Co. KG
Roeder Praezision GmbH
Rolls-Royce Deutschland GmbH
SFIM Industries Deutschland GmbH
Sitec Aerospace GmbH
STN Atlas Elektronik GmbH
TDW Gesellschaft für verteidigungstechnische Wirksystems mbH
Telair International GmbH
Teldix GmbH
Test Fuchs GmbH
Thales Electron Devices GmbH
Thyssen Umformtechnik Turbinen Komponenten GmbH
Tital –Titan-Aluminium Feinguss GmbH
Turbomeca GmbH
VCS Nachrichtentechnik GmbH
VIDAIR Avionics Aktiengesellschaft
Von Hoerner and Sulger GmbH
Wittenstein Motion Control GmbH
W.L. Gore & Associates GmbH
ZARM Technik GmbH
ZF Luftfahrttechnik GmbH
Z11 Imaging GmbH
Sponsoring members:
Aeronaval Ingenieurtechnik GmbH & Co. KG
AirTruck GmbH
Allianz Versicherungs-AG
Arianespace
BAE Systems Deutschland GmbH
Bayerische Landesbank Girozentrale
Bayerische Vereinsbank AG
BTG Messe - Spedition GmbH
Deutsche Bank AG
DLR Deutsches Zentrum für Luft- und Raumfahrt
Dresdner Bank AG
Gebrueder Krose
Edgar Hausmann GmbH
IBCOL Technical Services GmbH
Intospace GmbH

Ing. Horst Kegler GmbH
Lufthansa Consulting GmbH
Schenker-Eurocargo AG
Vereinigte Motor-Verlage GmbH & Co. KG
Walter Kostelezky GmbH & Co KG

UPDATED

Industry – Manufacturing

Access eV

(Materials and Processes)
Intzestrasse 5, D-52072 Aachen, Germany
Tel: (+49 241) 809 80 00
Fax: (+49 241) 385 78
e-mail: welcome@access.nwth-aachen.de
Web: http://www.access.nwth-aachen.de
Chairman: Andreas Bührig-Polaczck
Managing Director: Robert Guntlin
Project Engineer: Ruediger Tiefers
Chairman: Peter R Sahm

■ Development and construction of microgravity facilities. Casting processes, process simulation, multicomponent materials, microstructural modelling and metallic solidification.

Accreditations: ISO 9000ff

UPDATED

Aerodata AG

(ex-Aerodate Systems GmbH)
Herman Blenk Strasse 34-36, D-38108 Braunschweig, Germany
Tel: (+49 531) 235 90
Fax: (+49 531) 235 91 58
e-mail: mail@aerodata.de
Web: http://www.aerodata.de
President: Dr Hartmut Knigge
Director of Sales: Peter Kriening
Director of Sales: Bodo Gerlach
Director of Sales: Hubert Seher
Director for Strategy and Development: Manfred Haverland
Director of Competence Center: Friedhelm Schwarz
Director of Competence Center: George Edward Boag
Managing Director: Wolfgang Jöres
Director of Competence Center: Hans-Georg Niemöller
Director of Competence Center: Ms Mareile Langhorst
Contact: David Gondy

■ Development of applications in real-time data processing for aviation and land applications in the field of high-precision location and navigation. High quality solutions for the flight inspection of radio navigation aids, satellite navigation, airborne maritime patrol, atmospheric measurement technology, flight testing and comparable special systems are among these applications. The activities of Aerodata Systems are organised into five areas of business: flight inspection systems, modification of special mission aircraft, avionics systems, satellite navigation systems, integration of avionics systems into aircraft, research-oriented measurement technology.

UPDATED

Anite Systems GmbH

Robert-Bosch Straße 7 D-64293 Darmstadt, Germany
Tel: (+49 6151) 872 51 00
Fax: (+49 6151) 872 51 51
Web: http://www.anitesystems.de
Managing Director: Robert Rouse

UPDATED

Bayern-Chemie Gesellschaft für Flugchemische Antriebe mbH

(Bayern-Chemie GmbH)
PO Box 1131, D-84544 Aschau, am Inn, Germany
Tel: (+49 8638) 60 10
Fax: (+49 8638) 60 13 99
e-mail: info@bc.eads.de
Web: http://www.bayernchemie-protac.com
Managing Director: Dr Joachim Geisel
Tel: (+49 8638) 60 11 50
JV Marketing Activities: O Tréhu
Tel: (+33 2) 38 51 65 18
Technical Director: Dipl Ing H L Besser
Tel: (+49 8638) 60 12 70
Financial Director: D Fillard
Programmes Director: U Stoerchle
Key Account Manager: Dipl Ing Klaus Wölki
Tel: (+49 8638) 60 15 44
Key Account Manager: A Grelier

■ Development and manufacture of propulsions with solid propellant in steel or lightweight structures for: artillery rockets, guided missiles, multiple pulse motors, hypersonic missiles, side thrusters, mini rockets for model aeronautics and rockets for rescue systems. Programmes: Alarm, Crotale, DWS, Dar, Kormoran, Kriss, Mica, Magic, Otomat, Pars, Patriot, Martel, Super 530, Shahine. Throttleable and non-throttleable air breathing propulsions (ram rocket motors) with high energy propellant (boron). Applications on medium and long range missiles. Programmes for A3M, Aramis and Meteor. Components and propellants: Helical wound motor structures and flow turned technology. Non-asbestos thermal insulations. Solid propellant grains of extruded double base and cast composite propellants. Gas generators and pyrotechnics: gas generators with cool burning propellant for applications including ejection of sub-munitions, power supply for guided missiles, starter cartridges for turbines, inflation, deflation and fire extinguishing. Pyrotechnical devices and ignition mixtures for applications including valves, actuators, infra-red radiators and igniters. Chemical and physical analysis, environmental testing and simulation, x-ray of large objects, component testing and quality management. Consulting and sale of machinery for propellants. Environmental testing, laboratory analysis and modifications of production machines. Design and adaptation of special equipment. Research and development on hypervelocity propulsion (Mach 5), multiple pulse motors, ram rockets, throttleable and non-throttleable. Nozzleless booster (propellant technology).

UPDATED

Carl Zeiss

Carl Zeiss Strasse 22, D-73447 Oberkochen, Germany
Tel: (+49 7364) 200
Fax: (+49 7364) 20 68 08
(+49 7364) 68 08
e-mail: info@zeiss.de
Web: http://www.zeiss.de

■ Technological solutions and products in the areas of: semiconductor and optoelectronic technology, life sciences and health care, eye care, and industrial solutions.

UPDATED

Chemetall GmbH

(Aerospace Technologies)
(ex-Polymer Systems)
Trakehnerstrasse 3, PO Box 60441, D-60441 Frankfurt am Main, Germany
Tel: (+49 69) 716 50
Fax: (+49 69) 71 65 30 18
e-mail: www.aerospace.chemetall.com
Web: http://www.chemetall.com
Head of Department: Bruno Proefrock
Group Marketing Manager: Adrian Whyle
Tel: (+49 697) 165 32 13
Mobile: (+) 07740 662 727
Public Relations Manager: Guido Kruppa
Tel: (+49 697) 16 50

■ Aircraft sealants, surface treatments, adhesives.

UPDATED

Diehl & Eagle-Picher GmbH (D&EP)

(Batterie – Systeme)
Fischbachstrasse 20, D-90552 Röthenbach, Niemcy, Germany
Tel: (+49 911) 957 21 00
Fax: (+49 911) 957 30 70
e-mail: diehl-ep@battery.de
Web: http://www.battery.de
Managing Director: Peter F Weidner
Tel: (+49 911) 957 24 29
Sales Manager: Thomas Kromer
Tel: (+49 911) 957 24 29

■ Produces thermal batteries, lithium thionylchloride primary cells, nickel hydrogen secondary cells, silver zinc secondary cells, silver zinc batteries, lithium thionylchloride remotely activated batteries, battery-packs in all chemistries.

UPDATED

Drägerwerk AG & Co

Moislinger Allee 53-55, PO Box 1339, D-23542 Lübeck, Germany
Tel: (+49 451) 88 20
Fax: (+49 451) 882 20 80
e-mail: info@draeger.com
Web: http://www.draeger.com
Chairman: Dr Dieter Feddersen
President and Chief Executive Officer: Dipl Theo Dräger
Financial Officer: Hans-Oskar Sulzer
Press and Public Relations Manager: Dr Welf Böttcher
Public Relations Assistant: Ms Claudia Rohn

■ Development, production and product support of breathing equipment, gas protective equipment, compressed air systems, gas and particle filter systems, gas detecting and monitoring systems and medical rescue equipment. Aviation oxygen systems, portable oxygen units, air supply systems, liquid oxygen system and solid oxygen unit.

UPDATED

EADS Astrium

(German Headquarters)
D-81663 München, Germany
Tel: (+49 89) 60 70
Fax: (+49 89) 60 72 64 81
Web: http://www.astrium.eads.net

■ Production and development of LEO communications satellite platforms, navigation systems, avionics and electronics; solar arrays and antennas and provision of telecommunications operations and services.

UPDATED

EADS Astrium
(Friedrichshafen Facility)
(ex-Dornier Satellitensysteme GmbH)
An der Bundestrasse 31, D-88039 Friedrichshafen, Germany
Tel: (+49 75) 458 00
Fax: (+49 75) 45 44 11

■ Production and development of Earth observation and science satellites, microwave instruments, thermal components and power electronics.

UPDATED

EADS Defence & Security Systems Division

(Headquarters)
(ex-Daimlerbenz Aerospace AG)
(ex-DaimlerChrysler Aerospace AG)
(ex-Deutsche Aerospace AG)
(ex-MBB)
Willy Messerschmittstrasse, Postfach 801109, D-81663 München, Germany
Tel: (+49 89) 60 70
Fax: (+49 89) 60 72 64 81
Web: http://www.eads.net
Vice President, Corporate Media Relations: Dr-jur Rainer Ohler
Tel: (+49 89) 60 73 42 35
Mobile: (+49) 17 13 72 31 97
Fax: (+49 89) 60 73 42 39
Director, Corporate Communications: Christian Poppe
Tel: (+49 89) 60 73 42 50
Fax: (+49 89) 60 73 42 84
PR for Defence: Bernd Stuerzl

UPDATED

ESG Elektroniksystem und Logistik GmbH

Einsteinstrasse 174, D-81675 München, Germany
Tel: (+49 89) 92 16 27 45
(+49 89) 921 60
(+49 89) 92 16 26 16
Fax: (+49 89) 92 16 26 31
(+49 89) 92 16 27 36
Telex: 522594
e-mail: info@esg.de
Web: http://www.esg.de
Head of Corporate Communications: Jorg Riedle

■ Development of complex electronic and information systems, command and control systems. Cost-reducing logistics procedures.

UPDATED

Eurockot Launch Services GmbH

(ex-Rockot)
Hünefeldstraße 1-5 Postfach 10 59 46, Bremen, Germany
Tel: (+49 421) 539 65 01
(+49 421) 539 65 10
Fax: (+49 421) 539 65 00
Chief Executive Officer: Dr Matthias Oehm

Sales Director: Peter Freeborn
Tel: (+49 421) 539 65 08
Technical Director: Dr Mark Kinnersley
Tel: (+49 421) 539 65 08
■ Launch service provider.

UPDATED

Friemann & Wolf Batterietechnik GmbH (FRIWO)

(ex-Silberkraft Leichtakkumulatoren GmbH)
Meidericherstrasse 6-8, PO Box 100552, D-47005 Duisburg, Germany
Tel: (+49 203) 300 20
Fax: (+49 203) 30 02 40
e-mail: friemann-wolf@friwo-batterien.de
Web: http://www.friwo-batterien.de
Managing Director and Technical Director: J Cornelissen
Financial Director: H Plank
Sales Manager: H Kickel
■ Manufacture of batteries such as traditional nickel-cadium storage, lithium, silver-zinc, special batteries and battery systems with a heavy-duty capability.

UPDATED

Jena-Optronik GmbH (DJO)

Prüssingstrasse 41, D-07745 Jena, Germany
Tel: (+49 3641) 20 01 10
Fax: (+49 3641) 20 01 59
(+49 3641) 20 02 22
e-mail: info@jena-optronik.de
Web: http://www.jena-optronik.de
President and Chief Executive Officer: Dietmar Ratzsch
Tel: (+49 3641) 20 01 11
Director, Commerce and Finances: Dr Elke Schmiedeknecht
Tel: (+49 3641) 20 01 09
Director, Marketing and Programmes: Dr Reiner Strobel
Tel: (+49 3641) 20 01 31
Director, Engineering and Projects: Dr Frank Döngi
Tel: (+49 3641) 20 01 06
Public Relations: Sven Knuth
Tel: (+49 3641) 20 01 59
■ Development and production of components and systems for aerospace and security. Development, manufacture, assembly and test of opto-electronic sensors and instruments; optical, electrical and mechanical ground support equipment (hardware and software); software for satellite on-board data handling and instrument data processing.

UPDATED

Kayser-Threde GmbH (KT)

Wolfratshauser Strasse 48, D-81379 München, Germany
Tel: (+49 89) 72 49 50
Fax: (+49 89) 72 49 52 91
(+49 89) 72 49 52 32
e-mail: jm@kayser-threde.de
press@kayser-threde.de
Web: http://www.kayser-threde.com
Managing Director: Dipl.-Ing. Reiner Klett
Managing Director: Dr Klaus Eusslin
Management Space Department: Eckart Wulf
Secretary to Vice President of Space Strategy: Ms Alexandra Dietz
Tel: (+49 89) 72 49 52 37
■ Specialise in the design and development of high-technology solutions for the aerospace, industrial and scientific sectors. Solutions include applications in manned and unmanned space missions, data measurement and management, telematics, optics and precision control.

UPDATED

Lewicki Microelectronic GmbH

Allee 35, D-89610 Oberdischingen, Germany
Tel: (+49 7305) 960 20
Fax: (+49 7305) 86 01
e-mail: info@lewicki-gmbh.de
Web: http://www.lewicki-gmbh.de
Managing Director: Dr Edmund Rickus
Sales and Marketing Director: Max Sailer
Commercial Manager: Reinhold Fessler
■ Thickfilm hybrid ICs custom design and series production of highly miniaturised, high reliability hybrids for space, aerospace, industrial, military and medical applications.

UPDATED

Litef GmbH

Loerracherstrasse 18, D-79115 Freiburg im Breisgau, Germany
Tel: (+49 761) 490 10
Fax: (+49 761) 490 14 80
e-mail: info@litef.de
Web: http://www.litef.de
■ Development, production and support of inertial navigation systems, strapdown heading and attitude reference systems (AHRS), gyros and accelerometers, mobile digital high performance computers. Development of special electronic units and systems, application of specific integrated circuits (ASICs) for airborne and space applications, on-board ships and land vehicles.

UPDATED

MAN Technologie AG (MT)

PO Box 10 22 35, D-86012 Augsburg, Germany
Tel: (+49 82) 15 05 01
Fax: (+49 82) 15 05 10 00
e-mail: spacetransport@mt.man.de
Web: http://www.man-technologie.de
Executive Board, Chairman: Dr Wolfgang Brunn
Executive Board, Aerospace Sales, Marketing, Projects, Engineering: Walter Köppel
Executive Board, Business Administration, Personnel: Carl-Ferdinand Kolbow
External Relations: Gerhard Alex
Aerospace Production: Horst Choinowski
Press and Public Relations: Klaus-Dieter Naumann
Tel: (+49 82) 15 05 25 21
Fax: (+49 82) 15 05 19 07
Business Development and Public Relations: Ms Maria Zauninger
Tel: (+49 82) 15 05 14 80
Fax: (+49 82) 15 05 19 07
■ Development and manufacture of lightweight structures, tanks and propulsion components for space transportation systems (e.g. the European ARIANE, the Japanese H2A launcher), structural components and thermal protection systems for re-usable space vehicles, high-pressure vessels, fuel tanks and structural components for satellites.

UPDATED

MAN Mobile Bridges GmbH

(Ground Infrastructure Systems & Mobile Bridges)
(MAN Technologie AG, Military Bridges)
(ex-MAN Technologie AG)
PO Box 3480, D-55024 Mainz, Germany
Fax: (+49 6131) 215 53 83
Technical Director, Bridge Systems: Wilfried Mathes
Tel: (+49 61 34) 215 52 88
■ Ground systems and facilities used by Ariane, the European launcher; ground facilities used in satellite-based communication systems and for telescopes conducting extra-terrestrial research. Production of bridging systems including Leguan 26m wheeled vehicle comprising a Man 8 x 8 launcher vehicle with shifting frame capable of laying a 26m long MLC 70 bridge in seven minutes, the system can be automatically launched from such tank chassis as M1A1/A2 Wolverine heavy assault bridge, Leopard 1 and M60; can be fully automated, controlled by remote control or from inside the vehicle. The Leguan bridge can be provided with pontoons, featuring an integrated pump jet drive, and hydraulically adjusted ramps to enable it to float and be used as a ferry. Man 22m medium assault bridge and the Leguan modular bridge system (PSB2) on Leopard 2 containing 3 bridge modules each 9.7m which allow several combinations of bridge to be laid.

UPDATED

Orbitale Hochtechnologie Bremen System AG (OHB System AG)

Universitätsallee 27-29, D-28359 Bremen, Germany
Tel: (+49 421) 202 08
Fax: (+49 421) 202 07 00
e-mail: ohb@ohb-system.de
Web: http://www.ohb-systeme.de
Managing Director and Chief Executive Officer: Prof Manfred Fuchs
Chief Financial Officer: Mrs Christa Fuchs
Communications and Public Relations: Danela Sell
Tel: (+49 421) 202 06 20
■ Design and development of mechanical and electronic equipment (including MGSE); microgravity systems, experiment facilities and equipment; manned and unmanned space systems and mission operation; re-entry technology and arodynamics/aerothermodynamics; remote and in situ pollution sensors; small satellites for science, earth observation and telecommunication/reconnaissance.

UPDATED

PSM Drucklufttechnik Vertrieb GmbH

Fraunhoferstrasse 11 A, Postfach 1407, D-82152 München Martinsried, Germany
Tel: (+49 89) 899 62 10
Fax: (+49 89) 856 13 01
e-mail: psm.planegg@t-online.de
Web: http://www.psm-munich.de
Managing Director: W Schlembach
Technical Director (GFu): H März
Sales and Marketing Director: Wolfram Müller
Purchasing Manager: A Hausmann
Secretary: Ms Renate Naar
■ Manufacture of electronically controlled fastening systems, automotive industry, drilling techniques, aviation.
Representing:
FIAM, Italy
GSE – Techmotive, USA
Hios, Japan

UPDATED

RAM System GmbH

Daimlerstrasse 11, D-85521 Ottobrunn, Germany
Tel: (+49 89) 608 00 30
e-mail: info@ramsys.de
Contact: Andrea Trenkel
■ Consortium formed for the production of the RAM Rolling Airframe Missile ship-to-air missile system.
Member companies
Diehl BGT Defence GmbH & Co. KG
Diehl Stiftung & Co. KG
EADS Deutschland GmbH

UPDATED

Raufoss Metall GmbH

An der Schleuse 8, D-5870 Hemer-Becke, Germany
Tel: (+49 237) 29 19 75
Fax: (+49 237) 21 35 77

UPDATED

Rheinmetall Defence Electronics GmbH

(Headquarters)
Department MSA, Breggweg 54, D-28309 Bremen, Germany
Tel: (+49 160) 745 90 92
(+49 421) 457 41 99
Fax: (+49 421) 457 29 00
e-mail: marketing@rheinmetall-de.com
Web: http://www.rheinmetall-de.com
■ Systems for land forces include: command, control, communication, intelligence and reconnaissance systems, fire control systems and periscope and night sight units, air defence systems, artillery systems and test systems. Systems for air forces include aircraft simulators. Provide maintenance and logistics services for army, naval and air force related products.
Representing:
benntec Systemtechnik GmbH, Germany
Eurofighter Simulation Systems GmbH (ESS), Germany
Company divisions:
Airborne systems
Land systems
Naval and process simulation unit
Technical publication/logistics engineering
Vehicle and battlefield simulation unit
Flight simulation unit
Oy Finnish Defence Powersystems Ab (FDPS)
Rheinmetall Hellas SA

UPDATED

Teldix GmbH

PO Box 105608, D-69046 Heidelberg, Germany
Tel: (+49 6221) 51 20
Fax: (+49 6221) 51 23 05
e-mail: pr@teldix.de
operations@teldix.de
computer@teldix.de
space@teldix.de
Web: http://www.teldix.com
Managing Director: Dr Egon Tyssen
Deputy Managing Director: Rainer Götting
Engineering Director: Dr Oliver Stucky
Tel: (+49 6221) 51 22 70
Operations Director: Thomas Welsch
Tel: (+49 6221) 51 22 19
Financial Director: David Martin
Sales and Marketing Director, Computers and Displays: Uwe Kroen
Tel: (+49 6221) 51 24 32
Director of Computer Products: Wolfgang Becker
Tel: (+49 6221) 51 23 67

Public Relations Manager: Armin Göckel
Tel: (+49 6221) 51 22 34
Fax: (+49 6221) 51 25 58

■ Research, development and production of avionics computers and displays for cockpit/mission management, missile control, interface management systems; helmet displays, head-up displays, multi-function displays and GPS receivers; repair and overhaul of navigation systems for aircraft, ships and land vehicles; map displays for aircraft and land vehicles; high precision mechanisms for satellites including ball bearing momentum, reaction wheels, solar array drive assemblies (SADA); reference projects for: Tornado (head-up display HUD, fuel flow indicator, missile control unit, launcher decoder unit, weapon interface unit), EF 2000 head-up display control panel, cockpit interface unit, attack/navigation computer, interface processor unit, defensive aids computer; remote terminal unit, doppler velocity sensor and integrated helmet system for UHT Tiger.

UPDATED

Terma GmbH
(ex-Terma Elektronik (Deutschland) GmbH)
Weiterstadt Park, Brunnenweg 19, D-64331 Weiterstadt, Germany
Tel: (+49 6150) 105 90
Fax: (+49 6150) 10 59 99
e-mail: terma.de@terma.com
Web: http://www.terma.com

UPDATED

Thales ATM Navigation GmbH
(Thales Air Traffic Management GmbH)
(ex-Alcatel Air Navigation Systems GmbH)
Lilienthalstrasse 2, D-70825 Korntal-Münchingen, Germany
Tel: (+49 711) 82 14 50 23
Fax: (+49 711) 82 13 44 05
Web: http://www.thalesgroup.com
General Manager: Yvon Peresse
Finance Director: Ms Maria-Luise Teipel-Stephan
Tel: (+49 711) 82 13 46 35
Technical Director: Jean-Marc Klaus
Operations Director: Gerold Viems
Communications Manager: Ms Ellen Schellinger
Purchasing Manager: Volker Weis

■ Conventional navigation and ATC systems: VOR/DVOR/ILS Cat.I-III/DME;TACAN/ELTA/NDB; RMMC (remote maintenance and monitory concept); Mobile VOR/NDB/TACAN; Mobile Control Towers (2-4 operators). Factory and on-the-job training. New navigation and ATS systems: MLS Cat. I-III (azimuth-elevation-DME/D); DGPS (ground reference and monitor station); STREAMS (Surface Traffic Enhancement and Automation System).

UPDATED

Thyßen Krupp VDM GmbH
PO Box 1820, D-58778 Werdohl, Germany
Tel: (+49 2392) 550
Fax: (+49 2392) 55 22 17
e-mail: info@tbs-vdm.thyßenkrupp.com
Web: http://www.thyßenkruppvdm.com
Chairman: Rolf-Dieter Großkopf
Tel: (+49 2392) 55 22 13
Board of Management: Burkhard Becker
Tel: (+49 2392) 55 22 33
Board of Management: Ulrich W Leggewie
Tel: (+49 2392) 55 22 25
Marketing Director: Heinz-Jürgen Moller
Tel: (+49 2392) 55 25 01
Purchasing Manager: Dietmar Kronsfeld
Tel: (+49 2392) 55 25 76
Marketing Services Manager: Wolfgang Kämper
Tel: (+49 2392) 55 24 93

■ Development and production of high-performance materials for the aerospace industries: flat products (hot and cold-rolled sheet and plate, strip); rod and bars made of nickel-base alloys, cobalt alloys and demanding special stainless steels.

UPDATED

VCS Aktiengesellschaft
((SpaceCom))
(ex-VCS Nachrichtentechnik GmbH)
Borgmannstrasse 2, D-44894 Bochum, Germany
Tel: (+49 234) 925 81 15
Fax: (+49 234) 925 81 90
Web: http://www.vcs.de
Managing Director: Prof Dr Klaus G Meng
General Manager, SpaceCom: Dr Horst Wulf
Marketing and Sales, Remote Sensing Technologies: Dr Peter Scheidgen
Marketing and Sales, Spacecom: Ulli Leibnitz
Marketing and Sales: Christoph Wichmann

■ High-availability solutions for satellite communications, earth observation, and media broadcasting.

UPDATED

Industry – Service

Axel Neumann Versicherungsmakler GmbH (ANV)
Hauptstrasse 19, D-72124 Pliezhausen, (OT Rübgarten), Germany
Tel: (+49 7127) 975 40
Fax: (+49 7127) 97 54 44
e-mail: info@axelneumann.de
Web: http://www.axelneumann.de
Managing Director (GFu): Axel Neumann

■ National and international aviation and space insurance.

UPDATED

CAM Computer Anwendung fur Management GmbH
Oberföhringerstrasse 2, D-81679 München, Germany
Tel: (+49 8105) 372 31 02
Fax: (+49 8105) 37 23 11
Web: http://www.cam-comp.de
Contact: Ms Birgit Grundmann

■ Control centre software.

UPDATED

Energia Deutschland GmbH
Wolfratshauser Strasse 48, D-81379 München, Germany
Tel: (+49 89) 72 49 50
Fax: (+49 89) 72 49 52 91
e-mail: info@kayser-threde.de
Web: http://www.kayser-threde.de
Managing Director: Reiner Klett
Contracts Director: Mr Braeunig
Secretary to Dr. Wulf, Vice-President, Space: Ms Alexandra Dietz
Tel: (+49 89) 72 49 52 37
Fax: (+49 89) 72 49 52 32

■ Marketing and distribution of Earth observation data.

UPDATED

Eumetstat GmbH
Am Kavalleriesand 31, D-64295 Darmstadt, Germany
Tel: (+49 6151) 80 77
Fax: (+49 6151) 80 75 55
Web: http://www.eumetsat.de
Director-General: Dr Tillmann Mohr
Council Chairman: Dr Jorma Riissanen
Head, Technical Department: Alain Datiev
Head, Administration Department: Dr Volker Thiem
Head, Strategy and International Relations: Dr David Williams
Geostationary Programme Manager: Dr Anthony Harrison
EPS Programme Manager: Michel Langevin

UPDATED

Eurockot Launch Services GmbH
Airport Centre, Flughafenallee 26, D-28199 Bremen, Germany
Tel: (+49 421) 539 65 19
(+49 421) 539 65 08
Fax: (+49 421) 539 65 00
e-mail: eurockot@space.eads.net
Web: http://www.eurockot.com
Chief Executive Officer: Dr Matthias Oehm
Sales Director: Peter Freeborn
Technical Director: Dr Mark Kinnersley
Contact: Ms Anna Zorina

■ Commercial launch provider for Leo Satellite Industry.

UPDATED

European Space Agency (ESA)
(European Astronauts Centre)
Linder Höhe, D-51147 Köln, Germany
Tel: (+49 2203) 600 10
Fax: (+49 2203) 60 01 66
Web: http://www.estec.esa.int/spaceflight/astronaut

■ ESA centre for astronaut selection and training.

UPDATED

Industrieanlagen-Betriebs GmbH (IABG)
Einsteinstraße 20 D-85521 Ottobrunn, Munich, Germany
Tel: (+49 89) 60 88 20 30
(+49 89) 608 80
Fax: (+49 89) 60 88 40 00
(+49 89) 60 88 22 10
Telex: 524001
e-mail: iabg@iabg.de
Web: http://www.iabg.de
Managing Director: Dr Rudolf Falter
Chief Executive Officer: Prof. Dr Rudolf Schwarz
Senior Vice President: Wolfgang Raasch
Senior Vice President: Gerhard Vodermeier
Senior Vice President: Reinhard Hutter
Senior Vice President: Peter Rinnelt
Senior Vice President: Manfred Braitiuger
Purchasing Manager: Gerhard Gruber
Marketing Manager, Test and Analysis Centre: Dr Florian Halcour
Marketing Manager, Space Test Centre: Christian Henjes
Marketing Manager, Environment: Dr Evi Schuster
Marketing Manager, Weapon System Analysis: Ernst Budde
Marketing Manager, Information Systems and Communication: Klaus Peter Aupperle
Public Relations Manager: Wolfgang Mohr
Head of Corporate Communications: Ms Monica Amler

■ Scientific and technical services including systems engineering, operations research studies, logistics, aircraft and spacecraft technology. Qualification and acceptance testing, fatigue and static strength tests, environmental tests, realistic field tests, defence weapons systems simulation.

UPDATED

RapidEye AG
Wolfratshauser Strasse 48, D-81379 München, Germany
Tel: (+49 89) 72 49 52 83
Fax: (+49 89) 72 49 52 91
e-mail: info@rapideye.de
Web: http://www.rapideye.de

UPDATED

UKW Technik Jaeger
(ex-UKW-Technik Electronic GmbH)
PO Box 60, D-91081 Baiersdorf, Germany
Tel: (+49 9133) 779 40
Fax: (+49 9133) 77 94 30
e-mail: info@ukwtechnik.com
Web: http://www.ukwtechnik.com
Sales Director: Bernd Weitzmann

■ Production and sales of weather satellite receivers and image processing systems.

UPDATED

VSAT Satellitenkommunikation GmbH (VSAT Group)
Anzingerstrasse 5, D-85560 München, Ebersberg, Germany
Tel: (+49 80) 928 25 80
(+49 80) 92 82 58 22

Fax: (+49 80) 92 82 58 11
e-mail: vsat@vsat-sat.com
Web: http://www.vsat-sat.com
Chief Executive Officer: Dipl-Ing. Josef A H Geiger
Technical Director: Joachim Haberecht-Kaiser
Tel: (+49 8092) 82 58 33
Sales Director: Thomas Seguer
Projects Manager: N Tairi
Tel: (+49 8092) 82 58 55

UPDATED

Wilhelm-Foerster-Sternwarte e V mit Zeiss-Planetarium Berlin

Munsterdamm 90, D-12169 Berlin, Germany
Tel: (+49 30) 790 09 30
Fax: (+49 30) 79 00 93 12
e-mail: wfs@wfs.be.schule.de
Web: http://www.planetarium-am-insulaner.de
President: Dipl Ing Rolf Preuschmann
Scientific Director: Dipl Phys Jochen Rose
Tel: (+49 30) 79 00 93 13
Technical Director: Dip Ing Gerold Fass
Tel: (+49 30) 79 00 93 14
Librarian: Wolfgang Meyer
Tel: (+49 30) 79 00 93 30

■ Information on astronomic and space research.

UPDATED

Industry – Distributor

Autoflug Libelle GmbH

Industriestrasse 10, D-25462 Rellingen, Germany
Tel: (+49 4101) 30 75 04
Fax: (+49 4101) 30 75 09
e-mail: sales@autofluglibelle.com
Web: http://www.autofluglibelle.com
Managing Director: Dipl.-Ing. Mario Maevert
Tel: (+49 4101) 30 72 50
Fax: (+49 4101) 30 71 10
Product Manager: Rainer Stuwe
Tel: (+49 4101) 30 75 02
Sales Administrator: Ms Elke Tiedemann

■ Development, manufacture and distribution of Anti-G Protective Systems.

UPDATED

Hong Kong

Industry – Service

Asia Satellite Telecommunications Co. Ltd (ASIASAT)

23rd Floor, East Exchange Tower, 38–40 Leighton Road, Hong Kong
Tel: (+852) 25 00 08 88
(+852) 25 00 08 80
Fax: (+852) 25 00 08 95
e-mail: as-mkt@asiasat.com
Web: http://www.asiasat.com
Chief Executive Officer: Peter Jackson
General Manager, Marketing: Sabrina Cubbon

■ Operates AsiaSat regional satellite systems.

UPDATED

Hungary

Centres of Learning & Research

Institute for Geodesy, Cartography & Remote Sensing

Bosnyák tér 5, H-1149 Budapest, Hungary
Tel: (+36 1) 222 51 01
Fax: (+36 1) 222 51 12
Web: http://www.fomi.hu
Director: Dr Szabolcs Mihály
Tel: (+36 1) 222 51 11

UPDATED

Institute for Geodesy, Cartography & Remote Sensing (FÖMI)
(Remote Sensing Centre)
H-1149 Budapest, Hungary
Tel: (+36 1) 163 66 69
Fax: (+36 1) 252 82 82
e-mail: h3874win@ella.hu
Deputy General Director: Péter Winkler

UPDATED

India

Government – Ministry/Agency

Indian Space Research Organisation (ISRO)

Antariksh Bhavan, New BEL Road, Bangalore, 560094, India
Tel: (+91 80) 341 52 75
(+91 80) 341 53 57
Fax: (+91 80) 341 22 53
(+91 80) 341 53 28
Telex: 8452326
8452499
Cable: ISRO
Web: http://www.isro.org
Prime Minister and Minister of Space: Atal Behari Vajpayee
Tel: (+91 80) 301 23 12
Minister of State (Space): Vasundhra Raje
Chairman: Dr K Kasturirangan
Tel: (+91 80) 341 52 41
Director, Earth Observation System: Dr V Jayaraman
Tel: (+91 80) 341 63 58
Fax: (+91 80) 341 38 06
Director, Launch Vehicles: D Narayanamoorthi
Tel: (+91 80) 341 64 05
Fax: (+91 80) 341 52 97
Director, Technology Transfer and IC/Antrix: K R Sridharamurthi
Tel: (+91 80) 341 62 73
Fax: (+91 80) 341 89 81
Director, BEA: H N Madhusudhana
Tel: (+91 80) 341 26 77
Programme Director, INSAT/Director, SCP/FM: Dr K N Shankara
Tel: (+91 80) 341 53 01
Fax: (+91 80) 341 21 41
Assistant Scientific Secretary/Director, INSES: Rajeev Lochan
Tel: (+91 80) 351 18 21
Deputy Director, International Co-operation: Jacob Ninan
Tel: (+91 80) 341 63 61
Fax: (+91 80) 341 35 80
Executive Director, Space Industries Development: M N Sathyanavayan
Tel: (+91 80) 341 62 79
Director, Publications and Public Relations: S Krishnamurthy
Scientific Secretary: V Sundararamaiah
Tel: (+91 80) 341 63 56

■ Engaged in the design, development and operation of satellites for telecommunication, television broadcasting, meteorology, disaster management, resources survey and management. Design and development of launch vehicles including the Polar Satellite Launch Vehicle (PSLV) for launching IRS class of remote sensing satellites and Geo-synchrous Satellite Launch Vehicle (GSLV) to launch INSAT class of communications satellites. Conducts space science research. Products are available to international customers through Antrix Corporation.

UPDATED

Indian Space Research Organisation (ISRO)
(Development and Educational Communication Unit)
Ambavadi 4 Vistar Post Office, Jodhpur Tekra, Ahmedabad 380015, India
Tel: (+91 79) 677 31 24
(+91 79) 677 31 22
(+91 79) 677 25 35
(+91 79) 677 31 28
Fax: (+91 79) 676 85 56
(+91 79) 674 67 15
Director: B S Bhatia
Tel: (+91 79) 676 39 54
(+91 79) 677 31 20

■ Development of educational and social development software. Conducts research in relation to the impact of space applications.

UPDATED

Indian Space Research Organisation (ISRO)
(Inertial Systems Unit)
Vattiyoorkavu Post, Thiruvananthapuram (Trivandrum), 695013, India
Tel: (+91 471) 36 18 97
(+91 471) 36 30 92
(+91 471) 36 32 88
(+91 471) 36 32 89
(+91 471) 36 32 82
(+91 471) 36 32 94
Fax: (+91 471) 36 19 73
(+91 471) 36 18 13
Director: P S Veeraghavan

■ Design and development of inertial systems for satellites and launch vehicles.

UPDATED

Indian Space Research Organisation (ISRO)
(Liquid Propulsion Systems Centre)
Valiamala P. O., Thiruvananthapuram (Trivandrum), 695547, India
Tel: (+91 471) 56 77 15
(+91 471) 56 72 64
(+91 471) 56 72 95
(+91 80) 527 22 40
(+91 471) 56 77 36
Fax: (+91 471) 56 72 96
(+91 80) 526 20 85
(+91 471) 56 74 92
(+91 80) 526 04 15
(+91 471) 46 26 86
(+91 80) 526 20 47
(+91 471) 56 72 42
Telex: 8452469
Cable: PROPULSION
Director: N Vedachalam
Tel: (+91 80) 527 46 61
(+91 471) 56 72 57

■ Design and development of liquid propulsion systems for satellites and launch vehicles.

UPDATED

Indian Space Research Organisation (ISRO)
(Satellite Centre)
Airport Road, Vimanapura Post, Bangalore, 560017, India
Tel: (+91 80) 508 21 02
(+91 80) 508 21 08
(+91 80) 526 62 51
(+91 80) 527 45 16
(+91 80) 508 21 29
Fax: (+91 80) 508 21 10
(+91 80) 508 21 99
(+91 80) 526 94 90
(+91 80) 526 64 28
(+91 80) 526 36 22
(+91 80) 526 85 44
(+91 80) 526 90 21
(+91 80) 508 21 98
(+91 80) 508 23 21
Cable: UPAGRAM
Director: Dr P S Goel
Tel: (+91 80) 527 47 70

■ Design and development of communication, remote sensing and scientific satellites.

UPDATED

Indian Space Research Organisation (ISRO)
(SHAR Centre)
Sriharikota Range, Nellore District, Andhra Pradesh, 524124, India
Tel: (+91 8623) 450 55
(+91 8623) 450 43
Fax: (+91 8623) 451 54
(+91 8623) 451 59
(+91 8623) 450 67
(+91 8623) 451 52
(+91 8623) 450 02
(+91 8623) 451 53
Telex: 416723
Cable: SPACE
Director: K Narayana
Tel: (+91 8623) 450 50
(+91 44) 536 44 50

■ Launch station providing infrastructure and facilities for testing of large solid propellant motors.

UPDATED

Indian Space Research Organisation (ISRO)
(Space Applications Centre)
Ambavadi Vistar Post Office, Jodhpur Tekra, Ahmedabad, 380015, India
Tel: (+91 79) 674 20 10
(+91 79) 674 88 07
(+91 79) 676 93 91
(+91 79) 676 34 42
(+91 79) 676 55 32
(+91 79) 674 19 39
Fax: (+91 79) 676 80 73
(+91 79) 676 85 56
(+91 79) 676 17 60
(+91 79) 676 54 10
Cable: SPACE
ANTARIX
Director: Dr A K S Gopalan
■ Conducts space applications activities including development of satellite payloads for communications and remote sensing.

UPDATED

Indian Space Research Organisation (ISRO)
(Telemetry, Tracking & Command Network)
A1-6 Peenya Industrial Estate, Peenya, Bangalore, 560058, India
Tel: (+91 80) 839 42 61
(+91 80) 839 42 30
(+91 80) 837 63 16
(+91 80) 837 63 11
Fax: (+91 80) 839 45 15
(+91 80) 809 40 03
(+91 80) 809 40 21
(+91 80) 839 82 57
Telex: 8455010 trac
Director: S K Shivakumar
Tel: (+91 80) 809 45 81
(+91 80) 839 53 95
■ Provides telemetry, tracking and command network support for satellite and launch vehicle missions.

UPDATED

Indian Space Research Organisation (ISRO)
(Vikram Sarabhai Space Centre)
ISRO P.O, Thumba, Thiruvananthapuram (Trivandrum), 695022, India
Tel: (+91 471) 41 58 20
(+91 471) 56 41 23
Fax: (+91 471) 41 51 76
(+91 471) 41 53 58
Telex: 435201
435202
Cable: SPACE
Director: Dr G Madhavan Nair
Tel: (+91 471) 41 55 12
(+91 471) 56 55 67
■ Design and development of launch vehicles and related technologies.

UPDATED

National Remote Sensing Agency (NRSA)
Balanagar, Hyderabad, 500037, India
Tel: (+91 40) 387 87 88
(+91 40) 387 96 77
Fax: (+91 40) 387 72 10
(+91 40) 387 88 65
Telex: 4256522
Cable: REMOSEN
e-mail: sales@nrsa.gov.in
Web: http://www.nrsa.gov.in
Director: Dr R R Navalgund
Tel: (+91 40) 387 83 60
■ Remote sensing satellite data reception, processing and dissemination.

UPDATED

Association

Society of Indian Aerospace Technologies & Industries (SIATI)
Aeronautical Society Building, Suranjandas Road, Off Old Madras Road, Bangalore, 560 075, India
Tel: (+91 80) 521 99 51
(+91 80) 527 52 62
Fax: (+91 80) 529 24 40
e-mail: siati@vsnl.com
Web: http://www.siatiaero.com
President: Dr C G Krishnadas Nair
Honorary Secretary General: Air Cdre J Varkey
■ The encouragement of interaction between research, design, development and industry within India, and the encouragement of overseas collaborations. To promote aerospace technologies and industries in India, encourage and assist potential entrepreneurs with technology to set up medium/small scale industries for manufacture/maintenance of aerospace components, systems and materials. To provide a forum for interaction of industries and users, research and development in institutions and universities. To conduct seminars/workshops and participate in national and international air shows, exhibitions and seminars. To encourage education and research, institute scholarships/fellowships in the area of aerospace.
Members interests:
Aircraft, helicopter, satellite/launch vehicle structures; design, aircraft engines, starters and APUs; aircraft seats, systems, equipment and accessories; avionics and radars; airline operators; electrical connectors, cables and batteries; electrical components and systems; precision machined parts; rubber polymer, FRP parts; standard parts and fasteners; aircraft filters; ground support equipment; jigs, fixtures gauges and tools; metallic and non-metallic materials; castings and forgings; consumables; machines and special purpose machines; research and development and testing laboratories; training and educational institutions; government organisations; professional bodies and associations.

UPDATED

Industry – Manufacturing

Bharat Electronics Ltd (BEL)
(International Marketing)
2nd Floor Shankarnarayan Building, 25 M G Road, Bangalore, 560 001, India
Tel: (+91 80) 558 38 51
(+91 80) 559 50 17
Fax: (+91 80) 558 49 11
e-mail: imd@bell-india.com
Web: http://www.bel-india.com
Chairman and Managing Director: Dr V K Koshy
Tel: (+91 80) 226 26 26
Director, Research and Development: S K Mehta
Tel: (+91 80) 225 24 20
Director, Commercial and Management Services: S C Khanna
Tel: (+91 80) 225 11 07
Director, Finance: P R K Haragopal
Tel: (+91 80) 225 21 33
Director, Personnel: V Ammineedu
Tel: (+91 80) 228 39 18
Director, Bangalore Complex: Mr Basavarajaiah
Tel: (+91 80) 838 23 23
Director, Other Units: Y Gopala Rao
Tel: (+91 80) 225 86 72
■ Manufacture of products in the areas of defence communications, radars and sonars; telecommunications, sound and vision broadcasting, opto and tank electronics, solar systems and components. Products include: Secondary Surveillance Radar (SSR), Monopulse Secondary Surveillance Radar (MSSR), Airport Surveillance Radar and Automatic Visual Range Assessor (AVRA).
Accreditations: ISO 9001, ISO 9002

UPDATED

Bharat Electronics Ltd (BEL)
(Bangalore Unit)
Jalahalli Post, Bangalore, Karnataka, 560 013, India
Tel: (+91 80) 838 26 26
Fax: (+91 80) 838 01 60
(+91 80) 838 01 35
(+91 80) 838 18 79
(+91 80) 838 20 67
Director, Bangalore Unit: Mr Basavarajaiah
Public Relations: P G Kulkarni
Tel: (+91 80) 838 02 88
■ Defence and satellite communications, sound and vision broadcasting, radars, simulators, systems and components.

UPDATED

Bharat Electronics Ltd (BEL)
(Ghaziabad Unit)
Bharatnagar Post, Ghaziabad, Utter Pradesh, 201 010, India
Tel: (+91 120) 461 95 00
Fax: (+91 120) 477 07 20
Head of Unit: N K Sharma
Tel: (+91 120) 277 65 71
■ Radars, antennas. Satcom (defence) and microwave components.

UPDATED

Electronics Corporation of India Ltd (ECIL)
(Corporate Business Development)
ECIL Post Office, Hyderabad, Andhra Pradesh, 500 062, India
Tel: (+91 40) 712 34 09
(+91 40) 27 12 06 71
(+91 40) 712 01 31
(+91 40) 712 37 25
(+91 40) 712 23 49
Fax: (+91 40) 27 12 18 02
(+91 40) 712 25 35
(+91 40) 712 18 02
e-mail: cbdg@ecil.co.in
Web: http://www.ecil-india.com
Director, Technical: Wg Cdr K S Chandrasekhar
Senior Manager, Exports: A Chiramjeevi Rao
Senior Manager: T R Rao
■ Air traffic management systems, battlefield C3I systems, radar data processing and multi-sensor tracking; switching, transmission and access products, V-Sat networks, radio communication and EW products.

UPDATED

HBL NIFE Power Systems Ltd
(ex-HBL Limited)
(ex-Hyderabad Batteries Ltd)
8-2-601 Road No. 10, Banjara Hills, Hyderabad, 500034, India
Tel: (+91 8418) 24 46 42
Fax: (+91 40) 23 35 31 89
(+91 40) 23 35 50 48
e-mail: contact@hblnife.com
Web: http://www.hblnife.com
Chairman and Managing Director: Dr A J Prasad
Tel: (+91 8418) 24 46 51
Managing Director: A J Prasad
Finance Director: P Satish Kumar
Tel: (+91 98) 48 05 31 68
Executive Director: Ashok Nagarkatti
Tel: (+91 98) 48 01 25 91
Sales Director: M S Srinath
Tel: (+91 98) 48 01 25 80
Marketing Director, Exports: Ms Kavita Prasad
Tel: (+91 98) 48 01 28 01
Operations Director: L C Gangrade
Tel: (+91 98) 48 01 25 92
Technical Director, Defence Products Division: G K Rao
Tel: (+91 8418) 446 27
General Manager, Corporate Marketing: Zubain Hussain
Tel: (+91 98) 48 14 86 67
Purchasing Manager: P Srinivasa Rao
Tel: (+91 8418) 446 40
Executive Secretary: P Vijaya Lakshmi
■ Manufacture of silver, zinc and nickel-cadmium batteries for aircraft, torpedoes and UPS applications. Lithium, thermal, valve regulated lead acid, tubular and fuse batteries. Supplier of sealed maintenance-free lead acid generator start batteries and maintenance-free, rechargeable VRLA monoblock batteries for main battle tanks. Supplier for internal and export markets. Battery monitoring system for submarines.

UPDATED

Hindustan Aeronautics Ltd (HAL)
(Corporate Headquarters)
15/1 Cubbon Road, PO Box 5150, Bangalore, 560 001, India
Tel: (+91 80) 22 86 69 03
(+91 80) 22 86 69 04
(+91 80) 22 86 69 05
(+91 80) 22 86 69 06
(+91 80) 22 86 69 07
Fax: (+91 80) 22 86 71 40
(+91 80) 22 86 75 33
e-mail: marketing@hal-india.com
Web: http://www.hal-india.com
■ Design, development and manufacture of fixed and rotary-wing aircraft, avionics and accessories. Manufacture, maintenance, repair and overhaul of fighter, transport and trainer aircraft, helicopters, aero-engines, avionics, accessories, and ground support equipment. Structural components for satellites and launch vehicles, software development for aerospace applications, ground support equipment and design consultancy.

UPDATED

Hindustan Aeronautics Ltd (HAL)
(Aerospace Division)
New Thippasandra, PO Box 7502, Bangalore, 560075, India
Tel: (+91 80) 524 14 03
Fax: (+91 80) 524 16 51
General Manager: D K Naithani
■ Light-alloy structures for space launch vehicles.

UPDATED

Kaveri Telecoms Limited
Kaveri Industrial Complex, Plot No. 31 to 36, 1st Floor, 1st Main, 2nd Stage, Arakere Mico Layout, Bannerghatta Road, Bangalore, 560076, India
Tel: (+91 80) 658 05 18
(+91 80) 658 08 63
(+91 80) 658 13 01
(+91 80) 658 30 90
(+91 80) 658 75 43
Fax: (+91 80) 658 55 01
e-mail: kaveritl@vsnl.net
mktg@kaveritelecoms.com
Web: http://www.kaveritelecoms.com
General Manager: L Nicholas
■ Manufacture of antennas and RF products catering to civilian, military and satellite applications up to 18 GHz frequency range.
Accreditations: ISO 9001

UPDATED

Larsen & Toubro Limited
(Heavy Engineering Division)
Saki-Vihar Road, Powai, Mumbai (Bombay), 400 072, India
Tel: (+91 22) 28 58 10 91
(+91 22) 28 58 26 11
Fax: (+91 22) 28 58 10 76
e-mail: ranga9022@hed.ltindia.com
Web: http://www.lnthed.com
Assistant General Manager, Business Development: Nandkumar M Athawale
Tel: (+91 22) 28 58 14 01
Contact: S Ranganathan
Tel: (+91 22) 28 58 10 73
Fax: (+91 22) 28 58 10 70
(+91 22) 28 58 16 11

UPDATED

Remote Sensing Instruments
2-2-18/18/3/16, Bagh Amberpet, Hyderabad, Andhra Pradesh, 500 013, India
Tel: (+91 40) 23 07 85 66
(+91 40) 23 07 97 83
Fax: (+91 40) 23 07 95 66
e-mail: contact@rsi.bz
Web: http://www.rsi.bz
Managing Director: T Sesha Rao
Sales and Marketing Director: T Gangadhar Rao
Operations and Technical Director: G S Kumar
Purchasing Manager: Ramana Kumar
■ Remote sensing and geographic information systems (GIS). Application of GIS and digitisation.

UPDATED

Industry – Service

Antrix Corp. Ltd
Antariksh Complex, Near New Bel Road, Bangalore, 560094, India
Tel: (+91 80) 341 62 73
(+91 80) 341 62 74
Fax: (+91 80) 341 89 81
e-mail: krs@antrix.org
Web: http://www.antrix.org
■ Markets space products and services from Indian Space Research Organisation including satellite systems and subsystems (telecommunication, remote sensing and scientific), TTC and mission support, launch services and value added services based on satellite applications.

UPDATED

Speck Systems Pvt Ltd
B-49 Electronics Complex, Kushaiguda, Hyderabad, 500062, India
Tel: (+91 40) 712 53 05
(+91 40) 712 53 06
Fax: (+91 40) 712 24 89
e-mail: internationaldesk@specksystems.com
■ Digital image processing systems and electronic visual interpretation aids.

UPDATED

Indonesia

Pasifik Satelit Nusantara (PSN)
Kawasan Karyadeka Pancamurni, Blok-A Kav, 3 Lemah Abang, 17550 Bekasi, Indonesia
Tel: (+62 21) 89 90 81 11
Fax: (+62 21) 89 90 81 10
Web: http://www.psn.co.id
President Director: Adi Rahman Adiwoso
Vice President of Operations: Andreas K Peranginangin
Vice President of Information Technology: Yanuar Arif Machmud
Vice President of Legal: Ino A Simatupang
Director of Operations: Rian Alisjahbana
Director of Finance: Anggarini Surjaatmadja
Director of Regulatory Affairs and Administration: L Woerfiendarti Soedarmoro
Director of Technology: Achadiat Djajawinata
Head of Public Relations Department: Dewi Aryani
Contact: Niyta Korompis
■ Provider of a range of satellite-based telecommunications services, including both mobile and fixed telephony systems.

UPDATED

Industry – Sales Agent/Office

Radyne Comstream Inc.
(Indonesia)
(ex-Comstream Corporation)
Jalan MT Haryono, Kav 25, 12820, Jakarta, Indonesia
Tel: (+62 21) 521 37 33
Fax: (+62 21) 252 01 42
e-mail: asiapacsales@radn.com
Web: http://www.radn.com
Representing: Radyne Comstream Inc., US

UPDATED

Ireland

Industry – Manufacturing

Farran Technology Ltd (FTL)
Ballincollig, County Cork, Ireland
Tel: (+353 21) 487 28 14
Fax: (+353 21) 487 38 92
e-mail: sales@farran.com
Web: http://www.farran.com
Managing Director: Tony McEnroe
Financial Director: Gerry Landers
Technical Director: Dr Brendan Lyons
Marketing Manager: David R Vizard
Tel: (+353 21) 487 17 20
Sales/Marketing Department: Miss Claire Brun
■ Waveguide mixers, frequency multipliers for commercial and defence applications. MM and SUBMM components (mixers, detectors, frequency sources, diodes.) Low noise and medium power amplifiers. Custom MMIC-based subsystems in frequency range from 10GHz to 100 GHz for use in radar, wireless communications and EW applications. Amplifiers, transceivers, passive receivers, up-converters, down-converters, multipliers and mixers.

UPDATED

FLS Aerospace
(ex-Team Air Lingus)
Dublin Airport, MD 108, Dublin, Ireland
Tel: (+353 1) 705 33 06
Fax: (+353 1) 705 64 73
Communications: Ms Regina Ryan
Tel: (+353 1) 886 26 58
Fax: (+353 1) 888 88 89

UPDATED

L-3 Communications ESSCO Collins Ltd
Kilkishen, County Clare, Ireland
Tel: (+353 61) 36 72 44
Fax: (+353 61) 31 10 44
e-mail: info.esscogl-3com.com
Web: http://www.l-3com.com/essco
Director of International Marketing: Michael O'Connell
Finance Manager: John Hackett
■ Design and manufacture of ground-based, shipboard radomes (metal space frames, sandwich, solid laminate, and RF composite structures) for the military, air traffic control, weather radar, secure communications, satcom and remote sensing market. Design and manufacture of high-end millimetre wave antenna systems (aircraft, communications, radar) for specialised RF application, including satellite communications, radio astronomy, precision pointing and tracking applications. Design studies, structural and electromagnetic analysis, manufacture, installation and field support of all company products.

UPDATED

Israel

Association

Department of Geophysics & Planetary Sciences
Tel-Aviv University, IL-69978 Tel-Aviv, Israel
Tel: (+972 3) 640 69 72
Fax: (+972 3) 640 92 82
e-mail: akivab@luna.tau.ac.il
Chairman: Prof David Abir
Tel: (+972 9) 774 76 58

UPDATED

National Committee for Space Research
Israel Academy of Sciences and Humanities, PO Box 39040, IL-69978, Tel-Aviv, Israel
Tel: (+972 3) 640 69 72
Fax: (+972 3) 640 92 82
e-mail: akivab@luna.tau.ac.il
Secretary: Prof Akiva Bar-Nun

UPDATED

Industry – Manufacturing

Controp Precision Technologies Ltd
PO Box 611, IL-45105, Hod Hasharon, Israel
Tel: (+972 9) 744 06 61
(+972 9) 760 49 71
Fax: (+972 9) 744 06 62
(+972 9) 744 16 86
e-mail: cntrpnir@netvision.net.il
Web: http://www.controp.co.il
http://www.controp.com
General Manager: Shlomo Nir
Business Development Director: Eli Ben-Aharon
International Marketing Director: Meir Giber
Tel: (+972 9) 744 06 61
Marketing Communications Director: Ms Lori Erlich
Tel: (+972 9) 744 06 61
Marketing Director: Simi Golani
Tel: (+972 9) 744 06 61
■ Specialise in the development and production of electro-optical and precision motion control systems. Expertise in electro-optics, electronics, servo-control, precision mechanics and software. Main product lines include: high performance stabilised observation payloads used for day and night surveillance on board UAVs, helicopters and light aircraft; high performance wide area coverage (real time) EO stabilised payloads for day and night reconnaissance on board UAVs and light aircraft; intruder detection systems for land-based applications; directional antenna pedestals for UAV applications; stabilized antenna pedestals installed on aerial vehicles for satellite communication relay applications; special aerial reconnaissance film cameras for UAV applications; thermal imaging cameras with zoom lenses.
Accreditations: ISO 9001

UPDATED

Elbit Systems Ltd
(ex-Elbit Ltd)
Advanced Technology Center, PO Box 539, IL-31053 Haifa, Israel
Tel: (+972 4) 831 53 15
Fax: (+972 4) 855 00 02
e-mail: marcom@elbit.co.il
elbit-systems@elbit.co.il

Web: http://www.elbit.co.il
Chairman of the Board: Michael Federmann
Tel: (+972 3) 520 24 30
President and Chief Executive Officer: Joseph Ackerman
Tel: (+972 4) 831 55 26
Vice President and Chief Financial Officer: Joseph Gaspar
Tel: (+972 4) 831 64 04
Corporate Vice President, Business Development and Marketing: Zeev Gofer
Tel: (+972 4) 831 54 68
Vice President, International Marketing: Jacob Alon
Tel: (+972 8) 938 64 21
Vice President, Business Development and Marketing: Ran Galli
Tel: (+972 4) 831 52 64
Vice President, Finance: Ilan Pacholder
Tel: (+972 4) 831 63 19
Corporate Communications Director: Ms Dalia Rosen
Tel: (+972 4) 831 67 84
Fax: (+972 4) 831 54 20
Marketing Communications Director: Ms Ilana Gelfer
Tel: (+972 4) 831 50 56
Fax: (+972 4) 831 79 02
Vice President, Manufacturing and Purchasing: Marco Rosenthal
Tel: (+972 4) 831 64 65

■ International defense electronics company engaged in a wide range of defense-related programs throughout the world, in the areas of aerospace, land and naval systems, command, control, communications, computers, intelligence, surveillance and reconnaissance ("C4ISR"), advanced electro-optic and space technologies. The Company focuses on the upgrade of existing military platforms and development of new technologies for defense and homeland security applications.

UPDATED

ELOP Electro-Optics Industries Ltd

Advanced Technology Park, Kiryat Weizmann, PO Box 1165, IL-76111, Rehovot, Israel
Tel: (+972 8) 938 62 11
(+972 8) 938 66 06
Fax: (+972 8) 938 62 37
(+972 8) 938 61 25
e-mail: cus-service@elop.co.il
info@elop.co.il
Web: http://www.el-op.co.il
Managing Director: Haim Rousso
Tel: (+972 8) 938 66 44
Vice President, Marketing and Sales: Jacob Alon
Vice President, Financial: Mordechai Livne
Tel: (+972 8) 938 66 61
Marketing Communications Manager: Mrs Hagit Markel-Fried
Tel: (+972 8) 938 64 33

■ Research, development engineering and production of thermal imaging systems; laser systems, display systems, optronic stabilised payloads, space and airborne imagery inteligence systems; ground integrated sights, electro-optics countermeasures and homeland defence systems.

UPDATED

ELTA Systems Ltd. (IAI/ELTA)

(ex-Elta Electronics Industries Ltd)
PO Box 330, IL-77102 Ashdod, Israel
Tel: (+972 8) 857 23 12
(+972 8) 857 21 55
(+972 8) 857 24 14
(+972 8) 857 23 12
(+972 8) 857 25 43
Fax: (+972 8) 856 18 72
(+972 8) 856 45 68
e-mail: market@elta.co.il
Managing Director: Israel Livnat
Tel: (+972 8) 857 26 00
Deputy Managing Director for Technologies: Eli Golan
Tel: (+972 8) 857 23 22
Fax: (+972 8) 857 20 80
Deputy Managing Director for Finance: Haim Amitai
Tel: (+972 8) 857 25 23
Fax: (+972 8) 857 20 11
Director, Marketing and Sales: Pnina Joseph
Director of Purchasing and Subcontracting: Zvi Klagsburn
Tel: (+972 8) 857 25 36
Fax: (+972 8) 857 28 89
Deputy Director, Marketing, Sales and Communications: Baruch Reshef
Tel: (+972 8) 857 21 55
Mobile: (+972) 58 74 03 99

UPDATED

Israel Aircraft Industries Ltd (IAI)

(Headquarters)
Ben Gurion International Airport, IL-70100, Israel
Tel: (+972 3) 935 33 43
Fax: (+972 3) 935 82 78
Web: http://www.iai.co.il

■ Research, design, development, integration, testing, manufacture, upgrading, support and service of land, sea, air systems equipment and space technologies. Military and civil aircraft, UAVs, missiles, hydraulic, electro-mechanical, electronic and optronic systems, ordnance and navigation equipment, patrol boats, mine clearance devices, C3I systems, airborne and ground-based radars, software.

Accreditations: ISO 9001
Subsidiaries
Israel Aircraft Industries-Europe, France
Israel Aircraft Industries International Inc., US

UPDATED

Israel Aircraft Industries Ltd (IAI)

(MBT Division)
Yehud Industrial Zone, PO Box 105, IL-56000 Yehud, Israel
Tel: (+972 3) 531 40 01
(+972 3) 531 55 55
(+972 3) 531 41 84
(+972 3) 531 40 05
Fax: (+972 3) 536 33 76
(+972 3) 536 52 05
(+972 3) 531 41 80
(+972 3) 531 41 30
Telex: 341450 mbt
e-mail: pbichman@mbt.iai.co.il
Web: http://www.iai.co.il
General Manager: I Nissan

■ Design and development of satellites, full satellite integration and testing, launching and in-orbit operations. Also provides mission control centres, tracking and multi-satellite remote sensing stations. Design and manufacture of smart bombs, fire control systems, offensive and defensive precision and tactical battlefield missiles. Products include Nimrod, Gabriel, harpy air to surface missiles; Griffin and Guillotine bombs.

UPDATED

Israel Aircraft Industries Ltd (IAI)

(MLM Division)
PO Box 45, IL-70100 Beer Yaakov, Israel
Tel: (+972 8) 927 25 20
(+972 8) 972 24 25
Fax: (+972 8) 927 28 90
Telex: 381014
381033 israv
e-mail: mlm_marketing@mlm.iai.co.il
General Manager: Y Ramati
Tel: (+972 8) 927 24 36
Contact: Haim Berger
Tel: (+972 8) 927 22 07
MLM Marketing: Yoram Shimoni

■ Prime activites include: Anti-ballistic missile defence systems such as the Arrow, launchers and space systems including the Shavit small satellite launchers; advanced avionics, training and simulation systems including the EHUD autonomous ACMI for fighter aircraft and similar tailored systems for trainers, helicopters and situational awareness systems (SAW). Information Technology (IT) and command, control, communication and intelligence (C3I) systems. Communication, telemetry and test/launch range equipment and systems.

UPDATED

Israel Aircraft Industries Ltd (IAI)

(Tamam Division)
(ex-Tamam Precision Instrument Industries)
Yehud Industrial Zone, PO Box 75, IL-56100 Yehud, Israel
Fax: (+972 3) 531 51 40
Contact: Uri Bar-Zeev

■ Design and manufacture of inertial measurement, stabilisation, navigation and optronic systems and components for land, sea and air, manned and unmanned platforms, specialised civil space components.

UPDATED

Israel Military Industries Ltd (IMI)

(ex-TAAS Israel Industries Ltd)
64 Bialik Boulevard, PO Box 1044 or 6604, IL-47100 Ramat Hasharon, Israel
Tel: (+972 3) 548 56 17
(+972 3) 548 56 19
Fax: (+972 3) 540 69 08
e-mail: imimrktg@imi-israel.com
imi.mktg@netvision.net.il
Web: http://www.imi-israel.com
Chairman of the Board: Brig Gen Arie Mizrachi
President and Chief Executive Officer: Dr Ehud Udi Ganani
Vice President, Finance & Control: Ronen Wolfman
Vice President, Research and Development: Dr Avi Galor
Vice President, Marketing: Avinoam Zafir
Vice President, Human Resources: Reuven Chen

■ Manufacture of small arms, armoured fighting vehicles, ammunition, electronics systems and equipment, command and control equipment, airborne systems, countermeasures, rockets, unmanned aerial vehicles, personal protection equipment and security systems. Ammunition produced includes small arms: 5.56 x 45 mm M193, 7.62 x 39 mm M1943, 7.62 x 51 mm,.30-06, .50" BMG. Mortar: 51 mm, 52 mm, 81 mm and 120 mm. Rifle grenades. Tank: 60 mm APFSDS, 105 mm APFSDS, TPFSDS-T, I-HEAT-T M152/3, APAM, LAHAT and 120 mm APFSDS, LAHAT. Artillery: 130 mm HE, 152 mm ICM, 155 mm HE M107, Smk WP M110AS, Smk HC M116A1, Illg M485A2, 155 mm HE DPICM M395, 175 mm Cargo M366. Rockets 160 mm, 240 mm, 350 mm.

Air launched weapons include Delilah, STAR-1 ITALD, MAPATS and MSOV air to surface missiles; ATAP-300/-500/-1000, MSOV, Condib 70/120, M-2000, PB-500A1, IBDU-33, Mk 81, 82, 83, 84 GP and RAM bombs; Negev 5.56 mm LMG guns.

Operating units:
-Aircraft Systems
-Weapons & Armoured Fighting Vehicles
-Systems Group
-Advanced Systems Division
-Rocket Systems Division
-Airborne Electronic Systems
-Systems Engineering Division
-IMI Telecom
-Integrated Security Systems Group
-Small Arms
-Small Arms Ammunition
-Academy for Advanced Training
-Security Services Centre
-Heavy Ammunition Complex
-Haifa Metal Components
-Zoref Assembly Plant
-El-see Explosives and Chemicals
-Central Laboratory
-Test and Analysis Systems
-Ashot Ashkelon
-Re'em Electronics

UPDATED

Rafael Armament Development Authority Ltd

(Corporate Headquarters)
PO Box 2250 (19), IL-31021 Haifa, Israel
Tel: (+972 4) 879 12 44
(+972 4) 879 40 57
Fax: (+972 4) 879 46 57
(+972 4) 879 24 13
Web: http://www.rafael.co.il
President: Giora Shalgi
Tel: (+972 4) 879 47 77
Vice President, Research and Development: Dr Dan Lesham
Vice President, Research and Development: Dr Avi Grisburg
Vice President, Marketing and Business Development: Dr Eitan Yudelevich
Tel: (+972 4) 879 47 14
Vice President, Finance and Economics: Michael Weiner
Managing Director, Systems Division: David Zait
Operations Director: Rehuven Gal
Spokesman and Public Relations Director: Gila Harel
Strategic Programmes and Human Resources: Igal Srebro
Propulsion and Explosive Systems Division: Menachem Kidron
Deputy General Manager, Marketing and Business Development: Dror Shachnai
Marketing, Systems Division: Mrs Aviva Wolk
Acoustic Systems Directorate Contact: Giyora Saar
Contact: Jacob Grinstein

■ Design, development, manufacture and supply of a range of advanced defence systems. Products include: naval, air and ground precision guided weapons; electro-optical systems; Electronic Warfare (EW) systems; Command, Control, Communications, Computers and Intelligence (C4I) systems; acoustic defence systems and armoured protection systems. The company has also formed partnerships with civilian counterparts to develop commercial applications based on its proprietary technology.

Products include Shafrir 2, Python3, Python 4/5 and Derby air-to-air missiles; AGM-142 Popeye and NT-D air-to-surface missiles; TAL 1/2 Cluster, Barad, Pyramid and Spice Bombs.

Sub-divisions
Missiles Division
Ordnance Systems Division
Propulsion and Explosives Division
Systems Division

Subsidiaries
Rafael Development Corp. Ltd
Semi-Conductor Devices
Semiconductor Engineering Laboratories Ltd (SELA)

UPDATED

Rafael Armament Development Authority Ltd
(Propulsion and Explosive Systems Division)
PO Box 2250/ M1, IL-31021 Haifa, Israel
Manager, Expendable Decoys Directorate: Arnold Segal
Marketing Manager: Dor Fuchs
Marketing Department: Mrs Tova Tauber

■ Design, development and manufacture of rocket propulsion, warhead and pyrotechnic components for missile systems. Space propulsion systems and components, remote breaching devices and naval EW decoy rockets and launchers.

UPDATED

Tadiran Spectralink Ltd

29 Hamerkava Street, PO Box 150, IL-58101 Holon, Israel
Tel: (+972 3) 557 31 02
e-mail: info@tadspec.com
Web: http://www.tadiran-spectralink.com
President and Chief Executive Officer: Itzhak Beni
Tel: (+972 3) 557 31 01
Vice President, Finance and Control: Yehudit Rozenberg
Tel: (+972 3) 557 31 04
Vice President, Search & Rescue Systems & Guided Weapons: Zvi Krepel
Vice President, Data Link Systems for UAVs: Reuven Ron
Vice President, Operation & Customer Support: Itzhak Mor
Vice President Sales: Avi Peleg
Tel: (+972 3) 557 31 02

■ Supplier of data link systems for a variety of airborne platforms and ground installations including UAV's, guided weapons, C3I networks, TT&C for space vehicles and search and rescue systems. Datalink design, software development, systems integration, testing and manufacture of hardware, technical support and global marketing. Product lines include: digital processing techniques including direct sequence, spread spectrum, frequency hopping, error detection and correction codes, data interleave and deinterleave and burst synchronisation. Communication and network management protocols, real-time video compression and expansion, frequency bands from UHF, L, S, C, X, up to Ku.

Accreditations: ISO 9001:2000

UPDATED

The Elisra Group

(Corporate Headquarters)
(ex-Elisra Electronic Systems Ltd)
48 Mivtza Kadesh Street, IL-51203 Bene Beraq, Israel
Tel: (+972 3) 617 55 22
(+972 3) 617 51 11
Fax: (+972 3) 617 58 50
e-mail: marketing@elisra.com
Web: http://www.elisra.com
Chairman and Chief Executive Officer: Danny Bian
Tel: (+972 3) 617 55 60
President and Chief Executive Officer, Tadiran Electronic Systems and Tadiran Spectralink: Itzhak Ben
Senior Vice President, Marketing: Dov Avron
Tel: (+972 3) 617 54 66
Senior Vice President and Chief Technology Officer: Menachem Oren
Tel: (+972 3) 617 57 55
Vice President and General Manager, Microwave Division and Operations: Arie Dolev
Tel: (+972 3) 617 52 50
Vice President Finance: Ran Maidan
Tel: (+972 3) 617 59 09
Vice President and General Manager, Engineering Division: Moti Peleg
Tel: (+972 3) 617 57 47
Vice President and General Manager, Military Systems: Shlomo Livne
Tel: (+972 3) 617 51 89
Press Relations Contact: Ms Oreet Berr Yaacov
Tel: (+972 3) 570 65 27
Fax: (+972 3) 570 65 26

■ Military, security and civilian solutions for land, air and sea applications, providing defense and intelligence solutions for the armed and security forces of over 25 countries worldwide. The Group offers tailor-made integrated mission systems across the following fields Electronic Warfare (EW) and Defencive Air Suites (DAS); Command, Control, Communications, Computing and Intelligence, (C4I); Homeland Security (HLS); Search and Rescue (SAR), spectrum control and microwave components.

Products include new generation self-protection suites (DAS) for fighters, helicopters, transport MPA and commercial aircraft that include Radar Warning Receiver (RWR), Infra Red missile approach and laser warning systems, and the integration of chaff and flare dispensing systems; Electronic Surveillance Measures (ESM)/ Electronic Intelligence (ELINT), Communications Intelligence (COMINT), direction finding (DF) and Communications Jamming (COMJAM); C4I solutions for all battlefield command chains via IP technology; battlefield management and control; theater missile defense including test bed; artillery C4I; advanced SAR Systems; data links for UAVs & guided weapons.

The Elisra Group comprises Elisra Electronic Systems Ltd, Tadiran Electronic Systems Ltd. and Tadiran Spectralink Ltd with activities spanning most aspects of EW activity.

UPDATED

Industry – Service

Gilat Satellite Networks Ltd

(Headquarters)
21 Yegia Kapayim Street, Kiryat Arye, IL-49130 Petah Tikvah, Israel
Tel: (+972 3) 925 22 01
Fax: (+972 3) 925 22 52
Web: http://www.gilat.com
Chairman and Chief Executive Officer: Shlomo Rodav
Vice President, Marketing and Business Development: Tal Meirzon
Vice President, Finance: Ms Tal Payne
Vice President, Technology: Acik Keshet
Vice President, Sales: Avihu Bergman
Vice President, Information Systems: Amiram Beilis
Vice President of Operations: Rami Miron
Vice President, Customer Services: Doron Ron
Vice President, Human Resources: Ms Rachel Hermelin
Vice President, Treasury: Yaron Suher
Director, Corporate Marketing: Barry Spielman
Tel: (+972 3) 925 22 52

UPDATED

Images Sat International NV

(Tel-Aviv Office)
2 Kaufman Street, 17th Floor, IL-61500 Tel Aviv, Israel
Tel: (+972 3) 796 06 00
Fax: (+972 3) 516 34 30
e-mail: information@imagesatintl.com
Web: http://www.imagesatintl.com
Chief Executive Officer: Menashe Broder
President: Dr Moshe Bar-Lev
Senior Vice President and Chief Financial Officer: Ori Ben-Amotz
Senior Vice President, Business Development: Dave Krueger
Tel: (+972 602) 819 16 17
Vice President, Marketing and Sales: Noam Zafrir
Tel: (+972 3) 796 06 24
Vice President, Corporate Communication: Ms Karen Gold Anisfeld
Tel: (+972 3) 796 06 04
Mobile: (+972) 54 20 95 78

■ Commercial provider of high-resolution satellite imagery principally for national security, intelligence and homeland defence applications worldwide as well as for civilian applications.

UPDATED

Oreet International Media Ltd

(ex-Oreet-Marketing Communications)
15 Kineret Street, IL-51201, Bene Barak, Israel
Tel: (+972 3) 570 65 27
Fax: (+972 3) 570 65 26
e-mail: anat@oreet-marcom.com
liat_h@oreet-marcom.com
Web: http://www.oreet-marcom.com
Managing Director: Ms Oreet Ben-Yaacov
Press Relations: Efi Beck

■ Public relations and marketing communications company.

Representing:
Elbit Systems Ltd, Israel
Elisra Electronic Systems Ltd, Israel
ELOP Electro-Optics Industries Ltd, Israel
Silver Arrow, Israel
Tadiran Electronic Systems Ltd, Israel
Tadiran Spectralink, Israel

UPDATED

Paz Logistics Ltd

10 Yad Harutzim Street, PO Box 2242, IL-44425, Kfar-Saba, Israel
Tel: (+972 9) 766 15 55
Fax: (+972 9) 766 15 50
e-mail: info@pazlog.co.il
Web: http://www.pazlog.co.il
President: Ofer Paz
Technical Director: Ofer Ran
Sales Director: Roni Zmini
Operations and Purchasing Manager: Anat Paz

■ High level logistic and technical consulting. Specialise in weapons, guns, ammunition and related accessories, optical systems, night vision equipment, communication systems, GPS and satellite communication, navigation, compass systems, police and military equipment and textiles, automative and retrofit projects. Representation of foreign firms in Israel.

UPDATED

Spacecom Satellite Communication Services Ltd

Suite 1017 Twin Towers I, 33, Jabotinsky Street, IL-52511 Ramat Gan, Israel
Tel: (+972 3) 613 47 20
Fax: (+972 3) 613 47 23
e-mail: info@spacecom.co.il
Web: http://www.spacecom.co.il
Chief Financial Officer: Itzhak Shnaiberg
Marketing Director: Jacob Keret
Sales Director: Eyal Rosenfarb
Broadland Services Director: Erez Yaat
Technical Director: Avi Shaviv
Vice President, Business Development: Hanoch Dombek
Vice President, Contracts: Gil Lotan

■ Satellite communication services

UPDATED

Italy

Government – Intergovernmental

European Space Agency (ESA)

(European Space Research Institute)
Via Galileo Galilei, PO Box 64, I-00044 Frascati, Roma, Italy
Tel: (+39 06) 94 18 01
(+39 06) 94 18 03 50
(+39 06) 94 18 09 51
Fax: (+39 06) 94 18 02 80
Web: http://www.esa.int
Head of E.O Applications Department: Stephen Briggs
Tel: (+39 06) 94 18 04 01
Head of ESRIN Public and Institutional Relations Office: Ms Simmoneta Cheli

UPDATED

European Space Agency (ESA)
(ESRIN Earthnet Program Office)
Via Galileo Galilei, PO Box 64, I-00044 Roma, Italy
Tel: (+39 06) 940 11
(+39 06) 94 18 01
(+39 06) 94 18 03 50
(+39 06) 94 18 09 51
Fax: (+39 06) 940 13 61
(+39 06) 94 18 02 80
Head of EO Applications Department: Stephen Briggs
Tel: (+39 06) 94 18 04 01
Head of ESRIN Public and Institutional Relations Office: Ms Simonetta Cheli
■ Remote sensing satellite data acquisition preprocessing and distribution.
Publications: Earthnet Review

UPDATED

Centres of Learning & Research

Consiglio Nazionale delle Ricerche (CNR)
(Istituto di Radioastronomia)
Viap P Gobetti 101, I-40129 Bologna, Italy
Tel: (+39 051) 639 93 85
Fax: (+39 051) 639 94 31
Director: Prof Gianni Tofani
Tel: (+39 051) 639 94 60

UPDATED

Association

Italian Small and Medium Enterprise Aerospace Association
Via del Tempio 1, Roma, Italy
Tel: (+39 06) 686 92 22
Fax: (+39 06) 686 92 22
Web: http://www.aipas.it
Vice President: Lanfranco Zucconi
Tel: (+39 02) 38 04 82 12
Fax: (+39 02) 38 04 82 83

UPDATED

Industry – Manufacturing

Aero Sekur SpA
(ex-Irvin Aerospace SpA)
(ex-Irvin Manufatture Industriall SpA)
Via Delle Valli, I-04011 Aprilia, Latina, Italy
Tel: (+39 06) 92 01 61
Fax: (+39 06) 92 72 71 65
e-mail: commerciale@aerosekur.it
Web: http://www.aerosekur.it
Marketing Manager: Ing Mauro de Leonardis
Mobile: (+39 348) 654 19 95
■ Parachutes for personnel, cargo, weapon and aircraft retardation. Recovery systems for missiles/space vehicles and cargo delivery systems. Safety harnesses and air drop sequence control devices. Inflatable survival products and flexible fuel tanks. Nuclear, biological and chemical equipment. Camouflage, lifevests, aerial delivery platforms and field hospital equipment (air transportable). Flexible fuel tanks for aircraft and military vehicles.

UPDATED

Alenia Aeronautica SpA
Via Campania 45, I-00187 Roma, Italy
Tel: (+39 06) 42 08 81
Fax: (+39 06) 42 82 45 28
e-mail: addeo@alenia.finmeccanica.it
Web: http://www.alenia-aeronautica.it
Chairman and Chief Executive Officer: Giorgio Zappa
Press Officer: Ms Antonello De Montis
Tel: (+39 06) 41 72 39 56
■ Organised into different business lines: Combat Aircraft, Military Transport Aircraft, Commercial Aircraft (Regional and Aerostructures), Modification and Overhaul through Aeronavali Company.
Design and manufacture of combat and transport aircraft such as Eurofighter Typhoon, Tornado, AMX, ATR42 MP Surveyor and C-27J Spartan.
Commercial aircraft sector activities include the production of the ATR turbo-prop aircraft family jointly developed with EADS.
Alenia co-operates with Boeing participating in the manufacture of B767, B777, B717, B757 and B7E7. It also co-operates with Airbus by manufacturing aerostructures for the A321 and A380 and supplies parts for the A319/320/321/330, A300/310 and A340-500/600. Collaborates with Dassault in the production of the Falcon 2000 and 900EX.
In the modification and maintenance field Alenia Aeronautica, through Aeronavali, operates on many of the important commercial and military aircraft. It has full capability of design, production, installation and test of complex parts and systems.

UPDATED

Alenia Spazio SpA
(Headquarters)
Via Saccomuro 24, I-00131 Roma, Italy
Tel: (+39 06) 415 11
Fax: (+39 06) 419 06 75
e-mail: communications@roma.alespazio.it
Web: http://www.aleniaspazio.com
■ Space company involved in the construction of 200 satellites. Four operating areas: telecommunications, remote sensing, orbital infrastructures and scientific satellites.
Subsidiaries:
Laben
QSW
SSI

UPDATED

Avio SpA
(Headquarters)
(ex-Fiat Avio SpA)
Viale I° Maggio 99, I-10040 Rivalta, Italy
Tel: (+39 011) 685 81 11
Fax: (+39 011) 685 98 32
Cable: FIATAVIOTORINO
Web: http://www.fiatavio.it
President: Ing. Paolo Torricelli
Chief Executive and Operating Officer: Ing. Saverio Strati
Vice President, Business Development and Strategies: Ing. Giorgio Abrate
Vice President, Engineering and Quality: Ing. Leonard Caroni
Vice President, Procurement and Industrial Operations: Ing. F Canna
Vice President, Repair and Overhaul Unit – Civil Engines: Ing. U Catani
Vice President, Personnel and Organisation: Dr Roberto Ciervo
Director of Finance: V Rabajoli
Director of Aviation Business Unit – Civil Engines: Ing. Franco Rodi
Director of Aviation Business Unit – Military Engines: Ing. F Martini
Director of Space Business Unit: B Lasagni
Communications and Image Manager: Ms Silvia Maoli
■ Design, development and production of complete engines (both turbofan and turboprop) and engine mechanical components, accessory gear boxes and power transmissions, low pressure turbines, auxiliary power units for military and commercial aircraft and helicopters, compressors, turbine blades and vanes, discs and shafts, combustion chambers and afterburners. Design, development and production of helicopter engines and power transmissions, marine propulsion turbines. Assembly and testing of military engines and gas turbines for marine propulsion. Design and production of auxiliary power units. Space propulsion: liquid oxygen turbopumps for cryogenic propulsion, space solid propulsion for missile application and rockets. Liquid, solid, gas and arcjet propulsion for satellites, separation motors, orbital propulsion systems, light launchers missile and rocket engines. Repair, overhaul, technical assistance and logistic support of military/civil engines for aircraft, helicopters, marine and industrial applications. Design, development and production of advanced electronic automation and instrumentation systems for naval, defence, space, energy, railways and security applications. Marine and industrial engines and helicopter engines.

UPDATED

Avio SpA
(Colleferro (Roma) Plant)
Corso Garibaldi 22, I-00034 Colleferro, Roma, Italy
Tel: (+39 06) 972 91
Fax: (+39 06) 97 29 22 99
■ Space products: solid propellant apogee motors. Ariane 4 and 5 boosters and separation motors. Liquid propellant and cold gas propulsion systems. Magneto plasma dynamic electric propulsion. Small launcher (VEGA). Defence products: powders and propellants for missile motors.

UPDATED

Carlo Gavazzi Space SpA (CGS)
Via Gallarate 150, I-20151 Milan, Italy
Tel: (+39 02) 38 04 81
Fax: (+39 02) 308 64 58
e-mail: cgs@cgspace.it
Web: http://www.cgspace.it
President: Prof Manfred Fuchs
Vice President: Dr Marco Fuchs
Managing Director: Lanfranco Zucconi
Marketing Director: Roberto Aceti
Executive Assistant: Ms Sabrina Riccomini
Tel: (+39 02) 38 04 82 13
■ Facilities and payloads for microgravity applications, control electronics for on-board instrumentation and subsystems with associated test sets, GSE with related software, ground control station subsystems. Turn-key systems for small satellite missions, studies and research for facilities and experiments in microgravity, and scientific instruments for fundamental physics and microgravity.

UPDATED

Datamat Rome SpA
(ex-Datamat Ingegneria dei Sistemi SpA)
Via Laurentina 760, I-00143 Roma, Italy
Tel: (+39 06) 502 71
Fax: (+39 06) 50 51 14 07
Web: http://www.datamat.it
President: Dott. Gustavo Greco
Managing Director and General Manager: Ing. Giancarlo Giglio
Marketing Director: Ing. Franco Olivieri
Finance Director: Francesco Mannoni
Defence Business Development Director: Ing. Enrico Cuturi
Marketing Director, Defence Systems Division: Ing. Gabriele Pitzurra
Sales and Technical Director: Ing. E Cambise
Quality Assurance Director: A Aquili
Budget Control Director: C Carrara
Personnel and Organisation Director: G De Leonardis
■ C3I systems, training simulators, satellite ground stations, software technologies, system integration, logistic support.

UPDATED

Elettronica SpA
Via Tiburtina Valeria km 13.700, Loc Settecamini I-00131 Roma, Italy
Tel: (+39 06) 415 40 00
Fax: (+39 06) 415 49 24
e-mail: info@elt.it
■ Manufacture of RWR, ESM, ECM, ELINT and EWOS equipment for airborne, naval and ground applications. Participation in NH-90, EF-2000, Mirage 2000-9, Horizon and EH-101 programmes.

UPDATED

Eurimage SpA
Via E D'Onofrio 212, I-00155 Roma, Italy
Tel: (+39 06) 40 69 42 22
Fax: (+39 06) 40 69 42 32
e-mail: info@eurimage.com
Web: http://www.eurimage.com
Application Provider Network Manager: Ms Martina Wriessnegger
Tel: (+39 06) 40 69 43 34
■ Provides global satellite data and imagery.

UPDATED

Finmeccanica
Piazza Montegrappa 4, I-00195 Roma, Italy
Tel: (+39 06) 32 47 34 97
Fax: (+39 06) 32 65 71 70
e-mail: news@finmeccanica.com
Web: http://www.finmeccanica.it
Chairman: Pier Francesco Guarguaglini
■ Design and manufacture of aircraft, helicopters, satellites, missile systems, radar equipment, energy plants, trains, process and service automation systems. Active in the fields of aerospace, defence, energy, transportation and automation.

UPDATED

Industrie Cometto SpA
Via Cuneo No. 20, I-12011 Borgo San Dalmazzo CN, (Cuneo), Italy
Tel: (+39 0171) 26 33 00
Fax: (+39 0171) 26 63 35
e-mail: cometto@comettoind.com
Web: http://www.comettoind.com
Managing Director: Ing. Italo Vione
Export Area Manager: Stefano Tondelli
Mobile: (+39) 171 26 63 35
- Low-bed units for the transportation of heavy equipment; tanks and special loads in terms of weight and dimensions. Self-propelled units for various purposes including a NASA space shuttle SPU transporter and shipyard elevating transporters.

UPDATED

Laben SpA
SS Padana Superiore 290, I-20090 Vimodrone, Italy
Tel: (+39 02) 25 07 51
Fax: (+39 02) 250 55 15
e-mail: info@laben.it
Web: http://www.laben.it
Chairman: Dr Paolo Micheletta
Chief Executive Officer and Managing Director: Ing. Marco Pascucci
Human Resources and Organisation: Roberto De Franchis
- Payload data handling and spacecraft command and control subsystems for military surveillance satellites. Telemetry decommutation and processing (fixed and mobile stations). Flight test instrumentation for aircraft (PCM telemetry systems). On-board instrumentation for platforms, command, data handling, scientific payloads, remote sensing payloads, microgravity payloads. Space transportation electronics, on-board software and electrical ground support equipment. On-board processors for TDMA (Time Division Multiple Access) telecommunications payloads. Aircraft in-flight test instrumentation and ground stations (fixed and mobile).

UPDATED

Laben SpA
(Proel Tecnologie Division)
(ex-Laben Proel Tecnologie Division)
Viale Macchiavelli 31, I-50125 Florence, Italy
Tel: (+39 055) 22 87 31
Fax: (+30 055) 02 29 82 28
Managing Director: M Pascucci
Tel: (+39 02) 25 07 51
Financial and Operations Director: M Iacopini
Sales and Marketing Director: A Beretta
Operative Manager: Ing. Giovanni Matticari
Programs and Systems Manager: Giovanni E Noci
Purchasing Manager: E Roncada
- Electron sources and accelerators, plasma sources, ion thrusters, plasma contactors, E-beam cured composite materials.

UPDATED

Microtecnica-Torino Srl
Piazza Arturo Graf, PO Box 1210, I-10126 Turin, Italy
Tel: (+39 011) 693 21
Fax: (+39 011) 693 22 42
Telex: 221190 micro
President and Chief Executive Officer: Alberto Suppi
Financial Director: Roberto Hofman
Supply Chain Director: Enzo Bottero
Sales and Marketing Director: Carlo Marello
Deputy General Manager: Mario Botto Micca
Purchasing Manager: Giovanni Cerruti
Advertisement: Maurizio Bello
- Design, development and servicing of navigation and guidance systems for aircraft, helicopters and missiles. Primary and secondary flight controls, drive, control and actuation systems, air conditioning, pressurisation and heat transfer systems. Power generation and control. On-board power units for missiles, test stands.

UPDATED

Moog Italiana Srl
Via dei Tre Corsi, Zona Industriale sud D1,I-21046, Malnate, Italy
Tel: (+39 0332) 42 11 11
Fax: (+39 0332) 42 92 33
Telex: 380582
Chairman: Steve Huckvale
Managing Director: Ing Emanuele V Minneci
General Manager: Walter Terramant
- Flow servovalves, brushless servomotors and electronics. Propellant valves, redundant valves and servoactuators.

Representing: Moog Inc., US

UPDATED

Quadrics
(Rome Office)
Via Marcellina 11, I-00131 Roma, Italy
Tel: (+39 06) 41 23 86 15
Fax: (+39 06) 419 16 94
e-mail: sales@quadrics.com
Web: http://www.quadrics.com
- Manufacture high performance computer hardware.

UPDATED

Terma Srl
(ex-Terma Elektronik (Italia) Srl)
Via Milano 9 – 1 Piano, I-21023 Besozzo, Italy
Tel: (+39 033) 277 39 80
Fax: (+39 033) 277 39 81
e-mail: terma.it@terma.com
Web: http://www.terma.com

UPDATED

Thales Italia SpA
(Communications Division)
(ex-Alcatel Telettra)
(ex-Thales Communications SpA)
(ex-Thomson-CSF Electronic Systems Italia SpA)
Via Vincenzo Bellini 24, I-00198 Roma, Italy
Tel: (+39 06) 85 37 22 32
Fax: (+39 06) 85 37 22 50
- Development and manufacture of tactical communication systems and equipment.

UPDATED

Industry – Service

Space Software Italia (SSI)
Viale del Lavoro 101, I-74100 Taranto, Italy
Tel: (+39 099) 470 11 01
Fax: (+39 099) 470 17 77
e-mail: info@ssi.it
Web: http://www.ssi.it
- Space engineering systems and software.

UPDATED

Telespazio SpA
(Headquarters)
(ex-Nuova Telespazio SpA)
Via Tiburtina 965, I-00156 Roma, Italy
Tel: (+39 06) 407 91
Fax: (+39 06) 40 79 37 60
e-mail: comunicazione@telespazio.it
Web: http://www.telespazio.it
Chairman: Dott. Giorgio Zappa
Vice President: Ing. Alessandro Bellini
Chief Executive Officer: Ing. Enrico Saggese
Managing Director: Ing. Claudio Chierici
Director: Ing. Cosimo La Rocca
External Relations Manager: Dott. Domenico Primi
- Design, implementaion, operation and service of satellite communications systems, spatial services, closed user networks, mobile systems, in-orbit operation and control, and remote sensing with data processing and distribution.

UPDATED

The Institute of Information Science and Technologies (ISTI)
(ex-CNUCE)
(Istituto di Scienza e Tecnologies)
Consiglio Nazionale delle Ricerche, Area della Ricarca di Pisa, Via G. Moruzzi, I-56124 Pisa, Italy
Tel: (+39 050) 315 24 03
Fax: (+39 050) 315 28 10
Web: http://www.isti.cnr.it
External Relations Officer: Ms Claudia Raviolo

UPDATED

Vitrociset SpA
(Head Office and Plant)
Via Salaria 1027, I-00138 Roma, Italy
Tel: (+39 06) 882 01
Fax: (+39 06) 88 20 22 00
(+39 06) 88 20 25 75
(+39 06) 88 20 22 30
Web: http://www.vitrociset.it
Business Area Manager, Civil: Mauro Carosi
- Technical servicing in Italy and overseas for automated air traffic control (ATC) centres, ATC radar systems, air navigation, radio aids (including ILS, VOR, DME), communications systems, auxiliary systems. Overhaul, repair and calibration of test and measuring instruments. In flight radio-measurements. Servicing of Nike and Hawk missile systems in addition to Sparrow missile. Test engineering for SPADA missile system. Logistic engineering and technical servicing for radars and missile systems.

UPDATED

Industry – Sales Agent/Office

SIME sas di Paolo De Gaetano & Co.
Via F S Benucci 9, I-00149 Roma, Italy
Tel: (+39 06) 55 27 08 29
(+39 06) 55 27 30 49
Fax: (+39 06) 551 56 62
Web: http://www.simesas.com
General Partner: Paolo De Gaetano
- Numerical simulation, virtual prototype software, antennas satellite earth stations, telescopic hangars for ships and helicopter landing systems. Shipborne helicopter handling.

Representing:
Engineering Systems International, Netherlands
ESI Software Group, Paris
Indal Technologies Inc., Canada
Jered Industries, Inc., US
Vertex RSI, US

UPDATED

Japan

Government – Ministry/Agency

Japan Aerospace Exploration Agency (JAXA)
(Tokyo Office)
(ex-National Space Development Agency of Japan)
World Trade Centre Building 28F, 2-4-1 Hamamatsu-cho, Minato-ku, Tokyo, 105-8060, Japan
Tel: (+81 3) 34 38 60 00
(+81 3) 34 38 60 35
(+81 3) 34 38 61 11
Fax: (+81 3) 54 02 65 13
(+81 3) 54 02 65 12
e-mail: proffice@jaxa.jp
Web: http://www.jaxa.jp
President: Shuichiro Yamanouchi
Tel: (+81 3) 34 38 60 01
Vice President: Kaoru Mamiya
Tel: (+81 3) 34 38 60 02
Executive Director, Strategic Planning and Management Department, Industrial Collaboration Department and International Relations Department: Kiyoshi Higuchi
Tel: (+81 3) 34 38 60 04
Executive Director, General Affairs, Human Resources, Finance, Contracts, Assessment and Audit Office, Public Affairs, Safety and Mission Assurance, Information Systems, and Management: Kazuo Yoshikawa
Tel: (+81 3) 34 38 60 03
Executive Director, Office of Space Flight and Operations: Koji Yamamoto
Tel: (+81 3) 34 38 60 05
Executive Director, Office of Space Applications: Yoji Furuhama
Tel: (+81 3) 34 38 60 07
Executive Director, Office of Space Technology and Operations: Tsukasa Mito
Tel: (+81 3) 34 38 60 06
Executive Director. Institute of Space Technology and Aeronautics: Susumu Toda
Tel: (+81 29) 86 85 07
General Auditor: Shinji Nio
Tel: (+81 42) 759 80 07
General Auditor: Norio Kaya
Executive Director, Institute of Space and Astronautical Science: Koichiro Tsuruda

Tel: (+81 43) 759 80 07
Public Affairs Officer: Yasunori Matogawa
Tel: (+81 3) 34 38 60 10
Public Relations Department: Munetoh Wakana
■ Established as a corporate entity charged with developing launchers and applications satellites. Undertakes development and launch of vehicles development and operation of earth-observation satellites; tracking and control of satellites. Launching services, research and development and rockets. Space science, aircraft, earth observation and satellite data aquisition.
Overseas offices
Bangkok, Bonn, Houston, Kennedy Space Center, Los Angeles, Paris, Washington DC,

UPDATED

Japan Aerospace Exploration Agency (JAXA)
(Earth Observation Center)
(ex-National Space Development Agency of Japan)
1401 Aza-Numanoue, Oaza-Oohashi, Hiki-gun, Saitama, 350-0302, Japan
Tel: (+81 492) 98 12 00
Fax: (+81 492) 96 02 17
■ Satellite data acquisition and earth observation.

UPDATED

Japan Aerospace Exploration Agency (JAXA)
(ex-National Space Development Agency of Japan)
Triton Square X-23F, 1-8-10 Harmui, Chuo-ku, Tokyo 104-6023, Japan
Tel: (+81 3) 62 21 90 00
Fax: (+81 3) 62 21 91 91
Librarian: Mie Yukimura
■ Earth observation, research and development and study services.

UPDATED

Japan Aerospace Exploration Agency (JAXA)
(Katsuura Tracking and Communication Station)
1-14 Hanatateyama, Haga, Katsuura-shi, Chiba-ken, 299-5213, Japan
Tel: (+81 470) 73 06 54
Fax: (+81 470) 70 70 01
■ Tracking and control.

UPDATED

Japan Aerospace Exploration Agency (JAXA)
(Masuda Tracking and Communication Station)
(ex-National Space Development Agency of Japan)
1897, Masuda, Nakatane-machi, Kumage-gun, Kagoshima, 891-3603, Japan
Tel: (+81 9972) 719 90
Fax: (+81 9972) 420 00
■ Tracking and control.

UPDATED

Japan Aerospace Exploration Agency (JAXA)
(Nagoya Liaison Office)
(ex-National Space Development Agency of Japan)
Kanayama Sougo Building 10F, 1-12-14, Kanayama, Naka-ku, Nagoya-shi, Aichi 460-0022, Japan
Tel: (+81 52) 332 32 51
Fax: (+81 52) 339 12 80

UPDATED

Japan Aerospace Exploration Agency (JAXA)
(Ogasawara Downrange Station)
(ex-National Space Development Agency of Japan)
Kuwanokiyama Chichijima, Ogasawara–mura, Tokyo, 100-2101, Japan
Tel: (+81 4998) 225 22
Fax: (+81 4998) 223 60
■ Tracking and control.

UPDATED

Japan Aerospace Exploration Industry (JAXA)
(Tanegashima Space Center)
Aza Mazu, Minamitane-machi, Kumage-gun, Kagoshima-Ken, 891-3703, Japan
Tel: (+81 9972) 621 11
Fax: (+81 9972) 440 04
■ Rockets, launchign service and space vehicles.

UPDATED

Transport Administration – Space

Communications Research Laboratory

(Strategic Planning Division)
4-2-1 Nukui-Kitamachi, Koganei, Tokyo, 184-8795, Japan
Tel: (+81 42) 327 53 92
Fax: (+81 42) 327 75 87
Web: http://www.crl.go.jp/overview/index.html

UPDATED

Centres of Learning & Research

Institute of Space and Astronautical Science (ISAS)

3-1-1 Yoshinodai, Sagamihara, Kanagwa, 229-8510, Japan
Tel: (+81 42) 751 39 11
(+81 42) 759 82 38
Fax: (+81 42) 759 42 51
Web: http://www.isas.ac.jp
Research Associate: Yoshitaka Saito
■ Government agency responsible for scientific research of space.

UPDATED

Japan Aerospace Exploration Agency (JAXA)
(ex-National Aerospace Laboratory)
World Trade Center Building 2-4-1, Hamamatsu-cho, Minato-ku, Tokyo, 105-8060, Japan
Tel: (+81 3) 34 38 61 11
(+81 3) 34 38 60 73
(+81 3) 34 38 66 50
Web: http://www.jaxa.jp/index_e.html
Contact: Takashi Akuto
■ Government establishment for research and development in aeronautical and space technology.

UPDATED

National Astronomical Observatory (NAOJ)

(Kokuritsu Tenmondai)
Osawa Mitaka-shi, Tokyo, 181-8588, Japan
Tel: (+81 422) 34 36 00
Fax: (+81 422) 34 36 90
Telex: 2822307 taomtk
Web: http://www.nao.ac.jp/
Public Relations Director: Dr Toshio Fukushima
Division Chief, Public Information Office: Jun-ichi Watanabe
Tel: (+81 422) 34 36 88

UPDATED

Association

CBA International

Naka, PO Box 12, Yokohama, 231-0057, Japan
Tel: (+81 45) 664 24 84
Cable: CBAI
Executive Director: Yusuke J Matsumura
Research Director: S Fujii
■ Non-profit research organisation, technical investigation for radar/visual UFO reports by civilian and military aviation pilots.
Publications: Aerospace UFO News (Quarterly)
Divisions
Aerospace UFO Research Center
HQ International UFO Observer Corps

UPDATED

Japan Aerospace Exploration Agency (JAXA)

(Tsukuba Space Center)
(ex-National Space Development Agency of Japan)
2-1-1 Sengen, Tsukuba-shi, Ibaraki-ken, 305-8505, Japan
Tel: (+81 29) 868 50 00
Fax: (+81 29) 868 59 88
■ Research and devlopment, study services, satellites, tracking and control, satellite data acquisition and spacecraft.

UPDATED

Japan Aerospace Exploration Agency (JAXA)

(Okinawa Tracking and Communication Station)
(ex-National Space Development Agency of Japan)
1712, Kinrabaru, Aza Afuso, Onna–son, Kunigami–gun, Okinawa-Ken, 904-0402, Japan
Tel: (+81 989) 67 82 11
Fax: (+81 989) 83 30 01
■ Tracking and control.

UPDATED

The Society of Japanese Aerospace Companies Inc. (SJAC)

2nd Floor, Toshin-Tameike Building, 1–14 Akasaka 1–chome, Minato-ku, Tokyo, 107-0052, Japan
Tel: (+81 3) 35 85 05 11
Fax: (+81 3) 35 85 05 41
Web: http://www.sjac.or.jp
Chairman, Mitsubishi Heavy Industries, Ltd: Takashi Nishioka
Vice Chairman, Kawaski Heavy Industries Ltd: Masamato Tazaki
Vice Chairman, Shimadzu Corporation: Hidetoshi Yajima
Vice Chairman, Mitsubishi Electric Corporation: Ichiro Taniguchi
President: Takatoshi Hosoya
Senior Vice President: Toshitsugu Tanaka
Senior Vice President: Hidejiro Yamada
Vice President: Kazuyuki Hirose
Vice President, International: Shuichi Miwa
■ Non-profit trade association acting as the representative body of the Japanese aerospace industry. Members are mainly Japanese enterprises engaged in the manufacture of aircraft, engines, avionics, space vehicles and related equipment and products. The object of the society is to contribute to upgrading industrial activity as well as national welfare by encouraging the manufacture and expansion of the international trade of aerospace and its related equipment.

Associate members
Asashi Air Supply Inc.
BAE Systems
Barco Co. Ltd
Corus Aluminium Japan Ltd
CSP Japan, Inc.
Fuji Industries Co. Ltd
Fuji Research Institute Corporation
Galaxy Express Corporation
High-Reliability Components Corporation
International Aircraft Development Fund
International Task Force Ltd
The Ishida Corporation
Itochu Aviation Co., Ltd
Itochu Corporation
JGC Corporation
Jupitor Corporation
Kanematsu Corporation
Kawasho Corporation
Kyokuto Boeki Kaisha Ltd
Mainami Boeki Co. Ltd
Marubeni Aerospace Corporation
Marubeni Corporation
Mikuni Shoko Corporation
Mitsubishi Corporation
Mitsubishi Research Institute Inc.
Mitsui & Co. Ltd
Mitsui Chemicals, Inc.
Morimura Bros. Inc.
Nichimen Corporation
Nippon Aircraft Supply Co. Ltd
Nissho Iwai Corporation
NTK International Corporation
Rocket System Corporation
Shintoa Corporation
Space Engineering Development Co. Ltd
Sumitomo Corporation
The Tokyo Marine and Fire Insurance Co. Ltd
Yamada Corporation

Companies represented
Aero Asahi Corporation
All Nippon Airways Co. Ltd
AMS Corporation
APC Aerospeciality Inc.
Chiyoda Corporation
Chubu Nihon Maruko Co., Ltd
Commercial Airplane Company
Daicel Chemical Industries Ltd
Daido Steel Co. Ltd
Daikin Industries Ltd
Denso Corporation
DIC Hexcel Co., Ltd
Eagle Industry Co. Ltd
Fuji Bellows Co. Ltd
Fuji Heavy Industries Ltd
Fujikin International Inc.
Fujitsu Ltd
Fujiwara Co. Ltd
The Furukawa Battery Co. Ltd
The Furukawa Electric Co. Ltd
Furuno Electric Co. Ltd
Hitach Kokusai Electric Inc.
Hitachi Ltd
Hitachi Metals Ltd
Hoden Seimitsu Kako Kenkyusho Co. Ltd
Honda Motor Co. Ltd
IHI Aerospace Co., Ltd
Ishikawa Seisakusho Ltd
Ishikawajima-Harima Heavy Industries Co. Ltd
Jamco Corp.
Japan Air Lines Co. Ltd

Japan Air System Co. Ltd
Japan Aircraft Development Corp.
Japan Aircraft Manufacturing Co. Ltd
Japan Aviation Electronics Industry Ltd
Japan Radio Co. Ltd
Japanese Aero Engines Corp.
Kaiyo Denshi Kogyo K K (KDK)
Kanto Aircraft Instrument Co. Ltd
Kashifuji Work Ltd
Kawada Industries Inc.
Kawanishi Aero-Parts Products Co. Ltd
Kawasaki Heavy Industries Ltd
Kayaba Industries Co. Ltd
Kitamura Machinery Co. Ltd
Kobe Steel Ltd
Koito Industries Ltd
Komatsu Zenoah Co.
Koyo Seiko Co. Ltd
Machida Endoscope Co. Ltd
Matsuura Machinery Corp.
Meira Corporation
Minebea Co. Ltd
Mitsubishi Cable Industries Ltd
Mitsubishi Chemicals Inc.
Mitsubishi Electric Corp.
Mitsubishi Heavy Industries Ltd
Mitsubishi Materials Corp.
Mitsubishi Precision Co. Ltd
Mitsubishi Rayon Co. Ltd
Mitsubishi Space Software Co. Ltd
Murata Machinery Ltd
Nac Image Technology Inc.
Nachi-Fujikoshi Corp.
NEC Aerospace Systems Ltd
NEC Corporation
NEC Network and Sensor Systems, Ltd
NEC Toshiba Space Systems, Ltd
NGK Spark Plug Co. Ltd
Nihon Pall Ltd
Nihon Tokushu Toryo Co. Ltd
Nikkiso Co. Ltd
Nippon Avionics Co. Ltd
NOF Corporation
NSK Ltd
NTN Corporation
NTT Comware Corporation
NTT Data Corporation
OKI Electric Industry Co. Ltd
Owari Precise Products Co. Ltd
Sakura Rubber Co. Ltd
Sampa Kogyo K K
Shimadzu Corporation
Shinko Electric Co. Ltd
Shinmaywa Industry Co. Ltd
Shounan Seiki Co. Ltd
Showa Aircraft Industry Co. Ltd
Sogo Spring Mfg Co. Ltd
STS Corporation
Sumijyu Precision Forging Co., Ltd
Sumitomo 3M Ltd
Sumitomo Electric Industries Ltd
Sumitomo Light Metal Industries Ltd
Sumitomo Metal Industries Ltd
Sumitomo Precision Products Co. Ltd
Tamagawa Seiki Co. Ltd
Tatsumi Manufacturing Co. Ltd
Teijin Seiki Co. Ltd
Terauchi Mfg Co. Ltd
The Japan Steel Works Ltd
Toho Texas Co. Ltd
Tokimec Inc.
Tokyo Aircraft Instrument Co. Ltd
Toray Industries, Inc.
Toshiba Corporation
Toshiba Electronic Systems Co. Ltd
Toyo Communication Equipment Co. Ltd
Yokogawa Denshikiki Co. Ltd
Yokogawa Electric Corporation
The Yokohama Rubber Co. Ltd
Yoshimitsu Industries Inc.

UPDATED

Industry – Manufacturing

Fuji Heavy Industries Ltd (FHI)
(Fuji Jukogyo KK)
1-7-2 Nishi-Shinjuku, Shinjuku-ku, Tokyo, 160-8316, Japan
Tel: (+81 3) 33 47 25 25
(+81 3) 33 47 21 11
Web: http://www.fhi.co.jp

UPDATED

Fujitsu Ltd
(Kawasaki Office)
4-1-1 Kamikodanaka, Nakahara-ku, Kawasaki, Kanagawa, 211-8588, Japan
Tel: (+81 44) 777 11 11
Web: http://www.fujitsu.co.jp
http://www.fujitsu.com

■ Spaceborne electronics including remote sensing radiometers, telemetry systems, microwave amplifiers, small earth stations, data processing systems and satellite networks.

UPDATED

High-Reliability Components Corp.
BIF, Shiba Masuda Building, 14-4 Shiba 1-Chome Minato-ku, Tokyo, 105-0014, Japan
Tel: (+81 3) 54 44 76 91
Fax: (+81 3) 54 44 76 95
Web: http://www.hirec.co.jp
President: Joji Tabata

■ Quality assurance and distribution of high reliability components for space applications.

UPDATED

Hitachi Ltd
6 Kanda-Surugadai 4-chome, Chiyoda-Ku, Tokyo, 101-8010, Japan
Tel: (+81 3) 32 58 11 11
Fax: (+81 3) 32 58 52 30
Web: http://www.hitachi.co.jp
Chief Operating Officer of Transportation Systems: Gaku Suzuki
General Manager, Transportation Systems Sales Division: Chiaki Ueda
General Manager, Overseas Marketing Department: Jun Kataoka
Contact: Yaeko Higaki

■ Communications equipment, computers, software, electronic components and computer circuits.

UPDATED

IHI Aerospace Co., Ltd
(Head Office)
Tomioka Plant, 900 Fiujik, Tomioka–Shi, Gunina Ken, 370-2398, Japan
Tel: (+81 3) 32 44 59 13
Fax: (+81 3) 32 44 59 18
Web: http://www.ihi.co.jp/ia
Chairman: Teiichi Tamaki
President: Akio Shinohara
Senior Managing Director: Tetsuo Chikara
Managing, Technical and Operations Director: Tatsuro Asai
Financial Director: Koichi Hoshi
Sales and Marketing Director: Satoshi Suzuki
Public Relations Director: Yoshinori Nanamori
Administration Department Manager: Yoshifumi Abe
General Manager, Purchasing Control Department: Hayato Ochiai

■ Design, development, production and sales of space equipment systems, defence rocket systems and other aerospace related products.

UPDATED

IHI Aerospace Co., Ltd
(Tomioka Plant)
(ex-Nissan Motor Co. Ltd)
900 Fujiki, Tomioka–Shi, Gunma–Ken, 370-2398, Japan
Tel: (+81 274) 62 41 23
(+81 274) 62 76 11
Fax: (+81 274) 62 77 11
Chairman: Teiichi Tamaki
Senior Managing Director: Tetsuo Chikata
Managing Director: Tatsuro Asai
Sales, Marketing and Managing Director: Satoshi Suzuki
Director: Yoshiaki Ano
Director: Yoshinori Nanamori
Financial Director: Koichi Hoshi
Director: Hitoshi Sato
General Manager, No.1 Space Systems: Kazunori Kawasaki
General Manager, Administration Department: Takahisa Iwane
General Manager, Sales and Marketing: Takao Endo
General Manager, Purchasing Control Department: Hayato Ochiai
Manager, Administration Department: Yoshifumi Abe

■ Design, development, test and manufacture of launch vehicles, space equipment systems, defence rocket systems and other aerospace products.

UPDATED

Ishikawajima-Harima Heavy Industries Co. Ltd (IHI)
(Head Office)
Shin–Ohtemachi Building 2–1, Ohtemachi 2–chome, Chiyoda-Ku, Tokyo 100-8182, Japan
Tel: (+81 3) 32 44 53 33
(+81 3) 32 44 57 72
(+81 3) 32 44 51 11
Fax: (+81 3) 32 44 53 98
(+81 3) 32 44 56 68
(+81 3) 32 44 51 31
Telex: 22232 ihihqt
Cable: IHICOTOKYO
Web: http://www.ihi.co.jp
President and Chief Operating Officer: Mototsugu Ito
Executive Vice President and General Manager, Corporate Planning Division: Eiji Inoue
Executive Vice President and General Manager, Business Administration: Eiichiro Iwamoto
Executive Vice President: Tadaaki Yamazaki
Managing Director: Nobuhiro Shimizu
Managing Director and General Manager, Procurement Strategics: Koki Kinoshita
Managing Director and Deputy General Manager, Business Administration: Kazuo Kanaya
Managing Director: Naoteru Tsuda
Managing Director and General Manager, Energy and Plant, Deputy General Manager, Procurement Strategics: Reiji Ishimoto
Managing Director and General Manager, Research and Development: Jyunichi Hamanaka
Managing Director: Hiroshi Katayama
Managing Director and General Manager, Industrial Machinery: Yoshikazu Kobayashi
Managing Director and General Manager, Aero-engine & Space Operation: Susumu Nagano
Managing Director and General Manager, Logistics Systems and Structures: Isao Nakao
Public Relations Director: Koichi Ozawa
Manager, Public Relations and Advertising Division: Yoshiaki Yamada
Tel: (+81 3) 32 44 53 42
Fax: (+81 3) 32 44 51 39
General Manager, Corporate Planning Division: E Inoue
Manager, Business Planning Group: Hiroshi Suzuki
Director and Senior Counsellor: Kosaku Inaba
Director and Deputy General Manager: Yasuo Shinohara
Director and Deputy General Manager, Aero-engine and Space Operation: Teiichi Tamaki
Director and Deputy General Manager, Energy and Plant: Takayasu Kato
Director and General Manager, Finance and Accounting Division: Hiroyoshi Hiraga
Director and Deputy General, Research and Development: Yukiya Nakagawa
Director and Deputy General Manager, Industrial Machinery: Hiromasa Ohmura
Director and Deputy General Manager, Energy and Plant: Seiji Hatae
Director and Deputy General Manager, Research and Development: Yasuhiro Inagawa
Director and General Manager, Corporate Business Development Division: Morihiko Kawabe
Director and Deputy General Manager, Logistics Systems and Structures: Tsugharu Asayama
Director: Fumio Sato
Director: Naoteru Tsuda
Corporate Auditor: Koichi Ichida
Corporate Auditor: Katsuji Minato
Corporate Auditor: Koichiro Ejiri
Corporate Auditor: Sugiichiro Watari

■ Construction of ships for the naval and civilian markets. Engine manufacture for ships, aircraft, rockets and missiles. Training and simulation equipment. Other products include: boilers for power plants, LNG and LPG storage tanks, oil and gas processing plants, container cranes, bridges, gates, shield tunneling machines, hot presses, pumps, and turbochargers.

UPDATED

Ishikawajima-Harima Heavy Industries Co. Ltd (IHI)
(Aero Engine and Space Operations Division)
Shin-Ohtemachi Building, 2-2-1Ohtemachi, Chiyoda-ku, Tokyo 100-8182, Japan
Tel: (+81 3) 32 44 53 33
Fax: (+81 3) 32 44 53 98
Managing Director and President of Operations: Susumu Nagano
Board Director and Vice President of Operations: Teiichi Tamaki
Associates Board Director and Vice President of Operations: Yusuke Tasaka
Manager, Business Planning Group and Control: Hiroshi Suzuki

■ Manufacture of turbofan and turbojet engines. Parts, accessories, fuel control systems, overhaul and repair. Space systems; rocket engines and ceramic bearings.

UPDATED

Japan Radio Co. Ltd (JRC)

Nittochi Nishi-Shinjuku Building, 10-1 Nishi-Shinjuku 6 Chome, Shinjuku-Ku, Tokyo 160-8328, Japan
Tel: (+81 3) 33 48 38 83
Fax: (+81 3) 33 48 39 37
Web: http://www.jrc.co.jp/eng/index-e.html
President: Tadahiro Muta
Executive Managing Director: Shinji Takeuchi
Manager, Overseas Project Group: Yoshihiko Yashiro
General Manager, Maritime Sales Department: Hideaki Takahashi
■ Manufacture of communications systems and equipment. Radio equipment, navigation systems and simulators.

UPDATED

Kawasaki Heavy Industries Ltd (KHI)

(Tokyo Head Office)
(Kawasaki Jukogyo)
World Trade Centre Building, 4-1 Hamamatsu-cho, 2 Minato-ku, Tokyo 105-6116, Japan
Tel: (+81 3) 34 35 21 11
Web: http://www.khi.co.jp
President: Masamoto Tazaki
Contact: Tohru Yanaka
Contact, Marketing and Sales Division: Ken Takeda
■ Main products: aircraft and jet engines, spacecraft and equipment, ships, marine engineering, rolling stock, plant and factory automation systems, construction machinery, small engines, motorcycle and jet-ski personal watercraft. Aerospace products: aircraft; fuselage of Boeing B777 and B767, fuselage panel of Airbus A321, wing rib of B737, outboard flap for B747, flap hinge fairing of MD80 and wing components ERJ-170. T-4 medium jet trainer and T-4 aerobatics trainer (Blue Impulse), wing, tail and after fuselage of F-15, fuselage of F-2. Rotary-Wing: BK117 (joint development with MBB (ECD), Germany), OH-1 Observation Helicopter, CH-47J, OH-6.

Space-related activities: payload fairing for H-2 rocket, geological experiment satellite, engineering test satellite VII, H-2 orbiting plane (in design) and automatic landing flight experiment/hypersonic flight experiment, Japanese experiment module (part of space station Alpha), astronaut training facility.

Engines: T53 and T55-K-712 turboshaft engines, V2500, RB211/Trent, PW4000 and auxiliary power units (international joint programmes), transmission for OH-1, BK117 and MD900, Air-Turbo-Ramjet engine (under development).

Other products: construction of airport facilities for domestic and international airports, maintenance hangars, docks, passenger bridges, aircraft cleaning systems, engine run-up noise suppressors, air cargo terminals, cargo storage and distribution systems. Hang-glider, motorcycle and missile simulators.

Overseas offices
Bangkok, Thailand
Beijing, China
Jakarta, Indonesia
Kuala Lumpur, Malaysia
Manila, Philippines
Seoul, South Korea
Shanghai, China
Sydney, Australia
Taipei, Taiwan

Subsidiaries
Canadian Kawasaki Motors Inc., US
Glory Kawasaki Motors Co. Ltd, Thailand
Hainan Sundiro-Kawasaki Engine Co. Ltd, China
Kawasaki Construction Machinery Corp. of America, US
Kawasaki do Brasil Indústria e Comércio Ltda, Brazil
Kawasaki Gas Turbine Europe GmbH, Germany
Kawasaki Heavy Industries (Europe) BV, Netherlands
Kawasaki Heavy Industries (HK) Ltd, Hong Kong
Kawasaki Heavy Industries (Singapore) Pte Ltd, Singapore
Kawasaki Heavy Industries (UK) Ltd, UK
Kawasaki Heavy Industries (USA) Inc., US
Kawasaki Heavy Industries GmbH, Germany
Kawasaki Machine Systems Korea Ltd, Korea, South
Kawasaki Motoren GmbH, Germany
Kawasaki Motors Corp. USA, US
Kawasaki Motors Enterprise (Thailand) Co. Ltd, Thailand
Kawasaki Motors Holding (Malaysia) Sdn Bhd, Malaysia
Kawasaki Motors Manufacturing Corp. USA, US
Kawasaki Motors Netherlands NV, Netherlands
Kawasaki Motors Pty Ltd, Australia
Kawasaki Motors (Phils.) Corp., Philippines
Kawasaki Motors (UK) Ltd, UK
Kawasaki Precision Machinery (UK) Ltd, UK
Kawasaki Rail Car Inc., US
Kawasaki Robotics GmbH, Germany
Kawasaki Robotics (UK) Ltd, UK
Kawasaki Robotics (USA) Inc., US
KHI Design & Technical Service Inc., Philippines
KHI (Dalian) Computer Technology Co. Ltd, China
Nantong COSCO KHI Ship Engineering Co. Ltd,
PT Kawasaki Motor Indonesia, Indonesia
Shanghai Cosco Kawsaki Heavy Industries Steel Structure Co. Ltd, China
Tiesse Robot SpA, Italy
Wuhan Kawasaki Marine Machinery Co. Ltd, China

UPDATED

Kawasaki Heavy Industries Ltd (KHI)
(Aerospace Company)
World Trade Centre Building, 4-1 Hamamatsu-cho, 2-chome, Minato-ku, Tokyo 105-6116, Japan
Tel: (+81 3) 34 35 21 11
Fax: (+81 3) 35 78 15 73
e-mail: webtekkou@khi.co.jp
Web: http://www.khi.co.jp
Control and Planning Division: Kazumasa Inukai
■ Manufacture, repair and overhaul of equipment for commercial aircraft, anti-submarine aircraft, fighters, trainers, helicopters, missiles and space systems.

UPDATED

Mitsubishi Corporation

(Aerospace Division)
2-6-3 Marunouchi, Chiyoda-ku, Tokyo, 100-8086, Japan
Tel: (+81 3) 32 10 46 04
(+81 3) 32 10 47 11
(+81 3) 32 10 21 21
Fax: (+81 3) 32 10 47 83
Web: http://www.mitsubishi.co.jp
President: Mikio Sasaki
Executive Vice President: Takeshi Hashimoto
Vice President and General Manager, Aerospace Division and General Manager of Space Transport Unit (Acting): J Fujikawa
General Manager, Electronic Systems and General Manager, New Business Initiative: M Nagase
General Manager, Aeronautics: M Tanaka
General Manager, Space Systems Unit and Space Business Development Unit (Acting): Y Kamiyama
General Manager, Airline Business Unit: K Yasuno
General Manager, Spatial IT Business Unit: A Yamaura
Assistant to General Manager: Katsutoyo Kaida
■ Commercial and military helicopters, military aircraft, aircraft engines, guided missiles and defence electronics. Satellite and space systems and aircraft interior equipment including galley, lavatory fittings and seats. Aircraft financing and leasing.

Representing
ADI Limited, Australia
Aerojet Gen Corp., US
Arch Chemicals Inc., US
Atlantic Research Corp., US
Aurora Flight Sciences Corp., US
Control Flow Inc., US
Del Mar Avionics, US
EMS Technologies Canada Ltd, Canada
Ferrostaal AG, Germany
GE-P&W Engine Alliance LLC, US
Honeywell Inc., US
Indal Technologies Inc., Canada
International Aero Engines AG, US
Jena-Optronik, Germany
L-3 Communications Corp., US
Lockheed Martin Corporation, US
Longbow International, US
Macdonald Dettwiler, Canada
NASAM Inc., US
Northrop Grumman Corporation, US
Pratt & Whitney Canada, Canada
Pratt & Whitney, US
Raytheon Company, US
Saab Ericsson Space AB, Sweden
Saft, France
Science Applications International Corp., US
Sermatech International Inc., US
Sikorsky Aircraft Corporation, US
Space Bridge Network Corp., US
Spacehab Inc., US
Stephen Ward, Australia
Strachen & Henshaw Ltd, UK
Thales Avionics In-Flight Systems LLC, US
The Aerospace Corp., US
The Boeing Company, US
Vertical Circuit Inc., US

UPDATED

Mitsubishi Electric Corporation (MELCO)

(Head Office)
Mitsubishi Denki Building, 2-3 Marunouchi 2-chome, Chiyoda-Ku, Tokyo, 100-8310, Japan
Tel: (+81 3) 32 18 34 30
Fax: (+81 3) 32 18 28 95
(+81 3) 32 18 90 48
Web: http://www.mitsubishielectric.com
President: Tamotsu Nomakuchi
Executive General Manager, Public Utility Systems Group: Takaaki Kijima
International Public Relations Manager: Frank Hashimoto
Section-C, Transportation Overseas Marketing Department: Taro Okabe
■ Missile systems, communication equipment, combat direction systems, EW equipment, radars, fire control systems, signal processing equipment, power equipment, computers, semiconductors and satellites.

UPDATED

Mitsubishi Heavy Industries Ltd (MHI)

(Head Office)
(Mitsubishi Jukogyo Kabushiki Kaisha)
16-5 Konan, 2-Chome, Minato-ku, Tokyo, 108-8315, Japan
Tel: (+81 3) 67 16 31 11
Fax: (+81 3) 67 16 58 65
Web: http://www.mhi.co.jp
Acting General Manager: Kenichi Hori
Manager, Public Relations Section: Wataru Ichikawa
■ Fuselage panels and doors for Boeing 777/767, wings for Bombardier Global Express and parts for the PW4000 turbofan engine. Flaps for the Boeing 737/747 and cargo doors for the Airbus A380.

Shipbuilding; steel structures; power systems; nuclear energy; machinery; general machinery.

UPDATED

Mitsubishi Heavy Industries Ltd (MHI)

(Nagoya Guidance and Propulsion Systems Works)
1200 Higashi-Tanaka, Komaki-shi, Aichi Ken, 485-8561, Japan
Tel: (+81 568) 79 21 13
Fax: (+81 568) 78 25 52
Web: http://www.mhi.co.jp
Director and General Manager: Minoru Takakuru
■ Manufacture, repair, testing and calibration of missiles and aerospace engines and control equipment. Type 93 air-to-ship guided missiles, Type 90 air-to-air guided missiles, Patriot surface-to-air missile systems. Rocket engines LE-5B and LE-7A.

UPDATED

Mitsubishi Precision Co. Ltd (MPC)

(Head Office)
3-13-16 Mita, Minato-ku, Tokyo, 108-0073, Japan
Tel: (+81 3) 34 53 64 21
Fax: (+81 3) 34 53 64 34
Web: http://www.mpcnet.co.jp/e/index.html
President and Chief Executive Officer: Takehiko Tatsuko
Senior Vice President, Kamakura Operations: Yuzo Kawata
Senior Vice President: Yoshiaki Hayakawa
Vice President, Corporate Planning Division: Junichi Matsuo
Vice President, Defense and Space Systens: Michikiro Yamao
Vice President, Social Solution Systems: Masato Hasegawa
Vice President, Kamakura Operations: Kazuo Tomiyama
Vice President, Kamakura Operations: Masaaki Nakanishi
■ Design, manufacture, repair, maintenance and sale of space, aviation and vehicular equipment and systems. Simulator, trainer and visual systems, information processing systems, parking management systems, traffic control systems.

UPDATED

Moog Japan Ltd

1-8-37 Nishi Shindo, Hiratsuka, Kanagwa, 254-0019, Japan
Tel: (+81 463) 55 36 15
Fax: (+81 463) 54 47 09
e-mail: sales@moog.co.jp
Web: http://www.moog.com
President: Sean Gartland

Tel: (+81 463) 55 02 93
Financial Director: Ikuo Yanaga
Sales Director: Roy Itoh
Marketing Director: Ashis Bhattacharya
Technical Director: Yas Sakurya
Operations Director: Mitsuo Ikeda
Public Relations Director: Tom Harada
Purchasing Manager: Hiroshi Hase
■ Manufacture of precision control components and systems, used in a wide range of high performance aircraft, strategic and tactical missiles, space vehicles and military ground vehicles.
Representing: Moog Inc., US

UPDATED

NEC Corporation

(Headquarters)
7-1 Shiba 5-chome, Minato-Ku, Tokyo, 108-8001, Japan
Tel: (+81 3) 34 54 11 11
Fax: (+81 3) 37 98 15 10
(+81 3) 37 98 15 11
(+81 3) 37 98 15 12
e-mail: webmaster@nec.co.jp
Web: http://www.nec-global.com
http://www.nec.co.jp
Senior Manager, Aerospace and Defence Business: Yasuharu Kurosu
■ Solutions: supercomputers, computers, PCs, printers. Networks: network systems and equipment, mobile communications and related software and services. Electron Devices: compound semiconductors, display modules, rechargeable batteries, capacitors.
Group Divisions
NEC Electron Devices
NEC Networks
NEC Solutions

UPDATED

Nippi Corporation

(ex-Japan Aircraft Manufacturing Co. Ltd)
3175 Showa-machi, Kanazawa-ku, Yokohama, 236-8540, Japan
Tel: (+81 45) 773 51 18
Fax: (+81 45) 773 51 01
(+81 45) 771 12 53
e-mail: webchief@mail.nippi.co.jp
Web: http://www.nippi.co.jp
Chairman of the Board of Directors: Takashi Sugoh
President and Director: Toshihiko Tahara
Managing Director: Shigenori Tomita
Managing Director and General Manager, Aerospace Division: Kazo Furusawa
Director and General Manager, Aircraft Maintenance Division: Ryota Morikawa
Director: Masayuki Kobayashi
Senior Manager, International Sales Department, Aerospace: Mitsuo Oyama
Senior Manager, Sales Department, Aircraft Maintenance Division: Norimitsu Kurumada
Tel: (+81 46) 265 21 80
Senior Manager, General Affairs Department: Tsuneo Tomita
Contact, General Affairs Department: Takero Kasaya
■ Manufacture of aircraft parts, rocket parts and space equipment. Maintenance and modification of aircraft. Manufacture of non-destructive inspection systems, industrial fans and marine equipment.

UPDATED

Oki Electric Industry Co. Ltd

7-12 Toranomon 1-chome, Minato-ku, Tokyo, 105-8460, Japan
Tel: (+81 3) 35 01 31 11
Fax: (+81 3) 35 81 55 22
Web: http://www.oki.co.jp
■ Sonobuoys, sonars, computer systems, communications equipment and electronic devices.

UPDATED

Sumitomo Heavy Industries Ltd (SHI)

(Head Office)
Sumitomo Heavy Industries Building, 9-11 Kitashinagawa 5-chome, Shinagawa-Ku, Tokyo, 141-8686, Japan
Tel: (+81 3) 54 88 80 00
Fax: (+81 3) 54 88 80 56
Web: http://www.shi.co.jp

UPDATED

Toshiba Corp.

(Headquarters)
1-1-1 Shibaura, Minato-ku, Tokyo, 105-8001, Japan
Tel: (+81 3) 33 57 84 15
(+81 3) 34 57 30 50
(+81 3) 34 57 30 89
(+81 3) 34 57 32 45
(+81 3) 34 57 49 24
Fax: (+81 3) 34 56 16 31
(+81 3) 34 57 83 85
(+81 3) 54 44 92 18
(+81 3) 54 44 93 33
Telex: 22587 toshiba
Cable: TOSHIBA
Web: http://www.toshiba.co.jp
Chairman: Taizo Nishimuro
President and Chief Executive Officer: Tadashi Okamura
Senior Executive Vice President: Takashi Lida
Senior Executive Vice President: Yasuo Morimoto
Senior Executive Vice President: Makoto Nakagawa
Executive Vice President: Yuji Kiyokawa
Executive Vice President: Tadashi Matsumoto
Executive Vice President: Atsutoshi Nishida
Director, New Business Development: Jun Ichinose
Tel: (+81 3) 34 57 30 78
Fax: (+81 3) 34 56 03 59
Corporate Communications Office: Ms Naomi Furuyu
■ Close range SAM (SAM-2, Keiko SAM, SAM-3) for air-to-air and shipborne applications. Short range SAM (SAM-1 TAN SAM, SAM 1-C improved TAN SAM), base air defence. Medium range SAM, hawk and main firing radar. Long range SAM, PATRIOT, module, power plant, radar set and transmitter. Avionics for P-3C, HRN 107 ILS receiver, HSN-3 INS, AN/APS-115C search radar, AN/ASQ-212(J) data processing system. Avionics for SH-60J, HRQ-109 VOR receiver, HYQ-1 HCDS (Helicopter Combat Direction System). Avionics for fighter aircraft F-15 and F-2, AN/ASN-109 INS, lead computing gyro unit, map generator. Avionics for C-1 cargo aircraft, J/AYN-1 map navigation display system, J/APN-93 navigation weather radar. Avionics for AH-1S attack helicopter, FCC Fire Control Computer, JARN-A10 VOR/ILS receiver. Avionics for U-125 rescue aircraft, J/APQ-2 search radar. Avionics for E-2C and E-767 AWACS. Field artillery systems including FADS Field Artillery Data-Processing System, JMPQ-P13 mortar locating radar, JTPS-P16 artillery locating radar. Navigation support systems, GCA Ground Controlled Approach, RIPS Radar Information Processing Systems, ILS Instrument Landing System, relocatable control tower. Weather radars: J/FPH-6(); J/FPH-8() and LPN-10, NLPN-8(). Advanced bridge systems. Information and communication systems, electronic devices, medical equipment, power plant systems, transportation equipment, consumer products.
Subsidiaries with defence business
Toshiba Electro-Wave Products Corporation

UPDATED

Toyo Communication Equipment Co. Ltd (TOYOCOM)

484 Tsukagoshi 3-chome
Kawasaki-shi, Kanagwa, 212 8513, Japan
Tel: (+81 44) 542 65 00
Fax: (+81 3) 542 63 54
Web: http://www.toyocom.co.jp/english/index.html

UPDATED

Industry – Service

Broadcasting Satellite System Corporation

Parkside-Yamamotokan, 3rd Floor, 1-16-4 Tomigaya, Shibuya-ku, Tokyo, 151-0063, Japan
Tel: (+81 3) 54 53 57 07
(+81 3) 54 53 65 21
Fax: (+81 3) 54 53 65 24
President: Tatsuo Gunji
Managing Director: Kisaburo Kobayashi
Managing Director: Akihiro Takahashi
Managing Director: Kimio Aoyagi
Auditor: Iwao Obata
Contact: Tomohiko Daio
■ Broadcast satellite procurement, transfer and/or leasing of broadcast satellite transponders, control and management of satellites, commissioned broadcasting operations.

UPDATED

Earth Remote Sensing Data Analysis Center (ERSDAC)

(ex-Earth Resources Satellite Data Analysis Center)
Forefront Tower, 3-12-1 Kachidoki, Chuo-Ku, Tokyo, 104-0054, Japan
Tel: (+81 3) 35 33 93 10
Fax: (+81 3) 35 33 93 83
e-mail: ersdesk@ersdac.or.jp
Web: http://www.ersdac.or.jp
http://www.gds.aster.ersdac.or.jp
President: Kengo Ishii
Managing Director: Hiroji Tsu
Executive Managing Director: Osamu Nozaki
General Manager, Administration Department: Kyuji Yasuda
General Manager, Program Management Department: Yuichi Maruyama
General Manager, Technical Department and ASTER GDS Project Manager: Dr Hiroshi Watanabe
Tel: (+81 3) 35 33 93 80
Project Management Department: Akio Ogiso
Project Management Department: Masafumi Sawa
■ Research and development of image processing and analysis techniques of satellite remote sensing to non-renewable resources, development, environment protection and global monitoring. Dissemination of remote sensing technology. Research co-operation and information exchange with foreign organisations on remote sensing technology.

UPDATED

Jane's Information Group (JIG)

(Japan)
Palaceside Building, 1-1-1 Hitotsubashi, Chiyoda-ku, 5F, Palaceside Building, 5F, Tokyo, 100-0003, Japan
Tel: (+81 3) 52 18 76 82
Fax: (+81 3) 52 22 12 80
e-mail: japan@janes.com
Web: http://www.janes.com
Information Consultant, Japan: Norihisa Fukuyama
Representing: Jane's Information Group, UK

UPDATED

JSAT Corporation

(ex-Japan Satellite Systems Inc.)
Pacific Century Place Marunouchi 17-18F, 1-11-1 Marunouchi, Chiyoda-Ku, Tokyo, 100-6216, Japan
Tel: (+81 3) 52 19 77 78
Fax: (+81 3) 52 19 78 78
e-mail: info@jsat.net
Web: http://www.jsat.net
Chairman: Tetsuo Morimoto
President and Chief Executive Officer: Takuya Yoshida
Executive Vice President: Yoshio Miwa
Executive Vice President: Yoshihiro Imal
Executive Vice President: Koichi Narikawa
Executive Vice President: Hiroshi Tomomori
Managing Director: Yoshiro Aisaka
Director: Iwao Nakatani
Director: Hiroyuki Shinoki
Director: Masayuki Nomura
Director: Masao Nito
Senior Executive Officer: Takehiko Kinoshita
Executive Officer: Yutaka Nagai
Executive Officer: Naoshi Suzuki
Executive Officer: Yoichi Ibuka
Executive Officer, Corporate Planning and Communications Division: Yasuo Okuyama
General Manager: Toru Mizoguchi
Corporate Planning and Communications Division: Hideki Nakagawa
■ Provide telecommunication and broadcast satellite services.

UPDATED

Meteorological Satellite Center (MSC)

(ex-Japan Meteorological Agency)
3-235 Nakakiyoto, Kiyose-Shi, Tokyo, 204-0012, Japan
Tel: (+81 4) 24 93 49 70
Fax: (+81 4) 24 92 24 33
e-mail: syskan@msc.kishou.go.jp
Head, Systems Engineering Division: Hiroshi Fujimura

UPDATED

Nippon Hoso Kyokai (NHK)

(Japan Broadcasting Corporation)
2-2-1 Jinnan, Shibuya-ku, Tokyo, 150-8001, Japan
Tel: (+81 3) 34 65 11 11
Fax: (+81 3) 34 69 81 10
e-mail: intl@pr.nhk.or.jp

Web: http://www.nhk.or.jp
President: Katsuji Ebisawa
General Managing Director/Executive Director-General of Engineering: Hiroshi Nakamura
Associate Director, International Public Relations: Masahiko Asahina
■ Operates two terrestrial TV channels, three satellite TV channels including one HDTV channel (all satellite channels are transmitted in digital and analog), three radio channels and three world-wide services (two TV and one radio). Broadcasts three channels through BSAT-1a and BSAT-1b and will in future be using BSAT-2a and BSAT-2b for its broadcasting.

UPDATED

Nissho Iwai Corporation
(Aircraft & Advanced Technology Department)
3-1 Daiba 2-Chome, Minato-ku, 135-8655, Japan
Tel: (+81 3) 55 20 32 43
Fax: (+81 3) 55 20 32 46
Web: http://www.nisshoiwai.co.jp
President and Chief Executive Officer: Hidetoshi Nishimura
General Manager, Military Aircraft and Defence Electronics Department: Masafumi Doi
Tel: (+81 3) 55 20 32 50
■ Marketing consultant for military and aerospace companies.

UPDATED

Space Activities Promotion Council Keidanren
(Technology Group)
9-4 Ohtemachi 1-chome, Chiyodu-Ku, Tokyo, 100-8188, Japan
Tel: (+81 3) 32 79 14 11
Fax: (+81 3) 52 55 62 57
Telex: 2223188 kdrtok
Chairman: Ichiro Taniguchi

UPDATED

Industry – Sales Agent/Office

Arianespace SA
(Japan Liaison Office, Tokyo)
Kasumigaseki Building, 31st Floor, 3-2-5 Kasumigasui, Chiyoda-Ku, Tokyo, 100-6031, Japan
Tel: (+81 3) 35 92 27 66
Fax: (+81 3) 35 92 27 68
Representing: Arianespace SA, France

UPDATED

Kyokuto Boeki Kaisha Ltd (KBK)
(Head Office)
7th Floor New Otemachi Building, 2-1 Otemachi, 2-Chome, Chiyoda-Ku, Tokyo, 100-8670, Japan
Tel: (+81 3) 32 44 37 00
Fax: (+81 3) 32 46 27 65
Web: http://www.kbk.co.jp
Chairman and Chief Executive Officer: Hisago Miyoshi
President and Chief Operating Officer: Motoo Imakita
Executive Vice President: Takashi Hiraishi
Executive Managing Director: Kunihiko Irie
Managing Director: Masayoshi Fujimoto
Managing Director: Takehiko Takatori
Managing Director: Shinya Araki
Managing Director: Ryuji Sato
Director: Kosaku Inaba
Director: Kazutake Tsuji
Director: Hiroshi Inoue
Director: Masami Kurihara
General Manager, Aerospace and Defence Systems: Susumu Matsui
Corporate Auditor: Hiroshi Sato
Corporate Auditor: Hideyuki Aiso
Corporate Auditor: Kouzo Fujita
■ Aircraft and aerospace related products including engines, communications, navigation, ground support equipment, satellite equipment, launch related equipment. Global positioning systems, electronics, measuring and control technologies.
Representing:
L-3 Communications Corp., US
Rockwell Collins, Inc., US

UPDATED

Korea, South

Industry – Manufacturing

Korea Aerospace Industries, Ltd (KAI)
(Head Office)
(ex-Daewoo Heavy Industries)
(ex-Hyundai Space and Aircraft)
(ex-Samsung Aerospace Industries Ltd)
135 Seosomun-dong, Jung-gu, Seoul, 100-737, Korea, South
Tel: (+82 220) 20 01 31 14
Fax: (+82 220) 20 01 30 11
(+82 220) 20 01 30 12
Web: http://www.koreaaero.com
President and Chief Executive Officer: Hyong-Bo Kil
Executive Vice President: Taek-Soon Ahn
■ Production of satellites; fixed-wing aircraft include Korea Fighter Program KF-16, KT-1, T-50 and UAVs; rotary-wing aircraft include SB427, SOKOL, Korean Multipurpose Helicopter(KMH) and upgrade of ROKN Lynx helicopters; aerostructures including aircraft fuselages, wing assemblies and helicopter dynamic components. Development of composites and defence electronics including F-5E and M109 self propelled howizer simulators.

UPDATED

Korea Aerospace Industries, Ltd (KAI)
(Changwon Plant)
Seongju-dong, Changwon, Gyeongsangnam-do, 641-120, Korea, South
Tel: (+82 55) 280 61 14
(+82 55) 280 65 01
Fax: (+82 55) 285 23 83

UPDATED

Korea Aerospace Industries, Ltd (KAI)
(Sacheon Plant)
Sanam-Myeon, Sacheon, Gyeongsangnam-do, 664-942, Korea, South
Tel: (+82 55) 851 10 00
Fax: (+82 55) 851 10 05

UPDATED

Korea Aerospace Industries, Ltd (KAI)
(Sacheon Plant 2)
Sacheon-Eup, Sacheon, Gyeongsangnam-do, 664-802, Korea, South
Tel: (+82 55) 851 29 99
Fax: (+82 55) 854 86 38

UPDATED

Korea Aerospace Industries Ltd (KAI)
(Seoul Office)
(ex-Daewoo Heavy Industries Ltd)
(ex-Korea Aerospace Industries, Ltd)
Seosomun-dong, Jung-gu, Seoul, 100-737, Korea, South
Tel: (+82 2) 20 01 31 14
Fax: (+82 2) 20 01 30 11
Managing Director: Ki-Chui Moon
Tel: (+82 2) 726 15 30
Director, Military: In-Kil Kim
Tel: (+82 2) 726 30 24
Deputy General Manager: Won-Cheoi Seo
■ Manufacture of aerostructures, systems integration of KTX-1 basic trainer, design and development of UAVs and manufacture of satellite platforms. Prime contractor for Korean Basic Trainer (KTX-1), Korean Light Helicopter (KLH) and major partner in Korean Fighter Programme (KFP) supplying components including F-16 centre fuselages, side panels and ventral fins. Supplier of Boeing B-747 wings and fuselage, construction of Fairchild Dornier 328 fuselage shells and supply of complete wings for BAe Hawk and Lockheed Martin P-3B/C outer wings. Agreement with PZL-Swidnik (Poland) for worldwide marketing of W-3A Sokol helicopter, marketing alliance with Asiana Airline. Research and development activities include UAV program for reconnaissance and surveillance, co-axial rotating unmanned helicopter program for agriculture and other services. Participant member of Korean Commercial Aircraft Development Consortium (KCDC) for mid-sized commercial aircraft.

UPDATED

Korea Aerospace Industries, Ltd (KAI)
(Space Development & Research Center)
48-6, Moonpyung-Dong, Daeduck-Ku, Daejun, 306-220, Korea, South
Tel: (+82 42) 939 35 19
Fax: (+82 42) 939 35 00

UPDATED

Industry – Sales Agent/Office

Chang Woo, Inc.
Yeoeuido, PO Box 747, Seoul, 150-607, Korea, South
Tel: (+82 2) 782 90 56
Fax: (+82 2) 782 90 58
e-mail: salescwinc@hanafos.com
Web: http://www.changwooinc.com

UPDATED

Luxembourg

Industry – Service

SES Global
(ex-Société Européenne des Satellites)
Château de Betzdorf, L-6815 Betzdorf, Luxembourg
Tel: (+352) 710 72 51
Fax: (+352) 710 72 52 27
(+352) 710 72 53 24
Web: http://www.ses-global.com
Chairman of the Board of Directors: René Steichen
Vice Chairman of the Board of Directors: John F Connelly
Vice Chairman of the Board of Directors: Gerd Tenzer
Vice Chairman of the Board of Directors: Jean-Paul Zens
Member of the Board: Charles Alexander
Member of the Board: Wolfgang A Baertz
Member of the Board: Ernst Wilhelm Contzen
Member of the Board: Jean-Claude Finck
Member of the Board: Richard Goblet d'Alviella
Member of the Board: Raymond Kirsch
Member of the Board: Dr Joachim Kröske
Member of the Board: Dr Raphael Kübler
Member of the Board: Hadelin de Liedekerke Beaufort
Member of the Board: Denis J Nayden
Member of the Board: Gaston Reinesch
Member of the Board: Victor Rod
Member of the Board: Luis Sanchez-Merlo Ruiz
Member of the Board: Christian Schaack
Member of the Board: Georges Schmit
Member of the Board: Gaston Schwertzer
Member of the Board: François Tesch
General Counsel and Secretary to the Board of Directors: Roland Jaeger
President and Chief Executive Officer: Romain Bausch
President and Chief Executive Officer of SES Astra: Ferdinand Kayser
President and Chief Executive Officer, SES Americom: Dean Olmstead
Executive Vice President, Corporate Development: Robert Bednarek
Vice President, Corporate Communications: Jean-Paul Hoffmann
Vice President, Investor Relations: Mark Roberts
Chief Financial Officer, SES Global: Jürgen Schulte
Market Communications and Public Relations Manager: Ms Alison Black
Press Relations Manager: Yves Feltes
■ Operate a satellite services network providing seamless communications.

UPDATED

Netherlands

Transport Administration – Space

European Space Agency (ESA)
(European Space Research & Technology Centre)
PO Box 299, NL-2200 AG, Noordwijk, Netherlands

Tel: (+31 7) 15 65 65 65
Fax: (+31 7) 15 65 60 40
Web: http://www.estec.esa.nl
Head of Establishment and Director of Technical and Operational Support: Gaele Winters
Systems Engineering: Niels Jensen
Product Assurance/Safety: Alex Soons
Testing: Peter Brinkman
Communications: Ms Heidi Graf
Tel: (+31 71) 565 30 06
Fax: (+31 71) 565 57 28
■ Technical centre responsible for the design and execution of scientific and application satellite programmes plus the human space flight program. Also holds a large test centre for space craft.

UPDATED

Netherlands Industrial Space Organisation (NISO)

PO Box 32070, NL-2303 DB Leiden, Netherlands
e-mail: info@niso.nl
Web: http://www.niso.nl
■ Comprised of Dutch industries, research organisations and users in the space field with the aims of: defining common interests and strategies; promoting industrial involvement; stimulating the use/application of space facilities and products by industrial enterprises and institutes; broadening the social base of space; and promoting technology spin-offs.

UPDATED

Centres of Learning & Research

National Aerospace Laboratory of the Netherlands (NLR)

(Nationaal Luchten Ruimtevaartlaboratorium)
PO Box 90502, NL-1006 BM Amsterdam, Netherlands
Tel: (+31 20) 511 31 13
Fax: (+31 20) 511 32 10
e-mail: info@nlr.nl
Web: http://www.nlr.nl
Chairman: J van Houwelingen
General Director: Ir. F Holwerda
Technical Director: Ir F J Abbink
Marketing Director: Ir U Posthuma de Boer
Head of Flight Division: Prof. P G A M Jorna
Head of Air Transport Division: Ir H A Z M Offerman
Training Development and Concept Evaluation: J M Hoekstra
Purchasing Manager: G S Wijdeveld
Air Traffic Management: J C Terlouw
Engineering and Technical Services: Ir J van Twisk
Space Division: Ir B J P Van der Peet
Avionics Division: Ir M A G Peters
Fluid Dynamics Division: Dr B Oskam
Information and Communication Technology Division: Ir F J Heerema
General Secretary: E Folkers
Flight Mechanics: W P de Boer
Helicopters: L T Renirie
Operations Research: G J Alders
Flight Test and Safety: M A Piers
Airports: Ir A D J Rutten
Transport and Environmental Studies: Ms M E S Vogels
■ Independent non-profit research institute conducting contract research for national and international customers. The NLR owns and operates several research facilities. Seven main divisions of research: Fluid Dynamics, Structures and Materials, Space, Aircraft, Air Transport, Information and Communication Technology, Avionics. Research and test facilities wind tunnels, ATC simulators, research aircraft, supercomputers and environmental test facilities. Participation in DNW and ETW. Member of AEREA (Association of European Research Establishments) in Aeronautics.
Accreditations: ISO 9001

UPDATED

National Aerospace Laboratory of the Netherlands (Library)
PO Box 905 02, NL-1006 BM Amsterdam, Netherlands
Tel: (+31 527) 24 84 44
(+31 527) 24 83 17
Fax: (+31 527) 24 82 10
Chief Librarian: R Lammers

UPDATED

National Aerospace Laboratory of the Netherlands (Noordoostpolder Laboratory)
PO Box 153, NL-8300 AD Emmeloord, Netherlands
Tel: (+31 527) 24 84 44
Fax: (+31 527) 24 82 10
■ Aeronautical research and development in computational and experimental fluid dynamics, aeroacoustics, structures and materials, flight dynamics and operations research. Space research in the fields of fluid dynamics, stability and control, robotics environmental simulation, materials and structures, remote sensing, micro-gravity. Extensive facilities and additional equipment for experimental research and development in all fields.

UPDATED

Netherlands Agency for Aerospace Programs

(Nederlands Instituut voor Vliegtuigontwikkeling en Ruimtevaart (NIVR))
Kluyverweg 1,
PO Box 35, NL-2600 AA Delft, Netherlands
Tel: (+31 15) 278 80 25
Fax: (+31 15) 262 30 96
Web: http://www.nivr.nl
Chairman: Lieut Gen Ben A C Droste
General Director: Mr G M Driedonks
Head Aircraft Division: Henk J M van Leeuwen
Tel: (+31 15) 278 73 34
Head of Finance and General Affairs Division: Hendrik Jan Tiecken
Secretary General: Ms Marloes de Vogel
Head, In- and External Affairs: Ms Christine Windmeijer
■ Independent agency designated by the Government for the promotion of aerospace projects and research.

UPDATED

Space Research Organization of the Netherlands (SRON)

(Stichting Ruimteonderzoek Nederland)
Sorbonnelaan 2, NL-3584 CA Utrecht, Netherlands
Tel: (+31 30) 253 56 00
Fax: (+31 30) 254 08 60
e-mail: info@srou.nl
Web: http://www.sron.nl
Chairman: Dr A P Baede
General Director: Prof. Dr J A M Bleeker
Deputy Director: Dr R Gathier
Public Relations Director: Dr J A Wamsteker
Tel: (+31 30) 253 85 93
■ National agency for scientific space research, development and production of instrumentation for space research in astrophysics and earth observation.

UPDATED

Space Research Organization of the Netherlands (SRON)
(Groningen Office)
Postbus 800, NL-9700 AV Groningen, Netherlands
Tel: (+31 50) 363 40 74
Fax: (+31 50) 363 40 33
e-mail: secr-g@sron.nl

UPDATED

TNO Space

PO Box 477, NL-2600 AL Delft, Netherlands
Tel: (+31 15) 269 21 61
Fax: (+31 15) 269 21 11
e-mail: woenzel@tpd.tno.nl
Web: http://www.tnospace.tno.nl

UPDATED

TNO TPD

(ex-TNO Insitiute of Applied Physics)
PO Box 155, NL-2600 AD Delft, Netherlands
Tel: (+31 15) 269 20 00
Fax: (+31 15) 269 21 11
e-mail: maas@tpd.tno.nl
Web: http://www.tpd.tno.nl
Managing Director: Ir D Ph. Schmidt
Tel: (+31 15) 269 22 66
Head of Space: Dr A A M Maas
Tel: (+31 15) 269 21 86
Deputy Director: Ir A J Dalhuysen
Commercial Manager: Ing M R van der Kraan
Tel: (+31 15) 269 22 70
Purchasing Manager: R Grutter
Office Manager: Ms Carola Barth
■ The realisation of scientific instruments for use in space to observe the Earth and for other scientific purposes. The development of techniques for aperture synthesis for astronomical applications. The calibration under space conditions (spectral and radiometric) of optical instruments and diffusers for Earth observation. The realisation of precision mechanisms for use in space. The realisation of sensors for navigating and controlling the position of satellties.

UPDATED

Association

Netherlands Aerospace Group (NAG)

Boerhaavelaan 40, PO Box 190, NL-2700 AD Zoetermeer, Netherlands
Tel: (+31 79) 353 13 56
Fax: (+31 79) 353 13 65
e-mail: nag@fme.nl
Web: http://www.nag.aero
■ National association of aerospace companies and organisations engaged in education, research and development, engineering, manufacturing, maintenance repair and the overhaul of civil and military systems and equipment. The association facilitates the setting up of national pavilions at airshows and conferences, organises international trade missions, courses, seminars and meetings on specific areas of interest.

UPDATED

Netherlands Astronautical Society

(Nederlandse Vereniging v. Ruimtevaart)
Zonnenburg 2, NL-3512 NL Utrecht, Netherlands
Tel: (+31 30) 231 13 60
Fax: (+31 30) 234 28 52
e-mail: info@ruimtevaart-nvr.nl
Web: http://www.ruimtevaart-nvr.nl
President: Ir D de Hoop
Financial Director: Ir J van Casteren
Administration Manager: Dr M M A Drummen
Secretary: Ir H M Sanders
■ The NVR, member of the International Astronautical Federation (IAF), was established with the purpose to inform individuals interested in space research and space technology and to bring them together, by organising lectures, excursions, symposia and film-shows.
Publications: Ruimtevaart

UPDATED

Industry – Manufacturing

Bradford Engineering BV

De Wijper 26, NL-4726 TG, Heerle (NB), Netherlands
Tel: (+31 165) 30 51 00
Fax: (+31 165) 30 44 22
e-mail: info@bradford-space.com
Web: http://www.bradford-space.com
■ Development for satellites, space shuttle and ISS.

UPDATED

Brandt Fijnmechanische Industrie BV

De Stubbenweg 15, NL-1327 GA Almere, Netherlands
Tel: (+31 036) 523 13 98
Fax: (+31 036) 533 26 33
e-mail: info@brandtfmi.nl
Web: http://www.brandtfmi.nl
Managing Director: B M M Filart
Operations Director: T Bosma
Administration: A Sas
■ Manufacture of precision components, tools and assemblies. Repair and overhaul of mechanical components.
Representing:
Scheer & Co. GmbH, Germany

UPDATED

CHL Netherlands BV

(ex-Christiaan Huygens Laboratorium BV)
PO Box 3072, NL-2220 CB Katwijk, Netherlands
Tel: (+31 71) 402 55 14
Fax: (+31 71) 402 50 78
e-mail: marketing@chl.nl
Web: http://www.chl.nl
Marketing and Sales Manager: R Lens

- Design and production of complete radar antenna systems. Production of Artemis microwave range-bearing position fixing system, radar based traffic detection systems. Consultancy on radar and microwave based systems.

UPDATED

Dutch Space BV (DS)

(ex-Fokker Space BV)
PO Box 32070, NL-2303 DB Leiden, Netherlands
Tel: (+31 71) 524 50 00
Fax: (+31 71) 524 59 99
e-mail: info@dutchspace.nl
Web: http://www.dutchspace.nl
President: Dr B M Spee
Vice President: P Pieterse van Wijck
Quality: F Meiboom
Business Development: J de Kam
Marketing Co-ordination: H Mawira
Strategy and Communications: Dr G Mennenga
Tel: (+31 71) 524 51 20
Mobile: (+31) 653 22 02 77
Fax: (+31 71) 524 52 90

UPDATED

European Aeronautic Defence and Space Company NV (EADS NV)

Le Carré, Buchavinue 130-132, NL-1119 PR Schiphol, Rijk, Netherlands
Web: http://www.eads.net
President and Chief Executive Officer of Elbe Flugzeugwerke, Aeronautics Division: Horst Emker
President and Chief Executive Officer, EADS North America: Ralph D Crosby
Chairman and Chief Executive Officer of Sogerma, Aeronautics Division: Yves Richard
Co-Chairman of the Board: Dr Manfred Bischoff
Co-Chairman of the Board: Arnaud Lagardere
Co-Chief Executive Officer: Philippe Camus
Co-Chief Executive Officer: Rainer Hertrich
Chief Executive Officer, EADS Astrium, EADS Space Division: Antoine Bouvier
Chief Executive Officer of EADS Space Transportation, EADS Space Division: Hervé Guillou
President and Chief Executive Officer of Socata, Aeronautics Division: Stephane Mayer
President of EADS Space Tranportation, EADS Space Division: Josef Kind
Senior Vice President, Strategy and Planning, Strategic Co-ordination: Reinhold Lutz
Senior Vice President, Mergers and Acquisitions, Strategic Co-ordination: Pierre-Henn Ricaud
Senior Vice President, Finance/Treasury, Finance Division: Gérard Adsuar
Senior Vice President, Investor Relations, Finance Division: Pierre de Bausset
Senior Vice President, Accounting/Tax, Finance Division: Joachim Feyel
Senior Vice President, Controlling Finance Division: Christian Pechmann
Senior Vice President, Controlling Finance Division: Ms Brigitta Klasén
Senior Vice President, Purchasing: Hans-Erich Mundt
Senior Vice President, Legal Affairs: Eric Thomas
Senior Vice President, Political Affairs, France: Denis Verret
Senior Vice President, Political Affairs, Germany: Wolf-Peter Denker
Senior Vice President, Political Affairs, Spain: Carlos Grandal
Senior Vice President, Human Resources, France: Jacques Massot
Senior Vice President, Human Resources, Germany: Reinhard Havers
Senior Vice President, Human Resources, Spain: Mariano Alonso Romero
Senior Vice President, Management Development: Birgit Quecke
Senior Vice President, Communications: Christian Poppe
Tel: (+49 89) 60 73 42 50
Fax: (+49 89) 60 73 42 84
Head of Strategic Co-ordination: Jean-Louis Gergorin
Head of EADS International: Jean-Paul Gut
Head of Airbus Division/President and Chief Executive Officer of Airbus: Noël Forgeard
Head of Aeronautics Division: Dr Dietrich Russell
Head of ATR, Aeronautics Division: Jean-Michel Léonard
Head of Space Division: François Auque
Head of Defence and Security Systems Division: Dr Thomas Enders
Head of Eurocopter, Aeronautics Division: Fabrice Brégier
Head of Military Transport Aircraft Division and Chairman of the Board of CASA: Francisco Fernandez Sainz
Head of Human Resources: Jussi Itavuori
Chief Executive Officer of Defence and Communications Systems, Defence and Security Systems Division: Stefan Zoller
Head of CASA Espacio, EADS Space Division: Pedro Méndez
Head of Military Aircraft, Defence and Security Division: Aloysius Rauen
Head of Missiles and Chief Executive Officer of MBDA, Defence and Security Systems Division: Marwan Lahoud
Head of EADS Defence Electronics, Defence and Security Systems Division: Werner Heitzmann
Head of LFK, Defence and Defence Systems Division: Werner Kaltenegger
Head of Space Services, EADS Space Division: Eric Beranger
Head of SODERN, EADS Space Division: Henrie Dugre
Head of CILAS, EADS Space Division: Jacques Battistella
Head of EADS Services, Defence and Security Systems Division: Jacques Vannier
Chief Operating Officer, Defence and Communications Systems, Defence and Security Systems Division: Patrick Jourdan
Chief Financial Officer: Hans Peter Ring
Chief Operating Officer, Airbus: Dr Gustav Humbert
Chief Technology Officer, Strategic Co-ordination: Daniel Deviller
Corporate Secretary: Laurence Dors

- Manufacture of military and civil airborne, spaceborne and land-based platforms, systems, equipment and components.

Operations and activities of the five business units:
Airbus: manufacture of Airbus A300-600F, A300-600ST, A310, A318, A319, A319CJ, A320, A321, A330, A340, A380 and A380F.
Military Transport Aircraft Division: manufacture of C-212, C-295, CN-235 and A400M. Manufacture and supply of special mission aircraft for duties such as maritime surveillance and anti-submarine warfare.
Aeronautics Division: manufacture of military aircraft (Eurofighter Typhoon, Mako, C-101 Aviojet), helicopters (Eurocopter NH 90, Tiger, EC 120, EC 135, EC 155), regional aircraft (ATR 42-500, ATR 72-500), light aircraft (Socata TBGT, TBM 700, Epsilon) and aircraft conversion and maintenance (Sogerma, Elbe Flugzeugwerke GmbH). Technical and logistic support for F/A-18 Hornet, MiG-29, Tornado, F-4 Phantom, Mirage F-1, C-130 Hercules, Transall C-160, P-3 Orion and E-3A AWACS.
Space Division: design, development and production of satellites (broadcast satellites for Intelsat, Eutelsat and Inmarsat), earth observation and scientific satellite systems for civil and military applications (Envisat, Metop, Spot 5, Helios, XMM), the European navigation systems (Galileo), orbital infrastructure (space laboratory Columbus and the ATV for the international Space Station) and launchers (Arianespace heavy-lift launchers, Starsem medium-lift launchers, Eurockot small-lift launchers). Optronics and laser technologies through the subsidiaries of Sodern and Cilas.
Defence and Civil Systems Division: manufacture of missiles and missile systems, defence electronics, telecommunications and services. The Missiles and Missile Systems sector comprises MBDA and LFK-Lenflugkörpersysteme GmbH. Products include: Meteor, Aster, Exocet, Kormoran, Roland, Milan, HOT, Mistral, ASRAAM, Mica, Seawolf, RAM, Patriot, Stinger, Taurus, Trigat MR and Trigat LR, Polyphem and Scalp EG/Storm Shadow. The Defence Electronics sector produces C3I systems, reconnaissance and surveillance systems, airborne multi-mode radars and electronic warfare units. The Civil Communications sector supplies internet protocol networks, high-speed and long-distance networks, switching products, local loop equipment and fibre-optics and cellular telecommunications networks.
Corporate companies:
Airbus
Astrium
CELERG
EADS ATR
EADS CASA
EADS Dornier
EADS Eurocopter
EADS Socata
EDSN
Sogerma
Shareholder structure:
DaimlerChrysler: 30.3%
SOGEADE (Lagardère, together with French financial institutions and the French state holding company Sogepa): 30.3%
SEPI (Sociedad Estatal de Participaciones Industriales), Spain: 5.5%
Free Float: 33.9% (including EADS employees and about 3 percent held directly by DaimlerChrysler and the French state)

UPDATED

Fokker Elmo BV

PO Box 75, NL-4630 AB Hoogerheide, Netherlands
Tel: (+31 164) 61 76 44
Fax: (+31 164) 61 76 37
e-mail: jn11175@elmo.nl
Web: http://www.elmo.nl

UPDATED

Fokker Special Products BV

PO Box 59, NL-7900 AB Hoogeveen, Netherlands
Tel: (+31 528) 28 50 00
Fax: (+31 528) 28 50 10
e-mail: info@fokkersp.stork.com
Web: http://www.fokker.com

UPDATED

Philips Aerospace

(ex-Philips ETG)
(ex-Philips Machinefabrieken Nederland BV)
Zwaanstraat 1, Postbus 80004, NL-5600 JC Eindhoven, Netherlands
Tel: (+31 40) 278 82 54
Fax: (+31 40) 278 77 48
Managing Director: Geoff Bertenshaw
Tel: (+31 40) 278 96 44
Business Manager Aviation: Harold F Bousché

- Focus on design-to-manufacturing and production of turbomachinery, sheet metal, airframe components and other products. High speed and electro chemical machining. Products include blisks, impellers, exhaust nozzles and aerostructures.

Accreditations: ISO 9002, Mil-Q-9858A, Mil-I-45208A

UPDATED

RDM Technology BV (RDM-T)

PO Box 1039, NL-3000 BA Rotterdam, Netherlands
Tel: (+31 10) 487 91 11
Fax: (+31 10) 487 22 99
Sales Support: Björn de Lange
Tel: (+31 10) 487 21 38
Mobile: (+31 6) 51 91 52 95

- Lightdeck landing grid systems for shipborne helicopter operations. Secure helicopter restraints. Modification programs for M114/39 155 mm, M101/33 105 mm and M109L47 and M109L52 155 mm howitzers. Modification program of Leopard 1 and 2 Main Battle Tank. Modification, overhaul and maintenance of M109 155 mm howitzer. AMX-13 family modernisation program. Bore cleaning tool for maintenance of rifled and smooth barrels of both artillery and battle tanks. Mobile artillery systems on 4x4 and 6x6 army trucks. Mine clearing systems.

UPDATED

Stork NV

(Head Office)
PO Box 5004, NL-1410 AA Naarden, Netherlands
Tel: (+31 35) 695 74 11
Fax: (+31 35) 694 11 84
Web: http://www.storkgroup.com

- Supply high-level technological products and services to the industrial community.

UPDATED

Terma B V

(ex-Terma Elektronik Netherlands BV)
Schuttersveld 9, NL-2316 XG Leiden, Netherlands
Tel: (+31 71) 524 08 00
Fax: (+31 71) 514 32 77
e-mail: terma.nl@terma.com
Web: http://www.terma.com
Public Affairs Assistant: Ms Christina K Petersen

UPDATED

TNO PML

(Prinz Mauritz Laboratory)
PO Box 45, NL-2280 AA, Rijswijk ZH, Netherlands
Tel: (+31 15) 284 28 42

Fax: (+31 15) 284 39 91
e-mail: info@pml.tno.nl
Web: http://www.pml.tno.nl
Director: P A O G Korting
Tel: (+31 75) 284 35 18
Director of Research: Dr D W Hoffmans
Information Officer: F Dijkstra
■ Research and development into Chemical and Biological protection.

UPDATED

Industry – Service

Astrium NJRS
PO Box 38, NL-5340 AA Oss, Netherlands
Tel: (+31 24) 641 99 29
e-mail: gkspace@zonnet.nl
Web: http://http://www.njrs.nl
Editor-in-Chief: G Keyzers
■ Space magazine published in Dutch.

UPDATED

Omnistar BV
Postbus 113, NL-2260 AC Leidschendam, Netherlands
Tel: (+31 70) 317 09 00
Fax: (+31 70) 317 09 19
e-mail: dgps@omnistar.nl
Web: http://www.omnistar.nl
Managing Director: Erik Hammega
Financial Director: Gerard Scholtes
Director, Sales and Distribution and Public Relations Manager: Martin de Kievit
Director, Technical: Hans Visser
Purchasing Manager: Cor Landsman
Area Manager, Germany: Tony Ten Wolde
Area Manager, Middle East: Dave Hampshire

UPDATED

New Zealand

Centres of Learning & Research

Manaaki Whenua Landcare Research
(Corporate Office)
Canterbury Agriculture & Science Centre, Gorald Street, PO Box 40, 8152 Lincoln, New Zealand
Tel: (+64 3) 325 67 00
Fax: (+64 3) 325 21 27
Web: http://www.landcareresearch.co.nz
Chief Executive: Dr Andrew Pearce
Chief Operating Officer, Investments: Mark Cleaver
Chief Financial Officer: John Tan
Chief Operating Officer, Research: Ian Whitehouse
Manager, Human Resources: Michael Johnson
Manager, Treaty Responsibilities: Rauru Kirikiri
Manager, Operations: Dr Maggie Lawton
Research Manager: Dr David Penman
Public Relations and Communications Manager: Ms Judy Grindell
■ Development of satellite imagery, remote sensing equipment, radio telemetry equipment and geographic information systems for use with resource management applications.
Accreditations: ISO 14004
Representing
Sirtrack Ltd, Manaaki Whenua Press, International Business Group.

UPDATED

Industry – Manufacturing

Sirtrack Ltd
Private Bag 1403, Havelock North, New Zealand
Tel: (+64 6) 877 77 36
Fax: (+64 6) 877 54 22
e-mail: sirtrack@landcareresearch.co.nz
Web: http://www.sirtrack.com
General Manager: Dave Ward
Chief Engineer: Kevin Lay
Marketing Director: Rowan Calder
■ Design and manufacture of VHF and UHF tracking equipment, mainly for wildlife research, certified by Argos.
Accreditations: ISO 9001

UPDATED

Norway

Government – Ministry/Agency

Norwegian Meteorological Institute
PO Box 43 Blindern, N-0313 Oslo, Norway
Tel: (+47) 22 96 30 00
Fax: (+47) 22 96 30 50
e-mail: met.inst@dnmi.no

UPDATED

Centres of Learning & Research

Norwegian Defence Research Establishment
(Försvarets forskningsinstitutt)
PO Box 25, N-2027 Kjeller, Norway
Tel: (+47) 63 80 70 00
Fax: (+47) 63 80 71 25
e-mail: ffi@ffi.no
Web: http://www.ffi.no
Director-General: Dr Paul Narum
Tel: (+47 63) 80 71 00
Chief of Staff: Svein Rollvik
Director, Division for Systems Analysis: Ragnvald H Solstrand
Tel: (+47) 63 80 77 00
Fax: (+47) 63 80 77 15
Director, Division for Electronics: Johnny Bardal
Tel: (+47) 63 80 72 00
Fax: (+47) 63 80 72 12
Director, Division for Protection and Materiel: Jan Ivar Botnan
Tel: (+47) 63 80 75 00
Fax: (+47) 63 80 75 08
Financial Director: Jo Liseth
Tel: (+47 63) 80 71 10
Technical and Research Director: Jon Skjervold
Tel: (+47 63) 80 71 05
Librarian: Synnøve Eifring
Tel: (+47) 63 80 71 28
Contact: Bente Stern Bergersen
■ Defence research and technology establishment.
Programme areas:
Basic research
Command and control systems
Electronic warfare
Environmental measures in the armed forces
Operations and cost analysis
Protection against NBC weapons
Strategic analysis
Surveillance systems
Tactical underwater systems
Weapon effectiveness, vulnerability and protection
Weapon guidance and weapon control systems
Weapon technology

UPDATED

Association

Norwegian Industrial Forum for Space Activities (NIFRO)
Middelthunsgate 27, PO Box 5250 Majorstua, N-0303 Oslo, Norway
Tel: (+47) 23 08 83 08
Fax: (+47) 23 08 80 18
e-mail: nifro@nho.no
Web: http://www.nifro.no
■ A grouping of 17 Norwegian research and development and manufacturing companies and research institutes promoting members' participation in space activities. Areas of interest include: ground segment and support data systems, space electronics, materials technology, manfacturing techniques, space transportation technology, remote sensing, telecommunications and quality assurance/control and consulting. Integrated remote sensing data interpretation from multispectral and non-optical sources.

UPDATED

Norwegian Space Centre (NSC)
(Norsk Romsenter)
PO Box 113 Skoyen, N-0212 Oslo, Norway
Tel: (+47 22) 51 18 00
Fax: (+47 22) 51 18 01
e-mail: spacecentre@spacecentre.no
Web: http://www.spacecentre.no
Managing Director: Rolf Skår
Tel: (+47 22) 51 18 10
Director of Finance and Administration: Knut Myrvang
Tel: (+47 22) 51 18 37
Director of Applications Development: Jostein Ronnenberg
Tel: (+47 22) 51 18 23
Director of Industrial Development: Geir Hovmork
Tel: (+47 22) 51 18 13
Director of Information: Per Torbo
Tel: (+47 22) 51 18 17
Director of Space and Earth Sciences: Bo Andersen
Tel: (+47 22) 51 18 30
Director of Infrastructure and Business Development: Rolf Skatteboe
Tel: (+47 22) 51 18 28
Director of Applications and Direction: Jostein Rønneberg
Tel: (+47 22) 51 18 23
Managing Director (Andøya Rocket Range): Kolbjørn Adolfsen
Tromsø Satellite Station: Jan Petter Pederson
Industry: R Ericksen
Space Station Development: M Vinje Tantillo
Earth Observation: P E Skrovseth
Information Officer: Ms Ann-Lisbeth Ruud
■ Co-ordination of national and international space activities, development of resources, infrastructure and management of ESA participation.

UPDATED

Industry – Manufacturing

Alcatel Space Norway A/S
Knudsrødveien 7, PO Box 138, N-3191 Horten, Norway
Tel: (+47 33) 03 27 00
Fax: (+47 33) 03 28 00
e-mail: asn@alcatel.no
Web: http://www.asn.no
Managing Director: Sverre Bisgaard
Commercial Manager: Jorgen Galteland
Business Development Manager: Øyvind Andreassen
Communications Co-ordinator: Ms Kirsten Klostermann
Tel: (+47 33) 03 27 73
Fax: (+47 33) 03 28 01
■ Manufacture of on-board space electronics for satellites and launchers.

UPDATED

Eidsvoll Electronics A/S (Eidel)
Nedre Vilberg vei 8, N-2080 Eidsvoll, Norway
Tel: (+47) 63 95 97 21
(+47) 63 95 97 00
Fax: (+47) 63 95 97 10
e-mail: in@eidel.no
eidel@eidel.no
Web: http://www.eidel.no
Director: Tore Havstein
Technical Manager: Øystein Sandholt
Product Manager, Telemetry: Ivar Nordby
■ Analogue and digital designs with telemetry applications for aerospace and defence. Military control systems.

UPDATED

Jotron Electronics A/S
Kirkestien 1, PO Box 54, N-3280 Tjodalyng, Norway
Tel: (+47) 33 13 97 04
Fax: (+47) 33 12 67 80
e-mail: salesair@jotron.com
salesmar@jotron.com
Web: http://www.jotron.com
Area Sales Manager, Aeronautical Products: Jarle Skudem
Tel: (+47 33) 13 97 56
Marketing Manager, Ground to Air: Ole-Steinar Nordhaug
Sales Manager (UK), Ground to Air: Richard M Lorrison
Sales Manager (UK), Ground to Air: Stuart Ferguson
Sales Manager, Ground to Air: Rolf G Halvorsen
Sales Manager, Ground to Air: Harald Ibsen
In-House Sales Support: Ms Lise Groestad
■ Design and manufacture of ground-to-air radio communications equipment. VHF 117-137 MHz

AM base stations, UHF 225-400 MHz AM/FM base stations, remote control systems, hand-held transceivers, data transceivers, base station change-over units.

UPDATED

Kongsberg Defence & Aerospace AS (KDA)
(Headquarters)
(ex-Kongsberg Aerospace)
PO Box 1003, N-3601 Kongsberg, Norway
Tel: (+47) 32 28 82 00
Fax: (+47) 32 28 82 01
Web: http://www.kongsberg.com
■ Activities are focused on anti-ship missiles that can be launched from ships, fighter aircraft and helicopters. Also development and manufacture of military products based on related propulsion navigation and construction technologies.

UPDATED

Nammo A/S
(Head Office)
PO Box 142, N-2831 Raufoss, Norway
Tel: (+47 61) 15 36 00
Fax: (+47 61) 15 36 20
e-mail: nammo@nammo.com
Web: http://www.nammo.com
President: Edgar Fossheim
Chairman: Karl Glad
Senior Vice President, Chief Financial Officer: Jan Eystein Sabo
Senior Vice President, Human Resources: Bertil Palsrud
Executive Vice President, Demilitarisation: Reijo Perälä
Executive Vice President, Large Calibre Division: A Erland Paulsrud
Executive Vice President, Medium Calibre Division: Kjell I Kringsjaa
Executive Vice President, Small Calibre Division: Jan Koivurinta
Executive Vice President, Missile Product Division: Ola Skriverik
Vice President, Business Development: Johs Norheim
Marketing Director, Medium Calibre Division: Per Lillehaug
Corporate Communication Co-ordinator: Sissel Solum Belguendouz
Tel: (+47 61) 15 22 42
Mobile: (+47) 95 05 12 20
Corporate Communication Co-ordinator: Ragnhild Waldemar
■ Manufacture of small arms ammunition; medium calibre 12.7mm to 40mm; anti-armour weapons; ammunition artillery, ground based, APFSDS-T (30x173mm); mortar bombs; pyrotechnical products; 155 mm shell bodies; 40 mm brass cases; specialised aluminium components for the defence industry; rockets; rocket motors; warheads; space products; composite materials and pressure vessels; aerospace products include boosters and components, missile components and propulsion.

UPDATED

Nammo Raufoss A/S
(Head Office)
(ex-Raufoss ASA)
PO Box 162, N-2831 Raufoss, Norway
Tel: (+47) 61 15 36 50
Fax: (+47) 61 15 36 60
Web: http://www.nammo.com
■ Manufacture of Small Arms Ammunition; Cannon 20 x 102 mm M39, 20 x 110RB, 25 x 137 mm KBA, 25 mm Aden, 27 x 145 mm Mauser, 30 mm Aden, 30 mm DEFA; Mortar 81 mm; Air defence 40 mm L/60 MPT, 40 mm L/70 MPT; Naval 76 mm ASP; 105 mm ASP, 127 mm ASP; Field 155 mm HE NM28. Research, development and production in the space and automotive sectors. Ammunition and aerospace defence products.
Associated companies/subsidiaries
Raufoss Technology AS
Raufoss Service AS
Raufoss Metall GmbH, Germany
Raufoss France SARL, France
Raufoss do Brasil Ltda, Brasil
Kongsberg-Raufoss AS

UPDATED

Nera ASA
Kokstadveien 23, PO Box 7090, N-5020 Bergen, Norway
Tel: (+47) 55 22 51 00
Fax: (+47) 55 22 52 99
e-mail: webmaster@nera.no
Web: http://www.nera.no
Contact: Terje Hjelm
■ Microwave radio relay systems. Provision of equipment, systems and turnkey solutions for all Inmarsat analogue and digital land-earth stations used in maritime, land-based and aeronautical applications. Shipboard and land-based mobile terminals.

UPDATED

Norsk Elektro Optikk A/S (NEO)
PO Box 384, N-1471 Loerenskog, Norway
Tel: (+47 67) 97 47 00
Fax: (+47 67) 97 49 00
e-mail: neo@neo.no
sales@neo.no
support@neo.no
Web: http://www.neo.no
Managing Director: Peter Kaspersen

UPDATED

Industry – Service

Andøya Rocket Range
PO Box 54, N-8483 Andenes, Norway
Tel: (+47) 76 14 44 00
Fax: (+47) 76 14 44 01
e-mail: info@rocketrange.no
Web: http://www.rocketrange.no
Managing Director: Kolbjørn Adolfsen
Tel: (+47 76) 14 44 10
Head of Operational Services: Kjell Bøen
Tel: (+47 76) 14 44 21
■ Provides products, services and infrastructure with regard to space science and environmentally related science and surveillance. Launch of sounding rockets and balloons for investigation of upper atmosphere at high latitudes. Recovery of payloads. Users from more than 70 institutes and universities in Europe, USA, Canada andJapan. Owns and operates 'SvalRak' a non-permanent launching facility at Svalbard.

UPDATED

Det Norske Veritas (DNV)
Veritasveien 1, N-1322 Høvik, Norway
Tel: (+47) 67 57 99 00
Fax: (+47) 67 57 91 60
Telex: 76192
Web: http://www.dnv.com
■ Independent provider of safety, quality and environmental management services for industry and shipping. Research and development services. Certification services.

UPDATED

Kongsberg Spacetec A/S
PO Box 6244, N-9292 Tromsø, Norway
Tel: (+47) 77 66 08 00
Fax: (+47) 77 65 58 59
e-mail: marketing@spacetec.no
Web: http://www.spacetec.no
Managing Director: Bjorn A Kanck
Financial Director: Knut H Johannesen
Sales and Marketing Director: Harald Lauknes
Executive Secretary: Ms Astrid Movik
■ Satellite data receiving station equipment.

UPDATED

Norut Information Technology Ltd
N-9291 Tromsø, Norway
Tel: (+47) 77 62 94 00
Fax: (+47) 77 62 94 01
e-mail: itek@norut.no
Web: http://www.norut.no/itek/
Managing Director: Jan Borre Rydingen
Tel: (+47 77) 62 94 32
Information Officer: Sigurd Sjursen
Tel: (+47 77) 62 94 30
■ Research and technological development.

UPDATED

Pakistan

Centres of Learning & Research

SUPARCO
(SGSC Division)
G-8/1 Mauve Area, PO Box 1271, Islamabad, Pakistan
Tel: (+92 5777) 61 17 92
e-mail: sgs@comsats.net.pk
Web: http://www.suparco.gov.pk
General Manager: Imran Iqbal

UPDATED

Poland

Transport Administration – Space

Polish Space Office
ul. Bartycka 18a, PL-00-716 Warszawa, Poland
Tel: (+48 22) 851 18 12
e-mail: spaceoffice@cbk.waw.pl
Web: http://www.cbk.waw.pl/wsc2002

UPDATED

Association

Polish Astronautical Society (PAS)
(Polskie Towarzystwo Astronautyczne)
Space Research Centre, Bartycka str. 18A room 137, PL-00-716 Warszawa, Poland
Tel: (+48 22) 840 37 66
Fax: (+48 22) 840 31 31
e-mail: poczta@ptastronaut.org.pl
Web: http://www.ptastronaut.org.pl
President: Prof Zbigniew Klos
Vice President: PhD Col Maciej Mroczkowski
Vice President: Prof Rudolf Klemens
Vice President: M Eng Sc Marek Sadowski
General Secretary: M S Andrzej Kotarski
Treasurer: Koetowski Mirostaw
■ Society formed for the promotion of the studies in space sciences and space flight.
Publications: Astronautyka
Progress in Astronautics

UPDATED

Industry – Service

Geoinformation Laboratory & Consultancy
Bartycka 18a St., PL-00-716 Warszawa, Poland
Tel: (+48 22) 851 11 66
e-mail: office@geosystems.com.pl
Web: http://www.geosystems.com.pl
Librarian: Ms Krystyna Szewczyk

UPDATED

Polspace Ltd
Ul. Bartycka 18A, PL-00-716 Warszawa, Poland
Tel: (+48 22) 840 37 66
Fax: (+48 22) 840 31 31
e-mail: bbuszke@cbk.waw.pl
Web: http://www.polspace.pl
Managing Director: Bartosz Buszke
Tel: (+48 22) 840 37 66
■ Consultancy services in space policy and space business. Lobbying, brokering services, support to Polish SMEs and start-ups and promotion of technology transfer.

UPDATED

Space Research Centre
(Polish Academy of Sciences)
Bartycka 18A, PL-00-716 Warszawa, Poland
Tel: (+48 22) 840 37 66
e-mail: cbk@cbk.waw.pl
Web: http://www.cbk.waw

UPDATED

Portugal

NAV Portugal
(Communication and Image Office)
Rua D – Edificio 121, Aeroporto de Lisboa P-1700 -008 Lisboa, Portugal

Tel: (+351 218) 55 31 43
(+351 218) 55 35 07
(+351 218) 49 50 87
(+351 296) 88 62 08
Fax: (+351 218) 55 31 47
Telex: 12936 cctal
AFTN: lpppyoyc
lpppzqzx
lppozqzx
RadioCode: 132.3mhz
e-mail: gabcim@nav.pt
desica@nav.pt
acclis@nav.pt
dinis.resendes@nav.pt
Web: http://www.nav.pt
President: Carlos Alberto Gonçalves da Costa
Director: Alexandre Ulrich Kühl de Oliveira
Director: Antonio Barreira Paulino
Director: José Carlos Costa Infante de La Cerda
Director: Rui Pedro Correia Cabaço Gomes
Financial Director: Dr Mrs Teresa Gomes
Operations Director: Eng Mrs Conceiçào Lobào Ferreira
Marketing, Public Relations, Communication and Image Director: Dr Paulo Lagarto
Tel: (+351 21) 855 31 46
Contact: Pedro Gaspar
■ Responsibility for the management of air traffic in the Lisbon and Santa Maria flight information regions, under the jurisdiction of the Portugese State, in full compliance with the provisions of international agreements and those of the international civil aviation organisations to which Portugal subscribes or of which Portugal is a member state.

UPDATED

Romania

Government – Ministry/Agency

Romanian Space Agency (ROSA)
21-25 Mendeleev Street, Sector1, R-701681 Bucharest, Romania
Tel: (+40 21) 212 87 22
Fax: (+40 21) 312 88 04
e-mail: asr@rosa.ro
Web: http://www.rosa.ro
President: Dr Dumitru Dorin Prunariu
Chief Executive Officer: Dr Marius-Ioan Piso
General Director National Programs Division: Dr Dumitru Ion
Contact: Ms Irina Petra Manciu

UPDATED

Centres of Learning & Research

Commission d'Astronautique de l'Academie de Roumanie
125 Calea Victoriei, R-71102 Bucharest, 1, Romania
Tel: (+40 1) 650 76 80
Fax: (+40 1) 312 02 09
Telex: 11907 acad
President: Prof Dr N N Patraulea
Vice President: Prof Dr Ing Florin Zaganescu
Tel: (+40 44) 14 55 88
■ Main commission of the Romanian Academy, joining approximately 90 scientist-members, devoted to the promotion of non-governmental original scientific works in space related fields.

UPDATED

Russian Federation

Government – Ministry/Agency

International Organization of Space Communications (Intersputnik)
(Headquarters)
2 Smolensky per, 1/4, 121099 Moskva, Russian Federation
Tel: (+7 095) 244 03 33
Fax: (+7 095) 253 99 06
e-mail: dir@intersputnik.com
Web: http://www.intersputnik.com
Director General: Gennady G Kudryavtsev
Deputy Director General: Josef Dolecki
Tel: (+7 095) 244 05 03
Sales Director: Stefan Kollar
Tel: (+7 095) 241 72 77
Operations and Development Director: Victor Romantsov
Tel: (+7 095) 244 04 56
Director of International and Legal Department: Victor Veshchunov
Tel: (+7 095) 241 45 75
Finance and Strategic Planning Department Director: Svetlomir Stoitchev
Tel: (+7 095) 241 46 97
Marketing Communications Manager: Ms Olga Vasilieva
Tel: (+7 095) 244 06 98
Administrative Manager: Sergey Beliaev
Tel: (+7 095) 241 68 85
Chief Accountant: Zhanna Panina
Tel: (+7 095) 241 47 30
■ Direct access global organisation which can be used by licensed users of satellite services. Offers all satellite-based communication services (voice, data, TV and audio broadcasting) through LMI-1 and Express-A geostationary satellites together with full range of project engineering services.

UPDATED

Transport Administration – Space

Federal Space Agency
(ex-Russian Aviation and Space Agency)
(ex-Russian Space Agency)
42 Schepkin Street, 107996 Moskva, Russian Federation
Tel: (+7 095) 631 94 44
Fax: (+7 095) 288 90 63
(+7 095) 975 44 67
Web: http://www.rosaviakosmos.ru
Chief of Federal Space Agency: Anatoliy Nikolayevich Perminov
Deputy Chief of Federal Space Agency: Nikolai Fedorovich Moiseyev
Deputy Chief of Federal Space Agency: Alexander Ivanovich Medvedchikov
Tel: (+7 095) 971 91 76
(+7 095) 975 41 67
■ Responsible for the implementation of state policy in the field of exploration and use of outer space for peaceful purposes and for the general management of the development of new aeronautics and space technologies and products.

UPDATED

Centres of Learning & Research

Academy of Sciences
(Institute of Astronomy)
76 – A Khoroshevskoye Shosse, 123007 Moskva, Russian Federation
Tel: (+7 095) 195 23 63
(+7 095) 195 15 00
(+7 095) 195 63 35
Fax: (+7 095) 195 22 53
(+7 095) 195 15 00
Director: Anatoli I Grigoriev
First Deputy Director: Victor M Baranov

UPDATED

Babakin Space Research Center
Leningradskaya Street 24, Moscow Region, 141400 Khimki, 2, Russian Federation
Director General: Konstantin Mikhailovich Pichkhadze
Tel: (+7 095) 573 25 84
Fax: (+7 095) 573 25 84
Deputy Director General: Igor Petrovich Zaitsev
Tel: (+7 095) 573 90 56
Fax: (+7 095) 913 59 92
Foreign Trade Department Chief: Yaroslav Georgievich Podobedov
Tel: (+7 095) 745 22 85
Fax: (+7 095) 913 59 92
■ Development of automatic scientific spacecrafts.

UPDATED

Central Research Institute for Machine Building
(TSNIIMASH)
4 Pionerskaya Street, 141070 Korolev, Moscow Region, Russian Federation
Tel: (+7 095) 513 50 00
(+7 095) 187 03 22
e-mail: corp@tsniimash.ru
Web: http://www.tsniimash.ru
Director General: Nikolay A Anfimov
Deputy Director General: Vasiliy I Luckyatchenko

UPDATED

Central Scientific and Research Institute Elektroprobor (CSRI Elektroprobor)
30 Malaya Posadskaya Street, 197046 St Petersburg, Russian Federation
Tel: (+7 812) 232 59 15
Fax: (+7 812) 232 33 76
e-mail: elprib@online.ru
elprib@telros.net
Web: http://www.elektropribor.spb.ru
Head of Institute: Vladimir G Peshekhonov
Head of International Business & Marketing Department: Nikolai Minaev
■ Production and development of integrated satellite/inertial navigation systems, mobile airborne marine gravimeter and Periscopes.
Accreditations: ISO 9000-9003
Publications: Gyroscopy and Navigation, Annual

UPDATED

Gromov Flight Research Institute (Gromov FRI)
(High Computer Technologies, Computer Based Training (CBT) and Training Methodology Laboratory)
Zhukovsky 2, 140160 Moskva, Russian Federation
Tel: (+7 095) 556 56 07
Fax: (+7 095) 556 53 34
Web: http://www.lii.ru
■ Research and production of Computer Based Training (CBT) systems.

UPDATED

Institute for Biomedical Problems (RFSSC-IMBP, RAS)
(State Scientific Center Institute of Biomedical Problems Russian Academy of Sciences)
(Institut Mediko-Biologicheskikh Problemi)
Khoroshovskoye Shosse 76a, 123007 Moskva, Russian Federation
Tel: (+7 095) 195 23 63
Fax: (+7 095) 195 22 53
Telex: 411048
417508
Director: Anatoli I Grigoryev
First Deputy Director: Viktor M Bosanov
Deputy Director: Lieut Gen Yuvenali M Volynkin
Deputy Director: Yevgeni A Ilyin
Tel: (+7 095) 195 02 23
Deputy Director: Valery V Bogomolov
Marketing Director: Mark S Belakovski
■ Research in space biomedicine. Specific areas include the effect of stress on personnel under certain conditions – naval personnel on board atomic-powered submarines, on remote sites and in space.

UPDATED

International Science and Technology Center (ISTC)
Luganskaya Ulitsa 9, 115516 Moskva, Russian Federation
Tel: (+7 501) 797 60 10
Fax: (+7 501) 797 60 47
e-mail: istcinfo@istc.ru
Web: http://www.istc.ru
Executive Director: Prof. Dr Michael Kröning
Public Information Officer: Alexander Ivanchenko
■ Provide peaceful research opportunities to weapons scientists in CIS countries.

UPDATED

State Governmental Scientific-Testing Area of Aircraft Systems (FKP (GkNIPAS))
(ex-State Scientific-Testing Area of Aircraft Systems)
140250 Belozersky, Voskresensky District, Russian Federation
Tel: (+7 095) 556 07 09
Fax: (+7 095) 556 07 40

Web: http://www.aha.ru/~leokon

- Ground testing of aviation, technology products and armament. Research and development, study services in the field of aviation and space.

UPDATED

State Research & Production Space Rocket Center

(GNPRKTs TsSKB-Progress)
Pskovskaya Street 18, 443009 Samara, Russian Federation
Tel: (+7 8462) 24 28 14
(+7 8462) 92 65 31
Fax: (+7 8462) 92 65 18
Telex: 214153 nika ru

- Design and manufacture of launch vehicles such as Molniya, Soyuz-2, and spacecraft such as Photon and Resurs-DK. Development, manufacture and launch of middle-class LV Soyuz and Soyuz-2, SC for Earth remote sensing, research in the field of biology and material study.

UPDATED

Industry – Manufacturing

Academician M.F. Reshetnev's Nauchno-Proizvodstvennoe Obiedinenie Priklandnoi Mechaniki (NPO PM)

(Scientific Production Association of Applied Mechanics)
52 Lenin Street, Zheleznogozsk, Krasnoyarsk 26, 662972 Krashoyarsky Krai, Russian Federation
Tel: (+7 391) 972 80 08
(+7 391) 973 46 17
Fax: (+7 391) 972 26 35
(+7 391) 972 58 33
(+7 391) 972 36 13
e-mail: postmaster@npo-pm.krasnoyarsk.ru
Web: http://www.npopm.ru
General Designer and General Director: Albert G Kozlov
Head Patent and Information Department: Roman P Turkenich
First Deputy General Designer and Technical Director: Vladimir A Bartenev
Tel: (+7 391) 972 37 58

- Design of communications systems, TV broadcasting, navigation and geodite satellites. Manufacture of ground antennas for satellite communications. Projects include Express-AK, Express-2000, Express-K2 (Troika project) and SESAT communications satellites and Gonets low earth orbit communications satellite.

Representing:
Aerospatiale, France
Alcatel Espace, France
European Telecommunications Satellite Organization, France
Russian Satellite Communications Company, Russia
Russian Space Agency, Russia
SRI Precise Instruments, Russia

UPDATED

Aeropribor-Voskhod JS

(Aeropribor-Voskhod Joint Stock Company)
19 Tkatskaya Street, 105318 Moskva, Russian Federation
Tel: (+7 095) 363 23 01
Fax: (+7 095) 363 23 43
Managing Director: V Belotelov
General Director and Chief Designer: Vladimir G Kravtsov
Head of Department, Advertising and Exhibitions: I V Plyashkevich

- Manufacture of information systems, air data measurement sets for military and civil aircraft, back-up height/velocity measuring instruments, pitot/static tubes, flight emergency mode warning systems, automatic chute deployment devices, instrumentation for light and super-light aircraft, automatic devices for aerial/space vehicle life support systems, wide scope of precision pressure sensors and sensor-based test equipment, variety of induction pressure meters, electronic barometric altimeters with the function of FL deviation alert system. Helmet data displays, digital air pressure meters.

UPDATED

Arsenal Design Bureau named after M.V. Frunze Federal State Unitary Enterprise

(KB Arsenal)
Komsonmola Street 1-3, 195009 St Petersburg, Russian Federation
Tel: (+7 812) 542 29 73
(+7 812) 248 98 42
Fax: (+7 812) 542 20 60
Telex: 122774 grot
e-mail: kbarsenal@peterlink.ru
Director General and General Designer: Boris I Poletaev
First Deputy Director General and General Designer: Leonid D Fedotov
Tel: (+7 812) 542 31 11
Deputy General Designer, Marketing and Communications: Vladimir I Sapozhnikov

- Research, development and creation of space platforms, various spacecraft and space systems.

UPDATED

Fakel Experimental Design Bureau

(ex-OKB Fakel)
181 Moscovskiy Prospekt, 236001 Kaliningrad, Russian Federation
Tel: (+7 112) 46 19 64
Fax: (+7 112) 46 17 62
Telex: 262 104 zenit
Web: http://www.rosaviakosmos.ru/org/fakel_e.html
Manager and Chief Designer: A S Bobyor
First Deputy Chief Designer: Dr Nikolai A Maslennikov
Manager for the External Economic Relations Department, Responsible for Foreign Relations: Krasnoslav S Petrov

- Development and manufacture of electric thrust units and plasma sources for different applications. Attitude and orbit control systems of different spacecraft, full-scale production of stationary plasma thrusters (SPT) and SPT systems.

UPDATED

Federal State Unitary Enterprise 'Russian Institute of Space Device Engineering' (RISDE)

53 Aviamotornaya Street, 111250 Moskva, Russian Federation
Tel: (+7 095) 509 12 01
(+7 095) 509 12 02
Fax: (+7 095) 509 12 00
e-mail: contact@rniikp.ru
Web: http://www.rniikp.ru

- Development and production of ground stations for satellite data reception, satellite control and orientation systems; remote sensing and data transmission satellite payloads; and passive microwave units.

UPDATED

Geofizika-Cosmos NPP

(Geokos Special Design Bureau (SKB)
(Scientific Production Enterprise Geofizika-Cosmos)
11/17 Irkutskaya Street, 107497 Moskva, Russian Federation
Tel: (+7 095) 462 03 43
Fax: (+7 095) 462 13 14
e-mail: info@geofizika-cosmos.ru
Web: http://www.geofizika-cosmos.ru
Director: Anatoly N Egupov
First Deputy Director and Chief Designer: Mikhail G Pirogov
Deputy Chief Designer for International Co-operation: Alexander P Dmitriev

- Design and manufacture of opto-electronics including automatic opto-electronic instruments for altitude measurement using the sun, stars and planets, optical visualisation instruments for use by astronauts during manual manoeuvering in daylight and night conditions, instruments for studying the Earth's natural resources and atmosphere surrounding space vehicles (radiometers, photometers, spectroradiometers, spectrophotometers, actiometers, lightmeters and laser measuring devices), ground-based instruments for orientating heliostats towards the sun, star and Earth simulators, radiometers for star simulator calibration, simulators for the visualisation of external effects during training of astronauts, high precision optical angle-code encoders. Test facilities include precise opto-mechanical stands equipped with sun, star and planet simulators, test centre for vibration, shock, vacuum, acoustic and climatic environment, graduation complex for calibration of spectrometric instrumentation.

UPDATED

Joint Stock Company Research and Production Association Molniya (JSK NPO Molniya)

(ex-Molniya Scientific and Industrial Enterprise)
(OAO Nauchno-Proizvodstvennoye Obedinenie Molniya)
6 Novoposelkovaya Street, 123459 Moskva, Russian Federation
Tel: (+7 095) 493 50 53
(+7 095) 497 47 60
Fax: (+7 095) 497 47 23
(+7 095) 497 59 30
e-mail: molniya@dol.ru
Web: http://www.buran.ru
Managing Director: Dr Alexander S Bashilov
Deputy General Director and Financial Director: Valery Y Vorobjev
Tel: (+7 095) 497 49 53
Deputy General Director and Marketing Director: Prof Michail J Gofin
Operations Director: Yuri V Matveykin
Chief Engineer: Gennadiy G Kryuchkov
Tel: (+7 095) 497 55 78
Chief Marketing Department: Dr Vladimir I Fishelovich
Chief of Public Relations: Ms Nina V Andreeva

- Design and manufacture of re-usable orbital vehicle "Buran". Multipurpose aerospace systems, passenger, business and cargo aircraft (triplane) with payload from 500 to 450,000 kg.

UPDATED

JSC Motorostroitel

(ex-Motorostroitel Samara JSC)
29 Zavodskoje Shosse, 443009 Samara, Russian Federation
Tel: (+7 8462) 55 16 12
Fax: (+7 8462) 55 16 00
e-mail: marketing@motor-s.ru
General Director: Igor Leonidovich Shitarev

- Manufacture of engines for rockets (Vostok, Sojuz, Progress). Manufacture engines for MiG 15, Tu-114, Tu-144, An-22 and gas turbine engines for gas pumping and power industry.

UPDATED

JSC ND Kuznetsoc STC

(ex-Trud NPO)
Sergeya Lazo Street, 443026 Samara, Russian Federation
Tel: (+7 8462) 50 02 28
Fax: (+7 8462) 50 12 11
Deputy Chief Designer: L M Shirkin

- Design and production of rocket engines.

UPDATED

Kayser-Threde Moscow

Technical Center 'Splav', ulitsa Baikalskaya, 9, 107 497 Moskva, Russian Federation
Tel: (+7 95) 462 46 88
Fax: (+7 95) 462 40 35
e-mail: ktmoskau@glasnet.ru
ktmoskau@online.ru

UPDATED

Keldysh Research Centre (KeRC)

8 Onezhskaya Street, 125438 Moskva, Russian Federation
Tel: (+7 095) 456 46 08
Fax: (+7 095) 456 82 28
e-mail: kerc@elnet.msk.ru
Web: http://www.kerc.msk.ru
Director: Anatoly S Korotev
Supervisor for Foreign Economic Relations: Mrs Lorsana A Besedina
Tel: (+7 095) 456 87 56

- Development, manufacture and testing of rocket engines, space propulsion systems, high energy beam generators, particle accelerators, and lasers. Energy saving developments in the interests of ecology include: purification of natural and sewage water and water desalination, power supply on the basis of multi-stage processing of solid propellants and wastes, elimination of balneological infectious wastes, ozone-friendly refrigerant C1-freon substitute, elimination of oil spots by laser beam at the water surface, metallic filters, and vibrodampers.

UPDATED

Khrunichev State Research and Production Space Centre

(Gosudarstvennyi Kosmicheskii Naucho-Proizvodstvennyi Tsentr Imeni M V Krunicheva)
18 Novozavodskaya Street, 121309 Moskva, Russian Federation

Tel: (+7 095) 145 92 10
Fax: (+7 095) 142 59 00
Web: http://www.khrunichev.com
Chief, Public Relations Department: Alexander Bobrenev
■ Production of the launch-vehicle Proton, Angara and Rockot. Also produces multi-purpose space stations and orbital modules.

UPDATED

Komsomolsk-on-Amur Aircraft Production Association

1 Sovetskaya Street, 681018 Komsomolsk, on Amur, Russian Federation
Tel: (+7 42172) 636 09
(+7 42172) 635 89
Fax: (+7 42172) 298 51
(+7 42172) 634 51
e-mail: knaapo@kmscom.ru
Web: http://www.ceebd.co.uk/knaapo
■ Military aviation and equipment, ultralight and space technology.

UPDATED

Konstruktorskoe Buro Mashynostroyeniya (KBM)

(Design Bureau of Machine-Building)
42 Oksky Avenue, 140402 Kolomna, Moscow Region, Russian Federation
Tel: (+7 095) 26 16 33 01
(+7 095) 66 16 33 44
(+7 095) 26 16 34 68
Fax: (+7 095) 66 13 30 61
(+7 095) 26 13 30 64
Telex: 846521ogon
e-mail: kbm@kolomna.ru
Web: http://www.kbm.ru
Department Head Foreign Economic Activities: Georgy N Kuzyk
■ Production of short and medium range guided surface-to-air missile systems including Strela-2 and Igla as well as anti-tank weapon systems. Also manufactures explosive reactive armour (ERA) for the protection of armoured vehicles against kinetic and explosive weapons.

UPDATED

Lavochkin Scientific Production Association

Leningradskaya Street 24, Mascow Region, 141400 Khimki, 2, Russian Federation
Tel: (+7 095) 251 67 44
(+7 095) 573 56 75
Fax: (+7 095) 573 35 95
Web: http://www.laspace.ru
■ Design, development, testing and manufacturing of satellite systems for scientific research and space-borne telecopes as well as small multi-purpose satellites.

UPDATED

Moscow Scientific Research Institute of Instrument Engineering

(NPO Vega-M MNIIP)
(ex-Scientific & Research Institute of Instrumental Engineering)
34 Kutuzovsky Prospect, 121170 Moskva, Russian Federation
Tel: (+7 95) 249 76 10
Fax: (+7 95) 148 79 96
Telex: 412268 vega
■ Research, develop and manufacture radar early warning systems, surface surveillance radar and control systems.

UPDATED

NPO Energomash

1, Burdenko Street, 141400 Moskva, Russian Federation
Fax: (+7 095) 251 75 04
e-mail: energo@online.ru
First Deputy General Director: Nikolay A Pirogov
Chief of Information Department: Vladimir S Sudakov
Tel: (+7 095) 572 76 49

UPDATED

NPO Lavotchkin

(Babakin Engineering Research Centre)
24 Leningradskaya Street, 141400 Khimki, Moscow Oblast, Russian Federation
Tel: (+7 095) 573 90 56
(+7 095) 573 55 65
(+7 095) 573 86 92
Fax: (+7 095) 573 35 95
(+7 095) 573 27 86
Telex: 911721 irbis
e-mail: vkudz@mouse.berc.tssi.zu
Designer and Director General: Stanislav D Koulikov
Deputy Director General, Foreign Trade: Igor P Zaitsev
Foreign Trade Department Chief: Yaroslav Podobedon
Tel: (+7 95) 745 22 84
(+7 95) 745 21 12
■ Planetary and deep-space craft, science and communication satellites.

UPDATED

NPO Mashinostroyenia

33 Gagarin Street, 143966 Reutov, Moscow Oblast, Russian Federation
Tel: (+7 095) 528 30 18
Fax: (+7 095) 302 20 01
e-mail: fnpc@npomash.ru
Web: http://www.npom.ru
General Director: Dr Herbert A Yefremov
Director of Export Department: Vladimir Telkov
■ Development and manufacture of intercontinental ballistic missiles, heavy space launch vehicles, earth monitoring satellites, orbital manned and automatic space stations.

UPDATED

NPO Prikladnoi Mekhaniki (NPO PM)

(Scientific Production Association of Applied Mechanics)
52 Lenin Street, Zheleznogorsk, 662972 Krasnoyarsk, 26, Russian Federation
Tel: (+7 39197) 346 17
(+7 39197) 280 08
Fax: (+7 39197) 226 35
(+7 39197) 258 33
e-mail: gonti@npo-pm.krasnoyarsk.su
postmaster@npo.pm.krasnoyarek.ru
Web: http://www.npopm.ru
General Designer and General Director: Albert G Kozlov
Deputy General Designer: Vladimir A Bartenev
Tel: (+7 391) 972 37 58
Head, Patents and Information: Roman P Tourkenitch
■ Design of communications systems, TV broadcasting, navigation and geodetic satellites. Manufacture and operation of satellites. Manufacture of ground antennas for satellite communication.

UPDATED

Polyot Research and Production Company

(GUP NPP Polyot)
GSP-462, 603950 Nizhny Novgorod, Russian Federation
Tel: (+7 8312) 42 21 04
Fax: (+7 8312) 42 31 57
e-mail: polyot@nnov.rfnet.ru
Web: http://www.polyot.nnov.rfnet.ru
General Director: Y Belousov
Deputy General Director, Foreign Relations and Co-operation: Valentin A Kovalev
Tel: (+7 83 12) 42 21 64
Deputy Director for Foreign Relations: Ms Tatyana A Kholodova
■ Development of airborne radio communication equipment. Development and manufacture of aviation communication systems and devices, airborne and ground-based communication complexes for military and civil applications, automated receive-transmit centres for ATC purposes, antenna-feeders and aircraft-based ecological monitoring systems.

UPDATED

Samara Space Centre (SSC)

(State Research & Production Space Rocket Centre TsKB-Progress)
(ex-Central Specialised Design Bureau)
(ex-National Scientific Manufacturing Aerospace Center)
(ex-Samara Progress Plant)
18 Ulitsa Pskovskaya, 443009 Samara, Russian Federation
Tel: (+7 8462) 55 13 61
(+7 8462) 92 65 29
Fax: (+7 8462) 92 65 18
(+7 8462) 28 61 30
(+7 8462) 97 18 86
Telex: 214153 nika
e-mail: csdb@mail.samtel.ru
mail@progress.samara.ru
General Director: Alexander N Kirilin
First Deputy General Director, General Designer TsSKB Chief: Gennadi Anshakov
First Deputy General Director for Economics and Finances: Boris Melioransky
Tel: (+7 8462) 92 65 23
First Deputy General Director, Chief Engineer: Nikolai Rodin
Tel: (+7 8462) 97 18 43
Deputy General Director for Foreign Economic Activity: Vladimir Kozlov
Tel: (+7 8462) 92 64 90
Deputy General Director for Strategic Development, Enterprise Management and Labour Organisation: Anatoly Petrushin
Deputy General Director for Commercial Affairs: Dmitri Nosov
Deputy General Director for Production Operation: Vladimir Pashistov
■ Development and manufacturing of automated spacecraft for remote sensing, bio and microgravity experiments. Launchers and ground control complexes.

UPDATED

Tupolev PSC

(A.N. Tupolev Scientific and Technical Design Complex)
(Aviation Scientific Technical Complex Named for A N Tupolev JSC)
(ex-OKB-156)
(ex-Tupolev Design Bureau)
(ex-Tupolev JSC)
17 Academician Tupolev Embankment, Moskva, Russian Federation
Tel: (+7 095) 267 25 33
Fax: (+7 095) 267 27 33
(+7 095) 261 08 68
(+7 095) 261 71 41
e-mail: tu@tupolev.ru
Web: http://www.tupolev.ru
President, General Designer and Chairman of the Board: Igor Shevchuk
First Vice President: Alexander P Polyakov
Senior Vice President: Gennadiy I Korotkov
Vice President: Grigoriy Y Goutzenko
Deputy General Director, Manufacture: Oleg Y Alasheev
Executive Director: Seguey F Ilyushenkov
Corporation Development Director: Elionora P Romanova
■ Design and manufacture of civil and military aircraft and space projects. Activities involve the Tu-334 passenger aircraft and the Burlak air launched space booster. Design and production of Tu-22M and Tu-160 supersonic heavy bombers.

UPDATED

Zvezda-Strela State Research and Production Centre

7 Ilyicha Street, Korolev, 140070 Moskva, Region, Russian Federation
Tel: (+7 095) 519 20 02
Fax: (+7 095) 511 94 39
■ Development and production of airborne, naval and army tactical missiles.

UPDATED

Zvezda Joint Stock Company
123 Babushkina Street, 192012 St Petersburg, Russian Federation
Tel: (+7 812) 103 00 70
(+7 812) 262 07 47
Fax: (+7 812) 267 37 76
e-mail: office@zvezda.spb.ru
Web: http://www.zvezda.spb.ru
General Director and General Designer: Pavel Plavnik
Head of Public Relations Department: Yulia Grebneva

UPDATED

Industry – Service

Myasishchev Design Bureau (MDB)

(Eksperimentalnyi Mashinostroitelnyi Zavod Imeni V M Myasishcheva)
Zhukovsky 5, 140180 Moskva, Region, Russian Federation
Tel: (+7 095) 556 78 29
(+7 095) 912 60 41
Fax: (+7 095) 728 41 30
(+7 095) 556 55 83
(+7 095) 556 76 94
e-mail: mdb@mail.sitek.ru

Web: http://www.corbina.net
Commercial Director: Alexander Gorbunov
Tel: (+7 095) 556 71 49
Operations Director: Oleg Sergeivich Dolgikn
Public Relations Director: Stanislav Gavrilovich Smirnov
Deputy General Designer: Valery Andreevich Shirinyants
Deputy General Designer: Alexander Brook
Tel: (+7 095) 556 50 12
■ Aircraft design. Activities include defence and civil aircraft, manufacture of aerospace equipment and space technology. Company divided into four divisions: design bureau, experimental manufacturing base, flight test base and aviation transportation squad.

UPDATED

Sovinformsputnik

47 Leningradskiy Prospekt, 125167 Moskva, Russian Federation
Tel: (+7 095) 943 07 57
Fax: (+7 095) 943 05 85
e-mail: common@sovinformsputnik.com
Web: http://www.sovinformsputnik.com
General Director: Mikhail M Fomtchenko
Deputy General Director: Alexey Movlyav
■ Provide satellite imaging services.

UPDATED

Saudi Arabia

Arab Satellite Communications Organisation (ARABSAT)

PO Box 1038, 11431 Riyadh, Saudi Arabia
Tel: (+966 1) 482 00 00
Fax: (+966 1) 488 79 99
Web: http://www.arabsat.com
■ Provide satellite communications services to the Arab League and neighbouring countries.

UPDATED

Singapore

Centres of Learning & Research

National University of Singapore

Block SOC-1 Level 2, Singapore, 119260 Singapore
Tel: (+65) 68 74 63 96
Fax: (+65) 67 75 77 17
e-mail: crisp@nus.edu.sg
Web: http://www.crisp.nus.sg
Administration Manager: Ms Karen Wong

UPDATED

Industry – Manufacturing

Agilis Communication Technologies Pte Ltd

Singapore Technologies Building, 100 Jurong East Street 21, Level 4, Singapore, 609602, Singapore
Tel: (+65) 65 67 67 91
Fax: (+65) 65 68 74 02
e-mail: mktg@agilis.st.com.sg
Web: http://www.agilis.st.com.sg
General Manager and Vice President: Tang Kum Chuen
Tel: (+65) 65 68 71 83
Marketing Executive: Claris Lam
Tel: (+65) 65 68 74 17
Fax: (+65) 65 67 63 70
■ Manufacture and design of microwave and satellite components and subsystems, eg VSAT ODUs, microwave sensors, radar systems and microwave digital radios.

UPDATED

Honeywell Aerospace, Electronic Systems

(Singapore)
(ex-AlliedSignal Aerospace, Avionics and Lighting)
(ex-AlliedSignal Inc., Electronic & Avionics Inc.)
512 Chai Chee Lane #05-01/06, Singapore, 469028, Singapore
Tel: (+65) 64 48 61 22
Fax: (+65) 64 48 02 73
Managing Director: Michael Penafort
Tel: (+65) 62 49 91 81
Director, Quality Assurance: P H Chang
Tel: (+65) 249 91 25
Manager, Materials: Lai Yew Chin
Tel: (+65) 62 49 91 05
Manager, Test Engineering: Tan Yong Chuang
Program Manager: Chua Yak Chye
Tel: (+65) 62 49 91 33

UPDATED

Industry – Service

Jane's Information Group (JIG)

(Asia)
78 Shenton Way, #10-02, UIC Building, Singapore 079120, Singapore
Tel: (+65) 63 25 08 66
Fax: (+65) 62 26 11 85
e-mail: asiapacific@janes.com
Web: http://www.janes.com
Group Business Manager, Asia: David Fisher
Tel: (+65) 63 25 08 61
Information Manager, Asia: Andrew Vasko
Tel: (+65) 63 25 08 62
Regional Executive: Neo Kee Khoon
Marketing Specialist, Asia Pacific: Ms Patricia Pereira
Tel: (+65) 63 25 08 65
Representing:
Jane's Information Group, UK

UPDATED

ST Aviation Services Co. Pte Ltd (SASCO)

8 Changi North Way, Singapore, 499611, Singapore
Tel: (+65) 65 40 56 13
(+65) 65 45 09 88
(+65) 65 40 57 39
Fax: (+65) 65 45 67 57
(+65) 65 42 07 31
SitaCode: sinascr
Web: http://www.stengg.com/aero
Vice President and General Manager: Stephen Low
Tel: (+65) 65 40 56 18
(+65) 65 40 56 00
■ Specialise in maintenance, repair and overhaul of a wide range of commercial wide body and narrow body aircraft.

UPDATED

Industry – Sales Agent/Office

Arianespace SA

(Singapore Liaison Office)
Shenton House #25-06, 3 Shenton May, Singapore, 68805 Singapore
Tel: (+65) 62 23 64 26
Fax: (+65) 62 23 42 68
Representing:
Arianespace SA, France

UPDATED

Radyne Comstream Inc.

(Singapore)
(ex-Comstream Corporation)
15 McCallum Street, #12-04 Natwest Centre, 069045, Singapore
Tel: (+65) 62 25 40 16
Fax: (+65) 63 25 19 50
Web: http://www.radn.com
Representing:
Radyne Comstream Inc., US

UPDATED

South Africa

Industry – Service

CSIR

(Satellite Applications Centre)
PO Box 395, 0001 Pretoria, South Africa
Tel: (+27 12) 334 50 00
Fax: (+27 12) 334 50 01
Web: http://www.sac.co.za
■ Provision of spacecraft tracking, telemetry and command services to launchers, operators and owners. Remote sensing and information services.

UPDATED

Spain

Government – Ministry/Agency

Centre for the Development of Industrial Technology (CDIT)

Cid 4, E-28001 Madrid, Spain
Tel: (+34 91) 581 55 00
Fax: (+34 91) 581 55 94
e-mail: info@cdti.es
Web: http://www.cdti.es
Director General, Head of Spanish Delegation to ESA: Vicente Gómez
Director of Strategic Programmes: Emilia Buergo
Head of ESA Programmes Department: Roberto Molind
Head of Space Technology and Applications Department: J M Leceta

UPDATED

Government – Intergovernmental

European Union Satellite Centre

(ex-Western European Union)
(Union de l'Europe Occidentale Satellite Centre Satellitaire)
Apartado de Correos N° 511, E-28850 Torrejón de Ardoz, Madrid, Spain
Tel: (+34 91) 678 60 00
Fax: (+34 91) 678 60 06
e-mail: info@weusc.es
Web: http://www.weu.int
Director: Fernando Davara
Deputy Director: Brian Routledge
Head of Research Division: Giulio Prisco
Head of Technical Division: Juan Giner
Head of Administration/Personnel: Robert de Meyier

UPDATED

Villafranca Satellite Tracking Station

PO Box 50727, E-28080 Madrid, Spain
Tel: (+34 91) 401 96 61
Fax: (+34 91) 402 45 58
Telex: 42555
e-mail: info@esa.int
vilspaweb@esa.int
Head of Station: V Claros

UPDATED

Association

Instituto Iberoamericano de Derecho Aeronautico y del Espacio y de la Aviacion Comercial

Plaza del Cardenal Cisneros 3, E-28040 Madrid, Spain
Tel: (+34 91) 336 63 74
Fax: (+34 91) 543 98 59
e-mail: en201067@encomix.net
President: Javier Aparicio
Secretary-General: Cayetano Aguayo
■ Congress and publications on aeronautical and space law.

UPDATED

Industry – Manufacturing

Alcatel Espacio SA

(Alcatel Space Business Line)
c/o Einstein 7, E-28760 Madrid, Spain
Tel: (+34 91) 807 79 00
Fax: (+34 91) 807 79 99
e-mail: comunicacion.espacio@alcatel.es
Web: http://www.alcatel.es/espacio
General Director: Luis García Echegoyen
Commercial and Marketing Director: Jose Antonio Alvarez
Industrial Operations Director: Luis Cervera
Finance Director: Eduardo Terrón
Human Resources Director: Lorenzo Largacha
Quality Director: Antonio Arguello Berrocal
Planning and Information Systems Director: Loic Hamard
Logistics and Procurements Manager: Ms Flavia Buergo Bandera
■ Design, development and manufacture of satellite communications equipment including on board equipment and subsystems for space vehicle applications. Products include transponders and transceivers, TTC subsystems, amplifiers, transmitters, modulators, multiplexers, filters, diplexers and other passive devices. Digital processing units and on-board data handling and signal processing equipment for advanced communication payload.

UPDATED

Computadoras, Redes e Ingeniería SA (CRISA)

C/o Torres Quevedo, 9 (PTM), E-28760 Tres Cantos, Madrid, Spain
Tel: (+34 91) 806 86 00
Fax: (+34 91) 806 02 35
e-mail: mkt@crisa.es
Web: http://www.crisa.es
■ Develop and manufacture electronics and software products for aerospace, defence and telecommunications sectors.
Accreditations: ISO 9001: 2000; AQAP 110; ESA PSS-01-XXX

UPDATED

EADS CASA

(CASA Espacio)
Avenida de Aragón 404, E-28022 Madrid, Spain
Tel: (+34 91) 586 37 00
Fax: (+34 91) 747 47 99
Telex: 48540 casa e
e-mail: espacio@casa-de.es
Managing Director: Pedro Méndez
Programmes Director: Vicente Badenes
Commercial Director: Miguel Angel Llorca
Operations Director: Joaquin Iglesias
Business Development Manager: Gonzalo Galipienso
Commercial Logistics Manager: J F Lechon
■ Design, development, production and testing of space frames for satellites and launchers.

UPDATED

GMV S A

(ex-Grupo de Mecanica del Vuelo SA)
c/o Isaac Newton 11, Parque Technologico de Madrid, Tres Cantos, E-28760 Madrid, Spain
Tel: (+34 91) 807 21 00
Fax: (+34 91) 807 21 99
e-mail: info@gmv.es
Web: http://www.gmv.es
Executive President: Ms Monica Martinez Walter
General Secretary: Dr Enrique Revilla
Tel: (+34 1) 807 21 04
Space and Defence Director: Jesús Serrano Martinez
Marketing and Communications Department: Pedro Schoch Benlloch
Cost Control and Administration Department: F J Martinez
Quality Department: M Diaz
Department of Technologies: F J Campos
Space Engineering Business Unit: N A Molina
Business Development and Key Accounts Department: Jorge Potti
Ground Systems Business Unit: J Noguero
Space Engineering Business Unit: M A Molina
GNSS Business Unit: J Cosmen
Defence and Security Business Unit: M Perez
Marketing Assistant: Rebeca López
Tel: (+34 91) 807 21 80
Contracts Department: J C Carmona
■ Provide consulting and engineering services, software development, and turnkey systems integration to the aerospace and defence markets. Activities include: engineering and mission analysis, satellite control centres, satellite navigation systems, back-end information systems and defence systems.

UPDATED

Sener Grupo de Ingenieria

4 Severo Ochoa, Parque Technológico de Madrid, E-28760 Tres Cantos, Madrid, Spain
Tel: (+34 918) 07 70 00
Fax: (+34 918) 07 72 01
e-mail: dep.aeroespacial@sener.es
dir.exterior@sener.es
Web: http://www.sener.es
President: Dr Ing Jorge Sendagorta Gomendio
Tel: (+34 91) 807 70 00
Managing Director, Aerospace and Systems Department: Dr Ing Alvaro Azcárraga Arana
Tel: (+34 91) 807 71 60
Managing Director: Dr Ing Jorge Unda
Tel: (+34 94) 481 76 78
Director, Control and Actuation Systems: Dr Ing Rafael Quintana Alonso
Financial Director: Ignacio Mataix
Tel: (+34 91) 807 71 30
Corporate Communication: Ms Begona Francoy
Tel: (+34 91) 807 72 41
Corporate Communications: Ms Mercedes Dominguez Baron
Tel: (+34 91) 807 71 29
■ Provide aerospace engineering, systems integration and actuation control systems, specialising in critical satellite systems, such as precision mechanisms and attitude and orbit control systems, and the development and production of actuation control systems. Products for applications in the electronics, mechanisms and structures; guidance and control; software, aerothermodynamics, advanced navigation systems, airport systems and electro-optics markets.
Participates in European Space Agency (ESA) missions including Envisat, Meteosat, Cluster, XMM and Integral; also a collaborator in Galileo, Ariane, the ISS and the EJ 200 Eurofighter Typhoon engine.
Shareholdings
ITP (53.2%)
Arianespace (0.57%)
Galileo Sistemas y Servicios (14.3%)
Galileo Industries (2%)
Hisdesat (5%)
Bóreas (100%)

UPDATED

Sener Ingenieria y Sistemas SA

(ex-Sener Ingenieria y Sistemas)
4 Severo Ochoa, Parque Technológico de Madrid, E-28760 Tres Cantos, Madrid, Spain
Tel: (+34 91) 807 71 21
Fax: (+34 91) 80 72 09
e-mail: dep.aerospacial@sener.es
Web: http://www.sener.es
Managing Director, Aerospace and Systems Department: Dr-Ing. Alvaro Azcárraga Arana
Corporate Communication Department: Ms Mercedes Domínguez Barón
Tel: (+34 91) 807 71 29
Fax: (+34 91) 807 72 12
■ Rocket and satellite launching systems, air launched weapon systems, space mechanisms, booms, antennas, control electronics, military software systems analysis, ship design, ground support and test facilities, jet engine design and development. Design authority of the control section of the IRIS-T programme.

UPDATED

Industry – Service

Indra

(Corporate Headquarters)
(Indra Sistemas, SA)
(ex-Ceselsa and Inisel)
Arroyd de la Vega, Avenida de Bruselas 35, Alcobendas, E-28108 Madrid, Spain
Tel: (+34 91) 480 50 00
(+34 91) 396 33 00
Fax: (+34 91) 480 50 80
(+34 91) 396 31 31
e-mail: indra@indra.es
Web: http://www.indra.es
President: Javier Monzon
Vice Chairman: Humberto Figarola
Managing Director: Regino Moranchel
Communication Director: Ms Susana Velascu
General Manager: Juan Carlos Baena
General Manager: Angel Lucio
General Manager: Josep M Vila
EW Division Manager: Joaquin Uguet
Export Manager: Antonio Santamaría
Marketing Communication: Ms Ana Llaguno
Contact: Mar Guillém
Tel: (+34 91) 480 97 25
Fax: (+34 91) 480 97 32
Contact: Ms Silvia Valdesueiro
■ Radars, automatic test systems, air defence, simulation, avionics, air traffic management, radar display consoles, electronic warfare, command and control, fire control systems, satellite communications, earth observation systems, maintenance and logistic support.
Member companies:
Indra ATM
Indra EMAC
Indra Espacio SA

UPDATED

Industry – Sales Agent/Office

EADS CASA

(Headquarters)
(ex-Construcciones Aeronáuticas SA)
Avenida de Aragon 404, E-28022 Madrid, Spain
Tel: (+34 915) 85 70 00
(+34 915) 85 77 89
Fax: (+34 915) 85 76 68
(+34 915) 85 72 64
e-mail: communications@casa.eads.net
Web: http://www.eads.net
Chairman and Chief Executive Officer: Francisco Fernández Sainz
Executive Vice President, Commercial: Pablo de Bergia
Finance, Administration and Control Director: Enrique Gonzalez Tanarro
Legal Affairs Director: Pedro Montoya
Political Affairs Director: Carlos Grandal
Space Division Director: Pedro Mendez
Human Resources Director: Mariano Alonso
Commercial Communications Director: Miguel Sánchez
Communications: José A Barragán
■ Design, development and production of medium and light military transport aircraft, C-212, CN-235 and C-295; development of maritime patrol versions; development of a mission system of FITS (Fully Integrated Tactical System); responsible for some packages and the final assembly line of the A400M military aircraft; other programs including the transformation of A310 and A330 platforms into VIP and MRTT (Multi Role Tanker and Transport) versions; avionics update for old C-130 Hercules aircraft; development, manufacturing and delivery of advanced aerostructures. Military Aircraft concentrates its activities on the project, development, integration, production, product support, sales and marketing of fighters, trainers and helicopters; maintenance and modernisation; participates in the Eurofighter consortium. The space area develops structural, thermal and power distribution sub-systems for launchers, satellites and orbital infrastructures; participation in ESA programs and member and first level contractor of Arianespace.

UPDATED

Sweden

Transport Administration – Space

Swedish National Space Board (SNSB)
(Rymdstyrelsen)
Box 4006, SE-17104 Solna, Sweden
Tel: (+46 8) 627 64 80
Fax: (+46 8) 627 50 14
Web: http://www.snsb.se
Chairman and Director General: Per Tegnér
Tel: (+46 8) 627 64 86
Director of Earth Observation: Göran Boberg
Director of Industrial Affairs: Thorwald Larsson
Head of International Affairs Division: Silja Strömberg
Head of Public Relations: Johan Marcopoulos
Head of Science Division: Lennart Nordh
Administrative Officer: Ms Elsy Bergstedt
■ State authority for space and remote sensing activities and general space programme administration.

UPDATED

Centres of Learning & Research

Swedish Institute of Space Physics
(Institutet för rymdfysik)
Box 812, SE-981 28 Kiruna, Sweden
Fax: (+46 980) 790 91
Web: http://www.irf.se
Research and Development Officer: Dr Rick McGregor
Tel: (+46 980) 791 78
■ Government research institute whose primary task is to carry out basic research, education and associated observatory activities in space and atmosphere physics. IRF has participated in several sounding rocket and satellite investigations primarily targeted at ionosphere and magnetosphere studies. A programme for atmospheric research using ground-based and balloon experiments was established in 1996. Recently developed space instruments include hot plasma and wave experiments for the Swedish satellites Viking and Freja, spherical Langmuir probe for the ESA/NASA Cassini mission and ion mass spectrometers for the two Russian Interball spacecraft. Instruments for Swedish Astrid 2, Japanese mission to Mars (Nozomi) and ESA Rosetta, Cluster 2 and Mars Express missions.
Offices in:
Lund
Umea
Uppsala

UPDATED

Industry – Manufacturing

Saab AB
(Head Office)
SE 581 88 Linköping, Sweden
Tel: (+46 8) 463 00 00
(+46 8) 463 01 52
e-mail: info@saab.se
Web: http://www.saab.se
Chairman: Anders Scharp
Deputy Chairman: Marcus Wallenberg
President and Chief Executive Officer: Bengt Halse
Tel: (+46 13) 18 71 00
Executive Vice President: Göran Sjöblom
Group Senior Vice President: Ingemar Andersson
Group Senior Vice President: Dave Hewitson
Senior Vice President, Legal: Per Erlandsson
Senior Vice President, Human Resources: Mats Lindman
Senior Vice President, Financial Control: Lars Wahlund
Senior Vice President, Treasury: Peter Sandehed
Group Senior Vice President: Jan Nygren
Tel: (+46 13) 18 19 99
Director, Representing Engineering Workers: Ragnar Ludvigsson

UPDATED

Saab Ericsson Space AB
SE-405 15, Göteborg, Sweden
Tel: (+46 31) 735 00 00
Fax: (+46 31) 735 40 00
Web: http://www.space.se
President and Chief Executive Officer: Bengt Mortberg
Executive Vice President: Peter Möller
Financial Director: Adam Fritz Laurén
Commercial Director: Per-Gunnar Lundh
Director of Communication and Public Affairs: Lars Nordfeldt

UPDATED

SaabTech
(ex-Ericsson Saab Avionics AB)
(ex-Saab Avionics AB)
(ex-Saabtech Electronics)
Nettovägen 6, SE-175 88 Jarfalla, Sweden
Tel: (+46 8) 58 08 40 00
Fax: (+46 8) 58 03 22 44
e-mail: info@saabtech.se
Web: http://www.saabtech.se
■ Development and production of EW systems, cockpit displays, tactical reconnaissance systems, IFF radar, electromagnetic technology services, advanced electronics and mechanical equipment. Supply of avionics systems for the Gripen multi-role aircraft.

UPDATED

Telewide AB
Norrbölegatan 66, SE-931 37 Skellefteå, Sweden
Tel: (+46 910) 173 40
Fax: (+46 910) 173 48
e-mail: info@telewide.se
Web: http://www.telewide.se
Chief Executive Officer: Ulf Grape
Tel: (+46 910) 173 42

UPDATED

Volvo Aero Corp.
SE-461 81 Trollhättan, Sweden
Tel: (+46 520) 940 00
Fax: (+46 520) 340 10
Telex: 42040volfa
Web: http://www.volvoaero.com
President: Fred Bodin
Corporate Communications: Fredrik Fryklund
Tel: (+46 520) 944 01
Fax: (+46 520) 985 00
■ Development, manufacture and maintenance of commercial and military aircraft engines. Activities include operations development and production of key components for the Ariane space program.

UPDATED

Industry – Service

Metria
(ex-Metria Satellus)
(ex-Satellus)
PO Box 820, SE-981 28, Kiruna, Sweden
Tel: (+46 980) 670 00
Fax: (+46 980) 670 67
e-mail: lantmateriet@lm.se
satellite@lm.se
Web: http://www.lantmateriet.se
President: Jorgen Forsgren
Project Manager Market Communications: Thomas Karlsen
■ Space information retrieval services.

UPDATED

Swedish Space Corporation (SSC Solna)
(Metria Miljöanalys Office)
PO Box 4207, SE-171 04 Solna, Strandvag 86, Sweden
Tel: (+46 8) 627 62 00
Fax: (+46 8) 98 70 69
e-mail: info@ssc.se
Web: http://www.ssc.se
President (acting) and Strategic Development: Claes-Göran Borg
Vice President, Telecom: Per Zetterquist
Vice President, Technology: Sven Grahn
Tel: (+46 8) 627 62 05
Head, Small Satellite Projects Department: Fredrik von Scheele
Tel: (+46 8) 627 63 16
Head, Communications and Control Systems Department: Bengt Holmqvist
Tel: (+46 8) 627 62 13
Head, Propulsion Research and Development Department: Kjell Anflo
Tel: (+46 8) 627 63 05
Head, Space Avionics Department: Gunnar Andersson
Tel: (+46 8) 627 62 86
Head, Space Systems Integration Department: Conny Svensk
Tel: (+46 8) 627 63 59
Head, Microgravity and Payloads Department: Bengt Larsson
Tel: (+46 8) 627 62 33
General Manager, Space Systems Division: Gierth Olsson
General Manager, Esrange Division: Jan Englund
General Manager, Satellite Operations: Jonny Järnmark
General Manager, Airborne Systems: Olov Fäst
Controller: Bo C Johanson
Controller: Mrs Mari Eriksson
Tel: (+46 8) 627 62 12
Contracts Officer: Claes Hansen
Tel: (+46 8) 627 63 23
Public Relations and Information: Ms Helen Westman
Secretary to the General Manager: Mrs Ulla Robertsson
Tel: (+46 8) 627 62 92
Marketing Assistant: Mrs Yvonne Donlemar
■ Technical implementation of Sweden's space and remote sensing programmes. Design, development, operation and control of satellites and equipment for experiments on sounding rockets and balloons. Launchings of sounding rockets and balloons. Telecommunications services, data reception from remote sensing and other satellites. Processing, archiving and distribution of data from remote sensing satellites. Creation and distribution of environmental data sets. Value-added products and mapping services based on satellite images and geographical information; studies, remote sensing systems and methodology development. Airborne maritime surveillance systems.
Subsidiary
SSC Satellitbild, Kiruna

UPDATED

Swedish Space Corporation (SSC)
(Esrange)
PO Box 802, SE-981 28 Kiruna, Sweden
Tel: (+46 980) 720 00
Fax: (+46 980) 128 90
Web: http://www.ssc.se/esrange
Head of Esrange: Dr Olle Norberg
Public Relations and Information: Ms Johanna Bergstrom-Roos
Tel: (+46 980) 720 24

UPDATED

Systecon AB
PO Box 5205, SE-102 45 Stockholm, Sweden
Tel: (+46 8) 459 07 50
Fax: (+46 8) 459 07 80
e-mail: systecon@systecon.se
Web: http://www.systecon.se
Chairman: Hans Ebenfelt
Managing Director: Sven-Erik Westlund
Marketing Director: Olof Waak
Financial Director: Mrs Monica Schiffer
Purchasing Manager: Mrs Birgit Grandell
Military Activities: Bengt Thunstedt
Civil Aviation and Space Activities: Lennart Nyh
■ Systems and logistics engineering consultancy in aerospace, military, civil and space. Operations research model development. Computer programs for analysis and optimisation of logistic support systems. Professional training.

UPDATED

Switzerland

Government – Ministry/Agency

Swiss Space Office (SSO)
Hallwylstrasse 4, CH-3003 Bern, Switzerland
Tel: (+41 31) 324 10 74
Fax: (+41 31) 324 10 73
Web: http://www.sso.admin.ch
Contact: Ms Margaretha Hofer
■ The administrative unit charged with planning and implementing Swiss space policy, as defined by the Federal Council. The SSO is the executive body of the Federal Space Affairs Commission

(CFAS) and chairs the Interdepartmental Committee for Space Affairs (IKAR). The secretariat of both these groups is located at the SSO.

Swiss space policy is part of Swiss science and technology policy, as well as Swiss foreign policy in and outside Europe. The national space policy is determined by the Federal Council (Government) with the advice of a 20-member Federal Space affairs commission (CFAS). The Swiss Space Office (SSO) of the Federal Department of Home Affairs has the main responsibility within the Federal Administration for the implementation of this policy. The Interdepartmental Committee for Space Affairs (IKAR) assures co-ordination of interests and activities in space matters within the seven Departments. IKAR is chaired by the SSO. The Committee on Space Research of the Swiss Academy of Sciences promotes and coordinates space research in Switzerland.

Space research in Switzerland is carried out at various laboratories of the universities, at Federal Institutes of Technology and by Industry. Research is supported by the Swiss National Science Foundation (fundamental research), by the Commission for Technology and Innovation (applied research) as well as by the federal and cantonal governments.

The SSO has prime responsibility for Swiss participation in ESA programmes and activities and oversees the budget. The SSO looks after Swiss interests in ESA regarding political, institutional, legal, financial and industrial aspects of scientific, applications and infrastructure programmes. Under the SSO's guidance, the Federal Office for Professional Training and Technology (OFFT), the Federal Finance Administration (AFF) and the Foreign Ministry's Political Division V (DP V/DFAE) give their support on technological, financial and political aspects, respectively.

The SSO leads the Swiss Delegation to ESA and chairs the Delegation's coordination meetings and assures that Swiss space policy is consistent among international organisations for the utilisation of satellite operations, e.g. EUTELSAT, INTELSAT, IMSO and EUMETSAT as well as in the respective bodies of the United Nations and the European Union where Switzerland is represented. As there is no national space programme, Swiss space involvement is almost totally routed through ESA.

UPDATED

Association

Swiss Astronautics Association

(Association Suisse d'Astronautique)
(Schweizerische Raumfahrt-Vereinigung)
(ex-Schweizerische Vereinigung f. Weltraumtechnik (SVWT))

c/o Hesso-EIVD, Route de Cheseaux 1, CH-1400 Yverdon-les-Bains, Switzerland
Tel: (+41 24) 423 23 12
Fax: (+41 24) 423 22 05
e-mail: office@srv-ch.org
Web: http://www.srv-ch.org
President: Jean-Daniel Dessimoz

■ The organisation's primary activity is astronautics which involves the promotion of space related activities in Switzerland. Organisation of visits, forums and meetings. Careers advice for those interested in space.

UPDATED

Industry – Manufacturing

Ferriere Cattaneo SA

Via Ferriere 12, Casella Postale, CH-6512 Giubiasco, Ticino, Switzerland
Tel: (+41 91) 850 91 91
Fax: (+41 91) 850 91 92
e-mail: fcsa@ferrierecattaneo.ch
Web: http://www.ferrierecattaneo.ch
Chief Executive Officer: Aleardo Cattaneo
Technical Director: Ing Hans Tandetzki
Engineering Director: Ing Eugenio Moro
Bookkeeping Department: Fabrizio Redaelli
Engineering: Ing Erik Fregni

■ Capabilities include: freight wagons, workshop-machining, frameworks, engineering, research and development.

UPDATED

Mécanex SA

Zone Industrielle Vuarpillière 29, CH-1260 Nyon, Switzerland
Tel: (+41 22) 361 77 33
Fax: (+41 22) 361 67 52
e-mail: info@mecanex.ch
Web: http://www.mecanex.ch
President: Mrs Nicola Thibaudeau
Vice President and Operations Director: Volker Gass
Technical Director: Mrs Marie-Thérese Ivorra
Marketing Manager: Jean-Daniel Zimmermann

■ Research, development and production of slip ring-brush assemblies. Design and manufacture of high precision mechanisms and pointing mechanisms.

UPDATED

Oerlikon Contraves AG

Birchstrasse 155, CH-8050 Zürich, Switzerland
Tel: (+41 1) 316 22 11
(+41 1) 316 42 18
Fax: (+41 1) 311 31 54
e-mail: czjag@ocag.ch
info@ocag.com
Web: http://www.oerlikoncontraves.com
http://www.oerlikoncontraves.ch
Information Officer: Ernst Jäggi

■ Main products and activities:

Defence: Skyshield 35-Ahead gun and missile; Skyguard twin gun and missile air defence systems; Ahead system; Gun-King multi-divergence laser sight system; Shorar/Pagoda search and acquisition radar; Gun-Star compact fire control system; Seaguard CIWS system; ADATS missile systems; training systems and simulators. Automatic cannon for air defence, vehicle and naval armament.

Manufacture ammunition in the calibre range 12.7 to 35 mm. Cannon ammunition includes: 20 x 128 mm KAA, 20 x 139 mm HS820, 25 x 137 mm KBA, 23 mm VYa FAPDS-T, 25 x 184 mm KBB, 30 mm KCB, 35 x 228 mm, 35 mm AHEAD, 35 mm break-up, 40 mm L/60 break-up, 40 mm L/70 break-up. Production machinery and ballistic measuring systems. Facilities and equipment for disposal of ammunition.

Subsidiaries:
I.L.E.E. AG
Oerlikon Contraves GmbH
Oerlikon Contraves S.p.A
Contraves Advanced Devices Sdn.Bhd.
Oerlikon Singapore Pte. Ltd
Oerlikon Contraves Inc.
Oerlikon Transtec Inc.

UPDATED

RUAG Aerospace

(Systems Division)

PO Box 301, CH-6032 Emmen, Switzerland
Tel: (+41 412) 68 41 11
Fax: (+41 412) 60 25 88
e-mail: marketing.aerospace@ruag.com
Web: http://www.ruag.com

■ Integrated solution provider in aerospace core activities including development, manufacture and assembly of structural modules; systems integration, upgrading and maintenance of aircraft, helicopters, drones, missiles, anti-aircraft and related command and communication systems.

UPDATED

Stesalit AG

Fabrikstraße 54, CH-4234 Basel, Zullwil, Switzerland
Tel: (+41 61) 795 06 01
Fax: (+41 61) 795 06 04
e-mail: info@stesalit.com
Web: http://www.stesal.com

■ Development and manufacture of prepreg systems based on phenol, epoxy, BMI, Cyanatester resins for aerospace, space, rail and ship building industries. Structural components and laminates for aerospace and space applications. Manufacture of carbon epoxy based pultrusion profiles for civil engineering applications. NATO cage code: SH034

UPDATED

Syderal SA (SYD)

(ex-Alcatel Space Switzerland SA)
(ex-Compagnie Industrielle Radioélectrique SA (CIR))

Neuenburgstrasse 7, CH-2076 Gals, BE, Bern, Switzerland
Tel: (+41 32) 338 98 00
Fax: (+41 32) 338 99 34
(+41 32) 338 99 36
e-mail: info@syderal.ch
Managing Director: Vincent Pointet
Financial Director: Jean-Luc Rey
Technical Director: Olivier Moulin
Purchasing Manager: Olivier Henin
Marketing and Sales Manager: Arnaud Gisiger

■ Space and military electronic equipment.

UPDATED

Vectronix AG

(ex-Leica Geosystems AG, BU Defense and Special Projects)
(ex-Leica Vectronix AG)

Max-Schmidheinistrasse 202, CH-9435 Heerbrugg, Switzerland
Tel: (+41 71) 727 47 47
Fax: (+41 71) 727 46 79
e-mail: info@vectronix.ch
Web: http://www.vectronix.ch
President: Dr Joerg Wullschleger
International Sales International: Leonhard Gabathuler
Communications Manager: Kurt Hegetschweiler
Tel: (+41 71) 727 34 64

■ High-precision optronic instruments for military survey and artillery applications, digital compass, diode laser rangefinder. Day and night observation.

UPDATED

Industry – Service

Tecnavia SA

Via Cadepiano 28, Barbengo, CH-6917 Lugano, Switzerland
Tel: (+41 91) 993 21 21
Fax: (+41 91) 993 22 23
e-mail: info@tecnavia.com
Web: http://www.tecnavia.com
Managing Director: Bruno Rimoldi
Director: Giancarlo Vanoni

■ Manufacture of complete line of fixed and portable (land, aero, marine) stations for reception and display of images and data from meteorological satellites.

UPDATED

Industry – Sales Agent/Office

Cryophysics SA

Rue Rothschild 66, CH-1202 Genève, Switzerland
Tel: (+41 22) 732 95 20
Fax: (+41 22) 738 52 46
Cable: CRYOPHYSICS
e-mail: cryophysicsch@compuserve.com
President: K A Geiger
Operations Manager: René Koch
Purchasing Manager: Y Philipona

■ Cryopumps, cryogenic systems, cryogenic instruments.

Representing:
CTI Cryogenics, US
Lake Shore Cryotronics, US
Magnetic Solutions Ltd, Ireland
Magnex Scientific Ltd, UK
Sumitomo Heavy Industries (Cryogenics), Japan

UPDATED

Taiwan

Government – Ministry/Agency

National Space Program Office (NSPO)

8th Floor, 9 Prosperity 1st Road, Science–Based Industrial Park, 30077 Hsin-Chu, Taiwan
Tel: (+886 3) 578 42 08
Fax: (+886 3) 578 42 34
(+886 3) 578 42 10
e-mail: service@nspo.gov.tw
Web: http://www.nspo.gov.tw
Director: Dr Lou-Chuang Lee
Contact: Wan-Hua Sun

■ Development and operation of the national space programme including space technology, Earth observation and telecommunications.

UPDATED

Thailand

Japan Aerospace Exploration Agency (JAXA)
(Bangkok Office)
(ex-National Space Development Agency of Japan)
B B Building, Floor 13, Room 1305, 54 Asoke Road, Sukhumvit 21, Bangkok, 10110, Thailand
Tel: (+66 2) 260 70 26
Fax: (+66 2) 260 70 27

UPDATED

Ministry of Science, Technology and Environment
196 Paholyotin Road, Chatuchak, Bangkok, Thailand
Tel: (+66 2) 579 13 70
(+66 2) 579 13 79
Fax: (+66 2) 561 30 35
(+66 2) 561 30 49
(+66 2) 326 91 50
Minister of Science, Technology and Environment: Khunplum Sonthaya

UPDATED

Industry – Service

Geo-Informatics and Space Technology Development Agency (GISTDA)
(ex-Thailand Remote Sensing Centre)
196 Phahonyothin Road, Chatuchak, Bangkok, 10900, Thailand
Tel: (+66 2) 579 03 45
(+66 2) 940 64 20
(+66 2) 940 64 21
(+66 2) 940 64 22
(+66 2) 940 64 23
(+66 2) 940 64 24
(+66 2) 940 64 25
(+66 2) 940 64 26
(+66 2) 940 64 27
(+66 2) 940 64 28
(+66 2) 940 64 29
(+66 2) 940 63 45
Fax: (+66 2) 561 30 35
(+66 2) 562 04 29
(+66 2) 326 42 91
(+66 2) 326 42 83
(+66 2) 561 30 35
(+66 2) 561 24 72
Web: http://www.gistda.or.th
Director: Dr Suvit Vibulsresth
Tel: (+66 2) 940 56 53
Deputy Director: Dr Darasri Dowreang
Tel: (+66 2) 940 56 61
Head, Space Technology Centre: Chanchai Peanvijarnpong
Tel: (+66 2) 940 56 28
Head, Business Development Office: Mrs Praneet Ditsariyakul
Head, Geo-Informatics Centre: Dr Surachai Ratanasermpong
Tel: (+66 2) 579 01 16
Head, Administration Office: Ms Premchit Padamachitra
Tel: (+66 2) 940 56 59
Chief, International Relations Affairs: Ms Thanomsri Rangsikanbhum
Tel: (+66 2) 940 56 62
■ Development of space technology and geo-informatics applications, application of satellite imagery, and development of satellite database.

UPDATED

Ukraine

Centres of Learning & Research

National Space Agency of Ukraine (NKAU)
Bozhenka 11 Street, 03022 Kyiv, Ukraine
Head of Department for Space Programs: Vladimir M Shkapa
Tel: (+380 44) 227 87 37
Fax: (+380 44) 269 50 58
Chief of Directorate for Space Programs and Scientific Research: Dr Oleg P Fedorov
Tel: (+380 44) 261 54 22

UPDATED

Industry – Manufacturing

Marlin Yug Ltd
2 Kapitanskaya Strasse, 99000 Sevastopol, Ukraine
Tel: (+380 692) 54 04 50
Fax: (+380 692) 54 04 50
e-mail: marlin@stel.sebastopol.ua
Web: http://marlin.stel.sebastopol.ua
Director: Dr Sergey V Motyzhev
■ Manufacture of the WOCE SVP-B Drifter deployable autonomous buoy and the Diving Drifter surface buoy. Also manufacture the Platform Transmitter Terminal (PTT) and testing device for the data platform.

UPDATED

Yuzhnoye Design Office
(SDO Yuzhnoye)
3 Krivorozhskaya Street, 49 008 Dnepropetrovsk, Ukraine
Tel: (+380 56) 242 00 22
Fax: (+380 56) 770 01 25
(+380 56) 292 50 41
Telex: 143294 deviz
e-mail: kbu@public.ua.net
Web: http://www.yuzhnoye.dp.ua
■ Design and preparation of rocket and space complexes, launch vehicles (Cyclone, Zenit, Dnepr), spacecraft (Sich, Ocean) and production of their components. Provision of launch services. Testing: strength, pneumohydraulic, electrical, exposure to various environment conditions (temperature, humidity), radio frequency and firing tests of rotation motors.

UPDATED

United Arab Emirates

Industry – Service

Jane's Information Group (JIG)
(Middle East/Dubai)
PO Box 502138, Dubai, United Arab Emirates
Tel: (+971 4) 390 23 35
(+971 4) 390 23 36
Fax: (+971 4) 390 88 48
e-mail: mideast@janes.com
Web: http://www.janes.com
Representing:
Jane's Information Group, UK

UPDATED

United Kingdom

Government – Ministry/Agency

Defence Geographic and Imagery Intelligence Agency Headquarters (DGIA)
Room 17, Watson Building, Elmwood Avenue, Feltham, Middlesex TW13 7AH, United Kingdom
Tel: (+44 20) 88 18 24 22
Fax: (+44 20) 88 18 22 46
Chief Executive: Brig A P Walker
Commander of Geographic Engineer Group: Col N Rigby
Director, Defence Geographic Centre: Mrs M Jacobs
Officer Commanding Joint Air Reconnaissance Intelligence Centre: Gp Capt M Hallam
Corporate Planning: Miss J Hodson
Tel: (+44 20) 88 18 21 36
■ Provide imagery intelligence and geographic support to defence policy, planning, operations and training.

UPDATED

Centres of Learning & Research

Institute of Sound & Vibration Research (ISVR)
University of Southampton, Highfield, Southampton, SO17 1BJ, United Kingdom
Tel: (+44 23) 80 59 22 94
Fax: (+44 23) 80 59 31 90
e-mail: mzs@isvr.soton.ac.uk
Web: http://www.isvr.soton.ac.uk
Director: Prof P A Nelson
Technical Director: S J C Dyne
■ Teaching, research and consulting centre for the study of all aspects of sound and vibration. Defence and aerospace activities include work on spacecraft, satellites, aircraft, data processing recording systems, active control, automotive engineering, subjective acoustics and environmental noise.

UPDATED

Jodrell Bank Observatory (JBO)
(ex-Nuffield Radio Astronomy Laboratory)
University of Manchester, Macclesfield, Cheshire SK11 9DL, United Kingdom
Tel: (+44 1477) 57 13 21
Fax: (+44 1477) 57 16 18
Web: http://www.jb.man.ac.uk
Managing Director: Prof A G Lyne
Vice President and Member UAI Protection of Existing and Potential Observatory Commission: Dr Raymond J Cohen
Chief Engineer: J A Battilana
Tel: (+44 1477) 57 26 01
Librarian: Mrs S M Morris
■ Radio astronomy teaching and research.

UPDATED

Rutherford Appleton Laboratory (RAL)
(Central Laboratory of the Research Councils)
(ex-Appleton Laboratory)
(ex-Rutherford Lab)
Chilton, Didcot, Oxfordshire OX11 0QX, United Kingdom
Tel: (+44 1235) 82 19 00
Fax: (+44 1235) 44 58 48
Web: http://www.sstd.rl.ac.uk
Head of Technology Exploitation for Space: John Ellis
Tel: (+44 1235) 44 59 23
Mobile: (+44 77) 70 65 25 08

UPDATED

Space Research Centre
(Department of Physics and Astronomy)
(ex-Leicester University Space Research Centre)
University of Leicester, University Road, Leicester, LE1 7RH, United Kingdom
Tel: (+44 116) 252 35 52
Fax: (+44 116) 252 24 64
Web: http://www.src.le.ac.uk
Space Centre Director: Prof A Wells
Head of Department of Physics and Astronomy: Prof R S Warwick
Space Centre Manager: Dr J P Pye

UPDATED

University of Dundee
(NERC Satellite Receiving Station)
Department of Electronic Engineering and Physics, Perth Road, Dundee, DD1 4HN, United Kingdom
Tel: (+44 1382) 34 51 94
(+44 1382) 34 44 09
Fax: (+44 1382) 20 25 75
Web: http://www.sat.dundee.ac.uk
Principal of the University of Dundee: Sir Alan Langlands
Director: Dr Steve Parkes
Manager: Neil Lonie
■ Data acquisition and distribution services funded by the UK Natural Environment Research Council. 12,400 students and over 3,000 staff in the entire university.

UPDATED

University of London Observatory (ULO)

553 Watford Way, Mill Hill, London, NW7 2QS, United Kingdom
Tel: (+44 20) 89 59 04 21
Fax: (+44 20) 82 38 88 72
e-mail: vmp@star.ucl.ac.uk
Web: http://www.ulo.ucl.ac.uk
Director: Dr Mike Dworetsky
Tel: (+44 20) 82 38 88 56
Mobile: (+44) 79 39 03 30 47
■ Research activities: physics of the interstellar medium, physics of stellar atmospheres. Teaching of astronomy.

UPDATED

Association

Association in Scotland to Research into Astronautics Ltd (ASTRA)

(ASTRA)
Flat 65, Dalriada House, 56 Blythswood Court, Glasgow G2 7PE, United Kingdom
Tel: (+44 141) 221 76 58
Web: http://www.astra.org.uk
Treasurer: Duncan Lunan
■ Lectures, exhibitions, radio research, amateur astronomy and rocketry, observatory management and waverider re-entry vehicle research and development. Meetings in Airdrie (weekly) and Glasgow (monthly). Operates Airdrie public observatory for North Lanarkshire District Council.
Publications: Asgard
International Spacereport
Spacereport

UPDATED

The Association of Specialist Technical Organisations for Space (ASTOS)

c/o Nohmia Limited, 79 Larksway, Bishops Stortford, Hertfordshire, CM23 4DG, United Kingdom
Tel: (+44 870) 765 65 44
e-mail: enquiries@astos.org.uk
Web: http://www.astos.org.uk
Chairman: James Barrington-Brown
Treasurer: Paul Taylor
Secretary: Mark Allery
■ Association representing SMEs (small/medium enterprises) specialist skills and services to the space industry.
Members
Actheric Engineering
AEA Technology
Aegis Systems
Analyticon Ltd
ARC
ComDev Europe
Exemplar Associates
Grafton Technology
IGG Component Technology
ISP International Space Prolulsion
JRA Aerospace & Technology
L-3 Storm Control Systems
Moltek Consultants Ltd
Moreton Hall Associates
Mullard space Science Labarotary
Nohmia
Polyflex
RDSL
RPC Telecommunications Ltd
Starchaser
Spur Electron Ltd
Surrey Satellite Technology Ltd
Tri Polus

UPDATED

Industry – Manufacturing

Advanced Products (Seals & Gaskets) Ltd

Unit 25c, Number One Industrial Estate, Consett, County Durham DH8 6SR, United Kingdom
Tel: (+44 1207) 50 03 17
Fax: (+44 1207) 50 12 10
Regional Sales Manager (North): Gary Cordner
■ Manufacture of seals and gaskets.

UPDATED

Advent Communications Ltd

Preston Hill House, Nashleigh Hill, Chesham, Buckinghamshire HP5 3HE, United Kingdom
Tel: (+44 1494) 77 44 00
Fax: (+44 1494) 79 11 27
e-mail: sales@advent-comm.co.uk
Web: http://www.advent-comm.co.uk
Director: Steve McGuinness
Director: David Garrood
Director: Roger Crawshaw
■ Design and manufacture of satellite earth stations for fixed transportable and flyaway applications in commercial, government and military markets.

UPDATED

AEA Technology plc

Harwell, Didcot, Oxfordshire, OX11 0QJ, United Kingdom
Tel: (+44 870) 190 29 21
(+44 870) 112 10 24
Fax: (+44 870) 190 29 20
(+44 1235) 43 29 16

UPDATED

Aircraft Tanks Ltd

Spring Vale Works, Middleton, Manchester, M24 2HS, United Kingdom
Tel: (+44 161) 643 24 62
Fax: (+44 161) 643 34 90
e-mail: info@msmgroup.org
Web: http://www.msmgroup.org
Chairman: F Pedley
Managing Director: John D Pedley
Financial Director: James Barnes
Sales and Marketing Director: Michael Pedley
Purchasing Manager: Ms Sandra Cannon
■ Maintenance, repair and overhaul of aircraft and aircraft powerplants. Manufacture of fuel tanks for fighting vehicles and aircraft and aircrew seats. Sheet metal work and component manufacture. Subcontract work on missile programmes and Ariane space launcher.

UPDATED

AIT Recorders Ltd

(ex-Walton Radar Systems Ltd)
Bermuda House, 45B High Street, Hampton Wick, Surrey, KT1 4DG, United Kingdom
Tel: (+44 208) 255 06 66
Fax: (+44 208) 977 69 93
e-mail: enquiries@ait_recorders.com
Web: http://www.ait_recorders.com
Managing Director and Chief Executive Officer: Michael King
Tel: (+44 1256) 33 96 05
■ Recording systems for surveillance, archival and training use, datalogging and real-time data processing systems and graphics. Design, development and production of specialist electronic systems.

UPDATED

Ametek Aircontrol Technologies

(ex-Howden Aircontrol Ltd)
Hawthorne Road, Staines, Middlesex TW18 3AY, United Kingdom
Tel: (+44 1784) 46 61 66
Fax: (+44 1784) 46 58 94
e-mail: mail@aircontroltechnologies.co.uk
Web: http://www.aircontroltechnologies.co.uk
Managing Director: Jim Lambert
Sales Manager: Martin Hart
Financial Director: Roger Shipton
Engineering Director: Patrick Allan
Sales Manager: Paul Hammond
Operations Manager: Barry Hawyes
■ Design and manufacture of military and aerospace air and liquid conditioning systems including vapour cycle or air cycle cooling, NBC filtration equipment, engine cooling systems, ground support equipment and environmental control systems.

UPDATED

Andrew Ltd

ILEX Building, Mulberry Business Park, Fishponds Road, PO Box 134, Wokingham, Berkshire, RG41 2GY, United Kingdom
Tel: (+44 118) 936 67 00
(+44 118) 977 68 86
Fax: (+44 118) 936 67 77
(+44 118) 979 40 05
Web: http://www.andrew.com
Executive: Simon Lawrence
■ Antenna systems for military microwave and HF radio links. Coaxial cable systems, fibre-optic gyros. EW Signit equipment, tracking systems. Electro-optical communications systems equipment.
Representing:
Andrew Corporation, US

UPDATED

Avica

Boundary Way, Hemel Hempstead, Hertfordshire, HP2 7SL, United Kingdom
Tel: (+44 1442) 26 47 11
Fax: (+44 1442) 23 00 35
Managing Director: Kay Juss
Sales Manager: Timothy Figg
Purchasing Manager: Nigel Halls
■ Design, development and manufacture of a range of products and systems for aircraft, space vehicles, nuclear, cryogenic and marine applications. These include metallic ducting systems, gimbals, flexible joints, flexible metallic hoses, bellows, flanges, clamps and total ducting design service.

UPDATED

BAE Systems

(Corporate Headquarters)
(ex-British Aerospace plc)
(ex-Marconi Electronic Systems)
Warwick House, Farnborough Aerospace Centre, PO Box 87, Farnborough, Hampshire, GU14 6YU, United Kingdom
Tel: (+44 1252) 37 32 32
Fax: (+44 1252) 38 30 00
Web: http://www.baesystems.com
Executive Chairman: Dick Olver
Chief Executive Officer: Michael Turner
Chief Operating Officer, BAE Systems plc: Chris Geoghegon
Group Legal Director: Michael Lester
Chief Operating Officer: Steve Mogford
Chief Operating Officer and President, BAE Systems, North America: Mark Ronald
Group Finance Director: George Rose
Group Managing Director, Network Enabled Capability: Ms Alison Wood
Group Human Resources Director: Alastair Imrie
Chief of Staff: George Mayhew
Group Communications Director: Ms Charlotte Lambkin
Director of Audit: Grenville Hodge
Group Marketing Director: Mike Rouse
Chief Operating Officer, BAE Systems Programmes: Steve Mogford
President, Technology Solutions: Bruce Hamilton
President, Platform Solutions: Jim Scanlon
President, Information Solutions: Marshall Banker
Group Communications Director: Hugh Colver
Tel: (+44 1252) 38 32 80
Group Strategy Director: David Singleton
Group Managing Director, Major Programmes Assurance: Brian Phillipson
Group Managing Director, Customer Solutions & Support: Ian King
Group Managing Director, Avionics: Nick Franks
Group Managing Director, Shared Services & Ventures: Alistair Imrie
Group Managing Director, International Partnerships: Stephen Henwood
Group Managing Director, Air: Nigel Whitehead
Group Managing Director, C4ISR: Phill Blundell
Managing Director, Submarines: Murray Eastern
Managing Director, Naval Ships: Vic Emery
Managing Director, Underwater Systems: Andy Williams
Managing Director, Military Air Solutions Support: Kevin Taylor
Managing Director, International Programmes: Peter Wilson
Managing Director, Customer Solutions & Support, Australia: Harry Bradford
Managing Director, Training Solutions: Andrew Davis
Managing Director, Land & Sea: Rory Fisher
Managing Director, Operational Services: Neil Mason
Managing Director, Operations: Mike O'Callaghan
Managing Director, CFV Prime Contract: Nigel Stuart
Managing Director, RO Defence: Clive Richardson
Managing Director: Terence Black
President, Information and Electronic Systems: Galen Ho
Chief Executive Officer, BAE Systems Australia: Jim McDowell
Head of News: Richard Coltart
Tel: (+44 1252) 38 48 75
Director of Media Relations: Phil Soucy
Tel: (+44 1252) 38 47 97

Internal Communications: Stephen Windsor-Lewis
Tel: (+44 1252) 38 34 31
Chairman: Sir Richard H Evans
■ Global systems, defence and aerospace company. Prime contractor for the design and manufacture of civil and military aircraft, surface ships, submarines, space systems, radar, avionics, communications, electronics, guided weapon systems and a range of other defence products.

Corporate structure
BAE Systems, Future Systems
BAE Systems, Programmes
-Air Systems
-Eurofighter Typhoon
-Hawk
-JSF
-Nimrod MRA4
-New Business Group
-Naval Ships
-Submarines
-Underwater Systems
BAE Systems, C4ISR
BAE Systems, Customer Solutions & Support
-Customer Solutions & Support Australia
-Customer Solutions &Support International Programmes
-Customer Solutions & Support Land & Sea
-Customer Solutions & Support Military Air Solutions & Support
-Customer Solutions & Support Operational Services
-Customer Solutions & Support Training Solutions
BAE Systems, North America
-Information Systems Sector
-Information and Electronic Systems Integration Sector
-Platform Solutions Sector
-Technology Solutions Sector
BAE Systems, Operational, Shared Services, Capital
-International Partnerships
-Airbus (BAE Systems 20%)
-AMS (BAE Systems 50%)
-BAeHAL Software Limited (BAE Systems 40%)
-Gripen International (BAE Systems 50%)
-MBDA (BAE Systems 37.5%)
-SAAB AB (BAE Systems 35%)
-STN Atlas Elektronik (BAE Systems 49%)
-Regional Aircraft
-Avionics Group
-BAE Systems, Australia
-BAE Systems, South Africa
-Aerostructures
-Future Carrier
-Shared Services
-Advanced Technology Centre

UPDATED

Balzers Ltd
(ex-Balzers High Vacuum Ltd)
Bradbourne Drive, Tilbrook, MK7 8AZ, United Kingdom
Tel: (+44 1908) 37 72 77
Fax: (+44 1908) 36 13 62
e-mail: info.uk@balzers.com
Web: http://www.balzers.co.uk
Managing Director: P J Stokley
Marketing Director: Ms Joyce Atkinson
Sales and Marketing Co-ordinator: Ben Morling
■ Thin film components and vacuum components. The company also supplies PVD coatings for engineering tools and precision components.

UPDATED

Black Sky Technologies Ltd
Stonefield Park, Chilbolton, Hampshire, SO20 6BR, United Kingdom
Tel: (+44 1264) 86 11 00
Fax: (+44 1264) 86 00 33
Director: Andrew Cox
Director: Andrew Elson
■ Design and manufacture of high altitude platforms and aircraft.

UPDATED

Bristol Industrial & Research Associates Ltd (BIRAL)
PO Box 2, Portishead BS20 7JB, United Kingdom
Tel: (+44 1275) 84 77 87
Fax: (+44 1275) 84 73 03
e-mail: info@biral.com
Web: http://www.biral.com
Managing Director: Lance Croutear
Technical Director: Ian Rothwell
Director of Business Development: Charles Haythornthwaite
Mobile: (+44 7990) 57 72 40
Sales Manager: John Dimmock
Meteorologist: James Squires
Sales and Marketing Assistant: Ms Sarah Higgins
■ Meteorological systems and sensors.

UPDATED

Celab Ltd
25 Woolmer Way, Bordon, Hampshire, GU35 9QE, United Kingdom
Tel: (+44 1420) 47 70 11
Fax: (+44 1420) 47 80 43
Web: http://www.celab.co.uk
Director of Sales: Steve Pellett
Mobile: (+44 7768) 39 78 30
■ Design and manufacture of high density, switching regulator military power supplies for rigorous applications in airborne, ship and ground environments with AC and DC inputs. Multiple rail modular systems with full BITE facilities. Comprehensive sub-contract manufacturing facility including post design.

UPDATED

Centronic Ltd
Centronic House, King Henry's Drive, New Addington, Croydon, Surrey, CR9 0BG, United Kingdom
Tel: (+44 1689) 80 80 00
Fax: (+44 1689) 84 51 17
Cable: CENTRONIC London
e-mail: eosales@centronic.co.uk
Web: http://www.centronic.co.uk
Managing Director: Ian Crawford
Managing Director: Neil Foreman
Financial Director: Ms Ursula Olafisoye
Technical Director: Dr Leslie Howard
Purchasing Manager: Ian Emm
Sales Manager: D Fairweather
■ Manufacture of electro-optic components for commercial, military and space applications.

UPDATED

COM DEV Europe Ltd (CDE)
Triangle Business Park, Stoke Mandeville, Aylesbury, Buckinghamshire, HP22 5SX, United Kingdom
Tel: (+44 1296) 61 64 00
Fax: (+44 1296) 61 65 00
e-mail: info@comdev.co.uk
Web: http://www.comdev.co.uk
Managing Director: Sunil Chavda
Managing Director and Director of International Business Development: John R Stuart
■ Manufacture and design of microwave and electronic equipment for communications and earth observation, remote sensing and defence satellites. Products include: ferrite switches, isolators and circulators, input and output multiplexers, filters, couplers and waveguides, detector and preamplification electronics, ASICs design.
Accreditations: ISO 9001

UPDATED

Daco Scientific Ltd
Vulcan House, Calleva Industrial Park, Aldermaston, Berkshire, RG7 8PB, United Kingdom
Tel: (+44 118) 981 73 11
Fax: (+44 118) 981 99 63
e-mail: sales@daco.co.uk
Web: http://www.daco.co.uk
Managing Director: Roy Engel
Engineering Director: Chris Veel
Sales Co-ordinator: Ms Mahendra Patel
■ Design and manufacture of trackballs, control grips and joysticks for defence and aerospace related applications.
Accreditations: ISO 9001

UPDATED

Darchem Engineering Ltd
Stillington, Stockton-on-Tees, Cleveland, TS21 1LB, United Kingdom
Tel: (+44 1740) 63 04 61
Fax: (+44 1740) 63 05 29
e-mail: sales@darchem.co.uk
Web: http://www.darchem.co.uk
■ High temperature insulation materials and systems. Precision fabrications and honeycomb core structures in stainless steel, other alloys and components for aerospace and other industries. Marine gas turbine intake and exhaust systems. Sheet metal profiling. Specialist forming techniques. Heat shields.

UPDATED

Draeger Safety UK Limited
(ex-Draeger Ltd)
Ullswater Close, Kitty Brewster Industrial Estate, Blyth, Northumberland, NE24 4RG, United Kingdom
Tel: (+44 1670) 35 28 91
Fax: (+44 1670) 35 62 66
e-mail: marketing@draeger.ltd.uk
Web: http://www.draeger-safety.co.uk
Managing Director: Jim Varah
Operations and Technical Director: Vince Smith
Purchasing Manager: Peter Kirby
Marketing Communications Manager: Richard P Beckwith
Marketing Assistant: Ms Kerri Hepple
■ Manufacture and supply of breathing apparatus and gas detection systems for professional firefighters and industry worldwide.

UPDATED

EADS Astrium
(Stevenage Office)
(ex-Astrium Ltd)
(ex-Matra Marconi Space UK Ltd)
Gunnels Wood Road, Stevenage, Hertfordshire, SG1 2AS, United Kingdom
Tel: (+44 1438) 31 34 56
Fax: (+44 1438) 77 36 37
Web: http://www.astrium.eads.net
■ Production and development of scientific and earth observation satellites; civil and military communications satellites; antennas, filters, structures, cryocoolers, microwave products and ground communication systems.

UPDATED

EADS Astrium
(Portsmouth Division)
Anchorage Road, Portsmouth, Hampshire, PO3 5PU, United Kingdom
Tel: (+44 23) 92 70 57 05
Fax: (+44 23) 92 70 57 06

UPDATED

ETS-Lindgren Ray Proof Ltd
(ex-Lindgren-Rayproof)
(ex-Ray Proof Ltd)
Boulton Road, Pin Green Industrial Area, Stevenage, Hertfordshire, SG1 4TH, United Kingdom
Tel: (+44 1438) 73 07 00
Fax: (+44 1438) 73 07 50
e-mail: info@ets-lingren.co.uk
Web: http://www.ets-lindgren.com
Managing Director and Marketing and Sales Director: B Lawrence
Accounting Manager: Terry Hawkes
Operations Manager: A Smith
Technical Manager: C Castro
Sales Engineer, Fibre Optics: Stuart Levene
Buyer: Ms Helen Murphy
■ Specialise in RFI/EMC/NEMP and Tempest protection systems offering shielded enclosures, anechoic chambers, filters, fibre optics and suppression equipment. System design, manufacture, installation and testing are available on a worldwide basis.

UPDATED

E2V Technologies
(ex-EEV Ltd)
(ex-English Electric Valve Co. Ltd)
(ex-Marconi Applied Technologies)
106 Waterhouse Lane, Chelmsford, Essex, CM1 2QU, United Kingdom
Tel: (+44 1245) 49 34 93
Fax: (+44 1245) 49 24 92
e-mail: enquiries@e2vtechnologies.com
Web: http://www.e2vtechnologies.com
■ Electron tubes for radar, communications and EW, IR devices, night vision, TV camera tubes, image intensifiers, display devices including LCDs, hydrogen thyratrons, microwave tubes, travelling wave tubes, frequency agile magnetrons, receiver protection. CCD sensors and cameras for industrial, scientific, security purposes. Slapper detonators for electronic safety and arming circuitry. UV detectors. Caesium arc lamps.

UPDATED

Farsound Engineering Ltd (FEL)
Unit 3, Highams Park Industrial Estate, Jubilee Avenue, London, E4 9JD, United Kingdom
Tel: (+44 20) 84 98 38 88
Fax: (+44 20) 84 98 38 87
e-mail: sales@farsound.co.uk
info@farsound.co.uk

Web: http://www.farsound.co.uk
Managing Director: Alan Gardner
Financial Director: David Bartlett
Production Director: Melvin R Atkinson
Sales and Marketing Manager: Dave Skinner
Purchasing Manager: Ms Jacqui Jones
Marketing Manager: James Gulperin
■ Design, manufacture and distribution of a range of components for the aerospace, transport, marine, medical and petrochemical industries, utilising a range of engineering skills including: sheet metal fabrication, presswork, machining, welding and assembly, waterjet and laser cutting, overhaul and kitting solutions and NDT approved inspection and testing facilities.

UPDATED

Firetrench
(ex-Integrated Secure Information Consortium Limited)
Monks Farm, St James Road, All Saints South Elmham, Halesworth, Suffolk, IP19 0HG, United Kingdom
Fax: (+44 1986) 825 25
e-mail: editors@firetrench.com
sales@firetrench.com
Web: http://www.firetrench.com
Chairman: Ian M Johnstone-Bryden
Tel: (+44 1986) 78 25 47
■ Integrated, electronic data processing and communications systems for standard and secure environments together with associated consultancy and training services and avionics simulation systems.

UPDATED

Firth Rixson plc
Firth Rixson House, 25 Carbrook Hall Road, Sheffield, S9 2EJ, United Kingdom
Tel: (+44 114) 261 22 22
Fax: (+44 114) 261 22 23
e-mail: info@firthrixson.com
■ Forgings and rolled rings for the aerospace and defence industries. Steel, nickel and titanium alloys, which include turbine and compressor discs, gears, shafts, bearing rings, undercarriage components, missile and rocket engine forgings and engine parts. Cast and wrought superalloys. Steel and alloy iron castings. Components.

UPDATED

Flir Systems Ltd
(ex-AGEMA Infrared Systems Ltd)
(ex-Broadcast & Surveillance Systems Ltd)
(ex-Inframetrics Inc)
2 Kings Hill Avenue, KingsHill, West Malling, Kent, ME19 4AQ, United Kingdom
Tel: (+44 1732) 22 12 44
Fax: (+44 1732) 22 00 14
e-mail: marketing@flir.uk.com
Senior Vice President: Andrew C Teich
Managing Director: Andrew Philips
Technical Director: Neil Bertram
■ Design and manfuacture of thermal imaging systems for air, land, sea and security applications for the military, law enforcement and civilian sectors.
Accreditations: MoD

UPDATED

Forward Industries Ltd
(ex-VFP Fluid Power Ltd)
Hunts Rise, South Marston Industrial Park, Swindon, Wiltshire, SN3 4TQ, United Kingdom
Tel: (+44 1793) 82 32 41
Fax: (+44 1793) 82 84 74
e-mail: enquiries@forward-industries.co.uk
Finance Director: S Thomas
Business Manager, Hydraulics: D P Drudy
■ Provision of power, motion and control systems; design and manufacture of sophisticated electro-hydraulic installations for a variety of applications. Supplies a specialised range of hydraulic pumps, valves, actuators and systems for on-board or support functions, weapons loading, general services and simulation. Hangar ring main and mobile systems for servicing aircraft hydraulics, pneumatics and fuel systems.
Accreditations: ISO 9001

UPDATED

Francis & Lewis International Ltd (F&L)
Waterwells Drive, Waterwells Business Park, Gloucester, GL2 4AA, United Kingdom
Tel: (+44 1452) 72 22 00
Fax: (+44 1452) 72 22 44
e-mail: postmaster@fli.co.uk
Web: http://www.fli.co.uk
■ Wide product range of towers and masts for telecommunications, broadcasting, radar and cellular applications.

UPDATED

GKN Aerospace Transparency Systems Kings Norton Ltd
(ex-Pilkington Aerospace Ltd)
(ex-Triplex Aircraft & Special Products Ltd)
Triplex House, Eckersall Road, Kings Norton, Birmingham, B38 8SR, United Kingdom
Tel: (+44 121) 606 41 00
Fax: (+44 121) 606 41 91
e-mail: pilkaero@aol.com
Managing Director: John Saliture
Financial Director: Philip Mills
Tel: (+44 121) 606 41 51
Technical Director: Mark Burgess
Tel: (+44 121) 606 41 18
Sales and Marketing Director: Robert Harper
Tel: (+44 121) 606 41 94
General Manager: R A Harper
Purchasing Manager: Andy D Tither
Tel: (+44 121) 606 41 32
Publicity Co-ordinator: Ms Louise A Meredith
Tel: (+44 121) 606 41 44
■ Windows and transparencies for all aircraft, helicopters, hydrofoils, ships, tanks, fighting vehicles and space applications. De-icing/de-misting systems, anti-static coatings and bird/hail impact protection.

UPDATED

Goodrich Actuation Systems
(Group Headquarters)
Stafford Road, Fordhouses, Wolverhampton, West Midlands, WV10 7EH, United Kingdom
Tel: (+44 1902) 62 46 44
Fax: (+44 1902) 78 77 20
President and General Manager, Actuation Systems: Brian Gora
Tel: (+44 1902) 62 47 74
■ Providing a full range of actuation systems for airborne and land based platforms. Products include primary and secondary flight controls; helicopter main and tail rotor, engine and nacelle, utility, precision weapon and land vehicle actuators.
Provision of complete systems including hydraulic, fly-by-wire and power-by-wire controls using technologies such as hydraulic, rotary geared, electro-hydrostatic and electro-mechanical actuators.

UPDATED

Goodrich Optical & Space Systems (OSSD)
(Surveillance & Reconnaissance Systems, Malvern)
Malvern Hills Science Park, Geraldine Road, Malvern, Worcester, WR14 3SZ, United Kingdom
Tel: (+44 1684) 58 52 25
Fax: (+44 1684) 58 53 67
Web: http://www.oss.goodrich.com/surveilla-nceandreconnaissancesystems.shtml
■ Development and production of high performance cameras and ground stations for capturing and interpreting reconnaissance and surveillance data. Produce imagery products for intelligence, surveillance and reconnaissance (ISR) missions, sensors for day, night, and multi spectral image capture, along with the ground exploitation and data handling systems.

UPDATED

Holscot Fluoroplastics
(ex-Holscot Industrial Linings Limited)
Alma Park Road, Alma Park Industrial Estate, Grantham, Lincolnshire, NG31 9SE, United Kingdom
Tel: (+44 1476) 57 47 71
Fax: (+44 1476) 56 35 42
e-mail: sales@holscot.com
Web: http://www.holscot.com
Managing Director: David Joyce
Sales Director: Martin Daff
Financial Director: Ian Cripps
■ Specialists in melt processible fluoroplastics, FEP, PFA, PVDF or ECTFE. Production of fluoroplastic items for a wide range of industries and applications.

UPDATED

Insys Limited
Reddings Wood, Ampthill, Bedfordshire, MK45 2HD, United Kingdom
Tel: (+44 1525) 84 10 00
Fax: (+44 1525) 40 58 61
e-mail: marketing@insys-ltd.co.uk
Web: http://www.insys-ltd.co.uk
Head of Marketing and Public Relations: Chris I P Martin
Tel: (+44 1525) 84 37 60
■ Development and large scale production (as prime or sub contractor) of weapons and communication systems for HM Government and approved overseas customers.
Sister companies
Euro-Shelter Limited, France
Insys Limited, STL Pershore
Insys Limited, CBDE Porton Down

UPDATED

J S Chinn & Co. Ltd
Coventry Road, Exhall, Coventry, West Midlands, CV7 9FT, United Kingdom
Tel: (+44 24) 76 36 94 00
Fax: (+44 24) 76 36 80 00
e-mail: jschinn@intrnet-uk.net
Web: http://www.jschinn.co.uk
Managing Director: J K Atkinson
Project Director: Mark Cunnington
Tel: (+44 24) 76 36 94 21
Company Secretary and Financial Director: R P Hipkins
■ Precision fabrication and machining for aircraft, marine and industrial gas turbine industries. Intricate fabrications for airframes and aero-engines in stainless steel, heat resisting and hi-nickel alloys, aluminium and titanium. Vacuum brazing. CNC milling, turning and boring for six axis control and pipe bending. Repair and overhaul. Prototype development production programmes. Precision CNC CAD/CAM machining and tool making.

UPDATED

J S Chinn Holdings Ltd
Coventry Road, Exhall, Coventry, West Midlands, CV7 9FT, United Kingdom
Tel: (+44 24) 76 36 94 00
Fax: (+44 24) 76 36 80 00
e-mail: jschinn@internet-uk.net
Web: http://www.jschinn.co.uk
Chief Executive: J K Atkinson
Finance Director and Company Secretary: R P Hipkins
■ Subcontractors in precision fabrication and machining of components and complete assemblies for aero engines, airframe, satellites, missiles and underwater weapons and a wide range of aerospace applications.
Subsidiaries
A.O. Henton Engineering Co. Ltd
Colledge & Morley (Gears) Ltd
J S Chinn & Co. Ltd
J S Chinn Engineering Co. Ltd
J S Chinn Project Engineering Ltd

UPDATED

KEC Limited
(ex-Kern Electrical Components Ltd)
Orpheus House, Calleva Park, Reading, Berkshire, RG7 8TA, United Kingdom
Tel: (+44 118) 981 15 71
Fax: (+44 118) 981 15 70
e-mail: sales@kec.co.uk
Web: http://www.kec.co.uk
Managing Director: David J Dyson
Administration Director: Ms Cheryl Watson
Sales Manager: Don MacLean
Purchasing Manager: Ms Tracey Murray
Technical Manager: David Pike
Technical Engineer: Charlie Fleming
■ Electrical connector accessories for EMC applications. Through bulkhead fittings for EMC applications. Cable and enclosure systems specialising in EMI/RFI/EMP and Tempest areas and conduit system.
Representing:
Fastener Specialty Inc., US
Inotec GmbH, Germany
Uponor GmbH, Germany

UPDATED

Lobo Systems Ltd
Riverside Park, Raynesway, Derby, DE21 7RW, United Kingdom
Tel: (+44 1332) 68 00 28
Fax: (+44 1332) 66 80 68

e-mail: sales@lobosystems.com
Web: http://www.lobosystems.com
Sales Director: Robert Bokros
Financial Director: Ms Adele Harris
Technical Director: David Seale
Public Relations Director: Christine Chell
■ Manufacture of fast assembly access systems, work platforms and scaffolding.

UPDATED

Marshall of Cambridge Aerospace Ltd

(Marshall Aerospace)
(ex-Marshall of Cambridge (Engineering) Ltd)
The Airport, Cambridge, CB5 8RX, United Kingdom
Tel: (+44 1223) 37 37 37
Fax: (+44 1223) 32 10 32
Telex: 81208
Cable: MARFLY
e-mail: enquiries@marshallaerospace.com
Web: http://www.marshallaerospace.com
Chairman: Michael Marshall
Chief Executive Officer: Martin Broadhurst
Deputy Managing Director: Grahame Nix
Financial Director: Glen Clark
Engineering Director: R E Ward
Marketing Director: M E Milne
Commercial Sales Manager: M Chandler
Public Relations Assistant: Ms Sarah Beavis
■ OEM capabilities. Activities include conversion, modification and maintenance of military and civil aircraft; international engineering support services undertaken at the Cambridge facility or on site with the customer. UK design authority for the RAF C130 and sister design authority to Boeing on the E3D AWACS aircraft. Design authority for the L1011 Tristar Tanker responsible for the design and conversion of the aircraft. L1011 freighter conversions; B747 and MD11 interior reconfigurations; B767 maintenance. Approved Citation service centre.
Accreditations: ISO 9001, UK MoD PE design organisation approval, UK CAA ANO approval for design and manufacturing, UK CAA ANO approval for maintenance, JAR-145 approval CAA 00031, FAA FAR-145 approval.

UPDATED

Martin-Baker Aircraft Co. Ltd

Higher Denham, Uxbridge, Middlesex, UB9 5AJ, United Kingdom
Tel: (+44 1895) 83 22 14
Fax: (+44 1895) 83 25 87
e-mail: information@martin-baker.co.uk
Web: http://www.martin-baker.com
Joint Managing Director: John S Martin
Tel: (+44 1895) 83 65 21
Joint Managing Director: James W Martin
Tel: (+44 1895) 83 65 21
Engineering Director: Robert J Martin
Financial Controller and Company Secretary: Paul W Badman
Head of Sales: Ian Harbottle
Head of Quality: Tony Tofield
Business Development Manager: Andrew Martin
Tel: (+44 1895) 83 66 58
Head of Operations: Peter Lewis
IPT Manager (New Projects) and JSF Program Manager: Geoff Barnes
Human Resources Manager: Bob Blandamer
Tel: (+44 1835) 83 65 87
Design Support Manager: Richard J Wood
Purchasing Manager: Andrew Patrick
Product Support and Security Controller: Eric Thomas
Tel: (+44 1895) 83 65 35
Site Manager: Graham Dumbleton
Business Development Executive: Francisco de Florio
Chief Systems Engineer: Dr Kevin Hawkins
■ Advanced escape systems, ejection seats. Safety equipment, automatic life jacket inflation units, crashworthy crew and troop seats.

UPDATED

MB Aerospace Ltd

Logons Road, PO Box 4, Motherwell, ML1 3NP, United Kingdom
Tel: (+44 1698) 26 22 77
Fax: (+44 1698) 27 54 87
e-mail: sales@mbaerospace.com
Web: http://www.motherwellbridge.com
Managing Director: Bernie Phimister
Tel: (+44 1698) 24 23 80
Sales and Marketing Director: Norman Darroch
Tel: (+44 1698) 24 23 87
Operations Manager: Mike Kennedy
Tel: (+44 1698) 24 24 64
■ Highly specialised sub-contract machine company, supplying critical quality components to the aerospace and defence industries.

UPDATED

MBDA

(Headquarters)
(ex-British Aerospace (Dynamics) Ltd)
(ex-Matra BAE Dynamics UK Ltd)
11 Strand, London, WC2N 5RJ, United Kingdom
Tel: (+44 20) 74 51 60 00
Fax: (+44 20) 74 51 60 01
(+44 20) 74 51 60 89
Web: http://www.mbda.net
Chief Executive Officer: Marwan Lahoud
Chief Operating Officer: Guy Griffiths
Chief Operating Officer: Mario Di Donato
Sales and Business Development Director: Jean-Pierre Talamoni
MBDA UK, Managing Director: Steve Wadey
MBDA Italia, Managing Director: Sandro Pazzini
MBDA France, Managing Director: Didier Evrard
Chief Financial Officer: Julian Whitehead
Technical Director: Jean-Paul Genest
Engineering and Operations Director: Denis Ballet
Procurement Director: Franco Lucidi
Strategy and Planning Director: Guiseppe Snider
Communications Director: Pierre Bayle
■ Design, development, manufacture and supply of guided weapons and countermeasures systems and equipment.
Air Launched Weapons: Meteor, ASRAAM, Mica, Mistral Atam, Storm Shadow/Scalp EG, Brimstone, PGM, Diamond Back, BANG.
Countermeasures: Spectra, Eclair-M, DDM, Saphir 400, Lea/Spider.
Ground Based Air Defence: Rapier FSC/Jernas, Rapier B1X, SPADA 2000, Mistral 2, Mistral Atlas, Mistral Manpads, Mistral Co-oridnation post (MCP), Mistral Albi, VL Mica, Aster 30 (SAMP-T).
Battlefield Engagement: ERYX, Milan, HOT, GMLRS, DEADALE, SOUVIM.
Naval Systems: Vertical Launch Seawolf, Sea Skua, Albatros, Otomat Mk2-Block IV, Aster 15 (SAAM), Aster 15 and 30 (PAAMS), Marte Mk2/S, Milas, Simbad, Tetral, Sadral, VL Mica, Exocet AM39, SM39 and MM40.

UPDATED

McMurdo Ltd

(ex-McMurdo Ltd (Marine))
(ex-Nova Marine Systems Ltd)
Silver Point, Airport Service Road, Portsmouth, Hampshire, PO3 5PB, United Kingdom
Tel: (+44 23) 92 62 39 00
Fax: (+44 23) 92 62 39 98
e-mail: sales@mcmurdo.co.uk
Web: http://www.mcmurdo.co.uk
Managing Director: G J Mullins
Financial Director: Mike Kelly
Sales and Marketing Director: Kevin Robertson
Technical Director: C Hoffman
International Sales Manager: Richard Peacock
Purchasing Manager: Steve Newport
Marketing Communications Manager: Ms Corina Canning
■ Manufacture of advanced marine products including EPIRBs, GPS EPIRBs, Personal Location Beacons, a personal MOB wristwatch system, UAIS, DSC VHF radio, range of emergency lights for liferafts, lifejackets and lifebuoys. Supply of pyrotechnics for both commercial and leisure users.

UPDATED

McMurdo Ltd

(McMurdo Pains Wessex)
Silver Point, Airport Service Road, Portsmouth, Hampshire, PO3 5PB, United Kingdom
Tel: (+44 23) 92 62 39 00
Fax: (+44 23) 92 62 39 96
(+44 23) 92 62 39 98
e-mail: sales@mcmurdo.co.uk
Web: http://www.mcmurdo.co.uk
http://www.pwss.com
Vice President, McMurdo Pains Wessex Inc.: James Chandler
Managing Director: Gary Mullins
Tel: (+44 23) 92 62 39 02
Technical Director: Chris Hoffman
Tel: (+44 23) 92 62 39 03
Operations Director: Steve Lucas
Operations Manager: K Rough
Financial Director: Mike Kelly
■ Manufacture of marine and aviation safety, search and rescue equipment.

UPDATED

Metrum Information Storage Ltd

(Metrum Ltd)
2 Weller Drive, Finchampstead, Wokingham, Berkshire, RG40 4QZ, United Kingdom
Tel: (+44 118) 973 30 00
Fax: (+44 118) 973 43 63
e-mail: enquiries@metrum.co.uk
Web: http://www.metrum.co.uk
Managing Director: C G Beeton
Director, Sales: Stephen Wensley
Business Development Manager, Defence: G C Fabian
Marketing Co-ordinator: Ms Melany George
■ Design and manufacture of military magnetic tape data recorders.
Accreditations: ISO 9000

UPDATED

Microfiltrex

(ex-Fairey Microfiltrex Ltd)
Fareham Industrial Park, Fareham, Hampshire, PO16 8XG, United Kingdom
Tel: (+44 1329) 28 56 16
Fax: (+44 1329) 82 24 42
e-mail: info@porvairfiltration.com
Managing Director: John Sexton
Financial Director: James Robinson
Operations/Technical Director: David Amey
International Sales Manager: L Lee
Purchasing Manager: Les White
Group Marketing Manager: Mrs Helen Heiford
■ Design and manufacture of specialist filtration equipment for fuel, hydraulic, lubrication and air systems. Supply filters that protect vital subsystems for flight controls, fuel management, thrust reversers, breaking and steering, power generation and air intakes in aircraft, helicopters, military vehicles and space.
Accreditations: ISO 9001
JAR 145
JAR21 subpart G
BCAR a8.8
Defstan 05-91

UPDATED

Missiles & Space Batteries Ltd (MSB)

(ex-MSA (Britain) Limited, Catalyst Division)
Hagmill Road, East Shawhead, Coatbridge, Scotland, ML5 4UZ, United Kingdom
Tel: (+44 1236) 43 77 75
Fax: (+44 1236) 43 66 50
e-mail: msbscotland@aol.com
Operations Director: Nicolas Boudet
General Manager: Alistair Greig
Tel: (+44 1236) 43 84 50
Sales Manager: Gerry Smart
Financial Manager: Stephen Duffy
Sales and Marketing Manager: Ian Marshall
Purchasing Manager: R M Dick
■ Design and manufacture of thermal batteries.

UPDATED

Moog Controls Ltd

Ashchurch, Tewkesbury, Gloucestershire, GL20 8NA, United Kingdom
Tel: (+44 1684) 29 66 00
Fax: (+44 1684) 29 67 60
Web: http://www.moog.com
Managing Director: Colin Lewis
Operations Director: Nigel Cottell
Director of Aerospace Sales and Marketing: Andrew Yuill
Director of Aerospace Engineering: Steve Burks
■ Electro-hydraulic/pneumatic and electro-mechanical servo control systems and components, valves, actuators, motors and gassing systems and fuel control valves.

UPDATED

MPE Ltd

(ex-Dublier Limited)
(ex-Wego Condenser Co. Ltd)
PO Box 11, Liverpool, L33 7UL, United Kingdom
Tel: (+44 151) 632 91 00
Fax: (+44 151) 632 91 12
e-mail: sales@mpe.co.uk
Web: http://www.mpe.co.uk
Managing Director: Peter W Cotterill
Tel: (+44 151) 632 91 57
Financial Director: David Seabury
Technical Director: Jan Nalborczyk
Tel: (+44 151) 632 91 44
Sales and Marketing: Stephen Judge
Tel: (+44 151) 632 91 65
■ Provide EMC/RFI electrical filters, specialist plastic film and ceramic filters. EMC/RFI system analysis, test and solutions. Specialise in NEMP/

HEMP/EMPP/TEMPEST filter solutions, MilStud 188 125 filter solutions, shelter/rapid response equipment filter solutions, fighting vehicle filter solutions and telecommunications filter solutions.

UPDATED

Muirhead Aerospace Limited
(ex-Muirhead Vactric Components Ltd)
Oakfield Road, Penge, London, SE20 8EW, United Kingdom
Tel: (+44 20) 86 59 90 90
Fax: (+44 20) 86 59 99 06
Web: http://www.muirheadaerospace.com
Managing Director: Ken Smart
Financial Director: Pradeep Sharma
Sales and Marketing Director: Philip A Bowker
Technical Director: Colin Gambrill
Operations Director: Mark Evans
Marketing Support Manager: David Borley
Tel: (+44 20) 86 76 52 11
Mobile: (+44 7798) 50 47 18
Fax: (+44 20) 87 78 88 85
Purchasing Manager: Mark Pugh
■ Design and manufacture of high performance motion technology. Products include: servo components and control systems, synchros/resolvers, AC and DC motors, engine control components, torque motors, shaft encoders, gearheads, stepping motors, hybrids, brakes, flight controls, thrust reversal and fuel metering. Repair and overhaul of avionics and accessories, civil/military aircraft, fixed- and rotary-wing, including support chain management, ILS.
Accreditations: Lloyd's Register Quality Assurance Ltd to ISO 9001 Certificate No. 912120, CAA, DAI/4376/54, JAR145 repair station, CAA 00477, FAA, FAR145 repair station, M2UY050N, MoD approved No. IMRM06, Honeywell authorised warranty repair station.

UPDATED

NMB Minebea UK Ltd
(ex-Rose Bearings Ltd)
Doddington Road, Lincoln, LN6 3RA, United Kingdom
Tel: (+44 1522) 50 09 33
General Manager and Director: Mark N Stansfield
Sales Manager: Alan Morton
Mobile: (+44) 79 71 79 83 43
■ Rod end and spherical bearings to aerospace, railway, automotive and power generation industries. All types of military standard to customer bespoke. Specialist in coupling bearings for articulated PSV.

UPDATED

Octec Ltd
12 The Western Centre, Western Road, Bracknell, Berkshire, RG12 1RW, United Kingdom
Tel: (+44 1344) 46 52 00
Fax: (+44 1344) 46 52 01
e-mail: octec@octec.co.uk
sales@octec.co.uk
Web: http://www.octec.com
Managing Director: Gordon A Cain
Contact: Roger Joel
■ Automatic video target tracking and position analysis systems.

UPDATED

Parthus UK Ltd
Mansard Close, Westgate, Northampton, Northamptonshire, NN5 5DL, United Kingdom
Tel: (+44 1604) 58 55 88
Fax: (+44 1604) 58 55 99
Web: http://www.symmetricom.com
Managing Director: M Duckham
Technical Director: Tony Haddrell
Accounting Manager: D Jenkinson
Business Development Manager: Dr Clive De La Fuente
Purchasing Manager: Ms Michel Metcalfe
IS & CS Administrator: Ms Gill Rice
■ Production of GPS receivers for military and survey applications with centimetre accuracy.

UPDATED

Pascall Electronics Ltd
Westbridge Business Park, Cotheyway, Ryde, Isle of Wight, PO33 1QT, United Kingdom
Tel: (+44 1983) 81 73 00
Fax: (+44 1983) 56 47 08
e-mail: enquiries@pascall.co.uk
Web: http://www.pascall.co.uk
Chairman: Ian D Brodie
Managing Director: Alan Wainwright
Financial Director: Len Down
Marketing Controller: Ms Janice Gibson
Tel: (+44 1983) 81 74 12
■ Manufacture of amplifiers, attenuators, oscillators, signal processing components, descriminators and digital phase detectors.
Accreditations: BS EN ISO 9001

UPDATED

PW Defence Ltd
Wilne Mill, Draycott, Derbyshire, DE72 3QJ, United Kingdom
Tel: (+44 1332) 87 11 00
Fax: (+44 1332) 87 30 46
e-mail: info@pwdefence.com
Web: http://www.pwdefence.com
Managing Director: Paul A Greenwood
Tel: (+44 1980) 62 45 06
Sales Director: James Young
Tel: (+44 1332) 87 11 47
Mobile: (+44) 77 40 72 50 06
Fax: (+44 1332) 87 55 62
Regional Sales Manager: D Howell
UK Sales Manager: Lee Fuller
Regional Sales Manager: Ms Sharon Jones
Operations Director: Guy Wood
Research and Development Manager: S Quinby
■ A complete range of service pyrotechnics including products for illumination, signalling, search and rescue (SAR), training, law enforcement, shaped charges and battlefield simulation. Production of a range of 66 and 76 mm grenades that are launched from most standard grenade dischargers. Variants include multi-band infra-red screening smokes and low toxicity training smokes.

UPDATED

Quadrics
(Bristol Office)
One Bridewell Street, Bristol BS1 2AA, United Kingdom
Tel: (+44 117) 907 53 75
Fax: (+44 117) 907 53 95
e-mail: sales@quadrics.com
Web: http://www.quadrics.com
■ Manufacture of high performance computer hardware.

UPDATED

Racal Antennas Ltd
First Avenue, Millbrook Trading Estate, Southampton, SO15 0LJ, United Kingdom
Tel: (+44 23) 80 70 57 05
Fax: (+44 23) 80 70 11 22
e-mail: sales@racalantennas.com
Web: http://www.racalantennas.com
Managing Director: Derek McClelland
■ Design and manufacture of tactical antennas and telescopic masts for HF, VHF and UHF applications. Antennas are suitable for use with manpack radios and vehicle born installations in either a mobile or static role.

UPDATED

Raufoss Couplings UK Ltd
12 Brocks Business Park, Hodgson Way, Wickford, Essex, SS11 8YN, United Kingdom
Tel: (+44 1268) 56 29 00
Fax: (+44 1268) 76 61 13
e-mail: info@raufoss.co.uk
Web: http://www.raufoss.com

UPDATED

Reaction Engines Ltd (REL)
Building D5, Culham Science Centre, Abingdon, Oxon, OX14 3DB, United Kingdom
Tel: (+44 1865) 40 83 14
Fax: (+44 1865) 40 83 01
e-mail: reactionengines@aol.com
Web: http://www.gbnet.com/orgs/skylon
Managing Director: Alan Bond
Technical Director: Richard Varvill

UPDATED

Reynolds Industries Ltd
Navigation House, Canal View Road, Newbury, Berkshire, RG14 5UR, United Kingdom
Tel: (+44 1635) 311 37
Fax: (+44 1635) 52 19 36
e-mail: rey@reynoldsindustries.ltd.uk
Web: http://www.reynoldsindustries.ltd.uk
Managing Director: Nick S Bance
Tel: (+44 1635) 26 22 01
Engineering Manager: Paul I Belsey
Tel: (+44 1635) 26 22 18
Manager, Microwave Division: Brian H Huntsman
Tel: (+44 1635) 26 22 12
Sales Manager: Neil MacLean Smithers
Tel: (+44 1635) 26 22 07
Purchasing Manager: Malcolm Horne
Tel: (+44 1635) 26 22 11
■ Cable and wire products for space and aerospace. High voltage connectors, high and low voltage cable assemblies for the aerospace industry. Microwave cable assemblies (flexible and semi-rigid to 40 GHz). High voltage capacitors, transient protection devices used in military and airborne electronic equipment. Custom built electro-mechanical and electro-optical interconnect products for a wide range of space and aerospace applications.

UPDATED

Rotadata Ltd
Bateman Street, Derby DE23 8JQ, United Kingdom
Tel: (+44 1332) 34 80 08
Fax: (+44 1332) 33 10 23
e-mail: sales@rotadata.com
Web: http://www.rotadata.com
Managing Director: John Taylor
Sales Director: Mike Heyworth
Marketing Director: Derek Taylor
Operations Director: Graham Barsby
Technical Director: Brian Edmundson
Administrator: Mrs Amanda Marshall
■ Fixed and traversing aerodynamic probes for aircraft and aircraft engines. Tip clearance measurement equipment. Application of sensors to engine test components. Design and manufacture of turbo-machinery test rigs and wind tunnel models. High speed sliprings and telemetry.

UPDATED

Safe Training Systems Ltd (STS)
(Company Headquarters)
Holly House, Maidenhead Road, Wokingham, Berkshire, RG40 5RR, United Kingdom
Tel: (+44 1344) 48 35 63
Fax: (+44 1344) 48 51 75
e-mail: sales@safetrainingsystems.com
Web: http://www.safetrainingsystems.com
Managing Director: David R Ward
■ Design, manufacture and sales of simulators for hazardous materials monitors – radiation, chemical warfare agents and toxic chemicals.

UPDATED

Saft Ltd
(Alcad Ltd)
6th Floor, Westgate House, West Gate, Harlow, Essex, CM20 1JN, United Kingdom
Tel: (+44 1279) 77 25 50
Fax: (+44 1279) 42 09 09
Web: http://www.saftbatteries.com
Contact: Ms Sarah Carter
Tel: (+44 1279) 77 25 50
■ Manufacture of a range of vented nickel-cadmium batteries to match specific operating conditions. Ni-Cd and Ni-Mh cells and battery packs and Lithium primary and secondarycells and battery packs.
Representing:
Saft, France

UPDATED

Sea Tel Inc.
(European Office)
Unit 1, Orion Industrial Centre, WideLane, Swaythling, S018 2HJ, United Kingdom
Tel: (+44 2380) 67 11 55
Fax: (+44 2380) 67 11 66
e-mail: europe@seatel.com
Web: http://www.seatel.com
Public Relations Department: James Boulton

UPDATED

Serco Europe Ltd
5th Floor, Kempton Point, 68 Staines Road West, Sunbury-on-Thames, Middlesex, TW16 7AX, United Kingdom
Tel: (+44 1932) 73 30 00
Fax: (+44 1932) 73 30 99
Marketing Manager: James Smith
Business Development Support Manager: Ms Louise Greasley
Tel: (+44 1932) 73 30 11

UPDATED

Sonic Communications (Int) Ltd

Birmingham International Park, Starley Way, Bickenhill, Birmingham, West Midlands, B37 7HB, United Kingdom
Tel: (+44 121) 781 40 00
Fax: (+44 121) 781 44 02
e-mail: sales@sonic-comms.com
Web: http://www.sonic-comms.com
Managing Director: David John Bryan
Sales and Marketing Director: Neil Barker
Operations Director: Adrian Wright
Purchasing Manager: David Sutton
Regional Sales Manager: Eric Davidson
Regional Sales Manager: John Bellamy
Regional Sales Manager: Marcus Twomlow
Administration Secretary: Ms Lisa Barliss
■ Voice activated transmission (VOX) systems, intended to reject background noise and only transmit speech, for applications such as police firearms teams, SAR winchmen and armoured vehicles.

UPDATED

Spearhead Machinery Ltd

Pershore Trading Estate, Pershore, Worcestershire, WR10 2DD, United Kingdom
Tel: (+44 1386) 55 67 48
Fax: (+44 1386) 56 13 98
e-mail: enquiries@spearhead.uk.com
Web: http://www.spearhead.uk.com
Executive: William Davies
Sales Director: Jock McLaren
■ Design and manufacture of a range of tractor powered mowing equipment in use with civil and military airfields.

UPDATED

Specac Ltd

(ex-Graseby Specac Ltd)
River House, 97 Cray Avenue, St Mary Cray, Orpington, Kent, BR5 4HE, United Kingdom
Tel: (+44 1689) 87 31 34
Fax: (+44 1689) 87 85 27
e-mail: sales@specac.co.uk
Web: http://www.specac.com
■ Infra-red and laser optical components including polarisers and thermal imaging optics, FIR interferometers and laser beam splitters, optical filters, optical assemblies for astronomical telescopes and accessories for FT-IR spectroscopy.
Overseas agents
Specac Inc., US

UPDATED

Stop-choc Ltd

Banbury Avenue, Slough, Berkshire, SL1 4LR, United Kingdom
Tel: (+44 1753) 53 32 23
Fax: (+44 1753) 69 37 24
e-mail: sales@stop-choc.co.uk
Web: http://www.stop-choc.co.uk
Managing Director: J M H Dunn
Chief Executive, Sales: R W Steer
Sales Manager: Nigel Care
■ Design and manufacture of metal and elastomeric anti-vibration mountings, helicopter blade lead/lag dampers, avionics and military racking and accessories, illuminated instrument cockpit panels with integrated electronics, pushbutton controls and alpha-numeric displays.
Accreditations: ISO 9001

UPDATED

Sun Microsystems Inc

(European Headquarters, OEM Division)
Sparc House, Guillemont Park, Minley Road, Blackwater, Camberley, Surrey, GU17 9QG, United Kingdom
Tel: (+44 1252) 42 00 00
Fax: (+44 1252) 42 04 29
e-mail: info-oem@sun.com
Web: http://www.sun.com

UPDATED

Superform Aluminium

Cosgrove Close, Worcester, WR3 8UA United Kingdom
Tel: (+44 1905) 87 43 00
Fax: (+44 1905) 87 43 01
e-mail: sales@superform-aluminium.com
Web: http://www.superform-aluminium.com
General Manager: Frank K Holliday
Financial Manager: John Adams
Sales Manager: A M Lowerson
Technical Manager: Russell Stracey
■ Manufacture and distribution of superplastic components in aluminium alloys. Pressings and fabrications both complex and deep drawn.

UPDATED

T S Space Systems

Unit V, Rose Business Estate, Marlow Bottom, Marlow, Buckinghamshire, SL7 3ND, United Kingdom
Tel: (+44 1628) 47 40 40
Fax: (+44 1628) 47 78 00
Web: http://www.ts-space.demon.co.uk
Project Manager: K Howell

UPDATED

Thales Avionics Ltd

(ex-Racal Avionics Ltd)
86-88 Bushey Road, Raynes Park, London, SW20 0JH, United Kingdom
Tel: (+44 20) 89 46 80 11
Fax: (+44 20) 89 46 10 07
Web: http://www.thales-avionics.com
Managing Director: Paul Kahn
Operations Director: Rainer Koll
Marketing Director: John Mountifield
Tel: (+44 20) 89 71 51 20
Senior Marketing Manager: Denis McCallum
Sales Manager: Bruce Wright
Tel: (+44 20) 89 71 51 06
Procurement Manager: J Mansi
Tel: (+44 20) 89 71 51 72
Senior Publicity Manager: Guy Douglas
Tel: (+44 193) 282 48 32
■ Design of civil and military airborne navigation, flight/mission management and aeronautical satellite communications systems. Systems engineering/integration. Prime contractorship.
Accreditations: ISO 9001, MoD, CAA

UPDATED

Thales Defence Ltd

(Group Headquarters)
(ex-Racal Electronics plc)
(ex-Thomson Racal Defence)
2 Dashwood Lang Road, The Bourne Business Park, Addlestone, Near Weybridge, Surrey, KT15 2NX, United Kingdom
Tel: (+44 1344) 38 70 00
(+44 1344) 48 12 22
Fax: (+44 1344) 38 74 03
(+44 1344) 45 41 19
Web: http://www.thalesgroup.co.uk
Chairman: Sir Ernest T Harrison
Chief Executive: David C Elsbury
■ Principal activities: radio communications, telecommunications, data communications, radar defence systems, command, control, communications and intelligence systems, avionics, airborne navigation, precise positioning and energy technology. Other activities include: satellite communications, acoustics, antennas, automatic and diagnostic test systems, data and communications recording, electronic instrumentation, health and safety, intruder detection, logistic support; microwave components and systems.

UPDATED

Thales Sensors
(Thales Airborne Systems)
(ex-Redifon MEL Ltd)
Manor Royal, Crawley, West Sussex, RH10 2PZ, United Kingdom
Tel: (+44 1293) 52 87 87
Fax: (+44 1293) 54 28 18
Web: http://www.thales-defence.co.uk
■ Defence radar systems. ELINT, ESM, ECM and RWR systems. CCIS and C3I systems. Miss distance indicators; surveillance and target cueing radar. Non-communication electronic warfare systems and radar displays for fixed- and rotary-wing aircraft, surface ships and submarines. High resolution colour display systems, colour map display systems, ground-based ELINT systems, battlefield ESM and ELINT collection systems. Action information systems, SWSE/SWSA, high resolution display systems. Manpackable satellite communications.

UPDATED

Thompson Valves

(FCX Thompson Valves)
(ex-NEI Thompson Valves)
17 Balena Close, Creekmoor, Poole, Dorset, BH17 7EF, United Kingdom
Tel: (+44 1202) 69 75 21
Fax: (+44 1202) 60 53 85
e-mail: sales@fcx-thompson-valves.com
Web: http://www.fcx-thompson-valves.com
Managing Director: Bob Bowser
Marketing and Sales Director: Mark Harris
Business Development Director: Brian J Ellis
Technical Director: Steve Rawson
Operations Director: David Whitfield
Financial Director: Gary Cowen
Purchasing Manager: J Clarke
■ Solenoid valves for cryogenic solenoid fuel control for the second stage of the Ariane Space vehicle. Pressure regulators and solenoid valves for ground based support and service/test equipment. Valves for special development, high pressure air and fluids, containment and regulation. Solenoid valves.

UPDATED

Timet UK Ltd

PO Box 704, Witton, B6 7UR, United Kingdom
Tel: (+44 121) 356 11 55
Fax: (+44 121) 356 54 13
e-mail: eurosales@timet.com
Web: http://www.timet.com
■ Titanium and titanium alloys, ingots, bars, billet, sheet, rod, tube, extrusions, wire and castings. Superconducting composites wires and cables.

UPDATED

Timken Coventry

(Nadella UK Sales Office)
(ex-Nadella UK Ltd)
Progress Close, Leofrie Business Park, Binley, Coventry, West Midlands, CV3 2TF, United Kingdom
Tel: (+44 24) 76 29 69 00
Fax: (+44 24) 76 29 69 93
Web: http://www.timken.com
Managing Director: D G Barker
Sales Manager: John D Pugh
Tel: (+44 24) 76 29 69 21
Fax: (+44 24) 76 29 69 92
Sales Correspondent: R Gardner
Tel: (+44 24) 76 29 69 62
■ Manufacture of Torrington and Nadella airframe needle roller bearings. Fafnir aircraft control bearings and SNR mainshaft jet engine, gearbox, auxilliary drives, spacecraft and helicopter transmission bearings.

UPDATED

TRL Technology Ltd

Shannon Way, Tewkesbury, Gloucestershire, GL20 8ND, United Kingdom
Tel: (+44 1684) 27 87 00
Fax: (+44 1684) 85 04 06
e-mail: info@trltech.co.uk
Web: http://www.trltech.co.uk
■ Specialise in the design, development and integration of hardware and software products with a particular interest in radio and satellite communications systems for government and defence applications. Also specialise in digital processing and radio frequency engineering, real-time and off-line analysis software tools, radio electronic warfare systems, radio receivers, jammers, signal analysers and robust data links. Radio surveillance systems and radio direction finding systems. Bespoke DSP hardware and software design, real time/time critical software systems, and long-range optical surveillance equipment. Specialise in the design and implementation of secure private networks. Key products include the Tarax Lightweight EW system – a versatile EW tool that can be deployed in a wide cross-section of scenarios and applications, and a portfolio of satellite monitoring solutions which enable the identification of transmissions of interest, accurate and near real-time location of Inmarsat terminals under scrutiny, and traffic analysis.

UPDATED

Turbocam Europe Ltd

Unit 1, 155 Highlands Road, Fareham, Hampshire, PO15 6JR, United Kingdom
Tel: (+44 1329) 84 58 00
Fax: (+44 1329) 84 60 00
e-mail: tel@turbocam.com
Web: http://www.turbocam.com
Managing Director: Robin Taylor
Sales Manager: Onno Weststrate
Production Manager: Neil Lightfoot
Purchasing, Quality and Accounts Manager: Derek Hawkins

■ Specialist machining for turbomachinery, products include propellers and impellers for marine applications; aircraft engine components, impellors for spaceborne platforms and rail locomotive engine components.
Accreditations: ISO 9001; ISO 9002; AS9100A

UPDATED

Ultra Electronics Limited
(PMES)
(ex-Power Magnetics and Electronic Systems Ltd)
(ex-Thorn Automation)
(ex-Ultra Electronics Limited)
Armitage Road, Rugeley, Staffordshire, WS15 1DR, United Kingdom
Tel: (+44 1889) 50 33 00
Fax: (+44 1889) 57 29 29
e-mail: enquiries@ultra-pmes.com
Web: http://www.ultra-pmes.com
Managing Director: A M Freer
Engineering Director: J P Wardale
Marketing Director: J S Greenhalgh
Sales Manager, Transit Systems: T W Boston
Public Relations Officer: Mrs Lindsey M John
■ Military power conversion systems for ships and submarines. Rectifiers for civil applications (railways and metro systems). Data distribution systems, local area networks, acoustic ranges and intercept sonar.

UPDATED

Vitec Group Communications
(ex-Drake Electronics Ltd)
7400 Beach Drive, Cambridge Research Park, Cambridgeshire, CB5 9TP, United Kingdom
Tel: (+44 1223) 81 50 20
Fax: (+44 1223) 81 50 99
(+44 1223) 81 50 01
e-mail: vgc.uk@vitecgroup.com
Web: http://www.vitecgroupcomms.com
Managing Director: Daniel Burns
Sales and Marketing Director: Irwin D Nathanson
Tel: (+44 1223) 81 50 05
Technical Director: Steve Farmer
Operations Director: Terry Boon
Deputy Sales Director: David Bartlett
Marketing Communications Manager: Ms Suzanne Sutton
Sales Manager: George Batrouny
Sales Manager: Bruce Green
Sales Manager: Roger Mawditt
Systems and Applications Manager: Peter Stallard
Product Manager: Simon Browne
Purchasing Manager: Davinder Lachhar
■ Design, manufacture and supply of intercom systems for television, radio and satellite broadcasting, and voice communications solutions for air traffic control, space operations and simulation and training applications.

UPDATED

Industry – Service

Aetheric Engineering Ltd
Broadlaw House, Broadlaw Walk, Fareham, Hampshire, PO14 1LE, United Kingdom
Tel: (+44 1329) 82 35 83
Fax: (+44 1329) 28 86 75
e-mail: sales@aetheric.co.uk
Web: http://www.aetheric.co.uk
Principal Consultant: Peter Milne
■ Consultancy in all aspects of satellite communication systems, including system planning and operation, earth station implementation and satellite procurement. Services include feasibility and design studies; preparation of specifications and procurement documentation; evaluation of proposals and tenders; project management; supervision and witnessing of in-factory activities, including manufacture, test and integration; supervision and witnessing of on-site activities, including installation, test and commissioning; system operation; and frequency co-ordination and access planning. Provision of support to the establishment of effective communication regulatory organisations.

UPDATED

Aon Group Ltd
(Aviation Information Resources)
(ex-Leslie & Godwin Aviation Holdings Ltd)
(ex-Nicholson Leslie Aviation Ltd)
8 Devonshire Square, London EC2M 4PL, United Kingdom
Tel: (+44 20) 72 47 44 66
Fax: (+44 20) 72 77 65 09
Telex: 929464
Web: http://www.aor.com
Managing Director: William B B Smith
Operations Director: Mick Higgins
■ Specialise in risk management and insurance brokerage services for all satellite related risks, as well as major aerospace and aviation companies.

UPDATED

Barr Associates Ltd
Units 3 and 4 Home Farm Business Units, Yattendon, Newbury, Berkshire, RG18 0XT, United Kingdom
Tel: (+44 1635) 20 13 17
Fax: (+44 1635) 20 20 30
e-mail: info@barr-associates-uk.com
Web: http://www.barr-associates-uk.com
Managing Director: Alan Whatley
Operations and Technical Director: Dr Colin Cole
Contact: Rob Cassidy
Contact: Philip Gardner
Contact: Terry Gibbs
■ Advanced optical filters. Infra-red coating services. Consultancy on most aspects of infra-red optical design. Filters have been supplied for the IBSS as part of the SDI project.

UPDATED

British Sarozal Ltd
Handrail House, Maygrove Road, London, NW6 2EG, United Kingdom
Tel: (+44 20) 73 28 21 11
Fax: (+44 20) 76 24 51 44
Cable: SAROZAL
e-mail: sales@bslexpress.com
Web: http://www.bslexpress.com
Managing Director: J M Treister
Operations and Technical Director: Leonard Simmonds
Government Contracts Superintendent: Ms J Scott
■ Solid state electronic valves replacing thermionic valves for communications and radar installations. Supply of spare parts for British and American civil and military electronic equipment. Installation and maintenance of electronic systems for ATC. Manufacture of image intensifiers, synchros tacko-generators, servo motors, avionic tubes, display units (radar), avionic tubes, military tubes, transformers and cable assemblies. Pressure gauges, power supply units, integrated circuits, military tubes, image intensifiers, synchros, tacko generators and avionic tubes.

UPDATED

CAESAR Consultancy
35 Millington Road, Cambridge, CB3 9HW, United Kingdom
Tel: (+44 1223) 35 38 39
Fax: (+44 1223) 30 38 39
Web: http://www.caesarconsultancy.com
http://www.mjrycroft.com
Proprietor: Prof Michael J Rycroft
■ Provide advice on scientific, technical and commercial topics on atmospheric, environmental and space activities and research.

UPDATED

Commercial Space Technologies Ltd (CST)
67 Shakespeare Road, Hanwell W7 1LU, United Kingdom
Tel: (+44 20) 88 40 10 82
Fax: (+44 20) 88 40 77 76
e-mail: cst@commercialspace.co.uk
Web: http://www.commercialspace.co.uk
General Director: Gerry M Webb
Tel: (+44 20) 88 40 54 26
Financial Director: S T Perera
Technical Director: K Milayev
Director, Moscow Office: Ms Nina Pestmal
Tel: (+7 095) 415 77 32
■ Assessment studies of space activities (both technical and commercial) and assistance with the instigation or development of client's business activities in the space sector. Contacts with organisations in the Russian Federation and associated states.

UPDATED

Cospas-Sarsat
99 City Road, London, EC1Y 1AX, United Kingdom
Tel: (+44 20) 77 28 13 91
Fax: (+44 20) 77 28 11 70
e-mail: cospas_sarsat@imso.org
Web: http://www.cospas-sarsat.org
Principal Operations Officer: Ms Cheryl Bertoia
Tel: (+44 20) 77 28 10 66
Technical Officer: Wayne Carney
Tel: (+44 20) 77 28 14 44
Deputy Head: Bill Ruark
■ Satellite systems for distress alert and location data to assist search and rescue (SAR) operations.

UPDATED

Cranfield Aerospace Ltd
Cranfield, Bedfordshire, MK43 0AL, United Kingdom
Tel: (+44 1234) 75 40 46
Fax: (+44 1234) 75 11 81
e-mail: cae@cranfieldaerospace.com
Web: http://www.cranfield.ac.uk/cae
http://www.cranfieldaerospace.com

UPDATED

Earth Resource Mapping
Blenheim House, Crabtree Office Village, Eversley Way, Egham, Surrey, TW20 8RY, United Kingdom
Tel: (+44 1784) 43 06 91
Fax: (+44 1784) 43 06 92
Web: http://www.ermapper.com
http://www.earthetc.com
■ Image processing and internet distribution software development.

UPDATED

ERA Technology Ltd
Cleeve Road, Leatherhead, Surrey, KT22 7SA, United Kingdom
Tel: (+44 1372) 36 70 00
Fax: (+44 1372) 36 70 99
e-mail: info@era.co.uk
Web: http://www.era.co.uk
Managing Director: Fred J Cahill
Marketing Manager: Neil Gardner
Marketing Assistant: Ms Sasha Vigus-Ruffler
Tel: (+44 1372) 36 72 44
■ Design, development, testing, assessment and consultancy for communications, aerospace, defence, information technology, manufacturing, transport, electronics and energy.

UPDATED

Esys plc
1 Occam Court, Surrey Research Park, Guildford, Surrey, GU2 7HJ, United Kingdom
Tel: (+44 1483) 30 45 45
Fax: (+44 1483) 30 38 78
e-mail: info@esys.co.uk
Web: http://www.esys.co.uk
Director, Business Development: Andrew Davies
Director, Operations: Jim Boucher
Contact: Ms Abi Crowther
Tel: (+44 1483) 40 29 30
■ Divided between four consulting practices serving space, telecoms, defence and education, and research. Each of these practices offers a portfolio of consulting services, namely strategy and policy, business assessment, project management, technology evaluations and system engineering.

UPDATED

Flying Pictures Ltd
Stonefield Park, Chilbolton, Hampshire, SO20 6BR, United Kingdom
Tel: (+44 1264) 86 11 00
Fax: (+44 1264) 86 00 33
e-mail: info@flyingpictures.com
Web: http://www.flyingpictures.com
Chief Executive: Colin Prescoi
Managing Director, Filming: Marc Wolff
Financial Director: James Hunter Smart
Marketing Director: David Fish
■ Provide aviation publicity services, aerial stunts marketing and bespoke hot air balloons for marketing purposes; live and recorded aerial filming services using rotary wing aircraft and airships.

UPDATED

Flying Pictures Space Ltd
Stonefield Park, Chilbolton, Hampshire, SO20 6BR, United Kingdom
Tel: (+44 1264) 86 11 00
Fax: (+44 1264) 86 00 33
Director: Andrew Cox

Director: Andrew Elson
■ Design and manufacture of high altitude platforms and aircraft.

UPDATED

Hindustan Aeronautics Ltd (HAL)
(European Liason Office)
Room No. 602, India House, High Commission of India, Aldwych, London, WC2B 4NA, United Kingdom
Tel: (+44 20) 74 97 23 60
Fax: (+44 20) 74 97 83 98
e-mail: marketing@hal-india.com
Web: http://www.hal-india.com
Resident Manager: N M Yadav
■ Liaison work for design, development, manufacture and servicing of fixed and rotary-wing aircraft, engines, systems, accessories, equipment and avionics. Light alloy structures for space and ground handling equipment.
Representing:
Hindustan Aeronautics Ltd, India

UPDATED

IGG Component Technology Ltd
Waterside House, Waterside Gardens, Fareham, Hampshire, PO16 8RR, United Kingdom
Tel: (+44 1239) 82 93 11
Fax: (+44 1239) 82 93 12
e-mail: sales@igg.co.uk
Web: http://www.igg.co.uk
Chairman: Kelvin Harrison
Managing Director: Richard G Matthews
Sales and Marketing Director: Graham C Peters
Technical Director: Geoffrey K Penhaligon
Sales and Marketing Manager: Roger Buckley
■ Parts engineering of electronic components and centralised parts procurement. Procurement and testing of components for complete satellites. Also provides engineering and consultancy, obsolescence management, customer training, upscreening, destructive physical analysis, failure analysis, evaluation of new technologies and materials, and electronic component kitting.

UPDATED

Infoterra Ltd
(ex-National Remote Sensing Centre Ltd)
Delta House, Southwood Crescent, South wood, Farnborough, Hampshire, GU14 0NL, United Kingdom
Tel: (+44 1455) 84 92 29
Fax: (+44 1455) 84 17 85
e-mail: info@infoterra-global.com
Web: http://www.infoterra-global.com
■ Leading supplier of data and information products and services from Earth observation satellites, aircraft and ground observation.

UPDATED

Infoterra Ltd
(Barwell Office)
(ex-National Remote Sensing Centre Ltd)
Arthur Street, Barwell, Leicestershire, LE9 8GZ, United Kingdom
Tel: (+44 1455) 84 92 29
(+44 1455) 84 92 03
Fax: (+44 1455) 84 17 85
e-mail: info@infoterra-global.com
Web: http://www.infoterra-global.com
Managing Director: David Fox
Sales Manager: Andy Wells
Tel: (+44 1455) 84 92 42
Information Manager: John Shannon
■ Deliver advanced geo-information solutions to DGIA, European defence mapping agencies, JARIC, ESA, ordnance survey and the UK Armed services.

UPDATED

Inmarsat
99 City Road, London, EC1Y 1AX, United Kingdom
Tel: (+44 20) 77 28 10 00
(+44 20) 77 28 17 77
Fax: (+44 20) 77 28 10 44
(+44 20) 77 28 11 42
e-mail: customer_care@inmarsat.com
Web: http://www.inmarsat.com
Chairman: Richard Vos
Deputy Chairman: John Rennocks
President and Chief Executive Officer: Michael Storey
Director: Artur Schechtman
Director: Dr Henry Chasia
Director: Edward Berger
Director: Raymond Ch'ien
Director: Raynald Leconte
Director: Dick Hoefsloot
Director: Ms Britt-Carina Horncastle
Director: Bo Ake Lerenius
Director: Philip Permut
General Manager of Aeronautical Business: Robert Johnson
Director, Marketing and Communications: Steven Rogers
Press Enquiries: Dominic Cook
Tel: (+44 20) 77 28 12 56
■ Provision of global high bandwidth multimedia mobile communications solutions to corporations around the globe for maritime, air or land based applications.

UPDATED

Intelligence and Presentations Aviation and Aerospace Practice
(ex-J & H Marsh & McLennan Ltd)
(ex-Johnson & Higgins Ltd)
Marsh Specialty Operations Ltd, No.1 The Marsh Centre, London, E1 8DX, United Kingdom
Tel: (+44 20) 73 57 10 00
Fax: (+44 20) 79 29 27 05
Vice President: Rob Gardiner
■ Provide insurance, reinsurance broking and space insurance services.

UPDATED

Jane's Information Group (JIG)
(Europe, Middle East and Africa)
Sentinel House, 163 Brighton Road, Coulsdon, Surrey, CR5 2YH, United Kingdom
Tel: (+44 20) 87 00 37 00
Fax: (+44 20) 87 63 10 06
e-mail: info.uk@janes.com
Web: http://www.janes.com
Managing Director: Alfred Rolington
Tel: (+44 20) 87 00 37 01
Chief Sales and Marketing Officer: Ms Jo Moon
Tel: (+44 20) 87 00 37 52
Financial and Commercial Director: Mike Staton
Tel: (+44 20) 87 00 37 06
Strategy and Planning Director: Ian Kay
Tel: (+44 20) 87 00 37 96
Publisher (News Division): Sean Howe
Head of Advertising Sales, UK/RoW: Ms Janine Boxall
Tel: (+44 20) 87 00 38 52
Head of Advertising Sales, US: Ms Katie Taplett
Tel: (+1 703) 683 37 00
Group Communications Manager: Ms Claire Brunavs
Tel: (+44 20) 87 00 37 03
■ Jane's Information Group is the leading supplier of professional information to the defence and aerospace communities and a major information provider to selected areas of transportation and law enforcement worldwide. All Jane's publications, including magazines, are available on CD-ROM for personal computers and many UNIX platforms. All publications are now available ON-LINE (http://www.janes.com), updated a minimum of every three months. Jane's publishes Special Reports and provides consultancy services, data services and an extensive image library.
Air/Space Publications:
Jane's Aero-Engines
Jane's Aircraft Component Manufacturers
Jane's Aircraft Upgrades
Jane's Air-Launched Weapons
Jane's Air-Launched Weapons Image Library on CD-ROM
Jane's All The World's Aircraft
Jane's Avionics
Jane's Helicopter Markets & Systems
Jane's Military Aircraft Image Library on CD-ROM
Jane's Military Aircraft Markings Database
Jane's Space Directory
Jane's Unmanned Aerial Vehicles and Targets
Jane's World Air Forces
Air and Space Special Reports
Defence Markets Publications:
Jane's Defence Industry (newsletter)
Jane's International ABC Aerospace Directory
Jane's International Defence Directory
Jane's World Defence Industry
Defence Markets Special Reports
Land Forces Publications:
Jane's Ammunition Handbook
Jane's Armour and Artillery
Jane's Armour and Artillery Upgrades
Jane's Infantry Weapons
Jane's Land-Based Air Defence
Jane's Land Systems Image Library on CD-ROM
Jane's Military Biographies
Jane's Military Vehicles and Logistics
Jane's Mines and Mine Clearance
Jane's NBC Defence
Jane's World Armies
Land Forces Special Reports
Law Enforcement Publications:
Jane's Police and Security Equipment
Law Enforcement Special Reports
Other Police Review Titles
Police Review (weekly magazine)
Practical Police Management
Magazines and Newsletters:
Jane's Foreign Report
Jane's Defence Industry
Jane's Defence Weekly
Jane's Intelligence Review
Jane's International Defense Review
Jane's Missiles and Rockets
Jane's Navy International
Military Systems Publications:
Jane's C4I Systems
Jane's Electronic Mission Aircraft
Jane's Electro-Optic Systems
Jane's Military Communications
Jane's Radar and Electronic Warfare Systems
Jane's Simulation and Training Systems
Jane's Strategic Weapon Systems
Military Systems Special Reports
Naval/Maritime Publications:
Jane's Amphibious and Special Forces
Jane's Fighting Ships
Jane's Marine Propulsion
Jane's Merchant Ships
Jane's Naval Construction and Retrofit Markets
Jane's Naval Weapon Systems
Jane's Underwater Technology
Jane's Underwater Warfare Systems
Warships Image Library on CD-ROM
Naval and Maritime Special Reports
Risk and Security Publications:
Jane's Chem-Bio Handbook
Jane's Chemical-Biological Defense Guidebook
Jane's Counter Terrorism
Jane's Crisis Communications Handbook
Jane's Facility Security Handbook
Jane's Intelligence Digest
Jane's Intelligence Review
Jane's Intelligence Watch Report
Jane's Islamic Affairs Analyst
Jane's Mass Casualty - Pre hospital
Jane's Mass Casualty - Hospital
Jane's School Safety Handbook
Jane's Sentinel Security Assessments
Jane's Terrorism and Insurgency Centre
Jane's Terrorism and Security Monitor
Jane's Terrorism Watch Report
Jane's Unconventional Weapons Response Handbook
Jane's Workplace Security Handbook
Jane's World Insurgency and Terrorism
Geopolitical Special Reports
Transport Publications:
Jane's Air Traffic Control
Jane's Airport Review (Magazine)
Jane's Airports and Handling Agents
Jane's Airports, Equipment and Services
Jane's High Speed Marine Transport
Jane's Urban Transport Systems
Jane's World Airlines
Jane's World Railways
Jane's Transport Finance
Transport Special Reports

UPDATED

Jane's Information Group (JIG)
(London)
1st Floor, The Quadrangle,180 Wardour Street, London W1A 4YG, United Kingdom
Tel: (+44 20) 87 00 37 00
Fax: (+44 20) 72 87 47 65
(+44 20) 72 87 77 65

UPDATED

JRA Aerospace Ltd
(ex-JRA Aerospace & Technology Ltd)
JRA House, Marlow, Buckinghamshire, SL7 1PR, United Kingdom
Tel: (+44 1628) 89 11 05
Fax: (+44 1628) 81 65 81
Web: http://www.jra-aero.com
Managing Director: Peter Moores
Tel: (+44 1628) 81 65 83
Director, Project Support: R N Logan

UPDATED

Kal-Gard UK Ltd
(KG Coating Ltd)
(ex-Scanwel Scientific Ltd)
Canalwood Industrial Estate, Chirk, Clwyd, LL14 5RL, United Kingdom
Tel: (+44 1691) 77 20 70
(+44 1691) 77 80 70
Fax: (+44 1691) 77 83 03

Web: http://www.kal-gard.co.uk
Chairman and Managing Director: Adrian D Thompson
Commercial Manager: A V Barwood
■ Sub-contract vacuum coating, IVD aluminium and cadmium, anti-corrosion, low friction and specialised coating service. BAe, CAA and Lucas approved.

UPDATED

Kongsberg Defence & Aerospace AS (KDA)
(ex-Kongsberg Gruppen ASA)
4 Stanstead Close, Rowlands Castle, PO9 6AN, United Kingdom
Tel: (+44 23) 92 41 28 09
Fax: (+44 23) 92 41 38 09
UK Representative and Defence Consultant: Capt M J M Wilkin
■ Main areas of activity include anti-ship missiles for fighter aircraft, helicopters and naval vessels; missile launchers; missile control and test systems; ground based systems; naval defence systems; tactical communications; trainers and simulators; satellite-based surveillance and special electronics. Rocket analysis.
Representing:
Kongsberg Defence & Aerospace AS, Norway

UPDATED

Kudos Satellite Communications
60 Middle Watch, Swavesey, Cambridge, CB4 5RN, United Kingdom
Tel: (+44 1954) 23 08 24
Fax: (+44 1954) 20 07 21
e-mail: office@kudos-satellite.co.uk
Web: http://www.kudos-satellite.co.uk
Managing Director: Dr Michael J S Quigley
■ Consultancy in engineering and business planning of satellite projects. Provision of links for satellite internet access and broadband communications.

UPDATED

Marsh Space Projects Ltd
(ex-Sedgwick Space Services)
No.1, The Marsh Centre, London, E1 8DX, United Kingdom
Tel: (+44 20) 73 57 52 74
(+44 20) 73 57 10 00
Fax: (+44 20) 73 57 52 78
(+44 20) 79 29 27 05
Web: http://www.marsh.com
Chairman: Brian G Moore
Tel: (+44 20) 73 57 52 70
Assistant Vice President: Stephen Monks
Tel: (+44 20) 73 57 52 68
Managing Director: Yamin A Mustafa
■ Spacecraft risk management – manufacture, prelaunch, launch, in-orbit and Third Party Liability insurance.

UPDATED

Moreton Hall Associates (MHA)
Morar House, Altwood Close, Maidenhead, Berkshire, SL6 4PP, United Kingdom
Tel: (+44 1628) 78 34 55
Fax: (+44 1628) 63 75 86
Managing Director: Geoffrey E Hall
■ Consultancy, information services and support for projects in space, computer applications and risk management. Fields include product assurance, systems engineering, operations, software, technology and insurance.
Publications: Monthly bulletin: Projected Future Launches
Annual bulletin: Space Launches Pocket Book

UPDATED

Nallatech
Boolean House, One Napier Park, Cumbernauld, Glasgow, G68 0BH, United Kingdom
Tel: (+44 1236) 78 95 51
Fax: (+44 1236) 78 95 99
Web: http://www.nallatech.com
Systems Applications Engineer: James Alexander

UPDATED

Nigel Press Associates Ltd (NPA)
Crockham Park, Edenbridge, Kent, TN8 6SR, United Kingdom
Tel: (+44 1732) 86 50 23
Fax: (+44 1732) 86 65 21
e-mail: info@npagroup.com
Web: http://www.npagroup.com
Managing Director: Nigel Press
Managing Director: David Morten
Operations and Technical Director: Richard Chiles
Sales and Marketing Director: Ren Capes
Secretary: Ms Carol Howes
■ Digital and photographic enhancement of satellite imagery. Official spot image, radarsat and space imaging distribution in UK. Production and interpretive facilities.

UPDATED

Paradigm Secure Communications
Astrium Site, Gunnels Wood Road, Stevenage, Hertfordshire, SG1 2AS, United Kingdom
Tel: (+44 1438) 28 22 44
Fax: (+44 1438) 28 25 00
e-mail: contact@paradigmsecure.com
Web: http://www.paradigmsecure.com
Managing Director: Malcolm Peto
Business Development Director: Paul Millington
Tel: (+44 1438) 28 22 40
Marketing Development Manager: Pete Singleton
Tel: (+44 1438) 77 89 73
Personal Assistant to the Business Development Director: Mrs Helen Connelly

UPDATED

Paradigm Services
Astrium Site, Gunnels Wood Road, Stevenage, Hertfordshire, SG1 2AS, United Kingdom
Tel: (+44 1438) 28 22 44
Fax: (+44 1438) 28 25 00
e-mail: contact@paradigmservices.com
Web: http://www.paradigmservices.com
Managing Director: Peter Bruton
Business Development Director: Ken Hadfield
Services Management and Marketing Officer: Ms Emily Milsom
Tel: (+44 1438) 28 20 66

UPDATED

Plextek Ltd
London Road, Great Chesterford, Essex, CB10 1NY, United Kingdom
Tel: (+44 1799) 53 32 00
Fax: (+44 1799) 53 32 01
e-mail: defence@plextek.co.uk
Web: http://www.plextek.com/defence.htm
Director of Sales and Marketing: David Taylor
Mobile: (+44) 78 01 39 87 13
Marketing Co-ordinator: Ms Johanna Hughes
Senior Project Consultant: Mark Radford
Sales and Marketing Engineer: Robin Ireland
■ Consultancy service for the design and development of communications/electronics technology products and systems. Product design capability includes RF chip design on Si and GaAs, radio and synthesiser design, digital signal processing (DSP) using microprocessors, DSPs and FPGAs, electro-optic design, embedded and windows application software. Projects include: military airbourne IFF transponder design, VHF and UHF military transmitter/receiver design, microwave and milimetric frequency sources for airbourne radar applications, electronics for laser radar, DSP for various radar and multi-channel receiver applications, various RF units for operation under hostile conditions, portable electronic-scan radar, low noise fast switching frequency synthesisers, data links and receivers.

UPDATED

PMD (UK) Limited
Broad Lane, Coventry, CV5 7AY, United Kingdom
Tel: (+44 2476) 46 66 91
Fax: (+44 2476) 47 30 34
e-mail: sales@pmdgroup.co.uk
Web: http://www.pmdgroup.co.uk
Managing Director: A M T Naylor
Operations Director: Aynsley S Thom
Personal Assistant to Managing Director: Ms Georgina Butler
■ Gold plating service for satellites and aerospace applications.

UPDATED

QinetiQ
(Company Headquarters)
Cody Technology Park, Farnborough, Hampshire, GU14 0LX, United Kingdom
Tel: (+44 8700) 10 09 42
Fax: (+44 1252) 39 33 99
Web: http://www.qinetiq.com
Chief Executive: Sir John A R Chisholm
Chief Financial Officer: Graham Love
Managing Director, Through-Life Services Division: Dr Mike Goodfellow
Managing Director, Future Systems Technology Division: Dr Derek Barnes
Managing Director, Sensors and Electronics Division: Simon Bennett
Managing Director, Knowledge and Information Systems Division: Andrew Sleigh
Commercial Director: Hal Kruth
Managing Director, Defence: Graeme Ferrero
Managing Director, MoD Channel: Ms Mary Carver
Managing Director, QinetiQ Inc.: Dr David Anderson
Naval Programmes Director: Paul Wrobel
Marketing Director: Ms Brenda Jones
Technical Director: Adrian Mears
Business Continuity Director: Graham Coley
Human Resources Director: Stephen Luckhurst
Director, Sales, Sensors and Electronics: Jeremy Attree
Press Office, Head of Media & Corporate Affairs: Ms Nicky Louth-Davies
Tel: (+44 1252) 39 28 09
Manager of Media Relations: Douglas Millard
Tel: (+44 1252) 39 46 11
■ Science, technological, research, development and evaluation organisation. Non-nuclear research, technology and test evaluation services of the UK Ministry of Defence. Operational studies and analysis from basic to applied research. Consultancy services for procurement and evaluation requirements during development and operational services. Test facilities include indoor and outdoor weapon ranges, underwater target ranges, marine testing facilites, automotive test tracks and climate testing laboratories. Health care.

UPDATED

Q-Par Angus Ltd
Barons Cross Laboratories, Hereford, HR6 8RS, United Kingdom
Tel: (+44 1568) 61 21 38
(+44 1568) 61 49 13
Fax: (+44 1568) 61 63 73
e-mail: enquires@q-par.co.uk
Web: http://www.q-par.com
Operations, Technical and Managing Director: Dr H R Holliday
Sales and Financial Director: Mrs Jennifer B Holliday
Chief Engineer: Dr D C Hawkins
Purchasing Manager: Miss L Jones
Marketing: Julian Robbins
■ Microwave design and consultancy, including analysis and research, development through to manufacture, test and field trials.

UPDATED

Radarsat International (RSI)
(European Office)
21-22 Old Steyne, Brighton, Surrey, BN1 1EL, United Kingdom
Tel: (+44 1273) 64 83 46
Fax: (+44 1273) 64 83 49

UPDATED

Satcom Distribution Ltd
Unit 3, The Woodford Centre, Salisbury, Wiltshire, SP4 6BU, United Kingdom
Tel: (+44 1722) 41 08 00
Fax: (+44 1722) 41 07 77
e-mail: enquiries@satcomdistribution.com
Web: http://www.satcomdistribution.com
■ Worldwide supplier of satellite communications equipment and airtime.

UPDATED

Science Systems (Space & Defence) Ltd
(ex-Science Systems (Space) Ltd)
Methuen Park, Chippenham, Wiltshire, SN14 0GB, United Kingdom
Tel: (+44 1249) 46 64 66
Fax: (+44 1249) 46 66 61
e-mail: marketing@scisys.co.uk
Web: http://www.scisys.co.uk
Marketing Co-ordinator: Ms Antonia Orchard
■ Spacecraft monitor and control, flight dynamics support, simulation, mission planning, data processing, system studies and consultancy.

UPDATED

SEA (Group) Ltd

(ex-Systems Engineering & Assessment Ltd)
Beckington Castle, PO Box 800, Frome, BA11 6TB, United Kingdom
Tel: (+44 1373) 85 20 00
Fax: (+44 1373) 83 11 33
e-mail: info@sea.co.uk
Web: http://www.sea.co.uk
Chief Executive Officer: Ian Dale-Staples
Managing Director of Space Systems Division: Rob Andrewartha
Managing Director of Marine Systems Division: Phil Lock
Managing Director of Transport System Division: David Guy
Tel: (+44 1373) 85 21 49
Finance Director: Tim Marvell
Battlespace Marketing Manager: Bob O'Meara
■ A systems and software house with wide-ranging system and product development, systems engineering and software capabilities. Clients are blue-chip organisations principally in the marine, space, battlespace and transport domains, where the objective is to apply advanced technology to the solution of customers' most challenging problems. Assessed and approved to undertake design and development systems, feasibility studies and system specification, signal and image processing systems development, management systems and services, technical publications and manufacturing. Active in four broad domains: space, battlespace, marine and transport.
Accreditations: ISO 9001:2000 and ISO 9000-3 (TickIT)

UPDATED

SEA (Group) Ltd
(Bristol Office)
(ex-Systems Engineering & Assessment Ltd)
SEA House, PO Box 800, Fishponds, Bristol, BS16 1SU, United Kingdom
Tel: (+44 1373) 85 20 00
Fax: (+44 1373) 83 11 33
Chairman and Managing Director: Ian Dale-Staples

UPDATED

Sira Technology Ltd (SIRA)

(ex-Sira Electro-Optics Ltd)
South Hill, Chislehurst, Kent, BR7 5EH, United Kingdom
Tel: (+44 20) 84 67 26 36
Fax: (+44 20) 84 68 17 71
e-mail: marketing@siraeo.co.uk
info@siraeo.co.uk
Web: http://www.siraeo.co.uk
Managing Director: Steve Pickering
Manager, Business, Defence and Security: G M Bryan
Contact: Mrs Linda Lynn
■ Independent research and development organisation specialising in the design and custom manufacture of electro-optical systems. Reflectometers, thermal imager test equipment, high-speed IR spectro radiometers, helmet-mounted display test equipment, laser scanners. Special purpose CCD cameras, scanning sub-systems for laser radars, robust sensors, special purpose microscopes and telescopes. Satellite payloads including hyperspectral imagers, startrackers and optical intersatellite data link terminals.

UPDATED

Spur Electron Ltd

Hayward House, Havant, Hampshire, PO9 2NL, United Kingdom
Tel: (+44 23) 92 45 55 64
Fax: (+44 23) 92 47 08 74
Web: http://www.spurelectron.com
Financial Director: Ms Sadie Snowdon
Operations Director: John Moth
Laboratory Services: Ms Cathy Barnes
Tel: (+44 23) 92 45 72 26
Purchasing Manager: Ms Lorraine Harvey
■ Procurement of EEE parts, component engineering and laboratory services. Manufacturing to ESA/NASA and IPC standards, cleanroom assembly facility. Software development and technical publishing.

UPDATED

Surrey Satellite Technology Ltd (SSTL)

Surrey Space Centre, Guildford, Surrey, GU2 7XH, United Kingdom
Tel: (+44 1483) 68 92 78
Fax: (+44 1483) 68 95 03
e-mail: sstl@sstl.co.uk
Web: http://www.sstl.co.uk
Chief Executive Officer: Prof Sir Martin Sweeting
Public Relations Manager: Ms Audrey Nice
Ground Station Manager: Chris Jackson
■ Microsatellite and nanosatellite spacecraft and subsystem engineering and consultancy.

UPDATED

Timestep Weather Systems Ltd

(Timestep Electronics Ltd)
PO Box 2001, Dartmouth, Devon, TQ6 9QN, United Kingdom
Tel: (+44 1803) 83 33 66
e-mail: information@time-step.com
Web: http://www.time-step.com
Managing Director: David J Cawley
Company Secretary: Ms Jill Cawley

UPDATED

Vega Group plc

(Head Office)
2 Falcon Way, Shire Park, Welwyn Garden City, Hertfordshire, AL7 1TW, United Kingdom
Tel: (+44 1707) 39 19 99
Fax: (+44 1707) 39 39 09
e-mail: info@vega.co.uk
defence@vega.co.uk
Web: http://www.vega.co.uk
Chairman: Andy Roberts
Group Managing Director and Chief Executive: Phil Cartmell
Operations Director: Alan Gaby
Director, Business Development Space: Dave Whittle
Director, Human Resources: Martin Blomley
Director, Marketing: Peter Young
Director, Business Development Defence: Trevor Filtness
Director, Government: Paul Gibson
Director, Defence Business Development, Sea: Andre Knight
Director, Defence Business Development, Land Systems: Andy Allman
Defence Business Development, Land Consulting: Gary Smith
Marketing Manager, Defence: Mark Gunning
Tel: (+44 1707) 39 19 99
■ Consulting and technology for the implementation of international business and government organisation. Integrated training solutions for land, air and maritime acquisitions. Support for the UK Ministry of Defence in the acquisition and security of command, control, communication, computers and information systems.

UPDATED

Vega Group plc
(Bristol Office)
360 Bristol Business Park, Bristol BS16 1EJ, United Kingdom
Tel: (+44 117) 988 00 33
Fax: (+44 117) 988 00 34
General Manager: Mark Kirwan
■ Consulting and training systems.

UPDATED

Vega Group plc
(Training Systems Group)
2 Falcon Way, Shire Park, Welwyn Garden City, Hertfordshire, AL7 1TW, United Kingdom
Tel: (+44 1707) 39 19 99
Fax: (+44 1707) 39 39 09
e-mail: defence@vega.co.uk
Web: http://www.vega.co.uk
Chief Executive: Phil Cartmell
Chief Operating Officer: Alan Gaby
Business Development Director, Defence: Trevor Filtness
Marketing Director: Peter Young
Media Relations, Defence: Mark Gunning
■ Consulting and technology company that implements strategy for international businesses and government organisations. Integrated training in the defence market. Aquisition and security of command, control, communication, computers and information systems.

UPDATED

Industry – Sales Agent/Office

Astronautics Corporation of America

(European Operations)
28 Tekels Avenue, Camberley, Surrey GU15 2LB, United Kingdom
Tel: (+44 1276) 67 73 65
Fax: (+44 1276) 67 73 66
General Manager: David J Chalklin
■ Marketing office.
Representing:
Astronautics Corporation of America, US

UPDATED

Codan (UK) Ltd

Gostrey House, Union Road, Farnham, Surrey, GU9 7PT, United Kingdom
Tel: (+44 1252) 71 72 72
Fax: (+44 1252) 71 73 37
e-mail: uksales@codan.com.au
Web: http://www.codan.com.au
Managing Director: Mike Heard
General Manager: John Chaplain
■ Communications equipment: HF radio, microwave systems and satellite communications equipment.

UPDATED

Explorocean Technology Ltd (ETL)

Unit 1A/B, Thorpe Industrial Estate, Crabtree Road, Thorpe, Surrey, TW20 8RW, United Kingdom
Tel: (+44 1784) 47 21 30
Fax: (+44 1784) 47 30 32
e-mail: sales@explorocean.com
Web: http://www.explorocean.com
Director: Derek Noble
Purchasing Manager: Miss Jackie Holland
■ Oceanographic, hydrographic, geophysical and meteorological equipment sale and hire.

UPDATED

Leach International UK Ltd

(ex-LRE Relays & Electronics Ltd)
West Links, Chandler's Ford, Hampshire, SO53 3TG, United Kingdom
Tel: (+44 1962) 73 44 33
Fax: (+44 1962) 73 46 85
General Manager: Tim Whelan
■ Components, modules and equipment, central warning systems (software configurable), land light control boxes and lightweight 200 contractors. Supply of thin integrated keyboard systems, advisory and central warning panels, illuminated push button switches, anti-icing controllers and electronics. Supply of relays, contactors, time delay relays, solid state intelligent switching devices, knobs and fasteners.
Representing:
Leach International North America, US

UPDATED

Radyne Comstream Inc.

(European/Middle East/Africa Office)
8 Oriel Court, Alton, GU34 2YT, United Kingdom
Tel: (+44 1420) 54 42 00
e-mail: eurosales@radn.com
Web: http://www.radn.com
Representing:
Radyne Comstream Inc., US

UPDATED

Telonic Instruments Ltd

Toutley Industrial Estate, Wokingham, Berkshire, RG41 1QN, United Kingdom
Tel: (+44 118) 978 69 11
Fax: (+44 118) 979 23 38
e-mail: info@telonic.co.uk
Web: http://www.telonic.co.uk
Managing Director: R V (Bob) Lovell
■ Sale of oscilloscopes, recorders, oscillators, power supplies, RF filters, attenuators, amplifiers and quartz crystal microbalances for satellite applications.
Representing:
American Reliance Inc., US
ET Testsysteme Gmbh, Germany
E.P.S. (Takosago), Japan
Heinz-Gunter Lau GmbH, Germany
Kikusui Electronics Corp., Japan
QCM Research Inc., US
Telonic Berkeley Inc., US

UPDATED

United States

Government – Ministry/Agency

Defense Advanced Research Projects Agency (DARPA)

3701 North Fairfax Drive, Arlington, Virginia, 22203-1714, United States
Web: http://www.darpa.mil
Public Information Officer: Ms Jan Walker
Tel: (+1 703) 696 24 04

UPDATED

Department of Energy (DoE)

(Headquarters)
1000 Independence Avenue SW, Washington, District of Columbia, 20585, United States
Tel: (+1 202) 586 50 00
Freephone: (+1 800) DIAL DOE
Fax: (+1 202) 586 44 06
(+1 202) 586 84 03
Web: http://www.energy.gov
Secretary of Energy: Spencer Abraham

UPDATED

Japan Aerospace Exploration Agency (JAXA)

(Washington Office)
(ex-National Space Development Exploration Agency of Japan)
2020 K Street NW, Suite 325, Washington, District of Columbia, 20006, United States
Tel: (+1 202) 333 68 44
Fax: (+1 202) 333 68 45
Director: Masato Koyama

UPDATED

Japan Aerospace Exploration Agency (JAXA)
(Houston Office)
(ex-National Space Development Agency of Japan)
Cyberonics Building 16511, Space Center Boulevard, Suite 201, Houston, Texas, 77058, United States
Tel: (+1 281) 280 02 22
Fax: (+1 281) 486 10 24
(+1 281) 228 04 89

UPDATED

National Space Development Agency of Japan (JAXA-KSC)
(KSC Office)
O&C Building, Room 1014, Code: NASDA-KSD, John F Kennedy Space Center, Florida, 32899, United States
Tel: (+1 321) 867 38 79
(+1 321) 867 32 95
Fax: (+1 321) 452 96 62

UPDATED

National Aeronautics & Space Administration (NASA)

(Headquarters)
300 E Street SW, Washington, District of Columbia, 20546-0001, United States
Tel: (+1 202) 358 00 00
Fax: (+1 202) 358 32 51
e-mail: info-center@hq.nasa.gov

UPDATED

National Aeronautics & Space Administration (NASA)
(Flight Simulation Laboratories)
Ames Research Center, Mail Stop 243-1, Moffett Field, California, 94035-1000, United States
Tel: (+1 650) 604 32 71
Fax: (+1 650) 604 39 52
e-mail: sims@artemis.arc.nasa.gov
Web: http://www.simlabs.arc.nasa.gov
Chief, Aerospace Simulation Operations: Barry Sullivan
Tel: (+1 650) 604 67 56
Chief, Simulation Planning Office: Thomas 'Tom' Alderete
NASA Ames Public Affairs: Mike Mewhinney
Tel: (+1 650) 604 39 37
Future Flight Central Deputy Manager: Ms Debbi Ballinger
Tel: (+1 650) 604 06 34
Contact: Ms Kathleen Starmer
Contact: Ms Julie Mikula

UPDATED

National Aeronautics & Space Administration (NASA)
(Dryden Flight Research Center)
PO Box 273, Edwards, California, 93523-0273, United States
Tel: (+1 805) 258 34 49
Web: http://www.dfrc.nasa.gov
Public Affairs: Alan Brown

UPDATED

National Aeronautics & Space Administration (NASA)
(Goddard Space Flight Center)
Greenbelt, Maryland, 20771, United States
Tel: (+1 301) 286 20 00
Web: http://www.gsfc.nasa.gov
News Chief: Edward Campion
Head of Public Affairs: Mark Hess

- A national resource for the pursuit of exploration of the Earth and space. Provides customer centered leadership to implement the goals of NASA and its six enterprises with principal responsibilities in support of the Earth Science Enterprise and the Space Science Enterprise and secondary roles in support of the other four NASA enterprises.

UPDATED

National Aeronautics & Space Administration (NASA)
(John C Stennis Space Center)
NASA Public Affairs, PA00, Mississippi, 39529-6000, United States
Tel: (+1 228) 688 18 80
Web: http://www.ssc.nasa.gov

UPDATED

National Aeronautics & Space Administration (NASA)
(John F Kennedy Space Center)
Kennedy Space Center, Florida, 32899, United States
Tel: (+1 321) 867 44 44
Web: http://www.ksc.nasa.gov

UPDATED

National Aeronautics & Space Administration (NASA)
(Johnson Space Center)
Automation, Robotics and Simulation Division, Mail Code ER, Houston, Texas, 77058, United States
Tel: (+1 281) 483 15 55
Web: http://vesuvius.jsc.nasa.gov/er_er/html/robonaut/ robonaut.html

UPDATED

National Aeronautics & Space Administration (NASA)
(Langley Research Center)
100 NASA Road, Hampton, Virginia, 23681-0001, United States
Tel: (+1 757) 854 85 54
Web: http://www.larc.nasa.gov
Head, Public Services Office: Ms Karen R Credeur

- Research in atmospheric flight supports programs in earth and space science, space access and exploration, aerospace vehicle systems technology, airspace systems, and aviation safety and security. Specialising in aerodynamics, aerothermodynamics, acoustics, aerospace systems concepts and analysis, airborne systems, atmospheric sciences, structures and materials, and systems engineering.

UPDATED

National Aeronautics & Space Administration (NASA)
(Marshall Space Flight Center)
Huntsville, Alabama, 35812, United States
Web: http://www.msfc.nasa.gov
Contact: Ms Michelle Schneider
Contact: Ms Jody L Minor
Tel: (+1 256) 544 40 41
Fax: (+1 256) 544 88 07

UPDATED

National Aeronautics & Space Administration (NASA)
(Public Affairs Office)
(ex-National Space Technology (NSTL))
Stennis Space Center, St Louis, Mississippi, 39529, United States
Centre Director: Roy Estess
Contact: Paul Foerman

UPDATED

National Imagery and Mapping Agency (NIMA)

4600 Sangamore Road, Bethesda, Maryland, 20816-5003, United States
Tel: (+1 301) 227 20 57
Fax: (+1 301) 227 39 20
e-mail: queries@nima.mil
Web: http://www.nima.mil

- A combat support agency of the Department of Defense and a member of the intelligence community. Provides imagery, imagery intelligence and geospatial information to support US military forces and civilians.

UPDATED

United States International Trade Commission (USITC)

(US Int'l Trade Commission)
500 East Street, South West, Washington, District of Columbia, 20436, United States
Tel: (+1 202) 205 20 00
Fax: (+1 202) 205 27 98
Web: http://www.usitc.gov
Chairman: Ms Deanna Tanner Okun
Tel: (+1 202) 708 54 82
International Trade Analyst, Aerospace: Peder A Andersen
Tel: (+1 202) 205 33 88
Air Transportation Analyst: Ms Joanne Tortorice
Tel: (+1 202) 205 30 32
Public Information Officer: Ms Margaret O'Laughlin
Tel: (+1 202) 205 18 19

- Independent, nonpartisan and quasi-judicial federal agency providing trade expertise to the legislative and the executive branches of government. Also determines the impact of imports on US industries and directs actions against unfair trade practices such as patent, trademark and copyright infringement.

UPDATED

Defence Force – Army

US Army Space & Missile Defense Command (USASMDC)

(Public Affairs Office)
PO Box 15280, Arlington, Virginia, 22215-0280, United States
Tel: (+1 703) 607 20 39
(+1 703) 607 24 45
(+1 703) 607 18 74
Fax: (+1 256) 955 12 14
Web: http://www.smdc.army.mil
Commanding General: Maj Gen P Larry Dodgen
Tel: (+1 703) 607 18 73
Deputy Commanding General: Maj Gen John M Urias
Tel: (+1 256) 955 35 01
Deputy Commanding General-SMDC/ARSTRAT: Brig Gen Robert P Lennox
Tel: (+1 719) 554 19 01
Chief of Staff: Col David G Farrisee
Tel: (+1 703) 607 18 95
Deputy Chief of Staff, Engineer: Col J C Smith
Tel: (+1 256) 955 12 88
Deputy Chief of Staff, G-6 (IMO): Robert G Connell
Tel: (+1 256) 955 35 90
Deputy Chief of Staff, G-8: Col Mark Glynn
Tel: (+1 256) 955 31 00
Deputy Chief of Staff: Lieut Col Craig N Robinson
Deputy Chief of Staff, G-4: Lieut Col Robert B Steward
Tel: (+1 256) 955 44 12
Deputy Chief of Staff for Operations and Plans, G-3: Col Patrick Rayermann
Tel: (+1 703) 607 19 55
Deputy Chief of Staff, G-2: Col Antoinette Smart
Tel: (+1 703) 607 20 53
Deputy Chief of Staff, G-1: Lt Col Edward Mason
Command Sergeant Major, SMDC: Cmmd Sgt Maj David L Lady
Tel: (+1 703) 607 19 57
Commander US Army Kwajalein Atoll: Col Jerry Brown
Tel: (+1 805) 355 14 01
Inspector General: Col David J Thomas
Tel: (+1 703) 602 27 79
Chief Scientist: Dr Henry Dubin
Tel: (+1 703) 607 18 76
Space and Missile Defense Battle Laboratory: Laurence H Burger
Tel: (+1 205) 955 30 70
High Energy Laser System Test Facility Directorate Director: Tom Hodge
Tel: (+1 505) 679 56 05

Missile Defense and Technology Centre: Jess Granone
Tel: (+1 256) 955 35 20
Principle Assistant for Contracting: Mark J Lumer
Tel: (+1 256) 955 34 10
Public Affairs Officer (Located in Huntsville): William M Congo
Tel: (+1 256) 955 21 71
Protocol Officer: Ms Joy Ganus
Tel: (+1 256) 955 35 13
Public Affairs Specialist (Located in Washington): Cali Coulthard
Tel: (+1 703) 607 24 45
Public Affairs Specialist (Located in Colorado Springs, CO): Mike Howard
Tel: (+1 719) 554 19 83
Executive Officer: Lt Col Mike Zarbo
Tel: (+1 256) 955 35 02
Command Historian: James A Walker
Tel: (+1 205) 955 47 78

UPDATED

US Army Space & Missile Defense Command (USASMDC)

(Public Affairs Office)
PO Box 1500, Huntsville, Alabama, 35807-3801, United States
Tel: (+1 256) 955 34 12
(+1 256) 955 35 01
Fax: (+1 256) 955 10 56
Contact: Marco A Morales

UPDATED

Defence Force – Navy

Naval Research Laboratory (NRL)

4555 Overlook Avenue SW, Washington, District of Columbia, 20375, United States
Tel: (+1 202) 767 25 41
Fax: (+1 202) 767 69 91
e-mail: nrl1030@ccf.nrl.navy.mil
Web: http://www.nrl.navy.mil
Commanding Officer: Capt David Schubert
Inspector General: Capt John Horsman
Director of Research: Dr John Montgomery
Associate Director of Research (Acting): D Therning
Associate Director Research (Systems Directorate): Dr R LeFande
Tel: (+1 202) 767 33 24
Associate Director of Research (Materials Science and Component Technology Directorate): Dr B B Rath
Tel: (+1 202) 767 35 66
Chief Scientist and Director (Fluid Dynamics/Comp) Physics Laboratory: Dr Jay P Boris
Tel: (+1 202) 767 30 55
Director Naval Center for Space Technology: P G Wilhelm
Tel: (+1 202) 767 65 47
Associate Director of Research Ocean and Technology Directorate: Dr E O Hartwig
Tel: (+1 202) 404 86 90
Chesapeake Bay: Ms T Downing
Financial Management: Steven Birk
Contracting: J C Ely
Tel: (+1 202) 767 52 27
Supply Officer: Ms C Hartman
Tel: (+1 202) 767 34 46
Technical Information (acting): Jim Lucas
Research and Development Support: S Harrison
Tel: (+1 202) 767 33 71
Human Resources: Ms B Duffield
Tel: (+1 202) 767 34 21
Superintendent Radar (Acting): P K Hughes
Superintendent Information Technology (Acting): Dr J D McLean
Superintendent Optical Sciences: Dr T G Giallorenzi
Tel: (+1 202) 767 31 71
Tactical Electronic Warfare (Acting): Dr F Klemm
Superintendent Chemistry: Dr J S Murday
Tel: (+1 202) 767 30 26
Superintendent Material Science and Technology: Dr D U Gubser
Tel: (+1 202) 767 29 26
Superintendent Plasma Physics: Dr S Ossakow
Tel: (+1 202) 767 27 23
Superintendent Electronics Science and Technology: Dr G M Borsuk
Tel: (+1 202) 767 35 25
Superintendent Space Systems Development: R E Eisenhauer
Tel: (+1 202) 767 04 10
Superintendent Spacecraft Engineering (Acting): H E Senasack
Tel: (+1 202) 767 64 07
Superintendent Acoustics: Dr E R Franchi
Tel: (+1 202) 767 34 82
Superintendent Remote Sensing: Dr P Schwartz
Tel: (+1 202) 767 34 21
Superintendent Marine Geosciences Division (Stennis Space Center): Dr H C Eppert
Tel: (+1 601) 688 46 50
Superintendent Marine Meteorological Division (Monterey): Dr P E Merilees
Superintendent Space Science Division: Dr H Gursky
Tel: (+1 202) 767 63 43
Public Affairs Specialist: Ms Donna McKinney
Tel: (+1 202) 767 25 41
Public Affairs Officer: Richard Thompson

UPDATED

Naval Space Command HQ

5280 Fourth Street, Dahlgren, Virginia, 22448-5300, United States
Tel: (+1 540) 653 61 11
(+1 540) 653 61 00
Fax: (+1 540) 653 61 08
Web: http://www.navspace.navy.mil/leadership/biotitle.htm
Commander: Rear Adm John P Cryer
Deputy Commander: Col Raymond Adamiec
Chief of Staff: Cdr Cheryl Blake
Procurement and Logistics Director: Cdr Keith Chapman
Operations Director: Capt Mary McLendon
Technical Director: Pieter J Traas
Command Master Chief: CMDMC(SS) Alan P Steiner
Public Affairs Officer: Gary R Wagner

UPDATED

Defence Force – Air Force

Air Force Research Laboratory (AFRL/PA)

(Public Affairs Office)
1864 4th Street, Suite 1, Ohio, 45433-7132, United States
e-mail: afrl.pa.dl.all@wpafb.af.mil
Contact: Lieut J Elaine Hunnicut

UPDATED

Air Force Space Command (AFSPC)
(Headquarters)
150 Vandenberg Suite 1105, Peterson AFB, Colorado Springs, Colorado, 80914-4500, United States
Tel: (+1 719) 554 37 31
Fax: (+1 719) 554 60 13
e-mail: afspc/pa@peterson.af.mil
Web: http://www.peterson.af.mil/hgafspc
Commander: Gen Lance W Lord
Vice Commander: Lieut Gen Daniel Leaf
Operations Director: Brig Gen Douglas M Fraser
Mission Support Team Chief: Brig Gen Robert M Worley
Logistics Director: Brig Gen Richard E Webber
Plans: Brig Gen Roosevelt Mercer
Requirements: Brig Gen John T 'Tom' Sheridan
Chief of Staff: Col Christian Daehnick
Comptroller: Col Oswaldo Y Mullins
Public Affairs Director: Col David K Cannon
Civil Engineer: Col Marvin Fisher
Personnel: Col Jack F Anthony
Chief of Media Relations: Lieut Col Andy Roake
Tel: (+1 719) 554 38 42
Media Officer: Mike Kucharek
■ The primary missions of Air Force Space Command are: space and intercontinental ballistic missile operations, space warfare centre, including ballistic missile warning, space control spacelift and satellite operations. To integrate space capabilities and techniques into daily AF operations.
Major units include
Fourteenth Air Force (14AF)
-21st Space Wing – Ballistic Missile Warning Space Surveillance, Peterson AFB
-30th Space Wing – DoD, Civil and Commercial Spacelift, Vandenberg AFB
-45th Space Wing – DoD, Civil and Commercial Spacelift, Patrick AFB and Cape Canaveral AS
-50th Space Wing – Satellite Operations, Schriever AFB
-460th Air Base Wing – Missile Warning, Buckley AFB
Twentieth Air Force (20AF)
-90th Space Wing – Peacekeeper and Minuteman III, FE Warren AFB
-91st Space Wing – Minuteman III, Minot AFB
-341st Space Wing – Minuteman III, Malmstrom AFB
-721st Space Group – Mission Support Cheyenne Mountain AS

UPDATED

Air Force Space Command (AFSPC)
(North Dakota Office)
Grand Forks AFB, North Dakota, United States
Tel: (+1 701) 747 30 00
Web: http://www.grandforks.af.mil
■ Air Combat Command Base.

UPDATED

Air Force Space Command (AFSPC)
(Public Affairs Office)
150 Vandenberg Street, Suite 1105, Peterson, Force Base, Colorado, 80914-4500, United States
Chief of Staff: Col Christian Daehnick
Director of Logistics: Brig Gen Richard E Webber
Chief of Media Relations: Lieut Col Andy Roake
Deputy Chief, Media Relations: Mike Kucharek

UPDATED

Fourteenth Air Force (14AF)
747 Nebraska Avenue, Vandenberg AFB, California, 93437-6268, United States
Tel: (+1 805) 606 38 88
Fax: (+1 805) 605 88 56
Web: http://www.vafb.af.mil
Commander: Maj Gen Michael Hamel
■ Operators of the Air Force Space Command's space forces.

UPDATED

Air Force Space Command (AFSPC)
(21st Space Wing)
Peterson AFB, Colorado Springs, Colorado, 80914-1294, United States
Tel: (+1 719) 556 78 25
Fax: (+1 719) 556 78 48
Web: http://www.spacecom.af.mil/21sw
Commander: Brig Gen Duane W Deal
Command & Control unit at
Cheyenne Mountain AS
Space warning units at
Beale AFB
Cape Cod AS
Cavalier AS
Clear ASN
Thule AB (Greenland)
Woomera AS (Australia)
Surveillance units at
Diego Garcia
Eglin AFB
Maui
Misawa AB (Japan)
Peterson AFB
RAF Feltwell (UK)

UPDATED

Air Force Space Command (AFSPC)
(30th Space Wing)
747 Nebraska Avenue, Vandenberg AFB, California, 93437-6267, United States
Tel: (+1 805) 606 35 95
(+1 805) 606 58 14
(+1 805) 606 58 15
Fax: (+1 805) 606 83 03
Web: http://www.vandenberg.af.mil
Commander: Col Frank Gallegos
Commander, 30th Operations Group: Col Stephen Tanous
Commander, 2nd SLS: Lieut Col Clint Crosier
Director, Public Affairs: Maj Stacee Bako
■ The unit's mission is to spacelift satellites to polar orbit, operate and maintain the western range, support ICBM test launches, provide host base support and support commercial space launch.
Launch unit at
Vandenberg AFB
Tracking stations at
Kaena Point
Pillar Point

UPDATED

Air Force Space Command (AFSPC)
(45th Space Wing)
Patrick AFB, Florida 32925-6655, United States
Tel: (+1 407) 494 59 33
(+1 407) 494 11 10

Fax: (+1 407) 494 73 02
Web: http://www.patrick.af.mil
Commander: Brig Gen John G Pavlovich
Commander, 45th Range Squadron: Lieut Col Cynthia Grey
Commander, 45th Weather Squadron: Col Neil Wyse
Launch unit at
Cape Canaveral AS
Tracking stations at
Antigua AS
Ascension Aux Field (St Helena)
Cape Canaveral AS
Patrick AFB

UPDATED

Air Force Space Command (AFSPC)
(50th Space Wing)
Schriever AFB, Colorado, 80912-5000, United States
Tel: (+1 719) 567 50 40
Fax: (+1 719) 567 53 06
Web: http://www.schriever.af.mil
Commander: Col Suzanne M Vautrinot
Commander, 1 SOPS: Lieut Col Steven Lootens
Commander, 2 SOPS: Lieut Col Scott Henderson
Commander, 3 SOPS: Lieut Col Anthony Hinson
Commander, 4 SOPS: Lieut Col Ronald Huntley
Commander, 30th Space Group: Col Roscoe Griffin
Vice Commander: Col Mike Selva
Satellite operations units at
Cape Canaveral AS
Onizuka AS
Schriever AFB
Satellite remote tracking stations at
Andersen AFB
Diego Garcia
Kaena Point
New Boston AS
RAF Oakhanger (UK)
Schriever AFB
Thule AB (Greenland)
Vandenberg AFB

UPDATED

Air Force Space Command (AFSPC)
(21st Space Operations Squadron)
(ex-Air Force Satellite Test Center)
1080 Lockheed Martin Way,
Box 129, Sunnyvale, California, 94089-1235, United States
Tel: (+1 408) 752 40 35
Fax: (+1 408) 751 60 82
Web: http://www.schriever.af.mil/fact_sheets/21sops
http://www.onizuka.af.mil
Commander: Lieut Col Sandra Finan
Public Affairs Officer: Ms Valerie Joseph

UPDATED

Air Force Space Command (AFSPC)
(90th Space Wing)
5305 Randall Avenue, FE Warren Air Force Base, Wyoming, 82005-2266, United States
Tel: (+1 307) 773 20 74
Fax: (+1 307) 773 33 81
Web: http://www.warren.af.mil
Wing Commander: Col Thomas D Shearer
Vice Commander: Col Barry Kistler
Logistics Group Commander: Col Jack Weinstein
Support Group Commander: Col Craig Whitehead
Command Chief Master Sergeant: Chief Master Sgt Vincent Dicks
Chief, Public Affairs: Capt Stacy Vaughn
Tel: (+1 307) 773 33 83

UPDATED

Air Force Space Command (AFSPC)
(91st Space Wing)
201 Summit Drive, Minot Air Force Base, 58705-2000, United States
Tel: (+1 701) 723 62 12
Fax: (+1 701) 723 65 34
Web: http://www.minot.af.mil

UPDATED

Air Force Space Command (AFSPC)
(341st Space Wing)
2177th Street North, Room 150A, Malmstrom Air Force Base, 59402-7538, United States
Tel: (+1 406) 731 40 50
Fax: (+1 406) 731 40 48
e-mail: 3415wpa@nalmstrom.af.mil
Web: http://www.malmstrom.af.mil
Deputy Chief, Public Affairs: Shannon Nyberg

UPDATED

Air Force Space Command (AFSPC)
(721st Support Group)
Cheyenne Mountain Air Force Station, Colorado, United States
Tel: (+1 719) 554 73 21

UPDATED

Space Warfare Center (SWC)
730 Irwin Avenue Suite 83, Schriever Air Force Base, 80912, United States
Tel: (+1 719) 554 37 31
Fax: (+1 719) 554 83 47
Web: http://www.schriever.af.mil/swc

UPDATED

Department of the Air Force
(Public Affairs Resource Library)
1690 Air Force Pentagon, Washington, District of Columbia, 20330-1690, United States
Web: http://www.af.mil

UPDATED

Department of the Air Force
(Space & Missile Systems Center)
2420 Vela Way, El Segundo, Los Angeles, California, 90245, United States
Tel: (+1 310) 363 00 30
Fax: (+1 310) 363 25 49
Web: http://www.losangeles.af.mil/SMC/ds

UPDATED

US Air Force Materiel Command (AFMC)
(Headquarters)
Public Affairs Office, 4375 Childlaw Road, Building 262, Room N-152, Wright-Patterson AFB, Ohio, 45433-5001, United States
Tel: (+1 937) 257 76 20
e-mail: webmaster@wpafb.af.mil
Web: http://www.afmc-mil.wpafb.af.mil

- Management of research development, test, acquisition and support for Air Force systems.
- Organisationally this branch of the US Air Force is the Directed Energy Directorate at Kirtland Air Force Base in New Mexico. The branch is one of nine directorates of the Air Force Base research laboratory whose headquarters are at Wright-Patterson Air Force Base in Ohio. The Laboratory is under Air Force Materiel command whose headquarters are also at Wright-Patterson AFB.

Major AFMC installations (A-K)
Arnold AFB: (Arnold Engineering Development Center) Flight simulation test complex.
Brooks AFB: (Armstrong Laboratory and Human Systems Center) Concentrates on human aspects of USAF weapon systems and facilities.
Edwards AFB: (Air Armament Center) Flight test centre for all USAF and NASA aircraft.
Eglin AFB: (Air Force Developement Test Center) Tests and evaluates munitions (non-nuclear), electronic combat systems and navigation/guidance systems.
Griffiss AFB: (Rome Laboratory) Research and development for CSI.
Hanscom AFB: (Electronic Systems Center) Development and acquisition of C3I systems.
Hill AFB: (Ogden Air Logistics Center) Maintenance for ICBM, F-16, Maverick, GBU-15 and laser-guided bombs.
Kelly AFB: (San Antonio Air Logistics Center) Maintenance for C-17, T-37 and T-38 aircraft.
Kirtland AFB: (Phillips Laboratory) Research and development for space and missile related technology.
Major AFMC installations (L-Z)
Los Angeles AFB: (Space and Missile Systems Center) Development and acquisition of launch and space systems for military communications, navigation, weather and surveillance applications.
McClellan AFB: (Sacramento Air Logistics Center) Maintenance for all A-10, F-117A and F-22 aircraft.
Robins AFB: (Warner-Robins Air Logistics Center) Maintenance for F-15, C-141 and C-130 aircraft.
Tinker AFB: (Oklahoma City Air Logistics Center) Maintenance for B1-B, B-2, B-52, C-5, E-3, E-4 and KC-135 series aircraft.
Wright-Patterson AFB: (Aeronautical Systems Center and Wright Laboratory) Research, development and acquisition of aerospace systems, electronics, materials and manufacturing technologies.

UPDATED

US Air Force Materiel Command (AFMC)
(Directed Energy Directorate)
(ex-Air Force Research Laboratory)
(ex-Phillips Laboratory)
3550 Aberdeen Avenue SE, Kirkland AFB, New Mexico, 87117-5776, United States
Tel: (+1 505) 846 19 11
Fax: (+1 505) 846 04 23
e-mail: afrl.public.affairs@kirtland.af.mil
Web: http://www.de.afrl.af.mil
Managing Director: Dr L Bruce Simpson
Deputy Director: Col Thomas A Buter
Finance Director: Col Andrew L Detrick

- Conducts research in high-energy lasers, high power microwave technology, optics and beam control.

UPDATED

Transport Administration – Space

California Space Authority, Inc. (CSA)
(Corporate Office)
3201 Airpark Drive, Suite 204, Santa Maria, California, 93455, United States
Tel: (+1 805) 349 26 33
(+1 916) 551 15 43
Fax: (+1 805) 349 26 35
(+1 916) 551 15 79
Web: http://www.californiaspaceauthority.org
Executive Director: Ms Andrea Seastrand
Director, Member Services and Events: Ms Elizabeth Burkhead
Financial Director and Chief Business Officer: Pete Guernsey
Chief Information Officer: Jamie Foster
Chief Programme Office: Ms Vicki Conner

UPDATED

Dynspace
11710 Plaza American Drive, Reston, Virginia, 20190-6013, United States
Tel: (+1 703) 261 47 46
Fax: (+1 703) 261 54 20
Web: http://www.dyncorp.com/dynspace
President: Robert Parker

- High technology company specialising in the launching of commercial satellites into low earth orbits.

UPDATED

European Space Agency (ESA)
(Washington Office)
955 L'Enfant Plaza SW, Suite 7800, Washington, District of Columbia, 20024, United States
Tel: (+1 202) 488 41 58
Fax: (+1 202) 488 49 30
Web: http://www.esa.int
Liaison Officer: Frederic Nordlund

UPDATED

Florida Space Authority
(ex-Spaceport Florida Authority)
(ex-State of Florida Spaceport Florida Authority)
100 Spaceport Way, Cape Canaveral, Florida, 32920-4003, United States
Tel: (+1 321) 730 53 01
Fax: (+1 321) 730 53 07
e-mail: news@floridaspaceauthority.com
Web: http://www.floridaspaceauthority.com
Executive Director: Capt Winston Scott
Tel: (+1 321) 730 53 01
Deputy Director: Glenn Vera
Tel: (+1 321) 730 53 01
Director, Planning: E Keith Witt
Tel: (+1 321) 730 53 01
Director, Business Development and International Affairs: Ms Debra Holliday
Tel: (+1 321) 730 53 01
Director, Safety, Security and Environmental Compliance: Kenneth 'Pete' Gunn
Tel: (+1 321) 730 53 01
Manager, Media Affairs: Ms Margo F Witcher
Tel: (+1 321) 730 53 01
Manager, Finance: Ms Denise G Swanson
Tel: (+1 321) 730 53 01
Program Management Analyst: Ms Bonnie Sperry
Tel: (+1 321) 730 53 01
HQ Administrative: Ms Carmen Velazquez
Tel: (+1 321) 730 53 01
Executive Assistant: Ms Jacqueline Martin
Tel: (+1 321) 730 53 01
Programme Administrator: Ms Zorita Young
Tel: (+1 321) 730 53 01

- Provide economic development for space-related businesses.

UPDATED

Florida Space Research Institute (FSRI)

Building M6-306, Room 9030, Kennedy Space Centre, Florida, 32899, United States
Tel: (+1 321) 452 26 53
Fax: (+1 321) 456 99 61
e-mail: marketing@fsri.org
Web: http://www.fsri.org
Executive Director: Dr Samuel T Durrance
Director, Finances: Larry E Hughes

■ Centre for space research, leveraging academic and economic development resources with those of the industry, NASA and the military to expand and diversify Florida's space-related enterprise.

UPDATED

Gulf Coast Regional Spaceport

201 East Myrtle, Suite 139, Angleton, Texas, 77515, United States
Tel: (+1 979) 848 05 60
e-mail: gcspaceport@tgn.net
Web: http://www.gulfcoastspaceport.com

UPDATED

Pecos County/West Texas Spaceport Development Corporation

1000 Railroad Avenue, Fort Stockton, Texas, 79735, United States
Tel: (+1 915) 336 22 64
Fax: (+1 915) 336 61 14
e-mail: fortstocktontx@ftstockton.net

UPDATED

SMC Detachment 12

3548 Aberdeen Avenue SE, Kirtland AFB, New Mexico, 87117-5778, United States
Tel: (+1 505) 846 09 49
Fax: (+1 505) 846 13 49
Web: http://www.te.plk.af.mil
Chief, Commanders Action Group: Lieut David L Drummond

UPDATED

Willacy County Development Corporation for Spaceport Facilities

190 North 3rd Street, Raymondville, Texas, 78580, United States
Tel: (+1 956) 689 33 93
(+1 956) 689 48 17
e-mail: county.judge@willacycounty.org
Web: http://southtexasspaceport.org
Chair: Judge Simon Salinas

UPDATED

30th Space Wing

(Customer Support Office)
806 13th Street, Vandenberg, AFB, California, 93437-5244, United States
Fax: (+1 805) 606 79 79
Contact: Paul Klock
Tel: (+1 805) 605 72 10
Contact: Ms Rosemarie Leventis

UPDATED

Centres of Learning & Research

Center for Space and Planetary Sciences

(Arkansas – Oklahoma Center for Space and Planetary Sciences)
Chemistry Building, University of Arkansas, Fayetteville, Arkansas, 72701, United States
Fax: (+1 479) 575 78 88
e-mail: csaps@uark.edu
Web: http://www.uark.edu/csaps
Web: http://www.uark.edu/misc/csaps
Director: Derek Sears
Tel: (+1 479) 575 52 04
Fax: (+1 479) 575 77 78
Co-Director: Stephen McKeever
Tel: (+1 405) 744 58 02
Fax: (+1 405) 744 68 11

UPDATED

Institute for the Social Science Study of Space (ISSSS)

Fairmont State University, 1201 Locust Avenue, Fairmont, West Virginia, 26554, United States
Tel: (+1 304) 367 46 74
Fax: (+1 304) 367 47 85
Web: http://www.fscww.edu/users/mfulda
President: Dr Michael Fulda
Web: http://www.sm.bmstu.ru/spacepolicy

■ Co-ordinating institution for common interest among professionals in sixteen social science disciplines in space use and exploration.

UPDATED

International Space University (ISU)

(North America Office)
409 3rd Street, Suite 206, Washington, District of Columbia, 20024, United States
Tel: (+1 703) 522 67 31
Fax: (+1 703) 533 94 42
Web: http://www.isunet.edu
Vice President, North America Operations: Ms Margaret G Finarelli

UPDATED

Jet Propulsion Laboratory, NASA (JPL)

4800 Oak Grove Drive, Pasadena, California, 91109, United States
Tel: (+1 818) 354 43 21
Web: http://www.jpl.nasa.gov

■ Operated by the California Institute of Technology for NASA, with planetary exploration as its primary mission. The laboratory also carries out research for the US Department of Defence and other federal agencies. Working on the Mars Pathfinder mission. In addition, JPL manages the worldwide Deep Space Network, which communicates with spacecraft and conducts scientific investigations from its complexes in California's Mojave Desert near Goldstone, near Madrid, Spain and near Canberra, Australia.

UPDATED

Johns Hopkins University

(Applied Physics Laboratory)
11100 Johns Hopkins Road, Laurel, Maryland, 20723-6099, United States
Tel: (+1 301) 953 50 00
Web: http://www.jhuapl.edu
Manager, Satellite Communications Facility: Glen E Baer

UPDATED

Johns Hopkins University
(Chemical Propulsion Information Agency)
10630 Little Patuxent Parkway, Suite 202, Columbia, Maryland, 21044-3204, United States
Tel: (+1 410) 992 73 01
Fax: (+1 410) 730 49 69
Business Manager: Ms Karen L Brown

UPDATED

Los Alamos National Laboratory (LANL)

PO Box 1663, Los Alamos, New Mexico, 87545, United States
Web: http://www.lanl.gov

UPDATED

NASA Institute for Advanced Concepts (NIAC)

75 5th Street, North West, Suite 318, Atlanta, Georgia, 30308, United States
Tel: (+1 404) 347 96 33
Fax: (+1 404) 347 96 38
Web: http://www.niac.usra.edu
Director: Robert A Cassanova
Associate Director: Ms Patricia Russell

UPDATED

National Center for Microgravity Research (NCMR)

Case Western Reserve University, Glennan 312, 10900 Euclid Avenue, Cleveland, Ohio, 44106-7074, United States
Tel: (+1 216) 368 07 48
Fax: (+1 216) 368 07 18
Web: http://www.ncmr.org
Administrative Assistant: Ms Cindy Rosenberger

■ Perform critical-path microgravity research on fluids and combustion, supporting NASA missions, pursue knowledge transfer, outreach and education. Work to enhance the pool of microgravity research scientists, develop concepts leading to experiments on the International Space Station, and assist NASA in developing strategies for future microgravity research efforts.

UPDATED

National Research Council (NRC)

(Associateship Programs)
2101 Constitution Avenue NW, TJ2114, Washington, District of Columbia, 20418, United States
Tel: (+1 202) 334 27 60
Fax: (+1 202) 334 27 59
e-mail: rap@nas.edu
Web: http://www.national-academies.org/rap

UPDATED

New Mexico State University

(Physical Science Laboratory)
1050 Stewart Street, Box 30002, Las Cruces, New Mexico, 88003-8002, United States
Tel: (+1 505) 522 91 00
(+1 505) 522 94 21
(+1 505) 522 91 00
Fax: (+1 505) 522 94 93
Web: http://www.psl.nmsu.edu
Electromagnetics Leader: Dr Bruce Blevins
Financial Director: Brent Howe
Marketing Director: Ms Kathrine Hansen
Technical Director: Vicente Garcia
Operations Director: Scott Witt

■ UAVs, missile and rocket telemetry, information operations research, depleted uranium mitigation studies, electronic warfare and high altitude ballooning (NSBF). Rocket research development and production of missile and satellite antennas. Antenna products and measurement facilities. Environmental and atmosphere sciences and engineering.

UPDATED

Niagara University, Space Settlement Studies Project (3SP)

(Department of Sociology)
Niagara University, New York, 14109, United States
Tel: (+1 716) 286 80 94
Fax: (+1 716) 286 85 81
Director: Dr Stewart B Whitney
Operations Director: Ms Margaret Read
Public Relations Officer: Ms Barbara King

■ Research and study of social and cultural aspects of long-term space habitation.

UPDATED

Orbital Technologies Corporation (ORBITEC)

(Space Center)
1212 Fourier Drive, Madison, Wisconsin, 53717, United States
Tel: (+1 608) 827 50 00
Fax: (+1 608) 827 50 50
Web: http://www.orbitec.com
Vice President and Chief Executive Officer: Thomas M Crabb
Fax: (+1 608) 827 50 50

UPDATED

Penn State College of Engineering

(ex-Pennsylvania State University)
101 Hammond, University Park, Pennsylvania, 16802, United States
Tel: (+1 814) 865 75 37
(+1 814) 865 55 44
Fax: (+1 814) 863 47 49
e-mail: news@engr.psu.edu
Web: http://www.engr.psu.edu/news
Dean: David N Wormley
Public Relations Representative: Curtis Chan
Tel: (+1 814) 863 21 32

UPDATED

Penn State College of Engineering
(Communications & Space Sciences Laboratory)
318 EE East, University Park, Pennsylvania, 16802, United States
Tel: (+1 814) 865 23 54
Fax: (+1 814) 863 84 57
Web: http://www.engr.psu.edu/cssl
Director: John D Mathews

■ Current research interests focus on the upper atmosphere, weather, lightning and atmospheres of other planets and electromagnetic related problems such as antennas, microwaves, plasma diagnostics, scattering and shielding.

UPDATED

Penn State College of Engineering
(Propulsion Research Center)
(ex-NASA Center for Space Propulsion Engineering)
240 Research Building East, University Park, Pennsylvania, 16802, United States

Tel: (+1 814) 863 12 85
Fax: (+1 814) 865 33 89
Web: http://www.psu.edu/dept/perc
Director: Robert J Santoro
■ The research programme emphasises five major areas: chemical propulsion, electric nuclear propulsion; advanced propulsion concepts; diagnostics and material compatibility/reliabilty.

UPDATED

Philips Research
(Headquarters)
(ex-Philips Laboratories)
Briarcliff Manor, New York, NY 10510, United States
Tel: (+1 914) 945 60 00
Fax: (+1 914) 945 63 75
e-mail: info@philabs.research.philips.com
Senior Vice President Philips Research: Dr Barry M Singer
Public Relations Director: Ms Betsy McIlvaine

UPDATED

Research Institute for Advanced Computer Science (RIACS)
NASA Ames Research Center, Mail Stop T35B-1, Moffett Field, California 94035-1000, United States
Tel: (+1 650) 604 54 02
Fax: (+1 650) 604 49 54
e-mail: info@riacs.edu
Web: http://www.riacs.edu
Managing Director: Barry M Leiner
Deputy Director: Serdar Uckun
Programme Administration: Ms Diana Martinez
■ Human centered computing, high performance computing and autonomous reasoning.

UPDATED

Sandia National Laboratories (SNL)
(Sandia Corporation)
PO Box 5800, Albuquerque, New Mexico, 87185, United States
Tel: (+1 505) 845 00 11
(+1 505) 844 34 41
Fax: (+1 505) 844 63 67
e-mail: webmaster@sandia.gov
Web: http://www.sandia.gov
President and Laboratories Director: Dr C Paul Robinson
Executive Vice President and Chief Operating Officer: Dr Joan Woodard
Public Relations and Communications Director: Don Carson
Financial Director: Francisco Figueroa
Technical Director: Dr Alton D Romig
Operations Director: Ms Lynne Jones
Purchasing Manager: Dave Palmer
Community Involvement and Issues Management: Ginni Edmund
Media Relations: Ms Iris Aboytes
■ Operates Sandia National Laboratories on behalf of the US Department of Energy.

UPDATED

Sandia National Laboratories
(California Office)
PO Box 969, Livermore, California, 94551, United States
Tel: (+1 925) 294 20 65
Media Relations: Barry Schrader

UPDATED

SETI Institute
2035 Landings Drive, Mountain View, California, 94043, United States
Tel: (+1 650) 961 66 33
Fax: (+1 650) 961 70 99
Web: http://www.seti.org

UPDATED

The George Washington University
(Space Policy Institute)
Elliott School of International Affairs, 1957 E Street NW, Washington, District of Columbia, 20052, United States
Tel: (+1 202) 994 72 92
Fax: (+1 202) 994 16 39
Web: http://www.gwu.edu/~spi
Director: Prof John M Logsdon

UPDATED

University of Alaska Fairbanks
(Geophysical Institute)
Poker Flat Research Range (PFRR), PO Box 757320, Fairbanks, Alaska, 99775-7320, United States
Tel: (+1 907) 455 21 10
(+1 907) 474 70 15
Fax: (+1 907) 455 21 20
(+1 907) 474 57 05
e-mail: info@gi.alaska.edu
Web: http://www.pfrr.alaska.edu
Geophysical Institute Director: Roger Smith
Tel: (+1 907) 474 72 82
Assistant Director: Douglas H Christensen
Tel: (+1 907) 474 74 26
Range Manager: Gregory Walker
Tel: (+1 907) 455 21 02
Operations/Controller: Mrs Kathe Rich
Tel: (+1 907) 455 21 03
Public Information Officer, Geophysical Institute: Ms Amy Hartley
Tel: (+1 907) 474 77 98
Editorial Assistant: Ms Lynda McGilvary
Launch Officer: Ray Martinez
Tel: (+1 907) 455 21 04
■ University owned sounding rocket launch facility. Facilities to handle complete NASA sounding rocket inventory.

UPDATED

University of Colorado
(Center for Astrophysics and Space Astronomy)
389 UCB, Boulder, Colorado, 80309-0389, United States
Tel: (+1 303) 492 40 50
Fax: (+1 303) 492 71 78
Web: http://casa.colorado.edu
Managing Director: James Green
Administrative Assistant: Ms Susan Barnes
■ An astronomy and astrophysics centre that offers an academic program leading to the PhD degree in a variety of areas of astrophysics and planetary sciences.

UPDATED

University of Houston Clear Lake
2700 Bay Area Boulevard, Houston, Texas, 77058-1098, United States
Tel: (+1 281) 283 31 22
Fax: (+1 281) 283 31 23
Professor of Marketing and Director, Center for Advanced Management Programs School of Business and Public Administration: Sam J. Bruno
Professor of Management and Director, MBA Programme: Robert McGlashan
Tel: (+1 281) 283 31 24
Fax: (+1 281) 283 31 23

UPDATED

Association

Aerospace Industries Association of America, Inc. (AIA)
1000 Wilson Boulevard, Suite 1700, Arlington, Virginia, 22209-3901, United States
Tel: (+1 703) 358 10 00
Fax: (+1 703) 358 10 11
(+1 703) 358 10 12
(+1 703) 358 10 13
e-mail: aia@aia-aerospace.org
Web: http://www.aia-aerospace.org
Secretary-Treasurer: George F Copsey
Tel: (+1 202) 371 84 03
Vice President, Government Division: Terry Marlow
Tel: (+1 202) 371 85 25
Vice President, Supplier Management: William Lewandowski
Tel: (+1 202) 371 84 32
Assistant Vice President, Policy and Planning: Ms Sandra Carney-Talley
Tel: (+1 202) 371 85 03
Director, Communications Council: Alexis Allen
Tel: (+1 202) 371 85 44
Director, Space Policy: Bruce Mahone
Tel: (+1 202) 371 84 62
Director, Environmental, Safety and Health: Robert Peters
■ Trade association representing US manufacturers of commercial, military and business aircraft, helicopters, aircraft engines, missiles, spacecraft and related components and equipment.

Associate member companies
Abelconn, LLC
ADI American Distributors, Inc.
Advanced Technical Products, Inc.
Aeroflex Inc.
Air Industries Machining Corp.
Allen Aircraft Products, Inc.
Ambel Precision Manufacturing Corp.
American Data & Computer Products
AMI Metals, Inc.
Arkwin Industries, Inc.
Arrow Gear Company
Arrow/Zeus Electronics
AUSCO, Inc.
Auto-Valve, Inc.
AV Chem, Inc.
Avexus, Inc.
Avionics Specialities, Inc.
Avnet Electronics Marketing
Banneker Industries, Inc.
Berkshire Industries, Inc.
Bill-Jay Machine Tool Corporation
Brek Manufacturing Company
BTC Electronic Components, Inc.
Burton Industries Aerospace Heat Treating
California Screw Products
Chandler/May, Inc.
Cherokee Nation Distributors
Cincinnati Machine
Circle Seal Controls, Inc.
Cohesia Corporation
Compass Aerospace Corporation
Composites Atlantic Limited
Consolidated-PAC Foundries
CPI Aerostructures, Inc.
Cristek Interconnects, Inc.
Cytec Engineered Materials
Dy 4 Systems, Inc.
Dynabil Industries, Inc.
East West Technology
Eaton Aerospace
Electro-Methods, Inc.
Electronic/Fasteners, Inc.
EMS Technologies, Inc.
ENSCO, Inc.
Ensign-Bickford Aer. & Defense Co.
Envision, LLC
F.A.G. Bearings Limited
Faber Enterprises, Inc.
Fastener & Hose Technology, Inc.
Fortner Aerospace Manufacturing
Frontier Electroni Systems Corp.
G.S. Precision, Inc.
Genmech Aerospace Corporation
GEAR Software
Greene, Tweed & Company
Hagemeyer North America
Hamby Corporation
Hangsterfer's Laboratories, Inc.
Hartwell Corporation
Hi-Tech Aero Spares
Hi-Temp Insulation Inc.
Hobart Machined Products, Inc.
Holaday Circuits, Inc.
H & S Swanson's Tool Company
Hughes Bros. Aircrafters, Inc.
Hughes-Treitler Manufacturing Corp.
Industrial Precision, Inc.
Integrated Aerospace
ISPA, Inc.
JAMCO Aerospace
JCM Engineering Corp.
John Hassall, Inc.
Kennebec Tool & Die Co., Inc.
Kulite Semiconductor Products, Inc.
Lefiell Manufacturing Company
Lewis & Saunders
Lilly Software Associates, Inc.
LMI Aerospace, Inc
M/A-Com, Inc.
Magnetico, Inc.
Manufacturers' Services Control US Operations
Manugistics
Manzi Metals, Inc.
Marotta Controls, Inc.
McCann Aerospace Machining Corp.
Meyer Tool Inc.
Micro-Coax, Inc.
Millitech, L.L.C.
Morris Machine Company, Inc.
MPC Products Corporation
Natel Engineering Co. Inc.
National Machine Group (National Aviation Products, National Machine Company)
Navigant Consulting
Neuvant Aerospace
Norfil Manufacturing
Northwest Composites, Inc.
Omni International
Onboard Sofware, Inc.
Paravant, Inc.
Park Engineering & Manufacturing Co., Inc.
PC Guardian
Perkinelmer Fluid Sciences
Plymouth Extruded Shapes
Precision Machine & Manufacturing Co.

Precision Machine Works Inc.
Precision Plymouth Tube Bending
Precision Tube Bending
Primus International
PRMS Incorporated
Pro Fab
Production Engineering Corp.
PTI Technologies, Inc.
Quick-Wright Associates, Inc.
Radant Technologies, Inc.
Ram Manufacturing Inc.
REMEC Microwave, Inc.
Ryder Systems, Inc.
Safe Flight Instrument Corporation
Sechan Electronics, Inc.
Service Steel Aerospace
Servotronics, Inc.
Sparton Corporation
Spectra Lux Corporation
Spirit Electronics, Inc.
Spring Board technology Corp.
STADCO
Sunshine Metals
Sypris Electronics, LLC
Texas Composites, Inc.
Thayer Aerospace
The Deutsch Company
The Ferco Group
Therm, Inc.
Thermal Solutions, Inc.
Thomas James International
TMX Aerospace
Transtar Metals, Inc.
Tru Circle Aerospace Corp.
Trylon Machine Company
Tyco Electronics
Tyco Printed Circuit Group L.P.
UFC Aerospace Corp.
Uni-Tek, LLC
Vaupell Industrial Plastics
Viking Metallurical (Firth Rixson)
Welding Metallurgy, Inc.
Wems Electronics
West Cobb Eng. & Tool Co. Inc.
Williams International
Windings, Inc.
Xelus

Member companies
AAI Corporation
The Aerostructures Corporation
Alliant Techsystems Inc.
American Pacific Corporation
Analytical Graphics Inc.
Areté Associates
Argo-Tech Corporation
Astro Vision International, Inc.
Atlantic Research Corporation
Aviall, Inc.
BAE Systems, North America
Ball Aerospace & Technologies Corp.
Barnes Aerospace
B.H. Aircraft Company, Inc.
The Boeing Company
Celestica Corporation
Computer Sciences Corporation
Cordiem, LLC
Crane Aerospace
Cubic Corporation
Curtiss-Wright Corporation
Curtiss-Wright Flight Systems, Inc. Metal Improvement Company
Dassault Falcon Jet Corporation
DRS Technologies, Inc.
Ducommun Incorporated
Dupont Company
EDO Corporation
EFW Inc.
Embraer Aircraft Holding Inc.
ESIS, Inc.
Esterline Technologies
Exostar LLC
Federation Inc.
General Atomics Aeronautical Systems, Inc.
General Dynamics Corporation
General Electric Company
GKN Aerospace Services
Goodrich Corporation, Airframe Systems, Electronic Systems, Engine Systems
Groen Brothers Aviation, Inc.
Harris Corporation
HEICO Corporation
Hexcel Corporation
Hitco Carbon Composites, Inc.
Honeywell
ITT Industries Defense and Electronics
Jedco Inc.
Kaman Aerospace Corporation
Kistler Aerospace Corporation
L-3 Communications Holdings, Inc.
Lockheed Martin Corporation
Martin-Baker America Inc.
Matrixone, Inc.
MD Helicopters, Inc.
MOOG Inc.
Northrop Grumman Corporation, Mission Systems, Space Technology
Omega Air, Inc.
Orbital Sciences Corporation, Advanced Systems Division
Parker Aerospace
The Purdy Corporation
Raytheon Company
Remmele Engineering, Inc.
Rockwell Collins, Inc.
Rolls-Royce North America Inc.
RTI International Metals, Inc.
SiliconGraphics, Inc.
Smiths Aerospace Actuation Systems Los Angeles
Spectrum Astro, Inc.
Stellex Aerostructures, Inc.
Swales Aerospace
Teleflex Inc./TFX Sermatech Mal Toll & Engineering
Textron Inc.
Titan Coporation, Advanced Systems Division
Triumph Group, Inc.
United Defense
United Technologies Corporation
(Hamilton Sundstrand, Pratt & Whitney, Sikorsky)
Vought Aircraft Industries, Inc.
W.L. Gore & Associates
Woodward Governor Company

UPDATED

Aerospace Medical Association (ASMA)

320 S Henry Street, Alexandria, Virginia, 22314-3579, United States
Tel: (+1 703) 739 22 40
Fax: (+1 703) 739 96 52
Web: http://www.asma.org
Executive Director: Russell B Rayman
Tel: (+1 703) 739 22 40
Managing Editor: Ms Pamela Day
Tel: (+1 703) 739 22 40

■ Non-profit association of physicians, physiologists, nurses and other professionals specialising in maintaining the health and efficiency of man in atmospheric and space flight.

Publications: Aviation, Space & Environmental Medicine

UPDATED

American Astronautical Society (AAS)

6352 Rolling Mill Place #102, Springfield, Virginia, 22152-2354, United States

■ Scientific and technical society devoted to space activities.

Publications: AAS Directory (every 2 years)
AAS Newsletter (bi-monthly),
Journal of the Astronautical Sciences (quarterly).
Science & Technology series, Advances in the Astronautical Sciences, History series bound books)

UPDATED

American Institute of Aeronautics & Astronautics (AIAA)

1801 Alexander Bell Drive, Suite 500, Reston, Virginia, 20191-4344, United States
Tel: (+1 703) 264 75 00
Freephone: (+1 800) 639 24 22
Fax: (+1 703) 264 75 51
Web: http://www.aiaa.org
President: Alan Mulally
Executive Director: Cort Durocher
Treasurer and Secretary: David Quackenbush
Tel: (+1 703) 264 75 14
Sales and Marketing Director: Howard O'Brien
Tel: (+1 703) 264 75 35

■ Main purpose is to advance the arts, sciences and technology of aeronautics and astronautics. US representative on the International Astronautical Federation and the International Council on the Aeronautical Sciences.

Publications: Aerospace America

UPDATED

Council of Defense & Space Industry Associations (CODSIA)

2111 Wilson Boulevard, Suite 400, Arlington, Virginia, 22201, United States
Tel: (+1 703) 247 94 90
Fax: (+1 703) 243 85 39
e-mail: codsia@ndia.org
Web: http://www.codsia.org

■ Provides a central channel of communications for improving industry-wide consideration of many policies, regulations, implementation problems, procedures and questions involved in federal procurement actions.

Member associations
Aerospace Industries Association
American Electronics Association
American Shipbuilding Association
Contract Services Association
Electronic Industries Alliance
Manufacturers Alliance/MAPI
National Defense Industrial Association
Professional Services Council

UPDATED

National Air and Space Museum (NASM)

(Smithsonian Institution)
6th and Independence Avenue SW, Washington, District of Columbia, 20560, United States
Tel: (+1 202) 357 27 00
Fax: (+1 202) 633 81 74
Web: http://www.nasm.si.edu
Chairman, Aeronautics: Dominick A Pisano
Chairman, Space History: Allan A Needell
Chairman, Earth and Planetary Studies: Bruce Campbell
Director: Gen John R Dailey
Financial Director: Ms Liz Scheffler
Public Relations Director: Ms Claire H Brown
Editor, Air and Space/Smithsonian: George Larson
Public Affairs Specialist: Peter Golkin

■ Research, education, exhibits, collections management, public service, library and photographic collection and publications.

Publications: Air & Space/Smithsonian

UPDATED

Satellife

30 California Street, Watertown, Massachusetts, 02472, United States
Tel: (+1 617) 926 94 00
Fax: (+1 617) 926 12 12
e-mail: hnet@healthnet.org
Web: http://www.healthnet.org
Chairman and Founder: Dr Bernard Lown
Executive Director: Ms Holly D Ladd
Technical Director: Alexander Fraser
Development Director: Ms Melissa Boudreault
Senior Development Officer: Ms Mary Halpin

■ Design and dvelopment of information services for health care workers, governments, medical schools and health researchers in the developing world.

UPDATED

Space Transportation Association (STA)

2800 Shirlington Road, Suite 405A, Arlington, Virginia, 22206, United States
Tel: (+1 703) 671 41 16
Fax: (+1 703) 931 64 32
e-mail: spacetra@erols.com
sta4space@aol.com
Web: http://www.spacetransportation.org

■ Represents organisations engaged in developing, building, operating and using space transportation vehicles, systems and services to provide access to space for government, civil and military users.

UPDATED

Spaceweek

1110 NASA Road One, Suite 102, Houston, Texas, 77058, United States
Tel: (+1 281) 333 36 27
Fax: (+1 281) 335 02 29
Web: http://www.spaceweek.org
Director of Public Relations: Ms Victoria Koepsell

UPDATED

Universities Space Research Association (USRA)

(Headquarters)
10227 Wincopin Circle, Suite 500, Columbia, Maryland, 21044-3432, United States
Tel: (+1 410) 730 26 56
Fax: (+1 410) 730 34 96
e-mail: info@hq.usra.edu
Web: http://www.usra.edu
President and Chief Executive Officer: David C Black
Executive Director: David Cummings
Director of University Relations, Space Technology: Hussein Hussein
Director of University Relations, Space Sciences: Jacques L'Heureux
Controller: Ms Susan Ballard
Chief Financial Officer: Robert Senter
Manager, Human Resources: Ms Ginny Bachmann
Manager, Contracts: Ron Deacon
Manager, Information Systems: Ms Nancy Jenkins
Assistant Manager, Information Systems: Ms Juanita Harrington
Manager, Payroll/Personnel Operations: Ms Ruth Tutin
Senior Contract Specialist: Ms Debbie Devin
Chief Engineer: Lewis Peach
Contract Specialist: Ed Abrams
Contract Specialist: Ms Michelle Giller
Contract Specialist: Brian Deaton
Contract Assistant: Ernest Mabrey
Financial Analyst: Ms Shelley Hayes
Staff Accountant/Internal Auditor: Jeff Roddy
Staff Accountant: Ms Roberta Gloger
Subcontracts Specialist: Ms Sophia Hill
Payroll/Personnel Operations Assistant: Ms Tina Osborne
Accounts Payable Technician: Ms Kathy Smith
Accounts Payable Technician: Shenee Taylor
Systems Administrator: Matthew Taylor
Administrative Assistant: Ms Maria Turner
Computer Assistant: Steve Kirschbaum
Receptionist: Ms Susan Hartman

UPDATED

Universities Space Research Association (USRA)
(Huntsville Program Office)
6700 Odyssey Drive, Suite 203, Huntsville, Alabama, 35806, United States
Tel: (+1 256) 971 02 40
(+1 256) 971 02 42
Fax: (+1 256) 971 02 41
e-mail: debra@space.hsv.usra.edu

UPDATED

Universities Space Research Association (USRA)
(Washington Office)
1101 17th Street NW, Suite 1004, Washington, District of Columbia, 20036, United States
Tel: (+1 202) 689 12 90
Fax: (+1 202) 689 12 97

UPDATED

US Air Force Museum (USAFM)

1100 Spaatz Street, Wright-Patterson AFB, Dayton, Ohio, 45433-7102, United States
Tel: (+1 937) 255 32 86
(+1 937) 253 IMAX
Fax: (+1 937) 255 05 23
Web: http://www.wpafb.af.mil/museum
Director: Maj Gen Charles D Metcalf
Tel: (+1 937) 255 72 04
Curator: Terrill Aitken
Tel: (+1 937) 255 72 04
Air Force Museum Foundation Executive Secretary: Richard Johnson
Tel: (+1 937) 258 12 18
Chief, Public Affairs: Ms Diana Bachert
Tel: (+1 937) 255 47 04
Public Affairs Specialist: Chris McGee
Tel: (+1 937) 255 47 04
Media Representative: Robert Bardua
- Display over 300 aircraft and missiles plus many artifacts.

UPDATED

Industry – Manufacturing

Advanced Products Co.

33 Defco Park Road, PO Box 296, North Haven, Connecticut 06473, United States
Tel: (+1 203) 239 33 41
Freephone: (+1 800) 243 60 39
Fax: (+1 203) 234 72 33
Cable: ADVPRODCO
e-mail: sales@advpro.com
Web: http://www.advpro.com
Managing Director and Chief Operating Officer: Edward C Randall
Executive Secretary: Ms Susan Doolan
- Static metal gasket: C/E/O/V-rings, wire rings. Spring-energized PTFE seals (dynamic and static).

Subsidiaries
Advanced Products France
Advanced Products NV, Belgium
Advanced Products Dichtungen, Germany
Advanced Products (Seals & Gaskets) Ltd, UK

UPDATED

Advanced Technical Enterprises, Inc. (ATE)

1800 Penn Street, Suite 4A, Melbourne, Florida, 32901, United States
Tel: (+1 877) 700 94 26
Fax: (+1 321) 952 79 44
e-mail: postmaster@ate.cc
Web: http://www.ate.cc
President: Chuck Kohfeldt
Tel: (+1 877) 700 94 26

UPDATED

AEC-Able Engineering Co. Inc.

7200 Hollister Avenue, Goleta, California, 93117, United States
Tel: (+1 805) 685 22 62
Fax: (+1 805) 685 13 69
General Manager: Mark De Witt

UPDATED

AEDC Group

877 Avenue E, Tennessee, 37389-5051, United States
Tel: (+1 931) 454 70 63
Fax: (+1 931) 454 61 88
Programme Manager: Lex C Hutcheson

UPDATED

AeroAstro Corp.

20145 Ashbrook Place, Ashburn, Virginia, 20147, United States
Tel: (+1 703) 723 98 00
Fax: (+1 703) 723 98 50
e-mail: info@aeroastro.com
Web: http://www.aeroastro.com
Chief Executive Officer and President: Dr Rick Fleeter
Sales and Marketing Director: Ms Kara Campbell
Operations Director: Peter Goldberg
Technical marketing Manager: David Goldstein
- Design and manufacture of small, low cost space systems and components for small satellites.

UPDATED

Aerojet

(Headquarters)
Highway 50 and Aerojet Road, PO Box 13222, Sacramento, California 95813-6000, United States
Tel: (+1 916) 351 86 50
Fax: (+1 916) 351 86 67
Cable: AEROGENCOR
e-mail: comments@aerojet.com
Web: http://www.aerojet.com
President: Michael Martin
Tel: (+1 916) 351 85 55
Vice President, Human Resources: Bryan Ramsey
Tel: (+1 916) 355 20 72
Vice President, Legal and Contracts: Brian Sweeney
Tel: (+1 916) 351 85 88
Vice President, Controller: Frank Fogarty
Tel: (+1 916) 351 85 78
Vice President, Business Development: Ron Samborsky
Tel: (+1 916) 355 36 10
Vice President, Programmes: Dale Adams
Tel: (+1 916) 355 52 90
Vice President, Operations: Scott Neish
- Specialise in missile and space propulsion, defence, and armaments.

UPDATED

Aerojet
(Alabama Office)
1500 Perimeter Parkway, Suite 315, Huntsville, Alabama, 35806, United States
Tel: (+1 256) 837 33 40
Fax: (+1 256) 837 38 69
Contact: Craig Landrith
Contact: Claude Baldwin

UPDATED

Aerojet
(Heavy Metals)
1367 Old State Route 34, Jonesborough, Tennessee, 37659, United States
Tel: (+1 615) 753 12 00
Fax: (+1 615) 753 86 45

UPDATED

Aerojet
(Munition Loading and Packing)
604 Spring Street, Socorro, New Mexico, 87801, United States
Tel: (+1 505) 835 20 70
Fax: (+1 505) 835 49 11

UPDATED

Aerojet
(New Jersey Office)
100 Stierli Court, Mount Arlington, New Jersey, 07856, United States
Tel: (+1 201) 770 32 40
Fax: (+1 201) 770 32 87

UPDATED

Aerojet
(Propulsion Systems)
PO Box 13222, Sacramento, California, 95813, United States
Tel: (+1 916) 355 10 00
Fax: (+1 916) 351 86 67

UPDATED

Aerojet
(Tennessee Operations)
Old State Route 34, Jonesborough, Tennessee, 37659, United States
Tel: (+1 615) 753 12 00

UPDATED

AGM Container Controls, Inc.

PO Box 40020, Tucson, Arizona, 85717-0020, United States
Tel: (+1 520) 881 21 30
Freephone: (+1 800) 995 55 90
Fax: (+1 520) 881 49 83
e-mail: sales@agmcontainer.com
Web: http://www.agmcontainer.com
President: Howard N Stewart
Financial Director: Ms Ellen Howlett
Purchasing Manager: Ron Corbin
Sales Manager: Jim McCorry
Marketing Manager: Ms Cindy Ramos
- Manufacture of environmental control hardware for moisture-sensitive electronic and optical equipment, mobile electronic shelters, re-usable shipping and storage containers: breather valves, desiccators, desiccant, humidity indicators, record holders, shock overload indicators, tie down straps and tie down shelving.

Overseas agents/representatives
Adtech Inc., Japan
Airsec SA, France
Brownell LTD, UK
Diatero, France
Karsel LTD, Turkey
Sued-Chemie Espana, Spain
Sued-Chemie UK Ltd
Sued-Chemie AG, Germany
Yail-Noa Ltd Agencies, Israel

UPDATED

AIT-Atlanta Inc.

(Advanced Interconnection Technology-Atlanta Inc.)
3289 Montreal Industrial Way, Tucker, Georgia, 30084, United States
Tel: (+1 770) 938 93 87
Fax: (+1 770) 934 98 62

Web: http://www.ait-atlanta.com
National Sales Manager: Kenneth Giery
Tel: (+1 770) 938 93 87
■ Design and manufacture of multiwire and microwire circuit boards.

UPDATED

Aitech Defense Systems Inc.

3080 Olcott Street, Suite 105A, Santa Clara, California, 95054, United States
Tel: (+1 408) 980 62 00
Fax: (+1 408) 980 61 61
e-mail: sales@rugged.com
Web: http://www.rugged.com
Sales Manager: Peter J Sasure
■ Supply of Mil-Spec rugged computer systems for harsh environments offering VME for defence, space, and video applications. Products include: VMEbus boards, enclosures and systems; and provide hardware system integration services.

UPDATED

Alliance Spacesystems Inc.

1250 Lincoln Avenue, Suite 100, Pasadena, California, 91103-2466, United States
Tel: (+1 626) 296 13 73
Fax: (+1 626) 296 00 48
Web: http://www.asi-space.com
Director, Advanced Programmes: Scott Stanley

UPDATED

Allied Aerospace

(ex-Dynamic Engineering, Inc)
703 Middle Ground Boulevard, Newport News, Virginia, 23606, United States
Tel: (+1 757) 873 13 44
Fax: (+1 757) 873 21 83
Web: http://www.dynamic-eng.com
President: David H Mullins
Vice President, Director Marketing: Richard L Best
Operations and Technical Director: Jim Brunges
Marketing and Communications Director: Mark Fingar
Purchasing Manager: Ms Debbie Ross
■ Research and development of prototype hardware, precision metals and composite fabrication in support of research in aircraft, rotorcraft, missile systems, test stands, propulsion systems and remotely piloted vehicles.
Accreditations: ISO 9001

UPDATED

American Aerospace Controls Inc. (AAC)

570 Smith Street, Farmingdale, New York, 11735-1115, United States
Tel: (+1 631) 694 51 00
Freephone: (+1 888) 873 85 59
Fax: (+1 631) 694 67 39
e-mail: aac@rcn.com
Web: http://www.a-a-c.com

UPDATED

American Fuel Cell and Coated Fabrics Co. (AMFUEL)

601 Firestone Drive, Magnolia, Arkansas, 71753, United States
Tel: (+1 870) 235 72 12
Fax: (+1 870) 235 72 70
e-mail: amfuel@ipa.net
Web: http://www.amfuel.com
Contact: Bill Mow
■ Collapsible water drums constructed of water-resistant synthetic rubber-coated fabric.

UPDATED

Amptek, Inc.

6 De Angelo Drive, Bedford, Massachusetts, 01730-2204, United States
Tel: (+1 781) 275 22 42
Fax: (+1 781) 275 34 70
e-mail: sales@amptek.com
Web: http://www.amptek.com
President: John A Pantazis
Vice President: Alan C Huber
Advertising Director: Ms Linda Hantzis
Sales Director: Niki Pantazis
■ Development and manufacture of low-power, space qualified hybrid electronic circuits and scientific payload spacecraft instruments. Products include radiation hardenedhybrid amplifiers and preamplifiers which have been used on Galileo, NEAR and Mars Pathfinder; custom hybrid circuits, space instrumentation. Design, fabrication, calibration, integration and launch support of space plasma and energetic particle instruments, including instruments for the operational DMSP program, the CRRES mission and the Space Shuttle program. Has developed and built the Compact Environmental Anomoly Sensor (CEASE).

UPDATED

Andrew Corporation

(Corporate Offices)
10500 West 153rd Street, Orland Park, Illinois, 60462-3099, United States
Tel: (+1 708) 349 33 00
Web: http://www.andrew.com
■ Microwave antennas and antenna systems, telemetry and missile tracking antennas, scatter antennas and reflectors. Rigid and flexible transmission lines and waveguides. Antennas (airborne, microwave, telemetry, broadcast earth station; transmission lines). Defence electronics.

Branches in
Argentina, Australia, Austria, Brazil, Canada, China, France, Germany, India, Indonesia, Italy, Japan, Malaysia, Mexico, Netherlands, Philippines, Russia, South Africa, Spain, Sweden, Switzerland, Taiwan, Thailand

UPDATED

Applied Dynamics International (ADI)

3800 Stone School Road, Ann Arbor, Michigan, 48108, United States
Tel: (+1 734) 973 13 00
Fax: (+1 734) 668 00 12
e-mail: info@adi.com
Web: http://www.adi.com
President and Chief Executive Officer: J D McIntosh
Senior Vice President, Sales and Marketing: Charles R Moores
Senior Vice President: E J Fadden
Asia Sales Manager: Maurice F Snyder
■ Manufacture of digital computer systems for simulating models of dynamic, continuous systems. The SIM integrated system, provides an environment for real-time simulation of technologically advanced equipment such as aircraft, turbines, missiles and satellites. Beacon code generator for embedded microprocessors for safety critical applications such as Fadec engine control systems. Beacon offers graphical controller definition with C, Fortran, or Ada outputs and also generates unit test vectors.

Overseas representatives
Anacon Technology, Sweden
Anawell Corporation, Korea
Applied Dynamics International, UK
China National Aero-Technology International Supply Corp., China
Hinditron Infosystems (Pvt) Ltd, India
Kyokuto Boeki Kaisha Ltd, Japan
Otopos, Italy

UPDATED

Arch Chemicals, Inc.

(ex-Olin Chemicals)
PO Box 5204, Norwalk, Connecticut, 06856-5204, United States
Tel: (+1 203) 229 29 00
Fax: (+1 203) 229 26 13
Web: http://www.archchemicals.com
Director, Corporate Communications: Will Demers
Business Manager: Chris Cice
Tel: (+1 203) 229 34 34
Manager, Corporate Communications: Dale N Walter

UPDATED

Arde Inc.

(Head Office and Plant)
500 Walnut Street, Norwood, New Jersey, 07648, United States
Tel: (+1 201) 784 98 80
Fax: (+1 201) 784 97 10
e-mail: ardeinc@optonline.net
President: Mel Goldstein
Vice President and Technical Director: Kirk Sneddon
Production Director: Jeff Ehalt
Business Development Manager: Robert E Holland
■ Research, development and manufacturing of lightweight metal and metal/fibre composite pressurant and propellant tanks, with and without propellant management system and ASME code.

UPDATED

Arrowhead Products

4411 Katella Avenue, Los Alamitos, California, 90720, United States
Tel: (+1 714) 828 77 70
Web: http://www.arrowheadproducts.net
Vice President, Marketing: Andy Mead
North American Sales Director: Ed Espinola
European Business Director: Richard Mullings
East Coast Sales Manager: Chris Buntin

UPDATED

Astech Engineered Products Inc.

(ex-Astech, Inc.)
(ex-Astech Manufacturing Inc.)
3030 Red Hill Avenue, Santa Ana, California, 92705-5866, United States
Tel: (+1 949) 250 10 00
(+1 949) 250 49 82
(+1 949) 250 21 20
Fax: (+1 949) 250 21 35
e-mail: enquiries@chem-tronics.com
Web: http://www.chem-tronics.com
Director, Operations: Jody Callihan
Director, Marine and Space Applications: Al Dillingham
Tel: (+1 949) 250 21 42
Mobile: (+1) 94 96 36 53 95
Purchasing Manager: Dutch Ludt
■ Engineering, design, development, manufacture and test of aerostructures. Acoustic panels.

UPDATED

Astronautics Corporation of America

PO Box 523, Milwaukee, Wisconsin, 53201-0523, United States
Tel: (+1 414) 449 40 00
Fax: (+1 414) 447 82 31
Web: http://www.astronautics.com
Chairman Emeritus: N K Zelazo
Tel: (+1 414) 449 41 20
President: Michael Russek
Chief Executive Officer: Dr R E Zelazo
Tel: (+1 973) 785 60 00
Vice President, Administration: H Russek
Tel: (+1 414) 449 41 23
Vice President, Finance: Steve Givant
Tel: (+1 973) 785 67 35
Vice President, Engineering: Daniel Olson
Tel: (+1 414) 449 40 40
Vice President, Material and Production: Joe Potts
Tel: (+1 414) 449 45 20
Business Development Director: Daniel Wade
Tel: (+1 414) 449 42 42
Director, Quality Assurance: D Young
Director, Programmes and Contracts: Walter Bouwens
Tel: (+1 414) 449 42 50
Administrative Assistant: Jacky Schmitt
Tel: (+1 414) 449 42 43
■ Advanced colour and monochrome (AMLCD/CRT) displays and aircraft instruments including electronic and electromechanical EFI, HSI, ADI, BDHI and EICAS displays. Air data computers, airborne server units, multi-functions displays, mission and display processors, EFIS/FMS cockpit integration, flap control systems and autopilots. Mechanical and ring laser gyros, spaceborne, shipborne, airborne and ground-based inertial navigation systems. Synchros, resolvers, brushless motors, actuators and cryogenic refrigeration.

Joint venture
Astronautics Kearfott Electroautomatica (AKE), Russia

Subsidiaries
Astronautics C. A. Ltd
Bnei-Brak, Israel
Kearfott Guidance & Navigation Corp. Wayne, N.J.

UPDATED

ATI Engineering Inc.

227J Gateway Drive, Bel Air, Maryland, 21014, United States
Tel: (+1 410) 420 27 73
Fax: (+1 410) 420 27 75
e-mail: atie@ati-engineering.com
Web: http://www.ati-engineering.com
Vice President Marketing: Ronald P Leary
■ Research, development and manufacture of wireless and microwave RF systems for the military, mobile phone and satellite and wireless communication industries.

UPDATED

ATI Wah Chang (WC)
(ex-Oremet-Wah Chang)
(ex-Wah Chang)
1600 North East Old Salem Road, PO Box 460, Albany, Oregon, 97321, United States
Tel: (+1 541) 926 42 11
Freephone: (+1 888) 926 42 11
Fax: (+1 541) 967 69 94
e-mail: custservice@wahchang.com
Web: http://www.alleghenytechnologies.com/wahchang
http://www.wahchang.com
President: Lynn Davis
Vice President, Commercial: Parry Walborn
Financial Vice President: Joe Frazier
Sales Director: Gary Kneisel
Marketing Director: Andy Nichols
Operations Director: Carl Califano
Public Relations Director: Jim Denham
Purchasing Manager: Joe Michl
Marketing Co-ordinator: Ms Sheryl Renzoni
■ Manufacture of speciality metals and chemicals, used in energy production, chemical and mineral processing, aerospace, medical, research and consumer products. Materials include hafnium, niobium, titanium, vanadium, zirconium, silicon tetrachloride, and zirconium and hafnium chemicals.
European operations
Liechtenstein

UPDATED

ATK Composites
(ex-Alliant Aerospace Composite Structures Company)
Utah Composites Center, Freeport Center, MS YC14, PO Box 160433, Clearfield, Utah, 84016-0433, United States
Tel: (+1 801) 775 10 82
Fax: (+1 801) 775 12 07
e-mail: composite_structures@atk.com
Web: http://www.atk.com/composites
President: Travis E Campbell
Director of Business Development: Mike Blair
Tel: (+1 801) 775 17 22
■ Supplier of composite structures for space launch vehicles and military and commercial aircraft. Full service design and manufacturing capability, including fibre-placed filament-wound, hand lay-up and precision bonded assemblies.

UPDATED

ATK Thiokol Propulsion
(ex-Thiokol Propulsion)
PO Box 707, Brigham City, Utah, 84302-0707, United States
Tel: (+1 801) 863 35 11
Fax: (+1 801) 863 23 91
■ Manufacture, delivery, stacking, retrieval and refurbishment of space shuttle solid rocket booster propulsion.

UPDATED

Atlantic Research Corp. (ARC)
(Corporate Headquarters)
5945 Wellington Road, Gainesville, Virginia, 20155, United States
Tel: (+1 703) 754 50 00
(+1 703) 754 54 31
Fax: (+1 703) 754 54 96
(+1 703) 754 54 41
Web: http://www.atlanticresearchcorp.com
Chairman and Chief Executive Officer: John J Quicke
Vice President and General Manager, ARC Propulsion: B Frank Rohrback
Vice President and Chief Financial Officer: Patrick J Jenkins
Vice President, Human Resources and Administration: Ms Patty Park
Vice President and General Counsel: Paul Barchie
Vice President, Business Development: Richard T Yezzi
Director, Business Development: Craig W Clauss
Tel: (+1 703) 754 51 92
Manager, Publications and External Relations: Ms Libby Masaitis
Corporate Secretary: Ms Barbara F Biendl
Executive Assistant to the Chief Executive Officer: Ms Judy Briggs
■ High technology company with activities in both solid and liquid rocket propulsion, advanced composite materials and inflation systems for the automotive airbag industry.

UPDATED

Atlantic Research Corp. (ARC)
(ARC Liquid Propulsion)
6686 Walmore Road, Niagara Falls, New York, 14304, United States
Tel: (+1 716) 731 60 00
Fax: (+1 716) 731 62 81
Vice President and General Manager: Gerald L Greene
Marketing Director: Craig Clauss
■ Liquid rocket engines for space propulsion.

UPDATED

Atlantic Research Corp. (ARC)
(Arkansas Operations)
PO Box 1036, Camden, Arkansas, 71711-1036, United States
Tel: (+1 870) 574 31 24
Fax: (+1 870) 574 25 09
Vice President: Robert E Shenton
■ High volume production of solid rocket propellant. Large rocket motors. Advanced technology, analysis, testing, quality assurance and production.

UPDATED

Atlantic Research Corp. (ARC)
(Defense Products Group)
(ex-Solid Propulsion Division)
5945 Wellington Road, Gainesville, Virginia, 155-1699, United States
Tel: (+1 703) 754 50 00
Fax: (+1 703) 754 54 41
Vice President and General Manager, ARC Propulsion: B Frank Rohrback
■ Design, development and production of small tactical motors and gas generators. Advanced tactical and strategic propulsion systems.

UPDATED

Avalon Electronics Inc.
100 Bartow Municpal Airport, Bartow, Florida, 33830, United States
Tel: (+1 863) 519 09 05
Fax: (+1 863) 519 07 63
e-mail: us-info@avalon-electronics.com
Web: http://www.avalon-electronics.com
President: F Thames
■ Manufacture single and multi-channel S-VHS based analogue and video cassette recorders with 12 MHz bandwidth for airborne data acquisition, aerospace, satellite and high speed communications applications. Ruggedised disk-based analogue and digital recorders, single and multi-channel, recording up to 75 MHz, 1 Gigabits. Recording systems, SIGINT.

UPDATED

Aydin Displays, Inc.
1 Riga Lane, Birdsboro, Pennsylvania, 19508, United States
Tel: (+1 610) 404 53 53
Fax: (+1 610) 404 81 90
e-mail: sales@aydindisplays.com
Web: http://www.aydindisplays.com
Sales Director: Henry Casey
Tel: (+1 610) 404 53 48
Marketing Assistant: Ms Debbie Baker
■ Design and manufacture command and control equipment and systems, air traffic control systems and air defence consoles.

UPDATED

Ball Aerospace & Technologies Corp. (BATC)
1600 Commerce Street, PO Box 1062, Boulder, Colorado, 80301, United States
Tel: (+1 303) 939 40 00
(+1 303) 939 61 00
Fax: (+1 303) 939 61 04
Telex: 45605
e-mail: info@ball.com
Web: http://www.ballaerospace.com
http://www.ball.com/aerospace
President and Chief Executive Officer: David L Taylor
Vice President, Finance and Administration: Ms Lisa A Pauley
Vice President, Strategic Development: M Andrew Crouch
Senior Director, Human Resources: Joseph L Winslow
Public Affairs: Ms Jennifer Meyer
Director, Communication Service: Ms Rachelle S Wood
■ Provide systems engineering services. Design and manufacture of complete spacecraft and space systems, space and scientific sensors, cryogenic subsystems, antenna systems and video products for commercial and government customers.

UPDATED

Ball Aerospace & Technologies Corp. (BATC)
(Washington DC Office)
2200 Clarendon Boulevard, Suite 1202, Arlington, Virginia, 22201, United States
Tel: (+1 703) 284 54 00
Fax: (+1 703) 284 54 49

UPDATED

BBN Technologies
(ex-BBN Systems & Technologies)
10 Moulton Street, Cambridge, Massachusetts, 02138, United States
Tel: (+1 617) 873 80 00
Web: http://www.bbn.com
■ Research and product development in the areas of speech and language, communications and logistics for first responders; infrastructure security, information security solutions; unstructured data; wireless networking solutions.

UPDATED

Boeing Satellite Systems Inc. (BSS)
(ex-Hughes Space and Communications Company)
PO Box 92919, Los Angeles, California, 90009, United States
Tel: (+1 310) 364 60 00
(+1 310) 662 90 00
Fax: (+1 310) 364 49 11
Web: http://www.boeing.com/satellite
President of Boeing Satellite Systems: Randy Brinkley
Principal Deputy: Alexis Livanos
Chief Financial Officer: David Ramey
Vice President and General Manager: Dave Ryan
Senior Vice President of Programmes: Jack Wormington
Senior Vice President, Operations: Mark Weltman
Vice President for Business Development: John Konrad
Vice President of Communications: George Torres
General Manager, Government Business Development: Fred MacGregor
Public Relations Manager: Ms Diana Ball
Tel: (+1 310) 662 74 73
Fax: (+1 310) 364 63 97
Purchasing Manager: Daniel Bridleman
■ Development and production of space and communication systems for military, commercial and scientific uses. These systems supply communications and meteorological observations for domestic and international customers and meet many of the military and scientific space requirements of the U.S. government.

UPDATED

Brush Wellman Inc.
(Engineered Materials)
14710 West Portage River South Road, Elmore, Ohio, 43416, United States
Tel: (+1 419) 862 27 45
Fax: (+1 419) 862 41 74
e-mail: beproducts@brushwellman.com
Web: http://www.brushwellman.com
President: Gordon Harnett
Tel: (+1 216) 383 40 01
Managing Director: Mike Anderson
Tel: (+1 216) 383 40 87
Sales and Marketing Director: Lawrence Ryczek
Tel: (+1 419) 862 43 41
Financial Director: Bill Halm
Tel: (+1 419) 862 41 96
Technical Director: Donald Kaczynski
Operations Director: Keith Smith
Tel: (+1 419) 862 42 49
Public Relations Director: Pat Carpenter
Tel: (+1 216) 383 68 35
Purchasing Manager: William Smith
Tel: (+1 419) 862 43 29
Marketing Assistant: Ms Julie Humbarger
Tel: (+1 419) 862 41 73
■ Beryllium based materials including: beryllium metal matrix composites, aluminium beryllium alloys, beryllium oxide ceramics and beryllium copper. These materials are utilised in aerospace and defence applications where low density, high rigidity, high thermal conductivity and low coefficient of thermal expansion are useful in design.

UPDATED

Brush Wellman Inc.
(Electrofusion Products)
44036 South Grimmer Boulevard, Fremont, California, 94538-6346, United States

Tel: (+1 510) 623 15 00
Fax: (+1 510) 623 76 00
Telex: 3731106
President: Ed Hefter
Tel: (+1 510) 661 97 47
Sales and Marketing: Christopher Dorn
Tel: (+1 510) 623 76 00
■ Specialised metal fabrication; beryllium, tungsten, tantalum, moly. Beryllium X-Ray window, IF-1 beryllium foil

UPDATED

CIC International Ltd

5 Marine View Plaza, Hoboken, New Jersey, 07030, United States
Tel: (+1 201) 792 18 00
Fax: (+1 201) 792 57 55
Telex: 237593
Cable: COMMINTAL
e-mail: 103227.250@compuserve.com
cicinternational@compuserve.com
Web: http://www.cic-international.com
Chairman and President: Satiris G Fassoulis
Executive Vice President: Larry Thompson
Tel: (+1 304) 264 47 29
Vice President, Purchasing: Robert C Thompson
Vice President, Operations: Robert Volvo
Vice President, Sales and Marketing: Robert C Perry
Vice President: Ms Caron Zaroff
Vice President: Terry George
Vice President: M A Couvaras
Vice President: Robert Spitz
Operations Director: James Chladek
Financial Director: Robert Fassoulis
Sales and Marketing Director: Joseph Ceva
Public Relations Director: Ms Carole Dean
Advertising: J Gordon
■ Supplier of defence products to various defence establishments and munitions manufacturers. Principal products include a full range of shells and ammunition, howitzers, motor gun carriages and military vehicles. Supplier of aircraft parts, armament and avionics. The company maintains spare parts inventories for military vehicles, performs tank retrofits and acts as a consultant in modernising military vehicles.
Affiliated Companies
CIC Canada (Reg.)
CIC Iberia SA
CIC Task Ltd

UPDATED

Cicoil Corporation

24960 Avenue Tibbitts, Valencia, California, 91355, United States
Tel: (+1 661) 295 12 95
Fax: (+1 661) 295 08 13
e-mail: sales@cicoil.com
Web: http://www.cicoil.com
Chief Executive Officer: Peter G Nogradi
President: Thomas A Nogradi
Operations Director and Quality Assurance: Sam Palahnuk
Engineering Manager: Eric Scott
National Sales and Marketing Manager: James Alderson
Manufacturing Manager: Ms Tereasa De la Torres
Administration Manager: Ms Julie Tortelli
■ Flat, flexible, silicone encapsulated cables. Applications include: aerospace and weapons, space and satellite, medical, robotics and automation.

UPDATED

Circle Seal Controls, Inc.

2301 Wardlow Circle, PO Box 3300, Corona, California, 92878, United States
Tel: (+1 909) 270 62 00
Fax: (+1 909) 270 62 01
Web: http://www.circle-seal.com
■ Valves and controls for industrial and aerospace/military applications. Products range from discrete components to integrated assemblies to complete fluid control systems. Products include check, relief, shutoff, solenoid, cryogenic, fuse, and motor operated valves, pressure regulators, tyre fill valve gauges, restrictors and manifold systems.

UPDATED

Codan US, Inc.

10660 Wakeman Court, Manassas, Virginia, 20110, United States
Tel: (+1 703) 361 27 21
Fax: (+1 703) 361 38 12

UPDATED

COM DEV

(Sales and Marketing Office)
(ex-Com Dev Space)
4712 Admiralty Way, Marina Del Rey, California, 90292, United States
Tel: (+1 310) 667 88 23
Fax: (+1 310) 667 08 02

UPDATED

Computer Sciences Corporation (CSC)

(Federal Sector)
3170 Fairview Park Drive, Falls Church, Virginia, 22042, United States
Tel: (+1 703) 876 10 00
(+1 703) 641 22 22
Fax: (+1 703) 849 10 00
(+1 703) 204 83 55
Web: http://www.csc.com
President: Paul M Cofoni
Director of Communication: James Sullivan

UPDATED

Conduant Corporation

1501 South Sunset Street, Suite C, Longmont, Colorado, 80501, United States
Vice President of Business Development: Tom Skrobacz
Tel: (+1 303) 485 27 21
Fax: (+1 303) 485 51 04

UPDATED

Cubic Corporation

PO Box 85587, San Diego, California, 92186-5587, United States
Tel: (+1 858) 505 22 37
Freephone: (+1 800) 818 83 03
Fax: (+1 619) 277 18 78
e-mail: cubicinfo@cubic.com
Web: http://www.cubic.com
Chairman of the Board, Vice President and Chief Financial Officer: Walter J Zable
Chairman and Director, Cubic Transportation Systems: Raymond L Dekozan
Vice President and Director: William W Boyle
Vice President and Corporate Controller: Thomas A Baz
Vice President, International Business Development: Michael W David
Vice President of the Group and Defence: Gerry R Dinkel
Vice President, Financial Planning and Accounting: Mark A Harrison
Vice President and General Counsel: Kenneth F Kopf
Vice President, Human Resources: Bernard A Kulchin
Vice President, Information Technologies: John Minteer
Vice President and Secretary: William C Stewart
Tel: (+1 858) 505 22 76
Vice President, Finance and Treasurer: John D Thomas
Tel: (+1 858) 505 29 89
Manager of Investor Relations: Diane L Dyer
Corporate Communications Manager: Ms Jan Stevens
■ Training systems, tracking systems, data link, air and ground combat manoeuvering ranges, personnel location systems, aerospace products, operation and maintenance.

UPDATED

Custom Electronics Inc.

Browne Street, Oneonta, New York, 13820, United States
Tel: (+1 607) 432 38 80
Freephone: (+1 877) 735 92 34
Fax: (+1 607) 432 39 13
e-mail: ceisales@customelec.com
Web: http://www.customelec.com
President and Chief Executive Officer: P S Dokuchitz
Director, Financial: Michael Pentaris
Director, Human Relations: Ms Grace Alidi
Manager, Senior Sales: Michael Schulte
Manager, Quality: Ralph Ritton
Manager, Purchasing: Ms Carol Barrett
■ Manufacture of high voltage reconstituted mica paper capacitors for use in radar power supplies, ECM power supplies, high-voltage transmitters for missile applications, TWT power supplies for satellites, detonation systems, ignition systems and many other high voltage applications.

UPDATED

Dallas Remote Imaging Group (DRIG)

4209 Meadowdale Drive, Carrollton, Texas, 75010, United States
Tel: (+1 972) 898 35 63
e-mail: consulting@drig.com
Web: http://www.drig.com
President: Dr Jeff Wallach
■ Provides consulting and educational services on weather satellite imagery, satellite communications and tracking, and the use of image acquisition and processing in education. Provides remote imaging and space related content on a wide variety of platforms.

UPDATED

David Clark Company Incorporated

360 Franklin Street, PO Box 15054, Worcester, Massachusetts, 01615-0054, United States
Tel: (+1 508) 751 58 00
Fax: (+1 508) 753 58 27
e-mail: sales@davidclark.com
Web: http://www.davidclark.com
President: Robert A Vincent
Director, Sales and Marketing: Richard M Urella
Product Manager, Aviation: Dennis J Buzzell
Product Manager, Airlines: Robert Daigle
International Sales Manager: James Comer
■ Vehicle and mobile intercom systems, noise attenuating headsets with noise-cancelling microphones for airline ground support, de-icing and maintenance operations, noise attenuating headsets and adaptors for two-way radios.

UPDATED

DRS Technologies, Inc.

(Corporate Headquarters)
(ex-Diagnostic/Retrieval Systems, Inc.)
5 Sylvan Way, Parsippany, New Jersey, 07054, United States
Tel: (+1 973) 898 15 00
Freephone: (+1 866) 898 15 00
Fax: (+1 973) 898 47 30
e-mail: info@drs.com
Web: http://www.drs.com
Chairman, President and Chief Executive Officer: Mark S Newman
Tel: (+1 973) 898 60 10
Executive Vice President, Chief Operating Officer: Paul G Casner
Tel: (+1 973) 898 15 00
Executive Vice President, Chief Financial Officer and Treasurer: Richard A Schneider
Tel: (+1 973) 898 60 21
Senior Vice President, Marketing: Chris M Toffales
Tel: (+1 973) 898 60 30
Senior Vice President, Operations: Robert Russo
Tel: (+1 973) 898 60 19
Senior Vice President, Business Operations and Strategy: Robert F Mehmel
Tel: (+1 973) 898 60 06
Senior Vice President, Washington Operations: Michael Bowman
Tel: (+1 703) 416 80 00
Vice President, Business Planning: Christopher Wright
Tel: (+1 973) 451 35 10
Vice President, Contracts and Compliance: Alan Gross
Tel: (+1 973) 898 60 35
Vice President, Business Service: Al Moran
Vice President, Chief Information Officer: Louis J Belsito
Tel: (+1 973) 898 60 40
Vice President, Human Resources: Ms Andrea J Hayward
Tel: (+1 973) 898 60 15
Vice President, Corporate Communications and Investor Relations: Ms Patricia M Williamson
Tel: (+1 973) 898 60 25
Head of C4I Group: Steven T Schorer
Head of Surveillance and Reconnaissance Group: Fred L Marion
Corporate Counsel: Audrey Stern
Tel: (+1 973) 898 60 07
Corporate Controller: Thomas P Crimmins
Tel: (+1 973) 898 60 32
Director: Ira Albom
Director: Donald C Fraser
Director: William F Heitmann
Director: Steven S Honigman
Director: C Shelton James
Director: Mark N Kaplan
Director: Stuart F Platt
Director: Ms Nina Laserson Dunn
Tel: (+1 973) 898 60 20
Director: Dennis J Reimer
Director: Eric J Rosen
Public and Investor Relations Manager: James M Ryan

Tel: (+1 973) 451 35 30
- Provide products and services to defence, government, intelligence and industrial customers. Development and manufacture of mission critical systems for air, sea, subsurface and land military platforms.

Accreditations: ISO 9000; ISO 9001; AS-9000
Corporate Divisions
C4I Group
-C3 Business
-DRS Broadcast Technology
-DRS Communications Company, LLC
-DRS Electronic Systems, Inc.
-DRS Flight Safety and Communications, Carleton Place
-DRS Flight Safety and Communications, Kanata
-DRS Laurel Technologies
-DRS Surveillance Support Systems, Inc.
-DRS Technical Services
-DRS Technical Services, Inc.
-DRS Weather Systems, Inc.
-Intelligence Technologies Business
-DRS Engineering Development Labs, Inc.
-DRS Signal Recording Technologies, Inc.
-DRS Signal Solutions (West)
-DRS Signal Solutions, Inc.
-DRS Signal Technologies, Inc.
-Power Systems
-DRS Electric Power Technologies, Inc.
-DRS Power & Control Technologies, Inc. Danbury
-DRS Power & Control Technologies, Inc. Milwaukee
-DRS Power Technology, Inc.
-Tactical Computing Business
-DRS Tactical Systems (West), Inc.
-DRS Tactical Systems Ltd.
-DRS Tactical Systems, Inc.
Surveillance and Reconnaissance Group
-Reconnaissance, Surveillance and Target Acquisition (RSTA) Business
-DRS Data & Imaging Systems Inc.
-DRS Data & Imaging Systems Ltd.
-DRS Infrared Technologies, LP
-DRS Nytech Imaging Systems, Inc.
-DRS Optronics, Inc. Palm Bay
-DRS Optronics, Inc. Melbourne
-DRS Sensors & Targeting Systems, Inc. Anaheim
-DRS Sensors & Targeting Systems, Inc. Torrance
-DRS Unmanned Technologies, Inc.
-Training Systems Business
-DRS EW & Network Systems (Canada)
-DRS EW & Network Systems, Inc.
-DRS Training & Control Systems, Inc.
-Electronic Test Business
-DRS Test & Energy Management, Inc.

UPDATED

Dutch Space BV
(Washington Office)
(ex-Fokker Space BV)
1729 Abbey Oak Drive, Vienna, Virginia, 22182, United States
Tel: (+1 703) 319 01 03
Fax: (+1 703) 319 05 28
Director: Eduard van der Noordaa

UPDATED

Dynamic Instruments Inc.
3860 Calle Fortunada, San Diego, California, 92123, United States
Tel: (+1 619) 278 49 00
Freephone: (+1 800) 793 33 58
Fax: (+1 619) 278 67 00
e-mail: diinfo@dynamicinst.com
Managing Director: Hugh Ness
Marketing Director: Paul F Whitten
Advertising/Public Relations Manager: Holly Grover

UPDATED

Ecliptic Enterprises Corp.
398 West Washington Boulevard, #100 Pasadena, California, 91103-2000, United States
Tel: (+1 626) 798 24 36
Fax: (+1 626) 798 22 51
Web: http://www.eclipticenterprises.com
Business Development: James Cantrell
Mobile: (+1 435) 757 99 40

UPDATED

EDO Corporation
60 East 42nd Street, 42nd Floor, New York, New York, 10165, United States
Tel: (+1 212) 716 20 00
Fax: (+1 212) 716 20 50
Web: http://www.edocorp.com
Chairman, President and Chief Executive Officer: James M Smith
Executive Vice President and Chief Operating Officer: Frank W Otto
Vice President – Washington Operations: Jon A Anderson
Vice President – Finance, Chief Financial Officer and Treasurer: Frederic B Bassett
Vice President – Human Resources: Ms Patricia D Comiskey
Vice President – Administration, Shareholder Relations and Secretary: William J Frost
Vice President and General Counsel: Ms Lisa M Palumbo
Director, Business Development: Ed Mygland
Director – Investor Relations and Corporate Communications: William A Walkowiak
Counsellor: Ms Effie Pavlou
- Provides military and commercial products and professional services, with core competencies in a wide range of critical defence areas, including: Aircraft armament, defence electronics, undersea warfare, Command, Control, Communications, Computers and Intelligence (C4I); and integrated composite structures.

UPDATED

Elite Electronics, Inc.
1408 Summit Avenue, Suite 5, Plano, Texas, 75074-8115, United States
Tel: (+1 972) 881 00 26
Freephone: (+1 800) 553 43 76
Fax: (+1 972) 422 97 07
e-mail: sales@elitelex.com
Web: http://www.elitelex.com
President and Managing Director: Stephen Kaplan
Operations Director: Richard L Hill
Purchasing Manager: Ms Lisa Watson
- Aligned with a fibre and copper assembly/solution manufacturer that specialises in military (MIL-STD-790 facility and certified by DSCC), aerospace (M-5015, M-24308, M-38999 and ARINC 404/600), industrial including oil seismic and mining assemblies and custom engineering interconnect applications.

Accreditations: ISO 9001-2000 (tooling only)

UPDATED

EMS Technologies Inc.
(Corporate Headquarters)
(ex-Electromagnetic Sciences Inc.)
660 Engineering Drive, Technology Park/Atlanta, Norcross, Georgia, 30092, United States
Tel: (+1 770) 263 92 00
Fax: (+1 770) 447 43 97
e-mail: pr@ems-t.com
Web: http://www.ems-t.com
President and Chief Executive Officer: Al Hansen
Executive Vice President and Chief Financial Officer: Don Scartz
Tel: (+1 770) 729 65 10
Chairman of the Board: John B Mowell
Chief Information Officer: John Dunbar
Director, Advanced Technology for Space & Technology/Atlanta: James Chu
- Design and manufacture of wireless and satellite solutions.

Company divisions
EMS Wireless
LXE
SATCOM
Space and Technology
Customers include
Alcatel, Bombardier, Boeing, The Canadian Space Agency, Global Express, Gulfstream Aerospace Corp., Harris Hughes, Lockheed Martin, Matra Marconi Space, McHugh Software International, NEC, Nortel, PacBell, Raytheon, SES Astra, Space Systems/Loral, TRW and Unilever.

UPDATED

EMS Technologies, Inc.
(Space & Technology Group, Atlanta)
660 Engineering Drive,
PO Box 770, Norcross, Georgia, 30092, United States
Tel: (+1 770) 263 92 00
Fax: (+1 770) 446 57 39
e-mail: marketing@ems-t.com
Web: http://www.ems-t.com/seg
Group President: Gerry Bush
Senior Vice President and General Manager: Jay Grove
Vice President, Sales: Bill Reiner
Vice President, Business Development: Ron H Miyakawa
- Design and development of satellite-based terminals and antennas for the aeronautical, land mobile and search and rescue markets. Developed a steerable antenna system to provide live television to commercial aircraft by a wireless link to a direct broadcast satellite.

UPDATED

Engineered Arresting Systems Corporation (ESCO)
(Military Products)
(ex-Engineered Systems Co.)
2550 Market Street, Aston, Pennsylvania, 19014, United States
Tel: (+1 610) 494 80 00
Fax: (+1 610) 494 89 89
Web: http://www.esco-usa.com
http://www.arrestinggear.com
President and Chief Executive Officer: Spencer M Hoos
Vice President, Marketing: T Ladson Webb
Vice President, EMAS: P T Mahal
Tel: (+1 610) 595 28 40
Fax: (+1 610) 595 28 42
Engineering Director: Richard L Orner
Financial Director: Leonard Schlack
Operations Director: J Guida
Technology and Product Development Director: Edmond S Lopez
Purchasing Manager: E Trupkovich
- Manufacture of aircraft arresting systems and special material handling systems.

UPDATED

Entran Devices Inc.
10 Washington Avenue, Fairfield, New Jersey, 07004, United States
Tel: (+1 973) 227 10 02
Freephone: (+1 800) 635 06 50
(+1 888) 8 ENTRAN
Fax: (+1 973) 227 68 65
e-mail: sales@entran.com
Web: http://www.entran.com
Marketing Assistant: Ms Dorris E Raleigh
- Semiconductor accelerometers, pressure transducers, pressure transmitters, load cells, strain gauges, instrumentation modules and digital pressure meters.

UPDATED

ERG Materials and Aerospace Corp.
(ex-ERG Inc.)
900 Stanford Avenue, Oakland, California, 94608, United States
Tel: (+1 510) 658 97 85
Fax: (+1 510) 658 74 28
e-mail: sales@ergaerospace.com
Web: http://www.ergaerospace.com
- Manufacture of Duocel, an open celled, reticulated, rigid foam material of aluminium, carbon, ceramics and composites. Design and manufacture of high performance components such as heat exchangers, lightweight structures, energy absorbers, baffles, diffusers, optics and filters for the aerospace, semiconductor, energy, nuclear, military, optical, defence, biomedical and other high-tech industries.

UPDATED

Essex Corporation
9150 Guildford Road, Columbia, Maryland, 21046-2306, United States
Tel: (+1 301) 939 70 00
Fax: (+1 301) 953 78 80
Web: http://www.essexcorp.com
President and Chief Executive Officer: Leonard E Moodispaw
Senior Vice President and Chief Financial Officer: Joseph Kurry
Senior Vice President and Chief Financial Officer: Terry Turpin
Vice President, Business Development: Michael Piacenza
Chief Administrative Officer and Corporate Secretary: Ms Kimberley J Dechello
- Optical and communications engineering.

UPDATED

Essex Industries Inc.
(Corporate Headquarters)
7700 Gravois Avenue, St Louis, Missouri, 63123-4728, United States
Tel: (+1 314) 832 45 00
Fax: (+1 314) 832 16 33
e-mail: corp@essexind.com
Web: http://www.essexind.com
- Design, development and manufacture of life-saving, high technology components, systems and custom-design applications for the military,

aerospace, medical, automotive and general industry fields.

Subsidiary companies
Alar Parmeko Inc.
Alar Products Inc.
Dolphin Systems Corp.
Essex Cryogenics of Missouri
Essex Fluid Controls
Essex Manufacturing Division
Essex Medical Products
Essex Portable Breathing & Rescue Products
Essex Precision Controls
Essex Screw Products
Propellex Corp.

UPDATED

ETS – Lindgren LP

1301 Arrow Point Drive, Cedar Park, Texas, 78613, United States

■ Specialists in RFI/EMC/NEMP and Tempest protection systems offering shielded enclosures, anechoic chambers, filters, fibre optics and suppression equipment. system design, manufacture, installation and testing are available on a worldwide basis.

UPDATED

Fairchild Imaging

(ex-Lockheed Martin Fairchild Systems)
(ex-Loral Fairchild Imaging Sensors)
1801 McCarthy Boulevard, Milpitas, California, 95035, United States
Tel: (+1 408) 433 25 03
Freephone: (+1 800) 325 69 75
Fax: (+1 408) 435 73 52
Web: http://www.fairchildimaging.com/main/contact.htm
Vice President of Sales/Marketing: William Walker
Marketing Administrator: Sally McAuliffe

■ Design, manufacture and test of high performance, high-end charge coupled devices (CCD)s and cameras for both military and commercial applications.

UPDATED

Firstmark Aerospace Corporation

921 Holloway Street, Durham, North Carolina, 27701, United States
Tel: (+1 919) 956 42 00
Freephone: (+1 888) 832 77 22
Fax: (+1 919) 682 37 86
e-mail: info@firstmarkaerospace.com
repair@firstmarkaerospace.com
Web: http://www.firstmarkaerospace.com

■ Design, manufacture and repair of precision electronic, electromagnetic, and mechanical components and systems for aerospace and defence industries.

UPDATED

Fisher Space Pen Co.

711 Yucca Street, Boulder City, Nevada, 89005, United States
Tel: (+1 702) 293 30 11
Freephone: (+1 800) 634 34 94
Fax: (+1 702) 293 66 16
e-mail: fisher@spacepen.com
Web: http://www.spacepen.com
President and Owner: Paul C Fisher
Finance Director: Cary Fisher
Sales Director: Tim Lawson
Operations/Technical Director: Donald Wong
Marketing Co-ordinator: Ms Sharon Forbes
Vice President: Scott Fisher
Sales Director: Morgan Fisher

■ Manufacture of pens that write in extreme temperatures and at any angle. Can be used over grease, under water, through dirt, oil and in other harsh conditions. The full line includes a variety of colours and styles. Used by NASA and all manned space flights.

UPDATED

Fluid Regulators Company

313 Gillett Street, Painesville, Ohio, 44077, United States
Tel: (+1 440) 352 61 82
Fax: (+1 440) 354 29 12
e-mail: rdb@fluidreg.com

■ Fluid control and measurement components. Custom designed aerospace hydraulic and pneumatic control valves. Specialists in integrating sensing and fluid controls in modular sub-systems.

UPDATED

Garmin International Inc.

1200 East 151 Street, Olathe, Kansas, 66062, United States
Tel: (+1 913) 397 82 00
Freephone: (+1 800) 800 10 20
Fax: (+1 913) 397 82 82
Web: http://www.garmin.com
President: Gary Burrell
Vice President: Min Kao
Marketing Director: Gary Kelley
Senior Media Relations Specialist: Pete Brumbaugh
Tel: (+1 913) 397 82 00
Executive Assistant: Ms Debra Pollard
Media: Jim Stothard
Contact: Ms Andrea McCausland

■ Designs, manufactures and markets navigation and communications electronics worldwide for consumer, business and military applications.

UPDATED

Gencorp Inc.

(Headquarters)
(ex-Tire & General Rubber Company)
PO Box 537012, Sacramento, California, 95853-7012, United States
Tel: (+1 916) 355 40 00
Fax: (+1 916) 351 86 68
Web: http://www.gencorp.com
Chairman: Robert A Wolfe
Chief Executive Officer and President: Terry L Hall
Senior Vice President, Law: Gregory K Scott
Senior Vice President and Chief Financial Officer: Ms Yasmin R Seyal
Vice President and President, GDX Automotive: Michael Bryant
Vice President and President, Aerojet Fine Chemicals: Joseph Carleone
Vice President and President, Aerojet-General Corporation: Michael F Martin
Vice President and President, Gencorp Real Estate: William A Purdey
Vice President Environmental Health and Safety: Chris W Conely
Vice President, Corporate Communications: Ms Linda Cutler
Vice President and Controller: Douglas Jeffries
Vice President, Treasurer: Ms Kari Van Gundy
Assistant Secretary: Ms Margaret Hastings
Assistant Secretary: Robert G Hall

■ Technology-based manufacturing company within the automotive, aerospace and defence industries. Products include liquid and solid propulsion systems for strategic and tactical defence systems and manned and unmanned space launch missions.

Business units
Aerojet
Aerojet Fine Chemicals
GDX Automotive

UPDATED

General Atomics (GA)

PO Box 85608, San Diego, California, 92186-5608, United States
Tel: (+1 858) 455 30 00
Freephone: (+1 800) 854 22 33
Fax: (+1 858) 455 36 21
Telex: 695065
e-mail: info@ga.com
Web: http://www.ga.com
Chairman and Chief Executive Officer: J Neal Blue
Tel: (+1 858) 455 21 52
Senior Vice President: David I Roberts
Tel: (+1 858) 455 29 55
Marketing Communications Manager: Charles S Luby
Tel: (+1 858) 455 36 88
Business Development Manager: Michael R Reed
Tel: (+1 858) 455 24 46
Communications Manager: Doug Fouquet
Marketing Communication Manager: Ms Nancy Hitchcox
Tel: (+1 858) 455 39 51
Fax: (+1 858) 455 39 80
Contact: Ms Virginia Degutis Wolfe

■ Research and development of electromagnetic systems, advanced materials, robotic systems, chemical weapon demilitarisation, signature control, automated non-destructive inspection, repair technologies, hazardous waste elimination, unmanned air vehicles, synthetic aperture radar, electronics, superconducting magnets and materials, nuclear spacepower and advanced computing.

UPDATED

General Dynamics Corp.

2941 Fairview Park Drive, Suite 100, Falls Church, Virginia, 22042-4513, United States
Tel: (+1 703) 876 30 00
Fax: (+1 703) 876 31 25
Web: http://www.gd.com
Chairman and Chief Executive Officer: Nicholas D Chabraja
Tel: (+1 703) 876 32 51
Executive Vice President and Group Executive – Information Systems and Technology: Kenneth C Dahlberg
Tel: (+1 703) 876 33 56
Executive Vice President and Group Executive – Marine Systems and President, Electric Boat: Michael W Toner
Tel: (+1 860) 433 19 85
Executive Vice President and Group Executive – Combat Systems and Resources: Arthur J Veitch
Tel: (+1 703) 876 35 38
Senior Vice President – Planning and Development: David K Heebner
Tel: (+1 703) 876 35 67
Senior Vice President and Chief Financial Officer: Michael J Mancuso
Tel: (+1 703) 876 30 30
Senior Vice President – Human Resources and Administration: Walter M Oliver
Tel: (+1 703) 876 34 15
Senior Vice President and General Counsel, Secretary: David A Savner
Tel: (+1 703) 876 30 10
Vice President, Government Relations: Cordis Colburn
Tel: (+1 703) 876 30 34
Vice President and President, Gulfstream Aerospace: Brian T Moss
Tel: (+1 912) 965 35 00
Vice President – Communications: Kendell Pease
Tel: (+1 703) 876 30 93
Vice President – International: William O Schmieder
Tel: (+1 703) 876 35 77

■ The company is involved in mission-critical information systems and technologies, land and amphibious combat systems, shipbuilding and marine systems and business aviation.

UPDATED

General Dynamics Land Systems (GDLS)

(California Technical Center)
(ex-Delco Defense Systems Operations)
7410 Hollister Avenue, Goleta, California, 93117, United States
Tel: (+1 805) 961 74 77
Fax: (+1 805) 961 50 14
Web: http://www.gdls-ctc.com
http://www.delcosystems.com
Managing Director: Hugo Croft
Marketing and Sales Director: Hung Tran
Marketing Co-ordinator: P Penniman

■ Conducts research, design development and manufacture of turreted weapons systems and advanced land combat electronic systems.

UPDATED

General Dynamics Corp.

(Network Systems)
(ex-General Dynamics Worldwide Telecommunication Systems)
(ex-GTE Government Systems Corp.)
77 A Street, Needham, Massachusetts, 02494, United States
Tel: (+1 781) 449 20 00
Freephone: (+1 888) 483 00 22
Fax: (+1 781) 455 52 55
Web: http://www.gd-ns.com
President: Michael E Chandler
Tel: (+1 781) 455 25 00

■ Provide telecom, information and intelligence systems and is a supplier of customised systems for defence, government and industry in the US and abroad. Provides and manages integrated systems solutions tailored for information processing and communications needs. Capabilities include information resource management systems; software development; weather-related services for commercial, military and general aviation, telephone and call management systems.

Key technologies include advanced switching, information systems, network management and control, data systems, cellular systems, voice processing, secure voice and data, industry and government standards, imagery systems, multimedia systems, broad band networks, computer-based training systems, expert and

neural network applications, rapid prototyping and Human Machine Interface (HMI).
Design, development and production of tactical and strategic state-of-the-art C3 systems for governments, military forces and commercial organisations worldwide. Capabilities involve systems engineering and architecture, software-intensive systems design, utilisation of artificial intelligence and the total integration of information systems, hardware design, manufacturing, installation, training, logistics support, operation and maintenance.

UPDATED

General Dynamics C4 Systems
(Scottsdale)
(ex-General Dynamics, Decision Systems)
(ex-Motorola Integrated Information Systems Group)
8201 East McDowell Road, Scottsdale, Arizona, 85257, United States
Tel: (+1 480) 441 30 33
Freephone: (+1 877) 449 06 00
(+1 877) 466 94 67
(+1 877) 449 05 99
Fax: (+1 480) 726 29 71
e-mail: info@gdds.com
Web: http://www.gdds.com
President: Mark A Fried
Tel: (+1 480) 441 30 33
Manager, External Communications: Ms Diane Mitchell
Tel: (+1 480) 441 64 96
Fax: (+1 480) 441 50 67
Public Affairs Manager: Ms Fran Jacques
Tel: (+1 480) 441 28 85
■ Prime systems integrator of technologies, products and systems for government, defence and industrial customers worldwide, specialising in command and control, information assurance and interoperable communication solutions.
Accreditations: ISO 9001; CMMI Level 5; OSHAS 14001/18001

UPDATED

General Dynamics C4 Systems
(Radio Products)
(ex-General Dynamics Decision Systems)
(ex-Motorola Integrated Information Systems Group)
(ex-Motorola Systems and Technology Group)
(ex-Motorola Systems Solutions Group)
8201 East McDowell Road, Scottsdale, Arizona, 85257, United States
Tel: (+1 877) 449 06 00
(+1 480) 441 30 33
e-mail: info@gdds.com
Web: http://www.gdds.com
Public Affairs Director: Ms Fran Jacques
Tel: (+1 480) 441 28 85
Manager, External Communications: Ms Diane Mitchell
Tel: (+1 480) 441 64 96
Product Marketing Manager: Frances Connors
Tel: (+1 480) 441 40 79
Marketing Manager: Ms Liz Cress
Tel: (+1 480) 441 14 00
Sales Manager: Robert Boyle
Contact: Ms Carol Taska Smith
■ Turnkey ATC communications systems including: VHF and UHF receivers, transmitters and transceivers, emergency transceivers, antennas, multicouplers, remote maintenance and monitoring systems, remote-control systems, communication racks and switches.

UPDATED

Gentex Corp.
(Helmet Systems)
PO Box 315, Carbondale, Pennsylvania, 18407, United States
Tel: (+1 570) 282 35 50
Fax: (+1 570) 282 85 55
e-mail: info@gentexcorp.com
Web: http://www.gentexcorp.com
President: L P Frieder
Chief Financial Officer: Ms Heather Acker
Research and Development Director: Harvey Jagoe
Technical Director: George Hedges
Operations Director: John Pullo
Marketing Director: Frederick W D White
Purchasing Manager: Jim Simpson
Special Projects Manager: Charles Risio
Product Manager, Aircrew Helmets: Alexander R Clawson
Director of Marketing: Charles Rudolf
■ Manufacture of military life support systems, oxygen masks, air, ground and space shuttle helmets. Manufacture microphones, optical and opthalmic products and laser protective optics. Aluminised, engineered, ballistic chemical and biological defence textiles and chemical and biological casualty care systems.

UPDATED

Goodrich Corporation
(Electronc Systems Segment Headquarters)
Four Coliseum Centre, 2730 West Tyvola Road, Charlotte, North Carolina, 28217-4578, United States
Tel: (+1 704) 423 70 00
Fax: (+1 704) 423 70 02

UPDATED

Goodrich Corporation
(Corporate Headquarters)
(ex-BFGoodrich Aerospace)
Four Coliseum Centre, 2730 West Tyvola Road, Charlotte, North Carolina, 28217-4578, United States
Tel: (+1 704) 423 70 00
Fax: (+1 704) 423 70 02
e-mail: corporate.communications@goodrich.com
Web: http://www.goodrich.com
Chairman, President and Chief Executive Officer: Marshall O Larsen
President, Optical & Space Systems Business Unit: Ken Luczaj
Tel: (+1 203) 797 53 93
President, Aircraft Wheels & Brakes Business Unit: Brian Brandewie
President, Engineered Polymer Products Business Unit: Edward Hart
Tel: (+1 904) 696 41 01
Senior Vice President, Strategic and Business Development: Stephen R Huggins
Tel: (+1 704) 423 55 60
Senior Vice President, Technology and Innovation: Jerry S Lee
Tel: (+1 704) 423 55 20
Senior Vice President, Human Resources and Administration General Counsel: Terrence G Linnert
Tel: (+1 704) 423 75 62
Senior Vice President and Chief Financial Officer: Ulrich Schmidt
Tel: (+1 704) 423 75 62
Senior Vice President, Human Resources: Ms Jennifer Pollino
Segment President, Engine Systems Segment: John J Carmola
Tel: (+1 704) 423 34 84
Segment President, Airframe Systems Segment: John J Grisik
Tel: (+1 440) 262 14 40
Segment President, Electronic Systems Segment: Ms Cynthia M Egnotovich
Tel: (+1 704) 423 86 80
Vice President, Business Development and Tax: Joseph F Andolino
Tel: (+1 704) 423 71 28
Vice President and Controller: Scott E Kuechle
Tel: (+1 704) 423 55 35
Vice President, Airbus Customer Services: Graham Bailey
Tel: (+33 561) 30 53 56
Vice President, Corporate Communications: Ms Lisa Bottle
Tel: (+1 704) 423 70 60
Vice President, Government Programs: Jerry LaReau
Tel: (+1 703) 558 82 31
Director, Media Relations: Ms Gail K Warner
Tel: (+1 704) 423 70 48
Organised into three business segments:
-Airframe Systems
-Goodrich Actuation Systems
-Goodrich Aviation Technical Services
-Goodrich Engineered Polymer Products
-Goodrich Landing Gear
-Goodrich Wheels & Brakes
-Electronic Systems
-Goodrich Aircraft Interior Products
-Goodrich Propulsion Systems
-Goodrich De-Icing & Speciality Systems
-Goodrich Fuel & Utility Systems
-Goodrich Hella Aerospace Lighting Systems
-Goodrich Optical & Space Systems
-Goodrich Electro-Optical Systems
-Goodrich Space Flight Systems
-Goodrich Surveillance & Reconnaissance Systems
-Goodrich Power Systems
-Goodrich Sensor Systems
-Engine Systems
-Goodrich Aerostructures
-Goodrich Cargo Systems
-Goodrich Engine Control Systems
-Goodrich Turbine Fuel Technologies
-Goodrich Turbomachinery Products

UPDATED

Goodrich Actuation Systems
(Vermont Office)
(ex-Simmonds Precision, Aircraft Integrated Systems)
100 Panton Road, Vergennes, Vermont, 05491, United States
Tel: (+1 802) 877 29 11
Fax: (+1 802) 877 41 13
■ Fuel measurement/management systems, computer-based performance management systems, engine instruments, centre of gravity controls, speed and torque sensors, position sensors.

UPDATED

Goodrich Aircraft Interior Products (AIP)
(Propulsion Systems, Arizona)
(ex-Universal Propulsion Company, Inc.)
25401 North Central Avenue, Phoenix, Arizona, 85027-7899, United States
Tel: (+1 623) 516 33 40
Fax: (+1 623) 516 33 64
Web: http://www.upco.goodrich.com
■ Design, develop and manufacture emergency ballistic devices and systems, ejection seats; also produce highly engineered propellants, for rocket catapults, rocket motors, initiators, gas generators, drogue guns, thrusters, ballistic systems.

UPDATED

Goodrich Optical & Space Systems (OSSD)
(Group Headquarters)
(ex-Barnes Detectors & Sensors)
(ex-EDO Barnes)
100 Wooster Heights Road, Danbury, Connecticut, 06810-7589, United States
Tel: (+1 203) 797 50 00
Fax: (+1 203) 797 54 18
e-mail: eos@goodrich.com
Web: http://www.oss.goodrich.com
President: Ken Luczaj
Tel: (+1 203) 797 53 92
Vice President and General Manager Electro Optical Systems: David Olson
Tel: (+1 203) 797 59 69
Vice President and General Manager, Surveillance and Reconnaissance Systems: Thomas Bergeron
Tel: (+1 978) 967 21 29
Vice President and General Manager, Space Flight Systems: Chris Holmes
Tel: (+1 505) 938 51 00
■ Design and manufacture of high performance custom engineered electronics, optics and electro-optical products and services for defence, scientific, and commercial applications.
Accreditations: ISO 9001
Operating Units
-Electro-Optical Systems
-Space Flight Systems
-Surveillance and Reconnaissance Systems

UPDATED

Goodrich Optical & Space Systems (OSSD)
(Electro-Optical Systems, Danbury)
100 Wooster Heights Road, Danbury, Connecticut, 06810-7589, United States
Tel: (+1 203) 797 50 00
Fax: (+1 203) 797 54 18
Web: http://www.oss.goodrich.com/electroopticalsystems
Laser Warning and Counter Measures: Ronald Dale
Tel: (+1 203) 797 59 75
Advanced Windows, Electro-Optical and Opto-Mechanical Assemblies; Optical Coatings and Filters: David Imbrogno
Tel: (+1 203) 797 63 23
Lightweight Precision Optics: Matthew Magida
Tel: (+1 203) 797 63 23
Large Optics; Space Optics and Optical Systems; Telescopes and Beam Control: Stanley Kishner
Tel: (+1 203) 796 71 85
Attitude Determination and Control: Ms Celeste McGorty
Tel: (+1 203) 797 52 10
Commercial Detectors: Ms Marie Nicolosi
Tel: (+1 203) 926 41 49
Remote and Space Sensors: Bruce Boyce
Tel: (+1 203) 926 41 49
Space Observatories: Stanley Kishner
Tel: (+1 203) 796 71 65
Special Programmes: Andreas Nonnenmacher
Tel: (+1 203) 797 52 48
■ Design, development, and manufacture of precision optics and electro-optical systems and

sensors for defense, aerospace, scientific, and commercial applications.

The two main business segments, Sensor Systems and Optical Systems are active in the Chandra X-ray Observatory and the Hubble Space Telescope programmes; as well as development of laser threat warning systems for aircraft and armored vehicles, fine guidance sensors, satellite attitude determination systems, and Earth resources satellite optics.

UPDATED

Goodrich Optical & Space Systems (OSSD)
(Electro-Optical Systems, Ithaca)
950 Danby Road, Suite 100, Ithaca, New York, 14850-5720, United States
Tel: (+1 607) 272 76 40
Fax: (+1 607) 272 08 04

UPDATED

Goodrich Optical & Space Systems (OSSD)
(Space Flight Systems)
6600 Gulton Court, NE, Albuquerque, New Mexico, 87109, United States
Tel: (+1 505) 345 90 31
Fax: (+1 505) 344 98 79
Web:http://www.oss.goodrich.com/spaceflightsystems.shtml
Launch Vehicle, Spacecraft and Missile Mission Critical Systems: Jim Hanchett
Tel: (+1 505) 938 50 10

■ Applied engineering and manufacture of in-flight space data and control electronics space-qualified electronic systems and actuator products. Design and production of satellite subsystem architecture incorporating, attitude control, spacecraft thermal control, and power regulation and distribution systems and integrated 32-bit processors for space and other harsh environments.

UPDATED

Goodrich Optical & Space Systems (OSSD)
(Surveillance & Reconnaissance Systems, Beale)
2801 C Street, Beale Air Force Base, California, 95903-1517, United States
Tel: (+1 530) 788 92 14
Fax: (+1 530) 788 92 22

UPDATED

Hamilton Sundstrand

(Headquarters)
(ex-Hamilton Standard)
One Hamilton Road, Windsor Locks, Connecticut, 06096-1010, United States
Tel: (+1 860) 654 60 00
Fax: (+1 860) 654 37 73
Web: http://www.hamiltonsundstrand.com
President: Ron McKenna
President, Engine and Control Systems: Joseph Triompo
President, Aerospace Power Systems: Dave Hess
President, Flight Systems and Customer Service: Jim Gingrich
Vice President and Chief Financial Officer: David Nord
Vice President and General Counsel: Mike Monts
Vice President, Quality: Mike Ratchford
Vice President, Human Resources: Tatsuo Shirane
Vice President, Operations: John Boyd
Contact: Ms Peg Hashem
Tel: (+1 860) 654 34 69

■ A wide variety of high technology products and systems for the aerospace business market. Electric power generation systems, engine controls, environmental control systems, propeller systems, flight control systems, actuation systems and auxiliary power units.

International subsidiaries
Hamilton Standard Marston Aerospace, Ltd, UK
Hamilton Standard-Nauka, Russia
Microtecnica, Italy
Nord-Micro, Germany
Ratier-Figeac, France

UPDATED

Hamilton Sundstrand
(Flight and Undersea Systems)
4747 Harrison Avenue, PO Box 7002, Rockford, Illinois, 61125-7002, United States
Tel: (+1 860) 654 60 00
Fax: (+1 815) 226 72 88

■ Aircraft pumps, electrical generating systems, gearboxes, actuation systems and components, generators, electric motors, starting systems, pneumatic conditioning and control systems, fans, refrigeration systems, missile power units and emergency turbine systems.

UPDATED

Hamilton Sundstrand
(Power Systems)
4400 Ruffin Road,
PO Box 85757, San Diego, California, 92186-5757, United States
Tel: (+1 858) 627 65 27
Fax: (+1 860) 660 41 85
e-mail: businessdev@hs.utc.com
Web: http://www.hs-powersystems.com
General Manager Power Systems: Bob Ridgeway
Director, Engineering: Jerry Swadley
Director, Operations: Mike Blomgren
General Manager, Customer Service: Marty Kessell
Manager, Finance and Administration: Brian Roy
Manager, Marketing Services: Hilli M Christopherson
Vice President and General Manager Power Systems: Tim Morris
General Manager, Customer Service: Bill Worsham

■ Small gas turbine engines, generator sets and APU's for commercial and military aircraft and vehicles. Fans and vapour cycle systems. Propulsion engines for RPVs.

UPDATED

Harris Corporation

(Headquarters)
1025 West NASA Boulevard, Melbourne, Florida, 32919, United States
Tel: (+1 321) 727 92 07
(+1 321) 727 92 07
Freephone: (+1 800) 4 HARRIS
Fax: (+1 321) 727 96 46
Web: http://www.harris.com
Media Relations Director: Tom Hausman

UPDATED

Harris Corporation

(Government Communications Systems Division)
PO Box 37, Melbourne, Florida, 32902, United States
Tel: (+1 321) 727 40 00
Fax: (+1 321) 727 57 89
Web: http://www.govcomm.harris.com

UPDATED

Harris Corporation
(RF Communications Division)
1680 University Avenue, Rochester, New York, 14610-1887, United States
Tel: (+1 585) 244 58 30
Fax: (+1 585) 244 29 17
Web: http://www.rfcomm.harris.com
Senior Marketing Communications Specialist: Ms Harriet J Wado
Tel: (+1 585) 241 81 90

■ Secure wireless radio communication products, systems and services for defence, government and law enforcement markets. Multiband (HF-VHF-UHF) radios are provided in manpack and vehicular applications.

UPDATED

Harris Corporation
(Government Communications Systems Division)
PO Box 37, Melbourne, Florida, 32902-0037, United States
Tel: (+1 321) 727 40 00
Web: http://www.govcomm.harris.com
http://www.harris.com

■ Design, development and integration of assured communications and information systems design for US government customers including aerospace, defence and intelligence.

UPDATED

Hernandez Engineering, Inc. (HEI)

16055 Space Center Boulevard, Suite 725, Houston, Texas, 77062, United States
Tel: (+1 281) 280 51 59
Fax: (+1 281) 480 75 25
Web: http://www.hernandezengineering.com

■ Capabilities include: engineering analysis, operations, and science support; design and system engineering; system safety, reliability and quality engineering, product assurance, and independent assessment; documentation and information management; industrial and occupational safety; and multi-media computer-based training.

UPDATED

Hitco Carbon Composites, Inc.

1600 West 135th Street, PO Box 1097, Gardena, California, 90249, United States
Freephone: (+1 800) 421 54 44
Fax: (+1 310) 516 57 14
e-mail: refrasil@hitco.com
Web: http://www.hitco.com

■ Supply of composite structures, rocket nozzles and high temperature materials to aerospace and industrial markets. Design and manufacture of specialty insulation components fabricated from a range of high temperature fabrics, non-wovens and metal foils. The design of insulation products can be tailored to customer specifications. Markets served include: aircraft, chemical processing, personnel safety, foundry and forging, heavy industrial, power generating, automotive and basic metal processing.

Accreditations: ISO 9001

UPDATED

Honeywell Aerospace

(Headquarters)
(ex-AlliedSignal Aerospace)
(ex-AlliedSignal Inc., Aerospace)
1944 East Sky Harbor Circle, Buildings, 2101-2107 (Area 21), Phoenix, Arizona, 85034, United States
Tel: (+1 480) 592 50 00
(+1 602) 231 10 00
(+1 602) 365 21 80
Freephone: (+1 800) 421 21 33
Telex: chdgrcr
Web: http://www.honeywellaerospace.com
President and Chief Operating Officer, Honeywell: David M Cole
President and Chief Operating Officer, Honeywell Aerospace: Robert D Johnson
President Aerospace Electronics: Dean H Flatt
President, Commercial Avionics Electrical Systems: Frank Daly
President, Engines Systems and Services: Steven R Loranger
President, Federal Manufacturing and Technologies: David Douglass
Senior Vice President and Chief Information Officer, Group: Larry E Kittelberger
Senior Vice President and Chief Financial Officer: Richard F Wallman
Senior Vice President, Human Resources, Group: Thomas W Weidenkorpf
Senior Vice President, General Counsel: Peter H Kreindley
Vice President and General Manager, Commercial Aerospace: Ms Lynn Brubaker
Vice President, General Manager, Aircraft Landing Systems: Ms Margaret (Peg) Bilson
Vice President, Contracts and Pricing: Harry Borenstein
Vice President and Chief Legal Officer: Mike Denton
Vice President, Chief Information Officer: Ms Rita Heise
Vice President, Communications: William H Biateil

■ Manufacture aircraft guidance, navigation, communication and identification systems. Flight controls and cockpit instrumentation: weather radar, fuel controls and ignition systems. Commercial and military aircraft wheels, brakes and struts. Gas turbine engines, APUs, turboprop and turbofan engines. Environmental systems, cabin pressurisation systems, systems and controls for aircraft, missiles and undersea applications. Provide technical services and spacecraft and satellite tracking support to DoD, DoE and NASA.

UPDATED

Honeywell Aerospace, Aerospace Electronic Systems
(Fort Lauderdale)
(ex-Bendix/King Air Transport Avionics)
13350 US Highway 19 North, Clearwater, Florida, 33764, United States
Tel: (+1 727) 531 46 11
Fax: (+1 727) 539 34 47
Contact: Lee Williams

■ Design and production of precision components and systems for stabilisation, control, navigation and guidance of military and space vehicles.

UPDATED

Honeywell Aerospace, Aerospace Electronic Systems
(Glendale)
19019 North 59th Avenue, Glendale, Arizona, 85308, United States
Tel: (+1 602) 822 31 58
Fax: (+1 602) 822 36 04
Chief Systems Engineer: Mitch Fletcher

UPDATED

Honeywell Data Control
(Commercial Electronic Systems)
(ex-AlliedSignal Data Control)
(ex-Sundstrand Data Control)
15001 NE 36th Street, PO Box 97001, Redmond, Washington, 98073-9701, United States
Tel: (+1 425) 885 37 11
Fax: (+1 425) 885 20 61
Web: http://www.honeywellaerospace.com
http://www.egpws.com

■ Components and subsystems such as avionics systems for flight safety, accelerometers, thermal products, navigation and flight management equipment, telecommunication and satellite communication programmes, airborne communications systems and data management equipment.

UPDATED

Honeywell Aerospace, Engines, Systems & Services
(Headquarters)
(ex-Honeywell Aerospace, Engines & Systems)
PO Box 52181, Phoenix, Arizona, 85072-2181, United States
Tel: (+1 602) 231 10 00
Fax: (+1 602) 231 57 13
Web: http://www.honeywellaerospace.com
President, Engines, Systems and Services: Steven R Loranger
Vice President and General Manager, AS 900 Program: Michael Redenbaugh
Tel: (+1 602) 231 23 61
Chief Financial Officer and Vice President of Finance and Administration: Dominic Romeo
Tel: (+1 602) 365 38 66
Vice President and General Manager, Engine Systems and Accessories: Adrian Brown
Tel: (+1 480) 592 37 00
Vice President and General Manager, Airframe Systems: Roger E Wolfe
Vice President and General Manager, Asia-Pacific: Carl Baerst
Vice President, Quality Assurance: F R Pocock
Tel: (+1 602) 231 33 66
Vice President, Human Resources: Steve Yamasti
Tel: (+1 602) 231 20 47
Vice President, Supply Chain: Peter Riley
Tel: (+1 602) 231 56 00
General Counsel and Vice President: Gjon Nivica
Tel: (+1 602) 231 42 24

■ Gas turbine engines for aircraft and helicopters including: turboprop, turbofan turboshaft and auxiliary power units. Power management generation systems, engine systems and accessories and environmental control systems.

UPDATED

Honeywell Aerospace, Engines, Systems & Services
(Airframe Systems)
(ex-AlliedSignal Aerospace, Engines & Systems)
2525 West 190th Street, Torrance, California, 90504-6099, United States
Tel: (+1 310) 323 95 00

UPDATED

Honeywell Aerospace, Engines, Systems & Services
(Engine Systems and Accessories)
(ex-Honeywell Aerospace, Engines & Systems)
1300 West Warner Road, Tempe, Arizona, 85284, United States
Tel: (+1 480) 592 50 00
Vice President and General Manager, Engine Systems and Accessories: Ms Adrian Brown

UPDATED

Honeywell Aerospace, Engines, Systems & Services
(Manufacturing Center)
(ex-AlliedSignal Aerospace, Engines & Systems)
(ex-Honeywell Aerospace, Engines & Systems)
PO Box 52187, Phoenix, Arizona, 85034, United States
Tel: (+1 602) 231 10 00

UPDATED

HR Textron
(ex-Textron Motion Control)
25200 West Rye Canyon Road, Santa Clarita, California, 91355, United States
Tel: (+1 661) 294 60 00
Fax: (+1 661) 259 96 22
Web: http://www.hrtextron.com
President: Richard Hillman
Vice President, Aircraft Products: Joseph Klocko
Director, Business Development: Richard (Rick) Bartz

■ Flight control systems and components for high-performance aircraft, space launch vehicles, helicopters, missiles and turbine engines; servovalves, fuel and pneumatic systems components, automatic test equipment and product support. Land vehicle turret control and stabilisation systems.

UPDATED

Hydro Fitting Manufacturing Corp.
733 East Edna Place, PO Box 1558, Covina, California, 91722, United States
Tel: (+1 626) 967 51 51
Fax: (+1 626) 339 44 54
President: Seth D Schwartz
Financial Director: Ms Johanne Schwartz
Sales Director: Ms Susan Buchanan
Purchasing Manager: Ms Polly Teague-Edgell

■ Fittings, AN, MS, NAS, NSA approved Airbus. Valves, air charging; MIL-SPEC to order; 100% CNC shop. Covers, engine magnesium inspection plates.

UPDATED

ILC Dover Inc.
One Moonwalker Road, Frederica, Delaware, 19946-2080, United States
Tel: (+1 302) 335 39 11
(+1 800) 631 95 67
Fax: (+1 302) 335 07 62
(+1 303) 335 46 31
(+1 302) 335 13 20
e-mail: lawrer@ilcdover.com
Web: http://www.ilcdover.com
President: William J Wallach
Executive Vice President: Carl E Zlock
Vice President, Operations: Bradford S Walters
Director, Marketing: Douglas A Durney
Advertising Manager: Ronald W Lawrence
Administrative Assistant, Marketing: Ms Juanda Duley
Contact: Ms Kay Emery

■ Space suits and accessories, aircrew equipment, aerostats, airships and balloons. NBC masks, weapon decelerators, impact attenuation airbags, pressure suits and inflatable space structures.

UPDATED

Inspec Foams Inc.
(ex-IMI-Tech Corp.)
101 East Park Boulevard, Suite 201, Plano, Texas, 75074, United States
Tel: (+1 972) 516 07 02
Freephone: (+1 800) 666 76 10
Fax: (+1 972) 516 06 24
e-mail: inspecfoams@airmail.net
Web: http://www.inspecfoams.com
Operations Director: Dan Sutton
Marketing and Sales Director: Ms Betty Hartman
General Manager: Robert Tait
Technical Manager: Dr Scott Snider

■ Manufacture of Solimide polyimide flexible foams used for fire protection, thermal and acoustic insulation in marine and aircraft applications. Cushioning for aircraft and space applications.

UPDATED

Interpoint
(ex-Integrated Circuits Incorporated)
10201 Willows Road, PO Box 97005, Redmond, Washington, 98073-9705, United States
Tel: (+1 425) 882 31 00
Freephone: (+1 800) 822 87 82
Fax: (+1 425) 882 19 90
e-mail: power@intp.com
Web: http://www.interpoint.com
unknown: Winnie Chen
Fax: (+1 425) 882 19 90
President: Lou Bieck
Vice President of Finance and Administration: Robert Sorakubo
Executive Director, North American Sales: Vince Piscopo
Executive Director, International Sales: Simon Abel
Director of Space Products: Harvey Stanley
Tel: (+1 425) 895 40 52
Senior Tech Writer and Illustrator: Kenilee Campbell

■ Manufacture of high-density, high-reliability, microelectronics and power conversion products for the aerospace, space, military, medical, industrial and commercial markets. Products qualified to MIL-PRF-38534 Class H or K and space products available to radiation level R. Provides turn-key, co-design and build-to-print options specialising in light weight, high-reliability microelectronic assembly and test. Capabilities include BGA, flip-chip, COB, COF and hermetically sealed devices.

UPDATED

ITT Industries Inc.
(World Headquarters)
4 West Red Oak Lane, White Plains, New York, 10604, United States
Tel: (+1 914) 641 20 00
Fax: (+1 914) 696 29 50
Web: http://www.itt.com
Chairman of the Board: Louis J Giuliano
President and Chief Executive Officer: Steven R Loranger
President, Electronic Components and Senior Vice President ITT Industries: Ms Brenda Reichelderfer
Senior Vice President and Chief Financial Officer: Edward W Williams
Senior Vice President, ITT Industries and President, Fluid Technology: Robert L Ayers
Senior Vice President and Director of Human Resources: Scott A Crum
Senior Vice President, ITT Industries and President, Defense: Henry J Driesse
Senior Vice President, Treasurer and Director of Taxes: Donald E Foley
Senior Vice President and General Counsel: Vincent A Maffeo
Senior Vice President and Director of Corporate Relations: Thomas R Martin
Tel: (+1 914) 641 21 57
Fax: (+1 914) 696 29 77
Vice President and Chief Information Officer: Steven A Faas
Vice President, Secretary and Associate General Counsel: Ms Kathleen S Stolar
Vice President and Associate General Counsel: Lawrence J Swire
Vice President and Associate General Counsel Director, Environment, Safety and Health: Ms Usha Wright
Vice President and Corporate Controller: Mark E Lang
News and Public Affairs: Thomas Glover
Tel: (+1 914) 641 21 60
Purchasing Manager: Jay Furbay

UPDATED

ITT Industries Inc.
(Aerospace/Communications)
(ex-ITT Aerospace/Communications)
1919 West Cook Road, PO Box 3700, Fort Wayne, Indiana, 46801-3700, United States
Tel: (+1 219) 451 60 00
Fax: (+1 260) 451 50 92
(+1 260) 451 61 26
Web: http://www.acd.itt.com
Marketing Operations Manager: John T Kirkwood
Tel: (+1 260) 451 63 84

■ Design, development and production of forward area battlefield, tactical communications, signal processing equipment and systems. Voice and data communications receiving and transmitting systems; key management systems, surveillance and reconnaissance receiver systems; voice and data secure communications equipment and meteorological and navigation spaceborne payloads. Qualified for ground, airborne and space applications. Featuring microprocessor control and standard digital interface.

UPDATED

ITT Industries, Defense
(ex-ITT Defense & Electronics)
1650 Tysons Boulevard, Suite 1700, Mclean, Virginia, 22102, United States
Tel: (+1 703) 790 63 00
Fax: (+1 703) 790 63 60
Web: http://www.itt.com
President: Henry J Driesse

Vice President and General Counsel: Peter Johnsen
Vice President, International Marketing: Bruce Scott
Vice President, Government Relations: Jim Crumley
Acting Vice President and Controller: Mark Lang
Director, Corporate Operations and Office Support: Vince Thomas
Director of Communications: Ms Millie Rogers
Congressional Liaison and Program Development Director: Maj Gen Harry W Jenkins

■ Supply of advanced technology command, control and communications; electronic defence, electro-optical and surveillance systems; satellite imaging systems and information services; navigation and meteorological space payloads; related operations and management services. Primary business markets include avionics and data communications for military, space and transportation customers.

Group Companies
ITT Industries, Advanced Engineering and Sciences
ITT Industries, Aerospace/Communications
ITT Industries, Avionics
ITT Industries, Defence International
ITT Industries, Night Vision
ITT Industries, Space Systems
ITT Industries, Systems
ITT Industries, Gilfillan

UPDATED

Johns-Manville

(ex-Schuller International)
PO Box 5108, Denver, Colorado, 80217-5108, United States
Tel: (+1 303) 978 47 47
Freephone: (+1 800) 654 31 03
Fax: (+1 303) 978 20 00
Web: http://www.jm.com
Business Director: William A Riaski

■ Acoustical and thermal insulation for aircraft, Q-Fiber component of space shuttle tiles and blankets.

UPDATED

Joyce Telectronics Corp.

2049 Range Road, Clearwater, Florida, 33765-2124, United States
Tel: (+1 727) 461 35 25
Fax: (+1 727) 461 61 72
e-mail: joytc@msn.com
Web: http://www.joycetelectronics.com
President: Peter Joyce
Vice President: Michael Joyce

■ Aircraft and shipboard headset microphones, amplifiers, telephone handsets, headset adapters, ground communication cables and switches used in airlines, helicopters, military and space applications. Telephone, telegraph and aircraft equipment.

UPDATED

Kaiser Electro-Optics, Inc. (KEO)

2752 Loker Avenue West, Carlsbad, California, 92008, United States
Tel: (+1 760) 438 92 55
Fax: (+1 760) 438 68 75
e-mail: mktginfo@keo.com
Web: http://www.keo.com
President: Jerry T Carollo
Vice President, Business Development: Ben J Mall
Tel: (+1 619) 438 92 55
Vice President, Displays: Bill Maffucci
Vice President, Optics: Ms Janet Zeidler
Senior Sales Engineer, Optical Products: Chris Voita
Tel: (+1 760) 438 92 55
Marketing Manager, Displays: Dan Lowe
Displays Marketing Manager: Dave Kavahele
Chief Scientist: Dr Warren Smith

■ Design, development and manufacture of optical, opto-mechanical, electro-optical and display products for demanding aerospace, medical, defence, industrial, training and simulation applications. Products include: custom and commercial off-the-shelf optical systems for remote sensing, reconnaissance surveillance space instruments, space star tracker, countermeasures and laser communications; head-mounted, head-up, head-down displays for virtual flight simulators; tank sight display systems; land warrior monocular displays and complete virtual reality systems.

Accreditations: ISO 9001

UPDATED

Kaman Aerospace Corporation (KAC)

(Bloomfield Office)
Old Windsor Road, PO Box 2, Bloomfield, Connecticut, 06002-0002, United States
Tel: (+1 860) 243 73 36
(+1 860) 242 44 61
(+1 860) 243 71 00
(+1 860) 769 77 79
Fax: (+1 860) 243 70 43
(+1 860) 243 78 52
Telex: 99326
Web: http://www.kamanaero.com
Vice President, Finance: Dean Geoffrey

■ Helicopters (SH-2, K-Max), aircraft components, remotely piloted vehicles, electro-optical systems, precision measuring equipment, memory systems and fusing and safety devices.

UPDATED

Kaman Aerospace Corporation (KAC)
(Electro-Optics Development Center)
3480 East Britannia Drive, Suite 120, Tucson, Arizona, 85706-5007, United States
Tel: (+1 520) 295 21 61
Fax: (+1 602) 889 02 11
Vice President, EODC: Dr Bobby Ulich
Director, Programmes: Mark Yokley
Manager, Programme Development: Harold E Linton

UPDATED

Kayser-Threde NA Inc.

8469 South Saginaw Street, Grand Blanc, Michigan, 48439, United States
Tel: (+1 810) 695 69 10
Fax: (+1 810) 695 61 30
e-mail: support@kt-na.com
Web: http://www.kayser-threde-na.com

UPDATED

Kearfott Guidance & Navigation Corp. (KGN)

150 Totowa Road, Wayne, New Jersey, 07474-0946, United States
Tel: (+1 973) 785 60 00
(+1 973) 785 61 83
Fax: (+1 973) 785 59 05
(+1 973) 785 60 25
Telex: 133440
e-mail: marketing@kearfott.com
Web: http://www.kearfott.com
President and Chief Executive Officer: Dr Ronald E Zelazo
Tel: (+1 973) 785 68 00
Vice President, Business Development and Management: Dr James Gardner
Tel: (+1 973) 785 61 11
Vice President, Finance: S Givant
Tel: (+1 973) 785 67 35
Vice President, Engineering: Thomas Shanahan
Tel: (+1 973) 785 65 27
Vice President and General Manager, NJ Operations: Walter Peters
Business Development and Management Director: John Kelsey
Technical Marketing and Marketing Services Director: James Sweetser
Tel: (+1 973) 785 64 70
Purchasing Director: Richard Varina
Tel: (+1 973) 785 58 99
Manager, Marketing and Administrative Services: Joseph A Smalz
Tel: (+1 973) 785 60 61
Manager, Space Applications Business Development: Gerald Gilbert
Business Development Manager: James Fake
Tel: (+1 973) 785 62 10
Business Development Manager: Edward Kelly
Tel: (+1 973) 785 66 89
Business Development Manager: Robert Reisinger
Tel: (+1 973) 785 61 49
Business Development Manager: William Alameda
Tel: (+1 973) 785 69 96

■ Advanced inertial and stellar-inertial navigation and guidance using ring laser and tuned rotor gyroscopes. Design, manufacture and integration of systems for navigation, guidance and control of aircraft, space, land and marine vehicles, inertial components including gyroscopes, platforms, accelerometers, servo mechanisms, pilots displays and control panels, GSE, hydraulic and electro mechanical controls, motors, synchros, resolvers and actuators.

UPDATED

Kelly Space & Technology

294 South Leland Norton Way, Suite 3, San Bernardino, California, 92408, United States
Tel: (+1 909) 382 20 10
Fax: (+1 909) 382 20 12
e-mail: katadmin@kellyspace.com
Web: http://www.kellyspace.com
Facility Security Officer and Office Manager: Ms Donna Eatinger
Tel: (+1 909) 382 56 59
Fax: (+1 909) 382 2012

UPDATED

Kinesix Corporation

7700 San Felipe, Suite 200, Houston, Texas, 77063, United States
Tel: (+1 713) 953 83 00
Freephone: (+1 800) 953 53 30
Fax: (+1 713) 953 83 06
e-mail: info@kinesix.com
Web: http://www.kinesix.com
Chief Executive Officer: Mike Teague
General Manager: Paul Thiebaut
Product Marketing Manager: Ty Reid
Tel: (+1 713) 953 83 50

UPDATED

Liquidmetal Technologies

(Head Office)
100 North Tampa Street, Suite 3150, Tampa, Florida, 33602, United States
Tel: (+1 813) 314 02 80
Fax: (+1 813) 314 02 70
e-mail: information@liquidmetal.com
defense@liquidmetal.com
space@liquidmetal.com
Web: http://www.liquidmetal.com
Chairman of the Board: James Kang
Chief Executive officer and President: John Kang
Chief Financial Officer and Executive Vice President: Brian McDougall
Chief Strategy Officer and Executive Vice President: Scott Wiggins
Chief Technology Officer: Neil Paton

■ Development of alloy composites for use in environmentally safe Kinetic Energy Penetrators (KEP) and space borne applications.

UPDATED

Liquidmetal Technologies

(Lake Forest)
25800 Commercentre Drive, Suite 100, Lake Forest, California, 92630, United States
Tel: (+1 949) 206 80 00
Fax: (+1 949) 206 80 08

UPDATED

Lista International Corp.

106 Lowland Street, Holliston, Massachusetts, 01746-2094, United States
Tel: (+1 508) 429 13 50
Freephone: (+1 800) 722 30 20
Fax: (+1 508) 626 03 53
e-mail: sales@listaintl.com
Web: http://www.listaintl.com
President: Donald Brown
Financial Director: Debera Falcone
Sales Director: John Alfieri
Marketing Manager: Ms Anne Smagorinsky
Purchasing Manager: Greg Johnson
Marketing Assistant: Ms Kim Lucchesi

■ Design and Manufacture of modular storage and workspace systems-drawer storage cabinets, industrial and technical electronic workbenches. Lista International sells in the United States, Canada, Mexico and Latin America.

UPDATED

Lithion

82 Mechanic Street, Pawcatuck, Connecticut, 06379-2154, United States
Tel: (+1 860) 599 11 00
Fax: (+1 860) 599 51 22
e-mail: vyevoli@lithion.com
Web: http://www.lithion.com
Vice President and General Manager: Vincent A Yevoli
Tel: (+1 860) 599 11 00
Manager, Sales/Marketing: Robert Hellen
Tel: (+1 860) 599 11 00
Administrative Assistant: Ms Tracy Anderson
Tel: (+1 860) 599 11 00
Contact: Kris Johanessen

■ Manufacture of lithium-ion batteries.

UPDATED

Lockheed Martin Corporation
(ex-Lockheed Corporation)
(ex-Martin Marietta Corporation)
6801 Rockledge Drive, Bethesda, Maryland, 20817, United States
Tel: (+1 301) 897 60 00
Fax: (+1 301) 897 62 52
Web: http://www.lockheedmartin.com
Chairman and Chief Executive Officer: Vance D Coffman
President, LMC Properties, Inc.: Thomas J Quinn
President and Chief Executive Officer, Investment Management Co.: Anthony G Van Schaick
President and Chief Operating Officer: Robert J Stevens
Senior Vice President and General Counsel: Frank H Menaker
Executive Vice President, Technology Services: Michael F Camardo
Executive Vice President, Systems Integration: Robert B Coutts
Executive Vice President, Lockheed Martin Aeronautics Company: Dain M Hancock
Executive Vice President, Space Systems: Albert E Smith
Corporate Vice President, Washington Operations: Brian D Dailey
Corporate Chief Information Officer: Joseph R Cleveland
Vice President and Chief Executive Officer, Lockheed Martin Global Telecommunications: John V Sponyoe
Vice President, Corporate Shared Services: Ms Marillyn A Hewson
Vice President, Best Practices: Michael Joyce
Vice President, Ethics and Business Conduct: Ms Nancy Higgins
Vice President, Strategic Development: Arthur E Johnson
Vice President and Chief Technical Officer: Malcom O'Neill
Vice President, Corporate Secretary & Associate General Counsel: Ms Lillian M Trippett
Vice President, Corporate Communications: Dennis Boxx
Tel: (+1 301) 897 65 43
Vice President, Corporate Business Development: Robert H Trice
Vice President, Finances: Thomas F Kinstle
Vice President, Investor Relations: James R Ryan
Vice President and Controller: Rajeev Bhalla
Vice President, Contracts: Ms Eleanor R Spector
Vice President, Internal Audit: Ms Kimberly Gavaletz
Vice President, Corporate Energy, Environment, Safety and Health: Jim O'Brien
Vice President, Human Resources: Terry Powell
Vice President and Treasurer: Ms Janet L McGregor
Vice President, Financial Strategies: John E Montague
Director, Media Relations: Jim Fertig
Fax: (+1 301) 897 62 89
Director Public Relations: Hugh Burns
Tel: (+1 301) 897 63 08
Chief Financial Officer: Christopher E Kubasik
Contact: Eric Sollinger
■ A global enterprise engaged in the research, design, development, manufacture and integration of advanced-technology systems, products and services for government and commercial customers. The core businesses span systems integration, aeronautics, space technology services and global telecommunications.

Operating Units
-Aeronautics Business Area
-Aeronautics
-Electronic Systems Business Area
-Lockheed Martin Canada
-Maritime Systems & Sensors
-Missiles & Fire Control
-Simulation, Training & Support
-Systems Integration – Owego
-Transportation & Security Solutions
-Information and Technology Services Business Area
-Aircraft & Logistics Centers
-Information Technology
-KAPL, Inc. (Knolls Atomic Power Laboratory)
-Sandia Corporation (Sandia National Laboratory)
-Space Operations
-Systems Management
-Technical Operations
-Technology Ventures
-Integrated Systems & Solutions Business Area
-Integrated Systems & Solutions
-Lockheed Martin Orincon
-Space Systems Business Area
-Space Systems

UPDATED

Lockheed Martin Missiles and Fire Control
(Dallas Office)
(ex-Lockheed Martin Vought Systems)
PO Box 650003 M/S PT-42, Dallas, Texas, 75265-0003, United States
Tel: (+1 972) 603 14 01
(+1 972) 603 10 00
Fax: (+1 972) 603 10 09
Web: http://www.missilesandfirecontrol.com
President, Missiles and Fire Control: James F Berry
President, Missiles and Fire Control – Orlando: Stanley Arthur
Vice President, Finance: Mike Beck
Vice President, International Business Development: John Ward
Tel: (+1 972) 603 11 14
Vice President, Production Operations: Jeffrey Wilson
Vice President, Communications: Ms Joy Sabol
Tel: (+1 407) 356 22 07
Senior Manager, Media and Trade Relations: Craig Vanbebber
Tel: (+1 972) 603 14 01
Manager, International Business Development Programmes: Jon C Schreyach
Tel: (+1 972) 603 76 29
Fax: (+1 972) 603 02 77
■ Development, manufacture and support of advanced combat, missile, rocket and space systems.

UPDATED

Lockheed Martin Space Systems Company
(Headquarters)
12257 S Wadsworth Boulevard, Littleton, Colorado, 80125-8500, United States
Tel: (+1 303) 977 30 00
■ Provide space launch services, production and services relating to commercial satellites, government satellites and strategic missiles.

UPDATED

Lockheed Martin Space & Strategic Missiles (LMSS)
(Lockheed Martin Space Systems Company)
Mail Stop DC 1020, PO Box 179, Denver, Colorado, 80201, United States
Fax: (+1 303) 971 49 02
President: G Thomas Marsh
Deputy and Vice President, Space Transportation: Michael C Gass
Deputy and Vice President, Military Space: Leonard F Kwiatkowski
Vice President, Business Development: Daniel J Crowley
■ Design, development, testing and manufacture of a variety of advanced technology systems for the space and defence sectors. Chief projects include Titan II and Titan IV; Atlas, Centaur, Atlas V, crew return vehicle for use by International Space Station astronauts, Athenai I and II launch vehicles; Multi-Service Launch Systems (MSLS) and Target Delivery Vehicle (TDV).

UPDATED

Lockheed Martin Space Systems
(ex-Lockheed Martin Missiles & Space)
1111 Lockheed Martin Way, PO Box 3504, Sunnyvale, California, 894089, United States
Tel: (+1 408) 742 71 51
(+1 408) 742 66 88
Fax: (+1 408) 756 86 54
(+1 408) 742 84 84
e-mail: lmms.communications@lmco.com
Web: http://lmms.external.lmco.com
President: Anthony Tuffo
President, Lockheed Martin Special Programmes: G Thomas Marsh
Communications Department: Dave Waller
Tel: (+1 408) 742 16 06
Contact: Stephen O Tatum
■ Design, development, test, manufacture and operation of advanced technology systems for military, civil, and commercial customers. Products include: remote sensing and communications satellites for commercial and government customers, space observatories and interplanetry spacecraft, fleet ballistic missiles and missile defence systems.

UPDATED

Malin Space Science Systems
PO Box 910148, San Diego, California, 92191-0148, United States
Tel: (+1 858) 552 26 50
Fax: (+1 858) 458 05 03
e-mail: info@msss.com
Web: http://www.msss.com
President and Chief Scientist: Dr Michael C Malin
Contact: Ms Diana Michna
Tel: (+1 858) 552 26 50

UPDATED

Marathon Power Technologies
PO Box 8233, Waco, Texas, 76714-8233, United States
Tel: (+1 254) 776 06 50
Fax: (+1 254) 776 65 58
e-mail: marathon@mptc.com
Web: http://www.mptc.com

UPDATED

Marvin Engineering Co. Inc.
260 West Beach Avenue, Inglewood, California, 90302, United States
Tel: (+1 310) 674 50 30
Fax: (+1 310) 673 94 72
e-mail: sales@marvineng.com
Web: http://www.marvineng.com/mec
■ Manufacture of weapon pylons and ejectors for fixed- and rotary-wing aircraft including rail launchers for LAU-127, -128, -129 and -139; LAU-7 sidewinder, LAU-117 Maverick, M65 TOW, M272 Hellfire; ejector racks for TER-9A, MAU-12, B-1B Bomber 14-inch and 30-inch Ejector Racks; BRU-12, BRU-14 Rack, BRU-15A and SUU-20. Production of bomb hoists, missile and space launcher components.

UPDATED

McCormick Selph, Inc. (MSI)
(ex-Teledyne McCormick Selph)
3601 Union Road, PO Box 6, Hollister, California, 95024-0006, United States
Tel: (+1 831) 637 37 31
Fax: (+1 831) 637 14 50
Web: http://www.mcselph.com
Vice President of Business Development: Ray D Weiss
Vice President of Aerospace Programmes: Ben Griffin
Tel: (+1 831) 637 37 31
Vice President of Engineering: Mike Abbett
Tel: (+1 831) 637 37 31
Vice President of Operations: Dick Glover
Tel: (+1 831) 637 37 31
Chief Financial Officer: John Davis
Tel: (+1 831) 637 37 31
Public Relations Manager: Ms Verna Blaine
Purchasing Manager: Jerry Lundy
■ Design, development and manufacture of controlled explosive products for the aerospace, automotive and petroleum industries. Products include linear explosives, ordnance systems and components.

UPDATED

MCL, Inc.
501 South Woodcreek, Bolingbrook, Illinois, 60440-4999, United States
Tel: (+1 630) 759 95 00
Fax: (+1 630) 759 50 18
Web: http://www.mcl.com
■ Manufacture of high power microwave oscillators, amplifiers and severe environment ECM power supplies. Amplifiers for satellites and communications ground stations.

UPDATED

McMurdo Pains-Wessex Inc.
(ex-Pains-Wessex Safety Systems Inc.)
Suite 102, 200 Congress Park Drive, Delray Beach, Florida, 33445, United States
President: David Evans
Vice President: James Chandler
Vice President: Paul Rayner
Director: Gary Mullins
Treasurer and Secretary: Lawrence M D'Andrea
■ Manufacture of marine safety equipment.

UPDATED

MEDA Inc.
22611 Markey Court, Suite 114, Dulles, Virginia, 20166, United States
Tel: (+1 703) 471 14 45

Fax: (+1 703) 471 91 30
e-mail: sales@meda.com
Web: http://www.meda.com
President: Steven A Macintyre
Financial Director: Ms Patricia C K Macintyre
Operations Director: Robert E Kibler
■ Manufacture of TAM-1 and TAM-2 space qualified magnetometers for satellite attitude determination and control applications.

UPDATED

Metal Mart Inc.
12225 Coast Drive, Whittier, California, 90601, United States
Tel: (+1 562) 692 90 81
Fax: (+1 562) 699 68 68
Web: http://www.metal-mart.com
President: William Lippman
Chief Executive Officer: Andrew Lippman

UPDATED

Michigan Aerospace Corporation
1777 Highland Drive, Suite B, Ann Arbor, Michigan, 48108-2285, United States
Tel: (+1 734) 975 87 77
Fax: (+1 734) 975 02 39
Web: http://www.michiganaero.com
Chief Executive Officer: Carl Nardell
Executive Vice President: Pete Tchoryk
Tel: (+1 734) 975 87 77
Controller: Ms Nancy D Weber
Business Manager: Teri Julien
Tel: (+1 734) 975 87 77

UPDATED

Miltope Corp.
(Corporate Headquarters)
3800 Richardson Road South, Hope Hull, Alabama, 36043, United States
Tel: (+1 334) 284 86 65
Freephone: (+1 800) MILTOPE
Fax: (+1 334) 613 63 02
e-mail: info@miltope.com
custserv@miltope.com
Web: http://www.miltope.com

UPDATED

Missionspace Corporation
4519 Castle Road, La Canada, California, 91011, United States
Tel: (+1 818) 542 35 80
(+1 818) 542 35 81
Fax: (+1 818) 541 00 67
Web: http://www.missionspace.com
Chief Technology Officer: Randall E Reed
Manager of Business Development: John Eagles

UPDATED

Miteq Inc.
100 Davids Drive, Hauppauge, New York, 11788-2034, United States
Tel: (+1 516) 436 74 00
Fax: (+1 516) 436 74 31
(+1 516) 436 74 30
Web: http://www.miteq.com
Director, Corporate Communications: Hillar E Kiiss
Tel: (+1 631) 439 94 70
■ Manufacture of high performance components and systems for the microwave electronics industry. Products include satellite communications earth station equipment and space qualified equipment such as amplifiers, mixers, synthesizers and super components. Recent space platforms include P-97, TOPEX, SPINSAT, SEAWINDS, GEOSAT, SEASAT, SSMIS and AMSV-B.

UPDATED

Mobile Telesystems Inc.
205 Perry Parkway, Suite 14, Gaithersburg, Maryland, 20877, United States
e-mail: info@mti-usa.com
Web: http://www.mti-usa.com
General Manager: Doug Thomas
Tel: (+1 301) 590 85 22
Sales Manager: Eugene Lloyd
Tel: (+1 301) 590 85 29
■ Development, manufacture and distribution of mobile satellite communication systems for commercial and government military users.

UPDATED

Modular Devices Inc. (MDI)
One Roned Road, Shirley, New York, 11967, United States
Tel: (+1 516) 345 31 00
Fax: (+1 516) 345 31 06
e-mail: sales@mdipower.com
Web: http://www.mdipower.com
Outside Sales Manager: Bill Henry

UPDATED

Moog Inc.
(Headquarters)
PO Box 18, East Aurora, New York, 14052-0018, United States
Tel: (+1 716) 652 20 00
Fax: (+1 716) 687 44 57
Telex: 415239
Web: http://www.moog.com
Chairman, President and Chief Executive Officer: Robert T Brady
Executive Vice President and Chief Administrative Officer: Joe C Green
Executive Vice President and Chief Operating Officer: Robert H Maskrey
Executive Vice President and Chief Financial Officer: Robert R Banta
Vice President of Contracts and Pricing: Phil H Hubbell
Vice President, International: Stephen A Huckvale
Vice President, Strategy and Technology: Richard A Aubrecht
Vice President and General Manager, Systems Group: Jay Hennig
Vice President and General Manager, Industrial Controls Division: Martin J Berardi
Vice President and General Manager, Aircraft Group: Warren C Johnson
Treasurer: Timothy P Balkin
Secretary: John B Drenning
Controller: Donald R Fishback
■ Design and manufacture of electrohydraulic, electromechanical and electropneumatic controls, components and systems to the defence and aerospace industries.

UPDATED

Moog Inc.
(Chatsworth Operations)
(ex-Schaeffer Magnetics Inc.)
21339 Nordhoff Street, Chatsworth, California, 91311, United States
Tel: (+1 818) 341 51 56
Fax: (+1 818) 341 38 84
e-mail: sales.smo@moog.com
General Manager: Jerry Kraschinsky
Director of Finances: David Bailey
Sales and Marketing Director: Karl Anderson
Operations Director: John Nguyen
■ Supplier of spaceflight quality motors and actuators.

UPDATED

Motorola Inc.
(Corporate Office)
1303 East Algonquin Road, Schaumburg, Illinois, 60196, United States
Tel: (+1 847) 538 76 60
Freephone: (+1 800) 247 23 46
Fax: (+1 847) 576 49 59
(+1 847) 576 76 53
Web: http://www.motorola.com

UPDATED

Neptec
Suite 105, 16850 Saturn Lane, Houston, Texas, 77058, United States
Tel: (+1 281) 280 85 55
Fax: (+1 281) 280 86 16
Web: http://www.neptec.com
Administrative Assistant: Ms Kay Turner
Tel: (+1 281) 280 85 55

UPDATED

Northrop Grumman Corporation
(Corporate Headquarters)
1840 Century Park East, Los Angeles, California, 90067-2199, United States
Tel: (+1 310) 553 62 62
Web: http://www.northropgrumman.com
Vice President, Communications: Ms Roseanne O'Brien
■ Design, manufacture and systems integration: military surveillance, combat aircraft, defence electronics and systems, airspace management and information systems, marine systems, precision weapons, space systems and commercial and military aerostructures.

UPDATED

Northrop Grumman Electronic Systems
(Defensive Systems Division)
600 Hicks Road, Rolling Meadows, Illinois, 60009-1098, United States
Tel: (+1 847) 259 96 00
Fax: (+1 847) 870 57 05
Web: http://www.dsd.es.northropgrumman.com/

UPDATED

Northrop Grumman Corporation
(Electronic Systems)
1580-A West Nursery Road, MS A255, Baltimore, Maryland, 21090, United States
Freephone: (+1 800) 443 92 19
Web: http://www.sensor.northgrum.com
■ Production of radars and electronics for combat aircraft and battlespace management systems including the F-22 and F-16 fighters, B1-B bomber, Apache/Longbow attack helicopter, C-130 transport, E-3 AWACS and E-8 Joint stars as well as military space and undersea systems, electronic countermeasures, precision weapons, space systems, logistics systems and automatic standard intervention systems.

UPDATED

Northrop Grumman Corporation
(Electronic Systems)
(ex-Northrop Grumman Electronic Sensors & Systems International)
(ex-Westinghouse Electronic Systems International Co.)
Air Traffic Management Systems, PO Box 17320 B545, Baltimore, Maryland, 21203-7320, United States
Tel: (+1 410) 765 10 00
Fax: (+1 410) 993 87 71
Web: http://www.sensor.northgrum.com
http://www.northropgrumman.com
Director, Sector Communications: Robert Koch
Media Relations Manager: Debbi E McCallam
Tel: (+1 410) 765 15 21
Fax: (+1 410) 981 48 03
■ Aerostat radars, airborne fire control radars, air traffic control radars, worldwide air defence systems, command and control systems, electronic countermeasures, electronic support centres, integrated logistics support, shipboard radar systems and tactical surveillance radars.

UPDATED

Northrop Grumman Corporation
(Electronics and Systems Integration Division)
1111 Stewart Avenue, Bethpage, New York, 11714-3580, United States
Tel: (+1 516) 575 51 19
Fax: (+1 516) 575 36 91
Vice President and General Manager: John E Harrison
Vice President and Deputy General Manager: Max T Weiss

UPDATED

Northrop Grumman Sperry Marine
(ex-Sperry Marine Inc.)
1070 Seminole Trail, Charlottesville, Virginia, 22901-2891, United States
Tel: (+1 1434) 974 20 00
Fax: (+1 1434) 974 22 59
Web: http://www.sperry-marine.com
■ Design, development and manufacture of marine electronic products and systems for defence and commercial markets. Products include: integrated ship systems, ship bridge consoles, inertial navigation systems, gyrocompasses, steering systems, gyrofin stabilisers, marine surface search radars, airborne search radars, vessel traffic systems, dopplar speed logs, GMDSS communication systems, heading repeaters and marine navigation software.
Accreditations: ISO 9001

UPDATED

Northrop Grumman Electronic Systems
(Advanced Systems)
(ex-Litton Advanced Systems)
5225 Hellyer Avenue, Suite 100, San Jose, California, 95150-7012, United States
Tel: (+1 408) 365 47 47
Fax: (+1 408) 365 49 35
e-mail: info@littonas.com
Web: http://www.littonas.com

■ Manufacture of threat warning, advanced missile approach warning, electronic support measures, EW collection systems, active decoys, advanced self protection integrated EW suites, military and space computers, simulation, test and training systems, EW subsystem components, installation services.

UPDATED

Northrop Grumman Electronic Systems

(Navigation Systems Division)
(ex-Litton Advanced Systems)
(ex-Litton Industries Inc.)
21240 Burbank Boulevard, Woodland Hills, California, 91367-6675, United States
Tel: (+1 818) 715 40 11
(+1 818) 715 40 40
(+1 818) 715 24 70
Fax: (+1 818) 715 50 98
(+1 818) 715 45 40
e-mail: customerservice.nsd@northropgrumman.com
Web: http://www.ngnavsys.com
Director, Business Development: Joseph Geiger
Marketing Communications Specialist: Don Barteld
Contact: Ms Reyna D Gaar

UPDATED

Northrop Grumman, Electronic Systems
(ex-Denro Inc.)
(ex-Litton Advanced Systems)
(ex-Litton Denro)
9318 Gaither Road, Gaithersburg, Maryland, 20877-1441, United States
Tel: (+1 301) 527 79 00
Fax: (+1 301) 216 19 87
Site Director: Dave Wolt
Tel: (+1 301) 527 78 71
Director of Business Development: John Berthold
Tel: (+1 301) 527 78 89
Director of Business Management: Dan Roth
Tel: (+1 301) 527 79 60
Contact: Ms Liz Evans
Tel: (+1 301) 527 79 92
■ Supplier of voice switching and remote control systems used in air traffic control, shipboard communication and air defence applications. Products include: the model 400/ETVS, model 3080/RDVS and the model STVS (a small VCS). The Shipboard Integrated Voice Communication System (SIVCS) for ATC and internel shipboard communications on aircraft carriers and other vessels. The RCE 2001 Remote Control Equipment enables remote monitoring and control of radios over standard grade 4 wire telephone lines.

UPDATED

Nu-Cast, Inc.

29 Grenier Field Road, Londonderry, New Hampshire, 03053, United States
Tel: (+1 603) 432 16 00
Fax: (+1 603) 432 07 24
e-mail: nci@grolen.com
Web: http://www.nu-cast.com

UPDATED

Orbital Sciences Corp.

(ex-Orbital Image Corp.)
21700 Atlantic Boulevard, Dulles, Virginia, 20166, United States
Tel: (+1 703) 406 50 00
Web: http://www.orbital.com
Manager, Advanced Programmes, Launch Systems Group: Warren E Frick
Tel: (+1 703) 406 52 34
Fax: (+1 703) 404 80 42

UPDATED

Orbital Sciences Corporation

21839 Atlantic Boulevard, Dulles, Virginia, 20166, United States
Tel: (+1 703) 406 50 00
e-mail: public.relations@orbital.com
Web: http://www.orbital.com
Chairman of the Board, President and Chief Executive Officer: David W Thompson
President and Chief Operating Officer: James R Thompson
Executive Vice President and General Management, Advanced Programmes Group: Antonio L Elias
Executive Vice President and Chief Financial Officer: Garrett E Pierce
Executive Vice President and General Counsel: Ms Leslie C Seeman
Executive Vice President and General Manager, Launch Systems Group: Ronald L Grabe
Senior Vice President and General Manager, Space Systems Group: Jack Danko
Vice President of Investor Relations and Public Relations: Barron S Beneski
Tel: (+1 703) 406 55 28
Senior Vice President: Mark E Bitterman
■ Development and manufacture of space systems for commercial, civil government and military customers. Primary products include low-orbit, geosynchronous and planetary spacecraft for communications, scientific and remote sensing missions; ground- and air-launched rockets that deliver satellites into orbit; missile defence boosters for use as target and interceptor vehicles. Provides space-related technical services to government agencies, develops and builds satellite-based transportation management systems for public transit agencies and private vehicle fleet operators.

UPDATED

Orbital Sciences Corporation
(Chandler Operations)
3380 South Price Road, Chandler, Arizona, 85248, United States
Tel: (+1 602) 899 60 00
Fax: (+1 602) 899 88 00
Executive Vice President and General Manager, Launch Systems Group: Ronald Grabe

UPDATED

Para-Flite, Inc.

5800 Magnolia Avenue, Pennsauken, New Jersey, 08109-1399, United States
Tel: (+1 856) 663 12 75
Fax: (+1 856) 663 30 28
e-mail: marketing@paraflite.com
sales@paraflite.com
Web: http://www.paraflite.com
President: Elek Puskas
Vice President, Marketing: Joseph A Andrzejewski
Tel: (+1 856) 663 12 75
Operation Director: Randy Coperhaver
Materials Manager: Dan Fischer
Chief Engineer: John-Christophe Berland
Military Systems Technician: Mike Murphy
Human Resources: Ed Callahan
Master Rigger: Kenny George
Engineering Technician: Giorgio Piatti
■ Tactical gliding parachute systems, automatic aerial cargo delivery systems.

UPDATED

Parker Aerospace Group

(Group Headquarters)
(Parker Hannifin)
14300 Alton Parkway, Irvine, California, 92618, United States
Tel: (+1 949) 833 30 00
Fax: (+1 949) 851 32 77
e-mail: csd@parker.com
Web: http://www.parker.com
President: Bob Barker
Group Vice President, Marketing: Ed Arnold
Group Vice President, Flight Control and Hydraulic Systems: Joel Benkie
Group Vice President, Supplier Management and Integrity: Dick Hall
Group Vice President, Fluid Management and Control Systems: Frank Nichols
Group Vice President, Human Resources: Ms Linda Walker
Group Vice President, Customer Support: Mike Romito
■ Design, manufacture and service of hydraulic, fuel and pneumatic components, systems and related electronic controls for aerospace and other high-technology markets.

UPDATED

Parker Aerospace Group
(Air & Fuel Division, Irvine Plant)
16666 Von Karman Avenue, Irvine, California, 92606-4917, United States
Tel: (+1 949) 833 30 00
Fax: (+1 949) 851 33 41
Web: http://www.parker.com/airfuel
Marketing Director: Dale Elmore
General Manager: John Meston
■ Manufacture of aircraft fuel system components, aerial refuelling equipment, high-temperature bleed air valves, turbine clearance control valves, air turbine starters and aircraft fire-suppression systems.
Accreditations: ISO 9001

UPDATED

Parker Aerospace Group
(Control Systems Division – Military)
14300 Alton Parkway, Irvine, California, 92618-1898, United States
Tel: (+1 949) 833 30 00
Fax: (+1 949) 586 84 56
Group Vice President and General Manager: Bob Barker
Vice President Marketing: Jim Ryder
General Manager: Joel Benkie
■ Manufacture of hydraulic systems, primary flight control actuation and engine control systems, integrated fly-by-wire electrohydraulic servomodules, electrically controlled actuators, electromechanical actuators, flight control components and ballistic-tolerant hydraulic systems and components.
Accreditations: ISO 9000

UPDATED

Parker Aerospace Group
(Electronic Systems Division)
(ex-Gull Electronic Systems Division, Parker Aerospace)
300 Marcus Boulevard, PO Box 9400, Smithtown, New York, 11787-9400, United States
Tel: (+1 631) 231 37 37
Fax: (+1 631) 434 81 52
Group Vice President and General Manager: Stephen Long
Director of Marketing: Martin Hewitt
Director Operations: Thomas Renken
Marketing Representative: John Bunting
■ Fuel and oil management systems, fuel and oil quantity gauging systems, flight deck displays, flight inspection and aircraft-control systems.
Accreditations: ISO 9000

UPDATED

Parker Hannifin Corporation

6035 Parkland Boulevard, Cleveland, Ohio, 44124-4141, United States
Tel: (+1 216) 896 30 00
Freephone: (+1 800) 272 75 37
Fax: (+1 216) 746 11 48
(+1 440) 266 74 00
e-mail: c-parker@parker.com
Web: http://www.parker.com
President and Chief Executive Officer: Don Washkenicz
Chairman of the Board: Duane E Collins
Aerospace Group President: Stephen L Hayes
Executive Vice President: Dennis W Sullivan
Vice President and Technical Director: Claus Beneker
Climate and Industrial Controls Group President: Lynn M Cortright
Seal Group President: Heinz Droxner
Executive Vice President, Finance and Administration and Chief Financial Officer: Michael J Hiemstra
Vice President, Corporate Information Officer: William Eline
Vice President, Human Resources: Daniel T Garey
Vice President, General Counsel and Secretary: Thomas A Piraino
Vice President: John D Myslenski
Vice President: Nickolas W Vande Steeg
Vice President, Corporate Communications: Lorrie Paul Crum
Controller: Dana A Dennis
Treasurer: Timothy K Pistell
Corporate Marketing Services: Mike Marvin
■ Manufacture of motion control components and systems for industrial, automotive, aviation, space and marine markets.
Operation with defence business
Parker Aerospace

UPDATED

Perkinelmer Fluid Sciences (PKI)

(Belfab Products)
(ex-EG&G Belfab)
(ex-John Crane Belfab)
305 Fentress Boulevard, Daytona Beach, Florida, 32114, United States

Fax: (+1 386) 257 01 22
Web: http://www.perkinelmer.com/fluidsciences
■ Design and manufacture of welded metal bellows assemblies.

UPDATED

PerkinElmer Optoelectronics

(Headquarters)
44370 Christy Street, Freemont, California, 94538-3180, United States
Tel: (+1 510) 979 65 00
Freephone: (+1 800) 775 OPTO
Fax: (+1 510) 687 11 40
e-mail: opto@perkinelmer.com

UPDATED

Photo-Sonics Inc.

820 South Mariposa Street, Burbank, California, 91506-3196, United States
Tel: (+1 818) 842 21 41
(+1 818) 531 32 19
Fax: (+1 818) 842 26 10
e-mail: mail@photosonics.com
info@photosonics.com
Web: http://www.photosonics.com
President: John Kiel
Vice President: Roy Edwards
Vice President Operations: Philip Kiel
Fax: (+1 818) 842 26 10
Sales Manager: Graham Jones
Tel: (+1 818) 842 21 41
■ 16 mm gun sight, HUD cameras and airborne video recorder for military aircraft. 16, 35 and 70 mm and HS video photo instrumentation cameras for military and aerospace applications. Tracking mount mobile and fixed, instrumentation cameras. Automatic film and video analysis equipment.
Subsidiary
ARAN Electronics, Israel

UPDATED

Pioneer Aerospace Corp.

(Headquarters)
45 South Satellite Road, PO Box 207, South Windsor, Connecticut, 06074-0207, United States
Tel: (+1 860) 528 00 92
Fax: (+1 860) 528 81 22
e-mail: sales@pioneeraero.com
Web: http://www.pioneeraero.com
President: Ernest de Haas
Sales Development and Contracts Director: Chris Christensen
Tel: (+1 860) 528 00 92
Operations Director: Chris Powell
Contracts Manager: Ms Caren Wollenburg
Tel: (+1 860) 528 00 92
Purchasing Manager: Jim Edwards
Engineering Manager: John Smith
Tel: (+1 860) 528 00 92
Webmaster: Greg White
Controller: Tom Owens
Sales Administrator: John Whittles
■ Engineering and manufacture of parachutes and related items. Analysis, design, development, testing, manufacturing, integration and support of advanced recovery and retrieval systems for aerospace vehicles and components.
Accreditations: Mil-Q-9858A, Mil-45208A, ISO 9002

UPDATED

Pratt & Whitney (P&W)

400 Main Street, East Hartford, Connecticut, 06108, United States
Tel: (+1 203) 565 43 21
Web: http://www.pratt-whitney.com
President: Louis Chênevert
President, Pratt & Whitney Canada: Allain M Bellemare
President, Space Propulsion and Russian Operations: Walter Bylciw
President, Military Engines: Thomas E Farmer
Senior Vice President, Engineering: Arthur W Lucas
Senior Vice President, Operations: Larry O Moore
Senior Vice President, Finance: Jothi Purushotaman
Executive Vice President, Chief Operating Officer and President, Large Commercial Engines: Robert F Leduc
Vice President, Human Resources and Organisation: J Thomas Bowler
Vice President, General Manager, Power Systems: Peter Christman
Vice President, Environment, Health and Safety: Ms Claudia Coplein
Vice President, Quality: Brian G Cousins
Vice President and Chief Information Officer: Ms Nancy Davis
Vice President and General Counsel: Peter Gutermann
Vice President, Government and Business Development: David C Manke
Vice President, UTC Aerospace: Alfred J Mulvey
Vice President, Communications: Ms Dionn M Tron
Media Inquiries: Mark Sullivan
Tel: (+1 860) 565 96 00
Chairman, Pratt & Whitney Canada: Giles Ouimet
■ Design, develop and support of market propulsion systems. Installers of fighter, transport and surveillance aircraft. Rocket engines and propulsion systems for various space programmes.
Subsidiary company
Pratt & Whitney Canada Inc.

UPDATED

Pratt & Whitney (P&W)
(Military Engines)
(ex-Government Engines & Space Propulsion)
400 Main Street 181-36, East Hartford, Connecticut, 06108, United States
Tel: (+1 860) 565 96 00
(+1 860) 565 43 21
(+1 860) 565 01 40
e-mail: help24@pweh.com
Web: http://www.pw.utc.com
President, Military Engines: Thomas E Farmer
Senior Vice President, Military Customer Requirements & Support: Lloyd Newton
Vice President Operational Military Engine Programs: Warren M Boley
Vice President, Finance, Military Engines: John S Canzio
Vice President, Military Customer Support: Larry W Jones
Vice President, Human Resources: Michael D Smith
Vice President, Counsel, Military Engines: David K Ware
Director, Advanced Programs: Ms Simeon H Austin
Director, F119/F/A-22 Engine Program: Bennett Croswell
Director, F135 Program: William J Gostic
Director, F135 International Programs: Edward O'Donnell
Director, Military Aftermarket: Richard O'Donnell
Director, Mobility Programs: Steve Perry
Director, International Business Development and Programs: Raul Pereda
Director, Military Domestic Business Development: Brett D Rhodes
General Manager, Small Military Engines: James K Farrell
Manager, Military Derivative Programs: Jimmy B Reed
Manager, Communications: Laurie J Tardif
Assistant to the President: John P Wagner

UPDATED

Pratt & Whitney Space Propulsion

PO Box 109600, West Palm Beach, Florida, 33410-9600, United States
Tel: (+1 561) 796 20 00
Fax: (+1 561) 796 92 21
President, Space & Russian Operations: Walter N Bylcin
President, RD AMROSS LLC: Robert L Monaco
General Manager, San Jose, California: Mark Mounsey
General Manager, West Palm Beach, Florida: Thomas J Haack
Division Counsel: Ms Donna Clayton
Director, Strategic Planning: Don McMonagle
Director, Finance: Thomas H Wonnell
Director, Human Resources: Arnold P Mumford
Director, Engineering: Rich Pugh
Manager, Communications: Patrick W Louden
■ Solid and liquid rocket propulsion.

UPDATED

Programmed Composites Inc.

250 Klug Circle, Carona, California, 92880, United States
Tel: (+1 909) 520 73 00
Fax: (+1 909) 520 73 73
Web: http://www.pciaerospace.com/pci/default.htm

UPDATED

Puroflow Corp.

10616 Lanark Street, Sun Valley, California, 91352, United States
Tel: (+1 818) 504 40 00
Fax: (+1 818) 779 39 02
e-mail: sales@puroflow.com
Web: http://www.puroflow.com

UPDATED

Radio Research Instrument Co Inc

584 North Main Street, Waterbury, Connecticut, 06704-3506, United States
Tel: (+1 203) 753 58 40
Fax: (+1 203) 754 25 67
e-mail: radiores@prodigy.net
Web: http://www.radioresearch.thomasregister.com
President: P J Plishner
Executive Vice President: E B Doyle
Operations Director: Donald Smith
■ Radar systems and spares, threat emitters, high-power RF sources, antennas, azimuth-elevation pedestals and microwave tubes.

UPDATED

Radyne Comstream Inc.

(ex-Comstream Corporation)
3138 East Elwood Street, Phoenix, Arizona, 85034, United States
Tel: (+1 602) 437 96 20
Fax: (+1 602) 437 48 11
e-mail: sales@radn.com
Web: http://www.radn.com
Vice President of Sales and Marketing: Glen Green
■ Satellite-based communications technology, supplies interactive and broadcast networks and digital componentry. Systems include: SCPC earth stations, modems, receivers, uplinks, network management systems and custom chips. Point-to-multipoint voice, data, imaging and fax transmission. Point-to-point, full mesh or "star" connectivity. Uplink only for low cost network hubs. Applications include: mesh networking for private business and government services. Long distance telephone trunking terminals and voice backbone networks. LAN interconnection. Video conferencing. Data acquisition by a central site polling many remote locations (SCADA). Digital Data Applications. Real time financial market information. News/weather gathering and distribution. Computer file transfer. Photograph or image transmission. Paging networks. Digital Audio Applications. Radio programme distribution. Remote audio broadcasting. News gathering. Remote studio recording. Major systems designs and implementation of large turnkey satellite communication systems.

UPDATED

Radyne Comstream Inc.
(San Diego Office)
6340 Sequence Drive, San Diego, California, 92121, United States
Tel: (+1 858) 458 18 00
Fax: (+1 858) 657 54 03

UPDATED

Radyne Comstream Inc.
(Latin America)
7700 Congress Avenue, Suite 3206, Boca Ratón, Florida, 33487, United States
Tel: (+1 561) 988 12 10
Fax: (+1 561) 988 82 90
e-mail: lasales@radynecomstream.com

UPDATED

Raymond Engineering Operations (REO)

217 Smith Street, Middletown, Connecticut, 06457, United States
Tel: (+1 860) 632 45 82
Fax: (+1 860) 632 43 29
Web: http://www.raymond-engrg.com
Vice President: Harry J Hutchins
Purchasing Manager: William Houlberg
Marketing: James J Krupa
■ Electronic systems, both ground-based and airborne. Safety products for advanced tactical missiles, penetration fusing devices for major devices for major programmes such as AMRAAM, Phoenix, Hawk and Standard Missile; sensing systems for target recognition and data acquisition; complete testing facilities including sled testing and environmental laboratories for vibration, shock, temperature, firing and special purpose testing. Ruggedised mass memory products for weapons platforms, space satellites and the space shuttle.

UPDATED

Raytheon Company
(Corporate Headquarters)
870 Winter Street, Waltham, Massachusetts, 02451 – 1449, United States
Tel: (+1 617) 860 24 14
(+1 781) 862 66 00
Fax: (+1 617) 860 25 20
(+1 781) 860 25 20
e-mail: corpcom@raytheon.com
Web: http://www.raytheon.com
President and Chief Executive Officer: William H Swanson
President, Raytheon Technical Services: Bryan Even
President, Missile Systems: Ms Louise L Francesconi
President, Integrated Defence Systems: Charles E Franklin
President, Intelligence and Information Systems: Michael D Keebaugh
President, Network Centric Systems: Colin Schottlaender
President, Space and Airborne Systems: Jack R Kelble
President, Homeland Security: Dr Hugo B Poza
Senior Vice President and Chief Financial Officer: Edward S Pliner
Senior Vice President, Human Resources: Keith J Peden
Senior Vice President and General Counsel: Jay B Stephens
Executive Vice President, Business Development: Thomas M Culligan
Executive Vice President Chairman and Chief Executive Officer, Raytheon Aircraft Company: James Schustell
Executive Vice President, Raytheon, Leader Discounted Operations: Frank S Manchilean
Vice President, Contracts: John D Harris
Vice President and Secretary: John W Kapples
Vice President and Treasurer: Richard A Goglia
Vice President and Treasurer: Biggs Porter
Vice President and Chief Information Officer: Ms Rebecca Rhoads
Vice President, Corporate Affairs and Communication: Ms Phyllis J Piano
Vice President, Media Relations: James Fetig
Tel: (+1 781) 522 51 11
Fax: (+1 781) 522 58 78
Contact: Mr Sabrina Steele
Tel: (+1 781) 522 51 27
Fax: (+1 781) 522 58 78
Business Groups
Raytheon Aircraft
Raytheon Homeland Security
Raytheon Integrated Defense Systems
Raytheon Intelligence & information Systems
Raytheon Missile Systems
Raytheon Network Centric Systems
Raytheon Space & Airborne Systems
Raytheon Technical Services Company

UPDATED

Raytheon Company
(Santa Barbara Remote Sensing)
75 Coromar Drive, Goleta, California, 93117, United States
Tel: (+1 805) 968 35 11
(+1 805) 562 41 27
Fax: (+1 805) 685 82 27

UPDATED

Rockwell Collins
(Display Systems)
(ex-Kaiser Electronics)
2701 Orchard Parkway, San Jose, California, 95134, United States
Tel: (+1 408) 532 48 41
Fax: (+1 408) 954 10 42
e-mail: info@kaisere.com
Manager, Communications: Ms Leanne Kelsey
Tel: (+1 408) 532 45 16

■ Manufacture of advanced head-up displays, liquid crystal multifunction displays, helmet displays and display processors for combat aircraft.

UPDATED

Rockwell Collins, Inc.
(Headquarters)
(ex-Collins Avionics and Communications Division)
400 Collins Road NE, Cedar Rapids, Iowa, 52498, United States
Tel: (+1 319) 295 10 00
(+1 319) 295 51 00
(+1 800) 321 22 23
Fax: (+1 319) 295 47 77
e-mail: collins@rockwellcollins.com
Web: http://www.rockwellcollins.com
Chairman, President and Chief Executive Officer, Rockwell Collins: Clayton M Jones
Executive Vice President and Chief Operating Officer, Rockwell Collins Commercial Systems: Robert M Chiusano
Executive Vice President and Chief Operating Officer, Rockwell Collins Government Systems: Greg S Churchill
Senior Vice President and General Manager, Collins Aviation Services: Harry L Gregory
Senior Vice President & Chief Financial Officer: Patrick E Allen
Senior Vice President, General Counsel and Secretary: Gary R Chadwick
Senior Vice President, Corporate Development: Dr Barry M Abzug
Senior Vice President, Human Resources: Ron Kirchenbauer
Senior Vice President, Operations: Kent L Statler
Senior Vice President, e-Business and Lean Electronics: John-Paul Besong
Senior Vice President, Engineering and Technology: Ms Nan Mattai
Vice President & General Manager, Kaiser Aerospace & Electronics: John W Borghese
Vice President and General Manager, Integrated Systems: Walter S Hogle
Vice President and General Manager, Communication Systems: Bruce M King
Vice President and General Manager, International Subsidiaries: Bernard Loth
Vice President and General Manager, Sensor Systems: Ronald R Hornish
Vice President and General Manager, Passenger Systems: Glen T Dodson
Vice President and General Manager, Business and Regional Systems: Dennis W Helgeson
Vice President and General Manager, Air Transport: Robert K Ortberg
Senior Communications Specialist: Ms Teather Merritt
Media Contact: Pam J Tvrdy
Tel: (+1 319) 295 05 91
Fax: (+1 319) 295 43 33
Media Contact: Ms Nancy Welsh
Tel: (+1 319) 295 21 23
Fax: (+1 319) 295 43 33

■ Design, production and support of communications and aviation electronics solutions for government and commercial customers worldwide.

Domestic service centres in
Atlanta, Georgia
Boston, Massachusetts
Cedar Rapids, Iowa
Charlotte, North Carolina
Chicago, Illinois
Dallas/Fort Worth, Texas
Detriot, Michigan
Honolulu, Hawaii
Houston, Texas
Los Angeles, California
Miami, Florida
Minneapolis, Minnesota
Newark, New Jersey
New York, New York
Philadelphia, Pennsylvania
Pittsburg, Pennsylvania
Pomona, California
Raleigh, North Carolina
St Louis, Missouri
Salt Lake City, Utah
San Francisco, California
San Jose, California
Seattle, Washington
Tulsa, Oaklahoma
Washingtron, District of Colombia
Wichita, Kansas
International service centres in
Amsterdam, Netherlands
Auckland, New Zealand
Bangkok, Thailand
Beijing, China
Brisbane, Australia
Buenos Aires, Argentina
Frankfurt, Germany
Hong Kong, China
Johannesburg, South Africa
London, UK
Manchester, UK
Manila, Philippines
Melbourne, Australia
Moscow, Russia
Narita, Japan
Osaka, Japan
Paris, France
Reading, England
Rome, Italy
San Jose, Brazil
San Juan, Puerto Rico
Santiago, Chile
Shanghai, China
Singapore
Sydney, Australia
Toulouse, France
Zurich, Switzerland
Operations in
Ann Arbor, Michigan
Bellevue, Iowa
Carlsbad, California
Cedar Rapids, Iowa
Charlotte, North Carolina
Colchester, Vermont
Coralville, Iowa
Cypress, California
Decorah, Iowa
Irvine, California
Kirkland, Washington
Manchester, Iowa
Melbourne, Florida
Mexicali, Mexico
Pomona, California
Portland, Oregon
Richardson, Texas
San Jose, California
Tustin, California
White Marsh, Maryland

UPDATED

Rockwell Collins, Inc.
(Commercial Systems)
(ex-Air Transport Systems)
(ex-Business and Regional Systems)
(ex-Flight Dynamics)
(ex-Kaiser Electro Precision)
(ex-Passenger Systems)
400 Collins Road NE, Cedar Rapids, Iowa, 52498, United States
Tel: (+1 319) 295 10 00
Freephone: (+1 800) COLLINS
Fax: (+1 319) 295 51 00
e-mail: collins@rockwellcollins.com
mktgsvcs@rockwellcollins.com
Web: http://www.rockwellcollins.com
Executive Vice President and Chief Operating Officer, Commercial Systems: Robert M Chiusano
Vice President and General Manager, Air Transport Systems: Robert K Ortberg
Vice President and General Manager, Passenger Systems: Jonny Dodson
Vice President and General Manager, Business and Regional Systems: Denny W Helgeson
Vice President, Marketing and Strategic Management: Stephen L Belland
Vice President, Engineering: Steve J Nieuwsma
Vice President and Controller, Finance: Gary Widen
Vice President, Sales: Randy P Lincoln

■ Provide avionics systems products, and support the world's airlines, airframe manufactures, regional airlines and turbine-powered business aircraft. The company is directed at moving the aviation industry toward the satellite-based CNS/ATM free flight environment. Avionics solutions meeting those requirements include Pro Line 21 flight deck; Collins communication, navigation, situational awareness, display and automatic flight control systems; Integrated Information System (I2S); Rockwell Collins Flight Dynamics Head-up Guidance System (HGS) and Surface Guidance System (SGS) technology. Additionally produces cabin management and in-flight entertainment systems that are now flown with over 100 airlines around the world.

UPDATED

Rockwell Collins, Inc.
(Government Systems)
(ex-Collins Avionics Communications Division)
(ex-Government Systems)
350 Collins Road, NE, Cedar Rapids, Iowa, 52498-0120, United States
Tel: (+1 319) 295 10 00
(+1 319) 295 51 00
Freephone: (+1 800) COLLINS
Fax: (+1 319) 295 47 77
(+1 319) 295 54 29
e-mail: collins@rockwellcollins.com
mktgsvcs@rockwellcollins.com
Web: http://www.rockwellcollins.com
Executive Vice President and Chief Operating Officer, Government Systems: Greg S Churchill
Vice President and General Manager, Communication Systems: Bruce M King
Vice President and General Manager, Integrated Applications and Navigation Systems: Ron R Hornish
Vice President and General Manager, Kaiser Aerospace & Electronics: John W Borghese
Vice President and General Manager, International Subsidiaries: Bernard Loth

Vice President, Business Development: Ken A Peterman
Vice President, Strategy Development: Ms Meg Thompson
Vice President, Engineering: Nan Mattai
Vice President and Controller, Finance: Chuck A Harmeyer
Vice President, Government Operations: Mike K McDonald
■ Supply of defence electronics products and systems, including communications, navigation and integrated systems for airborne, ground and shipboard applications. Customers include the US Department of Defense, foreign militaries, government agencies and manufacturers of military aircraft and helicopters. Provide open systems architecture and commercial off-the-shelf technology solutions offering the growth and flexibility needed to address the emerging Global Air Traffic Management (GATM) requirements. Communication and navigation solutions meeting these mandates include the Joint Tactical Radio System (JTRS) Cluster 1 Radios, Collins Flight2 products and systems, GPS receivers, data links, flight management systems, communications systems and aviation electronics systems.

UPDATED

Sage Laboratories Inc.
11 Huron Drive, Natick, Massachusetts, 01760-1314, United States
Tel: (+1 508) 653 08 44
Fax: (+1 508) 653 56 71
e-mail: info@sagelabs.com
Web: http://www.sagelabs.com
Component Business Manager: J Andrew Funk
■ Microwave components operating DC-606 Hz (passive). Space qualified.

UPDATED

Sanmina-SCI Corporation
(ex-SCI Systems Inc.)
(ex-SCI Technology, Inc.)
2700 North First Street, San Jose, California, 95134, United States
Tel: (+1 256) 882 45 77
(+1 256) 882 41 99
Fax: (+1 256) 882 46 52
e-mail: products@sanmina-sci.corp
Web: http://www.sanmina-sci.corp
Advanced Product Support: Ms Carolyn Smith
■ Militarised voice recognition systems, communications terminals, control panels, data switches and avionics processors.

UPDATED

Sanmina-SCI Corporation
(Huntsville Office)
(ex-SCI Systems Inc.)
(ex-SCI Technology, Inc.)
8600 South Memorial Parkway, PO Box 1000, Huntsville, Alabama, 35802, United States
Tel: (+1 256) 882 48 00
Fax: (+1 256) 882 46 52
Web: http://www.sanmina-scl.com
Advanced Product Development Support: Ms Carolyn V Smith
Tel: (+1 256) 882 45 83
■ Design and manufacture of electronic products for the aerospace, defence, telecommunications industries and the US Government.

UPDATED

Scaled Composites, LLC
(Head Office)
1624 Flight Line, Mojave, California, 93501-1663, United States
Tel: (+1 661) 824 45 41
Fax: (+1 661) 824 41 74
e-mail: info@scaled.com
Web: http://www.scaled.com
President: Burt (Elbert L) Rutan
Vice President and General Manager: Michael W Melvill
Vice President, Corporate Secretary: Ms Kaye LeFebvre
■ Design, build and flight testing of prototype composite aircraft and spacecraft.

UPDATED

Schonstedt Instrument Co.
100 Edmond Road, Kearneysville, West Virginia, 25430, United States
Tel: (+1 304) 724 47 00
Freephone: (+1 304) 725 10 95
e-mail: info@schonstedt.com
Web: http://www.schonstedt.com
Chief Executive Officer: Mike Head
Customer Service: Ms Barb Mabry
■ Design, manufacture, marketing and servicing of instrumentation for domestic utilities, construction, excavation and exploration. The main lines of instrumentation include magnetic locating gradiometers capable of locating ferrous objects up to 16 feet underground, and multi-frequency pipe and cable locators used to find underground utilities for repair and damage avoidance during excavation.

UPDATED

Sea Tel Inc.
4030 Nelson Avenue, Concord, California, 94520, United States
Tel: (+1 925) 798 79 79
Fax: (+1 925) 798 79 86
e-mail: sales@seatel.com
service@seatel.com
Web: http://www.seatel.com
Vice President of Sales and Marketing: Richard D Stallbaum
Marketing: Ms Judy Bright
VSAT System Product Manager: Wes Schenck
Service Manager: Darren Manning
Human Resources Manager: Jim Lynch
■ Design, manufacture and marketing of shipboard stabilised antenna systems from 0.5 to 4.5 meters in diameter for VSAT and satellite television applications.

UPDATED

Sequa Corporation
(Corporate Headquarters)
200 Park Avenue, New York, New York, 10166, United States
Tel: (+1 212) 986 55 00
Fax: (+1 212) 370 19 69
Web: http://www.sequa.com
Chairman of the Board and Chief Executive Officer: Norman E Alexander
President and Chief Operating Officer: John J Quicke
Senior Vice President, Finance: Howard M Lettner
Vice President, Human Resources: Jesse Battino
Vice President and Treasurer: Kenneth A Drucker
Vice President, Corporate Communications: Ms Linda G Kyriakou
Vice President, Corporate Development and Strategy: Robert D DeVito
Corporate Secretary and Senior Associate General Counsel: Ms Diane Bunt
■ Manufacture of aircraft engine components, solid rocket propellant motors for missiles and space applications.
Subsidiaries with defence business
Atlantic Research Corp.
Chromalloy Gas Turbine Corp.

UPDATED

Shock-Tech Inc.
360 Route 59, Monsey, New York, 10952, United States
Tel: (+1 845) 368 86 00
Fax: (+1 845) 368 87 99
e-mail: info@shocktech.com
Web: http://www.shocktech.com
President: Kevork Kayayan
Chairman: Serge Seguin
Technical Director: Gary Melone
Sales Director: Paul Lorentzen
Purchasing Manager: Bill deYoung
■ Manufacture of shock and vibration isolators, cabinets and cases.
Representing:
SMAC S.A.S, France
Socitec International, France
Vibtech Ltd, Brazil

UPDATED

Smiths Aerospace, Electronic Systems – Grand Rapids
3290 Patterson Avenue, SE, Grand Rapids, Michigan, 49512-1991, United States
Tel: (+1 616) 241 70 00
Fax: (+1 616) 241 75 33
Web: http://www.smiths-aerospace.com
President: Robert F Ehr
Vice President, Finance: Ron Albrecht
Vice President, Human Resources: D M Carvalho
Vice President, Combat Fixed Wing – North America: James O Sweeney
Vice President, Combat Fixed Wing – ROW: N A McKeever
Vice President, Engineering Development: A Dimitriou
Vice President, Air Transport: D M Grady
Vice President, Rotorcraft: W Turner
Vice President and General Manager, Grand Rapids and Clearwater: D S Miller
Vice President, Manufacturing: W Plouff
Communications Manager: Ms Jennifer Villarreal
Tel: (+1 616) 241 86 43
Fax: (+1 616) 241 87 30
■ Provision of aviation electronics technology with systems capability in flight management, data management, mission systems and navigation/guidance. Instruments, systems and components for aircraft reference and navigation, missile guidance, anti-submarine warfare control, weapon systems operations, commercial flight management, aerospace ground support, self-contained navigation systems, weapon management systems, communication control systems, air data computers, data and video recording and land navigation systems.

UPDATED

Smiths Industries Aerospace
(Electronic Systems – Clearwater)
14180 Roosevelt Boulevard, Clearwater, Florida, 34622, United States
Tel: (+1 727) 531 77 81
Fax: (+1 727) 538 75 22
Web: http://www.smiths-aerospace.com
■ High-performance cockpit displays, high-resolution mission data displays, and control display units; graphics data processing involving high-performance computation of video, visual, and other mission data; fuel management and measurement systems; logistics support of products produced by Smiths Aerospace. Manufacture of electronic systems and circuit cards and systems including management systems, engine fuel displays and standby flight displays.

UPDATED

Soloy Corporation
450 Pat Kennedy Way South West, Olympia, Washington, 98501-7298, United States
Tel: (+1 360) 754 70 00
Fax: (+1 360) 943 76 59
e-mail: sales@soloy.com
Web: http://www.soloy.com
President: Dave A Stauffer
■ Manufacture of Soloy Dual Pac power plant, a single propeller twin engine powerplant, for general aviation aircraft. Turbine conversion for fixed wing aircraft, single or twin, utilising the 400-735 ship Soloy Turbine Pac. Engineering modifications and services. FAA repair station services. Custom subcontract sheet metal fabrication.

UPDATED

Southern Avionics Co. (SAC)
PO Box 5345, Beaumont, Texas, 77726-5345, United States
Tel: (+1 409) 842 17 17
Fax: (+1 409) 842 29 87
(+1 409) 842 13 24
Telex: 779450
Cable: SAC
Web: http://www.southernavionics.com
President: Brooks Goodhue
Tel: (+1 409) 842 17 17
Chief Financial Officer: Ms Jennifer Manriquez
Vice President, Marketing: Tyler Callis
Vice President, Engineering: Tim Hamilton
Vice President, Operations: Gary Hamel
■ Manufacture of stationary and portable non-directional beacons (NDB), marine band differential GPS transmitters (DGPS), antenna tuning units (ATU), antenna systems, monitor alarm and receivers, remote-control units (RCU), remote status indication, equipment shelters and battery-standby systems.

UPDATED

Space Dev (SPDV)
13855 Stowe Drive, Poway, California, 92064, United States
Tel: (+1 858) 375 20 30
Fax: (+1 858) 375 10 00
e-mail: info@spacedev.com
sales@spacedev.com
Web: http://www.spacedev.com
Financial Director: Richard Slanski

Sales Director: Kevin Heath
Tel: (+1 858) 375 20 12
Marketing and Public Relations Director: Ms Susan Benson
Tel: (+1 858) 375 20 75
Technical and Operations Director: Dan DaPra
Purchasing Manager: Charles Williams
Tel: (+1 858) 375 20 17

UPDATED

Space Systems/Loral (SS/L)

3825 Fabian Way, G-60, Palo Alto, California, 94303-4604, United States
Tel: (+1 650) 852 40 00
Freephone: (+1 800) 332 64 90
Fax: (+1 650) 852 47 88
Web: http://www.ssloral.com
Chairman: Robert E Berry
Tel: (+1 650) 852 69 92
President: C Patrick DeWitt
Tel: (+1 650) 852 54 57
Executive Vice President, Business: Ron Haley
Senior Vice President, Worldwide Marketing and Sales: Arnold Friedman
External Communications Director: Ms Cathy Crockett
Tel: (+1 650) 852 42 53

■ Produce commercial communications and weather satellites.

UPDATED

Space Vector Corp. (SVC)

9223 Deering Avenue, Chatsworth, California, 91311, United States
Tel: (+1 818) 734 26 00
Web: http://www.spacevector.com
President: Phil Panzera
Vice President, Finance and Administration: L A Kameen
Marketing Administrator: Ms Nancy Roche

UPDATED

Spectrolab, Inc.

12500 Gladstone Avenue, Sylmar, California, 91342-5373, United States
Tel: (+1 818) 365 46 11
Fax: (+1 818) 365 76 80
(+1 818) 361 51 02
(+1 818) 365 77 71
Telex: twx 9104961750 spectillm sylm
Web: http://www.spectrolab.com
President: Dr David R Lillington
Manager, Advanced Programmes: Dr Nasser H Karam
Business Manager: Ron Diamond
Communications and Advertising: M Kalachian

■ Helicopter and aircraft searchlights, ground based searchlights, solar cells and optoelectronic products.

UPDATED

Spectrum Astro Inc.

1440 North Fiesta Boulevard, Gilbert, Arizona, 85233, United States
Tel: (+1 480) 892 82 00
Fax: (+1 480) 892 29 49
Web: http://www.specastro.com
President and Chief Executive Officer: W David Thompson
Senior Vice President, Contracts and Finance and Chief Financial Officer: Ms Patricia Oleseon
Vice President: Scott Yeakel
Director, Business Development: Dan Toomey
Advertising Manager: Ms Maripat Meer

■ Active in the development of high performance, lower cost spacecraft for defence, scientific and commercial opportunities.

UPDATED

Stanford Mu Corporation

24003 Frampton Avenue, Harbor City, California, 90710-2101, United States
Tel: (+1 310) 325 63 00
Fax: (+1 310) 325 63 52
Web: http://www.stanfordmu.com
Director of Programs and Contracts: Lynn J Price

UPDATED

Swe-Dish Satellite Systems Inc.

1634 I Street North West, Suite 605, Washington, District of Columbia, 20006, United States
Tel: (+1 202) 393 49 40
Fax: (+1 202) 393 49 63
e-mail: us-sales@swe-dish.com
sales@swe-dish.se
Web: http://www.swe-dish.com

■ Supplier of mobile satellite communication equipment and related services for broadband applications.

UPDATED

Synthecon Incorporated

8054 El Rio, Houston, Texas, 77054, United States
Tel: (+1 713) 741 25 82
(+1 800) 853 07 40
Fax: (+1 713) 741 25 88
Web: http://www.synthecon.com
Regional Sales Manager: Fred Dinges

UPDATED

Sypris Data Systems

(ex-Datatape Inc.)
(ex-Metrum)
(ex-Metrum – Datatape Inc.)
160 East Via Verde, San Dimas, California, 91773, United States
Tel: (+1 909) 962 94 00
(+1 877) 797 74 78
Fax: (+1 909) 962 94 01
Web: http://www.syprisdatasystems.com
Marketing Communications: Ms Alicia Jones

■ Manufacture of data storage systems, products and services for various military/aerospace and commercial applications.

UPDATED

Taber Industries

(ex-Teledyne Taber)
455 Bryant Street, PO Box 164, North Tonawanda, New York, 14120-0164, United States
Tel: (+1 716) 694 40 00
Freephone: (+1 800) 333 53 00
Fax: (+1 716) 694 14 50
Telex: twx 7102621264
e-mail: sales@taberindustries.com
Web: http://www.taberindustries.com
President: Kenneth L Slawson
Senior Vice President: B W Daube
Sales Engineer: Tom Karlinski

■ Manufacture of testing equipment, machine tools and stainless steel bonded strain gauge pressure transducers. Users include Insat, Arabsat, tethered satellite, EOS, Landsatt and numerous commercial communications satellites and launch vehicles.

UPDATED

Tayco Engineering Inc.

10874 Hope Street, PO Box 6034, Cypress, California, 90630, United States
Tel: (+1 714) 952 22 40
Fax: (+1 714) 952 20 42
e-mail: sales@taycoeng.com
Web: http://www.taycoeng.com
President: Dr Jay Chung
Financial Director: Sheri Nik
Sales Director: Hamid Sadeghzadeh
Purchasing Manager: Don Hambly
Director of Engineering: Ron G Wilkes

■ Electrical heaters, temperature sensors, temperature controllers for satellites, infra-red devices, pressure switches and flexible cable (space qualified).

UPDATED

Taylor Devices, Inc.

90 Taylor Drive, PO Box 748, North Tonawanda, New York, 14120-0748, United States
Tel: (+1 716) 694 08 00
Fax: (+1 716) 695 60 15
e-mail: taylordevi@aol.com
Web: http://www.shockandvibration.com
http://www.taylordevices.com
President: Douglas P Taylor
Vice President: Richard G Hill
Treasurer: Mark McDonough
Aerospace/Defence Products Division: John R Mayfield
Advertising Director: Ms Sandi Taylor
Industrial Products Division: Robert Schneider
Industrial Products Division: Craig Winters
Purchasing Agent: Dennis Lis

■ Liquid springs, shock absorbers, shock isolators, energy absorbers, single and double acting dampers and snubbers, hydraulic components produced on a custom, design to order basis.

UPDATED

TCI

(ex-Technology for Communications International)
47300 Kato Road, Fremont, California, 94538, United States
Tel: (+1 510) 687 61 00
Fax: (+1 510) 687 61 01
Web: http://www.tcibr.com
Vice President, Business Development: Jim Sullivan
Director, Domestic Sales Signal Processing Division: Neal Rachlin
Tel: (+1 510) 687 62 10
Data Publications Specialist: Ms Phyllis Lopez Magana
Tel: (+1 510) 687 62 31

■ Provide solutions for communications and broadcasting applications and design direction finding technology. Produces high-power MF and HF broadcasting antennas as well as HF antennas for military and civilian communications.

UPDATED

Teledyne Brown Engineering (TBE)

300 Sparkman Drive North West, Cummings Research Park, PO Box 070007, Huntsville, Alabama, 35807-7007, United States
Tel: (+1 256) 726 13 19
(+1 256) 726 55 55
Freephone: (+1 800) 933 20 91
Fax: (+1 256) 726 34 34
(+1 256) 726 55 56
Web: http://www.tbe.com
Director, External Communications: Ms Eileen Heaton
Mobile: (+1 256) 25 65 20 77 02

■ Provide systems engineering and technology systems to the defence and space industries.

Business Units
Systems Group
Technologies Group
Advanced Programs
Teledyne Solutions, Inc.

UPDATED

Teledyne Brown Engineering (TBE)
(Albuquerque Office)
557 Windsong Lane, Corrales, New Mexico, 87048, United States
Tel: (+1 505) 792 44 18
Fax: (+1 505) 792 44 27
Contact: Steve R Booher

UPDATED

Teledyne Brown Engineering (TBE)
(Colorado Springs Office)
1330 Inverness Drive, Suite 350, Colorado Springs, Colorado, 80910, United States
Tel: (+1 719) 574 72 70
Fax: (+1 719) 574 63 29
Director: Gordon Bate

UPDATED

Teledyne Brown Engineering (TBE)
(Dayton Office)
4027 Colonel Glenn Highway, Suite 101, Beavercreek, Ohio, 45431-1670, United States
Tel: (+1 937) 427 95 54
Fax: (+1 937) 427 12 11

UPDATED

Teledyne Brown Engineering (TBE)
(Edgewood Project Office)
1310 Woodbridge Station Way, Suite 101, Edgewood, Maryland, 21040-3830, United States
Tel: (+1 410) 676 61 95
Fax: (+1 410) 676 62 12

UPDATED

Teledyne Brown Engineering (TBE)
(Gulf Office)
51 Third Street, Shalimar, Florida, 32579, United States
Tel: (+1 850) 651 34 43
Fax: (+1 850) 651 46 48

UPDATED

Teledyne Brown Engineering (TBE)
(Houston Aerospace Office)
13100 Space Center Boulevard, Mail Stop HB3-40, Houston, Texas, 77059, United States
Tel: (+1 281) 226 42 48
Fax: (+1 281) 226 61 33

UPDATED

Teledyne Brown Engineering (TBE)
(Knoxville Office)
2508 Quality Lane, Knoxville, Tennessee, 37931-3133, United States
Tel: (+1 865) 690 68 19
Fax: (+1 201) 690 61 87

UPDATED

Teledyne Brown Engineering (TBE)
(Los Angeles Field Office)
21221 Western Avenue, Suite 220, Room 235, Torrance, California, 90501, United States
Tel: (+1 310) 320 84 22
Fax: (+1 310) 320 84 20

UPDATED

Teledyne Brown Engineering (TBE)
(Oklahoma Office)
2701 Liberty Parkway, Suite D, Midwest City, Oklahoma, 73110, United States
Tel: (+1 405) 739 09 52
Fax: (+1 405) 739 09 47

UPDATED

Teledyne Brown Engineering (TBE)
(San Diego Office)
5045 Caminito Exquisito, San Diego, California, 92130-2851, United States
Tel: (+1 858) 509 36 80
Fax: (+1 858) 259 69 26

UPDATED

Teledyne Brown Engineering (TBE)
(Warner Robins Office)
PO Box 98568, Robins AFB, Georgia, 21098, United States
Tel: (+1 478) 929 23 91
Fax: (+1 478) 922 38 66

UPDATED

Teledyne Brown Engineering (TBE)
(Washington Operations)
2111 Wilson Boulevard, Suite 900, Arlington, Virginia, 22201-3052, United States
Tel: (+1 703) 276 46 00
Fax: (+1 703) 276 40 63
Director of Washington Operations: John A Braun

UPDATED

Telegenix Inc.
(ex-Telegenix/Grim Corp.)
PO Box 5550, Cherry Hill, New Jersey, 08034, United States
Tel: (+1 856) 424 52 20
Fax: (+1 856) 424 08 89
Telex: 6851048 induct
e-mail: sales@telegenix.com
Web: http://www.telegenix.com
President: John O'Meara
Operations Director: Tim Phelps
Manager, Business Development: John Peebles
Fax: (+1 334) 793 69 64
Purchasing Manager: Ms Joan Wivel
■ Manufacture of communications and control electronics systems and electronic graphic displays for use in air traffic, vessel and command and control applications. Programmable communications voice switches for internal and external communications, signalling interface to radio, telephone and intercom channels. Communications consoles for data, voice or telephone communications between central operator and remote operators. Intelligent voice and data acquisition control and transmission systems used for remote radio control using voice and data on a single telephone line. Contract manufacturing.

UPDATED

Texas Composite Inc.
1281 North Main Street, Boerne, Texas, 78006, United States
Tel: (+1 830) 249 33 99
Fax: (+1 830) 249 32 75
e-mail: mail@texascomposite.com
Web: http://www.texascomposite.com
President: Adair McOran-Campbell
Vice President: Charles Bamford
Purchasing Manager: Ms Carol Ford
■ Manufacture engineered composite structures for military and commercial aerospace. Turbine engine parts and assemblies, Bypass Vanes or Fan OGV's, fan ducts and assemblies. Wing and fuselage fairings, control surfaces, winglets, radomes and gear doors. Space rated structures for human space flight and satellite applications. Missile and launch vehicle structures.

UPDATED

Textron Systems (TS)
(Wilmington)
201 Lowell Street, Mail Stop 1110, Wilmington, Massachusetts, 01887-2941, United States
Tel: (+1 978) 657 21 00
Fax: (+1 978) 657 66 44
Web: http://www.systems.textron.com
■ Provide technology solutions to meet the needs of aerospace and defence industries. Support military precision engagement and dominant manoeuvre with strike weapons, mobility and surveillance systems. Weapons, surveillance systems, HR Textron aircraft control components, specialty marine craft, Cadillac Gage armoured vehicles, as well as turret control and stabilisation systems.

UPDATED

The Boeing Company
(World Headquarters)
100 North Riverside Plaza, Chicago, Illinois, 60606, United States
Tel: (+1 312) 544 20 00
(+1 312) 544 20 02
Web: http://www.boeing.com
Chairman: Lewis E Platt
Senior Vice President, Chief Technology Officer and Office of the Chairman: David O Swain
Senior Vice President, Chief People and Administration Officer and Office of the Chairman: Ms Laurette Koellner
Senior Vice President and General Council: Douglas Bain
Senior Vice President, International Relations: Thomas R Pickering
Vice President, Ethics and Business Conduct: Ms Gale C Andrews
Vice President and General Manager, Boeing Satellite Systems: David Ryan
■ Operates primarily in four market segments: the development, production and marketing of commercial and military aircraft; development and production of commercial and defence missiles and space exploration equipment; development and production of defence electronic systems; computer services and large-scale information networks.

Products
Commercial jet aircraft
Defence and space systems
Defence electronics
Guided missiles
Military aircraft
Rotorcraft
Space vehicles

UPDATED

The Boeing Company
(Integrated Defense Systems, Headquarters)
(Boeing Integrated Defense Systems)
(ex-Boeing Military Aircraft & Missile Systems)
(ex-McDonnell Douglas Corp.)
PO Box 516, St Louis, Missouri, 63166-0516, United States
Tel: (+1 314) 232 02 32
Fax: (+1 562) 797 20 20
(+1 314) 234 82 96
e-mail: susan.m.schantz@boeing.com
Web: http://www.boeing.com/ids
President and Chief Executive Officer: Jim Albaugh
Senior Vice President and General Manager, Naval Systems: John Lockard
Senior Vice President and General Manager, Air Force Systems: George Muellner
Senior Vice President, Army Systems: Roger Krone
Vice President and General Manager, NASA Systems: Mike Mott
Senior Vice President and General Manager, Space and Intelligence Systems: Roger Roberts
Vice President and General Manager, Missile Defense Systems: Jim Evatt
Vice President and General Manager, Homeland Security: Rick Stephens
Vice President and General Manager, Aerospace Support: Pat Finneran
Director, External Communications: Walt Rice
Tel: (+1 314) 234 21 49
Media Relations Manager: Jim Keler
Tel: (+1 314) 233 02 06
Manager, Media/International Programmes: Ms Mary Ann Brett
Tel: (+1 314) 234 71 11
Media Relations: Ms Susan M Schantz
Tel: (+1 314) 233 59 57
■ Provide defence, intelligence, communications and space capabilities combining sophisticated communications networks with air, land, sea and space-based platforms. Programs include B737-700 Airborne Early Warning and Control system (AEWA&C); B767 Tanker/Transport; AH-64D Apache Longbow; Airborne laser, AV-8B Harrier II Plus; Boeing 376, 601 and 702 Satellites; C-17 Globemaster III; C-32A Executive Transport; C-40A Military Transport; C-40B Executive Transport; CH-47 Chinook; Delta family of rockets (Delta II, III, IV); Explosive Detection System machines (EDS) and Explosive Trace Detection Machines (ETD); F/A-18 E/F Super Hornet; F-15 Eagle; F-22 Raptor; Future Combat Systems (FCS); Future Imagery Architecure; Global Positioning Satellites; Ground-based Midcourse Defense; International Space Station; Joint Tactical Radio Systems; Military Aerospace Support; munitions including Standoff Land Attack Missile Expanded Response (SLAM ER), Conventional Air-Launched Cruise Missile (CALCM), Joint Direct Attack Munition (JDAM), Harpoon; RAH-66 Comanche; Sea Launch Company, LLC; Space Shuttle; T-45 Goshawk; Unmanned Combat Air Vehicle; V-22 Osprey.

UPDATED

The Boeing Company
(Electronic Systems and Missile Defence)
PO Box 240002, Huntsville, Alabama, 35824-6402, United States
Tel: (+1 205) 461 28 05
(+1 205) 461 21 21
Fax: (+1 205) 561 22 52
■ Responsible for Boeing's partcipation in a variety of US DoD and NASA programmes, including the Avenger air defence system, Space Station Freedom and the Inertial Upper Stage.

UPDATED

The Recorder Co.
PO Box 8, San Marcos, Texas, 78667, United States
Tel: (+1 830) 629 14 00
Fax: (+1 572) 353 53 33
e-mail: recorder-78667@yahoo.com
Web: http://www.recordercompany.com
President: Floyd MacKenzie
■ CAD/CAM systems and equipment for design engineers. Strip chart recorders.

UPDATED

The Titan Corp.
(Corporate Headquarters)
3033 Science Park Road, San Diego, California, 92121, United States
Tel: (+1 858) 552 95 00
Fax: (+1 858) 552 96 45
Web: http://www.titan.com
■ Provide satellite communications systems, information technology solutions, sterilisation systems and services for commercial and governmental customers worldwide.

Subsidiary companies
Atlantic Aerospace Corporation
DBA Systems
Delfin Systems
Horizons Technology
Linkabit
Systems Resources Corporation
Titan Beta
Titan Eldyne
Titan Environmental
Titan PSI
Titan Scan
Titan Software Systems
Titan Systems
Unidyne
Validity
Visicom

UPDATED

Thomas Electronics Inc.
100 Riverview Drive, Wayne, New Jersey, 07470, United States
Tel: (+1 973) 696 52 00
Fax: (+1 973) 696 82 98
e-mail: sales@thomaselectronics.com
Web: http://www.thomaselectronics.com
Vice President, Business Development: Dennis Young
Vice President, Sales: Bruce Piaget

UPDATED

Turbocam Inc.
5 Faraday Drive, Dover, New Hampshire, 03820-4352, United States
Tel: (+1 603) 749 60 66
Fax: (+1 603) 749 24 04
e-mail: sales@turbocam.com
Web: http://www.turbocam.com
President: Marian B Noronha

Vice President, Engineering: John L Bressoud
Vice President, Manufacturing: Rob A Bujeaud
Sales Manager: William J R Cole
Operations Manager: William J Doyle
Purchasing Manager: Jonas P Zoller
Quality Manager: David P Romanie
Marketing Manager: David R Pincince
Controller/Accounting: David A Zelie
■ Specialist machining for turbomachinery, products include propellers and impellers for marine applications; aircraft engine components, impellers for spaceborne platforms and rail locomotive engine components.
Accreditations: ISO 9001; ISO 9002; AS9100A

UPDATED

United Technologies Corp. (UTC)
United Technologies Building, 1 Financial Plaza, Hartford, Connecticut, 06101, United States
Tel: (+1 860) 728 70 00
(+1 860) 728 79 21
Fax: (+1 860) 493 41 69
(+1 860) 493 41 65
Telex: 99212 utchfda
Web: http://www.utc.com
Chairman, Chief Executive Officer and President: George David
Senior Vice President, International Affairs and Government Relations: Ms Ruth Harkin
Senior Vice President and Chief Financial Officer: Stephen Page
Senior Vice President, Human Resources and Organisation: William L Bucknall
Vice President, Corporate Strategy and Development: Todd Kallman
Vice President, Industrial Relations: Jack P Leary
Director, Worldwide Director, Public Relations: Peter Murphy
Secretary, Worldwide Director, Public Relations: Ginger Schulman
■ Diversified manufacture of products for the aerospace and building industries.
Divisions and subsidiaries
Carrier Corp.
Hamilton Sundstrand
UTC Fuel Cells
Otis Elevator Co.
Pratt & Whitney
Pratt & Whitney Canada Corp.
Sikorsky Aircraft
United Technologies Research Center

UPDATED

Universal Avionics Systems Corporation
(Corporate Headquarters)
(ex-Universal Navigation Corp)
3260 East Universal Way, Tucson, Arizona, 85706, United States
Tel: (+1 520) 295 23 00
Freephone: (+1 800) 321 52 53
Fax: (+1 520) 295 23 95
e-mail: info@www.uasc.com
marketing@uasc.com
Web: http://www.uasc.com
President and Chief Executive Officer: Hubert L Naimer
Corporate Executive Vice President: Charles H Edmondson
Senior Vice President: Joachim (Ted) Naimer
Vice President/General Manager Marketing and Product Support Division: Donald D Berlin
Director, Corporate Finance: Michael Delgado
Director, Marketing Administration: David Upchurch
Director, Customer Services: Grady Dees
Director, Quality and Reliability: Ken Wise
Director, North American Marketing: Paul De Herrera
Director, International Marketing: Rolf Bickel
Director, Airline and Military Marketing: Dan Reida
General Manager, Research/Development/ Engineering Division: Frank Hummel
General Manager, Manufacturing Division: Steve Pagnucco
Manager of Marketing Communications: Stephen B Scolnik
Manager, European Marketing: Daniel Wildi
■ Products include: flight management and navigation management systems, long- and short-range navigation sensors, air-to-ground two way datalink, cockpit voice recorder and cabin display systems.

UPDATED

Vacco Industries
10350 Vacco Street, South El Monte, California, 91733, United States
Tel: (+1 626) 443 71 21
Fax: (+1 626) 442 69 43
e-mail: navy@vacco.com
Web: http://www.vacco.com
President: A E Gonzalez
Senior Vice President Navy Sales: R J McCredie
Executive Vice President: Alan Keskinen
Regional Sales Manager: Richard K Clark
Tel: (+1 626) 450 64 56
■ Specialised flow control valves, filters and filtration systems, photochemical machining.

UPDATED

Vertex RSI
(Santa Clara Facility)
(ex-TIW Systems Inc.)
(ex-Vertex Antenna Systems, LLC)
(ex-Vertex Satcom Systems, Inc.)
2211 Lawson Lane, Santa Clara, California, 95054, United States
Tel: (+1 408) 654 56 00
Fax: (+1 408) 654 56 13
(+1 408) 654 56 14
e-mail: info@tiw.com
Web: http://www.tripointglobal.com
President: Bernard Cahlander
Tel: (+1 408) 654 56 00
Director of Sales: Rob Wellins
Tel: (+1 408) 980 45 20
Manager of Microwave Engineering: Ken Goudey
Tel: (+1 408) 654 56 00
■ Antenna systems for Telemetry, Tracking and Control (TT&C) of satellites. Radar system and antennas for radio telescopes.

UPDATED

Vertex RSI
(Antenna Products Division)
(ex-Vertex Communications Corporat)
2600 North Longview Street, Kilgore, Texas, 75662, United States
Tel: (+1 903) 984 05 55
Fax: (+1 903) 984 18 26
Web: http://www.tripointglobal.com
http://www.vertexcomm.com
http://www.rsicom.com
■ Design, manufacture and installation of earth station antennas.

UPDATED

Viasat Inc.
6155 El Camino Real, Carlsbad, California, 92009-1699, United States
Tel: (+1 760) 476 22 00
Fax: (+1 760) 476 47 03
Web: http://www.viasat.com
President: Mark Dankberg
Marketing Communications Manager, GSD: Felice Chiapperini
Tel: (+1 760) 476 24 76
■ Digital communications company specialising in high-performance networking hardware and software for satellite and other wireless applications. Products for government and commercial applications. Satellite communication products include terminals and modems for UHF, Ku-, C- and Ka-band. Information security products include inline network encryptors that can secure communication on public networks up to top secret levels. Tactical communications products include MIDS/Link-16 terminals and advanced data controllers for error-free communication of images and email on military channels. Other products include multiple emitter C3 and EW simulators and satellite simulator/training terminals.

UPDATED

Voss Industries Inc.
2168 West 25th Street, Cleveland, Ohio, 44113-4172, United States
Tel: (+1 216) 771 76 55
Fax: (+1 216) 771 28 87
e-mail: voss@vossind.com
sales@vossind.com
Web: http://www.vossind.com
President: Daniel Sedor
Vice President, Engineering: John Fritskey
Vice President, Operations: Mark Schodowski
Vice President, Development: Nick Antonelli
Purchasing Manager: Bill Pugh
Sales Manager: David Kleinpeter
■ Standard and special T-bolt clamps, V-band couplings and mating flanges, strap assemblies, rigid V couplings and machined flanges, CNC machining, stampings, light fabrications and hydraulic bulge-formed shapes for aerospace and defence applications. Mil-Spec painting, spot welding and heliarc welding.

UPDATED

Votaw Precision Technologies Inc.
13153 Lakeland Road, Santa Fé Springs, California, 90670, United States
Tel: (+1 562) 944 06 61
Fax: (+1 562) 944 06 75
Web: http://www.votawprecision.com
■ Manufacture of very large precision tooling for the aerospace industry. Performs machining fabrication and tooling of aerospace components. Shuttle, Titan, Delta II, Atlas, EELV, JSF and F-22.

UPDATED

Wear-Cote International Inc.
PO Box 4177, Rock Island, Illinois, 61204-4177, United States
Tel: (+1 309) 793 12 50
Fax: (+1 309) 786 65 58
e-mail: webmaster@wear-cote.com
Web: http://www.wear-cote.com
Chief Executive Officer: Jim Henry
■ Electroless coatings for ferrous/non-ferrous metals. Electroless nickel is employed in Shuttle main engine ignition and mechanical systems and is now applied to other cryogenic propellant systems and ground handling equipment.

UPDATED

Zin Technologies
3000 Aerospace Parkway, Cleveland, Ohio, 44142, United States
Tel: (+1 216) 977 04 22
Fax: (+1 216) 977 06 26
Web: http://www.zin-tech.com
Creative Design Co-ordinator: Ms Dawn Wright

UPDATED

Industry – Service

Acme Rocket Company
930 South Dobson, Unit 8, Mesa, Arizona, 85202, United States
Tel: (+1 480) 461 83 26
President: Grant Boyd
■ Plans/dimensioned drawings and blueprints of rockets, missiles and launch vehicles. Intended for historians, researchers, model makers and space buffs. Large selection of non-classified authentic scale data includes international and obscure vehicles of all eras.

UPDATED

Airborne Air and Fuel Products
711 Taylor Street, Elyria, Ohio, 44035, United States
Tel: (+1 440) 937 13 15
Freephone: (+1 800) 382 84 22
Fax: (+1 440) 937 54 09
e-mail: techhelp@parker.com
Web: http://www.parker.com/airborne

UPDATED

Ambitech International
(Axiom Operations)
(ex-Axiom Electronics, Inc.)
14924 NW Greenbrier Parkway, Beaverton, Oregon, 5733, United States
Tel: (+1 503) 350 49 06
Freephone: (+1 800) 643 66 01
Fax: (+1 503) 350 43 66
Web: http://www.axiomsmt.com
Vice President and General Manager: Robert Toppel
Sales Manager: Doug Robertson
Engineering Manager: Bill King
Purchasing Manager: Roger Petit
■ Services include complete procurement service, surface mount assembly, mixed technology assembly, precision rework of existing products and complete final assembly and unit integration services.
Accreditations: ISO 9002, IPC-A-610, J-STD-001, NASA STD-8739.3, UL, Boeing QMS, FDA, ITAR

UPDATED

Americom Government Services (AGS)
2 Research Way, Princeton, New Jersey, 08540, United States
Tel: (+1 609) 987 45 00
Fax: (+1 609) 987 44 11
e-mail: ags-info.americom@americom-gs.com
Web: http://www.americom-gs.com
Chief Executive Officer: David Helfgott

Advertising and Promotion Manager: Andrew A Tytla
Tel: (+1 609) 987 42 68
Fax: (+1 609) 987 43 12

UPDATED

Analex Corporation
2001 Aerospace Parkway, Brook Park, Ohio, 44142, United States
Tel: (+1 216) 925 11 71
Fax: (+1 216) 925 10 40
Web: http://www.analex.com
Operations Engineer: Ms Angela M Ruwadi

UPDATED

Analytical Graphics Inc. (AGI)
220 Valley Creek Boulevard, Exton, Pennsylvania, 19341, United States
Tel: (+1 610) 981 80 00
Fax: (+1 610) 981 80 01
e-mail: info@stk.com
Web: http://www.stk.com
- Developers of the Satellite Tool Kit (STK) family of products, an interactive, graphical software package that helps analysts, developers, operators and users of satellite systems access, manage, display and manipulate their data. Users can analyse aerospace systems both graphically and numerically. Operates on UNIX platforms, Windows 95/98, Windows 2000 and Windows NT. Programmer's Library and 3D visualisation option available. The satellite orbit analysis software addresses market segments including: design, operations, communications, navigation, remote sensing, science and defence.

UPDATED

Ardak Corp.
1800 Old Meadow Drive, Suite 114, Mclean, Virginia, 22102, United States
Tel: (+1 703) 893 38 28
Fax: (+1 703) 893 40 10
Web: http://www.ardak.com

UPDATED

Astrotech Space Operations
(Headquarters)
1515 Chaffee Drive, Titusville, Florida, 32780, United States
Tel: (+1 407) 268 38 30
Fax: (+1 407) 268 38 34
Senior Vice President and General Manager: John B Satrom
Vice President General Manager: Jim Royston
- Pre-launch payload processing at Cape Canaveral and Kennedy Space Center; aerospace engineering services.

UPDATED

Astrotech
(Vandenberg Office)
Vandenberg Air Force Base, PO Box 5097, Vandenberg, 934387, United States
Tel: (+1 805) 734 11 02
Fax: (+1 805) 734 25 51

UPDATED

Atec Inc.
12600 Executive Drive, Stafford, Texas, 77477-3604, United States
Tel: (+1 281) 276 27 00
Freephone: (+1 866) 753 23 84
Fax: (+1 281) 240 26 82
e-mail: sales@atec.com
Web: http://www.atec.com
Programme Manager: David L Knape
Tel: (+1 281) 276 26 90
Fax: (+1 281) 276 26 79
- Diagnostic age and instrumentation systems for jet engine and aircraft systems testing at flight line, intermediate and depot maintenance levels.

UPDATED

Aviation Week & Space Technology
Two Penn Plaza, 5th floor, New York, New York, 10121-2298, United States
Tel: (+1 212) 904 20 00
Fax: (+1 212) 904 60 68
Web: http://www.aviationweek.com
Executive Vice President and Publisher: Kenneth E Gazzola
Director of Business Development: Mark Lipowitz
Sales Director: Kenneth F Luken
Promotions Director: Ms Marilyn J Zack
Managing Editor: David Hughes
Editor-in-Chief: David M North
Office Manager: Ms Mary Francis Koerner
Publications: Aviation Week and Space Technology

UPDATED

AXA Space
(ex-International Technology Underwriters)
4800 Montgomery Lane, 11th Floor, Bethesda, Maryland, 20814, United States
Tel: (+1 301) 654 85 85
Fax: (+1 301) 654 75 69
President and Chief Executive Officer: Frederick H Hauck
Senior Vice President, Space Underwriting: Patrick Rivalan
Director, Space Systems Engineering: Russell Fillers

UPDATED

Barrios Technology, Inc.
2525 Bay Area Boulevard, Suite 300, Houston, Texas, 77058-1556, United States
Tel: (+1 281) 280 19 00
Fax: (+1 281) 280 19 01
e-mail: info@activeiweb.com
Web: http://www.barrios.com
http://www.activeiweb.com
- Products include Active iWeb, a solution to information challenges that provides the tools necessary to publish information to the web in a secure environment and facilitate communications throughout an organisation. It allows administrators to post content without the need for extensive web development knowledge or training. It has an interface that is customisable allowing different user groups to see only the information they need.

UPDATED

BES Engineering
11200 Westheimer, Suite 630, Houston, Texas, 77042, United States
Tel: (+1 713) 780 75 27
Fax: (+1 713) 780 37 32
Vice President: Ms Liz Nevils

UPDATED

Blackhawk Management Corporation
1335 Regents Park Drive, Suite 130, Houston, Texas, 77058, United States
Tel: (+1 281) 286 57 51
Fax: (+1 281) 286 57 52
Web: http://www.blackhawkmgmt.com
President and Chief Operating Officer: Joseph D Di Iorio
Mobile: (+1 713) 703 94 16
Senior Director: James Wilcox
Tel: (+1 281) 461 17 73

UPDATED

Booz, Allen & Hamilton Inc.
(Headquarters & Worldwide Technology Business)
8283 Greensboro Drive, Mclean, Virginia, 22102, United States
Tel: (+1 703) 902 50 00
Fax: (+1 703) 902 33 33
Chairman and Chief Executive Officer: Ralph W Shrader
Contact: Ms Patricia Hardee
Tel: (+1 703) 902 50 21
- International management and technology consulting firm with major services in strategy, operations, technology and systems consulting.

UPDATED

Booz, Allen & Hamilton Inc.
(Colorado Springs Office)
1050 S Academy Boulevard, Suite 148, Colorado Springs, Colorado, 80910, United States
Tel: (+1 719) 637 68 80
Fax: (+1 719) 597 81 31
Contact: Christopher T Kucera
Mobile: (+1 301) 379 61 95

UPDATED

CD Adapco Group
1516 East Palm Valley Boulevard, Suite B-1, Round Rock, Texas, 78664, United States
Tel: (+1 631) 549 23 00
e-mail: starinfo@adapco.com
Web: http://www.cd-adapco.com
Technical Sales Manager: Ms Cathy Apple
Tel: (+1 512) 733 10 70
Mobile: (+1 512) 589 64 45
Fax: (+1 512) 244 29 66

UPDATED

Cimarron
1115 Gemini, Houston, Texas, 77058-3584, United States
Tel: (+1 281) 226 51 00
Fax: (+1 281) 226 51 90
e-mail: cimarron@cimarroninc.com
sales@cimarroninc.com
Web: http://www.cimarroninc.com

UPDATED

Command and Control Technologies Corporation (CCT)
1425 Chaffee Drive, Suite 1, Titusville, Florida, 32780, United States
Tel: (+1 321) 264 11 93
Fax: (+1 321) 383 50 96
e-mail: info@cctcorp.com
Web: http://www.cctcorp.com
President: Peter C Simons
Vice President, Business Development: Kevin Brown
Technical Director: Rodney Davis

UPDATED

Computer Sciences Corp. (CSC)
(Corporate Headquarters)
2100 East Grand Avenue, El Segundo, California, 90245, United States
Tel: (+1 310) 615 03 11
Web: http://www.csc.com
- Design and development of system software and application programs, system engineering, facilities management, systems integration, remote computing services and communication systems.

UPDATED

Comsol Inc.
8 New England Executive Park, Suite 310, Burlington, Massachusetts, 01803, United States
Tel: (+1 781) 273 33 22
Fax: (+1 781) 273 66 03
Web: http://www.comsol.com
Regional Sales Manager: Adam DeCruz

UPDATED

Dage Corporation
1011 High Ridge Road, Stamford, Connecticut, 06905, United States
Tel: (+1 203) 461 90 00
Fax: (+1 203) 461 97 95
e-mail: emailus@dage.com
Web: http://www.dage.com
President: Peter Sterling
Executive Vice President: John Everett
Contact: Ms Barbara Buchman
- Full service export management. Supplies custom designed, catalogue and spare parts components and subsystems for the commercial and military markets: communications of all types, cellular, line-of-sight, mobile earth stations, satellites. Radars: ground-based, fixed

and mobile, shipborne, missile and space. Electronic warfare system components. TV, radio transmitter and receiver parts. Linear and ring accelerator components. Industrial heating products.

Representing
Antelope Valley Microwave
Atlantic Microwave Corporation (Antennas)
Custom Microwave
DB Products
EMF Systems
EMR Corporation
MacTechnology
Mega Industries
Microwave Development Co. (MDC)
Microwave Development Labs. (MDL)
Microwave DeviceTechnology
Microwave Resources Inc. (MRI)
Microwave Resources Corporation (MRC)
Omniyig
Pendulum Electromagnetics
Reactel
Resin Systems
Res-Net Microwave
Sonoma Scientific
Telgaas Inc.

UPDATED

Digital Diecutting, Inc.

1405 16th Street #4002, PO Box 1824, Racine, 53403, United States
Tel: (+1 800) 331 96 63
Fax: (+1 888) 703 03 83
e-mail: info@digitaldiecutting.com
Web: http://www.digitaldiecutting.com
President: Greg Klema

UPDATED

Elsevier Science

(United States & Canada Regional Sales Office)
Customer Support Department, PO Box 945, New York, NewYork, 10159-0945, United States
Tel: (+1 212) 633 37 30
Fax: (+1 212) 633 36 80
e-mail: usinfo-f@elservier.com
Web: http://www.elsevier.com
■ Publisher of scientific, technical and medical information.

UPDATED

E-Systems Inc.

(Headquarters)
11 Penn Plaza, 5th Floor, New York, New York, 10001-2606, United States
Tel: (+1 212) 946 26 79
Fax: (+1 800) 613 74 56
e-mail: info@esystemsinc.com
Web: http://www.esystemsinc.com

UPDATED

Fluent

10 Cavendish Court, Centerra Resource Park, Lebanon, New Hampshire, 03766-1442, United States
Tel: (+1 603) 643 26 00
Fax: (+1 603) 643 39 67
Web: http://www.fluent.com
Account Manager: Christopher LaRiviere

UPDATED

Fluor Corporation

(Headquarters)
One Fluor Daniel Drive, Aliso Viejo, California, 92656-2606, United States
Tel: (+1 949) 349 20 00
Fax: (+1 949) 349 25 85
Web: http://www.fluordaniel.com
Chairman and Chief Executive Officer: Alan L Boeckmann
Senior Vice President and Chief Financial Officer: D Michael Steuert
Senior Vice President, Law and Secretary: Lawrence N Fisher
Vice President, Investor Relations: Ms Lila J Churney
Vice President, Corporate Communications: Lee Tashjian
Vice President, Corporate & Public Affairs: J Robert Fluor
Group Executive Sales, Marketing and Strategic Planning: John L Hopkins
Group Executive, Strategic Operations: Ronald W Oakley
Senior Executive, Human Resources: H Steven Gilbert
■ Provides worldwide engineering, construction, procurement, project management, operations and maintenance services to defence, aerospace, and government agencies. Services include project management, feasibility and engineering studies, planning, engineering studies, planning engineering, design, reliability, availability and maintainability (RAM) analysis, environmental studies and design, construction and construction management, procurement, telecommunications design, studies, and project management. Operations and maintenance.

UPDATED

Futron Corporation

7315 Wisconsin Avenue, Suite 900W, Bethesda, Maryland, 20814, United States
Tel: (+1 281) 333 01 90
e-mail: busdev@futron.com
Web: http://www.futron.com
■ Specialise in space and telecommunications analysis, risk management, aerospace safety and dependability, space remote sensing, advanced communication, education and training.

UPDATED

Glocom Inc.

20010 Century Boulevard, Germantown, Maryland, 20874, United States
Tel: (+1 301) 916 21 00
Fax: (+1 301) 916 94 38
e-mail: info@glocom-us.com
Web: http://www.glocom-us.com

UPDATED

Integral Systems Inc.

5000 Philadelphia Way, Lanham, Maryland, 20706-4417, United States
Tel: (+1 301) 731 42 33
Fax: (+1 301) 731 96 06
Web: http://www.integ.com
Satellite Systems Engineer: Dilene Nelson
Tel: (+1 301) 731 42 33

UPDATED

Intelsat Global Service Corporation

4455 Connecticut Avenue North West, Suite 2400, Washington, District of Columbia, 20008, United States
Tel: (+1 202) 243 40 00
e-mail: government.solutions@intelsat.com
Chairman: John Sponyoe
Chief Executive Officer: Conny Kullman
Executive Vice President and Chief Financial Officer: Joseph Corbett
Chief Operating Officer: Ramu Potarazu
Senior Vice President and Chief Administrative Officer: Tony A Trujillo
Corporate Communications: Jeff Hardy
Tel: (+1 202) 944 79 26
■ Provide telephony corporate network, video and internet solutions around the globe via capacity on 26 geosynchronous satellites in prime orbital locations.

UPDATED

Intergraph Corporation

PO Box 240000, Huntsville, Alabama, 35824, United States
Tel: (+1 256) 730 20 00
Fax: (+1 256) 730 20 48
Web: http://www.intergraph.com
■ Produce PC-based computers, workstations and image visualisation tools. Z/I Imaging Corporation is a joint venture of Intergraph and Carl Zeiss. It is involved in photogrammetry and products include aerial cameras, image processing software, and workstations.

UPDATED

International Launch Services (ILS)

(Headquarters)
1660 International Drive, McLean, Virginia, 22102-3900, United States
Tel: (+1 571) 633 74 00
Fax: (+1 571) 633 75 00
Web: http://www.ilslaunch.com
Director, Media Relations: Ms Fran Slimmer
Tel: (+1 571) 633 74 62
Mobile: (+1 646) 229 48 01
Fax: (+1 571) 633 75 37
■ Provide launch services.

UPDATED

International Space Brokers (ISB)

1300 Wilson Boulevard, Suite 990, Rosslyn, Virginia, 22209, United States
Tel: (+1 703) 841 13 34
Fax: (+1 703) 841 05 25
Web: http://www.isbworld.com
President/Chief Executive Officer: John W Vinter
Executive Vice President: John P Cozzi
Senior Vice President: Glen E Hogan
Executive Vice President: Stephen C Leonard
Senior Vice President: Richard E Rankin
Assistant Vice President: Ms Christi Chao
Contact: Ms Madeleine Jennings

UPDATED

Jackson and Tull

(Aerospace Engineering Division)
7375 Executive Place, Suite 200, Seabrook, Maryland, 20706, United States
Tel: (+1 301) 805 45 45
Fax: (+1 301) 805 45 38
Web: http://www.jnt.com
Director of Special Programme Development: Ms Kathy J Nado
Tel: (+1 301) 805 45 45

UPDATED

Jacobs Sverdrup, Inc. (JS)

(ex-Sverdrup Technology Inc.)
600 William Northern Boulevard, PO Box 884, Tullahoma, Tennessee, 37388, United States
Tel: (+1 931) 455 64 00
Fax: (+1 931) 393 63 89
e-mail: webmaster@jacobssverdrup.com
Web: http://www.jacobssverdrup.com
President: R F Starr
Senior Vice President, Business Development: Kenny Frame
Tel: (+1 931) 393 66 94
Region Controller: Richard Coe
Tel: (+1 931) 393 66 33
Marketing Director: Ms Lisa Wright
Tel: (+1 931) 393 66 31
Operations Director: Tom Mahler
Tel: (+1 931) 393 66 97
Office Manager of Hampton, VA Office of the Technology Group: Rick Wineman
Tel: (+1 757) 827 17 86
Marketing Specialist: Ms Beverly Charlton
■ Provide scientific and engineering services in support of the design, construction, operations, maintenance, testing and evaluation of advanced facilities and systems for customers including DoD, NASA and a diversity of commercial clients.

UPDATED

Jane's Information Group (JIG)

(North/Central/South America)
110 N Royal Street, Suite 200, Alexandria, Virginia, 22314, United States
Tel: (+1 703) 683 37 00
Freephone: (+1 800) 824 07 68
Fax: (+1 703) 836 02 97
(+1 800) 836 02 97
e-mail: info.us@janes.com
Web: http://www.janes.com
Sales Director: Bob Loughman
■ Visit Jane's on the net at http://www.janes.com.

UPDATED

Kamatics Corporation

PO Box 3, Bloomfield, Connecticut, 06002-0003, United States
Tel: (+1 860) 243 97 04
Fax: (+1 860) 243 79 93
Web: http://kamnrkt-kam@kaman.com
President: Dr John Kornegay
Vice President, Sales: Robert Paterson
Tel: (+1 860) 243 32 20

UPDATED

Keiser Engineering Inc.

2046 Carrhill Road, Vienna, Virginia, 22181-2917, United States
Tel: (+1 703) 281 95 82
Fax: (+1 703) 281 95 82
President: Bernhard E Keiser
■ Consulting in satellite communications systems engineering, feasibility studies, interface problems, applications and seminars.

UPDATED

KPMG Peat Marwick
(Commercial Space Group)
2001 M Street North West, Washington, District of Columbia, 20036-3310, United States
Tel: (+1 202) 467 30 00
(+1 202) 467 30 85
Fax: (+1 202) 293 54 37
Telex: 440477
■ Business advisory services on space transportation, materials processsing, remote sensing, communications, life sciences and industrial services.

UPDATED

Mackay Communications Inc.
(Headquarters)
3691 Trust Drive, Raleigh, North Carolina, 27616-2955, United States
Tel: (+1 919) 850 30 00
Fax: (+1 919) 954 17 07
Telex: 4612018 mackay
e-mail: info@mackaycomm.com
Web: http://www.mackaycomm.com
Co President, Sales and Marketing: Ben Pratt
Tel: (+1 919) 850 30 29
Financial Director: Dave Eckstine
Tel: (+1 919) 850 30 46
Director, Government Sales: Terry Croutharmel
Tel: (+1 904) 724 61 01
Sales Director: Patrick Fisher
Tel: (+1 919) 850 31 64
Technical Director: Bill Jackson
Tel: (+1 919) 850 30 32
Marketing Director: Ms Pamela McClure
Tel: (+1 919) 850 30 14
■ Offer a range of mobile satellite communications products and services, with products that have been designed to withstand rugged environments. Installation and service of shipboard and terrestrial-based Inmarsat satellite systems. Inmarsat A, B, C, M, Mini-M and Fleet F products and airtime provide reliable voice, fax, 64/128 kbps digital data, and video conferencing with ISDN connectivity. Also offer network and data application engineering services for new or existing applications, in addition to a full line of marine navigation and below-deck electronic equipment from manufacturers.

UPDATED

Major Tool & Machine, Inc.
1458 East 19th Street, Indianapolis, Indiana, 46218, United States
Tel: (+1 317) 636 64 33
Fax: (+1 317) 634 94 20
Web: http://www.majortool.com
Programme Manager: Michael H Kramer
Tel: (+1 317) 917 26 26

UPDATED

Mobile Satellite Ventures (MSV)
(Reston Office)
(ex-TMI Communications and Motient Corporation)
10802 Parkridge Boulevard, Reston, Virginia, 20191, United States
Tel: (+1 703) 390 27 00
Fax: (+1 703) 390 27 70
President and Chief Operating Officer: Dr Carson E Agnew
Vice President and Chief Financial Officer: Eric Swank
Vice President and Chief Technical Officer: Dr Peter Karabinis
Vice President, MSAT Operations: Ms Elizabeth Tasker
Manager, Services Marketing: Ms Linda Gustus
Tel: (+1 613) 742 41 68
■ Provide satellite communication services. Services include a choice of wireless data, voice, fax and dispatch radio services via the company's two MSAT satellites.

UPDATED

Nerac, Inc.
One Technology Drive, Tolland, Connecticut, 06084, United States
Tel: (+1 860) 872 70 00
Fax: (+1 860) 875 60 62
e-mail: info@nerac.com
Web: http://www.nerac.com
President: Kevin A Bouley
Vice President of Finance: John E Ruest
Vice President of Client Services: Gerri B Potash
Manager, Communications: James Kieffer
Marketing and Communicatons: Ms Suzi Craig-Edwards
■ Bibliographic reviews of aerospace technology and companies, technology tracking services and document retrieval. Information on aerospace engineering, agriculture and plant biology, automotives, biochemistry, biomedical and medical devices, business combined with management and marketing, coatings and adhesives, computers and software, electrical engineering, energy and power, environmental biology, good chemistry, inorganic chemistry, mechanical engineering, metallurgy, military and govenment, molecular and cell biology, optics combined with imaging and lasers, organic chemistry-synthesis, packaging, patent law, pharmaceuticals, polymer chemistry and textiles combined with paper and pigments.

UPDATED

Oologah Technologies, Inc. (OTI)
PO Box 2670, Bay St Louis, Mississippi, 39521, United States
Tel: (+1 228) 688 73 86
Fax: (+1 228) 688 73 86
e-mail: oti@oologahtech.com
Web: http://www.oologahtech.com
Chief Executive Officer: Joe M Stewart

UPDATED

PanAmSat Corporation
20 Westport Road, Wilton, Connecticut, 06897, United States
Tel: (+1 203) 210 80 00
Fax: (+1 203) 210 80 01
e-mail: corpcomm@panamsat.com
Web: http://www.panamsat.com
Contact: Philip Robertson
■ Provide global video and data broadcasting services via satellite. The company builds, owns and operates networks that deliver entertainment and information to cable television systems, TV broadcast for direct-to-home TV operators, internet service providers, telecommunications companies and corporations. The company currently has 20 satellites in orbit with plans to expand the fleet to 22.

UPDATED

Satellite Engineering Research Corporation
7701 Woodmount Avenue, Suite 208, Bethesda, Maryland, 20814, United States
Tel: (+1 301) 657 96 41
Fax: (+1 301) 657 96 42
Web: http://www.satellitecorp.com
President: Dr Robert A Nelson
■ Provides engineering and research services in satellite communication systems engineering, including RF link analysis, interface analysis, orbital mechanisms, constellation design, spacecraft design, GPS applications, theoretical analysis of high precision time transfer, and technical legal assistance. Conducts professional training courses in satellite communication.

UPDATED

Science Applications International Corporation (SAIC)
(Headquarters)
10260 Campus Point Drive, San Diego, California, 92121, United States
Tel: (+1 505) 842 78 91
(+1 858) 826 60 00
Web: http://www.saic.com
Public Affairs Director: Ron Zollars
Tel: (+1 858) 826 78 96
Fax: (+1 858) 826 66 34
Contact: James J Karns
■ Research and engineering services in technology development and analysis, computer systems, computer hardware and computer software. Applications in the areas of national and international security, information technology, transportation, telecommunications and environmental systems and engineering. Offices in over 150 cities worldwide.

UPDATED

Sermatech Power Solutions
PO Box 451047, Houston, Texas, 77245-1047, United States
Tel: (+1 832) 485 80 51
Fax: (+1 713) 413 10 09
Web: http://www.sermatech.com
IGT Solutions Manager: Bruce E Mizell
Mobile: (+1) 71 37 24 15 07

UPDATED

Simulogix
(Top-Down Engineering)
555 Veterans Boulevard, Suite 123, Reduced City, California, 94063, United States
Business Development Consultant: Leslie Schlom
Mobile: (+1 914) 799 11 55

UPDATED

Smith Semiconductor Inc.
3185 Tozer Road, North Branch, Michigan, 48461, United States
Tel: (+1 810) 688 11 74
Fax: (+1 810) 688 47 38
e-mail: jeremy@smithsemi.com
Web: http://www.smithsemi.com
■ Procurement and supply of obsolete and diminished military, aerospace and space electronic components.

UPDATED

Smiths Aerospace
(US Government Relations Office)
1225 Jefferson Davis Highway, Suite 1100, Arlington, Virginia, 22202-4301, United States
Tel: (+1 703) 416 94 00
Fax: (+1 703) 416 94 04
Web: http://www.smiths-aerospace.com
■ Focus on government relations and sales.

UPDATED

Space Data Resources/ Information (SDR/I)
PO Box 3438, Boulder, Colorado, 80307-3438, United States
Tel: (+1 303) 494 76 77
e-mail: newsspace@aol.com
President and Chief Executive Officer: Leonard David
Administrator: Ms Barbara Sprungman
■ Provide research services regarding the world's space programmes. Maintain databases on space applications programmes as well as space science projects. Freelance writing and press relations are also offered.

UPDATED

Space Machine Advisors Inc.
50 Washington Street, 11th Floor, Norwalk, Connecticut, 06854, United States
Tel: (+1 203) 661 02 02
Fax: (+1 203) 661 52 43
Chief Executive Officer: Alden M Richards

UPDATED

Space Technology Commerce & Communications
17 Heritage Drive, Lexington, Massachusetts, 02420, United States
Tel: (+1 781) 862 71 74
Fax: (+1 781) 862 22 45
e-mail: tfassoc@tiac.net
President: Ms Roseann Tully
■ Conference and exposition management; advertising and marketing.

UPDATED

Spacecoast Aeromedical Institute (SAMI)
1698B West Hibiscus Boulevard, Melbourne, Florida, 32901, United States
Tel: (+1 321) 676 32 00
Fax: (+1 321) 674 00 20
Web: http://www.aeromedicalinstitute.com
Director of Marketing and Development: Robert E Salonen

UPDATED

Spacehab, Inc.
(World Headquarters)
12130 Highway 3, Building 1, Webster, Texas, 77598, United States
Tel: (+1 713) 558 50 00
Freephone: (+1 888) 647 95 43
Fax: (+1 713) 558 59 60
Web: http://www.spacehab.com
Chairman and Chief Executive Officer: Dr Shelley A Harrison
President and Chief Executive Officer: Michael E Kearney
Senior Vice President, Space Flight Services: Dan Bland
Senior Vice President and Chief Financial Officer: Brian Harrington

Vice President, Spacehab Government Services: Jay Sauerwein
Marketing Director: Ms Kimberly Campbell
■ Provide habitation and logistics services to NASA. Provision of commercial access to space and offers payload processing services for manned and unmanned vehicles.

UPDATED

Spacehab, Inc.
(Payload Processing Facility)
620 Magellan Road, Cape Canaveral, Florida, 32920, United States
Tel: (+1 321) 868 74 11

UPDATED

Spacenet, Inc.
1750 Old Meadow Road, McLean, Virginia, 22102, United States
Tel: (+1 703) 848 10 00
Fax: (+1 703) 848 10 10
e-mail: info@spacenet.com
Web: http://www.spacenet.com
President and General Manager: Nick Supron

UPDATED

Spaceport Systems International (SSI)
(Commercial Spaceport)
3769-C Constellation Road, Lompoc, California, 93436, United States
Tel: (+1 805) 733 73 70
Fax: (+1 805) 733 73 72
Web: http://www.calspace.com
President and Chief Executive Officer: Orlando C Severo
Financial Director: Ms Cheryl M Samarco
Operations Director: William H Anders

UPDATED

Spot Image Corp. (SICORP)
14595 Avion Parkway, Suite 500, Chantilly, Virginia, 20151, United States
Tel: (+1 703) 715 31 00
Freephone: (+1 800) ASK SPOT
Fax: (+1 703) 715 31 20
Web: http://www.spot.com
President: Neal Carney
Tel: (+1 703) 715 31 10
Chief Operations Officer: Ms Diana Clemente
Director, Technical Operations: Ms Joann Barton
Tel: (+1 703) 715 31 54
Director, Corporate Communications and Marketing: Clark Nelson
Tel: (+1 703) 715 31 31
■ Marketing and distribution of the SPOT satellite system. Remotely sensed data includes high resolution imagery, rapid revist capability, worldwide coverage and digital elevation models.

UPDATED

Swales Aerospace
5050 Powder Mill Road, Beltsville, Maryland, 20705, United States
Fax: (+1 301) 902 41 14
Web: http://www.swales.com
Business Development: Darrell V Warner
Tel: (+1 301) 902 45 17

UPDATED

Technology Service Corp. (TSC)
1900 South Sepulveda Boulevard, Suite 300, Los Angeles, California, 90025-5659, United States
Tel: (+1 310) 954 22 00
Fax: (+1 310) 477 21 96
e-mail: info@tsc.com
Web: http://www.tsc.com
President: Robert Graziano
Vice President: Dr Russell Lefevre
Manager, Operations: Eric Wilen
■ Defence related services: radar, sonar, electro-optical and electronic warfare.

UPDATED

Terra-Mar Resource Information Services Inc.
(Terra-Mar Resource Inf. Serv. Inc.)
PO Box 369, Mountain Ranch, California, 95246, United States
Tel: (+1 209) 754 35 36
Fax: (+1 209) 754 35 37
e-mail: info@terra-mar.net
Web: http://www.terra-mar.net

■ Orthophoto maps. A selection of airborne, real-time mapping systems geared toward tactical emergency response and resource monitoring.
Representing:
Spot Image Corp., US

UPDATED

The Aerospace Corp.
(Corporate Headquarters)
PO Box 92957, Los Angeles, California, 90009-2957, United States
Tel: (+1 310) 336 50 00
Fax: (+1 310) 336 70 55
e-mail: corpcomm@aero.org
Web: http://www.aero.org
Chairman, Board of Trustees: Dr Bradford W Parkinson
Vice Chair of Board of Trustees: Gen Howell M Estes
President and Chief Executive Officer: Dr William F Ballhaus
Tel: (+1 310) 336 29 34
Executive Vice President: Dr Joseph M Straus
Tel: (+1 310) 336 69 33
Senior Vice President, Engineering and Technology Group: Dr Wanda M Austin
Tel: (+1 310) 336 89 24
Senior Vice President, National Systems Group: Jon H Bryson
Tel: (+1 703) 633 51 00
Senior Vice President, Space Systems Group: John R Parsons
Tel: (+1 310) 336 52 93
Senior Vice President, General Counsel and Secretary: Gordon J Louttit
Tel: (+1 310) 336 25 28
Senior Vice President, Systems Planning and Engineering: Dr Rodney C Gibson
Tel: (+1 310) 336 22 72
Vice President, Chief Financial Officer and Treasurer: Dale E Wallis
Tel: (+1 310) 336 11 07
Assistant Secretary: Ms Roberta L Ackley
■ An independent, non-profit corporation supporting the US government in planning, research, development, acquisition and operation of space launch and associated ground systems and in science and technology important to national security. Provide engineering support to Air Force space programmes and national security space systems. Co-operate with NASA and other civil agencies in order to apply space-related skills to critical issues.

UPDATED

The Aerospace Corp.
(Program Executive Office for Space Support)
1500 Wilson Boulevard, Suite 515, Arlington, Virginia, 22209-2404, United States
Tel: (+1 703) 526 17 40
Fax: (+1 703) 526 17 56

UPDATED

The Aerospace Corp.
(Washington D.C. Office)
1000 Wilson Boulevard, Suite 2600, Arlington, Virginia, 22209-3988, United States
Tel: (+1 703) 812 06 00
Fax: (+1 703) 812 94 15

UPDATED

The Aerospace Corp.
(Albuquerque Office)
PO Box 9045, Albuquerque, New Mexico, 87119-9045, United States
Tel: (+1 505) 872 62 00
Fax: (+1 505) 872 62 13

UPDATED

The Aerospace Corp.
(Aurora Office)
c/o 2SWS (Stop 69), 18300 East Crested Butte Avenue, Buckley Air Force Base, Aurora, Colorado, 80011-9518, United States
Tel: (+1 303) 677 55 17
Fax: (+1 303) 677 55 14

UPDATED

The Aerospace Corp.
(Omaha Office)
Defense Meteorological Satellite Program, PO Box 819, Bellevue, Nebraska, 68005-0819, United States
Tel: (+1 402) 292 50 80
Fax: (+1 402) 291 31 37

UPDATED

The Aerospace Corp.
(Eastern Range Directorate-Cape Canaveral)
Kennedy Space Center, PO Box 21205, Cape Canaveral, Florida, 32815-0205, United States
Tel: (+1 321) 853 66 66
Fax: (+1 321) 853 36 90

UPDATED

The Aerospace Corp.
(Chantilly Office)
15049 Conference Center Drive, Suite 600, Chantilly, Virginia, 20151-3824, United States
Tel: (+1 703) 633 50 00
Fax: (+1 703) 633 50 13

UPDATED

The Aerospace Corp.
(Colorado Springs Office)
1150 Academy Park Loop, Suite 136, Colorado Springs, Colorado, 80910, United States
Tel: (+1 719) 573 24 40
Fax: (+1 719) 573 24 97

UPDATED

The Aerospace Corp.
(Columbia Office)
8840 Stanford Boulevard, Suite 4400, Columbia, Maryland, 21045-5852, United States
Tel: (+1 410) 312 14 00
Fax: (+1 410) 312 29 15

UPDATED

The Aerospace Corp.
(Falls Church-Skyline)
5107 Leesburg Pike, Suite 103, Falls Church, Virginia, 22041-3234, United States
Tel: (+1 703) 575 21 80
Fax: (+1 703) 575 27 83

UPDATED

The Aerospace Corp.
(Houston Office)
16511 Space Center Boulevard, Suite 203, Houston, Texas, 77058, United States
Tel: (+1 281) 280 80 14
Fax: (+1 281) 280 81 88

UPDATED

The Aerospace Corp.
(Houston-Johnson Space Center Office)
2101 NASA Road One, Johnson Space Center, M/C NQ112, Houston, Texas, 77058-3696, United States
Tel: (+1 281) 280 80 14
Fax: (+1 281) 280 81 88

UPDATED

The Aerospace Corp.
(Peterson Air Force Base)
HQ AFSPC/DR-(A), 150 Vandenberg Street, Suite 1105, Peterson, AFB, Colorado, 80914-4680, United States
Tel: (+1 719) 554 58 12
Fax: (+1 719) 554 58 76

UPDATED

The Aerospace Corp.
(Schriever Air Force Base)
442 Discoverer Street, Suite 71, Schriever, AFB, Colorado, 80912-4471, United States
Tel: (+1 719) 567 25 64
Fax: (+1 719) 567 35 78

UPDATED

The Aerospace Corp.
(Suitland Office)
FB4, Room 3316, 4700 Silver Hill Road, Suitland, Maryland, 20746, United States
Tel: (+1 301) 457 51 54
Fax: (+1 301) 457 57 13

UPDATED

The Aerospace Corp.
(Sunnyvale-Onizuka Air Force Station)
1080 Lockheed Martin Way,
Box 018, Sunnyvale, California, 94089-1232, United States
Tel: (+1 408) 744 64 84
Fax: (+1 408) 744 66 68

UPDATED

The Aerospace Corp.
(Western Range Directorate Vandenberg AFB)
PO Box 5068, Vandenberg, AFB, California, 93437-0068, United States
Tel: (+1 805) 606 77 22
(+1 805) 606 20 03

UPDATED

The Charles Stark Draper Laboratory Inc. (CSDL)
(Headquarters)
(Draper Lab.)
555 Technology Square, Cambridge, Massachusetts, 02139-3563, United States
Tel: (+1 617) 258 17 19
(+1 617) 258 10 00
(+1 617) 258 29 17
Fax: (+1 617) 258 21 21
(+1 617) 258 13 33
e-mail: busdev@draper.com
communications@draper.com
Web: http://www.draper.com
■ Research and development of autonomous and highly automated vehicle systems; communication and intelligence systems; guidance, navigation and control systems; spacecraft systems; ocean systems; tactical systems; information systems.

UPDATED

The Space Store
1400 NASA Parkway, Suite D, Houston, Texas, 77058, United States
Freephone: (+1 877) 742 23 11
Fax: (+1 281) 333 38 80
Web: http://www.thespacestore.com
President: Ms Dayna Justiz

UPDATED

The Titan Corp.
(Communications and Electronic Warfare Division)
(Titan Wireless Inc.)
(ex-Linkabit)
(ex-Titan Systems Corporation)
3033 Science Park Road, San Diego, California, 92121, United States
Tel: (+1 858) 552 95 00
Fax: (+1 858) 552 99 09
Web: http://www.titan.com

UPDATED

Tietronix
1331 Gemini Avenue, Suite 300, Houston, Texas, 77058, United States
Tel: (+1 281) 461 92 00
Fax: (+1 281) 461 93 50
e-mail: emeimoun@tietronix.com
Vice President, Marketing: Eric Meimoun
Tel: (+1 281) 461 93 00
Mobile: (+1 832) 771 29 18
Fax: (+1 281) 461 93 50

UPDATED

Univelt Inc.
PO Box 28130, San Diego, California, 92198-0130, United States
Tel: (+1 760) 746 40 05
Fax: (+1 760) 746 31 39
e-mail: 76121.1532@compuserve.com
Web: http://www.univelt.com
President: Robert H Jacobs
Office Manager: Ms Madeleine Bera
■ Publisher in the field of astronautics and communications for the American Astronautical Society. Distribute books and microforms and provide services for technical associations.

UPDATED

Univelt Inc.
(Escondido Office)
740 Metcalf Street 13, Escondido, California, 92025, United States
Tel: (+1 760) 746 40 05
Fax: (+1 760) 746 31 39

UPDATED

W L Pritchard & Co. L.C.
4405 East West Highway, Suite 501, Bethesda, Maryland, 20814, United States
Tel: (+1 301) 654 11 44
Fax: (+1 301) 654 18 44
e-mail: wlpco@ix.netcom.com
Web: http://www.wlpco.com
President: Ms Ellen D Hoff
Vice President, Engineering: Jack L Dicks
Office Manager: Ms Joy Kelly

UPDATED

Willis Inspace
(ex-Willis Corroon Inspace)
6700 – A Rockledge Drive, Bethesda, Maryland, 20817, United States
Tel: (+1 301) 530 50 50
Fax: (+1 301) 897 88 17
Web: http://www.willis.com
Senior Vice President: Mark A Quinn
■ Provide international space insurance brokerage services. Provide services including customised risk management program design, risk management consulting and insurance consulting. Offers coverage for space industry companies worldwide.

UPDATED

Wyle Laboratories, Inc.
128 Maryland Street, El Segundo, California, 90245, United States
Tel: (+1 310) 322 17 63
Fax: (+1 310) 322 36 03
(+1 310) 322 07 81
e-mail: service@wylelabs.com
Web: http://www.wylelabs.com
Managing Director: John Shimudu
Technical Director: Gordon Bakken
■ Specialise in engineering, testing and support services, launch support, life sciences systems and services, small payload integration, and small experiment payload engineering.

UPDATED

Zero Gravity Corporation (ZERO-G)
254 Horizon Avenue, Venice, California, 90291, United States
Tel: (+1 310) 399 19 76
Fax: (+1 413) 604 01 42
Web: http://www.zerogcorp.com
Vice President, Marketing: George T Whitesides
Mobile: (+1 626) 833 58 69
■ Provide zero gravity parabolic flights for scientific, training and recreational purposes.

UPDATED

Industry – Distributor

Dspace Inc.
28700 Cabot Drive, Suite 1100, Novi, Michigan, 48377, United States
Tel: (+1 248) 567 13 00
Fax: (+1 248) 567 01 30
Web: http://www.dspaceinc.com
Aerospace Technical Sales Engineer: Zach Lott
Tel: (+1 248) 567 12 33
Fax: (+1 248) 567 01 31

UPDATED

Sheldahl Inc.
1150 Sheldahl Road, Northfield, Minnesota, 55057, United States
Tel: (+1 507) 663 80 00
Fax: (+1 507) 663 85 45
e-mail: sheldahl.info@sheldahl.com
Web: http://www.sheldahl.com
■ Supplier to the military and aerospace industries. Product line includes: flexible circuitry, passive thermal control for spacecraft and satellites, radar absorbing materials, engineered flexible structures.

UPDATED

Space Imaging Inc.
12076 Grant Street, Thornton, Colorado, 80241, United States
Tel: (+1 303) 254 20 00
(+1 301) 552 05 37
Freephone: (+1 800) 425 29 97
Fax: (+1 303) 254 22 11
(+1 301) 552 37 62
(+1 303) 254 22 15
e-mail: info@spaceimaging.com
custserv@spaceimaging.com
Web: http://www.spaceimaging.com
President and Chief Executive Officer: Robert Dalal
Chief Financial Officer: Chuck Stees
Executive Director, Government Affairs and Corporate Communication, and Public Relations Director: Mark Brender
Executive Vice President, Solutions: Gene Colabatistto
Executive Vice President, Satellite Access Systems: Scott Smith
Director, Customer Services: Howard Klayman
Public Relations Manager: Gary Napier
Public Relations Co-ordinator: Ms Val Webb
■ Supply of visual information products and services derived from space imagery and aerial photography.

UPDATED

Industry – Sales Agent/Office

Brüel & Kjaer North America Inc.
(Headquarters)
2815 Colonnades Court, Norcross, Georgia, 30071-1588, United States
Tel: (+1 770) 209 69 07
Fax: (+1 770) 448 32 46
e-mail: bkinfo@bksv.com
Web: http://www.bkhome.com
■ Instrumentation for sound, vibration, illumination, thermal environment and signal analysis.
Representing:
Bruel & Kjaer, Denmark

UPDATED

China Great Wall Industry Corp. (CGWIC)
(California Office)
21515 Hawthorne Boulevard, Suite 1065, Torrance, California, 90503, United States
Tel: (+1 310) 540 77 06
Fax: (+1 310) 540 34 75
e-mail: gwahuang@aol.com
Chief Representative: Ruofei Gao
Representing:
China Great Wall Industry Corp., China

UPDATED

Knight Electronics
(ex-Knight Electronics International)
31 Myrtle Street, Manhasset, New York, 11030, United States
Tel: (+1 516) 365 38 30
Fax: (+1 516) 365 20 54
Web: http://www.knight-intl.net
Managing Director: Chad Whitman
■ Broadline distributor in the electronics' procurement market servicing manufacturers, research and development facilities, and repair shops. Distributor of 1N, 2N & 4N; JAN, JANTX, JANTXV and JANS level semiconductor products for commercial, industrial, military, and space-level applications.
Also house integrated circuits including memory products, connectors/electromechanical, relays, switches, passive components, computer processors, computuer storage (HDD/SCSI Drives), development tools and compilers.
Accreditations:
EOS/ESD compliant
Representing:
3M, US
AMD, US
American Power Devices, US
AMP, US
Amphenol, US
Analog Devices, US
API, US
Arcotronics Italia SpA, Italy
Areva, France
ATC, US
Augat, US
AVX, US
BC Comp, US
Berg, France
Beyschlag, US
BKC, US
Bourns, US
Burndy, France
Burr-Brown, US
Cal-Chip, US
Central Semiconductor, US
Centralab, US
Chemicon, Japan
Cherry Semiconductor, US

Cirrus Logic, US
Compensated Devices, US
Conexant, US
Cornel-Dubilier, US
Cypress Semiconductor, US
Dale, US
Draloric, US
Epcos, Germany
Erie Technology, US
ERNI, Germany
ETC, US
Evox Rifa, Finland
Fairchild Semiconductor, US
FCI, France
Film Microelectronics, US
Germanium Power Devices, US
Glenair, US
Hirose, Japan
Hitachi, Japan
Honeywell Bendix/King, US
IDT, US
Integrated Device Technology, US
Intel, US
International Rectifier, US
Intersil, US
Isocom Components, US
ITT Cannon, US
Johnson Components, US
JST, US
Kemet, US
Kings, US
Knox Semiconductor, US
KOA Speer, US
Kyocera, Japan
Lattice Semiconductor, US
Linfinity Microelectronics, US
Mallory Sonalert, US
Maxim, US
Microdot, US
Micron, US
Micropac Industries, US
Microsemi, US
Mikro Elektronika, Serbia and Montenegro
Mill-Max, US
Molex, US
Motorola Inc, US
Murata, Japan
National Semiconductor, US
NIC, US
Nichicon, Japan
Nippon Chemicon, Japan
Novacap, US
Omnirel LLC, US
On Semiconductor, US
Paccom, US
Panasonic, Japan
Panduit Corp., US
Pericom, US
Philips Semiconductor, Netherlands
Powerex, US
Presidio, US
Protek, US
Raychem Corporation, US
Renesas, Japan
Roederstein, US
Rohm, Japan
Rubycon, Japan
Samsung, South Korea
Sanyo, US
Saraonix, US
Semicoa, US
Semicon Inc, US
Semiconductor Technology, US
Semtech, US
Sensitron, US
Sharp, US
Siemens, Germany
Signetics Inc., US
Silicon General, US
Siliconix, US
Solitron Devices, US
Sourio, US
Spectrum Control, US
Sprague, US
Syfer, UK
Taiyo Yuden, Germany
Tansistor, US
TDK, Japan
Telefunken, US
Texas Instruments, Inc., US
Toshiba, US
Tyco Electronics, US
United Chemicon, Japan
Vishay, US
Vitramon, US
Xilinx, US
Yageo, China
Zetex, US
Zicor, US
Zilog, US

UPDATED

Morcom International Inc

PO Box 220824, Chantilly, Virginia, 20153-0824, United States
Tel: (+1 703) 263 93 05
Fax: (+1 703) 263 93 08
Web: http://www.morcom.com
President: Manuel A Ojeda
Contact: Jacob Wertz

■ Supply of communications and weather systems.

UPDATED

SPACE LOGS

CONTENTS

SPACE LOGS

CHRONOLOGY NOTES

NOTES: 1. All times and dates are in GMT. **2.** Satellite masses are based on available information; estimates should be within 10% of the actual figure. **Launch sites: CC** Cape Canaveral Air Station (US); **EAFB** Edwards Air Force Base (US); **IS** Yevne (Israel); **JQ** Jiuquan (People's Republic of China; US Space Command designator is Shuang Cheng Tzu); **KA** Kagoshima (Japan); **KO** Kourou (French Guiana); **KY** Kapustin Yar (Russia); **KSC** Kennedy Space Center (US); **PL** Plesetsk (Russia); **SM** San Marco platform (Indian Ocean); **SR** Sriharikota (India); **TA** Tanegashima (Japan); **TT** Tyuratam/Baikonur (USSR/Kazakhstan); **TY** Taiyuan (People's Republic of China; US Space Command designator is Wuzhai); **WR** Vandenberg AFB (US); **XI** Xichang (People's Republic of China).

Table 1: 1996–2002 Chronology

1996

11 Jan: The year's 1st Shuttle mission, STS-72, is launched. It retrieved Japan's Space Flyer Unit and deployed/retrieved Spartan. See the Shuttle entry for mission description.

12 Jan: Ariane 44L (V82) is launched from Kourou's ELA-2 pad at 23:10 GMT into 200 × 35,992 km 6.99° GTO carrying PanAmSat's 2,918 kg PAS 3R and Malaysia's 1,450 kg MEASAT 1 GEO telecom satellites. 36 primary satellites remain contracted for Ariane launches. PAS 3R is the Hughes-built replacement for PAS 3, lost in Ariane's 1 Dec 1994 failure, to provide Latin America DTH from 43°W. Malaysia's first satellite was commercially provided by Hughes, but a national microsatellite built with India's assistance is in development for 1997 launch. See 5 Feb.

14 Jan: Korea's second national telecom satellite, KoreaTelecom's 1,459 kg Mugunghwa 2/Koreasat 2, is launched by McDonnell Douglas Delta 7925 from Cape Canaveral pad 17B at 11:10 GMT towards 116°E GEO. The orbit achieved after stage 3 firing was 1,358 × 35,421 km, 21.0° GTO. Koreasat's apogee kick motor was fired 16 Jan to attain GEO altitude. Matra Marconi Space delivered the payload Mar 1995 fully tested and ready for integration into its Lockheed Martin 3000-series bus at the Princeton, NJ facility. Koreasat 1 was left in a low orbit 5 Aug 1995 when one of Delta's strap-ons failed to separate.

16 Jan: Cosmos 2327 is launched by SL-8 Cosmos 3M from Plesetsk at 15:34 GMT into 952 × 1,021 km, 82.98° GTO. The orbital plane suggests that it is a Parus military navigation satellite, intended to replace 1993's C2266. The original replacement, C2321 (6 Oct 1995), was stranded in a low orbit.

25 Jan: Gorizont 31 is placed by a Proton SL-12 into geosynchronous orbit (1.5° inc) following launch from Tyuratam at 09:55 GMT. The satellite is positioned at 40°E 2 Feb, replacing G22. The launch announcement indicated there would be no more Gorizonts, but see 25 May.

26 Jan: Mikhail F Reshetnev, Director General and General Designer of NPO Prikladnoi Mekhaniki, dies. NPO PM is responsible for almost 900 satellites, including the Express, Gals, Ekran, Gorizont, Raduga and most other major telecom satellites; also the Strela 3 military sextets, Glonass and Cosmos navigation and GEO-IK geodetic satellites. Reshetnev founded NPO PM in 1959 as a design bureau under Sergei Korolev, specialising in missiles, before becoming independent in the mid-1960s.

1 Feb: Indonesia's 2,989 kg Palapa C1 telecom satellite is launched by Atlas 2AS AC-126 at 01:15:01 GMT from Cape Canaveral pad 36B towards 113°E GEO, replacing Palapa B2P (launched Mar 1987).

5 Feb: Ariane 44P+3 (V83) is launched from Kourou's ELA-2 pad at 07:19:38 GMT into 200 × 35,952 km 7.00° GTO carrying NTT's 3,420 kg NStar b to provide domestic Japanese telecom services from 136°E. 39 primary satellites remain contracted for Ariane. See 14 Mar.

8 Feb: A 182 min EVA by Mir's Gidzenko/Reiter recovers ESA experiments from Spektr.

11 Feb: NASDA's J1 launcher successfully debuts at 23:00 GMT from Tanegashima. The 2-stage suborbital version carried the 1,040 kg Hyflex hypersonic lifting body demonstrator to 110 km for a M14.4 re-entry. The chute deployed at 694 s and the flotation bag inflated, but Hyflex sank in 6 km-deep water after apparently breaking free of the bag. The thermal protection system could not be examined, but 12 of the 14 test objectives were achieved. The planned HOPE unmanned spaceplane will draw on Hyflex technology.

14 Feb: China's maiden CZ-3B is launched at 19:01 GMT from Xichang, carrying Intelsat's 4,576 kg Intelsat 708 towards 50°W. However, the inertial platform veered after 2 s and CZ crashed after 25 s into a hillside 1.5 km from the pad. The fault was determined to be caused by poor workmanship. The official death toll was six plus 57 injured, although unofficially it was claimed as up to 100. 250 homes were damaged or destroyed. The launch was reportedly insured for US$204.7 million at 13 per cent (the low rate reflecting its part in a 10-launch package).

17 Feb: NASA's first Discovery mission, the 805 kg NEAR (Near-Earth Asteroid Rendezvous), is launched by Delta 7925–8 at 20.43 GMT from Cape Canaveral's pad 17. NEAR, built by Johns Hopkins Univ's Applied Physics Lab, was due to fly out to the main asteroid belt (approaching to within 1,200 km of Mathilde Jun 1997) returning to Earth for a 415 km swing by 22 Jan 1998, setting up arrival at Eros Jan 1999 (entering orbit 6 Feb). NEAR is the first asteroid orbiter, coming within 24 km during its 1 yr of primary investigations.

19 Feb: Three 225 kg Strela 3 Cosmos satellites, 2328–2330, and three Gonets-D are launched aboard a single SL-14 Tsyklon 3 from Plesetsk at 00:58 GMT into orbits around 1,410 km, 82.6°. Strela are apparently part of a military tactical comms system, coplanar with 1994's C2299–2304, suggesting replacement. The Smolsat consortium plans a constellation of 36 Gonets satellites for E-mail store/forward and realtime bulk data relay.

19 Feb: Raduga 36 is launched by a Proton SL-12 from Tyuratam at 08:18 GMT towards geosynchronous orbit but is left in GTO when Block-DM2 stage 4 explodes, producing about 200 fragments. DM's burn 2 should have raised GTO into GSO but ignition failed; Raduga was released automatically soon after into 242 × 36,502 km, 48.60° GTO. The review board concluded that the Triethyl Aluminium ignition hypergol had not reached the gas generator or main chamber because of reduced pressure from a leaking joint in the pipe from the TEA bottle. The fitting nut had a broken lockwire, probably from poor installation, allowing it to back off during burn 1. Corrective actions include adding a second lockwire.

21 Feb: Soyuz-TM 23 is launched at 12:34 GMT by SL-4 Soyuz-U2 from Tyuratam site 2 carrying Commander Yuri Onufrienko and engineer Yuri Usachev as Mir's 21st main crew. Launch had been planned for 25 Dec 1995 but funding problems meant that its launcher was not ready in time, requiring extension of TM22's occupation. TM23 docked with Kvant 1's port 14:21 GMT 23 Feb.

22 Feb: The year's 2nd Shuttle mission, STS-75, is launched carrying USMP 3 and TSS 1R. The tether broke 26 Feb at about 19 km, leaving TSS as a separate satellite. See the Shuttle entry for mission description.

24 Feb: NASA's 1,258 kg Polar ionospheric research satellite is launched by Delta 7925–10 from Vandenberg pad 2W at 11:24 GMT into initial 186 × 50,495 km, 85.98°; perigee was raised 9 Mar to 5,141 km.

9 Mar: Orbital Sciences Corp's 10th Pegasus launcher is released from its Lockheed L-1011 carrier aircraft at 01:33 GMT at 11.94 km altitude, some 105 km out over the Pacific after taking off from Vandenberg AFB at 00:38 GMT. Stage 2 separated at 608 s instead of the expected 529 s as a result of high headwinds at the drop site; the autopilot ordered the delay to gain height before stage 3 ignition. Pegasus achieved 805 × 836 km 89.96° (target 821 × 834 km), releasing the 113 kg USAF REX II satellite to investigate radio propagation. This was the first successful Pegasus XL after two failures. REX was planned for a standard Pegasus but it switched to requalify XL; USAF still paid only US$6 million.

14 Mar: Intelsat's 4,175 kg 707 satellite, built by Space Systems/Loral, is launched by Ariane 44LP (V84) from Kourou's ELA-2 pad at 07:11 GMT into 200 × 35,913 km, 7.04° GTO. 707 entered service at 1°W GEO. It can handle up to 90,000 telephone circuits and three TV channels. Stationkeeping propellant should be sufficient for 15 yr. 38 primary satellites remain contracted for Ariane. See 20 Apr.

14 Mar: The Cosmos 2331 4th gen photorecon satellite is launched by SL-4 Soyuz-U from Plesetsk pad 43 at 17:40 GMT into 164 × 358 km, 67.1° for a Yantar close look mission. The descent capsule carried the USAF Space Test Program's deployable Binrad experiment to collect Be-7 samples. Returned to Earth 11 Jun after record 89 days.

15 Mar: 351 min EVA 1 starting 01:04 GMT by Mir cosmonauts Onufrienko/Usachev installs a crane on Mir's core module for moving the solar wing packages delivered Nov 1995 by STS-74 to Kvant 1. See 20 May.

21 Mar: India's 922 kg IRS P3 Earth observation satellite is launched by PSLV-D3 from Sriharikota at 04:53 GMT into 819 × 821 km, 98.8°. This was the second successful PSLV. Germany's stereo scanner (first activated 22 Mar on rev 15) is optimised for ocean studies and India's WiFS wide field sensor (first activated 23 Mar on rev 29) for vegetation dynamics. P3 also carries the XAP X-ray astronomy payload. India's space agency now plans to commercialise PSLV. Programme cost was Rs4,150 million through the first two launches; unit vehicle cost is Rs450 million.

22 Mar: The year's 3rd Shuttle mission, STS-76, is launched. It docks with Mir 25 Mar, delivering Shannon Lucid to join the station's main crew until STS-79's return Sep 1996. See the Shuttle entry for mission description.

22 Mar: Former NASA astronaut Bob Overmyer is killed while testing an experimental Cirrus VK-30 aircraft near Duluth International Airport. Overmyer retired from NASA in 1986 after flying the STS-5/51B missions. He was originally selected as an astronaut for the AF Manned Orbiting Laboratory.

28 Mar: Rockwell's 16th GPS Block 2A satellite, Navstar 2–25 (USA 117; GPS 38), is launched by McDonnell Douglas Delta 7925 from Cape Canaveral complex 17 at 00:21 GMT into the system's C2 slot. The Thiokol Star 37F solid propellant apogee motor was fired some 44 h later to establish the final orbit, joining 24 other operational Block 2/2A satellites. A further three 2As are available for launches at 2 months' notice to guarantee the service until Block 2R appears in 1997.

3 Apr: Inmarsat's 2,068 kg initial third-generation satellite is launched by Atlas Centaur 2A (AC-122) from Cape Canaveral's pad 36A at 23:01 GMT. The payload was activated 13 Apr for testing at 28°E from Fucino; it replaced Inmarsat 2-F1 at 64°E in mid-May. 2-F1 moved to 65°E as backup. 3-F2 is planned for Aug aboard Proton.

8 Apr: Luxembourg's 3,010 kg Astra 1F DTH satellite is launched at 23:09 GMT from Tyuratam by Proton SL-12 towards 19.2°E. It is Proton's first commercial launch. Block DM injected Astra into a GTO of 12,100 × 36,000 km, 7.0° using two burns. Astra separated after 6 h 41 min. It is Hughes' 29th HS-601 model to be launched.

20 Apr: TMI Communication's 2,855 kg MSAT 1 is launched by Ariane 42P (V85) from Kourou's ELA-2 pad at 22:36:00 GMT into 199.5 × 35,923 km, 7.00° GTO. MSAT began initial services at 106.5°W providing mobile phone services over N America. It is Hughes' 30th HS-601 model to be launched. The identical US AMSC 1 appeared 7 Apr 1995. 47 primary satellites remain contracted for Ariane. Ariane has now successfully launched 112 primary and 26 auxiliary payloads.

23 Apr: Russia's 19.7 t Priroda remote sensing module is launched by Proton from Tyuratam at 11:48 GMT to complete the Mir station. It docked with Mir's main axial face 12.43 GMT 26 Apr and transferred to the portside position 27 Apr. Cargo included 1 t of US equipment. Without solar wings, Priroda relied on battery power.
24 Apr: BMDO's US$800 million MSX Midcourse Sensor Expt mission is launched at 12:27 GMT by Delta 7920–10 from Vandenberg's pad 2W into 897 × 906 km, 99.4° Sun-synchronous. The 2,680 kg MSX is demonstrating early warning sensors for discriminating and tracking mid-course targets. The main instrument's nominal life is up to 20 months, limited by the cryogenic coolant supply, but the other sensors will operate for 5 yr. MSX is also gathering global environmental data, such as ozone levels, in co-operation with NASA.
24 Apr: Cosmos 2332 is launched at 13:03 GMT by SL-8 Cosmos 3M from Plesetsk into 295 × 1,565 km, 82.96°. A 2 m. {dia} sphere (500kg?) for tracking for radar calibration and passive monitoring of upper atmoshpere.
24 Apr: The 7th Titan 4 Centaur is launched from Canaveral's ETR 41 at 23:37:01 GMT carrying the classified USA 118 payload. Titan's core was #K16. Observers saw Titan heading due east (100.4° aximuth), indicating a 28.5° orbit preparatory to departure for GEO. Centaur's employment and the lack of other information suggest a GEO elint satellite for the National Reconnaissance Office. See 14 May 1995 and 12 May 1996.
30 Apr: The 1.4 t Italian-Dutch SAX X-ray satellite is launched by Atlas 1 from Canaveral's pad 36B at 04:31 GMT into 583 × 603 km, 3.96°. This was the 100th Atlas Centaur mission and the penultimate Atlas 1.

5 May: Progress-M 31 is launched by Soyuz-U from Tyuratam site 2 at 07:04 GMT, docking with Mir's forward port 08.54 GMT 7 May.
12 May: A Titan 4 Centaur is launched from Vandenberg's SLC 4E at 21:32 GMT carrying the classified USA 119 payload. The mission was identified as ocean surveillance when three subsatellites were designated (USA-120/121/122) and, uniquely, a 53 kg tether experiment was released 20 Jun into 1,023 km, 63.4°. The TiPS Tether Physics and Survivability experiment was provided by the US Naval Center for Space Technology, with the US$4 million total cost (including US$1.9 million for 2 yr of tracking) funded by the National Reconnaissance Office. The 4.0 km tether and its two end masses (9 kg + 41 kg) unreeled in 42 min; it became passive along the gravity gradient after its battery was exhausted. TiPS is being tracked by radar and laser to model its motion and longevity. Titan vehicle processing required 112 days; Titan's core was #K22. See 24 Apr.
14 May: A Kometa 4th generation topographic/mapping satellite is lost when the payload fairing of its Soyuz-U launcher fails after 49 s. Launch was at 08:55 GMT from Tyuratam site 31. In addition to its military mission, Kometa was to return 2/10 m-resolution film imagery after 45 days under Sovinformsputnik's SPIN-2 (Space Information 2 m) agreement with three US companies (Aerial Images, Inc; Lambda Tech International; Central Trading Systems). The commercial element was insured with Moscow's Megaruss Co for US$2.7 million. A second SPIN-2 mission is reportedly planned for early 1997. Sovinformsputnik planned to fill the current image orders using archived material and other military satellites.
16 May: Ariane 44L (V86) is launched from Kourou's ELA-2 pad at 00:56:29 GMT into 299 × 35,934 km, 4.00° GTO carrying Israel's 996 kg Amos 1 and Indonesia's 2,989 kg Palapa C2. Amos will provide mainly TV distribution from 4°E, and PT Satelindo's Palapa will replace Palapa B2R at 108°E. It is Hughes' 31st HS-601 model to be launched. 45 primary satellites remain contracted for Ariane. Ariane has now successfully launched 114 primary and 26 auxiliary payloads.
17 May: The 211 kg BMDO/USAF MSTI 3 is launched aboard Orbital Sciences Corp's 11th Pegasus launcher, released from its Lockheed L-1011 carrier aircraft at 02:44 GMT at 11.59 km altitude, some 96 km out over the Pacific after taking off from Vandenberg AFB. Pegasus achieved 296 × 361 km 97°; MSTI then manoeuvred into 425 km. MSTI is mapping Earth's IR background for a full season to help design future early warning sensors. This was the last standard Pegasus launch; future versions will be the extended XL.
19 May: The year's 4th Shuttle mission, STS-77, is launched. It deploys a 12 m inflatable antenna and carries Spacehab 4. See the Shuttle entry for mission description.
20 May: 320 min EVA 2, beginning 22:50 GMT, by Mir cosmonauts Onufrienko/Usachev attaches to Kvant 1 the Russian/US solar wing package delivered Nov 1995 by STS-74. They also filmed footage of a large inflated Pepsi Cola can for a reported fee of >US$1 million. See 24 May.
24 May: Hughes Communications' 1,397 kg Galaxy 9 is launched towards 123°W by a Delta 7925 from Canaveral's pad 17B at 01:10 GMT. Galaxy will primarily distribute cable TV programming.
24 May: 343 min EVA 3, beginning 20:47 GMT, by Mir cosmonauts Onufrienko/Usachev handcranks out the solar wing installed during 24 May EVA, adding 6 kW to Mir's power supply. Space Station Alpha will use the same US solar cells. See 30 May.
25 May: Gorizont 32 is placed by a Proton SL-12 into geosynchronous orbit (1.5° inc) following launch from Tyuratam at 02:05 GMT. The satellite is heading for 52°E.
30 May: 260 min EVA 4, beginning 18:20 GMT, by Mir cosmonauts Onufrienko/Usachev mounts Germany's MOMS-02P multi-spectral scanner on Mir's Priroda module.

Ariane 5 Europe's main contender for the heavy-lift market, Ariane 503 is launched from Kourou in 1998 on its third flight attempt (ESA)
0137119

4 Jun: ESA's maiden Ariane 5 breaks up after a software problem loses attitude control. The main Vulcain engine ignited at 12:33:59 GMT on Kourou's ELA 3 pad for launch 7.5 s later when the P230 solid boosters ignited. Flight was normal until 30 s after launch, at M0.7 and 3,500 m altitude, when both booster nozzles swivelled within 2 s to their extreme positions, followed by Vulcain. The intense aerodynamic loads broke up the vehicle and the self-destruct system was triggered by rupture of the electrical links between the boosters and core. An independent inquiry board reported 23 Jul that both inertial reference systems had failed simultaneously because of software specification and design errors. Flight 502 was planned for Sep/Oct 1996, but is now probable for mid-1997, followed by a new test flight 4Q 1997. 501 was to release four 1.17 t Cluster satellites into 10° GTO; they would have then manoeuvred into 25,513 × 140,318 km, 90° to study the 3D structure of Earth's plasma environment. ESA decided in Jul to build the single Phoenix replacement satellite from spare Cluster hardware, and will decide in Nov 1996 whether to fund another three.
6 Jun: Onifrienko and Usachyov began EVA 5 at 16:56 GMT during which they spent 3 h 34 min replacing experiment cassettes and installing American-made micrometeoroid devices.
13 Jun: Russian cosmonauts Onifrienko and Usachyov began their mission's EVA 6 at 12:45 GMT which lasted 5 h 42 min to assemble and erect a four-section truss to the side of Kvant module and deploy a large antenna on the Priroda module.
15 Jun: Intelsat's 3,420 kg 709 satellite, built by Space Systems/Loral, is launched by Ariane 44P (V87) from Kourou's ELA-2 pad at 06:55:09 GMT into 200 × 35,955 km, 6.99 GTO towards 18°W GEO. It can handle up to 90,000 telephone circuits and three TV channels. Stationkeeping propellant should be sufficient for 15 yr. 44 primary satellites remain contracted for Ariane. Ariane has now successfully launched 115 primary and 26 auxiliary payloads.
27 Jun: NASA's Galileo spacecraft conducted a fly-by of Ganymede when it passed within 835 km.
27 Jun: Canada's Industry Minister John Manly announced far reaching expansion of his country's participation in the International Space Station. The Canadian Space Agency reached agreement with NASA for three new positions for Canadian astronauts involving Julie Payette, Steve MacLean and Bjarni Tryggvason.

2 Jul: At 07:48 GMT Pegasus-XL F-12 dropped from a converted L-1011 aircraft and launched the 295 kg TOMS-EP (Total Ozone Mapping Spectrometer-Earth Probe) from WTR. This is the first satellite dedicated to mapping the Earth's ozone layer and is instrumented to also monitor sulphur dioxide levels emitted by volcanic eruptions. TOMS-EP was placed in a 494 × 511 km, 97.44° orbit with a period of 94.67 min.
3 Jul: At 00:31 GMT Titan 4 405 K-2 (45H-1) launched USA-125, the 700 kg SDS-B4 (SDS-2 series) satellite from LC-40 at the ETR. The satellite was placed in a 292 × 319 km, 54.99° orbit with a period of 90.64 min.
3 Jul: At 10:47 GMT APStar 1A, a satellite for the Asia-Pacific Telecommunications Satellite Co Ltd based in Hong Kong was launched by CZ3–10 from China's Xi Chang launch site to a 35,785 × 35,791 km, 0.07°, geostationary orbit. The 1,400 kg satellite was located at 133.9°E to replace the APStar 2 satellite lost during launch in January 1995.
9 Jul: At 22:24 GMT Ariane 44L V89 launched the 2,617 kg Arabsat 2A and 1,743 kg Turksat 1C from Kourou, French Guiana. Arabsat 2A is a telecommunications and direct broadcast satellite for the Arab League group of countries and was placed in a 35,782 × 35,794 km, 0.08°, geostationary orbit. BOL mass is 1,570 kg and the Spacebus 3000 satellite was located over 25° E. A telecommunications and direct broadcast satellite, the Turksat 1C Spacebus 2000 satellite was launched to a 35,781 × 35,791 km, 0.02°, geostationary orbit at 42° E with a BOL mass of 1,078 kg. Turksat 1C was initially located over 31° E and moved later to its present location.
10 Jul: Canada signed agreements with the European Space Agency extending co-operation into the fields of remote sensing and telecommunications.
12 Jul: NASA decides to replace STS-79's solid rocket boosters with new ones when examination of the boosters recovered from STS-78 indicated gas penetration down into the joint where the O-rings seal the pressure and prevent blow-through.
16 Jul: At 00:50 GMT Delta 7925A (237) launched from LC17A, ETR, carried the 1,881 kg (930 kg dry) Navstar 26 (SVN40) to a 20,138 × 20,224 km, 55.04°, orbit with a period of 717.93 min. Satellite also known as USA-126 was to be operated in plane E slot 3 of the GPS network
21 Jul: The first public acknowledgement of an accidental collision between two objects in space was made when a piece of debris from the Ariane 1 launcher that carried SPOT 1 into orbit in 1986 collided with CERISE launched July 1995, breaking off a section of boom making it a third object in orbit.
22 Jul: Rockwell and NPO Yuzhnoye concluded an agreement under which Rockwell's Space Systems Division would market commercial launch services for the Ukraine's Cyclone rocket. Derived from the two-stage SS-9 ICBM, with a third stage Cyclone can lift 3,600 kg to low Earth orbit.
25 Jul: At 12:42 GMT an Atlas 2 (AC-125 IABS/Centaur II) launched the 3,020 kg UHF F7 (UHF FO F-7-EHF) satellite from LC36A, ETR, for the DoD. Also known as UFO 7, it is an HS-601 communications and was placed in a geosynchronous orbit of 34,579 × 36,562 km with an inclination of 5.07° and a period of 1,425.08 min.
31 Jul: Progress M-32 was launched by Soyuz U (11A511U) from LC1, Baikonur, at 20:06 GMT to a 293 × 333 km, 51.65° orbit with a period of 90.78 min. This was the first successful launch of a Soyuz U vehicle in the last three attempts. It docked with the Mir forward port (–X) at 22:03:40 GMT 2 Aug, undocked at 09:33:45 GMT Aug 18 and redocked to the rear docking port (+X) at 09:35:00 GMT Sep 3. M-32 finally undocked at 19:51:20 GMT 20 Nov and broke up on re-entry at 22:42:25 GMT. Total docked time was 93.91 days, free time 2.2 days. Total mission duration was 96.11 days.
31 Jul: The McDonnell Douglas Clipper Graham VTOL test launcher was destroyed when it toppled and burst into flames after a landing leg failed to deploy on touchdown at the end of its fourth flight this summer.

6 Aug: NASA announced that a team of scientists from the Johnson Space Center had concluded that a meteorite (ALH84001) found in Antarctica in 1984 probably contained the tracks of primitive life forms and that isotopic analysis of the 4.5 billion year old rock indicated that it came from Mars.
6 Aug: Starsem S.A. is founded as a joint Russian-European venture to market industrial and commercial opportunities for the Soyuz-Icare series of launchers capable of lifting up to 5 tonnes to low orbit.
8 Aug: Ariane 44L V90 launched from ELA2 at Kourou at 22:49 GMT and carried the 1,990 kg Italsat 2 and 2,260 kg Telecom 2D satellites to geostationary orbits. Italsat 2 is a telecommunications satellite used for data transmission and digital TV with ESA's EMS package for mobile communications. It was placed in a preliminary orbit of 34,911 × 35,791 km with a period of 1,413.9 min drifting west to its planned position at 10.2° E. BOL mass was about 1,200 kg. The Eurostar 3000 Telecom 2D is used for telephone and TV communications and French government communications. It was placed in a 35,768 × 35,795 km geostationary orbit at 3° E. BOL mass was 1,400 kg.
14 Aug: Molniya-M launcher 8K78M/ML lifted off from LC43/3, Plesetsk, at 22:21 GMT carrying the 1,600 kg Molniya-1T/89 telecommunications satellite, replacing Molniya-1T/85.
17 Aug: Japan's ADEOS (ADvanced Earth Observing Satellite) and Fuji 3 amateur radio satellite were launched from Tanegashima LC-Y by H-2 (H-II-4F) at 01:53 GMT. The 3,560 kg ADEOS was placed in a 797 × 799 km, 98.62° orbit with a period of 100.83 min and is designed to monitor the Earth's surface, atmosphere and oceans to measure environmental changes. Designated JAS 2 prior to launch, Fuji 3 was placed in a 802 × 1,323 km, 98.58° orbit with a period of 106.46 min.
17 Aug: The Soyuz TM-24 manned spacecraft (TM 11F732 s/n 73) was launched at 13:18 GMT from LC1 at Baikonur by a Soyuz U (11A511U) launch vehicle. Designated Mir Expedition EO-22 the Soyuz TM-24 was placed in a 375 × 390, 51.65° orbit with a period of 92.2 min. It carried V. G. Korzun (Commander), A. Y. Kaleri (Flight Engineer) and C. Andre-Deshays, the first French woman in space. Soyuz TM-24 docked to the front longitudinal port (+X) on Mir at 14:50:21 GMT 19 Aug, allowing cosmonauts Korzun, Kaleri and Andre-Deshays to enter the space station.
18 Aug: At 10:27 GMT a CZ-3 (CZ3–11) launched the 1,200 kg Zhongxing 7 satellite from the LC1 complex at the Xi Chang launch site. An HS-376 telecommunications satellite for domestic internal use, it was stranded in a 200 × 17,229 km, 27.25° orbit with a period of 307.53 min when the second stage engine stopped firing 48 s prior to the planned time.
20 Aug: Maritime satellite Marecs A was retired from service.
21 Aug: A Pegasus-XL (s/n F13) launched at 09:47 GMT carried the FAST (Fast Auroral Snapshot Explorer) to an orbit of 351 × 4,165 km with an inclination of 82.98° and a period of 133.12 min. FAST is a 180 kg scientific satellite developed by NASA-GSFC for investigating plasma phenomena in aurora and general magnetospheric phenomena.
29 Aug: Launched at 05:22 GMT by Molniya-M (8K78M/2BL) from LC43/3 at the Plesetsk launch facility, the Interbal 2, Magion 5 and Musat science satellites were placed in elliptical orbits. Provided by the Instituto Aeronautico de Cordoba, Argentina, to demonstrate low-cost technologies, the 30 kg Musat was placed in a 236 × 1,171 km, 62.79° orbit with a period of 98.84 min. The 62 kg Magion 5 (Magnetosphere — Ionosphere) was provided by the Czech Republic to operate with Magion 4 launched in 1995 and Interball satellites to study energy transport mechanisms between the solar wind and the Earth's magnetosphere. It was placed in a 791 × 19,196 km, 62.8° orbit with a period of 347.46 min. Interball 2, also called Auroral Probe, is the second flight of the Prognoz-M2 science bus, weighs about 1,300 kg and was placed in a 769 × 19,211 km, 62.77°, orbit with a period of 347.34 min.

2 Sep: At 04:20:00 GMT the Soyuz TM-23 spacecraft carrying cosmonauts Y. Onufrienko, Y. Usachoyov and C. Andre-Deshays undocked from Mir and

Port of Long Beach, California, hosts the Sea Launch ship and launch platform before sailing to the Pacific Ocean for a satellite launch (ILS) 0137125

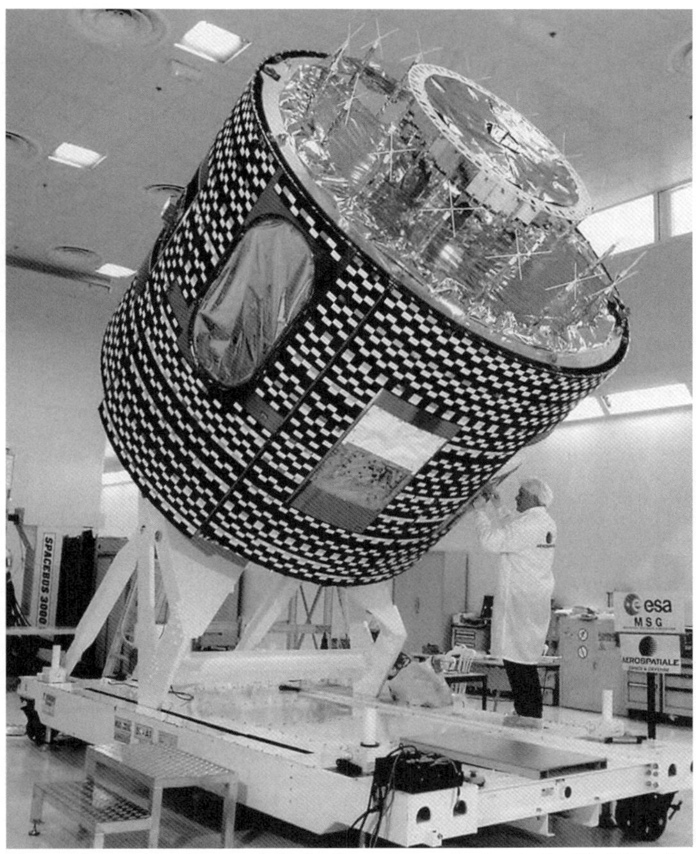

Second Generation Meteosat built for Europe's contribution to the World Weather Watch (ESA) 0137126

performed a small separation burn at 04:24:40 prior to the deorbit burn at 06:47:20 GMT. The modules separated at 07:14:36 with parachute deployment at 07:26 GMT followed by a landing at 07:41:40 100 km southwest of Akmola in Kazakhstan.
4 Sep: At 09:01 GMT the 3,250 kg Cosmos 2333 was carried by Zenit-2 from LC45L, Baikonur, to a 849 × 852 km, 71.01° 101.94 min, orbit. Cosmos 2333 was a Tselina-2 ELINT satellite.
5 Sep: A Parus military navigation satellite, Cosmos 2334 was launched from LC132/1, Plesetsk, at 12:47 GMT by Cosmos-3M (Kosmos 11K65M) launcher to a 970 × 1,000 km, 82.94° 104.8 min orbit. The navigation satellite was positioned in plane 1 of the constellation and operated at radio frequencies of 150.03 MHz and 400.08 MHz. Also launched with Cosmos 2334 was UNAMSAT 2, a science satellite from the Autonomous University of Mexico designed to study micrometeorites. It replaces the satellite of this type lost on the first flight of the Start launcher on 25 March 1995 and was placed in a 968 × 1,011 km, 82.93° 104.89 min, orbit.
6 Sep: At 17:37 GMT a Proton-K (8K82K/DM-1; s/n 375-01/DM-1 s/n 1L) launched Inmarsat-3 F2 into a 35,958 × 36,136 km, 2.65° orbit from LC81L, Baikonur. This was the second commercial launch for the Proton-K but the second Inmarsat-3 satellite did not need an apogee kick motor. The 1,144 kg (launch mass) satellite was temporarily positioned at 28° E on 18 Sep, but on 27 Sep it was moved further west.
6 Sep: The NASA Jupiter spacecraft Galileo performed a fly by of Ganymede at a distance of just 262 km.
8 Sep: An Atlas 2A (IIA (1N); s/n AC-123) launched from LC36B carried the 2,764 kg GE 1 telecommunications satellite to a 35,788 × 35,792 km supersynchronous orbit inclined 0.01° with a period of 1,436.27 min. Launched for GE American Communications (Americom), the satellite was located over 257° E.
11 Sep: An Ariane 42P (V91) launched from ELA2, Kourou, at 00:00 GMT carried the 2,865 kg (BOL 1,600 kg) EchoStar 2 direct broadcast satellite to a 35,679 × 35,787 km, 0.04° orbit for the EchoStar Communications Corporation. The satellite design was based on the Aerospatiale AS 7000 bus.
12 Sep: At 08:49 GMT a Delta 7925 (7925A; s/n 238) carried the 1,881 kg (launch mass) Navstar 27 (SVN30) navigation satellite from LC17A, ETR, to a 20,057 × 20,307 km, 54.72° orbit with a period of 717.98 min.
16 Sep: STS-79 (Atlantis) was launched from LC-39A at 08:54 GMT, the fourth Shuttle-Mir mission, carrying six astronauts and more than 2,100 kg of supplies to the Russian space station. Astronaut Blaha replaced Shannon Lucid as the resident NASA astronaut aboard Mir. See Shuttle log and the Human Space Flight entry for details.
26 Sep: At 17:50 GMT a Proton-K (8K82K; s/n 375-01) carried the 2,500 kg Ekspress telecommunications satellite from LC200L, Baikonur, to a 35 m 838 × 35,908 km, 0.21° orbit with a period of 1,440.53 min. This was the second launch of the new generation Ekspress and the satellite was stationed at 80° E.
28 Sep: With the de-orbiting of Cosmos 2320, launched 29 Sep 1996, there was no Russian military reconnaissance satellite currently in orbit.

20 Oct: At 07:20 GMT a Chinese CZ-2D (s/n CZ2D-3) launched from Jiuquan complex carried the 2,600 kg FSW2–3 (Fanhui Shi Weixing) military reconnaissance satellite to a 171 × 342 km, 63.04° orbit with a period of 89.64 min. The recoverable module was returned to Earth 4 Nov and landed in the Sichuan Province.
24 Oct: At 11:37 GMT a Molniya-M (8K78M) launch vehicle carried the 1,750 kg Molniya-3 48 telecommunications satellite from LC43/4 Plesetsk to a 610 × 39,768 km, 62.83° orbit with a period of 718.27 min coplanar with Molniya-3 36.

4 Nov: Pegasus XL (s/n F14) was launched from the L-1011 carrier aircraft at 17:08 GMT carrying the SAC-B (Satellite de Aplicaciones Cientificas) and HETE (High Energy Transient Experiment) satellites. The 183 kg SAC-B was built by INVAP of Argentina with both Argentinian and NASA instruments. The 118 kg HETE was a US science satellite. The Pegasus XL stage performed as planned but the satellite separation system malfunctioned and failed to release either payload. The solar panels on SAC-B were deployed but the combined third stage (176 kg) and satellites caused it to tumble and the batteries could not be recharged. The solar panels on HETE could not be deployed because it remained encapsulated. Initial orbit of the assembly was 488 × 556 km at 37.97° with a period of 95.07 min.
4 Nov: NASA's Galileo spacecraft conducted a fly-by of Jupiter's moon Callisto, passing within 1,104 km of the surface.
7 Nov: At 17:00 GMT a Delta 7925A (s/n 239) carried the 1,060 kg Mars Global Surveyor (MGS) spacecraft from LC-17A to a transplanetary trajectory via heliocentric orbit. Spacecraft carried a camera, laser altimeter, electron reflectometer, thermal emission spectrometer, Mars relay radio system and magnetometer. MGS was scheduled to arrive at Mars 11 September 1997 for a planned two years of operations from Mars orbiter, achieved through aerobraking techniques.
13 Nov: The 2,661 kg Arabsat 2B and the 1,512 kg MEASAT 2 telecommunications satellites were launched by Ariane 44L (s/n V92) from ELA-2 at French Guiana. Arabsat 2B was placed in a geostationary orbit over 32° E and MEASAT, launched for Binariang (Kuala Lumpur, Malaysia), operated from 148° E.
16 Nov: At 20:48 GMT a Proton K launcher (8K82K s/n 392-02) carried the 6,825 kg Mars 96 spacecraft to an initial Earth orbit of 160 km at 51.53° inclination. The Block D-2 fourth stage was then to have put Mars 96 in an elliptical 160 × 100,000 km Earth orbit. After separating from the D-2 stage, Mars 96 should have executed an escape manoeuvre to a heliocentric Mars transfer orbit. An attitude error put the D-2/Mars 96 combination in the wrong attitude for elliptical transfer and a single burn of two D-2 ullage motors resulted in a 146 × 171 km Earth orbit. Controlled by timer, the spacecraft separated and performed its own burn, putting it in a 87 × >1,500 km orbit from where it decayed early 17 November followed by the D-2 next day. The spacecraft carried two 90 kg surface landers to have been deployed before Mars orbit insertion and two 75 kg surface penetrometers released after orbit insertion.
19 Nov: Shuttle Columbia was launched at 19:55 GMT from LC-39B at the start of the STS-80 carrying five astronauts including Mission Specialist Story Musgrave, at 61 years the oldest astronaut to date. Problems with a stuck hatch prevented planned EVAs. See Shuttle flight log and Human Space Flight section for full details.
19 Nov: Progress M-33 was launched from Baikonur complex LC1 at 23:20 GMT and placed in a 371 ×390 km, 51.65° orbit with a period of 92.16 min. The 7,190 kg cargo tanker was launched by Soyuz U(11A511U) and docked with Mir (–X port) at 01:01:30 GMT on 22 November, undocking at 12:13:53 GMT on 6 February 1997 after 76.47 days. Continued in independent flight in 377 × 395 km orbit for 35.7 days until destroyed on re-entry at 03:23:37 GMT 12 March 1997 following a failed re-docking attempt 4 March.
20 Nov: NASA astronauts aboard the Shuttle Columbia on the STS-80 mission released the 3,573 kg ORFEUS-SPAS 2 (Orbiting Retrievable Far and Extreme Ultraviolet Spectrometer – Shuttle PAllet Satellite) to a 349 × 356 km orbit. From there it conducted studies of the magnetosphere and energy from very hot and very cold matter in the universe. It was retrieved by Columbia on 4 December and returned to Earth.
21 Nov: An Atlas 2A (AC-124) carrying the 2,912 kg Hot Bird 2 telecommunications satellite for Eutelsat was launched from LC-36A, ETR, at 20:47 GMT. The satellite was placed in a geostationary orbit at 13° E.
22 Nov: Astronauts aboard the Shuttle Columbia released the 2,109 kg Wake Shield Facility into a 347 × 359 km orbit from where it conducted microgravity experiments. The WSF was retrieved and re-berthed aboard Columbia on 26 November.

2 Dec: Russian cosmonauts Korzun and Kaleri perform a 5 h 57 min EVA to begin the installation of solar array cables and connectors and a new solar panel to augment power supplies.
4 Dec: A Delta 7925 (s/n 240) lifted off from LC-17B, ETR, at 06:58 GMT carrying the 890 kg Mars Pathfinder spacecraft to a heliocentric Mars transfer orbit. The lander would use a new means of safely reaching the surface, impact energy cushioned by inflated spheres. The lander spacecraft encapsulated a rover, Sojourner, designed to operate independently on the surface.
5 Dec: In a transaction valued at around US$8.1 billion, Boeing acquired the aerospace and defence interests of Rockwell, including Rocketdyne, while moving to acquire McDonnell Douglas.
9 Dec: Russian cosmonauts Korzun and Kaleri performed an EVA lasting more than 6 hours to install more electrical power cables for the new solar battery and attach an antenna on the docking unit to improve the final approach for Shuttle Atlantis.
11 Dec: A Tsyklon-2 launch vehicle carrying the 3,150 kg Cosmos 2335 EORSAT (ELINT Ocean Reconnaissance SATellite) was launched at 12:00 GMT from the Baikonur LC90L complex carrying its payload to a 403 × 419 km, 65.04° orbit with a period of 92.79 min. Cosmos 2335 operates with Cosmos 2313 and Cosmos 2326. It is virtually coplanar with Cosmos 2313, crossing the equator 120° radially ahead of it.
11 Dec: Equatorial Guinea joined Intelsat becoming its 140th member with an investment of 0.05 per cent, more than 11 years after it began using Intelsat services.
18 Dec: An Atlas 2A (AC-129) carrying the 2,074 kg Inmarsat-3 maritime communications satellite operated by Inmarsat was launched at 01:57 GMT from LC-36A to a 35,698 × 35,877 km geostationary orbit over 157° E. The AS4000 satellite has a BOL mass of 1,100 kg and a dry mass of 860 kg.
20 Dec: A Cosmos-3M (11K65M) launched at 06:43 GMT from LC132/1, Plesetsk, carried the 825 kg Cosmos 2336 Parus military navigation satellite to a 979 × 1,012 kg, 82.95°, 105.03 min, orbit in plane 4 of the constellation. Parus operated at 149.97 MHz and 399.92 MHz.
20 Dec: Titan 4 404A (s/n K-13) launched at 18L04 GMT carried the 19,600 kg USA 129 payload comprising a KH-12 (No. 4) Advanced Crystal military reconnaissance satellite to a 153 × 949 km, 97.9° orbit from SLC4E, WTR.
23 Dec: US DoD communications satellite FltSatCom 1 was retired from duty and moved off station to 344° E.
24 Dec: A Soyuz-U (11A511U) carrying the 6,000 kg Bion 11 biological satellite to a 217 × 379 km, 62.80°, 90.48 min, orbit was launched at 13:50 GMT from LC43/4, Plesetsk. Carrying two macaque monkeys, newts, snails, drosophilia flies and a variety of insects and bacteria, this was the 12th in the series. Bion 11 was returned to Earth, 130 km north of Kustani, on 7 January 1997.

1997

9 Jan: OSC announces that it will use a Taurus launch vehicle to place 420 kg Korean KOMPSAT into orbit during mid-1999.
10 Jan: Surge in electromagnetic radiation following an event on the Sun on 6 Jan resulted in serious damage to the Telstar 401 satellite four days later when that radiation reached the Earth's environment. Electronics on board the satellite appear to have short-circuited.
12 Jan: Launch of shuttle orbiter Atlantis on STS-81/SMM-5 mission from KSC: docks with Mir 15 Jan, undocked 20 Jan, landed KSC 22 Jan. Returned J Blaha from Mir, left J Linenger aboard the station.
14 Jan: Reported that GE purchases 17.25 per cent stake in Argentina's Nahuel-1 satellite.
16 Jan: India announces that two advanced remote sensing satellites, IRS-P4 and IRS-P5 are to be launched aboard PSLVs before the end of the century. Approval also given for R&D of a third generation earth observation satellite series.
17 Jan: First launch failure of a Delta-2 (7925); SRB anomaly just after launch from Cape Canaveral led to the vehicle self-destructing 21 seconds into the flight. Navstar 28 payload was the first Block 2R satellite to be launched.
22 Jan: Delta-2 second stage from the MSX launch re-entered the atmosphere and was almost intact until impact in Georgetown TX (US).
23 Jan: Chinasat authorises SS/Loral to build Chinasat 8 with 16 Ku-band and 32 C-band channels. Launch planned for late 1998.
30 Jan: Ariane-44L launched from Kourou; successfully places GE 2 (US) and Nahuel 1A (Argentina) comsat into orbit.
SS/Loral signs US$100 million contract with Kistler Aerospace for 10 COMSAT launches aboard the K-1 vehicle between 1999 and 2002.
JCSAT selects Atlas-2AS for the launch of JCSAT 6 in mid-1998.

3 Feb: An Australian businessman from Korea and several South Korean companies will put up A$160 million from a required A$200 million for a proposed launch site at Temple Bay, Queensland. Russian launch vehicles could start orbital missions from there around 2000.
6 Feb: Progress-M 33 undocked from rear port of Mir: planned re-docking on 4 Mar failed and the spacecraft was de-orbited 12 Mar.
7 Feb: Soyuz-TM 24 undocked from the front longitudinal port of the Mir Complex and re-docked at the back, in advance of the TM25 launch.
10 Feb: Soyuz-TM 25 launched from Baikonur (previously known a Tyuratam) using a Soyuz-U: carried two Russians and German astronauts to Mir: German returned to Earth aboard TM24 (see blow).
11 Feb: Discovery shuttle orbiter launched from KSC on the second mission to service Hubble Space Telescope. HST captured 13 Feb and two pairs of astronauts complete a total of five EVAs, undertaking repairs and replacing new equipment for old; released 19 Feb and Discovery landed KSC 21 Feb.

12 Feb: Japanese Haruka (Muses-B) launched using M-5 from Kagoshima: radio telescope with an 8 m diameter dish.
14 Feb: Russia launches three Gonets-D and three Strela-3/Cosmos communications satellites from Plesetsk using a three-stage Tsyklon vehicle.

SS/Loral books satellite launches aboard five Delta-3 vehicles; this gives 18 satellites booked to fly aboard the new launch vehicle, due to debut in 1998.
17 Feb: Japanese JCSAT 4 comsat launched from Cape Canaveral using an Atlas-2AS.
20 Feb: Galileo performs fly-by of jovian satellite Europa, minimum distance 587 km.

Reports that Malaysia is to fly Tiung (= mynah bird) satellite – built by Surrey Satellite Technology – aboard Russian launcher.

Wall Street Journal reports that Teledesic communications satellite system could become operational around 2002 with only 288 satellites operating in 12 orbital planes: full system is planned to have 840 satellites in orbit.
23 Feb: A solid-fuel oxygen-generating candle exploded in the Kvant 1 module, causing a fire which was brought under control by the crew on board Mir.
24 Feb: Maiden flight of the Titan-4B using an IUS upper stage assembly launches DSP 18 (USA 130) early warning satellite. Delay of six months in the ISS launch schedule is confirmed. First launch – Russian FGB module aboard a three-stage Proton-K – now scheduled for around June 1998. Delay caused by slippage of work on second Russian module, the Service Module, which cannot be launched until around November 1998.

1 Mar: INTELSAT 801 comsat launched by Ariane-44P from Kourou.
2 Mar: Soyuz-TM 24 undocks from Mir, returning to Earth with the original crew plus German astronaut launched aboard TM25.

Pioneer 10 celebrates 25 years in space; the spacecraft was the first to penetrate the asteroid belt and perform a fly-by of Jupiter.

Reports that Iran plans to manufacture communications satellites in the country's 'new year' which begins 21 March. First to fly would be an educational satellite in around three years.
4 Mar: Maiden launch from new Russian cosmodrome at Svobodny using a Start-1, places small Zeya satellite into orbit.
8 Mar: US Tempo 2 comsat launched from Cape Canaveral using Atlas-2A.
10 Mar: Carlos Varotto, president of Argentinean space agency, stated that a launch vehicle capable of placing around one tonne into a 1,000 km orbit is currently being designed with the backing of Argentina and Brazil. The programme is to replace Condor-2 which is to be cancelled.

Indonesia plans the launch of a replacement for Palapa-B 2R before the end of 1998.
12 Mar: NASDA selects Thiokol (US) for the supply of the solid-propellant strap-on boosters for the H-2A launch vehicle. Thiokol will supply Castor 4A- XL strap-ons to Mitsubishi Heavy Industries.
14 Mar: Russian government spokesman told ITAR-TASS that the first launch of the Sea Launch programme will take place in late 1998; a Zenit-2 launch vehicle will be used, although geosynchronous orbit missions require the still-unflown Zenit-3 version, carrying a Block DM third stage based upon the fourth stages used by the Proton-K launcher.
15 Mar: Mars Pathfinder overtakes Mars Global Surveyor en route to Mars.
16 Mar: Indonesia wants Lockheed Martin to build Telkon-1 COMSAT for launch aboard an Ariane in 1998–1999. Satellite will be of the same generation as Palapa-B.
18 Mar: Korea Telecom signs US$108 million contract with Lockheed Martin for Koreasat 3 (aka Mugunghwa), to be launched in August 1999 aboard an Ariane.
20 Mar: SS/Loral to build two INTELSAT 'follow-on series 2' comsats, for launch in Summer and Autumn 2000.
22 Mar: NASA celebrates having astronauts in orbit for a full 12 months, due to the visits to the Mir station. Russia has had a permanent presence in space since September 1989.
25 Mar: Official at KB Polyot indicates that the Kapustin Yar launch site will return to orbital launches around February 1999 with the launch of German Abrixas satellite aboard a Cosmos-3M launcher; last orbital launch from the site was in 1987.
31 Mar: Final signals received from Pioneer 10 after more than 20 years of communications.

2 Apr: MARISAT 1 manoeuvred off-station in geosynchronous orbit and retired after 21 years of communications service.

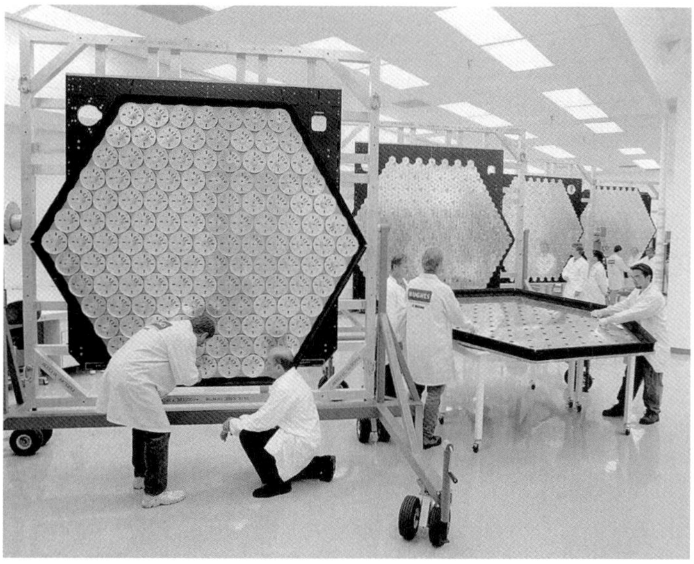

Antenna assembly line at the satellite production facility for the reconstituted NEW ICO (Boeing Company)
0137128

4 Apr: US DMSP-2 9 (USA 131) military metsat launched from Western Range (Vandenberg) using Titan-23G.

Columbia shuttle orbiter launched on STS-83 Microgravity Science Laboratory mission: planned 16-day mission curtailed due to fuel cell problems and shuttle landed KSC 8 Apr. Mission reflown in July as STS-94.
5 Apr: Galileo performs fly-by of jovian satellite Ganymede, minimum distance 3,102 km.
6 Apr: Launch of Progress-M 34 cargo freighter from Baikonur using Soyuz-U; docked with Mir 8 Apr, undocked 24 Jun and following a collision with Spektr module (see 25 Jun below) was de-orbited 2 Jul.
9 Apr: Launch of Cosmos 2340, Oko early warning satellite, from Plesetsk using Molniya-M.
16 Apr: Ariane-44LP from Kourou successfully launches THAICOM 3 (Thailand) and B-SAT 1A (Japan) comsats.
17 Apr: Cosmos 2341, Parus military navigation satellite, launched from Plesetsk using a Cosmos-3M. Planned launch of US FaiSat 2 aboard the same vehicle was cancelled, reportedly because documentation was not complete.
18 Apr: India plans to fly orbital mission using two-stage liquid oxygen/kerosene launcher by 2000.
21 Apr: Spanish MINISAT 01 launched from Gando AFB on the Canary Islands using US L-1011/Pegasus-XL system. Pegasus-XL third stage also carried Celestis 1 payload, comprising partial cremated ashes of 24 personalities.
22–23 Apr: Cosmos 2313 ELINT Ocean Reconnaissance Satellite performs end-of-life manoeuvre and begins to decay from orbit. Satellite disintegrates in orbit 26 Jun, marking the first EORSAT break-up since Novemeber 1987.
24 Apr: NASA withdraws from planned primate experiments aboard Russian Bion 12 because of unacceptable mortality risk to the primates. Criticism of primates aboard Bion missions ignores the fact, that primates were originally introduced to the programme at NASA's request.

Five candidates for the next one or two Discovery missions were approved; missions included flights to Mercury, Venus, the martian satellites, comets or the Sun. Detailed plans to be submitted by 15 Aug, launches to take place no later than 30 Sept 2002 and cost no more than US$183 million (1997 rate) over three years.
25 Apr: Final Atlas-1 launched from Cape Canaveral, carrying GOES 10 (GOES K before launch), meteorological satellite.
26 Apr: Xinhua news agency reports that India is to build a launch facility near the town of Irian Jaya in Indonesia: agreement with LAPAN calls for launches of Indonesian rockets, training from Indian personnel and TTC facilities.

Reports that Taiwan's first commercial satellite, ST-1, costing US$100 million, is to be launched towards geosynchronous orbit aboard an Ariane-5 in February 1998.
29 Apr: Russian Vasily Tsibliyev and NASA astronaut Jerry Linenger complete EVA outside Mir to install equipment outside the station.

RKA head Yuri Koptev said that a consortium of banks would offer an Rb80 billion (US$140 million) loan to permit work on the country's ISS commitment to continue.
30 Apr: Israel selects a shuttle astronaut candidate – identified only as 'Lt Col A' – to fly a mission during 1998 to mark the country's 50th anniversary. Since serving Israeli officers cannot be named under the country's security laws, the fighter pilot and electrical engineer will have to 'resign' from the military forces before his flight takes place.

Loral to build two digital radio comsats for CD Radio, launches due in 1999. Agreement signed in July calls for the satellites to be launched on separate Ariane-5 missions.

5 May: Return to flight for Delta-2 launch vehicle from Western Range when 7920 variant places first cluster of five Iridium satellites into orbit; full Iridium system calls for 60 operating satellites and six spare satellites spread around six orbital planes.
7 May: Galileo performs fly-by of Jovian satellite Ganymede, minimum distance 1,580 km.
9 May: London Daily Telegraph reports that launch platform (a re-fitted oil rig) for the Sea Launch programme has completed testing at Stavanger, Norway and is to process to Russia for further installations.
11 May: China launches second DFH-3 comsat from Xi Chang using CZ-3A launcher; first DFH-3 launched in 1993 suffered attitude control problems and had to be written off.
13 May: Amateur 'Rockoon' launched by Huntsville L5, a chapter of the National Space Society in the US, reached a peak altitude of 70 km. High Altitude Lift-Off Mission carried to an altitude of 15 km by a balloon and then a hybrid propulsion system fired to reach a higher altitude. Orbital mission planned by later version of the launch system.
14 May: Launch of Cosmos 2342, Oko early warning satellite, from Plesetsk using Molniya-M.

Reports that NASA and ISAS (Japan) are to co-operate on the Muses-C mission to collect asteroid samples. Due for launch using an M-5 in Jan 2002. Spacecraft to land on asteroid Nereus Sep 2003 and samples should return to Earth Jan 2006.
15 May: Launch of shuttle orbiter Atlantis on STS-84/SMM-6 mission from KSC; docks with Mir 17 May, undocked 22 May, landed KSC 24 May. Returned J Linenger from Mir, left M Foale aboard the station.

Russia launches Cosmos 2343 photo-reconnaissance satellite from Baikonur using Soyuz-U. This marks the first Russian photorecon satellite in orbit since 28 Sep 1996 when Cosmos 2320 was de-orbited – an unprecedented break in the programme since the first photorecon satellite was orbited in Apr 1962.
17 May: NASA confirms that Ukrainian cosmonaut Leonid Kadenyuk will fly a shuttle mission in October 1997.
20 May: Zenit-2 explodes seconds after launch from Baikonur; Tselina-2 ELINT satellite carried. Norwegian Thor 2 comsat launched from Cape Canaveral using Delta-2 (7925).
23 May: Kelly Space and Technology reported that it has been granted patent protection for the Eclipse tow-launch technique.
24 May: Russian four-stage Proton-K successfully launches US Telstar 5 comsat from Baikonur.
25 May: Inaccurate press reports suggest that a Chinese launch had placed a military communications satellite into orbit 'last week'; these are garbled reports of the successful second DFH-3 launch on 11 May.

3 Jun: Successful launch of INMARSAT-3 4 (UK) and INSAT 2D (India) comsats from Kourou using Ariane-44L.
6 Jun: Russian Proton-K launched from Baikonur carrying Cosmos 2344, said to be a new class of optical intelligence-gathering satellite. Orbit of

63.4°, 1,509–2,732 km is unlike any other known optical reconnaissance satellite. ESA considering a plan for the launch of a Mars sample-return mission in 2003 using an Ariane- 5.

10 Jun: China successfully launches second Feng Yun-2 #1 geosynchronous orbit metsat from Xi Chang using CZ-3; first FY-2 lost in 1994 pre-launch accident.

13 Jun: Arianespace announces that it has signed four new launch contracts: Sirius 3 will fly aboard an Ariane-4 in 1998, Telstar 6 aboard an Ariane-5 in late 1998 and two launches for an undisclosed customer.

18 Jun: Second cluster of Iridium satellites launched on four-stage Proton-K from Baikonur; seven satellites launched on this flight.

19 Jun: PANAMSAT selects Arianespace for the launch of its second HS-702 high-power satellite in 1999. Sydney Morning Herald reports that a deal has been signed with Starsem for the launch of Soyuz-class vehicles from Cape York by 2000.

21 Jun: China Great Wall Industry Corporation signs a deal with Hughes Space & Communications for the launch of 10 Hughes satellites aboard Chinese launchers during the period late 1998 to the end of 2006.

25 Jun: INTELSAT 802 comsat launched from Kourou using Ariane-44P.

Galileo performs fly-by of Jovian satellite Callisto, minimum distance 415 km.

Because of the possibility that ERS 1 might collide with Cosmos 614, the orbit of ERS 1 was raised slightly to ensure that a miss would take place and then it was lowered within 24 hours to the original altitude.

Manual re-docking of Progress-M 34 with the Mir Complex fails when the cargo freighter missed the rear of the station and performed a series of collisions with the Spektr module, damaging some of the solar panels and puncturing the module at least once. Two cosmonauts and NASA astronaut on Mir shut the hatch connecting Spektr to the Mir core module as the module loses pressure. Mir is powered down to conserve power while repairs are planned.

27 Jun: NEAR spacecraft performs a fly-by of asteroid 253 Mathilde: small manoeuvre 3 Jul put spacecraft on an Earth-fly-by trajectory (fly-by due Jan 1998) in advance of encounter with asteroid 433 Eros in early 1999.

30 Jun: Following a power loss over a period of a week, contact was lost with Japanese Midori (ADEOS 1).

1 Jul: Columbia shuttle orbiter launched on STS-94 Microgravity Science Laboratory mission – reflight of curtailed STS-83 mission; first time that a complete crew has been re-cycled to repeat a mission and the fastest return-to-orbit for the crew. Landed KSC 17 Jul. Arianespace signs an agreement for the launch of Mexico's Morelos 3 comsat; HS-601 satellite scheduled for launch at the end of 1998.

3 Jul: Gyrodynes on board Mir fail, causing the station to lose its lock on the Sun. Attitude control was later restored.

4 Jul: Mars Pathfinder spacecraft successfully lands on Mars in the Ares Vallis region. Spacecraft renamed Carl Sagan Memorial Station in honour of astronomer who died in Dec 1996. Sojourner rover deployed on martian surface 6 Jul. Both base station and rover still operating at end of August, far in excess of the planned mission.

5 Jul: Progress-M 35 cargo freighter launched to Mir Baikonur using Soyuz-U. Docked with Mir 7 Jul.

9 Jul: Third launch of Iridium satellite cluster; five satellites launched from Western Range using Delta-2 (7920); one of the satellites fails after orbital injection.

10 Jul: Indian government approves proposal for the development and launch of two advanced remote sensing satellites with an estimated cost of Rs3,900 million (US$120.4 million). CartoSat 1, a cartography satellite with a 2.5 metres ground resolution, and ResourceSat will both be launched using the PSLV.

14 Jul: OSC will pay US$12 million cash and refinance about US$25 million of debts to acquire the satellite manufacturing and communications service business of CTA Inc.

16 Jul: Cosmonaut aboard Mir inadvertently disconnects power/computer cable, resulting in the station losing its lock on the Sun. Systems restored within 24 hours.

21 Jul: Russia announces that the next Soyuz-TM launch to the Mir Complex will carry only the two Russians planned: the planned flight of a French spationaut is delayed until the following Soyuz-TM launch, due in early 1998.

22 Jul: Japan cancels the HOPE unmanned spaceplane programme in favour of a cheaper reusable 'rocket plane'.

23 Jul: Navstar satellite launched from Cape Canaveral using Delta-2 (7925); first Block 2R satellite to reach orbit. Japan delays the launch of COMETS aboard H-2 pending the results of the investigation into the Midori loss; launch now scheduled for early 1998.

24 Jul: Xinhua news agency reports that China will launch its first 'ocean exploration satellite' in 1999.

28 Jul: Japan's Superbird C comsat launched from Cape Canaveral using Atlas-2AS.

Norway's HS-376 Thor 3 comsat will be launched using a Delta-2 (7925) in July 1998.

1 Aug: L-1011/Pegasus-XL launches OrbView 2 remote sensing satellite from Vandenberg.

5 Aug: Two-man Soyuz-TM 26 launched from Baikonur using Soyuz-U: docks with Mir 7 Aug. Undocked 15 Aug from the rear port and re-docked at the front port in anticipation of the Progress-M 35 re-docking.

OSC will act as the prime contractor for 17 satellites which will form the Ellipso system.

7 Aug: Discovery shuttle orbiter launched from KSC for the STS-85 mission. CRISTA SPAS remote sensing payload deployed and re-captured before shuttle's return to Earth 19 Aug, landing KSC.

8 Aug: US PAS 6 comsat launched from Kourou using Ariane- 44P.

14 Aug: Soyuz-TM 25 with original two Russian cosmonauts undocks from Mir and makes a successful landing, marred by the solid propellant cushioning-rockets not firing, thus making the landing rougher than normal.

Cosmos 2345 launched from Baikonur using four-stage Proton-K; Prognoz early warning satellite.

18 Aug: Progress-M 35 re-docks with Mir after a delay of 24 hours. Just before docking the Mir computer system crashed, causing the Progress-M docking system to shut down; docking completed manually, but Mir lost its alignment with the Sun for about a day.

19 Aug: Chinese CZ-3B (currently the most powerful Chinese launch vehicle) launched from Xi Chang and successfully places Philippines Agila 2 comsat into orbit. South Africa plans to develop Overberg, Western Cape Province as a satellite launch site.

21 Aug: Fourth launch of Iridium satellite cluster: five satellites launched from Western Range using Delta-2 (7920). First Delta-2 launch since McDonnell Douglas merges with Boeing.

22 Aug: Mir cosmonauts Solovyov and Vinogradov perform IVA (intra-vehicular activity) using their EVA suits while NASA astronaut Foale waited in the Soyuz-TM 26 spacecraft. Russians isolated Spektr from the remaining Mir modules and working in the depressurised multiple docking adapter were able to reconnect cables from the Spektr module's solar panels via a new Spektr hatch attachment to the Mir power system. Vinogradov tried without success to spot where Spektr had been punctured.

23 Aug: LMLV-1 (Lockheed-Martin Launch Vehicle) successfully launches Lewis remote sensing satellite, but satellite is found to be tumbling after orbital injection.

24 Aug: Belgian company Verhaert Design and Development proposes to build 60 kg PROBA satellite for ESA for remote sensing and educational work; could be launched as a piggy-back payload aboard PSLV in mid-1999.

25 Aug: NASA's Advanced Composition Explorer (ACE) launched from Cape Canaveral using Delta-2 (7920). Mir briefly suffers problems with primary and back- up oxygen systems.

28 Aug: US PAS 5 comsat launched from Baikonur using four- stage Proton-K.

29 Aug: FORTE (Fast On-orbit Recording of Transient Events) launched from Western Range using L-1011/Pegasus-XL.

1 Sep: At 14:00 GMT two 689 kg Iridium Mass Frequency Simulator satellites were by a CZ-2C/SD launcher (s/n CZ2C-15) from the T'ai Yuan site in China to a 623 × 632 km, 86.34° orbit with a period of 97.26 min. These dummy satellites used the new CZ-2C Smart Dispenser launcher combination with MFS-1 in ascending node 271.9° and MFS-2 in ascending node 272°, both in plane 5 of the projected constellation.

2 Sep: At 22:21 GMT an Ariane 44LP (V99) carried the 2,915 kg (1,715 kg BOL) Hot Bird 3 communications satellite and the 703 kg (281 kg BOL) Meteosat 7 meterology satellite. They were launched from the ELA-2 launch pad at Kourou to geostationary orbits over 13.3° E and 349° E, respectively. Meteosat 7 was eventually located over the prime meridian.

4 Sep: From LC-36A, ETR, at 12:03 GMT an Atlas 2AS (AC-146) launched the 2,750 kg GE 3 telecommunications satellite to a geostationary orbit at 276° E, moved later to 272° E. Owned by GE American Communications, GE 3 is a Lockheed Martin AS2100 satellite.

14 Sep: A Proton-K (8K82K) launcher carried seven 689 kg Iridium satellites from complex LC91L, Baikonur, to medium orbit, the fifth Iridium cluster launch and the second to use the four-stage Proton-K. Launch came at 01:36 GMT. One (Iridium 27) was placed in a 542 × 553 km, 86.7° orbit and failed but six (Iridium 28–33 inclusive) were placed in a 776 × 779 km, 86.4° orbit conforming to plane 3 of the planned constellation.

23 Sep: At 16:44 GMT a Cosmos-3M (11K65M) launcher carried Cosmos 2346, a 795 kg Parus military navigation satellite from complex LC132, Plesetsk, to a 939 × 996 km, 82.92° orbit with a period of 104.42 min putting the satellite co-planar with Cosmos 2334. Also carried aloft was the 120 kg Faisat 2V store- forward communications satellite for Final Analysis, Inc., placed in a similar orbit.

23 Sep: Launched at 23:58 GMT, a Ariane 42L (V100) carried the 3,455 kg Intelsat 803 telecommunications satellite from the ELA-2 complex at Kourou to a geostationary orbit over 333.5° E. Satellite is built on a Lockheed Martin AS7000 bus.

24 Sep: At 21:32 GMT a Molniya-M launcher (8K78M) carried the 1,600 kg Molniya-1 90 telecommunications satellite from complex LC43/4 at Plesetsk to a 449 × 39,906 km, 62.84° orbit with a period of 717.78 min.

26 Sep: At 02:34 GMT the Shuttle Atlantis was launched from LC-39A at the start of STS-86, the last NASA mission to the Mir space station carrying seven astronauts (including one cosmonaut) and NASA crewmember Wolff who replaced Foale aboard the Russian facility. See Shuttle Log and Human Space Flight sections for further details.

27 Sep: At 01:23 GMT a Delta 7925 carried four Iridium satellites (19 and 34–37) from SLC2W at the WTR to an orbit of 775 × 790 km, inclination 86.4° with a period of 100.4 min. This was the sixth launch of Iridium series satellites, the fourth to use Delta.

28 Sep: At 04:47 GMT a PSLV (PSLVC-1) launched from India's Sriharikota launch site carried the 1,250 kg IRS 1D (Indian Remote Sensing 1D) satellite to a 308 × 822 km, 98.64° orbit with a period of 95.96 min. Following a malfunction of the fourth stage IRS 1D was left in an eccentric rather than circular (820 km) orbit. Onboard propulsion was used to change this orbit to a more useable 739 × 826 km orbit with the resultant reduction in lifetime to less than three years.

5 Oct: At 15:08 GMT a Soyuz-U (11A511U) carried the 7,195 kg Progress M-36 (s/n 237) from complex LC1 to a 382 × 391 km, 51.65° orbit with a period of 92.29 min. Unmanned supply vehicle docked with Mir –X port at 17:07:09 GMT, 8 Oct, undocked at 06:01:53 GMT 17 Dec after 69.54 days and was destroyed on re-entry at 13:20:01 GMT 19 Dec. Total free time 5.39 days.

5 Oct: The 2,900 kg (1,700 kg BOL) EchoStar 3 direct broadcast telecommunications satellite was launched at 21:01 GMT by Atlas 2AS from LC-36B, ETR, to a geostationary orbit at 298.6° E.

9 Oct: At 18:00 GMT a Soyuz-U (11A511U) launcher carried the 5,800 kg Foton 8 (F 11) satellite from LC43/3 at Baikonur to a 218 × 375 km, 62.81° orbit with a period of 90.45 min. Foton F11 carried microgravity experiments. The descent module landed 23 October 170 km northeast of Orsk in the Orenberg Region followed by the 150 kg MIRCA (Micro Re-entry Capsule) built by Kayser-Threde which had been attached to the exterior until after the de-orbit burn. It landed 110 km south-east of Orsk.

15 Oct: The 5,712 kg Cassini-Huygens spacecraft was launched at 08:43 GMT from LC-40, ETR, by direct ascent on a Titan 4B (401B/Centaur s/n 4B-33) to heliocentric orbit designed to facilitate two fly-bys of Venus, one fly-by of Earth and an encounter with Saturn in November 2004 where it will conduct observations from orbit and deploy the 320 kg Huygens probe to the surface of Titan. See Cassini under Planetary and Space Science section.

16 Oct: A CZ-3B (s/n CZ3B-3) launched from LC2 at the Xichang complex at 19:13 GMT carried the 3,745 kg APStar 2R telecommunications satellite to a geostationary orbit over 76° E. Launched as a replacement for APStar 2 lost shortly after lift off on 25 January 1995 the satellite is based on the FS-1300 bus from Space Systems/Loral.

22 Oct: The 395 kg STEP 4 (Space Test Experiment Platform 4) satellite was launched from a Pegasus XL (s/n F18) dropped from a Lockheed L-1011 carrier at 14:13 GMT. Placed in a 434 × 501 km, 44.96° orbit with a period of 93.95 min, STEP 4 carried a Digital Ion Drift Meter (DIDM), an ElectroMagnetic

Propagation Experiment (EMPE) and a Orbiting Ozone and Aerosol Measurement (OOAM) instrument.
24 Oct: Under the designation USA 133, at 02:32 GMT a Titan 4 (403A s/n 4A-18) carried the 14,500 kg Lacrosse 3 high-resolution radar imaging satellite from SLC4E at the WTR into a 666 × 679 km, 57° orbit for the National Reconnaissance Office. Lacrosse 1 was de-orbited in March 1997 but Lacrosse 2 was still in orbit at this date.
25 Oct: At 00:46 GMT an Atlas 2A (AC-131) carried the 1,040 kg USA 134, DSCS-3 (B13) military communications satellite from LC-36A to a geostationary orbit and deployed the USAF Falcon Gold technology satellite to a 151 × 34,457 km, 26.2° orbit with a period of 185.8 min. Falcon Gold was designed to demonstrate the use of GPS navigation technique in geosynchronous orbit.
30 Oct: The second, and first successful, Ariane 5 (s/n 502-V101) launched two satellites from ELA3 at Kourou at 13:43 GMT. The 2,650 kg MAQSAT-H/Teamsat satellite was placed in a 533 × 26,635 km, 7.79° orbit with a period of 467.8 min. The 1,400 kg MAQSAT-B/EPS satellite remained attached to the second stage and was placed in a 533 × 26,569 km, 7.78° orbit with a period of 466.64 min. The 60 kg Young Engineers Satellite (YES) was to have been unreeled from Teamsat but the tether was deleted and it became an independent satellite in a 540 × 26,626 km, 7.79° orbit with a period of 467.77 min. A malfunction in the second stage prevented the two MAQSAT satellites from being placed in elliptical orbit with apogee at geosynchronous altitude, about 9,000 km lower than planned.

2 Nov: Launched at 12:25 GMT from the Alacantara facility in Brazil, the first VLS launcher failed to reach orbit and was destroyed 65 s after liftoff when one of the four strap-on boosters while carrying the 115 kg SCD 2A data collection satellite.
6 Nov: At 00:30 GMT a Delta 7925 (s/n 249) carried the 1,881 kg Navstar 38 satellite from LC-17A, ETR, to a 19,910 × 20,449 km, 54.9° orbit with a period of 718 min. This was the last launch of the Block 2A series GPS series.
8 Nov: At 02:05 GMT a Titan 4 (401A/Centaur s/n 4A-17) carried the 5,000 kg Trumpet 3 Elint satellite from LC-41 at the ETR to a 1,100 × 39,060 km, 63.6° orbit with a period of 713.8 min. Under the designation USA 136, Trumpet 3 is the third such satellite after USA 103 (1994) and USA 112 (1995) launched for the NRO and the NSA.
9 Nov: a Delta 7925 (7920-10C s/n 250) from SLC2W at the WTR carried five 689 kg Iridium satellites (38–41 and 43) to a 775 × 779 km, 86.4° orbit with a period of 100.3 min. This is the seventh Iridium cluster to launch and was coplanar with the third sequence launched in July 1997.
12 Nov: At 17:00 GMT a Proton-K (8K82K) launched from LC200L at Baikonur carried the 2,500 kg Kupon satellite to a geostationary orbit at 55° E. Kupon (K95K) is the first in a series of four satellites built by NPO Lavochkin for the Russian Federation Central Bank. The series will relay information throughout the Bankir system.
12 Nov: An Ariane 44L (V102) launched at 21:48 GMT carried two communications satellites from ELA2 at the Kourou site to geostationary orbits. The 2,920 kg Sirius 2 telecommunications satellite launched for NSAB, Stockholm, was positioned over 5° E and the 1,385 kg (802 kg BOL).
18 Nov: A Soyuz U (11A511U) launched from Plesetsk carried the first 6,300 kg Resurs F-1M remote sensing satellite to a 180 × 236 km, 82.33° orbit with a period of 88.66 min. Derived from the old Zenit/Vostok design, the F-1M changed to a 238 × 270 km orbit with a period of 89.59 min.
19 Nov: At 19:46 GMT the Shuttle Columbia was launched from LC-39B, KSC, to a 281 × 286 km, 28.46° orbit with a period of 90.18 min at the start of the STS-87 mission to deploy a Spartan free-flying platform and conduct an EVA. For more details see Shuttle Log and the appropriate Human Space Flight section.
27 Nov: A Japanese H-2 (s/n H-II-6F) launched at 21.27 GMT carried the 3,620 kg TRMM (Tropical Rainfall Measuring Mission) satellite from complex Y at Tanegashima to a 367 × 385 km, 35° orbit with a period of 92.07 min. Manufactured by NASA's Goddard Space Flight Center, TRMM is designed to monitor rainfall conditions in the tropics. It also deployed the 2,900 kg Kiku-7 two-part satellite designed to demonstrate automated rendezvous and docking sequences.

2 Dec: At 22:52 GMT an Ariane 44P (V103) carried the 2,982 kg (1,819 kg BOL) JCSAT telecommunications satellite from ELA2 at the Kourou site to a geostationary orbit over 139° E and the 230 kg satellite Equator-S science satellite. Operated by the Max Planck Institute fur Extraterrestische Physik in Germany it was placed in a 496 × 67,232 km, 3.98° orbit with a period of 1,339.02 min. JCSAT was moved later to 150° E.
2 Dec: At 23:10 a Proton-K (9K82K) launched the 3,300 kg Astra 1G (HS 601) direct broadcast satellite for SES from complex LC81L at Baikonur to a geostationary position over 23–24°, moved later to 19.2° E.
8 Dec: A CZ-2C/SD (CZ2C-16) carried two 689 kg Iridium satellites (42 and 44) from China's T'ain Yuan launch site to a 775 × 779 km, 86.39° orbit with a period of 100.4 min. This was the first operational use of this launcher following a successful mass simulator launch on 1 Sep. They joined the cluster 3 orbit plane.
8 Dec: An Atlas 2AS (AC-149) launched from LC-36B at the ETR at 23:52 GMT carried the 3,000 kg class Galaxy 8I (HS 601) telecommunications satellite to a geostationary position above 280° E. Galaxy 8I was originally known as Galaxy 3R.
9 Dec: A Russian ELINT satellite, US-P EORSAT (Cosmos 2347) was carried to a 403 × 429 km, 65.04° orbit with a period of 92.79 min after launch at 07:17 GMT on a Tsyklon-M (Tysklon 2) from LC90 at the Baikonur complex.
15 Dec: Russia launched a Yantar-4K, fourth generation, military reconnaissance satellite on a Soyuz U (11A511U) at 15:40 GMT from the Plesetsk complex. Designated Cosmos 2348, it was placed in a 165 × 345 km, 67.15° orbit with a period of 89.6 min.
20 Dec: The Progress M-37 logistics vehicle was launched by Soyuz U (11A511U) at 08:45 GMT from complex LC1 at Baikonur. It docked with the –X port on Mir's Kvant 1 module at 10:22:20 GMT 22 Dec, undocked at 12:00:00 GMT 30 Jan 1998, re-docked at 09:42:28 GMT 23 Feb, undocked again at 19:16:01 15 Mar and was destroyed on re-entry at 23:04:00 the same day. Total docked time 59.47 days.
20 Dec: At 13:16 GMT a Delta 7925 (s/n 251) launched from SLC2W at the WTR carried five Iridium satellites (45–49) to a 775 × 779 km, 86.4° orbit, the ninth such cluster launched and the sixth by Delta 2.
22 Dec: The 3,455 kg (2,079 kg BOL) Intelsat 804 telecommunications satellite was launched at 00:16 GMT by Ariane 42L (V104) from the ELA 2 complex at Kourou to a geostationary orbit over 46° E, later moved to 64° E.
23 Dec: The first eight of 24 Orbcomm satellites (FM5-FM12 or A1 to A8 in orbit) were launched at 19:11 GMT by a Pegasus XL-HAPS (s/n F19). The 43 kg satellites were placed in closely spaced orbits of approximately 815 × 823 km, incln 45° with an orbital period of 101 min. Three 45° orbital planes of eight satellites each will be supplemented by 2–4 satellites in 70° orbits to cover high latitude requirements.
24 Dec: At 13:32 GMT a Start 1 launcher carried the 300 kg EarlyBird 1 commercial remote sensing satellite from the Svobodny launch site to a 480 × 488 km, 97.3° orbit with a period of 94.28 min. The satellite had been manufactured for EarthWatch of Longmont, USA, by Orbital Sciences Corp.
24 Dec: At 23:19 GMT the 3,410 kg (2,534 kg BOL) AsiaSat 3 was launched by Proton-K (8K82K/DM3) from LC81L at Baikonur to a geosynchronous transfer orbit of 361 × 35,999 km with an inclination of 51.1°. Asiasat 3 is a Hughes HS601 launched for the Asia Satellite Telecommunications Co of Hong Kong. The Proton fourth stage (DM3) failed 1 s into a planned second burn of 110 s and was unable to place the satellite in the planned orbit of 10,000 × 36,000 km, incln 12° leaving it stranded. Following an insurance settlement the satellite reverted to Hughes and a complex series of on-board manoeuvres was made to carry it around the moon and bring it to geostationary orbit.

1998 (orbital details can be found in the Satellite Launches 1996–1998 log)
7 Jan: NASA's Lunar Prospector, the third Discovery mission) is launched toward the moon at 02:28 GMT with lunar orbit injection four days later into a 100 km orbit. Prospector supported a geochemical study of the moon and exercised the first use of Athena 2 launcher.
10 Jan: The UK's Skynet 4D military communications satellite was launched by Delta 2 at 00:32 GMT. BOL mass is 850 kg.
22 Jan: Israel launched its fifth Ofeq military surveillance satellite (Ofeq 4) at 12:56 GMT but a malfunction in the Shavit launcher prevented it from reaching orbit.
23 Jan: NASA launched Shuttle Endeavour on STS-89, the eighth Mir visit and the first by this Orbiter to the space station, at 02:48 GMT returning to Earth 31 Jan after leaving astronaut Thomas and returning Wolff to Earth.
28 Jan: Russia launches Soyuz TM-27 with three cosmonauts on board including French cosmonaut L. Eyharts.
29 Jan: Under the designation USA 137, National Reconnaissance Office flies the Capricorn intelligence data relay satellite to a highly eccentric orbit following a launch at 18:37 GMT from the ETR.

4 Feb: At 23:29 GMT an Ariane 44LP (V105) launched the Brasilsat B3 telecommunications satellite located over 295° E (BOL mass is 1,052 kg) and the Inmarsat-3 F5 maritime mobile communications satellite (BOL mass 1,149 kg) over 25° E.
10 Feb: At 13:20 GMT the first launch of the full Taurus launcher from the WTR carried the GFO (Geosat Follow-On) and two Orbcomm satellites for Orbcomm Global LP into orbit.
14 Feb: The first four satellites in the planned constellation of 54 for Globalstar Ltd Partnership were launched by Delta 2 from the ETR at 14:34 GMT. The complete network would comprise six satellites and one spare in eight orbit planes. This was the first launch of the MedLite Delta 7420 with four solid strap-ons.
17 Feb: Following launch at 10:30 GMT a Soyuz U launcher (11A511U) carried Cosmos 2349 into orbit, a Yantar-1KFT/Kometa military topographic and mapping satellite with a planned lifetime of 45 days. The descent module was recovered late 2 April.
18 Feb: Four Iridium satellites (50–54) were launched at 13:58 GMT by Delta 2 (7920).
21 Feb: An H-2 launched from Tanegashima at 07:55 GMT placed the Kakehashi communications and broadcast engineering test satellite into orbit. Premature shutdown of the second stage prevented the satellite from reaching geosynchronous transfer orbit.
26 Feb: SNOE (Student Nitric Oxide Explorer), a science satellite operated by the University of Colorado, was launched at 07:07 GMT by Pegasus XL. It is designed to measure the density of nitric acid as a function of altitude and was the first satellite to be launched under the STudent Explorer Demonstration Initiative (STEDI). The final stage of Pegasus carried the second cluster of capsules in the Celestis series carrying human remains into space. A second satellite, Teledisc T-1 was released to a separate orbit where it would test micro-comsat technology.
27 Feb: Ariane 42P V106 launched at 22:38 GMT carrying the Hot Bird 4 telecommunications satellite for Eutelsat, placed at 13° E. Built on the Matra Marconi Eurostar 2000 satellite bus, it had a BOL mass of 1,770 kg.
28 Feb: Based on the Lockheed Martin (formerly RCA) AS7000 bus, the Intelsat 806 telecommunications satellite was launched by Atlas 2AS from LC-36B at the ETR at 00:21 GMT to a geostationary orbit over 40.5° W.

5 Mar: During a press conference at NASA's Ames Research Centre scientists from the Lunar Prospector programme announced that instruments aboard the moon probe had evidence for the existence of water ice on the lunar surface.
10 Mar: Operations with NASA's Mars Pathfinder ceased when the lander refused to acknowledge signals from the Deep Space Network's 34 m antenna at Goldstone, Calif.
12 Mar: The X-38 precursor Crew Return Vehicle for possible utilisation with the International Space Station made its first free-flight descent from a B-52 launched out of Edwards AFB, California. The origin of the X-38 design owes much to the evolution of M2, HL-10 and X-24 series lifting bodies.
14 Mar: The Progress M-38 cargo-tanker was launched at 22:46 GMT to a docking with the Mir space station just over two days later. It carried to orbit the Vynosnaya Dvigatel'naya Ustanovka (VDU) external engine unit to supplement attitude control equipment aboard Mir. M-38 was de-orbited 15 May.
16 Mar: Designated USA 138, the UFO 8 (UHF Follow-On) US Navy communications satellite was launched to geostationary orbit at 188° E by Atlas 2 at 21:32 GMT.
24 Mar: The 2,755 kg SPOT 4 remote sensing satellite was launched at 01:46 GMT by Ariane 40 (V107).
25 Mar: Two Iridium satellites (51 and 61) were launched by China's CZ-2C/SD into plane A of the system at 17:01 GMT.
30 Mar: Five Iridium satellites were launched into plane D of the system by Delta 2 (7920) which lifted off at 06:02 GMT.

2 Apr: At 02:42 GMT a Pegasus XL launched NASA's TRACE (Transition Region And Coronal Explorer) to study ultraviolet radiation from the Sun as a means of discriminating between the solar surface and features in the photosphere, the chromosphere and the corona. TRACE was the third in NASA's Small Explorer series.

7 Apr: Seven Iridium satellites (62–68) were launched by Russia's Proton-K at 02:13 GMT for the sixth and final orbit plane. It was the last cluster to be launched by this vehicle.
17 Apr: NASA's Shuttle Columbia was launched from KSC at 18:19 on the STS-90 mission to study a variety of neurological effects on living things including humans. See Shuttle Log and the Human Space Flight section. Columbia returned to Earth on 3 May.
24 Apr: Four Globalstar satellites (5–8) were launched by Delta 2 (7420) at 22:38 GMT to an orbit plane of 52° displaced approximately 49° to that of the first cluster.
26 Apr: NASA's Saturn-bound Cassini spacecraft performed a fly- by of Venus at a distance of only 284 km. It was expected to fly past Venus again 24 June 1999.
28 Apr: At 22:53 GMT Nilesat 1 became the first satellite to be launched for an African country when Ariane 44P (V108) carried Egypt's first telecommunication satellite into geostationary orbit over 353° E. Also launched was B-SAT 1B for the B-SAT Corporation, Japan, placed over 110° E.
29 Apr: Russia launched Cosmos 2350 at 04:37 GMT, a Geizer data relay satellite in the Potok system, to a geostationary orbit over 73° E.

2 May: Two Iridium satellites (69 and 71) were launched by China's CZ-2C/SD at 09:16 GMT and placed into plane B of the constellation.
7 May: Under the guise of Cosmos 2351 an Oko early warning satellite was launched at 08:53 GMT for the Russian Ministry of Defence intio an orbit plane 80° E of its companion Cosmos 2340.
7 May: At 23:45 GMT a Proton-K from Baikonur launched the EchoStar 4 direct broadcast satellite to a geostationary orbit at 233° E. Operated by the EchoStar Communications Company, the satellite was built around the Lockheed Martin AS2100 bus.
8 May: Renamed HGS 1 (Hughes Global Services), the AsiaSat 3 satellite launch 24 December 1997 was boosted toward the Moon using its onboard propulsion. It passed within 6,250 km of the Moon on 13 May and performed a fly-by of Earth at 36,000 km. A second lunar fly-by was made on 6 June at 34,354 km. HGS arrived in geosychronous drift orbit on 14 June with an inclination of 8.7° over 207° E.
9 May: USA 139, an Advanced Orion 2 ELINT satellite for the National Reconnaissance Office was launched at 01:38 GMT to a geostationary orbit by Titan 4B/Centaur.
13 May: NOAA 15 was carried into orbit by a Titan 23G launched at 15:52 GMT, the first of a new generation of meteorological satellite for the National Oceanic and Atmospheric Administration. The satellite carried a Thiokol Star-37XFP solid rocket motor which fired to take over from the two Titan stages and place it in orbit.
14 May: Progress M-39 was launched at 22:12 GMT for a rendezvous with Mir and a docking at the –X port two days later. M-39 de-orbited 29 Oct.
16 May: Technicians at the European Space Agency's ground station at Villafranca, Spain, turned off the Infra-red Space Observatory following more than two years of outstanding science studying the sources of infra-red energy in the universe.
17 May: Five Iridium satellites (70 and 72–75) were launched at 21:16 GMT by Delta 2 to plane A of the constellation.
30 May: The Zhongwei 1 (China Star) telecommunications satellite (a Lockheed Martin AS2100) was launched by CZ-3B at 10:00 GMT. It was located over 87.5° E. The satellite and liquid hydrogen upper stage were initially placed in a 216 × 85,035 km supersynchronous orbit.

2 Jun: At 22:06 GMT the Shuttle Discovery was launched from LC-39A at the start of STS-91, the ninth and last Shuttle-Mir mission. Six astronauts were launched but seven came back when Thomas was returned to Earth from Mir. Discovery landed on 12 Jun.
10 Jun: The Thor 3 (HS 376) Norwegian telecommunications satellite was launched at 00:35 GMT by Delta 2 and through a unique sequence of six burns by three motors was propelled to geostationary orbit. The first burn of the second stage placed the assembly in low orbit, the second burn raised apogee to 1,400 km, the third burn circularised the orbit prior to separation of the stage from the satellite. The fourth burn of the second stage de-orbited it into the atmosphere and the third stage fired to place the satellite on a geosynchronous transfer path with a 1,400 km perigee. The satellite fired its own Star 30 apogee kick motor for geostationary orbit over 0.8° W.
15 Jun: Six Strela 3 store-dump military communications satellites (Cosmos 2352–2357) were launched at 22:58 GMT by Tsyklon. They were delivered to the same orbital plane as the previous cluster of Gonets D1 and Strela 3 satellites in early 1997. The final stage of Tsyklon, however, failed to place them in the correct orbit and they gradually drifted out of phase with the previous clusters.
18 Jun: An Atlas 2AS launched at 22:48 GMT carried the Intelsat 805 satellite from LC-36A at the ETR to a geosynchronous orbit above 304.5° E.
24 Jun: A two-day ESA Council meeting in Brussels concludes with approval for the long-term programme establishing the initial step toward a Global Navigation Satellite System (GNSS), a vigorous Earth observation programme, first steps toward a more powerful Ariane 5 and development of the small VEGA launcher.
24 Jun: A Soyuz U launched at 18:30 GMT carried a fourth generation Yantar-4K2 Kobalt close-look military reconnaissance satellite (Cosmos 2358) to orbit.
25 Jun: A Yantar fifth generation photo-reconnaissance satellite was launched by Soyuz U at 14:00 GMT.

1 Jul: Molniya 3–49 was launched at 00:48 GMT from Plesetsk into a co-planar orbit with Molniya 3–39.
3 Jul: Japan's Mars probe Planet-B (Nozomi after launch) was placed on an interim Earth orbit of 146 × 417 km followed by ignition of the fourth (KM-V1) stage putting the spacecraft in a highly elliptical 359 × 401,491 km circumlunar orbit performing multiple Earth fly-bys and lunar fly-bys on 24 Sep and 18 Dec.
6 Jul: One component of Japan's technology satellite Kiku 7 satellite, called Orihime, launched 27 Nov 1997 separated from the other component, called Hikoboshi, and separated by 2 metres prior to re-docking after 21 min of free flight.
7 Jul: In the first orbital launch from a submerged submarine, Tubsat N and N1 were launched from the Delfin class submarine Novomoskovsk attached to Russia's Northern Fleet operating in the Barent Sea at 69.3° N × 35.3° E. The Shtil-1 launcher is a modified R-29RM ICBM known to NATO as the SS-N-23 Skiff.
10 Jul: At 06:30 GMT a Zenit-2 launched six separate technology and science satellites for Russia, Chile, Thailand and Australia.
18 Jul: A telecommunications satellites for the SINO Satellite Communications Company, SINOSAT 1 was launched at 09:20 from the Xi Chang launch complex on a CZ-3B. The Alcatel Spacebus 3000 satellite had been built in Cannes and was jointly owned by Daimler-Benz Aerospace and the Chinma Aerospace Corporation. Located at 110° E the satellite was handed over to its operators at the end of a checkout session.
28 Jul: A Zenit 2 launched Cosmos 2360, Tselina 2 ELINT satellite at 09:15 GMT from Baikonur.

2 Aug: Eight Orbocomm satellites (13–20) were launched by a Pegasus XL at 16:24 GMT in a cluster 120° to the east of the cluster orbited 23 Dec 1997.
6 Aug: Second undocking of the two elements of Japan's Kiku 7 satellite when Hikobashi separated from Orihime to a distance of 525 metres. Following several separate re-rendezvous and docking attempts during which maximum separation became 5 km, re-docking was achieved on 27 Aug.
12 Aug: A Titan 4A/Centaur launched the advanced Vortex Mercury ELINT 3 satellite from the ETR at 11:30 GMT but a power interrupt pitched the launcher into the sea at 40 s.
13 Aug: Soyuz TM-28 was launched by Soyuz U from Baikonur at 09:43 GMT, docked with Mir 12 Aug on a crew exchange mission, the spacecraft returning to earth 28 Feb 1999.
19 Aug: Two Iridium satellites (3 and 76) were orbited by CZ- 2C/SD, launched at 23:01 GMT, the first mission to replace in-orbit failures.
25 Aug: An Ariane 44P (V109) launched at 23:07 GMT carried the Singapore-Taiwan ST 1 telecommunications satellite to a geostationary orbit at 88° E.
27 Aug: Galaxy 10, launched for the PanAmSat Corporation by the first Delta 3 lifted off at 01:17 GMT but was destroyed when a guidance failure caused erratic flight and the guidance officer destroyed the vehicle at 75 s.
30 Aug: A Proton-K4 launched at 00:31 GMT carried the Astra 2A direct broadcast satellite to a geostationary orbit at 28.2° E where it was to be operated by SES.
31 Aug: Although initially calling it a success, North Korea suffered a failure with its first orbital attempt using the Taepo Dong 1 launcher which lifted off at 03:07 GMT.

8 Sep: Five Iridium satellites (77 and 79–82) were launched by Delta 2 as replacement satellites for plane F of the constellation.
9 Sep: A cluster of 12 Globalstar satellites were lost when the Zenit-2 that launched them at 20:29 GMT was destroyed after a failure in the second stage guidance at 272 s.
16 Sep: An Ariane 44LP launched the PAS 7 direct broadcast satellite at 06:31 GMT. Satellite was eventually located over 68.5° E from where it was operated by PanAmSat.
23 Sep: Eight Orbcomm satellites (22–28) were launched by Pegasus XL at 05:06 GMT, the third cluster to be oribited.
28 Sep: Molniya 1T telecommunications satellites was sent into the same plane as Molniya 1–86 by a Molniya-M launched at 23:41 GMT.

3 Oct: Sponsored by the US National Reconnaissance Office, the STEX (Space Technology Experiment) satellite was launched on a Taurus at 10:04 GMT. STEX incorporated a TAL-D55 plasma engine and a tether experiment which partly failed in flight and had to be jettisoned.
5 Oct: The Eutelsat W2 and Sirius 3 telecommunications satellites were launched by Ariane 44L (V111) at 22:51 GMT. Eutelsat W2 was the first of the new widebeam series and was located over 1° E. Sirius 3 was launched for the Nordic Television Company and was placed initially at 23° E prior to taking up its operating position at 16° E.
9 Oct: Hot Bird 5 was launched at 22:50 GMT by an Atlas 2A and placed over 13° E.
20 Oct: Launched as USA 140, the UFO 9 (HS 601) military communications satellite lifted off at 07:19 GMT and was located over 186° E.
21 Oct: The third qualification flight of the Ariane 5 launcher (s/n 503) began at 16:37 GMT with the launch of the Atmospheric Re-entry Demonstrator and the MAQSAT 3, a Eutelsat W mockup.
23 Oct: At 00:02 GMT the second SCD (Satelite de Coleta de Dados/SCD Data Collection Satellite) was launched by Pegasus to a medium orbit where it would be used as a Brazilian environmental data relay satellite.
24 Oct: NASA's Deep Space 1 technology spacecraft was launched at 12:08 GMT on a Delta 7326 from the ETR. The prime objective was to test new technologies for deep-space and interplanetary spacecraft, including an ion propulsion motor which would perform trajectory adjustment allowing encounters with an asteroid and a comet. As a secondary payload, SEDSAT 1 built by the Students foir the Exploration and Development of Space was released from the second stage of the Delta 7326 which was also making its first flight.
25 Oct: Progress M-40 was launched by Soyuz U at 04:14 GMT, docking at the Mir –X port two days later. It was de-orbited 5 Feb.
28 Oct: AfriStar, a broadcast communications satellite operated by the WorldSpace Corporation, and GE 5, a direct broadcast satellite for GE American Communications (Americom) were launched at 22:16 GMT. AfriStar was the first WorldSpace satellite and would broadcast 24–96 digital radio channels to hand-held L-band receivers throughout Africa and the Middle East from a position over 21° E. Originally the backup for Argentina's Nahuelsat, GE 5 was an Alcatel Euostar 2000 bus.
29 Oct: Shuttle Discovery was launched at 19:19 GMT carrying seven crewmembers including the first US astronaut to orbit the Earth, John Glenn, on a life sciences flight. Designated STS-95 payloads included a small telecommunications satellite called PANSAT and the Spartan free-flying platform released on 1 Nov and retrieved two days later prior to re-entry 7 Nov.

4 Nov: PAS 8, a direct broadcast satellite for the PanAmSat Corporation was launched by Proton-K from Baikonur at 05:12 GMT.
6 Nov: A Delta 2 (7920) launched from the WTR at 13:37 GMT placed five Iridium satellites (2 and 83–86) into plane E of the Iridium constellation.
20 Nov: The first element in the much delayed and re-designed International Space Station was launched from Baikonur at 06:40 GMT when a Proton-K carried the 20,700 kg Zarya FGB to Earth orbit. Funded by NASA and built by Khrunichev in Moscow under contract to Boeing, Zarya carried a multiple docking unit at one end and incorporated a pressurised section, a propulsion and instrument section and a rear docking port.
23 Nov: Owned by a Russian media consortium, the Bonum 1 telecommunciations satellite was launched at 23:54 GMT by Delta 2 (7925) from the ETR to geostationary orbit at 36° E. A HS 376, Bonum 1 is capable of broadcasting 50 channels to western Russia.

4 Dec: NASA's Shuttle Endeavour was launched at 08:35 GMT with a crew of six astronauts on

the STS-88 mission to begin assembly of the International Space Station by attaching the Unity module to the multiple docking adaptor on Zarya. The mission was a complete success and the Shuttle returned to Earth 16 Dec.
6 Dec: An Ariane 42L (V114) launched the SATMEX 5 telecommunications satellite at 00:43 GMT to a geostationary position over 243.2° E from where it would serve Mexico.
6 Dec: Another in the NASA Small Explorer series, the SWAS (Submillimeter Wave Astronomy Satellite) was launched from the WTR by Pegasus at 00:57 GMT.
10 Dec: At 11:57 GMT a Cosmos 3M launcher carried the civilian navigation satellite Nadezhda 5 and the Swedish particles and fields research satellite Astrid 2 into orbit from Plesetsk.
11 Dec: NASA launched the Mars Climate Orbiter to a heliocentric trajectory at 18:45 GMT. First of the two Mars '98 missions it is expected to arrive at Mars 23 Sep 1999.
19 Dec: Two Iridium satellites (11A and 20A) were launched by CZ-2C/SD from China's T'ai Yuan launch site at 11:30 GMT.
22 Dec: The PAS 6B telecommunications satellite was launched on an Ariane 42L at 01:08 GMT from Kourou, French Guiana.
24 Dec: Cosmos 2361, a navigation satellite in the Parus series, was launched at 20:03 GMT into a coplanar orbit with Cosmos 2233.
29 Dec: Launched 3 Jul, Japan's Nozomi Mars probe passed within 1,000 km of Earth at the end of multiple Earth-Moon fly-bys and performed a 7 min propulsive burn. The combined energies were to have placed the spacecraft on a heliocentric orbit for encounter with Mars on Oct 11, 1999. Course corrections on 21 December left Nozomi with too little propellant to reach Mars and the spacecraft was placed on a trajectory for Mars encounter Dec 2003.
30 Dec: Three Uragan navigation satellites in the Glonass series (Cosmos 2362–2364) were launched by Proton-K from the Baikonur LC200L complex.

1999
3 Jan: Second of the two Mars '98 planetary spacecraft scheduled for this launch window, Mars Polar Lander was launched at 20:21 GMT on a Delta 2 (7425) from the ETR. The spacecraft, also carrying two microprobes which will be released from the cruise stage just prior to entry into the Mars atmosphere on 3 Dec.
27 Jan: An Athena 1 launched from the ETR at 00:34 GMT placed Taiwan's first satellite (Chunghua or ROCSAT – Republic of China satellite) in orbit. It carries an ocean colour imager for fisheries and Earth resource evaluation.

7 Feb: Fourth mission in NASA's Discovery programme, the Stardust spacecraft was launched by Delta 2 (7426) at 21:04 GMT into a heliocentric orbit to fly by the comet Wild 2 in Jan 2004, capture primordial comet material and return it to Earth in Jan 2006.
9 Feb: Four Globalstar communications satellites (23, 36, 38 and 40) were launched by Soyuz U/Ikar at 03:54 GMT marking the first flight of this vehicle.
15 Feb: The Telstar 6 telecommunications satellite was launched at 05:12 GMT by Proton-K and placed in a geostationary orbit at 267° E.
16 Feb: An Atlas 2AS launched the JC-SAT 6 telecommunications satellite at 01:45 GMT for Japan Satellite Systems and placed it in geostationary orbit over 120° E.
20 Feb: Three cosmonauts from Russia, France and Slovakia were launched aboard the Soyuz TM-29 spacecraft at 04:18 GMT for a rendezvous with the +X Mir port two days later. The Slovak cosmonaut Bella returned with Padalka in Soyuz TM-28 on 28 Feb, but the Russian Afanasyev and the Frenchman Hagnere remained aboard Mir with cosmonaut Avdeyev. They were scheduled to return in late August as the last Mir crew.
23 Feb: A Delta 2 (7920) launched at 10:30 GMT carried three satellites to medium orbit: the joint-services ARGOS (Advanced Research and Global Observation Satellite) for near-Earth physics experiments; the Danish Ørstad microsatellite operated by the Danmarks Meterologiske Institut for mapping Earth's magnetic and particle environment; and Sunsat, South Africa's first satellite for Earth imaging and digital communications.
26 Feb: An Ariane 44L launched from Kourou at 22:44 GMT carried the ARABSAT 3A telecommunications satellite to a geostationary orbit at 26° E and the UK Skynet 4E military communications satellite to a position over the Gulf of Guinea at 6° E.
28 Feb: A Proton-4 launched at 04:00 GMT carried the Raduga-1 4 Globus telecommunications satellite into geostationary orbit.

5 Mar: At 02:56 GMT a Pegasus rocket launched the fifth of NASA Small Explorer satellites, the WIRE (Wide-Field Infra-Red Explorer) satellite designed to examine the origin and evolution of galaxies but technical problems prevented the equipment from obtaining data when an instrument cover was ejected prematurely.
15 Mar: Four Globalstar satellites (22, 37, 41 and 46) were launched at 03:06 GMT, the second flight of the Soyuz U/Ikar.
21 Mar: AsiaSat 3S, a HS 601P telecommunications satellite for the Asia Satellite Telecommunications Co Ltd of Hong Kong, was launched by Proton-K at 00:09 GMT and placed in geostationary transfer ellipse. From there, using its onboard R4D motor (the same as that used for attitude control on the Apollo spacecraft) it would be positioned for operations over 105.5° E. Asiasat 3S is the replacement for Asiasat 3 (now HGS), lost when the DM-SL stage of an earlier Proton failed dumping it in the wrong orbit.
28 Mar: A Zenit-3SL performed the first flight in the Sea Launch programme when, at 01:30 GMT it lifted off from the mobile launch platform named Odyssey located at 0° latitude by 154° W carrying a dynamic simulator model of a Hughes HS 702 satellite to geosynchronous transfer orbit. The DM-SL stage fired itself out of orbit.

2 Apr: The Progress M-41 cargo-tanker was launched at 11:28 GMT for a rendezvous with the Mir station two days later. Its propulsion system was also used to boost the orbit of Mir to inhibit premature orbital decay.
2 Apr: Ariane 44P (V117) was launched from the ELA-2 complex at Kourou carrying the Insat 2E communications and meteorology satellite.
2 Apr: RKA, the Russian Space Agency, announced that it had arranged finance to keep Mir in operation beyond August 1999 and NASA announced it would pay the Russians US$100 million to buy an additional Soyuz TMA to operate as an interim rescue vehicle.
9 Apr: A failure in the Inertial Upper Stage (IUS) of a Titan 4B/IUS left the DSP-19 early warning satellite in a useless orbit of 188 × 38,000 km. Launched at 17:01 GMT the first two stages of Titan placed the IUS/DSP-19 assembly in low Earth orbit. The first stage of the IUS fired as planned for geosynchronous transfer ellipse. However one or more connectors failed to disengage and prevented total separation of the lower stage and when the upper stage fired it caused the assembly to tumble.
12 Apr: An Atlas 2AS launched at 22:50 GMT carried Eutelsat W3, a Spacebus 3000B2 class telecommunications satellite, to a geostationary position over 7° E.
15 Apr: Four L-band Globalstar satellites (19, 42, 44 and 45) built by Space Systems/Loral were launched at 00:46 GMT by a Soyuz U/Ikar from Baikonur.
15 Apr: The Landsat 7 remote sensing satellite was launched by Delta 2 (7920) at 18:32 GMT into a 668 × 698 km orbit. It will be operated by NASA's Goddard Space Flight Center until control is transferred to the US Geological Survey in October 2000.
16 Apr: At 10:30 GMT an OSCAR amateur radio satellite was released from the Mir space station and designated Sputnik-99.
21 Apr: UOSAT-12 was launch at 04:59 GMT aboard the first Dnepr launch vehicle adapted from the R-36M2 ICBM. The third stage manoeuvering bus was used to propel UOSAT-12 into its proper orbit.
27 Apr: Ikonos 1, the first high-resolution civilian Earth observation satellite was launched at 18: 22 GMT but failed to reach orbit when the nose fairing failed to separate four minutes into flight. The added weight prevented the terminal stage on the Athena 2 launcher from reaching orbital velocity or position and the assembly re-entered over the Pacific.
28 Apr: A Cosmos 3M launched at 20:30 GMT carried the Abrixas X- ray astronomy satellite and the 35 kg Megsat 0 communications satellite into medium altitude orbits of 544 × 603 km.
30 Apr: The first in a series of upgraded Milstar 2 (F1) military communications satellites was stranded in a useless orbit when computer software in the Titan 4B/Centaur that launched it (at 16:30 GMT) commanded engine firings at the wrong times. Instead of three sequential Centaur burns over 6 h placing Milstar in parking orbit, geosynchronous transfer orbit and final geostationary orbit the three burns took place on the first orbit leaving the satellite in a 740 × 5,000 km path. Milstar was moved to a 953 × 4,475 km orbit where all propellant was depleted and it was declared dead.
30 Apr: Orbital Sciences Corporation rolled out the first of three X-34 spaceplane technology demonstrators. The X-34 would first fly a series of demonstration drops from a B-52 but eventually reach altitudes of 23,225 m and speeds of around Mach 8.

5 May: Following launch at 01:00 GMT from LC-17B at the ETR, the second Delta 3 failed to achieve its objective when the nozzle extender failed to operate and the engine burned for only 1 s, leaving the Orion 3 telecommunications satellite in a useless orbit of 162 × 1,378 km.
10 May: At 01:33 GMT a CZ-4A launched from China's T'ai Yuan complex carried the Fen Yung 1C meteorological satellite and the Shi Jian 5 magnetospheric satellite into orbit.
13 May: The US House of Representative's Science Committee voted to cancel the Triana project inspired by Vice President Al Gore conceived as a means of placing an imaging satellite between the Earth and Moon to provide continuous images of the sunlit Earth for access through the Internet.
13 May: AeroAstro Inc announces it has been authorised to build the first commercial interplanetary spacecraft designed to carry pictures, messages and human DNA from several million people worldwide on a gravity assist slingshot at Jupiter out of the Solar system following launch aboard an Ariane 5 in late 2002.
16 May: Canada's first contribution to the International Space Station, the Space Station Remote Manipulator System (SSRMS) arrived at NASA's Kennedy Space Center ready for launch in July 2000 aboard the STS-100 mission.
18 May: Launched at 05:10 GMT, a Pegasus XL carried two small satellites to orbit. One, the Terriers (Tomographic Experiment using Radiative Recombinative Ionospheric and Extreme UV and Radio Sources), was the second STEDI (Student Explorer Demonstration Initiative) and would study the Earth's upper atmosphere and ionosphere. The second, MUBLCOM (Multiple Path Beyond Line of Sight Communications) was operated by the Defense Advanced Research Projects Agency (DARPA) to research digital voice and data communications between handheld terminals without using any other terrestrial infrastructure.
20 May: Telesat Canada's first direct broadcast satellite, with 32 Ku-band transponders, the Lockheed Martin A2100 Nimiq, was launched by Proton-K from the Baikonur cosmodrome to a geostationary orbit over 91° W.
22 May: A Titan 4B launched from the ETR carried a classified military payload to orbit for the National Reconnaissance Office.
26 May: At 06:22 GMT In India's first commercial flight a PSLV-C2 launcher carried the 1,050 kg Oceansat 1 (IRS-P4) satellite into orbit along with the 107 kg Kitsat 3 for South Korea's Advanced Institute of Science and Technology and the 45 kg Tubsat-DLR for Germany's space agency. This was also the first time India had carried three satellites on one launcher.
27 May: NASA's Shuttle Discovery was launched at 10:49 GMT at the start of the STS-96 mission to carry about 1,800 kg of supplies and materials to the two docked International Space Station modules and perform three EVA operations to install equipment and prepare the assembly for later visits. Discovery returned to Earth 6 June.

1 Jun: Scheduled to fly in February 2000, the LE-7A engine for Japan's new H-IIA launcher performed its first static firing which had to be curtailed when inlet pressure fell and the gimbal system failed.
7 Jun: Sea Launch announced that its first commercial launch from the special equatorial ocean launch platform will take place in August or September. A Zenit 3SL will lift the Hughes built DirecTV-1R broadcast satellite to a geostationary position at 101° W replacing the 5.5 year old DBS-1 which began to fail last year.
10 Jun: A Delta 2 (7920) carried four Space Systems/Loral Globalstar communication satellites (25, 47, 49 and 52) from the ETR to orbit.
11 Jun: Two Iridium satellites (14A and 21A) were launched by China's CZ-2C/SD from the T'ai Yuan launch complex.
16 Jun: Sea Launch announced four more launch contracts for the period 2002–2003 bringing firm launch commitments to 19 satellites. A Zenit 3SL costs US$70–90 million.
18 Jun: A Proton-K/DM-3 launched from Baikonur at 01:49 GMT carried the Hughes Astra 1H (HS 601P) direct broadcast satellite into a geostationary orbit at 19.2° E for SES.
18 Jun: In a major challenge to the US Navstar GPS navigation programme, the 15 EU ministers of transport agreed in Brussels to initiate funding for the Galileo programme of European navigation

satellites, approving an initial payment of US$42 million. ESA has already spent US$62 million on the project.
20 Jun: A Titan 23G launched at 02:15 GMT from the WTR carried Nasa's Ball Aerospace QuickScat remote sensing satellite into a near-polar, 800 km, orbit. The satellite will provide detailed data on ocean winds to study their effects on oceans, weather and climate.
21 Jun: Iridium announced massive price cuts for its global mobile telephone service and shed 15 per cent of its 500 employee base in efforts to increase the number of subscribers. Company stock fell to US$5.375 a share compared with nearly US$71 a share in May 1998. Iridium had until 30 Jun to hit a total 27,000 subscribers.
24 Jun: A Delta 2 (the first 7320) launched from the ETR at 15:44 GMT carried NASA's 1,350 kg FUSE (Far Ultraviolet Spectroscopic Explorer) astronomy satellite, built by the Johns Hopkins Applied Physics Laboratory into orbit.
24 Jun: NASA's Cassini spacecraft, bound for Saturn, completed the second gravity-assist fly-by of Venus at 20:30 UT passing within 600 km of the planet. A 1,166 km fly-by of Earth 18 Aug will be followed by a fly-by of Jupiter on 30 December 2000 and a rendezvous with Saturn 1 Jul 2004.
29 Jun: NASA cancelled the Champollion mission designed to land a probe on the nucleus of comet P/Tempel 2 in late 2005.

5 Jul: Launched from Baikonur at 13:32 UT, a Proton launch vehicle carrying the new Breeze-M terminal stage and a Raduga military communications satellite failed when a second stage engine malfunctioned destroying the vehicle scattering debris over a wide area. Reacting to a potential threat to the local populace, Kazakh authorities temporarily banned all launches from Baikonur.
8 Jul: At 08:46 UT a Molniya-M launched the 52nd Molniya 3 communications satellite from Plesetsk to an elliptical 12-hour orbit.
9 Jul: A joint venture between Microsoft founder Bill Gates and Craig McCaw, Teledesic announced contracts with Motorola for the constellation of multimedia satellites and with Lockheed Martin for three Proton-M and three Atlas 5 launches.
10 Jul: A Boeing Delta 7420 launched from Cape Canaveral at 08:45 UT carried four 450 kg Globalstar satellites to medium orbits, the fourth quartet in a planned constellation of 48.
16 Jul: Rescinding a ban on launches from Baikonur following the Proton failure on 5 July, Kazakh authorities allowed the launch of the Progress M 42 cargo tanker which took place at 16:36 UT. Two days later the spacecraft docked with Mir and would remain attached when the current crew vacated Mir in late August.
17 Jul: A Zenit-2 launched from Baikonur at 06:38 UT carried the 4,360 kg Okean-O satellite into a low earth orbit. First in a new generation of oceanographic satellites carrying side-looking radar, visible and infra-red scanners and radiometers, it was a co-operative venture between the Russian Space Agency and the Ukrainian Space Agency.
19–30 Jul: Organised by the UN Office of Outer Space Affairs, the Unispace III exhibition and conference was held in Vienna, Austria, adopting as its theme "Space Benefits for Humanity in the 21st Century". Attended by delegates from 99 countries and comprising 40 workshops and symposia, there was unanimous agreement to allocate 4–10 October each year as World Space Week marking the anniversary of the world's first artificial satellite on 4 October 1957.
20 Jul: On the day NASA celebrated the 30th anniversary of the first manned landing on the moon, divers recovered the Mercury MR-4 spacecraft "Liberty Bell 7" which sank shortly after splashdown on 21 July 1961 at the end of Gus Grissom's 15 minute flight.
23 Jul: Shuttle Columbia was launched 04:31 UT with five crew members and the Chandra Advanced X-ray Astrophysical Facility (AXAF). At 122,534 kg this was the heaviest orbiter to date carrying, at 19,736 kg, the heaviest payload yet. During ascent two of the three main engine controllers failed and a leaking hydrogen coolant line raised operating temperatures on one main engine close to red-lines. Chandra was deployed at 11:47 UT prior to separate orbit-raising burns from the Inertial Upper Stage which left the observatory 900 km too low. A sequence of orbital adjustments was performed by Chandra's onboard propulsion system. The Shuttle returned to earth on 28 July but all four Shuttle orbiters were stood down for coolant line inspections.
25 Jul: Launched from Cape Canaveral at 07:46 UT a Delta 7925 carried a fifth quartet of Globalstar personal communication satellites to medium orbit.
11 Aug: Astronomers on the ground and the Mir cosmonauts in orbit observed a full eclipse of the sun visible along a line from southern England across southern Europe and on to India.
12 Aug: An Ariane 42P launched the Telkom 1 communications satellite from Kourous, French Guiana, to a geosynchronous transfer orbit from where the satellite's apogee motor positioned it at 108°E.
13 Aug: Iridium filed voluntary bankruptcy protection under Chapter 11 in efforts to stave off collapse of the company which had already launched 66 low earth orbit satellites for personal communication purposes but had been unable to sign up sufficient subscribers to meet revenue targets.
17 Aug: Delta 7925 launched from Cape Canaveral at 04:37 UT carried the fifth quartet of Globalstar satellites into a medium orbit.
18 Aug: The Cassini-Huygens spacecraft conducted a 1,172 km fly-by of planet Earth to gather energy for its final fling to Saturn where it will arrive in July 2004.
18 Aug: A Soyuz launched the Yantar-4K1 fourth-generation military reconnaissance satellite from Plesetsk at 18:00 UT. Equipped with high-resolution cameras, film was returned in two small SpK capsules with the main capsule landing in Russia on 15 December 1999.
26 Aug: Launched at 12:03 UT a Kosmos-3M carried Cosmos 2366, a navigation satellite in the Parus programme, from Plesetsk to a medium altitude orbit.
27 Aug: Following a series of financial problems ICO Global Communications filed Chapter 11 bankruptcy protection in the hope of resurrecting full funding for the constellation of 48 medium-orbit satellites for personal communications.
27 Aug: The last full day of activity aboard the Russian Mir space station, permanently occupied for almost 10 years during which 22,000 science experiments had been conducted for 20 research programmes involving 240 pieces of equipment weighing 14 tonnes. In all, 77 EVAs had been conducted and repairs on 1,500 technical incidents or malfunctions carried out. Russian cosmonauts Afanasyev and Avdeyev and French cosmonaut Haigneré returned to earth on 28 August.

4 Sep: An Ariane 42P launched from Kourou, French Guiana at 22:34 UT carried the Koreasat 3 communications satellite to a geosynchronous transfer orbit. A Lockheed Martin AS2100, the satellite was positioned at 112°E.
6 Sep: The first dual launch on a Proton took place at 16:36 UT from Baikonur carrying 1,360 kg Yamal 101 and 102 communication satellites. The DM-2M upper stage made two successful burns to place the two satellites in geostationary orbits, the first time two Russian geosynchronous satellites had been delivered by a single launcher.
7 Sep: Soyuz launch at 18:00 UT carried the Foton 12 spacecraft, with 650 kg of European microgravity experiments, from Plesetsk to a low earth orbit. The descent module returned to earth on 24 September 1999 about 130 km northwest of Orenberg.
9 Sep: Germany's DASA and the Pacific Century Group of Hong Kong in association with Pacific Century Matrix Ltd formed a joint venture to set up end-to-end broadband connectivity utilising space systems and terrestrial nets first through leased capacity and, from late 2002, via advanced Ka-band and X-band satellites.
20–23 Sep: SatCom Europe '99 was held in London during which global market opportunities were presented for satellite services and technologies.
22 Sep: SPOT Image and Orbital Imaging Corporation (Orbimage) signed an agreement for exclusive distribution in Europe of high-resolution imagery from OrbView 3 and OrbView 4 remote sensing satellites. Both satellites are planned for launch in 2000.
22 Sep: A Soyuz launched at 14:33 UT carried four Globalstar satellites to orbit from Baikonur, the fourth commercial Soyuz-Ikar flight.
23 Sep: At 06:02 UT an Atlas 2AS carried the 3,500 kg Echostar 5 communications satellite from Cape Canaveral to a supersynchronous transfer orbit of 131 × 45,526 km from where the satellite's apogee motor placed it at a geostationary position at 110°W.
23 Sep: Communication with Mars Climate Orbiter was lost when it burned up in the atmosphere of Mars due to a trajectory calculation error caused by misunderstanding on the unit of measurement being used by ground controllers.
24 Sep: The 1 m resolution Ikonos earth imaging photographic satellite was launched by Athena 2 from a specially modified site at Vandenberg Air Force Base at 18:21 UT.
25 Sep: An Ariane 44LP launched from Kourou, French Guiana, at 06:29 UT carried the 3,790 kg Telstar 7 C-band/Ku-band telecommunications satellite to a geosynchronous path from where it was placed at 129°W.
26 Sep: LM1 (Lockheed Martin Intersputnik), the first telecommunications satellite operated by a joint Russia-:US consortium, was launched by commercial Proton from Baikonur at 22:30 UT. The satellite was placed in geostationary orbit at 75°E.
28 Sep: At 11:00 UT a Soyuz launch vehicle carried the 6,300 kg Resurs F-1M earth observation satellite from Plesetsk to a low earth orbit. Equipped with high-resolution cameras, the satellite released its recoverable capsule for a landing on 22 October.

4–8 Oct: The 50th International Astronautical Federation was hosted at the RAI Centre in Amsterdam, Netherlands. Attended by some 1,400 people, the Congress supported some 850 papers.
7 Oct: A Delta 7925 launched at 12:51 UT carried a Navstar Block 2 navigation satellite from Cape Canaveral to a medium altitude orbit, consolidating the Global Positioning System network.
10 Oct: A Zenit 3SL launched from the Seal Launch Odyssey platform at 03:38 UT put the DirecTV1R satellite into geosynchronous transfer orbit from where it was placed in geostationary orbit at 101°W. This was the second successful Zenit launch from the Odyssey platform at 154°W on the equator.
11–13 Oct: The European Space Operations Centre at Darmstadt hosted an international meeting on the increasing problem of space debris.
14 Oct: China launched a CZ-4B from Tai Yuau at 03:15 UT carrying the SACI 1 technology satellite for Spain and Zi Yuan 1, a China-Brazil technology and remote sensing satellite, into sun-synchronous orbits. Technical problems prevented SACI 1 communicating with the ground.
18 Oct: At 13:22 UT a Soyuz-Ikar launched four Globalstar satellites to medium altitude orbits.
18 Oct: Aerospatiale Matra, DaimlerChrysler Aerospace and Marconi Electronic Systems announced formal agreement to form the first trinational space company. It would be called Astrium and employ 8,000 people with annual revenues of €2.25 billion and, when approved by the European Commission, would be the biggest space manufacturing unit in Europe.
19 Oct: At 06:22 UT an Ariane 44LP launched from Kourou, French Guiana, carried the Orion 2 satellite to geosynchronous orbit from where it was eventually positioned at 15°W. Renamed Telstar 12, it would augment the Loral Skynet constellation.
21 Oct: The Galileo spacecraft conducted the first close fly-by of Jupiter's moon Io, passing within 611 km of its surface.
27 Oct: Virtual Geosatellite Holdings Inc., a sister company of Ellipso, was granted patent to develop a constellation of 15 satellites each occupying a unique path to achieve total global coverage from low altitude in much the way geostationary satellites are able to achieve from an altitude of 36,200 km.
27 Oct: A Proton-K launched from Baikonur at 16:16 UT failed to achieve orbit and its payload, an Ekspress A1 communications satellite, was destroyed. Kazakhstan temporarily banned launch operations from Baikonur.

1 Nov: ICO Global Communications received US$1.2 billion from a group of international investors led by Teledesic and entrepreneur Craig McCaw.
4–5 Nov: Presentations were made by Iridium, Globalstar, ICO, Inmarsat Eutelsat among others at a mobile satellite services forum held in London.
10 Nov: European Space Agency approved the Beagle 2 Mars project proposed by the Open University of the UK. Beagle 2 is a joint project involving the OU, University of Leicester in the UK, Matra Marconi and Martin-Baker and is expected to search for microbial life on Mars following a landing scheduled for December 2003.
13 Nov: An Ariane 44LP launched at 22:55 UT carried the GE4 satellite into orbit where it would provide C-band and Ku-band services for GE American Communications. It would be stationed at 101°W.
15 Nov: A failure in the first stage propulsion system of the seventh H-2 launched at 07:29 UT from Tanegashima, Japan, destroyed the launch vehicle and its payload, a 1,200 kg FS-1300 satellite named MTSAT designed to provide communications and air traffic control and meteorological data. The failure resulted in NASDA abandoning the H-2 in favour of the H-2A.
19 Nov: China launched its first prototype spacecraft designed to carry humans into space when a CZ-2F lifted off its Shenzou launch pad at 22:30 UT. The

spacecraft was a close copy of Russia's Soyuz spacecraft and landed back on earch 110 km north-west of Wuhai, Inner Mongolia, at 19:41 UT on 20 November after completing 14 orbits.
22 Nov: At 16:20 UT a Soyuz-Ikar launched four Globalstar satellites from Baikonur to a medium altitude orbit.
23 Nov: An Atlas 2A launched from Cape Canaveral at 04:06 UT carried the UHF F/0 F10 US Navy HS-601 communications satellite to a geosynchronous transfer orbit from where it was placed in geostationary position at 170°W.
26 Nov: The Galileo spacecraft performed its second encounter with Io, innermost of the five major Jovian moons, passing within a mere 300 km of its surface.

1 Dec: Inmarsat Board of Directors approved a US$1.4 billion investment in the fourth generation Inmarsat satellite system for operation in late 2004 offering personal mobile communications compatible with cellular systems.
3 Dec: Contact with the NASA Mars Polar Lander was lost after it cut communications with earth to prepare for the landing sequence. Communications were never re-established and no signal was ever received from the Deep Space 2 probes. This and the loss of Mars Climate on 23 September brought a reassessment of NASA's Mars exploration programme resulting in the postponement of the 2001 Lander mission.
4 Dec: At 18:53 UT a Pegasus XL launched another seven Orbcomm satellites for global messaging and positioning services bringing the total to date to 35 microsatellites.
7 Dec: ESA and the European Commission signed a study contract for the proposed GalileoSat navigation satellite programme. At least 21 medium orbit satellites would operate with geosynchronous platforms to provide an advanced European navigation position system starting operations in 2005.
9 Dec: Eutelsat began delivering routine TV broadcasts to North America via the Atlantic Gate. The existing service using Eutelsat 1-F5 would be supplemented by Atlantic Bird 1 from mid-2001.
10 Dec: At 14:32 UT the fourth Ariane 5 was launched from Kourou, French Guiana, carrying the 3,764 kg XMM observatory to a highly elliptical orbit. This was the first commercial payload for Ariane 5.
11 Dec: At 18.30 UT the second flight of Brazil's VLS launcher was unsuccessful when the second stage failed to ignite destroying the SACI-2 satellite and denying Brazil indigenous launch capability.
12 Dec: At 17:38 UT a USAF Titan 23G carried the DMSP 5DF-15 meteorological satellite to a suborbital trajectory from where the satellite's own Star 37S kick motor fired for orbit insertion as planned.
18 Dec: At 18:57 UT an Atlas 2AS carried the Terra multispectral earth imaging satellite to a sun-synchronous earth orbit from Vandenberg AFB. Terra is the first launch in NASA's Earth Observing System series of satellites and carries multispectral imaging equipment, CO and methane detectors and instruments for cloud top and vegetation monitoring.
20 Dec: At 00:50 UT the Shuttle Discovery carrying seven crew members and repair equipment for the Hubble Space Telescope was launched from the Kennedy Space Center on an eight-day mission to repair the observatory and extend its life. The mission involved three highly successful EVAs.
21 Dec: At 07:12 UT a Faurus launcher carried the Acrimsat solar energy monitoring satellite, the Celestis 03 cremated remains satellite and the Kompsat Korean multipurpose satellite to orbit from Vandenberg AFB.
22 Dec: An Ariane 44L launched at 00:50 UT carried the Galaxy 11 satellite, first of the new-generation Hughes 702 series, to geosynchronous transfer orbit from where it was positioned first at 99°W and then 91°W. The world's largest telecommunications satellite, it carries 64 active C-band and Ku-band transponders.
26 Dec: Cosmos 2367 was launched from Baikonur at 08:00 UT by a Tsyklon 2 launcher and placed in an initial 147 × 442 km orbit. A passive electronic intelligence satellite for the Russian Navy, Cosmos 2367 replaced the only previous remaining satellite which ended its operations the previous month.
27 Dec: At 19:20 UT a Molniya-M was launched from Plesetsk carrying Cosmos 2368, an Oko early warning satellite placed in a highly elliptical 12-hour orbit with apogee of 39,138 km.

2000
21 Jan: Atlas 2A was launched for the USAF from LC-36A, CCAFS, Florida, carrying a DSCS-3 military communication satellite.
24 Jan: Ariane 42L flight 126 was launched from the ELA-2, Kourou, French Guiana, at 01:04 UT carrying the Galaxy 10R communications satellite.
26 Jan: China Great Wall Industry Corp launched a CZ-3A from the Xichang facility at 16:45 UT putting the Zhingxing-22 communications satellite in orbit.
27 Jan: An OSC Minotaur was launched from the Commercial Spaceport, Vandenberg AFB, California, at 03:03 UT carrying five satellites. These comprised JAWSAT, FalconSat and ASUSAT 1 and STENSAT and MASAT dispensed from the OPAL automated carrier).

1 Feb: The Progress M1-1 cargo-tanker was launched by Soyuz-U from Baikonur at 06:47 UT and docked with Mir two days later. The Progress M1 variant was originally developed for the ISS but the inaugural flight was assigned to Mir.
3 Feb: A Zenit 3 carrying the Tselina-2 signal intelligence satellite (Cosmos 2369) was launched from the Baikonur cosmodrome at 09:26 UT.
3 Feb: An Atlas 2AS was launched from Cape Canaveral at 23:30 UT carrying the Hispasat-1C communications satellite.
8 Feb: The Starsem Soyuz-Fregat made its first flight at 23:20 UT from Complex 31 at the Baikonur cosmosdrome. It was carrying the Argentinian satellite Microsat-2 and the IRDT (Inflatable Re-entry and Descent Technology) satellite.
10 Feb: The ASTRO-E x-ray astronomy satellite was launched from the Kagoshima centre at 01:30 UT by ISAS M-5 launcher.
11 Feb: The STS-99 mission began with the launch of the Shuttle orbiter Endeavour from LC-39A at the KSC, Fla., on the Shuttle Radar Topography Mission. The 11 day mission ended at the KSC runway.
12 Feb: An ILS Proton DM carrying the Garuda-1 communications satellite for the Asian Cellular Satellite System was launched from Baikonur at 09:11 UT.
18 Feb: An Arianespace Ariane 44LP launched from ELA-2 at the Kourou launch facility, Fr Guiana, at 01:04 UT carried Superbird 4 into a GTO for Space Communications Corporation of Japan.

12 Mar: A Russian Proton launched the Intersputnik Express 6A communications satellite from Baikonur at 04:07 UT.
12 Mar: An OSC Taurus T5 was launched from Vandenberg Air Force Base at 09:29 UT carrying the MTI (Multispectral Thermal Imager) satellite.
12 Mar: At 14:49 UT a Sea Launch Zenit 3SL carrying the first ICO satellite was launched at 14:49 UT from the Odyssey launch platform located on the equator at 154 deg west.
20 Mar: In a qualification flight, the second Starsem Soyuz-Fregat carried replicas of two Cluster 2 replacement satellites into orbit after launch at 18:28 UT from Complex 31 at the Baikonur cosmodrome.
21 Mar: The second commercial flight of Ariane 5 began at 23:28 UT from ELA-3 at Koutou, Fr Guiana, with the successful launch of the Isat 3B and Asiastar communication satellites. This was the fifth launch of Ariane 5.
25 Mar: NASA's Imager for Magnetropause-to-Aurora Global Exploration (IMAGE) was launched by Delta 7326 from SLC-2West at Vandenberg Air Force Base, Calif., at 20:35 UT.

4 Apr: A Soyuz-U carried the Soyuz TM-30 spacecraft into orbit following launch at 05:01 UT from Baikonur cosmodrome. On board were cosmonauts Zalyotin and Kaleri. TM-30 docked to the Mir space station two days later for the last period of human habitation.
17 Apr: Eutelsat's SESAT communications satellite was launched by Proton from Baikonur cosmodrome at 21:06 UT. The Gals/Express satellite carries 18 Ku-band transponders with telecoms package from Alcatel on a bus from NPO PM of Krasnoyarsk for broadcasts into Siberia.
19 Apr: An Ariane 42L (flight 129) was launched at 00:29 UT from ELA-2 at Kourou, Fr Guiana, carrying PanAmSat's Galaxy 4R communications satellite, a HS-601 bus destined for geosynchronous orbit at 99 deg west replacing Galaxy 4H which failed in May 1998.
25 Apr: The Progress M1-2 cargo-tanker was launched from Baikonur cosmodrome by Soyuz-U at 20:07 UT and docked with the Mir space station on 27 May.
26 Apr: Progress M1-1 undocked from the Kvant module at 16:33 UT and was deorbited over the Pacific Ocean at 19:27 UT, vacating that port for Progress M1-2.

3 May: An Atlas 2A launched at 0707 UT from LC-36A at the CCAFS, Fla., carried the GOES-L meteorological satellite into GTO orbit, renamed GOES 11 on station.
3 May: A Soyuz-U launched at 13:25 UT from Baikonur lifted an advanced Neman-class imaging reconnaissance satellite (Cosmos 2370) derived from the Yantar 4KS2 design into an initial orbit of 183 × 277 km, raised to 240 × 300 km next day. Utilising digital relay via orbiting geostationary communication satellites, it is the latest in a series first launched as Cosmos 1731 in February 1986. Satellites of this class usually operate for about a year.
8 May: A Titan 4B launched from Complex, CCAFS, Fla., at 16:10 UT carried the DSP-20 early warning satellite into geostationary orbit. It was the first fully successful Titan 4 launch since 9 May 1998 which was followed by three failures and a partial success on 22 May 1999.
11 May: A USAF Delta 7925 launched from LC-17A at 01:48 UT carried the Navstar GPS 2R-4 satellite into medium altitude orbit.
16 May: Eurockot's Rockot launched from Complex 133, Plesetsk, at 08:28 UT carried two dummy satellites, Simsat-1 and Simsat-2.
19 May: Shuttle Atlantis lifted off from LC-39A at the KSC, Fla., at the start of STS-101, an ISS supply flight (assembly flight 2A.2a) carrying a Spacehab Double Module and Integrated Cargo Carrier in addition to five astronauts including cosmonaut Yuri Usachev. The mission lasted almost 10 days and included one EVA lasting 6 hr 30 min before a landing at the KSC on 29 May.
24 May: The first flight of Atlas 3A began at 23: 10 UT from LC-36B, CCAFS, Fla., carrying the Eutelsat W5 communications satellite. Propelled by an Energomesh RD-180 propulsion system with 4,152 kN thrust (3,570 kN thrust for launch) through two throttleable combustion chambers, which replaced the MA-5 system comprising single sustainer and two booster engines. Atlas 3A also flew the single-engine Centaur 3A powered by an RL-10-A-4-1B.
30 May: Galileo conducted a close flyby of Jupiter's Galilean moon Ganymede at an altitude of 808 km during the perijove pass of the Jupiter system which brought the spacecraft within 479,000 km of the planet's atmosphere. Galileo then began a highly eccentric path which would carry it to an apojove 20.7 million kn from Jupiter in September 2000.

4 Jun: The highly successful Compton Gamma Ray Observatory impacted to destruction in the Pacific southeast of Hawaii at 06:18 UT. A controlled deorbit on 31 May 2000 put Compton in a 362 × 474 km orbit, lowerd by a second burn to 237 × 471 km on 1 June, and by a third burn to a perigee of 148 km on 4 June. The final burn came one revolution later putting it in a 28 × 470 km orbit.
6 Jun: The first successful launch of the Proton/Breeze-M was launched at 02:59 UT from Baikonur carrying the Gorizont 45 telecmmunications satellite, believed to be the last in the series.
7 Jun: A Pegasus launched by OSC at 13:20 UT from the L-1011 Stargazer carrier aircraft flying in the drop box at 36.0 deg north × 123.0 deg west put the USAF TSX-5 (Tri-Service Experiment-5) in a 403 × 1,704 km orbit. TSX-5 is the fifth STEP satellite and uses the LEOStar bus.
24 Jun: A Proton K-DM launched at 00:28 UT from Baikonur cosmodrome carried the Express-3A communications satellite to a low parking orbit at 51.6 deg inclination from where two additional Block DM burns placed it on geosynchronous drift.
25 Jun: China's second Fengyun-2 meteorological satellite was launched by CZ-3 from Xichang at 11:50 UT and placed in a GTO from where a jettisonable on-board apogee motor placed it in a geosynchronous drift orbit at an inclination of 1.1 deg. This satellite replaced the first FY-2 which had completed its three-year mission in two months before.
28 Jun: A Nadezhda COSPAS-Sarsat search and rescue satellite was launched by Cosmos-3M at 10:38 UT from Plesetsk to a 684 × 708 km, 98.1 deg orbit, the first sun-synchronous orbit from this launch site. From orbit two subsatellites were ejected: Tsinghua-1 with imager and communications payloads and SNAP-1 built by Surrey Satellite Technologies Ltd carrying imager and propulsion equipment for rendezvous tests with Tsinghua-1.
30 Jun: The first Advanced TDRS satellites (TDRS-H) was launched by Atlas 2A at 12:56 UT from LC-36A at CCAFS, Fla. To a geosynchronous orbit utilising the Centaur upper stage. This is the first TDRS built by Hughes (now Boeing), all previous TDRS launched 1983–1995 having been built by TRW.
30 Jun: At 22:09 UT a Proton K/Block DM launched from Baikonur carried Sirius 1, the first satellite for Sirius Satellite Radio (formerly CD Radio), to a 6,166 × 47,110 km orbit. Later, the onboard R4D liquid apogee engine placed Sirius 1 in a 24,388 × 47,097 km, 63.3 deg, elliptical orbit. This maintains

the satellite between 60 deg west and 140 deg west in its 24 hr orbit.

4 Jul: A Russian Geyser data relay satellite (Cosmos 2371) was launched from Baikonur at 23:44 UT on a Proton and placed in a geosynchronous orbit two days later.
12 Jul: A three-stage Proton K launched at 04:56 UT carried the Zvezda Service Module from Complex 81 at Baikonur to a 365 × 372 km orbit from where it would rendezvous and dock with the International Space Station. Zvezda was manufactured as the successor to Mir 1 and would been the core module for a Mir 2 which never flew.
14 Jul: An Atlas 2AS (AC-161) launched from LC-36B, CCAFS, Fla., at 05:21 UT carried the SS/L Echostar 6 communications satellite to GTO. The satellite provides direct-broadcast service for the DISH network.
15 Jul: A Cosmos-3M launched from Plesetsk at 12:00 UT carried Germany's CHAMP minisatellite, Italy's MITA experimental microsatellite and Rubin, a microsatellite developed by students at the Hochschule, Bremen, Germany, to a medium earth orbit.
16 Jul: A Delta 7925 launched for the USAF at 09:17 UT from CCAFS, Fla., carried the Navstar GPS 2R-5 navigation satellite to a medium earth orbit.
16 Jul: A Starsem Soyuz-Fregat launched from Complex 31, Baikonur, at 12:39 UT carried the first two replacement ESA Cluster satellites Samba and Salsa to an intermediate orbit of 250 × 18,072 km at 64.7 deg inclination. Following separation the two satellites utilised onboard liquid propulsion systems to reach a 90 deg polar orbit of 18,000 × 121,000 km.
19 Jul: An OSC Minotaur launched from Vandenberg AFB, Calif., at 20:09 UT lifted the USAF Mightysat 2.1 (Sindri) together with DARPA's Picosat 2 to a 547 × 581 km, 97.8 deg, orbit.
26 Jul: At 00:45 UT the Zvezda Service Module docked to the International Space Station, releasing for flight the next series of US manned missions to achieve initial operating capability and permanent habitation predicated by the launch of Expedition 1.
28 Jul: A Sea Launch Zenit-2SL launched from the Odyssey platform moored on the equator at 22:42 UT lifted the PanAmSat PAS-9 telecommunications satellite to a GTO. PAS-9 was positioned at 58 deg west to replace PAS-5.

6 Aug: A Soyuz-U launched from Baikonur at 18:27 UT carried the Progress M1-3 cargo-tanker to the International Space Station, docking to the rear Zvezda port at 20:13 UT on 8 Aug.
9 Aug: A Starsem Soyuz-Fregat launched the second pair of ESA's Cluster-2 satellites from Baikonur at 11:13 UT. Named Rumba and Tango through two burns with the Fregat stage they were dispensed to an orbit of 251 × 18,050 km at 64.8 deg inclination. After five onboard propulsion manoeuvres the two satellites had joined the first pair of Cluster-2 satellites in an orbit of 17,200 × 120,600 km at an inclination of 90 deg.
17 Aug: An Ariane 44LP (flight 131) was launched at 23:16 UT from ELA-2 at Kourou, Fr. Guiana, carrying Brasilsat B4 for Embratel and Nilesat 102 for the Egyptian telecommunications company Nilesat SA joining Nilesat 101 providing Ku-band services.
17 Aug: a USAF Titan 4B launched from LC-4E, Vandenberg AFB, Calif., at 23:45 UT carried a classified payload for the NRO to an intermediate orbit of 572 × 675 km at 68 deg inclination. After separation the payload, believed to be an Onyx radar imaging satellite, manoeuvred to a more nearly circular orbit of 681 × 695 deg at 68.1 deg.
19 Aug: NASA's Wind spacecraft performed its 32nd flyby of the moon at a distance of 7,600 km deflecting it to a 2 million km apogee, a trajectory adjusted on 26 Aug to a 567,000 × 1.62 million orbit.
23 Aug: At 11:05 UT the third Delta 3 was launched successfully from LC-17B, CCAFS, Fla., with a dummy payload on a simulated GTO demonstration flight aimed at restoring commercial confidence following two failed flights.
28 Aug: The Globus 1–5 communications satellite was launched by Proton K from Baikonur at 20:08 UT.

1 Sep: At 03:25 UT the Zi Yuan 2 remote sensing satellite was launched by CZ-4B from the T'ai Yuan launch facility. Whereas its predecessor launched in 1999 was a co-operative venture with Brazil this satellite was an independent Chinese project.
5 Sep: A Proton K launched at 09:44 UT from Baikonur carried the Sirius-2 digital mobile radio broadcast satellite to an elliptical orbit of 6,192 × 47,057 km at 63.4 deg inclination. The mission is similar to that of Sitius-1 (see 30 June).
6 Sep: An Ariane 44LP launched from ELA-2 at Kourou, Fr Guiana, at 22:33 UT carried the Eutelsat W1 communications satellite with 28 Ku-band transponders to GTO from where it was positioned at 10 deg east.
8 Sep: The STS-106 ISS 2A.2b logistics and outfitting mission utilising a Spacehab Double Module and Integrated Cargo Carrier was launched from LC-39B at 12:46 UT with a crew of seven. Atlantis docked to the PMA-2 adapter at 05:51 UT two days later. Astronaut Lu and cosmonaut Malenchenko conducted the 50th EVA in Shuttle history lasting 6 hr 14 min on 11 September and installed cables between Zvezda and Zarya before the Shuttle returned to earth on 20 September after a 11 dat 19 hr flight.
10 Sep: The YZ-2 remote sensing satellite was launched by CZ-4B from China's Taiyuan launch facility at 03:25 UT.
14 Sep: At 22:54 UT Ariane 506 flight 130, the sixth Ariane 5, lifted off from ELA-3 at Kourou, Fr. Guiana, carrying the Astra-2B satellite for SES and GE-7 for GE Americom.
20 Sep: Gherman Titov, the second Russian and fourth human in space died. Selected as a cosmonaut in early 1960, Titov was the pilot of Russia's second manned space flight, Vostok 2, launched on its 25 hr 18 min mission 6 August 1961.
21 Sep: A Titan 2 launched from LC-4W at Vandenberg AFB, Calif., at 10:22 UT carried NOAA-L, an Advanced Tiros meteorological satellite to a 2,500 × 800 km orbit at 98 deg from where the satellite's Thiokol Star 37 solid motor placed it in a sun-synchronous orbit at 800 km.
25 Sep: At 10:10 UT a Zenit -2 launched from Baikonur lifted the Cosmos 2372 military imaging satellite into a low earth orbit. Dubbed Yenisey, the satellite is possibly an improved Orlets first launched under the designation Cosmos 2290.
26 Sep: The second Dnepr launcher lifted off from Baikonur at 10:05 UT with five satellites. These were Saudisat 1A and Saudisat 1B (both amateur communications store-forward satellites, Tiungsat-1 (a Malaysian imaging satellite built by Surrey Technologies), Megsat-1 (Italian research satellite) and Unisat (Italian space debris monitoring satellite).
29 Sep: A Russian Soyuz-U carrying the Cosmos 2373 mapping satellite was launched from Baikonur at 09:30 UT. The satellite is the 14th Kometa (succeeding six Siluet satellite) and carried a recovery sphere which was returned to earth on 14 November. The first in this series was launched in February 1981.

1 Oct: At 22:00 a ILS Proton K was launched from Baikonur carrying the GE-1A communications satellite for GE Americom and was eventually located at 108 deg east providing broadcast services to eastern Asia.
6 Oct: Ariane 42L flight 133 launched at 23:00 UT from ELA-2 at Kourou, Fr. Guiana, lifted the Japanese N-Sat 110 (SuperBird 5/J-110) to GTO from where it was eventually manoeuvred to geosynchronous orbit.
9 Oct: The HETE-2 gamma ray burst satellite was launched at 05:38 UT by Orbital Sciences Pegasus XL from the drop zone at 7.65 deg N × 167.7 deg E. The L-1011 Stargazer launch aircraft operated out of Kawjalein Atoll for the first time. HETE-2 replaced its predecessor lost when its Pegasus launch adapter failed in November 1996. HETE-2 was put together from leftover spares and managed by NASA GSFC as an Explorer of opportunity. It was placed in a 595 × 636 km orbit at 2 deg inclination and carries a French gamma telescope as its primary instrument.
11 Oct: The 100th Space Shuttle mission began when Discovery lifted off from LC-39A at 23:17 UT on the STS-92/ISS 3A assembly flight carrying the Z1 truss, PMA-3, supplementary cargo and seven crewmembers. Four EVAs of 6 hr 28 min, 7 hr 7 min, 6 hr 48 min and 6 hr 56 min were conducted before Discovery returned to earth on 24 October.
13 Oct: Three Russian Uragan navigation satellites (no's 83, 87 and 88) were launched for the Glonass system at 14:13 UT by Proton K.
15 Oct: Progress M1-2 undocked from the Kvant module on Mir and was de-orbited over the Pacific Ocean.
16 Oct: The Progress M-43 cargo-tanker was launched from Baikonur at 21:27 UT and docked with the Kvant module on the Mir space station at 21:16 UT on 20 October.
20 Oct: The DSCS 2 B-11 military communications satellite was launched by Atlas 2A from LC-36A at the CCAFS, Fl., at 00:40 UT.
21 Oct: Sea Launch conducted a Zenit-3SL flight from the Odyssey platform located on the equator at 154 deg west at 05:52 UT carrying the Thuraya-1 satellite for L-band telephone services.
29 Oct: At 05:59 UT a Ariane 44LP was launched from ELA-2 at Kourou, Fr. Guiana, carrying the Europe*Star 1 communication satellite.
30 Oct: China launched its first experimental navigation satellite, Beidou, on a CZ-3A which lifted off from the Xichang launch facility at 16:02 UT.
31 Oct: At 07:52:47 UT the Soyuz TM-31 carrying NASA astronaut Bill Shepherd and cosmonauts Yuri Gidzenko and Sergei Krikalev was launched from Baikonur cosmodrome carrying the Expedition 1 crew to the International Space Station at the start of what is planned to be the permanent habitation of the ISS. TM-31 docked to Zvezda's rear port at 09:21 UT on 2 November. On the ISS Shepherd became the mission commander of Expedition 1.

1 Nov: Progress M1-3 was undocked from the ISS Zvezda port and de-orbited over the Pacific Ocean at 07:05 UT the same day.
2 Nov: NASA's X-38 precursor CRV test vehicle was drop-tested to the Edwards AFB, Calif., lakebed. The first flight into space was scheduled for 2002.
10 Nov: A Delta 7925 launched for the USAF from the CCAFS, Fla., at 17:14 UT carried Navstar GPS satellite 2R-6, arriving in medium earth orbit three days later.
16 Nov: The seventh launch of an Ariane 5 began when vehicle 507 lifted away from ELA-3 at Kourou, Fr. Guiana, at 01:07 UT. It carried the PanAmSat PAS 1R communications satellite, the AMSAT Phase 3-D (Oscar 40) amateur radio satellite (at 397 kg, the largest yet) and STRV-1c and STRV-1d both attached to the ASAP-5 secondary payload structure.
16 Nov: The Progress M1-4 cargo supply ship was launched to the ISS as assembly flight 2P at 01:33 UT from Baikonur and manually docked to Zarya's nadir port at 03:48 UT on 18 November.
20 Nov: A Russian Cosmos 3-M launched at 23:00 UT was unable to lift into the correct orbit the 1 metre resolution earth imaging satellite QuickBird 1when the second stage failed to restart and the terminal stage and satellite fell back into the atmosphere at first perigee.
21 Nov: NASA's Earth Observing-1 and Argentina's SAC-C were launched by Delta 7320 launched at 18:24 UT from LC-2 at Vandenberg-West, Calif. EO-1 is one of the New millenium series and demonstrates technology for next-generation Landsat satellites and SAC-C carries earth observing instruments.
21 Nov: At 23:56 UT an Ariane 44L (flight 136) launched from ELA-2 at Kourou, Fr. Guiana, carrying Anik F-1 with 36 C-band and 48 Ku-band transponders into GTO.
30 Nov: An ILS funded Proton K launched from Baikonur at 19:59 UT carried the Sirius-3 CD Radio satellite into an elliptical 63 deg orbit similar to its predecessors.

1 Dec: Progress M1-4 undocked from the Zarya nadir docking port at 16:23 UT and redocked to the ISS on 26 December.
1 Dec: Space Shuttle Endeavour and its five crewmembers was launched at 03:06 UT from LC-39B at the KSC on the STS-97/ISS-4A assembly flight delivering the P6 solar panel truss and array assembly. Endeavour docked to the station's PMA-3 port at 19:59 UT the following day and three EVAs were conducted lasting 7 hr 33 min, 6 hr 37 min and 5 hr 10 min before returning to earth after almost 10 days 20 hr.
5 Dec: Israel's Earth Remote Observation Satellite EROS A1 was launched from the Svobodny facility by Start-1 at 12:32 UT. To a 491 × 506 km, 97.3 deg orbit.
6 Dec: An Atlas 2AS launched from the CCAFS, Fla., at 02:47 UT carried a classified satellite into an initial parking orbit of 176 × 831 km for the NRO, then to a GTO prior to geosynchronous orbit where it serves as a data relay satellite.
20 Dec: Ariane 508, the first generic growth Ariane 5G, launched from ELA-3, Kourou, Fr. Guiana, at 00:26 UT carried the Astra 2D HS-376HP broadcast satellite into GTO for SES and the GE-8 communication satellite into GTO for GE Americom. The LDREX technology demonstration satellite was also lofted.
20 Dec: At 16:20 UT a CZ-3A lifted off from the Xichang complex carrying the second Beidou navigation satellite to GTO from where it was placed in geostationary orbit.
27 Dec: Believed to be Strela and Gonets types, six small Russian satellites were destroyed in a Tsyklon 3 launch vehicle failure following liftoff from Plesetsk at 18:56 UT.

2001
9 Jan: Launched at 17:00 UT, the second unmanned flight of China's Shenzhou spacecraft carried a monkey, a dog and a rabbit to test elements of

the environmental control system. The forward orbital module separated from the crew module on 16 January in preparation for the return to earth and after retrofire executed by the service module the crew module separated for atmospheric re-entry and a landing at 11:22 UT. With its own solar cell arrays, some experiments were conducted by the forward orbital module which remained in space in an orbit of 329 × 345 km.
21 Jan: An Ariane 44P carried the Turksat 2A (Eurasiasat 1) communications satellite into a geostationary transfer orbit from where it was positioned to GEO at 42° east replacing the ageing Turksat 1C.
18 Jan: Mir space station experiences a major electrical power failure delaying the launch of Progress M1-5 which was to deorbit the station. A standby crew was appointed to fly to Mir in the event the upcoming M1-5 failed to dock automatically.
24 Jan: Progress M1-5 was launched by Soyuz from Baikonur cosmodrome at 04:28 UT. The spacecraft performed a fuel efficient, three-day, rendezvous with Mir docking to the +X Kvant port at 05:33 UT on 27 January. Progress M1-5 carried 2,677 kg of fuel.
30 Jan: Navstar GPS-54 was launched by Delta 7925 from the CCAFS at 07:55 UT, the seventh Navstar 2R satellite launched and the 28th in the second generation series first flown in 1997.

7 Feb: An Ariane 44L carried the Italian Sicral defence communication satellite and the UK MOD Skynet 4F military communications satellite into a GTO following launch at 23:06 UT from Kourou, French Guiana. Sicral was parked at 16.2° east with Skynet 4F positioned at 1° east.
7 Feb: The 102nd Shuttle mission began at 23:13 UT from LC-39A at the KSC, Florida, when Atlantis carried five crewmembers on mission STS-98/ISS 5A to deliver the US Destiny experiment module and the PMA-2 module to the ISS. Atlantis docked to the ISS at 16:51 UT 9 February and undocked at 14:06 UT 16 February prior to landing at 20:33 UT on 20 February at EAFB, California, following three successive days delay due to bad weather. During the flight three successful EVAs were conducted on 10, 12 and 14 February.
20 Feb: Sweden's Odin science satellite used to study submillimetre wave astronomy and outer atmospheric phenomena was launched by Start-1 from Svobodniy to a 622 km circular orbit at an inclination of 97.8°.
26 Feb: At 08:09 UT a Soyuz launcher lifted off from Baikonur carrying the Progress M-44 spacecraft to a rendezvous orbit with the ISS. Carrying 2.5 tonnes of food, water, oxygen and fuel, it docked to the -Y Zvezda port at 09:47 UT 28 February after Soyuz TM-31, launched 31 October 2000, had been re-positioned to a port on the Zarya module.
27 Feb: The second Milstar Block 2 communications satellite was launched by Titan 4B from the CCAFS at 21:20 UT. Equipped with UHF, EHF and SHF transmitters, it is designed to provide secure communications for the DoD and was the first in the series with higher data transmission rates and increased resistance to jamming.

8 Mar: Launched at 11:42 UT from LC-39B at the KSC Shuttle Discovery carried six STS-102/ISS 5A.1 crewmembers to the ISS, three of which comprised the Expedition 2 crew Usachev, Helms and Voss. It also carried the Leonardo Multi-Purpose Logistics Module with 16 racks of equipment including the Human Research Facility rack. Discovery docked with the ISS PMA-2 port at 05:34 UT 10 March and two EVA were conducted on 11 and 13 March. PMA-3 had to be moved from its position at the -Z nadir port on Unity to make a docking port available for Leonardo, docked for a while to facilitate cargo transfer before it was re-berthed back in Discovery. Undocking took place at 04:32 UT 19 March and returned to a landing at KSC at 07:31 UT 21 March bringing back the Expedition 1 crew of Shepherd, Krikalev and Gidzenko.
8 Mar: Launched by Ariane 5 from Kourou, French Guiana, the Eurobird communications satellite was launched for Eutelsat to a GTO along with BSAT 2a for Japan. Operated from 29° east, Eurobird replaced the Copernicus (DFS 3) satellite. BSAT 2a was the second satellite in the Orbital STAR class and was to operate from 110° east.
18 Mar: The XM-2 Rock radio satellite was launched by Zenit 3SL from Sea Launch to a GTO. Located at 114.9° west it was to be accompanied by XM-1 Roll launched on 8 May 2001.
19 Mar: Toward the end of the STS-102 mission at the ISS, the Expedition 2 crew of Usachev, Helms and Voss formally assumed command of the space station beginning the second phase of permanent habitation.

23 Mar: The first of two small propulsion manoeuvres to de-orbit the Mir space station took place at 00:33 UT when the attached Progress M1-5 lowered the orbit from 212 × 218 km to 190 × 218 km. At 02:01 UT a second small burn changed the orbit to 150 × 215 km and the main de-orbit burn at 05:07 UT lowered perigee to less than 80 km. Observers on Fiji reported bight re-entry fragments illuminating the sky. Impact was around 160° west by 40° south.

7 Apr: A UHF TV satellite, Ekran M-18 was launched by Proton M from Baikonur to a GTO from where it was manoeuvred to a parking orbit at 99° east. Re-placement for Ekran M-15 operating since October 1992.
7 Apr: NASA's Mars Odyssey spacecraft was launched by Delta 7925 from SLC-17A at the CCAFS, Florida, at 15:02 UT, the first spacecraft in the newly configured Mars Exploration Programme following the failures of Mars Climate Orbiter and Mars Polar Lander. After entering a 195 × 215 km interim earth parking orbit, Odyssey drifted for 12 minutes before the Delta third stage fired again placing the spacecraft in a heliocentric orbit of AU 1.384 × 0.982 AU at 3.05°.
16 Apr: Having been docked to the ISS since two days after launch on 26 February the Progress M-44 undocked from the Zvezda aft port at 08:48 UT and deorbited over the Pacific Ocean at 13:23 UT.
18 Apr: At 10:13 UT India's first GSLV launch vehicle ascended from its Sriharkota pad to place the 1,500 kg GSat-1 scaled down test model (Insat 2) satellite in a GTO. The new cryogenic upper stage underperformed by 0.5 per cent and shut down early leaving the satellite short of its planned elliptical orbit. On board propulsion pushed Gsat-1 into a 35,665 × 33,806 km orbit with a period of 23 hours and a drift rate of 13°/day. The ITU required ISRO to switch off the otherwise functioning transponders and transmitters to prevent frequency pollution with operable satellites in their appropriate orbits.
19 Apr: Shuttle Endeavour was launched at 18:40 UT on the STS-100/ISS 6A with a crew of four Americans, one Russian, one Canadian and one Italian. Mission to install 1,700 kg Space Station Remote Manipulator Systems (SSRMS or Canadarm 2) and to deliver 4,500 kg of supplies and equipment transported to the ISS in the Rafaello cargo container. Two EVA successfully carried out on 22 and 24 April prior to return to earth. Endeavour undocked at 17:34 UT on 29 April with a landing at EAFB, California, at 16:11 UT on 1 May.
28 Apr: Soyuz TM-32/ISS-2S launched at 07:37 UT carrying Russian cosmonauts Musabayev and Baturin and US civilian Dennis Tito who had paid for his trip. TM-32 docked to the -Z port on Zarya at 07:58 UT 30 April after Shuttle Endeavour had departed and seat liners were transferred from that spacecraft to Soyuz TM-31, docked to the ISS since 2 November 2000, in which they returned to earth leaving TM-32 as the emergency evacuation vehicle. TM-31 returned to earth with Musabayev, Baturin and Tito on 6 May at 05:41 UT.

8 May: XM-1 Roll was launched by Zenit 3SL from the Sea Launch facility at 22:10 UT. Designed to provide radio services to North America, Roll and its companion Rock, launched 18 March, is a Boeing 702 series satellite.
15 May: PAS-10 was launched by Proton from Baikonur at 01:11 UT as replacement for PAS-4 and is parked at 68° east for DTH services to Europe, the Middle East and South Africa.
18 May: A Delta 7925 launched the NRO satellite GeoLITE spacecraft from the CCAFS, Florida, at 17:45 UT. A TRW T-310 class satellite carries an experimental laser communications and an operational UHD data relay payload.
20 May: Progress M 1-6 was launched by Soyuz from Baikonur at 22:33 UT on the ISS 4P supply mission, docking at the -Y aft Zvezda port at 00:24 UT on 23 May. Carrying 2,500 kg of food, water, life support material and fuel it also brought spare computer equipment to the US Destiny module.
29 May: A Russian Yantar 4K1 surveillance satellite, Cosmos 2377 was launched by Soyuz from Plesetsk at 17:55 UT. A Kobalt class imaging satellite, it returned to earth on 10 October.

8 Jun: At 14:20 UT ISS crewmembers Usachev and Voss depressurised the Zvezda transfer section so they could remove the flat hatch cover from the nadir docking port allowing them to install the Zvezvda docking cone removed from that port after docking with Zarya a year ago. Fellow ISS Expedition 2 crewmember Susan Helms remained in the pressurised Zarya.
8 Jun: A Parus navigation satellite, Cosmos 2378 was launched by Kosmos launcher from Plesetsk at 15:08 UT, the first flight with this launcher since a failure on 20 November 2000 prevented QuickBird 1 reaching orbit.
9 Jun: An Ariane 44L launched from Kourou, French Guiana, carried the Intelsat 901 satellite to GTO from where it was placed in a GEO at 18° west. Of the 54 Intelsat satellites launched since Early Bird 1 in 1965, 19 were currently operational.
16 Jun: Astra 2C, a Boeing 601HP class satellite, was launched by Proton K from Baikonur at 01:49 UT and moved to a parking position at 28.2° east.
19 Jun: ICO-2, a Boeing 601M class communications satellite for New ICO (formerly ICO Global), was launched by Atlas 2AS from LC-36B, CCAFS, at 04:41 UT. This class of Boeing 601 eliminated the R4D apogee motor and had a larger payload section. Two burns of the Centaur stage placed ICO-2 in a 10,100 km orbit.
30 Jun: The Microwave Anisotropy Probe (MAP) was launched by Delta 7925 from LC-17B, CCAFS, at 19:46 UT to a highly elliptical orbit facilitating lunar fly-by on fourth apogee at 16:39 UT on 30 July. The gravitational slingshot sent it to L2, the earth-moon Lagrangian point 1.65 million km from earth, where it arrived three months later in a Lissajous orbit about that point. From there it was to measure fluctuations in the cosmic microwave background.

12 Jul: NASA's Shuttle Atlantis was launched on mission STS-104/ISS 7A from LC-39B, KSC, at 09:04 UT carrying a crew of five and the Quest Joint Airlock. ISS docking at 03:08 UT on 14 July. EVA conducted on 15, 18 and 21 July. Crew undocked at 04:55 UT, 22 July, landing at KSC at 03:39 UT on 25 July.
12 Jul: The tenth Ariane 5 lifted off from Kourou, French Guiana, carrying the 3,100 kg Artemis ESA communications technologies test satellite and the 1,300 kg BSAT-2b TV broadcast satellite for Japan's B-SAT company. An early cutoff plus low thrust on the EPS upper stage left the BSAT-2b in an unusable elliptical orbit of 592 × 17,528 km from where it could not reach GEO. Artemis was able to use on-board propulsion to reach GEO.
20 Jul: Molniya 3K was launched from Plesetsk by a Molniya launcher at 00:17 UT and placed in a 407 × 40,831 km, 62.9°, orbit. Molniya 3K is an improved version of the Molniya 3 military communications satellite.
20 Jul: Russian Navy launched a R-29R Volna three-stage missile from the submarine Borisoglebsk in the Barents Sea in a suborbital test of the solar sail project managed by the Planetary Society and the NPO Lavochkin Babakin Centre. The sail failed to deploy from the third stage.
23 Jul: The GOES-M (GOES 12 in orbit) meteorological satellite was launched from LC-36A, CCAFS, by Atlas 2A at 07:23 UT.
31 Jul: A Tsyklon 3 launched from Plesetsk carried the 2,260 kg Koronas-F solar observation satellite to a sun-synchronous orbit of 499 × 544 km at 82.5°.

6 Aug: A Titan 4B carried the 2,300 kg DSP-21 infrared missile early warning satellite from LC-40, CCAFS, to a near geosynchronous earth orbit.
8 Aug: NASA's Genesis solar wind sample collector was launched from LC-17A by Delta 7925 at 16:13 UT and in three upper stage burns was placed in a highly elliptical trajectory with an apogee of 1.5 million km at the Lagrangian 1 point between earth and sun. By the first week in November Genesis had arrived at Lagrangian 1 around which it would perform five revolutions in just over 29 months before manoeuvring back in toward earth for a lunar fly-by and earth return trajectory lasting 5.3 months.
10 Aug: The STS-105/ISS 7A.1 mission began with the launch of Shuttle Discovery from LC-39A, KSC, at 21:10 UT. The 15,100 kg payload comprised the Leonardo MPLM module for delivering new science equipment and returning to earth with trash. The seven crewmembers included the ISS Expedition 3 crew of Culbertson, Dezhurov and Turin, returning with Expedition 2 crewmembers Usachev, Voss and Helms. Discovery docked with the PMA 2 port at 18:42 UT on 12 August. The Leonardo module was lifted out of the Discovery cargo bay and docked to the Unity nadir port at 15:54 UT next day following which 3,300 km of supplies were transferred before 1,700 kg of trash was loaded aboard. Leonardo was undocked by the Shuttle RMS and re-berthed in the cargo bay at 19:17 UT on 19 August. Two EVAs were conducted on 16 and 18 August. Discovery undocked at 14:52 UT on 20 August and landed at KSC at 18:23 UT, 22 August, returning the Expedition 2 crew to earth after 167 days in space.
20 Aug: A small astronomical test satellite, Simplesat was released from a GAS canister aboard the Shuttle Discovery on mission STS-105 but failed to operate after separation.

21 Aug: The Progress M-45 logistics carrier was launched from Baikonur by Soyuz to a docking with the ISS aft Zvezda port at 09:51 UT on 23 August.
22 Aug: The Progress M 1–6 cargo supply ship separated from the ISS -Y Zvezda port at 16:01 UT and deorbited at about 19:00 UT this day making way for Progress M-45.
24 Aug: An early warning geosynchronous satellite designated Cosmos 2379 was launched by Proton K from Baikonur at 20:34 UT.
29 Aug: First launch of Japan's new H-2A launch vehicle took place from Tanegashima at 07:00 UT carrying the LRE geodesic test satellite to a simulated GTO of 271 × 36,214 km and the VEP-2 test package which remained attached to the second stage.
30 Aug: Ariane 44L carried the Intelsat 902 communications satellite to a GTO from Kourou, French Guiana, following launch at 06:46 UT. Intelsat 902 has 44 C-band and 12 Ku-band transponders.

7 Sep: Picosat 7 and Picosat 8 were ejected as planned from the Mightysat 2.1 satellite launched 19 July 2000. The 250 g satellites were in an orbit of 511 × 539 km at 97.8°.
8 Sep: Designated USA 160, a new form of naval signals intelligence gathering satellites from the NRO was launched by Atlas 2AS from SLC-3E, VAFB, at 15:25 UT. As successor to the NOSS series, it was placed in a 1,100 km circular orbit at 63° inclination.
14 Sep: Progress M-S01 consisting of service module and Pirs docking and airlock module instead of standard cargo and fuel sections (equivalent to re-entry and orbital modules on Soyuz spacecraft) was launched by Soyuz launcher from Baikonur at 23:35 UT. Docking at the Zvezda nadir port occurred at 01:05 UT on 17 September facilitating three planned EVAs from that location for Expedition 3 crewmembers.
21 Sep: Orbview-4, QuickTOMS, SBD and Celestis 4 payloads were launched by Taurus at 18:49 UT from VAFB, California. A failure in the ascent phase caused all four to re-entry on the first orbit after separating from the final stage.
22 Sep: NASA's Deep Space 1 technology demonstration spacecraft passed within 2,200 km of Comet Borrelly at 22:30 UT obtaining the best pictures yet taken of a comet nucleus.
25 Sep: Atlantic Bird 2 communications satellite for Eutelsat was launched by Ariane 44P from Kourou, French Guiana, at 23:21 UT, replacing Telecom 2A at 8° west.
26 Sep: The Progress M-S01 service module undocked from the Pirs docking and airlock module at 15:36 UT and deorbited over the Pacific Ocean at 23:30 UT.
30 Sep: An Athena-1 launched at 02:40 UT from the Kodiak site carried the Starshine 3 satellite to a circular orbit of 472 km and STP P-97-1 Picosat, PCSat and Sapphire to a circular orbit of 794 km.

5 Oct: A Titan 4B launched from SLC-4E from VAFB at 21:20 UT carried an Improved Crystal imaging satellite to a sun-synchronous orbit of 150 × 1,050 km at 97.9°. This was the first NRO satellite launched after the 11 September terrorist attacks on New York and Washington DC.
6 Oct: A Raduga-1 class military communications satellite was launched from Baikonur by Proton K at 16:45 UT and placed in GEO by a second burn of the final stage at 23:18 UT.
8 Oct: Russian cosmonauts Dezhurov and Turin performed the first EVA from the Pirs docking and airlock module carried to the ISS by Progress M-S01.
11 Oct: An Atlas 2AS launched from CCAFS, Florida, at 02:32 UT carried what was believed to be an imaging relay satellite to a GTO of 274 × 37,538 km.
18 Oct: A Delta 7925 launched from SLC-2W, VAFB, at 18:51 UT carried the QuickBird 2 high-resolution earth imaging satellite to a 461 × 465 km orbit inclined at 97.2°. Reportedly capable of taking images with a resolution of 0.6 m the satellite was expected to begin operations after a few months of calibration and ground-truth tests.
19 Oct: ISS Expedition 3 crew entered Soyuz TM-32 and undocked from -Z port on Zarya and redocked at the Pirs nadir port at 11:04 UT freeing it for the arrival of Soyuz TM-33.
21 Oct: Launched from Baikonur on a Soyuz at 08:59 UT, Soyuz TM-33 was placed on a rendezvous trajectory to the ISS carrying Russian cosmonauts Afanasyev and Kozeev and French cosmonaut Claudie Haignere. It docked with the ISS at 10:00 UT on 23 October. Crew couch liners were exchanged with those in Soyuz TM-32 so the latter could be used to return the short-stay crew to earth eight days later. Carrying Afanasyev, Kozeev and Haignere, TM-32 undocked at 01:39 UT on October 31 and landed at 04:58 UT.
22 Oct: ISRO launched a PSLV from Sriharikota at 04:53 UT carrying a TES surveillance satellite, a PROBA technology development satellite for ESA, and the German BIRD research satellite to a 551 × 580 km, 97.8° orbit.
24 Oct: NASA's Mars Odyssey arrived at the planet Mars and with a 20 minute 19 second main propulsion burn starting at 02:18 UT entered an initial orbit of 272 × 26,818 km at 93.42°. Two days later 76 days of aerobraking began settling the spacecraft into a 400 km circular orbit with an operational sun-synchronous period of 2 hours. Odyssey was to conduct a 917 day mapping programme and serve as relay for 2003–2004 landers.
25 Oct: A Molniya 3 class military communications satellite was launched from Plesetsk by a Molniya launcher at 11:34 UT. It was placed in a 615 × 40,659 km at 62.8° with apogee over the northern hemisphere.

22 Nov: Having delivered 2,500 kg of supplies, the Progress M-45 logistics carrier was undocked from the ISS and deorbited.
26 Nov: A Soyuz FG launched at 18:24 UT from Baikonur carried the Progress M 1-7 logistics spacecraft to a docking with the ISS at 19:43 UT on 28 November. Although the docking probe retracted the eight latches would not secure a hard-dock condition and an EVA on 3 December cleared rubber seals left on the latches by the Progress M 45 flight and a hard dock was achieved.
27 Nov: An Ariane 44LP launched from Kourou, French Guiana, carried the DirecvTV-4S TV broadcast satellite to a GTO from where it was manoeuvred to a position over 101° west.

1 Dec: Cosmos 2382, the first of the Uragan-M improved Glonass navigation satellites, was launched by a Proton K from Baikonur at 18:04 UT also carrying two other Glonass satellites Cosmos 2380 and 2381.
5 Dec: NASA's Shuttle STS-108/ISS UF-1 mission began at 22:19 UT with the launch of Endeavour carrying seven crewmembers including Onufrienko, Bursch and Walz, the three crewmembers of ISS Expedition 4. Endeavour docked with the ISS at 20:03 UT on 7 December carrying the the Raffaello MPLM 2 logistics module. Rafaello was berthed to the Unity module at 17:55 UT on 8 December and transferred back to Endeavour at 21:18 UT on 14 December. Endeavour undocked from the ISS at 17:28 UT, 15 December, and landed back at the KSC at 17:55 UT on 15 December.
7 Dec: A Delta 7925 launched from SLC-2W, VAFB, at 15:07 UT carried the Jason 1 ionospheric research satellite to a 1,320 × 1,330 km, 66°, orbit and the TIMED solar terrestrial probe to a 627 × 640 km, 74.1°, orbit.
10 Dec: A Zenit 2 launched from Baikonur at 17:19 UT carried five satellites into 996 × 1,015 km, 99.7° orbits. These included Russia's Meteor 3M weather satellite, Pakistan's Badr B surveillance satellite, Russia's Kompass geophysical satellite, Morocco's Maroc-Turbsat store-forward communications satellite and Russia's Reflektor laser imaging calibration reflector.
18 Dec: NASA's Deep Space 1 technology demonstration spacecraft was retired after having successfully conducted close encounters with asteroid Braille in 1999 and comet Borrelly in September 2001.
21 Dec: A Tsyklon 3 launcher lifted off from Plesetsk at 03:24 UT carrying six Strela-3 military communication satellites and three Gonets-D1 civilian communications satellites into approximately 1,400 km circular orbits at 83° inclination. Because of a launcher failure in 2000 these were the first Strela series satellites orbited since 1998. The three Gonets were deployed last with perigee of the last one 15 km higher than the first.

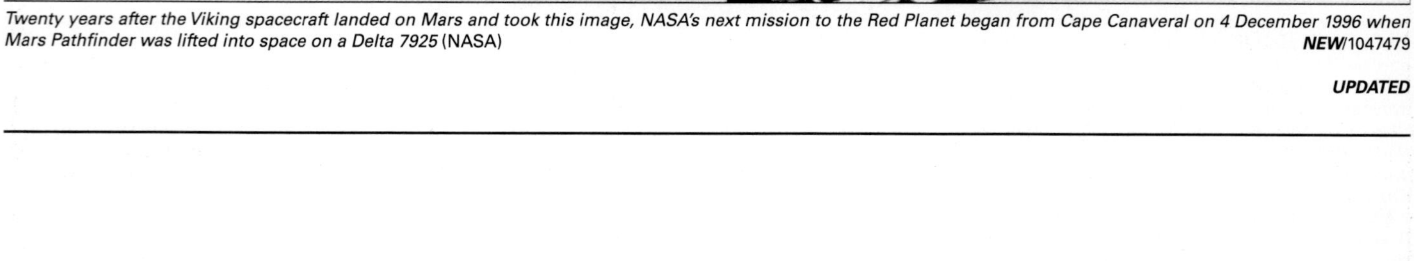

Twenty years after the Viking spacecraft landed on Mars and took this image, NASA's next mission to the Red Planet began from Cape Canaveral on 4 December 1996 when Mars Pathfinder was lifted into space on a Delta 7925 (NASA) ***NEW***/1047479

UPDATED

Table 2: Satellite Launches 1996

Desig	*Name*	*Date*	*Site*	*Vehicle*	*Mass (kg)*	*Perigee (km)*	*Apogee (km)*	*Period (min)*	*Inc (deg)*	*Notes*
1A	*Endeavour* (STS-72)	11 Jan	KSC	Shuttle	98,430	459	466	93.8	28.4	Duffy, Jett, Chiao, Scott, Wakata, Barry
						302	311	90.7	28.5	retrieved Japan's Space Flyer Unit 13 Jan and deployed/ retrieved Spartan 206/OAST-Flyer; landed KSC 20 Jan
1B	Spartan 206			Shuttle	1,198	303	312	90.7	28.5	Autonomous technology payload retrieved after 2 days
2A	PAS 3R	12 Jan	KO	Ar 44L	2,918	geostationary above 43° W				Atlantic telecommunications
2B	Measat 1				1,450	geostationary above 91.5° E				Malaysian telecommunications
3A	Koreasat 2	14 Jan	CC	Delta-2 (7925)	1,459	geosynchronous above 116° E				Korean domestic telecommunications
4A	Cosmos 2327	16 Jan	PL	Cosmos-3M	825	952	1,021	104.8	83.0	Parus military navigation, replacing C2266
5A	Gorizont 31	25 Jan	TT	Proton-K + DM-2	2,125	geosynchronous above 40° E				CIS civil telecommunications
6A	Palapa C1	1 Feb	CC	Atlas 2AS	2,989	geostationary above 113° E				Indonesian telecommunications
7A	NStar b	5 Feb	KO	Ar 44P	3,420	geostationary above 136° E				Domestic Japan telecommunications
–	Hyflex	11 Feb	TA	J1	1,040					Spaceplane test; J1 debut (2-stage suborbital)
–	Intelsat 708	14 Feb	XI	CZ-3B	4,576	failed to achieve orbit				Inertial platform failed within seconds
8A	NEAR	17 Feb	CC	Delta-2 (7925)	805	heliocentric				NASA Near-Earth Asteroid Rendezvous flew by 253 Mathilde 27 Jun 1997, due to encounter 433 Eros early 1999
9A	Gonets-D1	19 Feb	PL	Tsyklon	225	1,400	1,414	113.9	82.6	Gonets civil and Strela 3 military
9B	Gonets-D2					1,407	1,414	114.0	82.6	tactical communications
9C	Gonets-D3					1,410	1,417	114.1	82.6	
9D	Cosmos 2328					1,411	1,415	114.1	82.6	
9E	Cosmos 2329					1,412	1,422	114.2	82.6	
9F	Cosmos 2330					1,412	1,428	114.2	82.6	
10A	Raduga 36	19 Feb	TT	Proton-K + DM-2	2,000					released into GTO after stage 4 failed; CIS government/military comms
11A	Soyuz-TM 23	21 Feb	TT	Soyuz-U	7,150	200	234	88.8	51.6	Onufrienko/Usachov to Mir,
						391	398	92.5	51.6	docked 23 Feb. Undocked and landed 2 Sep, also carrying André-Deshays launched on TM24
12A	*Columbia* (STS-75)	22 Feb	KSC	Shuttle	103,370	297	303	90.5	28.5	Allen, Horowitz, Hoffman, Cheli, Nicollier,
						279	322	90.5	28.5	Chang-Diaz, Guidoni deployed/lost TSS 1R and operated USMP 3 µg payload. Returned KSC 9 Mar
12B	TSS 1R				518	316	414	91.9	28.5	Italy's tethered satellite lost when tether snapped 26 Feb; re-entered 19 Mar
13A	Polar	1 Nov	WR	Delta-2 (7925)	1,258	186	50,495	937.0	86.0	NASA fields/particles in near-Earth space
						5,141	50,605	1,051.4	86.0	
14A	REX 2	9 Mar	VAFB	L-1011/ Pegasus XL	113	804	832	101.3	90.0	USAF investigation of radio propagation; first successful Pegasus XL (third attempt)
15A	Intelsat 707	14 Mar	KO	Ar 44LP	4,175	geostationary above 1° W				International telecommunications
16A	Cosmos 2331	14 Mar	PL	Soyuz	6,500	164	358	89.7	67.1	Yantar 4th gen photoreconnaissance; recovered 11 Jun after record 89 days. Carried USAF Space Test Program Binrad experiment
17A	IRS P3	21 Mar	SR	PSLV D3	922	818	848	101.4	98.8	Indian Earth observation
18A	*Atlantis* (STS-76)	22 Mar	KSC	Shuttle	111,740	158	293	89.0	51.6	Chilton, Searfoss, Sega, Clifford, Godwin,
						337	349	91.4	51.6	Lucid docked with Mir 24 Mar, undocked 29 Mar (leaving behind Lucid to return STS-79 Aug 1996); landed EAFB 31 Mar
19A	Navstar 2–25	28 Mar	CC	Delta-2 (7925)	930	20,070	20,285	717.8	54.7	GPS Block 2A navigation satellite (USA 117)
20A	Inmarsat 3 F1	3 Apr	CC	Atlas 2A	2,068	geostationary above 64° E				International mobile telecommunications
21A	Astra 1F	8 Apr	TT	Proton-K + DM-3	3,010	geosynchronous above 19.2° E				European direct TV
22A	MSAT 1	20 Apr	KO	Ar 42P	2,855	geostationary above 106.5° W				N American mobile telecommunications
23A	Priroda	23 Apr	TT	Proton-K	19,700	214	328	89.9	51.7	Mir remote sensing module, docked 26 Apr,
						391	396	92.4	51.7	completed Mir complex after 10 yr
24A	MSX	24 Apr	WR	Delta-2 (7920)	2,680	897	906	103.0	99.4	US BMDO Midcourse Space Experiment

Desig	*Name*	*Date*	*Site*	*Vehicle*	*Mass (kg)*	*Perigee (km)*	*Apogee (km)*	*Period (min)*	*Inc (deg)*	*Notes*
25A	Cosmos 2332	24 Apr	PL	Cosmos-3M	500	295	1,565	103.6	83.0	2 m {dia} sphere for radar calibration and passive atmospheric monitoring
26A	USA 118	24 Apr	CC	Titan 401	?	geosynchronous?				Advanced Orion sigint?
27A	SAX	30 Apr	CC	Atlas 1	1,400	583	603	96.5	4.0	Italian-Dutch X-ray observatory
28A	Progress-M 31	5 May	TT	Soyuz-U	7,250	186	227	88.6	51.6	Mir supply ferry, docked 7 May, undocked and de-orbited 1 Aug
						391	396	92.4	51.6	
29A	USA 119	12 May	WR	Titan 4		1,023	1,023		63.4	probably PARCAE ocean surveillance. Three subsats designated (USA 120/121/122) and 53 kg TiPS Tether Physics & Survivability expt released 20 Jun into orbit typical for PARCAE
–	Cosmos	14 May	TT	Soyuz-U	6,500	failed to reach orbit				Kometa 4th gen topographic/mapping photoreconn, including commercial SPIN-2 for US companies, but shroud failed at 49 s
30A	Palapa C2	16 May	KO	Ar 44L	2,989	geostationary above 108° E				Indonesian telecommunications
30B	Amos 1				996	geostationary above 4° W				Israel/Middle East telecommunications
31A	MSTI 3	17 May	VAFB	L-1011/ Pegasus XL	211	291	365	91.1	97.0	BMDO sensor technology demonstration; last standard Pegasus
						422	435	93.1	97.1	
32A	*Endeavour* (STS-77)	19 May	KSC	Shuttle	115,600	280	290	90.2	39.0	Casper, Brown, Thomas, Bursch, Runco, Garneau operated Spacehab 4; deployed/retrieved Spartan 207/Inflatable Antenna Expt; deployed PAMS; landed KSC 29 May
32B	Spartan 207/IAE				1,296	282	294	90.3	39.0	Autonomous payload deployed 20 May, inflated 14 m {dia} antenna (released after 90 min), Spartan retrieved 21 May
32D	PAMS				52	278	288	90.2	39.0	Passive Aerodynamically Stabilised Magnetically-Damped Satellite, ejected 22 May
33A	Galaxy 9	24 May	CC	Delta-2 (7925)	1,397	geosynchronous above 123° W				US Hughes commercial communications
34A	Gorizont 32	25 May	TT	Proton-K	2,125	geosynchronous above 52° E				CIS civil telecommunications
–	Cluster F1	4 Jun	KO	Ar 5	1,183	failed to reach orbit				Maiden flight of ESA's Ariane 5 ended after control software problem caused severe attitude failure from 37 s and vehicle breakup. Satellites were to be released into 10° GTO before raising themselves to 25,513 × 140,318 km, 90° to study 3D structure of Earth's plasma environment
	Cluster F2				1,169					
	Cluster F3				1,171					
	Cluster F4				1,184					
35A	Intelsat 709	15 Jun	KO	Ar 44P	3,420	geostationary above 18° W				International telecommunications
36A	*Columbia* (STS-78)	20 Jun	KSC	Shuttle	116,200	273	286	90.1	39.0	Henricks, Kregel, Helms, Linnehan, Brady, Favier Thirsk undertook record-duration LMS Life & Microgravity Sciences Spacelab mission landed KSC, 7 Jul
						268	278	90.0	39.0	
–	Cosmos	20 Jun	PL	Soyuz-U	6,500	failed to reach orbit				Yantar 4th gen photoreconnaissance, but Soyuz shroud failed after 49 s
37A	TOMS-EP	2 Jul	VAFB	L-1011/ Pegasus XL	294	345	953		97.4	NASA's Total Ozone Mapping Spectrometer
						502	502		97.4	
38A	USA 125	3 Jul	CC	Titan 4	?	orbital data not available; inc >55°				DoD classified; trade reports suggest advanced SDS Satellite Data System (orbit about 400 × 40,000 km, 63°) for relaying imagery from photorecon satellites. Amateur observations showed manoeuvres into a high orbit
39A	APStar 1A	3 Jul	XI	CZ-3	1,400	geostationary above 77° E				Hong Kong regional telecommunications
40A	ARABSAT 2A	9 Jul	KO	Ar 44L	2,617	geostationary above 26° E				Arab League communications
40B	TURKSAT 1C				1,743	geostationary above 42° E				Turkish Telecom communications
41A	Navstar-2 26	16 Jul	CC	Delta-2 (7925)	1,881	20,138	20,224	717.93	55.0	GPS Block 2A navigation: USA 126

Desig	Name	Date	Site	Vehicle	Mass (kg)	Perigee (km)	Apogee (km)	Period (min)	Inc (deg)	Notes
42A	UFO 7	25 Jul	CC	Atlas-2	3,020	geostationary above 23° W				US Navy communications: USA 127
43A	Progress-M 32	31 Jul	TT	Soyuz-U	7,250?	186	229	88.7	51.6	Mir supply ferry: docked 2 Aug, undocked 18 Aug,
						375	391	92.2	51.6	redocked 3 Sep, undocked and de-orbited 21 Nov
44A	ITALSAT 2	8 Aug	KO	Ar 44L	1,990	geostationary above 10° E				Italian communications, carries EMS package
44B	Telecom 2D				2,260	geostationary above 5° W				French civil and military communications
45A	Molniya-1 89	14 Aug	PL	Molniya-M	1,600?	464	39,887	717.7	62.8	Orbita system communications
46A	Midori 1	17 Aug	TA	H-2	3,560?	797	799	100.8	98.6	Japanese remote sensing; ADEOS 1 before launch; power failure June 1997
47A	Soyuz-TM 24	17 Aug	TT	Soyuz-U	7,150?	233	292	89.8	51.6	Korzun, Kaleri, André-Deshays (CNES); docked with Mir
						375	390	92.2	51.6	19 Aug; André-Deshays returned to Earth aboard TM23 2 Sep; Korzun, Kaleri and Linenger undocked from Mir 7 Feb 1997 and re-docked at the opposite end of Mir the same day; Korzun, Kaleri and Ewald undocked and returned to Earth 2 Mar 1997
48A	Zhongxing 7	18 Aug	XI	CZ-3	1,200?	200	17,229	307.5	27.3	China Star communications; third stage shut down early, mission abandoned
49A	FAST	21 Aug	WR	L-1011/ Pegasus-XL	180	351	4,165	133.1	83.0	Fast Auroral Snapshot Explorer, science
50A	MUSAT	29 Aug	PL	Molniya-M	30	236	1,171	98.8	62.8	Argentine technology satellite
50B	MAGION 5				62	791	19,196	347.5	62.8	Czech Republic solar/ magnetosphere interaction research
50C	Interball 2				1,300?	769	19,211	347.3	62.8	Auroral Probe, international magnetosphere, plasmasphere research satellite
51A	Cosmos 2333	4 Sep	TT	Zenit-2	3,250?	849	852	101.9	71.0	Tselina-2 ELINT
52A	Cosmos 2334	5 Sep	PL	Cosmos-3M	825?	972	1,001	104.8	82.9	Parus navigation; co-planar with cosmos 2327
52B	UNAMSAT 2				17	968	1,011	104.9	82.9	Mexican micrometeorite research satellite
53A	INMARSAT-3 2	6 Sep	TT	Proton-K + DM-1	1,144	geostationary above 15° W				Maritime communications
54A	GE 1	8 Sep	CC	Atlas-2A	2,764	geostationary above 103° W				GE Americom communications
55A	EchoStar 2	11 Sep	KO	Ar 42P	2,865	geostationary above 119° W				EchoStar Communications Corporation communications
56A	Navstar 27	12 Sep	CC	Delta-2 (7925)	1,881	20,057	20,307	718.0	54.7	GPS Block 2A navigation: USA 128
57A	*Atlantis* (STS-79)	16 Sep	KSC	Shuttle	113,105	372	378	92.1	51.6	Readdy, Wilcutt, Apt, Akers, Walz and Blaha on fourth Shuttle-Mir mission: Blaha remained on Mir until STS-81, Lucid returned aboard Shuttle. Docked with Mir 19 Sep, undocked 23 Sep, landed KSC 26 Sep
58A	Ekspress 2	26 Sep	TT	Proton-K + DM-2	2,500	geostationary above 80° E				Communications
59A	FSW-2 3	20 Oct	JQ	CZ-2D	2,600?	171	342	89.6	63.0	Microgravity, remote sensing; capsule recovered 4 Nov, bus decayed
60A	Molniya-3 48	24 Oct	PL	Molniya-M	1,750?	610	39,768	718.3	62.8	Orbita system communications
61A	SAC-B + HETE + Pegasus 3rd stage	4 Nov	WI	L-1011/ Pegasus-XL	480?	488	556	95.1	38.0	SAC-B Argentine, HETE US: both satellites failed to separate from Pegasus-XL final stage once in orbit
62A	Mars Global Surveyor	7 Nov	CC	Delta-2 (7925)	1,060	heliocentric orbit				Planned to reach Mars orbit 11 Sep 1997
63A	ARABSAT 2B	13 Nov	KO	Ar 44L	2,661	geostationary above 30° E				Arab League communications
63B	MEASAT 2				1,512	geostationary above 148° E				Binariang, Malaysia, communications
64	Mars 8 (Mars-96)	16 Nov	TT	Proton-K + D-2	6,825	139	155	87.4	51.5	Mars probe, failed to leave Earth orbit; first orbit
						87?	1,500?	100.7?	51.6?	is for the fourth stage, second one for the spacecraft (approximate orbit); spacecraft not assigned an International Designator
65A	*Columbia* (STS-80)	19 Nov	KSC	Shuttle	102,946	347	358	91.6	28.5	Cockrell, Rominger, Jernigan, Jones and Musgrave; landed KSC 7 Dec
65B	ORFEUS-SPAS 2				3,573	349	356	91.6	28.5	Astrophysics payload, deployed 20 Nov, recaptured 4 Dec

Desig	*Name*	*Date*	*Site*	*Vehicle*	*Mass (kg)*	*Perigee (km)*	*Apogee (km)*	*Period (min)*	*Inc (deg)*	*Notes*
65C	WSF 3				2,109	347	359	91.6	28.5	Wake Shield Facility microgravity payload, deployed 22 Nov, re-captured 26 Nov
66A	Progress-M 33	19 Nov	TT	Soyuz-U	7,150?	250	309	90.1	51.6	Cargo freighter to Mir, docked 22 Nov, undocked 6 Feb 1997,
						371	390	92.2	51.6	re-docking 4 Mar 1997 failed and was de-orbited 12 Mar 1997
67A	Hot Bird 2	21 Nov	CC	Atlas-2A	2,912	geostationary above 13° E				Eutelsat communications
68A	Mars Pathfinder	4 Dec	CC	Delta-2 (7925)	890	heliocentric orbit				Landed on Mars 4 Jul 1997, deployed Sojourner rover
69A	Cosmos 2335	11 Dec	TT	Tsyklon-M	3,000?	403	419	92.8	54.0	EORSAT
70A	INMARSAT-3 3	18 Dec	CC	Atlas-2A	2,074	geostationary above 178° E				Maritime communications
71A	Cosmos 2336	20 Dec	PL	Cosmos-3M	825?	979	1,012	105.0	83.0	Parus navigation: co-planar with Cosmos 2173
72A	USA 129	20 Dec	WR	Titan-4	13,000?	orbital data classified				Advanced Crystal photoreconnaissance
73A	Bion 11	24 Dec	PL	Soyuz-U	6,000?	217	379	90.5	62.8	Biological research with two monkeys: US experiments carried: recovered 7 Jan 1997

Table 2: Satellite Launches 1997

Desig	Name	Date	Site	Vehicle	Mass (kg)	Perigee (km)	Apogee (km)	Period (min)	Inc (deg)	Notes
1A	*Atlantis* (STS-81)	12 Jan	KSC	Shuttle	113,370	370	390	92.2	51.6	Baker, Jett, Wisoff, Grunsfeld, Ivins and Linenger on fifth Shuttle-Mir mission: Linenger remained on Mir until STS-84, Blaha returned to Earth aboard the Shuttle. Docked with Mir 15 Jan, undocked 20 Jan, landed at KSC 22 Jan
–	Navstar 28	17 Jan	CC	Delta-2 (7925)	2,032	failed to reach orbit				First launch of a GPS Block 2R: launch vehicle self-destructed 21 seconds after launch (first Delta-2 to fail to reach orbit)
2A	GE 2	30 Jan	KO	Ar 44L	2,469	geostationary above 85° W				GE Americom communications
2B	Nahuel 1A				1,790	geostationary above 72° W				Nahuelsat, Argentina, communications
3A	Soyuz-TM 25	10 Feb	BA	Soyuz-U	7,150?	188	231	88.7	51.6	Tsibliev, Lazutkin and Ewald (Germany): docked with Mir
						378	394	92.3	51.6	12 Feb: Ewald returned to Earth aboard TM24, Tsibliev and Lazutkin aboard TM25 in August
4A	*Discovery* (STS-82)	11 Feb	KSC	Shuttle	96,758	589	594	96.5	28.5	Bowersox, Horowitz, Tanner, Hawley, Harbaugh, Lee and Smith: second Hubble Space Telescope servicing mission. HST captured 13 Feb, released 19 Feb: shuttle returned to KSC 21 Feb
5A	Haruka	12 Feb	KA	M-5	823	247	21,409	374.4	31.3	Radio telescope; space segment of VSOP
	(MUSES-B)					573	21,402	379.6	31.3	
6A	Gonets-D 4	14 Feb	PL	Tsyklon	225?	1,413	1,423	114.2	82.6	Civil communications; it is not certain which of the six satellites
6B	Gonets-D 5				225?	1,413	1,429	114.2	82.6	are the Gonets payloads and which are the Cosmos
6C	Gonets-D 6				225?	1,409	1,414	114.0	82.6	Strela-3 payloads
6D	Cosmos 2337				225?	1,402	1,414	114.0	82.6	Strela-3 military communications
6E	Cosmos 2338				225?	1,413	1,415	114.1	82.6	
6F	Cosmos 2339				225?	1,412	1,415	114.1	82.6	
7A	JCSAT 4	17 Feb	CC	Atlas-2AS	3,105	geostationary above 124° E				Japan Satellite Systems Inc communications
8A	DSP 18	24 Feb	CC	Titan-4B + IUS	2,400?	geostationary orbit				Early warning satellite; orbital data classified; maiden flight of Titan-4B; USA 130
9A	INTELSAT 801	1 Mar	KO	Ar 44P	3,420	geostationary above 62° E				International telecommunications
10A	Zeya	4 Mar	SV	Start-1	87	467	480	94.1	97.3	Geodetic satellite; maiden launch from Svobodny site
11A	Tempo 2	8 Mar	CC	Atlas-2A	3,561	geostationary above 119° W				Tele-Communications Inc communications
12A	DMSP-2 9	4 Apr	WR	Titan-23G	823	844	855	101.9	98.9	Block 5D-2 military meteorological satellite
13A	*Columbia* (STS-83)	4 Apr	KSC	Shuttle	96,643	299	304	90.5	28.5	Halsell, Still, Voss, Gernhardt, Thomas, Crouch and Linteris; first Microgravity Science Laboratory mission; flight curtailed because of fuel cell problems and re-flown as STS-94 in July; landed KSC 8 Apr
14A	Progress-M 34	6 Apr	BA	Soyuz-U	7,156	187	227	88.6	51.6	Cargo freighter; docked with Mir 8 Apr, undocked 24 Jun: re-docking attempt 25 Jun failed and spacecraft collided with Spektr module, the module's vanes of solar cells and puncturing its shell; Progress de-orbited 2 Jul
15A	Cosmos 2340	9 Apr	PL	Molniya-M	1,250?	541	39,815	717.8	62.9	Oko early warning
16A	THAICOM 3	16 Apr	KO	Ar 44LP	2,650	geostationary above 78.5° E				Shinawatra Satellite, Thailand, communications
16B	B-SAT 1A				1,236	geostationary above 110° E				B-SAT, Japan, communications
17A	Cosmos 2341	17 Apr	PL	Cosmos-3M	825?	978	1,014	105.0	82.9	Parus navigation, co-planar with Cosmos 2310
18A	MINISAT 01	21 Apr	GA	L-1011/ Pegasus-XL	209	563	582	96.1	151.0	Spanish technology satellite: Pegasus third stage carried Celestis 1 payloads, carrying cremated ashes of 24 people: first launch from Gando AFB, Canary Islands
19A	GOES 10	25 Apr	CC	Atlas-1	2,105	geostationary above 105° W				NOAA meterological satellite: final launch of Atlas-1
20A	Iridium 8	5 May	WR	Delta-2 (7920)	657	773	783	100.4	86.4	First launch within the Iridium communications constellation; Iridium 1–3 were non-flight vehicles
20B	Iridium 7				657	773	782	100.4	86.4	
20C	Iridium 6				657	771	783	100.4	86.4	
20D	Iridium 5				657	769	787	100.4	86.4	
20E	Iridium 4				657	771	784	100.4	86.4	
21A	Zhongxing 8	11 May	XI	CZ-3A	2,230?	geostationary above 125° E				Second DFH-3 communications
22A	Cosmos 2342	14 May	PL	Molniya-M	1,250?	525	39,825	717.7	62.8	Oko early warning
23A	*Atlantis* (STS-84)	15 May	KSC	Shuttle	100,284	371	396	92.2	51.6	Precourt, Collins, Clervoy, Noriega, Lu, Kondakova and Foale on sixth Shuttle-Mir mission; Foale remained on board Mir until STS-86, Linenger returned to Earth aboard the Shuttle; docked with Mir 17 May, undocked 22 May, landed KSC 24 May

Desig	*Name*	*Date*	*Site*	*Vehicle*	*Mass (kg)*	*Perigee (km)*	*Apogee (km)*	*Period (min)*	*Inc (deg)*	*Notes*
24A	Cosmos 2343	15 May	BA	Soyuz-U	7,000?	197	292	89.4	64.9	Photoreconnaissance, probably sixth generation type
						203	343	90.0	64.9	
–	Cosmos	20 May	BA	Zenit-2	3,250	failed to reach orbit				Tselina-2 ELINT: first stage engine shut down 85 seconds after launch
25A	Thor 2	21 May	CC	Delta-2 (7925)	1,300?	geostationary above 1° W				Telenor Satellite Services AS, Norway, communications
26A	Telstar 5	24 May	BA	Proton-K + DM4-1L	3,650	geostationary above 97° W				Loral Skynet communications
27A	INMARSAT-3 4	3 Jun	KO	Ar 44L	1,999	geostationary above 54° W				Maritime communications
27B	INSAT 2D				2,079	geostationary above 74° E				ISRO communications
28A	Cosmos 2344	6 Jun	BA	Proton-K + DM-5	5,000?	1,509	2,732	130.0	63.4	New class of optical reconnaissance platform
29A	Feng Yun-2 1R	10 Jun	XI	CZ-3	1,380	geostationary above 105° E				First Chinese geosynchronous meteorological satellite: first FY-2 1 lost in pre-launch accident, Apr 1994
30A	Iridium 14	18 Jun	BA	Proton-K + DM-5	657	555	560	95.8	86.4	Second launch of Iridium cluster, first to use Proton-K: orbits are as at the end of Jun 1997
30B	Iridium 12				657	608	616	96.9	86.4	
30C	Iridium 10				657	555	558	95.8	86.4	
30D	Iridium 9				657	560	561	95.9	86.4	
30E	Iridium 13				657	580	583	96.3	86.4	
30F	Iridium 16				657	585	591	96.4	86.4	
30G	Iridium 11				657	670	676	98.2	86.4	
31A	INTELSAT 802	25 Jun	KO	Ar 44P	3,435	geostationary above 174° E				International communications
32A	*Columbia* (STS-94)	1 Jul	KSC	Shuttle	117,801	300	304	90.6	28.5	Halsell, Still, Voss, Gernhardt, Thomas, Crouch and Linteris refly Microgravity Sciences Laboratory returned to earth early on STS-8 in Apr 1997. Landed 17 Jul
33A	Progress-M 35	5 Jul	TT	Soyuz-U	7,250	386	392	92.3	51.6	Mir supply
34A	Iridium 15	9 Jul	VAFB	Delta 2	657	774	780	100.4	86.4	Third Iridium cluster launch
34B	Iridium 17				657	775	780	100.4	86.4	
34C	Iridium 18				657	774	779	100.4	86.4	
34D	Iridium 20				657	772	782	100.4	86.4	
34E	Iridium 21				657	629	643	97.4	86.4	Failed to reach op orbit
35A	USA 132	23 Jul	CC	Delta 2	2,032	20,124	20,247	718.1	54.9	First successful Navstar Block 2R
36A	Superbird C	28 Jul	CC	Atlas AS2	3,130	geostationary above 143° E				HS 601 for Space Communication Corp of Japan
37A	Seastar	1 Aug	VAFB	L-1011/ Pegasus-XL	309	297 707	313 709	90.6 98.9	98.3 98.2	Remote sensing ocean satellite with SeaWifs sensor
38A	Soyuz-TM 26	5 Aug	TT	Soyuz-U	7,150	386	392	92.3	51.6	Solovyov and Vinogradov to Mir, relieving Tsibliev and Lazutkin who returned to earth in Soyuz-TM 25 leaving Foale on board
39A	*Discovery* (STS-85)	7 Aug	KSC	Shuttle	98,953	290	301	90.4	57.0	Brown, Reminger, Davis, Curbean, Robinson, Tryggvanson on earth science and ISS support mission. Landed 19 Aug
39B	CHRISTA-SPAS 2				3,504	291	301	90.4	57.0	Platform with cryogenic IR spectrometer deployed 7 Aug recaptured 16 Aug
40A	PAS 6	8 Aug	KO	Ar44P	3,420	geostationary above 316.5° E				DBS for PanAmSat
41A	Cosmos 2345	14 Aug	TT	Proton-K	2,500	geostationary above 336° E				Prognoz early warning
42A	Mabuhay Agila	19 Aug	XI	CZ-3B	2,560	geostationary above 147° E				Communications sat for Mabuhay Philippine Satellite Corporation
43A	Iridium 26	21 Aug	VAFB	Delta 2	657	765	763	100.2	86.4	Fourth Iridium cluster launched
43B	Iridium 25				657	775	780	100.4	86.4	
43C	Iridium 24				657	774	781	100.4	86.4	
43D	Iridium 23				657	775	780	100.4	86.4	
43E	Iridium 22				657	775	780	100.4	86.4	
44A	Lewis (SSTI)	23 Aug	VAFB	LMLV-1	386	298	320	90.7	97.6	Experimental TRW Small Spacecraft Technology Initiative satellite
45A	ACE	25 Aug	CC	Delta 2	752	177	1,367,769	95,140	28.7	Advanced Composition Explorer to measure heliospheric matter
46A	PAS 5	28 Aug	TT	Proton-K	3,720	geostationary over 302° E				HS 601P for PanAmSat
47A	FORTE (P94-1)	29 Aug	VAFB	L-1011/ Pegasus-XL	215	800	834	101.2	79.0	Detects and measures radio frequency pulses for US Dept of Energy
48A	Iridium MFS 1	1 Sep	TY	CZ-2C/SD	650	623	632	97.3	86.4	Test flight of new upper stage
48B	Iridium MFS 2				650	623	633	97.3	86.4	
49A	Hot Bird 3	2 Sep	KO	Ar44LP	2,915	geostationary over 13° E				DBS satellite for Eutelsat
49B	METEOSAT 7				703	geostationary over 349° E				Eumetsat satellite
50A	GE 3	4 Sep	CC	Atlas 2AS	2,750	geostationary over 276° E				For GE American Communications
51A	Iridium 29	14 Sep	TT	Proton-K	657	776	779	100.4	86.4	Fifth Iridium cluster
51B	Iridium 32				657	775	779	100.4	86.4	
51C	Iridium 33				657	776	779	100.4	86.4	
51D	Iridium 27				657	554	558	95.8	86.6	
51E	Iridium 28				657	775	780	100.4	86.4	
51F	Iridium 30				657	774	780	100.4	86.4	
51G	Iridium 31				657	776	779	100.4	86.4	
52A	Cosmos 2346	23 Sep	PL	Cosmos-3M	795	939	996	104.4	82.9	Parus military navigation satellite
52B	FAISAT 2V				120	939	996	104.4	82.9	Store-forward communication satellite
53A	Intelsat 803	24 Sep	KO	Ar42L		geostationary over 333.5° E				Intelsat telecommunications satellite
54A	Molniya-1 90	24 Sep	PL	Molniya-M	3,455	449	39,906	717.8	0.1	Molniya-IT series; coplanar with Molniya-1 84
55A	*Atlantis* (STS-86)	26 Sep	KSC	Shuttle	114,184	256	298	90.1	51.7	Wetherbee, Bloomfield, Titov Parazynski, Chretien, Lawrence, Wolf on 7th Mir visit. Wolf replaced Foale on Mir who returned with the STS-86 crew 6 Oct
56A	Iridium 19	27 Sep	VAFB	Delta 2	657	775	779	100.4	86.4	Sixth Iridium launch
56B	Iridium 37				657	774	779	100.4	86.4	
56C	Iridium 36				657	641	652	97.7	86.4	
56D	Iridium 35				657	774	781	100.4	86.4	
56E	Iridium 34				657	775	780	100.4	86.4	

Desig	Name	Date	Site	Vehicle	Mass (kg)	Perigee (km)	Apogee (km)	Period (min)	Inc (deg)	Notes
57A	IRS 1D	28 Sep	SR	PSLV	1,250	308	822	95.9	98.6	ISRO remote sensing but spacecraft left in wrong orbit due to stage 4 malfunction. Perigee raised using satellite propellant as shown
						739	826	100.5		
58A	Progress-M 36	4 Oct	TT	Soyuz-U	7,250	382	392	92.3	51.6	Mir cargo supply flight
58C	Sputnik model				50	385	389	92.3	51.6	Sputnik model launched from Progress
59A	EchoStar 3	4 Oct	CC	Atlas 2AS	2,900	geostationary over 298° E				DBS for EchoStar Communications Corp
60A	Foton 8 (F11)	9 Oct	PL	Soyuz-U	5,800	218	375	90.4	62.8	Microgravity research. Capsule with samples landed 23 Oct
61A	Cassini/Huygens	15 Oct	KSC	Titan 4B/ Centaur	5,712	Heliocentric				Saturn orbiter by way of two fly-bys of Venus and one fly-by of Earth, Jupiter and Phoebe arriving Nov 2004
62A	APStar 2R	16 Oct	XI	CZ-3B	3,747	geostationary over 76° E				Replaced APStar 2 for APT Satellite Co.
63A	STEP M4	22 Oct	WI	L-1011/ Pegasus XL	395	434	501	93.5	45.0	Near-earth space properties satellite. Failed when solar panels did not deploy
64A	USA 133	24 Oct	VAFB	Titan 4A	14,500	673	674	98.2	57.0	Third Lacrosse/Vega launch. Orbit shown is unofficial
65A	USA 134	25 Oct	CC	Atlas 2A	1,040	geostationary				DSCS-3B 5, improved type
66A	MAQSAT-H/ TEAMSAT	30 Oct	KO	Ar 5 (502)	2,650	533	26,635	467.8	7.8	Technology and debris monitor but prevented from achieving 36,000 km apogee by second stage malfunction
66B	MAQSAT-B/EPS				4,100	533	26,569	466.6	7.8	Remained attached to second stage
66C	YES				60	540	26,626	467.8	7.8	Young Engineers Satellite planned for tether deploy to become independent satellite
67A	USA 135	6 Nov	CC	Delta 2	1,881	19,912	20,449	358.3	54.9	NAVSTAR 38
68A	USA 136	8 Nov	CC	Titan 4A/ Centaur	5,000	1,100	39,060	713.8	63.6	ELINT Trumpet satellite
69A	Iridium 43	9 Nov	VAFB	Delta 2	657	774	780	100.4	86.4	Seventh cluster launch
69B	Iridium 41				657	772	782	100.4	86.4	
69C	Iridium 40				657	768	770	100.2	86.4	
69D	Iridium 39				657	773	780	100.4	86.4	
69E	Iridium 38				657	768	771	100.2	86.4	
70A	Kupon 1	12 Nov	TT	Proton-K	2,500	geostationary above 55° E				First satellite for Bankir system
71A	Sirius 2	12 Nov	KO	Ar44L	2,920	geostationary above 5° E				Communications satellite for NSAB, Stockholm
71B	Cakrawarta				1,385	geostationary above 107.7° E				DBS for Media Citra, Indonesia
72A	Resurs-F 21	18 Nov	PL	Soyuz-U	6,300	180	236	88.7	82.3	First flight of improved F-1M
						238	270	88.6	82.3	
73A	*Columbia* (STS-87)	20 Nov	KSC	Shuttle	102,717	281	286	90.2	28.5	USMP-4 microgravity payload with Kregel, Lindsey, Chawla, Scott, Doi and Kadenyuk. Landed 5 Dec
73B	Spartan 201–04				1,352	280	285	90.2	28.5	Deployed 21 Nov, retrieved 22 Nov
74A	TRMM	27 Nov	TA	H-2	3,620	367	385	92.7	35.0	NASA Tropical Rainfall Mission
74B	Hikobashi/Orihame				2,900	379	538	93.8	35.0	Kiku-7 (ETS 7 before launch)
75A	JCSAT 5	2 Dec	KO	AR44P	2,982	geostationary over 139° E				Telecommunications satellite for Japan Satellite Systems Inc. To operate from 150° E
75B	Equator-S				230	496	67,232	1,339	4.0	Science satellite from Max Planck Inst, Germany
76A	Astra 1G	2 Dec	TT	Proton-K	3,300	geostationary over 19.2° E				DBS satellite for SES, Luxembourg
77A	Iridium 42	8 Dec	TY	CZ-2C/SD	657	775	779	100.4	86.4	Eighth Iridium cluster
77B	Iridium 44				657	775	779	100.4	86.4	
78A	Galaxy 8I	8 Dec	CC	Atlas 2AS	3,000	geostationary over 280° E				Telecommunications satellite operated by Hughes Communications Inc
79A	Cosmos 2347	9 Dec	TT	Tsyklon-M	3,000	403	419	92.8	65.0	ELINT ocean reconnaissance satellite
80A	Cosmos 2348	15 Dec	PL	Soyuz-U	6,600	165	345	89.6	67.1	Yantar-4K fourth-generation reconnaissance satellite
81A	Progress-M 37	20 Dec	TT	Soyuz-U	7,195	378	389	92.3	51.7	Mir cargo-tanker supply ship, docked 22 Dec. Re-entered 15 Mar 1998
82A	Iridium 45	20 Dec	VAFB	Delta 2	657	775	778	100.4	86.4	Ninth Iridium cluster launch
82B	Iridium 46				657	775	780	100.4	86.4	
82C	Iridium 47				657	776	778	100.4	86.4	
82D	Iridium 48				657	775	779	100.4	86.4	
82E	Iridium 49				657	716	720	100.4	86.4	
83A	Intelsat 804	22 Dec	Ko	Ar42L	3,455	geostationary over 64° E				Telecommunications satellite for Intelsat
84A	Orbcomm FM 8	23 Dec	WI	L-1011/Pegasus	43	815	823	101.3	45	First cluster for Orbcomm Global LP
84B	Orbcomm FM 10				43	814	823	101.3	45	
84C	Orbcomm FM 11				43	815	823	101.3	45	
84D	Orbcomm FM 12				43	815	823	101.3	45	
84E	Orbcomm FM 9				43	815	823	101.3	45	
84F	Orbcomm FM 5				43	816	827	101.3	45	
84G	Orbcomm FM 6				43	815	826	101.3	45	
84H	Orbcomm FM 7				43	816	826	101.3	45	
85A	Early Bird 1	24 Dec	SV	Start-1	300	480	488	101.3	97.3	Commercial remote sensing satellite for EarthWatch Inc.
86A	AsiaSat 3	24 Dec	TT	Proton-K	3,401	301	35,999	638.3	51.1	Telecommunications satellite for Asia Satellite Telecommunications Co Ltd, Hong Kong

Table 2: Satellite Launches 1998

Desig	Name	Date	Site	Vehicle	Mass (kg)	Perigee (km)	Apogee (km)	Period (min)	Inc (deg)	Notes
01A	Lunar Prospector	7 Jan	CC	Athena-2	233	translunar/heliocentric				Geochemical lunar orbiter
02A	Skynet 4D	10 Jan	CC	Delta 2	1,493	geostationary over 6° E				UK military communications satellite
03A	*Endeavour* (STS-89)	23 Jan	KSC	Shuttle	114,130	378	387	92.2	51.7	Wilcutt, Edwards, Reilly, Anderson, Dunbar, Sharipov and Thomas, replacing Wolff on Mir. Landed 31 Jan
04A	Soyuz-TM 27	28 Jan	TT	Soyuz-U	7,150	379	385	92.2	51.7	Musabayev, Budarin and Eyharts, the latter returning with Solovyov and Vinogradov in Soyuz-TM 26
05A	USA 137	29 Jan	CC	Atlas 2A	2,500	200	39,500		63.0	Capricorn classified successor to the DSP data relay satellites for the National Reconnaissance Office
06A	Brasilsat B3	4 Feb	KO	Ar44LP	1,780	geostationary over 295° E				Telecommunications satellite for EBRATEL (Brazil)
06B	Inmarsat-3 5				2,000	geostationary over 25° E				Mobile communications satellite for Inmarsat
07A	GFO	10 Feb	VAFB	Taurus	347	780	878	101.5	108.0	Geosat Follow-On radar altimeter satellite operated by Naval Meteorology and Oceanographic Command
07B	Orbcomm FM 3				43	781	876	101.5	108.0	Orbcomm operated by Orbcomm Global LP
07C	Orbcomm FM 4				43	781	874	101.5	108.0	
08A	Globalstar 1	14 Feb	CC	Delta 2	450	1,244	1,259	110.5	52.0	First launch for Globalstar Ltd Partnership
08B	Globalstar 2				450	1,237	1,254	110.4	52.0	
08C	Globalstar 3				450	1,240	1,260	110.5	52.0	
08D	Globalstar 4				450	1,244	1,256	110.5	52.0	
09A	Cosmos 2349	17 Feb	TT	Soyuz-U	6,600	212	278	89.4	70.4	Yantar-1KFT/Kometa topographic and mapping photo-reconnaissance satellite
10A	Iridium 50	18 Feb	VAFB	Delta 2	657	723	726	99.3	86.5	Tenth Iridium cluster
10B	Iridium 51				657	768	773	100.2	86.4	
10C	Iridium 52				657	699	694	98.6	86.5	
10D	Iridium 53				657	766	771	100.2	86.4	
10E	Iridium 54				657	774	780	100.4	86.4	
11A	Kakehashi	21 Feb	TA	H-2	3,960	247	1,883	106.5	30.1	Broadcast research and development test satellite
12A	SNOE	26 Feb	VAFB	Pegasus-XL	132	535	581	95.8	97.8	Student Nitric Oxide Explorer operated by University of Colorado
12B	Teledisc T-1				45	535	580	95.8	97.7	Test satellite for Teledisc communications satellite system
13A	Hot Bird 4	27 Feb	KO	Ar42P	2,885	geostationary orbit at 13° E				Telecommunications satellite for Eutelsat
14A	Intelsat 806	28 Feb	CC	Atlas 2AS	3,400	geostationary orbit at 304° E				Telecommunications satellite for Intelsat
15A	Progress-M 38	14 Mar	TT	Soyuz-U	7,200	376	383	92.1	51.7	Cargo freighter to Mir
16A	USA 138	16 Mar	CC	Atlas 2	3,000	geostationary orbit at 188° E				UHF Follow-On military communications satellite
17A	SPOT 4	24 Mar	KO	Ar40	2,755	824	826	101.4	98.7	CNES remote sensing satellite
18A	Iridium 51	25 Mar	TY	CZ-2C/SD	657	775	780	100.4	86.4	Eleventh Iridium cluster launch
18B	Iridium 61				657	687	692	98.6	86.5	
19A	Iridium 55	30 Mar	VAFB	Delta 2	657	728	731	99.4	86.5	Twelfth Iridium cluster launch
19B	Iridium 57				657	665	668	98.1	86.6	
19C	Iridium 58				657	666	668	98.1	86.6	
19D	Iridium 59				657	687	696	98.6	86.6	
19E	Iridium 60				657	675	676	98.3	86.6	
20A	TRACE	2 Apr	VAFB	Pegasus-XL	250	599	641	97.1	97.8	Transition Regional & Coronal Explorer
21A	Iridium 62	7 Apr	TT	Proton-K	657	776	779	100.4	86.4	Thirteenth Iridium cluster of satellites
21B	Iridium 63				657	776	779	100.4	86.4	
21C	Iridium 64				657	775	779	100.4	86.4	
21D	Iridium 65				657	775	778	100.4	86.4	
21E	Iridium 66				657	772	782	100.4	86.4	
21F	Iridium 67				657	774	780	100.4	86.4	
21G	Iridum 68				657	775	778	100.4	86.4	
22A	*Columbia* (STS-90)	17 Apr	KSC	Shuttle	105,500	257	286	89.5	39.0	Searfoss, Altman, Linnehan, Hire, Williams, Buckey and Pawelczyk on Neurolab mission. Landed 3 May
23A	Globalstar 5	24 Apr	CC	Delta 2	450	1,258	1,265	110.7	52.0	Second launch of Globalstar cluster
23B	Globalstar 6				450	1,256	1,270	110.8	52.0	
23C	Globalstar 7				450	1,260	1,263	110.8	52.0	
23D	Globalstar 8				450	1,259	1,269	110.8	52.0	
24A	Nilesat 101	29 Apr	KO	Ar44P	1,840	geostationary orbit at 353° E				First satellite to be launched for an African country, Nilesat of Egypt
24B	B-Sat 1B				1,230	geostationary orbit at 110° E				Telecommunications satellite for B-SAT Corporation of Japan
25A	Cosmos 2350	29 Apr	TT	Proton-K	2,500	geostationary orbit at 80° E				Data relay satellite
26A	Iridium 69	2 May	TY	CZ-2C/SD	657	625	641	97.4	86.4	Fourteenth cluster of Iridium satellites
26B	Iridium 71				657	774	780	100.4	86.4	
27A	Cosmos 2351	7 May	PL	Molniya-M	1,250	525	39,829	717.7	63.0	Oko early warning satellite

Desig	Name	Date	Site	Vehicle	Mass (kg)	Perigee (km)	Apogee (km)	Period (min)	Inc (deg)	Notes
28A	EchoStar 4	7 May	TT	Proton-K	2,900	geostationary orbit at 233° E				DBS for EchoStar Communications Corp. Operated over 212° E
29A	USA 139	9 May	CC	Titan 4B/ Centaur	4,500	geostationary orbit				Advanced Orion ELINT satellite
30A	Noaa 15	13 May	VAFB	Titan 23G	2,234	808	823	101.2	98.7	First of new generation meteorological satellite for NOAA
31A	Progress-M 39	14 May	TT	Soyuz-U	7,135	371	379	92.1	51.7	Mir cargo freighter
32A	Iridium 70	17 May	VAFB	Delta 2	657	775	780	100.4	86.4	Fifteenth cluster of Iridium satellites
32B	Iridium 72				657	773	780	100.4	86.4	
32C	Iridium 73				657	775	779	100.4	86.4	
32D	Iridium 74				657	774	778	100.4	86.4	
32E	Iridium 75				657	692	695	98.6	86.4	
33A	Zhongwei 1	30 May	XI	CZ-3B	2,984	35,761	84,991	2,860.7	1.3	China Star telecommunications satellite operated for China Orient Telecom Satellite Company. Eventually operated in eccentric geosynchronous orbit at 87.5° E
34A	*Discovery* (STS-91)	2 Jun	KSC	Shuttle	117,860	370378	92.0	51.7		Precourt, Gorie, Chang-Diaz, Lawrence, Kavandi and Ryumin on ninth and last Mir visit. Returned 12 Jun
35A	Thor 3	10 Jun	CC	Delta 2	1,300	geostationary orbit at 358° E				Telecommunications satellite for Telenor, Norway
36A	Cosmos 2352	15 Jun	PL	Tsyklon	225	1,310	1,875	118.0	82.6	First of six Strela military store-dump
36B	Cosmos 2353				225	1,300	1,870	117.9	82.6	communications satellites. Failure in final
36C	Cosmos 2354				225	1,307	1,872	118.0	82.6	stage attitude system left satellites in eccentric
36D	Cosmos 2355				225	1,302	1,868	117.9	82.6	rather than near-circular orbits of 1,400–1,415 km
36E	Cosmos 2366				225	1,298	1,867	117.8	82.6	Strela 3 satellites launched into same orbit
36F	Cosmos 2367				225	1,294	1,863	117.7	82.6	plane as Gonets-D1, 4–6 and Cosmos 2337–2339 cluster
37A	Intelsat 805	18 Jun	CC	Atlas 2AS	3,524	geostationary at 304.5° E				Telecommunications satellite for Intelsat
38A	Cosmos 2358	24 Jun	PL	Soyuz-U	6,600	167	334	89.5	67.1	Yantar-4K2 Kobalt fourth-generation military reconnaissance satellite
39A	Cosmos 2359	25 Jun	TT	Soyuz-U	7,000	240	303	89.9	64.9	Yantar fifth-generation photo-reconnaissance satellite
40A	Molniya-3 49	1 Jul	P	Molniya-M	1,750	432	39,943	718.2	62.8	Telecommunications satellite co-planar with Molniya-3 39
41A	Nozomi	3 Jul	KA	M-5	536	1,421	516,130	22,646.8	25.4	Planet-B Mars spacecraft in earth-moon orbit with lunar fly-bys
42A	Tubsat N	7 Jul	BS	Shtil-1	8.5	401	777	96.5	78.9	Sensor and transceiver test platform. First launch from a submarine, located in Barents Sea at 69.3° north × 35.3° east
42B	Tubsat N 1				3	400	771	96.4	78.9	Store-forward transceiver communications
43A	Resurs-02 1	10 Jul	TT	Zenit 2	2,800	817	818	101.2	98.8	First Resurs-02 class. Belgian/German IRIS
43B	Fasat-Bravo				50	817	819	101.2	98.8	Chilean Air Force to replace Sich 1, 1995–46
43C	Tmsat				50	817	818	101.2	98.8	Thai remote sensing satellite sensor payloads
43D	Techsat 1B				50	816	818	101.2	98.8	Replaces failed Techsat 1B, March 1995
43E	Westpac				24	817	819	101.2	98.8	Australian-Russian geodetic satellite
43F	Safir 2				60	815	819	101.2	98.8	Successor to Safir R1, 1994–97
44A	Sinosat 1	18 Jul	XI	CZ-3B	2,840	geostationary at 87.5° E				Telecommunications satellite for Sino Satellite Communications Company
45A	Cosmos 2360	28 Jul	TT	Zenit 2	3,250	848	854	101.9	71.0	Tselina-2 ELINT satellite
46A	Orbcomm FM 13	2 Aug	WI	L-1011/Pegasus XL + HAPS	43	813	825	101.3	45.0	Orbcomm Global LP communications
46B	Orbcomm FM 14				43	813	826	101.3	45.0	
46C	Orbcomm FM 15				43	813	826	101.3	45.0	
46D	Orbcomm FM 16				43	813	827	101.3	45.0	
46E	Orbcomm FM 20				43	816	828	101.3	45.0	
46F	Orbcomm FM 19				43	816	828	101.3	45.0	
46G	Orbcomm FM 18				43	815	828	101.3	45.0	
46H	Orbcomm FM 17				43	816	828	101.3	45.0	
47A	Soyuz-TM 28	13 Aug	TT	Soyuz-U	7,150	364	374	91.9	51.7	Padalka, Avdeyev, who remained aboard Mir, and Baturin, who returned to earth in Soyuz-TM 27
48A	Iridium 3	19 Aug	TY	CZ-2C/SD	657	776	780	100.4	86.4	Sixteenth Iridium cluster launched
48B	Iridium 76				657	775	779	100.4	86.4	
49A	ST 1	25 Aug	KO	AR44P	3,255	geostationary orbit at 88° E				Telecommunications satellite for Singapore Telecom and Chunghwa Telecom International
50A	Astra 2A	30 Aug	TT	Proton-K	3,500	geostationary orbit at 28° E				Broadcast satellite for SES, Luxembourg
51A	Iridium 82	8 Sep	VAFB	Delta 2	657	703	707	98.9	86.5	Seventeenth Iridium cluster launch
51B	Iridium 81				657	776	779	100.4	86.4	
51C	Iridium 80				657	776	779	100.4	86.4	
51D	Iridium 79				657	704	708	98.9	86.4	
51E	Iridium 77				657	704	708	98.9	86.4	
52A	PAS 7	16 Sep	KO	Ar44LP	3,838	geostationary at 68.5° E				Broadcast satellite for PanAmSat

Desig	*Name*	*Date*	*Site*	*Vehicle*	*Mass (kg)*	*Perigee (km)*	*Apogee (km)*	*Period (min)*	*Inc (deg)*	*Notes*
53A	Orbcomm-FM 21	23 Sep	WI	L-1011/Pegasus XL + HAPS	43	811	822	101.2	45.0	Third cluster launch for Orbcomm Global LP
53B	Orbcomm-FM 22				43	810	818	101.2	45.0	
53C	Orbcomm-FM 23				43	812	821	101.2	45.0	
53D	Orbcomm-FM 24				43	812	821	101.2	45.0	
53E	Orbcomm-FM 25				43	809	819	101.2	45.0	
53F	Orbcomm-FM 26				43	809	819	101.2	45.0	
53G	Orbcomm-FM 27				43	809	819	101.2	45.0	
53H	Orbcomm-FM 28				43	810	819	101.2	45.0	
54A	Molniya-1 91	28 Sep	PL	Molniya-M	1,600	420	40,657	732.5	62.8	Telecommunications satellite launched to the same orbital plane as Molniya-1 86
55A	Stex	3 Oct	VAFB	Taurus	540	697	716	98.9	85.0	Space Technology Experiment satellite for National Reconnaissance Office
56A	Eutelsat W2	5 Oct	KO	Ar44L	2,950	geostationary orbit at 16° E				First of new widebeam Eutelsat telecommunications satellite
56B	Sirius 3				1,465	geostationary orbit at 5° E				Broadcast satellite for Nordic Television Co in Sweden
57A	Hot Bird 5	9 Oct	CC	Atlas 2A	2,900	geostationary orbit at 13° E				Telecommunications satellite for Eutelsat
58A	USA 140	20 Oct	CC	Atlas 2A	3,208	geostationary orbit at 186° E				UHF Follow-On (UFO) 9 for US Navy
59A	Maqsat 3	21 Oct	KO	Ar5	6,000	1,003	35,494	641.0	7.0	Eutelsat W communications satellite mock-up
60A	SCD 2	23 Oct	CC	Pegasus	115	744	768	99.9	25.0	Brazilian data collection satellite
61A	Deep Space 1	24 Oct	CC	Delta 2	486	heliocentric orbit				First of NASA's New Millenium Programme satellites; flew by asteroid 9969 Braille on 28 Jul 1999 at 15 km distance
61B	Sedsat 1				40	547	1,079	101.1	31.4	Student microsatellite with imaging camera
62A	Progress-M 40	25 Oct	TT	Soyuz-U	7,150	355	365	91.7	51.7	Cargo freighter to Mir
63A	AfriStar	28 Oct	KO	Ar44L	2,739	geostationary orbit at 21° E				Radio broadcast satellite for WorldSpaceCorp
63B	GE 5				1,698	geostationary orbit 281° E				Broadcast satellite for GE Americom
64A	*Discovery* (STS-95)	29 Oct	KSC	Shuttle	130,973	552	561	95.8	28.5	Brown, Lindsey, Robinson, Parazynski, Duque, Mukai, Glenn (the first US astronaut to orbit the earth) on life sciences mission. Returned 7 Nov
64B	Pansat				70	554	558	95.8	28.5	US Navy student telecommunications satellite
64C	Spartan 201–5				1,269	550	560	95.7	28.5	Reflight of Spartan 201–4. Retrieved 3 Nov
65A	PAS 8	4 Nov	TT	Proton-K	3,800	geostationary orbit 166° E				Broadcast satellite for PanAmSat
66A	Iridium 2	6 Nov	VAFB	Delta 2	657	566	579	96.1	86.1	Eighteenth Iridium cluster launch
66B	Iridium 86				657	517	533	96.1	86.1	
66C	Iridium 85				657	514	534	95.3	86.1	
66D	Iridium 84				657	702	706	98.8	86.1	
66E	Iridium 83				657	515	533	95.1	86.1	
67A	Zarya	20 Nov	TT	Proton-K	20,700	388	401	92.4	51.6	First element in the orbital assembly of the International Space Station
68A	Bonum 1	23 Nov	CC	Delta 2	1,425	geostationary orbit at 36° E				Telecommunications satellite owned by Russian Bonum-1 news group
69A	*Endeavour* (STS-88)	4 Dec	KSC	Shuttle	119,717	383	395	92.3	51.6	Cabana, Sturckow, Ross, Currie, Newman and Krikalev. Second element in the orbital assembly of the International Space Station, Unity module attached to Zarya. Landed 16 Dec
69B	SAC A				268	383	400	92.4	51.6	Argentine satellite
69C	MightySat 1				320	382	399	92.4	51.6	Developed by USAF Phillips Laboratory
69D	Slidewire				3.9	382	394	92.3	51.6	Carrier lost overboard from Endeavour
69E	Socket				1.5	382	394	92.3	51.6	Lost overboard during mission
69F	Unity				11,600	383	395	92.3	51.6	Interconnecting ISS module
70A	Satmex 5	6 Dec	KO	Ar42L	4,144	geostationary at 243.2° E				Telecommunications satellite owned by Satmex S.S. de C.V., Mexico
71A	Swas	6 Dec	VAFB	L-1011/ Pegasus-XL	283	637	653	97.6	69.9	Submillimetre Wave Astronomy Satellite
72A	Nadezhda 5	10 Dec	PL	Cosmos-3M	825	977	1,013	105.0	82.9	Tsikada system civilian navigation satellite
72B	Astrid 2				30	979	1,013	105.0	82.9	Swedish science payload
73A	MCO	11 Dec	CC	Delta 2	629	heliocentric orbit				NASA Mars Climate Orbiter crashed into Mars September 1999
74A	Iridium 11A	19 Dec	TY	CZ-2C/SD	657	688	713	98.8	86.5	Nineteenth Iridium cluster launch
74B	Iridium 20A				657	686	713	98.8	86.5	
75A	PAS 6B	22 Dec	KO	Ar42L	3,600	geostationary at 313° E				Broadcast telecommunications satellite for PanAmSat. Operated from 317° E
76A	Cosmos 2361	24 Dec	PL	Cosmos-3M	795	969	1,013	104.9	82.9	Parus military navigation satellite
77A	Cosmos 2362	30 Dec	TT	Proton-K	1,300	19,125	19,129	675.6	64.8	Three Uragan navigation satellites
77B	Cosmos 2362				1,300	19,126	19,129	675.6	64.8	
77C	Cosmos 2363				1,300	19,066	19,200	675.8	64.8	

UPDATED

Table 2: Satellite Launches 1999

Desig	*Name*	*Date*	*Site*	*Vehicle*	*Mass (kg)*	*Perigee (km)*	*Apogee (km)*	*Period (min)*	*Inc (deg)*	*Notes*
01A	MPL + DS2	3 Jan	CC	Delta 2	583	heliocentric orbit				Mars Polar Lander + 2 DS penetrators arrived at Mars 3 Dec but failed to land intact
02A	Chunghua 1	27 Jan	CC	Athena-1	410	595	605	96.7	35.0	First satellite launched for Taiwan, a marine resources satellite
03A	Stardust	7 Feb	CC	Delta 2	385	heliocentric orbit				Fourth in NASA's Discovery series, to retrieve cometary material following 100 km fly-by of P/Comet Wild 2, returning to earth Jan 2006
04A	Globalstar M36	9 Feb	TT	Soyuz-U/Ikar	450	1,337	1,356	112.6	52.0	Fourth cluster launch for Globalstar and first use of Soyuz-U/Ikar launcher
04B	Globalstar M23				450	1,323	1,338	112.2	52.0	
04C	Globalstar M38				450	1,343	1,404	113.2	52.0	
04D	Globalstar M40				450	1,005	1,032	105.5	52.0	
05A	Telstar 6	15 Feb	TT	Proton-K	3,650	geostationary orbit at 267° E				Loral Skynet high power telecommunications satellite
06A	JC-Sat	16 Feb	CC	Atlas-2AS	3,000	geostationary orbit at 120° E				Telecommunications satellite owned by Japan Satellite Systems
07A	Soyuz-TM 29	20 Feb	TT	Soyuz-U	7,120	346	364	91.6	51.7	Afanasyev, Hagnere and Bella to dock with Mir. Bella returned to earth with Padalka in Soyuz-TM 28
08A	Argos	23 Feb	VAFB	Delta 2	2,700	831	838	101.7	98.7	Joint-services advanced research and global observation satellite
08B	Ørsted				62	647	864	99.9	96.5	Danish magnetic and particle science satellite
08C	Sunsat				61	647	865	99.9	96.5	South Africa's first satellite carries earth imager and voice/digital communications
09A	Arabsat 3A	26 Feb	KO	Ar44L	2,708	geostationary at 26° E				First of new generation Arab broadcast and telecommunications satellite
09B	Skynet 4E				1,490	geostationary at 6° E				UK military communications satellite
10A	Raduga-1 4	28 Feb	TT	Proton-K	2,000	geostationary				Globus military and governmental communications satellite
11A	Wire	5 Mar	VAFB	L-1011/ Pegasus-XL	270	539	594	96.0	97.5	Wide-Field Infra-Red Explorer to study the origin and evolution of galaxies
12A	Globalstar M22	15 Mar	TT	Soyuz-U/Ikar	450	1,409	1,417	114.1	52.0	Fifth cluster launch for Globalstar Ltd
12B	Globalstar M41				450	1,413	1,414	114.1	52.0	
12C	Globalstar M46				450	1,411	1,416	114.1	52.0	
12D	Globalstar M37				450	1,413	1,415	114.1	52.0	
13A	AsiaSat 3S	21 Mar	TT	Proton-K	3,465	geostationary orbit at 98° E				Telecommunications satellite for Asia Satellite Telecommunications Co, Hong Kong
14A	Sea Launch	28 Mar	PO	Zenit-3SL	4,500	639	36,064	645.0	1.2	HS 702 mass model in first test of Sea Launch Odyssey platform at 0° north × 154° west
15A	Progress-M 41	2 Apr	TT	Soyuz-U	7,180	339	355	91.5	51.7	Unmanned freighter carrying 2,438 kg cargo to Mir
15B	Sputnik 99				3	338	352	91.4	51.7	Amateur radio satellite released by Mir crew
16A	Insat 2E	2 Apr	KO	Ar42P	2,600	geostationary orbit at 83° E				Indian telecommunications and meteorology satellite
17A	USA 142	9 Apr	CC	Titan-4B/IUS	2,400	188	38,000			Defense Support Program 19 satellite stranded in GTO by IUS burn timing failure
18A	Eutelsat-W 3	12 Apr	CC	Atlas-2AS	3,183	geostationary orbit at 7° E				Second wideband Eutelsat satellite
19A	Globalstar M19	15 Apr	TT	Soyuz-U/Ikar	450	902	944	103.5	51.9	Sixth cluster launch for Globalstar Partnership Ltd
19B	Globalstar M42				450	904	946	103.5	51.9	
19C	Globalstar M44				450	931	977	103.5	51.9	
19D	Globalstar M45				450	1,412	1,415	114.1	52.0	
20A	Landsat 7	15 Apr	VAFB	Delta 2	2,170	685	702	98.6	98.2	NASA, NOAA, USGS remote sensing satellite
21B	Uosat 12	21 Apr	TT	Dniepr	320	649	652	97.7	64.6	Surrey Satellite Technology Ltd satellite with mobile radio experiment, GPS receiver and imaging cameras. Intl designation is correct following an officially incorrect designation of Sputnik 99 as 21A
22A	Abrixas	28 Apr	KY	Cosmos-3M	470	554	603	96.2	48.4	Broadband X-ray survey astronomy satellite
22B	Megsat-0				35	549	603	96.2	48.4	Italian technology development satellite
23A	USA 143	30 Apr	CC	Titan-4B/Centaur	4,700	147?	740?	147.5	28.0	Milstar-2 DoD communications satellite stranded in wrong orbit by Centaur
24A	Orion 3	5 May	CC	Delta 3	4,300	422	1,317	102.3	29.1	Loral Orion Inc telecommunications satellite stranded in useless orbit by Delta second stage
25A	Fen Young-1 3	10 May	TY	CZ-4B	958	849	868	102.1	98.8	Third launch of FY-1 meteorological satellite
25B	Shi Jian 5				298	846	865	102.0	98.8	Test of new satellite bus
26A	Terriers	18 May	VAFB	Pegasus-XL + HSPS	123	542	554	95.6	97.7	NASA Student Explorer Demonstration Initiative satellite
26B	Mublcom				45	773	776	100.3	97.7	US Army mobile communications test satellite
27A	Nimiq 1	20 May	TT	Proton-K	3,646	geostationary at 269° E				Broadcast satellite for Telesat Canada
28A	USA 144	22 May	VAFB	Titan-4B						Classified payload
29A	Kitsat 3	26 May	SR	PSLV	107	711	737	99.3	98.4	Second operational PSLV launch with earth imaging satellite for South Korea
29B	DLR-Tubsat				45	716	737	99.3	98.4	High resolution earth observation satellite for DLR and University of Berlin
29C	Oceansat				1,048	718	739	99.4	98.4	ISRO ocean resources satellite (IRS-P4)

Desig	*Name*	*Date*	*Site*	*Vehicle*	*Mass (kg)*	*Perigee (km)*	*Apogee km)*	*Period (min)*	*Inc (deg)*	*Notes*
30A	*Discovery* (STS-96)	27 May	KSC	Shuttle	100,236	324342	91.2	51.6		Rominger, Husband, Jernigan, Ochoa,Barry, Payette, and Tokarev on second Shuttle flight to ISS Unity/Zarya modules
30B	Starshine 1				160	380	397	97.3	51.6	US Navy/University of Utah atmospheric
31A	Globalstar M52	10 Jun	CC	Delta 2	450	1,360	1,380	113.1	52.0	Seventh cluster for Globalstar Partnership
31B	Globalstar M49				450	1,413	1,414	114.1	52.0	
31C	Globalstar M25				450	1,413	1,414	114.1	52.0	
31D	Globalstar M47				450	1,411	1,416	114.1	52.0	
32A	Iridium 14A	11 Jun	TY	CZ-2C/SD	667	709	713	99.0	86.5	Twentieth Iridium cluster launch
32B	Iridium 21A				667	746	752	99.8	86.4	
33A	Astra 1H	18 Jun	TT	Proton-K	3,728	geostationary at 19.2° E				Broadcast satellite for SES, Luxembourg
34A	Quikscat	20 Jun	VAFB	Titan-23G	870	280	813	95.6	98.6	NASA remote sensing technology satellite with in-orbit perigee boost for circularisation
						792	821	101.0	98.6	
35A	Fuse	24 Jun	CC	Delta 2	1,335	754	770	100.1	25.0	NASA Far Ultraviolet Spectroscopic Explorer
36A	Molniya-3 50	8 Jul	PL	Molniya-M	1,750	40,811	468	736.6	62.8	Molniya plane C
37A	Globalstar 32	9 Jul	ETR	Delta 7420	450	1,415	1,412	114.1	52.0	Eighth Globalstar cluster
37B	Globalstar 30				450	1,414	1,411	114.1	52.0	
37C	Globalstar 35				450	1,414	1,411	114.1	52.0	
37D	Globalstar 51				450	1,414	1,411	114.1	52.0	
38A	Progress-M 42	16 Jul	TT	Soyuz-U	7,200	353	346	91.5	51.7	Freighter docked to Mir-X port
39A	Okean 5	17 Jul	TT	Zenit-2	6,300	663	661	97.8	98.1	First of new generation oceansat
40A	Columbia	23 Jul	KSC	Shuttle	105,000	302	289	90.4	28.5	Eileen Collins, Ashby, Hawley Coleman, Tognini on Chandra delivery mission
40B	Chandra				5,865	140,012	10,037	3,852.3	28.5	Chandra X-ray observatory
41A	Globalstar 26	25 Jul	CC	Delta 7420	450	1,370	1,368	113.1	52.0	Ninth Globalstar cluster launch
41B	Globalstar 28					1,376	1,364	113.3	52.0	
41C	Globalstar 43					1,412	1,410	114.0	52.0	
41D	Globalstar 48					1,414	1,396	113.9	52.0	
42A	Telekom 1	12 Aug	KO	Ar42P	2,655	35,804	35,778	1,436.3	0.1	Indonesian communications
43A	Globalstar 24	17 Aug	CC	Delta 7420	450	1,414	1,411	114.1	52.0	Tenth Globalstar cluster
43B	Globalstar 27				450	1,415	1,411	114.1	52.0	
43C	Globalstar 53				450	1,413	1,412	114.1	52.0	
43D	Globalstar 54				450	1,415	1,411	114.1	52.0	
44A	Cosmos 2365	18 Aug	PL	Soyuz-U	6,600	343	167	89.6	67.1	Fourth gen Yantar-4K2 reconsat
45A	Cosmos 2366	26 Aug	PL	Cosmos-3M	795	1,008	963	104.8	82.9	Parus military navigation satellite
46A	Mugunghwa 3	4 Sep	KO	Ar42P	2,800	geostationary at 116° E				Koreasat 3 telecoms-sat
47A	Yamal 101	6 Sep	TT	Proton-K	1,360	geostationary at 340° E				Communications; first dual launch of Russian sats to geostationary
47B	Yamal 102				1,360	geostationary at 75° E				
48A	Foton 12	7 Sep	PL	Soyuz-U	6,200	384	217	90.5	62.8	Ninth launch of Foton microgravity research satellite with experiments from Russia, and CNES. Module landed 24 Sep
49A	Globalstar 33	22 Sep	TT	Soyuz-U	450	957	901	103.6	51.9	Eleventh Globalstar cluster
49B	Globalstar 50				450	1,413	1,408	114.0	52.0	
49C	Globalstar 55				450	957	901	103.6	52.0	
49D	Globalstar 58				450	1,416	1,411	114.1	52.0	
50A	EchoStar 5	23 Sep	CC	Atlas 2AS	3,500	geostationary at 260° E				Telecoms for EchoStar Corp
51A	Ikonos	24 Sep	WTR	Athena-2	726	682	678	98.4	98.2	First successful launch of 1 m commercial earth imaging satellite
52A	Telstar 7	25 Sep	KO	Ar44LP	3,790	geostationary at 231° E				Telecoms for Loral Skynet
53A	LM1	26 Sep	TT	Proton-K	2,600	geostationary at 75° E				First US-Russia telecoms satellite
54A	Resurs-F22	28 Sep	PL	Soyuz-U	6,300	230	222	89.0	82.3	Second flight of improved performance Resurs-F1M remote sensing satellite
55A	Navstar 46	7 Oct	CC	Delta 7925		20,073	20,091		53.0	Third Block 2R GPS satellite
56A	DirecTV 1R	10 Oct	PO	Zenit-3SL		geostationary at 101° W				Second successful sea launch
57A	SAC 1	14 Oct	TY	CZ-4B		744	733		98.6	Contact lost in mid-October
57B	Zi Yuan 1					774	773		98.6	China-Brazil earth resources
58A	Globalstar 31	18 Oct	TT	Soyuz-Ikar	450	1,414	1,412		52.0	Twelfth Globalstar cluster
58B	Globalstar 56					1,414	1,412		52.0	
58C	Globalstar 57					1,414	1.412		52.0	
58D	Globalstar 59					948	x896		52.0	
59A	Orion 2	19 Oct	KO	Ar44LP		geostationary at 15° W				Compliment to Telstar for Loral Skynet
60A	GE4	13 Nov	KO	Ar44LP		geostationary at 101° W				C-band and Ku-band comsat for GE Americom
61A	Shenzou	19 Nov	XI	CZ-2F		324	196		42.6	Unmanned prototype flight of China's first manned spacecraft
62A	Globalstar 29	22 Nov	TT	Soyuz-Ikar	450	941	897		52.0	Thirteenth Globalstar cluster
62B	Globalstar 34				450	1,414	1,412		52.0	
62C	Globalstar 39				450	942	900		52.0	
62D	Globalstar 61				450	1,415	1,411		52.0	
63A	UHF FO F10	23 Nov	CC	Atlas 2A		geostationary at 170° W				UHF/EHF coms for global broadcast service TV for US Navy
64A	Clementine	3 Dec	KO	Ar40		664	646		98.1	French military electronic intelligence technology satellite
64B	Helios 1B					681	679		98.1	French optical military military reconnaissance satellite
65A	Orbcomm FM30	4 Dec	W	Pegasus XL	43	834	824		45.0	Fourth cluster launch for Orbcom
65B	Orbcomm FM31					834	824		45.0	
65C	Orbcomm FM32					834	824		45.0	
65D	Orbcomm FM33					834	824		45.0	
65E	Orbcomm FM34					834	824		45.0	
65F	Orbcomm FM35					834	824		45.0	
65G	Orbcomm FM36					834	824		45.0	
66A	XMM	10 Dec	KO	Ar5		113,678	7,417		38.8	ESA XMM X-ray multi-mirror multi-mirror space observatory
67A	DMSP 5D-3 F-15	12 Dec	VAFB	Titan 2		851	837		98.9	First launch of Block 5D-3 military weather satellite
68A	Terra	18 Dec	VAFB	Atlas 2AS		684	654		98.2	First launch in NASA's Earth Observing System with multi-spectral imagers, radiation budget instrument and CO and methane detectors

Desig	*Name*	*Date*	*Site*	*Vehicle*	*Mass (kg)*	*Perigee (km)*	*Apogee (km)*	*Period (min)*	*Inc (deg)*	*Notes*
69A	Discovery	20 Dec	KSC	Shuttle	112,565	610	591		28.5	Hubble Space Telescope repair mission 3A with Brown, Scott, Kelly, Steven Smith, Grunsfeld, Foale, Nicollier, Clervoy
70A	Acrimsat	21 Dec	VAFB	Taurus		724	683		98.3	Solar energy output monitor
70B	Celestis-03					723	683		98.3	Burial satellite with cremated human remains
70C	Kompsat					722	690		98.3	Korean multipurpose satellite
71A	Galaxy 11	22 Dec	KO	Ar44L		38,900	278		5.4	First HS 702, carrying 64 transponders for PanAmSat Corp
72A	Cosmos 2367	26 Dec	TT	Tsyklon-2		418	404		65.0	Passive naval intelligence satellite
73A	Cosmos 2368	27 Dec	PL	Molniya-M		39,138	551		62.9	Early warning satellite

UPDATED

Table 2: Satellite Launches 2000

Desig	*Name*	*Date, Time*	*Site*	*Vehicle*	*Mass (kg)*	*Perigee (km)*	*Apogee (km)*	*Period (min)*	*Inc (deg)*	*Notes*
001A	DSCS 3B F7	21 Jan 01:03	CC	Atlas 2A	1,175	Elements not available				Defense Satellite Communications System
002A	Galaxy 10R	25 Jan 01:04	KO	Ariane 42L	3,651	35,785	35,788	1,436.1	0.0	Telecommunications for PanAmSat. Stationed above 122° W
003A	Xhongzing 22	25 Jan 16:45	XI	CZ 3A	2,300	35,774	35,803	1,436.2	0.0	'China Star' communications satellite. Located above 98° E
004A	Jawsat	27 Jan 03:03	VAFB	Minotaur	64	748	800	100.3	100.2	Joint Air Force Academy/Weber State satellite carried student experiments. Maiden flight of Minotaur, booster derived from Minuteman stages
004B	OCSE	27 Jan	VAFB	Minotaur	22	597	612	96.8	100.2	Optical Calibration Sphere Experiment used to calibrate ground-based lasers. Decayed 5 Mar 01
004C	Opal	27 Jan	VAFB	Minotaur	13	750	803	100.4	100.2	Orbiting Picosat Automated Launcher, deployed series of small satellites
004D	Falconsat	27 Jan	VAFB	Minotaur	52	750	805	100.4	100.2	Developed by US Air Force Academy Students to provide operational experience
004E	Asusat	27 Jan	VAFB	Minotaur	5	750	804	100.4	100.2	Earth imaging, amateur radio
004H	Picosat 1&2	27 Jan	VAFB	Minotaur	0.5	747	795	100.3	100.2	Tethered pair deployed from OPAL 15 Jul
004J	Picosat 3	27 Jan	VAFB	Minotaur	0.5	747	798	100.3	100.2	Deployed from OPAL 1 Feb. Also called Thelma
004K	Picosat 4	27 Jan	VAFB	Minotaur	0.5	746	800	100.3	100.2	Deployed from OPAL 12 Feb. Also called Louise
004L	Picosat 5	27 Jan	VAFB	Minotaur	0.5	745	798	100.3	100.2	Deployed from OPAL 11 Feb. Also called JAX
004M	Picosat 6	27 Jan	VAFB	Minotaur	0.2	747	801	100.3	100.2	Deployed from OPAL 11 Feb. Also called Stensat
005A	Progress-M11	1 Feb 06:47	TT	Soyuz-U	7,200	320	302	90.7	51.6	Resupply Mir. Docked 3 Feb, undocked and re-entered 26 Apr 99
006A	Cosmos 2369	3 Feb 09:26	TT	Zenit 2	3,250	845	857	102.0	71.0	Elint
007A	Hispasat 1C	3 Feb 23:30	CC	Atlas 2AS	3,112	35,779	35,796	1,436.1	0.0	Telecommunications for Spain. Located above 30° W
008A	Globalstar 63	8 Feb 21:24	CC	Delta 2	450	911	933	103.5	52.0	Continued series of clusters launched to populate commsat constellation
008B	Globalstar 62	8 Feb	CC	Delta 2	450	1,412	1,415	114.1	52.0	Communications for individual users
008C	Globalstar 60	8 Feb	CC	Delta 2	450	920	933	103.6	52.0	Communications for individual users
008D	Globalstar 64	8 Feb	CC	Delta 2	450	911	930	103.4	52.0	Communications for individual users
009A	Dummy Satellite	8 Feb 23.20	TT	Soyuz-U	966	575	608	96.5	64.9	Maiden flight of Soyuz-U with Fregat 4th stage
009B	Fregat/IRDT	8 Feb	TT	Soyuz-U	1,100	581	604	96.5	64.8	Fregat 4th stage housed Inflatable Re-entry and Descent Technology payload, which separated before re-entry. Both vehicles re-entered and landed 9 Feb. IRDT recovered
010A	STS-99	11 Feb 17:43	CC	Endeavour	116,221	239	230	89.2	57.0	Crew: Kevin Kregel, Dom Gorie, Gerhard Thiele, Janet Kavandi, Janice Voss, Mamoru Mohri. Shuttle Radar Topography Mission, mapped land surface of earth. Returned 22 Feb 99 after 11 days, 5 h, 38 min
011A	Garuda 1	12 Feb 09:10	TT	Proton-K	4,500	35,731	35,843	1,436.1	2.3	Telecommunications for Asia Cellular Satellite.Located above 123° E
012A	Superbird 4	18 Feb 01:04	KO	Ariane 44LP	4,057	35,780	35,794	1,436.1	0.0	Telecommunications for Japan. Stationed above 162° E
013A	Ekspress 2A	12 Mar 04:07	TT	Proton-K	2,600	35,777	35,795	1,436.1	0.1	Telecommunications. Stationed above 80° E
014A	MTI	12 Mar 09:29	VAFB	Taurus	1,610	567	600	96.4	97.4	Multispectral Thermal Imager, US Dept of Energy Experiments
015A	Dumsat 2/ Fregat	20 Mar 18:28	TT	Soyuz-U	3,530	294	17,867	318.8	64.8	Test flight of launch vehicle. Dummy payload Simulated future ESA Cluster satellites, remained attached to Fregat 4th stage
016A	Asiastar	21 Mar 23:28	KO	Ariane 5	2,777	35,771	35,803	1,436.1	0.0	Digital radio broadcasting for Worldspace. Located above 105° E
016B	Insat 3B	21 Mar	KO	Ariane 5	2,070	35,767	35,807	1,436.1	0.0	Telecommunications for India. Stationed above 83° E
017A	Image	25 Mar 20:34	VAFB	Delta 2	494	1,526	45,349	853.9	89.3	Imager for Magnetopause-to-Aurora Global Exploration. Space environment research
018A	Soyuz-TM 30	4 Apr 05:01	TT	Soyuz-U	7,150	330	327	91.0	51.6	Crew: Sergei Zaletin, Alexander Kaleri. Mission to repopulate Mir funded by Western interests. Docked 6 Apr. Returned 16 Jun, concluding a 72 day, 19 h, 43 min flight
019A	Sesat	17 Apr 21:06	TT	Proton-K	2,500	35,772	35,800	1,436.1	0.0	Siberia-Europe SATellite. Communications. Positioned above 36° E
020A	Galaxy IVR	19 Apr 00:29	KO	Ariane 42L	3,668	35,781	35,793	1,436.1	0.0	Telecommunications for PanAmSat. Stationed above 90° W
021A	Progress-M1 2	25 Apr 20:08	TT	Soyuz-U	7,280	332	328	91.1	51.6	Resupply Mir. Docked 27 Apr. Undocked and re-entered 15 Oct
022A	GOES 11	3 May 07:07	CC	Atlas 2A	2,217	35,765	35,810	1,436.1	0.0	Geostationary Operational Environmental Satellite. Meteorological. On orbit spare parked above 105° W
023A	Cosmos 2370	3 May 13:25	TT	Soyuz-U	7,000	247	300	90.0	64.7	Photoreconn
024A	DSP 20	8 May 16:01	CC	Titan 4B	2,400	Elements not available				Defense Support Program satellite. Early warning
025A	GPS 47	11 May 01:48	CC	Delta 2	2,032	20,108	20,256	718.0	55.1	Global Positioning System satellite. Navigation
026A	Simsat 1	16 May 08:27	PL	Rokot	657	539	555	95.6	86.4	Dummy payload flown in flight test of commercial version of Rokot booster
026B	Simsat 2	16 May	PL	Rokot	660	541	551	95.6	86.4	Dummy payload

Desig	*Name*	*Date, Time*	*Site*	*Vehicle*	*Mass (kg)*	*Perigee (km)*	*Apogee (km)*	*Period (min)*	*Inc (deg)*	*Notes*
027A	STS-101	19 May 10:11	CC	Atlantis	118,925	381	352	91.8	51.5	Crew: James Halsell, Scott Horowitz, Mary Weber, Jeff Williams, James Voss, Susan Helms, Yuri Usachev. Mission to service International Space Station. Docked 21 May. Undocked 26 May, returned 29 May after 9 days, 20 h, 10 min flight
028A	Eutelsat W4	24 May 23:10	CC	Atlas 3A	3,190	35,735	35,826	1,435.8	0.0	Telecommunications. Positioned above 36° E. First Atlas 3A flight
029A	Gorizont 33	6 Jun 02:59	TT	Proton-K	2,125	35,780	35,796	1,436.1	1.0	Telecommunications. Stationed above 145° E
030A	TSX 5	7 Jun 13:19	VAFB	L1011/ Pegasus XL	250	405	1,689	106.1	69.0	Tri-Service Experiments Mission. Space technology, space environment research
031A	Ekspress 3A	24 Jun 00:28	TT	Proton-K	2,600	35,778	35,793	1,436.1	0.1	Telecommunications. Positioned above 11° W
032A	Feng Yun-2 2	25 Jun 11:50	XI	CZ-3A	1,400	35,781	35,791	1,436.1	0.7	'Wind and Cloud' meteorological satellite. Located above 105° E
033A	Nadezhda 6	2 Jun 10.38	PL	Cosmos 3M	800	681	706	98.6	98.2	Navigation. Houses search and rescue transponders for international emergency operations
033B	Tsinghua 1	28 Jun	PL	Cosmos 3M	50	682	707	98.7	98.2	Imaging technology demonstration. Sponsored by Chinese University
033C	Snap 1	28 Jun	PL	Cosmos 3M	6.5	684	707	98.6	98.1	Nanosatellite housing vision system designed to inspect other orbiters
034A	TDRS 8	30 Jun 12:56	CC	Atlas 2A	3,180	35,682	35,891	1,436.1	6.7	Tracking and Date Relay Satellite. Positioned above 189° E
035A	Sirius 1	30 Jun 22.08	TT	Proton-K	3,800	24,916	46,658	1,436.1	63.2	First in series of new digital radio broadcast satellites. Not to be confused with Swedish Sirius series already on orbit
036A	Cosmos 2371	4 Jul 23:44	TT	Proton-K	2,300	35,771	35,802	1,436.1	1.1	Data relay, military communications. Located above 80° E
037A	Zvezda	12 Jul 28:03	TT	Proton-K	19,710	367	351	91.7	51.8	Accommodations component of ISS. Second Russian-supplied module. Docked 28 Jul
038A	Echostar 6	14 Jul 06:21	CC	Atlas 2AS	3,700	35,782	35,791	1,436.1	0.0	Telecommunications. Positioned above 119° W
039A	MITA	15 Jul 12:00	PL	Cosmos 3M	170	382	429	92.7	87.3	Italian experimental microsatellite, observing space environment
039B	Champ	15 Jul	PL	Cosmos 3M	550	412	467	93.4	87.3	German geophysics research
039C	BIRD-Rubin	15 Jul	PL	Cosmos 3M	37	389	434	92.8	87.3	Student-built microsatellite remained attached to 2nd stage. Measure launch vehicle parameters
040A	GPS 48	16 Jul 09:17	CC	Delta 2	2,032	20,054	20,310	718.0	55.2	Global Positioning System satellite
041A	Cluster II-7 Samba	16 Jul 12:39	TT	Soyuz-U/ Fregat	1,200	18,188	119,589	3,425.1	89.6	One of four ESA satellites studying solar wind and magnetosphere
041B	Cluster II-6 Salsa	16 Jul	TT	Soyuz-U/ Fregat	1,200	18,438	119,341	3,425.1	89.6	Formation flying with 3 companions, studying space environment in multiple dimensions
042A	Mightysat	19 Jul 20:09	VAFB	Minotaur	119	534	568	95.7	97.8	Hyperspectral imaging, satellite technology experiments for USAF. Also carries an undeployed tethered experiment.
043A	PAS 9	28 Jul 22:42	OP	Zenit 3SL	3,659	35,782	35,793	1,436.1	0.1	Telecommunications for PanAmSat. Boosted from Sea Launch platform at 0° E, 154° N. In orbit above 58° W
044A	Progress-M1 3	6 Aug 18:27	TT	Soyuz-U	7,300	364	348	91.6	51.5	Resupply ISS. 1st Progress to dock with international station. Brought provisions for future crew. Docked 8 Aug. Undocked and re-entered 1 Nov
045A	Cluster II-5 Rhumba	9 Aug 11:13	TT	Soyuz-U/ Fregat	1,200	18,407	119,370	3,425.0	89.6	2nd set of satellites launched to study magnetosphere, solar wind with companions on orbit since 16 Jul
045B	Cluster II-8 Tango	9 Aug	TT	Soyuz-U/ Fregat	1,200	18,303	119,475	3,425.1	89.6	Formation flying with 3 companions
046A	Brazilsat B4	17 Aug 23:16	KO	Ariane 44LP	1,757	35,777	35,797	1,436.1	0.1	Telecommunications for Brazil. Located above 92° W
046B	Nilesat 102	17 Aug	KO	Ariane 44LP	1,827	35,764	35,808	1,436.1	0.0	Direct broadcast for Egypt. Stationed above 7° W
047A	US 152	17 Aug 23:45	VAFB	Titan 4B		Elements not available				
048A	DM 3	23 Aug 11:05	CC	Delta 3	213	19,506	343.2	27.6		Dummy payload flown to demonstrate Delta 3 capabilities. Used as optical target for USAF imaging studies
049A	Raduga-1 5	28 Aug 20:08	TT	Proton-K	2,000	35,722	35,841	1,435.9	1.7	Military/government communications Stationed above 49° E
050A	Zi Yuan 2	1 Sep 03:25	TY	CZ-4B	1,500	486	496	94.4	97.4	Remote sensing
051A	Sirius 2	5 Sep 09:43	TT	Proton-K	3,765	24,508	47,065	1,436.1	63.4	Digital radio broadcast. 2nd in series, not related to Swedish Sirius
052A	Eutelsat W1R	6 Sep 22:33	KO	Ariane 44P	3,250	35,774	35,803	1,436.1	0.0	Telecommunications. Replaces Eutelsat W1 lost in ground fire. Stationed above 10° E
053A	STS-106	8 Sep 12:45	CC	Atlantis	115,259	386	375	92.1	51.5	Crew: Terrence Wilcutt, Scott Altman, Edward Lu, Richard Mastracchio, Daniel Burbank, Yuri Malenchenko, Boris Morukov. ISS mission, preparing station for habitation. Docked 10 Sept. Undocked 18 Sept, returned 20 Sept. Mission duration: 11 days, 19 h, 11 min
054A	Astra 2B	14 Sep 22:54	KO	Ariane 5	3,315	35,365	36,207	1,436.1	0.1	Telecommunications. Positioned above 28.2° E
054B	GE 7	14 Sep	KO	Ariane 5	1,935	35,783	35,789	1,436.1	0.0	Telecommunications. Stationed above 146° W
055A	NOAA 16	21 Sep 10:22	VAFB	Titan 23G	2,234	848	864	102.1	98.8	Meteorological for National Oceanic and Atmospheric Administration
056A	Cosmos 2372	25 Sep 10:09	TT	Zenit 2	12,000?	208	305	89.6	64.8	Photoreconn
057A	Tiungsat 1	26 Sep	TT	Dneipr	52	632	651	97.6	64.6	Remote sensing, amateur radio
057B	Megsat 1	26 Sep	TT	Dneipr	56	631	650	97.5	64.6	Houses transmitters for commercial use
057C	Unisat	26 Sep	TT	Dneipr	10	637	664	97.7	64.6	Space experiments for educational purposes

Desig	Name	Date, Time	Site	Vehicle	Mass (kg)	Perigee (km)	Apogee (km)	Period (min)	Inc (deg)	Notes
057D	Saudisat 1A	26 Sep 10:05	TT	Dneipr	10	636	655	97.6	64.6	Amateur store/forward communications
057E	Saudisat 1B	26 Sep	TT	Dneipr	10	637	660	97.7	64.6	Amateur store/forward communications
058A	Cosmos 2373	29 Sep 09:30	TT	Soyuz-U	6,600	279	212	89.4	70.3	Photoreconn. Re-entered 14 Nov
059A	GE 1A	1 Oct 22:00	TT	Proton-K	3,593	35,783	35,792	1,436.1	0.1	Telecommunications, for Americom Asia-Pacific. Located above 108.2° E
060A	Nsat 110	6 Oct 23:01	KO	Ariane 42L	3,531	35,784	35,790	1,436.1	0.0	Telecommunications for JSAT Corp. Stationed Above 110° E
061A	HETE 2	9 Oct 05:38	KWI	L1011/ Pegasus	124	593	634	97.0	2.0	High Energy Transient Experiment replacing platform that failed to separate from booster in 1996. First orbital launch commencing from Kwajalein Missile Range
062A	STS-62	11 Oct 23:17	CC	Discovery	114,974	379	390	92.2	51.5	Crew: Brian Duffy, Pamela Melroy, Koichi Wakata, Peter Wisoff, Leroy Chiao, William McArthur, Michael Lopez-Alegria. ISS assembly flight. Returned 24 Oct after 12 days, 21 h, 43 min
063A	Cosmos 2374	13 Oct 14:12	TT	Proton-K	1,400	19,125	19,134	675.7	64.8	Navigation, part of Glonass system
063B	Cosmos 2375	13 Oct	TT	Proton-K	1,400	19,115	19,139	675.7	64.8	Navigation, part of Glonass system
063C	Cosmos 2376	13 Oct	TT	Proton-K	1,400	19,123	19,230	675.7	64.7	Navigation, part of Glonass system
064A	Progress-M 43	16 Oct	TT	Soyuz-U	6,860	336	325	91.1	51.6	Resupply Mir (unoccupied). Docked 20 Oct
065A	DSCS 3B F8	20 Oct 00:40	CC	Atlas 2A	1,173	Elements not available				Communications for US military
066A	Thuraya 1	21 Oct 05:52	OP	Zenit 3SL	5,108	35,753	35,819	1,436.1	6.2	Mobile communications for United Arab Emirates. Positioned above 44° E. Sea Launch from Pacific Ocean site
067A	GE 6	21 Oct 22:00	TT	Proton-K	3,800	35,781	35,793	1,436.1	0.0	Telecommunications for GE Americom. Located above 72° W
068A	EuropeStar 1	29 Oct 05:59	KO	Ariane 44LP	4,167	35,772	35,799	1,436.1	0.0	Telecommunications and digital relay. Stationed above 45° E
069A	Bei Dou 1	3 Oct 16:02	XI	CZ-3A	2,300	35,771	35,801	1,436.1	0.1	'North Dipper' navigation satellite. Stationed above 140° E
070A	Soyuz-TM 31	31 Oct 07:52	TT	Soyuz-U	7,150	361	373	91.9	51.6	Crew: Yuri Gidzenko, Sergei Krikalev, William Shepard. First residents of ISS. Docked 2 Nov. Returned 21 Mar, aboard STS-102 after 14 days, 23 h, 38 min mission
071A	GPS 49	10 Nov 17:14	CC	Delta 2	2,032	20,106	20,272	718.3	55.1	Global Positioning System. Navigation
072A	PAS 1R	16 Nov 01:07	KO	Ariane 5	4,793	34,816	36,779	1,436.1	0.1	Telecommunications for PanAmSat. Located above 45° W
072B	Amsat P3D	16 Nov	KO	Ariane 5	630	344	58,982	1,134.9	6.0	Amateur radio communications
072C	STRV 1C	16 Nov	KO	Ariane 5	100	680	39,790	720.1	6.3	Space environment studies
072D	STRV 1D	16 Nov	KO	Ariane 5	100	603	39,263	707.9	6.2	Space environment studies
073A	Progress-M14	16 Nov 01:39	TT	Soyuz-U	7,300	363	374	91.9	51.6	Supply ISS. Docked 18 Nov. Undocked and re-entered 08 Feb
075A	EO 1	21 Nov 18:24	CC	Delta 2	573	701	703	98.8	98.2	Earth Observing satellite. Remote sensing
075B	SAC C	21 Nov	CC	Delta 2	475	671	707	98.6	98.3	Satellite de Aplicationes Cientificas. Joint remote sensing mission for US, Argentina, with additional international participation
075C	Munin	21 Nov	CC	Delta 2	6	698	1,800	110.5	95.4	Aurora studies
076A	Anik-F1	21 Nov 23:56	KO	Ariane 44L	4,859	33,448	38,270	1,439.8	0.1	Telecommunications for Telesat Canada. Stationed above 107° W
077A	Sirius 3R	30 Nov 19:59	TT	Proton-K	3,765	24,475	47,090	1,435.9	63.4	Digital radio broadcast
078A	STS-97	1 Dec 23:03	CC	Endeavour	102,461	374	380	92.0	51.5	Crew: Brent Jett, Michael Bloomfield, Joseph Tanner, Marc Garneau, Carlos Noriega. ISS assembly mission. Docked 2 Dec. Undocked 9 Dec, returned 11 Dec after 10 days, 19 h, 58 min
079A	EROS A1	5 Dec 12:32	S	Start 1	250	488	502	94.5	97.3	Commercial high-resolution imaging
080A	USA 155	6 Dec 02:47	CC	Atlas 2AS		Elements not available				
081A	Astra 2D	20 Dec 00:26	KO	Ariane 5	1,414	35,887	35,895	1,441.4	0.3	Telecommunications. Stationed above 28.2° E
081B	GE 8	20 Dec	KO	Ariane 5	2,015	35,780	35,794	1,436.1	0.1	Telecommunications. Located above 139° W
081C	LDREX Antenna	20 Dec	KO	Ariane 5	140	213	35,581	627.3	1.9	Large Scale Deployable Reflector Experiment. A 1/2 scale antenna planned for use on future NASDA craft separated from satellite/2nd stage to test deployment
081D	LDREX/ Ariane Stage	20 Dec	KO	Ariane 5	1,232	215	36,037	636.2	2.0	LDREX satellite remained attached to Ariane 5 second stage
082A	Bei Dou 2	20 Dec 16:20	XI	CZ-3A	2,300	35,776	35,797	1,436.1	0.1	Second 'North Dipper' navigation satellite. Positioned above 80° E

UPDATED

Table 2: Satellite Launches 2001

Desig	Name	Date, Time	Site	Vehicle	Mass (kg)	Perigee (km)	Apogee (km)	Period (min)	Inc (deg)	Notes
001A	Shen Zhou 2	9 Jan 17:00	J	CZ-2F	7,600	330	346	91.2	42.5	Anticipated crewed vehicle. Craft carried an assortment of animals during a week-long test flight. Returned 16 Jan, descending by parachute into Inner Mongolia
002A	Turksat 2A	10 Jan 22:9	KO	Ariane 44P	3,535	35,767	35,803	1,436.1	0.1	Telecommunications for Eurasiasat. Stationed above 42° E
003A	Progress-M1 5	24 Jan 04:28	TT	Soyuz-U	7,300	285	306	90.4	51.6	Unmanned cargo vehicle, dispatched to Mir to provide control during station's planned de-orbit. Decayed with Mir 23 Mar
004A	GPS 50	30 Jan 07:55	CC	Delta 2	2,032	20,122	20,242	718.0	55.0	Global Positioning System satellite
005A	Sicral 1	7 Feb 23:06	KO	Ariane 44L	2,596	35,783	35,789	1,436.1	0.1	Military telecommunications for Italy, Positioned above 16.2° E
005B	Skynet 4F	7 Feb	KO	Ariane 44L	1,489	35,718	35,798	1,436.1	3.4	Military telecommunications for UK. Located above 4° W
006A	STS-98	7 Feb 23:13	CC	Atlantis	115,376	369	387	92.1	51.5	Crew: Kenneth Cockrell, Mark Polansky, Robert Curbeam, Marsha Ivins, Thomas Jones. ISS assembly mission; transported Destiny module and Power Data Grapple Fixture. Docked 9 Feb; undocked 16 Feb. Landed at Edwards AFB 20 Feb after a 12 day, 21 h, 20 min flight
006B	Destiny	7 Feb	CC	Atlantis						Laboratory module. Docked with ISS 10 Feb. Power grapple fixture installed on module for later use
007A	ODIN	20 Feb	S	Start 1	250	605	613	96.9	97.8	Astronomical, atmospheric research
008A	Progress-M 44	26 Feb 08:09	TT	Soyuz-U	7,100	373	385	92.1	51.5	Resupply ISS. Docked 28 Feb; undocked and re-entered 16 Apr
009A	Milstar 2 F2	2 Feb 21:20	CC	Titan 4B		Elements not available				Military EHF communications
010A	STS-102	8 Mar 11:42	CC	Discovery	89,924	362	379	91.9	51.5	Crew: James Kelly, Andrew Thomas and 2nd ISS resident team Paul Richards, Yuri Usachev, James Voss, Susan Helms. Also carried Leonardo Multipurpose Logistics Module. Docked 10 Mar; undocked 19 Mar. Returned with 1st ISS residents (Shepard, Gidzenko, Krikalev) 21 Feb. STS-102 flight time 12 days, 19 h, 49 min. 1st ISS resident crew flight time: 140 days, 23 h, 38 min
011A	Eurobird	8 Mar 22:51	KO	Ariane 5	3,050	35,758	35,814	1,436.1	0.0	Telecommunications for Eutelsat. Operational above 28.5° E
011B	B-Sat 2A	8 Mar	KO	Ariane 5	1,317	35,771	35,804	1,436.1	0.6	Telecommunications for B-Sat Corp. Stationed above 110° E
012A	XM-2	18 Mar 22:33	OP	Zenit	4,450	32,785	38,787	1,436.1	0.1	Digital radio for XM Satellite Radio, also called 'Rock'. Located above 85° W. Boosted from Sea Launch platform in Pacific Ocean
013A	Mars Odyssey	7 Apr 15:02	CC	Delta 2	725	Orbit		Mars		Entered Mars orbit 24 Oct. Slated for two-year mission to study surface chemical composition (seeking water ice) and radiation levels
014A	Ekran 21	7 Apr 03:47	TT	Proton-M	1,900	35,773	35,797	1,436.0	1.6	Direct broadcast, last of Ekran series.Located above 99° E
015A	GSAT 1	18 Apr 10:13	SR	GSLV	1,540	33,825	35,768	1,385.8	0.6	Experimental communications targeting remote areas of India. Maiden flight of Geostationary Launch Vehicle. Premature engine shutdown left satellite in orbit lower than intended
016A	STS-100	19 Apr 18:40	CC	Endeavour	103,369	390	392	92.3	51.5	Crew: Kent Rominger, Jeffrey Ashby, Chris Radfield, Scott Parazynski, John Phillips, Umberto Guidoni, Yuri Lonchakov. ISS construction, install Canadian remote manipulator system. Docked 21 Apr; undocked 27 Apr. Returned 5 Jan after 11 days, 21 h, 30 min
017A	Soyuz-TM 32	28 Apr 07:37	TT	Soyuz-U	7,150	385,	397	92.4	51.6	Crew: Talgat Musabayev, Yuri Baturin, Dennis Tito (space tourist). Taxi mission to ISS. Docked 30 Apr. Crew returned aboard Soyuz-TM 31 6 May after 7 day, 22 h, 4 min flight. Soyuz-TM 32 returned with TM-33 crew, 31 Oct
018A	XM-1	8 May 22:10	OP	Zenit 3SL	4,682	35,786	35,788	1,436.1	0.0	Digital audio broadcast satellite. Also called 'Roll' (companion to 'Rock'). Positioned above 85° W

Desig	*Name*	*Date, Time*	*Site*	*Vehicle*	*Mass (kg)*	*Perigee (km)*	*Apogee (km)*	*Period (min)*	*Inc (deg)*	*Notes*
019A	PAS 10	15 May 01:11	TT	Proton-K	3,772	35,774	35,799	1,436.1	0.0	Telecommunications for PamAmSat Corp. Operational above 68.5° E
020A	GeoLITE	18 May 17:45	CC	Delta 2	1,870	Elements not available				Geosynchronous Lightweight Technology Experiment satellite. Built by TRW for NRO
021A	Progress-M1 6	21 May 22:33	TT	Soyuz-FG	7,200	375	399	92.2	51.5	Resupply ISS. Docked 23 May; undocked and re-entered 22 Aug. 1st flight of improved Soyuz vehicle
022A	Cosmos 2377	29 May 17:55	PL	Soyuz-U	6,600	165	358	89.7	67.1	Photoreconn. Reentered 10 Oct
023A	Cosmos 2378	8 Jun 15:08	PL	Cosmos-3M	795	964	1,010	104.8	82.9	Military navigation
024A	Intelsat 901	9 Jun 6:45	KO	Ariane 44L	4,723	35,660	35,661	1,429.7	0.0	Telecommunications. Located above 94° W
025A	Astra 2C	16 Jun 01:49	TT	Proton-K	3,643	35,768	35,805	1,436.0	0.1	Telecommunications for Societe Europeenne des Satellites. Positioned above 19° E
026A	ICO F2	19 Jun 04:41	CC	Atlas 2AS	2,750	10,383	10,392	360.1	44.9	Mobile communcations owned by NEW ICO. First on-orbit entrant in planned constellation of 12 satellites
027A	MAP	30 Jun 19:46	CC	Delta 2	840	L2 point		Enroute to		Microwave Anisotropy Probe. Stationed at L2, 1.5 million km from earth, to produce a full sky picture of cosmic microwave background radiation
028A	STS-104	12 Jul 09:04	CC	Atlantis	116,975	380	398	92.3	51.5	Crew: Steve Lindsay, Charles Hobaugh, Michael Gemhardt, Janet Kavandi, James Reilly. Delivered Quest airlock to ISS. Docked 14 Jul; undocked 22 Jul. Landed at KSC 25 Jul after 12 days, 18 h, 35 min
N/A	Quest	12 Jul	CC	Atlantis	6,064					Joint Airlock attached to ISS 15 Jul
029A	Artemis	12 Jul 21.38	KO	Ariane 5	3,105	30,894	30,957	1,195.1	0.9	Experimental telecommunications. Left in low orbit by launch vehicle malfunction. Attempts made to boost orbit using onboard fuel and experimental electric propulsion system
029B	BSat 2B	12 Jul	KO	Ariane 5	1,298	616	17,444	317.3	3.0	Direct broadcast for Japan. Slated to fly above 110° E
030A	Molniya-3 51	20 Jul 00:17	PL	Molniya-M	1,700	575	39,779	717.8	62.9	Domestic communications
031A	GOES 12	23 Jul 07:23	CC	Atlas 2A	2,279	35,784	35,794	1,436.2	0.2	Geostationary Operational Environmental satellite. Meteorology. Stationed above 30° W
032A	Koronas F	31 Jul 08:00	PL	Tsyklon	2,260	478	522	94.6	82.4	Solar science
033A	DSP 21	6 Aug 7:28	CC	Titan 4B	2,400	Elements not available				Defense Support Program
034A	Genesis	8 Aug 16:13	CC	Delta 2	636	Eccentric	Earth	Highly	Orbit	Stationed near L1 point, 1.5 million km from earth. Collecting samples of solar wind material. Slated to return to earth in 2004
035A	STS-105	10 Aug 21:10	CC	Discovery	116,760	373	402	92.3	51.6	Crew: Scott Horowitz, Fred Sturckow, Patrick Forrester, Daniel Barry, and ISS Crew #3: Frank Culbertson, Mikhail Tyurin, Vladimir Dezhurov. Ferried Leonardo Multi-Purpose Logistics Module and deployed one small satellite. Docked with ISS, 12 Aug; undocked 20 Aug. Returned with ISS Crew #2, who concluded a 167 day, 6 h, 41 min mission. STS 105 landed at KSC, 22 Aug after 11 days, 21 h, 13 min
035B	Simplesat	30 Aug 18:30		STS	52	385	403	92.4	51.6	Microsatellite hosting optical Telescope for science, engineering tests. Decayed 30 Jan 02
036A	Progress-M 45	21 Aug 09:24	TT	Soyuz-U	7,100	385	397	92.4	51.6	Resupplied ISS. Docked 23 Aug. Undocked and reentered 22 Nov
037A	Cosmos 2379	24 Aug 20:34	TT	Proton-K	2,500	35,817	36,081	1,444.4	2.2	Early warning. Positioning above 80° E
038A	LRE	29 Aug 07:00	TA	H-2A	86	279	36,110	638.9	28.6	Laser Ranging Experiment, carried 126 laser reflectors for geodesy
039A	Intelsat 902	30 Aug 06:46	KO	Ariane-44L	4,723	35,778	35,795	1,436.1	0.1	Telecommunications. Stationed Above 63° E
040A	USA 160	8 Sep 15:25	VAFB	Atlas 2AS		Elements not available				
041A	Pir/ Progress-M Module	14 Sep 23:35	TT	Soyuz-U	6,900	329	335	91.1	51.6	Pirs, a docking module carrying an airlock and docking system and the Progress-M standard instrument module were launched together. The unit docked with ISS 17 Sep. Instrument module separated and deorbited 26 Sep
	OrbView 4	21 Sept 18:49	VAFB	Taurus	368					Fail. Vehicle veered off course and then recovered following 1st stage separation; upper stages fired but without sufficient force to achieve orbit

Desig	*Name*	*Date, Time*	*Site*	*Vehicle*	*Mass (kg)*	*Perigee (km)*	*Apogee (km)*	*Period (min)*	*Inc (deg)*	*Notes*
	QuikTOMS	21 Sep	VAFB	Taurus	168					Fail. Carried Total Ozone Mapping Spectrometer
	SBD/Celestis	21 Sep	VAFB	Taurus	73					Fail. Special Bus Design test flight. Slated to remain with 3rd stage, along with Celestis burial canister
042A	Atlantic Bird 2	25 Sep 23:21	KO	Ariane-44P	3,150	35,770	35,799	1,436.3	0.1	Telecommunications. Positioned above 48° W
043A	Starshine 3	30 Sep	KOD	Athena 1	90	193	466	93.9	67.1	Geodetic. 1st orbital launch from Alaskan facility
043B	STP P97-1	30 Sep	KOD	Athena 1	68	790	798	100.8	67.0	Space Test Program, also called Picosat 9, Monitoring ionosphere
043C	PCSat	30 Sep	KOD	Athena 1	10	793	891	100.8	67.1	Amateur radio satellite
043D	Sapphire	30 Sep	KOD	Athena 1	16	791	800	100.8	67.1	Student experiments including infrared sensors, voice synthesizer and digital camera
044A	US 161	5 Oct 21:20	VAFB	Titan 4B		Elements not available				
045A	Raduga 1-6	6 Oct 16:45	TT	Proton-K	2,000	35,784	35,791	1,435.9	1.4	Military, government communications
046A	US 162	11 Oct 02:32	CC	Atlas 2AS		Elements not available				
047A	Quickbird 2	18 Oct 18:51	VAFB	Delta 2	1,028	462	466	93.8	97.2	Commerical imaging
048A	Soyuz-TM 33	21 Oct 08:59	TT	Soyuz-U	7,150	388	399	92.4	51.6	Crew: Viktor Afanaseev, Konstantin Kozeev, Claudia Haignere. Taxi flight to ISS. Docked 23 Oct. Crew returned aboard Soyuz-TM 32, 31 Oct 02 after 9 day, 20 h mission
049A	TES	22 Oct 04:53	SR	PSLV	1,108	551	580	95.9	97.7	Technology Experiment Satellite

Alenia Spazio is the prime contractor for ESA's Integral science satellite delayed beyond its planned launch date in 2002 (ESA) 0137124

Canadarm 2 is the next logical step for Canada's unique role in the Shuttle and ISS, alone among partner nations in designing and erecting the main manipulator systems for both programmes (Canadian Space Agency) 0137139

Milstar F5 antenna element in pre-assembly testing for Boeing Satellite Systems (Boeing Company)
0137127

Desig	*Name*	*Date, Time*	*Site*	*Vehicle*	*Mass (kg)*	*Perigee (km)*	*Apogee (km)*	*Period (min)*	*Inc (deg)*	*Notes*
049B	Proba	22 Oct	SR	PSLV	94	553	677	97.0	97.9	Technology development minisat. Houses infrared spectrometer, space debris detector, earth imager
049C	Bird 2	22 Oct	SR	PSLV		551	581	95.9	97.7	Bispectral IR Detector for earth imaging
050A	Molniya-3 52	25 Oct	PL	Molniya-M	1,700	612	39,885	720.6	62.8	Domestic communications for military, government use
051A	Progress-M1 7	26 Nov 18:24	TT	Soyuz-U	7,200	230	244	89.2	51.6	Resupply ISS. Attempted docking 28 Nov but docking port obstruction delayed completion of manoeuvre until 3 Dec when impediment was removed during EVA
052A	DirecTV 4S	27 Nov	KO	Ariane 44LP	4,300	35,781	35,783	1,435.8	0.0	Direct broadcast for DirecTV. Stationed above 1° W
053A	Cosmos 2380	1 Dec 18:04	TT	Proton-K	1,400	19,115	19,144	675.7	64.8	Glonass navigation system
053B	Cosmos 2381	1 Dec	TT	Proton-K	1,400	19,069	19,191	675.7	64.8	Glonass navigation system
053C	Cosmos 2382	1 Dec	TT	Proton-K	1,400	19,113	19,147	675.7	64.8	Glonass navigation system
054A	STS-108	5 Dec 22:19	KSC	Endeavour	116,000?	388	390	92.3	51.6	Crew: Dominic Gorie, Mark Kelly, Linda Goodwin, Daniel Tani, and ISS-4 crew Yury Onufrienko, Carl Walz, Daniel Bursch. Docked 7 Dec. Ferred Raffaello cargo module; deployed one payload. Returned, 17 Dec, with ISS-3 crew, who concluded a 128 day, 20 h, 45 min flight. STS-103 flight time was 11 days, 19 h, 36 min
054B	Starshine II	16 Dec 15:02		Endeavour	39	355	382	91.9	51.6	
055A	Jason	7 Dec 15:07	VAFB	Delta 2	500	1,319	1,333	112.6	66.0	Ocean monitoring, follow-on to Topex Poseidon. Houses altimeter and microwave radiometer. Joint NASA/JPL/CNES mission
055B	Timed	7 Dec	VAFB	Delta 2	587	627	629	97.2	74.0	Thermosphere-Ionosphere-Mesosphere Energetics and Dynamics. Monitoring solar, auroral and atmospheric properties
056A	Meteor-3M1	10 Dec	TT	Zenit-2	2,500	996	1,016	105.3	99.7	Meteorology, housing visible light, IR sensors, and NASA SAGE 3 instrument for ozone studies
056B	Kompas	10 Dec	TT	Zenit-2	80	987	1,014	105.1	99.7	Geophysics studies
056C	BADR 2	10 Dec	TT	Zenit-2	70	986	1,014	105.1	99.6	Earth imaging
056D	Maroc-Tubsat	10 Dec	TT	Zenit-2	45	985	1,014	105.1	99.7	Earth imaging, store/dump communications
056E	Reflector	10 Dec	TT	Zenit-2	6	986	1,014	105.1	99.7	Calibration sphere for laser imaging, optical sensing tests
057A	Cosmos 2383	21 Dec	TT	Tsyklon-M	3,000	405	417	92.8	65.0	Elint
058A	Cosmos 2384	28 Dec 03:24	PL	Tsyklon	225	1,417	1,432	114.3	82.5	Store/dump communications for military use
058B	Cosmos 2385	28 Dec	PL	Tsyklon	225	1,417	1,426	114.2	82.5	Store/dump communications for military use
058C	Cosmos 2386	28 Dec	PL	Tsyklon	225	1,415	1,419	114.6	82.5	Store/dump communications for military use
058D	Gonets-D1 10	28 Dec	PL	Tsyklon	225	1,412	1,418	114.1	82.5	Store/dump communications for civil use
058E	Gonets-D1 11	28 Dec	PL	Tsyklon	225	1,417	1,418	114.1	82.5	Store/dump communications for civil use
058F	Gonets-D1 12	28 Dec	PL	Tsyklon	225	1,404	1,418	114.0	82.5	Store/dump communications for civil use

Table 2: Satellite Launches 2002

Design	Name	Date, Time	Site	Vehicle	Mass (kg)	Perigee (km)	Apogee (km)	Period (min)	Inc (deg)	Notes
001A	Milstar2 F5	16 Jan 00:30	CC	Titan 4B	4,545	elements not available				Military EHF communications, fourth flight vehicle of five
002A	Insat 3C	23 Jan 23:47	KO	Ariane 44L	2,750	geosynchronous orbit				India's S-band/C-band communications satellite
003A	MDS-1	4 Feb 02:45	T	H-2A	480	500	35,696	635	28/5	Mission Demonstration Satellite 1 named Tsubasa to test electronic components
003B	DASH	4 Feb 02:45	T	H-2A	89	474	35,740	636	28.5	Technology demonstration satellite for ISAS
004A	HESSI	5 Feb 20:58	CC	Pegasus XL	293	579	607	96.5	38.0	High Energy Solar Spectroscopic Imager for NASA
005A	Iridium 91	11 Feb 17:44	VAFB	Delta	657	660	660	98.0	86.6	Iridium LEO satellite constellation
005B	Iridium 90									
005C	Iridium 94									
005D	Iridum 95									
005E	Iridum 96									
006A	Echostar 7	21 Feb 12:43	CC	Atlas 3B	4,020	geosynchronous orbit				Telecommunications for DTH and data
007A	Intelsat 904	23 Feb 02:00	KO	Ariane 44L	4,700	geosynchronous orbit				C-band/Ku-band telecommunications
008A	Cosmos 2387	25 Feb 17:25	PL	Soyuz-U		176	369	89.6	67.1	Russian military reconnaissance
009A	Envisat 1	1 Mar 01:07	KO	Ariane 5	8,100	785	791	100.6	98.6	ESA environmental and remote-sensing monitoring satellite
010A	STS-109	1 Mar 11:22	KSC	Columbia	116,989	486	578	95.3	28.5	Crew: Scott Altman, Duane Carey, John Grunsfeld, Nancy Currie, Richard Linnehan, James Newman, Michael Massimino. Fourth servicing mission to HST. Total flight time 10 days, 22 h, 9 min 51 s
011A	TDRS-9	8 Mar 22:59	CC	Atlas 2A	1,781	geosynchronous orbit				Second of three second-generation TDRS satellite
012A	GRACE 1	17 Mar 09:21	PL	Rockot	500	483	508	94.5	89.0	One of a pair of earth research satellite for NASA-DLR, named Tom
012B	GRACE 2	17 Mar 09:21	PL	Rockot	500	483	508	94.5	89.0	Second NASA-DLR, named Gerry
013A	Progress M1-8	21 Mar 20:13	TT	Soyuz-U	7,100	389	394	92.4	51.6	Re-supplied ISS. Docked 24 Mar
014A	Shenzhou 3	25 Mar 14:00	J	CZ-2F	7,600	332	337	91.2	42.4	Unmanned test of China's prospective manned spacecraft. Landed 1 Apr
015A	JCSat 8	29 Mar 01:29	KO	Ariane 44L	2,500	geosynchronous orbit				DTH TV satellite for JCSat
015B	Astra 3A	29 Mar 01:29	KO	Ariane 44L	1,500	geosynchronous orbit				TV broadcast satellite for SES-Astra
016A	Intelsat 903	30 Mar 17:25	TT	Proton-K	4,700	geosynchronous orbit				C-band/Ku-band telecommunications
017A	Cosmos 2388	1 Apr 22:07	PL	Molniya-M		518	39,582	718	62.9	Russian military communications satellite
018A	STS-110	8 Apr 20:44	KSC	Atlantic	116,611	115	225	88.3	51.6	Crew: Michael Bloomfield, Stephen Frick, Rex Walheim, Ellen Ochoa, Lee Morin, Jerry Ross, Steven Smith. ISS S0 truss delivery. Docked 9 Apr, undocked 17 Apr. Total flight duration 10 days 19 h, 42 min, 38 s
019A	NSS7	16 Apr 23:01	KO	Ariane 44LP	4,700	geosynchronous orbit				New Skies Satellite C-band/Ku-band
020A	Soyuz TM-34	25 Apr 06:26	TT	Soyuz-U	7,150	383	393	92.4	51.6	Crew: Yuri Gidzenko, Mark Shuttleworth (space tourist), Roberto Vittori. Soyuz replacement mission delivered Soyuz TM-34 to ISS. Docked 27 Apr. Crew returned in TM-33 on 5 May
021A	SPOT 5	4 May 00:31	KO	Ariane 42P	3,024	825	826	101.4	98.8	CNES earth imaging satellite
022A	Aqua	4 May 09:55	VAFB	Delta 2	2,929	673	686	98.4	98.2	NASA hydrology survey satellite previously known as EOS PM-1
023A	DirecTV5	7 May 17:00	TT	Proton-K	3,633	geosynchronous orbit				Ku-band digital TV broadcast
024A	Haiyang 1	15 May 01:50	TY	CZ-4B	428	851	872	102.2	98.8	Chinese earth observation satellite
024B	Fengyun 1D	15 May 01:50	TY	CZ-4B	360	851	872	102.2	98.8	Chinese oceanographic imaging satellite
025A	Ofeq 5	28 May 14.55	PB	Shavit		369	770	96	143.5	Israeli technology development and observation satellite
026A	Cosmos 2389	28 May 18.15	PL	Cosmos 3-M		950	1,017	105	83	Russian military communications satellite
027A	Intelsat 905	5 Jun 06:44	KO	Ariane 44L	4,700	geosynchronous orbit				Telecommunications satellite for Intelsat
028A	STS-111	5 Jun 21:23	KSC	Endeavour	116,520	349	387	91.9	51.6	Crew: Kenneth Cockrell, Paul Lockhart Philippe Perrin, Franklin Chang-Diaz, Peggy Whitson Valery Korzun, Sergei Treschev. Rotation of Expedition 4 and 5 crews, returning with Carl Walz, Daniel Bursch and Yury Onu Frienko, leaving Whitson, Korzun and Treschev. Total flight time 13 days, 20 h, 34 min, 52 s
029A	Express 4A	10 Jun 01:15	TT	Proton-K	1,600	geostationary orbit				Telecommunications Satellite
030A	Galaxy 3C	15 Jun 22.39	OP	Zenit 3SL	2,800	geostationary orbit				C-band/Ku-band telecommunications
031A	Iridium 97	20 Jun 09:34	PL	Rockot	690	658	670	98.0	86.6	LEO communications satellites for mobile telephone users
032A	NOAA-17	24 Jun 18:23	VAFB	Titan 2	1,500	807	823	101.2	98.8	US Meteorological satellite
033A	Progress M46	26 Jun 06:36	TT	Soyuz-U	7,100	193	245	88.6	51.7	Resupplied ISS. Docked 29 Jun
034A	CONTOUR	3 Jul 06:47	CC	Delta 2	970	183	108,498	2,482	30.2	Heliospheric Comet Nucleus Tour spacecraft destined for comet Encke in November 2003 and Schwassmann-Wachmann 3 in 2006
035A	STELLAT 5	5 Jul 23:22	KO	Ariane 5	4,042	geostationary orbit				C-band/Ku-band telecommunications for Stellat in France
035B	N-Star 3	5 Jul 23:22	KO	Ariane 5	1,625	geostationary orbit				Japanese mobile telecoms and maritime communications satellite
036A	Cosmos 2390	8 Jul 06:36	PL	Cosmos 3M	227	1,467	1,507	115.7	82.5	Strela-3 class military communications
036B	Cosmos 2391	8 Jul 06:36	PL	Cosmos 3M	227	1,467	1,507	115.7	82.5	Strela-3 class military communications
037A	Cosmos 2391	25 Jul 15:13	TT	Proton-K		1,507	1,834	120	63.5	Arkan-2 military reconnaissance satellite
038A	Hot Bird 6	21 Aug 22:05	CC	Atlas 5	4,900	geostationary orbit				Eutelsat communications satellite
039A	Echostar 8	22 Aug 05:15	TT	Proton-K	4,700	geostationary orbit				US telecommunications satellite

Design	*Name*	*Date, Time*	*Site*	*Vehicle*	*Mass (kg)*	*Perigee (km)*	*Apogee (km)*	*Period (min)*	*Inc (deg)*	*Notes*
040A	Atlantic Bird	28 Aug 22:05	KO	Ariane 5	2,700	geostationary orbit				Eutelsat telecommunications satellite
040B	MSG1	28 Aug 22:05	KO	Ariane 5	2,000	geostationary orbit				Meteosat Second Generation for ESA stationed at 0 deg longitude
041A	Intelsat 906	6 Sep 06:44	KO	Ariane 44L	4,700	geostationary orbit				Intelsat telecommunications satellite
042A	USERS	10 Sep 08:20	T	H2A	1,700	504	519	94.9	30.9	Japanese microgravity experimental satellite (Unmanned Space Experiment Recovery System)
042B	DRTS	10 Sep 08:20	T	H2A	2,800	geostationary orbit				Japanese (Data Relay Transponder Satellite) communications
043A	METSAT 1	12 Sep 10:24	S	PSLV-C4	1,050	geostationary orbit				Indian meteorological satellite
044A	Hispasat 1D	18 Sep 22:04	CC	Atlas 2AS	3,300	geostationary orbit				Telecommunications satellite
045A	Progress M1-9	25 Sep 16:58	TT	Soyuz-U	7,100	281.5	323.5	90.6	51.64	ISS resupply flight. Docked 29 Sep
046A	Nadezhda 7	26 Sep 15:30	PL	Cosmos 3M		987	1,022	102	83	Russian navigation satellite
047A	STS-112	7 Oct 19:46	KSC	Atlantis	116,538	273	405	91.3	51.6	Crew: Jeffrey Ashnby, Pamela Melroy, David Wolff, Sandra Magnus, Piers Sellers, Fyodor Yurchikhin. ISS assembly flight delivering S-1 truss. Total flight time 10 days, 19 h, 57 min, 49 s
048A	INTEGRAL	17 Oct 04.41	TT	Proton-K	4,000	639	153,000	3,960	51.7	ESA International Gamma Ray Astrophysics Laboratory
049A	JB-3 2	27 Oct 03:17	TY	CZ-4B		470	483	94.1	97.4	Chinese remote-sensing satellite
050A	Soyuz TMA-1	30 Oct 03:11	TT	Soyuz-U		278	295	90.2	51.62	Crew: Sergei Zalyotin, Yuri Lonchakov, Frank DeWinne (ESA). Soyuz replacement. Docked 1 Nov. Crew returned in TM-34 9 Nov
051A	W5	20 Nov 22:39	CC	Delta 4	3,000	geostationery orbit				Eutelsat telecommunication satellite on inaugural Delta 4 launch vehicle
052A	STS-113	24 Nov 00:50	KSC	Endeavour	116,460	379	379	92.3	51.6	Crew: Jim Wetherbee, Paul Lockhart, Michael Lopez-Alegria, John Herrington, Ken Bowersox, Nikolai Budarin, Donald Pettit. Crew replacement and P1 truss delivery flight. Returned to earth with ISS Expedition 5 crew Korzun, Whitson and Treschev leaving Bowersox, Budarin and Pettiton Expedition 6. Total flight time 13 days, 18 h, 47 min, 25 s
053A	Astra 1K	25 Nov 23:04	TT	Proton-K	5,000	geostationary orbit				Telecommunication satellite for SES-Astra
054A	Alsat 1	28 Nov 06:07	PL	Cosmos-3M	90	700	700	98.5	98.0	Algerian imaging minisatellite
054B	Mozhayets	28 Nov 06:07	PL	Cosmos-3M	90	700	700	98.5	98.0	Russian experimental minisatellite
055A	TDRS-10	5 Dec 14:42	CC	Atlas 2A	3,200	geostationary orbit				NASA TDRS satellite
056A	Adeos 2	14 Dec 01:31	T	H2	3,700	806	807	101	98.7	Japanese remote-sensing and global climate research satellite
056B	Fedsat	14 Dec 01:31	T	H2	50	793	806	100.9	98.7	Australian technology test satelllite
056C	WEOS	14 Dec 01:31	T	H2		791	805	100.8	98.7	Whale Ecology Observation Satellite for Japan
056D	Micro-Labsat	14 Dec 01:31	T	H2		789	805	100.8	98.7	Japanese technology experimental satellite
057A	NSS-6	20 Dec 23:04	KO	Ariane 4	4,500	geostationary orbit				Netherlands communication satellite
058A	Rubin 2	20 Dec 17:00	TT	Dnepr		635	679	97.9	64.6	German microsatellite operated via internet link from a PC using Orbcomm mobile managing satellite network
058B	Latinsat B	20 Dec 17:00	TT	Dnepr	12	632	702	98.1	64.6	Argentinian picosatellite to monitor mobile goods for transport industry
058C	Saudisat 1C	20 Dec 17:00	TT	Dnepr		633	690	98.0	64.6	Saudi Arabian picosatellite
058D	Unisat 2	20 Dec 17:00	TT	Dnepr	12	636	667	97.8	64.6	Italian picosatellite with cameras for debris/aerosol detection
058E	Trailblazer	20 Dec 17:00	TT	Dnepr	100	639	645	97.6	64.6	Mockup of possible lunar orbiters and landers carrying cremated remains
058H	Latinsat A	20 Dec 17:00	TT	Dnepr	12	932	702	98.1	64.6	Argentinian picosat to monitor mobile goods for transport industry
059A	Cosmos 2393	24 Dec 12:20	PL	Molniya-M		517	39,187	704.6	62.8	Russian military reconnaissance satellite
060A	Cosmos 2394	25 Dec 07:38	TT	Proton-K		19,127	19,137	676	64.8	Glonass navigation satellite
060B	Cosmos 2395	25 Dec 07:38	TT	Proton-K		19,127	19,137	676	64.8	Glonass navigation satellite
060C	Cosmos 2396	25 Dec 07:38	TT	Proton-K		19,127	19,137	676	64.8	Glonass navigation satellite
061A	Shenzhon 4	29 Dec 16:40	J	CZ-2F	7,600	196	329	89.8	42.4	Fourth test flight of China's manned spacecraft landed 5 Jan 2003
062A	Nimiq 2	29 Dec 23:17	TT	Proton-M	3,600	geostationary orbit				Canadian communications satellite in Ku-band

Table 3: Solar System Exploration Chronology

The following table presents a chronological listing of known and, in some cases, surmised planetary missions. Speculative material on early Soviet failures abounds; the flights included here are the more probable ones. Beginning in 1964, Cosmos numbers were assigned to Soviet payloads stranded in Earth parking orbit. US Apollo manned lunar and Helios solar missions are also included.

	Launch	*Nation*	*Launcher*	*Mission Type*	*Comments*
1958					
Pioneer 0	Aug 17	US	Thor-Able	Lunar orbiter	Stage 1 explosion
Luna-1958A	Sep 23	USSR	Vostok	Lunar impact	Structural failure T+92 s
Pioneer 1	Oct 11	US	Thor-Able	Lunar orbiter	Failed to reach Earth escape velocity; attained 113,854 km during 43 h 17 m flight. 38 kg
Luna-1958B	Oct 12	USSR	Vostok	Lunar impact	Launcher exploded T+100 s
Pioneer 2	Nov 8	US	Thor-Able	Lunar orbiter	Stage 3 failure. 39.5 kg
Luna-1958C	Dec 4	USSR	Vostok	Lunar impact	Stage 1 failure T+245 s
Pioneer 3	Dec 6	US	Juno II	Lunar flyby	Failed to reach Earth escape velocity, re-entered from 102,333 km apogee. 5.9 kg
1959					
Luna 1	Jan 2	USSR	Vostok	Lunar impact	First lunar flyby (5,995 km)
Pioneer 4	Mar 3	US	Juno II	Lunar flyby	Partial success; 59,983 km flyby. 5.9 kg
Luna-1959A	Jul 18	USSR	Vostok	Lunar impact	Failed to reach Earth orbit
Luna 2	Sep 12	USSR	Vostok	Lunar impact	First lunar impact
Luna 3	Oct 4	USSR	Vostok	Lunar flyby	First far side images (barycentric)
Atlas-Able 4	Nov 26	US	Atlas-Able	Lunar orbiter	Launcher failure
1960					
Pioneer 5	Mar 11	US	Thor-Able	Solar monitor	0.81 × 0.99 AU solar orbit; returned solar wind data to 26 Jun 1960. 43 kg
Luna-1960A	Apr 15	USSR	Vostok	Lunar flyby	Failed to reach Earth orbit (premature stage 2 cutoff); Luna 3-type
Luna-1960B	Apr 16	USSR	Vostok	Lunar flyby	Strap-on failed to ignite, broke up near pad; Luna 3-type
Atlas-Able 5A	Sep 25	US	Atlas-Able	Lunar orbiter	Launcher failure
Mars-1960A	Oct 10	USSR	Molniya	Mars flyby	Failed to reach Earth orbit
Mars-1960B	Oct 14	USSR	Molniya	Mars flyby	Failed to reach Earth orbit
Atlas-Able 5B	Dec 15	US	Atlas-Able	Lunar orbiter	Launcher failure
1961					
Venera-1961A	Feb 4	USSR	Molniya	Venus impact	Failed to leave Earth orbit
Venera 1	Feb 12	USSR	Molniya	Venus impact	Contact lost en route; 100,000 km flyby. 644 kg
Ranger 1	Aug 23	US	Atlas-Agena B	Lunar test	High-altitude lunar test; failed to leave low orbit
Ranger 2	Nov 18	US	Atlas-Agena B	Lunar test	As Ranger 1
1962					
Ranger 3	Jan 26	US	Atlas-Agena B	Lunar lander	Launch error, lunar flyby
Ranger 4	Apr 23	US	Atlas-Agena B	Lunar lander	Spacecraft failure; first US lunar impact
Mariner 1	Jul 22	US	Atlas-Agena B	Venus flyby	Launcher failure; error in flight guidance equations
Venera-1962A	Aug 25	USSR	Molniya	Venus capsule	Failed to leave Earth orbit
Mariner 2	Aug 27	US	Atlas-Agena B	Venus flyby	First successful Venus flyby; 34,830 km after 109 days. Surface T far higher than expected
Venera-1962B	Sep 1	USSR	Molniya	Venus capsule	Failed to leave Earth orbit
Venera-1962C	Sep 12	USSR	Molniya	Venus flyby	Failed to leave Earth orbit
Ranger 5	Oct 18	US	Atlas-Agena B	Lunar lander	Spacecraft failure; close lunar flyby
Mars-1962A	Oct 24	USSR	Molniya	Mars flyby	Final stage exploded in Earth orbit. 893.5 kg
Mars 1	Nov 1	USSR	Molniya	Mars flyby	Contact lost 21 Mar 1963; 200,000 km flyby
Mars-1962B	Nov 4	USSR	Molniya	Mars lander?	Failed to leave Earth orbit
1963					
Luna-1963A	Jan 4	USSR	Molniya	Lunar lander	Failed to leave Earth orbit
Luna-1963B	Feb 2	USSR	Molniya	Lunar lander	Failed to reach Earth orbit
Luna 4	Apr 2	USSR	Molniya	Lunar lander	Lunar miss (8,529 km)
Cosmos 21	Nov 11	USSR	Molniya	Venus test	Failed to leave Earth orbit
1964					
Ranger 6	Jan 30	US	Atlas-Agena B	Lunar impact	Cameras failed
Venera-1964A	Feb 19	USSR	Molniya-M	Venus flyby	Failed to reach Earth orbit
Venera-1964B	Mar 1	USSR	Molniya-M	Venus flyby	Tentative identification; launcher failure?
Venera-1964C	Mar 21	USSR	Molniya-M	Venus flyby	Failed to reach Earth orbit
Cosmos 27	Mar 27	USSR	Molniya-M	Venus flyby	Failed to leave Earth orbit
Zond 1	Apr 2	USSR	Molniya-M	Venus flyby	Contact lost May; 100,000 km flyby 19 Jul. 825 kg
Luna-1964B	Apr 20	USSR	Molniya-M	Lunar lander	Failed to reach Earth orbit
Ranger 7	Jul 28	US	Atlas-Agena B	Lunar impact	Successful image return
Mariner 3	Nov 5	US	Atlas-Agena D	Mars flyby	Payload shroud failed to separate. Intended 13,840 km flyby
Mariner 4	Nov 28	US	Atlas-Agena D	Mars flyby	First successful Mars flyby (14 Jul 1965; 9,844 km) + imaging. 21 TV images (1% surface) showed arid, lunar-like surface
Zond 2	Nov 30	USSR	Molniya	Mars flyby	Contact lost May 1965; 1,500 km flyby 6 Aug 1965. 1,145 kg
1965					
Ranger 8	Feb 17	US	Atlas-Agena B	Lunar impact	Successful image return
Cosmos 60	Mar 12	USSR	Molniya	Lunar lander	Failed to leave Earth orbit
Ranger 9	Mar 21	US	Atlas-Agena B	Lunar impact	Successful image return
Luna-1965A	Apr 10	US	Molniya	Lunar lander	Failed to reach Earth orbit
Luna 5	May 9	USSR	Molniya-M	Lunar lander	Retro failed; lunar impact
Luna 6	Jun 8	USSR	Molniya-M	Lunar lander	Missed Moon (161,000 km)
Zond 3	Jul 18	USSR	Molniya	Lunar flyby	Photographed lunar far side; Mars probe test. 1,145 kg
Luna 7	Oct 4	USSR	Molniya	Lunar lander	Failed landing sequence, crashed
Venera 2	Nov 12	USSR	Molniya-M	Venus flyby	Contact lost en route; 23,810 km flyby. 963 kg
Venera 3	Nov 16	USSR	Molniya	Venus capsule	Contact lost before entry. 960 kg
Cosmos 96	Nov 23	USSR	Molniya-M	Venus capsule	Failed to leave Earth orbit
Luna 8	Dec 3	USSR	Molniya-M	Lunar lander	Retro delay; lunar impact
Pioneer 6	Dec 16	US	Delta	Solar orbiter	0.81 × 0.985 AU orbit; still operable for brief periods
1966					
Luna 9	Jan 31	USSR	Molniya-M	Lunar lander	First semi-soft landing; first surface images
Cosmos 111	Mar 1	USSR	Molniya-M	Lunar orbiter?	Failed to leave Earth orbit

	Launch	Nation	Launcher	Mission Type	Comments
Luna 10	Mar 31	USSR	Molniya-M	Lunar orbiter	First lunar orbiter; in contact 56 days
Luna-1966A	Apr 30	USSR	Molniya-M	Lunar orbiter	Failed to reach Earth orbit
Surveyor 1	May 30	US	Atlas-Centaur	Lunar lander	First true soft landing
Explorer 33	Jul 1	US	Delta	Lunar orbiter	Entered high Earth orbit
Lunar Orbiter 1	Aug 10	US	Atlas-Agena D	Lunar orbiter	Second lunar orbiter; photographic survey
Pioneer 7	Aug 17	US	Delta	Solar monitor	1.012 × 1.125 AU orbit; still operable for brief periods
Luna 11	Aug 24	USSR	Molniya-M	Lunar orbiter	Third lunar orbiter; possibly carried failed TV system
Surveyor 2	Sep 20	US	Atlas-Centaur	Lunar lander	Impacted Moon
Luna 12	Oct 22	USSR	Molniya-M	Lunar orbiter	Fourth lunar orbiter; TV imaging
Lunar Orbiter 2	Nov 6	US	Atlas-Agena D	Lunar orbiter	Fifth lunar orbiter; photographic survey
Luna 13	Dec 21	USSR	Molniya-M	Lunar lander	Third lunar landing
1967					
Lunar Orbiter 3	Feb 5	US	Atlas-Agena D	Lunar orbiter	Sixth lunar orbiter; photographic survey
Surveyor 3	Apr 17	US	Atlas-Centaur	Lunar lander	Fourth lunar lander
Lunar Orbiter 4	May 4	US	Atlas-Agena D	Lunar orbiter	Seventh lunar orbiter; photographic survey
Venera 4	Jun 12	USSR	Molniya-M	Venus capsule	First successful Venus atmospheric probe (18 Oct). 1,106 kg (383 kg capsule); tx to 27 km altitude
Mariner 5	Jun 14	US	Atlas-Agena D	Venus flyby	Second successful Venus flyby; 3,990 km 19 Oct
Cosmos 167	Jun 17	USSR	Molniya-M	Venus capsule	Failed to leave Earth orbit
Surveyor 4	Jul 14	US	Atlas-Centaur	Lunar lander	Soft-landed? Contact lost
Explorer 35	Jul 19	US	Delta	Lunar orbiter	Eighth lunar orbiter
Lunar Orbiter 5	Aug 1	US	Atlas-Agena D	Lunar orbiter	Ninth lunar orbiter; photographic survey
Surveyor 5	Sep 8	US	Atlas-Centaur	Lunar lander	Fifth lunar lander
Zond-1967A	Sep 28	USSR	Proton-K + Block D	Lunar test	Stage 1 failure; Zond 4-type mission
Surveyor 6	Nov 7	US	Atlas-Centaur	Lunar lander	Sixth lunar lander
Zond-1967B	Nov 22	USSR	Proton-K + Block D	Lunar test	Stage 2 engine failure; Zond 4-type mission
Pioneer 8	Dec 13	US	Delta	Solar monitor	0.990 × 1.087 AU orbit; still operable for brief periods
1968					
Surveyor 7	Jan 7	US	Atlas-Centaur	Lunar lander	Seventh lunar lander
Luna 1968A	Feb 7	USSR	Molniya-M	Lunar orbiter	Failed to reach Earth orbit
Zond 4	Mar 2	USSR	Proton-K + Block D	Lunar distance	Test of manned craft
Luna 14	Apr 7	USSR	Molniya-M	Lunar orbiter	Tenth lunar orbiter
Zond-1968A	Apr 23	USSR	Proton-K + Block D	Lunar test	Control system short circuit shut down stage 2 at 260 s
Zond 5	Sep 14	USSR	Proton-K + Block D	Lunar flyby	Earth return; test of manned craft
Pioneer 9	Nov 8	US	Delta	Solar monitor	Successful; 0.756 × 0.990 AU orbit
Zond 6	Nov 10	USSR	Proton-K + Block D	Lunar flyby	Earth return; test of manned craft
Apollo 8	Dec 21	US	Saturn 5	Lunar orbiter	First manned lunar orbiter
1969					
Venera 5	Jan 5	USSR	Molniya-M	Venus capsule	Atmospheric entry (16 May), tx to 25 km altitude. 1,130 kg (405 kg capsule)
Venera 6	Jan 10	USSR	Molniya-M	Venus capsule	Atmospheric entry (17 May), tx to 12 km altitude. 1,130 kg (405 kg capsule)
Zond-1969A	Jan 20	USSR	Proton-K + Block D	Lunar test	One stage 2 engine shut down 25 s early
Luna-1969A	Feb 19	USSR	Proton-K + Block D	Lunar rover	Lunokhod. Exploded at 40 s
Zond L1S	Feb 21	USSR	SL-15/N-1	Circumlunar	Exploded T+70 s. N1 test but Zond cameras to image potential manned landing sites. Mockup lander
Mariner 6	Feb 25	US	Atlas-Centaur	Mars flyby	Imaging flyby; 3,412 km 31 Jul, 24 near-encounter images concentrating on equatorial region
Mariner 7	Mar 27	US	Atlas-Centaur	Mars flyby	Imaging flyby; 3,524 km 5 Aug, 33 near-encounter images. Confirmed ice cap mainly CO_2 + some H_2O ice
Mars-1969A	Mar 27	USSR	Proton-K + Block D	Mars lander?	Stage 3 exploded at 438 s
Mars-1969B	Apr 2	USSR	Proton-K + Block D	Mars lander?	Early stage 1 failure, crashed nearby pad
Apollo 10	May 18	US	Saturn 5	Lunar orbiter	Second manned lunar orbiter
Luna-1969B	Jun 14	USSR	Proton-K + Block D	Lunar sampler	Launcher failure
Zond L1S	Jul 3	USSR	SL-15/N-1	Circumlunar	Exploded T+5 s. N1 test but Zond cameras to image potential manned landing sites. Mockup lander
Luna 15	Jul 13	USSR	Proton-K + Block D	Lunar sampler	11th unmanned lunar orbiter; crashed landing attempt (Luna 16 type)
Apollo 11	Jul 16	US	Saturn 5	Lunar lander	First manned lunar landing (eighth in all)
Zond 7	Aug 7	USSR	Proton-K + Block D	Lunar flyby	Earth return; test of manned craft
Cosmos 300	Sep 23	USSR	Proton-K + Block D	Lunar sampler	Failed to leave Earth orbit
Cosmos 305	Oct 22	USSR	Proton-K + Block D	Lunar sampler	Failed to leave Earth orbit
Apollo 12	Nov 14	US	Saturn 5	Lunar lander	Second manned landing (ninth in all)
1970					
Luna-1970A	Feb 6	USSR	Proton-K + Block D	Lunar sampler	Failed to reach Earth orbit. Luna 16-type
Apollo 13	Apr 11	US	Saturn 5	Lunar lander	Manned Earth return/flyby after craft failure
Venera 7	Aug 17	USSR	Molniya-M	Venus capsule	23 min data after landing 15 Dec. 1,180 kg (500 kg capsule)
Cosmos 359	Aug 22	USSR	Molniya-M	Venus capsule	Failed to leave Earth orbit
Luna 16	Sep 12	USSR	Proton-K + Block D	Lunar sampler	13th unmanned lunar orbiter; first sample return (=~100 g)
Zond 8	Oct 20	USSR	Proton-K + Block D	Lunar flyby	Earth return; test for manned craft
Luna 17	Nov 10	USSR	Proton-K + Block D	Lunar rover	14th unmanned orbiter; first rover: Lunokhod 1 operated for 11 months, travelled 10,540 m, returned >22,000 TV images
1971					
Apollo 14	Jan 31	US	Saturn 5	Lunar lander	Third manned landing, 12th in all
Mariner 8	May 8	US	Atlas-Centaur	Mars flyby	Booster failure; Centaur autopilot failure
Cosmos 419	May 10	USSR	Proton-K + Block D	Mars orbiter	Stage 4 failed to ignite to leave Earth orbit because of incorrect timer setting. Orbiter-only to beat Mariner 8/9 as first Mars orbiter
Mars 2	May 19	USSR	Proton-K + Block D	Mars orbit/land	Second in Mars orbit; 27 Nov landing failed during dust storm. 4,650 kg
Mars 3	May 28	USSR	Proton-K + Block D	Mars orbit/land	Third in Mars orbit; 2 Dec lander failed at 45°S/158°W. Claimed 20 s of TV signals received, but dust storm in progress; claimed in 1993 there were no signals – a propaganda lie to claim first 'successful' landing. Also carried a small tethered rover
Mariner 9	May 30	US	Atlas-Centaur	Mars orbiter	First in Mars orbit, 13 Nov; mapping survey (7,329 images) in 349 days' operations. Mapped most of surface at 1–2 km resolution (2% at 100–300 m)
Apollo 15	Jul 26	US	Saturn 5	Lunar lander	Fourth manned landing; 13th in all
Apollo 15 subsat	Aug 4	US	Apollo 15	Lunar orbiter	Released from Apollo 15
Luna 18	Sep 2	USSR	Proton-K + Block D	Lunar sampler	Failed landing attempt
Luna 19	Sep 28	USSR	Proton-K + Block D	Lunar orbiter	Returned images from orbit; contact >1 yr
1972					
Luna 20	Feb 14	USSR	Proton-K + Block D	Lunar sampler	Second successful unmanned sample return (=~50 g?); 14th landing in all
Pioneer 10	Mar 2	US	Atlas-Centaur	Jupiter flyby	First jovian flyby; still operating (contact to 67 AU in 1997 projected)
Venera 8	Mar 27	USSR	Molniya-M	Venus capsule	Successful atmospheric probe; tx for 50 min after touchdown 22 Jul. 1,184 kg
Cosmos 482	Mar 31	USSR	Molniya-M	Venus capsule	Failed to leave Earth orbit

	Launch	Nation	Launcher	Mission Type	Comments
Apollo 16	Apr 16	US	Saturn 5	Lunar lander	Fifth manned landing; 15th successful in all
Apollo 16 subsat	Apr 24	US	Apollo	Lunar orbiter	Released from Apollo 16
Soyuz L3	Nov 23	USSR	SL-15/N-1	Lunar distance	Exploded T+107 s. N1 test away from Moon, with orbiter + model lander
Apollo 17	Dec 7	US	Saturn 5	Lunar lander	Sixth manned landing; 16th in all
1973					
Luna 21	Jan 8	USSR	Proton-K + Block D	Lunar rover	17th lunar landing; second rover, Lunokhod 2 covered 37 km, returned >80,000 TV images
Pioneer 11	Apr 5	US	Atlas-Centaur	Jup/Sat flyby	First Saturn, second Jupiter flyby; last contact 30 Sep 1995 at 42 AU
Explorer 49	Jun 10	US	Delta	Lunar orbiter	Successful
Mars 4	Jul 21	USSR	Proton-K + Block D	Mars orbiter	Retro failure; 2,200 km flyby 10 Feb 1974 (returned one image swath). 4,385 kg
Mars 5	Jul 25	USSR	Proton-K + Block D	Mars orbiter	Fourth Mars orbiter; operated 'a few days' returning S hemisphere images, 4,385 kg
Mars 6	Aug 5	USSR	Proton-K + Block D	Mars lander	Contact lost shortly before landing 12 Mar 1974 at 24°S/25°W; first atmosphere data
Mars 7	Aug 9	USSR	Proton-K + Block D	Mars lander	Missed planet 9 Mar 1974 by 1,300 km (attitude or retro failure)
Mariner 10	Nov 4	US	Atlas-Centaur	Merc/Venus flyby	First Mercury flyby; first Mercury/Venus imaging (6,800 images). Venus 5,760 km 5 Feb 1974. Mercury 271 km 29 Mar 1974 (2,300 images), 48,069 km 21 Sep 1974 (1,000), 319 km 16 Mar 1975 (349; 1 km res); in total covered 57%
1974					
Luna 22	May 29	USSR	Proton-K + Block D	Lunar orbiter	Imaging survey. Contact >15 months
Luna 23	Oct 28	USSR	Proton-K + Block D	Lunar sampler	18th successful landing; sampler damaged (abandoned after 3 days)
Helios 1	Dec 10	US/FRG	Titan-Centaur	Solar probe	Approached to within 47 million km of Sun
1975					
Venera 9	Jun 8	USSR	Proton-K + Block D	Venus orbit/lander	First Venus orbiter; first surface image; operated for 53 min after landing, 4,936 kg (660 kg lander)
Venera 10	Jun 14	USSR	Proton-K + Block D	Venus orbit/lander	Second Venus orbiter; second surface image; operated for 65 min after landing. 5,033 kg
Viking 1	Aug 20	US	Titan-Centaur	Mars orbit/lander	First Mars soft landing; first surface images; fifth orbiter (2,325 kg; operated to 7 Aug 1980). 1,067 kg lander 20 Jul 1976, operated to Nov 1982
Viking 2	Sep 9	US	Titan-Centaur	Mars orbit/lander	Second Mars soft landing; second surface images; sixth orbiter (2,325 kg; operated to 24 Jul 1980). 1,067 kg lander 3 Sep 1976, operated to 11 Apr 1980
Luna-1975A	Oct 16	USSR	Proton-K + Block D	Lunar sampler	Failed to reach Earth orbit
1976					
Helios 2	Jan 16	US/FRG	Titan-Centaur	Solar probe	Approached to within 43 million km of Sun
Luna 24	Aug 9	USSR	Proton-K + Block D	Lunar sampler	19th landing; ninth sample return in all
1977					
Voyager 2	Aug 20	US	Titan-Centaur	Multiple flyby	Fourth Jupiter, third Saturn, first Uranus/Neptune flybys; still operating
Voyager 1	Sep 5	US	Titan-Centaur	Jup/Sat flyby	Third Jupiter, second Saturn flybys; still operating
1978					
Pioneer Venus 1	May 20	US	Atlas-Centaur	Venus orbiter	Third Venus orbiter; op until burned up 8 Oct 1992
Pioneer Venus 2	Aug 8	US	Atlas-Centaur	Venus capsules	Five successful atmospheric probes
ISEE-3/ICE	Aug 12	US	Delta	Interplanetary monitor	First cometary encounter
Venera 11	Sep 9	USSR	Proton-K + Block D	Venus lander	No surface images, but op for 95 min after landing 23 Dec. 4,450 kg
Venera 12	Sep 14	USSR	Proton-K + Block D	Venus lander	No surface images, but op for 110 min after landing 21 Dec. 4,450 kg
1981					
Venera 13	Oct 30	USSR	Proton-K + Block D	Venus lander	Third surface images (first colour); first soil analysis. Op for 127 min after 1 Mar 1982 landing. 4,363 kg (760 kg lander)
Venera 14	Nov 4	USSR	Proton-K + Block D	Venus lander	Fourth surface images (second colour), second soil analysis. Op for 57 min after 3 Mar 1982 landing. 4,364 kg (760 kg lander)
1983					
Venera 15	Jun 2	USSR	Proton-K + Block D	Venus orbiter	Radar mapper (1–2 km res); fourth Venus orbiter 10 Oct 1983. 5,250 kg
Venera 16	Jun 7	USSR	Proton-K + Block D	Venus orbiter	Radar mapper (1–2 km res); fourth Venus orbiter 16 Oct 1983. 5,300 kg
1984					
Vega 1	Dec 15	USSR	Proton-K + Block D	Venus lander/ comet flyby	First Venus balloon 11 Jun 1985 (46 h flt), lander survived 56 min (no camera); first Halley flyby 6 Mar 1986, 8,890 km. 4,920 kg (21 kg balloon)
Vega 2	Dec 21	USSR	Proton-K + Block D	Venus lander/ comet flyby	Second Venus balloon 15 Jun 1985 (461/2 h flt), lander survived 57 min (no camera); second Halley flyby 9 Mar 1986, 8,030 km. 4,920 kg (21 kg balloon)
1985					
Sakigake	Jan 7	Japan	M-3SII	Halley flyby	Distant flyby successful; last Earth approach of 4,000R_E Jul 1994; contact lost 15 Nov 1995
Giotto	Jul 2	Europe	Ariane 1	Halley flyby	First comet coma encounter, Mar 86; Earth swingby Jul 90; 2nd comet flyby Jul 92; Earth flyby due Jul 1999; dormant
Suisei	Aug 18	Japan	M-3SII	Halley flyby	Flyby successful; mission ended Feb 1991
1988					
Phobos 1	Jul 7	USSR	Proton-K + Block D	Mars orbiters,	Contact lost Sep 1988 en route
Phobos 2	Jul 12	USSR	Proton-K + Block D	Phobos flyby/ landings	Seventh Mars orbiter; contact lost Mar 1989
1989					
Magellan	May 4	US	Shuttle/IUS	Venus orbiter	Radar, altimeter, gravity mapper; sixth Venus orbiter (10 Aug 1990); last contact 12 Oct 1994, burned up 13 Oct
Galileo	Oct 18	US/FRG	Shuttle/IUS	Jupiter orbiter/ atmos probe	First Jupiter orbiter, Dec 1995; 16,000 km Venus flyby 10 Feb 90; Earth flyby 8 Dec 90; first asteroid (Gaspra) 29 Oct 91; Earth flyby 8 Dec 92; second asteroid (Ida) 28 Aug 93; 57 min probe data 7 Dec 1995; 1st orbiter 8 Dec 1995; began 2 yr orbital tour with 835 km Ganymede flyby 27 Jun 1996; still operating
1990					
Hiten/Hagomoro	Jan 24	Japan	M-3SII	Lunar flyby orbiter	Hagomoro subsat injected into lunar orbit 19 Mar during parent Hiten's geocentric flyby; Hiten also became lunar orbiter 15 Feb 1992, impacted 10 Apr 1993

	Launch	Nation	Launcher	Mission Type	Comments
Ulysses	Oct 6	ESA/US	Shuttle/IUS	High ecliptic/ solar polar	Jupiter swingby 8 Feb 1992 for solar polar overflights 1994/95 + 2000/01; still operating
1992					
Mars Observer	Sep 25	US	Shuttle/TOS	Mars orbiter	Mars arrival due 24 Aug 1993 for global survey but last contact 22 Aug; fate unknown
1994					
Clementine	Jan 25	US	Titan 2G	Lunar orbiter	In lunar orbit 19 Feb–4 May; 31 Aug flyby of asteroid 1620 Geographos cancelled after attitude propellant exhausted 7 May while in Earth orbit; perturbed from Earth-Moon system early Sep 1994, entering 1.0243 × 1.0561 AU heliocentric orbit. Intermittent contact made in 1995
1996					
NEAR	Feb 17	US	Delta	Asteroid	Near Earth Asteroid Rendezvous planned as first asteroid orbiter; flew to within 1,200 km of C-class asteroid 253 Mathilde on 27 Jun 1997, Earth on 23 Jan 1998 and to within 3,830 km of 433 Eros on 23 Dec 1998 when 222 pictures were sent back. Began year-long orbit of 433 Eros in Feb 2000
MGS	Nov 8	US	Delta	Mars orbiter	Mars Global Surveyor arrived in Mars orbit 11 Sep 1997 but aerobraking extended by problem with solar panel strut and mapping operations only started 9 Mar 1999. Continued in initial phase until 31 Jan 2001 at an average altitude of 400 km. Extended mission maintains MGS operations at least through the end of 2002
Mars-8	Nov 16	Russia	Proton-K	Mars orbiter W/2 landers	Failed to leave Earth orbit
Mars Pathfinder	Dec 4	US	Delta	Mars lander	Landed 4 Jul 1997 and returned copious data and images from lander base and Sojourner roving vehicle for three months compared to predictions of one month and one week respectively
1997					
Cassini/Huygens	Oct 15	US	Titan 4	Saturn orbiter	Performed fly-bys of Venus on 26 Apr 1998 and 24 Jun 1999 and Earth on 18 Aug 1999. Jupiter swing-by on 30Dec 2000 at a distance of 9.7 million km. Will reach Saturn in Jun 2004, releasing the Huygens probe to Titan six months later
1998					
Lunar Prospector	Jan 7	US	LMLV-2	Lunar orbiter	Began geochemical mapping operations from 100 km orbit from 15 Jan 1998 until start of extended mission at 30 km altitude from Jan 1999. Impacted on the moon 31 Jul 1999
Nozomi	Jul 3	Japan	M-5	Mars orbiter	Aeronomy orbiter given insufficient energy during multiple earth-moon swing-bys for planned arrival in Oct 1999. Spacecraft trajectory error prevented it reaching Mars as planned in January 2004
Deep Space 1	Oct 24	US	Delta	Asteroid flyby	Demonstrated 12 key technologies for future missions including ion propulsion. Passed within 15 km of asteroid 9969 Braille on 29 Jul 1999. Began extended mission Sep 1999 and performed close approach of comet Borrelly 22 Sep 2001. Retired 18 Dec 2001
MCO	Dec 11	US	Delta	Mars orbiter	Mars Climate Orbiter was second flight in Mars Surveyor programme and was expected to enter Mars polar orbit 23 Sep 1999 for aerobraking prior to climate/weather studies. However, human error in calculating the trajectory put MCO into the atmosphere of Mars at a close approach of just 60 km where it was destroyed
1999					
MPL/DS2	Jan 3	US	Delta	Mars lander	Carries two Deep Space 2 Microprobes released prior to landing on 3 Dec 1999 at polar site to search for frozen water. All contact with MCO lost during descent. Review Board suggested several areas of weak engineering design probably caused its destruction
Stardust	Feb 7	US	Delta	Comet sampler	Flew within 100 km of comet 81P/Wild-2 in Jan 2004 and expected return samples to earth in Jan 2006
2001					
Mars Odyssey	Apr 7	US	Delta		Arrived in Mars orbit 24 Oct 2001 for extensive two-year survey of surface composition

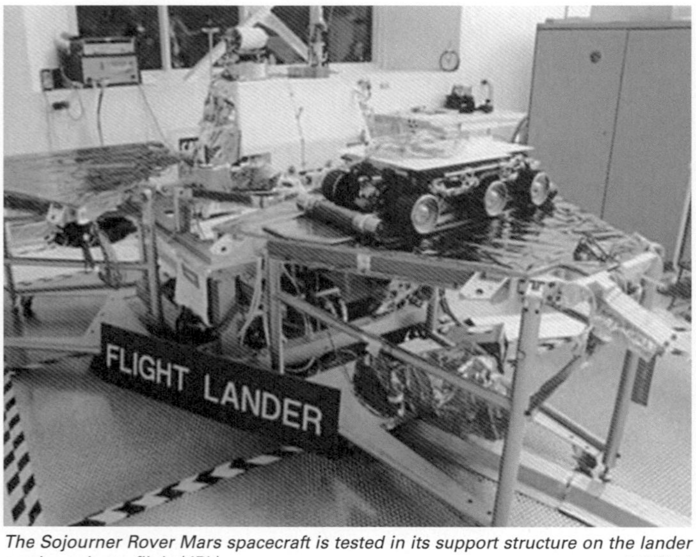

The Sojourner Rover Mars spacecraft is tested in its support structure on the lander portion prior to flight (JPL) 0137144

Viewed by the Ranger spacecraft rushing to destruction, a breathtaking view of a cratered surface sent live to earth prior to impact (JPL) 0137145

Viking Lander 1 images survey the surrounding terrain marked by basaltic rocks and sand dunes (JPL) 0137142

A mockup of NASA's Phoenix Mars lander due for launch in 2007 (Lockheed Martin) 0572996

Technicians prepare a lunar-bound Ranger spacecraft with survivable hard-landing capsule built from layers of balsa wood (JPL) 0572998

Ranger 4 is launched by Atlas-Agena in April 1962 and, although failing in its primary mission, becomes the first US spacecraft to hit the moon (JPL) 0137146

Pioneer 3 is scrutinised before launch in one of NASA's earliest, unsuccessful, attempts at a lunar fly-by in December 1958, a programme adopted from the US a programme adopted from the US Army (US Army) 0137163

NASA's Pioneer-Venus orbiter at the planet in an artists impression. (NASA) 0573000

NASA's first planetary launch, Mariner 1, lifts off from Cape Canaveral on 22 July 1962. Intended for a fly-by of Venus, the spacecraft was destroyed when the launch vehicle failed due to an error in flight guidance equations (NASA/JPL) 0137168

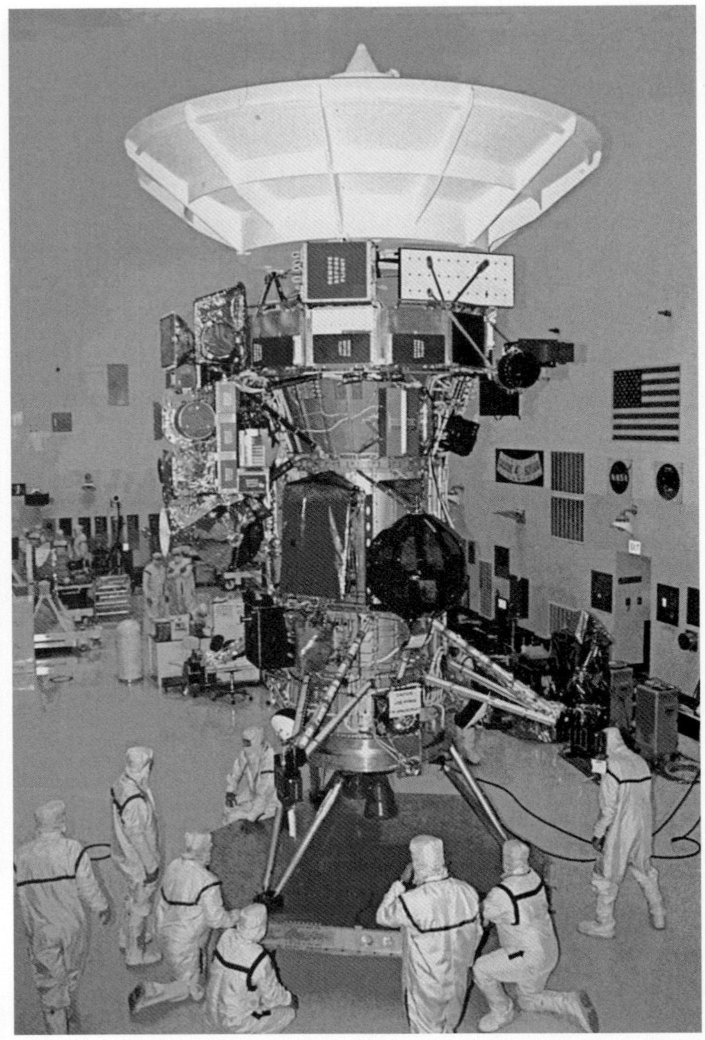

Destined for Saturn, NASA's Cassini spacecraft is checked out prior to launch (NASA) 0572498

A correct pan image taken by Viking Lander 2 revealing a littered surface, rocks possibly transported by water (JPL) 0137143

An artists view of ESA's Beagle 2 lander on the surface of Mars (ESA) 0573206

Pioneer 3 is installed in the top of the Juno 2 launch vehicle prior to its launch attempt (US Army)
0137164

UPDATED

Table 4: US Manned Flights (Mercury to Apollo-Soyuz)

Mission	Launch Date	Crew	Revs	Duration (d, h, m)
Mercury MR-3	05.05.61	Alan Shepard	Sub	00.00.15
1st American in space; Freedom 7 attained suborbital 187.5 km, 8,336 km/h				
Mercury MR-4	21.07.61	Virgil Grissom	Sub	00.00.16
Second suborbital flight, attained 190.4 km. Liberty Bell 7 sank				
Mercury MA-6	20.02.62	John Glenn	3	00.04.55
1st American in orbit, Friendship 7 attained 161 × 261 km. Telemetry falsely indicated heatshield unlatched				
Mercury MA-7	24.05.62	M Scott Carpenter	3	00.04.56
MA-6 repeat, Aurora 7 overshot landing site by 420 km				
Mercury MA-8	03.10.62	Walter Schirra	6	00.09.13
Engineering evaluation, conserving propellant. Sigma 7 landed 7.24 km from target				
Mercury MA-9	15.05.63	L Gordon Cooper	22	01.10.20
1st US 1-day mission; performed manual re-entry in Faith 7 after automatic stabilisation & control system failed, landing 6.44 km from recovery carrier. In-orbit mass 1,373 kg, carrying additional supplies for long flight				
Gemini 3	23.03.65	Virgil Grissom John Young	3	00.04.53
Test flight; first manned orbit changes; first computer, allowing onboard calculation of manoeuvres				
Gemini 4	03.06.65	James McDivitt Edward White	62	04.01.56
White made 1st US EVA (21 min); 1st US 4-day flight. Manual entry made after computer failed				
Gemini 5	21.08.65	L Gordon Cooper Charles Conrad	120	07.22.56
Endurance record (sufficient for return lunar mission)				
Gemini 7	04.12.65	Frank Borman James Lovell	206	13.18.35
Longest US flight for 8 yr, record until Soyuz 9. See G6				
Gemini 6	15.12.65	Walter Schirra Thomas Stafford	16	01.01.51
1st manned rendezvous, to within 1.8 m of Gemini 7 (as planned Agena was lost)				
Gemini 8	16.03.66	Neil Armstrong David Scott	6	00.10.41
1st space docking (with Agena); emergency re-entry after Gemini thruster jammed on and re-entry thrusters had to be activated to regain control				
Gemini 9	03.06.66	Thomas Stafford Eugene Cernan	45	03.00.21
127 min EVA by Cernan, rendezvous but no docking with target. Landed 800 m from recovery ship				
Gemini 10	18.07.66	John Young Michael Collins	43	02.22.47
Docked with Agena 10 and used engine to attain record 763 km; rendezvous with Agena 8, Collins retrieved package in 39 min EVA				
Gemini 11	12.09.66	Charles Conrad Richard Gordon	44	02.23.17
Agena engine used to attain record 1,369 km; Gordon 163 min SEVA/EVA. Gordon connected Gemini and Agena by tether; when undocked, Gemini initiated cartwheel motion, generating 0.00015 g. First automatic, computer-guided re-entry				
Gemini 12	11.11.66	James Lovell Eugene Aldrin	59	03.22.34
Agena docking; SEVA/EVA totalling 330 min by Aldrin. Automatic computer-guided re-entry				
Apollo 7	11.10.68	Walter Schirra Donn Eisele Walter Cunningham	163	10.20.09
CSM demonstration, 1st manned Apollo, 1st US 3-man flight				
Apollo 8	21.12.68	Frank Borman James Lovell William Anders	10 lunar	06.03.00
1st manned lunar mission, 1st manned Saturn 5				
Apollo 9 *Gumdrop* *Spider*	03.03.69	James McDivitt David Scott Russell Schweickart	151	10.01.01
1st manned Lunar Module (Earth orbit); 56 min Schweickart EVA tested lunar suit				
Apollo 10 *Charlie Brown* *Snoopy*	18.05.69	Thomas Stafford Eugene Cernan John Young	31 lunar	08.00.03
Descent to within 14 km of Moon in landing dress rehearsal. Holds manned speed record: 11.0825 km/s at atmospheric entry				
Apollo 11 *Columbia* *Eagle*	16.07.69	Neil Armstrong Eugene Aldrin Michael Collins	31 lunar	08.03.18
1st manned lunar landing; 151 min EVA, 21 kg samples, 21 h 36 min on Moon				
Apollo 12 *Yankee Clipper* *Intrepid*	14.11.69	Charles Conrad Alan Bean Richard Gordon	49 lunar	10.04.36
2 lunar EVAs totalling 465 min; 34 kg samples; 31 h 31 min on Moon				
Apollo 13 *Odyssey* *Aquarius*	11.04.70	James Lovell John Swigert Fred Haise		05.22.55
Mission aborted following Service Module oxygen tank explosion; circumlunar return. Holds manned altitude record: 400,187 km above Earth's surface				
Apollo 14 *Kitty Hawk* *Antares*	31.01.71	Alan Shepard Edgar Mitchell Stuart Roosa	34 lunar	09.00.42
2 lunar EVAs, total 563 min; 43 kg samples; 33 h 31 min on Moon				
Apollo 15 *Endeavour* *Falcon*	26.07.71	David Scott James Irwin Alfred Worden	74 lunar	12.07.12
3 lunar EVAs, total 19 h 8 min; 77 kg samples; 1st lunar rover; 66 h 54 min on Moon. 38 min Worden EVA. Subsatellite released				

On board the ISS the androgynous docking probe is removed and re-installed during routine preparations configuring the modules for habitation (NASA) 0137136

A fish-eye lens view of the control displays of NASA's two-man Gemini spacecraft (NASA) 0573214

Mission Control-Houston during the Gemini programme 1965–66 (JSC) 0573213

Mission	Launch Date	Crew	Revs	Duration (d, h, m)
Apollo 16 *Casper* *Orion*	16.04.72	John Young Charles Duke Tom Mattingly	64 lunar	11.01.51

3 lunar EVAs, total 20 h 14 min; 94 kg samples; 71 h 14 min on Moon. Subsatellite released

Mission	Launch Date	Crew	Revs	Duration (d, h, m)
Apollo 17 *America* *Challenger*	07.12.72	Eugene Cernan Jack Schmitt Ron Evans	75 lunar	12.13.51

3 lunar EVAs, total 22 h 4 min; 110 kg samples; 74 h 59 min on Moon. 66 min Evans EVA

Mission	Launch Date	Crew	Revs	Duration (d, h, m)
Skylab 1	14.05.73	Unmanned	34,981	

Space station, suffered launch damage. Hosted 3 crews. Re-entered 11.07.79

Mission	Launch Date	Crew	Revs	Duration (d, h, m)
Skylab 2	25.05.73	Charles Conrad Paul Weitz Joe Kerwin	404	28.00.50

Endurance record. Repaired Skylab, deploying 6.7 × 7.3 m parasol through 20 cm science airlock on 7.6 m boom and releasing jammed solar wing. 81 h of solar observations and 11 Earth resources passes (9,850 images); 2 EVAs + SEVA

Mission	Launch Date	Crew	Revs	Duration (d, h, m)
Skylab 3	28.07.73	Alan Bean Owen Garriott Jack Lousma	858	59.11.09

Endurance record. Erected new sunshade, tested EVA manoeuvring units, 305 h solar observations (24,940 frames), 39 Earth resources passes, welding & materials processing experiments; 3 EVAs

Mission	Launch Date	Crew	Revs	Duration (d, h, m)
Skylab 4	16.11.73	Gerald Carr Edward Gibson William Pogue	1214	84.01.15

Endurance record. Observed Comet Kohoutek (2,500 images) and returned 73,400 solar + 19,400 Earth images; 4 EVAs

Mission	Launch Date	Crew	Revs	Duration (d, h, m)
ASTP	15.07.75	Thomas Stafford Vance Brand Deke Slayton	138	09.01.28

1st US-Soviet joint flight, 46 h 44 min docked over two periods for activities with Leonov/Kubasov in Soyuz 19. 28 US expts (including 5 with Soyuz). Docked mass 1,500 kg (Apollo 12,700 kg). US cost $218 million

The above manned flights totalled 189,985,890 km (118,077,569 miles).

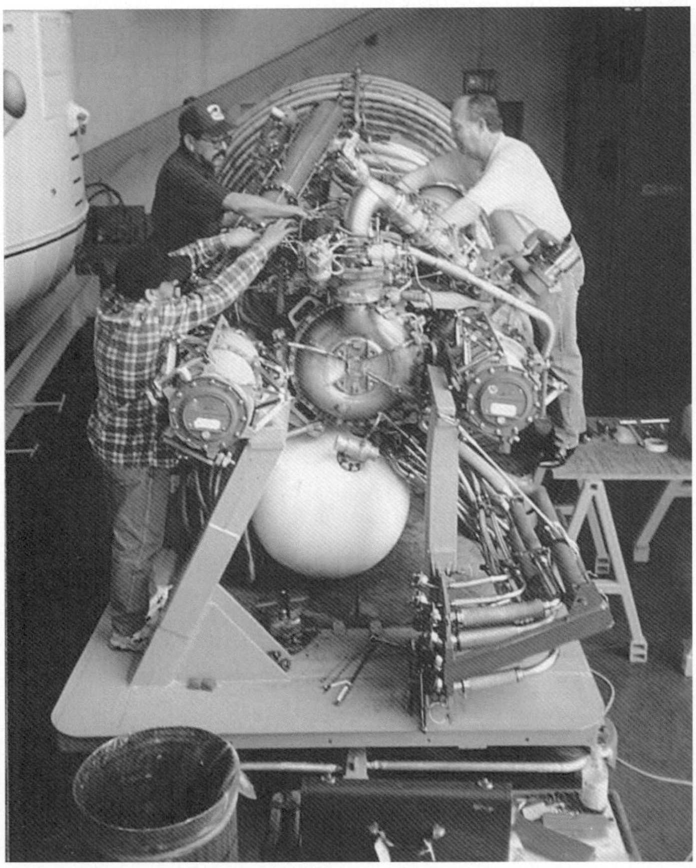

Columbia, the Apollo 11 CSM, is fitted atop the Spacecraft Launch Vehicle Adapter in July 1969 (NASA) 0137176

The last man on the moon during the Apollo era, astronaut Eugene Cernan salutes from the valley of Taurus-Littrow on the Apollo 17 mission, December 1972 (NASA) 0137170

A Gemini-Titan launch vehicle lifts off from Launch Complex 19 at Cape Canaveral Air Force Station (NASA) 0137152

NASA's first group of astronauts selected for the Mercury programme in 1959 (NASA) 0573203

A technician checks deployment characteristics of the Lunar Roving Vehicle (NASA)
0573207

Astronaut Grissom is recovered from the sea as his Mercury spacecraft sinks in the Atlantic Ocean (NASA)
0573202

Following the first US manned flight into space during a ballistic flight on 5 May 1961, astronaut Alan Shepard meets President Kennedy at the White House (NASA)
0573205

An Apollo spacecraft as seen from the Lunar Module during the mission of Apollo 10 (NASA)
0569812

Armstrong, Collins and Aldrin in the hatch of their Apollo 11 command module prior to the launch of the first manned lunar landing mission (NASA)
0573201

A Redstone missile adapted for the Mercury programme lifts off from Cape Canaveral (NASA) 0572999

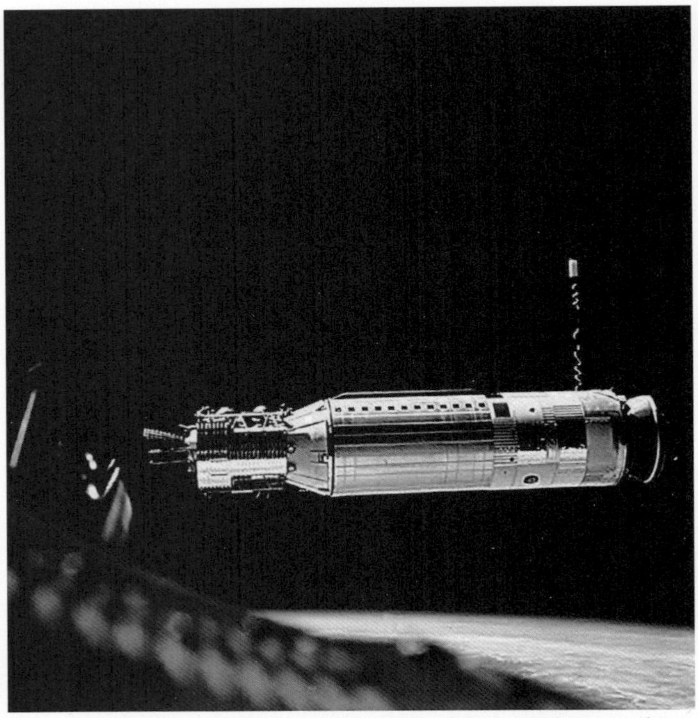

Agena Target Vehicle previously launched by an Atlas launch vehicle as seen from a Gemini spacecraft (NASA) 0137175

Astronaut Charlie Duke busy on lunar geologic tasks during the Apollo 16 mission (NASA) 0573204

The site for man's last landing on the moon-Apollo 17, December 1972 (NASA) 0573209

Table 5: US Space Shuttle Flights

Flight	Launch date	Orbiter	Crew	Duration (d, h, m)	Incl	Notes
01 **STS-1**	12.04.81	Columbia	Young, Crippen	02.06.21	40.4°	*Shuttle debut mission; some thermal tiles lost*
02 **STS-2**	12.11.81	Columbia	Engle, Truly	02.06.13	38.0°	*Tested RMS arm; 5-day mission halved by fuel cell fault*
03 **STS-3**	22.03.82	Columbia	Lousma, Fullerton	08.00.05	38.0°	*Third test; White Sands landing delayed 1 day by storm*
04 **STS-4**	27.06.82	Columbia	Mattingly, Hartsfield	07.01.10	28.45°	*Final test; first concrete runway landing. SRBs lost. Cirris DoD payload failure*
05 **STS-5**	11.11.82	Columbia	Brand, Overmyer, Allen, Lenoir	05.02.14	28.45°	*First operational flight. SBS 3/Anik C3 deployed. EVA cancelled*
06 **STS-6**	05.04.83	Challenger	Weitz, Bobko, Peterson, Musgrave	05.00.23	28.45°	*Challenger debut. TDRS 1 deployed on IUS. First Shuttle EVA*
07 **STS-7**	18.06.83	Challenger	Crippen, Hauck, Fabian, Ride, Thagard	06.02.24	28.45°	*First US woman (Ride). First deploy/retrieve (SPAS). Anik C2/Palapa B deployed*
08 **STS-8**	30.08.83	Challenger	Truly, Brandenstein, Bluford, Gardner, W Thornton	06.01.07	28.45°	*First night launch/landing. India's Insat 1B telecom sat deployed*
09 **STS-9**	28.11.83	Columbia	Young, Shaw, Garriott, Parker, Lichtenberg, Merbold	10.07.47	57.0°	*Spacelab 1 demonstration; 90% science achievd. First 6-man flight*
10 **41B**	03.02.84	Challenger	Brand, Gibson, McCandless, Stewart, McNair	07.23.17	28.45°	*First MMU tests + SMM EVA repair rehearsal. Westar 6/Palapa B2 lost. 1st KSC landing*
11 **41C**	06.04.84	Challenger	Crippen, Scobee, G Nelson, van Hoften, Hart	06.23.40	28.45°	*LDEF deploy. First satellite retrieval/repair (Solar Max)*
12 **41D**	30.08.84	Discovery	Hartsfield, Coats, Resnik, Hawley, Mullane, C Walker	06.00.56	28.45°	*Discovery debut. SBS 4/Leasat 2/Telstar 3C deploy. Solar panel deploy test. First commercial Payload Specialist*
13 **41G**	05.10.84	Challenger	Crippen, McBride, Sullivan, Ride, Leestma, Garneau, Scully-Power	08.05.23	57.0°	*OSTA-3, ERBS, Large Format Camera. Firsts: 7 crew, US woman EVA, 2-woman flight. 2nd KSC landing*
14 **51A**	08.11.84	Discovery	Hauck, D Walker, A Fisher, D Gardner, Allen	07.23.45	28.45°	*First sat recovery/return: Palapa B2/Westar 6. Anik D/Leasat 1 deployed. 3rd KSC landing*
15 **51C**	24.01.85	Discovery	Mattingly, Shriver, Onizuka, Buchli, Payton	03.01.33	28.45°	*First dedicated DoD flight: Magnum elint on IUS. 100th manned orbital mission. 4th KSC landing*
16 **51D**	12.04.85	Discovery	Bobko, Williams, Seddon, Hoffman, Griggs, C Walker, Garn	06.23.56	28.45°	*Anik C/Leasat 3 deployed; EVA failed to repair Leasat 3. 5th KSC landing*
17 **51B**	29.04.85	Challenger	Overmyer, Gregory, Lind, Thagard, W Thornton, van den Berg, Wang	07.00.08	57.0°	*Spacelab 3 science mission; released small Nusat satellite*
18 **51G**	17.06.85	Discovery	Brandenstein, Creighton, Lucid, Fabian, Nagel, Baudry, Al Saud	07.01.38	28.45°	*Spartan deploy/retrieve; Morelos/Arabsat/Telstar 3D deployments*
19 **51F**	29.07.85	Challenger	Fullerton, Bridges, Musgrave, Henize, England, Acton, Bartoe	07.22.45	49.5°	*Spacelab 2 astronomy/science mission (no pressurised module)*
20 **51I**	27.08.85	Discovery	Engle, Covey, van Hoften, Lounge, W Fisher	07.02.18	28.45°	*MSL; Aussat/ASC/Leasat 4 deployed; Leasat 3 EVA repair*
21 **51J**	03.10.85	Atlantis	Bobko, Grabe, Hilmers, Stewart, Pailes	04.01.44	28.5°	*DoD DSCS 3 pair on IUS. Atlantis debut*
22 **61A**	30.10.85	Challenger	Hartfield, Nagel, Dunbar, Buchli, Bluford, Messerschmid, Furrer, Ockels	07.00.44	57.0°	*Spacelab D1: First foreign (West German) Shuttle flight; First 8-man crew*
23 **61B**	27.11.85	Atlantis	Shaw, O'Connor, Cleave, Ross, Spring, C Walker, Vela	06.21.05	28.45°	*Morelos B/Satcom K2/Aussat 2 deploy, Ease/Access First Space Station assembly demo*
24 **61C**	12.01.86	Columbia	Gibson, Bolden, G Nelson, Hawley, Chang-Diaz, Cenker, W Nelson	06.02.04	28.45°	*Satcom K1 deploy; SDI experiments*
25 **51L**	28.01.86	Challenger	Scobee, Smith, Resnik, McNair, Onizuka, McAuliffe, Jarvis	00.00.01		*Vehicle exploded after 73 s; crew killed*
26 **STS-26**	29.09.88	Discovery	Hauck, Covey, Hilmers, Lounge, Nelson	04.01.00	28.5°	*TDRS 3 deployed on IUS stage; 11 mid-deck secondary payloads*
27 **STS-27**	02.12.88	Atlantis	R Gibson, G Gardner, Mullane, Ross, Shepherd	04.09.06	57.0°	*First DoD Lacrosse imaging radar satellite for all-weather day/night reconnaissance*
28 **STS-29**	13.03.89	Discovery	Coats, Blaha, Buchli, Springer, Bagian	04.23.39	28.45°	*TDRS 4 on IUS; five mid-decK+two cargo bay secondary payloads*
29 **STS-30**	04.05.89	Atlantis	D Walker, Grabe, Thagard, Cleave, Lee	04.00.58	28.85°	*Magellan Venus orbiter on IUS stage; arrived Venus Aug 1990*

New and improved Block 2 Space Shuttle Main Engine elements are installed in Atlantis in the Orbiter Processing Facility at NASA's Kennedy Space Center, Florida (NASA) 0137134

Flight	Launch date Orbiter	Crew	Duration (d, h, m) Incl
30 **STS-28**	08.08.89 Columbia	Shaw, Richards, Leestma, Adamson, Brown	05.01.00 57.00°
Deployed military satellite into Molniya-type orbit for relaying imaging reconnaissance satellite imagery? Flight similar to STS-53			
31 **STS-34**	18.10.89 Atlantis	Williams, McCulley, Lucid, Baker, Chang-Diaz	04.23.39 34.30°
Galileo Jupiter orbiter on IUS stage; arrive Jupiter Dec 1995			
32 **STS-33**	23.11.89 Discovery	Gregory, Blaha, Musgrave, K Thornton, Carter	05.00.07 28.45°
DoD mission; deployed second Magnum signal intelligence satellite?			
33 **STS-32**	09.01.90 Columbia	Brandenstein, Wetherbee, Dunbar, Low, Ivins	10.21.01 28.50°
Deployed USN Leasat 5 telecom satellite, retrieved Long Duration Exposure Facility			
34 **STS-36**	28.02.90 Atlantis	Creighton, Casper, Hilmers, Mullane, Thuot	04.10.18 62.00°
DoD mission; deployed AFP-731 reconnaissance satellite			
35 **STS-31**	24.04.90 Discovery	Shriver, Bolden, Hawley, McCandless, Sullivan	05.01.16 28.45°
Deployed Hubble Space Telescope; Shuttle altitude record (619 km)			
36 **STS-41**	06.10.90 Discovery	Richards, Cabana, Shepherd, Melnick, Akers	04.02.10 28.45°
Ulysses solar probe on IUS stage			
37 **STS-38**	15.11.90 Atlantis	Covey, Culbertson, Springer, Meade, Gemar	04.21.54 28.45°
DoD mission; deployed third Magnum signal intelligence satellite? 6th KSC landing			
38 **STS-35**	02.12.90 Columbia	Brand, G Gardner, Lounge, Hoffman, Parker, Parise, Durrance	08.23.05 28.45°
Carried Astro-1/BBXRT-1 ultraviolet/X-ray telescopes			
39 **STS-37**	05.04.91 Atlantis	Nagel, Cameron, Ross, Godwin, Apt	05.23.33 28.45°
Deployed Gamma Ray Observatory; 2 EVAs (first since 1985)			
40 **STS-39**	28.04.91 Discovery	Coats, Hammond, Harbaugh, McMonagle, Bluford, Veach, Hieb	08.07.22 57.0°
DoD AFP-675/IBSS IR Background Signature Survey platforms. 7th KSC landing			
41 **STS-40**	05.06.91 Columbia	O'Connor, Gutierrez, Bagian, Jernigan, Seddon, Gaffney, Hughes-Fulford	09.02.14 39.0°
Spacelab Life Sciences (SLS-1); first dedicated life sciences research; 12 GAS canisters			
42 **STS-43**	02.08.91 Atlantis	Blaha, Baker, Adamson, Low, Lucid	08.21.21 28.45°
Deployment of TDRS 5 on IUS stage; Lucid first woman to make 3 flights. 8th KSC landing			
43 **STS-48**	12.09.91 Discovery	Creighton, Reightler, Buchli, Brown, Gemar	05.08.28 57.0°
Deployment of UARS Upper Atmosphere Research Satellite			
44 **STS-44**	24.11.91 Atlantis	Gregory, Henricks, Jim Voss, Musgrave, Runco, Hennen	06.22.51 28.5°
Unclassified DoD deployment of DSP early warning satellite on IUS			
45 **STS-42**	22.01.92 Discovery	Grabe, Oswald, Thagard, Readdy, Hilmers, Bondar, Merbold	08.01.15 57.0°
Spacelab International Microgravity Lab (IML-1)			
46 **STS-45**	24.03.92 Atlantis	Bolden, Duffy, Foale, Leestma, Sullivan, Lichtenberg, Frimout	08.22.09 57.0°
Atmospheric Lab for Applications & Science (Atlas-1). 9th KSC landing			

The Shuttle lifts off at the start of an ISS re-supply flight, one of many scheduled before the ISS is fully configured for its international crew (NASA) 0137137

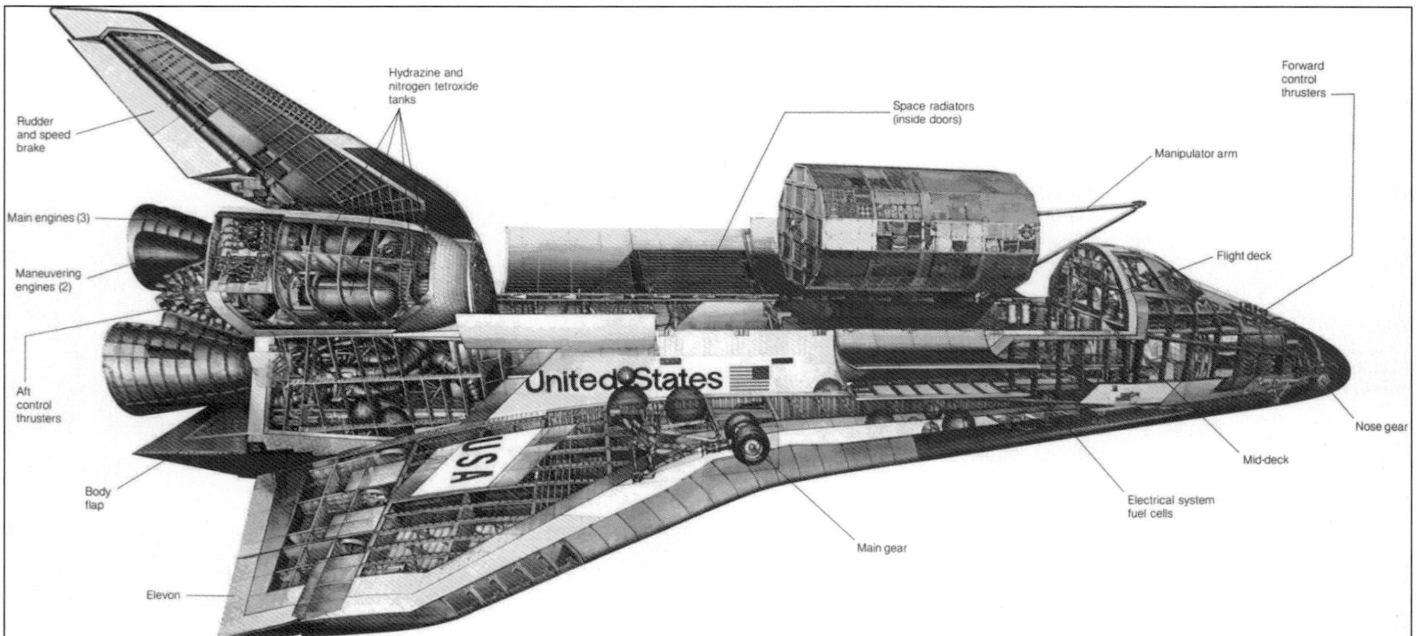

Cutaway view of the Shuttle Orbiter with remote manipulator and typical payload above the cargo bay (Rockwell International) 0137158

Flight	Launch date Orbiter	Crew	Duration (d, h, m) Incl
47 **STS-49**	07.05.92 Endeavour	Brandenstein, Chilton, Melnick, Thuot, Hieb, K Thornton, Akers	08.21.18 28.35°
Endeavour debut; Intelsat 603 recovery/repair. 4 EVAs by 4 crew, total 60.1 man-h (including 1st 3-man)			
48 **STS-50**	25.06.92 Columbia	Richards, Bowersox, Dunbar, Baker, Meade, DeLucas, Trinh	13.19.31 28.5°
US Microgravity Lab (USML 1) Spacelab. 1st EDO; Shuttle duration record. 10th KSC landing			
49 **STS-46**	31.07.92 Atlantis	Shriver, Allen, Chang-Diaz, Ivins, Hoffman, Nicollier, Malerba	07.23.15 28.5°
Deployment of ESA's Eureca 1; deployment/retrieval of Italy's Tethered Satellite. 11th KSC landing			
50 **STS-47**	12.09.92 Endeavour	Gibson, Brown, Lee, Apt, Davis, Jemison, Mohri	07.22.30 57.0°
Japan-sponsored Spacelab-J: 43 materials/life sciences expts. 12th KSC landing			
51 **STS-52**	22.10.92 Columbia	Wetherbee, Baker, Veach, Shepherd, Jernigan, MacLean	09.20.56 28.45°
Deployment of Italy's Lageos 2. USMP 1 US Microgravity Payload. 13th KSC landing			
52 **STS-53**	02.12.92 Discovery	D Walker, Cabana, Bluford, Jim Voss, Clifford	07.07.19 57.00°
Deployment of DoD 1 satellite, purpose unknown, into Molniya-type orbit possibly to relay reconnaissance satellite imagery. Flight similar to STS-28			
53 **STS-54**	13.01.93 Endeavour	Casper, McMonagle, Runco, Harbaugh, Helms	05.23.38 28.45°
Deployment of TDRS 6 on IUS. 268 min EVA. 14th KSC landing			
54 **STS-56**	08.04.93 Discovery	Cameron, Oswald, Foale, Cockrell, Ochoa	09.06.08 57.00°
Atmospheric Lab for Applications & Science (Atlas 2); Spartan 201 deploy/retrieve. 15th KSC landing			
55 **STS-55**	26.04.93 Columbia	Nagel, Henricks, Ross, Precourt, Harris, Walter, Schlegel	09.23.40 28.46°
Spacelab D2: 90 materials/life sciences, Earth observation & astronomy expts. Shuttle programme exceeded 1 yr aggregate flight time			
56 **STS-57**	21.06.93 Endeavour	Grabe, Duffy, Low, Jan Voss, Sherlock, Wisoff	09.23.45 28.45°
Eureca 1 retrieval; Spacehab 1 with 21 expts. Generic EVA. 16th KSC landing			
57 **STS-51**	12.09.93 Discovery	Culbertson, Readdy, Bursch, Newman, Walz	09.20.11 28.45°
ACTS deploy, German Orfeus-SPAS deploy/retrieve, generic EVA. 17th KSC landing (1st night)			
58 **STS-58**	18.10.93 Columbia	Blaha, Searfoss, Seddon, Lucid, McArthur, Wolf, Fettman	14.00.13 39.0°
Spacelab Life Sciences (SLS-2) dedicated life sciences research; 2nd EDO (record duration)			

Flight	Launch date Orbiter	Crew	Duration (d, h, m) Incl
59 **STS-61**	02.12.93 Endeavour	Covey, Bowersox, Musgrave, Akers, Hoffman, K Thornton, Nicollier	10.19.59 28.47°
First Hubble servicing mission; 5 EVAs for 4 crew totalled 70.93 man-h. 18th KSC landing			
60 **STS-60**	03.02.94 Discovery	Bolden, Reightler, Davis, Sega, Chang-Diaz, Krikalev	08.07.10 57.0°
Spacehab 2; employed Wake Shield Facility; first Russian aboard US craft. 19th KSC landing			
61 **STS-62**	04.03.94 Columbia	Casper, Allen, Thuot, Gemar, Ivins	13.23.17 39.0°
USMP 2 US Microgravity Payload; OAST 2; 3rd EDO. 20th KSC landing			
62 **STS-59**	09.04.94 Endeavour	Gutierrez, Chilton, Godwin, Jones, Apt, Clifford	11.05.49 57.0°
Space Radar Lab 1 mapped 20% Earth, 133 h data. Endeavour made record 412 manoeuvres			
63 **STS-65**	8.07.94 Columbia	Cabana, Halsell, Hieb, Chiao, D Thomas, Walz, Mukai	14.17.55 28.5°
Spacelab International Microgravity Lab IML-2; 4th EDO (record duration). 21st KSC landing			
64 **STS-64**	09.09.94 Discovery	Richards, Hammond, Helms, Lee, Linenger, Meade	10.22.50 57.0°
LITE-1 Lidar In-space Technology Expt; deploy/retrieve Spartan 201–02; EVA			
65 **STS-68**	30.09.94 Endeavour	Baker, Wilcutt, Jones, Bursch, Smith, Wisoff	11.05.46 57.0°
Space Radar Lab SRL-2			
66 **STS-66**	03.11.94 Atlantis	McMonagle, Brown, Ochoa, Parazynski, Tanner, Clervoy	10.22.34 57.0°
Atmospheric Lab for Applications & Science (Atlas-3); deploy/retrieve Germany's Crista-SPAS 1			
67 **STS-63**	03.02.95 Discovery	Wetherbee, Collins, Foale, Harris, Jan Voss, Titov	08.06.28 51.6°
Rendezvous with Mir; Spacehab 3; deploy/retrieve Spartan 204. 1st female pilot (Collins); 2nd Russian aboard US craft (Titov); generic EVA. 22nd KSC landing			
68 **STS-67**	02.03.95 Endeavour	Oswald, W Gregory, Jernigan, Grunsfeld, Lawrence, Durrance, Parise	16.15.10 28.5°
Astro-2 UV telescope array; 5th EDO (record duration)			
69 **STS-71**	27.06.95 Atlantis	Gibson, Precourt, Baker, Harbaugh, Dunbar, Solovyov, Budarin	09.19.22 51.6°
1st docking with Mir, delivering station's 19th main crew (Solovyov/Budarin), returning with 18th crew (Dezhurov/Strekalov) + Soyuz-TM 21's Dr Norman Thagard (aboard since Mar 1995). Spacelab carried. 23rd KSC landing			
70 **STS-70**	13.07.95 Discovery	Henricks, Kregel, Sherlock, D Thomas, Weber	08.22.20 28.5°
Deploy TDRS 7 on IUS. 24th KSC landing			

Italy's Leonardo module provides resupply container space for essential ISS equipment (ESA) 0137140

Astronaut Sellers in the cargo bay of the Shuttle during a scheduled EVA (NASA) 0573536

Flight	Launch date Orbiter	Crew	Duration (d, h, m) Incl
71 **STS-69**	7.09.95 Endeavour	D Walker, Cockrell, Jim Voss, Newman, Gernhardt	10.20.29 28.45°
Deploy/retrieve WSF 2 Wake Shield Facility; deploy/retrieve Spartan 201–03; generic EVA. 25th KSC landing			
72 **STS-73**	20.10.95 Columbia	Bowersox, Rominger, K Thornton, Coleman, Lopez-Alegria, Leslie, Sacco	15.21.52 39.0°
Spacelab USML 2 US Microgravity Lab; 6th EDO; 26th KSC landing			
73 **STS-74**	12.11.95 Atlantis	Cameron, Halsell, McArthur, Ross, Hadfield	08.04.31 51.64°
#2 Mir docking. Delivered docking unit for future Shuttle missions and new solar arrays. 27th KSC landing			
74 **STS-72**	11.01.96 Endeavour	Duffy, Jett, Barry, Chiao, Scott, Wakata	08.22.01 28.45°
Retrieve Japan's SFU Space Flyer Unit; deploy/retrieve Spartan OAST-Flyer; 2 generic EVAs. 28th KSC landing			
75 **STS-75**	22.02.96 Columbia	Allen, Horowitz, Chang-Diaz, Cheli, Hoffman, Nicollier, Guidoni	15.17.40 28.45°
USMP 3 US Microgravity Payload; deploy/retrieve Italy's TSS-1R tethered satellite (TSS lost: tether snapped); 7th EDO. 29th KSC landing			
76 **STS-76**	22.03.96 Atlantis	Chilton, Searfoss Lucid, Godwin, Clifford, Sega	09.05.16 51.6°
#3 Mir docking, with Spacehab short module. Delivered logistics and Lucid for Mir stay (returning STS-79). Godwin/Clifford EVA mounted expts on Mir's docking module			
77 **STS-77**	19.05.96 Endeavour	Casper, Brown, Bursch, Runco, A Thomas, Garneau	10.00.39 28.5°
Spacehab 4; deploy/retrieve Spartan 207 (inflatable antenna expt); deploy PAMS Passive Aerodynamically Stabilised Magnetically Damped Satellite. 30th KSC landing			
78 **STS-78**	20.06.96 Columbia	Henricks, Kregel, Helms, Linnehan, Brady, Favier, Thirsk	16.21.48 39.0°
Life & Microgravity Science Spacelab; 8th EDO. 31st KSC landing			
79 **STS-79**	16.09.96 Endeavour	Readdy, Wilcutt, Apt, Akers Walz, Blaha (up), Lucid (down)	10.03.18 51.6°
#4 Mir docking with Spacehab double module. Delivered supplies to Mir, left Blaha on board for medium-length stay, returned with Lucid (mission duraction 188.04.00) from STS-76			
80 **STS-80**	19.11.96 Columbia	Cockrell, Rominger, Jernigan, Jones, Musgrave	17.15.53 28.5°
Deployment and retrievel of Wake Shield Facility 3 and ORFEUS-SPAS 2; 9th EDO; planned EVAs by Jernigan and Jones cancelled			
81 **STS-81**	12.01.97 Atlantis	Baker, Jett, Wilsoff, Grunsfeld, Ivins, Linenger (up), Blaha (down)	10.04.56 51.6°
#5 Mir docking with Spacehab double module. Delivered supplies to Mir, left Linenger on board for medium-length stay, returned with Blaha from STS-79 (mission duration 128.05.27)			
82 **STS-82**	11.02.97 Discovery	Bowersox, Horowitz, Tanner Hawley, Harbaugh, Lee, Smith	9.23.37 28.5°
Second servicing mission for the Hubble Space Telescope: new instruments installed, older ones returned to Earth, repairs to insulation; five EVAs completed, Lee/Smith 3, Harbaugh/Tanner 2			
83 **STS-83**	04.04.97 Columbia	Halsell, Still, Voss, Gernhardt Thomas, Crounch, Linteris	3.23.13 28.5°
Microgravity Science Laboratory 1 primary payload for 10th EDO: mission curtailed due to fuel cell problems and re-flown as STS-94			
84 **STS-84**	15.05.97 Atlantis	Precourt, Collins, Clervoy, Noriega Lu, Kondakova, Foale (up), Linenger (down)	9.05.21 51.6°
#6 Mir docking mission with Spacehab-DM module. Delivered supplies to Mir, left Foale on board for medium-length stay, returned with Linenger from STS-81 (mission duration 132.03.50)			

Flight	Launch date Orbiter	Crew	Duration (d, h, m) Incl
85 **STS-94**	01.07.97 Columbia	Halsell, Still, Voss, Gernhardt, Thomas, Crouch, Linteris	15.16.45 28.5°
Reflight of STS-83, Microgravity Science Laboratory 1 primary payload for 11th EDO: fastest return-to-orbit for the crewmembers			
86 **STS-85**	07.08.97 Discovery	Brown, Rominger, Davis, Curbeam, Robinson, Tryggvason	11.20.28 57.0°
CRSTA-SPAS deployment, test of Japanese remote manipulator system			
87 **STS-86**	25.09.97 Atlantis	Wetherbee, Bloomfield, Titov, Parazynski, Chretien, Lawrence, Wolf (up), Foale (down)	10.19.22 51.6°
#7 Mir docking; Lawrence originally planned to remain aboard Mir but replaced by Wolf because her height precludes use of Russian EVA suit. EVA with Parazynski and Titov, first to involve non-US crewmember			
88 **STS-87**	19.11.97 Columbia	Kregel, Lindsey, Chawla, Scott, Doi and Kadenyuk	15.16.35 29.45°
US microgravity payload #4, deployment and retrieval of SPARTAN 201–04. 12th EDO. Scott and Doi performed two EVAs, 1 to manually capture SPARTAN			
89 **STS-89**	22.01–98 Endeavour	Wilcutt, Edwards, Reilly, Anderson, Dunbar, Sharipov, Thomas (up) and Wolf (down)	08.19.48 51.6°
#8 Mir docking. Thomas replaced Wolf			
90 **STS-90**	17.04.98 Columbia	Searfoss, Altman, Hire, Linnehan, Williams, Buckey and Pawelczyk	15.21.51 39.0°
US Neurolab mission. 13th and last planned EDO before ISS operations for long duration missions			
91 **STS-91**	02.06.98 Discovery	Precourt, Gorie, Chang-Diaz, Lawrence, Kavandi, Ryumin and Thomas (down)	09.19.55 51.6°
#9 and last Mir docking, first with Discovery. Cosmonaut Ryumin conducted Mir inspection to determine suitability for sustained operations. Thomas returned to Earth			
92 **STS-95**	29.10.98 Discovery	Brown, Lindsey, Parazynski, Robinson, Duque, Mukai and Glenn	08.21.45 28.45°
Life sciences mission including world's oldest astronaut, John Glenn age 77. Deployed and retrieved SPARTAN 201–05 and tested Hubble Space Telescope orbital systems test platform			
93 **STS-88**	04.12.98 Endeavour	Cabana, Sturckow, Ross, Currie, Newman and Krikalev	11.19.53 51.6°
Carried Unity module to Zarya in first docking to begin assembly of ISS. Ross and Newman made three EVAs to attach cables and deploy fixtures to ISS exterior			
94 **STS-96**	27.05.99 Discovery	Rominger, Husband, Jernigan, Ochoa, Barry, Payette and Tokarev	09.19.14 51.6°
First Shuttle ISS servicing mission to docked Zarya/Unity modules. Released Starshine satellite			
95 **STS-93**	23.07.99 Columbia	FUTURE MISSIONS (as of May 2002) Collins, Ashby, Hawley, Tognini	04.22.50 28.45°
Launch of Chandra X-ray observatory			
96 **STS-103**	19.12.99 Discovery	Brown, Kelly, Clervoy, Smith Foale, Grunsfeld, Nicollier	07.23.11 28.45°
Third Hubble Space Telescope servicing mission			
97 **STS-99**	11.02.00 Endeavour	Kregel, Gorie, Thiele, Kavandi, Voss, Mohri	11.05.38 57°
Radar Topography Mission			
98 **STS-101**	19.05.00 Atlantis	Halsell, Horowitz, Weber, Williams, Voss, Helms, Usachev	09.20.10 51.6°
Third Shuttle ISS flight (2A.2a) with double Spacehab module			

Flight	Launch date Orbiter	Crew	Duration (d, h, m) Incl
99	08.09.00	Wilcutt, Altman, Lu, Mastracchio,	11.19.11
STS-106	Atlantis	Burbank, McArthur, Lopez-Alegria	51.6°
Fourth Shuttle ISS flight (2A.2b) with double Spacehab module			
100	11.10.00	Duffy, Melroy, Wakata, Wisoff,	12.21.44
STS-92	Discovery	Chiao, McArthur, Lopez-Alegria	51.6°
Fifth Shuttle ISS flight (3A) delivering Z1 truss and PMA-3			
101	30.11.00	Jett, Bloomfield, Garneau, Tanner,	10.19.58
STS-97	Endeavour	Noriega	51.6°
Sixth Shuttle ISS flight (4A) delivering P6 solar array			
102	07.02.01	Cockrell, Polansky, Ivins, Jones,	12.21.20
STS-98	Atlantis	Curbeam	51.6°
Seventh Shuttle ISS flight (5A) carrying US Destiny laboratory			
103	08.03.01	Weatherbee, Kelly, Thomas, Richards, Usachev (up), Helms (up) Voss (up), Shepherd (down), Krikalev	12.19.49
STS-102	Discovery	(down), Gidzenko (down)	51.6°
Eighth Shuttle ISS flight (5A.1) with Leonardo MPLM, Expedition 2 crew			
104	19.04.01	Rominger, Ashby, Hadfield, Phillips	11.21.30
STS-100	Endeavour	Parazynski, Guidoni, Lonchakov	51.6°
Ninth Shuttle ISS flight (6A) with Rafaello MPLM and SSRMS			
105	12.07.01	Lindsay, Hobaugh, Gernhardt,	12.18.35
STS-104	Atlantis	Kavandi, Reilly	51.6°
Tenth Shuttle ISS flight (7A) delivers Quest airlock module			
106	10.08.01	Horowitz, Sturckow, Forrester, Culbertson (up), Tyurin (up), Dezhurov (up), Voss (down), Helms	11.21.13
STS-105	Discovery	(down), Usachev (down)	51.6°
Eleventh Shuttle ISS flight (7A.1) with Leonardo MPLM, Expedition 3 crew			
107	05.12.01	Gorie, Kelly, Goodwin, Tani, Onufrienko (up), Bursch (up),	11.19.36
STS-108		Walz (up), Culbertson (down), Endeavour Dezhurov (down), Tyurin (down)	51.6°
Twelfth Shuttle ISS flight (UF1) with Rafaello MPLM, Expedition 4 crew			
108	01.03.02	Altman, Carey, Grunsfeld, Currie	10.22.10
STS-109	Colombia		28.45°
Fourth Hubble Space Telescope servicing mission			
109	04.08.02	Bloomfield, Frick, Walhelm, Ochoa, Morin, Ross, Smith	10.19.43
STS-110	Atlantis		51.6°
Thirteenth Shuttle ISS flight (8A) with S0 truss and Mobile Transporter			
110	06.02.02	Cockrell, Lockhart, Perrin, Chang-Diaz, Whitson (up)	13.20.35
STS-111	Endeavour	Korzun (up), Treschev (up), Walz (down), Bursch (down), Onufrienko (down)	51.6°
Fourteenth Shuttle ISS flight (UF-2) with Leonardo MPLM, Expedition 5 crew			
111	10.07.02	Ashby, Melroy, Wolf, Magnus, Sellers, Yurchikhin	10.19.58
STS-112	Atlantis		51.6°
Fifteenth Shuttle ISS flight (9A) with S1 truss			
112	11.23.02	Wetherbee, Lockhart, Lopez-Alegria, Herrington, Bowersox (up), Budarin (up), Petit (up)	13.18.47
STS-113		Korzun (down), Whitson (down), Treschev (down)	51.6°
Sixteenth Shuttle ISS flight (11A) with P1 truss, Expedition 6 crew			
113	01.16.03	Husband, McCool, Brown, Chawla, (to destruction) Anderson, Clark,	15.22.10
STS-107		Ramon	39°
Microgravity research with Spacehab-DM			

* rev on which landing occurred (equals completed revs for Kennedy Space Center landings). Durations are in days/hours/minutes. **Shuttle Orbiters:** *Challenger* 099, *Enterprise* 101, *Columbia* 102, *Discovery* 103, *Atlantis* 104, *Endeavour* 105. EDO = Extended Duration Orbiter

Clearly seen in the overhead view of the Shuttle on its mobile launch platform, the two tail service umbilical shrouds through which fluids and electrical power flow to the stacked vehicle prior to lift-off (NASA) ***NEW***/1047474

Shuttle moves to the pad up the crawlerway ramp at Complex 39 (NASA)
NEW/1047473

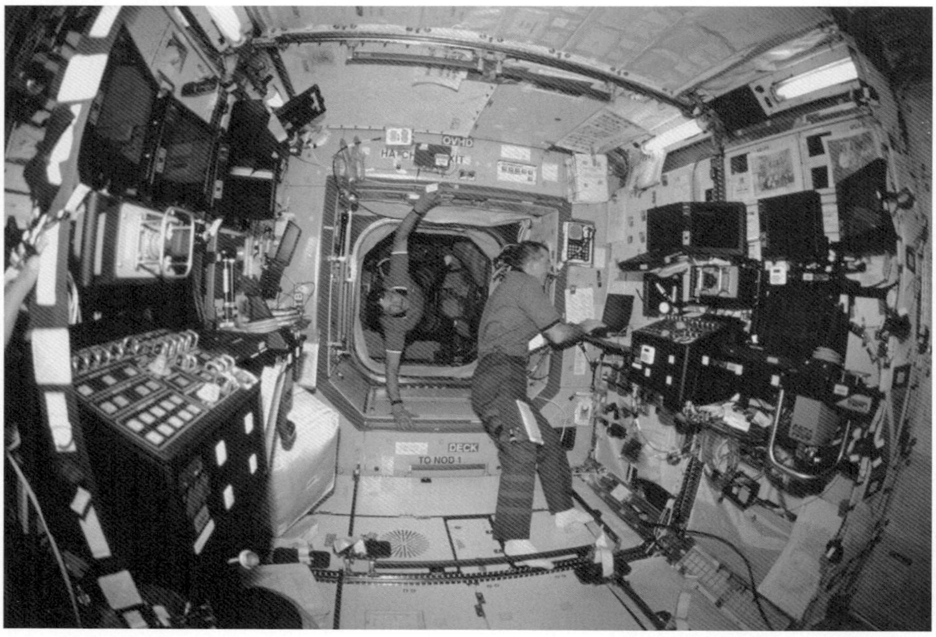

Shuttle deliveries of crew and cargo to the International Space Station facilitates extension of experiments formerly conducted from inside the Orbiter (NASA)
NEW/1047477

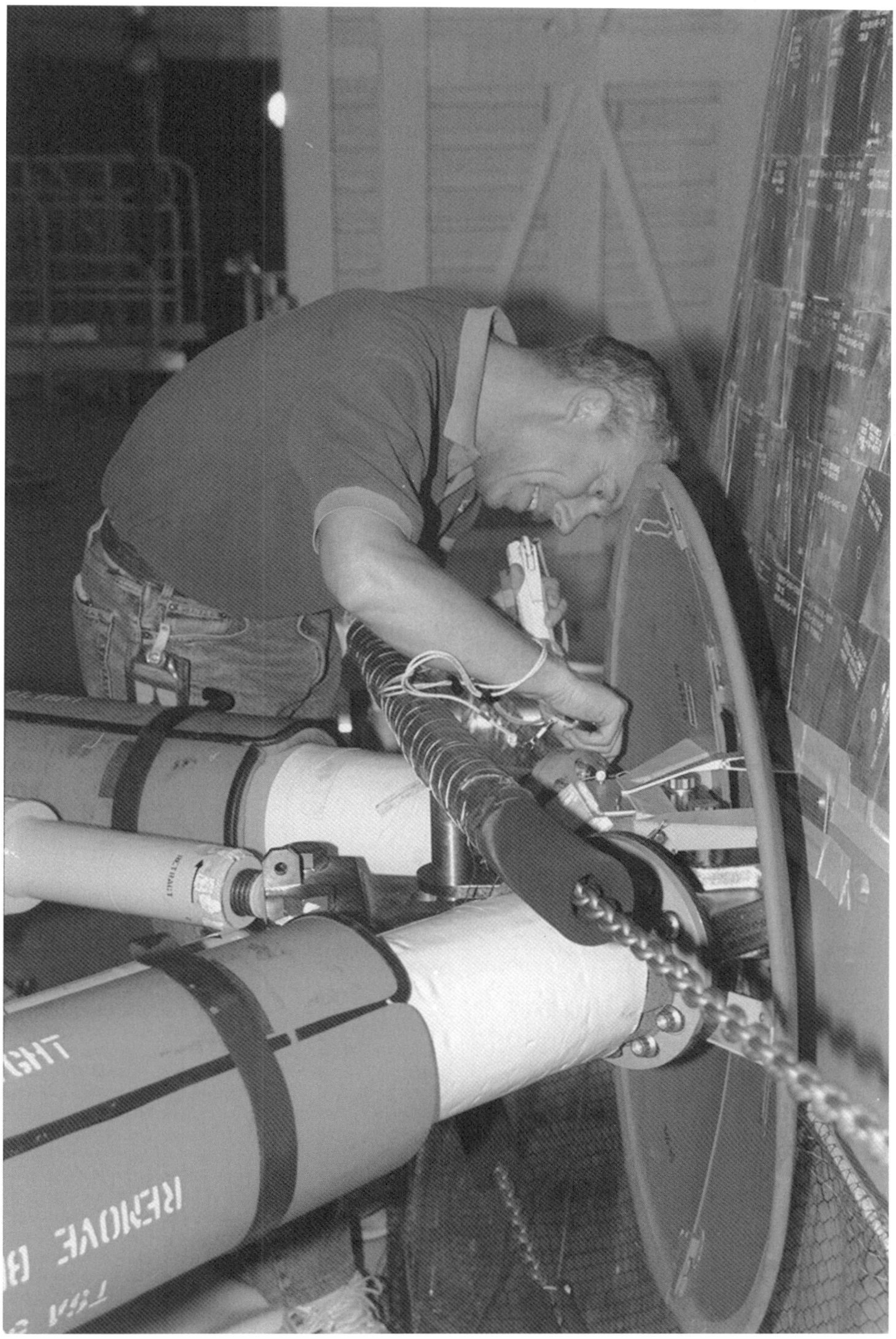

The forward bipod connecting the External Tank to the underside of the Orbiter forward fuselage (NASA)
NEW/1047478

The Payload Changeout Room, or PCR, is seen clearly in the view of the Shuttle at Launch Complex 39 (NASA)
NEW/1047499

One of the new Block 2 Space Shuttle Main Engines to be fitted to Atlantis for STS-104 launched December 2001 (Boeing)
NEW/1047500

Atlantis is moved on its flatbed transporter at the Kennedy Space Center (NASA)
NEW/1047504

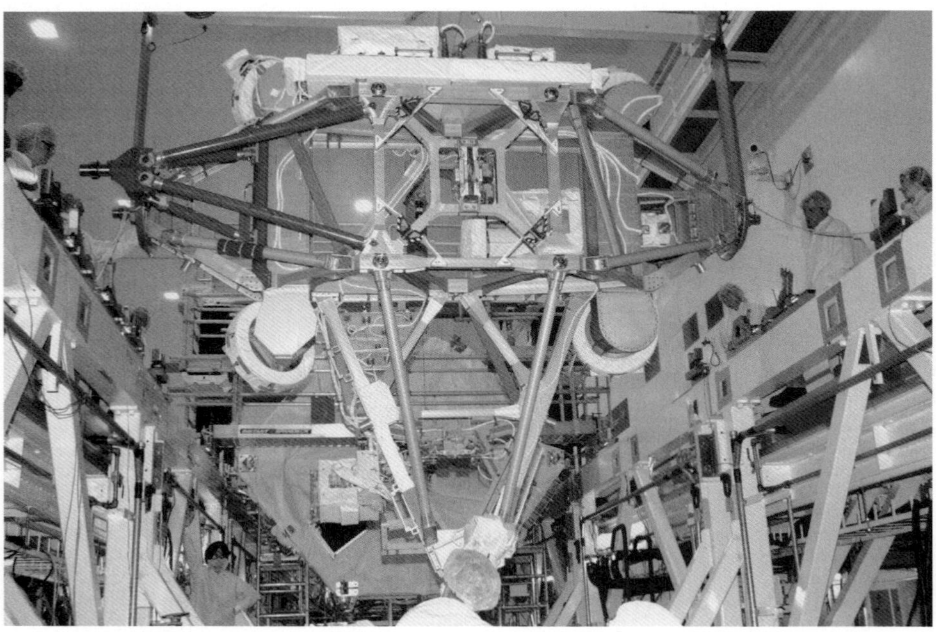

The Base system for the ISS is placed in the payload bay of the Shuttle in readiness for STS-111 (NASA)
NEW/1047512

UPDATED

Table 6: General Human Space Flight Statistics to 1 February 2003

Shuttle mission STS-51L is not considered a space flight in this list.

	STS-113	*STS-107*
Total Fliers	426	430
Nations	30	31
Male	387	390
Female	39	40
Total Tickets	946	953
United States	268	271
United States men	237	239
United States Women	31	32
USSR	72	72
USSR Men	70	70
USSR Women	2	2
CIS	27	27
CIS Men	26	26
CIS Women	1	1
Non US/Russian	59	60

	STS-113	*STS-107*
Men	54	55
Women	5	5
Men with 7 flights	2	2
Men with 6 flights	4	4
Women/6	0	0
Men/5	15	15
Women/5	6	6
Men/4	47	47
Women/4	4	4
Men/3	70	70
Women/3	7	7
All/2	97	100
All/1	174	175

Early stages in the assembly of the International Space Station provided the first true measure of international co-operation on a scale never before attempted in space (NASA) 0573539

Artist's impression of Europe's Automated Transfer Vehicle carrying cargo, gases and fluids to the International Space Station (ESA) 0137121

The International Space Station embodied the national aspirations of the USA, Europe, Japan, Canada and Russia and underpins further expansion in human space flight (NASA) 0573533

Attached to the International Space Station, Canadarm 2 extends the options available for moving modules and equipment around the exterior of the facility (NASA)
NEW/1047476

UPDATED

Table 7: US/Russian Space Endurance Records

Single Flight Individual Records

Astronaut/Cosmonaut	*Days*	*Flights*
Valery Polyakov	438	Mir-17
C Walz/D Bursch	196	ISS-4

Cumulative Records

World Rank	*NASA*	*Days*	*Flights*
1	Avdeyev	748	3
2	Polyakov	679	2
3	Solovyev	652	5
4	Krikalev	625	5
5	Afanasyev	556	4
6	Usachev	553	4
7	Manarov	541	2
8	Viktorenko	489	4
9	Romanenko	430	3
10	Kaleri	416	3
11	Volkov	392	3
12	Onufrienko	389	2
13	Titov	387	4
14	Tsibliev	383	2
15	Korzun	382	2
US Endurance Data			
30	Walz	231	4
31	Bursch	227	4
32	Lucid	223	5
36	Helms	211	5
38	Voss, J	202	5
42	Whitson	185	1
48	Foale	179	5
51	Thomas	164	3
52	Blaha	161	5
53	Shepherd	159	4
54	Wolf	153	3
58	Culbertson	144	3
59	Linenger	143	2

The Shuttle edges its way to the launch pad more than 5 km from the giant assembly building where moon-bound Saturn Vs were once assembled (NASA) 0137131

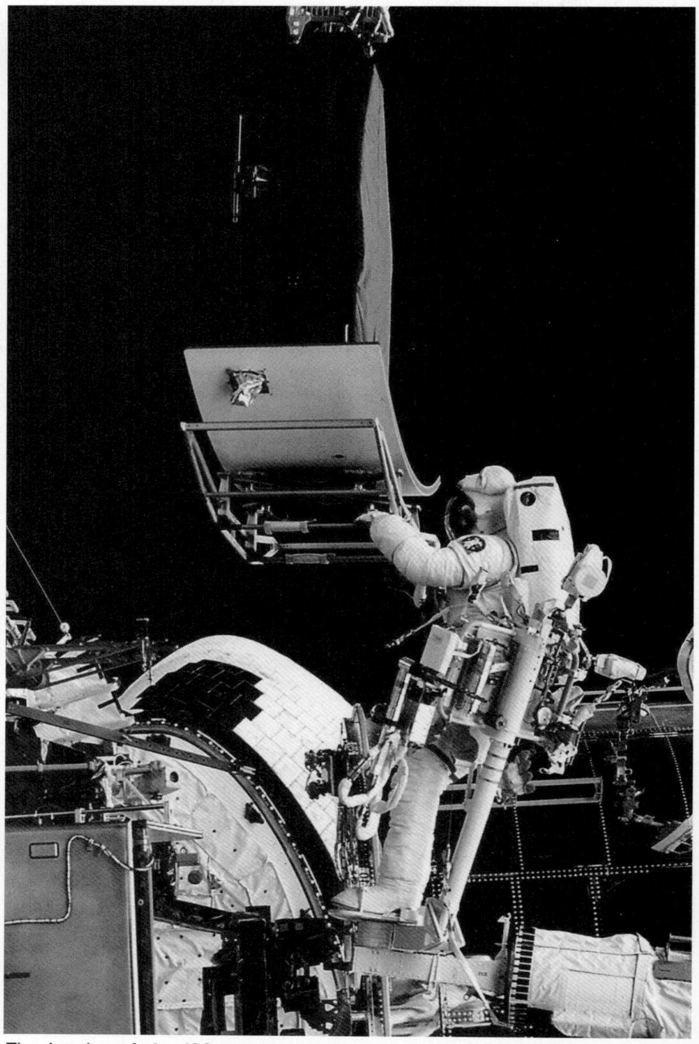

The interior of the ISS provides freedom for busy crewmembers performing housekeeping and science tasks (ESA) 0137161

Table 8: NASA Phase One Mir Missions and Flight Durations

Astronaut	Up	Date (GMT)	Down	Date (GMT)	Days	Mir Crew
Norm Thagard	SOY	14/03/95	71	7/7/95	115.4	Mir-18
Shannon Lucid	76	22/03/96	79	26/9/96	188.2	Mir-21/22
John Blaha	79	16/09/96	81	22/1/97	128.2	Mir-22
Jerry Linenger	81	12/01/97	84	24/5/97	132.2	Mir-22/23
Mike Foale	84	15/05/97	86	6/10/97	144.6	Mir-23/24
David Wolf	86	26/09/97	89	31/1/98	127.8	Mir-24/25
Andrew Thomas	89	22/01/98	91	12/6/98	140.6	Mir-25

TOTAL US PHASE ONE TIME IN SPACE 977 days

Table 9: US EVA Durations

General EVA STATS (through STS-113)

	Expedition EVAs	HH	MM
0	Expedition 1 EVAs	00	00
1	Expedition 2 EVAs (2001)	00	19
4	Expedition 3 EVAs (2001)	18	40
3	Expedition 4 EVAs (2001–2002)	17	49
2	Expedition 5 EVAs (2002)	04	25
0	Expedition 6 EVAs (2002–2003)	00	00
10	TOTAL Expedition EVAs	41	13

	Expedition EVAs	HH	MM
25	Total STS-based ISS EVAs	167	15
24	Total ISS-based ISS/STS EVAs	138	05
	Longest STS EVA (STS-102)	08	56
	Shortest STS EVA (STS-41C)	03	05
	Most STS EVA time (STS-109)	35	55

UPDATED

Table 9A: US/Russian Spacewalk Endurance Records

	Astronaut/Cosmonaut	EVAs	HH	MM
1	Anatoly Solovyov	16	77	41
2	Jerry Ross	9	58	18
3	Steven Smith	7	49	48
4	Nikolai Budarin	9	46	14
5	James Newman	6	43	13
6	Talgat Musabayev	8	43	02
7	Yury Onufrienko	8	42	52
8	Sergei Avdeyev	10	41	59
9	Victor Afanasyev	7	38	33
10	John Grunsfeld	5	38	02
11	Vladimir Dezhurov	9	37	56
12	Sergei Krikalev	7	36	00
13	Musa Manarov	7	34	32

Table 10: ISS Assembly EVAs

No	*Mission*	*EVA Date*	*HH*	*MM*	*Notes*
3	STS-88	07/12/98	07	21	ISS-2A: Node 1 connection to Zarya (Ross, Newman)
		09/12/98	07	02	ISS-2A
		12/12/98	06	59	ISS-2A
					STS-88 TOTAL: 21:22
1	STS-96	29/05/99	07	55	ISS-2A 1: Station outfitting (Jernigan, Barry)
1	STS-101	21/05/00	06	44	ISS-2A 2a: Station outfitting (Voss, Williams)
1	STS-106	11/09/00	06	14	ISS-2A 2b: Zvezda connections (Lu, Malenchenko)
4	STS-92	15/10/00	06	28	ISS-3A: Z1, PMA-3 (Chiao, McArthur)
		16/10/00	07	07	ISS-3A (Wisoff, Lopez-Alegria)
		17/10/00	06	48	ISS-3A (Chiao, McArthur)
		18/10/00	06	56	ISS-3A (Wisoff, Lopez-Alegria)
					STS-92 TOTAL: 27:19
3	STS-97	04/12/00	07	33	ISS-4A: P6 solar array truss (Tanner, Noriega)
		05/12/00	06	37	ISS-4A
		07/12/00	05	10	ISS-4A
					STS-97 TOTAL: 19:20
3	STS-98	10/02/01	07	34	ISS-5A: Destiny lab module (Jones, Curbeam)
		12/02/01	06	50	ISS-5A
		14/02/01	05	25	ISS-5A
					STS-98 TOTAL: 19:49
2	STS-102	11/03/01	08	56	ISS-5A 1: Station outfitting (Voss and Helms)
		13/03/01	06	21	ISS-5A 1 (Thomas and Richards)
					STS-102 TOTAL: 15:17
2	STS-100	22/04/01	07	10	ISS-6A: SSRMS installation (Hadfield, Parazynski)
		24/04/01	07	40	ISS-6A
					STS-100 TOTAL: 14:50
1	ISS-1/E2	08/06/01	00	19	E2 crew: Zvezda hatch move (Usachev, Voss)
2	STS-104	14/07/01	05	59	ISS-7A: Joint airlock (Gernhardt, Reilly)
		17/07/01	06	29	ISS-7A
					STS-104 TOTAL: 12:47
1	ISS-2/7A	20/07/01	04	02	ISS-7A/Quest 0 (conducted from new airlock)
2	STS-105	16/08/01	06	16	ISS-7A 1: Ammonia servicer; MISSE (Barry, Forrester)
		18/08/01	05	29	ISS-7A 1
					STS-105 TOTAL: 15:47
1	ISS-3/E3	08/10/01	04	58	E3 crew: Pirs module outfitting (Dezhurov and Tyurin)
1	ISS-4/E3	15/10/01	05	52	E3 crew: Experiments (Dezhurov and Tyurin)
1	ISS-5/E3	12/11/01	05	04	E3 crew: Pirs outfitting (Culbertson and Dezhurov)
1	ISS-6/E3	03/12/01	02	46	E3 crew: Debris removal (Dezhurov and Tyurin)
					ISS/E3 TOTAL: 18:40
1	STS-108	10/12/01	04	12	ISS-UF1: P6 blanket installation (Godwin and Tani)
1	ISS-7/E4	14/01/02	06	03	E4 crew: Strela relocation (Onufrienko and Walz)
1	ISS-8/E4	25/01/02	05	59	E4 crew: Plume deflectors (Onufrienko and Bursch)
1	ISS-9/E4	20/02/02	05	47	E4/Quest 1: 8A get-ahead tasks (Walz and Bursch)
					ISS/E4 TOTAL: 17:49
4	ISS-10/8A	11/04/02	07	48	ISS-8A/Quest 2: S0 installation (Smith and Walheim)
	ISS-11/8A	13/04/02	07	30	ISS-8A/Quest 3: S0 installation (Ross and Morin)
	ISS-12/8A	14/04/02	06	27	ISS-8A/Quest 4: SSRMS config (Smith and Walheim)
	ISS-13/8A	16/04/02	06	37	ISS-8A/Quest 5: Get-ahead (Ross and Morin)
					STS-110 TOTAL: 28:22
3	ISS-14 UF2	01/01/04	07	14	ISS-UF2/Quest 7 (Chang-D and Perrin)
	ISS-15 UF2	01/01/04	05	00	ISS-UF2/Quest 8: (Chang-Diaz and Perrin)
	ISS-16 UF2	01/01/04	07	17	ISS-UF2/Quest 9: SSRMS joint (Chang-D and Perrin)
					STS-111 TOTAL: 19:31

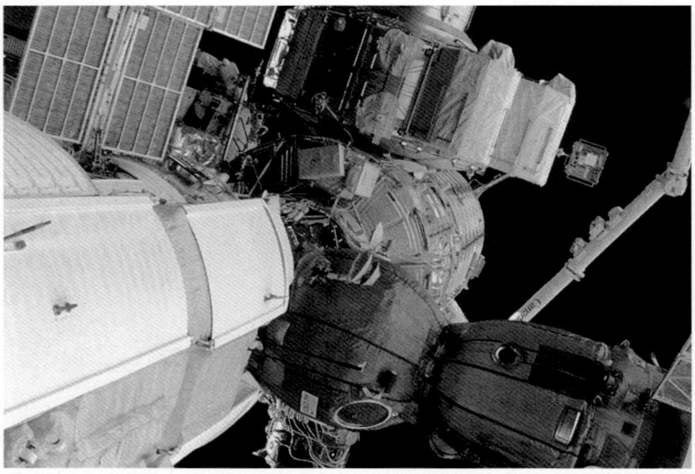

The ISS Expedition 3 crew take over their long duration shift and capture this evocative image with Russia's Progress cargo tanker in the foreground shrouded in black thermal protection (NASA) 0137135

The S0 truss in an essential element of the ISS and marks the beginning of the construction and assembly of the long girder keel to which the large solar arrays will be attached (NASA) 0137138

No	Mission	EVA Date	HH	MM	Notes
1	ISS-17 (E5)	16/08/02	04	25	E5 crew: SM debris shields (Korzun, Whitson)
1	ISS-18 (E5)	23/08/02	05	21	E5 crew: Ham antennas, handrails (Korzun, Treschev)
					ISS/E5 TOTAL: 09:46
3	ISS-19 (9A)	10/10/02	07	01	ISS-9A/Quest 10: S1 installation (Wolf and Sellers)
	ISS-20 (9A)	12/10/02	06	04	ISS-9A/Quest 11: Wolf and Sellers
	ISS-21 (9A)	14/10/02	06	36	ISS-9A/Quest 12: Wolf and Sellers
					STS-112 TOTAL: 19:41
3	ISS-22 (11A)	26/11/02	06	45	ISS-11A/Quest 13: P1 installation
	ISS-21 (11A)	28/11/02	06	10	ISS-11A/Q 14: Lopez-A, Herrington
	ISS-22 (11A)	30/11/02	07	00	ISS-11A/Q 15: Lopez-A, Herrington
					STYS-113 TOTAL: 19:55

49	ISS EVAs
Total EVA time	305:20
Nationalities	35 US, 1 Canada, 1 France, 7 Russia

Build-up of the International Space Station involved high work loads on EVA at a level not experienced since the Gemini and Apollo days in the USA and the Mir days in Russia (NASA-JSC) 0573542

Developed in Canada, the Shuttle Remote Manipulator System played a vital role in assisting astronauts on EVA Space Station assembly tasks (NASA) 0573540

Docked elements of the International Space Station following the first launches by Russia and the USA in 1998 (NASA) 0573538

Technology developed for the ISS reached new levels of sophistication and included the development of some of the largest solar arrays ever flown in space (NASA) 0573541

Astronauts Lopez and Herrington conduct EVA operations from the ISS during the STS-113 mission (NASA) 0573537

Astronaut Sellers on the Space Station Remote Manipulator System (NASA-JSC)
0573535

Viewed alongside the Shuttle RMA attachment point, astronaut Sellers in the Shuttle cargo bay (NASA)
0573534

Table 11: Active NASA EVA Astronauts

	EVA Missions (EVAs/HH:MM)
Daniel Barry	STS-72: 1/6:09; STS-96: 2/07:55; STS-105: 2/11:41
Franklin Chang-Diaz	STS-111: 3/28:22
Leroy Chiao	STS-72: 2/13:02; STS-92: 2/13:16
Robert Curbeam	STS-98: 3/19:49
Takao Doi	STS-87: 2/12:42
Michael Foale*	STS-63: 1/6:39; STS-103: 1/8:10
Michael Forrester	STS-105: 2/11:41
Michael Gernhardt	STS-69: 1/6:46; STS-104: 3/16:30
Linda Godwin	STS-76: 1/6:02; STS-108: 1/04:12
John Grunsfeld	STS-103: 2/16:23; STS-109: 3/21:09
Chris Hadfield	STS-100: 2/14:50
Gregory Harbaugh	STS-54: 1/4:28; STS-82: 2/14:00
Susan Helms	STS-102: 1/8:56
John Herrington	STS-113: 3/19:55
Tamara Jernigan	STS-96: 1/07:55
Thomas Jones	STS-98: 3/19:49
Mark Lee	STS-64: 1/6:51; STS-82: 3/19:10
Richard Linnehan	STS-109: 3/21:09
Michael Lopez-Alegria	STS-92: 2/14:13; STS-113: 3/19:55
Edward Lu	STS-106: 1/6:14
Michael Massimino	STS-109: 2/14:46
William McArthur	STS-92: 2/13:16
Lee Morin	STS-110: 2/14:07
James Newman	STS-51: 1/7:05; STS-88: 3/21:22; STS-109: 2/14:46
Claude Nicollier	STS-103: 1/8:10
Carlos Noriega	STS-97: 3/19:20
Scott Parazynski	STS-86: 1/5:01; STS-100: 2/14:50
Philippe Perrin	STS-111: 3/28:22
James Reilly	STS-104: 3/16:30
Paul Richards	STS-102: 1/6:21
Jerry Ross	STS-37: 2/10:29; STS-61B: 2/12:20; STS-88: 3/21:22; STS-110: 2/14:07
Mario Runco	STS-54: 1/4:28
Piers Sellers	STS-112: 3/19:41
Steven Smith	STS-82: 3/19:10; STS-103: 2/16:23; STS-110: 2/14:15
Daniel Tani	STS-108: 1/04:12
Joseph Tanner	STS-82: 2/14:00; STS-97: 3/19:20
Andrew Thomas	STS-102: 1/06:21
James Voss	STS-69: 1/6:46; STS-101: 1/06:44; STS-102: 1/8:56
Rex Walheim	STS-110: 2/14:15
Carl Walz	STS-51: 1/7:05
Peggy Whitson	ISS-5: 1/4:25
Jeffrey Williams	STS-101: 1/06:44
Jeff Wisoff	STS-57: 1/5:50; STS-92: 2/14:13
David Wolf	STS-112: 3/19:41

* Does not count a Mir spacewalk

Table 12: World Astronaut and Cosmonaut List

STS-51L is not considered a space flight

Name	Flts	Sex	Nation	Missions
Acton, Loren	1	M	US	STS-51F
Adamson, James	2	M	US	STS-28 STS-44
Afanasyev, Viktor	4	M	USSR	TM-11 TM-8 TM-29 TM-33
Akers, Thomas	4	M	US	STS-41 STS-49 STS-61 STS-79
Akiyama, Toyohiro	1	M	Japan	TM 11
Aksenov, Vladimir	2	M	USSR	Soyuz 22, Soyuz T2
Aldrin, Edwin	2	M	US	Gemini 12 Apollo 11
Alexandrov, Alexander	1	M	Bulgaria	TM 5
Alexandrov, Alexander	2	M	USSR	Soyuz T9 TM 3
Allen, Andrew	3	M	US	STS-46 STS-62 STS-75
Allen, Joseph	2	M	US	STS-5 STS-51A
AlSaud, Sultan	1	M	S Arabia	STS-51G
Altman, Scott	3	M	US	STS-90 STS-106 STS-109
Anders, William	1	M	US	Apollo 8
Anderson, Michael	1	M	US	STS-89
Apt, Jay	4	M	US	STS-37 STS-47 STS-59 STS-79
Armstrong, Neil	2	M	US	Gemini 8 Apollo 11
Artsebarsky, Anatoly	1	M	USSR	TM 12
Artyukhin, Yuri	1	M	USSR	Soyuz 14
Ashby, Jeff	3	M	US	STS-93 STS-100 STS-112
Atkov, Oleg	1	M	USSR	Soyuz T10
Aubakirov, Toktar	1	M	USSR	TM 13
Avdeyev, Sergei	3	M	CIS	TM 15 TM 22 TM 28
Bagian, James	2	M	US	STS-29 STS-40
Baker, Ellen	3	F	US	STS-34 STS-50 STS-71
Baker, Michael	4	M	US	STS-43 STS-52 STS-68 STS-81
Balandin, Alexander	1	M	USSR	TM 9
Barry, Daniel	3	M	US	STS-72 STS-96 STS-105
Bartoe, John-David	1	M	US	STS-51F
Baturin, Yuri	2	M	CIS	TM 28 STS-32
Baudry, Patrick	1	M	France	STS-51G
Bean, Alan	2	M	US	Apollo 12 Skylab 3
Bella, Ivan	1	M	Slovakia	TM 29
Belyayev, Pavel	1	M	USSR	Voskhod 2
Beregovoy, Georgi	1	M	USSR	Soyuz 3
Berezovoy, Anatoly	1	M	USSR	Soyuz T5-7
Blaha, John	5	M	US	STS-29 STS-33 STS-43 STS-58 79/Mir3/81
Bloomfield, Michael	2	M	US	STS-86 STS-97 STS-110
Bluford, Guion	4	M	US	STS-8 STS-61A STS-39 STS-53
Bobko, Karol	3	M	US	STS-6 STS-51D STS-51J
Bolden, Charles	4	M	US	STS-61C STS-31 STS-45 STS-60
Bondar, Roberta	1	F	Canada	STS-42
Borman, Frank	2	M	US	Gemini 7 Apollo 8
Bowersox, Kenneth	5	M	US	STS-50 STS-61 STS-73 STS-82
Brady, Charles	1	M	US	STS-78
Brand, Vance	4	M	US	ASTP STS-5 STS-41B STS-35
Brandenstein, Daniel	4	M	US	STS-8 STS-51G STS-32 STS-49
Bridges, Roy	1	M	US	STS-51F
Brown, Curt	6	M	US	STS-47 STS-66 STS-77 STS-85 STS-95 STS-103
Brown, Mark	2	M	US	STS-28 STS-48
Buchli, James	4	M	US	STS-51C STS-61A STS-29 STS-48
Buckey, Jay	1	M	US	STS-90
Budarin, Nikolai	3	M	CIS	STS-71/Mir19 TM-27/Mir25 STS-113/ISS-6
Burbank, Daniel	1	M	US	STS-106
Bursch, Daniel	4	M	US	STS-51 STS-68 STS-77 108/ISS-4/111
Bykovsky, Valery	3	M	USSR	Vostok 5 Soyuz 22 Soyuz 31
Cabana, Robert	4	M	US	STS-41 STS-53 STS-65 STS-88
Cameron, Kenneth	3	M	US	STS-37 STS-56 STS-74
Carey, Duane	1	M	US	STS-109
Carpenter, Scott	1	M	US	Mercury 7
Carr, Gerald	1	M	US	Skylab 4
Carter, Manley	1	M	US	STS-33
Casper, John	4	M	US	STS-36 STS-54 STS-62 STS-77
Cenker, Robert	1	M	US	STS-61C
Cernan, Eugene	3	M	US	Gemini 9 Apollo 10 Apollo 17
Chang-Diaz, Franklin	7	M	US	STS-61C STS-34 STS-46 STS-60 STS-75 STS-91 STS-107
Chawla, Kalpana	1	F	US	STS-87
Cheli, Maurizio	1	M	Italy	STS-75
Chiao, Leroy	3	M	US	STS-65 STS-72 STS-92
Chilton, Kevin	3	M	US	STS-49 STS-59 STS-76
Chretien, Jean-Loup	3	M	France	Soyuz T6 TM 7 STS-86
Cleave, Mary	2	F	US	STS-61B STS-30
Clervoy, Jean-Francois	3	M	France	STS-66 STS-84 STS-103
Clifford, Richard	3	M	US	STS-53 STS-59 STS-76
Coats, Michael	3	M	US	STS-41D STS-29 STS-39
Cockrell, Kenneth	5	M	US	STS-56 STS-69 STS-80 STS-98 STS-111
Coleman, Catherine	2	F	US	STS-73 STS-93
Collins, Eileen	3	F	US	STS-63 STS-84 STS-93
Collins, Michael	2	M	US	Gemini 10 Apollo 11
Conrad, Charles	4	M	US	Gemini 5 Gemini 11 Apollo 12 Skylab 2
Cooper, Gordon	2	M	US	Mercury 9 Gemini 5
Covey, Richard	4	M	US	STS-51I STS-26 STS-38 STS-61
Creighton, John	3	M	US	STS-51G STS-36 STS-48
Crippen, Robert	4	M	US	STS-1 STS-7 STS-41C STS-41G

Name	*Flts*	*Sex*	*Nation*	*Missions*
Crouch, Roger	2	M	US	STS-83 STS-94
Culbertson, Frank	3	M	US	STS-38 STS-51 ISS-3
Cunningham, Walter	1	M	US	Apollo 7
Curbeam, Robert	2	M	US	STS-85 STS-98
Currie, Nancy	4	F	US	STS-57 STS-70 STS-88 STS-109
Davis, Jan	3	F	US	STS-47 STS-60 STS-85
DeLucas, Larry	1	M	US	STS-50
Demin, Lev	1	M	USSR	Soyuz 15
Dezhurov, Vladimir	2	M	CIS	TM21/STS-71 ISS-3
Dobrovolsky, Georgy	1	M	USSR	Soyuz 11
Doi, Takao	1	M	Japan	STS-87
Duffy, Brian	4	M	US	STS-45 STS-57 STS-72 STS-92
Duke, Charles	1	M	US	Apollo 16
Dunbar, Bonnie	5	F	US	STS-61A STS-32 STS-50 STS-71 STS-89
Duque, Pedro	1	M	Spain	STS-95
Durrance, Samuel	2	M	US	STS-35 STS-67
Dzhanibekov, Vladimir	5	M	USSR	Soyuz-27 Soyuz-39 Soyuz-T6 Soyuz-T12 Soyuz-T13
Edwards, Joseph	1	M	US	STS-89
Eisele, Donn	1	M	US	Apollo 7
England, Anthony	1	M	US	STS-51F
Engle, Joseph	2	M	US	STS-2 STS-51I
Evans, Ronald	1	M	US	Apollo 17
Ewald, Reinhold	1	M	Germany	TM-25
Eyharts, Leopold	1	M	France	TM-27
Fabian, John	2	M	US	STS-7 STS-51G
Faris, MA	1	M	Syria	TM-3
Farkas, Bertalan	1	M	Hungary	Soyuz 36
Favier, Jean-Jacques	1	M	France	STS-78
Feoktistov, Konstantin	1	M	USSR	Voskhod 1
Fettman, Martin	1	M	US	STS-58
Filipchenko, Anatoly	2	M	USSR	Soyuz 7 Soyuz 16
Fisher, WIlliam	1	M	US	STS-51I
Fisher, Anna	1	F	US	STS-51A
Flade, Klaus-Dietrich	1	M	Germany	TM-14
Foale, Michael	5	M	US	STS-45 STS-56 STS-63 84/Mir5/86 STS-103
Forrester, Patrick	1	M	US	STS-105
Frick, Stephen	1	M	US	STS-110
Frimout, Dirk	1	M	Belgium	STS-45
Fullerton, Gordon	2	M	US	STS-3 STS-51F
Furrer, Reinhard	1	M	Germany	STS-61A
Gaffney, Andrew	1	M	US	STS-40
Gagarin, Yuri	1	M	USSR	Vostok 1
Gardner, Dale	2	M	US	STS-8 STS-51A
Gardner, Guy	2	M	US	STS-27 STS-35
Garn, Jake	1	M	US	STS-51D
Garneau, Marc	3	M	Canada	STS-41G STS-77 STS-97
Garriott, Owen	2	M	US	Skylab 3 STS-9
Gemar, Charles	3	M	US	STS-38 STS-48 STS-62
Gernhardt, Michael	4	M	US	STS-69 STS-83 STS-94 STS-104
Gibson, Edward	1	M	US	Skylab 4
Gibson, Robert	5	M	US	STS-41B STS-61C STS-27 STS-47 STS-71
Gidzenko, Yuri	2	M	CIS	TM 22 ISS-1
Glazkov, Yuri	2	M	USSR	Soyuz 24 TM-34 (ISS taxi)
Glenn, John	2	M	US	MA- 6 STS-95
Godwin, Linda	4	F	US	STS-37 STS-59 STS-76 STS-108
Gorbatko, Viktor	3	M	USSR	Soyuz 7 Soyuz 24 Soyuz 37
Gordon, Richard	2	M	US	Gemini 11 Apollo 12
Gorie, Dominic	3	M	US	STS-91 STS-99 STS-108
Grabe, Ronald	4	M	US	STS-51J STS-30 STS-42 STS-57
Grechko, Georgy	3	M	USSR	Soyuz 17 Soyuz 26 Soyuz T14
Gregory, Frederick	3	M	US	STS-51B STS-33 STS-44
Gregory, William	1	M	US	STS-67
Griggs, David	1	M	US	STS-51D
Grissom, Virgil	2	M	US	Mercury 4 Gemini 3
Grunsfeld, John	4	M	US	STS-67 STS-81 STS-103 STS-109
Gubarev, Alexei	2	M	USSR	Soyuz 17 Soyuz 28
Guidoni, Umberto	2	M	Italy	STS-75 STS-100
Gurragcha, J	1	M	Mongolia	Soyuz 39
Gutierrez, Sidney	2	M	US	STS-40 STS-59
Hadfield, Chris	2	M	Canada	STS-74 STS-100
Haigere, Jean-Pierre	2	M	France	TM 17 TM 29
Haignere, Claudie	2	F	France	TM 24 TM 33
Haise, Fred	1	M	US	Apollo 13
Halsell, James	5	M	US	STS-65 STS-74 STS-83 STS-94 STS-101
Hammond, Blaine	2	M	US	STS-39 STS-64
Harbaugh, Gregory	4	M	US	STS-39 STS-54 STS-71 STS-82
Harris, Bernard	2	M	US	STS-55 STS-63
Hart, Terry	1	M	US	STS-41C
Hartsfield, Henry	3	M	US	STS-4 STS-41D STS-61A
Hauck, Frederick	3	M	US	STS-7 STS-51A STS-26
Hawley, Stephen	5	M	US	STS-41D STS-61C STS-31 STS-82 STS-93
Helms, Susan	5	F	US	STS-54 STS-64 STS-78 STS-101 ISS-2
Henize, Karl	1	M	US	STS-51F
Hennen, Thomas	1	M	US	STS-44
Henricks, Terence	4	M	US	STS-44 STS-55 STS-70 STS-78
Hermaszewski, M	1	M	Poland	Soyuz 30
Herrington, John	1	M	US	STS-113
Hieb, Richard	3	M	US	STS-39 STS-49 STS-65
Hilmers, David	4	M	US	STS-51J STS-26 STS-36 STS-42
Hire, Kay	1	F	US	STS-90
Hobaugh, Charles	1	M	US	STS-104

Name	Flts	Sex	Nation	Missions
Hoffman, Jeffrey	5	M	US	STS-51D STS-35 STS-46 STS-61 STS-75
Horowitz, Scott	4	M	US	STS-75 STS-82 STS-101 STS-105
Hughes-Fulford, Millie	1	F	US	STS-40
Husband, Rick	1	M	US	STS-96
Irwin, James	1	M	US	Apollo 15
Ivanchenkov, Alexander	2	M	USSR	Soyuz 29 Soyuz T-6
Ivanov, Georgy	1	M	Bulgaria	Soyuz 33
Ivins, Marsha	5	F	US	STS-32 STS-46 STS-62 STS-81 STS-98
Jaehn, Sigmund	1	M	Germany	Soyuz 31
Jemison, Mae	1	F	US	STS-47
Jernigan, Tamara	5	F	US	STS-40 STS-52 STS-67 STS-80 STS-96
Jett, Brent	3	M	US	STS-72 STS-81 STS-97
Jones, Thomas	4	M	US	STS-59 STS-68 STS-80 STS-98
Kadenyuk, Leonid	1	M	CIS	STS-87
Kaleri, Alexander	3	M	CIS	TM 14 TM 24 Mir-28
Kavandi, Janet	3	F	US	STS-91 STS-99 STS-104
Kelly, James M	1	M	US	STS-102
Kelly, Mark	1	M	US	STS-108
Kelly, Scott	1	M	US	STS-103
Kerwin, Joseph P	1	M	US	Skylab 2
Khrunov, Yevgeny	1	M	USSR	Soyuz 4-5
Kilrain, Susan	2	F	US	STS-83 STS-94
Kizim, Leonid	3	M	USSR	Soyuz T3 Soyuz T10 Soyuz T15
Klimuk, Pyotr	3	M	USSR	Soyuz 13 Soyuz 18 Soyuz 30
Komarov, Vladimir	2	M	USSR	Voskhod 1 Soyuz 1
Kondakova, Elena	2	F	CIS	TM 20 STS-84
Korzun, Valery	2	M	CIS	TM 24 STS-111/ISS-5/STS-113
Kovalenok, Vladimir	3	M	USSR	Soyuz 25 Soyuz 29 Soyuz T4
Kozeev, Konstantin	1	M	CIS	TM 33
Kregel, Kevin	4	M	US	STS-70 STS-78 STS-87 STS-99
Krikalev, Sergei	5	M	USSR	TM 7 TM 12 STS-60 STS-88 ISS-1
Kubasov, Valery	3	M	USSR	Soyuz 6 Soyuz 19ASTP Soyuz 36
Laveikin, Alexander	1	M	USSR	TM 2
Lawrence, Wendy	3	F	US	STS-67 STS-86 STS-91
Lazarev, Vasily	2	M	USSR	Soyuz 12 Soyuz 18A
Lazutkin, Alexander	1	M	CIS	TM-25
Lebedev, Valentin	2	M	USSR	Soyuz 13 Soyuz T5
Lee, Mark	4	M	US	STS-30 STS-47 STS-54 STS-82
Leestma, David	3	M	US	STS-41G STS-28 STS-45
Lenoir, William	1	M	US	STS-5
Leonov, Alexei	2	M	USSR	Voskhod 2 Soyuz 19
Leslie, Fred	1	M	US	STS-73
Levchenko, Anatoly	1	M	USSR	TM 4
Lichtenberg, Byron	2	M	US	STS-9 STS-45
Lind, Donald	1	M	US	STS-51B
Lindsey, Steven	3	M	US	STS-87 STS-95 STS-104
Linenger, Jerry	2	M	US	STS-64 81/Mir4/84
Linnehan, Richard	3	M	US	STS-78 STS-90 STS-109
Linteris, Gregory	2	M	US	STS-83 STS-94
Lockhart, Paul	2	M	US	STS-111 STS-113
Lonchakov, Yuri	1	M	CIS	STS-100
Lopez-Alegria, Michael	3	M	US	STS-73 STS-92 STS-113
Lounge, Michael	3	M	US	STS-51I STS-26 STS-35
Lousma, Jack	2	M	US	Skylab 3 STS-3
Lovell, James	4	M	US	Gemini 7 Gemini 12 Apollo 8 Apollo 13
Low, David	3	M	US	STS-32 STS-43 STS-57
Lu, Edward	2	M	US	STS-84 STS-106
Lucid, Shannon	5	F	US	STS-51G STS-34 STS-43 STS-58 76/Mir2/79
Lyakhov, Vladimir	3	M	USSR	Soyuz 32 Soyuz T9 TM 6
MacLean, Steve	1	M	Canada	STS-52
Makarov, Oleg	4	M	USSR	Soyuz 12 Soyuz 18A Soyuz 27 Soyuz T3
Malenchenko, Yuri	2	M	CIS	TM 19 STS-106
Malerba, Franco	1	M	Italy	STS-46
Malyshev, Yuri	2	M	USSR	Soyuz T2 Soyuz T11
Manakov, Gennady	2	M	USSR	TM 10 TM 16
Manarov, Musa	2	M	USSR	TM 4 TM 11
Massimino, Michael	1	M	US	STS-109
Mastracchio, Richard	1	M	US	STS-106
Mattingly, Thomas	3	M	US	Apollo 16 STS-4 STS-51C
McArthur, William	3	M	US	STS-58 STS-74 STS-92
McBride, Jon	1	M	US	STS-41G
McCandless, Bruce	2	M	US	STS-41B STS-31
McCulley, Michael	1	M	US	STS-34
McDivitt, James	2	M	US	Gemini 4 Apollo 9
McMonagle, Donald	3	M	US	STS-39 STS-54 STS-66
McNair, Ronald	1	M	US	STS-41B
Meade, Carl	3	M	US	STS-38 STS-50 STS-64
Melnick, Bruce	2	M	US	STS-41 STS-49
Melroy, Pamela	2	F	US	STS-92 STS-112
Mendez, Arnaldo	1	M	Cuba	Soyuz 38
Merbold, Ulf	3	M	Germany	STS-9 STS-42 TM 20
Messerschmid, Ernst	1	M	Germany	STS-61A
Mitchell, Edgar	1	M	US	Apollo 14
Mohmand, A	1	M	Afghanistan	TM 6
Mohri, Mamoru	2	M	Japan	STS-47 STS-99
Morin, Lee	1	M	US	STS-110
Morukov, Boris	1	M	CIS	STS-106
Mukai, Chiaki	2	F	Japan	STS-65 STS-95
Mullane, Michael	3	M	US	STS-41D STS-27 STS-36
Musabayev, Talgat	3	M	CIS	TM 19 TM 27/Mir25 TM 32
Musgrave, Story	6	M	US	STS-6 STS-51F STS-33 STS-44 STS-61 STS-80

Name	*Flts*	*Sex*	*Nation*	*Missions*
Nagel, Steven	4	M	US	STS-51G STS-61A STS-37 STS-55
Nelson, George	3	M	US	STS-41C STS-61C STS-26
Nelson, William	1	M	US	STS-61C
Neri, Rodolfo	1	M	Mexico	STS-61B
Newman, James	4	M	US	STS-51 STS-69 STS-88 STS-109
Nicollier, Claude	4	M	Switzerland	STS-46 STS-61 STS-75 STS-103
Nikolayev, Andrian	2	M	USSR	Vostok 3 Soyuz 9
Noriega, Carlos	2	M	US	STS-84 STS-97
Ochoa, Ellen	4	F	US	STS-56 STS-66 STS-96 STS-110
Ockels, Wubbo	1	M	Netherlands	STS-61A
O'Connor, Bryan	2	M	US	STS-61B STS-40
Onizuka, Ellison	1	M	US	STS-51C
Onufrienko, Yuri	2	M	CIS	Mir21 108/ISS-4/111
Oswald, Stephen	3	M	US	STS-42 STS-56 STS-67
Overmyer, Robert	2	M	US	STS-5 STS-51B
Padalka, Gennady	1	M	CIS	TM 28
Pailes, William	1	M	US	STS-51J
Parazynski, Scott	4	M	US	STS-66 STS-86 STS-95 STS-100
Parise, Ronald	2	M	US	STS-35 STS-67
Parker, Robert	2	M	US	STS-9 STS-35
Patsayev, Viktor	1	M	USSR	Soyuz 11
Pawelczyk, James	1	M	US	STS-90
Payette, Julie	1	F	Canada	STS-96
Payton, Gary	1	M	US	STS-51C
Perrin, Philippe	1	M	France	STS-111
Peterson, Donald	1	M	US	STS-6
Pettit, Donald	1	M	US	STS-113/ISS-6
Phillips, John	1	M	US	STS-100
Pogue, William	1	M	US	Skylab 4
Polansky, Mark	1	M	US	STS-98
Poleshchuk, Alexander	1	M	CIS	TM 16
Polyakov, Valery	2	M	USSR	TM 6TM 18
Popov, Leonid	3	M	USSR	Soyuz 35 Soyuz 40 SoyuzT7
Popovich, Pavel	2	M	USSR	Vostok 4 Soyuz 14
Precourt, Charles	4	M	US	STS-55 STS-71 STS-84 STS-91
Prunariu, D	1	M	Romania	Soyuz 40
Readdy, William	3	M	US	STS-42 STS-51 STS-79
Reightler, Kenneth	2	M	US	STS-48 STS-60
Reilly, James	2	M	US	STS-89 STS-104
Reiter, Thomas	1	M	Germany	TM 22
Remek, Vladimir	1	M	Czech	Soyuz 28
Resnik, Judith	1	F	US	STS-41D
Richards, Paul	1	M	US	STS-102
Richards, Richard	4	M	US	STS-28 STS-41 STS-50 STS-64
Ride, Sally	2	F	US	STS-7 STS-41G
Robinson, Stephen	2	M	US	STS-85 STS-95
Romanenko, Yuri	3	M	USSR	Soyuz 26 Soyuz 38TM 2
Rominger, Kent	5	M	US	STS-73 STS-80 STS-85 STS-96 STS-100
Roosa, Stuart	1	M	US	Apollo 14
Ross, Jerry	7	M	US	STS-61B STS-27 STS-37 STS-55 STS-74 STS-88 STS-110
Rozhdestvensky, Valery	1	M	USSR	Soyuz 23
Rukavishnikov, Nikolai	3	M	USSR	Soyuz 10 Soyuz 16 Soyuz 33
Runco, Mario	3	M	US	STS-44 STS-54 STS-77
Ryumin, Valery	4	M	USSR	Soyuz 25 Soyuz 32 Soyuz 35 STS-91
Sacco, Albert	1	M	US	STS-73
Sarafanov, Gennadi	1	M	USSR	Soyuz 15
Savinykh, Viktor	3	M	USSR	SoyuzT4 SoyuzT13TM 5
Savitskaya, Svetlana	2	F	USSR	SoyuzT7 SoyuzT12
Schirra, Walter	3	M	US	Mercury 8 Gemini 6 Apollo 7
Schlegel, Hans	1	M	Germany	STS-55
Schmitt, Harrison	1	M	US	Apollo 17
Schweickart, Russell	1	M	US	Apollo 9
Scobee, Francis	1	M	US	STS-41C
Scott, David	3	M	US	Gemini 8 Apollo 9 Apollo 15
Scott, Winston	2	M	US	STS-72 STS-87
Scully-Power, Paul	1	M	US	STS-41G
Searfoss, Richard	3	M	US	STS-58 STS-76 STS-90
Seddon, Rhea	3	F	US	STS-51D STS-40 STS-58
Sega, Ronald	2	M	US	STS-60 STS-76
Sellers, Piers	1	M	US	STS-112
Serebrov, Alexander	4	M	USSR	SoyuzT7 SoyuzT8TM 8TM 17
Sevastianov, Vitaly	2	M	USSR	Soyuz 9 Soyuz 18
Sharipov, Salizhan	1	M	CIS	STS-89
Sharma, Rakesh	1	M	India	SoyuzT11
Sharman, Helen	1	F	UK	TM-12
Shatalov, Vladimir	3	M	USSR	Soyuz 4 Soyuz 8 Soyuz 10
Shaw, Brewster	3	M	US	STS-9 STS-61B STS-28
Shepard, Alan	2	M	US	Mercury 3 Apollo 14
Shepherd, William	4	M	US	STS-27 STS-41 STS-52 ISS-1
Shonin, Georgy	1	M	USSR	Soyuz 6
Shriver, Loren	3	M	US	STS-51C STS-31 STS-46
Shuttleworth, Mark	1	M	SA	TM-34 (ISS taxi)
Slayton, Donald	1	M	US	ASTP
Smith, Steven	4	M	US	STS-68 STS-82 STS-103, STS-110
Solovyev, Anatoly	5	M	USSR	TM 5TM 9TM 15 71/Mir19TM 26
Solovyov, Vladimir	2	M	USSR	SoyuzT10 SoyuzT15
Spring, Sherwood	1	M	US	STS-61B
Springer, Robert	2	M	US	STS-29 STS-38
Stafford, Thomas	4	M	US	Gemini 6 Gemini 9 Apollo 10 ASTP
Stewart, Robert	2	M	US	STS-41B STS-51J
Strekalov, Gennady	5	M	USSR	SoyuzT3 SoyuzT8 SoyuzT11TM 10 Mir18/STS-71
Sturckow, Frederick	2	M	US	STS-95 STS-105

Name	*Flts*	*Sex*	*Nation*	*Missions*
Sullivan, Kathryn	3	F	US	STS-41G STS-31 STS-45
Swigert, John	1	M	US	Apollo 13
Tani, Daniel	1	M	US	STS-108
Tanner, Joseph	3	M	US	STS-66 STS-82 STS-97
Tereshkova, Valentina	1	F	USSR	Vostok 6
Thagard, Norman	5	M	US	STS-7 STS-51B STS-30 STS-42 TM 21/Mir1/71
Thiele, Gerhard	1	M	Germany	STS-99
Thirsk, Robert	1	M	Canada	STS-78
Thomas, Andrew	3	M	US	STS-77 89/Mir7/91 STS-102
Thomas, Donald	4	M	US	STS-65 STS-70 STS-83 STS-94
Thornton, Kathryn	4	F	US	STS-33 STS-49 STS-61 STS-73
Thornton, William	2	M	US	STS-8 STS-51B
Thuot, Pierre	3	M	US	STS-36 STS-49 STS-62
Tito, Dennis	1	M	US	Soyuz TM 32
Titov, Gherman	1	M	USSR	Vostok 2
Titov, Vladimir	4	M	USSR	Soyuz T8 Soyuz TM 4 STS-63 STS-86
Tognini, Michel	2	M	France	TM 15 STS-93
Tokarev, Valery	1	M	CIS	STS-96
Treschev, Sergei	1	M	CIS	STS-111/ISS-5/STS-113
Trinh, Eugene	1	M	US	STS-50
Truly, Richard	2	M	US	STS-2 STS-8
Tryggvason, Bjarni	1	M	Canada	STS-85
Tsibliyev, Vasily	2	M	CIS	TM 17 TM 25
Tuan, Pham	1	M	Vietnam	Soyuz 37
Turin, Mikhail	1	M	CIS	ISS-3
Usachev, Yuri	4	M	CIS	TM 18 Mir21 STS-101 ISS-2
van den Berg, Lodewijk	1	M	US	STS-51B
van Hoften, James	2	M	US	STS-41C STS-51I
Vasyutin, Vladimir	1	M	USSR	Soyuz T14
Veach, Charles	2	M	US	STS-39 STS-52
Viehboeck, Franz	1	M	Austria	TM 13
Viktorenko, Alexander	4	M	USSR	TM 3 TM 8 TM 14 TM20
Vinogradov, Pavel	1	M	CIS	TM 26
Vittori, Roberto	1	M	Italy	TM-24 (ISS taxi)
Volk, Igor	1	M	USSR	Soyuz T12
Volkov, Alexander	3	M	USSR	Soyuz T14 TM 7 TM 13
Volkov, Vladislav	2	M	USSR	Soyuz 7 Soyuz 11
Volynov, Boris	2	M	USSR	Soyuz 5 Soyuz 21
Voss, James	5	M	US	STS-44 STS-53 STS-69 STS-101 ISS-2
Voss, Janice	5	F	US	STS-57 STS-63 STS-83 STS-94 STS-99
Wakata, Koichi	2	M	Japan	STS-72 STS-92
Walheim, Rex	1	M	US	STS-110
Walker, Charles	3	M	US	STS-41D STS-51D STS-61B
Walker, David	4	M	US	STS-51A STS-30 STS-53 STS-69
Walter, Ulrich	1	M	Germany	STS-55
Walz, Carl	4	M	US	STS-51 STS-65 STS-79 108/ISS-4/111

The spacious interior of the ISS with numerous location points for experiments and equipment (NASA) 0573543

Name	*Flts*	*Sex*	*Nation*	*Missions*
Wang, Taylor	1	M	US	STS-51B
Weber, Mary Ellen	2	F	US	STS-70 STS-101
Weitz, Paul	2	M	US	Skylab 2 STS-6
Wetherbee, James	6	M	US	STS-32 STS-52 STS-63 STS-86 STS-102 STS-113
White, Edward	1	M	US	Gemini 4
Whitson, Peggy	1	F	US	STS-111/ISS-5/STS-113
Wilcutt, Terrence	4	M	US	STS-68 STS-79 STS-89 STS-106
Williams, Dafydd	1	M	Canada	STS-90
Williams, Donald	2	M	US	STS-51D STS-34
Williams, Jeffrey	1	M	US	STS-101
Wisoff, Jeff	4	M	US	STS-57 STS-68 STS-81 STS-92
Wolf, David	3	M	US	STS-58 86/Mir6/89 STS-112
Worden, Alfred	1	M	US	Apollo 15
Yegorov, Boris	1	M	USSR	Voskhod 1
Yeliseyev, Alexei	3	M	USSR	Soyuz 4 Soyuz 8 Soyuz 10
Young, John	6	M	US	Gemini 3 Gemini 10 Apollo 10 Apollo 16 STS-1 STS-9
Yurchikhin, Fyodor	1	M	CIS	STS-112
Zalyotin, Sergei	1	M	CIS	Mir-28
Zholobov, Vitaly	1	M	USSR	Soyuz 21
Zudov, Vyacheslav	1	M	USSR	Soyuz 23

Table 13: Astronaut and Cosmonaut Fatalities

GENERAL STATISTICS

NASA Astronauts Killed in Space Flight	14
Russian Cosmonauts Killed in Space Flight	4
Total Individuals in Space	385
Total Space Fatalities	18

IN-FLIGHT FATALITIES

Komarov, Vladimir	24/04/67	Soyuz 1 parachute fails during entry
Dobrovolsky, Georgy	29/06/71	Soyuz 11 depressurized during entry
Patsayev, Victor	29/06/71	Soyuz 11 depressurized during entry
Volkov, Vladislav	29/06/71	Soyuz 11 depressurized during entry
Scobee, Francis "Dick"	28/01/86	SRB booster failure, mission STS-51L
Smith, Michael	28/01/86	SRB booster failure, mission STS-51L
Resnik, Judith	28/01/86	SRB booster failure, mission STS-51L
Onizuka, Ellison	28/01/86	SRB booster failure, mission STS-51L
McNair, Ronald	28/01/86	SRB booster failure, mission STS-51L
Jarvis, Gregory	28/01/86	SRB booster failure, mission STS-51L
McAuliffe, Christa	28/01/86	SRB booster failure, mission STS-51L
Husband, Rick	01/02/03	re-entry failure, mission STS-107
McCool, William	01/02/03	re-entry failure, mission STS-107
Brown, David	01/02/03	re-entry failure, mission STS-107
Chawla, Kalpana	01/02/03	re-entry failure, mission STS-107
Anderson, Michael	01/02/03	re-entry failure, mission STS-107
Clark, Laurel	01/02/03	re-entry failure, mission STS-107
Ramon, Ilan	01/02/03	re-entry failure, mission STS-107

ACTIVE-DUTY FATALITIES

Freeman, Theodore	31/10/64	T-38 jet crash in Houston
Bassett, Charles	28/02/66	T-38 jet crash in St Louis
See, Elliott	28/02/66	T-38 jet crash in St Louis
Grissom, Virgil "Gus"	27/01/67	Apollo 1 launch pad fire
White, Edward	27/01/67	Apollo 1 launch pad fire
Chaffee, Roger	27/01/67	Apollo 1 launch pad fire
Givens, Edward	06/06/67	Houston car crash
Williams, Clifton	15/10/67	airplane crash near Tallahassee, Fla
Robert Lawrence	??/06/67	F-104 crash (MOL Air Force astronaut)
Gagariin, Yuri	27/03/68	MiG jet trainer crash near Star City
Belyayev, Pavel	10/01/70	died during surgery
Thorne, Stephen	24/05/86	private plane crash near Houston
Levchenko, Anatoly	06/08/88	inoperable brain tumor
Shchukin, Alexander	18/08/88	experimental plane crash
Griggs, David	17/06/89	plane crash
Carter, Manley "Sonny"	04/05/91	commuter plane crash in Georgia
Veach, Charles Lacy	03/10/95	cancer

Table 14: Space Shuttle Orbiter Flight Histories and Statistics to 31 March 2002

Shuttle	*Days*	*Hours*	*Minutes*	*Seconds*	*Flights*	*Most Recent*
Challenger	062	07	56	22	10	STS-51L: 28/01/86
Columbia	284	19	19	01	27	STS-109: 01/03/02
Discovery	241	23	19	54	30	STS-105: 10/08/01
Atlantis	219	21	27	17	26	STS-112: 07/10/02
Endeavour	206	14	12	17	19	STS-113: 23/11/02
Program Total	1,015	14	14	51	112	

SPACE SHUTTLE FLIGHT HISTORIES

NOTE: FRF: Flight Readiness Firing (engine test firing); RSLS: Redundant Set Launch Sequencer Abort (on-pad engine shutdown prior to launch); ATO: Abort to Orbit Detailed mission reports are available from World Spaceflight News.

Challenger (OV-099)

OV	*Number*	*STS*	*Launch*	*Days*	*Hours*	*Minutes*	*Seconds*	*Notes*
		FRF	18/12/82	00	00	00	00	FRF-1
		FRF	25/01/83	00	00	00	00	FRF-2
1	06	06	04/04/83	05	00	23	42	TDRS-1
2	07	07	18/06/83	06	02	23	59	Three comsats
3	08	08	30/08/83	06	01	08	43	Insat, CFES
4	10	41B	02/02/84	07	23	15	55	Westar, Palapa; MMU EVA
5	11	41C	06/04/84	06	23	40	07	Solar Max repair; MMU EVA
6	13	41G	05/10/84	08	05	23	38	ERBS, SIR-B; female EVA
7	17	51B	29/04/85	07	00	08	46	Spacelab-3
		RSLS	12/07/85	00	00	00	00	SSME2 CCV slow closing
8	19	51F	29/07/85	07	22	45	26	Spacelab-2; RSLS, ATO
9	22	61A	30/10/85	07	00	44	53	Spacelab D1
10	25	51L	28/01/86	00	00	01	13	SRB failure at T+73
TOTAL				**62**	**07**	**56**	**22**	

Columbia (OV-102)

Flow Director: Grant Cates

OV	*Number*	*STS*	*Launch*	*Days*	*Hours*	*Minutes*	*Seconds*	*Notes*
		FRF	28/02/81	00	00	00	00	FRF
1	01	01	12/04/81	02	06	20	53	First shuttle flight
2	02	02	12/11/81	02	06	13	11	Fuel cell/MDM flight
3	03	03	22/03/82	08	00	04	46	White Sands landing
4	04	04	27/06/82	07	01	09	31	Final test flight
5	05	05	11/11/82	05	02	14	26	First satellite launches
6	09	09	28/11/83	10	07	47	24	First Spacelab flight
7	24	61C	12/01/86	06	02	03	51	Last pre-Challenger flight
8	30	28	08/08/89	05	01	00	08	DOD
9	33	32	09/01/90	10	21	00	36	LDEF recovery
10	38	35	02/12/90	08	23	05	08	ASTRO Spacelab mission
11	41	40	05/06/91	09	02	14	20	Life Science Spacelab
12	48	50	25/06/92	13	19	30	04	USML-1
13	51	52	22/10/92	09	20	56	13	LAGEOS-2, USMP-1
		RSLS	22/03/93	00	00	00	00	ME-3 LOX preburner valve
14	55	55	26/04/93	09	23	39	59	Spacelab D2
15	57	58	22/10/93	14	00	12	32	SLS-2
16	61	62	04/03/94	13	23	16	41	USMP-2, OAST-2
17	63	65	08/07/94	14	17	55	00	IML-2
18	72	73	20/10/95	15	21	52	21	USML-2
19	75	75	22/02/96	15	17	40	21	TSS-2, USMP-3
20	78	78	20/06/96	16	21	47	36	LMS (record duration)
21	80	80	19/11/96	17	15	53	18	ORFEUS-SPAS, WSF-3
22	83	83	04/04/97	03	23	12	39	MSL-1
23	85	94	01/07/97	15	16	44	34	MSL-1 reflight
24	88	87	19/11/97	15	16	34	04	USMP, Spartan-201
25	90	90	17/04/98	15	21	49	59	Neurolab
26	95	93	23/07/99	04	22	49	35	Chandra X-ray Observatory
27	109	109	01/03/02	10	22	09	51	Hubble Space Telescope Servicing Mission 3B
TOTAL				**284**	**19**	**19**	**01**	

Discovery (OV-103)

Flow Director: Scott Cilento

OV	Number	STS	Launch	Days	Hours	Minutes	Seconds	Notes
		FRF	02/06/84	00	00	00	00	FRF
		RSLS	26/06/84	00	00	00	00	ME-3 MFV actuator
1		41D	30/08/84	06	00	56	04	SBS, Syncom, Telstar
2	14	51A	07/11/84	07	23	44	56	Westar, Palapa retrieve
3	15	51C	24/01/85	03	01	33	23	DOD (Magnum)
4	16	51D	12/04/85	06	23	55	23	Telesat, Syncom; unplanned EVA
5	18	51G	17/06/85	07	01	38	52	Morelos, Arabsat, Telstar
6	20	51I	27/08/85	07	02	17	42	ASC, Aussat, Syncom
		FRF	10/08/88	00	00	00	00	FRF
7	26	26	29/09/88	04	01	00	11	TDRS-3 (return to flight)
8	28	29	13/03/89	04	23	38	50	TDRS-4
9	32	33	22/11/89	05	00	06	48	DOD
10	35	31	24/04/90	05	01	16	06	Hubble Space Telescope
11	36	41	06/10/90	04	02	10	04	Ulysses Solar Probe
12	40	39	28/04/91	08	07	22	23	DOD/SDI (unclassified)
13	43	48	12/09/91	05	08	27	38	UARS
14	45	42	22/01/91	08	01	14	44	IML-1
15	52	53	02/12/92	07	07	19	47	DOD-1 (classified payload)
16	54	56	08/04/93	09	06	08	24	ATLAS-2
		RSLS	12/08/93	00	00	00	00	ME-2 fuel flow meter fails
17	57	51	12/09/93	09	20	11	11	ACTS, SPAS
18	60	60	03/02/94	08	07	09	22	WSF-1 /Russian MS
19	64	64	09/09/94	10	22	49	57	LITE, SAFER, SPIFEX; EVA
20	67	63	03/02/95	08	06	28	15	Mir-1, Spartan, EVA
21	70	70	13/07/95	08	22	20	07	TDRS-G
22	82	82	11/02/97	09	23	37	09	HST Servicing Mission
23	86	85	07/08/97	11	20	26	59	CRISTA-SPAS, MFD
24	91	91	02/06/98	09	19	53	57	Mir Docking No 9
25	92	95	29/10/98	08	21	43	57	Spartan 201R; John Glenn flight
26	94	96	27/05/99	09	19	13	01	ISS 2A 1
27	96	103	19/12/99	07	23	10	47	HST SM-3A
28	100	92	11/10/00	12	21	42	41	ISS 3A
29	103	102	08/03/01	12	19	49	32	ISS 5A 1
30	106	105	10/08/01	11	21	12	44	ISS 7A 1
TOTAL				**241**	**23**	**19**	**54**	

Atlantis (OV-104)

Flow Director: TBD

OV	Number	STS	Launch	Days	Hours	Minutes	Seconds	Notes
		FRF	12/09/85	00	00	00	00	FRF
1	21	51J	03/10/85	04	01	44	38	DOD
2	23	61B	26/11/85	06	21	04	49	3 comsats EASE/ACCESS EVA
3	27	27	02/12/88	04	09	05	37	DOD (Lacrosse?)
4	29	30	04/05/89	04	00	56	27	Magellan Venus Probe
5	31	34	18/10/89	04	23	39	21	Galileo Jupiter Probe
6	34	36	28/02/90	04	10	18	22	DOD
7	37	38	15/11/90	04	21	54	31	DOD
8	39	37	05/04/91	05	23	32	44	Gamma Ray Observatory
9	42	43	02/08/91	08	21	21	25	TDRS-5
10	44	44	19/11/91	06	22	50	44	DSP
11	46	45	24/03/92	08	22	09	28	ATLAS-1
12	49	46	31/07/92	07	23	15	03	TSS; EURECA deployment
13	66	66	03/11/94	10	22	34	02	ATLAS-3
14	69	71	27/06/95	09	19	22	17	Mir Docking No 1
15	73	74	12/11/95	08	04	30	46	Mir Docking No 2
16	76	76	22/03/96	09	05	15	53	Mir Docking No 3
17	79	79	16/09/96	10	03	19	28	Mir Docking No 4
18	81	81	12/01/97	10	04	55	21	Mir Docking No 5
19	84	84	15/05/97	09	05	19	56	Mir Docking No 6
20	87	86	25/09/97	10	19	20	50	Mir Docking No 7
21	98	101	19/05/00	09	20	09	08	ISS 2A 2a (Zarya repair)
22	99	106	08/09/00	11	19	11	01	ISS 2A 2b (Zvezda outfitting)
23	102	98	07/02/01	12	21	20	03	ISS 5A (Destiny lab)
24	105	104	12/07/01	12	18	34	56	ISS 7A (Airlock)
25	109	110	08/04/02	10	19	42	38	ISS 8A (S0 truss)
26	110	112	07/10/02	10	19	57	49	ISS 9A (S1 truss)
TOTAL				**219**	**21**	**27**	**17**	

Endeavour (OV-105)

Flow Director: Nichole Passino

OV	*Number*	*STS*	*Launch*	*Days*	*Hours*	*Minutes*	*Seconds*	*Notes*
		FRF	06/04/92	00	00	00	00	FRF
1	47	49	07/05/92	08	21	17	38	Intelsat rescue
2	50	47	12/09/92	07	22	30	23	Spacelab-J
3	53	54	13/01/93	05	23	38	19	TDRS-6, DXS
4	56	57	21/06/93	09	23	44	54	Eureca retrival, Spacehab-1
5	60	61	02/12/93	10	19	58	37	HST repair
6	62	59	09/04/94	11	05	49	30	Shuttle Radar Laboratory 1
		RSLS	18/08/94	00	00	00	00	ME-3 HPOT discharge temp
7	65	68	30/09/94	11	05	46	08	Shuttle Radar Laboratory 2
8	68	67	02/03/95	16	15	08	48	ASTRO-2 UV astronomy mission
9	71	69	07/09/95	10	20	28	56	WSF-2/Spartan-201
10	74	72	11/01/96	08	22	00	45	SFU retrieve; OAST deploy
11	77	77	19/05/96	10	00	39	24	Inflatable antenna; PAM-STU
12	89	89	22/01/98	08	19	46	54	Mir Docking No 8
13	93	88	04/12/98	11	19	17	55	ISS 2A
14	97	99	11/02`/00	11	05	38	43	SRTM
15	101	97	30/11/00	10	19	57	24	ISS-4A: P6 solar array
16	104	100	19/04/01	11	21	30	00	ISS-6A: SSRMS
17	108	108	05/12/01	11	19	35	42	ISS-UF1: Crew rotation
18	108	111	05/06/02	13	20	34	52	ISS-UF2: Crew rotation
19	112	113	23/11/02	13	18	47	25	ISS-11A: Crew rotation; P1 truss
TOTAL				**206**	**14**	**12**	**17**	

SPACE SHUTTLE LAUNCH STATISTICS

	LC-39A	*LC-39B*	*Total*
Night	16	12	28
Daylight	47	37	84
Total	63	49	112

Inclinations:

28.5	54
34.3	1
38	2
39	7
40.3	1
49.5	1
51.6	26
57	19
62	1

NASA's Shuttle gets away from Launch Complex 39 at the Kennedy Space Centre, Florida, raising wild life from the swamps (NASA) 0137130

SPACE SHUTTLE LANDING STATISTICS

	KSC	*EAFB*	*WS*	*Total*
Night	14	5	0	19
Daylight	47	44	1	92
Total	61	49	1	111
Most Recent	07/12/02	19/06/02	30/03/82	

Landing site	*Runway*	*Surface*	*Landings*	*Most recent*
KSC	33	Concrete	TBD	07/12/02
KSC	15	Concrete	TBD	17/12/01
White Sands	17	Lakebed	01	30/03/82
Edwards AFB	22	Concrete	27	19/06/02
Edwards AFB	04	Concrete	02	20/09/94
Edwards AFB	23	Lakebed	08	04/03/90
Edwards AFB	15	Lakebed	01	24/06/83
Edwards AFB	17	Lakebed	08	13/08/89
Edwards AFB	33	Lakebed	01	11/04/91
Edwards AFB	05	Lakebed	01	01/12/91

SPACE SHUTTLE ABORTS

NOTE: RSLS: Redundant Set Launch Sequencer Abort (on pad engine shutdown); SSME: Space Shuttle Main Engine; ATO: Abort to Orbit

Shuttle	*Date*	*Time*	*Abort*	*Notes*
Discovery	26/06/84	–3 s	RSLS-1	SSME 3 main fuel valve
Challenger	12/07/85	–3 s	RSLS-2	SSME 2 chamber coolant valve
Challenger	29/07/85	+5:45	ATO-1	SSME 2 shutdown
Columbia	22/03/93	–3 s	RSLS-3	SSME oxidizer valve
Discovery	12/08/93	–3 s	RSLS-4	SSME 2 fuel flow meter
Endeavour	18/08/94	–1.9 s	RSLS-5	SSME high 02 pump temps

Seven crewmembers gather for a photo-call during the STS-42 mission in January 1992 carrying the International Microgravity Laboratory (NASA) 0137157

MINIMUM MISSION

NOTE: shortened due to technical problems in orbit

Flight	*Cause*
STS-2	Fuel cell failure. 5-day flight cut to MET 2/06:13
STS-44	IMU failure. 10-day flight cut to MET 6/22:52
STS-83	Fuel cell NO 2 failure. 16-day flight cut to MET 3/23:12:39

FLIGHT READINESS FIRINGS

NOTE: 23-second main engine test firings at the launch pad

Orbiter	*No*	*Date*	*SSMEs*	*Time*	*Notes*
Columbia OV-102	1	28/02/81	2007, 2006, 2005	23 s	Successful
Challenger OV-099	1	18/12/82	2011, 2015, 2012	23 s	Leaks
	2	25/01/83	2011, 2015, 2012	23 s	Cracks found
Discovery OV-103	1	02/06/84	2021, 2018, 2017	23 s	Successful
	2	10/08/88	2019, 2022, 2028	23 s	Successful
Atlantis OV-104	1	12/09/85	2011, 2019, 2017	23 s	Successful
Endeavour OV-105	1	06/04/92	2035, 2033, 2034	23 s	Successful

The Michoud Assembly Facility has been home to Shuttle External Tank production since the late 1970s. Note the spray-on foam insulation for thermal protection (NASA) 0137159

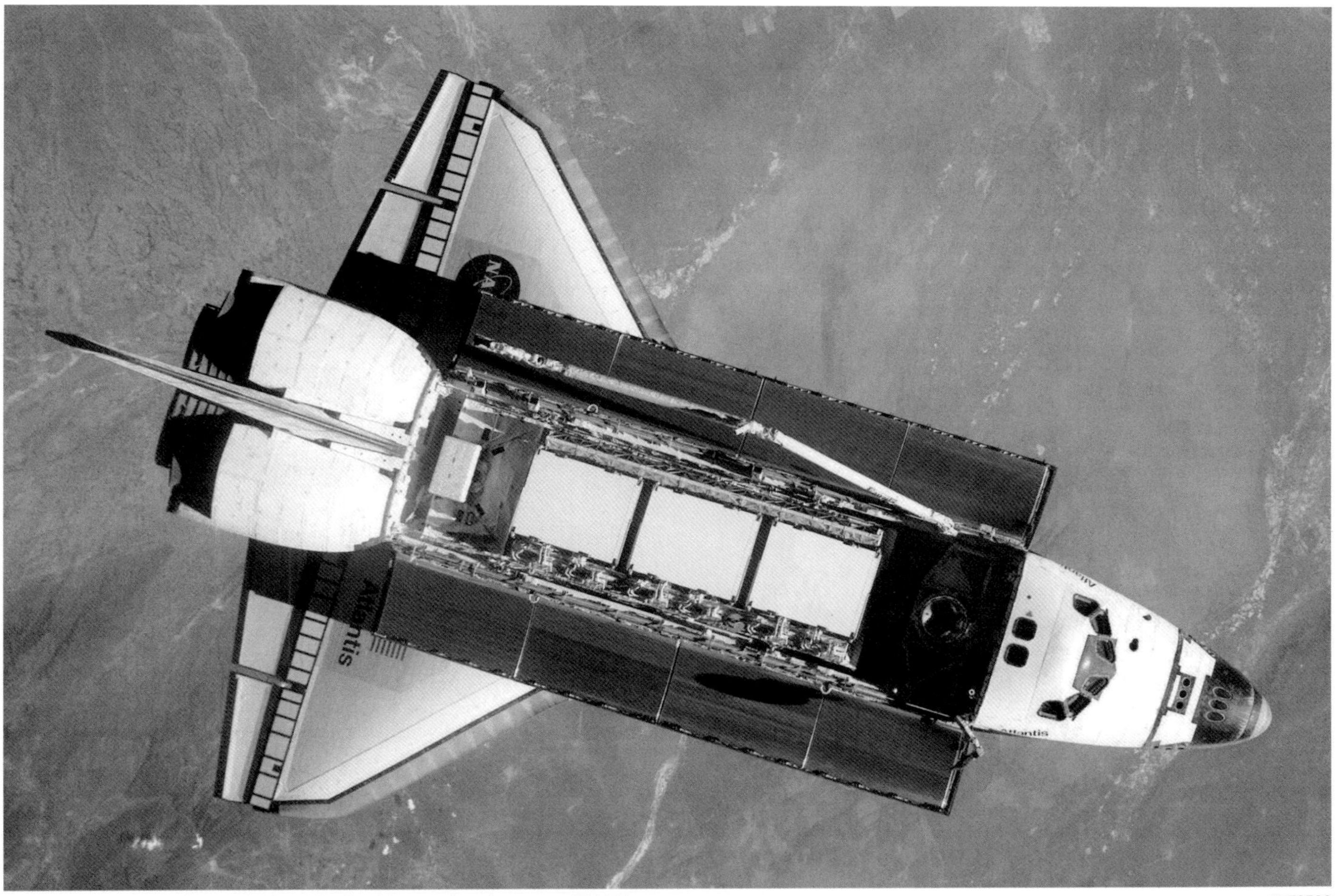

An impressive view of Shuttle Orbiter Atlantis with cargo bay doors open (NASA) 0573532

UPDATED

Table 15: US EVA Record (excluding lunar and ISS)

Mission	EVA date	Astronaut	h, min
Gemini 4	03.06.65	White	00.36
Gemini 9	05.06.66	Cernan	02.08
Gemini 10	19.07.66	Collins (SEVA)	00.49
	20.07.66	Collins	00.39
Gemini 11	13.09.66	Gordon	00.33
	14.09.66	Gordon (SEVA)	02.08
Gemini 12	12.11.66	Aldrin (SEVA)	02.29
	13.11.66	Aldrin	02.06
	14.11.66	Aldrin (SEVA)	00.55
Apollo 9	06.03.69	Schweickart	00.37
	06.03.69	Scott (SEVA)	00.46
Apollo 15	05.08.71	Worden	00.38
	05.08.71	Irwin (SEVA)	00.38
Apollo 16	25.04.72	Mattingly	01.13
	25.04.72	Duke (SEVA)	01.13
Apollo 17	17.12.72	Evans	01.06
	17.12.72	Schmitt (SEVA)	01.06
Skylab 2	25.05.73	Weitz (SEVA)	01.15
	07.06.73	Conrad	03.30
	07.06.73	Kerwin	03.30
	19.06.73	Conrad	01.44
	19.06.73	Weitz	01.44
Skylab 3	06.08.73	Garriott	06.31
	06.08.73	Lousma	06.31
	24.08.73	Garriott	04.30
	24.08.73	Lousma	04.30
	22.09.73	Bean	02.45
	22.09.73	Garriott	02.45
Skylab 4	22.11.73	Gibson	06.33
	22.11.73	Pogue	06.33
	25.12.73	Carr	07.01
	25.12.73	Pogue	07.01
	29.12.73	Carr	03.28
	29.12.73	Gibson	03.28
	03.02.74	Carr	05.19
	03.02.74	Gibson	05.19
STS-6	07.04.83	Musgrave	03.54
	07.04.83	Peterson	03.54
STS-41B	08.02.84	McCandless	05.35*
	08.02.84	Stewart	05.35
	10.02.84	McCandless	06.02
	10.02.84	Stewart	06.02*
STS-41C	09.04.84	Nelson	02.59*
	09.04.84	van Hoften	02.59
	11.04.84	Nelson	07.07
	11.04.84	van Hoften	07.07*
STS-41G	12.10.84	Sullivan	03.29
	12.10.84	Leestma	03.29
STS-51A	13.11.84	Allen	06.13*
	13.11.84	Gardner	06.13
	15.11.84	Allen	06.01
	15.11.84	Gardner	06.01*
STS-51D	16.04.85	Hoffman	03.10
	16.04.85	Griggs	03.10
STS-51I	31.08.85	van Hoften	07.20
	31.08.85	W. Fisher	07.20
	01.09.85	van Hoften	04.31
	01.09.85	W. Fisher	04.31
STS-61B	29.11.85	Spring	05.34
	29.11.85	Ross	05.34
	01.12.85	Spring	06.46
	01.12.85	Ross	06.46
STS-37	07.04.91	Ross	04.31
	07.04.91	Apt	04.31
	08.04.91	Ross	05.58
	08.04.91	Apt	05.58
STS-49	10.05.92	Thuot	03.53
	10.05.92	Hieb	03.53
	11.05.92	Thuot	05.38
	11.05.92	Hieb	05.38
	13.05.92	Thuot	08.30
	13.05.92	Hieb	08.30
	13.05.92	Akers	08.30
STS-51	14.05.92	Thornton	07.47
	14.05.92	Akers	07.47
STS-54	17.01.93	Runco	04.28
	17.01.93	Harbaugh	04.28
STS-57	25.06.93	Low	05.50
	25.06.93	Wisoff	05.50
	16.09.93	Walz	07.05
	16.09.93	Newman	07.05
STS-61	05.12.93	Musgrave	07.54
	05.12.93	Hoffman	07.54
	06.12.93	Akers	06.36
	06.12.93	Thornton	06.36
	07.12.93	Musgrave	06.47
	07.12.93	Hoffman	06.47
	08.12.93	Akers	06.50
	08.12.93	Thornton	06.50
STS-63	09.12.93	Musgrave	07.21
	09.12.93	Hoffman	07.21
STS-64	16.09.94	Lee	06.51
	16.09.94	Meade	06.51
	09.02.95	Foale	04.38
	09.02.95	Harris	04.38
STS-69	16.09.95	Voss	06.46
	16.09.95	Gernhardt	06.46
STS-72	15.01.96	Chiao	06.09
	15.01.96	Barry	06.09
	17.01.96	Chiao	06.54
	17.01.96	Scott	06.54
STS-76	27.03.96	Clifford	06.03
	27.03.96	Godwin	06.03
STS-82	14.02.97	Lee	06.42
	14.02.97	Smith	06.42
	15.02.97	Harbaugh	07.27
	15.02.97	Tanner	07.27
	16.02.97	Lee	07.11
	16.02.97	Smith	07.11
	17.02.97	Harbaugh	06.34
	17.02.97	Tanner	06.34
	18.02.97	Lee	05.17
	18.02.97	Smith	05.17
STS-86	01.10.97	Parazynski	05.01
STS-87	24.11.97	Scott	07.43
	24.11.97	Doi	07.43
	02.12.97	Scott	05.00
	02.12.97	Doi	05.00
STS-88	07.12.98	Ross	07.21
	07.12.98	Newman	07.21
	09.12.98	Ross	07.02
	09.12.98	Newman	07.02
	12.12.98	Ross	06.59
	12.12.98	Newman	06.59
STS-96	30.05.99	Jernigan	07.55
	30.05.99	Barry	07.55
STS-103	22.12.99	Smith	08.15
		Grunsfeld	08.15
	23.12.99	Foale	08.10
		Nicollier	08.10
	24.12.99	Smith	08.08
		Grunsfeld	08.08
STS-109	04.03.02	Grunsfeld	07.01
		Linnehan	07.01
	05.03.02	Massimino	07.16
		Newman	07.16
	06.03.02	Grunsfeld	06.48
		Linnehan	06.48
	07.03.02	Massimino	07.30
		Newman	07.30
	08.03.02	Grunsfeld	07.20
		Linnehan	07.20

Hubble gets a service call from STS-61 astronauts in December 1993 in which a record five space walks are concluded (JSC) *0137155*

* = includes untethered flight with Manned Manoeuvring Unit (total 10.22); SEVA = Stand-Up EVA, ie astronaut stood in hatchway but did not leave craft. Times given are depressurisation of spacecraft; actual time spent in space will be less. Planned EVAs for STS-80 by Jernigan and Jones were cancelled because of faulty hatch.

Servicing of the NASA Hubble Space Telescope in orbit allowed successive upgrades and improvements with repair and replacement a functional part of the mission (NASA) 0573544

Table 16: US Lunar EVA Record

Mission	*EVA date*	*Astronaut*	*h, min*
Apollo 11	21.07.69	Armstrong	02.31
	21.07.69	Aldrin	02.31
Apollo 12	19.11.69	Conrad	03.56
	19.11.69	Bean	03.56
	20.11.69	Conrad	03.49
	20.11.69	Bean	03.49
Apollo 14	05.02.71	Shepard	04.48
	05.02.71	Mitchell	04.48
	06.02.71	Shepard	04.35
	06.02.71	Mitchell	04.35
Apollo 15	30.07.71	Scott (SEVA)	00.33
	31.07.71	Scott	06.33
	31.07.71	Irwin	06.33
	01.08.71	Scott	07.12
	01.08.71	Irwin	07.12
	02.08.71	Scott	04.50
	02.08.71	Irwin	04.50
Apollo 16	21.04.72	Young	07.11
	21.04.72	Duke	07.11
	22.04.72	Young	07.23
	22.04.72	Duke	07.23
	23.04.72	Young	05.40
	23.04.72	Duke	05.40
Apollo 17	11.12.72	Cernan	07.12
	11.12.72	Schmitt	07.12
	12.12.72	Cernan	07.37
	12.12.72	Schmitt	07.37
	13.12.72	Cernan	07.16
	13.12.72	Schmitt	07.16
Totals	**EVA**	**28**	**161.06**
	SEVA	**1**	**00.33**

Apollo 16 Commander John Young with the LRV at the Descartes landing site on NASA's penultimate visit to the launch surface (NASA) 0137169

Table 17: Russian Human Space Flights

Mission	Launch Date	Crew	Duration (d, h, m)
Vostok 1	12.04.61	Yuri Gagarin	00.01.48
1st manned spaceflight, single rev			
Vostok 2	06.08.61	Gherman Titov	01.01.18
1st full day in space			
Vostok 3	11.08.62	Andrian Nikolayev	03.22.22
1st double flight (with Vostok 4)			
Vostok 4	12.08.62	Pavel Popovich	02.22.57
1st double flight (with Vostok 3)			
Vostok 5	14.06.63	Valeri Bykovsky	04.23.06
2nd double flight (with Vostok 6)			
Vostok 6	16.06.63	Valentina Tereshkova	02.22.50
1st woman in space; joint flight with Vostok 5			
Voskhod 1	12.10.64	Vladimir Komarov Konstantin Feoktistov Boris Yegorov	01.00.17
Modified Vostok; 1st 3-man flight			
Voskhod 2	18.03.65	Pavel Belyayev Alexei Leonov	01.02.02
Modified Vostok; 1st EVA (Leonov) via inflatable airlock			
Soyuz 1	23.04.67	Vladimir Komarov	01.02.48
Komarov killed after descent parachute snarled			
Soyuz 3	26.10.68	Georgi Beregovoi	03.22.51
Rendezvous with unmanned Soyuz 2			
Soyuz 4	14.01.69	Vladimir Shatalov	02.23.21
1st docking of 2 manned craft; Khrunov/Yeliseyev returned in Soyuz 4			
Soyuz 5	15.01.69	Boris Volynov Yevgeni Khrunov Alexei Yeliseyev	03.00.54
1st manned docking; Khrunov/Yeliseyev transferred by EVA to Soyuz 4			
Soyuz 6	11.10.69	Georgi Shonin Valeri Kubasov	04.22.42
Rendezvous with Soyuz 7/8 (intended Soyuz 7 docking); welding experiment			
Soyuz 7	12.10.69	Anatoli Filipchenko Vladislav Volkov Viktor Gorbatko	04.22.41
Triple rendezvous with Soyuz 6/8 (intended Soyuz 6 docking)			
Soyuz 8	13.10.69	Vladimir Shatalov Alexei Yeliseyev	04.22.51
Triple rendezvous with Soyuz 6/7 (remains only triple flight)			
Soyuz 9	01.06.70	Andrian Nikolayev Vitali Sevastyanov	17.16.59
Endurance record for solo craft remains			
Salyut 1	19.04.71	Unmanned space station	–
First space station; re-entered 11 Oct 1971			
Soyuz 10	23.04.71	Vladimir Shatalov Alexei Yeliseyev Nikolai Rukavishnikov	01.23.46
Docked twice with Salyut 1 but no boarding made			
Soyuz 11	06.06.71	Georgi Dobrovolsky Vladislav Volkov Viktor Patsayev	23.18.22
23 days aboard Salyut 1; Soyuz pressure loss killed crew before re-entry			
Soyuz 12	27.09.73	Vasili Lazarev Oleg Makarov	01.23.16
Tested post-Soyuz 11 modifications			
Soyuz 13	18.12.73	Pyotr Klimuk Valentin Lebedev	07.20.55
Solo flight with Orion ultraviolet telescope			
Salyut 3	25.06.74	Unmanned space station	–
2nd station, occupied by Soyuz 14; re-entered 24 Jan 1975			
Soyuz 14	03.07.74	Pavel Popovich Yuri Artyukhin	15.17.30
Sole occupation of Salyut 3			
Soyuz 15	26.08.74	Gennadi Sarafanov Lev Demin	02.00.12
Failed to dock with Salyut 3 because of Soyuz systems failures			
Soyuz 16	02.12.74	Anatoli Filipchenko Nikolai Rukavishnikov	05.22.24
Solo rehearsal for joint Apollo-Soyuz mission			
Salyut 4	26.12.74	Unmanned space station	–
3rd station; occupied by 2 crews for 93 days; re-entered 3 Feb 1977			
Soyuz 17	11.01.75	Alexei Gubarev Georgi Grechko	29.13.20
1st Salyut 4 occupation; Soviet endurance record			
Soyuz 18A	05.04.75	Vasili Lazarev Oleg Makarov	00.00.21
Launch to Salyut 4 aborted when stage 1 failed to separate; 14 g re-entry			
Soyuz 18B	24.05.75	Pyotr Klimuk Vitali Sevastyanov	62.23.20
2nd Salyut 4 occupation; Soviet endurance record			
Soyuz 19	15.07.75	Alexei Leonov Valeri Kubasov	05.22.31
1st joint US/USSR flight, docked with Apollo for 2 days			
Soyuz 20	17.11.75	Unmanned test	–
Resupply ferry rehearsal to Salyut 4			
Salyut 5	22.06.76	Unmanned space station	–
Occupied by 2 crews for 65 days; re-entered 8 Aug 1977			
Soyuz 21	06.07.76	Boris Volynov Vitali Zholobov	49.06.23
1st Salyut 5 occupation (48 days), acrid fumes forced return			
Soyuz 22	15.09.76	Valeri Bykovsky Vladimir Aksyonov	07.21.52
Solo Earth observation mission using backup ASTP craft			
Soyuz 23	14.10.76	Vyacheslav Zudov Valeri Rozhdestvensky	02.00.06
Salyut 5 docking cancelled after launch error (almost causing ascent abort) and Igla automatic approach system failure 200 m out left only enough propellant for one docking attempt and one de-orbit burn; mission rules required enough for two de-orbits. 1st Soviet splashdown after emergency return, into Lake Tengiz at night with –22°C outside. Frost was forming inside when rescuers found them after 9 h (presuming them dead because of no radio contact) and helicopter pulled them to shore			
Soyuz 24	07.02.77	Viktor Gorbatko Yuri Glazkov	17.17.26
2nd Salyut 5 occupation			
Salyut 6	29.09.77	Unmanned space station	–
Occupied for 676 days by 5 long stay + 11 visiting crews. Re-entered 28 Jul 1982 guided by C1267			
Soyuz 25	09.10.77	Vladimir Kovalyonok Valeri Ryumin	02.00.45
Failed to convert soft docking with Salyut 6 into hard docking (see S26)			
Soyuz 26	10.12.77	Yuri Romanenko Georgi Grechko	96.10.00
1st Salyut 6 occupation, endurance record, returned in Soyuz 27. Grechko EVA examined forward docking unit for reason for S25 docking failure, none apparent			
Soyuz 27	10.01.78	Vladimir Dzhanibekov Oleg Makarov	05.22.59
1st station dual occupancy; returned in Soyuz 26			
Soyuz 28	02.03.78	Alexei Gubarev Vladimir Remek	07.22.16
1st international flight (Czech), to Salyut 6			
Soyuz 29	15.06.78	Vladimir Kovalyonok Alexander Ivanchenkov	139.14.18
1st 100+ day flight, Salyut 6; returned in Soyuz 31			
Soyuz 30	27.06.78	Pyotr Klimuk Miroslaw Hermaszewski	07.22.03
Polish Intercosmos visit to Salyut 6			
Soyuz 31	26.08.78	Valeri Bykovsky Sigmund Jahn	07.20.49
E German Intercosmos Salyut 6 visit; returned in Soyuz 29			

Mission	Launch Date	Crew	Duration (d, h, m)
Soyuz 34	06.06.79	Unmanned	73.18.24
Delivered for Soyuz 32 crew return to compensate for Soyuz 33 failure			
Soyuz-T 1	16.12.79	Unmanned	100.09.20
T test, docked with Salyut 6			
Soyuz 35	09.04.80	Leonid Popov Valeri Ryumin	184.20.11
Endurance record, 4th Salyut 6 long stay, returned in Soyuz 37			
Soyuz 36	26.05.80	Valeri Kubasov Bertalan Farkas	07.20.46
Hungarian Salyut 6 visit; returned in Soyuz 35			
Soyuz-T 2	05.06.80	Yuri Malyshev Vladimir Aksyonov	03.22.19
1st manned T flight, Salyut 6 visit			
Soyuz 37	23.07.80	Viktor Gorbatko Pham Tuan	07.20.42
N Vietnamese Intercosmos Salyut 6 visit; returned in Soyuz 36			
Soyuz 38	18.09.80	Yuri Romanenko Arnaldo Tamayo Mendez	07.20.43
Cuban Intercosmos visit to Salyut 6			
Soyuz-T 3	27.11.80	Leonid Kizim Oleg Makarov Gennadi Strekalov	12.19.08
Resumption of 3-man flights; Salyut 6 repair work			
Soyuz-T 4	12.03.81	Vladimir Kovalyonok Viktor Savinykh	74.17.38
Last Salyut 6 long-stay; 50th Soviet/100th spaceman			
Soyuz 39	22.03.81	Vladimir Dzhanibekov Jugderdemidiyn Gurragcha	07.20.43
Mongolian Intercosmos visit to Salyut 6			
Soyuz 40	14.05.81	Leonid Popov Dumitru Prunariu	07.20.42
Romanian Intercosmos Salyut 6 visit; last flight of old Soyuz design			
Salyut 7	19.04.82	Unmanned space station	–
Occupied for 812 days by 10 crews; 13 EVAs. Re-entered 7 Feb 1991 with C1686 attached			
Soyuz-T 5	13.05.82	Anatoli Berezovoi Valentin Lebedev	211.09.05
1st Salyut 7 occupation + long stay; returned in T7			
Soyuz-T 6	24.06.82	Vladimir Dzhanibekov Alexander Ivanchenkov Jean-Loup Chrétien	07.21.51
French visiting mission to Salyut 7			
Soyuz-T 7	19.08.82	Leonid Popov Alexander Serebrov Svetlana Savitskaya	07.21.52
Salyut 7 visit, 2nd Soviet woman, returned in T5			
Soyuz-T 8	20.04.83	Vladimir Titov Gennadi Strekalov Alexander Serebrov	02.00.18
Failed to dock with Salyut 7 for long stay. Radar failed (possibly torn off during launch) and manual approach aborted			
Soyuz-T 9	27.06.83	Vladimir Lyakhov Alexander Alexandrov	149.10.46
Salyut 7 long stay; 2 EVAs added 2 solar panels			
Soyuz-T 10A	26.09.83	Vladimir Titov Gennadi Strekalov	
Launchpad fire, 1st use of escape tower. Descent module re-used by T15			
Soyuz-T 10B	08.02.84	Leonid Kizim Vladimir Solovyev Oleg Atkov	236.22.50
1st long stay triple crew, Salyut 7; 6 EVAs totalling 22 h 56 min			
Soyuz-T 11	03.04.84	Yuri Malyshev Gennadi Strekalov Rakesh Sharma	07.21.40
Indian visit to Salyut 7; returned in Soyuz-T10B			
Soyuz-T 12	17.07.84	Vladimir Dzhanibekov Svetlana Savitskaya Igor Volk	11.19.15
Salyut 7 visit, 1st EVA by a woman			
Soyuz-T 13	06.06.85	Vladimir Dzhanibekov Viktor Savinykh	112.03.12 168.03.51
Reactivation of Salyut 7; Savinykh stayed with T14 crew			
Soyuz-T 14	17.09.85	Vladimir Vasyutin Georgi Grechko Alexander Volkov	64.21.52 08.21.13 64.21.52
Salyut 7, 1st mission to be ended by illness (Vasyutin); Grechko returned with T13			
Mir	20.02.86	Unmanned space station	–
New-generation space station with multiple ports			
Soyuz-T 15	13.03.86	Leonid Kizim Vladimir Solovyev	125.00.01
1st Mir occupation; excursion to Salyut 7 5 May–26 Jun. 2 Salyut EVAs 8 h 50 min. Re-used descent module from T10A abort			
Soyuz-TM 1	21.05.86	Unmanned test	08.22.27
TM demonstration; docked to Mir for 6 days			
Soyuz-TM 2	05.02.87	Yuri Romanenko Alexander Laveikin	326.11.18 174.03.26
2nd Mir long stay; Romanenko new endurance record; Romanenko returned TM3			
Kvant 1	31.03.87	Unmanned Mir module	–
Astrophysics module docked with Mir 09.04.87			
Soyuz-TM 3	22.07.87	Alexander Viktorenko Alexander Alexandrov Mohammed Faris	07.23.05 160.07.17 07.23.05
Syrian Mir visit, Alexandrov replacing Laveikin			
Soyuz-TM 4	21.12.87	Vladimir Titov Musa Manarov Anatoli Levchenko	365.22.39 365.22.39 07.21.58
Titov/Manarov swapped with Romanenko/Alexandrov for Mir's 3rd long stay occupation: endurance record, 3 EVAs totalling 13 h 40 min, returned in TM6			
Soyuz-TM 5	07.06.88	Anatoli Solovyov Viktor Savinykh Alexander Alexandrov	09.20.10
Bulgarian Mir visit, all returning in Soyuz-TM 4			
Soyuz-TM 6	29.08.88	Vladimir Lyakhov Valeri Poliakov Abdol Mohmand	08.20.27 240.22.36 08.20.27
Afghani Mir visit, physician Poliakov remained with Titov/Manarov			
Soyuz-TM 7	26.11.88	Alexander Volkov Sergei Krikalev Jean-Loup Chrétien	151.11.09 151.11.09 24.18.07
French Mir visit, Volkov/Krikalev swapped with Titov/Manarov. Chrétien EVA, returned in TM6, Volkov/Krikalev in TM7			
Soyuz-TM 8	05.09.89	Alexander Viktorenko Alexander Serebrov	166.06.58
5th Mir long stay: received Kvant 2, 5 EVAs including 2 with manoeuvring backpack			
Kvant 2	26.11.89	Unmanned Mir module	–
Radial expansion module, docked with Mir 06.12.89			
Soyuz-TM 9	11.02.90	Anatoli Solovyov Alexander Balandin	179.01.19
6th Mir long stay: received Kristall, 2 EVAs			
Kristall	31.05.90	Unmanned Mir module	–
2nd radial expansion (technology) module, docked with Mir 10.06.90			
Soyuz-TM 10	01.08.90	Gennadi Manakov Gennadi Strekalov	130.20.36
7th Mir long stay: emphasised Kristall materials processing, 1 EVA			
Soyuz-TM 11	02.12.90	Viktor Afanasyev Musa Manarov Toyohiro Akiyama	175.01.52 175.01.52 07.21.15
8th Mir long stay: 4 EVAs; Akiyama (Japan) returned in Soyuz-TM 10			
Soyuz-TM 12	18.05.91	Anatoli Artsebarsky Sergei Krikalev Helen Sharman	144.15.22 311.20.01 07.21.15
9th Mir long stay; 1st UK cosmonaut (Sharman) returned in Soyuz-TM 11, Artsebarsky in TM12, Krikalev in TM13. 6 EVAs totalling 31 h 58 min			
Soyuz-TM 13	02.10.91	Alexander Volkov Toktar Aubakirov Franz Viehböch	175.02.52 07.22.13 07.22.13
10th Mir long stay; 1st Austrian- + Kazakh-sponsored cosmonauts (1st mission with 2 researchers), returned in TM12, Volkov in TM13. 1 EVA			
Soyuz-TM 14	17.03.92	Alexander Viktorenko Alexander Kaleri Klaus-Dietrich Flade	145.14.11 145.14.11 07.21.57
11th Mir long stay; 1st German cosmonaut, returned in TM13, Viktorenko/Kaleri in TM14. 123 min EVA			
Soyuz-TM 15	27.07.92	Anatoli Solovyov Sergei Avdeyev Michel Tognini	188.21.40 188.21.40 13.18.57
12th Mir long stay; 2nd French cosmonaut (3rd French visit mission), returned in TM14; Solovyov/Avdeyev in TM15. 4 EVAs totalling 18 h 21 min			

Mission	Launch Date	Crew	Duration (d, h, m)
Soyuz-TM 16	24.01.93	Gennadi Manakov	179.00.44
		Alexander Poleshchuk	179.00.44
13th Mir long stay; 1st docking with Kristall androgynous port. 2 EVAs to prepare for move of Kristall solar arrays to Kvant 1			
Soyuz-TM 17	01.07.93	Vasili Tsebliyev	196.17.45
		Alexander Serebrov	196.17.45
		Jean-Pierre Haignere	20.16.09
14th Mir long stay; 3rd French cosmonaut (4th French visit mission), returned in TM16; Tsebliyev/Serebrov in TM17. 5 EVAs to test new support truss and monitor ageing Mir			
Soyuz-TM 18	08.01.94	Viktor Afanasyev	182.00.27
		Yuri Usachev	182.00.27
		Valeri Poliakov	437.17.58
15th Mir long stay; physician Poliakov remained aboard for record duration, returning with TM20 (record unlikely to be broken for many years)			
Soyuz-TM 19	01.07.94	Yuri Malenchenko	125.22.53
		Talgat Musabayev	125.22.53
16th Mir long stay. 2 EVAs inspected damage from TM17/Progress-M 24 impacts			
Soyuz-TM 20	03.10.94	Alexander Viktorenko	169.05.22
		Yelena Kondakova	169.05.22
		Ulf Merbold	31.12.36
17th Mir long stay; ESA researcher Merbold returned with TM19			
Soyuz-TM 21	03.03.95	Vladimir Dezhurov	115.08.44
		Gennadi Strekalov	115.08.44
		Norman Thagard	115.08.44
18th Mir long stay; NASA researcher Thagard; all crew returning on NASA Shuttle STS-71 7 Jul 1995. 3 EVAs transferred one Kristall solar array to Kvant 1			
Spektr	20.05.95	Unmanned Mir module	–
3rd radial module (remote sensing), docked with Mir 01.06.95			
NASA Shuttle STS-71	27.06.95		
Delivery of 19th Mir main crew, Cdr Anatoli Solovyov + engineer Nikolai Budarin; return TM21 crew. Solovyov/Budarin returned in TM21. 3 EVAs partially deployed jammed Spektr solar wing and installed MIRAS spectrometer			
Soyuz-TM 22	03.09.95	Yuri Gidzenko	179.01.42
		Sergei Avdeyev	179.01.42
		Thomas Reiter	179.01.42
20th Mir long stay; ESA researcher Reiter performed 2 EVAs			
NASA Shuttle STS-74	12.11.95		
Attached docking module to Kristall, delivered solar arrays to increase Mir's power, delivered supplies and returned with experimental results			
Soyuz-TM 23	21.02.96	Yuri Onufrienko	193.19.08
		Yuri Usachev	193.19.08
21st Mir long stay			
NASA Shuttle STS-76	22.03.96		
Shannon Lucid remained on Mir, returning aboard STS-79 in Sep 1996. NASA's Clifford/Godwin exited docking adapter hatch to mount three experiments on Mir's docking module			
Priroda	23.04.96	Unmanned Mir module	–
4th radial module (remote sensing), docked with Mir 26 Apr Planned:			
Soyuz-TM 24	17.08.96	Valeri Korzun	196.17.26
		Alexandr Kaleri	196.17.26
		Claudie André-Deshays	15.18.24
22nd long-stay Russian mission to Mir; French 'Cassiopeia' mission (5th French visit)			
NASA Shuttle STS-79	16.09.96		188.04.00
Delivered Blaha to Mir, returned Lucid			
NASA Shuttle STS-81	22.12.97		128.05.27
Delivered Linenger to Mir, returned with Blaha			
Soyuz-TM 25	10.02.97	Vasily Tsibilev	184.22.07
		Aleksandr Lazutkin	184.22.07
		Reinhold Ewald	19.16.35
23rd long-stay Russian mission to Mir; second German visit to the station. Accident-prone residency saw fire aboard Mir during the hand-over period with the Soyuz-TM 24 crew, collison of Progress-M 34 with Spektr module and the brief accidental disconnection of the computer system			
NASA Shuttle STS-84	15.05.97		132.03.59
Launched delivered Foale to Mir, returned Linenger			
Soyuz-TM 26	05.08.97	Anatoli Solovyov	197.17.34
		Pavel Vinogradov	197.17.34
24th long-stay Russian mission to Mir: planned flight of a French spationaut delayed until Soyuz-TM 27 because of power problems on Mir following the Progress-M 34 collision with Spektr			
NASA Shuttle STS-86	26.09.97		144.13.48
Delivered Wolf to Mir, returned with Linenger			
NASA Shuttle STS-89	23.01.98		127.10.01
Delivered Thomas, final US astronaut to Mir, returned Wolf			
Soyuz-TM 27	29.01.98	Talgat Musabeyev	207.12.50
		Nikolai Budarin	207.12.50
		Leopold Eyharts (CNES)	20.16.37
25th long-stay aboard Mir carrying French cosmonaut Eyharts			
NASA Shuttle STS-91	02.06.98		140.15.12
Returned Thomas, last US astronaut to Mir			
Soyuz-TM 28	13.08.98	Gennady Padalka	198.04.31
		Sergei Avdeyev	188.20.16
		Yuri Baturin	11.07.40
26th long-stay aboard Mir and with seats on the next and last, manned flight to Mir pre-sold to Slovak and French cosmonauts, their flight plan was being constantly revised			
Soyuz-TM 29	20.02.99	Afanasyev	188.20.16
		Haignere	188.20.16
		Bella	07.21.56
27th long-stay aboard Mir with Slovak and French cosmonauts and Avdeyev the sole Russian on board			
Soyuz-TM 30	04.04.00	Zalyotin	73.19.32
		Kaleri	73.19.32
Reopened uninhabited Mir for engineering evaluation			
Soyuz-TM 31	31.10.00	Krikalev	140.23.39
		Gidzenko	140.23.39
		Shepard	140.23.39
Expedition 1 crew for ISS, returned in STS-102			
Soyuz-TM 32	28.04.01	Musabeyev	07.22.04
		Baturin	07.22.04
		Tito	07.22.04
Taxi flight to ISS carrying private US citizen Tito, returned in TM 31			
Soyuz-TM 33	21.10.01	Afanaseyev	09.19.59
		Kozeev	09.19.59
		Haignere	09.19.59
Taxi flight to ISS, returned in TM 32			

Table 18: Russian Cosmonauts in Flight Order

No	Name	Total Flts	Total Duration (d, h, m)	Missions
1	Y Gagarin	1	00.01.48	V1
2	G Titov	1	01.01.18	V2
3	A Nikolayev	2	21.15.25	V3, S9
4	P Popovich	2	18.16.29	V4, S14
5	V Bykovsky	3	20.17.49	V5, S22, S31
6	V Tereshkova	1	02.22.50	V6
7	V Komarov	2	02.02.54	Vd1, S1
8	K Feoktistov	1	01.00.17	Vd1
9	B Yegorov	1	01.00.17	Vd1
10	P Belyayev	1	01.02.02	Vd2
11	A Leonov	2	07.00.33	Vd2, S19
12	G Beregovoi	1	03.22.51	S3
13	V Shatalov	3	09.21.59	S4, S8, S10
14	B Volynov	2	52.07.20	S5, S21
15	A Yeliseyev	3	08.22.24	S5, S8, S10
16	Y Khrunov	1	01.23.49	S5
17	G Shonin	1	04.22.42	S6
18	V Kubasov	3	18.17.19	S6, S19, S36
19	A Filipchenko	2	10.21.05	S7, S16
20	V Volkov	2	28.17.03	S7, S11
21	V Gorbatko	3	30.12.31	S7, S24, S37
22	V Sevastyanov	2	80.16.19	S9, S18B
23	N Rukavishnikov	3	09.22.10	S10, S16, S33
24	G Dobrovolsky	1	23.18.22	S11
25	V Patsayev	1	23.18.22	S11
26	V Lazarev	2	01.23.16	S12, S18A
27	O Makarov	4	20.18.28	S12, S18A, S27, T3
28	P Klimuk	3	78.18.19	S13, S18B, S30
29	V Lebedev	2	219.05.00	S13, T5
30	Y Artyukhin	1	15.17.30	S14
31	G Sarafanov	1	02.00.12	S15
32	L Demin	1	02.00.12	S15
33	A Gubarev	2	37.11.37	S17, S28
34	G Grechko	3	133.20.33	S17, S26, T14
35	V Zholobov	1	49.06.24	S21
36	V Aksyonov	2	11.20.35	S22, T2
37	V Zudov	1	02.00.06	S23
38	V Rozhdestvensky	1	02.00.06	S23
39	Y Glazkov	1	17.17.23	S24
40	V Kovalyonok	3	216.10.17	S25, S29, T4
41	V Ryumin	4	371.12.19	S25, S32, S35, STS-91
42	Y Romanenko	3	430.18.12	S26, S38, TM2
43	V Dzhanibekov	5	147.15.59	S27, S39, T6, T12, T13
44	A Ivanchenkov	2	147.12.39	S29, T6
45	V Lyakhov	3	333.06.49	S32, T9, TM6
46	L Popov	3	199.01.32	S35, S30, T7
47	Y Malyshev	2	11.20.32	T2, T11
48	L Kizim	3	374.17.59	T3, T10B, T15
49	G Strekalov	6	268.22.37	T3, T8, T10A, T11, TM10, TM21
50	V Savinykh	3	196.17.00	T4, T13, TM5
51	A Berezovoi	1	211.08.05	T5
52	A Serebrov	4	372.22.53	T7, T8, TM8, TM17
53	S Savitskaya	2	19.17.07	T7, T12
54	V Titov	4	375.05.25	T8, T10A, TM4, STS-63
55	A Alexandrov	2	309.17.03	T9, TM3
56	V Solovyev	2	360.22.51	T10B, T15
57	O Atkov	1	236.22.50	T10B
58	I Volk	1	11.19.14	T12
59	V Vasyutin	1	64.21.52	T14
60	A Volkov	3	391.11.53	T14, TM7, TM13
61	A Laveikin	1	174.03.26	TM2
62	A Viktorenko	4	489.01.36	TM3, TM8, TM14, TM20
63	M Manarov	2	541.00.31	TM4, TM11
64	A Levchenko	1	07.21.58	TM4
65	A Solovyov	5	651.00.04	TM5, TM9, TM15, STS-71, TM26
66	V Poliakov	2	678.16.34	TM6, TM18
67	S Krikalev	5	627.30.34	TM7, TM12, STS-60, STS-88, TM31
68	A Balandin	1	179.01.19	TM9
69	G Manakov	2	309.21.20	TM10, TM16
70	V Afanasyev	4	555.18.35	TM11, TM18, TM29, TM33
71	A Artsebarsky	1	144.15.22	TM12
72	A Kaleri	3	415.27.09	TM14, TM24, TM30
73	S Avdeyev	3	747.14.14	TM15, TM22, TM28
74	A Poleshchuk	1	179.00.44	TM16
75	V Tsibliyev	2	381.15.52	TM17, TM25
76	Y Usachev	2	376.19.35	TM18, TM23
77	Y Malenchenko	1	125.22.53	TM19
78	T Musabayev	3	340.33.47	TM19, TM27, TM32
79	Y Kondakova	2	178.10.43	TM20, STS-84
80	V Dezhurov	1	115.08.44	TM21
81	N Budarin	2	283.00.11	STS-71, TM27
82	Y Gidzenko	2	319.25.21	TM22, TM31
83	Y Onufrienko	1	194.19.08	TM23
84	V Korzun	1	196.17.26	TM24
85	A Lazutkin	1	184.22.07	TM25
86	P Vinogradov	1	197.17.34	TM26
87	G Padalka	1	198.16.31	TM28
88	S Zalyotin	1	73.19.32	TM30
89	K Kozeev	1	09.19.59	TM33

Notes: Some cosmonauts have flown aboard shuttle missions. Dezhurov and Strekalov were launched aboard Soyuz-TM 21 but returned to Earth aboard Atlantis/STS-71; A Solovyov and Budarin were launched on Atlantis/STS-71 and returned to Earth aboard Soyuz-TM 21. Other cosmonaut shuttle missions did not involve a switch between launch and landing craft. T10A was a launch pad abort and is not included in the totals. V = Vostok, Vd = Voskhod, S = Soyuz, T = Soyuz-T, TM = Soyuz-TM.

Table 19: CIS/USSR EVA Record

Mission	EVA date	Cosmonaut	h, min
Voskhod 2	18.03.65	Leonov	00.24
Soyuz 4	16.01.69	Yeliseyev	01.00
Soyuz 5	16.01.69	Khrunov	01.00
Soyuz 26	20.12.77	Grechko	01.28
		Romanenko[1]	
Soyuz 29	29.07.78	Kovalyonok	02.05
		Ivanchenkov	02.05
Soyuz 32	15.08.79	Lyakhov	01.23
		Ryumin	01.23
Soyuz-T 5	30.07.82	Berezovoi	02.33
		Lebedev	02.33
Soyuz-T 9	01.11.83	Lyakhov	02.50
		Alexandrov	02.50
Soyuz-T 9	03.11.83	Lyakhov	02.55
		Alexandrov	02.55
Soyuz-T 10B	23.04.84	Kizim	04.15
		Solovyev	04.15
Soyuz-T 10B	26.04.84	Kizim	05.00
		Solovyev	05.00
Soyuz-T 10B	29.04.84	Kizim	02.45
		Solovyev	02.45
Soyuz-T 10B	04.05.84	Kizim	02.45
		Solovyev	02.45
Soyuz-T 10B	18.05.84	Kizim	03.05
		Solovyev	03.05
Soyuz-T 12	25.07.84	Dzhanibekov	03.35
		Savitskaya	03.35
Soyuz-T 10B	08.08.84	Kizim	05.00
		Solovyev	05.00
Soyuz-T 13	02.08.85	Dzhanibekov	05.00[2]
		Savinykh	05.00[2]
Soyuz-T 15	28.05.86	Kizim	03.50
		Solovyev	03.50
Soyuz-T 15	31.05.86	Kizim	05.00
		Solovyev	05.00
Soyuz-TM 2	12.04.87	Romanenko	03.40
		Laveikin	03.40
Soyuz-TM 2	12.06.87	Romanenko	01.53
		Laveikin	01.53
Soyuz-TM 2	16.06.87	Romanenko	03.15
		Laveikin	03.15
Soyuz-TM 4	26.02.88	Titov	04.25
		Manarov	04.25
Soyuz-TM 4	30.06.88	Titov	05.10
		Manarov	05.10
Soyuz-TM 4	20.10.88	Titov	04.12
		Manarov	04.12
Soyuz-TM 7	09.12.88	Volkov	05.57
		Chrétien	05.57
Soyuz-TM 8	08.01.90	Viktorenko	02.56
		Serebrov	02.56
Soyuz-TM 8	11.01.90	Viktorenko	02.54
		Serebrov	02.54
Soyuz-TM 8	26.01.90	Viktorenko	03.02
		Serebrov	03.02
Soyuz-TM 8	01.02.90	Viktorenko	04.59
		Serebrov[3]	04.59
Soyuz-TM 8	05.02.90	Viktorenko[4]	03.45
		Serebrov	03.45
Soyuz-TM 9	17.07.90	Solovyov	07.16
		Balandin	07.16
Soyuz-TM 9	26.07.90	Solovyov	03.31
		Balandin	03.31
Soyuz-TM 10	29.10.90	Manakov	03.45
		Strekalov	03.45
Soyuz-TM 11	07.01.91	Afanasyev	05.18
		Manarov	05.18
Soyuz-TM 11	23.01.91	Afanasyev	05.33
		Manarov	05.33
Soyuz-TM 11	26.01.91	Afanasyev	06.20
		Manarov	06.20
Soyuz-TM 11	25.04.91	Afanasyev	03.34
		Manarov	03.34
Soyuz-TM 12	24.06.91	Artsebarsky	04.58
		Krikalev	04.58
Soyuz-TM 12	28.06.91	Artsebarsky	03.24
		Krikalev	03.24
Soyuz-TM 12	15.07.91	Artsebarsky	05.56
		Krikalev	05.56
Soyuz-TM 12	19.07.91	Artsebarsky	05.28
		Krikalev	05.28
Soyuz-TM 12	23.07.91	Artsebarsky	05.34
		Krikalev	05.34
Soyuz-TM 12	27.07.91	Artsebarsky	06.49
		Krikalev	06.49
Soyuz-TM 13	20.02.92	Volkov	04.12
		Krikalev	04.12
Soyuz-TM 14	08.07.92	Viktorenko	02.03
		Kaleri	02.03
Soyuz-TM 15	03.09.92	Solovyov	03.56
		Avdeyev	03.56
Soyuz-TM 15	07.09.92	Solovyov	05.08
		Avdeyev	05.08
Soyuz-TM 15	11.09.92	Solovyov	05.44
		Avdeyev	05.44
Soyuz-TM 15	15.09.92	Solovyov	03.33
		Avdeyev	03.33
Soyuz-TM 16	19.04.93	Manakov	05.25
		Poleshchuk	05.25
Soyuz-TM 16	18.06.93	Manakov	04.33
		Poleshchuk	04.33
Soyuz-TM 17	16.09.93	Tsebliyev	04.18
		Serebrov	04.18
Soyuz-TM 17	20.09.93	Tsebliyev	03.13
		Serebrov	03.13
Soyuz-TM 17	28.09.93	Tsebliyev	01.52
		Serebrov	01.52
Soyuz-TM 17	22.10.93	Tsebliyev	00.38
		Serebrov	00.38
Soyuz-TM 17	29.10.93	Tsebliyev	04.12
		Serebrov	04.12
Soyuz-TM 19	09.09.94	Malenchenko	05.04
		Musabayev	05.04
Soyuz-TM 19	13.09.94	Malenchenko	06.01
		Musabayev	06.01
Soyuz-TM 21	12.05.95	Dezhurov	06.15
		Strekalov	06.15
Soyuz-TM 21	17.05.95	Dezhurov	06.52
		Strekalov	06.52
Soyuz-TM 21	22.05.95	Dezhurov	05.15
		Strekalov	05.15
'STS 71'	14.07.95	Solovyov	05.34
		Budarin	05.34
'STS 71'	19.07.95	Solovyov	03.08
		Budarin	03.08
'STS 71'	21.07.95	Solovyov	05.35
		Budarin	05.35
Soyuz-TM 22	20.10.95	Avdeyev	05.16
		Reiter	05.16
Soyuz-TM 22	08.02.96	Gidzenko	03.02
		Reiter	03.02
Soyuz-TM 23	15.03.96	Onufrienko	05.51
		Usachev	05.51
Soyuz-TM 23	20.05.96	Onufrienko	05.20
		Usachev	05.20
Soyuz-TM 23	24.05.96	Onufrienko	05.43
		Usachev	05.43
Soyuz-TM 23	30.05.96	Onufrienko	04.20
		Usachev	04.20
Soyuz-TM 23	06.06.96	Onufrienko	03.34
		Usachev	03.34
Soyuz-TM 24	02.12.96	Korzun	05.57
		Kaleri	05.57
Soyuz-TM 24	09.12.96	Korzun	06.36
		Kaleri	06.36
Soyuz-TM 25	29.04.97	Tsibilev	04.58
		Linenger	04.58
Soyuz-TM 26	22.08.97	Solovyov	06.00
		Vinogradov	06.00
Soyuz-TM 26	20.10.97	Solovyov	06.38
		Vinogradov	06.38
Soyuz-TM 26	06.11.97	Solovyov	06.17
		Vinogradov	06.17
Soyuz-TM 26	14.01.98	Solovyov	03.52
		Wolf	03.52
Soyuz-TM 27	01.04.98	Musabayev	06.40
		Budarin	06.40
Soyuz-TM 27	06.04.98	Musabayev	04.23
		Budarin	04.23
Soyuz-TM 27	11.04.98	Musabayev	06.25
		Budarin	06.25
Soyuz-TM 27	17.04.98	Musabayev	06.33
		Budarin	06.33
Soyuz-TM 27	22.04.98	Musabayev	06.21
		Budarin	06.21
Soyuz-TM 28	10.11.98	Padalka	05.54
		Avdeyev	05.54
Soyuz-TM 29	16.04.99	Afanasyev	06.19
		Haignere	06.19
Soyuz-TM 29	23.07.99	Afanasyev	06.07
		Avdeyev	06.07
Soyuz-TM 29	28.07.99	Afanasyev	05.22
		Avdeyev	05.22

[1] Unauthorised, duration unknown.
[2] Approximate time.
[3] first test of manoeuvring unit,
[4] second test of manoeuvring unit. Linenger was the visiting NASA astronaut at the time.

Table 20: Chinese Satellite Launches

List of Chinese Satellite Launches, 1970–2001 (including known test flights)

Launch Date		*Satellite*	*Mass (kg)*	*Vehicle*	*Site*	*Orbit Class*
10 January 1970				CZ-1	J	Test flight
24 April 1970		DFH 1	173	CZ-1	J	E-LEO
3 March 1971		SJ 1	221	CZ-1	J	E-LEO
10 August 1972				FB-1	J	Test flight
18 September 1973		JSSW	1,100 ?	FB-1	J	Failure
14 July 1974		JSSW	1,100 ?	FB-1	J	Failure
4 November 1974		FSW-0	1,800 ?	CZ-2A	J	Failure
26 July 1975		JSSW 1	1,107	FB-1	J	LEO
26 November 1975		FSW-0 1	1,790	CZ-2C	J	LEO (3)
16 December 1975		JSSW 2	1,109	FB-1	J	LEO
30 August 1976		JSSW 3	1,108	FB-1	J	E-LEO
10 November 1976		JSSW	1,100 ?	FB-1	J	Failure
7 December 1976		FSW-0 2	1,812	CZ-2C	J	LEO (3)
14 September 1977				FB-1	J	Test flight
26 January 1978		FSW-0 3	1,810	CZ-2C	J	LEO (3)
16 April 1978				FB-1	J	Test flight
28 July 1979		SJ	770 ?	FB-1	J	Failure
19 September 1981		SJ 2	257	FB-1	J	E-LEO
		SJ 2A	483			E-LEO
		SJ 2B	28			E-LEO
9 September 1982		FSW-0 4	1,783	CZ-2C	J	LEO (5)
19 August 1983		FSW-0 5	1,842	CZ-2C	J	LEO (5)
29 January 1984		SW (DFH-2)	1,000	CZ-3	XC	GEO fail
8 April 1984		STTW A (DFH-2)	1,000	CZ-3	XC	GEO
12 September 1984		FSW-0 6	1,809	CZ-2C	J	LEO (5)
21 October 1985		FSW-0 7	1,820	CZ-2C	J	LEO (5)
1 February 1986		STTW 1 (DFH-2)	1,000	CZ-3	XC	GEO
6 October 1986		FSW-0 8	1,820	CZ-2C	J	LEO (5)
5 August 1987		FSW-0 9	1,820	CZ-2C	J	LEO (5)
9 September 1987		FSW-1 1	2,100	CZ-2C	J	LEO (8)
7 March 1988		STTW 2 (DFH-2A)	1,025	CZ-3	XC	GEO
5 August 1988		FSW-1 2	2,100	CZ-2C	XC	LEO (8)
6 September 1988		FY-1 1	757	CZ-4A	T	SSO
22 December 1988		STTW 3 (DFH-2A)	1,025	CZ-3	XC	GEO
4 February 1990		STTW 4 (DFH-2A)	1,024	CZ-3	XC	GEO
7 April 1990	*	AsiaSat 1 (Hong Kong)	1,442	CZ-3	XC	GEO
16 July 1990	*	Badr 1 (Pakistan)	52	CZ-2E/PKM	XC	LEO
		AUSSAT-B (model only)	2,700 ?		XC	GTO fail
3 September 1990		FY-1 2	881	CZ-4A	T	SSO
		QQW 1	4 ?			SSO
		QQW 2	4 ?			SSO
5 October 1990		FSW-1 3	2,080	CZ-2C	J	LEO (8)
28 December 1991		STTW 5	1,025 ?	CZ-3	XC	GEO fail
9 August 1992		FSW-2 1	2,400	CZ-2D	J	LEO (16)
13 August 1992	*	Optus-B 1 (Australia)	3,000 ?	CZ-2E/STAR-63F	XC	GEO
6 October 1992	*	Freja (Sweden)	259	CZ-2C	J	E-LEO
		FSW-1 4	2,100			LEO (7)
21 December 1992	*	Optus-B 2 (Australia)	3,000 ?	CZ-2E/STAR-63F	XC	GEO fail (satellite exploded)
8 October 1993		FSW-1 5	2,099	CZ-2C	J	LEO
8 February 1994		SJ 4	410	CZ-3A	XC	GTO
3 July 1994		FSW-2 2	2,600 ?	CZ-2D	J	LEO (15)
21 July 1994	*	APStar 1	1,383	CZ-3	XC	GEO
27 August 1994	*	OPTUS-B 3	3,000 ?	CZ-2E/STAR-63F	XC	GEO
29 November 1994		Zhongxing 6	2,230	CZ-3A	XC	GEO
25 January 1995	*	APStar 2 (Hong Kong)	3,000 ?	CZ-2E/STAR-63F	XC	Failure (satellite exploded ?)
29 May 1995			?	CZ-1D	J	Failure
25 November 1995	*	AsiaSat 2 (Hong Kong)	3,379	CZ-2E/EPKM	XC	GEO
28 December 1995	*	EchoStar 1 (USA)	3,288	CZ-2E/EPKM	XC	GEO
14 February 1996	*	INTELSAT 708 (USA)	4,576	CZ-3B	XC	Failure
4 July 1996	*	APStar 1A (Hong Kong)	1,385 ?	CZ-3	XC	GEO
18 August 1996		Zhongxing 7	1,400 ?	CZ-3	XC	GEO fail
20 October 1996		FSW-2 3	2,600 ?	CZ-2D	J	LEO (15)
11 May 1997		Zhongxing 8	2,230 ?	CZ-3A	XC	GEO
10 Jun 1997		Feng Yun-2 1R	1,380	CZ-3	XC	GEO
19 Aug 1997		Mabuhay Ayilu 2	2,560	CZ-3B	XC	GEO
1 Sept 1997		Iridium mFS-1	650	CZ-2C/5D	T	LEO
		Iridium mFS-2	650			LEO
22 Oct 1997		APStar 2R	3,747	CZ-3B	XC	GEO
30 Mar 1998		Iridium 51	657	CZ-2C/5D	T	LEO
		Iridium 61	657			LEO
2 May 1998		Iridium 69	657	CZ-2C/5D	T	LEO
		Iridium 71	657			LEO
30 May 1998		Zhongwei 1	2,984	CZ-3B	XC	GEO
18 July 1998		SINOSAT 1	2,840	CZ-3B	XC	GEO
19 Aug 1998		Iridium 3	657	CZ-2C/5D	T	LEO
		Iridium 76	657			LEO
19 Dec 1998		Iridium 11A	657	CZ-2C/SD	T	LEO
		Iridium 20A	657			
10 May 1999		FY-1 3	958	CZ-4B	T	LEO
		Shi Jian 5	298			
11 Jun 1999		Iridium 14A	667	CZ-2C/SD	T	LEO
		Iridium 21 A	667			
14 Oct 1999		Zi Yuan 1	1,450	CZ-4B1	T	LEO
		SACI-1	60			
19 Nov 1999		Shen Zhou 1	9,000	CZ-2F	J	LEO
25 Jan 2000		Zhongxing 22	2,300	CZ-3A	XC	GEO
25 Jun 2000		Feng Yun-2 2	1,400	CZ-3	XC	GEO

Launch Date	*Satellite*	*Mass (kg)*	*Vehicle*	*Site*	*Orbit Class*
1 Sep 2000	Zi Yuan 2	1,500	CZ-4B	T	LEO
30 Oct 2000	Bei Dou	2,300	CZ-3A	XC	E-LEO
20 Dec 2000	Bei Dou 2	2,300	CZ-3A	XC	E-LEO
9 Jan 2001	Shen Zhou 2	9,600	CZ-2F	J	LEO

Sub-orbital test flights are shown as 'Test flight'. Abbreviations of satellite names are as follows. DFH – Dong Fang Hong (The East is Red). SJ – Shi Jian (Practice). JSSW – Ji Shu Shiyan Weixing (technical experimental satellite). FSW – Fanhui Shi Weixing (recoverable experimental satellite). SW – Shiyan Weixing (experimental satellite). STTW A – Shiyan Tongbu Tongxin Weixing – (experimental stationary communications satellite). STTW 1–5 – Shiyong Tongbu Tongxin Weixing (operational stationary communications satellite). FY – Feng Yun (wind and cloud). QQW – Qi Qui Weixing (atmospheric observation satellite). An asterisk indicates a non-Chinese satellite, with the owning country shown under the satellite's name. Launch sites are J – Jiuquan, XC – Xi Chang and T – T'ai Yuan. Orbits are indicated in general terms: LEO – low Earth orbit, E-LEO – eccentric low Earth orbit, GEO – geosynchronous Earth orbit, GTO – geosynchronous transfer orbit and SSO – Sun-synchronous orbit. 'Failure' indicates that nothing reached orbit: 'GEO fail' indicates a mission intended to reach GEO which failed after the initial orbital injection. For the recoverable FSW satellites the lifetime in days is shown in parentheses following 'LEO'.

Table 21: Chinese Launch Failures

List of Chinese Launch Failures

Launch Date	*Satellite*	*Comments*	
18 September 1973	Ji Shu Shiyan Weixing?	Rumoured launch failure	(FB-1)
14 July 1974	Ji Shu Shiyan Weixing?	Second stage engine lost thrust	(FB-1)
5 November 1974	Fanhui Shi Weixing -0	First stage control system failure	(CZ-2A)
10 November 1976	Ji Shu Shiyan Weixing?	Rumoured launch failure	(FB-1)
28 July 1979	Shi Jian	Second stage thrusters failure	(FB-1)
29 January 1984	Shiyan Weixing	Third stage failed to re-ignite after parking orbit injection	(CZ-3)
16 July 1990	Badr + AUSSAT model	Chinese solid-propellant PKM failed to place the AUSSAT model in an eccentric Orbit: Badr was deployed as planned	(CZ-2E/PKM)
28 December 1991	Shiyong Tongbu Tongxin Weixing 5	Third stage failed to re-ignite after parking orbit injection	(CZ-3)
21 December 1992	Optus-B 2	Explosion of the satellite shortly after launch	(CZ-2E/STAR-63F)
25 January 1995	APStar 2	Possible explosion of the satellite shortly after launch	(CZ-2E/STAR-63F)
29 May 1995	No details of failure (CZ-1D)		
14 February 1996	INTELSAT 708	Attitude control lost within seconds of launch	(CZ-3B)
18 August 1996	Zhongxing 7	Third stage shutdown during burn to GTO	(CZ-3)

Notes: In April 1994 the first Feng Yun-2 geosynchronous orbit satellite was destroyed during propellant loading: since it had not been attached to its CZ-3 launch vehicle, this cannot be classed as a launch failure.

Table 22: World Orbital Launch Rates

(First digit indicates total known launches; second, the number of failures)

	1983	1984	1985	1986	1987	1988	1989	1990	1991
BRAZIL									
VLS	-	-	-	-	-	-	-	-	-
CHINA									
CZ-2C	1	1	1	1	2	1	-	1	-
CZ-2D	-	-	-	-	-	-	-	-	-
CZ-2E	-	-	-	-	-	-	-	1	-
CZ-3	-	2/1*	-	1	-	2	-	2	1/1
CZ-3A	-	-	-	-	-	-	-	-	-
CZ-3B	-	-	-	-	-	-	-	-	-
CZ-3C	-	-	-	-	-	-	-	-	-
CZ-4A	-	-	-	-	-	1	-	1	-
* CZ-2E attained orbit, but without payload; CZ-3 partial failure									
EUROPE									
Ariane 1	2	2	1	1	-	-	-	-	-
Ariane 2	-	-	-	1/1	1	3	2	-	-
Ariane 3	-	2	3/1	1	1	2	1	-	-
Ariane 40	-	-	-	-	-	-	-	1	1
42P	-	-	-	-	-	-	-	1	-
44P	-	-	-	-	-	-	-	-	2
42L	-	-	-	-	-	-	-	-	-
44LP	-	-	-	-	-	2	2	1	1
44L	-	-	-	-	-	-	2	3/1	4
Ariane 5	-	-	-	-	-	-	-	-	-
Ariane 5E	-	-	-	-	-	-	-	-	-
INDIA									
SLV-3	1	-	-	-	-	-	-	-	-
ASLV	-	-	-	-	1/1	1/1	-	-	-
PSLV	-	-	-	-	-	-	-	-	-
GSLV	-	-	-	-	-	-	-	-	-
ISRAEL									
Shavit	-	-	-	-	-	1	-	1	-
JAPAN									
J-1*	-	-	-	-	-	-	-	-	-
M-3S	-	1	-	-	-	-	-	-	-
M-3SII	-	-	2	-	1	-	1	1	1
M-5	-	-	-	-	-	-	-	-	-
N-2	2	2	-	1	1	-	-	-	-
H-1	-	-	-	1	1	2	1	2	1
H-2	-	-	-	-	-	-	-	-	-
H-2A	-	-	-	-	-	-	-	-	-
* J-1 launch was sub-orbital									
KOREA									
Taepo Dong			1/1						
UNITED STATES									
Atlas E	3	3	2	1	1	2	-	2	2
Atlas H	2	1	-	1	1	-	-	-	-
Atlas 1*	1	1/1	3	1	1/1	-	1	1	1/1
Atlas 2	-	-	-	-	-	-	-	-	1
Atlas 2A	-	-	-	-	-	-	-	-	-
Atlas 2AS	-	-	-	-	-	-	-	-	-
Atlas 2AR	-	-	-	-	-	-	-	-	-
Conestoga	-	-	-	-	-	-	-	-	-
Delta 3910	1	-	-	-	-	-	-	-	-
3914	2	1	1/1	-	-	-	-	-	-
3920	-	1	1	-	2	1	1	-	-
3920/PAM	3	1	-	-	-	-	-	-	-
3924	2	1	-	-	-	-	-	-	-
4925	-	-	-	-	-	-	1	1	-
5920	-	-	-	-	-	-	1	-	-
6920	-	-	-	-	-	-	-	2	-
6925	-	-	-	-	-	-	5	7	1
7326	-	-	-	-	-	-	-	-	-
7424	-	-	-	-	-	-	-	-	-
7920	-	-	-	-	-	-	-	-	-
7925	-	-	-	-	-	-	-	1	4
3	-	-	-	-	-	-	-	-	-
LMLV 1 Athena 1	-	-	-	-	-	-	-	-	-
LMLV 2 Athena 2	-	-	-	-	-	-	-	-	-
Pegasus	-	-	-	-	-	-	-	1	1
Pegasus XL	-	-	-	-	-	-	-	-	-
Scout G-1	1	1	2	1	1	4	-	1	1
Space Shuttle	4	5	9	2/1	-	2	5	6	6
Start 1									
Taurus	-	-	-	-	-	-	-	-	-
Titan 2G	-	-	-	-	-	1	1	-	-
3B	1	1	-	-	-	-	-	-	-
34B	1	1	1	-	1	-	-	-	-
3D	1	-	-	-	-	-	-	-	-
34D	-	5/1	1/1	1/1	2	2	2	-	-
4	-	-	-	-	-	-	1	2	2
Titan 3	-	-	-	-	-	-	-	3/1	-
* Atlas 1 failed to attain operational orbit; payload failed to provide orbital injection after separation from Titan 2G									
RFAS/USSR									
Vostok	4	-	1	-	-	2	-	-	1
Soyuz-class	43/1	44	40	37/1	44/1	45/3	38	32/2	24
Molniya-m	11	11	16	14/1	4	11	6	12/1	5

	1983	*1984*	*1985*	*1986*	*1987*	*1988*	*1989*	*1990*	*1991*
Kosmos 3M	21/1	17	12/1	15	13	7	9	10	12/1
Tsyklon 2/m	2	4	5	5	6	3	3	4	1
Proton-K*	11	13	9	7	11/2	13/2	10	10/1	8
Proton-K	1	-	1	2/1	2	-	1	1	1
Tsyklon	5	7/1	12	12/1	11	10	8/1	8	8
Zenit-2	-	-	4/3	2	5	2	-	2/1	1/1
Zenit-3	-	-	-	-	-	-	-	-	-
Energia	-	-	-	-	1	1	-	-	-
Shtil									
Start-1	-	-	-	-	-	-	-	-	-
Start	-	-	-	-	-	-	-	-	-
Rockot	-	-	-	-	-	-	-	-	-

Note: some failures might not be included in the Chinese and RFAS/USSR analyses. * Four stage variant

	1992	*1993*	*1994*	*1995*	*1996*	*1997*	*1998*	*1999*	*2000*	*2001*
BRAZIL										
VLS	-	-	-	-	-	1/1		1/1	-	-
CHINA										
CZ-2C	1	1	-	-	-	2	4	1	-	-
CZ-2D	1	-	1	-	1				-	-
CZ-2E	2*	-	1	3/1	-			-	-	
CZ-2F	-	-	-	-	-	-	-	1	-	1
CZ-3	-	-	1	-	2/1	1				
CZ-3A	-	-	2	-	-	1			4	-
CZ-3B	-	-	-	-	1/1	2	2			
CZ-3C	-	-	-	-	-					
CZ-4A	-	-	-	-	-					
CZ-4B	-	-	-	-	-	-	-	2	1	
* CZ-2E attained orbit, but without payload;										
CZ-3 partial failure										
EUROPE										
Ariane 1	-	-	-	-	-					
Ariane 2	-	-	-	-	-					
Ariane 3	-	-	-	-	-					
Ariane 40	-	1	-	2	-		1	1		
42P	3	1	2/1	1	2		1	3		
42L	-	1	1	3	-	2	2		3	
44LP	1	2	3/1	2	1	2	2	2	3	1
44L	3	2	2	1	5	3	2	2	1	3
44P	-	-	-	2	2	4	2	2	1	2
Ariane 5	-	-	-	-	1/1	1	1	1	4	2
Ariane 5E	-	-	-	-	-					
INDIA										
SLV-3	-	-	-	-	-					
ASLV	1	-	1	-	-					
PSLV	-	1/1	1	-	1	1/1		1	-	1
GSLV	-	-	-	-	-					1
ISRAEL										
Shavit	-	-	-	1	-		1/1			
JAPAN										
J-1*	-	-	-	-	1					
M-3S	-	-	-	-	-					
M-3SII	-	1	-	1/1**	-					
M-5	-	-	-	-	-	1	1		1	
N-2	-	-	-	-	-					
H-1	1	-	-	-	-					
H-2	-	-	2	1	1	1				1
H-2A	-	-	-	-	-	1				
* J-1 launch was sub-orbital										
** payload left in untenable low orbit										
KOREA										
Taepo Dong										
UNITED STATES										
Atlas E	-	1	2	1	-					
Atlas H	-	-	-	-	-					
Atlas 1	2/1	2/1*	1	1	1					
Atlas 2	2	2	-	3	1		1			
Atlas 2A	1	-	2	3	4	2	3	1	4	1
Atlas 2AS	-	1	1	4	1	5	2	4	3	3
Atlas 2AR	-	-	-	-	-					
Atlas 3A	-	-	-	-	-	-	-		1	
Conestoga	-	-	-	1/1	-					
Delta 3910	-	-	-	-	-					
3914	-	-	-	-	-					
3920	-	-	-	-	-					
3920/PAM	-	-	-	-	-					
3924	-	-	-	-	-					
4925	-	-	-	-	-					
5920	-	-	-	-	-					
6920	1	-	-	-	-					
6925	1	-	-	-	-					
7326	-	-	-	-	-		1			
7420	-	-	-	-	-	-	-	4	-	-
7424	-	-	-	-	-		2			
7425								2		
7920	-	-	-	2	-	7	5	2		
7925	9	7	3	1	3	3/1	3	1	7	7
3	-	-	-	-	-			1/1	1	
LMLV 1 Athena 1	-	-	-	1/1	-	1	1			1

	1992	1993	1994	1995	1996	1997	1998	1999	2000	2001
LMLV 2 Athena 2	-	-	-	-	-			1	2	
Minotaur	-	-	-	-	-	-	-	2		
Pegasus	-	2	2	1	1	5	6	1		
Pegasus XL	-	-	1/1	1/1	3			3	1	
Scout G-1	2	1	1	-	-					
Space Shuttle	8	7	7	7	7	8	5	3	5	6
Start 1						2	1	1		
Taurus	-	-	1	-	-		2	1	1	1
Titan 2G	1	1/1*	1	-	-	1	1	2	1	
3B	-	-	-	-	-					
34B	-	-	-	-	-					
3D	-	-	-	-	-					
34D	-	-	-	-	-					
4	1	1/1	4	4	4	4	2	3/2		
4B	-	-	-	-	-	-	-		2	3
Titan 3	1	-	-	-	-					
* Atlas 1 failed to attain operational orbit; payload failed to provide orbital injection after separation from Titan 2G										
RFAS/USSR										
Vostok	-	-	-	-	-					
Soyuz-class	24	17	15	12	4/2	10	8		12	9
Molniya-m	8	8	3	4	3	3	3	2		2
Kosmos 3M	7	6	5	5/1	4	2	2	2	3/1	1
Tsyklon 2/m	-	4	1	2	1	1		1	1/1	3
Proton-K*	8	6/1	13	6	9/2**	9	7	9/2	12	6
Proton-K	-	-	-	1	1					1
Tsyklon	5	4	7/1	1	1	1	1			
Zenit-2	3/1	2	4	1	1	1/1	3	1	2	1
Zenit-3	-	-	-	-	-			2	3	2
Energia	-	-	-	-	-					
Shtil							1			
Start-1	-	1	-	-	-				1	1
Start	-	-	-	1/1	-					
Rockot	-	-	1	-	-				1	
Dneipr	-	-	-	-	-	-	-	1	1	

Note: some failures might not be included in the Chinese and RFAS/USSR analyses. * Four stage variant ** In-orbit failures

Table 23: World Launch Log 1995–1999 Inclusive

See *jsd.janes.com* for World Launch Log 1957–1994

Launch	*Launch Date*	*COSPAR*	*PL Name*	*Orig PL Name*	*SATCAT*	*LV Type*	*LV S/N*	*Launch Site*	*Ref*
1995-001	1995 Jan 10 0618	1995-001A	Intelsat 704	Intelsat 704	S23461 Intelsat	Atlas IIAS	AC-113	CC LC36B	Wire
1995-U01	1995 Jan 15 1345	1995-U01	EXPRESS	EXPRESS	A04472 DARA	Mu-3S-II	M-3S2-8	KASC M1	JCM
		1995-U01	EXPRESS RV	EXPRESS RV	A04473 DARA				
1995-002	1995 Jan 24 0354:22	1995-002A	Tsikada	Tsikada	S23463 MO RF	Kosmos 11K65M	–	GIK-1 LC132/1	NK9501-24
		1995-002B	Astrid	Astrid	S23464 SSC				
		1995-002C	FAISAT	FAISAT	S23465 FAI				
1995-F01	1995 Jan 25 1926	1995-F01	Apstar 2	Apstar 2	F01134 APT	Chang Zheng 2E	CZ2E-5	XSC LC2	JCM
1995-003	1995 Jan 29 0125	1995-003A	UHF F/O F4	UFO F4 EHF	S23467 HCI	Atlas II	AC-112	CC LC36A	Wire
1995-004	1995 Feb 3 0522:04	1995-004A	Discovery	OV-103	S23469 NASA JSC	Space Shuttle	STS-63	KSC LC39B/ MLP2	STSJSC
		1995-004	Spacehab SH-03	Spacehab	A04477 Spacehab				
		1995-004B	Spartan-204	Spartan-204	S23470 NRL				
		1995-004	CGP/ODERACS	HH-M	A04479 NASA GSFC				
		1995-004C	ODERACS 2A	ODERACS	S23471 NASA JSC				
		1995-004D	ODERACS 2B	ODERACS	S23472 NASA JSC				
		1995-004E	ODERACS 2C	ODERACS	S23473 NASA JSC				
		1995-004F	ODERACS 2D	ODERACS	S23474 NASA JSC				
		1995-004G	ODERACS 2E	ODERACS	S23475 NASA JSC				
		1995-004	ODERACS 2F	ODERACS	A04474 NASA JSC				
1995-005	1995 Feb 15 1648:28	1995-005A	Progress M-26	Progress 7K-TGM No. 226	S23477 MOM	Soyuz 11A511U	Ya15000-641	GIK-5 LC1	NK9807-46
1995-006	1995 Feb 16 1739:59	1995-006A	Foton	Foton No. 10	S23497 MOM	Soyuz 11A511U	–	GIK-1 LC43/4?	NK9721-34
1995-007	1995 Mar 2 0638:13	1995-007A	Endeavour	OV-105	S23500 NASA JSC	Space Shuttle	STS-67	KSC LC39A/ MLP1	STSJSC
		1995-007	ASTRO-2 Fwd	ASTRO-2 PLT + Igloo	A04483 NASA MSFC				
		1995-007	ASTRO-2 Aft	ASTRO-2 PLT	A04484 NASA MSFC				
		1995-007	EDO	EDO	A04486 NASA JSC				
1995-008	1995 Mar 2 1300:00	1995-008A	Kosmos-2306	Romb	S23501 MO RF	Kosmos 11K65M	–	GIK-1 LC132/1	NK9505-47
		1995-008G	Kosmos-2306 SS 1	ESO	S24781 MO SSSR				
		1995-008H	Kosmos-2306 SS 2	ESO	S24782 MO SSSR				
		1995-008J	Kosmos-2306 SS 3	ESO	S24783 MO SSSR				
		1995-008K	Kosmos-2306 SS 4	ESO	S24784 MO SSSR				
		1995-008L	Kosmos-2306 SS 5	ESO	S24785 MO SSSR				
		1995-008M	Kosmos-2306 SS 6	ESO	S24789 MO SSSR				
		1995-008N	Kosmos-2306 SS 7	ESO	S26584 MO SSSR				
		1995-008P	Kosmos-2306 SS 8	ESO	S26585 MO SSSR				
		1995-008Q	Kosmos-2306 SS 9	ESO	S26586 MO SSSR				
		1995-008R	Kosmos-2306 SS 10	ESO	S26587 MO SSSR				
		1995-008S	Kosmos-2306 SS 11	ESO	S26588 MO SSSR				
		1995-008T	Kosmos-2306 SS 12	ESO	S26589 MO SSSR				
		1995-008U	Kosmos-2306 SS 13	ESO	S26592 MO SSSR				
		1995-008V	Kosmos-2306 SS 14	ESO	S26593 MO SSSR				
		1995-008W	Kosmos-2306 SS 15	ESO	S26594 MO SSSR				
		1995-008X	Kosmos-2306 SS 16	ESO	S26595 MO SSSR				
		1995-008Y	Kosmos-2306 SS 17	ESO	S26596 MO SSSR				
		1995-008Z	Kosmos-2306 SS 18	ESO	S26597 MO SSSR				
		1995-008AA	Kosmos-2306 SS 19	ESO	S26598 MO SSSR				
		1995-008	Kosmos-2306 SS 20	ESO	A04500 MO SSSR				
		1995-008	Kosmos-2306 SS 21	ESO	A04501 MO SSSR				
		1995-008	Kosmos-2306 SS 22	ESO	A04502 MO SSSR				
		1995-008	Kosmos-2306 SS 23	ESO	A04503 MO SSSR				
		1995-008	Kosmos-2306 SS 24	ESO	A04504 MO SSSR				
1995-009	1995 Mar 7 0923:44	1995-009A	Kosmos-2307	Uragan No. 65L	S23511 MOM	Proton-K/ DM-2	370-02	GIK-5 LC200/39	NK9505-47
		1995-009B	Kosmos-2308	Uragan No. 66L	S23512 MOM				
		1995-009C	Kosmos-2309	Uragan No. 67L	S23513 MOM				
1995-010	1995 Mar 14 0611:34	1995-010A	Soyuz TM-21	Soyuz 7K-STM No. 70	S23519 MOM	Soyuz 11 A511U2	–	GIK-5 LC1	AVMN
1995-011	1995 Mar 18 0801	1995-011A	SFU	SFU	S23521 NASDA	H-II	H-II-3F	TNSCY	NK9506-46
		1995-011B	Himawari-5	GMS-5	S23522 NASDA				
1995-012	1995 Mar 22 0409:03	1995-012A	Kosmos-2310	Parus	S23526 MO RF	Kosmos 11K65M	–	GIK-1 LC132/1	NK9910-34
1995-013	1995 Mar 22 0618	1995-013A	Intelsat 705	Intelsat 705	S23528 Intelsat	Atlas IIAS	AC-115	CC LC36B	Wire
1995-014	1995 Mar 22 1644:59	1995-014A	Kosmos-2311	Kobal't	S23530 MOM	Soyuz 11A511U	–	GIK-1 LC43/3	AVMN
		1995-014	Spuskaemiy Kapsula	SpK	A04510 MOM				
		1995-014	Spuskaemiy Kapsula	SpK	A04514 MOM				
1995-015	1995 Mar 24 1405	1995-015A	DMSP 24547	DMSP S-13	S23533 USAF	Atlas E	45E	V SLC3W	Wire
1995-F02	1995 Mar 28 1000	1995-F02	TECHSAT-1	Gurwin-1	F01138 AMSAT-IL	Start	–	GNIIP LC158	NK9507-24
		1995-F02	ENB	ENB UNAMSAT-1	F01139 UNAM/SAI				
		1995-F02	EKA-2	EKA-2	F01140 MO RF				
1995-016	1995 Mar 28 2314:19	1995-016B	Hot Bird 1	Eutelsat HB1	S23537 Eutelsat	Ariane 44LP	V71	CSG ELA2	CNES32
		1995-016A	Brasilsat B2	Brasilsat B2	S23536 Telebras				

Launch	*Launch Date*	*COSPAR*	*PL Name*	*Orig PL Name*	*SATCAT*	*LV Type*	*LV S/N*	*Launch Site*	*Ref*
1995-017	1995 Apr 3 1348	1995-017A	Orbcomm F1	Orbcomm FM1	S23545 Orbcomm	Pegasus H	F8	V RW30/12 PAWA	Wire
		1995-017B	Orbcomm F2	Orbcomm FM2	S23546 Orbcomm				
		1995-017C	OrbView-1	Microlab 1	S23547 Orbimage				
1995-018	1995 Apr 5 1116	1995-018A	'Ofeq-3	'Ofeq-3	S23549 ISA	Shaviyt 1	3	PALB	JCM
1995-019	1995 Apr 7 2347	1995-019A	AMSC 1	AMSC 1	S23553 AMSC	Atlas IIA	AC-114	CC LC36A	Wire
1995-020	1995 Apr 9 1934:12	1995-020A	Progress M-27	Progress 7K-TGM No. 227	S23555 MOM	Soyuz 11A511U	–	GIK-5 LC1	NK9807-46
		1986-017JE	GFZ-1	GFZ-1	S23558 GFZ				
1995-021	1995 Apr 21 0144:02	1995-021A	ERS-2	ERS-2	S23560 ESA	Ariane 40	V72	CSG ELA2	CNES32
1995-022	1995 May 14 1345:00	1995-022A	USA 110	Adv ORION 1	S23567 NRO/CIA	Titan 401A/Centaur	K-23 (45E-8)	CC LC40	Wire
1995-023	1995 May 17 0634:00	1995-023A	Intelsat 706	Intelsat 706	S23571 Intelsat	Ariane 44LP	V73	CSG ELA2	CNES32
1995-024	1995 May 20 0333:22	1995-024A	Spektr	TsM-O No. 17301	S23579 MOM	Proton-K	378-02	GIK-5 LC81/23	NK9510-30
1995-025	1995 May 23 0552:02	1995-025A	GOES 9	GOES J	S23581 NOAA	Atlas I	AC-77	CC LC36B	Wire
1995-026	1995 May 24 2010:09	1995-026A	Kosmos-2312	Oko	S23584 MOM	Molniya 8K78M	–	GIK-1 LC16/2	NK9715-48
1995-027	1995 May 31 1527:01	1995-027A	UFO 5	UHF F/O F5-EHF	S23589 USN	Atlas II	AC-116	CC LC36A	Wire
1995-028	1995 Jun 8 0443:00	1995-028A	Kosmos-2313	US-P	S23596 MO RF	Tsiklon-2	–	GIK-5 LC90/20	AVMN
1995-029	1995 Jun 10 0024:00	1995-029A	DBS 3	DBS 3	S23598 HCI	Ariane 42P	V74	CSG ELA2	CNES32
1995-F03	1995 Jun 22 1958	1995-F03	STEP M3	STEP 3	F01141 USAF STP	Pegasus XL	F9	V RW30/12 PAWA	Wire
1995-030	1995 Jun 27 1932:18	1995-030A	Atlantis	OV-104	S23600 NASA JSC	Space Shuttle	STS-71	KSC LC39A/ MLP3	STSJSC
		1995-030	External Airlock/ODS	EAL/ODS	A04519 NASA JSC				
		1995-030	Spacelab-Mir LM	Spacelab Long Module	A04521 NASA JSC				
1995-031	1995 Jun 28 1825:00	1995-031A	Kosmos-2314	Kobal't	S23601 MOM	Soyuz 11A511U	–	GIK-1 LC43/3	AVMN
		1995-031	Spuskaemiy Kapsula	SpK	A04523 MOM				
		1995-031	Spuskaemiy Kapsula	SpK	A04530 MOM				
1995-032	1995 Jul 5 0309:03	1995-032A	Kosmos-2315	Nadezhda-M	S23603 MO RF	Kosmos 11K65M	–	GIK-1 LC132/1	AVMN
1995-033	1995 Jul 7 1623:34	1995-033A	Helios 1A	Helios 1A	S23605 CNES/DGA	Ariane 40	V75	CSG ELA2	CNES32
		1995-033B	CERISE	CERISE	S23606 CNES/DGA				
		1995-033C	UPM/SAT 1	UPM/LBSAT	S23607 UPM				
1995-034	1995 Jul 10 1238:00	1995-034A	USA 112	TRUMPET 2	S23609 USAF/NSA	Titan 401A/ Centaur	K-19 (45E-5)	CC LC41	Rigg
1995-035	1995 Jul 13 1341:55	1995-035A	Discovery	OV-103	S23612 NASA JSC	Space Shuttle	STS-70	KSC LC39B/ MLP2	STSJSC
		1995-035B	TDRS 7	TDRS G	S23613 NASA GSFC				
1995-036	1995 Jul 20 0304:41	1995-036A	Progress M-28	Progress 7K-TGM No. 228	S23617 MOM	Soyuz 11A511U	–	GIK-5 LC1	NK9807-46
1995-037	1995 Jul 24 1552:10	1995-037A	Kosmos-2316	Uragan No. 80L	S23620 MOM	Proton-K/DM-2	374-01	GIK-5 LC200/39	NK9515-28
		1995-037B	Kosmos-2317	Uragan No. 81L	S23621 MOM				
		1995-037C	Kosmos-2318	Uragan No. 85L	S23622 MOM				
1995-038	1995 Jul 31 2330:00	1995-038A	USA 113	DSCS III B-7	S23628 USAF	Atlas IIA	AC-118	CC LC36A	Rigg
1995-039	1995 Aug 2 2359:11	1995-039A	Interbol-1	SO-M2 No. 511	S23632 MOM	Molniya 8K78M	N15000-294 10M1	GIK-1 LC43-3	NK9516-38
		1995-039F	Magion-4	Magion-4	S23646 Czech				
1995-040	1995 Aug 3 2358:00	1995-040A	PAS 4	Panamsat K3	S23636 Panamsat	Ariane 42L	V76	CSG ELA2	JCM
1995-041	1995 Aug 5 1110:00	1995-041A	Mugunghwa	Koreasat 1	S23639 Korea Tel	Delta 7925	D228	CC LC17B	Rigg
1995-042	1995 Aug 9 0121:00	1995-042A	Molniya-3	Molniya-3 No. 59	S23642 MOM	Molniya 8K78M	PVB77031-674	GIK-1 LC43-3	NK9815-26
1995-F04	1995 Aug 15 2230	1995-F04	Gemstar DSS-1	Vitasat 1	F01143 CTA/VITA	LLV-1	DLV	V SLC6	JCM
1995-043	1995 Aug 29 0053:01	1995-043A	JCSAT 3	JCSAT 3	S23649 JSAT	Atlas IIAS	AC-117	CC LC36B	Wire
1995-044	1995 Aug 29 0641:00	1995-044A	N-STAR a	N-STAR a	S23651 NTT	Ariane 44P	V77	CSG ELA2	CNES32
1995-045	1995 Aug 30 1933:00	1995-045A	Kosmos-2319	Geizer No. 20L	S23653 MOM	Proton-K/DM-2	369-02	GIK-5 LC200/39	NK9518-35
1995-046	1995 Aug 31 0649:59	1995-046A	Sich-1	Okean-O1 No. 8 (NKhM 10)	S23657 NKAU	Tsiklon-3	801	GIK-1 LC32/2	NK9518-36
		1995-046	Fasat-Alfa	Fasat-Alfa	A04532 FACh				
1995-047	1995 Sep 3 0900:23	1995-047A	Soyuz TM-22	Soyuz 7K-STM No. 71	S23665 MOM	Soyuz 11A511U2	–	GIK-5 LC1	AVMN
1995-048	1995 Sep 7 1509:00	1995-048A	Endeavour	OV-105	S23667 NASA JSC	Space Shuttle	STS-69	KSC LC39A/ MLP1	STSJSC
		1995-048C	Wake Shield Facility	WSF	S23669 SII				
		1995-048B	Spartan 201	Spartan 201	S23668 NASA GSFC				
		1995-048	GBA-8/CAPL	GBA	A04537 NASA GSFC				
		1995-048	IEH-1	IEH-1/HH-M	A04538 NASA GSFC				
1995-049	1995 Sep 24 0006:00	1995-049A	Telstar 402R	Telstar 402R	S23670 AT&T	Ariane 42L	V78	CSG ELA2	CNES32
1995-050	1995 Sep 26 1120:00	1995-050A	Resurs-F2	Resurs-F2 No. 10	S23672 MOM	Soyuz 11A511U	–	GIK-1 LC43-4	AVMN
1995-051	1995 Sep 29 0425:00	1995-051A	Kosmos-2320	Neman	S23674 MOM	Soyuz 11A511U	–	GIK-5 LC31	AVMN
1995-052	1995 Oct 6 0323:10	1995-052A	Kosmos-2321	Parus	S23676 MO RF	Kosmos 11K65M	–	GIK-1 LC132/1	NK9910-34

Launch	*Launch Date*	*COSPAR*	*PL Name*	*Orig PL Name*	*SATCAT*	*LV Type*	*LV S/N*	*Launch Site*	*Ref*
1995-053	1995 Oct 8 1850:40	1995-053A	Progress M-29	Progress 7K-TGM No. 229	S23678 MOM	Soyuz 11A511U	V15000-645	GIK-5 LC1	NK9807-46
1995-054	1995 Oct 11 1626:00	1995-054A	Luch-1	Gelios	S23680 MOM	Proton-K/DM-2	386-01	GIK-5 LC81/23	NK9521-36
1995-055	1995 Oct 19 0038:00	1995-055A	Astra 1E	Astra 1E	S23686 SES	Ariane 42L	V79	CSG ELA2	CNES32
1995-056	1995 Oct 20 1353:00	1995-056A	Columbia	OV-102	S23688 NASA JSC	Space Shuttle	STS-73	KSC LC39B/ MLP3	STSJSC
		1995-056	Spacelab USML-2	Spacelab Long Module	A04548 NASA MSFC				
		1995-056	EDO	EDO	A04549 NASA JSC				
1995-057	1995 Oct 22 0800:02	1995-057A	UFO F6	UHF F/O F6-EHF	S23696 USN	Atlas II	AC-119	CC LC36A	Rigg
1995-F05	1995 Oct 23 2203	1995-F05	Meteor SM	Meteor SM	F01145 CTA	Conestoga 1620	F1	WI LA0	Wire
		1995-F05	Meteor RV	Meteor RV	F01146 EER				
1995-058	1995 Oct 31 2019:00	1995-058A	Kosmos-2322	Tselina-2	S23704 MO RF	Zenit-2	–	GIK-5 LC45L	AVMN
1995-059	1995 Nov 4 1422:00	1995-059A	Radarsat	Radarsat	S23710 CSA	Delta 7920-10	D229	V SLC2W	Rigg
		1995-059B	SURFSAT	SURFSAT	S23711 NASA/JPL				
1995-060	1995 Nov 6 0515:00	1995-060A	Milstar DFS 2	Milstar DFS 2	S23712 USAF	Titan 401A/ Centaur	K-21 (45E-7)	CC LC40	Rigg
1995-061	1995 Nov 12 1230:43	1995-061A	Atlantis	OV-104	S23714 NASA JSC	Space Shuttle	STS-74	KSC LC39A/ MLP2	STSJSC
		1995-061	External Airlock/ ODS	EAL/ODS	A04555 NASA JSC				
		1995-061	Stikovochniy Otsek	DM 316GK No. 1	A04553 RKA				
1995-062	1995 Nov 17 0120:00	1995-062A	ISO	ISO	S23715 ESA	Ariane 44P	V80	CSG ELA2	CNES32
1995-063	1995 Nov 17 1425:00	1995-063A	Gals 2	Gals No. 12L	S23717 MOM	Proton-K/DM-2	384-01	GIK-5 LC200/39	NK9523-59
1995-064	1995 Nov 28 1130	1995-064A	Asiasat 2	Asiasat 2	S23723 Asiasat	Chang Zheng 2E	CZ2E-6	XSC LC2	JCM
1995-065	1995 Dec 2 0808:01	1995-065A	SOHO	SOHO	S23726 ESA	Atlas IIAS	AC-121	CC LC36B	Wire
1995-066	1995 Dec 5 2118:00	1995-066A	USA 116	Improved Crystal	S23728 NRO/CIA	Titan 404A	K-15 (45J-3)	V SLC4E	Rigg
1995-067	1995 Dec 6 2323:16	1995-067B	Insat 2C	Insat 2C	S23731 ISRO	Ariane 44L	V81	CSG ELA2	CNES32
		1995-067A	Telecom 2C	Telecom 2C	S23730 France Tel				
1995-068	1995 Dec 14 0610:31	1995-068A	Kosmos-2323	Uragan No. 82L	S23734 MOM	Proton-K/DM-2	378-01	GIK-5 LC200/39	NK9525-30
		1995-068B	Kosmos-2324	Uragan No. 78L	S23735 MOM				
		1995-068C	Kosmos-2325	Uragan No. 76L	S23736 MOM				
1995-069	1995 Dec 15 0023	1995-069A	Galaxy 3R	Galaxy 3R	S23741 HCG	Atlas IIA	AC-120	CC LC36A	Wire
1995-070	1995 Dec 18 1431:35	1995-070A	Progress M-30	Progress 7K-TGM No. 230	S23744 MOM	Soyuz 11A511U	647	GIK-5 LC1	NK9807-46
1995-071	1995 Dec 20 0052:14	1995-071A	Kosmos-2326	US-P	S23748 MO RF	Tsiklon-2	–	GIK-5 LC90/20	AVMN
1995-072	1995 Dec 28 0645:18	1995-072A	IRS-1C	IRS-1C	S23751 ISRO	Molniya 8K78M	–	GIK-5 LC31	AVMN
		1995-072B	Skipper	Skipper	S23752 BMDO				
1995-073	1995 Dec 28 1150	1995-073A	Echostar 1	Echostar 1	S23754 TCI	Chang Zheng 2E	CZ2E-7	XSC LC2	WSL
1995-074	1995 Dec 30 1348:00	1995-074A	Rossi X-ray Timing Explorer	XTE	S23757 NASA GSFC	Delta 7920-10	D230	CC LC17A	Wire
1996-001	1996 Jan 11 0941:00	1996-001A	Endeavour	OV-105	S23762 NASA JSC	Space Shuttle	STS-72	KSC LC39B/MLP1	STSJSC
		1996-001B	OAST-Flyer	Spartan 206	S23763 NASA GSFC				
		1996-001	SLA-1/GAS	SLA-1/GAS	A04565 NASA GSFC				
1996-002	1996 Jan 12 2310:00	1996-002B	Measat 1	HS376 Measat 1	S23765 Binariang	Ariane 44L	V82	CSG ELA2	CNES32
		1996-002A	PAS 3R	Panamsat K4	S23764 Panamsat				
1996-003	1996 Jan 14 1110	1996-003A	Koreasat 2	Koreasat 2	S23768 Korea Tel	Delta 7925	D231	CC LC17B	Wire
1996-004	1996 Jan 16 1533:45	1996-004A	Kosmos-2327	Parus	S23773 MO RF	Kosmos 11K65M	–	GIK-1 LC132/1	NK9602-23
1996-005	1996 Jan 25 0956:00	1996-005A	Gorizont	Gorizont No. 43L	S23775 MOM	Proton-K/DM-2	374-02	GIK-5 LC200/39	NK9602-24
1996-006	1996 Feb 1 0115:01	1996-006A	Palapa C1	Palapa C1	S23779 Satelindo	Atlas IIAS	AC-126	CC LC36B	Wire
1996-007	1996 Feb 5 0719:38	1996-007A	N-Star b	N-Star b	S23781 NTT	Ariane 44P	V83	CSG ELA2	CNES32
1996-F01	1996 Feb 14 1901	1996-F01	Intelsat 708	Intelsat 708	F01151 Intelsat	Chang Zheng 3B	CZ3B-1	XSC LC2	JCM
1996-008	1996 Feb 17 2043:27	1996-008A	NEAR	NEAR	S23784 NASA GSFC	Delta 7925-8	D232	CC LC17B	Wire
1996-009	1996 Feb 19 0058:25	1996-009A	Gonets-D1	Gonets-D1 No. 1	S23787 RKA	Tsiklon-3	–	GIK-1 LC32/1	NK9701-40
		1996-009B	Gonets-D1	Gonets-D1 No. 2	S23788 RKA				
		1996-009C	Gonets-D1	Gonets-D1 No. 3	S23789 RKA				
		1996-009D	Kosmos-2328	Strela-3	S23790 MO RF				
		1996-009E	Kosmos-2329	Strela-3	S23791 MO RF				
		1996-009F	Kosmos-2330	Strela-3	S23792 MO RF				
1996-010	1996 Feb 19 0819:00	1996-010A	Raduga	Gran' No. 44L	S23794 MOM	Proton-K/DM-2	383-02	GIK-5 LC200/39	NK9604-45
1996-011	1996 Feb 21 1234:05	1996-011A	Soyuz TM-23	Soyuz 7K-STM No. 72	S23798 MOM	Soyuz 11A511U	651	GIK-5 LC1	KIAM
1996-012	1996 Feb 22 2018:00	1996-012A	Columbia	OV-102	S23801 NASA JSC	Space Shuttle	STS-75	KSC LC39B/ MLP3	STSJSC
		1996-012	TSS-1R Deployer	TSS-1R Deployer	A04569 NASA MSFC				
		1996-012B	TSS-1R	TSS-1R	S23805 ASI				
		1996-012	TSS-1R MPESS	TSS-1R MPESS	A04570 NASA MSFC				
		1996-012	USMP-3 Fwd	USMP-3	A04571 NASA MSFC				
		1996-012	USMP-3 Aft	USMP-3	A04572 NASA MSFC				
		1996-012	EDO	EDO	A04573 NASA JSC				

Launch	*Launch Date*	*COSPAR*	*PL Name*	*Orig PL Name*	*SATCAT*	*LV Type*	*LV S/N*	*Launch Site*	*Ref*
1996-013	1996 Feb 24 1124:00	1996-013A	Polar	Polar	S23802 NASA GSFC	Delta 7925-10	D233	V SLC2W	Rigg
1996-014	1996 Mar 9 0153	1996-014A	REX II	REX II	S23814 USAF STP	Pegasus XL	F10	V RW30/12 PAWA	Wire
1996-015	1996 Mar 14 0711:01	1996-015A	Intelsat 707	Intelsat 707	S23816 Intelsat	Ariane 44LP	V84	CSG ELA2	CNES32
1996-016	1996 Mar 14 1740:00	1996-016A	Kosmos-2331	Kobal't	S23818 MOM	Soyuz 11A511U	–	GIK-1 LC43-4	NK9701-40
		1996-016	Spuskaemiy Kapsula	SpK	A04588 MOM				
		1996-016	Spuskaemiy Kapsula	SpK	A04597 MOM				
1996-017	1996 Mar 21 0453	1996-017A	IRS-P3	IRS-P3	S23827 ISRO	PSLV	PSLV-D3	SHAR PSLV	JCM
1996-018	1996 Mar 22 0813:04	1996-018A	Atlantis	OV-104	S23831 NASA JSC	Space Shuttle	STS-76	KSC LC39B/ MLP2	STSJSC
		1996-018	External Airlock/ODS	EAL/ODS	A04578 NASA JSC				
		1996-018	Spacehab-SM	Spacehab	A04580 NASA JSC				
1996-019	1996 Mar 28 0021:00	1996-019A	Navstar GPS 33	Navstar SVN 33	S23833 USAF	Delta 7925	D234	CC LC17B	Rigg
1996-020	1996 Apr 3 2301:01	1996-020A	Inmarsat III F-1	Inmarsat III F-1	S23839 Inmarsat	Atlas IIA	AC-122	CC LC36A	Wire
1996-021	1996 Apr 8 2309:01	1996-021A	Astra 1F	Astra 1F	S23842 SES	Proton-K/ DM-2M	390-01	GIK-5 LC81/23	NK9608-29
1996-022	1996 Apr 20 2236:00	1996-022A	MSAT-1	MSAT-1	S23846 TMI	Ariane 42P	V85	CSG ELA2	CNES32
1996-023	1996 Apr 23 1148:50	1996-023A	Priroda	TsM-I 77KSI No. 17401	S23848 RKA	Proton-K	385-01	GIK-5 LC81/23	NK9609-13
1996-024	1996 Apr 24 1227:40	1996-024A	MSX	MSX	S23851 BMDO	Delta 7920-10	D235	V SLC2W	Wire
1996-025	1996 Apr 24 1300:01	1996-025A	Kosmos-2332	Yug	S23853 MO RF	Kosmos 11K65M	–	GIK-1 LC132/1	NK9701-40
1996-026	1996 Apr 24 2337:00	1996-026A	USA 118	MERCURY 2	S23855 USAF/NSA	Titan 401A/ Centaur	K-16 (45E-4)	CC LC41	Rigg
1996-027	1996 Apr 30 0431:01	1996-027A	BeppoSAX	SAX	S23857 ASI	Atlas I	AC-78	CC LC36B	Wire
1996-028	1996 May 5 0704:18	1996-028A	Progress M-31	Progress 7K-TGM No. 231	S23860 MOM	Soyuz 11A511U	–	GIK-5 LC1	NK9807-46
1996-029	1996 May 12 2132:00	1996-029D	USA 122	NOSS B-4	S23862 NRO/NRL	Titan 403A	K-22 (45F-11)	V SLC4E	Rigg
		1996-029A	USA 119	SSU	S23893 NRO/NRL				
		1996-029B	USA 120	SSU	S23907 NRO/NRL				
		1996-029C	USA 121	SSU	S23908 NRO/NRL				
		1996-029E	USA 123	TIPS Ralph	S23936 NRO/NRL				
		1996-029F	USA 124	TIPS Norton	S23937 NRO/NRL				
		1996-029	TLD	TLD	A05013 NRO/NRL				
1996-F02	1996 May 14 0855:00	1996-F02	–	Kometa No. 18	F01154 MOM	Soyuz 11A511U	PVB78051-368	GIK-5 LC31	NK96-10-32
1996-030	1996 May 16 0156:29	1996-030A	Palapa C2	Palapa C2	S23864 PSN	Ariane 44L	V86	CSG ELA2	CNES32
		1996-030B	AMOS 1	AMOS 1	S23865 IAI/Mabat				
1996-031	1996 May 17 0244	1996-031A	MSTI 3	MSTI 3	S23868 BMDO	Pegasus H	F11	V RW30/12 PAWA	Wire
1996-032	1996 May 19 1030:00	1996-032A	Endeavour	OV-105	S23870 NASA JSC	Space Shuttle	STS-77	KSC LC39B/ MLP1	STSJSC
		1996-032	Spacehab 4	Spacehab 4	A04601 Spacehab				
		1996-032B	Spartan 207	Spartan 207	S23871 JPL				
		1996-032C	IAE	Inflatable Antenna Expt.	S23872 JPL				
		1996-032	TEAMS	TEAMS	A04603 NASA GSFC				
		1996-032	GBA-9	GBA	A04604 NASA GSFC				
		1996-032D	PAMS	PAMS	S23876 NASA GSFC				
1996-033	1996 May 24 0109:59	1996-033A	Galaxy 9	Galaxy 9	S23877 HCI	Delta 7925	D236	CC LC17B	Rigg
1996-034	1996 May 25 0205	1996-034A	Gorizont	Gorizont No. 44L	S23880 MOM	Proton-K/DM-2	379-01 LC200/39	GIK-5	NK9611-38
1996-F03	1996 Jun 4 1234:06	1996-F03	Cluster F3	Cluster F3	F01157 ESA	Ariane 5G	V88 (501)	CSG ELA3	ESAPR
		1996-F03	Cluster F4	Cluster F4	F01158 ESA				
		1996-F03	Cluster F1	Cluster F1	F01160 ESA				
		1996-F03	Cluster F2	Cluster F2	F01161 ESA				
1996-035	1996 Jun 15 0655:09	1996-035A	Intelsat 709	Intelsat 709	S23915 Intelsat	Ariane 44P	V87	CSG ELA2	CNES32
1996-036	1996 Jun 20 1449:00	1996-036A	Columbia	OV-102	S23931 NASA JSC	Space Shuttle	STS-78	KSC LC39B/ MLP3	STSJSC
		1996-036	Spacelab LMS	Spacelab Long Module	A04611 NASA MSFC				
		1996-036	EDO	EDO	A04613 NASA JSC				
1996-F04	1996 Jun 20 1845	1996-F04	[Kosmos]	Kobal't	F01163 MOM	Soyuz 11A511U	–	GIK-1 LC16	NK96-12
		1996-F04	Spuskaemiy Kapsula	SpK	F01166 MOM				
		1996-F04	Spuskaemiy Kapsula	SpK	F01168 MOM				
1996-037	1996 Jul 2 0748	1996-037A	TOMS-EP	TOMS-EP	S23940 NASA-GSFC	Pegasus XL	F12	V RW30/12 PAWA	Wire
1996-038	1996 Jul 3 0031	1996-038A	USA 125	SDS B-4	S23945 NRO/USAF	Titan 405A	K-2 (45H-1)	CC LC40	Wire
1996-039	1996 Jul 3 1047	1996-039A	Apstar 1A	Apstar 1A	S23943 APT	Chang Zheng 3	CZ3-10	XSC LC1	JCM
1996-040	1996 Jul 9 2224:55	1996-040B	Turksat 1C	Turksat 1C	S23949 Turk Telecom	Ariane 44L	V89	CSG ELA2	CNES32
		1996-040A	Arabsat IIA	Arabsat 2A	S23948 Arabsat				
1996-041	1996 Jul 16 0050	1996-041A	Navstar SVN 40	Navstar SVN 40	S23953 USAF	Delta 7925	D237	CC LC17A	Wire
1996-042	1996 Jul 25 1242	1996-042A	UFO F7	UHF F/O F7-EHF	S23967 HCI	Atlas II	AC-125	CC LC36A	Wire

Launch	Launch Date	COSPAR	PL Name	Orig PL Name	SATCAT	LV Type	LV S/N	Launch Site	Ref
1996-043	1996 Jul 31 2000:06	1996-043A	Progress M-32	Progress 7K-TGM No. 232	S24071 MOM	Soyuz 11A511U	–	GIK-5 LC1	NK9807-46
1996-044	1996 Aug 8 2249:00	1996-044A	Italsat F2	Italsat F2	S24208 ASI	Ariane 44L	V90	CSG ELA2	PR/AE960809
		1996-044B	Telecom 2D	Telecom 2D	S24209 France Tel				
1996-045	1996 Aug 14 2220:59	1996-045A	Molniya-1T	Molniya-1T	S24273 MOM	Molniya 8K78M	–	GIK-1 LC43/3	NK9701-40
1996-046	1996 Aug 17 0153	1996-046A	Midori	ADEOS	S24277 NASDA	H-II	H-II-4F	TNSC Y	WWW-NASDA
		1996-046B	JAS-2	JAS-2	S24278 JARL				
1996-047	1996 Aug 17 1318:03	1996-047A	Soyuz TM-24	Soyuz 7K-STM No. 73	S24280 RKA	Soyuz 11A511U	–	GIK-5 LC1	MIR.323
1996-048	1996 Aug 18 1027	1996-048A	Zhongxing 7	Chinasat 7	S24282 Chinasat	Chang Zheng 3	CZ3-11	XSC LC1	Wire
1996-049	1996 Aug 21 0947:26	1996-049A	FAST	FAST	S24285 NASA-GSFC	Pegasus XL	F13	V RW30/12 PAWA	WWW-FAST
1996-050	1996 Aug 29 0522:00	1996-050A	Victor	Microsatelite	S24291 Cordoba	Molniya 8K78M	–	GIK-1 LC43-3	AVMN
		1996-050B	Interbol-2	SO-M2 No. 512	S24292 RKA				
		1996-050C	Magion-5	Magion-5	S24293 Czech				
1996-051	1996 Sep 4 0901:00	1996-051A	Kosmos-2333	Tselina-2	S24297 MO RF	Zenit-2	–	GIK-5 LC45L	NK9701-40
1996-052	1996 Sep 5 1247:39	1996-052A	Kosmos-2334	Parus	S24304 MO RF	Kosmos 11K65M	–	GIK-1 LC132/1	NK9618-23
		1996-052B	UNAMSAT-B	UNAMSAT-B	S24305 UNAM				
1996-053	1996 Sep 6 1737:39	1996-053A	Inmarsat III F-2	Inmarsat III F-2	S24307 INMARSAT	Proton-K/DM-2	375-01	GIK-5 LC81/23	NK9618-26
1996-054	1996 Sep 8 2149	1996-054A	GE 1	GE 1	S24315 GE Americom	Atlas IIA	AC-123	CC LC36B	Wire
1996-055	1996 Sep 11 0000:59	1996-055A	Echostar II	Echostar II	S24313 Echostar	Ariane 42P	V91	CSG ELA2	PC/ Ransom 960912
1996-056	1996 Sep 12 0849	1996-056A	Navstar SVN 30	Navstar SVN 30	S24320 USAF	Delta 7925	D238	CC LC17A	Wire
1996-057	1996 Sep 16 0854:49	1996-057A	Atlantis	OV-104	S24324 NASA JSC	Space Shuttle	STS-79	KSC LC39A/ MLP1	STSJSC
		1996-057	External Airlock/ODS	Orbiter Docking System	A04628 NASA JSC				
		1996-057	Spacehab Double Module	Spacehab FU2/STA	A04630 NASA JSC				
1996-058	1996 Sep 26 1750:53	1996-058A	Ekspress	Ekspress No. 12L	S24435 AO Inform	Proton-K/ DM-2M	379-02	GIK-5 LC200/39	NK9810-25
1996-059	1996 Oct 20 0720	1996-059A	FSW-2 No. 3	FSW-2 No. 3	S24634 CASC	Chang Zheng 2D	CZ2D-3	JQ	Wire
		1996-059	FSW-2 RV	FSW-2 RV	A05014 CASC				
1996-060	1996 Oct 24 1137:00	1996-060A	Molniya-3	Molniya-3 No. 62	S24640 MOM	Molniya 8K78M	PVB71612-697	GIK-1 LC43-4	NK9701-40
1996-061	1996 Nov 4 1708:56	1996-061A	SAC-B	SAC-B	A04634 CONAE	Pegasus XL	F14	WI RW04/22? DZWI	Wire
		1996-061A	HETE	HETE	S24645 NASA-GSFC				
1996-062	1996 Nov 7 1700:49	1996-062A	Mars Global Surveyor	MGS	S24648 NASA/JPL	Delta 7925	D239	CC LC17A	Wire
1996-063	1996 Nov 13 2240:00	1996-063A	Arabsat IIB	Arabsat 2B	S24652 Arabsat	Ariane 44L	V92	CSG ELA2	Wire
		1996-063B	Measat 2	Measat 2	S24653 Binariang				
1996-064	1996 Nov 16 2048:53	1996-064	Mars-8	M1 No. 520 (Mars-96)	A04638 RKA	Proton-K/D-2	392-02	GIK-5 LC200/39	NK9622-25
		1996-064	MAS 1	MAS No. 520/1	A04640 RKA				
		1996-064	MAS 2	MAS No. 520/2	A04641 RKA				
		1996-064	Penetrator 1	PN No. 520/4	A04642 RKA				
		1996-064	Penetrator 2	PN No. 520/5	A04643 RKA				
1996-065	1996 Nov 19 1955:47	1996-065A	Columbia	OV-102	S24660 NASA JSC	Space Shuttle	STS-80	KSC LC39B/ MLP3	MSFCFLASH
		1996-065B	ORFEUS-SPAS	ASTRO-SPAS	S24661 DLR				
		1996-065C	Wake Shield Facility	WSF	S24662 SII				
		1996-065	EDO	EDO	A04646 NASA JSC				
1996-066	1996 Nov 19 2320:38	1996-066A	Progress M-33	Progress 7K-TGM No. 233	S24663 RKA	Soyuz 11A511U	–	GIK-5 LC1	NK9807-46
1996-067	1996 Nov 21 2047	1996-067A	Hot Bird 2	Eutelsat HB2	S24665 Eutelsat	Atlas IIA	AC-124	CC LC36A	Wire
1996-068	1996 Dec 4 0658:07	1996-068	MPF Cruise Stage	MPF Cruise Stage	A04654 NASA/JPL	Delta 7925	D240	CC LC17B	Wire
		1996-068A	Sagan Memorial Station	MPF Lander	S24667 NASA/JPL				
		1996-068	Sojourner	MFEX Rover	A04656 NASA/JPL				
1996-069	1996 Dec 11 1200:00	1996-069A	Kosmos-2335	US-P	S24670 MO RF	Tsiklon-2	–	GIK-5 LC90/19	NK96-25-41
1996-070	1996 Dec 18 0157	1996-070A	Inmarsat III F3	Inmarsat III F3	S24674 Inmarsat	Atlas IIA	AC-129	CC LC36B	Wire
1996-071	1996 Dec 20 0643:58	1996-071A	Kosmos-2336	Parus	S24677 MO RF	Kosmos 11K65M	–	GIK-1 LC132/1	NK9626-30
1996-072	1996 Dec 20 1804	1996-072A	USA 129	Improved CRYSTAL 4	S24680 NRO/USAF	Titan 404A	K-13 (45J-5)	V SLC4E	Wire
1996-073	1996 Dec 24 1350:00	1996-073A	Bion No. 11	Bion No. 11	S24701 RKA	Soyuz 11A511U	PVB15000-050	GIK-1 LC43-4	NK96-26-34
1997-001	1997 Jan 12 0927:23	1997-001A	Atlantis	OV-104	S24711 NASA JSC	Space Shuttle	STS-81	KSC LC39B/ MLP2	MSFCFLASH
		1997-001	External Airlock/ODS	EAL/ODS	A04670 NASA JSC				
		1997-001	Spacehab Double Module	Spacehab DM	A04672 NASA JSC				
1997-F01	1997 Jan 17 1628	1997-F01	GPS SVN 42	GPS SVN 42	F01173 USAF	Delta 7925	D241	CC LC17A	Wire
1997-002	1997 Jan 30 2204:00	1997-002A	GE 2	GE 2	S24713 GE	Ariane 44L Americom	V93	CSG ELA2	Wire
		1997-002B	Nahuel 1A	Nahuel 1A	S24714 Nahuelsat				
1997-003	1997 Feb 10 1409:30	1997-003A	Soyuz TM-25	Soyuz 7K-STM No. 74	S24717 RKA	Soyuz 11A511U	–	GIK-5 LC1	AVMN

Launch	*Launch Date*	*COSPAR*	*PL Name*	*Orig PL Name*	*SATCAT*	*LV Type*	*LV S/N*	*Launch Site*	*Ref*
1997-004	1997 Feb 11 0855:17	1997-004A	Discovery	OV-103	S24719 NASA JSC	Space Shuttle	STS-82	KSC LC39A/ MLP1	MSFCFLASH
		1997-004	External Airlock	EAL	A04674 NASA JSC				
		1997-004	SAC	SAC	A04675 NASA JSC				
		1997-004	ORUC	ORUC	A04676 NASA JSC				
		1997-004	FSS	FSS	A04677 NASA JSC				
1997-005	1997 Feb 12 0450	1997-005A	Haruka	MUSES-B	S24720 ISAS	M-V	M-V-1	KASC M-V	Wire
1997-006	1997 Feb 14 0347:22	1997-006D	Kosmos-2337	Strela-3	S24728 MO RF	Tsiklon-3	–	GIK-1 LC32/1	NK9704-54
		1997-006E	Kosmos-2338	Strela-3	S24729 MO RF				
		1997-006F	Kosmos-2339	Strela-3	S24730 MO RF				
		1997-006A	Gonets-D1	Gonets-D1 No. 4	S24725 RKA				
		1997-006B	Gonets-D1	Gonets-D1 No. 5	S24726 RKA				
		1997-006C	Gonets-D1	Gonets-D1 No. 6	S24727 RKA				
1997-007	1997 Feb 17 0142:02	1997-007A	JCSAT 4	JCSAT 4	S24732 JSAT	Atlas IIAS	AC-127	CC LC36B	Wire
1997-008	1997 Feb 23 2020	1997-008A	DSP F18	DSP 20	S24737 USAF	Titan 402B/IUS	4B-24 (K-24, 45	CC LC40	Wire
1997-009	1997 Mar 1 0107:42	1997-009A	Intelsat 801	Intelsat 801	S24742 Intelsat	Ariane 44P	V94	CSG ELA2	Wire
1997-010	1997 Mar 4 0200:02	1997-010A	Zeya	Zeya	S24744 MO RF	Start-1.2	–	GIK-2 LC5	NK9705-11
1997-011	1997 Mar 8 0601	1997-011A	Tempo 2	Tempo 2	S24748 TCI	Atlas IIA	AC-128	CC LC36A	Wire
1997-012	1997 Apr 4 1647	1997-012A	DMSP 5D-2 F-14	DMSP 5D-2	S24753 USAF S-14	Titan II SLV	23G-6	V SLC4W	Wire
1997-013	1997 Apr 4 1920:32	1997-013A	Columbia	OV-102	S24755 NASA JSC	Space Shuttle	STS-83	KSC LC39A/ MLP3	MSFCFLASH
		1997-013	Spacelab MSL-1	Spacelab Long Module 1	A04684 NASA MSFC				
		1997-013	EDO	EDO	A04685 NASA JSC				
1997-014	1997 Apr 6 1604:05	1997-014A	Progress M-34	Progress 7K-TGM No. 234	S24757 RKA	Soyuz 11A511U	–	GIK-5 LC1	NK9807-46
1997-015	1997 Apr 9 0858:44	1997-015A	Kosmos-2340	Oko	S24761 MO RF	Molniya 8K78M	PVB76032-647	GIK-1 LC16-2	NK9708-30
1997-016	1997 Apr 16 2308:44	1997-016A	Thaicom 3	Thaicom 3	S24768 Shinawatra	Ariane 44LP	V95	CSG ELA2	Wire
		1997-016B	BSAT 1a	BSAT 1a	S24769 BSAT				
1997-017	1997 Apr 17 1303:21	1997-017A	Kosmos-2341	Parus	S24772 MO RF	Kosmos 11K65M	–	GIK-1 LC132/1	NK9708-36
1997-018	1997 Apr 21 1159	1997-018B	Celestis	CPAC	S24780 Celestis	Pegasus XL	F15	GAN RW03/ 21 DZGC	Wire
		1997-018A	Minisat-01	Minisat-01	S24779 INTA				
1997-019	1997 Apr 25 0549	1997-019A	GOES 10	GOES K	S24786 NOAA	Atlas I	AC-79	CC LC36B	Wire
1997-020	1997 May 5 1455:28	1997-020E	Iridium 4	Iridium SV004	S24796 Iridium	Delta 7920-10C	D242	V SLC2W	AWST970512-25
		1997-020D	Iridium 5	Iridium SV005	S24795 Iridium				
		1997-020C	Iridium 6	Iridium SV006	S24794 Iridium				
		1997-020B	Iridium 7	Iridium SV007	S24793 Iridium				
		1997-020A	Iridium 8	Iridium SV008	S24792 Iridium				
1997-021	1997 May 11 1617	1997-021A	Zhongxing 6	DFH-3	S24798 Chinasat	ChangZheng 3A	CZ3A-3	XSC LC2	Stein
1997-022	1997 May 14 0033:57	1997-022A	Kosmos-2342	Oko	S24800 MO RF	Molniya 8K78M	–	GIK-1 LC43-4	NK9710-34
1997-023	1997 May 15 0807:48	1997-023A	Atlantis	OV-104	S24804 NASA JSC	Space Shuttle	STS-84	KSC LC39A/ MLP2	MSFCFLASH
		1997-023	External Airlock/ODS	EAL/ODS	A04696 NASA JSC				
		1997-023	Spacehab Double Module	Spacehab Double Module	A04698 NASA JSC				
1997-024	1997 May 15 1210:00	1997-024A	Kosmos-2343	Don	S24805 MO RF	Soyuz 11A511U	–	GIK-5 LC31	NK9710-35
		1997-024	Spuskaemiy Kapsula	SpK	A04660 MO RF				
		1997-024	Spuskaemiy Kapsula	SpK	A04661 MO RF				
		1997-024	Spuskaemiy Kapsula	SpK	A04662 MO RF				
		1997-024	Spuskaemiy Kapsula	SpK	A04663 MO RF				
		1997-024	Spuskaemiy Kapsula	SpK	A04664 MO RF				
		1997-024	Spuskaemiy Kapsula	SpK	A04665 MO RF				
		1997-024	Spuskaemiy Kapsula	SpK	A04666 MO RF				
1997-F02	1997 May 20 0707:00	1997-F02	–	Tselina-2	F01179 MO RF	Zenit-2	–	GIK-5 LC45L	NK9711-24
1997-025	1997 May 20 2239	1997-025A	Thor II	Thor 2A	S24808 Telenor	Delta 7925	D243	CC LC17A	Wire
1997-026	1997 May 24 1700:00	1997-026A	Telstar 5	Telstar 5	S24812 Loral Skynet	Proton-K/ DM-2M	380-02	GIK-5 LC81/23	NK9711-18
1997-027	1997 Jun 3 2320:06	1997-027B	Insat 2D	Insat 2D	S24820 ISRO	Ariane 44L	V97	CSG ELA2	Wire
		1997-027A	Inmarsat III F4	Inmarsat III F4	S24819 Inmarsat				
1997-028	1997 Jun 6 1656:54	1997-028A	Kosmos-2344	11F664 No. 1	S24827 MO RF	Proton-K/17S40	380-01	GIK-5 LC200/39	NK9712-20
1997-029	1997 Jun 10 1201	1997-029A	Feng Yun 2	FY-2	S24834 CASC	Chang Zheng 3	CZ3-12	XSC LC1	Wire
1997-030	1997 Jun 18 1402:45	1997-030C	Iridium 9	Iridium SV009	S24838 Iridium	Proton-K/17S40	390-02	GIK-5 LC81/23	NK9713-34
		1997-030D	Iridium 10	Iridium SV010	S24839 Iridium				
		1997-030G	Iridium 11	Iridium SV011	S24842 Iridium				
		1997-030B	Iridium 12	Iridium SV012	S24837 Iridium				
		1997-030E	Iridium 13	Iridium SV013	S24840 Iridium				
		1997-030A	Iridium 14	Iridium SV014	S24836 Iridium				
		1997-030F	Iridium 16	Iridium SV016	S24841 Iridium				

Launch	*Launch Date*	*COSPAR*	*PL Name*	*Orig PL Name*	*SATCAT*	*LV Type*	*LV S/N*	*Launch Site*	*Ref*
1997-031	1997 Jun 25 2344:00	1997-031A	Intelsat 802	Intelsat 802	S24846 Intelsat	Ariane 44P	V96	CSG ELA2	Wire
1997-032	1997 Jul 1 1802:00	1997-032A	Columbia	OV-102	S24849 NASA JSC	Space Shuttle	STS-94	KSC LC39A/ MLP1	MSFCFLASH
		1997-032	Spacelab Long Module 1	Spacelab Long Module 1	A04713 NASA MSFC				
		1997-032	EDO	EDO	A04714 NASA JSC				
1997-033	1997 Jul 5 0411:54	1997-033A	Progress M-35	Progress 7K-TGM No. 235	S24851 RKA	Soyuz 11A511U	–	GIK-5 LC1	NK9807-46
1997-034	1997 Jul 9 1304	1997-034A	Iridium 15	Iridium SV015	S24869 Iridium	Delta 7920-10C	D244	V SLC2W	Wire
		1997-034B	Iridium 17	Iridium SV017	S24870 Iridium				
		1997-034D	Iridium 18	Iridium SV018	S24872 Iridium				
		1997-034C	Iridium 20	Iridium SV020	S24871 Iridium				
		1997-034E	Iridium 21	Iridium SV021	S24873 Iridium				
1997-035	1997 Jul 23 0343	1997-035A	GPS SVN 43	GPS SVN 43	S24876 USAF	Delta 7925	D245	CC LC17A	Wire
1997-036	1997 Jul 28 0115	1997-036A	Superbird C	Superbird C	S24880 SCC	Atlas IIAS	AC-133	CC LC36B	Wire
1997-037	1997 Aug 1 2020	1997-037A	OrbView-2	Seastar	S24883 Orbimage	Pegasus XL	F16	V RW30/12 PAWA	Wire
1997-038	1997 Aug 5 1535:53	1997-038A	Soyuz TM-26	Soyuz 7K-STM No. 75	S24886 RKA	Soyuz 11A511U	–	GIK-5 LC1	AVMN
1997-039	1997 Aug 7 1441:00	1997-039A	Discovery	OV-103	S24889 NASA JSC	Space Shuttle	STS-85	KSC LC39A/ MLP3	MSFCFLASH
		1997-039B	CRISTA-SPAS	ASTRO-SPAS	S24890 DLR				
		1997-039	MFD	MFD	A04722 NASDA				
		1997-039	TAS-1	TAS-1	A04723 NASA GSFC				
		1997-039	IEH-2	IEH-2	A04724 NASA GSFC				
1997-040	1997 Aug 8 0646:00	1997-040A	PAS 6	PAS 6	S24891 Panamsat	Ariane 44P	V98	CSG ELA2	Wire
1997-041	1997 Aug 14 2049:14	1997-041A	Kosmos-2345	Prognoz	S24894 MO RF	Proton-K/DM-2	381-01	GIK-5 LC200/39	NK9713-57
1997-042	1997 Aug 19 1750	1997-042A	Agila 2	Mabuhay	S24901 MPSC	Chang Zheng 3B	CZ3B-2	XSC LC2	Wire
1997-043	1997 Aug 21 0038:40	1997-043E	Iridium 22	Iridium SV022	S24907 Iridium	Delta 7920-10C	D246	V SLC2W	Seesat-L
		1997-043D	Iridium 23	Iridium SV023	S24906 Iridium				
		1997-043C	Iridium 24	Iridium SV024	S24905 Iridium				
		1997-043B	Iridium 25	Iridium SV025	S24904 Iridium				
		1997-043A	Iridium 26	Iridium SV026	S24903 Iridium				
1997-044	1997 Aug 23 0651:01	1997-044A	Lewis	SSTI/Lewis	S24909 TRW/ NASA HQ	LMLV-1	LM-002	V SLC6	AWST970901-28
1997-045	1997 Aug 25 1439	1997-045A	ACE	Advanced Composition Expl	S24912 NASA GSFC	Delta 7920-8	D247	CC LC17A	Wire
1997-046	1997 Aug 28 0033:30	1997-046A	PAS 5	PAS 5	S24916 Panamsat	Proton-K/ DM-2M	387-02	GIK-5 LC81/23	NK9721-38
1997-047	1997 Aug 29 1502:22	1997-047A	FORTE	FORTE	S24920 USAF STP	Pegasus XL	019/F17	V RW30/12 PAWA	Wire
1997-048	1997 Sep 1 1400:15	1997-048A	Iridium MFS 1	Iridium MFS 1	S24925 CASC	Chang Zheng 2C-III/SD	CZ2C-15	TYSC LC1	McDonald-B-priv
		1997-048B	Iridium MFS 2	Iridium MFS 2	S24926 CASC				
1997-049	1997 Sep 2 2221:07	1997-049A	Hot Bird 3	Hot Bird 3	S24931 Eutelsat	Ariane 44LP	V99	CSG ELA2	Wire
		1997-049B	Meteosat 7	MTP 1	S24932 Eumetsat				
1997-050	1997 Sep 4 1203	1997-050A	GE 3	GE 3	S24936 GE XAmericom	Atlas IIAS	AC-146	CC LC36A	Wire
1997-051	1997 Sep 14 0136:54	1997-051D	Iridium 27	Iridium SV027	S24947 Iridium	Proton-K/17S40	391-01	GIK-5 LC81/23	NK9718-55
		1997-051E	Iridium 28	Iridium SV028	S24948 Iridium				
		1997-051A	Iridium 29	Iridium SV029	S24944 Iridium				
		1997-051F	Iridium 30	Iridium SV030	S24949 Iridium				
		1997-051G	Iridium 31	Iridium SV031	S24950 Iridium				
		1997-051B	Iridium 32	Iridium SV032	S24945 Iridium				
		1997-051C	Iridium 33	Iridium SV033	S24946 Iridium				
1997-052	1997 Sep 23 1644:51	1997-052A	Kosmos-2346	Parus	S24953 MO RF	Kosmos 11K65M	–	GIK-1 LC132/1	NK9910-34
		1997-052B	FAISAT-2V	FAISAT-2V	S24954 FAI				
1997-053	1997 Sep 23 2358	1997-053A	Intelsat 803 (NSS 803)	Intelsat 803	S24957 Intelsat	Ariane 42L	V100	CSG ELA2	Wire
1997-054	1997 Sep 24 2130:59	1997-054A	Molniya-1T	Molniya-1T	S24960 MO RF	Molniya 8K78M	–	GIK-1 LC43-4	AVMN
1997-055	1997 Sep 26 0234:19	1997-055A	Atlantis	OV-104	S24964 NASA JSC	Space Shuttle	STS-86	KSC LC39A/ MLP2	Wire
		1997-055	External Airlock/ODS	EAL/ODS	A04747 NASA JSC				
		1997-055	Spacehab Double Module	Spacehab Double Module	A04749 NASA JSC				
1997-056	1997 Sep 27 0123:36	1997-056A	Iridium 19	Iridium SV019	S24965 Iridium	Delta 7920-10C	D248	V SLC2W	Wire
		1997-056E	Iridium 34	Iridium SV034	S24969 Iridium				
		1997-056D	Iridium 35	Iridium SV035	S24968 Iridium				
1997-056C		Iridium 36	Iridium SV036	S24967 Iridium					
		1997-056B	Iridium 37	Iridium SV037	S24966 Iridium				
1997-057	1997 Sep 29 0447	1997-057A	IRS-1D	IRS-1D	S24971 ISRO	PSLV	PSLV-C1	SHAR PSLV	Wire
1997-058	1997 Oct 5 1508:57	1997-058A	Progress M-36	Progress 7K-TGM No. 237	S25002 RKA	Soyuz 11A511U	–	GIK-5 LC1	NK9807-46
		1997-058C	Spoutnik-40	RS-17	S24958 RKA				
		1997-058D	X-Mir Inspector	Inspector	S25100 DASA				
1997-059	1997 Oct 5 2101	1997-059A	Echostar 3	Echostar 3	S25004 Echostar	Atlas IIAS	AC-135	CC LC36B	Wire
1997-060	1997 Oct 9 1759:59	1997-060A	Foton	Foton No. 11	S25006 RKA	Soyuz 11A511U	–	GIK-1 LC43/3	NK9721-34
		1997-060	Mirka	Mirka	A04762 DLR				
1997-061	1997 Oct 15 0843	1997-061A	Cassini	Cassini	S25008 NASA/JPL	Titan 401B/ Centaur	4B-33 (K-33, 45)	CC LC40	Wire

Launch	*Launch Date*	*COSPAR*	*PL Name*	*Orig PL Name*	*SATCAT*	*LV Type*	*LV S/N*	*Launch Site*	*Ref*
		1997-061	Huygens	Huygens	A04759 ESA				
1997-062	1997 Oct 16 1913	1997-062A	Apstar 2R	Apstar 2R	S25010 APT	Chang Zheng 3B	CZ3B-3	XSC LC2	Wire
1997-063	1997 Oct 22 1313	1997-063A	STEP M4	STEP M4	S25013 USAF STP	Pegasus XL	F18	WI RW04/ 22? DZWI	Wire
1997-064	1997 Oct 24 0232	1997-064A	USA 133	LACROSSE 3	S25017 NRO/CIA	Titan 403A	4A-18 (K-18, 45	V SLC4E	Wire
1997-065	1997 Oct 25 0046	1997-065A	USA 135	DSCS III B-13	S25019 USAF	Atlas IIA	AC-131	CC LC36A	Wire
		1997-065B	Falcon Gold	Falcon Gold	A04764 USAF				
1997-066	1997 Oct 30 1343	1997-066B	Maqsat-B	Maqsat-B	S25024 ESA	Ariane 5G	V101 (502)	CSG ELA3	Wire
		1997-066A	Maqsat-H	Maqsat-H	A04766 ESA				
		1997-066A	TEAMSAT	TEAMSAT	S25023 ESA				
		1997-066C	YES	Young Engineers Satellite	S25025 ESA				
1997-F03	1997 Nov 2 1225:00	1997-F03	SCD-2A	SCD-2A	F01181 INPE	VLS-1	V01	ALCA VLS	Wire
1997-067	1997 Nov 6 0030:00	1997-067A	GPS SVN 38	GPS SVN 38	S25030 USAF	Delta 7925	D249	CC LC17A	Wire
1997-068	1997 Nov 8 0205	1997-068A	USA 136	TRUMPET 3	S25034 USAF/NSA	Titan 401A/ Centaur	4A-17 (K-20, 45	CC LC41	Wire
1997-069	1997 Nov 9 0134:26	1997-069E	Iridium 38	Iridium SV038	S25043 Iridium	Delta 7920-10C	D250	V SLC2W	Wire
		1997-069D	Iridium 39	Iridium SV039	S25042 Iridium				
		1997-069C	Iridium 40	Iridium SV040	S25041 Iridium				
		1997-069B	Iridium 41	Iridium SV041	S25040 Iridium				
		1997-069A	Iridium 43	Iridium SV043	S25039 Iridium				
1997-070	1997 Nov 12 1700:00	1997-070A	Kupon	K95K	S25045 TsBank	Proton-K/ DM-2M	382-01	GIK-5 LC200/39	NK9723-36
1997-071	1997 Nov 12 2148	1997-071B	Cakrawarta 1	Indostar 1	S25050 Indostar	Ariane 44L	V102	CSG ELA2	Wire
		1997-071A	Sirius 2	Sirius 2	S25049 NSAB				
1997-072	1997 Nov 18 1114:59	1997-072A	Resurs F-1M	Resurs F-1M	S25059 RKA	Soyuz 11A511U	–	GIK-1	Wire
1997-073	1997 Nov 19 1946:00	1997-073A	Columbia	OV-102	S25061 NASA JSC	Space Shuttle	STS-87	KSC LC39B/ MLP1	MSFCFLASH
		1997-073B	Spartan 201	Spartan 201	S25062 NASA GSFC				
		1997-073	USMP-4 Forward	USMP-4	A04774 NASA MSFC				
		1997-073	USMP-4 Aft	USMP-4	A04775 NASA MSFC				
		1997-073	EDO	EDO	A04776 NASA JSC				
		1997-073	AERCam/Sprint	Sprint	A04788 NASA JSC				
1997-074	1997 Nov 27 2127	1997-074A	TRMM	TRMM	S25063 NASA GSFC	H-II	H-II-6F	TNSCY	WWW-NASDA
		1997-074B	Hikoboshi	ETS-7	S25064 NASDA				
		1997-074E	Orihime	ETS-7 Target	S25424 NASDA				
1997-075	1997 Dec 2 2252	1997-075A	JCSAT 5	JCSAT 5	S25067 JSAT	Ariane 44P	V103	CSG ELA2	Wire
		1997-075B	Equator-S	Equator-S	S25068 MPE				
1997-076	1997 Dec 2 2310:37	1997-076A	Astra 1G	Astra 1G	S25071 SES	Proton-K/ DM-2M	382-02	GIK-5 LC81/23	NK9725-35
1997-077	1997 Dec 8 0716	1997-077A	Iridium 42	Iridium SV042	S25077 Iridium	Chang Zheng 2C-III/SD	CZ2C-16	TYSC LC1	Wire
		1997-077B	Iridium 44	Iridium SV044	S25078 Iridium				
1997-078	1997 Dec 8 2352	1997-078A	Galaxy 8i	Galaxy 8-i	S25086 Panamsat	Atlas IIAS	AC-149	CC LC36B	Wire
1997-079	1997 Dec 9 0717:20	1997-079A	Kosmos-2347	US-P	S25088 MO RF	Tsiklon-2	–	GIK-5 LC90/19	Wire
1997-080	1997 Dec 15 1540:00	1997-080A	Kosmos-2348	Kobal't	S25095 MO RF	Soyuz 11A511U	–	GIK-1	TLE
		1997-080	Spuskaemiy Kapsula	SpK	A04798 MO RF				
		1997-080	Spuskaemiy Kapsula	SpK	A04821 MO RF				
1997-081	1997 Dec 20 0845:02	1997-081A	Progress M-37	Progress 7K-TGM No 236	S25102 RKA	Soyuz 11A511U	–	GIK-5 LC1	NK9807-46
1997-082	1997 Dec 20 1316	1997-082A	Iridium 45	Iridium SV045	S25104 Iridium	Delta 7920-10C	D251	V SLC2W	Wire
		1997-082B	Iridium 46	Iridium SV046	S25105 Iridium				
		1997-082C	Iridium 47	Iridium SV047	S25106 Iridium				
		1997-082D	Iridium 48	Iridium SV048	S25107 Iridium				
		1997-082E	Iridium 49	Iridium SV049	S25108 Iridium				
1997-083	1997 Dec 22 0016	1997-083A	Intelsat 804	Intelsat 804	S25110 Intelsat	Ariane 42L	V104	CSG ELA2	Wire
1997-084	1997 Dec 23 1911:42	1997-084A	Orbcomm A1	Orbcomm FM5	S25112 Orbcomm	Pegasus XL/ HAPS	F19	WI RW04/ 22? DZWI	Rosenberg
		1997-084B	Orbcomm A2	Orbcomm FM6	S25113 Orbcomm				
		1997-084C	Orbcomm A3	Orbcomm FM7	S25114 Orbcomm				
		1997-084D	Orbcomm A4	Orbcomm FM8	S25115 Orbcomm				
		1997-084E	Orbcomm A5	Orbcomm FM9	S25116 Orbcomm				
		1997-084F	Orbcomm A6	Orbcomm FM10	S25117 Orbcomm				
		1997-084G	Orbcomm A7	Orbcomm FM11	S25118 Orbcomm				
		1997-084H	Orbcomm A8	Orbcomm FM12	S25119 Orbcomm				
1997-085	1997 Dec 24 1332:52	1997-085A	EarlyBird	EarlyBird	S25123 EarthWatch	Start-1	–	GIK-2 LC5	Lissov
1997-086	1997 Dec 24 2319	1997-086A	HGS-1	Asiasat 3	S25126 Asiasat	Proton-K/ DM-2M	394-01	GIK-5 LC81/23	NK9726-45
1998-001	1998 Jan 7 0228:44	1998-001A	Lunar Prospector	LP	S25131 NASA ARC	Athena-2	LM-004	SPFL SLC46	Wire
1998-002	1998 Jan 10 0032:01	1998-002A	Skynet 4D	Skynet 4D	S25134 UK MOD	Delta 7925-9.5	D252	CC LC17B	Wire
1998-F01	1998 Jan 22 1256	1998-F01	'Ofeq-4	'Ofeq-4	F01183 ISA	Shaviyt 1	4	PALB	Wire
1998-003	1998 Jan 23 0248:15	1998-003A	Endeavour	OV-105	S25143 NASA JSC	Space Shuttle	STS-89	KSC LC39A/ MLP3	MSFCFLASH
		1998-003	External Airlock/ODS	EAL/ODS	A04809 NASA JSC				

Launch	*Launch Date*	*COSPAR*	*PL Name*	*Orig PL Name*	*SATCAT*	*LV Type*	*LV S/N*	*Launch Site*	*Ref*
		1998-003A	Spacehab Double Module	Spacehab Double Module	A04807 NASA JSC				
1998-004	1998 Jan 29 1633:42	1998-004A	Soyuz TM-27	Soyuz 7K-STM No. 76	S25146 RKA	Soyuz 11A511U	–	GIK-5 LC1	NK9902-45
1998-005	1998 Jan 29 1837	1998-005A	CAPRICORN	CAPRICORN	S25148 NRO	Atlas IIA	AC-109	CC SLC36A	Wire
1998-006	1998 Feb 4 2329	1998-006A	Brasilsat B3	Brasilsat B3	S25152 Embratel	Ariane 44LP	V105	CSG ELA2	Wire
		1998-006B	Inmarsat 3 F5	Inmarsat 3 F5	S25153 Inmarsat				
1998-007	1998 Feb 10 1320	1998-007D	Celestis-02	CPAC	S25160 Celestis	Taurus 2210	T2 2210	V 576E	Wire
		1998-007B	Orbcomm G1	Orbcomm FM3	S25158 Orbcomm				
		1998-007C	Orbcomm G2	Orbcomm FM4	S25159 Orbcomm				
		1998-007A	GFO	GFO	S25157 USN				
1998-008	1998 Feb 14 1434	1998-008A	Globalstar FM1	Globalstar FM1	S25162 Globalstar	Delta 7420-10C	D253	CC LC17A	Wire
		1998-008B	Globalstar FM2	Globalstar FM2	S25163 Globalstar				
		1998-008C	Globalstar FM3	Globalstar FM3	S25164 Globalstar				
		1998-008D	Globalstar FM4	Globalstar FM4	S25165 Globalstar				
1998-009	1998 Feb 17 1035:00	1998-009A	Kosmos-2349	Kometa No. 19	S25167 RKA	Soyuz 11A511U	–	GIK-5 LC31	NK9902-45
1998-010	1998 Feb 18 1358:09	1998-010D	Iridium 50	Iridium SV050	S25172 Iridium	Delta 7920-10C	D254	V SLC2W	Wire
		1998-010A	Iridium 52	Iridium SV052	S25169 Iridium				
		1998-010E	Iridium 53	Iridium SV053	S25173 Iridium				
		1998-010C	Iridium 54	Iridium SV054	S25171 Iridium				
		1998-010B	Iridium 56	Iridium SV056	S25170 Iridium				
1998-011	1998 Feb 21 0755	1998-011A	Kakehashi	COMETS	S25175 NASDA	H-II	H-II-5F	TNSCY	WWW-NASDA
1998-012	1998 Feb 26 0707	1998-012A	SNOE	SNOE	S25233 NASA GSFC	Pegasus XL	F20	V RW30 PAWA	Wire
		1998-012B	T1	BATSAT	S25234 Teledesic				
1998-013	1998 Feb 27 2238	1998-013A	Hot Bird 4	Hot Bird 4	S25237 Eutelsat	Ariane 42P	V106	CSG ELA2	Wire
1998-014	1998 Feb 28 0021	1998-014A	Intelsat 806 (NSS 806)	Intelsat 806	S25239 Intelsat	Atlas IIAS	AC-151	CC SLC36B	Wire
1998-015	1998 Mar 14 2245:55	1998-015A	Progress M-38	Progress 7K-TGM No 240	S25256 RKA	Soyuz 11A511U	–	GIK-5 LC1	NK9807-46
1998-016	1998 Mar 16 2132	1998-016A	UHF F/O F8	UHF F/O F8	S25258 USN	Atlas II	AC-132	CC SLC36A	Wire
1998-017	1998 Mar 24 0146	1998-017A	SPOT-4	SPOT-4	S25260 CNES	Ariane 40	V107	CSG ELA2	Wire
1998-018	1998 Mar 25 1701:06	1998-018A	Iridium 51	Iridium SV051	S25262 Iridium	Chang Zheng 2C-III/SD	CZ2C-17	TYSC LC1	CALT-WWW
		1998-018B	Iridium 61	Iridium SV061	S25263 Iridium				
1998-019	1998 Mar 30 0602:46	1998-019A	Iridium 55	Iridium SV055	S25272 Iridium	Delta 7920-10C	D255	V SLC2W	Rigg
		1998-019B	Iridium 57	Iridium SV057	S25273 Iridium				
		1998-019C	Iridium 58	Iridium SV058	S25274 Iridium				
		1998-019D	Iridium 59	Iridium SV059	S25275 Iridium				
		1998-019E	Iridium 60	Iridium SV060	S25276 Iridium				
1998-020	1998 Apr 2 0242:39	1998-020A	TRACE	TRACE	S25280 NASA GSFC	Pegasus XL	F21	V RW30/12 PAWA	Wire
1998-021	1998 Apr 7 0213:03	1998-021A	Iridium 62	Iridium SV062	S25285 Iridium	Proton-K/17S40	391-02	GIK-5 LC81/23	NK9809-11
		1998-021B	Iridium 63	Iridium SV063	S25286 Iridium				
		1998-021C	Iridium 64	Iridium SV064	S25287 Iridium				
		1998-021D	Iridium 65	Iridium SV065	S25288 Iridium				
		1998-021E	Iridium 66	Iridium SV066	S25289 Iridium				
		1998-021F	Iridium 67	Iridium SV067	S25290 Iridium				
		1998-021G	Iridium 68	Iridium SV068	S25291 Iridium				
1998-022	1998 Apr 17 1819:00	1998-022A	Columbia	OV-102	S25297 NASA JSC	Space Shuttle	STS-90	KSC LC39B/MLP2	MSFCFLASH
		1998-022A	Neurolab	Spacelab Long Module 2	A04835 NASA MSFC				
		1998-022	EDO	EDO	A04838 NASA JSC				
1998-023	1998 Apr 24 2238:34	1998-023A	Globalstar FM6	Globalstar FM6	S25306 Globalstar	Delta 7420-10C	D256	CC LC17A	Wire
		1998-023B	Globalstar FM8	Globalstar FM8	S25307 Globalstar				
		1998-023C	Globalstar FM14	Globalstar FM14	S25308 Globalstar				
		1998-023D	Globalstar FM15	Globalstar FM15	S25309 Globalstar				
1998-024	1998 Apr 28 2253	1998-024A	Nilesat 101	Nilesat 101	S25311 Nilesat	Ariane 44P	V108	CSG ELA2	Wire
		1998-024B	BSAT 1b	BSAT 1b	S25312 BSAT				
1998-025	1998 Apr 29 0436:54	1998-025A	Kosmos-2350	Prognoz	S25315 RVSN	Proton-K/DM-2	384-02	GIK-5 LC200/39	NK9810-25
1998-026	1998 May 2 0916:53	1998-026A	Iridium 69	Iridium SV069	S25319 Iridium	Chang Zheng 2C-III/SD	CZ2C-18	TYSC LC1	CALT-WWW
		1998-026B	Iridium 71	Iridium SV071	S25320 Iridium				
1998-027	1998 May 7 0853:22	1998-027A	Kosmos-2351	Oko	S25327 MO RF	Molniya 8K78M	–	GIK-1 LC16-2	NK9811-14
1998-028	1998 May 7 2345:00	1998-028A	Echostar 4	Echostar 4	S25331 Echostar	Proton-K/DM-2M	393-02	GIK-5 LC81/23	NK9811-15
1998-029	1998 May 9 0138:01	1998-029A	USA 139	Adv ORION 2	S25336 NRO/NSA	Titan 401B/Centaur	4B-25 (K-25)	CC LC40	Wire
1998-030	1998 May 13 1552:04	1998-030A	NOAA 15	NOAA K	S25338 NOAA	Titan II SLV	23G-12	V SLC4W	Wire
1998-031	1998 May 14 2212:59	1998-031A	Progress M-39	Progress 7K-TGM No 238	S25340 RKA	Soyuz 11A511U	–	GIK-5 LC1	NK9811-03
1998-032	1998 May 17 2116:56	1998-032A	Iridium 70	Iridium SV070	S25342 Iridium	Delta 7920-10C	D257	V SLC2W	Wire
		1998-032B	Iridium 72	Iridium SV072	S25343 Iridium				
		1998-032C	Iridium 73	Iridium SV073	S25344 Iridium				
		1998-032D	Iridium 74	Iridium SV074	S25345 Iridium				
		1998-032E	Iridium 75	Iridium SV075	S25346 Iridium				
1998-033	1998 May 30 1000:04	1998-033A	Zhongwei 1	Chinastar 1	S25354 China Orient	Chang Zheng 3B	CZ3B-4	XSC LC2	CALT-WWW

Launch	*Launch Date*	*COSPAR*	*PL Name*	*Orig PL Name*	*SATCAT*	*LV Type*	*LV S/N*	*Launch Site*	*Ref*
1998-034	1998 Jun 2 2206:24	1998-034A	Discovery	OV-103	S25356 NASA JSC	Space Shuttle	STS-91	KSC LC39A/ MLP1	MSFCFLASH
		1998-034	External Airlock/ODS	EAL/ODS	A04854 NASA JSC				
		1998-034A	Spacehab	Spacehab FU1	A04851 Spacehab				
		1998-034A	AMS	Alpha Magnetic Spectromet	A04852 CERN				
1998-035	1998 Jun 10 0035	1998-035A	Thor III	Thor 3	S25358 Telenor	Delta 7925-9.5	D258	CC LC17A	Wire
1998-036	1998 Jun 15 2258:05	1998-036A	Kosmos-2352	Strela-3	S25363 MO RF	Tsiklon-3	-	GIK-1 LC32/1	NK9814-08
		1998-036B	Kosmos-2353	Strela-3	S25364 MO RF				
		1998-036C	Kosmos-2354	Strela-3	S25365 MO RF				
		1998-036D	Kosmos-2355	Strela-3	S25366 MO RF				
		1998-036E	Kosmos-2356	Strela-3	S25367 MO RF				
		1998-036F	Kosmos-2357	Strela-3	S25368 MO RF				
1998-037	1998 Jun 18 2248	1998-037A	Intelsat 805	Intelsat 805	S25371 Intelsat	Atlas IIAS	AC-153	CC SLC36A	Wire
1998-038	1998 Jun 24 1829:58	1998-038A	Kosmos-2358	Kobal't	S25373 MO RF	Soyuz 11A511U	–	GIK-1 LC43/3	NK9814-10
		1998-038	Spuskaemiy Kapsula	SpK	A04862 MO RF				
		1998-038	Spuskaemiy Kapsula	SpK	A04867 MO RF				
1998-039	1998 Jun 25 1400	1998-039A	Kosmos-2359	Neman	S25376 MO RF	Soyuz 11A511U	–	GIK-5 LC31	NK9814-10
1998-040	1998 Jul 1 0048:01	1998-040A	Molniya-3	Molniya-3 No. 61	S25379 MOM	Molniya 8K78M	–	GIK-1 LC43/3	NK9815-26
1998-041	1998 Jul 3 1812	1998-041A	Nozomi	Planet B	S25383 ISAS	M-V	M-V-3	KASC M-V	Wire
1998-042	1998 Jul 7 0315:00	1998-042A	Tubsat-N	Tubsat-N	S25389 TUB	Shtil'-1	–	BLA, K-407	NK9815-15
		1998-042B	Tubsat-N1	Tubsat-N1	S25390 TUB				
		1998-042C	Shtil'-1	Shtil'-1 instr. package		S25391 Makeev			
1998-043	1998 Jul 10 0630:00	1998-043A	Resurs-O1	Resurs-O1 No. 4L	S25394 RKA	Zenit-2	–	GIK-5 LC45L	NK9815-19
		1998-043B	Fasat-Bravo	Fasat-Bravo	S25395 FACh				
		1998-043C	TM-SAT	TM-SAT	S25396 FACh				
		1998-043D	Gurwin Techsat 1B	Techsat 1B	S25397 Technion				
		1998-043E	WESTPAC	WESTPAC	S25398 WPLTN				
		1998-043F	SAFIR-2	SAFIR-2	S25399 DLR				
1998-044	1998 Jul 18 0920	1998-044A	Sinosat	Sinosat	S25404 Eurasspace	Chang Zheng 3B	CZ3B-5	XSC LC2	CALT-WWW
1998-045	1998 Jul 28 0915:00	1998-045A	Kosmos-2360	Tselina-2	S25406 MO RF	Zenit-2	–	GIK-5 LC45L/1	NK9815-25
1998-046	1998 Aug 2 1624	1998-046H	Orbcomm B1	Orbcomm FM13	S25420 Orbcomm	Pegasus XL/ HAPS	F22	WI DZWI	Wire
		1998-046G	Orbcomm B2	Orbcomm FM14	S25419 Orbcomm				
		1998-046F	Orbcomm B3	Orbcomm FM15	S25418 Orbcomm				
		1998-046E	Orbcomm B4	Orbcomm FM16	S25417 Orbcomm				
		1998-046A	Orbcomm B5	Orbcomm FM17	S25413 Orbcomm				
		1998-046B	Orbcomm B6	Orbcomm FM18	S25414 Orbcomm				
		1998-046C	Orbcomm B7	Orbcomm FM19	S25415 Orbcomm				
		1998-046D	Orbcomm B8	Orbcomm FM20	S25416 Orbcomm				
1998-F02	1998 Aug 12 1130:01	1998-F02	USA	MERCURY 3	F01185 USAF/NSA	Titan 401A/ Centaur	4A-20 (K-17)	CC LC41	ISIR980831-3
1998-047	1998 Aug 13 0943:11	1998-047A	Soyuz TM-28	Soyuz 7K-STM No. 77	S25429 RKA	Soyuz 11A511U	–	GIK-5 LC1	NK9902-45
1998-048	1998 Aug 19 2301:46	1998-048A	Iridium 3	Iridium SV078	S25431 Iridium	Chang Zheng 2C-III/SD	CZ2C-19	TYSC LC1	CALT-WWW
		1998-048B	Iridium 76	Iridium SV076	S25432 Iridium				
1998-049	1998 Aug 25 2307	1998-049A	ST-1	Singapore-Taiwan 1	S25460 Singapore T	Ariane 44P	V109	CSG ELA2	Wire
1998-F03	1998 Aug 27 0117	1998-F03	Galaxy X	Galaxy 10	F01225 Panamsat	Delta 8930	D259	CC LC17B	Wire
1998-050	1998 Aug 30 0031:00	1998-050A	Astra 2A	Astra 2A	S25462 SES	Proton-K/ DM-2M	383-01	GIK-5 LC81/23	NK9817-28
1998-F04	1998 Aug 31 0307:00	1998-F04	Kwangmyongsong 1	–	F01187 CMIK	Taepodong 1	–	MUSU	NYT980901-1
1998-051	1998 Sep 8 2113	1998-051E	Iridium 77	Iridium SV077	S25471 Iridium	Delta 7920-10C	D260	V SLC2W	Wire
		1998-051D	Iridium 79	Iridium SV079	S25470 Iridium				
		1998-051C	Iridium 80	Iridium SV080	S25469 Iridium				
		1998-051B	Iridium 81	Iridium SV081	S25468 Iridium				
		1998-051A	Iridium 82	Iridium SV082	S25467 Iridium				
1998-F05	1998 Sep 9 2029:00	1998-F05	Globalstar FM5	Globalstar FM5	F01194 Globalstar	Zenit-2 11K77.05	22D (67047801)	GIK-5 LC45L/1	Wire
		1998-F05	Globalstar FM7	Globalstar FM7	F01195 Globalstar				
		1998-F05	Globalstar FM9	Globalstar FM9	F01196 Globalstar				
		1998-F05	Globalstar FM10	Globalstar FM10	F01197 Globalstar				
		1998-F05	Globalstar FM11	Globalstar FM11	F01198 Globalstar				
		1998-F05	Globalstar FM12	Globalstar FM12	F01199 Globalstar				
		1998-F05	Globalstar FM13	Globalstar FM13	F01200 Globalstar				
		1998-F05	Globalstar FM16	Globalstar FM16	F01201 Globalstar				
		1998-F05	Globalstar FM17	Globalstar FM17	F01202 Globalstar				
		1998-F05	Globalstar FM18	Globalstar FM18	F01203 Globalstar				
		1998-F05	Globalstar FM20	Globalstar FM20	F01204 Globalstar				
		1998-F05	Globalstar FM21	Globalstar FM21	F01205 Globalstar				
1998-052	1998 Sep 16 0631	1998-052A	PAS 7	PAS 7	S25473 Panamsat	Ariane 44LP	V110	CSG ELA2	Wire
1998-053	1998 Sep 23 0506	1998-053A	Orbcomm C1	Orbcomm FM21	S25475 Orbcomm	Pegasus XL/ HAPS	F23	WI DZWI	Wire
		1998-053B	Orbcomm C2	Orbcomm FM22	S25476 Orbcomm				
		1998-053C	Orbcomm C3	Orbcomm FM23	S25477 Orbcomm				
		1998-053D	Orbcomm C4	Orbcomm FM24	S25478 Orbcomm				
		1998-053E	Orbcomm C5	Orbcomm FM25	S25479 Orbcomm				

Launch	*Launch Date*	*COSPAR*	*PL Name*	*Orig PL Name*	*SATCAT*	*LV Type*	*LV S/N*	*Launch Site*	*Ref*
		1998-053F	Orbcomm C6	Orbcomm FM26	S25480 Orbcomm				
		1998-053G	Orbcomm C7	Orbcomm FM27	S25481 Orbcomm				
		1998-053H	Orbcomm C8	Orbcomm FM28	S25482 Orbcomm				
1998-054	1998 Sep 28 2341:27	1998-054A	Molniya-1T	Molniya-1T	S25485 MOM	Molniya 8K78M	–	GIK-1 LC43/3	NK9821-10
1998-055	1998 Oct 3 1004:49	1998-055A	STEX	STEX	S25489 NRO	ARPA Taurus	T3 1110	V 576E	Rigg
		1998-055C	ATEX	ATEX	S25615 NRO/NRL				
1998-056	1998 Oct 5 2251	1998-056B	Sirius 3	Sirius 3	S25492 NSAB	Ariane 44L	V111	CSG ELA2	Wire
		1998-056A	Eutelsat W2	Eutelsat W2	S25491 Eutelsat				
1998-057	1998 Oct 9 2250	1998-057A	Hot Bird 5	Eutelsat HB5	S25495 Eutelsat	Atlas IIA	AC-134	CC SLC36B	Wire
1998-058	1998 Oct 20 0719	1998-058A	UHF F/O F9	UHF F/O F9	S25501 USN	Atlas IIA	AC-130	CC SLC36A	Wire
1998-059	1998 Oct 21 1637:21	1998-059A	Maqsat-3	Maqsat-3	S25503 ESA	Ariane 5G	V112 (503)	CSG ELA3	Wire
1998-060	1998 Oct 23 0002	1998-060A	SCD-2	SCD-2	S25504 INPE	Pegasus H	F24/P-33	CC RW30/ 12 MFWA	Wire
1998-061	1998 Oct 24 1208:00	1998-061B	SEDSAT	SEDSAT	S25509 SEDS	Delta 7326-9.5	D261	CC SLC17A	Wire
		1998-061A	Deep Space 1	Deep Space 1	S25508 NASA/JPL				
1998-062	1998 Oct 25 0414:57	1998-062A	Progress M-40	Progress 7K-TGM No 239	S25512 RKA	Soyuz 11A511U	660	GIK-5 LC1	NK9823-01
		1998-062C	Spoutnik-41	RS-18	S25533 ACF/RuAF				
1998-063	1998 Oct 28 2215:00	1998-063B	GE 5	GE 5	S25516 GE	Ariane 44L	V113	CSG ELA2	JCM-P
		1998-063A	Afristar	Afristar	S25515 Worldspace				
1998-064	1998 Oct 29 1919:34	1998-064A	Discovery	OV-103	S25519 NASA JSC	Space Shuttle	STS-95	KSC LC39B/ MLP2	MSFCFLASH
		1998-064A	Spacehab	Spacehab FU1	A04876 Spacehab				
		1998-064C	Spartan 201	Spartan 201	S25521 NASA GSFC				
		1998-064B	PANSAT	PANSAT	S25520 USN PGS				
1998-065	1998 Nov 4 0512:00	1998-065A	PAS 8	PAS 8	S25522 Panamsat	Proton-K/DM-2M	395-02	GIK-5 LC81/23	NK9823-30
1998-066	1998 Nov 6 1337:52	1998-066A	Iridium 2	Iridium SV087	S25527 Iridium	Delta 7920-10C	D262	V SLC2W	Wire
		1998-066E	Iridium 83	Iridium SV083	S25531 Iridium				
		1998-066D	Iridium 84	Iridium SV084	S25530 Iridium				
		1998-066C	Iridium 85	Iridium SV085	S25529 Iridium				
		1998-066B	Iridium 86	Iridium SV086	S25528 Iridium				
1998-067	1998 Nov 20 0640:00	1998-067A	Zarya	77KM No. 17501	S25544 NASA	Proton-K	395-01	GIK-5 LC81/23	NK9901-02
1998-068	1998 Nov 22 2354	1998-068A	BONUM-1	BONUM-1	S25546 Telenor	Delta 7925-9.5	D263	CC SLC17B	Wire
1998-069	1998 Dec 4 0835:34	1998-069A	Endeavour	OV-105	S25549 NASA JSC	Space Shuttle	STS-88	KSC LC39A/ MLP3	MSFCFLASH
		1998-069F	Unity	Node 1	S25575 NASA JSC				
		1998-069F	PMA-1	PMA-1	A04890 NASA JSC				
		1998-069F	PMA-2	PMA-2	A04891 NASA JSC				
		1998-069B	SAC-A	SAC-A	S25550 CONAE				
		1998-069C	Mightysat 1	Mightysat 1	S25551 USAF				
1998-070	1998 Dec 6 0043	1998-070A	Satmex 5	Satmex 5	S25558 Satmex	Ariane 42L	V114	CSG ELA2	Wire
1998-071	1998 Dec 6 0057:54	1998-071A	SWAS	SWAS	S25560 NASA GSFC	Pegasus XL	F25	V RW30/ 12 PAWA	Wire
1998-072	1998 Dec 10 1157:07	1998-072A	Nadezhda	Nadezhda	S25567 MO RF	Kosmos 11K65M	–	GIK-1 LC132/1	SGrahn
		1998-072B	Astrid-2	Astrid 2	S25568 SSC				
1998-073	1998 Dec 11 1845:51	1998-073A	Mars Climate Orbiter	MCO	S25571 JPL	Delta 7425-9.5	D264	CC SLC17A	Wire
1998-074	1998 Dec 19 1139:44	1998-074A	Iridium 11A	Iridium SV088?	S25577 Iridium	Chang Zheng 2C-III/SD	CZ2C-20	TYSC LC1	CALT-WWW
		1998-074B	Iridium 20A	Iridium SV089?	S25578 Iridium				
1998-075	1998 Dec 22 0108	1998-075A	PAS 6B	PAS 6B	S25585 Satmex	Ariane 42L	V115	CSG ELA2	Wire
1998-076	1998 Dec 24 2002:19	1998-076A	Kosmos-2361	Parus	S25590 MO RF	Kosmos 11K65M	–	GIK-1 LC132/1	NK9902-17
1998-077	1998 Dec 30 1835:46	1998-077A	Kosmos-2362	Uragan No. 86L	S25593 MOM	Proton-K/DM-2	385-02	GIK-5 LC200/39	WWW-CFSCIC
		1998-077B	Kosmos-2363	Uragan No. 84L	S25594 MOM				
		1998-077C	Kosmos-2364	Uragan No. 79L	S25595 MOM				
1999-001	1999 Jan 3 2021:10	1999-001A	Mars Polar Lander	MPL	S25605 JPL	Delta 7425-9.5	D265	CC SLC17B	Wire
		1999-001	Deep Space 2 Microprobe 1	Mars Microprobe 1	A04907 JPL				
		1999-001	Deep Space 2 Microprobe 2	Mars Microprobe 2	A04908 JPL				
1999-002	1999 Jan 27 0034:02	1999-002A	ROCSAT-1	ROCSAT-1	S25616 NSPO	Athena-1	LM-006	SPFL SLC46	Wire
1999-003	1999 Feb 7 2104:15	1999-003A	Stardust	Stardust	S25618 JPL	Delta 7426-9.5	D266	CC SLC17A	Wire
		1999-003	Sample Return Capsule	SRC	A04909 JPL				
1999-004	1999 Feb 9 0353:59	1999-004A	Globalstar FM36	Globalstar FM36	S25621 Globalstar	Soyuz 11A511U	S15000-058 ST01	GIK-5 LC1	NK9903-32
		1999-004B	Globalstar FM23	Globalstar FM23	S25622 Globalstar				
		1999-004C	Globalstar FM38	Globalstar FM38	S25623 Globalstar				
		1999-004D	Globalstar FM40	Globalstar FM40	S25624 Globalstar				
1999-005	1999 Feb 15 0512:00	1999-005A	Telstar 6	Telstar 6	S25626 Loral Skynet	Proton-K/ DM-2M	396-01	GIK-5 LC81/23	NK9904-28
1999-006	1999 Feb 16 0145:26	1999-006A	JCSAT 6	JCSAT 6	S25630 JSAT	Atlas IIAS	AC-152	CC SLC36A	Wire
1999-007	1999 Feb 20 0418:01	1999-007A	Soyuz TM-29	Soyuz 7K-STM No. 78	S25632 RKA	Soyuz 11A511U	M15000-662	GIK-5 LC1	NK9904-03
1999-008	1999 Feb 23 1029:55	1999-008A	ARGOS	P91-1	S25634 USAF SMC/TEV	Delta 7920-10	D267	V SLC2W	Wire

Launch	*Launch Date*	*COSPAR*	*PL Name*	*Orig PL Name*	*SATCAT*	*LV Type*	*LV S/N*	*Launch Site*	*Ref*
		1999-008B	Sunsat	Sunsat	S25635 Stellenbosch				
		1999-008C	Orsted	Orsted	S25636 DMI				
1999-009	1999 Feb 26 2244	1999-009B	Skynet 4E	Skynet 4E	S25639 UK MoD	Ariane 44L	V116	CSG ELA2	Wire
		1999-009A	Arabsat 3A	Arabsat 3A	S25638 Arabsat				
1999-010	1999 Feb 28 0400:00	1999-010A	Raduga-1	Globus No. 14	S25642 MO RF	Proton-K/DM-2	387-01	GIK-5 LC81/23	NK9904-38
1999-011	1999 Mar 5 0256	1999-011A	WIRE	WIRE	S25646 NASA GSFC	Pegasus XL	F26/M-22	V RW30/ 12 PAWA	Wire
1999-012	1999 Mar 15 0306:00	1999-012A	Globalstar FM22	Globalstar FM22	S25649 Globalstar	Soyuz 11A511U	#NAME?	GIK-5 LC1	NK9905-21
		1999-012B	Globalstar FM41	Globalstar FM41	S25650 Globalstar				
		1999-012C	Globalstar FM46	Globalstar FM46	S25651 Globalstar				
		1999-012D	Globalstar FM37	Globalstar FM37	S25652 Globalstar				
1999-013	1999 Mar 21 0009:30	1999-013A	Asiasat 3S	Asiasat 3S	S25657 Asiasat	Proton-K/ DM-2M	388-01	GIK-5 LC81/23	NK0003-17
1999-014	1999 Mar 28 0129:59	1999-014A	DemoSat	DemoSat	S25661 SeaLaunch	Zenit-3SL	–	KLA, SL Odyssey	Wire
1999-015	1999 Apr 2 1128:43	1999-015A	Progress M-41	Progress 7K-TGM No 241	S25664 RKA	Soyuz 11A511U	–	GIK-5 LC1	NK9905-01
		1999-015C	Sputnik-99	Sputnik-99	S25685 AMSAT-F				
1999-016	1999 Apr 2 2203	1999-016A	Insat 2E	Insat 2E	S25666 ISRO	Ariane 42P	V117	CSG ELA2	Wire
1999-017	1999 Apr 9 1701:00	1999-017A	DSP F19 (USA 142)	DSP 18?	S25669 USAF	Titan 402B/IUS	4B-27 (K-32)	CC LC41	Wire
1999-018	1999 Apr 12 2250	1999-018A	Eutelsat W3	Eutelsat W3	S25673 Eutelsat	Atlas IIAS	AC-154	CC SLC36A	Wire
1999-019	1999 Apr 15 0046:00	1999-019A	Globalstar FM19	Globalstar FM19	S25676 Globalstar	Soyuz 11A511U	S15000-060 ST03	GIK-5 LC1	NK9906-12
		1999-019B	Globalstar FM42	Globalstar FM42	S25677 Globalstar				
		1999-019C	Globalstar FM44	Globalstar FM44	S25678 Globalstar				
		1999-019D	Globalstar FM45	Globalstar FM45	S25679 Globalstar				
1999-020	1999 Apr 15 1832:00	1999-020A	Landsat 7	Landsat 7	S25682 NASA/GSFC	Delta 7920-10	D268	V SLC2W	Wire
1999-021	1999 Apr 21 0459:12	1999-021A	UoSAT-12	UoSAT-12	S25693 SSTL	Dnepr	67035 42509	GIK-5 LC109/95	NK9906-16
1999-F01	1999 Apr 27 1822:01	1999-F01	Ikonos 1	Ikonos 1	F01208 SpaceIm	Athena-2	LM-005	V SLC6	Wire
1999-022	1999 Apr 28 2030:00	1999-022A	ABRIXAS	ABRIXAS	S25721 DLR	Kosmos 11K65M	65036-413	GTsMP -4 LC107	NK9906-1
		1999-022B	Megsat-0	Megsat-0	S25722 MegSat				
1999-023	1999 Apr 30 1630	1999-023A	USA 143	Milstar-2 F1	S25724 USAF	Titan 401B/ Centaur	4B-32 (K-26)	CC LC40	Wire
1999-024	1999 May 5 0100	1999-024A	Orion 3	Orion 3	S25727 Loral Orion	Delta 8930	D269	CC SLC17B	Wire
1999-025	1999 May 10 0133:00	1999-025A	Feng Yun	Feng Yun 1C	S25730 CASC	Chang Zheng 4B	CZ4B-1	TYSC LC1	CGW-WWW
		1999-025B	Shi Jian 5	SJ-5	S25731 CASC				
1999-026	1999 May 18 0509:36	1999-026B	MUBLCOM	MUBLCOM	S25736 Orbcomm	Pegasus XL/ HAPS	F27	V RW30/12 PAWA	Wire
		1999-026A	TERRIERS	TERRIERS	S25735 NASA GSFC				
1999-027	1999 May 20 2230:00	1999-027A	Nimiq	Telesat DTH-1	S25740 Telesat	Proton-K/ DM-2M	396-02	GIK-5 LC81/23	NK9907-42
1999-028	1999 May 22 0936	1999-028A	USA 144	NRO X5?	S25744 NRO	Titan 404B	4B-12	V SLC4E	Wire
1999-029	1999 May 26 0522	1999-029A	Oceansat	IRS-P4	S25756 ISRO	PSLV	PSLV-C2	SHAR PSLV	Wire
		1999-029B	KITSAT-3	KITSAT-3	S25757 KAIST				
		1999-029C	DLR-TUBSAT-C	DLR-TUBSAT-C	S25758 DLR				
1999-030	1999 May 27 1049:42	1999-030A	Discovery	OV-103	S25760 NASA JSC	Space Shuttle	STS-96	KSC LC39B/ MLP2	MSFCFLASH
		1999-030A	Spacehab Double Module	Spacehab Double Module	A04924 NASA JSC				
		1999-030A	ICC	Integrated Cargo Carrier	A04925 NASA JSC				
		1999-030B	Starshine	Starshine	S25769 NASA GSFC				
1999-031	1999 Jun 10 1348:43	1999-031A	Globalstar M052	Globalstar M052	S25770 Globalstar	Delta 7420-10C	D270	CC SLC17B	Wire
		1999-031B	Globalstar M049	Globalstar M049	S25771 Globalstar				
		1999-031C	Globalstar M025	Globalstar M025	S25772 Globalstar				
		1999-031D	Globalstar M047	Globalstar M047	S25773 Globalstar				
1999-032	1999 Jun 11 1715:33	1999-032A	Iridium 14A	Iridium SV090?	S25777 Iridium	Chang Zheng 2C-III/SD	CZ2C-21	TYSC LC1	CGW-WWW
		1999-032B	Iridium 20A	Iridium SV091?	S25778 Iridium				
1999-033	1999 Jun 18 0149:30	1999-033A	Astra 1H	Astra 1H	S25785 SES	Proton-K/ DM-2M	397-02	GIK-5 LC81/23	NK9908-12
1999-034	1999 Jun 20 0215:00	1999-034A	QuikScat	QuikScat	S25789 NASA GSFC	Titan II SLV	23G-7	V SLC4W	Wire
1999-035	1999 Jun 24 1544:00	1999-035A	FUSE	Far UV Spec. Explorer	S25791 NASA GSFC	Delta 7320-10	D271	CC SLC17A	Wire
1999-F02	1999 Jul 5 1332:00	1999-F02	Raduga	Gran' No. 45	F01213 MO RF	Proton-K/Briz-M	389-01	GIK-5 LC81/24	NK9909-26
1999-036	1999 Jul 8 0845:06	1999-036A	Molniya-3	Molniya-3 No. 63?	S25847 MOM	Molniya 8K78M	–	GIK-1 LC43/3	NK9909-32
1999-037	1999 Jul 10 0845:37	1999-037A	Globalstar M032	Globalstar M032	S25851 Globalstar	Delta 7420-10C	D272	CC SLC17B	Wire
		1999-037B	Globalstar M030	Globalstar M030	S25852 Globalstar				
		1999-037C	Globalstar M035	Globalstar M035	S25853 Globalstar				
		1999-037D	Globalstar M051	Globalstar M051	S25854 Globalstar				
1999-038	1999 Jul 16 1637:33	1999-038A	Progress M-42	Progress 7K-TGM No 242	S25858 RAKA	Soyuz 11A511U	667	GIK-5 LC1	NK9909-05
1999-039	1999 Jul 17 0638:00	1999-039A	Okean-O	Okean-O No. 1	S25860 RAKA	Zenit-2	17L	GIK-5 LC45L/1	NK9909-34
1999-040	1999 Jul 23 0431:00	1999-040A	Columbia	OV-102	S25866 NASA JSC	Space Shuttle	STS-93	KSC LC39B/ MLP1	JCM-P
		1999-040B	Chandra X-ray Observatory	AXAF	S25867 NASA MSFC				

Launch	*Launch Date*	*COSPAR*	*PL Name*	*Orig PL Name*	*SATCAT*	*LV Type*	*LV S/N*	*Launch Site*	*Ref*
1999-041	1999 Jul 25 0746:03	1999-041A	Globalstar M026	Globalstar M026	S25872 Globalstar	Delta 7420-10C	D273	CC SLC17A	Wire
		1999-041B	Globalstar M028	Globalstar M028	S25873 Globalstar				
		1999-041C	Globalstar M043	Globalstar M043	S25874 Globalstar				
		1999-041D	Globalstar M048	Globalstar M048	S25875 Globalstar				
1999-042	1999 Aug 12 2252	1999-042A	Telkom 1	Telkom 1	S25880 PT Telkom	Ariane 42P	V118	CSG ELA2	Wire
1999-043	1999 Aug 17 0437:31	1999-043A	Globalstar M024	Globalstar M024	S25883 Globalstar	Delta 7420-10C	D274	CC SLC17B	Wire
		1999-043B	Globalstar M027	Globalstar M027	S25884 Globalstar				
		1999-043C	Globalstar M053	Globalstar M053	S25885 Globalstar				
		1999-043D	Globalstar M054	Globalstar M054	S25886 Globalstar				
1999-044	1999 Aug 18 1800	1999-044A	Kosmos-2365	Kobal't	S25889 MO RF	Soyuz 11A511U	–	GIK-1 LC43/3	NK9910-32
		1999-044	Spuskaemiy Kapsula	SpK	A04938 MO RF				
		1999-044	Spuskaemiy Kapsula	SpK	A04949 MO RF				
1999-045	1999 Aug 26 1202:15	1999-045A	Kosmos-2366	Parus	S25892 MO RF	Kosmos 11K65M	–	GIK-1 LC132/1	NK9910-34
1999-046	1999 Sep 4 2234	1999-046A	Mugunghwa 3	Koreasat 3	S25894 Korea Tel	Ariane 42P	V120	CSG ELA2	Wire
1999-047	1999 Sep 6 1636:00	1999-047A	Yamal 101	Yamal 101	S25896 AO Gazcom	Proton-K/ DM-2M	388-02	GIK-5 LC81/23	NK9911-04
		1999-047B	Yamal 102	Yamal 102	S25897 AO Gazcom				
1999-048	1999 Sep 9 1800:00	1999-048A	Foton	Foton No. 12	S25902 MOM	Soyuz 11A511U	-	GIK-1 LC43/4	NK9911-08
1999-049	1999 Sep 22 1433:00	1999-049A	Globalstar M033	Globalstar FM33	S25907 Globalstar	Soyuz 11A511U	S15000-061 ST04	GIK-5 LC1	NK9911-15
		1999-049B	Globalstar M050	Globalstar FM50	S25908 Globalstar				
		1999-049C	Globalstar M055	Globalstar FM55	S25909 Globalstar				
		1999-049D	Globalstar M058	Globalstar FM58	S25910 Globalstar				
1999-050	1999 Sep 23 0602	1999-050A	Echostar 5	Echostar 5	S25913 Echostar	Atlas IIAS	AC-155	CC SLC36A	Wire
1999-051	1999 Sep 24 0621:08	1999-051A	Ikonos	Ikonos 2	S25919 SpaceIm	Athena-2	LM-007	V SLC6	Wire
1999-052	1999 Sep 25 0629	1999-052A	Telstar 7	Telstar 7	S25922 Loral Skynet	Ariane 44LP	V121	CSG ELA2	Wire
1999-053	1999 Sep 27 2230:00	1999-053A	LMI 1	LMI 1	S25924 LMI	Proton-K/ DM-2M	398-02	GIK-5 LC81/23	NK9911-20
1999-054	1999 Sep 28 1100:05	1999-054A	Resurs F-1M	Resurs F-1M	S25929 RKA	Soyuz 11A511U	–	GIK-1 LC43/4	NK9911-24
1999-055	1999 Oct 7 1251:01	1999-055A	GPS SVN 46	GPS SVN 46	S25933 USAF	Delta 7925-9.5	D275	CC SLC17A	Wire
1999-056	1999 Oct 10 0328:00	1999-056A	DirecTV-1R	DirecTV-1R	S25937 DirecTV	Zenit-3SL	–	KLA, SL Odyssey	Wire
1999-057	1999 Oct 14 0315	1999-057A	Zi Yuan 1	CBERS	S25940 CASC/INPE	Chang Zheng 4B	CZ4B-1	TYSC LC1	Wire
		1999-057B	SACI 1	SACI 1	S25941 INPE				
1999-058	1999 Oct 18 1322:00	1999-058A	Globalstar M031	Globalstar FM31	S25943 Globalstar	Soyuz 11A511U	M15000 -062 ST05	GIK-5 LC1	NK9912-10
		1999-058B	Globalstar M056	Globalstar FM56	S25944 Globalstar				
		1999-058C	Globalstar M057	Globalstar FM57	S25945 Globalstar				
		1999-058D	Globalstar M059	Globalstar FM59	S25946 Globalstar				
1999-059	1999 Oct 19 0622	1999-059A	Orion 2	Orion 2	S25949 Loral Orion	Ariane 44LP	V122	CSG ELA2	Wire
1999-F03	1999 Oct 27 1616:00	1999-F03	Ekspress A1	Ekspress A No. 1	F01219 AO Inform	Proton-K/DM-2	386-02	GIK-5 LC200/39	Kosmodrom99-11
1999-060	1999 Nov 13 2254	1999-060A	GE 4	GE 4	S25954 GE Americom	Ariane 44LP	V123	CSG ELA2	Wire
1999-F04	1999 Nov 15 0729	1999-F04	MTSAT	MTSAT	F01221 NASDA	H-II	H-II-8F	TNSCY	WWW-NASDA
1999-061	1999 Nov 19 2230	1999-061A	Shenzhou	Shenzhou	S25956 CASC	Chang Zheng 2F	CZ2F-1	JQ SLS	CALT-WWW
1999-062	1999 Nov 22 1620:00	1999-062A	Globalstar M029	Globalstar FM29	S25961 Globalstar	Soyuz 11A511U	#NAME?	GIK-5 LC1	NK0003-17
		1999-062B	Globalstar M034	Globalstar FM34	S25962 Globalstar				
		1999-062C	Globalstar M039	Globalstar FM39	S25963 Globalstar				
		1999-062D	Globalstar M061	Globalstar FM61	S25964 Globalstar				
1999-063	1999 Nov 23 0406	1999-063A	UHF F/O F10	UHF F/O F10	S25967 USN	Atlas IIA	AC-136	CC SLC36B	Wire
1999-064	1999 Dec 3 1622:46	1999-064A	Helios 1B	Helios 1B	S25977 CNES/DGA	Ariane 40	V124	CSG ELA2	Wire
		1999-064B	Clementine	Clementine	S25978 CNES/DGA				
1999-065	1999 Dec 4 1853	1999-065A	Orbcomm D1	Orbcomm FM30	S25980 Orbcomm	Pegasus XL/ HAPS	F28	WI DZWI	Wire
		1999-065B	Orbcomm D2	Orbcomm FM31	S25981 Orbcomm				
		1999-065C	Orbcomm D3	Orbcomm FM32	S25982 Orbcomm				
		1999-065D	Orbcomm D4	Orbcomm FM33	S25983 Orbcomm				
		1999-065E	Orbcomm D5	Orbcomm FM34	S25984 Orbcomm				
		1999-065F	Orbcomm D6	Orbcomm FM35	S25985 Orbcomm				
		1999-065G	Orbcomm D7	Orbcomm FM36	S25986 Orbcomm				
1999-066	1999 Dec 10 1432:07	1999-066A	XMM	XMM	S25989 ESA	Ariane 5G	V119 (504)	CSG ELA3	Wire
1999-067	1999 Dec 12 1738:01	1999-067A	DMSP 5D-3 F-15	DMSP 5D-3 S-15	S25991 USAF	Titan II SLV	23G-8	V SLC4W	Wire

Launch	*Launch Date*	*COSPAR*	*PL Name*	*Orig PL Name*	*SATCAT*	*LV Type*	*LV S/N*	*Launch Site*	*Ref*
1999-F05	1999 Dec 12 1940	1999-F05	SACI-2	SACI-2	F01223 INPE	VLS-1	V02	ALCA VLS	AEB-WWW
1999-068	1999 Dec 18 1857:39	1999-068A	Terra	EOS AM-1	S25994 NASA GSFC	Atlas IIAS	AC-141	V SLC3E	Wire
1999-069	1999 Dec 20 0050:00	1999-069A	Discovery	OV-103	S25996 NASA JSC	Space Shuttle	STS-103	KSC LC39B/ MLP2	Wire
		1999-069	ORUC	ORUC	A04965 NASA JSC				
		1999-069	FSS	FSS	A04966 NASA JSC				
1999-070	1999 Dec 21 0713	1999-070C	Celestis-03	CPAC	A04967 Celestis	Taurus 2110	T4 2110	V 576E	Wire
		1999-070A	KOMPSAT	KOMPSAT	S26032 KARI				
		1999-070B	ACRIMSAT	ACRIMSAT	S26033 JPL				
1999-071	1999 Dec 22 0050	1999-071A	Galaxy 11	Galaxy 11	S26038 Panamsat	Ariane 44L	V125	CSG ELA2	Wire
1999-072	1999 Dec 26 0800	1999-072A	Kosmos-2367	US-P	S26040 MO RF	Tsiklon-2	801 (45082801)	GIK-5 LC90/20	NK0002-42
1999-073	1999 Dec 27 1912:44	1999-073A	Kosmos-2368	Oko	S26042	Molniya MO RF	– 8K78M	GIK-1 LC16/2	NK0002-46

Table 23: World Launch Log 2000–2005

Launch	*Launch Date*	*COSPAR*	*PL Name*	*Orig PL Name*	*SATCAT*	*LV Type*	*LV S/N*	*Launch Site*	*Ref*
2000-001	2000 Jan 21 0103	2000-001A	USA 148	DSCS III B-8	S26052 USAF	Atlas IIA	AC-138	CC SLC36A	Wire
2000-002	2000 Jan 25 0104	2000-002A	Galaxy 10R	Galaxy 10R	S26056 Satmex	Ariane 42L	V126	CSG ELA2	Wire
2000-003	2000 Jan 25 1645:05	2000-003A	Zhongxing 22	DFH	S26058 Chinasat	Chang Zheng 3A	CZ3A-4	XSC LC2	CALT-PR
2000-004	2000 Jan 27 0303:06	2000-004A	JAWSAT	JAWSAT	S26061 -	Minotaur	1	V CLF	ASUSAT-WWW
		2000-004E	ASUSAT	ASUSAT	S26065 -				
		2000-004C	OPAL	OPAL	S26063 -				
		2000-004B	OCS	OCSE	S26062 -				
		2000-004D	Falconsat	Falconsat I	S26064 -				
		2000-004L	JAK	Artemis	S26093 -				
		2000-004M	STENSAT	STENSAT	S26094 -				
		2000-004H	Picosat 1	MEMS 1	S26080 -				
		2000-004H	Picosat 2	MEMS 2	S26080 -				
		2000-004J	Thelma	Artemis	S26091 -				
		2000-004K	Louise	Artemis	S26092 -				
2000-005	2000 Feb 1 0647:23	2000-005A	Progress M1-1	Progress 7K-TGM No 250	S26067 RAKA	Soyuz 11A511U	A15000 -669	GIK-5 LC1	NK0004-14
2000-006	2000 Feb 3 0926:00	2000-006A	Kosmos-2369	Tselina-2	S26069 MO RF	Zenit-2	45025801	GIK-5 LC45L	NK0004-17
2000-007	2000 Feb 3 2339	2000-007A	Hispasat 1C	Hispasat 1C	S26071 Hispasat	Atlas IIAS	AC-158	CC SLC36B	Wire
2000-008	2000 Feb 8 2124	2000-008A	Globalstar M060	Globalstar M060	S26081 Globalstar	Delta 7420-10C	D276	CC SLC17B	Wire
		2000-008B	Globalstar M062	Globalstar M062	S26082 Globalstar				
		2000-008C	Globalstar M063	Globalstar M063	S26083 Globalstar				
		2000-008D	Globalstar M064	Globalstar M064	S26084 Globalstar				
2000-009	2000 Feb 8 2320:00	2000-009A	Dummy satellite	-	S26086 Lavochkin	Soyuz 11A511U	A15000 -079 ST07	GIK-5 LC31	NK0004-24
		2000-009	IRDT	Mission 2000	A04971 ESA				
2000-F01	2000 Feb 10 0130	2000-F01	ASTRO E	ASTRO E	F01226 ISAS	M-V	M-V-4	KASC M-V	Wire
2000-010	2000 Feb 11 1743:40	2000-010A	Endeavour	OV-105	S26088 NASA JSC	Space Shuttle	STS-99	KSC LC39A/ MLP3	Wire
		2000-010	SRL-3	SRL PLT	A04974 JPL/DARA				
		2000-010	SRTM Mast	SRTM Mast	A04978 -				
2000-011	2000 Feb 12 0910:54	2000-011A	Garuda 1	Garuda 1	S26089 ACES	Proton-K/ DM-2M	399-02	GIK-5 LC81/23	NK0004-32
2000-012	2000 Feb 18 0104	2000-012A	Superbird 4	Superbird 4	S26095 SCC	Ariane 44LP	V127	CSG ELA2	Wire
2000-013	2000 Mar 12 0407:00	2000-013A	Ekspress 6A	Ekspress A No. 2	S26098 AO Inform	Proton-K/ DM-2M	399-01	GIK-5 LC200/39	NK0005-20
2000-014	2000 Mar 12 0929	2000-014A	MTI	MTI	S26102 DOE	Taurus 1110	T5 1110	V 576E	Wire
2000-F02	2000 Mar 12 1449:15	2000-F02	ICO F-1	ICO-F1	F01228 ICO	Zenit-3SL	- Odyssey	KLA, SL	Wire
2000-015	2000 Mar 20 1828:30	2000-015A	Dumsat	-	S26106 Starsem	Soyuz 11A511U		GIK-5 LC31	NK0005-27
2000-016	2000 Mar 21 2328:19	2000-016B	Insat 3B	Insat 3B	S26108 ISRO	Ariane 5G	V128 (505)	CSG ELA3	AE-TV
		2000-016A	Asiastar	Asiastar	S26107 Worldspace				
2000-017	2000 Mar 25 2034:43	2000-017A	IMAGE	IMAGE	S26113 NASA GSFC	Delta 7326-9.5	D277	V SLC2W	Wire
2000-018	2000 Apr 4 0501:29	2000-018A	Soyuz TM-30	Soyuz 7K-STM No. 204	S26116 RKA	Soyuz 11A511U	-	GIK-5 LC1	NK0006-08
2000-019	2000 Apr 17 2106:00	2000-019A	Sesat	Sesat	S26243 Eutelsat	Proton-K/ DM-2M	397-01	GIK-5 LC200/39	NK0006-26
2000-020	2000 Apr 19 0029	2000-020A	Galaxy 4R	Galaxy 4R	S26298 Satmex	Ariane 42L	V129	CSG ELA2	Wire
2000-021	2000 Apr 25 2008:02	2000-021A	Progress M1-2	Progress 7K-TGM No 252	S26301 RAKA	Soyuz 11A511U	-	GIK-5 LC1	NK0006-16
2000-022	2000 May 3 0707	2000-022A	GOES 11	GOES L	S26352 NOAA	Atlas IIA	AC-137	CC SLC36A	Wire
2000-023	2000 May 3 1325	2000-023A	Kosmos-2370	Neman	S26354 MO RF	Soyuz 11A511U	A15000-649	GIK-5 LC1	NK0007-32
2000-024	2000 May 8 1601	2000-024A	DSP F20 (USA 149)	DSP 20	S26356 USAF	Titan 402B/ IUS	4B-29	CC LC40	Wire
2000-025	2000 May 11 0148	2000-025A	GPS SVN 51	GPS SVN 51	S26360 USAF	Delta 7925-9.5	D278	CC SLC17A	Wire
2000-026	2000 May 16 0827:41	2000-026A	Simsat-1	Simsat-1	S26365 Krunichev	Rokot	-	GIK-1 LC133	Eurockot-WWW
		2000-026B	Simsat-2	Simsat-2	S26366 Krunichev				
2000-027	2000 May 19 1011:10	2000-027A	Atlantis	OV-104	S26368 NASA JSC	Space Shuttle	STS-101	KSC LC39A/ MLP1	Wire
		2000-027	External Airlock/ODS	EAL/ODS	A04995 NASA JSC				
		2000-027	Spacehab Double Module	Spacehab Double Module	A04999 NASA JSC				
		2000-027	ICC	Integrated Cargo Carrier	A05000 NASA JSC				
2000-028	2000 May 24 2310	2000-028A	Eutelsat W4	Eutelsat W4	S26369 Eutelsat	Atlas 3A	AC-201	CC SLC36B	Wire
2000-029	2000 Jun 6 0259:00	2000-029A	Gorizont	Gorizont No. 45L	S26372 MOM	Proton-K/ Briz-M	392-01	GIK-5 LC81/24	NK0008-16
2000-030	2000 Jun 7 1319:30	2000-030A	TSX-5	STEP M5	S26374 USAF STP	Pegasus XL	F29 12 PAWA	V RW30/	Wire
2000-031	2000 Jun 24 0028:00	2000-031A	Ekspress 3A	Ekspress A No. 3	S26378 AO Inform	Proton-K/ DM-2M	394-02	GIK-5 LC200/39	NK0008-23
2000-032	2000 Jun 25 1150	2000-032A	Feng Yun 2	FY-2	S26382 CASC	Chang Zheng 3	CZ3-13	XSC LC1	CALT-WWW
2000-033	2000 Jun 28 1037:42	2000-033A	Nadezhda	Nadezhda	S26384 MO RF	Kosmos 11K65M	-	GIK-1 LC132/1	NK0008-25
		2000-033B	Tsinghua	-	S26385 Tsinghua				
		2000-033C	SNAP	-	S26386 SSTL				

Launch	*Launch Date*	*COSPAR*	*PL Name*	*Orig PL Name*	*SATCAT*	*LV Type*	*LV S/N*	*Launch Site*	*Ref*
2000-034	2000 Jun 30 1256	2000-034A	TDRS 8	TDRS H	S26388 NASA	Atlas IIA	AC-139	CC SLC36A	Wire
2000-035	2000 Jun 30 2208:47	2000-035A	Sirius 1	Sirius 1	S26390 Loral	Proton-K/ DM-2M	400-01	GIK-5 LC81/24	NK0008-35
2000-036	2000 Jul 4 2344:00	2000-036A	Kosmos-2371	Geizer	S26394 MO RF	Proton-K/ DM-2	389-02	GIK-5 LC200/39	NK0009-24
2000-037	2000 Jul 12 0456:36	2000-037A	Zvezda	17KSM No. 128-01	RAKA	Proton-K	398-01	GIK-5 LC81/23	NK0009-02
2000-038	2000 Jul 14 0521	2000-038A	Echostar VI	Echostar VI	S26402 Echostar	Atlas IIAS	AC-161	CC SLC36B	Wire
2000-039	2000 Jul 15 1200:00	2000-039B	CHAMP	-	S26405 DLR	Kosmos 11K65M	47136-414	GIK-1 LC132/1	NK0009-29
		2000-039A	MITA	-	S26404 ASI				
		2000-039	BIRD	-	A05007 DLR				
2000-040	2000 Jul 16 0917:00	2000-040A	GPS SVN 48	GPS SVN 48	S26407 USAF	Delta 7925-9.5	D279	CC SLC17A	Wire
2000-041	2000 Jul 16 1239:34	2000-041A	Samba	Cluster FM7	S26410 ESA	Soyuz 11A511U	A15000-069 ST09	GIK-5 LC31	NK0009-35
		2000-041B	Salsa	Cluster FM6	S26411 ESA				
2000-042	2000 Jul 19 2009:00	2000-042A	Sindri	Mightysat 2.1	S26414 -	Minotaur	2	V CLF	Wire
		-	Picosat 1	MEMS 1	-				
		-	Picosat 2	MEMS 2	-				
2000-043	2000 Jul 28 2242:00	2000-043A	PAS 9	PAS 9	S26451 DirecTV	Zenit-3SL	-	KLA, SL Odyssey	NK0009-45
2000-044	2000 Aug 6 1826:42	2000-044A	Progress M1-3	Progress 7K-TGM No 251	S26461 RAKA	Soyuz 11A511U	K15000-668	GIK-5 LC1	NK0010-17
2000-045	2000 Aug 9 1113:35	2000-045A	Rumba	Cluster FM5	S26463 ESA	Soyuz 11A511U	A15000-070 ST10	GIK-5 LC31	NK0010-02
		2000-045B	Tango	Cluster FM8	S26464 ESA				
2000-046	2000 Aug 17 2316:00	2000-046B	Nilesat 102	Nilesat 102	S26470 Nilesat	Ariane 44LP	V131	CSG ELA2	Wire
		2000-046A	Brasilsat B4	Brasilsat B4	S26469 Embratel				
2000-047	2000 Aug 17 2345:01	2000-047A	USA 152	ONYX 4	S26473 NRO/CIA	Titan 403B	4B-28	V SLC4E	Wire
2000-048	2000 Aug 23 1105:00	2000-048A	DM-F3	DM-F3	S26475 Boeing/HB	Delta 8930	D280	CC SLC17B	Wire
2000-049	2000 Aug 28 2008	2000-049A	Kosmos-2372	Globus	S26477 MO RF	Proton-K/ DM-2	401-02	GIK-5 LC81/24	NK0010-13
2000-050	2000 Sep 1 0325	2000-050A	ZiYuan 2	ZY-2	S26481 CASC	Chang Zheng 4B	CZ4B-2	TYSC LC1	Wire
2000-051	2000 Sep 5 0943:58	2000-051A	Sirius 2	Sirius 2	S26483 Loral	Proton-K/ DM-2M	400-02	GIK-5 LC81/23	NK0011-29
2000-052	2000 Sep 6 2233	2000-052A	Eutelsat W1	Eutelsat W1	S26487 Singapore T	Ariane 44P	V132	CSG ELA2	Wire
2000-053	2000 Sep 8 1245:47	2000-053A	Atlantis	OV-104	S26489 NASA JSC	Space Shuttle	STS-106	KSC LC39B/ MLP2	Wire
		2000-053	External Airlock/ODS	EAL/ODS	A05049 NASA JSC				
		2000-053	Spacehab Double Module	Spacehab Double Module	A05053 NASA JSC				
		2000-053	ICC	Integrated Cargo Carrier	A05054 NASA JSC				
2000-054	2000 Sep 14 2254:07	2000-054A	Astra 2B	Astra 2B	S26494 SES	Ariane 5G	V130 (506)	CSG ELA3	Wire
		2000-054B	GE 7	GE 7	S26495 GE				
2000-055	2000 Sep 21 1022	2000-055A	NOAA 16	NOAA L	S26536 NOAA	Titan II SLV	23G-13	V SLC4W	Wire
2000-056	2000 Sep 25 1010:00	2000-056A	Kosmos-2372	Yenisey	MO RF	Zenit-2	-	GIK-5 LC45L	NK0011-36
2000-057	2000 Sep 26 1005:00	2000-057A	Tiungsat	-	ATSB	Dnepr	-	GIK-5 LC109/95	NK0011-40
2000-058	2000 Sep 29 0930:15	2000-058A	Kosmos-2373	Kometa No. 20?	S26552 RKA	Soyuz 11A511U	-	GIK-5 LC31	NK0011-46
2000-059	2000 Oct 1 2200	2000-059A	GE-1A	GE-1A	S26554 LMMS	Proton-K/ DM-2M	401-01	GIK-5 LC81/23	Wire
2000-060	2000 Oct 6 2300	2000-060A	N-SAT-110	Superbird 5	S26559 SCC	Ariane 42L	V133	CSG ELA2	Wire
2000-061	2000 Oct 9 0538:18	2000-061A	HETE 2	HETE 2	S26561 MIT	Pegasus H	F30/P-35	KMR RW06/ 24 DZKMR	Wire
2000-062	2000 Oct 11 2317:00	2000-062A	Discovery	OV-103	S26563 NASA JSC	Space Shuttle	STS-92	KSC LC39A/ MLP3	Wire
		-	PMA-3	PMA-3	NASA JSC				
		-	Z1						
2000-063	2000 Oct 13 1412:45	2000-063A	Kosmos-2374	Uragan No.	S26564 MOM	Proton-K/ DM-2	-	GIK-5 LC200/39?	CSIC-WWW
		2000-063B	Kosmos-2375	Uragan No.	S26565 MOM				
		2000-063C	Kosmos-2376	Uragan No.	S26566 MOM				
2000-064	2000 Oct 16 2127:06	2000-064A	Progress M-43	Progress 7K-TGM No 243	S26570 RAKA	Soyuz 11A511U	-	GIK-5 LC1	Wire
2000-065	2000 Oct 20 0040	2000-065A	USA 153	DSCS III B-11	S26575 USAF	Atlas IIA	AC-140	CC SLC36A	Wire
2000-066	2000 Oct 21 0552	2000-066A	Thuraya 1	Thuraya 1	S26578 Thuraya	Zenit-3SL	-	KLA, SL Odyssey	Wire
2000-067	2000 Oct 21 2200:00	2000-067A	GE-6	GE-6	S26580 LMMS	Proton-K/ DM-2M	-	GIK-5 LC81/23	Wire
2000-068	2000 Oct 29 0559	2000-068A	Europe* Star F1	Europe* Star FM1	S26590 EuropeStar	Ariane 44LP	V134	CSG ELA2	Wire
2000-069	2000 Oct 30 1602	2000-069A	Beidou	Beidou	S26599 CNSA?	Chang Zheng 3A	CZ3A-5	XSC LC2	Wire
2000-070	2000 Oct 31 0752:47	2000-070A	Soyuz TM-31	Soyuz 7K-STM No. 205	S26603 RKA	Soyuz 11A511U	-	GIK-5 LC1	Wire
2000-071	2000 Nov 10 1714:02	2000-071A	GPS SVN 41	GPS SVN 41	S26605 USAF	Delta 7925-9.5	D281	CC SLC17A	Wire
2000-072	2000 Nov 16 0107:07	2000-072A	PAS 1R	PAS 1R	S26608 Panamsat	Ariane 5G	V135 (507)	CSG ELA3	AE-TV
		2000-072B	Amsat P3D	Amsat P3D	Amsat				
		2000-072C	STRV-1c	STRV-1c	DERA				
		2000-072D	STRV-1d	STRV-1d	DERA				
2000-073	2000 Nov 16 0132:36	2000-073A	Progress M1-4	Progress 7K-TGM No 253?	S26615 RAKA	Soyuz 11A511U	-	GIK-5 LC1	AVMN
2000-074	2000 Nov 20 2300	2000-074A	QuickBird	QuickBird	S26617 EWatch	Kosmos 11K65M	-	GIK-1 LC132/1	Wire
2000-075	2000 Nov 21 1824:25	2000-075A	EO-1	EO-1	-	Delta 7320-10	D282	V SLC2W	Wire
		2000-075B	SAC-C	-	-				
		2000-075C	Munin	-	-				
2000-076	2000 Nov 21 2356	2000-076A	Anik F1	Anik F1	S26624 Panamsat	Ariane 44L	V136	CSG ELA2	Wire

Launch	*Launch Date*	*COSPAR*	*PL Name*	*Orig PL Name*	*SATCAT*	*LV Type*	*LV S/N*	*Launch Site*	*Ref*
2000-077	2000 Nov 30 1959:47	2000-077A	Sirius 3	Sirius 3	S26626 Loral	Proton-K/ DM-2M	402-02	GIK-5 LC81/23	Wire
2000-078	2000 Dec 1 0306:01	2000-078A	Endeavour	OV-105	S26630 NASA JSC	Space Shuttle	STS-97	KSC LC39B/ MLP1	Wire
2000-079	2000 Dec 5 1232	2000-079A	EROS A1	EROS A1	S26631 ImageSat	Start-1	-	GIK-2 LC5	Wire
2000-080	2000 Dec 6 0247	2000-080A	USA 155	NRO satellite	S26635 NRO	Atlas IIAS	AC-157	CC SLC36A	Wire
2000-081	2000 Dec 20 0026	2000-081D	Astra 2D	Astra 2D	S26641 -	Ariane 5G	V138 (508)	CSG ELA3	AE-TV
		2000-081B	GE 8	-	S26639 -				
		2000-081C	LDREX	-	S26640 -				
2000-082	2000 Dec 20 1620	2000-082A	Beidou	Beidou	S26643 CNSA?	Chang Zheng 3A	CZ3A-6	XSC LC2	Wire
2000-F03	2000 Dec 27 0956:31;	2000-F03;	Kosmos;	Strela-3;	F01232 MO RF;	Tsiklon-3	-	GIK-1 LC32/1	NK0102-35
		2000-F03	Kosmos	Strela-3	F01233 MO RF				
		2000-F03	Kosmos	Strela-3	F01234 MO RF				
		2000-F03	Gonets-D1	Gonets-D1 No. 7	F01235 RKA				
		2000-F03	Gonets-D1	Gonets-D1 No. 8	F01236 RKA				
		2000-F03	Gonets-D1	Gonets-D1 No. 9	F01237 RKA				
2001-001	2001 Jan 9 1700:03	2001-001A	Shenzhou	Shenzhou	S26664 CASC	Chang Zheng 2F	CZ2F-2	JQ SLS	Wire
2001-002	2001 Jan 10 2209	2001-002A	Turksat 2A	Eurasiasat 1	S26666 -	Ariane 44P	V137	CSG ELA2	Wire
2001-003	2001 Jan 24 0428:42	2001-003A	Progress M1-5	Progress 7K-TGM No 254	S26688 RAKA	Soyuz 11A511U	K15000-673	GIK-5 LC1	NK0103-25
2001-004	2001 Jan 30 0755	2001-004A	GPS SVN 54	GPS SVN 54	S26690 USAF	Delta 7925-9.5	D283	CC SLC17A	Wire
2001-005	2001 Feb 7 2306	2001-005B	Skynet 4F	Skynet 4F	S26695 UK MoD	Ariane 44L	V139	CSG ELA2	Wire
		2001-005A	Sicral	Sicral	S26694 -				
2001-006	2001 Feb 7 2313:02	2001-006A	Atlantis	OV-104	S26698 NASA JSC	Space Shuttle	STS-98	MLP2,KSC LC39A	Wire
		2001-006B	Destiny	Destiny	S26700 NASA MSFC				
2001-007	2001 Feb 20 0848:27	2001-007A	Odin	-	S26702 SSC	Start-1	-	GIK-2 LC5	Wire
2001-008	2001 Feb 26 0809:35	2001-008A	Progress M-44	Progress 7K-TGM No 244	S26713 RAKA	Soyuz 11A511U	670	GIK-5 LC1	NK0104-29
2001-009	2001 Feb 27 2120	2001-009A	USA	Milstar-2 F2	S26715 USAF	Titan 401B/ Centaur	4B-41 (K-30)	CC SLC40	
2001-010	2001 Mar 8 1142:09	2001-010A	Discovery	OV-103	S26718 NASA JSC	Space Shuttle	STS-102	MLP3,KSC LC39B	Wire
		-	Leonardo	MPLM	NASA JSC				
2001-011	2001 Mar 8 2251	2001-011A	Eurobird	Eurobird	S26719 -	Ariane 5G	V140 (509)	CSG ELA3	Wire
		2001-011B	BSAT 2a	-	S26720 -				
2001-012	2001 Mar 18 2233:30	2001-012A	XM Radio-2	XM Radio 2	S26724 XM Radio	Zenit-3SL	-	ODYSSEY,KLA	Wire
2001-013	2001 Apr 7 0347:00	2001-013A	Ekran	Ekran	S26736 -	Proton-M/ Briz-M	535-01	GIK-5 LC81/24	NK0106-38
2001-014	2001 Apr 7 1502:22	2001-014A	2001 Mars Odyssey	Mars Odyssey	S26734 -	Delta 7925-9.5	D284	CC SLC17A	Wire
2001-015	2001 Apr 18 1013	2001-015A	GSAT-1	-	S26745 -	GSLV	GSLV-D1	SHAR PSLV	Wire
2001-016	2001 Apr 19 1840:42	2001-016A	Endeavour	OV-105	S26747 NASA JSC	Space Shuttle	STS-100	MLP1,KSC LC39A	Wire
2001-017	2001 Apr 28 0737:20	2001-017A	Soyuz TM-32	Soyuz 7K-STM No. 206	S26749 RKA	Soyuz 11A511U	674	GIK-5 LC1	NK0106-30
2001-018	2001 May 8 2210:29	2001-018A	XM Radio-1	XM Radio 1	S26761 XM Radio	Zenit-3SL	-	ODYSSEY,KLA	Wire
2001-019	2001 May 15 0111:30	2001-019A	PAS 10	PAS 10	S26766 -	Proton-K/ DM-2M	403-01	GIK-5 LC81/23	NK0107-22
2001-020	2001 May 18 1745:00	2001-020A	GeoLITE	GeoLITE	S26770 -	Delta 7925-9.5	D285	CC SLC17B	Wire
2001-021	2001 May 20 2232:40	2001-021A	Progress M1-6	Progress 7K-TGM No 255	S26773 RAKA	Soyuz-FG	-	GIK-5 LC1	NK0107-09
2001-022	2001 May 29 1755:00	2001-022A	Kosmos-2377	Kobal't	S26775 MO RF	Soyuz 11A511U	-	GIK-1 LC43/4	NK0107-26
2001-023	2001 Jun 8 1508:42	2001-023A	Kosmos-2378	Parus	S26818 MO RF	Kosmos 11K65M	-	GIK-1 LC132/1	NK0108-21
2001-024	2001 Jun 9 0645	2001-024A	Intelsat 901	-	S26824 -	Ariane 44L	V141	CSG ELA2	Wire
2001-025	2001 Jun 16 0149:00	2001-025A	Astra 2C	Astra 2C	S26853 -	Proton-K/ DM-2M	403-02	GIK-5 LC81/23	NK0108-23
2001-026	2001 Jun 19 0441:02	2001-026A	ICO 2	-	S26857 -	Atlas IIAS	AC-156	CC SLC36B	Wire
2001-027	2001 Jun 30 1946:46	2001-027A	MAP	MAP	S26859 -	Delta 7425-10	D286	CC SLC17B	Wire
2001-028	2001 Jul 12 0903:59	2001-028A	Atlantis	OV-104	S26862 NASA JSC	Space Shuttle	STS-104	KSC LC39B	Wire
		-	Quest	-	NASA JSC				
2001-029	2001 Jul 12 2158	2001-029A	Artemis	-	S26863 -	Ariane 5G	V142 (510)	CSG ELA3	Wire
		2001-029B	BSAT 2b	-	S26864 -				
2001-030	2001 Jul 20 0017:00	2001-030A	Molniya-3	Molniya-3K	S26867 MOM	Molniya 8K78M	-	GIK-1 LC43/4	NK0109-38
2001-031	2001 Jul 23 0723	2001-031A	GOES 12	GOES M	S26871 NOAA	Atlas IIA	AC-142	CC SLC36A	Wire
2001-032	2001 Jul 31 0800:00	2001-032A	Koronas-F	AUOS-SM-KF-IK?	S26873 RKA	Tsiklon-3	-	GIK-1 LC32/1	NK0109-45
2001-033	2001 Aug 6 0728	2001-033A	DSP F21	-	S26880 USAF	Titan 402B/IUS	4B-31	CC SLC40	Wire
2001-034	2001 Aug 8 1613:40	2001-034A	Genesis	Genesis	S26884 -	Delta 7326-9.5	D287	CC SLC17A	Wire
2001-035	2001 Aug 10 2110:14	2001-035A	Discovery	OV-103	S26888 NASA JSC	Space Shuttle	STS-105	KSC LC39A	Wire
		2001-035B	Simplesat	-	S26889 -				
2001-036	2001 Aug 21 0923:54	2001-036A	Progress M-45	Progress 7K-TGM No 245	S26890 RAKA 11A511U	Soyuz	-	GIK-5 LC1	NK0110-18
2001-037	2001 Aug 24 2035	2001-037A	Kosmos-2379	-	S26892 MO RF	Proton-K/ DM-2	404-01	GIK-5 LC81/24	NK0110-37
2001-038	2001 Aug 29 0700:00	2001-038A	LRE	LRE	S26898 NASDA	H-IIA 202	H-IIA-1F	TNSC Y	NASDARept112
2001-039	2001 Aug 30 0646	2001-039A	Intelsat 902	-	S26900 -	Ariane 44L	V143	CSG ELA2	Wire
2001-040	2001 Sep 8 1525:05	2001-040A	NRO	-	S26905 -	Atlas IIAS	AC-160	V SLC3E	Wire
2001-041	2001 Sep 14 2334:55	2001-041A	Progress SO-1	Progress	S26908 RAKA	Soyuz 11A511U	677	GIK-5 LC1	NK0111-06
		2001-041	Pirs	-	A05160 RAKA				

Launch	*Launch Date*	*COSPAR*	*PL Name*	*Orig PL Name*	*SATCAT*	*LV Type*	*LV S/N*	*Launch Site*	*Ref*
2001-F01	2001 Sep 21 1849	-	Celestis-04	CPAC	Celestis	Taurus 2110	T6	V576E	Wire
		2001-F01	Orbview-4	-	F01239 -				
		2001-F01	QuikTOMS	-	F01240 -				
		-	SBD	-	-				
2001-042	2001 Sep 25 2321	2001-042A	Atlantic Bird 2	-	S26927 -	Ariane 44P	V144	CSG ELA2	Wire
2001-043	2001 Sep 30 0240	2001-043A	Starshine 3	-	S26929 LMA	Athena-1	LM-001	KLC	Wire
		2001-043B	Picosat	-	S26930 LMA				
		2001-043C	PCSat	-	S26931 LMA				
		2001-043D	SAPPHIRE	-	S26932 LMA				
2001-044	2001 Oct 5 2121:01	2001-044A	USA 161	Improved CRYSTAL 5	S26934 NRO/ USAF	Titan 404B	4B-34	V SLC4E	Wire
2001-045	2001 Oct 6 1645:00	2001-045A	Raduga-1	Globus	S26936 MO RF	Proton-K/ DM-2	-	GIK-5 LC81/24	NK0112-37
2001-046	2001 Oct 11 0301	2001-046A	USA 162	NRO satellite	S26948 NRO	Atlas IIAS	AC-162	CC SLC36B	Wire
2001-047	2001 Oct 18 1851:26	2001-047A	QB 2	QB	S26953 -	Delta 7320-10	D288	V SLC2W	spaceflightnow .com
2001-048	2001 Oct 21 0859:35	2001-048A	Soyuz TM-33	Soyuz 7K-STM No. 207	S26955 RKA	Soyuz 11A511U	672	GIK-5 LC1	NK0112-03
2001-049	2001 Oct 22 0453:00	2001-048B	TES	-	S26956 ISRO	PSLV	PSLV-C3	SHAR PSLV	
			2001-049C	BIRD	-	S26959 -			
		2001-049B	PROBA	-	S26958 -				
2001-050	2001 Oct 25 1134:13	2001-050A	Molniya-3	Molniya-3	S26970 MOM	Molniya 8K78M	77013-687	GIK-1 LC43/3	NK-0112-45
2001-051	2001 Nov 26 1824:12	2001-051A	Progress M1-7	Progress 7K-TGM No 256	S26983 RAKA	Soyuz-FG	-	GIK-5 LC1	rosaviakosmos.ru
2001-052	2001 Nov 27 0035	2001-052A	DirecTV-4S	-	S26985 DirecTV	Ariane 44LP	V146	CSG ELA2	Wire
2001-053	2001 Dec 1 1804	2001-053A	Kosmos-2380	Uragan No.	S26987 MOM	Proton-K/ DM-2	-	GIK-5 LC81	Wire
		2001-053B	Kosmos-2381	Uragan No.	S26988 MOM				
		2001-053C	Kosmos-2382	Uragan No.	S26989 MOM				
2001-054	2001 Dec 5 2219:28	2001-054A	Endeavour	OV-105	S26995 NASA JSC	Space Shuttle	STS-108	MLP1,KSC LC39B	Wire
2001-055	2001 Dec 7 1507:35	2001-055B	TIMED	-	S26998 -	Delta 7920-10	D289	V SLC2W	spaceflightnow .com
2001-055A			Jason	Jason	S26997 -				
2001-056	2001 Dec 10 1718:56	2001-056A	Meteor-3M	-	S27001 -	Zenit-2	-	GIK-5 LC45/1	rosaviakosmos.ru
		2001-056C	Badr B	-	S27003 -				
		2001-056D	Maroc-Tubsat	-	S27004 -				
		2001-056B	Kompas	-	S27002 -				
		2001-056E	Reflektor	-	S27005 -				
2001-057	2001 Dec 21 0400	2001-057A	Kosmos-2383	US-P	S27053 MO RF	Tsiklon-2	-	GIK-5 LC90/20?	
2001-058	2001 Dec 28 0324:24	2001-058A	Kosmos	Strela-3	S27055 MO RF	Tsiklon-3	-	GIK-1 LC32/1?	rosaviakosmos.ru
		2001-058B	Kosmos	Strela-3	S27056 MO RF				
		2001-058C	Kosmos	Strela-3	S27057 MO RF				
		2001-058D	Gonets-D1	Gonets-D1 No. 10	S27058 RKA				
		2001-058E	Gonets-D1	Gonets-D1 No. 11	S27059 RKA				
		2001-058F	Gonets-D1	Gonets-D1 No. 12	S27060 RKA				
2002-001	2002 Jan 16 0030	2002-001A	USA	Milstar-2 F2	S27168 USAF	Titan 401B/ Centaur	4B-38	CC SLC40	Wire
2002-002	2002 Jan 23 2347	2002-002B	Insat 3C	-	S27299 ISRO	Ariane 42L	V147	CSG ELA2	Wire
2002-003	2002 Feb 4 0245	-	DASH	DASH	ISAS	H-IIA 2024	H-IIA-2F	TNSCY	Wire
		2002-003B	VEP-3	-	S27368 NASDA				
		2002-003A	MDS-1	-	S27367 NASDA				
2002-004	2002 Feb 5 2058:12	2002-004A	HESSI	-	S27370 OSC	Pegasus XL	F31	L-1011, CC RW30	Wire
2002-005	2002 Feb 11 1743:44	2002-2005A	Iridium	-	S27372	Delta 7920-10	CD290	V SLC2W	Spacelightnow .com
		2002-005B	Iridium	-	S27373				
		2002-005C	Iridium	-	S27374				
		2002-005D	Iridium	-	S27375				
		2002-005E	Iridium	-	S27376				
2002-006	2002 Feb 21 1243:00		Atlas 3B	AC-204	CC SLC36B				Wire
2002-007	2002 Feb 23 0659	2002-007A	Intelsat 904	-	S27380	Ariane 44L	V148	CSG ELA2	Wire
2002-008	2002 Feb 25 1726	1996-007B	Kosmos-2377	Kobal't	S23782 MO RF	Soyuz-U	-	GIK-1 LC43/4	Wire
2002-009	2002 Mar 1 0107:59	-	Envisat	-	-	Ariane 5G	V145 (511)	CSG ELA3	Wire
2002-010	2002 Mar 1 1122:02	2002-010A	Columbia	OV-102	S27388 NASA JSC	Space Shuttle	STS-109	KSC LC39A	Wire
2002-011	2002 Mar 8 2259	2002-011A	TDRS 9	TDRS	S27389	Atlas IIA	AC-143	CC SLC36A	Wire
2002-012	2002 Mar 17 0921:17	2002-012A	Grace-1	Grace-1	S27391 Krunichev	Rokot	-	GIK-1 LC133	Wire
		2002-012B	Grace-2	Grace-2	S27392 Krunichev				
2002-013	2002 Mar 21 2013:36	2002-013A	Progress M1-8	Progress 7K-TGM No 257	S27395 RAKA	Soyuz-FG	-	GIK-5 LC1	rosaviakosmos. ru
2002-014	2002 Mar 25 1415	2002-014A	Shenzhou	Shenzhou	S27397 CASC	Chang Zheng 2F	CZ2F-3 (66)	JQ SLS	Wire
2002-015	2002 Mar 29 0129	2002-015A	Astra 3A	-	S27399	Ariane 44L	V149	CSG ELA2	Wire
		2002-015B	JCSAT 8	-	S27400				
2002-016	2002 Mar 30 1725:00	2002-016A	Intelsat 903	Intelsat 903	S27403	Proton-K/ DM-2M	-	GIK-5 LC81/23	Spaceflightnow. com

Table 24: World Launch Log 1957–2001 Inclusive

Number of orbital launches by site

Country	*Launch site*	*Latitude*	*Longitude*	*1957–87*	*1988*	*1989*	*1990*	*1991*	*1992*	*1993*	*1994*	*1995*	*1996*	*1997*	*1998*	*1999*	*2000*	*2001*	*1957–01*
RFAS/USSR	Tyuratam (Baikonur)	45.6°N	63.4°E	775	43	29	27	22	21	21	30	19	15	18	16	17	30	15	1,098
	Barent Sea	69.3°N	35.3°	0	0	0	0	0	0	0	0	0	0	0	1	0	0		1
	Kapustin Yar	48.4°N	45.8°E	83	0	0	0	0	0	0	0	0	0	0	0	1	0		84
	Plesetsk	62.8°N	40.1°E	1,159	47	45	48	37	33	26	18	13	10	9	7	4	5	6	1,467
	Svobodny	51.37°N	128.3°E	0	0	0	0	0	0	0	0	0	0[1]	2	0	0	1	1	4
USA	Cape Canaveral/ KSC (ER)	28.5°N	81.0°W	376	4	16	23	12	24	20	19	23	24	24	23	16	18	16	638
	EdwardsAFB/B52	35°N	118°W	0	0	0	1	1	0	1	2	0	-	1	0	1	0		7
	Pacific Sea Launch			0	0	0	0	0	0	0	0	0	0	0	0	2	3	2	7
	Vandenberg AFB (WR)	34.4°N	120.35°W	475	7	2	3	5	4	2	5	4	8	11	11	8	8	5	558
	Wallops Island	37.8°N	75.5°W	19	0	0	0	0	0	0	0	0	1	2	2	1	0		25
	Kwajalein	8.9917°N	167.72°W	0	0	0	0	0	0	0	0	0	0	0	0	0	1		1
	Kodiak																	1	1
France	Hammaguir	31.0°N	8.0°W	4	0	0	0	0	0	0	0	0	0	0	0	0	0		4
Italy	San Marco platform	2.9°S	40.3°E	8	1	0	0	0	0	0	0	0	0	0	0	0	0		9
Australia/UK	Woomera	31.1°S	136.8°E	2	0	0	0	0	0	0	0	0	0	0	1	0	0		3
Japan	Kagoshima	31.2°N	131.1°E	17	0	1	1	1	0	1	0	1	0	1	0	0	1		24
	Tanegashima	30.4°N	131.0°E	17	2	1	2	1	1	0	2	1	1	1	1	1	0	1	32
France/Europe	Kourou	5.2°N	52.8°W	22	7	7	5	8	7	7	6	11	10	12	11	9	12	7	141
China	Jiuquan	40.6°N	99.9°E	16	1	0	1	0	2	1	1	0	1	0	0	1	0	1	25
	Taiyuan/Wuzhai	37.5°N	112.6°E	0	1	0	1	0	0	0	0	0	0	2	4	2	1		11
	Xichang	28.25°N	102.0°E	3	2	0	3	1	2	0	4	2	2	4	2	0	3		28
India	SHAR, Sriharikota	13.9°N	80.4°E	3	0	0	0	0	1	0	2	0	1	1	0	0	0	2	10
Israel	Palmachim/Yavne	31.5°N	34.5°E	0	1	0	1	0	0	0	0	1	0	0	0	0	0		3
Brazil	Alcantara	2.3°S	44.4°W	0	0	0	0	0	0	0	0	0	0	0	0	0	0		0
Spain	Torrejon AB	40.488°N	3.457°W	0	0	0	0	0	0	0	0	0	0[2]	0	0	0	0		0
South Africa	OTB	34.583°S	20.317°E	0	0	0	0	0	0	0	0	0	0	0	0	0	0		0
			Total Launches	**2,979**	**116**	**101**	**116**	**88**	**95**	**79**	**89**	**75**	**73**	**88**	**79**	**63**	**83**		**4,181**

[1]First launch 4 March 1997
[2]First launch 21 April 1997

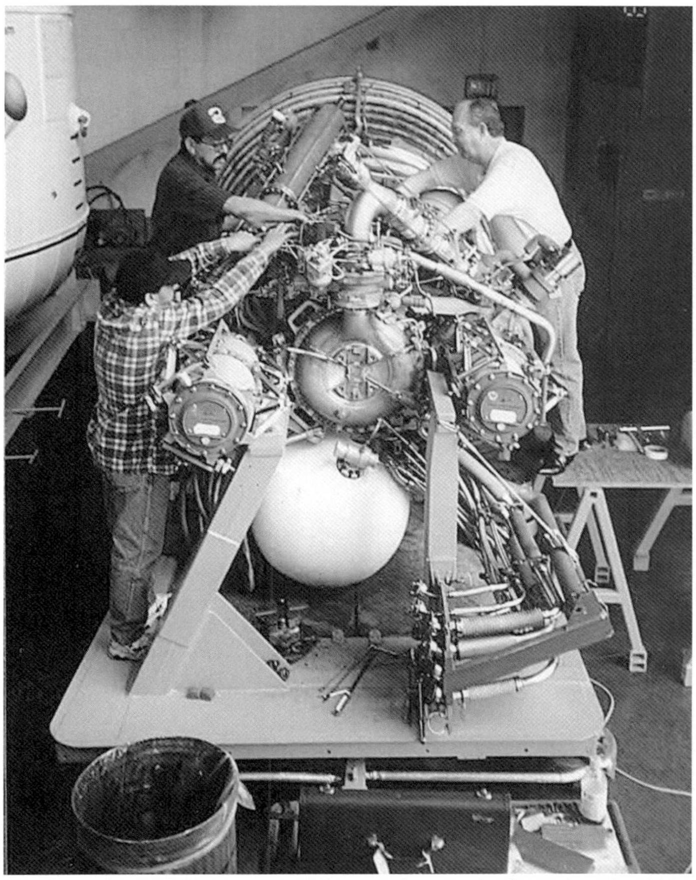

The J-2 engine, NASA's most powerful cryogenic motor for Saturn launch vehicles and pivotal to providing the energy necessary to launch Apollo to the moon (MSFC)
0137177

Ariane 5 is transported to its launch complex at Kourou in French Guiana (ESA)
0137120

PAS-9 gets a boost into space from the Sea Launch Odyssey platform located in the Pacific Ocean close to the equator (Sea Launch) 0137129

Europe's largest science satellite, the XMM X-ray observatory launched in 1999 is the most powerful of its type yet sent into space (ESA) 0137133

The Apollo-era Mobile Launch Platform supports External Tank, Solid Rocket Boosters and Orbiter as the Shuttle is carried to its launch pad (NASA) 0137132

Mercury Atlas 2 launched in February 1961 qualifies a new 'belly-band' securing the Mercury spacecraft to the launcher in a high-velocity sub-orbital test of high heat rate during a simulated abort (NASA) 0137147

Launched by Atlas MA-3 in April 1961, Mercury spacecraft 8 was a quick test of an Atlas abort qualifying the spacecraft for human survivability (NASA) 0137148

Viewed from the last departing Apollo spacecraft in 1974 Skylab displays its sole remaining solar panel, the other having been ripped off during launch in 1973 (MSFC) 0137149

The crew compartment of the two-man Gemini spacecraft displays the console and crew instruments in this fish-eye view (NASA) 0137150

Operational from 1965, the Mission Control Centre at NASA's Manned Spacecraft Center gears up for two-man Gemini flights (JSC) 0137151

Commanded by veteran astronaut Pete Conrad, the first manned visit to Skylab gets under way from Cape Canaveral in May 1973 (NASA) 0137153

NASA's High Energy Astronomy Observatory is encapsulated prior to launch (MSFC) 0137154

Shuttle Orbiter gets a workout in the Mated Vertical Vibration Test Facility designed to shake, rattle and roll the assembly looking for weak points (MSFC) 0137156

Pioneer 10 in its assembly rig prior to tests which will qualify it for the first flight to Jupiter (NASA) 0137162

The mirror for NASA's Chandra in its test cell during pre-assembly testing. Launched in July 1999 Chandra is an advanced x-ray facility (NASA) 0137160

Developed by the US Army from the Jupiter A ballistic missile, Juno 2 was readily adapted into a satellite launcher carrying early probes into space (US Army) 0137165

The first of the Mariners, Mariner 1 departs Cape Canaveral in July 1962 (NASA) 0137166

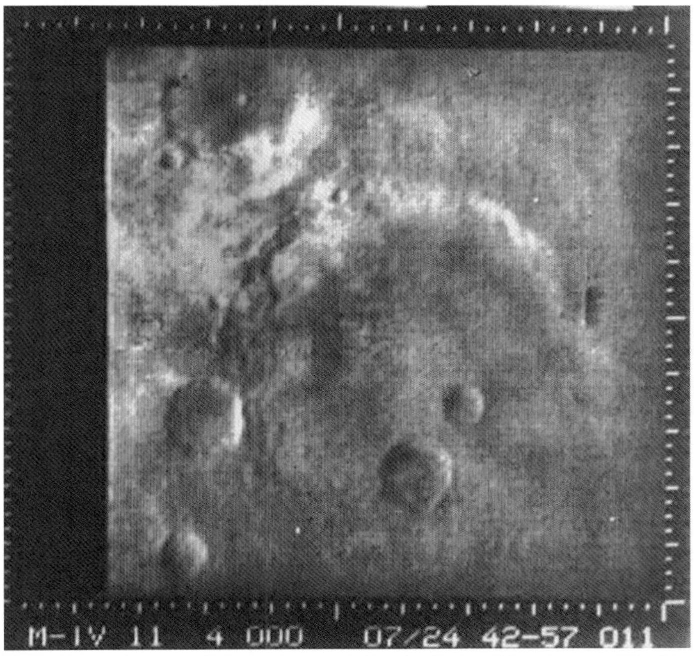

One of the 21 images of Mars returned during the fly-by of Mariner 4 in July 1965 revealing a lunar-like landscape of craters and archaic terrain (JPL) 0137167

Rock sampling provided earth bound geologists with hands-on science when Apollo crewmembers returned to earth with their valuable cargo (NASA) 0137172

An ethereal scene of desolation, the Apollo 17 site, location of NASA's last Apollo landing (NASA) 0137171

At the third attempt, astronauts –Schirra and Stafford launch from Cape Canaveral to a rendezvous with Borman and Lovell aboard Gemini 7 already in space, December 1965 (NASA) 0137173

A scale model of Gemini and its Titan launcher are prepared for test in a wind tunnel (NASA) 0137174

Reflecting the International nature of launch vehicle marketing services, models of (left to right): Ariane 4, Soyuy and Ariane 5
0084625

Crucial to moving astronauts several kilometres across the lunar surface, the Lunar Roving Vehicle built by Boeing undergoes a deployment test prior to its first flight on Apollo 15 in 1971 (NASA)
0137178

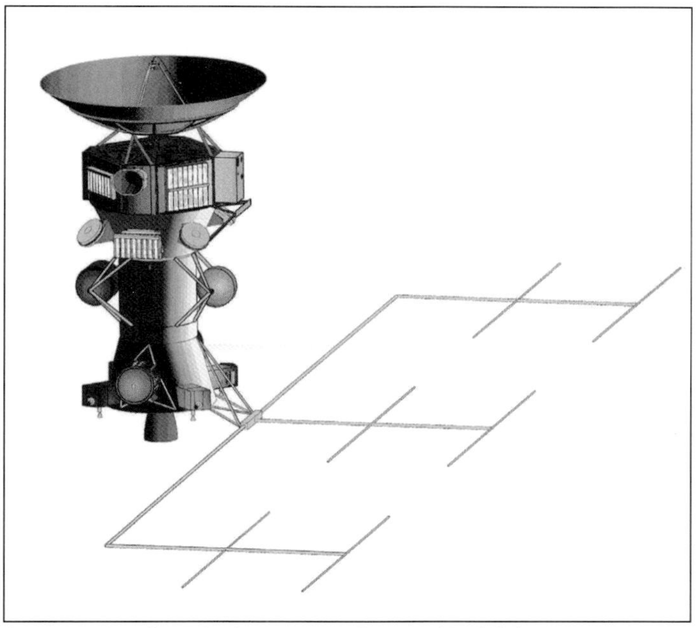

A conceptual configuration for a projected Europa orbiter (ESA) ***NEW***/1047486

ESA's Mars Express in Earth orbit prior to ignition of its injection stage (ESA) ***NEW***/1047485

Separated by seven years of development, the diminutive Mars Pathfinder (launched 1996) compares in scale with the Mars Exploration Rover spacecraft (launched in 2003) (JPL) ***NEW***/1047528

Elements of one of two Mars Exploration Rover spacecraft come together. Note the white vertical mast carrying TV cameras and the solar panels (JPL) ***NEW***/1047529

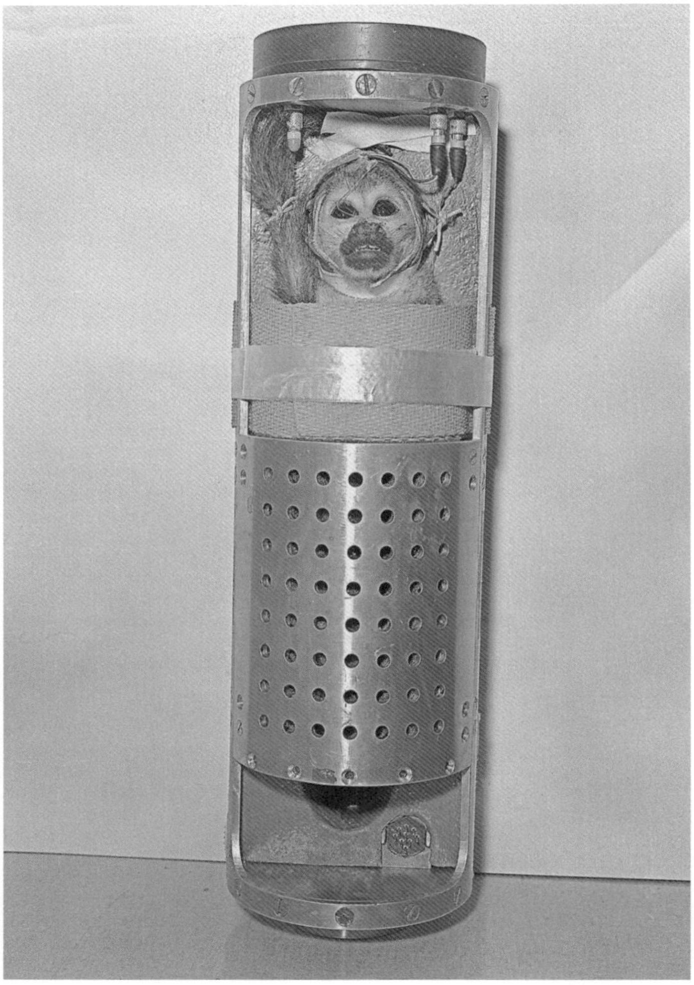

A female squirrel monkey, Baker, is prepared for a space flight on 28 May 1959. Baker lived until 1984 (NASA)
NEW/1047487

Europe's Rosetta comet chaser in assembly building prior to tests (ESA)
NEW/1047534

Crater Endurance on Mars imaged by the MER rover showing layered outcrops (JPL)
NEW/1047543

ESA's SMART-1 engineering test and qualification satellite launched 27 September 2003 and entered lunar orbit 15 November 2004 (ESA)
NEW/1047535

Table 25: Geosynchronous Satellite Listing

Geosynchronous satellite catalogue for 22 March 2001
Number of satellites: 783

Cat No	*International Designator*	*Common Name*	*Source*	*Apogee (km)*	*Perigee (km)*	*Inclination (°)*
553	1963-004A	SYNCOM 1	US	37015	34181	33.5
634	1963-031A	SYNCOM 2 (A 26)	US	35833	35753	24.9
858	1964-047A	SYNCOM 3	US	35858	35743	8.6
1317	1965-028A	INTELSAT 1-F1	ITSO	35836	35760	12.5
2207	1966-053A	OPS 9381 (GGTS)	US	33899	33624	9.1
2215	1966-053B	OPS 9311 (IDSCS 1)	US	33880	33666	9.1
2216	1966-053C	OPS 9312 (IDSCS 2)	US	36136	31432	8.9
2217	1966-053D	OPS 9313 (IDSCS 3)	US	33954	33672	9.2
2218	1966-053E	OPS 9314 (IDSCS 4)	US	34018	33693	9.3
2219	1966-053F	OPS 9315 (IDSCS 5)	US	34105	33698	9.5
2220	1966-053G	OPS 9316 (IDSCS 6)	US	34209	33714	9.6
2221	1966-053H	OPS 9317 (IDSCS 7)	US	34375	33695	9.6
2222	1966-053J	TITAN 3CTRANSTAGE R/B	US	34710	33432	9.8
2608	1966-110A	ATS 1	US	35797	35766	13.2
2639	1967-001A	INTELSAT 2-F2	ITSO	35865	35761	13.8
2645	1967-003A	OPS 9321 (IDSCS 8)	US	33848	33500	9.2
2649	1967-003B	OPS 9322 (IDSCS 9)	US	33833	33532	9.3
2650	1967-003C	OPS 9323 (IDSCS 10)	US	33847	33548	9.2
2651	1967-003D	OPS 9324 (IDSCS 11)	US	33893	33560	9.5
2652	1967-003E	OPS 9325 (IDSCS 12)	US	33962	33571	9.5
2653	1967-003F	OPS 9326 (IDSCS 13)	US	34041	33589	9.3
2654	1967-003G	OPS 9327 (IDSCS 14)	US	34127	33621	9.6
2655	1967-003H	OPS 9328 (IDSCS 15)	US	34243	33642	9.7
2660	1967-003J	TITAN 3CTRANSTAGE R/B	US	33796	33561	11.4
2717	1967-026A	INTELSAT 2-F3	ITSO	35893	35697	13.2
2862	1967-066A	OPS 9331 (IDSCS 16)	US	33480	33067	12.2
2863	1967-066B	OPS 9332 (IDSCS 17)	US	33518	33047	12.2
2864	1967-066C	OPS 9333 (IDSCS 18)	US	33532	33096	12.2
2865	1967-066D	OPS 9334 (IDSCS 19)	US	33537	33169	12.3
2866	1967-066E	LES 5	US	33594	33211	12.3
2867	1967-066F	DODGE 1	US	33654	33273	12.4
2868	1967-066G	TITAN 3CTRANSTAGE R/B	US	33833	33222	12.4
2969	1967-094A	INTELSAT 2-F4	ITSO	35823	35707	13.6
3029	1967-111A	ATS 3	US	35859	35713	13.9
3284	1968-050A	OPS 9341 (IDSCS 20)	US	33844	33729	10.1
3286	1968-050C	OPS 9343 (IDSCS 22)	US	33900	33725	10.1
3287	1968-050D	OPS 9344 (IDSCS 23)	US	33939	33744	10
3288	1968-050E	OPS 9345 (IDSCS 24)	US	34065	33699	10.3
3289	1968-050F	OPS 9346 (IDSCS 25)	US	34119	33741	10.2
3290	1968-050G	OPS 9347 (IDSCS 26)	US	34265	33722	10.3
3291	1968-050H	OPS 9348 (IDSCS 27)	US	34393	33728	10.6
3292	1968-050J	TITAN 3CTRANSTAGE R/B	US	35054	33661	10.5
3428	1968-081A	OV2-5	US	35762	35098	12.7
3430	1968-081C	ERS 21 (OV5-4)	US	35823	35765	13.6
3431	1968-081D	LES 6	US	35796	35733	13.2
3432	1968-081E	TITAN 3CTRANSTAGE R/B	US	35827	35053	12.8
3623	1968-116A	INTELSAT 3-F2	ITSO	37129	35965	15
3674	1969-011A	INTELSAT 3-F3	ITSO	35803	35767	6.1
3691	1969-013A	OPS 0757 (TACSAT)	US	35772	35766	7
3692	1969-013B	TITAN 3CTRANSTAGE R/B	US	35804	35694	13.7
4068	1969-069A	ATS 5	US	36012	36000	14.9
4250	1969-101A	SKYNET 1	UK	35876	35696	14.3
4297	1970-003A	INTELSAT 3-F6	ITSO	36139	36064	14.7
4353	1970-021A	NATO 2A	NATO	35801	35767	14.1
4376	1970-032A	INTELSAT 3-F7	ITSO	35791	35775	0.9
4478	1970-055A	INTELSAT 3-F8	ITSO	36598	33877	12.8
4881	1971-006A	INTELSAT 4-F2	ITSO	36250	36136	15.4
4902	1971-009A	NATO 2B	NATO	35824	35748	15
5587	1971-095A	OPS 9431 (DSCS II-1)	US	35808	35765	15
5588	1971-095B	OPS 9432 (DSCS II-2)	US	35808	35762	14.9
5589	1971-095C	TITAN 3CTRANSTAGE R/B	US	37375	35973	16
5709	1971-116A	INTELSAT 4-F3	ITSO	36030	35907	14.1
5775	1972-003A	INTELSAT 4-F4	ITSO	35943	35884	13.6
6052	1972-041A	INTELSAT 4-F5	ITSO	35857	35808	14.2
6278	1972-090A	ANIK A1 (TELESAT 1)	CA	36277	36116	14.2
6437	1973-023A	ANIK A2 (TELESAT 2)	CA	35966	35876	13.5
6796	1973-058A	INTELSAT 4-F7	ITSO	36148	36064	13.7
6973	1973-100A	OPS 9433 (DSCS II-3)	US	36675	36399	16.2
6974	1973-100B	OPS 9434 (DSCS II-4)	US	36873	36279	15.6
6976	1973-100D	TITAN 3CTRANSTAGE R/B	US	38597	36068	17
7229	1974-017A	COSMOS 637	CIS	35791	35494	14.5
7250	1974-022A	WESTAR 1	US	35910	35867	13.2
7298	1974-033A	SMS 1	US	36314	36202	17.3
7318	1974-039A	ATS 6	US	35433	35194	14.7
7324	1974-039C	TITAN 3CTRANSTAGE R/B	US	35758	35591	15.1
7392	1974-060A	MOLNIYA 1-S	CIS	35824	35722	15
7466	1974-075A	WESTAR 2	US	35937	35874	13
7544	1974-093A	INTELSAT 4-F8	ITSO	35961	35881	12.4
7547	1974-094A	SKYNET 2B	UK	35872	35768	14.3
7578	1974-101A	SYMPHONIE 1	FGER	35895	35847	14.9
7648	1975-011A	SMS 2	US	36059	35943	14.7
7790	1975-038A	ANIK A3 (TELESAT-3)	CA	35865	35824	12.5
7815	1975-042A	INTELSAT 4-F1	ITSO	36120	36023	12.4
8132	1975-077A	SYMPHONIE 2	FGER	35894	35846	14.9
8330	1975-091A	INTELSAT 4A-F1	ITSO	35915	35850	12.4
8357	1975-097A	COSMOS 775	CIS	35837	35696	15.1
8366	1975-100A	GOES 1 (SMS-C)	US	35792	35755	14.8

Cat No	International Designator	Common Name	Source	Apogee (km)	Perigee (km)	Inclination (°)
8476	1975-117A	SATCOM 1	US	36054	35899	12.5
8513	1975-123A	RADUGA 1	CIS	35822	35759	15
8585	1976-004A	CTS	CA	35876	35735	15
8620	1976-010A	INTELSAT 4A-F2	ITSO	35967	35932	12.5
8697	1976-017A	MARISAT 1	US	36152	36023	13.8
8746	1976-023A	LES 8	US	35824	35755	12
8747	1976-023B	LES 9	US	35879	35695	12
8751	1976-023F	TITAN 3CTRANSTAGE R/B	US	36913	35807	12.3
8774	1976-029A	SATCOM 2	US	36517	35993	12.3
8808	1976-035A	NATO 3A	NATO	35991	35826	13.2
8832	1976-023J	TITAN 3C DEB	US	36927	35793	12.3
8838	1976-042A	COMSTAR 1	US	35927	35896	12.4
8882	1976-053A	MARISAT 3	US	37583	36534	14.7
9009	1976-066A	PALAPA 1	INDO	35870	35810	12.2
9047	1976-073A	COMSTAR 2	US	35892	35840	12.2
9416	1976-092A	RADUGA 2	CIS	35874	35706	15.1
9478	1976-101A	MARISAT 2	US	35802	35772	13
9503	1976-107A	EKRAN 1	CIS	36087	35487	15
9785	1977-005A	NATO 3B	NATO	37436	37063	13.6
9852	1977-014A	KIKU 2 (ETS 2)	JPN	35881	35833	14.5
9862	1977-018A	PALAPA 2	INDO	35890	35821	11.4
10000	1977-034A	OPS 9437 (DSCS II-7)	US	36879	36774	15.6
10001	1977-034B	OPS 9438 (DSCS II-8)	US	37379	37025	15.3
10002	1977-034C	TITAN 3CTRANSTAGE R/B	US	38491	35831	15.9
10024	1977-041A	INTELSAT 4A-F4	ITSO	36070	35972	11.6
10061	1977-048A	GOES 2	US	35797	35723	13.9
10143	1977-065A	HIMAWARI 1 (GMS 1)	JPN	36153	36000	14.1
10159	1977-071A	RADUGA 3	CIS	35848	35767	15
10294	1977-080A	SIRIO	IT	35892	35658	12.6
10365	1977-092A	EKRAN 2	CIS	35899	35679	14.9
10489	1977-108A	METEOSAT 1	ESA	35842	35717	14.5
10516	1977-118A	SAKURA 1A (CS-1A)	JPN	36203	36139	13.8
10557	1978-002A	INTELSAT 4A-F3	ITSO	35911	35872	11.1
10637	1978-012A	IUE	US	42219	29461	38.3
10778	1978-035A	INTELSAT 4A-F6	ITSO	35773	35757	11.1
10792	1978-039A	YURI (BSE)	JPN	35854	35745	14.4
10855	1978-044A	OTS 2	ESA	36118	36095	12.8
10953	1978-062A	GOES 3	US	35816	35754	13.1
10975	1978-068A	COMSTAR 3	US	36156	36023	11.1
10981	1978-071A	GEOS 2	ESA	36045	36032	14.5
10987	1978-073A	RADUGA 4	CIS	35829	35769	14.7
11115	1978-106A	NATO 3C	NATO	36304	36284	11.4
11144	1978-113A	OPS 9441 (DSCS II-11)	US	37758	37505	13.6
11145	1978-113B	OPS 9442 (DSCS II-12)	US	36350	36265	13.3
11147	1978-113D	TITAN 3CTRANSTAGE R/B	US	38920	36516	15.5
11153	1978-116A	ANIK B1 (TELESAT-4)	CA	35946	35880	10.6
11158	1978-118A	GORIZONT 1	CIS	49748	21816	25.3
11256	1979-007A	SCATHA	US	43137	27737	15.2
11261	1979-009A	AYAME 1 (ECS 1)	JPN	37278	29407	5.6
11273	1979-015A	EKRAN 3	CIS	35944	35659	14.5
11343	1979-035A	RADUGA 5	CIS	35793	35767	14.5
11353	1979-038A	OPS 6392 (FLTSATCOM 2)	US	36365	36192	13.2
11440	1979-062A	GORIZONT 2	CIS	35803	35788	14.3
11484	1979-072A	WESTAR 3	US	35889	35863	9.7
11561	1979-087A	EKRAN 4	CIS	35879	35721	14.4
11567	1974-017F	SL-12 R/B(2)	CIS	35763	35401	14.4
11568	1975-123F	SL-12 R/B(2)	CIS	35806	35647	14.9
11569	1976-107F	SL-12 R/B(2)	CIS	35520	35392	14.7
11570	1977-071F	SL-12 R/B(2)	CIS	36579	36444	15.6
11571	1977-092G	SL-12 R/B(2)	CIS	35569	35443	14.6
11581	1977-092H	EKRAN 2 DEB	CIS	35861	35718	14.9
11621	1979-098A	OPS 9443 (DSCS II-13)	US	37210	37088	13.2
11622	1979-098B	OPS 9444 (DSCS II-14)	US	36367	36302	12.9
11623	1979-098C	TITAN 3CTRANSTAGE R/B	US	38612	35858	14.6
11648	1979-105A	GORIZONT 3	CIS	35842	35747	14.1
11669	1980-004A	OPS 6393 (FLTSATCOM 3)	US	35914	35708	12
11676	1975-097F	SL-12 R/B(2)	CIS	35947	35724	15.1
11684	1979-105E	SL-12 R/B(2)	CIS	36296	36180	14.4
11708	1980-016A	RADUGA 6	CIS	35819	35791	14.2
11715	1980-018A	AYAME 2 (ECS-2)	JPN	36839	32784	1.4
11728	1980-016D	SL-12 R/B(2)	CIS	36597	36494	14.8
11841	1980-049A	GORIZONT 4	CIS	36260	36246	14.1
11862	1980-049F	SL-12 R/B(2)	CIS	36590	36318	14.4
11890	1980-060A	EKRAN 5	CIS	35868	35723	11.3
11926	1978-118C	SL-12 R/B(2)	CIS	49088	21747	24.8
11941	1978-073E	SL-12 R/B(2)	CIS	36648	36474	15.4
11964	1980-074A	GOES 4	US	36241	35920	12.9
12003	1980-081A	RADUGA 7	CIS	35799	35739	13.9
12065	1980-091A	SBS 1	US	35962	35865	10.3
12089	1980-098A	INTELSAT 502	ITSO	36218	36141	9
12120	1980-104A	EKRAN 6	CIS	35802	35746	13.8
12309	1981-018A	COMSTAR 4	US	35792	35782	11
12351	1981-027A	RADUGA 8	CIS	36119	35398	13.9
12447	1980-081F	SL-12 R/B(2)	CIS	35880	35858	14
12471	1980-104E	SL-12 R/B(2)	CIS	35638	35340	13.6
12472	1981-049A	GOES 5	US	35805	35787	10.5
12474	1981-050A	INTELSAT 501	ITSO	36208	36167	9.5
12544	1981-057A	METEOSAT 2	ESA	36338	36113	10.6
12545	1981-057B	APPLE	IND	35972	35739	13.2
12564	1981-061A	EKRAN 7	CIS	35796	35765	13.6
12618	1981-069A	RADUGA 9	CIS	35790	35736	13.5
12635	1981-073A	FLTSATCOM 5	US	36323	36196	13.9
12677	1981-076A	HIMAWARI 2 (GMS 2)	JPN	36018	35963	12.6
12850	1981-069F	SL-12 R/B(2)	CIS	36630	36416	14.1
12851	1981-061F	SL-12 R/B(2)	CIS	35604	35559	13.4
12855	1981-096A	SBS 2	US	35857	35781	9.5
12897	1981-102A	RADUGA 10	CIS	35811	35759	13.4

Cat No	*International Designator*	*Common Name*	*Source*	*Apogee (km)*	*Perigee (km)*	*Inclination (°)*
12967	1981-114A	SATCOM 3R	US	35874	35776	7.5
12994	1981-119A	INTELSAT 503	ITSO	36120	35908	8.7
13010	1981-122A	MARECS A	ESA	37834	36807	10.3
13035	1982-004A	SATCOM 4	US	36004	35955	6.9
13056	1982-009A	EKRAN 8	CIS	35972	35779	13.4
13069	1982-014A	WESTAR 4	US	35942	35914	7
13083	1982-017A	INTELSAT 504	ITSO	36228	35937	8.7
13092	1982-020A	GORIZONT 5	CIS	36425	36140	13.5
13129	1982-031A	INSAT 1A	IND	35882	35701	13
13177	1982-044A	COSMOS 1366	CIS	35816	35748	13.3
13269	1982-058A	WESTAR 5	US	36139	36030	6.7
13431	1982-082A	ANIK D1 (TELESAT 6)	CA	35847	35801	7.2
13554	1982-093A	EKRAN 9	CIS	35860	35688	12.8
13595	1982-097A	INTELSAT 505	ITSO	36437	36197	8.3
13624	1982-103A	GORIZONT 6	CIS	35769	35745	12.5
13630	1982-103E	SL-12 R/B(2)	CIS	35823	35686	12.5
13631	1982-105A	SATCOM C5	US	35804	35768	7.3
13636	1982-106A	DSCS II-15	US	37331	37278	10.7
13643	1982-106D	TITAN 34D IUS R/B(2)	US	36232	35843	11.8
13651	1982-110B	SBS 3	US	35931	35837	7
13652	1982-110C	ANIK C3 (TELESAT-5)	CA	35910	35873	7.1
13669	1982-113A	RADUGA 11	CIS	36688	36354	12.4
13753	1976-023K	LES 8,9/SOL 11A,B DEB	US	35506	35470	14.1
13782	1983-006A	SAKURA 2A (CS-2A)	JPN	36048	36013	10
13878	1983-016A	EKRAN 10	CIS	37464	37183	13.9
13899	1982-020F	SL-12 R/B(2)	CIS	36381	36123	13.6
13900	1979-015D	SL-12 R/B(2)	CIS	35521	35461	14.3
13907	1977-108D	METEOSAT 1 AKM	ESA	36896	36089	16
13954	1982-113F	SL-12 R/B(2)	CIS	36646	36476	12.5
13969	1983-026B	TDRS 1	US	35804	35768	11
13974	1983-028A	RADUGA 12	CIS	35808	35777	12
13983	1983-028F	SL-12 R/B(2)	CIS	35968	35735	12
13984	1983-030A	SATCOM 1R	US	35970	35836	6.1
14005	1979-062D	SL-12 R/B(2)	CIS	36555	36509	14.9
14050	1983-041A	GOES 6	US	35794	35757	9.5
14077	1983-047A	INTELSAT 506	ITSO	36107	36091	7.5
14086	1983-016F	SL-12 R/B(2)	CIS	35600	35511	12.7
14114	1982-044F	SL-12 R/B(2)	CIS	35870	35701	13.3
14115	1982-093F	SL-12 R/B(2)	CIS	35532	35493	12.7
14117	1982-009F	SL-12 R/B(2)	CIS	35761	35417	13.1
14128	1983-058A	EUTELSAT 1-F1 (ECS 1)	EUTE	36239	36130	8.5
14133	1983-059B	NAHUEL I2 (ANIK C2)	ARGN	36295	35930	7
14134	1983-059C	PALAPA B1	INDO	35846	35772	7.9
14158	1983-065A	GALAXY 1	US	35826	35816	5.4
14160	1983-066A	GORIZONT 7	CIS	36364	36305	11.9
14193	1980-060F	SL-12 R/B(2)	CIS	35488	35346	13.7
14194	1981-027F	SL-12 R/B(2)	CIS	36621	36448	14.5
14195	1981-102F	SL-12 R/B(2)	CIS	35883	35761	13.4
14234	1983-077A	ARABSAT 1DR (TELSTAR 3A)	AB	36115	35878	5.5
14248	1983-081A	SAKURA 2B (CS-2B)	JPN	36234	36166	9.5
14307	1983-088A	RADUGA 13	CIS	36465	36305	12
14318	1983-089B	INSAT 1B	IND	35788	35723	8.4
14328	1983-094A	SATCOM 2R	US	36124	35932	4.7
14333	1983-088F	SL-12 R/B(2)	CIS	36633	36462	12.1
14365	1983-098A	GALAXY 2	US	35847	35771	5.2
14377	1983-100A	EKRAN 11	CIS	35796	35769	12.2
14394	1983-100F	SL-12 R/B(2)	CIS	35672	35469	12.1
14421	1983-105A	INTELSAT 507	ITSO	35995	35903	7.8
14532	1983-118A	GORIZONT 8	CIS	36477	36237	11.6
14548	1983-118F	SL-12 R/B(2)	CIS	35968	35620	11.3
14659	1984-005A	YURI 2A (BS-2A)	JPN	36162	36103	9.4
14725	1984-016A	RADUGA 14	CIS	35809	35751	11.4
14783	1984-022A	COSMOS 1540	CIS	35825	35746	12.4
14786	1984-023A	INTELSAT 508	ITSO	36747	36540	7.4
14821	1984-028A	EKRAN 12	CIS	37077	36944	13
14867	1984-031A	COSMOS 1546	CIS	35868	35708	11.2
14899	1984-035A	STTW-T2	PRC	35856	35767	10.1
14940	1984-041A	GORIZONT 9	CIS	35826	35773	11.1
14943	1984-041D	SL-12 R/B(2)	CIS	36326	36183	11.3
14948	1984-022F	SL-12 R/B(2)	CIS	36018	35786	12.5
14951	1984-031F	SL-12 R/B(2)	CIS	36090	35961	11.4
14985	1984-049A	CHINASAT 5 (SPACENET 1)	PRC	35889	35838	3.9
15057	1984-063A	RADUGA 15	CIS	35815	35766	11.1
15139	1984-028F	SL-12 R/B(2)	CIS	35545	35388	12.1
15141	1983-066F	SL-12 R/B(2)	CIS	36606	36487	12.1
15144	1984-078A	GORIZONT 10	CIS	35809	35772	10.9
15152	1984-080A	HIMAWARI 3 (GMS 3)	JPN	35962	35860	9.2
15158	1984-081A	EUTELSAT 1-F2 (ECS 2)	EUTE	36243	36167	7.8
15159	1984-081B	TELECOM 1A	FR	36473	36164	7.8
15181	1984-078F	SL-12 R/B(2)	CIS	35866	35669	10.8
15219	1984-090A	EKRAN 13	CIS	37099	36944	12.3
15235	1984-093B	SBS 4	US	35796	35778	5.9
15236	1984-093C	LEASAT 2	US	36719	36472	12.5
15237	1984-093D	TELSTAR 302	US	35956	35891	4.7
15308	1984-101A	GALAXY 3	US	35948	35869	4.7
15383	1984-113B	ARABSAT 1D (ANIK D2)	AB	36288	36060	5.8
15384	1984-113C	LEASAT 1	US	36390	36164	7
15385	1984-114A	SPACENET 2	US	35988	35875	3.1
15386	1984-114B	MARECS B2	ESA	35801	35772	9.2
15391	1984-115A	NATO 3D	NATO	35808	35765	6.2
15484	1985-007A	GORIZONT 11	CIS	35789	35743	10.5
15487	1985-007D	SL-12 R/B(2)	CIS	35089	34974	10
15560	1985-015A	ARABSAT 1A	AB	35772	35728	7.5
15561	1985-015B	BRAZILSAT 1	BRAZ	35791	35780	4.9
15574	1985-016A	COSMOS 1629	CIS	35808	35760	10.7
15581	1985-016F	SL-12 R/B(2)	CIS	36135	35925	10.8
15626	1985-024A	EKRAN 14	CIS	37459	37331	12.1

Cat No	*International Designator*	*Common Name*	*Source*	*Apogee (km)*	*Perigee (km)*	*Inclination (°)*
15629	1985-025A	INTELSAT 510	ITSO	36487	36217	6.4
15630	1985-024D	SL-12 R/B(2)	CIS	35582	35459	11.1
15642	1985-028B	NAHUEL I1 (ANIK C1)	ARGN	35792	35782	3.2
15643	1985-028C	LEASAT 3	US	37060	36421	13.9
15677	1985-035A	GSTAR 1	US	35824	35749	3.8
15678	1985-035B	TELECOM 1B	FR	35817	35752	9.5
15693	1984-063F	SL-12 R/B(2)	CIS	35029	34892	10.6
15824	1985-048B	MORELOS 1	MEX	36039	35986	5.3
15825	1985-048C	ARABSAT 1B	AB	35802	35706	6.9
15826	1985-048D	TELSTAR 303	US	35913	35884	4.4
15873	1985-055A	INTELSAT 511	ITSO	35812	35760	6
15946	1985-070A	RADUGA 16	CIS	35777	35748	10.3
15963	1985-070F	SL-12 R/B(2)	CIS	36556	36429	10.7
15993	1985-076B	OPTUS A1 (AUSSAT 1)	AUS	35953	35903	6.1
15994	1985-076C	ASC 1	US	35811	35767	5
15995	1985-076D	LEASAT 4	US	36523	36467	9
16101	1985-087A	INTELSAT 512	ITSO	36126	36102	5.3
16199	1985-102A	COSMOS 1700	CIS	35813	35751	10
16214	1985-102D	SL-12 R/B(2)	CIS	35786	35582	9.9
16250	1985-107A	RADUGA 17	CIS	35882	35772	10.1
16274	1985-109B	MORELOS 2	MEX	35794	35778	2.1
16275	1985-109C	OPTUS A2 (AUSSAT 2)	AUS	35794	35783	5.7
16276	1985-109D	SATCOM K2	US	35802	35773	3.1
16339	1985-107F	SL-12 R/B(2)	CIS	36681	36486	10.5
16482	1986-003B	SATCOM K1	US	36015	35962	3.4
16497	1986-007A	RADUGA 18	CIS	36504	35896	10.2
16526	1986-010A	STTW-1	PRC	35782	35757	9.1
16597	1986-016A	BS-2B (YURI 2B)	JPN	36139	35990	7.7
16649	1986-026A	GSTAR 2	US	35957	35881	4.5
16650	1986-026B	BRAZILSAT 2	BRAZ	35792	35782	3.6
16667	1986-027A	COSMOS 1738	CIS	35866	35703	10.1
16676	1986-027F	SL-12 R/B(2)	CIS	36696	36356	10.5
16729	1986-038A	EKRAN 15	CIS	36918	36811	11
16732	1986-038D	SL-12 R/B(2)	CIS	35581	35381	10.3
16769	1986-044A	GORIZONT 12	CIS	35794	35779	9.6
16797	1986-044F	SL-12 R/B(2)	CIS	36582	36481	9.9
16870	1986-007F	SL-12 R/B(2)	CIS	36633	36354	10.4
17046	1986-082A	RADUGA 19	CIS	36360	36243	9.5
17065	1986-082F	SL-12 R/B(2)	CIS	36681	36425	9.7
17083	1986-090A	GORIZONT 13	CIS	36868	36754	9.6
17125	1986-090D	SL-12 R/B(2)	CIS	35840	35735	9.2
17561	1987-022A	GOES 7	US	35790	35774	6.3
17611	1987-028A	RADUGA 20	CIS	37177	36901	10.2
17705	1987-028D	SL-12 R/B(2)	CIS	36033	35771	9.8
17706	1987-029A	AGILA 1 (PALAPA B2P)	RP	35921	35861	4
17872	1976-092F	SL-12 R/B(2)	CIS	35869	35696	15
17873	1979-035E	SL-12 R/B(2)	CIS	35938	35694	14.5
17874	1984-016F	SL-12 R/B(2)	CIS	35948	35641	11.3
17875	1984-090F	SL-12 R/B(2)	CIS	35599	35420	11.4
17939	1979-087C	SL-12 R/B(2)	CIS	35861	35598	14.3
17969	1987-040A	GORIZONT 14	CIS	36662	36409	11
17972	1987-040D	SL-12 R/B(2)	CIS	35080	34988	10.4
18316	1987-070A	KIKU 5 (ETS 5)	JPN	36105	35997	7.1
18328	1987-073A	EKRAN 16	CIS	36918	36848	9.9
18331	1987-073D	SL-12 R/B(2)	CIS	35564	35389	9.3
18350	1987-078A	OPTUS A3 (AUSSAT 3)	AUS	35807	35766	3.6
18351	1987-078B	EUTELSAT 1-F4 (ECS 4)	EUTE	35805	35765	5.8
18384	1987-084A	COSMOS 1888	CIS	35805	35764	8.6
18387	1987-084D	SL-12 R/B(2)	CIS	35974	35735	8.6
18443	1987-091A	COSMOS 1894	CIS	35831	35804	8.7
18446	1987-091D	SL-12 R/B(2)	CIS	35836	35670	8.7
18570	1987-095A	TVSAT 1	GER	36449	35892	8.8
18575	1987-096A	COSMOS 1897	CIS	35841	35751	8.5
18578	1987-096D	SL-12 R/B(2)	CIS	35788	35613	8.4
18631	1987-100A	RADUGA 21	CIS	35815	35766	8.7
18634	1987-100D	SL-12 R/B(2)	CIS	34991	34872	8.4
18715	1987-109A	EKRAN 17	CIS	37247	36880	8.7
18718	1987-109D	SL-12 R/B(2)	CIS	35901	35360	8.2
18877	1988-012A	CS 3A	JPN	36429	36363	3.8
18922	1988-014A	STTW-2	PRC	35808	35747	5.8
18951	1988-018A	SPACENET 3R	US	35979	35941	1.3
18952	1988-018B	TELECOM 1C	FR	36880	36025	4.4
19017	1988-028A	GORIZONT 15	CIS	36637	36335	8.7
19020	1988-028D	SL-12 R/B(2)	CIS	36606	36391	8.7
19073	1988-034A	COSMOS 1940	CIS	35746	35601	8.3
19076	1988-034D	SL-12 R/B(2)	CIS	35953	35726	8.3
19090	1988-036A	EKRAN 18	CIS	37359	37215	9.5
19094	1988-036E	SL-12 R/B(2)	CIS	35658	35446	8.9
19121	1988-040A	INTELSAT 513	ITSO	35800	35772	4.4
19215	1988-051A	METEOSAT 3	ESA	36784	36706	6.6
19217	1988-051C	PAS 1	US	35799	35776	0.1
19330	1988-063A	INSAT 1C	IND	35842	35750	8.3
19331	1988-063B	EUTELSAT 1-F5 (ECS 5)	EUTE	36459	36317	5.1
19344	1988-066A	COSMOS 1961	CIS	35809	35757	7.9
19347	1988-066D	SL-12 R/B(2)	CIS	36404	36085	8.1
19397	1988-071A	GORIZONT 16	CIS	35942	35792	8
19400	1988-071D	SL-12 R/B(2)	CIS	35793	35631	7.9
19483	1988-081A	GSTAR 3	US	35820	35753	10.1
19484	1988-081B	SBS 5	US	35943	35897	0.9
19508	1988-086A	CS 3B	JPN	36106	36059	2.9
19548	1988-091B	TDRS 3	US	35802	35771	5.8
19550	1988-091D	IUS R/B(2)	US	35787	35669	7.5
19596	1988-095A	RADUGA 22	CIS	35824	35784	7.8
19621	1988-098A	TDF 1	FR	36591	35564	3.8
19683	1988-108A	EKRAN 19	CIS	36954	36710	8
19686	1988-108D	SL-12 R/B(2)	CIS	35509	35375	7.7
19687	1988-109A	SKYNET 4B	UK	35964	35943	6.6

Cat No	International Designator	Common Name	Source	Apogee (km)	Perigee (km)	Inclination (°)
19688	1988-109B	ASTRA 1A	LUXE	35799	35774	0
19710	1988-111A	STTW-3	PRC	35799	35743	4.5
19765	1989-004A	GORIZONT 17	CIS	36227	36028	7.6
19772	1989-006A	INTELSAT 515	ITSO	35800	35776	2.9
19776	1989-004F	SL-12 R/B(2)	CIS	36554	36319	7.8
19777	1988-095F	SL-12 R/B(2)	CIS	36502	36404	8
19874	1989-020A	JCSAT 1	JPN	36024	35967	3
19876	1989-020B	METEOSAT 4 (MOP 1)	ESA	36785	36608	5.5
19883	1989-021B	TDRS 4	US	35801	35770	3.5
19913	1989-021D	IUS R/B(2)	US	35811	35578	9.6
19919	1989-027A	TELE-X	SWED	36106	36081	3.3
19928	1989-030A	RADUGA 23	CIS	35896	35712	7.4
19931	1989-030D	SL-12 R/B(2)	CIS	36534	36380	7.7
20040	1989-041A	SUPERBIRD A	JPN	35948	35918	7.5
20041	1989-041B	DFS 1	GER	35706	35649	4.1
20083	1989-048A	RADUGA 1-1	CIS	36321	36141	7.4
20086	1989-048D	SL-12 R/B(2)	CIS	36578	36357	7.5
20107	1989-052A	GORIZONT 18	CIS	36164	35903	7.2
20110	1989-052D	SL-12 R/B(2)	CIS	35146	34898	6.9
20122	1989-053A	OLYMPUS 1	ESA	35519	35445	6.5
20168	1989-062A	TVSAT 2	GER	35951	35860	2.2
20193	1989-067A	SIRIUS (MARCOPOLO 1)	SWED	35799	35774	1.3
20217	1989-070A	HIMAWARI 4 (GMS 4)	JPN	36814	36402	5.2
20263	1989-081A	GORIZONT 19	CIS	35772	35759	7.1
20266	1989-081D	SL-12 R/B(2)	CIS	35830	35561	7
20315	1989-087A	INTELSAT 602	ITSO	35790	35782	0
20317	1989-070C	HIMAWARI 4 AKM	JPN	37292	35138	6.8
20367	1989-098A	RADUGA 24	CIS	35779	35756	7.1
20370	1989-098D	SL-12 R/B(2)	CIS	36559	36396	7.3
20391	1989-101A	COSMOS 2054	CIS	35810	35740	6.9
20394	1989-101D	SL-12 R/B(2)	CIS	36404	36321	7.1
20401	1990-001A	SKYNET 4A	UK	35792	35783	4.7
20402	1990-001B	JCSAT 2	JPN	35794	35780	0
20410	1990-002B	LEASAT 5	US	35793	35780	4.1
20473	1990-011A	STTW-4	PRC	35826	35777	4.3
20499	1990-016A	RADUGA 25	CIS	35836	35780	6.8
20502	1990-016D	SL-12 R/B(2)	CIS	36018	35685	6.8
20523	1990-021A	INTELSAT 603	ITSO	35826	35748	0
20558	1990-030A	ASIASAT 1	AC	35791	35782	1.6
20570	1990-034A	PALAPA B2R	INDO	35798	35774	0.8
20643	1990-051A	INSAT 1D	IND	35842	35729	2.6
20659	1990-054A	GORIZONT 20	CIS	35812	35763	6.6
20662	1990-054D	SL-12 R/B(2)	CIS	35808	35644	6.5
20667	1990-056A	INTELSAT 604	ITSO	35789	35783	0
20693	1990-061A	COSMOS 2085	CIS	35804	35774	6.5
20696	1990-061D	SL-12 R/B(2)	CIS	35937	35676	6.5
20705	1990-063A	TDF 2	FR	36421	36098	2
20706	1990-063B	DFS 2	GER	35795	35776	0.3
20762	1990-074A	THOR 1 (MARCOPOLO 2)	NOR	35794	35777	0
20771	1990-077A	BS-3A (YURI 3A)	JPN	36229	36175	2.4
20776	1990-079A	SKYNET 4C	UK	35799	35774	3.4
20777	1990-079B	EUTELSAT 2-F1	EUTE	35804	35767	1.9
20799	1977-048G	GOES 2 AKM	US	36623	34778	14.6
20800	1989-020E	METEOSAT 4 AKM	ESA	36296	35168	7.5
20801	1978-062D	GOES 3 AKM	US	36482	35538	14.8
20835	1975-011F	SMS 2 AKM	US	36708	35822	15.5
20836	1974-060F	SL-12 R/B(2)	CIS	35910	35711	15
20837	1981-057F	METEOSAT 2 AKM (MAGE 1)	ESA	36364	35715	13.4
20872	1990-091A	SBS 6	US	35792	35780	0
20873	1990-091B	GALAXY 6	US	35795	35778	0
20918	1990-093A	INMARSAT 2-F1	IM	35804	35770	1.8
20923	1990-094A	GORIZONT 21	CIS	35829	35803	6.3
20926	1990-094D	SL-12 R/B(2)	CIS	35778	35465	6.3
20945	1990-100A	SATCOM C1	US	35802	35771	0
20946	1990-100B	GSTAR 4	US	35793	35780	0
20953	1990-102A	GORIZONT 22	CIS	35819	35803	6.2
20962	1975-100F	GOES 1 AKM	US	36478	34176	14.7
21016	1990-112A	RADUGA 26	CIS	35812	35735	6.2
21019	1990-112D	SL-12 R/B(2)	CIS	35976	35741	6.2
21038	1990-116A	RADUGA 1-2	CIS	35800	35745	6.2
21041	1990-116D	SL-12 R/B(2)	CIS	36616	36286	6.3
21046	1990-102D	SL-12 R/B(2)	CIS	36555	36394	6.4
21047	1991-001A	NATO 4A	NATO	35799	35775	3.2
21052	1969-069D	ATS 5 AKM	US	36927	35840	15.2
21055	1991-003A	ITALSAT 1	IT	35903	35832	3.2
21056	1991-003B	EUTELSAT 2-F2	EUTE	35810	35763	1.1
21111	1991-010A	COSMOS 2133	CIS	35812	35754	5.1
21129	1991-010F	SL-12 R/B(2)	CIS	35894	35754	5.2
21132	1991-014A	RADUGA 27	CIS	35802	35763	6.4
21135	1991-014D	SL-12 R/B(2)	CIS	35042	34805	6.2
21139	1991-015A	ASTRA 1B	LUXE	35799	35774	0
21140	1991-015B	METEOSAT 5 (MOP 2)	ESA	35793	35777	4.1
21149	1991-018A	INMARSAT 2-F2	IM	35804	35769	1.7
21222	1991-026A	ANIK E2	CA	35791	35783	0
21227	1991-028A	SPACENET 4 (ASC 2)	US	35792	35780	0
21392	1991-037A	AURORA 2	US	36152	36133	0
21533	1991-046A	GORIZONT 23	CIS	36190	36157	5.9
21536	1991-046D	SL-12 R/B(2)	CIS	35662	35541	5.7
21639	1991-054B	TDRS 5	US	35791	35777	2.6
21641	1991-054D	IUS R/B(2)	US	35887	35627	7.4
21648	1989-101G	COSMOS 2054 DEB	CIS	36579	36380	7.1
21653	1991-055A	INTELSAT 605	ITSO	35793	35780	0
21668	1991-060A	BS-3B (YURI 3B)	JPN	36243	36173	2.4
21702	1991-064A	COSMOS 2155	CIS	35837	35803	5.8
21703	1991-064B	SL-12 R/B(2)	CIS	35920	35860	5.8
21726	1991-067A	ANIK E1	CA	35793	35780	0
21759	1991-074A	GORIZONT 24	CIS	36379	36234	5.6

Cat No	*International Designator*	*Common Name*	*Source*	*Apogee (km)*	*Perigee (km)*	*Inclination (°)*
21762	1991-074D	SL-12 R/B(2)	CIS	35963	35929	5.6
21765	1991-075A	INTELSAT 601	ITSO	35817	35756	0.1
21789	1991-079A	COSMOS 2172	CIS	35797	35773	5.5
21792	1991-079D	SL-12 R/B(2)	CIS	36294	36217	5.7
21803	1991-083A	EUTELSAT 2-F3	EUTE	35803	35769	0.9
21813	1991-084A	TELECOM 2A	FR	35807	35765	0.1
21814	1991-084B	INMARSAT 2-F3	IM	35820	35752	1.2
21821	1991-087A	RADUGA 28	CIS	35819	35748	5.5
21824	1991-087D	SL-12 R/B(2)	CIS	36523	36337	5.6
21893	1992-010A	SUPERBIRD B1	JPN	35792	35781	0
21894	1992-010B	INSAT 2R (ARABSAT 1C)	IND	36004	35566	0.1
21904	1991-015E	METEOSAT 5 AKM	ESA	36498	35155	6.1
21906	1992-013A	GALAXY 5	US	35792	35780	0
21922	1992-017A	GORIZONT 25	CIS	35803	35774	5.2
21925	1992-017D	SL-12 R/B(2)	CIS	35665	35455	5.2
21939	1992-021A	TELECOM 2B	FR	35792	35781	0.3
21940	1992-021B	INMARSAT 2-F4	IM	35796	35779	2.5
21964	1992-027A	PALAPA B4	INDO	35790	35786	0
21989	1992-032A	INTELSAT K	ITSO	35805	35769	0
22027	1992-041A	INSAT 2A	IND	35875	35746	3
22028	1992-041B	EUTELSAT 2-F4	EUTE	35824	35748	0
22041	1992-043A	GORIZONT 26	CIS	35791	35784	5
22044	1992-043D	SL-12 R/B(2)	CIS	36593	36376	5.1
22087	1992-054A	OPTUS B1 (AUSSAT B1)	AUS	35801	35772	0
22096	1992-057A	SATCOM C4	US	35795	35776	0
22112	1992-059A	COSMOS 2209	CIS	35812	35778	5
22115	1992-059D	SL-12 R/B(2)	CIS	35957	35884	5.1
22116	1992-060A	HISPASAT 1A	SPN	35794	35781	0
22117	1992-060B	SATCOM C3	US	35797	35775	0.1
22175	1992-066A	DFS 3	GER	35961	35614	0
22205	1992-072A	GALAXY 7	US	36023	35866	0.3
22210	1992-074A	EKRAN 20	CIS	35802	35771	4.8
22213	1992-074D	SL-12 R/B(2)	CIS	35648	35441	4.7
22245	1992-082A	GORIZONT 27	CIS	35798	35773	4.8
22248	1992-082D	SL-12 R/B(2)	CIS	36483	36375	4.9
22253	1992-084A	SUPERBIRD A1	JPN	35805	35768	0
22266	1984-080E	N-2 R/B(3)	JPN	36554	35313	10.6
22269	1992-088A	COSMOS 2224	CIS	35791	35759	3.8
22272	1992-088D	SL-12 R/B(2)	CIS	35890	35793	3.8
22314	1993-003B	TDRS 6	US	35797	35777	1.9
22316	1993-003D	IUS R/B(2)	US	36139	35508	4.9
22557	1993-013A	RADUGA 29	CIS	35796	35772	4.5
22563	1993-015A	UFO 1 (USA 98)	US	36112	36037	22
22624	1993-013D	SL-12 R/B(2)	CIS	36516	36346	4.6
22653	1993-031A	ASTRA 1C	LUXE	35799	35774	0
22694	1993-039A	GALAXY 4	US	35811	35760	2.3
22723	1993-048A	HISPASAT 1B	SPN	35794	35779	0
22724	1993-048B	INSAT 2B	IND	35816	35760	0.4
22796	1993-058B	ACTS	US	35808	35765	2.1
22836	1993-062A	RADUGA 30	CIS	35796	35771	4.1
22839	1993-062D	SL-12 R/B(2)	CIS	35843	35718	4.1
22871	1993-066A	INTELSAT 701	ITSO	35799	35773	0
22880	1993-069A	GORIZONT 28	CIS	35803	35776	4.1
22883	1993-069D	SL-12 R/B(2)	CIS	35767	35693	4.1
22907	1993-072A	GORIZONT 29	CIS	35792	35782	4
22910	1993-072D	SL-12 R/B(2)	CIS	35083	34905	3.8
22911	1993-073A	SOLIDARIDAD 1	MEX	35798	35771	0.4
22912	1993-073B	METEOSAT 6	ESA	35802	35772	1.1
22921	1993-076A	NATO 4B	NATO	35798	35775	1.6
22927	1993-077A	TELSTAR 401	US	35816	35750	3.3
22930	1993-078A	DBS 1	US	35796	35777	0
22931	1993-078B	THAICOM 1	THAI	35790	35779	0.1
22963	1994-002A	GALS 1	CIS	35798	35768	2.5
22966	1994-002D	SL-12 R/B(2)	CIS	36204	35823	5
22981	1994-008A	RADUGA 1-3	CIS	35816	35790	3.9
22984	1994-008D	SL-12 R/B(2)	CIS	36549	36340	4
23010	1994-012A	RADUGA 31	CIS	35833	35760	3.8
23013	1994-012D	SL-12 R/B(2)	CIS	36567	36240	3.9
23016	1994-013A	GALAXY 1R	US	35790	35782	0
23051	1994-022A	GOES 8	US	35797	35779	0.2
23108	1994-030A	GORIZONT 30	CIS	35801	35778	3.8
23111	1994-030D	SL-12 R/B(2)	CIS	34996	34633	3.6
23118	1993-073E	METEOSAT 6 AKM	ESA	36177	35559	4.2
23124	1994-034A	INTELSAT 702	ITSO	35801	35771	0
23168	1994-038A	COSMOS 2282	CIS	35870	35761	2.7
23171	1994-038D	SL-12 R/B(2)	CIS	36138	35909	2.7
23175	1994-040A	PAS 2	US	35797	35776	0
23176	1994-040B	BS-3N	JPN	35794	35782	0.1
23185	1994-043A	APSTAR 1	PRC	35790	35783	0.1
23192	1994-047A	DBS 2	US	35795	35778	0
23199	1994-049A	BRAZILSAT B1	BRAZ	35795	35778	0
23200	1994-049B	TURKSAT 1B	TURK	35796	35778	0
23227	1994-055A	OPTUS B3	AUS	35813	35760	0
23267	1994-060A	COSMOS 2291	CIS	35828	35763	3.5
23270	1994-060D	SL-12 R/B(2)	CIS	35886	35799	3.4
23305	1994-064A	INTELSAT 703	ITSO	35813	35758	0
23313	1994-065A	SOLIDARIDAD 2	MEX	35799	35773	0
23314	1994-065B	THAICOM 2	THAI	35818	35756	0.1
23319	1994-067A	EXPRESS 1	CIS	35801	35772	0.6
23322	1994-067D	SL-12 R/B(2)	CIS	35826	35737	4.5
23327	1994-069A	ELEKTRO (GOMS)	CIS	35814	35762	3.7
23330	1994-069D	SL-12 R/B(2)	CIS	35941	35730	3.6
23331	1994-070A	ASTRA 1D	LUXE	35799	35772	0.1
23413	1994-079A	ORION 1	US	35795	35780	0
23415	1994-080A	DFH-3 1	PRC	35971	35214	4.4
23426	1994-082A	LUCH	CIS	35810	35751	2.4
23429	1994-082D	SL-12 R/B(2)	CIS	35769	35507	2.3

Cat No	International Designator	Common Name	Source	Apogee (km)	Perigee (km)	Inclination (°)
23448	1994-087A	RADUGA 32	CIS	35790	35776	3.2
23451	1994-087D	SL-12 R/B(2)	CIS	35903	35775	3.2
23461	1995-001A	INTELSAT 704	ITSO	35840	35732	0
23522	1995-011B	GMS 5	JPN	35793	35786	0.7
23524	1995-011D	GMS 5 AKM	JPN	36406	34621	3.4
23528	1995-013A	INTELSAT 705	ITSO	35802	35772	0
23536	1995-016A	BRAZILSAT B2	BRAZ	35795	35780	0
23537	1995-016B	HOT BIRD 1	EUTE	35800	35770	0
23553	1995-019A	AMSC 1	US	35796	35777	0
23571	1995-023A	INTELSAT 706	ITSO	35798	35777	0
23581	1995-025A	GOES 9	US	35831	35741	0.6
23598	1995-029A	DBS 3	US	35796	35778	0
23613	1995-035B	TDRS 7	US	35797	35773	4.3
23615	1995-035D	IUS R/B(2)	US	35768	35671	5.9
23636	1995-040A	PAS 4	US	35799	35774	0
23639	1995-041A	KOREASAT 1	KOR	35906	35666	1.6
23649	1995-043A	JCSAT 3	JPN	35800	35773	0
23651	1995-044A	NSTAR 1	JPN	35799	35775	0
23653	1995-045A	COSMOS 2319	CIS	35810	35763	2.7
23656	1995-045D	SL-12 R/B(2)	CIS	35899	35723	2.7
23670	1995-049A	TELSTAR 402R	US	35797	35778	0
23680	1995-054A	LUCH 1	CIS	35828	35744	1.5
23683	1995-054D	SL-12 R/B(2)	CIS	35894	35679	1.5
23686	1995-055A	ASTRA 1E	LUXE	35799	35774	0
23715	1995-062A	ISO	ESA	70217	578	4.6
23716	1995-062B	ARIANE 44P+3 R/B	ESA	73763	903	6.6
23717	1995-063A	GALS 2	CIS	35792	35775	0.5
23720	1995-063D	SL-12 R/B(2)	CIS	35942	35592	3.8
23723	1995-064A	ASIASAT 2	AC	35796	35777	0.1
23730	1995-067A	TELECOM 2C	FR	35800	35773	0
23731	1995-067B	INSAT 2C	IND	35814	35759	0.3
23741	1995-069A	GALAXY 3R	US	35796	35777	0
23754	1995-073A	ECHOSTAR 1	US	35798	35774	0
23764	1996-002A	PAS 3R	US	35796	35778	0
23765	1996-002B	MEASAT 1	MALA	35791	35783	0
23768	1996-003A	KOREASAT 2	KOR	35792	35782	0
23775	1996-005A	GORIZONT 31	CIS	35789	35777	2.4
23778	1996-005D	SL-12 R/B(2)	CIS	36583	36477	2.5
23779	1996-006A	ANATOLIA 1 (PALAPA C1)	TURK	35901	35670	0
23781	1996-007A	NSTAR 2	JPN	35796	35778	0
23816	1996-015A	INTELSAT 707	ITSO	35802	35770	0
23839	1996-020A	INMARSAT 3-F1	IM	35801	35771	0
23842	1996-021A	ASTRA 1F	LUXE	35799	35774	0
23846	1996-022A	MSAT M1	CA	35801	35772	0
23864	1996-030A	PALAPA C2	INDO	35791	35783	0
23865	1996-030B	AMOS	ISRA	35806	35766	0
23877	1996-033A	GALAXY 9	US	35792	35780	0
23880	1996-034A	GORIZONT 32	CIS	35791	35787	2.2
23883	1996-034D	SL-12 R/B(2)	CIS	36606	36295	2.3
23915	1996-035A	INTELSAT 709	ITSO	35791	35784	0
23943	1996-039A	APSTAR 1A	PRC	35793	35782	0
23948	1996-040A	ARABSAT 2A	AB	35789	35782	0
23949	1996-040B	TURKSAT 1C	TURK	35802	35770	0.1
24208	1996-044A	ITALSAT 2	IT	35832	35740	0.1
24209	1996-044B	TELECOM 2D	FR	35813	35760	0
24282	1996-048A	CHINASAT 7	PRC	46626	21548	28
24307	1996-053A	INMARSAT 3-F2	IM	35807	35766	0.1
24313	1996-055A	ECHOSTAR 2	US	35796	35776	0
24315	1996-054A	GE 1	US	35797	35776	0
24435	1996-058A	EXPRESS 2	CIS	35801	35777	2.4
24438	1996-058D	SL-12 R/B(2)	CIS	35850	35740	3.2
24652	1996-063A	ARABSAT 2B	AB	35801	35769	0
24653	1996-063B	MEASAT 2	MALA	35793	35780	0
24665	1996-067A	HOT BIRD 2	EUTE	35800	35772	0.1
24674	1996-070A	INMARSAT 3-F3	IM	35801	35772	0.1
24700	1995-062C	ISO DEB	ESA	70362	1219	6.6
24713	1997-002A	GE 2	US	35802	35772	0
24714	1997-002B	NAHUEL 1A	ARGN	35796	35777	0.1
24732	1997-007A	JCSAT 4	JPN	35797	35777	0
24742	1997-009A	INTELSAT 801	ITSO	35803	35770	0
24748	1997-011A	TEMPO 2	US	35796	35776	0
24768	1997-016A	THAICOM 3	THAI	35815	35758	0
24769	1997-016B	B-SAT 1A	JPN	35796	35778	0.1
24786	1997-019A	GOES 10	US	35799	35776	0.2
24798	1997-021A	DFH-3 2	PRC	35811	35766	0.2
24808	1997-025A	THOR 2A	NOR	35800	35772	0
24812	1997-026A	TELSTAR 5	US	35796	35778	0
24819	1997-027A	INMARSAT 3-F4	IM	35922	35650	0.1
24820	1997-027B	INSAT 2D	IND	35971	33160	2.7
24834	1997-029A	FENGYUN 2A	PRC	35788	35785	0.9
24846	1997-031A	INTELSAT 802	ITSO	35801	35771	0
24880	1997-036A	SUPERBIRD C	JPN	35793	35781	0
24891	1997-040A	PAS 6	US	35791	35784	0
24894	1997-041A	COSMOS 2345	CIS	36444	35081	1.5
24897	1997-041D	SL-12 R/B(2)	CIS	37339	34375	1.5
24901	1997-042A	AGILA 2	RP	35806	35769	0
24916	1997-046A	PAS 5	US	35794	35781	0
24931	1997-049A	HOT BIRD 3	EUTE	35808	35765	0.1
24932	1997-049B	METEOSAT 7	ESA	35793	35781	0.2
24936	1997-050A	GE 3	US	35800	35773	0
24957	1997-053A	INTELSAT 803	ITSO	35801	35773	0
25000	1968-081G	TITAN 3C TRANSTAGE DEB	US	36046	35100	12.8
25001	1968-081H	TITAN 3C TRANSTAGE DEB	US	35676	35098	12.7
25004	1997-059A	ECHOSTAR 3	US	35813	35760	0
25010	1997-062A	APSTAR 2R	PRC	35804	35769	0
25021	1997-065C	IABS R/B	US	35612	35523	2.6
25045	1997-070A	KUPON	CIS	35803	35760	2.4

Cat No	International Designator	Common Name	Source	Apogee (km)	Perigee (km)	Inclination (°)
25048	1997-070D	SL-12 R/B(2)	CIS	36036	35894	2.6
25049	1997-071A	SIRIUS 2	SWED	35792	35780	0
25050	1997-071B	INDOSTAR 1	INDO	35798	35778	0.1
25067	1997-075A	JCSAT 5	JPN	35793	35781	0
25068	1997-075B	EQUATOR S	GER	67057	622	8.6
25071	1997-076A	ASTRA 1G	LUXE	35799	35774	0
25086	1997-078A	GALAXY 8	US	35797	35777	0
25110	1997-083A	INTELSAT 804	ITSO	35803	35769	0
25126	1997-086A	HGS-1 (ASIASAT 3)	US	35988	35585	6.6
25134	1998-002A	SKYNET 4D	UK	35795	35780	2.2
25152	1998-006A	BRAZILSAT B3	BRAZ	35801	35772	0.1
25153	1998-006B	INMARSAT 3-F5	IM	35819	35752	0.7
25237	1998-013A	HOT BIRD 4	EUTE	35845	35726	0.1
25239	1998-014A	INTELSAT 806	ITSO	35804	35770	0
25311	1998-024A	NILESAT	EGYP	35803	35770	0.1
25312	1998-024B	B-SAT 1B	JPN	35794	35782	0
25315	1998-025A	COSMOS 2350	CIS	35823	35752	0.3
25318	1998-025D	SL-12 R/B(2)	CIS	35992	35703	0.3
25331	1998-028A	ECHOSTAR 4	US	35792	35780	0.1
25339	1996-053D	SL-12 R/B(2)	CIS	36609	36210	0.9
25353	1997-049E	METEOSAT 7 AKM	ESA	35968	35393	1
25354	1998-033A	CHINASTAR 1	PRC	35803	35771	0
25358	1998-035A	THOR 3	NOR	35792	35780	0.1
25371	1998-037A	INTELSAT 805	ITSO	35798	35776	0
25404	1998-044A	SINOSAT 1	PRC	35798	35774	0.1
25460	1998-049A	ST-1	STCT	35800	35772	0
25462	1998-050A	ASTRA 2A	LUXE	35820	35750	0.1
25473	1998-052A	PAS 7	US	35800	35772	0
25491	1998-056A	EUTELSAT W2	EUTE	35812	35760	0
25492	1998-056B	SIRIUS 3	SWED	35792	35779	0
25495	1998-057A	HOT BIRD 5	EUTE	35822	35751	0
25515	1998-063A	AFRISTAR	US	35803	35768	0
25516	1998-063B	GE 5	US	35796	35778	0
25522	1998-065A	PAS 8	US	35801	35772	0
25546	1998-068A	BONUM 1	CIS	35791	35782	0
25558	1998-070A	SAT MEX 5	MEX	35790	35783	0
25585	1998-075A	PAS 6B	US	35793	35782	0
25611	1997-029C	FENGYUN 2A AKM	PRC	36067	34076	1.8
25626	1999-005A	TELSTAR 6	US	35795	35778	0.1
25630	1999-006A	JCSAT 06	JPN	35792	35783	0
25638	1999-009A	ARABSAT 3A	AB	35800	35771	0
25639	1999-009B	SKYNET 4E	UK	35814	35760	2.6
25642	1999-010A	RADUGA 1-4	CIS	35795	35784	1.6
25645	1999-010D	SL-12 R/B(2)	CIS	36500	36345	1.6
25657	1999-013A	ASIASAT 3S	AC	35803	35772	0
25666	1999-016A	INSAT 2E	IND	35805	35770	0
25673	1999-018A	EUTELSAT W3	EUTE	35826	35746	0
25740	1999-027A	NIMIQ 1	CA	35792	35781	0
25785	1999-033A	ASTRA 1H	LUXE	35799	35774	0
25869	1999-040D	CXO R/B(2)	US	71420	1935	33.9
25880	1999-042A	TELKOM 1	INDO	35798	35776	0
25894	1999-046A	KOREASAT 3	KOR	35801	35775	0
25896	1999-047A	YAMAL 101	CIS	36302	35495	1.2
25897	1999-047B	YAMAL 102	CIS	35801	35771	0.2
25900	1999-047E	SL-12 R/B(2)	CIS	36288	35345	1.2
25913	1999-050A	ECHOSTAR 5	US	35794	35779	0
25922	1999-052A	TELSTAR 7	US	35796	35776	0
25924	1999-053A	LMI 1	CIS	35789	35784	0
25937	1999-056A	DIRECTV 1-R	US	35790	35783	0
25949	1999-059A	ORION 2	US	35787	35786	0
25954	1999-060A	GE 4	US	35788	35785	0
25967	1999-063A	UFO 10 (USA 146)	US	35800	35773	5.2
26038	1999-071A	GALAXY 11	US	35788	35785	0
26056	2000-002A	GALAXY 10R	US	35792	35780	0
26058	2000-003A	ZHONGXING-22	PRC	35805	35769	0.2
26071	2000-007A	HISPASAT 1C	SPN	35802	35772	0
26089	2000-011A	GARUDA 1	INDO	35798	35775	2.2
26095	2000-012A	SUPERBIRD 4	JPN	35795	35779	0
26098	2000-013A	EXPRESS 2A	CIS	35795	35777	0.1
26101	2000-013D	SL-12 R/B(2)	CIS	35693	35608	0.7
26107	2000-016A	ASIASTAR	US	35803	35771	0
26108	2000-016B	INSAT 3B	IND	35808	35766	0
26243	2000-019A	SESAT	EUTE	35795	35778	0
26246	2000-019D	SL-12 R/B(2)	CIS	36036	35939	0.8
26298	2000-020A	GALAXY 4R	US	35790	35784	0
26352	2000-022A	GOES 11	US	35806	35772	0.1
26369	2000-028A	EUTELSAT W4	EUTE	35805	35768	0
26372	2000-029A	GORIZONT 33	CIS	35805	35765	0.8
26373	2000-029B	BREEZE-M R/B	CIS	36826	34727	1.1
26378	2000-031A	EXPRESS 3A	CIS	35797	35776	0.1
26381	2000-031D	SL-12 R/B (2)	CIS	36089	35952	0.4
26382	2000-032A	FENGYUN 2B	PRC	35791	35783	0.5
26388	2000-034A	TDRS 8	US	35884	35686	6.5
26390	2000-035A	SIRIUS-1	US	47096	24475	63.3
26394	2000-036A	COSMOS 2371	CIS	35799	35771	0.9
26397	2000-036D	SL-12 R/B(2)	CIS	35908	35699	0.9
26402	2000-038A	ECHOSTAR 6	US	35793	35779	0
26451	2000-043A	PAS 9	US	35795	35779	0

Cat No	*International Designator*	*Common Name*	*Source*	*Apogee (km)*	*Perigee (km)*	*Inclination (°)*
26460	2000-032C	FENGYUN 2B AKM	PRC	35901	35716	0.6
26469	2000-046A	BRAZILSAT B4	BRAZ	35808	35766	0.1
26470	2000-046B	NILESAT 102	EGYP	35804	35769	0
26477	2000-049A	RADUGA 1-5	CIS	35792	35781	1
26480	2000-049D	SL-12 R/B(2)	CIS	36463	36298	1.1
26483	2000-051A	SIRIUS-2	US	47030	24562	63.4
26487	2000-052A	EUTELSAT W1	EUTE	35894	35679	0
26494	2000-054A	ASTRA 2B	LUXE	35841	35729	0
26495	2000-054B	GE 7	US	35801	35770	0.1
26554	2000-059A	GE 1A	US	35793	35782	0
26559	2000-060A	NSAT 110	JPN	35790	35784	0
26578	2000-066A	THURAYA 1	UAE	35800	35771	6
26580	2000-067A	GE 6	US	35794	35781	0
26590	2000-068A	EUROPE STAR F1	ESA	35803	35768	0.1
26599	2000-069A	BEIDOU 1	PRC	35813	35760	0.1
26608	2000-072A	PAS 1R	US	35788	35787	0.1
26624	2000-076A	ANIK F1	CA	35789	35784	0
26626	2000-077A	SIRIUS 3	US	47161	24411	63.5
26638	2000-081A	ASTRA 2D	LUXE	35803	35769	0.1
26639	2000-081B	GE 8	US	35802	35771	0
26643	2000-082A	BEIDOU 1B	PRC	35809	35763	0.1
26666	2001-002A	TURKSAT-2A	TURK	35796	35775	0.1
26694	2001-005A	SICRAL 1	IT	35834	35736	0.1
26695	2001-005B	SKYNET 4F	UK	35795	35778	3.9
26719	2001-011A	EUROBIRD	EUTE	35787	35787	0.1
26720	2001-011B	B-SAT 2A	JPN	35827	35715	0.2
26724	2001-012A	XM-Rock	US	35786	35785	0.03
26738	2001-013C	Briz-M	CIS	37222	35669	1.33
26736	2001-013A	Ekran-M	CIS	35805	35773	1.50
26745	2001-015A	GSAT-1	US	35759	33836	0.47
26761	2001-018A	XM-Roll	US	35788	35784	0.04
26766	2001-019A	PAS 10	US	35799	35773	0.06
26770	2001-020A	GeoLITE	US	(Classified mission)		
26824	2001-024A	Intelsat 901	ITSO	35801	35772	0.06
26853	2001-025A	Astra 2C	LUXE	35804	35768	113.1
26883	2001-033D	IUS-16 SRM-2	US	(Classified mission)		
26880	2001-033A	DSP F21 (USA 159)	US	(Classified mission)		
26871	2001-031A	GOES 12	US	(Classified mission)		
26895	2001-037D	Clok M-2 No.99L	CIS	35913	35705	2.13
26892	2001-037A	Kosmos-2379	CIS	36078	35812	2.09
26900	2001-039A	Intelsat 902	ITSO	35786	35770	97.8
26927	2001-042A	Atlantic Bird 2	US	35798	35774	73.9
26939	2001-045D	Blok DM2	CIS	35927	35681	1.39
26936	2001-045A	Raduga-1	CIS	35796	35777	1.35
26948	2001-046A	USA 162	US	(Classified NRO mission)		
26985	2001-052A	DirecTV-4S	US	35789	35779	0.04

CZ-2E Parking Orbits for HS-601 Satellites

Payload	*Inclination (deg)*	*Period (min)*	*Perigee (km)*	*Apogee (km)*
AUSSAT-B model	28.49	96.6	204	990
Optus-B 1	27.99	97.1	203	1,032
Optus-B 2	28.15	96.9	208	1,010
Optus-B 3	27.85	97.5	190	1,086
APStar 2	Failed to reach orbit			

AUSSAT-B was a model with a Chinese third stage, but was flown to simulate the stack for operational launches. From the data in this table, it would appear that the CZ-2E performed successfully in placing the third stage and payload into the planned parking orbit on all four orbital missions.

INDEXES

COMPANY INDEX

Germany

Hungary

India

Indonesia

International

United States

Venezuela

GENERAL INDEX

To help users of this title evaluate the published data, *Jane's Information Group* has divided entries into three categories.
[V] VERIFIED The editor has made a detailed examination of the entry's content and has checked its relevancy and accuracy for publication in the new edition to the best of his ability.
[U] UPDATED During the verification process, significant changes to content have been made to reflect the latest position known to *Jane's* at the time of publication.
[N] NEW ENTRY Information on new equipment and/or system appears for the first time in the title.

J

K

T

U

NOTES

NOTES

NOTES

NOTES

NOTES